D1564706

THE
MASTER
SEMICONDUCTOR
REPLACEMENT
HANDBOOK

Listed by Manufacturer's Number

by TAB Editorial Staff

D1564705

TAB TAB BOOKS Inc.
BLUE RIDGE SUMMIT, PA. 17214

FIRST EDITION

FIRST PRINTING

Copyright © 1982 by TAB BOOKS Inc.

Printed in the United States of America

Library of Congress Cataloging in Publication Data

Main entry under title:

The Master semiconductor replacement handbook—
    listed by manufacturer's number.

    Includes index.
    1. Semiconductors—Catalogs. I. Tab Books.
TK7871.85.M367   621.3815′2′0216   82-5670
ISBN 0-8306-1471-0   AACR2

Preface

Over the years it has been a constant challenge to keep up with the never-ending flood of semiconductor devices on the consumer market. Simply knowing the equipment manufacturer's part number has not been enough to identify the device because a universal standardization of semiconductors has yet to be realized.

In the past few years, leading semiconductor suppliers—Sylvania, RCA, General Electric, Radio Shack, and Motorola—have individually introduced a "general replacement" line of semiconductors for wide ranges of applications that meet the needs of electronic servicemen, hobbyists, experimenters, hams, and engineers. It is these wide-range semiconductor lines that have been cross-referenced to thousands and thousands of Industry Standard numbers.

Indexed by each manufacturer's part numbers, this list covers over 140,000 diodes, transistors, and ICs. This list will shorten the time it takes to find a replacement device in most cases. Especially when used in conjunction with TAB book No. 1470, *The Master Semiconductor Replacement Handbook—Listed by Industry Standard Number*, this book will soon become one of your most worthwhile references.

Contents

Introduction

The information that begins on page 9 is unique. We believe it has never been published before. For the first time, anyone in electronics can find an Industry Standard number for a semiconductor by looking it up under a "general replacement" number assigned by a particular semiconductor distributor. As has been the case since more than one manufacturer of semiconductors existed, different semiconductor distributors speak different semiconductor languages; in other words, they use different part numbers for the same semiconductor in many cases.

The list that follows is divided into sections by semiconductor distributor. It is simple to use this list. For example, if you have a Radio Shack semiconductor part number, but you want to know what Industry Standard part number it replaces, simply turn to the Radio Shack section of the list and look up the Radio Shack number. You will find the Industry Standard number immediately to the right of the Radio Shack number in the list. If you want to carry the procedure one step further—for example, by trying to find the General Electric semiconductor replacement number for that Industry Standard number—you will need a copy of TAB book No. 1470, *The Master Semiconductor Replacement Handbook—Listed by Industry Standard Number*. Look up the Industry Standard number in the list of book No. 1470, and you will find the General Electric part number under the GE column. It's that easy!

ECG	Industry Standard No.	ECG	Industry Standard No.	ECG	Industry Standard No.	ECG	Industry Standard No.	ECG	Industry Standard No.	ECG	Industry Standard No.
019	43200103	100	GA52829	100	R258	100	T1902	100	2D021		
100	A1243	100	GA53149	100	R488	100	T2038	100	2D021-11		
100	A1474-3	100	GA53242	100	R506	100	T2039	100	2D021-56		
100	A1474-39	100	GC1159	100	RE1	100	T2040	100	2D021-B		
100	A1488C	100	GC1302	100	RS-1539	100	T2091	100	2G1024		
100	A1488C9	100	GC181	100	RS-1550	100	T2122	100	2G1026		
100	A160(JAPAN)	100	GC182	100	RS-2683	100	T2172	100	2G138		
100	A167	100	GC31	100	RS-2684	100	T2173	100	2G139		
100	A168(JAPAN)	100	GC32	100	RS-2685	100	T2256	100	2G140		
100	A168A	100	GC33	100	RS-2686	100	T2257	100	2G383		
100	A169	100	GC34	100	RS-2687	100	T2258	100	2G394		
100	A170(JAPAN)	100	GC35	100	RS-2688	100	T2259	100	2G396		
100	A171(JAPAN)	100	GC360	100	RS-2690	100	T2260	100	2G397		
100	A172	100	GC4022	100	RS-2691	100	T2261	100	2G524		
100	A172A	100	GC532	100	RS-2692	100	T50944	100	2G525		
100	A204	100	GC60	100	RS-2694	100	T52147	100	2G526		
100	A205	100	GC61	100	RS-2695	100	T52147Z	100	2G527		
100	A206	100	GE-1	100	RS-2696	100	T52148Z	100	2G577		
100	A207	100	GET880	100	RS-3277	100	T52149	100	2G601		
100	A208(JAPAN)	100	GET881	100	RS-3278	100	T52149Z	100	2G602		
100	A210(JAPAN)	100	GET882	100	RS-3279	100	TA-1575B	100	2G603		
100	A211(JAPAN)	100	GET887	100	RS-3288	100	TA-1655B	100	2G604		
100	A212	100	GET888	100	RS-3309	100	TA-1704	100	2G605		
100	A217	100	GET889	100	RS-3867	100	TA-1763	100	2N107		
100	A248(JAPAN)	100	GET890	100	RS-3868	100	TA-1763A	100	2N114		
100	A26	100	GET891	100	RS-3907	100	TA-1778	100	2N1171		
100	A277(JAPAN)	100	GET892	100	RS-3913	100	TA-1782	100	2N1176		
100	A278(JAPAN)	100	GET895	100	RS-3929	100	TA-1783	100	2N1176A		
100	A279(JAPAN)	100	GET896	100	RS-5104	100	TA1704	100	2N1176B		
100	A280	100	GET897	100	RS-5105	100	TA1763	100	2N1185		
100	A281	100	GI1	100	RS-5106	100	TA1763A	100	2N1186		
100	A283	100	GT11	100	RS-5401	100	TA1778	100	2N1187		
100	A284	100	GT12	100	RS-5504	100	TA1782	100	2N1188		
100	A304	100	GT13	100	RS-5511	100	TA1783	100	2N1264		
100	A305	100	GT153	100	RS-5540	100	TI-363	100	2N1280		
100	A311(JAPAN)	100	GT1604	100	RS2690	100	TI-364	100	2N1281		
100	A350A	100	GT1605	100	RS2691	100	TIA03	100	2N1282		
100	A391	100	GT1606	100	RS2692	100	TIA05	100	2N1284		
100	A392	100	GT1607	100	RS2696	100	TIA05A	100	2N1307		
100	A393	100	GT269	100	RS3281	100	TIX895	100	2N1309		
100	A394	100	GT2694	100	RS3287	100	TIXA-03	100	2N1316		
100	A395	100	GT348	100	RS3892	100	TIXA-04	100	2N1317		
100	A414	100	GT5153	100	RS3914	100	TIXA-05	100	2N1318		
100	A415(JAPAN)	100	GT760	100	RS3915	100	TIXA01	100	2N1319		
100	A514-027662	100	GT761	100	RS5104	100	TIXA02	100	2N1344		
100	A514-032815	100	GT761R	100	RS5105	100	TIXA03	100	2N1345		
100	A74-3-3A9G	100	GT762	100	RS5302	100	TIXA04	100	2N1346		
100	A74-3-70	100	GT762R	100	RS5303	100	TIXA05	100	2N1347		
100	A74-3-705	100	GT764	100	RS5402	100	TM100	100	2N1349		
100	A74-3A9G	100	GT766	100	RS5403	100	TNJ-60610	100	2N135		
100	A88C-70	100	GT766A	100	RS5504	100	TNJ-60611	100	2N1350		
100	A88C-705	100	GT83	100	RS5511	100	TNJ-60612	100	2N1351		
100	A88C19G	100	GT832	100	RS5540	100	TNJ60608	100	2N1354		
100	AA1	100	GT87	100	RS5743.3	100	TO-101	100	2N1355		
100	ACR810-104	100	GT88	100	RS682	100	TO-102	100	2N1356		
100	ACR810-105	100	GTE1	100	S74-3-A-3P	100	TQ5020	100	2N1357		
100	ACR810-106	100	GTE2	100	S88C-1-3P	100	TR-05	100	2N136		
100	ACR83-1004	100	GTV	100	SC43	100	TR-06	100	2N1361		
100	ACR83-1005	100	HA-00102	100	SC44	100	TR-044	100	2N1361A		
100	ACR83-1006	100	HA-12	100	SC46	100	TR-044A	100	2N137		
100	AF-101	100	HA00052	100	SF.T171	100	TR-045A	100	2N1393		
100	AF138/290	100	HA00053	100	SF.T172	100	TR05C	100	2N1395		
100	APZ23	100	HA15	100	SF.T173	100	TR07C	100	2N1404A		
100	ALZ10	100	HA202	100	SF.T174	100	T811	100	2N1469		
100	ASI76	100	HA49	100	SFT-307	100	TR310015	100	2N1470		
100	AST77	100	HEP-G0005	100	SFT-319	100	TR310161	100	2N1471		
100	AST80	100	HEP-G0006	100	SFT223	100	TR321(HPGH1)	100	2N1570		
100	AT-15	100	HEP-G0007	100	SFT226	100	TR482	100	2N1581		
100	AT-5	100	HJ41	100	SFT227	100	TR482A	100	2N1583		
100	B290	100	HM-00049	100	SFT228	100	TR53	100	2N1584		
100	B291	100	MA389	100	SFT229	100	TR55	100	2N1664		
100	B292	100	MA100	100	SFT237	100	TR760	100	2N1684		
100	B292A	100	MA286	100	SFT251	100	TRC44	100	2N1729		
100	B392	100	MA287	100	SFT252	100	TRC44A	100	2N1731		
100	B393	100	MA288	100	SFT253	100	TRC45	100	2N1940		
100	B394	100	MA901	100	SFT288	100	TRC45A	100	2N1969		
100	B395	100	NK74-3A19	100	SK3005	100	TS-601	100	2N1997		
100	B401	100	NK88C119	100	SM-217	100	TS-602	100	2N1998		
100	B402	100	NKT128	100	ST-28B	100	T8669A	100	2N2171		
100	B403	100	NKT129	100	ST-28C	100	T8669B	100	2N2172		
100	B416	100	NKT141	100	ST-37C	100	T8669D	100	2N2648		
100	B417	100	NKT142	100	ST-37C	100	T8669E	100	2N27		
100	B74-3-A-21	100	NKT143	100	ST-37D	100	T8669F	100	2N271		
100	B74-3A21	100	NKT144	100	SYL-105	100	TV24152	100	2N271A		
100	B75A	100	NKT162	100	SYL-106	100	TV4152	100	2N30		
100	B88C-1-21	100	NKT163	100	SYL-160	100	TVS-28A171	100	2N3075		
100	BE6	100	NKT16325	100	SYL-1608	100	TVS-28S172A	100	2N308		
100	BE6A	100	NKT164	100	SYL-2248	100	UPI1345	100	2N309		
100	C1437	100	NKT16425	100	SYL-2249	100	UPI1347	100	2N31		
100	C73	100	NKT203	100	SYL-2250	100	V10/2S	100	2N311		
100	C75	100	NKT204	100	SYL105	100	V10/2SJ	100	2N315		
100	C76	100	NKT205	100	SYL1588	100	V13/11	100	2N316		
100	CK14	100	NKT206	100	SYL160	100	VFQ-2745P	100	2N317		
100	CK14A	100	NKT207	100	SYL1608	100	VFY-2745E	100	2N317A		
100	CK16	100	NKT22281	100	SYL1690	100	VL/8RJ	100	2N3216		
100	CK17	100	NKT222282	100	SYL1697	100	VSP2745	100	2N327		
100	CK17A	100	NKT243	100	SYL1717	100	W1	100	2N327A		
100	CK25	100	NKT261	100	SYL2120	100	WTVB6	100	2N394		
100	CK25A	100	NKT262	100	SYL2247	100	X78	100	2N394A		
100	CK26	100	NKT263	100	SYL2250	100	001-01202-1	100	2N395		
100	CK26A	100	NKT264	100	T-109	100	001-012021	100	2N396		
100	CK27	100	NKT42	100	T-116	100	001-01203-1	100	2N396A		
100	CK27A	100	NKT43	100	T-152148	100	001-012031	100	2N397		
100	CK661	100	NKT62	100	T-1877	100	1-21-100	100	2N403		
100	CK662	100	NKT63	100	T-2038	100	1-21-102	100	2N404		
100	CK759	100	NKT64	100	T-2039	100	1-21-103	100	2N413		
100	CK759A	100	NKT72	100	T-2040	100	1-21-104	100	2N413A		
100	CK760	100	NKT73	100	T-2091	100	1-21-105	100	2N414		
100	CK760A	100	NKT74	100	T-2439	100	1-21-128	100	2N414A		
100	CK761	100	OC-130	100	T-2440	100	1-21-161	100	2N415		
100	CK768	100	OC-140	100	T-2441	100	1-21-162	100	2N415A		
100	CK776	100	OC-410	100	T-46	100	1-21-179	100	2N416		
100	CK776A	100	OC-44	100	T-47	100	1-21-180	100	2N417		
100	D019	100	OC-45	100	T-48	100	1-21-186	100	2N426		
100	D078	100	OC-46	100	T-52148Z	100	1-21-234	100	2N427		
100	DS-28(DELCO)	100	OC-47	100	T-52149	100	1-21-235	100	2N428		
100	D821	100	PF-530A	100	T-52149Z	100	1-21-256	100	2N428A		
100	D822	100	PTC102	100	T-78	100	1-21-240	100	2N481		
100	D823	100	Q-1A	100	TO-101	100	1-21-241	100	2N482		
100	D853	100	R-119	100	TO-102	100	1-21-254	100	2N483		
100	E2412	100	R-163	100	T0101	100	1-21-273	100	2N484		
100	ED52	100	R-186	100	T0102	100	1-21-275	100	2N485		
100	ED53	100	R-227	100	T03323	100	1-21-289	100	2N486		
100	ED54B	100	R-244	100	T1251	100	1-21-73	100	2N486B		
100	EK159	100	R-424	100	T1289	100	1-21-74	100	2N487		
100	E0105	100	R-425	100	T1291	100	1-21-75	100	2N505		
100	E065	100	R-488	100	T1312	100	1-21-76	100	2N518		
100	E066	100	R-506	100	T1322	100	1-21-78	100	2N519		
100	E067	100	R119	100	T1326	100	1-21-83				
100	E068	100	R163	100	T1474	100	1-21-91				
100	E825	100	R186	100	T1510	100	1-21-92				
100	E826	100	R227	100	T152148	100	1-21-93				
100	FV2747C	100	R244	100	T1877	100	2B				
						100	2C				

ECG	Industry Standard No.	ECG	Industry Standard No.	ECG	Industry Standard No.	ECG	Industry Standard No.	ECG	Industry Standard No.
100	2N519A	100	2V631	100	48-134466	100	86X0014-001	100	310-189
100	2N520	100	2V632	100	48-134468	100	86X3	100	324-0090
100	2N520A	100	2V633	100	48-134469	100	88C	100	350
100	2N521	100	4-74-3A7-1	100	48-134470	100	088C-12	100	365T1
100	2N521A	100	4-88077-1	100	48-134471	100	088C-12-7	100	412
100	2N522	100	4JD1A17	100	48-134472	100	88C-70	100	421-9682
100	2N522A	100	4JX1A520D	100	48-134473	100	88C-70-12	100	474-3A5
100	2N523	100	4JX1A520E	100	48-134474	100	88C-70-12-7	100	488C15
100	2N523A	100	4JX1C850A	100	48-134475	100	8800	100	520T1
100	2N529	100	4JX13B21	100	48-134476	100	8801	100	521T1
100	2N530	100	4JX2A60	100	48-134477	100	88C10	100	674-3A5L
100	2N531	100	6A12678	100	48-134494	100	88C10R	100	688C-1-5L
100	2N532	100	8-0060	100	48-134495	100	8811	100	690V047H56
100	2N533	100	9-5120A	100	48-134496	100	88C119	100	690V047H57
100	2N572	100	011-H01	100	48-134499	100	8812	100	1145
100	2N578	100	12-1-100	100	48-134500	100	88C121	100	1146
100	2N579	100	12-1-102	100	48-134501	100	8813	100	1340
100	2N580	100	12-1-103	100	48-134509	100	88C13P	100	1350
100	2N581	100	12-1-104	100	48-134510	100	8814	100	1390
100	2N582	100	12-1-105	100	48-134512	100	88C14-7	100	1400
100	2N583	100	12-1-128	100	48-134538	100	88C14-7B	100	1410
100	2N586	100	12-1-161	100	48-134539	100	8815	100	1474-3
100	2N59	100	12-1-162	100	48-134540	100	88C15L	100	1474-3-12
100	2N592	100	12-1-179	100	48-134541	100	8816	100	1474-3-12-8
100	2N593	100	12-1-180	100	48-134542	100	88C16D	100	1488C
100	2N597	100	12-1-186	100	48-134543	100	8817	100	1488C-12
100	2N598	100	12-1-234	100	48-134544	100	88C17-1	100	1488C-12-8
100	2N599	100	12-1-235	100	48-134553	100	8818	100	1874-3
100	2N60	100	12-1-236	100	48-134554	100	88C182	100	1874-3-12
100	2N600	100	12-1-240	100	48-134555	100	8819	100	1874-3-127
100	2N61	100	12-1-241	100	48-134556	100	88C19G	100	1874-3L
100	2N614	100	12-1-254	100	48-134557	100	8802	100	1874-3L8
100	2N615	100	12-1-273	100	48-134558	100	8803	100	1888C
100	2N616	100	12-1-275	100	48-134559	100	8804	100	1888C-12
100	2N617	100	12-1-289	100	48-134562	100	8805	100	1888C-127
100	2N658	100	12-1-73	100	48-134563	100	8806	100	1888CL
100	2N659	100	12-1-74	100	48-134564	100	8807	100	1888CL8
100	2N662	100	12-1-75	100	48-134565	100	8808	100	2093A9-3
100	2N674	100	12-1-76	100	48-134567	100	8809	100	2093A9-4
100	2N801	100	12-1-78	100	48-134603	100	88C0-70-12	100	2904-038H05
100	2N802	100	12-1-83	100	48-134604	100	93A9-3	100	3425
100	2N809	100	12-1-91	100	48-134610	100	93A9-4	100	3434
100	2N810	100	12-1-92	100	48-134625	100	94T1	100	3435
100	2N811	100	12-1-93	100	48-134626	100	95-120A	100	3500
100	2N812	100	012-H02	100	48-134631	100	99BA6	100	3746
100	2N83	100	12A6240	100	48-134636	100	99BB6	100	3852
100	2NJ5A	100	13-18944-1	100	48-134637	100	998003	100	3970
100	2NJ6	100	13-18944-2	100	48-134641	100	101-12	100	4451
100	2NJ8A	100	13-50484-1	100	48-134655	100	112-001	100	4484
100	2N111	100	13-50486-1	100	48-134656	100	120-001192	100	4485
100	2S12	100	13-50631-1	100	48-134657	100	121-128	100	4486
100	2S13	100	13-50944-1	100	48-134956	100	121-1330	100	6445
100	2S155	100	019-003324	100	48-56P1	100	121-1350	100	12110-0
100	2S159	100	019-003342	100	48-63029A92	100	121-1360	100	12110-3
100	2S160	100	019-003343	100	48-63075A76	100	121-1390	100	12110-4
100	2S167	100	19-020-033	100	48-63077A03	100	121-1400	100	12110-5
100	2S174	100	19A115208	100	48-63078A59	100	121-145	100	12112-8
100	2S178	100	19A115208-P1	100	48-63078A60	100	121-146	100	12116-1
100	2S25	100	19A115208-P2	100	48-63078A61	100	121-147	100	12116-2
100	2S30	100	19A115301-P1	100	48-63078A64	100	121-160	100	12117-9
100	2S31	100	19A115301-P2	100	48-63082A15	100	121-205	100	12118-5
100	2S45	100	19B2000129-P1	100	51D188	100	121-206	100	12118-6
100	2S49	100	21A051-000	100	51D189	100	121-207	100	12123-4
100	2S51	100	21M007	100	57A6-6A	100	121-208	100	12123-5
100	2S52	100	24MW11	100	57A6-6B	100	121-209	100	12123-6
100	2S53	100	24MW77	100	57A6-6C	100	121-210	100	12124-0
100	2S60	100	27T401	100	5706-6A	100	121-211	100	12124-1
100	2S91	100	27T402	100	5706-6B	100	121-212	100	12125-4
100	2S92	100	29V008M01	100	5706-6C	100	121-213	100	12127-3
100	2S92A	100	29V011H01	100	57D180	100	121-219	100	12127-5
100	2S93	100	29V012H01	100	57D184	100	121-220	100	12128-9
100	2S93A	100	31-0253	100	62-18415	100	121-221	100	12173
100	2SA167	100	43P1	100	62-18416	100	121-222	100	12174
100	2SA168	100	43P3	100	62-18417	100	121-225	100	12175
100	2SA168A	100	43P6C	100	62-18423	100	121-234	100	12176
100	2SA169	100	45X2	100	62-18424	100	121-235	100	12185
100	2SA170	100	48-124307	100	63-10200	100	121-236	100	12191
100	2SA171	100	48-124308	100	63-11585	100	121-254	100	12192
100	2SA172	100	48-124314	100	63-18416	100	121-354	100	12193
100	2SA172A	100	48-124315	100	63-25946	100	121-397	100	18529
100	2SA182	100	48-124327	100	63-26849	100	121-53	100	020156
100	2SA204	100	48-124328	100	63-27278	100	121-80	100	27125-360
100	2SA205	100	48-124343	100	63-27279	100	121-81	100	30208
100	2SA206	100	48-124344	100	63-27280	100	121-82	100	30231
100	2SA207	100	48-124345	100	63-27367	100	121-830	100	34119
100	2SA211	100	48-124357	100	63-29662	100	121-83	100	34219
100	2SA212	100	48-124358	100	63-29663	100	121-84	100	34220
100	2SA217	100	48-124359	100	63-29664	100	121-85	100	34221
100	2SA26	100	48-124370	100	63-29863	100	121-86	100	36816
100	2SA277	100	48-124371	100	63-7547	100	121-87	100	37833
100	2SA278	100	48-124378	100	64T1	100	121-88	100	38209
100	2SA279	100	48-124379	100	070-020	100	121-89	100	40269
100	2SA280	100	48-124380	100	74-3	100	121-90	100	671193-82
100	2SA281	100	48-124389	100	74-3-70	100	121-91	100	72784-21
100	2SA283	100	48-124398	100	74-3-70-12	100	121-92	100	81502-0
100	2SA284	100	48-124443	100	74-3-70-12-7	100	121-93	100	81502-0A
100	2SA304	100	48-124444	100	74-30	100	121-94	100	81502-0B
100	2SA305	100	48-124446	100	74-31	100	122-229	100	81502-1
100	2SA322K	100	48-125229	100	74-32	100	129-11	100	81502-1A
100	2SA391	100	48-125230	100	74-33	100	174-3A82	100	81502-1B
100	2SA392	100	48-125231	100	74-34	100	188C-1-82	100	81502-5A
100	2SA393	100	48-125232	100	74-35	100	235	100	81502-5B
100	2SA394	100	48-125237	100	74-36	100	260P21001	100	81502-7
100	2SA395	100	48-125238	100	74-37	100	260P21002	100	81502-7A
100	2SA414	100	48-125239	100	74-38	100	297V008H01	100	81502-7B
100	2SA415	100	48-125240	100	74-39	100	297V011H02	100	81502-7C
100	2SB290	100	48-125242	100	74-3A	100	297V012H02	100	81502-8
100	2SB291	100	48-125296	100	74-3A0	100	297V012H03	100	81502-8A
100	2SB292	100	48-128303	100	74-3A0R	100	297V012H05	100	81502-8B
100	2SB292A	100	48-134415	100	74-3A1	100	297V012H06	100	81502-8C
100	2SB392	100	48-134416	100	74-3A19	100	297V012H08	100	81503-6
100	2SB393	100	48-134417	100	74-3A2	100	297V012H09	100	81503-6A
100	2SB394	100	48-134418	100	74-3A21	100	297V017H01	100	81503-6B
100	2SB395	100	48-134419	100	74-3A3	100	297V017H02	100	81503-6C
100	2SB401	100	48-134420	100	74-3A3P	100	297V019B01	100	81503-7
100	2SB402	100	48-134421	100	74-3A4	100	297V019H01	100	81503-7A
100	2SB403	100	48-134422	100	74-3A4-7	100	297V020H02	100	81503-7B
100	2SB416	100	48-134423	100	74-3A4-7B	100	297V020M01	100	81503-7C
100	2SB417	100	48-134424	100	74-3A5	100	297V021H01	100	81503-8
100	2SH203	100	48-134425	100	74-3A5L	100	297V021H02	100	81504-1
100	2TN15	100	48-134426	100	74-3A6	100	297V021H03	100	81504-1A
100	2TN32	100	48-134427	100	74-3A6-3	100	297V022H01	100	81504-1B
100	2TN48	100	48-134428	100	74-3A7	100	297V026H03	100	81504-1C
100	2TN49	100	48-134432	100	74-3A7-1	100	297V038H01	100	81504-3
100	2TN52	100	48-134433	100	74-3A8	100	297V038H05	100	81504-3A
100	2TN53	100	48-134443	100	74-3A82	100	297V038H09	100	81504-3B
100	2V464	100	48-134444	100	74-3A9	100	297V043H02	100	81504-3C
100	2V465	100	48-134445	100	74-3A9G	100	297V044H01	100	81506-5
100	2V466	100	48-134446	100	77	100	297V054C01	100	81506-5A
100	2V467	100	48-134450	100	78BLK	100	297V054C02	100	81506-5B
100	2V482	100	48-134459	100	78GRN	100	297V055C01	100	81506-5C
100	2V483	100	48-134462	100	78RED	100	297V065H01	100	81506-6
100	2V484	100		100	78YEL	100	297V065H02	100	81506-6A
100	2V486	100		100		100	297V065H03	100	81506-6B

ECG	Industry Standard No.
100	81506-6C
100	81510-5
100	81510-0
100	81511-5
100	81511-6
100	81511-7
100	95120A
100	116084
100	117616
100	117617
100	121153
100	121154
100	223372
100	223473
100	223475
100	224584
100	226181
100	226338
100	227752
100	231588
100	233507
100	233345
100	244007
100	256126
100	257470
100	266702
100	310223
100	454549
100	573356
100	610035-1
100	610036-1
100	610036-2
100	610036-3
100	610059-2
100	610074-2
100	801507
100	815020
100	815020A
100	815020B
100	815021
100	815021A
100	815021B
100	815025
100	815025A
100	815025B
100	815027
100	815027A
100	815027B
100	815027C
100	815028
100	815028A
100	815028B
100	815028C
100	815036
100	815036A
100	815036B
100	815036C
100	815037
100	815037A
100	815037B
100	815037C
100	815041
100	815041A
100	815041B
100	815041C
100	815043
100	815043A
100	815043B
100	815043C
100	815055
100	815056
100	815057
100	815065
100	815065A
100	815065B
100	815065C
100	815066
100	815066A
100	815066B
100	815066C
100	815101
100	815103
100	815105
100	815107
100	815108
100	815109
100	815115
100	815116
100	815117
100	815308A
100	922896
100	980136
100	980516
100	980426
100	980432
100	980434
100	980438
100	980439
100	1221648
100	1221649
100	2001653-22
100	2001812-65
100	2091211-0014
100	2091241-0013
100	2091241-0014
100	2091241-0015
100	2091241-1
100	2091241-10
100	2091241-11
100	2091241-12
100	2091241-13
100	2091241-13A
100	2091241-14
100	2091241-15
100	2091241-15A
100	2091241-2
100	2091241-3
100	2091241-4
100	2091241-6
100	2091241-7
100	2091241-8
100	2320194
100	2320512
100	2320514-1
100	2970038H05
100	4036612-P1
100	4036612-P2
100	4036707-P1
100	4036707-P2
100	4036937-P1
100	4036937-P2
100	4037993-P1
100	4037993-P2
100	4038260-P1
100	4038260-P2
100	6100724-2
100	7276211
100	7278421
100	7279379
100	7279782
100	7279788
100	7279789
100	7279940
100	7281307
100	7281308
100	7281309
100	7281891
100	8010520
100	8526849-1
100	27125240
100	62087684
100	62752319
100	62752327
100	62771364
100	120001192
100	8014712
1001	4001(IC)
1002	GEIC-69
1002	LA1111
1002	LA1111P
1002	09-308050
1002	1111P
1003	A1200
1003	A1201
1003	A1201B
1003	A1201C
1003	A1201T
1003	BA16X35
1003	BA3282
1003	BA33X8367
1003	G09-015-B
1003	G09015
1003	GEIC-43
1003	LA1200
1003	LA1201
1003	000LA1201B
1003	LA1201C
1003	LA1201C-W
1003	LA1201T
1003	LA1201W
1003	LA1202
1003	RE335-IC
1003	RT7399
1003	8-759-812-01
1003	09-308063
1003	14IG007
1003	19-076001
1003	24MW1028
1003	410-407
1003	46-1343-3
1003	307-005-9-001
1003	740-8160-190
1003	905-30B
1003	917-1201-0
1003	1077-2382
1003	003501
1003	61443
1003	171179-036
1003	172252
1003	7910112
1003	7910112-01
1003	7910122-01
1004	A1364N
1004	B0306000
1004	B0306004
1004	GEIC-149
1004	GEIC-36
1004	GEIC-96
1004	HA1126D
1004	LA1364N
1004	PTC754
1004	RE325-IC
1004	REN1004
1004	HH-1X0020CEZZ
1004	RH-1X0020CEZZ
1004	TA7070P
1004	TA7070PPA-1
1004	TA7070PGL
1004	09-308089
1004	46-1357-3
1004	46-1369-3
1004	200X2100-022
1004	905-105
1004	2360401
1004	5351361
1004	2002100022
1004	4206002400
1004	4206004400
1005	A3300
1005	BA33X8356
1005	GEIC-42
1005	HC1000403
1005	LA3300
1005	RE336-IC
1005	SVI-LA3300
1005	09-308058
1005	051-0038-00
1005	740-5853-300
1005	740-8120-160
1005	754-5853-300
1005	2091-50
1006	740940
1006	A3301
1006	G09-009-A
1006	GEIC-38
1006	LA3301
1006	PC-20007
1006	RE326-IC
1006	REN1006
1006	RVILA3301
1006	09-308034
1006	09-308064
1006	21M532
1006	87-0004
1006	740-5903-301
1006	740-9003-301
1006	1077-2390
1006	2091-49
1006	000074010
1006	171179-051
1006	172272
1006	741098
1006	916072
1006	16233010
1006	9020090
1009	A4030P
1009	GEIC-77
1009	LA4030P
1009	46-1131-3
101	A121-1410
101	A121-15
101	A121-16
101	A121-17
101	A121-21
101	A121-50
101	A121-762
101	A127-7
101	A129-30
101	A13-86420-1
101	A13-87433-1
101	A1396
101	A1465-4
101	A1465-49
101	A1858
101	A198794-1
101	A2039-2
101	A2092418
101	A2092418-0711
101	A3607
101	A3609
101	A3T201
101	A3T202
101	A3T203
101	A46-8614-3
101	A4700
101	A48-124216
101	A48-124217
101	A48-124218
101	A48-124220
101	A48-124221
101	A48-125233
101	A48-125234
101	A48-125235
101	A48-125236
101	A48-128239
101	A48-134520
101	A48-134700
101	A48-134931
101	A4JD381
101	A514-023553
101	A65-4-70
101	A65-4-705
101	A65-4A90
101	A86-10-2
101	A86-44-2
101	A95115
101	A95211
101	A99307
101	A98K7
101	AA2
101	AC130
101	AC187/01
101	AF192
101	A07
101	AS3428
101	ASY74
101	ASY75
101	AST86
101	AST87
101	AST88
101	AST89
101	AT52
101	AT521
101	AT53
101	AT551
101	AT71
101	AT72
101	AT75
101	AT75R
101	AT76R
101	AT77
101	B65-4-4-21
101	B65-4A21
101	B92-1-A-21
101	C128
101	C129
101	C13(TRANSISTOR)
101	C14
101	C173
101	C175
101	C176
101	C177
101	C178
101	C36(TRANSISTOR)
101	C50(TRANSISTOR)
101	C50A
101	C60(TRANSISTOR)
101	C71
101	C72
101	C73(JAPAN)
101	C75(JAPAN)
101	C76(JAPAN)
101	C77
101	C77C
101	C78
101	C89
101	C90
101	CK261
101	CK262
101	D-P1
101	D101
101	D19
101	D20
101	D21
101	D22
101	D23
101	D44
101	DP1
101	D811
101	D812
101	DS2
101	DS4
101	DS5
101	DS6
101	DS7
101	DS8
101	DS9
101	E2427
101	E2428
101	E2429
101	E4002
101	ES5
101	ET8
101	ET9
101	G101079
101	G16506
101	GA53270
101	GC1034
101	GC1035
101	GC1036
101	GC452
101	GC453
101	GC454
101	GE-5
101	GE-6
101	GE-7
101	GI5
101	GI6
101	GI6506
101	GT1200
101	GT1202
101	GT1608
101	GT1609
101	GT167
101	GT229
101	GT2765
101	GT2766
101	GT2767
101	GT2884
101	GT2886
101	GT2888
101	GT2906
101	GT3150
101	GT792
101	GT904
101	GT905
101	GT905R
101	GT947
101	GT948
101	GT949R
101	HA5001
101	HA5002
101	HA5003
101	HA5005
101	HA5009
101	HA5011
101	HA5012
101	HA5014
101	HA5020
101	HA5021
101	HA5022
101	HA5023
101	HA5024
101	HA5025
101	HA5026
101	HC-00730
101	HD-187
101	HEP-00011
101	HEP641
101	M2N168A
101	M4700
101	M8120
101	MHT2002
101	MHT2003
101	MHT2004
101	MHT2008
101	MHT2009
101	MHT2010
101	MIS-14150-18A
101	MIS14150-18A
101	NA20
101	NK65-4A19
101	NKT734
101	NKT736
101	NKT753
101	NR-10
101	NR05
101	NR10
101	NR30
101	NR5
101	NR700
101	PTC108
101	Q-2
101	Q-3
101	Q-5
101	Q-9
101	R-125
101	R-135
101	R-136
101	R-137
101	R-1533
101	R-202
101	R-203
101	R-33
101	R-34
101	R-62
101	R-63
101	R117
101	R12
101	R125
101	R135
101	R136
101	R137
101	R14
101	R1533
101	R202
101	R203
101	R41
101	R592
101	RS-104
101	RS-1536
101	RS-1537
101	RS-1538
101	RS-1545
101	RS-1547
101	RS-1553
101	RS-2001
101	RS-2359
101	RS-2360
101	RS-2364
101	RS-2365
101	RS-2366
101	RS-2373
101	RS-2374
101	RS-2375
101	RS1513
101	RS1530
101	RS1531
101	RS1532
101	RS1534
101	RS1536
101	RS1537
101	RS1538
101	RS1547
101	RS1553
101	RS2356
101	RS2359
101	RS2360
101	RS2364
101	RS2365
101	RS2366
101	RS2375
101	RS3306
101	S028
101	S65-4-A-3P
101	SA-7
101	SA7
101	SF.T184
101	SFT-298
101	SFT259
101	SFT260
101	SFT261
101	SFT298
101	SK-7
101	SK3011
101	SK7
101	S860
101	SQ7
101	ST-172
101	ST172
101	STL-101
101	STL-102
101	STL-1297
101	STL-1310
101	STL-1311
101	STL-1987
101	STL-2130
101	STL-2131
101	STL-2132
101	STL-2245
101	STL-2246
101	STL101
101	STL102
101	STL1279
101	STL1310
101	STL1311
101	STL1312
101	STL1313
101	STL1326
101	STL1327
101	STL1380
101	STL1408
101	STL1454
101	STL1537
101	STL1591
101	STL1617
101	STL1750
101	STL1941
101	STL1987
101	STL2130
101	STL2131
101	STL2132
101	STL2245
101	STL2246
101	STL4339
101	T59276
101	T59277
101	TA-1759
101	TA-1767
101	TA-1771
101	TA-1772
101	TA1759
101	TA1767
101	TA1771
101	TA1772
101	TP70
101	TP71
101	TP72
101	TIX896
101	TK33C
101	TM101
101	TN616671
101	TP4274
101	TQ5031
101	TQ5032
101	TQ5039
101	TR-08
101	TR-08C
101	TR-09
101	TR-09C
101	TR-10
101	TR-10C
101	TR-159(OLSON)
101	TR-160(OLSON)
101	TR04
101	TR07
101	TR08
101	TR08C
101	TR09C
101	TR10
101	TR10C
101	TR167
101	TR182
101	TR183
101	TR184
101	TR193
101	TR194
101	TR211
101	TR212
101	TR213
101	TR214
101	TR216
101	TR335
101	TR336
101	TR337
101	TY828C647A
101	W2
101	WTV-L6
101	WTVSA7
101	WTVSK7
101	WTVSQ7
101	XA701
101	XA702
101	XA703
101	ZEN315
101	001-01101-0
101	001-011010
101	20339A
101	2M78
101	2N100
101	2N1000
101	2N1012
101	2N102
101	2N103
101	2N105B
101	2N1086
101	2N1086A
101	2N1087
101	2N1090
101	2N1091
101	2N1102/5
101	2N1112
101	2N1114
101	2N1121
101	2N1198
101	2N1217
101	2N124
101	2N125

ECG	Industry Standard No.	ECG	Industry Standard No.	ECG	Industry Standard No.	ECG	Industry Standard No.	ECG	Industry Standard No.
101	2N12b	101	28D-F1	101	998K5	101	62050217	102	MA113
101	2N127	101	28D101	101	998K7	101	62087617	102	MA114
101	2N1288	101	28D19	101	998Q7	1010	G8IC-37	102	MA115
101	2N1289	101	28D20	101	120-004888	1010	LA4031P	102	MA116
101	2N1299	101	28D21	101	121-100	1011	G8IC-79	102	MA117
101	2N1302	101	28D22	101	121-1410	1011	LA4032P	102	MA1700
101	2N1304	101	28D23	101	121-15	102	A174	102	MA1703
101	2N1306	101	28D44	101	121-16	102	A397	102	MA1704
101	2N1308	101	28DP1	101	121-17	102	A398	102	MA1705
101	2N1310	101	2T513	101	121-21	102	A399	102	MA1706
101	2N1311	101	2T52	101	121-22	102	A42X00286-01	102	MA1707
101	2N1366	101	2T520	101	121-24	102	ACY40	102	MA1708
101	2N1367	101	2T521	101	121-25	102	ACY41	102	MA894
101	2N145	101	2T524	101	121-26	102	AFB-11H-1008	102	MA895
101	2N146	101	2T53	101	121-302	102	B100	102	MA896
101	2N147	101	2T54	101	121-50	102	B101	102	MA897
101	2N1473	101	2T55	101	121-51	102	B1022	102	MA898
101	2N148	101	2T551	101	121-70	102	B103	102	MA899
101	2N150	101	2T56	101	121-71	102	B1058	102	MA900
101	2N150A	101	2T57	101	121-762	102	B1154	102	NKT133
101	2N1510	101	2T58	101	124-N16	102	B161	102	NKT221
101	2N1585	101	2T650	101	124N1	102	B163	102	NKT222
101	2N1605	101	2T67	101	127-7	102	B165	102	NKT223
101	2N1605A	101	2T71	101	130-40089	102	B199	102	NKT224
101	2N1622	101	2T72	101	151-0040	102	B202	102	NKT225
101	2N1624	101	2T73	101	151-0040-00	102	B218	102	NKT226
101	2N164	101	2T73R	101	151-0238	102	B219	102	NKT227
101	2N164A	101	2T74	101	165-4A82	102	B220	102	NKT228
101	2N165	101	2T75	101	394-3102-1	102	B220A	102	NKT231
101	2N166	101	2T75R	101	465-4A5	102	B221	102	NKT232
101	2N167	101	2T76	101	665-4A5L	102	B221A	102	NKT242
101	2N167A	101	2T76R	101	965-4A6-2	102	B222	102	NKT245
101	2N168	101	2T77	101	1034-43	102	B223	102	NKT247
101	2N1685	101	2T77R	101	1102-63	102	B224(JAPAN)	102	NS121
101	2N168A	101	2T78	101	1344-3767	102	B225	102	NS32
101	2N169	101	2T78R	101	1396	102	B328	102	OC32
101	2N1694	101	2T82	101	1465-4	102	B335	102	OC41
101	2N169A	101	2T83	101	1465-4-12	102	B34	102	P6460006
101	2N170	101	3R22	101	1465-4-12-8	102	B34N	102	P6460037
101	2N172	101	3N23	101	1515(PNP)	102	B350	102	PBX103
101	2N1730	101	3N23A	101	1865-4	102	B377B	102	PBX113
101	2N1732	101	3N23B	101	1865-4-12	102	B379	102	PXB103
101	2N1779	101	3N23C	101	1865-4-127	102	B379-2	102	PXB113
101	2N1780	101	3N29	101	1865-4L	102	B379A	102	PXC101
101	2N1781	101	3N30	101	1865-4L8	102	B379B	102	PXC101A
101	2N1783	101	3N31	101	2039-2	102	B38	102	PXC101AB
101	2N1808	101	3N36	101	3607	102	B380	102	Q-1
101	2N184	101	3N37	101	3609	102	B380A	102	Q-4
101	2N1891	101	3T201	101	4700	102	B396	102	QN2613
101	2N1994	101	3T202	101	38199	102	B422	102	R2SB492
101	2N1995	101	3T203	101	38200	102	B423	102	R46
101	2N1996	101	4-65-4A7-1	101	43992-2	102	B48	102	RS-2004
101	2N2085	101	4JD3B1	101	45495-2	102	B486	102	RS-2005
101	2N2426	101	4JX2A801	101	46490-2	102	B52	102	RS3276
101	2N253	101	6-89X	101	46631-2	102	B53	102	RS5825
101	2N254	101	9-5112	101	46774-1	102	B72	102	RS6843
101	2N2699	101	9-5113	101	46775-2	102	B98	102	RS6846
101	2N28	101	9-5114	101	49138-2	102	CK721	102	S413796
101	2N29	101	12AA2	101	61012-4-1	102	CK722	102	SB-5-0819
101	2N292	101	13-86420-1	101	70598-1	102	CK725	102	SF.T222
101	2N292A	101	13-87433-1	101	81502-6	102	CK727	102	SF.T227
101	2N293	101	19A115103-P1	101	81502-6A	102	CK751	102	SF.T318
101	2N312	101	19A115201-P1	101	81502-6B	102	CK754	102	SFT221
101	2N313	101	19A115201-P2	101	81502-6C	102	CK790	102	SFT222
101	2N314	101	19A115546-P1	101	81502-6D	102	CK791	102	SMB447610
101	2N35	101	19A115546-P2	101	91021	102	CK793	102	SMB447610A
101	2N356	101	19A115673-P1	101	95112	102	CK794	102	SMB454549
101	2N357	101	19A115673-P2	101	95113	102	CK871	102	SMB621960
101	2N358	101	19B200065-P1	101	95114	102	CK872	102	SYL107
101	2N377	101	19B200065-P2	101	95115	102	CK875	102	SYL108
101	2N377A	101	2TT408	101	95117	102	CK878	102	SYL1430
101	2N385	101	46-8614-3	101	95211	102	CK879	102	SYL1535
101	2N385A	101	48-124216	101	101678	102	CK882	102	SYL1583
101	2N388	101	48-124217	101	103443	102	CK888	102	SYL1655
101	2N388A	101	48-124218	101	122061	102	DE-3181	102	SYL1665
101	2N440	101	48-124220	101	122111	102	DRC-81252	102	SYL1668
101	2N444	101	48-124221	101	122112	102	EE100	102	SYL2248
101	2N445	101	48-125233	101	198794-1	102	ET3	102	SYL2249
101	2N446	101	48-125234	101	219016	102	ET4	102	SYL2300
101	2N447	101	48-125235	101	221601	102	ET5	102	SYL3615
101	2N448	101	48-125236	101	221924	102	FBN-CP2293	102	T-2122
101	2N449	101	48-128239	101	223367	102	FD-1029-BG	102	T50931B
101	2N556	101	48-134520	101	223368	102	FD-1029-EE	102	T51573A
101	2N557	101	48-134700	101	223370	102	FD1029EE	102	TIAO4
101	2N558	101	48-134931	101	223684	102	G0006	102	TK40
101	2N576	101	65-4	101	226441	102	GE-2	102	TK40C
101	2N576A	101	65-4-70	101	230209	102	GET898	102	TK41
101	2N587	101	65-4-70-12	101	230256	102	GP20	102	TK42
101	2N594	101	65-4-70-12-7	101	232949	102	GP21	102	TK45C
101	2N595	101	65-40	101	236265	102	GP32	102	TM102
101	2N596	101	65-41	101	256127	102	GPT3008/40	102	TNJ70634
101	2N634	101	65-42	101	257385	102	GI2	102	TNJ70635
101	2N634A	101	65-43	101	258993	102	GI4	102	TQ8A0-222
101	2N635	101	65-44	101	260468	102	GT122	102	TR-045
101	2N635A	101	65-45	101	270781	102	GT123	102	TR-070
101	2N636	101	65-46	101	300486	102	GT20	102	TR-071
101	2N636A	101	65-47	101	300536	102	GT222	102	TR-072
101	2N679	101	65-48	101	300542	102	GT2693	102	TR310225
101	2N78	101	65-49	101	300774	102	GT2695	102	TR320
101	2N78A	101	65-4A	101	581024	102	GT2696	102	TR320A
101	2N821	101	65-4A0	101	650860	102	GT2883	102	TR321
101	2N822	101	65-4AOR	101	723000-18	102	GT2885	102	TR323
101	2N823	101	65-4A1	101	723001-19	102	GT2887	102	TR323A
101	2N824	101	65-4A19	101	815026	102	GT34	102	TR332
101	2N94A	101	65-4A2	101	815026A	102	GT74	102	TR383
101	2N955	101	65-4A21	101	815026B	102	GT75	102	TR383(HGFH-2)
101	2N955A	101	65-4A3	101	815026C	102	GT758	102	TR508
101	2N97A	101	65-4A3P	101	815026D	102	GT759	102	TR508A
101	2N98	101	65-4A4	101	2092418-0711	102	GT81	102	TR5R26
101	2N98A	101	65-4A4-7	101	4036749-P1	102	GT82	102	TR650
101	2N99	101	65-4A4-7B	101	4037289-P1	102	HEP252	102	TR653
101	2SC128	101	65-4A5	101	4037289-P2	102	HEP629	102	TR6R26
101	2SC129	101	65-4A5L	101	4037839-P1	102	HEP630	102	TR721
101	2SC13	101	65-4A6	101	4037839-P2	102	HEP631	102	TR722
101	2SC14	101	65-4A6-2	101	4038264-P1	102	HEP632	102	TR763
101	2SC173	101	65-4A7	101	4038264-P2	102	HEP633	102	TR8007
101	2SC175	101	65-4A7-1	101	5492653-P1	102	HS-15	102	TRC70
101	2SC175B	101	65-4A8	101	5492653-P2	102	HS-22D	102	TRC71
101	2SC176	101	65-4A82	101	5492655-P1	102	HS170	102	TRC72
101	2SC177	101	65-4A9	101	5492655-P2	102	HS23D	102	TS-164
101	2SC178	101	65-4A9Q	101	5492655-P3	102	HS29D	102	TS13
101	2SC050	101	86-10-2	101	5492655-P5	102	M-P3D	102	TS14
101	2SC050A	101	86-11-2	101	5492655-P6	102	M4327	102	TS15
101	2SC71	101	86-12-2	101	5492659-P2	102	M4482	102	TV-65
101	2SC72	101	86-26-2	101	8510744-1	102	M4483	102	TV4825B172
101	2SC73	101	86-31-2	101	8512001-2	102	M4567	102	TVSSA71B
101	2SC75	101	86-4-2	101	8521502-1	102	M4596	102	UPI1303
101	2SC76	101	86-44-2	101	8521502-2	102	M4597	102	UPI1305
101	2SC77	101	95-112	101	8521502-4	102	M4597GRN	102	UPI1307
101	2SC77C	101	95-113	101	8524402-1	102	M4597RED	102	UPI1309
101	2SC78	101	95-114	101	8524402-4	102	M4607	102	V10/15A
101	2SC89	101	97N2	101	8975103-2	102	M4627	102	V10/30A
101	2SC90	101	99K7			102	M9148	102	V10/50A
101	2SC91	101	99807			102	MA112	102	VS2SA385

ECG	Industry Standard No.	ECG	Industry Standard No.	ECG	Industry Standard No.	ECG	Industry Standard No.	ECG	Industry Standard No.
102	VS28B171	102	2N159	102	2N60R	102	6D0000105	102	48-869282
102	VS28B172	102	2N1681	102	2N610	102	8-0062	102	48-869475
102	VS28B176	102	2N1705	102	2N611	102	12A7239P1	102	48-869475A
102	VS28B178	102	2N1706	102	2N612	102	12A9275	102	48-97046A53
102	VS28B178A	102	2N1707	102	2N613	102	12A9275-1	102	48-97127A32
102	WRT1114	102	2N189	102	2N61A	102	13-14888-3	102	48A124327
102	WTVB6A	102	2N1892	102	2N61B	102	13-15805-1	102	48K134482
102	XB1	102	2N190	102	2N61C	102	13-15836-1	102	48K134483
102	XB2	102	2N191	102	2N631	102	16A787	102	48B869250
102	XB3	102	2N192	102	2N632	102	19-020-015	102	48X97238A05
102	XB3B	102	2N195	102	2N633	102	19A115077-P1	102	51D170
102	XB3C	102	2N196	102	2N633B	102	19A115077-P2	102	53P15T
102	XJ13	102	2N197	102	2N65	102	19A115281-P1	102	57A1-11
102	ZEN303	102	2N198	102	2N650	102	19C300073-P1	102	57A126
102	ZEN306	102	2N199	102	2N650A	102	19C300073-P2	102	57A143
102	001-012011	102	2N200	102	2N651	102	19C300073-P3	102	57A169
102	001-01206-0	102	2N2001	102	2N651A	102	19C300073-P4	102	57A170
102	001-012060	102	2N204	102	2N652	102	19C300073-P5	102	57A189
102	2D013	102	2N205	102	2N652A	102	19C300073-P6	102	57A6(PNP)
102	2D013-109	102	2N207	102	2N653	102	19C300074-P2	102	57A6-1
102	2D013-13	102	2N207A	102	2N654	102	19C300128-P1	102	57A6-22
102	2D013-160	102	2N207B	102	2N655	102	19C300128-P2	102	57B6
102	2D013-54	102	2N207BLU	102	2N655GRN	102	19C300128-P3	102	57C6-1
102	2D016	102	2N2209	102	2N655RED	102	19C300128-P4	102	57C6-22
102	2D016-45	102	2N2271	102	2N71	102	19C300128-P5	102	57D1-11
102	2D016-54	102	2N23	102	2N73	102	19C300128-P6	102	57D1127
102	2D023	102	2N237	102	2N74	102	19C300128-P7	102	57D1143
102	201027	102	2N2374	102	2N76	102	19C300128-P8	102	57D126
102	2G108	102	2N2375	102	2N79	102	19C300138-P4	102	57D143
102	2G270	102	2N2376	102	2N799	102	19C300138-P8	102	57D156
102	2G271	102	2N24	102	2N80	102	21-34	102	57D169
102	2G302	102	2N2468	102	2N81	102	21-36	102	57D170
102	2G306	102	2N2469	102	2N82	102	21-37	102	57D189
102	2G319	102	2N25	102	2N825	102	22-002006	102	57D3-6
102	2G320	102	2N2564	102	2N826	102	022-3504-040	102	57D68
102	2G321	102	2N2565	102	2N96	102	022-3504-060	102	57P1
102	2G322	102	2N26	102	2N9A	102	022-3505-910	102	610002-1
102	2G323	102	2N265	102	2N9D	102	022-3504-040	102	610003-1
102	2G324	102	2N266	102	2N8121	102	022-3504-060	102	63P3
102	2G384	102	2N270	102	2N831	102	022-3505-910	102	86-156-2A
102	2G385	102	2N270-5M	102	2N832	102	24MW107	102	86X00011-001
102	2G386	102	2N270A	102	2814	102	24MW115	102	93A9
102	2G387	102	2N282	102	2815	102	24MW116	102	95-11414000
102	2G395	102	2N2930	102	2815A	102	24MW132	102	96X26053-09N
102	2G508	102	2N302	102	2816	102	24MW15	102	96X26053-27N
102	2G509	102	2N303	102	28179	102	24MW16	102	97P1
102	2N1127	102	2N310	102	2822	102	24MW263	102	99A76
102	2N1008	102	2N315A	102	2824	102	24MW29	102	99B5
102	2N1008A	102	2N315B	102	2832	102	24MW43	102	100B63
102	2N1008B	102	2N316A	102	2833	102	24MW60	102	112-003
102	2N1009	102	2N319	102	2834	102	24MW69	102	112-004
102	2N1045	102	2N32	102	2837	102	24MW70	102	120-00419
102	2N1056	102	2N321	102	2838	102	24MW78	102	120-001195
102	2N1057	102	2N322	102	2839	102	24MW83	102	120-001795
102	2N109/5	102	2N323	102	2840	102	24MW84	102	121-1032
102	2N1093	102	2N324	102	2843	102	25T1	102	121-1034
102	2N1097	102	2N32A	102	2844	102	026-100005	102	121-106
102	2N1098	102	2N3427	102	2846	102	026-100012	102	121-107
102	2N111	102	2N3428	102	2847	102	026-100018	102	121-266
102	2N111A	102	2N359	102	2854	102	26T1	102	121-267
102	2N124	102	2N360	102	2856	102	27T403	102	121-27
102	2N125	102	2N361	102	28A174	102	27T404	102	121-372
102	2N126	102	2N362	102	28A396	102	27T405	102	121-734
102	2N112A	102	2N362B	102	28A397	102	29V0038H03	102	121-95
102	2N144	102	2N363	102	28A398	102	31-0247	102	121-96
102	2N145	102	2N367	102	28A399	102	31-025	102	129-17
102	2N174	102	2N368	102	28B100	102	33-1000-00	102	129-18
102	2N175	102	2N369	102	28B101	102	33-1001-00	102	154A3675-105
102	2N175A	102	2N381	102	28B103	102	35P1	102	154A3676-205
102	2N189	102	2N382	102	28B161	102	35P2	102	154A3679
102	2N190	102	2N383	102	28B163	102	35P2C	102	154A8681
102	2N191	102	2N39	102	28B165	102	35P1	102	173A9970
102	2N192	102	2N40	102	28B199	102	36P1	102	231-0009
102	2N193	102	2N402	102	28B218	102	36P1C	102	297V003H03
102	2N194	102	2N404A	102	28B219	102	36P1F	102	297V003M01
102	2N1265	102	2N41	102	28B220	102	36P2F	102	297V003M07
102	2N1265A	102	2N414B	102	28B220A	102	36P3	102	297V004H10
102	2N273	102	2N414C	102	28B221	102	36P3A	102	297V004H14
102	2N1273BLU	102	2N42	102	28B221A	102	36P3C	102	297V004H15
102	2N1273GRN	102	2N422	102	28B222	102	36P4	102	297V004H16
102	2N1273ORN	102	2N422A	102	28B223	102	36P4C	102	297V004M01
102	2N1273RED	102	2N425	102	28B225	102	36P5	102	297V021H14
102	2N1273YEL	102	2N43	102	28B328	102	36P5C	102	297V025H02
102	2N1274	102	2N43A	102	28B335	102	36T1	102	297V025H04
102	2N1274BLU	102	2N44	102	2834	102	38T1	102	297V025H05
102	2N1274BRN	102	2N46	102	28B34N	102	39A9	102	297V025H15
102	2N1274ORN	102	2N460	102	28B350	102	39T1	102	297V027H01
102	2N1274GRN	102	2N461	102	28B377	102	40-601	102	297V033H01
102	2N1274PUR	102	2N462	102	28B377B	102	40P1	102	297V033H02
102	2N1274RED	102	2N464	102	28B379	102	40P2	102	297V037H01
102	2N1287	102	2N465	102	28B379-2	102	42-17143	102	297V037H02
102	2N1287A	102	2N466	102	28B379A	102	42X233	102	297V038H07
102	2N130	102	2N467	102	28B379B	102	42X309	102	297V040H01
102	2N1303	102	2N468	102	2838	102	42X311	102	297V040H08
102	2N1305	102	2N47	102	28B380	102	43P2	102	297V040H10
102	2N130A	102	2N48	102	28B380A	102	43P4	102	297V040H11
102	2N131	102	2N483-6M	102	28B396	102	43P4C	102	297V040H12
102	2N131A	102	2N483B	102	28B422	102	43P6	102	297V040H13
102	2N132	102	2N48A	102	28B423	102	43P6A	102	297V040H16
102	2N133	102	2N49	102	2848	102	43P7	102	297V042C01
102	2N1348	102	2N50	102	2849	102	43P7A	102	297V042C02
102	2N1352	102	2N508	102	2850	102	43P7C	102	297V042C03
102	2N1353	102	2N508A	102	2851	102	45X1A502C	102	297V042C04
102	2N1370	102	2N51	102	28B52	102	48-10073A02	102	297V042H01
102	2N1371	102	2N52	102	28B53	102	48-10074A02	102	297V042H02
102	2N1372	102	2N524	102	28B72	102	48-124219	102	297V043H01
102	2N1373	102	2N524A	102	28B98	102	48-124286	102	297V050H03
102	2N1374	102	2N525	102	2T2001	102	48-124297	102	297V051C03
102	2N1375	102	2N525A	102	2T230	102	48-124303	102	297V051C04
102	2N1376	102	2N526	102	2T231	102	48-124304	102	297V051B01
102	2N1377	102	2N526A	102	2TN45A	102	48-124306	102	297V051H03
102	2N1378	102	2N527	102	2TN56	102	48-124309	102	297V052C01
102	2N1379	102	2N527A	102	2TN95	102	48-124318	102	297V052H02
102	2N1380	102	2N53	102	2TN95A	102	48-124319	102	297V053C01
102	2N1381	102	2N54	102	2V362	102	48-124353	102	297V076B01
102	2N1382	102	2N55	102	2V363	102	48-124354	102	297V076C01
102	2N1383	102	2N56	102	03-0020-0	102	48-124355	102	301
102	2N1392	102	2N564	102	03-0022	102	48-134573	102	302
102	2N1404	102	2N566	102	03-0023-0	102	48-37T199	102	310-188
102	2N1413	102	2N568	102	03-57-301	102	48-35P1	102	324-0041
102	2N1414	102	2N570	102	03-57-304	102	48-36P1	102	324-0088
102	2N1415	102	2N573	102	003-H03	102	48-36P3	102	324-0133
102	2N1432	102	2N573BRN	102	3B15	102	48-39P3	102	324-0139
102	2N1446	102	2N573ORN	102	3B15-1	102	48-43P3	102	324-0140
102	2N1447	102	2N573RED	102	04-57-303	102	48-43P4	102	324-0142
102	2N1448	102	2N58	102	4JD1A73	102	48-63029A18	102	324-0143
102	2N1449	102	2N59A	102	4JX1A520	102	48-63029A19	102	324-0144
102	2N1451	102	2N59B	102	4JX1A520B	102	48-63029A91	102	324-0146
102	2N1452	102	2N59C	102	4JX1A520C	102	48-63029A93	102	324-144
102	2N1478	102	2N59D	102	4JX101224	102	48-63029A94	102	325-1442-8
102	2N151	102	2N609	102	4JX1C850	102	48-869148	102	354-3052
102	2N1561	102	2N60A	102	6-0000155A	102	48-869249	102	394-3074-2
102	2N1562	102	2N60B	102	6-000105	102	48-869250		
		102	2N60C	102	6-155	102	48-869253		

ECG	Industry Standard No.
102	394-3074-5
102	394-3097-1
102	394-3097-2
102	421-13C
102	421-19
102	473B6-2
102	473B6-2A
102	473B6-4
102	473B6-5
102	473B6-7
102	650-105
102	650-106
102	650-107
102	650-108
102	660B
102	690V034H39
102	690V040H61
102	690V040H62
102	690V047H58
102	690V047H60
102	690V052H23
102	690V052H24
102	690V054H21
102	690V056H90
102	690V057H27
102	690V057H28
102	690V061H98
102	690V061H99
102	690V063H16
102	690V063H17
102	690V063H51
102	690V066H46
102	690V066H47
102	690V068H30
102	690V068H31
102	690V077H37
102	690V080H39
102	690V080H44
102	690V086H39
102	720-35019
102	750-137
102	750-138
102	750-139
102	750-140
102	792-286
102	792-287
102	792-288
102	792-289
102	792-290
102	815-181D
102	964-17142
102	965T1
102	987T1
102	988T1
102	989T1
102	990T1
102	991T1
102	1001-7663
102	1021-17
102	1023-17
102	1032
102	1033-5
102	1033-7
102	1113-2875
102	1166-7821
102	1178-3453
102	1180-0182
102	1192
102	1241A
102	1320
102	1321-7724
102	1321-7732
102	1329
102	1330
102	1344-7321
102	1360
102	1413-160
102	1413-175
102	1459
102	1510-2718
102	1850-0040
102	1850-0040-1
102	1850-0060
102	1850-0062
102	1850-0062-1
102	1850-0101
102	1850-0184
102	1850-0184-1
102	2012
102	2057A2-206
102	2057B2-32
102	2057B2-34
102	2057B2-4
102	2057B2-43
102	2057B2-44
102	2057B2-45
102	2057B2-49
102	2057B2-57
102	2057B2-72
102	2057B2-78
102	2057B2-83
102	2057B2-86
102	2061A45-47
102	2093A38-23
102	2093A41-40
102	2093A41-41
102	2781
102	3600
102	3907
102	4313
102	4315
102	4348
102	4349
102	4398
102	4450
102	4462
102	44660RN
102	4468BRN
102	4469RED
102	44710RN
102	4471YEL
102	4472GRN
102	4474YEL
102	4475GRN
102	4476BLU
102	4477PUR
102	4553BLU
102	4553BRN
102	4553GRN
102	45530RN
102	4553RED
102	4553YEL
102	4562
102	4563
102	4564
102	4565
102	4596
102	4607
102	4627
102	5052
102	5766-25
102	6440
102	6452
102	7239
102	9390
102	9391
102	16598
102	16958
102	17047-1
102	17142
102	17143
102	18109
102	18530
102	18601
102	18611
102	18731
102	27910-12153
102	30204
102	30206
102	30207
102	30208-2
102	30216
102	30218
102	30244
102	34493P
102	34871
102	35045
102	35086
102	35454
102	35820
102	35950
102	35952
102	35953
102	35954
102	35955
102	36534
102	36557
102	36558
102	37677
102	38177
102	40403
102	42324
102	44616-1
102	44967-2
102	46776-2
102	47394-2
102	47737-2
102	49139-2
102	64071-1
102	67193-85
102	68895-13
102	69107-45
102	75960CH
102	86452
102	86832
102	86842
102	95201
102	95203
102	95204
102	95208
102	95209
102	95212
102	95213
102	95214
102	95217
102	95218
102	95219
102	95222-1
102	95255-000
102	99217
102	99218
102	107274
102	111001
102	111011
102	111012
102	111013
102	112071
102	177105
102	300008
102	537790
102	581005
102	581042
102	610036-4
102	610036-6
102	610036-7
102	650196
102	650859-1
102	650859-2
102	650859-3
102	651012
102	651236
102	801500
102	801501
102	801509
102	801510
102	801511
102	802560
102	815033
102	815228A01
102	982820
102	984746
102	985217
102	992289
102	2047102
102	3130011
102	3130025
102	3130060
102	3404114-1
102	3404114-2
102	5496665-P1
102	5496665-P2
102	5496665-P3
102	5496665-P4
102	5496665-P5
102	5496665-P6
102	5496666-P1
102	5496666-P2
102	5496666-P3
102	5496666-P4
102	5496666-P5
102	5496666-P6
102	5496666-P7
102	5496666-P8
102	5496667-P1
102	5496667-P2
102	5496774-P1
102	5496774-P2
102	5496774-P3
102	5496774-P4
102	5496774-P6
102	5496839-P1
102	6460006
102	6460037
102	7210027-003
102	7210036-001
102	8516861-1
102	8516986
102	8516986-1
102	8516986-2
102	8524440-2
102	8945077-12
102	10017663
102	11132875
102	11667821
102	11783453
102	11800182
102	11858909
102	13217724
102	13217732
102	13443767
102	13447321
102	15102718
102	54967774-P5
102	62052570
102	62054336
102	62078502
102	62088502
102	62380020
102	62713755
102	2791012153
102$	A417
102$	A446
102$	2SA417
102$	2SA446
1022	15-31015-1
1024	GE-720
1024	PC-20006
1024	STK-011
1024	24MW997
1024	16740135809
1025	GE-722
1025	STK-020
1025	STK-020B
1025	STK-020D
1025	STK-020E
1025	STK-020F
1025	STK-020K
1025	STK-020L
1027	GE-721
1027	PC-20005
1027	STK-015
1027	06008
1027	415200050
1028	GE-724
1028	STK-056
1029	GEIC-162
1029	HA1306
1029	000HA1306U
1029	HA1306W
102A	A-514-027662
102A	A059-106
102A	A059-107
102A	A059-116
102A	A061-114
102A	A061-115
102A	A065-106
102A	A069-105
102A	A115(JAPAN)
102A	A116(JAPAN)
102A	A12
102A	A12A
102A	A12B
102A	A12C
102A	A12D
102A	A12H
102A	A12V
102A	A13
102A	A138(JAPAN)
102A	A14-1004
102A	A14-1005
102A	A14-1006
102A	A14-1007
102A	A14-1008
102A	A14-1009
102A	A14-1010
102A	A141
102A	A142
102A	A143
102A	A1466-1
102A	A1466-19
102A	A1466-19-
102A	A1466-2
102A	A1466-29
102A	A1466-3
102A	A1466-39
102A	A148P-2
102A	A148P-29
102A	A148P2
102A	A148P2-29
102A	A149L-4
102A	A149L-49
102A	A15(TRANSISTOR)
102A	A15-1004
102A	A15-1005
102A	A15BK
102A	A15BL
102A	A15BLU
102A	A15H
102A	A15K
102A	A15R
102A	A15V
102A	A15VR
102A	A15Y
102A	A16
102A	A17
102A	A173
102A	A173B
102A	A17H
102A	A182
102A	A187TV
102A	A18H
102A	A198
102A	A203
102A	A203(PNP)
102A	A203AA
102A	A203B
102A	A203P
102A	A248
102A	A25-1004
102A	A25-1005
102A	A25-1006
102A	A282
102A	A2SA666PQR
102A	A30
102A	A302
102A	A303
102A	A31
102A	A311
102A	A312
102A	A32
102A	A321
102A	A322
102A	A33
102A	A35
102A	A36
102A	A374
102A	A40
102A	A406
102A	A407
102A	A49
102A	A50(TRANSISTOR)
102A	A52
102A	A53
102A	A538
102A	A595
102A	A615-1010
102A	A615-1011
102A	A64
102A	A65
102A	A66
102A	A66-1-70
102A	A66-1-705
102A	A66-1A9G
102A	A66-2-70
102A	A66-2-705
102A	A66-2A9G
102A	A66-3-3A9G
102A	A66-3-70
102A	A66-3-705
102A	A66-3A9G
102A	A67
102A	A78
102A	A78B
102A	A78C
102A	A78D
102A	A79
102A	A8P-2-70
102A	A8P-2-705
102A	A8P-2A9G
102A	A909-1011
102A	A909-1012
102A	A909-1013
102A	A9L-4-70
102A	A9L-4-705
102A	A9L-4A9G
102A	AC-107
102A	AC-113.
102A	AC-113A
102A	AC-114
102A	AC-116
102A	AC-117
102A	AC-117A
102A	AC-117P
102A	AC-121IV
102A	AC-123
102A	AC-125
102A	AC-126
102A	AC-128
102A	AC-132
102A	AC-150
102A	AC-151
102A	AC-152
102A	AC-154
102A	AC-155
102A	AC-156
102A	AC-161
102A	AC-162
102A	AC-165
102A	AC-166
102A	AC-167
102A	AC-168
102A	AC-169
102A	AC-N7B
102A	AC105
102A	AC106
102A	AC108
102A	AC109
102A	AC110
102A	AC113
102A	AC114
102A	AC115
102A	AC116
102A	AC117
102A	AC117A
102A	AC117B
102A	AC117P
102A	AC118
102A	AC119
102A	AC120
102A	AC121
102A	AC121-IV
102A	AC121-V
102A	AC121-VI
102A	AC121-VII
102A	AC121V
102A	AC121VI
102A	AC121VII
102A	AC122
102A	AC122-30
102A	AC122GRN
102A	AC122RED
102A	AC122YEL
102A	AC123
102A	AC124
102A	AC125
102A	AC131
102A	AC132
102A	AC132-01
102A	AC133A
102A	AC134
102A	AC135
102A	AC136
102A	AC137
102A	AC150
102A	AC150GRN
102A	AC150YEL
102A	AC151
102A	AC151-IV
102A	AC151-RIV
102A	AC151-RV
102A	AC151-RVI
102A	AC151-V
102A	AC151-VI
102A	AC151-VII
102A	AC151IV
102A	AC151R
102A	AC151RIV
102A	AC151RV
102A	AC151RVI
102A	AC151V
102A	AC151VI
102A	AC151VII
102A	AC152
102A	AC152-IV
102A	AC152-V
102A	AC152-VI
102A	AC152IV
102A	AC152V
102A	AC152VI
102A	AC155
102A	AC156
102A	AC160
102A	AC160A
102A	AC160B
102A	AC160GRN
102A	AC160RED
102A	AC160YEL
102A	AC161
102A	AC162
102A	AC163
102A	AC165
102A	AC168
102A	AC170
102A	AC171
102A	AC173
102A	AC182
102A	AC184
102A	AC191
102A	AC192
102A	ACY-32
102A	ACY16
102A	ACY23VI
102A	ACY27
102A	ACY28
102A	ACY29
102A	ACY30
102A	ACY31
102A	ACY32
102A	ACY32-V
102A	ACY32-VI
102A	ACY32V
102A	ACY32VI
102A	ACY34
102A	ACY35
102A	ACY36
102A	ACY38
102A	ACY41-1
102A	ACY44
102A	ACY44-1
102A	AE-50
102A	AF187
102A	AF188
102A	APB-11A-1008
102A	APB-11H-1010
102A	AR-102
102A	AR-103
102A	AR-104
102A	AS33867
102A	AS33868
102A	ASY90
102A	ASY91
102A	AT-50
102A	AT-6A
102A	AT100H
102A	AT100M
102A	AT100N
102A	AT10H
102A	AT10M
102A	AT10N
102A	AT20H
102A	AT20M
102A	AT20N
102A	AT30H
102A	AT30M
102A	AT30N
102A	AT50
102A	AT6
102A	AT6A
102A	AT74
102A	AT74S
102A	AT874
102A	ATAF1
102A	ATAF2
102A	ATGP
102A	B-1058
102A	B-22-3
102A	B-22-4
102A	B-23
102A	B-23-1
102A	B-23-2
102A	B-24-1
102A	B-26
102A	B-26-1
102A	B-F1A
102A	B1022-1
102A	B1058-1
102A	B111K
102A	B112
102A	B113
102A	B114
102A	B115
102A	B1154-1
102A	B116
102A	B117
102A	B117K
102A	B120
102A	B134(JAPAN)
102A	B134-D
102A	B134-E
102A	B135
102A	B135B
102A	B135C
102A	B135E
102A	B136
102A	B136-2
102A	B136-3
102A	B136A
102A	B136B
102A	B136C
102A	B136U
102A	B153
102A	B154
102A	B155
102A	B155A
102A	B155B
102A	B156
102A	B156A
102A	B156AA

ECG	Industry Standard No.	ECG	Industry Standard No.	ECG	Industry Standard No.	ECG	Industry Standard No.	ECG	Industry Standard No.	ECG	Industry Standard No.
102A	B156AB	102A	B302	102A	B56A	102A	CTP2076-1011	102A	FPR40-1004	102A	HB-0054
102A	B156AC	102A	B303-0	102A	B56B	102A	CTP2076-1012	102A	FPR40-1005	102A	HB-00056
102A	B156B	102A	B303A	102A	B56C	102A	D-F1A	102A	FPR50-1005	102A	HB-00156
102A	B156C	102A	B303B	102A	B57	102A	D008	102A	FPR50-1006	102A	HB-00171
102A	B156D	102A	B303C	102A	B5A	102A	D018	102A	G0007	102A	HB-00172
102A	B156P	102A	B303H	102A	B60	102A	D021	102A	G004	102A	HB-00173
102A	B168	102A	B303K	102A	B601-1009	102A	D030	102A	G11	102A	HB-00175
102A	B170	102A	B315	102A	B60A	102A	D031	102A	G12	102A	HB-00176
102A	B171	102A	B316	102A	B61	102A	D038	102A	G14	102A	HB-00178
102A	B171(JAPAN)	102A	B317	102A	B65	102A	D043	102A	GC1097	102A	HB-00186
102A	B171B	102A	B32	102A	B66	102A	D059	102A	GC1134	102A	HB-00187
102A	B171TB	102A	B32-0	102A	B66-1-A-21	102A	D101B	102A	GC1136	102A	HB-172
102A	B172	102A	B32-1	102A	B66-1A21	102A	D105B	102A	GC1143	102A	HB-173
102A	B172A	102A	B32-2	102A	B66-2-A-21	102A	D135	102A	GC1145	102A	HB-85
102A	B172AP	102A	B32-4	102A	B66-2A21	102A	DC-10	102A	GC1150	102A	HB156
102A	B172B	102A	B321	102A	B66-3-A-21	102A	DC-9	102A	GC1183	102A	HB156C
102A	B172C	102A	B322	102A	B66-3A21	102A	D8-14	102A	GC1184	102A	HB171
102A	B172D	102A	B323	102A	B66H	102A	D8-19	102A	GC1186	102A	HB172
102A	B172E	102A	B326	102A	B71	102A	D8-8	102A	GC1187	102A	HB175
102A	B172F	102A	B327	102A	B73	102A	D813	102A	GC1257	102A	HB176
102A	B172R	102A	B329	102A	B73A	102A	D816	102A	GC1422	102A	HB178
102A	B173	102A	B329K	102A	B73B	102A	D826	102A	GC250	102A	HB186
102A	B173A	102A	B32N	102A	B73C	102A	D829	102A	GC345	102A	HB187
102A	B173B	102A	B33	102A	B73GR	102A	D83	102A	GC408	102A	HB263
102A	B173C	102A	B33-4	102A	B75	102A	DU3	102A	GC4144	102A	HB270
102A	B173L	102A	B336	102A	B75AH	102A	DU5	102A	GC464	102A	HB32
102A	B174	102A	B33C	102A	B75B	102A	E-044A	102A	GC466	102A	HB33
102A	B175	102A	B33D	102A	B75C	102A	E-070	102A	GC5000	102A	HB459
102A	B175A	102A	B33B	102A	B75F	102A	E-2465	102A	GC520	102A	HB54
102A	B175B	102A	B33P	102A	B75H	102A	E132	102A	GC521	102A	HB55
102A	B175C	102A	B345	102A	B75LB	102A	E158	102A	GC551	102A	HB56
102A	B175E	102A	B346	102A	B76	102A	E181	102A	GC552	102A	HB75
102A	B176	102A	B346K	102A	B77	102A	E181A	102A	GC578	102A	HB75C
102A	B176-0	102A	B346Q	102A	B77(B)	102A	E181B	102A	GC579	102A	HB77
102A	B176-P	102A	B347	102A	B77A	102A	E181C	102A	GC580	102A	HB77B
102A	B176-PR	102A	B348	102A	B77AA	102A	E181D	102A	GC581	102A	HB77C
102A	B176B	102A	B348Q	102A	B77AB	102A	E241	102A	GC588	102A	HC-00176
102A	B176M	102A	B348R	102A	B77AC	102A	E24104	102A	GC639	102A	HEP-00004
102A	B176P	102A	B349	102A	B77AD	102A	E24106	102A	GC640	102A	HEP-0004
102A	B176PRC	102A	B37	102A	B77AH	102A	E241A	102A	GC680	102A	HEP250
102A	B176R	102A	B370	102A	B77AP	102A	E241B	102A	GC681	102A	HEP251
102A	B177	102A	B370A	102A	B77B	102A	E2445	102A	GC682	102A	HEP280
102A	B177(JAPAN)	102A	B370AA	102A	B77B-11	102A	E2448	102A	GC896	102A	HEP281
102A	B178	102A	B370AB	102A	B77C	102A	E2453	102A	GC864	102A	HEP634
102A	B178(JAPAN)	102A	B370AC	102A	B77D	102A	E2465	102A	GB-52		
102A	B178-0	102A	B370AHA	102A	B77H	102A	E2467	102A	GB-X9		
102A	B178-S	102A	B370AHB	102A	B77V	102A	E2476	102A	GER-A		
102A	B178A	102A	B370B	102A	B77VRED	102A	E2480	102A	GET103		
102A	B178C	102A	B370C	102A	B78	102A	E2481	102A	GET113		
102A	B178D	102A	B370D	102A	B79	102A	E2482	102A	GET113A		
102A	B178M	102A	B370P	102A	B89	102A	EA0009	102A	GET114		
102A	B178N	102A	B370PB	102A	B89A	102A	EA1081	102A	GP139		
102A	B178T	102A	B370V	102A	B89AH	102A	EA1346	102A	GP139A		
102A	B178U	102A	B371	102A	B89H	102A	EA15X164	102A	GP139B		
102A	B178V	102A	B371D	102A	B8P-2-A-21	102A	EA15X19	102A	GT109		
102A	B178X	102A	B376	102A	B8P-2A21	102A	EA15X2	102A	GT109R		
102A	B178Y	102A	B376Q	102A	B90	102A	EA15X203	102A	GT1223		
102A	B183	102A	B377	102A	B92	102A	EA15X207	102A	GT132		
102A	B184	102A	B378	102A	B94	102A	EA15X212	102A	GT14		
102A	B185	102A	B378A	102A	B95	102A	EA15X23	102A	GT14H		
102A	B185(0)	102A	B37A	102A	B97	102A	EA15X25	102A	GT1665		
102A	B185(O)	102A	B37B	102A	B9L-4-A-21	102A	EA15X28	102A	GT18		
102A	B185AA	102A	B37C	102A	B9L-4A21	102A	EA15X3	102A	GT20H		
102A	B185P	102A	B37E	102A	BCM1002-4	102A	EA15X36	102A	GT20R		
102A	B185P	102A	B37F	102A	BCM1002-5	102A	EA15X4	102A	GT31		
102A	B186	102A	B381	102A	BP1	102A	EA15X67	102A	GT32		
102A	B186(0)	102A	B382	102A	BP1A	102A	EA15X6840	102A	GT33		
102A	B186-1	102A	B383	102A	BP3	102A	EA15X7	102A	GT34HV		
102A	B186-K	102A	B383-1	102A	BZX070	102A	EA15X8	102A	GT40		
102A	B186A	102A	B383-2	102A	C10227	102A	EA15X8442	102A	GT41		
102A	B186AQ	102A	B384	102A	C10230-3	102A	EA15X8444	102A	GT42		
102A	B186B	102A	B385	102A	C11021	102A	EA2134	102A	GT43		
102A	B186BY	102A	B386	102A	C175B	102A	EA2135	102A	GT44		
102A	B186G	102A	B387	102A	C202	102A	EA2176	102A	GT45		
102A	B186H	102A	B389	102A	C58	102A	EB0001	102A	GT46		
102A	B186L	102A	B39	102A	CA2D2	102A	EB0003	102A	GT47		
102A	B186Y	102A	B40	102A	CB161	102A	ED55	102A	GT751		
102A	B187	102A	B400	102A	CB248	102A	ED56	102A	GT759R		
102A	B187AA	102A	B4004B	102A	CB249	102A	ED57	102A	GT760R		
102A	B187B	102A	B400A	102A	CB0360/7839	102A	EK136	102A	GT763		
102A	B187C	102A	B400B	102A	CB0361/7839	102A	E044A	102A	GT81H8		
102A	B187D	102A	B400K	102A	CB0362/7839	102A	EQG-15	102A	GT81R		
102A	B187G	102A	B408	102A	CB213811	102A	EQ-9	102A	GT33231		
102A	B187K	102A	B427	102A	CGB-52	102A	ER15X17	102A	GT33232		
102A	B187R	102A	B428	102A	CJ5204	102A	ER15X18	102A	H10		
102A	B187RED	102A	B43	102A	CK13	102A	ER15X22	102A	HA1360		
102A	B187S	102A	B439	102A	CK13A	102A	ER15X23	102A	HA54		
102A	B187Y	102A	B439A	102A	CK22	102A	ER15X7	102A	HA56		
102A	B187YBL	102A	B43A	102A	CK22A	102A	ER15X9	102A	HA4-1		
102A	B188	102A	B44	102A	CK22B	102A	ES15X100				
102A	B200	102A	B440	102A	CK22C	102A	ES15X31				
102A	B200A	102A	B443	102A	CK64	102A	ES15X32				
102A	B201	102A	B443A	102A	CK64A	102A	ES15X4				
102A	B22	102A	B443B	102A	CK64B	102A	ES15X49				
102A	B22-3	102A	B459	102A	CK64C	102A	ES15X50				
102A	B22-4	102A	B459-0	102A	CK65	102A	ES15X53				
102A	B224	102A	B459A	102A	CK65A	102A	ES15X55				
102A	B225	102A	B459B	102A	CK65B	102A	ES15X63				
102A	B226	102A	B459C	102A	CK65C	102A	ES15X72				
102A	B227	102A	B459D	102A	CK66	102A	ES15X8				
102A	B22A	102A	B46	102A	CK66A	102A	ES17(ELCOM)				
102A	B22B	102A	B460	102A	CK66B	102A	ES19				
102A	B22L	102A	B460A	102A	CK66C	102A	ES2(ELCOM)				
102A	B22R	102A	B460B	102A	CK67	102A	ES23				
102A	B22Y	102A	B47	102A	CK67A	102A	ES26(ELCOM)				
102A	B23	102A	B470	102A	CK67B	102A	ES3(ELCOM)				
102A	B23-1	102A	B475	102A	CK67C	102A	ES3120				
102A	B23-2	102A	B475A	102A	CK891	102A	ES3121				
102A	B24	102A	B475B	102A	CK892	102A	ES3122				
102A	B24-1	102A	B475D	102A	CQ1	102A	ES3123				
102A	B257	102A	B475E	102A	CT1009	102A	ES3124				
102A	B261	102A	B475F	102A	CTP-2001-1001	102A	ES3125				
102A	B262	102A	B475G	102A	CTP-2001-1002	102A	ES3126				
102A	B263	102A	B475P	102A	CTP-2001-1003	102A	ET15111				
102A	B264	102A	B475Q	102A	CTP-2001-1004	102A	ET15X1				
102A	B265	102A	B482	102A	CTP-2001-1009	102A	ET15X25				
102A	B266	102A	B49	102A	CTP-2006-1001	102A	ET15X31				
102A	B266P	102A	B496	102A	CTP-2006-1002	102A	ET15X32				
102A	B266Q	102A	B497	102A	CTP-2006-1003	102A	ETTB-2SB176				
102A	B267	102A	B498	102A	CTP1032	102A	ETTB-2SB176A				
102A	B268	102A	B5	102A	CTP1033	102A	ETTB-2SB176B				
102A	B269	102A	B516C	102A	CTP1034	102A	ETTB-2SB176R				
102A	B270	102A	B516CD	102A	CTP1035	102A	ETTB-75LB				
102A	B270A	102A	B516D	102A	CTP1036	102A	EX15X25				
102A	B270B	102A	B516P	102A	CTP2076-1001	102A	F20-1006				
102A	B270C	102A	B534	102A	CTP2076-1002	102A	F20-1007				
102A	B270D	102A	B534A	102A	CTP2076-1003	102A	F20-1008				
102A	B270E	102A	B54	102A	CTP2076-1004	102A	F20-1009				
102A	B271	102A	B54B	102A	CTP2076-1005	102A	F215-1006				
102A	B272	102A	B54E	102A	CTP2076-1006	102A	F215-1007				
102A	B273	102A	B54F	102A	CTP2076-1007	102A	F215-1008				
102A	B293	102A	B54Y	102A	CTP2076-1008	102A	F215-1009				
102A	B299	102A	B55(TRANSISTOR)	102A	CTP2076-1009	102A	FB420				
		102A	B56	102A	CTP2076-1010	102A	FB421				

ECG	Industry Standard No.	BCG	Industry Standard No.	ECG	Industry Standard No.	ECG	Industry Standard No.	ECG	Industry Standard No.
102A	HJ15	102A	M75517-1	102A	OC-71A	102A	R-24-1001	102A	R8707
102A	HJ17	102A	M75517-2	102A	OC-71N	102A	R-24-1002	102A	R8883
102A	HJ17D	102A	M75517-7	102A	OC-72	102A	R-242	102A	R8884
102A	HJ22	102A	M8062A	102A	OC-73	102A	R-245	102A	R8885
102A	HJ22D	102A	M8062B	102A	OC-74	102A	R-291	102A	R8886
102A	HJ23	102A	M8062C	102A	OC-75	102A	R-28B186	102A	R8971
102A	HJ23D	102A	M8604	102A	OC-75N	102A	R-28B187	102A	R98
102A	HJ43	102A	M8604A	102A	OC-77	102A	R-28B303	102A	RCA34101
102A	HJ50	102A	M9198	102A	OC-79	102A	R-530	102A	RCA34106
102A	HJ51	102A	MA1318	102A	OC-81DD	102A	R-593	102A	RCA35517
102A	HJ54	102A	MA206	102A	OC110	102A	R-608A	102A	RCA35953
102A	HJ60	102A	MA240	102A	OC120	102A	R-64	102A	RCA35954
102A	HJ606	102A	MA393	102A	OC123	102A	R-66	102A	RCA3858
102A	HJ62	102A	MA393A	102A	OC302	102A	R-67	102A	RCA40395
102A	HJX2	102A	MA393B	102A	OC303	102A	R-83	102A	RCA40396P
102A	HM-08014	102A	MA393C	102A	OC304	102A	R06-1007	102A	R84
102A	HR-1	102A	MA393E	102A	OC304-1	102A	R06-1008	102A	R8-1192
102A	HR-2	102A	MA393G	102A	OC304-2	102A	R06-1009	102A	R8-1540
102A	HR-3	102A	MA393R	102A	OC304-3	102A	R06-1010	102A	R8-1541
102A	HR-39	102A	MA815	102A	OC304N	102A	R100-1	102A	R8-1542
102A	HR-4	102A	MA881	102A	OC305	102A	R100-8	102A	R8-1543
102A	HR-4A	102A	MA882	102A	OC305-1	102A	R100-9	102A	R8-1544
102A	HR-5	102A	MA883	102A	OC305-2	102A	R101-2	102A	R8-1546
102A	HR-6	102A	MA884	102A	OC306	102A	R101-3	102A	R8-1548
102A	HR-7	102A	MA885	102A	OC306-1	102A	R101-4	102A	R8-1555
102A	HR-7A	102A	MA886	102A	OC306-2	102A	R120	102A	R8-2350
102A	HR-8	102A	MA887	102A	OC306-3	102A	R1273	102A	R8-2351
102A	HR-8A	102A	MA888	102A	OC307	102A	R1274	102A	R8-2352
102A	HR-9	102A	MA889	102A	OC307-1	102A	R152	102A	R8-2353
102A	HR-9A	102A	MA890	102A	OC307-2	102A	R1540	102A	R8-2354
102A	HR53	102A	MA891	102A	OC307-3	102A	R1541	102A	R8-2355
102A	HR61	102A	MA892	102A	OC308	102A	R1542	102A	R8-2367
102A	HS102	102A	MA893	102A	OC309	102A	R1543	102A	R8-2675
102A	HS17D	102A	MA902	102A	OC309-1	102A	R1544	102A	R8-2677
102A	HS5	102A	MA903	102A	OC309-2	102A	R1546	102A	R8-2689
102A	HT1001510	102A	MA904	102A	OC309-3	102A	R1548	102A	R8-2697
102A	HT200540	102A	MA909	102A	OC318	102A	R1555	102A	R8-3275
102A	HT200540A	102A	MA910	102A	OC33	102A	R16	102A	R8-3276
102A	HT200541	102A	MD501	102A	OC330	102A	R164	102A	R8-3282
102A	HT200541A	102A	MD501B	102A	OC34	102A	R23-1003	102A	R8-3283
102A	HT200541B	102A	MM1151	102A	OC340	102A	R23-1004	102A	R8-3284
102A	HT200541B-0	102A	MM1152	102A	OC350	102A	R2350	102A	R8-3285
102A	HT200561	102A	MM1153	102A	OC360	102A	R2351	102A	R8-3286
102A	HT200561A	102A	MM1154	102A	OC364	102A	R2352	102A	R8-3289
102A	HT200561B	102A	MM1742	102A	OC38	102A	R2353	102A	R8-3299
102A	HT200561C	102A	MN52	102A	OC41A	102A	R2355	102A	R8-3301
102A	HT200561C-0	102A	MN53	102A	OC41N	102A	R2366	102A	R8-3308
102A	HT200751B	102A	MN53BLU	102A	OC42N	102A	R2367	102A	R8-3310
102A	HT200770B	102A	MN53GRN	102A	OC45	102A	R2373	102A	R8-3316
102A	HT200771	102A	MN53RED	102A	OC46	102A	R24-1001	102A	R8-3316-1
102A	HT200771B	102A	MN60	102A	OC47	102A	R24-1002	102A	R8-3316-2
102A	HT200771C	102A	MP1014-1	102A	OC56	102A	R24-1003	102A	R8-3318
102A	HT201721A	102A	MP1014-4	102A	OC601	102A	R24-1004	102A	R8-3904
102A	HT201721D	102A	MP1014-5	102A	OC602	102A	R242	102A	R8-3925
102A	HT201725A	102A	MP1014-6	102A	OC6028P	102A	R245	102A	R8-406
102A	HT201782A	102A	N57B2-15	102A	OC6028Q	102A	R2482-i	102A	R8-5008
102A	HT201861A	102A	N57B2-25	102A	OC603	102A	R255	102A	R8-5206
102A	HT201871L	102A	N57B2-3	102A	OC604	102A	R2675	102A	R8-5311
102A	HT203701	102A	N57B2-6	102A	OC6048P	102A	R2677	102A	R8-5476
102A	HT203701B	102A	N57B2-7	102A	OC65	102A	R2689	102A	R8-5502
102A	HV0000102	102A	NA-1114-1004	102A	OC66	102A	R2749	102A	R8-5505
102A	HV0000202	102A	NA-1114-1005	102A	OC70	102A	R2749M	102A	R8-5506
102A	HV12	102A	NA-1114-1006	102A	OC70N	102A	R289	102A	R8-5530
102A	HV16	102A	NA-1114-1007	102A	OC71	102A	R290	102A	R8-5531
102A	HV17	102A	NA-1114-1008	102A	OC711	102A	R291	102A	R8-5532
102A	HV17B	102A	NA-1114-1009	102A	OC71A	102A	R324	102A	R8-5533
102A	HV19	102A	NA-1114-1010	102A	OC71N	102A	R3275	102A	R8-5534
102A	IP-65	102A	NA-1114-1011	102A	OC72	102A	R3276	102A	R8-5535
102A	IRTR85	102A	NA1022-1007	102A	OC73	102A	R3280(RCA)	102A	R8-5536
102A	I8BP	102A	NA5018-1013	102A	OC74	102A	R3282	102A	R8-5541
102A	J241164	102A	NA5018-1014	102A	OC74N	102A	R3284	102A	R8-5542
102A	J241190	102A	NA5018-1015	102A	OC75	102A	R3286	102A	R8-5544
102A	J24626	102A	NA5018-1016	102A	OC75N	102A	R3299	102A	R8-5551
102A	J24639	102A	NAP-T-Z-10	102A	OC76	102A	R3301	102A	R8-5552
102A	J24869	102A	NAP-TZ-10	102A	OC77	102A	R3578-1	102A	R8-5553
102A	J24870	102A	NC30	102A	OC77M	102A	R3598-2	102A	R8-5554
102A	J24934	102A	N8269	102A	OC78	102A	R364	102A	R8-5555
102A	J310159	102A	NJ181B	102A	OC79	102A	R428	102A	R8-5556
102A	J310224	102A	NK66-1A19	102A	OC80	102A	R4348	102A	R8-5557
102A	J310252	102A	NK66-2A19	102A	OC81	102A	R4349	102A	R8-5558
102A	J5063	102A	NK66-3A19	102A	OC810	102A	R5051	102A	R8-5602
102A	J5064	102A	NK8P-2A19	102A	OC81D	102A	R5052	102A	R8-5704
102A	JP5063	102A	NK9Z-4A19	102A	OC81DD	102A	R5055	102A	R8-57062
102A	JP5064	102A	NKT11	102A	OC81DN	102A	R5097	102A	R8-57062
102A	JR15	102A	NKT12	102A	OF-129	102A	R5098	102A	R8-5708
102A	JR5	102A	NKT123	102A	OF129	102A	R5099	102A	R8-5708-2
102A	KD2101	102A	NKT126	102A	P1L	102A	R5100	102A	R8-5709
102A	KG81000	102A	NKT208	102A	P1L4956	102A	R5181	102A	R8-5711
102A	KR-Q0001	102A	NKT223A	102A	P30	102A	R52	102A	R8-5717
102A	KR-Q0002	102A	NKT224A	102A	PA9156	102A	R530	102A	R8-5717-3
102A	KR-Q0004	102A	NKT225A	102A	PA9157	102A	R537	102A	R8-5717-6
102A	KR-Q1010	102A	NKT226A	102A	PBE3162-1	102A	R5523	102A	R8-5720
102A	KR-Q1011	102A	NKT227A	102A	PBB3162-2	102A	R5524	102A	R8-5731
102A	KR-Q1012	102A	NKT231A	102A	PTO-139	102A	R5525	102A	R8-5733
102A	KV-1	102A	NKT232A	102A	PT530A	102A	R558(T.I.)	102A	R8-5734
102A	KV-4	102A	NKT24	102A	PTC109	102A	R56	102A	R8-5736
102A	KV1	102A	NKT244	102A	PT0139	102A	R563(T.I.)	102A	R8-5737
102A	KV2	102A	NKT246	102A	PXB-103	102A	R5708	102A	R8-5742
102A	KV4	102A	NKT247A	102A	PXB-113	102A	R579(T.I.)	102A	R8-5743
102A	L5021	102A	NKT25	102A	PXC-101	102A	R60-1004	102A	R8-5743-1
102A	L5022	102A	NKT273	102A	PXC-101AB	102A	R60-1005	102A	R8-5743-2
102A	L5022A	102A	NKT278	102A	Q-16	102A	R60-1006	102A	R8-5743-3
102A	L5025	102A	NKT303	102A	Q-6	102A	R608	102A	R8-5744
102A	L5025A	102A	NKT308	102A	Q-7	102A	R608A	102A	R8-5744-3
102A	M108	102A	NKT32	102A	Q-8	102A	R64	102A	R8-5749
102A	M4313	102A	NKT33	102A	Q0V60526	102A	R65	102A	R8-5852
102A	M4315	102A	NKT4	102A	Q0V60528	102A	R6553	102A	R8-5854
102A	M4398	102A	NKT5	102A	Q1	102A	R66	102A	R8-686
102A	M4462	102A	NKT52	102A	Q1-7C	102A	R67	102A	R8-697
102A	M4466	102A	NKT53	102A	Q11/6515	102A	R6922	102A	R81192
102A	M4468	102A	NKT54	102A	Q12/6515	102A	R7048	102A	R81540
102A	M4469	102A	OC-307	102A	Q2-7C	102A	R7124	102A	R81541
102A	M4470	102A	OC-308	102A	Q2N2428	102A	R7127	102A	R81542
102A	M4471	102A	OC-318	102A	Q2N2613	102A	R7164	102A	R81543
102A	M4472	102A	OC-330	102A	Q2N406	102A	R7166	102A	R81544
102A	M4474	102A	OC-34	102A	Q35218	102A	R7363	102A	R81545
102A	M4475	102A	OC-340	102A	Q4	102A	R7489	102A	R81546
102A	M4476	102A	OC-341	102A	Q40263	102A	R7490	102A	R81548
102A	M4477	102A	OC-342	102A	Q7/6515	102A	R7491	102A	R81555
102A	M4553	102A	OC-343	102A	Q8/6515	102A	R7612	102A	R82003
102A	M4553BLU	102A	OC-350	102A	Q00076	102A	R7888	102A	R82350
102A	M4553BRN	102A	OC-351	102A	Q0V60526	102A	R7889	102A	R82351
102A	M4553GRN	102A	OC-360	102A	Q0V60528	102A	R8121	102A	R82352
102A	M4553RN	102A	OC-362	102A	Q0V60538	102A	R83	102A	R82353
102A	M4553RED	102A	OC-363	102A	QQ061210	102A	R8310	102A	R82354
102A	M4553YEL	102A	OC-364	102A	QQV60528	102A	R8311	102A	R82355
102A	M4565	102A	OC-38	102A	QQV60539	102A	R868	102A	R82367
102A	M4564	102A	OC-602	102A	QR2378	102A	R8687	102A	R82373
102A	M4565	102A	OC-604	102A	R-120	102A	R8688	102A	R82374
102A	M4573	102A	OC-65	102A	R-152	102A	R8695	102A	R82675
102A	M4595	102A	OC-66	102A	R-16	102A	R8697	102A	R82677
102A	M5285	102A	OC-70	102A	R-164	102A	R87	102A	R82689
102A		102A	OC-71	102A	R-23-1003	102A	R8706	102A	R82697
102A		102A		102A	R-23-1004	102A		102A	R82867

ECG	Industry Standard No.	ECG	Industry Standard No.	ECG	Industry Standard No.	ECG	Industry Standard No.	ECG	Industry Standard No.
102A	R83211	102A	RT3098	102A	SM843	102A	T1342	102A	TNJ72289
102A	R83275	102A	RT3229	102A	S0-25	102A	T1346	102A	TO-005
102A	R83280	102A	RT3230	102A	S0-88	102A	T1352	102A	TO-04
102A	R83282	102A	RT3231	102A	S025	102A	T1363	102A	TO-041
102A	R83283	102A	RT3363	102A	S065	102A	T1364	102A	TO-103
102A	R83284	102A	RT3364	102A	S088	102A	T1546	102A	TO-104
102A	R83285	102A	RT3365	102A	SR1	102A	T1559	102A	TO101
102A	R83286	102A	RT3449	102A	SS0001	102A	T1573	102A	TO102
102A	R83289	102A	RT3467	102A	SS0001A	102A	T1574	102A	TO103
102A	R83293	102A	RT3468	102A	SS0002	102A	T1577	102A	TO104
102A	R83299	102A	RT3564	102A	SS0002A	102A	T1583	102A	TQ5023
102A	R83301	102A	RT3566	102A	SS0003A	102A	T1593	102A	TQ5025
102A	R83308	102A	RT3568	102A	SS0004	102A	T1594	102A	TQ5026
102A	R83310	102A	RT4624	102A	SS0004A	102A	T1595	102A	TQ5051
102A	R83316	102A	RT5468	102A	SS0005	102A	T1596	102A	TQ5061
102A	R83316-1	102A	RT5521	102A	SS0005A	102A	T1597	102A	TR-04
102A	R83316-2	102A	RT5522	102A	ST-122	102A	T1598	102A	TR-14
102A	R83318	102A	RT5637	102A	ST-123	102A	T1599	102A	TR-14C
102A	R83717	102A	RT61015	102A	ST-301	102A	T160	102A	TR-15
102A	R83726	102A	RT61016	102A	ST-302	102A	T1740	102A	TR-15C
102A	R83857	102A	RT6205	102A	ST-304	102A	T1905	102A	TR-169(OLSON)
102A	R83866	102A	RT6604	102A	ST-332	102A	T1904	102A	TR-2R2641C
102A	R83880	102A	RT6734	102A	ST122	102A	T1961	102A	TR-320
102A	R83897	102A	RT6736	102A	ST123	102A	T2159	102A	TR-320A
102A	R83904	102A	RT6990	102A	ST28A	102A	T2439	102A	TR-321
102A	R83913	102A	RT7401	102A	ST28B	102A	T2440	102A	TR-321A
102A	R83925	102A	RT8442	102A	ST28C	102A	T2441	102A	TR-323A
102A	R83926	102A	RT8602	102A	ST301	102A	T2515	102A	TR-482A
102A	R8406	102A	RV1180	102A	ST302	102A	T2517	102A	TR-508A
102A	R85008	102A	RV1475	102A	ST303	102A	T282(SEARS)	102A	TR-5R26
102A	R85102	102A	S-1348	102A	ST304	102A	T3005	102A	TR-6R26
102A	R85103	102A	S-1349	102A	ST332	102A	T3321	102A	TR04C
102A	R85202	102A	S-1639	102A	ST370	102A	T3322	102A	TR14
102A	R85203	102A	S-95201	102A	ST37C	102A	T3321	102A	TR14C
102A	R85243-2	102A	S-95204	102A	ST37D	102A	T3323	102A	TR15
102A	R85401	102A	S025	102A	ST37E	102A	T39	102A	TR15C
102A	R85406	102A	S065A	102A	ST382	102A	T45	102A	TR2R2614C
102A	R85502	102A	S088	102A	STX0096	102A	T46	102A	TR310011
102A	R85503	102A	S1348	102A	STX0099	102A	T47	102A	TR310012
102A	R85505	102A	S1349	102A	STX0104	102A	T48	102A	TR310017
102A	R85506	102A	S1639	102A	STX0105	102A	T50	102A	TR310018
102A	R85507	102A	S2041635	102A	STX0110	102A	T50339A	102A	TR310026
102A	R85530	102A	S4248	102A	STX0114	102A	T50631	102A	TR310075
102A	R85531	102A	S66-1-A-3P	102A	STX0123	102A	T52150	102A	TR310107
102A	R85532	102A	S66-2-A-3P	102A	STX0260	102A	T52150Z	102A	TR310125
102A	R85533	102A	S66-3-A-3P	102A	STX0263	102A	T52151	102A	TR310136
102A	R85534	102A	S685	102A	STX0264	102A	T52151Z	102A	TR310149
102A	R85535	102A	S686	102A	STX0265	102A	T52159	102A	TR310153
102A	R85536	102A	S687	102A	STX0268	102A	T59	102A	TR310159
102A	R85541	102A	S8P-2-A-3P	102A	STX0269	102A	T59247	102A	TR310164
102A	R85542	102A	S95201	102A	SYL-107	102A	T59249	102A	TR310227
102A	R85543	102A	S95203	102A	SYL-108	102A	T60	102A	TR310235
102A	R85544	102A	S95204	102A	SYL-1583	102A	T61	102A	TR310251
102A	R85545	102A	S95206	102A	SYL-1668	102A	T72	102A	TR310252
102A	R85551	102A	S95207	102A	SYL107A	102A	T74	102A	TR310255
102A	R85552	102A	S95214	102A	SYL108A	102A	T77	102A	TR320AN
102A	R85553	102A	S95218	102A	SYL1583A	102A	T78	102A	TR353AN
102A	R85554	102A	S99201	102A	SYL1655A	102A	T82	102A	TR38117
102A	R85555	102A	S99203	102A	SYL1665A	102A	T83	102A	TR43
102A	R85556	102A	S99218	102A	SYL1668A	102A	T84	102A	TR44
102A	R85557	102A	S9L-4-A-3P	102A	SYL2248A	102A	T87	102A	TR45
102A	R85558	102A	SA128	102A	SYL2249A	102A	T95	102A	TR508AN
102A	R85563	102A	SA128-1	102A	SYL2300A	102A	T99	102A	TR54
102A	R85564	102A	SA15V	102A	SYL3613A	102A	TA-1575	102A	TR650A
102A	R85565	102A	SA197	102A	T-00014	102A	TA-1697	102A	TR653A
102A	R85566	102A	SA197-1	102A	T-126	102A	TA-1706	102A	TR71
102A	R85567	102A	SA197-2	102A	T-129	102A	TA-1730	102A	TR72
102A	R85568	102A	SA197-3	102A	T-130	102A	TA-4	102A	TR721A
102A	R85602	102A	SA204	102A	T-23	102A	TA1575	102A	TR722A
102A	R85603	102A	SA205BLU	102A	T-3321	102A	TA1575B	102A	TR763A
102A	R85605	102A	SA205BRN	102A	T-3322	102A	TA1655B	102A	TR81
102A	R85607	102A	SA205GRN	102A	T-3321	102A	TA1697	102A	TRA-32
102A	R85608	102A	SA205ORN	102A	T-3323	102A	TA1706	102A	TRA-33
102A	R85610	102A	SA205RED	102A	T-39	102A	TA1730	102A	TRA32
102A	R85704	102A	SA205VIO	102A	T-52150	102A	TC3123041557	102A	TRA33
102A	R85704-2	102A	SA205WHT	102A	T-52150Z	102A	TF-30	102A	TS-1
102A	R857042	102A	SA205YEL	102A	T-52151	102A	TF-65	102A	TS-1007
102A	R857062	102A	SA240	102A	T-52151Z	102A	TF-66	102A	TS-1266
102A	R85708	102A	SA33	102A	T-95	102A	TF30	102A	TS-13
102A	R85708-2	102A	SA33BRN	102A	T00014	102A	TF49	102A	TS-14
102A	R85709	102A	SA33RED	102A	T0003	102A	TF65	102A	TS-15
102A	R85711	102A	SA529	102A	T0004	102A	TF65/30	102A	TS-162
102A	R85717	102A	SA565	102A	T0005	102A	TF65/M	102A	TS-163
102A	R85717-1	102A	SA646	102A	T0012	102A	TF65/S/30	102A	TS-165
102A	R85717-3	102A	SA853	102A	T0014	102A	TF66	102A	TS-166
102A	R85717-6	102A	SB168	102A	T0015	102A	TF66/30	102A	TS-1727
102A	R85720	102A	SB169	102A	T0031	102A	TF66/60	102A	TS-1728
102A	R85731	102A	SC-12	102A	T0033	102A	TF75	102A	TS-2
102A	R85732	102A	SC-63	102A	T0038	102A	TF77	102A	TS-603
102A	R85733	102A	SC-66	102A	T0039	102A	TF80/302	102A	TS-604
102A	R85734	102A	SC-68	102A	T0040	102A	THU60U	102A	TS-616
102A	R85735	102A	SC-69	102A	T0041	102A	TIA-01	102A	TS-617
102A	R85736	102A	SC-73	102A	T0051	102A	TIA01	102A	TS-618
102A	R85737	102A	SC12	102A	T100	102A	TIX90	102A	TS-627
102A	R85738	102A	SC45	102A	T1001	102A	TK-23C	102A	TS-629
102A	R85740	102A	SC63	102A	T10010	102A	TK-40C	102A	TS-739
102A	R85740-1	102A	SC66	102A	T1002	102A	TK-41C	102A	TS-739B
102A	R85742	102A	SC68	102A	T1002A	102A	TK-42C	102A	TS-740
102A	R85743	102A	SC69	102A	T1003	102A	TK-45C	102A	TS-765
102A	R85743-1	102A	SC73	102A	T1004	102A	TK1228-1002	102A	TS1792
102A	R85743-2	102A	SP.T237	102A	T1005	102A	TK1228-1003	102A	TV-61A
102A	R85743-3	102A	SP.T251	102A	T1006	102A	TK1228-1004	102A	TV24115
102A	R85744	102A	SP.T252	102A	T1007	102A	TK1228-1005	102A	TV24154
102A	R85744-3	102A	SP.T253	102A	T1008	102A	TK1228-1006	102A	TV24156
102A	R85745	102A	SP.T306	102A	T10085	102A	TK1228-1007	102A	TV24189
102A	R85746	102A	SP.T337	102A	T1009	102A	TK23C	102A	TV24194
102A	R85747	102A	SPT-306	102A	T101	102A	TK40A	102A	TV2428
102A	R85748	102A	SPT-322	102A	T1010	102A	TK40CA	102A	TV2429
102A	R85749	102A	SPT-323	102A	T1013	102A	TK41A	102A	TV2434
102A	R85750	102A	SPT-327	102A	T1023	102A	TK42A	102A	TV24370
102A	R85751	102A	SPT-337	102A	T102A	102A	TK49C	102A	TV24599
102A	R85752	102A	SPT-352	102A	T1036	102A	TM102A	102A	TV24945
102A	R85765	102A	SPT-353	102A	T1037	102A	TN591	102A	TV24984
102A	R85766	102A	SPT151	102A	T1042	102A	TNJ-60070	102A	TVS-28A385
102A	R85767	102A	SPT152	102A	T1043	102A	TNJ-60074	102A	TVS-28A385A
102A	R85768	102A	SPT221A	102A	T1046	102A	TNJ-60079	102A	TVS-28A385L
102A	R85852	102A	SPT222A	102A	T1047	102A	TNJ-60282	102A	TVS-28A71B
102A	R85854	102A	SPT232	102A	T1076	102A	TNJ-60283	102A	TVS-28B171
102A	R86840	102A	SPT306	102A	T108	102A	TNJ-60728	102A	TVS-28B172
102A	R86843A	102A	SPT321	102A	T109	102A	TNJ60070	102A	TVS-28B172F
102A	R86846A	102A	SPT322	102A	T116	102A	TNJ60074	102A	TVS-28B176
102A	R87568	102A	SPT323	102A	T11618	102A	TNJ60079	102A	TVS-28B234
102A	R88421	102A	SPT327	102A	T1202	102A	TNJ60282	102A	TVS28B171A
102A	R88444	102A	SPT337	102A	T1203	102A	TNJ60283	102A	TVS28B171B
102A	RT-185	102A	SPT337B	102A	T126	102A	TNJ60365	102A	TVS28B324
102A	RT185	102A	SPT337V	102A	T127	102A	TNJ60610	102A	UPI1352
102A	RT2230	102A	SPT351	102A	T129	102A	TNJ60611	102A	UPI1355
102A	RT2329	102A	SPT352	102A	T130	102A	TNJ60728	102A	V10/15A18
102A	RT2330	102A	SPT353	102A	T1300	102A	TNJ61222	102A	V10/30A18
102A	RT2331	102A	SPT526	102A	T13000	102A	TNJ61282	102A	V10/50A18
102A	RT2709	102A	SK3003	102A	T1310	102A	TNJ72278	102A	V51
102A	RT3097	102A	SK3004	102A	T1327	102A	TNJ72283	102A	V6/2RC
102A		102A	SMO843	102A	T1328	102A	TNJ72285	102A	V6/2RJ
102A		102A	SM8341	102A	T1334	102A	TNJ72287	102A	

ECG	Industry Standard No.	ECG	Industry Standard No.	ECG	Industry Standard No.	ECG	Industry Standard No.	ECG	Industry Standard No.
102A	V6/4RC	102A	2N109M1	102A	2SA12D	102A	2SB175E	102A	2SB32N
102A	V6/4RJ	102A	2N109M2	102A	2SA12H	102A	2SB176	102A	2SB33
102A	V6/6RJ	102A	2N109WHT	102A	2SA12V	102A	2SB176-0	102A	2SB33-4
102A	V6/7RC	102A	2N109YEL	102A	2SA13	102A	2SB176-P	102A	2SB336
102A	VB11	102A	2N1115	102A	2SA138	102A	2SB176-PR	102A	2SB33C
102A	VFQ2745F	102A	2N1130	102A	2SA14	102A	2SB176B	102A	2SB33D
102A	VFB-2745	102A	2N132A	102A	2SA15	102A	2SB176M	102A	2SB33F
102A	VFB-2745J	102A	2N133A	102A	2SA15BK	102A	2SB176N	102A	2SB345
102A	VFT-2745H	102A	2N138A	102A	2SA15BL	102A	2SB176PRC	102A	2SB346
102A	VM-30244	102A	2N138B	102A	2SA15BLU	102A	2SB176R	102A	2SB346K
102A	VM30244	102A	2N1416	102A	2SA15H	102A	2SB177	102A	2SB346Q
102A	VS-28B171	102A	2N1753	102A	2SA15K	102A	2SB177A	102A	2SB347
102A	VS-28B172	102A	2N180	102A	2SA15R	102A	2SB177D	102A	2SB348
102A	VS-28B172FN	102A	2N181	102A	2SA15U	102A	2SB177B	102A	2SB348Q
102A	VS-28B176	102A	2N185BLU	102A	2SA15V	102A	2SB177M	102A	2SB348R
102A	VS-28B178	102A	2N186	102A	2SA15VR	102A	2SB177R	102A	2SB349
102A	VS-28B178A	102A	2N186A	102A	2SA15Y	102A	2SB178	102A	2SB37
102A	VS-28B524	102A	2N187	102A	2SA16	102A	2SB178-0	102A	2SB370
102A	W3	102A	2N187A	102A	2SA17	102A	2SB178-8	102A	2SB370A
102A	WC19863	102A	2N188	102A	2SA173	102A	2SB178A	102A	2SB370AA
102A	WC19864	102A	2N188A	102A	2SA173B	102A	2SB178C	102A	2SB370AC
102A	WTV-BMC	102A	2N1924	102A	2SA17H	102A	2SB178D	102A	2SB370AHA
102A	WTV15VMG	102A	2N1925	102A	2SA18	102A	2SB178M	102A	2SB370AHB
102A	WTV15VMQ	102A	2N1926	102A	2SA187TV	102A	2SB178N	102A	2SB370B
102A	WTV20MG	102A	2N1954	102A	2SA18H	102A	2SB178S	102A	2SB370C
102A	WTV20VH6	102A	2N1955	102A	2SA19S	102A	2SB178T	102A	2SB370D
102A	WTV20VHG	102A	2N1956	102A	2SA24S	102A	2SB178U	102A	2SB370P
102A	WTV20VMG	102A	2N1957	102A	2SA282	102A	2SB178V	102A	2SB370PB
102A	WTV30VHG	102A	2N2000	102A	2SA30	102A	2SB178X	102A	2SB370V
102A	WTV30VLG	102A	2N206	102A	2SA302	102A	2SB178Y	102A	2SB371D
102A	WTV30VMG	102A	2N215	102A	2SA303	102A	2SB183	102A	2SB376
102A	WTVA56	102A	2N217	102A	2SA31	102A	2SB184	102A	2SB376Q
102A	WTVB5A	102A	2N217A	102A	2SA311	102A	2SB185	102A	2SB378
102A	WTVBA6A	102A	2N217RED	102A	2SA312	102A	2SB185(0)	102A	2SB378A
102A	WTVBB6A	102A	2N217WHT	102A	2SA32	102A	000028B185-0	102A	2SB37A
102A	X-78	102A	2N217YEL	102A	2SA321	102A	002SB18500	102A	2SB37B
102A	X101644	102A	2N220	102A	2SA322	102A	0028SB185AA	102A	2SB37C
102A	X45C-H06	102A	2N222	102A	2SA33	102A	2SB185P	102A	2SB37E
102A	XA122	102A	2N222A	102A	2SA35	102A	2SB185P	102A	2SB37F
102A	XB102	102A	2N223	102A	2SA36	102A	00002SB186	102A	2SB381
102A	XB103	102A	2N224	102A	2SA374	102A	2SB186(0)	102A	2SB382
102A	XB104	102A	2N225	102A	2SA375	102A	0002SB186-0	102A	2SB383
102A	XB112	102A	2N226	102A	2SA40	102A	2SB186-1	102A	2SB383-1
102A	XB113	102A	2N227	102A	2SA406	102A	2SB186-X	102A	2SB383-2
102A	XB114	102A	2N238	102A	2SA407	102A	2SB186A	102A	2SB384
102A	XB13	102A	2N238-ORN	102A	2SA49	102A	2SB186AG	102A	2SB385
102A	XB2A	102A	2N238U	102A	2SA50	102A	2SB186B	102A	2SB386
102A	XB3A	102A	2N238B	102A	2SA52	102A	2SB186BY	102A	2SB387
102A	XB3BN	102A	2N238F	102A	2SA53	102A	2SB186C	102A	2SB387A
102A	XB3C-1	102A	2N241	102A	2SA538	102A	2SB186G	102A	2SB389
102A	XB4-1	102A	2N241A	102A	2SA64	102A	2SB186H	102A	2SB39
102A	XC101	102A	2N242B	102A	2SA65	102A	2SB186L	102A	2SB40
102A	XC121	102A	2N2429	102A	2SA66	102A	2SB186O	102A	2SB400
102A	XC131	102A	2N2447	102A	2SA67	102A	2SB186Y	102A	2SB400A
102A	XC171	102A	2N2448	102A	2SA78	102A	000002SB187	102A	2SB400B
102A	XG8	102A	2N2449	102A	2SA78B	102A	2SB187(1)	102A	2SB400BL
102A	XJ13-1	102A	2N2450	102A	2SA78C	102A	00002SB187(RED)	102A	2SB400K
102A	Y363	102A	2N249	102A	2SA78D	102A	2SB187AA	102A	2SB408
102A	YV1	102A	2N2613	102A	2SA79	102A	2SB187B	102A	2SB427
102A	YV1A	102A	2N2614	102A	2SB-P1A	102A	2SB187C	102A	2SB428
102A	YV2	102A	2N269	102A	2SB110	102A	2SB187D	102A	2SB43
102A	ZA105604	102A	2N272	102A	2SB111	102A	2SB187G	102A	2SB439
102A	ZEN309	102A	2N273	102A	2SB111K	102A	2SB187H	102A	2SB439A
102A	ZEN310	102A	2N279	102A	2SB112	102A	2SB187K	102A	2SB43A
102A	ZTR-B54	102A	2N280	102A	2SB113	102A	2SB187N	102A	2SB44
102A	ZTR-B56	102A	2N281	102A	2SB114	102A	2SB187RED	102A	2SB440
102A	01-1201-0	102A	2N283	102A	2SB115	102A	2SB187S	102A	2SB443
102A	1-21-106	102A	2N284	102A	2SB116	102A	2SB187X	102A	2SB443A
102A	1-21-107	102A	2N284A	102A	2SB117	102A	2SB187YEL	102A	2SB443B
102A	1-21-120	102A	2N291	102A	2SB117K	102A	2SB188	102A	2SB459
102A	1-21-148	102A	2N2953	102A	2SB120	102A	2SB200	102A	2SB459-0
102A	1-21-164	102A	2N320	102A	2SB134	102A	2SB200A	102A	2SB459A
102A	1-21-184	102A	2N34	102A	2SB134-D	102A	2SB201	102A	2SB459B
102A	1-21-191	102A	2N34A	102A	2SB134-E	102A	2SB202	102A	2SB459C
102A	1-21-192	102A	2N36	102A	2SB134C	102A	2SB22	102A	2SB459D
102A	1-21-225	102A	2N38A	102A	2SB134E	102A	2SB22-0	102A	2SB46
102A	1-21-226	102A	2N405	102A	2SB135	102A	2SB224	102A	2SB460
102A	1-21-227	102A	2N406	102A	2SB135A	102A	2SB226	102A	2SB460A
102A	1-21-232	102A	2N406BLU	102A	2SB135B	102A	2SB227	102A	2SB460B
102A	1-21-246	102A	2N406BRN	102A	2SB135C	102A	2SB22A	102A	2SB47
102A	1-21-266	102A	2N406GRN-YEL	102A	2SB135D	102A	2SB22B	102A	2SB470
102A	1-21-267	102A	2N406GRN	102A	2SB135E	102A	2SB22H	102A	2SB475
102A	1-21-272	102A	2N406RED	102A	2SB135F	102A	2SB22I	102A	2SB475A
102A	1-21-274	102A	2N407BLK	102A	2SB136	102A	2SB22R	102A	2SB475B
102A	1-21-95	102A	2N407GRN	102A	2SB136-2	102A	2SB22Y	102A	2SB475C
102A	1-21-96	102A	2N407J	102A	2SB136-3	102A	2SB23	102A	2SB475D
102A	1-52221011	102A	2N407RED	102A	2SB136A	102A	2SB24	102A	2SB475E
102A	1-522210111	102A	2N407WHT	102A	2SB136B	102A	2SB257	102A	2SB475F
102A	1-522211200	102A	2N407YEL	102A	2SB136C	102A	2SB261	102A	2SB475G
102A	1-522211328	102A	2N408	102A	2SB136D	102A	2SB262	102A	2SB475Q
102A	1-522216500	102A	2N408J	102A	2SB136U	102A	2SB263	102A	2SB482
102A	1-6207190405	102A	2N408WHT	102A	2SB153	102A	2SB264	102A	2SB486
102A	1-801-005-23	102A	2N427A	102A	2SB154	102A	2SB265	102A	2SB496
102A	1-801-006-12	102A	2N44A	102A	2SB155	102A	2SB266	102A	2SB497
102A	1-801-006-14	102A	2N45	102A	2SB155A	102A	2SB266P	102A	2SB498
102A	1-801-308	102A	2N450	102A	2SB155B	102A	2SB266Q	102A	2SB516
102A	1-801-308-24	102A	2N45A	102A	2SB156	102A	2SB267	102A	2SB516C
102A	1-TR-111	102A	2N534	102A	2SB156/7825B	102A	2SB268	102A	2SB516CD
102A	1A0055	102A	2N535	102A	2SB156A	102A	2SB269	102A	2SB516D
102A	1A0056	102A	2N535A	102A	2SB156AA	102A	2SB270	102A	2SB516F
102A	002-005100	102A	2N535B	102A	2SB156AC	102A	2SB270A	102A	2SB534
102A	002-006600	102A	2N536	102A	2SB156B	102A	2SB270B	102A	2SB534A
102A	002-006800	102A	2N563	102A	2SB156C	102A	2SB270C	102A	2SB54
102A	002-006900	102A	2N565	102A	2SB156D	102A	2SB270D	102A	2SB54B
102A	002-007900	102A	2N567	102A	2SB156P	102A	2SB270E	102A	2SB54F
102A	002-008400	102A	2N571	102A	2SB168	102A	2SB271	102A	2SB54Y
102A	002-011000	102A	2N584	102A	2SB169	102A	2SB272	102A	2SB55
102A	002-011900	102A	2N591	102A	2SB170	102A	2SB273	102A	2SB56
102A	002-11800	102A	2N591/5	102A	2SB171	102A	2SB293	102A	2SB56A
102A	002-11900	102A	2N591A	102A	2SB171A	102A	2SB294	102A	2SB56B
102A	2D036	102A	2N62	102A	2SB171B	102A	2SB299	102A	2SB56C
102A	2D039	102A	2N63	102A	2SB172	102A	2SB302	102A	2SB57
102A	2Q303	102A	2N64	102A	2SB172A	102A	00002SB303	102A	2SB60
102A	2Q304	102A	2N680	102A	2SB172AF	102A	00028SB303-0	102A	2SB60A
102A	2Q308	102A	2N77	102A	2SB172C	102A	2SB303A	102A	2SB61
102A	2Q309	102A	2N800	102A	2SB172D	102A	2SB303B	102A	2SB65
102A	2Q344	102A	2N807	102A	2SB172E	102A	2SB303C	102A	2SB66
102A	2Q345	102A	2N808	102A	2SB172P	102A	2SB303H	102A	2SB66H
102A	2Q371	102A	2N813	102A	2SB172H	102A	2SB303K	102A	2SB71
102A	2Q371A	102A	2N814	102A	2SB172P	102A	2SB315	102A	2SB73
102A	2Q374	102A	2N815	102A	2SB172R	102A	2SB316	102A	2SB73A
102A	2Q374A	102A	2N816	102A	2SB173	102A	2SB32	102A	2SB73B
102A	2Q376	102A	2N819	102A	2SB173B	102A	2SB32-0	102A	2SB73C
102A	2Q377	102A	2N820	102A	2SB173C	102A	2SB32-1	102A	2SB73GR
102A	2Q381	102A	2PB187	102A	2SB173L	102A	2SB32-2	102A	2SB73S
102A	2Q381A	102A	2N189	102A	2SB174	102A	2SB32-4	102A	2SB75
102A	2Q382	102A	2S273	102A	2SB175	102A	2SB321	102A	2SB75A
102A	2N104	102A	2SA12	102A	2SB175A	102A	2SB322	102A	2SB75AH
102A	2N105	102A	2SA12A	102A	2SB175B	102A	2SB323		
102A	2N106	102A	2SA12B	102A	2SB175C	102A	2SB326		
102A	2N108	102A	2SA12C			102A	2SB329		
102A	2N109					102A	2SB329K		
102A	2N109BLU								
102A	2N109GRN								

ECG	Industry Standard No.
102A	2SB75B
102A	2SB75C
102A	2SB75F
102A	2SB75H
102A	2SB75LB
102A	2SB76
102A	2SB77
102A	2SB77(B)
102A	2SB77A
102A	2SB77AA
102A	2SB77AB
102A	2SB77AC
102A	2SB77AD
102A	2SB77AH
102A	2SB77AP
102A	2SB77B
102A	2SB77B-11
102A	2SB77C
102A	2SB77D
102A	2SB77H
102A	2SB77V
102A	2SB77VRED
102A	2SB78
102A	2SB79
102A	2SB89
102A	2SB89A
102A	2SB89AH
102A	2SB89H
102A	2SB90
102A	2SB91
102A	2SB92
102A	2SB94
102A	2SB95
102A	2SB97
102A	2SBF1
102A	2SBF1A
102A	2SBF2
102A	2SD-F1A
102A	2T3
102A	03-156B
102A	03-57-302
102A	3E-27
102A	3SB347
102A	4-66-1A7-1
102A	4-66-2A7-1
102A	4-66-3A7-1
102A	4-686163-3
102A	4-686195-3
102A	4-686196-3
102A	4-68681-2
102A	4-68681-3
102A	4-8P-2A7-1
102A	4-9L-4A7-1
102A	4JX1D925
102A	6-13
102A	6-53
102A	6-5363
102A	6-53A
102A	6-53F
102A	6-60
102A	6-60B
102A	6-60D
102A	6-60P
102A	6-61
102A	6-61B
102A	6-61D
102A	6-61P
102A	6-62
102A	6-62A
102A	6-62B
102A	6-62D
102A	6-63
102A	6-63A
102A	6-63T
102A	6-65
102A	6-65T
102A	6-66
102A	6-66T
102A	6-67
102A	6-67T
102A	6-87
102A	6-88
102A	6A10624
102A	6A11665
102A	6A11668
102A	6A12516
102A	6A12684
102A	6A12685
102A	6A12989
102A	6A12990
102A	6D122
102A	6D122R
102A	6D122T
102A	6D122TC
102A	6D122TH
102A	6D122U
102A	6D122V
102A	6D122W
102A	6D122Y
102A	6L122
102A	6X97047A02
102A	07-07119
102A	07-1075-01
102A	07-1075-02
102A	07-1156-03
102A	07-2012-04
102A	07-3012-04
102A	07-3015-05
102A	7-59-0293477
102A	7-59-0603477
102A	07-6015-16
102A	8-0205400
102A	8-0205600
102A	8-0222631U
102A	8-0236400
102A	8-0236430
102A	8-0243900
102A	8-619-030-007
102A	8-619-030-008
102A	8-619-030-009
102A	8-619-030-014
102A	8-619-030-016
102A	8-619-030-017
102A	8-697-020-567
102A	8-697-020-568
102A	8-697-020-569
102A	8-905-605-016
102A	8-905-605-030
102A	8-905-605-032
102A	8-905-605-050
102A	8-905-605-051
102A	8-905-605-075
102A	8-905-605-090
102A	8-905-605-091
102A	8-905-605-230
102A	8-905-605-232
102A	8-905-605-234
102A	8-905-605-292
102A	8-905-605-305
102A	8-905-606-750
102A	8-905-606-800
102A	8-905-606-815
102A	8-905-606-817
102A	8-905-606-885
102A	8-905-613-010
102A	8-905-613-640
102A	8-905-613-710
102A	8-905-613-955
102A	8-905-615-156
102A	08P-12-12
102A	08P-2
102A	08P-2-12-7
102A	8P-2-70
102A	8P-2-70-12
102A	8P-2-70-12-7
102A	8P-20
102A	8P-21
102A	8P-22
102A	8P-23
102A	8P-24
102A	8P-25
102A	8P-26
102A	8P-27
102A	8P-28
102A	8P-29
102A	8P-2A
102A	8P-2A0
102A	8P-2A0R
102A	8P-2A1
102A	8P-2A19
102A	8P-2A2
102A	8P-2A21
102A	8P-2A3
102A	8P-2A3P
102A	8P-2A4
102A	8P-2A4-7
102A	8P-2A4-7B
102A	8P-2A5
102A	8P-2A5L
102A	8P-2A6
102A	8P-2A6-2
102A	8P-2A7
102A	8P-2A7-1
102A	8P-2A8
102A	8P-2A82
102A	8P-2A9
102A	8P-2A9G
102A	09-300017
102A	09-301001
102A	09-301002
102A	09-301002-6
102A	09-301003
102A	09-301004
102A	09-301005
102A	09-301006
102A	09-301007
102A	09-301008-18
102A	09-301009
102A	09-301016
102A	09-301020
102A	09-301023
102A	09-301025
102A	09-301025-6
102A	09-301026
102A	09-301032
102A	09-301036
102A	09-301048
102A	09-30104B
102A	09-301056
102A	09-301072
102A	09-301074
102A	09-30126
102A	09-30313
102A	09-305131
102A	9-511410100
102A	9-511410200
102A	9-511410900
102A	9-511413500
102A	9-5201
102A	9-5203
102A	9-5204
102A	9-5208
102A	9-5209
102A	9-5212
102A	9-5213
102A	9-5214
102A	9-5217
102A	9-5218
102A	9-5222-1
102A	9-5224-1
102A	9-9104
102A	9-9201
102A	9-9202
102A	9-9203
102A	9L-4
102A	09L-4-12
102A	09L-4-12-7
102A	9L-4-70
102A	9L-4-70-12
102A	9L-4-70-12-7
102A	9L-40
102A	9L-41
102A	9L-42
102A	9L-43
102A	9L-44
102A	9L-45
102A	9L-46
102A	9L-47
102A	9L-48
102A	9L-49
102A	9L-4A
102A	9L-4A0
102A	9L-4A0R
102A	9L-4A1
102A	9L-4A19
102A	9L-4A2
102A	9L-4A21
102A	9L-4A3
102A	9L-4A3P
102A	9L-4A4
102A	9L-4A4-7
102A	9L-4A4-7B
102A	9L-4A5
102A	9L-4A5L
102A	9L-4A6
102A	9L-4A6-1
102A	9L-4A7
102A	9L-4A7-1
102A	9L-4A8
102A	9L-4A82
102A	9L-4A9
102A	9L-4A9G
102A	9S011
102A	10-2SB54
102A	10-2SB56
102A	12-1-106
102A	12-1-107
102A	12-1-120
102A	12-1-148
102A	12-1-164
102A	12-1-184
102A	12-1-191
102A	12-1-226
102A	12-1-227
102A	12-1-232
102A	12-1-246
102A	12-1-266
102A	12-1-267
102A	12-1-272
102A	12-1-274
102A	12-1-95
102A	12-1-96
102A	13-087005
102A	13-14085-10
102A	13-14085-18
102A	13-14085-23
102A	13-14085-35
102A	13-14085-60
102A	13-14085-71
102A	13-14085-9
102A	13-18032-1
102A	13-18033-1
102A	13-18304-1
102A	13-18671-1
102A	13-18671-1A
102A	13-18671-1B
102A	13-18671-1C
102A	13-23785-1
102A	13-35792-1
102A	13-67599-3
102A	13-94096-2
102A	14-557-10
102A	14-577-10
102A	14-584-01
102A	14-602-04
102A	14-602-05
102A	14-602-05A
102A	14-602-10
102A	14-602-15
102A	14MW69
102A	15-2221011
102A	15-22210111
102A	15-22211200
102A	15-22211328
102A	15-22216500
102A	16-207190405
102A	18P-2A82
102A	19-020-003
102A	19-020-007
102A	19-020-034
102A	19-020-035
102A	19-020-036
102A	19A115674-P1
102A	19A115674-P2
102A	19AR14-1
102A	19AR14-2
102A	19AR16-1
102A	19AR16-2
102A	19AR19-1
102A	19AR19-2
102A	19AR25
102A	19AR26
102A	19AR27
102A	19AR32
102A	19AR7-1
102A	19AR7-2
102A	19B2000132-P1
102A	19B2000132-P2
102A	19B2000132-P3
102A	19B2000132-P4
102A	19B200054-P1
102A	19B200061-P1
102A	19B200061-P2
102A	19B200061-P3
102A	19B200061-P4
102A	19B200063-P1
102A	19B200210-P1
102A	19B200210-P2
102A	19B200210-P3
102A	19L-4A82
102A	20A0007
102A	20A0009
102A	20A0015
102A	20C71
102A	20C72
102A	20V-HG
102A	21A005-000
102A	21A015-001
102A	21A015-006
102A	21A038-000
102A	21A039-000
102A	21A040-000
102A	21A040-005
102A	21A040-014
102A	21A040-021
102A	21A040-022
102A	21A040-057
102A	21A040-058
102A	21A040-079
102A	21A040-36
102A	21A053-000
102A	21A054-000
102A	21A055-000
102A	21A074-000
102A	22-002007
102A	23-5014
102A	23-5017
102A	24MW1083
102A	24MW1084
102A	24MW111
102A	24MW1115
102A	24MW178
102A	24MW179
102A	24MW185
102A	24MW187
102A	24MW27
102A	24MW28
102A	24MW34
102A	24MW370
102A	24MW384
102A	24MW441
102A	24MW598
102A	24MW599
102A	24MW600
102A	24MW601
102A	24MW608
102A	24MW613
102A	24MW614
102A	24MW615
102A	24MW777
102A	24MW780
102A	24MW781
102A	24MW782
102A	24MW789
102A	24MW799
102A	24MW824
102A	24MW853
102A	24MW856
102A	24MW857
102A	24MW892
102A	24MW893
102A	025-10031
102A	25B378
102A	26MW613
102A	29V12H01
102A	30V-H6
102A	30V-HG
102A	31-0006
102A	31-0008
102A	31-0018
102A	31-0025
102A	31-0026
102A	31-0033
102A	31-0053
102A	31-0070
102A	31-0107
102A	31-0148
102A	31-0153
102A	31-0172
102A	31-0188
102A	31-0189
102A	31-0205
102A	31-22005400
102A	33-1009-01
102A	33-1019-00
102A	33-1020-00
102A	33-1021-00
102A	34-6000-15
102A	34-6000-27
102A	34-6000-28
102A	34-6000-29
102A	34-6000-30
102A	34-6000-31
102A	34-6000-32
102A	34-6000-33
102A	34-6000-34
102A	34-6000-4
102A	34-6000-5
102A	34-6000-6
102A	34-6000-7
102A	34-6000-8
102A	34-6000-83
102A	34-6000-84
102A	34-6000-85
102A	34-6001-10
102A	34-6001-11
102A	34-6001-12
102A	34-6001-13
102A	34-6001-14
102A	34-6001-16
102A	34-6001-17
102A	34-6001-18
102A	34-6001-19
102A	34-6001-20
102A	34-6001-21
102A	34-6001-22
102A	34-6001-23
102A	34-6001-26
102A	34-6001-29
102A	34-6001-30
102A	34-6001-31
102A	34-6001-33
102A	34-6001-41
102A	34-6001-42
102A	34-6001-43
102A	34-6001-44
102A	34-6001-47
102A	34-6001-7
102A	34-6001-72
102A	34-6001-76
102A	34-6001-8
102A	34-6001-9
102A	34-6008
102A	34-6009
102A	34-6015-44A
102A	34-6016-11
102A	34-6016-50
102A	36J003-1
102A	36P6C
102A	36P7
102A	36P7C
102A	36P7T
102A	36P8
102A	36P8C
102A	42-19671
102A	42-19863
102A	42-19863A
102A	42-19864
102A	42-19864A
102A	42-20222
102A	42-21406
102A	42-22534
102A	42-23622
102A	42X230
102A	45X1A520C
102A	46-163-3
102A	46-840-3
102A	46-8610-3
102A	46-8611
102A	46-8611-3
102A	46-86163-3
102A	46-86165-3
102A	46-86195-3
102A	46-86256-3
102A	46-8631-3
102A	46-86373-3
102A	46-86430-3
102A	46-8660-3
102A	46-8664-3
102A	46-8665-3
102A	46-8666-3
102A	46-8668-3
102A	46-8679-1
102A	46-8679-2
102A	46-8679-3
102A	46-8680-1
102A	46-8680-2
102A	46-8680-3
102A	46-8681-1
102A	46-8681-2
102A	46-869-3
102A	48-10073A01
102A	48-10074A01
102A	48-10074A03
102A	48-124158
102A	48-124159
102A	48-124175
102A	48-124373
102A	48-12443
102A	48-125282
102A	48-125285
102A	48-125294
102A	48-134408
102A	48-134572
102A	48-134621
102A	48-134632
102A	48-171-A06
102A	48-17162A06
102A	48-17162A17
102A	48-17162A22
102A	48-17271A03
102A	48-21598B01
102A	48-43351A01
102A	48-63044A05
102A	48-63084A03
102A	48-63084A04
102A	48-63084A05
102A	48-644678
102A	48-869001
102A	48-869198
102A	48-97046A02
102A	48-97046A03
102A	48-97046A10
102A	48-97046A32
102A	48-97046A33
102A	48-97046A34
102A	48-97046A54
102A	48-97046A55
102A	48-97046A56
102A	48-97046A57
102A	48-97127A20
102A	48-97127A22
102A	48-97127A30
102A	48-97127A31
102A	48-97162A11
102A	48-97162A16
102A	48-97162A18
102A	48-97162A19
102A	48-97162A20
102A	48-97162A24
102A	48-97162A25
102A	48-97221A01
102A	48-97221A02
102A	48-97221A03
102A	48-97221A04
102A	48-97221A05
102A	48-97238A05
102A	48-97238A07
102A	48-97271A01
102A	48-97271A02
102A	48-97271A03
102A	48-97271A04
102A	48-97271A2
102A	48-97271A3
102A	48-97271A4
102A	48-97271A5
102A	48-97271A6
102A	48P-2A5
102A	48P1
102A	48R134621
102A	48R134632
102A	48X97046A53
102A	48X97046A54
102A	48X97046A55
102A	48X97162A06
102A	48X97162A07
102A	49L-4A5
102A	50B173-C
102A	50B173-S
102A	50B175A
102A	50B175B
102A	50B175C
102A	50B423
102A	50B54
102A	50BU75-C
102A	51D176
102A	52C04
102A	55-1016
102A	56-8091
102A	56-8091A
102A	56-8091B
102A	56-8091C
102A	56-8091D
102A	56-8092
102A	56-8092A
102A	56-8092B
102A	57-4015-27
102A	57A1-104
102A	57A1-105
102A	57A1-106
102A	57A1-111
102A	57A1-116
102A	57A1-117
102A	57A1-118
102A	57A1-121
102A	57A1-14
102A	57A1-23
102A	57A1-26
102A	57A1-27
102A	57A1-28
102A	57A1-34
102A	57A1-40
102A	57A1-43
102A	57A1-44
102A	57A1-53
102A	57A1-58
102A	57A1-59
102A	57A1-66
102A	57A1-68
102A	57A1-70
102A	57A1-71
102A	57A1-79
102A	57A1-8
102A	57A1-82
102A	57A1-83
102A	57A1-90
102A	57A1-91
102A	57A1-92
102A	57A1-93
102A	57A1-95

ECG	Industry Standard No.
102A	57A1-97
102A	57A100-7
102A	57A1127
102A	57A1143
102A	57A148-1
102A	57A2-15
102A	57A2-23
102A	57A2-24
102A	57A2-29
102A	57A2-32
102A	57A2-33
102A	57A2-34
102A	57A2-36
102A	57A2-39
102A	57A2-43
102A	57A2-44
102A	57A2-45
102A	57A2-46
102A	57A2-52
102A	57A2-60
102A	57A2-72
102A	57A2-78
102A	57A2-83
102A	57A2-88
102A	57A2.4
102A	57A3-4
102A	57A3-5
102A	57A3-6
102A	57A5-3
102A	57A5-5
102A	57A6-25
102A	57A68
102A	57B100-7
102A	57B1127
102A	57B1143
102A	57B168
102A	57B169
102A	57B170
102A	57B2-12
102A	57B2-23
102A	57B2-24
102A	57B2-29
102A	57B2-32
102A	57B2-34
102A	57B2-36
102A	57B2-39
102A	57B2-43
102A	57B2-44
102A	57B2-45
102A	57B2-52
102A	57B2-6
102A	57B2-60
102A	57B2-72
102A	57B2-78
102A	57B2-83
102A	57B2-88
102A	57B3-4
102A	57B3-5
102A	57B3-6
102A	57C5-3
102A	57C5-5
102A	57C6-25
102A	57C68
102A	57D1-104
102A	57D1-105
102A	57D1-106
102A	57D1-111
102A	57D1-116
102A	57D1-117
102A	57D1-118
102A	57D1-121
102A	57D1-14
102A	57D1-23
102A	57D1-26
102A	57D1-27
102A	57D1-28
102A	57D1-34
102A	57D1-40
102A	57D1-43
102A	57D1-44
102A	57D1-53
102A	57D1-58
102A	57D1-59
102A	57D1-66
102A	57D1-68
102A	57D1-70
102A	57D1-71
102A	57D1-79
102A	57D1-8
102A	57D1-82
102A	57D1-83
102A	57D1-90
102A	57D1-91
102A	57D1-92
102A	57D1-93
102A	57D1-95
102A	57D1-97
102A	57L1-6
102A	57L1-8
102A	57M1-5
102A	57M1-6
102A	59P2C
102A	61-1130
102A	61-1131
102A	61-1215
102A	61-1907
102A	61-1934
102A	61-1935
102A	61-607
102A	61-608
102A	61-654
102A	61-655
102A	61-656
102A	61-928
102A	61-929
102A	61B0015-1
102A	61B004-1
102A	61B005-1
102A	61B006-1
102A	61B009-1
102A	61B015-1
102A	61B016-1
102A	61B017-1
102A	61B018-1
102A	61B019-1
102A	61B020-1
102A	61B021-1
102A	61B022-2
102A	61B022-3
102A	61B023-1
102A	61B026-1
102A	61B45-14
102A	61J004-1
102A	62-13258
102A	62-13494
102A	62-16918
102A	62-18420
102A	62-18421
102A	62-18430
102A	63-10037
102A	63-10038
102A	63-10147
102A	63-10151
102A	63-10152
102A	63-10153
102A	63-10154
102A	63-10156
102A	63-10158
102A	63-10159
102A	63-10408
102A	63-11144
102A	63-11474
102A	63-11586
102A	63-11661
102A	63-12316
102A	63-12317
102A	63-12876
102A	63-12880
102A	63-12881
102A	63-13840
102A	63-16918
102A	63-18420
102A	63-18421
102A	63-18430
102A	63-23041
102A	63-25179
102A	63-25180
102A	63-25181
102A	63-25182
102A	63-25281
102A	63-25282
102A	63-25720
102A	63-25727
102A	63-25728
102A	63-25729
102A	63-25942
102A	63-25944
102A	63-26851
102A	63-27281
102A	63-28390
102A	63-28399
102A	63-29665
102A	63-29666
102A	63-7247
102A	63-7396
102A	63-7397
102A	63-7398
102A	63-7399
102A	63-7420
102A	63-7564
102A	63-7596
102A	63-7871
102A	63-7872
102A	63-7873
102A	63-8380
102A	63-8703
102A	63-8704
102A	63-9519
102A	63-9520
102A	63-9521
102A	63-9659
102A	65T1
102A	66-1
102A	066-1-12
102A	066-1-12-7
102A	66-1-70
102A	66-1-70-12
102A	66-1-70-12-7
102A	66-10
102A	66-11
102A	66-12
102A	66-13
102A	66-14
102A	66-15
102A	66-16
102A	66-17
102A	66-18
102A	66-19
102A	66-1A
102A	66-1A0
102A	66-1A0R
102A	66-1A1
102A	66-1A19
102A	66-1A2
102A	66-1A21
102A	66-1A3
102A	66-1A3P
102A	66-1A4
102A	66-1A4-7
102A	66-1A4-7B
102A	66-1A5
102A	66-1A5L
102A	66-1A6
102A	66-1A6-3
102A	66-1A7
102A	66-1A7-1
102A	66-1A8
102A	66-1A82
102A	66-1A9
102A	66-1A9G
102A	66-2
102A	066-2-12
102A	066-2-12-7
102A	66-2-70
102A	66-2-70-12
102A	66-2-70-12-7
102A	66-20
102A	66-21
102A	66-22
102A	66-23
102A	66-24
102A	66-25
102A	66-26
102A	66-27
102A	66-28
102A	66-29
102A	66-2A
102A	66-2A0
102A	66-2A0R
102A	66-2A1
102A	66-2A19
102A	66-2A2
102A	66-2A21
102A	66-2A3
102A	66-2A3P
102A	66-2A4
102A	66-2A4-7
102A	66-2A4-7B
102A	66-2A5
102A	66-2A5L
102A	66-2A6
102A	66-2A6-4
102A	66-2A7
102A	66-2A7-1
102A	66-2A8
102A	66-2A82
102A	66-2A9
102A	66-2A9G
102A	66-3
102A	066-3-12
102A	066-3-12-7
102A	66-3-70
102A	66-3-70-12
102A	66-3-70-12-7
102A	66-30
102A	66-31
102A	66-32
102A	66-33
102A	66-34
102A	66-35
102A	66-36
102A	66-37
102A	66-38
102A	66-3A
102A	66-3A0
102A	66-3A0R
102A	66-3A1
102A	66-3A19
102A	66-3A2
102A	66-3A21
102A	66-3A3
102A	66-3A3P
102A	66-3A4
102A	66-3A4-7
102A	66-3A4-7B
102A	66-3A5
102A	66-3A5L
102A	66-3A6
102A	66-3A6C
102A	66-3A7
102A	66-3A7-1
102A	66-3A8
102A	66-3A82
102A	66-3A9
102A	66-3A9G
102A	66-6023
102A	66-6023-00
102A	66-6024-00
102A	66-6025-00
102A	66-6026-00
102A	66-6027-00
102A	66-6028-00
102A	66-6033
102A	68P-2A5L
102A	69L-4A5L
102A	75-461
102A	77-271025-1
102A	77-271026-1
102A	77-271027-1
102A	77-271036-1
102A	77-271037-1
102A	77-271039-1
102A	79P1
102A	80-205400
102A	80-205600
102A	80-2226314
102A	80-236400
102A	80-236430
102A	80-243900
102A	80P1
102A	81-46125002-9
102A	81-46125003-7
102A	81-46125004-5
102A	81-46125009-4
102A	81-46125010-2
102A	81-46125011-0
102A	81-46125029-2
102A	86-103-2
102A	86-114-2
102A	86-115-2
102A	86-126-2
102A	86-128-2
102A	86-129-2
102A	86-130-2
102A	86-131-2
102A	86-132-2
102A	86-133-2
102A	86-152-2
102A	86-156-2
102A	86-159-2
102A	86-16-2
102A	86-169-2
102A	86-172-2
102A	86-176-2
102A	86-21-2
102A	86-22-2
102A	86-23-2
102A	86-249-2
102A	86-27-2
102A	86-28-2
102A	86-283-2
102A	86-29-2
102A	86-295-2
102A	86-297-2
102A	86-30-2
102A	86-300-2
102A	86-303-2
102A	86-304-2
102A	86-305-2
102A	86-32-2
102A	86-33-2
102A	86-39-2
102A	86-392-2
102A	86-419-2
102A	86-421-2
102A	86-45-2
102A	86-46-2
102A	86-47-2
102A	86-476-2
102A	86-48-2
102A	86-49-2
102A	86-497-2
102A	86-50-2
102A	86-5000-2
102A	86-5001-2
102A	86-5006-2
102A	86-5027-2
102A	86-5042-2
102A	86-5063-2
102A	86-5067-2
102A	86-509-2
102A	86-5091-2
102A	86-54-2
102A	86-59-2
102A	86-60-2
102A	86-61-2
102A	86-72-2
102A	86-73-2
102A	86-74-2
102A	86-75-2
102A	86-77-2
102A	86-78-2
102A	86-79-2
102A	86-80-2
102A	86-82-2
102A	86-83-2
102A	86-84-2
102A	86-93-2
102A	86-95-2
102A	86-98-2
102A	86A318
102A	86X0017-001
102A	86X0018-001
102A	86X0037-002
102A	87-0018
102A	87-0019
102A	87-0020
102A	87-0021
102A	089-222
102A	90-58
102A	93A9-1
102A	93A9-2
102A	95-11410100
102A	95-11410200
102A	95-11410900
102A	95-11413500
102A	95-201
102A	95-203
102A	95-204
102A	95-208
102A	95-209
102A	95-212
102A	95-213
102A	95-214
102A	95-217
102A	95-218
102A	95-222-1
102A	95-224-1
102A	96-5032-01
102A	96-5033-01
102A	96-5033-02
102A	96-5033-03
102A	96-5033-04
102A	96-5085-02
102A	96-5098-01
102A	96-5101-01
102A	96-5102-01
102A	96XZ6053/27N
102A	96XZ6053/51N
102A	96XZ801/50N
102A	98-1
102A	98-2
102A	98-3
102A	98-4
102A	98-5
102A	98-6
102A	98-7
102A	98P-2A6-2
102A	99-104
102A	99-201
102A	99-202
102A	99-203
102A	99L-4A6-1
102A	99P3
102A	99P3AA
102A	99S002
102A	99S004A
102A	99S005
102A	99S010
102A	99S010A
102A	99S011
102A	99S011A
102A	101-15
102A	0102-0371
102A	0000103
102A	0000104
102A	110-01563-00
102A	112-034923
102A	112-200525
102A	116-091
102A	116-201
102A	116-203
102A	116-206
102A	116-685
102A	116-686
102A	116-757
102A	116-997
102A	120-00195
102A	120-002012
102A	120-002013
102A	120-002014
102A	120-002521
102A	120-002748
102A	120-004727
102A	120-004729
102A	120-01193
102A	120-190
102A	121-10
102A	121-11
102A	121-12
102A	121-14(ZENITH)
102A	121-148
102A	121-151
102A	121-152
102A	121-163
102A	121-164
102A	121-170
102A	121-184
102A	121-19
102A	121-190
102A	121-191
102A	121-192
102A	121-193
102A	121-226
102A	121-227
102A	121-239
102A	121-240X
102A	121-245
102A	121-246
102A	121-272
102A	121-273
102A	121-274
102A	121-275
102A	121-291
102A	121-300
102A	121-301
102A	121-305
102A	121-306
102A	121-307
102A	121-309
102A	121-310
102A	121-311
102A	121-311B
102A	121-314
102A	121-319
102A	121-320
102A	121-327
102A	121-328
102A	121-33
102A	121-34
102A	121-347
102A	121-348
102A	121-374
102A	121-375
102A	121-395
102A	121-396
102A	121-399
102A	121-408
102A	121-409
102A	121-420
102A	121-43
102A	121-437
102A	121-46
102A	121-47
102A	121-490
102A	121-52
102A	121-543
102A	121-544
102A	121-61
102A	121-632
102A	121-633
102A	121-634
102A	121-635
102A	121-636
102A	121-64
102A	121-640
102A	121-68
102A	121-69
102A	121-72
102A	121-9
102A	125A134
102A	129-31
102A	129-32
102A	129-8
102A	129-8-1
102A	129-8-1A
102A	129-8-2
102A	130-40095
102A	130-40236
102A	130-40352
102A	0131-000100
102A	0131-000101
102A	0131-000102
102A	0131-000563
102A	0131-001419
102A	0131-001426
102A	0131-002656
102A	132-001
102A	132-010
102A	132-090
102A	132-90
102A	148P-2
102A	148P-2-12
102A	148P-2-12-8
102A	149L-4
102A	149L-4-12
102A	149L-4-12-8
102A	152-221011
102A	166-1A82
102A	166-2A82
102A	166-3A82
102A	173A4348
102A	173A4349
102A	173A4389-1
102A	173A4390
102A	175-006-9-001
102A	175-007-9-001
102A	175-008-9-001
102A	188P-2
102A	188P-2-12
102A	188P-2-127
102A	188P-2L
102A	188P-2L8
102A	189L-4
102A	189L-4-12
102A	189L-4-127
102A	189L-4L
102A	189L-4L8
102A	201-15
102A	205A2-210
102A	207A3
102A	207A7
102A	231-000-001
102A	247-623
102A	260D00401
102A	260D00402
102A	260D00403
102A	260D00404
102A	260D02501
102A	260D02601
102A	260D08514
102A	260D09413
102A	260D13704
102A	260P2100
102A	264P06301
102A	296-18-9
102A	296-19-9
102A	296-60-9
102A	296-62-9
102A	297L001M02
102A	297L002H01
102A	297L005H01
102A	297V003H09
102A	297V004H01
102A	297V004H03
102A	297V004H04
102A	297V004H06
102A	297V004H08
102A	297V004H09
102A	297V004H11
102A	297V005H01
102A	297V018H01
102A	297V037B02
102A	297V050002
102A	297V051H04
102A	297V052H01
102A	297V052H04
102A	297V053H01
102A	297V053H02
102A	297V081001

ECG	Industry Standard No.
102A	309-327-931
102A	322T1
102A	323T1
102A	324
102A	324-0029
102A	324-0074
102A	324-0089
102A	324-0091
102A	324-0092
102A	324-0093
102A	324-0100
102A	324T1
102A	324T2
102A	325-0025-329
102A	325-0025-330
102A	325-0025-331
102A	325-0028-79
102A	325-0028-80
102A	325-0028-81
102A	325-0028-83
102A	325-0030-315
102A	325-0030-317
102A	325-0030-318
102A	325-0030-319
102A	325-0036-536
102A	325-0047-516
102A	325-0054-310
102A	325-0054-311
102A	325-0670
102A	325-0670-1
102A	325-0670-7
102A	325-0670A
102A	325-1370-18
102A	325-1376-53
102A	325-1376-56
102A	325-1376-57
102A	325-1376-58
102A	325-1378-20
102A	325-1378-22
102A	325T1
102A	326T1
102A	353-9012-001
102A	353-9312-001
102A	410-012-0150
102A	410-013-0240
102A	417-103
102A	417-122
102A	417-17
102A	417-40
102A	417-41
102A	417-47
102A	417-48
102A	417-5
102A	417-52
102A	417-73
102A	417-74
102A	417-75
102A	417-78
102A	420T1
102A	421-10
102A	421-11
102A	421-11B
102A	421-12
102A	421-12B
102A	421-13
102A	421-13B
102A	421-14
102A	421-14B
102A	421-15
102A	421-15B
102A	421-7143
102A	421-8109
102A	421-9
102A	421-9671
102A	421-9862A
102A	421-9863
102A	421-9863A
102A	421-9864
102A	421-9864A
102A	421T1
102A	422-0222
102A	422-1406
102A	422-2535
102A	429-0092-2
102A	429-0092-3
102A	429-0910-51
102A	429-0910-52
102A	465-005-19
102A	465-036-19
102A	465-072-19
102A	465-073-19
102A	465-075-19
102A	465-080-19
102A	465-082-19
102A	465-115-19
102A	465-132-19
102A	465-163-19
102A	465-165-19
102A	465-191-15
102A	466-1A5
102A	466-2A5
102A	466-3A5
102A	473B5
102A	573-529
102A	576-0001-003
102A	576-0001-009
102A	576-0001-014
102A	576-0002-004
102A	576-0002-012
102A	576-001
102A	576-0040-253
102A	576-005
102A	602-040
102A	605-030
102A	606-020
102A	610-035
102A	610-035-1
102A	610-036
102A	610-036-1
102A	610-036-2
102A	610-036-3
102A	610-036-4
102A	610-036-5
102A	610-036-7
102A	610-036-8
102A	610-040
102A	610-040-1
102A	610-040-2
102A	610-043
102A	610-043-1
102A	610-043-2
102A	610-043-3
102A	610-043-4
102A	610-043-6
102A	610-043-7
102A	610-046-7
102A	610-079
102A	610-079-1
102A	610-080
102A	610-080-1
102A	614X1
102A	614X10
102A	614X2
102A	614X3
102A	614X4
102A	614X5
102A	614X6
102A	614X7
102A	614X9
102A	617-50
102A	617-52
102A	617-70
102A	642-117
102A	642-150
102A	656-136
102A	656-137
102A	656-139
102A	660-227
102A	666-1-A-5L
102A	666-2-A-5L
102A	666-3A-5L
102A	675-154
102A	675-155
102A	675-156
102A	690V034H30
102A	690V034H31
102A	690V047H59
102A	690V054H20
102A	690V056H33
102A	690V056H34
102A	690V059H20
102A	690V059H52
102A	690V059H55
102A	690V062H47
102A	690V063H50
102A	690V077H73
102A	690V084H61
102A	690V084H63
102A	690V088H52
102A	690V089H90
102A	690V094H18
102A	690V094H19
102A	690V094H20
102A	690V097H59
102A	690V104H53
102A	690V104H54
102A	690V118H62
102A	720-35019A
102A	750-35019
102A	750M63-115
102A	800-502-00
102A	800-505-00
102A	802-05400
102A	802-05600
102A	802-22631U
102A	802-36400
102A	802-36430
102A	802-43900
102A	808-304
102A	808-305
102A	808-306
102A	808-307
102A	808-308
102A	808-310
102A	808-311
102A	815-1810
102A	830
102A	901-000-6-51
102A	910X3
102A	910X4
102A	910X5
102A	910X6
102A	916-31001-1
102A	916-31001-1B
102A	916-31001-7B
102A	916-31003-5B
102A	916-31007-5
102A	916-31007-5B
102A	916-31012-6
102A	916-31012-6B
102A	916-31026-9B
102A	921-100B
102A	921-147B
102A	921-148B
102A	921-14B
102A	921-150B
102A	921-153B
102A	921-216B
102A	921-217B
102A	921-222B
102A	921-223B
102A	921-238B
102A	921-24B
102A	921-27
102A	921-273B
102A	921-274B
102A	921-27A
102A	921-318B
102A	921-319B
102A	921-35
102A	921-35A
102A	921-35B
102A	921-36B
102A	921-37B
102A	921-38
102A	921-38A
102A	921-38B
102A	921-39
102A	921-39A
102A	921-39B
102A	921-40
102A	921-40A
102A	921-40B
102A	921-41B
102A	921-45
102A	921-45A
102A	921-45B
102A	921-51B
102A	921-52B
102A	921-67
102A	921-67A
102A	921-67B
102A	921-68
102A	921-68A
102A	921-68B
102A	924-16598
102A	930X10
102A	930X7
102A	930X8
102A	930X9
102A	964-19862A
102A	964-19863
102A	964-19864
102A	966-1A6-3
102A	966-2A6-4
102A	966-3A6C
102A	972X10
102A	972X11
102A	972X12
102A	972X9
102A	991-00-1221
102A	991-00-1222
102A	991-01-1221
102A	991-01-1222
102A	991-01-1223
102A	991-01-1224
102A	991-011221
102A	991-011222
102A	991-011223
102A	992T1
102A	1001
102A	1002(SQUELCH)
102A	1002-05
102A	1003
102A	1005(JULIETTE)
102A	1005-17
102A	1006-93
102A	1007-17
102A	1009
102A	1009(SEARS)
102A	1013-16
102A	1019-74
102A	1052-17
102A	1057-17
102A	1060-17
102A	1102-17A
102A	1104-94
102A	1119-57
102A	1119-59
102A	1128-17
102A	1140-17
102A	1145-17
102A	1248
102A	1277-17
102A	1316-17
102A	1317-17
102A	1347-17
102A	1362-17
102A	1362-17A
102A	1364-17
102A	1436-17
102A	1466-1
102A	1466-1-12
102A	1466-1-12-8
102A	1466-2
102A	1466-2-12
102A	1466-2-12-8
102A	1466-3
102A	1466-3-12
102A	1466-3-12-8
102A	1476-17
102A	1476-17-6
102A	1777-17
102A	1827-17
102A	1866-1
102A	1866-1-12
102A	1866-1-127
102A	1866-1L
102A	1866-1L8
102A	1866-2
102A	1866-2-12
102A	1866-2-127
102A	1866-2L
102A	1866-2L8
102A	1866-3
102A	1866-3-12
102A	1866-3-127
102A	1866-3L
102A	1866-3L8
102A	1917-17
102A	1919-17
102A	1919-17A
102A	1946-17
102A	1954-17
102A	1960-17
102A	1973-17
102A	1974-17
102A	2042-17
102A	2047A2-288
102A	2057A100-21
102A	2057A100-24
102A	2057A100-55
102A	2057A100-8
102A	2057A100-9
102A	2057A2-147
102A	2057A2-148
102A	2057A2-165
102A	2057A2-210
102A	2057A2-241
102A	2057A2-288
102A	2057A2-329
102A	2057A2-988
102A	2057B100-1
102A	2057B100-13
102A	2057B100-6
102A	2057B100-8
102A	2057B100-9
102A	2057B168
102A	2057B169
102A	2057B2-107
102A	2057B2-124
102A	2057B2-129
102A	2057B2-135
102A	2057B2-137
102A	2057B2-142
102A	2057B2-206
102A	2057B2-23
102A	2057B2-28
102A	2057B2-29
102A	2057B2-52
102A	2057B2-94
102A	2057B2-99
102A	2057B206
102A	2057B45-14
102A	2106-119
102A	2106-121
102A	2106-122
102A	2106-123
102A	2112-17
102A	2402-453
102A	2402-454
102A	2402-455
102A	2402-457
102A	2405-453
102A	2405-454
102A	2405-455
102A	2405-456
102A	2405-457
102A	2408-326
102A	2408-328
102A	2490
102A	2490A
102A	2495-014
102A	2495-080
102A	2495-388
102A	2495-567-2
102A	2495-567-3
102A	2495-586-2
102A	2497-473
102A	2497-496
102A	2502(RCA)
102A	2603-180
102A	2603-181
102A	2603-182
102A	2606-286
102A	2606-287
102A	2606-291
102A	2612
102A	2703-384
102A	2703-385
102A	2703-386
102A	2704-384
102A	2704-385
102A	2704-386
102A	3004-856
102A	3009(SEARS)
102A	3010(SEARS)
102A	3014(SEARS)
102A	3112
102A	3507(RCA)
102A	3517
102A	3544
102A	3578-1
102A	3686(RCA)
102A	3970(G.E.)
102A	3970CL
102A	4001-224
102A	4001-225
102A	4001-226
102A	4004(PENNCREST)
102A	4014-000-10160
102A	4041-000-10160
102A	4041-000-20120
102A	4041-000-30180
102A	0004201
102A	0004202
102A	4473
102A	4477V10
102A	4553V10
102A	4686-163-3
102A	4686-195-3
102A	4686-196-3
102A	4686-81-2
102A	4686-81-3
102A	4800-200
102A	4800-220
102A	4800-221
102A	4800-222
102A	4822-130-40235
102A	4822-130-40236
102A	4907-976
102A	5464
102A	5612-370
102A	5612-370C
102A	5612-75C
102A	5612-77C
102A	7215-0
102A	8000-00001-068
102A	8000-00003-038
102A	8000-00003-039
102A	8000-00004-088
102A	8000-00004-P088
102A	8000-0004-P088
102A	8070-4
102A	8071-4
102A	8072-4
102A	8073-4
102A	8500-201
102A	8500-202
102A	8500-203
102A	8999-202
102A	9012HE
102A	9012HF
102A	9330-011-70112
102A	9400-8
102A	9400-9
102A	9401-7
102A	9403-7
102A	9403-9
102A	9564
102A	9920-4
102A	9920-5
102A	9921-7
102A	9921-8
102A	10032
102A	10036
102A	10037
102A	10038
102A	10039
102A	11609-1
102A	11620-3
102A	11620-6
102A	11668-5
102A	11668-6
102A	11675-7
102A	11699-7
102A	12110-6
102A	12110-7
102A	12112-0
102A	12114-8
102A	12116-4
102A	12118-4
102A	12119-1
102A	12119-2
102A	12122-5
102A	12122-6
102A	12122-7
102A	12123-2
102A	12124-6
102A	12126-6
102A	12126-7
102A	12127-2
102A	12127-4
102A	12195
102A	12196
102A	15009
102A	23114-061
102A	23785(SYLVANIA)
102A	24785
102A	25114-101
102A	25114-102
102A	25114-103
102A	25114-104
102A	25651-020
102A	25651-021
102A	25651-033
102A	25655-055
102A	25655-056
102A	25657-050
102A	27125-120
102A	27125-150
102A	27125-170
102A	27125-330
102A	27125-340
102A	27125-350
102A	27125-480
102A	27125-540
102A	27125-550
102A	30208-1
102A	30263
102A	30293
102A	30302(RCA)
102A	34262
102A	35590
102A	35628
102A	35677
102A	35678
102A	35792
102A	35816
102A	35817
102A	35824
102A	37549
102A	37550
102A	37551
102A	38057
102A	38176
102A	38685
102A	40034
102A	40034-1
102A	40034-2
102A	40034-3
102A	40034VM
102A	40035
102A	40035-1
102A	40035-2
102A	40035-3
102A	40036
102A	40036-1
102A	40036-2
102A	40036-3
102A	40038
102A	40038-1
102A	40038-2
102A	40038-3
102A	40038VM
102A	40253
102A	40263
102A	40329
102A	40359
102A	40395
102A	40396P
102A	40490
102A	40763
102A	40828
102A	40852(VM)
102A	41570
102A	42305
102A	42322
102A	49341
102A	53201-01
102A	53201-11
102A	59557-48
102A	59625-1
102A	59625-11
102A	59625-12
102A	59625-2
102A	59625-3
102A	59625-4
102A	59625-5
102A	59625-6
102A	59625-7
102A	59625-8
102A	59625-9
102A	59987-1
102A	61008-8
102A	61008-8-1
102A	61008-8-2
102A	61009-9-3
102A	62032
102A	65804-63
102A	66008-2
102A	68504-63
102A	71193-2
102A	72117
102A	72784-22
102A	72784-23
102A	72799-41
102A	72813-10
102A	72847-51
102A	72941-33
102A	73100-9
102A	77052-3
102A	77052-4
102A	77053-2
102A	77272-0
102A	77272-1
102A	77272-5
102A	77272-7
102A	77272-9
102A	77273-2
102A	77273-3
102A	77273-6
102A	77273-7
102A	78527-75-01
102A	78527-76-01
102A	78527-78-01
102A	78527-79-01
102A	080003
102A	080004
102A	080043
102A	080047
102A	080052
102A	80818VM
102A	081001
102A	081018
102A	081019
102A	081038
102A	081046
102A	081047
102A	081048

ECG	Industry Standard No.	ECG	Industry Standard No.	ECG	Industry Standard No.	ECG	Industry Standard No.	ECG	Industry Standard No.
102A	081049	102A	123806	102A	573001	102A	815030A	102A	1443200-3
102A	081050	102A	123809	102A	0573001-14	102A	815030B	102A	1471100-1
102A	081056	102A	123877	102A	0573001H	102A	815031	102A	1471100-8
102A	81404-4A	102A	124626	102A	0573003	102A	815031A	102A	1471100-9
102A	81500-3	102A	126093	102A	0573003H	102A	815031B	102A	1471101-15
102A	81501-5	102A	126093-1	102A	0573004	102A	815034A	102A	1471101-2
102A	81502-2	102A	126093-2	102A	0573005	102A	815034B	102A	1471101-3
102A	81502-2A	102A	126093-3	102A	0573005-14	102A	815034C	102A	1471101-4
102A	81502-2B	102A	126093-4	102A	0573005H	102A	815038	102A	1472482-1
102A	81502-3B	102A	126187	102A	0573011	102A	815038A	102A	1473578-1
102A	81502-4	102A	126276	102A	0573012	102A	815038B	102A	1473598-2
102A	81502-4A	102A	126697	102A	0573012H	102A	815038C	102A	1956016
102A	81502-4B	102A	126945	102A	0573018	102A	815058	102A	1960643
102A	81502-5	102A	127112	102A	0573018H	102A	815058A	102A	1961837
102A	81502-9	102A	127114	102A	0573022	102A	815058B	102A	2000287-28
102A	81502-9A	102A	127297	102A	0573022H	102A	815058C	102A	2000625-31
102A	81502-9B	102A	127303	102A	0573023	102A	815058X	102A	2000625-33
102A	81502-9C	102A	127589	102A	0573023H	102A	815069	102A	2000625-34
102A	81503-0	102A	127962	102A	0573024	102A	815069A	102A	2000646-108
102A	81503-0A	102A	128343	102A	0573024-14	102A	815069B	102A	2000804-9
102A	81503-0B	102A	128940	102A	0573025	102A	815069C	102A	2001653
102A	81503-1	102A	129286	102A	573029	102A	815070	102A	2001653-23
102A	81503-1A	102A	129802	102A	0573034	102A	815070A	102A	2001653-24
102A	81503-1B	102A	130200-00	102A	0573036	102A	815070B	102A	2001809-47
102A	81503-3B	102A	130200-02	102A	0573036H	102A	815070C	102A	2001809-48
102A	81503-3A	102A	130400-95	102A	0573056	102A	815070D	102A	2001809-48A
102A	81503-4	102A	130403-36	102A	573135	102A	815074	102A	2001809-48B
102A	81503-4A	102A	130403-52	102A	573110	102A	815082	102A	2002151-19
102A	81503-4B	102A	137093	102A	573114	102A	815083	102A	2002152-14
102A	81503-4C	102A	161705	102A	0573114H	102A	815104	102A	2002153-71
102A	81503-8C	102A	165976	102A	573117	102A	815114	102A	2002153-78
102A	81505-8	102A	166882	102A	0573117-14	102A	815118	102A	2002210-110
102A	81505-8A	102A	166883	102A	573119	102A	815120	102A	2002211-24
102A	81505-8B	102A	167679	102A	573125	102A	815120A	102A	2002211-25
102A	81505-8C	102A	167998	102A	0573131	102A	815120B	102A	2003073-14
102A	81505-8X	102A	167999	102A	0573142	102A	815120C	102A	2003073-15
102A	81506-9	102A	168907	102A	0573142H	102A	815120D	102A	2003073-8
102A	81506-9A	102A	168954	102A	0573152	102A	815120E	102A	2004358-123
102A	81506-9B	102A	168983	102A	573153	102A	815122	102A	2004358-168
102A	81506-9C	102A	169359	102A	0573153H	102A	815136	102A	2006226-14
102A	81507-0	102A	169360	102A	573184	102A	815139	102A	2006441-113
102A	81507-0A	102A	169361	102A	0573187	102A	815158	102A	2076945-0701
102A	81507-0B	102A	169773	102A	0573200	102A	815160	102A	2090924-0008
102A	81507-0C	102A	171016(SEARS)	102A	573328	102A	815160-I	102A	2090924-009
102A	81507-0D	102A	171017	102A	0573422	102A	815160-J	102A	2090924-6
102A	81507-4	102A	171018	102A	0573422H	102A	815160-K	102A	2090924-8A
102A	81511-0	102A	171026(SEARS)	102A	0573429	102A	815160-L	102A	2090924B
102A	81511-4	102A	171049	102A	573432	102A	815160-O	102A	2091241-0018
102A	81511-8	102A	171162-074	102A	0573742	102A	815160-P	102A	2091241-9
102A	81512-0	102A	171162-075	102A	574003	102A	815160-Q	102A	2091260-1(PNP)
102A	81512-0A	102A	171162-076	102A	576001	102A	815160A	102A	2091260-2(PNP)
102A	81512-0B	102A	171162-080	102A	576005	102A	815160B	102A	2091578-0702
102A	81512-0C	102A	171162-120	102A	605030	102A	815160C	102A	2091578-1
102A	81512-0D	102A	171162-121	102A	606112	102A	815160D	102A	2092693-1
102A	81512-0E	102A	171522	102A	610035	102A	815160E	102A	2092693-8
102A	81513-6	102A	171916	102A	610035-2	102A	815160F	102A	2243225-1
102A	81515-9	102A	171917	102A	610036	102A	815160H	102A	2320011
102A	81515-8	102A	175006-186	102A	610036-5	102A	815177	102A	2320261
102A	81516-0	102A	195601-6	102A	610036-8	102A	815178	102A	2320302
102A	81516-0A	102A	196064-3	102A	610040	102A	815179	102A	2320302H
102A	81516-0B	102A	196183-7	102A	610040-1	102A	815181	102A	2320422-1
102A	81516-0C	102A	200028-7-28	102A	610040-2	102A	815181A	102A	2320423
102A	81516-0D	102A	200062-5-31	102A	610043	102A	815181B	102A	2320492
102A	81516-0E	102A	200062-5-32	102A	610043-1	102A	815181C	102A	2320513
102A	81516-0F	102A	200062-5-33	102A	610043-2	102A	815181D	102A	2320515
102A	81516-0G	102A	200062-5-34	102A	610043-3	102A	815189	102A	2495014
102A	81516-0H	102A	200064-6-108	102A	610043-4	102A	815195	102A	2495080
102A	81516-0I	102A	200064-6-111	102A	610043-6	102A	815196	102A	2495388
102A	81516-0J	102A	218502	102A	610043-7	102A	815218-3	102A	2495388-1
102A	087003	102A	218503	102A	610052-1	102A	815228A	102A	2495388-2
102A	94000	102A	222509	102A	610079	102A	815228A1	102A	2495567-2
102A	94001	102A	223124	102A	610079-1	102A	815228B	102A	2495567-3
102A	94002	102A	223366	102A	610080	102A	815228B1	102A	2495568-2
102A	94006	102A	223971	102A	610080-1	102A	825065	102A	2497473
102A	94008	102A	223483	102A	610088	102A	910050-2	102A	2497473-1
102A	94014	102A	223484	102A	610088-1	102A	910062-1	102A	2497496
102A	94015	102A	223485	102A	610088-2	102A	910070-6	102A	2498897-4
102A	94016	102A	223486	102A	610099-3	102A	910094-4	102A	2851296-01
102A	94030	102A	223810	102A	610126-1	102A	980144	102A	3004856
102A	94037(EICO)	102A	224696	102A	611020	102A	980148	102A	3460679-1
102A	94038(EICO)	102A	224857	102A	660030	102A	980149	102A	3464482-1
102A	94059	102A	225593	102A	660059	102A	980153	102A	4036598-P2
102A	95172-2	102A	226924	102A	660060	102A	980375	102A	4037145-P1
102A	95173-1	102A	228287	102A	660072	102A	980376	102A	4037145-P2
102A	95224-1	102A	230253	102A	660082	102A	980508	102A	4037804-P1
102A	95224-3	102A	230259	102A	731009	102A	980510	102A	4037804-P2
102A	99201	102A	230525	102A	740417	102A	980511	102A	4907976
102A	99202	102A	231140-21	102A	770523	102A	980536	102A	4999887
102A	99203	102A	231706	102A	770524	102A	980837	102A	5073004
102A	99204	102A	234076	102A	770730	102A	980960	102A	5320011
102A	99205	102A	234630	102A	772718	102A	980961	102A	5320141
102A	100693	102A	235194	102A	772720	102A	981147	102A	5320295
102A	101973	102A	236709	102A	772721	102A	981148	102A	5320296
102A	101974	102A	238417	102A	772722	102A	981149	102A	5320305H
102A	103562	102A	238418	102A	772723	102A	981206	102A	5320485
102A	104444	102A	240003	102A	772724	102A	981672	102A	5320672
102A	110957	102A	240006	102A	772725	102A	981673	102A	5406665-P2
102A	110959	102A	244221	102A	772727	102A	981674	102A	5406665-P5
102A	111957	102A	243939	102A	772728	102A	981675	102A	5955748
102A	111959	102A	255728	102A	772729	102A	982151	102A	7030105
102A	112297	102A	257340	102A	772736	102A	982152	102A	7278422
102A	116091	102A	257473	102A	772737	102A	982244	102A	7278429
102A	116201	102A	262113	102A	800747	102A	982283	102A	7279941
102A	116203	102A	266686	102A	801520	102A	982284	102A	7281310
102A	116205	102A	269374	102A	802032-2	102A	982285	102A	7284751
102A	116206	102A	297240-1	102A	802032-4	102A	982375	102A	7294133
102A	116286	102A	300538	102A	802033-3	102A	982531	102A	7303105
102A	116628	102A	300540	102A	802054-0	102A	982532	102A	7851316-01
102A	116685	102A	300541	102A	802056-0	102A	982537	102A	7851317-01
102A	116686	102A	309421	102A	802189-7	102A	982538	102A	7852775-01
102A	116757	102A	310017	102A	802189-8	102A	983405	102A	7852776-01
102A	116996	102A	310055	102A	802263-0	102A	983406	102A	7852778-01
102A	116997	102A	310159	102A	802263-1	102A	983407	102A	7852779-01
102A	116998	102A	310160	102A	802389-2	102A	983408	102A	7853351-01
102A	117208	102A	310201	102A	802415-2	102A	983409	102A	7853352-01
102A	117209	102A	310225	102A	802439-0	102A	983411	102A	7853354-01
102A	117210	102A	322968-167	102A	814044A	102A	984160	102A	7853356-01
102A	117727	102A	322968-17	102A	815003	102A	984221	102A	7855297-01
102A	117728	102A	346016-1	102A	815015	102A	984228	102A	7910070-01
102A	120075	102A	346016-11	102A	815022	102A	985216	102A	7910071-01
102A	120143	102A	373003	102A	815022A	102A	985468	102A	7910588-01
102A	120144	102A	373117	102A	815022B	102A	985468A	102A	7910589-01
102A	120545	102A	373119	102A	815023	102A	985469	102A	7910801-01
102A	120546	102A	489751-045	102A	815023A	102A	985469A	102A	8020322
102A	120909-24.4	102A	489751-108	102A	815023B	102A	985470A	102A	8020324
102A	121151	102A	489751-109	102A	815024	102A	985609	102A	8020333
102A	121152	102A	489751-113	102A	815024A	102A	985610	102A	8020534
102A	122243	102A	489751-114	102A	815024B	102A	986302	102A	8020540
102A	122244	102A	510007	102A	815029	102A	986766	102A	8020560
102A	122901	102A	537200	102A	815029A	102A	995002	102A	8021897
102A	123379	102A	551015	102A	815029B	102A	995003	102A	8021898
102A	123791	102A	551051	102A	815029C	102A	1420427-1		
102A	123805	102A	0563012H	102A	815030	102A	1420427-2		
						102A	1420427-3		

ECG	Industry Standard No.	ECG	Industry Standard No.	ECG	Industry Standard No.	ECG	Industry Standard No.	ECG	Industry Standard No.
102A	8022630	103	D11	103	42X210	103A	D178	103A	PBE3020-2
102A	8022631	103	ET10	103	42X310	103A	D178A	103A	Q0V60527
102A	8023643	103	ET11	103	45N1	103A	D178Q	103A	Q0V60537
102A	8023892	103	GE-8	103	45N2	103A	D178T	103A	Q2N4105
102A	8024390(AIWA)	103	GE-X8	103	48-869092	103A	D186	103A	Q5039
102A	8024400	103	GIB	103	48-869093	103A	D186A	103A	5050
102A	8112023	103	GT364	103	48-869254	103A	D186B	103A	QOV60527
102A	8112027	103	GT365	103	48-869283	103A	D187	103A	QOV60537
102A	8112028	103	GT366	103	48-869476	103A	D187A	103A	QQV60527
102A	8112071	103	GT949	103	48-869476A	103A	D187R	103A	QQV61772
102A	8112090	103	HA5016	103	48R869092	103A	D187Y	103A	R-28D187
102A	8112146	103	ISD-162	103	48R869093	103A	D191	103A	R1530
102A	8112161	103	KD2124	103	57A5	103A	D192	103A	R1531
102A	8112162	103	M9092	103	57A6-6	103A	D193	103A	R1532
102A	8975158-1	103	M9093	103	57B2-5	103A	D194	103A	R1534
102A	9100502	103	NR20	103	5705	103A	D195	103A	R1537
102A	9100621	103	RB5	103	5706-6	103A	D195A	103A	R1538
102A	9100706	103	SA354B	103	86X037-001	103A	D30-O	103A	R1545
102A	9100944	103	SYL104	103	089-233	103A	D30-N	103A	R1547
102A	09301020	103	SYL1297	103	99L6	103A	D31	103A	R1549
102A	11220009/7825	103	SYL1329	103	122-1962	103A	D31D	103A	R1553
102A	11220018/7611	103	SYL1396	103	200A	103A	D33	103A	R177
102A	11220022/7611	103	SYL1524	103	202A	103A	D33C	103A	R2356
102A	11220076/7825	103	SYL1536	103	251M1	103A	D34	103A	R2359
102A	12090924-4	103	SYL1538	103	324-0134	103A	D35	103A	R2360
102A	13020000	103	SYL1539	103	650-109	103A	D352D	103A	R2364
102A	13020002	103	SYL1547	103	1035-6	103A	D352E	103A	R2365
102A	13040095	103	SYL2134	103	2057B2-46	103A	D352P	103A	R2374
102A	13040236	103	SYL2135	103	38175	103A	D36	103A	R2375
102A	13040352	103	SYL2136	103	46590-2	103A	D367	103A	R3293
102A	23114021	103	SYL2650	103	46591-2	103A	D367A	103A	R33
102A	23114061	103	SYL4315	103	46592-2	103A	D367B	103A	R34
102A	23114211	103	TA1620A	103	46593-2	103A	D367C	103A	R3573-1
102A	25114101	103	TA1620B	103	47645-2	103A	D367D	103A	R5050
102A	25114102	103	TM103	103	48385-2	103A	D367E	103A	R5054
102A	25114103	103	TR09	103	49058-2	103A	D367F	103A	R5056
102A	25114104	103	TR338	103	95202	103A	D367P	103A	R5179
102A	27125110	103	TV27	103	95222-2	103A	D37	103A	R5180
102A	27125120	103	WTVL6	103	4036754-P1	103A	D37A	103A	R62
102A	27125260	103	XB4	103	4036754-P2	103A	D37B	103A	R63
102A	27125330	103	XNC101	103	5492639-P1	103A	D37C	103A	R7362
102A	27125340	103	2-36	103	5492639-P2	103A	D38	103A	R79
102A	27125350	103	20339	103	62728574	103A	D43	103A	R80
102A	27125360	103	2N1095	1030	170964	103A	D43A	103A	RCA40231
102A	27125550	103	2N1096	1035	GEIC-167	103A	D61	103A	RCA40396N
102A	30000021	103	2N1101	1036	HA1314	103A	D62	103A	R86
102A	3122005A-00	103	2N1169	1036	GEIC-168	103A	D63	103A	R8-1524
102A	62013559	103	2N1170	1036	HA1316	103A	D65-1	103A	R81524
102A	62018526	103	2N1251	1037	GEIC-170	103A	D72	103A	R81533
102A	62034327	103	2N1312	1037	HA1322	103A	D72-2C	103A	R81549
102A	62041528	103	2N1591	1037	HA1322C	103A	D72-3C	103A	R83931
102A	62041536	103	2N1431	1037	051-0036-00	103A	D72-4C	103A	R88407
102A	62042591	103	2N1858	1037	051-0036-01	103A	D72A	103A	R88420
102A	62042664	103	2N193	1037	051-0036-02	103A	D72B	103A	R88441
102A	62045531	103	2N194	1037	051-0036-03	103A	D72C	103A	R88443
102A	62045930	103	2N194A	1037	740781	103A	D72RE	103A	R88445
102A	62049200	103	2N1993	1039	GEIC-159	103A	D75	103A	RT3096
102A	62087498	103	2N211	1039	HA1201	103A	D75A	103A	RT7440
102A	62118946	103	2N212	103A	A122-1962	103A	D75AH	103A	RT7944
102A	62135301	103	2N213	103A	A13-86416-1	103A	D75B	103A	S-95202
102A	62711728	103	2N214A	103A	A1465-29	103A	D75C	103A	S2042634
102A	62887676	103	2N216	103A	A14665-2	103A	D75H	103A	S65-2-A-3P
102A	80203350	103	2N229	103A	A16A1	103A	D77(TRANSISTOR)	103A	SC-56
102A	80205400	103	2N233	103A	A16A2	103A	D77A	103A	SO56
102A	80205600	103	2N233A	103A	A2T682	103A	D77AH	103A	SE-7001
102A	80226310	103	2N235A	103A	A42X210	103A	D77B	103A	SFI377
102A	80236400	103	2N2482	103A	A48-869254	103A	D77C	103A	SFP-184
102A	80243900	103	2N306	103A	A48-869283	103A	D77D	103A	SFT377
102A	92005600	103	2N356A	103A	A48-869476	103A	D77H	103A	SK3010
102A	95114101-00	103	2N357A	103A	A48-869476A	103A	D77P	103A	SP2158
102A	95114102-00	103	2N358A	103A	A4JX2A822	103A	D96	103A	SP2188
102A	95114109-00	103	2N364	103A	A4L	103A	E-2466	103A	SQ-7
102A	95114135-00	103	2N365	103A	A5705	103A	E24105	103A	SYL-103
102A	120001195	103	2N366	103A	A65-2-70	103A	E2447	103A	SYL-104
102A	120002013	103	2N438	103A	A65-2-705	103A	E2466	103A	SYL-1329
102A	120002014	103	2N438A	103A	A65-2A9G	103A	EA15X8443	103A	SYL-1396
102A	120002511	103	2N439	103A	AC-127	103A	EQ1-1	103A	SYL-152
102A	120002748	103	2N439A	103A	AC-157	103A	E815X48	103A	SYL-1524
102A	120004493	103	2N440A	103A	AC-172	103A	E815X71	103A	SYL-2134
102A	120004494	103	2N444A	103A	AC-175A	103A	E815X74	103A	SYL-2135
102A	120004495	103	2N445A	103A	AC-175B	103A	E838(ELCOM)	103A	SYL-2136
102A	152221011	103	2N446A	103A	AC-175P	103A	E85(ELCOM)	103A	SYL-4315
102A	1522210300	103	2N447A	103A	AC127	103A	E86(ELCOM)	103A	SYL1468
102A	152221200	103	2N447B	103A	AC127-01	103A	GC1137	103A	T-81
102A	152221221	103	2N515	103A	AC127-132	103A	GC1185	103A	T81
102A	1522211328	103	2N516	103A	AC141	103A	GC1423	103A	TA-1620A
102A	1522216500	103	2N517	103A	AC141B	103A	GC148	103A	TA-1620B
102A	2004305600	103	2N625	103A	AC141K	103A	GC285	103A	TM103A
102A	2004305618	103	2N648	103A	AC172	103A	GC286	103A	TNJ61671(2SD72)
102A	312205400	103	2N797	103A	AC175	103A	GC463	103A	TNJ61734
102A	4102100220	103	2N94	103A	AC176	103A	GC465	103A	TNJ72284
102A	4202104770	103	2N97	103A	AC179	103A	GC467	103A	TQ5044
102A	9511410100	103	2SD11	103A	AC181	103A	GC608	103A	TQ5050
102A	9511410200	103	2S51	103A	AC181K	103A	GC609	103A	TQ5062
102A	9511410900	103	2S522	103A	AC183	103A	GT1658	103A	TR310160
102A	9511413500	103	2S523	103A	AC185	103A	GT2768	103A	TR310236
102A	16201190022	103	2S552	103A	AC186	103A	GT336	103A	TV-27
102A	400100010160	103	2S61	103A	AC187	103A	GT35	103A	TV24143
102A	400100020120	103	2S62	103A	AC187K	103A	GT903	103A	TV24983
102A	933001170112	103	2S63	103A	AC187R	103A	HA5010	103A	W4
102A$	A3773	103	2S64	103A	AC194	103A	HD-00072	103A	WC19862
102A$	A478	103	2S64R	103A	AC194K	103A	HT400721A	103A	XG28
102A$	A479	103	2S65	103A	B65-2-A-21	103A	HT400721B	103A	XG29
102A$	ACY-17	103	2S65R	103A	B65-2A21	103A	HT400721C	103A	XG33
102A$	ACY-18	103	2S66	103A	BD-00072	103A	HT400721D	103A	XS26
102A$	ACY-19	103	2S66R	103A	BTX071	103A	HT400721E	103A	1-801-309
102A$	ACY-20	103	2S681	103A	C179	103A	HT400723	103A	1-T8-112
102A$	ACY-21	103	2S682	103A	C181	103A	HT400723A	103A	002-011700
102A$	ACY-22	103	2S69	103A	C277C	103A	HT400723B	103A	002-11700
102A$	ACY-23	103	2S84	103A	C34(TRANSISTOR)	103A	HT400770B	103A	2N1010
102A$	ACY17	103	2S85	103A	C35(TRANSISTOR)	103A	HT403523A	103A	2N1059
102A$	ACY17-1	103	2S85A	103A	C60	103A	IP-850	103A	2N102
102A$	ACY18	103	2S86	103A	C81759	103A	IP20-0008	103A	2N1173
102A$	ACY18-1	103	2S89	103A	D083	103A	IP20-0076	103A	2N1173W
102A$	ACY19	103	4JX1E850	103A	D085	103A	IP20-0160	103A	2N148A
102A$	ACY19-1	103	4JX2816	103A	D100A	103A	J241185	103A	2N148B
102A$	ACY20	103	4JX2825	103A	D104	103A	J24868	103A	2N148C/D
102A$	ACY20-1	103	4JX2A601	103A	D105	103A	K4-501	103A	2N149
102A$	ACY21	103	4JX2A816	103A	D127	103A	N57B2-8	103A	2N149A
102A$	ACY21-1	103	4JX2A822	103A	D127A	103A	NA30	103A	2N1672
102A$	ACY22	103	6A12993	103A	D128	103A	NC53	103A	2N1672A
102A$	ACY22-1	103	8-723-650	103A	D128A	103A	NK65-2A19	103A	2N182
102A$	ACY23	103	13-27050-1	103A	D162	103A	NKT701	103A	2N183
102A$	ACY23-V	103	13-86416-1	103A	D167	103A	NKT703	103A	2N213A
102A$	ACY23-VI	103	16A1	103A	D170	103A	NKT713	103A	2N214
102A$	ACY23V	103	16A2	103A	D170A	103A	NKT717	103A	2N228
102A$	28A373	103	019-003317	103A	D170AA	103A	NKT751	103A	2N2430
102A$	28A478	103	019-003318	103A	D170AB	103A	NKT752	103A	2N306A
102A$	28A479	103	019-003333	103A	D170AC	103A	NKT773	103A	2N507
102A(2)	297Y003H06	103	19A115129-2	103A	D170B	103A	NKT781	103A	2N585
102A/410	B189	103	19A115129-P1	103A	D170BC	103A	OC139	103A	2N646
103	AC157	103	24MW130	103A	D170C	103A	OC140	103A	2N647
		103	27T410	103A	D170PB	103A	OC141		

ECG	Industry Standard No.
103A	2N647/22
103A	2N649
103A	2N649/5
103A	2SC11
103A	2SC179
103A	2SC180
103A	2SC181
103A	2SC277C
103A	2SC34
103A	2SC35
103A	2SC36
103A	2SC60
103A	2SD100
103A	2SD100A
103A	2SD104
103A	2SD105
103A	2SD127
103A	2SD127A
103A	2SD128
103A	2SD128A
103A	2SD162
103A	2SD167
103A	2SD170
103A	2SD170A
103A	2SD170AA
103A	2SD170AB
103A	2SD170AC
103A	2SD170B
103A	2SD170BC
103A	2SD170C
103A	2SD170PB
103A	2SD178
103A	2SD178A
103A	2SD178Q
103A	2SD178T
103A	2SD186
103A	2SD186A
103A	2SD186B
103A	2SD187
103A	2SD187A
103A	2SD187R
103A	2SD187Y
103A	2SD191
103A	2SD192
103A	2SD193
103A	2SD194
103A	2SD195
103A	2SD195A
103A	2SD25
103A	2SD30
103A	2SD30-0
103A	2SD30-N
103A	2SD30G
103A	2SD31
103A	2SD31D
103A	2SD32
103A	2SD33
103A	2SD33C
103A	2SD34
103A	2SD35
103A	2SD352
103A	2SD352D
103A	2SD352E
103A	2SD352F
103A	2SD36
103A	2SD367
103A	2SD367A
103A	2SD367B
103A	2SD367C
103A	2SD367D
103A	2SD367E
103A	2SD367F
103A	2SD367P
103A	2SD37
103A	2SD37A
103A	2SD37B
103A	2SD37C
103A	2SD38
103A	2SD43
103A	2SD43A
103A	2SD61
103A	2SD62
103A	2SD63
103A	2SD64
103A	2SD65
103A	2SD65-1
103A	2SD66
103A	2SD72
103A	2SD72-2C
103A	2SD72-3C
103A	2SD72-4C
103A	2SD72A
103A	2SD72B
103A	2SD72BR
103A	2SD72C
103A	2SD72K
103A	2SD72P
103A	2SD72RE
103A	2SD75
103A	2SD75A
103A	2SD75AH
103A	2SD75B
103A	2SD75C
103A	2SD75H
103A	2SD77
103A	2SD77A
103A	2SD77AH
103A	2SD77B
103A	2SD77C
103A	2SD77D
103A	2SD77H
103A	2SD77P
103A	2SD96
103A	04-00072-01
103A	4-65-2A7-1
103A	4JX2A616
103A	6A12992
103A	07-07167
103A	8-0050300
103A	8-0050600
103A	8-0052800
103A	8-905-605-105
103A	8-905-605-108
103A	8-905-605-109
103A	8-905-605-110
103A	8-905-605-111
103A	8-905-605-112
103A	8-905-605-113
103A	8-905-605-365
103A	8-905-605-384
103A	8-905-605-390
103A	8-905-613-015
103A	8-905-613-062
103A	8Q-3-04
103A	09-033006
103A	09-303006
103A	09-303012
103A	09-303013
103A	09-303023
103A	09-303030
103A	09-309075
103A	9-5202
103A	9-5222-2
103A	9-5224-2
103A	13-14279-1
103A	13-18654-1
103A	14-602-21
103A	14-602-52
103A	19AR6-1
103A	19AR6-2
103A	19AR6-3
103A	21A015-005
103A	022-2876-002
103A	030-034-0
103A	34-6001-84
103A	42-19862
103A	42-21404
103A	45N2A
103A	46-86115-3
103A	46-86211-3
103A	46-8671-3
103A	55-1027
103A	55-1032
103A	56-8100
103A	56-8100A
103A	56-8100B
103A	56-8100C
103A	56-8100D
103A	57A1-3
103A	57A1-4
103A	57A1-5
103A	57A1-6
103A	57A1-78
103A	57A6-20
103A	57A6-21
103A	57A6-5
103A	57B2-8
103A	57C6-20
103A	57C6-21
103A	57C6-5
103A	57D1-3
103A	57D1-4
103A	57D1-5
103A	57D1-6
103A	57D1-78
103A	57M1-16
103A	57M1-18
103A	57M2-1
103A	57M2-2
103A	57M2-6
103A	57M2-9
103A	62-13259
103A	63-10383
103A	63-7248
103A	63-7549
103A	63-7565
103A	63-8473
103A	63-8705
103A	63-9340
103A	065-2
103A	065-2-12
103A	65-2-70
103A	65-2-70-12
103A	65-2-70-12-7
103A	65-20
103A	65-21
103A	65-22
103A	65-23
103A	65-24
103A	65-25
103A	65-26
103A	65-27
103A	65-28
103A	65-29
103A	65-2A
103A	65-2A0
103A	65-2A0R
103A	65-2A1
103A	65-2A19
103A	65-2A2
103A	65-2A21
103A	65-2A3
103A	65-2A3P
103A	65-2A4
103A	65-2A4-7
103A	65-2A4-7B
103A	65-2A5
103A	65-2A5L
103A	65-2A6
103A	65-2A6-1
103A	65-2A7
103A	65-2A7-1
103A	65-2A8
103A	65-2A82
103A	65-2A9
103A	65-2A9G
103A	80-050300
103A	80-050600
103A	80-052800
103A	86-13-2
103A	86-14-2
103A	86-24-2
103A	86-25-2
103A	86-301-2
103A	86-35-2
103A	86-5-2
103A	86-5003-2
103A	86-5004-2
103A	86-5005-2
103A	86-5007-2
103A	86-5008-2
103A	86-5011-2
103A	86-5012-2
103A	86-5013-2
103A	86-5015-2
103A	86-5016-2
103A	86-5017-2
103A	86-5026-2
103A	86-5029-2
103A	86-5034-2
103A	86-5047-2
103A	86-5048-2
103A	86-5060-2
103A	86-5061-2
103A	86-5062-2
103A	86-5080-2
103A	86-5086-2
103A	86-5087-2
103A	86-6-2
103A	86-76-2
103A	86-81-2
103A	95-202
103A	95-222-2
103A	95-224-2
103A	96-5205-01
103A	97N2U
103A	99SA7
103A	114N4U
103A	116-687
103A	121-237
103A	121-238
103A	121-247
103A	121-248
103A	121-557
103A	121-558
103A	121-59
103A	121-6
103A	121-60
103A	121-641
103A	121-7
103A	121-762CL
103A	121-8
103A	126N1
103A	126N2
103A	129-30
103A	130-40096
103A	130-40314
103A	130-40347
103A	165-2A82
103A	174-002-9-001
103A	247-256
103A	297L001H01
103A	297L001H02
103A	297L001M01
103A	297V002H03
103A	297V002H04
103A	297V002H05
103A	297V002M04
103A	297V002M05
103A	324-0122
103A	417-121
103A	421-9862
103A	465-2A5
103A	576-0002-013
103A	601-065
103A	614X8
103A	642-277
103A	660-228
103A	665-2A5L
103A	690V067H35
103A	690V081H96
103A	690V102H39
103A	800-50300
103A	800-506-00
103A	800-50600
103A	800-528
103A	800-528-00
103A	800-52800
103A	800-537-01
103A	808-309
103A	921-46B
103A	921-54B
103A	921-5A
103A	921-5B
103A	921-6A
103A	921-6B
103A	921-71
103A	921-71A
103A	921-71B
103A	921-7B
103A	965-2A6-1
103A	1040-80
103A	1104-95
103A	1119-58
103A	1349-17
103A	1371-17
103A	1465-2
103A	1465-2-12
103A	1465-2-12-8
103A	1865-2
103A	1865-2-12
103A	1865-2L
103A	1865-2L8
103A	1906-17
103A	1937-17
103A	1958-17
103A	2057A2-167
103A	2057B100-4
103A	4822-130-40096
103A	4822-130-40314
103A	5001-512
103A	5614-77C
103A	8000-00004-086
103A	8000-00011-086
103A	8000-00030-009
103A	8000-00032-027
103A	8000-0004-086
103A	8000-0004-P086
103A	11668-7
103A	27125-310
103A	27125-490
103A	030812-1
103A	37279
103A	37552
103A	40037
103A	40037-1
103A	40037-2
103A	40037-3
103A	40037VM
103A	40396N
103A	72191
103A	80817VM
103A	81507-5
103A	81507-6
103A	94023
103A	94029
103A	95224-2
103A	95224-4
103A	104080
103A	110495
103A	110958
103A	111958
103A	116204
103A	116687
103A	130400-96
103A	130403-47
103A	168953
103A	170783-3
103A	171005(TOSHIBA)
103A	223482
103A	224820
103A	225300
103A	226791
103A	231140-45
103A	243837
103A	250400
103A	269367
103A	279317
103A	0573037
103A	0573037H
103A	0573139
103A	601065
103A	610126-2
103A	815075
103A	815076
103A	815218-4
103A	985735
103A	985735A
103A	1473573-1
103A	1817017
103A	2002153-83
103A	2003073-16
103A	2010952-14
103A	2091260-1(NPN)
103A	2091260-2(NPN)
103A	2091260-3(NPN)
103A	2320331
103A	4038256-P1
103A	4038256-P2
103A	5320295H
103A	5320305
103A	5320306
103A	5320361
103A	5320475
103A	7269847
103A	7277066
103A	7851318-01
103A	7851319-01
103A	7851321-01
103A	7851467-01
103A	13040089
103A	13040096
103A	23114045
103A	27125310
103A	27125490
103A	62087609
103A	62119578
103A	80050300
103A	80052800
103A	2004503018
104	A059-115
104	A2SB240A
104	A2SB242A
104	A2SB248A
104	AD139
104	AD149
104	AR4
104	AR5
104	AR6
104	AR7
104	AUY21
104	AUY22
104	B10064
104	B10069
104	B123
104	B123A
104	B141
104	B142
104	B142B
104	B142C
104	B143
104	B143P
104	B144
104	B144P
104	B145
104	B146
104	B147
104	B149
104	B246
104	B247
104	B248
104	B248A
104	B250
104	B250A
104	B26
104	B26(JAPAN)
104	B26A
104	B27
104	B28(JAPAN)
104	B29
104	B2SB241
104	B2SB244
104	B30(JAPAN)
104	B31(TRANSISTOR)
104	B355
104	B356
104	B41
104	B83
104	BC1073
104	BC1073A
104	BC1274
104	BC1274A
104	BC1274B
104	CST1739
104	CST1740
104	CST1741
104	CST1742
104	CTP1104
104	CTP1106
104	CTP1108
104	CTP1109
104	CTP1117
104	CTP1119
104	DS-503
104	DS503
104	DT401
104	DU6
104	EQG-8
104	ES13
104	ES13(ELCOM)
104	ES15X45
104	ES15X51
104	ES15X78
104	ES18(ELCOM)
104	ES503(ELCOM)
104	ES7
104	ES9
104	ES9(ELCOM)
104	ET6
104	HEP-G6000
104	HEP-G6003
104	HEP200
104	HEP230
104	HJ35
104	HR101
104	JP40
104	JR40
104	K04774
104	K4-520
104	K4-521
104	K04774
104	LS52
104	M4463
104	M4570
104	M4619
104	M4620
104	M4649
104	M4727
104	M4974
104	M4974/P1R
104	M84
104	MN76
104	MP525
104	OC16
104	OC22
104	OC23
104	OC24
104	OC25
104	OC26
104	OC27
104	OC28
104	OC29
104	OC35
104	OC36
104	P1R
104	P2C
104	P2R
104	P3EBLK
104	P3EBLU
104	P3EGRN
104	P3ERED
104	P4D
104	P4L
104	P4N
104	P75534-2
104	P75534-3
104	PA10890
104	PADT50
104	PQ31
104	PT150
104	PT235A
104	PT236
104	PT236A
104	PT554
104	PTC114
104	R3515
104	R7167
104	R7620
104	R8313
104	RE7
104	RS-105
104	RS-2006
104	SP1013B
104	SP1108
104	SP1137
104	SP1323
104	SP1403
104	SP1600
104	SP1619
104	SP1651
104	SP176
104	SP1927
104	SP2234
104	SP2247
104	SP230
104	SP2358
104	SP2431
104	SP26
104	SP441D
104	SP441S
104	SP47
104	SP486
104	SP880
104	SP891
104	T1040
104	T1041
104	T1366
104	T1366A
104	T1367
104	T1367A
104	T1368
104	T1368A
104	T1369
104	T1369A
104	T1370
104	T1370A
104	T139
104	TF78
104	TF78/30
104	TI370
104	TI370A
104	TR-02
104	TR-16
104	TR-16C
104	TR-172(OLSON)
104	TR56
104	TS-610
104	TS-612
104	TS-613
104	TS-614
104	TX103-1
104	V30/30DP
104	VFG-274513
104	VFG2745B
104	VFP-2746C
104	W5
104	XB5
104	XB7
104	XC155
104	XC156
104	XN12A
104	XN12B
104	XN12C
104	XN12E
104	XN12F
104	Z20
104	ZEN325
104	001-01204-0
104	001-012040
104	001-01205-0
104	001-01205-1
104	001-012051
104	002-007000
104	002-008100
104	2G223
104	2N1007
104	2N1040
104	2N1136
104	2N1136A
104	2N1137

ECG	Industry Standard No.	ECG	Industry Standard No.	ECG	Industry Standard No.	ECG	Industry Standard No.	ECG	Industry Standard No.
104	2N1165	104	2841A	104	2090056-27	105	SPT-266	105	122-1028A
104	2N1227	104	2842	104	2090056-5	105	SPT-267	105	148-134622
104	2N1227-3	104	2SB123	104	2091859-0008	105	SPT264	105	417-177
104	2N1227-4	104	2SB123A	104	2091859-0025	105	ST-106	105	1417-177
104	2N1227-4R	104	2SB140	104	2091859-0720	105	ST-107	105	1477A
104	2N1227A	104	2SB141	104	2091859-10	105	ST-109	105	1477A-12
104	2N1245	104	2SB142	104	2091859-11	105	ST-110	105	1477A-12-8
104	2N1246	104	2SB142B	104	2091859-2	105	ST-111	105	1492-1
104	2N1314	104	2SB142C	104	2091859-4	105	ST-112	105	1492-1-12
104	2N1314R	104	2SB143	104	4082501-001	105	TA-3	105	1492-1-12-8
104	2N1501	104	2SB143P	104	7279039	105	TRO3	105	1877A
104	2N1502	104	2SB144	104	7279049	105	TRO3C	105	1877A-12
104	2N1668	104	2SB144P	104	7285774	105	TR334	105	1877A-127
104	2N1669	104	2SB145	104	7285778	105	TR57	105	1877AL
104	2N1755	104	2SB146	104	7289047	105	TS-609	105	1877AL8
104	2N1756	104	2SB147	104	7290594	105	WK5458	105	1892-1
104	2N1757	104	2SB149	104	7291252	105	WK5459	105	1892-1-12
104	2N1758	104	2SB246	104	7297043	105	10100	105	1892-1-127
104	2N1759	104	2SB247	104	7297092	105	2N1099	105	1892-1L
104	2N176	104	2SB248	104	7297093	105	2N1100	105	1892-1L8
104	2N176-1	104	2SB248A	104	78513322-01	105	2N1202	105	2009
104	2N1760	104	2SB250	104	23311006	105	2N1203	105	2010
104	2N1761	104	2SB250A	104	62051509	105	2N1261	105	4573
104	2N1762	104	2SB26	104	62742585	105	2N1262	105	4597
104	2N176A	104	2SB26A	1040	HA1211	105	2N1263	105	4597GRN
104	2N176G	104	2SB27	1041	GEIC-160	105	2N1433	105	4622
104	2N176W	104	2SB28	1041	HA-1202	105	2N1434	105	6818
104	2N178	104	2SB29	1041	HA1202	105	2N1435	105	7172-54
104	2N179	104	2SB30	1043	GEIC-169	105	2N1523	105	14573
104	2N2061	104	2SB31	1043	HA1319	105	2N2210	105	030531-1
104	2N2061A	104	2SB355	1045	8-759-101-60	105	2N2266	105	33901-0001
104	2N2062	104	2SB356	1045	09-308033	105	2N2267	105	217119
104	2N2062A	104	2SB41	1045	000074020	105	2N2268	105	217230
104	2N2063	104	2SB83	1045	171179-027	105	2N2269	105	226789
104	2N2063A	104	4-435	1046	GEIC-118	105	2N2728	105	229045
104	2N2064	104	8A12991	1046	KD6311	105	2N2730	105	233305
104	2N2064A	104	8L301V	1046	RE537-IC	105	2N2731	105	233307
104	2N2065	104	8P40	1046	SAJ72157	105	2N2732	105	233508
104	2N2065A	104	8P404	1046	TV8-MPC23C	105	2N2793	105	234078
104	2N2066	104	8P404P	1046	TV8-UPC23C	105	2SB236	105	243215
104	2N2066A	104	8P404RN	1046	TV8UPC23C	105	2SB237	105	243815
104	2N2067	104	09-301010	1046	09-308009	105	2SB237-12A	105	257004
104	2N2067B	104	14-586-01	1046	21A101-005	105	2SB237-12B	105	256480
104	2N2067G	104	14-601-08	1046	489751-175	105	2SB351	105	257242
104	2N2067W	104	34-6001-1	1047	37002001	105	3N50	105	257243
104	2N2137	104	34-6002-17	1047	37001003	105	3N51	105	257534
104	2N2137A	104	34-6002-18	1048	C51C	105	3N52	105	261401
104	2N2138	104	34-6002-18A	1048	37004001	105	4-77A17-1	105	261488
104	2N2138A	104	34-6002-20	1049	GEIC-121	105	4-92-1A7-1	105	262309
104	2N2139	104	34-6002-22	104MP	RE7MP	105	5AA5-1	105	660144
104	2N2139A	104	34-6002-22A	104MP	TM104MP	105	014-556	105	660144A
104	2N2140	104	48-134727	104MP	7-2(STANDEL)	105	19A115094-P1	105	980052
104	2N2140A	104	48-134731	104MP	32-16591	105	19A115487-P1	105	980052A
104	2N2141	104	48-134907	104MP	48-134747	105	19A115540-P1	105	980462
104	2N2141A	104	48-137025	104MP	57A3-10	105	27T407	105	980462A
104	2N2142	104	48-137026	104MP	57A3-11	105	48-134622	105	980463
104	2N2142A	104	48-137102	104MP	57A3-8	105	53N49	105	980463A
104	2N2143	104	48-137213	104MP	57A3-9	105	63N50	105	981969
104	2N2143A	104	48-137215	104MP	57B3-10	105	73N51	105	981969A
104	2N2144	104	48-137216	104MP	57B3-11	105	O77A	105	983945
104	2N2144A	104	48-137217	104MP	57B3-8	105	O77A-12	105	985432
104	2N2145	104	48-137218	104MP	57B3-9	105	O77A-12-7	105	988414
104	2N2145A	104	48-137219	104MP	800-196	105	77A-70-12	105	988477
104	2N2146	104	48-137220	104MP	99257	105	77A-70-12-7	105	989171
104	2N2146A	104	48-137329	104MP	4082501-0001	105	77A0	105	989615
104	2N2282	104	48-869087B	105	A1477A	105	77A1	105	989692
104	2N230	104	48-869093B	105	A1477A9	105	77A10	105	989693
104	2N242	104	57A124-10	105	A1492-1	105	77A10R	105	1221028
104	2N250	104	57B124-10	105	A1492-9	105	77A11	105	4036831-P1
104	2N250A	104	57I5-1	105	A77A-70	105	77A119	105	4036832-P1
104	2N255	104	57M3-7	105	A77A-705	105	77A12	105	5985945A
104	2N255A	104	57M3-8	105	A77A19G	105	77A121	105	5985432
104	2N256	104	62-18427	105	A92-1-70	105	77A13	105	5988414
104	2N256A	104	63-18427	105	A92-1-705	105	77A13P	105	5988977
104	2N257	104	76-11770	105	A92-1A9G	105	77A14	105	5989171
104	2N257A	104	97A83	105	AA5	105	77A14-7	105	5989615
104	2N257B	104	111P5C	105	ADZ11	105	77A15	105	5989692
104	2N257G	104	111P7C	105	ADZ12	105	77A15L	105	5989693
104	2N257W	104	129-6	105	B236	105	77A16	105	7276605
104	2N285	104	129-7	105	B237	105	77A16-1	105	7279005
104	2N285A	104	207A20	105	B237-12A	105	77A17	105	7279007
104	2N285B	104	207A20A	105	B237-12B	105	77A17-1	105	7279011
104	2N296	104	297V041H05	105	B351	105	77A18	105	7279017
104	2N307	104	297V041H07	105	B92-1A21	105	77A182	105	7279025
104	2N307A	104	417-62	105	C81F4	105	77A19	105	7279027
104	2N307B	104	576-0002-002	105	CTP1509	105	77A19G	105	7279033
104	2N325	104	642-206	105	CTP1512	105	77A2	105	7279073
104	2N350	104	642-264	105	D8-501	105	77A3	105	7279076
104	2N351	104	642-316	105	D8570	105	77A4	105	7279293
104	2N351A	104	690V081H97	105	DU47	105	77A5	105	7279298
104	2N375	104	800-329	105	DU7	105	77A7	105	7279566
104	2N376	104	964-16599	105	EA15X6	105	77A8	105	7279793
104	2N376A	104	964-17887	105	ES810	105	77A9	105	7280281
104	2N378	104	992-00-1192	105	ES21	105	83N52	105	7282315
104	2N380	104	1008-17	105	ES501	105	86-512-2	105	7288072
104	2N386	104	1024-17	105	EE7	105	092-1	105	7288073
104	2N387	104	1124C	105	GE-4	105	092-1-12	105	7288076
104	2N392	104	2057A2-302	105	GP1622	105	92-1-70-12	105	7288079
104	2N399	104	2347-17	105	HEP-06004	105	92-1-70-12-7	105	7299720
104	2N400	104	3107-204-90140	105	IO-100	105	92-10	105	57276605
104	2N401	104	3107-204-90190	105	JC100	105	92-11	105	57279005
104	2N419	104	4247	105	JEM1	105	92-12	105	57279007
104	2N456	104	4347	105	JEM2	105	92-13	105	57279009
104	2N456A	104	4822-130-40233	105	JEM3	105	92-14	105	57279011
104	2N456B	104	5253	105	JEM4	105	92-15	105	57279017
104	2N457	104	8999-115	105	JEM5	105	92-16	105	57279025
104	2N457A	104	12163	105	M4553PUR	105	92-17	105	57279027
104	2N457B	104	12178	105	M4622	105	92-18	105	57279033
104	2N458	104	15354-3	105	MHT180	105	92-19	105	57279073
104	2N458A	104	15927	105	MHT1801O	105	92-1A0	105	57279076
104	2N511	104	25661-022	105	MHT1807	105	92-1AOR	105	57279293
104	2N511A	104	30302	105	MHT1808	105	92-1A19	105	57279298
104	2N511B	104	34715	105	MHT1809	105	92-1A2	105	57279566
104	2N512	104	35084	105	MHT181	105	92-1A3	105	57279793
104	2N512A	104	35201	105	MHT230	105	92-1A3P	105	57280281
104	2N512B	104	35260	105	MHT2305	105	92-1A4	105	57282315
104	2N513	104	36896	105	MP503	105	92-1A4-7	105	57288072
104	2N513A	104	40022	105	MP503A	105	92-1A5	105	57288073
104	2N513B	104	40050	105	MP507	105	92-1A5L	105	57288076
104	2N539	104	40254	105	MP507A	105	92-1A6	105	57288079
104	2N539A	104	40462	105	POWER40	105	92-1A6-1	105	57299720
104	2N540	104	057040	105	POWER500	105	92-1A7	1050	GEIC-123
104	2N540A	104	60770	105	POWER60	105	92-1A7-1	1050	GEIC-133
104	2N637	104	94004	105	POWER80	105	92-1A8	1050	IC-142
104	2N637A	104	110515	105	PT-501	105	92-1A82	1050	M5191P
104	2N637B	104	175027-022	105	PT201	105	92-1A9	1050	MPC32C
104	2N638	104	218012D8	105	PTC106	105	92-1A90	1050	MPC562C
104	2N638A	104	0573166	105	QP8-6623N	105	119-0016	1050	RH-1X0025CEZZ
104	2N638B	104	610111-5	105	RE8	105	122-1028	1051	21A101-018
104	2N639	104	610111-7	105	R8-106			1052	G09-029-B
104	2N639A	104	610152-1	105	SPT-265			1052	GEIC-135
104	2N639B	104	801519					1052	RE327-IC
104	2N669	104	1221625					1052	TA70639
104	2841	104	2090056-1					1052	09-308076

ECG	Industry Standard No.	ECG	Industry Standard No.	ECG	Industry Standard No.	ECG	Industry Standard No.	ECG	Industry Standard No.
1052	61A001-12	106	2N2968	1068	2SA500-Y	107	BF218	107	C668E1
1052	740-9000-566	106	2N2969	1068	2SA741H	107	BF219	107	C668B2
1052	1077-2408	106	2N2970	106(4)	19AR21	107	BP220	107	C668EP
1052	003522	106	2N2971	106(4)	100-19B	107	BF229	107	C668EV
1052	26810-158	106	2N3209	1060	AN217	107	BF230	107	C668EX
1052	37510-166	106	2N3304	1060	AN217AA	107	BF253	107	C668P
1052	916084	106	2N3342	1060	AN217AB	107	BF329	107	C674
1054	AN203	106	2N5576	1060	AN217BA	107	BF332	107	C674(JAPAN)
1054	AN203AA	106	2N3915	1060	AN217BB	107	BF333	107	C674B
1054	AN203BA	106	2N4034	1060	AN217CA	107	BF333C	107	C674C
1054	AN203BB	106	2N4035	1060	AN217CB	107	BF333D	107	C674DV
1054	AN203U	106	2N4058	1060	AN217D	107	BF334	107	C674D
1054	GEIC-45	106	2N4059	1060	AN217PBB	107	BF335	107	C674E
1054	IC-554(ELCOM)	106	2N4060	1060	GEIC-50	107	BF813E	107	C674F
1054	09-308011	106	2N4061	1060	PC-20069	107	BF813P	107	C674G
1054	70270730	106	2N4062	1062	AN220	107	BF813G	107	C684
1054	4150002031	106	2N4451	1061	GEIC-143	107	BF814E	107	C684A
1055	AN210	106	2N4453	1061	GEIC-51	107	BF814P	107	C684B
1055	AN210A	106	2N4872	1061	HA1108	107	BF814G	107	C684BK
1055	AN210B	106	2N495	1062	GEIC-52	107	BF815E	107	C684F
1055	AN210C	106	2N495/18	1062	ULN2224A	107	BF815P	107	C705
1055	AN210D	106	2N496	1068	AN242	107	BF815G	107	C705B
1055	GEIC-47	106	2N5140	1069	AN288	107	BF816E	107	C705D
1056	AN211	106	2N5141	1069	GEIC-64	107	BF816P	107	C705E
1056	AN211A	106	2N5208	107	A-1854-0092-1	107	BF816G	107	C705F
1056	AN211AB	106	2N5228	107	A054-148	107	BF818	107	C705TV
1056	AN211B	106	2N5352	107	A054-170	107	BF818R	107	C717
1056	GEIC-48	106	2T23	107	A060-100	107	BF819	107	C717(FINAL IP)
1056	IC-552(ELCOM)	106	2T26	107	A061-105	107	BF819R	107	C717BE
1056	REN1056	106	2T27	107	A061-106	107	BF820	107	C717BK
1056	1018-25	106	2N864A	107	A061-108	107	BF820R	107	C717BLK
1056	741853	106	2N865	107	A061-109	107	BFX32	107	C717C
1058	AN214	106	2N865A	107	A061-112	107	BFY47	107	C717E
1058	AN214P	106	2N869	107	A069-101	107	BFY48	107	C722
1058	AN214PQR	106	2N869A	107	A069-102	107	BFY49	107	C835
1058	AN214Q	106	2N978	107	A069-103	107	BFY69A	107	C836M
1058	AN214R	106	2N995	107	A069-114	107	BO-71	107	C837
1058	909-013-A	106	2N995A	107	A069-116	107	B8V53P	107	C837P
1058	JEIC-49	106	2N996	107	A069-119	107	B8V54P	107	C837H
1058	RE338-IC	106	2SA524	107	A121-585	107	C1023(JAPAN)	107	C837K
1058	REN1056	106	3N112	107	A121-585B	107	C1023-0	107	C837L
1058	88-18920	106	3N113	107	A121-687	107	C1023-Y	107	C837WP
1058	8000-00032-030	106	13-34369-1	107	A1G	107	C1023G	107	C922
1058	18600-156	106	14-807-12	107	A1G-1	107	C1026	107	C922A
1058	741854	106	14-861-12	107	A1M-1	107	C1026Q	107	C922B
1058	5350231	106	15-008-1	107	A1P	107	C1026Y	107	C922C
106	A-1853-0009-1	106	21A062-000	107	A1P-1	107	C1032	107	C922K
106	A-1853-0010-1	106	25-001528	107	A1P-1A	107	C1032G	107	C922L
106	A-1853-0034-1	106	34-1011	107	A1P/4922	107	C1032Y	107	C922M
106	A499	106	40C2PW8V18P	107	A1P/4923	107	C1047	107	C924
106	A499-0	106	48-137066	107	A1R	107	C1047B	107	C924E
106	A499-R	106	48-3003A04	107	A1R-2	107	C1047C	107	C924F
106	A499-Y	106	48-97127A015	107	A1R/4924	107	C1047D	107	C924M
106	A500	106	48-97127A15	107	A1R/4925	107	C1047E	107	C929
106	A500-0	106	57A1-52	107	A1R/4926	107	C1293B	107	C929-0
106	A500-R	106	57A108-1	107	A2057B2-115	107	C1342	107	C929B
106	A500-Y	106	57A108-2	107	A2M	107	C1342A	107	C929C
106	BC205L	106	57A108-3	107	A2M-1	107	C1342B	107	C929C1
106	BCW29	106	57A108-4	107	A2N	107	C1342C	107	C929D
106	BCW29R	106	57A108-5	107	A2N-1	107	C1390A	107	C929D1
106	BCW30	106	57A108-6	107	A2N-2	107	C1417	107	C929DB
106	BCW30R	106	57A108-7	107	A2P	107	C1417C	107	C929DP
106	BF315	106	57A108-8	107	A2Y	107	C1417D	107	C929DU
106	BF316	106	57D1-52	107	A417-154	107	C1417D(U)	107	C929DV
106	BF339	106	62-22529	107	A417-190	107	C1417DU	107	C929ED
106	BFX12	106	121-615	107	A417-205	107	C1417E	107	C929EP
106	BFX13	106	121-711	107	A46-86101-3	107	C1417G	107	C929F
106	BFX48	106	0131-001438	107	A46-86109-3	107	C1417H	107	C929FK
106	BI-82	106	136P1	107	A46-86110-3	107	C1417U	107	C930
106	B8X29	106	151-0417-00	107	A46-86133-3	107	C1417V	107	C930B
106	CS1124G	106	154A5947-7732	107	A46-86301-3	107	C1417VW	107	C930BK
106	D14	106	210BWTP421	107	A46-86302-3	107	C1417W	107	C930BV
106	D8-68	106	297L012C0-01	107	A4E	107	C155	107	C930C
106	D868	106	352-0950-010	107	A4Y-1	107	C156	107	C930CK
106	D896	106	352-0950-020	107	A6Y	107	C1674	107	C930CL
106	EA15X69	106	417-102	107	A772B1	107	C1674K	107	C930CB
106	EA15X70	106	549-1	107	A772BH	107	C1674L	107	C930D
106	EA15X71	106	576-0003-012	107	A772FE	107	C1674M	107	C930DE
106	FI-1019	106	921-103B	107	AR200(GREEN)	107	C186	107	C930DH
106	FK2894	106	921-112B	107	AR21YY	107	C187	107	C930DP
106	FM2894	106	921-333B	107	AR220(YELLOW)	107	C351	107	C930DP-2
106	FT1702	106	921-333P	107	AR220GY	107	C351(PA)	107	C930DX
106	FV2894	106	1853-0069	107	AR222(BLUE)	107	C375	107	C930DZ
106	GMB0404	106	2057A2-200	107	AR222(YELLOW)	107	C375-Y	107	C930E
106	GMB0404-2	106	2057A2-203	107	AR222BY	107	C381	107	C930EP
106	GMB404-1	106	3C12	107	AR224	107	C381-0	107	C930ET
106	HP-47	106	4484-1	107	AR224(WHITE)	107	C381-0	107	C930EV
106	H8-40053	106	4484-2	107	AR224(YELLOW)	107	C381-R	107	C930EX
106	I50865	106	4485-1	107	AT310	107	C381BN	107	C930P
106	K071687	106	4590	107	AT311	107	C381R	107	CDC12050B
106	RR53	106	4822-130-40369	107	AT312	107	C384	107	CDC5000
106	8500	106	4822-130-40477	107	AT313	107	C384-0	107	CS-6225E
106	8501	106	4822-130-40508	107	AT314	107	C384Y	107	CS1585H
106	SA495	106	4822-130-40614	107	AT315	107	C385A	107	CT1012
106	SA495A	106	6855K90	107	AT316	107	C386	107	CT1013
106	SA496	106	7340	107	AT318	107	C386A-0(TV)	107	D006
106	SA496A	106	95229	107	AT319	107	C535	107	D10B1051
106	SA496B	106	97680	107	AT321	107	C535A	107	D10B1055
106	SA537	106	105468	107	AT322	107	C535B	107	D10G1051
106	SA538	106	115270-101	107	AT323	107	C535C	107	D10G1052
106	SA539	106	124634	107	AT324	107	C535G	107	D160
106	SA540	106	131262	107	AT325	107	C544	107	D1G6
106	SAC040	106	135347	107	AT326	107	C544C	107	D16K1
106	SAC040A	106	150865	107	AT327	107	C544D	107	D16K2
106	SAC040B	106	160196	107	AT328	107	C544E	107	D16K3
106	SAC42	106	405192	107	AT330	107	C545	107	D16K4
106	SAC42A	106	531298-001	107	AT33B	107	C545A	107	D1T
106	SAC42B	106	547684	107	A2Y	107	C545B	107	D26B1
106	SAC44	106	610102-1	107	B-75568-2	107	C545C	107	D26B2
106	SL200	106	610136-1	107	B-T1000-139	107	C545D	107	D26C1
106	SL201	106	741050	107	B22	107	C545E	107	D26C2
106	SP12271	106	986015	107	BC121	107	C56	107	D26C3
106	ST2-2517	106	1473651-1	107	BC122	107	C629	107	D26E2
106	T2357	106	2621811	107	BC123	107	C657	107	D26G1
106	T40	106	22115192	107	BC155	107	C658	107	DDBI269001
106	TM-22	106	43027619	107	BC156	107	C658A	107	DDBI295002
106	TM106	106	50211300-00	107	BC188	107	C659	107	DS-71
106	TW135	106	50211300-10	107	BC189	107	C668-0	107	DS-72
106	W7	106	50211300-10	107	BC442	107	C668A	107	DS-74
106	002-0105-00	106	50211600-01	107	BC510C	107	C668B1	107	DS-781
106	2H1256	106	50211600-02	107	BCM1002-2	107	C668BC2	107	DS71
106	2H1257	106	50211600-10	107	BF173	107	C668C	107	DS72
106	2H1024	106	50211600-12	107	BF121	107	C668C1	107	DS73
106	2N1238	106	50211610-10	107	BF123	107	C668CD	107	DS74
106	2N1239	106	50211610-12	107	BF125	107	C668D	107	DS75
106	2N1240	106	62539313	107	BF127	107	C668D0	107	E1A
106	2N1241	106S	2SA499	107	BF194B	107	C668D1	107	EA15X134
106	2N1640	106S	2SA499-0	107	BF195C	107	C668DC	107	EA15X135
106	2N1641	106S	2SA499-R	107	BF195D	107	C668D0	107	EA15X239
106	2N1642	106S	2SA499-Y	107	BF197	107	C668DV	107	EA15X4064
106	2N2175	106S	2SA500	107	BF216	107	C668DX	107	EA15X7117
106	2N2176	106S	2SA500-0	107	BF217	107	C668DZ	107	EA15X7125
106	2N2177	106S	2SA500-R			107	C668E		
106	2N2395								

ECG	Industry Standard No.	ECG	Industry Standard No.	ECG	Industry Standard No.	ECG	Industry Standard No.	ECG	Industry Standard No.
107	EA15X7141	107	N-EA15X134	107	TX-107-3	107	28C545D	107	27919
107	EA15X7215	107	N-EA15X135	107	V118	107	28C545E	107	3L4-6007-17
107	EA15X7231	107	NCS9018D	107	V129	107	28C56	107	3L4-6007-21
107	EA15X7264	107	NJ100A	107	V143	107	28C629	107	3L4-6007-23
107	EA2131	107	NJ202B	107	V828C684	107	28C657	107	3L4-6007-35
107	EA2132	107	NL100B	107	W29	107	28C658	107	3N71
107	EA2494	107	NR421	107	X0371	107	28C658A	107	3N72
107	EA2496	107	NR421DG	107	X840	107	28C659	107	3N73
107	EA2812	107	NR461AF	107	01-030930	107	28C668	107	04-00461-02
107	ED1502B	107	NS1510	107	1-801-304-15	107	28C668A-0	107	04-00535-02
107	ED1502C	107	NS3039	107	1-TR-048	107	28C668A	107	04-01585-08
107	ED1502D	107	NS3040	107	1A0044	107	28C668B	107	04-15850-06
107	ED1502E	107	NS3041	107	1S535B	107	28C668B1	107	4-850
107	EL434	107	PL1024	107	2N1586	107	28C668BC2	107	7-16
107	EN5172	107	PL1025	107	2N1589	107	28C668C	107	7-26
107	EQ2-18	107	PL1026	107	2N1592	107	28C668C1	107	7-39
107	ES10186	107	PL1113	107	2N162A	107	28C668CD	107	7A30
107	ES10188	107	PM194	107	2N2610	107	28C668D0	107	7A31
107	ES15046	107	PM195	107	2N2715	107	28C668D1	107	7A32
107	ES15047	107	PT4816	107	2N2716	107	28C668DB	107	8-722-923-00
107	ES15X1D2	107	PT4830	107	2N3082	107	28C668DQ	107	8-722-034-00
107	ES15X66	107	Q-01115C	107	2N3083	107	28C668DV	107	8-729-671-14
107	ES15X67	107	Q1/6515	107	2N3662	107	28C668DX	107	8-729-803-04
107	ES15X80	107	Q2/6515	107	2N3663	107	28C668DZ	107	9-003
107	ES15X81	107	Q301	107	2N3709	107	28C668E	107	09-302014
107	ES15X82	107	Q4/6515	107	2N3710	107	28C668E1	107	09-302024
107	ES15X97	107	Q6/6515	107	2N3846	107	28C668E2	107	09-302044
107	ET15X2	107	R-28C535	107	2N3855A	107	28C668EP	107	09-302063
107	ET15X3	107	R-28C772	107	2N3983	107	28C668EV	107	09-302072
107	EW165	107	R-28C858	107	2N3984	107	28C668EX	107	09-302075
107	EW164	107	R118	107	2N3985	107	28C668F	107	09-302079
107	EW165	107	R8529	107	2N4254	107	28C674	107	09-302092
107	FI 1023	107	RE5001	107	2N4255	107	28C674C	107	09-302114
107	FS1308	107	RE5002	107	000028606	107	28C674CK	107	09-302129
107	G04-041B	107	R89	107	2882	107	28C674CL	107	09-302138
107	G04041B	107	R87143	107	28C1023	107	28C674CV	107	09-302142
107	G05-050-C	107	R87222	107	28C1023-0	107	28C674CZ	107	09-302143
107	GE-11	107	R87523	107	28C1023-Y	107	28C674D	107	09-302151
107	GE129	107	R89510	107	28C1023G	107	28C674E	107	09-302152
107	GN770	107	R89511	107	28C1023Y	107	28C674F	107	09-302203
107	HC-00380	107	R89512	107	28C1026	107	28C674G	107	09-302216
107	HC-00784	107	RT112	107	0002801026A	107	28C674V	107	09-305033
107	HC-00829	107	RT5464	107	0002801026B	107	28C684	107	09-305041
107	HC-00920	107	RT5465	107	0002801026C	107	28C684A	107	09-309069
107	HC-00930	107	RT6601	107	28C1026G	107	28C684B	107	09-309072
107	HC-01047	107	RT6602	107	28C1026Y	107	28C684BK	107	09-309672
107	HC-01359	107	RT6602Y	107	28C1032	107	28C684R	107	9TR1
107	HC-01417	107	RT7704	107	0002801032A	107	28C684F	107	9TR11001-01
107	HC380	107	RT8668	107	0002801032B	107	28C705	107	10-280380
107	HC381	107	RT8669	107	00028C1032C	107	28C705B	107	12-23163-3
107	HC394	107	S1041	107	28C1032G	107	28C705C	107	13-10321-29
107	HC454	107	S1041-16GN	107	28C1032Y	107	28C705D	107	13-10321-31
107	HC460	107	S1122	107	28C1047	107	28C705E	107	13-10321-32
107	HC461	107	S1126	107	28C1047B	107	28C705F	107	13-14085-16
107	HC535	107	S1308	107	28C1047BC	107	28C705TV	107	13-14085-17
107	HC535A	107	S1897	107	28C1047BCD	107	28C717	107	13-14085-3
107	HC535B	107	S2131	107	28C1047C	107	28C717B	107	13-14085-4
107	HC537	107	S2132	107	28C1047D	107	28C717BK	107	13-23163-2
107	HC545	107	S2133	107	28C1047E	107	28C717BLK	107	13-67583-6/3464
107	HC784	107	S2134	107	28C1215	107	28C717C	107	13-67583-4/3464
107	HB-00829	107	S2164B	107	28C1215R	107	28C717R	107	14-602-77
107	HEP721	107	S40545	107	28C1293	107	28C722	107	14-602-77B
107	HEP731	107	SB2020	107	28C1293A	107	28C723	107	14-603-05
107	HEP732	107	SB2397	107	28C1293C	107	28C738A	107	14-603-05-2
107	HEP734	107	SK3356	107	28C1293D	107	28C739A	107	14-603-06
107	HR79	107	SKA1416	107	28C1320K	107	28C763(C)	107	15-05001-00
107	HR80	107	SM-A-595830-12	107	28C1342	107	28C763A	107	15-088003
107	HT303711A	107	SPS-856	107	28C1342A	107	0000280772	107	15-166N
107	HT303711B	107	SPS-860	107	28C1342B	107	0002807720	107	16GN
107	HT303711C	107	SPS83787	107	28C1342C	107	0000280829	107	16164
107	HT303710	107	SPS84143	107	28C1417	107	280835	107	018B
107	HT303801	107	SPS84168	107	28C1417D	107	280836M	107	19-020-071
107	HT303801J-B	107	SPS43-1	107	28C1417D(U)	107	280837	107	19QC17
107	HT303801A	107	SPS856	107	28C1417DU	107	280837P	107	020-00024
107	HT303801A0	107	SPS860	107	28C1417F	107	280837H	107	020-00025
107	HT303801B	107	ST11	107	28C1417G	107	280837K	107	020-00028
107	HT303801B-0	107	ST1502C	107	28C1417H	107	280837L	107	020-1112-004
107	HT303801B0	107	ST1502D	107	28C1417U	107	280837WP	107	21A040-023
107	HT303801C	107	ST6510	107	28C1417V	107	280922	107	21A040-024
107	HT303801C0	107	T-203	107	28C1417VW	107	280922A	107	21A040-054
107	HT303451A	107	T-Q5079	107	28C1417W	107	280922B	107	21A040-055
107	HT307720B	107	T-Q5106	107	28C155	107	280922J	107	21A040-063
107	HT307721C	107	T1886	107	28C156	107	280922K	107	21A112-007
107	HT307721D	107	T3568	107	28C674	107	280922T	107	21MM04
107	HT308291B0	107	T3601(RCA)	107	28C1674K	107	280922M	107	21M140
107	HT308291B	107	TQ28C1293	107	28C1674L	107	280924	107	21M151
107	HT309301D	107	TQ28C927A	107	28C1674M	107	280924E	107	21M152
107	HV0000302	107	TQ28C927C	107	28C1730	107	280924F	107	21M153
107	HX50003	107	TI-3016	107	28C1730L	107	280924M	107	21M154
107	J107	107	TIS818	107	28C1789	107	280929	107	21M178
107	J241177	107	TIS412	107	28C187	107	280929-0	107	21M179
107	J241188	107	TIM107	107	28C187R	107	280929A	107	21M182
107	J241189	107	TNJ-60605	107	28C187Y	107	280929C	107	21M188
107	J24561	107	TNJ60069(28C74)	107	28C1906	107	280929C1	107	22-1
107	J24562	107	TNJ60447	107	28C1990B	107	280929D	107	24MW 656
107	J24563	107	TNJ60448	107	28C2012	107	280929D1	107	24MW1057
107	J24701	107	TNJ60449	107	28C351	107	280929DE	107	24MW1058
107	J24903	107	TNJ60604	107	28C351(PA)	107	280929DP	107	24MW1081
107	K121J688-1	107	TNJ60605	107	28C375	107	280929DU	107	24MW1106
107	K2001	107	TNJ60606	107	28C375-Y	107	280929DV	107	24MW287
107	K2501	107	TNJ60607	107	28C381	107	280929EB	107	24MW361
107	K2502	107	TNJ61217	107	28C381-0	107	280929ED	107	24MW595
107	K2503	107	TNJ61218	107	28C381-Q	107	280929P	107	24MW593
107	K2509	107	TNJ61679	107	28C381-R	107	280929PK	107	24MW594
107	K4002	107	TNJ61172(28C722)	107	28C381BN	107	280929PR	107	24MW595
107	KLE4792	107	TNJ61730	107	28C381R	107	280929NP	107	24MW596
107	LM1110B	107	TNJ61731	107	28C384	107	280930	107	24MW597
107	LM1120B	107	TNJ71629	107	28C384-0	107	280930B	107	24MW653
107	LM1120C	107	TNJ71937	107	28C384T	107	280930BB	107	24MW656
107	LTI1016	107	TNJ71963	107	28C385	107	280930BK	107	24MW673
107	M-128J2510-1	107	TNJ72277	107	28C385A	107	280930BV	107	24MW700
107	M140-1	107	TNJ72279	107	28C386	107	280930C	107	24MW727
107	M1400-1	107	TR-8010	107	28C386A-0(TV)	107	280930CK	107	24MW738
107	M4757	107	TR-8043	107	28C387	107	280930CL	107	24MW793
107	M4789	107	TR228735046011	107	28C387A	107	280930CS	107	24MW805
107	M4825	107	TV24210	107	28C387G	107	280930D	107	24MW813
107	M4840A	107	TV24580	107	000028C460	107	280930DE	107	24MW813
107	M4904	107	TV24382	107	000028C535	107	280930DH	107	24MW814
107	M9032	107	TV24383	107	28C535(B)	107	280930DS	107	24MW815
107	M91	107	TV24385	107	28C535A	107	280930DT	107	24MW863
107	ME3002	107	TV24438	107	28C535B	107	280930DT-2	107	24MW953
107	ME3011	107	TV24806	107	28C535C	107	280930DX	107	24T-011-003
107	ME9002	107	TV57	107	28C535G	107	280930DZ	107	24T-016
107	ME9003	107	TV58	107	28C544	107	280930E	107	24T-016-010
107	ME9022	107	TV60	107	28C544D	107	280930EP	107	24T013003
107	MP1161	107	TVS-28288A	107	28C544E	107	280930ET	107	24T013005
107	MP1162	107	TVS-28C429A	107	28C545	107	280930EV	107	24T016
107	MP1163	107	TVS-28C469A	107	28C545A	107	280930EX	107	24T016001
107	MP1164	107	TVS-28C644	107	28C545B	107	280930J	107	24T016005
107	MP28823	107	TVS-28C645	107	28C545C	107	280930L	107	025-100026
107	MP28894	107	TVS-28C645A			107	280930NP	107	25T-002
107	MP86528	107	TVS-28C645B			107	28D829	107	28-819-172
107	MP86529	107	TVS-28C645C					107	31-0048
107	N-EA15X130	107	TV828C829B						
		107	TX-107-12						

ECG	Industry Standard No.
107	31-0049
107	31-0050
107	31-0054
107	31-0098
107	31-0206
107	34-3015-46
107	34-3015-47
107	34-3015-49
107	34-6015-50
107	34-6015-8
107	36-1
107	36-6015-46
107	42-22532
107	42-23960
107	42-23960P
107	42-23961
107	42-23961P
107	42-23962
107	42-23962P
107	42-23963
107	42-23963P
107	46-06311-3
107	46-86101-3
107	46-86109-3
107	46-86110-3
107	46-86120-3
107	46-86126-3
107	46-86127-3
107	46-86133-3
107	46-86172-3
107	46-86207-3
107	46-86209-3
107	46-86239-3
107	46-86251-3
107	46-86265-3
107	46-8629
107	46-86295-3
107	46-86299-3
107	46-86301-3
107	46-86302-3
107	46-86352-3
107	46-86354-3
107	46-86435-3
107	46-8647
107	46-96434-3
107	48-01-017
107	48-01-031
107	48-134789
107	48-134837
107	48-134845
107	48-134904
107	48-134904A10
107	48-134904P
107	48-134922
107	48-134923
107	48-134924
107	48-134925
107	48-134926
107	48-134945
107	48-134961
107	48-134962
107	48-134963
107	48-134964
107	48-134965
107	48-134966
107	48-134981
107	48-137040
107	48-137059
107	48-137071
107	48-137075
107	48-137076
107	48-137077
107	48-137400
107	48-137612
107	48-3003A03
107	48-41815J02
107	48-41816J01
107	48-41816J02
107	48-43351A02
107	48-63076A81
107	48-63076A82
107	48-63078A52
107	48-63078A54
107	48-97046A05
107	48-97046A06
107	48-97046A07
107	48-97046A17
107	48-97046A20
107	48-97046A21
107	48-97127A06
107	48-97162A26
107	48P65146A61
107	50C380-0
107	50C380-OR
107	50C394-0
107	50C394-R
107	50C784-R
107	051-0049
107	56-8086
107	56-8086A
107	56-8086B
107	56-8086C
107	56-8087
107	56-8087B
107	56-8087C
107	56-8088
107	56-8088A
107	56-8088C
107	57A126-12
107	57A21
107	57A21-12
107	57A21-13
107	57A21-14
107	57A21-45
107	57B21-9
107	57C21
107	57C21-5
107	62-19581
107	65(TRANSISTOR)
107	70N3
107	81-46125007-8
107	81-46125012-8
107	81-46125013-6
107	81-46125030-0
107	81-46125032-6
107	81-46125033-4
107	86-525-2
107	86-596-2
107	86-597-2
107	86X0060-001
107	86X0061-001
107	86X0062-001
107	86X7-6
107	90T2
107	100-4107
107	0103-0531/4460
107	0103-0568
107	0103-0568/4460
107	105-00107-09
107	105-02005-07
107	105-02006-05
107	105-02008-01
107	121-480
107	121-723
107	121-735B
107	121-849
107	121J688-1
107	121J688-2
107	130
107	130-112
107	1300RN
107	151-0259-00
107	151-0427-00
107	151-0471-00
107	176-025-9-002
107	176-073-9-001
107	201-254343-13
107	201-254343-28
107	201-254343-30
107	201-254343-33
107	207A25
107	229-0190-31
107	229-0191-29
107	229-0191-30
107	229-0210-19
107	229-0260-18
107	260D05704
107	260D08013
107	260P05402
107	260P05402A
107	260P1760
107	297L011C01
107	325-1378-18
107	325-1378-19
107	325-1771-16
107	344-6015-8
107	344-6015-9
107	352-0630
107	352-0630-010
107	352-0653-010
107	352-0653-020
107	417-125-12903
107	417-154
107	417-19
107	417-190
107	417-205
107	429-0981-12
107	429-0985-12
107	465-181-19
107	522(ZENITH)
107	535A
107	576-0003-029
107	642-229
107	642-230
107	642-254
107	642-260
107	642-268
107	642-269
107	642-270
107	642-274
107	642AS4076-101
107	690V088H46
107	690V088H47
107	690V088H49
107	690V089H46
107	690V089H86
107	690V103H28
107	690V109H46
107	690V116H20
107	702-0002
107	715FB
107	753-2000-460
107	753-4000-668
107	753-4000-929
107	753-5751-359
107	753-5851-359
107	753-9000-922
107	772A
107	772B
107	772B1
107	772BJ
107	772BL
107	772BM
107	772BN
107	772BY
107	772C
107	772CC
107	772D
107	772D1
107	772DC
107	772DG
107	772E
107	772EH
107	772F
107	772FE
107	772G
107	909(RCA)
107	909-27125-160
107	916-31024-3B
107	916-31024-5
107	916-31025-5B
107	921-102
107	921-102A
107	921-102B
107	921-141B
107	921-142B
107	921-143B
107	921-145B
107	921-1528
107	921-176B
107	921-177B
107	921-204B
107	921-210B
107	921-211B
107	921-213B
107	921-226B
107	921-232B
107	921-233B
107	921-235B
107	921-326B
107	921-59B
107	921-62
107	921-62A
107	921-62B
107	921-63
107	921-63A
107	921-63B
107	921-64
107	921-64A
107	921-64B
107	921-64C
107	921-84
107	921-84A
107	921-84B
107	921-85
107	921-85A
107	921-85B
107	921-86
107	921-86A
107	921-86B
107	930DZ
107	991-01-1316
107	1000-135
107	1002A(JULIETTE)
107	1003-01
107	1039-0482
107	1100-9446
107	1100-9453
107	1373-17A
107	1420-1-1
107	1455-7-4
107	1585H
107	1687-17
107	1854-0417
107	1931-17
107	1998-17
107	1999-17
107	2004-01
107	2057A100-53
107	2057A2-109
107	2057A2-110
107	2057A2-116
107	2057A2-117
107	2057A2-120
107	2057A2-128
107	2057A2-216
107	2057A2-217
107	2057A2-218
107	2057A2-219
107	2057A2-220
107	2057A2-221
107	2057A2-237
107	2057A2-259
107	2057A2-304
107	2057A2-305
107	2057A2-322
107	2057A2-323
107	2057A2-325
107	2057A2-326
107	2057A2-331
107	2057A2-356
107	2057A2-395
107	2057A2-466
107	2057A2-501
107	2057A2-503
107	2057A2-539
107	2057A2-540
107	2057B2-101
107	2057B2-102
107	2057B2-114
107	2057B2-115
107	2057B2-116
107	2057B2-117
107	2057B2-125
107	2057B2-143
107	2197-17
107	2213-17
107	2284-17
107	2291-17
107	3003(SEARS)
107	3004(SEARS)
107	3029
107	3107-204-90100
107	3301(SEARS)
107	3507(SEARS)
107	3514(SEARS)
107	3560-1(RCA)
107	4041-000-40270
107	4041-000-40300
107	4041-000-60170
107	4041-000-60200
107	4789
107	4822-130-40214
107	4822-130-40215
107	4822-130-40216
107	4822-130-40311
107	4822-130-40312
107	4822-130-40313
107	4822-130-40317
107	4822-130-40318
107	4851
107	5613-460
107	5613-460A
107	5613-460B
107	5613-460C
107	5613-535
107	5613-535A
107	5613-535B
107	5613-535C
107	8000-00006-005
107	8000-0005-002
107	8000-0005-003
107	8503
107	8840-162
107	9330-229-60112
107	9330-229-70112
107	18410-142
107	18600-151
107	0023645
107	0023829
107	28810-172
107	37986-3563
107	37986-4040
107	37986-4046
107	040001
107	58810-161
107	58840-192
107	75568-3
107	88510-172
107	116119
107	118822
107	119412
107	122902
107	122904
107	124024
107	124623
107	124624
107	125329
107	125389
107	125390
107	125390(RCA)
107	125394
107	125474
107	125475
107	125994
107	125995
107	129392
107	129393
107	129394
107	129571
107	129573
107	129574
107	129979
107	130793
107	165392
107	165931
107	165932
107	166906
107	167263
107	168567
107	169194
107	169196
107	169505
107	169574
107	170294
107	170388
107	170794
107	171009
107	171029(TOSHIBA)
107	171031(TOSHIBA)
107	171034
107	171915
107	171983
107	183015
107	183016
107	183018
107	183019
107	489751-052
107	489751-173
107	514045
107	515041S
107	0573485
107	0573507
107	0573508
107	0573509H
107	0573510H
107	0573511H
107	0573607
107	610100-2
107	740949
107	740950
107	740951
107	741726
107	741855
107	741862
107	742537
107	742547
107	742549
107	742728
107	749002
107	749014
107	760249
107	760253
107	772738
107	772739
107	916049
107	916068
107	960201
107	960202
107	972306
107	972307
107	972417
107	972418
107	972419
107	972420
107	983233
107	983234
107	983235
107	983742
107	984876
107	984877
107	984878
107	985442
107	985444
107	985619
107	986576
107	986693
107	986694
107	1223911
107	1472636-1
107	1473529-1
107	1473544-1
107	1473558-1
107	2003342-244
107	2003779-22
107	2003779-23
107	2003779-24
107	2003779-25
107	2006513-39
107	2006681-93
107	2006681-94
107	2006681-95
107	2008292-56
107	2008299-1
107	2092417-0017
107	2092417-0018
107	2092417-0019
107	2092693-0724
107	2092693-0725
107	2093308-0704
107	2093308-0705
107	2093308-0706
107	2093308-0725
107	2320031
107	2320042
107	2320043
107	2320471-1
107	2320471H
107	2320981
107	2498456-2
107	2498482-2
107	2498508-2
107	2498508-3
107	2498902-1
107	2498902-2
107	2498903-1
107	2498903-3
107	3596440
107	3597114
107	5320326H
107	5320861
107	5321431
107	06120009
107	06120015
107	7305468
107	7910781-01
107	7914009-01
107	7914010-01
107	8031837
107	8036683
107	8037722
107	10027973-101
107	13038115
107	23114126
107	23114157
107	23114165
107	23126289
107	23126290
107	23126291
107	43021067
107	43022860
107	43027618
107	44089001
107	55440111-001
107	62105038
107	62105291
107	62695919
107	62695927
107	62798564
107	70260450
107	80366830
107	80366840
107	83093005
107	93078420
107	93082920
107	93939440
107	1760609001
107	1760609002
107	1760829001
107	2003038092
107	2003190604
107	4100907633
107	4104204602
107	4104204603
107	4104204612
107	16103190930
107	16112190710
1071	AN342
1071	GEIC-67
1072	AN253
1072	GEIC-59
1073	AN277
1073	AN277AB
1073	AN277B
1073	AN277BA
1073	GEIC-63
1074	AN260
1074	GEIC-60
1076	37007001
1077	37006001
1078	GEIC-136
1078	MPC571C
1078	RVIUPC22C
1079	RH-1X0023CEZZ
108	A-1854-0485
108	A-1854-JBD1
108	A054-157
108	A054-158
108	A054-159
108	A054-163
108	A054-164
108	A054-470
108	A068-111
108	A068-112
108	A1109
108	A1170
108	A121-480
108	A124623
108	A124624
108	A125329
108	A129509
108	A129510
108	A129511
108	A129512
108	A129513
108	A129571
108	A129572
108	A129573
108	A129574
108	A1300RN
108	A130V100
108	A14-602-63
108	A14-603-05
108	A14-603-06
108	A1462
108	A1518
108	A1519
108	A1520
108	A1521
108	A154(NPN)
108	A155(NPN)
108	A164(NPN)
108	A165(NPN)
108	A176-025-9-002
108	A19-020-072
108	A1E
108	A1G-1A
108	A1K
108	A1P-/4923
108	A1P-5
108	A1P/4923-1
108	A1R-1/4925
108	A1R-1A
108	A1R-2/4926
108	A1R-24926
108	A1R-2A
108	A1R-5
108	A1R/4925A
108	A1R/4926A
108	A2006681-95
108	A2092693-0724
108	A2092693-0725
108	A24
108	A245
108	A245(AMC)
108	A2498
108	A24MW594
108	A24MW595
108	A24MW596
108	A24MW597
108	A24T-016-016
108	A2C
108	A2D
108	A2F
108	A2G
108	A2H
108	A2L
108	A2N-2A
108	A2P-5
108	A2T

ECG	Industry Standard No.	ECG	Industry Standard No.	ECG	Industry Standard No.	ECG	Industry Standard No.	ECG	Industry Standard No.
108	A2T919	108	BCY89	108	C81227	108	E815X30	108	HEP723
108	A2V	108	BD71	108	C81227D	108	E815X6	108	HEP727
108	A2W	108	BP-115	108	C81227E	108	E815X60	108	HEP733
108	A31-0206	108	BP162	108	C81227F	108	E815X69	108	HEP80020
108	A32-2809	108	BP163	108	C81227G	108	E83266	108	HR58
108	A3A	108	BP164	108	C81238	108	ET15X18	108	HR59
108	A3C	108	BP165	108	C81238G	108	ET15X19	108	HR60
108	A3D	108	BP173A	108	C81238H	108	ET15X21	108	H3-1225
108	A3H	108	BP176	108	C81238I	108	ET15X23	108	H3-1226
108	A3N71	108	BP187	108	C81243E	108	ET15X30	108	H3-1227
108	A3N72	108	BP188	108	C81243H	108	ET15X7	108	H3-40017
108	A3N73	108	BP194A	108	C81244H	108	EU15X1	108	H3-40020
108	A3P	108	BP223	108	C81244J	108	EU15X2	108	H3-40045
108	A3R	108	BP233-4	108	C81252B	108	EU15X3	108	H3-40047
108	A38B	108	BP235	108	C81252C	108	EU15X6	108	H3-40055
108	A418	108	BP236	108	C81293	108	EW162	108	HS40049
108	A419	108	BP262	108	C81330A	108	F15810	108	HT3037010
108	A420	108	BP263	108	C81330B	108	F15835	108	HT3037I1A-Q
108	A427	108	BP264	108	C81330C	108	F15841	108	HT3037201
108	A427(JAPAN)	108	BP357	108	C81340B	108	P20-1001	108	HT3037201A
108	A4789	108	BP817	108	C81340E	108	P20-1002	108	HT3037201B
108	A48-134789	108	BP817R	108	C81340F	108	P20-1003	108	HT3037I20A
108	A48-134837	108	BPV83	108	C81340G	108	P20-1004	108	HT3039I41
108	A48-134845	108	BPV85A	108	C81340H	108	P20-1005	108	HT3039I41A
108	A48-134902	108	BPV85D	108	C81350	108	P215-1001	108	HT3039I41B
108	A48-134904	108	BPV85E	108	C81351	108	P215-1002	108	HT3045540A0
108	A48-134922	108	BPV85F	108	C81359	108	P2427	108	HT3046010C0
108	A48-134923	108	BPV85G	108	C81360	108	P2450	108	HT30461IB
108	A48-134924	108	BPV64	108	C81361E	108	P24T-016-024	108	HT3039551C0
108	A48-134925	108	BPX18	108	C81361P	108	P2633	108	HT306451
108	A48-134926	108	BPX19	108	C81386H	108	P2634	108	HT306451B
108	A48-134945	108	BPX20	108	C81508G	108	P2636	108	HT3066810
108	A48-134961	108	BPX21	108	C81509B	108	P5530	108	HT306962A-Q
108	A48-134962	108	BPX45	108	C81509P	108	P4706	108	HT3039291C
108	A48-134963	108	BPY-37	108	C81518E	108	P9623(G.E.)	108	HX500001
108	A48-134964	108	BPY-47	108	C81555	108	P9625	108	IRTR95
108	A48-134965	108	BPY-48	108	C81661	108	P96N	108	J108
108	A48-134966	108	BPY78	108	C81834	108	P05006	108	J187
108	A48-134981	108	BPY90B	108	C8184J	108	PC81168E	108	J24596
108	A48-137071	108	B8V35	108	C82006G	108	PC81168E641	108	J24635
108	A48-137075	108	B8V35B	108	C82008	108	PC81225E	108	J24636
108	A48-137076	108	B8V35C	108	C82008G	108	PC81227B814	108	J24637
108	A48-137077	108	B8V35D	108	C82008H	108	PC81227F	108	J24813
108	A48-137197	108	B8V52	108	C82008H552	108	PC81227F743	108	J24814
108	A48-40247901	108	B8V52R	108	C8429J	108	PC81227G	108	J24852
108	A48-43351A02	108	B8X12	108	C8430H	108	PC81270810	108	J24863
108	A48-63076A82	108	B8X26	108	C8461P	108	PC89018D	108	J24904
108	A48-97046A05	108	B8X27	108	C8469P	108	PC89018E	108	J24905
108	A48-97046A06	108	B8X35	108	C86225F	108	PC89018G	108	J24915
108	A48-97046A07	108	B8X92	108	C86226F	108	PC89018H	108	J24921
108	A48-97127A06	108	B8X93	108	C86227E	108	PC89066	108	J24923
108	A48-97127A12	108	B8Y-62	108	C86227F	108	PK2369A	108	J24933
108	A48-97127A18	108	B8Y-72	108	C89001	108	PK2484	108	J8P7005
108	A480	108	B8Y-73	108	C89018	108	PK3014	108	J8P7006
108	A4851	108	B8Y-74	108	C89018/3490	108	PK3299	108	K4-510
108	A497	108	B8Y-80	108	C89018D	108	PK3300	108	KD2119
108	A4B-5	108	B8Y-95	108	C89018E	108	PK914	108	LM1110A
108	A4G	108	B8Y22	108	C89018F	108	PK918	108	LM1123H
108	A4T	108	B8Y23	108	C89018P	108	FM1613	108	LM1138G/H
108	A4Y-1A	108	C1023	108	C89018F/3490	108	FM1711	108	LT1016
108	A57A144-12	108	C111B	108	C89018PG	108	FM2368	108	LT1016E
108	A57A145-12	108	C185V	108	C890210-I	108	FM2369	108	LT1016G
108	A5J	108	C2475078-3	108	C89124-C2	108	FM2846	108	LT81016(G.E.)
108	A62-19581	108	C2485078-1	108	C89124B1	108	FM3014	108	LTH1016(G.E.)
108	A642-254	108	C2485079-1	108	C89125-B1	108	FM709	108	M012
108	A642-260	108	C271	108	CT1500	108	FM709	108	M024
108	A642-268	108	C272	108	CT1500	108	FM720A	108	M4709
108	A667RED	108	C289	108	D058	108	FM870	108	M4733
108	A6B	108	C3123	108	D069	108	FM871	108	M4820
108	A6E	108	C387A(FA-3)	108	D087	108	FM910	108	M4826
108	A6F	108	C387G	108	D088	108	FM911	108	M4837
108	A6P	108	C39A	108	D1666	108	FM914	108	M4845
108	A6T	108	C40	108	D24A3394	108	FPR40-1003	108	M4855
108	A6U	108	C561	108	D562	108	FPR50-1003	108	M4857
108	A6V-5	108	C63(JAPAN)	108	DX1018	108	FPR50-1004	108	M75547-1
108	A715FB	108	C662	108	E2434	108	FS3266	108	M75547-2
108	A772738	108	C684(JAPAN)	108	E2435	108	FS326690	108	M9010
108	A772739	108	C684A(JAPAN)	108	E629	108	FS35529	108	M9482
108	A7720I	108	C735(FA-3)	108	EA0013	108	FS3683	108	M9575
108	A7A30	108	C740	108	EA0091	108	FSE1001	108	MB5001
108	A7A31	108	C79	108	EA0093	108	FSE3001	108	MB8201
108	A7A32	108	C828	108	EA0094	108	FS85002	108	MB9021
108	A7N	108	CC82006D	108	EA0095	108	FSP-1	108	MH1001
108	A7P	108	CC82008FP015	108	EA1343	108	FSP-164	108	MIM1101
108	A7U	108	CC89016D	108	EA1562	108	FSP-165	108	MI1536
108	A7V	108	CC89016E	108	EA1563	108	FSP-166	108	MJE9411T
108	A7W	108	CC89016F	108	EA15X113	108	FSP-166-1	108	MM1367/280684
108	A88	108	CC89018P	108	EA15X131	108	FSP-215	108	MM1382
108	A909-27125-160	108	CDC12112C	108	EA15X132	108	FSP-242-1	108	MM1387
108	A90T2	108	CDC12112D	108	EA15X48	108	FSP-270-1	108	MM1945
108	A916-31025-58	108	CDC12112E	108	EA15X49	108	FSP-289-1	108	MM709
108	A916-31025-5B	108	CDC12112F	108	EA15X50	108	FSP-42	108	MP835S6
108	A921-59B	108	CDC5000-1B	108	EA15X51	108	FSP-42-1	108	MP835S3
108	A921-62B	108	CDC5030A-1A	108	EA15X54	108	FT1315	108	MP8607
108	A921-63B	108	CDC5071A	108	EA15X55	108	FT1324B	108	MP86511
108	A921-64B	108	CDC5075B	108	EA15X7113	108	FT1324C	108	MP86511-S
108	A991-01-1316	108	CE4010D	108	EA15X7140	108	FT709	108	MP86568
108	A9A	108	CE4010E	108	EA15X7177	108	FV2369A	108	MP8918
108	A9D	108	CF1	108	EA15X7228	108	FV2484	108	MP89423F
108	AR200	108	CIL511	108	EA15X7243	108	FV3014	108	MP89423G
108	AR200W	108	CIL512	108	EA15X7263	108	FV3299	108	MP89423H
108	AR201	108	CIL513	108	EA15X7587	108	FV3300	108	MP89423I
108	AR201(YELLOW)	108	CIL521	108	EA15X7722	108	FV914	108	MP89601
108	AR201Y	108	CIL523	108	EA15X8589	108	FV918	108	MP89604PG
108	AR202	108	CIL531	108	EA15X860B	108	G05-004A	108	MP8H07
108	AR202(GREEN)	108	CIL532	108	EA15X8609	108	G05004A	108	MP8H08
108	AR202G	108	CIL533	108	EA15X8610	108	GB-61	108	M8701T
108	AR212	108	CS-2004C	108	EA15X94	108	GMO380	108	M875018
108	AR221	108	CS-2007G	108	EA1733	108	GME3001	108	M875017
108	AR25(ORANGE)	108	CS-2007H	108	EA1793	108	GME3002	108	M875028
108	AR25(WHITE)	108	CS-2008F	108	EA2493	108	GME9001	108	M87502T
108	AR313	108	CS-3001B	108	EA2495	108	GME9002	108	M8R7502
108	AT520	108	CS-4618	108	EA2600	108	GME9021	108	M8R7501
108	AX91770	108	CS-6270G	108	EA2601	108	GME9022	108	M8R7502
108	B9426	108	CS-9018	108	EA2602	108	H1V	108	M8R7501
108	BC112	108	CS-9018E	108	EA2603	108	H442	108	MT100
108	BC155A	108	CS-9018F	108	EA2604	108	H9625	108	MT101
108	BC156A	108	CS-9018G	108	EA2605	108	HC-00668	108	MT102
108	BC194	108	CS-9018H	108	EA3406	108	HC-00772	108	MT106
108	BC194B	108	C81014	108	ED592M	108	HC-00839	108	MT107
108	BC195	108	C81014A	108	EF718A	108	HC206	108	MT743
108	BC195CD	108	C81014E	108	EN916	108	HC645	108	MT744
108	BC295	108	C81014F	108	EN918	108	HC668	108	MT753
108	BC6500	108	C81018	108	EP15X55	108	HC772	108	N-EA15X131
108	BCW31	108	C81168E	108	EQS-21	108	HC829	108	N-EA15X132
108	BCW31R	108	C81225D	108	E810187	108	HEP-80016	108	NPC173
108	BCW32	108	C81225E	108	E815I0	108	HEP-80021	108	NPC188
108	BCW32R	108	C81225F	108	E815X18	108	HEP56	108	N81356
108	BCW71	108	C81226	108	E815X19	108	HEP718	108	N83300
108	BCW71R	108	C81226E	108	E815X2	108	HEP720	108	N8381
108	BCW72	108	C81226F	108	E815X3	108	HEP722	108	N8382
108	BCY87	108	C81226G							108	N86112
108	BCY88	108	C81226H							108	N86113

ECG	Industry Standard No.	ECG	Industry Standard No.	ECG	Industry Standard No.	ECG	Industry Standard No.	ECG	Industry Standard No.
108	N87261	108	R87334	108	S81219	108	T1XM17	108	TR210
108	N87267	108	R87511	108	SE3001R	108	T2634	108	TR228755045311
108	N89710	108	R87512	108	SE3001Y	108	T308	108	TR228735048617
108	NTC-5	108	R87520	108	SE3003	108	T3530	108	TR228735048618
108	P346	108	R87522	108	SE5006	108	T3535	108	TR22C
108	PA-10556	108	R87524	108	SE5006Q	108	T3536	108	TR24
108	PBC182	108	R87532	108	SE5010	108	T3539	108	TR28C371
108	PEP1001	108	R87533	108	SE5015	108	T3568(RCA)	108	TR28C372
108	PET-101-1	108	RT2915	108	SE5029	108	T386(SEARS)	108	TR28C384
108	PET1075	108	RT3069	108	SE5030	108	T9011CD	108	TR310230
108	PET3001	108	RT3070	108	SE5031	108	T9011EF	108	TR310244
108	PET8201	108	RT3095	108	SE5036	108	T9011G	108	TR310249
108	PET8250	108	RT3225	108	SE504	108	T9011GEF	108	TR310250
108	PET8251	108	RT3226	108	SE5040	108	T9011GH	108	TR38
108	PET8500	108	RT3227	108	SE5056	108	T9011H	108	TR8004
108	PL1051	108	RT3232	108	SE21	108	T9011HEF	108	TR8042
108	PL1053	108	RT5061	108	SK-31024-3	108	T9011J	108	TS9013
108	PL1055	108	RT5200	108	SK1320	108	T9016F	108	TSC614
108	PL1061	108	RT5201	108	SK3019	108	T9016H	108	TT-204
108	PL1062	108	RT5205	108	SK7181	108	TA-7	108	TT-204A
108	PL1063	108	RT5900	108	SKA4074	108	TA2401	108	TT-204AB
108	PL1064	108	RT5902	108	SKA4075	108	TA2503	108	TT-204B
108	PL1065	108	RT5903	108	SKA4076	108	TA7319	108	TT-204C
108	PL1081	108	RT5904	108	SKA4525	108	TC-0918	108	TT204
108	PL1082	108	RT6157	108	SKA9013	108	TC0914	108	TT204A
108	PM195A	108	RT6158	108	SL-100	108	TC0918	108	TT204AB
108	PMT1767	108	RT6159	108	SL100	108	TC2369A	108	TT204B
108	Q-00284R-3	108	RT6160	108	SM5796	108	TC2483	108	TT204C
108	Q-00384R-3	108	RT6203	108	SP4168	108	TC2484	108	TV-15
108	Q-0064R-1	108	RT6600	108	SP8-1351	108	TE2484	108	TV-15B
108	Q-005R8R-3	108	RT6991	108	SP8-1352	108	TE2715	108	TV-22
108	QSE3001	108	RT7320	108	SP8-1353	108	TE2716	108	TV-7
108	R-280545	108	RT7321	108	SP8-1473	108	TE3707	108	TV1000
108	R-280668	108	RT7323	108	SP8-2111	108	TE3708	108	TV115
108	R-3530-1	108	RT7324	108	SP8-2320	108	TE3709	108	TV17A
108	RO6-1001	108	RT9308	108	SP8-4145	108	TE3710	108	TV18
108	RO6-1002	108	RV1467	108	SP81351	108	TE3711	108	TV2403
108	RO6-1003	108	RV1468	108	SP81352	108	TE706	108	TV2404
108	RO6-1004	108	RV1469	108	SP81353	108	TI-410	108	TV24102
108	RO6-1005	108	RV1470	108	SP820	108	TI-417	108	TV24148
108	RO6-1006	108	RVS280645	108	SP82110	108	TI-420	108	TV24203
108	R2473	108	RVTCS81384	108	SP82111	108	TI-430	108	TV24204
108	R2476	108	S-1037	108	SP82224	108	TI-431	108	TV24313
108	R2477	108	S-1041	108	SP82265	108	TI-474	108	TV24387
108	R62194	108	S-1058	108	SP82265-2	108	TI-490	108	TV24573
108	R810	108	S-1059	108	SP82266	108	TI-495	108	TV24574
108	RB1001	108	S-1060	108	SP83003	108	TI25A	108	TV24589
108	RB1002	108	S-1062	108	SP83370	108	TI25B	108	TV24684
108	RR8070	108	S-1078	108	SP83929	108	TI3016	108	TV32
108	RR8116	108	S-1079	108	SP83937	108	TI407	108	TV55
108	RR8118	108	S-1153	108	SP83948	108	TI408	108	TVS-1818
108	RR8119	108	S-1227	108	SP83952	108	TI409	108	TVS-28C185A
108	RR8989	108	S-1276	108	SP83968	108	TI410	108	TVS-28C208
108	RS-109	108	S-1286	108	SP83971	108	TI431	108	TVS-28C287A
108	RS-2015	108	S-1296	108	SP84	108	TI824	108	TVS-28C446
108	RS-7102	108	S-1316	108	SP840	108	TI862	108	TVS-28C605
108	RS-7104	108	S-1317	108	SP84002	108	TI863	108	TVS-28C606
108	RS-7106	108	S-1318	108	SP84005	108	TI864	108	TVS-TI818
108	RS-7107	108	S-1360	108	SP84008	108	TI885	108	TVS28C288A
108	RS-7108	108	S-1361	108	SP84016	108	TI898A	108	TVS28C466
108	RS-7109	108	S-1362	108	SP84030	108	TIX876	108	TVS28C538
108	RS-7110	108	S-1408	108	SP84043	108	TIX880	108	TVS28C645
108	RS-7113	108	S-1409	108	SP84050	108	TIX809	108	TVS28C645A
108	RS-7114	108	S-532E	108	SP84051	108	TIX810	108	TVS28C645C
108	RS-7115	108	S-5670-E	108	SP84068	108	TIX828	108	TVS28C683
108	RS-7202	108	S-95125	108	SP84079	108	TIX829	108	TVS28C762
108	RS-7512	108	S-95125A	108	SP84080	108	TIX830	108	TVS28C828Q
108	RS1726	108	S-95126	108	SP84091	108	TIX831	108	TVS28C920-0Q
108	RS6523	108	S-95126A	108	SP84145	108	TM108	108	V120PH
108	RS7101	108	S0016	108	SP84167	108	TMT-2427	108	V220
108	RS7102	108	S0020	108	SP84288	108	TMT2427	108	V221
108	RS7104	108	S0021	108	SP84399	108	TMT696	108	V222
108	RS7106	108	S1009	108	SP84423	108	TMT697	108	V405
108	RS7107	108	S1019	108	SP84610	108	TMT839	108	V415
108	RS7109	108	S1037	108	SR130-1	108	TMT840	108	V417
108	RS7110	108	S1044	108	ST10	108	TMT841	108	V435
108	RS7112	108	S1058	108	ST1026	108	TMT842	108	VS-28C206
108	RS7113	108	S1059	108	ST1050	108	TMT843	108	VS-28C208
108	RS7115	108	S1062	108	ST1051	108	TNJ-60066	108	VS-28C288A
108	RS7116	108	S1076	108	ST1J	108	TNJ60066	108	VS-28C563
108	RS7117	108	S1078	108	ST13	108	TNJ61729	108	VS-28C565
108	RS7118	108	S1079	108	ST1336	108	TNJ70478	108	VS-28C645
108	RS7119	108	S1142	108	ST14	108	TNJ70478-1	108	VS-28C645A
108	RS7120	108	S1153	108	ST15	108	TNJ70479	108	VS-28C684
108	RS7123	108	S1227	108	ST29	108	TNJ70480	108	VS-28C762
108	RS7123	108	S1276	108	ST30	108	TNJ70484	108	VS28C1855//-1
108	RS7124	108	S1296	108	ST31	108	TNJ71173	108	W8
108	RS7125	108	S1313	108	ST32	108	TNJ7149B	108	X81
108	RS7126	108	S1316	108	ST33	108	TNT-839	108	X814
108	RS7128	108	S1317	108	ST34	108	TNT-840	108	X815
108	RS7135	108	S1318	108	ST35	108	TNT-841	108	X82
108	RS7138	108	S1360	108	ST40	108	TNT839	108	X83
108	RS7139	108	S1361	108	ST41	108	TNT840	108	X84
108	RS7140	108	S1362	108	ST415	108	TNT843	108	X86
108	RS7141	108	S1408	108	ST42	108	TP4275	108	XT15X3
108	RS7142	108	S1409	108	ST43	108	TQ1	108	ZA100962B
108	RS7144	108	S1636	108	ST44	108	TQ2	108	ZDT-30
108	RS7145	108	S1674	108	ST45	108	TQ3	108	ZDT-31
108	RS7161	108	S1674A	108	ST60	108	TQ5	108	ZDT10
108	RS7163	108	S1682	108	ST61	108	TQ5049	108	ZDT11
108	RS7163	108	S2159	108	ST6110	108	TQ6	108	ZDT20
108	RS7164	108	S2224	108	ST6120	108	TQ7	108	ZDT21
108	RS7165	108	S2805	108	ST62	108	TQ8	108	ZDT30
108	RS7166	108	S2617	108	ST70	108	TQ9	108	ZDT31
108	RS7167	108	S3019	108	ST71	108	TR-016	108	ZEN104
108	RS7168	108	S3020	108	ST72	108	TR-01B	108	ZEN108
108	RS7169	108	S32669	108	ST80	108	TR-163(OLSON)	108	ZEN109
108	RS7170	108	S326690	108	ST82	108	TR-1R31	108	ZEN118
108	RS7174	108	S5327E	108	ST9	108	TR-1R35	108	ZJ40
108	RS7175	108	S5328E	108	ST903	108	TR-24	108	Z8C555B
108	RS7176	108	S5670E	108	ST904	108	TR-2R31	108	ZZ2475
108	RS7177	108	S95125	108	ST904A	108	TR-2R33	108	ZZ2857
108	RS7201	108	S95125A	108	ST905	108	TR-2R35	108	ZT3269A
108	RS7202	108	S95126	108	ST910	108	TR-28C371	108	ZT3600
108	RS7209	108	S95126A	108	ST400	108	TR-28C372	108	ZT709
108	RS7210	108	SATCS2339	108	SX-3825	108	TR-28C384	108	ZT917
108	RS7211	108	SC12277P	108	SX3001	108	TR-3R31	108	ZT918
108	RS7212	108	SC12279	108	SX3825	108	TR-3R33	108	ZTX320
108	RS7214	108	SE-1019	108	SX3827	108	TR-3R35	108	ZTX321
108	RS7215	108	SE-1044	108	SYL-2300	108	TR-4R35	108	01-031855
108	RS7216	108	SE-1419	108	SYL-4131	108	TR-8004-4	108	01-031906
108	RS7217	108	SE-3001	108	SYL4131	108	TR-8004-5	108	1-041/2207
108	RS7218	108	SE-3002	108	T-483	108	TR-8028	108	01-117005
108	RS7219	108	SE-3005	108	T-484	108	TR-8029	108	01-117006
108	RS7221	108	SE-5019	108	T-486	108	TR-8030	108	1-801-003-12
108	RS7225	108	SE-5001	108	T-H28C313	108	TR-8031	108	1-801-003-13
108	RS7227	108	SE-5002	108	T-H28C387	108	TR-8032	108	1-801-003-14
108	RS7228	108	SE-5005	108	T-Q5055	108	TR-8058	108	1-801-003-15
108	RS7229	108	SE-5006-14	108	T1003-521	108	TR01026	108	1-801-305-13
108	RS7230	108	SE-5020	108	T1003521	108	TR010602-1	108	1-801-306
108	RS7231	108	SE-5021	108	T1394	108	TR112	108	1-801-306-13
108	RS7237	108	SE-5050	108	T1898	108	TR1512-80	108	1-801-306-14
108	RS7333	108	SE1019	108	T1828			108	1-801-306-15
		108	SE1044	108	T1XM15			108	01-9016-42221-3

ECG	Industry Standard No.	ECG	Industry Standard No.	ECG	Industry Standard No.	ECG	Industry Standard No.	ECG	Industry Standard No.
108	01-9018-62221-3	108	09-305069	108	19A123160-1	108	48-134787	108	57A143-2
108	1-TR-046	108	09-305070	108	19A123160-2	108	48-134800	108	57A143-3
108	002-009600	108	09-305071	108	020-00186	108	48-134806	108	57A143-4
108	002-009601	108	09-305072	108	020-00027	108	48-134814	108	57A143-5
108	002-9601	108	09-305074	108	20-00229-001	108	48-134818	108	57A143-6
108	002-9601-12	108	09-309007	108	20-00444-001	108	48-134820	108	57A143-7
108	2N1005	108	09-309013	108	20-1	108	48-134821	108	57A143-8
108	2N1060	108	09-309024	108	21A015-004	108	48-134826	108	57A146-11
108	2N2032	108	09-309027	108	21A015-014	108	48-134827	108	57A146-12
108	2N2197	108	09-309028	108	21A015-016	108	48-134828	108	57A151-6
108	2N2475	108	09-309032	108	21A040-003	108	48-134855	108	57A152-12
108	2N2615	108	09-309073	108	21A040-004	108	48-134879	108	57A160-1
108	2N2616	108	09-321124	108	21A040-007	108	48-134891	108	57A160-2
108	2N2711	108	9-5125	108	21A040-010	108	48-134892	108	57A160-3
108	2N2784	108	9-5126	108	21A040-016	108	48-134893	108	57A160-4
108	2N2784/52	108	9-5127	108	21A040-017	108	48-134902	108	57A20-1
108	2N3010	108	9-5128	108	21A040-019	108	48-134908	108	57A21-1
108	2N3035	108	9-5129	108	21A050-004	108	48-134937	108	57A21-10
108	2N3562	108	9-5130	108	21A105-001	108	48-134946	108	57A21-15
108	2N3563	108	9-5131	108	21A112-086	108	48-134948	108	57A21-2
108	2N3563-1	108	9-5223	108	21A112-087	108	48-134949	108	57A21-3
108	2N3682	108	10-2SC080	108	21M476	108	48-134950	108	57A21-4
108	2N4292	108	10-2SC094	108	21M481	108	48-134960	108	57A21-5
108	2N4293	108	10B1051	108	22-001002	108	48-134979	108	57A21-6
108	2N4996	108	10B1055	108	22-001003	108	48-134983	108	57A21-7
108	2N4997	108	10B551	108	22-001004	108	48-134985	108	57A21-9
108	2N5770	108	10B553	108	22-001005	108	48-137004	108	57A27-2
108	2N709	108	10B556	108	022-3640-080	108	48-137006	108	57A5-6
108	2N709/52	108	100573	108	022_3640-080	108	48-137033	108	57A5-7
108	2N709A	108	100574	108	23-P2275-122	108	48-137055	108	57A5-8
108	2N717A	108	10G1051	108	24-3564	108	48-137104	108	57A7-1
108	2N849	108	10G1052	108	24A1	108	48-137105	108	57A7-2
108	2N850	108	10H1051	108	24MW103B	108	48-137126	108	57A7-3
108	2N851	108	10H1053	108	24MW1082	108	48-137136	108	57A7-4
108	2N852	108	10H551	108	24MW654	108	48-137140	108	57A7-5
108	2N914/51	108	10H553	108	24MW657	108	48-137144	108	57A7-6
108	2N918	108	11B1052	108	24MW675	108	48-137158	108	57A7-7
108	2SC006	108	11B1055	108	24MW724	108	48-137166	108	57B101-4
108	2SC148	108	11B551	108	24MW725	108	48-137190	108	57B134-12
108	2SC1636	108	11B552	108	24MW739	108	48-137191	108	57B141-1
108	2SC39	108	11B554	108	24MW827	108	48-137194	108	57B141-2
108	2SC39A	108	11B555	108	24MW852	108	48-137196	108	57B141-3
108	2SC40	108	11C1051	108	24MW865	108	48-137339	108	57B142-1
108	000028C460C	108	11C1053	108	24T-002	108	48-137352	108	57B142-2
108	000028C461	108	11C1057	108	24T-011-008	108	48-137355	108	57B142-3
108	2SC467	108	110551	108	24T-013-005	108	48-137372	108	57B143-1
108	2SC561	108	110553	108	24T-016-001	108	48-137375	108	57B143-10
108	2SC63	108	110557	108	24T-016-005	108	48-137376	108	57B143-11
108	2SC662	108	13-0321-14	108	24T-016-013	108	48-137388	108	57B143-2
108	000028C668D	108	13-0321-15	108	24T-016-015	108	48-137483	108	57B143-3
108	2SC740	108	13-0321-16	108	24T-016-016	108	48-43351A03	108	57B143-4
108	2SC79	108	13-0321-17	108	24T011-008	108	48-43351A04	108	57B143-5
108	2SCF1	108	13-0321-21	108	025-100003	108	48-43351A05	108	57B143-6
108	2SC8184J	108	13-1032-5	108	025-100004	108	48-43992J01	108	57B143-8
108	2SC8429J	108	13-10321-1	108	025-100009	108	48-63026A46	108	57B143-9
108	2SC8430H	108	13-10321-10	108	025-100013	108	48-63077A29	108	57B151-6
108	2SC8461F	108	13-10321-11	108	025-100014	108	48-65112A65	108	57B152-12
108	2SC8469F	108	13-10321-12	108	25A	108	48-65112A67	108	57B160-1
108	2SD562	108	13-10321-14	108	25A1	108	48-65113A88	108	57B160-2
108	2SB629	108	13-10321-16	108	25A1262-005	108	48-65118A64	108	57B160-3
108	2SQ371	108	13-10321-17	108	25A1281-001	108	48-65123A67	108	57B160-4
108	03-460C	108	13-10321-2	108	25AM624	108	48-65123A95	108	57B160-5
108	03-461B	108	13-10321-20	108	25B	108	48-65144T2	108	57B160-6
108	03-535A	108	13-10321-26	108	25B-1	108	48-90232A03	108	57B160-7
108	31A-6007-51	108	13-10321-30	108	25B1	108	48-90232A04	108	57B160-8
108	04-00535-06	108	13-10321-43	108	25C206	108	48-90232A10	108	57B166-12
108	4-0485	108	13-10321-5	108	25R	108	48-90232A19	108	57B21
108	4-3022861	108	13-10321-51	108	31-0051	108	48-97046A04	108	57B21-1
108	4-3025763	108	13-10321-6	108	31-0097	108	48-97046A18	108	57B21-12
108	4-3025764	108	13-10321-67	108	31-0103	108	48-97046A51	108	57B21-13
108	4-3025765	108	13-10321-7	108	31-0242	108	48-97127A02	108	57B21-14
108	4-3025767	108	13-10321-77	108	31-0243	108	48-97127A03	108	57B21-16
108	4-399	108	13-10321-79	108	33H50	108	48-97162A01	108	57B21-18
108	4-400	108	13-10321-8	108	34-6001-3	108	48-97162A02	108	57B21-2
108	4-433	108	13-10321-9	108	34-6001-6	108	48-97177A04	108	57B21-3
108	4-434	108	13-14085-1	108	34-6015-49	108	48-97177A07	108	57B21-5
108	4-443	108	13-14085-2	108	34-6016-17	108	48-97177A08	108	57C10-1
108	4-684120-3	108	13-14085-24	108	34E31	108	48-97M1A04	108	57C10-2
108	4-685285-3	108	13-14085-27	108	34M31	108	48-97762A02	108	57C020-1
108	4-686107-3	108	13-14085-74	108	41M	108	48-K869575	108	57C027-2
108	4-686109-3	108	13-14085-75	108	41N2	108	48P63082A45	108	57C5-6
108	4-686112-3	108	13-14085-76	108	41N2A	108	48P63082A71	108	57C5-7
108	4-686114-3	108	13-14085-77	108	41N2AA	108	48P65146A63	108	57C7-1
108	4-686119-3	108	13-15810-1	108	41N2B	108	48P65175A78	108	57C7-2
108	4-686119-3	108	13-15841-1	108	41R3	108	488134902	108	57C7-3
108	4-686126-3	108	13-16744-1	108	42-19683	108	488134946	108	57C7-4
108	4-686127-3	108	13-18949-1	108	42-22785	108	488134960	108	57C7-5
108	4-686131-3	108	13-18950-1	108	42-28203	108	488134970	108	57C7-6
108	4-686140-3	108	13-23323-6	108	42-28204	108	488134979	108	57C7-7
108	4-686169-3	108	13-28584	108	42-28206	108	488137006	108	57D107-8
108	4-686171-3	108	13-28584-1	108	43-022861	108	48843991J01	108	57D24-1
108	4-686172-3	108	13-29392-2	108	43-025763	108	48843992J01	108	57D24-2
108	4-686207-3	108	13-31013-4	108	43-025764	108	48X97046A51	108	57D24-3
108	4-686208-3	108	13-32366-2	108	43-025765	108	48X97162A01	108	61B007-1
108	4-686209-3	108	13-34045-1	108	43-025767	108	48X97162A02	108	61B007-2
108	4-686224-3	108	13-34045-2	108	045-1(SYLVANIA)	108	48X97162A04	108	66X0007-104
108	4-686228-3	108	13-35550	108	045-2(SYLVANIA)	108	48X97162A09	108	69N1
108	4-686244-3	108	13-55020-1	108	45NP	108	48X97162A10	108	72N1
108	4-686251-3	108	13-55063-1	108	46-85120-3	108	49-1	108	72N2
108	4-686695-3	108	13-55065-1	108	46-85285-3	108	50-40101-04	108	74
108	4-JBD1	108	13-67583-5	108	46-86140-3	108	50-40101-05	108	76-13570-39
108	6A12677	108	14-32430	108	46-86208-3	108	50C1047	108	76-13570-59
108	6A12679	108	14-602-41	108	46-86244-3	108	50CT84	108	76-13866-17
108	7-4	108	14-603-12	108	46-86262-3	108	50C829	108	76-13866-18
108	7-44	108	14-609-49A	108	46-86269-3	108	500829B	108	76-13866-19
108	7-59-0193477	108	015	108	46-86314-3	108	500829C	108	76-13866-20
108	7-59-0203477	108	15-088004	108	46-86314-3A	108	51	108	76-13866-59
108	7-59-0213477	108	15-088000	108	46-86357-3	108	54BLK	108	76-13866-62
108	7-59-0233477	108	16-736	108	46-86397-3	108	54BRN	108	78001
108	7-59-0233477	108	16J1	108	46-864-3	108	54ORN	108	78002
108	7-6	108	16J2	108	46-8677-2	108	54RED	108	80-0X5600
108	7-7	108	16K1	108	47-2(BRADFORD)	108	56-234	108	80-338030
108	7-8	108	16K2	108	48-01-004	108	57A10-1	108	80-338040
108	8-0024-1	108	16K3	108	48-01-010	108	57A10-2	108	80-339430
108	8-0024-2	108	16L22	108	48-124804	108	57A101-4	108	80-339440
108	8-0053600	108	16L23	108	48-124805	108	57A107-1	108	80-383840
108	8-0339030	108	16L3	108	48-124808	108	57A107-3	108	80-383930
108	8-0339040	108	16L5	108	48-134706	108	57A107-4	108	81-46125006-0
108	8-0339430	108	019-003929	108	48-134709	108	57A107-5	108	86-138-2
108	8-0383840	108	19-020-047	108	48-134713	108	57A107-6	108	86-243-2
108	8-0383930	108	19-020-052	108	48-134717	108	57A107-8	108	86-244-2
108	09-302005	108	19-020-44	108	48-134719	108	57A10A-8-6	108	86-245-2
108	09-302009	108	19-19420	108	48-134724	108	57A134-12	108	86-386-2
108	09-302010	108	19A115249-1	108	48-134725	108	57A139-4-6	108	86-416-2
108	09-302017	108	19A115342-1	108	48-134772	108	57A141-1	108	86-417-2
108	09-302060	108	19A115440-1	108	48-134773	108	57A141-3	108	86-418-2
108	09-302115	108	19A115440-2	108	48-134774	108	57A142-1	108	86-488-2
108	09-302141	108	19A115441-1	108	48-134777	108	57A142-3	108	86-490-2
108	09-302149	108	19A115666-1	108	48-134779	108	57A143-1	108	86-491-2
108	09-302162	108	19A115925-1	108	48-134780	108	57A143-10	108	86X0007-004
108	09-302190			108	48-134783	108	57A143-11		
108	09-302241			108	48-134784				
				108	48-134786				

ECG	Industry Standard No.
108	86X0007-204
108	86X0038-001
108	86X0043-001
108	86X6029-001
108	86X7-6013
108	87-0027
108	089-214
108	089-215
108	089-216
108	91A
108	91B
108	91BGRN
108	91F
108	92N1
108	92N1B
108	95-125
108	95-126
108	95-127
108	95-128
108	95-129
108	95-130
108	95-131
108	95-223
108	96-056-234
108	96-138-2
108	96N(AIRLINE)
108	96N927
108	96N932
108	96NPT
108	99S016
108	99S016-1
108	99S017
108	99S018
108	99S018A
108	99S019
108	99S019A
108	99S019B
108	99S037
108	99S090-1
108	100W1
108	0101-0060A
108	0101-0531
108	101-2(ADMIRAL)
108	101-3(ADMIRAL)
108	101-4(ADMIRAL)
108	0103-0060
108	0103-0060B
108	0103-0191
108	0103-0389
108	0103-0521
108	0103-0521B
108	0103-0531
108	0103-389
108	103-4
108	0103-9531/4460
108	105-001-08
108	105-00106-00
108	105-00108-07
108	105-005-12
108	105-02004-09
108	105-06004-00
108	105-24191-04
108	112-520
108	112-521
108	112-522
108	113-398
108	113-938
108	114-118
108	114-267
108	116-073
108	116-079
108	116-080
108	116-082
108	116-083
108	116-198
108	116-199
108	116-200
108	118-1
108	118-2
108	118-3
108	118-4
108	120-004496
108	120-004497
108	120-004723
108	120-004724
108	120-004725
108	120-004881
108	120-005291
108	120-005292
108	120-005293
108	120-005294
108	120-005295
108	120-005296
108	120-005297
108	120-005298
108	121-113
108	121-303
108	121-316
108	121-317
108	121-318
108	121-318L
108	121-321
108	121-345
108	121-453
108	121-472
108	121-481
108	121-482
108	121-483
108	121-520
108	121-546B
108	121-547
108	121-551
108	121-560
108	121-612
108	121-612-16
108	121-613
108	121-613-16
108	121-614
108	121-614-9
108	121-616
108	121-630
108	121-637
108	121-638
108	121-638B
108	121-735
108	121-742
108	121-753
108	121-819
108	121-827
108	121-834
108	121-835
108	121-841
108	121-846
108	121-848
108	121-851
108	121-857
108	121-869
108	121-884
108	121-898
108	121-899
108	121-900
108	121-925
108	121-932
108	121-968
108	121-974
108	122-A484
108	128
108	128N4
108	130-138
108	130-40304
108	130-40362
108	130-40421
108	130-40459
108	131(ARVIN)
108	131(SEARS)
108	132-015
108	139-4
108	0142
108	142N6
108	151-0138
108	151M11
108	151N1
108	151N11
108	151N116
108	161T2
108	162T2
108	173A04490-1
108	173A04490-2
108	176-003
108	176-003-9-001
108	176-004-9-001
108	176-005
108	176-005-9-001
108	176-006
108	176-006-9-001
108	176-007
108	176-007-9-001
108	189
108	200-007
108	200-010
108	200-015
108	200-055
108	200-056
108	200X3190-604
108	201-25-4343-12
108	201-254323-12
108	201-254323-13
108	201-254343-12
108	207A9
108	217-1
108	220-001011
108	220-001012
108	223
108	229-0151-3
108	229-0180-124
108	229-0180-149
108	229-0180-34
108	229-0185-2
108	229-0185-3
108	229-0190-29
108	229-0192-19
108	229-0204-23
108	229-0204-4
108	229-0210-14
108	229-0214-40
108	229-0220-19
108	229-0220-9
108	229-0248-45
108	229-0250-10
108	229-5100-15U
108	229-5100-15V
108	229-5100-224
108	229-5100-225
108	229-5100-226
108	229-5100-228
108	229-5100-33V
108	232N2
108	247-016-013
108	260D05701
108	260D05707
108	260P05801
108	260P06901
108	260P06902
108	260P06903
108	260P07004
108	260P07901
108	260P08001
108	260P08401
108	260P10403
108	260P10501
108	260P10502
108	260P1060
108	260P10602
108	260P11101
108	260P11101A
108	260P16301
108	260P16302
108	260P17201
108	260P17602
108	260P17603
108	260P36501
108	260P70403
108	260P70501
108	260P70502
108	260Z00109
108	260Z00209
108	260Z00309
108	297V070H49
108	297V072C01
108	297V072C03
108	297V072C04
108	297V074C09
108	297V078C01
108	297V078C02
108	324-0149
108	324-0150
108	324-1
108	325-0028-84
108	330-1304-8
108	344-6000-3
108	344-6000-3A
108	344-6015-10
108	344-6015-11
108	344-6015-7
108	344-6015-7A
108	344-6017-6
108	366-1(SYLVANIA)
108	366-2(SYLVANIA)
108	386-7118P1
108	386-7188P1
108	396-7178P1
108	417-124
108	417-125
108	417-83
108	421-9683
108	422-1401
108	422-1402
108	422-2532
108	429-0986-12
108	430-22861
108	430-25763
108	430-25764
108	430-25765
108	430-25767
108	488-2(SEARS)
108	490-2(SEARS)
108	491-2(SEARS)
108	499-1
108	501ES001M
108	515-521
108	537FS
108	546
108	551
108	573-472
108	573-474
108	573-474A
108	573-475
108	573-491
108	573-494
108	573-495
108	573-507
108	573-509
108	576-0001-006
108	576-0003-001
108	576-0003-002
108	576-0003-003
108	576-0003-004
108	576-0003-005
108	576-0003-006
108	576-0003-007
108	576-0003-018
108	576-0003-020
108	576-0003-021
108	576-0003-027
108	576-0003-028
108	576-0006-011
108	576-0036-918
108	576-0036-919
108	600X0092-086
108	601-113
108	602-113
108	602-61
108	603-113
108	604-113
108	605-113
108	610-041
108	610-041-1
108	610-041-2
108	610-041-3
108	610-042
108	610-042-1
108	610-045
108	610-045-1
108	610-045-2
108	610-069
108	610-069-1
108	610-072
108	610-072-1
108	610-072-2
108	610-073
108	610-073-1
108	612-16(ZENITH)
108	612-16A
108	613(ZENITH)
108	613-72
108	614(ZENITH)
108	614-12
108	653-202
108	660-127
108	668CS
108	690V010H41
108	690V028H28
108	690V028H48
108	690V028H69
108	690V028H89
108	690V02H69
108	690V049H81
108	690V060H58
108	690V060H59
108	690V070H49
108	690V070H98
108	690V075H68
108	690V081H07
108	690V084H94
108	690V084H95
108	690V084H96
108	690V086H52
108	690V086H87
108	690V086H96
108	690V088H44
108	690V088H45
108	690V088H48
108	690V103H23
108	690V103H24
108	690V103H25
108	690V103H26
108	690V103H27
108	690V110H30
108	690V110H31
108	690V110H32
108	690V110H33
108	690V114H29
108	690V114H31
108	690V116H19
108	690V118H59
108	0703
108	750D858-213
108	753-0101-047
108	753-2000-710
108	773RED
108	7740RN
108	775BRN
108	779BLU
108	783RED
108	7840RN
108	786
108	787BLU
108	800-536-00
108	800-53600
108	803-38030
108	803-38040
108	803-39430
108	803-39440
108	803-83840
108	803-83930
108	916-31024-5B
108	921-119B
108	921-129B
108	921-158B
108	921-170B
108	921-171B
108	921-172B
108	921-173B
108	921-174B
108	921-20
108	921-20A
108	921-20B
108	921-21
108	921-212B
108	921-21A
108	921-21B
108	921-21BK
108	921-22
108	921-22A
108	921-22B
108	921-23
108	921-23A
108	921-23B
108	921-264B
108	921-265
108	921-265B
108	921-266
108	921-266B
108	921-267
108	921-267B
108	921-30
108	921-30A
108	921-30B
108	921-31
108	921-313B
108	921-31A
108	921-31B
108	921-32
108	921-325B
108	921-32A
108	921-32B
108	921-33
108	921-334B
108	921-335B
108	921-336B
108	921-338B
108	921-33B
108	921-34
108	921-34A
108	921-34B
108	921-428
108	921-72B
108	921-97B
108	921-98B
108	930X4
108	930X5
108	947-1(SYLVANIA)
108	1002-02A
108	1002-04-1
108	1004-17
108	1011-11(R.F.)
108	1016-83
108	1016-84
108	1027(G.E.)
108	1062-0615
108	1106-97
108	1123-55
108	1123-56
108	1123-57
108	1123-58
108	1123-59
108	1229H
108	1284
108	1373-17AL
108	1634-17-14A
108	1761-17
108	1792-17
108	1852-17
108	1880-17
108	1881-17
108	1890-17
108	1923-17
108	1923-17-1
108	1931-17A
108	1983-17
108	2028
108	2028-00
108	2032-33
108	2032-34
108	2057A-120
108	2057A-429
108	2057A2-119
108	2057A2-127
108	2057A2-163
108	2057A2-179
108	2057A2-201
108	2057A2-224
108	2057A2-309
108	2057A2-310
108	2057A2-311
108	2057A2-313
108	2057A2-314
108	2057A2-342
108	2057A2-386
108	2057A2-392
108	2057A2-393
108	2057A2-394
108	2057A2-402
108	2057A2-448
108	2057A2-465
108	2057A2-504
108	2057A2-505
108	2057A2-509
108	2057B-113
108	2057B100-12
108	2057B2-103
108	2057B2-108
108	2057B2-109
108	2057B2-110
108	2057B2-111
108	2057B2-112
108	2057B2-119
108	2057B2-120
108	2057B2-127
108	2057B2-128
108	2057B2-14
108	2057B2-160
108	2057B2-161
108	2057B2-162
108	2057B2-64
108	2057B2-85
108	2057B2-87
108	2093A2-289
108	2180-151
108	2180-152
108	2427(RCA)
108	2445
108	2450(RCA)
108	2473(RCA)
108	2476(RCA)
108	2477(RCA)
108	2495-166-1
108	2495-166-4
108	2495-166-8
108	2495-166-9
108	2495-520
108	2495-521
108	2495-522-1
108	2495-523-1
108	2498-507-2
108	2498-507-3
108	2498-508-2
108	2498-508-3
108	2498-903-2
108	2498-903-3
108	2606-294
108	2633(RCA)
108	2634(RCA)
108	2634-1
108	2636
108	2900-007
108	2904-033
108	3002
108	3012(NPN)
108	3018
108	3020
108	3021
108	3028
108	3107-204-90080
108	3227-E
108	3370
108	3508(WARDS)
108	3509
108	3510(SEARS)
108	3510(WARDS)
108	3511
108	3516(WARDS)
108	3524(RCA)
108	3524-1
108	3524-1(RCA)
108	3524-2(RCA)
108	3527(RCA)
108	3530(RCA)
108	3537(RCA)
108	3539(RCA)
108	3539-307-001
108	3539-307-002
108	3568(RCA)
108	3568(WARDS)
108	3572-3
108	3576(RCA)
108	3598(RCA)
108	3603(RCA)
108	3604(RCA)
108	3610(RCA)
108	3618(RCA)
108	3646-2(RCA)
108	3693(ARVINE)
108	4167(SEARS)
108	4168(SEARS)
108	4169(SEARS)
108	4473-1
108	4473-11
108	4473-2
108	4473-3
108	4473-6
108	4473-7
108	4473-8
108	4490-1
108	4587
108	4684-120-3
108	4685-285-3
108	4686-107-3
108	4686-108-3
108	4686-112-3
108	4686-114-3
108	4686-118-3
108	4686-119-3
108	4686-120-3
108	4686-126-3
108	4686-127-3
108	4686-131-3
108	4686-140-3
108	4686-169-3
108	4686-171-3
108	4686-172-3
108	4686-207-3
108	4686-208-3
108	4686-209-3
108	4686-224-3
108	4686-228-3
108	4686-244-3
108	4686-251-3
108	4686-95-3
108	4706
108	4709
108	4801-0000-035
108	4802-00002
108	4820
108	4821
108	4826
108	4837
108	4845
108	4855
108	5001-510
108	5093
108	5313-461B
108	5613-46B
108	6158
108	6158-3
108	6185-3
108	6507(AIRLINE)
108	7117(GE)
108	7118
108	7122
108	7123
108	7124
108	7125
108	7126
108	7127
108	7128
108	7131
108	7132
108	7133
108	7134

ECG	Industry Standard No.	ECG	Industry Standard No.	ECG	Industry Standard No.	ECG	Industry Standard No.	ECG	Industry Standard No.
108	7173	108	61015-9-1	108	171162-131	108	824960-0	108	2093308-2
108	7174	108	61015-0-1	108	171206-1	108	910799	108	2093308-3
108	7175	108	61133	108	171206-2	108	916029	108	2320062
108	7177	108	61661	108	171206-4	108	916060	108	2320062
108	7178	108	61663	108	171206-5	108	916069	108	2320073
108	7214	108	62449	108	171207-1	108	964634	108	2321511
108	7215	108	67802	108	171207-2	108	964715	108	2495166-1
108	7216	108	70167-8-00	108	175006-187	108	965074	108	2498507-1
108	7217	108	70231	108	175043-062	108	965633	108	2498507-2
108	7218	108	70260-11	108	175043-063	108	965634	108	2498507-3
108	7219	108	70260-12	108	175043-064	108	970046	108	2596071
108	7220	108	70260-13	108	175043-100	108	970046-1	108	3181972
108	7221	108	72949-10	108	175043-107	108	970046-2	108	3596067
108	7232	108	72951-95	108	181003-7	108	970046A	108	3596068
108	7234	108	72951-96	108	181003-8	108	970244	108	3596069
108	7235	108	72979-80	108	181003-9	108	970245	108	3596070
108	7237	108	75616-6	108	181503-6	108	970249	108	3596071
108	7238	108	75810-17	108	181503-7	108	970309	108	3596072
108	7262	108	79855	108	181503-9	108	970309-1	108	3596260
108	7425	108	79856	108	181504-1	108	970309-12	108	3596261
108	7426	108	080006	108	181504-7	108	970309-2	108	3597103
108	7427	108	080021	108	181506-7	108	970309-3	108	3597104
108	7593-2	108	080022	108	200064-6-103	108	970309-4	108	3597260
108	7642	108	080023	108	200064-6-105	108	970309-5	108	3597261
108	7810	108	080041	108	209417-0714	108	970310	108	5320328
108	7811	108	080042	108	227000	108	970310-1	108	6212922
108	7812	108	080059	108	229392	108	970310-2	108	7026011
108	7813	108	080060	108	231140-01	108	970310-3	108	7026012
108	7814	108	82716	108	231140-07	108	970310-4	108	7026013
108	7815	108	94044	108	231140-23	108	970310-5	108	7295195
108	8606	108	95125	108	231140-31	108	970332	108	7295196
108	8607	108	95126	108	231140-34	108	970332-12	108	7910780-01
108	8609	108	95127	108	231140-44	108	970911	108	8031825
108	8611	108	95128	108	232840	108	980138	108	8031836
108	8710-162	108	95129	108	236251	108	980139	108	8033803
108	90164	108	95130	108	236706	108	982268	108	8033804
108	9018F	108	95131	108	236907	108	982269	108	8033945
108	9018G	108	95242-1	108	237020	108	982521	108	8037723
108	9300	108	101434	108	237021	108	982815	108	8556188
108	9300A	108	111193	108	237024	108	982816	108	13040304
108	9300B	108	112355	108	237026	108	982817	108	13040362
108	9300Z	108	113938	108	237785	108	982818	108	13040421
108	9314	108	114143-1	108	237840	108	982819	108	13040459
108	9426B	108	114525	108	241249	108	983095	108	19901806
108	9426C	108	115440	108	241778	108	983096	108	19901807
108	9513	108	115910	108	241960	108	984156	108	22901513
108	9600C	108	115925	108	242590	108	984158	108	22902046
108	9600F	108	116073	108	242960	108	984159	108	23114001
108	9600G	108	116079	108	243318	108	984194	108	23114023
108	9600H	108	116080	108	243645	108	984195	108	23114031
108	9601	108	116082	108	245078-3	108	984577	108	23114034
108	9601-12	108	116083	108	257540	108	984743	108	23114036
108	9604F	108	116198	108	260565	108	984744	108	23114043
108	9623F	108	116199	108	265074	108	984851	108	23114056
108	9623G	108	116200	108	265241	108	984852	108	23114057
108	9623H	108	117823	108	267197	108	984853	108	23114060
108	9625F	108	119414	108	304900	108	984875	108	23114082
108	9625H	108	122517	108	346015-15	108	985096	108	23114109
108	9630C	108	122518	108	346015-16	108	985097	108	23114127
108	11252-0	108	123160	108	346015-17	108	985215	108	23114171
108	11252-1	108	123429	108	346015-18	108	985442A	108	23114172
108	11252-2	108	123430	108	346015-19	108	985443A	108	23114180
108	11339-8	108	123431	108	346015-20	108	985443AA	108	23114181
108	11393-8	108	124263	108	346015-21	108	986634	108	23114238
108	11426-7	108	124412	108	346015-22	108	986635	108	23124037
108	11607-5	108	125137	108	346015-25	108	988000	108	23126620
108	11607-9	108	125138	108	346015-37	108	988001	108	25114161
108	11608-0	108	125263	108	489751-027	108	988002	108	26010056
108	11608-2	108	125264	108	489751-131	108	988985	108	27125210
108	11608-3	108	125392	108	489751-137	108	988986	108	27126220
108	11619-8	108	125475-14	108	489751-143	108	988987	108	30200091
108	11619-9	108	125944	108	489751-145	108	988988	108	43027614
108	11620-0	108	125994-14	108	489751-147	108	988989	108	43027615
108	11620-1	108	126023	108	489751-148	108	994634	108	43027616
108	16190	108	126024	108	489751-162	108	1222463	108	43027617
108	16194	108	126670	108	489751-165	108	1223781	108	44007301
108	19420	108	126698	108	489751-167	108	1408615-1	108	44008401
108	23114-056	108	127529	108	489751-168	108	1408640-1	108	55440011-001
108	23114-057	108	127693	108	489751-169	108	1471115-13	108	55440023-001
108	23114-060	108	127792	108	489751-171	108	1471115-14	108	62048158
108	23114-078	108	127794	108	489751-206	108	1472450-1	108	62104078
108	23114-104	108	129050	108	573101	108	1472634-1	108	62506296
108	23125-077	108	129144	108	573472	108	1473524-2	108	62506318
108	25114-121	108	129392-14	108	573494	108	1473530-1	108	62537124
108	30292	108	129393-14	108	0573495	108	1473532-1	108	62539305
108	35004	108	129394-14	108	0573506	108	1473533-1	108	62539321
108	35449	108	130278	108	0573506H	108	1473537-1	108	62541555
108	36212	108	130403-04	108	0573507H	108	1473568-1	108	62543892
108	36212V1	108	130403-62	108	0573511	108	1473603-1	108	62543906
108	36578	108	130404-21	108	0573570	108	1473604-3	108	62563303
108	36581	108	130404-59	108	601113	108	1473606-1	108	62563346
108	36847	108	131221	108	602113	108	1473617-1	108	62563354
108	36918	108	131544	108	603113	108	1473652-1	108	62565122
108	36919	108	131545	108	604113	108	1810037	108	62605465
108	37383	108	131648	108	610041	108	1810038	108	62691271
108	37384	108	131844	108	610041-1	108	1810039	108	62695595
108	37694A	108	134142	108	610041-3	108	1815036	108	62695943
108	37694B	108	134144	108	610042-0	108	1815037	108	62713747
108	38207	108	134263	108	610042-1	108	1815039	108	62766239
108	38208	108	134419	108	610045	108	1815045	108	62789263
108	38246A	108	147245-0-1	108	610045-1	108	1815047	108	80055600
108	38511	108	147356-9-1	108	610045-2	108	1815067	108	80338030
108	38511A	108	147357-2-1	108	610046-77	108	1815068	108	80338040
108	38785	108	147357-9-1	108	610069	108	1817004	108	80339430
108	38786	108	148751-147	108	610069-1	108	1817005-3	108	80339440
108	38920	108	156931	108	610072	108	1817006-3	108	120004496
108	38921	108	162002-090	108	610072-1	108	1817008	108	120004497
108	39331	108	165995	108	610072-2	108	1817045	108	120004880
108	39730	108	166272	108	610075	108	1819045	108	120004881
108	39731	108	168657	108	610091	108	2000646-105	108	120004882
108	39789	108	168658	108	610091-1	108	2000757-80	108	226021014
108	40413	108	168659	108	610091-2	108	2000804-7	108	229018032
108	40414	108	169195	108	610092	108	2000804-8	108	229018033
108	40470	108	171003(SEARS)	108	610092-1	108	2002332-53	108	229018034
108	40480	108	171009(SEARS)	108	610092-2	108	2002332-54	108	229020423
108	40482	108	171028	108	610096	108	2002332-55	108	229021014
108	41689	108	171029(SEARS)	108	610096-1	108	2002332-56	108	229025010
108	41694	108	171030(SEARS)	108	610100	108	2002620-18	108	229510015V
108	043001	108	171031(SEARS)	108	610100-3	108	2002620-19	108	229510031V
108	43022-860	108	171032	108	610107-1	108	2003342-109	108	229510032V
108	50957-03	108	171033	108	610128-4	108	2004746-114	108	229510033V
108	57000-5452	108	171045	108	610139-2	108	2004746-115	108	450010201
108	60048	108	171048	108	610142-6	108	2006431-44	108	2003185506
108	60314	108	171052	108	610150	108	2006513-19	108	2290180119
108	61009-1	108	171090-1	108	610174-1	108	2091859-0711	108	2295100224
108	61009-1-1	108	171139-1	108	613112	108	2092418-0715	108	2295100225
108	61009-1-2	108	171140-1	108	815164	108	2092418-0724	108	2295100226
108	61009-2	108	171141-1	108	815165	108	2093308-070	108	2295100228
108	61009-2-1	108	171162-027	108	815170	108	2093308-0700	108	4108296047
108	61009-6	108	171162-128	108	815172	108	2093308-0704A	108	16100190668
108	61009-6-1	108	171162-129	108	815172A	108	2093308-0705A	108	16102190929
108	61010-0	108	171162-130	108	815173A	108	2093308-0706A	108	16103190668
108	61010-0-1			108	815173C	108	2093308-1	108	16104191226
108	61010-7-1			108	815173F			108	16104191226
				108	815209				

ECG	Industry Standard No.	ECG	Industry Standard No.	ECG	Industry Standard No.	ECG	Industry Standard No.	ECG	Industry Standard No.
108	310720490080	109	AAY15	109	EA16X97	109	GD556	109	NGP3002
108	933022960012	109	AAY18	109	EA2502	109	GD5E	109	NTC-13
108	933022970112	109	AAY22	109	EA3127	109	GD663	109	NTC-14
1080	GBIC-41	109	AAY27	109	EA3718	109	GD6R	109	NU34
1080	GEIC-98	109	AAY30	109	ED-46	109	GD72E/3	109	OA-90
1080	HA1140	109	AAY33	109	ED-60	109	GD72E/4	109	OA-90(G)
1080	LA1355	109	AAY46	109	ED12(ELCOM)	109	GD72E/5	109	OA-90G
1080	RE328-IC	109	AAZ10	109	ED219464	109	GD73E/4	109	OA-91
1080	REN1080	109	AA218	109	ED4(ELCOM)	109	GD73E/5	109	OA134Q
1080	TA7075	109	ACR810-107	109	ED46	109	GD74E/3	109	OA150
1080	TA7075P	109	ACR83-1007	109	ED6(ELCOM)	109	GD74E/4	109	OA159
1080	46-13101-3	109	AFS-160-1017	109	ED60	109	GD74E/5	109	OA160
1080	2002120033	109	B-28	109	ED9(ELCOM)	109	GD8E	109	OA161
1080	4206002600	109	B-30	109	EDG-0003	109	GE-X66	109	OA172
1080	4206003900	109	B-30P	109	EDG-0006	109	G83658	109	OA174
1082	DDEY064001	109	B-3P	109	EDG-1	109	GED05B850	109	OA47
1082	GEIC-140	109	B28	109	EDG-3	109	G00-003-A	109	OA50
1082	MB3202	109	B30	109	EDG-4	109	GP-354	109	OA541
1082	MPC577H	109	B601-1011	109	EI16X3	109	GP2354	109	OA6
1082	MX-3389	109	B692X13	109	ES10189	109	GPM1NA	109	OA7
1082	NJM2201	109	C21480	109	ES10224	109	GPM1NB	109	OA70
1082	RE341-M	109	C5005	109	ES10225	109	GPM2NA	109	OA71
1082	REN1082	109	C60(DIODE)	109	ES10231	109	G9T3E/3	109	OA71C
1082	09-308052	109	CB106	109	ES15054	109	H091	109	OA72
1082	61A001-10	109	CB163	109	ES16X103	109	H316	109	OA73
1082	307-112-9-007	109	CB393	109	ES16X12	109	HD-1000101	109	OA73C
1082	2056-04	109	CD-0000	109	ES16X14	109	HD10-001-01	109	OA74
1082	28810-175	109	CD-0000N	109	ES16X2	109	HD10000101	109	OA74A
1082	741852	109	CD-0014N	109	ES16X3	109	HD1000001	109	OA79
1082	916085	109	CD000	109	ES16X5	109	HD10000302	109	OA81
1085	GEIC-104	109	CD0000/7825B	109	ES16X6	109	HD1000101-0	109	OA81C
1085	HC1000505	109	CD101	109	ES16X7	109	HD10001010	109	OA85
1085	TA7122AP	109	CE0495/7839	109	ET16X1	109	HD10001050	109	OA85C
1085	TA7122AP-D	109	CG12E	109	ET16X19	109	HD1000301	109	OA9
1085	TA7122AR	109	CG64H	109	ET16X20	109	HD1000302	109	OA90
1085	TA7122P-B	109	CG65H	109	ET16X21	109	HD1468	109	OA909
1085	15-144471-1	109	CG66H	109	ET41X37	109	HD2149	109	OA90G
1086	37009006	109	CG74H	109	ETD-1N60	109	HD2155	109	OA90GA
1087	G09-008-B	109	CG86H	109	ETD-8D46	109	HD4000109	109	OA90LF
1087	G09-008-C	109	CGD1029	109	EU16X1	109	HE-10001	109	OA90M
1087	G09-008-E	109	CGD462	109	EU16X2	109	HE-10003	109	OA90MLF
1087	G09-017-B	109	CGD591	109	EW166	109	HE-10024	109	OA90Z
1087	G09-017-C	109	CGD685	109	EW167	109	HE-10025	109	OA90ZA
1087	G09-017-D	109	CK705	109	EYV420D1R5JB	109	HE-10027	109	OA91
1087	GEIC-103	109	CK706	109	FH36	109	HE-10040	109	OA92
1087	GEIC-171	109	CK706A	109	F20-1010	109	HE-1024	109	OA99
1087	HA1406	109	CK706P	109	F20-1012	109	HE-1N34	109	OA9D
1087	HA1406-2	109	CK715	109	F20-1013	109	HE-1N34A	109	OA9Z
1087	HA1406-3	109	CR0000	109	F20-1014	109	HE-1N60	109	OP173
1087	HA1406-4	109	CR101/6515	109	F20303	109	HE-1N60P	109	ON67A
1087	TA7120	109	CR102/6515	109	F215-1010	109	HE-18188	109	0870
1087	TA7120B	109	CT2002	109	F215-1012	109	HE-18426	109	OSS-16685
1087	TA7120P	109	CT2007	109	F215-1013	109	HE-18446	109	OS816308
1087	051-0020-00	109	CT461	109	F215-1014	109	HE-DA90	109	OS816685
1087	051-0020-01	109	CTP-2001-1010	109	FA-1(DIODE)	109	HEP-R9134	109	P1O155
1087	051-0035-01	109	CTP-2006-1004	109	FB1043	109	HEP-R9135	109	PBE5322
1087	051-0035-02	109	CTP461	109	FD1980	109	HEP134	109	P093
1087	051-0035-04	109	CTP573	109	FPR40-1006	109	HEP135	109	PTC206
1087	588-40-203	109	CV425	109	FS19	109	HP-10024	109	PTC207
1087	740-2007-120	109	CV442	109	FV-23	109	HP-20008	109	Q-02115C
1087	740-9037-120	109	CX0036	109	FV23	109	HG1090	109	Q49
1087	25810-166	109	CX0041	109	G00-003-A	109	HG5002	109	Q50
1087	25840-166	109	D-00169C	109	G00-004-A	109	HG5004	109	Q51
1087	58810-172	109	D-00204R	109	G00-008-A	109	HG5006	109	QD-G1N60PKT
1087	58840-203	109	D-00269C	109	G00-009-A	109	HG5007	109	QD-G1N60XXT
1087	171179-045	109	D-00669C	109	G00-013-B	109	HG5008	109	QD-G1852XXT
1087	0207120	109	D-2	109	G00003A	109	HG5009	109	QVD1KP114
1087	742563	109	D093	109	G00009A	109	HG5078	109	R-18188
1087	5350251	109	D1-2	109	G0100	109	HG5079	109	R-7051
1089*	37003001	109	D286	109	G1010	109	HG5085	109	R1106
109	A054-103	109	D2R31	109	G1288	109	HG5088	109	R1107
109	A054-105	109	D4R	109	G156	109	HG5808	109	R1109
109	A054-107	109	D8410	109	G157	109	HN-00003	109	R1667
109	A066-120	109	DA90	109	G158	109	IC743050	109	R1889
109	A066-121	109	DAAY001002	109	G159	109	IN60-1	109	R2164
109	A068-101	109	DANZ0060000	109	G198	109	INJ60284	109	R2334
109	A069-109	109	DANZ0060000	109	G199	109	INJ61675	109	R5096
109	A069-115	109	DC-13	109	G1HA	109	IP20-0015	109	R5522
109	A07	109	DDAY001001	109	G200	109	IP20-0016	109	R60-1007
109	A090	109	DDAY001002	109	G297	109	IP20-0020	109	R7028
109	A15-1007	109	DDAY001004	109	G409	109	IP20-0060	109	R7029
109	OA180	109	DDAY001010	109	G498	109	IP20-0283	109	R7743
109	A20371	109	DDAY001022	109	G580	109	IR5JA	109	R7892
109	A2419	109	DDMV-1	109	G5C	109	IT22	109	R7893
109	A2420	109	DDMV-2	109	G5F	109	IT23	109	R8060
109	A2473	109	DG1N60	109	G5K	109	IT23G	109	R8061
109	A2476	109	DG1834	109	G766	109	ITT102	109	R8219
109	A25-1007	109	DGM-2	109	G769	109	ITT301	109	R8257
109	A30(DIODE)	109	DGM-3	109	G770	109	ITT718	109	R8475
109	A36508	109	DI-1	109	G788	109	J241245	109	R8887
109	A42X00340-01	109	DI-2	109	G789	109	J242	109	R8970
109	A48-63078A52	109	DI-8(DIODE)	109	G790	109	J243	109	R9590
109	A556-142	109	DIJ61224	109	G7D	109	J2441	109	RE47
109	A615-1012	109	DIJ70542	109	G7E	109	J24911	109	RP18111
109	A692X13-4	109	DIJ70645	109	G7F	109	J24913	109	RP3550-1
109	OA7	109	DIJ70646	109	G7G	109	J24914	109	RP60034
109	A7001800	109	DIJ71776	109	G814	109	J320041	109	RPJ60614
109	OA81	109	DIJ72294	109	G815	109	J685	109	RL232G
109	OA9	109	DIJ72349	109	G816	109	JT-E1014	109	RL246
109	OA90	109	DK19	109	G820	109	JT-E1031	109	RL252
109	A909-1015	109	DK20	109	G821	109	K115J511-2	109	RL31
109	A909-1017	109	DK21	109	G822	109	K3	109	RL32
109	OA9OLF	109	DR291	109	G823	109	K4-550	109	RL32G
109	OA9OZ	109	DR351	109	G824	109	K52	109	RL34
109	OA91	109	DR352	109	G825	109	K6	109	RL34G
109	AA111	109	DR365	109	G844	109	K60	109	RL41
109	AA112	109	DR385	109	G845	109	K882	109	RL41G
109	AA112P	109	DR426	109	G846	109	KD27	109	RL42
109	AA113	109	DR434	109	G847	109	KDD-0013	109	RL52
109	AA114	109	DR449	109	G868	109	KGE41959	109	RS1811
109	AA116	109	DR464	109	G869	109	KR-Q0005	109	RS2801
109	AA117	109	DS-18(DELCO)	109	GB-1	109	M34A	109	RT-2016
109	AA118	109	D827	109	GC5012	109	M51	109	RT1008
109	AA119	109	D833(DELCO)	109	GD-25	109	M60	109	RT1106
109	AA120	109	DS410(G.E.)	109	GD-26	109	M8489-A	109	RT1108
109	AA121	109	DS410R(G.E.)	109	GD-29	109	M95	109	RT1184
109	AA123	109	DS816685	109	GD-30	109	MA3	109	RT2334
109	AA130	109	DX-0161	109	GD1001	109	MA51A	109	RT2452
109	AA131	109	DX-0162	109	GD11E	109	MA55	109	RT2694
109	AA132	109	DX-0241	109	GD12E	109	MA8	109	RT3072
109	AA134	109	DX-0725	109	GD13E	109	MA90	109	RT3099
109	AA135	109	DX6873	109	GD1E	109	MC2526	109	RT3233
109	AA136	109	E0018	109	GD1P	109	MC308	109	RT3336
109	AA137-M	109	E21430	109	GD3658-00	109	MD-34	109	RT3469
109	AA138	109	EA1123	109	GD400	109	MD34	109	RT4293
109	AA139	109	EA16X1	109	GD402	109	MD34A	109	RT4644
109	AA140	109	EA16X11	109	GD403	109	MD46	109	RT5213
109	AA142	109	EA16X140	109	GD404	109	MD60	109	RT5214
109	AA143	109	EA16X22	109	GD405	109	MD604A	109	RT5908
109	AA143B	109	EA16X27	109	GD406	109	MN34A	109	RT5939
109	AA144	109	EA16X48	109	GD409	109	MN51	109	RT61012
109	AA218	109	EA16X5	109	GD4E	109	MN60(DIODE)	109	RT6119
109	AA779	109	EA16X9	109	GD5004	109	N2A(DIODE)	109	RT6179
109	AAY139	109		109		109	N48	109	RT6180
						109	NC29		

ECG	Industry Standard No.	ECG	Industry Standard No.	ECG	Industry Standard No.	ECG	Industry Standard No.	ECG	Industry Standard No.
109	RT6181	109	TV80A70	109	1N295X	109	1N636	109	00001S4460
109	RT6182	109	TV80A90	109	1N296	109	1N63A	109	00001S446D
109	RT6183	109	TV80A91	109	1N297	109	1N64	109	1S447
109	RT6184	109	UP-SD1	109	1N297A	109	1N64A	109	1S447P
109	RT6189	109	V-10916-3	109	1N298	109	1N64B	109	1S448
109	RT6619	109	V-210C	109	1N298A	109	1N64G	109	1S449
109	RT7330	109	V10916-3	109	1N304	109	1N64GA	109	1S451
109	RT7636	109	V115	109	1N305	109	1N65	109	1S452
109	RT7689	109	V117	109	1N306	109	1N65A	109	1S453
109	RT8671	109	V135	109	1N307	109	1N66	109	1S454
109	RV1479	109	V50260-16	109	1N308	109	1N66A	109	1S455
109	RVD1K110	109	V50A260-36	109	1N309	109	1N67	109	1S466
109	RVD1N34A	109	V74	109	1N3110	109	1N67A	109	1S467R
109	RVD2-1K110	109	VD11	109	1N312	109	1N68	109	1S5454
109	S-21271	109	VD12	109	1N3121	109	1N68A	109	1S60P
109	OS-90	109	VD13	109	1N3122	109	1N69	109	1S72
109	S1820	109	VHD1N60-1	109	1N3125	109	1N695	109	1S73
109	S21271	109	VHD1N60////-1	109	1N314	109	1N695A	109	1S74
109	S30300	109	VHD1834///-1	109	1N3146	109	1N698	109	1S75
109	S35770	109	V8-0A70	109	1N31A	109	1N69A	109	1S76
109	S36030	109	WD1	109	1N3204	109	1N70	109	1S78
109	S3776B	109	WX6	109	1N3287	109	1N70A	109	1S78B
109	S3838GA	109	X16	109	1N3287W	109	1N71	109	1S79
109	S3885G	109	X18	109	1N3287N	109	1N72	109	1S80
109	SC-6	109	YAAD001	109	1N34	109	1N72G	109	1S82
109	SC54	109	YAAD009	109	1N3465	109	1N73	109	1S88
109	SD-12	109	YEAD1N60P	109	1N3466	109	1N74	109	1S926G
109	SD-13(PHILCO)	109	YSG-V139-2-2	109	1N3467	109	1N75A	109	1S990AM
109	SD-14	109	YSG-V139-22	109	1N3468	109	1N76	109	1Z188
109	SD-150	109	ZC1N34A	109	1N3469	109	1N76A	109	1Z213
109	SD-16	109	ZE-1.5	109	1N3470	109	1N76C	109	1Z22
109	SD-1N60S	109	ZEN430	109	1N3483	109	1N76G	109	1Z22AJ
109	SD-46	109	ZTR-1N60	109	1N3484	109	1N770	109	1Z22AJ
109	SD-46-2	109	001-0000-00	109	1N34A	109	1N771	109	1Z22B
109	SD-56	109	001-0010-00	109	1N34A-Z	109	1N771A	109	1Z22G
109	SD-60	109	001-0022-00	109	1N34AM	109	1N771B	109	1Z23
109	SD12	109	001-0081	109	1N34AS	109	1N772A	109	1Z231
109	SD12B	109	001-01501-0	109	1N34G	109	1N773	109	1Z236
109	SD12E	109	001-015010	109	1N34GA	109	1N773A	109	1Z236
109	SD12M	109	001-015011	109	1N35	109	1N774	109	1Z238
109	SD14	109	1-017	109	1N355	109	1N774A	109	1Z23G
109	SD15	109	1-037/2207	109	1N3564	109	1N775	109	1Z23J
109	SD16	109	1-12689	109	1N3592	109	1N776	109	1Z23M
109	SD21A	109	1-425-636	109	1N36	109	1N777	109	1Z24V
109	SD34	109	1-DI-009	109	1N367	109	1N781	109	1Z240A
109	SD46	109	1A11306	109	1N367B	109	1N805	109	1Z26-2
109	SD46-2	109	1A12689	109	1N3753	109	1N81	109	1Z261
109	SD56	109	1A14384	109	1N3773	109	1N81A	109	1Z262
109	SD60	109	1C0029	109	1N38	109	1N835	109	1V9002
109	SDH-2	109	1C0039	109	1N38A	109	1N84	109	02-1001-1221-3
109	SPD107	109	1DG2	109	1N38B	109	1N86	109	2MW665
109	SPD112	109	1G02	109	1N3991	109	1N86AG	109	2XAA111
109	SPT104	109	1GD2	109	1N40	109	1N87	109	2XAA112
109	SPT108	109	1G04	109	1N408B	109	1N87A	109	2XAA113
109	SI820	109	1GD6	109	1N41	109	1N87G	109	03-0009-0
109	SK-1W80	109	1K110	109	1N417	109	1N87GA	109	003-005400
109	SK3087	109	1K261	109	1N418	109	1N87S	109	003-006700
109	SK3088	109	1K34A	109	1N419	109	1N87T	109	003-009000
109	SK3090	109	1K60	109	1N43	109	1N88	109	03-160
109	SK5091	109	1K60A	109	1N435	109	1N89	109	03-931051
109	SQ46	109	1N10	109	1N44	109	1N90	109	03-931771
109	SS0007	109	1N100	109	1N447	109	1N909	109	3A90
109	SS0008	109	1N100A	109	1N449	109	1N90G	109	3E-4
109	SU-31	109	1N103	109	1N4502	109	1N900A	109	3I4-2001-1
109	SV-31	109	1N104	109	1N452	109	1N910	109	3I4-2001-1A
109	SV-31(DIODE)	109	1N105	109	1N4523	109	1N911	109	3I4-2003-1
109	SV30	109	1N107	109	1N454	109	1N949	109	3I4-2003-3
109	SVDOA79	109	1N108	109	1N455	109	1N95	109	3I4-2003-4
109	SVD2OA70	109	1N109	109	1N46	109	1N96A	109	4-2020
109	SVD2OA79	109	1N1093	109	1N46A	109	1N97	109	4-2020-03500
109	SVDMA26-1	109	1N111	109	1N47	109	1N97A	109	4-2020-03600
109	SVDOA70	109	1N112	109	1N476	109	1N98	109	4-2020035000
109	SVDOA90	109	1N113	109	1N477	109	1N98A	109	4-2020035571
109	SVDS020	109	1N114	109	1N478	109	1N99	109	4-2021-05870
109	SX-990	109	1N115	109	1N479	109	1N994	109	4-2021A16
109	T-B1014	109	1N116	109	1N48	109	1N996	109	4-282
109	T-B1031	109	1N116A	109	1N480	109	1N99A	109	4-852
109	T11	109	1N117	109	1N48A	109	1NA4	109	4-853
109	T12	109	1N117A	109	1N497	109	1NA4G	109	4-854
109	T12G	109	1N118	109	1N498	109	1NJ33233	109	4-855
109	T13	109	1N118A	109	1N50	109	1NJ60284	109	4-857
109	T13G	109	1N119	109	1N500	109	1NJ61224	109	48D46-2
109	T14	109	1N119A	109	1N51	109	1NJ61675	109	05-00060-00
109	T14G	109	1N120	109	1N52	109	1NJ70973	109	05-000060-00
109	T17	109	1N120A	109	1N527	109	1NJ71185	109	05-00060-01
109	T18	109	1N125	109	1N52A	109	1P541	109	05-00160-01
109	T1G	109	1N126	109	1N54	109	1P542	109	05-170034
109	T20	109	1N126A	109	1N54A	109	1S-188AM	109	05-170060
109	T20G	109	1N128	109	1N54G	109	1S-446D	109	05-180034
109	T21	109	1N128A	109	1N54GA	109	1S1007	109	05-180034
109	T21237	109	1N132	109	1N56	109	1S1007S	109	05-180188
109	T21238	109	1N133	109	1N564	109	1S127	109	05-610046
109	T21271	109	1N134	109	1N569	109	1S13	109	05-931771
109	T21313	109	1N134GA	109	1N56A	109	1S15	109	05-932510
109	T21G	109	1N139	109	1N57	109	1S1690	109	05A03
109	T22	109	1N1391	109	1N571	109	1S17D1	109	7-0005
109	T22G	109	1N140	109	1N57A	109	1S5	109	7-0006
109	T23	109	1N142	109	1N58	109	1S6	109	07-5134-14
109	T23G	109	1N143	109	1N58A	109	1S186(FM)	109	07-5134-14A
109	T24G	109	1N144	109	00001N60	109	1S187	109	07-5134-14B
109	T26G	109	1N145	109	1N60(TV)(PA-1)	109	0001S188	109	07-5134-14C
109	T27G	109	1N148	109	1N60-M3	109	1S188A	109	7-59-0013477
109	T2G(DIODE)	109	1N1561	109	1N60-P	109	1S188AM	109	08-08111
109	T3G	109	1N1562	109	1N60-S	109	1S188BFM	109	8-619-030-011
109	T7	109	1N191	109	1N60-T	109	1S188BFM-1	109	8-697-020-571
109	T8G	109	1N192	109	1N60-Z	109	1S188BFM1	109	8-719-026-11
109	T9	109	1N198	109	1N60/3490	109	1S188BFM1A	109	8-719-422-21
109	T9G	109	1N198A	109	1N60/4454C	109	1S188BFMA	109	8-905-305-007
109	TC311200600	109	1N198B	109	1N60A	109	1S188G	109	8-905-305-020
109	TC3112006000	109	1N22	109	1N60AM	109	1S188FM	109	8-905-305-055
109	TB1014	109	1N265	109	1N60C	109	1S188S	109	8-905-305-318
109	TB1031	109	1N266	109	1N60D	109	1S188TV	109	8-905-305-327
109	TB1098	109	1N267	109	1N60P	109	1S19	109	8-905-305-330
109	TB1105	109	1N268	109	1N60PD1	109	1S20	109	8-905-305-336
109	TG-28	109	1N270	109	1N60PM	109	1S318	109	8-905-305-338
109	TG-48	109	1N273	109	1N60PMX	109	1S32	109	8-905-305-339
109	TP34	109	1N276	109	1N60Q	109	1S33	109	8-905-305-342
109	TP34A	109	1N277	109	1N60QA	109	1S34	109	8-905-305-348
109	TR0575002	109	1N278	109	1N60QB	109	1S34A	109	8-905-305-405
109	TR12001-4	109	1N279	109	1N60M	109	1S348	109	8-905-305-555
109	TR28736002003	109	1N2801	109	1N60P	109	1S35	109	8-905-305-561
109	TR320008	109	1N281	109	1N60S	109	1S354	109	8-905-305-580
109	TR320039	109	1N282	109	1N60QD60	109	1S355	109	8-905-305-635
109	TR320041	109	1N283	109	1N60TV	109	1S357	109	8-905-313-010
109	TR320048	109	1N285	109	1N60TVGL	109	1S3585	109	8-905-313-011
109	TR48	109	1N287	109	1N617	109	1S426	109	8-905-313-100
109	TVS-0A70	109	1N288	109	1N618	109	1S4266FM	109	8-905-313-101
109	TVS-0A81	109	1N289	109	1N62	109	1S426FM	109	8-905-313-120
109	TVS-0A90	109	1N290	109	1N63	109	1S426G	109	8-905-405-077
109	TVS-0A91	109	1N292	109	1N631	109	1S428	109	8-905-405-838
109	TVS-0A95	109	1N294	109	1N632	109	1S441	109	8A01
109	TVS-0A70	109	1N294A			109	1S446	109	09-306002
109	TVS-0A90	109	1N295					109	09-306009
109	TVS-0A91	109	1N295A					109	09-306010
		109	1N295S					109	09-306012

ECG	Industry Standard No.
109	09-306020
109	09-306024
109	09-306036
109	09-306037
109	09-306040
109	09-306047
109	09-306049
109	09-306051
109	09-306061
109	09-306064
109	09-306091
109	09-306093
109	09-306107
109	09-306108
109	09-306222
109	09-306290
109	09-306331
109	09-306334
109	09-306336
109	09-306339
109	09-306349
109	09-306370
109	9D12
109	9D16
109	9DI2
109	9DT1100310
109	10-085001
109	10-085025
109	11-085001
109	11-085004
109	11-085005
109	11-085007
109	11-085008
109	11-085015
109	11-085022
109	12-085005
109	12-085006
109	12-085009
109	12-085029
109	12-085034
109	12-085035
109	12-085038
109	12-087003
109	13-0004
109	13-004
109	13-085012
109	13-12001-0
109	13-12002-0
109	13-12003-0
109	13-14094-1
109	13-14094-11
109	13-14094-15
109	13-14094-2
109	13-14094-3
109	13-14094-5
109	13-14094-9
109	13-14890-1
109	13-16235-8
109	13-30281
109	13-35621-1
109	13-55046-1
109	13-55166-1
109	13-55166-1/3464
109	13-85962-1
109	13Z-1005
109	14-10
109	14-504-01
109	14-510-01
109	14-511-01
109	14-512-01
109	14-513-01
109	14-514-01
109	14-514-01/51
109	14-514-05
109	14-514-05/55
109	14-514-06
109	14-514-08
109	14-514-08/58
109	14-514-09
109	14-514-10
109	14-514-10/60
109	14-514-21
109	14-514-22
109	14-514-22/72
109	14-514-55
109	14-514-61
109	14-514-72
109	14-515-06/66
109	15-085002
109	15-085003
109	15-085009
109	15-085027
109	15-085032
109	15-085037
109	15-085061
109	15-20A-90M
109	019-001918
109	019-001980
109	019-002718
109	019-00301980
109	019-005043
109	19-080-009
109	19-085005
109	19-085018
109	019-301980
109	19A115086-P1
109	19AR3
109	020-00030
109	20-1680-175
109	20A-70
109	20A-90
109	20A-90H
109	20A-90M
109	20A70
109	20A79
109	20A90
109	20A90LF
109	20A90M
109	20A9M
109	21A009-000
109	21A020-005
109	21A040-001
109	21A103-010
109	21A103-011
109	21A103-017
109	21A103-019
109	21A103-022
109	21A103-046
109	21A103-052
109	21A103-055
109	21A109-001
109	21A109-002
109	21A109-022
109	21A119-005
109	21K60
109	21M288
109	21M288(DIODE)
109	21M323
109	21M432
109	21M594
109	22-004003
109	022-2823-003
109	022-2823-006
109	022-2823-007
109	022-2823-008
109	022-3901-001
109	022-3902-001
109	022.3901-001
109	022.3902-001
109	24MW1029
109	24MW1030
109	24MW1043
109	24MW1051
109	24MW1067
109	24MW1092
109	24MW199
109	24MW243
109	24MW603
109	24MW771
109	24MW860
109	24MW87
109	24MW967
109	025-100027
109	27P1
109	28P1
109	29-505
109	30P1
109	31-0039
109	32-0000
109	32-0001
109	32-0002
109	32-0003
109	32-0004
109	32-0008
109	32-0013
109	32-0023
109	32-0029
109	32-0036
109	32-18537
109	034-001-0
109	34-028-0
109	34-2001-1
109	34-8002-1
109	34-8002-2
109	34-8002-3
109	34-8002-4
109	34-8002-5
109	34-8002-6
109	34-8002-7
109	34-8022-1
109	34-8022-2
109	34-8022-3
109	34-8022-4
109	34-8022-5
109	34-8022-6
109	34-8022-7
109	34-8022-77
109	34-8057-23
109	34-8057-25
109	34-8057-26
109	34-8057-29
109	34-8057-30
109	34-8057-52
109	36E004-1
109	42-22537
109	42-22539(DIODE)
109	42-22755
109	42-23969
109	42-27543
109	42-28199
109	42A14
109	44T-300-97
109	46-8619-3
109	46-86214-3
109	46-8623-3
109	46-86253-3
109	46-86266-3
109	46-8644-3
109	46-8646-3
109	46-86484-3
109	48-03005A03
109	48-06-001
109	48-134387
109	48-134537
109	48-134587
109	48-134588
109	48-137299
109	48-137495
109	48-137497
109	48-155039
109	48-155061
109	48-155114
109	48-355009
109	48-41768G01
109	48-60022A97
109	48-60077A06
109	48-61074B01
109	48-61767B01
109	48-62334A02
109	48-63006A56
109	48-63029A20
109	48-63075A78
109	48-63077A11
109	48-63077A32
109	48-63084A06
109	48-63590A01
109	48-644587
109	48-644681
109	48-647311
109	48-647313
109	48-65837A02
109	48-65937A02
109	48-67020A11
109	48-67120A09
109	48-711052
109	48-739300
109	48-741280
109	48-82139G01
109	48-82139G02
109	48-82178A01
109	48-82178A02
109	48-82178A03
109	48-82178A04
109	48-82178A05
109	48-82178A06
109	48-82178A07
109	48-82178A08
109	48-82178A09
109	48-82178A10
109	48-82178A11
109	48-82178A12
109	48-82178A13
109	48-82292A01
109	48-82292A02
109	48-82292A03
109	48-82292A04
109	48-82292A05
109	48-86168-3
109	48-8619-3
109	48-86200-3
109	48-863030
109	48-86343-3
109	48-867716
109	48-90210A01
109	48-90222A08
109	48-90233A01
109	48-90233A06
109	48-90233A08
109	48-90343A66
109	48-97048A02
109	48-97048A06
109	48-97048A19
109	48-97168A01
109	48-97168A04
109	48-97168A07
109	48-97168A09
109	48-97168A13
109	48-97177A15
109	48-97222A01
109	48-97239A01
109	48-97270A02
109	48B41768G01
109	48C61074B01
109	48C65837A02
109	48K644681
109	48K863030
109	48K867716
109	48S10346A02
109	48S137299
109	48S155039
109	48S155061
109	48S155078
109	48X97048A06
109	48X97048A19
109	48X97168A01
109	48X97168A03
109	48X97168A04
109	48X97239A01
109	51-04001-01
109	51IN60P
109	51IS34S
109	51S1S34
109	52-050-021-0
109	53-1519
109	53A001-1
109	53A006-1
109	53A008-1
109	53A0081
109	53A022-2
109	53B001-1
109	53B001-2
109	53B001-3
109	53B004-1
109	53B005-2
109	53B007-1
109	53C001-1
109	53C006-1
109	53C006-2
109	53C006-52
109	53C009-2
109	53C020-1
109	53CD01-1
109	53D002-1
109	53D002-2
109	53E001-1
109	53E003-1
109	53E006-1
109	53H001-2
109	53K001-2
109	53K001-5
109	53L001-10
109	53N001-7
109	53N003-1
109	53N003-2
109	53N004-11
109	53N004-14
109	53N004-5
109	53N004-6
109	53T001-1
109	53T001-4
109	53T001-5
109	53T001-6
109	53U001-1
109	53V001-2
109	54V001-2
109	56-1
109	56-10
109	56-11
109	56-2
109	56-20
109	56-26
109	56-3
109	56-4
109	56-4886
109	56-8
109	56-8093
109	57A1-1
109	57A1-2
109	57A1-54
109	57A1-62
109	57A1-64
109	57D1-1
109	57D1-2
109	57D1-54
109	57D1-62
109	57D1-64
109	61-259
109	61-59395
109	62-10234
109	62-10655
109	62-12034
109	62-15318
109	62-16769
109	62-16841
109	62-19846
109	63-10739
109	63-11074
109	63-11879
109	63-12158
109	63-12607
109	63-12645
109	63-12754
109	63-12755
109	63-12756
109	63-12757
109	63-13080
109	63-13842
109	63-22724
109	63-25933
109	63-26382
109	63-28250
109	63-28888
109	63-8381
109	63-8955
109	63-9523
109	065-013
109	65-085002
109	65-085003
109	65-085010
109	65-085012
109	66X0020-000
109	66X0020-001
109	66X0039-001
109	66X0047-001
109	66X0047-901
109	66X0049-002
109	66X0051-001
109	66X20
109	69-2922
109	76-12965-26
109	77-271031-1
109	77-271032-1
109	78-271199-1
109	78-271228-1
109	78-273002-1
109	79F015
109	80-60-1
109	81-27123150-8
109	81-46123001-3
109	81-46123006-2
109	81-46123013-8
109	81-46123015-3
109	83B38-1
109	86-0002
109	86-0008
109	86-001
109	86-0513
109	86-10-1
109	86-125-1
109	86-146-1
109	86-20-1
109	86-22-1
109	86-45-1
109	86-49-1
109	86-5007-3
109	86-60-1
109	86-64-1
109	86-88-3
109	87-10-0
109	089-236
109	089-248
109	089-293
109	090A64-1
109	91-3
109	91-46
109	92-1001
109	93.24.401
109	93.24.601
109	93.24.604
109	93A105-1
109	93A110-1
109	93A25-1
109	93A25-3
109	93A27-1
109	93A33-1
109	93A38-1
109	93A41-5
109	93A77-1
109	93A8-1
109	93B25-1
109	93B25-2
109	93B25-3
109	93B27-1
109	93B38-1
109	93B38-5
109	93B41
109	93B41-1
109	93B41-3
109	93B8-1
109	93C218
109	93C25-3
109	93C55-1
109	93C7-1
109	93C8-1
109	094-014
109	94-42-9
109	96-0008
109	96-5007-01
109	96-5059-01
109	96-5087-01
109	96-5363-01
109	96X2077844N
109	96XZ778/21N
109	96XZ778/44N
109	100-00340-00
109	100-0051
109	100-00910-07
109	100-0124
109	100-0125
109	100-12
109	100-136
109	100-160
109	100-180
109	100-181
109	100-215
109	100-436
109	102-02
109	102-207
109	103-114
109	103-19
109	103-202
109	103-22
109	103-23-01
109	103-271
109	103-31
109	103-34
109	103-44
109	103-73
109	103-74
109	103-79
109	103-87
109	105-02
109	105-03
109	106-008
109	108(PARFISA)
109	109-036500
109	110-763
109	0112-0019
109	0112-0026
109	0112-0028
109	0112-0028/4460
109	0112-0037
109	0112-0046
109	0112-0073
109	0112-0082
109	120-004498
109	120-004499
109	120-004730
109	120-005299
109	121-31
109	123-013
109	123-015
109	130-30281
109	130-30301
109	130-40229
109	137-824
109	141-003
109	142-011
109	150-001-005
109	150-001-9-005
109	150-001-9-007
109	150-002-9-001
109	150-004-9-001
109	150-005-9-001
109	150-006-9-001
109	150-013-9-001
109	150-014-9-001
109	185-013
109	186-013
109	200X8000-026
109	201X2000-118
109	209-31
109	229-0182-65
109	229-5100-231
109	230-0006
109	260-10-025
109	260-10-048
109	260D00507
109	264D00612
109	264D00701
109	264D00901
109	264D01001
109	264P00401
109	264P0080
109	264P00801
109	264P01301
109	264P01305
109	264P01306
109	264P03802
109	264Z00701
109	294-42-9
109	295V005H03
109	296-42-9
109	296V002H01
109	296V002H02
109	296V002H05
109	296V002H06
109	296V002H07
109	296V002H08
109	296V002M01
109	296V006H02
109	296V007H02
109	296V012H01
109	296V015B01
109	296V015H01
109	309-324-616
109	309-327-916
109	324-0014
109	324-0035
109	324-0037
109	324-0049
109	324-0057
109	324-0105
109	324-0107-01
109	324-0108
109	324-0141
109	324-0160
109	325-0025-327
109	325-0028-86
109	325-0028-87
109	325-0031-335
109	325-0036-562
109	325-1376-60
109	339-529-001
109	344-6006-6
109	400A
109	400D
109	403-1
109	429-0092-56
109	429-0910-54
109	464-100-19
109	464-103-19
109	464-106-19
109	464-111-19
109	464-113-19
109	503-T21271
109	503-T21472
109	510A90
109	510ED46
109	510IN60
109	510IS34
109	521-145
109	523-1000-067
109	523-1000-294
109	523-1000-295
109	523-1002-326
109	523-1500-067
109	523-2003-0-1
109	523-2003-001
109	524-457
109	525-877
109	0575-005
109	600X0096-066
109	600X0097-066
109	601X0150-066
109	601X0151-066
109	617-15
109	617-156
109	617-17
109	630-002
109	630-079
109	642-028
109	642-102
109	642-119
109	642-132
109	642-199
109	642-221
109	650
109	656-142
109	675-158

ECG	Industry Standard No.
109	690V034H32
109	690V034H74
109	690V040H63
109	690V047H61
109	690V052H50
109	690V052H68
109	690V059H63
109	690V066H48
109	690V067H09
109	690V068H32
109	690V073H60
109	690V083H89
109	690V092H85
109	690V098H52
109	690V103H54
109	690V68H32
109	690V73H60
109	690V037H91
109	742
109	754-2000-009
109	754-4000-088
109	754-4000-188
109	754-4000-410
109	754-5900-090
109	754-9000-460
109	792-292
109	800-003-00
109	800-005-00
109	800-020-00
109	800-022-00
109	800-039-00
109	800-517-00
109	805-1060
109	805-10600
109	808-312
109	903-00390
109	903-103B
109	903-108
109	903-108B
109	903-10B
109	903-113B
109	903-114B
109	903-115B
109	903-12B
109	903-168B
109	903-16B
109	903-18B
109	903-212
109	903-23
109	903-23A
109	903-23B
109	903-23C
109	903-23D
109	903-23E
109	903-27
109	903-27B
109	903-29B
109	903-30B
109	903-34
109	903-34A
109	903-34B
109	903-34C
109	903-34D
109	903-34E
109	903-37B
109	903-43B
109	903-45B
109	903-51B
109	903-54B
109	903-65B
109	903-83B
109	903-92B
109	903-9B
109	904-97B
109	908D1
109	914-000-4-00
109	916-32000-7
109	916-32003-2
109	916-32006-2
109	919-01-0867
109	919-010867
109	1000-131
109	1000-17
109	1001-10
109	1002-08
109	1002-09
109	1002-17
109	1002-17(DIODE)
109	1004(DIODE)
109	1006(DIODE)
109	1006-9292
109	1007-17(DIODE)
109	1010-145
109	1011
109	1012-17
109	1013
109	1016-77
109	1019-6699
109	1030-17
109	1033-1916
109	1040-08
109	1041-65
109	1042-12
109	1042-13
109	1042-3
109	1048-9870
109	1063-3553
109	1063-8591
109	1074-124
109	1074-24
109	1077-2325
109	1077-2760
109	1077-9296
109	1118-17
109	1119-17
109	01122-0073
109	1207-17
109	1410-171
109	1489-17
109	1512
109	1550
109	1778-17
109	1956-17
109	1977-17
109	1980-17
109	2000-301
109	2000-318
109	2022-07
109	2093A25-3
109	2093A33-1
109	2093A38-1
109	2093A38-10
109	2093A38-14
109	2093A38-15
109	2093A38-21
109	2093A38-27
109	2093A38-30
109	2093A38-31
109	2093A38-32
109	2093A38-33
109	2093A38-34
109	2093A38-40
109	2093A38-5
109	2093A41
109	2093A41-14
109	2093A41-141
109	2093A41-148
109	2093A41-154
109	2093A41-167
109	2093A41-169
109	2093A41-181
109	2093A41-187
109	2093A41-196
109	2093A41-2
109	2093A41-38
109	2093A41-50
109	2093A41-59
109	2093A41-68
109	2093A41-92
109	2093A77-1
109	2093A8-1
109	2093B41-11
109	2102-010
109	2102-025
109	2102-028
109	2106-124
109	2122-17
109	2151-17
109	2196-17
109	2232-17
109	2279-13
109	2280-13
109	2281-13
109	2282-13
109	2282-17
109	2283-13
109	2284-13
109	2290-13
109	2402-459
109	2405-458
109	2408-330
109	2510-31
109	2603-186
109	2606-296
109	2703-389
109	2704-388
109	2789
109	003016
109	3322-6
109	3505
109	03571
109	4001-230
109	4002(SEARS)
109	4003(SEARS)
109	4005(SEARS)
109	4006(SEARS)
109	4007(SEARS)
109	4008(SEARS)
109	4009(SEARS)
109	4010(PENNCREST)
109	4010(SEARS)
109	4011(SEARS)
109	4012(SEARS)
109	4041-000-1018
109	4041-200-10180
109	4041-200-40100
109	4354
109	04770
109	4801-00628
109	4801-00629
109	4822-130-30281
109	4822-130-30301
109	4822-130-30311
109	4822-130-30312
109	4822-130-40229
109	4828-4
109	5001-080
109	5001-134
109	5001-141
109	5001-161
109	5101834
109	5120A90
109	5631-1N34A
109	5631-1N60
109	5631-1N60P
109	8000-00003-045
109	8000-00004-045
109	8000-00004-060
109	8000-00004-063
109	8000-00005-015
109	8000-00005-016
109	8000-00005-017
109	8000-00005-018
109	8000-00005-023
109	8000-00006-007
109	8000-00011-046
109	8000-00011-060
109	8000-00038-009
109	8000-0004-042
109	8000-0004-063
109	8000-0004-P060
109	8000-0004-P063
109	8000-00041-015
109	8000-00041-016
109	8000-0005-015
109	8000-0005-016
109	8000-0005-017
109	8000-0005-018
109	8000-0005-023
109	8010-53
109	8710-53
109	8840-54
109	9861B-43
109	10181
109	11252
109	0012060
109	12808
109	12850
109	15027(DIODE)
109	18410-42
109	18600-53
109	25201-001
109	25810-53
109	25810-55
109	25840-53
109	25840-55
109	26810-51
109	27840-42
109	28287
109	031033
109	031040
109	36508
109	42020
109	43959
109	45810-53
109	48287
109	50505-01
109	53092-1
109	55166
109	55810-51
109	55810-52
109	057001
109	057001H
109	057005
109	057005H
109	58810-83
109	059395
109	59395(RCA)
109	60215-1
109	67590
109	68504-77
109	71778
109	72013
109	72060
109	72080
109	000072090
109	72128A
109	72129
109	72129A
109	000072160
109	075005
109	76675
109	76675A
109	76675B
109	79985
109	085002
109	085004
109	085006
109	085016
109	86001
109	95000
109	95001
109	95004
109	95007
109	95008
109	95014
109	95017
109	95018
109	100017
109	100844
109	102989
109	104152
109	105517
109	110610
109	111207
109	111605
109	112330
109	112526
109	112529
109	115101
109	116048
109	116273
109	117659
109	117730
109	117760
109	119199
109	119919
109	120617
109	122166
109	122520
109	126177
109	127017
109	127784
109	129028
109	129157
109	129158
109	129474
109	132912
109	134180
109	139634
109	143595
109	144585
109	146575
109	161006
109	161016
109	161038
109	162002-039
109	162002-39
109	165572
109	166273
109	167562
109	168910
109	169115
109	169362
109	169501
109	170373
109	171162-042
109	171162-269
109	175043-066
109	175043-068
109	190716
109	195617
109	200648-26
109	202315
109	216001
109	225410
109	226334
109	226344
109	245517
109	320007
109	489752-001
109	489752-003
109	489752-031
109	489752-042
109	489752-049
109	489752-076
109	489850-004
109	500009G
109	0517022
109	0517826
109	0517828
109	0517829
109	0525002
109	0525002H
109	0526224
109	530015-2
109	530063-10
109	530063-4
109	530065-1
109	530065-1002
109	530065-1002A
109	530065-2
109	530065-3
109	530072-7
109	530072-8
109	530085-2
109	530092-1
109	530092-2
109	530127-1
109	0537820
109	0570519
109	0575001
109	0575001H
109	0575002
109	0575002H
109	0575004
109	0575005
109	0575005H
109	0575007
109	575009
109	0575067
109	575091
109	0575099
109	576063
109	0577001
109	601030
109	615010
109	740402
109	740952
109	740954
109	741866
109	742004
109	742730
109	755722
109	760101-0005
109	760101-0006
109	765722
109	771909
109	771910
109	771911
109	772712
109	785278-01
109	801722
109	817032
109	817077
109	817125
109	817158
109	817159
109	817194
109	817199
109	871125
109	922021
109	922604
109	972258-6
109	972258-8
109	972258-9
109	972259-8
109	980514
109	981150
109	981153
109	981207
109	981522
109	981676
109	982065
109	982275
109	982290
109	982822
109	983099
109	983239
109	984163
109	984200
109	984226
109	984666
109	988994
109	988997
109	988998
109	992143
109	1121008/7611
109	1223770
109	1223931
109	1408649-1
109	1471872-15
109	1471872-17
109	1476179-001
109	1476179-1
109	1800002
109	2000433-150
109	2000648-26
109	2000757-18
109	2001786-134
109	2002151-020
109	2002151-20
109	2002336-20
109	2004357-106
109	2006422-132
109	2006431-50
109	2006441-122
109	2006582-20
109	2092055-0001
109	2092055-0010
109	2092055-0713
109	2092055-1
109	2092055-7
109	2095083
109	2485080(DIODE)
109	2495083-2
109	2495084
109	2495380
109	2495383
109	2496436
109	2498530
109	3596062
109	5330331
109	5330332
109	5330335
109	5330721
109	5330731
109	5330732
109	5491236-P1
109	5491236-P2
109	06200001
109	06200002
109	06200017
109	7100460
109	7129386-P2
109	7279893
109	7282358
109	7570014-02
109	7570016
109	7570016-02
109	7570016-03
109	7777146-P2
109	7777146-P23
109	7777146-P3
109	7777146-P4
109	7851654-01
109	7851655-01
109	7851947-01
109	7852223-01
109	7852223-01A
109	7852438-01
109	7852782-01
109	7852902-01
109	7853098-01
109	7853357-01
109	7855282
109	7855282-01
109	7855352-01
109	7855357-01
109	8051060
109	8052446
109	8121002
109	8121014
109	11210016/7825
109	13030301
109	13030312
109	13040229
109	17101881
109	20001786-134
109	20115070
109	22115181
109	22115182
109	23115070
109	23115088
109	23115115
109	23115129
109	25115102
109	27123150
109	27123240
109	27123270
109	30600010
109	30600020
109	37000918
109	41027612
109	41027613
109	41027992
109	41029009
109	41029290
109	41527380
109	41624836
109	62094613
109	62119551
109	62134496
109	62134941
109	62543299
109	62761261
109	62784822
109	80510600
109	80521881
109	87100600
109	87100605
109	87201881
109	87204260
109	231150050
109	600000060
109	1522270100
109	1522270101
109	2008000001
109	2008000026
109	2008000064
109	2012000092
109	2012000100
109	2012000118
109	4120100600
109	4120129000
109	4129100600
109	4202003500
109	4202008600
109	6612004000
109	6612009000
109	9511510200
109	16400690060
109	16401190188
109	16411190188
109	16415490000
109	16419990032
109	134800847274
109	134800855216
109	134800863030
109	134800867716
109	134882139701
109	134882139702
109	134882178101
109	134882178102
109	134882178103
109	134882178104
109	134882178105
109	134882178106
109	134882178107
109	134882178108
109	134882178110
109	134882178111
109	134882178112
109	134882921701
109	134882921702
109	134882921704
109	134882921705
109	301720490201
109	404120010180
109	404120030100
109	404120040100
109	2025911018810
1090	GE-723
1090	STK-025
1090	STK-025G
1090	STK-032
1091*	37008002
1092	GEIC-130
1092	RE342-M
1092	09-308096
1092	916063
1092	916092
1096	M5134P
1100	BA402
1100	DDEY004001
1100	EA33X8500
1100	GEIC-92
1100	HC10002050
1100	TA7061
1100	TA7061(AP)
1100	TA7061AP
1100	TA7061P
1100	09-308041
1100	09-308059
1100	90-36
1100	90-73
1100	260-10-036
1100	307-007-9-001
1100	307-020-9-001
1100	307-029-1-001

ECG	Industry Standard No.
1100	588-40-201
1100	1021-25
1100	5002-007
1100	8710-171
1100	8840-170
1100	8910-147
1100	18600-154
1100	25810-164
1100	37510-164
1100	58810-170
1100	58840-201
1100	88060-147
1100	88510-177
1101	GEIC-95
1101	23119011
1101	23119022
1102	GEIC-93
1102	IP20-0014
1102	TA7062P
1102	90-35
1102	90-72
1102	260-10-035
1102	307-007-9-002
1102	307-009-9-002
1102	26010035
1103	EICM-14
1103	G09-007-A
1103	G09-007-B
1103	G09-029-C
1103	G09-029-D
1103	G09007
1103	G09007B
1103	GEIC-94
1103	MPC566HB
1103	MPC566HC
1103	MPC566HD
1103	RE329-IC
1103	TA-7063P
1103	TA7063
1103	TA7063P
1103	TA7063P-B
1103	TA7063P-C
1103	TA7063P-D
1103	TA7063P-O
1103	09-308036
1103	09-308043
1103	051-0011-00
1103	051-0011-00-04
1103	051-0011-01
1103	051-0011-02
1103	051-0011-03
1103	051-0011-04
1103	57A32-19
1103	266P30706
1103	740583
1103	916067
1104	BA401
1104	GEIC-91
1104	Q-01369C
1104	TA7060
1104	TA7060P
1104	TA7060P-10
1104	TA7060PR
1104	TA7060PRW
1104	TA7060PW
1104	90-37
1104	90-37(IC)
1104	90-74
1104	8000-00005-014
1104	916105
1105	GEIC-101
1105	RH-1X0024CEZZ
1105	TA7102P
1105	TA7102P(FA-1)
1105	TA7102P(FA-2)
1105	23119016
1105	23119028
1107	09-308067
1107	051-0011-00-05
1107	740502
1109	GEIC-99
1109	IC-545(ELCOM)
1109	TA7076P
1109	TA7076P(FA-1)
1109	TA7076P(FA-6)
1109	TA7076P(FA-7)
1109	21A120-002
1109	46-1370-3
1109	23119013
1109	23119023
1109	23119030
1109	23119031
1109	23119032
1109	23119033
1109	40306604
1109	4206105370
110MP	A054-226
110MP	A069-111
110MP	A069-118
110MP	A909-1016
110MP	OA95
110MP	AFS-160-1020
110MP	CIJ70645
110MP	CT2005
110MP	CT2008
110MP	CX0042
110MP	D010
110MP	D1R35
110MP	D1R39
110MP	DIJ70543
110MP	DIJ70644
110MP	DIJ71778
110MP	DS39
110MP	E21135
110MP	E2484
110MP	EA2137
110MP	EA2606
110MP	EP16X21
110MP	ES16X21
110MP	ES16X70
110MP	EU16X19
110MP	EW168
110MP	G00004A
110MP	H8287
110MP	H8287-4
110MP	HD1000105
110MP	HD1000105-O
110MP	HD1000303
110MP	HE-10002
110MP	HE-10044
110MP	HF-20088
110MP	INJ61224
110MP	J241
110MP	J24567
110MP	J24628
110MP	J24643
110MP	J24820
110MP	JP575005
110MP	JP575995
110MP	MD-60A
110MP	MD60A
110MP	N-EA16X27
110MP	OA-95
110MP	OA90FM
110MP	OA95
110MP	RE86
110MP	RT1667
110MP	RT2451
110MP	RT4880
110MP	RT5212
110MP	RT5379
110MP	RT5470
110MP	RT5738
110MP	RT5912
110MP	RT6728
110MP	RT6731
110MP	RT7538
110MP	S1384
110MP	SD-1N60
110MP	SD020
110MP	SD46(4)
110MP	SD46R
110MP	S046
110MP	SYL128
110MP	T-E1105
110MP	T-E1177
110MP	T4590
110MP	TR0575005
110MP	TR228736002004
110MP	TR320007
110MP	TV241013
110MP	TV24122
110MP	TV24273
110MP	VHD1N34A///-1
110MP	XD2A
110MP	YEAD032
110MP	001-0020-00
110MP	1N34M
110MP	1N541
110MP	1N542
110MP	1N60-5
110MP	1N60/7825B
110MP	1N60MP
110MP	1S128FM
110MP	1S186FM
110MP	1S188FMI
110MP	1S188MPX
110MP	1S188P
110MP	1S50
110MP	1S60
110MP	1T23S
110MP	2A119
110MP	2AA119
110MP	20A90
110MP	003-004200
110MP	003-007500
110MP	4-2020-05600
110MP	4-2020-08600
110MP	07-5160-15
110MP	08-08112
110MP	09-306019
110MP	09-306057
110MP	09-306229
110MP	09-306335
110MP	09-30636
110MP	10-085004
110MP	12-085041
110MP	13-085029
110MP	13-14094-8
110MP	14-514-11
110MP	14-514-13
110MP	14-514-13/63
110MP	15-20A70
110MP	15-20A90
110MP	020-00011
110MP	020-00012
110MP	20A90MLF
110MP	20A90Z
110MP	21A009-002
110MP	21A103-006
110MP	21A103-016
110MP	21A103-048
110MP	21A109-003
110MP	21M289
110MP	21M325
110MP	21M568
110MP	24MW122
110MP	24MW244
110MP	24MW665
110MP	24MW820
110MP	32-0007
110MP	32-0039
110MP	32-18539
110MP	34-8057-3
110MP	42-19681
110MP	42-21362
110MP	42-23972
110MP	46-8616-3
110MP	46-86200-3
110MP	48-134954
110MP	48-355008
110MP	48-60154A01
110MP	48-62334A01
110MP	48-97048A05
110MP	48-97168A03
110MP	53A001-2
110MP	53B010-7
110MP	53N003-3
110MP	53N004-10
110MP	56-8095
110MP	065-014
110MP	66X0049-100
110MP	76-14196-1
110MP	86-14-1
110MP	86-15-1
110MP	86-48-1
110MP	87-10-1
110MP	91N1
110MP	93A25-2
110MP	93A41-2
110MP	93A83-1
110MP	93B27-3
110MP	93C25-2
110MP	100-00914-10
110MP	103-102
110MP	103-23
110MP	103-90
110MP	120-001301
110MP	150-015-9-001
110MP	264D00801
110MP	264D0701
110MP	324-0107
110MP	354-9001-001
110MP	354-9101-002
110MP	429-0910-53
110MP	464-110-19
110MP	464-119-19
110MP	754-1003-030
110MP	800-002-00
110MP	903-167B
110MP	903-8B
110MP	914-001-7-00
110MP	1007-1124
110MP	1010-8173
110MP	1048-9888
110MP	1206-17
110MP	1947-17
110MP	2093A38-22
110MP	2093A38-35
110MP	2093A41-29
110MP	2093B11-21
110MP	2093B41-29
110MP	5101N60
110MP	5361-1N60P
110MP	5631-20A90
110MP	8910-53
110MP	011119
110MP	28810-64
110MP	37510-53
110MP	46287-4
110MP	48287-4
110MP	58840-114
110MP	68504-76
110MP	72128
110MP	085005
110MP	085026
110MP	88060-53
110MP	88510-53
110MP	112524
110MP	161001
110MP	161015
110MP	167572
110MP	171162-270
110MP	216003
110MP	489752-036
110MP	489752-125
110MP	500001
110MP	511534S
110MP	530063-1
110MP	530063-14
110MP	530065-1003
110MP	530072-1001
110MP	530105-1001
110MP	0575019
110MP	0575019H
110MP	610030
110MP	614020
110MP	772740
110MP	817160
110MP	817177
110MP	982270
110MP	982271
110MP	984881
110MP	985621
110MP	1471822-11
110MP	2092055-0007
110MP	2495083
110MP	06200003
110MP	7116060
110MP	17101882
110MP	41522875
110MP	62154543
110MP	62381620
110MP	71773660
110MP	87500620
110MP	4120200602
110MP	4129100602
110MP	4129200602
110MP	4202005600
1110	M5115P
1115	CA810M
1115	GEIC-278
1115	TBA810AS
1115	TBA810DS
1115	TBA810S
1115	88-0550
1115	544-2006-001
1115	3130-3248-801
1115	6001
1115	3597049
1115A	CA810Q
1115A	RH-1X1020AFZZ
1116	GEIC-279
1116	TBA800
1116	56A42-1
112	AA015
112	ET16X14
112	EU16X14
112	EU16X4
112	G01A
112	HEP-R0700
112	JT-1601-41
112	K3E
112	M8482
112	MBD101
112	RB48
112	SD-404
112	SD-51
112	SD51
112	SD82
112	SD82A
112	SD82AG
112	SP82A
112	TH-1S750
112	TV24103
112	TV24103A
112	TV24103B
112	TV24103C
112	TV24103D
112	TV24159
112	TV24182
112	TVS-182G
112	TVS-1N82G
112	TVS-1S750
112	TVS-82G
112	TVS-S82
112	TVS-SD82A
112	1N82A
112	1N82AG
112	1N82G
112	1S1922
112	1S1925
112	1S1926
112	1S1926K
112	1S2198
112	1S750
112	1SS16
112	1T13
112	1T13A
112	1T13B
112	4JB2C6
112	4JBC12
112	07-5160-15A
112	07-5160-15B
112	07-5160-15C
112	09-306077
112	09-306089
112	09-306209
112	09-306210
112	09-306216
112	13-10321-42
112	13-31014-4
112	15-085018
112	19A115322
112	19A115322-P1
112	23-PT275-124
112	24E-002
112	24E-022
112	025-100008
112	46-861-3
112	46-8613
112	46-86134-3
112	46-8625
112	46-8643
112	46-8643-3
112	46-8688-3
112	47-2
112	47-4
112	48-65112A73
112	48-65113A84
112	48-674297U
112	48-674970
112	48-742970
112	48-90234A66
112	48P65112A73
112	57D1-65
112	62-17232
112	62-19260
112	63-250128-2
112	76-13570-65
112	76-13848-23
112	86-12-1
112	93A43-1
112	93A43-2
112	93A59-1
112	93D112-57
112	103-49
112	103-60
112	103-61
112	103-65
112	229-0240-20
112	264P03401
112	2090A43-1
112	2093A43-2
112	2093A59-1
112	8000-00012-038
112	107729
112	116314
112	119662
112	127532
112	129556
112	131214
112	134074
112	134264
112	143162
112	147922-2
112	489752-020
112	489752-052
112	489752-089
112	489752-090
112	575037
112	970047
112	970759
112	1442415-2
112	1442415-3
112	1471922-1
112	2092055-0708
112	23115113
112	23115121
112	62034297
112	62139455
1123	AN236
1123	GEIC-281
1123	IC-535(ELCOM)
1123	51-90305A20
1123	144011
1127	09-308084
1128	A0311400
1128	B0311400
1128	B0311402
1128	GEIC-105
1128	REN1128
1128	TA7124P
1128	46-1396-3
1128	1061-9153
1128	23119981
1128	23119990
1128	6644001100
1130	B0313800
1130	GEIC-111
1130	REN1130
1130	TA7150P
1130	46-1393-3
1130	1061-9666
1130	23119993
1131	B0313600
1131	GEIC-109
1131	TA7148P
1131	46-1394-3
1131	23119995
1131	6644001700
1132	B0313700
1132	GEIC-110
1132	TA7149P
1132	46-1397-3
1132	23119994
1132	6644001800
1133	B0313400
1133	GEIC-107
1133	REN1133
1133	TA7146P
1133	46-1398-3
1133	1061-9856
1133	23119988
1133	6644001500
1134	B0313300
1134	GEIC-106
1134	REN1134
1134	TA7145P
1134	46-1395-3
1134	1061-9161
1134	23119989
1134	6644001400
1135	BA301
1135	REN1135
1135	09-308062
113A	A40-6704
113A	A40-6722
113A	A4201
113A	A42946
113A	A42946B
113A	A86-9-1
113A	A95-5280
113A	A95-5297
113A	ALC1
113A	ALC1A
113A	B522-893
113A	B527-062
113A	C05-03C
113A	C08P1
113A	C10
113A	C10-02A
113A	C10-13B
113A	C10-18B
113A	C10-1B
113A	C10-31A
113A	C10-38C
113A	C10-47B
113A	CM
113A	D16U3
113A	D16U4
113A	D1B
113A	D1C
113A	D4
113A	DD04
113A	DIC
113A	E1176ALC1A
113A	EP-2798
113A	EP16X6
113A	ES57X6
113A	FSA1177
113A	FSA1178
113A	G5019
113A	GE-6GD1
113A	H615
113A	HD2000710
113A	K112
113A	K112C
113A	K115J510-1
113A	K115J510-2
113A	K117J460-1
113A	K117J460-2
113A	K122
113A	K122C
113A	K1615
113A	K8533137
113A	M109474
113A	M8534992
113A	P15
113A	R109328
113A	R109474
113A	R3057
113A	R3314
113A	R3533
113A	R4409
113A	R4666
113A	R8474
113A	RE87
113A	REJ701148-1
113A	REJ70148
113A	REJ70148A
113A	RER023
113A	RF33426-7
113A	RF5464
113A	RF5464-1P
113A	RF5465
113A	RF5465-1P
113A	RFJ60313
113A	RFJ70148
113A	RSLNB0001CEZZ
113A	SDD-C10
113A	SDD4
113A	SELEN-26
113A	SELEN-38
113A	SI-RECT-174
113A	SLEN-26
113A	SR-0004
113A	SR20
113A	SR29
113A	SR6
113A	SR9002
113A	SRR13
113A	T-E1086A
113A	TV24226
113A	TVM-526
113A	TVM554
113A	TVS-K112C
113A	1NJ70972
113A	003-002200
113A	4-2020-0500
113A	4-2020-05000
113A	4-2020-05400
113A	6G4
113A	6GC1
113A	6GC1BY1
113A	7K705M
113A	7VM705M
113A	09-306101
113A	09-306193
113A	9IR2
113A	9IR2-1
113A	9IR2-24
113A	9IR2-3
113A	9IR2-S
113A	9IR2-S1
113A	9IR21
113A	10DC05N
113A	12/3-04
113A	13-31014-2
113A	13-85943-1
113A	13-85943-2
113A	13-85943-3
113A	14-501-01
113A	14-501-02
113A	15-085038
113A	15-085039
113A	15-085047
113A	15-108002
113A	15-108009
113A	19AR29

ECG	Industry Standard No.	ECG	Industry Standard No.	ECG	Industry Standard No.	ECG	Industry Standard No.	ECG	Industry Standard No.
113A	19AR29-1	114	TC0.09M21/3	1155	26810-159	116	A1F5	116	A600
113A	21A002	114	TVH-526	1155	45810-170	116	A1F9	116	A600(RECTIFIER)
113A	21A002-000	114	TVM526	116	0207205	116	A1G1	116	A692275
113A	21A024	114	TVMTC00921-3	116	741687	116	A1G5	116	A692514-0
113A	24MW1124	114	TVS-TC009M21/3	116	742564	116	A1G9	116	A692116-0
113A	27-226	114	WT-16X9	116	916110	116	A200	116	A7102001
113A	027-300226	114	Y100	116	1223910	116	A23	116	A75-68-500
113A	27-C226	114	6GD1	116	GEIC-71	116	A2421	116	A75092201
113A	32-C062	114	9LR2-2	116	LA1366N	116	A2422	116	A7568500
113A	33059019	114	012-1022-002	116	4-2060-07500	116	A2460	116	A7572100
113A	34-8034-7	114	14-503-01	116	4206007500	116	A2461	116	A7572200
113A	34-8037	114	14-503-02	116	GEIC-72	116	A2462	116	A7C
113A	34-8037-1	114	14-503-03	116	LA1367	116	A2481	116	A7D
113A	34-8037-2	114	14-503-04	116	4-2060-05200	116	A2485	116	A7E
113A	34-8037-3	114	14-503-08	116	4206005200	116	A25-1008	116	A7G
113A	34-8037-4	114	14-504-04	116	.7E05	116	A2A1	116	A909-1018
113A	34-9037-1	114	19AR4	116	.7E1	116	A2A4	116	A909-1019
113A	46-86220-3	114	19AR4-1	116	.7E2	116	A2A5	116	A95-5281
113A	46-86332-3	114	21/3	116	.7E3	116	A2A9	116	A95-5289
113A	46-86336-3	114	21/3.92	116	.7E4	116	A2B1	116	AA100
113A	48-90235A01	114	46-86304-3	116	.7E5	116	A2B4	116	AA200
113A	488134916	114	48-741752	116	.7E6	116	A2B5	116	AA300
113A	53B01O-1	114	48-754153	116	.7J05	116	A2B9	116	AA400
113A	53B01O-2	114	65-744238	116	.7J1	116	A2C1	116	AA50
113A	62-16712	114	86-3-1	116	.7J2	116	A2C4	116	AA500
113A	62-18337	114	930267	116	.7J3	116	A2C5	116	AA600
113A	62-19734	114	9305-3	116	.7J5	116	A2C9	116	AAY-22
113A	66XO024-000	114	93051-3	116	.7J6	116	A2D1	116	AA218D
113A	66XO025-000	114	513-891	116	A-04	116	A2D4	116	ACR81-100B
113A	66XO025-000-001	114	1616C	116	A-04049-B	116	A2D5	116	ACR83-100B
113A	66XO025-001	114	2093A5-2	116	A-04091-A	116	A2D9	116	AD-1UF
113A	66X21	114	489752-017	116	A-04092	116	A2E1	116	AD10
113A	66X218	114	530045-2	116	A-04092-B	116	A2E4	116	AD100
113A	66X25	114	530045-3	116	A-04093	116	A2E5	116	AD200
113A	66X25-0	114	530045-4	116	A-04093A	116	A2E9	116	AD4001
113A	66XZ18	114	530093-3	116	A-04212-A	116	A2P1	116	AD50
113A	86-18-1	114	654032	116	A-04212-B	116	A2P4	116	AG100D
113A	86-18-1A	114	817962	116	A-04226	116	A2P5	116	AG1000
113A	86-9-1	114	84226	116	A-042313	116	A2P9	116	AG100J
113A	86-97-1	114	1045494-1	116	A-04242	116	A201	116	AJ-30
113A	93A5-10	114	20001786-139	116	A-04901A	116	A2G4	116	AJ10
113A	93A5-2	1140	GEIC-138	116	A-1.5-01	116	A205	116	AJ15
113A	93A5-9	1140	MP0575C2	116	A-100(RECT.)	116	A209	116	AJ20
113A	93B5-1	1140	001-0091	116	A-10105	116	A300	116	AJ25
113A	93B5-10	1140	36-0083	116	A-10113	116	A3A1	116	AJ30
113A	93B5-3	1140	57502	116	A-10118	116	A3A3	116	AJ35
113A	93B5-3-6	1140	1001-0091/4460	116	A-1946	116	A3A5	116	AJ40
113A	93B5-3-8	1140	90200100	116	A-95-5281	116	A3A9	116	AJ5
113A	93B5-3-9	1140	EA33X8372	116	A-95-5289	116	A3B1	116	AJ50
113A	93B5-4	1140	GEIC-128	116	A0377	116	A3B3	116	AJ60
113A	93B5-5	1142	K24154	116	A04	116	A3B5	116	AM-010
113A	93B5-6	1142	MP0554	116	A04049B	116	A3C1	116	AM-020
113A	93B5-8	1142	MP0554C	116	A04091A	116	A3C3	116	AM-025
113A	93B5-9	1142	PC554	116	A04092	116	A3C5	116	AM-030
113A	9305-10	1142	09-308038	116	A04092A	116	A3C9	116	AM-035
113A	9305-5	1142	21M485	116	A04092B	116	A3D1	116	AM-060
113A	9305-6	1142	36-0041	116	A04093	116	A3D3	116	AM-22
113A	9305-7	1142	57A132-29	116	A04210A	116	A3D5	116	AM-35
113A	9305-8	1142	57A32-2	116	A04212-B	116	A3D9	116	AM-G-5
113A	9305-9	1142	57A32-29	116	A04212B	116	A3E1	116	AM-6-5
113A	93059	1142	61A001-11	116	A042313	116	A3E3	116	AM-G-10
113A	93K2-1	1142	588-40-202	116	A04233	116	A3E5	116	AM-G-22
113A	96XZ778/27W	1142	740-9000-554	116	A04242	116	A3E9	116	AM-G-5
113A	103-20	1142	2057A32-26	116	A04331-021	116	A3F1	116	AM-022
113A	103-32	1142	8710-172	116	A04331-023	116	A3F3	116	AMO05
113A	103-43	1142	8910-148	116	A04331-043	116	A3F5	116	AMO10
113A	131A	1142	25840-165	116	A04350-022	116	A3F9	116	AMO20
113A	165	1142	28810-165	116	A04731	116	A3G1	116	AMO25
113A	227-200001	1142	58810-171	116	A04901A	116	A3G3	116	AMO30
113A	264P00501	1142	000074030	116	A0491A	116	A3G5	116	AMO35
113A	264P00506	1142	88510-178	116	A05	116	A3G9	116	AMO40
113A	269V004-H01	1142	916070	116	A054-150	116	A400(RECTIFIER)	116	AMO50
113A	296V004H01	115	C08P1R	116	A054-230	116	A4212-A	116	AMO60
113A	420-2005-000	115	D6	116	A059-114	116	A422-A	116	AM13
113A	977-14B	115	D7	116	A06	116	A42XO0269-01	116	AM23
113A	1615C	115	DD06	116	A061-118	116	A42XO0374-01	116	AM3
113A	2093A5-10	115	FSA1169	116	A065-110	116	A4A1	116	AM33
113A	2113	115	FSA1202	116	A066-124	116	A4A5	116	AM405
113A	10031	115	K118J966-2	116	A068-100	116	A4A9	116	AM410
113A	12871	115	K118J966-4	116	A068-112	116	A4B1	116	AM415
113A	53093-1	115	K118J9663	116	A069-112	116	A4B5	116	AM420
113A	72053	115	K1617	116	A100(RECTIFIER)	116	A4B9	116	AM425
113A	72148	115	P17	116	A101-A(RECT.)	116	A4C1	116	AM43
113A	100471	115	RB89	116	A10105	116	A4C5	116	AM430
113A	100581	115	RPJ70643	116	A10113	116	A4C9	116	AM435
113A	103872	115	SDD6	116	A10118	116	A4D1	116	AM440
113A	107474	115	SI-RECT-178	116	A10142	116	A4D5	116	AM445
113A	109328	115	6GX1	116	A10164	116	A4D9	116	AM450
113A	109474	115	6GX1BY1	116	A10165	116	A4E1	116	AM460
113A	489752-044	115	9LR2-4	116	A10169	116	A4E5	116	AM53
113A	489765-005	115	12/1N	116	A10A	116	A4E9	116	AM63
113A	530127-5	115	86-0007	116	A10B	116	A4F1	116	AM65
113A	607101	115	166	116	A10C	116	A4F5	116	AM66
113A	611132	115	1617C	116	A10D	116	A4F9	116	AN-1
113A	616010	115	5203RNI	116	A10E	116	A4G1	116	AM-G-5B
113A	633977	115	107268	116	A10M	116	A4G5	116	AQ2(PHILCO)
113A	661010	115	107628	116	A123-7	116	A4G9	116	AQ5(PHILCO)
113A	700055-00	115	1107832-6	116	A13(RECTIFIER)	116	A50(RECTIFIER)	116	AR16
113A	744002	115	2001786-139	116	A132	116	A514-023626	116	AR17
113A	744006	115	23115078	116	A132-1	116	A514-025607	116	AR18
113A	817074	1150	GEIC-286	116	A13A2	116	A514-027757	116	AR19
113A	817126	1150	LA4050P	116	A13AA2	116	A514-028072	116	AR20
113A	817127	1150	LA4051P	116	A13B2	116	A514-028073	116	AR21
113A	982361	1153	GEIC-182	116	A13C2	116	A514-033903	116	AR22
113A	1107832-10	1153	TA7204P	116	A13D2	116	A514-0339903	116	AR22(RECTIFIER)
113A	1107832-11	1154	GEIC-287	116	A13E2	116	A5A1	116	AR70
113A	1107832-7	1154	TA7203	116	A13P2	116	A5A2	116	AS-14
113A	1107832-8	1154	TA7203P	116	A13M2	116	A5A5	116	AS-15
113A	1107832-9	1155	BO319200	116	A14B	116	A5A9	116	AS-2
113A	2004107-40	1155	EA33X8389	116	A14C	116	A5B1	116	AS-3
113A	2006512-79	1155	EA33X8396	116	A14D	116	A5B2	116	AS-4
113A	2006512-80	1155	EICM-0060	116	A14E	116	A5B5	116	AS-5
113A	23115079	1155	ETI-23	116	A14E2	116	A5B9	116	AS11
113A	62522682	1155	GEIC-179	116	A14F	116	A5C1	116	AS14
113A	4202005000	1155	IP20-0161	116	A14M	116	A5C2	116	AS15
113A	4202007900	1155	KIA7205AP	116	A15-1008	116	A5C5	116	AS2
114	A86-4-1	1155	KIA7205P	116	A1946	116	A5C9	116	AS3
114	B1045494P1	1155	QQ-MO7205AT	116	A1A1	116	A5D1	116	AS4
114	D5	1155	QQMO7205AT	116	A1A5	116	A5D2	116	AS5
114	DD05	1155	REN1155	116	A1A9	116	A5D5	116	AS6
114	EU16X8	1155	SK3231	116	A1B1	116	A5D9	116	B-1501U
114	K112D	1155	TA7205	116	A1B5	116	A5E1	116	B-31
114	K118J966-1	1155	TA7205A	116	A1B9	116	A5E2	116	B01-02
114	K122D	1155	TA7205AP	116	A1C1	116	A5E5	116	B0102
114	K1616	1155	TA7205P	116	A1C5	116	A5E9	116	B12-02
114	P16	1155	TM1155	116	A1C9	116	A5F1	116	B1A1
114	RB88	1155	TVCM-81	116	A1D1	116	A5F2	116	B1A5
114	RP5794	1155	02-257205	116	A1D5	116	A5F5	116	B1A9
114	RNVTC00921-3	1155	44T-100-120	116	A1D9	116	A5F9	116	B1B
114	SDD5	1155	44T-300-102	116	A1E1	116	A5G1	116	B1B1
114	SR10	1155	051-0055-02	116	A1E5	116	A5G2	116	B1B5
114	SR14	1155	051-0055-03	116	A1E9	116	A5G5	116	B1B9
114	SR15	1155	740-9007-205	116	A1F1	116	A5G9	116	B1C1
		1155	740-9607-205					116	B105
		1155	5002-031					116	B109

ECG	Industry Standard No.	ECG	Industry Standard No.	ECG	Industry Standard No.	ECG	Industry Standard No.	ECG	Industry Standard No.
116	B1D5	116	BA119	116	CD1124	116	CTP-2001-1011	116	DR4
116	B1D9	116	BA127	116	CD1125	116	CTP-2001-1012	116	DR400
116	B1D9	116	BA128	116	CD1126	116	CX0037	116	DR427
116	B1E1	116	BA129	116	CD1127	116	CX0039	116	DR435
116	B1E5	116	BA130	116	CD1142	116	CX0040	116	DR5
116	B1E9	116	BA153	116	CD1143	116	CX0047	116	DR500
116	B1F1	116	BAX12	116	CD1147	116	CX0048	116	DR5101
116	B1F5	116	BAY44	116	CD1148	116	CX0049	116	DR5102
116	B1F9	116	BAY64	116	CD1149	116	CX9001	116	DR600
116	B101	116	BAY86	116	CD13532	116	CY40	116	DR668
116	B105	116	BAY87	116	CD13333	116	CY50	116	DR669
116	B109	116	BB-2	116	CD13335	116	D-00384R	116	DR670
116	B200C40	116	BB-68	116	CD13336	116	D-05	116	DR671
116	B250C100	116	BB107	116	CD13337	116	D004	116	DR695
116	B250C100TD	116	BB117	116	CD13338	116	D01-100	116	DR698
116	B250C125	116	BB127	116	CD13339	116	D028	116	DR699
116	B250C125K4	116	BB1A	116	CD37A2	116	D1-528	116	DR826
116	B250C125N2	116	BB2A184	116	CD4	116	D1-7	116	DR848
116	B250C125X4	116	BC-207	116	CDG005	116	D100	116	DR863
116	B250C150	116	BC-307	116	CDR-2	116	D101167	116	DR8
116	B250C150K4	116	BD-107	116	CDR-4	116	D10167	116	DR8102
116	B250C75	116	BD-107(RECT.)	116	CR0398/7839	116	D10168	116	DR8104
116	B250C75K4	116	BE107	116	CE502	116	D1201P	116	DR8106
116	B250C75K41	116	BE117	116	CE504	116	D1448	116	D8-0065
116	B250C75K45	116	BE127	116	CE506	116	D15A	116	D8-13
116	B250C75K5	116	BH481	116	CE06050	116	D15C	116	D8-13(COURIER)
116	B250C75K4S	116	BR42	116	CER500	116	D1E	116	D8-130
116	B294(RECTIFIER)	116	BR44	116	CER500A	116	D1H	116	D8-130B
116	B2A1	116	BR46	116	CER500B	116	D1J	116	D8-130C
116	B2A5	116	BR47	116	CER500C	116	D1L	116	D8-130E
116	B2A9	116	BR48	116	CE667	116	D2-1	116	D8-130YB
116	B2B1	116	BR51400-1	116	CER670	116	D220	116	D8-130YE
116	B2B5	116	BR51401-2	116	CER670A	116	D220M	116	D8-131
116	B2B9	116	BR52	116	CER670B	116	D227(DIODE)	116	D8-131A
116	B2C1	116	B51	116	CER670C	116	D25	116	D8-131B
116	B2C5	116	B52	116	CER67A	116	D25A	116	D8-132
116	B2C9	116	BTM50	116	CER67B	116	D25B	116	D8-132A
116	B2D1	116	BV25	116	CER67C	116	D25C	116	D8-132B
116	B2D5	116	BY101	116	CER68	116	D2600EF	116	D8-13A(SANYO)
116	B2D9	116	BY102	116	CER680	116	D28	116	D8-13B(SANYO)
116	B2E1	116	BY106	116	CER680A	116	D2H(DIODE)	116	D8-14(DIODE)
116	B2E5	116	BY107	116	CER680B	116	D2J	116	D8-16A
116	B2E9	116	BY111	116	CER680C	116	D2X4	116	D8-16B(SANYO)
116	B2F1	116	BY112	116	CER68A	116	D3R	116	D8-16C(SANYO)
116	B2F5	116	BY113	116	CER68B	116	D3R38	116	D8-16D
116	B2F9	116	BY114	116	CER68C	116	D3R39	116	D8-16B(SANYO)
116	B2G1	116	BY115	116	CER69	116	D3U	116	D8-16B(SANYO)
116	B2G5	116	BY116	116	CER690	116	D3V	116	D8-16NY
116	B2G9	116	BY117	116	CER690A	116	D3Z	116	D8-16YA
116	B300250	116	BY121	116	CER690B	116	D400(RECT.)	116	D8-17-6A
116	B300250-1	116	BY124	116	CER690C	116	D45C	116	D8-18
116	B300350-1	116	BY125	116	CER69A	116	D45CZ	116	D8-1M
116	B300500	116	BY126	116	CER69B	116	D48	116	D8-1P
116	B300600	116	BY130	116	CER69C	116	D4M	116	D8-79
116	B300600CB	116	BY134	116	CER63B	116	D4R26	116	D8-79(DELCO)
116	B31	116	BY135	116	CER670	116	D4R39	116	D8-1M
116	B31(RECTIFIER)	116	BY141	116	CER700	116	D500	116	D81
116	B350600	116	BY153	116	CER700A	116	D5R35	116	D8130
116	B36564	116	BYX22/200	116	CER700B	116	D5R39	116	D8130B
116	B3A1	116	BYX22/400	116	CER700C	116	D65C	116	D8130B
116	B3A5	116	BYX22/600	116	CER70A	116	D6623	116	D8130B
116	B3A9	116	BYX36-300	116	CER70C	116	D6623A	116	D8130ND
116	B3B1	116	BYX36-600	116	CER71	116	D6624	116	D8130Y
116	B3B5	116	BYX36/150	116	CER710	116	D6624A	116	D8130YC
116	B3B9	116	BYX36/300	116	CER710A	116	D6625	116	D8130YE
116	B3C1	116	BYX36/600	116	CER710C	116	D6625A	116	0000DB131
116	B3C5	116	BYX60-100	116	CER71A	116	D68Z	116	D8131A
116	B3C9	116	BYX60-200	116	CER71B	116	DA000	116	D8131B
116	B3D1	116	BYX60-300	116	CER71C	116	DA001	116	D8132A
116	B3D5	116	BYX60-400	116	CP102DA	116	DA002	116	D8160(G.E.)
116	B3D9	116	BYX60-50	116	CH119D	116	DAAY002001	116	D816B
116	B3E1	116	BYX60-500	116	CL010	116	DD-000	116	D816N
116	B3E5	116	BYX60-600	116	CLO25	116	DD-003	116	D816NB
116	B3E9	116	BYY-31	116	CLO5	116	DD-006	116	D816NC
116	B3F1	116	BYY-32	116	CL1	116	DD-007	116	D816ND
116	B3F5	116	BYY-35	116	CL1.5	116	DD056	116	D816NE
116	B3F9	116	BYY31	116	CL2	116	DD175C	116	D817
116	B3G1	116	BYY32	116	CL3	116	DD177C	116	D817(ADMIRAL)
116	B3G5	116	BYY33	116	CL4	116	DD2066	116	D817N
116	B3G9	116	BYY34	116	CL5	116	DD2320	116	D818
116	B4A1	116	BYY35	116	CL6	116	DD2321	116	D818N
116	B4A5	116	BYY36	116	CL7	116	DD236	116	D81K
116	B4A9	116	BYY89	116	CL8	116	DD266	116	D81K7
116	B4B1	116	C10110	116	CLM05	116	DDAY002001	116	D81N
116	B4B5	116	C10159	116	CLM1	116	DDAY002002	116	D81P
116	B4B9	116	C10176	116	CO49	116	DDAY103001	116	D82K
116	B4C1	116	C1181C1E1C	116	COD1531	116	DDBY002001	116	D82N
116	B4C5	116	C1B	116	COD1532	116	DDE-201	116	D838(CRAIG)
116	B4C9	116	C1H	116	COD1533	116	DE14	116	D838(SANYO)
116	B4D1	116	C21382	116	COD1534	116	DE14A	116	D8430
116	B4D5	116	C248507	116	COD1535	116	DE16	116	D858(SANYO)
116	B4D9	116	C2A102	116	COD1536	116	DE16A	116	D870
116	B4E1	116	C83-829	116	COD1551	116	DE201	116	DU400
116	B4E5	116	C83-880	116	COD1552	116	DG1PR	116	DU600
116	B4E9	116	CA10	116	COD1553	116	DH-001	116	DX-0099
116	B4F1	116	CA100	116	COD1554	116	DH14	116	DX-0445
116	B4F5	116	CA100A	116	COD1555	116	DH14A	116	DX-0475
116	B4F9	116	CA102BA	116	COD1556	116	DH16	116	DX520
116	B4G1	116	CA102DA	116	COD16047	116	DH16A	116	E-0704W
116	B4G5	116	CA102FA	116	COD1611	116	DH4R2	116	E-075L
116	B4G9	116	CA102HA	116	COD1612	116	DI-1649	116	EO3155-001
116	B50	116	CA102MA	116	COD1613	116	DI-1728	116	EO3155-002
116	B51	116	CA150	116	COD1614	116	DI-428	116	EO788C
116	B59	116	CA20	116	COD1615	116	DI-46	116	E1
116	B5A1	116	CA200	116	COD1616	116	DI-528	116	E1011
116	B5A5	116	CA250	116	CODI11556	116	DI-55	116	E10116
116	B5A9	116	CA50	116	CODI1531	116	DI-56	116	E10157
116	B5B1	116	CB10	116	CODI15524	116	DI-645	116	E10171
116	B5B5	116	CB150	116	CODI15531	116	DI-646	116	E10172
116	B5B9	116	CB20	116	CODI15534	116	DI-647	116	E1018N
116	B5C1	116	CB200	116	CODI15561	116	DI-648	116	E102(ELCOM)
116	B5C5	116	CB250	116	CODI15564	116	DI-649	116	E106(ELCOM)
116	B5C9	116	CB5	116	CODI6045	116	DI-7	116	E1124
116	B5D1	116	CB50	116	CODI6047	116	DI-705	116	E125C200
116	B5D5	116	CC102BA	116	CP102BA	116	DI-71	116	E13-020-00
116	B5D9	116	CC102DA	116	CP102DA	116	DI-728	116	E13-20-00
116	B5E1	116	CC102FA	116	CP102FA	116	DIE	116	E135
116	B5E5	116	CC102HA	116	CP102HA	116	DIJ	116	E1410
116	B5E9	116	CC102KA	116	CP102KA	116	DIJ70488	116	E1411
116	B5F1	116	CC102MA	116	CP102MA	116	DIJ70544	116	E1412
116	B5F5	116	CD-2N	116	CR/2	116	DIJ71958	116	E1413
116	B5F9	116	CD-4	116	CR1034	116	DIJ71959	116	E1415
116	B601	116	CD-860037	116	CR1035	116	DIJ72168	116	E143
116	B605	116	CD05	116	CS131D(AXIAL)	116	DIL	116	E1440
116	B609	116	CD1111	116	CS16E	116	DI8-18	116	E146
116	B601-1012	116	CD1112	116	CT100	116	DR1	116	E140350
116	B91	116	CD1113	116	CT200	116	DR100	116	E150L
116	BA-100	116	CD1114	116	CT300	116	DR1100	116	E2
116	BA-104	116	CD1115	116	CT3003	116	DR1PR	116	E21
116	BA-142-01	116	CD1116	116	CT3005	116	DR2	116	E24100
116	BA100	116	CD1117	116	CT600	116	DR200	116	E25C5
116	BA104	116	CD1121	116	CTN200	116	DR3	116	E3
116	BA105	116	CD1122			116	DR300	116	E3006
116	BA108	116	CD1123					116	E300L

ECG	Industry Standard No.	ECG	Industry Standard No.	ECG	Industry Standard No.	ECG	Industry Standard No.	ECG	Industry Standard No.
116	E41	116	EP57X1	116	FW100	116	HD20007030	116	JC-SG005
116	E4676B	116	EP57X12	116	FW400	116	HD2000903	116	JC00049
116	E5	116	EP535	116	FW500	116	HD2001310	116	JCN1
116	E6	116	EP600	116	FW600	116	HD200207	116	JCN2
116	E650L	116	ER1	116	FW600A	116	HD200301	116	JCN3
116	E750(ELCOM)	116	ER101	116	FWL100	116	HD6147	116	JCN4
116	E752(ELCOM)	116	ER102D	116	FWL200	116	HD6865	116	JCN5
116	E756(ELCOM)	116	ER103D	116	FWL300	116	HE-20011	116	JCM6
116	EA0015	116	ER103E	116	G00-502A	116	HB-8D1	116	JCV-2
116	EA0016	116	ER104D	116	G00-534-A	116	HEP154	116	JCV-3
116	EA0031	116	ER105D	116	G00-535-B	116	HEP156	116	JD-00040
116	EA005	116	ER106D	116	G00-535A	116	HEP157	116	JD-BB1A
116	EA010	116	ER11	116	G00-536-A	116	HEP158	116	JDBD1D
116	EA020	116	ER12	116	G00-536A	116	HP-08W05	116	JT-B1024D
116	EA030	116	ER181	116	G00-543-A	116	HP-20042	116	JT-B1064
116	EA040	116	ER182	116	G00-551-A	116	HP-20047	116	K1.3G22A
116	EA050	116	ER183	116	G0055	116	HP-20050	116	K1A5
116	EA060	116	ER184	116	G0055A	116	HP-20052	116	K1B5
116	EA1072	116	ER185	116	G01	116	HP-20067	116	K1C5
116	EA1448	116	ER2	116	G01211	116	HP-20083	116	K1D5
116	EA15X14	116	ER201	116	G02	116	HP-20084	116	K1E5
116	EA1672	116	ER21	116	GOQ-535-B	116	HP20066	116	K1P5
116	EA16X2	116	ER22	116	01	116	HPOW05	116	K1Q5
116	EA16X21	116	ER301	116	G100G	116	HPB8005	116	K1H5
116	EA16X30	116	ER31	116	G100D	116	HGR-10	116	K200
116	EA16X33	116	ER381	116	G100G	116	HGR-20	116	K2A5
116	EA16X34	116	ER401	116	G100J	116	HGR-30	116	K2B5
116	EA16X55	116	ER41	116	G1O119	116	HGR-40	116	K2C5
116	EA16X71	116	ER42	116	G1242	116	HGR-5	116	K2D5
116	EA16X8	116	ER501	116	G2	116	HGR-60	116	K2E5
116	EA16X92	116	ER51	116	G296	116	HGR1	116	K2P5
116	EA2140	116	ER57X2	116	G2A	116	HGR2	116	K2G
116	EA2499	116	ER57X3	116	G3	116	HGR3	116	K2Q5
116	EA2501	116	ER57X4	116	G4	116	HGR4	116	K3B5
116	EA2741	116	ER601	116	G5	116	HIPI	116	K3C5
116	EA3827	116	ER61	116	G6	116	HN-00008	116	K3D5
116	EA3989	116	ER62	116	G657	116	HN-00018	116	K3P5
116	EA5711	116	ERB11-01	116	G659	116	HN-00029	116	K3Q5
116	EA57X1	116	ERB22-15	116	G700	116	HN-00032	116	K4-555
116	EA57X10	116	ERB24(GE)	116	G701	116	HP-5A	116	K4-557
116	EA57X11	116	ERD300	116	G702	116	HP205	116	K4A5
116	EA57X14	116	ERD400	116	GD12	116	HR-05A	116	K4B5
116	EA57X3	116	ERV-02P2150	116	GE-1N5061	116	HR-5A	116	K4C5
116	EA57X8	116	ES10233	116	GE-5A	116	HR-5AX2	116	K4D5
116	EA75X1	116	ES15056	116	GE-5AX2	116	HR-5B	116	K4E5
116	EC401	116	ES16X13	116	GE-X36	116	HR10	116	K4P5
116	EC402	116	ES16X25	116	GE42-7	116	HR11	116	K4Q5
116	EDR-600-2	116	ES47X1	116	GE6366	116	HR13	116	K5A5
116	ED-4	116	ES57X1	116	GI-1N4385	116	HR5A	116	K5B5
116	ED-5	116	ES57X2	116	GI-3008	116	HR5A8E	116	K5C5
116	ED-6	116	ES57X4	116	GI-300D	116	HR5B	116	K5D5
116	ED1804	116	ES57X5	116	GI-P100-D	116	HS1001	116	K5E5
116	ED1892	116	ESA-10C	116	GI08B	116	HS1002	116	K5P5
116	ED2106	116	ESA-10N	116	GI3002	116	HS1003	116	K5Q5
116	ED2107	116	ESK1/06	116	GI3992-17	116	HS1007	116	KB533058-1
116	ED2108	116	ESKE400C500	116	GI411	116	HS1008	116	KB-182
116	ED2109	116	ET200	116	GI420	116	HS1009	116	KB265A(RECT)
116	ED2110	116	ET400	116	GJ4M	116	HS1010	116	KC0-0691L/8
116	ED224548	116	ET51X25	116	G00535	116	HS1012	116	K00.8
116	ED224550	116	ET52X25	116	G00-502-A	116	HS1020	116	KC06911
116	ED2842	116	ET55-25	116	GP-1	116	HS3103	116	KC06B11/8
116	ED2843	116	ET55X25	116	GP05A	116	HS3104	116	KC08C1110
116	ED2844	116	ET57X25	116	GPO8B	116	HSFD-1A	116	KC08C215
116	ED2845	116	ET57X30	116	GPO8D	116	HS8000710	116	KC08C221
116	ED2846	116	ET57X33	116	GP230	116	HV-100	116	KC1-.3G
116	ED2914	116	ET57X35	116	GP250	116	HV-26	116	KC1-.3G
116	ED2915	116	ET600	116	GSM482	116	HV-26G	116	KC1-.3G12/1X2
116	ED2916	116	ETD-10D1	116	GSM483	116	HV0000105	116	KC1-.3G22/12
116	ED2917	116	ETD-10D2	116	GSM51	116	HV00000105-0	116	KC13C221
116	ED2918	116	ETD-VO6C	116	GSM52	116	HV0000406	116	KC2B922/1B
116	ED2919	116	EU16X20	116	GSM53	116	HV0000705	116	KC2D221
116	ED2920	116	EU57X40	116	GSM54	116	IC743048	116	KC2DP
116	ED2921	116	EYY-420D1R5JA	116	H100	116	INJ61726	116	KC2DP12/1N
116	ED3000	116	F-05	116	H200	116	IP20-0022	116	KC2DP121N
116	ED3000A	116	F-14C	116	H300	116	IP20-0024	116	KC2DP122
116	ED3000B	116	F1	116	H400	116	IP20-0025	116	KC2DP221
116	ED3001	116	F10124	116	H50	116	IP20-0054	116	KC2DP221B
116	ED3001A	116	F10148	116	H500	116	IP20-016J	116	KC0-8CP
116	ED3001B	116	F10180	116	H585	116	IR10E6J	116	KD2103
116	ED3002	116	F14A	116	H600	116	IR1D	116	KD2104
116	ED3002A	116	OP164	116	H616	116	IR20	116	KDD0032
116	ED3002B	116	P2	116	H617	116	IR2A	116	KB-262
116	ED3003	116	P20-1015	116	H618	116	IR2E	116	KGE41007
116	ED3003A	116	P20-1016	116	H619	116	IT10D4K	116	KUB4567
116	ED3003B	116	P215-1016	116	H620	116	ITT350	116	KS-05
116	ED3003B	116	P215-1017	116	H625	116	ITT402	116	KS-05X
116	ED3004	116	P3	116	H626	116	ITT992	116	KSKE400C200
116	ED3004A	116	P4	116	H7126-3	116	J100	116	KSKE400C500
116	ED3004B	116	P5	116	H781	116	J101183	116	L32H
116	ED3005	116	P6	116	H783	116	J20437	116	L62H
116	ED3005A	116	PA4	116	HA100	116	J241100	116	LA300
116	ED3005B	116	PA6	116	HA200	116	J241102	116	LA600
116	ED3006	116	PD-1029-DP	116	HA300	116	J241142	116	LA800
116	ED3006A	116	PD-1029-DG	116	HA400	116	J241209	116	LAA300
116	ED3006B	116	PD1599	116	HA50	116	J241210	116	LAA600
116	ED329128	116	PD3	116	HA500	116	J241211	116	LAA800
116	ED329130	116	PD3389	116	HA600	116	J241214	116	LL-2
116	ED494583	116	PD6	116	HAR10	116	J241232	116	LM-1158
116	ED511097	116	PDH400	116	HAR15	116	J241260	116	LM-1160
116	ED7	116	PG-2NA	116	HAR20	116	J241271	116	LM-1862
116	EDJ-363	116	FM1J2	116	HB2	116	J24570	116	LM1862
116	EDS-0002	116	PO5	116	HB3	116	J24630	116	LP1H
116	EDS-0004	116	FPR50-1011	116	HC-30	116	J24645	116	LP2H
116	EDS-0017	116	FR-1	116	HC500	116	J24647	116	LP5H
116	EDS-0024	116	FR-1H	116	HC67	116	J24756	116	LP4H
116	EDS-17	116	FR-1H(M)	116	HC670	116	J24871	116	LRR-100
116	EDS-24	116	FR-1M	116	HC68	116	J24877	116	LRR-200
116	EDS-4	116	FR-1MD	116	HC680	116	J24920	116	LRR-300
116	ED100	116	FR-1N	116	HC69	116	J24935	116	LRR-400
116	ED100H	116	FR-1P	116	HC700	116	J24939	116	LRR-50
116	EM1021	116	FR-2	116	HC71	116	J24940	116	LRR-500
116	EM1J2	116	FR-202	116	HC710	116	J320020	116	M-0027
116	EM401	116	FR-2P	116	HC80	116	JAMT02C	116	M-31
116	EM402	116	FR1M	116	HCV	116	JB-00030	116	M12
116	EM403	116	FR1MB	116	HD-1	116	JB-BB1A	116	M12.4J779-1
116	EM404	116	FR2	116	HD-2000308	116	JC-00012	116	M14
116	EM405	116	FR2(S1B1)	116	HD-3000301	116	JC-00014	116	M150
116	EM406	116	FR2-02	116	HD20-003-01	116	JC-00028	116	M172A
116	EM407	116	FR2-02C	116	HD2000	116	JC-00032	116	M1H
116	EM408	116	FR2-02	116	HD2000-301	116	JC-00033	116	M22
116	EM410	116	OOOOFR202	116	HD20000703	116	JC-00035	116	M2497
116	EM501	116	FR2P	116	HD20000903	116	JC-00037	116	M3016
116	EM502	116	FRH-101	116	HD2000110	116	JC-00044	116	M41223-2
116	EM503	116	FRH101	116	HD2000110-0	116	JC-00047	116	M42
116	EM504	116	OOOOOOOOFRI	116	HD2000110-0	116	JC-00049	116	M4HZ
116	EM505	116	PSP-288-1	116	HD20000301	116	JC-00051	116	M500
116	EM506	116	PST2	116	HD20000301-0	116	JC-00055	116	M500A
116	E0704	116	PST3	116	HD20003010	116	JC-10D1	116	M500B
116	EP1259-2	116	FT-1	116	HD2000307	116	JC-DS16E	116	M500C
116	EP1428-2H	116	FT14A	116	HD2000413	116	JC-KS05	116	M60GHT
116	EP16X13	116	FT1N	116	HD2000501	116	JC-8D-1X	116	MC2
116	EP200	116	FU1H	116	HD2000510	116	JC-8D-12	116	M67
116	EP3149	116	FU1U	116	HD2000510			116	M670
116	EP400			116	HD2000703			116	M670A

ECG	Industry Standard No.	ECG	Industry Standard No.	ECG	Industry Standard No.	ECG	Industry Standard No.	ECG	Industry Standard No.
116	M670B	116	MR2065	116	P32H	116	PS2412	116	RD-31903P
116	M670C	116	MR2261	116	P3A5	116	PS2413	116	RD-3472
116	M67A	116	MR9600	116	P3B5	116	PS2415	116	R8250
116	M67B	116	MR9601	116	P305	116	PS405	116	RD26235-1
116	M67C	116	MR9602	116	P3D5	116	PS415	116	RD29799P
116	M68	116	MS11H	116	P400	116	PS425	116	RD31903P
116	M680	116	MS12H	116	P4A5	116	PS440	116	RDA2
116	M680A	116	MS13H	116	P4B5	116	PS450	116	R9037
116	M680B	116	MS14H	116	P4C5	116	PS4559	116	RE49
116	M680C	116	MS1H	116	P4D5	116	PS4560	116	RE70643
116	M68A	116	MS2H	116	P5A5	116	PS460	116	RE70931
116	M68B	116	MS35H	116	P5B5	116	PS4725	116	RE820
116	M68C	116	MS36H	116	P5C5	116	PS5300	116	R8-3160
116	M69	116	MS3H	116	P5D5	116	PS5301	116	RF-32101-8
116	M690	116	MS4H	116	P6/2H	116	PS5302	116	RF-32101R
116	M690A	116	MS5	116	P600	116	PS603	116	RF-6235-1
116	M690B	116	MS5H	116	P62H	116	PS604	116	RF26231-1
116	M690C	116	MSR-500	116	P6A5	116	PS605	116	RF26234-1
116	M69A	116	MSR-V5	116	P6B5	116	PS609	116	RF26235-1
116	M69B	116	MSR500	116	P6C5	116	PS611	116	RF26235-2
116	M69C	116	MSB-1000	116	P6D5	116	PS615	116	RF26235-5
116	M68IZ	116	MT021	116	P7A5	116	PS616	116	RF29799P
116	M70	116	MT021A	116	P7B5	116	PS617	116	R9160
116	M700	116	MT022	116	P705	116	PS621	116	RF31903P
116	M700A	116	MT022A	116	P7D5	116	PS622	116	RF32101-8
116	M700B	116	MT14	116	P8/2H	116	PS623	116	RF32101-9
116	M700C	116	MT24	116	P82H	116	PS627	116	RF32101R
116	M701B	116	MT44	116	P9459	116	PS628	116	RF32645
116	M70A	116	MT64	116	PA-069	116	PS629	116	RF33976
116	M70B	116	MV-5	116	PA-3	116	PS632	116	RF34383
116	M70C	116	MV11	116	PA-320	116	PS633	116	RF3472
116	M71	116	MV3(RECTIFIER)	116	PA-320A	116	PS636	116	RFA70597
116	M710	116	MVA-05A	116	PA-320B	116	PS637	116	RFA70600
116	M710A	116	MY-1	116	PA069	116	PT-3	116	RFC61197
116	M710B	116	N-02	116	PA070	116	PT-510	116	RFJ-30704
116	M710C	116	N-41	116	PA071	116	PT-530	116	RFJ-31218
116	M71A	116	N-EA16X30	116	PA10556	116	PT-550	116	RFJ-31362
116	M71B	116	NA-22	116	PA10887	116	PT-560	116	RFJ-31363
116	M71C	116	NA-25	116	PA200	116	PT-5B	116	RFJ-33292
116	M8222	116	NA-33	116	PA3	116	PT3	116	RFJ-60366
116	M8399	116	NA-36	116	PA300	116	PT5	116	RFJ30704
116	M91A01	116	NA-46	116	PA305	116	PT505	116	RFJ31218
116	M91A02	116	NA13	116	PA305A	116	PT510	116	RFJ31362
116	M91A03	116	NA22	116	PA310	116	PT520	116	RFJ31363
116	M9206	116	NA25	116	PA310A	116	PT525	116	RFJ33292
116	M9312	116	NA32	116	PA315	116	PT530	116	RFJ60286
116	M9314	116	NA33	116	PA315A	116	PT540	116	RFJ60366
116	M9317	116	NA35	116	PA320	116	PT550	116	RFJ60869
116	M9319	116	NA36	116	PA320A	116	PT560	116	RFJ6134
116	MA101	116	NA42	116	PA320B	116	PT5B	116	RFJ70432
116	MA102	116	NA45	116	PA325	116	PT72130	116	RFJ70487
116	MA110	116	NA46	116	PA325A	116	PTG202	116	RFJ70703
116	MA2	116	NF500	116	PA325B	116	PU6022	116	RFJ70931
116	MA203	116	NF501	116	PA330	116	PV-8	116	RFJ70970
116	MA211	116	NF506	116	PA330A	116	PV8	116	RFJ70974
116	MA215	116	NF550	116	PA330B	116	PY-5	116	RFJ70977
116	MA242	116	NL-10	116	PA340	116	Q-20115C	116	RFJ71122
116	MA350	116	NL10	116	PA340A	116	Q-26115C	116	RFJ71122
116	MA351	116	NL15	116	PA340B	116	Q1B	116	RFJ72360
116	MB01	116	NL20	116	PA350	116	Q1H	116	RFJ72787
116	MB244	116	NL25	116	PA350A	116	Q3/2	116	RFJZO432
116	MB257	116	NL30	116	PA360	116	Q32	116	RFL-30596
116	MB258	116	NL40	116	PA360A	116	Q4B	116	RFL30596
116	MB269	116	NL5	116	PA400	116	Q52	116	RFM-33160
116	MB270	116	NL50	116	PA600	116	Q53	116	RFM33160
116	MC010	116	NL60	116	PA7615	116	Q54	116	RFP-33118
116	MC015	116	NN50	116	PA8645	116	Q55	116	RFP33118
116	MC020	116	NP50A	116	PA9160	116	Q56	116	RFV60500
116	MC020A	116	NP60A	116	PC4004	116	Q57	116	RG1004
116	MC021	116	NPC0010	116	PD101	116	Q58	116	RG100B
116	MC021A	116	NPC0050	116	PD102	116	Q59	116	RG100D
116	MC022	116	NPC0100	116	PD103	116	Q6/2	116	RG100G
116	MC022A	116	NTC-19	116	PD104	116	Q60	116	RG100J
116	MC023	116	NU398B	116	PD105	116	Q62	116	RG1127
116	MC023A	116	O101	116	PD106	116	Q8/2	116	RGP-10D
116	MC025	116	O234	116	PD107	116	Q82	116	RH-DX00038EZZ
116	MC030	116	OA127	116	PD107A	116	QD-SS1885XT	116	RH-DX0003TAZZ
116	MC030A	116	OA128	116	PD108	116	QD-SSR1KX4P	116	RH-DX0008CEZZ
116	MC030B	116	OA129	116	PD110	116	QD-SV06CXXB	116	RH-DX0025CEZZ
116	MC035	116	OA130	116	PD111	116	R-1	116	RH-DX0026AGZZ
116	MC040	116	OA131	116	PD122	116	R-106379	116	RH-DX00038CEZZ
116	MC040A	116	OA132	116	PD125	116	R-113321	116	RH-DX00041CEZZ
116	MC1521	116	OA180	116	PD129	116	R-113392	116	RH-DX00042CEZZ
116	MC170	116	OA210	116	PD130	116	R-154B	116	RH-DX00055TAZZ
116	MC19	116	OA211	116	PD131	116	R-1A	116	RH-DX00056CEZZ
116	MC456	116	OA214	116	PD132	116	R-2-02	116	RH-DX00059TAZZ
116	MCV	116	OA8	116	PD133	116	R-3-1720	116	RH-DX00064CEZZ
116	MD04	116	OF160	116	PD134	116	R-81264	116	RH-DX00068TAZZ
116	MD134	116	OF164	116	PD135	116	R-81347	116	RH-DX00069TAZZ
116	MD135	116	OG-30L125	116	PD154	116	R1035	116	RH-DX00072CEZZ
116	MD136	116	08-16308	116	PD155	116	R106379	116	RH-DX0081TAZZ
116	MD137	116	0816308	116	PD910	116	R10D1	116	RH-DX1005AFZZ
116	MD138	116	08S-16308	116	PB-401	116	R1ODC	116	RH-DX0003SEZZ
116	MH500	116	08S-36885	116	PE401	116	R113321	116	RHDX0043TAZZ
116	MH67	116	08S36503	116	PE401N	116	R113392	116	RLP1G
116	MH670	116	08S836685	116	PE402	116	R122C	116	RM-1V
116	MH68	116	08S36885	116	PE403	116	R1329	116	RM1A
116	MH680	116	OY-5061	116	PE404	116	R1B	116	RM1ZM
116	MH70	116	OY-5062	116	PE405	116	R1K	116	RM1ZV
116	MH700	116	OY101	116	PE406	116	R2159	116	RM26
116	MH71	116	OY5061	116	PB502	116	R2252	116	RRB24-06
116	MH710	116	OY5062	116	PB504	116	R2442	116	R8-1264
116	MI-15R	116	OY5063	116	PB506	116	R3/2H	116	R8-1347
116	MI-15S	116	OY5064	116	PH-10B	116	R3285	116	R8-3570
116	MJR1C	116	OY5065	116	PH1021	116	R4A	116	R8-6344
116	MMO	116	OY5066	116	PH204	116	R5970	116	R8-6461
116	MM2	116	P-10115	116	PH208	116	R5971	116	R8-6471
116	MM3	116	P100	116	PH25C22	116	R6/2H	116	R810
116	MM4	116	P100A	116	PH25C22/1	116	R6048	116	R81264
116	MM5	116	P100B	116	PH404	116	R6110	116	R81720
116	MM6	116	P100D	116	PH9DS22	116	R6422	116	R81749
116	MP-01	116	P100G	116	PB204	116	R7162	116	R81805
116	MP-5115	116	P100J	116	PB-225	116	R7248	116	R81832
116	MP100	116	P10115	116	PB-035	116	R7271	116	RS220AF
116	MP1003-1	116	P10156	116	PB-040	116	R7682	116	RS230AF
116	MP1003-2	116	P150A	116	PB-060	116	R7954	116	R83570
116	MP1003-4	116	P150B	116	PB-120	116	R8024	116	R83727
116	MP225	116	P150D	116	PB005	116	R8470	116	R86344
116	MP300	116	P150G	116	PB015	116	R8473	116	R86461
116	MP400	116	P150J	116	PB025	116	R855-2	116	R86471
116	MP500(RECT.)	116	P1A5	116	PB040	116	R9470	116	R86705
116	MP5113	116	P1B5	116	PB050	116	R9597	116	R84430
116	MP651	116	P1C5	116	PB060	116	R9A	116	RT-2669
116	MP9602	116	P1D5	116	PB105	116	RA-1	116	RT-3858
116	MP89602	116	P20	116	PB125	116	RA-1ZC	116	RT-4232
116	MPX-25	116	P200	116	PB140	116	RA132BA	116	RT1595
116	MPX215	116	P21316	116	PB150	116	RA1B	116	RT1840
116	MQ32	116	P21317	116	PB160	116	RA1Y	116	RT213
116	MQ62	116	P21443	116	PB2207	116	RA1Z	116	RT215
116	MQ82	116	P2A5	116	PB2208	116	RA1ZC	116	RT3443
116	MR-150-01	116	P2B5	116	PB2209	116	RCC-7022	116	RT3858
116	MR12573L	116	P2C5	116	PB2247	116	RD-26235-1	116	RT3981
116	MR1M	116	P2D5	116	PB2249	116	RD-29799P	116	RT4050
116	MR2064	116	P3/2H	116	PB2411	116	RD-3		

ECG	Industry Standard No.	ECG	Industry Standard No.	ECG	Industry Standard No.	ECG	Industry Standard No.	ECG	Industry Standard No.
116	RT4232	116	S203	116	SCA116	116	SG-105	116	SL833A
116	RT4764	116	S204	116	SCA05	116	SG-1198	116	SL91
116	RT5070	116	S205	116	SCA1	116	SG-205	116	SL92
116	RT5472	116	S206	116	SCA1103	116	SG-305	116	SL93
116	RT5911	116	S20ND400	116	SCA2	116	SG-805	116	SLA-445
116	RT6322	116	S20NH400	116	SCA3	116	SG105	116	SLA1095
116	RT6332	116	S21	116	SCA4	116	SG1198	116	SLA1096
116	RT6605	116	S217	116	SCA5	116	SG323	116	SLA1100
116	RT6791	116	S218	116	SCA6	116	SG3400	116	SLA1101
116	RT7648	116	S219	116	SCB1	116	SG505	116	SLA1102
116	RT7849	116	S22	116	SCB2	116	SGR100	116	SLA1104
116	RT7850	116	S220	116	SCB4	116	SH-1	116	SLA1105
116	RT8231	116	S221	116	SCB6	116	SH-1A	116	SLA11AB
116	RT8340	116	S222	116	SC05	116	SH-1DE	116	SLA11C
116	RT8840	116	S223	116	SC05E	116	SH1	116	SLA12AB
116	RT8841	116	S224	116	SC05E	116	SH15	116	SLA12C
116	RV-2289	116	S22A	116	SD-02	116	SH1A	116	SLA13AB
116	RV06	116	S23	116	SD-1-211B	116	SH4D05	116	SLA13C
116	RV06/7825B	116	S230	116	SD-1-211C	116	SH4D1	116	SLA1487
116	RV1189	116	S232	116	SD-101	116	SH4D2	116	SLA1488
116	RV1424	116	S233	116	SD-15	116	SH4D3	116	SLA1489
116	RV147B	116	S234	116	SD-16A	116	SH4D4	116	SLA1490
116	RV2072	116	S235	116	SD-16D	116	SH4D6	116	SLA1491
116	RV220	116	S238	116	SD-18	116	SHAD-1	116	SLA1492
116	RV2250	116	S239	116	SD-1A	116	SHAD1	116	SLA14AB
116	RV2289	116	S23A	116	SD-1C-4P	116	SI-RECT-044	116	SLA140
116	RV2327	116	S240	116	SD-1C-UF	116	SI-RECT-100	116	SLA15AB
116	RVD08C22/1A	116	S241	116	SD-1CUF	116	SI-RECT-102	116	SLA15C
116	RVD10D1	116	S243	116	SD-1HF	116	SI-RECT-122	116	SLA1692
116	RVD10DC1	116	S250	116	SD-1HP	116	SI-RECT-144	116	SLA1693
116	RVD10DC1R	116	S251	116	SD-1L	116	SI-RECT-154	116	SLA1694
116	RVD10E1	116	S252	116	SD-1LA	116	SI-RECT-155	116	SLA1695
116	RVD10E1LF	116	S253	116	SD-1UP	116	SI-RECT-156	116	SLA1696
116	RVD18854	116	S254	116	SD-1Z	116	SI-RECT-2	116	SLA1697
116	RVD2DF	116	S255	116	SD-2	116	SI-RECT-218	116	SLA16A7
116	RVD2DP221P	116	S256	116	SD-201	116	SI-RECT-222	116	SLA160
116	RVD2P22/1B	116	S26	116	SD-80	116	SI-RECT-226	116	SLA17AB
116	RVD4B265J2	116	S262	116	SD-91	116	SI-RECT-25	116	SLA17C
116	RVDDS-410	116	S2A06	116	SD-91A	116	SI-RECT-27	116	SLA2610
116	RVDDP05A	116	S2A10	116	SD-918	116	SI-RECT-33	116	SLA2611
116	RVDKB16205	116	S2AR1	116	SD-92	116	SI-RECT-34	116	SLA2612
116	RVD8D-1	116	S2AR2	116	SD-92A	116	SI-RECT-37	116	SLA2613
116	RVD8D-1U	116	S2C30	116	SD-928	116	SI-RECT-39	116	SLA2614
116	RVD8D-1Y	116	S2C40	116	SD-93	116	SI-RECT-48	116	SLA2615
116	RVD8R3AM2N	116	S2C40A	116	SD-93A	116	SI-RECT-49	116	SLA3194
116	S-05	116	S2E20	116	SD-94	116	SI-RECT-53	116	SLA3195
116	S-05-005	116	S2B60	116	SD-94A	116	SI-RECT-59	116	SLA440
116	S-05-01	116	S2B60-1	116	SD-94AB	116	SI-RECT-69	116	SLA440B
116	S-050	116	S30	116	SD-95	116	SI-RECT-73	116	SLA441
116	S-0501	116	S31	116	SD-95A	116	SI-RECT-74	116	SLA441B
116	S-1.5	116	S33	116	SD-Y	116	SI-RECT-75	116	SLA442
116	S-1.5-0	116	S35	116	SD040	116	SI-RECT-77	116	SLA442B
116	S-10	116	S36	116	SD05	116	SI-RECT-84	116	SLA443
116	S-17	116	S3A06	116	SD07	116	SI-RECT-92	116	SLA443B
116	S-17A	116	S3AR1	116	SD1	116	SI-RECT-94	116	SLA444
116	S-2	116	S3MX	116	SD1-1	116	SI100E	116	SLA444B
116	S-262	116	S40	116	SD1-211B	116	SI50B	116	SLA445
116	S-3MX	116	S4001	116	SD1-211C	116	SI91G	116	SLA445B
116	S-500B	116	S40A	116	SD102	116	SIB-01-02	116	SLA536
116	S-500C	116	S42B	116	SD103	116	SIB-01-022	116	SLA537
116	S-5277B	116	S43	116	SD104	116	SIBO-1	116	SLA538
116	S-58	116	S431	116	SD13	116	SIB01	116	SLA539
116	S-58R	116	S44	116	SD18	116	SIB01-06	116	SLA540
116	S05	116	S46	116	000000SD1AB	116	SIB01-06B	116	SLA547
116	S0501	116	S47	116	SD1CUF	116	SIB02-03CR	116	SLA599
116	S1-1	116	S48	116	SD1DM-4	116	SIB02-CR	116	SLA599A
116	S1-B01-02	116	S49	116	SD1HP	116	SIB0201CR	116	SLA600
116	S1-RECT-102	116	S4A06	116	SD1L	116	SIBOL	116	SLA600A
116	S1-RECT-154	116	S4AR1	116	SD1LA	116	SIBOL-02	116	SLA601A
116	S1-RECT-155	116	S4AR2	116	SD1X	116	SID01E	116	SLA602
116	S1.5	116	S4AR30	116	000000SD1T	116	SID01L	116	SLA602A
116	S1.5-01	116	S4FN300	116	SD1Z	116	SID02E	116	SLA603
116	S10	116	S5089-A	116	SD1ZHF	116	SID02L	116	SLA603A
116	S101	116	S5AR1	116	SD201	116	SIG1/200	116	SLA604
116	S102	116	S5AR2	116	SD202	116	SIG1/400	116	SLA604A
116	S103	116	S58R	116	SD23	116	SIG1/600	116	SLA605
116	S104	116	S6005	116	SD2A	116	SIL-200	116	SLA605A
116	S105	116	S6AR1	116	SD2B	116	SIL200	116	SLA606A
116	S106	116	S6AR2	116	SD4	116	SIR-80	116	SM-1
116	S107	116	S72	116	SD45	116	SIR-RECT-44	116	SM-1-005
116	S108	116	S73	116	SD470	116	SIRECT-102	116	SM-1-47
116	S10A	116	S75	116	SD5	116	SIRECT-2	116	SM-10
116	S115	116	S77	116	SD500C	116	SIRECT-36	116	SM-1K
116	S1243N	116	S79	116	SD6	116	SIRECT-48	116	SM1-02
116	S129	116	S81	116	SD600C	116	SIRECT-59	116	SM10
116	S13	116	S82	116	SD80	116	SIRECT-92	116	SM105
116	S14	116	S83	116	SD93B	116	SI8D-1X	116	SM11
116	0000000S15	116	S84	116	SD94AB	116	SI8D-K	116	SM110
116	S16	116	S85	116	SD94B	116	SI8M-150-01	116	SM120
116	S1600	116	S86	116	SD94S	116	SI8W-05-02	116	SM130
116	S16A	116	S91	116	SD95	116	SI8W-0502	116	SM140
116	S16B	116	S91-A	116	SD950	116	SJ-570	116	SM160
116	S17	116	S91H	116	SD95A	116	SJ051E	116	SM20
116	S17A	116	S92	116	SD96	116	SJ051P	116	SM205
116	S18	116	S92-A	116	SD96A	116	SJ052E	116	SM210
116	S1801-02	116	S92H	116	SD96B	116	SJ052P	116	SM220
116	S18A	116	S93	116	SD8-113	116	SJ101F	116	SM230
116	S18B	116	S93A	116	SD8113	116	SJ102F	116	SM240
116	S19	116	S93H	116	SE-0.5B	116	SJ201P	116	SM250
116	S191G	116	S94	116	SE-05	116	SJ202P	116	SM260
116	S19A	116	S95	116	SE-05-01	116	SJ301P	116	SM30
116	S1A	116	SA2	116	SE-05X	116	SJ302P	116	SM31
116	S1A06	116	SA2B	116	SE-2	116	SJ401P	116	SM40
116	S1A060	116	SA3B	116	SE-5	116	SJ402P	116	SM483
116	S1A60	116	SB-1	116	SE05	116	SJ501P	116	SM486
116	S1AR1	116	SB-3	116	SE05B	116	SJ601P	116	SM487
116	S1AR2	116	SB-3-02	116	SE05D	116	SJ60P	116	SM488
116	S1B	116	SB-309A	116	SE058	116	SK-1B	116	SM5
116	S1B-01-02	116	SB-309C	116	SE0588	116	SK-218	116	SM50
116	0000081B01	116	SB-3P01	116	SE08-01	116	SK1PM	116	SM505
116	S1B01-01	116	SB-3N	116	SE30B26A	116	SK1K-2	116	SM51
116	S1B01-02	116	SB01	116	SE46	116	SK3016	116	SM510
116	S1B01-06	116	SB1-01-04	116	SE6	116	SK3017A	116	SM512
116	S1B01101CR	116	SB302	116	SELEN-70	116	SK3030	116	SM513
116	S1B0102	116	SB315	116	SELEN-701	116	SK3031	116	SM514
116	S1B02	116	SB332	116	SFB6183	116	SK3174	116	SM515
116	S1B02-06CE	116	SB333	116	SFR135	116	SK3311	116	SM516
116	S1B02-06CRE	116	SB393	116	SFR151	116	SK3312	116	SM60
116	S1B02-C	116	SBR-260	116	SFR152	116	SK3313	116	SM645
116	S1B0201CR	116	SC-110	116	SFR153	116	SL-030	116	SM646
116	S1BD1-02	116	SC-16	116	SFR154	116	SL-030T	116	SM705
116	S1CN1	116	SC05	116	SFR155	116	SL-2	116	SM710
116	S1D51C052-19	116	SC05E	116	SFR156	116	SL-3	116	SM720
116	S1D510169-1	116	SC1	116	SFR164	116	SL-4	116	SM730
116	S1RC20	116	SC110	116	SFR251	116	SL-433	116	SM740
116	S1RC20R	116	SC1414	116	SFR252	116	SL-833A	116	SM750
116	S18M-150-01	116	SC1431	116	SFR253	116	SL030/3490	116	SM760
116	S18M-150-02	116	SC1631	116	SFR254	116	SL0308	116	SN-1
116	S18W-05-02	116	SC1631(GE)	116	SFR255	116	SL030T	116	SN-1Z
116	S200	116	SC2	116	SFR256	116	SL103	116	SN0303
116	S201	116	SC23-3	116	SFR264	116	SL2		
116	S202	116	SC23-9	116	SFR266	116	SL5		
		116	SC305	116	SG-005	116	SL833		
		116	SC4						

ECG	Industry Standard No.	ECG	Industry Standard No.	ECG	Industry Standard No.	ECG	Industry Standard No.	ECG	Industry Standard No.
116	SN1	116	SR22	116	T-E1050	116	TI55	116	TVS-DS-1K
116	S05	116	SR23	116	T-E1064	116	TI56	116	TVS-DS-1M
116	S0501	116	SR2301	116	T-E1078	116	TI57	116	TVS-DS1K
116	SOD200D	116	SR2301A	116	T-E1078A	116	TI58	116	TVS-DS1M
116	SP-1	116	SR24	116	T-E1080	116	TI59	116	TVS-DS2K
116	SP1	116	SR27	116	T-E1089	116	TI60	116	TVS-ET1P
116	SP1K-1	116	SR28	116	T-E1090	116	TI71	116	TVS-FR-1P
116	SP1K-2	116	SR2A-1	116	T-E1097	116	TIRO1	116	TVS-FR-1P(FR1P)
116	SP1E-2	116	SR2A-2	116	T-E1102	116	TIRO2	116	TVS-FR-2PC
116	SP1E-01	116	SR2A-4	116	T-E1102A	116	TIRO3	116	TVS-FR1-PC
116	SR-05K-2	116	SR2A1	116	T-E1124	116	TIRO4	116	TVS-FR1MD
116	SR-1	116	SR2A12	116	T-E1133	116	TIRO5	116	TVS-FR1PC
116	SR-101-1	116	SR2A2	116	T-E1138	116	TIRO6	116	TVS-FR2M
116	SR-101-2	116	SR2A4	116	T-E1144	116	TJ-5A	116	TVS-FT-1P
116	SR-112	116	SR2A8	116	T-E1148	116	TJ10A	116	TVS-PU1N
116	SR-120	116	SR3	116	T-E1155	116	TJ15A	116	TVS-HP-8D-12
116	SR-120-1	116	SR30	116	T-E1157	116	TJ20A	116	TVS-HPSD1Z
116	SR-130-1	116	SR3010	116	T-E1171	116	TJ25A	116	TVS-KC2-LP
116	SR-131-1	116	SR35	116	T-E1176	116	TJ30A	116	TVS-KC2OP12/1
116	SR-132-1	116	SR3582	116	T0150	116	TJ35A	116	TVS-KC2OP12/1
116	SR-136	116	SR390	116	T065	116	TJ40A	116	TVS-KC2OP12/2
116	SR-13H	116	SR3943	116	T075	116	TJ5A	116	TVS-OV-02
116	SR-14	116	SR5BM-6	116	T10144	116	TJ60A	116	TVS-PC02P11/2
116	SR-1849-1	116	SR4	116	T10175	116	TK-10	116	TVS-PCD2P11/2
116	SR-1K	116	SR40	116	T10185	116	TK-30	116	TVS-S1B02-03C
116	SR-1K-2	116	SR401	116	T10453	116	TK-40	116	TVS-S1B02-03CR
116	SR-1K2	116	SR405	116	T1085	116	TK-41	116	TVS-SD1A
116	SR-1Z	116	SR5	116	T1450	116	TK10	116	TVS-TC009M11/1Q
116	SR-22	116	SR50	116	T56	116	TK11	116	TVS-UPSD-1
116	SR-23	116	SR500	116	T53	116	TK20	116	TV80A71
116	SR-24	116	SR500B	116	T1J6G	116	TK21	116	TV810D
116	SR-27	116	SR50411-1	116	T200	116	TK30	116	TV810DC4
116	SR-28	116	SR60	116	T21312	116	TK400	116	TV810DC4R
116	SR-3	116	SR605	116	T21333	116	TK5	116	TV81850
116	SR-30	116	SR6154	116	T21507	116	TK50	116	TV81N4002
116	SR-390	116	SR6324	116	T21602	116	TK60	116	TV81181850
116	SR-401	116	SR6325	116	T21649	116	TK600	116	TV81181906
116	SR-5	116	SR6385	116	T21679	116	TK61	116	TVS550
116	SR-846-2	116	SR6415	116	T3/2	116	TKP10	116	TVSBEE
116	SR-889	116	SR6560	116	T300	116	TKP20	116	TVSD1K
116	SR-IK-2	116	SR6567	116	T30155-001	116	TKP40	116	TVSD52K
116	SRO5K-2	116	SR6617	116	T30155-1	116	TKP5	116	TVSDG1NR
116	SR1-K2	116	SR6723	116	T400	116	TKP60	116	TVSD32M
116	SR100	116	SR6724	116	T42692-001	116	TL1	116	TVSBBA06
116	SR101-1	116	SR76	116	T42692-1R	116	TL12	116	TVSFR1P
116	SR101-2	116	SR806-126	116	T450	116	TM-33	116	TVSFR1PC
116	SR1024	116	SR846-2	116	T500	116	TM-43	116	TVSFR2-06
116	SR105	116	SR846-3	116	T550	116	TM33	116	TVSFT1P
116	SR1104	116	SR851	116	T600	116	TM43	116	TVSF01N
116	SR112	116	SR851-121	116	T650	116	TM62	116	TVSHFSD-1A
116	SR114	116	SR889	116	T8/2	116	TM63	116	TVSHFSD12C
116	SR120	116	SR9005	116	TA100	116	TM65	116	TVS1181850
116	SR1266	116	SRIK-2	116	TA1062	116	TM66	116	TVSJL41A
116	SR131-1	116	SRK-2	116	TA1063	116	TMD41	116	TVSN1-02
116	SR135-1	116	SRK1	116	TA1064	116	TMD42	116	TV80A71
116	SR1378-1	116	SRLFM-1	116	TA200	116	TMD45	116	TVSPCD2P11/2
116	SR1378-3	116	SS-1	116	TA300	116	TP101	116	TVSRMP5020
116	SR15H	116	SS00010	116	TA400	116	TP201	116	TVSS1P20
116	SR144	116	SS00010	116	TA50	116	TP302	116	TVSS1R8D
116	SR145	116	SS00009	116	TA500	116	TP402	116	TVSS3-2
116	SR1493	116	SS321	116	TA600	116	TP4067-409	116	TVSS34RECT
116	SR151	116	SS322	116	TA7802	116	TR-02E	116	TVSS4C
116	SR152	116	SS324	116	TA7803	116	TR-2880	116	TVSSA2B
116	SR1549	116	SS334	116	TA7804	116	TR-77	116	TVSSD1A
116	SR1668	116	SS337	116	TA7996	116	TR1N4002	116	TVSSV02
116	SR1692	116	SS455	116	TC0.2P11/2	116	TR2327041	116	TVSUPSD1P
116	SR1693	116	SSD974	116	TC02P112	116	TR2880	116	TVSWP2
116	SR1694	116	ST-12	116	TC02P112	116	TR2A	116	TW10
116	SR1695	116	ST-14	116	TC0P11/2	116	TR320020	116	TW20
116	SR17	116	ST-2040P	116	TC136	116	TR320022	116	TW3
116	SR1731-1	116	ST16	116	TC3112319300	116	TR330027	116	TW30
116	SR1731-2	116	ST2040P	116	TE1010	116	TS-2A	116	TW40
116	SR1731-3	116	STB0L-02	116	TE1011	116	TS05	116	TW5
116	SR1731-4	116	SV-01A	116	TE1024C	116	TS1	116	TW50
116	SR1731-5	116	SV-01B	116	TE1024D	116	TS2	116	TW60
116	SR1742	116	SV-05	116	TE1029	116	TS2A	116	TWV
116	SR1766	116	SV-12388	116	TE1042	116	TS6	116	TX1N3190
116	SR1984	116	SV-1238E	116	TE1050	116	TSB-1000	116	TX1N3191
116	SR1A-1	116	SV01A	116	TE1078	116	TSB-245	116	TX1N645
116	SR1A-2	116	SV02A	116	TE1080	116	TSB245	116	TX1N647
116	SR1A-4	116	SV1238	116	TE1088	116	TS0159	116	U
116	SR1A-8	116	SV12388	116	TE1089	116	TT66X26	116	U-2400-03
116	SR1A1	116	SV12388E	116	TE1090	116	TUS-185D	116	U-633
116	SR1A12	116	SV1238I	116	TE1097	116	TV-24104	116	U06E
116	SR1A2	116	SVD10D-1	116	TE1108	116	TV-24125	116	U13033801
116	SR1A4	116	SVD13I850	116	TP20	116	TV-24266	116	U18102
116	SR1A8	116	SVDVD1223	116	TP21	116	TV-2496	116	U212
116	SR1D1M	116	SW-05-005	116	TP22	116	TV24104	116	U212-25
116	SR1DM	116	SW-05-02	116	TP23	116	TV241073	116	U213
116	SR1DM-1	116	SW-05V	116	TFR-120	116	TV241074	116	U214
116	SR1DM-4	116	SW-1A	116	TFR120	116	TV24125	116	U2400-03
116	SR1DM1	116	SWO.5A	116	TG-11	116	TV24136	116	U633
116	SR1DMX	116	SW05	116	TG-12	116	TV24155	116	UPO1
116	SR1E	116	SW05-01	116	TG-21	116	TV2419	116	UPSD-1
116	SR1EM	116	SW05-02	116	TG-22	116	TV24191	116	UPSD-1A
116	SR1EM-1	116	SW05A	116	TG-31	116	TV24193	116	UR105
116	SR1EM-2	116	SW05B	116	TG-32	116	TV24200	116	UR110
116	SR1EM-X	116	SW05S	116	TG-41	116	TV24222	116	UR115
116	SR1EM1	116	SW05SS	116	TG-42	116	TV24224	116	UR120
116	SR1EM2	116	SW05V	116	TG-51	116	TV24232	116	UR125
116	SR1FM	116	SX-642	116	TG-52	116	TV24234	116	USPD-1
116	SR1FM-1	116	SX623	116	TG-61	116	TV24266	116	USPD-1A
116	SR1FM-4	116	SX631	116	TG-62	116	TV24282	116	UT-112
116	SR1FM10	116	SX633	116	TG12	116	TV24283	116	UT-16
116	SR1FM12	116	SX641	116	TG20A	116	TV24292	116	UT-234
116	SR1FM2	116	SX642	116	TG21	116	TV24298	116	UT-258
116	SR1FM4	116	SX643	116	TG22	116	TV24582	116	UT11
116	SR1FM6	116	SX644	116	TG31	116	TV24586	116	UT111
116	SR1FM8	116	SX645	116	TG32	116	TV24803	116	UT112
116	SR1FMA	116	T-0150	116	TG41	116	TV24941	116	UT113
116	SR1K	116	T-065	116	TG42	116	TV2496	116	UT114
116	SR1K-1	116	T-075	116	TG51	116	TV24979	116	UT115
116	SR1K-1K	116	T-100	116	TG52	116	TV34232	116	UT116
116	SR1K-2	116	T-13	116	TG61	116	TV4	116	UT117
116	SR1K-2/494	116	T-13G	116	TG62	116	TVC-3	116	UT14
116	SR1K-4	116	T-14	116	TQ12	116	TVD81M	116	UT15
116	SR1K-8	116	T-14G	116	TH18557	116	TVM-511	116	UT16
116	SR1K-Z	116	T-200	116	TH400	116	TVM-EH2C	116	UT17
116	SR1K/494	116	T-22G	116	TH50	116	TVM-EH2C11	116	UT18
116	SR1K08	116	T-26G	116	TH600	116	TVM-M204B	116	UT21
116	SR1K1	116	T-300	116	TH801	116	TVM-PH9D22/1	116	UT22
116	SR1K2	116	T-400	116	TH802	116	TVM-TC0.2P11/2	116	UT221
116	SR1K8	116	T-450	116	TH803	116	TVM35	116	UT222
116	SR1T	116	T-4590	116	TH804	116	TVM550	116	UT223
116	SR1Z	116	T-50	116	TH805	116	TVM56	116	UT224
116	SR200	116	T-500	116	TH806	116	TVM563	116	UT225
116	SR200B	116	T-550	116	TH8105	116	TVMH8151B	116	UT226
116	SR205	116	T-600	116	TI-53	116	TVML00.09M1115	116	UT227
116	SR2121	116	T-650	116	TI-55	116	TVMM204B	116	UT228
		116	T-R01029D	116	TI-71	116	TVMPH9DZ2/1	116	UT229
		116	T-E1011	116	TI152	116	TVS-185D	116	UT23
		116	T-E1024	116	TI52	116	TVS-181850	116	UT231
		116	T-E1024C	116	TI53	116	TVS-181906	116	UT232
		116	T-E1024D	116	TI54	116	TVS-DG-1N-R	116	UT233

ECG	Industry Standard No.	ECG	Industry Standard No.	ECG	Industry Standard No.	ECG	Industry Standard No.	ECG	Industry Standard No.
116	UT24	116	XS17	116	001-02603-0	116	1N1564A	116	1N3278
116	UT25	116	XS17A	116	1-101	116	1N1565A	116	1N3279
116	UT26	116	XS18	116	1-20-001-890	116	1N1566A	116	1N3279A
116	UT27	116	XS22(RECT.)	116	01-2405-0	116	1N1567A	116	1N3298
116	V-06B	116	XS23	116	01-2405-1	116	1N1568A	116	1N354
116	V-06C	116	XS23A	116	01-2406-1	116	1N58	116	1N3544
116	V-270D1	116	XS31	116	1-530-012-11	116	1N1617	116	1N3545
116	V-442	116	XS40A	116	1-531-027	116	1N1618	116	1N3546
116	VO-6	116	XU604	116	1-531-105	116	1N1619	116	1N3547
116	VO-6B	116	YAAD004	116	1-531-10513	116	1N1620	116	1N3548
116	VO-6-401	116	YAAD007	116	1-531-106-13	116	1N169	116	1N3549
116	VO3G	116	YAAD019	116	1-531-106-17	116	1N1692	116	1N3559A
116	VO6	116	YAAD020	116	1-531-5-11	116	1N1693	116	1N360
116	VO6-C	116	YAAD022	116	1-534-105-13	116	1N1694	116	1N3560A
116	VO6A	116	YBAD009	116	1-650119001G	116	1N1695	116	1N3611
116	VO6ZK4	116	YBAD030	116	1-8259	116	1N1696	116	1N3612
116	VO6C	116	YR-011	116	1-RE-004	116	1N1697	116	1N3613
116	VO6E	116	YR011	116	1A10425	116	1N1701	116	1N3561A
116	V10158	116	Y3G-V47-1-3	116	1A11184	116	1N1702	116	1N362
116	V11189-1	116	Y3G-V47-7-51-1	116	1A11671	116	1N1703	116	1N362A
116	V15920	116	Y3G-V47-7-51-2	116	1A12214	116	1N1704	116	1N363
116	V15C200/80-VF	116	Z330611	116	1A12407	116	1N1705	116	1N3639
116	V17L	116	ZCOM-5683-0	116	1A12690	116	1N1706	116	1N3639A
116	V210C	116	ZCOM3679	116	1A13219	116	1N1707	116	1N3640
116	V270-D1	116	ZJ252B	116	1A13719	116	1N1708	116	1N3641
116	V3074A20	116	ZR-1025	116	1A13720	116	1N1709	116	1N3669
116	V3074A21	116	ZR-1031	116	1A15790	116	1N1710	116	1N3754
116	V442	116	ZR-1035	116	1A16550	116	1N1711	116	1N3895
116	V66	116	ZR-1076	116	1A50	116	1N1712	116	1N400
116	V6C	116	ZR-500	116	1B-2C1	116	1N1763	116	1N4002
116	V9446-4	116	ZR-590	116	1B05J20	116	1N1764	116	1N4003
116	VAMV-4	116	ZR-590A	116	1B05J40	116	1N1907	116	1N4003GP
116	VB-11	116	ZR-61	116	1B10J20	116	1N1908	116	1N4004
116	VB-400	116	ZR-63	116	1C0009	116	1N1909	116	1N4005
116	VB-600	116	ZR1025	116	1C0025	116	1N1911	116	1N400B
116	VB100	116	ZR1031	116	1C0026	116	1N1912	116	1N4245
116	VB300	116	ZR1035	116	1C0031	116	1N1913	116	1N4246
116	VB400	116	ZR1076	116	1D261	116	1N2069	116	1N4247
116	VB500	116	ZR15	116	1D281	116	1N2069A	116	1N4364
116	VB600A	116	ZR500	116	1DC1	116	1N2070	116	1N4365
116	VBH600	116	ZR590A	116	1E05	116	1N2070A	116	1N4366
116	VO6E	116	ZR60	116	1E1	116	1N2071	116	1N4367
116	VD-121C	116	ZR61	116	1E2	116	1N2071A	116	1N4368
116	VD6	116	ZR62	116	1E3	116	1N2072	116	1N4369
116	VFA-2745C	116	ZR63	116	1E4	116	1N2073	116	1N4383
116	VFA2745C	116	ZR64	116	1E5	116	1N2074	116	1N4384
116	VHD181885-1	116	ZR66	116	1E6	116	1N2075	116	1N4385
116	VHD181885//-1	116	ZS-10B	116	1ET02	116	1N2077	116	1N440
116	VHD182250//1B	116	ZS-20A	116	1ET05	116	1N2078	116	1N440B
116	VM-PH11D522/1	116	ZS-20B	116	1ET1	116	1N2079	116	1N441
116	VM-PH9D522/1	116	ZS-21	116	1ET2	116	1N2080	116	1N441B
116	VM-TC02P11/2	116	ZS-23	116	1ET3	116	1N2081	116	1N442
116	VMPH11D522-1	116	ZS-24	116	1ET4	116	1N2082	116	1N442B
116	VO-5X	116	ZS-25	116	1ET5	116	1N2083	116	1N443
116	VO-6A	116	ZS-30A	116	1ET6	116	1N2084	116	1N443B
116	VO-6C	116	ZS-30B	116	1F05	116	1N2085	116	1N444
116	VO6-C	116	ZS-31A	116	1F14A	116	1N2086	116	1N444B
116	VO6B	116	ZS-31B	116	1F2	116	1N2088	116	1N445
116	VO6C	116	ZS-32A	116	1FM2	116	1N2089	116	1N445B
116	VOO	116	ZS-32B	116	1G2C1	116	1N2090	116	1N448
116	VS-1	116	ZS-33A	116	1G2Z1	116	1N2091	116	1N450
116	VS-102	116	ZS-33B	116	1GA	116	1N2092	116	1N451
116	VS-DG1NR	116	ZS-34A	116	1HY40	116	1N2093	116	1N453
116	VS-FR-1	116	ZS-34B	116	1HT50	116	1N2094	116	1N480B
116	VS-FR-1P	116	ZS-50	116	M8513A	116	1N2095	116	1N4937
116	VS-FR1	116	ZS-52	116	1N-4002	116	1N2096	116	1N503
116	VS-FR1P	116	ZS-53	116	1N1008	116	1N2103	116	1N504
116	VS-FT-1N	116	ZS-73	116	1N1028	116	1N2104	116	1N505
116	VS-PH9D522/1	116	ZS100	116	1N1029	116	1N2105	116	1N506
116	VS-SD-1Z	116	ZS101	116	1N1030	116	1N2106	116	1N507
116	VS-TO0-2P11/2	116	ZS102	116	1N1031	116	1N2107	116	1N508
116	VS-TCO2P11/2	116	ZS103	116	1N1032	116	1N2108	116	1N511
116	VS1	116	ZS104	116	1N1033	116	1N2115	116	1N512
116	VS120	116	ZS108	116	1N1052	116	1N2116	116	1N513
116	VS202	116	ZS10A	116	1N1053	116	1N2117	116	1N514
116	VS89-0001-911	116	ZS10B	116	1N1081	116	1N2323	116	1N515
116	VS89-0002-911	116	ZS120	116	1N1081A	116	1N2482	116	1N516
116	VS89-0005-911	116	ZS121	116	1N1082	116	1N2483	116	1N519
116	VS89-0006-911	116	ZS122	116	1N1083	116	1N2484	116	1N520
116	VS89-0007-911	116	ZS123	116	1N1084	116	1N2485	116	1N521
116	VSFR1	116	ZS124	116	1N1095	116	1N2486	116	1N5211
116	VSFR1P	116	ZS173	116	1N1096	116	1N2487	116	1N5212
116	VSFT1N	116	ZS174	116	1N10D-4F	116	1N2488	116	1N5215
116	VSG-20024	116	ZS174B	116	1N1100	116	1N2489	116	1N5216
116	VSSD1B	116	ZS20A	116	1N1101	116	1N2609	116	1N5217
116	VSSD1Z	116	ZS20B	116	1N1102	116	1N2610	116	1N522
116	VSTCO2P11/2	116	ZS21	116	1N1103	116	1N2611	116	1N523
116	WO3A	116	ZS22	116	1N1104	116	1N2612	116	1N524
116	WO3B	116	ZS23	116	1N1105	116	1N2613	116	1N530
116	WO6	116	ZS24	116	1N122A	116	1N2614	116	1N531
116	WO6A	116	ZS25	116	1N1169	116	1N2615	116	1N532
116	WO6B	116	ZS30	116	1N1169A	116	1N2858	116	1N533
116	WO6C	116	ZS30A	116	1N1217	116	1N2859	116	1N534
116	W4002	116	ZS30B	116	1N1217A	116	1N2860	116	1N535
116	WC-14020	116	ZS31	116	1N1217B	116	1N2861	116	1N536
116	WC-14027	116	ZS31A	116	1N1218	116	1N2862	116	1N537
116	WC120	116	ZS31B	116	1N1218A	116	1N2863	116	1N538
116	WC14020	116	ZS32	116	1N1218B	116	1N2864	116	1N539
116	WC14027	116	ZS32A	116	1N1219	116	1N3094	116	1N540
116	WC19865	116	ZS32B	116	1N1219A	116	1N3315	116	1N547
116	WDO01	116	ZS34	116	1N1219B	116	1N3315A	116	1N596
116	WDO02	116	ZS34A	116	1N1220	116	1N3316	116	1N599
116	WDO03	116	ZS34B	116	1N1220A	116	1N3160	116	1N599A
116	WDO04	116	ZS50	116	1N1221	116	1N3316A	116	1N600
116	WDO05	116	ZS51	116	1N1221A	116	1N3317	116	1N600A
116	WDO06	116	ZS52	116	1N1224	116	1N3317A	116	1N601
116	WDO07	116	ZS53	116	1N1224A	116	1N3318	116	1N601A
116	WDO08	116	ZS7	116	1N1251	116	1N3189	116	1N602
116	WDO09	116	ZS70	116	1N1252	116	1N3318A	116	1N602A
116	WDO10	116	ZS71	116	1N1253	116	1N3319	116	1N603
116	WDO11	116	ZS72	116	1N1254	116	1N3190	116	1N603A
116	WDO12	116	ZS73	116	1N1255	116	1N3191	116	1N604A
116	WDO13	116	ZS74	116	1N1255A	116	1N3193	116	1N605
116	WDO14	116	ZS74B	116	1N1256	116	1N3194	116	1N605A
116	WDO15	116	ZS76	116	1N1257	116	1N3195	116	1N606
116	WO-6A	116	ZS8	116	1N1337-5	116	1N3319A	116	1N606A
116	WO6B	116	ZS90	116	1N1406	116	1N3203	116	1N645
116	WR-013	116	ZS91	116	1N1415	116	1N3320A	116	1N645A
116	WR-200	116	ZS92	116	1N1439	116	1N321	116	1N645B
116	WRO06	116	ZS94	116	1N1440	116	1N3323A	116	1N646
116	WR100	116	ZTR-WO6B	116	1N1441	116	1N324	116	1N647
116	WR200	116	ZTR-WO6C	116	1N1442	116	1N3324A	116	1N648
116	WR300	116	ZTR-WO6B	116	1N1486	116	1N325	116	1N649
116	WR400	116	ZW2	116	1N1487	116	1N3253	116	1N673
116	WRE-981	116	001-0077-00	116	1N1488	116	1N3254	116	1N676
116	WR8981	116	001-0153-00	116	1N1489	116	1N3255	116	1N677
116	X5M6	116	001-02405-0	116	1N1490	116	1N3255A	116	1N678
116	XA121	116	001-02405-1	116	1N1491	116	1N326	116	1N679
116	XS-10	116	001-02405-2	116	1N1492	116	1N3326A	116	1N681
116	XS-31	116	001-02405-1	116	1N151	116	1N327	116	1N682
116	XS10	116	001-024051	116	1N152	116	1N3277		
116	XS16	116	001-024052	116	1N153				
116	XS16A	116	001-02601-0	116	1N1563A				

ECG	Industry Standard No.	ECG	Industry Standard No.	ECG	Industry Standard No.	ECG	Industry Standard No.	ECG	Industry Standard No.
116	1N683	116	181885-3	116	2X3A	116	05-931601	116	09-306214
116	1N684	116	181886	116	2W4A	116	05-931971	116	09-306224
116	1N685	116	181887	116	2W5A	116	5A-D	116	09-306245
116	1N686	116	181888	116	2X9A116	116	05A07	116	09-306249
116	1N687	116	181906	116	003-001	116	5A1	116	09-306250
116	1N689	116	181941	116	03-0018-0	116	5A2	116	09-306254
116	1N692	116	181 9413	116	003-009400	116	5A3	116	09-306255
116	1N819	116	181942	116	003-009900	116	5A4	116	09-306263
116	1N846	116	181943	116	03-3016	116	5A4D	116	09-306264
116	1N847	116	181209	116	03-931601	116	5A4D-C	116	09-306285
116	1N848	116	18204	116	03-931609	116	5A5	116	09-306300
116	1N849	116	18205	116	03-931971	116	5A5D	116	09-306303
116	1N850	116	18206	116	3A152	116	5A6	116	09-306312
116	1N851	116	18207	116	3A154	116	5A6D	116	09-306315
116	1N852	116	18208	116	3A156	116	5A6D-C	116	09-306323
116	1N857	116	182080	116	3A200	116	5B-15H	116	09-306333
116	1N858	116	182081	116	3A252	116	5B-2	116	09-306341
116	1N859	116	18209	116	3A254	116	5B-2-H5W	116	09-306350
116	1N860	116	182230	116	3A256	116	5B3	116	09-306353
116	1N861	116	182310	116	3A81	116	5D1	116	09-306365
116	1N862	116	182313	116	3A82	116	5D2	116	09-306376
116	1N863	116	182351	116	3B81	116	5E1	116	09-306384
116	1N868	116	182352	116	3B82	116	5E2	116	09-306389
116	1N869	116	182356	116	3C81	116	5E4	116	09-306394
116	1N870	116	182357	116	3C82	116	5E5	116	09-306417
116	1N871	116	182361	116	3D81	116	5GA	116	09-306421
116	1N872	116	182362	116	3D82	116	5GB	116	09-306422
116	1N873	116	182363	116	3E-64	116	5GD	116	09-306427
116	1N874	116	182367	116	3E-65	116	5GFH	116	09-306432
116	1N879	116	182372	116	3E81	116	50JFR1N	116	09-306433
116	1N880	116	182373	116	3E82	116	5GL	116	09-307043
116	1N881	116	182374	116	3F81	116	5H	116	09-307084
116	1N882	116	182375	116	3F82	116	5H4D1	116	9D13
116	1N883	116	182376	116	3G152	116	5H750M	116	9RE1
116	1N884	116	182401	116	3G154	116	5J-P1	116	10-012
116	1N885	116	182402	116	3G156	116	5MA2	116	10-085006
116	1N91	116	182404	116	3G252	116	5MA4	116	10-085009
116	1N92	116	182462	116	3G254	116	5MA5	116	10-085010
116	1N93	116	182606	116	3G256	116	5MA6	116	10-085026
116	1N93A	116	1826BT	116	3G8	116	5MF1	116	10-12
116	1N947	116	18301 6(RECT)	116	3GA	116	5MB10	116	10-42
116	1N998	116	18309	116	3GB1	116	5MB20	116	10-7
116	1NC61684	116	18310	116	3GB2	116	5MB30	116	10-D1
116	1NJ61676	116	18312	116	3H81	116	5MB40	116	10A590B
116	1NJ61726	116	18313	116	3H82	116	5MB5	116	10AG2
116	1R01	116	18314	116	3L4-3001-5	116	5MB50	116	10AG4
116	1R1K	116	18315	116	3L4-3001-8	116	5N1	116	10AG6
116	1R2A	116	18358	116	3M810	116	05V-50	116	10AL2
116	1R2D	116	18358(S)	116	3M820	116	6-59010	116	10AL6
116	1R2E	116	18358B	116	3M830	116	608	116	10A8
116	1R3G	116	18395	116	3M840	116	60A1750	116	10AT2
116	1R3J	116	18396	116	3M85	116	60A175D	116	10AT4
116	1R5A	116	18399	116	3M850	116	6M4	116	10AT6
116	1R5B	116	18400	116	38B-B732	116	6M404-1	116	10B-2
116	1R5G	116	18456	116	38B629	116	6M404-2	116	10B-Y
116	1R5H	116	18457	116	5T501	116	6M404-3	116	10B1
116	1R9	116	18458	116	5T502	116	6M404-4	116	10B2
116	1R90	116	18459	116	5T503	116	6M404-5	116	10B3
116	1R91	116	0001849	116	5T504	116	6M404-6	116	10B4
116	1R9E	116	18558	116	5T505	116	6M404-7	116	10B5
116	1R9J	116	18559	116	5T506	116	6R522PC7BAD1	116	10B6
116	1R0F	116	18588	116	004-002000	116	6RW62EY	116	10C
116	1R0H	116	18685	116	004-002700	116	7-0002	116	10005
116	18005	116	1871	116	004-002800	116	7-0004	116	10C1
116	18100	116	1881	116	004-003000	116	7-0008	116	10C2
116	181004	116	1883	116	004-003300	116	7D	116	10C3
116	18101	116	18844	116	004-005400	116	7MA60	116	10C4
116	18102	116	18846	116	004-003500	116	008-024-00	116	10C4D
116	18103	116	18849	116	004-003600	116	08-08122	116	10C5
116	18104	116	18849,R	116	004-003900	116	08-0821	116	10C6
116	18105	116	18885	116	004-004000	116	8-22	116	10D
116	18106	116	1890	116	004-004100	116	8-25	116	10D-02
116	181061	116	1891	116	004-03500	116	8-38	116	10D-05
116	181062	116	1892	116	004-03500	116	8-619-030-012	116	10D-06
116	181063	116	1893	116	004-03600	116	8-639-001-095	116	10D-1
116	181064	116	1893/SGJ	116	004-03700	116	8-710-222-21	116	10D-2B
116	18106A	116	1894	116	4-1807	116	8-719-205-10	116	10D-2B(-4)
116	181096	116	1894/4454C	116	4-2020-03173	116	8-719-900-63	116	10D-4
116	18110	116	189413	116	4-2020-03200	116	8-719-901-02	116	10D-6
116	18110A	116	1895	116	4-2020-05200	116	8-719-901-13	116	10D-V
116	18111	116	18B01-02	116	4-2020-07300	116	8-719-901-92	116	10D05
116	18112	116	18D-2	116	4-2020-07600	116	8-719-908-03	116	10D1
116	18113	116	18D2	116	4-2020-07700	116	8-905-013-752	116	10D2
116	18113A	116	18IZ09	116	4-2020-08500	116	8-905-013-759	116	10D2L
116	18114	116	18L1885	116	4-2020-14500	116	8-905-013-760	116	10D3
116	18115	116	18030	116	4-2021-04970	116	8-905-198-001	116	10D3G
116	18116	116	18031	116	4-2021-09370	116	8-905-198-004	116	10D4
116	18119	116	18032	116	4-2021 04170	116	8-905-198-005	116	10D4C
116	18120	116	18034	116	4-202R101	116	8-905-198-007	116	10D4D
116	18121	116	18036	116	4-3033	116	8-905-198-008	116	10D4E
116	18121(RECT)	116	18054	116	4-3540012	116	8-905-198-010	116	10D4L
116	18122	116	181K	116	4-50(SEARS)	116	8-905-198-034	116	10D5C
116	181221	116	18854	116	4-686149-3	116	8-905-305-400	116	10D5B
116	181222	116	1T2011	116	04-8054-3	116	8-905-405-002	116	10D5F
116	181224	116	1T2012	116	04-8054-4	116	8-905-405-026	116	10D6
116	18123	116	1T2013	116	04-8054-7	116	8-905-405-069	116	10D6D
116	181230	116	1T2014	116	4AJ4DX520	116	8-905-405-134	116	10D6R
116	181231	116	1T2015	116	4AJ4DX52D	116	8-905-405-146	116	10DBF
116	181232	116	1T2016	116	4D4	116	8-905-405-206	116	10DC
116	18124	116	1T378	116	4D6	116	8-905-413-092	116	10DC-1R
116	18125	116	1T501	116	4D8	116	8A11667	116	10DC-2C
116	18126	116	1T502	116	4GA	116	8CG15	116	10DC-2J
116	18134	116	1T503	116	4I39104002	116	8CG15RE	116	10DC-4
116	181341	116	1T504	116	4JA10DX3	116	8D4	116	10DC-4R
116	181342	116	1T505	116	4JA10DX32	116	8D6	116	10DC05R
116	181343	116	1T506	116	4JA10EX3	116	09-306-083	116	10DC05(RED)
116	181344	116	1805	116	4JA211A	116	09-306033	116	10DCO5R
116	181346	116	1V3074A20	116	4JA2FX355	116	09-306034	116	000001ODC1
116	18136	116	1V3074A21	116	4JA2X355	116	09-306046	116	10DC22P
116	18147	116	1VA10	116	4JA4DR700	116	09-306050	116	10DC4
116	18148	116	1WM6	116	4JA4DX520	116	09-306054	116	10DC4R
116	18149	116	2-55R	116	4JA6MR700	116	09-306059	116	10DCO5R
116	18150	116	2D-02	116	4T501	116	09-306063	116	10DC0B
116	181622	116	2E05	116	4T502	116	09-306088	116	10DC0H
116	181623	116	2E4	116	4T503	116	09-306100	116	10D0
116	181624	116	2F4	116	4T504	116	09-306103	116	10D0.5
116	181625	116	2G13	116	4T505	116	09-306104	116	10DRV
116	181664	116	2G8	116	4T506	116	09-306112	116	10DY
116	181666	116	2G805	116	05-03016-01	116	09-306114	116	10DZ2
116	181668	116	2GA	116	05-04001-02	116	09-306115	116	10DZ
116	181668F	116	28-16E	116	5-30086.1	116	09-306119	116	10E-1
116	181691	116	281K	116	5-30088.1	116	09-306125	116	10E-2
116	181693	116	28B-C731	116	5-30094.1	116	09-306138	116	10E-4D
116	181697	116	28J2A	116	5-30095.1	116	09-306149	116	10E1
116	181698	116	28J4A	116	5-30098.1	116	09-306160	116	10E1LF
116	18180/5GB	116	28J60A	116	5-30099.1	116	09-306162	116	10E6
116	18185	116	28B1K	116	5-30099.4	116	09-306169	116	10H
116	181851	116	2T501	116	5-30106.1	116	09-306172	116	10J2
116	181851R	116	2T502	116	5-30109.1	116	09-306176	116	10J2F
116	181850R	116	2T503	116	5-30113.1	116	09-306177	116	10K
116	181881AM	116	2T504	116	5-30120.1	116	09-306192	116	10K-1
116	181885	116	2T505	116	05-540001	116	09-306205	116	10M
116		116	2T506	116	05-750010	116	09-306213	116	10N1
116		116		116		116		116	10R1B

ECG	Industry Standard No.	ECG	Industry Standard No.	ECG	Industry Standard No.	ECG	Industry Standard No.	ECG	Industry Standard No.
116	1OR2B	116	19AR34	116	28-25-01	116	46-86139-3	116	48-4488790
116	1OR3B	116	19AR5	116	28-254566-1	116	46-86148-3	116	48-57120A01
116	1OR4B	116	19B200011-P5	116	28-29-01	116	46-86149-3	116	48-60022A98
116	1OR5B	116	190300076-P1	116	28-6-01	116	46-86177-3	116	48-63086A16
116	1OR6B	116	190300076-P2	116	28-65-01	116	46-86199-3	116	48-64169
116	11-0429	116	190300076-P3	116	28-7-01	116	46-86212-3	116	48-64954
116	11-0769	116	190300076-P4	116	28J2	116	46-86261-3	116	48-647829
116	11-0771	116	190300076-P5	116	30A8	116	46-86267-3	116	48-651145A74
116	11-085003	116	190300076-P6	116	30B5	116	46-86271-3	116	48-660370A05
116	11-085013	116	19P2	116	30B8	116	46-86307-3	116	48-66037A03
116	11-085024	116	20-1680-143	116	30C	116	46-86321-3	116	48-66037A04
116	11-102001	116	20-22-08	116	30H	116	46-86322-3	116	48-66037A05
116	11-120007	116	20A0054	116	30K	116	46-86326-3	116	48-66037A08
116	11-1592	116	20A8	116	30M	116	46-86339-3	116	48-66037A10
116	11/15	116	20B8	116	31-195	116	46-86351-3	116	48-66037A12
116	11J2	116	20C	116	32-0026	116	46-86355-3	116	48-66629A02
116	11J2P	116	20H	116	32-0037	116	46-86358-3	116	48-66629A03
116	12-085031	116	20K	116	32-0038	116	46-86364-3	116	48-66629A05
116	12-085040	116	20M	116	32-0042	116	46-86365-3	116	48-66629A06
116	12-100001	116	20N1	116	32-0045	116	46-86402-3	116	48-66654A02
116	12-100003	116	21-810	116	32-0046	116	46-86416-3	116	48-67120A01
116	12-100008	116	21-810-2	116	32-0047	116	46-86420-3	116	48-67120A06
116	12-102001	116	21A006-000	116	32-0050	116	46-86433-3	116	48-67120A07
116	0012-911	116	21A007-000	116	32-0059	116	46-86437-3	116	48-67926A01
116	12B-2	116	21A008-000	116	32-0060	116	46-86438-3	116	48-68688A79
116	12B-2B1P-M	116	21A008-001	116	32-0061	116	46-86443-3	116	48-68688A79B
116	1202	116	21A008-003	116	33-0002	116	46-86647-3	116	48-733746
116	12J2	116	21A020-001	116	33-0006	116	46-86497-3	116	48-746831
116	12J2P	116	21A020-006	116	33-0023	116	46-86503-3	116	48-752497
116	13-0015	116	21A040-44	116	33-0024	116	46-8661-3	116	48-82095001
116	13-085039	116	21A079-000	116	33-0026	116	46-8687-3	116	48-82095002
116	13-087027	116	21A102-001	116	33-0029	116	46-8689-3	116	48-82095003
116	13-10102-1	116	21A103-007	116	33-0030	116	46-8697-3	116	48-82095064
116	13-10321-3	116	21A103-018	116	33G59024	116	46AR1	116	48-8240006
116	13-14094-12	116	21A103-021	116	33G59121	116	46AR10	116	48-8240008
116	13-14094-16	116	21A103-044	116	33G59122	116	46AR11	116	48-82466H01
116	13-14094-17	116	21A103-050	116	34-8003	116	46AR12	116	48-82466H02
116	13-14094-24	116	21A103-058	116	34-8034-1	116	46AR2	116	48-82466H03
116	13-14094-38	116	21A103-070	116	34-8034-2	116	46AR21	116	48-82466H04
116	13-14094-39	116	21A110-001	116	34-8034-3	116	46AR27	116	48-82466H06
116	13-14094-42	116	21A110-002	116	34-8034-4	116	46AR28	116	48-82466H07
116	13-14094-54	116	21A110-003	116	34-8036-1	116	46AR29	116	48-86148
116	13-14261-1	116	21A110-006	116	34-8036-2	116	46AR3	116	48-90229A01
116	13-14261-3	116	21A110-007	116	34-8036-3	116	46AR35	116	48-90233A04
116	13-14627-1	116	21A110-010	116	34-8036-4	116	46AR4	116	48-90343A01
116	13-14627-4	116	21A6	116	34-8042-0	116	46AR5	116	48-90343A64
116	13-16104-8	116	21A7	116	34-8042-1	116	46AR6	116	48-90343A91
116	13-16104-9	116	21B-14	116	34-8042-2	116	46AR7	116	48-90343A92
116	13-16247-3	116	21B-17	116	34-8042-3	116	46AR8	116	48-90343A93
116	13-17557-1	116	21M248	116	34-8048-1	116	46AR9	116	48-90420A79
116	13-17025-1	116	21M283	116	34-8048-2	116	46AX1	116	48-90420A80
116	13-18458-1	116	21M302	116	34-8048-5	116	46AX10	116	48-97048A04
116	13-22452-0	116	21M312	116	34-8050-14	116	46AX16	116	48-97048A07
116	13-22463-1	116	21M315	116	34-8050-2	116	46AX17	116	48-97048A10
116	13-31013-6	116	21M316	116	34-8050-7	116	46AX19	116	48-97048A18
116	13-31014-1	116	21M386(PWR)	116	34-8051	116	46AX2	116	48-97127A01
116	13-31014-3	116	21M416	116	34-8054-1	116	46AX21	116	48-97168A02
116	13-33376-1	116	21M417	116	34-8054-10	116	46AX3	116	48-97168A06
116	013-339	116	21M419	116	34-8054-11	116	46AX30	116	48-97168A10
116	13-34057-1	116	21M433	116	34-8054-12	116	46AX34	116	48-97168A11
116	13-59860-1	116	21M434	116	34-8054-13	116	46AX4	116	48-97168A14
116	13-43766-1	116	21M435	116	34-8054-14	116	46AX5	116	48-97172A01
116	13-55029-1	116	21M436	116	34-8054-15	116	46AX52	116	48-97222A02
116	0013-911	116	21M437	116	34-8054-18	116	46AX54	116	48-97270A01
116	13D4	116	21M469	116	34-8054-2	116	46AX7	116	48A41508A01
116	13J2	116	21M487	116	34-8054-23	116	46AX8	116	48B43265G01
116	13J2P	116	21M519	116	34-8054-27	116	46AX9	116	48C-40235-602
116	13P1	116	21M545	116	34-8054-4	116	46BD1	116	48D40235G01
116	14J2	116	21M590	116	34-8054-5	116	46BD11	116	48D67120A02
116	14J2P	116	022-2823-011	116	34-8054-7	116	46BD12	116	488134921
116	14F2	116	022-3905-001	116	34-8055-3	116	46BD14	116	488134939
116	015-002	116	022.3905-001	116	34-8057-11	116	46BD19	116	488155037
116	015-006	116	022D	116	34-8057-18	116	46BD2	116	488155040
116	15-085006	116	23-0004	116	34-8057-28	116	46BD25	116	488155041
116	15-085007	116	23-0010	116	34-8057-45	116	46BD5	116	488155054
116	15-085040	116	23-0017	116	35-003-001	116	46BD8	116	488155063
116	15-085043	116	23-0018	116	35-1004	116	46BD9	116	488155079
116	15-100001	116	23B8C101	116	35-1005	116	46BR10	116	488155099
116	15-100002	116	24-198	116	38	116	46BR11	116	488155100
116	15-100004	116	24DP1	116	39-02	116	46BR15	116	488155106
116	15-108003	116	24MW1066	116	40A8	116	46BR17	116	488155126
116	15-108004	116	24MW1071	116	40B5	116	46BR5	116	48B191A02
116	15-108005	116	24MW1108	116	40B8	116	46BR7	116	48B191A07
116	15-108006	116	24MW1113	116	40C	116	46BR9	116	48A40235G02
116	15-108010	116	24MW1123	116	40D6665A03	116	46BX2	116	48X97048A10
116	15-108011	116	24MW1144	116	40H	116	46BX3	116	48X97172A01
116	15-108016	116	24MW1146	116	40J2	116	48-01	116	49-1042
116	15-108020	116	24MW1162	116	40JZ	116	48-05-001	116	49-3112
116	15-108021	116	24MW197	116	40K	116	48-10001-A01	116	50A
116	15-108022	116	24MW267	116	40KR	116	48-10001-A030-1	116	50A8
116	15-108036	116	24MW268	116	40M	116	48-1005	116	50B5
116	15-108037	116	24MW269	116	40N1	116	48-10062A01	116	50C
116	15-108049	116	24MW619	116	40NJ	116	48-10062A01A	116	50D2
116	15-108050	116	24MW669	116	40Y3P	116	48-10062A02	116	50D4
116	15-123105	116	24MW671	116	41-J2	116	48-10062A04	116	50E05
116	0015-911	116	24MW721	116	42-14027	116	48-10062A05	116	50E1
116	15B2	116	24MW768	116	42-17443	116	48-10062A05A	116	50E2
116	15J2	116	24MW772	116	42-17443A	116	48-134769	116	50E3
116	15J2P	116	24MW851	116	42-19865	116	48-134790	116	50E4
116	15M1	116	24MW862	116	42-21400	116	48-134958	116	50E5
116	16-2	116	24MW867	116	42-21408	116	48-134959	116	50E6
116	1604	116	24MW871	116	42-21866	116	48-134990	116	50M
116	1605	116	24MW995	116	42-22835	116	48-137029	116	051-0006
116	16J2(DIODE)	116	025-100016	116	42-23350	116	48-137074	116	51-03007-06
116	16J2P	116	025-100028	116	42-23350A	116	48-137143	116	52A011
116	16X10	116	025-100029	116	42-23975	116	48-137198	116	52BBIA
116	17-10	116	025-100055	116	42-27202	116	48-137208	116	52DS-18
116	17-410	116	25-5	116	42-27278	116	48-137291	116	52BD1
116	018-00006	116	26D00505	116	42-28058	116	48-155063	116	53-0051-2
116	018-00007	116	027-000296	116	42-28202	116	48-155126	116	53-0086-1
116	018-00008	116	027-000306	116	42A11	116	48-155136	116	53-0088-1
116	018-00009	116	027-000312	116	42A23	116	48-155193	116	53-0094-1
116	18-085001	116	28-13-01	116	42B16	116	48-155235	116	53-0098-1
116	18-22-17	116	28-14-01	116	42B2	116	48-155236	116	53-0099-1
116	18P2	116	28-15-01	116	42J2	116	48-191807	116	53-0099-3
116	019-002935	116	28-15-02	116	42X244	116	48-191A01	116	53-0099-4
116	019-003420	116	28-19-01	116	42X244B	116	48-191A01-9	116	53-0106-1
116	019-003870-013	116	28-20-01	116	42X245	116	48-191A02	116	53-0109-1
116	019-003870-020	116	28-20-02	116	42X245B	116	48-191A03	116	53-0113-1
116	19-040-002	116	28-21-01	116	42X25	116	48-191A04	116	53-0120-1
116	19-040-003	116	28-22-01	116	42X32	116	48-191A07	116	53-1086
116	19-040-004	116	28-22-02	116	43-540012	116	48-191A08	116	53A001-12
116	19-080-002	116	28-22-03	116	44-590	116	48-191A11	116	53A001-3
116	19-085010	116	28-22-04	116	45A2FX355	116	48-355016	116	53A001-35
116	19-085022	116	28-22-05	116	46-16261	116	48-355023	116	53A001-9
116	19A115024-P4	116	28-22-06	116	46-16261-3	116	48-355025	116	53A002-1
116	19A115100-P1	116	28-22-07	116	46-34-5	116	48-355049	116	53A010-1
116	19A115145-P3	116	28-22-10	116	46-61249-3	116	48-40235G01	116	53A022-4
116	19A115145-P4	116	28-22-12	116	46-67120A13	116	48-40739P01	116	53B001-6
116	19A115569-P2	116	28-22-14	116	46-86-3	116	48-41508A02	116	53B003-1
116	19AR11	116	28-22-15	116	46-8601-3	116	48-43265G01	116	53B010-3
116	19AR12	116	28-22-17	116	46-8611-4			116	53B010-4
116	19AR17	116	28-22-21	116	46-861148-3			116	53B011-2
116	19AR2								

ECG	Industry Standard No.	ECG	Industry Standard No.	ECG	Industry Standard No.	ECG	Industry Standard No.	ECG	Industry Standard No.
116	53B019-2	116	66X0037-001	116	92B12-2	116	103-261-0	116	212-61
116	53B020-2	116	66X0048-001	116	93-302	116	103-261-01	116	212-62
116	53C003-1	116	66X053-001	116	93A12-1	116	103-29	116	212-64
116	53C005-3	116	66X23	116	93A1D-1	116	103-315-03A	116	212-65
116	53C007-1	116	66X24	116	93A3D-2	116	103-76	116	212-70
116	53C009-1	116	66X26	116	93A4-2	116	103-82	116	212-71
116	53C012-1	116	67-1000-00	116	93A42-7	116	103B6	116	212-76-02
116	53C014-1	116	67-1003-00	116	93A51-3	116	10303125	116	212-77
116	53C015-1	116	69-2246	116	93A53-3	116	10303125A	116	212-79
116	53C016-1	116	070-019	116	93A6-1	116	104B6	116	212-92
116	53C017-1	116	70P40	116	93A6-2	116	105B6	116	212-94
116	53C022-1	116	70840	116	93A60-14	116	106	116	212-94B
116	53D001-7	116	70T40	116	93A60-2	116	106-011	116	215-51
116	53D003-1	116	72-111	116	93A60-8	116	106-1	116	215-58
116	53D003-2	116	72-15	116	93A60-9	116	106-111	116	215-76(GE)
116	53E001-1	116	72Z	116	93A67-1	116	106B6	116	220-003001
116	53F001-1	116	75	116	93A78-1	116	107B6	116	229-1054-5
116	53F002-1	116	75D1	116	93A97-2	116	108	116	229-1054-82
116	53H001-1	116	75D2	116	93A97-3	116	108A4	116	229-1054-9
116	53K003-1	116	75B1	116	93A97-37	116	108B4	116	229-5100-232
116	53K001-11	116	75E2	116	93B1-1	116	108B6	116	229-5100-233
116	53K001-6	116	75E3	116	93B1-10	116	108E-E2	116	232-0001
116	53K001-7	116	75E4	116	93B1-11	116	109B6	116	232-0006
116	53K001-8	116	75E5	116	93B1-12	116	0110	116	232-1006
116	53N001-2	116	75E6	116	93B1-13	116	0110-0011	116	232-1009
116	53N001-3	116	75P05	116	93B1-14	116	0110-0141	116	232-1011
116	53N002-2	116	75R1B	116	93B1-15	116	0110-0141/4460	116	0234
116	53N004-7	116	75R2B	116	93B1-16	116	0110-0209	116	0244
116	53N004-8	116	75R3B	116	93B1-17	116	110-629	116	247-255
116	53N004-9	116	75R4B	116	93B1-18	116	110-635	116	247-621
116	53T001-3	116	75R5B	116	93B1-2	116	110-672	116	260-10-033
116	53T001-1	116	75R6B	116	93B1-3	116	110-684	116	264D00505
116	56-4839	116	7601450J	116	93B1-4	116	0111	116	264D01112
116	56-52	116	77-271374-1	116	93B1-5	116	111-1	116	264P00601
116	56-8097	116	078-0016	116	93B1-6	116	111-4-2020-0800	116	264P01011
116	56-8198	116	078-1696	116	93B1-7	116	0112	116	264P01012
116	56-8199	116	078-2400	116	93B1-8	116	112-601-0-102	116	264P02001
116	57-0006	116	78-254566-1	116	93B1-9	116	112-826	116	264P02402
116	57-1	116	78-254566-4	116	93B12-1	116	113-039	116	264P03001
116	57-12	116	78-271145-1	116	93B12-2	116	113-321	116	264P03607
116	57-13	116	78-272160-1	116	93B12-3	116	113-592	116	264P04301
116	57-15	116	78-273008-1	116	93B24-2	116	113A7739	116	264P04402
116	57-17	116	78-273085	116	93B24-3	116	114-013	116	264P04702
116	57-2	116	078-5001	116	93B27-2	116	114-42020-14500	116	264P04703
116	57-20	116	81-27123100-3	116	93B30-1	116	115-059	116	264P04801
116	57-21	116	81-27123300-3	116	93B30-3	116	115-559	116	264P04901
116	57-22	116	81-27123307-3	116	93B41-12	116	115-599	116	264P05001
116	57-23	116	81-46123004-5	116	93B41-14	116	116-052	116	264P05002
116	57-24	116	81-46123004-7	116	93B41-4	116	119-6511	116	264P06601
116	57-27	116	81-46123005	116	93B41-6	116	120-001-300	116	264P06606
116	57-28	116	81-46123011-2	116	93B41-8	116	120-001300	116	264P08002
116	57-31	116	81-46123012-2	116	93B42-1	116	120-003148	116	264P08801
116	57-33	116	81-46123014-6	116	93B42-10	116	120-004503	116	264P09801
116	57-46	116	81-46123018-7	116	93B42-11	116	120-004878	116	264P10103
116	57-58	116	81-46123022-9	116	93B42-12	116	122-80	116	264P10105
116	57-59	116	81-46123043-5	116	93B42-2	116	123-021	116	264P20021
116	57-6	116	82-4	116	93B42-3	116	124-0178	116	288(SEARS)
116	57-60	116	83-829	116	93B42-4	116	124J490	116	291-04
116	57-65	116	83-880	116	93B42-6	116	126-4	116	291-20
116	60C	116	85-5	116	93B42-7	116	126-7(ARVIN)	116	292-10
116	60D	116	86-0006	116	93B42-8	116	130-30192	116	295L001H01
116	60H	116	86-0010	116	93B45-1	116	130-30313	116	295L001M01
116	60J2	116	86-0012	116	93B45-2	116	130-338-00	116	295L001M02
116	60M	116	86-0013	116	93B47-1	116	130-398-99	116	295L003M01
116	61-1320	116	86-0016-01	116	93B51-3	116	130A17268	116	295V022H01
116	61-1765	116	86-0511	116	93C1-20	116	130TB	116	295V008H01
116	61-7728	116	86-0516	116	93C1-21	116	0131-0026	116	295V006H01
116	61J2	116	86-1-3	116	93C16-2	116	0131-0035	116	295V006H02
116	62-13261	116	86-111-3	116	93C19-1	116	0131-0044	116	295V006H03
116	62-13477	116	86-139-3	116	93C24-2	116	0131-053	116	295V006H05
116	62-15483	116	86-147-1	116	93C30-1	116	0131-026	116	295V006H07
116	62-16711	116	86-147-3	116	93C50-3	116	0131-053	116	295V006H09
116	62-18135	116	86-21-1	116	93C40-2	116	135-3	116	295V007H02
116	62-18431	116	86-22-3	116	93C42-7	116	137-718	116	295V008H01
116	62-18434	116	86-28-3	116	93ERH-DX0156//	116	140-013	116	295V012H02
116	62-18435	116	86-3-3	116	93L101-2	116	0140-8	116	295V012H03
116	62-18436	116	86-30-3	116	93L102-2	116	0140-9	116	295V012H06
116	62-18438	116	86-32-1	116	93L103-4	116	142-013	116	295V014H01
116	62-19115	116	86-34-3	116	93L107-2	116	142-014	116	295V014H04
116	62-19749	116	86-35-3	116	93L5-1	116	145A9254	116	295V015H02
116	62-19814	116	86-4-1	116	93L5-4	116	145A9786	116	295V016H01
116	62-21369	116	86-40-3	116	93L5-6	116	150	116	295V017H01
116	62A01	116	86-42-3	116	93L5-7	116	151-011-9-011	116	295V020B01
116	62A02	116	86-46-3	116	094-010	116	151-018-9-001	116	295V020H01
116	62A04	116	86-49-3	116	94-1066-1	116	151-023-9-001	116	295V028H01
116	62A05	116	86-50-3	116	96-5022-01	116	151-029-9-003	116	295V028B01
116	62J2	116	86-5000-3	116	96-5022-1	116	151-030-9-005	116	295V027001-1
116	63-10064	116	86-5001-3	116	96-5023-01	116	151-040-7-003	116	295V028001
116	63-10709	116	86-5002-3	116	96-5046-01	116	151-040-9-003	116	295V028002
116	63-11291	116	86-5005-3	116	96-5082-01	116	151-045-9-001	116	295V028002
116	63-11881	116	86-5006-3	116	96-5088-01	116	151-046-9-001	116	295V029001
116	63-11957	116	86-5009-3	116	96-5096-01	116	151-1	116	295V029002
116	63-13903	116	86-5012-3	116	96-5103-01	116	152-047	116	295V035001
116	63-13919	116	86-5024-3	116	96-5109-01	116	154A3992	116	296V020H03
116	63-14195	116	86-5029-3	116	96-5109-02	116	162-005A	116	296V020B02
116	63-15483	116	86-51-3	116	96-5113-01	116	162J2	116	297V027001
116	63-26597	116	86-52-1	116	96-5196-01	116	163J2	116	0300
116	63-27622	116	86-54-3	116	96-5333-01	116	164J2	116	0304
116	63-7433	116	86-57-3	116	98-301	116	165J2	116	307B
116	63-8685	116	86-58-3	116	98A12518	116	16J2	116	307C
116	63-8824	116	86-59-1	116	0100	116	173A3981	116	307D
116	63J2	116	86-59-3	116	100-10121-05	116	173A3981-1	116	307E
116	64-J2	116	86-5911-3	116	100-10132-00	116	173A4393	116	307F
116	65(DIODE)	116	86-62-3	116	100-138	116	174-1	116	307K
116	065-015	116	86-63-3	116	100-15	116	174-2	116	307M
116	65-085013	116	86-65-3	116	100-162	116	174-3	116	309-327-910
116	65J2	116	86-65-9	116	100-217	116	180-1	116	309-327-927
116	65P117	116	86-67-0	116	100-438	116	185-014	116	309-327-932
116	65P124	116	86-67-9	116	100-520	116	185-6(RCA)	116	310
116	65P124-1	116	86-7-1	116	100-527	116	186-014	116	310-4
116	65P124-2	116	86-75-3(SEARS)	116	100A	116	246-654	116	0311
116	65P133	116	86-80-1	116	100O-4R	116	2013120-255	116	0312
116	65P155	116	86-80-3	116	100R1B	116	203	116	0314
116	65P206	116	86-84-1	116	100R2B	116	212-18	116	0320
116	65P284	116	86-85-3	116	100R3B	116	212-192	116	320A
116	65P297	116	86-89-1	116	100R4B	116	212-21	116	320B
116	66-2246	116	86-89-3	116	100R5B	116	212-23	116	320C
116	66-6030-00	116	86-9-3	116	100R6B	116	212-25	116	320D
116	66-6031-00	116	87-67-3	116	101B6	116	212-254	116	320F
116	66P001	116	88-125	116	0102	116	212-27	116	320H
116	66P001-1	116	88-3	116	102B6	116	212-33	116	320K
116	66J2	116	88-831	116	102D	116	212-35	116	320M
116	66X0023-001	116	88-832	116	103-145-01	116	212-36	116	0321
116	66X0023-002	116	88-833	116	103-191	116	212-37	116	0322
116	66X0023-003	116	91A01	116	103-203	116	212-39	116	322-0147
116	66X0023-004	116	91A01B	116	103-216	116	212-40	116	324-0117
116	66X0023-005	116	91A02	116	103-228	116	212-41	116	324-0119
116	66X0023-006	116	91A03	116	103-245	116	212-42	116	324-0120
116	66X0023-007	116	91A05	116	103-254	116	212-47	116	324-0135
116	66X0023-008	116	91A06	116	103-254-01	116	212-48	116	324-0147
116	66X0023-1	116	91A11	116	103-261	116	212-49	116	324-0162
116	66X0028-001	116	92A11-1	116		116	212-50	116	325-0028-89
116	66X0033-000	116		116		116	212-51	116	
116	66X0033-001	116		116		116	212-58	116	

ECG	Industry Standard No.	ECG	Industry Standard No.	ECG	Industry Standard No.	ECG	Industry Standard No.	ECG	Industry Standard No.
116	325-0031-338	116	538J2F	116	919-007394	116	2061A45-72	116	5301-06-1
116	325-0047-517	116	539J2F	116	919-009459RA	116	2061B45-35	116	5301-09-1
116	325-0054-312	116	540-010	116	919-01-0459	116	2093A12-1	116	5301-13-1
116	325-0076-315	116	540J2F	116	919-01-0623	116	2093A38-13	116	5301-20-1
116	325-0081-110	116	0557-010	116	919-01-1211	116	2093A38-28	116	5380-21
116	325-0135-B	116	575-02B	116	919-01-1212	116	2093A41-103	116	5416
116	325-0141-23	116	575-042	116	919-011212	116	2093A41-104	116	5601-MS
116	325-0670-16	116	575-04B	116	919-013044	116	2093A41-108	116	5601-NI
116	325-1441-10	116	575-050	116	919-013072	116	2093A41-110	116	5632-81B01-02
116	325-1441-11	116	580-029	116	936-10	116	2093A41-115	116	5632-WO3B
116	325-1446-29	116	600	116	936-20	116	2093A41-116	116	5632-WO6A
116	354-9101-006	116	600X0099-066	116	964-17443	116	2093A41-126	116	5632-WO6B
116	354-9102-001	116	601	116	964-174443	116	2093A41-131	116	5641-MV11
116	354-9110-001	116	601X0048-066	116	964-19865	116	2093A41-139	116	6171-28
116	359A	116	601X0152-066	116	964-21866	116	2093A41-151	116	7211-9
116	359B	116	601X0224-066	116	972D7	116	2093A41-152	116	7212-3
116	359C	116	601X0227-066	116	977-10B	116	2093A41-173	116	7212-3A
116	359D	116	602X0019-000	116	977-11B	116	2093A41-182	116	7213(DIODE)
116	359F	116	604B	116	977-13B	116	2093A41-185	116	7213-7
116	359H	116	606-113	116	977-18B	116	2093A41-189	116	7314
116	359K	116	607-113	116	977-19	116	2093A41-197	116	7568
116	359M	116	608	116	977-19B	116	2093A41-45	116	7962
116	369-2	116	608-030	116	977-20B	116	2093A41-49	116	8000-00003-044
116	369-3	116	608-101	116	977-22B	116	2093A41-51	116	8000-00004-044
116	380H61	116	608-113	116	977-23B	116	2093A41-54	116	8000-00004-064
116	380K62	116	609-030	116	977-25B	116	2093A41-55	116	8000-00004-068
116	380M63	116	609-113	116	977-27B	116	2093A41-57	116	8000-00005-022
116	384A	116	610	116	977-28B	116	2093A41-62	116	8000-00005-152
116	384B	116	610-001-103	116	977-2B	116	2093A41-66	116	8000-00006-201
116	384C	116	610C	116	977-8B	116	2093A41-75	116	8000-00006-231
116	384D	116	612	116	991-00-7776	116	2093A41-78	116	8000-00011-041
116	384F	116	614	116	992-17	116	2093A41-81	116	8000-00011-044
116	384K	116	616	116	1000-129	116	2093A41-86	116	8000-00011-104
116	385A	116	617-162	116	1001-11	116	2093A41-88	116	8000-00030-010
116	385B	116	617-163	116	1002-0219	116	2093A41-89	116	8000-00035-001
116	385C	116	617-46	116	1003-13	116	2093A42-7	116	8000-00035-002
116	385G	116	617-53	116	1005-20	116	2093B38-14	116	8000-0004-P067
116	385P	116	618	116	1006-24	116	2093B4-6	116	8000-0004-P068
116	385H	116	620	116	1007-0951	116	2093B41-10	116	8000-00049-010
116	385K	116	622	116	1009-09	116	2093B41-12	116	8000-00005-022
116	385M	116	624	116	1010(JULIETTE)	116	2093B41-18	116	8003-115
116	386-1AY	116	624-0009	116	1010-9486	116	2093B41-20	116	8020-203
116	386-1CY	116	624-0010	116	1010-9494	116	2093B41-21	116	08050
116	386-1FY	116	629A02	116	1011(VO-6C)	116	2093B41-22	116	8300-8
116	386AK	116	630-052	116	1016-17	116	2093B41-28	116	8500-206
116	386AW	116	642-011	116	1016-78	116	2093B41-34	116	8710-55
116	386AX	116	650-110	116	1016-79	116	2093B41-35	116	8800-201
116	386AY	116	656-141	116	1018-6963	116	2093B41-37	116	8800-206
116	386BW	116	660-230	116	1020	116	2093B41-6	116	8840-56
116	386BY	116	684-652423-1	116	1022-5548	116	2093B41-8	116	8864-1
116	386CW	116	690V031H53	116	1024	116	2093B41-9	116	8910-55
116	386CX	116	690V039H52	116	1033-17	116	2102	116	8999-201
116	386CY	116	690V041H08	116	1033-8	116	2102-014	116	8999-205
116	386DW	116	690V053H57	116	1038-1697	116	2102-017	116	9100-1
116	386DY	116	690V069H39	116	1040	116	2102-074	116	9300-5
116	386FW	116	690V080H47	116	1040-10	116	2110N-42	116	9300-6
116	386FX	116	690V080H52	116	1040-81	116	2116-17	116	9300-7
116	386FY	116	690V080H91	116	1040-932	116	2148-17	116	9301-1
116	386K	116	690V086H91	116	1041-109	116	2152	116	9302-3
116	386KX	116	690V089H91	116	1041-63	116	2182-17	116	9330-006-11112
116	386KY	116	690V098H53	116	1041-64	116	2198-17	116	9330-229-20112
116	386MW	116	690V109H44	116	1042-17	116	2199-17	116	9348-5
116	386MY	116	690V114H36	116	1043-7382	116	2200-17	116	9650-001
116	0400	116	690V116H41	116	1044	116	2206	116	10180
116	4000-11958	116	690V118H58	116	1045-0518	116	2206-17	116	10909
116	0401	116	690V119H14	116	1045-0534	116	2211B	116	11282-6
116	0402	116	0700	116	1048-9946	116	2233-17	116	11303-9
116	0411	116	700-137	116	1048-9995	116	2283-17	116	11332-1
116	421-4027	116	700A858-285	116	1049-3435	116	2285-17	116	11339-2
116	421-7443	116	700A858-286	116	1050	116	2302	116	11401-3
116	421-7443A	116	0701	116	1050-64	116	2315-046	116	11503-9
116	421-9865	116	0702	116	1059-7961	116	2319-17	116	11555-9
116	422-1362	116	0704	116	1060	116	2400-23	116	11559-9
116	422-1866	116	710	116	1061-8379	116	2400-27	116	11605-2
116	422-2540	116	0727-50	116	1061-8916	116	2402	116	11746
116	429-0989-68	116	750-141	116	1061-8924	116	2402-461	116	12602
116	430-31	116	750A858-285	116	1063-79	116	2402-462	116	12720
116	435-40012	116	750M63-149	116	1065-9928	116	2402-463	116	12736
116	0450	116	751-2001-212	116	1074-119	116	2405-459	116	12746
116	454-A2534-1	116	751-5300-001	116	1077-0261	116	2405-462	116	12768
116	454-A2534-10	116	751-9001-124	116	1077-2366	116	2405-463	116	12786
116	454425	116	754-2000-002	116	1077-3836	116	2410-17	116	12788
116	0460	116	754-2000-005	116	1080-08	116	2411	116	12803
116	460-1013	116	754-5000-021	116	1095J2F	116	2510-32	116	12837
116	464-285-15	116	754-5750-284	116	1096J2F	116	2522	116	12844
116	471-010	116	754-900-124	116	1103-88	116	2523	116	13782
116	474-004	116	754-9000-953	116	1104-96	116	2606-299	116	14027
116	474-025	116	754-9001-124	116	1106-29	116	2704-389	116	14126-1
116	486-40235-002	116	756	116	1106-36	116	2762	116	14588-9
116	511-898	116	792-238	116	1108-73	116	2763	116	16681
116	515-299	116	800-006-00	116	1110-86	116	2766	116	17002
116	517-0021	116	800-013-00	116	1115-16	116	2777	116	17443
116	517-0025	116	800-014-00	116	1116-42	116	2779	116	19042
116	517-0031	116	800-014-01	116	1117-76	116	2784	116	19865
116	517-0033	116	800-024-00	116	1118-20	116	2786	116	20810-21
116	518-499	116	800-032-00	116	1120-18	116	2794	116	20810-22
116	521-094	116	800-038-00	116	1121-17	116	3001A	116	021154
116	522-726	116	800-041-00	116	1123(JULIETTE)	116	3003007	116	24198(RECT.)
116	523-0001-001	116	808-206	116	1138	116	5069(ARVIN)	116	025026
116	523-0001-002	116	822-1	116	1139-17	116	3074A20	116	025056
116	523-0001-003	116	822-2	116	1263A	116	3074A21	116	25115-115
116	523-0001-004	116	822-3	116	01339	116	3377	116	25202-002
116	523-0001-005	116	822-4	116	1405-17	116	3529	116	25810-51
116	523-0001-006	116	822-5	116	1410-167	116	3700-153	116	25840-51
116	523-0013-002	116	851-0372-130	116	1412-170	116	4013(RECTIFIER)	116	25840-52
116	523-0501-002	116	854-0372-020	116	1488(SEARS)	116	4014	116	26235-1
116	523-0501-003	116	880-207-000	116	1562	116	4041-000-10150	116	27113-100
116	523-1000-001	116	900-00394	116	1611-17	116	04049B	116	27123-050
116	523-1000-881	116	903-104B	116	1674-17	116	4051-300-10150	116	27123-070
116	523-1000-882	116	903-105B	116	1713-17	116	4101-685	116	27123-100
116	523-1000882	116	903-117B	116	1848-17	116	4403	116	27123-120
116	523-1500-002	116	903-14B	116	1849	116	004567	116	27840-41
116	523-1500-881	116	903-156B	116	1854-17	116	4686-149-3	116	28810-461
116	525-24	116	903-15B	116	1855-17	116	4802-2000-012	116	031034
116	525-498	116	903-164B	116	1859R	116	4802-2000-019	116	34174
116	526-376	116	903-169B	116	1885-17(RECT.)	116	4806-0000-004	116	34394
116	530-086-1	116	903-197	116	1941-17	116	4822-130-30192	116	35287
116	530-088-1	116	903-28B	116	1949-17	116	4822-130-30256	116	35287A
116	530-094-1	116	903-303	116	1950-17	116	4822-130-30259	116	35289
116	530-095-1	116	903-330	116	1970-16	116	4822-130-50221	116	35306
116	530-098-1	116	903-49B	116	1970-17	116	5001-117	116	35500
116	530-099-1	116	903-52B	116	1982-17	116	5001-129	116	36201
116	530-099-3	116	903-69B	116	2000-304	116	5001-135	116	36549
116	530-099-4	116	903-79B	116	2000-309	116	5001-163	116	36554
116	530-106-1	116	919-00-2440-1	116	2000-320	116	5051-300-10150	116	36555
116	530-109-1	116	919-00-2440-2	116	2010-02	116	5205NI	116	36564
116	530-111-1	116	919-00-7766	116	2017-114	116	5300-86-1	116	37126-1
116	530-113-1	116	919-00-7776	116	2021-05	116	5300-88-1	116	37680
116	530-120-1	116	919-004799	116	2022-0B	116	5300-94-1	116	37987
116	532-341	116	919-007109	116	2027	116	5300-95-1	116	38174
116	532-341A	116	919-007109RA	116	2031-17	116	5300-98-1	116	40024
116	536J2F			116	2045-17	116	5300-99-1	116	40024VM
116	537J2F			116	2052	116	5300-99-3	116	40143
				116	2060-024	116	5300-99-4	116	40265

ECG	Industry Standard No.	ECG	Industry Standard No.	ECG	Industry Standard No.	ECG	Industry Standard No.	ECG	Industry Standard No.
116	40383	116	110636	116	218612	116	530082-1003	116	922567
116	41001-4	116	110873	116	219935	116	530082-1004	116	922799
116	41616	116	111086	116	221128	116	530084-4	116	922860
116	42221	116	111516	116	222611	116	530086-1	116	922969
116	440088(RCA)	116	111642	116	223215	116	530088-1003	116	924605-3
116	44465-3	116	111776	116	223216	116	530088-1004	116	924801-5
116	44465-4	116	111820	116	223358	116	530088-4	116	924805-5
116	44465-5	116	112017	116	223462	116	530094-1	116	924805-8
116	44465-6	116	112018	116	223467	116	530095-1	116	945820-4
116	44937	116	112329	116	223724	116	530098-1	116	972216
116	44938	116	112531	116	223753	116	530098-1001	116	972571-4
116	45810-51	116	112826	116	224159	116	530099-1	116	973935-20
116	46140-2	116	113039	116	224597	116	530099-3	116	973936-1
116	47126	116	113392	116	224774	116	530099-4	116	973936-10
116	47126-003	116	113398	116	225200	116	530109-1	116	973936-11
116	47126-12	116	113998	116	225592	116	530111-1	116	973936-12
116	47126-1	116	115039	116	226058	116	530111-1001	116	973936-13
116	47126-1A	116	115559	116	226182	116	530116-1003	116	973936-14
116	47126-2	116	115559A	116	226237	116	530124-1	116	973936-15
116	47126-3	116	115599	116	226546	116	530126-1	116	973936-16
116	47126-3A	116	116052	116	226788	116	530127-4	116	973936-17
116	47126-4	116	116054	116	227348	116	530127-6	116	973936-18
116	47126-4A	116	117145	116	227565	116	530135-1003	116	973936-2
116	47126-6	116	117145A	116	227720	116	530135-3	116	973936-3
116	47126-7	116	118825	116	229522	116	530162-1	116	973936-4
116	47126-9	116	118873	116	230218	116	530171-1	116	973936-5
116	47127-3	116	120503	116	231339	116	530171-1001	116	973936-6
116	50745-01	116	120544	116	231665	116	530171-1002	116	973936-7
116	50745-02	116	121468	116	231669	116	530171-1003	116	973936-8
116	50745-06	116	121680	116	231923	116	530171-3	116	973962-1
116	50745-07	116	122129	116	232203	116	530180-1001	116	980143
116	50745-08	116	122788	116	233011	116	533038	116	980164
116	53099-1	116	123004	116	233561	116	535151-1001	116	980540
116	53099-3	116	123296	116	235597	116	0557640	116	980964
116	53300-41	116	123804-1	116	234552	116	551026	116	981445
116	055210H	116	124098	116	234565	116	0552006	116	981739
116	58810-80	116	124812	116	234611	116	0552006H	116	981952
116	58810-85	116	125105	116	234761	116	0552010	116	981953
116	58840-111	116	125127	116	235313	116	0552010H	116	981954
116	58840-116	116	125458	116	235543	116	575028	116	981955
116	59557-49	116	125471	116	235546	116	575042	116	981956
116	59844-1	116	125549	116	237227	116	575051	116	982214
116	59844-3	116	125835	116	237453	116	0576054	116	982253
116	6101-2	116	125964	116	237929	116	580029	116	982577
116	66682-42	116	125993	116	239221	116	603114	116	982826
116	68177	116	126131	116	240456	116	606113	116	982823
116	68177A	116	126148	116	240594	116	608030	116	983101
116	68504-78	116	126176	116	240603	116	608101	116	983413
116	68643-03	116	126320	116	242029	116	608113	116	984182
116	69111	116	126855	116	242226	116	609030	116	984183
116	69213-78	116	126861	116	243364	116	609113	116	984184
116	70064-7	116	127102	116	248917	116	610020	116	984189
116	70066-3	116	127176	116	249217	116	610112	116	984254
116	70064-4	116	127396	116	252817	116	611111	116	984522
116	70270-23	116	127695	116	258882	116	612112	116	984594
116	70372-2	116	127992	116	259315	116	612130	116	984713
116	70432-16	116	128256	116	259368	116	612132	116	984882
116	71006-28	116	129029	116	259878	116	613020	116	985105
116	71006-30	116	129213	116	260429	116	613130	116	985218
116	71588-1	116	129334	116	261463	116	614010	116	985472
116	71588-2	116	129759	116	261596	116	619130	116	986414
116	71588-5	116	130338-01	116	261898	116	651030	116	986935
116	71588-6	116	130389-00	116	261975	116	651038	116	988051
116	71779	116	130607	116	262112	116	652092	116	992171
116	000072020	116	131245	116	262872	116	652615	116	1043176-1
116	72041	116	131501	116	265164	116	700043-00	116	1043176-2
116	000072050	116	131950	116	265235	116	700063-00	116	1043176-3
116	72058	116	132149	116	266534	116	700647	116	1043176-4
116	72058A	116	132814	116	276097	116	700663	116	1043176-5
116	72097	116	133266	116	280217	116	700664	116	1045013
116	72101	116	133950	116	300233	116	740183	116	1045154-1
116	72103	116	135281	116	300315	116	740289	116	1045154-2
116	72109	116	135284	116	300524	116	740570	116	1045154-3
116	72113	116	136635	116	300550	116	740630	116	1223772
116	72113A	116	137652	116	301586	116	741101	116	1223928
116	72119	116	138172	116	330003	116	741740	116	1471872-6
116	72123	116	138173	116	330018	116	741865	116	1472171-32
116	72123A	116	139605	116	330019	116	742008	116	1472460-16
116	72130-1	116	141849	116	348048-2	116	742009	116	1474778-10
116	72145	116	141872-6	116	348053-3	116	746004	116	1474778-11
116	72158	116	144052	116	348054-1	116	752309	116	1474778-2
116	72171-1	116	144860	116	348054-10	116	760202-0003	116	1474778-3
116	72176	116	145839	116	348054-2	116	761113	116	1474778-7
116	74200-8	116	161029	116	348054-5	116	771907	116	1474788-10
116	74200-9	116	161030	116	348054-9	116	772213	116	1476171-28
116	75230-9	116	161031	116	348055-2	116	801707	116	1476171-29
116	75700-14-02	116	161037	116	348055-3	116	801711	116	1476171-31
116	75702-15-21	116	165739	116	464010	116	801714	116	1800006
116	77190-7	116	166593	116	464070	116	801716	116	1800018
116	77271-3	116	166726	116	489752-005	116	815138	116	1956486
116	78894	116	166881	116	489752-013	116	815142	116	1961843
116	080040	116	166920	116	489752-015	116	817042	116	1962594
116	080050	116	166922	116	489752-022	116	817043	116	1966808
116	81513-8	116	167034	116	489752-025	116	817044	116	1967813
116	81514-2	116	167413	116	489752-027	116	817053	116	1969497
116	83008	116	167543	116	489752-028	116	817064	116	2000625-36
116	88641	116	167544	116	489752-035	116	817066	116	2000626-32
116	90203-6	116	168339	116	489752-043	116	817067	116	2000646-115
116	91001	116	169113	116	489752-050	116	817068	116	2000646-119
116	93005	116	169116	116	489752-051	116	817068P	116	2000646-120
116	93006	116	169363	116	489752-073	116	817079	116	2000648-120
116	93007	116	169565	116	489752-075	116	817088	116	2001786-141
116	93011	116	169590	116	489752-092	116	817104	116	2002332-57
116	93012	116	169765	116	489752-094	116	817109	116	2002332-58
116	93022	116	170193	116	489752-097	116	817111	116	2002536-115
116	93022A	116	170297	116	489752-108	116	817112	116	2002402-29
116	93023	116	170750	116	500859	116	817114	116	2003069
116	93028	116	170856	116	501010	116	817117	116	2003073-67
116	95015	116	170965	116	501152	116	817121	116	2003073-68
116	99023-3	116	170970-1	116	0510006	116	817122	116	2004358-142
116	99203-005	116	171019	116	0517550-3	116	817128	116	2006422-133
116	99203-006	116	171149-024	116	0517750	116	817130	116	2006436-38
116	0099203-007	116	171162-252	116	0517750H	116	817133	116	2006441-91
116	99203-5	116	171416	116	530051-2	116	817134	116	2006463-89
116	99203-6	116	171561	116	530051-1	116	817135	116	2006512-40
116	99203-6(RCA)	116	171657	116	530063-11	116	817138	116	2006582
116	99203-9	116	172551	116	530063-12	116	817140	116	2006623-49
116	100412	116	185704	116	530063-6	116	817141	116	2008302-41
116	100520	116	195648-6	116	530065-7	116	817143	116	2010957-49
116	100617	116	196184-3	116	530071-2	116	817148	116	2013019-117
116	100624	116	196259-4	116	530071-3	116	817156	116	2057062-0702
116	101403	116	200062-5-36	116	530072-1	116	817157	116	2060041
116	103318	116	200062-6-32	116	530072-1017	116	817161	116	2327031
116	104081	116	200064-6-115	116	530072-1019	116	817164	116	2327041
116	104213	116	200064-6-119	116	530072-11	116	817166	116	2330020
116	104273	116	200064-6-120	116	530072-14	116	817167	116	2330251H
116	104325	116	202463	116	530072-15	116	817180	116	2330256
116	104615	116	214105	116	530072-2	116	817195	116	2330362
116	107540	116	215669	116	530072-4	116	852158-7.1	116	2330553
116	110033	116	216014	116	530072-5	116	921608	116	2330561
116	110388	116	216020	116	530072-6	116	922092	116	2330512
116	110496	116	216449	116	530072-9	116	922094	116	2330721
116	110629	116	216817	116	530082-1	116	922183	116	2331142
						116	922311	116	2332141

ECG	Industry Standard No.
116	2337071
116	2347021
116	2486836
116	2498513
116	3430063
116	4037325-P1
116	4101685
116	5300113-1
116	5301710-001
116	5330031
116	5330041
116	5330041H
116	5330042
116	5330101
116	5330101H
116	5330102
116	5330102H
116	5330336
116	5330341
116	5330371
116	5330381
116	5330431
116	5340021
116	5340022
116	5490415-P4
116	5490459-P1
116	5490459-P2
116	5490459-P3
116	5490459-P4
116	5490804-P1
116	5490804-P2
116	5490804-P3
116	5490804-P4
116	5490804-P5
116	5490804-P6
116	5494922-P1
116	5494922-P2
116	5494922-P3
116	5494922-P4
116	5494922-P5
116	5494922-P6
116	5494922-P7
116	5494922-P8
116	5955749
116	6200039
116	6846503
116	7023416-00
116	7027002
116	7027005
116	7027023
116	7027031
116	7027038
116	7027039
116	7043216
116	7043261
116	7070692
116	7100628
116	7100630
116	7200953
116	7279497
116	7299325
116	7570014-05
116	7570015-34
116	7570024-02
116	7570215-24
116	7570215-34
116	7851441-01
116	7851657-01
116	7852264-01
116	7852907-01
116	7853358-01
116	7855284-01
116	7910076-01
116	7910805-01
116	7910872-01
116	7910873-01
116	7914011-01
116	8120037
116	8120067
116	8120133
116	8521587-1
116	8981399-1
116	8981399-2
116	9246053
116	9246055
116	9248015
116	9248055
116	9248058
116	9739636-20
116	11210024/7825
116	13030192
116	13030313
116	13033801
116	13038900
116	17210010
116	19080002
116	20001786-141
116	20001786-142
116	22115183
116	22115405
116	23115042
116	23115046
116	23115085
116	23115140
116	23115142
116	23115185
116	23115199
116	23115250
116	23115263
116	23115273
116	23115294
116	23115300
116	23115337
116	25115108
116	25115115
116	26010033
116	27113100
116	27123050
116	27123070
116	27123100
116	27123100A
116	27123120
116	27123220
116	30600040
116	32510001
116	37568800
116	41013566
116	41020618
116	41021007
116	41023224
116	41023225
116	41025849
116	41027063
116	41029499
116	41520419
116	43540012
116	43600101
116	62047119
116	62052791
116	62085053
116	62087447
116	62103264
116	62103272
116	62140844
116	62256877
116	62360755
116	62522674
116	62522739
116	62562226
116	62727772
116	62744162
116	62749067
116	62749558
116	62760087
116	62761334
116	0063050118
116	70270050
116	70270390
116	70270720
116	70370050
116	76276722
116	80001300
116	80001400
116	80002400
116	87096010
116	87219410
116	87400020
116	87500720
116	90270060
116	91190032
116	120001300
116	134804204-101
116	134804596-301
116	440008401
116	602000002
116	1348042041
116	1348045963
116	1523270105
116	2008100027
116	2008100130
116	2008130084
116	2009120101
116	2009120187
116	2012100126
116	2013120208
116	4120910010
116	4129104002
116	4202003173
116	4202006800
116	4202007600
116	4202007802
116	4202008000
116	4202008500
116	4202014500
116	4202017600
116	4202018500
116	4202021000
116	4202023100
116	4202023400
116	4202108270
116	4202109370
116	6611001700
116	6611001800
116	6611007300
116	13488240306
116	13488240308
116	16501090016
116	16501190016
116	16505090005
116	16629291210
116	134804204101
116	134804596301
116	134882095301
116	134882095302
116	134882095303
116	134882095304
116	134882466801
116	134882466802
116	134882466803
116	134882466804
116	134882466806
116	134882466807
116	404130010150
116	505130010150
116	933000611112
116	20225230001701
116(2)	B-1881U
116(2)	B-1882U
116(2)	BD-1A
116(2)	CD2
116(2)	CR501/6515
116(2)	EW169B
116(2)	INJ61227
116(2)	JC-V03C
116(2)	K8532799
116(2)	KC2DP12/1
116(2)	M702C
116(2)	RQ-409S
116(2)	RT6729
116(2)	RVDC08P1
116(2)	S1B02-0
116(2)	S1B02-CR
116(2)	S2VC
116(2)	S5S
116(2)	SN0303
116(2)	TV24278
116(2)	TVS1S1950
116(2)	WR013
116(2)	1B2C1
116(2)	1C05
116(2)	1C1
116(2)	1C2
116(2)	1C4
116(2)	1D1
116(2)	1D2
116(2)	1D6
116(2)	00001S1849
116(2)	00001S1849R
116(2)	1S1850
116(2)	1S1850R
116(2)	1S849R
116(2)	1T2C1
116(2)	1T2Z1
116(2)	2AA113
116(2)	10B-2-N1W
116(2)	10DC-1
116(2)	10DC-2B
116(2)	10DC-2F
116(2)	24MW208
116(2)	24MW227
116(2)	24MW241
116(2)	46-86264-3
116(2)	46-86303-3
116(2)	53B001-5
116(2)	53B003-2
116(2)	81-46123034-4
116(2)	81-46123035-1
116(2)	96-5119-01
116(2)	96-5121-01
116(2)	96-5166-01
116(2)	617-62
116(2)	617-66
116(2)	1849R
116(2)	1850
116(2)	1850R
116(2)	2408-331
116(2)	2703-390
116(2)	140693
116(2)	140694
116(2)	171305
116(2)	699739
116(2)	2008292-89
116(2)	5330372
116(2)	7852439-01
116(2)	13050221
116(2)	23115019
116(2)	27123290
116(2)	202530001710
116(3)	H891
116(3)	KC-1.3C3X11/1
116(3)	KC0.8CP11/H12/1
116(3)	KC08CP111
116(3)	KC08CP121
116(3)	KC2DP11/1412/1
116(3)	RF32412-3
116(3)	T-E1088
116(3)	TVM511
116(3)	15-108015
116(3)	21A001-000
116(3)	33-0031
116(3)	34-8055-2
116(3)	46-8619
116(3)	46-86270-3
116(3)	46-86290-3
116(3)	48-66653A001
116(3)	48-66653A002
116(3)	53B011-1
116(3)	57-38
116(3)	93B1-20
116(3)	93B1-21
116(3)	93B2-1
116(3)	93B53-1
116(3)	229-5100-234
116(3)	113321
116(3)	199985
116(3)	282601
116(3)	489752-066
116(3)	530122-1
116(3)	2001786-207
116(3)	23115052
116(4)	AFS-160-1021
116(4)	B60C300
116(4)	DDAY042001
116(4)	JB16C4
116(4)	KC-08C221
116(4)	KC2DP22/1
116(4)	M9235
116(4)	MRB-20C
116(4)	RCC7225
116(4)	RH-DX0066TAZZ
116(4)	RH-DX0073TAZZ
116(4)	S10149
116(4)	S2VB
116(4)	SID51C169
116(4)	TVM-PH9D522/11
116(4)	1S2371A
116(4)	5B-1
116(4)	10DB
116(4)	12CZP-114
116(4)	16C-4
116(4)	28-18-01
116(4)	41-032
116(4)	75W-005
116(4)	93A104-1
116(4)	212-72
116(4)	212-76
116(4)	800-017-00
116(4)	916-33003-2
116(4)	1002-6219
116(4)	118244
116(4)	120231
116(4)	123702
116(4)	530124-3
116(4)	2006582-23
116(4)	2006582-24
116(4)	2206582-22
116(5)	114013
1160	GEIC-196
1160	GEIC-197
1160	TM1160
1161	AN239
1161	AN239Q
1161	AN239QA
1161	AN239QB
1161	GEIC-300
1162	AN241
1162	AN241D
1162	AN241P
1162	AN241PD
1162	51-13753A04
1162	51M33801A01
1163	AN343
1164	AN245
1164	AN246
1164	GEIC-301
1164	51-13753A10
1164	51S13753A02
1164	51S13753A10
1164	1052-6408
1165	BA511A
1166	BA521
1166	BA521A
1166	QQ-MBA521AX
1167	MC145109
1167	PLL02
1167	PLL02A
1167	REN1167
1167	TC9100P
1167	TM1167
1167	02-360002
1168	AN331
1168	GEIC-66
1168	51S13753A08
1169	HA1139A
1169	HA1339
1169	HA1339A
117	BCF5
117	BD3A-184
117	BF26235-5
117	C934(RECTIFIER)
117	DS89
117	EER600-2
117	ER102
117	ET16X16
117	ITT73
117	J241212
117	J24919
117	JB-00036
117	JC-0025
117	MR2262
117	NA62
117	NA63
117	NA65
117	NA66
117	P7776
117	R-7096
117	R2460-1
117	R2460-4
117	R8721
117	RE50
117	RE504
117	RPS61436
117	RS1234
117	RS1823
117	RT5216
117	RT8199
117	RT8839
117	SC6
117	SE-05-2
117	SK3017
117	SR16
117	SR1A-12
117	SR390-2
117	SR3AM-8
117	SR50517
117	SVDMA26
117	T21638
117	T30155
117	TVSSA-2H
117	TVSSB-2T
117	V148
117	V171
117	V78
117	Y0D4
117	001-0081-00
117	1EA10A
117	1EA20A
117	1EA30A
117	1EA40A
117	1EA50A
117	1EA60A
117	1N1222B
117	1N359
117	1N5198
117	1N5412
117	1S1472
117	1S1665
117	1S1692
117	1S1694
117	1S208/2SJ2A
117	1S209/2SJ4A
117	1S311
117	1S54
117	1S58
117	1S963
117	4-686116-3
117	4-686147-3
117	4-686148-3
117	4-686150-3
117	4-686151-3
117	4-686177-3
117	4-68687-3
117	5-30088.2
117	5-30111.1
117	5-30122.1
117	13-88302
117	24MW175
117	24MW196
117	24MW974
117	24MW975
117	32-0048
117	40-0502
117	46-6661-2
117	46-86116-3
117	46-86150-3
117	46-86151-3
117	46-86807-3
117	48-90234A02
117	53-0082-1003
117	53-0088-2
117	53-0111-1
117	53-0122-1
117	53J002-2
117	53K001-9
117	56-33
117	56-78
117	57-25
117	57-26
117	57-29
117	57-43
117	57-49
117	57A1-63
117	57D1-63
117	61-756
117	61-820
117	61-926
117	63-11215
117	63-12077
117	63-18135
117	63-27483
117	63-29383
117	63-8819
117	63-9787
117	77-270993-1
117	78-271030-1
117	089-252
117	93A15-1
117	93A2
117	93B20-1
117	93B20-2
117	93B20-3
117	93B41-20
117	93L3-1
117	93L3-2
117	93L5-2
117	93M8-1
117	98-302
117	113-998
117	115-867
117	135P4
117	151-025-9-001
117	200-6582-22
117	295L002M03
117	295V006H06
117	295V006H08
117	324-0102
117	464-280-19
117	464-285-19
117	527-798
117	528-325
117	530-082-1003
117	530-088-2
117	530-122-1
117	642-304
117	690V092H88
117	903-13B
117	903Y00212
117	919-00-5045
117	919-00-7109
117	919-007776RA
117	919-01-0829
117	919-01-0829-1
117	919-01-1339
117	991-00-1449
117	991-00-1449-1
117	991-00-2440
117	991-00-2440-1
117	991-00-2440-2
117	1851-17
117	2093A3D-2
117	2093A4-2
117	2093A6-1
117	2093A6-2
117	4686-116-3
117	4686-147-3
117	4686-148-3
117	4686-150-3
117	4686-151-3
117	4686-177-3
117	4686-87-3
117	5300-82-1003
117	5300-88-2
117	5301-11-1
117	5301-22-1
117	7213-0
117	9302-2
117	9302-2A
117	11399-8
117	11586-7
117	013339
117	36535
117	36537
117	055210
117	75702-15-24
117	93027
117	95016
117	106379
117	147478-8-2
117	172721
117	219245
117	223489
117	223720
117	225267
117	227015
117	265236
117	266583
117	348054-11
117	348054-14
117	348054-6
117	0552007H
117	988049
117	1472460-1
117	1472460-4
117	1472460-5
117	7201212
117	7302340
117	402003173
117	2295100232
117	2295100234
117	2295100235
117	13480084523
117	13480084555
117	16516390010
117(4)	BU029
1170	DDEY123001
1170	HC10001200
1170	M51521L
1170	51-42211P01
1170	003526
1170	916106
1171	AMLM201
1171	AMLM301
1171	AMLM301A
1171	CA201AT
1171	CA201T
1171	CA301AT
1171	CA748CT
1171	LM201AH
1171	LM201H
1171	LM301AH
1171	LM301AL
1171	LM748CH
1171	ML201AT
1171	ML301T
1171	RC748T
1171	SG201AT
1171	SG201T
1171	SG301AT
1171	SN72301AL
1171	SN72748L
1171	SSS201AJ
1171	SSS301AJ
1171	748BE
1171	748CE
1173	AN247
1173	AN247P
1173	GEIC-302
1173	M51247
1173	51-13753A18
1173	51S13753A06
1173	1052-6416
1173	1065-2055
1175	CA3134EM
1175	CA3134QM
1176	CA3137E
1178	LA1369
1179	GEIC-75
1179	LA3201
1179	4152032010
118	A-95-5296

ECG	Industry Standard No.	ECG	Industry Standard No.	ECG	Industry Standard No.	ECG	Industry Standard No.	ECG	Industry Standard No.
118	A04735-A	119	34-8056-1	121	A2091859-0025	121	A8Z15	121	CQT1111A
118	A95-5296	119	34-8057-15	121	A2091859-0720	121	A8Z16	121	CQT1112
118	B66X0035-001	119	42-22540	121	A2091859-10	121	A8Z17	121	CQT1129
118	EE57X32	119	46-86105-3	121	A2091859-11	121	A8Z18	121	CQT940A
118	EE58X32	119	61-8968	121	A218012DS	121	ATC-TR-14	121	CQT940B
118	EU57X32	119	66X0036-002	121	A30302	121	ATC-TR-5	121	CQT940BA
118	FR-1035	119	86-45-3	121	A34-6001-1	121	ATC-TR-6	121	CRT1544
118	GECR-1	119	93C40	121	A34-6002-17	121	AU102	121	CRT1545
118	HS-7/1	119	93C40-1	121	A34715	121	AUY-21	121	CRT1252
118	HS-8/1	119	264P01508	121	A35084	121	AUY10	121	CRT1553
118	H81/7	119	295V031B	121	A35201	121	AUY19	121	CRT1602
118	H81/9	119	295V034C01	121	A35260	121	AUY20	121	CRT3602A
118	H89/1	119	690V039H17	121	A36896	121	AUY21A	121	CST1743
118	JT-E1095	119	800-011-00	121	A416	121	AUY22A	121	CST1744
118	K122J176-1	119	70581-1	121	A417-62	121	AUY31	121	CST1745
118	K122J176-2	119	72111	121	A4247	121	AUY33	121	CST1746
118	PTC208	119	97251-3	121	A4347	121	AV105	121	CT1122
118	R105064	119	113391	121	A48-134727	121	B-1511	121	CT1124
118	RE167	119	126862	121	A48-134731	121	B-1914	121	CT1124A
118	RF32103-1	119	530096-1	121	A48-134907	121	B10163	121	CT1124B
118	RF32103R	119	607113	121	A48-137102	121	B1017	121	CTP1111
118	8913	119	817124	121	A48-137213	121	B10474	121	CTP1124
118	8926	119	972571-2	121	A48-137214	121	B10475	121	CTP1133
118	SR-32	119	972571-3	121	A48-137215	121	B107	121	CTP1135
118	SR9000	119	1476183-8	121	A48-137216	121	B107A	121	CTP1136
118	T-E1095	119	4202009400	121	A48-137217	121	B1085	121	CTP1137
118	TV6.5	1192	8000-00058-009	121	A48-137218	121	B10912	121	CTP1265
118	3MA	1194	1156	121	A48-137219	121	B10913	121	CTP1266
118	004-003200	1194	003536	121	A48-137220	121	B1151	121	CTP1306
118	004-03200	1196	HA1177	121	A48-63076A81	121	B1151A	121	CTP1307
118	5-30132.1	120	A04727	121	A4A-1-70	121	B1151B	121	CTP1500
118	6R81BPH110BEB1	120	B66X0041-001	121	A4A-1-705	121	B1503	121	CTP1503
118	6R81BPH110BHB1	120	E13-013-03	121	A4A-1A9G	121	B1152A	121	CTP1504
118	6R81BPH110BMB1	120	E13-013-04	121	A5253	121	B1152B	121	CTP1508
118	6R87PH130BCB1	120	EE57X26	121	A57B124-10	121	B1181	121	CTP1511
118	6R87PH130BCB1	120	EU57X38	121	A57L5-1	121	B119	121	CTP1513
118	13-16106-1	120	GECR-3	121	A57M3-7	121	B119A	121	CTP1514
118	15-108012	120	GI-TVO3	121	A57M3-8	121	B122	121	CTP1550
118	15-108017	120	M105064	121	A62-18427	121	B126	121	CTP1551
118	19AR30	120	PTC404	121	A63-18427	121	B126A	121	CTP1728
118	21A3	120	R-3	121	A65C-19G	121	B126P	121	CTP1729
118	23E210679-1	120	R3	121	A65C-70	121	B126V	121	CTP1730
118	28-23-01	120	R8560	121	A65C-705	121	B127	121	CTP1731
118	34-8053-2	120	RCC7022	121	A65C19G	121	B1274	121	CTP1732
118	34-8053-3	120	RE91	121	A660097	121	B1274A	121	CTP1733
118	34-8053-4	120	RSLNA0004CEZZ	121	A690V081H97	121	B1274B	121	CTP1735
118	46-E6180-3	120	SELEN-42	121	A7279039	121	B127A	121	CTP1736
118	48-66653A003	120	SR-97	121	A7279049	121	B128V	121	CTP1739
118	48-66653A005	120	SR-9005	121	A7285774	121	B131(JAPAN)	121	CTP3500
118	48-66653APT015	120	TV24237	121	A7285778	121	B131A	121	CTP3503
118	48-66653APT1003	120	1TV24237	121	A7289047	121	B132(JAPAN)	121	CTP3504
118	53B011-5	120	004-003700	121	A7290594	121	B132A	121	CTP3508
118	61-8969	120	6R8510X12	121	A7291252	121	B134	121	DS-515
118	66X0035-001	120	10B-4	121	A7297043	121	B134A	121	DS515
118	83-43-5	120	10B-4-C4	121	A7297092	121	B134C	121	DS520
118	86-44-3	120	13-17596-1	121	A7297093	121	B136A	121	DT1040
118	93A57-1	120	15-108007	121	A76-11770	121	B1368A	121	DT41
118	93B57-1	120	15-108038	121	A77C-70	121	B1368B	121	DTG110
118	110BH1	120	21A005	121	A77C-705	121	B1368C	121	DTG1011
118	212-85	120	21A008-016	121	A77C19G	121	B1368D	121	DTG1040
118	212-85B	120	21A500	121	A84A-70	121	B1368F	121	DTG110
118	295V003H01	120	21A500-000	121	A84A-705	121	B1368P	121	EA1082
118	295V003M01	120	28-26-01	121	A84A19G	121	B137	121	EA1341
118	295V033B01	120	34-8058	121	A8P404-ORN	121	B138	121	EA15X10
118	800-012-00	120	34-8058-1	121	A8P404-ORN	121	B14A-1-21	121	EA15X15
118	2093A57-1	120	34-8058-7	121	A8P404F	121	B179	121	EA15X173
118	23115-022	120	41-001	121	A94004	121	B1904	121	EA15X26
118	72110	120	48-66653A02	121	A964-177887	121	B215	121	EA15X33
118	126860	120	48-90068A01	121	A971B3	121	B216	121	EA15X35
118	530096	120	510X14	121	A992-00-1192	121	B216A	121	EA15X38
118	530132	120	510X3	121	AA28B240A	121	B217	121	EA15X53
118	530132-1	120	66X0041-001	121	AA4	121	B217A	121	EA15X88
118	605131	120	66X41-1	121	AC148	121	B217G	121	EA1700
118	700021-00	120	86-37-3	121	AD-140	121	B217U	121	EQ0-6
118	817123	120	86-55-3	121	AD-148	121	B25	121	ER15X10
118	983995	120	86-56-3	121	AD-149	121	B25B	121	ER15X17
118	1440977-1	120	86-56-3C	121	AD-150	121	B283	121	ER15X43
118	62605618	120	86-56-3F	121	AD-159	121	B284	121	ES18
118	EA33X8363	120	93A5-2	121	AD130	121	B2W	121	ES21(ELCOM)
1180	GEIC-80	120	93A75-1	121	AD130-III	121	B337	121	ES503
1180	LA4100	120	93B19-1	121	AD130-IV	121	B337A	121	ET15X17
1180	LA4101	120	93B53-2	121	AD130-V	121	B337B	121	ET15X14
1181	AN271	120	9301-2021	121	AD131	121	B337BK	121	ET15X43
1181	AN271B	120	9305-2	121	AD131-III	121	B337H	121	ET15X5
1181	GEIC-61	120	173A4424	121	AD131-IV	121	B337HA	121	F67B
1181	14LN034	120	188-70-48	121	AD131-V	121	B337HB	121	FD-1029-ET
1182	AN289P	120	212-46	121	AD132	121	B338	121	FD-1029ET
1183	M5192P	120	212-63	121	AD138	121	B338H	121	G04-701-A
1184	GEIC-137	120	295V031B02	121	AD138/50	121	B338HA	121	G19
1185	GEIC-139	120	295V031C02	121	AD13850	121	B338HB	121	G6013
1185	61K001-13	120	297V050H01	121	AD140	121	B407	121	GC4045
1186	87-0217	120	297V050H02	121	AD143	121	B407-0	121	GC4062
1186	BPR4X10	120	301-1	121	AD143B	121	B407V	121	GC4087
1186	GEIC-175	120	324-0015	121	AD143R	121	B42	121	GC4097
1186	TVSMPC595C	120	690V037H92	121	AD145	121	B424	121	GC4111
1186	TVSUPC595C	120	690V039H54	121	AD149-01	121	B425	121	GC4125
1186	51-90305A04	120	2090A53-2	121	AD149-02	121	B426	121	GC4156
1187	GEIC-176	120	2093A75-1	121	AD149-IV	121	B426BL	121	GC4251
1187	RH-IX0001PAZZ	120	72098	121	AD149-V	121	B426R	121	GC4267-2
1187	TVSMPC596C2	120	105064	121	AD149B	121	B426Y	121	GC641
1187	TVSUPC596C2	120	489752-072	121	AD149C	121	B449	121	GC691
1188	GEIC-82	120	530087-2	121	AD150	121	B449F	121	GC692
1188	TA70CEM	120	530097-2	121	AD150-IV	121	B449P	121	GE-16
119	A-95-5295	120	575995	121	AD150-V	121	B471	121	GB-3
119	A04710	120	817149	121	AD153	121	B471-2	121	GBT572
119	B66X0036-001	120	1470990-1	121	AD159	121	B471A	121	GM428
119	DBCZ0373000	120	1887048	121	ADY-27	121	B471B	121	GP1432
119	EE57X31	120	62047259	121	ADY22	121	B4A-1-A-21	121	GP1493
119	EU57X31	120	62542704	121	ADY23	121	B65C-1-21	121	GP1494
119	GECR-2	1205	GRIC-178	121	ADY24	121	B69	121	GP1882
119	PTC209	1205	TA7055P	121	ADY27	121	B77B-1-21	121	GP420
119	R2	1206	TA7157P	121	ADY28	121	B84	121	GPT-16
119	RF32102R	121	A059-10B	121	AF280	121	B84A-1-21	121	GTZ001
119	SELEN-48	121	A1124C	121	AH101	121	BDY62	121	H034
119	SR-31	121	A12163	121	AR-10	121	BP5	121	HEP-06005
119	SR9001	121	A1217B	121	AR-11	121	C50BA042	121	HEP-06013
119	T-E1107	121	A14-586-01	121	AR-12	121	CDT1309	121	HEP232
119	TVM-537	121	A1414A	121	AR-13	121	CDT1310	121	HEP623
119	2MA	121	A1414A9	121	AR-14	121	CDT1311	121	HEP624
119	004-003100	121	A144A-1	121	AR-4	121	CDT1319	121	HEP628
119	004-03100	121	A144A-19	121	AR-5	121	CDT1320	121	HP19
119	4-2020-06300	121	A1465C	121	AR-6	121	CDT1321	121	HP20
119	4-2020-09400	121	A146509	121	AR-7	121	CDT1349	121	HR-101A
119	6R356PH13BJK1	121	A1477C	121	AR-9	121	CDT1349A	121	HR102
119	6R356PHL3BLJ1	121	A1477O9	121	AR10	121	CDT1350	121	HR102C
119	6R356PHL3BCJ1	121	A1484A	121	AR11	121	CDT1350A	121	HR103
119	6R86PH13BJJ1	121	A1484A9	121	AR12	121	CM2550	121	HR105
119	6R86PH13BJJ1	121	A14A-70	121	AR13	121	CQT1075	121	HR105A
119	6R86PH13BKJ1	121	A14A-705	121	AR14	121	CQT1076	121	HR1053A
119	6R86PH13BLJ1	121	A14A10G	121	AR8	121	CQT1077	121	I472446-I
119	6R86PH13BMJ1	121	A15927	121	AR8P404R	121	CQT1110	121	KT1017
119	13-16105-3	121	A2090056-1	121	AR9	121	CQT1110A	121	L-417-29BLK
119	21A4	121	A2090056-27	121	A8215	121	CQT1111	121	L-417-29GRN
119	33C59113	121	A2090056-5	121		121		121	

ECG	Industry Standard No.	ECG	Industry Standard No.	ECG	Industry Standard No.	ECG	Industry Standard No.	ECG	Industry Standard No.
121	L-417-29WHT	121	P1KBRN	121	S15556-2	121	SP880-3	121	V30/20DP
121	L-417-60	121	P1KGRN	121	S4A-1-A-3P	121	SP891B	121	V60/10DP
121	M4331	121	P1KORN	121	S65C-1-3P	121	SP891BLU	121	V60/10P
121	M4582	121	P1KRED	121	S67809	121	SP891G	121	V60/20DP
121	M4582BRN	121	P1KYEL	121	S77C-1-3P	121	SP891GRN	121	V60/20P
121	M4583	121	P1T	121	S84A-1-3P	121	SP891R	121	V60/30DP
121	M4583RED	121	P2D	121	S95253	121	SP891W	121	V60/30P
121	M4584	121	P2DBLU	121	S95253-1	121	SP891WHT	121	VFP-6537C
121	M4584GRN	121	P2DBRN	121	SC-70	121	SP891	121	VFP2746C
121	M4606	121	P2DGRN	121	SC70	121	SS1606A	121	VFP65370
121	M4608	121	P2DORN	121	SDT-3048	121	ST-235	121	VFU-2746B
121	M4619RED	121	P2DRED	121	SE-40022	121	ST235	121	VFU2746B
121	M4620GRN	121	P2DYEL	121	SE40022	121	STL109	121	VFU65326B
121	M4722	121	P31898	121	SFT190	121	T-101	121	VM-30203
121	M4722BLU	121	P4M	121	SFT191	121	T-127	121	VM30203
121	M4722GRN	121	P6480001	121	SFT192	121	T-235	121	V8-28B126
121	M4722PUR	121	P75534	121	SFT212	121	T1167	121	V828B126
121	M4722RED	121	P75534-1	121	SFT213	121	T1168	121	V828B126F
121	M4722YEL	121	P75534-4	121	SFT214	121	T13029	121	V828B126V
121	M4730	121	P75534-5	121	SFT238	121	T142	121	V9
121	M4766	121	P8870	121	SFT239	121	T1601	121	WTV12PWR
121	M4767	121	P8890	121	SFT240	121	TA-1614	121	WTV199PWR
121	M4888	121	P8890A	121	SFT250	121	TA-1682	121	WTV25PWR
121	M4888A	121	P8890L	121	SK3009	121	TA-1682A	121	WTV40PWR
121	M4888B	121	PA-10889-1	121	SK3014	121	TA-1705	121	WTV6PWR
121	M501	121	PA-10889-2	121	SP-108	121	TA-1765	121	WTV99PWR
121	M7031	121	PA-10890	121	SP-148-3	121	TA-1766	121	X1005
121	M84B	121	PA-10890-1	121	SP-1482-5	121	TA-1773	121	XB-5
121	M9141	121	PA10890-1	121	SP-1483	121	TA-1794	121	XB-7
121	M9142	121	PAR-12	121	SP-1484	121	TA-1881	121	XB14
121	M9202	121	PAR12	121	SP-1556-2	121	TA-1890	121	XC141
121	M9237	121	PB110	121	SP-1603	121	TA-1891	121	XC142
121	M9241	121	PO3004	121	SP-1603-1	121	TA-2	121	Y410
121	M9255	121	PIK	121	SP-1603-2	121	TA1614	121	ZEN326
121	M9265	121	POWER12	121	SP-404T	121	TA1682A	121	ZEN330
121	M9342	121	POWER25	121	SP-441	121	TA1705	121	ZEN331
121	M9436	121	POWER299	121	SP-485	121	TA1765	121	1-21-270
121	M9550	121	POWER99	121	SP-486	121	TA1766	121	1-21-271
121	MA4670	121	PS-1	121	SP-486W	121	TA1773	121	002-008800
121	MF-55-62	121	PS1	121	SP-634	121	TA1794	121	002-009701
121	M4194	121	PT-12	121	SP-649	121	TA1881	121	002-010100
121	MN22	121	PT-150	121	SP-649-1	121	TA1890	121	002-012700
121	MN23	121	PT-155	121	SP-834	121	TA1891	121	002-9700
121	MN24	121	PT-176	121	SP-880	121	TA2672	121	2AD140
121	MN25	121	PT-234	121	SP-880-1	121	TP-80/30	121	2D001
121	MN26	121	PT-235	121	SP-880-3	121	TT78/30Z	121	2D004
121	MN29	121	PT-235A	121	SP-891	121	TT78/60	121	2D004-9
121	MN29BLK	121	PT-236	121	SP-891B	121	TP80/30	121	2D015
121	MN29GRN	121	PT-236A	121	SP-891W	121	TP80/30Z	121	2G210
121	MN29PUR	121	PT-236B	121	SP1013A	121	TI-266A	121	2G220
121	MN29WHT	121	PT-242	121	SP111B	121	TI-269	121	2G222
121	MN32	121	PT-25	121	SP271	121	TI-366	121	2G224
121	MN46	121	PT-255	121	SP148-3	121	TI-366A	121	2G225
121	MN49	121	PT-256	121	SP1481	121	TI-367	121	2G240
121	MN73	121	PT-285	121	SP1481-1	121	TI-367A	121	2N1020
121	MN73BLK	121	PT-285A	121	SP1481-3	121	TI-368	121	2N1033
121	MN75WHT	121	PT-301	121	SP1481-4	121	TI-368A	121	2N1046B
121	MP1014	121	PT-301A	121	SP1481-5	121	TI-369	121	2N1168
121	MP1509-1	121	PT-307	121	SP1482	121	TI-369A	121	2N1292
121	MP1509-2	121	PT-307A	121	SP1482-3	121	TI-370	121	2N1293
121	MP1509-3	121	PT-3A	121	SP1482-4	121	TI-370A	121	2N1296
121	MP2060	121	PT-40	121	SP1482-5	121	TI266A	121	2N1359
121	MP2060-1	121	PT-50	121	SP1482-6	121	TI269	121	2N1364
121	MP2061	121	PT-554	121	SP1482-7	121	TI3012	121	2N1365
121	MP2062	121	PT-555	121	SP1483	121	TI3027	121	2N1430
121	MP2137A	121	PT-6	121	SP1483-1	121	TI3028	121	2N1530A
121	MP2138A	121	PT06	121	SP1483-2	121	TI3029	121	2N1533
121	MP2139A	121	PT12	121	SP1483-3	121	TI366	121	2N1535A
121	MP2142A	121	PT155	121	SP1556-3	121	TI366A	121	2N1535B
121	MP2134	121	PT176	121	SP1556	121	TI367	121	2N1536A
121	MP2144A	121	PT235	121	SP1556-1	121	TI367A	121	2N1538
121	MP3611	121	PT236B	121	SP1556-2	121	TI368	121	2N1538A
121	MP3612	121	PT236C	121	SP1556-3	121	TI368A	121	2N155
121	MP3613	121	PT242	121	SP1563-2	121	TI369	121	2N157
121	MP3614	121	PT25	121	SP1595BLK	121	TI369A	121	2N157A
121	MP3615	121	PT255	121	SP1595BLU	121	TM121	121	2N1667
121	MP3617	121	PT256	121	SP1595GRN	121	TNJ60454	121	2N176-1BLU
121	MP825	121	PT285	121	SP1595RED	121	TNJ72318	121	2N176-1WHT
121	N57B4-2	121	PT285A	121	SP1596BLK	121	TO-012	121	2N176-1YEL
121	N57B4-4	121	PT30	121	SP1596BLU	121	TO-015	121	2N176-3PUR
121	NA P-T2-8	121	PT301	121	SP1596GRN	121	TQ-5064	121	2N176-4PUR
121	NK14A119	121	PT301A	121	SP1596RED	121	TQ5036	121	2N176-5WHT
121	NK4A-1A19	121	PT307	121	SP1603	121	TQ5064	121	2N176-6WHT
121	NK650119	121	PT307A	121	SP1603-1	121	TR-01A	121	2N176BLK
121	NK77C119	121	PT366B	121	SP1603-2	121	TR-01C	121	2N176BLU
121	NK84A119	121	PT3A	121	SP1603-3	121	TR-178(OLSON)	121	2N176GRN
121	NKT-401	121	PT40	121	SP1657	121	TR-43B	121	2N176R
121	NKT-402	121	PT50	121	SP1801	121	TR-5	121	2N176RN
121	NKT-403	121	PT501	121	SP1844	121	TR-8006	121	2N176PUR
121	NKT-404	121	PT555	121	SP1938	121	TR010	121	2N176RED
121	NKT-405	121	PT6	121	SP1950	121	TRO2	121	2N176WHT
121	NKT-415	121	PTO105	121	SP2045	121	TRO2C	121	2N176YEL
121	NKT-416	121	PTO-6	121	SP2046	121	TR16C	121	2N1971
121	NKT-451	121	R102-41	121	SP2048	121	TR353	121	2N2147
121	NKT-452	121	R2446	121	SP2072	121	TR35144	121	2N2148
121	NKT-453	121	R2446-1	121	SP2076	121	TR35144A	121	2N2212
121	NKT-454	121	R265A	121	SP2094	121	TR35524	121	2N2287
121	NKT-501	121	R3512-1	121	SP2155	121	TR8006	121	2N2293
121	NKT-503	121	R3515(RCA)	121	SP2341	121	TRA-7R	121	2N2294
121	NKT-504	121	R516	121	SP2361	121	TRA-7RM	121	2N2295
121	NKT401	121	R7253	121	SP2361BLU	121	TRA-8R	121	2N2296
121	NKT402	121	R8659	121	SP2361BRN	121	TRA7R	121	2N234
121	NKT403	121	R811	121	SP2361GRN	121	TRA8R	121	2N234A
121	NKT404	121	RE11MP	121	SP2361ORN	121	T8-1657	121	2N235
121	NKT405	121	R8-3858-1	121	SP2361RED	121	T8-175	121	2N2357
121	NKT406	121	R8-5613	121	SP2361YEL	121	T8-176	121	2N235B
121	NKT415	121	R8-5835	121	SP2395	121	T81657	121	2N235A
121	NKT416	121	R8-5855	121	SP2493	121	T8173	121	2N235B
121	NKT450	121	R83358-1	121	SP2541	121	T8176	121	2N236
121	NKT454	121	R83359-1	121	SP354	121	T8610	121	2N236A
121	NKT501	121	R83858	121	SP404	121	T8612	121	2N236B
121	NKT503	121	R83858-1	121	SP404T	121	T8613	121	2N251
121	NKT504	121	R83959	121	SP441G	121	T8614	121	2N251A
121	OC-16	121	R83959-1	121	SP485	121	TT-1083	121	2N2836
121	OC-22	121	R85612	121	SP485B	121	TT1083	121	2N2870
121	OC-23	121	R85613	121	SP485BLK	121	TV24337	121	2N297
121	OC-24	121	R85614	121	SP485BLU	121	TV24678	121	2N297A
121	OC-25	121	R85616	121	SP485BRN	121	TV28B126F	121	2N301
121	OC-28	121	R85835	121	SP485W	121	TV8-28B126	121	2N301A
121	OC-29	121	S-1556-2	121	SP485WHT	121	TV8-28B126F	121	2N301B
121	OC-35	121	S-39T	121	SP486W	121	TV8-28B449F	121	2N301W
121	OC-36	121	S-40T	121	SP486WHT	121	TV825126F	121	2N3212
121	0019	121	S-40TB	121	SP534	121	TV828B126	121	2N3213
121	OC20	121	S-41T	121	SP649	121	TV828B449	121	2N3214
121	P-31898	121	S-42T	121	SP649-1	121	ULTRA-7RM	121	2N3215
121	P-D-30	121	S-43T	121	SP744	121	V145	121	2N352
121	P1A	121	S-46T	121	SP819R	121	V15/10DP	121	2N353
121	P1E-1	121	S-48T	121	SP834	121	V15/20DP	121	2N420
121	P1G	121	S-49T	121	SP875	121	V15/30DP	121	2N420A
121	P1K	121	S-58TB	121	SP880-1	121	V152A	121	2N4241
121	P1KBLK	121	S-95253	121		121	V162A	121	2N458B
121	P1KBLU	121	S-95253-1	121		121	V30/10DP	121	2N538
121		121	S14A-1-3P	121		121		121	2N538A
121		121		121		121		121	2N553

ECG	Industry Standard No.	ECG	Industry Standard No.	ECG	Industry Standard No.	ECG	Industry Standard No.	ECG	Industry Standard No.
121	2N554	121	4A-1A7-1	121	31-0192	121	48R869241	121	084A-12-7
121	2N555	121	4A-1A8	121	31-0196	121	48R869255	121	84A-70
121	2N589	121	4A-1A82	121	31-0240	121	48R869436	121	84A-70-12
121	2N663	121	4A-1A9	121	33-1004-00	121	48R869550	121	84A-70-12-7
121	2N665	121	4A-1A9G	121	34-6001-79	121	49P1C	121	84A0
121	2826	121	04A1	121	34-6002-10	121	52P153	121	84A1
121	2826A	121	04A1-12	121	34-6002-11	121	57A1-119	121	84A10
121	28A416	121	4JX8D404	121	34-6002-13	121	57A4-1	121	84A10R
121	28B107	121	4JX8P404	121	34-6002-14	121	57A4-2	121	84A12
121	28B107A	121	4JX8P409	121	34-6002-19	121	57A4-4	121	84A121
121	28B119	121	6-0000015B	121	34-6002-2	121	57A6-12	121	84A13
121	28B119A	121	6-158	121	34-6002-3	121	57A6-2	121	84A13P
121	28B122	121	06P1C	121	34-6002-34	121	57A6-23	121	84A14
121	28B126	121	7-1(STANDEL)	121	34-6002-4	121	57A6-3	121	84A14-7
121	28B126A	121	8-0050700	121	34-6002-5	121	57A6-8	121	84A14-7B
121	28B126B	121	8-905-605-624	121	34-6002-6	121	57A9-2	121	84A15
121	28B126C	121	8-905-605-635	121	34-6002-7	121	57B4-1	121	84A15L
121	28B126D	121	8-905-605-636	121	34-6002-8	121	57B4-2	121	84A16
121	28B126E	121	8-905-605-637	121	34-6002-9	121	57B4-4	121	84A16B
121	28B126F	121	8-905-606-720	121	38P1	121	5706-12	121	84A17
121	28B126G	121	8-905-613-210	121	38P1C	121	5706-2	121	84A17-1
121	28B126H	121	8-905-613-215	121	39P1	121	5706-23	121	84A18
121	28B126V	121	8-905-613-250	121	39P1C	121	5706-3	121	84A182
121	28B127	121	8H303	121	42-16599	121	5706-8	121	84A19
121	28B127A	121	8L201	121	42-23968	121	5709-2	121	84A19G
121	28B127M	121	8L201B	121	42-23968P	121	57D1-119	121	84A2
121	28B131	121	8L201C	121	44A-1A5	121	57D4-1	121	84A3
121	28B131A	121	8L201R	121	46-8136-3	121	57D4-2	121	84A4
121	28B132	121	8L201V	121	46-8615-3	121	57D6-12	121	84A5
121	28B132A	121	8L404	121	46-8617-3	121	57D9-2	121	84A6
121	28B137	121	8P404B	121	46-8621-3-3	121	57D9-23	121	84A7
121	28B138	121	8P404M	121	46-8634-3	121	57D0-32	121	84A8
121	28B215	121	8P404N	121	46-8638-3	121	61-782	121	84A9
121	28B216	121	8P404R	121	47P1	121	610005-1	121	84AA1
121	28B216A	121	8P404V	121	48-124204	121	62-18428	121	84AA19
121	28B217	121	8P415C	121	48-124246	121	63-10378	121	84B
121	28B217A	121	8P416C	121	48-124247	121	63-29451	121	85-370-2 BLU
121	28B217G	121	8PC60	121	48-124285	121	63-29459	121	86-120-2
121	28B217U	121	8PS60	121	48-124302	121	63-8590	121	86-127-2
121	28B25	121	09-300037A	121	48-124332	121	63-8706	121	86-141-2
121	28B25B	121	09-301052	121	48-124356	121	0650-12	121	86-142-2
121	28B283	121	9-51141400	121	48-125204	121	0650-12-7	121	86-146-2
121	28B284	121	9-5250	121	48-125208	121	650-70	121	86-147-2
121	28B337	121	9-5251	121	48-125267	121	650-70-12	121	86-173-2
121	28B337A	121	11-0399	121	48-125288	121	650-70-12-7	121	86-173-9
121	28B337B	121	11-0400	121	48-125332	121	6500	121	86-19-2
121	28B337BK	121	12-1-270	121	48-129934	121	65C10	121	86-230-2
121	28B337H	121	12-1-271	121	48-129935	121	65C10R	121	86-231-2
121	28B337HA	121	12M2	121	48-129936	121	65C11	121	86-232-2
121	28B337HB	121	13-14735	121	48-129937	121	65C119	121	86-235-2
121	28B338	121	13-14735-1	121	48-134302	121	65C12	121	86-248-2
121	28B338H	121	13-14735A	121	48-134447	121	65C121	121	86-313-2
121	28B338HA	121	13-15806-1	121	48-134448	121	65C13	121	86-317-2
121	28B338HB	121	13-18034	121	48-134449	121	65C13P	121	86-319-2
121	28B391	121	13-18034-1	121	48-134463	121	65C14	121	86-353-2
121	28B407	121	13-18034A	121	48-134487	121	65C14-7	121	86-354-2
121	28B407-0	121	13-22741	121	48-134488	121	65C14-7B	121	86-370-2TEL
121	28B407TV	121	14-574-10	121	48-134493	121	65C15	121	86-5039-2
121	28B42	121	14-574-10	121	48-134519	121	65C15L	121	86-5043-2
121	28B424	121	14-578-10	121	48-134560	121	65C16	121	86-5057-2
121	28B425	121	14-579-10	121	48-134570	121	65C16-4	121	86-5058-2
121	28B425Y	121	14-589-01	121	48-134574	121	65C17	121	86-5083-2
121	28B426	121	14-590-01	121	48-134575	121	65C17-1	121	86-5088-2
121	28B426BL	121	14-601-01	121	48-134582	121	65C18	121	86-5089-2
121	28B426R	121	14-601-03	121	48-134583	121	65C182	121	86-5090-2
121	28B426Y	121	14-601-04	121	48-134592	121	65C19	121	86-5113-2
121	28B449	121	14-601-05	121	48-134606	121	65C19G	121	86-5125-2
121	28B449A	121	14-601-06	121	48-134611	121	6502	121	86-5943-2
121	28B449B	121	14-601-07	121	48-134612	121	6503	121	86-62-2
121	28B449C	121	14-601-09	121	48-134613	121	6504	121	86-63-2
121	28B449D	121	14-601-11	121	48-134634	121	6505	121	86-8-2
121	28B449E	121	14-604-07	121	48-134638	121	6506	121	86X0009-001
121	28B449F	121	14-604-08	121	48-134639	121	6507	121	863001S-001
121	28B449L	121	14A	121	48-134644	121	6508	121	094-013
121	28B449M	121	014A-12	121	48-134645	121	6509	121	94A-1A6-4
121	28B449P	121	014A-12-7	121	48-134646	121	67P1	121	95-250
121	28B471	121	14A0	121	48-134647	121	67P1C	121	95-251
121	28B471-2	121	14A1	121	48-134649	121	67P2	121	96-5026-01
121	28B471A	121	14A1-A82	121	48-134651	121	67P2C	121	96-5045-01
121	28B471B	121	14A10	121	48-134670	121	67P3	121	96-5064-01
121	28B69	121	14A10R	121	48-134672	121	67P3C	121	96-5081-01
121	28B84	121	14A11	121	48-134696	121	68P1	121	96-5086-02
121	28B95	121	14A12	121	48-134722	121	68P1B	121	96-5100-01
121	28P212	121	14A13	121	48-134725	121	77-270877-2	121	96-5100-03
121	2T3011	121	14A13P	121	48-134730	121	77-270878-2	121	96-5125-01
121	2T3021	121	14A14	121	48-134738	121	77-271491-1	121	96-5143-01
121	2T3022	121	14A14-7	121	48-134744	121	77C	121	96-5143-02
121	2T3030	121	14A14-7B	121	48-134746	121	077C-12	121	96-5143O-02
121	2T3031	121	14A15	121	48-134750	121	077C-12-7	121	96-5145-01
121	2T3032	121	14A15L	121	48-134751	121	77C-70	121	96-5155-01
121	2T3033	121	14A16	121	48-134757	121	77C-70-12	121	96-5192-01
121	2T3041	121	14A16-5	121	48-134758	121	77C-70-12-7	121	96-5378-01
121	2T3042	121	14A17	121	48-134759	121	77C0	121	96XZ801/06N
121	2T3043	121	14A17-1	121	48-134763	121	77C1	121	96XZ801/10N
121	03-57-501	121	14A18	121	48-134764	121	77C10R	121	96XZ801/34X
121	4-14A17-1	121	14A19	121	48-134766	121	77C119	121	998001
121	4-4A-1A7-1	121	14A19G	121	48-134767	121	77C12	121	998014
121	4-65017-1	121	14A2	121	48-134888	121	77C121	121	998015
121	4-68621-3-3	121	14A3	121	48-134930	121	77C13	121	106P1
121	4-77017-1	121	14A4	121	48-134938	121	77C13P	121	106P1AG
121	004-8000	121	14A5	121	48-134947	121	77C14	121	106P1T
121	4-88A17-1	121	14A6	121	48-134974	121	77C14-7	121	112-202147
121	4A-1	121	14A7	121	48-134977	121	77C14-7B	121	112-524
121	04A-1-12-7	121	14A8	121	48-137031	121	77C15	121	114A-1-82
121	4A-1-70	121	14A9	121	48-137118	121	77C15L	121	115-063
121	4A-1-70-12	121	17A4422-1	121	48-137119	121	77C16	121	115-268
121	4A-1-70-12-7	121	18AA-1-82	121	48-137120	121	77C16-3	121	115-269
121	4A-1-4A-7B	121	19A115101	121	48-137122	121	77C17	121	121-1134
121	4A-10	121	19A115101-P1	121	48-137123	121	77C17-1	121	121-1134
121	4A-11	121	19A115184-P1	121	48-137124	121	77C18	121	121-171
121	4A-12	121	19A115267P1	121	48-137978	121	77C182	121	121-270
121	4A-13	121	19A115268	121	48-39P1	121	77C19	121	121-271
121	4A-14	121	19A115341P1	121	48-57B2	121	77C19G	121	121-308
121	4A-15	121	19A115361-P1	121	48-57B42	121	77C2	121	121-363
121	4A-16	121	19A115376	121	48-869141	121	77C3	121	121-371
121	4A-17	121	19A115385-P1	121	48-869142	121	77C4	121	121-382
121	4A-18	121	19A115561	121	48-869182	121	77C5	121	121-389
121	4A-19	121	19A115561-1	121	48-869202	121	77C6	121	121-398
121	4A-1A	121	19AR31	121	48-869237	121	77C7	121	121-793
121	4A-1AO	121	19C300113-P1	121	48-869241	121	77C8	121	122-1625
121	4A-1AOR	121	20A0017	121	48-869255	121	77C9	121	124-1
121	4A-1A1	121	20A0041	121	48-869436	121	78-272212-1	121	129-10
121	4A-1A19	121	20A0042	121	48-869550	121	78-5009	121	129-13
121	4A-1A2	121	20A0074	121	48K134583	121	80-050700	121	129-5
121	4A-1A21	121	21-28	121	48K134584	121	81-27126130-7	121	129-9
121	4A-1A3	121	21A064-000	121	48K869342	121	81-27126130-7A	121	130-104
121	4A-1A3P	121	21A007-000	121	48NBP1035	121	81-27126130-7B	121	0131-000192
121	4A-1A4	121	022-3640-050	121	48R134582	121	81-46125028-4	121	0131-000336
121	4A-1A-7	121	022-3640-050	121	48R134606	121	83-1056	121	0131-000337
121	4A-1A5	121	23-5042	121	48R869141	121	84	121	131-000562
121	4A-1A5L	121	23B-210-025	121	48R869142	121	84A	121	0131-001425
121	4A-1A6	121	026-100003	121	48R869202	121	084A-12	121	144A-1
121	4A-1A6-4	121	026-100028	121	48R869205	121		121	144A-1-12
121	4A-1A7	121	27T406	121		121		121	

ECG	Industry Standard No.
121	2057B2-133
121	2243
121	2446-1(RCA)
121	2577
121	2780
121	2780-4
121	2780-5
121	2901-010
121	2904-008
121	2904-014
121	3107-204-90070
121	3512
121	3514
121	3515(RCA)
121	3618-1
121	4082-501-0001
121	4331
121	4463
121	4570
121	4582BRN
121	4583RED
121	4584GRN
121	4608
121	4619RED
121	4620GRN
121	4649
121	4686-213-3
121	4722
121	4722BLU
121	4722GRN
121	4722ORN
121	4722PUR
121	4722RED
121	4722YEL
121	4727
121	4730
121	4801-1100-011
121	4888A
121	4888B
121	8883-2
121	9005-0
121	9403-2
121	9404-0
121	9925-0
121	11252-4
121	11506-3
121	11526-8
121	11526-9
121	12127-0
121	12127-1
121	014382
121	15024
121	15027
121	16599
121	16959
121	17887
121	17945
121	23311-006
121	25661-020
121	27126-090
121	30203
121	30211
121	30215(RCA)
121	30216(RCA)
121	30246
121	30246A
121	33989-2069
121	34022
121	34298
121	34315
121	34425
121	34526
121	35044
121	35144
121	35231
121	35349
121	35728
121	35885A
121	35885B
121	35951
121	36203
121	36303
121	36304
121	36304-4
121	36312
121	36395
121	36477
121	36687
121	36800-2
121	36800-3
121	36800-4
121	36800-5
121	36800-6
121	36800-7
121	36910
121	36971
121	39893
121	40051
121	40051-2
121	40421
121	40612
121	40626
121	43046
121	50447-4
121	50477-4
121	51650
121	59990-1
121	61010-6
121	61010-6-1
121	62177
121	66009-5
121	66010-3
121	70434
121	71448
121	71448-1
121	71448-2
121	71448-3
121	71448-4
121	71448-5
121	71448-6
121	71448-7
121	71488-4
121	71488-5
121	71488-6
121	72856-63
121	75700-03-01
121	080048
121	80416C
121	81513-7
121	084001C
121	88832
121	90050
121	094013
121	94024
121	94025
121	94026
121	94032
121	94034
121	94040
121	95250
121	95250-1
121	95251
121	99250
121	115268
121	115269
121	115281
121	115282
121	115283
121	115284
121	116093
121	119721
121	121243
121	122792
121	123792
121	125703
121	125761
121	145134-526
121	147351-5-1
121	162002-033
121	162002-062
121	162002-062A
121	162002-095
121	170307-1
121	170376
121	170376-1
121	170407-1
121	170479-1
121	171004(SEARS)
121	171162-082
121	171162-086
121	175043-023
121	175043-81
121	190425
121	190425A
121	194474-8
121	196058-4
121	196148-0
121	196183-5
121	196501-7
121	196607-9
121	209185-962
121	214396
121	216986
121	217892
121	219301
121	219361
121	219440
121	219940
121	221602
121	221605
121	221940
121	221941
121	222915
121	223365
121	223490
121	223576
121	224503
121	224873
121	225595
121	225596
121	225925
121	225927
121	226634
121	226999
121	227566
121	227804
121	228229
121	228230
121	228558
121	228559
121	230208
121	230523
121	231140-11
121	231140-33
121	231672
121	231797
121	232194
121	232674
121	232675
121	233509
121	234077
121	234178
121	234566
121	235312
121	236935
121	237452
121	242183
121	242838
121	256068
121	256071
121	257341
121	257403
121	257536
121	258990
121	261970
121	262114
121	262370
121	263856
121	270744
121	270745
121	270746
121	270780
121	270785
121	275612
121	309412
121	322968-140
121	0573040
121	0573205
121	602032
121	603031
121	610039
121	610039-1
121	610049-1
121	610067
121	610067-1
121	610067-2
121	610067-D
121	610068
121	610106
121	610106-1
121	617871-1
121	618139-1
121	650970
121	651202
121	652085
121	652086
121	655319
121	660077
121	660094
121	660095
121	660097
121	660103
121	702885
121	702885-00
121	801518
121	801522
121	815246-2
121	980132
121	980134
121	980135
121	980155
121	980437
121	983036
121	983795
121	983874
121	983975
121	984261
121	984431
121	985036
121	985431
121	985443
121	985447
121	985449
121	985453
121	985455
121	985686
121	988080
121	988336
121	988413
121	988468
121	989387
121	995001
121	995014
121	995015
121	1221615
121	1407205-1
121	1407206-1
121	1471036-14
121	1471036-20
121	1471102-41
121	1472446
121	1472446-1
121	1473501-1
121	1473515-1
121	1960584
121	1961479
121	1961480
121	1961835
121	1965017
121	1966079
121	2000646-113
121	2006607-59
121	2091858-0712
121	2091858-11
121	2091859-0011
121	2091859-0712
121	2091859-0713
121	2091859-0714
121	2091859-0715
121	2091859-0716
121	2091859-0717
121	2091859-0718
121	2091859-0723
121	2091859-16
121	2091859-25
121	2091859-6
121	2091859-8
121	2091859-9
121	2091959-16
121	2904014
121	3130006
121	3460553-2
121	3460553-4
121	3462221-1
121	3462306-1
121	3700085
121	3731313-1
121	4036598-P1
121	4036733-P1
121	4037594-P1
121	4037607-P1
121	4082501-0001A
121	4999774
121	5493158-1
121	5496663-P1
121	5496939-P1
121	5496939-P2
121	6480001
121	6480004
121	7274653
121	7285663
121	7285776
121	7287110
121	7289041
121	7290593
121	7292690
121	7292955
121	7294796
121	7298079
121	7570003
121	7570003-01
121	7910072-01
121	20918596-2
121	23111006
121	23114011
121	23114033
121	23311066
121	27126090
121	27126130
121	27126130-12
121	43022577
121	62081579
121	62084103
121	62084936
121	62087633
121	62371566
121	62736976
121	80050700
121	95114140-00
121	120004887
121	131000562
121	134804290-101
121	1348042901
121	4803101109-02
121	134804290101
121	310720490070
121$	B295
121$	B472
121$	B472A
121$	B472B
121$	2SB295
121$	2SB472
121$	2SB472A
121$	2SB472B
1211	LA4420
1213	GEIC-150
1213	HA11228
1213	HA1144
1213	61L001-3
1213	2360151
1215	GEIC-74
1215	LA3155
1217	GEIC-76
1217	LA3350
1217	051-0088-00
1217	003516
121MP	A13-14604-1A
121MP	A13-14604-1B
121MP	A13-14604-1C
121MP	A13-14604-1D
121MP	A13-14604-1E
121MP	A13-14777-1
121MP	A13-14777-1A
121MP	A13-14777-1B
121MP	A13-14777-1C
121MP	A13-14777-1D
121MP	A13-14778-1A
121MP	A13-14778-1B
121MP	A13-14778-1C
121MP	A13-14778-1D
121MP	A13-22741-2
121MP	A146B-3
121MP	A146B-39
121MP	A1477B
121MP	A1477B9
121MP	A168P1
121MP	A48-10075A01
121MP	A48-10075A02
121MP	A48-10075A03
121MP	A48-10075A04
121MP	A48-10075A05
121MP	A48-10075A06
121MP	A48-10075A07
121MP	A48-10075A08
121MP	A48-10103A01
121MP	A48-10103A02
121MP	A48-10103A03
121MP	A48-10103A04
121MP	A48-10103A05
121MP	A48-10103A06
121MP	A48-10103A07
121MP	A48-10103A08
121MP	A48-10103A09
121MP	A48-10103A10
121MP	A48-10103A11
121MP	A48-64978A10
121MP	A48-64978A11
121MP	A48-64978A24
121MP	A642-271
121MP	A660031
121MP	A6B-3-70
121MP	A6B-3-705
121MP	A6B-3A9G
121MP	A77B-70
121MP	A77B-705
121MP	A77B19G
121MP	A815203-5
121MP	A86X0030-100
121MP	B6B-3-A-21
121MP	B6B-3A21
121MP	B77C-1-21
121MP	NK6B-3A19
121MP	NK77B119
121MP	RS5855
121MP	S6B-3-A-3P
121MP	S77B-1-3P
121MP	SK3013
121MP	SK3015
121MP	TM121MP
121MP	W9MP
121MP	2AD149
121MP	4-6B-3A7-1
121MP	4-77B17-1
121MP	6B-3
121MP	06B-3-12
121MP	06B-3-12-7
121MP	6B-3-70
121MP	6B-3-70-12
121MP	6B-30
121MP	6B-31
121MP	6B-32
121MP	6B-33
121MP	6B-34
121MP	6B-35
121MP	6B-36
121MP	6B-37
121MP	6B-38
121MP	6B-39
121MP	6B-3A
121MP	6B-3A0
121MP	6B-3A1
121MP	6B-3A19
121MP	6B-3A2
121MP	6B-3A21
121MP	6B-3A3
121MP	6B-3A3P
121MP	6B-3A4
121MP	6B-3A4-7
121MP	6B-3A4-7B
121MP	6B-3A5
121MP	6B-3A5L
121MP	6B-3A6
121MP	6B-3A7
121MP	6B-3A7-1
121MP	6B-3A8
121MP	6B-3A82
121MP	6B-3A9
121MP	6B-3A9G
121MP	6B-3A0R
121MP	9-5250-1
121MP	9-5257
121MP	13-14604-1
121MP	13-14604-1A
121MP	13-14604-1B
121MP	13-14604-1C
121MP	13-14604-1D
121MP	13-14604-1E
121MP	13-14777-1
121MP	13-14777-1A
121MP	13-14777-1B
121MP	13-14777-1C
121MP	13-14777-1D
121MP	13-14778-1
121MP	13-14778-1A
121MP	13-14778-1B
121MP	13-14778-1C
121MP	13-14778-1D
121MP	13-22739-1
121MP	13-22741-1
121MP	13-22741-2
121MP	16B-3A82
121MP	026-100004
121MP	026-100020
121MP	32-16599
121MP	42-21443
121MP	42-22834
121MP	46B-3A5
121MP	48-10075A01
121MP	48-10075A02
121MP	48-10075A03
121MP	48-10075A04
121MP	48-10075A05
121MP	48-10075A06
121MP	48-10075A07
121MP	48-10075A08
121MP	48-10103A01
121MP	48-10103A02
121MP	48-10103A03
121MP	48-10103A04
121MP	48-10103A05
121MP	48-10103A07
121MP	48-10103A08
121MP	48-10103A09
121MP	48-10103A10
121MP	48-64978A10
121MP	48-64978A11
121MP	48-64978A24
121MP	57A3-12
121MP	57B3-12
121MP	66B-3A5L
121MP	077B
121MP	077B-12
121MP	077B-12-7
121MP	77B-70
121MP	77B-70-12
121MP	77B-70-12-7
121MP	77B0
121MP	77B1
121MP	77B10
121MP	77B10R
121MP	77B11
121MP	77B119
121MP	77B121
121MP	77B13
121MP	77B13P
121MP	77B14
121MP	77B14-7
121MP	77B14-7B
121MP	77B15
121MP	77B15L
121MP	77B16
121MP	77B16-2
121MP	77B17
121MP	77B17-1
121MP	77B18
121MP	77B182
121MP	77B19
121MP	77B19G
121MP	77B2
121MP	77B3
121MP	77B4
121MP	77B5
121MP	77B6
121MP	77B7
121MP	77B8
121MP	77B9
121MP	86X0030-001
121MP	86X0030-100
121MP	95-250-1
121MP	95-257
121MP	96B-3A65
121MP	998013
121MP	998013A
121MP	115-281
121MP	115-282
121MP	115-283
121MP	115-284
121MP	146B-3
121MP	146B-3-12
121MP	146B-3-12-8
121MP	168P1
121MP	177B-1-82
121MP	177C-1-82
121MP	186B-3
121MP	186B-3-12
121MP	186B-3-127
121MP	186B-3L
121MP	186B-3L8
121MP	297V041H01
121MP	353-9201-001
121MP	0418
121MP	422-1443
121MP	477B15
121MP	477C15
121MP	642-271
121MP	677B-1-5L
121MP	677C-1-5L
121MP	800-253
121MP	977B1-6-2
121MP	977C1-6-3
121MP	992-008-890
121MP	1477B
121MP	1477B-12
121MP	1477B-12-8
121MP	1477C
121MP	1477C-12
121MP	1477C-12-8
121MP	1877B
121MP	1877B-12
121MP	1877B-127
121MP	1877BL
121MP	1877BL8
121MP	1877C
121MP	1877C-12
121MP	1877C-127
121MP	1877CL
121MP	1877CL8
121MP	2780(AIRLINE)
121MP	2780-3
121MP	11528-1
121MP	11528-2
121MP	11528-3
121MP	11528-4
121MP	36359-4
121MP	38094
121MP	40623
121MP	67085-0
121MP	67085-0-1
121MP	115063
121MP	170666-1

ECG	Industry Standard No.	ECG	Industry Standard No.	ECG	Industry Standard No.	ECG	Industry Standard No.	ECG	Industry Standard No.
121MP	170668-1	123	2N541	123	576-0036-847	123A	A054-114	123A	A4R
121MP	170850-1	123	2N542A	123	605	123A	A054-115	123A	A4U
121MP	560004	123	2N552	123	617-87	123A	A054-155	123A	A4V
121MP	610067-3	123	2N696	123	660-134	123A	A054-173	123A	A4V-2
121MP	610068-1	123	2N701	123	690V080H41	123A	A054-195	123A	A54-96-001
121MP	660031	123	2N728	123	800-250-102	123A	A054-221	123A	A54-96-002
121MP	670850	123	2N70	123	800-53001	123A	A054-222	123A	A567
121MP	670850-1	123	2N71	123	1004	123A	A054-225	123A	A593
121MP	815137	123	2N772	123	1524	123A	A054-233	123A	A5C
121MP	815203-3	123	2N773	123	2044-17	123A	A054-234	123A	A5E
121MP	815203-5	123	2N774	123	2057B2-59	123A	A069-109	123A	A5K
121MP	7299771	123	2N775	123	2057B2-63	123A	A06-1-12	123A	A5L
121MP	7299803	123	2N776	123	2057B2-73	123A	A065-102	123A	A5M
121MP	43020418	123	2N777	123	2204-17	123A	A065-103	123A	A5N
121MP	62084212	123	2N778	123	2584	123A	A065-104	123A	A5P
121MP	62084200	123	2N789	123	2904-034	123A	A065-108	123A	A5R
121MP	310720440170	123	2N790	123	2904-035	123A	A065-109	123A	A58
121MP	310720490190	123	2N791	123	4464	123A	A065-110	123A	A5T2222
1222	SK3730	123	2N792	123	4465	123A	A065-113	123A	A5T3903
1223	AN360	123	2N793	123	4470	123A	A066-109	123A	A5T3904
1223	GEIC-295	123	2S701	123	4470-31	123A	A066-112	123A	A5U
1223	87-0246	123	2S702	123	4470-33	123A	A066-133	123A	A5W
1227	GEIC-70	123	2S703	123	4470M-32	123A	A068-108	123A	A641(NPN)
1227	LA1222	123	2S741	123	4594	123A	A068-113	123A	A649L
1228	LA4102	123	2S74	123	4624	123A	A069-102/103	123A	A6498
123	A-184/5	123	2SC116	123	4630	123A	A069-104	123A	A670720K
123	A1472-19	123	2SC150	123	4705	123A	A069-104/106	123A	A670722D
123	AQ4	123	2SC151	123	4714	123A	A069-106	123A	A6HD
123	AQ6	123	2SC151H	123	15809-1	123A	A069-120	123A	A6J
123	B8780010	123	2SC152	123	15820-1	123A	A069-122	123A	A6K
123	BF321A	123	2SC152A	123	15835-1	123A	A106(JAPAN)	123A	A6R
123	BPX96	123	2SC152B	123	17144	123A	A108	123A	A68
123	BIP7201	123	2SC152C	123	17444	123A	A108A	123A	A748B
123	C151	123	2SC152H	123	20738	123A	A108B	123A	A749B
123	C157	123	2SC57	123	27910-12150	123A	A10005-010-A	123A	A7A
123	C158	123	2SC58	123	30210	123A	A10005-011-A	123A	A7R(TRANSISTOR)
123	C28	123	2SC028	123	30219	123A	A10005-015-D	123A	A7R
123	C29	123	2SC029	123	30224	123A	A111	123A	A78
123	C702	123	2SC74	123	30226	123A	A11414257	123A	A7T
123	CT4	123	2SC8183E	123	30248	123A	A12-1-70	123A	A7Y
123	CDQ10035	123	2SC8184E	123	30259	123A	A12-1-705	123A	A88
123	CDQ10036	123	2SD134	123	37585	123A	A12-1A9G	123A	A88(JAPAN)
123	C8183E	123	2T402	123	38178	123A	A128	123A	A8B
123	C8184E	123	8-0024-3	123	40517	123A	A128A	123A	A8G
123	D4D24	123	13-18363	123	40518	123A	A137	123A	A8L
123	D4D25	123	13-18563-1A	123	57000-5503	123A	A137(NPN)	123A	A937
123	D4D26	123	13-31013-1	123	86812	123A	A1379	123A	A937-1
123	D926640-1	123	022-006500	123	86822	123A	A1380	123A	A937-3
123	DRC-87540	123	23-5052	123	88686	123A	A13N1	123A	A9B
123	DT1610	123	026-100017	123	95216	123A	A1412-1	123A	A9E
123	E846	123	27T409	123	95216RED	123A	A153	123A	A9G
123	GC783	123	277411	123	95216YEL	123A	A156	123A	A9H
123	GC784	123	31-0187	123	95225	123A	A1567	123A	A9J
123	IRTB86	123	34-6001-71	123	95226-2	123A	A1567-1	123A	A9N
123	KGB41055	123	42-17444	123	99109-1	123A	A157	123A	A9R
123	MT014	123	42-18511	123	99109-2	123A	A157A	123A	A9U
123	MT55565-1	123	42-19644	123	116148	123A	A157B	123A	A9W
123	M9170	123	42-19670	123	121664	123A	A157C	123A	A9Y
123	M9282	123	42-20738	123	215074	123A	A158	123A	ALD-3141
123	M9475	123	42-21407	123	328785	123A	A158A	123A	ALD-35
123	M9491	123	43X3	123	379102	123A	A158B	123A	ALB-8922
123	RB12	123	43N6	123	567312	123A	A158C	123A	AN
123	RT6921	123	43X16A567	123	600096-413	123A	A159	123A	AR-107
123	SC785	123	53P151	123	851881	123A	A159A	123A	AR-108
123	SQD-2170	123	53P158	123	2622284	123A	A159B	123A	AR-200
123	ST3030	123	53P159	123	2640830-1	123A	A159C	123A	AR-201
123	ST3031	123	53P161	123	14500022-001	123A	A168	123A	AR-202
123	STL152	123	53P162	123	24562100	123A	A1B	123A	AR107
123	TK128B-1008	123	53P163	123	26501505	123A	A1F	123A	AR108
123	TK128B-1009	123	53P165	123	55440007-001	123A	A1H	123A	AR204
123	TM123	123	57A6-27	123	55440043-001	123A	A1J	123A	AR205
123	TVB28C968	123	57A6-29	123	62042583	123A	A1L	123A	AR206
123	W10	123	57A6-30	123	62260408	123A	A1T	123A	AR208
123	001-02011	123	57A6-32	123	2791012150	123A	A1T-1	123A	AR306
123	002-006500	123	57A6-7	123	5700045452	123A	A1V	123A	AR306(BLUE)
123	002-012000	123	57D6-19	1231	TDA1190Z	123A	A1VB	123A	AR306(ORANGE)
123	2N1006	123	610001-1	1232	AB920	123A	A1W	123A	AT329
123	2N1082	123	70X2	1232	CA2002	123A	A2019ZC	123A	AT335
123	2N1139	123	77N3	1232	LM383	123A	A20372	123A	AT336
123	2N1199	123	86-389-2	1232	LM383T	123A	A2410	123A	AT337
123	2N1199A	123	86-5065-2	1232	NA2002	123A	A2411	123A	AT347
123	2N1200	123	86X0012-001	1232	NAM383	123A	A2412	123A	AT348
123	2N1201	123	86X0025-001	1232	TDA2002	123A	A2413	123A	AT349
123	2N1205	123	93039+11	1232	544-2006-011	123A	A2434	123A	AT370
123	2N1247	123	998020	1232	640000003	123A	A246	123A	AT400
123	2N1248	123	113-118	1234	TA7130P	123A	A246(AMC)	123A	AT401
123	2N1249	123	115-1	1234	TA7130PB	123A	A2466	123A	AT402
123	2N1252	123	115-4	1234	TA7130PC	123A	A2468	123A	AT403
123	2N1253	123	121-276	1234	02-257130	123A	A2469	123A	AT404
123	2N1267	123	121-277	1234	51844789J02	123A	A2470	123A	AT405
123	2N1268	123	121-278	1236	BO316403	123A	A248(AMC)	123A	AT406
123	2N1269	123	121-422	1236	TA7176P	123A	A249	123A	AT407
123	2N1270	123	129-14	1236	23119978	123A	A2499	123A	AT420
123	2N1271	123	129-15	1237	HA1151	123A	A25A509-016-101	123A	AT421
123	2N1272	123	200-016	1237	THHA1151	123A	A2B	123A	AT422
123	2N1386	123	297V049H01	1237	415A011510	123A	A2BRN	123A	AT423
123	2N1387	123	297V049H03	1239	HA1342A	123A	A2FGRN	123A	AT424
123	2N1390	123	297V049H04	1239	51-44837J04	123A	A2J	123A	AT425
123	2N1409	123	297V059H03	123A	A-11095924	123A	A280538PQR	123A	AT426
123	2N1417	123	297V059H02	123A	A-11237336	123A	A301	123A	AT427
123	2N1418	123	297V059H03	123A	A-1141 6062	123A	A306	123A	AT490
123	2N1528	123	297V061C03	123A	A-120278	123A	A307	123A	AT491
123	2N1663	123	297V061C04	123A	A-125332	123A	A323	123A	AT492
123	2N1682	123	297V061C06	123A	A-1379	123A	A344	123A	AT493
123	2N1763	123	297V074C02	123A	A-1380	123A	A345	123A	AT494
123	2N2234	123	297V074C03	123A	A-156	123A	A346	123A	AT495
123	2N2235	123	297V074C04	123A	A-1567	123A	A3E	123A	WH-24
123	2N2240	123	310-187	123A	A-158B	123A	A3F	123A	B 722246-2
123	2N2241	123	324-0151	123A	A-158C	123A	A3G	123A	B-1338
123	2N2309	123	324-0152	123A	A-168	123A	A3N	123A	B-1421
123	2N2330	123	324-0154	123A	A-1854-0003-1	123A	A3R	123A	B-1433
123	2N2389	123	325-0031-304	123A	A-1854-0019-1	123A	A3T	123A	B-1666
123	2N2390	123	325-0031-305	123A	A-1854-0025-1	123A	A3T2221	123A	B-169
123	2N2396	123	325-1446-26	123A	A-1854-0027-1	123A	A3T2221A	123A	B-1842
123	2N2397	123	325-1446-27	123A	A-1854-0071-1	123A	A3T2222	123A	B-1872
123	2N335A	123	325-1446-28	123A	A-1854-0094-1	123A	A3T2222A	123A	B-75583-1
123	2N392A	123	325-1513-29	123A	A-1854-0099-1	123A	A3T3011	123A	B-75583-2
123	2N431	123	325-1513-30	123A	A-1854-0201-1	123A	A3T929	123A	B-75583-I02
123	2N432	123	417-127	123A	A-1854-0215-1	123A	A3T930	123A	B-75589-13
123	2N433	123	511-515	123A	A-1854-0241-1	123A	A3W	123A	B-75589-3
123	2N470	123	511-519	123A	A-1854-0246-1	123A	A3Z	123A	B-75608-3
123	2N471	123	515	123A	A-1854-0251-1	123A	A415	123A	B12-1-A-21
123	2N471A	123	516	123A	A-1854-0255-1	123A	A42X00434-01	123A	B12-1A21
123	2N473	123	517-518	123A	A-1854-0354-1	123A	A43021415	123A	B133578
123	2N474	123	576-0001-004	123A	A-1854-0408-1	123A	A454	123A	B169
123	2N474A	123	576-0001-005	123A	A-1854-0434-1	123A	A481A002B	123A	B169(JAPAN)
123	2N476			123A	A-1854-0471-1	123A	A481A0031	123A	B1K
123	2N477			123A	A-1854-0492-1	123A	A494(JAPAN)	123A	B1N
123	2N478			123A	A-1854-0541-1	123A	A4A	123A	B1P7201
123	2N479			123A	A-1854-0554-1	123A	A4M	123A	B1W
123	2N479A			123A	A-567A	123A	A4N	123A	B2D
123	2N5135			123A	AO-54-195	123A	A4P	123A	BA67
123	2N5136			123A	A054-108			123A	BA71
123	2N5183								

ECG	Industry Standard No.	ECG	Industry Standard No.	ECG	Industry Standard No.	ECG	Industry Standard No.	ECG	Industry Standard No.
123A	BACSB2M1	123A	BC238B	123A	BFX92A	123A	BSY20	123A	C2538-11
123A	BACSB2M2	123A	BC238C	123A	BFX93	123A	BSY21	123A	C26
123A	BACSB2M3	123A	BC239C	123A	BFX94	123A	BSY24	123A	C267A
123A	BACS2P	123A	BC267	123A	BFX95	123A	BSY25	123A	C27
123A	BB71	123A	BC268	123A	BFX95A	123A	BSY26	123A	C281
123A	BC-107	123A	BC269	123A	BFY	123A	BSY27	123A	C281A
123A	BC-1072	123A	BC270	123A	BFY-22	123A	BSY28	123A	C281B
123A	BC-107A	123A	BC280A	123A	BFY-23	123A	BSY29	123A	C281C
123A	BC-108	123A	BC280B	123A	BFY-23A	123A	BSY34	123A	C281C-EP
123A	BC-1082	123A	BC280C	123A	BFY-24	123A	BSY38	123A	C281D
123A	BC-1086	123A	BC282	123A	BFY-29	123A	BSY39	123A	C281EP
123A	BC-108B	123A	BC284	123A	BFY-30	123A	BSY48	123A	C281H
123A	BC-1096	123A	BC284A	123A	BFY-39	123A	BSY49	123A	C281HA
123A	BC-109B	123A	BC284B	123A	BFY10	123A	BSY58	123A	C281HB
123A	BC-114	123A	BC289	123A	BFY11	123A	BSY59	123A	C281HC
123A	BC-121	123A	BC289A	123A	BFY12	123A	BSY61	123A	C282
123A	BC-122	123A	BC289B	123A	BFY18	123A	BSY62	123A	C282H
123A	BC-123	123A	BC290	123A	BFY19	123A	BSY62A	123A	C282HA
123A	BC-148A	123A	BC290B	123A	BFY22	123A	BSY63	123A	C282HB
123A	BC-148B	123A	BC290C	123A	BFY23	123A	BSY70	123A	C282HC
123A	BC-148C	123A	BC377	123A	BFY23A	123A	BSY72	123A	C283
123A	BC-167-B	123A	BC378	123A	BFY24	123A	BSY73	123A	C284
123A	BC-1690	123A	BC382	123A	BFY25	123A	BSY74	123A	C284H
123A	BC-169B	123A	BC383	123A	BFY28	123A	BSY75	123A	C284HA
123A	BC-169C	123A	BC384	123A	BFY29	123A	BSY76	123A	C284HB
123A	BC-71	123A	BC408	123A	BFY30	123A	BSY80	123A	C300
123A	BC107	123A	BC409	123A	BFY33	123A	BSY89	123A	C301
123A	BC107A	123A	BC456	123A	BFY37	123A	BSY91	123A	C302(JAPAN)
123A	BC107B	123A	BC507A	123A	BFY371	123A	BSY93	123A	C31
123A	BC108	123A	BC507B	123A	BFY39	123A	BSY95	123A	C315
123A	BC108A	123A	BC508A	123A	BFY39/1	123A	BSY95A	123A	C316
123A	BC108B	123A	BC508B	123A	BFY39/2	123A	BT-94	123A	C317
123A	BC109	123A	BC508C	123A	BFY39/3	123A	BT67	123A	C317C
123A	BC109A	123A	BC509B	123A	BFY39I	123A	BT71	123A	C318
123A	BC109B	123A	BC509C	123A	BFY391	123A	BTX-070	123A	C318(JAPAN)
123A	BC110	123A	BC510B	123A	BFY72	123A	BTX-094	123A	C318A
123A	BC113	123A	BC583	123A	BFY73	123A	BTX-095	123A	C318A(JAPAN)
123A	BC114	123A	BC71	123A	BFY74	123A	BTX-096	123A	C32
123A	BC114TR	123A	BCW54	123A	BFY75	123A	BTX-2567B	123A	C321
123A	BC115	123A	BCW56	123A	BFY76	123A	BTX068	123A	C321H
123A	BC118	123A	BCW60A	123A	BFY77	123A	BTX2367B	123A	C321HA
123A	BC125	123A	BCW60AA	123A	BG-66	123A	BU67	123A	C321HB
123A	BC125A	123A	BCW83	123A	BG-94	123A	BU71	123A	C321HC
123A	BC125B	123A	BCY-50	123A	BH71	123A	BUC 97704-2	123A	C323
123A	BC129	123A	BCY-58	123A	BI71	123A	BV67	123A	C324
123A	BC130	123A	BCY13	123A	BN-66	123A	BV71	123A	C324A
123A	BC131	123A	BCY15	123A	BN7517	123A	BW67	123A	C324H
123A	BC132	123A	BCY16	123A	BN7518	123A	BW71	123A	C324HA
123A	BC134	123A	BCY36	123A	BP67	123A	BX67	123A	C335
123A	BC135	123A	BCY42	123A	BQ-94	123A	BX71	123A	C337B
123A	BC136	123A	BCY43	123A	BQ67	123A	BY67	123A	C348
123A	BC147	123A	BCY50	123A	BR-66	123A	BY71	123A	C350H
123A	BC147B	123A	BCY501	123A	BR67	123A	B267	123A	C352
123A	BC147A	123A	BCY51	123A	BS-66	123A	B271	123A	C352(JAPAN)
123A	BC147B	123A	BCY51I	123A	BS-94	123A	COO 68602300	123A	C352A
123A	BC148	123A	BCY56	123A	BS475	123A	C1007	123A	C360
123A	BC148A	123A	BCY57	123A	BS67	123A	C10279-1	123A	C360D
123A	BC148B	123A	BCY58B	123A	BS810	123A	C10279-3	123A	C37
123A	BC148C	123A	BCY58D	123A	BS821	123A	C103	123A	C37(TRANSISTOR)
123A	BC149	123A	BCY58VII	123A	BS826	123A	C103(JAPAN)	123A	C38(TRANSISTOR)
123A	BC149A	123A	BCY58VIII	123A	BSV35A	123A	C104	123A	C39-207
123A	BC149B	123A	BCY59	123A	BSV40	123A	C104A	123A	C395
123A	BC149C	123A	BCY59A	123A	BSV41	123A	C105	123A	C395A
123A	BC149Q	123A	BCY59B	123A	BSV53	123A	C1071	123A	C395R
123A	BC150	123A	BCY59D	123A	BSV54	123A	C111	123A	C400
123A	BC151	123A	BCY59VII	123A	BSV59	123A	C111E	123A	C400-0
123A	BC152	123A	BCY59VIII	123A	BSV88	123A	C120	123A	C400-GR
123A	BC155B	123A	BCY69	123A	BSV89	123A	C1244	123A	C400-R
123A	BC156B	123A	BCY84A	123A	BSV90	123A	C127	123A	C400-T
123A	BC168B	123A	BE-66	123A	BSV91	123A	C131	123A	C401(JAPAN)
123A	BC169	123A	BF-214	123A	BSW11	123A	C132	123A	C402(JAPAN)
123A	BC169A	123A	BF-215	123A	BSW12	123A	C133	123A	C405
123A	BC169B	123A	BF-226	123A	BSW19	123A	C134	123A	C406
123A	BC169GL	123A	BF183A	123A	BSW33	123A	C134B	123A	C423B
123A	BC170	123A	BF189	123A	BSW34	123A	C135	123A	C423C
123A	BC170A	123A	BF224J	123A	BSW39	123A	C136	123A	C423D
123A	BC170B	123A	BF225J	123A	BSW41	123A	C1390I	123A	C423E
123A	BC170C	123A	BF248	123A	BSW42	123A	C1390J	123A	C423F
123A	BC171	123A	BF250	123A	BSW42A	123A	C1390K	123A	C424
123A	BC171A	123A	BF291	123A	BSW43	123A	C1390V	123A	C424(JAPAN)
123A	BC171B	123A	BF291A	123A	BSW43A	123A	C1390W	123A	C424D
123A	BC172	123A	BF291B	123A	BSW51	123A	C1390WH	123A	C425B
123A	BC172A	123A	BF293	123A	BSW52	123A	C1390WI	123A	C425C
123A	BC172B	123A	BF293A	123A	BSW53	123A	C1390WX	123A	C425D
123A	BC172C	123A	BF293D	123A	BSW82	123A	C1390WY	123A	C425E
123A	BC173	123A	BF321B	123A	BSW83	123A	C1390X	123A	C425F
123A	BC173A	123A	BF321C	123A	BSW84	123A	C1390XJ	123A	C444
123A	BC173B	123A	BF321D	123A	BSW85	123A	C1390XK	123A	C45
123A	BC173C	123A	BF321E	123A	BSW88	123A	C1390YM	123A	C450
123A	BC175	123A	BF321F	123A	BSW89	123A	C1416	123A	C456
123A	BC180	123A	BF71	123A	BSW92	123A	C1416BL	123A	C468(LGR)
123A	BC180B	123A	BFR11	123A	BSW19	123A	C15(TRANSISTOR)	123A	C468A
123A	BC182	123A	BFR16	123A	BSX24	123A	C15-1	123A	C472Y
123A	BC182KA	123A	BFR26	123A	BSX25	123A	C15-2	123A	C478
123A	BC182KB	123A	BFS31P	123A	BSX30	123A	C15-3	123A	C478(D)
123A	BC182L	123A	BFS36A	123A	BSX38	123A	C16	123A	C478-4
123A	BC183	123A	BFS36B	123A	BSX38A	123A	C160	123A	C478D
123A	BC183KB	123A	BFS36C	123A	BSX38B	123A	C1639	123A	C52(TRANSISTOR)
123A	BC183KC	123A	BFS38	123A	BSX44	123A	C166	123A	C53
123A	BC184	123A	BFS38A	123A	BSX48	123A	C167	123A	C537GP
123A	BC184B	123A	BFS42	123A	BSX49	123A	C16A	123A	C538
123A	BC184K	123A	BFS42A	123A	BSX51	123A	C17	123A	C538A
123A	BC184KB	123A	BFS42B	123A	BSX51A	123A	C170	123A	C538AQ
123A	BC185	123A	BFS42C	123A	BSX52	123A	C171	123A	C538P
123A	BC190B	123A	BFS43	123A	BSX52A	123A	C172	123A	C538Q
123A	BC197	123A	BFS43A	123A	BSX53	123A	C172A	123A	C538R
123A	BC197A	123A	BFS43B	123A	BSX54	123A	C1739	123A	C538S
123A	BC197B	123A	BFS43C	123A	BSX66	123A	C174	123A	C538T
123A	BC198	123A	BFT55	123A	BSX67	123A	C17A	123A	C539
123A	BC199	123A	BFV83B	123A	BSX68	123A	C18	123A	C539K
123A	BC207	123A	BFV83C	123A	BSX69	123A	C191	123A	C539L
123A	BC207A	123A	BFV85	123A	BSX75	123A	C192	123A	C539R
123A	BC207B	123A	BFV85A	123A	BSX76	123A	C193	123A	C539S
123A	BC207BL	123A	BFV85B	123A	BSX77	123A	C194	123A	C54
123A	BC208	123A	BFV85C	123A	BSX78	123A	C1945295DY1	123A	C55
123A	BC208A	123A	BFV87	123A	BSX79	123A	C195	123A	C564
123A	BC208AL	123A	BFV88	123A	BSX80	123A	C196	123A	C564A
123A	BC208B	123A	BFV88A	123A	BSX81	123A	C197	123A	C564P
123A	BC208BL	123A	BFV88C	123A	BSX87	123A	C199	123A	C564Q
123A	BC208C	123A	BFW29	123A	BSX88	123A	C201(JAPAN)	123A	C564R
123A	BC208CL	123A	BFW32	123A	BSX89	123A	C202(JAPAN)	123A	C564S
123A	BC209	123A	BFW46	123A	BSX90	123A	C203	123A	C564T
123A	BC209A	123A	BFW58	123A	BSX91	123A	C204	123A	C587
123A	BC209B	123A	BFW60	123A	BSX94A	123A	C205	123A	C587A
123A	BC209BL	123A	BFW68	123A	BSX97	123A	C230	123A	C588
123A	BC209C	123A	BFX43	123A	BSY10	123A	C231	123A	C593
123A	BC209CL	123A	BFX44	123A	BSY11	123A	C232	123A	C594
123A	BC210	123A	BFX59P	123A	BSY165	123A	C233	123A	C595
123A	BC220	123A	BFX92	123A	BSY168	123A	C237	123A	C602E
123A	BC222			123A	BSY17	123A	C238	123A	C62(TRANSISTOR)
123A	BC223A			123A	BSY18	123A	C239	123A	C622
123A	BC223B			123A	BSY19	123A	C248	123A	C63
123A	BC233A					123A	C250	123A	C648
123A	BC238A								

ECG	Industry Standard No.	ECG	Industry Standard No.	ECG	Industry Standard No.	ECG	Industry Standard No.	ECG	Industry Standard No.
123A	C648H	123A	CI2712	123A	C82219	123A	DDBY233001	123A	EA1703
123A	C649	123A	CI2713	123A	C82221	123A	DN	123A	EA1716
123A	C650	123A	CI2714	123A	C82222	123A	DS-47	123A	EA1718
123A	C650B	123A	CI2923	123A	C82369	123A	DS-66	123A	EA1735
123A	C66-P11111-0001	123A	CI2924	123A	C82481	123A	DS-66L	123A	EA1872
123A	C6862400	123A	CI2925	123A	C82711	123A	DS-67	123A	EA2271
123A	C694D	123A	CI2926	123A	C82712	123A	DS-76	123A	EA2489
123A	C735-R	123A	CI3390	123A	C82713	123A	D81B	123A	EA2490
123A	C7350RN	123A	CI3391	123A	C82714	123A	D844	123A	EA2739
123A	C796	123A	CI3391A	123A	C82922	123A	D845	123A	EA2740
123A	C847	123A	CI3392	123A	C82923	123A	D846	123A	EA2770
123A	C848	123A	CI3393	123A	C82924	123A	D847	123A	EA3149
123A	C849	123A	CI3395	123A	C82925	123A	D866	123A	EA112
123A	C850	123A	CI3396	123A	C83390	123A	D867	123A	ED-1402
123A	C87	123A	CI3397	123A	C83391	123A	D867W	123A	ED1402A
123A	C896	123A	CI3398	123A	C83391A	123A	D876	123A	ED1402B
123A	C899	123A	CI3402	123A	C83392	123A	D877	123A	ED1502
123A	C899K	123A	CI3403	123A	C83393	123A	DT161	123A	ED1702L
123A	C907	123A	CI3404	123A	C83394	123A	DW6034/M	123A	EDC-Q10-1
123A	C907A	123A	CI3405	123A	C83395	123A	E13-000-03	123A	EDO 219
123A	C907AC	123A	CI3414	123A	C83396	123A	E13-000-04	123A	ED8-100
123A	C907AD	123A	CI3415	123A	C83397	123A	E13-002-03	123A	EL232
123A	C907AH	123A	CI3416	123A	C83398	123A	E13-003-00	123A	EMB-73500
123A	C907C	123A	CI3417	123A	C83402	123A	E13-003-01	123A	EN2219
123A	C907D	123A	CI3900	123A	C83403	123A	E13-005-02	123A	EN2222
123A	C907H	123A	CI3900A	123A	C83404	123A	E24103	123A	EN3009
123A	C907HA	123A	CI3901	123A	C83405	123A	E2430	123A	EN3011
123A	C923E	123A	CI4256	123A	C83414	123A	E2431	123A	EN3013
123A	C934C	123A	CI4424	123A	C83415	123A	E2436	123A	EN3014
123A	C934D	123A	CI4425	123A	C83416	123A	E2444	123A	EN3903
123A	C934E	123A	CJ-5206	123A	C83417	123A	E2452	123A	EN3904
123A	C934F	123A	CJ-5207	123A	C83560	123A	E2454	123A	EN697
123A	C934G	123A	CJ-5208	123A	C83605	123A	E2455	123A	EN706
123A	C934P	123A	CJ5201	123A	C83606	123A	E2459	123A	EN708
123A	C943	123A	CJ5202	123A	C83607	123A	E2461	123A	EN744
123A	C943A	123A	CJ5203	123A	C83843	123A	E2497	123A	EN914
123A	C943B	123A	CJ5211	123A	C83844	123A	E2499	123A	EN930
123A	C943C	123A	CJ5212	123A	C83845	123A	EA0092	123A	EN956
123A	C9604	123A	CK419	123A	C83854	123A	EA1080	123A	EP15X47
123A	C966	123A	CK420	123A	C83854A	123A	EA1128	123A	EP15X48(NPN)
123A	C967	123A	CK421	123A	C83855	123A	EA1129	123A	EP15X49
123A	C968	123A	CK422	123A	C83855A	123A	EA1135	123A	EP15X7
123A	C968P	123A	CK474	123A	C83859	123A	EA1145	123A	EQB-0100
123A	C98	123A	CK475	123A	C83859A	123A	EA1344	123A	EQB-0196
123A	C99	123A	CK476	123A	C83860	123A	EA1345	123A	EQB-0198
123A	C993	123A	CK477	123A	C83900	123A	EA1406	123A	EQB-100
123A	CA-9011H	123A	CMO334-423	123A	C83900A	123A	EA1407	123A	EQB-1-5
123A	CAM-12	123A	CS-1235F	123A	C83901	123A	EA1408	123A	EQB-22
123A	CB246	123A	CS-1386E	123A	C83903	123A	EA1451	123A	EQB-5
123A	CC1168F	123A	CS-460B	123A	C83904	123A	EA1452	123A	EQB-61
123A	CC81259G	123A	CS-616B0	123A	C84003	123A	EA1499	123A	EQB-9
123A	CC82004	123A	CS-616BH	123A	C84007	123A	EA1564	123A	ES10222
123A	CC84004	123A	CS-6225G	123A	C84021	123A	EA1578	123A	ES10223
123A	CC86168	123A	CS-9011	123A	C84060	123A	EA1581	123A	ES10232
123A	CC86168F	123A	CS-9011G	123A	C84061	123A	EA15X1	123A	ES15050
123A	CC89016R	123A	CS-9011L	123A	C84193	123A	EA15X101	123A	ES15X1
123A	CD0014NA	123A	CS-9014	123A	C84194	123A	EA15X103	123A	ES15X11
123A	CD0014NG	123A	CS-9014B	123A	C84424	123A	EA15X111	123A	ES15X14
123A	CD0015N	123A	CS-9014D	123A	C84425	123A	EA15X112	123A	ES15X16
123A	CD0021	123A	CS-9104	123A	C85088	123A	EA15X136	123A	ES15X20
123A	CD12000	123A	CS-9125B	123A	C85369	123A	EA15X137	123A	ES15X23
123A	CD38	123A	CS1166	123A	C8616BF	123A	EA15X142	123A	ES15X24
123A	CD446	123A	CS1166D	123A	C86229F	123A	EA15X143	123A	ES15X37
123A	CD6019	123A	CS1166E	123A	C86229G	123A	EA15X153	123A	ES15X42
123A	CD6150	123A	CS1166F	123A	C8696	123A	EA15X157	123A	ES15X58
123A	CD6157	123A	CS1166G	123A	C8697	123A	EA15X162	123A	ES15X62
123A	CD6775	123A	CS1166H	123A	C8706	123A	EA15X163	123A	ES15X64
123A	CD8000	123A	CS1166H/F	123A	C8718	123A	EA15X167	123A	ES15X68
123A	CD8000-1	123A	CS1168G	123A	C8718A	123A	EA15X168	123A	ES15X7
123A	CD9525	123A	CS1168H	123A	C8720A	123A	EA15X18	123A	ES15X70
123A	CDC-13000-1	123A	CS1229	123A	C87229G	123A	EA15X180	123A	ES15X76
123A	CDC-8001	123A	CS1229A	123A	C89011	123A	EA15X189	123A	ES15X83
123A	CDC12000-1C	123A	CS1229B	123A	C89011/3490	123A	EA15X20	123A	ES15X84
123A	CDC12018C	123A	CS1229C	123A	C89011D	123A	EA15X22	123A	ES15X85
123A	CDC12013C	123A	CS1229D	123A	C89011E	123A	EA15X24	123A	ES20(ELCOM)
123A	CDC12077F	123A	CS1229E	123A	C89011F	123A	EA15X240	123A	ES46(ELCOM)
123A	CDC13000-1	123A	CS1229F	123A	C89011G	123A	EA15X272	123A	ES53(ELCOM)
123A	CDC13000-18	123A	CS1229G	123A	C89011G/3490	123A	EA15X31	123A	ES95(ELCOM)
123A	CDC13000-1B	123A	CS1229H	123A	C89011H	123A	EA15X330	123A	ET15X10
123A	CDC13000-1C	123A	CS1229J	123A	C89011I	123A	EA15X331	123A	ET15X11
123A	CDC13000-1D	123A	CS1235E	123A	C89015	123A	EA15X367	123A	ET15X12
123A	CDC13016A	123A	CS1235G	123A	C89015/3490	123A	EA15X37	123A	ET15X13
123A	CDC13500-1	123A	CS1236C	123A	C89013A	123A	EA15X370	123A	ET15X14
123A	CDC2010	123A	CS1236H	123A	C89013B	123A	EA15X371	123A	ET15X15
123A	CDC2010C	123A	CS1236P	123A	C89013C	123A	EA15X373	123A	ET15X20
123A	CDC2010D	123A	CS1238P	123A	C89013D	123A	EA15X379	123A	ET15X24
123A	CDC25100-6	123A	CS1245F	123A	C89013F	123A	EA15X412	123A	ET15X27
123A	CDC25100	123A	CS1245H	123A	C89013F	123A	EA15X44	123A	ET15X37
123A	CDC25100-G	123A	CS1245J	123A	C89013H	123A	EA15X45	123A	ET15X41
123A	CDC25100	123A	CS1245F	123A	C89013HE	123A	EA15X52	123A	ET15X42
123A	CDC430	123A	CS1250B	123A	C89013HF	123A	EA15X56	123A	ET15X45
123A	CDC4306813	123A	CS1257	123A	C89013HG	123A	EA15X58	123A	ET234843
123A	CDC745(ZENITH)	123A	CS1258	123A	C89013HG/3490	123A	EA15X59	123A	ET238894
123A	CDC746	123A	CS1259	123A	C89013HH	123A	EA15X63	123A	ET368021
123A	CD08000	123A	CS1283A	123A	C89014	123A	EA15X68	123A	ET412626
123A	CD08000-1B	123A	CS1288	123A	C89014/3490	123A	EA15X7112	123A	ET8-068
123A	CD08001	123A	CS1289	123A	C89014A	123A	EA15X7115	123A	ETT-CDC-12000
123A	CD08011B	123A	CS1295E	123A	C89014B	123A	EA15X7118	123A	ETTC-458LG
123A	CD08021	123A	CS1295G	123A	C89014C	123A	EA15X7119	123A	ETTC-CD12000
123A	CD8054	123A	CS1401	123A	C89014C/3490	123A	EA15X7120	123A	ETTC-CD13000
123A	CD86X7-5	123A	CS1340I	123A	C89014D	123A	EA15X7175	123A	ETTC-CD8000
123A	CDQ10001	123A	CS1344	123A	C89101B	123A	EA15X7176	123A	ETX18
123A	CDQ10002	123A	CS1345	123A	C89125B	123A	EA15X72	123A	EW165V
123A	CDQ10003	123A	CS1348	123A	C89126	123A	EA15X7232	123A	EW181
123A	CDQ10004	123A	CS1349	123A	C89259	123A	EA15X73	123A	EW182
123A	CDQ10005	123A	CS1353	123A	C89600-4	123A	EA15X75	123A	EY2P-632
123A	CDQ10006	123A	CS13610	123A	C89600-5	123A	EA15X7514	123A	EY2P-791
123A	CDQ10007	123A	CS1362	123A	CTP-2001-1007	123A	EA15X7517	123A	F121-453804
123A	CDQ10008	123A	CS1363	123A	CTP-2001-1008	123A	EA15X7586	123A	F15840
123A	CDQ10009	123A	CS1368	123A	D048	123A	EA15X76	123A	F15840-1
123A	CDQ10010	123A	CS1368A	123A	D053	123A	EA15X7638	123A	F222
123A	CDQ10016	123A	CS1368B	123A	D16E7	123A	EA15X7643	123A	F2443
123A	CDQ10017	123A	CS1368C	123A	D16E9	123A	EA15X7	123A	F2448
123A	CDQ10018	123A	CS1368D	123A	D16EC18	123A	EA15X83	123A	F2584
123A	CDQ10019	123A	CS1370	123A	D1A	123A	EA15X84	123A	F302-1
123A	CDQ10020	123A	CS1371	123A	D1838	123A	EA15X85	123A	F302-1532
123A	CDQ10021	123A	CS1372	123A	D294	123A	EA15X8502	123A	F302-2
123A	CDQ10022	123A	CS1383	123A	D2R38	123A	EA15X8511	123A	F302-2532
123A	CDQ10023	123A	CS1420	123A	D328	123A	EA15X8529	123A	F306-001
123A	CDQ10024	123A	CS1453E	123A	D33028	123A	EA15X86	123A	F306-022
123A	CDQ10025	123A	CS1463A	123A	D342	123A	EA15X89	123A	F5519
123A	CDQ10026	123A	CS1585	123A	D372EL	123A	EA15X9	123A	F5532
123A	CDQ10027	123A	CS1585E/F	123A	D912	123A	EA15X91	123A	F5569
123A	CDQ10028	123A	CS1585G	123A	D917254-2	123A	EA15X96	123A	F5571
123A	CDQ10052	123A	CS1625	123A	D921881-1	123A	EA15X98	123A	F366
123A	CE40001B	123A	CS1665	123A	D928121	123A	EA1628	123A	F572-1
123A	CE40025E	123A	CS2001	123A	DBCZ083905	123A	EA1629	123A	F587
123A	CP-2	123A	CS2004	123A	DBCZ083906	123A	EA1630	123A	F75116
123A	CP2	123A	CS2004D	123A	DBCZ136406	123A	EA1638	123A	F9600
123A	CP5	123A	CS2006	123A	DDBY209003	123A	EA1695	123A	F9623
123A	CQ1	123A	CS2218	123A	DDBY222002	123A	EA1696	123A	FA-1(SEARS)
123A	CI2711					123A	EA1697	123A	FB6853
								123A	FBC237

ECG	Industry Standard No.	ECG	Industry Standard No.	ECG	Industry Standard No.	ECG	Industry Standard No.	ECG	Industry Standard No.
123A	PCS1168F813	123A	HC372	123A	HX-50063	123A	M54A	123A	MPS65615
123A	PCS11168G	123A	HC373	123A	HX-50072	123A	M54B	123A	MPS6590
123A	PCS11168G704	123A	HC458	123A	HX-50113	123A	M54BLK	123A	MPS9185
123A	PCS1229F	123A	HC539	123A	HX50002	123A	M54BLU	123A	MPS89423
123A	PCS1229G	123A	HC561	123A	HT3045801C	123A	M54BRN	123A	MPS8933
123A	FD-1029-JA	123A	HCL-29	123A	I9A115728-2	123A	M54C	123A	MPS89433J
123A	FD-1029-LL	123A	HCL-6066	123A	IC743042	123A	M54D	123A	MPS89434J
123A	FD-1029-NG	123A	HD-00227	123A	IP20-0001	123A	M54E	123A	MPS89434K
123A	FD-1029-PP	123A	HEP-80004	123A	IP20-0003	123A	M54GRN	123A	MPS89600
123A	FD-1029-PT	123A	HEP-80011	123A	IP20-0006	123A	M54ORN	123A	MPS89600-5
123A	FMPEA20	123A	HEP-80022	123A	IP20-0029	123A	M54RED	123A	MPS89600F
123A	FPR40-1001	123A	HEP-80030	123A	IP20-0032	123A	M54WHT	123A	MPS89600G
123A	FPR50-1001	123A	HEP50	123A	IP20-0039	123A	M54YEL	123A	MPS89600G/H
123A	FPR50-1002	123A	HEP53	123A	IP20-0040	123A	M671	123A	MPS89600H
123A	FS1221	123A	HEP54	123A	IP20-0122	123A	M7003	123A	MPS89604D
123A	FS1974	123A	HEP55	123A	IRTR62	123A	M7015	123A	MPS89604B
123A	FS36999	123A	HEP724	123A	IRTR63	123A	M7033	123A	MPS89604I
123A	FT005	123A	HEP725	123A	J139A	123A	M7108	123A	MPS89604R
123A	FT006	123A	HEP728	123A	J241054	123A	M7108/A5N	123A	MPS89611-5
123A	FT008	123A	HEP729	123A	J241099	123A	M7109	123A	MPS89616
123A	FT008A	123A	HEP735	123A	J241290	123A	M7109/A5P	123A	MPS89616A
123A	FT023	123A	HEP738	123A	J241251	123A	M7171	123A	MPS89616J
123A	FT024	123A	HP-40	123A	J24458	123A	M773	123A	MPS89618
123A	FT025	123A	HP2	123A	J24564	123A	M773RED	123A	MPS89618H
123A	FT026	123A	HP3	123A	J24565	123A	M774	123A	MPS89618I
123A	FT053	123A	HP4	123A	J24624	123A	M774ORN	123A	MPS89618J
123A	FT3643	123A	HP5	123A	J24625	123A	M775	123A	MPS89623G
123A	G005-036C	123A	HP6	123A	J24641	123A	M775BRN	123A	MPS89623E
123A	G005-036E	123A	HP7	123A	J24658	123A	M776	123A	MPS89623F
123A	G05-010-A	123A	HP8	123A	J24752	123A	M776GRN	123A	MPS89623G
123A	G05-011-A	123A	HKK-158	123A	J24753	123A	M779BLU	123A	MPS89623G/H
123A	G05-015-D	123A	HKT-161	123A	J24817	123A	M780WHT	123A	MPS89623H/I
123A	G05-015C	123A	HR-11	123A	J24855	123A	M783	123A	MPS89623I/J
123A	G05-034-D	123A	HR-11A	123A	J24874	123A	M783RED	123A	MPS89626
123A	G05-035-D	123A	HR-11B	123A	J24878	123A	M784	123A	MPS89630
123A	G05-055E	123A	HR-13	123A	J24906	123A	M784ORN	123A	MPS89630H
123A	G05-036-C	123A	HR-13A	123A	J24907	123A	M785	123A	MPS89630I
123A	G05-036-E	123A	HR-14A	123A	J24909	123A	M785YEL	123A	MPS89630J
123A	G05-036D	123A	HR-15A	123A	J24916	123A	M787BLU	123A	MPS89631
123A	G05-036E	123A	HR-16	123A	J310249	123A	M791	123A	MPS89651I
123A	G05-037B	123A	HR-16a	123A	J310250	123A	M8105	123A	MPS89651J
123A	G05-064-A	123A	HR-17	123A	J9618(G.E.)	123A	M818	123A	MPS89651K
123A	G05010C	123A	HR-17A	123A	JA-H	123A	M818WHT	123A	MPS89651T
123A	G05035E	123A	HR-18	123A	JA-L	123A	M822	123A	MPS89632
123A	G05036	123A	HR-19	123A	JA1200	123A	M8221	123A	MPS89632H
123A	G05036B	123A	HR-19A	123A	JE9011	123A	M822A	123A	MPS89632I
123A	G05036C	123A	HR-32	123A	JLM-20	123A	M822A-BLU	123A	MPS89632J
123A	G05036D	123A	HR-36	123A	JN271	123A	M822B	123A	MPS89632K
123A	G05037B	123A	HR-37	123A	JSP7001	123A	M823	123A	MPS89632T
123A	G05059	123A	HR-38	123A	J8P7001B	123A	M823B	123A	MPS8966H
123A	G395967	123A	HR-48	123A	JT-16GN-40	123A	M823WHT	123A	MPS87000
123A	G395967-2	123A	HR36	123A	K4-506	123A	M827	123A	MPS87000B
123A	G9600	123A	HR62	123A	KB8339	123A	M827BRN	123A	MPS87000D
123A	G9600(G.E.)	123A	HR63	123A	KD2102	123A	M828GRN	123A	MFX9623H/I
123A	G9629	123A	HR64	123A	KIAH1422	123A	M847BLK	123A	MFX96301
123A	G9696	123A	HR66	123A	KIA704	123A	M9095	123A	MQ1
123A	GC1144	123A	HR84(NPN)	123A	KPG6682	123A	M9159	123A	MQ2
123A	GE-10	123A	HS-1168	123A	KR-Q1013	123A	M91A	123A	MR3932
123A	GE-17	123A	HS-1229	123A	KTR0710C	123A	M91B	123A	MR9604
123A	GE-20	123A	HS-40037	123A	KTR08150	123A	M91C	123A	MS2223
123A	GE-210	123A	HS-40044	123A	KTR0859C	123A	M91CM624	123A	MST502R
123A	GE-211	123A	HS-40046	123A	KTR16870	123A	M91D	123A	MST503R
123A	GE-3265	123A	HS40046	123A	LLB-23	123A	M91E	123A	MSR7503
123A	GE-81	123A	HT303620B	123A	LM-1129	123A	M91F	123A	MT104
123A	GE-X16A1938	123A	HT303620	123A	LM-1130	123A	M91FM624	123A	MT4101
123A	GET2221	123A	HT303711AO	123A	LM-1132	123A	M9226	123A	MT4102
123A	GET2222	123A	HT303711B-0	123A	LM-1133	123A	M9525	123A	MT4102A
123A	GET2369	123A	HT303711BO	123A	LM-1147	123A	M9532	123A	MT4103
123A	GET3013	123A	HT303721-0	123A	LM-1148B	123A	M9563	123A	MT6001
123A	GET5014	123A	HT303721O	123A	LM-1155	123A	M9568	123A	MT6002
123A	GET5646	123A	HT303721-0	123A	LM1090B	123A	M9570	123A	MT6003
123A	GET706	123A	HT303730	123A	LM1090F	123A	MA4101	123A	MT696
123A	GET708	123A	HT303730A	123A	LM1090G	123A	MA4103	123A	MT697
123A	GET914	123A	HT303730O	123A	LM1117D	123A	MA6001	123A	MT706
123A	GI10	123A	HT303731O	123A	LM1403	123A	MA6002	123A	MT706A
123A	GI2711	123A	HT304531	123A	LM141-6	123A	MA6003	123A	MT706B
123A	GI2712	123A	HT304531A	123A	LM1415-6	123A	MA6102	123A	MT707
123A	GI2713	123A	HT304531B	123A	LM1415-7	123A	MA9426	123A	MT708
123A	GI2714	123A	HT304540BQ	123A	LM1540	123A	MA77786	123A	MT9001
123A	GI2715	123A	HT304580	123A	LM1566F	123A	MC9427	123A	MT9002
123A	GI2716	123A	HT304580A	123A	LM1614D	123A	ME-1	123A	N-EA15X136
123A	GI2921	123A	HT304580B	123A	LM1614M	123A	ME-2	123A	N-EA15X137
123A	GI2922	123A	HT304580C0	123A	LM1818	123A	ME-3	123A	N-EA15X138
123A	GI2923	123A	HT304580K	123A	LRQ849	123A	ME1001	123A	0N047204-2
123A	GI2924	123A	HT304580YO	123A	LS-0085-01	123A	ME1002	123A	N201AY
123A	GI3641	123A	HT304580Z	123A	L83705	123A	ME2001	123A	N3565
123A	GI3643	123A	HT304581	123A	M-1002-2	123A	ME2002	123A	OM47204-1
123A	GI3704	123A	HT304581A	123A	M-75557-1	123A	ME213	123A	N4T
123A	GI3705	123A	HT304581B	123A	M-75557-2	123A	ME213A	123A	NC207AL
123A	GI3706	123A	HT304581B-0	123A	M-75557-3	123A	ME216	123A	NJ100B
123A	GI3707	123A	HT304581C	123A	M-75557-4	123A	ME217	123A	NJ102C
123A	GI3708	123A	HT304861B	123A	M-75557-5	123A	ME4001	123A	NKT2-1A19
123A	GI3709	123A	HT305361E	123A	M-75557-6	123A	ME4002	123A	NKT10339
123A	GI3710	123A	HT305361O	123A	M-8641	123A	ME4003	123A	NKT10419
123A	GI3711	123A	HT305371E	123A	M140-3	123A	ME4003C	123A	NKT10439
123A	GMB1001	123A	HT306441	123A	M24	123A	ME4101	123A	NKT10519
123A	GMB1002	123A	HT306441A	123A	M24A	123A	ME4102	123A	NKT12329
123A	GME2001	123A	HT306441B-0	123A	M24B	123A	ME4103	123A	NKT12429
123A	GME2002	123A	HT306441BO	123A	M25	123A	ME4104	123A	NKT13329
123A	GMB4001	123A	HT306441C	123A	M25A	123A	ME6001	123A	NKT13429
123A	GMB4002	123A	HT306441C-0	123A	M25B	123A	ME6002	123A	NL-102
123A	GMB6003	123A	HT307321A	123A	M25B2	123A	ME6003	123A	NP0737
123A	GMB6003	123A	HT307321B-0	123A	M31001	123A	ME900	123A	NR-071AU
123A	G05-010-A	123A	HT307322A	123A	M3519	123A	ME900A	123A	NR-431AG
123A	G05-011-A	123A	HT307331B	123A	M4464	123A	ME901	123A	NR-431A8
123A	G05056	123A	HT307331C	123A	M4465	123A	ME901A	123A	NR-461A8
123A	G05059	123A	HT307331CO	123A	M447	123A	MFP-25	123A	NR041
123A	G89018F	123A	HT307341B	123A	M4594	123A	MI9623	123A	NR041B
123A	G89023K	123A	HT307341O-0	123A	M4624	123A	MI9623	123A	NR071AU
123A	GV6063	123A	HT308281B	123A	M4630	123A	MI9630	123A	NR091BT
123A	GVL 20077	123A	HT308281C	123A	M4705	123A	MI9623	123A	NR201AY
123A	H102	123A	HT308281G	123A	M4714	123A	MI9630	123A	NR261A8
123A	H1567	123A	HT308281O	123A	M4732	123A	MM1755	123A	NR271AY
123A	H931	123A	HT308282A	123A	M4734	123A	MM1756	123A	NR461
123A	H933	123A	HT308282A-0	123A	M4737	123A	MM1757	123A	NR461BH
123A	H934	123A	HT308291A	123A	M4739	123A	MM1758	123A	NS1500
123A	H9423	123A	HT308291A-0	123A	M4765	123A	MM3903	123A	NS1972
123A	H9618	123A	HT308291B	123A	M4768	123A	MM3904	123A	NS1973
123A	H9623	123A	HT308291B-0	123A	M4821	123A	MP1014-2	123A	NS1974
123A	H9696	123A	HT308291BO	123A	M4834	123A	MPB 9623G	123A	NS1975
123A	HC-00372	123A	HT308291C	123A	M4840	123A	MPS2926BRN	123A	NS3903
123A	HC-00537	123A	HT308291DO	123A	M4841	123A	MPS2926GRN	123A	NS3904
123A	HC-00693	123A	HT308291C	123A	M4842	123A	MPS2926ORN	123A	NS475
123A	HC-00735	123A	HT308291DO	123A	M4842A	123A	MPS2926RED	123A	NS476
123A	HC-00828	123A	HT309842A-0	123A	M4842C	123A	MPS2926YEL	123A	NS478
123A	HC-00871	123A	HT400	123A	M4844	123A	MPS2992	123A	NS479
123A	HC-00921	123A	HT401	123A	M4852	123A	MPS6351	123A	NS6114
123A	HC-00924	123A	HT800011F	123A	M4854	123A	MPS6413	123A	NS6115
123A	HC-00945	123A	HT800011H	123A	M4898	123A	MPS6552	123A	NS6207
123A	HC-01820	123A	HT800011H	123A	M4906	123A	MPS6553	123A	NS6210
123A	HC-56	123A	HT800011K	123A	M4926	123A	MPS6554	123A	NS7262
123A	HC00838	123A	HT800181O	123A	M4933	123A	MPS6556	123A	NS731
123A	HC371	123A	HV25	123A	M4935				
				123A	M4937				
				123A	M54				

ECG	Industry Standard No.	ECG	Industry Standard No.	ECG	Industry Standard No.	ECG	Industry Standard No.	ECG	Industry Standard No.
123A	N8731A	123A	Q3/6515	123A	R85856	123A	RV2249	123A	S2984
123A	N8733	123A	Q35242	123A	R85857	123A	RVTC81381	123A	S2985
123A	N8733A	123A	Q5053	123A	R87103	123A	RVTC81383	123A	S2989
123A	N8734	123A	Q5123B	123A	R87105	123A	RVT822410	123A	S2996
123A	N8734A	123A	Q5123F	123A	R87108	123A	RYN121105	123A	S2997
123A	N8949	123A	QA-12	123A	R87111	123A	RYN121105-3	123A	S2998
123A	09-309060	123A	QA-13	123A	R87121	123A	RYN121105-4	123A	S2999
123A	0N47204-2	123A	QA-13	123A	R87127	123A	S-1061	123A	S34540
123A	0N271	123A	QA-14	123A	R87129	123A	S-1065	123A	S36999
123A	0N274	123A	QA-15	123A	R87132	123A	S-1066	123A	S85369
123A	0N47204-1	123A	QA-16	123A	R87133	123A	S-1068	123A	S6801
123A	P-8393	123A	QA-19	123A	R87136	123A	S-1128	123A	S95202
123A	P/N10000020	123A	Q00254	123A	R87160	123A	S-1143	123A	S9631
123A	P04-44-0028	123A	Q0V60529	123A	R87225	123A	S-1221	123A	SC-4044
123A	P04-45-0014-P2	123A	Q0V60530	123A	R87224	123A	S-1221A	123A	SC-4244
123A	P04-45-0014-P5	123A	Q8054	123A	R87226	123A	S-1245	123A	SC-65
123A	P04440028-001	123A	Q8C380	123A	R87232	123A	S-1331W	123A	SC-832
123A	P04440028-009	123A	Q8E1001	123A	R87234	123A	S-1363	123A	SC1001
123A	P04440028-014	123A	QT-C0372XAT	123A	R87235	123A	S-1564	123A	SC1168G
123A	P04440028-8	123A	QT-C0710XBE	123A	R87236	123A	S-1403	123A	SC1168H
123A	P04440032-001	123A	QT-00829XAN	123A	R87241	123A	S-1512	123A	SC1229Q
123A	P15153	123A	QT-C0829XBN	123A	R87242	123A	S-1533	123A	SC950
123A	P1901-50	123A	QT-C01687XAN	123A	R87405	123A	S-1559	123A	SC1010
123A	P480A0028	123A	QT-C0839XDA	123A	R87406	123A	S001466	123A	SC4044
123A	P480A0029	123A	R-280537	123A	R87407	123A	S0015	123A	SC65
123A	P5152	123A	R3273-P1	123A	R87408	123A	S0022	123A	SC832
123A	P5153	123A	R3273-P2	123A	R87409	123A	S0025	123A	SC842
123A	P533567	123A	R3283	123A	R87410	123A	S022010	123A	SCDT323
123A	P64447	123A	R3293(GE)	123A	R87411	123A	S022011	123A	SD-109
123A	P8393	123A	R34-6016-58	123A	R87412	123A	S024428	123A	SD109
123A	P8394	123A	R340	123A	R87413	123A	S024987	123A	SDD3000
123A	P9623	123A	R4057	123A	R87415	123A	S025232	123A	SDD421
123A	PA7001/0001	123A	R582	123A	R87421	123A	S025289	123A	SDB821
123A	PA9006	123A	R7163	123A	R87504	123A	S031A	123A	SE-0566
123A	PEP2	123A	R7165	123A	R87510	123A	S037	123A	SE-1002
123A	PEP5	123A	R7249	123A	R87512	123A	S0704	123A	SE-1331
123A	PEP6	123A	R7343	123A	R87513-15	123A	S1016	123A	SE-2001
123A	PEP7	123A	R7359	123A	R87514	123A	S1061	123A	SE-3646
123A	PEP8	123A	R7360	123A	R87515	123A	S1065	123A	SE-4001
123A	PEP9	123A	R7361	123A	R87516	123A	S1066	123A	SE-4002
123A	PET1002	123A	R7582	123A	R87517	123A	S1068	123A	SE-4010
123A	PET2001	123A	R7887	123A	R87517-19	123A	S1069	123A	SE-5006
123A	PET2002	123A	R7953	123A	R87518	123A	S1074	123A	SE-6001
123A	PET5704	123A	R8066	123A	R87519	123A	S1074R	123A	SE-6002
123A	PET5705	123A	R8067	123A	R87521	123A	S1128	123A	SE1331
123A	PET5706	123A	R8068	123A	R87525	123A	S1143	123A	SE2401
123A	PET4001	123A	R8070	123A	R87526	123A	S12-1-A-3P	123A	SE2402
123A	PET4002	123A	R8115	123A	R87527	123A	S1221	123A	SE4001
123A	PET6001	123A	R8116	123A	R87528	123A	S1221A	123A	SE4002
123A	PET6002	123A	R8117	123A	R87529	123A	S1226	123A	SE4010
123A	PET8000	123A	R8118	123A	R87530	123A	S1241	123A	SE4020
123A	PET8002	123A	R8119	123A	R87542	123A	S1242	123A	SE4172
123A	PET8003	123A	R8120	123A	R87543	123A	S1243	123A	SE5-0128
123A	PET8004	123A	R8223	123A	R87544	123A	S1245	123A	SE5-0253
123A	PET9002	123A	R8224	123A	R87555	123A	S1272	123A	SE5-0274
123A	PL1052	123A	R8225	123A	R87606	123A	S1307	123A	SE5-0367
123A	PL1054	123A	R8243	123A	R87607	123A	S1309	123A	SE5-0567
123A	PM1121	123A	R8244	123A	R87609	123A	S133-1	123A	SE5-0608
123A	PN107	123A	R8259	123A	R87610	123A	S1331	123A	SE5-0848
123A	PRT-101	123A	R8260	123A	R87611	123A	S1331N	123A	SE5-0854
123A	PRT-104	123A	R8261	123A	R87612	123A	S1331W	123A	SE5-0855
123A	PRT-104-1	123A	R8305	123A	R87613	123A	S1363	123A	SE5-0887
123A	PRT-104-2	123A	R8312	123A	R87614	123A	S1364	123A	SE5-0888
123A	PRT-104-3	123A	R8528	123A	R87620	123A	S1369	123A	SE5-0938-54
123A	PS209800	123A	R8530	123A	R87621	123A	S1373	123A	SE5030A
123A	PT1558	123A	R8543	123A	R87622	123A	S1374	123A	SE5030B
123A	PT1559	123A	R8551	123A	R87623	123A	S1403	123A	SE5151
123A	PT1610	123A	R8552	123A	R87624	123A	S1405	123A	SE6010
123A	PT1835	123A	R8553	123A	R87625	123A	S1419	123A	SE6420
123A	PT1836	123A	R8554	123A	R87626	123A	S1420	123A	SE8040
123A	PT1837	123A	R8555	123A	R87627	123A	S1429-3	123A	SF1001
123A	PT2760	123A	R8556	123A	R87628	123A	S1432	123A	SF1713
123A	PT2896	123A	R8557	123A	R87634	123A	S1443	123A	SF1714
123A	PT3141	123A	R8620	123A	R87635	123A	S1453	123A	SF1726
123A	PT3141A	123A	R8646	123A	R87636	123A	S1475	123A	SF1730
123A	PT3141B	123A	R8647	123A	R87637	123A	S1476	123A	SF7713
123A	PT3151A	123A	R8648	123A	R87638	123A	S1487	123A	SF7714
123A	PT31513	123A	R8658	123A	R87639	123A	S1502	123A	SH1064
123A	PT31510	123A	R8889	123A	R87640	123A	S1510	123A	SJ570
123A	PT3500	123A	R8914	123A	R87641	123A	S1526	123A	SK1640A
123A	PT4-7158	123A	R8916	123A	R87642	123A	S1527	123A	SK1641
123A	PT4-7158-012	123A	R8963	123A	R87643	123A	S1529	123A	SK3020
123A	PT4-7158-013	123A	R8964	123A	R87814	123A	S1530	123A	SK3038
123A	PT4-7158-01A	123A	R8965	123A	R88442	123A	S1559	123A	SK3046
123A	PT4-7158-021	123A	R8966	123A	R88567	123A	S1649	123A	SK3434A
123A	PT4-7158-022	123A	R9004	123A	R88605T332	123A	S1568	123A	SK5801
123A	PT4-7158-023	123A	R9005	123A	RT-100	123A	S1570	123A	SK5915
123A	PT4-7158-02A	123A	R9006	123A	RT-929-H	123A	S1619	123A	SK8215
123A	PT4800	123A	R9025	123A	RT-929H	123A	S1620	123A	SK8251
123A	PT627	123A	R9071	123A	RT-930H	123A	S1629	123A	SKA1080
123A	PT703	123A	R9384	123A	RT100	123A	S1697	123A	SKA1117
123A	PT720	123A	R9385	123A	RT114	123A	S169N	123A	SKA1395
123A	PT851	123A	R9483	123A	RT2016	123A	S1761	123A	SKA4141
123A	PT886	123A	R813	123A	RT2332	123A	S1761A	123A	SKB8359
123A	PT887	123A	RH120	123A	RT2914	123A	S1761B	123A	SL300
123A	PT897	123A	RR7504	123A	RT3053	123A	S1761C	123A	SL7990
123A	PT898	123A	RR8068	123A	RT3064	123A	S1764	123A	SM-4508-B
123A	PTC115	123A	RR8914	123A	RT3228	123A	S1765	123A	SM-5564
123A	PTC136	123A	RS-107	123A	RT3565	123A	S1766	123A	SM-5643
123A	Q-00269	123A	RS-108	123A	RT3567	123A	S1768	123A	SM-716
123A	Q-00269A	123A	RS-2009	123A	RT4760	123A	S1770	123A	SM-7815
123A	Q-00269B	123A	RS-2013	123A	RT5202	123A	S1772	123A	SM-7836
123A	Q-00269C	123A	RS-2016	123A	RT5206	123A	S1784	123A	SM-A-726655
123A	Q-00369	123A	RS-5851	123A	RT5207	123A	S1785	123A	SM-A-726664
123A	Q-00369A	123A	RS-5853	123A	RT5551	123A	S1835	123A	SM-B-610342
123A	Q-00369B	123A	RS-5856	123A	RT5901	123A	S1871	123A	SM-B-686767
123A	Q-00369C	123A	RS-5857	123A	RT6732	123A	S1891	123A	SM-C-583256
123A	Q-00484R	123A	RS-7103	123A	RT6733	123A	S1891A	123A	SM2700
123A	Q-00569	123A	RS-7105	123A	RT69221	123A	S1891B	123A	SM2701
123A	Q-00569A	123A	RS-7409	123A	RT697M	123A	S1955	123A	SM3104
123A	Q-00569B	123A	RS-7411	123A	RT6989	123A	S1993	123A	SM3117A
123A	Q-00669	123A	RS-7412	123A	RT7322	123A	S2034	123A	SM3505
123A	Q-00669A	123A	RS-7413	123A	RT7325	123A	S2043	123A	SM3986
123A	Q-00669C	123A	RS-7504	123A	RT7326	123A	S2044	123A	SM4508-B
123A	Q-00684R	123A	RS-7511	123A	RT7327	123A	S2121	123A	SM5379
123A	Q-0115C	123A	RS-7606	123A	RT7511	123A	S2122	123A	SM5564
123A	Q-02115C	123A	RS-7607	123A	RT7514	123A	S2123	123A	SM5643
123A	Q-03115C	123A	RS-7609	123A	RT7515	123A	S2124	123A	SM576-1
123A	Q-04115C	123A	RS-7610	123A	RT7517	123A	S2171	123A	SM576-2
123A	Q-05115C	123A	RS-7611	123A	RT7518	123A	S2172	123A	SM5981
123A	Q-06115C	123A	RS-7612	123A	RT7528	123A	S2225	123A	SM5773
123A	Q-07115C	123A	RS-7613	123A	RT7577	123A	S22543	123A	SM716
123A	Q-10115C	123A	RS-7614	123A	RT7838	123A	S2397	123A	SM7545
123A	Q-14115C	123A	RS-7622	123A	RT7845	123A	S24591	123A	SM7815
123A	Q-15115C	123A	RS-7623	123A	RT7943	123A	S24596	123A	SM7836
123A	Q-16115C	123A	R81049	123A	RT8143	123A	S2581	123A	SM8112
123A	Q-35	123A	R81059	123A	RT8195	123A	S2582	123A	SM8113
123A	Q-RF-2	123A	R8128	123A	RT8197	123A	S2590	123A	SM8978
123A	Q-SE1201	123A	R8136	123A	RT8198	123A	S2593	123A	SM900B
123A	Q0V60529	123A	R81048	123A	RT8201	123A	S2635	123A	SM9135
123A	Q0V60530	123A	R88914	123A	RT8230	123A	S2636	123A	SM9253
123A	Q0V60538	123A	R85851	123A	R8929H	123A	S2935	123A	SPC040
123A		123A	R85853	123A	RV1471	123A	S2944	123A	SPC42
123A		123A		123A	RV1474	123A	S29445	123A	

ECG	Industry Standard No.	ECG	Industry Standard No.	ECG	Industry Standard No.	ECG	Industry Standard No.	ECG	Industry Standard No.
123A	SP050	123A	ST04	123A	T459(SEARS)	123A	TIS71	123A	TR8025
123A	SP051	123A	ST05	123A	T460	123A	TIS72	123A	TR8028
123A	SP052	123A	ST06	123A	T461-16	123A	TIS90-2	123A	TR8029
123A	SP8-1475	123A	ST1242	123A	T461-16(SEARS)	123A	TIS92	123A	TR8030
123A	SP8-1475YT	123A	ST1243	123A	T462	123A	TIS92-BLU	123A	TR8031
123A	SP8-1476	123A	ST1244	123A	T462(SEARS)	123A	TIS92-GRN	123A	TR8034
123A	SP8-4075	123A	ST1290	123A	T472	123A	TIS92-GRY	123A	TR8035
123A	SP8-41	123A	ST150	123A	T472(SEARS)	123A	TIS92-VIO	123A	TR8040
123A	SP8-952	123A	ST1506	123A	T485(SEARS)	123A	TIS92-YEL	123A	TR8043
123A	SP8-952-2	123A	ST151	123A	T484(SEARS)	123A	TIX94	123A	TR9100
123A	SP81045	123A	ST152	123A	T485(SEARS)	123A	TIX712	123A	TRA-34
123A	SP81475	123A	ST153	123A	T486(SEARS)	123A	TIX812	123A	TRA-36
123A	SP82225	123A	ST154	123A	T59235A	123A	TIX813	123A	TRA-4
123A	SP82270	123A	ST155	123A	T615A002	123A	TK1228-1010	123A	TRA-4A
123A	SP83015	123A	ST156	123A	T615A006-1	123A	TK1228-1011	123A	TRA-4B
123A	SP83735	123A	ST157	123A	T6565	123A	TK1228-1012	123A	TRA-9R
123A	SP83751	123A	ST160	123A	T76	123A	TM123A	123A	TRA34
123A	SP83900	123A	ST1607	123A	T9011A1C	123A	TM1613	123A	TRA36
123A	SP83907	123A	ST161	123A	T9011A1G	123A	TM1711	123A	TRA4
123A	SP83908	123A	ST162	123A	T9011AZ	123A	TM2613	123A	TRA4A
123A	SP83909	123A	ST163	123A	TA-6	123A	TM2711	123A	TRA4B
123A	SP83915	123A	ST1694	123A	TA198030-4	123A	TMT-1543	123A	TRA9R
123A	SP83923	123A	ST175	123A	TA6	123A	TMT1543	123A	TRAPLC711
123A	SP83925	123A	ST176	123A	TA7	123A	TN237	123A	TRAPLC871
123A	SP83926	123A	ST177	123A	TAC-047	123A	TN53	123A	TRAPLC871A
123A	SP83930	123A	ST178	123A	TA0047	123A	TN55	123A	TRB0147B
123A	SP83936	123A	ST180	123A	TC3123036722	123A	TN56	123A	T82221
123A	SP83938	123A	ST181	123A	TC3123036900	123A	TN59	123A	T82222
123A	SP83940	123A	ST182	123A	TC3123037111	123A	TN60	123A	TSC499
123A	SP83951	123A	ST250	123A	TC3123037222	123A	TN61	123A	TSC695
123A	SP83957C	123A	ST251	123A	TC3123037412	123A	TN62	123A	T8T705899A
123A	SP83967	123A	ST25A	123A	TE1420	123A	TN63	123A	TT-1097
123A	SP83972	123A	ST25C	123A	TE2369	123A	TN64	123A	TT1097
123A	SP83973	123A	ST403	123A	TE3414	123A	TN80	123A	TV-18
123A	SP83999	123A	ST50	123A	TE3415	123A	TN561689	123A	TV-21
123A	SP84003	123A	ST501	123A	TE3605	123A	TN061702	123A	TV-23
123A	SP84004	123A	ST502	123A	TE3605A	123A	TNJ-60076	123A	TV-40
123A	SP84006	123A	ST503	123A	TE3607	123A	TNJ-60606	123A	TV-46
123A	SP84009	123A	ST504	123A	TE3704	123A	TNJ-60607	123A	TV-51
123A	SP84017	123A	ST5060	123A	TE3705	123A	TN60076	123A	TV-52
123A	SP84020	123A	ST51	123A	TE3903	123A	TNJ61219	123A	TV-53
123A	SP84029	123A	ST53	123A	TE3904	123A	TNJ61220	123A	TV-56
123A	SP84032	123A	ST54	123A	TE3906	123A	TNJ70479-1	123A	TV-58
123A	SP84034	123A	ST55	123A	TE4123	123A	TNJ70537	123A	TV-6
123A	SP84037	123A	ST56	123A	TE4124	123A	TNJ70539	123A	TV-60
123A	SP84039	123A	ST57	123A	TE4424	123A	TNJ70637	123A	TV-65
123A	SP84040	123A	ST58	123A	TE4951	123A	TNJ70638	123A	TV-66
123A	SP84041	123A	ST59	123A	TE4952	123A	TNJ70639	123A	TV-68
123A	SP84042	123A	ST63	123A	TE4953	123A	TNJ70640	123A	TV-84
123A	SP84044	123A	ST64	123A	TE4954	123A	TNJ71036	123A	TV-92
123A	SP84045	123A	ST6511	123A	TE5309A	123A	TNJ72280	123A	TV17
123A	SP84049	123A	ST6512	123A	TE5311A	123A	TNJ72281	123A	TV241077
123A	SP84052	123A	SX3709	123A	TE5368	123A	TNJ72783	123A	TV241078
123A	SP84053	123A	SX3711	123A	TE5369	123A	TNJ72784	123A	TV24215
123A	SP84055	123A	SX55	123A	TE5370	123A	TNT-843	123A	TV24216
123A	SP84059	123A	SYL-1182	123A	TE5371	123A	TNT842	123A	TV24281
123A	SP84060	123A	SYL1182	123A	TE5376	123A	TO-033	123A	TV24372
123A	SP84061	123A	SYL5460	123A	TE5377	123A	TO-038	123A	TV24453
123A	SP84062	123A	T-1416	123A	TE5449	123A	TO-039	123A	TV24454
123A	SP84063	123A	T-255	123A	TE5450	123A	TO-040	123A	TV24458
123A	SP84066	123A	T-256	123A	TE5451	123A	TO1-101	123A	TV24576
123A	SP84067	123A	T-399	123A	TE697	123A	TO1-104	123A	TV24655
123A	SP84069	123A	T-H280536	123A	TF-78	123A	TO1-105	123A	TV28C208
123A	SP84074	123A	T-H280693	123A	TG28C1175-D	123A	TP4123	123A	TV38
123A	SP84075	123A	T-H280715	123A	TG28C1175-E	123A	TP4124	123A	TV39
123A	SP84077	123A	T-Q5053	123A	TH28C1175C	123A	TP86512	123A	TV46
123A	SP84081	123A	T-Q5053C	123A	TH28C596	123A	TP86513	123A	TV48
123A	SP84083	123A	T-Q5073	123A	TH28C693	123A	TP86514	123A	TV56
123A	SP84084	123A	T01-013	123A	TH28C715	123A	TP86515	123A	TV57A
123A	SP84085	123A	T01-014	123A	TI-412	123A	TP86520	123A	TV58A
123A	SP84088	123A	T01-101	123A	TI-413	123A	TP86521	123A	TV59A
123A	SP84089	123A	T01-104	123A	TI-415	123A	TQ-5052	123A	TV60A
123A	SP84095	123A	T01-105	123A	TI-422	123A	TQ-5053	123A	TV65
123A	SP84169	123A	T1-1A6	123A	TI-423	123A	TQ-5054	123A	TV71
123A	SP84199	123A	T1004671	123A	TI-432	123A	TQ-5060	123A	TV92
123A	SP84303	123A	T1008-834	123A	TI-433	123A	TQ4	123A	TVS-2C8645A
123A	SP84313	123A	T1008834	123A	TI-714	123A	TQ5052	123A	TVS-28C538
123A	SP84345	123A	T1340A31	123A	TI-714A	123A	TQ5053	123A	TVS-28C538A
123A	SP84347	123A	T1340A3I	123A	TI-751	123A	TQ5054	123A	TVS-28C828
123A	SP84356	123A	T1340A3J	123A	TI-806G	123A	TQ5060	123A	TVS-28C828A
123A	SP84359	123A	T1340A3K	123A	TI-907	123A	TR-1033-1	123A	TVS-28C828Q
123A	SP84360	123A	T1413	123A	TI-908	123A	TR-1033-2	123A	TVS-828A
123A	SP84363	123A	T1414	123A	TI-92	123A	T-1347	123A	TVS-C81255H
123A	SP84367	123A	T1415	123A	TI1A6	123A	TR-162(OLSON)	123A	TVS-C81255HF
123A	SP84368	123A	T1416	123A	TI24A	123A	TR-1R33	123A	TV828C538A
123A	SP84382	123A	T1417	123A	TI24B	123A	TR-21	123A	TV828C644
123A	SP84446	123A	T143	123A	TI411	123A	TR-21-6	123A	TV828C645B
123A	SP84450	123A	T1495	123A	TI415	123A	TR-21C	123A	TV828C684
123A	SP84451	123A	T157	123A	TI416	123A	TR-22	123A	TV828C828A
123A	SP84453	123A	T158	123A	TI417	123A	TR-22C	123A	TV828C828P
123A	SP84455	123A	T1642B	123A	TI419	123A	TR-24(PHILCO)	123A	TV828C828R
123A	SP84456	123A	T170	123A	TI420	123A	TR-28C567	123A	TVSJA1200
123A	SP84457	123A	T171	123A	TI422	123A	TR-28C373	123A	TX-100-1
123A	SP84459	123A	T1746	123A	TI424	123A	TR-28C735	123A	TX-100-2
123A	SP84472	123A	T1746A	123A	TI430	123A	TR-3R38	123A	TX-101-12
123A	SP84476	123A	T1746B	123A	TI432	123A	TR-4R33	123A	TX-107-1
123A	SP84478	123A	T1746C	123A	TI480	123A	TR-5R33	123A	TX-107-10
123A	SP84491	123A	T1748	123A	TI482	123A	TR-5R35	123A	TX-107-16
123A	SP84493	123A	T1748A	123A	TI483	123A	TR-5R38	123A	TX-107-4
123A	SP84494	123A	T1748B	123A	TI484	123A	TR-6R33	123A	TX-107-5
123A	SP84920	123A	T1748C	123A	TI485	123A	TR-7R35	123A	TX-107-6
123A	SP84942	123A	T1802	123A	TI492	123A	TR-8004	123A	TX-108-1
123A	SP85000	123A	T1802A	123A	TI493	123A	TR-8014	123A	TX-112-1
123A	SP85006	123A	T1802B	123A	TI494	123A	TR-8025	123A	TX-119-1
123A	SP85006-1	123A	T1804	123A	TI495	123A	TR-8035	123A	TX100-1
123A	SP85006-2	123A	T1805	123A	TI496	123A	TR-8039	123A	TX100-2
123A	SP85457	123A	T1810	123A	TI54A	123A	TR-8042	123A	TX100-3
123A	SP86111	123A	T1810B	123A	TI54B	123A	TR-8835	123A	TX101-12
123A	SP86112	123A	T185	123A	TI54C	123A	TR-9100-18	123A	TX102-1
123A	SP86113	123A	T1909	123A	TI54E	123A	TR-BC147B	123A	TX102-2
123A	SP86571	123A	T1855	123A	TI714	123A	TR-BRC149C	123A	TX107-1
123A	SP87652	123A	T235A013-2	123A	TI751	123A	TR-RR38	123A	TX107-10
123A	SP8817	123A	T237	123A	TI802B	123A	TR-TR38	123A	TX107-12
123A	SP8817N	123A	T2446	123A	TI803B	123A	TR01037	123A	TX107-16
123A	SP8868	123A	T255	123A	TI810B	123A	TR0573491	123A	TX107-3
123A	SP79844	123A	T256	123A	TI904	123A	TR1011	123A	TX107-4
123A	S81-145128	123A	T277	123A	TI907	123A	TR1031	123A	TX107-5
123A	S82508	123A	T291	123A	TI908	123A	TR1033	123A	TX107-6
123A	S82504	123A	T327	123A	TIA06	123A	TR1033-3	123A	TX108-1
123A	S83694	123A	T327-2	123A	TIA102	123A	TR1993-2	123A	TX112-1
123A	ST-1242	123A	T328	123A	TI8113	123A	TR21	123A	TI585E
123A	ST-1243	123A	T339	123A	TI8114	123A	TR281570LH	123A	UPI2222
123A	ST-1244	123A	T342	123A	TI822	123A	TR28C3677	123A	UPI2222B
123A	ST-1290	123A	T3565	123A	TI823	123A	TR28C373	123A	UPI4046
123A	ST-MP89433	123A	T55A-5	123A	TI844	123A	TR28C735	123A	UPI4046-46
123A	ST-MP89700D	123A	T3601	123A	TI845	123A	TR302	123A	UPI4047-46
123A	ST-MP89700F	123A	T386	123A	TI846	123A	TR310231	123A	UPI706
123A	ST.082.112.005	123A	T399	123A	TI847	123A	TR310243	123A	UPI706A
123A	ST.082.114.015	123A	T416-16(SEARS)	123A	TI848	123A	TR310245	123A	UPI706B
123A	ST/217/Q	123A	T417	123A	TI849	123A	TR4010-2	123A	UP2718A
123A	ST01	123A	T457-16	123A	TI851	123A	TR601	123A	V119
123A	ST02	123A	T457-16(SEARS)	123A	TI852	123A	TR8004-4	123A	V169
123A	ST03	123A	T458-16	123A	TI855	123A	TR8014	123A	
123A		123A		123A		123A	TR8021	123A	

ECG	Industry Standard No.
123A	V297
123A	VM-30209
123A	VM-30241
123A	VM-30242
123A	VM30209
123A	VM30241
123A	VM30242
123A	VS-2SC-458
123A	VS-2SC324H
123A	VS-2SC458
123A	VS-2SC538
123A	VS2SB324
123A	VS2SC206
123A	VS2SC208
123A	VS2SC288A
123A	VS2SC324H
123A	VS2SC458
123A	VS2SC645A
123A	VS2SD227V-1
123A	VS9-0005-913
123A	VS9-0006-913
123A	W20
123A	W24
123A	WRR1952
123A	WRR1953
123A	WRR1954
123A	X16A1938
123A	X16A545-7
123A	X16E3860
123A	X16E3960
123A	X16N1485
123A	X19001-A
123A	X6584-C
123A	X735-41
123A	XA-1071
123A	XA-1139
123A	XC372
123A	XC373
123A	XC374
123A	XEJ040017
123A	XG30
123A	XN-400-318-P1
123A	XS21
123A	XS22
123A	Y49001-21
123A	Y56001-86
123A	Y56601-08
123A	Y56601-45
123A	Y56601-49
123A	Y56601-51
123A	Y56601-73
123A	Y56601-75
123A	Y56601-80
123A	Y56601-86-AD
123A	Y56601-93
123A	YEAN2SC941
123A	Z-28058-1
123A	Z4MW333
123A	ZDT
123A	ZEN100
123A	ZEN102
123A	ZEN103
123A	ZEN110
123A	ZEN111
123A	ZEN112
123A	ZEN113
123A	ZEN114
123A	ZEN115
123A	ZEN119
123A	ZEN120
123A	ZEN127
123A	ZT-110
123A	ZT-62
123A	ZT-82
123A	ZT111
123A	ZT112
123A	ZT113
123A	ZT114
123A	ZT116
123A	ZT117
123A	ZT118
123A	ZT119
123A	ZT1420
123A	ZT1708
123A	ZT20
123A	ZT20-1
123A	ZT20-12
123A	ZT20-55
123A	ZT202
123A	ZT203
123A	ZT204
123A	ZT20A
123A	ZT20B
123A	ZT20C
123A	ZT21
123A	ZT21-1
123A	ZT21-12
123A	ZT21-55
123A	ZT21A
123A	ZT21B
123A	ZT21C
123A	ZT22
123A	ZT22-1
123A	ZT22-12
123A	ZT22-55
123A	ZT2205
123A	ZT2206
123A	ZT22A
123A	ZT22B
123A	ZT22C
123A	ZT23
123A	ZT23-1
123A	ZT23-12
123A	ZT2368
123A	ZT2369
123A	ZT2369A
123A	ZT23A
123A	ZT23B
123A	ZT23C
123A	ZT24
123A	ZT24-1
123A	ZT24-12
123A	ZT24-55
123A	ZT2476
123A	ZT2477
123A	ZT24A
123A	ZT24B
123A	ZT24C
123A	ZT2938
123A	ZT40
123A	ZT402
123A	ZT403
123A	ZT404
123A	ZT406
123A	ZT41
123A	ZT42
123A	ZT43
123A	ZT44
123A	ZT50
123A	ZT60
123A	ZT60-1
123A	ZT60-12
123A	ZT60-55
123A	ZT60A
123A	ZT60B
123A	ZT60C
123A	ZT61
123A	ZT61-1
123A	ZT61-12
123A	ZT61-55
123A	ZT61A
123A	ZT61B
123A	ZT61C
123A	ZT62-1
123A	ZT62-12
123A	ZT62-15
123A	ZT62C
123A	ZT63
123A	ZT63-1
123A	ZT63-12
123A	ZT63-55
123A	ZT63A
123A	ZT63B
123A	ZT63C
123A	ZT64
123A	ZT64-1
123A	ZT64-12
123A	ZT64-5
123A	ZT64-55
123A	ZT64A
123A	ZT64B
123A	ZT64C
123A	ZT66
123A	ZT68
123A	ZT696
123A	ZT697
123A	ZT706
123A	ZT706A
123A	ZT708
123A	ZT80
123A	ZT81
123A	ZT83
123A	ZT84
123A	ZT87
123A	ZT89
123A	001-00
123A	1-0006-0021
123A	1-0006-0022
123A	1-001-003-15
123A	001-02020
123A	001-02101-0
123A	001-02101-1
123A	001-02102-0
123A	001-02103-0
123A	001-02104-0
123A	001-02105-0
123A	001-02106-0
123A	001-02107-0
123A	001-02108-0
123A	001-02109-0
123A	001-02110-0
123A	001-02111-1
123A	001-02113-2
123A	001-02113-3
123A	001-02113-4
123A	001-02113-5
123A	001-02121-0
123A	001-021210
123A	01-030458
123A	01-031175
123A	01-031364
123A	01-031687
123A	1-042/2207
123A	1-044/2207
123A	1-21-276
123A	1-21-277
123A	1-21-278
123A	1-21-279
123A	01-2101
123A	01-2101-0
123A	001-21011
123A	01-2102
123A	01-2104
123A	01-2105
123A	01-2106
123A	01-2107
123A	01-2108
123A	01-2109
123A	1-522223720
123A	01-571941
123A	1-6147191229
123A	1-6171191368
123A	1-801-003
123A	1-801-004
123A	1-801-004-17
123A	1-801-314
123A	1-801-314-15
123A	1-801-314-16
123A	01-9011-52221-3
123A	01-9013-72221-3
123A	01-9014-22221-3
123A	1A0034
123A	1A0035
123A	1A0043
123A	1A0063
123A	1A0079
123A	1A0080
123A	1A0081
123A	1A4757-1
123A	1C3576
123A	1J1
123A	1U585F
123A	1U585F/7825B
123A	1W8358
123A	1W9723
123A	1W9787
123A	002-009500
123A	002-009501
123A	002-009502
123A	002-009900
123A	002-010400
123A	002-010800
123A	002-03
123A	02-1078-01
123A	002-12000
123A	2-8454-031
123A	2C8900
123A	2D002-168
123A	2D002-169
123A	2N1103
123A	2N1104
123A	2N1140
123A	2N1276
123A	2N1277
123A	2N1278
123A	2N1279
123A	2N1388
123A	2N1389
123A	2N1409A
123A	2N1410A
123A	2N1505
123A	2N1506
123A	2N1613/46
123A	2N1644A
123A	2N1674
123A	2N1704
123A	2N1708
123A	2N1708A
123A	2N1711/46
123A	2N1840
123A	2N1944
123A	2N1945
123A	2N1946
123A	2N1947
123A	2N1948
123A	2N1949
123A	2N1950
123A	2N1951
123A	2N1952
123A	2N1962
123A	2N1963
123A	2N1964
123A	2N1965
123A	2N1992
123A	2N2096A
123A	2N2097A
123A	2N2205
123A	2N2206
123A	2N2220
123A	2N2221
123A	2N2221A
123A	2N2222
123A	2N2222A
123A	2N2224
123A	2N2236
123A	2N2237
123A	2N2242
123A	2N2244
123A	2N2245
123A	2N2246
123A	2N2247
123A	2N2248
123A	2N2249
123A	2N2250
123A	2N2253
123A	2N2254
123A	2N2255
123A	2N2256
123A	2N2257
123A	2N2272
123A	2N2314
123A	2N2315
123A	2N2318
123A	2N2319
123A	2N2320
123A	2N2331
123A	2N2349
123A	2N2368
123A	2N2368A
123A	2N2369
123A	2N2413
123A	2N2417
123A	2N2427
123A	2N2432
123A	2N2476
123A	2N2477
123A	2N2481
123A	2N2483
123A	2N2484
123A	2N2501
123A	2N2514
123A	2N2515
123A	2N2523
123A	2N2524
123A	2N2529
123A	2N2530
123A	2N2531
123A	2N2532
123A	2N2533
123A	2N2534
123A	2N2539
123A	2N2540
123A	2N2569
123A	2N2570
123A	2N2571
123A	2N2572
123A	2N2586
123A	2N2645
123A	2N2651
123A	2N2656
123A	2N2673
123A	2N2674
123A	2N2675
123A	2N2676
123A	2N2677
123A	2N2678
123A	2N2692
123A	2N2693
123A	2N2719
123A	2N2790
123A	2N2791
123A	2N2792
123A	2N2831
123A	2N2845
123A	2N2846
123A	2N2847
123A	2N2848
123A	2N2883
123A	2N2884
123A	2N2926-6
123A	2N2926G
123A	2N2926GRN
123A	2N2926ORN
123A	2N2938
123A	2N2951
123A	2N2952
123A	2N2958
123A	2N2959
123A	2N2960
123A	2N2961
123A	2N3009
123A	2N3011
123A	2N3013
123A	2N3014
123A	2N3015
123A	2N3115
123A	2N3116
123A	2N3210
123A	2N3211
123A	2N3241
123A	2N3241A
123A	2N3242
123A	2N3242A
123A	2N3246
123A	2N3247
123A	2N3261
123A	2N3268
123A	2N3299
123A	2N3300
123A	2N3301
123A	2N3302
123A	2N3310
123A	2N332
123A	2N332A
123A	2N333
123A	2N333A
123A	2N334
123A	2N3340
123A	2N334A
123A	2N335
123A	2N335B
123A	2N336
123A	2N336A
123A	2N337
123A	2N337A
123A	2N338
123A	2N338A
123A	2N3415
123A	2N3462
123A	2N3463
123A	2N3508
123A	2N3509
123A	2N3510
123A	2N3511
123A	2N3512
123A	2N3564
123A	2N3565
123A	2N3566
123A	2N3567
123A	2N3641
123A	2N3642
123A	2N3643
123A	2N3646
123A	2N3647
123A	2N3648
123A	2N3688
123A	2N3689
123A	2N3690
123A	2N3691
123A	2N3692
123A	2N3693
123A	2N3694
123A	2N3725
123A	2N3946
123A	2N3947
123A	2N4013
123A	2N4014
123A	2N4046
123A	2N4072
123A	2N4074
123A	2N4138
123A	2N4140
123A	2N4141
123A	2N4227
123A	2N4274
123A	2N4275
123A	2N4432
123A	2N4432A
123A	2N4436
123A	2N4437
123A	2N4450
123A	2N472
123A	2N472A
123A	2N475
123A	2N475A
123A	2N480
123A	2N480A
123A	2N4966
123A	2N4968
123A	2N4969
123A	2N4970
123A	2N5081
123A	2N5082
123A	2N5107
123A	2N5127
123A	2N5128
123A	2N5129
123A	2N5134
123A	2N5137
123A	2N5186
123A	2N5187
123A	2N5417
123A	2N543
123A	2N543A
123A	2N5449
123A	2N546
123A	2N548
123A	2N550
123A	2N551
123A	2N619
123A	2N620
123A	2N621
123A	2N622
123A	2N702
123A	2N703
123A	2N706
123A	2N706/46
123A	2N706A
123A	2N706B
123A	2N706C
123A	2N707A
123A	2N708
123A	2N708/46
123A	2N708A
123A	2N715
123A	2N717
123A	2N718
123A	2N718A
123A	2N729
123A	2N742
123A	2N742A
123A	2N743
123A	2N744
123A	2N745
123A	2N746
123A	2N747
123A	2N748
123A	2N749
123A	2N751
123A	2N753
123A	2N753/46
123A	2N754
123A	2N756
123A	2N756A
123A	2N757
123A	2N757A
123A	2N758
123A	2N758A
123A	2N758B
123A	2N759
123A	2N759A
123A	2N759B
123A	2N760
123A	2N760A
123A	2N760B
123A	2N761
123A	2N762
123A	2N780
123A	2N783
123A	2N784
123A	2N784A
123A	2N834
123A	2N834A
123A	2N835
123A	2N839
123A	2N840
123A	2N841
123A	2N842
123A	2N843
123A	2N844
123A	2N866
123A	2N867
123A	2N909
123A	2N913
123A	2N914
123A	2N914A
123A	2N915
123A	2N915A
123A	2N916
123A	2N916A
123A	2N919
123A	2N920
123A	2N921
123A	2N922
123A	2N929
123A	2N929/46
123A	2N929A
123A	2N930
123A	2N930/46
123A	2N930A
123A	2N930A/46
123A	2N930B
123A	2N947
123A	2N956
123A	2N957
123A	2N988
123A	2N989
123A	2S001
123A	2S002
123A	2S003
123A	2S004
123A	2S005
123A	2S101
123A	2S102
123A	2S103
123A	2S104
123A	2S131
123A	2S501
123A	2S502
123A	2S503
123A	2S512
123A	2S711
123A	2S712
123A	2S731
123A	2S732
123A	2S733
123A	2S741A
123A	2S744A
123A	2S95A
123A	2SC1007
123A	2SC103
123A	2SC103A
123A	2SC104
123A	2SC104A
123A	2SC105
123A	2SC1071
123A	2SC110
123A	2SC111
123A	2SC111D
123A	2SC120
123A	2SC1244
123A	2SC127
123A	2SC131
123A	2SC132
123A	2SC133
123A	2SC134
123A	2SC134B
123A	2SC135
123A	2SC136
123A	2SC137
123A	2SC1380
123A	2SC1380A
123A	2SC1390
123A	2SC13901
123A	2SC1390I
123A	2SC1390J
123A	2SC1390K
123A	2SC1390L
123A	2SC1390V
123A	2SC1390W
123A	2SC1390WH
123A	2SC1390WI
123A	2SC1390WX
123A	2SC1390X
123A	2SC1390XJ
123A	2SC1390XK
123A	2SC1390Y
123A	2SC1390YM
123A	2SC1416
123A	2SC1416BL
123A	2SC15
123A	2SC15-1
123A	2SC15-2
123A	2SC15-3
123A	2SC159
123A	2SC16
123A	2SC160
123A	2SC1639
123A	2SC1641
123A	2SC1641Q

ECG	Industry Standard No.
123A	2SC1641R
123A	2SC166
123A	2SC167
123A	2SC16A
123A	2SC17
123A	2SC170
123A	2SC170R
123A	2SC171
123A	2SC172
123A	2SC172A
123A	2SC1739
123A	2SC174
123A	2SC17A
123A	2SC18
123A	2SC191
123A	2SC192
123A	2SC193
123A	2SC194
123A	2SC195
123A	2SC196
123A	2SC197
123A	2SC199
123A	2SC200
123A	2SC201
123A	2SC202
123A	2SC204
123A	2SC204F
123A	2SC205
123A	2SC237
123A	2SC238
123A	2SC239
123A	2SC248
123A	2SC250
123A	2SC26
123A	2SC27
123A	2SC281
123A	2SC281A
123A	2SC281B
123A	2SC281C
123A	2SC281C-EP
123A	2SC281D
123A	2SC281EP
123A	2SC281H
123A	2SC281HA
123A	2SC281HB
123A	2SC281HC
123A	2SC282
123A	2SC282H
123A	2SC282HA
123A	2SC282HB
123A	2SC282HC
123A	2SC283
123A	2SC284
123A	2SC284H
123A	2SC284HA
123A	2SC284HB
123A	2SC284HC
123A	2SC300
123A	2SC300B
123A	2SC300C
123A	2SC300D
123A	2SC300E
123A	2SC301
123A	2SC301B
123A	2SC301C
123A	2SC301D
123A	2SC301E
123A	2SC302
123A	2SC302B
123A	2SC302C
123A	2SC302D
123A	2SC302E
123A	2SC315
123A	2SC316
123A	2SC317
123A	2SC317C
123A	2SC318
123A	2SC318A
123A	2SC321
123A	2SC321H
123A	2SC321HA
123A	2SC321HB
123A	2SC321HC
123A	2SC323
123A	2SC324
123A	2SC324A
123A	2SC324H
123A	2SC324HA
123A	2SC337B
123A	2SC348
123A	2SC350
123A	2SC350H
123A	2SC352
123A	2SC352A
123A	2SC360
123A	2SC360B
123A	2SC360D
123A	2SC37
123A	2SC38
123A	2SC395
123A	2SC395A
123A	2SC395R
123A	2SC3T2Y
123A	2SC400
123A	2SC400-O
123A	2SC400-GR
123A	2SC400-R
123A	2SC400-Y
123A	2SC405
123A	2SC406
123A	2SC423
123A	2SC423B
123A	2SC423C
123A	2SC423D
123A	2SC423E
123A	2SC423F
123A	2SC424
123A	2SC424D
123A	2SC425
123A	2SC425B
123A	2SC425C
123A	2SC425D
123A	2SC425E
123A	2SC425F
123A	2SC45
123A	2SC468
123A	2SC468(LGR)
123A	2SC468A
123A	2SC468B
123A	2SC472Y
123A	2SC474
123A	2SC478
123A	2SC478(D)
123A	2SC478-4
123A	2SC478D
123A	2SC52
123A	2SC53
123A	0002SC531F
123A	2SC538
123A	2SC538A
123A	2SC538AQ
123A	2SC538P
123A	2SC538Q
123A	2SC538R
123A	2SC538S
123A	2SC538T
123A	2SC539
123A	2SC539K
123A	2SC539L
123A	2SC539R
123A	2SC539S
123A	2SC54
123A	2SC55
123A	2SC564
123A	2SC564A
123A	2SC564P
123A	2SC564Q
123A	2SC564R
123A	2SC564S
123A	2SC564T
123A	2SC587
123A	2SC587A
123A	2SC587B
123A	2SC587C
123A	2SC588
123A	2SC593
123A	2SC594
123A	2SC595
123A	2SC596
123A	2SC62
123A	2SC622
123A	2SC626
123A	2SC648
123A	2SC648H
123A	2SC649
123A	2SC649A
123A	2SC649B
123A	2SC650
123A	2SC650A
123A	2SC650B
123A	0002SC710B
123A	0002SC710C
123A	2SC796
123A	2SC847
123A	2SC847A
123A	2SC847B
123A	2SC847C
123A	2SC848
123A	2SC848A
123A	2SC848B
123A	2SC848C
123A	2SC849
123A	2SC849A
123A	2SC849B
123A	2SC849C
123A	2SC850
123A	2SC850A
123A	2SC850B
123A	2SC850C
123A	2SC87
123A	00002SC870A
123A	00002SC870B
123A	2SC896
123A	2SC899
123A	2SC899K
123A	2SC906
123A	2SC906F
123A	2SC907
123A	2SC907A
123A	2SC907AC
123A	2SC907AD
123A	2SC907AH
123A	2SC907C
123A	2SC907D
123A	2SC907H
123A	2SC907HA
123A	2SC934
123A	2SC934C
123A	2SC934D
123A	2SC934E
123A	2SC934F
123A	2SC934G
123A	2SC934P
123A	2SC943
123A	2SC943A
123A	2SC943B
123A	2SC943C
123A	2SC963
123A	2SC964
123A	2SC965
123A	2SC966
123A	2SC967
123A	2SC968
123A	2SC98
123A	2SC99
123A	2SC993
123A	2SCF-2
123A	2SCF2
123A	2SCF5
123A	2SG536GN
123A	2T172
123A	2T202
123A	2T2708
123A	2T2785
123A	2T2857
123A	2T40
123A	2T403
123A	2T404
123A	2T41
123A	2T42
123A	2T43
123A	2T44
123A	2T918
123A	3-0033
123A	03-1585/G
123A	03A03
123A	03A05
123A	3E-1
123A	3E-2
123A	3E-3
123A	3L4-6007-02
123A	3L4-6007-03
123A	3L4-6007-04
123A	3L4-6007-08
123A	3L4-6007-09
123A	3L4-6007-1
123A	3L4-6007-2
123A	3L4-6007-3
123A	3L4-6007-37
123A	3L4-6007-38
123A	3L4-6010-03
123A	3L4-6010-6
123A	3L4-6015-01
123A	3N35
123A	3S004
123A	004-00
123A	04-00460-03
123A	04-01585-06
123A	04-01585-07
123A	04-02090-02
123A	4-12-1A7-1
123A	4-1545
123A	4-1790
123A	4-3023212
123A	4-3023221
123A	4-3025766
123A	4-47(SEARS)
123A	4-5145
123A	4-686132-3
123A	4-686143-3
123A	4-686144-3
123A	4-686173-3
123A	4-686183-3
123A	4-686231-3
123A	4-686257-3
123A	4-68682-3
123A	4-851
123A	4D20
123A	4D21
123A	4D22
123A	4D24
123A	4D25
123A	4D26
123A	4JX16A567
123A	4JX16A667
123A	4JX16A667G
123A	4JX16A6670
123A	4JX16A667R
123A	4JX16A667Y
123A	4JX16A668
123A	4JX16A668G
123A	4JX16A6680
123A	4JX16A668Y
123A	4JX16A669
123A	4JX16A669G
123A	4JX16A669Y
123A	4JX16A670
123A	4JX16A670G
123A	4JX16B670B
123A	4JX16B670G
123A	4JX16B670R
123A	4JX16B670Y
123A	4JX16E3860
123A	4JX16E3890
123A	4JX16E3960
123A	4JX1E596
123A	4JX7A972
123A	005-02
123A	5-70004503
123A	5-70005452
123A	5-70005503
123A	5-7000901504
123A	5-8
123A	006-0000134
123A	6-04
123A	6-0451
123A	6-0452
123A	6-04GRN
123A	6-04ORN
123A	6-04S1
123A	6-04S2
123A	6-05
123A	6-05F
123A	6-05YEL
123A	6-11
123A	6-19
123A	6-2708
123A	6-30
123A	6-4
123A	006-6450032
123A	6-6450036
123A	6-90
123A	6-9029-15D
123A	6-9029-15E
123A	6-93
123A	6A10227
123A	6A10422
123A	6A10423
123A	6A10520
123A	6A10851
123A	6A10855
123A	6A11180
123A	6A12681
123A	6A12682
123A	6A12683
123A	6A12725
123A	6A12788
123A	6A12789
123A	6A16399
123A	6E4850/56-0001
123A	6X97047A01
123A	7-0015
123A	07-07124
123A	07-07125
123A	07-07139
123A	07-07156
123A	7-1(SARKES)
123A	07-1458-85
123A	7-15(SARKES)
123A	7-16(SARKES)
123A	7-17
123A	7-17(SARKES)
123A	7-18(SARKES)
123A	7-19(SARKES)
123A	7-2(SARKES)
123A	7-20(SARKES)
123A	7-3(SARKES)
123A	7-4(SARKES)
123A	7-5
123A	7-5(SARKES)
123A	7-59-0243477
123A	7-59-068
123A	7-6(SARKES)
123A	7-7(SARKES)
123A	007-74655-02
123A	007-74655-06
123A	007-74661-01
123A	007-7466101
123A	7-8(SARKES)
123A	7A30(SHERWOOD)
123A	7A31(SHERWOOD)
123A	8-00243
123A	8-0050100
123A	8-0051500
123A	8-005202
123A	8-0052102
123A	8-0052302
123A	8-0052600
123A	8-0053001
123A	8-0053400
123A	8-0318250
123A	8-0337390
123A	8-0383940
123A	8-0389910
123A	8-0389930
123A	8-0421980
123A	8-2409501
123A	8-4(BENDIX)
123A	8-697-020-570
123A	8-724-733-30
123A	8-726-357-10
123A	8-81250109
123A	8-902-0706-071
123A	8-905-014-017
123A	8-905-705-112
123A	8-905-705-403
123A	8-905-705-405
123A	8-905-706-104
123A	8-905-706-201
123A	8-905-706-202
123A	8-905-706-203
123A	8-905-706-206
123A	8-905-706-208
123A	8-905-706-211
123A	8-905-706-215
123A	8-905-706-235
123A	8-905-706-236
123A	8-905-706-238
123A	8-905-706-239
123A	8-905-706-240
123A	8-905-706-242
123A	8-905-706-244
123A	8-905-706-245
123A	8-905-706-246
123A	8-905-706-250
123A	8-905-706-257
123A	8-905-706-260
123A	8-905-706-263
123A	8-905-706-336
123A	8-905-706-606
123A	8-905-707-254
123A	8-905-707-265
123A	8-905-707-313
123A	8A12789
123A	08A8300-2
123A	09-002012
123A	9-006
123A	09-302007
123A	09-302012
123A	09-302033
123A	09-302034
123A	09-302039
123A	09-302045
123A	09-302045-12
123A	09-302054
123A	09-302058
123A	09-302062
123A	09-302074
123A	09-302078
123A	09-302101
123A	09-302106
123A	09-302118
123A	09-302124
123A	09-302131
123A	09-302140
123A	09-302148
123A	09-302153
123A	09-302165
123A	09-302172
123A	09-302175
123A	09-302189
123A	09-302204
123A	09-302215
123A	09-302227
123A	09-303025
123A	09-304044
123A	09-304045
123A	09-304058
123A	09-305034
123A	09-305062
123A	09-305063
123A	09-305064
123A	09-305065
123A	09-305066
123A	09-305067
123A	09-305068
123A	09-305077
123A	09-305139
123A	09-305148
123A	09-305152
123A	09-309006
123A	09-309012
123A	09-309023
123A	09-309049
123A	09-309050
123A	09-309064
123A	09-309076
123A	9-5216
123A	9-5221
123A	9-5225
123A	9-5226-2
123A	9-5227
123A	9-5296
123A	9-9109-1
123A	9-9109-2
123A	9GR2
123A	09N1
123A	9S037
123A	9TR10
123A	9TR2
123A	9TR21001-02
123A	9TR31001-03
123A	9TR7
123A	9TRZ1001-02
123A	10-080010
123A	11-085010
123A	012-1-12
123A	012-1-12-7
123A	12-1-276
123A	12-1-277
123A	12-1-278
123A	12-1-279
123A	12-1-70
123A	12-1-70-12
123A	12-1-70-12-7
123A	12-10
123A	12-101001
123A	12-11
123A	12-12
123A	12-13
123A	12-14
123A	12-15
123A	12-16
123A	12-17
123A	12-18
123A	12-19
123A	12-1A
123A	12-1A0
123A	12-1A0R
123A	12-1A1
123A	12-1A19
123A	12-1A2
123A	12-1A21
123A	12-1A3
123A	12-1A3P
123A	12-1A4
123A	12-1A4-7
123A	12-1A4-7B
123A	12-1A5
123A	12-1A5L
123A	12-1A6
123A	12-1A6A
123A	12-1A7
123A	12-1A7-1
123A	12-1A8
123A	12-1A82
123A	12-1A9
123A	12-1A9G
123A	12-4
123A	12CLN
123A	012E
123A	12X047
123A	13-0021
123A	13-0022
123A	13-0024
123A	13-0041
123A	13-0048
123A	13-0058
123A	13-0321-10
123A	13-0321-11
123A	13-0321-12
123A	13-0321-5
123A	13-0321-6
123A	13-0321-7
123A	13-0321-8
123A	13-0321-81
123A	13-0321-9
123A	13-14085-15
123A	13-14085-34
123A	13-14085-50
123A	13-14085-6
123A	13-14085-7
123A	13-14085-72
123A	13-14085-83
123A	13-14085-84
123A	13-14085-85
123A	13-14085-89
123A	13-14606-1
123A	13-15804-1
123A	13-15840-1
123A	13-15840-2
123A	13-15865-1
123A	13-16769-1
123A	13-17-6(SEARS)
123A	13-18087-1
123A	13-18087-2
123A	13-18158-1
123A	13-1836-1
123A	13-18363-1
123A	13-18364-1
123A	13-18927-1
123A	13-22581
123A	13-22581-1
123A	13-23160-4
123A	13-23309-5
123A	13-23324-6
123A	13-23916-1
123A	13-27404-2
123A	13-27433-1
123A	13-29432-1
123A	13-33350-1
123A	13-33595-1
123A	13-33595-2
123A	13-33595-3
123A	13-35807-2
123A	13-55061-1
123A	13-55061-2
123A	13-55066-1
123A	13-55066-2
123A	13-55067-1
123A	13-55068-1
123A	13-67583-6
123A	13-67585-4
123A	13-67585-5
123A	13-67585-5/3464
123A	13-67586-3
123A	13-67586-3/3464
123A	13-68617-1
123A	14 806 12
123A	14-0104-7
123A	14-1
123A	14-2
123A	14-3
123A	14-575-10
123A	14-583-01
123A	14-601-28
123A	14-602-01
123A	14-602-02
123A	14-602-03
123A	14-602-12
123A	14-602-13
123A	14-602-14
123A	14-602-16
123A	14-602-17
123A	14-602-22
123A	14-602-23
123A	14-602-26
123A	14-602-35
123A	14-602-48
123A	14-602-50
123A	14-602-55A
123A	14-602-61
123A	14-602-62
123A	14-602-69
123A	14-602-78
123A	14-602-80
123A	14-602-81
123A	14-602-87
123A	14-602-89
123A	14-603-03

ECG	Industry Standard No.
123A	14-603-04
123A	14-603-10
123A	14-603-11
123A	14-651-12
123A	14-655-13
123A	14-656-21
123A	14-659-12
123A	14-660-12
123A	014-680
123A	014-686
123A	014-698
123A	014-784
123A	14-800-32
123A	14-802-12
123A	14-805-12
123A	14-806-12
123A	14-806-23
123A	14-809-23
123A	14-809-32
123A	14-851-32
123A	14-853-23
123A	14-854-12
123A	14-858-12
123A	14-862-23
123A	14-862-32
123A	14-864-12
123A	14-865-12
123A	14-866-32
123A	15-01999
123A	15-03014-00
123A	15-03100
123A	15-05302
123A	15-05369
123A	15-05650
123A	15-082019
123A	15-09338
123A	15-09980
123A	15-1
123A	15-2
123A	15-22223720
123A	15-875-075-003
123A	16-147191229
123A	16-171191368
123A	16A1938
123A	16A545-7
123A	016B12
123A	016B810
123A	016B812
123A	16E1330(GE)
123A	16G2
123A	16L42
123A	16L43
123A	16L44
123A	16L62
123A	16L63
123A	16U1(HEATH KIT)
123A	16X1
123A	16X2
123A	17-451
123A	17-457
123A	017B824
123A	018-00003
123A	18-148A
123A	019-00009
123A	019-00010
123A	019-003675-196
123A	019-003675-203
123A	019-003675-207
123A	019-003675-246
123A	019-003932
123A	019-003934
123A	019-004111
123A	019-004428-002
123A	019-005006
123A	019-005021
123A	19-020-043
123A	19-020-043A
123A	19-020-058
123A	19-020-067
123A	19-020-073
123A	19-020-074
123A	19-020071
123A	19-2-02616
123A	19A115061-P1
123A	19A115061-P2
123A	19A115102-P1
123A	19A115108-P1
123A	19A115108-P2
123A	19A115123-2
123A	19A115123-P1
123A	19A115123-P2
123A	19A115142-P1
123A	19A115142-P2
123A	19A115157-1
123A	19A115167-2
123A	19A115245-P1
123A	19A115245-P2
123A	19A115253-P1
123A	19A115253-P2
123A	19A115315-P1
123A	19A115315-P2
123A	19A115328-1
123A	19A115330
123A	19A115342-P1
123A	19A115342-P2
123A	19A115359-P1
123A	19A115359-P2
123A	19A115362-P1
123A	19A115362-P2
123A	19A115410-P1
123A	19A115410-P2
123A	19A115552P1
123A	19A115552P2
123A	19A115591P1
123A	19A115591P2
123A	19A115720-1
123A	19A115720-2
123A	19A115728-1
123A	19A115728-2
123A	19A115786
123A	19A115786A
123A	19A115910P1
123A	19A115944P1
123A	19A115944P2
123A	19A116631P1
123A	19A116755P1
123A	19A116774-P1
123A	19A116865
123A	19A129207P1
123A	19AR20
123A	19AR36
123A	19C300114-P1
123A	19C300114-P2
123A	19C300114P1

ECG	Industry Standard No.
123A	19C300114P2
123A	19C300114P3
123A	020-1112-001
123A	20A0053
123A	20A0073
123A	20A10849
123A	021-0121-00
123A	21-1
123A	21-1L
123A	21A015-013
123A	21A015-020
123A	21A015-027
123A	21A040-020
123A	21A040-032
123A	21A040-033
123A	21A040-033A
123A	21A040-034
123A	21A040-037
123A	21A040-056
123A	21A040-077
123A	21A040-078
123A	21A040-092
123A	21A040-37
123A	21A112-013
123A	21A112-015
123A	21A112-017
123A	21A112-018
123A	21A112-020
123A	21A112-050
123A	21A112-062
123A	21A112-063
123A	21A112-085
123A	21A112-088
123A	21A112-089
123A	21A112-090
123A	21A112-091
123A	21A112-092
123A	21A112-101
123A	21A112-102
123A	21A112-104
123A	21M084
123A	21M085
123A	21M086
123A	21M122
123A	21M123
123A	21M125
123A	21M139
123A	21M146
123A	21M149
123A	21M150
123A	21M186
123A	21M200
123A	21M205
123A	21M366
123A	21M488
123A	21M520
123A	21M563
123A	21M578
123A	21M579
123A	21M60
123A	21M605
123A	22-001001
123A	22-001006
123A	22-001007
123A	022-2876-003
123A	23
123A	23-5033
123A	23-PT274-121
123A	23B114044
123A	23E001-1
123A	24-0003714-1
123A	24-000451
123A	24-000457
123A	24-000653-1
123A	24-001327-1
123A	24-002
123A	24-602-25
123A	24A
123A	24B
123A	24B1
123A	24MW1023
123A	24MW1024
123A	24MW1059
123A	24MW1068
123A	24MW1069
123A	24MW1089
123A	24MW1096
123A	24MW1120
123A	24MW1141
123A	24MW1147
123A	24MW119
123A	24MW333
123A	24MW372
123A	24MW454
123A	24MW458
123A	24MW460
123A	24MW461
123A	24MW609
123A	24MW655
123A	24MW658
123A	24MW659
123A	24MW676
123A	24MW740
123A	24MW760
123A	24MW773
123A	24MW774
123A	24MW775
123A	24MW776
123A	24MW790
123A	24MW795
123A	24MW796
123A	24MW797
123A	24MW801
123A	24MW807
123A	24MW808
123A	24MW809
123A	24MW817
123A	24MW818
123A	24MW823
123A	24MW854
123A	24MW855
123A	24MW874
123A	24MW899
123A	24MW954
123A	24MW961
123A	24MW988
123A	24MW992
123A	25-0060-4
123A	025-100018
123A	025-100030
123A	025-100040
123A	025-10030
123A	25A1273-001
123A	25A2

ECG	Industry Standard No.
123A	025B-YEL
123A	25C858LGBM
123A	25H2
123A	026-100026
123A	31-0007
123A	31-0009
123A	31-0068
123A	31-0069
123A	31-0080
123A	31-0081
123A	31-0082
123A	31-0084
123A	31-0085
123A	31-0104
123A	31-0106
123A	31-0115
123A	31-0116
123A	31-0177
123A	31-0230
123A	31-0239
123A	31-0246
123A	31-058
123A	31-1
123A	31-16
123A	031A
123A	32-13843-2
123A	32-20738
123A	33-00706A
123A	33-070
123A	33-0706
123A	33-071
123A	34-1010
123A	34-34-6015-43
123A	34-6000-64
123A	34-6000-69
123A	34-6000-70
123A	34-6000-71
123A	34-6001-34
123A	34-6001-48
123A	34-6001-49
123A	34-6001-5
123A	34-6001-52
123A	34-6001-53
123A	34-6001-54
123A	34-6001-55
123A	34-6001-56
123A	34-6001-57
123A	34-6001-58
123A	34-6001-60
123A	34-6001-61
123A	34-6001-62
123A	34-6001-63
123A	34-6001-69
123A	34-6001-70
123A	34-6001-73
123A	34-6001-74
123A	34-6001-77
123A	34-6007-1
123A	34-6007-2
123A	34-6007-3
123A	34-6015-1
123A	34-6015-10
123A	34-6015-11
123A	34-6015-13
123A	34-6015-14
123A	34-6015-2
123A	34-6015-21
123A	34-6015-3
123A	34-6015-4
123A	34-6015-41
123A	34-6015-42A
123A	34-6015-43A
123A	34-6015-44
123A	34-6015-5
123A	34-6015-54
123A	34-6015-6
123A	34-6015-60
123A	34-6015-63
123A	34-6015-7
123A	34-6015-80
123A	34-6015-9
123A	34-6016-14
123A	34-6016-16
123A	34-6016-18
123A	34-6016-19
123A	34-6016-2
123A	34-6016-24
123A	34-6016-25
123A	34-6016-26
123A	34-6016-3
123A	34-6016-4
123A	34-6016-49
123A	34-6016-49A
123A	34-6016-63
123A	34-6016-7
123A	34-6016-8
123A	35-39306001
123A	35-39306002
123A	35-39306003
123A	037
123A	041
123A	41-0499
123A	042
123A	42-19840
123A	42-21234
123A	42-22158
123A	42-22533
123A	42-22786
123A	42-22787
123A	42-22809
123A	42-22811
123A	42-22812
123A	42-22847
123A	42-23348
123A	42-23349
123A	42-23542
123A	42-23964
123A	42-23964P
123A	42-23966
123A	42-23966P
123A	42-28056
123A	42-28205
123A	42-28207
123A	42-28210
123A	42-9029-31M
123A	42-9029-31R
123A	42-9029-31V
123A	42-9029-40C
123A	42-9029-40L
123A	42-9029-40Y
123A	42-9029-60A
123A	42-9029-60C
123A	42-9029-70C
123A	42-9029-70F

ECG	Industry Standard No.
123A	42-9029-70P
123A	43-023212
123A	43-023221
123A	43-025766
123A	43A128340-1
123A	43A128340-2
123A	43A128340-3
123A	43A128340-4
123A	43A128342-1
123A	43A128342-2
123A	43A128342-3
123A	43A128342-4
123A	43A128342-5
123A	43A128342-6
123A	43A128342-7
123A	43A167851
123A	43A168016P1
123A	43A180002-P1
123A	43A180002-P2
123A	43B140883-1
123A	43B168610
123A	43C168567
123A	044-9667-02
123A	44A333463-001
123A	44A390247
123A	44A390249
123A	44A390251
123A	44A390251-001
123A	44A395994-001
123A	44B311097
123A	45AN4AA
123A	45N2M
123A	45N3
123A	45N4
123A	45N4M
123A	46-8257-3
123A	46-86121-3
123A	46-86122-3
123A	46-86143-3
123A	46-86144-3
123A	46-86145-3
123A	46-86152-3
123A	46-86169-3
123A	46-86171-3
123A	46-86192-3
123A	46-86228-3
123A	46-86231-3
123A	46-8624-3
123A	46-86247-2
123A	46-86247-3
123A	46-86257-3
123A	46-86268-3
123A	46-86274-3
123A	46-86310-3
123A	46-86375-3
123A	46-86376-3
123A	46-86378-3
123A	46-86404-3
123A	46-86407-3
123A	46-86408-3
123A	46-86409-3
123A	46-86419-3
123A	46-86485-3
123A	46-8682-2
123A	46-8695-3
123A	48-01-005
123A	48-123802
123A	48-123803
123A	48-134173
123A	48-134464
123A	48-134465
123A	48-134654
123A	48-134664
123A	48-134665
123A	48-134666
123A	48-134667
123A	48-134668
123A	48-134669
123A	48-134673
123A	48-134674
123A	48-134675
123A	48-134690
123A	48-134691
123A	48-134703
123A	48-134705
123A	48-134714
123A	48-134718
123A	48-134720
123A	48-134721
123A	48-134726
123A	48-134732
123A	48-134733
123A	48-134733A
123A	48-134734
123A	48-134734A
123A	48-134737
123A	48-134739
123A	48-134765
123A	48-134768
123A	48-134775
123A	48-134776
123A	48-134782
123A	48-134785
123A	48-134791
123A	48-134801
123A	48-134804
123A	48-134807
123A	48-134808
123A	48-134809
123A	48-13481
123A	48-134811
123A	48-134817
123A	48-134822
123A	48-134824
123A	48-134839
123A	48-134840
123A	48-134841
123A	48-134842
123A	48-134844
123A	48-134847
123A	48-134848
123A	48-134852
123A	48-134854
123A	48-134889
123A	48-134894
123A	48-134895
123A	48-134896
123A	48-134897
123A	48-134899
123A	48-134903
123A	48-134905
123A	48-134906
123A	48-134918
123A	48-134928

ECG	Industry Standard No.
123A	48-134929
123A	48-134933
123A	48-134933E
123A	48-134935
123A	48-134942
123A	48-134952
123A	48-134970
123A	48-134980
123A	48-134988
123A	48-134992
123A	48-134994
123A	48-134996
123A	48-136665
123A	48-137003
123A	48-137007
123A	48-137010
123A	48-137013
123A	48-137014
123A	48-137019
123A	48-137022
123A	48-137043
123A	48-137044
123A	48-137047
123A	48-137056
123A	48-137057
123A	48-137072
123A	48-137073
123A	48-137083
123A	48-137089
123A	48-137096
123A	48-137101
123A	48-137106
123A	48-137107
123A	48-137108
123A	48-137109
123A	48-137110
123A	48-137111
123A	48-137115
123A	48-137137
123A	48-137138
123A	48-137139
123A	48-137171D
123A	48-137172
123A	48-137174
123A	48-137192
123A	48-137206
123A	48-137257
123A	48-137260
123A	48-137265
123A	48-137336
123A	48-137350
123A	48-137353
123A	48-137354
123A	48-137373
123A	48-137374
123A	48-137377
123A	48-137378
123A	48-137384
123A	48-137398
123A	48-137399
123A	48-137498
123A	48-137500
123A	48-137509
123A	48-137530
123A	48-137543
123A	48-137855
123A	48-137998
123A	48-155088
123A	48-3003A05
123A	48-3003A11
123A	48-3003A12
123A	48-355002
123A	48-355052
123A	48-40171G01
123A	48-40246G02
123A	48-40606J01
123A	48-43354A81
123A	48-43354A82
123A	48-44885G02
123A	48-60022A13
123A	48-63005A66
123A	48-63005A72
123A	48-63026A47
123A	48-63026A48
123A	48-63076A52
123A	48-63076A83
123A	48-63077A10
123A	48-63077A30
123A	48-63077A31
123A	48-63078A70
123A	48-63078A71
123A	48-63079A97
123A	48-63082A25
123A	48-63082A26
123A	48-63082A27
123A	48-63082A45
123A	48-63082A71
123A	48-65123A94
123A	48-65147A72
123A	48-69394A01
123A	48-83750G01
123A	48-86376-3
123A	48-869226-0
123A	48-869248
123A	48-869312
123A	48-869325
123A	48-869329
123A	48-869444
123A	48-869525
123A	48-869563
123A	48-869568
123A	48-869570
123A	48-869767
123A	48-90172A01
123A	48-90232A05
123A	48-90232A11
123A	48-90232A13
123A	48-97046A22
123A	48-97046A23
123A	48-97046A24
123A	48-97046A28
123A	48-97046A42
123A	48-97046A43
123A	48-97046A46
123A	48-97046A50
123A	48-97046A52
123A	48-971-A95
123A	48-97127A012
123A	48-97127A013
123A	48-97127A018
123A	48-97127A12
123A	48-97127A13
123A	48-97127A18
123A	48-97127A19

ECG	Industry Standard No.
123A	48-97127A24
123A	48-97127A29
123A	48-97127A33
123A	48-97162A04
123A	48-97162A05
123A	48-97162A09
123A	48-97162A12
123A	48-97162A15
123A	48-97162A21
123A	48-97162A23
123A	48-97162A33
123A	48-97177A09
123A	48-97177A12
123A	48-97177A13
123A	48-971A05
123A	48-97238A04
123A	48-P02597A
123A	48B407624P1
123A	48P63078A71
123A	48R869312
123A	48R869325
123A	48R869329
123A	48R869444
123A	48R869525
123A	48R869563
123A	48R869568
123A	48R869570
123A	48R869767
123A	48S134903
123A	48S134933
123A	48S134997
123A	48S137107
123A	48S137110
123A	48S137115
123A	48S137171
123A	48S137172
123A	48S137174
123A	48S137300
123A	48S137315
123A	48S137476
123A	48S137498
123A	48S137530
123A	48S137543
123A	48S155088
123A	48S44885G01
123A	48S44885G02
123A	48X97046A60
123A	48X97046A61
123A	48X97046A62
123A	48X97162A05
123A	48X97162A21
123A	48X97238A04
123A	50-40102-04
123A	50-40105-08
123A	50-40201-08
123A	50-40201-09
123A	50-40201-10
123A	50C371
123A	50C372
123A	50C373
123A	50C374
123A	50C538
123A	50C644
123A	50C828
123A	50C838
123A	50CJ139
123A	051-0046
123A	051-0047
123A	051-0155
123A	51-47-23
123A	51-47-24
123A	52-020-108-0
123A	53-1110
123A	054
123A	54-1
123A	54A
123A	54B
123A	54BLU
123A	54C
123A	54D
123A	54F
123A	54GRN
123A	54WHT
123A	54YEL
123A	055
123A	55-1026
123A	55-1034
123A	55-1082
123A	56-35
123A	56-8089
123A	56-8089A
123A	56-8089C
123A	56-8090
123A	56-8090A
123A	56-8090C
123A	56A22-1
123A	56B22-1
123A	057
123A	57-0004503
123A	57-0005452
123A	57-0005503
123A	57-000901504
123A	57-01491-B
123A	57A1-123
123A	57A1-124
123A	57A1-51
123A	57A1-75
123A	57A105-12
123A	57A11-1
123A	57A117-9
123A	57A118-12
123A	57A119-12
123A	57A120-12
123A	57A121-9
123A	57A125-9
123A	57A135-12
123A	57A136-12
123A	57A140-12
123A	57A15-1
123A	57A15-2
123A	57A15-3
123A	57A15-4
123A	57A152-1
123A	57A152-10
123A	57A152-11
123A	57A152-2
123A	57A152-3
123A	57A152-4
123A	57A152-5
123A	57A152-6
123A	57A152-7
123A	57A152-8
123A	57A152-9
123A	57A153-1
123A	57A153-2
123A	57A153-3
123A	57A153-4
123A	57A153-5
123A	57A153-6
123A	57A153-7
123A	57A153-8
123A	57A153-9
123A	57A156-9
123A	57A16-1
123A	57A166-12
123A	57A181-12
123A	57A184-12
123A	57A191-12
123A	57A199-4
123A	57A2-101
123A	57A2-102
123A	57A2-103
123A	57A2-113
123A	57A2-116
123A	57A2-126
123A	57A2-153
123A	57A2-192
123A	57A2-27
123A	57A2-28
123A	57A2-59
123A	57A2-62
123A	57A2-63
123A	57A2-64
123A	57A2-73
123A	57A2-85
123A	57A2-87
123A	57A2-97
123A	57A200-12
123A	57A201-13
123A	57A202-13
123A	57A203-14
123A	57A204-14
123A	57A21-8
123A	57A24-1
123A	57A24-2
123A	57A24-3
123A	57A24-4
123A	57A252-1
123A	57A253-14
123A	57A265-4
123A	57A268-9
123A	57A27-1
123A	57A282-12
123A	57A6-11
123A	57A6-4
123A	57A6-9
123A	57A7-10
123A	57A7-15
123A	57A7-17
123A	57A7-18
123A	57A7-20
123A	57A7-9
123A	57B105-12
123A	57B107-8
123A	57B117-9
123A	57B118-12
123A	57B119-12
123A	57B120-12
123A	57B121-9
123A	57B125-9
123A	57B126-12
123A	57B129-9
123A	57B135-12
123A	57B136-12
123A	57B140-12
123A	57B143-12
123A	57B144-12
123A	57B146-12
123A	57B152-1
123A	57B152-10
123A	57B152-11
123A	57B152-2
123A	57B152-3
123A	57B152-4
123A	57B152-5
123A	57B152-6
123A	57B152-7
123A	57B152-8
123A	57B152-9
123A	57B153-1
123A	57B153-2
123A	57B153-3
123A	57B153-4
123A	57B153-5
123A	57B153-6
123A	57B153-7
123A	57B153-8
123A	57B153-9
123A	57B156-9
123A	57B182-12
123A	57B184-12
123A	57B191-12
123A	57B194-11
123A	57B2-101
123A	57B2-102
123A	57B2-103
123A	57B2-113
123A	57B2-116
123A	57B2-126
123A	57B2-153
123A	57B2-192
123A	57B2-27
123A	57B2-28
123A	57B2-59
123A	57B2-62
123A	57B2-63
123A	57B2-64
123A	57B2-73
123A	57B2-85
123A	57B2-87
123A	57B2-97
123A	57B200-12
123A	57B202-13
123A	57B253-14
123A	57B282-12
123A	57C11-1
123A	57C121-9
123A	57C15-1
123A	57C15-2
123A	57C15-3
123A	57C15-4
123A	57C156-9
123A	57C16-1
123A	57C24-1
123A	57C24-2
123A	57C24-3
123A	57C24-4
123A	57C27-1
123A	57C6-11
123A	57C6-4
123A	57C6-9
123A	57C7-10
123A	57C7-15
123A	57C7-17
123A	57C7-18
123A	57C7-20
123A	57C7-9
123A	57D1-123
123A	57D1-124
123A	57D1-51
123A	57D1-75
123A	57D136-12
123A	57D14-1
123A	57D14-2
123A	57D14-3
123A	57D6-4
123A	57L2-2
123A	57L3-1
123A	57L3-4
123A	57M1-14
123A	57M1-15
123A	57M1-19
123A	57M1-20
123A	57M1-23
123A	57M1-24
123A	57M1-26
123A	57M1-27
123A	57M1-28
123A	57M1-29
123A	57M1-30
123A	57M1-31
123A	57M1-32
123A	58-1(TRUETONE)
123A	61-1400
123A	61-1401
123A	61-1402
123A	61-1403
123A	61-1404
123A	61-1763
123A	61-746
123A	61-751
123A	61-754
123A	61-755
123A	61-814
123A	61-815
123A	61J001-1
123A	61J002-1
123A	61J003-1
123A	62-16905
123A	62-17550
123A	62-18425
123A	62-18641
123A	62-18642
123A	62-18643
123A	62-18828
123A	62-19280
123A	62-19516
123A	62-19548
123A	62-19837
123A	62-19838
123A	62-20155
123A	62-20240
123A	62-20241
123A	62-20242
123A	62-20243
123A	62-20360
123A	62-21496
123A	62-22038
123A	62-22039
123A	62-22250
123A	62-22251
123A	62-7567
123A	62A11868
123A	63-10188
123A	63-10377
123A	63-10708
123A	63-10725
123A	63-10732
123A	63-10733
123A	63-10734
123A	63-10735
123A	63-10736
123A	63-10737
123A	63-10860
123A	63-11025
123A	63-11143
123A	63-11289
123A	63-11468
123A	63-11469
123A	63-11470
123A	63-11471
123A	63-11472
123A	63-11660
123A	63-11757
123A	63-11758
123A	63-11759
123A	63-11831
123A	63-11832
123A	63-11833
123A	63-11916
123A	63-11934
123A	63-11935
123A	63-11937
123A	63-12003
123A	63-12004
123A	63-12062
123A	63-12272
123A	63-12605
123A	63-12608
123A	63-12609
123A	63-12641
123A	63-12642
123A	63-12696
123A	63-12697
123A	63-12706
123A	63-12707
123A	63-12750
123A	63-12751
123A	63-12752
123A	63-12753
123A	63-12874
123A	63-12875
123A	63-12877
123A	63-12878
123A	63-12879
123A	63-12933
123A	63-12940
123A	63-12941
123A	63-12942
123A	63-12943
123A	63-12946
123A	63-12948
123A	63-12949
123A	63-12950
123A	63-12951
123A	63-12952
123A	63-12953
123A	63-13419
123A	63-13438
123A	63-13440
123A	63-13441
123A	63-13864
123A	63-13927
123A	63-14032
123A	63-14051
123A	63-14052
123A	63-14057
123A	63-18643
123A	63-19280
123A	63-19282
123A	63-29461
123A	63-7421
123A	63-7567
123A	63-7670
123A	63-8555
123A	63-8701
123A	63-8702
123A	63-9337
123A	63-9338
123A	63-9339
123A	63-9341
123A	63-9516
123A	63-9518
123A	63-9829
123A	63-9830
123A	63-9831
123A	63-9832
123A	63-9833
123A	63-9847
123A	065-004
123A	65-1
123A	66-127119
123A	66-F29-1
123A	66A00008A
123A	66A00010A
123A	66F027-1
123A	66F028-1
123A	66F029-1
123A	66F057-1
123A	66F057-2
123A	66M
123A	68A7380-1
123A	68A7715P1
123A	68A8321
123A	069
123A	70-943-722-001
123A	70-943-754-002
123A	70-943-762-001
123A	70-943-772-002
123A	70.01.704
123A	70N1
123A	70N4
123A	71-126268
123A	71N1B
123A	72N2B
123A	73B-140-003-5
123A	73C182081-31
123A	73N1B
123A	74N1
123A	75N5AA
123A	76N1
123A	76N1M
123A	76N2
123A	76N2369-000
123A	76N2369-001
123A	77-271453-1
123A	77-271819-1
123A	77-271967-1
123A	77-273001-2
123A	77N1
123A	77N2
123A	77N4
123A	77N5
123A	77N6
123A	78N1
123A	78N2B
123A	80-050100
123A	80-051500
123A	80-052102
123A	80-052202
123A	80-052302
123A	80-052600
123A	80-053001
123A	80-053400
123A	80-308-2
123A	80-318250
123A	80-337390
123A	80-383940
123A	80-389910
123A	80-389930
123A	80-421980
123A	81-27125140-7
123A	81-27125140-7A
123A	81-27125140-7B
123A	81-27125160-5
123A	81-27125160-5A
123A	81-27125160-5B
123A	81-27125270-2
123A	81-27125270-2A
123A	81-27125270-2B
123A	81-27125300-7
123A	81-46125016-9
123A	81-46125019-3
123A	81-46125026-8
123A	81-46125027-6
123A	82-409501
123A	85004
123A	86 A 86A327
123A	86-0007-004
123A	86-0022-001
123A	86-0029-001
123A	86-0031-001
123A	086-005132-02
123A	86-100005
123A	86-100008
123A	86-110-2
123A	86-119-2
123A	86-123-2
123A	86-139-2
123A	86-1392
123A	86-143-2
123A	86-144-2
123A	86-155-2
123A	86-158-2
123A	86-166-2
123A	86-171-2
123A	86-175-2
123A	86-182-2
123A	86-188-2
123A	86-189-2
123A	86-190-2
123A	86-191-2
123A	86-192-2
123A	86-193-2
123A	86-194-2
123A	86-195-2
123A	86-196-2
123A	86-197-2
123A	86-198-2
123A	86-199-2
123A	86-201-2
123A	86-202-2
123A	86-237-2
123A	86-238-2
123A	86-247-2
123A	86-250-2
123A	86-255-2
123A	86-256-2
123A	86-264-2
123A	86-265-2
123A	86-277-2
123A	86-291-9
123A	86-293-2
123A	86-308-2
123A	86-309-2
123A	86-310-2
123A	86-323-2
123A	86-324-2
123A	86-327-2
123A	86-328-2
123A	86-339-2
123A	86-339-9
123A	86-342-2
123A	86-359-2
123A	86-362-2
123A	86-365-2
123A	86-379-2
123A	86-390-2
123A	86-391-2
123A	86-399-1
123A	86-399-2
123A	86-399-9
123A	86-403-2
123A	86-42-1
123A	86-420-2
123A	86-445-2
123A	86-457-2
123A	86-458-2
123A	86-460-2
123A	86-461-2
123A	86-462-2
123A	86-472-2
123A	86-481-1
123A	86-481-2
123A	86-483-2
123A	86-483-3
123A	86-484-2
123A	86-485-2
123A	86-486-2
123A	86-493-2
123A	86-494-2
123A	86-495-2
123A	86-496-2
123A	86-5018-2
123A	86-502-2
123A	86-5040-2
123A	86-5041-2
123A	86-5044-2
123A	86-5045-2
123A	86-5046-2
123A	86-5049-2
123A	86-5050-2
123A	86-5051-2
123A	86-5055-2
123A	86-5056-2
123A	86-5081-2
123A	86-5097-2
123A	86-5103-2
123A	86-5110-2
123A	86-5111-2
123A	86-5114-2
123A	86-5117-2
123A	86-514-2
123A	86-515-2(SEARS)
123A	86-520-2
123A	86-534-2
123A	86-539-2
123A	86-548-2
123A	86-551-2
123A	86-554-2
123A	86-559-2
123A	86-560-2
123A	86-561-2
123A	86-565-2
123A	86-573-2
123A	86-58-2
123A	86-595-2
123A	86-598-2
123A	86-599-2
123A	86-646-2
123A	86-655-2
123A	86-661-2
123A	86-702-2
123A	86A334
123A	86A336
123A	86A350
123A	86X0006-001
123A	86X0007-001
123A	86X0007-104
123A	86X0008-001
123A	86X0022-001
123A	86X0029-001
123A	86X0031-001
123A	86X0031-002
123A	86X0031-003
123A	86X0032-001
123A	86X0035-001
123A	86X0040-00
123A	86X0040-001
123A	86X0045-001
123A	86X0048-001
123A	86X0050-001
123A	86X0051-001
123A	86X0054-001
123A	86X0058-001
123A	86X0058-002
123A	86X0058-003
123A	86X006-001
123A	86X0063-001
123A	86X007-004

ECG	Industry Standard No.
123A	86X007-034
123A	86X0079-001
123A	86X0090-001
123A	86X34-1
123A	86X6
123A	86X6-1
123A	86X6-4-518
123A	86X7-2
123A	86X7-3
123A	86X7-4
123A	86X8-1
123A	86X8-2
123A	86X8-3
123A	86X8-4
123A	87-0014
123A	88-1250109
123A	089-223
123A	089-226
123A	90-2213-00-18
123A	90-30
123A	90-32
123A	90-452
123A	90-453
123A	90-457
123A	90-48
123A	90-57
123A	90-601
123A	90-602
123A	90-61
123A	90-612
123A	90-65
123A	90-69
123A	90-71
123A	91C
123A	91D
123A	91E
123A	92-30942
123A	93A39-15
123A	94N1
123A	94N1B
123A	94N1R
123A	94N2
123A	95-216
123A	95-221
123A	95-225
123A	95-226-2
123A	95-227
123A	95-296
123A	96-5080-02
123A	96-5115-01
123A	96-5115-02
123A	96-5115-03
123A	96-5115-04
123A	96-5115-05
123A	96-5152-01
123A	96-5152-03
123A	96-5153-01
123A	96-5153-03
123A	96-5177-01
123A	96-5187-01
123A	96-5213-01
123A	96-5220-01(NPN)
123A	96-5221-01
123A	96-5228-01
123A	96-5229-01
123A	96-5237-01
123A	96-5255-01
123A	96-5257-01
123A	96-5281-01
123A	96-5290-01
123A	96-5314-01
123A	96XZ6052/52N
123A	96XZ6053/11N
123A	96XZ6053/35N
123A	96XZ6053/36N
123A	96XZ801/14N
123A	99-109-1
123A	99-109-2
123A	998012
123A	998012A
123A	998012E
123A	998025
123A	998025A
123A	998033A
123A	998035
123A	998036
123A	998038
123A	998085
123A	0101-0491
123A	0101-0540
123A	0103-0014
123A	0103-0014/4460
123A	0103-0088
123A	0103-0088H
123A	0103-0088R
123A	0103-0088S
123A	103-0227-18
123A	0103-0473
123A	0103-0482
123A	0103-0491
123A	0103-0491/4460
123A	0103-0504
123A	0103-0540
123A	0103-94
123A	105(ADMIRAL)
123A	105-003-06
123A	105-003-09
123A	105-006-08
123A	105-060-09
123A	105-06007-05
123A	105-08243-05
123A	105-085-33
123A	105-085-54
123A	105-12
123A	105-28196-07
123A	107-8
123A	107BRN
123A	107N2
123A	109
123A	109-1(RCA)
123A	112-000088
123A	112-1A82
123A	112-523
123A	115-225
123A	115-875
123A	116-074
123A	116-078
123A	116-085
123A	116-092
123A	116-588
123A	116-875
123A	119-0054
123A	119-0056
123A	120-004480

ECG	Industry Standard No.
123A	120-004482
123A	120-004483
123A	120-004880
123A	120-004882
123A	120-004883
123A	120-1
123A	120-2
123A	120-3
123A	120-7
123A	120-8
123A	120-8A
123A	120BLU
123A	121(SEARS)
123A	121-195B
123A	121-364
123A	121-365
123A	121-366
123A	121-367
123A	121-369
123A	121-404
123A	121-423
123A	121-448
123A	121-450
123A	121-505501
123A	121-5065
123A	121-581
123A	121-610
123A	121-629
123A	121-660
123A	121-662
123A	121-678
123A	121-701
123A	121-706
123A	121-711
123A	121-730
123A	121-737
123A	121-751
123A	121-764
123A	121-767CL
123A	121-768CL
123A	121-773
123A	121-836
123A	121-837
123A	121-856
123A	121-889
123A	121G3019
123A	121G3020
123A	122-1
123A	122-2
123A	122-6
123A	123-004
123A	123-010
123A	124N16
123A	0125
123A	125B132
123A	125C211
123A	0126
123A	126-12
123A	127
123A	128WHT
123A	129-16
123A	129-21
123A	129-33(PILOT)
123A	129BRN
123A	129WHT
123A	130-40214
123A	130-40215
123A	130-40294
123A	130-40311
123A	130-40312
123A	130-40313
123A	130-40317
123A	130-40318
123A	130-40357
123A	130-40896
123A	130-40922
123A	0131
123A	0131-000473
123A	0131-000704
123A	0131-001417
123A	0131-001418
123A	0131-001421
123A	0131-001422
123A	0131-001423
123A	0131-001424
123A	0131-001464
123A	0131-001864
123A	0131-004323
123A	132-002
123A	132-004
123A	132-005
123A	132-011
123A	132-017
123A	132-018
123A	132-021
123A	132-023
123A	132-026
123A	132-030
123A	132-041
123A	132-042
123A	132-050
123A	132-051
123A	132-054
123A	132-055
123A	132-057
123A	132-062
123A	132-063
123A	132-069
123A	132-075
123A	132-077
123A	132-501
123A	132-502
123A	132-503
123A	132-504
123A	132-539
123A	132-540
123A	0134
123A	135C44322-542
123A	135ORN
123A	136RED
123A	138-4
123A	140-0007
123A	0140-6
123A	142-001
123A	142-3
123A	142-4
123A	142N3T
123A	142N4
123A	142N5
123A	146N3
123A	146N5
123A	148N2
123A	148N212
123A	150N2

ECG	Industry Standard No.
123A	151-0103-00
123A	151-0127-00
123A	151-0190-00
123A	151-0223-00
123A	151-0224-00
123A	151-0302-00
123A	151-0424-00
123A	151N2
123A	151N4
123A	151N5
123A	154A5941
123A	154A5946
123A	156
123A	156WHT
123A	158(SEARS)
123A	161-016K
123A	165A4383
123A	167N1
123A	167N2
123A	168N1
123A	173A4057
123A	173A4399
123A	173A4416
123A	173A4470-11
123A	173A4470-13
123A	173A4470-32
123A	176-008-9-001
123A	176-014-9-001
123A	176-017-9-001
123A	176-024-9-001
123A	176-042-9-002
123A	176-047-9-002
123A	176-047-9-003
123A	176-054-9-001
123A	185-003
123A	185-004
123A	185-009
123A	186-002
123A	186-007
123A	200-057
123A	200-058
123A	200-846
123A	200-862
123A	200-863
123A	207A10
123A	207A29
123A	207A31
123A	207A35
123A	209-846
123A	209-862
123A	209-863
123A	211AVPF3415
123A	211AVTE4275
123A	212-695
123A	218-22
123A	218-23
123A	218-24
123A	218-25
123A	220-001001
123A	220-001002
123A	221(SEARS)
123A	226-1(SYLVANIA)
123A	229-0050-13
123A	229-0050-14
123A	229-0050-15
123A	229-0180-123
123A	229-0190-90
123A	229-1200-36
123A	229-1301-24
123A	247-257
123A	247-629
123A	247AS-C1249-001
123A	250-0712
123A	260-10-020
123A	260-10-20
123A	260D07412
123A	260D08801
123A	260D09001
123A	260D09314
123A	260D106A1
123A	260D13702
123A	260P02903
123A	260P02903A
123A	260P02908
123A	260P04001
123A	260P04002
123A	260P04003
123A	260P04004
123A	260P04502
123A	260P04503
123A	260P04505
123A	260P06904
123A	260P07001
123A	260P07002
123A	260P07301
123A	260P07701
123A	260P07702
123A	260P07703
123A	260P07704
123A	260P07705
123A	260P07707
123A	260P08801
123A	260P08801A
123A	260P09902
123A	260P11302
123A	260P11303
123A	260P11304
123A	260P11305
123A	260P11502
123A	260P11503
123A	260P11504
123A	260P11505
123A	260P12001
123A	260P12002
123A	260P14101
123A	260P14102
123A	260P14103
123A	260P14105
123A	260P141103
123A	260P17104
123A	260P17105
123A	260P17106
123A	260P17501
123A	260P17503
123A	260P19101
123A	260P19103
123A	260P19501
123A	260P4002
123A	260Z00402
123A	270-950-030
123A	281
123A	284HC
123A	296-50-9
123A	296-51-9

ECG	Industry Standard No.
123A	296-56-9
123A	296-59-9
123A	296-77-9
123A	296-98-9
123A	297L006H01
123A	297L006H02
123A	297L007C02
123A	297L007H01
123A	297L007H02
123A	297L013B01
123A	297V049H05
123A	297V061C07
123A	297V061H01
123A	297V061H02
123A	297V061H03
123A	297V072C06
123A	297V074C06
123A	297V074C07
123A	297V074C08
123A	297V083C02
123A	297V085C01
123A	297V085C02
123A	297V085C03
123A	297V085C04
123A	297V086C02
123A	297V086C03
123A	306-1
123A	309-327-926
123A	314-6007-1
123A	314-6007-2
123A	314-6007-3
123A	317-8504-001
123A	319C
123A	322(CATALINA)
123A	324-6005-5
123A	325-0031-303
123A	325-0031-310
123A	325-0042-351
123A	325-0076-306
123A	325-0076-307
123A	325-0076-308
123A	325-0081-100
123A	325-0081-101
123A	325-0574-30
123A	325-0574-31
123A	325-1370-19
123A	325-1370-20
123A	325-1771-15
123A	344-6000-2
123A	344-6000-4
123A	344-6000-5
123A	344-6000-5A
123A	344-6002-3
123A	344-6005-1
123A	344-6005-2
123A	344-6005-5
123A	344-6017-2
123A	344-6017-3
123A	344-6017-5
123A	352-0195-000
123A	352-0197-000
123A	352-0206-001
123A	352-0316-00
123A	352-0318-00
123A	352-0318-001
123A	352-0319-000
123A	352-0322-010
123A	352-0349-000
123A	352-0365-000
123A	352-0400-00
123A	352-0400-010
123A	352-0400-030
123A	352-0433-00
123A	352-0477-00
123A	352-0506-000
123A	352-0519-00
123A	352-0546-00
123A	352-0569-00
123A	352-0569-010
123A	352-0569-020
123A	352-0579-00
123A	352-0579-010
123A	352-0579-020
123A	352-0596-010
123A	352-0596-020
123A	352-0596-030
123A	352-0629-010
123A	352-0661-010
123A	352-0661-020
123A	352-0667-010
123A	352-0675-010
123A	352-0675-020
123A	352-0675-030
123A	352-0675-040
123A	352-0675-050
123A	352-0680-010
123A	352-0680-020
123A	352-0713-030
123A	352-0809
123A	352-7500-010
123A	352-7500-450
123A	352-8000-010
123A	352-8000-020
123A	352-8000-030
123A	352-8000-040
123A	352-9036-00
123A	352-9079-00
123A	352-9103-000
123A	353-9306-001
123A	353-9306-002
123A	353-9306-003
123A	353-9306-004
123A	353-9306-005
123A	353-9310-001
123A	353-9314-001
123A	353-9315-001
123A	353-9319-001
123A	353-9319-002
123A	354-3127-1
123A	355D9
123A	365-1
123A	375-1005
123A	386-1102-P1
123A	386-1102-P2
123A	386-1102-P3
123A	386-7178P1
123A	386-7185P1
123A	394-3003-1
123A	394-3003-3
123A	394-3003-7
123A	394-3003-9
123A	400-1371-101
123A	400-2023-101
123A	400-2023-201

ECG	Industry Standard No.
123A	404-2(SYLVANIA)
123A	412-1A5
123A	417-105
123A	417-106
123A	417-108
123A	417-109-13163
123A	417-110
123A	417-110-13163
123A	417-114-13163
123A	417-118
123A	417-129
123A	417-134
123A	417-134-13271
123A	417-135
123A	417-155-13163
123A	417-171-13163
123A	417-172
123A	417-172-13271
123A	417-185
123A	417-192
123A	417-197
123A	417-213
123A	417-217
123A	417-226-1
123A	417-228
123A	417-229
123A	417-233-13163
123A	417-244
123A	417-67
123A	417-69
123A	417-7
123A	417-77
123A	417-801-12903
123A	417-91
123A	417-92
123A	417-93
123A	417-93-12903
123A	417-94
123A	421-7444
123A	421-8111
123A	421-9644
123A	421-9670
123A	421-9840
123A	422-0738
123A	422-1234
123A	422-1407
123A	422-2533
123A	422-2534
123A	429-0958-42
123A	430(ZENITH)
123A	430-10034
123A	430-10034-06
123A	430-10053-0
123A	430-10053-0A
123A	430-1044-0A
123A	430-23212
123A	430-23221
123A	430-25766
123A	430-86
123A	430-87
123A	430CL
123A	433(ZENITH)
123A	433CL
123A	436-403-001
123A	444(SEARS)
123A	447(ZENITH)
123A	450-1167-2
123A	450-1261
123A	462-0119
123A	462-1000
123A	462-1009-01
123A	462-1061
123A	462-1063
123A	462-2002
123A	465-106-19
123A	472-0491-001
123A	472-1198-001
123A	483-3141
123A	486-1551
123A	499(ZENITH)
123A	509(SEARS)
123A	512RED
123A	515ORN
123A	519-1(RCA)
123A	524WHT
123A	536-2(RCA)
123A	536D
123A	536D9
123A	536F
123A	536FS
123A	536FU
123A	536G(WARDS)
123A	537D
123A	537E
123A	537FV
123A	537FY
123A	000546-1
123A	550-026-00
123A	560-2
123A	565-074
123A	567-0003-011
123A	570-004503
123A	570-005503
123A	572-683
123A	573-469
123A	573-479
123A	573-480
123A	573-481
123A	576-0001-008
123A	576-0001-012
123A	576-0001-018
123A	576-0002-006
123A	576-0003-011
123A	576-0004-010
123A	576-0036-916
123A	576-0036-917
123A	576-0036-920
123A	576-0036-921
123A	586-2
123A	590-593031
123A	593D742-1
123A	595-1
123A	595-1(SYLVANIA)
123A	595-2
123A	595-2(SYLVANIA)
123A	600-188-1-13
123A	600-188-1-20
123A	600-188-1-23
123A	600-301-801
123A	600X0091-086
123A	601-0100793
123A	601-1
123A	601-1(RCA)
123A	601-2

ECG	Industry Standard No.
123A	601X0149-086
123A	602X0018-000
123A	604
123A	604(SEARS)
123A	606-9601-101
123A	606-9602-101
123A	607-030
123A	609-112
123A	610-045-3
123A	610-045-4
123A	610-070
123A	610-070-1
123A	610-070-2
123A	610-070-3
123A	610-076
123A	610-076-1
123A	610-076-2
123A	610-077
123A	610-077-1
123A	610-077-2
123A	610-077-3
123A	610-077-4
123A	610-077-5
123A	610-078
123A	610-078-1
123A	612-1A5L
123A	614-3
123A	617-10
123A	617-161
123A	617-29
123A	617-63
123A	617-64
123A	617-67
123A	617-68
123A	617-71
123A	626
123A	626-1
123A	630-076
123A	638
123A	639(ZENITH)
123A	642-174
123A	642-242
123A	642-246
123A	642-319
123A	000653
123A	660-126
123A	660-131
123A	660-220
123A	660-221
123A	660-222
123A	660-225
123A	690V0103H27
123A	690V047H97
123A	690V084H62
123A	690V086H51
123A	690V086H88
123A	690V088H50
123A	690V088H51
123A	690V089H89
123A	690V092H52
123A	690V092H54
123A	690V092H81
123A	690V092H84
123A	690V092H96
123A	690V092H97
123A	690V094H21
123A	690V097H62
123A	690V098H48
123A	690V098H49
123A	690V098H50
123A	690V099H79
123A	690V102H71
123A	690V103H31
123A	690V103H33
123A	690V114H30
123A	690V114H33
123A	690V116H21
123A	693EP
123A	693FS
123A	693G
123A	693GT
123A	694D
123A	694E
123A	699
123A	700A858-318
123A	700A858-328
123A	703 056 (4)
123A	715EN
123A	737(ZENITH)
123A	739H01
123A	748(ZENITH)
123A	750C838-123
123A	750C838-124
123A	750C838-125
123A	750D858-212
123A	750M63-119
123A	750M63-120
123A	750M63-146
123A	753-1372-100
123A	753-1828-001
123A	753-2000-003
123A	753-2000-004
123A	753-2000-008
123A	753-2000-009
123A	753-2000-011
123A	753-2000-711
123A	753-2000-735
123A	753-2000-870
123A	753-2000-871
123A	753-2100-001
123A	753-2100-008
123A	753-4000-011
123A	753-4000-101
123A	753-4000-537
123A	767(ZENITH)
123A	767CL
123A	772-110
123A	773(ZENITH)
123A	0776-0160
123A	776-151
123A	776-183
123A	776-2(PHILCO)
123A	776GRN
123A	780WHT
123A	785YEL
123A	791
123A	800-001-034
123A	800-101-101-1
123A	800-101-102-1
123A	800-501-00
123A	800-501-01
123A	800-501-02
123A	800-501-03
123A	800-501-04
123A	800-501-11
123A	800-501-22
123A	800-50100
123A	800-508-00
123A	800-509-00
123A	800-514-00
123A	800-51500
123A	800-521-01
123A	800-52102
123A	800-522-01
123A	800-522-02
123A	800-522-04
123A	800-52202
123A	800-52302
123A	800-526-00
123A	800-52600
123A	800-529-00
123A	800-530-00
123A	800-530-01
123A	800-534-00
123A	800-534-01
123A	800-53400
123A	800-538-00
123A	800-544-00
123A	800-544-10
123A	800-544-20
123A	800-544-30
123A	800-548-00
123A	801B
123A	803-18250
123A	803-37390
123A	803-83940
123A	803-89910
123A	803-89930
123A	804
123A	804-21980
123A	818WHT
123A	822
123A	822A
123A	822ABLU
123A	822B
123A	823B
123A	823WHT
123A	824-09501
123A	827BRN
123A	828GRN
123A	828S
123A	834-6066
123A	847BLK
123A	853-0300-632
123A	853-0300-900
123A	853-0300-923
123A	853-0373-110
123A	858GS
123A	880-250-102
123A	880-250-108
123A	880-250-109
123A	881-250-102
123A	881-250-108
123A	881-250-109
123A	903-3
123A	903-3G
123A	903Y002149
123A	904-95
123A	904-95A
123A	904-96B
123A	909-27125-140
123A	912-1A6A
123A	916-31024-3
123A	916-31025-5
123A	916-31026-8B
123A	921-109B
123A	921-117B
123A	921-120B
123A	921-123B
123A	921-124B
123A	921-125B
123A	921-128B
123A	921-159B
123A	921-161B
123A	921-189B
123A	921-191B
123A	921-195B
123A	921-200B
123A	921-20BK
123A	921-214B
123A	921-215B
123A	921-225B
123A	921-228B
123A	921-229B
123A	921-22BG
123A	921-234B
123A	921-237B
123A	921-23BK
123A	921-252B
123A	921-255B
123A	921-26
123A	921-268B
123A	921-269B
123A	921-26A
123A	921-272B
123A	921-275R
123A	921-27B
123A	921-28
123A	921-28A
123A	921-28B
123A	921-28BLU
123A	921-305B
123A	921-314B
123A	921-345B
123A	921-369
123A	921-43B
123A	921-46
123A	921-46A
123A	921-46BK
123A	921-47
123A	921-47A
123A	921-47BL
123A	921-48B
123A	921-49B
123A	921-50B
123A	921-7
123A	921-73B
123A	921-77B
123A	921-8
123A	921-93B
123A	921-99B
123A	926C193-1
123A	926C193-2
123A	926C193-P1M4165
123A	930X1
123A	930X2
123A	930X3
123A	935-1
123A	964-17444
123A	964-20738
123A	964-2073B
123A	964-24584
123A	991-00-1219
123A	991-00-2232
123A	991-00-2248
123A	991-00-2298
123A	991-00-2356
123A	991-00-2356/K
123A	991-00-2873
123A	991-00-3144
123A	991-00-3304
123A	991-00-8393
123A	991-00-8393A
123A	991-00-8393M
123A	991-00-8394
123A	991-00-8394A
123A	991-00-8394AH
123A	991-00-8395
123A	991-01-1219
123A	991-01-1220
123A	991-01-1306
123A	991-01-1312
123A	991-01-1318
123A	991-01-1705
123A	991-01-3044
123A	991-01-3056
123A	991-01-3057
123A	991-01-3068
123A	991-01-3544
123A	991-01-3683
123A	991-01-3740
123A	991-011219
123A	991-011220
123A	991-011306
123A	991-011312
123A	991-011313
123A	991-011318
123A	991-015587
123A	992-00-2298
123A	992-00-3144
123A	992-01-3738
123A	998-0061114
123A	998-0200816
123A	1000-136
123A	1000-137
123A	1001(JULIETTE)
123A	1001-02
123A	1001-03
123A	1001-04
123A	1001-05
123A	1001-06
123A	1002-03
123A	1002-04
123A	1004(2SC537)
123A	1004-03
123A	1005
123A	1005(2SC537)
123A	1005-03
123A	1008-02
123A	1009-02
123A	1009-02-16
123A	1009-17
123A	1010-7928
123A	1010-8082
123A	1010-8090
123A	1019-3852
123A	1020-17
123A	1023G
123A	1023G(GE)
123A	1024G(GE)
123A	1026G
123A	1026G(GE)
123A	1028(G.E.)
123A	1028G
123A	1028G(GE)
123A	1029(G.E.)
123A	1029G
123A	1029G(GE)
123A	1034-17
123A	1038-1
123A	1038-1-10
123A	1038-10
123A	1038-15
123A	1038-15CL
123A	1038-18
123A	1038-18CL
123A	1038-21
123A	1038-23
123A	1038-23CL
123A	1038-24
123A	1038-6
123A	1038-6CL
123A	1038-8
123A	1039-01
123A	1040-01
123A	1040-03
123A	1040-155
123A	1040-2
123A	1041-72
123A	1041-73
123A	1041-75
123A	1042-04
123A	1042-07
123A	1042-7
123A	1043-07
123A	1045-2951
123A	1048-9912
123A	1049-1744
123A	1063-5381
123A	1077-07
123A	1080-03
123A	1080-20
123A	1080-6396
123A	1081-3301
123A	1081-3319
123A	1081-3475
123A	1081-9464
123A	1087-02
123A	1113-03
123A	1119-8132
123A	1123-60
123A	1203(GE)
123A	1203-169
123A	1205(GE)
123A	1208(GE)
123A	1210-17B
123A	1227-17
123A	1228-17
123A	1236-3776
123A	1272
123A	1315
123A	1316
123A	1368C/D
123A	1374-17AC
123A	1402B
123A	1402C
123A	1402E
123A	1412-1
123A	1412-1-12
123A	1412-1-12-8
123A	1415
123A	001422
123A	1424
123A	1431 8349
123A	1463
123A	1465
123A	1471-4778
123A	1482
123A	1493-17
123A	1540
123A	1567
123A	1567-0
123A	1567-2
123A	1705-4834
123A	1711
123A	1711-17
123A	1711MC
123A	1712-17
123A	1723-17
123A	1751G036
123A	1799-17
123A	1800-17
123A	1812-1
123A	1812-1-12
123A	1812-1-127
123A	1812-1L
123A	1812-1L8
123A	1841-17
123A	1854-0003
123A	1854-0005
123A	1854-0033
123A	1854-0353
123A	1854-0432
123A	1866-17
123A	1879-17
123A	1882-17
123A	1883-17
123A	1884-17
123A	1893-17
123A	1915-17
123A	1929-17
123A	1932-17
123A	1961-17
123A	1966-17
123A	1984-17
123A	1999
123A	2000-210
123A	2001-17
123A	2003-17
123A	2004-03
123A	2004-04
123A	2004-05
123A	2004-14
123A	2006
123A	2008-17
123A	2017-115
123A	2018-01
123A	2020-06
123A	2022-05
123A	2022-244
123A	2026
123A	2026-00
123A	2057A10-64
123A	2057A2-103
123A	2057A2-113
123A	2057A2-121
123A	2057A2-122
123A	2057A2-131
123A	2057A2-145
123A	2057A2-146
123A	2057A2-152
123A	2057A2-153
123A	2057A2-154
123A	2057A2-155
123A	2057A2-184
123A	2057A2-208
123A	2057A2-209
123A	2057A2-215
123A	2057A2-222
123A	2057A2-225
123A	2057A2-226
123A	2057A2-264
123A	2057A2-276
123A	2057A2-278
123A	2057A2-279
123A	2057A2-280
123A	2057A2-281
123A	2057A2-285
123A	2057A2-289
123A	2057A2-294
123A	2057A2-296
123A	2057A2-297
123A	2057A2-300
123A	2057A2-303
123A	2057A2-306
123A	2057A2-316
123A	2057A2-324
123A	2057A2-332
123A	2057A2-341
123A	2057A2-374
123A	2057A2-387
123A	2057A2-390
123A	2057A2-396
123A	2057A2-398
123A	2057A2-399
123A	2057A2-401
123A	2057A2-412
123A	2057A2-433
123A	2057A2-434
123A	2057A2-449
123A	2057A2-452
123A	2057A2-463
123A	2057A2-464
123A	2057A2-479
123A	2057A2-487
123A	2057A2-510
123A	2057A2-511
123A	2057A2-559
123A	2057A2-62
123A	2057A2-64
123A	2057B-85
123A	2057B101-4
123A	2057B102-4
123A	2057B103-4
123A	2057B117-9
123A	2057B118-12
123A	2057B119-2
123A	2057B120-12
123A	2057B125-9
123A	2057B141-4
123A	2057B142-4
123A	2057B143-12
123A	2057B146-12
123A	2057B151-6
123A	2057B152-12
123A	2057B2-113
123A	2057B2-121
123A	2057B2-122
123A	2057B2-123
123A	2057B2-130
123A	2057B2-152
123A	2057B2-153
123A	2057B2-154
123A	2057B2-155
123A	2057B2-38
123A	2057B2-62
123A	2057B2-69
123A	2057B2-97
123A	2063-17-12
123A	2064(CROWN)
123A	2065-03
123A	2132(GE)
123A	2158-1541
123A	2180-153
123A	2180-154
123A	2263
123A	2270
123A	2275-17
123A	2290-17
123A	2320-17
123A	2321-17
123A	2337-17
123A	2443(RCA)
123A	2446(RCA)
123A	2447(RCA)
123A	2448-17
123A	2472-5632
123A	2475(RCA)
123A	2495(RCA)
123A	2495-166-2
123A	2495-522-4
123A	2495-529
123A	2496-125-2
123A	2510-104
123A	2546(RCA)
123A	2603-184
123A	2787
123A	2854
123A	2904-045
123A	2904-054
123A	2925
123A	3005(SEARS)
123A	3005-861
123A	3006(SEARS)
123A	3007(SEARS)
123A	03008-1
123A	3011
123A	3026
123A	3107-204-9000
123A	3107-204-90010
123A	3107-204-90020
123A	3111
123A	3113
123A	3505(RCA)
123A	3506(RCA)
123A	3507(WARDS)
123A	3508(RCA)
123A	3509(SEARS)
123A	3509(WARDS)
123A	3510(RCA)
123A	3513(RCA)
123A	3514(WARDS)
123A	3519(RCA)
123A	3521(SEARS)
123A	3523(SEARS)
123A	3525(RCA)
123A	3526(RCA)
123A	3532(RCA)
123A	3538(RCA)
123A	3541(RCA)
123A	3543(RCA)
123A	3544-1
123A	3546(RCA)
123A	3546-1(RCA)
123A	3546-2(RCA)
123A	3548(RCA)
123A	3551(RCA)
123A	3551A(RCA)
123A	3554(RCA)
123A	3555(RCA)
123A	3556-1
123A	3558(RCA)
123A	3560(RCA)
123A	3560-2(RCA)
123A	3561
123A	3561(RCA)
123A	3561-1(RCA)
123A	3569(RCA)
123A	3571(RCA)
123A	3571R
123A	3572(RCA)
123A	3577(RCA)
123A	3577-1
123A	3586(RCA)
123A	3588
123A	3589
123A	3601(RCA)
123A	3601-1
123A	3614-1
123A	3614-3
123A	3625(RCA)
123A	3626
123A	3631-1
123A	3634.0011
123A	3706
123A	3867
123A	3999
123A	4002(PACE)
123A	4002(PENNCREST)
123A	4003E
123A	4010
123A	4021
123A	4046(SEARS)
123A	4057
123A	4085
123A	4150-01
123A	4309(AIRLINE)
123A	4322-542
123A	4473-12

ECG	Industry Standard No.
123A	4473-4
123A	4473-5
123A	4473-5X
123A	4473-9
123A	4686-132-3
123A	4686-143-3
123A	4686-144-3
123A	4686-173-3
123A	4686-183-3
123A	4686-231-3
123A	4686-257-3
123A	4686-82-3
123A	4732
123A	4733
123A	4734
123A	4737
123A	4765
123A	4768
123A	4801-0000-003
123A	4801-0000-010
123A	4801-0000-016
123A	4802-00003
123A	4802-00006
123A	4802-00009
123A	4802-00012
123A	4802-00015
123A	4822-130-40184
123A	4822-130-40343
123A	4822-130-40354
123A	4822-130-40361
123A	4822-130-40454
123A	4839
123A	4840
123A	4841
123A	4842
123A	4852
123A	4854
123A	4856-0101
123A	4856-0107
123A	4856-0109
123A	4856-0110
123A	5001-002
123A	5001-014
123A	5001-020
123A	5001-069
123A	5001-072
123A	5001-542
123A	5065
123A	5226-2
123A	5380-71
123A	5380-72
123A	5613-1335
123A	5613-1335D
123A	5613-4581C
123A	5613-558C
123A	5613-711
123A	5613-711E
123A	5613-870
123A	5613-870F
123A	5721
123A	6136
123A	06246-00
123A	6343-1
123A	6367
123A	6367-1
123A	6514
123A	6854K90-074
123A	6954K90-074
123A	7005G(LOWREY)
123A	7113
123A	7122-5
123A	7129
123A	7171
123A	7171(GE)
123A	7172
123A	7176
123A	7306
123A	7306-1
123A	7306-4
123A	7306-5
123A	7318
123A	7318-2
123A	7340-2
123A	7398-6117P1
123A	7398-6118P1
123A	7398-6119P
123A	7398-6119P1
123A	7429
123A	7430
123A	7431
123A	7432
123A	7506
123A	7515
123A	7516
123A	7517
123A	7518
123A	7585
123A	7586
123A	7586(GE)
123A	7587
123A	7587(GE)
123A	7588
123A	7588(GE)
123A	7589
123A	7590
123A	7590(GE)
123A	7591(GE)
123A	7637
123A	7641
123A	7675
123A	7676
123A	7816
123A	7817
123A	7818
123A	7992
123A	8000-00003-033
123A	8000-00004-079
123A	8000-00004-082
123A	8000-00004-243
123A	8000-00004-85
123A	8000-00004-P079
123A	8000-00005-002
123A	8000-00005-007
123A	8000-00005-055
123A	8000-00006-003
123A	8000-00006-280
123A	8000-00009-174
123A	8000-00009-280
123A	8000-00011-004
123A	8000-00011-048
123A	8000-00012-039
123A	8000-00028-206
123A	8000-00029-006
123A	8000-00029-007
123A	8000-00030-007
123A	8000-00032-025
123A	8000-0004-P079
123A	8000-0004-P082
123A	8000-0004-P085
123A	8000-00049-055
123A	8000-00049-057
123A	8000-0005-007
123A	8000-00005-009
123A	8000-0009-089
123A	8003-114
123A	8010-176
123A	8020-205
123A	8074-4
123A	8075-4
123A	8210-1203
123A	8281
123A	8281-1
123A	8302
123A	8440-122
123A	8440-123
123A	8440-124
123A	8504
123A	8509
123A	8600
123A	8602
123A	8614 007 0
123A	8710-163
123A	8710-164
123A	8710-165
123A	8710-166
123A	8710-167
123A	8710-168
123A	8800-202
123A	8800-203
123A	8800-204
123A	8840-163
123A	8840-164
123A	8840-165
123A	8840-166
123A	8840-167
123A	8840-168
123A	8868-8
123A	8886-2
123A	8910-143
123A	8910-145
123A	9013H
123A	9013HF
123A	9013HG
123A	9013HH
123A	9014D
123A	9033
123A	9033(SYLVANIA)
123A	9033-1
123A	9033BROWN
123A	9033G(SYLVANIA)
123A	9410A
123A	09502-8
123A	9600
123A	9600-5
123A	9920-6-1
123A	9920-6-2
123A	9920-7-2
123A	10226/2
123A	10416-009
123A	11252-3
123A	11522-5
123A	11587-5
123A	11607-4
123A	11607-8
123A	11608-5
123A	11609-2
123A	11658-8
123A	11687-5
123A	012013-1
123A	12112-C
123A	12112-D
123A	12112-E
123A	12112-F
123A	12127-6
123A	12127-7
123A	12127-8
123A	12127-9
123A	12593
123A	14303
123A	15840-1
123A	15841-1
123A	17412-5
123A	18410-143
123A	18410-145
123A	18410-146
123A	18509
123A	18555
123A	18600-152
123A	18600-153
123A	19645
123A	22158
123A	22635-002
123A	22635-003
123A	22810-173
123A	23114-046
123A	23114-053
123A	23114-054
123A	23114-082
123A	23114-095
123A	23115-057
123A	23115-058
123A	23316
123A	0023828
123A	25114-116
123A	25114-130
123A	25114-161
123A	25810-161
123A	25810-162
123A	25810-163
123A	25840-161
123A	26810-154
123A	27125-080
123A	27125-090
123A	27125-140
123A	27125-160
123A	27125-270
123A	27125-300
123A	27125-370
123A	27125-500
123A	27125-530
123A	27127-550
123A	27840-161
123A	27840-162
123A	030010
123A	030010-1
123A	30227
123A	30228
123A	30229
123A	30235
123A	30241
123A	30242
123A	30243
123A	30253
123A	30268
123A	30269
123A	30289
123A	030512-1
123A	030512-2
123A	030515
123A	030515-4
123A	030527
123A	030536
123A	030537
123A	030537-1
123A	030537-2
123A	030538
123A	030542
123A	030542-1
123A	030543
123A	030543-1
123A	030543-2
123A	31001
123A	31003
123A	31009
123A	33563
123A	36580
123A	36917
123A	36920
123A	36921
123A	37510-162
123A	37510-163
123A	37884
123A	38283
123A	38510-163
123A	38510-166
123A	38510-167
123A	38788
123A	39034
123A	39096
123A	40084
123A	40217
123A	40218
123A	40219
123A	40220
123A	40221
123A	40222
123A	40283
123A	40397
123A	40398
123A	40399
123A	40400
123A	40405
123A	40432
123A	40456(RCA)
123A	40473
123A	40474
123A	40477
123A	40500
123A	40519
123A	40577
123A	40637
123A	41051
123A	41176
123A	42464
123A	43021-017
123A	43044
123A	43045
123A	43139
123A	45184
123A	45810-162
123A	45810-163
123A	45810-164
123A	48004-07
123A	48004-08
123A	50202-1
123A	50202-13
123A	50202-14
123A	50202-2
123A	50202-23
123A	51213
123A	51213-01
123A	51213-02
123A	51213-03
123A	51213-2
123A	51428-01
123A	51429-02
123A	51429-03
123A	51429-3
123A	51441
123A	51441-01
123A	51441-02
123A	51441-03
123A	51442
123A	51442-01
123A	51442-02
123A	51442-03
123A	51545
123A	51547
123A	53200-22
123A	53200-23
123A	53200-51
123A	53400-01
123A	053492
123A	55606
123A	58215-01
123A	58810-162
123A	58810-163
123A	58810-165
123A	58810-166
123A	58810-168
123A	58840-193
123A	58840-194
123A	58840-195
123A	58840-196
123A	58840-197
123A	58840-198
123A	58840-199
123A	60395
123A	61009-4
123A	61009-4-1
123A	61010-7-2
123A	61011-3-2
123A	61013-2-1
123A	61049
123A	67586
123A	68617
123A	70023-0-00
123A	70023-1-00
123A	70260-14
123A	70260-15
123A	70260-16
123A	70260-20
123A	70511
123A	71226-10
123A	71266-4
123A	71819-1
123A	72114
123A	72115
123A	72116
123A	72151
123A	72204
123A	72206
123A	72207
123A	72874-52
123A	72963-14
123A	000073090
123A	000073100
123A	000073120
123A	000073130
123A	000073230
123A	000073231
123A	000073290
123A	000073310
123A	000073332
123A	000073333
123A	000073390
123A	000073391
123A	74651-02
123A	75561-16
123A	75561-28
123A	75561-3
123A	75561-33
123A	75614-1
123A	75700-04
123A	75700-04-01
123A	75700-05
123A	75700-05-01
123A	75700-05-02
123A	75700-05-03
123A	75700-08
123A	75700-08-02
123A	75700-09-01
123A	75700-09-21
123A	76236
123A	80540
123A	80544
123A	80545
123A	80813VM
123A	80814VM
123A	80815VM
123A	80816VM
123A	81170-6
123A	81513-3
123A	082006
123A	082019
123A	85549
123A	86287
123A	87757
123A	88060-143
123A	88510-173
123A	88510-175
123A	88687
123A	88688
123A	88862
123A	90209-172
123A	90209-182
123A	90326-001
123A	90429
123A	91605
123A	94027
123A	94047
123A	94048
123A	95223
123A	96457-1
123A	99206-1
123A	99206-2
123A	99207-2
123A	100092
123A	100093
123A	101119
123A	101185
123A	102002
123A	103521
123A	104389
123A	105432
123A	110697
123A	111303
123A	112356
123A	112357
123A	112358
123A	112359
123A	112520
123A	112521
123A	112522
123A	112523
123A	113348
123A	113438
123A	113524
123A	115167
123A	115225
123A	115720
123A	115728
123A	115875
123A	116074
123A	116076
123A	116077
123A	116078
123A	116085
123A	116092
123A	116588
123A	116875
123A	118200
123A	118713
123A	119232-001
123A	119258-001
123A	119636
123A	119724
123A	119725
123A	119726
123A	119982
123A	120073
123A	120074
123A	120085
123A	120481
123A	120482
123A	120483
123A	121655
123A	121658
123A	121660
123A	121661(RCA)
123A	121662
123A	121663
123A	122074
123A	122664
123A	122665
123A	123807
123A	124557
123A	124759
123A	125135
123A	125139
123A	125140
123A	125143
123A	125389-14
123A	126150
123A	126156
123A	126331
123A	126525
123A	126526
123A	126699
123A	126702
123A	126704
123A	126706
123A	126708
123A	126711
123A	126713
123A	126714
123A	127263
123A	129147
123A	129949
123A	130403-13
123A	130403-17
123A	130403-18
123A	130536
123A	130537
123A	131240
123A	131243-12
123A	133743
123A	137383
123A	137614
123A	138378
123A	138789-1
123A	138789-2
123A	138789-3
123A	140622
123A	140858-12
123A	141558
123A	146144-2
123A	146153-1
123A	147115
123A	147115-5
123A	147115-6
123A	147357-1-1
123A	147357-7-1
123A	147363-1
123A	147513
123A	147555-1
123A	150117
123A	150741
123A	150763
123A	150768
123A	157008
123A	161918-28
123A	165668
123A	165827
123A	165828
123A	167569
123A	167688
123A	169197
123A	169679
123A	169680
123A	170967-1
123A	170968-1
123A	171003(TOSHIBA)
123A	171009(TOSHIBA)
123A	171026(TOSHIBA)
123A	171027
123A	171030(TOSHIBA)
123A	171040(SEARS)
123A	171040(TOSHIBA)
123A	171044(TOSHIBA)
123A	171046
123A	171162-005
123A	171162-006
123A	171162-008
123A	171162-009
123A	171162-119
123A	171162-132
123A	171162-143
123A	171162-161
123A	171162-162
123A	171162-188
123A	171162-190
123A	171162-191
123A	171162-202
123A	171162-286
123A	175007-275
123A	175007-276
123A	175027-021
123A	175043-058
123A	175043-059
123A	175043-060
123A	181023
123A	181214
123A	181504-2
123A	183017
123A	183030
123A	183031
123A	187218
123A	190426
123A	190428
123A	190715
123A	196023-1
123A	196023-2
123A	198003-1
123A	198003-2
123A	198007-3
123A	198013-P1
123A	198023-1
123A	198023-3
123A	198023-4
123A	198023-5
123A	198030
123A	198030-2
123A	198030-3
123A	198030-4
123A	198030-6
123A	198030-7
123A	198031-1
123A	198031-2
123A	198042-2
123A	198042-3
123A	198051-1
123A	198051-2
123A	198051-3
123A	198051-4
123A	198067-1
123A	198581-1
123A	198581-2
123A	198581-3
123A	200064-6-107

ECG	Industry Standard No.
123A	200076
123A	200251-5377
123A	202609-0713
123A	202862-947
123A	202907-047P1
123A	202914-417
123A	202915-627
123A	202922-237
123A	204210-002
123A	204969
123A	210074
123A	215072
123A	215081
123A	216445-2
123A	221600
123A	221857
123A	221897
123A	221918
123A	222131
123A	224506
123A	229017
123A	231140-15
123A	231374
123A	232017
123A	232678
123A	234612
123A	234758
123A	234763
123A	235192
123A	235205
123A	235206
123A	236285
123A	236286
123A	237025
123A	237223
123A	238368
123A	239970
123A	240401
123A	242758
123A	242759
123A	244817
123A	256217
123A	256517
123A	256917
123A	262066
123A	265240
123A	266685
123A	267898
123A	267899
123A	268044L
123A	270819
123A	275131
123A	276331
123A	279917
123A	282317
123A	290458LGD
123A	299371-1
123A	300113
123A	301591
123A	302342
123A	304581B
123A	308449
123A	309442
123A	313309-1
123A	0320031
123A	320529
123A	321517
123A	321573
123A	325079
123A	330803
123A	333241
123A	334724-1
123A	335288-4
123A	335774
123A	346015-24
123A	346016-14
123A	346016-18
123A	346016-19
123A	346016-25
123A	346016-26
123A	00352080
123A	379101K
123A	388060
123A	405457
123A	433836
123A	00444028-010
123A	00444028-014
123A	489751-025
123A	489751-026
123A	489751-029
123A	489751-030
123A	489751-040
123A	489751-041
123A	489751-107
123A	489751-122
123A	489751-125
123A	489751-166
123A	489751-172
123A	514023
123A	5150438
123A	5150458
123A	533802
123A	552308
123A	570000-5452
123A	570000-5503
123A	570004-503
123A	570005-452
123A	570005-503
123A	570009-01-504
123A	572683
123A	0573066
123A	0573202
123A	0573418
123A	0573430
123A	0573460
123A	573467
123A	0573468
123A	0573469
123A	0573469H
123A	573479
123A	0573479H
123A	0573481H
123A	0573490
123A	0573523
123A	0573529
123A	0573556
123A	0573981
123A	581034A
123A	581054
123A	581055
123A	600080-413-001
123A	600080-413-002
123A	600098-413-001
123A	601122
123A	602113(SHARP)
123A	602909-2A
123A	603122
123A	604122
123A	609112
123A	610045-3
123A	610045-4
123A	610045-5
123A	610070
123A	610070-1
123A	610070-2
123A	610070-3
123A	610070-4
123A	610076
123A	610076-1
123A	610076-2
123A	610077
123A	610077-1
123A	610077-2
123A	610077-3
123A	610077-4
123A	610077-5
123A	610077-6
123A	610078
123A	610078-1
123A	610078-2
123A	610094
123A	610094-1
123A	610094-2
123A	610132
123A	610132-1
123A	610142-2
123A	610142-3
123A	610142-4
123A	610142-5
123A	610142-7
123A	610143-3
123A	610146-3
123A	610146-5
123A	610147-1
123A	610148-2
123A	610148-2A
123A	610150-2
123A	610151-2
123A	610151-4
123A	610165-1
123A	610165-2
123A	610167-1
123A	610167-2
123A	610168-1
123A	610168-2
123A	610232-1
123A	611428
123A	615093-2
123A	615179-1
123A	615179-2
123A	618072
123A	618126-1
123A	618217-2
123A	618810-2
123A	619006
123A	619006-7
123A	651955
123A	651995-1
123A	651995-2
123A	651995-3
123A	656204
123A	700047-47
123A	700047-49
123A	700181
123A	700230-00
123A	700231-00
123A	702884
123A	720236
123A	720240
123A	740437
123A	757008-02
123A	760142
123A	760236
123A	760239
123A	760251
123A	785278-101
123A	800132-001
123A	803182-5
123A	803369-6
123A	803372-0
123A	803373-0
123A	803733-0
123A	803733-3
123A	803735-3
123A	810000-373
123A	815133
123A	815134
123A	815171
123A	815171D
123A	815174
123A	815174L
123A	815182
123A	815183
123A	815184
123A	815184E
123A	815186
123A	815186C
123A	815186L
123A	815190
123A	815191
123A	815198
123A	815201
123A	815202
123A	815210
123A	815227
123A	815233
123A	815237
123A	845050
123A	848082
123A	883802
123A	900552-20
123A	900552-30
123A	900552-6
123A	900552-8
123A	911743-1
123A	916009
123A	916028
123A	916030
123A	916031
123A	916050
123A	916059
123A	916091
123A	928103-1
123A	930236
123A	941295-2
123A	941295-3
123A	943720-001
123A	954330-2
123A	959492-2
123A	961544-1
123A	965000
123A	970247
123A	970250
123A	970252
123A	970659
123A	970660
123A	970661
123A	970662
123A	970916
123A	970916-6
123A	972155
123A	972156
123A	972214
123A	972215
123A	980147
123A	980440
123A	982231
123A	982510
123A	982511
123A	982512
123A	983097
123A	983743
123A	984197
123A	984198
123A	984222
123A	984224
123A	984286
123A	984590
123A	984591
123A	984593
123A	984686
123A	984687
123A	984745
123A	984854
123A	984879
123A	985098
123A	985099
123A	985100
123A	985101
123A	985102
123A	985543
123A	986542
123A	986636
123A	987010
123A	988003
123A	988991
123A	992052
123A	992129
123A	995016
123A	995017
123A	995870-1
123A	995870-3
123A	996746
123A	1022612
123A	1127859
123A	1221962
123A	1222123
123A	1222133
123A	1222424
123A	1223782
123A	1223912
123A	1223916
123A	1223920
123A	1261915-383
123A	1320135
123A	1320135A
123A	1320135BC
123A	1320135C
123A	1417302-1
123A	1417306-2
123A	1417306-4
123A	1417306-5
123A	1417312-1
123A	1417312-2
123A	1417318-1
123A	1417340-2
123A	1471113-2
123A	1471113-3
123A	1471115-1
123A	1471115-12
123A	1471120-15
123A	1471120-7
123A	1471120-8
123A	1471120-8-9
123A	1472495-1
123A	1473500-1
123A	1473505-1
123A	1473519-1
123A	1473527-1
123A	1473532-1
123A	1473536-001
123A	1473536-1
123A	1473538-1
123A	1473539-1
123A	1473546-1
123A	1473546-2
123A	1473546-3
123A	1473548-1
123A	1473550-1
123A	1473551-1
123A	1473554-1
123A	1473555-1
123A	1473555-2
123A	1473556-1
123A	1473557-1
123A	1473560-002
123A	1473569-1
123A	1473572-1
123A	1473582-1
123A	1473586-2
123A	1473589-1
123A	1473595-1
123A	1473601-001
123A	1473601-1
123A	1473601-2
123A	1473614-1
123A	1473614-3
123A	1473622-1
123A	1473626-1
123A	1473631-1
123A	1476188-1
123A	1501883
123A	1522237-20
123A	1563295-101
123A	1596408
123A	1611708-2
123A	1617510-1
123A	1690019-01
123A	1700008
123A	1700019
123A	1780145-1
123A	1780145-2
123A	1780145-2-001
123A	1780724-1
123A	1780738-1
123A	1815041
123A	1815042
123A	1815043
123A	1815054
123A	1815154
123A	1817005
123A	1817007
123A	1817108
123A	1840399-1
123A	1846282-1
123A	1851515
123A	1950039
123A	1960023
123A	1960177-2
123A	1968958
123A	2000646-103
123A	2000646-107
123A	2002153-77
123A	2002621-2
123A	2003073-0701
123A	2003073-10
123A	2003073-9
123A	2003168-135
123A	2003168-136
123A	2003229-25
123A	2006227-51
123A	2006334-115
123A	2006431-45
123A	2006431-46
123A	2006514-60
123A	2006582-25
123A	2006607-60
123A	2006613-77
123A	2006623-145
123A	2006681-96
123A	2010088-49
123A	2010499-52
123A	2041614
123A	2092055-0714
123A	2092417-0719
123A	2092417-0720
123A	2092417-0721
123A	2092417-0724
123A	2092417-0725
123A	2092417-17
123A	2092417-18
123A	2092417-19
123A	2092605-0705
123A	2092608-22
123A	2092609
123A	2092609-0001
123A	2092609-0002
123A	2092609-001
123A	2092609-0022
123A	2092609-0023
123A	2092609-0024
123A	2092609-0026
123A	2092609-0027
123A	2092609-0028
123A	2092609-0705
123A	2092609-0706
123A	2092609-0707
123A	2092609-0713
123A	2092609-0715
123A	2092609-0718
123A	2092609-0720
123A	2092609-0721
123A	2092609-1
123A	2092609-2
123A	2092609-3
123A	2092609-5
123A	2093308-0701
123A	2093308-0702
123A	2093308-0703
123A	2093308-0708
123A	2320022
123A	2320111
123A	2320123
123A	2320413
123A	2320441
123A	2320591-1
123A	2320696
123A	2320696-1
123A	2326953
123A	2327023
123A	2327122
123A	2327363
123A	2360924-5601
123A	2469749
123A	2469755
123A	2479692
123A	2479836
123A	2485078-1
123A	2485078-2
123A	2485078-3
123A	2485079-1
123A	2485079-2
123A	2485079-3
123A	2495166-2
123A	2495166-4
123A	2495166-8
123A	2495166-9
123A	2495521-1
123A	2495522-4
123A	2496125-2
123A	2498457-2
123A	2498904-3
123A	2498904-4
123A	2498904-6
123A	2505209
123A	2520063
123A	2530733
123A	2621567-1
123A	2621764
123A	2640843-1
123A	2712080
123A	3005861
123A	3068305-2
123A	3201104-10
123A	3457107-1
123A	3457632-5
123A	3468182-1
123A	3468182-2
123A	3468242-2
123A	3596117
123A	3596338
123A	3596339
123A	3596570
123A	3670724H
123A	3700072
123A	3700109
123A	3700171
123A	3700279
123A	3731132-1
123A	4002862-0001
123A	4017621-0701
123A	4036887-P8
123A	4036924-P1
123A	4036924-P2
123A	4037586-P1
123A	4037586-P2
123A	4037800-P1
123A	4037800-P2
123A	04440028-001
123A	004440028-002
123A	04440028-003
123A	04440028-006
123A	004440028-007
123A	004440028-008
123A	004440028-010
123A	04440028-013
123A	004440028-014
123A	04440052-001
123A	04440052-002
123A	04450002-001
123A	04450002-004
123A	04450002-005
123A	4813466
123A	4906071
123A	4906072
123A	4906073
123A	5294477-1
123A	5294477-2
123A	5320023H
123A	5320024
123A	5320026
123A	5320064H
123A	5320074
123A	05320074H
123A	5320241
123A	5320326
123A	5320372
123A	5320372H
123A	5320373
123A	5320851
123A	5321261
123A	6208839
123A	6212839
123A	6218945
123A	6984590
123A	6984600
123A	6993650
123A	7002453
123A	7011507
123A	7011507-00
123A	7026014
123A	7026015
123A	7026016
123A	7026020
123A	7071021
123A	7071031
123A	7121105-01
123A	7279779(G.M.)
123A	7284137
123A	7286858
123A	7287452
123A	7295197
123A	7296314
123A	7296811
123A	7297053
123A	7297054
123A	7303120
123A	7304380
123A	7314584
123A	7570004
123A	7570004-01
123A	7570005
123A	7570005-01
123A	7570005-02
123A	7570005-03
123A	7570008
123A	7570008-01
123A	7570008-02
123A	7570009-01
123A	7570009-21
123A	7576015-01
123A	7840540-1
123A	7851324-01
123A	7851325
123A	7851326
123A	7851327
123A	7851379-01
123A	7851380-01
123A	7851949-01
123A	7851950-01
123A	7851952-01
123A	7851953-01
123A	7852454
123A	7852454-01
123A	7852455-01
123A	7852459-01
123A	7852781-01
123A	7853092-01
123A	7853094-01
123A	7853463-01
123A	7853464-01
123A	7853465-01
123A	7855291-01
123A	7855292-01
123A	7855293-01
123A	7855294-01
123A	7910584-01
123A	7910585-01
123A	7910586-01
123A	7910587-01
123A	7910804-01
123A	7936256
123A	7936331
123A	8031839
123A	8033690
123A	8033696
123A	8033720
123A	8033730
123A	8033944
123A	8037330
123A	8037332
123A	8037333
123A	8037353
123A	8113034
123A	8113051
123A	8113052
123A	8113060
123A	8113134
123A	8114031
123A	8522468-1

ECG	Industry Standard No.
123A	9001630
123A	9002159
123A	9008964-01
123A	9176494
123A	10000020
123A	10015595
123A	10022104-101
123A	10106058
123A	10180722
123A	10545502
123A	10644433
123A	10849792
123A	10896074
123A	11198132
123A	11220046/7825
123A	11802400
123A	11802500
123A	12965471
123A	12994851
123A	13035807-1
123A	13037215
123A	13040216
123A	13040313
123A	13040317
123A	13040318
123A	13040357
123A	13083908
123A	14714760
123A	14714786
123A	15038433
123A	15039456
123A	15039464
123A	16270092
123A	16520001-1
123A	16797300
123A	16797301
123A	18179900
123A	20025153-77
123A	20030703-0701
123A	20052600
123A	23114046
123A	23114118
123A	23114119
123A	23114155
123A	23114212
123A	23114214
123A	23114255
123A	23114275
123A	23114276
123A	23114296
123A	23115057
123A	23115058
123A	24501000
123A	24553600
123A	24562000
123A	24562001
123A	24562101
123A	24562200
123A	25114116
123A	25114121
123A	26004001
123A	26010020
123A	27125080
123A	27125090
123A	27125140
123A	27125150
123A	27125160
123A	27125250
123A	27125270
123A	27125300
123A	27125370
123A	27125380
123A	27125470
123A	27125500
123A	30200053
123A	32600025-01-08A
123A	35393060-01
123A	35393060-02
123A	35393060-03
123A	36171100
123A	38970300
123A	43021017
123A	43021083
123A	43022861
123A	43023212
123A	43023221
123A	43023844
123A	43024972
123A	43025055
123A	43025056
123A	43025059
123A	43025972
123A	43027379
123A	43027620
123A	44011001
123A	44090004
123A	48134666
123A	48134842
123A	50210104
123A	50210300-00
123A	50210300-01
123A	50210300-11
123A	50210310-10
123A	50210310-11
123A	50210510
123A	50210800-00
123A	50210800-01
123A	50210800-02
123A	50210800-11
123A	51003059
123A	51003092
123A	51122245
123A	51565600 VF
123A	51581300
123A	55440048-001
123A	56301500
123A	57000901-504
123A	59700278
123A	62084960
123A	62105275
123A	62236302
123A	62379995
123A	62389699
123A	62506334
123A	62506377
123A	62537140
123A	62539283
123A	62563362
123A	62593687
123A	62593707
123A	62596279
123A	62604727
123A	62618663
123A	62638214
123A	62638230
123A	62638281
123A	62638303
123A	62638311
123A	62652381
123A	62675004
123A	62675012
123A	62691263
123A	62711914
123A	62733942
123A	62737409
123A	62737476
123A	62737484
123A	62737492
123A	62766220
123A	62785225
123A	62793163
123A	75857322
123A	80050100
123A	80051500
123A	80052102
123A	80052202
123A	80052302
123A	80052600
123A	80053001
123A	80053400
123A	80318250
123A	80383840
123A	80383930
123A	80383940
123A	80389910
123A	82409501
123A	83037204
123A	83073205
123A	83073206
123A	83073304
123A	83073305
123A	83073504
123A	83094502
123A	86401000
123A	88125010-9
123A	89942601
123A	89962306
123A	89962307
123A	89962308
123A	89962404
123A	89963008
123A	89963009
123A	93037230
123A	93037240
123A	93063240
123A	93063270
123A	93063280
123A	93063470
123A	93064450
123A	93073540
123A	93082830
123A	93082840
123A	93094502
123A	93938040
123A	94824101
123A	120004883
123A	229005013
123A	229005014
123A	229005015
123A	436005001
123A	485134922
123A	485134923
123A	485134924
123A	485134925
123A	485134926
123A	570004503
123A	570005452
123A	570005503
123A	632037218
123A	881250108
123A	881250109
123A	1522223720
123A	1611819064
123A	1760089001
123A	2003037227
123A	2003189025
123A	3539306001
123A	3539306002
123A	3539306003
123A	4104206440
123A	4108296238
123A	4109208284
123A	4360021001
123A	6621003100
123A	6621003200
123A	6621003400
123A	8001200001
123A	16102190693
123A	16102190930
123A	16104191168
123A	16105190536
123A	16106190537
123A	16109209536
123A	16116190634
123A	16147191229
123A	16171190693
123A	16171190858
123A	16171191368
123A	16172190693
123A	16172190858
123A	16307190632
123A	16377190632
123A	57000901504
123A	134800869525
123A	310720490000
123A	310720490010
123A	310720490020
123A	310720490100
123A	310720490150
123A	404100900160
123A	933022960112
123A$	C913
123A$	2SC1385H
123A$	2SC335
123A$	2SC913
123A$	2SC914
123A$	2SC915
123A(2)	M779
123A(2)	M780
123A(2)	M786
123A(2)	M787
123A(2)	R8900
123A(2)	86-5099-2
123A(2)	297L007H03
123A*	13-17-6
123AP	A5T3391
123AP	A5T3391A
123AP	A5T3392
123AP	A5T3565
123AP	A5T3707
123AP	A5T3708
123AP	A5T3709
123AP	A5T3710
123AP	A5T3711
123AP	A5T4123
123AP	A5T4124
123AP	A5T5172
123AP	A5T5219
123AP	A5T5220
123AP	A5T5223
123AP	A5T5225
123AP	A670729B
123AP	A6H
123AP	A8T3391
123AP	A8T3391A
123AP	A8T3392
123AP	A8T3706
123AP	A8T3707
123AP	A8T3708
123AP	A8T3709
123AP	A8T3710
123AP	A8T3711
123AP	A8T5172
123AP	B6P
123AP	BC109BP
123AP	BC167
123AP	BC168
123AP	BC183A
123AP	BC183B
123AP	BC237A
123AP	BC237B
123AP	BC239
123AP	BC239A
123AP	BC239B
123AP	BC317
123AP	BC317A
123AP	BC317B
123AP	BC318
123AP	BC318A
123AP	BC318B
123AP	BC318C
123AP	BC319
123AP	BC319B
123AP	BC319C
123AP	BC337-16
123AP	BC338
123AP	BC338-16
123AP	BC382B
123AP	BC382C
123AP	BC383B
123AP	BC383C
123AP	BC384B
123AP	BC413B
123AP	BC413C
123AP	BC414B
123AP	BC414C
123AP	BC547B
123AP	BC547C
123AP	BC548
123AP	BC582
123AP	BC582A
123AP	BC582B
123AP	BC583A
123AP	BC583B
123AP	BCW48A
123AP	BCW94
123AP	BCW94A
123AP	BCW94B
123AP	BCW94C
123AP	BCW94KA
123AP	BCW94KB
123AP	BCW94KC
123AP	BF237
123AP	BF238
123AP	BFS39
123AP	BLY27
123AP	BS9011G
123AP	BSV86
123AP	BSV87
123AP	BSW58
123AP	C1123
123AP	C1362
123AP	CCS2001H
123AP	CE4001B
123AP	CE4001C
123AP	CE4004C
123AP	CE4013E
123AP	CS2001H
123AP	CS2004C
123AP	DDBY270001
123AP	EA15X365
123AP	EA15X7262
123AP	EA15X7588
123AP	EA15X8518
123AP	EA15X8601
123AP	EA15X8602
123AP	ED1402C
123AP	ED1402D
123AP	ED1402E
123AP	ED1702M
123AP	EP15X1
123AP	EP15X2
123AP	EP15X8
123AP	EP15X86
123AP	EP15X88
123AP	EP15X9
123AP	FCS-9013F
123AP	FCS-9013G
123AP	FCS-9016G
123AP	FCS9011E
123AP	FCS9011F
123AP	FCS9011G
123AP	FCS9011H
123AP	FCS9013
123AP	FCS9013F
123AP	FCS9013G
123AP	FCS9013H
123AP	FCS9013HG
123AP	FCS9013HH
123AP	FCS9014
123AP	FCS9014B
123AP	FCS9014C
123AP	FCS9014D
123AP	FCS9016
123AP	FCS9016D
123AP	FCS9016E
123AP	FCS9016F
123AP	FCS9016G
123AP	GS9014
123AP	GS9014I
123AP	GS9014J
123AP	GS9014K
123AP	GS9023H
123AP	GS9023I
123AP	GS9023J
123AP	HEPS0015
123AP	HEPS0025
123AP	HEPS0030
123AP	HS-40014
123AP	HS-40030
123AP	HX-50092
123AP	HX-50097
123AP	HX-50161
123AP	IP20-0002
123AP	JA1350
123AP	JE9011G
123AP	JE9011H
123AP	KGE46338
123AP	KM917F
123AP	KM917G
123AP	LM2152
123AP	ME9001
123AP	MPS2369
123AP	MPS2711
123AP	MPS2712
123AP	MPS2713
123AP	MPS2714
123AP	MPS2715
123AP	MPS2716
123AP	MPS2923
123AP	MPS2924
123AP	MPS2925
123AP	MPS2926
123AP	MPS3392
123AP	MPS3393
123AP	MPS3394
123AP	MPS3395
123AP	MPS3396
123AP	MPS3398
123AP	MPS3642
123AP	MPS3693
123AP	MPS3694
123AP	MPS3704
123AP	MPS3705
123AP	MPS3706
123AP	MPS3707
123AP	MPS3708
123AP	MPS3709
123AP	MPS3710
123AP	MPS3711
123AP	MPS3721
123AP	MPS3826
123AP	MPS3827
123AP	MPS5172
123AP	MPS6512
123AP	MPS6513
123AP	MPS6514
123AP	MPS6515
123AP	MPS6520
123AP	MPS6521
123AP	MPS6530
123AP	MPS6531
123AP	MPS6532
123AP	MPS6544
123AP	MPS6545
123AP	MPS6555
123AP	MPS6560
123AP	MPS6561
123AP	MPS6564
123AP	MPS6565
123AP	MPS6566
123AP	MPS6571
123AP	MPS6573
123AP	MPS6574
123AP	MPS6575
123AP	MPS6576
123AP	MPS706
123AP	MPS706A
123AP	MPS8001
123AP	MPS8097
123AP	MPS8098
123AP	MPS8099
123AP	MPS834
123AP	MPS9418
123AP	MPS9418T
123AP	MPS9623
123AP	MPS9623H
123AP	MPS9623I
123AP	MPS9626G
123AP	MPS9626H
123AP	MPS9626I
123AP	MPS9630K
123AP	MPS9630T
123AP	MPS9631S
123AP	MPS9634
123AP	MPS9634C
123AP	MPS9634D
123AP	MPSA09
123AP	MPSA10
123AP	MPSA20
123AP	MPSD05
123AP	MPSD06
123AP	MPSH37
123AP	MPSK20
123AP	MPSK21
123AP	MPSK22
123AP	NPC069
123AP	NPC069-98
123AP	PE3001
123AP	PN2222
123AP	PN2369
123AP	PN2369A
123AP	PN2484
123AP	PN930
123AP	PTC121
123AP	Q-2N5225
123AP	QT-C0710XAE
123AP	QT-C0710XEE
123AP	SC147A
123AP	SC147B
123AP	SC148
123AP	SC148A
123AP	SC148B
123AP	SC148C
123AP	SC149
123AP	ST-LM2152
123AP	ST1402E
123AP	T9631
123AP	TIS125
123AP	TIS133
123AP	TIS134
123AP	TIS83
123AP	TIS97
123AP	TR01062-1
123AP	TV-32
123AP	UPI2222P
123AP	ZTX107
123AP	ZTX108
123AP	ZTX109
123AP	ZTX300
123AP	ZTX301
123AP	ZTX302
123AP	ZTX303
123AP	ZTX304
123AP	ZTX310
123AP	ZTX311
123AP	ZTX312
123AP	ZTX330
123AP	ZTX331
123AP	ZTX360
123AP	01-030710
123AP	01-030829
123AP	01-030945
123AP	01-031815
123AP	01-032076
123AP	01-349418
123AP	01-571821
123AP	01-680815
123AP	1A0025
123AP	1A0070
123AP	1A0076
123AP	1A0077
123AP	1A0078
123AP	2D002
123AP	2D002-170
123AP	2D002-175
123AP	2D002-41
123AP	2D026
123AP	2D026-274
123AP	2N1149
123AP	2N1150
123AP	2N1151
123AP	2N1152
123AP	2N1153
123AP	2N117
123AP	2N118
123AP	2N118A
123AP	2N119
123AP	2N120
123AP	2N1587
123AP	2N1588
123AP	2N1590
123AP	2N1591
123AP	2N1593
123AP	2N1594
123AP	2N160
123AP	2N160A
123AP	2N161
123AP	2N161A
123AP	2N162
123AP	2N163
123AP	2N163A
123AP	2N2617
123AP	2N263
123AP	2N264
123AP	2N3128
123AP	2N3129
123AP	2N3130
123AP	2N3705
123AP	2N3706
123AP	2N3793
123AP	2N3794
123AP	2N3828
123AP	2N3903
123AP	2N3904
123AP	2N4123
123AP	2N4124
123AP	2N4264
123AP	2N4265
123AP	2N4400
123AP	2N4401
123AP	2N4418
123AP	2N4419
123AP	2N4420
123AP	2N4421
123AP	2N4422
123AP	2N4951
123AP	2N4952
123AP	2N4953
123AP	2N4994
123AP	2N4995
123AP	2N5088
123AP	2N5089
123AP	2N5131
123AP	2N5209
123AP	2N5210
123AP	2N5219
123AP	2N5220
123AP	2N5223
123AP	2N5224
123AP	2N5225
123AP	2N5368
123AP	2N5369
123AP	2N5370
123AP	2N5371
123AP	2N5380
123AP	2N5381
123AP	2N5769
123AP	2N5772
123AP	2N5810
123AP	2N5812
123AP	2N5824
123AP	2N5825
123AP	2N5826
123AP	2N5827
123AP	2N5827A
123AP	2N5828
123AP	2N5828A
123AP	2N5845
123AP	2N5845A
123AP	2N5961
123AP	2N6000
123AP	2N6002
123AP	2N6006
123AP	2SC1123
123AP	2SC1851
123AP	2SC1852
123AP	2SC235Y
123AP	03A12
123AP	8-729-663-47
123AP	13-15808-1
123AP	13-27404-1
123AP	13-27432-2
123AP	13-29033-1
123AP	13-29033-2
123AP	13-29033-3
123AP	13-29033-5
123AP	13-29033-6
123AP	13-29947-1
123AP	13-32362-1

ECG	Industry Standard No.
123AP	13-32630-1
123AP	13-32630-3
123AP	13-35226-1
123AP	13-39114-1
123AP	13-39114-2
123AP	13-43773-1
123AP	022-2844-002
123AP	022-2876-007
123AP	31-002-0
123AP	31-025-0
123AP	34-6015-12
123AP	48-137171
123AP	57A182-12
123AP	121-1004
123AP	121-1040
123AP	121-430
123AP	121-430B
123AP	121-430CL
123AP	121-433
123AP	121-433CL
123AP	121-434
123AP	121-434H
123AP	121-435
123AP	121-442
123AP	121-447
123AP	121-499
123AP	121-499-01
123AP	121-587
123AP	121-600
123AP	121-600(ZENITH)
123AP	121-639
123AP	121-671
123AP	121-675
123AP	121-677
123AP	121-695
123AP	121-744
123AP	121-745
123AP	121-748
123AP	121-767
123AP	121-768
123AP	121-825
123AP	121-850
123AP	121-862
123AP	121-863
123AP	121-877
123AP	121-881
123AP	121-888
123AP	121-895
123AP	121-895A
123AP	121-931
123AP	121-972
123AP	121-972-01
123AP	121-975
123AP	121-982
123AP	142N3
123AP	147-7031-01
123AP	200X3174-006
123AP	200X3174-014
123AP	200X3174-021
123AP	232N1
123AP	260P17101
123AP	260P17102
123AP	260P17103
123AP	417-109
123AP	417-126
123AP	417-171
123AP	417-801
123AP	700-134
123AP	700-135
123AP	700-156
123AP	921-01127
123AP	921-291B
123AP	921-303B
123AP	921-304B
123AP	921-306B
123AP	921-307
123AP	921-307B
123AP	921-309
123AP	921-309B
123AP	921-339B
123AP	921-349
123AP	921-351
123AP	921-351B
123AP	921-352B
123AP	921-353
123AP	921-353B
123AP	921-354B
123AP	921-355B
123AP	921-360B
123AP	921-407
123AP	921-408
123AP	921-449
123AP	921-450
123AP	921-462
123AP	921-463
123AP	921-464
123AP	921-470
123AP	947-1
123AP	991-012686
123AP	1100-9461
123AP	1100-9479
123AP	1414-173
123AP	2000-206
123AP	3536(RCA)
123AP	3536-1
123AP	3536-2
123AP	5613-710B
123AP	8000-00011-047
123AP	9011F
123AP	9011G
123AP	9014
123AP	9014B
123AP	9014C
123AP	026237
123AP	119635
123AP	123139
123AP	123941
123AP	124753
123AP	124754
123AP	124756
123AP	125141
123AP	126334
123AP	126712
123AP	126716
123AP	126717
123AP	126720
123AP	127354
123AP	127899
123AP	129897
123AP	129899
123AP	131140
123AP	131243
123AP	131311
123AP	132329
123AP	132823
123AP	132824
123AP	133178
123AP	133218
123AP	133249
123AP	133275
123AP	133690
123AP	134143
123AP	136165
123AP	136239
123AP	136430
123AP	137339
123AP	137875
123AP	138191
123AP	138763
123AP	139268
123AP	139362
123AP	141330
123AP	141331
123AP	142683
123AP	142684
123AP	142686
123AP	142711
123AP	143316
123AP	143792
123AP	143793
123AP	143794
123AP	143795
123AP	143796
123AP	143804
123AP	143805
123AP	145173
123AP	145395
123AP	145398
123AP	146141
123AP	146142
123AP	146484
123AP	147664
123AP	147665
123AP	166917
123AP	172761
123AP	185236
123AP	227517
123AP	610124-1
123AP	610232-2
123AP	658577
123AP	658578
123AP	2321541
123AP	5321291
123AP	06120005
123AP	1760579001
123AP	1760589001
123AP	1760609003
123AP	2003174014
123AP	4100900102
123AP	4100900103
123AP	4100900104
124	A1N
124	A2U
124	A3Y
124	A7B
124	AR-15
124	AR-18
124	C1059
124	C1102
124	C1102A
124	C1102B
124	C1102C
124	C1105
124	C1105A
124	C1105B
124	C1105C
124	C1168
124	C514
124	C515
124	C515A
124	C582
124	C582A
124	C582C
124	C635A
124	C685
124	C685A
124	C685B
124	C685GU
124	C685P
124	C685Y
124	C795
124	C795A
124	D11C201B20
124	D11C203B20
124	D11C205B20
124	D11C207B20
124	D11C210B20
124	D11C211B20
124	E2460
124	EP15X16
124	EP15X35
124	ES15X95
124	FBN-37605
124	FBN-38982
124	FBN-L108
124	FT300
124	GE-12
124	HC515
124	HEP-S5011
124	HR106
124	HR107
124	J24566
124	M4872
124	M4885
124	MJ2251
124	MJ2252
124	MJ3201
124	MJ3202
124	MJ400
124	MT3202
124	PTC104
124	Q34450
124	Q5075CLY
124	Q5075DLY
124	Q5075DXY
124	Q5075ELY
124	Q5075EXY
124	Q5075FXY
124	Q5075ZMY
124	Q5104Z
124	Q5113ZLM
124	Q5113ZMM
124	R2444
124	RE14
124	RS-7315
124	RS-7316
124	RS-7317
124	RS-7318
124	RS7310
124	RS7311
124	RS7312
124	RS7313
124	RS7315
124	RS7316
124	RS7317
124	RS7318
124	RS7320
124	RS7321
124	RS7327
124	RS7328
124	RS7329
124	RS7330
124	RS7365
124	RS7366
124	RS7367
124	RS7368
124	RT7311
124	S2059
124	SC-4004
124	SC-4131
124	SC-4131-1
124	SC-4167
124	SC-727
124	SC4004
124	SC727
124	SC777
124	SE7006
124	SE7020
124	SE7030
124	SJ1165
124	SJ1201
124	SJ1286
124	SJ805
124	SJ806
124	SK5184A
124	SP-2158
124	STX0015
124	T-Q5057
124	T-Q5075
124	T-Q5104
124	TA2509
124	TA2509A
124	TG2SD24Y
124	TNJ72147
124	TNJ72153
124	TR-23
124	TR-23C
124	TR-8005
124	TR1605LP
124	TR23C
124	TR262-2
124	TR266-2
124	TR8005
124	TRS1005
124	TRS1005LP
124	TRS1205
124	TRS1205LP
124	TRS1405
124	TRS1405LP
124	TRS1605
124	TRS1805
124	TRS1805LP
124	TRS2005
124	TRS2005LP
124	TRS2255
124	TRS2255LP
124	TRS2505
124	TRS2505LP
124	TRS2755
124	TRS2755LP
124	TRS3015LP
124	TRS3016LC
124	TRS4016LC
124	TRS5016LC
124	TRS6016LC
124	TV113
124	TV122
124	TX-100-3
124	TX-101-11
124	TX-101-8
124	TX-102-4
124	TX-107-13
124	TX-111-1
124	TX100-5
124	TX101-11
124	TX101-8
124	TX102-4
124	TX107-13
124	TX111-1
124	001-02115-0
124	001-021150
124	001-021151
124	001-02116-0
124	001-021160
124	001-021162
124	01-571921
124	1-801-301-13
124	1-801-301-14
124	1-801-301-15
124	1A0038
124	002-009100
124	2N2204
124	2N4298
124	2N4299
124	2N6425
124	2SC1059
124	2SC1102
124	2SC1105
124	2SC1105A
124	2SC1105B
124	2SC1105C
124	2SC1105K
124	2SC1105L
124	2SC1105M
124	2SC1168
124	2SC1168X
124	2SC1235AL
124	2SC1235AM
124	2SC1391
124	2SC1391VL
124	2SC1456
124	2SC1456L
124	2SC1456M
124	2SC514
124	2SC515
124	2SC515A
124	2SC515AM
124	2SC515AX
124	2SC515BK
124	2SC515Y
124	2SC582
124	2SC582A
124	2SC582B
124	2SC582BC
124	2SC582C
124	2SC685
124	2SC685A
124	2SC685B
124	2SC685BK
124	2SC685GU
124	2SC685P
124	2SC685S
124	2SC685Y
124	2SC795
124	2SC795A
124	2SD156
124	2SD156B
124	2SD156C
124	2SD156F
124	2SD157
124	2SD157A
124	2SD157B
124	2SD157C
124	2SD190
124	2SD24
124	2SD24B
124	2SD24C
124	2SD24CK
124	2SD24D
124	2SD24E
124	2SD24F
124	2SD24K
124	2SD24KC
124	2SD24KD
124	2SD24Y
124	2SD24YB
124	2SD24YD
124	2SD24YE
124	2SD24YK
124	2SD24YLC
124	2SD24YLD
124	2SD24YLE
124	2SD24YM
124	2SD257
124	2SD324
124	2SD326
124	03-0018-6
124	3-19
124	3L4-6006-1
124	7B1
124	7B13
124	7B2
124	7C1
124	7C2
124	7C3
124	7D1
124	7D2
124	7D3
124	7E1
124	7E2
124	7E3
124	7G1
124	7G2
124	7G3
124	7G4
124	8-760-343-10
124	09-302146
124	09-302156
124	09-302160
124	09-302185
124	09-303028
124	9-5252
124	9-5252-1
124	9-5252-2
124	9-5252-3
124	9-5252-4
124	11C10B1
124	11C11B1
124	11C1B1
124	11C3B1
124	11C3B3
124	11C5B1
124	11C7B1
124	13-14085-29
124	13-18282
124	13-18282-1
124	13-18359
124	13-18359-1
124	13-18359-3
124	13-18359A
124	13-23543-1
124	19A115623-P1
124	19A115623-P2
124	19AR35
124	21-35
124	21A112-025
124	21A112-029
124	21A112-033
124	34-6002-21
124	34-6002-26
124	34-6002-46
124	34-6002-62
124	42-18310
124	46-86384-3
124	46-86386-3
124	46-86439-3
124	48-134920
124	48-134972
124	48-137207
124	48-63026A45
124	48-90232A07
124	48G13001A01
124	48S137528
124	48S137535
124	57A158-10
124	57A192-10
124	57A195-10
124	57B158-1
124	57B158-2
124	57B158-3
124	57B158-4
124	57B158-5
124	57B158-6
124	57B158-7
124	57B158-8
124	57B158-9
124	57B195-10
124	57C6-10
124	57C6-14
124	57C6-24
124	57D6-10
124	059
124	71N1
124	71N1T
124	71N2
124	71N2T
124	86-227-2
124	86-236-2
124	86-259-2
124	86-260-2
124	86-261-2
124	86-275-2
124	86-487-2
124	86-487-3
124	86-624-2
124	86-624-9
124	86X0028-001
124	95-252
124	95-252-1
124	95-252-2
124	95-252-3
124	95-252-4
124	96-5132-01
124	96-5135-01
124	96N1
124	116-075
124	116-1
124	117-1
124	121-315
124	121-436
124	121-451
124	121-582
124	121-713
124	132-028
124	132-033
124	132-059
124	132-515
124	132-516
124	132-521
124	132-522
124	132-523
124	132-524
124	132-525
124	132-526
124	154A5943
124	154A5943-1
124	158-10
124	169-257
124	169-284
124	171-003-9-001
124	260P09402
124	260P15100
124	260P15108
124	260P16202
124	260P16208
124	260P21208
124	297V060H01
124	297V060H02
124	297V060H03
124	297V071C03
124	297V071H03
124	325-1442-9
124	421-18
124	421-8310
124	573-515
124	610-071
124	800-122
124	800-158
124	800-172
124	800-203
124	800-204
124	800-321
124	921-156B
124	921-61B
124	921-88
124	921-88A
124	921-88B
124	992-00-3172
124	1043-7309
124	2057A2-223
124	2057A2-81
124	2057A2-84
124	2057B123-10
124	2057B2-47
124	2057B2-58
124	2057B2-84
124	2417
124	2491
124	2491A
124	2491B
124	3520-1
124	3565
124	3566
124	4872
124	7311
124	7317
124	9925-2
124	16232
124	18310
124	19278
124	25114-143
124	30234
124	30245
124	30257
124	36634
124	37584
124	37730
124	40264
124	40328
124	40374
124	40422
124	40423
124	40425
124	40426
124	40427
124	40491
124	40546
124	40547
124	40850
124	95252
124	95252-1
124	95252-2
124	95252-3
124	95252-4
124	99252
124	99252-1
124	99252-3
124	116075
124	118279
124	118686
124	119650
124	123275
124	123275-14
124	123375
124	126138
124	126188
124	126722
124	126726
124	146286-01
124	237450
124	240388

ECG	Industry Standard No.	ECG	Industry Standard No.	ECG	Industry Standard No.	ECG	Industry Standard No.	ECG	Industry Standard No.	ECG	Industry Standard No.
124	244357	125	A759500	125	CODI1537	125	H-881	125	RM-2C		
124	263561	125	A800	125	CODI1538	125	H1000	125	RM2A		
124	263857	125	AA1000	125	CODI1617	125	H621	125	RM2C		
124	489751-043	125	AA800	125	CODI1618	125	H800	125	RT8665		
124	0573415	125	AD4002	125	CODI1537	125	HA1000	125	S-05/01		
124	573515	125	AD4003	125	CODI1617	125	HA800	125	S-500		
124	0573515H	125	AD4004	125	CP102	125	HC72	125	S0100		
124	610071	125	AD4005	125	CP102PA	125	HC720	125	S100		
124	610071-1	125	AD4006	125	CP102RA	125	HC73	125	S10AR1		
124	610071-2	125	AD4007	125	CP102VA	125	HC730	125	S10AR2		
124	699414-164	125	AH1005	125	CP103	125	HEP159	125	S1201F		
124	770763-3170756	125	AH1010	125	CP152VA	125	HEP160	125	S1B-0306		
124	815166	125	AH1015	125	CT100	125	HEP170	125	S1C		
124	815166-4	125	AH14	125	D1000	125	HF-20071	125	S1D		
124	815167-3	125	AH805	125	D105C	125	HS3108	125	S20		
124	815175	125	AH810	125	D108	125	HS3110	125	S208		
124	815175H	125	AH815	125	D1201A	125	IP20-0026	125	S210		
124	815180-3	125	AM-G-5A	125	D1201B	125	JCN7	125	S24		
124	815180-4	125	AM-G-5C	125	D1201D	125	JCV7	125	S257		
124	815180-7	125	AN-G5B	125	D1201M	125	K1K5	125	S258		
124	816135	125	AR23	125	D1201N	125	K1M5	125	S260		
124	970255	125	AR24	125	D1201P	125	K2H5	125	S28		
124	982576	125	B1H1	125	D1D	125	K2K5	125	S2H100		
124	984608	125	B1H5	125	D1K	125	K3H5	125	S2V		
124	984932	125	B1H9	125	D4L	125	K3K5	125	S2VC10R		
124	995030	125	B1K1	125	D800	125	K3M5	125	S394		
124	1018734-001	125	B1K5	125	D85C	125	K4H5	125	S61		
124	1471117-1	125	B1K9	125	D8HZ	125	K4K5	125	S62		
124	1473520-1	125	B1M1	125	DA05B	125	K4M5	125	S63		
124	1473567-1	125	B1M5	125	DA2068	125	K5H5	125	S750		
124	1473584-1	125	B1M9	125	DD05B	125	K5K5	125	S7AR1		
124	1473536-1	125	B24-06B	125	DD2068	125	K5M5	125	S8AK1		
124	1969113	125	B2H1	125	DD268	125	KSKE125C200	125	S8AR2		
124	2004746-87	125	B2H5	125	DER1	125	KSKE125C500	125	S9AR1		
124	2320221	125	B2H9	125	DI-650	125	L82H	125	SA2Z		
124	2320222	125	B2K1	125	DICR1	125	M102	125	SB-1000		
124	2320223	125	B2K5	125	DID	125	M702	125	SC10		
124	2320228	125	B2K9	125	DIK	125	M702B	125	SC10A		
124	2320931	125	B2M1	125	DP100	125	M72	125	SC8		
124	2321001	125	B2M5	125	DR1000	125	M720	125	SC8A		
124	2321403	125	B2M9	125	DR700	125	M720A	125	SCA10		
124	2327182	125	B3H1	125	DR800	125	M720B	125	SCA8		
124	23114200	125	B3H5	125	DR900	125	M720C	125	SCBR05F		
124	23114268	125	B3H9	125	DR8107	125	M72A	125	SCER1F		
124	23114277	125	B3K5	125	DR8108	125	M72B	125	SCBR6F		
124	23114315	125	B3K9	125	DS-16A(SANYO)	125	M72C	125	SC810		
124	25111143	125	B3M1	125	DS-1K	125	M73	125	SC8A		
124	30771100	125	B3M5	125	DS16NA	125	M730	125	SCT1		
124	62372540	125	B3M9	125	D82M	125	M730A	125	SCT2		
124	62563370	125	B4H1	125	DU1000	125	M730B	125	SCT3		
124	62734612	125	B4H5	125	DU800	125	M730C	125	SCT4		
124	62737417	125	B4H9	125	E10	125	M73A	125	SCT5		
124	62737646	125	B4K1	125	E108(ELCOM)	125	M73B	125	SD-6AUF		
124	94029220	125	B4K5	125	E1M3	125	M73C	125	SD1C		
124	6621001100	125	B4K9	125	E1M3	125	M82	125	SD2		
1242	AN366	125	B4M1	125	E758(ELCOM)	125	MBHZ	125	SD2C		
1242	87-0233	125	B4M5	125	E760(ELCOM)	125	M91A06	125	SD7		
1243	TA711592	125	B4M9	125	E8	125	MC070	125	SD8		
1243	09-308094	125	B5H1	125	E9	125	MC070A	125	SD800		
1243	6207159	125	B5H5	125	EA080	125	MC080	125	SD910		
1248	AN362	125	B5H9	125	EA100	125	MC080A	125	SD910A		
125	.7810	125	B5K1	125	EC100	125	MC090	125	SD9108		
125	.7R7	125	B5K5	125	ED2847	125	MC090A	125	SD98		
125	.7R8	125	B5K9	125	ED2848	125	MT72	125	SD98A		
125	.7J10	125	B5M1	125	ED2849	125	MT720	125	SD98A		
125	.7J7	125	B5M5	125	ED2910	125	MT730	125	SD98B		
125	.7J8	125	B5M9	125	ED2911	125	MM10	125	SE0		
125	A-118038	125	B7579500Q	125	ED2912	125	MM7	125	SE05C		
125	A04212-A	125	BAY15	125	ED2913	125	MM8	125	SE1730		
125	A059-118	125	BAY16	125	ED2922	125	MM9	125	SF1CN1		
125	A068-104	125	BAY23	125	ED2923	125	MR-1M	125	SF3CN1		
125	A1000	125	BAY90	125	ED2924	125	MR990	125	SF4		
125	A10N	125	BXY10	125	ED3007	125	MT84	125	SF4CN1		
125	A10P	125	BY100	125	ED3007A	125	NA104	125	SF5		
125	A114C	125	BY1001	125	ED3007B	125	NA105	125	SF6		
125	A14N	125	BY1002	125	ED3008	125	NA74	125	SFR258		
125	A14P	125	BY1008	125	ED3008A	125	NA75	125	SFR268		
125	A1H1	125	BY104	125	ED3008A	125	NA76	125	SH1B		
125	A1H5	125	BY105	125	ED3009	125	NA84	125	SH1C		
125	A1H9	125	BY108	125	ED3009A	125	NA85	125	SH4D8		
125	A1K1	125	BY109	125	ED3009B	125	NA86	125	SI-RECT-102A		
125	A1K5	125	BY1101	125	ED3010	125	NS81021	125	SI-RECT-136		
125	A1K9	125	BY1102	125	ED3010A	125	OT5067	125	SI-RECT-204		
125	A1M1	125	BY118	125	ED3010B	125	P1000	125	SI1000B		
125	A1M5	125	BY119	125	EP100	125	P10115A	125	SIB-0306		
125	A1M9	125	BY12 00	125	EM507	125	P10156A	125	SID01K		
125	A2H1	125	BY1201	125	EM508	125	P5100	125	SID02K		
125	A2H4	125	BY1202	125	EM510	125	P580	125	SID-1HF		
125	A2H5	125	BY127/500	125	EP100	125	P6RP10	125	SK3032		
125	A2H9	125	BY127/600	125	EP800	125	P6RP8	125	SK3033		
125	A2K1	125	BY127/700	125	ER1001	125	P800	125	SK3080		
125	A2K4	125	BY128	125	ER107D	125	PA380	125	SL608		
125	A2K5	125	BY129	125	ER108D	125	PD114	125	SL610		
125	A2K9	125	BYX10	125	ER186	125	PD115	125	SL708		
125	A2M1	125	BYX12/400	125	ER187	125	PD116	125	SL710		
125	A2M4	125	BYX13/600	125	ER308	125	PD913	125	SLA01		
125	A2M5	125	BYX22/800	125	ER310	125	PD914	125	SLA18AB		
125	A3H1	125	BYX60-700	125	ER801	125	PD915	125	SLA18C		
125	A3H5	125	BTY37	125	ER81	125	PD916	125	SLA19AB		
125	A3H9	125	BTY91	125	ERB12-02	125	PE408	125	SLA19C		
125	A3K1	125	C0410	125	ERC01-06	125	PE410	125	SLA2616		
125	A3K3	125	C1.0BO2	125	ERC04-10	125	PE508	125	SLA2617		
125	A3K5	125	CA102PA	125	ERC54-06	125	PE510	125	SLA3196		
125	A3K9	125	CA102RA	125	ERD1000	125	PH109	125	SLA560		
125	A3M1	125	CA102VA	125	ERD700	125	PS1140	125	SLA561		
125	A3M3	125	CA152VA	125	ERD800	125	PS2416	125	SLA01		
125	A3M5	125	CC102PA	125	ERD900	125	PS2417	125	SM-150		
125	A3M9	125	CC102RA	125	ESKE125C500	125	PT580	125	SM-150-005		
125	A4212A	125	CC102VA	125	F-14A	125	PTC203	125	SM-150-02		
125	A4H1	125	CC152VA	125	F10	125	PTC204	125	SM-150A		
125	A4H5	125	CB508	125	F11034	125	PTC205	125	SM-150B		
125	A4H9	125	CB510	125	F14B	125	QDSSR3AMBE	125	SM100		
125	A4K1	125	CER72	125	F14C	125	RO80	125	SM101		
125	A4K5	125	CER720	125	F14D	125	R5	125	SM103		
125	A4K9	125	CER720A	125	F14E	125	R5B	125	SM150		
125	A4M1	125	CER720B	125	F14F	125	R5C	125	SM150-01		
125	A4M5	125	CER720C	125	F14H	125	R6	125	SM150-02		
125	A4M9	125	CER72A	125	F14J	125	R8	125	SM150-11		
125	A5H1	125	CER72B	125	F8	125	R8/2H	125	SM150-6		
125	A5H2	125	CER72C	125	FA8	125	RA-1B	125	SM150A		
125	A5H5	125	CER72F	125	FR-10	125	RA-1C	125	SM150B		
125	A5H9	125	CER73	125	FR2-06	125	RA-2	125	SM150C		
125	A5K1	125	CER730	125	FT10	125	RA-2C	125	SM150D		
125	A5K2	125	CER730A	125	G10	125	RA2	125	SM150S		
125	A5K5	125	CER730B	125	G100K	125	RA2C	125	SM1508S		
125	A5K9	125	CER730C	125	G100M	125	RC080	125	SM170		
125	A5M1	125	CER73A	125	G8	125	R851	125	SM180		
125	A5M2	125	CER73B	125	GE-509	125	RE90	125	SM200		
125	A5M5	125	CER73C	125	GE-509A	125	RLO05	125	SM270		
125	A5M9	125	CER73D	125	GE-531	125	RLO10	125	SM280		
125	A725EH2AB1	125	CER73F	125	GE505	125	RLO20	125	SM300		
125	A7568700	125	CP102PA	125	GI237	125	RLO40	125	SM517		
125	A7580111	125	CP102RA	125	GM1J2	125	RLO60	125	SM518		
				125	GSR1	125	RM-2AV	125	SM520		

ECG	Industry Standard No.	ECG	Industry Standard No.	ECG	Industry Standard No.	ECG	Industry Standard No.	ECG	Industry Standard No.
125	SM70	125	1N1258	125	1897	125	182JP	125	75R10B
125	SM71	125	1N1259	125	1898	125	19-100001	125	75R7B
125	SM73(DIODE)	125	1N1260	125	1899	125	19A115024-P6	125	75R8B
125	SM770	125	1N1261	125	18058	125	21A103-012	125	75R9B
125	SM780	125	1N1407	125	18058	125	21A103-013	125	80-001300
125	SM80	125	1N1408	125	1T507	125	21A103-015	125	80-001400
125	SM800	125	1N1730	125	1T508	125	21A103-104	125	80A5
125	SM81	125	1N1730A	125	1T509	125	21A110-004	125	80A8
125	SM83	125	1N1914	125	2-64701508	125	21A110-012	125	80H
125	SO10G	125	1N1915	125	2HR3J	125	21A110-072	125	86-128-3
125	SOA	125	1N1916	125	2HR3M	125	21A119-030	125	86-67-2
125	SR-1K-2A	125	1N2616	125	2EO2	125	22-004004	125	86-72-3
125	SR-1K-2B	125	1N2617	125	2N8M-1	125	24MW246	125	088-2
125	SR-1K-2C	125	1N2878	125	28J8A	125	24MW602	125	91A04
125	SR-1K-2D	125	1N2879	125	2T507	125	24MW605	125	93A79-6
125	SR-2	125	1N2880	125	2T508	125	24MW779	125	93B45-3
125	SR-35	125	1N2881	125	2T509	125	24MW829	125	93B65-1
125	SR-9001	125	1N2882	125	2T510	125	24MW864	125	93C118-1
125	SR-9007	125	1N2883	125	2V6A	125	24MW924	125	93C118-2
125	SR150	125	1N3196	125	2W7A	125	025-10029	125	93C18-1
125	SR1PM-12	125	1N321A	125	2W9A	125	25D10	125	93C18-2
125	SR1PM20	125	1N322	125	3A1510	125	25K10	125	094-011
125	SR2A-12	125	1N3221	125	3A158	125	26-4701508	125	95-11511500
125	SR2A-8	125	1N322A	125	3A2510	125	30D2	125	96-5178-01
125	SR3AM	125	1N3256	125	3A258	125	31-194	125	96-5345-01
125	SR3AM2	125	1N328	125	3Q1510	125	34-8050-10	125	098GI219
125	SR3AM4	125	1N3280	125	30158	125	35-1000	125	100-13
125	ST18	125	1N3281	125	3Q2510	125	35-1003	125	100I0
125	SW05C	125	1N3282	125	3Q258	125	35-1029	125	100K10
125	SW05D	125	1N328A	125	3T507	125	42-051	125	100R10B
125	SW1C	125	1N329	125	3T508	125	42-19645	125	100R7B
125	SW1D	125	1N329A	125	3T509	125	46-86141-3	125	100R8B
125	SWC	125	1N3614	125	3T510	125	46-86146-3	125	100R9B
125	SWD	125	1N364	125	4-2020-03900	125	46-86147-3	125	110-636
125	T-1000	125	1N3642	125	4-2020-8000	125	46-86148	125	110B6
125	T-1000X	125	1N364A	125	4-2020-8700	125	46-86488-3	125	120-003149
125	T1000X	125	1N365	125	4-2021-04470	125	46-86553-3	125	120-004061
125	T800	125	1N365A	125	4-2021-04770	125	46-86562-3	125	124-0028
125	T800X	125	1N3929	125	4-686105-3	125	46AR13	125	130-30256
125	TA1000	125	1N4001	125	4-686106-3	125	46AR15	125	137-684
125	TA7805	125	1N4006	125	4-686139-3	125	46AR16	125	137-828
125	TA7806	125	1N4007	125	4-686179-3	125	46AR18	125	142-002(RECT.)
125	TA800	125	1N4011	125	4-686184-3	125	46AR50	125	167J2
125	TB1064	125	1N4248	125	4-686186-3	125	46AR52	125	201X3130-109
125	TPR105	125	1N4249	125	4-686189-3	125	46AR59	125	264-701508
125	TPR110	125	1N4250	125	4-686199-3	125	46AX12	125	264P00602
125	TH1040	125	1N4251	125	4-686201-3	125	46AX13	125	264P04303
125	TH1000	125	1N4361	125	4-686212-3	125	46AX14	125	264P09001
125	TH800	125	1N4818	125	4-68689-3	125	46AX55	125	264P14701
125	TH808	125	1N5054	125	4-68697-3	125	46AX56	125	290V034C01
125	TH810	125	1N5054A	125	4G8	125	46AX59	125	295V028004
125	TIR07	125	1N5059	125	4JA16MR700M	125	46AX70	125	0307
125	TIR08	125	1N5060	125	4T507	125	46AX82	125	309-327-803
125	TIR09	125	1N5061	125	4T508	125	46AX84	125	0317
125	TIR10	125	1N5062	125	4T509	125	46AX85	125	320P
125	TK1000	125	1N509	125	4T510	125	46BD101	125	320S
125	TK800	125	1N510	125	05-190061	125	46BD27	125	320Z
125	TKP100	125	1N517	125	5-30082.3	125	46BD30	125	325-4610-100
125	TKP80	125	1N518	125	5-30082.4	125	46BD32	125	0327
125	TM86	125	1N5214	125	5-30088.3	125	46BD33	125	359P
125	TR2327031	125	1N5218	125	5A10	125	46BD34	125	359B
125	TR8R3AM	125	1N525	125	5A10C	125	46BD38	125	359V
125	TS3	125	1N526	125	5A10D	125	46BD39	125	359Z
125	TS8	125	1N5392	125	5A10D-C	125	46BD52	125	384K
125	TV24221	125	1N548	125	5A8	125	46BR18	125	384M
125	TV24942	125	1N560	125	5A8D	125	46BR21	125	384P
125	TVB	125	1N561	125	5A8DC	125	46BR27	125	384S
125	TVB-10D8	125	1N597	125	5D10	125	46BR62	125	384Y
125	TVS-FR10	125	1N598	125	5D8	125	46BR63	125	384Z
125	TVS-FT-1N	125	1N853	125	5E6	125	46BR64	125	385P
125	TVS-FT10	125	1N854	125	5E8	125	46BR68	125	385S
125	TVS-88-2C	125	1N855	125	5G8	125	48-10577A04	125	385Z
125	TVS-SB-2C	125	1N856	125	5GF	125	48-137205	125	420-2003-173
125	TVS-SD-1B	125	1N864	125	50J	125	48-137212	125	420-2104-570
125	TVS-SD1B	125	1N865	125	5K10	125	48-137290	125	422-1408
125	TVS10D8	125	1N866	125	5MA10	125	48-137301	125	429-0093-71
125	TVS1P80	125	1N867	125	5MA8	125	48-137302	125	500B10
125	TVS00410	125	1N875	125	7D210	125	48-137316	125	500810
125	TVSD8-2K	125	1N876	125	7D210A	125	48-157083	125	523-0001-007
125	TVSMR-1M	125	1N877	125	8-0001300	125	48-44125A07	125	523-0001-008
125	TVSMR1C	125	1N878	125	8-0001400	125	48-67120A04	125	523-0001-009
125	TVSS3G4	125	1N886	125	8-905-405-105	125	48-67120A05	125	523-0001-010
125	TVSSA-2B	125	1N887	125	8-905-405-160	125	48-67120A10	125	525-212
125	TVSSP-1	125	1N888	125	8-905-405-170	125	48-82466H	125	530-082-3
125	TVSSID30-15	125	1N889	125	8D10	125	48-90343A62	125	530-082-4
125	TW100	125	1NJ27	125	8D8	125	48-97048A20	125	530-088-1002
125	TW80	125	1R5G261	125	8G7	125	48-97768A06	125	530-088-1003
125	U119	125	1R5G261PA-1	125	89A	125	488155083	125	530-088-1004
125	U120	125	1R96	125	09-306226	125	50D10	125	530-088-3
125	U361	125	1S1065	125	09-306274	125	50D8	125	530-106-1001
125	UPSD-18	125	1S1066	125	09-306311	125	50E10	125	530-111-1001
125	UT118	125	1S107	125	09-306424	125	50E7	125	530-136-T
125	UT345	125	1S108	125	9-511511500	125	50E8	125	540-014
125	VO-3C	125	1S109	125	10-102005	125	52A011-1	125	547(2P
125	VO-6A	125	1S117	125	10A010	125	53-0082-3	125	626(RECT.)
125	VO1G	125	1S1223	125	10AG10	125	53-0088-1002	125	750D858-211
125	VO3	125	1S1225	125	10AG8	125	53-0088-1003	125	903-36
125	VO3-C	125	1S1225A	125	10AT10	125	53-0088-1004	125	903-36A
125	VO3-E	125	1S1233	125	10AT8	125	53-0088-3	125	903-36B
125	VO3C	125	1S1234	125	10D10	125	53-0106-1001	125	903-36C
125	VO3E	125	1S1255A	125	10B8	125	53-0111-1001	125	903-36D
125	VO3O	125	1S1345	125	10C10	125	53-0136T	125	903-36F
125	VO6-B	125	1S1347	125	10C8	125	53A001-34	125	977-38
125	VO6-G	125	1S1348	125	10D-7K	125	53A001-4	125	1006-17
125	VO6B	125	1S138	125	10D10	125	60DB10	125	1061-8361
125	VO6G	125	1S1695	125	10D7	125	63-11762	125	1063-8971
125	VO-3C	125	1S1829	125	10D7F	125	63-12366	125	1070
125	VO3-C	125	1S210	125	10DC1R	125	66-8504	125	1070-0623
125	VO3C	125	1S211	125	10DCB	125	66X0023-009	125	1076-1674
125	VO3U	125	1S211/28J8A	125	10G4	125	66X0028-008	125	1080
125	V89-0003-911	125	1S2311	125	10GA	125	67J2	125	1084
125	V89-0004-911	125	1S2312	125	10I10	125	67J2A	125	1090
125	W06A	125	1S2314	125	10R10B	125	070-004	125	1971-17
125	ZV604	125	1S2315	125	10RT8	125	070-005	125	2084-17
125	Z1A103-018	125	1S2353	125	10RR8	125	070-006	125	2093A41-42
125	Z878	125	1S2354	125	10R9B	125	070-007	125	2093A41-43
125	Z878A	125	1S2364	125	13-0050	125	070-008	125	2093A41-6
125	Z878B	125	1S2365	125	13-085024	125	070-009	125	2093A41-65
125	001-0072-00	125	1S2403	125	13-085027	125	070-010	125	2093A41-77
125	1210	125	1S2405	125	13-14094-6	125	070-013	125	2093A52-1
125	187	125	1S2406	125	13-17174-1	125	070-014	125	2093A69-1
125	188	125	1S2608	125	13-17174-3	125	070-015	125	2093A71-1
125	18T10	125	1S2610	125	13-17174-5	125	070-016	125	2093A78-1
125	18T7	125	1S397	125	13-41122-1	125	070-017	125	2093A79-6
125	18T8	125	1S398	125	13-41122-4	125	070-030	125	2330-191
125	1P8	125	1S401	125	13-41123-2	125	070-032	125	2330-201
125	108	125	1S402	125	13-41123-4	125	070-033	125	2330-252
125	1HY100	125	1S557	125	13P2(RECTIFIER)	125	070-050	125	2495-489
125	1HY80	125	1S686	125	15-085033	125	70-270050	125	2498-513
125	1JZ61	125	1S687	125	15-085041	125	75D10	125	2802
125	1N1225	125	1S848	125	15-108046	125	75D8	125	4001-151
125	1N1225A	125	1S850	125	15-123243	125	75E10	125	4686-105-3
125	1N1226	125	1S96	125	16-501190016	125	75E7	125	4686-106-3
125	1N1226A			125	18J2	125	75E8	125	4686-139-3

ECG	Industry Standard No.	ECG	Industry Standard No.	ECG	Industry Standard No.	ECG	Industry Standard No.	ECG	Industry Standard No.
125	4686-179-3	125	2006582-21	126	A142(JAPAN)	126	A350T	126	AF126
125	4686-184-3	125	2330251	126	A142A	126	A350TY	126	AF127
125	4686-186-3	125	2330252	126	A142B	126	A350Y	126	AF129
125	4686-189-3	125	2330254	126	A142C	126	A351	126	AF130
125	4686-199-3	125	2331991	126	A143(JAPAN)	126	A351A	126	AF131
125	4686-201-3	125	5330001	126	A144	126	A351B	126	AF132
125	4686-212-3	125	5330041(HR-5A)	126	A144C	126	A352	126	AF133
125	4686-89-3	125	5330104	126	A145	126	A352A	126	AF134
125	4686-97-3	125	06200005	126	A145A	126	A352B	126	AF135
125	04970	125	06200009	126	A145C	126	A353	126	AF136
125	5001-089	125	7570011-01	126	A146	126	A353A	126	AF137
125	5300-82-3	125	7570011-02	126	A147	126	A353C	126	AF138
125	5300-82-4	125	13030256	126	A148	126	A354	126	AF144
125	5300-88-1002	125	20130109	126	A149	126	A354A	126	AFY10
125	5300-88-1003	125	23115012	126	A15-1001	126	A354B	126	AFY17
125	5300-88-1004	125	23115098	126	A15-1002	126	A355	126	AFY18
125	5300-88-3	125	23115117	126	A15-1003	126	A355A	126	AFY19
125	5301-06-1001	125	23115118	126	A150	126	A356	126	A0134
125	5301-11-1001	125	23115130	126	A151	126	A357	126	A01
125	5301-36-T	125	23115296	126	A152	126	A359	126	APB-11H-1007
125	5861	125	23115947	126	A153(JAPAN)	126	A364	126	AS210
125	6629	125	23115966	126	A154(PNP)	126	A365	126	AS211
125	6629-A05	125	23115994	126	A155(PNP)	126	A366	126	AS221
125	7214-6	125	41025773	126	A156(JAPAN)	126	A367	126	ATB30
125	7215-1	125	41027991	126	A157(JAPAN)	126	A368	126	ATRP1
125	7215-2	125	62047895	126	A160	126	A369	126	ATRP2
125	7701	125	95115115-00	126	A164(PNP)	126	A37	126	ATB13
125	7702-1A	125	96421866	126	A165(PNP)	126	A375	126	B157
125	7704-1	125	42020B700	126	A166	126	A376(JAPAN)	126	B158
125	7706-1	125	2008100093	126	A175	126	A38	126	B159
125	7708-1	125	2009130017	126	A176	126	A380	126	B160
125	7711-1	125	2009130075	126	A181	126	A381	126	B444
125	7712-1	125	2295100233	126	A183	126	A382	126	B444A
125	7713-1	125	4202007800	126	A188	126	A383	126	B444B
125	8000-00006-147	125	4202008700	126	A189	126	A384	126	C10291
125	8000-004-P061	125	4202017200	126	A19	126	A385	126	CB103
125	8000-004-P064	125	4202104470	126	A197	126	A385A	126	CB157
125	8000-004-P067	125	9511511500	126	A20	126	A385D	126	CB158
125	8000-004-P068	125	134882466808	126	A201	126	A39	126	CK28
125	9920-3-6	125(2)	1010	126	A201-0	126	A393A	126	CK28A
125	10100	125(2)	1D10	126	A201A	126	A401	126	CK28A
125	23115-042	125(2)	1D8	126	A201B	126	A408	126	CK4
125	23115-072	125(2)	KBF-02	126	A201B	126	A409	126	CK4A
125	025072	125(4)	7-11(STANDEL)	126	A201TV0	126	A412	126	CK762
125	35333	125(4)	7-15(STANDEL)	126	A202	126	A413	126	CK766
125	40808	125(4)	25-7	126	A202(JAPAN)	126	A428	126	CK766A
125	40809	125(4)	525-26	126	A202A	126	A43	126	D020
125	41020-618	1254	10-001	126	A202B	126	A436	126	D026
125	41023-224	126	A-102	126	A202C	126	A437	126	D134
125	41023-225	126	A059-104	126	A202D	126	A438	126	D65
125	42021	126	A059-105	126	A203A	126	A44	126	D66
125	44001	126	A061-107	126	A203AA(PNP)	126	A447	126	DAT1A
125	44003	126	A061-110	126	A203B(PNP)	126	A45	126	DAT2
125	44004	126	A061-111	126	A203P(PNP)	126	A45-1	126	D824
125	44005	126	A069-121	126	A21	126	A45-3	126	D825
125	44006	126	A1-3	126	A213	126	A453	126	D836
125	44007	126	A100A	126	A214	126	A454(JAPAN)	126	D838
125	53088-4	126	A100B	126	A215	126	A456	126	D851
125	53300-31	126	A100C	126	A216	126	A457	126	D852
125	53301-01	126	A101	126	A219	126	A466(JAPAN)	126	E-065
125	055228	126	A101AA	126	A221	126	A466-2	126	E-066
125	055228H	126	A101AY	126	A222	126	A466-3	126	E-067
125	62140	126	A101B	126	A225	126	A466BLK	126	E-068
125	70205-8A	126	A101BA	126	A234B	126	A466BLU	126	E105
125	70270-05	126	A101BB	126	A234C	126	A466YEL	126	E2451
125	70270-38	126	A101BC	126	A235A	126	A468	126	EA15X133
125	72172-1	126	A101BX	126	A235C	126	A469	126	EA15X140
125	75700-14-05	126	A101C	126	A236	126	A470	126	EA15X141
125	75700-24-02	126	A101CA	126	A237	126	A471	126	EA2133
125	75702-15-54	126	A101CV	126	A246(JAPAN)	126	A471-1	126	EA2491
125	77190-8	126	A101CX	126	A246V	126	A471-2	126	EA2497
125	77211-4	126	A101E	126	A247(JAPAN)	126	A471-3	126	EA2498
125	78524-39-01	126	A101QA	126	A25-1001	126	A472(JAPAN)	126	E070
125	99203-7	126	A101V	126	A25-1002	126	A472A	126	ES14
125	99203-8	126	A101X	126	A25-1003	126	A472B	126	ES15X61
125	115524	126	A101Y	126	A251	126	A472C	126	ES15X73
125	115529	126	A101Z	126	A252	126	A472D	126	ES3
125	120471	126	A102	126	A253(JAPAN)	126	A472E	126	ES3110
125	126885	126	A102(JAPAN)	126	A254	126	A474	126	ES3111
125	127379	126	A102A	126	A255	126	A476	126	ES3112
125	129348	126	A102AA	126	A256	126	A477	126	ES3113
125	130110	126	A102AB	126	A257	126	A51	126	ES3114
125	136605	126	A102B	126	A258	126	A517	126	ES3115
125	141489	126	A102BA	126	A259	126	A518	126	ES3116
125	144050	126	A102BN	126	A266	126	A55	126	ES41
125	144051	126	A102CA	126	A267	126	A56	126	ESA213
125	147015	126	A102TV	126	A268	126	A60	126	ESA233
125	147187-2-14	126	A103	126	A269	126	A65B-70	126	ET12
125	147477-8-10	126	A103A	126	A270(JAPAN)	126	A65B-705	126	G0008
125	147477-8-2	126	A103B	126	A271(JAPAN)	126	A65B19G	126	G0010
125	147477-8-3	126	A103C	126	A272(JAPAN)	126	A69	126	G010B1
125	147477-8-7	126	A103CA	126	A273(JAPAN)	126	A70	126	GET871
125	147477-8-8	126	A103CAK	126	A274(PNP)	126	A72	126	GET872
125	222867	126	A103CG	126	A275(JAPAN)	126	A72BLU	126	GET873
125	223357	126	A103DA	126	A28	126	A72BRN	126	GET873A
125	226922	126	A104	126	A288	126	A720RN	126	GET874
125	228560	126	A104(JAPAN)	126	A288A	126	A72WHT	126	GET875
125	234553	126	A104A	126	A289	126	A74	126	GET883
125	235382	126	A104B	126	A29	126	A75B	126	GET885
125	239429	126	A104D	126	A290	126	A76	126	G13
125	241295	126	A104P	126	A292	126	A77	126	GM380
125	241302	126	A104Y	126	A293	126	A77A	126	GT5116
125	242141	126	A105	126	A294	126	A77B	126	GT5117
125	270642	126	A106	126	A295	126	A77C	126	GT5148
125	275851	126	A107	126	A296	126	A77D	126	GT5149
125	348054-15	126	A109	126	A297	126	A80	126	GT5151
125	348054-7	126	A111(JAPAN)	126	A30(TRANSISTOR)	126	A82	126	GT66
125	348057-8	126	A112	126	A301(JAPAN)	126	A85	126	GTJ33141
125	348057-9	126	A113	126	A308	126	A87	126	GTJ33229
125	348058-2	126	A115	126	A309	126	A90(TRANSISTOR)	126	GTJ33230
125	500003	126	A116	126	A310(JAPAN)	126	A909-1009	126	HA-101
125	530073-4	126	A117	126	A313	126	A909-1010	126	HA-15
125	530082-1002	126	A118	126	A314	126	A92	126	HA-269
125	530082-2	126	A121	126	A315	126	A93	126	HA-350
125	530088-1	126	A122(JAPAN)	126	A316	126	A94	126	HA-350A
125	530088-2	126	A123	126	A324	126	AC164	126	HA-353
125	530111-1002	126	A124	126	A325	126	AC169	126	HA-353C
125	530113-2	126	A125	126	A326	126	ACB810-101	126	HA-354
125	530135-1	126	A126	126	A329	126	ACB810-102	126	HA-354B
125	552005	126	A131	126	A329A	126	ACB810-103	126	HA-49
125	0575050	126	A132(JAPAN)	126	A329B	126	ACB83-1001	126	HA102
125	765713	126	A133(JAPAN)	126	A330	126	ACB83-1002	126	HA103
125	801715	126	A134	126	A331	126	ACB83-1003	126	HA104
125	801723	126	A136	126	A335	126	AF105	126	HA201
125	817193	126	A137(JAPAN)	126	A337	126	AF10B	126	HA235
125	973936-20	126	A138	126	A338	126	AF114	126	HA330
125	991064	126	A139(JAPAN)	126	A341	126	AF115	126	HA342
125	991129	126	A14-1001	126	A342	126	AF116	126	HA471
125	991421	126	A14-1002	126	A344(JAPAN)	126	AF117	126	HA52
125	991422	126	A14-1003	126	A350	126	AF120	126	HA53
125	991429	126	A141(JAPAN)	126	A350C	126	AF124	126	HA422
125	1223929	126	A141B	126	A350H	126	AF125	126	HEP-G0001
125	2001786-142	126	A141B	126	A350R			126	HEP-G0008
125	2001786-142	126	A141C					126	HEP-G0009
125	2003069-1							126	HEP-G0010

ECG	Industry Standard No.	ECG	Industry Standard No.	ECG	Industry Standard No.	ECG	Industry Standard No.	ECG	Industry Standard No.
126	HEP1	126	RCA34099	126	TG28A201C	126	2N1636	126	2N501/18
126	HEP635	126	RCA34100	126	TK122B-1001	126	2N1637	126	2N501A
126	HEP636	126	RCA44098	126	TM126	126	2N1638	126	2N502
126	HEP638	126	RE15	126	TNJ60362	126	2N1639	126	2N502A
126	HEP639	126	RS-1554	126	TNJ60363	126	2N1683	126	2N502B
126	HEP640	126	RS-2003	126	TNJ60364	126	2N1699	126	2N503
126	HT10101IX	126	RS-3322	126	TNJ71248	126	2N1715	126	2N504
126	HT101021A	126	RS-3323	126	TR-11	126	2N1726	126	2N506
126	HT103501A	126	RS-3324	126	TR-168(OLSON)	126	2N1727	126	2N544
126	HT103531C	126	RS-3862	126	TR02012	126	2N1728	126	2N544/33
126	HT103541B	126	RS-3863	126	TR02063-1	126	2N1742	126	2N588
126	19A115180-2	126	RS-3866	126	TRT61	126	2N1743	126	2N588A
126	K417-68	126	RS-5107	126	TRT62	126	2N1744	126	2N591-6M
126	LU28544	126	RS-5108	126	TRA-10R	126	2N1745	126	2N602
126	M4363	126	RS-5109	126	TRA-11R	126	2N1746	126	2N602A
126	M4363BLU	126	RS-5201	126	TRA-12R	126	2N1747	126	2N603
126	M4363GRN	126	RS-5205	126	TRA-2	126	2N1748	126	2N603A
126	M4363ORN	126	RS-5207	126	TRA-22A	126	2N1748A	126	2N604
126	M4363WHT	126	RS-5209	126	TRA-22B	126	2N1752	126	2N604A
126	M4364	126	RS-5301	126	TRA-23	126	2N1754	126	2N605
126	M4365	126	RS-5305	126	TRA-23A	126	2N1785	126	2N606
126	M4366	126	RS-5306	126	TRA-23B	126	2N1786	126	2N607
126	M4367	126	RS-5312	126	TRA-24	126	2N1787	126	2N608
126	M4368	126	RS-5313	126	TRA-24A	126	2N185	126	2N623
126	M4388	126	RS-5752	126	TRA-24B	126	2N1853	126	2N624
126	M4454	126	RS-5753	126	V10/18	126	2N1854	126	2N640
126	M4456	126	RS-5754	126	V10/18J	126	2N1864	126	2N641REDM/P
126	M4457	126	RS-5755	126	V15/20R	126	2N1865	126	2N642
126	M4501	126	RS-5756	126	V6/2R	126	2N1868	126	2N643
126	M4509	126	RS-5757	126	V6/4R	126	2N2048	126	2N644
126	M4545	126	RS-5758	126	V6/8R	126	2N2048A	126	2N645
126	M4545BLU	126	RS-5759	126	VPI-2744K	126	2N2059	126	2N694
126	M4545WHT	126	RS-5760	126	VP83K	126	2N2083	126	2N695
126	M4586	126	RS-5761	126	VPW-27450	126	2N2089	126	2N705
126	M4589	126	RS-5762	126	VL18RJ	126	2N2090	126	2N705A
126	M4603	126	RS-5802	126	V8-28A358	126	2N2091	126	2N710
126	M4604	126	RS-5818	126	V8-28A378	126	2N2092	126	2N710A
126	M4605	126	RS-684	126	V8-28A379	126	2N2093	126	2N711A
126	M4605RED	126	RS-685	126	V8-28A385	126	2N2099	126	2N711B
126	M4621	126	RS5753	126	XT300	126	2N2168	126	2N72
126	M4632	126	RS5753-2	126	XT400	126	2N2169	126	2N725
126	M8116	126	RS5754	126	ZEN311	126	2N2170	126	2N740
126	MA820	126	RS5755	126	ZEN312	126	2N218	126	2N741
126	MA821	126	RS5756	126	ZEN313	126	2N2180	126	2N75
126	MA822	126	RS5757	126	ZEN314	126	2N2188	126	2N768
126	MA823	126	RS5758	126	1-21-138	126	2N2189	126	2N769
126	MD835	126	RS5759	126	1-21-189	126	2N2199	126	2N770
126	N-RA15X133	126	RS5760	126	1-21-242	126	2N2200	126	2N779A
126	NO20	126	RS5761	126	1-21-243	126	2N2207	126	2N781
126	N57B2-17	126	RS5762	126	1-21-244	126	2N2208	126	2N782
126	N57B2-18	126	RS5802	126	1-21-257	126	2N2225	126	2N794
126	N57B2-19	126	RS6821	126	002-006300	126	2N2258	126	2N795
126	N57B2-23	126	RS6822	126	002-011600	126	2N2259	126	2N796
126	NA5018-1002	126	RT-61014	126	2A	126	2N2273	126	2N827
126	NA5018-1003	126	RT3361	126	2E	126	2N231	126	2N828
126	NA5018-1004	126	RT3362	126	2G	126	2N231-YEL-RED	126	2N828A
126	NA5018-1005	126	RT5063	126	2G101	126	2N231BLU	126	2N829
126	NA5018-1006	126	RT5466	126	2G402	126	2N231TEL	126	2N837
126	NA5018-1007	126	RT5467	126	2G414	126	2N232	126	2N838
126	NA5018-1008	126	RT5520	126	2G415	126	2N2381	126	2N846A
126	NA5018-1009	126	RT61014	126	2G416	126	2N240	126	2N85
126	NA5018-1010	126	S-55TB	126	2G417	126	2N2400	126	2N86
126	NA5018-1011	126	S-70T	126	2MC	126	2N2401	126	2N87
126	NA5018-1022	126	S-80T	126	2N1003	126	2N2402	126	2N88
126	NA5018-1219	126	S-87TB	126	2N1004	126	2N2451	126	2N89
126	NA5018-1220	126	S-88TB	126	2N1042	126	2N247	126	2N90
126	NK65B159	126	S-95101	126	2N1043	126	2N247/33	126	2N933
126	NKT103	126	S-95102	126	2N1107	126	2N248	126	2N934
126	NKT127	126	S-95103	126	2N1108	126	2N2487	126	2N960
126	NKT131	126	S1332	126	2N1108RED	126	2N2488	126	2N961
126	NKT132	126	S65B-1-3P	126	2N1109	126	2N2489	126	2N962
126	NKT151	126	SA102	126	2N1110	126	2N2494	126	2N963
126	NKT152	126	SB-100	126	2N1111	126	2N252	126	2N964
126	NKT15325	126	SB100	126	2N1111A	126	2N2588	126	2N964A
126	NKT15425	126	SB200	126	2N1111RED	126	2N2621	126	2N965
126	NKT249	126	SB5122	126	2N1111B	126	2N2622	126	2N966
126	NKT252	126	SFT-315	126	2N1111M1	126	2N2623	126	2N967
126	NKT253	126	SFT-316	126	2N1111M2	126	2N2624	126	2N969
126	NKT254	126	SFT-317	126	2N112	126	2N2625	126	2N970
126	NKT255	126	SFT-320	126	2N112M1	126	2N2626	126	2N971
126	NKT265	126	SFT-357	126	2N1158	126	2N2627	126	2N972
126	NKT270	126	SFT307	126	2N1158A	126	2N2628	126	2N973
126	NKT618	126	SFT308	126	2N1180	126	2N2629	126	2N974
126	NKT674F	126	SFT317	126	2N1204	126	2N2630	126	2N975
126	NKT675	126	SFT318	126	2N1204A	126	2N2635	126	2N977
126	NKT676	126	SFT319	126	2N1213	126	2N2717	126	2N979
126	NKT677	126	SFT320	126	2N1214	126	2N275	126	2N980
126	NKT677P	126	SFT354	126	2N1215	126	2N2783	126	2N982
126	00-170	126	SFT357	126	2N1216	126	2N2795	126	2N983
126	00-171	126	SFT357P	126	2N1224	126	2N2796	126	2N984
126	00-390	126	SK3008	126	2N1225	126	2N2955	126	2N985
126	00-613	126	SM862	126	2N1226	126	2N2956	126	2N986
126	00-614	126	S065A	126	2N123A	126	2N2957	126	2N990
126	00-975	126	SP0871	126	2N1266	126	2N299	126	2N991
126	00130	126	ST-103	126	2N128	126	2N3318	126	2N992
126	00331	126	STX0033	126	2N129	126	2N33	126	2N993
126	00341	126	STX0034	126	2N1300	126	2N3320	126	2N994
126	00342	126	SWT1728	126	2N1313	126	2N3321	126	2SJ50
126	00343	126	T-1363	126	2N138	126	2N3322	126	2SJ51
126	00351	126	T-1364	126	2N1384	126	2N3323	126	2SJ52
126	00361	126	T-1460	126	2N139	126	2N3324	126	2SJ53
126	00362	126	T-348	126	2N1396	126	2N3325	126	2S109
126	00363	126	T-6028	126	2N1397	126	2N3400	126	2S110
126	00390	126	T-6029	126	2N1398	126	2N3412	126	2S112
126	0040	126	T-6030	126	2N1399	126	2N3444	126	2S141
126	0042	126	T-6031	126	2N1400	126	2N3443	126	2S142
126	0043	126	T-6032	126	2N1401	126	2N3449	126	2S143
126	0043N	126	T1388	126	2N1401A	126	2N345	126	2S145
126	0044	126	T1460	126	2N1402	126	2N346	126	2S146
126	0044N	126	T1524	126	2N1404M1	126	2N37	126	2S176
126	0045N	126	T1524BRN	126	2N1404M2	126	2N370	126	2S35
126	0046N	126	T1524BRN/RED	126	2N1425	126	2N370/33	126	2S36
126	0047N	126	T1654	126	2N1426	126	2N370A	126	2S58
126	0050	126	T1654BLU	126	2N1427	126	2N371	126	2S96
126	00612	126	T2322	126	2N1450	126	2N371/33	126	2S97
126	00613	126	T2323	126	2N1499	126	2N372	126	2S98
126	00614	126	T2324	126	2N1499A	126	2N372/33	126	2SA100
126	00615	126	T280(SEARS)	126	2N1499B	126	2N373	126	2SA100A
126	Q2R15226	126	T281(SEARS)	126	2N1500	126	2N374	126	2SA100B
126	Q9/6515	126	T449(SEARS)	126	2N1515	126	2N38	126	2SA100C
126	R-539	126	T50818	126	2N1516	126	2N3883	126	2SA101
126	R104-5	126	T52054	126	2N1517	126	2N393	126	2SA101A
126	R104-6	126	T6028	126	2N1524	126	2N407	126	2SA101AA
126	R104-7	126	T6029	126	2N1526	126	2N409	126	2SA101AY
126	R104-8	126	T6030	126	2N1527	126	2N410	126	2SA101B
126	R336	126	T6031	126	2N1531	126	2N411	126	2SA101BA
126	R337	126	T6032	126	2N1632	126	2N412	126	2SA101BB
126	R581	126	TA-1650A	126	2N1633	126	2N499	126	2SA101BC
126	R60-1001	126	TA-1755	126	2N1634	126	2N499A	126	2SA101BX
126	R684	126	TA-1756	126	2N1635	126	2N500	126	2SA101C
126	R714	126	TA-5			126	2N500BLU	126	2SA101CA
126	R715	126	TG28A201			126	2N500RED	126	2SA101CV
126	RCA34098	126	TG28A201-0			126	2N500WHT	126	2SA101CX
		126	TG28A201-N			126	2N501		

ECG	Industry Standard No.	ECG	Industry Standard No.	ECG	Industry Standard No.	ECG	Industry Standard No.	ECG	Industry Standard No.
126	2SA101E	126	2SA254	126	2SA472C	126	31-0135	126	57A5-4
126	2SA101QA	126	2SA255	126	2SA472D	126	31-0139	126	57A5-9
126	2SA101V	126	2SA256	126	2SA472E	126	31-0178	126	57A6-16
126	2SA101X	126	2SA259	126	2SA474	126	31-0184	126	57C16
126	2SA101T	126	2SA266	126	2SA476	126	31-0190	126	5705-1
126	2SA101Z	126	2SA267	126	2SA477	126	31-0191	126	5705-10
126	2SA102	126	2SA268	126	2SA51	126	31-0217	126	5705-4
126	2SA102A	126	2SA269	126	2SA517	126	31-0228	126	5705-9
126	2SA102AA	126	2SA270	126	2SA518	126	34-8001-43	126	5706-16
126	2SA102AB	126	2SA271	126	2SA55	126	37T1	126	57D130
126	2SA102B	126	2SA272	126	2SA56	126	42-21403	126	57D131
126	2SA102BA	126	2SA273	126	2SA60	126	42-23965	126	57D132
126	2SA102BN	126	2SA274	126	2SA69	126	44-13	126	57D168
126	2SA102CA	126	2SA275	126	2SA72	126	46-86102-3	126	57D186
126	2SA102TV	126	2SA28	126	2SA72BLU	126	46-8613-3	126	57D187
126	2SA103	126	2SA288	126	2SA72BLU-BLU	126	46-8625-1	126	57D5-1
126	2SA103A	126	2SA288A	126	2SA72BRN	126	46-8625-2	126	57D5-2
126	2SA103B	126	2SA289	126	2SA720RN	126	46-8625-3	126	57D5-4
126	2SA103C	126	2SA29	126	2SA72WHT	126	46-8625-4	126	61-260039
126	2SA103CA	126	2SA290	126	2SA74	126	46-8625-5	126	61-260039A
126	2SA103CAK	126	2SA291	126	2SA75	126	46-8625-6	126	61B003-1
126	2SA103CG	126	2SA292	126	2SA75B	126	46-86256-1	126	61P1
126	2SA103DA	126	2SA293	126	2SA76	126	46-86256-2	126	61P10
126	2SA103K	126	2SA294	126	2SA77	126	46-8636-3	126	065B-12
126	2SA104	126	2SA295	126	2SA77A	126	48-123522	126	065B-12-7
126	2SA104D	126	2SA296	126	2SA77B	126	48-123536	126	65B-70
126	2SA104P	126	2SA297	126	2SA77C	126	48-124255	126	65B-70-12
126	2SA105	126	2SA301	126	2SA77D	126	48-124256	126	65B-70-12-7
126	2SA106	126	2SA308	126	2SA80	126	48-124305	126	65B1
126	2SA107	126	2SA309	126	2SA82	126	48-124310	126	65B10
126	2SA108	126	2SA310	126	2SA83	126	48-124311	126	65B10R
126	2SA109	126	2SA313	126	2SA85	126	48-124312	126	65B11
126	2SA110	126	2SA314	126	2SA87	126	48-124316	126	65B119
126	2SA111	126	2SA315	126	2SA92	126	48-124346	126	65B12
126	2SA112	126	2SA316	126	2SA93	126	48-124347	126	65B121
126	2SA113	126	2SA324	126	2SA94	126	48-124349	126	65B13
126	2SA114	126	2SA324F	126	2SB157	126	48-124349	126	65B13P
126	2SA115	126	2SA325	126	2SB158	126	48-124350	126	65B14
126	2SA116	126	2SA326	126	2SB159	126	48-124351	126	65B14-7
126	2SA117	126	2SA329	126	2SB160	126	48-124352	126	65B14-7B
126	2SA118	126	2SA329A	126	2SB444	126	48-124360	126	65B15
126	2SA121	126	2SA329B	126	2SB444A	126	48-124364	126	65B15L
126	2SA122	126	2SA330	126	2SB444B	126	48-124365	126	65B16
126	2SA123	126	2SA331	126	2T14A	126	48-124366	126	65B16-2
126	2SA124	126	2SA335	126	03-57-001	126	48-124367	126	65B17
126	2SA125	126	2SA337	126	03-57-002	126	48-124377	126	65B17-1
126	2SA126	126	2SA338	126	03-57-100	126	48-124588	126	65B18
126	2SA131	126	2SA339	126	03-57-101	126	48-125228	126	65B182
126	2SA132	126	2SA344	126	03-57-201	126	48-125278	126	65B19
126	2SA133	126	2SA350	126	4-279	126	48-128093	126	65B190
126	2SA134	126	2SA350A	126	4-280	126	48-128095	126	65B3
126	2SA136	126	2SA350C	126	4-686256-1	126	48-128096	126	65B4
126	2SA137	126	2SA350H	126	4-686256-2	126	48-134101	126	65B5
126	2SA139	126	2SA350R	126	4-686256-3	126	48-134372	126	65B6
126	2SA141	126	2SA350T	126	6-50	126	48-134404	126	65B7
126	2SA141B	126	2SA350TY	126	6-62F	126	48-134405	126	65B8
126	2SA141C	126	2SA350TT	126	6-62P	126	48-134406	126	65B9
126	2SA142	126	2SA551	126	6-62X	126	48-134414	126	65B0
126	2SA142A	126	2SA551A	126	6-70	126	48-134434	126	70.00.730
126	2SA142B	126	2SA551B	126	6-71	126	48-134454	126	76
126	2SA142C	126	2SA551GR	126	6-72	126	48-134456	126	80-104900
126	2SA143	126	2SA552	126	6-89	126	48-134457	126	80-105200
126	2SA144	126	2SA552A	126	6A12680	126	48-134479	126	80-105300
126	2SA144C	126	2SA552B	126	07-33050-57	126	48-134480	126	81-46125001-1
126	2SA145	126	2SA553	126	7-3401A	126	48-134481	126	86-108-2
126	2SA145A	126	2SA553-AC	126	7-7340102	126	48-134508	126	86-109-2
126	2SA145C	126	2SA553A	126	8E(AUTOMATIC)	126	48-134514	126	86-111-2
126	2SA146	126	2SA553C	126	8F(AUTOMATIC)	126	48-134521	126	86-116-2
126	2SA147	126	2SA554	126	09-300002	126	48-134522	126	86-280-2
126	2SA148	126	2SA554A	126	09-300006	126	48-134536	126	86-281-2
126	2SA149	126	2SA554B	126	09-300007	126	48-134545	126	86-311-2
126	2SA150	126	2SA555	126	09-300011	126	48-134547	126	86-449-2
126	2SA151	126	2SA555A	126	09-300012	126	48-134561	126	86X0011-001
126	2SA152	126	2SA556	126	09-300015	126	48-134576	126	86X0013-001
126	2SA153	126	2SA557	126	09-300016	126	48-134577	126	87-0015
126	2SA154	126	2SA559	126	09-300027	126	48-134578	126	87-0016
126	2SA155	126	2SA564	126	09-300029	126	48-134591	126	87-0017
126	2SA156	126	2SA565	126	09-300078	126	48-134600	126	089-220
126	2SA157	126	2SA566	126	09-300079	126	48-134601	126	95-111
126	2SA159	126	2SA567	126	9-9101	126	48-134602	126	96X26051-28N
126	2SA160	126	2SA568	126	9-9102	126	48-134605	126	96X26051-35N
126	2SA161	126	2SA569	126	9-9103	126	48-134635	126	96X26053-16N
126	2SA162	126	2SA57	126	10-28A49	126	48-134680	126	96X26053-10N
126	2SA163	126	2SA576	126	12-1-189	126	48-134681	126	96X26053-24N
126	2SA164	126	2SA58	126	12-1-242	126	48-134682	126	99-101
126	2SA165	126	2SA380	126	12-1-243	126	48-134683	126	99-102
126	2SA166	126	2SA381	126	12-1-244	126	48-134684	126	99-103
126	2SA175	126	2SA382	126	12-1-257	126	48-134697	126	0000101
126	2SA176	126	2SA383	126	13-14085-31	126	48-134711	126	0101-0034
126	2SA180	126	2SA384	126	13-14085-32	126	48-134795	126	0101-0222
126	2SA181	126	2SA385	126	13-14085-33	126	48-134796	126	104-17
126	2SA183	126	2SA385A	126	13-14085-93	126	48-134797	126	104-19
126	2SA188	126	2SA385D	126	13-14886-1	126	48-134798	126	104-21
126	2SA189	126	2SA39	126	13-14887-1	126	48-134859	126	111-6935
126	2SA19	126	2SA393A	126	13-14889-1	126	48-134860	126	112-002
126	2SA20	126	2SA400	126	13-18951-1	126	48-134861	126	115-275
126	2SA197	126	2SA401	126	13-18951-2	126	48-134862	126	116-072
126	2SA201	126	2SA408	126	14-582-01	126	48-134880	126	116-756
126	2SA201-0	126	2SA409	126	019-003777	126	48-63029A90	126	120-00190
126	2SA201A	126	2SA412	126	19-020-001	126	48-63075A74	126	120-00213
126	2SA201B	126	2SA413	126	19-020-002	126	48-63078A65	126	120-002513
126	2SA201R	126	2SA427	126	19-020-005	126	48-63082A24	126	120-002515
126	2SA201N	126	2SA428	126	19A115087-P1	126	48-8613-3	126	120-002656
126	2SA201TV	126	2SA43	126	19A115098	126	48-97046A08	126	120-004048
126	2SA201TVO	126	2SA436	126	19A115098-P1	126	48-97046A09	126	120-004492
126	2SA202	126	2SA437	126	19A115099-P1	126	48-97046A16	126	121-150
126	2SA202A	126	2SA438	126	020-00023	126	48-97162A03	126	121-153
126	2SA202B	126	2SA44	126	21-32	126	48-97238A01	126	121-154
126	2SA202C	126	2SA447	126	21-33	126	48-97238A02	126	121-161
126	2SA202D	126	2SA45	126	21A040-031	126	48-97238A02	126	121-162
126	2SA203	126	2SA45-1	126	21M006	126	48X97238A01	126	121-186
126	2SA203A	126	2SA45-2	126	022-3516-380	126	48X97238A02	126	121-187
126	2SA203AA	126	2SA45-3	126	022-5311-770	126	48X97238A03	126	121-228
126	2SA203B	126	2SA453	126	022-5311-780	126	50A102	126	121-229
126	2SA203P	126	2SA454	126	022.3516-380	126	50A103K	126	121-230
126	2SA21	126	2SA455	126	24MW152	126	50P2	126	121-231
126	2SA213	126	2SA456	126	24MW205	126	50P3	126	121-232
126	2SA214	126	2SA457	126	24MW271	126	051-0079	126	121-233
126	2SA215	126	2SA466	126	24MW303	126	51P2	126	121-256
126	2SA216	126	2SA466-2	126	24MW352	126	51P4	126	121-257
126	2SA219	126	2SA466-3	126	24MW353	126	55P2	126	121-258
126	2SA220	126	2SA466BLK	126	24MW368	126	55P3	126	121-259
126	2SA221	126	2SA466BLU	126	24MW61	126	57A130	126	121-260
126	2SA222	126	2SA466YEL	126	24MW74	126	57A131	126	121-261
126	2SA223	126	2SA468	126	24MW816	126	57A132	126	121-262
126	2SA225	126	2SA469	126	24MW991	126	57A16	126	121-263
126	2SA236	126	2SA470	126	27T412	126	57A168	126	121-290
126	2SA237	126	2SA471	126	31-0005	126	57A186	126	121-292
126	2SA246	126	2SA471-1	126	31-0041	126	57A2-1	126	121-293
126	2SA246V	126	2SA471-2	126	31-0042	126	57A2-19	126	121-294
126	2SA247	126	2SA471-3	126	31-0065	126	57A5-1		
126	2SA251	126	2SA472	126	31-0123	126	57A5-10		
126	2SA252	126	2SA472A	126	31-0124	126	57A5-2		
126	2SA253	126	2SA472B	126	31-0132				
				126	31-0134				

ECG	Industry Standard No.	ECG	Industry Standard No.	ECG	Industry Standard No.	ECG	Industry Standard No.	ECG	Industry Standard No.
126	121-295	126	690V065H14	126	4595	126	225594A	127	AD167
126	121-296	126	690V065H15	126	4603	126	232681	127	A1100
126	121-297	126	690V066H44	126	4604	126	249588	127	A1102
126	121-298	126	690V066H45	126	4605	126	310157	127	A1103
126	121-299	126	690V066H49	126	4605RED	126	310158	127	AT200
126	121-360	126	690V068H29	126	4621	126	310221	127	AU101
126	121-381	126	690V073H89	126	4632	126	310224	127	AU103
126	121-44	126	690V073H95	126	4677	126	573303	127	AU104
126	121-45	126	690V077H34	126	4686-256-1	126	573329	127	AU105
126	121-491	126	690V077H35	126	4686-256-2	126	573330	127	AU106
126	121-492	126	690V077H36	126	4686-256-3	126	573366	127	AU107
126	121-493	126	690V080H37	126	6154	126	573402	127	AU108
126	121-494	126	690V080H40	126	6155	126	0573427	127	AU110
126	121-54	126	690V084H60	126	6162	126	0573471	127	AU111
126	121-554	126	690V085H42	126	9403-8	126	0573518	127	AU112
126	121-555	126	690V102H96	126	9510-3	126	601054	127	AU113
126	121-62	126	690V119H94	126	11527-5	126	604112	127	AUY28
126	121-65	126	800-50500	126	11607-2	126	610052	127	AUY38
126	121-66	126	801-04900	126	11675-6	126	610056	127	B10142
126	121-67	126	801-05200	126	12118-9	126	610074	127	B10142A
126	122-1648	126	801-05300	126	12124-2	126	740946	127	B10142B
126	145-21B	126	916-31019-3	126	12124-3	126	740947	127	B10143
126	151-045	126	916-31019-3B	126	12124-4	126	772768	127	B10143A
126	151-1002	126	921-43	126	12125-7	126	815064	127	B10143B
126	154A3676	126	921-65	126	12180	126	815064A	127	B102000
126	154A3677-5110	126	921-65A	126	18493	126	815064B	127	B102001
126	229-5100-227	126	921-65B	126	24198	126	815064C	127	B102002
126	260P1300	126	921-66	126	25566-01	126	815064D	127	B102003
126	260P13001	126	921-66A	126	25642-020	126	815068	127	B103000
126	297V011H01	126	921-66B	126	25642-030	126	815068A	127	B103001
126	297V012H01	126	930X6	126	25642-031	126	815068B	127	B103002
126	297V012H10	126	951-1(SYLVANIA)	126	25642-040	126	815068C	127	B103004
126	297V012H15	126	964-16598	126	25642-041	126	980372	127	B1178
126	297V026H01	126	965B16-2	126	25642-110	126	980373	127	B128
126	297V034H01	126	972X6	126	25642-115	126	980374	127	B128A
126	297V035H01	126	972X7	126	25642-120	126	980626	127	B129
126	297V036H01	126	972X8	126	30213	126	980833	127	B228
126	297V036H02	126	1003(JULIETTE)	126	30214	126	980834	127	B229
126	297V038H06	126	1007(JULIETTE)	126	30215	126	980835	127	B230
126	297V038H10	126	1010-89	126	30217	126	980958	127	B231
126	297V038H11	126	1033-1	126	30221	126	980959	127	B232
126	297V038H12	126	1033-2	126	30222	126	982150	127	B233
126	297V042H03	126	1033-3	126	30223	126	982267	127	B234
126	297V042H04	126	1033-4	126	30230	126	982289	127	B234N
126	297V045H01	126	1122-96	126	30247	126	983236	127	B251
126	297V045H02	126	1241-719	126	30274	126	983271	127	B251A
126	297V054H01	126	1465B	126	34048	126	983272	127	B252
126	297V054H02	126	1465B-12	126	55168	126	984685	127	B252A
126	297V055H01	126	1465B-12-8	126	55818	126	985445	127	B253
126	297V063001	126	1865B	126	56559	126	985446	127	B253A
126	297V065001	126	1865B-12	126	56560	126	985611	127	B274(JAPAN)
126	297V065002	126	1865B-127	126	56563	126	1061854-2	127	B275
126	297V065003	126	1865BL	126	58093	126	1471104-5	127	B276
126	297V070001	126	1865BL8	126	58933	126	1471104-6	127	B282
126	297V077001	126	1951-17	126	40261	126	1471104-7	127	B285
126	310-068	126	2015-00	126	40262	126	1471104-8	127	B296
126	310-123	126	2020-00	126	40268	126	2000646-106	127	B300
126	310-124	126	2021-00	126	40488	126	2000646-109	127	B301
126	310-139	126	2057A2-263	126	42302	126	2000648-23	127	B309
126	310-190	126	2057A2-37	126	42311	126	2001653-20	127	B310
126	310-191	126	2057A2-60	126	48937-2	126	2001653-21	127	B311
126	310-68	126	2057A2-65	126	48939-2	126	2002336-19	127	B312
126	324-0027	126	2057A2-66	126	61010-2-1	126	2002403-19	127	B313
126	324-0028	126	2057A2-80	126	61102-0	126	2008292-87	127	B318
126	324-0038	126	2057B2-104	126	69107-42	126	2076393	127	B319
126	324-0079	126	2057B2-105	126	69107-44	126	2091241-005	127	B320
126	324-0098	126	2057B2-118	126	72797-80	126	2091241-0719	127	B341
126	324-0099	126	2057B2-141	126	72879-39	126	2091241-5	127	B341H
126	324-0106	126	2057B2-35	126	72879-40	126	2091241-5A	127	B341V
126	324-0121	126	2057B2-37	126	080001	126	2495012	127	B342
126	324-0129	126	2057B2-41	126	080072	126	2495013	127	B343
126	324-0130	126	2057B2-42	126	080206	126	2495200	127	B357
126	324-0136	126	2057B2-48	126	080224	126	2495378	127	B358
126	324-0137	126	2057B2-50	126	080225	126	2495379	127	B359
126	324-0138	126	2057B2-51	126	080228	126	2495488-1	127	B360
126	324-0145	126	2057B2-60	126	080236	126	2495488-2	127	B361
126	325-0036-564	126	2057B2-65	126	080253	126	3404520-81	127	B362
126	325-0036-565	126	2057B2-67	126	080274	126	7279779	127	B366
126	325-1375-10	126	2057B2-68	126	080275	126	7287940	127	B375-2B
126	325-1375-11	126	2057B2-70	126	080276	126	7852899-01	127	B375-5B
126	325-1375-12	126	2057B2-71	126	080277	126	7910783-01	127	B375A-2B
126	325-1376-55	126	2057B2-77	126	81506-4	126	8010490	127	B375A-5B
126	353-9001-001	126	2057B2-79	126	81506-4A	126	8010530	127	B375A-NB
126	353-9001-002	126	2057B2-80	126	81506-4B	126	8510747-4	127	B375TV
126	353-9001-003	126	2057B2-81	126	81506-4C	126	8989457-1	127	B390
126	353-9002-002	126	2057B2-88	126	81506-7	126	18525200	127	B391
126	421-16	126	2058	126	81506-7A	126	27125230	127	B410
126	421-17	126	2478	126	81506-7B	126	30000010	127	B425Y
126	421-20	126	2478A	126	81506-7C	126	43024833	127	B432
126	421-20B	126	2478B	126	81506-8	126	43024834	127	B447
126	421-21	126	2488	126	81506-8A	126	61260039	127	B464
126	421-21B	126	2488A	126	81506-8B	126	61260039A	127	B465
126	421-22	126	2489	126	81506-8C	126	62042575	127	B468
126	421-22B	126	2495-012	126	94038	126	62044888	127	B468A
126	421-26	126	2495-013	126	95116	126	62047364	127	B468B
126	422-1403	126	2495-200	126	95118	126	62049626	127	B468C
126	429-0093-69	126	2603-183	126	95119	126	62049774	127	B468D
126	429-0094-39	126	2606-292	126	95120	126	62052821	127	B64
126	455-1	126	2791	126	95122	126	62067205	127	B85
126	576-0003-008	126	2904-016	126	95123	126	62087205	127	B87
126	576-0003-009	126	2904-029	126	99101	126	62111607	127	D080
126	576-0003-013	126	3008(CB)	126	99102	126	62119500	127	D081
126	576-0003-014	126	3009(MIXER)	126	99120	126	62123648	127	DTG1010
126	576-0003-015	126	3010	126	99121	126	62129788	127	DTG1110
126	576-0003-024	126	3010(IP)	126	100678	126	62140887	127	ES15252
126	576-2000-990	126	3019(EPJOHNSON)	126	101089	126	62140895	127	ES15X54
126	576-2000-993	126	3551	126	104009	126	62140933	127	ES15X77
126	601-094	126	3551A	126	104059	126	62530898	127	ES42(ELCOM)
126	610-052	126	3551A-BLU	126	112041	126	80050500	127	ET15X26
126	610-052-1	126	3551A-GRN	126	115775	126	80104900	127	ET15X40
126	610-056	126	3851	126	116072	126	80105200	127	GE-25
126	610-056-1	126	3907(2N404A)	126	116756	126	80105300	127	HEP-06007
126	610-056-2	126	3961	126	117724	126	1522210131	127	HEP-06008
126	610-056-3	126	4041-000-80100	126	117824	126	1522210921	127	HEP234
126	610-056-4	126	4363	126	119526	126	1522211021	127	HEP235
126	610-074	126	4363BLU	126	122725	126	1522214411	127	HO300
126	610-074-1	126	4363GRN	126	123244	126	1522214435	127	M4459
126	612-60039	126	4363ORN	126	123511	126	1522214821	127	M4623
126	612-60039A	126	4363WHT	126	124097	126	1522216600	127	M4652
126	617-56	126	4364	126	125330	126	16000190201	127	M7342
126	665B-1-5L	126	4365	126	126185	126S	A238	127	M7342/P4P
126	690V010H42	126	4366	126	128938	126S	A250	127	M9090
126	690V034H29	126	4367	126	162002-040	126S	A358	127	MN61
126	690V040H57	126	4454	126	162002-041	126S	28A238	127	MN63
126	690V040H58	126	4456	126	162002-042	126S	28A250	127	MN64
126	690V040H59	126	4457	126	162002-41	126S	28A358	127	MP1612
126	690V040H60	126	4501	126	165667	126D	HA1366	127	MP1612A
126	690V047H54	126	4509	126	166908	126S	AN262	127	MP1612B
126	690V047H55	126	4545	126	166909	126S	51-90305A21	127	MP1613
126	690V056H89	126	4545BLU	126	171005	126S	610001-4	127	MP3730
126	690V057H59	126	4545WHT	126	198010-1	126S	144012	127	MP3731
126	690V057H62	126	4586	126	221856	127	A339	127	P1F
		126	4589	126	223487	127	A340	127	P1Y
						127	AD166	127	P3H

ECG	Industry Standard No.	ECG	Industry Standard No.	ECG	Industry Standard No.	ECG	Industry Standard No.	ECG	Industry Standard No.
127	P3J	127	28B468B	128	A-1854-0087-1	128	BCW90KB	128	BSY82
127	P4F	127	28B468C	128	A-1854-0090-1	128	BCW90KC	128	BSY83
127	P4H	127	28B468D	128	AO-54-175	128	BCW91	128	BSY84
127	PTC122	127	28B64	128	A054-156	128	BCW91A	128	BSY85
127	Q5030	127	28B85	128	A054-160	128	BCW91B	128	BSY86
127	R-2001	127	28B87	128	A054-186	128	BCW95	128	BSY87
127	R2001	127	28S7341V	128	A054-206	128	BCW95A	128	BSY88
127	R2003	127	3-20	128	A059-110	128	BCW95B	128	BSY90
127	R2460-9	127	8-905-605-775	128	A1314	128	BCW95KA	128	BSY92
127	R2494-1	127	8-905-605-908	128	A1341	128	BCW95KB	128	C1008
127	R2590-1	127	8-905-613-232	128	A2471	128	BCY443	128	C108
127	R2964	127	8-905-613-295	128	A249(AMC)	128	BCY46	128	C109
127	R3514-1	127	09-301071	128	A25A305020101	128	BCY47	128	C109A
127	RE16	127	09-301073	128	A3011112	128	BCY48	128	C112(JAPAN)
127	SB-5-0399	127	09-301079	128	A32-2805-50-1	128	BCY49	128	C113
127	SB5-0399	127	13-16592-1	128	A322805-50-1	128	BCY65	128	C114
127	SB50399	127	13-16607-1	128	A466	128	BCY66	128	C115
127	SP.T212	127	13-16607-1	128	A47392R-0	128	BCY85	128	C118(JAPAN)
127	SP.T213	127	13-16607-1	128	A5B	128	BCY86	128	C119(JAPAN)
127	SP.T214	127	13-16607B	128	A5T2192	128	BF71	128	C121(TRANSISTOR)
127	SP.T238	127	13-16608-1	128	A5T4409	128	BF177	128	C121
127	SP.T239	127	13-16608-2	128	A5T5209	128	BF387	128	C124
127	SP.T240	127	13-17608B	128	A5T5210	128	BFR18	128	C130
127	SP.T250	127	13-17608C	128	A66-P11138-0001	128	BFR19	128	C147
127	SP1742	127	13-17609-1	128	A8D	128	BFR20	128	C150T
127	SP53	127	13-21606-1	128	AMP-2971-4	128	BFR22	128	C188
127	SP62	127	13-21606-1	128	AQ2	128	BFR36	128	C188A
127	TA1928A	127	13-55009-1	128	AQ3	128	BFR40TO5	128	C188AB
127	TA2083	127	13-55009-2	128	AQ5	128	BFR41TO5	128	C189
127	TA2188	127	19A115056-P1	128	AR203	128	BFR51	128	C19
127	TNJ-60075	127	19A11531-P1	128	AR203(RED)	128	BFR52	128	C190
127	TNJ-60080	127	31-0175	128	AR203R	128	BFR77	128	C20
127	TNJ60080	127	34-6002-24	128	AT-12	128	BFR78	128	C210
127	TQ5028	127	34-6002-31	128	AT-12(PHILCO)	128	BFR29	128	C211
127	TR-183(OLSON)	127	34-6002-49	128	AT-7	128	BF536	128	C215
127	TR16	127	46-86518-3	128	AT12	128	BFT30	128	C216
127	TR34	127	48-134659	128	AT339	128	BFT31	128	C217
127	TV-114	127	48-134623	128	AT380	128	BFT39	128	C218
127	TV106	127	48-134652	128	AT381	128	BFT40	128	C218A
127	TV111	127	48-134934	128	AT382	128	BFT41	128	C220
127	TV114	127	48-137001	128	AT383	128	BFT54	128	C221
127	TV24142	127	48-137178	128	AT384	128	BFX17	128	C222
127	TV24162	127	48-137234	128	AT385	128	BFX50	128	C226
127	TV24163	127	48-137235	128	AT386	128	BFX51	128	C227
127	TV24341	127	48-137342	128	AT387	128	BFX52	128	C228
127	TV24468	127	48-137342-P47	128	AT388	128	BFX61	128	C229
127	TV24568	127	48-137367	128	AT391	128	BFX68	128	C236
127	TV28B126	127	48-869090	128	AT392	128	BFX68A	128	C246
127	TV28B448	127	48B869090	128	AT393	128	BFX69	128	C2485076-3
127	TV8-28B126V	127	73B140-004	128	AT440	128	BFX69A	128	C2485077-2
127	TV8-28B448	127	86-292-2	128	AT441	128	BFX74	128	C249
127	TV8Z5B448	127	116-068	128	AT442	128	BFX84	128	C254
127	002-012700-12	127	116-086	128	AT443	128	BFX85	128	C30
127	2N1073B	127	116-087	128	AT444	128	BFX86	128	C306
127	2N1905	127	116-088	128	AT445	128	BFX96A	128	C307
127	2N2527	127	116-089	128	AT446	128	BFX97	128	C308
127	2N2528	127	121-370	128	AT470	128	BFX97A	128	C309
127	2N3730	127	324-0116	128	AT471	128	BFY13	128	C32A
127	2N3731	127	324-0128	128	AT472	128	BFY14	128	C352A(JAPAN)
127	2N3732	127	417-112	128	AT473	128	BFY15	128	C353
127	2N4346	127	608-112	128	AT474	128	BFY17	128	C353A
127	28B128	127	690V080H42	128	AT475	128	BFY27	128	C420
127	28B128A	127	690V080H43	128	AT476	128	BFY34	128	C425
127	28B128B	127	2057A2-265	128	AT477	128	BFY40	128	C426
127	28B128C	127	2057B2-134	128	AT478	128	BFY44	128	C443
127	28B128D	127	2057B2-156	128	AT479	128	BFY46	128	C46
127	28B128E	127	2496(RCA)	128	ATC-TR-13	128	BFY50	128	C47
127	28B128F	127	2500(RCA)	128	ATC-TR-4	128	BFY51	128	C479
127	28B128G	127	3514(RCA)	128	ATC-TR-7	128	BFY52	128	C48
127	28B128H	127	3583(RCA)	128	B274(SYLVANIA)	128	BFY53	128	C481
127	28B128V	127	3648(RCA)	128	B3746	128	BFY55	128	C482
127	28B129	127	4459	128	B87J0007	128	BFY56	128	C48C
127	28B152	127	4623	128	BC-119	128	BFY56A	128	C497
127	28B152A	127	4652	128	BC-138	128	BFY66	128	C497-0
127	28B228	127	11606-8	128	BC-140	128	BFY67	128	C497-R
127	28B229	127	11606-8	128	BC-140A	128	BFY67A	128	C497-Y
127	28B230	127	11608-7	128	BC-140B	128	BFY67C	128	C498
127	28B231	127	11608-8	128	BC-140C	128	BFY68	128	C498-0
127	28B232	127	11608-9	128	BC-140D	128	BFY68A	128	C498-R
127	28B233	127	17607-1	128	BC-141	128	BFY70	128	C498-Y
127	28B234	127	21606-1	128	BC-142	128	BFY80	128	C49X
127	28B234N	127	23114-097	128	BC103	128	BFY99	128	C503
127	28B251	127	40440	128	BC103C	128	BSB40	128	C503-0
127	28B251A	127	61218	128	BC117	128	BSB41	128	C503-Y
127	28B252	127	116068	128	BC119	128	BSV51	128	C503GR
127	28B252A	127	116086	128	BC120	128	BSV69	128	C504
127	28B253	127	116087	128	BC138	128	BSV84	128	C504-0
127	28B253A	127	116088	128	BC140	128	BSW10	128	C504-Y
127	28B274	127	116089	128	BC140-10	128	BSW26	128	C504GR
127	28B275	127	119722	128	BC140-16	128	BSW27	128	C51
127	28B276	127	119723	128	BC140-6	128	BSW28	128	C516
127	28B282	127	126701	128	BC140C	128	BSW29	128	C560
127	28B285	127	135744	128	BC140D	128	BSW35	128	C580
127	28B296	127	147351-4-1	128	BC141	128	BSW49	128	C61
127	28B300	127	171003	128	BC141-10	128	BSW65	128	C64
127	28B301	127	171004	128	BC141-16	128	BSW66	128	C696
127	28B309	127	200064-6-109	128	BC141-6	128	BSX12A	128	C696A
127	28B310	127	200064-6-110	128	BC142	128	BSX23	128	C696D
127	28B311	127	231140-04	128	BC144	128	BSX33	128	C696E
127	28B312	127	231140-09	128	BC174	128	BSX45	128	C696F
127	28B313	127	231140-26	128	BC211	128	BSX45-10	128	C696G
127	28B319	127	236288	128	BC216	128	BSX45-16	128	C696H
127	28B320	127	240403	128	BC216A	128	BSX45-6	128	C696I
127	28B341	127	243168	128	BC216B	128	BSX46	128	C708
127	28B341H	127	275845	128	BC254	128	BSX46-10	128	C708A
127	28B341S	127	322968-141	128	BC255	128	BSX46-16	128	C708AA
127	28B341V	127	489751-115	128	BC286	128	BSX46-6	128	C708AB
127	28B342	127	0573199	128	BC288	128	BSX59	128	C708AC
127	28B343	127	0573199H	128	BC301	128	BSX60	128	C708AH
127	28B357	127	0573212H	128	BC302-4	128	BSX61	128	C708AHA
127	28B358	127	608112	128	BC302-5	128	BSX62	128	C708AHB
127	28B359	127	802425-0	128	BC302-6	128	BSX62B	128	C708AHC
127	28B360	127	980150	128	BC310	128	BSX62C	128	C708B
127	28B361	127	1471125-3	128	BC340-10	128	BSX62D	128	C708C
127	28B362	127	1472494-1	128	BC340-16	128	BSX63B	128	C744
127	28B375	127	1472500-1	128	BC340-6	128	BSX63C	128	C774
127	28B375-2B	127	1473514-1	128	BC341-10	128	BSX70	128	C775
127	28B375-5B	127	1473648-1	128	BC341-6	128	BSX71	128	C781
127	28B375A	127	1962323	128	BC429	128	BSX72	128	C797
127	28B375A-2B	127	2000646-110	128	BC431	128	BSX95	128	C816
127	28B375A-5B	127	2314009	128	BC535	128	BSX96	128	C816K
127	28B375A-NB	127	2320201	128	BC537	128	BSY44	128	C826
127	28B375TV	127	2320803	128	BC538	128	BSY45	128	C827
127	28B390	127	7292684	128	BC682	128	BSY46	128	C875
127	28B410	127	7297348	128	BCW44	128	BSY51	128	C875-1C
127	28B411	127	8024250	128	BCW46	128	BSY52	128	C875-1D
127	28B432	127	23114004	128	BCW47	128	BSY53	128	C875-1E
127	28B447	127	23114009	128	BCW48	128	BSY54	128	C875-1F
127	28B464	127	23114026	128	BCW49	128	BSY55	128	C875-2
127	28B465	127	23114097	128	BCW90	128	BSY71	128	C875-2C
127	28B468	127	62051450	128	BCW90A	128	BSY77	128	C875-2D
127	28B468A	127	62141042	128	BCW90B	128	BSY78	128	C875-2E
		128	A-1141-5932	128	BCW90C	128	BSY81		
		128	A-120018	128	BCW90KA				

ECG	Industry Standard No.	ECG	Industry Standard No.	ECG	Industry Standard No.	ECG	Industry Standard No.	ECG	Industry Standard No.
128	C875-2F	128	D4D2O	128	HT104861	128	P04450026P5	128	S1864
128	C875-3	128	D4D21	128	HT104861A	128	P4069	128	S1874
128	C875-3C	128	D7A30	128	HT104861B	128	P480A0018	128	S19386
128	C875-3D	128	D7A31	128	HT304861	128	P4Z	128	S2118
128	C875-3B	128	D7A32	128	HT304861A	128	P633024	128	S21549
128	C875-3F	128	D911138-1	128	HT304971C	128	P633024G	128	S2209
128	C875BR	128	D911138-2	128	HT304971A	128	P6450026	128	S2369
128	C875C	128	D911138-3	128	HT304971AO	128	P6786	128	S2371
128	C875D	128	D911138-4	128	HT306441AO	128	P9962-1	128	S2400
128	C875B	128	D911138-5	128	HT307341	128	P9962-2	128	S2400A
128	C875F	128	D911138-6	128	HT307341A	128	P9962-5	128	S2400B
128	C876	128	D911138-7	128	HT307342B	128	PC04900-1	128	S2401
128	C876C	128	DDBY410002	128	HT307342C	128	PEE2001	128	S2401A
128	C876D	128	DN20-00453	128	HT309680B	128	PET9021	128	S2401B
128	C876B	128	DS-512	128	HT309714A-0	128	PET9022	128	S2401C
128	C876F	128	D8512	128	HT309841B0	128	P933024G	128	S2402
128	C876BV	128	DT1110	128	HT313181C	128	PIT-74	128	S2402A
128	C876BVD	128	DT1111	128	HT8001210	128	PL1083	128	S2402B
128	C876BVE	128	DT1112	128	HT8000310	128	PL1084	128	S2402C
128	C876BVEF	128	DT1120	128	I473608-2	128	PMC-Q8-0320	128	S2427
128	C88	128	DT1121	128	I473679-1	128	PMC-Q8-0400	128	S2459B
128	C88A	128	DT1122	128	I6191	128	PRT-104-4	128	S2461A
128	C934	128	DT1311	128	I9651	128	PT1544	128	S24616
128	C95	128	DT1321	128	INTROR-108	128	PT1545	128	S2487
128	C959A	128	DT1510	128	IR-TR53	128	PT3502	128	S2526
128	C959B	128	DT1511	128	IRTR87	128	PT612	128	S2648
128	C959C	128	DT1512	128	J241255	128	PT850	128	S27233
128	C959D	128	DT1520	128	J241256	128	PT850A	128	S2794
128	C959H	128	DT1521	128	K071961-001	128	PT888	128	S2992
128	C9598A	128	DT1522	128	KD2118	128	PT898	128	S409F
128	C9598B	128	DT1621	128	KGE41054	128	PTC101	128	S6004
128	C9598C	128	DTN206	128	KLR5807	128	Q-0-172	128	SC1229E
128	C9598D	128	DW6195	128	KS20180-L1	128	Q-00869	128	SCI44191005
128	C972	128	E-01381	128	L532 008 012	128	Q-00869A	128	SCI444103053
128	C972C	128	B-167-228	128	L532000162	128	Q-00869B	128	SDD1220
128	C972D	128	E-2491B	128	LDA404	128	Q-00969	128	SDD420
128	C972E	128	E2441	128	LDA405	128	Q-00969A	128	SE-8001
128	C984	128	E2449	128	LDA406	128	Q-00969B	128	SE-8010
128	C984A	128	E318-1	128	LDA408	128	Q-01169	128	SE5-0452
128	C984B	128	EA0081	128	LDB200	128	Q-01169A	128	SE5-0958
128	C984C	128	EA0090	128	LDB201	128	Q-01169B	128	S86001
128	C99CD	128	EA1549	128	LM2682	128	Q-01169C	128	S86006
128	C9FF-10A580	128	EA15X102	128	M-75536-1	128	Q-0172	128	S86020A
128	CC86168Q	128	EA15X144	128	M-75536-2	128	QA-10	128	S86021
128	CC86229H	128	EA15X57	128	M1X	128	QA-8	128	S86021A
128	CD6153	128	EA15X7655	128	M300-1300A	128	QA8	128	S86022
128	CD6153-2	128	EA1684	128	M4689	128	QQV60529	128	S86023
128	CDC120700	128	EA1698	128	M530	128	QRF-2	128	S87005
128	CDC5000B	128	EA1873	128	M9138	128	QT-0131BXDN	128	S87015
128	CDC5028A	128	EDC TR11-4	128	M9184	128	R-3552-1	128	S88001
128	CD08000-1	128	EL214	128	M9209	128	R-3553-1	128	S88002
128	CD08000-1C	128	EMS73278	128	M9221	128	R-3555	128	S88010
128	CD08000-CM	128	EMS73279	128	M9228	128	R-3580-1	128	S88012
128	CDC8002	128	EN1613	128	M9380	128	R123	128	S88041
128	CD08002-1	128	EN1711	128	M9519	128	R123-2	128	S88042
128	CDC9002-1B	128	EN870	128	M9562	128	R123-3	128	S88510
128	CDC9002-1C	128	EP15X33	128	MA6101	128	R123-4	128	S88520
128	CDQ10011	128	EP15X5	128	MA8001	128	R123-5	128	S88521
128	CDQ10012	128	EP16X7	128	MB1075	128	R15003	128	SPT443
128	CDQ10014	128	EQ8131	128	MB8001	128	R15003P1	128	SPT443A
128	CDQ10033	128	ES15226	128	MB8002	128	R2270-60106	128	SPT445
128	CDQ10048	128	ES15X93	128	MB8003	128	R2270-77873D	128	S05013
128	CDQ10051	128	ES22(ELCOM)	128	MH9410A	128	R227077873D	128	SJ2032
128	CDQ10052	128	ES51X65	128	MHT2414	128	R3508	128	SK3024
128	CDQ10053	128	ES56(ELCOM)	128	MHT2418	128	R3555-3	128	SK3047
128	CDQ10057	128	ES57(ELCOM)	128	MHT4401	128	R3593	128	SKA4616
128	CDZ15000	128	ES62(ELCOM)	128	MHT4411	128	R3608	128	SL3010
128	CE4003D	128	ESD918964P	128	MHT4412	128	R3608-1	128	SL301CE
128	CI3704	128	ET15X36	128	MHT4413	128	R3608-2	128	SM2716
128	CI3705	128	ET15X8	128	MHT4483	128	R3679	128	SM3978
128	CI3706	128	ETS-070	128	MHT4485	128	R5048	128	SM6251
128	CJ5210	128	EU15X27	128	MHT4511	128	R7613	128	SM7991
128	CJ5213	128	EU15X34	128	MHT4512	128	R8915	128	SMB-706009D
128	CJ5215	128	EX-141216	128	MHT4513	128	RC2270	128	SMC-583259
128	CM0770	128	F318-1	128	MHT7401	128	RCA1A01	128	SMC-620774-1
128	CN2484	128	F3560	128	MHT7411	128	RCA1A06	128	SN166
128	CP2357	128	F5561	128	MHT7412	128	RCA1A07	128	SN167
128	CS1129E	128	F5565	128	MHT7414	128	RCA1A17	128	SPD-80123
128	CS1225E	128	F5589	128	MHT7417	128	RCA1A18	128	SPS-0122
128	CS1225EF	128	F4709	128	MH79001	128	RE17	128	SPS-4077
128	CS1229K	128	F625-1	128	MH79002	128	R87O	128	SPS0122
128	CS1248	128	FBN-CP34634	128	MH79004	128	R8-2014	128	SPS3912
128	CS1248I	128	FBN-L109	128	MH79005	128	R8132	128	SPS3914
128	CS1248F	128	FBN-L113	128	MM1809	128	R87672	128	SPS4038
128	CS1250F	128	FBN-L115	128	MM1809A	128	R87678	128	SPS4300
128	CS1255B	128	FBN-L148	128	MM1810A	128	R88101	128	SPS4309
128	CS1295H	128	FBTX070	128	MM1943	128	R88103	128	SPS4311
128	CS1305	128	FC81229	128	MM2266	128	R88105	128	SPS4361
128	CS1352	128	FD-1029-FY	128	MM486	128	R88107	128	SPS4461
128	CS1453F	128	FD-1029-GE	128	MM487	128	R88109	128	SPS4490
128	CS1453G	128	FD-1029-GM	128	MM488	128	R88111	128	SPS4495
128	CS1462I	128	FD-1029-JN	128	MM511	128	R88113	128	S2849
128	CS1464E	128	FD-1029-PA	128	MM512	128	RT-141	128	SPS8909
128	CS1591LE	128	F81331	128	MM513	128	RT-188	128	S86124
128	CS1609F	128	F827233	128	MP4906063	128	RT141	128	ST 254 Q
128	CS1615	128	FT001	128	MP89410A	128	RT154	128	ST-201
128	CS1664	128	FT0019H	128	MP89410AJ	128	RT188	128	ST-213
128	CS1711	128	FT002	128	MP89410H	128	RT482	128	S-LM2682
128	CS1893	128	FT003	128	MPX9410H	128	RT483	128	ST213
128	CS1990	128	FT004	128	MR3933	128	RT484	128	ST4150
128	CS2484	128	FT004A	128	MS2991	128	RT5151	128	ST4201
128	CS3704	128	FT027	128	MST506H	128	RT5152	128	ST4202
128	CS3705	128	G05413A	128	MST506J	128	RT5203	128	ST4203
128	CS3706	128	G05413B	128	MS97506	128	RT5204	128	ST4204
128	CS4006	128	G23-46	128	MSK5405	128	RT5401	128	ST4341
128	CS5449	128	G5A7A66-2	128	MSP99905B-1	128	RT5402	128	ST6573
128	CS5450	128	GC1615-1	128	MST-10	128	RT5403	128	ST6574
128	CS5451	128	GB-18	128	MT1613	128	RT5404	128	STIA2880
128	CS6168Q	128	GB-243	128	MT1711	128	RT5906	128	T-04689
128	CS7229F	128	GET929	128	ONO20540	128	RT5907	128	T-291
128	CS9022LE	128	GRT930	128	N1X	128	RT6993M	128	T-339
128	CS89103B	128	G05-015-B	128	NCR046	128	RT6996M	128	T-Q5081
128	CS89103C	128	HC-00268	128	NCR047	128	RT7945	128	T-Q5099
128	CS8956	128	HC-00509	128	NJ107	128	RV1473	128	TO1-022
128	D11101OB1	128	HC-01209	128	NPC115	128	RYN12104	128	T13015
128	D11101CB1	128	HC-01317	128	NPC187	128	S001683	128	T1340A3H
128	D11C1B1	128	HC-01318	128	NP0189	128	S007220	128	T164215
128	D11C3B1	128	HEP-85014	128	NS2100	128	S1368	128	T1706
128	D1105B1	128	HEP243	128	NS2101	128	S1514	128	T1706A
128	D11C7B1	128	HEP736	128	NS9400	128	S1516	128	T1706B
128	D149	128	HP11(PHILCO)	128	NS9420	128	S1517	128	T1706C
128	D204	128	HP9	128	NS9500	128	S1523	128	T1811
128	D204L	128	HR12A	128	NS9540	128	S1525	128	T1811B
128	D215	128	HR12B	128	NS9728	128	S15660	128	T1811G
128	D219	128	HR12C	128	NS9729	128	S1642	128	T23-94
128	D221	128	HR12D	128	NS9730	128	S1644	128	T247
128	D233	128	HR12E	128	NS9731	128	S1671	128	T336-2
128	D3288	128	HR12F	128	OA-10	128	S1689	128	T452
128	D4C28	128	HR28	128	00V60529	128	S1762	128	TA198035-1
128	D4C29	128	HR29	128	ORF-2	128	S1773	128	TA6200
128	D4C30	128	HR81	128	P-11748-1	128	S1777	128	TB1990
128	D4C31	128	HS-40039	128	P-11903-1	128	S17900	128	TB3416
128		128	HS40026	128	P04-45-0026-P5	128	S18000	128	TF101-A
128		128		128		128	S18200	128	TF101-B

ECG	Industry Standard No.	ECG	Industry Standard No.	ECG	Industry Standard No.	ECG	Industry Standard No.	ECG	Industry Standard No.	ECG	Industry Standard No.
128	TP101-D	128	001-02119-0	128	2N2787	128	28C124	128	28C875-2E	128	28C875-2E
128	TI-424	128	001-021190	128	2N2788	128	28C130	128	28C875-2F	128	28C875-2F
128	TI-425	128	001-021290	128	2N2789	128	28C130L	128	28C875-3	128	28C875-3
128	TI-475	128	01-030734	128	2N2863	128	28C13838	128	28C875-3C	128	28C875-3C
128	TI-480	128	1-035J2207	128	2N2864	128	28C140	128	28C875-3D	128	28C875-3D
128	TI-482	128	01-2110	128	2N2868	128	28C147	128	28C875-3E	128	28C875-3E
128	TI-483	128	01-2111	128	2N2890	128	28C150	128	28C875-3F	128	28C875-3F
128	TI-484	128	01-2114	128	2N2891	128	28C150T	128	28C875BR	128	28C875BR
128	TI-496	128	1-2114-0	128	2N2895	128	28C154H	128	28C875C	128	28C875C
128	TI3015	128	1A0066	128	2N2897	128	28C188	128	28C875D	128	28C875D
128	TI481	128	1A34	128	2N2900	128	28C188A	128	28C875E	128	28C875E
128	TI64213	128	1A348	128	2N2909	128	28C188AB	128	28C875F	128	28C875F
128	TI8110	128	1A348R	128	2N2J324	128	28C189	128	28C876	128	28C876
128	TI8107	128	1A348	128	2N3020	128	28C189D	128	28C876D	128	28C876D
128	TI8110	128	1N3112	128	2N3056	128	28C189E	128	28C876E	128	28C876E
128	TI8111	128	1WB995A	128	2N3053	128	28C19	128	28C876F	128	28C876F
128	TI860A	128	002-010600	128	2N3056	128	28C190	128	28C876TV	128	28C876TV
128	TI860B	128	002-012500	128	2N3077	128	28C20	128	28C876TVD	128	28C876TVD
128	TI860C	128	2E1A20A22AAB	128	2N3078	128	28C210	128	28C876TVR	128	28C876TVR
128	TI860D	128	2N1051	128	2N3107	128	28C211	128	28C876TVEF	128	28C876TVEF
128	TI860E	128	2N1081	128	2N3108	128	28C212	128	28C88	128	28C88
128	TI860M	128	2N1092	128	2N3109	128	28C216	128	28C88A	128	28C88A
128	TI892M	128	2N1116	128	2N3110	128	28C217	128	28C95	128	28C95
128	TI898	128	2N1117	128	2N3122	128	28C218A	128	28C959	128	28C959
128	TI899	128	2N1338	128	2N3326	128	28C218A	128	28C959A	128	28C959A
128	TM128	128	2N1410	128	2N339	128	28C220	128	28C959B	128	28C959B
128	TNJ70482	128	2N1420	128	2N339A	128	28C221	128	28C959C	128	28C959C
128	TNJ71035	128	2N1420A	128	2N340	128	28C222	128	28C959D	128	28C959D
128	TNJ71234	128	2N1444	128	2N340A	128	28C225	128	28C959M	128	28C959M
128	TNJ72288	128	2N1472	128	2N342	128	28C226	128	28C959R	128	28C959R
128	TQ-5055	128	2N1491	128	2N342A	128	28C227	128	28C959SA	128	28C959SA
128	TQ5055	128	2N1492	128	2N342B	128	28C228	128	28C959SB	128	28C959SB
128	TQPD3053	128	2N1506A	128	2N343	128	28C234	128	28C959SC	128	28C959SC
128	TR-01B	128	2N1507	128	2N343A	128	28C236	128	28C959D	128	28C959D
128	TR-1000-3	128	2N1564	128	2N343B	128	28C247	128	00028C96P	128	00028C96P
128	TR-164(OLSON)	128	2N1565	128	2N3469	128	28C249	128	28C972	128	28C972
128	TR-25	128	2N1566	128	2N3498	128	28C291	128	28C972C	128	28C972C
128	TR-28C482	128	2N1566A	128	2N3568	128	28C30	128	28C972D	128	28C972D
128	TR-31B	128	2N1613	128	2N3569	128	28C306	128	28C972E	128	28C972E
128	TR-4R31	128	2N1613L	128	2N3665	128	28C306B	128	28C984	128	28C984
128	TR-5R31	128	2N1613B	128	2N3666	128	28C306C	128	28C984A	128	28C984A
128	TR-7R31	128	2N1615	128	2N3678	128	28C306D	128	28C984B	128	28C984B
128	TR-8021	128	2N1644	128	2N3723	128	28C306E	128	28C984C	128	28C984C
128	TR-8023	128	2N1700	128	2N3830	128	28C306F	128	28C984H	128	28C984H
128	TR-8024	128	2N1711	128	2N3861	128	28C307	128	28C995B	128	28C995B
128	TR-8036	128	2N1711A	128	2N4047	128	28C307B	128	28C995D	128	28C995D
128	TR01054-7	128	2N1711L	128	2N4237	128	28C307C	128	28D204	128	28D204
128	TR1001	128	2N1711B	128	2N4238	128	28C307D	128	28D204L	128	28D204L
128	TR1003	128	2N1714	128	2N4239	128	28C307E	128	28D215	128	28D215
128	TR1005	128	2N1716	128	2N4385	128	28C308	128	28D219	128	28D219
128	TR23	128	2N1764	128	2N4395	128	28C309	128	28D220	128	28D220
128	TR25	128	2N1837	128	2N480B	128	28C309B	128	28D221	128	28D221
128	TR28C482	128	2N1838	128	2N4943	128	28C309C	128	28D3288	128	28D3288
128	TR36643	128	2N1839	128	2N4944	128	28C309D	128	3I4-6007-4	128	3I4-6007-4
128	TR8036	128	2N1889	128	2N4946	128	28C309E	128	4-0498	128	4-0498
128	TR01054-7	128	2N1890	128	2N4960	128	28C31	128	4-1001	128	4-1001
128	TS2218	128	2N1941	128	2N4961	128	28C32	128	4-274	128	4-274
128	TS2219	128	2N1943	128	2N4962	128	28C32A	128	4-288	128	4-288
128	TSC-722	128	2N1953	128	2N4963	128	28C353	128	4-3023223	128	4-3023223
128	TSC722	128	2N1958	128	2N4964	128	28C353A	128	4-3023844	128	4-3023844
128	TV-26	128	2N1958A	128	2N497	128	28C443	128	4-397	128	4-397
128	TV-28	128	2N1959	128	2N497A	128	28C46	128	4-398	128	4-398
128	TV-41	128	2N1959A	128	2N498	128	28C47	128	4-686182-3	128	4-686182-3
128	TV-43	128	2N1972	128	2N5188	128	28C479	128	4C28	128	4C28
128	TV-45	128	2N1973	128	2N5211	128	28C479H	128	4C29	128	4C29
128	TV-59	128	2N1974	128	2N542	128	28C48	128	4C30	128	4C30
128	TV-67	128	2N1975	128	2N545	128	28C48C	128	4C31	128	4C31
128	TV-70B	128	2N1983	128	2N5450	128	28C49	128	4JX11C2848	128	4JX11C2848
128	TV23	128	2N1984	128	2N5451	128	28C497	128	5E4850/56-0002	128	5E4850/56-0002
128	TV41	128	2N1985	128	2N547	128	28C497-0	128	6-0000139	128	6-0000139
128	TV42	128	2N1986	128	2N549	128	28C497-R	128	6-139	128	6-139
128	TV45	128	2N1987	128	2N560	128	28C497-Y	128	6-456	128	6-456
128	TV51	128	2N1988	128	2N5681	128	28C498	128	6A10228	128	6A10228
128	TV52	128	2N1989	128	2N5814	128	28C498-0	128	6E4850/56-0002	128	6E4850/56-0002
128	TV53	128	2N1990	128	2N5816	128	28C498-R	128	6I0148-1	128	6I0148-1
128	TV59	128	2N2017	128	2N5818	128	28C498-Y	128	7-0002-00	128	7-0002-00
128	TV6080	128	2N2038	128	2N5820	128	28C49Y	128	7-0011-00	128	7-0011-00
128	TVS-28C582	128	2N2039	128	2N5822	128	28C501	128	007-0051	128	007-0051
128	TVS-28C582A	128	2N2040	128	2N5856	128	28C503	128	7-0051-00	128	7-0051-00
128	TVS28C1255	128	2N2041	128	2N6010	128	28C503-0	128	007-74659-01	128	007-74659-01
128	TVS28C1255HF	128	2N2049	128	2N6012	128	28C503-Y	128	007-74659-04	128	007-74659-04
128	TVS28C696	128	2N2094	128	2N6014	128	28C503GR	128	007-74659-06	128	007-74659-06
128	TVS28D968	128	2N2094A	128	2N655A	128	28C504	128	7A1011(GE)	128	7A1011(GE)
128	TX-100-4	128	2N2095A	128	2N656	128	28C504-0	128	7A30(GE)	128	7A30(GE)
128	TX100-4	128	2N2102L	128	2N657	128	28C504-Y	128	7A31(GE)	128	7A31(GE)
128	TX101-4	128	2N2102B	128	2N696A	128	28C504GR	128	7A32(SHERWOOD)	128	7A32(SHERWOOD)
128	UPI1613	128	2N2106	128	2N697	128	28C51	128	7A35(GE)	128	7A35(GE)
128	UPI2217	128	2N2107	128	2N697A	128	28C516	128	7A995(GE)	128	7A995(GE)
128	UPI2218	128	2N2108	128	2N697L	128	28C516A	128	7A995(SHERWOOD)	128	7A995(SHERWOOD)
128	UPI1956	128	2N2161	128	2N6978	128	28C560	128	8-0053300	128	8-0053300
128	V120RH	128	2N2192	128	2N730	128	28C580	128	8-905-705-41Q	128	8-905-705-41Q
128	V139	128	2N2192A	128	2N731	128	28C59	128	8Q-3-13	128	8Q-3-13
128	V146	128	2N2192B	128	2N734	128	28C61	128	09-302090	128	09-302090
128	V154	128	2N2193	128	2N735	128	28C64	128	09-302171	128	09-302171
128	V166	128	2N2193A	128	2N736	128	28C69	128	09-305076	128	09-305076
128	V172	128	2N2193B	128	2N736A	128	28C696	128	9-5226	128	9-5226
128	V177	128	2N2194	128	2N736B	128	28C696(D)	128	9-5226-004	128	9-5226-004
128	V8-081255H	128	2N2194A	128	2N752	128	28C696A	128	9-5226-4	128	9-5226-4
128	V8-081255HF	128	2N2194B	128	2N845	128	28C696B	128	9TR61001-07	128	9TR61001-07
128	V8C81255H	128	2N2195	128	2N870	128	28C696D	128	110-11536	128	110-11536
128	W14	128	2N2195A	128	2N871	128	28C696F	128	13-0096666-001	128	13-0096666-001
128	X19001-B	128	2N2195B	128	2N910	128	28C696G	128	13-105698-1	128	13-105698-1
128	X19001-C	128	2N2198	128	2N911	128	28C696H	128	13-14085-15A	128	13-14085-15A
128	X300-1300A	128	2N2217	128	2N912	128	28C696I	128	13-14085-91	128	13-14085-91
128	X3205099	128	2N2218	128	2N981	128	28C708	128	13-14605-1	128	13-14605-1
128	X3205111	128	2N2218A	128	28014	128	28C708A	128	13-15835-1	128	13-15835-1
128	XA-1095	128	2N2219	128	28017	128	28C708AA	128	13-18927-1A	128	13-18927-1A
128	XA1018	128	2N2219A	128	28018	128	28C708AB	128	13-23840-1	128	13-23840-1
128	XB12	128	2N2243A	128	28019	128	28C708AC	128	13-26666	128	13-26666
128	X830	128	2N2270	128	28020	128	28C708AH	128	13-28394-1	128	13-28394-1
128	Y56601-47	128	2N2270L	128	28742	128	28C708AHA	128	13-27443-1	128	13-27443-1
128	ZT1479	128	2N2270B	128	2742AA	128	28C708AHB	128	13-28394-1	128	13-28394-1
128	ZT1481	128	2N2297	128	28745	128	28C708AHC	128	13-28394-2	128	13-28394-2
128	ZT1613	128	2N2317	128	2745A	128	28C708B	128	13-28394-3	128	13-28394-3
128	ZT1700	128	2N2350	128	280-NJ107	128	28C708C	128	13-34371-1	128	13-34371-1
128	ZT1711-1	128	2N2350A	128	28C1008	128	28C797	128	13-55018-4	128	13-55018-4
128	ZT190	128	2N2351	128	28C108	128	28C802	128	13-55064-1	128	13-55064-1
128	ZT191	128	2N2352	128	28C108A	128	28C816	128	14-0104-4	128	14-0104-4
128	ZT192	128	2N2352A	128	28C108A-0	128	28C816K	128	14-602-18	128	14-602-18
128	ZT193	128	2N2353	128	28C108A-R	128	28C826	128	14-602-24	128	14-602-24
128	ZT2102	128	2N2353A	128	28C109	128	28C827	128	14-602-29	128	14-602-29
128	ZT2270	128	2N2380	128	28C109A	128	28C827A	128	14-602-30	128	14-602-30
128	ZT3512	128	2N2380A	128	28C109A-0	128	28C827C	128	14-602-43	128	14-602-43
128	ZT3866	128	2N2410	128	28C109A-R	128	28C875	128	14-602-49	128	14-602-49
128	ZT66A	128	2N2466	128	28C109A-Y	128	28C875-1	128	14-602-59	128	14-602-59
128	ZT66B	128	2N2479	128	28C112	128	28C875-1C	128	14-602-65	128	14-602-65
128	ZT860	128	2N2489A	128	28C113	128	28C875-1D	128	14-602-72	128	14-602-72
128	ZT86	128	2N2510	128	28C114	128	28C875-1E	128	14-603-01	128	14-603-01
128	ZT88	128	2N2511	128	28C115	128	28C875-1F	128	14-603-02A	128	14-603-02A
128	ZT90	128	2N2537	128	28C12	128	28C875-2	128	14-603-07	128	14-603-07
128	ZT93	128	2N2538	128	28C121	128	28C875-2C	128	14-654-21	128	14-654-21
128	ZT94	128	2N2594	128	28C122	128	28C875-2D	128	15-02155	128	15-02155
128	001-021110			128	28C123			128	15-09650	128	15-09650
128	001-021111										

ECG	Industry Standard No.
128	15-10062-0
128	17-443
128	18-3539-1
128	019-003349
128	019-003637
128	19-020-038
128	19-020-075
128	19-15840
128	19-3349
128	19-3692
128	19-3934-643
128	19A115238-2
128	19A115238-P1
128	19A115300
128	19A115300-1
128	19A115300-2
128	19A115304-2
128	19A115889-P1
128	19A115889-P2
128	19A115889-P3
128	19C300115-1
128	19C300115-P1
128	020-1110-008
128	020-1111-018
128	020-1111-038
128	20-1680-174
128	20A0055
128	20A0076
128	021
128	021-0137-00
128	21A015-018
128	21A015-019
128	21A015-026
128	21A040-051
128	21A105-004
128	21A105-006
128	21M185
128	21M387
128	21M448
128	21M455
128	23-5039
128	23B114053
128	23B114054
128	24-000452
128	24MW1161
128	24MW660
128	24MW663
128	24MW674
128	24MW677
128	24MW714
128	025-100015
128	026-100013
128	31-0083
128	33-048
128	34-6001-50
128	34-6001-51
128	34-6001-51MR-12
128	34-6001-78
128	34-6001-80
128	34-6001-82
128	34-6001-83
128	34-6001-85
128	34-6015-23
128	34-6015-24
128	34-6015-25
128	34-6015-30
128	34-6015-51
128	34-6016-22
128	34-6016-27
128	34-6016-30
128	34-6016-33
128	35-020-21
128	40-0068-2
128	41B581014
128	42-19642
128	42-21233
128	42-9029-40T
128	42-9029-70J
128	42-9029-70K
128	43-023223
128	43-023844
128	43-VD-09
128	43A126932
128	43A162455-1
128	43A223060-1
128	44A319819-1
128	44A359497-001
128	44A359497-002
128	44A390264-001
128	44A395992-001
128	44A395992-002
128	44A417063-001
128	46-86233-3
128	46-86371-3
128	46-86381
128	46-86381-3
128	46-86405-3
128	46-86426-3
128	46-86427-3
128	48-01-007
128	48-03-04046103
128	48-03-05013702
128	48-134689
128	48-134838
128	48-134941
128	48-134953
128	48-134982
128	48-137005
128	48-137041
128	48-137088
128	48-137307
128	48-137315
128	48-137988
128	48-40212
128	48-859248
128	48-869138
128	48-869170
128	48-869184
128	48-869221
128	48-869228
128	48-869263
128	48-869380
128	48-869464
128	48-869562
128	48-869599
128	48-869703-3
128	48-90232A08
128	48-97046A29
128	48-97046A40
128	48-97127A019
128	48-97127A04
128	48K869228
128	48R869170
128	48R869562
128	48S137041
128	48X97046A52
128	49-62139
128	51-47-20
128	53P169
128	55-1084
128	55-642
128	57-01491-C
128	57-01494C
128	57A104-8-6
128	57A109-9
128	57A113-9
128	57A12-4
128	57A128-9
128	57A129-9
128	57A131-10
128	57A14-1
128	57A14-2
128	57A14-3
128	57A156
128	57A2-38
128	57A2141-14
128	57A7-8
128	57B109-9
128	57B113-9
128	57B128-9
128	57B165-11
128	57B167-9
128	57B170-9
128	57B2-38
128	57C109-9
128	57C12-4
128	57C14-1
128	57C14-2
128	57C14-3
128	57C7-8
128	57L105-12
128	57L2-1
128	57M1-33
128	57M1-34
128	57M2-11
128	57M2-14
128	57M2-18
128	57M2-7
128	57M3-3
128	61-1764
128	61-309-458
128	61-309687
128	61-747
128	61-813
128	61C004-1
128	63-11936
128	63-11989
128	63-12989
128	63-12990
128	63-13214
128	63-13215
128	066
128	66A10298
128	66F025-1
128	66S00000A
128	68A7368
128	68A7380-2
128	72N1B
128	73B140385-001
128	76-14090-1
128	76N1B
128	76N2B
128	76N3B
128	76S1030-000
128	77-271490
128	77-271490-1
128	77-272999-1
128	77N2B
128	79F114-1
128	79F114-3
128	80-053300
128	81T2
128	86-161-2
128	86-170-2
128	86-207-2
128	86-208-2
128	86-210-2
128	86-211-2
128	86-234-2
128	86-266-2
128	86-267-2
128	86-273-2
128	86-291-2
128	86-330-2
128	86-336-2
128	86-393-2
128	86-428-9
128	86-440-2
128	86-441-2
128	86-452-2
128	86-463-2
128	86-463-3(SEARS)
128	86-5070-2
128	86-5073-2
128	86-5074-2
128	86-5093-2
128	86-510-2
128	86-543-2
128	86-567-2
128	86-650-2
128	86-674-2
128	86-675-2
128	94N1V
128	95-220
128	95-226-004
128	95-226-4
128	96-5107-01
128	96-5107-02
128	96-5170-01
128	96-5180-01
128	96-5180-02
128	96-5203-01
128	96-5204-01
128	96-5208-01
128	96-5231-01
128	96-5244-01
128	96-5252-01
128	96-5256-01
128	96-5364-01
128	96XZ6053/38N
128	99L6(SHARP)
128	99S074
128	100-4846-001
128	100-5338
128	100-5338-001
128	0103-0014R
128	0103-0014S
128	0103-0051
128	0103-0503
128	0103-0503S
128	0103-0607
128	0103-0616
128	0104-0013
128	104H01
128	107N1
128	112-203053
128	112-361
128	112-525
128	119-0077
128	121-431
128	121-639CL
128	121-676
128	121-722
128	121-737CL
128	121-766
128	121-773CL
128	121-844
128	121-878
128	123S-437
128	123S425
128	125-121
128	125A137
128	125A137A
128	125B139
128	128N2
128	130-191-00
128	0131-000561
128	0131-001429
128	0131-001430
128	131-005-807
128	131-005807
128	131-045-60
128	132-021B
128	132-022
128	132-038
128	132-066
128	132N1
128	148N1
128	148N3
128	151-0096-00
128	151-0136
128	151-0136-00
128	151-0136-02
128	151-0211
128	151-0211-00
128	151-0211-01
128	151-096-1C
128	151-150
128	151-150-1B
128	151-211
128	158-045-0027
128	170-9
128	173A-4490-5
128	173A4391
128	173A4489-2
128	173A4490-5
128	174-25566-21
128	174-25566-50
128	174-25566-63
128	174-25566-76
128	176-004
128	176-074-9-004
128	182B2003JDP1
128	200-011
128	200-12
128	218-26
128	226-4
128	231-0013
128	233(SEARS)
128	241B
128	247-625
128	247-626
128	248-38104-1
128	260D07901
128	260D08601
128	260D08701
128	260D09612
128	260P10003
128	260P10003A
128	260P10005
128	260P12401
128	260P12481
128	260P17002
128	270-950-037-02
128	296-58-9
128	296-81-9
128	297V049H06
128	297V061C05
128	297V062C06
128	297V072C05
128	297V074C10
128	297V074C12
128	297V082B01
128	297V083C04
128	300C043
128	324-019
128	324-6011
128	324-6013
128	325-1513-46
128	342-1
128	344-1
128	344-6001-1
128	344-6001-2
128	344-6011-1
128	344-6011-2
128	344-6013-1B
128	344-6013-4
128	344-6017-4
128	345-2
128	349-1
128	349-2
128	352-0092-020
128	352-0364-000
128	352-0364-010
128	352-043-010
128	352-0479-010
128	352-0766-010
128	352-0783-020
128	352-0816-010
128	352-9014-00
128	353-9301-001
128	360-1(RCA)
128	385-1
128	386-7181P2
128	386-7316-P2
128	394-3005-2
128	394-3127-1
128	394-3127-2
128	394-3127-3
128	394-3141-1
128	400-1362-101
128	400-1362-102
128	400-1362-201
128	411-237
128	417-100
128	417-114
128	417-115-13173
128	417-128
128	417-133
128	417-136
128	417-137
128	417-155
128	417-178
128	417-180
128	417-193
128	417-224
128	417-233
128	417-237
128	417-237-13163
128	417-247
128	417-250
128	417-257
128	417-49
128	417-59
128	417-821-13163
128	417-87
128	417-88
128	417-89
128	422-1233
128	422-1404
128	422-2158
128	430-10047-0C
128	430-23223
128	430-23844
128	448A662
128	454A104
128	462-1007
128	462-1016
128	462-1019
128	462-I007
128	469-646-3
128	472-0309-001
128	472-0445-001
128	481-201-A
128	481-201-B
128	491A948
128	520-301
128	536GU(WARDS)
128	542-1033
128	555-3(RCA)
128	555-4
128	559-1516-001
128	565-072
128	565-1
128	573-532
128	576-0004-02
128	576-0005-004
128	577R819H01
128	600-188-1-21
128	601-0100792
128	608-2
128	614-1
128	622-1
128	622-1(RCA)
128	622-2
128	625-1(RCA)
128	638-1
128	639CL
128	642-306
128	660-125
128	660-128
128	660-145
128	666-1(RCA)
128	686-0012
128	686-0112
128	686-0130
128	686-0165
128	686-229-0
128	690V075H62
128	690V086H90
128	690V103H30
128	690V110H34
128	690V110H36
128	690V116H24
128	691C844
128	700-110
128	703-1
128	703-2
128	703B
128	729-3
128	0731
128	753-2000-107
128	753-8510-470
128	763-1
128	772-120-00
128	772-121-00
128	773CL
128	800-001-106-1
128	800-101-101-2
128	800-512-00
128	800-513-00
128	800-515-00
128	800-521-02
128	800-521-03
128	800-522-03
128	800-53300
128	819-1
128	914F298-1
128	921-188B
128	921-236B
128	921-337B
128	921-60B
128	943-728-001
128	964-22158
128	964-24387
128	964-25046
128	991-00-2888
128	991-01-1305
128	991-01-1314
128	991-011305
128	991-011314
128	991-3N
128	999-4601
128	1001-01
128	1002A-2
128	1008
128	1009-03-17
128	1010-7993
128	1010-8041
128	1027G
128	1027G(GE)
128	1038(G.E.)
128	1040-11
128	1040-7
128	1044-9544
128	1045-3082
128	1099-0950
128	1106-99
128	1112-78
128	1116-6535
128	1157
128	1187
128	1204(GE)
128	1206(GE)
128	1207(GE)
128	1300-1
128	1300A
128	1414-157
128	1414-184
128	1414-186
128	1414-189
128	1473-4255
128	1476 1118
128	1483
128	1573-00
128	1573-01
128	1673-0475
128	1676-1991
128	1705-5351
128	1789-17
128	1854-0022-1
128	1854-0090
128	1854-0090-1
128	1854-0274
128	1854-0332
128	1854-0498
128	1879-17A
128	1933-17
128	1972-17
128	2000-209
128	2003JDP1
128	2057A100-17
128	2057A2-151
128	2057A2-156
128	2057A2-199
128	2057A2-230
128	2057A2-284
128	2057A2-295
128	2057A2-468
128	2057B100-17
128	2057B107-8
128	2057B109-9
128	2057B113-9
128	2057B126-12
128	2057B129-9
128	2057B144-12
128	2057B153-9
128	2057B156-9
128	2093A38-24
128	2208-17
128	2270-5
128	2495
128	2498-163
128	2904-032
128	3022
128	3034(RCA)
128	3202-51-01
128	3202-5H01
128	3223
128	3508
128	3532-1
128	3544(RCA)
128	3553-3
128	3555-1(RCA)
128	3555-3
128	3560-2
128	3561-1(GE)
128	3565(RCA)
128	3566(RCA)
128	3591
128	3601
128	3608(RCA)
128	3608-2(RCA)
128	3615(RCA)
128	3622(RCA)
128	3622-1
128	3622-2(RCA)
128	3625-1
128	3666(RCA)
128	4080-187-0507
128	4306
128	4473-M-12
128	4473-M3
128	4473-N
128	4483
128	4490
128	4686-182-3
128	4689
128	4822-130-40356
128	4824-33
128	5258
128	5380-73
128	5553
128	5613-1213D
128	5613-1788R
128	6151(RCA)
128	6284
128	6651-486
128	7325-1
128	7334
128	7344-1
128	7345-2
128	7349
128	7381-2
128	7507
128	7513
128	7909
128	8000-00004-P082
128	8000-0004-P083
128	8000-0004-P084
128	8000-0005-010
128	8000-0005-011
128	8304
128	8554-9
128	8868-6
128	8880-3
128	8883-4
128	9013G
128	9279
128	9367-1
128	9404-2
128	9405-0
128	9405-1
128	9617K
128	9696H
128	10416-010
128	11236-1
128	11252-5
128	012015
128	13213E4252

ECG	Industry Standard No.	ECG	Industry Standard No.	ECG	Industry Standard No.	ECG	Industry Standard No.	ECG	Industry Standard No.
128	14692	128	41053	128	139696	128	800019-001	128	7902310
128	14995	128	41502	128	140501	128	800073-6	128	8000736
128	14996-1	128	43021-198	128	141008	128	800073-7	128	8000737
128	14996-2	128	43082	128	141019	128	800946-001	128	8004265
128	15486	128	43088	128	141355	128	810002-736	128	8037343
128	16065	128	43117	128	141355	128	860001-8	128	8378759
128	16082	128	43122	128	141767	128	900201-104	128	8398315
128	16191	128	43165	128	141783	128	900201-105	128	8421133
128	16254	128	45354	128	143042	128	900201-167	128	8524457
128	018077	128	49092	128	143797	128	900201-81	128	10811788
128	20810-93	128	50137-2	128	143798	128	910088	128	11166595
128	21201	128	50200-12	128	145258	128	910634	128	11253441
128	21221	128	50202-12	128	147355	128	916034	128	11322253
128	21290	128	50202-24	128	150045	128	922114	128	11718319
128	21076A	128	51194	128	150046A	128	922125	128	13104560
128	23648	128	51194-01	128	150095	128	928291-101	128	17771400
128	25566-21	128	51194-02	128	150730	128	928291-102	128	18151417
128	25566-50	128	51194-03	128	150787	128	932017-0001	128	19901403
128	25566-63	128	53203-72	128	150796	128	932055-1	128	22130009
128	25566-76	128	57001-01	128	161919-29	128	960494-1	128	23114195
128	26664-1	128	59988-1	128	162002-081	128	960494-2	128	23114252
128	27125-110	128	60031	128	162002-082	128	970108	128	23200596-1
128	27125-380	128	60091	128	165735	128	982300	128	24551302
128	28977	128	60106	128	165736	128	984196	128	36001200
128	030011	128	60194	128	168660	128	984229	128	43020731
128	030011-1	128	60228	128	171039	128	987030	128	43021415
128	030011-2	128	60294	128	171044(SEARS)	128	988993	128	43023223
128	34044	128	60335	128	171162-163	128	993624-2	128	43024879
128	34588A	128	60417	128	175007-277	128	995928 1	128	43026284
128	35210	128	60423	128	181515-4	128	1288055	128	43040313
128	35212	128	60428	128	181515-6	128	1417306-1	128	50210800-10
128	35303	128	60597	128	181515-7	128	1417325-1	128	50211900
128	35383	128	60659	128	181515-9	128	1417338-5	128	51122870
128	35888	128	60677	128	186342A	128	1417342-1	128	51122880
128	36145	128	60680	128	194243	128	1417344-1	128	55440063-001
128	36387	128	60682	128	196779-9	128	1417345-2	128	62054298
128	36466	128	60697	128	196779-9-1	128	1417381-2	128	62104759
128	36579	128	60700	128	196780-1	128	1471120-14	128	62208945
128	36682	128	60703	128	198014-1	128	1471123-2	128	62389494
128	36748	128	60720	128	198020-1	128	1471123-3	128	62530915
128	37280	128	60994	128	198020-2	128	1471123-4	128	62638265
128	37287	128	61209	128	198020-3	128	1471123-5	128	62691298
128	37393	128	61219	128	198023-2	128	1473508-1	128	65372000
128	37445	128	61275	128	198035-1	128	1473545-1	128	67283700
128	37464	128	61359	128	198035-3	128	1473547-1	128	80052201
128	37649	128	61970/4560	128	198045-4	128	1473553-1	128	80052301
128	37694	128	61538	128	198047-1	128	1473555-3	128	80337390
128	37741	128	61562	128	198047-2	128	1473555-I	128	80421980
128	37767	128	61667	128	198047-3	128	1473560-1	128	89941008
128	37800	128	61733	128	198047-5	128	1473560-2	128	93073350
128	37806	128	61828	128	198047-6	128	1473561-1	128	93097120
128	37840	128	61841	128	198048-1	128	1473565-001	128	94742308
128	37847	128	61917	128	198048-2	128	1473565-1	128	95522800
128	37899	128	62019	128	198072-1	128	1473566-1	128	131000561
128	37975	128	62185	128	198077-1	128	1473573-2	128	346190001
128	37982	128	62192	128	200252	128	1473580-1	128	436006101
128	38045	128	62243	128	202913-057	128	1473593-1	128	485137002
128	38058	128	62398	128	204201-001	128	1473608-002	128	2003162700
128	38182	128	62404	128	215075	128	1473608-1	128	6622000100
128	38270	128	62446	128	218511	128	1473608-2	128	8001200004
128	38271	128	62452	128	221158	128	1473608-3	128	16105191229
128	38334	128	62540	128	230214	128	1473613-5	128	16108290536
128	38354	128	62612	128	230233	128	1473615-1	128	16156197229
128	38361	128	63900-229	128	231375	128	1473622	128	16356179229
128	38398	128	66007-4	128	232841	128	1473622-002	128(2)	LM1501H
128	38424	128	67001	128	233735	128	1473622-2	128(2)	023
128	38432	128	67003	128	233944	128	1473625-001	128/400	024
128	38468	128	70087-31	128	234024	128	1473625-1	128/400	28023
128	38475	128	000073110	128	236287	128	1473679-1	128/400	28024
128	38476	128	75145-3	128	239612	128	1502059	128/401	19-020-050
128	38495	128	75561-00	128	262417	128	1815153	128/427	280960
128	38497	128	75561-32	128	262417-2	128	1815154-9	128/427	280960A
128	38551	128	80902-1	128	267704	128	1815156	128/427	280960B
128	38588	128	80904-1	128	276160	128	1815157	128/427	280960B
128	38659	128	87532	128	276413	128	1815159	128/427	28D205
128	38716	128	88860-145	128	276415	128	1817006	128/427	28D219F
128	38725	128	88803	128	301606	128	1820829	128/427	28D220F
128	38735	128	88803-2-1	128	310110	128	1944313A1	128/427	Z3-5037
128	38736	128	88803-3-1	128	315930	128	1960083-1	1288	314-9020-1
128	38837	128	91271	128	315932	128	1967799	129	A-1853-0041-1
128	38869	128	91272	128	321145	128	1967799-1	129	A054-109
128	38916	128	91273	128	321166	128	1967801	129	A1016
128	39097	128	91411	128	325101	128	1968959	129	A116084
128	39231	128	94041	128	331383	128	1971489	129	A116284
128	39238	128	94042	128	340866-2	128	2000752-80	129	A119983
128	39248	128	94050	128	346015-23	128	2000334-155	129	A126724
128	39252	128	94051	128	346015-30	128	2006436-40	129	A1473549-1
128	39255	128	94052	128	346016-16	128	2006623-148	129	A1473616-1
128	39311	128	94066	128	346016-17	128	2097013-0702	129	A170
128	39329	128	94070	128	346016-27	128	2320233	129	A2057B110-9
128	39443	128	94855-145-00	128	348446-1	128	2320233	129	A2057B112-9
128	39462	128	95220	128	00351980	128	2320647-1	129	A2057B114-9
128	39485	128	95226-1	128	400108	128	2320946	129	A2057B115-9
128	39486	128	95235	128	400909	128	2327022	129	A2057B116-9
128	39561	128	96481	128	405965-8A	128	2327292	129	A2057B121-9
128	39587	128	101435	128	425411-01	128	2327293	129	A2057B122-9
128	39617	128	102209	128	0440002-003	128	2327332	129	A2057B145-12
128	39705	128	104719	128	450826-1	128	2327403	129	A2057B163-12
128	39713	128	105180	128	481335	128	2469936-1	129	A2482
128	39835	128	110669	128	489751-037	128	2485076-2	129	A297074C11
128	39842	128	110699	128	489751-038	128	2485005-3	129	A297V073C001
128	39863	128	11127B	128	489751-129	128	2485077-2	129	A297V073C002
128	39864	128	112360	128	489751-144	128	2485077-3	129	A297V082B03
128	39868	128	112361	128	500879	128	2495529(ARVIN)	129	A29V082B03
128	39920	128	113942	128	505287	128	2498163	129	A2K9
128	39940	128	115300-1	128	511806	128	2545989-2	129	A28A550P
128	40053	128	115810P2	128	516598	128	2605022	129	A28A564P
128	40309	128	119278	128	531972	128	2608169-1	129	A30278
128	40309V1	128	11922	128	0573480H	128	2666307-1	129	A3525
128	40309V2	128	123243	128	0575517	128	2777301	129	A3S-9008-001
128	40311	128	123703	128	0573527	128	2928054-1	129	A3533
128	40314	128	123940	128	573532	128	3130053	129	A3533-1
128	40315	128	126703	128	0573557	128	3130092	129	A3616-1
128	40317	128	126721	128	602122	128	3403866-3	129	A36577
128	40320	128	126725	128	602909-7A	128	3404520-301	129	A4037764-2
128	40323	128	127376	128	603165	128	3412907-1	129	A40410
128	40326	128	127845	128	608122	128	3457633-1	129	A417-138
128	40360	128	129051	128	610107-2	128	3457633-2	129	A417-170
128	40361	128	130172	128	610148-1	128	3458267-1	129	A417-234
128	40366	128	130174	128	610148-3	128	3463099-1	129	A417-43
128	40392	128	132175	128	610213-1	128	3596116	129	A4478
128	40407	128	132327	128	618136-1	128	3700162	129	A497(JAPAN)
128	40409	128	132328	128	618197	128	3731418-2	129	A498(JAPAN)
128	40457	128	132500	128	619006-1	128	4450026-P5	129	A498Y
128	40458	128	133177	128	651956	128	5049911	129	A501
128	40461-2	128	133576	128	652231	128	5320612	129	A503
128	40501	128	134155	128	653406	128	6480006	129	A503-0
128	40539	128	134989	128	660070	128	7281806	129	A503-R
128	40539L	128	136424	128	660074	128	7311074	129	A503-Y
128	40539B	128	136696	128	660100	128	7311350	129	A503GR
128	40578	128	137241	128	696575-198	128	7570009	129	A504
128	40611	128	137648	128	699410-140	128	7576015-01	129	A504-R
128	40616	128	138035-001	128	702407-00	128	7601010		
128	40635			128	723020-41	128	7851956-01		

ECG	Industry Standard No.	ECG	Industry Standard No.	ECG	Industry Standard No.	ECG	Industry Standard No.	ECG	Industry Standard No.
129	A504-Y	129	BCY30	129	M9520	129	TI861R	129	2N4235
129	A504GR	129	BCY31	129	MA0401	129	TI861M	129	2N4236
129	A527	129	BCY32	129	MA0402	129	TI893M	129	2N4314
129	A528	129	BCY33	129	MH9460A	129	TM129	129	2N4404
129	A532	129	BCY34	129	MJ4645	129	TQ63	129	2N4405
129	A532A	129	BCT67	129	MM4005	129	TQ63A	129	2N4407
129	A532B	129	BCY78A	129	MM4006	129	TQ64	129	2N4412
129	A532C	129	BCY78B	129	MM4008	129	TQ64A	129	2N4412A
129	A532D	129	BCY96	129	MM4009	129	TR-04C	129	2N4414
129	A532E	129	BCY96B	129	MPS9460A	129	TR-165(OLSON)	129	2N4414A
129	A532F	129	BCY97	129	MPS9460H	129	TR-28	129	2N4890
129	A537	129	BCY97B	129	M3934	129	TR-8020	129	2N4928
129	A537A	129	BDY70	129	M87506Q	129	TRO2054-7	129	2N5022
129	A537AA	129	BFS96	129	MBJ7505	129	TR1000	129	2N5023
129	A537AB	129	BFS97	129	P1M	129	TR1002	129	2N5040
129	A537AC	129	BFT60	129	PMC-Q8-0280	129	TR1004	129	2N5041
129	A537AH	129	BFT61	129	Q5100A	129	TR1012	129	2N5042
129	A537B	129	BFT62	129	QA-11	129	TR22	129	2N5110
129	A537C	129	BFT79	129	QA-17	129	TR22A	129	2N5111
129	A537H	129	BFT80	129	QA-9	129	TR2327743	129	2N5160
129	A546	129	BFT81	129	R2270-76963	129	TR23A	129	2N5242
129	A546A	129	BFW44	129	RCA1A02	129	TR8037	129	2N5243
129	A546B	129	BFW91	129	RCA1A05	129	TV-29	129	2N5322
129	A546E	129	BFX38	129	RCA1A08	129	TV29	129	2N5821
129	A546H	129	BFX39	129	RCA1A19	129	TV828A546	129	2N5834
129	A551	129	BFX40	129	R818	129	TV828C1256	129	2N5865
129	A551C	129	BFX41	129	R88100	129	TV828C1256HG	129	28021
129	A551D	129	BFX74A	129	R88102	129	V180	129	28022
129	A551E	129	BFX87	129	R88104	129	VS-C81256HG	129	28023
129	A552	129	BFX88	129	R88106	129	VS28A844-D/-1	129	28024
129	A560	129	BFY64	129	R88108	129	VS28C1741-1	129	2S3010
129	A571	129	BSB17	129	R88110	129	W15	129	2S302
129	A594	129	BSB18	129	R88112	129	XA-495C	129	2S302A
129	A594-0	129	BSB44	129	RT5230	129	XA492D	129	2S303
129	A594-R	129	BSV15	129	RV1472	129	XA495D	129	2S304
129	A594-Y	129	BSV16	129	RV2069	129	Y56601-46	129	2SA119
129	A595C	129	BSV82	129	S-320P	129	Y56601-76	129	2SA497
129	A604	129	BSV83	129	S-437	129	ZDP-D22-69 54	129	2SA497R
129	A606	129	BSW23	129	S-437P	129	ZEN101	129	2SA498
129	A606S	129	BSW40	129	S1430	129	ZT210	129	2SA498Y
129	A6532921	129	BSX40	129	S1431	129	ZT211	129	2SA501
129	A708	129	BSX41	129	S1520	129	ZT286	129	2SA503
129	A708A	129	BTX-071	129	S1698	129	ZTX500	129	2SA503-0
129	A708B	129	BTX-097	129	S1863	129	002-010300-6	129	2SA503-R
129	A708C	129	C106(PNP)	129	S1983	129	002-010700	129	2SA503-Y
129	A717	129	C112(PNP)	129	S2117	129	002-012600	129	2SA503GR
129	A756	129	C201	129	S2274	129	002-9800-12	129	2SA504
129	A75	129	C301A(PNP)	129	S2368	129	2A12	129	2SA504-0
129	A800-511-00	129	C302(PNP)	129	S2370	129	2N1034	129	2SA504-R
129	A800-516-00	129	C41001	129	S2398C	129	2N1035	129	2SA504-Y
129	A8015613	129	C9080	129	S24594	129	2N1036	129	2SA504GR
129	A815518S	129	C9081	129	S24597	129	2N1037	129	2SA532
129	A815185E	129	CDC-9000-1B	129	S24612	129	2N1197	129	2SA532A
129	A832P2B	129	CDC9002	129	S24612A	129	2N1228	129	2SA532B
129	A880-250-107	129	CB4002D	129	S24615	129	2N1229	129	2SA532C
129	A94065	129	CJ5209	129	S2771	129	2N1230	129	2SA532D
129	AEX-85715	129	CS-6228G	129	S2991	129	2N1231	129	2SA532E
129	AMP2970-2	129	CS1237	129	S2993	129	2N1232	129	2SA532F
129	AR304	129	CS1256H	129	S2994	129	2N1233	129	2SA537
129	AR304(RED)	129	CS1312G	129	S2995	129	2N1254	129	2SA537AA
129	AR308	129	CS1369	129	S3012	129	2N1255	129	2SA537AB
129	AT2848	129	CS1465R	129	S3586	129	2N1258	129	2SA537AH
129	AT394	129	CS7228G	129	S33886	129	2N1259	129	2SA537B
129	AT395	129	CS89021HF	129	S33886A	129	2N1275	129	2SA537H
129	AT396	129	CS9102B	129	S504-0	129	2N1429	129	2SA537C
129	AT397	129	DDBY008001	129	S80C121	129	2N1606	129	2SA537H
129	AT398	129	DS-86	129	SCI444204037	129	2N1623	129	2SA544
129	AT460	129	E2498	129	SCI444291004	129	2N1654	129	2SA546
129	AT461	129	EA0086	129	SDT3321	129	2N1922	129	2SA546A
129	AT462	129	EA0087	129	SDT3322	129	2N2591	129	2SA546B
129	AT463	129	EA15X194	129	SDT3501	129	2N2592	129	2SA546H
129	AT464	129	EL264	129	SDT3502	129	2N2593	129	2SA551
129	AT465	129	EN2905	129	SDT3503	129	2N2800	129	2SA551C
129	AT466	129	EN3502	129	SB8540	129	2N2801/46	129	2SA551D
129	AT467	129	EP15X4	129	SB8541	129	2N2904	129	2SA551E
129	AT468	129	ES51(ELCOM)	129	SB8542	129	2N2904A	129	2SA552
129	AT480	129	ET15X33	129	SFD-23	129	2N2905	129	2SA560
129	AT481	129	ET15X38	129	SJ2031	129	2N2905A	129	2SA581
129	AT482	129	ET15X39	129	SK16510006-2	129	2N3039	129	2SA594
129	AT483	129	ET8-069	129	SK16510006-4	129	2N3040	129	2SA594-0
129	AT484	129	ET8-071	129	SK3025	129	2N3120	129	2SA594-R
129	AT485	129	F3549	129	SKA1079	129	2N3133	129	2SA594-Y
129	AT5156	129	PC81795D	129	SKA4621	129	2N3134	129	2SA595C
129	B5493957-4	129	FD-1029-JP	129	SL3101	129	2N3202	129	2SA604
129	B5493957-5	129	FD-1029-ML	129	SL3111	129	2N3203	129	2SA606
129	B5493957-6	129	FD-1029-RB	129	SM2718	129	2N3208	129	2SA606S
129	BC139	129	G03-404-B	129	SN3987	129	2N3244	129	2SA612
129	BC143	129	G03-404-C	129	SM6728	129	2N3245	129	2SA708
129	BC160	129	GE-21A	129	SMC449077	129	2N327B	129	2SA708A
129	BC160-10	129	GE-244	129	S080121	129	2N328	129	2SA708B
129	BC160-16	129	HA7530	129	SPS-0121	129	2N328A	129	2SA708C
129	BC160-6	129	HA7531	129	SPS-29	129	2N328B	129	2SA717
129	BC161	129	HA7630	129	SPS-4076	129	2N329	129	2SA734
129	BC161-10	129	HA7631	129	SPS-4078	129	2N329A	129	2SA736
129	BC161-16	129	HA7632	129	SPS0121	129	2N329B	129	2SB810
129	BC161-6	129	HA9500	129	SPS2226	129	2N330	129	2SB510S
129	BC287	129	HA9501	129	SPS4010	129	2N330A	129	3-041
129	BC303	129	HA9502	129	SPS4301	129	2N3343	129	3-215
129	BC311	129	HEP-83012	129	SPS4310	129	2N3344	129	3-7
129	BC313	129	HEP-85013	129	SPS4312	129	2N3345	129	3L4-6010-4
129	BC560-10	129	HEP-85023	129	SPS4462	129	2N3467	129	3L4-6010-8
129	BC560-16	129	HEP242	129	SPS4477	129	2N3468	129	3L4-6011-02
129	BC560-6	129	HEP51	129	SPS4492	129	2N3494	129	4-30203845
129	BC361-10	129	HEP710	129	SPS4497	129	2N3503	129	4-3025222
129	BC361-6	129	HEPS0012	129	SP86125	129	2N3505	129	4-686170-3
129	BC396	129	HP10	129	SP-021660	129	2N3671	129	4-686229-3
129	BC404A	129	HP12(PHILCO)	129	ST72039	129	2N3677	129	4-686230-3
129	BC404V1	129	HS40032	129	ST72040	129	2N3719	129	4-686235-3
129	BC461-4	129	HT104971A-0	129	STC5610	129	2N3720	129	4-686238-3
129	BC461-5	129	HT104971A0	129	STC5611	129	2N3762	129	006-0004956
129	BC461-6	129	IP20-0217	129	STC5612	129	2N3763	129	006-0005191
129	BC477	129	IRTR88	129	STX0011	129	2N3764	129	6-9029-20J
129	BC477A	129	J241015	129	SX61	129	2N3774	129	7-0012-00
129	BC477V1	129	J24908	129	T-340	129	2N3775	129	7-14A
129	BC527	129	KO71818-001	129	T-482	129	2N3776	129	7-3(STANDEL)
129	BC528	129	KO71962-001	129	T1275	129	2N3779	129	7-4(STANDEL)
129	BC534	129	K1181	129	T1276	129	2N3780	129	8-0051600
129	BC727	129	KD2120	129	T1808	129	2N3782	129	8-0052700
129	BC728	129	KE1007-0004-00	129	T1808A	129	2N3867	129	8-2410300
129	BCW79-25	129	KLH5808	129	T1808B	129	2N3868	129	8-905-706-545
129	BCW80-25	129	LDA450	129	T1808C	129	2N4026	129	8-905-713-810
129	BCX10	129	LN76963	129	T1808D	129	2N4027	129	09-300043
129	BCY17	129	M447B	129	T1808E	129	2N4028	129	09-305075
129	BCY18	129	M652P1C	129	T246	129	2N4029	129	09-305134
129	BCY19	129	M75561-17	129	T354-2	129	2N4030	129	9-5226-003
129	BCY21	129	M75561-8	129	T459	129	2N4031	129	9-5226-1
129	BCY22	129	M828	129	T475	129	2N4032	129	9-5226-3
129	BCY23	129	M9145	129	TI808E	129	2N4033	129	12-100027
129	BCY24	129	M9257	129	TIS61A	129	2N4036	129	13-14085-92
129	BCY25	129	M9308	129	TIS61B	129	2N4037	129	13-28386-1
129	BCY26	129	M9400	129	TIS61C	129	2N4234	129	13-28394-4
129	BCY27	129	M9426	129	TIS61D	129		129	13-28394-4A
129	BCY28	129	M9432						
129	BCY29	129	M9435						

ECG	Industry Standard No.	ECG	Industry Standard No.	ECG	Industry Standard No.	ECG	Industry Standard No.	ECG	Industry Standard No.
129	13-28394-5	129	57B115-9	129	344-6012-1	129	36577	129	810002-733
129	13-28394-5A	129	57B115-9A	129	344-6012-3	129	37269	129	815185
129	13-28394-6	129	57B116-9	129	344-6014-1B	129	37664	129	815185E
129	13-4800869145	129	57B116-9A	129	353-9008-001	129	37740	129	970107
129	14-0086-1	129	57B148-1	129	353-9301-004	129	37764	129	970762-6
129	14-0104-3	129	57B148-10	129	353-9304-004	129	37793	129	986030
129	14-602-28	129	57B148-11	129	353-9317-001	129	37918	129	986931
129	14-602-28A	129	57B148-12A	129	380-0171-000	129	37966	129$	993570-4
129	14-602-60	129	57B148-2	129	386-7184P1	129	38388	129	126195-191
129	14-602-66	129	57B148-3	129	386-7254-P202	129	38458	129	1471112-12
129	14-602-73	129	57B148-4	129	393-1(SYLVANIA)	129	38496	129	1471112-3
129	14-602-79	129	57B148-5	129	404B(NCR)	129	38654	129	1471112-8-9
129	14-602-79A	129	57B148-6	129	417-111	129	38734	129	1473516-1
129	14-602-85	129	57B148-7	129	417-138	129	38737	129	1473592-1
129	14-607-29	129	57B148-8	129	417-170	129	38870	129	1473616-1
129	14-607-29A	129	57B148-9	129	417-181	129	39114	129	1473666-1
129	17-458	129	57B163-12	129	417-234	129	39250	129	1827322
129	19-020-100	129	57B163-12A	129	417-255	129	39440	129	1965016
129	19A115180-2	129	57B168-9	129	417-260	129	39618	129	2004746-116
129	19A115976P1	129	57B171-9	129	417-43	129	39619	129	2004746-117
129	020-1110-014	129	57C110-9	129	417-822-13262	129	39853	129	2006436-37
129	020-1111-017	129	57C120-9	129	422-0739	129	39865	129	2096700
129	20A0075	129	57C148-12	129	422-1232	129	40319	129	2096700-TM18
129	021-0224-00	129	57C148-12A	129	422-1405	129	40319B	129	2320242
129	21A040-049	129	57C23	129	422-2008	129	40362	129	2320243
129	21A040-050	129	57C23-1	129	430-203845	129	40406	129	2327282
129	21A112-002	129	57C23-2	129	430-23222	129	40406B	129	2327283
129	24MW727	129	57C23-3	129	433M852	129	40408	129	2412949-0001
129	026-100019	129	5706-15	129	461-1014	129	40410	129	3596340
129	29A	129	5706-26	129	461-1048	129	40537	129	3596341
129	030-007-0	129	5706-26A	129	610-083	129	40537L	129	3700163
129	32K64	129	5706-31	129	610-083-1	129	40537B	129	3700258
129	33-00742	129	5706-31A	129	610-083-2	129	40538	129	3731418-1
129	34-6001-86	129	5706-33	129	610-083-3	129	40538L	129	3731418-3
129	34-6016-12	129	57M1-13	129	631-1	129	40538B	129	3731418-4
129	34-6016-23	129	57M1-13A	129	631-3(SYLVANIA)	129	40595VX	129	04450016-003
129	34-6016-23A	129	57M1-21	129	690V086H89	129	40654	129	5320111
129	34-6016-32A	129	57M1-21A	129	690V110H55	129	41052	129	5321253
129	039	129	57M1-25	129	700-136	129	41503	129	7011515
129	41-0500	129	57M1-25A	129	762-105-00	129	43089	129	7012411
129	41-0500A	129	57M2-10	129	762-120	129	43107	129	7026019
129	42-19643	129	57M2-10A	129	774CL	129	43116	129	7570032-01
129	42-20739	129	57M2-15	129	800-101-114-1	129	59989-1	129	8015613
129	42-21292	129	57M2-15A	129	800-284	129	60154	129	8508309
129	42-9029-40U	129	61-309688	129	800-51-1-00	129	60701	129	9341834
129	42-9029-60W	129	000062	129	800-516-00	129	60947	129	13040429
129	42-9029-70Q	129	66-P11139	129	800-51600	129	61009-9	129	13104561
129	43-0203845	129	66-P11141	129	800-525-03	129	61009-9-2	129	24733362
129	43-025222	129	698P112	129	800-52700	129	61011-0	129	43021168
129	44A358624-003	129	74P1M	129	824-10300	129	61011-0-1	129	43023222
129	44A391505	129	77-271818-1	129	880-250-107	129	61013-4-1	129	43024880
129	46-86403-3	129	79P114-2A	129	881-250-107	129	61244	129	43026285
129	46-86406-3	129	79P114-4	129	921-110B	129	61371/4561	129	62208953
129	46-86425-3	129	79P114-4A	129	921-182B	129	61666	129	62278566
129	46-86517-3	129	80-051600	129	921-270B	129	61774	129	62393696
129	46-86565-3	129	80-052700	129	921-315B	129	61937	129	62593660
129	46-86574-3	129	80P2	129	921-47B	129	62204	129	62638257
129	48-03-10111102	129	80P2B	129	964-20739	129	62584	129	80050600
129	48-03-10744802	129	80P3	129	964-2200B	129	62708	129	80051600
129	48-134478	129	80P3B	129	978-1923	129	70158-9-00	129	80052700
129	48-134478A	129	81P3	129	991-01-1315	129	70260-19	129	82410300
129	48-134951	129	82-410300	129	991-01-2686	129	70260-29	129	89946008
129	48-134951A	129	83	129	991-2P	129	071818	129	131005808
129	48-65177A77	129	83P2B	129	1002A-1	129	75700-13-01	129	430203845
129	48-65177A77A	129	86-329-2	129	1004P	129	84001	129	436006201
129	48-869145	129	86-334-2	129	1011M57P01	129	84001A	129	65240011-00
129	48-869257	129	86-431-2	129	1011M62P01	129	84001B	129	16256197228
129	48-869308	129	86-431-9	129	1043-1278	129	86257	129$	A597
129	48-869400	129	86-5064-2	129	1080G	129	94063	129$	28A571
129	48-869426	129	86-5064-2A	129	1184G	129	94064	129$	28A597
129	48-869432	129	86-5082-2	129	1414-185	129	94067	129(2)	LM150ZH
129	48-869435	129	86-5082-2A	129	1414-187	129	94068	129/427	A547
129	48-869520	129	86-616-2	129	1428	129	95239-1	129/427	28A547
129	48-869681	129	95-226-003	129	1853-0041	129	111945	129/427	28A547A
129	48-90232A06	129	95-226-1	129	1853-0045	129	116084(RCA)	129/427	28A607A
129	48-90232A06A	129	95-226-3	129	1853-0215	129	116284	129/427	28A607B
129	48-90232A09	129	96-5115-01	129	1889-17	129	119298-001	129/427	28A607C
129	48-90232A09A	129	96-5176-01	129	2000-202	129	119983	129/427	28A607D
129	48-90232A12	129	96-5209-01	129	2057A2-150	129	123944	129/427	28A607K
129	48-90232A12A	129	96-5220-01(PNP)	129	2057B110-9	129	124616	129/427	28A607L
129	48-90232A15	129	96-5230-01	129	2057B112-9	129	125142	129/427	28A607M
129	48-90232A15A	129	99P117	129	2057B114-9	129	125707	129/427	23-5045
129	48-97046A36	129	99PJ	129	2057B115-9	129	126724	130	A-11166527
129	48-97046A39	129	99PJ3C	129	2057B116-9	129	130404-29	130	A-1203527
129	48R869145	129	998073	129	2057B121-9	129	131242-12	130	A-140605
129	48R869257	129	0101-439	129	2057B122-9	129	136423	130	A-1854-0291-1
129	48R869432	129	106K80	129	2057B145-12	129	147112-7	130	A-1854-0294-1
129	48R869520	129	106KBA	129	2057B163-12	129	147357-0-1	130	A-6-67703
129	48R969681	129	112-2	129	2482(RCA)	129	147357-4-1	130	A-6-67703-A-7
129	50-40205-09	129	112-2A	129	2502	129	147359-0-1	130	A054-154
129	50A103	129	121-1007	129	3222	129	157004	130	A08-105018
129	53D166	129	121-603	129	3523(RCA)	129	171162-193	130	A112363
129	53D170	129	121-746	129	3533(RCA)	129	188226	130	A13-0032
129	55-643	129	121-765	129	3533-1	129	198065-1	130	A13-17918-1
129	57A108-6-8	129	121-774CL	129	3574-1	129	198065-3	130	A13-23594-1
129	57A110-9	129	121-845	129	3590	129	198074-1	130	A13-33188-2
129	57A112-9	129	121-879	129	3592(RCA)	129	198078-1	130	A14-601-10
129	57A112-9A	129	121-952	129	3616-1(RCA)	129	202917-137	130	A14-601-12
129	57A114-9	129	0131-001427	129	3746-00	129	203718	130	A14-601-13
129	57A114-9A	129	0131-001428	129	3746-01	129	218537	130	A18-4
129	57A115-9	129	131-005-808	129	4219	129	233969	130	A241B
129	57A115-9A	129	131-005808	129	4367-001	129	236433	130	A2E
129	57A116-9	129	131-045-61	129	4478	129	240402	130	A2E-2
129	57A116-9A	129	131-04561	129	4686-170-3	129	241052	130	A2EBLK
129	57A122-9	129	132-007	129	4686-229-3	129	242422	130	A2EBRN
129	57A130-9	129	132-032	129	4686-230-3	129	242460	130	A2EBRN-1
129	57A132-10	129	132-039	129	4686-235-3	129	242958	130	A2B
129	57A148-10	129	134P1AA	129	4686-238-3	129	262638	130	A2B-3
129	57A148-11	129	13472	129	5680	129	268717	130	A3902441
129	57A148-12A	129	151-0208	129	6285	129	297074C11	130	A3I4-6001
129	57A148-2	129	151-0208-00-AA	129	7301-1	129	302865	130	A3I4-6001-01
129	57A148-3	129	154A5946-667	129	7508	129	319304	130	A3TB120
129	57A148-4	129	158P1M	129	8000-004-P089	129	325077	130	A3TB230
129	57A148-5	129	171-9	129	8303	129	335613	130	A3TB240
129	57A148-6	129	174-25566-01	129	8400-1	129	386726-1	130	A3TX003
129	57A148-7	129	174-25566-62	129	8400-1A	129	405965-35A	130	A3TX004
129	57A148-8	129	177-001	129	8400-1B	129	5140678	130	A3U
129	57A148-9	129	177-001-9-001	129	8624-003	129	5140728	130	A3U-4
129	57A23	129	177-025-9-002	129	9405-2	129	564671	130	A417033
129	57A23-1	129	183P1	129	09800-12	129	0573542	130	A43023843
129	57A23-2	129	226-3	129	10035-001	129	0573559	130	A4J
129	57A23-3	129	260P13704	129	10300-12	129	0573560	130	A4JBRN
129	57A6-15	129	297V073001	129	12048-0011	129	602909-3A	130	A4JRED-1
129	57A6-26	129	297V073C02	129	012099-1	129	610099	130	A4S
129	57A6-26A	129	297V074C11	129	16239	129	610099-1	130	A4B-1
129	57A6-31	129	297V080C01	129	018069	129	610099-2	130	A4Z
129	57A6-31A	129	297V082B02	129	20739	129	610099-5	130	A522
129	57A6-33	129	297V082B03	129	23114-050	129	610110	130	A522-3
129	57A6-33A	129	297V083003	129	23114-051	129	610129-1	130	A572
129	57B110-9	129	303-2	129	25566-62	129	610158-1	130	A572-1
129	57B112-9	129	317-0139-001	129	30278	129	650175	130	A6L
129	57B112-9A			129	31006	129	681266	130	A6LBLK
129	57B114-9			129	31032-0	129	681266-1	130	A6LBLK-1
129	57B114-9A			129	033571	129	723043-1		

ECG	Industry Standard No.	ECG	Industry Standard No.	ECG	Industry Standard No.	ECG	Industry Standard No.	ECG	Industry Standard No.
130	A6LBRN	130	D12	130	S2403B	130	2N6254	130	44A355565-001
130	A6LBRN-1	130	D146UK	130	S2403C	130	2N6571	130	44A355565001
130	A6LRED	130	D163	130	S2741	130	2N6569	130	44A390244-001
130	A6LRED-1	130	D164	130	S305	130	28033	130	44A417035-001
130	A6N	130	D172	130	S305A	130	28034	130	44A417716
130	A6N-6	130	D211	130	S305D	130	28C1667	130	44A417716-001
130	A7-12	130	D212	130	S355	130	28221	130	44-86588-3
130	A7-13	130	D3005VN	130	S356	130	28C240	130	48-03-04093403
130	A7M	130	D341	130	S3771	130	28C241	130	48-03-041840-2
130	A80052402	130	D341H	130	S93SE133	130	28C242	130	48-134701
130	A80414120	130	D41	130	S93SE165	130	28C242D	130	48-134701A
130	A80414130	130	D68	130	SDT9201	130	28C664	130	48-134715
130	A8P	130	D69	130	SDT9205	130	28C664B	130	48-134715A
130	A8U	130	DD-79D107-1	130	SDT9206	130	28C664C	130	48-134882
130	A8W	130	DP-2	130	SDT9210	130	28C7	130	48-134884
130	AEX79846	130	DS-509	130	SDT9261	130	28C793	130	48-134969
130	AEX9846	130	DS-514	130	SDT9301	130	28C793BL	130	48-137008
130	AM3235	130	DS509	130	SDT9302	130	28C793R	130	48-137008A
130	AMP-121	130	DS519	130	SDT9303	130	28C793T	130	48-137027
130	AMP104	130	DT4011	130	SDT9304	130	28C794R	130	48-137027A
130	AMP105	130	DT4110	130	SDT9305	130	28C889	130	48-137036
130	AMP115	130	DT4111	130	SDT9306	130	28I2	130	48-137036A
130	AMP116	130	DT4120	130	SDT9307	130	28D163	130	48-137053
130	AMP117	130	DT4121	130	SDT9308	130	28D164	130	48-137053A
130	AMP117A	130	EA15X100	130	SDT9309	130	28D172	130	48-137079
130	AMP118	130	EA1740	130	SE-3033	130	28D172A	130	48-137079A
130	AMP118A	130	EC961	130	SE3033	130	28D172B	130	48-137175
130	AMP119	130	ES316(ELCOM)	130	SE3035	130	28D172C	130	48-137175A
130	AMP119A	130	ES31(ELCOM)	130	SE3036	130	28D173	130	48-137180
130	AMP120	130	ET8-003	130	SE9002	130	28D173A	130	48-137180A
130	AMP120A	130	FBN-36220	130	SE9080	130	28D173B	130	48-137251
130	AMP201	130	FBN-36485	130	SES632	130	28D173C	130	48-137333
130	AMP210	130	FBN-36603	130	SES881	130	28D174	130	48-137344
130	AMP210A	130	FS2003-1	130	SJ1106	130	28D174P	130	48-137368
130	AMP2919-2	130	G23-45	130	SJ1470	130	28D175	130	48-15
130	AR15	130	G23-67	130	SJ2000	130	28D175B	130	48-869244
130	AR15-L8-0026	130	GE-14	130	SJ2008	130	28D175F	130	48-869259
130	A2-10	130	GB-19	130	SJ3464	130	28D175M	130	48-869278
130	A2-1856	130	GRA88-R2982	130	SJ3604	130	28D176	130	48-869302
130	A21856	130	HEP-87002	130	SJ3678	130	28D211	130	48-869321
130	A23260	130	HEP-87004	130	SJ619	130	28D211M	130	48-869515
130	ATC-TR-15	130	HEP247	130	SJ619-1	130	28D211Y	130	48K869515
130	B-12822-2	130	HEP704	130	SJ820	130	28D212	130	48K869244
130	B133550	130	HF22	130	SJ8701	130	28D341	130	48K869321
130	B133577	130	HT30494	130	SJ9190	130	28D341H	130	48K869302
130	B133684	130	HT304941	130	SK3027	130	28D41	130	48B137344
130	B133685	130	HT304941X	130	SK3510	130	28D68D	130	48B155053
130	B170000	130	HT304942X	130	SK5601-01	130	28D69	130	48B159067
130	B170000-ORG	130	HT401191	130	SP4231	130	3L4-6001-01	130	500401
130	B170000-RED	130	HT401191A	130	SPD-80062	130	3TE120	130	57-4018-60
130	B170000BLK	130	HT401191B	130	SPT3713	130	3TE140	130	57B175-9
130	B170000BRN	130	IR-TR59	130	STC-1035	130	3TE230	130	57B175-9A
130	B170001	130	K071964-001	130	STC-1035A	130	3TE240	130	57M3-1
130	B170001-BLK	130	K4-525	130	STC-1036	130	32X003	130	57M3-1A
130	B170001-BRN	130	KB-1007	130	STC-1036A	130	32X004	130	57M3-2
130	B170001-ORG	130	KS-19938	130	STC-1085	130	4-0294	130	57M3-4
130	B170001-RED	130	KSD1051	130	STC1035	130	4-0563	130	57M3-4A
130	B170001BLK	130	KSD1055	130	STC1035A	130	4-3023843	130	57M3-6
130	B170001BRN	130	KSD1056	130	STC1036	130	6-137	130	57M3-6A
130	B170002	130	LN75497	130	STC1036A	130	6-138(PWR)	130	61-309449
130	B170002-ORG	130	M18-12795B	130	STC1080	130	6-6490004	130	63-11991
130	B170002-RED	130	M4715	130	STC1081	130	007-0040-00	130	63-11991A
130	B170003	130	M4882	130	STC1082	130	7-12(STANDEL)	130	63-8707
130	B170004	130	M7543-1	130	STC1083	130	7-13(STANDEL)	130	63-8707A
130	B170005	130	M75549-2	130	STC1084	130	007-7450301	130	68A8319001
130	B170006	130	M9244	130	STC4252	130	8-0052402	130	070
130	B170007	130	M9259	130	STC4253	130	8-0414120	130	80-052402
130	B170009	130	M9278	130	STC4254	130	8-0414130	130	80-414120
130	B170010	130	M9302	130	STC4255	130	8-905-706-555	130	80-414130
130	B170011	130	M9321	130	STX0014	130	8-905-706-556	130	085
130	B170012	130	M9515	130	STX0027	130	8-905-706-557	130	86-5084-2
130	B170013	130	MHT7601	130	STX0032	130	8-905-713-101	130	86-5084-2A
130	B170014	130	MHT7602	130	T-Q5105	130	8-905-713-556	130	86-5101-2
130	B170015	130	MHT7607	130	TA2577A	130	09-302122	130	86-5101-2A
130	B170016	130	MHT7607	130	TA7068	130	9TR91001-09	130	86-5112-2
130	B170018	130	MHT7608	130	TA7069	130	10-13-002-003	130	86-665-2
130	B170019	130	MHT7609	130	TM130	130	10-13-002-004	130	86A332
130	B170020	130	MJ2801	130	TNJ72148	130	10-13-002-3	130	93-8B-124
130	B170021	130	MJ2802	130	TQ-PD-3055	130	10-13-002-4	130	93B0165133
130	B170022	130	MJ480	130	TR-1000-7	130	10-13002-003	130	93B0165133A
130	B170024	130	MJ481	130	TR-1039-4	130	10-13002-004	130	96-5117-01
130	B170025	130	MJE2940	130	TR-26	130	10-374101	130	96-5117-91
130	B17307	130	N-121122	130	TR1039-4	130	10-13002-004	130	96-5162-03
130	B5020	130	N121122	130	TR1039-6	130	12-100047	130	96-5162-04
130	B66X0040-006	130	P-10954-1	130	TR1077	130	13-0032	130	96-5164-02
130	BD111	130	P-10954-1	130	TR1490	130	13-17918-1	130	96-5164-03
130	BD130	130	P-11901-1	130	TR1491	130	13-23594-1	130	96-5201-01
130	BD142	130	P10619-1	130	TR1492	130	13-33188-2	130	96-5207-01
130	BD181	130	P2271	130	TR1493	130	14-40325A	130	96-5285-01
130	BD182	130	P3139	130	TR26	130	14-40363A	130	10082
130	BD183ELK	130	P50200-11	130	TR26C	130	14-40369A	130	100T2A
130	BDX10	130	P5034	130	TR271TR26	130	14-40421A	130	100X2
130	BDF-10	130	P5149-	130	TS-1193-736	130	14-40464A	130	100X6
130	BDY10	130	PP-AR15	130	TVS28C647	130	14-40465A	130	100X6A
130	BDY11	130	PMC-QP0010	130	W16	130	14-40466A	130	104T2
130	BDY17	130	PMC-QP0012	130	X194-3005829A	130	14-40471	130	104T2A
130	BDY20	130	PN350	130	XA-1078	130	14-40934-1	130	111N2C
130	BDY38	130	PP3000	130	XA-1161	130	14-601-10	130	111N4
130	BDY39	130	PP3003	130	XC723	130	14-601-12	130	111N4A
130	BDY53	130	PP3006	130	XI-548	130	14-601-13	130	111N4B
130	BDY73	130	PT1941	130	XT-548A	130	14-601-15	130	111N4C
130	BLY10	130	PTC119	130	ZT1487	130	14-601-15A	130	111N6
130	BLY11	130	QP-11	130	ZT1488	130	14-601-16	130	112-203055
130	BNT133	130	QP-12	130	ZT1489	130	14-601-16A	130	112-363
130	BN7214	130	QP-8	130	ZT1490	130	15-03068	130	121-726
130	BRO-116	130	QP001200A	130	ZT1702	130	17-50(FISHER)	130	121-726A
130	BUY10	130	QP8	130	2-002	130	19A116761P1	130	128-9050
130	BUY11	130	R135-1	130	2-003	130	19A126813	130	130-146
130	C1667	130	R2270-75497	130	001-021270	130	19A126813A	130	131-001-007
130	C21	130	R2270-78399	130	181-0336-00-A	130	20-1111-002	130	131-001007
130	C240	130	R22707-8399	130	2-G-3055	130	20-1111-003	130	0131-001597
130	C241	130	R227075497	130	2C0198B	130	020-1111-008	130	0131-002068
130	C242	130	R227077499	130	2C010	130	020-1111-019	130	0131-004367
130	C664	130	R2982	130	2G3055	130	23-5035	130	131-043-67
130	C664B	130	R4569	130	2N1069	130	23-5038	130	131-043A67
130	C664C	130	RC-1700	130	2N1070	130	23-5041	130	0132
130	C793	130	RC1700	130	2N1422	130	025	130	132-070
130	C793BL	130	RC8242	130	2N1423	130	33-052	130	132-085
130	C793R	130	RE19	130	2N1703	130	34-1000	130	132-541
130	C793T	130	RT-154	130	2N2305	130	34-1000A	130	140N1C
130	C794R	130	S-305	130	2N3055	130	34-1028	130	140N2
130	C889	130	S-305-PD	130	2N3055-1	130	34-1028A	130	151-0140
130	CD461	130	S-305A	130	2N3226	130	34-6002-32	130	151-0140-00
130	CD461-014-614	130	S-356	130	2N3232	130	34-6002-32A	130	151-0336-00
130	CII-025-Q	130	S1691	130	2N3233	130	41-0318	130	151-0336-00-A
130	CP400	130	S1692	130	2N3448	130	41-0318A	130	151-0337-00
130	CP401	130	S1865	130	2N3863	130	42-20961	130	151-0337-00-A
130	CP404	130	S1905	130	2N3864	130	42-20961A	130	152N2
130	CP405	130	S1905A	130	2N4395	130	43-023843	130	156-042
130	CP406	130	S1907	130	2N4396	130	43A165137P1	130	156-042A
130	CP407	130	S1977634	130	2N6253	130	43A165137P3	130	156-043
130	CP408	130	S2003-1	130		130	43A165137P4	130	156-043A
130	CP409	130	S2241	130		130	43A212067	130	156-053
130	D116	130	S2392	130		130		130	156-063
130		130		130		130		130	161-2NC

ECG	Industry Standard No.	ECG	Industry Standard No.	ECG	Industry Standard No.	ECG	Industry Standard No.	ECG	Industry Standard No.
130	161-1NC	130	3055-3	130	70260-18	130	19020095	131	TNJ70483
130	161N1C	130	3146-977	130	78399	130	23114070	131	TNJ70541
130	161N4	130	3152-170	130	80807	130	23114070A	131	TR-184(OLSON)
130	161N4C	130	3499	130	91274	130	23114100	131	TR50
130	172-003-9-001	130	3665-2	130	94065	130	23114108	131	W17
130	172-003-9-001A	130	3685-1	130	94065A	130	27126100	131	1R9.1
130	172-024-9-005	130	3714H1	130	94094(EICO)	130	30200101	131	2N2835
130	173A4491-2	130	3714H1A	130	113875	130	43023843	131	2N4078
130	173A4491-2A	130	4216	130	124511	130	43024216	131	2N5893
130	173A4491-4	130	4491-4	130	130014	130	50221300-01	131	2N5897
130	173A4491-7	130	4491-7	130	133684	130	51126850	131	2N8130
130	174-20989-22	130	4701	130	133685	130	55440087-001	131	2SB130A
130	176-040-9-001	130	4715	130	138194	130	55440091-001	131	2SB367
130	180T2	130	4715A	130	146466-1	130	62737425	131	2SB367(A)
130	180T2A	130	4802-00005	130	162002-101	130	74140226-001	131	2SB367A
130	180T2B	130	4802-0005A	130	162002-71	130	74140226-002	131	2SB367B
130	180T2C	130	4822-130-40132	130	188165	130	74140226-003	131	2SB367C
130	181T2A	130	4882	130	198034-1	130	80051001	131	2SB367H
130	181T2C	130	5001-508	130	198034-2	130	80052402	131	2SB443
130	211-40140-18	130	5565-001	130	198034-3	130	80041420	131	2SB448
130	211A6580-1	130	6580-1	130	198034-4	130	80414130	131	2SB458
130	211AE8U3055	130	7214(LOWREY)	130	198034-9	130	131001007	131	2SB458A
130	229-1301-64	130	7214A	130	198039-0507	130MP	E835(ELCOM)	131	2SB463
130	231-0004	130	7320-1	130	198039-506	130MP	GE-15MP	131	2SB463BL
130	260D09301	130	7356-2	130	198039-507	130MP	8938E140	131	2SB463E
130	260P2A008	130	7514	130	198039-6	130MP	SK3029	131	2SB463R
130	297V061001	130	8000-00004-241	130	198039-7	130MP	TM130MP	131	2SB463Y
130	297V061001A	130	8000-00006-006	130	198079-2	130MP	W16MP	131	2SB473
130	297V061002	130	8000-00006-190	130	202909-827	130MP	8-0053702	131	2SB473A
130	297V061002A	130	8000-0005-008	130	204211-001	130MP	80-053702	131	2SB473B
130	0301-3055-00	130	8319-001	130	231378	130MP	132-542	131	2SB473C
130	332-2911	130	11236-3	130	232359	130MP	161N2	131	2SB473D
130	352-0583-011	130	12536	130	252359A	130MP	800-537-02	131	2SB473E
130	352-0677-010	130	12536A	130	241657	130MP	800-53702	131	2SB473F
130	352-0677-011	130	13002-3	130	267791	130MP	2057A100-16	131	2SB473H
130	352-0677-020	130	13002-4	130	309449	130MP	2057A100-26	131	2SB473M
130	352-0677-021	130	13298	130	318835	130MP	2057B100-16	131	2SB481
130	352-0677-030	130	16001	130	395253-1	130MP	171174-1	131	2SB481A
130	352-0677-031	130	16083	130	445023-P1	130MP	4090187-0502	131	2SB481B
130	352-0677-041	130	16176	130	489751-119	130MP	62573741	131	2SB481D
130	352-0677-051	130	16201	130	502349	130MP	80053702	131	2SB481F
130	352-0677-40	130	16230	130	505198	131	A061-116	131	2SB481G
130	378-44	130	16234	130	570029	131	A065-111	131	2SB481H
130	378-44A	130	16235	130	0573562	131	AD-152	131	2SB481J
130	386-40	130	16240	130	601190	131	AD-156	131	2SB481K
130	386-7183P1	130	16261	130	600115-413-001	131	AD152	131	2SB481L
130	386-7183P1	130	16266	130	610111-2	131	AD155	131	2SB481M
130	394-3127-4	130	16287	130	610111-4	131	AD156	131	2SB481X
130	394-3135A	130	16292	130	610140-1	131	AD162	131	2SB63
130	403-009/07	130	16299	130	610161-2	131	AD164	131	2SB80
130	417-139	130	16319	130	610161-4	131	AD169	131	4-142
130	417-139A	130	16320	130	618955-2	131	AD262	131	6-88(AUTOMATIC)
130	417-162	130	16338	130	649002	131	AD263	131	6A10229
130	417-212	130	23114-070	130	659174	131	B130	131	8-619-030-015
130	417-212A	130	027762	130	700080	131	B130A	131	8-905-605-607
130	417-215	130	27126-100	130	70008OA	131	B367	131	8-905-605-650
130	417-215-13286	130	28474	130	700083	131	B367(A)	131	8-905-613-240
130	417-215A	130	30276	130	700083A	131	B367A	131	8-905-613-241
130	417-273	130	030539-1	130	740306	131	B367B	131	8-905-613-242
130	417-275-13286	130	33188-2	130	793556-1	131	B367C	131	8-905-613-245
130	422-0961	130	34208	130	801537	131	B367H	131	8-905-613-265
130	430-23843	130	35001	130	815246-1	131	B413	131	8-905-613-266
130	445-0023	130	36545	130	882028	131	B448	131	8-905-613-277
130	445-0023-P1	130	36846	130	891032	131	B458	131	8-905-613-282
130	445-0023-P3	130	36855	130	984259	131	B458A	131	8-905-613-283
130	445-0023-P4	130	36892	130	984259A	131	B463	131	8-905-613-284
130	472-0946-001	130	36946	130	1289050	131	B463BL	131	8-905-613-555
130	472-0946-002	130	36953	130	1417320	131	B463E	131	8A10522
130	514-054214	130	037085	130	1417320-1	131	B463R	131	8A10625
130	542-1034	130	37267	130	1417356-2	131	B463Y	131	8A11083
130	571-844	130	37475	130	1471135-001	131	B473	131	8A11721
130	571-844A	130	37563	130	1471135-1	131	B473D	131	8A12359
130	576-0002-003	130	37663	130	1473665-2	131	B473F	131	8A13164
130	576-0040-251	130	37888	130	1473683-1	131	B473H	131	8P2202
130	686-0243	130	38137	130	1701790-1	131	B481D	131	8P404T
130	686-0243-0	130	38138	130	1702601-1	131	B481E	131	8P505
130	686-143	130	38166	130	1968977	131	B63	131	8P508
130	686-143A	130	38272	130	2003073-91	131	B80	131	09-301024
130	690V094B17	130	38474	130	2003073-91A	131	BCM1002-6	131	09-301034
130	700-113	130	38494	130	2057199-0700	131	C465(I)	131	09-301075
130	74202030-020	130	38626	130	2057199-0701	131	E24107	131	13-26377-1
130	800-510-00	130	38731(KALOP)	130	2057199-0703A	131	EA15X139	131	21A015-003
130	800-510-01	130	38897	130	2057199-701	131	EA15X154	131	21A015-022
130	800-524-02	130	39127	130	2057323-0500	131	EB10110	131	22-002001
130	800-524-02A	130	39148	130	2057323-0501	131	ES29(ELCOM)	131	22-002008
130	800-524-03	130	39213	130	2327052	131	EB50(ELCOM)	131	22-002009
130	800-524-03A	130	39251	130	2327053	131	EB7(ELCOM)	131	24MW994
130	800-524-04A	130	39369	130	3130058	131	ETTB-367B	131	43-025834
130	800-52402	130	39414	130	3130091	131	G04-704-A	131	46-86125-3
130	800-525-04A	130	39492	130	3146977	131	G6016	131	46-86135-3
130	804-14120	130	39921	130	3146977A	131	GC4094	131	48-137214
130	804-14130	130	39954	130	3152170	131	GE-30	131	48-137267
130	866-6	130	40151	130	3152170A	131	GE-44	131	48-137268
130	866-6(BENDIX)	130	40251	130	3463101-1	131	HB367	131	48-137269
130	921-1011	130	40363	130	3700135	131	HEP-G6016	131	48-137270
130	991-01-1317	130	40369	130	3700164	131	HEP642	131	48-137271
130	991-01-3063	130	40464	130	3700219	131	HEP643	131	48-137308
130	992-00-2271	130	40465	130	3700228	131	HEPG6016	131	48-40172001
130	992-00-3139	130	40466	130	4080187-0504	131	HT20436A	131	48-97046M15
130	992-00-3139A	130	40934-1	130	4080320-0501	131	HT2046710	131	48-97046A15
130	992-00-4091	130	43060	130	4080320-0504	131	HT204756	131	48-97238A06
130	992-00-4092	130	43095	130	4080320-050B	131	IR-RE50	131	48813270
130	992-002271	130	43114	130	4080866-000A	131	MZ9	131	57A100-11
130	992-00271	130	43168	130	4080866-8006	131	NKT451	131	57B100-11
130	992-00271A	130	52215-00	130	4082886-001	131	NKT452	131	57M3-109
130	992-000139	130	52329	130	4082886-002	131	NKT452-S1	131	57M3-12
130	992-004091	130	52360	130	4082886-3	131	NKT455	131	57M3-9P
130	992-01-1317	130	60041	130	4361620	131	OC-30A	131	61-1906
130	992-017169	130	60046	130	4450023	131	OC30	131	86-480-9
130	992-02271	130	60047	130	04450023-001	131	OC30A	131	86X0053-001
130	1001-09	130	60085	130	04450023-006	131	OC30B	131	98P1?
130	1045-7844	130	60127	130	4450023-007	131	P3E	131	120-003150
130	1045-7851	130	60130	130	4450023-P3	131	P3R	131	120-004887
130	1064-6032	130	60205	130	4450023-P4	131	P3R-2	131	171-016-9-001
130	1071-3642	130	60243	130	4450023-P6	131	P3R-3	131	220-00201
130	1080-5564	130	60465	130	4450023P5	131	P3R-4	131	260P07502
130	1116-6527	130	60710	130	4822354	131	P3T	131	296-61-9
130	1471-4802	130	60837	130	4822354J	131	P3T-1	131	324-0126
130	1806-17	130	60944	130	6101161-1	131	P3T-2	131	417-160
130	1806-17A	130	60973	130	6480000	131	PC5010	131	430-25834
130	1854-0563	130	61012	130	7026018	131	PT32	131	465-137-19
130	2000-221	130	61019	130	7306982	131	PTC120	131	465-206-19
130	2003	130	61234	130	7309160	131	Q-01084R	131	480-9(SEARS)
130	2015-1	130	61367	130	7937586	131	RO092	131	642-152
130	2015-1A	130	61369/4367	130	8510694-1	131	RE20	131	642-217
130	2015-2	130	61868	130	9340311	131	RE20MP	131	690V098H51
130	2015-2A	130	62004	130	9340311-D5514	131	RT4762	131	753-2100-002
130	2015-3	130	62005-1	130	10646032	131	SC4274	131	753-4001-474
130	2015-3A	130	62143	130	10713642	131	SK3052	131	1008(OULLETTE)
130	2015-4	130	62282	130	10805364	131	SP2551	131	1008(POWER)
130	2015-5	130	62792	130	11069934A	131	TI-1A6	131	1840-17
130	2015-6	130	70008-0	130	11166627	131	TI-7A	131	1945-17
130	2015-7	130	70008-3	130	13104367	131	TM131		
130	3055-1			130	17809000				

ECG	Industry Standard No.	ECG	Industry Standard No.	ECG	Industry Standard No.	ECG	Industry Standard No.	ECG	Industry Standard No.
131	1955-17	134A	1BZ3.6	135A	1Z5.1T10	136A	OA225.6A	136A	WS-054
131	2057A2-211	134A	1N3822	135A	1ZF5.1T10	136A	OA2202	136A	WS-054A
131	2057B100-11	134A	1N4729	135A	1ZF5.1T20	136A	OAZ242	136A	WS-054B
131	2093A3D-20	134A	1N4729A	135A	1ZF5.1T5	136A	OZ5.6T5	136A	WS-054C
131	2402-454	134A	1N5227B	135A	1ZM5.1T10	136A	PD6008	136A	WS-055
131	0004203	134A	1N747A	135A	1ZM5-1T20	136A	PD6049	136A	WS-056B
131	4822-130-40213	134A	1N747B	136A	10-085027	136A	PD6049C	136A	WS-056C
131	7219-3	134A	182036A	136A	13-33187-12	136A	P88907	136A	WS-057
131	8000-00003-040	134A	123.6	136A	14-509-02	136A	PTC502	136A	WS-058
131	8500-204	134A	123.6A	136A	14-511-02	136A	QD-ZRD56EAA	136A	WS-058A
131	002481	134A	12O3.6	136A	14-515-15	136A	QD-ZRD56PAA	136A	WS-058B
131	25658-120	134A	12M3.6T10	136A	1626	136A	QZ5.6T10	136A	WS-058C
131	25658-121	134A	12M3.6T20	136A	1626A	136A	QZ5.6T5	136A	WT-054
131	27126-060	134A	12M3.6T5	136A	20B409	136A	RD-6A	136A	WT-054A
131	00071090	134A	1226F	136A	20B410	136A	RD5.6E	136A	WT-054B
131	72193	134A	29P1	136A	34-8057-37	136A	RD5.6E-B	136A	WT-054C
131	167285	134A	3226	136A	3626	136A	RD5.6EC	136A	WT-056
131	171162-083	134A	3226A	136A	3626A	136A	RD5.6EK	136A	WT-056B
131	171162-089	134A	3326	136A	48-134698	136A	RD5.6F	136A	WT-056C
131	171217-1	134A	3326A	136A	48-134991	136A	RD5.6FA	136A	WU-054
131	175043-065	134A	56-70	136A	56-51	136A	RD5.6FB	136A	WU-054A
131	215089	134A	93C39-1	136A	56-44	136A	RD6	136A	WU-054B
131	489751-163	134A	1102	136A	56-59	136A	RD6A	136A	WU-054C
131	0573030	134A	1103	136A	264P02501	136A	RD64M	136A	WU-056
131	0573030-14	134A	129903	136A	394-3155	136A	RE107	136A	WU-056B
131	0573212	135A	-.75N5.1	136A	523-2003-519	136A	RT1306	136A	WU-056C
131	605122	135A	AZ-050	136A	919-01-3058	136A	RT1306(Q.R.)	136A	WU-056D
131	880092	135A	A25.1	136A	2093A41-90	136A	RT5793	136A	WV-054
131	984521	135A	BZ5.1	136A	2774	136A	RV1181	136A	WV-054A
131	985103	135A	BZY9205V1	136A	4822-130-30264	136A	RZ5.6	136A	WV-054B
131	200651A-59	135A	CD3122	136A	5001-197	136A	RZZ5.6	136A	WV-054C
131	2320092	135A	CHA4Z5.1	136A	256817	136A	SD-32	136A	WV-056
131	7855295-01	135A	CHMZ5.1	136A	741738	136A	SPZ708	136A	WV-056B
131	8112143	135A	DX-0987	136A	760298	136A	SK5357	136A	WV-056C
131	13040349	135A	EA16X136	136A	986578	136A	SV123	136A	WV-058
131	62046287	135A	EP16X225	136A	7852904-01	136A	SZ5.6	136A	WW-054
131	62752300	135A	EQAO1-058	136A	.25T5.6	136A	TMDO2	136A	WW-054A
131	9511414000	135A	EQAO1-05T	136A	.25T5.6A	136A	TMDO2A	136A	WW-054B
131S	B368	135A	F25.1T10	136A	.25T5.6B	136A	T25.6	136A	WW-054C
131S	B368A	135A	F25.1T5	136A	.4T5.6	136A	VHRHZ6B3///1A	136A	WW-056
131S	B368B	135A	FZ901	136A	.4T5.6A	136A	VR5.6	136A	WW-056A
131S	B368H	135A	G5.1T10	136A	.4T5.6B	136A	VR5.6A	136A	WW-056B
131S	B462	135A	G5.1T20	136A	.75N5.6	136A	VR5.6B	136A	WW-056C
131S	28B368	135A	G5.1T5	136A	A7285900	136A	WM-054	136A	WW-056D
131S	28B368B	135A	GEZD-5.0	136A	AZ-052	136A	WM-054A	136A	WX-054
131S	28B368H	135A	GEZD-5.1	136A	AZ-054	136A	WM-054B	136A	WX-054A
131S	28B462	135A	G1A47	136A	AZ-056	136A	WM-054C	136A	WX-054B
131MP	GE-31MP	135A	G1A47A	136A	AZ5.6	136A	WM-054D	136A	WX-054C
131MP	J24366	135A	HEP-Z0406	136A	A2752A	136A	WM-056	136A	WX-054D
131MP	N-R-15X139	135A	H82051	136A	BZ-052	136A	WM-056A	136A	WX-056
131MP	TM131MP	135A	H87051	136A	BZ-054	136A	WM-056B	136A	WX-056A
131MP	W17MP	135A	IP20-0086	136A	BZ-056	136A	WM-056C	136A	WX-056B
131MP	4-1848	135A	K82047B	136A	BZ052	136A	WM-056D	136A	WX-056C
131MP	24MW618	135A	K82051A	136A	BZ5.6	136A	WN-054	136A	WX-058
131MP	33-1002-00	135A	K82051B	136A	BZX2905V6	136A	WN-054A	136A	WX-058B
131MP	48X97238A06	135A	K834A	136A	BZXA605V6	136A	WN-054B	136A	WX-058C
131MP	86-0023-007	135A	K834B	136A	BZX5505V6	136A	WN-054C	136A	WX-058D
131MP	96XZ6054/45X	135A	K834BF	136A	BZX7905V6	136A	WN-054D	136A	WY-054
131MP	465-166-19	135A	K835A	136A	BZX8305V6	136A	WN-056	136A	WY-054A
131MP	675-206	135A	K835AF	136A	BZY58	136A	WN-056A	136A	WY-054C
131MP	753-2000-463	135A	K8047A	136A	BZY85/05V6	136A	WN-056B	136A	WY-054D
131MP	4800-223	135A	K8047B	136A	BZY85/D5V6	136A	WN-056C	136A	WY-056
131MP	081042	135A	K8051A	136A	BZY8305V6	136A	WN-056D	136A	WY-056B
131MP	123808	135A	LPM4.7	136A	BZY85C5V6	136A	WN-058	136A	WY-056C
131MP	171162-090	135A	LPM4.7A	136A	BZY9205V6	136A	WN-058A	136A	WY-056D
131MP	175043-081	135A	LR47CH	136A	CDN125A	136A	WN-058B	136A	WZ060
131MP	215071	135A	LR51CH	136A	CD31-00007	136A	WN-058C	136A	XZ5.6
131MP	0573031	135A	LR56CH	136A	CD48	136A	WN-058D	136A	XZ523
131MP	740247	135A	M4Z4.7	136A	CHA4Z5.6	136A	WO-054	136A	XZ055
131MP	740443	135A	M4Z4.7-20	136A	CHMZ5.6	136A	WO-054A	136A	YZ058
131MP	740471	135A	M4Z4.7A	136A	D1M	136A	WO-054C	136A	Z-1006
134A	A23.6	135A	M425.1	136A	DDA7009001	136A	WO-054D	136A	Z-1104
134A	A2747A	135A	M425.1-20	136A	DIM	136A	WO-056	136A	Z-1140
134A	B23.6	135A	M425.1A	136A	D850	136A	WO-056A	136A	Z-1145
134A	BZY85C3V6	135A	MGLA47	136A	DX-0061	136A	WO-056B	136A	Z-1150
134A	BZY88C4V3	135A	MGLA51	136A	EA16X118	136A	WO-056C	136A	Z-1155
134A	CD31-00002	135A	MGLA51A	136A	EA26C8	136A	WO-056D	136A	Z-1160
134A	CHMZ3.6	135A	MGLA51B	136A	EDZ-19	136A	WO-058	136A	Z-1165
134A	F23.6T10	135A	MR51C-H	136A	EQAO105T	136A	WO-058A	136A	Z-1170
134A	P23.6T5	135A	MR51E-H	136A	ET16X17	136A	WO-058B	136A	Z-1240
134A	GEZD-3.6	135A	MZ1005	136A	EVR4	136A	WO-058C	136A	Z-1245
134A	G1A39	135A	MZ205	136A	EVR4A	136A	WP-054	136A	Z-1250
134A	G1A39A	135A	M24624	136A	EVR4B	136A	WP-054A	136A	Z-1255
134A	HEP-Z0402	135A	M24625	136A	EZ5R6(ELCOM)	136A	WP-054B	136A	Z-1260
134A	H82036	135A	MZ92-4.7A	136A	F25.6T10	136A	WP-054C	136A	Z-1265
134A	H82039	135A	MZ92-5.1A	136A	F25.6T5	136A	WP-054D	136A	Z-1270
134A	H97036	135A	OA201	136A	G5.6T10	136A	WP-056	136A	Z-1540
134A	H87039	135A	OA209	136A	G5.6T20	136A	WP-056A	136A	Z-1545
134A	K82039B	135A	OA2240	136A	G5.6T5	136A	WP-056B	136A	Z-1550
134A	K830A	135A	OA2241	136A	GEZD-5.6	136A	WP-056C	136A	Z-1555
134A	K830AF	135A	PD6006	136A	G1A56	136A	WP-056D		
134A	K830B	135A	PD6007	136A	G1A56A	136A	WP-058		
134A	K830BF	135A	PD6007A	136A	G1A56B	136A	WP-058B		
134A	K831A	135A	PD6048	136A	HD30042090	136A	WP-058C		
134A	K832A	135A	PR515	136A	HEP-Z0407	136A	WP-058D		
134A	K832AF	135A	PR605	136A	HEP603	136A	WQ-054		
134A	K832B	135A	PR804	136A	H82056	136A	WQ-054A		
134A	K832BF	135A	PR93017	136A	H87056	136A	WQ-054B		
134A	K8033A	135A	P86468	136A	J241186	136A	WQ-054C		
134A	K8033B	135A	PTC214	136A	J24186	136A	WQ-056		
134A	K8039A	135A	PTC215	136A	K8056A	136A	WQ-056A		
134A	K8039A	135A	QD-ZMZ205XB	136A	K82056A	136A	WQ-056B		
134A	K8039B	135A	QZ5.1T10	136A	K82056B	136A	WQ-056C		
134A	LPM3.6	135A	QZ5.1T5	136A	K836A	136A	WQ-056D		
134A	LPM3.6A	135A	RD-5A	136A	K836AF	136A	WQ-058		
134A	LR39CH	135A	RD5.1EB	136A	K836B	136A	WQ-058A		
134A	M423.5	135A	RD5B	136A	K836BF	136A	WQ-058B		
134A	M423.5A	135A	RH-EX0024TA2Z	136A	K8056A	136A	WQ-058C		
134A	M423.9	135A	SD-6	136A	K8056B	136A	WR-054		
134A	M423.9A	135A	SV122	136A	LPM5.6	136A	WR-054A		
134A	MGLA39	135A	SZ200	136A	LPM5.6A	136A	WR-054B		
134A	MGLA39B	135A	SZ200-5	136A	L45.6	136A	WR-054C		
134A	MR36H	135A	SZ4.7	136A	M425.6	136A	WR-056		
134A	MR390-H	135A	SZ5	136A	M425.6-20	136A	WR-056A		
134A	M24620	135A	SZ5.1	136A	M425.6A	136A	WR-056B		
134A	M2500-5	135A	TMD01	136A	M758	136A	WR-056C		
134A	MZ92-3.9A	135A	TMD01A	136A	MA1056	136A	WR-056D		
134A	PD6003	135A	TVSRD5A	136A	MC6607	136A	WR-058		
134A	PD6003A	135A	WP-052D	136A	MC6607A	136A	WR-058A		
134A	PD6004	135A	001-0163-02	136A	MC6107	136A	WR-058B		
134A	PD6044	135A	001-02303-0	136A	MC6107A	136A	WR-058C		
134A	PD6501	135A	001-023030	136A	MD752	136A	WR-058D		
134A	QZ3.6T10	135A	1D5.1A	136A	MD752A				
134A	QZ3.6T5	135A	1EZ5.1	136A	MEZ5.6T10				
134A	RE101	135A	1N1765	136A	MEZ5.6T5				
134A	R23.6	135A	1N3826	136A	MGLA56				
134A	R323.6	135A	1N3826A	136A	MGLA56A				
134A	S23.6	135A	1N4733A	136A	MGLA56B				
134A	TIXD747	135A	1N4733AA	136A	MR560-H				
134A	TVSRD4AM	135A	1N761A	136A	MT2607				
134A	1D3.6	135A	18320	136A	MT2607A				
134A	1D3.6A			136A	M24626				
134A	1D3.6B			136A	MZ54				
				136A	MZ92-5.6A				

ECG	Industry Standard No.	ECG	Industry Standard No.	ECG	Industry Standard No.	ECG	Industry Standard No.	ECG	Industry Standard No.
136A	Z-1560	136A	1T5.6	137A	HEP-Z040B	137A	WM-061C	137A	WV-061B
136A	Z-1565	136A	1T5.6B	137A	HEP103	137A	WM-061D	137A	WV-061C
136A	Z-1570	136A	1TA5.6	137A	HS2062	137A	WM-063	137A	WV-061D
136A	Z-5140	136A	1TA5.6A	137A	HST7062	137A	WM-063A	137A	WV-063
136A	Z-5145	136A	125.6	137A	HZ-6C	137A	WM-063B	137A	WV-063A
136A	Z-5150	136A	125.6A	137A	INJ61225	137A	WM-063C	137A	WV-063B
136A	Z-5155	136A	1Z05.6T10.5	137A	J241179	137A	WM-063D	137A	WV-063C
136A	Z-5160	136A	1ZF5.6T10	137A	J24631	137A	WM-065	137A	WV-063D
136A	Z-5165	136A	1ZF5.6T20	137A	KD2503	137A	WM-065A	137A	WV-065
136A	Z-5170	136A	1ZF5.6T5	137A	KS062A	137A	WM-065B	137A	WV-065A
136A	Z-5240	136A	1ZM5.6T10	137A	KS2062A	137A	WM-065C	137A	WV-065B
136A	Z-5245	136A	1ZM5.6T20	137A	KS2062B	137A	WM-065D	137A	WV-065C
136A	Z-5250	136A	1ZM5.6T5	137A	K837A	137A	WN-061	137A	WV-065D
136A	Z-5255	136A	02Z5.6	137A	K837AF	137A	WN-061A	137A	WW-061
136A	Z-5260	136A	02Z5.6A	137A	K8062A	137A	WN-061B	137A	WW-061A
136A	Z-5265	136A	3L4-3506-21	137A	LR62CH	137A	WN-061C	137A	WW-061B
136A	Z-5270	136A	3L4-3506-31	137A	M4Z6.2	137A	WN-061D	137A	WW-061C
136A	Z-5540	136A	4-4356(SBARS)	137A	M4Z6.2-20	137A	WN-063	137A	WW-061D
136A	Z-5545	136A	6V-200	137A	M4Z6.2A	137A	WN-063A	137A	WX-061
136A	Z-5550	136A	07-5331-86A	137A	MA1062	137A	WN-063B	137A	WX-061A
136A	Z-5555	136A	07-5331-86B	137A	MC6008	137A	WN-063C	137A	WX-061B
136A	Z-5560	136A	07-5331-86C	137A	MC6008A	137A	WN-063D	137A	WX-061C
136A	Z-5565	136A	07-5331-86D	137A	MC6108	137A	WN-065	137A	WX-061D
136A	Z-5570	136A	8-905-421-109	137A	MC6108A	137A	WN-065A	137A	WX-063
136A	Z0212	136A	09-306109	137A	MD753	137A	WN-065B	137A	WX-063A
136A	Z0407	136A	10-016	137A	MD753A	137A	WN-065C	137A	WX-063B
136A	Z0B5.6	136A	012-1023-007	137A	MGLA62	137A	WN-065D	137A	WX-063C
136A	Z0D5.6	136A	13-67544-1	137A	MGLA62A	137A	WO-061	137A	WX-063D
136A	Z0D5.6	136A	13-67544-1/3464	137A	MGLA62B	137A	WO-061A	137A	WX-065
136A	Z1104	136A	17Z6	137A	MR6Z0-H	137A	WO-061B	137A	WX-065A
136A	Z1104-C	136A	17Z6A	137A	MR62E-H	137A	WO-061C	137A	WX-065B
136A	Z1140	136A	17Z6AF	137A	MR62H	137A	WO-061D	137A	WX-065C
136A	Z1145	136A	17Z6F	137A	MTZ60B	137A	WO-063	137A	WX-065D
136A	Z1150	136A	21A119-041	137A	MTZ60BA	137A	WO-063A	137A	WY-061
136A	Z1155	136A	34-8057-32	137A	MZ4627	137A	WO-063B	137A	WY-061A
136A	Z1160	136A	37Z6	137A	MZ6	137A	WO-063C	137A	WY-061B
136A	Z1165	136A	37Z6A	137A	MZ6.2	137A	WO-063D	137A	WY-061C
136A	Z1170	136A	48-134993	137A	MZ6.2A	137A	WO-065	137A	WY-061D
136A	Z1250	136A	48-40458A04	137A	MZ6.2T5	137A	WO-065A	137A	WY-063
136A	Z1255	136A	48-83461E38	137A	MZ605	137A	WO-065B	137A	WY-063A
136A	Z1260	136A	48-97305A02	137A	MZ610	137A	WO-065C	137A	WY-063B
136A	Z1270	136A	48-97305A05	137A	MZ620	137A	WO-065D	137A	WY-063C
136A	Z1540	136A	61E38	137A	MZ640	137A	WO-61	137A	WY-063D
136A	Z1545	136A	73-31	137A	MZ9P-6.2A	137A	WP-061	137A	WY-065
136A	Z1555	136A	93A39-19	137A	NGP5002	137A	WP-061A	137A	WY-065A
136A	Z1560	136A	100-10	137A	OA2203	137A	WP-061B	137A	WZ-061
136A	Z1565	136A	100-135	137A	OA2210	137A	WP-061C	137A	WZ-061A
136A	Z1570	136A	264904005	137A	OA2243	137A	WP-061D	137A	WZ-063A
136A	Z1B5.6	136A	523-2003-569	137A	OA2270	137A	WP-063	137A	WZ-065A
136A	Z1D5.6	136A	65200	137A	OZ6.2T10	137A	WP-063A	137A	WZ.061
136A	Z2A56CP	136A	903-120B	137A	OZ6.2T5	137A	WP-063B	137A	WZ065
136A	Z2A56F	136A	903-333	137A	PA9267	137A	WP-063C	137A	WZ.6
136A	Z4B5.6	136A	919-00-1445	137A	PD6009	137A	WP-063D	137A	XBADAW01-06
136A	Z4D5.6	136A	2323-17	137A	PD6009A	137A	WP-065	137A	YZ065
136A	Z4X5.6	136A	2362-17	137A	PD6050	137A	WP-065A	137A	Z0214
136A	Z5.6	136A	4082-748-0002	137A	PS1325	137A	WP-065B	137A	Z0408
136A	Z5140	136A	4822-130-30193	137A	P88908	137A	WP-065C	137A	ZOB6.2
136A	Z5145	136A	5001-160	137A	PTC503	137A	WP-065D	137A	ZOD6.2
136A	Z5150	136A	67544	137A	QZ6.2T10	137A	WQ-061	137A	ZOD6.2
136A	Z5155	136A	99201-210	137A	QZ6.2T5	137A	WQ-061A	137A	Z1B6.2
136A	Z5160	136A	99210-210	137A	R-7093	137A	WQ-061B	137A	Z1C6.2
136A	Z5165	136A	132865	137A	RD-6L	137A	WQ-061C	137A	Z1D6.2
136A	Z5170	136A	134993	137A	RD6.2F	137A	WQ-061D	137A	Z2A62F
136A	Z5240	136A	169199	137A	RD6.2FA	137A	WQ-063	137A	Z4X16.2
136A	Z5250	136A	171842	137A	RD6.2FB	137A	WQ-063A	137A	Z4X16.2B
136A	Z5255	136A	225316	137A	RE109	137A	WQ-063B	137A	Z5B6.2
136A	Z5260	136A	348057-17	137A	RH-EX0024CEZZ	137A	WQ-063C	137A	Z5OB.2
136A	Z5265	136A	530073-31	137A	RT7539	137A	WQ-063D	137A	Z5OB.2
136A	Z5270	136A	530145-1569	137A	RV6.2	137A	WQ-065	137A	Z6.2
136A	Z5540	136A	530145-569	137A	RZ6.2	137A	WQ-065A	137A	Z801
136A	Z5545	136A	530157-569	137A	RZZ6.2	137A	WQ-065B	137A	ZB6.2
136A	Z5550	136A	741116	137A	S1A3	137A	WQ-065C	137A	ZB6.2A
136A	Z5555	136A	1471898-5	137A	S3004-1715	137A	WQ-065D	137A	ZB6.2B
136A	Z5560	136A	1642606B6	137A	S3004-1718	137A	WR-061	137A	ZD6.2A
136A	Z5565	136A	1800012	137A	SA821	137A	WR-061A	137A	ZD6.2B
136A	Z5570	136A	2010967-84	137A	SA821A	137A	WR-061B	137A	ZP6.2
136A	Z5B5.6	136A	2092055-0016	137A	SA823	137A	WR-061C	137A	ZM6.2B
136A	Z5D5.6	136A	4082748-0002	137A	SA823A	137A	WR-061D	137A	ZOB6.2
136A	Z5D5.6	136A	5330842	137A	SA825	137A	WR-063	137A	ZP6.2
136A	ZB5.6	136A	62522666	137A	SA825A	137A	WR-063A	137A	ZPD6.2
136A	ZB5.6A	136A	134883461538	137A	SA827	137A	WR-063B	137A	ZT6.2
136A	ZB5.6B	137A	.25F6.2	137A	SA827A	137A	WR-063C	137A	ZT6.2A
136A	ZD5.6A	137A	.25F6.2B	137A	SA829	137A	WR-063D	137A	ZT6.2B
136A	ZD5.6B	137A	.25F6.2B	137A	SA829A	137A	WR-065	137A	001-02303-3
136A	ZD56A	137A	.75N6.2	137A	SB821	137A	WR-065A	137A	001-023033
136A	ZEC5.6	137A	AZ-061	137A	SB821A	137A	WS-061	137A	10003B
136A	ZEC5.6	137A	AZ-063	137A	SB823	137A	WS-061A	137A	1D6.2
136A	ZP5.6	137A	AZ6.2	137A	SB823A	137A	WS-061B	137A	1D6.2A
136A	ZQ5.6	137A	AZ75.3A	137A	SB825	137A	WS-061C	137A	1D6.2B
136A	ZM5.6	137A	BZ-061	137A	SB825A	137A	WS-061D	137A	1D6.2SA
136A	ZM5.6B	137A	BZ-063	137A	SB827	137A	WS-063	137A	1D6.2BB
136A	ZOB5.6	137A	BZ6.2	137A	SB827A	137A	WS-063A	137A	1M6.2Z810
136A	ZR5.6	137A	BZX10	137A	SB829	137A	WS-063B	137A	1M6.2Z85
136A	ZR5.6	137A	BZX2906V2	137A	SB829A	137A	WS-063C	137A	1M753A
136A	ZR5.6A	137A	BZX46C6V2	137A	SC821	137A	WS-063D	137A	1N1485
136A	ZR5.6B	137A	BZX55C6V2	137A	SC821A	137A	WS-065	137A	1N1766A
136A	ZT5.6	137A	BZX79C6V2	137A	SC823	137A	WS-065A	137A	1N3443
136A	ZT5.6A	137A	BZX83C6V2	137A	SC823A	137A	WS-065B	137A	1N3828
136A	ZY5.6B	137A	BZY59	137A	SC825	137A	WS-065C	137A	1N3828A
136A	ZZ5.6	137A	BZY89C6V2	137A	SC825A	137A	WS-065D	137A	1N4499
136A	001-0163-04	137A	BZY85C6V2	137A	SC827	137A	WT-061	137A	1N4691
136A	1-210	137A	C4011	137A	SC827A	137A	WT-061A	137A	1N4735
136A	1D5.6	137A	CDO033	137A	SC829	137A	WT-061B	137A	1N4735A
136A	1D5.6A	137A	CD31-00008	137A	SC829A	137A	WT-061C	137A	1N5234B
136A	1D5.6B	137A	CHAZ6.2	137A	SK7058	137A	WT-061D	137A	1N5525B
136A	1E25.6	137A	CHAZ6.2A	137A	SUC650	137A	WT-063	137A	1N5850B
136A	1M5.6Z810	137A	D-00469C	137A	SUM6010	137A	WT-063A	137A	1N675
136A	1M5.6Z85	137A	D3H	137A	SUM6011	137A	WT-063B	137A	1N709
136A	1N1509	137A	D3Y	137A	SUM6020	137A	WT-063C	137A	1N709A
136A	1N1509A	137A	D5W	137A	SUM6021	137A	WT-063D	137A	1N753A
136A	1N1520	137A	D6.2	137A	SVC625	137A	WT-065	137A	1S136(ZENER)
136A	1N1520A	137A	DDAY008001	137A	SVC650	137A	WU-061	137A	1S1715
136A	1N1765A	137A	DIJ72293	137A	SVM6010	137A	WU-061A	137A	1S2062A
136A	1N1929A	137A	E21431	137A	SVM6011	137A	WU-061B	137A	1S2112A
136A	1N1983A	137A	E2486	137A	SVM6020	137A	WU-061C	137A	1S331
136A	1N2033A	137A	EA16X80	137A	SVM6021	137A	WU-061D	137A	1S331'A
136A	1N3827	137A	ED6.2EB	137A	SVM605	137A	WU-063	137A	1S331'AZ
136A	1N4690	137A	EQA01-06T	137A	SVM61	137A	WU-063A	137A	1T6.2
136A	1N4734	137A	EVR5	137A	SZ-RD6.2EB	137A	WU-063B		
136A	1N4734A	137A	EVR5A	137A	SZ-YZ063	137A	WU-063C		
136A	1N5232B	137A	EVR5B	137A	SZ6.2	137A	WU-063D		
136A	1N5248B	137A	F16H1	137A	SZ6.2A	137A	WU-065		
136A	1N5848B	137A	FV-22	137A	TIXU753	137A	WV-061		
136A	1N708A	137A	FV22	137A	TMD03	137A	WV-061A		
136A	1N762	137A	FZ6.2T10	137A	TMD03A				
136A	1N762a	137A	G01-036A	137A	TR2337123				
136A	1NJ61225	137A	G01036A	137A	THR6				
136A	1S2056A	137A	GARE	137A	WO-61				
136A	1S2111A	137A	GEZD-6.2	137A	WM-061				
136A	1S211A	137A	G1A62	137A	WM-061A				
136A	1S221	137A	G1A62A	137A	WM-061B				
136A	1S692	137A	G1A62B						
		137A	GZ6.2						

ECG	Industry Standard No.	ECG	Industry Standard No.	ECG	Industry Standard No.	ECG	Industry Standard No.	ECG	Industry Standard No.
137A	126.2B	138A	EP36X12	139A	AWO1-9	139A	MZ-209	139A	WN-088D
137A	1TA6.2	138A	EQA01-072	139A	AWO9	139A	MZ090	139A	WN-090
137A	1TA6.2A	138A	EQB01-07	139A	AZ-090	139A	MZ209	139A	WN-090A
137A	1Z6.2	138A	E857X7	139A	AZ-092	139A	MZ309	139A	WN-090B
137A	1Z6.2T10	138A	ETD-RD75	139A	AZ757A	139A	MZ509B	139A	WN-090C
137A	126.2T5	138A	FV21	139A	AZ9.1	139A	MZ409-02B	139A	WN-090D
137A	1Z8.2A	138A	GEZD-7.5	139A	AZ960B	139A	MZ92-9.1A	139A	WN-092
137A	1ZC6.2	138A	HD3000309	139A	B090	139A	MZ92-9.1B	139A	WN-092A
137A	1ZM6.2	138A	HE-20049	139A	B2090	139A	MZX9.1	139A	WN-092B
137A	1ZM6.2T10	138A	HE-7	139A	B66XO040-003	139A	NGP5007	139A	WN-092C
137A	1ZM6.2T20	138A	HE-7A	139A	BXO90	139A	OA2207	139A	WN-092D
137A	1ZM6.2T5	138A	HEP-Z0216	139A	BX909	139A	OA2212	139A	WN-094
137A	2MW742	138A	HEP-Z0410	139A	BXY63	139A	OA2247	139A	WN-094A
137A	02Z-6.2A	138A	IP20-0027	139A	BZ-090	139A	OA2272	139A	WN-094B
137A	02Z6.2	138A	MZ1007	139A	BZ-090(ZENER)	139A	P21544	139A	WN-094C
137A	02Z6.2W	138A	OSD-0033	139A	BZ-0900	139A	PD6013	139A	WN-094D
137A	02262A	138A	PTC504	139A	BZ-0901	139A	PD6013A	139A	WO-088
137A	4-2020-07500	138A	R7030	139A	BZ-092	139A	PD6054	139A	WO-088B
137A	4T6.2	138A	RD-7.5EB	139A	BZ-094	139A	P810019B	139A	WO-088C
137A	4T6.2A	138A	RD-7A	139A	BZ090	139A	P810063	139A	WO-088D
137A	4T6.2B	138A	RD7R5EB	139A	BZ090.1Z9	139A	P88912	139A	WO-090
137A	05Z-6.2L	138A	RE111	139A	BZ094	139A	PTC505	139A	WO-090A
137A	05Z6.2	138A	RVDEQA01078	139A	BZ1-9	139A	Q-25115C	139A	WO-090B
137A	6.28R1	138A	RVDRD7R5EB	139A	BZ9.1	139A	QD-ZMZ409BE	139A	WO-090C
137A	6.28R1A	138A	OSD-0033	139A	BZX14	139A	QD-2RD9EXAA	139A	WO-090D
137A	6.28R2	138A	SZ-7	139A	BZX29C9V1	139A	RD-9.1E	139A	WO-092
137A	6.28R2A	138A	T17-A	139A	BZX46C9V1	139A	RD-91	139A	WO-092A
137A	6.28R3	138A	TR-7GS	139A	BZX55C9V1	139A	RD-91E	139A	WO-092B
137A	6.28R3A	138A	TVS-RD7A	139A	BZX71C9V1	139A	RD-96	139A	WO-092C
137A	6.28R4	138A	Z-1010	139A	BZX79C9V1	139A	RD-9A	139A	WO-092D
137A	6.28R4A	138A	001-0099-01	139A	BZY63	139A	RD-9L	139A	WO-094
137A	07-5331-86	138A	001-0099-02	139A	BZY92C9V1	139A	RD9	139A	WO-094A
137A	09-306055	138A	1A7.5M	139A	C4015	139A	RD9.1E	139A	WO-094B
137A	09-306183	138A	1A7.5MA	139A	CA-092	139A	RD9.1EA	139A	WO-094D
137A	09-306377	138A	1C0020	139A	CD31-00012	139A	RD9.1EC	139A	WP-088
137A	1826	138A	1EZ-5210	139A	CD31-12019	139A	RD9.1ED	139A	WP-088A
137A	1826A	138A	1EZ7.5Z	139A	CD31-12039	139A	RD9.1EK	139A	WP-088B
137A	1826AF	138A	1N1768	139A	CD3112039	139A	RD9.1F	139A	WP-088C
137A	1826F	138A	1N2034-3	139A	CD3122055	139A	RD9.1FA	139A	WP-088D
137A	19A11552B-P1	138A	1N3017B	139A	CDZ-9V	139A	RD9.1FB	139A	WP-090
137A	3826	138A	1N3676	139A	CDZ-09V	139A	RD9A	139A	WP-090A
137A	3826A	138A	1N3676A	139A	CP-092	139A	RD9A(10)	139A	WP-090B
137A	46-86394-3	138A	1N4159	139A	CP3212055	139A	RD9A-N	139A	WP-090C
137A	48-10641D62	138A	1N4159A	139A	CZ-092	139A	RE114	139A	WP-090D
137A	48-137170	138A	1N4324	139A	CZ-92	139A	RFJ71480	139A	WP-092
137A	48-137210	138A	1N4324A	139A	CZO92	139A	RF-9A	139A	WP-092A
137A	48-137387	138A	1N4401	139A	CZ094	139A	RS1290	139A	WP-092B
137A	48-83461I36	138A	1N4629	139A	D-00569C	139A	RT8339	139A	WP-092D
137A	48810641D62	138A	1N4737	139A	D-18	139A	RV2213	139A	WP-094
137A	53E011-2	138A	1N4737A	139A	D1-8	139A	RVD1N4739	139A	WP-094A
137A	53J004-1	138A	1N5852	139A	D176	139A	RVDMZ209	139A	WP-094B
137A	61ES6	138A	182114A	139A	D10	139A	RVDRD11AN	139A	WP-094C
137A	66XO040-003	138A	18223	139A	D2Z	139A	RXO90	139A	WP-094D
137A	121-0041	138A	18332	139A	D3F	139A	RZ9.1	139A	WQ-088
137A	123B-005	138A	1T7.5	139A	DAAY010092	139A	RZZ9.1	139A	WQ-088A
137A	151-030-9-003	138A	1T7.5B	139A	DDAY010002	139A	S	139A	WQ-088B
137A	151-031-9-001	138A	1TA7.5	139A	DDAY010005	139A	S0702	139A	WQ-088C
137A	151-031-9-003	138A	1TA7.5A	139A	DDAY126001	139A	SA-93794	139A	WQ-088D
137A	152-019-9-001	138A	127.5	139A	DIJ72165	139A	SC91	139A	WQ-090
137A	152-047-9-001	138A	1ZM7.5T10	139A	DZ10	139A	SD-15	139A	WQ-090A
137A	152-047-9-004	138A	1ZM7.5T20	139A	EA16X62	139A	SD-27	139A	WQ-090B
137A	152-052-9-002	138A	02-165143	139A	EA16X74	139A	SD-632	139A	WQ-090C
137A	152-079-9-002	138A	02Z7.5A	139A	EA3866	139A	SD105	139A	WQ-092
137A	152-079-9-003	138A	314-3505-3	139A	EDZ-0045	139A	SD27	139A	WQ-092B
137A	260-10-046	138A	4-2021-05470	139A	EDZ-23	139A	SD632	139A	WQ-092C
137A	260-10-46	138A	05-110108	139A	EO771-3	139A	SD632(10)	139A	WQ-094
137A	429-0958-43	138A	09-306181	139A	EQ-09R	139A	SK3060A	139A	WQ-094A
137A	523-2003-629	138A	09-306197	139A	EQA-0109S	139A	S0-632	139A	WQ-094B
137A	65207	138A	09-307082	139A	EQA01-09	139A	SV-9(ZENER)	139A	WQ-094D
137A	1411-137	138A	13-14879-4	139A	EQAO1-09R	139A	SVD-181717	139A	WR-088
137A	2000-322	138A	14-515-04	139A	EQAO9R	139A	SVD02Z9.5A	139A	WR-088A
137A	2000-328	138A	14-515-13	139A	EQBO1-09	139A	SVZ181717	139A	WR-088B
137A	2000-329	138A	2026AF	139A	EQB01-908	139A	SZ-200-8	139A	WR-088C
137A	3701	138A	2026F	139A	ETD-RD9.1FB	139A	SZ-200-9V	139A	WR-088D
137A	3702	138A	21A103-049	139A	EVR9	139A	SZ-9	139A	WR-090
137A	3703	138A	24MW207	139A	EVR9A	139A	SZ9	139A	WR-090B
137A	3704	138A	33-0025	139A	EVR9B	139A	SZ9.1	139A	WR-090C
137A	5001-131	138A	34-8057-16	139A	EZ9(ELCOM)	139A	SZO9	139A	WR-092
137A	8000-00004-239	138A	34-8057-27	139A	G01-036-G	139A	SZT-9	139A	WR-092B
137A	8000-00005-020	138A	34-8057-9	139A	G01-036-H	139A	SZT9	139A	WR-092C
137A	8000-00006-146	138A	38A64C	139A	G01036	139A	T-R9B	139A	WR-092D
137A	8000-0005-020	138A	44T-300-94	139A	G01036G	139A	T21334	139A	WR-094
137A	9730-092-90112	138A	46-86225-3	139A	G9.1T10	139A	T21639	139A	WR-094A
137A	72168-3	138A	46-86456-3	139A	G9.1T20	139A	TMD07	139A	WR-094B
137A	99201-208	138A	53B015-1	139A	G9.1T5	139A	TMD07A	139A	WR-094C
137A	99201-211	138A	53J004-3	139A	GEZD-9.1	139A	TR-95	139A	WR-094D
137A	99201-212	138A	56-31	139A	GIA91	139A	TR-95(B)	139A	WS-088
137A	129938	138A	56-8096	139A	GIA91A	139A	TR-908	139A	WS-088A
137A	138974	138A	93C39-10	139A	GIA91B	139A	TR-98	139A	WS-088B
137A	161199	138A	96-5124-02	139A	G01036	139A	TR-98A	139A	WS-088C
137A	168653	138A	264P11003	139A	HD3000401	139A	TR-98B	139A	WS-088D
137A	302540	138A	1009-06	139A	HD30017090	139A	TR98	139A	WS-090A
137A	530073-1031	138A	1041-67	139A	HEP-Z0412	139A	TR98B	139A	WS-090B
137A	530073-5	138A	2000-324	139A	HEP104	139A	TVSRD9AL	139A	WS-090C
137A	530166-1004	138A	2000-325	139A	HP-18334	139A	TZ8.2	139A	WS-090D
137A	1471898-3	138A	2093A41-53	139A	HP-18339	139A	TZ9.1	139A	WS-092
137A	1642606B4	138A	2330-021	139A	HP-2004	139A	TZ9.1A	139A	WS-092A
137A	2002209-4	138A	4822-130-30287	139A	HP-20011	139A	TZ9.1B	139A	WS-092B
137A	4037413-P1	138A	05470	139A	HP-20065	139A	TZ9.1C	139A	WS-092C
137A	23115168	138A	7574	139A	HM9.1B	139A	U28709	139A	WS-092D
137A	23115955	138A	8000-00005-019	139A	HN-00012	139A	U28809	139A	WS-094
137A	26010029	138A	8000-00011-103	139A	HN-00024	139A	U28814	139A	WS-094A
137A	26010046	138A	72135	139A	HS2091	139A	UZ9.1	139A	WS-094B
137A	4202021700	138A	115123	139A	HS7091	139A	UZ9.1B	139A	WS-094C
137A	134883461536	138A	171162-172	139A	HW9.1	139A	VHEXZ-090-1	139A	WS-094D
137A	933009290112	138A	239219	139A	HW9.1A	139A	VR9.1	139A	WT-088
138A	.25T7.5	138A	281917	139A	HW9.1B	139A	VR9.1A	139A	WT-088B
138A	.25T7.5B	138A	983011	139A	ICF43047	139A	VR9.1B	139A	WT-088C
138A	.75N7.5	138A	984252	139A	IP20-0019	139A	VR9B	139A	WT-088D
138A	.7JZ7.5	138A	2005779-26	139A	J24632	139A	WD90	139A	WT-090
138A	.7Z7.5	138A	2006607-65	139A	KD2504	139A	WG91	139A	WT-090A
138A	.7Z7.5B	138A	2092055-5	139A	KS2091A	139A	WM-088	139A	WT-090B
138A	.7Z7.5C	138A	2330021	139A	KS2091B	139A	WM-088A	139A	WT-090C
138A	.7Z7.5D	138A	2330302H	139A	KS41A	139A	WM-088B	139A	WT-090D
138A	.7ZM7.5A	138A	23115984	139A	KS41AF	139A	WM-088C	139A	WT-092A
138A	.7ZM7.5B	138A	4202024200	139A	KS091A	139A	WM-088D	139A	WT-092B
138A	.7ZM7.5D	138A	4202105470	139A	LPM9-1	139A	WM-090A	139A	WT-092C
138A	A04332-007	138A	.25T9.1	139A	LPM9-1A	139A	WM-090B	139A	WT-092D
138A	A04344-007	139A	.25T9.1A	139A	LR91CH	139A	WM-090D	139A	WT-094
138A	A059-117	139A	.25T9.1B	139A	M4653	139A	WM-092	139A	WT-094A
138A	A21268	139A	.7JZ9.1	139A	M4Z9-1	139A	WM-092A	139A	WT-094B
138A	AZ-073	139A	.7Z9.1A	139A	M4Z9.1-20	139A	WM-092B	139A	WT-094C
138A	AZ-075	139A	.7Z9.1B	139A	M4Z9.1A	139A	WM-092C		
138A	AZ7.5	139A	.7Z9.1D	139A	M9Z	139A	WM-092D		
138A	AZ755A	139A	.72M9.1A	139A	MD757	139A	WM-094A		
138A	AZ95BB	139A	.72M9.1B	139A	MD757A	139A	WM-094B		
138A	B071	139A	.72M9.1C	139A	ME409-02B	139A	WM-094C		
138A	BZ-071	139A	.72M9.1D	139A	MGLA91	139A	WM-094D		
138A	BZ071	139A	A04234-2	139A	MGLA91A	139A	WN-088		
138A	CDZ-318-75	139A	A061-119	139A	MGLA91B	139A	WN-088A		
138A	D-00369C	139A	A068-103	139A	MR91C-H	139A	WN-088B		
138A	EA16X88	139A	AW-01-09	139A	MR91E-H	139A	WN-088C		
138A	EDZ-2	139A	AWO1-09	139A	MTZ613	139A			
138A	EP16X12	139A	AWO1-09	139A	MTZ613A	139A			

ECG	Industry Standard No.
139A	WT-094D
139A	WU-088
139A	WU-088A
139A	WU-088B
139A	WU-088C
139A	WU-088D
139A	WU-090
139A	WU-090A
139A	WU-090B
139A	WU-090C
139A	WU-090D
139A	WU-092
139A	WU-092A
139A	WU-092B
139A	WU-092C
139A	WU-092D
139A	WU-094
139A	WU-094A
139A	WU-094B
139A	WU-094C
139A	WU-094D
139A	WV-088
139A	WV-088A
139A	WV-088B
139A	WV-088C
139A	WV-088D
139A	WV-090
139A	WV-090A
139A	WV-090B
139A	WV-090C
139A	WV-090D
139A	WV-092
139A	WV-092A
139A	WV-092B
139A	WV-092C
139A	WV-092D
139A	WV-094
139A	WV-094A
139A	WV-094B
139A	WV-094C
139A	WV-094D
139A	WW-088
139A	WW-088A
139A	WW-088B
139A	WW-088C
139A	WW-088D
139A	WW-090
139A	WW-090A
139A	WW-090B
139A	WW-090C
139A	WW-090D
139A	WW-092
139A	WW-092A
139A	WW-092B
139A	WW-092C
139A	WW-092D
139A	WW-094
139A	WW-094A
139A	WW-094B
139A	WW-094C
139A	WW-094D
139A	WX-088
139A	WX-088A
139A	WX-088B
139A	WX-088C
139A	WX-088D
139A	WX-090
139A	WX-090A
139A	WX-090B
139A	WX-090C
139A	WX-090D
139A	WX-092
139A	WX-092A
139A	WX-092B
139A	WX-092C
139A	WX-092D
139A	WX-094
139A	WX-094A
139A	WX-094B
139A	WX-094C
139A	WX-094D
139A	WY-088
139A	WY-088A
139A	WY-088B
139A	WY-088C
139A	WY-088D
139A	WY-090
139A	WY-090A
139A	WY-090B
139A	WY-090C
139A	WY-090D
139A	WY-092
139A	WY-092A
139A	WY-092B
139A	WY-092C
139A	WY-092D
139A	WY-094
139A	WY-094A
139A	WY-094B
139A	WY-094C
139A	WY-094D
139A	WZ-090
139A	WZ-092
139A	WZ-092A
139A	WZ-094
139A	WZ-90
139A	WZ090
139A	WZ090A
139A	WZ094
139A	WZ9.1
139A	X092
139A	XS9.1B
139A	XZ-086
139A	XZ-092
139A	XZ090
139A	XZ092
139A	YEAD015
139A	Z0219
139A	Z0412
139A	Z0B9.1
139A	Z0C9.1
139A	Z0D9.1
139A	Z1109
139A	Z1B9.1
139A	Z2A91F
139A	Z4B9.1
139A	Z4X9.1
139A	Z4XL9.1
139A	Z4XL9.1B
139A	Z5B9.1
139A	Z5C9.1
139A	Z5D9.1
139A	Z694
139A	Z714
139A	ZA9.1
139A	ZA9.1A
139A	ZA9.1B
139A	ZB-1-9.5
139A	ZB1-9
139A	ZB1-9V
139A	ZB9.1
139A	ZB9.1A
139A	ZB9.1B
139A	ZBI-09
139A	ZD9.1A
139A	ZD9.1B
139A	ZF9.1
139A	ZG9.1
139A	ZH9.1
139A	ZH9.1A
139A	ZH9.1B
139A	ZM9.1B
139A	ZP9.1
139A	ZP9.1A
139A	ZP9.1B
139A	ZPD9.1
139A	ZQ9.1
139A	ZQ9.1A
139A	ZQ9.1B
139A	ZS9.1
139A	ZS9.1A
139A	ZS9.1B
139A	ZT9.1
139A	ZT9.1A
139A	ZT9.1B
139A	ZV9.1
139A	ZV9.1A
139A	ZV9.1B
139A	ZWO-9.1
139A	ZW9.1
139A	ZX9.1
139A	ZZ9.1
139A	001-0163-15
139A	1-20363
139A	1A12688
139A	1A9.1M
139A	1A9.1MA
139A	1E9.1Z
139A	1E9.1Z10
139A	1EZ9.1
139A	1M9.1Z
139A	1M9.1Z10
139A	1M9.1Z5
139A	1M9.1ZS10
139A	1M9.1ZS5
139A	1N1770
139A	1N1770A
139A	1N2035
139A	1N3019
139A	1N3019A
139A	1N3019B
139A	1N3678
139A	1N3678A
139A	1N3855A
139A	1N4161
139A	1N4161A
139A	1N4161B
139A	1N4326
139A	1N4326A
139A	1N4326B
139A	1N4403
139A	1N4403A
139A	1N4403B
139A	1N4631
139A	1N4739
139A	1N4739A
139A	1N5239B
139A	1N5529B
139A	1N5562
139A	1N5562A
139A	1N5855A
139A	1N5855B
139A	1N713A
139A	1N713B
139A	1N757A
139A	1N960
139A	1N960A
139A	1N960B
139A	1R9.1A
139A	1R9.1B
139A	1S130(ZENER)
139A	1S1717L
139A	1S1718
139A	1S1728
139A	1S1959
139A	1S2116A
139A	1S3009
139A	000001S334
139A	1S334K
139A	1S334N
139A	1S482
139A	1S757A
139A	1S758GS
139A	1T243M
139A	1T9.1
139A	1T9.1B
139A	1TA9.1
139A	1TA9.1A
139A	1Z9.1
139A	1Z9.1A
139A	1Z9.1B
139A	1Z9.1C
139A	1Z9.1D10
139A	1Z9.1D5
139A	1Z9.1T10
139A	1Z9.1T20
139A	1Z9.1T5
139A	1ZC9.1
139A	1ZC9.1T10
139A	1ZC9.1T5
139A	1ZF9.1T10
139A	1ZF9.1T20
139A	1ZF9.1T5
139A	1ZM9.1T10
139A	1ZM9.1T20
139A	1ZM9.1T5
139A	02Z-9.1A
139A	02Z102A
139A	02Z9.1A
139A	05-990094
139A	05Z9.1
139A	05Z9.1L
139A	08
139A	08-08125
139A	8-905-421-118
139A	8-905-421-228
139A	8C91
139A	09-306106
139A	09-306124
139A	09-306158
139A	09-306165
139A	09-306180
139A	09-306194
139A	09-306232
139A	09-306239
139A	09-306241
139A	09-306247
139A	09-306275
139A	09-306287
139A	09-306327
139A	09-306332
139A	09-306351
139A	09-306375
139A	09-306382
139A	09-306401
139A	9D141003-12
139A	9D15
139A	10-013
139A	13-085028
139A	13-085042
139A	19A115528-P3
139A	19B200379-P1
139A	21A103-064
139A	21A108-002
139A	21M214
139A	21M307
139A	21M584
139A	21M585
139A	022-2823-002
139A	22Z6AF
139A	22Z6F
139A	24MW331
139A	24MW998
139A	32-0005
139A	33R-09
139A	42-Z6A
139A	42Z6
139A	44T-300-92
139A	46-86284-3
139A	46-86421-3
139A	48-03073A09
139A	48-134653
139A	48-134957
139A	48-137130
139A	48-137164
139A	48-41763C01
139A	48-41763C03
139A	48-41873J02
139A	48-44080J05
139A	48-82256C22
139A	48-82256C43
139A	48-82256C56
139A	50Z9.1
139A	50Z9.1A
139A	50Z9.1B
139A	50Z9.1C
139A	051-0024
139A	52-053-013-0
139A	53A022-6
139A	53A022-7
139A	53K001-14
139A	53K001-18
139A	56-19
139A	56-46
139A	56-62
139A	75Z9.1
139A	75Z9.1A
139A	75Z9.1B
139A	75Z9.1C
139A	86-0005
139A	86-35-1
139A	86-37-1
139A	86-58-1
139A	93C39-3
139A	100-139
139A	100-14
139A	100-522
139A	100-523
139A	103-272
139A	103-279-18
139A	106-010
139A	0114-0017
139A	0114-0090
139A	120-004879
139A	123-020
139A	142-012
139A	145-118
139A	151-012-9-001
139A	152-008-9-001
139A	152-012-9-001
139A	152-051-9-001
139A	152-082-9-001
139A	185-015
139A	186-016
139A	230-0023
139A	523-2003-919
139A	523-2005-919
139A	690V105H32
139A	700-159
139A	754-9053-090
139A	903-119B
139A	903-179
139A	903Y00228
139A	919-01-1213
139A	919-01-3035
139A	1000-12
139A	1000-132
139A	1001-12
139A	1042-18
139A	1074-120
139A	1080-10
139A	1794-17
139A	1858A
139A	2000-308
139A	2000-327
139A	2023-41
139A	2041-06
139A	2079-93
139A	2087-46
139A	2093A41-159
139A	2093A41-16
139A	2093A41-163
139A	2093A41-186
139A	2093A41-24
139A	2093A41-24A
139A	2093B41-24
139A	2102-032
139A	2150-17
139A	2309
139A	2452-17
139A	2795A
139A	003102
139A	4653
139A	4801-00801
139A	5001-125
139A	5001-152
139A	6432-3
139A	7554
139A	8000-00004-065
139A	8000-00005-021
139A	8000-00006-009
139A	8000-00006-232
139A	8000-00011-043
139A	8000-0004-P065
139A	8000-00041-018
139A	8000-00043-021
139A	8000-00043-068
139A	8000-0005-021
139A	8000-00057-011
139A	8710-54
139A	8840-55
139A	000072150
139A	79408
139A	112530
139A	166921
139A	171162-272
139A	171162-292
139A	236266
139A	245217
139A	280317
139A	602081
139A	741739
139A	801726
139A	817197
139A	922358
139A	922524
139A	922603
139A	982254
139A	1223927
139A	1965019
139A	2002209-1
139A	2003779-26
139A	5330011
139A	5330012
139A	8120069
139A	23115076
139A	55310007-115
139A	70260540
139A	87500920
139A	87600920
139A	601100001
139A	134882256322
139A	134882256343
139A	134882256356
140A	.25T10
140A	.25T10.5
140A	.25T10.5A
140A	.25T10A
140A	.25T10B
140A	.4T10
140A	.4T10A
140A	.4T10B
140A	.7JZ10
140A	.7Z10A
140A	.7Z10B
140A	.7Z10C
140A	.7Z10D
140A	.7ZM10A
140A	.7ZM10B
140A	.7ZM10C
140A	.7ZM10D
140A	AW-01-10
140A	AW01-10
140A	AZ-094
140A	AZ-096
140A	AZ-098
140A	AZ-100
140A	AZ10
140A	AZ758A
140A	AZ961B
140A	B094
140A	B100(ZENER)
140A	B66X0040-005
140A	B094
140A	BZ-096
140A	BZ-098
140A	BZ100
140A	BZ79C10
140A	BZX29C10
140A	BZX46C10
140A	BZX55C10
140A	BZX79C10
140A	BZX83C10
140A	BZY83/C10
140A	BZY83/D10
140A	BZY83C10
140A	BZY85C10
140A	BZY92C10
140A	C4016
140A	CHZ10
140A	CHZ10A
140A	CX0051
140A	CZD010
140A	CZD010-5
140A	DZ10A
140A	E84
140A	EA16X150
140A	EA16X157
140A	EQA01-10S
140A	EQB01-10
140A	EQB01-10
140A	EZ10(ELCOM)
140A	G01-012-F
140A	GEZD-10
140A	GEZD-10-4
140A	GLA100
140A	GLA100A
140A	GLA100B
140A	HD30001200
140A	HEP-Z0413
140A	HEP101
140A	HM10B
140A	HN-00061
140A	HS2100
140A	HS7100
140A	HW10
140A	HW10A
140A	IP20-0233
140A	J24872
140A	KD2501
140A	KS100A
140A	KS100B
140A	KS2100A
140A	KS2100B
140A	KS42A
140A	KS42AF
140A	KS42B
140A	KS42BF
140A	KVR10
140A	LMZX-10
140A	LR100CH
140A	LZ10
140A	M10Z
140A	M4Z10
140A	M4Z10-20
140A	M4Z10A
140A	MA1100
140A	MC6014
140A	MC6014A
140A	MGLA100
140A	MGLA100A
140A	MGLA100B
140A	MR100C-H
140A	MR100H
140A	MTZ614
140A	MTZ614A
140A	MZ-210
140A	MZ-210B
140A	MZ10
140A	MZ1000-13
140A	MZ1010
140A	MZ10A
140A	MZ310A
140A	MZ92-10A
140A	02Z-10A
140A	OA12610
140A	OZ10T10
140A	OZ10T5
140A	PC1879-004
140A	PD6014
140A	PD6014A
140A	PD6055
140A	PS1511
140A	PS1512
140A	PS1513
140A	PS1514
140A	PS1515
140A	PS1516
140A	PS1517
140A	PS8913
140A	PTC506
140A	QZ10T10
140A	QZ10T5
140A	RD-10EB
140A	RD-11B
140A	RD-9E
140A	RD10EA
140A	RD10EB-2
140A	RD10F
140A	RD10FA
140A	RD10FB
140A	RD11A
140A	RE115
140A	RVD1N4740
140A	RZ10
140A	RZZ10
140A	S6-10
140A	S6-10A
140A	S6-10B
140A	S6-10C
140A	SK3061
140A	SV133
140A	SZ10
140A	SZL9
140A	TIXD758
140A	TMD08
140A	TMD08A
140A	TR228736003026
140A	TR330028
140A	TVSRD11
140A	TVSRD11A
140A	UZ8710
140A	UZ8810
140A	VHEWZ-100-1F
140A	VHEWZ-100//1F
140A	VR10
140A	VR10A
140A	VR10B
140A	WM-098
140A	WM-098A
140A	WM-098B
140A	WM-098C
140A	WM-098D
140A	WM-100
140A	WM-100A
140A	WM-100B
140A	WM-100C
140A	WM-100D
140A	WN-098
140A	WN-098A
140A	WN-098B
140A	WN-098C
140A	WN-098D
140A	WN-100
140A	WN-100A
140A	WN-100B
140A	WN-100C
140A	WN-100D
140A	WO-098
140A	WO-098A
140A	WO-098B
140A	WO-098C
140A	WO-098D
140A	WO-100A
140A	WO-100B
140A	WO-100C
140A	WO-100D
140A	WP-098
140A	WP-098A
140A	WP-098B
140A	WP-098C
140A	WP-098D
140A	WP-100
140A	WP-100A
140A	WP-100B
140A	WP-100C
140A	WP-100D
140A	WQ-098
140A	WQ-098A
140A	WQ-098B
140A	WQ-098C
140A	WQ-098D
140A	WQ-100
140A	WQ-100A
140A	WQ-100B
140A	WQ-100C
140A	WQ-100D
140A	WR-098
140A	WR-098A
140A	WR-098B

RCG	Industry Standard No.	ECG	Industry Standard No.	ECG	Industry Standard No.	ECG	Industry Standard No.	ECG	Industry Standard No.
140A	WR-098C	140A	1N1771A	140A	45810-54	142A	AW01-12C	142A	0A12612
140A	WR-098D	140A	1N1876	140A	58810-066	142A	AW01-12V	142A	0A2Z13
140A	WR-100	140A	1N1876A	140A	58840-117	142A	AX12	142A	0AZZT3
140A	WR-100A	140A	1N1932A	140A	93030	142A	AZ-120	142A	0Z12T10
140A	WR-100B	140A	1N1986A	140A	99201-216	142A	AZ12	142A	0Z12T5
140A	WR-100C	140A	1N2036	140A	99201-316	142A	OA2Z13	142A	PD6015
140A	WR-100D	140A	1N3020	140A	111994	142A	AZ759A	142A	PD6016A
140A	WS-098	140A	1N3020A	140A	130044	142A	AZ965B	142A	PD6057
140A	WS-098A	140A	1N3020B	140A	130047	142A	B66X0040-001	142A	PR617
140A	WS-098B	140A	1N3434	140A	147477-7-?	142A	BN7551	142A	P810022B
140A	WS-098C	140A	1N3446	140A	171560	142A	BZ-12	142A	P810066
140A	WS-098D	140A	1N3679	140A	190404	142A	BZ-120	142A	P88915
140A	WS-100	140A	1N3679A	140A	228007	142A	BZ120	142A	PTO507
140A	WS-100A	140A	1N4162	140A	262111	142A	BZX17	142A	QZ12T10
140A	WS-100B	140A	1N4162A	140A	262546	142A	BZX29C12	142A	QZ12T5
140A	WS-100C	140A	1N4327	140A	263424	142A	BZX46C12	142A	R-1348
140A	WS-100D	140A	1N4327A	140A	300732	142A	BZX55C12	142A	R-7103
140A	WT-098	140A	1N4404	140A	489752-040	142A	BZX79C12	142A	R2D
140A	WT-098A	140A	1N4404A	140A	489752-109	142A	BZX83C12	142A	R7894
140A	WT-098B	140A	1N4632	140A	530073-3	142A	BZY18	142A	R8364
140A	WT-098C	140A	1N4697	140A	530145-100	142A	BZT85C12	142A	RD11EC
140A	WT-098D	140A	1N4740	140A	530157-1100	142A	BZY85C12	142A	RD12
140A	WT-100	140A	1N4740A	140A	610122	142A	BZY92C12	142A	RD12EC
140A	WT-100A	140A	1N5240B	140A	741867	142A	C4018	142A	RD12F
140A	WT-100B	140A	1N5530B	140A	988050	142A	CD31-00015	142A	RD12FA
140A	WT-100C	140A	1N5563	140A	1223773	142A	CD31-12022	142A	RD12FB
140A	WT-100D	140A	1N5563A	140A	1800013	142A	CD4116	142A	RD13K
140A	WU-098	140A	1N5856B	140A	2002209-5	142A	CD4117	142A	RE118
140A	WU-098A	140A	1N714A	140A	2002209-9	142A	CD4118	142A	RH-EX0017CEZZ
140A	WU-098B	140A	1N754	140A	4036392-P2	142A	CD4121	142A	RH-EX0019TAZZ
140A	WU-098C	140A	1N758A	140A	5490307-P2	142A	CD4122	142A	RH-EX003GCEZZ
140A	WU-098D	140A	1N765-1	140A	42020041A	142A	CD2D12-5	142A	RH-EX0047CEZZ
140A	WU-100	140A	1N765A	140A	42020737	142A	D-00484R	142A	RS1348
140A	WU-100A	140A	1N961	140A	16602995856	142A	D2G-3	142A	RZ12
140A	WU-100B	140A	1N961A	140A	134882256328	142A	D2G-4	142A	RZZ12
140A	WU-100C	140A	1N961B	140A	134882256340	142A	D3W	142A	S1R12B
140A	WU-100D	140A	1R10	141A	.25Z11.5	142A	DS-108	142A	S1R13B
140A	WU2N1307	140A	1R10A	141A	.25Z11.5B	142A	DZ12A	142A	SD33
140A	VV-098	140A	1R10B	141A	.4T11.5	142A	EO771-7	142A	SD53
140A	VV-098A	140A	181718/4454C	141A	.4T11.5A	142A	E1852	142A	SPZ716
140A	VV-098B	140A	18195	141A	AV01-07	142A	E262	142A	SI-RECT-228
140A	VV-098C	140A	181960	141A	AZ-115	142A	EA1318	142A	SI-RECT-230
140A	VV-098D	140A	182100A	141A	D2G	142A	EA16X162	142A	SK3062
140A	VV-100	140A	182117	141A	D2G-1	142A	EA16X29	142A	SV1017
140A	VV-100A	140A	182117A	141A	D2G-2	142A	EA16X77	142A	SV135
140A	VV-100B	140A	18215	141A	D5N	142A	EA2500	142A	SV4012
140A	VV-100C	140A	18216	141A	D4N	142A	EA3719	142A	SV4012A
140A	VV-100D	140A	18225	141A	EA16X4	142A	EP16X36	142A	SWO1
140A	WW-098	140A	183010	141A	EA16X6	142A	EQA01-12	142A	SZ-EQBO1-12A
140A	WW-098A	140A	18335	141A	EVR11	142A	EQA01-12R	142A	SZ-RD12FB
140A	WW-098B	140A	1T10B	141A	EVR11A	142A	EQA01-12B	142A	SZ12
140A	WW-098C	140A	1TA10	141A	EVR11B	142A	EQA01-12Z	142A	SZ12.0
140A	WW-098D	140A	1TA10A	141A	FA8005	142A	EQBO1	142A	SZ671-B
140A	WW-100	140A	1Z10	141A	FA8006	142A	EQBO1-02R	142A	SZ671-G
140A	WW-100A	140A	1Z10A	141A	FA8007	142A	EQBO1-12	142A	SZ8
140A	WW-100B	140A	1Z10T10	141A	FA8008	142A	EQBO1-12A	142A	T-B1068
140A	WW-100C	140A	1Z10T20	141A	GEZD-11.5	142A	EQBO1-12B	142A	T-B1077
140A	WW-100D	140A	1Z10T5	141A	HEF604	142A	EQBO1-12BV	142A	T-B1106
140A	WX-098	140A	1Z010T5	141A	HW11B	142A	EQBO1-12R	142A	T-B1140
140A	WX-098A	140A	1ZC10	141A	LPM11	142A	EQBO1-12Z	142A	TE1068
140A	WX-098B	140A	1ZC10T10	141A	M4850	142A	EQO01-12A	142A	TB1077
140A	WX-098C	140A	1ZM10T10	141A	M4851	142A	ES10234	142A	TMD10
140A	WX-098D	140A	1ZM10T20	141A	MZ-110MA	142A	ET16X15	142A	TMD10A
140A	WX-100	140A	1ZM10T5	141A	SV11021	142A	EVR12	142A	TR-75
140A	WX-100A	140A	281718/4454C	141A	SVC1125	142A	EVR12A	142A	TR14002-6
140A	WX-100B	140A	02Z-10A	141A	SVC1150	142A	EVR12B	142A	TR-337
140A	WX-100C	140A	02Z10A-U	141A	SVM1020	142A	EE12(ELCOM)	142A	TVS-RD(M)(P)D
140A	WX-100D	140A	05Z-10	141A	SVM1105	142A	FZ12A	142A	TVS-RD13A
140A	WY-098	140A	05Z10	141A	SVM111	142A	FZ12T10	142A	TVS-RD13D
140A	WY-098A	140A	8-719-937-10	141A	SZ11.0	142A	FZ12T5	142A	TVS-RD13M
140A	WY-098B	140A	09-306228	141A	TVS-ZB1-11	142A	G01-037-A	142A	TVS-RD13F
140A	WY-098C	140A	09-306235	141A	TV81N4741A	142A	G12T10	142A	TV81N741A
140A	WY-098D	140A	10B62	141A	ZB1-11	142A	G12T20	142A	TV81N741H
140A	WY-100	140A	10BB1	141A	1B11Z	142A	G12T5	142A	TVSEQA01-125
140A	WY-100A	140A	10V	141A	1B11Z10	142A	GEZD-12	142A	TVSEQBO1-12
140A	WY-100B	140A	10VJ	141A	1B11A	142A	HC899	142A	TVSQA01-12B
140A	WY-100C	140A	13-0002	141A	1N492A	142A	HD3000101-0	142A	TZ12
140A	WY-100D	140A	13-33179-4	141A	1N492B	142A	HD3002409	142A	TZ12A
140A	WZ-088A	140A	13DD02P	141A	1N493	142A	HEP-ZO415	142A	TZ12B
140A	WZ-098A	140A	019-003411	141A	1N493A	142A	HEP105	142A	TZ120
140A	WZ-100	140A	19-090-008A	141A	1N493B	142A	HF-20041	142A	UZ-12B
140A	WZ-100A	140A	21M493	141A	18135	142A	HM12B	142A	UZ13B
140A	WZ096	140A	2326AP	141A	1Z11-5	142A	HN-0005	142A	UZ8712
140A	WZ10	140A	2326P	141A	1Z11.5B	142A	HS2120	142A	UZ8812
140A	WZ100	140A	24MW1065	141A	1TA11.5	142A	HS7120	142A	V160
140A	WZ533	140A	24MW1125	141A	1TA11.5A	142A	HW12	142A	VR12
140A	XZ-096	140A	24MW743	141A	14-515-18	142A	HW12A	142A	VR12A
140A	XZ-100	140A	32-0049	141A	48-134850	142A	HW12B	142A	VR12B
140A	XZ098	140A	4326	141A	48-134851	142A	HZ-212	142A	WM-120
140A	XZ102	140A	4326A	141A	48-137188	142A	J20438	142A	WM-120A
140A	Z0220	140A	44T-300-93	141A	48-82256C08	142A	KD2505	142A	WM-120B
140A	Z0413	140A	46-86323-3	141A	50Z11	142A	KS120A	142A	WM-120D
140A	Z0B10	140A	48-11-005	141A	50Z11A	142A	KS120B	142A	WN-120
140A	Z0C10	140A	48-41873J03	141A	50Z11B	142A	KS2120A	142A	WN-120A
140A	Z0D10	140A	48-82256C28	141A	50Z11C	142A	KS2120B	142A	WN-120C
140A	Z10	140A	48-82256C040	141A	56-53	142A	KS44A	142A	WN-120D
140A	Z10K	140A	55Z4	141A	75Z11	142A	KS44AF	142A	WO-120
140A	Z1110A	140A	56-67	141A	75Z11A	142A	KS44B	142A	WO-120A
140A	Z1110-C	140A	66X0040-005	141A	75Z11B	142A	KS44BF	142A	WO-120B
140A	Z1B10	140A	68X0040-005	141A	75Z11C	142A	LPM12	142A	WO-120C
140A	Z1C10	140A	070-011	141A	2112Z	142A	LPM12A	142A	WO-120D
140A	Z1D10	140A	070-024	141A	296Y019B04	142A	LPZ212	142A	WP-120
140A	Z5B10	140A	86-118-1	141A	297V019B04	142A	LR120CH	142A	WP-120A
140A	Z5C10	140A	93339-6	141A	3112Z	142A	M12Z	142A	WP-120B
140A	Z5D10	140A	96-5116-01	141A	412Z4	142A	M4Z12	142A	WP-120C
140A	ZB10	140A	145-100	141A	4850	142A	M4Z12-20	142A	WP-120D
140A	ZB10A	140A	157-100	141A	71411-1	142A	M4Z12A	142A	WQ-120
140A	ZB10B	140A	179-4	142A	.25Z12	142A	MA1120	142A	WQ-120A
140A	ZB10X	140A	187-6	142A	.25Z12A	142A	MC6016	142A	WQ-120B
140A	ZD10B	140A	264P09703	142A	.25Z12B	142A	MC6016A	142A	WQ-120C
140A	ZE10	140A	339-529-002	142A	.4T12A	142A	MC6116	142A	WQ-120D
140A	ZP10	140A	464-311-19	142A	.4T12B	142A	MC6116A	142A	WR-120
140A	Z010	140A	523-2001-100	142A	.75Z12	142A	MD759	142A	WR-120A
140A	ZM10B	140A	523-2003-100	142A	.7JZ12	142A	MD759A	142A	WR-120B
140A	ZP10	140A	523-2005-100	142A	.7Z12A	142A	MEZ12T10	142A	WR-120C
140A	Z010	140A	523-2505-100	142A	.7Z12B	142A	MEZ12T5	142A	WR-120D
140A	001-0127-00	140A	523-2509-100	142A	.7Z12D	142A	MR1200-H	142A	WS-120
140A	001-023035.	140A	623-2003-100	142A	.72M12A	142A	MTZ616	142A	WS-120A
140A	1-20398	140A	630-077	142A	.72M12B	142A	MTZ616A	142A	WS-120B
140A	1A10M	140A	903-337	142A	.72M12C	142A	MZ-11	142A	WS-120C
140A	1A10MA	140A	1006-5985	142A	.72M12D	142A	MZ-12	142A	WS-120D
140A	1E10Z	140A	1010	142A	A-1341M66-2	142A	MZ-212	142A	WT-120
140A	1B10Z10	140A	1040-09	142A	A054-151	142A	MZ1000-15	142A	WT-120A
140A	1M10Z	140A	1040-09(ZENER)	142A	A2474	142A	MZ1012	142A	WT-120B
140A	1M10Z10	140A	1065-6775	142A	A36939	142A	MZ12	142A	WT-120C
140A	1M10Z5	140A	1411-136	142A	A7287500	142A	MZ12A	142A	WT-120D
140A	1M10Z810	140A	2040-17	142A	A7287513	142A	MZ12B	142A	WU-120
140A	1M10Z85	140A	5632-HZ11A	142A	AU2012	142A	MZ212	142A	WU-120A
140A	1N1512	140A	5635-HZ11A	142A	AV2012	142A	MZ92-12A	142A	WU-120C
140A	1N1512A	140A	5635-ZB1-10	142A	AV5	142A	N-EA16X29	142A	WU-120D
140A	1N1523	140A	8000-00028-047	142A	AW-01-12	142A	NGP5010	142A	WU-125
140A	1N1523A	140A	8000-00058-004	142A	AW-01-12C	142A	02Z12A	142A	WU-125A
140A	1N1744	140A	26810-52			142A	OA126-12		
140A	1N1771	140A	38510-53			142A	OA126/12		
		140A	42020-737						

ECG	Industry Standard No.	ECG	Industry Standard No.	ECG	Industry Standard No.	ECG	Industry Standard No.	ECG	Industry Standard No.	ECG	Industry Standard No.
142A	WU-125B	142A	1M122	142A	48-82256054	143A	.7JZ13	143A	WQ-130	143A	1A13M
142A	WU-125C	142A	1M122I0	142A	48-97048A08	143A	.7Z13A	143A	WQ-130A	143A	1A13MA
142A	WU-125D	142A	1M122I0	142A	48-97048A16	143A	.7Z13B	143A	WQ-130B	143A	1M13Z810
142A	WU-20B	142A	1M122I10	142A	488137000	143A	.7Z13C	143A	WQ-130C	143A	1N1774
142A	WV-120A	142A	1M122S5	142A	488155103	143A	.7Z13D	143A	WQ-130D	143A	1N1774A
142A	WV-120B	142A	1N426	142A	502I2	143A	.7ZM13A	143A	WR-125	143A	1N1783
142A	WV-120C	142A	1N1515	142A	50Z12A	143A	.7ZM13B	143A	WR-125A	143A	1N1A13MA
142A	WV-120D	142A	1N1513A	142A	50Z12B	143A	.7ZM13C	143A	WR-125B	143A	1N2037
142A	WW-115	142A	1N1524	142A	50Z12C	143A	.7ZM13D	143A	WR-125C	143A	1N2037-2
142A	WW-115A	142A	1N1524A	142A	53A001-10	143A	AO54-231	143A	WR-125D	143A	1N3023
142A	WW-115B	142A	1N1773	142A	56-4857	143A	AWO1(RCA)	143A	WR-130	143A	1N3023A
142A	WW-115C	142A	1N1773A	142A	56-51	143A	AZ-125	143A	WR-130A	143A	1N3023B
142A	WW-115D	142A	1N1877	142A	56-57	143A	AZ-130	143A	WR-130B	143A	1N3682
142A	WW-120	142A	1N1877A	142A	63-9942	143A	AZ13	143A	WR-130C	143A	1N3682A
142A	WW-120A	142A	1N1960A	142A	65-085004	143A	AZ964B	143A	WR-130D	143A	1N3682B
142A	WW-120B	142A	1N2037-1	142A	75Z12	143A	BZ-125	143A	WS-125	143A	1N4165
142A	WW-120C	142A	1N2037A	142A	75Z12A	143A	BZ-130	143A	WS-125A	143A	1N4165A
142A	WW-120D	142A	1N3022	142A	75Z12B	143A	BZ130	143A	WS-125B	143A	1N4165B
142A	WX-120	142A	1N3022A	142A	75Z12C	143A	BZX29C13	143A	WS-125D	143A	1N4330
142A	WX-120A	142A	1N3022B	142A	78-271383-1	143A	BZX46C13	143A	WS-130	143A	1N4330A
142A	WX-120B	142A	1N3435	142A	86-052I	143A	BZX55C13	143A	WS-130A	143A	1N4330B
142A	WX-120C	142A	1N3447	142A	86-51-1	143A	BZX79C13	143A	WS-130C	143A	1N4407
142A	WX-120D	142A	1N3681	142A	86-61-1	143A	BZX83C13	143A	WS-130D	143A	1N4407A
142A	WY-120A	142A	1N3681A	142A	86-85-1	143A	BZY85C13V5	143A	WT-125	143A	1N4635
142A	WY-120C	142A	1N3681B	142A	93A39-12	143A	BZY92C13	143A	WT-125B	143A	1N4700
142A	WY-120D	142A	1N4164	142A	93A39-43	143A	D18	143A	WT-125C	143A	1N4743
142A	WZ12	142A	1N4164A	142A	93A80-1	143A	D12VIO	143A	WT-130	143A	1N4743A
142A	WZ120	142A	1N4164B	142A	93B39-12	143A	DIS	143A	WT-130A	143A	1N4743B
142A	WZ535	142A	1N4329	142A	93B39-13	143A	DZ12	143A	WT-130B	143A	1N5533
142A	WZ917	142A	1N4329A	142A	93C39-12	143A	DZ13	143A	WT-130C	143A	1N5566
142A	X330302	142A	1N4329B	142A	93C39-13	143A	EA16X81	143A	WT-130D	143A	1N5566A
142A	XZ-122	142A	1N4406	142A	93C39-2	143A	EQAO1-13	143A	WU-130	143A	1N5739B
142A	Z-1014	142A	1N4406A	142A	93C39-6	143A	EQAO1-13R	143A	WU-130A	143A	1N5959B
142A	Z-1112C	142A	1N4406B	142A	96-5091-01	143A	EQBO1-13	143A	WU-130B	143A	1N717A
142A	Z-963B	142A	1N4634	142A	96-5133-01-02	143A	EZ13(ELCOM)	143A	WU-130D	143A	1N766-2
142A	ZO-12	142A	1N4699	142A	96-5133-02	143A	GEZD-13	143A	WV-125	143A	1N766A
142A	ZO-12A	142A	1N4742	142A	96-5248-04	143A	HEP-Z0416	143A	WV-125A	143A	1R13
142A	ZO-12B	142A	1N4742A	142A	103-158	143A	HEP605	143A	WV-125B	143A	1R13B
142A	ZO222	142A	1N5242B	142A	103-279-01	143A	HW13	143A	WV-125C	143A	1S228
142A	ZO415	142A	1N5532B	142A	103-279-21	143A	HW13A	143A	WV-125D	143A	1S3013
142A	ZOB-12	142A	1N5565	142A	160	143A	HW13B	143A	WV-130	143A	1S338Q
142A	ZOB12	142A	1N5565A	142A	232-0009-31	143A	LPM13	143A	WV-130A	143A	1T12.8
142A	ZOC012	142A	1N5858A	142A	264P03303	143A	LPM13A	143A	WV-130B	143A	1T12.8B
142A	ZOD12	142A	1N5858B	142A	264P09301	143A	MA1130	143A	WV-130C	143A	1TA12.8
142A	Z1112	142A	1N665	142A	264P10502	143A	MMZ12(06)	143A	WV-130D		
142A	Z1112-C	142A	1N716A	142A	29GLOO3B01	143A	PTC50S	143A	WX-125		
142A	Z111Z	142A	1N716B	142A	296I017B01	143A	RD-13A	143A	WX-125A		
142A	Z1212	142A	1N759A	142A	296V018B01	143A	RD-13AK	143A	WX-125B		
142A	Z12A	142A	1N766-1	142A	523-2003-120	143A	RD-13AK-P	143A	WX-125C		
142A	Z12K	142A	1N963B	142A	600X0101-066	143A	RD-13AKP	143A	WX-125D		
142A	OZ12T10	142A	1R12	142A	690V101853	143A	RD-13E	143A	WY-125A		
142A	OZ12T5	142A	1R12A	142A	690V116IA0	143A	RD-13M	143A	WY-125B		
142A	Z1B12	142A	1R12B	142A	800-004-00	143A	RD13	143A	WY-125C		
142A	Z1C12	142A	1S1962	142A	800-023-00	143A	RD13A	143A	WY-125D		
142A	Z1D12	142A	1S197	142A	800-035-00	143A	RD13AK	143A	WY-130		
142A	Z2A120P	142A	1S2120A	142A	903-97B	143A	RD13AKP	143A	WY-130A		
142A	Z4B12	142A	1S227	142A	919-00-1445-002	143A	RD13AN	143A	WY-130B		
142A	Z4X12	142A	1S3012	142A	919-001-1445-2	143A	RD13AN-P	143A	WY-130C		
142A	Z4XL12	142A	1S337	142A	919-001445-2	143A	RD13B	143A	WY-130D		
142A	Z4XL12B	142A	1S337-Y	142A	919-013058	143A	RD13B-Z	143A	WZ12.8		
142A	Z5B12	142A	1S337A	142A	1002-4404	143A	RD13P	143A	WZ130		
142A	Z5B2	142A	1S337B	142A	1033	143A	RD13FA	143A	WZ13B		
142A	Z5C12	142A	1S337Y	142A	1043-1534	143A	RD13FB	143A	Z01-13		
142A	Z5D12	142A	1S696	142A	1411-135	143A	RE119	143A	Z0416		
142A	ZA12	142A	1T12	142A	1766	143A	RH-EXO011CEZZ	143A	Z4X13		
142A	ZA12A	142A	1T12B	142A	2093A39-12	143A	RZ13	143A	ZA13		
142A	ZA12B	142A	1TA12	142A	2093A41-56	143A	RZ213	143A	ZA13A		
142A	ZB1-10	142A	1TA12A	142A	2093A80-1	143A	SD-33	143A	ZA13B		
142A	ZB1-12	142A	1Z12	142A	3520(WARDS)	143A	SK3093	143A	ZB1-13		
142A	ZB12	142A	1Z12A	142A	004887	143A	SZ-200-13	143A	ZB13		
142A	ZB12A	142A	1Z12B	142A	5122	143A	SZ13	143A	ZB13B		
142A	ZB12B	142A	1Z12C	142A	5222	143A	SZ13.0	143A	ZBE506		
142A	ZOC012	142A	1Z12T10	142A	35219	143A	SZ961-V	143A	ZF13		
142A	ZOD012	142A	1Z12T20	142A	36539	143A	TVSRD13AL	143A	ZH13		
142A	ZD12A	142A	1Z12T5	142A	42025-850	143A	UZ8713	143A	ZH13A		
142A	ZD12B	142A	1ZB12	142A	59991-1	143A	UZ8813	143A	ZH13B		
142A	ZB12A	142A	1ZB12B	142A	71411-2	143A	VR13	143A	ZM13B		
142A	ZB12B	142A	1ZC12	142A	99201-110	143A	VR13A				
142A	ZBC12	142A	1ZC12T10	142A	99201-219	143A	VR13B				
142A	ZENNER-122	142A	1ZC12T20	142A	120504	143A	WL-125				
142A	ZF12	142A	1ZC12T5	142A	125126	143A	WL-125A				
142A	ZF12A	142A	1ZF12T10	142A	130328	143A	WL-125B				
142A	ZF12B	142A	1ZF12T20	142A	130380	143A	WL-125D				
142A	ZG12	142A	1ZM12T10	142A	132416	143A	WL-130				
142A	ZH12	142A	1ZM12T20	142A	136634	143A	WL-130A				
142A	ZH12A	142A	1ZM12T5	142A	141429	143A	WL-130B				
142A	ZH12B	142A	1ZT12	142A	147704-6-4	143A	WL-130D				
142A	ZJ12A	142A	1ZT12B	142A	165358	143A	WM-125				
142A	ZJ12B	142A	02I12A	142A	224780	143A	WM-125A				
142A	ZM12B	142A	003-009200	142A	317208	143A	WM-125B				
142A	ZO-12	142A	003-009700	142A	0330302	143A	WM-125D				
142A	ZO-12A	142A	003-010000	142A	489752-045	143A	WM-130				
142A	ZO-12B	142A	03-933943	142A	530073-1013	143A	WM-130A				
142A	ZOB12	142A	4-2020-12400	142A	530073-13	143A	WM-130B				
142A	ZP12	142A	4JX24X539	142A	530073-14	143A	WM-130D				
142A	ZP12A	142A	4J24X539	142A	530073-6	143A	WO-125				
142A	ZP12B	142A	4J24XL12	142A	530118-6	143A	WO-125A				
142A	ZQ12	142A	4WO1-13	142A	530145-120	143A	WO-125B				
142A	ZQ12A	142A	05212	142A	530163-120	143A	WO-125D				
142A	ZQ12B	142A	8-719-930-12	142A	760304	143A	WO-130				
142A	ZR50B793-1	142A	8-905-421-234	142A	801731	143A	WO-130A				
142A	ZS12	142A	8-905-421-319	142A	811155	143A	WO-130B				
142A	ZS12A	142A	09-306179	142A	817208	143A	WO-130D				
142A	ZS12B	142A	09-306391	142A	922693	143A	WO130				
142A	ZT12	142A	1205Y	142A	1444875-1	143A	WP-125				
142A	ZT12A	142A	12V	142A	1474777-1	143A	WP-125A				
142A	ZT12B	142A	13-14879-2	142A	1477046-4	143A	WP-125B				
142A	ZU12	142A	13-33179-6	142A	2002209-10	143A	WP-125D				
142A	ZU12A	142A	13-55332-1	142A	2327071	143A	WP-130				
142A	ZU12B	142A	14-509-01	142A	2330241	143A	WP-130A				
142A	ZV12	142A	14-515-01	142A	2331152	143A	WP-130B				
142A	ZV12A	142A	14-515-03	142A	2337101	143A	WP-130D				
142A	ZV12B	142A	14-515-25	142A	5490907-P3	143A	WQ-125				
142A	ZY12A	142A	1624	142A	7492377-5	143A	WQ-125A				
142A	ZY12B	142A	18-085002	142A	23115919	143A	WQ-125B				
142A	ZZ12	142A	21A037-003	142A	30600140	143A	WQ-125C				
142A	ZZ12M	142A	21A037-008	142A	62390174	143A	WQ-125D				
142A	1A12MA	142A	2526	142A	62564504						
142A	1AC12	142A	2526A	142A	62921967						
142A	1AC12A	142A	2526AF	142A	2012230109						
142A	1AC12B	142A	2526P	142A	2012230312						
142A	1C12I	142A	34-8057-14	142A	4129501200						
142A	1C12ZA	142A	34-8057-33	142A	134882256354						
142A	1D212	142A	34-8057-41	142A	.25T12.8						
142A	1E12Z	142A	34-8057-53	142A	.25T12.8A						
142A	1E12Z10	142A	36-6343	142A	.25T12.8B						
142A	1E12Z5	142A	39(SHARP)	142A	.25T13						
142A	1E225	142A	41-0905	142A	.25T13A						
142A	1EZ12	142A	42-27541	143A	.4T12-8						
		142A	46-86379-3	143A	.4T12-8A						
		142A	48-03073A08	143A	.4T13						
		142A	48-137177	143A	.4T13A						
		142A	48-137272	143A	.4T13B						
		142A	48-52000								

ECG	Industry Standard No.	ECG	Industry Standard No.	ECG	Industry Standard No.	ECG	Industry Standard No.	ECG	Industry Standard No.
143A	17A12.8A	144A	WQ-140B	145A	.25T15B	145A	Q21ST5	145A	XZ-152
143A	1Z13	144A	WQ-140C	145A	.4T15	145A	R2E	145A	Z-1016
143A	1Z13A	144A	WQ-140D	145A	.4T15B	145A	RD-15E	145A	Z-1016A
143A	1Z13T10	144A	WR-140	145A	.75N5	145A	RD-16H	145A	Z-1114
143A	1Z13T20	144A	WR-140A	145A	.7JZ15	145A	RD-16HA	145A	Z-15
143A	1Z13T5	144A	WR-140B	145A	.7Z14A	145A	RD15E	145A	Z-15E
143A	1ZC13T10	144A	WR-140C	145A	.7Z14C	145A	RD15EB	145A	20225
143A	1ZC13T5	144A	WR-140D	145A	.7Z15D	145A	RD15F	145A	Z0415B
143A	02Z13A	144A	WS-140	145A	.7ZM15A	145A	RD15FA	145A	ZOB-15
143A	05Z13	144A	WS-140A	145A	.7ZM15B	145A	RD15FB	145A	ZOB15
143A	09-506127	144A	WS-140B	145A	.7ZM15C	145A	RD158	145A	ZOC-15
143A	21A037-018	144A	WS-140C	145A	.7ZM15D	145A	RD16H	145A	ZOC15
143A	21A119-008	144A	WS-140D	145A	A04166-2	145A	RB121	145A	ZOD-15
143A	21M432(REG)	144A	WT-140	145A	AV-2015	145A	RT3671	145A	ZOD15
143A	2626AF	144A	WT-140A	145A	AV2015	145A	RT3671A	145A	Z1114
143A	2626P	144A	WT-140B	145A	AW01-15	145A	RZ-15AB	145A	Z1114-C
143A	42-19917	144A	WT-140C	145A	AZ-145	145A	RZ15	145A	Z1214
143A	46-86296-3	144A	WT-140D	145A	AZ-15	145A	RZ15A	145A	Z15
143A	48-137000	144A	WU-140	145A	AZ-150	145A	RZZ156	145A	Z15K
143A	48-137209	144A	WU-140A	145A	AZ15	145A	SK3063	145A	Z1B-15
143A	56-32	144A	WU-140B	145A	AZ95B	145A	SU5	145A	Z1B15
143A	75Z13	144A	WU-140C	145A	B-6002	145A	SV-1020A	145A	Z1C-15
143A	75Z13A	144A	WU-140D	145A	BZ-145	145A	SV-138A	145A	Z1C15
143A	75Z13B	144A	WV-140	145A	BZ-150	145A	SV-4015A	145A	Z1D-15
143A	75Z13C	144A	WV-140A	145A	BZ-X19	145A	SV1020	145A	Z1D15
143A	86-67-1	144A	WV-140B	145A	BZ150	145A	SV138	145A	Z2A-150P
143A	93A102-2	144A	WV-140C	145A	BZX-71C15	145A	SV4015	145A	Z2A150P
143A	93A39-31	144A	WV-140D	145A	BZX19	145A	SV4015A	145A	Z4B-15
143A	100-286	144A	WW-125	145A	BZX29C15	145A	SZ-150	145A	Z4B15
143A	103-301-17A	144A	WW-125A	145A	BZX46C15	145A	SZ-15B	145A	Z4X-15
143A	264P1T402	144A	WW-125B	145A	BZX55C15	145A	SZ-200-15	145A	Z4X15
143A	296V011H03	144A	WW-125C	145A	BZX83C15	145A	SZ-200-15A	145A	Z5B-15
143A	523-2003-130	144A	WW-125D	145A	BZY-19	145A	SZ15	145A	Z5C-15
143A	601X0226-066	144A	WW-130	145A	BZY-83D15	145A	SZ15.0	145A	Z5C15
143A	1550-17	144A	WW-130A	145A	BZY19	145A	SZ150	145A	Z5D-15
143A	2093A41-113	144A	WW-130B	145A	BZY83/C15	145A	SZ961-B	145A	ZA15
143A	3506	144A	WW-130C	145A	BZY83/D15	145A	TRI25B	145A	ZA15A
143A	99202-220	144A	WW-130D	145A	BZY83C15	145A	TRI4002-12	145A	ZA15B
143A	129940	144A	WW-140A	145A	BZY83C16	145A	TVS-ZB1-15	145A	ZA15V
143A	161022	144A	WW-140B	145A	BZY85C15	145A	TVSBQB01-15	145A	ZB-15A
143A	166985	144A	WW-140C	145A	BZY82C15	145A	TVSQA01-07R	145A	ZB1-15
143A	2002331-46	144A	WW-140D	145A	C4020	145A	TVSQA01-15RB	145A	ZB13A
143A	13030401	144A	WX-140	145A	CD31-00017	145A	TVSQB01-15ZB	145A	ZB15
143A	42023425	144A	WX-140A	145A	CD31-10365	145A	TZ-15A	145A	ZB15A
143A	42029566	144A	WX-140C	145A	CD31-12024	145A	TZ-15AB	145A	ZC-015
144A	.25T14	144A	WX-140D	145A	CZD015	145A	TZ-15BC	145A	ZC015
144A	.25T14A	144A	WY-140	145A	CZD015-5	145A	TZ-15C	145A	ZD-015
144A	.25T14B	144A	WY-140A	145A	D1ZBLU	145A	TZ15	145A	ZD-15
144A	.4T14	144A	WY-140B	145A	D1ZYBL	145A	TZ15A	145A	ZD-15B
144A	.4T14A	144A	WY-140C	145A	D5B	145A	TZ15B	145A	ZD015
144A	.7Z14	144A	WY-140D	145A	DHD805(ZENER)	145A	TZ15C	145A	ZD15A
144A	.7Z14B	144A	WZ14	145A	DI15A	145A	UZ-15C	145A	ZD15B
144A	.72M14A	144A	WZ140	145A	EA16X124	145A	UZ15	145A	ZE-15
144A	.72M14B	144A	WZ537	145A	EP16X5	145A	UZ8715	145A	ZE-15B
144A	.72M14C	144A	WZ919	145A	EQA01-15R	145A	UZ8815	145A	ZE15
144A	.72M14D	144A	Z0417	145A	EQB01-15	145A	V126	145A	ZE15A
144A	AZ-135	144A	Z4X14	145A	EQB01-15Z	145A	VR-14	145A	ZE15B
144A	AZ-140	144A	Z4X14A	145A	EQB01-15ZB	145A	VR15	145A	ZEN508
144A	AZ7	144A	Z4XL14	145A	EVR15	145A	VR15A	145A	ZF-15
144A	B140	144A	Z4XL14B	145A	EVR15A	145A	VR15B	145A	ZF-15A
144A	BZ-135	144A	ZB1-14	145A	EVR15B	145A	W-150	145A	ZF-15B
144A	BZ-140	144A	ZEN507	145A	EZ15(ELCOM)	145A	WL-150	145A	ZF-16
144A	BZ140	144A	ZM14B	145A	EZ150	145A	WL-150A	145A	ZF15
144A	CZD014	144A	ZXL-14	145A	FA8009	145A	WL-150C	145A	ZF15A
144A	CZD014-5	144A	ZXY-14	145A	FA8010	145A	WL-150D	145A	ZF15B
144A	D1T	144A	ZXY14B	145A	FA8011	145A	WM-150	145A	ZG-15
144A	D1ZRED	144A	1Z14Z	145A	FA8012	145A	WM-150A	145A	ZG-15B
144A	D4P	144A	1Z14Z10	145A	FZ15A	145A	WM-150C	145A	ZG15
144A	DIT	144A	1Z14Z5	145A	FZ15T10	145A	WM-150D	145A	ZG15B
144A	DS-110	144A	1M14Z	145A	FZ15T5	145A	WN-150	145A	ZH-15
144A	EO771-6	144A	1M14Z10	145A	G15T10	145A	WN-150A	145A	ZH-15A
144A	EQA01-14R	144A	1M14Z5	145A	G15T20	145A	WN-150D	145A	ZH-15B
144A	EQA01-14RD	144A	1N2037-3	145A	GEZD-15	145A	WO-150	145A	ZH15
144A	EQB01-14	144A	1N4701	145A	GT15T5	145A	WO-150A	145A	ZH15A
144A	EZ14(ELCOM)	144A	1N5244B	145A	HEP-Z0418	145A	WO-150C	145A	ZH15B
144A	FZ14T10	144A	1N5534B	145A	HEP607	145A	WO-150D	145A	ZJ-15
144A	FZ14T5	144A	1N5860B	145A	HM15B	145A	WP-150	145A	ZJ-15B
144A	GEZD-14	144A	1N766-3	145A	HS2150	145A	WP-150A	145A	ZJ15
144A	HD3001109-0	144A	181963	145A	HS7150	145A	WP-150C	145A	ZJ15A
144A	HD30011090	144A	18198	145A	HW15	145A	WP-150D	145A	ZO-15
144A	HD3002109	144A	18229	145A	HW15A	145A	WQ-150	145A	ZO-15A
144A	HEP-Z0417	144A	18338U	145A	HW15B	145A	WQ-150A	145A	ZO-15AC
144A	HEP606	144A	18484	145A	KS3150A	145A	WQ-150B	145A	ZO-15B
144A	M4552	144A	1T14	145A	KS3150B	145A	WQ-150C	145A	ZO-15BY
144A	M4659	144A	1T14B	145A	KS2150A	145A	WQ-150D	145A	ZOB-15
144A	MZ1014	144A	1TA14	145A	KS2150B	145A	WR-150	145A	ZOB15
144A	MZ14	144A	1TA14A	145A	KS46	145A	WR-150A	145A	ZOD-15
144A	OA126/14	144A	12F14T10	145A	KS46AF	145A	WR-150B	145A	ZOD15
144A	OA12614	144A	12F14T20	145A	KS46BF	145A	WR-150C	145A	ZP-15
144A	PTO511	144A	12F14T5	145A	LPM15	145A	WR-150D	145A	ZP-15A
144A	QZ14T10	144A	09-306278	145A	LPM15A	145A	WS-150	145A	ZP-15B
144A	QZ14T5	144A	27Z6	145A	LPZ15	145A	WS-150A	145A	ZP15
144A	RE120	144A	27Z6A	145A	LR150CH	145A	WS-150B	145A	ZP15A
144A	SK3094	144A	27Z6AF	145A	M15Z	145A	WS-150C	145A	ZP15B
144A	SV1019	144A	27Z6P	145A	M4Z15	145A	WS-150D	145A	ZQ-15
144A	SV13T	144A	34M14Z	145A	M4Z15-20	145A	WT-150	145A	ZQ-15A
144A	SV4014	144A	34M14Z10	145A	M4Z15A	145A	WT-150A	145A	ZQ-15B
144A	SV4014A	144A	34M14Z5	145A	MA1150	145A	WT-150B	145A	ZQ15
144A	SZ14	144A	34Z14D	145A	MC6018	145A	WT-150C	145A	ZQ15A
144A	SZ961-R	144A	34Z14D10	145A	MC6018A	145A	WT-150D	145A	ZQ15B
144A	SZ961-T	144A	34Z14D5	145A	MC6118	145A	WU-150	145A	ZQ15X
144A	TRZ33T103	144A	46-86341-3	145A	MC6118A	145A	WU-150A	145A	ZR50B921-3
144A	TVSQA01-14RD	144A	46-86401-3	145A	MEZ15T10	145A	WU-150C	145A	ZS-15
144A	UZ8714	144A	48-134552	145A	MEZ15T5	145A	WU-150D	145A	ZS-15B
144A	VR14	144A	48-134659	145A	MR150C-H	145A	WV-150	145A	ZS15
144A	VR14A	144A	48-137017	145A	MTZ618	145A	WV-150A	145A	ZS15A
144A	VR14B	144A	48-137048	145A	MTZ618A	145A	WV-150B	145A	ZS15B
144A	WL-140	144A	48-137298	145A	MZ1000-17	145A	WV-150C	145A	ZT-15
144A	WL-140A	144A	48-137365	145A	MZ15A	145A	WW-150	145A	ZT-15A
144A	WL-140B	144A	48-82256010	145A	MZ15T20	145A	WW-150A	145A	ZT-15B
144A	WL-140C	144A	50Z14	145A	MZ92-15A	145A	WW-150B	145A	ZT15
144A	WL-140D	144A	50Z14A	145A	021S110	145A	WW-150C	145A	ZT15A
144A	WM-140	144A	50Z14B	145A	021ST5	145A	WW-150D	145A	ZT15B
144A	WM-140A	144A	50Z14Z	145A	P38103/507-10	145A	WX-150	145A	ZU-15
144A	WM-140B	144A	53A001-33	145A	PAB261	145A	WX-150A	145A	ZU-15B
144A	WM-140C	144A	75Z14	145A	PD-6018	145A	WX-150B	145A	ZU15
144A	WM-140D	144A	75Z14A	145A	PD-6018A	145A	WX-150C	145A	ZU15A
144A	WN-140	144A	75Z14C	145A	PD-6059	145A	WX-150D	145A	ZU15B
144A	WN-140A	144A	523-2003-140	145A	PD6018	145A	WY-150	145A	ZV-15
144A	WN-140B	144A	540-013	145A	PD6018A	145A	WY-150A	145A	ZV-15A
144A	WN-140C	144A	1000-133	145A	PD6059	145A	WY-150C	145A	ZV-15B
144A	WN-140D	144A	4552	145A	PR-620	145A	WY-150D	145A	ZV15
144A	WO-140	144A	165740	145A	PR620	145A	WZ-15B	145A	ZV15A
144A	WO-140A	144A	530073-1023	145A	PS-10068	145A	WZ-920	145A	ZV15B
144A	WO-140B	144A	530073-23	145A	PS-8917	145A	WZ15	145A	ZY-15
144A	WO-140C	144A	1474777-11	145A	PS10024B	145A	WZ150	145A	ZY-15A
144A	WO-140D	144A	2330305	145A	PS10068	145A	WZ538		
144A	WP-140	144A	134882256310	145A	PS8917	145A	WZ920		
144A	WP-140A	144A	.25T15	145A	PTO509	145A	XB152		
144A	WP-140B	145A	.25T15	145A	QBO1-15ZB				
144A	WP-140C	145A	.25T15A	145A	QZ-15T10				
144A	WP-140D			145A	QZ-15T5				
144A	WQ-140A			145A	QZ15T5				
				145A	QZ15T10				

ECG	Industry Standard No.	ECG	Industry Standard No.	ECG	Industry Standard No.	ECG	Industry Standard No.	ECG	Industry Standard No.
145A	Z1-15B	145A	50215	146A	QZ27T10A	147A	.7J233	147A	50233
145A	ZY15A	145A	50215A	146A	QZ27T5	147A	.7233A	147A	50233A
145A	ZT15B	145A	50215B	146A	QZZ7T5A	147A	.7233B	147A	50233B
145A	ZZ-15	145A	50215C	146A	R227	147A	.7233D	147A	50233C
145A	ZZ15	145A	56-25	146A	R227A	147A	.72M33A	147A	52-053-005-0
145A	001-02303-4	145A	62-20643	146A	R227AC	147A	.72M33B	147A	56A7-1
145A	001-023034	145A	66200040-009	146A	R2Z27=12	147A	.72M33D	147A	5607-1
145A	1A15M	145A	070-021	146A	SV4027	147A	AA10	147A	75233
145A	1A15MA	145A	73-15	146A	SV4027-6	147A	AA10-1	147A	75233A
145A	1AC15	145A	75215	146A	SV4027A	147A	AN155	147A	75233B
145A	1AC15A	145A	75215A	146A	SV4027A-6	147A	AV10	147A	75233C
145A	1AC15B	145A	75215B	146A	TC27A5	147A	AV2033	147A	86-94-1
145A	1C15Z	145A	75215C	146A	TC27A5A	147A	AW-01-33	147A	93A39-40
145A	1C15ZA	145A	78-271383-3	146A	TC27A5A-5	147A	AWO1-33	147A	103-236
145A	1D08	145A	78-271383-4	146A	TX27A	147A	AZ973B	147A	152-006-9-001
145A	1E15Z	145A	93A39-24	146A	ZA27B	147A	BZ-310	147A	152-029-9-002
145A	1E15Z10	145A	96-5110-02	146A	1A27M	147A	BZX27	147A	264P11009
145A	1E15Z5	145A	96-5110-03	146A	1A27MA	147A	BZY29C33	147A	1081-3186
145A	1EZ15	145A	96-5248-06	146A	1AC27	147A	BZY29C33	147A	1133
145A	1M15Z	145A	11524	146A	1AC27A	147A	CD31-00025	147A	4663
145A	1M15Z10	145A	137-759	146A	1AC27B	147A	CD31-12032	147A	485B
145A	1M15Z5	145A	200X8220-531	146A	1C27Z	147A	D2E	147A	5132
145A	1M15Z810	145A	21524	146A	1C27ZA	147A	D4H	147A	5252
145A	1M15Z85	145A	264P04003	146A	1D27Z5	147A	D4J	147A	8000-00006-010
145A	1N1427	145A	264P10308	146A	1E27Z	147A	DZ33A	147A	17973A
145A	1N1514	145A	523-2003-150	146A	1E27Z10	147A	EQA01-32R	147A	0099202-128
145A	1N1514A	145A	525-2003-150	146A	1E227	147A	EQB01-33	147A	99202-228
145A	1N1525	145A	642-255	146A	1M27Z	147A	FZ33A	147A	136721
145A	1N1525A	145A	690V081H92	146A	1M27Z10	147A	GEZD-33	147A	141302
145A	1N1775	145A	1006-1737	146A	1M27Z5	147A	HEP-Z0426	147A	142670
145A	1N1775A	145A	1015	146A	1M27Z810	147A	H8Z330	147A	146138
145A	1N1878	145A	2396-17	146A	1M27Z85	147A	H3T330	147A	1478164-1
145A	1N1878A	145A	5124	146A	1N1430	147A	HW33	147A	2327076
145A	1N1934A	145A	5224(ZENER)	146A	1N1517	147A	HW33A	147A	62674997
145A	1N1988A	145A	99201-221	146A	1N1517A	147A	HW33B	148A	.25255
145A	1N2038-1	145A	126852	146A	1N1528	147A	LPM33	148A	.25255B
145A	1N3024	145A	127382	146A	1N1528A	147A	LPM33A	148A	.7J256
145A	1N3024A	145A	129946	146A	1N1781	147A	LPZT33	148A	.7256A
145A	1N3024B	145A	130762	146A	1N1781A	147A	M4663	148A	.7256C
145A	1N3436	145A	134444	146A	1N1881	147A	M4858	148A	.7256D
145A	1N3448	145A	140973	146A	1N1881A	147A	M4Z33	148A	AV2055
145A	1N3683	145A	171092-1	146A	1N1937A	147A	MC6026	148A	EVR56
145A	1N3683A	145A	233150	146A	1N1964A	147A	MC6026A	148A	EVR56A
145A	1N3683B	145A	489752-041	146A	1N1991A	147A	MC6126	148A	EVR56B
145A	1N4166	145A	530073-1017	146A	1N1991B	147A	MC6126A	148A	GEZD-55
145A	1N4166A	145A	530073-12	146A	1N3030	147A	MZ33A	148A	HW56B
145A	1N4166B	145A	530073-17	146A	1N3030A	147A	PTC512	148A	LPM56
145A	1N4331	145A	530073-9	146A	1N3030B	147A	RD-35A	148A	SV4055
145A	1N4331A	145A	1473777-2	146A	1N3439	147A	RE130	148A	SV4055A
145A	1N4331B	145A	1474717-2	146A	1N3451	147A	RH-EX0033CEZZ	148A	SZ1200
145A	1N4408	145A	1474777-002	146A	1N5689	147A	RH-IX0037CEZZ	148A	1A56M
145A	1N4408A	145A	1474777-2	146A	1N5689A	147A	R3-35	148A	1A56MA
145A	1N4408B	145A	1477046-1	146A	1N5689B	147A	SV4033	148A	1255
145A	1N4636	145A	1477081-501	146A	1N4172	147A	SV4033A	148A	1255B
145A	1N4702	145A	1945295A1	146A	1N4172A	147A	TC33A5A	148A	1TA55
145A	1N4744	145A	5300073-15	146A	1N4337	147A	TR-14002-10	148A	1TA55A
145A	1N4744A	145A	54903007-P4	146A	1N4337A	147A	TVS-RD29AN	148A	1255
145A	1N5245B	145A	62166916	146A	1N4337B	147A	TVSAN155	148A	1255A
145A	1N5535B	145A	62746483	146A	1N4414	147A	1A33M	148A	1255B
145A	1N5567	145A	4202012700	146A	1N4414A	147A	1A33MA	148A	48-134643
145A	1N5567A	145A	4202970330	146A	1N4414B	147A	1AC33	148A	48-134971
145A	1N5661B	145A	134882256359	146A	1N4642	147A	1AC33A	148A	75256
145A	1N666	146A	.25227	146A	1N4711	147A	1AC33B	148A	75256A
145A	1N718A	146A	.25227A	146A	1N4750	147A	1C33Z	148A	75256B
145A	1N718B	146A	.25227B	146A	1N4750A	147A	1C33ZA	148A	75256C
145A	1N767-1	146A	.4T27	146A	1N5254B	147A	1E33Z	148A	0086
145A	1N965B	146A	.4T27A	146A	1N5746B	147A	1E33Z10	148A	79949
145A	1R15	146A	.4T27B	146A	1N669	147A	1E33Z5	149A	130096
145A	1R15A	146A	.75N27	146A	1N724A	147A	1M33Z	149A	.25T62
145A	1R15B	146A	.7J227	146A	1N724B	147A	1M33Z10	149A	.25T62A
145A	1S137	146A	.7T27A	146A	1N971	147A	1M33Z85	149A	.25T62B
145A	1S1964	146A	.7227C	146A	1N971A	147A	1N1783A	149A	.75N62
145A	1S2121A	146A	.7227D	146A	1N971B	147A	1N1882	149A	.7J262
145A	1S2150A	146A	.72M27A	146A	1R27	147A	1N1882A	149A	.7262A
145A	1S230	146A	.72M27B	146A	1R27A	147A	1N1938A	149A	.7262B
145A	1S301.5	146A	.72M27C	146A	1R27B	147A	1N1965A	149A	.7262C
145A	1S697	146A	.72M27D	146A	1S2270A	147A	1N1992A	149A	.7262D
145A	1S990S	146A	AB2027	146A	1S239	147A	1N1992B	149A	.72M62A
145A	1S990S(ZENER)	146A	AV2027	146A	1S700	147A	1N3032	149A	.72M62B
145A	1T15	146A	AV2027A	146A	1T27	147A	1N3032A	149A	.72M62C
145A	1T15B	146A	AWO1-27	146A	1T27B	147A	1N3032B	149A	.72M62D
145A	1T19-15B	146A	AZ27	146A	1TA27	147A	1NZ440	149A	AV2062
145A	1TA15	146A	AZ27A	146A	1TA27A	147A	1N3453	149A	BZX29C62
145A	1ZA15A	146A	AZ971A	146A	1Z27	147A	1N3691	149A	BZX94C62
145A	1Z15	146A	BZ-260	146A	1Z27A	147A	1N3691A	149A	CD3171
145A	1Z15A	146A	BZX25	146A	1Z27B	147A	1N3691B	149A	GEZD-62
145A	1Z15B	146A	BZX25A	146A	1Z27C	147A	1N4174	149A	HEP-Z0433
145A	1Z15C	146A	BZX79C27	146A	1Z27D	147A	1N4174A	149A	HW62
145A	1Z15D	146A	BZX79C27A	146A	1Z27D10	147A	1N4174B	149A	HW62A
145A	1Z15D10	146A	BZY85C27	146A	1Z27D5	147A	1N4339	149A	HW62B
145A	1Z15D5	146A	BZY92C27	146A	1Z27T10	147A	1N4339A	149A	LPM62
145A	1Z15T10	146A	C4026	146A	1Z27T20	147A	1N4339B	149A	M4262
145A	1Z15T20	146A	CD31-00023	146A	1Z27T5	147A	1N4416	149A	M4262A
145A	1Z15T5	146A	CD31-10361	146A	1ZP27T10	147A	1N4416A	149A	PTC514
145A	1ZB15	146A	CD31-12030	146A	1ZP27T5	147A	1N4416B	149A	RE132
145A	1ZB15B	146A	D1W	146A	1ZP27T10	147A	1N4644	149A	SV4062
145A	1Z15	146A	D28-2	146A	1ZP27T20	147A	1N4714	149A	SV4062A
145A	1ZC15T10	146A	DZ27A	146A	1ZP27T5	147A	1N4752	149A	1A62M
145A	1ZC15T5	146A	EVR27	146A	1Z27	147A	1N4752A	149A	1A62MA
145A	1ZP15Z10	146A	EVR27A	146A	1Z27B	147A	1N5257B	149A	1AC62
145A	1ZP15Z20	146A	EVR27B	146A	25N27	147A	1N726A	149A	1AC62A
145A	1ZP15Z5	146A	FZ27A	146A	48-137034	147A	1N5973B	149A	1AC62B
145A	1ZM15Z10	146A	F227T10	146A	50227	147A	1R33	149A	1C62Z
145A	1ZM15Z20	146A	F227T5	146A	50227A	147A	1R33A	149A	1C62ZA
145A	1ZM15Z5	146A	G27T10	146A	50227C	147A	1R33B	149A	1B62Z
145A	1ZT15	146A	G27T20	146A	56-47	147A	1S241	149A	1B62Z10
145A	1ZT15B	146A	G27T5	146A	75227	147A	1T33	149A	1B62Z5
145A	02215A	146A	GEZD-27	146A	75227A	147A	1T33B	149A	1M62Z
145A	003-009100	146A	G227	146A	75227B	147A	1TA33	149A	1M62Z10
145A	314-3506-29	146A	HEP-Z0424	146A	75227C	147A	1TA33A	149A	1M62Z5
145A	4-2020-12700	146A	HM27B	146A	0087	147A	1Z33	149A	1M62Z810
145A	05215	146A	HS2270	146A	1027	147A	1Z33A	149A	1M62Z85
145A	08-08120	146A	HS7270	146A	1127	147A	1Z33B	149A	1N1790
145A	8-905-421-128	146A	HW27	146A	5130	147A	1Z33D	149A	1N3039
145A	8-905-421-239	146A	HW27A	146A	5230	147A	1Z33D5	149A	1N3039A
145A	8-905-421-715	146A	HW27B	146A	99202-226	147A	1Z33T10	149A	1N3039B
145A	09-306567	146A	LPM27	146A	110351	147A	1Z33T20	149A	1N3698
145A	13-14879-1	146A	LPM27A	146A	138174	147A	1Z33T5	149A	1N3698A
145A	13-14879-3	146A	LPZ227	146A	267272	147A	1ZP33T10	149A	1N3698B
145A	13-33179-2	146A	M2TZ	146A	530073-18	147A	1ZP33T20	149A	1N4181
145A	14-515-07	146A	M4227	146A	7570021-12	147A	1ZP33T5	149A	1N4181A
145A	14-515-09	146A	M4227-20	147A	.25233	147A	1Z33	149A	1N4181B
145A	14-515-11	146A	M4227A	147A	.25233A	147A	1Z33B	149A	1N4346
145A	14-515-73	146A	MC6024	147A	.25233B	147A	13-33186-1	149A	1N4346A
145A	25N15	146A	MC6024A	147A	.4T33	147A	48-134663	149A	1N4346B
145A	2826	146A	MC6124	147A	.4T33A	147A	48-134858	149A	1N4423
145A	2826A	146A	MC6124A	147A	.4T33B	147A	488137266	149A	1N4423A
145A	2826AF	146A	MZ227T10	147A	.75N33	147A	488137272	149A	1N4423B
145A	2826F	146A	MZE27T5					149A	1N4759
145A	34-8057-12	146A	MTZ624					149A	1N4759A
145A	46-86464-3	146A	MTZ624A					149A	1N5265B
145A	48-82256059	146A	MZ27A					149A	1N733
145A	488137330	146A	M292-27A					149A	1N733A
145A	488155104	146A	QZ27T10					149A	1N980B

ECG	Industry Standard No.	ECG	Industry Standard No.	ECG	Industry Standard No.	ECG	Industry Standard No.	ECG	Industry Standard No.
149A	1R62	150A	48-134699	152	B143018	152	D28A10	152	SDM345
149A	1R62A	150A	48-134704	152	B143019	152	D28A12	152	SDN345
149A	1R62B	150A	50282	152	B143026	152	D28A13	152	SD75102
149A	18250	150A	50282A	152	B143027	152	D28A2	152	SD75907
149A	1762	150A	50282B	152	B1D	152	D28A3	152	SDT6001
149A	1762B	150A	50Z82B	152	B3547	152	D28A4	152	SDT6011
149A	1T462	150A	75282	152	B3548	152	D28A6	152	SDT6013
149A	1TA62A	150A	75282A	152	B3550	152	D28A7	152	SDT6035
149A	1262	150A	75282B	152	B3551	152	D28A9	152	SDT6103
149A	1262A	150A	75Z82C	152	B3577	152	D28D1	152	SDT7511
149A	1262B	150A	187-1	152	B3578	152	D28D10	152	SDT7512
149A	1262C	150A	2093A41-150	152	B3580	152	D28D2	152	SDT7514
149A	1262D10	150A	4699	152	B3584	152	D28D3	152	SDT7515
149A	1262D5	150A	4704	152	B3585	152	D28D4	152	SDT9009
149A	48-134955	150A	5142	152	B3586	152	D28D5	152	SE5-0963
149A	50256	150A	5242	152	B3588	152	D28D7	152	SJB-513
149A	50256A	151A	.25T110	152	B3589	152	D313C	152	SJB-515
149A	50256B	151A	.25T110B	152	B5000	152	D313D	152	SJA42
149A	50256C	151A	.4T110	152	B5002	152	D315	152	SJE513
149A	50262	151A	.4T110A	152	B5021	152	D317	152	SJE515
149A	50262A	151A	.75N110	152	B5022	152	D317P	152	SJE783
149A	50262B	151A	.7Z110A	152	B5031	152	D317P	152	SK3041
149A	50262C	151A	.7Z110C	152	B5032	152	D318	152	SP8416
149A	66X0040-012	151A	.7Z110D	152	BD106	152	D325	152	SP8660
149A	75262	151A	.7ZM110A	152	BD106A	152	D325C	152	SP81436
149A	75262A	151A	.7ZM110B	152	BD106B	152	D325D	152	STC1300
149A	75262B	151A	.7ZM110C	152	BD107A	152	D325E	152	STC1850
149A	75262C	151A	.7ZM110D	152	BD107B	152	D330D	152	STC1860
149A	93A39-25	151A	AV2110	152	BD109	152	D343	152	T01-050
149A	5139	151A	EVR110A	152	BD124	152	D359	152	T1486
149A	5259	151A	EVR110B	152	BD162	152	D359C	152	T1487
149A	138107	151A	HEP-Z0439	152	BD163	152	D359C2	152	T1P29
149A	1474777-7	151A	HW110	152	BD220	152	D359D	152	T1P29X
149A	62746491	151A	HW110A	152	BD221	152	D359D1	152	T1P31
150A	.25282	151A	LPM110	152	BD222	152	D359D2	152	T611-1
150A	.25282A	151A	LPM110A	152	BD231A	152	D360	152	T612-1
150A	.25282B	151A	M4728	152	BD239	152	D360C	152	TA2911
150A	.75282	151A	M4Z110	152	BD239A	152	D365H	152	TA7156
150A	.7J282	151A	PTO516	152	BD241	152	D366-0	152	TA7262
150A	.7Z82A	151A	TC110A5B	152	BD243	152	D366P	152	TA7363
150A	.7Z82B	151A	ZH110	152	BD243B	152	D366Q	152	TA7554
150A	.7Z82C	151A	1A110M	152	BD271	152	D382	152	TA7555
150A	.7Z82D	151A	1A110MA	152	BD433	152	D382LM	152	TI486
150A	.7ZM82A	151A	1AC110	152	BD435	152	D389	152	TI487
150A	.7ZM82B	151A	1AC110A	152	BD437	152	D389-0	152	TIP-14
150A	.7ZM82C	151A	1AC110B	152	BD439	152	D389-0P	152	TIP-31
150A	.7ZM82D	151A	1C110Z	152	BD533	152	D389APP	152	TIP-31A
150A	AV2082	151A	1C110ZA	152	BD535	152	D389B	152	TIP24
150A	BZX29082	151A	1B110Z	152	BDX74	152	D389BL	152	TIP31
150A	CD5174	151A	1B110Z10	152	BDX75	152	D389LB	152	TIP31A
150A	D2824	151A	1B110D5	152	BDY12	152	D90	152	TM152
150A	EVR82	151A	1Z110D5	152	BDY13	152	D91	152	TR-19
150A	EVR82A	151A	1M110Z	152	BDY34	152	D91P	152	TR2527203
150A	EVR82B	151A	1M110Z10	152	BLY21	152	DDBY228001	152	TR28D330E
150A	FZ82A	151A	1M110Z5	152	BLY36	152	DDBY278001	152	TV-115
150A	G8ZD-82	151A	1M110Z85	152	BLY63	152	DDBY278002	152	TV-117
150A	HEP-Z0436	151A	1N1796	152	BLY88	152	DS-513	152	Y409
150A	HW82	151A	1N1796A	152	BLY89	152	DS513	152	V828C1173-Y-3
150A	HW82A	151A	1N3045	152	BR101B	152	E2496	152	VX3733
150A	HW82B	151A	1N3045A	152	BRO5296	152	EA15X327	152	W18
150A	LPM82	151A	1N3045B	152	BUY24	152	EA15X333	152	XA-1160
150A	LPM82A	151A	1N3704	152	C1060	152	EA15X8119	152	XB404
150A	M4699	151A	1N3704A	152	C1060A	152	EA15X99	152	XB408
150A	M4704	151A	1N3704B	152	C1060B	152	EA3716	152	XB476
150A	M4Z82	151A	1N4187	152	C1060BM	152	EA4055	152	ZT1483
150A	MZ82A	151A	1N4187B	152	C1060C	152	EA4085	152	ZT1484
150A	PS6525	151A	1N4352	152	C1060D	152	EP15X22	152	ZT1485
150A	PTC515	151A	1N4352A	152	C1061	152	EP15X68	152	ZT1486
150A	SV4082	151A	1N4352B	152	C1061A	152	EP3053	152	ZT1701
150A	SV4082A	151A	1N4429	152	C1061B	152	EQS-140	152	ZT2876
150A	TC82A5B	151A	1N4429B	152	C1061C	152	EQ8140	152	01-031173
150A	1A82M	151A	1N5272B	152	C1061D	152	ES15X12	152	01-040243
150A	1A82MA	151A	1N739	152	C1061T	152	ES15X86	152	01-572784
150A	1AC82	151A	1N739A	152	C1061T-B	152	ES80(ELCOM)	152	01-572791
150A	1AC82A	151A	1N986B	152	C1061TB	152	ETTD-235	152	01-572861
150A	1AC82B	151A	1R110	152	C1173	152	G05705	152	01-57291
150A	1C82Z	151A	1R110A	152	C1173-0	152	GB-66	152	00A058
150A	1C82ZA	151A	1R110B	152	C1173-GR	152	G05705	152	002-012400
150A	1B82Z	151A	18257	152	C1173-R	152	HC495	152	002D235RY
150A	1B82Z10	151A	1Z110	152	C1173-Y	152	HP57	152	2N5293
150A	1B82Z5	151A	1Z110A	152	C1173C	152	HT308301BO	152	2N5295
150A	1M82Z	151A	1Z110B	152	C1173R	152	HT311621B	152	2N5296
150A	1M82Z10	151A	1Z110C	152	C1173X	152	HT402352B	152	2N5298
150A	1M82Z5	151A	1Z110D	152	C1173XO	152	IP20-0007	152	2N6288
150A	1MB2ZS10	151A	1Z110D10	152	C1173Y	152	IP20-0036	152	2N6289
150A	1MB2ZS5	151A	1Z110D5	152	C1398	152	IP20-0323	152	2N6290
150A	1N1795	151A	000028A550	152	C1398Q	152	IRTR76	152	2SC1060
150A	1N1795A	151A	4-2020-14000	152	C1418	152	J241241	152	2SC1060A
150A	1N1887	151A	8P9253	152	C1418A	152	LS-0066	152	2SC1060B
150A	1N1887A	151A	17-1	152	C1418B	152	M75543-1	152	2SC1060BM
150A	1N1943A	151A	48-134728	152	C1418C	152	M9576	152	2SC1060C
150A	1N3042	151A	50Z110	152	C1418D	152	M9661	152	2SC1060D
150A	1N3042A	151A	50Z110A	152	C1419	152	MHT5906	152	2SC1061
150A	1N3042B	151A	50Z110B	152	C1419A	152	P/PTV/117	152	002801061A
150A	1N3458	151A	50Z110C	152	C1419B	152	P4J148	152	2SC1061B
150A	1N3701	151A	56-18	152	C1419C	152	PLE-48	152	2SC1061BT
150A	1N3701A	151A	56-29	152	C1419D	152	PN66	152	2SC1061C
150A	1N3701B	151A	56-48	152	C1429	152	PP5250	152	2SC1061D
150A	1N4184	151A	75Z110	152	C1429-1	152	PP5310	152	2SC1061KA
150A	1N4184A	151A	75Z110A	152	C1429-2	152	PP5312	152	2SC1061KB
150A	1N4184B	151A	75Z110B	152	C154	152	PT2635	152	2SC1061KC
150A	1N4349	151A	75Z110C	152	C154B	152	PT5693	152	2SC1061T
150A	1N4349A	151A	187-2	152	C325E	152	PT665	152	2SC1061T-B
150A	1N4349B	151A	4728	152	C36583	152	Q-11115C	152	2SC1061TB
150A	1N4426	151A	5145	152	C789	152	Q-12115C	152	2SC1173
150A	1N4426A	151A	5245	152	C789-0	152	QT-DO325XAC	152	2SC1173-0
150A	1N4426B	151A	5300073-1020	152	C789-0	152	R3611-1	152	2SC1173-Y
150A	1N4762	151A	1473590-1	152	C789-R	152	R3681-1	152	2SC1173A
150A	1N4762A	151A	1960642	152	C789-Y	152	R612-1	152	2SC1173C
150A	1N5268B	152	A-1854-0420-1	152	CGE-66	152	R621-1	152	2SC1173R
150A	1N736	152	A-1854-0464	152	D141	152	R632-1	152	2SC1173XO
150A	1N736A	152	A066-114	152	D141H01	152	R632-2	152	2SC1398
150A	1N985B	152	A272	152	D141H9Z	152	RCA29	152	2SC1398P
150A	1R82	152	A273	152	D154	152	RCA29/SDH	152	2SC1398Q
150A	1R82A	152	A28D2260P	152	D234	152	RCA29A	152	2SC1418
150A	1R82B	152	A417032	152	D234-0	152	RCA29A/SDH	152	2SC1418A
150A	18253	152	A54-3	152	D234-0	152	RCA40250	152	2SC1418B
150A	1T82	152	A5A-1B	152	D234-R	152	S-310E	152	2SC1418C
150A	1T82B	152	A5A-1B	152	D234-Y	152	S12020-04	152	2SC1418D
150A	1TA82	152	A68-23-560	152	D235	152	S1D153	152	2SC1419
150A	1TA82A	152	A9V	152	D235-0	152	S2042	152	2SC1419A
150A	1282	152	AR-17	152	D235-0	152	SB0319	152	2SC1419B
150A	1282A	152	AR17(GREY)	152	D235-R	152	SC4303	152	2SC1419C
150A	1282B	152	ATC-TR-19	152	D235-Y	152	SC4303-1	152	2SC1419D
150A	1282C	152	B-1823	152	D235D	152	SC4303-2	152	2SC1429
150A	1282D	152	B131	152	D235G	152	SC4308	152	2SC1429-1
150A	1282D10	152	B143004	152	D259GR	152	SDA345	152	2SC1429-2
150A	1282D5	152	B143011	152	D259R	152	SDA345	152	2SC789
150A	1Z82Z	152	B143012	152	D253B	152	SDB345	152	2SC789-0
150A	1Z82B			152	D27C1	152	SDJ345	152	2SC789-0
150A	1ZT82			152	D27C2	152	SDK345	152	2SC789-R
150A	1ZT82B			152	D27C3	152	SDI345	152	2SC789-Y
150A	8-905-421-315			152	D27C4			152	2SC789R
				152	D28A1			152	2SC789Y

ECG	Industry Standard No.	ECG	Industry Standard No.	ECG	Industry Standard No.	ECG	Industry Standard No.	ECG	Industry Standard No.
152	28C790	152	9TR8	152	149N2004	152	40629	152	9001324
152	28C790-0	152	10-26-123-313	152	149N201	152	40630	152	9001756
152	28C790Y	152	012-103002	152	149P1D	152	40631	152	9341510
152	28D141	152	13-14085-88	152	153N1C	152	41500	152	19020065
152	28D141H01	152	13-28469-2	152	153N2C	152	41504	152	23114253
152	28D141H9Z	152	13-2P64	152	153N4C	152	42942	152	23114254
152	28D154	152	13-2P64-1	152	153N5C	152	43163	152	26010034
152	28D234	152	13-32636-1	152	153N6	152	44208	152	26010059
152	28D234-0	152	13-32640-1	152	172-010-9-001	152	44699	152	28105203-001
152	28D234-0	152	13-39004-1	152	172-014-9-001	152	50200-1B	152	28105203-002
152	28D234-R	152	13-39004-2	152	172-014-9-003	152	50200-24	152	30200033
152	28D234-Y	152	13-39099-1	152	172-014-9-007	152	50200-9	152	43027213
152	28D234R	152	13-39884-1	152	172-024-9-004	152	50308-0100	152	43027214
152	28D235	152	13-39884-2	152	172-031-9-003	152	60175	152	43027987
152	28D235-0	152	13-4763	152	173A-4490-7	152	60216	152	50221800
152	28D235-0	152	13-4763-1	152	173A-4491-5	152	60408	152	50221800-01
152	28D235-R	152	13-4763-I	152	173A4490-7	152	60679	152	62593688
152	28D235-Y	152	14-0104-1	152	173A4491-5	152	60719	152	62734671
152	28D235D	152	14-609-00	152	173A4491-8	152	60835	152	62743182
152	28D235Q	152	14-609-05	152	176-042-9-007	152	60838	152	80053300
152	28D235QR	152	14-609-04	152	176-055-9-004	152	60886	152	430253843
152	28D235LBY	152	14-609-06	152	185-007	152	60966	152	6621007200
152	28D235R	152	14-609-08	152	195N1	152	60987	153	A-1853-0233-1
152	0028D235RY	152	14-902-23	152	195N1D	152	61102	153	A-1853-0234-1
152	28D235Y	152	14-905-23	152	195N5	152	61193	153	A-1853-0254-1
152	28D256	152	14-905-23	152	200-2076	152	61285	153	A473(JAPAN)
152	28D313	152	14-908-23	152	207A50	152	61286	153	A473-0R
152	28D313C	152	15-123065	152	207A53	152	61418	153	A473-0
152	28D313B	152	19-020-056	152	209-30	152	61636	153	A473-R
152	28D313K	152	19-020-066	152	211A6380-3	152	61772	153	A489(JAPAN)
152	28D313M	152	19A115200-P1	152	211A6381-2	152	61875	153	A489-0
152	28D314	152	19A116118-1	152	211A6582-2	152	61958	153	A489-R
152	28D314C	152	19A116118-2	152	216-001-001	152	61981	153	A489-Y
152	28D314D	152	19A116118-I	152	229-1301-35	152	62156	153	A490(JAPAN)
152	28D314B	152	19A116118P1	152	250-0359	152	62571	153	A490(POWER)
152	28D315	152	19A116118P2	152	260-10-024	152	62681	153	A670
152	28D315C	152	21A040-052	152	260-10-054	152	79992	153	A670A
152	28D315D	152	21A112-095	152	260P12701	152	95261-1	153	A670B
152	28D317	152	21A118-029	152	260P28401	152	99252-4	153	A670C
152	28D317P	152	21A118-049	152	296(REGENCY)	152	116118	153	A748
152	28D317P	152	24MW662	152	353-9203-001	152	11611B-2	153	A748Q
152	28D318	152	24MW778	152	353-9502-001	152	127798	153	A754
152	28D318(0)	152	24MW977	152	417-175-12993	152	128056	153	A754A
152	28D318B	152	34-1002	152	474A410BEP2	152	131075	153	A754B
152	28D318P	152	34-6002-50	152	474A410BW-2	152	131848(RCA)	153	A754C
152	28D318Q	152	34-6002-52	152	499(CHRYSLER)	152	131849	153	A754D
152	28D325	152	34-6002-56	152	617-117	152	132499	153	A755
152	28D325C	152	42-23459	152	623(RCA)	152	132573	153	A755A
152	28D325D	152	42-27539	152	640-1(SYLVANIA)	152	132697	153	A755B
152	28D325R	152	44A395986-001	152	690L-021H25	152	136648	153	A755C
152	28D330	152	44A417032-001	152	750A858-448	152	137527	153	A755D
152	28D330D	152	46-86317-3	152	753-2001-173	152	138192	153	AR-25
152	28D330E	152	46-86335-3	152	753-4001-932	152	138379	153	AR-50
152	28D343	152	46-86348-3	152	753-9010-235	152	142691	153	AR25(GREEN)
152	28D359	152	46-86374-3	152	770-045	152	146139	153	AR27(GREEN)
152	28D359C2	152	46-86400-3	152	800-533-00	152	153107	153	AR37(GREEN)
152	28D359D	152	46-86516-3	152	800-546-00	152	167691	153	AR44
152	28D359D1	152	48-13309X3	152	834-250-011	152	171162-265	153	B152
152	28D359D2	152	48-137396	152	921-1009	152	171162-291	153	B411
152	28D360	152	48-137437	152	991-013063	152	172463	153	B434
152	28D360C	152	48-137506	152	992-011317	152	172643	153	B434-0
152	28D360D	152	48-355040	152	995-01-6151	152	242102	153	B434-R
152	28D365	152	48-355059	152	1000-138	152	279517	153	B434-Y
152	28D365-0	152	48-43240001	152	1000-142	152	309441	153	B435
152	28D365B	152	48-44883001	152	1016-81	152	309459	153	B435-0
152	28D365H	152	48-869576	152	1041-74	152	489751-033	153	B435-R
152	28D365P	152	48-869661	152	1045-1286	152	489751-044	153	B435-Y
152	28D365Q	152	48-97046A30	152	1045-7828	152	610111-6	153	B507
152	28D366	152	48-97046A38	152	1069(GE)	152	610149-2	153	B508
152	28D366-0	152	488137309	152	1080-130	152	610153-1	153	B509
152	28D366P	152	488137311	152	1081-3368	152	610153-2	153	B511
152	28D366Q	152	488137569	152	1098-14	152	610153-3	153	B511C
152	28D389	152	488155013	152	1111(GE)	152	610153-4	153	B511D
152	28D389(LP)	152	488155062	152	1968-17	152	610153-5	153	B512
152	28D389(O)	152	48843240001	152	1969-17	152	610153-6	153	B513
152	28D389(P)	152	48843441001	152	2000-216	152	610162-4	153	B514
152	28D389-OP	152	48844885901	152	2017-109	152	610162-8	153	B515
152	28D389APO	152	57A250-14	152	2036-59	152	610195-3	153	BD223
152	28D389B	152	57A278-14	152	2057B2-151	152	702886	153	BD224
152	28D389BL	152	57A279-14	152	2082-6	152	717101	153	BD225
152	28D389BLB	152	57A286-10	152	2210-17	152	740856	153	BD240
152	28D389BLB-0	152	57B131-10	152	2582-17	152	996817	153	BD240A
152	28D389BLB-P	152	57B250-14	152	2408-17	152	1417358-1	153	BD242
152	28D389BP	152	57B278-14	152	2853-2	152	1471132-4	153	BD242B
152	28D389LB	152	57B279-14	152	2854-2	152	1473567-4	153	BD244
152	28D389LP	152	61-309689	152	2855-2	152	1473611	153	BD244A
152	28D389P	152	62B046-1	152	2856-2	152	1473611-1	153	BD244B
152	28D389Q	152	62B046-3	152	3107-204-90182	152	1473612	153	BD272
152	28D390	152	62B046-4	152	3612(RCA)	152	1473612-11	153	BD434
152	28D390(O)	152	73C180028-11	152	3621(RCA)	152	1473681-1	153	BD436
152	28D390P	152	73C180829-12	152	3631(RCA)	152	1473681-I	153	BD438
152	28D390Q	152	73C180829-12	152	3631-1(RCA)	152	1700036	153	BD440
152	28D477	152	77-272913-1	152	3843	152	1950160	153	BD534
152	28D762	152	77-273715-1	152	4490-7	152	2320083	153	BD536
152	28D880	152	77-273759-1	152	4491-8	152	2320482	153	BDX27
152	28D90	152	86-5102-2	152	5001-053	152	2320482H	153	BDX27-10
152	28D91	152	86-5107-2	152	6380-2	152	2320485	153	BDX27-6
152	28D91P	152	86-5107-2	152	6381-2	152	2320485	153	BDX28
152	3L4-6005-1	152	86-510B-2	152	6382-2	152	2320486	153	BDX28-10
152	3L4-6005-55	152	86-529-2	152	7215	152	2320602	153	BDX28-6
152	3L4-6005-55	152	86-544-2	152	7252	152	2320651	153	BDY82
152	3L4-6012-02	152	86-663-2	152	7358-1	152	2320652	153	BU183
152	3L4-6012-2	152	90-600	152	7414	152	2327206	153	C9B-69
152	3L4-6012-3	152	96-5225-01	152	8000-00011-050	152	2875493	153	D43C7
152	004-2	152	96-5348-01	152	8000-00028-041	152	3438095	153	EA15X328
152	4-464	152	96-5357-01	152	8020-206	152	3438867	153	EA15X334
152	6-0005193	152	099-1(PHILCO)	152	8800-205	152	3596446	153	EA15X311B
152	6-5193	152	099-1(SYL)	152	12020-02	152	3596447	153	EA15X8130
152	007-0112	152	998100-1	152	16113	152	3596448	153	EA3715
152	7-0112-00	152	102-1061-01	152	16164	152	3755171	153	EP15X15
152	007-0112-03	152	103-0235-85	152	16166	152	4080187-0502	153	EP15X25
152	7-0112-04	152	111N6C	152	16181	152	4080866-0013	153	EO81(ELCOM)
152	7-0112-05	152	121-1006	152	16182	152	4080866-006	153	EW183
152	007-112-04	152	121-719	152	16207	152	4080866-009	153	GE-26
152	8-1074	152	121-770	152	16241	152	4080866-4	153	HA-00699
152	8P345	152	121-804	152	16305	152	4080879-0001	153	HA505
152	8P73BLU	152	121-808	152	16334	152	4550106-001	153	HP5B
152	8P73GRN	152	121-874	152	16335	152	4550106-003	153	IRTR77
152	8P73YEL	152	121-887	152	16336	152	5320422H	153	L8-0067
152	09-302083	152	121-927	152	23754	152	5320432	153	M9348
152	09-302132	152	121-966	152	30294	152	5320433	153	NPS6518
152	09-302164	152	121-966-01	152	38733	152	5320492H	153	P1Y-4
152	09-302236	152	121-970-02	152	39302	152	5320671	153	P2B
152	09-303018	152	125-B415	152	39750	152	5321301	153	P2K
152	09-303022	152	129-33	152	39767	152	06120018	153	P4W
152	09-303031	152	0131-005352	152	39789(POWER)	152	06120021	153	P5H
152	09-303032	152	131N2G	152	39824	152	06120030	153	P5U
152	09-303033	152	132-3	152	39948	152	06120096	153	P8H
152	09-303318	152	132-4	152	39981	152	6902021H25	153	RCA30
152	09-304140	152	133-3	152	40613	152	7026024	153	RCA30A
152	09-305140	152	142-004	152	40618	152	7570031-01	153	S2041
152		152	142-008	152	40621	152	7855298-01	153	SCO T334
152		152	149N2002D	152	40622	152	8034903	153	SDA445
152		152		152		152	8113102	153	SDA445

ECG	Industry Standard No.	ECG	Industry Standard No.	ECG	Industry Standard No.	ECG	Industry Standard No.	ECG	Industry Standard No.
153	SDB445	153	3L4-6013-3	154	A3170717	154	C65Y	154	TIS102
153	SDI445	153	3L4-6013-5	154	A35(JAPAN)	154	C65YA	154	TIS103
153	SDJ445	153	3L4-6013-5	154	A3M	154	C65YB	154	TM154
153	SDK445	153	3L4-6013-56	154	A3MA	154	C65YTV	154	TNJ60072
153	SDL445	153	3L4-6013-58	154	A417-115	154	C66	154	TNJ72282
153	SDM445	153	8P445	154	A46-867-3	154	C686	154	TQ5063
153	SDN445	153	09-300090	154	A4648	154	C69	154	TR301
153	SDT-445	153	09-301077	154	A48-134819	154	C70	154	TRB100
153	SDT3509	153	13-39100-1	154	A48-134843	154	C788	154	TRB120
153	SDT3510	153	13-39819-1	154	A48-134853	154	C805	154	TRB140
153	SDT3513	153	13-39819-2	154	A48-134898	154	C818	154	TRB160
153	SDT3514	153	14-0104-2	154	A48-134919	154	C856	154	TRB180
153	SDT3702	153	19A116375	154	A48-134927	154	C856-02	154	TRB200
153	SDT3703	153	19A116375P1	154	A48-137002	154	C856C	154	TRB225
153	SDT3704	153	21A112-094	154	A48-137035	154	C857	154	TRB250
153	SDT3706	153	34-1003	154	A4819	154	C857H	154	TRB275
153	SDT3707	153	34-6002-57	154	A4838	154	C995	154	TRB301
153	SDT3708	153	42-27538	154	A4843	154	CDC744	154	TRB3011
153	SDT3709	153	46-86515-3	154	A4853	154	CDQ10013	154	TRB3012
153	SDT3710	153	48-137312	154	A4H	154	CDQ10015	154	TV-19
153	SDT3711	153	48-137507	154	A573501	154	CDQ10034	154	TV-49
153	SDT3712	153	48-137540	154	A57C12-1	154	CDQ10037	154	TV-70
153	SDT3715	153	48-137566	154	A57C12-2	154	CDQ10044	154	TV19
153	SDT3715	153	48-43241001	154	A57D1-122	154	CDQ10045	154	TV24164
153	SDT3716	153	48-44884001	154	A57M2-16	154	CDQ10046	154	TV24435
153	SDT3717	153	48-90420A06	154	A57M2-17	154	CDQ10047	154	TV24499
153	SDT3720	153	488137310	154	A610075-1	154	CDQ10049	154	TV70
153	SDT3721	153	488137312	154	A6181-1	154	CS1347	154	TV8-28C526
153	SDT3722	153	488137370	154	A63-18426	154	CS6776	154	V8-28C58
153	SDT3725	153	488155066	154	A7253	154	DT1003	154	TV8-28C58A
153	SDT3726	153	48844884001	154	A86-213-2	154	DT1602	154	V8-28C58
153	SDT3727	153	57A206-14	154	A86-214-2	154	DT1603	154	V8-28C58A
153	SDT3729	153	57A207-14	154	A86-215-2	154	DT1612	154	V8-28C58B
153	SDT3730	153	57B132-10	154	A86-316-2	154	DT1613	154	V8-28C58C
153	SDT3733	153	57B206-14	154	A8V	154	EP15X18	154	W28
153	SDT5112	153	57B207-14	154	A8VA	154	EP15X59	154	ZEN205
153	SE5-0964	153	61-309690	154	AT350	154	ES15X89	154	01-572831
153	SJ1152	153	730180830-11	154	AT351	154	ES32(ELCOM)	154	2N1052
153	SJ1171	153	73018O830-12	154	B5D	154	EP15X34	154	2N1053
153	SJ1284	153	77-272914-1	154	BC100	154	ETP2008	154	2N1054
153	SJE514	153	77-273716-1	154	BC285	154	ETP3923	154	2N1207
153	SPB1437	153	77-273739-1	154	BC394	154	FM1893	154	2N1335
153	ST27020	153	86-530-2	154	BP108	154	FT34C	154	2N1336
153	ST05202	153	96-5303-01	154	BP109	154	FT34D	154	2N1337
153	ST05203	153	96-5349-01	154	BP110	154	FT3641	154	2N1339
153	ST05205	153	96-5356-01	154	BP111	154	GB-40	154	2N1340
153	ST05206	153	998099-1	154	BP114	154	H932	154	2N1341
153	ST05303	153	100-1(PHILCO)	154	BP117	154	HEP-83033	154	2N1492
153	ST05805	153	121-803	154	BP118	154	HEP-83034	154	2N1572
153	ST05806	153	121-875	154	BP119	154	HEP-83035	154	2N1573
153	STX0029	153	121-886	154	BP140	154	HEP-85024	154	2N1574
153	STX0030	153	121-926	154	BP140A	154	HEP-85025	154	2N1893
153	TAT556	153	121-969-02	154	BP140R	154	HEP-85026	154	2N2008
153	TAT557	153	121-997	154	BP140S	154	HEP706	154	2N2006
153	TIP32	153	0131-002049	154	BP155R	154	HEP712	154	2N2443
153	TIP32A	153	0131-005353	154	BP155B	154	HEP713	154	2N2509
153	TM153	153	149P2001D	154	BP156	154	HEP85024	154	2N2618
153	TR-8019	153	149P2003	154	BP157	154	HT8000101	154	2N3114
153	TR2327723	153	177-023-9-001	154	BP157B	154	IRTR78	154	2N3388
153	TRB019	153	195P2	154	BP174	154	M4648	154	2N3389
153	TV-116	153	195P4	154	BP178	154	M4819	154	2N341
153	W19	153	229-1301-36	154	BP179	154	M4838	154	2N341A
153	01-010473	153	624(RCA)	154	BP179A	154	M4839	154	2N3526
153	01-572774	153	690L-021H26	154	BP179B	154	M4843	154	2N3700
153	002-012300	153	995-01-6130	154	BP179C	154	M4853	154	2N3701
153	2N4387	153	1045-1294	154	BP186	154	M4927	154	2N3712
153	2N4388	153	1045-7836	154	BP259	154	M7002	154	2N3742
153	2N6021	153	1071(GB)	154	BP292	154	M819	154	2N3743
153	2N6109	153	1113(GB)	154	BP292A	154	ME1110	154	2N3925
153	28A473	153	2036-58	154	BP292B	154	MIB14150/37	154	2N4068
153	28A473-GR	153	2057B2-150	154	BP292C	154	MJ420	154	2N4269
153	28A473-O	153	2081-6	154	BP294	154	MM2260	154	2N4270
153	28A473-R	153	2211-17	154	BP305	154	MM3000	154	2N4925
153	28A489	153	7420	154	BP336	154	MM3002	154	2N4926
153	28A489-O	153	31004-1	154	BP337	154	MM3009	154	2N4927
153	28A489-R	153	44209	154	BP338	154	MM3100	154	2N5058
153	28A489-Y	153	50201-4	154	BP355	154	MM3101	154	2N5059
153	28A490	153	75803-1	154	BPR57	154	MM7087	154	2N5176
153	28A490(POWER)	153	75803-2	154	BPR58	154	MM7088	154	2N5184
153	28A490-O	153	75803-3	154	BPW53	154	MT1893	154	2N5965
153	28A670	153	115792	154	BFW37	154	MT698	154	2N738
153	28A670A	153	128057	154	BPW45	154	MT699	154	2N740
153	28A670B	153	144076	154	BFX98	154	MT870	154	2N743
153	28A670C	153	146081	154	BPT441	154	MT971	154	2N743A
153	28A700	153	199063-1	154	BFY43	154	MT910	154	2N746
153	28A700B	153	489751-032	154	BFY45	154	MT911	154	2N746A
153	28A700Y	153	610112-1	154	BFY57	154	MT912	154	28C1012
153	28A748Q	153	610149-1	154	BFY65	154	N2XA	154	28C1012A
153	28A754	153	610149-3	154	BLT55	154	N26212	154	28C1012A
153	28A754A	153	610195-4	154	BN7253	154	PEE1075A	154	28C1048
153	28A754B	153	802037-001	154	BSW70	154	PRT101	154	28C1048B
153	28A754C	153	1471134	154	C1012	154	PTC117	154	28C1048C
153	28A754D	153	2320884	154	C1012A	154	R2474-2	154	28C1048D
153	28A755	153	2321281	154	C1048	154	R825	154	28C1048E
153	28A755A	153	3596451	154	C1048B	154	R8-2008	154	28C1048F
153	28A755B	153	3596452	154	C1048C	154	S1366	154	28C1048N
153	28A755C	153	3596453	154	C1048D	154	S1407	154	28C1056
153	28A755D	153	3596454	154	C1048E	154	S1769	154	28C1103
153	28A768	153	6902021H26	154	C1048F	154	S17862	154	28C1103(A)
153	28B434	153	7020203	154	C1056	154	S2986	154	28C1103A
153	28B434-0	153	9001757	154	C1103	154	S3002	154	28C1103B
153	28B434-R	153	28105249-001	154	C1103(A)	154	S3033	154	28C1103C
153	28B434-Y	153	62393661	154	C1103A	154	S3034	154	28C1103L
153	28B435	153	62734450	154	C1103B	154	S3035	154	28C154
153	28B435-0	153	62743794	154	C1103C	154	S40205	154	28C154B
153	28B435-O	153	2004047310	154	C1103L	154	SET001	154	28C154C
153	28B435-R	153	2004049081	154	C154C	154	SET002	154	28C1550
153	28B435-Y	153	6624003100	154	C273	154	SET010	154	28C273
153	0028B435RY	153	A111E2	154	C470	154	SET016	154	28C473
153	28B355Y	154	A116081	154	C49	154	SET017	154	28C500
153	28B507	154	A121-361	154	C500	154	SET050	154	28C500R
153	28B508	154	A12546	154	C500R	154	SET055	154	28C500Y
153	28B509	154	A126705	154	C500Y	154	SET056	154	28C505
153	28B511	154	A127712	154	C505	154	SPT186	154	28C505-0
153	28B511C	154	A130	154	C505-0	154	SPT187	154	28C505-R
153	28B511D	154	A130-ORN	154	C505-R	154	SK3040	154	28C506
153	28B511E	154	A130-V10	154	C506	154	SK8261	154	28C506-0
153	28B512	154	A1409	154	C506-0	154	SM6727	154	28C506-R
153	28B512P	154	A1A	154	C506-R	154	SPB400	154	28C507
153	28B513	154	A1M	154	C507	154	SPB401	154	28C507-0
153	28B513P	154	A18	154	C507-0	154	SS1912	154	28C507-R
153	28B513Q	154	A2057B104-8	154	C507-R	154	SS524	154	28C507-Y
153	28B513R	154	A2090	154	C507-Y	154	SS6111	154	28C526
153	28B514	154	A247	154	C526	154	STX0028	154	28C58
153	28B515	154	A247(AMC)	154	C589	154	SX60M	154	28C589
153	28B523	154	A2511A4130	154	C58A	154	SX-601	154	28C58A
153	3L4-6011-11	154	A25762-010	154	C59	154	T-481	154	28C590
153	3L4-6011-12	154	A25762-012	154	C590	154	T-Q5082	154	28C627
153	3L4-6011-14	154	A2620	154	C627	154	T481(SEARS)	154	28C65
153	3L4-6011-9	154	A2A	154	C64(JAPAN)	154	TA7292	154	28C65B
153	3L4-6013-02	154	A2K	154	C65	154	TA7293	154	28C65N
153	3L4-6013-15	154	A2Z	154	C65B	154	TG28C65	154	28C65Y
153	3L4-6013-2	154	A310	154	C65N	154	TG28C65Y		
						154	TI-722		

ECG	Industry Standard No.	ECG	Industry Standard No.	ECG	Industry Standard No.	ECG	Industry Standard No.	ECG	Industry Standard No.
154	28065XA	154	321-264	156	A6A5	156	GM-1Z	156	1N056
154	28065YB	154	352-0403-010	156	A6A9	156	GM-3	156	1N057
154	28065YTV	154	417-115	156	A6B1	156	GM-3Y	156	1N225B
154	28066	154	613-4	156	A6B5	156	GM-3Z	156	1N226B
154	28066B	154	656-4	156	A6B9	156	GM5Y	156	1N5170
154	28070	154	690Y080H38	156	A6C5	156	GP25G	156	1N5172
154	28C788	154	992-01-3684	156	A6C9	156	H109	156	1N5177
154	28C805	154	1018	156	A6D1	156	HC50	156	1N5178
154	28C818	154	1067	156	A6D5	156	HD20005100	156	1N5400
154	28C856	154	1067(GE)	156	A6D9	156	HEP-R0604	156	1N821
154	28C856-02	154	1076-1559	156	A6E1	156	HP8D12	156	1N5B261
154	28C856C	154	1804-17	156	A6E5	156	IP20-0164	156	181071
154	28C857	154	1885-17	156	A6E9	156	J10	156	181076
154	28C857H	154	2017-110	156	A6F1	156	JA-KCDP	156	181891
154	280995	154	2057A2-261	156	A6F5	156	JC-00059	156	181892
154	4-686145-3	154	2057B104-8	156	A6F9	156	KCDP12-1	156	181944
154	4-686232-3	154	2057B2-140	156	A6G1	156	M72D	156	182409
154	6A12988	154	2090(CROWN)	156	A6G5	156	MA242RC	156	1W810
154	09-302099	154	2114-0	156	A6G9	156	MR-1	156	2A100
154	010-694(AMPEX)	154	2474(RCA)	156	A6H1	156	MR-1C	156	2A200
154	12-1	154	2620	156	A6H5	156	MR1030A	156	2A30
154	13-15809-1	154	3545(RCA)	156	A6H9	156	MR1031A	156	2B10
154	13-23825-1	154	3552(RCA)	156	A6K1	156	MR1031B	156	2B6
154	13-28432-2	154	3553(RCA)	156	A6K5	156	MR1032A	156	2B8
154	13-55062-1	154	3582(RCA)	156	A6K9	156	MR1032B	156	2BM015
154	14-602-19	154	3687	156	A6M1	156	MR1C	156	3A1
154	14-602-36	154	3878	156	A6M5	156	MR2369	156	3A100
154	14-602-37	154	4648	156	A6M9	156	N82006	156	3AF2
154	19-020-046	154	4686-232-3	156	A7A1	156	N82007	156	3B261
154	34-6001-65	154	4819	156	A7A5	156	N82008	156	3D83
154	34-6015-59	154	4858	156	A7A9	156	N83006	156	5AD10
154	46-86173-3	154	4843	156	A7G1	156	N83007	156	5AD10C
154	46-86182-3	154	4853	156	A7G5	156	N83008	156	5AD10D
154	46-86183-3	154	6181-1	156	A7G9	156	P82346	156	5AD8
154	46-86210-3	154	7253(LOWREY)	156	A7H1	156	P82347	156	5AD8DC
154	46-86232-3	154	12546	156	A7H5	156	QD-81B953XA	156	5B2
154	46-86318-3	154	014558	156	A7H9	156	RO080	156	8-719-911-54
154	46-867-3	154	23114-052	156	A7K1	156	RO081	156	8-719-912-54
154	48-134648	154	25672-016	156	A7K5	156	RO082	156	09-306053
154	48-134819	154	37725	156	A7K9	156	RO086	156	09-306379
154	48-134843	154	381-20	156	A7M1	156	RO08A	156	012-1025-002
154	48-134853	154	38121	156	A7M5	156	RO090	156	13-17174-4
154	48-134898	154	38996	156	A7M9	156	RO091	156	13-26614-1
154	48-134919	154	40354	156	AB100	156	RO094	156	13-29165-1
154	48-134927	154	40355	156	AB1000	156	RO096	156	13-29165-2
154	48-137002	154	40459	156	AB200	156	RO097	156	022-2823-001
154	48-137035	154	60684	156	AB300	156	RO098	156	24MW1118
154	48-137364	154	94051(EICO)	156	AB400	156	R1000	156	28-22-11
154	48-137415	154	116081	156	AB50	156	R210	156	28-22-16
154	48-155006	154	126705	156	AB500	156	R250	156	030
154	488134919	154	126709	156	AB600	156	R250F	156	30D1
154	48B155006	154	126710	156	AB800	156	R350	156	30R10
154	57A1-122	154	127712	156	AC100	156	RA132AA	156	30R8
154	57A104-1	154	136066	156	AC200	156	RA132DA	156	031
154	57A104-2	154	140259	156	AC30	156	RA132MA	156	032
154	57A104-3	154	171162-124	156	AC50	156	RA132RA	156	035
154	57A104-4	154	171162-125	156	AB1A	156	RA132VA	156	46-86558-3
154	57A104-5	154	171162-126	156	AB1C	156	RE54	156	533010-5
154	57A104-6	154	231140-28	156	AB1D	156	RE92	156	53J001-1
154	57A104-7	154	256282	156	AB1B	156	RPJ71123	156	53J0002-1
154	57A104-8	154	0320051	156	AB1B	156	RH-DX0091CEZZ	156	57-42
154	57A12-1	154	540205	156	AB1F	156	RVDSD-1Z	156	57-42A
154	57A12-2	154	0573501	156	AB1G	156	RVDSG-5N	156	60LA
154	57A136-1	154	0573519	156	ATC-SR-3	156	RVDSG-5P	156	065-016
154	57A136-10	154	610075-1	156	B172-10	156	RZ-3	156	070-031
154	57A136-11	154	610135-1	156	B172-100	156	S1B02-03C	156	070-035
154	57A136-2	154	610144-3	156	B172-20	156	S3V20	156	070-041
154	57A136-3	154	652321	156	B172-5	156	S6BR2	156	96-5193-01
154	57A136-4	154	984227	156	B172-60	156	S7BR2	156	96-5194-01
154	57A136-5	154	1449098-1	156	B172-70	156	S8-06	156	106-013
154	57A136-6	154	1472474-2	156	B172-80	156	S8-10	156	150R10B
154	57A136-7	154	1473541-1	156	B172-90	156	S808	156	150R8B
154	57A136-8	154	1473552-1	156	B6A1	156	S8BR2	156	150R9B
154	57A136-9	154	1473656-4	156	B6A5	156	S9BR2	156	200-6582
154	57A160-8	154	1960085-2	156	B6A9	156	SA-2	156	200R10B
154	57A194-11	154	2320051	156	B6B1	156	SA-2B	156	200R1B
154	57A207-8	154	2320051H	156	B6B5	156	SA-2C	156	200R7B
154	57A261-10	154	2320191	156	B6B9	156	SA-2Z	156	200R8B
154	57B104-4	154	2320892	156	B6C5	156	SCBR10	156	200R9B
154	57B136-1	154	2321101	156	B6C9	156	SCBR6	156	024B
154	57B136-10	154	3170717	156	B6G1	156	SE15B	156	264P03703
154	57B136-11	154	3170757	156	B6G5	156	SE15C	156	503-210192
154	57B136-2	154	3450842-10	156	B6G9	156	SE15D	156	523-0017-001
154	57B136-3	154	3450842-20	156	B6H1	156	SG-5N	156	690V038H22
154	57B136-4	154	3450842-30	156	B6H5	156	SG-5P	156	690V008H54
154	57B136-5	154	06120001	156	B6H9	156	SI-RECT-208	156	800-018-00
154	57B136-6	154	14735521	156	B6K1	156	SI10A	156	903-17
154	57B136-7	154	23114007	156	B6K5	156	SI8A	156	903-17A
154	57B136-8	154	23114028	156	B6K9	156	SIDA05K	156	903-17B
154	57B136-9	154	23114052	156	B6M1	156	SIDA05N	156	903-17C
154	57B207-8	154	23114053	156	B6M5	156	SIDA05P	156	903-33A
154	57B283-11	154	23114054	156	B6M9	156	SJ1003F	156	919-002440
154	57C012-1	154	23114095	156	B7A1	156	SK3081	156	919-002440-1
154	57C012-2	154	23114125	156	B7A5	156	SR-499	156	919-011339
154	57D1-122	154	25114130	156	B7A9	156	SR154	156	919-013061
154	57M2-16	154	62051493	156	B7G1	156	SR1EM-4	156	919-013079
154	57M2-17	154	6211633I	156	B7G5	156	SR2B12	156	2000-348
154	62-18426	154	62543914	156	B7G9	156	SR2B16	156	3529A
154	63-28426	154	62563309	156	B7H1	156	SR2B20	156	3529B
154	730180497-4	154	62734639	156	B7H5	156	SR3AM-2	156	5529C
154	730180499-5	154	62737522	156	B7H9	156	SSIB0140	156	8000-00011-105
154	730180499-6	154	930Y5440	156	B7K1	156	SSIB0180	156	12550
154	730182080-33	154/427	0996	156	B7K5	156	SSIB0640	156	23115-085
154	730182088-31	154/427	28C996	156	B7K9	156	SSIB0680	156	47126-10
154	75N1	155	AD-157	156	B7M1	156	SSIC0840	156	47126-8
154	86-213-2	155	AD157	156	B7M5	156	SSIC1140	156	50745-03
154	86-214-2	155	AD161	156	B7M9	156	SSIC1180	156	50745-04
154	86-215-2	155	AD165	156	BH4R6	156	SSIC1740	156	50745-05
154	86-316-2	155	BDY15	156	BTB1000	156	SSIC1780	156	61436
154	86-612-2	155	BDY15B	156	BTM1000	156	SVDA82HB10	156	71794
154	86-629-2	155	BDY15B	156	BY172	156	SW-1	156	112528
154	998087-1	155	BDY15C	156	BY173	156	SW-1-01	156	127835
154	998101-1	155	BDY16A	156	BYX26-60	156	SW-1-02	156	131815
154	104-8	155	BDY16B	156	BYX38	156	SZA100	156	132815
154	1111T2	155	GE-43	156	BYY60	156	TM61	156	229805
154	121-241	155	IRTR91	156	BYY61	156	TM84	156	235541
154	121-361	155	PT4	156	CB100	156	TM85	156	237070
154	121-445	155	2N4077	156	CF102VA	156	TVSD81M	156	237421
154	121-473	155	8Q-3-23	156	CF152VA	156	TVSD82K	156	239097
154	121-743	155	13-26377-2	156	CT80	156	U05B	156	240055
154	121-776	155	21A015-021	156	DIJ72292	156	U05C	156	240076
154	121-777	155	57M3-10N	156	DR82	156	U05E	156	243115
154	121-777-01	155	57M3-11	156	DS-1.5-2	156	UO5E	156	489752-029
154	121-792	155	57M3-9N	156	DS-2N	156	UO-5E	156	530072-1014
154	121-843	155	94N2P	156	D81M	156	U05E	156	530082-4
154	133N	155	4822-130-40212	156	D88953	156	001-02406-0	156	530088-3
154	135N1M	156	A054-229	156	E24101	156	001-02406-1	156	5600208
154	144-4	156	A22B	156	EA16X149	156	1-506-319	156	817179
154	260-10-052	156	A22B1	156	ES16X33	156	1.5B05	156	817179A
154	260D07201	156	A22D	156	EW169	156	1.5C05	156	984794
154	260P10301	156	A22D1	156	GE-510	156	1AB027	156	984795
154	260P10801	156	A22M	156	GB-512	156	1AB029	156	1303256
154	260P22101	156	A2MA	156	GM-1	156	1LE11	156	1471405-1
154	260P22203	156	A6A1	156	GM-1A	156	1LF11	156	1472460-14

ECG	Industry Standard No.	ECG	Industry Standard No.	ECG	Industry Standard No.	ECG	Industry Standard No.	ECG	Industry Standard No.
156	1474778-013	158	AC193	158	HT203243A	158	SF.T322	158	28B324J
156	1474778-13	158	AC193K	158	HT204051	158	SF.T351	158	28B324K
156	1474778-4	158	ACY33	158	HT204051B	158	SF.T352	158	28B324L
156	1474778-8	158	ACY33-VI	158	HT204051C	158	SPT122	158	28B324N
156	1476171-21	158	ACY33-VII	158	HT204051D	158	SPT123	158	28B324P
156	1800017	158	ACY33-VIII	158	HT204051E	158	SPT124	158	28B324R
156	2330332	158	ACY33VI	158	HT204053	158	SPT125	158	28B324V
156	2330771	158	ACY33VII	158	HT204053A	158	SPT125P	158	28B564
156	2330773	158	AR102	158	HT204053B	158	SPT131	158	28B565
156	17240010	158	ASY12-1	158	HT20451A	158	SPT131P	158	28B565B
156	41020687	158	ASY12-2	158	HVO000405	158	SPT143	158	0000028B405
156	53015201	158	ASY13-1	158	IRTRB4	158	SPT144	158	28B405-2C
156	4202003200	158	ASY13-2	158	J241178	158	SPT145	158	28B405-3C
156(2)	4202021100	158	ASY14	158	K4-500	158	SPT146	158	28B405-4C
156(2)	100C-2	158	ASY14-1	158	M-8641A	158	SPT241	158	28B405A
156(4)	SI-RECT-110	158	AU100N	158	M4450	158	SPT242	158	28B405B
156(4)	SI-RECT-112	158	B-315-1	158	M44660RN	158	SPT243	158	28B405BR
156(4)	523-4001-001	158	B-324	158	M4468BRN	158	SPT267	158	28B405C
157	A3L	158	B105	158	M4469RED	158	SPT523	158	28B405D
157	BD127	158	B108	158	M4470GRN	158	SS0005	158	28B405F
157	BD157	158	B167	158	M4471YEL	158	STX0121	158	28B405G
157	BD158	158	B238	158	M4472GRN	158	STX0224	158	28B405H
157	BD215	158	B238-12A	158	M4473	158	0000008V31	158	28B405P
157	BD216	158	B238-12B	158	M4474YEL	158	T-131	158	28B405R
157	BF458	158	B238-12C	158	M4475GRN	158	T1000	158	28B405RE
157	BF459	158	B304	158	M4476BLU	158	T131	158	28B415
157	C1501Q	158	B304A	158	M4477PUR	158	T50933B	158	28B415A
157	C1501R	158	B324	158	M4510	158	T814	158	28B415B
157	E185B121712	158	B324A	158	M4562	158	T815	158	28B431
157	E839(ELCOM)	158	B324B	158	M8640	158	T65M	158	28B450
157	GB-232	158	B324D	158	M8640A	158	T048	158	28B450A
157	HEP-85015	158	B324E	158	M9002	158	TM158	158	28B451
157	HEP244	158	B324E-1	158	MA1702	158	TNJ60612	158	28B452
157	IR-TR60	158	B324F	158	N-EA2136	158	TNJ61674	158	28B452A
157	M4998	158	B324G	158	NC52	158	TNJ70688	158	28B453
157	M912	158	B324H	158	NKT102	158	TQ-5051	158	28B457
157	MJE340	158	B324I	158	NKT104	158	TQ-5061	158	28B457-Q
157	MJE341	158	B324J	158	NKT105	158	TR-157(OLSON)	158	28B457A
157	MJE344	158	B324K	158	NKT106	158	TR-158(OLSON)	158	28B457AQ
157	MJE440	158	B324L	158	NKT107	158	TR-170(OLSON)	158	28B494
157	MJ9742	158	B324N	158	NKT108	158	T81007	158	28B495
157	RB24	158	B324P	158	NKT109	158	T81266	158	28B495A
157	S5472	158	B324S	158	NKT153/25	158	T8163	158	28B495B
157	895252	158	B324V	158	NKT154/25	158	T8164	158	28B495C
157	899252	158	B364	158	NKT163/25	158	T8165	158	28B495D
157	SJE205	158	B365	158	NKT164/25	158	T8166	158	28B495T
157	SJE218	158	B365B	158	NKT211	158	T81727	158	28B535
157	SJE232	158	B405	158	NKT212	158	T81728	158	28B58
157	SJE290	158	B405-2C	158	NKT213	158	T8603	158	28B74
157	SJE9754	158	B405-3C	158	NKT214	158	T8604	158	2T11
157	SJB400	158	B405-4C	158	NKT215	158	T8616	158	2T12
157	ST-MJE9742	158	B405A	158	NKT216	158	T8617	158	2T13
157	T287	158	B405C	158	NKT217	158	T8618	158	2T14
157	TIP27	158	B405D	158	NKT218	158	T8619	158	2T15
157	TV-120	158	B405E	158	NKT219	158	T8627	158	2T16
157	TV8SJE5472-1	158	B405G	158	NKT271	158	T8629	158	2T17
157	2N4054	158	B405H	158	NKT272	158	T8739	158	2T21
157	2N4055	158	B405K	158	NKT274	158	T8739B	158	2T22
157	2N4056	158	B405R	158	NKT275	158	T8740	158	2T23
157	2N4057	158	B405RE	158	NKT275A	158	T8765	158	2T24
157	2N5655	158	B415	158	NKT275B	158	TV47	158	2T25
157	2N5656	158	B415A	158	NKT275J	158	TV61	158	2T26
157	28C1501	158	B415B	158	NKT281	158	TVS-28B171A	158	2T311
157	28C1501Q	158	B431	158	OC-304/1	158	TVS-28B324	158	2T312
157	28C1501R	158	B450	158	OC-304/2	158	TVS28B171	158	2T313
157	13-27974-1	158	B450A	158	OC-304/3	158	TX-104-3	158	2T314
157	13-29974-4	158	B451	158	OC-305/1	158	VP82745J	158	2T315
157	13-34089-2	158	B452	158	OC-305/2	158	W25	158	2T321
157	13-34089-4	158	B452A	158	OC-306/1	158	WC19862A	158	2T322
157	13-35257-1	158	B453	158	OC-306/2	158	WTV30VH6	158	2T323
157	13-35257-2	158	B457	158	OC-306/3	158	WTVBMC	158	2T324
157	34-1006	158	B457-C	158	OC83	158	XA151	158	2T325
157	46-86319-3	158	B457A	158	OC84	158	XA152	158	33-28
157	48-134998	158	B495	158	P3B	158	X032	158	33-29
157	51-4	158	B495A	158	PA10889-1	158	X8101	158	04-00156-03
157	52-4	158	B495C	158	PA10889-2	158	X8104	158	4-202104770
157	57A127	158	B495D	158	PA915B	158	X8121	158	6A10622
157	57D127	158	B495T	158	PBE3014-1	158	XT100	158	6A11501
157	66P020-1	158	B535	158	PBE3014-2	158	XT200	158	6A12515
157	66P020-2	158	B58	158	PBE3020-1	158	Y633	158	6A12517
157	84G01	158	B601-1010	158	PBB3162	158	ZC28B172	158	7-59-0103477
157	84G01	158	B74	158	PC10662	158	ZC28B172A	158	8-721-323-00
157	86-165-2	158	BA6	158	PC1067T	158	ZEN302	158	8-729-447-53
157	86-177-2	158	BA6A	158	PC1068T	158	ZEN304	158	8-905-605-120
157	86-228-2	158	BCM1002-18	158	PC3002	158	ZEN305	158	8-905-605-123
157	86-251-2	158	BCM1002-3	158	PC3003	158	ZEN307	158	8-905-605-124
157	86-287-2	158	BX-324	158	PC3005	158	ZEN308	158	8-905-605-125
157	86X0073-002	158	CE0363/7839	158	PC3006	158	ZJ13	158	8-905-605-126
157	089-4(SYLVANIA)	158	CGB-53	158	PC3007	158	Z838	158	8-905-605-127
157	121-712	158	CP800	158	PC3009	158	Z856	158	8-905-605-129
157	132-078	158	CP801	158	PD28	158	001-02010	158	8-905-605-250
157	297V084C01	158	CP802	158	PD29	158	1-801-005	158	8-905-605-260
157	417-159	158	CP803	158	PTO139	158	1-801-006	158	8-905-605-264
157	417-195	158	CS1758	158	Q2N4106	158	1-801-310	158	8-905-605-266
157	417-227	158	CT1017	158	Q6	158	1T495	158	8-905-605-268
157	974-1(SYLVANIA)	158	D156	158	Q7	158	2N149B	158	8-905-605-269
157	974-2(SYLVANIA)	158	D352	158	Q8	158	2N1614	158	8-905-613-070
157	974-3	158	DC-12	158	QQ061209	158	2N2095	158	8-905-613-071
157	974-3A6-3	158	DU4	158	QQV60526	158	2N2096	158	8-905-613-131
157	974-4(SYLVANIA)	158	E4	158	QQV60538	158	2N2431	158	8-905-613-132
157	3567-2(RCA)	158	EA15X257	158	R-28B405	158	2N2706	158	8-905-613-133
157	59810	158	EA15X326	158	R-56	158	2N2707	158	8-905-613-160
157	06120053	158	EA2136	158	R5101	158	2N4106	158	8A13718
157#	2801566	158	ES15X75	158	R5182	158	2N803	158	09-300005
157#	2802258	158	ES15X99	158	R9533	158	2N804	158	09-301008
158	A065-105	158	ES37(ELCOM)	158	R9534	158	2N805	158	09-301014
158	A065-112	158	ES4(ELCOM)	158	R9603	158	2N806	158	09-301019
158	A069-107	158	G0005	158	R9604	158	2N817	158	09-301022
158	A128(JAPAN)	158	G04-711-E	158	RE25	158	2N818	158	09-301027
158	A2414	158	G04-711-F	158	RB-102	158	20C72	158	09-301054
158	AC126	158	G04-711-G	158	RB-2007	158	28A128	158	09-301065
158	AC128	158	G04-711-H	158	RB3667	158	28A129	158	13-14085-11
158	AC128-01	158	GB-53	158	RB8406	158	28B105	158	13-14085-12
158	AC128/01	158	GETO-50P	158	RB8424	158	28B167	158	13-14085-25
158	AC12801	158	HB-00324	158	RB8446	158	28B238	158	13-67599-3/3464
158	AC128K	158	HB-00405	158	RT-61015	158	28B238-12A	158	14-602-06
158	AC131-30	158	HB-475	158	RT-61016	158	28B238-12B	158	14-602-07
158	AC138	158	HB324	158	RT121	158	28B238-07	158	14-602-08
158	AC139	158	HB365	158	RT4625	158	28B304	158	14-602-09
158	AC142	158	HB415	158	RT7558	158	28B304A	158	14-602-51
158	AC142K	158	HB475	158	RT7846	158	28B324	158	019-003415
158	AC153	158	HEP253	158	RT8842	158	28B324/4454C	158	019-003416
158	AC153K	158	HEP254	158	S065	158	28B324A	158	19-3415
158	AC154	158	HJ226	158	S1672	158	28B324B	158	19-3416
158	AC166	158	HJ228	158	SA29	158	28B324D	158	20-1680-189
158	AC167	158	HJ230	158	SA318-2	158	28B324E	158	21A040-036
158	AC176K	158	HJ315	158	SA318-3	158	28B324E-1	158	21A040-060
158	AC178	158	HJ71	158	SA681	158	28B324F	158	21A040-061
158	AC180	158	HJ72	158	SF.T124	158	28B324G	158	21A040-081
158	AC180K	158	HJ73	158	SF.T130	158	28B324H		
158	AC188	158	HJ74	158	SF.T131P				
158	AC188/01	158	HR30	158	SF.T221				
158	AC18801			158	SF.T223				
158	AC188K			158	SF.T321				

ECG	Industry Standard No.	ECG	Industry Standard No.	ECG	Industry Standard No.	ECG	Industry Standard No.	ECG	Industry Standard No.
159	A-1005-725	159	A402	159	A829B	159	B54731-30	159	BC214KA-1
159	A-113110	159	A4086	159	A829C	159	BC-261	159	BC214KB
159	A-120417	159	A4087	159	A829D	159	BC116	159	BC214KB-1
159	A-120526	159	A4126	159	A829E	159	BC116A	159	BC214KC
159	A-1853-0016-1	159	A41440	159	A833	159	BC126	159	BC214KC-1
159	A-1853-0020-1	159	A417-116	159	A8405	159	BC126-1	159	BC214L
159	A-1853-0027-1	159	A417-132	159	A8540	159	BC137	159	BC214L-1
159	A-1853-0036-1	159	A417-153	159	A8867	159	BC137-1	159	BC214LA
159	A-1853-0039-1	159	A417-176	159	A8T3702	159	BC153-1	159	BC214LA-1
159	A-1853-0049-1	159	A417-182	159	A8T3703	159	BC154-1	159	BC214LB
159	A-1853-0058-1	159	A417-184	159	A8T404A	159	BC157	159	BC214LB-1
159	A-1853-0062-1	159	A417-196	159	A8T405B	159	BC157-1	159	BC214LC
159	A-1853-0065-1	159	A417-200	159	A8T4059	159	BC157A	159	BC214LC-1
159	A-1853-0092-1	159	A417-201	159	A8T4060	159	BC157B	159	BC221
159	A-1853-0285-1	159	A417-235	159	A8T4061	159	BC158	159	BC221-1
159	A-1853-0321-1	159	A43023845	159	A8T4062	159	BC158-1	159	BC225-1
159	A-195C	159	A4310	159	A921-70B	159	BC158A	159	BC237
159	A054-223	159	A4442	159	A94037	159	BC158A-1	159	BC250
159	A066-113	159	A4745	159	A945-0	159	BC158B	159	BC250-1
159	A066-113A	159	A4801(JAPAN)	159	A95227	159	BC158B-1	159	BC250A
159	A066-113AB	159	A4802-00004	159	A95232	159	BC159	159	BC250A-1
159	A066-118	159	A4815	159	A970246	159	BC159-1	159	BC250B
159	A066-118A	159	A481A0030	159	A970248	159	BC159A	159	BC250B-1
159	A068-109	159	A482(JAPAN)	159	A970251	159	BC159A-1	159	BC250C
159	A068-109A	159	A4822-130-40548	159	A970254	159	BC159B	159	BC250C-1
159	A112-000172	159	A4844	159	A984193	159	BC159B-1	159	BC251
159	A112-000185	159	A489751-028	159	A991-01-0098	159	BC174A	159	BC251A
159	A112-000187	159	A489751-031	159	A991-01-1225	159	BC174B	159	BC251A-1
159	A1200482	159	A493-0	159	A991-01-1319	159	BC177-1	159	BC251B
159	A116078	159	A495-R	159	A991-01-3058	159	BC177A	159	BC251B-1
159	A118284	159	A500(JAPAN)	159	AC9082	159	BC177A-1	159	BC251C
159	A119730	159	A514-044910	159	AC9083	159	BC177B	159	BC251C-1
159	A120P1	159	A522(JAPAN)	159	AC9084	159	BC177B-1	159	BC252
159	A121-1	159	A5226-1	159	AC9085	159	BC177V	159	BC252-1
159	A121-1RED	159	A522A	159	ACD09000-1	159	BC177V-1	159	BC252A
159	A121-444	159	A530	159	AD29A-4	159	BC177V1	159	BC252A-1
159	A121-446	159	A530H	159	AD29A-5	159	BC177V1-1	159	BC252B
159	A121-495	159	A5320111	159	AD29A-6	159	BC178-1	159	BC252B-1
159	A121-496	159	A54-96-005	159	AD29A-9	159	BC178A	159	BC252C
159	A121-497	159	A544	159	AD29B-1	159	BC178A-1	159	BC252C-1
159	A121-497WHT	159	A548	159	AD29B-2	159	BC178B	159	BC253
159	A121-602	159	A550	159	AD29B10	159	BC178B-1	159	BC253-1
159	A121-603	159	A550A	159	AD29B4	159	BC178D	159	BC253A
159	A121-679	159	A550AQ	159	AD29B5	159	BC178D-1	159	BC253A-1
159	A121-699	159	A550Q	159	AD29B6	159	BC178V	159	BC253B
159	A121-746	159	A550R	159	AD29B7	159	BC178V-1	159	BC253B-1
159	A121-774	159	A550S	159	AD29B8	159	BC178V1	159	BC253C
159	A1214	159	A564-0	159	AD29B9	159	BC178V1-1	159	BC253C-1
159	A1214467	159	A565	159	AD30A1	159	BC179-1	159	BC256
159	A121659	159	A565A	159	AD30A2	159	BC179A	159	BC256-1
159	A122GRN	159	A565B	159	AD30A3	159	BC179A-1	159	BC256A
159	A122YEL	159	A565C	159	AD30A4	159	BC179B	159	BC256A-1
159	A124047	159	A565D	159	AD30A5	159	BC179B-1	159	BC256B
159	A124755	159	A565K	159	AF21490	159	BC181	159	BC256B-1
159	A12594	159	A567(JAPAN)	159	AF3570	159	BC181A	159	BC256C-1
159	A126524	159	A567A	159	AF3590	159	BC181A-1	159	BC257
159	A126700	159	A569J	159	AF699	159	BC186	159	BC257-1
159	A126707	159	A576-0001-002	159	AFC81170F	159	BC187	159	BC257VI
159	A126715	159	A576-0001-013	159	AFS24226	159	BC187-1	159	BC258
159	A126718	159	A592Y	159	AFT0019M	159	BC192	159	BC258-1
159	A126719	159	A59625-1	159	AFT052	159	BC192-1	159	BC258VI
159	A12888	159	A59625-10	159	AFT1341	159	BC196	159	BC259
159	A129-34	159	A59625-11	159	AFT17746	159	BC196-1	159	BC259-1
159	A129697	159	A59625-12	159	ANJ101	159	BC196A	159	BC25BB
159	A129699	159	A59625-2	159	AR304(GREEN)	159	BC196A-1	159	BC260
159	A130-149	159	A59625-3	159	AR308(VIOLET)	159	BC196B	159	BC260-1
159	A130-40315	159	A59625-4	159	AT-11	159	BC196B-1	159	BC260A
159	A130-40429	159	A59625-5	159	AT331	159	BC196V1	159	BC260A-1
159	A130159	159	A59625-6	159	AT331A	159	BC196V1-1	159	BC260B
159	A137(INP)	159	A59625-7	159	AT332	159	BC200	159	BC260B-1
159	A1471114-1	159	A59625-8	159	AT332A	159	BC200-1	159	BC260C
159	A1473563-1	159	A59625-9	159	AT333	159	BC201	159	BC260C-1
159	A1473570-1	159	A5T2604	159	AT335A	159	BC201-1	159	BC261
159	A1473574-1	159	A5T2605	159	AT410	159	BC202	159	BC261A
159	A1473590-1	159	A5T2907	159	AT410-1	159	BC202-1	159	BC261A-1
159	A1473591-1	159	A5T3504	159	AT412	159	BC203	159	BC261B
159	A1473597-1	159	A5T3505	159	AT412-1	159	BC203-1	159	BC261B-1
159	A1558-17	159	A5T3638	159	AT413	159	BC204-1	159	BC261C
159	A161	159	A5T3638A	159	AT413-1	159	BC204A	159	BC261C-1
159	A162	159	A5T3644	159	AT414	159	BC204A-1	159	BC262
159	A171	159	A5T3645	159	AT414-1	159	BC204V	159	BC262A
159	A177	159	A5T3905	159	AT415	159	BC204V-1	159	BC262A-1
159	A177(A)	159	A5T3906	159	AT415-1	159	BC204V1	159	BC262B
159	A177A	159	A5T4125	159	AT416	159	BC204V1-1	159	BC262B-1
159	A177AB	159	A5T4126	159	AT416-1	159	BC205	159	BC262C
159	A178A	159	A5T4248	159	AT417	159	BC205-1	159	BC263
159	A178AB	159	A5T4249	159	AT417-1	159	BC205A	159	BC263A
159	A178B	159	A5T4250	159	AT418	159	BC205A-1	159	BC263A-1
159	A178BA	159	A5T4402	159	AT418-1	159	BC205V	159	BC263B
159	A179A	159	A5T4403	159	AT419	159	BC205V-1	159	BC263B-1
159	A179AC	159	A5T5086	159	AT419-1	159	BC205V1	159	BC263C
159	A179B	159	A5T5087	159	AT430	159	BC205V1-1	159	BC263C-1
159	A179BB	159	A5T5221	159	AT430-1	159	BC206-1	159	BC266
159	A1844-17	159	A5T5226	159	AT431	159	BC206A	159	BC266-1
159	A1867-17	159	A603	159	AT431-1	159	BC206B-1	159	BC266A
159	A1901-5338	159	A608-F	159	AT432	159	BC212	159	BC266B
159	A190429	159	A610074-1	159	AT432-1	159	BC212-1	159	BC281
159	A200-052	159	A610083	159	AT433	159	BC212A	159	BC281-1
159	A20057073-0702	159	A610083-1	159	AT433-1	159	BC212B	159	BC281A
159	A2057013-0701	159	A610083-2	159	AT434	159	BC212K	159	BC281A-1
159	A2057013-0702	159	A610083-3	159	AT434-1	159	BC212K-1	159	BC281B
159	A2057013-0703	159	A610110-1	159	AT435	159	BC212KA	159	BC281B-1
159	A2057A2-198	159	A610120-1	159	AT435-1	159	BC212KA-1	159	BC281C
159	A2057B106-12	159	A617K	159	AT436	159	BC212KB	159	BC281C-1
159	A2057B108-6	159	A618K	159	AT436-1	159	BC212KB-1	159	BC283
159	A20K	159	A640(JAPAN)	159	AT437	159	BC212L	159	BC283-1
159	A20KA	159	A650232E	159	AT437-1	159	BC212LA	159	BC291
159	A22008	159	A669	159	AT438	159	BC212LA-1	159	BC291-1
159	A23114050	159	A672	159	AT438-1	159	BC212LB	159	BC291A
159	A23114051	159	A672A	159	AT451	159	BC212LB-1	159	BC291A-1
159	A23114550	159	A672B	159	AT451-1	159	BC212VI	159	BC291D
159	A2428	159	A672C	159	AT452	159	BC213	159	BC291D-1
159	A2448	159	A690D	159	AT452-1	159	BC213-1	159	BC292
159	A2498512	159	A701584-00	159	AT453	159	BC213A	159	BC292A
159	A2798	159	A701589-00	159	AT453-1	159	BC213B	159	BC292A-1
159	A297LC12C01	159	A71687-1	159	AT454	159	BC213K	159	BC292D
159	A297V073C03	159	A718	159	AT454-1	159	BC213K-1	159	BC292D-1
159	A297V073C04	159	A753-4004-248	159	AT455	159	BC213KA	159	BC297
159	A28A564PR	159	A7570013-01	159	AT455-1	159	BC213KA-1	159	BC307
159	A30270	159	A7576004-01	159	B-1426	159	BC213KB	159	BC307A
159	A30290	159	A759A	159	B-75561-31	159	BC213KB-1	159	BC307B
159	A3513	159	A759B	159	B1J	159	BC213KC	159	BC307C
159	A3540	159	A78331	159	B1N-1	159	BC213KC-1	159	BC307V1
159	A3549	159	A800-523-01	159	B1N-2	159	BC213L	159	BC308
159	A3559	159	A800-523-02	159	B266A-1	159	BC213L-1	159	BC308A
159	A3562	159	A800-527-00	159	B266B-1	159	BC213LA	159	BC308B
159	A3563	159	A815199	159	B2A	159	BC213LA-1	159	BC308C
159	A3574	159	A815199-6	159	B2B	159	BC213LB	159	BC308V1
159	A3581	159	A815211	159	B2G	159	BC213LB-1	159	BC309
159	A3T2894	159	A815213	159	B2M-1	159	BC213LC	159	BC309A
159	A3T2906	159	A815229	159	B2M-2	159	BC213LC-1	159	BC309B
159	A3T2906A	159	A815247	159	B2M-3	159	BC214		
159	A3T2907	159	A829	159	B2S	159	BC214-1		
		159	A829A	159	B2Y	159	BC214A		
						159	BC214B		
						159	BC214K		
						159	BC214K-1		
						159	BC214KA		

ECG	Industry Standard No.	ECG	Industry Standard No.	ECG	Industry Standard No.	ECG	Industry Standard No.	ECG	Industry Standard No.
159	BC309C	159	BCY95A	159	BJ2B	159	C89012	159	G89015I
159	BC315	159	BCY95B	159	BJ2C	159	C89012/3490	159	G89015J
159	BC320	159	BCY95B-1	159	BJ2D	159	C89012E	159	G89022F
159	BC325	159	BCY98	159	BJ3	159	C89012E-F	159	G89022G
159	BC325A	159	BCY98A	159	BJ3A	159	C89012F	159	GT1644
159	BC326	159	BCY98B	159	BJ3B	159	C89012FC	159	HA-00495
159	BC326A	159	BCY98B-1	159	BJ4	159	C89012PG	159	HA-00610
159	BC327	159	BCZ10	159	BJ4A	159	C89012H	159	HA9048
159	BC328	159	BCZ10A	159	BJ4B	159	C89012HE	159	HA9049
159	BC337	159	BCZ10B	159	BJ4C	159	C89012HF	159	HEP-80026
159	BC381	159	BCZ10C	159	BJ4D	159	C89012HG	159	HEP-80031
159	BC400	159	BCZ11	159	BJ5	159	C89012HG/3490	159	HEP-80032
159	BC405	159	BCZ11A	159	BJ5G	159	C89012HH	159	HEP52
159	BC405A	159	BCZ12	159	BJ5A	159	C89012HH/3490	159	HEP57
159	BC405B	159	BCZ12A	159	BJ5B	159	C89012I	159	HEP708
159	BC406B	159	BCZ12B	159	BJ6	159	C89015	159	HEP715
159	BC415A	159	BCZ12C	159	BJ6A	159	C89015B	159	HEP716
159	BC415B	159	BCZ13	159	BJ6B	159	C89015C	159	HEP717
159	BC416A	159	BCZ13A	159	BJ6C	159	C89015C2	159	HEP739
159	BC416B	159	BCZ13B	159	BJ6D	159	C89015D	159	HR-71
159	BC417	159	BCZ13C	159	BJ7	159	C89012HF	159	HR71
159	BC418	159	BCZ13D	159	BJ7A	159	C89020E	159	HR84(PNP)
159	BC419	159	BCZ13E	159	BJ7B	159	C89020F	159	HS-40031
159	BC432	159	BCZ13F	159	BJ7C	159	C89102	159	HS-40035
159	BC478	159	BCZ13G	159	BJ7D	159	C89128-B2	159	HS-40040
159	BC478A	159	BCZ13H	159	BJ8	159	C89128C1	159	HS-40050
159	BC478B	159	BCZ14	159	BJ8A	159	C89129	159	HS-90028
159	BC479	159	BCZ14A	159	BJ8B	159	C89129B	159	HT100
159	BC479B	159	BCZ14B	159	BJ8C	159	C89129B1	159	HT101
159	BC512	159	BCZ14C	159	BJ8D	159	C89129B2	159	HT104941B-Q
159	BC512A	159	BCZ14D	159	BJ9	159	D-50492-01	159	HT104941C-Q
159	BC513	159	BCZ14E	159	BJ9A	159	D29A4	159	HT104941CO
159	BC513A	159	BCZ14F	159	BJ9B	159	D29A5	159	HT104942A
159	BC514A	159	BF-832	159	BJ9C	159	D29A6	159	HT104951A-Q
159	BC556	159	BF249	159	BJ9D	159	D29A9	159	HT104951B
159	BC556A	159	BF340	159	BMT1991	159	D29B1	159	HT104951B-Q
159	BC557	159	BF340A	159	BMT2303	159	D29B2	159	HT104951C0
159	BC557A	159	BF340B	159	BMT2411	159	D29B4	159	HT105611
159	BC558	159	BF340C	159	BMT2412	159	D29B5	159	HT105611A
159	BC558B	159	BF340D	159	B08875/2	159	D29B6	159	HT105611B
159	BC559	159	BF341A	159	BR-82	159	D29E7	159	HT105611BO
159	BCW35	159	BF341B	159	BR-832	159	D29E8	159	HT105611C0
159	BCW37	159	BF341C	159	BSV21	159	D30A1	159	HT105612B
159	BCW37A	159	BF341D	159	BSV21A	159	D30A2	159	HT105621B-Q
159	BCW45	159	BF342	159	BSV43A	159	D30A3	159	HT105621B0
159	BCW56	159	BF342A	159	BSV44A	159	D30A5	159	HT106731B
159	BCW56A	159	BF342B	159	BSV45A	159	DE04191	159	HT600011F
159	BCW57	159	BF342C	159	BSV47A	159	D8-82	159	HT600011H
159	BCW57A	159	BF342D	159	BSV48A	159	D8-83	159	HT600210
159	BCW58	159	BF343	159	BSV49A	159	D882	159	HX-50094
159	BCW58A	159	BF440	159	BSV55A	159	D883	159	HX-50105
159	BCW58B	159	BF441	159	BSV55AP	159	E13-001-02	159	HX-50112
159	BCW59	159	BF540	159	BSV55P	159	E13-001-03	159	HX-50176
159	BCW59A	159	BF541	159	BSV96	159	E13-001-04	159	I81030
159	BCW61	159	BFS14A	159	BSV97	159	E13-006-02	159	I9680
159	BCW61C	159	BFS14B	159	BSV98	159	E213	159	IP20-0009
159	BCW61D	159	BFS14C	159	BSW-21A	159	EA15X233	159	IP20-0159
159	BCW62	159	BFS14D	159	BSW-22	159	EA3714	159	J241225
159	BCW62A	159	BFS16	159	BSW-22A	159	EKD1802M	159	J241226
159	BCW63	159	BFS16A	159	BSW-24	159	E13-006-02	159	J241253
159	BCW63A	159	BFS16B	159	BSW-44	159	EN1132	159	J241259
159	BCW64A	159	BFS16C	159	BSW-44A	159	EN2894	159	J24640
159	BCW69	159	BFS16D	159	BSW-45	159	EN2894A	159	J9680
159	BCW69R	159	BFS26	159	BSW-45A	159	EN2907	159	J9697
159	BCW79-10	159	BFS26A	159	BSW-72	159	EN3250	159	JA1050
159	BCW79-16	159	BFS26B	159	BSW-73	159	EN3504	159	JA1050G
159	BCW80-10	159	BFS26C	159	BSW-74	159	EN3905	159	JA1050QL
159	BCW80-16	159	BFS26D	159	BSW-75	159	EN3906	159	K4-505
159	BCW86	159	BFS26E	159	BSW21	159	EN3962	159	K9682
159	BCW93A	159	BFS26F	159	BSW21A	159	EN722	159	KLM746
159	BCW93KA	159	BFS26G	159	BSW22	159	EP15X17	159	KSA495Y
159	BCW93KB	159	BFS31	159	BSW22A	159	EP15X26	159	LJ-152
159	BCW96	159	BFS32	159	BSW24	159	EP15X48(PNP)	159	LJ152
159	BCW96A	159	BFS32P	159	BSW44	159	EP15X53	159	LJ152(0)
159	BCW96KA	159	BFS33	159	BSW44A	159	EP15X60	159	LJ152-0
159	BCW96KB	159	BFS33P	159	BSW45	159	EP15X90	159	LJ152B
159	BCW97	159	BFS34P	159	BSW45A	159	EPX15X17	159	LJ152G
159	BCW97A	159	BFS37	159	BSW72	159	ES15X101	159	LM-1149
159	BCW97KA	159	BFS37A	159	BSW73	159	ES15X107	159	LM-1150
159	BCW97KB	159	BFS40	159	BSW74	159	ES15X128	159	LM-151
159	BCY10	159	BFS40A	159	BSW75	159	ES15X9	159	LM-1153
159	BCY10A	159	BFS41	159	BSX36	159	ES34(ELCOM)	159	LM-2589
159	BCY11	159	BFS69	159	BSY-40	159	ES65(ELCOM)	159	LM1153
159	BCY11A	159	BFT70	159	BSY-41	159	EW202	159	LM1404
159	BCY12	159	BFV20	159	BSY40	159	EYZP-546	159	LM1795
159	BCY12A	159	BFV22	159	BSY41	159	EYZP-623	159	L8-0079-01
159	BCY35	159	BFV25	159	BT832	159	F209	159	L8-0079-02
159	BCY35A	159	BFV26	159	BYZP-546	159	F21490	159	M094-585-46
159	BCY37	159	BFV29	159	BYZP-623	159	F3559	159	M442
159	BCY37A	159	BFV30	159	C00686-0258-0	159	F3570	159	M446
159	BCY38	159	BFV33	159	C00686602720	159	F3590	159	M4525
159	BCY38A	159	BFV82	159	067302	159	F3597	159	M4590
159	BCY39	159	BFV82A	159	0673C	159	F543-1	159	M4745
159	BCY39A	159	BFV82B	159	0686-248-0	159	F699	159	M4815
159	BCY40	159	BFV82C	159	09082	159	FC8-9012-HH	159	M4815D
159	BCY40A	159	BFV86	159	09083	159	FC8-9012F	159	M4943
159	BCY54	159	BFV86A	159	09084	159	FC8-9012G	159	M4989
159	BCY54A	159	BFV86B	159	09085	159	FC81170F	159	M644
159	BCY58	159	BFV86C	159	CC82005B	159	FC89012	159	M65A
159	BCY70	159	BFW20	159	CC86228F	159	FC89012H	159	M65B
159	BCY70A	159	BFW22	159	CC8015	159	FC89012HE	159	M65C
159	BCY71	159	BFW31	159	CD10000-1B	159	FC89012HG	159	M65D
159	BCY71A	159	BFW37	159	CD437	159	FC89015B	159	M65E
159	BCY72	159	BFW89	159	CD445	159	FC89015C	159	M65F
159	BCY72A	159	BFW90	159	CDC-9000-1D	159	FC89015D	159	M7127/P28
159	BCY77VX	159	BFX29	159	CDC10000-1E	159	FD-1029-MB	159	M829A
159	BCY77VII	159	BFX30	159	CDC496	159	FI-1007	159	M829B
159	BCY77VIII	159	BFX35	159	CDC746(ZENITH)	159	FI-1008	159	M829C
159	BCY79	159	BFX65	159	CDC9000-1	159	FS24226	159	M829D
159	BCY79VII	159	BFY94	159	CDC9000-1D	159	FS24954	159	M829E
159	BCY90	159	BJ10	159	CE4005C	159	FTO019M	159	M829F
159	BCY90A	159	BJ11	159	CE40012D	159	FTO52	159	M833
159	BCY90B	159	BJ11A	159	CK942	159	FT1541	159	M9334
159	BCY90B-1	159	BJ11B	159	C8-2005B	159	FT1746	159	M9514
159	BCY91	159	BJ12	159	C8-2005C	159	FT3638	159	M9526
159	BCY91A	159	BJ12B	159	C8015	159	GO3-407-Y	159	M9527
159	BCY91B	159	BJ12C	159	C81170F	159	GO3007	159	M9531
159	BCY91B-1	159	BJ13	159	C81221F	159	GI3638	159	M9571
159	BCY92	159	BJ13A	159	C8122B	159	GE-21	159	M6649
159	BCY92A	159	BJ13B	159	C81228	159	GE-22	159	ME0404
159	BCY92B	159	BJ14	159	C81251E	159	GB-82	159	ME0404-1
159	BCY92B-1	159	BJ14A	159	C81294E	159	GET3638	159	ME0404-2
159	BCY93	159	BJ15	159	C81294H	159	GET3638A	159	ME501
159	BCY93A	159	BJ160	159	C8129B	159	GI3638	159	MM3504
159	BCY93B	159	BJ161	159	C81303	159	GI3638A	159	MM3726
159	BCY93B-1	159	BJ161A	159	C81308	159	GI3644	159	MM3905
159	BCY94	159	BJ161B	159	C81354	159	GI3703	159	MM3906
159	BCY94A	159	BJ161C	159	C81627	159	GM760	159	MM4048
159	BCY94B	159	BJ1A	159	C83702	159	GMB040-1	159	MM999
159	BCY94B-1	159	BJ1B	159	C83703	159	GO3007	159	MPS-A55
159	BCY95	159	BJ1C	159	C85447	159	GO3014	159	MPS-A56
159		159	BJ2	159	C85448	159	G89015H	159	MPS1572
159		159	BJ2A	159	C86228F	159		159	MPS3636

ECG	Industry Standard No.	ECG	Industry Standard No.	ECG	Industry Standard No.	ECG	Industry Standard No.	ECG	Industry Standard No.	ECG	Industry Standard No.
159	MP83638A	159	N8661	159	PN72	159	SK3118	159	SS1906A	159	TI-752
159	MP83640	159	N8662	159	PTC103	159	SK5797	159	SS2503	159	TI-743
159	MP83644	159	N8663	159	PTC131	159	SK5798	159	SS2503A	159	TI-744
159	MP83645	159	N8664	159	Q-00984R	159	SK6345	159	SSA43	159	TI-752
159	MP83702	159	N8665	159	Q-285226	159	SK6346	159	SSA43A	159	TI-890
159	MP83703	159	N8666	159	Q-36	159	SK6347	159	SSA43A-1	159	TI-905
159	MP8404	159	N8667	159	Q-36A	159	SK6347A	159	SSA46	159	TI-906
159	MP8404A	159	N8668	159	Q0-419	159	SK7664	159	SSA46A	159	TIS03
159	MP84554	159	N8732	159	Q0415	159	SKA1279	159	SSA48	159	TI743
159	MP84555	159	N8752A	159	Q5087Z	159	SKA4129	159	SSA48A	159	TI744
159	MP85086	159	00415	159	Q5135	159	SKWH07006	159	ST-MP89682J	159	TI752
159	MP86076	159	00200	159	QA-21	159	SM-A-726658	159	ST-MP89750D	159	TIS-03
159	MP86134	159	00201	159	QA-21A	159	SM-B-523974	159	ST.082.115.015	159	TI803
159	MP86516	159	00202	159	QQ61689	159	SM-B-574495	159	ST129-1	159	TI804
159	MP86517	159	00203	159	QQ616B9A	159	SM1507	159	ST16020	159	TIS104
159	MP86518	159	00204	159	Q8316	159	SM4547	159	ST8014	159	TIS112
159	MP86519	159	00205	159	R8967	159	SM4574A	159	ST8033	159	TI837
159	MP86522	159	00206	159	R8967A	159	SM4719	159	ST8033A	159	TI838
159	MP86523	159	00207	159	R8969	159	SN-400-319-P1	159	ST8034	159	TI850
159	MP86524	159	00430	159	R8969A	159	SNT204	159	ST8035	159	TI853
159	MP86533	159	00430K	159	R826	159	SNT204A	159	ST8035A	159	TI854
159	MP86533M	159	00440	159	RS-110	159	SP70	159	ST8036	159	TI861
159	MP86534	159	00440K	159	RS-2021	159	SP90	159	ST8036A	159	TI891
159	MP86534M	159	00443	159	RS-2022	159	SP8-1539	159	ST8065	159	TI893
159	MP86535	159	00443K	159	RS-2023	159	SP81097	159	ST8065A	159	TIS93-BLU
159	MP86535M	159	00445	159	RS-2024	159	SP81097A	159	ST8190	159	TIS93-GRN
159	MP86562	159	00445K	159	RS-7665A	159	SP812	159	ST8500	159	TIS93-GRY
159	MP86563	159	00449	159	R87665	159	SP81523	159	ST8509	159	TIS93-YIO
159	MP86580	159	00449K	159	R73065	159	SP81523A	159	ST8509A	159	TIS93-YEL
159	MP88000	159	00450	159	R73065A	159	SP822	159	STM73Q	159	TIX804
159	MP88598	159	00450K	159	RT3071	159	SP82269	159	SX3702	159	TIX805
159	MP88599	159	00460	159	RT3071A	159	SP82272	159	SX3702A	159	TIX890
159	MP89666	159	00460K	159	RT8895	159	SP82274	159	SX61M	159	TIX891
159	MP89680	159	00463	159	RV1059	159	SP82279	159	SX61MA	159	TM159
159	MP89680H	159	00463K	159	RV2260	159	SP83329	159	SY14275	159	TM1614
159	MP89680H/I	159	00465	159	RV2351	159	SP83724	159	T 112	159	TM1712
159	MP89680I	159	00465K	159	RVTC81382	159	SP83724A	159	T-246	159	TM2614
159	MP89680I/J	159	00466	159	RVTS22411	159	SP83786	159	T-251	159	TM2712
159	MP89680J	159	00466K	159	S-1367A	159	SP83786A	159	T-Q5077	159	TNC61690
159	MP89680T	159	00467	159	S0026	159	SP83924	159	T-Q5087	159	TNC61703
159	MP89681	159	00467K	159	S017446	159	SP83924A	159	T01-023	159	TNC61690
159	MP89681I	159	00468	159	S019843	159	SP83927	159	T1-503		
159	MP89681J	159	00468K	159	S022012	159	SP83927A	159	T1-503A		
159	MP89681K	159	00469	159	S023735	159	SP83931	159	T1-743		
159	MP89681T	159	00469K	159	S1047	159	SP83931A	159	T1-743A		
159	MP89682	159	00470	159	S1550	159	SP83987	159	T1-744		
159	MP89682-I	159	00470K	159	S1350A	159	SP83988	159	T1-752		
159	MP89682I	159	00700	159	S1367	159	SP83988A	159	T1-752A		
159	MP89682J	159	00700A	159	S18100	159	SP83990	159	T1-906		
159	MP89682K	159	00700B	159	S1889	159	SP83990A	159	T1602		
159	MP89682T	159	00702	159	S2091	159	SP84007	159	T1803		
159	MP89750D	159	00702A	159	S2128	159	SP84007A	159	T1803A		
159	MP89750P	159	00702B	159	S2129	159	SP84013	159	T1804		
159	MP89750Q	159	00704	159	S2A226	159	SP84013A	159	T1804A		
159	MPSA55	159	00714	159	S2525	159	SP84014	159	T1837A		
159	MPSA56	159	00740	159	S2645	159	SP84014A	159	T1858		
159	MPSA70	159	00740G	159	S2645A	159	SP84018	159	T1858A		
159	MPSD55	159	00740M	159	S2988	159	SP84018A	159	T1853		
159	MPSD56	159	00740O	159	S2988A	159	SP84019	159	T1853A		
159	MPSBB1	159	00742	159	S3004	159	SP84019A	159	T1854		
159	MPSK70	159	00742G	159	S3639	159	SP8401K	159	T185A		
159	MPSK71	159	00742M	159	S3640A	159	SP84025	159	T1861A		
159	MPSK72	159	00742O	159	S3655	159	SP84025A	159	T1891		
159	MPX9681J	159	P00347100	159	S3655A	159	SP84026	159	T1891A		
159	MPT7505	159	P00347101	159	S4249	159	SP84026A	159	T1893		
159	MS9667	159	P04-45-0015-P1	159	S520	159	SP84027	159	T1893A		
159	MS9681	159	P04-45-0016-P1	159	S608	159	SP84027A	159	T276		
159	MT0404	159	P04-45-0016-002	159	SA310	159	SP84028	159	T340		
159	MT0404-1	159	P04450016-004	159	SA310A	159	SP84028A	159	T5570		
159	MT0404-2	159	P1000A	159	SA311	159	SP84031	159	T5570A		
159	MT0411	159	P1901-70	159	SA311A	159	SP84031A	159	T460(SEARS)		
159	MT0412	159	P1B	159	SA312	159	SP84054	159	T475(SEARS)		
159	MT0413	159	P1C	159	SA312A	159	SP84054A	159	T482(SEARS)		
159	MT1131	159	P1CG	159	SA313	159	SP84056	159	T597-1		
159	MT1131A	159	P1D	159	SA314	159	SP84056A	159	T9681		
159	MT1132	159	P1H	159	SA314A	159	SP84064	159	TA198036-2		
159	MT1132A	159	P1J	159	SA315	159	SP84064A	159	TCB98		
159	MT1132B	159	P1N	159	SA315A	159	SP84072	159	TB3905		
159	MT1254	159	P1N-1	159	SA316	159	SP84072A	159	TB4125		
159	MT1255	159	P1N-2	159	SA316A	159	SP84073	159	TB4126		
159	MT1256	159	P1N-3	159	SA410	159	SP84073A	159	TB5365		
159	MT1257	159	P1P	159	SA410A	159	SP84076	159	TB5566		
159	MT1258	159	P1P-1	159	SA411	159	SP84078	159	TB5378		
159	MT1259	159	P1W	159	SA411A	159	SP84078A	159	TB5379		
159	MT1420	159	P2A	159	SA412	159	SP84082	159	TB5447		
159	MT1991	159	P2E	159	SA412A	159	SP84082A	159	TB5448		
159	MT2303	159	P2G	159	SA413	159	SP84086	159	TG28A608		
159	MT2411	159	P2GE	159	SA413A	159	SP84086A	159	TG28A608C		
159	MT2412	159	P2H	159	SA414	159	SP84087	159	TI-428		
159	MT726	159	P2L	159	SA414A	159	SP84087A	159	TI-429		
159	MT869	159	P2M-1	159	SA415	159	SP84090	159	TI-503		
159	NB121	159	P2M-2	159	SA415A	159	SP84090A	159	TI-743		
159	NJ101B	159	P2M-3	159	SA416	159	SP842	159	TI-744		
159	NKT20329	159	P2P	159	SA416A	159	SP842A	159	TI-752		
159	NKT20329A	159	P2S	159	SA50	159	SP84302	159	TI-890		
159	NKT20339	159	P2W	159	SA50A	159	SP84314	159	TI-905		
159	NR-601AT	159	P2Y	159	SA51	159	SP84314A	159	TI-906		
159	NR601BT	159	P5C	159	SA52	159	SP84348	159	TI503		
159	NR621AT	159	P5CA	159	SA52A	159	SP84348A	159	TI743		
159	NR621BU	159	P5Z	159	SA52AC	159	SP84354	159	TI744		
159	NR651AY	159	P480A0022	159	SA52B	159	SP84354A	159	TI752		
159	NS1000	159	P480A0023	159	SA52BC	159	SP84365	159	TIS-03		
159	NS1000A	159	P480A0027	159	SA53	159	SP84365A	159	TI803		
159	NS1001	159	P4C	159	SA53A	159	SP84452	159	TI804		
159	NS1001A	159	P4G	159	SA54	159	SP84452A	159	TIS104		
159	NS1672	159	P4K	159	SA55	159	SP84458	159	TIS112		
159	NS1672A	159	P4P	159	SA55A	159	SP84458A	159	TI837		
159	NS1673	159	P4R	159	SA56	159	SP84460	159	TI838		
159	NS1674	159	P4Y	159	SA56A	159	SP84460A	159	TI850		
159	NS1674A	159	P5B	159	SA70	159	SP84473	159	TI853		
159	NS1675	159	P5C	159	SA70A	159	SP84473A	159	TI854		
159	NS1675A	159	P5D	159	SE5-0370	159	SP84480	159	TI861		
159	NS1861	159	P67	159	SE5-0798	159	SP84480A	159	TI891		
159	NS1861A	159	PA1000	159	SE5-0831	159	SP84489	159	TI893		
159	NS1862	159	PA1001	159	SE5-0949	159	SP84489A	159	TIS93-BLU		
159	NS1863	159	PA1001A	159	SE5-1057	159	SP847	159	TIS93-GRN		
159	NS1863A	159	PI-10,131	159	SE5-1223	159	SP84813	159	TIS93-GRY		
159	NS1864	159	PIC	159	SP8014	159	SP85007	159	TIS93-YIO		
159	NS1864A	159	PIT-50	159	SHA7530	159	SP85007-1	159	TIS93-YEL		
159	NS3905	159	PIT-79	159	SHA7531	159	SP85007-1A	159	TIX804		
159	NS3906	159	PIT-81	159	SHA7532	159	SP85007-2	159	TIX805		
159	NS404	159	PL1031	159	SHA7533	159	SP85007-2A	159	TIX890		
159	NS6001	159	PL1033	159	SHA7534	159	SP85007A	159	TIX891		
159	NS6062	159	PL1034	159	SHA7536	159	SP85008	159	TM159		
159	NS6062A	159	PL1101	159	SHA7537	159	SP8514	159	TM1614		
159	NS6063	159	PL1102	159	SHA7538	159	SP8514A	159	TM1712		
159	NS6063A	159	PL1103	159	SK1639	159	SP85458	159	TM2614		
159	NS6064	159	PL1104	159	SK1639D	159	SP86109	159	TM2712		
159	NS6064A	159	PN	159	SK1640	159	SP86109A	159	TNC61690		
159	NS6065	159	PN2904	159	SK1856	159	SS1606	159	TNC61703		
159	NS6065A	159	PN2906	159	SK1856A	159	SS1906	159	TNC61690		
159	NS6211	159	PN2906A	159	S2604						
159	NS6211A	159	PN2907	159	SK2604A						
159	NS86241	159	PN2907A	159	SK3114						
		159	PN70								
		159	PN71								

ECG	Industry Standard No.	ECG	Industry Standard No.	ECG	Industry Standard No.	ECG	Industry Standard No.	ECG	Industry Standard No.
159	TNC61703	159	Y56601-84	159	2N1223	159	2N3451	159	2N938
159	TNJ71037	159	ZAG-9673	159	2N1256	159	2N3464	159	2N939
159	TNJ71173	159	ZEN106	159	2N1257	159	2N3485	159	2N940
159	TNJ71774	159	ZEN107	159	2N1428	159	2N3485A	159	2N941
159	TNJ72154	159	ZEN122	159	2N1439	159	2N3486	159	2N942
159	TP3638	159	ZT131	159	2N1440	159	2N3486A	159	2N943
159	TP4125	159	ZT131A	159	2N1441	159	2N3496	159	2N944
159	TP4126	159	ZT152	159	2N1442	159	2N3504	159	2N945
159	TP4257	159	ZT152A	159	2N1443	159	2N3505	159	2N946
159	TP4258	159	ZT153	159	2N1474	159	2N3527	159	02P1B
159	TP5142	159	ZT153A	159	2N1474A	159	2N3545	159	2S3020
159	TP86516	159	ZT154	159	2N1475	159	2N3546	159	2S3021
159	TP86517	159	ZT154A	159	2N1607	159	2N3547	159	2S3030
159	TP86518	159	ZT180	159	2N1608	159	2N3548	159	2S3040
159	TP86519	159	ZT180A	159	2N1643	159	2N3549	159	2S306
159	TP86522	159	ZT181	159	2N1676	159	2N3550	159	2S307
159	TP86523	159	ZT181A	159	2N1677	159	2N3579	159	2S321
159	TQ-63	159	ZT182	159	2N1919	159	2N3580	159	2S3210
159	TQ-64	159	ZT182A	159	2N1920	159	2N3581	159	2S322
159	TQ61	159	ZT183	159	2N1921	159	2N3582	159	2S3220
159	TQ61A	159	ZT183A	159	2N1991	159	2N3638	159	2S322A
159	TQ62	159	ZT184	159	2N2002	159	2N3638A	159	2S323
159	TQ62A	159	ZT184A	159	2N2003	159	2N3639	159	2S3230
159	TR-1000-2	159	ZT187	159	2N2004	159	2N3640	159	2S324
159	TR-1030-1	159	ZT187A	159	2N2005	159	2N3644	159	2S3240
159	TR-1030-2	159	ZT189	159	2N2006	159	2N3672	159	2S326
159	TR-1032-2	159	ZT280	159	2N2007	159	2N3673	159	2S327
159	TR-167(OLSON)	159	ZT280A	159	2N2104	159	2N3702	159	2S673C
159	TR-19A	159	ZT281	159	2N2105	159	2N3703	159	2SA-NJ101
159	TR-20	159	ZT281A	159	2N2121	159	2N3798	159	2SA402
159	TR-20A	159	ZT282	159	2N2162	159	2N3798A	159	2SA480
159	TR-30	159	ZT282A	159	2N2163	159	2N3799	159	2SA482
159	TR-30A	159	ZT283	159	2N2164	159	2N3799A	159	2SA522
159	TR-4R58	159	ZT283A	159	2N2165	159	2N3829	159	2SA522A
159	TR-6R55	159	ZT284	159	2N2166	159	2N3840	159	2SA530
159	TR-6R55A	159	ZT284A	159	2N2167	159	2N3857	159	2SA530H
159	TR-8007	159	ZT287	159	2N2178	159	2N3905	159	2SA548
159	TR-8007(FISHER)	159	ZT287A	159	2N2181	159	2N3906	159	2SA550A
159	TR-8026	159	ZTX501	159	2N2182	159	2N3913	159	2SA550AQ
159	TR-8057	159	ZTX502	159	2N2183	159	2N3914	159	2SA550AR
159	TR0055	159	ZTX503	159	2N2184	159	2N3930	159	2SA550A8
159	TRO1053-1	159	ZTX504	159	2N2185	159	2N3931	159	2SA550P
159	TRO2051-1	159	ZTX510	159	2N2186	159	2N3962	159	2SA550Q
159	TRO2051-3	159	ZTX530	159	2N2187	159	2N3963	159	2SA550R
159	TRO2051-5	159	ZTX530A	159	2N2274	159	2N3964	159	2SA550B
159	TRO2051-6	159	ZTX530C	159	2N2275	159	2N3977	159	2SA561
159	TRO2062-1	159	ZTX530D	159	2N2276	159	2N3978	159	2SA565
159	TRO2062-6	159	ZTX531	159	2N2277	159	2N3979	159	2SA565A
159	TRO2062-8	159	ZTX531A	159	2N2278	159	2N4121	159	2SA565B
159	TR1000A	159	ZTX531B	159	2N2279	159	2N4125	159	2SA565C
159	TR1030	159	1,000,111-00	159	2N2280	159	2N4126	159	2SA565D
159	TR1030-1	159	1-0006-0023	159	2N2281	159	2N4142	159	2SA565K
159	TR1030-2	159	01-010628	159	2N2299	159	2N4143	159	2SA567
159	TR1030A	159	01-010733	159	2N2303	159	2N4208	159	2SA567A
159	TR1032	159	01-010844	159	2N2332	159	2N4209	159	2SA567B
159	TR1032-1	159	001-02	159	2N2333	159	2N4228	159	2SA567C
159	TR1032A	159	001-021170	159	2N2334	159	2N4248	159	2SA592Y
159	TR1034	159	001-021171	159	2N2335	159	2N4249	159	2SA603
159	TR1034A	159	001-021172	159	2N2336	159	2N4250	159	000028A609
159	TR20	159	001-021173	159	2N2337	159	2N4250A	159	2SA617K
159	TR30	159	001-02201-0	159	2N2370	159	2N4257	159	2SA618K
159	TR31	159	001-022010	159	2N2371	159	2N4257A	159	2SA672
159	TR8007A	159	001-03	159	2N2372	159	2N4288	159	2SA672A
159	TR8020	159	01-030733	159	2N2373	159	2N4289	159	2SA672B
159	TR8026A	159	1-034/2207	159	2N2377	159	2N4290	159	2SA672C
159	TR8055	159	001-04	159	2N2378	159	2N4291	159	2SA690D
159	TS-21756640	159	1-043/2207	159	2N2393	159	2N4313	159	2SA701P0
159	TS2904	159	001-533-00	159	2N2394	159	2N4354	159	2SA718
159	TS2905	159	01-571591	159	2N2411	159	2N4355	159	2SA735
159	TS2906	159	01-571751	159	2N2412	159	2N4356	159	2SA888
159	TS2907	159	01-572588	159	2N2424	159	2N4359	159	2SA889
159	TS97-1	159	1B3096-1	159	2N2425	159	2N4589	159	2SA890
159	TT28A495-0-A	159	1V68611A47	159	2N2595	159	2N4402	159	2SA891
159	TT28A495-0-A	159	1V68611A47A	159	2N2596	159	2N4403	159	2SA945-0
159	TT28A495-Y-A	159	1W11700	159	2N2597	159	2N4411	159	2SA945Y
159	TV-44	159	1W11700A	159	2N2601	159	2N4413	159	2SANJ101
159	TV-47A	159	1W11702	159	2N2602	159	2N4413A	159	2SM610B
159	TV-57	159	1W11702A	159	2N2603	159	2N4415	159	3L4-6007-34
159	TV-87	159	1W11711	159	2N2604	159	2N4415A	159	3L4-6017-01
159	TV-95	159	1W8537	159	2N2605	159	2N4452	159	04-440032-002
159	TV24214	159	1W9148	159	2N2605A	159	2N4889	159	04-440032-008
159	TV24214A	159	1W9640	159	2N2695	159	2N4916	159	04-67000-01
159	TV24363	159	1W9640A	159	2N2696	159	2N4917	159	4JX29A826
159	TV24363A	159	1W9728	159	2N2709	159	2N4965	159	4JX29A829
159	TV24495	159	1W9728A	159	2N2800/46	159	2N4971	159	006-000135
159	TV24495A	159	1W9782	159	2N2801	159	2N4972	159	006-02
159	TV44	159	1W9782A	159	2N2837	159	2N4982	159	6-31
159	TV72	159	1W9810	159	2N2838	159	2N5086	159	6-31A
159	TVS-28A564	159	1W9810A	159	2N2861	159	2N5087	159	6-38
159	TVS-28A564A	159	1W98108	159	2N2862	159	2N5138	159	6-98A
159	TVS-28A564P	159	1W98108A	159	2N2894	159	2N5139	159	7-0014
159	TVS-28C564	159	002-009800	159	2N2894A	159	2N5142	159	007-74004-01
159	TV8-C81303	159	002-010300	159	2N2906	159	2N5143	159	007-74008-01
159	TV8-C81303A	159	002-010300A	159	2N2906A	159	2N5226	159	8-905-706-247
159	TV828A564	159	002-010500	159	2N2907	159	2N5227	159	8-905-706-251
159	TV828A564-0	159	002-010500A	159	2N2907A	159	2N5356	159	8-905-706-253
159	TV828A564A	159	002-010900	159	2N2927	159	2N5365	159	8-905-706-254
159	TV828A564C	159	002-012800	159	2N2927/46	159	2N5366	159	8-905-706-255
159	TV828A564P	159	002-012800A	159	2N2944	159	2N5367	159	8-905-706-256
159	TV828A564PY	159	002-9800-A	159	2N2944A	159	2N5372	159	8-905-706-280
159	TV828A564Q	159	2.01.03.02	159	2N2945	159	2N5373	159	8-905-706-286
159	TV828A607	159	2D017	159	2N2945A	159	2N5374	159	8-905-706-287
159	TV828A609	159	2D017-165	159	2N2946	159	2N5375	159	8-905-706-288
159	TV828C564	159	2D017-167	159	2N2946A	159	2N5378	159	8-905-706-289
159	TV828C564-0	159	2D017-169	159	2N3012	159	2N5379	159	8-905-706-290
159	TV828C5640	159	2D027	159	2N3058	159	2N5382	159	8-905-713-058
159	TV828C564R	159	2H1254	159	2N3059	159	2N5383	159	8C200
159	V152	159	2H1255	159	2N3062	159	2N5447	159	8C201
159	V162	159	2H1258	159	2N3072	159	2N5448	159	8C202
159	V410	159	2H1259	159	2N3073	159	2N5857	159	8C203
159	V410A	159	2N1025	159	2N3081	159	2N6005	159	8C204
159	V435A	159	2N1026	159	2N3121	159	2N6067	159	8C205
159	V761	159	2N1026A	159	2N3135	159	2N6076	159	8C206
159	V763	159	2N1027	159	2N3136	159	2N6222	159	8C207
159	V828A495-0/1B	159	2N1028	159	2N3217	159	2N721	159	8C430
159	V828A495-Y/1B	159	2N1118	159	2N3218	159	2N721A	159	8C430K
159	VSC81256HG	159	2N1118A	159	2N3219	159	2N722	159	8C440
159	W21	159	2N1119	159	2N3248	159	2N722A	159	8C440K
159	X29A829	159	2N1131A	159	2N3249	159	2N858	159	8C443
159	XA-1072	159	2N1132	159	2N3250	159	2N859	159	8C443K
159	XA-1140	159	2N1132/46	159	2N3250A	159	2N860	159	8C445
159	XA-495	159	2N1132A	159	2N3251	159	2N861	159	8C445K
159	XA494	159	2N1132A46	159	2N3251A	159	2N862	159	8C449
159	XA495	159	2N1132B	159	2N3305	159	2N863	159	8C449K
159	XA495(C)	159	2N1135	159	2N3306	159	2N864	159	8C450
159	XA495AC	159	2N1135A	159	2N3307	159	2N923	159	8C460
159	XA495C	159	2N1196	159	2N3308	159	2N924	159	8C460K
159	XN-400-319-P2	159	2N1219	159	2N3317	159	2N925	159	8C463
159	X819	159	2N1220	159	2N3318	159	2N926	159	8C463K
159	Y56601-50	159	2N1221	159	2N3319	159	2N927	159	8C465
159	Y56601-63	159	2N1222	159	2N3341	159	2N928	159	8C465K
159	Y56601-74			159	2N3346	159	2N935	159	8C466
159	Y56601-79			159	2N3401	159	2N936	159	8C466K
159	Y56601-82					-159	2N937	159	8C467

ECG	Industry Standard No.
159	8C467K
159	8C468
159	8C468K
159	8C469
159	8C469K
159	8C470
159	8C470K
159	8C700
159	8C700A
159	8C700B
159	8C702
159	8C702A
159	8C702B
159	8C704
159	8C740
159	8C7400
159	8C740G
159	8C740M
159	8C742
159	8C7420
159	8C742G
159	8C742M
159	8Q-3-11
159	8Q-3-14
159	09-300037
159	09-300059
159	09-300061
159	09-300062
159	09-300063
159	09-300074
159	09-300077
159	09-300307
159	09-30063
159	09-304012
159	09-304047
159	09-304049
159	09-304050
159	09-304051
159	09-305024
159	09-305073
159	09-305149
159	09-309038
159	09-309042
159	10P1
159	10P1A
159	11-691504
159	12-CAM
159	13-0006
159	13-0006A
159	13-0043
159	13-0043A
159	13-0044
159	13-0044A
159	13-006
159	13-0061
159	13-0061A
159	13-13532-1
159	13-14085-13
159	13-14085-87
159	13-16570-1
159	13-16570-1A
159	13-16570-2
159	13-16570-2A
159	13-19776-1
159	13-22582-1
159	13-22582-1A
159	13-23325-5
159	13-23826-1
159	13-23826-1A
159	13-23826-2
159	13-23826-2A
159	13-23826-3
159	13-23826-3A
159	13-26386-1
159	13-26386-1A
159	13-26386-2
159	13-26386-2A
159	13-26386-3
159	13-26386-4
159	13-28391-1
159	13-28391-1A
159	13-28391-2
159	13-28391-2A
159	13-28393-1
159	13-28393-1A
159	13-28393-2
159	13-28393-2A
159	13-28393-3
159	13-29776-1
159	13-29776-1A
159	13-29776-2
159	13-29776-3
159	13-31013-1/2
159	13-32364-1
159	13-32631-1
159	13-32631-3
159	13-34367-1
159	13-34367-3
159	13-34940-1
159	13-36386-1
159	13-39115-3
159	13-39970-1
159	13-40083-1
159	13-40083-2
159	13-43634-1
159	13-55069-1
159	13-55069-1A
159	14-602-11
159	14-602-11A
159	14-602-20
159	14-602-20A
159	14-602-32
159	14-602-32A
159	14-602-42
159	14-602-42A
159	14-602-44
159	14-602-44A
159	14-602-47
159	14-602-47A
159	14-602-54A
159	14-602-56
159	14-602-56A
159	14-602-58
159	14-602-580
159	14-602-58A
159	14-602-600
159	14-602-68
159	14-602-88
159	14-602-90
159	014-611
159	014-652
159	014-652C
159	014-772
159	14-803 12

ECG	Industry Standard No.
159	14-803-12
159	14-804-12
159	14-808-12
159	14-855-12
159	14-856-23
159	14-857-12
159	14-857-79
159	14-863-23
159	14-864-23
159	14-867-32
159	15-01742
159	15-01913-00
159	15-02762-00
159	15-02762-1
159	15-02762-2
159	15-02979
159	15-03099
159	15-03409-0
159	15-03409-02
159	15-03409-1
159	15-088002
159	15-088002A
159	15-09090-01
159	15-3
159	15-30
159	15-4
159	15-40
159	15-5
159	15-50
159	15-875-075-001
159	15Z3
159	17-459
159	17-459A
159	018-00001
159	018-00002
159	019-003675-231
159	019-003675-232
159	019-003675-234
159	019-003675-257
159	019-003931
159	019-004558
159	019-005010
159	019-005179
159	19-1
159	19-10
159	19-2
159	19-20
159	19-3
159	19-30
159	19A115178-P1
159	19A115178-P2
159	19A115458-P1
159	19A115458-P2
159	19A115562P2
159	19A115653-P1
159	19A115653-P2
159	19A115654-P1
159	19A115654-P2
159	19A115688-P1
159	19A115688-P2
159	19A115706-1
159	19A115706-2
159	19A115706-P1
159	19A115706-P2
159	19A115768-1
159	19A115768-2
159	19A115768-3
159	19A115768-P1
159	19A115768-P2
159	19A115779P1
159	19A115852P1
159	19A116223P1
159	19A116408-1
159	020-1110-004C
159	20-JLM
159	21A015-008
159	21A015-008A
159	21A015-009
159	21A015-009A
159	21A015-011
159	21A015-011A
159	21A015-012
159	21A015-012A
159	21A015-025
159	21A040-059
159	21A112-001
159	21A112-003
159	21A112-047
159	21A112-065
159	21A112-075
159	21A112-093
159	21A112-100
159	21M022
159	21M355
159	21M581
159	22-001010
159	022-1110-005C
159	022-2876-004
159	23-1
159	23-10
159	23-2
159	23-20
159	23-3
159	23-30
159	23-LLB
159	23-PT274-122
159	24-AWH
159	24MW1031
159	24MW1049
159	24MW1061
159	24MW661
159	24MW976
159	24T-011-011
159	25-000453
159	25-000462
159	25-MEF
159	27A10533
159	29-HCL
159	32-20739
159	33-016
159	33-086
159	34-1013
159	34-1022
159	34-143-12
159	34-3015-28
159	34-6001-15
159	34-6015-26
159	34-6015-42
159	34-6016-15
159	34-6016-15A
159	34-6016-32
159	34-6016-47
159	34-6016-60
159	34P1AA

ECG	Industry Standard No.
159	35(RCA)
159	35-ALD
159	0036-001
159	42-22008
159	42-22008A
159	42-22810
159	42-22810A
159	42-23541
159	42-23541A
159	42-27536
159	42-28208
159	42-28211
159	42-9029-40X
159	42-9029-60Q
159	42-9029-70D
159	42-9029-70E
159	43A145291-1
159	43A145291-2
159	43A167207P1
159	43A167207P2
159	43A168064-1
159	43A168064P1
159	43A176002
159	43B168450-1
159	43B168495-1
159	43B168566-P1
159	44A333464
159	44A333464-1
159	44A390248-001
159	44A390256-001
159	44A390261
159	44A397905
159	44A417031-001
159	44A418041-001
159	44B238203-1
159	44B238246
159	46-86170-3
159	46-86229-3
159	46-86229-3A
159	46-86230-3
159	46-86238-3
159	46-86238-3A
159	46-86283-3
159	46-86293-3
159	46-86377-3
159	46-86399-3
159	46-86412-3
159	46-86424-3
159	48-03-002
159	48-134525
159	48-134525A
159	48-134702
159	48-134702A
159	48-134745
159	48-134745A
159	48-134815
159	48-134815A
159	48-134829
159	48-134829A
159	48-134830
159	48-134833
159	48-134833A
159	48-134865
159	48-134865A
159	48-134866
159	48-134866A
159	48-134867
159	48-134867A
159	48-134868
159	48-134868A
159	48-134869
159	48-134869A
159	48-134870
159	48-134870A
159	48-134871
159	48-134871A
159	48-134909
159	48-134909A
159	48-134910
159	48-134910A
159	48-134910F
159	48-134911
159	48-134913
159	48-134913A
159	48-134914
159	48-134914A
159	48-134915
159	48-134915A
159	48-134940
159	48-134940A
159	48-134943
159	48-134943A
159	48-134967
159	48-134967A
159	48-134973
159	48-134973A
159	48-134975
159	48-134975A
159	48-134989
159	48-134989A
159	48-137020
159	48-137020A
159	48-137021
159	48-137032
159	48-137032A
159	48-137045
159	48-137045A
159	48-137046
159	48-137061
159	48-137067
159	48-137067A
159	48-137068
159	48-137068A
159	48-137069
159	48-137069A
159	48-137090
159	48-137090A
159	48-137127
159	48-137127A
159	48-137173
159	48-137173A
159	48-137176
159	48-137176A
159	48-137195
159	48-137318
159	48-137324
159	48-137366
159	48-137379
159	48-137380
159	48-137381
159	48-137382
159	48-137383
159	48-137391
159	48-137502

ECG	Industry Standard No.
159	48-137504
159	48-3003A06
159	48-355007
159	48-40118B01
159	48-40118B01A
159	48-42098B01
159	48-42098B01A
159	48-43238G01
159	48-64978A40
159	48-64978A40A
159	48-64978A41
159	48-64978A41A
159	48-869334
159	48-869413
159	48-869526
159	48-869571
159	48-869649
159	48-90165A01
159	48-90165A01A
159	48-97046A26
159	48-97046A27
159	48-97177A14
159	48-97177A14A
159	48R869334
159	48R869413
159	48R869526
159	48R869571
159	48R869649
159	48S134815
159	48S137032
159	48S137127
159	48S137173
159	50-40106-09
159	50-40204-10
159	051-0107
159	51-47-21
159	53-1516
159	55-1083
159	55-1083A
159	55-1085
159	55-1085A
159	55-152579
159	056
159	56-8098
159	56-8098A
159	56-8098B
159	56-8098C
159	57A1-76
159	57A1-76A
159	57A106-12
159	57A108-6A
159	57A122-9A
159	57A130-9A
159	57A133-12
159	57A137-12
159	57A137-12A
159	57A145-12
159	57A145-12A
159	57A147-12
159	57A147-12A
159	57A148-12
159	57A15-5
159	57A15-50
159	57A157
159	57A157-9
159	57A157-90
159	57A157-9A
159	57A159-12
159	57A159-12A
159	57A174-8
159	57A175-12
159	57A178-12
159	57A185-12
159	57A189-8
159	57A19
159	57A19-1
159	57A19-10
159	57A19-1A
159	57A19-2
159	57A19-20
159	57A19-3
159	57A19-30
159	57A197-12
159	57A2-70
159	57A2-70A
159	57A2-71
159	57A2-71A
159	57A201-14
159	57A215-12
159	57A216-12
159	57A235-12
159	57A258-8
159	57A305-12
159	57B106-12
159	57B108-6
159	57B108-6A
159	57B122-9
159	57B122-9A
159	57B130-9
159	57B130-9A
159	57B133-12
159	57B137-12
159	57B137-12A
159	57B145-12
159	57B145-12A
159	57B147-12
159	57B147-12A
159	57B157-9
159	57B157-9A
159	57B159-12
159	57B159-12A
159	57B175-12
159	57B178-12
159	57B185-12
159	57B189-8
159	57B197-12
159	57B2-70
159	57B2-70A
159	57B2-71
159	57B2-71A
159	57B201-14
159	57B216-12
159	57B235-12
159	57B258-8
159	57C15-5
159	57C15-50
159	57C157-9
159	57C157-90
159	57C19-1
159	57C19-1A
159	57D1-76
159	57D1-76A
159	57D19
159	57D19-1

ECG	Industry Standard No.
159	57D19-10
159	57D19-2
159	57D19-20
159	57D19-3
159	57D19-30
159	62-19452
159	62-20154
159	62-20154A
159	62-20244
159	62-20244A
159	62A11871
159	63-12154
159	63-12154A
159	63-12156
159	63-12156A
159	63-12157
159	63-12157A
159	63-13322
159	63-13322A
159	65A
159	65B
159	65C
159	65C1
159	65D
159	65D1
159	65E
159	65E1
159	65F
159	65F1
159	66-P11120
159	66F023-1
159	66F024-1
159	66F041-1
159	68 A 8318-P1
159	68-110-02
159	68A7370-1
159	68A7370-P3
159	68A7382-P1
159	68A7734P1
159	73B-140-005-1
159	73B-140005-4
159	73C180831-1
159	73C180831-2
159	73C182082-31
159	74P1
159	81-46125071-4
159	082.115.015
159	83P1
159	83P1A
159	83P1B
159	83P1BC
159	83P1M
159	83P1MC
159	83P2
159	83P2A
159	83P2AA
159	83P2AA1
159	83P2M
159	83P2M1
159	83P2N
159	83P3
159	83P3A
159	83P3AA
159	83P3AA1
159	83P3B
159	83P3B1
159	83P3M
159	83P3M1
159	83P4
159	83PS
159	86-0036-001
159	86-100009
159	86-178-2
159	86-178-20
159	86-183-2
159	86-183-20
159	86-216-2
159	86-217-2
159	86-217-20
159	86-218-2
159	86-218-20
159	86-219-2
159	86-233-2
159	86-246-2
159	86-246-20
159	86-251-2
159	86-251-20
159	86-276-2
159	86-276-20
159	86-286-2
159	86-286-20
159	86-294-2
159	86-294-20
159	86-298-2
159	86-298-20
159	86-340-2
159	86-340-20
159	86-406-2
159	86-407-2
159	86-459-2
159	86-475-2
159	86-482-2
159	86-501-2
159	86-527-2
159	86-528-2
159	86-533-2
159	86-547-2
159	86-552-2
159	86-570-2
159	86-600-2
159	86-622-2
159	86-669-2
159	86A335
159	86P1AA
159	86X0016-001
159	86X0016-001A
159	86X0036-001A
159	86X0041-001
159	86X0041-001A
159	86X0044-001
159	86X0044-001A
159	86X0046-001
159	86X0047-001
159	86X0066-001
159	86X0066-003
159	86X0072-001
159	86X46
159	86X47
159	93P1AA
159	96-5215-01
159	96-5282-01
159	96-5283-01
159	96-5365-01
159	98P1

ECG	Industry Standard No.	ECG	Industry Standard No.	ECG	Industry Standard No.	ECG	Industry Standard No.	ECG	Industry Standard No.
159	98P10	159	260P15201	159	991-011225	159	4856-0106	159	141345
159	99P1	159	260P15202	159	991-011319	159	5001-0468	159	141421
159	99P10	159	260P15203	159	991-012328	159	5001-066	159	141771
159	99P1M	159	260P16502	159	991-013058	159	5001-509	159	141738P63-1
159	99P5	159	260P16504	159	1005M19	159	5059-0236	159	142838
159	998039	159	294	159	1010-7738	159	5226-1	159	142839
159	998039A	159	297L012001	159	1012(G.B.)	159	5611-628	159	143791
159	998062-1	159	297L013B02	159	1012(GB)	159	5611-628P	159	143802
159	998084-1	159	297V073C03	159	1013(GE)	159	5611-673	159	143803
159	100-4790	159	297V073C04	159	1016(GE)	159	5611-673D	159	143806
159	100-0495-15	159	297V073001	159	1030	159	5701	159	143807
159	101P1	159	297V086C01	159	1040-9068	159	6201	159	143963
159	101P10	159	309(CATALINA)	159	1042-06	159	6854K90-062	159	143410
159	102P1	159	311D589-P2	159	1043-7374	159	7303-1	159	145776
159	102P10	159	317-0083-001	159	1044-0295	159	7363-1	159	147549-1
159	103P(AIRLINE)	159	344-6017-1	159	1061-8312	159	7503	159	147549-2
159	103P935	159	352-0219-000	159	1061-9068	159	7626	159	147663
159	103P935A	159	352-0551-010	159	1062-6018	159	8000-00004-P089	159	150742
159	103PA	159	352-0551-021	159	1063-5423	159	8000-00006-004	159	150753
159	104-17(RCA)	159	352-0610-030	159	1063-5431	159	8000-0004-P089	159	150758
159	104-170	159	352-0610-040	159	1063-5449	159	8000-00049-056	159	150762
159	105	159	352-0636-010	159	1063-6926	159	8501	159	150771
159	106-12	159	352-0636-020	159	1079-85	159	8405	159	167690
159	106-120	159	352-0754-020	159	1081-4000	159	8540	159	172336
159	106RED	159	352-0778-010	159	1081-4010	159	8601	159	181015
159	108-6	159	352-0848-020	159	1084-9784	159	8710-169	159	181030
159	108-60	159	352-0959-010	159	1089.6199	159	8867	159	181034
159	108GRN	159	352-0959-020	159	11124B	159	9015C	159	189052
159	110P1	159	352-0959-030	159	1125-2582	159	9330-767-60112	159	187217
159	110P1AA	159	353-9304-001	159	1147(GE)	159	9330-908-10112	159	188180
159	110P1M	159	364-1	159	1186(GE)	159	96528I	159	190429
159	112-000172	159	364-1(SYLVANIA)	159	1214	159	09800	159	198024
159	112-000185	159	386-1	159	1236-3750	159	10300	159	198056-1
159	112-000187	159	386-1(SYLVANIA)	159	1254(GE)	159	12594	159	198050
159	112-10	159	394-3145	159	1294(GE)	159	12888	159	200067
159	112-7	159	00415	159	1314	159	13162	159	200220
159	112-8	159	417-116	159	1314(GE)	159	17045	159	200433
159	119-0055	159	417-116-13165	159	1414-158	159	19680	159	202909-577
159	120-006604	159	417-132	159	1414-176	159	20011	159	202909-587
159	120P1	159	417-153	159	1479-8029	159	22008	159	202911-737
159	120P1M	159	417-168	159	1553-17	159	22595-000	159	203364
159	121-1	159	417-176	159	1582	159	22605-005	159	205052
159	121-1005	159	417-182	159	1679.7391	159	23826(SYLVANIA)	159	205048
159	121-1019	159	417-184	159	1844-17	159	26810-152	159	205049
159	121-1RED	159	417-196	159	1850-17	159	29076-023	159	205367
159	121-417	159	417-200	159	1853-0001-1	159	30270	159	210076
159	121-441	159	417-201	159	1853-0081	159	30290	159	232651
159	121-444	159	417-234-13165	159	1867-17	159	31005	159	241517
159	121-446	159	417-235	159	1940-17	159	33509-1	159	267838
159	121-495	159	417-235-13262	159	1979-808-10	159	37486	159	309684
159	121-496	159	417-242-8181	159	2004-06	159	38095	159	320280
159	121-4949	159	417-260-50127	159	2017-107	159	41177	159	321165
159	121-497WHT	159	430-20015-0B	159	2020-01	159	41440	159	324144
159	121-602	159	430-20018-0A	159	2020-07	159	43127	159	333060-1029
159	121-608	159	430-20021	159	2043-17	159	45122	159	337342
159	121-679	159	430-20023-0A	159	2057A100-51	159	45337-C	159	401003-001
159	121-699	159	430-20026-0	159	2057A2-182	159	50203-12	159	436119-002
159	121-699-02	159	436-404-002	159	2057A2-183	159	50203-8	159	489751-028
159	121-774	159	450-1167-1	159	2057A2-198	159	59625-10	159	489751-031
159	121-838	159	461-1006	159	2057A2-298	159	60719-1	159	489751-042
159	121-861	159	461-1055-01	159	2057A2-307	159	63282	159	489751-097
159	121-865	159	514-044910	159	2057A2-343	159	000071150	159	489751-124
159	121-875	159	549-2	159	2057A2-353	159	000071151	159	489751-130
159	121-933	159	550-027-00	159	2057A2-359	159	71687	159	489751-146
159	121-973	159	559-1	159	2057A2-397	159	71687-1	159	543995
159	121-978	159	570-1	159	2057A2-400	159	71687-101	159	552503
159	121-986	159	574	159	2057A2-403	159	71818-1	159	610074-1
159	1220RN	159	576-0001-013	159	2057A2-406	159	75561-1	159	610083
159	122YEL	159	576-0002-008	159	2057A2-450	159	75561-2	159	610083-3
159	123-006	159	576-0003-017	159	2057A2-457	159	75561-31	159	610093-1
159	0124(KNIGHT)	159	576-0003-019	159	2057A2-489	159	75617-1	159	610095-1
159	0124A	159	580R304H01	159	2057A2-561	159	75617-2	159	610099-6
159	125B133	159	597-1(RCA)	159	2057B106-12	159	77561-27	159	610101-1
159	125P1	159	600X0095-086	159	2057B108-6	159	78331	159	610110-1
159	125P116	159	601X0417-086	159	2057B147-12	159	83272	159	610110-2
159	125P1M	159	602-56	159	2057B159-12	159	87758	159	610120-1
159	125PI	159	602-60	159	215B-1558	159	87759	159	610125-1
159	129-20	159	620-1	159	2220-17	159	90330-001	159	610147-2
159	129-34	159	627-1	159	2269	159	90432	159	610158-2
159	129-34(PILOT)	159	635	159	2272	159	94037	159	610221-1
159	130-149	159	669	159	2300.036.096	159	95227	159	610225-1
159	130-40315	159	686-0325-0	159	2381-17	159	95227-1	159	610246-1
159	130-40429	159	686-2700	159	2448(RCA)	159	95232	159	615180-1
159	0131-000335	159	690V086HB6	159	2798	159	95240-1	159	615180-2
159	0131-001328	159	690V116H23	159	2904-038	159	96458-1	159	615180-3
159	0131-001329	159	690V118H60	159	3012(PNP)	159	101497	159	615180-4
159	0131-001420	159	690V118H61	159	3017	159	102001	159	650060
159	0131-001439	159	700-133	159	3019	159	102060	159	698941-1
159	0131-004746	159	753-2000-101	159	3513	159	102263	159	701584-00
159	0131-005351	159	753-4004-248	159	3513(SEARS)	159	113182	159	701599-00
159	0131-4328	159	755-422494	159	3513(WARDS)	159	115517-001	159	721272
159	132-056	159	774(ZENITH)	159	3522(SEARS)	159	116078(RCA)	159	760269
159	132-074	159	0776-0195	159	3524(SEARS)	159	118284	159	801540
159	0133	159	776-1(SYLVANIA)	159	3540(RCA)	159	119228-001	159	815199
159	134P1A	159	800-001-031-1	159	3540-1(GE)	159	119730	159	815199-6
159	134P1M	159	800-101-108-1	159	3549(RCA)	159	121467	159	815211
159	134P4	159	800-523-01	159	3549-1(RCA)	159	121693	159	815213
159	134P4A	159	800-525-02	159	3549-2(RCA)	159	123971	159	815229
159	134PM	159	800-525-04	159	3559(RCA)	159	123991	159	815236
159	137(ADMIRAL)	159	800-527-00	159	3559-1	159	124047	159	815247
159	147-7009-01	159	800-547-00	159	3562(RCA)	159	124755	159	858105
159	151-0087-00	159	826-1	159	3563(RCA)	159	126524	159	891008
159	151-0124-00	159	829	159	3570	159	126700	159	900552-17
159	151-0188-00	159	829A	159	3570(RCA)	159	126707	159	908864-2
159	151-0221-00	159	829B	159	3570-1	159	126715	159	916051
159	151-0221-02	159	829C	159	3570P	159	126718	159	916062
159	151-0325-00	159	829D	159	3574	159	126719	159	928408-101
159	151-0458-00	159	829E	159	3574(RCA)	159	123699	159	932040
159	151-0459-00	159	829F	159	3574-1(RCA)	159	130139	159	932107-1
159	157	159	833	159	3581(RCA)	159	130215	159	960106-3
159	157YEL	159	921-160B	159	3597	159	131241	159	970246
159	158P2	159	921-197B	159	3597-1	159	131242	159	970248
159	158PZM	159	921-254B	159	3597-1(RCA)	159	131647	159	970251
159	161-012J	159	921-296B	159	3597-2	159	132176	159	970254
159	173A4483-1	159	921-29B	159	3620(RCA)	159	132285	159	970663
159	173A4483-2	159	921-308A	159	3620-1	159	132498	159	970762
159	177-006-9-001	159	921-308B	159	3627	159	132850	159	971059
159	177-012-9-001	159	921-332B	159	3627(RCA)	159	133182	159	984193
159	177-019-9-003	159	921-348B	159	3627-1	159	133253	159	988990
159	200-052	159	921-405	159	3631	159	135286	159	1417330-3
159	200X4082-614	159	921-70	159	3634.2011	159	137155	159	1417330-4
159	200X4095-415	159	921-70A	159	4013	159	137340	159	1417339-1
159	200X4101-500	159	921-70B	159	4086	159	138376	159	1417347-1
159	207V073004	159	943-721-001	159	4087	159	139455	159	1417363-1
159	209-1	159	958-023	159	4151-01	159	140290	159	1471112-7
159	209P1	159	972-659E-0	159	4310 (AIRLINE)	159	140371	159	1471112-8
159	210ATTF3638	159	991-01-0098	159	4442	159	140572	159	1471114-1
159	211ATF83591	159	991-01-0462	159	004746	159	140623	159	1472501-1
159	212-699	159	991-01-1225	159	4801-0000-001	159	141018	159	1473501-1
159	223P1	159	991-01-1319	159	4801-0000-060	159	141227	159	1473523-1
159	246P1	159	991-01-2328	159	4802-00004	159	141343	159	1473540-1
159	260-10-016	159	991-01-3058	159	4815	159	141344	159	1473549-1
159	260P08201	159	991-01-3599	159	4822-130-40315	159		159	1473549-2
159	260P11403	159	991-010098	159	4844	159		159	1473559-001
159		159		159		159		159	1473559-1
159		159		159		159		159	1473562-1

ECG	Industry Standard No.	ECG	Industry Standard No.	ECG	Industry Standard No.	ECG	Industry Standard No.	ECG	Industry Standard No.
159	1473563-1	159	50210600-01	160	A347	160	AP107	160	D086
159	1473570-1	159	50210600-03	160	A348	160	AP109	160	D173
159	1473570-2	159	50210610-10	160	A349	160	AP109R	160	D174
159	1473574-1	159	50210610-13	160	A360	160	AP110	160	D175
159	1473581-1	159	50211210	160	A361	160	AP111	160	DS34
159	1473591-1	159	50211500	160	A376	160	AP112	160	DS35
159	1473597-1	159	50211500-01	160	A377	160	AP113	160	DS37
159	1473597-2	159	50211510-10	160	A378	160	AP114N	160	DS38(DELCO)
159	1473599-1	159	50211510-11	160	A379	160	AP115N	160	DS41
159	1473620-001	159	50212100	160	A403	160	AP116N	160	DS42
159	1473620-1	159	51003108	160	A404	160	AP117C	160	DS56
159	1473627-1	159	51161325	160	A405	160	AP117N	160	DS62
159	1503097-0	159	55430001-001	160	A41	160	AP118	160	DS63
159	1616226-1	159	55517007	160	A417(JAPAN)	160	AP119	160	DS64
159	1617032	159	56301600	160	A419(JAPAN)	160	AP121	160	DS65
159	1700001	159	62256893	160	A42	160	AP1218	160	DU1
159	1700034	159	62279664	160	A420(JAPAN)	160	AP122	160	DU12
159	1780142	159	62589702	160	A421	160	AP127/01	160	DU2
159	1780522-1	159	62438096	160	A422	160	AP128	160	E2438
159	1780522-2	159	62506288	160	A425	160	AP137A	160	E2439
159	1780522-2-001	159	62539291	160	A426	160	AP139	160	E2440
159	1861223-1	159	62563265	160	A430(JAPAN)	160	AP142	160	E2450
159	1945294	159	62563311	160	A431	160	AP143	160	E2462
159	1950052	159	62565965	160	A431A	160	AP146	160	E2474
159	1950056-1	159	62566369	160	A432	160	AP147	160	E2475
159	1969281	159	62590661	160	A432A	160	AP148	160	E2477
159	2003073-0702	159	62608498	160	A433	160	AP149	160	E2478
159	2056606-0701	159	62737433	160	A434	160	AP150	160	EA0002
159	2057013-0004	159	62737468	160	A435	160	AP164	160	EA0007
159	2057013-0007	159	62741457	160	A435A	160	AP165	160	EA0053
159	2057013-0008	159	62752742	160	A435B	160	AP166	160	EA1337
159	2057013-0012	159	62759720	160	A440	160	AP167	160	EA1338
159	2057013-0701	159	62766247	160	A440A	160	AP168	160	EA1339
159	2057013-0702	159	62766271	160	A451(JAPAN)	160	AP169	160	EA1340
159	2057013-0703	159	62806060	160	A460	160	AP170	160	EA1342
159	2092609-0025	159	72035900	160	A461	160	AP171	160	EA15X11
159	2092693-0734	159	91056140	160	A462	160	AP172	160	EA15X13
159	2132523-1	159	94650700-00	160	A463	160	AP178	160	EA15X27
159	2320161	159	94650700-01	160	A464	160	AP179	160	EA15X29
159	2320162	159	632049518	160	A506	160	AP180	160	EA15X30
159	2320671	159	910678870	160	A507	160	AP181	160	EA15X40
159	2320681	159	2004049514	160	A508	160	AP182	160	EA15X41
159	2327262	159	2004060858	160	A525	160	AP185	160	EA15X43
159	2327387	159	2004082606	160	A525A	160	AP186	160	EA15X5
159	2487340	159	2004082614	160	A525B	160	AP186G	160	EA15X66
159	2487341	159	2004356122	160	A57	160	AP186W	160	ED51
159	2487424(PNP)	159	3520743010	160	A58	160	AP193	160	ER15X11
159	2498512	159	6623001100	160	A59	160	AP200	160	ER15X12
159	2505734-105	159	6623002000	160	A61	160	AP200U	160	ER15X13
159	2521108-1	159	6623002100	160	A615-1008	160	AP201	160	ER15X14
159	2621570	159	6623002200	160	A615-1009	160	AP201C	160	ER15X15
159	2905993-1	159	6624002000	160	A65-1-1A9G	160	AP201U	160	ER15X19
159	2905993-I	159(2)	86-5079-2	160	A65-1-70	160	AP202	160	ER15X20
159	3468185-1	159(2)	A-1384	160	A65-1-705	160	AP202L	160	ER15X21
159	3468242-1	160	A01(TRANSISTOR)	160	A65-149G	160	AP202B	160	ER15X24
159	3596063	160	A059-100	160	A65A-70	160	AP239	160	ER15X25
159	3596118	160	A059-101	160	A65A-705	160	AP239S	160	ER15X26
159	3650258A	160	A059-102	160	A65A19G	160	AP240	160	ER15X4
159	3700144	160	A059-103	160	A70F	160	AP251	160	ER15X5
159	3700249	160	A107(JAPAN)	160	A70L	160	AP252	160	ER15X6
159	3731133-1	160	A108(JAPAN)	160	A70MA	160	AP253	160	ES1(ELCOM)
159	4031986-0701	160	A122	160	A71	160	AP256	160	ES11(ELCOM)
159	0444032-002	160	A1220	160	A71AB	160	AP267	160	ES14(ELCOM)
159	0444032-005	160	A131(TRANSISTOR)	160	A71AC	160	AP279	160	ES15(ELCOM)
159	0444032-004	160	A150(JAPAN)	160	A71BS	160	AP306	160	ES19(ELCOM)
159	0444032-005	160	A135	160	A71Y	160	AFY12	160	ES23(ELCOM)
159	0444032-006	160	A1377	160	A73	160	AFY15	160	ES25(ELCOM)
159	0444032-007	160	A1378	160	OA8-1	160	AFY16	160	ES41(ELCOM)
159	0444032-008	160	A14(TRANSISTOR)	160	OA8-1-12	160	AFY18C	160	ES8(ELCOM)
159	0444032-009	160	A1465-1	160	OA8-1-12-7	160	AFY18D	160	ET1
159	0450016-001	160	A1465-19	160	A8-1-70	160	AFY18E	160	ET15X29
159	0450016-002	160	A1465A	160	A8-1-70-1	160	AFY34	160	ET2
159	5320042H	160	A1465A9	160	A8-1-70-12	160	AFY37	160	ET3216
159	5320043H	160	A1465B	160	A8-1-70-12-7	160	AFY39	160	FB401
159	5321184	160	A1465B9	160	A8-1-A-4-7B	160	AFY40	160	FB402
159	6993400	160	A1488B	160	A8-10	160	AFY40K	160	FB403
159	6993650	160	A1488B9	160	A8-11	160	AFY40R	160	FB440
159	7020202	160	A14A8-1	160	A8-12	160	AFY41	160	FS2299
159	7302024	160	A14A8-19	160	A8-13	160	AFY42	160	G0002
159	7570013	160	A14A8-19G	160	A8-14	160	AFZ11	160	G13
159	7570013-01	160	A151(JAPAN)	160	A8-15	160	AFZ12	160	GC1003
159	7576004-01	160	A152(JAPAN)	160	A8-16	160	AL210	160	GC1004
159	7851652-01	160	A163	160	A8-17	160	APB-11H-11	160	GC1005
159	7851994-01	160	A218	160	A8-18	160	APB-11H-1001	160	GC1006
159	7932515	160	A224	160	A8-19	160	APB-11H-1004	160	GC1007
159	7939186	160	A226	160	A8-1A	160	AR103	160	GC1092
159	8111227	160	A227	160	A8-1A0	160	AR104	160	GC1093
159	8111230	160	A228	160	A8-1A0R	160	AR105	160	GC1093X3
159	8115009	160	A229	160	A8-1A1	160	AS34280	160	GC1142
159	9000940	160	A290	160	A8-1A19	160	ASZ20	160	GC1146
159	9001638	160	A233	160	A8-1A2	160	ASZ20N	160	GC1148
159	9004508-01	160	A233A	160	A8-1A21	160	AT-1	160	GC1149
159	9007038	160	A233B	160	A8-1A3	160	AT-14	160	GC1155
159	9340388	160	A233C	160	A8-1A3P	160	AT-2	160	GC1182
159	9341767	160	A234	160	A8-1A4	160	AT-3	160	GC1573
159	9342291	160	A234A	160	A8-1A4-7B	160	AT-4	160	GC282
159	10112562	160	A235	160	A8-1A5	160	AT-6	160	GC283
159	10182330	160	A235B	160	A8-1A5L	160	AT-8	160	GC284
159	10641140	160	A239	160	A8-1A6	160	AT-9	160	GC387
159	10669666	160	A240	160	A8-1A6-4	160	AT/RF1	160	GC388
159	10670399	160	A240A	160	A8-1A7	160	AT/RP2	160	GC389
159	11076437	160	A240B	160	A8-1A7-1	160	AT/813	160	GC460
159	12901503	160	A240B2	160	A8-1A8	160	AT13	160	GC461
159	14736221	160	A240BL	160	A8-1A82	160	AT14	160	GC462
159	14798029	160	A241	160	A8-1A9	160	AT15	160	GC630
159	17942800-01	160	A242	160	A8-1A9G	160	AT16	160	GC630A
159	17942800-801	160	A243	160	A84	160	AT17	160	GC631
159	19901503	160	A244(JAPAN)	160	A88-70	160	AT4	160	GE-208
159	20030073-0702	160	A245(JAPAN)	160	A88-705	160	AT5	160	GE-245
159	22114253	160	A261	160	A88B-70	160	B601-1006	160	GE-50
159	23114050	160	A262	160	A88B-705	160	B601-1007	160	GE-51
159	23114051	160	A263	160	A88B19G	160	B601-1008	160	GE-9
159	23114124	160	A264	160	A89	160	B65-1-A-21	160	GE-9A
159	23114156	160	A265	160	A909-1008	160	B65-1A21	160	GE-M100
159	23114158	160	A276(JAPAN)	160	AA5	160	B65A-1-21	160	GER-A-D
159	23114300	160	A285	160	AA8-1-70	160	B65B-1-21	160	GET5116
159	23114301	160	A286	160	AA8-1-705	160	B8BB-1-21	160	GET5117
159	23114302	160	A287	160	AA8-1A9G	160	BA8-1A-21	160	GET671
159	23114325	160	A298	160	AC107	160	BCM1002-1	160	GET672
159	23114550	160	A306(JAPAN)	160	AC107M	160	BF1371	160	GET672A
159	24559800	160	A307(JAPAN)	160	AC129	160	BP5263	160	GET673
159	24562300	160	A321(JAPAN)	160	ACY24	160	C10215-2	160	GET691
159	26004301	160	A322(JAPAN)	160	ACZ	160	C1025B	160	GET692
159	26004961	160	A323(JAPAN)	160	AF-105A	160	C10260	160	GET693
159	26005121	160	A324(JAPAN)	160	AF-109	160	C10261	160	GFT44
159	26010016	160	A332	160	AF-121	160	C10262	160	GFT45
159	26011310	160	A341-0A	160	AF-166	160	CB156	160	JGB-51
159	26316032	160	A341-0B	160	AF-182	160	CB244	160	JM0290
159	43022458	160	A342A	160	AP101	160	CB254	160	GM0375
159	43023845	160	A343	160	AP102	160	CGE-50	160	GM0376
159	48137195A	160	A345(JAPAN)	160	AP105A	160	CGE-51	160	GM0377
159	50210400-00	160	A346(JAPAN)	160	AP106	160	D063	160	GM290
159	50210400-00								
159	50210600-00 VP			160	AP106A	160	D079		

ECG	Industry Standard No.	ECG	Industry Standard No.	ECG	Industry Standard No.	ECG	Industry Standard No.	ECG	Industry Standard No.
160	GM290A	160	M75516-2R	160	R341	160	S877B	160	T2029
160	GM378	160	M755162-P	160	R424	160	S88B-1-3P	160	T2030
160	GM378A	160	M755162-R	160	R424-1	160	S88TB	160	T2191
160	GM378RED	160	M76	160	R425	160	S95101	160	T2364
160	GM656A	160	M77	160	R497	160	S95102	160	T2379
160	GM875	160	M78	160	R5102	160	S95103	160	T2384
160	GM876	160	M78A	160	R5103	160	S95104	160	T253
160	GM877	160	M78B	160	R515	160	S95106	160	T253 (SEARS)
160	GM878	160	M78BLK	160	R515A	160	S99101	160	T278
160	GM878A	160	M78C	160	R516(T.I.)	160	S99102	160	T279
160	GM878B	160	M78D	160	R516A	160	S99103	160	T280
160	GM2900	160	M78GRN	160	R539	160	S99104	160	T281
160	GM0375	160	M78RED	160	R558	160	SA8-1-A-3P	160	T282
160	GM0376	160	M78YEL	160	R563	160	SB101	160	T2878
160	GM0377	160	M8124	160	R564	160	SB102	160	T2896
160	GM0378	160	MA1	160	R565	160	SB103	160	T2945
160	HA-234	160	MC101	160	R579	160	SC-71	160	T2946
160	HA-268	160	MC103	160	R593	160	SC-72	160	T348
160	HA1040	160	MD420	160	R593A	160	SC-74	160	T367
160	HA12	160	MDS31	160	R60-1002	160	SC-78	160	T368
160	HA2190	160	MDS32	160	R60-1003	160	SC-79	160	T373
160	HA2556	160	MDS33	160	R7885	160	SC-80	160	T374
160	HA235A	160	MDS33A	160	R7886	160	SC1007	160	T449
160	HA240	160	MDS33C	160	R7891	160	SC71	160	T6058
160	HA266	160	MDS33D	160	R7962	160	SC72	160	T811
160	HA267	160	MDS34	160	R8240	160	SC74	160	TA-1628
160	HA30	160	MDS36	160	R8241	160	SC78	160	TA-1658
160	HA3210	160	MDS37	160	R8242	160	SC79	160	TA-1659
160	HA3480	160	MDS38	160	R8559	160	SC80	160	TA-1660
160	HA3670	160	MDS39	160	R8685	160	SE2400	160	TA-1662
160	HA4400	160	MDS40	160	R8686	160	SF.T163	160	TA-1731
160	HA525	160	MM1139	160	R8692	160	SF.T316	160	TA-1757
160	HA70	160	MM1199	160	R8693	160	SF.T317	160	TA-1796
160	HEP-2	160	MM2503	160	R8694	160	SF.T319	160	TA-1797
160	HEP-G0002	160	MM2550	160	R8703	160	SF.T320	160	TA-1798
160	HEP-G0003	160	MM2552	160	R8704	160	SF.T354	160	TA-1828
160	HEP3	160	MM2594	160	R8705	160	SF.T357	160	TA-1829
160	HEP637	160	MM2894	160	R8881	160	SF.T358	160	TA-1846
160	HP12H	160	MM380	160	R8882	160	SFT-163	160	TA-1847
160	HP12M	160	MM5000	160	R9531	160	SFT-358	160	TA-1860
160	HP12N	160	MM5001	160	R9532	160	SFT120	160	TA-1861
160	HP20B	160	MM5002	160	R9601	160	SFT162	160	TA1628
160	HP20M	160	MPS1097	160	R9602	160	SFT165	160	TA1650A
160	HP3H	160	MT102351A	160	RE27	160	SFT171	160	TA1658
160	HP3M	160	N-020	160	RE69	160	SFT172	160	TA1659
160	HP50H	160	N57B2-11	160	RS-101	160	SFT173	160	TA1660
160	HP50M	160	N57B2-13	160	RS-103	160	SFT174	160	TA1662
160	HP6H	160	N57B2-14	160	RS-2002	160	SFT268	160	TA1731
160	HP6M	160	N57B2-22	160	RS-3892	160	SFT315	160	TA1755
160	HJ15D	160	NA-1114-1001	160	RS-3898	160	SFT316	160	TA1756
160	HJ32	160	NA-1114-1002	160	RS-3900	160	SFT358	160	TA1757
160	HJ34	160	NA1022-1001	160	RS-3902	160	SK3006	160	TA1796
160	HJ34A	160	NA5018-1001	160	RS-3903	160	SK3007	160	TA1797
160	HJ37	160	NK1302	160	RS-3911	160	SM1297	160	TA1798
160	HJ55	160	NK1404	160	RS-5208	160	SM1600	160	TA1828
160	HJ56	160	NK65-1A19	160	RS1539	160	SM217	160	TA1829
160	HJ57	160	NK65A119	160	RS1550	160	SM2491	160	TA1830
160	HJ60A	160	NK88B119	160	RS1554	160	SM2492	160	TA1846
160	HJ60C	160	NKA8-1A19	160	RS2679	160	SM3014	160	TA1847
160	HJ70	160	NKT121	160	RS2680	160	SMB454760	160	TA1860
160	HR-20	160	NKT122	160	RS2683	160	S0-1	160	TA1861
160	HR-20A	160	NKT124	160	RS2684	160	S0-2	160	TA2222
160	HR-21	160	NKT125	160	RS2685	160	S0-3	160	TI-338
160	HR-21A	160	NKT251	160	RS2686	160	S0-65A	160	TI-387
160	HR-22	160	OC-169	160	RS2687	160	S01	160	TI-388
160	HR-22A	160	OC-615	160	RS2688	160	S02	160	TI-400
160	HR-22B	160	OC169	160	RS2694	160	S03	160	TI-403
160	HR-24	160	OC169R	160	RS2695	160	S8155	160	TI-403
160	HR-25	160	OC170	160	RS3277	160	ST-125	160	TI3010
160	HR-25A	160	OC170N	160	RS3278	160	STX0036	160	TI3011
160	HR-26	160	OC170R	160	RS3279	160	STX0085	160	TI338
160	HR-27	160	OC170V	160	RS3288	160	STX0087	160	TI363
160	HR-27A	160	OC171	160	RS3309	160	STX0089	160	TI364
160	HR24A	160	OC171N	160	RS3322	160	STX0090	160	TI365
160	HR40	160	OC171R	160	RS3323	160	SWT3588	160	TI387
160	HR40836	160	OC171V	160	RS3324	160	SYL2189	160	TI388
160	HR40837	160	OC320	160	RS3668	160	T-163	160	TI389
160	HR41	160	OC400	160	RS3862	160	T-2028	160	TI390
160	HR42	160	OC410	160	RS3863	160	T-2029	160	TI391
160	HR43	160	OO53	160	RS3864	160	T-2030	160	TI393
160	HR43835	160	OO54	160	RS3868	160	T-278	160	TI395
160	HR44	160	OO55	160	RS3898	160	T-279	160	TI396
160	HR448636	160	OC615N	160	RS3900	160	T-99	160	TI397
160	HR45	160	ON174	160	RS3901	160	T1011	160	TI398
160	HR45838	160	PA10880	160	RS3902	160	T1012	160	TI399
160	HR45910	160	PA9154	160	RS3903	160	T1028	160	TI400
160	HR45913	160	PA9155	160	RS3905	160	T1032	160	TI401
160	HR46	160	PADT20	160	RS3906	160	T1033	160	TI402
160	HR50	160	PADT21	160	RS3907	160	T1034	160	TI403
160	HR51	160	PADT22	160	RS3911	160	T1038	160	TIA02
160	HR52	160	PADT23	160	RS3912	160	T1166	160	TIM-01
160	HT102341	160	PADT24	160	RS3929	160	T1224	160	TIM-10
160	HT102341A	160	PADT25	160	RS3986	160	T1225	160	TIM-11
160	HT102341B	160	PADT26	160	RS3995	160	T1232	160	TIX-M01
160	HT102341C	160	PADT27	160	RS5101	160	T1233	160	TIX-M02
160	HT102351	160	PADT28	160	RS5106	160	T1250	160	TIX-M03
160	HT102351A	160	PADT30	160	RS5107	160	T1298	160	TIX-M04
160	HT103501	160	PADT351	160	RS5108	160	T1299	160	TIX-M05
160	HT200541C	160	PADT35	160	RS5109	160	T1305	160	TIX-M06
160	I12032	160	PADT40	160	RS5201	160	T1306	160	TIX-M07
160	J24620	160	PADT51	160	RS5204	160	T1314	160	TIX-M08
160	J24621	160	PIL/4956	160	RS5205	160	T1387	160	TIX-M11
160	J24622	160	PQ27	160	RS5206	160	T1389	160	TIX-M17
160	J24623	160	PQ30	160	RS5207	160	T1390	160	TIX-M201
160	J310251	160	PT2A	160	RS5208	160	T1391	160	TIX-M202
160	J5062	160	PT28	160	RS5209	160	T1400	160	TIX-M203
160	JP5062	160	PT855	160	RS5301	160	T1401	160	TIX-M204
160	JR10	160	PT856	160	RS5305	160	T1403	160	TIX-M205
160	JR100	160	PTC107	160	RS5306	160	T1454	160	TIX-M206
160	JR200	160	Q40359	160	RS5311	160	T1459	160	TIX-M207
160	JR30	160	Q5044	160	RS5312	160	T1461	160	TIX3016
160	JR30X	160	QA-1	160	RS5313	160	T1548	160	TIX3016A
160	K75508-1	160	R-28A222	160	RS5314	160	T1618	160	TIX3032
160	L2091241-2	160	R1539	160	RS5318	160	T163	160	TIX316
160	L2091241-3	160	R1550	160	S593	160	T1657	160	TIX91
160	L5108	160	R1554	160	S8684	160	T1690	160	TIX92
160	L5121	160	R2683	160	S8685	160	T1691	160	TIXM-201
160	L5122	160	R2684	160	S8686	160	T1692	160	TIXM-203
160	L5181	160	R2685	160	S8687	160	T1737	160	TIXM-205
160	M351	160	R2686	160	RT3466	160	T1738	160	TIXM-206
160	M4439	160	R2687	160	RT4525	160	T1788	160	TIXM01
160	M4484	160	R2688	160	RT6988	160	T1814	160	TIXM02
160	M4485	160	R2694	160	S-1640	160	T1831	160	TIXM03
160	M4486	160	R2695	160	S-371	160	T2015	160	TIXM04
160	M4504	160	R2696	160	S-875YB	160	T2016	160	TIXM05
160	M4506	160	R2697	160	S-874TB	160	T2017	160	TIXM06
160	M4507	160	R3277	160	S01	160	T2019	160	TIXM07
160	M4524	160	R3278	160	S02	160	T2020	160	TIXM08
160	M4526	160	R3279	160	S03	160	T2021	160	TIXM10
160	M4697	160	R3287	160	S1640	160	T2022	160	TIXM101
160	M4860	160	R3288	160	S65-1-A-3P	160	T2024	160	TIXM103
160	M75516-2	160	R3309	160	S65A-1-3P	160	T2025	160	TIXM104
160	M75516-2P	160	R338	160	S684	160	T2026		
		160	R339	160	S70T	160	T2028		

ECG	Industry Standard No.	ECG	Industry Standard No.	ECG	Industry Standard No.	ECG	Industry Standard No.	ECG	Industry Standard No.
160	TIXM105	160	TS-673A	160	2G413	160	2N700	160	2T201
160	TIXM106	160	TS-673B	160	2J72	160	2N700/18	160	2T203
160	TIXM107	160	TS630	160	2J73	160	2N700A	160	2T204
160	TIXM108	160	TS669C	160	2K48	160	2N700A/18	160	2T204A
160	TIXM11	160	TS672B	160	2N1017	160	2N741A	160	2T205
160	TIXM13	160	TS673A	160	2N1018	160	2N779B	160	2T205A
160	TIXM14	160	TS673B	160	2N1025	160	2N846	160	2T485
160	TIXM15	160	TV24137	160	2N1065	160	2N846B	160	2V559
160	TIXM16	160	TV24158	160	2N1066	160	2N960/46	160	2V560
160	TIXM17	160	TV24166	160	2N1094	160	2N964/46	160	2V561
160	TIXM18	160	TV24172	160	2N1115A	160	2N968	160	2V562
160	TIXM19	160	TV24229	160	2N1122	160	2N976	160	2V563
160	TIXM201	160	TV24230	160	2N1122A	160	2N987	160	03-57-003
160	TIXM202	160	TV24239	160	2N1177	160	28A130	160	03-57-102
160	TIXM203	160	TV24351	160	2N1178	160	28A135	160	03-57-200
160	TIXM204	160	TV455	160	2N1179	160	28A218	160	03-57-202
160	TIXM205	160	TV2479	160	2N1285	160	28A224	160	3MC
160	TIXM206	160	TVS-28A103	160	2N1301	160	28A226	160	3N54
160	TIXM207	160	TVS25A103	160	2N1309A	160	28A227	160	3N35A
160	TK1228-001	160	V120	160	2N1385	160	28A228	160	38A324
160	TK41C	160	V205	160	2N1403	160	28A229	160	4-2073
160	TK42C	160	V58	160	2N1405	160	28A230	160	4-432
160	TM160	160	V75	160	2N1406	160	28A233	160	4-44-0012-PT2
160	TNJ-60067	160	VFL2744K	160	2N1407	160	28A233A	160	4-65-1A7-1
160	TNJ-60068	160	VFW2745D	160	2N1408	160	28A233B	160	4-65A17-1
160	TNJ-60069	160	VS-28A103	160	2N1411	160	28A233C	160	4-65B17-1
160	TNJ-60071	160	VS-28A2385L	160	2N1436	160	28A234	160	4-88B17-1
160	TNJ-60073	160	VS-28A71	160	2N1495	160	28A234A	160	4A8-1A5
160	TNJ-60077	160	VS-28A71B	160	2N1500/18	160	28A234B	160	4A8-1A7-1
160	TNJ-60279	160	VS-28A71BS	160	2N1517A	160	28A234C	160	4JX1A813
160	TNJ-60280	160	VS-28C385L	160	2N1524-1	160	28A235	160	4JX1C07
160	TNJ-60281	160	VS28A103	160	2N1524-2	160	28A235A	160	6-1260039
160	TNJ-60362	160	VS28A378	160	2N1524/33	160	28A235B	160	6-1260039A
160	TNJ-60363	160	VS28A379	160	2N1525	160	28A235C	160	6-60A
160	TNJ-60364	160	VS28A71B	160	2N1526/33	160	28A239	160	6-60C
160	TNJ-60365	160	VS28A71BS	160	2N1637/33	160	28A240	160	6-60B
160	TNJ-60608	160	VS28C385L	160	2N1639/33	160	28A240A	160	6-60P
160	TNJ60063	160	V12	160	2N1646	160	28A240B	160	6-60T
160	TNJ60064	160	V22	160	2N1665	160	28A240B2	160	6-60X
160	TNJ60065	160	WTV12MC	160	2N1670	160	28A240BL	160	6-61A
160	TNJ60067	160	WTV20MC	160	2N1673	160	28A241	160	6-61B
160	TNJ60068	160	WTV3MC	160	2N1678	160	28A242	160	6-61P
160	TNJ60069	160	WTV6MC	160	2N1749	160	28A243	160	6-61T
160	TNJ60071	160	WTVAT6A	160	2N1750	160	28A244	160	6-61X
160	TNJ60073	160	WTVB5	160	2N1782	160	28A245	160	6-62C
160	TNJ60279	160	WTVBA6	160	2N1784	160	28A260	160	6-62E
160	TNJ60280	160	WTVBE6	160	2N1788	160	28A261	160	6-69
160	TNJ60281	160	X42	160	2N1789	160	28A262	160	6-69X
160	TNJ60450	160	XA101	160	2N1790	160	28A263	160	6-84F
160	TNJ60456	160	XA102	160	2N1853/18	160	28A264	160	6-85F
160	TNJ70641	160	XA105	160	2N1867	160	28A265	160	6A12889
160	TO-003	160	XA104	160	2N1960	160	28A276	160	6A8-1A5L
160	TO-004	160	XA111	160	2N1960/46	160	28A285	160	6MC
160	TQ-5034	160	XA112	160	2N1961	160	28A286	160	07-3080-06
160	TQ5021	160	XA123	160	2N1616	160	28A287	160	07-4233-19
160	TQ5022	160	XA124	160	2N1999	160	28A298	160	07-4235-13
160	TQ5034	160	XA126	160	2N2022	160	28A306	160	07-4235-73
160	TQ5035	160	XA131	160	2N2084	160	28A307	160	8-0050400
160	TQ5038	160	XA141	160	2N2098	160	28A323	160	8-0050500
160	TR-07	160	XA142	160	2N2100	160	28A332	160	8-0104900
160	TR-11C	160	XA145	160	2N2219	160	28A340	160	8-0105200
160	TR-12	160	XA161	160	2N2190	160	28A341	160	8-0105300
160	TR-12C	160	XB10	160	2N2191	160	28A341-0A	160	8-905-605-320
160	TR-13	160	XB8	160	2N2258	160	28A341-0B	160	8-905-606-001
160	TR-13C	160	XB9	160	2N2360	160	28A342	160	8-905-606-003
160	TR-161(OLSON)	160	XG1	160	2N2361	160	28A342A	160	8-905-606-007
160	TR-166(OLSON)	160	XG10	160	2N2562	160	28A343	160	8-905-606-008
160	TR-17	160	XG11	160	2N2363	160	28A345	160	8-905-606-010
160	TR-17A	160	XG12	160	2N2382	160	28A346	160	8-905-606-016
160	TR-17C	160	XG2	160	2N2398	160	28A347	160	8-905-606-051
160	TR-18	160	XG24	160	2N2399	160	28A348	160	8-905-606-075
160	TR-18C	160	XG3	160	2N2415	160	28A349	160	8-905-606-077
160	TR-1R26	160	XG5	160	2N2416	160	28A360	160	8-905-606-090
160	TR-2R26	160	XJ71	160	2N2455	160	28A361	160	8-905-606-105
160	TR-3R26	160	XJ72	160	2N2456	160	28A377	160	8-905-606-106
160	TR-4R26	160	XJ73	160	2N2495	160	28A378	160	8-905-606-120
160	TR-8001	160	ZC28A101	160	2N2496	160	28A379	160	8-905-606-142
160	TR-8002	160	ZC28A101BA	160	2N2512	160	28A403	160	8-905-606-152
160	TR-8003	160	ZC28A102CA	160	2N2587	160	28A404	160	8-905-606-153
160	TRO6C	160	ZC28A103	160	2N2654	160	28A405	160	8-905-606-154
160	TR110	160	ZC28A103CA	160	2N267	160	28A41	160	8-905-606-155
160	TR12	160	ZC28A377	160	2N2671	160	28A419	160	8-905-606-158
160	TR12C	160	ZC28A70	160	2N2671A	160	28A42	160	8-905-606-165
160	TR13	160	ZC28A700A	160	2N2672BLK	160	28A420	160	8-905-606-168
160	TR13C	160	ZC28A700B	160	2N2672GRN	160	28A421	160	8-905-606-180
160	TR17	160	ZC28A71	160	2N274	160	28A422	160	8-905-606-211
160	TR18	160	ZC28A71A	160	2N274BLU	160	28A425	160	8-905-606-225
160	TR18C	160	ZEN300	160	2N274WHT	160	28A426	160	8-905-606-241
160	TR1R26	160	ZEN301	160	2N276	160	28A430	160	8-905-606-255
160	TR2R26	160	ZJ72	160	2N2786	160	28A431	160	8-905-606-256
160	TR310019	160	ZJ73	160	2N2786A	160	28A431A	160	8-905-606-349
160	TR310025	160	1-21-135	160	2N2797	160	28A432	160	8-905-606-350
160	TR310065	160	1-21-137	160	2N2798	160	28A432A	160	8-905-606-351
160	TR310068	160	1-21-139	160	2N2799	160	28A433	160	8-905-606-352
160	TR310069	160	1-21-150	160	2N286	160	28A434	160	8-905-606-360
160	TR310123	160	1-21-157	160	2N2860	160	28A435	160	8-905-606-375
160	TR310124	160	1-21-190	160	2N2873	160	28A435A	160	8-905-606-390
160	TR310139	160	1-21-228	160	2N2928	160	28A435B	160	8-905-606-391
160	TR310147	160	1-21-229	160	2N2929	160	28A440	160	8-905-606-392
160	TR310150	160	1-21-230	160	2N2942	160	28A440A	160	8-905-606-405
160	TR310155	160	1-21-231	160	2N2943	160	28A460	160	8-905-606-419
160	TR310156	160	1-21-233	160	2N2996	160	28A461	160	8-905-606-420
160	TR310157	160	1-21-256	160	2N2997	160	28A462	160	8-905-606-423
160	TR310158	160	1-21-258	160	2N2998	160	28A463	160	8-905-706-790
160	TR310193	160	1-21-259	160	2N2999	160	28A464	160	8D
160	TR310224	160	1-21-260	160	2N300	160	28A506	160	8B
160	TR310232	160	1-522210131	160	2N3074	160	28A507	160	8F
160	TR331	160	1-522210300	160	2N3127	160	28A508	160	8L
160	TR3R26	160	1-522210921	160	2N3148	160	28A525	160	8P
160	TR4R26	160	1-522211021	160	2N3153	160	28A525A	160	09-300021
160	TR51	160	1-522211921	160	2N3267	160	28A525B	160	09-300024
160	TR52	160	1-522214400	160	2N3279	160	28A54	160	09-300028
160	TR62	160	1-522214411	160	2N3280	160	28A57	160	09-304011
160	TR8001	160	1-522214435	160	2N3281	160	28A58	160	9-5108
160	TR8002	160	1-522214821	160	2N3282	160	28A59	160	9-5110
160	TR8003	160	1-522214831	160	2N3283	160	28A61	160	9-5111
160	TRA10R	160	1-522216600	160	2N3284	160	28A70	160	9-5116
160	TRA11R	160	1-522217400	160	2N3285	160	28A70F	160	9-5117
160	TRA12R	160	1A8-1A82	160	2N3286	160	28A70L	160	9-5118
160	TRA22A	160	002-007100	160	2N331	160	28A70MA	160	9-5120
160	TRA22B	160	002-007200	160	2N3371	160	28A71	160	9-5121
160	TRA23	160	002-007400	160	2N3399	160	28A71A	160	9-5122
160	TRA23A	160	2P	160	2N3588	160	28A71AB	160	9-5123
160	TRA23B	160	2G102	160	2N3770	160	28A71AC	160	9-5124
160	TRA24C	160	2G103	160	2N3783	160	28A71B	160	9-9105
160	TS-615	160	2G104	160	2N3784	160	28A71BS	160	9-9106
160	TS-620	160	2G106	160	2N3785	160	28A71D	160	9-9107
160	TS-621	160	2G109	160	2N384	160	28A71Y	160	9-9108
160	TS-627A	160	2G110	160	2N384/33	160	28A73	160	9-9120
160	TS-627B	160	2G201	160	2N3995	160	28A81	160	9-9121
160	TS-630	160	2G202	160	2N509	160	28A84	160	9A8-1A64
160	TS-672A	160	2G301	160	2N537	160	28A88	160	10A
160	TS-672B	160	2G403	160	2N559	160	28A89	160	10B
		160	2G404			160	2T20		

ECG	Industry Standard No.	ECG	Industry Standard No.	ECG	Industry Standard No.	ECG	Industry Standard No.	ECG	Industry Standard No.
160	12-1-135	160	32-12066-10	160	57A1-16	160	57D1-120	160	65-1-70
160	12-1-137	160	34-119	160	57A1-22	160	57D1-13	160	65-1-70-12
160	12-1-138	160	34-220	160	57A1-24	160	57D1-15	160	65-1-70-12-7
160	12-1-139	160	34-221	160	57A1-25	160	57D1-16	160	65-10
160	12-1-150	160	34-298	160	57A1-30	160	57D1-22	160	65-11
160	12-1-157	160	34-6	160	57A1-31	160	57D1-24	160	65-12
160	12-1-190	160	34-6000-10	160	57A1-32	160	57D1-25	160	65-13
160	12-1-228	160	34-6000-11	160	57A1-33	160	57D1-30	160	65-14
160	12-1-229	160	34-6000-12	160	57A1-35	160	57D1-31	160	65-15
160	12-1-230	160	34-6000-13	160	57A1-36	160	57D1-32	160	65-16
160	12-1-231	160	34-6000-14	160	57A1-37	160	57D1-33	160	65-17
160	12-1-233	160	34-6000-16	160	57A1-41	160	57D1-35	160	65-18
160	12-1-256	160	34-6000-17	160	57A1-45	160	57D1-36	160	65-19
160	12-1-258	160	34-6000-18	160	57A1-46	160	57D1-37	160	65-1A
160	12-1-259	160	34-6000-19	160	57A1-47	160	57D1-41	160	65-1AO
160	12-1-260	160	34-6000-20	160	57A1-48	160	57D1-45	160	65-1AOR
160	12A9244-1	160	34-6000-25	160	57A1-49	160	57D1-46	160	65-1A1
160	12A9244-P2	160	34-6000-26	160	57A1-50	160	57D1-47	160	65-1A19
160	12MC	160	34-6000-3	160	57A1-57	160	57D1-48	160	65-1A2
160	12MZ	160	34-6000-58	160	57A1-61	160	57D1-49	160	65-1A21
160	13-14085-28	160	34-6000-59	160	57A1-67	160	57D1-50	160	65-1A3
160	13-14085-30	160	34-6000-60	160	57A1-69	160	57D1-57	160	65-1A3P
160	13-18946-1	160	34-6000-61	160	57A1-72	160	57D1-61	160	65-1A4-7
160	13-18946-2	160	34-6000-62	160	57A1-73	160	57D1-67	160	65-1A4-7B
160	13-18947-1	160	34-6000-63	160	57A1-74	160	57D1-69	160	65-1A5
160	13-18948-1	160	34-6000-66	160	57A1-80	160	57D1-72	160	65-1A5L
160	13-18948-2	160	34-6000-67	160	57A1-81	160	57D1-73	160	65-1A6-5
160	14-569-09	160	34-6000-68	160	57A1-84	160	57D1-74	160	65-1A7
160	14-580-01	160	34-6000-76	160	57A1-85	160	57D1-80	160	65-1A7-1
160	14-581-01	160	34-6000-77	160	57A1-86	160	57D1-81	160	65-1A8
160	14-585-01	160	34-6000-78	160	57A1-87	160	57D1-84	160	65-1A82
160	14-587-01	160	34-6000-79	160	57A1-89	160	57D1-85	160	65-1A9
160	14-588-01	160	34-6000-80	160	57A1-9	160	57D1-86	160	65-1A9G
160	14-591-01	160	34-6000-81	160	57A1-94	160	57D1-87	160	065A-12
160	14-600-01	160	34-6000-82	160	57A1-96	160	57D1-89	160	065A-12-7
160	14-600-02	160	34-6000-9	160	57A1-98	160	57D1-9	160	65A-70
160	14-600-04	160	34-6005-1	160	57A1-99	160	57D1-94	160	65A-70-12
160	14-600-11	160	34-6015-32	160	57A1130	160	57D1-96	160	65A-70-12-7
160	14-600-13	160	34-6015-33	160	57A1131	160	57D1-98	160	65AO
160	14-600-16	160	34-6015-34	160	57A1132	160	57D1-99	160	65A1
160	14-600-19	160	34-6015-35	160	57A1186	160	57D1130	160	65A10
160	14-600-20	160	34-6015-36	160	57A132-9	160	57D1131	160	65A1OR
160	14-600-22	160	34-6015-39	160	57A180	160	57D1132	160	65A119
160	14A8-1	160	34-6015-40	160	57A184	160	57D1186	160	65A12
160	14A8-1-12	160	34-6016-28	160	57A187	160	57D132-9	160	65A121
160	15-22210131	160	34-6016-29	160	57A188	160	57D9-1	160	65A13
160	15-22210300	160	36(SEARB)	160	57A2-104	160	57L1-1	160	65A13P
160	15-22210921	160	40D1547	160	57A2-105	160	57L1-10	160	65A14
160	15-22211021	160	42-19682	160	57A2-149	160	57L1-11	160	65A14-7
160	15-22211921	160	42-19792	160	57A2-157	160	57L1-12	160	65A14-7B
160	15-22214400	160	42-22778	160	57A2-158	160	57L1-2	160	65A15
160	15-22214411	160	42-22779	160	57A2-159	160	57L1-3	160	65A15L
160	15-22214435	160	42-22780	160	57A2-22	160	57L1-4	160	65A16
160	15-22214821	160	42-22781	160	57A2-26	160	57L1-9	160	65A16-3
160	15-22214831	160	42-22784	160	57A2-30	160	57M1-1	160	65A17
160	15-22216600	160	42-23965P	160	57A2-31	160	57M1-10	160	65A17-1
160	15-22217400	160	43A111449	160	57A2-35	160	57M1-11	160	65A18
160	18A8-1	160	44P1	160	57A2-37	160	57M1-12	160	65A182
160	18A8-1-12	160	46-8612-3	160	57A2-40	160	57M1-17	160	65A19
160	18A8-1-127	160	46-86123-3	160	57A2-41	160	57M1-2	160	65A19G
160	18A8-1L	160	46-862-3	160	57A2-42	160	57M1-3	160	65A2
160	18A8-1L8	160	46-86300-3	160	57A2-48	160	57M1-4	160	65A3
160	019-003315	160	46-865-3	160	57A2-50	160	57M1-9	160	65A4
160	019-00377B	160	46-866-3	160	57A2-51	160	58B2-14	160	65A5
160	19-020-031	160	46-868-3	160	57A2-65	160	61B002-1	160	65A6
160	19-020-032	160	48-10079A01	160	57A2-66	160	61P1D	160	65A7
160	19A115140-P1	160	48-10079A02	160	57A2-67	160	62-17390	160	65A8
160	19A115140-P2	160	48-124296	160	57A2-68	160	62-17591	160	65A9
160	19A115192-P1	160	48-124363	160	57A2-75	160	62-18418	160	070-001
160	19A115192-P2	160	48-124368	160	57A2-77	160	62-18419	160	74Q1262
160	19A115628-P1	160	48-128219	160	57A2-80	160	62-18422	160	77-271029-1
160	19A115628-P2	160	48-134411	160	57A2-89	160	62-26851	160	77-271029-2
160	19A115635-1	160	48-134412	160	57A2-90	160	62-8781	160	77-271029-1
160	19A115636-P1	160	48-134413	160	57A2-93	160	63-10035	160	77-271166-3
160	19A115665-P1	160	48-134439	160	57A9-1	160	63-10036	160	77-271166-2
160	19A115665-P2	160	48-134484	160	57B1130	160	63-10145	160	77-273001-3
160	19A126265-1	160	48-134486	160	57B1131	160	63-10146	160	80-050400
160	19A126265-2	160	48-134504	160	57B1186	160	63-10148	160	80-050500
160	19AR13-1	160	48-134506	160	57B180	160	63-10149	160	86-100-2
160	19AR13-2	160	48-134507	160	57B184	160	63-10150	160	86-101-2
160	19AR13-3	160	48-134524	160	57B187	160	63-10195	160	86-102-2
160	19AR13-4	160	48-134526	160	57B188	160	63-10196	160	86-107-2
160	19AR18	160	48-134579	160	57B2-1	160	63-10375	160	86-112-2
160	19AR24	160	48-134676	160	57B2-104	160	63-10376	160	86-117-2
160	19B2000130-P1	160	48-134678	160	57B2-105	160	63-11055	160	86-135-2
160	19B2000130-P2	160	48-134679	160	57B2-11	160	63-11496	160	86-136-2
160	19G300216-P1	160	48-134693	160	57B2-13	160	63-11582	160	86-149-2
160	19G300216-P2	160	48-134694	160	57B2-14	160	63-11584	160	86-150-2
160	20MC	160	48-63029A16	160	57B2-149	160	63-12610	160	86-151-2
160	21A045-000	160	48-63029A60	160	57B2-157	160	63-13025	160	86-162-2
160	21A048-000	160	48-63075A72	160	57B2-158	160	63-13839	160	86-163-2
160	21A049-000	160	48-63075A73	160	57B2-159	160	63-13899	160	86-164-2
160	21A050-000	160	48-63075A75	160	57B2-17	160	63-17390	160	86-179-2
160	21A050-001	160	48-63078A63	160	57B2-19	160	63-18418	160	86-18-2
160	022-3511-770	160	48-63081A82	160	57B2-2	160	63-18419	160	86-180-2
160	022-3511-780	160	48-644676	160	57B2-20	160	63-18423	160	86-181-2
160	022-3511-790	160	48-644677	160	57B2-22	160	63-18424	160	86-20-2
160	022-3511-770	160	48-645867	160	57B2-26	160	63-25726	160	86-253-2
160	022-3511-780	160	48-64978A27	160	57B2-30	160	63-26850	160	86-254-2
160	022-3511-790	160	48-64978A28	160	57B2-31	160	63-27366	160	86-278-2
160	23-PT284-122	160	48-64978A29	160	57B2-35	160	63-27500	160	86-279-2
160	23-PT284-123	160	48-65132A79	160	57B2-37	160	63-28548	160	86-29-2
160	24MW197	160	48-869040GP	160	57B2-40	160	63-28558	160	86-312-2
160	24MW351	160	48-971A203	160	57B2-41	160	63-29614	160	86-320-2
160	24MW44	160	48-97271A05	160	57B2-42	160	63-29819	160	86-321-2
160	24MW55	160	48-97271A06	160	57B2-48	160	63-29820	160	86-322-2
160	24MW59	160	48R134545	160	57B2-50	160	63-29821	160	86-347-2
160	31-0001	160	051-0062	160	57B2-51	160	63-29862	160	86-348-2
160	31-0002	160	051-0063	160	57B2-65	160	63-3954	160	86-36-2
160	31-0003	160	56P1	160	57B2-66	160	63-7538	160	86-363-2
160	31-0004	160	56P2	160	57B2-68	160	63-7541	160	86-366-2
160	31-0015	160	56P3	160	57B2-75	160	63-7548	160	86-367-2
160	31-0016	160	56P4	160	57B2-77	160	63-7579	160	86-368-2
160	31-0141	160	56P4P	160	57B2-80	160	63-7580	160	86-37-2
160	31-0150	160	57A1-100	160	57B2-89	160	63-7581	160	86-373-2
160	31-0161	160	57A1-101	160	57B2-9	160	63-7582	160	86-374-2
160	31-0163	160	57A1-102	160	57B2-90	160	63-7660	160	86-376-2
160	31-0165	160	57A1-103	160	57B2-93	160	63-8119	160	86-38-2
160	31-0166	160	57A1-107	160	57D1-10	160	63-8376	160	86-449-9
160	31-0168	160	57A1-108	160	57D1-100	160	63-8377	160	86-86-2
160	31-0170	160	57A1-109	160	57D1-101	160	63-8378	160	86-87-2
160	31-0171	160	57A1-110	160	57D1-102	160	63-8379	160	86-89-2
160	31-0180	160	57A1-112	160	57D1-103	160	63-8699	160	86-90-2
160	31-0181	160	57A1-113	160	57D1-107	160	63-8700	160	86-91-2
160	31-0241	160	57A1-114	160	57D1-108	160	63-8954	160	86-92-2
160	31-0241-1	160	57A1-115	160	57D1-109	160	63-9072	160	86-99-2
160	31-21004900	160	57A1-12	160	57D1-110	160	63-9517	160	088B
160	31-21007744	160	57A1-120	160	57D1-112	160	63-9644	160	088B-12
160	31-21024033	160	57A1-13	160	57D1-113	160	63-9665	160	088B-12-7
160	31-21024044	160	57A1-15	160	57D1-115	160	63-9876	160	88B-70
160	31-21047111			160	57D1-12	160	63-9877	160	88B-70-12
160	31-21047733					160	63-9941		
160	31-21050611					160	065-1-12		
160	31-21050622					160	065-1-12-7		

ECG	Industry Standard No.	ECG	Industry Standard No.	ECG	Industry Standard No.	ECG	Industry Standard No.	ECG	Industry Standard No.
160	88B-70-12-7	160	121-384	160	417-60	160	1119-56	160	61012-5-1
160	88B0	160	121-385	160	417-66	160	1301-1	160	65804-62
160	88B1	160	121-411	160	417-68	160	1501-2	160	68504-62
160	88B10	160	121-412	160	417-70	160	1449	160	69107-43
160	88B10R	160	121-413	160	417-71	160	1465-1	160	72797-81
160	88B11	160	121-414	160	417-72	160	1465-1-12	160	72923-08
160	88B119	160	121-415	160	417-76	160	1465-1-12-8	160	77271-8
160	88B12	160	121-415B	160	417-79	160	1465A	160	080026
160	88B121	160	121-426	160	421-6	160	1465A-12	160	080027
160	88B13	160	121-427	160	421-6B	160	1465A-12-8	160	080028
160	88B13P	160	121-428	160	421-7	160	1488B	160	080061
160	88B14	160	121-429	160	421-7B	160	1488B-12	160	080244
160	88B14-7	160	121-432	160	421-8	160	1488B-12-8	160	080245
160	88B14-7B	160	121-48	160	421-8B	160	1526	160	080258
160	88B15	160	121-49	160	421-9792	160	1865-1	160	080266
160	88B15L	160	121-538	160	422-2778	160	1865-1-12	160	080267
160	88B16	160	121-538B	160	422-2779	160	1865-1-127	160	080269
160	88B16B	160	121-539	160	422-2780	160	1865-1L	160	94007
160	88B17	160	121-540	160	429-0092-1	160	1865A	160	94028
160	88B17-1	160	121-540B	160	429-0910-50	160	1865A-12	160	94033
160	88B18	160	121-541	160	444-0012-PT2	160	1865A-127	160	94035
160	88B182	160	121-541B	160	444-012-P1	160	1865AL8	160	94036
160	88B19	160	121-542	160	465-032-19	160	1888B	160	95101
160	88B19Q	160	121-542B	160	465-042-19	160	1888B-12	160	95102
160	88B2	160	121-552	160	465-045-19	160	1888B-127	160	95103
160	88B3	160	121-553	160	465-049-19	160	1888BL	160	95107
160	88B4	160	121-601	160	465-061-19	160	1888BL8	160	95108
160	88B5	160	121-63	160	465-086-19	160	2015	160	95110
160	88B6	160	121-697	160	465-146-19	160	2020	160	95111
160	88B7	160	121-698	160	465-1A5	160	2021	160	95121
160	88B8	160	121-73	160	465-223-19	160	2057A2-159	160	99103
160	88B9	160	121-74	160	465A-15	160	2057A2-166	160	99104
160	90-54	160	121-75	160	465B-15	160	2057A2-205	160	99105
160	90-59	160	121-76	160	488B15	160	2057A2-231	160	99106
160	91-4	160	121-78	160	501T1	160	2057A2-232	160	99107
160	95-108	160	121-79	160	503T1	160	2057A2-252	160	99108
160	95-110	160	0131-000418	160	504T1	160	2057B1B6	160	101078
160	95-116	160	0131-000419	160	505T1	160	2057B2-149	160	101087
160	95-117	160	0131-000498	160	506T1	160	2057B2-157	160	111117
160	95-118	160	0131-000802	160	507T1	160	2057B2-158	160	111118
160	95-119	160	0131-000859	160	508T1	160	2057B2-159	160	111313
160	95-120	160	0131-000862	160	573-518	160	2057B2-89	160	111954
160	95-121	160	0131-000863	160	576-0003-010	160	2057B2-90	160	111955
160	95-122	160	0131-001314	160	576-0003-025	160	2057B2-93	160	111956
160	95-123	160	0131-001332	160	576-0003-009	160	2057B2A2-118	160	112001
160	95-124	160	0131-001433	160	601-040	160	2106-120	160	112002
160	96-5062-01	160	0131-001434	160	602-075	160	2215-17	160	112011
160	96-5095-01	160	0131-001435	160	603-020	160	2487B	160	112032
160	96-5099-01	160	0131-001436	160	603-030	160	2489A	160	112296
160	96-5138-01	160	0131-001697	160	603-040	160	2495-078	160	115227
160	96-5139-01	160	0131-003029	160	604-030	160	2495-079	160	115228
160	96-5140-01	160	132-019	160	604-080	160	2495-082	160	115229
160	96-5141-01	160	132-020	160	609-020	160	2495-376	160	115504
160	96X2801/37N	160	132-027	160	610-050	160	2495-377	160	116202
160	99-105	160	154T1	160	610-050-1	160	2495-378	160	116207
160	99-106	160	154T1A	160	610-050-2	160	2495-488-1	160	116208
160	99-107	160	154T1B	160	610-050-3	160	2495-488-2	160	116209
160	99-108	160	155T1	160	610-051	160	2606-295	160	116683
160	99-120	160	156T1	160	610-051-1	160	2700	160	116684
160	99-121	160	157T1	160	610-051-2	160	3008	160	117618
160	998006	160	159T1	160	610-051-4	160	3024	160	117658
160	998007	160	161T1	160	610-053	160	3025	160	117725
160	101A	160	162T1	160	610-053-1	160	345B	160	117726
160	101B	160	165-1A82	160	610-053-2	160	3504(RCA)	160	117866
160	101M	160	165A-182	160	610-055	160	3534(RCA)	160	119013
160	107A	160	165B-182	160	610-055-1	160	3750	160	124625
160	107B	160	188B-1-82	160	610-055-2	160	3961(G.E.)	160	125790
160	107M	160	201A	160	610-055-3	160	4001-222	160	125972
160	108-1	160	201B	160	617-54	160	4001-223	160	126184
160	108-2	160	207A	160	617-55	160	4368	160	126186
160	108-3	160	207A1	160	617-57	160	4822-130-40252	160	129389
160	108-4	160	207B	160	617-58	160	4822-130-40255	160	156766
160	112-000267	160	207M	160	642-116	160	4822-130-40441	160	171016
160	115-227	160	241-15A	160	642-147	160	5085	160	171039(SEARS)
160	115-228	160	296-46-9	160	642-173	160	6313	160	171039(TOSHIBA)
160	115-229	160	297V008M01	160	642-202	160	6990	160	171162-169
160	116-202	160	297V012H04	160	642-207	160	9403-3	160	175006-181
160	116-207	160	297V012H07	160	665-1A5L	160	9403-6	160	175006-182
160	116-208	160	297V012H11	160	665A-1-5L	160	9510-1	160	175006-183
160	116-209	160	297V012H12	160	688B-1-5L	160	9510-2	160	175006-184
160	116-683	160	297V012H15	160	690L297H01	160	9510-7	160	175006-185
160	116-684	160	297V020H01	160	690V052H63	160	11522-7	160	190427
160	120-001190	160	297V024H01	160	690V056H27	160	11522-8	160	200064-6-104
160	120-001190	160	297V024H03	160	690V056H29	160	11620-2	160	200064-6-106
160	120-002213	160	297V038H02	160	690V056H30	160	11620-7	160	223369
160	120-002214	160	297V038H03	160	690V056H31	160	11620-8	160	223474
160	120-002216	160	297V038H04	160	690V056H32	160	11620-9	160	224586
160	120-004722	160	297V064H01	160	690V057H25	160	11668-3	160	224587
160	120-02213	160	324-0016	160	690V066H89	160	11668-4	160	225311
160	121-101	160	324-0026	160	690V081H08	160	12113-5	160	225594
160	121-102	160	324-0077	160	690V119H95	160	12113-7	160	225600
160	121-103	160	324-0082	160	690V119H96	160	12113-8	160	229133
160	121-104	160	324-0083	160	690V119H97	160	12113-9	160	232676
160	121-105	160	324-0086	160	750M63-104	160	12115-0	160	232680
160	121-119	160	324-0087	160	750M63-116	160	12115-7	160	234015
160	121-132	160	324-0095	160	750M63-117	160	12119-0	160	234631
160	121-134	160	324-0110	160	800-504-00	160	12122-8	160	235200
160	121-135	160	324-0111	160	800-50400	160	12122-9	160	261586
160	121-136	160	324-0112	160	921-10B	160	12123-0	160	265771
160	121-137	160	324-0123	160	921-11B	160	12123-1	160	310030
160	121-138	160	324-0131	160	921-12B	160	12123-3	160	310132
160	121-139	160	324-0132	160	921-13B	160	12123-6	160	310162
160	121-177	160	324-0187	160	921-15B	160	12125-8	160	310204
160	121-179	160	324-132	160	921-16B	160	12125-9	160	346607-4
160	121-180	160	325-0028-85	160	921-17B	160	12126-0	160	454760
160	121-181	160	353-9301-002	160	921-1A	160	18540	160	0510079
160	121-185	160	367(SEARS)	160	921-1B	160	18541	160	0510079H
160	121-242	160	417-11	160	921-25B	160	30218(RCA)	160	0573335
160	121-243	160	417-12	160	921-26B	160	30238	160	573356
160	121-244	160	417-13	160	921-2A	160	30240	160	573371
160	121-268	160	417-143	160	921-2B	160	30273	160	573398
160	121-269	160	417-16	160	921-3A	160	34342	160	573405
160	121-284	160	417-2	160	921-3B	160	34389	160	573406
160	121-304	160	417-22	160	921-4A	160	34423	160	0573428
160	121-312	160	417-23	160	921-9B	160	34553	160	601032
160	121-313	160	417-25	160	955-1	160	34675	160	601040
160	121-329	160	417-26	160	955-2	160	34942	160	601052
160	121-330	160	417-27	160	955-3	160	35070	160	602075
160	121-331	160	417-31	160	965-1A6-5	160	35169	160	603020
160	121-332	160	417-35	160	965A16-3	160	35170	160	603030
160	121-333	160	417-36	160	972X1	160	35815	160	603040
160	121-334	160	417-37	160	972X2	160	37278	160	603112
160	121-335	160	417-38	160	972X3	160	38680	160	604030
160	121-336	160	417-39	160	972X4	160	38681	160	604040
160	121-349	160	417-50	160	972X5	160	39053	160	604080
160	121-350	160	417-52	160	1006-78	160	40004	160	605112
160	121-351	160	417-54	160	1010-78	160	40005	160	609020
160	121-352	160	417-56	160	1010-87	160	40006	160	610050
160	121-353	160	417-57	160	1040-59	160	40263(RCA)	160	610050-1
160	121-356	160	417-58	160	1111-17	160	40487	160	610050-2
160	121-357	160		160	1111-18	160	40489	160	610050-3
160	121-358	160		160	1113-13	160	49939-2	160	610051
160	121-359	160		160	1119-54	160		160	610051-1
160	121-383	160		160	1119-55	160		160	

ECG	Industry Standard No.	ECG	Industry Standard No.	ECG	Industry Standard No.	ECG	Industry Standard No.	ECG	Industry Standard No.
160	610051-2	160	4038359-P2	161	AT340	161	C927B	161	PL1111
160	610051-4	160	4038406-P1	161	AT341	161	C927C	161	PL1112
160	610053	160	4038406-P2	161	AT342	161	C927CJ	161	PTC126
160	610053-1	160	5495957-P1	161	AT343	161	C927CU	161	PTC132
160	610053-2	160	5495957-P2	161	AT344	161	C927CW	161	PTG132
160	610055	160	5495957-P3	161	AT345	161	C927D	161	Q35259
160	610055-1	160	5495957-P5	161	AT346	161	C927E	161	Q8B5020
160	610055-2	160	5495957-P6	161	B1R	161	C928	161	RCA40245
160	610055-3	160	7279780	161	BP155	161	C928B	161	RCA40246
160	610056-1	160	7279781	161	BP161	161	C928C	161	RB2001
160	610056-2	160	7284513	161	BP166	161	C928D	161	RB2002
160	610056-3	160	7292508	161	BP167	161	C928E	161	RB28
160	610056-4	160	7851323	161	BP168	161	C947	161	RB5001
160	610061-1	160	7852897-01	161	BP169	161	C948	161	RB5002
160	660064	160	7852900-01	161	BP173	161	C997	161	RP200
160	660084	160	8014711	161	BP175	161	CC86225P	161	RS-2011
160	660085	160	8024390	161	BP180	161	CC86226G	161	S1286
160	772716	160	8510671-1	161	BP181	161	CC86227P	161	S130-138
160	772717	160	8510671-2	161	BP182	161	CC89017	161	S130-251
160	772719	160	8510671-4	161	BP183	161	CC890170925	161	S15650
160	785897-01	160	8511759-1	161	BP184	161	CC89018H924	161	S15657
160	815067	160	8538640	161	BP185	161	CS1014G	161	S15658
160	815067A	160	8989441-2	161	BP200	161	CS1014H	161	S15659
160	815067B	160	31210049-00	161	BP206	161	CS1120C	161	S2002
160	815067C	160	31210077-44	161	BP207	161	CS1120D	161	S27604
160	815193	160	31210240-33	161	BP208	161	CS1120E	161	S4002
160	815197	160	31210240-44	161	BP209	161	CS1120F	161	S40204
160	815294	160	31210471-11	161	BP212	161	CS1120H	161	S5021
160	851759-3	160	31210506-11	161	BP213	161	CS1284B	161	SAB1044
160	853864-0	160	31210506-22	161	BP214	161	CS1284F	161	SAB3469
160	902521	160	43022055	161	BP215	161	CS1284G	161	SCA3244
160	980140	160	43025620	161	BP222	161	CS1284H	161	SD8240
160	980142	160	62034076	161	BP226	161	CS1330	161	SD8420
160	980146	160	62042648	161	BP232	161	CS1460E	161	SE-5023
160	980435	160	62042656	161	BP233-5	161	CS1460H	161	SE-5025
160	980441	160	62043679	161	BP251	161	CS1461J	161	SE1001
160	980505	160	62043695	161	BP252	161	CS1461X	161	SE1002
160	980506	160	62046406	161	BP260	161	CS1462F	161	SE1002-1
160	980507	160	62046414	161	BP261	161	CS1589E	161	SE1002-2
160	980509	160	62046422	161	BP270	161	CS1589F	161	SE1010
160	980514A	160	62087641	161	BP271	161	CS1589B	161	SE2001
160	980545A	160	62135182	161	BP273	161	CS1594E	161	SE2002
160	980636A	160	62139633	161	BP273C	161	CS1596E	161	SE3001
160	981143	160	62139668	161	BP273D	161	CS2715	161	SE3002
160	981144	160	62380004	161	BP274	161	CS2716	161	SE3005
160	981145	160	62593733	161	BP274B	161	CS3662	161	SE3019
160	981146	160	62596025	161	BP274C	161	CS3663	161	SE4021
160	981203	160	62614765	161	BP287	161	CS3707	161	SE4022
160	981204	160	62707666	161	BP288	161	CS3708	161	SE5-0249
160	981959	160	62714832	161	BP290	161	CS3709	161	SE5-0250
160	982522	160	62723645	161	BP302	161	CS3710	161	SE5001
160	982374	160	62726605	161	BP304	161	CS3711	161	SE5002
160	982497	160	80050400	161	BP306	161	CS89017F	161	SE5003
160	985445A	160	80146620	161	BP344	161	CS89017G	161	SE5004
160	985446A	160	80146630	161	BP345	161	CS89017H	161	SE5020
160	1222136	160	120001190	161	BP516	161	CS929	161	SE5021
160	1222314	160	120002213	161	BF818CA	161	CS990	161	SE5022
160	1222371	160	120002214	161	BF827B	161	D4D22	161	SE5023
160	2000646-104	160	120002216	161	BF827F	161	D9-78	161	SE5024
160	2000648-21	160	120002513	161	BF827G	161	D8-85	161	SE5032
160	2000648-22	160	120002515	161	BF862	161	D878	161	SE5035
160	2001653-58	160	120002520	161	BFW41	161	D881	161	SE5050
160	2001653-59	160	120002656	161	BFW63	161	D885	161	SE5051
160	2002151-18	160	120004492	161	BFX31	161	ES15X104	161	SE5052
160	2002151-18A	160	152221483	161	BFX60	161	ES15X105	161	SE6002
160	2002153-58	160	312104732	161	BFX62	161	ES15X122	161	SG887231
160	2002153-59	160	312104733	161	BFX75	161	ES15X123	161	SK3018
160	2002153-60	160	485134956	161	BFX77	161	ES15X56	161	SK3117
160	2002153-76	160	1522211921	161	BFX89	161	ES15X57	161	SKA4768
160	2003073-11	160	1522214400	161	BFY69B	161	ES15X65	161	SM-4304-8
160	2003073-12	160	1522214831	161	BFY79	161	ES15X79	161	SP4436
160	2003073-13	160	1522217400	161	BFY87	161	ES15X87	161	SR8343
160	2076403	160	2295100227	161	BFY87A	161	ES15X88	161	SS4042
160	2076403-0703	160	3121004900	161	C1035	161	ES15X96	161	ST5641
160	2091217-0014	160	3121007744	161	C1035C	161	ES54(ELCOM)	161	T-Q5071
160	2091241-0005	160	3121024033	161	C1035D	161	ES73(ELCOM)	161	T-Q5086
160	2091247-005	160	3121024044	161	C1035E	161	ES86(ELCOM)	161	T1202(GE)
160	2092417-005	160	3121047111	161	C1036	161	EW212	161	T1X-M14
160	2092417-0704	160	3121050611	161	C1044	161	F3535	161	T1X-M15
160	2092417-0707	160	3121050622	161	C1128	161	F3574	161	T1X-M16
160	2092417-0708	160$	A362	161	C1182B	161	F501	161	T381
160	2092417-0709	160$	A363	161	C1182C	161	F501(ZENITH)	161	T381(SEARS)
160	2092417-0710	160$	A372	161	C1182D	161	F501-16	161	T576-1
160	2092417-0717	160$	A450	161	C174A	161	F502	161	TA2554
160	2092417-1	160$	A450H	161	C206	161	F502(ZENITH)	161	TA2555
160	2092417-3	160$	A451	161	C208	161	F523	161	TG28C1293A
160	2092417-4	160$	A451H	161	C251	161	FS1682	161	TG28C927
160	2092417-5	160$	A452	161	C251A	161	FT118	161	TH-R28C313
160	2092417-6	160$	A452H	161	C252	161	FT45	161	TI-407
160	2092417-7	160$	28A362	161	C253	161	GE-59	161	TI-408
160	2092417-9	160$	28A363	161	C296	161	GE-60	161	TI-409
160	2092418-071	160$	28A372	161	C313(JAPAN)	161	G13793	161	TI-492
160	2092418-0710	160$	28A450	161	C313C	161	GM508	161	TI-493
160	2092418-0712	160$	28A450H	161	C313H	161	GM0580	161	TI-494
160	2092418-1	160$	28A451	161	C33	161	HEP709	161	TI856
160	2092418-10	160$	28A451H	161	C389	161	HEP719	161	TI857
160	2092418-11	160$	28A452	161	C389-0	161	HS40021	161	TIX-M14
160	2092418-2	160$	28A452H	161	C389R	161	HS40022	161	TIX-M15
160	2092418-5	160(2)	62087668	161	C463	161	HS40023	161	TIX-M16
160	2092418-6	161	A1U	161	C463H	161	HS40025	161	TM161
160	2092418-7	161	A2332	161	C477	161	HT8001610	161	TNJ11173
160	2092418-8	161	A244	161	C562	161	HT8001710	161	TNJ71964
160	2092693-2	161	A244(AMC)	161	C562O	161	IRTR70	161	TNJ72150
160	2092693-3	161	A2464	161	C562Y	161	IRTR71	161	TNJ72151
160	2092693-4	161	A2465	161	C563	161	K2115	161	TNJ72275
160	2092693-9	161	A2479	161	C563A	161	M-128J509-1	161	TNJ72368
160	2320161(RCA)	161	A2480	161	C663	161	M-128J511-3	161	TNJ72701
160	2320514	161	A2746	161	C682	161	M401	161	TR-171
160	2495078	161	A3T918	161	C682A	161	M4756	161	TR01042
160	2495079	161	A417-19	161	C682B	161	M546	161	TV-36
160	2495082	161	A429-0981-12	161	C683	161	M612	161	TV-37
160	2495376	161	A465-181-19	161	C683A	161	M613	161	TV-38
160	2495377	161	A481	161	C683B	161	M614	161	TV15
160	2498837	161	A482	161	C683V	161	M75545-1	161	TV15A
160	3460550-1	161	A484	161	C720	161	M9266	161	TV15B
160	3460550-3	161	A486	161	C761	161	M9450	161	TV15C
160	3460550-4	161	A489	161	C761I	161	M9481	161	TV20
160	4036715-P1	161	A490	161	C761Z	161	MM8006	161	TV24160
160	4036715-P2	161	A492	161	C762	161	MM8007	161	TV24161
160	4036923-P1	161	AIE	161	C786	161	MRP501	161	TV24209
160	4036923-P2	161	AR111	161	C786R	161	MRP502	161	TV24399
160	4036962-P1	161	AR209	161	C787	161	NKT16229	161	TV24436
160	4036962-P2	161	AR210	161	C80	161	NKT35219	161	TV24437
160	4036963-P1	161	AR211	161	C860	161	NPC167	161	TV24571
160	4036963-P1	161	AR213	161	C860C	161	NR461AA	161	TV33
160	4036965-P1	161	AR213(VIOLET)	161	C860D	161	NS345	161	TV55
160	4036965-P2	161	AR213V	161	C860E	161	NS406	161	TV50
160	4037410-P1	161	AR213(ORANGE)	161	C863	161	P/N14-603-02	161	TV54
160	4037410-P2	161	AR218(RED)	161	C864	161	PL1021	161	TVS-28C313
160	4037764-P1	161	AR218RO	161	C917K	161	PL1022	161	TVS-28C466
160	4037764-P2	161	AR219	161	C918	161	PL1023	161	TVS-28C562
160	4038359-P1	161	AR220	161	C927	161	PL1066	161	TVS-28C563
		161	AR222	161	C927A	161	PL1067	161	TVS-28C563A

ECG	Industry Standard No.	ECG	Industry Standard No.	ECG	Industry Standard No.	ECG	Industry Standard No.	ECG	Industry Standard No.
161	TVS-28C683	161	28C927B	161	48-869450	161	229-0190-30	161	40352
161	TVS-28C683V	161	28C927C	161	48-869481	161	229-0204-6	161	40472
161	TVS-28C684	161	28C927CJ	161	48-90232A01	161	229-0240-25	161	40475
161	TVS-28C762	161	28C927CT	161	48-90232A17	161	229-1301-22	161	40476
161	TVS-28C948	161	28C927CU	161	48-90232A18	161	229-5100-31V	161	40478
161	TVS28C562	161	28C927CW	161	48-97046A25	161	229-5100-32	161	40479
161	TVS28C565	161	28C927D	161	48-97046A45	161	229-5100-32V	161	40481
161	V8-28C446	161	28C927E	161	48-97162A28	161	260P03201	161	55810-161
161	V8-28C683	161	28C928	161	48-97162A30	161	260P03201A	161	55810-163
161	V828C717///+1	161	28C928B	161	48-97162A31	161	260P05901	161	55810-164
161	W23	161	28C928C	161	48-97162A32	161	260P05901A	161	61558
161	X856	161	28C928D	161	48-97177A02	161	260P09201	161	61755
161	X857	161	28C928E	161	48-97177A03	161	260P10601	161	119554
161	X858	161	28C947	161	48P65174A24	161	260P1601	161	119555
161	X859	161	28C948	161	48B134981	161	260P16101	161	119556
161	ZEN105	161	28C948E	161	51A180-4	161	260P17702	161	119557
161	Z22708	161	28C997	161	57A102-4	161	260P24901	161	119823
161	01-050682	161	3L4-6007-10	161	57A103-4	161	297V074C01	161	119824
161	01-571794	161	3L4-6007-11	161	57A119-2	161	352-0658-010	161	119825
161	002-011400	161	3L4-6007-12	161	57A139-1	161	352-0658-020	161	126025
161	002-011500	161	3L4-6007-13	161	57A139-2	161	352-0658-030	161	131543
161	2A0	161	3L4-6007-14	161	57A139-3	161	352-0658-040	161	137388
161	2AH	161	3L4-6007-19	161	57A141-4	161	352-0658-050	161	141402
161	2N2865	161	3L4-6007-20	161	57A142-7	161	355D6	161	1471155-9
161	2N3287	161	3L4-6007-22	161	57A164-4	161	355D8	161	147353-0-1
161	2N3288	161	6A11223	161	57A177-12	161	386-7243-P001	161	170753-1
161	2N3289	161	7-10	161	57A179-4	161	417-243	161	170756-1
161	2N3290	161	7-10(SARKES)	161	57A21-16	161	417-258	161	170906
161	2N3291	161	7-11(SARKES)	161	57A21-17	161	417-262	161	170906-1
161	2N3292	161	7-21(SARKES)	161	57A21-18	161	536-1(RCA)	161	171030
161	2N3293	161	7-22(SARKES)	161	57A24	161	576-0003-023	161	171031
161	2N3294	161	7-23(SARKES)	161	57A249-4	161	576-0003-026	161	171162-118
161	2N3337	161	7-24(SARKES)	161	57B102-4	161	576-1(RCA)	161	190714
161	2N3338	161	7-25(SARKES)	161	57B103-4	161	600X0093-086	161	231140-36
161	2N3339	161	7-9	161	57B119-2	161	600X0094-086	161	231140-37
161	2N3407	161	7-9(SARKES)	161	57B141-4	161	617-65	161	231140-43
161	2N3493	161	8-905-605-644	161	57B21-17	161	657-31	161	236039
161	2N3600	161	8-905-706-044	161	57B21-6	161	680-1(RCA)	161	489751-039
161	2N3683	161	8-905-706-055	161	57B21-7	161	690Y010H40	161	489751-047
161	2N3832	161	8-905-706-060	161	57B249-4	161	690V080H36	161	489751-121
161	2N3952	161	8-905-706-070	161	057B474H	161	690V086H94	161	489751-127
161	2N3953	161	8-905-706-071	161	570142-4	161	690V086H95	161	489751-128
161	2N4134	161	8-905-706-075	161	570164-4	161	800-557-00	161	489752-095
161	2N4135	161	8-905-706-080	161	57D24	161	822-1(SYLVANIA)	161	540204
161	2N4252	161	8-905-706-101	161	66P021-1	161	824-1(SYLVANIA)	161	05734A74
161	2N4253	161	8-905-706-110	161	66P022-1	161	916-31025-4	161	05734A74H
161	2N4259	161	8-905-706-730	161	66P042-1	161	916-31025-4B	161	05734A75
161	2N4433	161	09-302128	161	73N1	161	921-55B	161	610041-2
161	2N4434	161	09-302201	161	86-186-2	161	921-56B	161	610042
161	2N4435	161	09-302240	161	86-204-2	161	921-57B	161	610073-1
161	2N5031	161	09-305006	161	86-205-2	161	921-58B	161	610100-1
161	2N5032	161	09-305011	161	86-262-2	161	1004(G.E.)	161	610150-1
161	2N5126	161	09-305132	161	86-263-2	161	1006(G.E.)	161	610150-3
161	2N5130	161	10-2	161	86-289-2	161	1009(G.E.)	161	610181-2
161	2N5132	161	13-10320-14	161	86-290-2	161	1076-1377	161	610186-1
161	2N5133	161	13-10321-47	161	86-381-2	161	1420-2-2	161	701678-00
161	2N5180	161	13-10321-76	161	86-442-2	161	1801	161	815206
161	2N5181	161	13-10321-78	161	90-45	161	1843-17	161	965632
161	2N5182	161	13-1032176	161	91N1B	161	1845-17	161	972505
161	2N5230	161	13-15808-2	161	96-5131-01	161	1952-17	161	984191
161	2N5852	161	13-15835-1	161	96-5163-01	161	2020-02	161	984192
161	2N917	161	13-2384-2	161	96-5174-01	161	2057A2-157	161	1471115-10
161	2N917A	161	13-26009-1	161	96-5175-01	161	2057A2-158	161	1472633
161	2SC1035	161	13-26576-1	161	96-5198-01	161	2057A2-180	161	1473556-2
161	2SC1035C	161	13-26576-2	161	96-5199-01	161	2057A2-181	161	1473543-1
161	2SC1035D	161	13-26577-1	161	96-5235-01	161	2057A2-185	161	1473571-1
161	2SC1035E	161	13-26577-2	161	96-5236-01	161	2057A2-187	161	1473576
161	2SC1036	161	13-26577-3	161	96-5259-01	161	2057A2-192	161	1473577-1
161	2SC1044	161	13-31013-3	161	96-5260-01	161	2057A2-193	161	1473579-1
161	2SC1054	161	14-602-31	161	96-5334-01	161	2057A2-195	161	1473586
161	2SC1117	161	14-602-34	161	96XZ6050/25N	161	2057A2-196	161	2006582-101
161	2SC1117H	161	14-603-02	161	998031	161	2057A2-197	161	2092417-0711
161	2SC1180	161	14-603-08	161	998032	161	2057A2-202	161	2092417-0712
161	2SC1182B	161	14-603-09	161	998044	161	2057A2-204	161	2092417-0713
161	2SC1182C	161	14-603-13	161	998067-1	161	2057A2-251	161	2092417-0714
161	2SC1182D	161	14-652-12	161	100N1	161	2057A2-258	161	2092417-0715
161	2SC1254	161	14-653-21	161	100N1AS	161	2057B2-138	161	2092417-0716
161	2SC1547	161	14-661-21	161	100N1P	161	2057B2-139	161	2092418-0022
161	2SC174A	161	14-850-12	161	100N3P	161	2057B2-192	161	2092418-0023
161	2SC206	161	14-851-12	161	101-1(ADMIRAL)	161	2110N-132	161	2092418-0024
161	2SC208	161	16-21426	161	102-4	161	2110N-133	161	2092418-0716
161	2SC313	161	019-005157	161	121-283	161	3025	161	2092418-0717
161	2SC33C	161	19-020-044	161	121-377	161	3027	161	2092418-0718
161	2SC313H	161	19-020-048	161	121-378	161	3476(RCA)	161	2092418-0719
161	2SC33	161	19A115342-2	161	121-379	161	3508(SEARS)	161	2092418-0720
161	2SC389	161	020-1112-003	161	121-380	161	3516(RCA)	161	2092418-0721
161	2SC389-0	161	21A112-010	161	121-460	161	3518(RCA)	161	2316183
161	2SC389R	161	23-PT274-120	161	121-461	161	3569	161	23200041H
161	2SC397	161	23-PT274-123	161	121-462	161	3571-1	161	23201041H
161	2SC463	161	23-PT275-121	161	121-470	161	3579	161	2495520
161	2SC463H	161	23-PT283-122	161	121-471	161	3579(RCA)	161	2495521
161	2SC477	161	23-PT283-124	161	121-500	161	3680	161	2495522-1
161	2SC503	161	24MW957	161	121-501	161	3680-1	161	2495523-1
161	2SC562-0	161	24MW958	161	121-502	161	3881	161	2497094-1
161	2SC562Y	161	24T-011-001	161	121-510	161	4167(AIRLINE)	161	2497094-2
161	2SC563	161	24T-013-003	161	121-521	161	4167(PENNCREST)	161	3459332-1
161	2SC563(3RDIP)	161	24T011-012	161	121-523	161	4168(PENNCREST)	161	3463609-2
161	2SC563A	161	24T021	161	121-580	161	4168(WARDS)	161	3468068-1
161	2SC602	161	025-100012	161	121-692	161	4169(PENNCREST)	161	3468068-2
161	2SC618	161	025-100036	161	121-704	161	4169(WARDS)	161	3468068-3
161	2SC618A	161	025-100037	161	121-732	161	4756	161	3468068-4
161	2SC663	161	025-100038	161	121-760	161	4822-130-40304	161	3539307-001
161	2SC682	161	41N	161	121-761	161	8000-00032-026	161	3539307-002
161	2SC682A	161	41N2M	161	121-775	161	8010-174	161	3539307-002
161	2SC682B	161	42-21401	161	121-779	161	8010-175	161	4028839
161	2SC682E	161	42-21402	161	121-823	161	25671-020	161	5320051
161	2SC683	161	46-119-3	161	121-824	161	25671-021	161	7294910
161	2SC683A	161	46-86107-3	161	129-27	161	25671-023	161	7297980
161	2SC683B	161	46-86108-3	161	129B1	161	29076-005	161	7851650-01
161	2SC683R	161	46-86112-2	161	130-185	161	29076-006	161	7851651-01
161	2SC683V	161	46-86113-3	161	130-240	161	37334	161	7910108-01
161	2SC761	161	46-86114-3	161	132-008	161	38246	161	7939165
161	2SC761Y	161	46-86117-3	161	132-009	161	38787	161	8249600
161	2SC761Z	161	46-86118-3	161	132-076	161	40231	161	23114037
161	2SC762	161	46-86119-3	161	132-082	161	40232	161	23114164
161	2SC786	161	46-86224-3	161	132-087	161	40233	161	23126183
161	2SC786R	161	46-86285-3	161	132-185	161	40234	161	62539348
161	2SC787	161	46-8672-3	161	139N2	161	40235	161	62565160
161	2SC787A	161	48-01-027	161	141 402	161	40236	161	62593717
161	2SC80	161	48-134190A1G	161	142N1P	161	40237	161	62593725
161	2SC811	161	48-134756	161	145N1	161	40238	161	62636696
161	2SC860	161	48-137197	161	145N1P	161	40239	161	62748273
161	2SC860C	161	48-137491	161	150-18	161	40240	161	62771240
161	2SC860D	161	48-65112A68	161	150N1	161	40242	161	62789247
161	2SC860E	161	48-65146A61	161	150N3	161	40243	161	62789255
161	2SC863	161	48-65146A62	161	170(RCA)	161	40244	161	450010701
161	2SC863D	161	48-65146A63	161	186N1	161	40245	162	A1C
161	2SC864	161	48-65173A78	161	201-254343-22	161	40246	162	A6532841
161	2SC917	161	48-65174A24	161	201-254343-26	161	40259	162	A6738701
161	2SC917K	161	48-869266	161	207A27	161	40260	162	A8M
161	2SC918			161	217(RCA)	161	40294	162	A9K
161	2SC918A			161	229-0180-119	161	40350	162	A9P
161	2SC927			161	229-0180-32	161	40351	162	B170008
161	2SC927A			161	229-0180-33			162	B176001

ECG	Industry Standard No.	ECG	Industry Standard No.	ECG	Industry Standard No.	ECG	Industry Standard No.	ECG	Industry Standard No.
162	B176002	162	610064-1	164	C1184C	165	2SD380	166	R154B
162	B176003	162	1473564-1	164	C1184D	165	2SD418	166	R204B
162	B2H	162	1473637-1	164	C1184E	165	2SD577	166	RT7402
162	BD144	162	2320271	164	C642	165	2SD649	166	RVD12B-1
162	BDX12	162	2320273	164	C642A	165	2SD725	166	RVD2DP22/10
162	BLY49	162	7293818	164	C936	165	8-729-118-76	166	RVD2DP221B
162	BLY50	162	7293819	164	DTS013	165	8-729-341-34	166	RVD8MB4
162	BU107	162	7301664	164	DTS701	165	009-00	166	RVDD1245
162	BUY20	162	7301665	164	GB-37	165	09-302187	166	RVDD124B
162	BUY21	162	7301666	164	IRTR68	165	09-304056	166	RVD81543B
162	BUY21A	162	7302699	164	PTC130	165	13-37870-1	166	81B0201B
162	BUY22	162	7303304	164	RE31	165	13-37870-2	166	S1RB
162	C1185	162	7313063	164	SJ5525	165	13-43465-2	166	S1RB10
162	C1185A	162	7913605	164	T-Q5083	165	14-601-27	166	S2FB
162	C1185B	162	23114208	164	TNJ72146	165	15-123230	166	S6-3
162	C1185C	162	23114317	164	2SC1004	165	21A112-036	166	SB309A
162	C1185K	162	6621002300	164	2SC1004A	165	21A112-103	166	SB309C
162	C1185L	163A	A1D	164	2SC1045B	165	34-6002-59	166	SD-19
162	C1185M	163A	A1DJ	164	2SC1045C	165	34-6002-63	166	SELEN-44
162	C895	163A	A2417	164	2SC1045D	165	34-6002-64	166	SEN2A1
162	D285	163A	A3H-1	164	2SC1045E	165	46-86389-3	166	SIB020-1B
162	D353	163A	A6A	164	2SC1045R	165	46-86415-3	166	SIB50B794-1
162	D45	163A	A6M	164	2SC1101	165	46-86477-3	166	SIRB-10
162	D46	163A	A6Z	164	2SC1101A	165	46-86486-3	166	SIRB10
162	D59	163A	A6ZH	164	2SC1101B	165	48-137539	166	SR1B
162	D60	163A	A8T	164	2SC1101C	165	48-155189	166	SVD12B2B1P-M
162	DTS0710	163A	A9R	164	2SC1101D	165	48-155224	166	SVD81RB10
162	DTS3705	163A	B176000	164	2SC1101E	165	48-90343A52	166	T-E1042
162	DTS3705A	163A	B176004	164	2SC1101F	165	488137341	166	T10195
162	DTS3705B	163A	B176005	164	2SC1101L	165	488137539	166	T00.09M22/1
162	DS401	163A	B176006	164	2SC1151	165	488155006	166	TI-365A
162	E13-008-00	163A	B176007	164	2SC1151A	165	488155076	166	TV24285
162	G4N1	163A	B176009	164	2SC1171	165	53B012-1	166	TVM-PT6D22/1
162	GB-35	163A	B176010	164	2SC1184	165	57A186-11	166	TVM29
162	GB-72	163A	B176011	164	2SC1367	165	57A186-12	166	TVSW04
162	HEP-85020	163A	B176013	164	2SC1367A	165	57A198-11	166	TVSW04M
162	IR-TR61	163A	B176014	164	2SD642	165	57A213-11	166	W-005
162	M4900	163A	B176015	164	2SD642A	165	57B198-11	166	WRO11
162	M9408	163A	B176024	164	2SD936	165	57B199-11	166	WRO30
162	MJ3010	163A	B176025	164	2SD936A	165	57B213-11	166	WRO40
162	MJ3026	163A	B176026	164	2SD936BK	165	86-563-2	166	ZA150
162	MJ3027	163A	B176027	164	2SD199	165	86-563-9	166	1B05
162	MJB423	163A	B176028	164	2SD312	165	86-564-3	166	1B05J05
162	NO400	163A	B176029	164	09-302157	165	86-564-9	166	1B08T05
162	PTC118	163A	B1P	164	09-304057	165	86-626-9	166	1B1
162	Q50830	163A	B1P-1	164	21A112-023	165	86-633-2	166	1N1054
162	Q5119	163A	BU109	164	46-86390-3	165	998079-1	166	1RLD
162	Q5119D	163A	C41	164	121-758	165	121-1029	166	182371
162	Q5120P	163A	C41TV	164	121-758X	165	121-759	166	28B-8851
162	Q5120Q	163A	C42	164	121-821	165	121-759X	166	33B-20BO1
162	Q5120R	163A	C42A	164	123N1	165	121-831	166	4-2021-04170
162	RE29	163A	C43	164	260P09501	165	121-985	166	7-0003
162	SDT413	163A	C44	164	260P19108	165	260P08901	166	10B2-B1W
162	ST-2SC1106	163A	ES15X126	164	260P24308	165	260P09701	166	10DB1
162	STA9364	163A	ES84	164	260P33408	165	260P19209	166	11-102-001
162	TNJ72320	163A	GB-73	164	610123-1	165	260P19208	166	11-102003
162	TR1009	163A	IRTR67	164	2320281	165	260P21608	166	11-108002
162	TR68	163A	M4901	165	A6317800	165	260P33108	166	12C2P-114
162	TV121	163A	M4995	165	BU105	165	260P39008	166	13-67539-1
162	2N6211	163A	M7511	165	BU108	165	375-1	166	13-67539-1/3464
162	2N6215	163A	MJ3011	165	BU115	165	370-1	166	16C-4P
162	2S035	163A	PTC129	165	BU110	165	1010-8025	166	16C4B1P
162	2S036	163A	RE30	165	BU205	165	1062-7511	166	24MW192
162	2SC1152	163A	SDT-423	165	BU207	165	1115	166	46-8621-3
162	2SC1152F	163A	T-Q5019	165	C1046	165	1190	166	48-97305A03
162	2SC1152G	163A	TNJ72319	165	C270	165	3647-1	166	53B018-1
162	2SC1185	163A	TV108	165	D56W1	165	3649-1	166	53B011-1
162	2SC1185A	163A	TV118	165	D56W2	165	137607	166	86-68-3
162	2SC1185B	163A	TVS-2SC901	165	DTS702	165	139295	166	86-68-3A
162	2SC1185C	163A	2SC1027	165	DTS704	165	140976	166	130-50261
162	2SC1185L	163A	2SC41	165	DTS802	165	140977	166	325-0042-311
162	2SC1185M	163A	2SC41TV	165	DTS804	165	142689	166	429-0958-41
162	2SD895	163A	2SC42	165	E13-009-00	165	145671	166	642-216
162	2SD995	163A	2SC42A	165	EP15X126	165	610194-1	166	690V081H40
162	2SD283	163A	2SC43	165	EP15X28	165	610194-3	166	977-218
162	2SD285	163A	2SC44	165	EP15X45	165	610216-2	166	977-24B
162	2SD320	163A	2SC901	165	ES15X94	165	610216-3	166	977-4B
162	2SD353	163A	2SC901A	165	GB-237	165	610233-1	166	977-6B
162	2SD45	163A	09-302158	165	GB-238	165	610242-1	166	992-531-01
162	2SD46	163A	09-302176	165	GB-36	165	1417731-1	166	04170
162	2SD59	163A	09-302177	165	IRTR93	165	1417735-1	166	4822-130-30414
162	2SD60	163A	29V069002	165	MJ105	165	1417366-1	166	72174-1
162	09-302159	163A	34-6002-58	165	MJ8400	165	1417370-1	166	0112945
162	13-331182-1	163A	48-134901	165	NTC-12	165	1473647-1	166	0112945(ELGIN)
162	13-33182-1	163A	48-134995	165	Q5140ZP	165	1473649-1	166	125787
162	13-33182-2	163A	48-137134	165	Q5140ZI	165	1473669-1	166	126413
162	14-601-14	163A	48-137179	165	Q5140ZXP	165	1473669-2	166	489752-038
162	21A112-051	163A	48-137203	165	Q5140ZXQ	165	2320291	166	530088-1002
162	21A112-107	163A	48-869273C	165	Q5140ZXR	165	2320299	166	530120-1
162	34-6002-61	163A	488137524	165	Q5142	165	23114321	166	530152-1
162	46-86455-3	163A	63N1	165	RE52	165	23114343	166	0551029
162	46-86455-3	163A	86-221-2	165	SJ5526	165	62737441	166	0551029H
162	46-86462-3	163A	86-222-2	165	ST-BU208	165	62741449	166	771908
162	48-134900	163A	86-224-2	165	T-Q5084	165	2003189109	166	772714
162	48-137326	163A	86-225-2	165	TQ28C1046N-A	165	2003189304	166	984690
162	48-86475-3	163A	121-452	165	TQ28C1295	166	A04210-A	166	2330011
162	48-86279C	163A	121-996	165	TQ28C12950	166	A04231-A	166	2330011H
162	48-869357	163A	260P24408	165	TNJ72149	166	A04284-A	166	2330361
162	48-86940R	163A	417-239	165	TH67	166	A04716	166	7571329-01
162	488137548	163A	417-248	165	TV-119	166	B50C1000	166	7851329-01
162	488155105	163A	690V080H45	165	TV-125	166	BD3A-1B4	166	7852577-01
162	60P19009	163A	800-550-10	165	TV124	166	BDOA	166	7853099-01
162	63-12273	163A	1073	165	TVS28C647-0	166	BR-1	167	B30C250KP
162	63-13216	163A	1074	165	TVS28C647-P	166	BY122	167	B50050KP
162	64N1	163A	3634-1	165	TVS28C647B	166	BY164	167	B8-1
162	96-5284-01	163A	80249-910787	165	TVS28C647E	166	D256(RECT.)	167	BY123
162	121-449	163A	137352	165	TVS28C647Q	166	EB1(ELCOM)	167	E03090-002
162	121-829	163A	137718	165	TVS28C647R	166	EB10(ELCOM)	167	EB1(ELCOM)
162	132-025	163A	0573526	165	2SC1005	166	EB4(ELCOM)	167	EB4(ELCOM)
162	184T2C	163A	610063-1	165	2SC1005A	166	EB9(ELCOM)	167	FW200
162	200X3110-607	163A	610017-5	165	2SC1046	166	FW50	167	FWB5001
162	260P19009	163A	1473601	165	2SC1046K	166	HEP175	167	FWB5002
162	260P19909	163A	2320961	165	2SC1046N	166	JB00036	167	HEP176
162	260P21901	163A	3438854	165	2SC1099	166	K1.3922-1A	167	KBP02
162	260P21908	163A	7304149	165	2SC1295	166	K2CDP221B	167	MDA102
162	269P19009	163A	62734604	165	2SC1296	166	KBP005	167	MDA920-4
162	334-377	163A	2003055813	165	2SC1309	166	KCO8C2219	167	MDA942A-3
162	417-101	164	C1045	165	2SC1358	166	KC2AP221B	167	SD-W04
162	417-158	164	C1045C	165	2SC1358A	166	KC2D022/1	167	SEN2A2
162	417-204	164	C1045D	165	2SC1358B	166	KC2D022/1B	167	SH6G253-2
162	574-844	164	C1045E	165	2SC1358K2	166	KC2DP221C	167	WO4
162	637-1	164	C1045R	165	2SC1358K3	166	M604	167	ZEN433
162	800-550-00	164	C1101	165	2SC1358Y	166	MDA100	167	1-531-024
162	1061-6282	164	C1101A	165	2SC1358Q	166	MDA101	167	1B08T20
162	1063-8369	164	C1101B	165	2SC1358R	166	MDA920-1	167	1B2
162	2057B157-9	164	C1101C	165	2SC1413	166	MDA942A-1	167	2N6214
162	2057B175-9	164	C1101D	165	2SC1413A	166	MDA942A-2	167	10DB2A
162	41506	164	C1101E	165	2SC270	166	PD1011	167	18DB2A
162	126900	164	C1151	165	2SD350	166	PH25G221	167	18DB2A-C
162	135352	164	C1171	165	2SD350A	166	PH9-221	167	25P-B1F
162	255997	164	C1184	165	2SD350Q	166	PH9D5221	167	35B611
162	240404	164	C1184A	165	2SD350T	166	PH9D522M	167	35BL611
162	417214	164	C1184B			166	PH9D8221	167	63-11148
						166	PT6D22-1	167	63-12287

ECG	Industry Standard No.	ECG	Industry Standard No.	ECG	Industry Standard No.	ECG	Industry Standard No.	ECG	Industry Standard No.
167	93.20.709	171	57A295-8	172A	86-541-2	175	D226	175	TVS-28C840
167	93.20.714	171	57A312-11	172A	86-549-2	175	D226-0	175	TVS-28C840A
167	325-0081-109	171	57B193-11	172A	86X0089-001	175	D226A	175	TVS-28C226A
167	690V0R0H53	171	57B236-11	172A	113N1AQ	175	D226AP	175	TVS28D226
167	2093A41-112	171	66P074-1	172A	113N2	175	D226B	175	TVS28D226-0
167	4822-130-30261	171	66P074-2	172A	121-752	175	D226BP	175	TVS28D226A
167	126849	171	66P074-4	172A	123-017	175	D226P	175	TVS28D226B
167	199919	171	86-556-2	172A	132-052A	175	D226Q	175	TVS28D226C
167	1471393-4	171	86-628-2	172A	229-1301-27	175	D238	175	TVS28D226D
167	4202104170	171	86-628-9	172A	296	175	D238P	175	TVS28D226P
167	BY17G4	171	86-631-2	172A	417-161	175	D254	175	X7338934
168	EB11(ELCOM)	171	86-664-2	172A	417-222	175	D255	175	XI-549
168	EB2(ELCOM)	171	86-672-2	172A	576-0007-001	175	D29	175	XT-549A
168	EB6(ELCOM)	171	86X0065-001	172A	921-01129	175	D290	175	YAAN28D141
168	E8A-06	171	96-5005-01	172A	921-403	175	D290L	175	1A0013(YAMAHA)
168	ESA06	171	96-5219-01	172A	924-2209	175	D291	175	1A0027
168	FB-200	171	121-1037	172A	964-22009	175	D292	175	1A0027(YAMAHA)
168	FB200	171	121-822	172A	991-01-0461	175	D297	175	1A0048
168	FWB3003	171	121-868	172A	991-01-3543	175	D34014094	175	1A0048(YAMAHA)
168	FWB5004	171	121-868-01	172A	991-015663	175	D57	175	2N3483
168	GM037B	171	121-868-02	172A	2122-3	175	D58	175	2N4231
168	HEP177	171	121-989	172A	3367	175	D70	175	2N4232
168	KBP04	171	121-990	172A	22009	175	D71	175	2N4233
168	MDA920-6	171	132-003	172A	T2054-1	175	D71L	175	2N5598
168	MDA942A-5	171	132-014	172A	95296	175	DT3301	175	2N5600
168	PTC401	171	144N2	172A	130040	175	DT3302	175	2N5602
168	SEN2A4	171	417-245	172A	134442	175	E570022-01	175	2N5604
168	SK3106	171	1063-5142	172A	139618	175	EA15X121	175	2N6260
168	131	171	1063-7916	172A	53013-1	175	ES15X98	175	2N6263
168	7-8(STANDEL)	171	131139	172A	610113-1	175	ES36(ELCOM)	175	2N6264
168	10DB4A	171	133265	172A	610113-2	175	ES44(ELCOM)	175	2N6500
168	10DB6A	171	134772	172A	760012	175	ES45(ELCOM)	175	28C1024
168	10DB6A-C	171	135716	172A	801533	175	ETS-017	175	28C1024-D2
168	18DB4A	171	141295	172A	1471136-3	175	ETTC-28C490	175	28C1024B
168	18DB4A-C	171	142677	172A	1473604-1	175	EW183B	175	28C1024C
168	4822-130-50228	171	146826	172A	4082671-0002	175	FBN-35469	175	28C1024D
169	GEBR-600	171	610144-1	172A	62256885	175	FBN-35903	175	28C1024E
169	KBP06	171	610144-2	173BP	GB-305	175	FBN-36486	175	28C1024F
169	MDA920-7	171	801541	173BP	RE93P	175	FBN-36488	175	28C1025
169	PD1020	171	1417352-5	173BP	B5000	175	FRB-564	175	28C1025CTV
169	PD60	171	1473613-4	173BP	010	175	GB-23	175	28C1025D
169	PH9D522/1	171	1473628-2	173BP	010-6742	175	GE-246	175	28C1025E
169	PTC402	171	1473632-1	173BP	010-6744	175	HEP-85012	175	28C1025J
169	R145B	171	1473633-1	173BP	57-52	175	HEP-85019	175	28C1025MT
169	534	171	23114269	173BP	57-56	175	HEP241	175	28C1160
169	7B-531	171	23114250	173BP	63-19173	175	HEP703	175	28C1160K
169	18DB6A	171	23114251	173BP	120818	175	HEP85019	175	28C1160L
169	530002-1	171	62737395	173BP	135320	175	HR-107(PHILCO)	175	28C1161
169	530021-1	171$	62748451	173BP	135532	175	HR107H	175	28C487
169	48106	171$	28C2278	173BP	1471908-1	175	HT304911	175	28C488
169	71783	171/403	144BL	175	A-120304	175	HT304911A	175	28C489
169	129241	171/403	144N1G	175	A-1854-0449-1	175	HT304911B	175	28C489Y
169	7851328-01	171/403	610144-101	175	A054-224	175	HT401301B0	175	28C490
170	A-0205	172A	A9C	175	A068-114	175	HT403152A	175	28C491
170	AM-0-11	172A	BA1003	175	A14743	175	HT403152B	175	28C491BL
170	PD100	172A	C1280	175	A27(RCA)	175	I51-0141-00	175	28C491R
170	PD80	172A	C1280A	175	A2415	175	I6114	175	28C491Y
170	18DB10A	172A	C1280AS	175	A28D226PQ	175	INTRON-127	175	28C791
170	18DB8A	172A	C1280B	175	A6843401	175	IP20-0212	175	28C830
170	21A102-002	172A	C1472K	175	AD160	175	IR-TR57	175	28C830A
170	522-958	172A	C982	175	AR17A	175	J24123	175	28C830B
171	A278	172A	CB-3024	175	AR17B	175	J24642	175	28C830C
171	A279	172A	D16P2	175	B13823	175	M9225	175	28C840
171	A6519400	172A	D8-60	175	B13	175	M9274	175	28C840A
171	A673515F	172A	EA15X247	175	B2B	175	M9301	175	28C840AC
171	BCW50	172A	EA15X266	175	B5C	175	M9309	175	28C840H
171	BF899	172A	EL401	175	BDY16	175	M9316	175	28C840HP
171	BS819	172A	GE-64	175	BDY71	175	M9393	175	28C840P
171	BS820	172A	GET5305	175	BDY72	175	MJ4101	175	28C840PQ
171	D40N1	172A	GET5306	175	BLY15A	175	MJ4102	175	28D102
171	D40N3	172A	GET5307	175	BLY47A	175	MJ5202	175	28D102-0
171	D40N5	172A	GET5308	175	BLY48A	175	MJ5203	175	28D102-R
171	E13-007-00	172A	GET5308A	175	BLY49A	175	MJ5204	175	28D102-Y
171	EP15X227	172A	HEP-89100	175	BN7168	175	P3172	175	28D103
171	EP15X261	172A	IRTR69	175	C1024	175	P5148	175	28D103-0
171	ES89	172A	M433	175	C1024-D2	175	P6022A	175	28D103-R
171	GE-27	172A	MA10	175	C1024B	175	P6128	175	28D103-Y
171	IRTR79	172A	MP8-A12	175	C1024C	175	P6804	175	28D129
171	LM-1154	172A	M87504	175	C1024D	175	PF-AR18	175	28D129-BL
171	PL1085	172A	PEP95	175	C1024E	175	PN26	175	28D129-R
171	PT2525	172A	PTC153	175	C1024F	175	PTC112	175	28D129-Y
171	PT2524	172A	R832	175	C1025	175	Q-00769	175	28D130
171	PT2525	172A	89100	175	C1025CTV	175	Q-00769A	175	28D130-R
171	R9382	172A	TE2713	175	C1160	175	Q-00769B	175	28D130-Y
171	R9383	172A	TE2714	175	C1160K	175	Q-00769C	175	28D130BL
171	RCP111A	172A	TNO61688	175	C1160L	175	Q-01269	175	28D130Y
171	RCP111B	172A	TV62	175	C1161	175	Q-01269B	175	28D142
171	RCP111C	172A	2N5305	175	C487	175	Q-01269Y	175	28D142M
171	RCP111D	172A	2N5306	175	C488	175	Q5134Z	175	28D143
171	RCP113A	172A	2N5307	175	C489	175	R2096	175	28D144
171	RCP113B	172A	2N5308	175	C490	175	RE34	175	28D145
171	RCP113C	172A	2N5525	175	C491	175	S21520	175	28D150
171	RCP113D	172A	2N997	175	C491BL	175	S2321	175	28D155
171	RCP115	172A	2N998	175	C491R	175	S2486	175	28D155H
171	RCP115A	172A	2N999	175	C491Y	175	S2527	175	28D155K
171	RCP115B	172A	28C1280	175	C791	175	S306A	175	28D155L
171	RCP117	172A	28C1280A	175	C830	175	S354	175	28D226
171	RCP117B	172A	28C1280AS	175	C830A	175	SE9060	175	28D226-0
171	RBT3	172A	28C1280B	175	C830B	175	SE9061	175	28D226A
171	RT1116	172A	28C1472K	175	C830C	175	SE9062	175	28D226AP
171	RT1893	172A	28C982	175	C840	175	SE9063	175	28D226BP
171	SF.T187	172A	8A1002	175	C840A	175	SJ1172	175	28D226P
171	TQ-5063	172A	8A1003	175	C840AC	175	SJ2095	175	28D226Q
171	TR83014	172A	09-309029	175	C840H	175	SJ3408	175	28D226R
171	T28C3903-0-A	172A	11-0423	175	C840PQ	175	SJ3447	175	28D236
171	01-572088	172A	11-0775	175	0893-1007	175	SJ3648	175	28D238
171	2D002-171	172A	13-14085-14	175	D250	175	SJ3680	175	28D238F
171	2N6557	172A	13-29775-1	175	D102	175	SJ811	175	28D254
171	2N6558	172A	13-33175-1	175	D102-0	175	SK3026	175	28D255
171	28C1127	172A	13-33175-2	175	D102-R	175	SM-A-618687-1	175	28D28
171	28C1127E	172A	14-0104-5	175	D102-Y	175	STC-4401	175	28D29
171	28C1127H	172A	14-2000-01	175	D103	175	STC4401	175	28D290
171	28C1127JR	172A	14-2000-02	175	D103-0	175	STX0010	175	28D290L
171	007-00	172A	14-2000-03	175	D103-R	175	STX0016	175	28D291
171	8-729-322-78	172A	14-2000-04	175	D103-Y	175	T-Q5080	175	28D291(R)
171	8-905-706-010	172A	14-2000-05	175	D129	175	T271	175	28D292
171	8-905-706-067	172A	14-2000-23	175	D129-BL	175	TA2402	175	28D297
171	8-905-706-068	172A	14-2053-23	175	D129-R	175	TA2402A	175	28D57
171	09-304046	172A	16P2881	175	D129-Y	175	TG28C1025	175	28D58
171	13-33174-1	172A	16P3367	175	D130	175	TG28C1025D	175	28D556
171	13-33174-2	172A	23-5044	175	D130-R	175	TIP503	175	28D70
171	13-33176-1	172A	34-6017-3	175	D130-Y	175	TM175	175	28D71
171	14-609-02	172A	34-601703	175	D130BL	175	TNJ60451	175	28D71L
171	14-901-12	172A	42-22009	175	D142	175	TNJ60453	175	314-6004
171	34-6015-64	172A	48-137392	175	D142M	175	TNJ72286	175	4-005721-00
171	46-81187-3	172A	052	175	D143	175	TR-180(OLSON)	175	4-265
171	46-86327-3	172A	052A	175	D144	175	TR-188(OLSON)	175	4-686130-3
171	46-86344-3	172A	53-1173	175	D145	175	TR1591	175	4-686226-3
171	46-86346-3	172A	55-641	175	D150	175	TR1593	175	6-000140
171	48-355054	172A	66P059-1	175	D155	175	TR2327852	175	6-000555-2
171	57A193-11	172A	66P059-2	175	D155H	175	TR26-1	175	6-140
171	57A208-8	172A	86-5105-2	175	D155K	175	TV109	175	6-24
171	57A236-11			175	D155L	175	TV24211		
						175	TV24487		

ECG	Industry Standard No.	ECG	Industry Standard No.	ECG	Industry Standard No.	ECG	Industry Standard No.	ECG	Industry Standard No.
175	6-490001	175	0243-001	175	198039-1	176	2N1315	177	AM620A
175	6-5552	175	260P08601	175	198039-3	176	2N2173	177	AM626
175	6-6490001	175	260P16009	175	198039-501	176	2N2406	177	AM626A
175	007-0030	175	260P20101	175	198039-503	176	2N2541	177	AM632
175	7-0030-00	175	260P24803	175	198049-1	176	2N2672	177	AM632A
175	007-0074	175	314-6006	175	230084	176	2N660	177	AV0000105-0
175	7-29(STANDEL)	175	322-1	175	232268	176	2N661	177	B-1599
175	7L6-0105	175	352-0581-011	175	258208-002	176	2N672	177	B-1702
175	8-	175	352-0581-020	175	262116	176	28A208	177	B111
175	8-1(BENDIX)	175	352-0581-021	175	400127	176	28A209	177	B614 007 0
175	09-302218	175	352-0581-030	175	489751-174	176	28A210	177	BA114
175	09-303005	175	352-0581-031	175	505256	176	28A231	177	BA147
175	10-13159-002	175	352-0606-011	175	505257	176	28A232	177	BA167
175	10-3159-002	175	417-104	175	505434	176	28B372	177	BA168
175	10-3159-002	175	417-199	175	505469	176	28B373	177	BA170
175	13-0014	175	422-0960	175	532003	176	28B461	177	BA174
175	14-602-67	175	424-9001	175	0573525	176	28B476	177	BA187
175	141A-180	175	638H	175	619009-1	176	28B492	177	BA216
175	18-177-1	175	638HJ	175	619361-1	176	28B492B	177	BA219
175	19-00-3485	175	686-0210	175	700191	176	4-1546	177	BA244
175	019-003485	175	686-0210-0	175	700195	176	7-73004-02	177	BA316
175	19-020-045	175	750-045	175	705784-1	176	7-73004-03	177	BAW10TP20
175	19A115527	175	753-2000-006	175	723060-29	176	7-73004-04	177	BAW11TP21
175	19A115527-1	175	817-275	175	723423-16	176	7-73004-1	177	BAW12TP22
175	19A115527-P1	175	992-01-3705	175	723423-20	176	09-301015	177	BAW13TP23
175	19A115527-1	175	1040-02	175	723423-7	176	09-301031	177	BAW16
175	22-001008	175	1064-4417	175	723423-9	176	21A040-035	177	BAW17
175	22-001009	175	1065-9944	175	910807-11	176	22B89	177	BAW18
175	23-5031	175	1123-3355	175	932081-1	176	23-5034	177	BAW24
175	31-0010	175	1239 5782	175	982523	176	42-18109	177	BAW24A
175	33-00234-B	175	1507-7183	175	1417322-1	176	46-86198-3	177	BAW24B
175	33-050	175	1534-8931	175	1851516	176	48-134431	177	BAW25
175	34-1026	175	1548-17	175	1851518	176	48-134584	177	BAW25A
175	34-6002-1	175	1714-0402	175	1971296	176	48-134585	177	BAW25B
175	34-6002-28	175	1714-0602	175	2520084	176	48-134856	177	BAW45
175	34-6002-02	175	1714-0605	175	3130057	176	48-137039	177	BAW53
175	34-6005-2	175	1714-0802	175	3130093	176	48-869177	177	BAW59
175	34-6005-3	175	1714-0805	175	3152159	176	73A01	177	BAW59A
175	41-0909	175	1714-1002	175	3463100-1	176	73A02	177	BAW59B
175	41E581144-P001	175	1714-1005	175	3463604-1	176	74A01	177	BAW63A
175	42-00960	175	1854-0265	175	3463604-2	176	74A02	177	BAW63B
175	44A353980-002	175	1936-17	175	3468071-1	176	74A03	177	BAW99
175	44A354657-001	175	2008	175	3468841-1	176	129-4	177	BAX-16
175	44A390243-001	175	2057B155-10	175	4080627-0501	176	505B3105	177	BAX16
175	44B258208-001	175	2057B158-10	175	4080838-0001	176	512E8040P	177	BAX28
175	44B258208-002	175	2110N-134	175	4080838-0002	176	513E8045P	177	BAX30
175	44B258208002	175	2163-17	175	4080838-2	176	941T1	177	BAX74
175	46-86189-3	175	2180-155	175	4080838-3	176	2606-288	177	BAX87
175	46-86226-5	175	2295	175	4080866-0007	176	2703-387	177	BAX87A
175	46-86349-3	175	02375-A	175	4080866-1	176	3456	177	BAX88TP11
175	46-86349-9	175	2842-056	175	4080866-2	176	4467	177	BAX89A
175	46-86350-3	175	3054	175	4080879-0015	176	34966	177	BAY17
175	46-86476-5	175	3152-159	175	5320003	176	35218	177	BAY18
175	46-86496-5	175	3721	175	6490001	176	56340	177	BAY19
175	48-134936	175	3980-002	175	693243-001	176	56673	177	BAY20
175	48-355042	175	4080-838-0001	175	7570022-01	176	56695	177	BAY31
175	48-60022A14	175	4080-838-2	175	7852460-01	176	42321	177	BAY36
175	48-869225	175	4080-838-3	175	8002866	176	72003	177	BAY41
175	48-869274	175	4080-866-2	175	10644417	176	72004	177	BAY41A
175	48-869301	175	4686-130-3	175	10653086	176	72006	177	BAY41B
175	48-869309	175	4686-226-3	175	11233335	176	114504	177	BAY52
175	48-869393	175	4823-0018	175	12363800	176	114504A03	177	BAY68
175	48-90254A11	175	5320-003	175	15077183	176	127590	177	BAY73
175	48-90347A56	175	5380-73(POWER)	175	15348931	176	171162-025	177	B8A01
175	48K869309	175	6381-1	175	23114084	176	171162-108	177	B8A02
175	48K869225	175	7312	175	23114088	176	194086-3	177	B8A11
175	48K869301	175	7322	175	23114220	176	245568-2	177	BYX58-100
175	48B155005	175	8000-00005-008	175	23114221	176	524966	177	BYX58-200
175	48B155044	175	8000-00005-012	175	23114961	176	0573185	177	BYX58-50
175	051-0151	175	8620	175	50220601	176	620782	177	B210/244
175	57A12-3	175	9404-9	175	50220602	176	983012	177	C-4401
175	57A12-5	175	9409-4	175	51126540	176	2227367	177	C-505
175	57A125-10	175	12044-0021	175	258208001	176	5958539	177	C255110-011
175	57A155-10	175	012085	175	314745006	176	258208001	177	C6141990
175	57A2-4T	175	12538	175	2003102529	176	8511724-3	177	C490
175	57A2-58	175	13159-2	175(2)	GB-24MP	176	8511724-4	177	CD-OO-9
175	57A2-84	175	16114	176	A127364	176	20912578-1	177	CD-0021
175	57B123-10	175	30236	176	A208	176	27125540	177	CD-20
175	57B155-10	175	30254	176	A209	176	B0285400	177	CD-37
175	57B158-10	175	30256	176	A210	177	A-201125(DIODE)	177	CD-57A
175	57B2-47	175	35002	176	A231	177	A-1901-0025-1	177	CD-84857
175	57B2-58	175	35405	176	A232	177	A-1901-0033-1	177	CD0000NC
175	57B2-84	175	36274	176	A300043-06	177	A-1901-0053-1	177	CD0014(MORSE)
175	57012-3	175	36320	176	AOT39	177	A-1901-0096-1	177	CD1224
175	57012-5	175	36370-05490	176	A8477	177	A-1901-0150-1	177	CD57
175	57M3-5	175	37077	176	A8YB1	177	A-1901-0156-1	177	CD57A
175	62-16919	175	37484	176	B-5	177	A-1901-1067-1	177	CD5003
175	63-10062	175	37599	176	B372	177	A01	177	CDB457
175	63-11290	175	37763	176	B373	177	A01(MOTOROLA)	177	CDB6003
175	63-11878	175	37900	176	B461	177	A02	177	CDB60037
175	63-11938	175	37913	176	B476	177	A054-228	177	CDG-00
175	63-12954	175	38443	176	FBN-2N1183	177	A066-12	177	CDG-20
175	63-8512	175	39285	176	FBN-CF34759	177	A42X00390-01	177	CDG-20/494
175	63-8945	175	40310	176	GP290	177	A488-A0001	177	CDG-21
175	065	175	40310V1	176	HER-06011	177	A514-042791	177	CDG-22
175	76-0105	175	40312	176	IRTR82	177	A65-P11311-0001	177	CDG-24
175	86-271-2	175	40312V1	176	KD2121	177	A66X0043-001	177	CDG00
175	86-272-2	175	40316	176	M75537-2	177	A691M5	177	CDG025
175	86-412-2	175	40324	176	M9177	177	A691M5-2	177	CDG22
175	86-5085-2	175	40364	176	N4967	177	A7246602	177	CDG24
175	86-5106-2	175	40372	176	NC54	177	A7246711	177	CDG24/3490
175	86A338	175	40373	176	NKT224J	177	A7246727	177	CDG27
175	91AJ150	175	40424	176	NKT225J	177	A73-16-179	177	CDJ-00
175	96-5190-01	175	40664	176	NKT226J	177	ACA50(DIODE)	177	CE57
175	96-5191-01	175	40910	176	NKT247J	177	AD100(DIODE)	177	CG9-500
175	98-24320-2	175	42065	176	NKT618J	177	AD150(DIODE)	177	CX-0055
175	998047	175	42342	176	OC-304	177	AD200(DIODE)	177	D-00184R
175	100-5765-001	175	50280-3	176	P2F	177	AD30	177	D-12
175	119-0068	175	51300	176	R-28B492	177	AB10	177	D1-20
175	121-588	175	55810-166	176	R56	177	AE100	177	D2-77-1
175	125-410	175	60201	176	RS-5735	177	AE150	177	D3356
175	125A8251	175	60219	176	RS5715	177	AE200	177	D34005220-001
175	125B410	175	60632	176	RS5715-1	177	AE50	177	D5V
175	129-23	175	62142	176	RS5788	177	A050	177	D6462
175	0131-04560	175	62660	176	RSKK36	177	AM300	177	D6126
175	132-065	175	62763	176	0S492	177	AM300A	177	D77(DIODE)
175	151-0141	175	70019-1	176	SF.T353	177	AM301	177	D7E
175	151-0141-00	175	70019-5	176	SMB620782-1	177	AM301A	177	D7Z
175	151-0148	175	75700-22-01	176	ST-28A	177	AM302	177	DAAY003002
175	151-0149	175	79922	176	ST61583	177	AM302A	177	DAAY004001
175	151-0149-1A	175	88801-3-1	176	TIA35	177	AM303	177	DB457
175	151-0217	175	94049	176	TNJ61223	177	AM303A	177	DDAY004001
175	151-0217-1	175	94094	176	TRO31	177	AM304	177	DDAY048001
175	151-0I49	175	115527-1	176	TS1541	177	AM304A	177	DDAY048008
175	151-148-00-BC	175	131095-2	176	TS67BB	177	AM305	177	DFFY004002
175	185-756	175	133823	176	002-011800	177	AM305A	177	DFFY007001
175	200-018	175	139270	176	2N1038	177	AM306	177	DHD800
175	207A16	175	150060	176	2N1128	177	AM307	177	DHD8001
175	207A16A	175	1657383	176	2N1131	177	AM307A	177	DHD805
175	211A6381-1	175	167958	176	2N1174W	177	AM308	177	DHD806
175	21A6381-I	175	198058-1	176	2N1183	177	AM308A	177	DI-5
175	220	175	198058-3	176	2N1184	177	AM620	177	DIJ70486
175	231-0008	175	198058-4					177	DIJ70545
175	00234-B							177	DIJ71273

ECG	Industry Standard No.
177	DIJ71711
177	DIJ71960
177	DIJ72164
177	DND800
177	DRH575
177	DS-117
177	DS-410
177	DS-410(AMPEX)
177	DS-442
177	DS-442(SEARS)
177	DS-97
177	DS1-002-0
177	DS104
177	DS51
177	DS410
177	DS410(COURIER)
177	DS410(EMERSON)
177	DS410(PANON)
177	DS410(OLYMPIC)
177	DS410R
177	DS441
177	DS442
177	DS442PM
177	DS443
177	DS97
177	DS97(DELCO)
177	DSA150
177	DX-0270
177	E1121R
177	E13-017-01
177	EA1405
177	EA1661
177	EA16X101
177	EA16X110
177	EA16X134
177	EA16X146
177	EA16X1171
177	EA16X20
177	EA16X39
177	EA16X49
177	EA16X60
177	EA16X661
177	EA16X75
177	EA16X84
177	EA2607
177	EA3447
177	ED21(ELCOM)
177	ED31(ELCOM)
177	ED32(ELCOM)
177	ED514721
177	ED515790
177	ED516420
177	ED536062
177	ED560913
177	ED8-0001
177	ED8-0014
177	ED8-25
177	EH16X20
177	EP16X20
177	EP16X22
177	EP16X23
177	EP16X27
177	EP16X4
177	EP16X10
177	ES15057
177	ES1627
177	ES16X23
177	ES16X24
177	ES16X27
177	ES16X30
177	ES16X32
177	ES16X40
177	ES57X12
177	ETD-899150
177	ETD182788
177	ETDCDG21
177	EU16X11
177	EYV-320D1R2J
177	EY2P-384
177	FA111
177	FCD0003PC
177	FCD0014NCS
177	FD-1029-MC
177	FD1708
177	FD1843
177	FD200
177	FD222
177	FD6451
177	FD6489
177	FDH-9
177	FDH6229
177	FDH694
177	FDH900
177	FG-12377
177	FT1
177	FT1(SHARP)
177	G00-003-A
177	G00-012-A
177	G00-014-A
177	G00-803-A
177	G01-083-A
177	G01-209-B
177	G01-209B
177	G01-217-A
177	G01-803-A
177	G01-803-A
177	G1209
177	G01209B
177	G1803
177	G01803A
177	G1010A
177	GD3638
177	GE-300
177	GBH14
177	GB6063
177	G01209
177	G01211
177	G01803
177	GP2-345
177	GP2-345/MA161
177	GV5760
177	H623
177	H8513
177	H889
177	ED-2000106
177	HD2000206
177	HD2001105
177	HD2001105Q
177	HE6001
177	HE6125
177	HE-10030
177	HE-M8489
177	HF-20014
177	HP-20034
177	HP-20048
177	HP-20060
177	HP-20064
177	HP-20095
177	HP-DS410
177	HP-MV2
177	HP20032
177	HN-00002
177	HN-0047
177	HT-230
177	HV-25
177	HV-25(DIODE)
177	HV-25(RCA)
177	HV-27
177	HV00001050
177	HV25(HITACHI)
177	HV460R
177	I964(6
177	IC743051
177	INJ61433
177	INJ61677
177	INJ61725
177	IP20-0018
177	IP20-0023
177	IP20-0145
177	IP20-0184
177	ITT413
177	ITT7215
177	ITT921
177	J241182
177	J241213
177	J241234
177	J241235
177	J241242
177	J24755
177	J24912
177	JL40A
177	JM-40
177	JM40
177	JM401
177	K0120SA
177	K119J804-5
177	KB102
177	KD300A
177	KGE46109
177	KGE46465
177	KLH4763
177	KX-1
177	LM-1159
177	M1-301
177	M150-1
177	M26
177	M8640E(C-M)
177	MB513AR
177	MC2
177	MC2326
177	MC5321
177	MDP173
177	ME-4
177	MI-301
177	MI301
177	ML2812
177	MPC5500
177	MPN-5401
177	MPS9444
177	MPS9606
177	MPS9606H
177	MPS9606I
177	MPS9644
177	MPS9646
177	MPS964611
177	MPS9646G
177	MPS9646H
177	MPS9646I
177	MPS9646J
177	MPS9646M
177	MSS1000
177	MV3(DIODE)
177	MV4
177	MZ-00
177	MZ2360
177	MZ2361
177	N5406(RCA)
177	NE-446AQ
177	09-306113
177	09-306195
177	0A200
177	0A202
177	0A205
177	0F156
177	0F162
177	P-6006
177	P1172
177	P1172-1
177	P4326
177	P7394
177	PD137
177	PS700
177	PB720
177	PT4-2287-01
177	Q-21115C
177	Q-24115C
177	QD-SMA150XN
177	QD-SS1555XT
177	QD-SS855XXA
177	R-7026
177	R-7027
177	R-7092
177	R8022
177	R8023
177	RD1343
177	RE52
177	REJ71253
177	RP34661
177	RH-DX0033TAZZ
177	RH-DX0046CEZZ
177	RH-DX0054CEZZ
177	RH-DX0083CEZZ
177	RHDX003STAZZ
177	RS14281
177	RT1669
177	RT1689
177	RT2061(G.E.)
177	R2219
177	RT5217
177	RT5554
177	RT5909
177	RT7946
177	RV1017
177	RV1226
177	RV2071
177	RVDKB265J2
177	RVDVD1250M
177	S-3016R
177	S04
177	S04-1
177	S04A-1
177	S04B-1
177	S074-007-0001
177	S1-RECT-55
177	S11
177	S1428
177	S180
177	S1D50B851-A
177	S1R20
177	S2087G
177	S3004-1716
177	S3072C
177	S502
177	S502A
177	S502B
177	S506
177	S506A
177	S506B
177	S509
177	S509A
177	S509B
177	SA-93792
177	SB01-02
177	SC12(DIODE)
177	SC1431(GE)
177	SD-110
177	SD-1AUF
177	SD-34
177	SD-43
177	SD-5
177	SD-630
177	SD100
177	SD110
177	SD12(PHILCO)
177	SD165
177	SD500
177	SD600
177	SD630
177	SD701-02
177	SD974
177	SE5-0966
177	SPD43
177	SPD83
177	SG-9150
177	SG5182
177	SG5193
177	SG3198
177	SG3432
177	SG3516
177	SG5583
177	SG5028
177	SG5392
177	SG5400
177	SG9150
177	SI-RECT-152
177	SI-RECT-55
177	SIB-01-02
177	SID50894
177	SID50B851
177	SMB-541191
177	SP101
177	SR-34
177	STB01-02
177	STB576
177	SV-3
177	SV-3B
177	SV-3C
177	SV-8
177	SVDVD1121
177	SX780
177	T-E1098
177	T-E1118
177	T-E1119
177	T-E1121
177	TF44
177	TF44J
177	TI-51
177	TI-UG-1888
177	TI-UG1888
177	TI51
177	TR2083-42
177	TR2337011
177	TSC136
177	TV24554
177	TVB-181211
177	TVS-0A81
177	TVS-0A95
177	TVS182076
177	TVS181954
177	TVSB01-2
177	TVSBAX-13
177	TVSBAX13
177	TVSGP2-354
177	TVSMA26
177	TVSMA1A
177	TZ1153
177	UG1888
177	US1555
177	V1112
177	V50260-10
177	V50260-36
177	VAR
177	VAR-1R2
177	VD-1123
177	VD1127
177	VHD181553//-1
177	VHD181555-R-1
177	VHD181555//1A
177	VHD182076-1
177	VHEYZ-1001F
177	VHY181209-1
177	V89-0008-911
177	V89-0014-911
177	W1R
177	W1R(DIODE)
177	WA-26
177	WD4
177	WG-1010-A
177	WG-1010A
177	WG-1012
177	WG-10AS
177	WG-599
177	WG-713
177	WGOA-90
177	000WG1010
177	WG1010A
177	WG1010B
177	WG1012
177	WG1014A
177	WG1021
177	WG599
177	WG713
177	WG714
177	WS100
177	WS100A
177	WS100B
177	WS100C
177	WS200
177	WS200A
177	WS200B
177	WS200C
177	WS300
177	WS300A
177	WS300B
177	WS300C
177	WSD002C
177	X1022220-1
177	X72A42416
177	X925940-5018
177	X925940-501B
177	YAAD010
177	YAAD018
177	YDI121
177	Z-175-011
177	ZC183588
177	ZS142
177	ZS40
177	ZS4P
177	1-001/2207
177	001-0095-00
177	001-0125-00
177	1-014/2207
177	001-0151-00
177	001-0151-01
177	001-026010
177	001-026030
177	01-1501
177	1-16549
177	001-226010
177	01-2601
177	01-2601-0
177	01-2603-0
177	1-DI-007
177	1-A16549
177	1A6551
177	1C0017
177	1D098-001V-022
177	1N135
177	1N137A
177	1N137B
177	1N138
177	1N138A
177	1N138B
177	1N630
177	1N638
177	1N839
177	1N1840
177	1N1841
177	1N1843
177	1N1844
177	1N1845
177	1N1846
177	1N1847
177	1N94
177	1N194A
177	1N195
177	1N196
177	1N200
177	1N202
177	1N203
177	1N204
177	1N205
177	1N2075X
177	1N208
177	1N210
177	1N211
177	1N212
177	1N213
177	1N214
177	1N216
177	1N217
177	1N218
177	1N2473
177	1N251
177	1N251A
177	1N252
177	1N252A
177	1N300
177	1N300A
177	1N300B
177	1N301
177	1N301A
177	1N301B
177	1N303
177	1N303A
177	1N303B
177	1N3063
177	1N3064
177	1N3065
177	1N3066
177	1N3067
177	1N3068
177	1N3069
177	1N3069M
177	1N3070
177	1N3071
177	1N3123
177	1N3124
177	1N3147
177	1N3197
177	1N3206
177	1N3207
177	1N3223
177	1N3257
177	1N3258
177	1N331
177	1N3471
177	1N350
177	1N351
177	1N352
177	1N3550
177	1N3575
177	1N3576
177	1N3577
177	1N3593
177	1N3594
177	1N3598
177	1N3598
177	1N3599
177	1N3601
177	1N3602
177	1N3604
177	1N3605
177	1N3606
177	1N3607
177	1N3609
177	1N3609
177	1N3625
177	1N3653
177	1N3654
177	1N3666
177	1N3668
177	1N3722
177	1N380
177	1N381
177	1N382
177	1N383
177	1N386
177	1N3864
177	1N3872
177	1N3873
177	1N389
177	1N390
177	1N391
177	1N392
177	1N393
177	1N394
177	1N3953
177	1N3954
177	1N3956
177	1N4043
177	1N4087
177	1N4147
177	1N414A
177	1N4315
177	1N4318
177	1N432
177	1N4322
177	1N432A
177	1N432B
177	1N433
177	1N433A
177	1N433B
177	1N434
177	1N434A
177	1N434B
177	1N4363
177	1N4375
177	1N4392
177	1N4395
177	1N4395A
177	1N4450
177	1N4453
177	1N4455
177	1N4531
177	1N4532
177	1N4533
177	1N4534
177	1N4536
177	1N4547
177	1N4548
177	1N456
177	1N456A
177	1N457
177	1N457A
177	1N457M
177	1N458
177	1N458M
177	1N459
177	1N459A
177	1N459M
177	1N460
177	1N460A
177	1N460B
177	1N461
177	1N461A
177	1N462
177	1N462A
177	1N463
177	1N463A
177	1N464
177	1N464A
177	1N4726
177	1N4727
177	1N482
177	1N4827
177	1N4828
177	1N4829
177	1N482A
177	1N482B
177	1N482C
177	1N4830
177	1N483A
177	1N483AM
177	1N483B
177	1N483BM
177	1N483C
177	1N484
177	1N484A
177	1N484AB
177	1N484C
177	1N485
177	1N485A
177	1N485B
177	1N485C
177	1N486

ECG	Industry Standard No.	ECG	Industry Standard No.	ECG	Industry Standard No.	ECG	Industry Standard No.	ECG	Industry Standard No.
177	1N4861	177	1N903A	177	182186QR	177	09-307039	177	46-61267-3
177	1N4862	177	1N903AM	177	182276	177	09-307045	177	46-61307-3
177	1N4863	177	1N903M	177	182460	177	09-307055	177	46-80509-3
177	1N486A	177	1N904	177	182461	177	09-307075	177	46-836380-3
177	1N486B	177	1N904A	177	18273T	177	09-307080	177	46-861187-3
177	1N4938	177	1N904AM	177	182788	177	09-307081	177	46-86168-3
177	1N4949	177	1N904M	177	182788B	177	09-307083	177	46-86184
177	1N950	177	1N905	177	18306	177	09-307085	177	46-86184-3
177	1N5062(SEARS)	177	1N905A	177	18307	177	09-307089	177	46-86186-3
177	1N5194	177	1N905AM	177	183076	177	9D11	177	46-86187-3
177	1N5195	177	1N905M	177	18322	177	9D14	177	46-86250-3
177	1N5196	177	1N906	177	1838	177	9DI	177	46-86338-3
177	1N5208	177	1N906A	177	18444	177	9DI3	177	46-86343-3
177	1N5209	177	1N906M	177	18560	177	10-010	177	46-86422-3
177	1N5210	177	1N907	177	18642	177	10-085005	177	46-86428-3
177	1N5219	177	1N907A	177	1884	177	10-10112	177	46-86431-3
177	1N5220	177	1N907AM	177	1889	177	11-0430	177	46-86436-3
177	1N5315	177	1N907M	177	18920	177	11-0781	177	46-86481-3
177	1N5316	177	1N908	177	18921	177	11-085012	177	46-8676-3
177	1N5317	177	1N908A	177	18922(DIODE)	177	012-0121-001	177	047(DIODE)
177	1N5318	177	1N908AM	177	18923	177	12-087004	177	472102536-P1
177	1N5319	177	1N908M	177	18941	177	012-1020-005	177	48-03005A01
177	1N5320	177	1N914	177	18942	177	012-1024-001	177	48-03005A05
177	1N5413	177	1N914B	177	18951	177	12-200009	177	48-03005A07
177	1N5414	177	1N914M	177	18952	177	12A1027-P5	177	48-05-011
177	1N5426	177	1N915	177	18953	177	12P2	177	48-10577A01
177	1N5605	177	1N916A	177	18955	177	13-085015	177	48-14781
177	1N5606	177	1N916B	177	18983	177	13-085022	177	48-14816
177	1N5607	177	1N917	177	0000018990	177	13-085023	177	48-137385
177	1N5610	177	1N919	177	18990-AM	177	13-085026	177	48-137514
177	1N568	177	1N920	177	18990A	177	13-10321-54	177	48-137573
177	1N5711	177	1N921	177	189908(DIODE)	177	13-10321-55	177	48-155159
177	1N5712	177	1N922	177	18994A	177	13-14094-33	177	48-34816
177	1N5713	177	1N924	177	18994A	177	13-14097-7	177	48-355035
177	1N5719	177	1N925	177	1802367	177	13-17204-1	177	48-355036
177	1N5767	177	1N926	177	18853	177	13-17569-1	177	48-40738P01
177	1N619	177	1N927	177	18881	177	13-17569-2	177	48-42899J01
177	1N622	177	1N928	177	18243	177	13-22017-0	177	48-46669H01
177	1N625	177	1N930	177	18240	177	13-22606-0	177	48-67120A03
177	1N625A	177	1N931	177	1X9179	177	13-22609-0	177	48-67120A11
177	1N625M	177	1N932	177	1X9805	177	13-23917-1	177	48-67120A13
177	1N626	177	1N933	177	1X9809	177	13-29687-2	177	48-67120A15
177	1N626A	177	1N934	177	2-1860	177	13-29867-1	177	48-6712A11
177	1N626M	177	1N948	177	2D165	177	13-29867-2	177	48-6720A02
177	1N627	177	1N993	177	2P2150M	177	13-34056-1	177	48-82392B01
177	1N627A	177	1N995	177	2810032	177	13-43250-1	177	48-82392B02
177	1N628	177	1N997	177	03-0063-04	177	13-67590-1	177	48-82392B03
177	1N628A	177	1N999	177	03-931641	177	13-67590-1/3464	177	48-82392B04
177	1N629	177	1NJ61433	177	03-931642	177	13P2	177	48-82392B05
177	1N629A	177	1NJ61677	177	03-931645	177	14-507-01	177	48-82392B06
177	1N633	177	1NJ61725	177	3I4-2001-3	177	14-514-02	177	48-82392B07
177	1N643	177	1NJ70980	177	3I4-2001-4	177	14-514-17	177	48-82392B08
177	1N643A	177	1NJ71224	177	3I4-3001-1	177	14-514-19	177	48-82392B09
177	1N658	177	1R0	177	3I4-3001-7	177	14-514-70	177	48-82392B10
177	1N658A	177	1R10D3K	177	3I4-3002-10	177	0014-911	177	48-82392B11
177	1N658M	177	1R2	177	3I4-3002-25	177	15-085005	177	48-82392B12
177	1N659	177	1R3A	177	3I4-3002-31	177	15-085008	177	48-82392B13
177	1N659A	177	1R4	177	3I4-3002-32	177	15-108008	177	48-82392B14
177	1N660	177	1R0E	177	3I4-3002-7	177	15-108025	177	48-82392B15
177	1N660A	177	18-1555V	177	4-1724	177	15-108042	177	48-8240009
177	1N661	177	18-180	177	4-1726	177	15-123101	177	48-8240C13
177	1N661A	177	18-2144Z	177	4-2020-05800	177	15P2	177	48-8240C15
177	1N662	177	1S1001	177	4-2020-06100	177	16028	177	48-82420001
177	1N662A	177	1S1052	177	4-2020-06200	177	16028A	177	48-82420002
177	1N663	177	1S1053	177	4-2020-06400	177	16X39	177	48-82420005
177	1N663A	177	1S1124	177	4-2020-10100	177	17P2	177	48-86289-3
177	1N663M	177	1S1155	177	4-2020-15600	177	019-002964	177	48-97048A01
177	1N690	177	1S11941	177	4-2020-16200	177	19-080-001	177	48-97048A17
177	1N696	177	1S12	177	4-2021-07470	177	19-080-008	177	48-V34816
177	1N697	177	1S1213	177	4-2021-07670	177	19A115371-1	177	48D67
177	1N760	177	1S1214	177	4-202104970	177	19A116081	177	48D67120A11
177	1N778	177	1S1215	177	4-3034	177	19B200249-P1	177	48D67120A13
177	1N779	177	1S1216	177	05-02160-01	177	19B200249-P2	177	48D82420001
177	1N788	177	1S1217	177	05-110442	177	19P1	177	48D82420002
177	1N789	177	1S1218	177	05-180053	177	20-161002	177	48D82420005
177	1N789M	177	1S1218GR	177	05-181555	177	20A11	177	48810577A01
177	1N790	177	1S1219	177	05-330150	177	20A13	177	48810577A02
177	1N790M	177	1S1220	177	05-931642	177	21-606-0001	177	48810577A11
177	1N791	177	1S130	177	05-931645	177	21-606-0001-00H	177	48I34816
177	1N791M	177	1S1302	177	05-936470	177	21-609-3595-009	177	488137133
177	1N792	177	1S1303	177	6X97174A01	177	21A008-008	177	488155047
177	1N792M	177	1S1305	177	6X97174XA08	177	21A009	177	488155077
177	1N793	177	1S131	177	7-0013	177	21A009-008	177	49B1
177	1N793M	177	1S132	177	08-08117	177	21A009-009	177	051-0003
177	1N794	177	1S1420H	177	08-08119	177	21A103-065	177	051-0020
177	1N795	177	1S1473	177	8-719-815-55	177	21A108-001	177	51-13-14
177	1N796	177	1S1514	177	8-905-406-020	177	21A108-004	177	52A4
177	1N797	177	1S1515	177	8-905-421-300	177	21A110-001	177	53A001-32
177	1N798	177	1S1516	177	08A159-007	177	21M330	177	53A001-36
177	1N799	177	1S1532	177	08A165-001	177	21M415	177	53A019-1
177	1N800	177	1S1533	177	09-306060	177	21M562	177	53A020-1
177	1N801	177	1S155	177	09-306062	177	21M562(DIODE)	177	53A030-1
177	1N802	177	1S155-1	177	09-306110	177	022	177	53B001-7
177	1N803	177	1S1554	177	09-306111	177	022-2823-004	177	53B010-6
177	1N804	177	0001S1555	177	09-306113	177	23-Pt274-125	177	53B013-1
177	1N806	177	1S1555-8	177	09-306129	177	23-Pt283-125	177	53B014-1
177	1N808	177	1S1555-Z	177	09-306134	177	23J2	177	53B001-4
177	1N809	177	1S15551	177	09-306145	177	24-28201	177	53B001-8
177	1N810	177	1S1580	177	09-306148	177	24B-001	177	53B005-2
177	1N811	177	1S1585	177	09-306151	177	24B-006	177	53J003-2
177	1N811M	177	1S1586	177	09-306154	177	24J2	177	53B001-15
177	1N812	177	1S1587	177	09-306159	177	24MW1109	177	53B001-5
177	1N812M	177	1S1589	177	09-306161	177	24MW667	177	53B004-12
177	1N813	177	1S1621	177	09-306163	177	24MW744	177	53B001-7
177	1N813M	177	1S1621-0	177	09-306170	177	24MW825	177	53B001-2
177	1N814	177	1S1621-R	177	09-306171	177	24MW858	177	56-24
177	1N814M	177	1S1621-Y	177	09-306178	177	24MW861	177	56-27
177	1N815	177	1S1650	177	09-306195	177	24MW994	177	56-28
177	1N815M	177	1S1651	177	09-306198	177	24MW956	177	56-4852
177	1N817	177	1S17	177	09-306199	177	25J2	177	56-5
177	1N818	177	1S18	177	09-306202	177	26J2	177	56-73
177	1N818M	177	1S180	177	09-306206	177	27J2	177	56-93-2867
177	1N837	177	1S181	177	09-306211	177	31-011	177	59B001-1
177	1N837A	177	1S182	177	09-306219	177	31-093	177	62-18782
177	1N838	177	1S1825	177	09-306220	177	32-0022	177	62-20223
177	1N839	177	1S1993	177	09-306221	177	32-0057	177	62-20319
177	1N840	177	1S1994	177	09-306223	177	32-0063	177	62-20437
177	1N840M	177	1S1995	177	09-306231	177	34-0037-102	177	62-20597
177	1N841	177	1S2074H	177	09-306233	177	34-8057-40	177	62-21952
177	1N842	177	1S20751C	177	09-306236	177	34-8057-5	177	62-26597
177	1N843	177	1S2076	177	09-306248	177	34-8057-6	177	065-012
177	1N844	177	1S2076-TFI	177	09-306266	177	34P4	177	65-P11308-0001
177	1N845	177	1S2091	177	09-306276	177	35-1014	177	65-P11311-0001
177	1N890	177	1S2091(DIODE)	177	09-306285	177	35P4	177	66P016-1
177	1N891	177	1S2091-BK	177	09-306288	177	36D-32	177	66P016-2
177	1N892	177	1S2091-BL	177	09-306291	177	39A69-2	177	66P018
177	1N897	177	1S2091-W	177	09-306309	177	42-22538	177	66X0043-001
177	1N898	177	1S2092	177	09-306313	177	42-24387	177	66X0044-001
177	1N899	177	1S2097	177	09-306326	177	42-28201	177	66X0044-100
177	1N900	177	1S2098	177	09-306368	177	43A113534	177	66X0062-001
177	1N901	177	1S2144	177	09-306373	177	43A167229-01	177	68A7252
177	1N902	177	1S2144A	177	09-306390	177	44T-300-96	177	69A49728-P1
177	1N903	177	1S2144Z	177	09-306426	177	046-0134	177	070-022
						177	046-0909	177	070-047

ECG	Industry Standard No.
177	72-18
177	77A01
177	77A02
177	81-46123023-7
177	81-46123038-5
177	84A20
177	86-0515
177	86-0522
177	86-100013
177	86-100014
177	86-27-1
177	86-41-1(SEARS)
177	86-5010-3
177	86-5011-3
177	86-5015-3
177	86-5027-3
177	86-5037-3
177	86-62-1
177	86-65-4
177	86-70-1
177	86-74-1
177	86-74-9
177	86-76-1
177	86-77-1
177	86-78-1
177	87-190XX-001
177	88-77-1
177	089-241
177	93-13
177	93A27-8
177	93A31-1
177	93A3912
177	93A60-5
177	93A60-6
177	93A64-1
177	93A64-2
177	93A64-3
177	93A64-5
177	93A64-7
177	93A69-2
177	93B27-8
177	93B48-1
177	93B48-2
177	93B48-3
177	93B48-4
177	93B64-1
177	93B64-2
177	93C60-5
177	93C60-6
177	93C7-2
177	93C7-3
177	96-5118-01
177	96-5254-01
177	100-00310-09
177	100-011-20
177	100-01110-01
177	100-01120-09
177	100-11
177	100-120
177	100-125
177	100-130
177	100-137
177	100-161
177	100-184
177	100-216
177	100-435
177	102-339
177	103-131
177	103-142
177	103-142-01
177	103-145
177	103-159
177	103-178
177	103-178-01
177	103-222
177	103-240
177	103-42
177	103-51
177	106-009
177	112-500-0-50
177	112-500-0-501
177	118-030
177	120-003147
177	120-004877
177	120A02
177	120A11
177	120A13
177	123-016
177	123-022
177	123-025
177	123B-003
177	130-30189
177	130-30265
177	130-30266
177	130-30274
177	130-30702
177	141-078-0001
177	142-006(DIODE)
177	142-009
177	142-010
177	144-3
177	149-142-G1
177	150-040-9-002
177	150-1
177	151-001-9-001
177	151-002-9-001
177	151-011-9-001
177	151-014-9-1
177	151-021-9-001
177	151-022-9-001
177	151-022-9-002
177	151-024-9-001
177	151-029-9-001
177	151-029-9-002
177	151-030-9-001
177	151-030-9-006
177	151-030-9-009
177	151-032-9-001
177	151-032-9-002
177	151-032-9-004
177	151-034-9-001
177	151-035-0-003
177	151-035-9-003
177	151-040-9-001
177	151-040-9-002
177	151-042-9-001
177	151-045-9-002
177	151-049-9-001
177	151-049-9-002
177	151-066-9-001
177	151-067-9-001
177	151-267-9-001
177	152-0061-00
177	152-0242-00
177	165A4378P2
177	168-107-001
177	170-1
177	185-011
177	185-012
177	186-011
177	186-011(DIODE)
177	186-012
177	186-015
177	200X8010-165
177	204-1
177	209-32
177	229-1301-19
177	229-1301-20
177	230-0014
177	232-0006-02
177	260-10-011
177	260-10-032
177	260-10-047
177	260-10-049
177	260-61-011
177	260-61-047
177	260-61-067
177	264D00101
177	264D00209
177	265D00702
177	265Z00101
177	290-1003
177	296V020B01
177	296V024B02
177	300-0003-002
177	300-0003-003
177	311-0139-001
177	311D883-P01
177	332-4009
177	344-6005-6
177	351-3031
177	353-2575-00
177	353-2655-000
177	353-3083-000
177	353-3273-00
177	353-3339-000
177	360-32
177	394-1592-1
177	394-1602-1
177	400-1596
177	413(TRUETONE)
177	429-20004-0B
177	460-1009
177	479-0663-005
177	0500
177	514-042791
177	540-028
177	596-2
177	596-5
177	600C
177	600X0100-066
177	601-1(DIODE)
177	601C
177	604C
177	606-6021-101
177	610-017-706
177	612C
177	614-118
177	614C
177	616C
177	618C
177	620C
177	622C
177	642-126
177	642-275
177	642-281
177	690V088H20
177	690V102H40
177	690V118H57
177	690V119H15
177	700A858-322
177	754-2000-001
177	754-2000-011
177	754-2009-150
177	754-2509-150
177	754-2720-021
177	754-5750-283
177	800-016-00
177	800-021-00
177	800-036-00
177	800-040-00
177	800-042-00
177	863-254B
177	863-567B
177	863-776B
177	903-00393
177	903-100B
177	903-112B
177	903-116B
177	903-118B
177	903-121B
177	903-177B
177	903-178
177	903-20
177	903-25B
177	903-311
177	903-332
177	903-41B
177	903-42B
177	903-48B
177	903-58B
177	903-72B
177	903-82B
177	903-84B
177	903-95B
177	914-000-2-00
177	914-000-6-00
177	914-001-1-00
177	919-00-4326
177	919-00-4799
177	919-00-7394
177	919-00-9929
177	919-001172
177	919-001172-1
177	919-004326
177	919-01-0873
177	919-01-1172-1
177	919-01-1215
177	919-01-1307
177	919-01-3072
177	919-010873
177	919-011215
177	919-011307
177	919-013059
177	919-013060
177	919-013067
177	919-013081
177	919-013082
177	921-342B
177	931
177	939-12
177	941-026-0001
177	976-0036-921
177	977-1
177	991-00-1172
177	991-00-1172-1
177	1000-130
177	1003-11
177	1006-5977
177	1009-07
177	1009-08
177	1010-143
177	1018-3259
177	1018-9884
177	1033-0911
177	1033-0983
177	1033-0991
177	1040-07
177	1041-66
177	1042-16
177	1042-23
177	1043-0049
177	1043-10
177	1044-8983
177	1045-7802
177	1048-6421
177	1048-9987
177	1062GE
177	1076-1484
177	1077-2341
177	1079-01
177	1079-89
177	1092-16
177	1120-17
177	1410-102
177	1627 1843
177	1699-17
177	1750-103
177	1756-17
177	1818
177	1846-17
177	1872-6
177	1937-17(DIODE)
177	1981-17
177	2000-345
177	2004-67
177	2010-01
177	2010-03
177	2041-05
177	2061A45-38
177	2061A45-93
177	2065-17
177	2066-17
177	2076
177	2093A38-11
177	2093A38-37
177	2093A41-102
177	2093A41-105
177	2093A41-129
177	2093A41-155
177	2093A41-158
177	2093A41-164
177	2093A41-165
177	2093A41-171
177	2093A41-172
177	2093A41-76
177	2093B38-4
177	2101-03
177	2110N-41
177	2209-17
177	2304
177	2328
177	2328-17
177	2400-17
177	2403
177	2405
177	4001(MAGNAVOX)
177	4011(PENNCREST)
177	4014-200-30110
177	004763
177	4801-00154
177	4822-130-40182
177	5001-083
177	5001-107
177	5001-144
177	5001-145
177	5001-156
177	5001-162
177	5001-164
177	5631-MA150
177	6019
177	7112
177	7344
177	7550
177	8000-00003-046
177	8000-00004-061
177	8000-00004-062
177	8000-00004-066
177	8000-00004-067
177	8000-00004-184
177	8000-00006-008
177	8000-00010-109
177	8000-00011-042
177	8000-00028-045
177	8000-00038-008
177	8000-0004-066
177	8000-0004-P061
177	8000-0004-P062
177	8000-0004-P064
177	8000-0004-P066
177	8000-00041-019
177	8010-52
177	8020-202
177	8710-52
177	8840-53
177	8910-52
177	9128-1503-001
177	9330-228-60112
177	9440
177	9644
177	9646
177	9803
177	12011
177	12101
177	014002-1
177	15122
177	18600-52
177	25810-52
177	27840-43
177	28810-63
177	37510-52
177	38510-330
177	38510-54
177	39042
177	43062
177	50212-19
177	50212-28
177	50212-30
177	53016R
177	057293
177	58810-82
177	58840-113
177	59840
177	67055
177	70055
177	71119-1
177	71467-1
177	71667-1
177	000072130
177	72146
177	72147
177	72197
177	085003
177	87756
177	88060-52
177	88060-55
177	88510-52
177	88510-55
177	95003
177	95012
177	95015(EICO)
177	95332-1
177	99203-3
177	99203-5(DIODE)
177	100828-003
177	101154
177	102009
177	113397
177	116021(SEARS)
177	118527-04
177	119507
177	119596
177	119597
177	119697
177	119956
177	123805(DIODE)
177	123813
177	125397
177	125528
177	125588
177	126321
177	127474
177	127993
177	128474
177	129375
177	130045
177	131501(DIODE)
177	131502
177	132645
177	133390
177	135571
177	135872
177	136162
177	136163
177	136688
177	137028
177	137028-001
177	138196
177	139065
177	139328
177	139706
177	143837
177	161021
177	168706
177	169114
177	169117
177	169558
177	0170301
177	170370
177	170857
177	171149-016
177	171162-196
177	171162-197
177	171162-271
177	171840
177	171841
177	171843
177	172165
177	172201
177	172253
177	172547
177	181073
177	181681
177	183033
177	185411
177	193717
177	198809-1
177	199596
177	202315(THOMAS)
177	206617
177	240564
177	245192
177	248817
177	249508-3
177	267611
177	269922-001
177	281101-97
177	304281-P1
177	476690-2
177	501343
177	504720
177	0517132
177	0517133
177	0518926
177	0526232
177	529657
177	530063-13
177	530072-1002
177	530072-1008
177	530072-1009
177	530072-1010
177	530072-1011
177	530072-1015
177	530072-18
177	530082-3
177	530092-1002
177	530116-1001
177	530116-3
177	530144-1
177	530144-1002
177	530144-1003
177	530150-1
177	530151-1
177	530170-1
177	530179-1
177	530179-1001
177	530179-2
177	530179-3
177	530181-1
177	530181-1001
177	530181-1003
177	534001H
177	604407
177	615004-8
177	618150-3
177	650845
177	650854
177	701662-00
177	710838-1
177	740828
177	741100
177	741741
177	741864
177	801712
177	801728
177	817172
177	817190
177	860011
177	900546-20
177	908703-1
177	908721-1
177	922433
177	922873
177	925521-1B
177	925521-1C
177	925915-101
177	925939-101
177	925940-1B
177	925940-501B
177	942677-9
177	984162
177	984880
177	985106
177	986934
177	988995
177	988996
177	992150
177	1223930
177	1470872-6
177	1471072-4
177	1471872-1
177	1471872-12
177	1471872-13
177	1471872-14
177	1471872-16
177	1471872-18
177	1471872-2
177	1471872-4
177	1471872-7
177	1471872-8
177	1472460-2
177	1474778-16
177	1474872-1
177	1476171-33
177	1476690-2
177	1588035-42
177	1641141-101
177	1641141-102
177	1780169-1
177	1813913-1
177	1846794-1
177	1950060
177	2002402
177	2003069-2
177	2003069-6
177	2006623-48
177	2006627-54
177	2008292-88
177	2010967-83
177	2337011
177	2485080
177	2487305
177	2610032
177	2621499-1
177	2635012
177	2760438-1
177	3110883-P01
177	3195638
177	3596061
177	3596559
177	3755864
177	5001266-3
177	5330133
177	5330212H
177	5330261
177	5330261H
177	5330334
177	5340001
177	5340001H
177	5340022H
177	5340051
177	5340111
177	7012126
177	7141123-1
177	7570215-21
177	7855283-01
177	7914014-01
177	7932638
177	8130014
177	8131010
177	8505214-1
177	10041265-101
177	10180316
177	11176377
177	11290111
177	11803000
177	13030189
177	13030266
177	13030274
177	17200000
177	17200240
177	22692012
177	23115049
177	23115050
177	23115064
177	23115068
177	23115071
177	23115072
177	23115080
177	23115108
177	23115109
177	23115120(DIODE)
177	23115194
177	23115249
177	23115285
177	23115331
177	23115897
177	23115978
177	26006820
177	26006910

ECG	Industry Standard No.	ECG	Industry Standard No.	ECG	Industry Standard No.	ECG	Industry Standard No.	ECG	Industry Standard No.
177	26008020	178MP	12/1N10	179	QP-1A	179	2N1556A	179	992-008870
177	26010028	178MP	13-15465-1	179	QP-2	179	2N1557	179	1413-168
177	26010030	178MP	13-55335-1	179	QP-3	179	2N1557A	179	1413-172
177	26010032	178MP	14-514-20	179	QP-4	179	2N1558	179	4640
177	26010047	178MP	14-514-65	179	QP-5	179	2N1558A	179	4640P
177	26010057	178MP	15-100003	179	QP-6	179	2N1559	179	4702
177	28100812-001	178MP	15-108001	179	QP-7	179	2N1559A	179	4729
177	28102163-001	178MP	16P2	179	QP1	179	2N1560A	179	43074
177	37246602	178MP	21M586	179	QP1A	179	2N1651	179	94010
177	37246711	178MP	30-8057-13	179	QP2	179	2N1652	179	121244
177	37275350	178MP	34-8047-1	179	QP6	179	2N1666	179	127828
177	41622859	178MP	34-8057-34	179	QP7	179	2N1751	179	405965-30A
177	62468211	178MP	34-8057-8	179	RE36	179	2N1906	179	612020
177	62522720	178MP	46-86178-3	179	SP1029	179	2N1907	179	801525
177	62564539	178MP	46-86254-3	179	SP1650	179	2N1907A	179	801538
177	62697571	178MP	46-86380-3	179	SP1817	179	2N1908	179	1471124-5
177	62741406	178MP	48-134916	179	SP2077	179	2N1908A	179	1944748
177	62741422	178MP	48-134917	179	SP2708	179	2N2069	179	1960652
177	62761350	178MP	48-137167	179	SP838	179	2N2070	179	1962326
177	62761377	178MP	48-355014	179	SP838-1	179	2N2285	179	7279069
177	70270250	178MP	48-6712A02	179	T370	179	2N2286	179	7287107
177	87227880	178MP	48-741255	179	TI-3030	179	2N2288	179	7287112
177	87227881	178MP	48-741656	179	TI-3031	179	2N2289	179	7287117
177	87352650	178MP	48-741724	179	TG5050	179	2N2290	179	7289097
177	92115000-23	178MP	488134917	179	TR1036	179	2N2291	179	7292689
177	110654492-001	178MP	488137167	179	TR1038	179	2N2292	179	7297347
177	600000150	178MP	488155060	179	TR35	179	2N2423	179	7299780
177	2008010028	178MP	53A018-1	179	001-012050	179	2N2446	179	7301661
177	2008010094	178MP	53A022-3	179	2G226	179	2N2526	179$	B485
177	2008010110	178MP	53008-1	179	2G227	179	2N2612	179$	2SB485
177	2008010127	178MP	66P017	179	2G228	179	2N2636	179(2)	121-418
177	2008010131	178MP	73A64-1	179	2N1011	179	2N2637	18	131005807
177	2008010159	178MP	86-41-1	179	2N1014	179	2N2638	180	BD250
177	2012010165	178MP	86-574-2(DIODE)	179	2N1021	179	2N2668	180	BD250A
177	4120900010	178MP	92A64-1	179	2N1021A	179	2N2668A	180	BD250B
177	4129100000	178MP	93A64-2(APC)	179	2N1022	179	2N2691	180	BD250C
177	4202006100	178MP	103-101	179	2N1022A	179	2N2691A	180	EA151124
177	4202006200	178MP	103-142(DET)	179	2N1029	179	2N2832	180	E874(ELCOM)
177	4202010100	178MP	103-192	179	2N1029A	179	2N2869	180	GE-74
177	4202015600	178MP	120-004629	179	2N1029B	179	2N2869/2N301	180	HEP-87001
177	4202016200	178MP	146D1	179	2N1029C	179	2N3125	180	HF23
177	4202018700	178MP	146D1B112	179	2N1030	179	2N3132	180	HF25
177	4202021900	178MP	150-030-9-002	179	2N1030A	179	2N3611	180	M9344
177	4202107470	178MP	151-013-9-001	179	2N1030B	179	2N3612	180	M3559
177	4202109170	178MP	151-030-9-002	179	2N1030C	179	2N3613	180	MJ450
177	6611001200	178MP	205-142-G1	179	2N1031	179	2N3614	180	MJ4502
177	6612006000	178MP	601X0225-066	179	2N1031A	179	2N3615	180	P2J
177	8001100005	178MP	1011-0302	179	2N1031B	179	2N3616	180	R874
177	13488240509	178MP	1042-14	179	2N1031C	179	2N3617	180	S35486
177	13488240313	178MP	2102-029	179	2N1032	179	2N3618	180	SDT3760
177	13488240315	178MP	2180-41	179	2N1032A	179	2N379	180	SDT3764
177	16405990022	178MP	2231-17	179	2N1032B	179	2N418	180	SDT3765
177	62170355330	178MP	2289-13	179	2N1032C	179	2N459	180	SDT3766
177	134800843022	178MP	3505(AIRLINE)	179	2N1046	179	2N459A	180	SDT3826
177	134800854093	178MP	5205N1	179	2N1073	179	2N514	180	SDT3827
177	134882392202	178MP	72152	179	2N1073A	179	2N514A	180	SDT3875
177	134882392203	178MP	72163-1	179	2N1120	179	2N514B	180	SDT3876
177	134882392206	178MP	130046	179	2N1136B	179	2N561	180	SDT3877
177	134882392207	178MP	139605(APC)	179	2N1137A	179	2N627	180	S1272
177	134882392208	178MP	170733-1	179	2N1137B	179	2N628	180	SJ2023
177	134882392209	178MP	503146-1	179	2N1138	179	2N629	180	SJ2024
177	134882392210	178MP	530092-1001	179	2N1138A	179	2N630	180	SJ3507
177	134882392211	178MP	530093-1	179	2N1146	179	2N677	180	SJ3636
177	134882392212	178MP	530146-1	179	2N1146A	179	2N677A	180	SJ3637
177	134882392213	178MP	530146-2	179	2N1146B	179	2N677B	180	SJE264
177	134882392214	178MP	740629	179	2N1146C	179	2N677C	180	SJE764
177	134882392215	178MP	983744	179	2N1147	179	2N678	180	ST29045
177	134882393301	178MP	1421207-1	179	2N1147A	179	2N678A	180	ST29046
177	134882393305	178MP	1471872-10	179	2N1147B	179	2N678B	180	ST29047
177	134882420301	178MP	1471876-6	179	2N1147C	179	2N678C	180	TM180
177	134882420302	179	23115065	179	2N1159	179	2SB203AA	180	TR02060-7
177	134882420305	179	23115074	179	2N1160	179	2SB249	180	2N4398
177	4041200030110	179	AD103	179	2N1162	179	2SB249A	180	2N5621
177	2025953101010	179	AD104	179	2N1162A	179	2SB339	180	2N5625
177(2)	14-514-15	179	AD105	179	2N1163	179	2SB339H	180	2N5741
177(2)	24MW785	179	AD133	179	2N1163A	179	2SB340	180	2N5742
177(2)	34-8057-13	179	AD142	179	2N1164	179	2SB340H	180	2N5743
177(2)	48-67120A02	179	AL113	179	2N1164A	179	2SB483	180	2N5885
177(3)	KC020	179	AT1138	179	2N1165A	179	2SB484	180	2N5884
177(3)	11/1	179	AT1138A	179	2N1166	179	004-001	180	2N6329
178MP	A11	179	AT1138B	179	2N1166A	179	8P1555	180	13-40347-1
178MP	ALC-1A	179	AT1833	179	2N1182	179	13-18198-1	180	34-6002-36
178MP	B-46-110	179	AT1834	179	2N1295	179	13-18642-1	180	34-6002-58
178MP	C-10-20A	179	AUY29	179	2N1297	179	13-18642-2A	180	48-137049
178MP	C10-20A	179	AUY37	179	2N1360	179	13-18642-2D	180	51N3M
178MP	C10-22C	179	B113000	179	2N1362	179	13-18642-2E	180	152P1C
178MP	C1A	179	B203AA	179	2N1363	179	13-18642-2F	180	4419-4
178MP	CDB60011	179	B204	179	2N1419	179	13-18642-3	180	15428I
178MP	D30	179	B205	179	2N1529	179	13-18642-3A	180	171053-1
178MP	D1J71895-1	179	B206	179	2N1529A	179	13-18642-3B	180	559557
178MP	E1176R	179	B249	179	2N1530	179	14-601-02	180	1471139-1
178MP	EP16X1	179	B249A	179	2N1531	179	23-5009	181	A-1854-045B
178MP	IS-446D	179	B339	179	2N1531A	179	30-004-001	181	A08-1050115
178MP	M8569	179	B339H	179	2N1532	179	34-6002-33	181	A391591
178MP	PTC406	179	B340	179	2N1534	179	48-134640	181	A417014
178MP	PTC407	179	B340H	179	2N1534A	179	48-134692	181	A4I7014
178MP	RBR-025	179	B483	179	2N1536	179	48-134695	181	A580-040215
178MP	RP5246-7	179	B484	179	2N1537	179	48-134737	181	A580-040315
178MP	SIB02-CR1	179	DTG110A	179	2N1537A	179	48-134740	181	A580-040515
178MP	SIB02-CR1	179	DTG110B	179	2N1539	179	48-134741	181	A580-080215
178MP	SR-13	179	DTG1200	179	2N1539A	179	48-134742	181	A580-080315
178MP	SR-20	179	DTG2000	179	2N1540	179	48-134743	181	A580-080515
178MP	SR0004	179	DTG2000A	179	2N1540A	179	48-134749	181	A5V
178MP	TR-072I(DIODE)	179	DTG2100	179	2N1541	179	48-134752	181	A9N
178MP	TVM-530	179	DTG2100A	179	2N1541A	179	48-134753	181	AMP201B
178MP	TVM-K-112C	179	DTG2200	179	2N1542	179	48-134760	181	AMP201C
178MP	TVM554A	179	DTG400M	179	2N1542A	179	48-134761	181	AMP210B
178MP	Y8G-V81-2-3	179	DTG600	179	2N1543	179	48-134788	181	B-12822-4
178MP	EBN432	179	DTG601	179	2N1543A	179	48-137121	181	B0301-049
178MP	1N1842	179	DTG602	179	2N1544	179	48-137125	181	B170003-BLK
178MP	1N201	179	DTG603	179	2N1544A	179	48-869205	181	B170003-BRN
178MP	1N206	179	DTG603M	179	2N1545	179	48-869427	181	B170003-RED
178MP	1N207	179	GE-239	179	2N1545A	179	57A22-1	181	B170004-BLK
178MP	1N209	179	GE-76	179	2N1546	179	57A22-2	181	B170004-BRN
178MP	1N330	179	HEP-G6001	179	2N1546A	179	57022-1	181	B170004-RED
178MP	1N353	179	HEP-G6014	179	2N1547	179	57022-2	181	B170005-BRN
178MP	1N379	179	HEP-G6015	179	2N1547A	179	62-18429	181	B170005-RED
178MP	1N384	179	HEP-G6018	179	2N1548	179	86-370-2	181	B170006-BLK
178MP	1N385	179	HP-19D	179	2N1548A	179	86X2	181	B170006-BRN
178MP	1N387	179	HR-19E	179	2N1549	179	998057	181	B170006-BRN
178MP	1N588	179	M4701	179	2N1549A	179	112-7292955	181	B170007-BLK
178MP	1N4092	179	M4702	179	2N1550	179	121-406	181	B170007-BRN
178MP	1N4093	179	M9181	179	2N1550A	179	121-419	181	B170008-BRN
178MP	1N4389	179	MP110B	179	2N1551	179	231-0006-03	181	B170008-BLK
178MP	1N4951	179	MP2300A	179	2N1551A	179	324-0115	181	B170008-RED
178MP	1N4952	179	MP600	179	2N1552	179	417-113	181	B170017
178MP	1N5179	179	MP601	179	2N1552A	179	417-120	181	B170025
178MP	1N914A	179	MP602	179	2N1553	179	417-142	181	B170026
178MP	1N929	179	MP603	179	2N1553A	179	417-42	181	B177000
178MP	1S1579	179	P23	179	2N1554	179	992-00-8890	181	BD249
178MP	1S1701	179	QP-1	179	2N1554A	179	992-00-8890L		
178MP	4-2021-05000	179	QP-10	179	2N1555				
178MP	09-306168			179	2N1555A				
				179	2N1556				

ECG	Industry Standard No.	ECG	Industry Standard No.	ECG	Industry Standard No.	ECG	Industry Standard No.	ECG	Industry Standard No.
181	BD249A	181	44A-417014-001	181	170891-1	184	A5A-1	184	SJE583
181	BD249B	181	44A391593-001	181	198064-1	184	A5A-2	184	SJE634
181	BD249C	181	44A395909-1	181	198079-1	184	A5A-3	184	SJE669
181	BDX13	181	44A4417714	181	211040-1	184	A5A-4	184	SJE721
181	BDX61	181	44A4417714-001	181	236854	184	A5A-5	184	SJE724
181	BDY76	181	48-12091A	181	239713	184	A5G	184	SJE737
181	BUT43	181	48-137116	181	267878	184	A5Y	184	SJE769
181	BUY46	181	48-217241	181	291509	184	A6C	184	SJE781
181	D113	181	48-232796	181	505568	184	A6C-1	184	SJE784
181	D113-O	181	48-869480	181	983055	184	A6C-2	184	SJE785
181	D113-R	181	48-869628	181	1471132-6	184	A6C-3	184	SK3190
181	D113-Y	181	48-869639	181	1471135-2	184	A6C-4	184	SP8918
181	D114	181	48P217241	181	1851517	184	A6W	184	STX0013
181	D114-O	181	48P232796	181	1967784	184	A7Z	184	STX0026
181	D114-R	181	60-211040	181	3130090	184	B-1790	184	T-342
181	D114-Y	181	938E165	181	3130104	184	B1C	184	T-344
181	D232	181	96-5315-01	181	04450023-005	184	B1C-1	184	T344
181	D55	181	998103-1	181	04450037-001	184	B1C-2	184	T611-1(RCA)
181	D55A	181	998103-2	181	4832800	184	B1D-1	184	T612-1(RCA)
181	EA15X123	181	998103-3	181	9000630	184	B1F	184	TIP14
181	ES43(ELCOM)	181	116C3475	181	11041003-1	184	B1U148	184	TM184
181	FBN-36972	181	119-0075	181	11194628	184	B2J	184	TNJ70540
181	FBN-36973	181	151-0275	181	14317184	184	BD131	184	TB01057-3
181	FBN-38022	181	151-0275-00	181	45023190	184	BD135	184	V176
181	FD4500AL	181	152N2C	181	48869715	184	BD137	184	X713
181	G181-725-001	181	332-2912	181	94813000	184	BD153	184	YAANZ8C1096
181	GE-75	181	386-7270-P2	181	AR22(PHILCO)	184	BD154	184	ZEN202
181	HEF-87000	181	417-139-13286	182	B1A	184	BD155	184	ZEN210
181	HP24	181	417-214-13286	182	BD148	184	BD165	184	002-0012200
181	K8D2203	181	417-254	182	BD195	184	BD167	184	2N4921
181	K8D3055	181	430-23190	182	BD197	184	BD169	184	2N4922
181	LM511160	181	445-0034-1	182	BD205	184	BD175	184	2N4923
181	LM75116	181	860-022-01	182	BD207	184	BD177	184	2N5190
181	M9461	181	1092(GE)	182	BD461	184	BD179	184	2N5191
181	M9628	181	1119-4628	182	BD463	184	BD185	184	2N5192
181	M9659	181	1339(GE)	182	E13-012-00	184	BD187	184	2SC1449
181	M9715	181	1414-188	182	ES86(ELCOM)	184	BD189	184	2SC1449M
181	M33771	181	1431-7184	182	HEP-85001	184	BD561	184	2SD612
181	M33772	181	1561-0408	182	HEP-85004	184	C1162	184	2SD612K
181	MJ802	181	1561-0410	182	LM-1157	184	C1162WTD	184	3I4-6005-2
181	NO2820T	181	1561-0608	182	LM1157	184	C1386C	184	3I4-6005-3
181	P-11810-1	181	1561-0615	182	MJE182	184	C1449	184	8-905-706-801
181	P-11901-3	181	1561-0615	182	MJ204	184	C495-0	184	8-905-713-110
181	PO445-0034-1	181	1561-0808	182	MJE205	184	C496Y	184	11-0772
181	PO445-0034-2	181	1561-0810	182	MJE2801	184	D18A12	184	13-0049
181	PO4450034-1	181	1561-0815	182	MJE3055	184	EA15X244	184	13-14085-126
181	PO4450034-2	181	1561-1010	182	RS-2019	184	EA15X269	184	13-23507-0
181	PO44450057	181	1561-1015	182	SJE820	184	EA15X360	184	13-28336-2
181	P650OA	181	1723-1805	182	SJE5018	184	EAI-380	184	13-32638-1
181	PMC-QP0040	181	1723-1810	182	SJE5019	184	ES58(ELCOM)	184	13-32642-1
181	PP3001	181	1763-0415	182	SJE5020	184	G03-007C	184	13-33925-1
181	PP3004	181	1763-0420	182	SJE816	184	G03007C	184	14-601-17
181	PP3007	181	1763-0425	182	TA7311	184	G06-717-B	184	14-608-01
181	PT0116	181	1763-0615	182	TA7312	184	G06-717-C	184	14-608-02
181	R2270-75116	181	1763-0620	182	TA7313	184	G06-717-D	184	14-608-03
181	R837	181	1763-0625	182	TA7314	184	HC-00496	184	14-608-04
181	S55487	181	1763-0815	182	TA7315	184	HEP-85000	184	14-906-13
181	SDP9701	181	1763-0825	182	TA7316	184	HEP-85003	184	14-907-13
181	SDT9704	181	1763-1015	182	TA7318	184	HEP245	184	14-910-13
181	SDT9707	181	1763-1020	182	TR-1037-1	184	HEP701	184	14-911-13
181	SJ2047	181	1763-1025	182	TR-1037-2	184	HT104961C	184	21M466
181	SJ2064	181	1763-1215	182	TRO-1057-1	184	HT304961B	184	24MW1143
181	SJ3519	181	1763-1220	182	TRO-1057-3	184	HT304961C	184	24MW978
181	SK0296	181	1763-1225	182	TRO-1057-4	184	HT307902B	184	30-090
181	SK3511	181	1763-1415	182	314-6012-4	184	HT313681BO	184	33-090
181	SK3535	181	1763-1420	182	09-304055	184	IP20-0230	184	42-28212
181	SPD-80059	181	1763-1420	182	012-00	184	J241250	184	46-86234-3
181	SPD-80060	181	1854-0245	182	15-0097	184	M3567-2	184	48-137037
181	SPD-80061	181	1854-0490-1	182	13-28222-2	184	MB44	184	48-137095
181	ST101	181	2071	182	13-28222-4	184	MB52	184	48-137128
181	STC1094	181	2842-875	182	13-28532-1	184	M9556	184	48-137145
181	STC2220	181	3373	182	13-331188-1	184	M9618	184	48-137146
181	STC2221	181	3771	182	86X0059-002	184	MJE-200	184	48-137147
181	STC2224	181	3772-1	182	656-2	184	MJE-521	184	48-137148
181	STC2225	181	3772-2	182	3656-1	184	MJE180	184	48-137200
181	STC2228	181	5504	182	3656-2	184	MJE200	184	48-137211
181	STC2229	181	5505	182	3656-4	184	MJE482	184	48-137277
181	T841	181	005575	182	4802-3274-200	184	MJE483	184	48-137323
181	T842	181	5909-001	182	43118	184	MJE488	184	48-137473
181	T843	181	7885-1	182	43119	184	MJE520	184	48-137549
181	T844	181	7885-2	182	43120	184	MJE521	184	48-41784J03
181	TK9201	181	7885-3	182	140506	184	MJ89400	184	48-41884J03
181	TM181	181	8005(PENNCREST)	182	1473656-1	184	MOTMJE521	184	48-86161B
181	TR-176(OLSON)	181	10003	182	1473656-2	184	PLB52	184	48-97162A35
181	TR-8018	181	13030-4	182	04450037-002	184	PLZ52	184	48-97162A69
181	TRO1060-7	181	16267	183	AR23(PHILCO)	184	PTC110	184	48R869618
181	TR1009A	181	21280	183	BD196	184	Q2-14	184	56-86412-3
181	TR56	181	37476	183	BD200	184	R840	184	072
181	TR8018	181	37967	183	BD206	184	RS-2017	184	77-271798-1
181	TVB-280646	181	37974	183	BD208	184	RS-2020	184	77-271798-3
181	001-021280	181	38049	183	BD464	184	S-85	184	080
181	1A1123100-1	181	38267	183	D2E	184	S10153	184	081
181	2N3771	181	38268	183	ES67(ELCOM)	184	S8660	184	86-344-2
181	2N5301	181	38397	183	GE-34	184	S8660121-808	184	86-506-2
181	2N5302	181	38473	183	G356	184	SC4133	184	86-5100-2
181	2N5303	181	38491	183	HEP-85002	184	SJE-649	184	86X0059-001
181	2N5885	181	38965	183	HEP-85005	184	SJE100	184	998105-1
181	2N5886	181	39140	183	HEP-85008	184	SJE106	184	1148
181	2N6257	181	39196	183	HEP-85009	184	SJE115	184	121-708
181	2N6326	181	39455	183	HEP-85010	184	SJE133	184	121-710
181	2N6327	181	39465	183	HP50	184	SJE1519	184	0122(AIRLINE)
181	2N6528	181	39466	183	MJE104	184	SJE1520	184	0124
181	2SD113	181	59616	183	MJE105	184	SJB203	184	0124(HOFFMAN)
181	2SD113-O	181	59635	183	MJE2901	184	SJE211	184	0124(WARDS)
181	2SD113-R	181	59751	183	MJE2955	184	SJE222	184	132-046
181	2SD113-Y	181	40411	183	P5N-5	184	SJB228	184	132-072
181	2SD114	181	60076	183	P5UA	184	SJE229	184	132-080
181	2SD114-O	181	60115	183	P4B-1	184	SJE237	184	132-081
181	2SD114-R	181	60142	183	P4B-2	184	SJE242	184	149N1
181	2SD114-Y	181	60187	183	P4T	184	SJE244	184	149N2
181	2SD55	181	60234	183	P4V-2	184	SJE246	184	149N4
181	2SD55A	181	60339	183	R339	184	SJE248	184	162N2
181	3-30173	181	60350	183	R8-2027	184	SJE253	184	190N1C
181	4-245	181	60380	183	SJE517	184	SJE254	184	190N3
181	4-3025190	181	60810	183	TM183	184	SJE255	184	190N3C
181	4-458	181	60885	183	TR-1036-1	184	SJE261	184	314-6005-1
181	4-490	181	60991	183	TR-1036-2	184	SJE262	184	364-6004
181	7-7466201	181	61173	183	TRO-2057-1	184	SJE271	184	417-144
181	10-13030-004	181	61451	183	TRO-2057-3	184	SJE272	184	449
181	10-13030-005	181	61456	183	TRO-2057-4	184	SJE274	184	576-0002-011
181	11-11911-1	181	62287	183	TRO-2058-1	184	SJE278	184	642-261
181	13-34374-1	181	74662	183	TRO-2058-5	184	SJE280	184	690V116H26
181	13-34684-1	181	98484-001	183	TR1036-2	184	SJE284	184	753-4001-931
181	13-40346-1	181	101568	183	TR1036-3	184	SJE289	184	0772
181	14-601-26	181	113876	183	TR1037-1	184	SJE305	184	800-533-01
181	19A115818	181	131257	183	TR1038-6	184	SJE320	184	884-250-011
181	19A116753P1	181	133550	183	2N5986	184	SJE340	184	936 NPN
181	19A126826-P2	181	133923	183	314-6013-4	184	SJE401	184	1043-1328
181	34-6002-35	181	133925	183	13-28222-1	184	SJE402	184	01057-1
181	34-6002-37	181	134282	183	48-137472	184	SJE404	184	1132-2(RCA)
181	43-023190	181	140612	183	659141	184	SJE407	184	1269
181	43A167885-P1	181	150070	184	A4K	184	SJE527	184	2011
181	43A167885-P2	181	162002-071			184	SJE5402	184	2057A100-49
181	43A167885-P3							184	2085-17

ECG	Industry Standard No.	ECG	Industry Standard No.	ECG	Industry Standard No.	ECG	Industry Standard No.	ECG	Industry Standard No.
184	2312-17	185	P3M	185	417-225	186	G06714C	186	176-024-9-004
184	2904-057	185	P3N	185	624-1	186	GE-28	186	260P21102
184	3526(SEARS)	185	P3N-1	185	642-266	186	GE-X18	186	260P21308
184	4491-6	185	P3N-2	185	644-1	186	GEMR-6	186	260P28701
184	4491-9	185	P3N-3	185	6901116H25	186	HC-01096	186	361-1
184	4802-0000-002	185	P3N-4	185	0770	186	HC-01098	186	576-002-001
184	4822-130-40537	185	P3P	185	0773	186	HC-01226	186	623-1
184	5847(RCA)	185	P3P-1	185	936PNP	186	HR-69	186	753-0101-226
184	7413	185	P3P-2	185	02057-1	186	HT304961C-0	186	753-9000-096
184	8000-00004-P185	185	P3P-3	185	3525(SEARS)	186	KQE41061	186	903Y002152
184	12539	185	P3P-4	185	3682(RCA)	186	PA6900	186	921-163B
184	30272	185	P3P-5	185	7351	186	PPR1006	186	1097(GB)
184	39458	185	P3S	185	7359-1	186	PPR1008	186	1116(GB)
184	41178	185	P3S-1	185	9582	186	PT2620	186	1125(GE)
184	41344	185	P4E	185	30271	186	PT2640	186	1345(GE)
184	95262-1	185	P4S	185	41179	186	PT2660	186	15580R
184	95263-1	185	P4U	185	41342	186	PT3503	186	2132
184	127978	185	P4V	185	95262-2	186	PT4690	186	2245-17
184	132495	185	P4V-1	185	95263-2	186	PT600	186	2510-105
184	132776	185	P4W-1	185	132571	186	PT601	186	2853-1
184	134280	185	P4W-2	185	134279	186	PT6618	186	2853-3
184	137369	185	P5L	185	138193	186	PT6669	186	2854-1
184	144856	185	P5R	185	166919	186	PT6696	186	2854-3
184	166918	185	PLE37	185	570030	186	RB42	186	2855-1
184	570031	185	PTC111	185	657179	186	SD1023	186	2855-3
184	610162-2	185	QP-13	185	657181	186	SDT3326	186	2856-1
184	610162N2	185	R841	185	910952	186	SDT4455	186	2856-3
184	657180	185	RS-2025	185	1417359-1	186	SDT4483	186	3632-2(RCA)
184	916114	185	SB9570	185	1471141-1	186	SDT4551	186	8000-00004-185
184	1471132-2	185	SB9571	185	1473682-1	186	SDT4553	186	8000-0004-P185
184	1471140-1	185	SB9572	185	5320723	186	SDT4583	186	8710-170
184	1473567-2	185	SB9573	185	7570030-01	186	SDT4611	186	8840-169
184	2006623-88	185	SJ285	185	23114131	186	SDT4612	186	20810-94
184	2320845	185	SJE108	185	23114133	186	SDT4614	186	38448
184	2320846	185	SJE111	185	43024219	186	SDT4615	186	53810-169
184	4080873-0001	185	SJE112	185	62389478	186	SDT5011	186	58840-200
184	4082886-0002	185	SJE114	185	2004074304	186	SDT5511	186	60407
184	5320643	185	SJE151B	185	4104007380	186	SDT5901	186	60413
184	5320921	185	SJE202	185	310720490060	186	SDT5902	186	61242
184	7910073-01	185	SJE210	185#	A715	186	SDT5906	186	62950
184	23114132	185	SJE221	185#	A715A	186	SDT6101	186	000073280
184	23114134	185	SJE227	185#	A715B	186	SDT6102	186	000073320
184	43024218	185	SJE231	185#	A715C	186	SDT6104	186	000073380
184	62589486	185	SJE241	185#	A715D	186	SDT6105	186	88060-146
184	62766212	185	SJE243	185#	A715WBP	186	SDT6106	186	88510-176
184	2003121216	185	SJE245	185#	A715WTA	186	SDT9001	186	133573
184#	C1162A	185	SJE256	185#	A715WTD	186	SDT9002	186	741115
184#	C1162B	185	SJE257	185#	A738C	186	SDT9005	186	760275
184#	C1162C	185	SJE265	185#	A758D	186	SDT9005	186	916046
184#	C1162CP	185	SJE273	186	28A715	186	SDT9006	186	986932
184#	C1162D	185	SJE275	186	28A715A	186	SDT9007	186	1473632-2
184#	C1162MP	185	SJE276	186	28A715B	186	SDT9008	186	2006436-35
184#	C1162WB	185	SJE277	186	28A715C	186	SJE-5038	186	2006436-36
184#	C1162WBP	185	SJE279	186	28A715D	186	SJB-5402	186	2327152
184#	C1162WTB	185	SJE283	186	28A715WB	186	STC1800	186	2327153
184#	C1162WTC	185	SJE288	186	28A715WBP	186	STC1862	186	2327803
184#	C1368	185	SJE403	186	28A715WT	186	STT4451	186	4037647-P1
184#	C1368C	185	SJE405	186	28A715WTA	186	T1882	186	4037647-P2
184#	C1368D	185	SJE408	186	28A715WTB	186	TM186	186	5320642
184#	2SC1162	185	SJE584	186	28A715WTC	186	TNJ70450	186	22114225
184#	2SC1162A	185	SJE653	186	28A715WTD	186	TNJ72775	186	43027571
184#	2SC1162B	185	SJE723	186	28A738	186	TR-1037-3	186	16102190931
184#	2SC1162C	185	SJE736	186	28A758B	186	TRAPLC1013	186	16343190142
184#	2SC1162CP	185	SJET43	186	28A758C	186	TV-73	186A	C1095
184#	2SC1162D	185	SJE768	186	28A758D	186	TV-75	186A	C1095(6)
184#	2SC1162MP	185	SJE797	185$	2N4920	186	TV-82	186A	C1095L
184#	2SC1162WB	185	SJE799	186	A203(NPN)	186	VX3375	186A	C1095M
184#	2SC1162WBP	185	SK3191	186	A208(NPN)	186	YAAN28C1096K	186A	C1096
184#	2SC1162WT	185	SPS4237	186	AR24(PHILCO)	186	YAAN28C1096K	186A	C1096(M)
184#	2SC1162WTA	185	STX0020	186	XB401	186	YAAN28C1096L	186A	C1096-3ZM
184#	2SC1162WTB	185	T-345	186	B143000	186	YAAN28C1096M	186A	C1096-3ZZM
184#	2SC1162WTC	185	T-396	186	B143001	186	YAAN28C1096N	186A	C1096-4ZL
184#	2SC1162WTD	185	T345	186	B143003	186	YAAN28C1096	186A	C1096AZL
184#	2SC1368	185	T396	186	B143009	186	YAANZ	186A	C1096AZL
184#	2SC1368B	185	TA7520	186	B143010	186	YAANZ281096	186A	C1096B
184#	2SC1368C	185	TM185	186	B143015	186	YAANZ28C1096	186A	C1096B
184#	2SC1368D	185	TR-182(OLSON)	186	B143016	186	YAANZ28C1096K	186A	C1096C
184#	2SC1568	185	TR02057-3	186	B143024	186	YAANZ28C1096L	186A	C1096D
184#	2SC1568R	185	TR2327393	186	B143025	186	YAANZ28C1096M	186A	C1096K
184(2)	121-772X	185	ZEN203	186	B3531	186	ZT3375	186A	C1096L
185	A496(JAPAN)	185	ZEN211	186	B3537	186	ZT600	186A	C1096LM
185	A715WTB	185	002-012100	186	B3538	186	2N6551	186A	C1096M
185	A715WTC	185	2N4918	186	B3540	186	2N6554	186A	C1096N
185	A8C	185	2N4919	186	B3541	186	28C1098	186A	C1096W
185	B-1695	185	2N5193	186	B3542	186	28C1098(4)K	186A	C1098Q
185	BD-132	185	2N5194	186	B3570	186	28C1098(4)L	186A	C1226
185	BD132	185	2N5195	186	B3576	186	28C1098A	186A	C1226-0
185	BD136	185	2SB632	186	B3606	186	28C1098B	186A	C1226A
185	BD138	185	2SB632K	186	B3607	186	28C1098C	186A	C1226AC
185	BD140	185	2SB744	186	B3608	186	28C1098D	186A	C1226AO
185	BD151	185	4-283	186	B3609	186	28C1098L	186A	C1226AP
185	BD152	185	11-0770	186	B3610	186	28C1098M	186A	C1226AQ
185	BD156	185	11-0773	186	B3611	186	13-32634-1	186A	C1226AR
185	BD166	185	13-23508-0	186	B3612	186	13-34046-1	186A	C1226C
185	BD168	185	13-28536-1	186	B3613	186	13-34046-2	186A	C1226CP
185	BD176	185	21M465	186	B3614	186	13-34046-3	186A	C1226P
185	BD178	185	33-096	186	B3747	186	13-34046-4	186A	C1226Q
185	BD180	185	42-28213	186	B3748	186	13-34372-2	186A	C1226R
185	BD183	185	46-86235-3	186	B3750	186	13-39046-3	186A	DBBY003001
185	BD186	185	46-86411-3	186	B5001	186	13-41628-3	186A	DBBY003002
185	BD188	185	48-134987	186	BF351	186	13-43790-1	186A	DDBY227001
185	BD190	185	48-137153	186	BLY20	186	018-00005	186A	DDBY227004
185	BD462	185	48-137154	186	BLY37	186	19-020-101	186A	EQ8-89
185	BD562	185	48-137155	186	BLY38	186	19A115300-P1	186A	ET453611
185	EA15X243	185	48-137156	186	BLY55	186	19A115300-P2	186A	G05-416-C
185	EA15X270	185	48-137157	186	BLY62	186	19A115300-P3	186A	GE-247
185	EA15X8605	185	48-137256	186	BLY78	186	21A112-049	186A	KBC1096-0
185	ES60(ELCOM)	185	48-137258	186	BLY91	186	21M180	186A	KBC1096-Y
185	ES68(ELCOM)	185	48-137303	186	BR100B	186	21M181	186A	KTR1096C
185	G05-406-C	185	48-137304	186	C1098	186	21M183	186A	MJE200E
185	HA-00496	185	48-137501	186	C1098A	186	21M184	186A	NCBV14
185	HEP-85006	185	48-137550	186	C1098B	186	21M286	186A	PT-029
185	HEP-S5007	185	48-471185J03	186	C1098D	186	21M367	186A	TM186A
185	HEP246	185	48-869582	186	C1098L	186	21M556	186A	YAANL28C1096K
185	HEP700	185	48R869582	186	C1098M	186	31-0066	186A	YAANL28C1096L
185	HT104961B	185	63-11990	186	D4201	186	34-6002-41	186A	YAANL28C1096M
185	M9582	185	079	186	D42C2	186	046-1(SYLVANIA)	186A	01-031096
185	MJ3370	185	86-345-2	186	D42C3	186	46-86360-3	186A	28C1095
185	MJ3371	185	86-396-2	186	D42C4	186	48-155097	186A	28C1095(6)
185	MJ3492	185	86-507-2	186	D42C5	186	48-40382J01	186A	28C1095L
185	MJ3493	185	86004	186	D42C6	186	48840382J01	186A	28C1095M
185	MJ3510	185	114P1P	186	D42C7	186	53-1362	186A	28C1096
185	MJ3711	185	0121(AIRLINE)	186	D42C8	186	53-1967	186A	28C1096(M)
185	MJ9450	185	121-707	186	EA43X160	186	074-1(PHILCO)	186A	28C1096-3ZM
185	MOTMJE371	185	121-709	186	EA15X248	186	0103-0419	186A	28C1096-3ZZM
185	P1V	185	0123	186	EA2488	186	0103-0419A	186A	28C1096-4ZL
185	P1V-2	185	0123(WARDS)	186	EA3674	186	0103-512M	186A	28C1096AZL
185	P1V-3	185	132-024	186	EP100	186	0140-7	186A	28C1096AZL
185	P2T	185	132-079	186	EP15X25	186	172-006-9-001	186A	28C1096B
185	P2T-1	185	149P1	186	EP15X227			186A	28C1096B
185	P2T-2	185	149P3	186	ET495371			186A	28C1096C
185	P2T-3	185	162P1	186	G06-714C			186A	28C1096D
185	P2T-4	185	297V086C004					186A	28C1096K
		185	417-145					186A	28C1096LM

ECG	Industry Standard No.	ECG	Industry Standard No.	ECG	Industry Standard No.	ECG	Industry Standard No.	ECG	Industry Standard No.
186A	28C1096M	187	13-34047-3	188	87B02	189	121-1021	191	40459
186A	28C1096N	187	13-34373-2	188	96-5263-01	189	121-980	191	40459V1
186A	28C1096W	187	13-39047-3	188	999091-1	189	121-980-01	191	40459V2
186A	28C1226	187	13-43791-1	188	121-1020	189	0126(WARDS)	191	139017
186A	28C1226A	187	018-00004	188	133-1	189	134-1	191	237075
186A	28C1226AC	187	19-020-102	188	157N3	189	157P4	191	346016-63
186A	28C1226AP	187	21A112-048	188	258-1	189	202-1	191	610131-2
186A	28C1226AQ	187	21M025	188	297V087B02	189	202N1	191	1473613-2
186A	28C1226AR	187	21M026	188	471-1(PLASTIC)	189	202P1	191	1473613-3
186A	28C1226ARL	187	21M028	188	628-3	189	202P2	191	1473687-1
186A	28C1226B	187	21M345	188	632-1(SYLVANIA)	189	628-3	191	62638222
186A	28C1226C	187	21M395	188	742-1	189	629-3	191	93507676O112
186A	28C1226CF	187	21M443	188	2904-058	189	3629-3	192	A188103
186A	28C1226D	187	21M028	188	3628-3	189	4099A	192	A3T2484
186A	28C1226P	187	34-6002-42	188	3742	189	40362V1	192	A417O34
186A	28C1226L	187	34-6016-65	188	4442-3	189	40362V2	192	AR207
186A	28C1226Q	187	047-1(SYLVANIA)	188	7316	189	44764	192	A5T
186A	28C1226P	187	48-40383J01	188	7555	189	60458	192	BC226
186A	28C1226Q	187	48B40383J01	188	8471(SYLVANIA)	189	95258-2	192	BC510
186A	28C1226QR	187	0101-0448	188	36213	189	132447	192	BCW51
186A	28C1226R	187	0101-0448A	188	40311A	189	132448	192	BCW73-16
186A	28C1226RL	187	733-0100-699	188	40311B	189	132488	192	BCW74-16
186A	28C1226RLP	187	753-0100-699	188	40314V1	189	137065	192	BCW77-16
186A	28C1226RLQ	187	853-0300-634	188	40314V2	189	138121	192	BCW78-16
186A	28C1226RLR	187	903Y002151	188	40315V2	189	139267	192	BCW82
186A	28C1226C	187	1100(GE)	188	40317V1	189	145172	192	BCW82A
186A	28C1848	187	1285(GE)	188	40317V2	189	610157-4	192	BCW82B
186A	28C1848Q	187	61239	188	40319V1	189	610202-1	192	BCW83B
186A	28C1848R	187	000071120	188	40319V2	189	610202-2	192	BCW84
186A	04-11620-01	187	760276	188	40320V1	189	610227-1	192	BCW90K
186A	4ZL	187	986933	188	40320V2	189	1417317-1	192	BCW91K
186A	0V-07165	187	5320652	188	40325V1	189	1445829-501	192	BCW91KA
186A	09-302080	187	22114254	188	40323V2	189	1445829-502	192	BCW91KB
186A	09-302121	187S	28A779	188	40326V2	189	1471134-1	192	BCW94K
186A	09-302123	187S	28A779K	188	40327V2	189	1473629-1	192	BCW95K
186A	09-302126	187S	28A779KB	188	40360V1	189	1473629-3	192	BF829P
186A	09-309071	187S	28A779KC	188	40360V2	190	A5P	192	BF830
186A	10-009	187S	28A779KD	188	40361V2	190	BDY65	192	BF830P
186A	48-155042	187A	G05-406-C	188	40361V3	190	C1157	192	BFX53
186A	48-155074	187A	GE-248	188	40385V1	190	HEP-83019	192	BFY16
186A	48-355059	187A	K8A634-Y	188	40385V2	190	IRTR74	192	BLY28
186A	48-41784J04	187A	28A634	188	40457V1	190	LM2701	192	B8825
186A	488155042	187A	28A634A	188	40457V2	190	MPSU03	192	B8Y62B
186A	488155074	187A	28A634D	188	40594	190	MPSU04	192	C1347Q
186A	065-007	187A	28A634K	188	44763	190	SJE103	192	C1347R
186A	90-111	187A	28A634L	188	60457	190	SJE1032	192	C1347S
186A	90-176	187A	28A634M	188	95258-1	190	TR-8022	192	CDC-8000-1D
186A	90-38	187A	28A699	188	95285-1	190	28C1124	192	CDC8000-1A
186A	90-451	187A	28A699-0	188	112525	190	28C1124B	192	CDC9000-1B
186A	90-75	187A	28A699A	188	130474	190	28C1155	192	CGE-63
186A	106-004	187A	28A699A0	188	132445	190	28C1156	192	CJ5206A
186A	123-011	187A	28A699AP	188	132446	190	28C1157	192	CS-2143
186A	123-011A	187A	28A699AQ	188	134774	190	28C1663	192	CS1256HG
186A	123B-002	187A	28A699AR	188	137066	190	28C1663H	192	C82023
186A	142-006	187A	28A699P	188	139266	190	13-34372-1	192	C89417
186A	172-011-9-001	187A	28A699Q	189	610157-3	190	19-3935-641	192	D00D
186A	172-058-9-003	187A	28A699R	189	610228-1	190	57A172-8	192	D35324J1
186A	176-042-9-005	187A	04-07150-01	189	1417316-1	190	57A211-8	192	EA15X274
186A	185-001	187A	09-300073	189	1445829-503	190	57B211-8	192	EA15X349
186A	186-008	187A	48-41785J04	189	1445829-504	190	998102-1	192	EA15X4531
186A	186-009	187A	48-41788J04	189	1471135-1	190	121-1014	192	EA15X7519
186A	260P22801	187A	48-90A20A02	189	1473628-1	190	121-755	192	EA3990
186A	853-0301-096	187A	488155116	189	1473628-3	190	190N1	192	ED-1029-N8
186A	1039-0433	187A	260P26301	189	62786272	190	576-0004-035	192	GE-63
186A	1074-116	187A	000071130	189	BFW88	190	921-1020	192	HD-00261
186A	1080-06	187A	000071131	189	EA15X364	190	1041-76	192	HEP-80002
186A	2017-108	187A	146569	189	EA15X8122	190	1081-3350	192	HEP-80003
186A	2041-03	188	A5E	189	ES81	190	159044	192	HR-67
186A	2081-17	188	A6D	189	ES83(ELCOM)	190	142690	192	HR82
186A	2160-17	188	A6D-1	189	GE-218	190	146143	192	HR83
186A	2168-17	188	A6D-2	189	GE-84	190	610190-1	192	HS5810
186A	2202-17	188	A6D-3	189	HEP-85030	190	610190-4	192	HS5812
186A	2334-17	188	A6G	189	HEP-85031	191	A5T	192	HS5814
186A	5001-064	188	ABJ	189	HEP-85032	191	A5T-1	192	HS5816
186A	5001-075	188	AG6	189	IRTR73	191	A9M	192	HS5818
186A	8000-00041-043	188	B1M	189	M9641	191	B1E	192	HS5820
186A	8910-146	188	B2M	189	MPS-U51	191	B1B-1	192	HS5822
186A	18410-147	188	BC430	189	MPS-U55	191	B1G	192	HS6010
186A	38510-168	188	ES82(ELCOM)	189	MPS-U56	191	E13-010-00	192	HS6012
186A	000073581	188	GE-217	189	MPSU51	191	E13-011-00	192	HS6014
186A	532-125	188	GE-83	189	MPSU51A	191	E896	192	HS6016
186A	632000001	188	H7618	189	MPSU52	191	GE-224	192	HT514071Q
186A	632122617	188	HEP-85026	189	MPSU55	191	GE-249	192	HT404001E
187	AR25	188	IRTR72	189	MPSU56	191	HEP-83021	192	HX-50103
187	AR250	188	M9640	189	MU9660	191	HEP-83022	192	LM8000
187	C636	188	MPS-U01	189	MU9660B	191	IRTR75	192	M9521
187	D4301	188	MPSU01A	189	MU9660T	191	LM-1156	192	MH1501
187	D4302	188	MPSU02	189	MU9661	191	M7476	192	NN7000
187	D4303	188	MPSU05	189	MU9661T	191	MPS-U10	192	NN7001
187	D4304	188	MPSU06	189	P218-2	191	MPSU10	192	NN7002
187	D4305	188	MU9610	189	P2U	191	MPSU11	192	NN7003
187	D4306	188	MU9610P	189	P2U-1	191	K3613-3(RCA)	192	NN7004
187	D4308	188	MU9610T	189	P2V	191	RE77	192	NN7005
187	EP101A	188	MU9611	189	P3K	191	ST-2N6558	192	NR-141E8
187	EP15X24	188	MU9611Q	189	P3Y	191	TA2819	192	NR-141ET
187	ES15051	188	MU9611T	189	RE76	191	TV-25	192	N81355
187	G03703C	188	P218-1	189	SDT5505	191	TV22	192	N81900
187	GE-29	188	RE75	189	SDT5506	191	TV49	192	N81960
187	HA-00634	188	SPS1107	189	SDT5550	191	V82SC1921//1B	192	N8950
187	HA-00636	188	T422	189	SDT3552	191	28C1962	192	PA9483
187	HB-00564	188	TM188	189	SDT5553	191	4-686252-3	192	PBC184
187	HF51	188	28C2194	189	SF84099	191	4-8134842	192	PEE1001
187	HR-70	188	13-33180-1	189	SP8837	191	8-765-170-01	192	PT2540
187	HR104961C	188	13-36508-1	189	T423	191	09-304052	192	RB196
187	IR-TR56	188	13-37326-1	189	TM189	191	011-00	192	S-522
187	RE43	188	46-86583-3	189	TRO2053-5	191	13-29437-1	192	SE1012
187	RY2356	188	48-137091	189	TRO2053-7	191	14-609-01	192	SEC1078
187	SDT3325	188	48-137092	189	02P13O	191	14-609-01A	192	SEC1079
187	SDT3775	188	48-137149	189	28A706	191	14-609-02A	192	SEC1477
187	SDT3776	188	48-137169	189	28A962	191	14-609-05	192	SEC1479
187	SDT3778	188	48-137202	189	004-1	191	14-900-12	192	SF.T440
187	SDT9004	188	48-137319	189	13-34004-1	191	14-904-12	192	SF.T443
187	TA7741	188	48-137505	189	13-34373-1	191	34-6001-64	192	SF.T443A
187	TM187	188	48-869640	189	13-36509-1	191	34-6015-28	192	SF.T445
187	TNJ72774	188	48-97177A10	189	13-37527-1	191	48-137476	192	SF.T714
187	TR-1036-3	188	48B86964O	189	20A0059	191	48B137364	192	SKA4410
187	TV-74	188	488137169	189	20A0060	191	60P22204	192	SPA402
187	28A636	188	488137572	189	46-86584-3	191	120-004884	192	ST1504
187	28A636(4)K	188	57A214-12	189	48-137160	191	121-743-01	192	ST1505
187	28A636(4)L	188	66P058-2	189	48-137168	191	131N24	192	ST402
187	28A636A	188	66P069-1	189	48-137240	191	260P22204	192	ST5061
187	28A636B	188	81-23860400-3	189	48-137314	191	481-34842	192	STC1336
187	28A636C	188	81-23860400A	189	48-137320	191	687-1	192	TE5417
187	28A636D	188	81-23860400B	189	48-137610	191	1010-17	192	TE3859A
187	28A636K	188	81-27125530-9	189	48-869641	191	3613	192	TE4425
187	28A636M	188	81-27125530-9A	189	48-97177A11	191	3613-2(GE)	192	TI-485
187	09-300067	188	81-27125530-9B	189	48B869641	191	3613-3	192	TM192
187	09-300068	188	81-27126100-0	189	488137168	191	3679	192	TN79
187	09-302170	188	86-284-2	189	667069-2	191	4686-252-3	192	TN81
187	13-14085-86	188	86-422-3	189	86-423-3	191	40256	192	TNJ71252
187	13-26353-1	188	86-660-2	189	86-659-2	191	40346V1	192	TV21
187	13-34047-1	188	86X0073-001	189	96-5262-01	191	40390	192	TV26
187	13-34047-2			189	96-5320-01			192	TV28
				189	998092-1			192	TV3-CS1256HG

ECG	Industry Standard No.
192	TVS2SA543
192	X16E3890
192	ZT3053
192	2N5320HS
192	2N5820HS
192	2SC1406
192	2SC14070
192	00002SD261
192	4-1792
192	09-302075
192	09-302116
192	09-302226
192	09-303019
192	09-30319
192	09-309031
192	13-23327-4
192	13-32630-2
192	13-32630-4
192	13-39114-3
192	14-602-55
192	14-602-70
192	14-602-74
192	21-0101
192	21M192
192	21M193
192	21M606
192	24MW1152
192	31-0101
192	34-6016-51
192	34-6016-59(NPN)
192	44A417034
192	48-03-10744702
192	48-869521
192	48R869521
192	48S134988
192	48S155123
192	57A167-9
192	065-008
192	86-5075-2
192	86-601-2
192	87-0006
192	96-5161-01
192	114-1(PHILCO)
192	121-703
192	121-766-01
192	167-9
192	229-1301-37
192	260Z00401
192	261RAX
192	362A10
192	921-230B
192	1187(GE)
192	2246-17
192	4218
192	5613-1209
192	5613-1209C
192	8000-00012-040
192	12047-0023
192	25840-163
192	38510-165
192	39741
192	39876
192	39919
192	43115
192	45810-165
192	60172
192	62203
192	000073300
192	000073301
192	000073302
192	000073303
192	000073305
192	94062
192	95226-004
192	171162-026
192	172763
192	183034
192	870006
192	2320664
192	2487424(NPN)
192	2865101
192	3458573-1
192	8113024
192	8114024
192	22114210
192	23114314
192	23114994
192	43025539
192	62727128
192	62734647
192	84026103
192A	BCW83A
192A	C1330
192A	C1330A
192A	C1330B
192A	C1330C
192A	C1330D
192A	C1330L
192A	C1330R
192A	C1346
192A	C1346R
192A	C1346S
192A	C1347
192A	C814
192A	C853
192A	C853A
192A	C853B
192A	C853C
192A	C853KLM
192A	C853L
192A	C881
192A	C881A
192A	C881B
192A	C881C
192A	C881D
192A	C881K
192A	C881L
192A	D228
192A	D261
192A	D261A
192A	D261B
192A	D261C
192A	D261D
192A	D261E
192A	D261F
192A	D261L
192A	D261P
192A	D261R
192A	D261V
192A	D261W
192A	D33D21J1
192A	D33D22J1
192A	D33D25J1
192A	D33D26J1
192A	D33D27J1
192A	D33D29J1
192A	D33D30J1
192A	GE-88
192A	SK3137
192A	001-021232
192A	2N3402
192A	2N3403
192A	2N3404
192A	2N3405
192A	2N3417
192A	2N4425
192A	2SC1330
192A	2SC1330A
192A	2SC1330B
192A	2SC1330C
192A	2SC1330D
192A	2SC1330L
192A	2SC1330R
192A	2SC1346
192A	2SC1346Q
192A	2SC1346R
192A	2SC1346S
192A	2SC1347
192A	2SC1347Q
192A	2SC1347R
192A	2SC1347S
192A	2SC814
192A	2SC853
192A	2SC853A
192A	2SC853B
192A	2SC853C
192A	2SC853K
192A	2SC853KLM
192A	2SC853L
192A	2SC881
192A	2SC881A
192A	2SC881B
192A	2SC881C
192A	2SC881D
192A	2SC881K
192A	2SC881L
192A	2SC881M
192A	2SD261(R)
192A	2SD261(U)
192A	2SD261A
192A	2SD261B
192A	2SD261C
192A	2SD261D
192A	2SD261E
192A	2SD261F
192A	2SD261L
192A	2SD2610
192A	2SD261P
192A	2SD261Q
192A	2SD261R
192A	2SD261S
192A	2SD261V
192A	2SD261W
192A	13-34003-1
192A	71447-1
192A	71447-2
192A	71447-3
192A	801527
193	A3T2907A
193	BC370
193	BC406
193	BC512B
193	BC513B
193	BC514
193	BC514B
193	BCW52
193	BCW61A
193	BCW61B
193	BCW62B
193	BCW63B
193	BCW64
193	BCW64B
193	BCW75-10
193	BCW75-16
193	BCW76-10
193	BCW76-16
193	BCW92
193	BCW92K
193	BCW93
193	BCW93K
193	BCW96K
193	BCW97K
193	BCY78VII
193	BCY78VIII
193	BCY79VIII
193	BF834
193	BSV43B
193	BSV44B
193	BSV45B
193	BSV47B
193	BSV48B
193	BSV49B
193	CGE-67
193	CS-2142
193	CS1255HF
193	D29E10
193	D29E10J1
193	D29E1J1
193	D29E2J1
193	D29E4J1
193	D29E5J1
193	D29E6J1
193	D29E7J1
193	D29E9
193	D29E9J1
193	EA15X273
193	EP15X13
193	EP15X21
193	EP15X51
193	GE-67
193	GE-89
193	HA-00643
193	HR-72
193	HR105611C
193	HR72
193	HS5811
193	HS5813
193	HS5815
193	HS5817
193	HS5819
193	HS5821
193	HS5823
193	HS6011
193	HS6013
193	HS6015
193	HS6017
193	HX-50104
193	MH1502
193	NN7500
193	NN7501
193	NN7502
193	NN7503
193	NN7504
193	NN7505
193	NN7511
193	NR-671ET
193	P4B
193	RE197
193	SHA7520
193	SHA7521
193	SHA7522
193	SHA7523
193	SHA7524
193	SHA7526
193	SHA7527
193	SHA7528
193	SHA7597
193	SHA7598
193	SHA7599
193	SK3138
193	ST61000
193	TE3702
193	TE3703
193	TE5086
193	TE5087
193	TE5367
193	TI-891
193	TM193
193	TNJ70481
193	TNJ72152
193	TP3638A
193	TQ53
193	TQ54
193	TQ55
193	TQ57
193	TQ58
193	TQ59
193	TQ59A
193	TQ60
193	TQ60A
193	V654
193	V655
193	V741
193	001-02117-2
193	2N5821HS
193	2SH643
193	09-300072
193	09-302161
193	09-309030
193	13-32631-2
193	13-32631-4
193	13-33178-1
193	13-39115-2
193	14-602-54
193	14-602-71
193	14-602-75
193	21A112-004
193	21A118-032
193	21M027
193	21M459
193	31-0055
193	31-0102
193	34-6016-59(PNP)
193	36-6016-59
193	48-137321
193	48-355055
193	48S134989
193	57A168-9
193	86-5104-2
193	86-602-2
193	87-0029
193	0101-0439
193	121-765-01
193	168-9
193	178-1(PHILCO)
193	229-1301-28
193	229-1301-38
193	643RIX
193	853-0300-643
193	1072
193	1072K(GE)
193	1080G(GE)
193	1184G(GE)
193	5611-695
193	5611-695C
193	95226-003
193	183035
193	255821HS
193	5140688
193	988992
193	2320632
193	23114313
193	23114995
193	62711205
193	62734663
193	16009090545
193A	A545
193A	A545GRN
193A	A545K
193A	A545KLM
193A	A545L
193A	A545LM
193A	A643
193A	A643A
193A	A643B
193A	A643C
193A	A643D
193A	A643E
193A	A643F
193A	A643L
193A	A643R
193A	A643S
193A	A643V
193A	A643W
193A	A730
193A	A731
193A	A751R
193A	A751S
193A	BC257A
193A	BC257B
193A	BC258A
193A	BC258B
193A	BC259A
193A	BC259B
193A	2SA545
193A	2SA545GRN
193A	2SA545K
193A	2SA545KLM
193A	2SA545L
193A	2SA545LM
193A	2SA545M
193A	2SA643
193A	2SA643A
193A	2SA643B
193A	2SA643C
193A	2SA643D
193A	2SA643E
193A	2SA643F
193A	2SA643R
193A	2SA643V
193A	2SA643W
193A	2SA730
193A	2SA731
193A	2SB544P1
193A	2SB544P1D
193A	2SB544P1E
193A	2SB544P1F
193A	033589
194	A-1854-0358-1
194	A-1854-0365-1
194	A-1854-0474-1
194	A-1854-0533
194	A133
194	A86-565-2
194	A8N
194	A8Z
194	BC532
194	BC533
194	BSX21
194	BSY79
194	C499R
194	C499Y
194	C780
194	C780AG
194	C780AG-O
194	C780AG-R
194	C780G
194	ES97
194	HEP-S0001
194	HEP-S0005
194	HF40
194	MPSA05
194	PTC125
194	PTC127
194	RE66
194	RE78
194	SE5-0745
194	SPS5450
194	TR32
194	001-021100
194	2D020
194	2D020-173
194	2D020-174
194	2N1154
194	2N1155
194	2N1156
194	2N4410
194	2N4945
194	2N5175
194	2N719
194	2N719A
194	2N720
194	2N720A
194	2N750
194	2SC189F
194	2SC499-R(FA-1)
194	2SC499-RY
194	2SC499-Y(FA-1)
194	2SC499R
194	2SC499Y
194	2SC780
194	2SC780AG
194	2SC780AG-O
194	2SC780AG-R
194	2SC780AG-Y
194	2SC780G
194	09-302130
194	13-27432-1
194	13-28432-1
194	13-35550-1
194	13-35807-1
194	13-39851-1
194	020-1110-038
194	21A112-074
194	34-6016-41
194	34-6016-56
194	034A
194	48-137238
194	48-137278
194	48-137332
194	48-137386
194	48S137386
194	48S155094
194	48S155095
194	57B181-12
194	66X0049-001
194	68A7702P1
194	73B140585-21
194	73B140585-22
194	73B140585-23
194	73B140585-24
194	73B140585-25
194	73B140585-26
194	73B140585-27
194	86-515-2
194	86-521-2
194	86-550-2
194	86-557-2
194	86-566-2
194	86-611-2
194	86-615-2
194	86-630-2
194	86-649-2
194	86X0049-001
194	86X0071-001
194	99S034
194	99S040
194	99S060-1
194	99S061-1
194	99S070-1
194	99S077-1
194	590-591731
194	590-591811
194	4686-145-3
194	139269
194	140624
194	1417321-1
195A	A066-116
195A	A2416
195A	A270
195A	A67-15-280
195A	B3465
195A	B3468
195A	BC412
195A	BFS22
195A	BFS23
195A	BSX22
195A	C261
195A	C311
195A	C456-O
195A	C456A
195A	C456D
195A	C481X
195A	C482-GR
195A	C482-O
195A	C482-Y
195A	C482GR
195A	C482X
195A	C482Y
195A	C502
195A	C614
195A	C614D
195A	C614E
195A	C614F
195A	C614G
195A	C615
195A	C615A
195A	C615B
195A	C615C
195A	C615D
195A	C615E
195A	C615G
195A	C776
195A	C776(Y)
195A	C776Y
195A	C798
195A	C803
195A	CF6
195A	EQS-56
195A	EQS-57
195A	FS1978
195A	GE-219
195A	GE-320
195A	GE-45
195A	GR102
195A	HEP-S3001
195A	IC743038
195A	IC743039
195A	IP20-0004
195A	IP20-0048
195A	IRTR64
195A	IRTR65
195A	MM1810
195A	MM8004
195A	MRF8004
195A	PT1537
195A	PT2040A
195A	PT2677C
195A	PT3141C
195A	PT31961
195A	PT857
195A	PTC2677C
195A	Q-00869C
195A	Q-00969C
195A	Q-01069
195A	Q-01069A
195A	Q-01069B
195A	Q-01184R
195A	Q-01284R
195A	RE79
195A	RE79A
195A	S3001
195A	SS3935
195A	TIS109
195A	TIX888
195A	TR-65
195A	TR2SC671
195A	2N2540A
195A	2N2949
195A	2N3426
195A	2N3554
195A	2SC116T
195A	2SC1556
195A	2SC261
195A	2SC307T
195A	2SC311
195A	2SC456
195A	2SC456-O
195A	2SC456A
195A	2SC456D
195A	2SC481
195A	2SC481X
195A	2SC482
195A	2SC482-GR
195A	2SC482GR
195A	2SC482X
195A	2SC482Y
195A	2SC502
195A	2SC6
195A	2SC614
195A	2SC614C
195A	2SC614D
195A	2SC614E
195A	2SC614F
195A	2SC614G
195A	2SC615
195A	2SC615A
195A	2SC615B
195A	2SC615C
195A	2SC615D
195A	2SC615E
195A	2SC615F
195A	2SC615G
195A	2SC774
195A	2SC775
195A	2SC776
195A	2SC776(Y)
195A	2SC776Y
195A	2SC781
195A	2SC798
195A	2SC803
195A	2SCF6
195A	09-302015
195A	09-302030
195A	09-302035
195A	09-302055
195A	09-302068
195A	09-302081
195A	09-302082
195A	09-309062
195A	9TR11
195A	9TR4
195A	9TR6
195A	13-0028
195A	13-0035
195A	019-003691
195A	019-003692
195A	19-020-076
195A	19-020-077
195A	022-3640-081

ECG	Industry Standard No.	ECG	Industry Standard No.	ECG	Industry Standard No.	ECG	Industry Standard No.	ECG	Industry Standard No.
195A	022-3640-082	196	M9676/NPN	196	229-1301-34	197	TA8212	198	SC4411
195A	48-869209	196	QT-DO313XAC	196	239A7920	197	TIP5A	198	SM6814
195A	48-869491	196	R623-1	196	260P14202	197	TM197	198	ST-2801514
195A	48-869519	196	R640-1	196	260P34202	197	TV116	198	ST84027
195A	48B869209	196	RCA1005	196	331-1	197	XA1199	198	ST84028
195A	48B869491	196	RCA1010	196	417-175	197	ZN 35024712	198	ST84029
195A	48B869519	196	RCA1C14	196	417-203	197	01-020566	198	TG28C1756
195A	051-0156	196	RE21	196	417-4-00226	197	2N6106	198	TG28D386Y-D-A
195A	051-0157	196	32D153	196	514-047830	197	2N6107	198	TG28D386Y-E-A
195A	120-001798	196	SK3054	196	652-3	197	2N6108	198	TIP47
195A	120-003151	196	STX3526	196	991-01-3170	197	2N6110	198	TIP48
195A	130-245	196	T23-93	196	991-01-5001	197	2N6111	198	TIP49
195A	176-018-9-001	196	TA-7155	196	991-01-5063	197	2N6132	198	TIP50
195A	176-029-9-002	196	TA7155	196	1009-05A	197	2N6133	198	TN28C1507
195A	176-029-9-003	196	TA7562	196	1010-7936	197	2N6134	198	TN28C1507-K-A
195A	260-10-010	196	TA7782	196	1061-8338	197	28A764	198	TN28C1507-L-A
195A	386-7182P1	196	TA7783	196	1109(GE)	197	28A765	198	TN28C1507-M-A
195A	576-0004-013	196	TA7784	196	1384(GE)	197	28A769	198	TN28C1520-1
195A	576-0036-212	196	TA8231	196	3632-1	197	28B502	198	TN28C1520-K-1A
195A	576-0036-212	196	TA8232	196	3632-3	197	28B503	198	TN28C1520-L-1A
195A	576-0036-913	196	TA8233	196	3640(RCA)	197	28B512A	198	TN28C1520-M-3A
195A	1000-139	196	TH7251	196	3843(SEARS)	197	28B513A	198	TR81204
195A	1000-141	196	TIP-24	196	4080-866-0006	197	28B596	198	TR81404
195A	1001-07	196	TIP29XA	196	4080-879-0001	197	28B596-0	198	TR8140MP
195A	1004-02	196	TIP4	196	4442-566	197	28B683	198	TR8160A
195A	1005-02	196	TM196	196	4686-234-3	197	3-233	198	TR8160MP
195A	1009-03	196	TV-112	196	4745	197	3L4-6013-6	198	TR81804
195A	1009-04-17	196	TV112	196	5259	197	3L4-6013-8	198	TR8180MP
195A	1010-14	196	TV117	196	6099-2	197	13-283361-1	198	TR82004
195A	1011-01	196	X735-40C	196	06115	197	13-34839-1	198	TR82006
195A	1040-04	196	Z8D25SY	196	7364-1	197	44A417756-1	198	TR8200MP
195A	1041-77	196	01-040313	196	7920-1	197	48-137080	198	TR82254
195A	1096-11	196	01-040476	196	16029	197	48-137185	198	TR8225MP
195A	1098-15	196	002-1(SYLVANIA)	196	16059	197	48-137259	198	TR82504
195A	1100-75	196	002-2(SYLVANIA)	196	16115	197	48-137310	198	TR8250MP
195A	1112-79	196	002-2(SYLVANIA)	196	16165	197	48-869677	198	TR82754
195A	1183-17	196	2N5294	196	16165	197	48-869701	198	TR8275MP
195A	1482-17	196	2N5297	196	16277	197	57A245-14	198	TR82804S
195A	1487-17	196	2N5490	196	16341	197	57A251-14	198	TR828058
195A	1916-17	196	2N5491	196	020425-3	197	57B245-14	198	TR85006
195A	2027-00	196	2N5492	196	40624	197	57B251-14	198	TR83015
195A	2904-037	196	2N5493	196	40627	197	61-3096-90	198	TR8301LC
195A	3868	196	2N5494	196	40632	197	96-5316-01	198	TR8301MP
195A	4004	196	2N5495	196	40816(RCA)	197	998099	198	TR832048
195A	4004 (SEARS)	196	2N5496	196	50200-8	197	121-977	198	TR8205B
195A	4005 (CB)	196	2N5497	196	60237	197	121-988-02	198	TR83254
195A	4007	196	2N6291	196	60416	197	131-005-353-1	198	TR83255
195A	4009	196	2N6292	196	60678	197	131-005353-1	198	TR83742
195A	4802-00007	196	2N6293	196	60977	197	195P2C	198	TR84926
195A	4802-00017	196	28C1826P	196	61035	197	259A7921-1	198	TR84927
195A	7364-6053P1	196	28C1826Q	196	61252	197	417-289	198	TT28C983-0-A
195A	8000-00005-009	196	28C1826R	196	61534	197	977-64197	198	TT28C983-R-A
195A	8000-00005-010	196	28C296	196	61997	197	991-01-5000	198	TT28C983-Y-A
195A	8000-00006-001	196	28D257	196	62144	197	991-01-5062	198	TV28C1505
195A	8000-00011-051	196	28D288	196	62511	197	1385(GE)	198	TV28C1507
195A	8000-00011-151	196	28D288A	196	131161	197	4781	198	TV82801505
195A	8000-0004-P083	196	28D288B	196	131239	197	7921-1	198	TV82BC1520
195A	8000-0004-P084	196	28D288C	196	134771	197	16167	198	001-02114-0
195A	9404-3	196	28D288K	196	135735	197	16169	198	01-051507
195A	36213A2	196	28D288L	196	135739	197	16175	198	01-051756
195A	36913	196	28D289	196	140624	197	16279	198	2
195A	38122	196	28D289A	196	140979	197	16306	198	28C1505
195A	38789	196	28D289B	196	141020	197	16342	198	28C1505K
195A	082022	196	28D289C	196	171174-3	197	020426-3	198	28C1505L
195A	082033	196	28D517A	196	239517	197	41501	198	28C1505LA
195A	94043	196	28D517AP	196	309689	197	62277	198	28C1505M
195A	101436	196	28D565A	196	610111-8	197	62279	198	28C1506
195A	111279	196	28D589A	196	610155-1	197	62512	198	28C1507
195A	115304	196	28D589AF	196	610195-1	197	62759	198	28C1507A
195A	198005-1	196	28D589AFF	196	1417364-1	197	140625	198	28C1507H
195A	237028	196	28D589AP	196	1471132-002	197	147624-1	198	28C1507K
195A	321264-2	196	28D476	196	1471132-3	197	309690	198	28C1507L
195A	1223784	196	28D476A	196	1471132-5	197	610195-2	198	28C1507LM
195A	1223785	196	28D476B	196	1473612-1	197	779821	198	28C1507M
195A	1223786	196	28D476C	196	1473621-1	197	1473624-1	198	28C519
195A	1700037	196	28D476D	196	1473621-1	197	1835667	198	28C1520
195A	1700038	196	28D476YL	196	1473632-3	197	3596100	198	28C1520-1A
195A	2006514-61	196	28D570	196	1473640-1	197	3596101	198	28C1520-3A
195A	2006607-61	196	3596091	196	3596091	197	4082873-0001	198	28C1520I
195A	5320501	196	3596092	196	3596092	197	23114939	198	28C1520K
195A	26010010	196	3L4-6011-1	196	4080187-0506	197	35024712	198	28C1520KL
195A	2601005B	196	3L4-6012-06	196	4080187-0507	197	95349700	198	28C1520L
195A	120004884	196	3L4-6012-5	196	4080835-0002	197	131005353-1	198	28C1520M
195A	16304198000	196	3L4-6012-55	196	4080858-002	197	485137370	198	28C1905
196	A1Y	196	3L4-6012-56	196	4080866-0012	198	A6789760A	198	28C1929
196	A5A14-047830	196	3L4-6012-58	196	4080879-0006	198	A6789971C	198	28C1929Q
196	A5A	196	3L4-6012-6	196	4080879-0011	198	A6789971D	198	28C1929R
196	A8E	196	3L4-6012-8	196	23114923	198	BW866	198	2802231Y
196	A8K	196	38D313	196	23114969	198	BW867	198	8-729-372-31
196	ALB6494612	196	38D313C	196	26734596	198	C1505	198	09-302186
196	AR24(RED)	196	4-686234-3	197	A1853-0233-1	198	C1505K	198	09-303029
196	AR28	196	09-303021	197	A417756	198	C1505L	198	14-605-76
196	AR28(RED)	196	13-1847-1	197	A764	198	C1505LA	198	21A112-070
196	AR30	196	13-34002-1	197	AR27	198	C1505M	198	21A112-071
196	AR35	196	13-34002-2	197	AR37	198	C1506	198	21A112-098
196	AR38	196	13-34002-4	197	B23-79	198	C1507A	198	21A112-099
196	AR38(RED)	196	13-34389-1	197	B502	198	C1507K	198	21A118-008
196	B23-82	196	13-34635-1	197	B503	198	C1507L	198	27D127
196	B5E	196	14-608-02A	197	B512A	198	C1507LM	198	41-0609
196	BD273	196	14-609-03A	197	B513A	198	C1507M	198	46-86466-3
196	BD275	196	14-609-09	197	BD276	198	C1520	198	46-86469-3
196	BD278	196	15-00009-05	197	BD442	198	C1520-1A	198	46-86482-3
196	BD441	196	34-6002-43	197	BD538	198	C1520-3A	198	46-86506-3
196	BD537	196	46-86347-3	197	BRC6109	198	C1520K	198	46-86508-3
196	BDX70	196	46-86459-3	197	EP15X44	198	C1520KL	198	46-86509-3
196	BDX71	196	48-137309	197	ES104	198	C1520L	198	46-86556-3
196	BDX72	196	48-869676	197	GE-250	198	C1520M	198	46-86567-3
196	BDX73	196	488869676	197	16342	198	C1521L	198	48-155059
196	BRC-5496	196	488134936	197	KLH4781	198	C1521LM	198	48-155070
196	D257	196	488137323	197	KLH5353	198	C1569-0	198	48-155110
196	D288	196	57A244-14	197	MT310	198	D44R1	198	48-155130
196	D288A	196	57B244-14	197	M9677	198	D44R2	198	48-155153
196	D288B	196	61-1053-1	197	M9677/PNP	198	D44R3	198	48-155213
196	D288C	196	86-568-2	197	M9701	198	D44R4	198	48-355005
196	D288L	196	96-5232-01	197	P3A	198	D44R6	198	48-355038
196	D289	196	96-5232-02	197	P5U	198	EP15X50	198	48-355044
196	D289A	196	96-5232-03	197	PBB	198	GE-32	198	48-869265
196	D289B	196	96-5245-01	197	PIV	198	GE-325	198	48-869286
196	D289C	196	96-5267-01	197	PIV-1	198	J241233	198	48-869320
196	D389A	196	99-8-075	197	PIV-2	198	K8C1507	198	48-869465
196	DDBY407001	196	998075	197	PIV-3	198	K8C1520	198	48-90343A61
196	DDBY407004	196	998083-1	197	QP-31	198	M9320	198	488155034
196	EP15X11	196	998100	197	R264-1	198	M9465	198	488155059
196	EP15X14	196	111N8C	197	RCA1006	198	MJE2360	198	488155070
196	EP15X30	196	121-770CL	197	RCA1011	198	MJE2561	198	488155072
196	EP15X43	196	121-772CL	197	RE22	198	NTC-9	198	488155096
196	E103	196	121-853	197	SK3083	198	Q5138	198	488155110
196	GE-241	196	121-880	197	SK3084	198	Q5138K	198	86-648-2
196	HR-68	196	121-967	197	SK3085	198	Q5138L	198	200X3172-208
196	IRTR57	196	121-976	197	STH7251	198	Q5138M	198	200X3206-800
196	IRTR92	196	121-987-02	197	TA7742	198	Q5160	198	455-2452
196	J241252	196	132-5	197	TA7743	198	Q5160Y	198	462-1059
196	KLH4745	196	153N1	197	TA8210	198	RE191	198	1043-7358
196	M9676	196	155N1	197	TA8211			198	1061-8346
		196	161-5						

ECG	Industry Standard No.	ECG	Industry Standard No.	ECG	Industry Standard No.	ECG	Industry Standard No.	ECG	Industry Standard No.
198	1061-8353	199	C1313G	199	C858F	199	HX-50107	199	ST2T003
198	1061-8668	199	C1313H	199	C858FG	199	IP20-0034	199	ST53026
198	1061-8908	199	C1313Y	199	C858G	199	IRTR51	199	ST7100
198	1070-0631	199	C1313YF	199	C859	199	J24812	199	T-Q5093
198	7362-1	199	C1313YG	199	C859E	199	J24875	199	T01-047
198	22959	199	C1313YH	199	C859F	199	J24932	199	T1007(ZENITH)
198	91371	199	C1327	199	C859FG	199	JA1350B	199	T13415K
198	198075-1	199	C1327FS	199	C859G	199	JA1350W	199	T14402
198	618960-1	199	C1327U	199	C859GK	199	JA7010	199	TE2711
198	619010-1	199	C1328	199	C900	199	JE9014C	199	TE2712
198	1417362-1	199	C1328T	199	C900(L)	199	KGE46146	199	TE2921
198	1417362-4	199	C1328U	199	C900A	199	LM1117	199	TE2922
198	2321095	199	C1335A	199	C900B	199	LM1117C	199	TE2923
198	25114266	199	C1335B	199	C900C	199	LM1540C	199	TE2924
198	25114336	199	C1335C	199	C900D	199	LMT540C	199	TE2925
198	25114344	199	C1335D	199	C900E	199	LS-0031-AR-218	199	TE2926
198	25114941	199	C1335E	199	C900F	199	LS-0095-AR-213	199	TE3390
198	25114974	199	C1335F	199	C900L	199	LS-0095-AR-2I3	199	TE3391
198	25114975	199	C1344	199	C900M	199	M9197	199	TE3391A
198	25114982	199	C1344C	199	C900U	199	M9269	199	TE3392
198	25114993	199	C1344D	199	C923	199	M9293	199	TE3393
198	25114999	199	C1344E	199	C923A	199	M9329	199	TE3394
198	35050411	199	C1344F	199	C923B	199	M9338	199	TE3395
198	35050412	199	C1345	199	C923C	199	M9384	199	TE3396
198	35062712	199	C1345C	199	C923D	199	M9409	199	TE3397
198	2003172216	199	C1345F	199	C923F	199	M9416	199	TE3398
198	2003172305	199	C1537	199	C934	199	M9447	199	TE3844
198	2003206800	199	C1537-0	199	CC82004B	199	M9474	199	TE3845
198	6621001800	199	C1537B	199	CC82004D303	199	M9486	199	TE3854
198$	C1521K	199	C1537S	199	CD441	199	M9547	199	TE3854A
198$	28C1521	199	C1538	199	CD562	199	M9594	199	TE3855
198$	28C1521K	199	C1538A	199	CGB-62	199	M9597	199	TE3855A
198$	28C1521L	199	C1538B	199	DO57	199	MP89433K	199	TE3859
199	A-1853-0404-1	199	C1538SA	199	D308	199	MP89633	199	TE3860
199	A-1854-0023-1	199	C1542	199	D9634	199	MP89633C	199	TE3900
199	A-1854-0088-1	199	C1648	199	DBCZ037300	199	MP89633D	199	TE3900A
199	A-1854-0284-1	199	C1681BL	199	DBCZ073304	199	MP896330	199	TE3901
199	A066-143	199	C1685S	199	DBCZ094504	199	MP89634B	199	TE4256
199	A110	199	C368	199	DBCZ373000	199	NB013	199	TE5088
199	A110(JAPAN)	199	C368BL	199	DDBY224001	199	NR-421AS	199	TE5089
199	A1238	199	C368GR	199	DDBY224003	199	NTC-7	199	TE5249
199	A138	199	C368Y	199	DDBY224004	199	P1901-48	199	TE5309
199	A139	199	C369	199	DDBY224006	199	P69941	199	TE5310
199	A1460	199	C369BL	199	DDBY222001	199	PA8260	199	TE5311
199	A2570773	199	C369G	199	EA15X152	199	PA8543	199	TG280536
199	A280538R	199	C369G-BL	199	EA15X161	199	PA9004	199	TG280536-D-A
199	A3J	199	C369G-GR	199	EA15X213	199	PA9005	199	TG280536-D-B
199	A3K	199	C369G-V	199	EA15X241	199	PBC107	199	TG280536-E
199	A4B	199	C369GBL	199	EA15X245	199	PBC107A	199	TG280536-E-A
199	A4F	199	C369GGR	199	EA15X258	199	PBC107B	199	TG280536-E-B
199	A6642L(NPN)	199	C369GR	199	EA15X264	199	PBC108	199	TG280536-P
199	A6642R(NPN)	199	C369V	199	EA15X288	199	PBC108A	199	TG280536-P-A
199	A644L	199	C374	199	EA15X325	199	PBC108B	199	TG280536E
199	A644S	199	C374-BL	199	EA15X336	199	PBC108C	199	TI-415
199	A645L	199	C374-V	199	EA15X352	199	PBC109	199	TI-416
199	A645S	199	C374JA	199	EA15X353	199	PBC109B	199	TI-418
199	A667-GRN	199	C536	199	EA15X354	199	PBC109C	199	TI-419
199	A667-ORG	199	C536FS	199	EA15X355	199	PFT4005	199	TI-421
199	A667-RED	199	C536G	199	EA15X361	199	PIT-37	199	TI54D
199	A667-YEL	199	C536H	199	EA15X386	199	PS6010-1	199	TIB-94
199	A668-GRN	199	C537(F)	199	EA15X404	199	Q-00469	199	TIB-97
199	A668-ORG	199	C537(G)	199	EA15X7245	199	Q-00469A	199	TM199
199	A669-GRN	199	C537A	199	EA15X7583	199	Q-00469B	199	TN280945-R
199	A669-YEL	199	C537B	199	EA2429	199	Q-00469C	199	TNJ1034
199	A67-07-244	199	C537FV	199	EA2738	199	Q-00569C	199	TNJ70691
199	A67-33-340	199	C537G	199	EA2771	199	Q-08115C	199	TNJ71034
199	A675419H	199	C537GI	199	EA3763	199	Q-09115C	199	TNJ71271
199	A747B	199	C537H	199	EA4025	199	Q-13115C	199	TNJ71277
199	A748C	199	C537HT	199	EN2484	199	Q5053D	199	TNJ71965
199	A749C	199	C537W	199	EP15X3	199	Q5053E	199	TP4067-410
199	A76228	199	C644	199	EQ8-0061	199	Q5053F	199	TP4067-411
199	A8R	199	C644C	199	EQ8-131	199	Q5053G	199	TR 19A
199	A9-175	199	C644F	199	EQ8-78	199	Q5078E	199	TR-1993
199	B-1910	199	C644F/494	199	ES15049	199	Q5121	199	TR-8034
199	BC108C	199	C644FR	199	ES15052	199	Q51210	199	TR-8040
199	BC109C	199	C644FS	199	ES379262	199	Q51210	199	TR-BC1490
199	BC127	199	C644H	199	ET379462	199	Q51210	199	TRO1014
199	BC128	199	C644HR	199	ET380634	199	Q5121R	199	TRO1040
199	BC146	199	C644HS	199	ET398711	199	Q5180	199	TR105(SPRAGUE)
199	BC155C	199	C644P	199	ET398777	199	Q5183	199	TR2327363
199	BC156C	199	C644PJ	199	ET511263	199	Q5183P	199	TR2327443
199	BC168C	199	C644Q	199	ET517994	199	QR102	199	TR2327444
199	BC169C	199	C644R	199	ETTC-945	199	QR105	199	TR33
199	BC182K	199	C644RST	199	EW8-78	199	QT-C0828XAN	199	TR4010
199	BC183L	199	C644S	199	F079	199	QT-C0828XDN	199	T8C-499
199	BC184L	199	C644S/494	199	FSE4002	199	QT-C0900XBA	199	T80767
199	BC209/7825B	199	C644H	199	G05-012-G	199	QT-C0900XBD	199	V828C1335D/-1
199	BC20C	199	C693	199	G05-035-E	199	QT-C0900XCA	199	V828C1335D/1
199	BC384C	199	C693(JAPAN)	199	G05-036-B	199	QT-C0945ACA	199	V828C16819/1E
199	BC385B	199	C693A	199	G05-036-D	199	QT-C0945AGA	199	V828C371-R-1
199	BC386A	199	C693B	199	G05-706-D	199	QT-CBC546AA	199	V828C375-//1E
199	BC386B	199	C693C	199	G05-706-E	199	RE192	199	V828C375Q-1
199	BC408B	199	C693D	199	G05039D	199	RE4001	199	V828C574-B-1
199	BC408C	199	C693E	199	G212	199	RE4002	199	V828C585-W/1E
199	BC520	199	C693E(JAPAN)	199	GE-10A	199	RE401Q	199	V828C732-V1F
199	BC520B	199	C693EB	199	GE-212	199	RE64	199	V828C733B-1
199	BC520C	199	C693ET	199	GE-62	199	RE67	199	V828C945LK-1
199	BC521	199	C693F	199	G8-85	199	RLB-17	199	V828C945LK-1
199	BC521C	199	C693FC	199	HC-00373	199	RS-279US	199	V89-0003-913
199	BC521D	199	C693FL	199	HC-00536	199	RS-805US	199	X19001-D
199	BC522	199	C693FU	199	HC-00711	199	RT2309	199	ZEN116
199	BC522C	199	C693G	199	HC-00732	199	RT4761	199	Z930
199	BC522E	199	C693G(JAPAN)	199	HC-00900	199	RT5208	199	001-021070
199	BC523	199	C693GL	199	HC-00923	199	RT5435	199	01-030373
199	BC523B	199	C693GS	199	HC-01000	199	RT5905	199	01-030536
199	BC525C	199	C693GU	199	HC-01335	199	RT6737	199	01-030711
199	BC546	199	C693GZ	199	HE-00950	199	RT7559	199	01-030712
199	BC546A	199	C693H	199	HEP-80023	199	RT8047	199	01-030828
199	BC546B	199	C694	199	HEP-80024	199	RT8666	199	01-030900
199	BC737	199	C694E	199	HEP726	199	RT8863	199	01-031327
199	BC738	199	C694F	199	HEP730	199	RV2070	199	1A0024
199	BFR25	199	C694G	199	HEP737	199	RV2248	199	002-104-000
199	BFV89	199	C694Z	199	HP09S901E	199	RV2354	199	002-9501
199	BFV89A	199	C715(JAPAN)	199	HR-14	199	RVTC81473	199	002-9502
199	BFY39-1	199	C732	199	HR-15	199	S001465	199	002-9502-12
199	BFY39-2	199	C732BL	199	HR-47	199	S0023	199	2N2931
199	BFY39-3	199	C732GR	199	HR-75	199	S006793	199	2N2932
199	BTX-068	199	C732B	199	HR47	199	S006927	199	2N2933
199	C1000	199	C732V	199	HT306442A	199	S007764	199	2N2934
199	C1000-BL	199	C732Y	199	HT306442B	199	S24592	199	2N2935
199	C1000-GR	199	C733	199	HT30733100	199	SE5-0127	199	2N3117
199	C1000-Y	199	C733-0	199	HT308281D	199	SE5-0565	199	2N3390
199	C1000Y	199	C733-0	199	HT308281H	199	SE5-0569	199	2N3391
199	C1204D	199	C7355-BL	199	HT308282B	199	SE5-0938	199	2N3391A
199	C1215	199	C733BL	199	HT309301C	199	SE5-0938-55	199	2N3398
199	C1222	199	C739GR	199	HT309301E	199	SE5-0938-56	199	2N3901
199	C1222A	199	C733R	199	HT309301F	199	SE5-0938-57	199	2N4967
199	C1222B	199	C733Y	199	HT309451LO	199	S019806	199	2N5232
199	C1222C	199	C785	199	HT309451RO	199	S025094	199	2N5232A
199	C1222D	199	C828R/494	199	HT310002A	199	SP82271	199	2N5249
199	C1313	199	C858	199	HT313271T	199	SP84272	199	2N5249A
199	C1313F	199	C858K	199	HT313272B	199	SP84814	199	28C-NJ100
				199	HT36441B	199	ST-28C383W		
						199	ST.082.114.016		

ECG	Industry Standard No.
199	2SC1000
199	2SC1000-BL
199	2SC1000-GR
199	2SC1000-Y
199	2SC1000BL
199	2SC1000GR
199	2SC1000Y
199	2SC1204D
199	2SC1222
199	2SC1222A
199	2SC1222B
199	2SC1222C
199	2SC1222D
199	2SC1222E
199	2SC1222U
199	2SC1327
199	2SC1327FS
199	2SC1327S
199	2SC1327T
199	2SC1327TU
199	2SC1327TV
199	2SC1327U
199	2SC1328
199	2SC1328T
199	2SC1328U
199	2SC1335
199	2SC1335A
199	2SC1335B
199	2SC1335C
199	2SC1335D
199	2SC1335E
199	2SC1335F
199	2SC1344
199	2SC1344C
199	2SC1344D
199	2SC1344E
199	2SC1344F
199	2SC1345
199	2SC1345C
199	2SC1345D
199	2SC1345E
199	2SC1345F
199	2SC1537
199	2SC1537-O
199	2SC1537B
199	2SC1537S
199	2SC1538
199	2SC1538A
199	2SC1538S
199	2SC1538SA
199	2SC1542
199	2SC1570
199	2SC1570LH
199	2SC1571
199	2SC1571G
199	2SC1571L
199	2SC1637
199	2SC1647
199	2SC1647Q
199	2SC1647RY
199	2SC1648
199	2SC1648E
199	2SC1648S
199	2SC1648SH
199	2SC1681
199	2SC1681BL
199	2SC1681GR
199	2SC1681V
199	2SC1682
199	2SC1682-BL
199	2SC1682-GR
199	2SC1682V
199	2SC1684S
199	2SC1684T
199	2SC1685T
199	2SC1787
199	2SC2309
199	2SC368
199	2SC368BL
199	2SC368GR
199	2SC368V
199	2SC369
199	2SC369BL
199	2SC369G
199	2SC369G-BL
199	2SC369G-GR
199	2SC369G-V
199	2SC369GBL
199	2SC369GGR
199	2SC369GR
199	2SC369V
199	2SC374
199	2SC374-BL
199	2SC374-V
199	2SC374JA
199	2SC536
199	2SC537(G)
199	2SC537F-C7
199	2SC537FV
199	2SC537G
199	2SC537G2
199	2SC537GF
199	2SC537GI
199	2SC537H
199	2SC537HT
199	2SC537W
199	2SC644
199	2SC644C
199	2SC644F
199	2SC644FR
199	2SC644FS
199	2SC644H
199	2SC644HR
199	2SC644HS
199	2SC644P
199	2SC644PJ
199	2SC644Q
199	2SC644R
199	2SC644RST
199	2SC644S
199	2SC644T
199	2SC693
199	2SC693E
199	2SC693EB
199	2SC693ET
199	2SC693F
199	2SC693FC
199	2SC693FL
199	2SC693FU
199	2SC693G
199	2SC693GL
199	2SC693GS
199	2SC693GU
199	2SC693GZ
199	2SC693H
199	2SC694
199	2SC694E
199	2SC694F
199	2SC694G
199	2SC694Z
199	2SC732
199	2SC732B
199	2SC732BL
199	2SC732GR
199	2SC732GR/4454C
199	2SC732S
199	2SC732V
199	2SC732Y
199	2SC733
199	2SC733(GR)
199	2SC733-O
199	2SC733-0
199	2SC733-Y
199	2SC7335-BL
199	2SC733B
199	2SC733BL
199	2SC733GR
199	2SC733Q
199	2SC733R
199	2SC733V
199	2SC733Y
199	00002SC858
199	2SC858E
199	2SC858F
199	2SC858FG
199	2SC858G
199	2SC859
199	2SC859E
199	2SC859F
199	2SC859FG
199	2SC859G
199	2SC859GK
199	2SC900
199	2SC900(F)
199	2SC900(L)
199	2SC900A
199	2SC900B
199	2SC900C
199	2SC900D
199	2SC900E
199	2SC900F
199	2SC900L
199	2SC900M
199	2SC900U
199	2SC900VE
199	2SC923
199	2SC923A
199	2SC923B
199	2SC923C
199	2SC923D
199	2SC923E
199	2SC923F
199	2SCNJ100
199	2SD591
199	2SD591R
199	2SD599
199	2SE4002
199	3L4-6007-15
199	3L4-6007-16
199	4-1544
199	4-1791
199	4-46
199	07-07166
199	8-729-309-06
199	8-81250108
199	8Q-3-10
199	09-302038
199	09-302053
199	09-302085
199	09-302086
199	09-302093
199	09-302097
199	09-302107
199	09-302125
199	09-302127
199	09-302139
199	09-302194
199	09-30501
199	09-305048
199	09-305052
199	09-305123
199	09-305126
199	09-309059
199	09-309070
199	9TR3
199	10-003
199	11-27070
199	13-14085-41
199	13-14085-49
199	13-14085-95
199	13-14085-96
199	13-14085-97
199	13-18365-1
199	13-23338-3
199	13-23338-4
199	13-23339-2
199	13-23339-3
199	13-29033-4
199	13-34381-1
199	13-34381-2
199	13-40312-1
199	13-4085-121
199	13-4085-122
199	13-4085-41
199	14-602-46
199	14-602-46A
199	14-602-63
199	14-801-23
199	014-862
199	15-02757-00
199	15-09587
199	019-003675-205
199	019-004094
199	020-1110-012
199	020-1110-013
199	21A040-064
199	21A040-065
199	21A040-066
199	21A040-067
199	21A040-082
199	21A040-083
199	21A112-019
199	21A112-045
199	21A112-046
199	21A112-058
199	21A404-066
199	21M087
199	21M091
199	21M095
199	21M096
199	21M124
199	21M137
199	21M138
199	21M160
199	21M161
199	21M170
199	21M174
199	21M408
199	21M446
199	21M550
199	21M603
199	24-001326
199	24-001354
199	24M125
199	24MW1022
199	24MW1025
199	24MW1060
199	24MW826
199	24MW964
199	24MW965
199	24MW990
199	24T-009
199	025-100017
199	27A10489-101-11
199	31-0012
199	31-0013
199	31-0052
199	31-0099
199	31-0100
199	31-027-0
199	033A
199	34-1009
199	34-1019
199	37-193MP
199	37-21401
199	42-27277
199	42-27534
199	42-27535
199	42-30092
199	42-9029-31B
199	42-9029-31L
199	42-9029-31P
199	42-9029-31Q
199	42-9029-60D
199	43A162445P1
199	43A212090P1
199	43B168613-1
199	44T-300-106
199	44T-300-111
199	44T-300-112
199	46-86353-3
199	46-86434-3
199	46-86535-3
199	46-86536-3
199	46-86543-3
199	46-8682-3
199	48-01-049
199	48-134810
199	48-134813
199	48-134823
199	48-134997
199	48-137015
199	48-137300
199	48-137325
199	48-137390
199	48-155073
199	48-155154
199	48-355004
199	48-40170-G01
199	48-40170G01
199	48-40246-G01
199	48-40246G01
199	48-40247G02
199	48-40606J02
199	48-40607J01
199	48-44885G01
199	48-45N2
199	48-869197
199	48-869269
199	48-869293
199	48-869338
199	48-869384
199	48-869409
199	48-869416
199	48-869447
199	48-869474
199	48-869486
199	48-869497
199	48-869547
199	48-869594
199	48-90343A06
199	48-90343A73
199	48-90343A79
199	48K869269
199	48K869293
199	48K869409
199	48K869447
199	48K869474
199	48K869486
199	48R869197
199	48R869338
199	48R869384
199	48R869416
199	48R869547
199	48R869594
199	48S137015
199	48S155073
199	48S155089
199	48S40170G01
199	48S40246G01
199	48S40247G01
199	48S40606J02
199	48S40607J01
199	56-4827
199	56-4829
199	56-8196
199	56-8197
199	57A142-2
199	57A143-12
199	57A144-12
199	57A6(NPN)
199	57B203-14
199	57C6
199	61-309686
199	065-006
199	65A11573
199	66-P11112-0001
199	66F026-1
199	67A9060
199	68A7366-1
199	74-01-772
199	77-271798-2
199	77-27198-3
199	78E67-2
199	81-46125053-2
199	81-46125063-1
199	86-005135-2
199	86-400-2
199	86-444-2
199	86-526-2
199	86-607-2
199	86A327
199	87-0005
199	87-0009
199	87-0013
199	87-0023-T
199	87-0212-1
199	87-0218-U
199	87-0227
199	87-0230
199	87-0230-1
199	87-0231
199	87-0235B
199	87-0238
199	87-218-U
199	88-1250108
199	90-140
199	90-181
199	90-33
199	90-458
199	90-459
199	90-603
199	90-605
199	90-614
199	90-70
199	96-5346-01
199	99S033
199	102-0373-00
199	102-0732-28
199	102-0828-17
199	102-0945-16
199	102-0945-17
199	102-0945-38
199	102-0945-39
199	102-1335-04
199	0103-0088/4460
199	0103-0492
199	0103-93
199	106-003
199	112-203391
199	122-7
199	122-7(RCA)
199	123-005
199	123-007
199	127-115
199	130-40216
199	130-40883
199	130-40901
199	0131-004792
199	0131-005347
199	0131-005348
199	0131-005349
199	0131-005350
199	0140-5
199	142-002
199	142-003
199	147-7016-01
199	151-0341-00
199	151-0341-00-A
199	151-0456-00
199	176-006-9-002
199	176-025-9-001
199	176-031-9-002
199	176-037-9-003
199	176-037-9-004
199	176-042-9-003
199	176-060-9-003
199	176-062-9-001
199	185-005
199	185-010
199	186-003
199	186-004
199	186-005
199	195
199	207A17
199	215-37567
199	229-1301-23
199	231-0004-01
199	231-0004-03
199	250-0373
199	250-0711
199	250-1312
199	260-10-023
199	260-10-042
199	260D13701
199	260D15901
199	260P17502
199	260P17701
199	260P17704
199	260P19503
199	296-55-9
199	297L007C
199	297L007C03
199	297L015C01
199	311D916P01
199	325-0500-12
199	325-0500-13
199	352-0549-000
199	352-0638
199	353-9306-006
199	353-9306-007
199	353-9318-001
199	353-9318-002
199	386-7178-P001
199	400-1569-101
199	417-108-13163
199	417-126-12903
199	417-226-13163
199	417-244-12903
199	417-283-13271
199	434
199	462-1038-01
199	462-1066-01
199	462-2004
199	536F(JVC)
199	536GT
199	576-0003-022
199	599P3430
199	600-207-801
199	600-224-605
199	602X0008-002
199	602X0018-002
199	686-257-0
199	693GV
199	700-155
199	700A-858-318
199	700A-858-328
199	700A858-319
199	750A858-328
199	753-1644-100
199	753-2000-100
199	753-4000-010
199	753-6000-002
199	753-8400-230
199	753-8500-380
199	753-9000-839
199	772-101-00
199	853-0300-644
199	858
199	859GK
199	881-250108
199	902-000-2-04
199	902-002-3-06
199	902-003-3-17
199	902-003-6-06
199	903Y002150
199	904-95B
199	921-106B
199	921-111B
199	921-114B
199	921-115B
199	921-116B
199	921-127B
199	921-133
199	921-133B
199	921-152B
199	921-154B
199	921-155B
199	921-196
199	921-196B
199	921-202B
199	921-205B
199	921-206
199	921-206B
199	921-207
199	921-207B
199	921-208B
199	921-209B
199	921-239B
199	921-240B
199	921-257B
199	921-258B
199	921-281B
199	921-92B
199	929C
199	929CA
199	929CU
199	929D(JVC)
199	929DX
199	930(OV)
199	930B
199	930C
199	930C(WARDS)
199	930D
199	930D(WARDS)
199	930DU
199	930DX
199	930E
199	930E(JVC)
199	930E(WARDS)
199	930EX
199	943-742-002
199	991-002298
199	991-002356
199	991-002873
199	991-013544
199	1003-6754
199	1004-0780
199	1025G(GE)
199	1030-21
199	1035-80
199	1039-0961
199	1042-03
199	1048-9920
199	1049-0092
199	1049-0100
199	1050-21
199	1057-2071
199	1074-03
199	1074-115
199	1080-21
199	1096-12
199	1472 8349
199	1479 7963
199	1515(NPN)
199	1751-17
199	1835-17
199	1854-0387
199	1854-SRK1-1
199	2000-245
199	2020-05
199	2035-5100-53660
199	2035-5100-69372
199	2057A2-212
199	2057A2-249
199	2057A2-257
199	2057A2-260
199	2057A2-262
199	2057A2-272
199	2057A2-273
199	2057A2-274
199	2057A2-275
199	2057A2-290
199	2057A2-333
199	2057A2-334
199	2057A2-352
199	2057A2-370
199	2057A2-373
199	2057A2-385
199	2057A2-391
199	2057A2-404
199	2057A2-405
199	2057A2-428
199	2057A2-436
199	2057A2-454
199	2057A2-475
199	2057A2-502
199	2057A2-518
199	2057A2-542
199	2057A2-543
199	2057A2-560
199	2057B-59
199	2063-17
199	2065-54
199	2101
199	2121-17
199	2132-17
199	2195-17
199	2207-17

ECG	Industry Standard No.	ECG	Industry Standard No.	ECG	Industry Standard No.	ECG	Industry Standard No.	ECG	Industry Standard No.	ECG	Industry Standard No.
199	2212-17	199	171162-180	199	436004601	213	2N2076	219	TR2327841	221	M75561-23
199	2271	199	171162-204	199	436006401	213	2N2076A	219	TR29	221	M75561-23RN
199	2338-17	199	171162-247	199	485134997	213	2N2077	219	2N3789	221	MF83005
199	2449-17	199	171162-285	199	2003037309	213	2N2077A	219	2N3790	221	M08365
199	2510-103	199	171162-288	199	2003082823	213	2N2078	219	2N3791	221	SF8503
199	3107-204-90150	199	171553	199	2003082837	213	2N2078A	219	2N3792	221	TA7149
199	3391	199	171554	199	4100900116	213	2N2079	219	2N4901	221	TA7150
199	3391(SEARS)	199	171558	199	4104206442	213	2N2079A	219	2N4902	221	TA7151
199	3391A	199	171559	199	4104208282	213	2N2080	219	2N4903	221	TA7260(RCA)
199	3391A(SEARS)	199	171676	199	4104208283	213	2N2080A	219	2N4907	221	TR5453
199	003460	199	171677	199	4109213354	213	2N2081	219	2N4908	221	TR40603
199	003461	199	171678	199	4203970101	213	2N2081A	219	2N4909	221	V00236-001
199	3519-1(RCA)	199	181012	199	16108190536	213	2N2082	219	2N6246	221	V00236-00I
199	4039-00	199	200200	199	16109190536	213	2N2082A	219	2N6247	221	28N124
199	4039-01	199	200200-700	199	2035510069362	213	2N2152A	219	2N6469	221	3N140
199	4824-0014	199	202617	2011	ITT652	213	2N2153	219	2N6594	221	3N141
199	4824-0014-02	199	204117	2011	L201	213	2N2153A	219	28A663	221	38K35
199	5001-038	199	213217	2011	9665PC	213	2N2154A	219	28A663-R	221	38K35-BL
199	5001-043	199	257017	2012	ITT654	213	2N2155	219	28A744	221	38K35-GR
199	5001-074	199	268003	2012	L202	213	2N2155A	219	020-1111-009	221	38K35-Y
199	5001-505	199	281001-53	2012	9666PC	213	2N2156A	219	020-1111-080	221	38K35G
199	5001-511	199	281001-83	2013	ITT656	213	2N2157	219	34-1029	221	38K41
199	5001-545	199	325934	2013	L203	213	2N2157A	219	34-1029	221	38K41(L)
199	05206-00	199	506902	2013	9667PC	213	2N2158	219	48-137313	221	38K41C
199	5613-1327T	199	510584	2014	9668PC	213	2N2158A	219	111P5	221	38K41L
199	5613-1684T	199	530130-1	210	C1018B	213	2N2159	219	0137	221	38K41M
199	5613-8288	199	531841-002	210	D40D1	213	2N2159A	219	2549	221	38K45
199	5613-871	199	532775	210	D40D10	213	2N2490	219	5118	221	38K45-B-09
199	5613-871P	199	610079-2	210	D40D11	213	2N2491	219	12537	221	38K45B
199	7502	199	610128-1	210	D40D2	213	2N2492	219	99240-292	221	38K45B09
199	7509	199	610128-2	210	D40D3	213	2N2493	219	181038	221	7-117-02
199	7591	199	610128-5	210	D40D4	213	2N277	219	202925-047	221	045
199	8000-00004-089	199	610128-6	210	D40D5	213	2N278	219	332762	221	46-86396-3
199	8000-00005-003	199	610151-1	210	D40D7	213	2N290	219	760021	221	047
199	8000-00005-004	199	610151-3	210	D40D8	213	2N3311	219	9004017-01	221	86-540-2
199	8000-00005-005	199	610151-5	210	ET392927	213	2N3312	219	9006527-01	221	998045
199	8000-00009-089	199	610224-1	210	GE-252	213	2N3313	220	A054-142	221	0117-02
199	8000-00041-041	199	610226-1	210	HEP-85023	213	2N3314	220	DDCY103001	221	132-045
199	8000-00041-042	199	618165-1	210	HEP-85024	213	2N3315	220	HEP-F2005	221	132-047
199	8020-204	199	618181-1	210	HEP-85025	213	2N441	220	M75561-10RK	221	576-0006-221
199	8200-202	199	699291	210	HR-73	213	2N442	220	MF83004	221	635-1(RCA)
199	8200-203	199	740438	210	HR85	213	2N443	220	TA2840	221	2079-40
199	8394	199	740439	210	MB-5	213	2N575	220	W1E	221	3618
199	8440-126	199	740440	210	RE80	213	2N575A	220	W1R	221	6221
199	8868-7	199	740441	210	RS-2018	213	28B235	220	2N5823	221	40600
199	8910-144	199	740442	210	TM210	213	28B235A	220	3N128	221	40601
199	9033-2	199	740886	210	TV73	213	28B259	220	3N142	221	40602
199	9033-3	199	740887	210	TV75	213	28B260	220	3N143	221	40603
199	9033-4	199	741114	210	13-33185-1	213	28B331	220	3N152	221	40604
199	9033-5	199	741737	210	34-6016-31	213	28B531H	220	38K33	221	40841
199	9033GREEN	199	741857	210	34-6016-53	213	28B332	220	13-33173-1	221	60337
199	9033ORANGE	199	741861	210	86-673-2	213	28B332H	220	13-37900-1	221	135324
199	9033Q	199	742512	210	260D0B214	213	28B333	220	04B	221	135963
199	9033RED	199	742513	210	260P21106	213	28B352	220	48-137707	221	985175
199	9033WHITE	199	742548	210	964-27986	213	28B352D	220	48-137070	221	1473588-2
199	9330-688-30112	199	803696	210	3107-204-90180	213	28B353	220	123-002	221	1473635-2
199	09500	199	815173	210	171162-164	213	1415-159	220	132-048	221	2068491-704
199	09501	199	908844-1	210	171162-235	2138	B258	220	417-206	221	3404520-601
199	9502	199	916033	210	702415-00	2138	28B258	220	417-207	222	A498(F.E.T.)
199	16237	199	916052	2104	N4027	218	A566	220	921-157B	222	AR501
199	18410-144	199	916055	211	A645	218	A566A	220	38378	222	AR502
199	19500-253	199	964547-2	211	AR26	218	A566B	220	40467A	222	DDCY104001
199	20103	199	971460	211	AR29	218	A613	220	40468A	222	DDCY104003
199	20810-91	199	992066	211	D41D1	218	A614	220	40593A	222	D8-102
199	20810-92	199	1223921	211	D41D10	218	A616	220	60793	222	D8-105
199	22810-174	199	1417308-1	211	D41D11	218	HEP-85018	220	5320031		
199	25840-162	199	1417308-2	211	D41D2	218	J241258	220	5518017		
199	26810-155	199	1443024-1	211	D41D4	218	MJ2253	221	BF828		
199	27125-460	199	1471115-11	211	D41D5	218	MJ2254	221	BF828R		
199	27125-470	199	1471115-2	211	D41D7	218	MJ3701				
199	37510-161	199	1471115-3	211	D41D8	218	RB68				
199	38478	199	1471115-4	211	GE-253	218	SDT5575				
199	38510-164	199	1471120-11	211	HEP-83027	218	SJ2009				
199	38510-350	199	1471122	211	HEP-83028	218	SJ652				
199	41175	199	1471122-6	211	HEP-83029	218	SJ822				
199	43054	199	1471122-7	211	HR-74	218	2N4898				
199	43055	199	1472475-1	211	HR86	218	2N4899				
199	45337-A	199	1473506-1	211	P2U-2	218	2N4900				
199	45810-166	199	1473614-2	211	R881	218	2N6049				
199	51429	199	1612738-1	211	RS-2026	218	28A566				
199	55200-74	199	2125310	211	SPS4092	218	28A566A				
199	58810-164	199	2320598	211	TM211	218	28A566B				
199	58810-167	199	2498665-1	211	28A645	218	28A613				
199	71963-1	199	2498665-2	211	28A646	218	28A614				
199	000073070	199	2498665-3	211	3I4-6011-2	218	28A616				
199	000073080	199	2499950	211	3I4-6011-3	218	024				
199	000073350	199	2505207	211	3I4-6011-52	218	3638-1				
199	000073351	199	3755862	211	3I4-6011-53	218	135551				
199	000073360	199	4000921	211	34-6015-54	218	1473638-1				
199	000073361	199	4906093	211	921-1021	2188	A483				
199	000073373	199	5320813	211	13299	2188	A566C				
199	000073374	199	7576015-02	211	171162-195	2188	28A483				
199	75613-1	199	7576015-03	2118	A780AK	2188	28A566C				
199	75613-2	199	7576015-04	2118	A780AKA	218	28A566C				
199	81410-145	199	7576015-05	2118	A780AKB	219	A663				
199	88060-144	199	7852452-01	2118	A780AKC	219	A744				
199	88510-174	199	8113323	2118	A780AKD	219	BDX18				
199	90934-35	199	8113327	2118	28A780AK	219	BDX18N				
199	99240-269	199	8722248-2	2118	28A780AKA	219	B964(ELCOM)				
199	100292	199	10183001	2118	28A780AKB	219	B990				
199	122519	199	10545506	2118	28A780AKC	219	B990(ELCOM)				
199	127355	199	11706998-2	2118	28A780AKD	219	HEP-248				
199	129425	199	13094517	213	ADY26	219	HEP-705				
199	129509	199	20088508	213	B235	219	HEP-87003				
199	129510	199	23114216	213	B235A	219	KS-20033L2				
199	129511	199	23114217	213	B259	219	MJ2901				
199	129512	199	23114258	213	B260	219	MJ2955				
199	129513	199	23114349	213	B331	219	NKT4055				
199	136282	199	26010025	213	B332	219	P18				
199	138001	199	28102128-002	213	B353	219	P1E-1BLK				
199	138019-001	199	28102128-004	213	B552	219	P1E-1GRN				
199	138789-4	199	28105789-001	213	B553	219	P1E-1RED				
199	138789-5	199	43024859	213	B430	219	P1E-1VIO				
199	144034	199	43024873	213	B433	219	P1E-2BLK				
199	144035	199	43025538	213	0B-240	219	P1E-2BLU				
199	144040	199	43029483	213	2N1157	219	P1E-2GRN				
199	144044	199	44079004	213	2N1157A	219	P1E-2RED				
199	144858	199	50210700-00	213	2N1358	219	P1E-2VIO				
199	146512	199	50210700-01 VF	213	2N1358A	219	P1E-3BLK				
199	147115-7	199	50210700-10	213	2N1358M	219	P1E-3BLU				
199	147115-8	199	50210700-10 VF	213	2N1518	219	P1E-3GRN				
199	147122P7	199	62563273	213	2N1519	219	P1E-3RED				
199	150714-1	199	62563281	213	2N1520	219	P1E-3VIO				
199	162002-085	199	62638249	213	2N1521	219	P5W				
199	167541	199	62741465	213	2N1522	219	RB82				
199	167956	199	62744081	213	2A173	219	SJ2001				
199	167957	199	62748281	213	2N174	219	SJ3520				
199	168405	199	62766255	213	2N174A	219	SJ3679				
199	168651	199	62766263	213	2N174RED	219	SK173				
199	168716	199	80052101	213	2N1970	219	TM219				
199	171162-004	199	82073206	213	2N1980	219	TR-29				
199	171162-095	199	83073306	213	2N1981	219	TR0259-6				
199	171162-100	199	93064440	213	2N1982	219	TR1038-4				
199	171162-113	199	93073260	213	2N2075						
				213	2N2075A						

ECG	Industry Standard No.
222	DS-106
222	EA15X402
222	EA15X405
222	EF4(ELCOM)
222	EF5(ELCOM)
222	EP15X64
222	FT0601
222	GE-FET-4
222	HF200301E
222	IP20-0157
222	IP20-0218
222	K39Q(2 GATE)
222	K39R(2 GATE)
222	K40(2 GATE)
222	K45(2 GATE)
222	KD2130
222	MEM564C
222	MEM630
222	MEM680Y
222	MFE121
222	MFE130
222	MFE130-712
222	MFE131
222	MFE3006
222	MFE3007
222	MFE3008
222	MFP121
222	MPF-121
222	MPF121
222	R3651-1
222	RE199
222	RT180
222	SE5315
222	SFB8970
222	SFC8999
222	SFD2285
222	SFE303424
222	SFE425
222	SK3050
222	SK3065
222	SPF274
222	SPF512
222	T1731
222	T464
222	TA2644
222	TA7189
222	TA7274
222	TA7374
222	TA7399
222	TA7669
222	TA7684
222	TA8242
222	TR08004
222	TR2327431
222	W1U
222	W1U-1
222	YEAN3SK39Q
222	01-080045
222	03A10
222	3L4-6503-1
222	3L4-6503-2
222	3N187
222	3N201A
222	3N213
222	3SK34
222	3SK37
222	3SK39
222	3SK39E
222	3SK39P
222	3SK39Q
222	3SK39R
222	3SK44
222	3SK49
222	3SK49Q
222	3SK59
222	3SK59BL
222	3SK59GR
222	7-40
222	09-305040
222	13-0118
222	13-0165
222	13-10321-37
222	13-28583-1
222	022-2876-008
222	24MW1122
222	033-014-0
222	041A
222	46-8646
222	48-137488
222	48-137567
222	48-3003A10
222	48-64978A39
222	57A267-4
222	86-464-2
222	86-605-2
222	86-606-2
222	86-625-2
222	86-632-2
222	90-178
222	99S041
222	99S045A
222	99S046
222	121
222	121-1024
222	121-1030
222	121-783
222	121-784
222	121-785
222	121-786
222	121-787
222	121-826
222	121-953
222	182-038-9-001
222	182-138-9-001
222	201-283818-1
222	201-283818-2
222	201-283818-3
222	203-1
222	203-4
222	229-0192-18
222	355D7
222	576-0003-224
222	576-0006-222
222	576-0006-227
222	1009-01
222	1042-01
222	1251-1-1
222	2000-102
222	2000-103
222	2022-06
222	2065-55
222	2359-17
222	3588(RCA)
222	3635(RCA)
222	3635-1
222	5001-046
222	6227
222	7372-1
222	8000-00011-053
222	8000-00042-013
222	38563
222	40673
222	40823
222	95132
222	114267
222	127980
222	129980
222	134450
222	138946
222	141332
222	171206-6
222	258017
222	610166-1
222	610203-1
222	610203-2
222	610203-3
222	610203-4
222	610203-5
222	610203-6
222	610358-1
222	610358-2
222	610358-3
222	741860
222	760005
222	986930
222	1408694-1
222	1417372-1
222	1473618-1
222	1473635
222	1473635-1
222	2327431
222	3596401
222	3596402
222	4084114-0001
222	4084114-0002
224	A059-111
224	C312
224	RE202
224	SE-3034
224	SE3034
224	SM7989
224	2N2196
224	2SC312
224	019-003935
224	120-004885
224	120-004886
224	172-001
224	172-001-9-001
224	1068-17A
224	2904-030
224	40446
224	40582
224	112362
224	112527
224	120004885
224	120004886
225	A104-8
225	A12-1
225	A12-2
225	A13-15809-1
225	A13-28432-1
225	A13-55062-1
225	A14-602-19
225	A14-602-36
225	A14-602-37
225	A3170757
225	A3545
225	A3547
225	A3552
225	A3553
225	A3582
225	A48-134648
225	AA133
225	AA1M
225	AA2A
225	AA2K
225	AA2Z
225	AA310
225	AA3M
225	C522-R
225	C523-R
225	C524-R
225	GE-256
225	001-021221
225	12-2
225	3552
225	3553
225	3582
225	4686-210-3
225	25762-010
225	25762-012
225	40544
225	239103
226	B254
226	B255
226	B256
226	B414
226	B474
226	B474-2
226	B474-3
226	B474-4
226	B474-6D
226	B474MP
226	B474S
226	B474V10
226	B474V4
226	B474Y
226	B481
226	B62
226	D146
226	ES61(ELCOM)
226	GE-49
226	RE83
226	RE83MP
226	SK3082
226	TM226
226	2SB254
226	2SB255
226	2SB256
226	2SB414
226	2SB474
226	2SB474-2
226	2SB474-3
226	2SB474-4
226	2SB474-6D
226	2SB474MP
226	2SB474S
226	2SB474V10
226	2SB474V4
226	2SB474Y
226	2SB62
226	2SD146
226	8200-204
226$	B445
226$	B446
226$	B466
226$	B467
226$	2SB445
226$	2SB446
226$	2SB466
226$	2SB467
226MP	IRTR94
226MP	IRTR94MP
226MP	R-2SB474
226MP	SK3086
226MP	TM226MP
226MP	4-48
226MP	4-48(SEARS)
226MP	09-301030
226MP	65-080001
226MP	1042-10
226MP	5001-049
226MP	22881
228A	PTC124
228A	TA7739
228A	TA7740
228A	2N6175
228A	2N6176
228A	2N6177
228A	2SC2637
228A	2SD625
228A	13-35089-1
228A	13-35089-2
228A	13-35089-3
228A	13-35089-4
228A	13-45018-1
228A	86X0076-001
228A	089-2
228A	089-3
228A	40885
228A	40886
228A	40887
228A	41505
229	A066-111
229	A67-37-940
229	AR218
229	BC238
229	BD113
229	BF194
229	BF195
229	BF196
229	BF198
229	BF199
229	BF224
229	BF225
229	BF233-2
229	BF233-3
229	BF234
229	BF240
229	BF241
229	BF254
229	BF255
229	BF310
229	BF311
229	BF314
229	BF365
229	BF494
229	BF495
229	BF594
229	BF595
229	C1126
229	C1128(S)
229	C1128D
229	C1129
229	C1129(R)
229	C1187
229	C1393
229	C1394
229	C1686
229	C1687
229	C668
229	C772CS
229	C772R(JAPAN)
229	C7840
229	C785(O)
229	C785BN
229	C785D
229	C785E
229	C785R
229	C903D
229	C920
229	C9426
229	CE4008B
229	CE4008C
229	CE4008D
229	CS-9016
229	CS-9016D
229	CS-9016F
229	CS2006F
229	CS9016
229	CS9016/3490
229	CS9016D
229	CS9016E
229	CS9016EF
229	CS9016F
229	CS9016FG
229	CS9016G
229	CS9016G/3490
229	CS9016H
229	D009
229	D072
229	DBCZ039404
229	DDBY216002
229	DDBY219001
229	DDBY259001
229	DDBY259002
229	DDBY261002
229	DDBY287001
229	E9625
229	EA15X130
229	EA15X190
229	EA15X251
229	EA15X350
229	EA15X351
229	EA15X374
229	EA15X376
229	EA15X393
229	EA15X437
229	EA15X7233
229	EA15X7234
229	EA15X7235
229	EA15X7236
229	EA15X7244
229	EA3713
229	ED-1502C
229	ED1502
229	ED1502A
229	ED592K
229	EL75
229	EP15X6
229	EQS-139
229	ETTC-930D
229	F24T-011-015
229	FRS3693
229	G05-037-A
229	G05-037-B
229	G05-037-D
229	G05-063-R
229	G05-065-A
229	G05-066A
229	G9423
229	G9625
229	HC-00394
229	HC-00461
229	HC-00535
229	HC-00929
229	HC00930
229	HEP-S0008
229	HEP-S0010
229	HEP-S0020
229	HEP-S0033
229	HEPS0008
229	HEPS0010
229	HEPS0033
229	HR76
229	HR77
229	HR78
229	HR87
229	HS-40019
229	HS-40054
229	HT309002AO
229	HT313592B
229	HT316751MO
229	HX-50091
229	HX-50110
229	IC743040
229	IC743041
229	IE460B
229	IP20-0037
229	IP20-0038
229	IP20-0110
229	KSC1187R
229	KSC1674R
229	LM1138
229	LM1138E
229	LM1138E/F
229	LM1138F
229	LM1138G
229	LM1138G/F
229	LM1138H
229	LM1138H/I
229	LM1138I
229	LT1016H
229	LTE1016
229	LTH1016
229	ME1138
229	MPS-H17
229	MPS6539
229	MPS6540
229	MPS6541
229	MPS6542
229	MPS6543
229	MPS6546
229	MPS6547
229	MPS6548
229	MPS6567
229	MPS6568A
229	MPS6569
229	MPS6570
229	MPS7513
229	MPS9426
229	MPS9426A
229	MPS9426B
229	MPS9426BC
229	MPS9426C
229	MPS9427
229	MPS9427B
229	MPS9427C
229	MPS9604
229	MPS9625
229	MPS9625C
229	MPS9625D
229	MPS9625E
229	MPS9625F
229	MPS9625G
229	MPS9625H
229	MPSH02
229	MPSH10
229	MPSH11
229	MPSH17
229	MPSH19
229	MPSH30
229	MPSH31
229	MPSH32
229	Q-00284R
229	QT-C1047XAN
229	QT-C1047XBN
229	QT-C1359XAN
229	R9600
229	RQ-444S
229	RS7114
229	RS7173
229	RS7233
229	RT6201
229	RT6202
229	RT6787
229	RT7703
229	RT8333
229	RT8527
229	RV1068
229	SE-1001
229	SE1001-1
229	SE1001-2
229	SE50
229	SE5025
229	SX3826
229	SX408
229	T1408
229	T1S-18
229	T1S18
229	T9423
229	TG2SC2057-D
229	TIS105
229	TIS108
229	TIS84
229	TIS86
229	TIS87
229	TM229
229	TR0573507
229	TR2SC1342
229	TR2SC535
229	TR5320326
229	TV-48
229	VS2SC1674L/-1
229	VS2SC1675M-1
229	VS2SC394-O-1
229	VS2SC394-Y-1
229	VS2SC784-R1F
229	VS2SC784R1F
229	ZEN117
229	ZEN121
229	01-030380
229	01-030394
229	01-030763
229	01-030784
229	01-031047
229	01-031359
229	01-031674
229	01-031675
229	1E535A
229	1E535A/7825B
229	2N5222
229	2SC1126
229	2SC1128
229	2SC1128(M)
229	2SC1128(S)
229	2SC1128D
229	2SC1128H
229	2SC1129
229	2SC1129(M)
229	2SC1129(R)
229	2SC1129R
229	2SC1187
229	2SC1189L
229	2SC1393
229	2SC1393K
229	2SC1393M
229	2SC1394
229	2SC1686
229	2SC1686B
229	2SC1687
229	2SC1727
229	2SC1778
229	2SC1779
229	2SC1855
229	2SC1856
229	2SC1923
229	2SC1923A
229	2SC1923BN
229	2SC2009
229	2SC2057
229	2SC2057C
229	2SC2057D
229	2SC2057E
229	2SC2057F
229	2SC784R
229	003-00
229	003-01
229	3L4-6007-41
229	04-46000-02
229	07-07129
229	09-3002006
229	09-302002
229	09-302003
229	09-302006
229	09-302036
229	09-302037
229	09-302095
229	09-302206
229	09-302207
229	09-302224
229	09-302225
229	09-302242
229	09-304042
229	09-304043
229	09-305050
229	09-305051
229	09-305093
229	09-305094
229	09-305096
229	09-309065
229	10-002
229	13-0009
229	13-0010
229	13-0020
229	13-0040
229	13-0062
229	13-0062-1
229	13-0063
229	13-0065
229	13-10321-35
229	13-10321-36
229	13-10321-41
229	13-10321-46
229	13-10321-62
229	13-10321-66
229	13-10321-71
229	13-14085-26
229	13-23001-2
229	13-23002-2
229	13-23013-2
229	13-23160-5
229	13-23822-1
229	13-23824-1
229	13-32366-1
229	19-020-070
229	21A040-045
229	21A040-053
229	21A040-091
229	21A040-54
229	21A112-084
229	21M093
229	21M099
229	21M577
229	24T-011-013
229	25B2
229	34-6015-27
229	34-6015-46
229	34-6015-47
229	34-6015-48
229	34-6015-52
229	42-27529
229	42-27530
229	44T-300-104
229	46-86240-3
229	46-86512-3
229	48-134805
229	48-134825
229	48-134857
229	48-137351

ECG	Industry Standard No.	ECG	Industry Standard No.	ECG	Industry Standard No.	ECG	Industry Standard No.	ECG	Industry Standard No.
229	48-155087	229	2057A2-558	230	28P1188L	233	3576	234	HEP-80006
229	48-355053	229	2057A2-87	230	28P1188M	233	3676	234	HEP80031
229	48-90343A75	229	2057A42-477	230	28P1188N	233	3676(RCA)	234	HP47
229	48-90343A80	229	2065-04	230	28P1188R	233	8000-00003-034	234	HT105641C
229	488155087	229	2127-17	230	28P1188Y	233	12454	234	HT105641D
229	50-40102-05	229	2214-17	233	537040	233	61086-1	234	HT105641H
229	56-48826	229	2224-17	233	126898	233	082020	234	HT105642B
229	86-100002	229	2225-17	233	126899	233	137338(3RD IP)	234	HT107211E
229	86-100004	229	2226-17	233	131346	233	147676-1	234	HT305642B
229	86-100006	229	2904-053	233	132325	234	0573486	234	M9412
229	86-100007	229	005449	233	132326	234	0573486H	234	M9461
229	86-100003	229	3652-2	233	140763	234	0573487	234	MI813674/47
229	86-593-2	229	3657-1	233	1415762-1	234	0573487H	234	M283639
229	86-593-9	229	3657-2	231	28P1189	234	610145-1	234	NR-621AU
229	86-594-2	229	4825	231	28P1189A	234	0820220	234	P/N297L010001
229	86-619-2	229	4857	231	28P1189C	234	1473676-1	234	P12407-1
229	86-620-2	229	5001-032	231	28P1189D	234	1473676-1	234	PM1120
229	86-621-2	229	5001-037	231	28P1189F	234	2320041	234	Q5102
229	87-0002	229	5001-541	231	28P1189G	234	2320141	234	Q5102P
229	87-0002-1	229	5001-543	231	28P1189H	234	06120008	234	Q5102Q
229	87-0003	229	5001-544	231	28P1189K	234	23114044	234	Q5102R
229	87-0255	229	5613-1342	231	28P1189L	234	23114078	234	QRT106
229	87-0255-C	229	5613-1342C	231	28P1189M	234	23114104	234	R8193
229	87-0235A	229	5613-1359B	231	28P1189R	234	30200062	234	RE62
229	87-0255C	229	5613-461	231	28P1189Y	234	43029471	234	RV2353
229	87-593-2	229	5613-461C	231	131347	234	43029472	234	8026094
229	90-180	229	5710	231	140764	234	62506350	234	81990
229	90-604	229	7116	231	1415762-2	234	62734655	234	SE5-0909
229	998055	229	7233	232	ES107	234	2003038812	234	T309
229	998056	229	7233B	232	GB-258	234	2004700116	234	TM234
229	102-0394-25	229	7236	232	HEP-89120	234	6621005000	234	TRO2020-2
229	102-0461-02	229	7236B	232	MPS-A65	234	6621009100	234	TR106(SPRAGUE)
229	102-0535-02	229	7261	232	MPS-A66	234	A-1853-0050-1	234	TR28A763
229	102-1047-03	229	7264	232	921-404	234	A-1853-0066-1	234	VS9-0004-923
229	102-1342-02	229	8000-00004-298	233	A10(3RDIF)	234	A-1853-0077-1	234	VS9-0008-923
229	102-1675-11	229	8000-00009-177	233	A1U(LAST IP)	234	A-1853-0086-1	234	XA-1164
229	102-1675-12	229	8000-00028-037	233	A1Z	234	A-1853-0098-1	234	Y56601-44
229	106-001	229	8000-00035-003	233	A6708850	234	A-1853-0300-1	234	Y56601-48
229	106-002	229	8000-00041-040	233	C1128(3RD IP)	234	A2311	234	01-010564
229	106-351	229	8000-00041-046	233	C1293A(LAST IP)	234	A493GR	234	28A1017
229	121-498	229	8000-00043-019	233	C1293B(LAST IP)	234	A494	234	28A494
229	121-503	229	8000-00049-053	233	CF14	234	A494-GR	234	28A494-0R
229	121-504	229	8000-00049-054	233	CGE-61	234	A494-0	234	28A494-0
229	121-505	229	8910-142	233	DDBY277002	234	A494-Y	234	28A494-Y
229	121-506	229	9011E	233	E13-004-00	234	A564-0	234	28A564-0
229	121-508	229	9011H	233	ED15028	234	A564A8	234	28A564-Q
229	121-509	229	9016	233	EP15220	234	A564AT	234	28A564A8
229	121-524A	229	9016D	233	EP15X54	234	A564B	234	28A564AT
229	121-526	229	9016E	233	EQ8-0018	234	A564T	234	28A564B
229	121-546	229	9016F	233	E815X106	234	A565Q	234	28A564T
229	121-754	229	9018D	233	E815X127	234	A629	234	28A572
229	121-855	229	9018E	233	E815X222	234	A640	234	28A580Q
229	121-885	229	9618	233	G05-003-A	234	A640A	234	28A629
229	121-885	229	26810-151	233	G05-003-B	234	A640B	234	28A640
229	121-924	229	38510-162	233	HT304601B0	234	A640C	234	28A640A
229	121-950	229	55170-1	233	HT305351B0	234	A640D	234	28A640B
229	121-951	229	000073140	233	LM1153	234	A640L	234	28A640C
229	121-954	229	88060-142	233	M4937(3RD IP)	234	A640M	234	28A640D
229	121-983	229	95170-1	233	MPS-A10	234	A640S	234	28A640L
229	121-984	229	95170-2	233	NPC1075	234	A641	234	28A640M
229	139N1	229	125144	233	P85025	234	A641(JAPAN)	234	28A640S
229	139N1D	229	129145	233	Q-00584R	234	A641(PNP)	234	28A641
229	176-026-9-001	229	134417	233	Q-00584R	234	A641A	234	28A641A
229	176-037-9-001	229	134857	233	Q50787	234	A641B	234	28A641B
229	176-039-9-001	229	136168	233	Q50787Z	234	A641C	234	28A641C
229	176-042-9-001	229	136240	233	QT-00460CBB	234	A641D	234	28A641D
229	176-056-9-001	229	137127	233	R3576-1	234	A641L	234	28A641L
229	176-056-9-003	229	137338	233	R3676-1	234	A641M	234	28A641M
229	176-060-9-001	229	144582	233	RT6204	234	A641S	234	28A666S
229	176-060-9-002	229	171162-186	233	80002	234	A666S	234	28A721
229	176-065-9-001	229	171162-187	233	80024	234	A721	234	28A721B
229	176-074-9-001	229	171162-279	233	SK3132	234	A721B	234	28A721T
229	176-075-9-001	229	171162-280	233	T-Q5078	234	A721T	234	28A721U
229	176-075-9-003	229	212717	233	TG28C1293-A-A	234	A721U	234	28A722
229	185-002	229	228417	233	TG28C1293-B-A	234	A722	234	28A722B
229	186-001	229	233117	233	TG28C1293-C-A	234	A722B	234	28A722T
229	249-1L	229	256417	233	TG28C1293-D-A	234	A722T	234	28A722U
229	249N1	229	256417	233	TM233	234	A722U	234	28A763
229	250-0380	229	279417	233	TNJ72276	234	A763	234	28A763-W
229	260-10-026	229	0573509	233	TRO573486	234	A836	234	28A763-WL-3
229	260-10-040	229	0575510	233	TV-20	234	A836E	234	28A763-WL-4
229	260-10-051	229	610139-1	233	TV-54	234	A844	234	28A763-WL-5
229	260P17601	229	610249-1	233	TV-55	234	A844D	234	28A763-WL-6
229	296-86	229	741731	233	TV-77	234	A844E	234	28A763-WN
229	753-1303-801	229	992108	233	TV81	234	BC153	234	28A763-WN-3
229	753-2000-007	229	1223915	233	VS28C460B-1	234	BC154	234	28A763-WN-4
229	753-2000-535	229	1223919	233	01-031293	234	BC177	234	28A763-WN-5
229	753-5000-535	229	1473657-1	233	01-571804	234	BC177A	234	28A763-WN-6
229	753-5151359	229	1473657-2	233	28C1128(3RD IP)	234	BC177B	234	28A763-Y
229	753-9001-674	229	1700020	233	28C1128B(3RDIF)	234	BC178	234	28A763-YL
229	753-9001-675	229	1700032	233	28C372Y(3RD IP)	234	BC178A	234	28A763-YL-3
229	753-9020-784	229	1700033	233	28C536A(3RD IP)	234	BC178B	234	28A763-YL-4
229	753-9050-785	229	53207722	233	28CP14	234	BC179	234	28A763-YL-5
229	902-001-7-16	229	5321901	233	004-00(LAST IP)	234	BC179A	234	28A763-YL-6
229	902-003-0-12	229	06120073	233	09-302004	234	BC179B	234	28A763-YN
229	921-1014	229	7853091-01	233	09-302016	234	BC204	234	28A763-YN-3
229	921-181B	229	7853093-01	233	09-302200	234	BC204A	234	28A763-YN-4
229	921-301	229	19901606	233	13-23160-2	234	BC205B	234	28A763-YN-5
229	921-301B	229	19901607	233	13-23824-3	234	BC205P	234	28A763-YN-6
229	921-312B	229	23114280	233	21A040-046	234	BC206	234	28A774
229	921-349B	229	23114282	233	21A040-047	234	BC206B	234	28A836
229	921-350B	229	23114966	233	34-6000-72	234	BC206C	234	28A836D
229	921-379	229	26010026	233	34-6015-37	234	BC224	234	28A836E
229	921-350B	229	26010051	233	34-6015-38	234	BC225	234	28A836F
229	1002-02	229	83038004	233	34-6015-61	234	BC320A	234	28A841
229	1002-68	229	83078402	233	34-6015-62	234	BC320C	234	28A841-GR
229	1006-48	229	83167405	233	46-86531-3	234	BC321	234	28A842-BL
229	1007-3054	229	83167503	233	46-86531-3	234	BC321A	234	28A842-GR
229	1007-3062	229	83167505	233	48-134932	234	BC321B	234	28A844
229	1007-3088	229	89942702	233	48-4937	234	BC321C	234	28A844C
229	1013-15	229	89960404	233	488134932	234	BC322	234	28A844D
229	1026(GE)	229	632039418	233	571142-4	234	BC322B	234	28A844E
229	1041-71	229	4100208283	233	57A142-4(3RDIP)	234	BC322C	234	28A860
229	1049-9904	229	4100208292	233	57A180-4(3RDIP)	234	BC326A	234	28B642R
229	1049-0060	229	4100208293	233	57B142-4	234	BC326B	234	006-00055182
229	1080-01	229	4100204613	233	86-513-2	234	BC326C	234	6-138
229	1501	229	4108296255	233	86X0034-001	234	BCY78	234	6-49
229	1502B	229	4108296257	233	90-455	234	BF243	234	8Q-00003-11
229	1502D	229$	C1688	233	102-0460-02	234	BF65	234	09-300026
229	1925-17	229$	28C1688	233	105-001-07	234	BF357	234	09-300036
229	2000-205	229*	K3C11870	233	121-522	234	C532000585	234	09-305058
229	2000-213	229*	K3C16740	233	121-524*	234	C673	234	11-691501
229	2004-02	230	28P1188	233	176-072-9-005	234	CD500	234	11-691502
229	2022-03	230	28P1188A	233	180N1	234	E8112	234	12-21-050
229	2041-01	230	28P1188B	233	260P11102	234	ET350335	234	12-PG 01
229	2048-17	230	28P1188C	233	576-1	234	ET2P-80B	234	14-602-64
229	2057A2-432	230	28P1188D	233	676-1	234	FI-1021	234	014-754
229	2057A2-477	230	28P1188E	233	1010(GE)	234	FS1990	234	14-805-32
229	2057A2-478	230	28P1188F	233	1010-8066	234	GE-65	234	17-RL8
229	2057A2-485	230	28P1188G	233	1075	234	HA-00564	234	19Q019
229	2057A2-507	230	28P1188H	233	1373-17			234	020-1110-005
229	2057A2-508	230	28P1188K	233	2000-204			234	020-1110-027
229	2057A2-526			233	2302-17			234	21M604
229	2057A2-527								
229	2057A2-541								

ECG	Industry Standard No.
234	25-000456-1
234	34-1007
234	34-1008
234	34-6016-64
234	37-19201
234	40-09437
234	40-09952
234	40-11253
234	42-9029-40F
234	42-9029-40W
234	42-9029-60B
234	42-9029-70Q
234	46-86429-3
234	48-134830A
234	48-134831
234	48-134831A
234	48-134832
234	48-134832A
234	48-355006
234	48-86429-3
234	48-869412
234	48-869467
234	48-90234A38
234	48-90343A59
234	48R869412
234	48R869467
234	48S155045
234	51-47-34
234	66A10310
234	67A8926
234	68A7355P1
234	70-943-773-001
234	73C180831-3
234	86-608-2
234	96-5258-01
234	132-029
234	132-031
234	134P1
234	151-0219-00
234	151-0342-00
234	151-0410-00
234	151-0453-00
234	177-007-9-001
234	207A14
234	297L010C
234	297L010C01
234	352-0773-010
234	352-0773-020
234	352-0773-030
234	417-153-13431
234	417-196-13262
234	461-2001
234	601-0100810
234	762-110
234	0831
234	1002
234	1016
234	1038-9922
234	1063-8435
234	1152
234	1341
234	1574-01
234	1935-17
234	2000-218
234	2057A2-529
234	4823-0031-01
234	06102
234	11000
234	31015
234	57009
234	75561-18
234	75561-21
234	75596-1
234	75596-2
234	75596-3
234	75596-4
234	90209-246
234	130253
234	131710
234	136281
234	138049-001
234	138049-004
234	144030
234	144858
234	146902
234	165737
234	170128
234	171555
234	309683
234	309685
234	610083-1
234	610083-2
234	610083-4
234	610134-1
234	610134-2
234	610134-4
234	610134-5
234	610134-6
234	611233
234	741863
234	1471112-10
234	2321351
234	2321521
234	3457936-1
234	06100053
234	28101149
234	28105207-001
234	30682592-001
234	50211400-00
234	50211400-01
234	50211410-10
234	50211410-11
234	436005901
234 5	BTX-084
234$	2SA493GR
235	C1237
235	C1237E
235	C1909
235	C1975
235	D235Y
235	DBBY005001
235	DDBY230001
235	DDBY257001
235	DDBY289001
235	EA15X363
235	EA15X372
235	EA15X381
235	EA15X414
235	EQS-0160
235	ET517375
235	FT527
235	GE-322
235	GE-329
235	IC743045
235	IP20-0083
235	IP20-0135
235	IP20-0155
235	MDS1678
235	NCB35
235	NCBW35
235	QS1306
235	QT-C1306XZA
235	RE203
235	01-031306
235	01-031909
235	1C73045
235	2SC1237
235	2SC1237E
235	2SC1306
235	2SC1909
235	2SC1909K
235	2SC1975
235	2SC2020
235	2SC2029
235	2SC2029-1
235	2SC2029-B/10
235	2SC2029/3
235	2SC2029B
235	2SC2029C
235	04S06
235	09-302119
235	09-305095
235	09-305136
235	022-2876-011
235	31-069-0
235	44T-300-109
235	48-40004S06
235	48-40764P01
235	48-42885P01
235	90-450
235	90-609
235	102-1678-00
235	106-006
235	123-009
235	172-024-9-002
235	172-028-9-001
235	172-038-9-002
235	176-024-9-002
235	176-074-9-003
235	576-0004-105
235	1042-08
235	1042-8
235	1212-4
235	2000-208
235	2000-271
235	2022-04
235	2041-04
235	2065-06
235	2399-17
235	5001-050
235	5001-514
235	8000-00038-001
235	8000-00041-044
235	8000-00043-065
235	8000-00049-060
235	239917
235	279717
235	741732
235	741859
235	742970
235	916118
235	1223918
235	5321321
235	632000002
235	632197406
235$	2SC2092
236	A67-76-200
236	BLY15
236	C1307
236	C1377
236	C1678
236	C1679
236	C1964
236	DDBY231002
236	EQS-0159
236	EQS-141
236	ESQ-141
236	GE-332
236	IP20-0154
236	PL-172-013-9001
236	QT-C1307XZA
236	RE201
236	TM236
236	VS2SC1237-1
236	01-031964
236	2SC1307
236	2SC1307-1
236	2SC1377
236	2SC1678
236	2SC1678E
236	2SC1679
236	2SC1964
236	2SC1969
236	2SC1969B
236	2SC1969BH
236	2SC1969H
236	2SC1974
236	2SC2043
236	2SC2078
236	2SC2166
236	2SC2184
236	2SC9
236	09-302192
236	09-302193
236	09-305137
236	022-2876-012
236	31-091-3
236	065-210
236	90-610
236	123-008
236	172-013-9-001
236	172-024-9-003
236	172-029-9-001
236	176-024-9-003
236	176-044-9-001
236	176-073-9-002
236	176-073-9-012
236	260-10-043
236	260-10-053
236	260-10-055
236	1042-09
236	2000-211
236	2000-212
236	2437-17
236	5001-071
236	8000-00028-039
236	8000-00032-028
236	26010053
236$	C1816
236$	2SC1816
236$	2SC2075
237	A066-117
237	C-F8
237	C517
237	C517C
237	C525
237	C525-R
237	C608
237	C608E
237	C608T
237	C609
237	C609T
237	C777
237	C778
237	C778B
237	C799
237	C799K
237	CF8
237	EQS-60
237	EQS-86
237	IP20-0005
237	Q-01069C
237	Q-01384R
237	RE198
237	S20446
237	SC609
237	2SC-F8
237	2SC517
237	2SC517C
237	2SC525
237	2SC608
237	2SC608E
237	2SC608T
237	2SC609
237	2SC609F
237	2SC609T
237	2SC609Y
237	2SC777
237	2SC778
237	2SC778B
237	2SC778D
237	2SC799
237	2SC799K
237	2SC866
237	2SCF8
237	07-07141
237	07-07164
237	09-302050
237	09-302051
237	09-302056
237	09-302117
237	09-302169
237	09-305091
237	9TR5
237	19-020-078
237	90-110
237	90-175
237	172-007-9-001
237	172-008-9-001
237	172-009-9-001
237	1001-08
237	1009-04
237	1040-05
237	1068-17
237	2062-17
237	2062-17A
237	2167-17
237	2418-17
237	4802-00008
237	4802-00019
237	8000-00004-083
237	8000-00004-084
237	8000-00004-P083
237	8000-00004-P084
237	8000-00005-011
237	8000-00006-002
237	8000-00006-230
237	8000-00011-052
237	082025
237	082029
237	0573541
237	2006607-62
237	5320151
237	5320511
237	26010022
238	A6317900
238	A6771373
238	C1170A
238	C1325
238	CGE-79
238	EP15X12
238	GE-259
238	J241227
238	Q5111ZK
238	Q5161Z
238	Q5207
238	Q5207Z
238	TVS2SC1629A
238	2SC1172
238	2SC1172A
238	2SC1172B
238	2SC1308
238	2SC1308K
238	2SC1308N
238	2SC1325
238	2SC1325A
238	2SC1325AK
238	2SC1325AL
238	2SC1358L
238	2SC1894
238	2SC1895
238	2SC1896
238	2SD348
238	2SD368
238	2SD416
238	3L4-6020-01
238	8-729-372-52
238	09-302238
238	09-302245
238	13-43463-1
238	46-86461-3
238	46-86492-3
238	46-86557-3
238	48-155058
238	48-355043
238	48S155058
238	48S155090
238	48S155140
238	57A263-11
238	57B186-11
238	57B263-11
238	86-626-2
238	86-633-9
238	121-1003
238	121-1033
238	189N1
238	189N1G
238	203X3189-408
238	1039-1290
238	1061-6274
238	1074T
238	1081-3343
238	62547
238	145648
238	146823
238	610189-1
238	610189-2
238	23114323
238	23114945
238	23114962
238	2003064314
238	2003117044
238	6621002400
241	A8F
241	A8Y
241	B1U
241	BD-131
241	BD575
241	BD577
241	BD585
241	BD587
241	BD589
241	ES113
241	ES15X125
241	MJE205K
241	MJE2480
241	MJE2481
241	MJE2482
241	MJE2483
241	MJE2520
241	MJE2521
241	MJE2522
241	MJE2523
241	MJE3054
241	MJE520K
241	MJE521K
241	MJES191
241	SJE5439
241	SJE5441
241	TM241
241	002-3(SYLVANIA)
241	2SD526
241	2SD526-0
241	13-34002-3
241	13-34002-5
241	13-36314-1
241	48-137030
241	48-137311
241	48-137369
241	48-137526
241	86X0042-002
241	86X0080-002
241	149N2B
241	149N4B
241	576-0002-001
241	576-0002-026
241	576-0002-029
241	707
241	2026-4
241	2026-5
241	20294
241	20295
241	80131
241	171033-1
241	1223783
242	BD576
242	BD578
242	BD580
242	BD586
242	BD588
242	BD590
242	ES114
242	JSP6009
242	MJE105K
242	MJE2370
242	MJE2371
242	MJE2490
242	MJE2491
242	MJE370K
242	MJE371K
242	MJE3740
242	MJE3741
242	MJG5194
242	P3V
242	P4E-3
242	P4E-4
242	P4J
242	P5F
242	P5S
242	P6009
242	SJE5442
242	TM242
242	48-137331
242	48-137370
242	48-137527
242	48-137562
242	48-37312
242	86X0042-001
242	86X0080-001
242	149P1B
242	149P4B
243	MJ1000
243	MJ1001
243	MJ4000
243	RCA1000
243	RCA1001
243	2N6055
243	2N6056
243	2N6492
243	2N6493
243	2N6494
243	2SD692
243	1414-180
244	MJ4010
244	MJ4011
244	MJ900
244	MJ901
244	2N6053
244	2N6054
245	MJ3000
245	MJ3001
245	MJ3520
245	MJ3521
245	2N6383
245	2N6384
245	2N6385
246	MJ2500
246	MJ2501
246	RCA8350
246	RCA8350A
246	RCA8350B
246	SK3183
246	2N6648
246	2N6649
246	2N6650
247	B1R
247	2N6057
247	2N6058
247	2N6059
247	2N6357
247	2SD803
247	48S137512
248	2N6050
248	2N6051
248	2N6052
249	MJ4033
249	MJ4034
249	MJ4035
249	MJE3520
250	MJ4030
250	MJ4031
250	MJ4032
250	2N6285
251	2N6282
251	2N6283
251	2N6284
251	2N6355
251	2N6356
251	2N6358
251	1414-179
252	2N6286
252	2N6287
253	MJE800
253	MJE801
253	MJE802
253	MJE803
253	2N6037
253	2N6038
253	2N6039
254	BD262A
254	MJE700
254	MJE701
254	MJE702
254	MJE703
254	2N6034
254	2N6035
254	2N6036
257	MJE1100
257	MJE1101
257	MJE1102
257	MJE1103
258	MJE1090
259	MJE6043
259	MJE6044
259	MJE6045
259	2N6043
259	2N6044
260	MJE6040
260	MJE6041
260	MJE6042
260	2N6040
260	2N6041
261	MJE2100
261	MJE2101
261	MJE2103
261	RCA120
261	RCA121
261	TIP111
261	TIP120
261	TIP121
261	2N6045
261	2N6300
261	2N6301
261	2SD678
261	2SD679
262	MJE2090
262	RCA125
262	RCA126
262	RCA8203
262	TIP116
262	TIP125
262	TIP126
262	2N6042
262	2N6298
262	2N6299
262	2SB668
262	2SB669
263	D44E1
263	D44E2
263	D44E3
263	2N6386
263	2N6387
263	2N6388
263	2N6530
264	D45E1
264	D45E2
264	D45E3
264	RCA8203A
264	2N6666
264	2N6667
264	2N6668
265	D4001
265	D4004
265	EA15X332
265	13-35324-1
266	D4005
268	D40K2
270	TIP140
270	TIP141
270	TIP142
270	TIP640
270	TIP641
270	TIP642
271	TIP145
271	TIP146
271	TIP147
271	TIP645
271	TIP646
271	TIP647
272	MPSU45
273	MPSU95
274	2N6294
274	2N6295
275	2N6296
275	2N6297
277	C1034
277	D56
277	GE-260
277	2SC1034
277	2SC1316

ECG	Industry Standard No.	ECG	Industry Standard No.	ECG	Industry Standard No.	ECG	Industry Standard No.	ECG	Industry Standard No.
277	2SC867	280	2N5069	281	2SA648	282	2N2034	283	C940L
277	2SD56	280	2N5874	281	2SA648A	282	2N2405	283	C940M
278	BFW16A	280	2SC1051	281	2SA648B	282	2N2849	283	CGE-75
278	BFW17A	280	2SC1051D	281	2SA648C	282	2N2850	283	D171
278	BFW30	280	2SC1051E	281	2SA679	282	2N2851	283	D321
278	BFW47	280	2SC1051F	281	2SA679R	282	2N2852	283	D551
278	BFX55	280	2SC1051LC	281	2SA679Y	282	2N2853	283	D583
278	C412	280	2SC1051LD	281	2SA680	282	2N2853-1	283	D458
278	C651	280	2SC1051LE	281	2SA680R	282	2N2854	283	DTA303
278	EA15X356	280	2SC1051LF	281	2SA680Y	282	2N2854-1	283	DTA304
278	TA2710	280	2SC1079	281	2SA746	282	2N2855	283	DTA305
278	TA2800	280	2SC1079R	281	2SA753	282	2N2855-1	283	DTA306
278	2N3633	280	2SC1079Y	281	2SA753A	282	2N2856	283	DTS402
278	2N4251	280	2SC1080	281	2SA753B	282	2N2856-1	283	DTS403
278	2N5108	280	2SC1080R	281	2SA753C	282	2N3262	283	DTS413
278	2N5109	280	2SC1080Y	281	2SA758	282	2N3506	283	DTS423M
278	2N5200	280	2SC1115	281	2SA758A	282	2N3507	283	DTS424
278	2N5200	280	2SC1343	281	2SA758B	282	2N3631	283	DTS425
278	2N709/46	280	2SC1343A	281	2SA758C	282	2N4001	283	DTS431
278	2N709A46	280	2SC1343B	281	2SB531	282	2N4877	283	DTS431M
278	2SC1324	280	2SC1343C	281	2SB532	282	2N4924	283	GE-36
278	38281	280	2SC1343E	281	2SB532P	282	2N5189	283	MJ3030
278	40404	280	2SC1343HA	281	2SB532Q	282	2N5262	283	MJ413
278	40608	280	2SC1343HB	281	2SB532R	282	2N5320	283	MJ431
278$	2SC1365	280	2SC1402	281	2SB528	282	2N5321	283	MJ9000
278$	2SC824	280	2SC1402C	281	2SB555	282	2N5682	283	RCA1B04
280	BD141	280	2SC243	281	2SB555-0	282	2N5784	283	RCA1B05
280	BDX11	280	2SC492	281	2SB556	282	2N5785	283	RCA1B09
280	BDX24	280	2SC586	281	2SB557	282	2N5846	283	SDP-413
280	BDY54	280	2SC687	281	2SB558	282	2N2054	283	SPC431
280	BDY56	280	2SC897	281	34-6002-30	282	2SC213	283	ST8402
280	BDY65	280	2SC897A	282	A574410	282	2SC214	283	ST8403
280	BLY47	280	2SC897B	282	BDY60	282	2SC215	283	ST8409
280	BLY48	280	2SC897C	282	BFS22A	282	2SC223	283	2N1490
280	BUY44	280	2SC898	282	BFT33	282	2SC224	283	2N1702
280	C1030P	280	2SC898A	282	BFT34	282	2SC229	283	2N3902
280	C1051	280	2SC898B	282	BFT42	282	2SC230	283	2N5241
280	C1051C	280	2SC898C	282	BFT43	282	2SC231	283	2N5466
280	C1051D	280	2SD110	282	BFW25	282	2SC232	283	2N5467
280	C1051E	280	2SD110-R	282	BFW26	282	2SC233	283	2N5804
280	C1051F	280	2SD110-Y	282	BFX34	282	2SC292	283	2N5805
280	C1051LC	280	2SD111	282	BSS14	282	2SC293	283	2N5869
280	C1051LD	280	2SD111-R	282	BSS15	282	2SC298	283	2N5870
280	C1051LE	280	2SD111-Y	282	BSS16	282	2SC298-4	283	2N6306
280	C1051LF	280	2SD118	282	BSS42	282	2SC310	283	2N6307
280	C1079	280	2SD118BL	282	BSS43	282	2SC310B	283	2N6308
280	C1079R	280	2SD118R	282	BSV60	282	2SC310C	283	2N6510
280	C1079Y	280	2SD118Y	282	BSW67	282	2SC310D	283	2N6511
280	C1080	280	2SD119	282	BSW68	282	2SC310E	283	2N6512
280	C1080R	280	2SD119BL	282	BSX63	282	2SC756	283	2N6513
280	C1080Y	280	2SD119R	282	C213	282	2SC756-1	283	2N6514
280	C1115	280	2SD119Y	282	C214	282	2SC756-1-1	283	2N6544
280	C1343	280	2SD189	282	C215	282	2SC756-1-2	283	2SC1433
280	C1343A	280	2SD189A	282	C224	282	2SC756-1-3	283	2SC1477
280	C1343B	280	2SD189R	282	C225	282	2SC756-1-4	283	2SC1617
280	C1343C	280	2SD213	282	C234	282	2SC756-2	283	2SC2122
280	C1343E	280	2SD214	282	C235	282	2SC756-2-1	283	2SC681
280	C1343HA	280	2SD316	282	C235-0	282	2SC756-2-3	283	2SC681A
280	C1343HB	280	2SD322	282	C292	282	2SC756-2-4	283	2SC681AYL
280	C1402	280	2SD322A	282	C293	282	2SC756-2-5	283	2SC681B
280	C243	280	2SD322B	282	C310	282	2SC756-3	283	2SC681YL
280	C586	280	2SD322C	282	C756	282	2SC756-3-1	283	2SC806
280	C687	280	2SD323	282	C756-1	282	2SC756-3-2	283	2SC806A
280	C897	280	2SD323A	282	C756-1-2	282	2SC756-3-3	283	2SC940
280	C897A	280	2SD323B	282	C756-1-2	282	2SC756-3-4	283	2SC940L
280	C897B	280	2SD323C	282	C756-1-3	282	2SC756-4	283	2SC940M
280	C897C	280	2SD334	282	C756-1-4	282	2SC756-4-1	283	2SC940Q
280	C898	280	2SD334A	282	C756-2	282	2SC756-4-2	283	2SD321
280	C898A	280	2SD334R	282	C756-2-1	282	2SC756-4-3	283	2SD551
280	C898B	280	2SD379	282	C756-2-2	282	2SC756-4-4	283	2SD583
280	C898C	280	2SD379P	282	C756-2-3	282	2SC756A	283	2SD458
280	D110	280	2SD379Q	282	C756-2-4	282	2SC756A-1	283	2SD533
280	D110-R	280	2SD379R	282	C756-2-5	282	2SC756A-2	283	2SD640
280	D110-Y	280	2SD425	282	C756-3	282	2SC756A-3	283	13-43633-1
280	D111	280	2SD4250	282	C756-3-1	282	2SC756A-4	283	15-123100
280	D111-R	280	2SD427	282	C756-3-3	282	2SC756C	283	21A116-124
280	D111-Y	280	2SD67	282	C756-3-4	282	2SC756D	283	21A118-124
280	D118	280	2SD67B	282	C756A	282	2SC756E	283	48-137524
280	D118BL	280	2SD67C	282	C756A-1	282	2SC756H	283	488155118
280	D118R	280	2SD67D	282	C756A-2	282	2SC92	283	86-564-2
280	D118Y	280	2SD67E	282	C756A-3	282	2SC93	283	122R2
280	D119	280	96-5310-01	282	C756C	282	2SC94	283	260P08908
280	D119BL	280	181T2B	282	C756D	282	2SD78	283	410
280	D119R	280	610217-1	282	C94	282	2SD78B	283	413
280	D119Y	280	610217-2	282	D121	282	09-302166	283	938-3
280	D124B	280	610245-1	282	D121A	282	52-010-151-0	283	40852
280	D189	280	2321652	282	D121B	282	142-007	283	40853
280	D189A	280MP	GE-262MP	282	D121H	282	172-014-9-002	283	610122-2
280	D322	281	A648	282	D121HA	282	185-008	283	2321121
280	D322A	281	A648A	282	D121HB	282	186-010	283	2321241
280	D322B	281	A648B	282	D78	282	501-068	283	06120012
280	D322C	281	A648C	282	D78A	282	1080-05	284	BD117
280	D323	281	A679	282	D78B	282	5001-068	284	BD184
280	D323A	281	A679R	282	D78C	282	2SC106	284	BDX23
280	D323B	281	A679Y	282	D78D	282/427	C299	284	BDX40
280	D323C	281	A680	282	DBBY001003	282/427	C697A	284	BDX41
280	D334	281	A680R	282	M1113HA	282/427	C697FP	284	BDX50
280	D334A	281	A680Y	282	MM3005	282/427	2SC1239	284	BDX51
280	D334R	281	A746	282	Q-18115C	282/427	2SC299	284	BDX60
280	D379	281	A747	282	RCA1A03	282/427	2SC524	284	BDT18
280	D379P	281	A747A	282	SDT3421	282/427	2SC697	284	BDT19
280	D379Q	281	A753	282	SDT3422	282/427	2SC697A	284	BDT55
280	D379R	281	A753A	282	SDT3423	282/427	2SC697B	284	BDT74
280	D379B	281	A753B	282	SDT3424	282/427	2SC697D	284	BDT77
280	D425	281	A753C	282	SDT3425	282/427	2SC697E	284	C1116-0
280	D4250	281	A758	282	SDT3426	282/427	2SC697F	284	D287
280	D67	281	A758A	282	SDT3427	282/427	2SC697H	284	D319
280	D67B	281	A758B	282	SDT3428	282/427	2SC697J	284	D424
280	D67C	281	A758C	282	SDT4301	282/427	2SC895	284	D424-R
280	D67D	281	B532	282	SDT4302	283	B1T	284	DTS104
280	D67E	281	B532P	282	SDT4304	283	B2L	284	FD-1029-LU
280	D73D	281	B532Q	282	SDT4305	283	BDY24	284	GE-265
280	GE-262	281	B532R	282	SDT4307	283	BDY25	284	KSD9701
280	HT401193A0	281	B528	282	SDT4308	283	BDY26	284	KSD9701A
280	KSD1052	281	B555	282	SDT4310	283	BDY27	284	KSD9704
280	KSD1057	281	B555-0	282	SDT4311	283	BDY28	284	KSD9707
280	MJ6302	281	BDY69	282	SDT5001	283	BU106	284	MJ2840
280	RCA1B06	281	BDY69B	282	SDT5501	283	BU110	284	MJ2841
280	S124AHB	281	FD-1029-LW	282	SDT5503	283	BU111	284	RCA1801
280	SAB-1	281	GE-263	282	SDT5506	283	BU120	284	SDT9202
280	SE3032	281	MJ490	282	SDT5507	283	BU208	284	SDT9203
280	TM280	281	MJ491	282	SDT5508	283	BUY35	284	SDT9204
280	2N1487	281	R861	282	SDT7401	283	BUY69B	284	SDT9207
280	2N1488	281	SAB-2	282	SDT7402	283	BUY69C	284	SDT9208
280	2N1489	281	TM281	282	SDT7411	283	C1477	284	SDT9209
280	2N3234	281	TR1038-5	282	SDT7412	283	C681	284	SJ2519
280	2N4347	281	2N3447	282	SDT7414	283	C681A	284	SK5563
280	2N4348	281	2N4904	282	SDT7415	283	C681B	284	TM284
280	2N4913	281	2N4905	282	SK3512	283	C681YL	284	2SA442
280	2N4914	281	2N4906	282	UPD611	283	C806	284	2N3772
280	2N4915	281	2N5738	282	1A0013	283	C806A	284	2N3773
280	2N5067	281	2N5871	282	2N2033	283	C940	284	2N5629
280	2N5068								

ECG	Industry Standard No.	ECG	Industry Standard No.	ECG	Industry Standard No.	ECG	Industry Standard No.	ECG	Industry Standard No.
284	2N5630	287	C728	289A	BC182LA	289A	C735(0)	289A	001-021135
284	2N5651	287	GB-220	289A	BC183IA	289A	C735-0	289A	01-030454
284	2N5878	287	GB-222	289A	BC252A	289A	C735-0	289A	01-030509
284	2N5970	287	GE-235	289A	BC252B	289A	C755B	289A	01-030735
284	2N6259	287	G89019H	289A	BSW32	289A	C755F	289A	01-031213
284	2N6262	287	HX-50108	289A	C1166	289A	C755FA3	289A	01-031317
284	2N6302	287	KSD4152I	289A	C1166-0	289A	C755GR	289A	01-7572814
284	2N6359	287	MPS-A05	289A	C1166D	289A	C755H	289A	1A0021
284	2N6360	287	MPS-A06	289A	C1166GR	289A	C755J	289A	1A0022
284	28C1116	287	MPSA05	289A	C1166O	289A	C755K	289A	1A0032
284	28C1116-0	287	MPSA06	289A	C1166R	289A	C755L	289A	1A0033
284	28C1584	287	MPSA42	289A	C1166Y	289A	C755Y	289A	1A0037
284	28D208	287	MPSLO1	289A	C1175	289A	C815	289A	1A0045
284	28D287	287	TG28C222B	289A	C1175C	289A	C815(M)	289A	1A0051
284	28D519	287	TIS100	289A	C1175D	289A	C815A	289A	1A0067
284	28DA24	287	TIS101	289A	C1175E	289A	C815B	289A	1A0085
284	28DA24-0	287	TN28C1941	289A	C1175F	289A	C815C	289A	1A0084
284	28DA24-R	287	01-030983	289A	C1204	289A	C815F	289A	2D033
284	28D582	287	01-051921	289A	C1204B	289A	C815K	289A	2N2712
284	28D582A	287	2N2310	289A	C1204C	289A	C815L	289A	2N2713
284	28D675	287	2N2312	289A	C1205	289A	C815M	289A	2N2714
284	28D676	287	2N2316	289A	C1205A	289A	C815S	289A	2N2921
284	13-36440-1	287	2N2437	289A	C1205B	289A	C815SA	289A	2N2922
284	13-37708-1	287	2N2438	289A	C1205C	289A	C815SC	289A	2N2923
284	13-41738-1	287	2N2439	289A	C1213AA	289A	C838	289A	2N2924
284	96-5370-01	287	2N2520	289A	C1213AB	289A	C838(H)	289A	2N2925
284	40656	287	2N2521	289A	C1213AC	289A	C838(J)	289A	2N2926
284	43104	287	2N2522	289A	C1213AD	289A	C838(K)	289A	2N3392
284	138195	287	2N3057	289A	C1213AK	289A	C838(M)	289A	2N3393
284$	28D665	287	2N5550	289A	C1213AKA	289A	C838A	289A	2N3394
285	B506C	287	2N5551	289A	C1213AKB	289A	C838B	289A	2N3395
285	B506D	287	2N5850	289A	C1213AKC	289A	C838C	289A	2N3396
285	B539	287	28C1033	289A	C1213AKD	289A	C838D	289A	2N3397
285	B539R	287	28C1033A	289A	C1213B	289A	C838E	289A	2N3414
285	B554	287	28C1279	289A	C1213BC	289A	C838F	289A	2N3416
285	B554-R	287	28C1279B	289A	C1213C	289A	C838H	289A	2N3605
285	BDX20	287	28C727	289A	C1213CD	289A	C838J	289A	2N3605A
285	GE-266	287	28C727A	289A	C1213D	289A	C838K	289A	2N3606
285	MJ2267	287	28C727B	289A	C1214	289A	C838L	289A	2N3606A
285	MJ2268	287	28C728	289A	C1214A	289A	C838M	289A	2N3607
285	SJ2520	287	28C728A	289A	C1214B	289A	C838R	289A	2N3707
285	TIP544	287	8-729-213-01	289A	C1214C	289A	C894	289A	2N3708
285	2N5876	287	13-23510-4	289A	C1214D	289A	C933	289A	2N3711
285	2N5879	287	13-37869-1	289A	C1220E	289A	C933BB	289A	2N3721
285	2N5880	287	46-86345-3	289A	C1317	289A	C933C	289A	2N3825
285	2N6029	287	46-86591-3	289A	C1317P	289A	C933D	289A	2N3826
285	2N6030	287	46-86630-3	289A	C1317Q	289A	C933E	289A	2N3827
285	2N6031	287	46-86647-3	289A	C1317R	289A	C933F	289A	2N3843
285	2N6226	287	46-86678-3	289A	C1317S	289A	C933FP	289A	2N3843A
285	2N6227	287	48-134846	289A	C1317T	289A	C933FPC	289A	2N3844
285	2N6228	287	48-155051	289A	C1318	289A	C933FPD	289A	2N3844A
285	2N6229	287	48-355012	289A	C1318Q	289A	C933FPB	289A	2N3845
285	2N6231	287	54E	289A	C1318R	289A	C933FPP	289A	2N3845A
285	28B539	287	86-651-2	289A	C1318S	289A	C958	289A	2N3854
285	28B552	287	87-0239-A	289A	C1335	289A	C958A	289A	2N3854A
285	28B554	287	121-1017	289A	C1361	289A	C958B	289A	2N3855
285	28B554-R	287	200X3192-101	289A	C1363	289A	C958C	289A	2N3856
285	28B655	287	200X3222-907	289A	C1364	289A	C941	289A	2N3856A
285	28B656	287	200X3223-025	289A	C1364A	289A	C941-0	289A	2N3858
285	96-5371-01	287	200X3224-007	289A	C1372Y	289A	C941-0	289A	2N3858A
286	B1V	287	260P35501	289A	C1684	289A	C941-R	289A	2N3859A
286	B1Y	287	417-294	289A	C361	289A	C941-T	289A	2N3860
286	B2P	287	1573PQ	289A	C362	289A	C941R	289A	2N3877
286	C1104	288	2041-02	289A	C370	289A	DBCZ073503	289A	2N3877A
286	C1104A	288	95226-4	289A	C370P	289A	DBC2073504	289A	2N3900
286	C1104B	288	5321214	289A	C370G	289A	DDBY273001	289A	2N3900A
286	C1104C	288	2004546826	289A	C370H	289A	DDBY283001	289A	2N4286
286	C1304	288	A549A	289A	C370J	289A	DS-94	289A	2N4286
286	C14508	288	A549AH	289A	C370K	289A	EA15X249	289A	2N4287
286	C680	288	A5T4059	289A	C371	289A	EA15X256	289A	2N4424
286	C680A	288	A5T4060	289A	C371-(Q)	289A	EA15X267	289A	2N4954
286	C680R	288	A637	289A	C371-0	289A	EL403	289A	2N5172
286	C779-R	288	A638	289A	C371-R	289A	EQ8-0192	289A	2N5174
286	C779-Y	288	A778	289A	C371-R-1	289A	ET234854	289A	2N5309
286	C782	288	A778AK	289A	C371B	289A	ET329218	289A	2N5310
286	C783	288	A778K	289A	C371G	289A	ET352146	289A	2N5311
286	C825	288	BP299	289A	C371R	289A	FC880500C	289A	2N5418
286	D159	288	GE-221	289A	C378	289A	G05-055-C	289A	2N5419
286	EP15X34	288	GE-223	289A	C394	289A	G05-055-D	289A	2N5420
286	FD-1029-PB	288	G89022H	289A	C394-0	289A	G05-055-E	289A	2N5998
286	GE-267	288	G89022I	289A	C394-0	289A	G05-413A	289A	28C006
286	SK3558	288	HX-50109	289A	C394GR	289A	GE-268	289A	28C1166
286	2N2251	288	Q6521	289A	C394R	289A	HC-01390	289A	28C1166-0
286	2N2252	288	TIS138	289A	C394W	289A	HD-00471	289A	28C1166-0
286	2N3583	288	2N1132B46	289A	C394Y	289A	HT31213100	289A	28C1166-Y
286	2N3585	288	2N1655	289A	C401	289A	HT313171R	289A	28C1166D
286	2N3738	288	2N1917	289A	C402	289A	HT313172A	289A	28C1166GR
286	2N3739	288	2N1918	289A	C402A	289A	IC743044	289A	28C1166O
286	2N4240	288	2N2598	289A	C403(C)	289A	IP20-0041	289A	28C1166R
286	2N4296	288	2N2599	289A	C403B(SONY)	289A	IP20-0165	289A	28C1166Y
286	2N4297	288	2N2599A	289A	C403C	289A	IP20-0179	289A	28C1175
286	28C1104	288	2N2600	289A	C404	289A	IP20-0191	289A	28C1175C
286	28C1304BK	288	2N2600A	289A	C455	289A	IP20-0231	289A	28C1175D
286	28C1450	288	2N3841	289A	C509	289A	KSCB15-0	289A	28C1175E
286	28C14508	288	2N5400	289A	C509(0)	289A	KSC945Y	289A	28C1175F
286	28C680	288	2N5401	289A	C509G	289A	M9389	289A	28C1204
286	28C680A	288	2SA1018	289A	C509Y	289A	MP89696	289A	28C1204B
286	28C680R	288	2SA657	289A	C536D	289A	MP89696F	289A	28C1204C
286	28C779	288	2SA778	289A	C536E	289A	MP89696G	289A	28C1205
286	28C779R	288	2SA778AK	289A	C537	289A	MP89696I	289A	28C1205A
286	28C782	288	2SA778K	289A	C537-01	289A	PBC183	289A	28C1205B
286	28C783	288	2SA912	289A	C537C	289A	Q-00784R	289A	28C1205C
286	28C825	288	2SA912Q	289A	C537D2	289A	QT-00755XBT	289A	28C1213
286	28C825A	288	2SA912R	289A	C537E	289A	SK3122	289A	28C1213AA
286	28C825B	288	2SA912B	289A	C537EF	289A	SK3124	289A	28C1213AB
286	28C825C	288	8-762-020-00	289A	C537EH	289A	TG28C1175	289A	28C1213AC
286	28D158	288	13-59115-1	289A	C537EJ	289A	TI860	289A	28C1213AD
286	28D158A	288	48-355058	289A	C537EK	289A	TI890	289A	28C1213AK
286	28D158B	288	52-010-109-0	289A	C537F	289A	TM289	289A	28C1213AKA
286	28D158C	288	121-1016	289A	C537P1	289A	TN28C945-Q	289A	28C1213AKB
286	28D158F	288	161-015G	289A	C537P2	289A	TN28C945R	289A	28C1213AKC
286	28D159	288	260P26201	289A	C537PC	289A	TR320063	289A	28C1213AKD
286	28D159A	288	95226-3	289A	C590Y	289A	TR327293	289A	28C1213B
286	28D159B	288	2321111	289A	C716	289A	TR327333	289A	28C1213BC
286	28D159C	288	2004085415	289A	C716B	289A	VS28C1213-C/1A	289A	28C1213C
286	28D159F	288$	2004356220	289A	C716C	289A	VS28C372-Y/1E	289A	28C1213CD
286	09-302188	288$	A778A	289A	C716E	289A	VS28C454-B/1E	289A	28C1213D
286	46-86150-3	288	A840	289A	C716F	289A	VS28C454-C/1A	289A	28C1214
286	48-137528	288$	28A778A	289A	C725	289A	VS28C454-C/1B	289A	28C1214A
286	121-1028	289A	A066-115	289A	C725-0	289A	VS28C454-C/3A	289A	28C1214C
286	121-993	289A	A514-033338	289A	C734	289A	VS28C458-C/1E	289A	28C1214D
286	40522	289A	A67-33-540	289A	C734-0	289A	VS28C735-Y-1	289A	28C1220
286	23114038	289A	A673354K	289A	C734-0	289A	VS28D467-C/-1	289A	28C1220-003
287	A5T5058	289A	A673355H	289A	C734-R	289A	001-021010	289A	28C1220E
287	A5T5059	289A	A673355K	289A	C734-Y	289A	001-021020	289A	28C1247
287	A673518A	289A	A7T3391	289A	C734GR	289A	001-021030	289A	28C1317
287	BP297	289A	A7T3391A	289A	C734Y	289A	001-021040	289A	28C1317B
287	BP298	289A	A7T3392	289A	C735	289A	001-021050	289A	28C1317D
287	C1033	289A	A7T5172	289A		289A	001-021060	289A	28C1317Q
287	C1033A	289A	BC167A	289A		289A	001-021080	289A	28C1317R
287	C1279	289A	BC167B	289A		289A	001-021090	289A	
287	C1279B	289A	BC168A	289A		289A	001-021132	289A	
287	C727	289A	BC182A	289A		289A	001-021133	289A	
287		289A		289A		289A	001-021134	289A	

ECG	Industry Standard No.
289A	2SC1317S
289A	2SC1317T
289A	2SC1318
289A	2SC1318A
289A	2SC1318P
289A	2SC1318Q
289A	2SC1318QR
289A	2SC1318R
289A	2SC1318S
289A	2SC1318X
289A	2SC1361
289A	2SC1362
289A	2SC1363
289A	2SC1364
289A	2SC1364-6
289A	2SC1364A
289A	2SC1372Y
289A	2SC1627
289A	2SC1627-0
289A	2SC1627AY
289A	2SC1627Y
289A	2SC1684
289A	2SC1684BL
289A	2SC1684P
289A	2SC1684Q
289A	2SC1684R
289A	2SC1788
289A	2SC1788R
289A	2SC2120
289A	2SC2120Y
289A	2SC2308
289A	2SC361
289A	2SC362
289A	2SC370
289A	2SC370F
289A	2SC370G
289A	2SC370H
289A	2SC370J
289A	2SC370K
289A	2SC371
289A	2SC371(O)
289A	2SC371-O
289A	2SC371-O
289A	2SC371-R
289A	2SC371-R-1
289A	2SC371B
289A	2SC371G
289A	2SC371O
289A	2SC371R
289A	2SC378
289A	2SC378-O
289A	2SC378-R
289A	2SC378-Y
289A	2SC394
289A	2SC394-O
289A	2SC394-O
289A	2SC394GR
289A	2SC394R
289A	2SC394W
289A	2SC394X
289A	2SC394Y
289A	2SC401
289A	2SC401A
289A	2SC402
289A	2SC402A
289A	2SC403
289A	2SC403(C)
289A	2SC403A
289A	2SC403C
289A	2SC404
289A	2SC455
289A	2SC509
289A	2SC509(O)
289A	2SC509-O
289A	2SC509G
289A	2SC509Y
289A	2SC537
289A	2SC537(F)
289A	2SC537-01
289A	2SC537C
289A	2SC537D
289A	2SC537D2
289A	2SC537E
289A	2SC537EF
289A	2SC537EH
289A	2SC537EJ
289A	2SC537EK
289A	2SC537F
289A	2SC537F1
289A	2SC537F2
289A	2SC537FC
289A	2SC590Y
289A	2SC716
289A	2SC716B
289A	2SC716C
289A	2SC716D
289A	2SC716E
289A	2SC716F
289A	2SC725
289A	2SC725-O
289A	2SC734
289A	2SC734-O
289A	2SC734-O
289A	2SC734-R
289A	2SC734-Y
289A	2SC734GR
289A	2SC734O
289A	2SC734Y
289A	2SC735
289A	2SC735(FA-3)
289A	2SC735(O)
289A	2SC735-O
289A	2SC735-O
289A	2SC735-Y
289A	2SC735/4454C
289A	002SC7350Y
289A	2SC7354
289A	2SC735B
289A	2SC735F
289A	2SC735G
289A	2SC735GR
289A	2SC735H
289A	2SC735J
289A	2SC735K
289A	2SC735L
289A	2SC735ORN
289A	2SC735R
289A	2SC735Y
289A	2SC735Y/4454C
289A	2SC815
289A	2SC815(M)
289A	2SC815A
289A	2SC815B
289A	2SC815C

ECG	Industry Standard No.
289A	2SC815F
289A	2SC815K
289A	2SC815L
289A	2SC815M
289A	2SC815S
289A	2SC815SA
289A	2SC815SC
289A	2SC838
289A	2SC838(H)
289A	2SC838(J)
289A	2SC838(K)
289A	2SC838(L)
289A	2SC838(M)
289A	2SC838O
289A	2SC838A
289A	2SC838B
289A	2SC838C
289A	2SC838D
289A	2SC838E
289A	2SC838F
289A	2SC838H
289A	2SC838J
289A	2SC838K
289A	2SC838L
289A	2SC838M
289A	2SC838S
289A	2SC894
289A	2SC903
289A	2SC904
289A	2SC933
289A	2SC933BB
289A	2SC933C
289A	2SC933D
289A	2SC933E
289A	2SC933F
289A	2SC933FP
289A	2SC933FPC
289A	2SC933FPD
289A	2SC933FPE
289A	2SC933FPF
289A	2SC938
289A	2SC938A
289A	2SC938B
289A	2SC938C
289A	2SC941
289A	2SC941-O
289A	2SC941-R
289A	2SC941-Y
289A	2SC941R
289A	2SC941Y
289A	00002SC945
289A	2SD228
289A	2SD228A
289A	2SD228B
289A	2SD228C
289A	2SD228D
289A	2SD228E
289A	07-07163
289A	09-302020
289A	09-302040
289A	09-302173
289A	09-309061
289A	9TR9
289A	10-008
289A	11-0422
289A	11-0774
289A	11-0778
289A	13-14085-54
289A	13-23323-4
289A	21A118-031
289A	23-5020
289A	23-5021
289A	23-5022
289A	23-5023
289A	23-5024
289A	23-5025
289A	23-5026
289A	23-5027
289A	23-5029
289A	42-27537
289A	44-44886G01
289A	46-86132-3
289A	46-86252-3
289A	46-86513-3
289A	46-86572-3
289A	46-86589-3
289A	48-155119
289A	48-44886G01
289A	48-869389
289A	48-90234A13
289A	48-90343A77
289A	48-90343A83
289A	48R869389
289A	48S44886G01
289A	57A219-14
289A	57A241-14
289A	57A280-14
289A	57B219-14
289A	57B241-14
289A	57B280-14
289A	76-042-9-006
289A	86-422-2
289A	87-0203-1
289A	90-31
289A	90-66
289A	102-0454-02
289A	102-0735-25
289A	102-1317-18
289A	123-012
289A	142-005
289A	142N1
289A	161-001I
289A	161-011J
289A	161-014H
289A	161-016H
289A	176-016-9-001
289A	176-029-9-001
289A	176-031-9-001
289A	176-042-9-006
289A	176-043-9-002
289A	176-047-9-001
289A	176-048-9-002
289A	176-049-9-002
289A	185-006
289A	186-006
289A	250-1213
289A	260-10-021
289A	260D15902
289A	260P28107
289A	260P31303
289A	514-033338
289A	600-229-201
289A	853-0301-317
289A	902-000-8-04

ECG	Industry Standard No.
289A	921-1013
289A	921-1017
289A	921-1022
289A	921-275B
289A	921-276B
289A	991-002232
289A	991-003304
289A	991-008393
289A	991-010462
289A	991-013044
289A	991-013056
289A	991-013057
289A	991-013068
289A	991-015615
289A	991-016274
289A	1007-3153
289A	1039-0441
289A	1042-05
289A	1043-1229
289A	1043-1260
289A	1063-8963
289A	1071-4913
289A	1080-07
289A	1374-17
289A	1374-17A
289A	1414-174
289A	1414-183
289A	1710
289A	2000-203
289A	2057A2-427
289A	2361-17
289A	2796
289A	2797
289A	4802-00014
289A	5001-021
289A	5001-070
289A	5001-506
289A	5613-458
289A	5613-458B
289A	5613-458C
289A	5613-458D
289A	5613-458LGC
289A	7501
289A	7505
289A	7519
289A	8000-00003-035
289A	8000-00003-036
289A	8000-00004-085
289A	8000-00004-242
289A	8000-00004-P085
289A	8000-00004-P086
289A	8000-00011-049
289A	8000-00028-038
289A	26810-155
289A	061366
289A	71226-1
289A	71226-15
289A	71226-2
289A	71226-3
289A	71226-4
289A	71226-5
289A	71226-6
289A	71412-4
289A	71412-5
289A	082028
289A	95171-1
289A	95171-2
289A	95171-3
289A	95171-4
289A	95221
289A	123274
289A	125519
289A	129146
289A	132642
289A	132643
289A	144039
289A	146570
289A	167540
289A	169771
289A	170308
289A	170967
289A	508762
289A	0573480
289A	0573481
289A	0573491
289A	0573491H
289A	0573492
289A	610142-1
289A	651891
289A	651955-1
289A	651955-2
289A	651955-3
289A	652072
289A	652091
289A	652230
289A	654000
289A	656524
289A	656719
289A	656746
289A	658657
289A	740461
289A	740462
289A	740463
289A	740466
289A	740470
289A	740857
289A	801512
289A	801513
289A	801514
289A	801515
289A	801516
289A	801517
289A	801524
289A	801529
289A	801530
289A	801532
289A	801534
289A	801536
289A	801543
289A	801729
289A	2320591
289A	2320595
289A	2320596
289A	2320643
289A	2320644
289A	2320646
289A	2320647
289A	3673351K
289A	3673354G
289A	5320004
289A	5320022
289A	5320023
289A	05320064
289A	5320067

ECG	Industry Standard No.
289A	5320613
289A	5320651
289A	06100007
289A	06120063
289A	7852457-01
289A	13073525
289A	23114015
289A	23114017
289A	23114259
289A	23114260
289A	23114261
289A	23114297
289A	23114347
289A	26010021
289A	62771210
289A	2003073517
289A	2003174006
289A	4100217172
289A	4104213173
289A	4109213174
289A	6621001000
289A	6621003500
289A	6621008100
289A	6624003200
289MP	GE-268MP
29	131848
290A	A0666-118
290A	A1030
290A	A467
290A	A467-O
290A	A467-Y
290A	A467G
290A	A467G-O
290A	A467G-R
290A	A467G-Y
290A	A493
290A	A493Y
290A	A495(JAPAN)
290A	A495-G
290A	A495-O
290A	A495-Y
290A	A495A
290A	A495D
290A	A495G-O
290A	A495W
290A	A495Y
290A	A496
290A	A502
290A	A539
290A	A539L
290A	A539S
290A	A561
290A	A561-O
290A	A561-R
290A	A561-Y
290A	A562
290A	A562-O
290A	A562-O
290A	A562-R
290A	A562-Y
290A	A562G
290A	A562R
290A	A562Y
290A	A564
290A	A564A
290A	A564ABQ
290A	A564AO
290A	A564AP
290A	A564AQ
290A	A564F
290A	A564FQ
290A	A564FR
290A	A564J
290A	A564P
290A	A564POR
290A	A564Q
290A	A564QHD
290A	A564QR
290A	A564R
290A	A568
290A	A569
290A	A575K
290A	A575L
290A	A610
290A	A610B
290A	A611
290A	A611-4E
290A	A642
290A	A642(JAPAN)
290A	A642A
290A	A642B
290A	A642C
290A	A642D
290A	A642E
290A	A642F
290A	A642L
290A	A642W
290A	A65-09-220
290A	A650235G
290A	A650238A
290A	A650372D
290A	A650923F
290A	A650925H
290A	A659
290A	A659C
290A	A659D
290A	A659E
290A	A659F
290A	A661
290A	A661R
290A	A661Y
290A	A666
290A	A666A
290A	A666H
290A	A666HR
290A	A666QRS
290A	A666R
290A	A673
290A	A673A
290A	A673AA
290A	A673AB
290A	A673AC
290A	A673AD
290A	A673AE
290A	A673AS
290A	A673ASC
290A	A673B
290A	A673C
290A	A673D
290A	A675
290A	A675A
290A	A675B
290A	A675C
290A	A677

ECG	Industry Standard No.
290A	A678
290A	A705
290A	A719
290A	A719P
290A	A719Q
290A	A719R
290A	A719RS
290A	A719S
290A	A720
290A	A720P
290A	A720Q
290A	A720R
290A	A720S
290A	A723
290A	A723A
290A	A723B
290A	A723C
290A	A723D
290A	A723E
290A	A723F
290A	A733
290A	A733A
290A	A733B
290A	A733C
290A	A733D
290A	A733E
290A	A733F
290A	A733H
290A	A733I
290A	A733P
290A	A733Q
290A	A733R
290A	A825
290A	BC231A
290A	BC231B
290A	DDBY003001
290A	DDBY004001
290A	DDCY007002
290A	EA15X242
290A	EA15X395
290A	EP15X48
290A	EQR-0038
290A	ES15X90
290A	FCS8550
290A	FCS8550C
290A	GE-269
290A	HA00562
290A	HS-40057
290A	HT106731BO
290A	IC743043
290A	IP20-0046
290A	IP20-0192
290A	IP20-0213
290A	Q5116C
290A	Q5116CA
290A	Q5205
290A	QT-A0719XAN
290A	QT-A0719XCN
290A	QT-A0719XHN
290A	RE63
290A	RT8670
290A	RV2355
290A	TG2SA608-D
290A	TG2SA608-E
290A	TM290
290A	TN2SA733-Q
290A	TN2SA733-R
290A	TT2SA495-O
290A	TT2SA495-Y
290A	VS2SA562-Y/1E
290A	VS2SA673-B/1E
290A	VS2SA673-C/1A
290A	VS2SA673-C/1E
290A	VS2SA854-Q/1E
290A	VS2SB561-C/-1
290A	01-010495
290A	01-010562
290A	01-010673
290A	01-010719
290A	001-022020
290A	01-572811
290A	2N5221
290A	2N5354
290A	2N5355
290A	2N5999
290A	2SA1015-O
290A	2SA1015Y
290A	2SA1029
290A	2SA1029C
290A	2SA467
290A	2SA467-O
290A	2SA467-Y
290A	2SA467G
290A	2SA467G-O
290A	2SA467G-R
290A	2SA467G-Y
290A	2SA493
290A	2SA493Y
290A	2SA495
290A	2SA495-O
290A	2SA495-G
290A	2SA495-O
290A	2SA495-R
290A	2SA495-Y
290A	2SA495A
290A	2SA495D
290A	2SA495G-O
290A	2SA495GR
290A	2SA495H
290A	2SA495J
290A	2SA495O
290A	2SA495R
290A	2SA495W
290A	2SA495X
290A	2SA495Y
290A	2SA502
290A	2SA539
290A	2SA539K
290A	2SA539L
290A	2SA539M
290A	2SA539S
290A	2SA556
290A	2SA561-O
290A	2SA561-R
290A	2SA561-Y
290A	2SA561R
290A	2SA561Y
290A	2SA562
290A	2SA562-O
290A	2SA562-O
290A	2SA562-Y
290A	2SA562G
290A	2SA562R
290A	2SA562Y

ECG	Industry Standard No.
290A	2SA564
290A	2SA564-0
290A	2SA564A
290A	2SA564ABQ
290A	2SA564AO
290A	2SA564AP
290A	2SA564AQ
290A	2SA564AR
290A	2SA564D
290A	2SA564F
290A	2SA564FQ
290A	2SA564FR
290A	2SA564G
290A	2SA564H
290A	2SA564J
290A	2SA564P
290A	2SA564POR
290A	2SA564Q
290A	2SA564QGD
290A	2SA564QHD
290A	2SA564QR
290A	2SA564R
290A	2SA564Y
290A	2SA568
290A	2SA568B
290A	2SA568C
290A	2SA568D
290A	2SA568E
290A	2SA568F
290A	2SA568G
290A	2SA568GN
290A	2SA568H
290A	2SA568J
290A	2SA568K
290A	2SA568L
290A	2SA568M
290A	2SA568R
290A	2SA568X
290A	2SA568Y
290A	2SA569
290A	2SA569J
290A	2SA570
290A	2SA575K
290A	2SA575L
290A	2SA610
290A	2SA610B
290A	2SA611
290A	2SA611-4E
290A	2SA642
290A	2SA642A
290A	2SA642B
290A	2SA642C
290A	2SA642D
290A	2SA642E
290A	2SA642F
290A	2SA642L
290A	2SA642R
290A	2SA642S
290A	2SA642V
290A	2SA642W
290A	2SA642Y
290A	2SA659
290A	2SA659A
290A	2SA659B
290A	2SA659C
290A	2SA659D
290A	2SA659E
290A	2SA659F
290A	2SA659G
290A	2SA659L
290A	2SA659P
290A	2SA659R
290A	2SA659Y
290A	2SA661
290A	2SA661-0
290A	2SA661R
290A	2SA661Y
290A	2SA666
290A	2SA666A
290A	2SA666H
290A	2SA666HR
290A	2SA666Q
290A	2SA666QRS
290A	2SA666R
290A	2SA673(B)
290A	2SA673A
290A	2SA673AA
290A	2SA673AB
290A	2SA673AC
290A	2SA673AD
290A	2SA673AE
290A	2SA673AK
290A	2SA673AS
290A	2SA673ASC
290A	2SA673B
290A	2SA673C
290A	2SA673D
290A	2SA675
290A	2SA675A
290A	2SA675B
290A	2SA675C
290A	2SA677
290A	2SA677HL
290A	2SA678
290A	2SA678(C)
290A	2SA678E
290A	2SA685
290A	2SA704
290A	2SA705
290A	2SA719
290A	2SA719K
290A	2SA719P
290A	2SA719PQR
290A	2SA719Q
290A	2SA719QR
290A	2SA719R
290A	2SA719RS
290A	2SA719S
290A	2SA720
290A	2SA720A
290A	2SA720P
290A	2SA720Q
290A	2SA720R
290A	2SA720S
290A	2SA723
290A	2SA723A
290A	2SA723B
290A	2SA723C
290A	2SA723D
290A	2SA723E
290A	2SA723F
290A	2SA723R
290A	2SA728
290A	2SA728A

ECG	Industry Standard No.
290A	2SA733
290A	2SA733A
290A	2SA733AP
290A	2SA733H
290A	2SA733I
290A	2SA733IQ
290A	2SA733P
290A	2SA733Q
290A	2SA733QP
290A	2SA733R
290A	2SA825
290A	2SA825Q
290A	2SA825R
290A	2SA826
290A	2SA826P
290A	2SA826Q
290A	2SA826R
290A	2SA826RY
290A	2SA854
290A	2SA854Q
290A	2SA950
290A	2SA950-0
290A	2SA952
290A	2SB641
290A	2SB641S
290A	2SB642
290A	2SB642Q
290A	2SB643
290A	2SB643Q
290A	2SB643R
290A	2SB643S
290A	2SB774
290A	03A04
290A	07-07113
290A	09-300064
290A	09-300070
290A	09-300080
290A	09-300081
290A	19-020-114
290A	46-86546-3
290A	46-86551-3
290A	46-86578-3
290A	48-155035
290A	48-155045
290A	48-155065
290A	48-155095
290A	48-155156
290A	48-355057
290A	48-90234A14
290A	48-90343A68
290A	48S155035
290A	57A220-14
290A	57A240-14
290A	57A281-14
290A	57B220-14
290A	57B240-14
290A	57B281-14
290A	86-555-2
290A	86-610-9
290A	87-423-2
290A	100-0673-04
290A	177-006-9-002
290A	177-025-9-001
290A	260-10-027
290A	260-10-039
290A	260P16503
290A	260P16603
290A	260P36001
290A	509R
290A	509Y
290A	921-1016
290A	991-015614
290A	1039-0060
290A	1061-8320
290A	1061-8807
290A	2000-201
290A	2181-17
290A	2799
290A	5001-540
290A	8000-00003-037
290A	144031
290A	256317
290A	741729
290A	742546
290A	1223914
290A	2320631
290A	2320637
290A	5320592
290A	5320593
290A	06100035
290A	11056225
290A	23114081
290A	23114137
290A	23114262
290A	23114293
290A	26010027
290A	62771224
290A	81073304
290A	2004056215
290A	2004067350
290A	4100006285
290A	4104007193
290A	6623001900
290A	6623002300
290A$	A495G-GR
290A$	A495G-R
290A$	A495G-Y
290A$	A561GR
290A$	A562GR
290A$	A661GR
290A$	A673C2
290A$	A673WT
290A$	2SA495G
290A$	2SA495G-GR
290A$	2SA495G-R
290A$	2SA495G-Y
290A$	2SA561GR
290A$	2SA562GR
290A$	2SA661GR
290A$	2SA673C2
290A$	2SA673WT
290A$	2SB726S
291	C1107Q
291	D49
291	D92
291	D92D
291	EP15X19
291	RCA1C03
291	RCA1C12
291	RCA29B
291	RCA29C
291	T1P31A
291	TIP-29
291	TIP-31B

ECG	Industry Standard No.
291	TIP29
291	TIP29A
291	TIP29B
291	TIP29C
291	TIP31B
291	1A0046
291	1A0059
291	2N6473
291	2N6474
291	2SC1107
291	2SC1107Q
291	2SC1625
291	2SC1625YLBGL1
291	2SC1827
291	2SD258
291	2SD259
291	2SD356
291	2SD358
291	2SD381
291	2SD382
291	2SD49
291	2SD92
291	2SD92D
291	13-34046-5
291	13-39074-1
291	13-41628-2
291	13-41628-5
291	13-43005-1
291	46-86500-3
291	57A187-12
291	86-613-2
291	121-1008
291	121-970
291	121-987
291	121-992
291	121-992-01
291	260P37108
291	2321302
291	2003162538
291	2004547806
292	B537LM
292	BD274
292	RCA1C04
292	RCA1C13
292	RCA30B
292	SPS5451
292	TIP-30
292	TIP30
292	TIP30A
292	TIP30B
292	TIP32B
292	TVS2SB546
292	2N3741
292	2N3741A
292	2N6022
292	2N6023
292	2N6024
292	2N6025
292	2N6026
292	2N6467
292	2N6468
292	2N6475
292	2N6476
292	2SA490-0
292	2SA775
292	2SA775C
292	2SA814
292	2SA815
292	2SB526
292	2SB528
292	2SB536
292	2SB537
292	2SB537L
292	2SB537LM
292	2SB566
292	2SB566A
292	2SB566C
292	2SB566D
292	13-34047-4
292	13-41629-4
292	57A188-12
292	57B188-12
292	57B205-14
292	86-614-2
292	121-1009
292	121-969
292	121-988
292	121-994
292	260P34008
292	2320541
292	2321381
292	06100008
292	2004356801
293	C1383
293	C1383P
293	C1383Q
293	C1383R
293	C1383S
293	C1383X
293	C1384
293	C1384Q
293	C1384R
293	C1384S
293	CS1909B
293	D400
293	D471
293	D471L
293	DDBY272001
293	DDBY410001
293	EA15X250
293	EA15X397
293	EA15X413
293	EA15X8517
293	ED1702N
293	G05-415-B
293	G05415
293	GE-47
293	G05415
293	HT313831X
293	HT313832C
293	HT313841R
293	IP20-0214
293	KGE41414
293	MPS9466A
293	RV2068
293	T9418
293	TIS95
293	TIS96
293	TM293
293	2N3973
293	2N3974
293	2N3975
293	2N3976
293	2SC1383

ECG	Industry Standard No.
293	2SC1383P
293	2SC1383Q
293	2SC1383R
293	2SC1383RS
293	2SC1383S
293	2SC1383X
293	2SC1384
293	2SC1384Q
293	2SC1384R
293	2SC1384S
293	2SC1518
293	2SC2060
293	2SC2060Q
293	2SC235
293	2SC235-0
293	2SD468
293	2SD468A
293	2SD468B
293	2SD468C
293	2SD468D
293	2SD471
293	2SD471K
293	2SD471L
293	2SD471M
293	2SD545
293	2SD545F
293	2SD571K
293	03A02
293	8-760-335-10
293	09-302155
293	13-0178
293	46-86550-3
293	48-3003A02
293	48S155121
293	50-40306-07
293	56-4830
293	86-100010
293	87-0236-Q
293	87-0236-R
293	87-0236-S
293	87-0236R
293	102-1384-17
293	147N1
293	1049-0167
293	1702M
293	2000-276
293	2057A2-484
293	8000-00049-059
293	8000-00049-061
293	8517
293	742729
293	2321881
293MP	GE-47MP
293MP	TM293MP
294	A683
294	A684
294	A8T4026
294	A8T4028
294	B544
294	DBAZ073304
294	DDBY003002
294	DDBY003003
294	DDBY104002
294	EA15X185
294	EA15X268
294	EA15X385
294	EA15X396
294	EA15X7592
294	EA15X7639
294	ED1802N
294	EQR-0016
294	ET539122
294	G03-017-B
294	G03014
294	GE-334
294	GE-48
294	HA-00733
294	HT107331Q0
294	JE9015B
294	NB021
294	Q5209
294	QT-A0733XAA
294	T9468
294	2N3765
294	2SA683
294	2SA683P
294	2SA683Q
294	2SA683R
294	2SA683S
294	2SA684
294	2SA684Q
294	2SA684R
294	2SA966
294	2SB544
294	2SB544E
294	2SB544F
294	2SB562
294	2SB562B
294	2SB562C
294	2SB564
294	2SB564L
294	2SB621
294	2SB621R
294	2SB684
294	03A06
294	8-729-468-43
294	8-760-514-10
294	09-300076
294	46-86514-3
294	48-90420A05
294	48S155122
294	56-4831
294	86-100011
294	86-423-2
294	86-511-9
294	87-0237-R
294	87-0237-S
294	147P2
294	177-018-9-001
294	177-020-9-001
294	921-292B
294	1060-6564
294	1062-6414
294	1076-0999
294	1097-85
294	2032-35
294	7638
294	7639
294	95241-1
294	95241-3
294	145840
294	1224031
294	2321891
294	5321252

ECG	Industry Standard No.
295	A671656K
295	C495
295	C495-R
295	DDBY256001
295	GE-270
295	NCB14
295	01-031957
295	2SC1449(CB)
295	2SC1957
295	2SC2024
295	2SC2314
295	2SC2314D
295	2SC495
295	2SC495-0
295	2SC495-R
295	2SC495-Y
295	2SC495T
295	2SC496
295	2SC496-0
295	2SC496-R
295	2SC496-Y
295	09-302136
295	09-302212
295	10-006
295	44T-300-107
295	44T-300-108
295	48-40734P01
295	48-42884P01
295	106-005
295	176-029-9-004
295	176-060-9-004
295	2000-270
295	2065-07
295	8888-00005-013
295	167542
295	916119
295	5321311
295#	2SC1846
295#	2SC1846B
295#	2SC1846P
295#	2SC1846Q
295#	2SC1846R
295#	2SC1846S
295#	2SC2091
297	A67-70-960
297	BFR40
297	BFR41
297	C1318C
297	C1509
297	C1509P
297	C1509Q
297	C1509R
297	C1509S
297	CS8050
297	D336
297	D336R
297	D336Y
297	D33D21
297	D33D22
297	D33D24
297	D33D25
297	D33D26
297	D33D27
297	D33D29
297	D33D30
297	EA15X378
297	EA15X408
297	EA15X8521
297	EQS-0165
297	GE-271
297	HS-40016
297	MPS9416
297	MPS9416A
297	MPS9416AT
297	MPS9416ST
297	MPS9417A
297	MPS9417AT
297	MPS9417T
297	MPS9418S
297	NB211
297	NB211EI
297	Q5124
297	QSC509
297	QT-C1318XAN
297	QT-C1318XDN
297	TM297
297	VS2SC1166-0-1
297	VS2SC1166-Y-1
297	VS2SC1166Y-1
297	01-031166
297	01-031318
297	2N3704
297	2N5233
297	2N5234
297	2N5235
297	002SC1209C
297	2SC1509
297	2SC1509P
297	2SC1509Q
297	2SC1509R
297	2SC1509S
297	2SD336
297	2SD336R
297	2SD336Y
297	09-302222
297	09-305092
297	09-309063
297	13-15842-1
297	13-23505-2
297	21M541
297	46-86540-3
297	48-155046
297	48-355037
297	48S155046
297	48S155117
297	86-609-2
297	86-609-9
297	102-1166-25
297	121-1035
297	176-056-9-005
297	260-10-041
297	260P31402
297	318-2
297	1039-0458
297	2057A100-67
297	2057A2-524
297	5001-539
297	7318-1
297	171557
297	231017
297	248017
297	741856
297	1223913
297	1417318-2

ECG	Industry Standard No.	ECG	Industry Standard No.	ECG	Industry Standard No.	ECG	Industry Standard No.	ECG	Industry Standard No.
297	1417349-2	302	28C1728	311	C441	312	EA15X169	312	NKT80213
297	8114057	302	28C1728-3	311	C442	312	EA15X192	312	NKT80214
297	23114915	302	28C1728D	311	C628	312	EA15X193	312	NPC512N
297	26010041	302	09-302219	311	C652	312	EA15X394	312	NRT916
297	26010052	302	8000-00043-064	311	C654	312	EA15X400	312	NS316
297	43029484	302	610228-3	311	C844	312	EA15X401	312	PL1091
297	4100905554	302	16304190031	311	C991	312	EA15X446	312	PL1092
298	A707	3028	CL110	311	C992	312	EA3278	312	PL1093
298	A707V	3028	OP130	311	C84001	312	EP1(ELCOM)	312	PL1094
298	A777	3028	OP131	311	GE-277	312	EP2(ELCOM)	312	PTC151
298	A777R	3028	TIL31	311	HEP-75	312	EF3	312	PTC152
298	B560	3028	TIL34	311	HEP-83013	312	EL131	312	Q-00169
298	BC527-16	3032	TIL81	311	MM1803	312	EQP-0009	312	Q-00169A
298	BC528-16	3034	L14H1	311	MM1941	312	EQP-4	312	Q-00169B
298	EA15X8522	3035	2N5777	311	MPS3646	312	ES15X92	312	Q-00169C
298	EA15X8524	3035	2N5778	311	S3013	312	ET491051	312	Q-00184R
298	FD-1029-JB	3035	2N5779	311	2N2369A	312	F1462	312	QA-18
298	GE-272	3035	2N5780	311	2N3137	312	F1463	312	QA-20
298	H8-40027	3036	L14P1	311	2N3227	312	FE100A	312	QKT0033XBE
298	IP20-0211	3040	MCT26	311	2N3544	312	FE102	312	QRF-3
298	MPS9466AT	3040	OPI2151	311	2N3687A	312	FE102A	312	QRF3
298	MPS9467A	3040	OPI2152	311	2N3724	312	FE104A	312	QRG-3
298	MPS9467T	3040	OPI2252	311	2N3862	312	FE402	312	QT-K0023AAS
298	MPS9468	3040	4N25	311	2N3866	312	FE402A	312	QT-K0033XBE
298	MPS9468B	3040	4N27	311	2N3869	312	FE404A	312	R845
298	MPS9468T	3040	4N28	311	2N4073	312	FF400	312	R846
298	MPS9476AT	3041	H11A1	311	2N4137	312	FT743	312	R8-2028
298	2SA707	3041	H11A3	311	2N4449	312	G08-005L	312	R87916
298	2SA707V	3041	H11B2	311	2N4873	312	G08-007-B	312	RT175
298	2SA777	3041	IL1	311	2N4956	312	G08005L	312	RT8351
298	2SA777P	3041	TIL113	311	2N5272	312	GE-FET-1	312	RT8667
298	2SA777Q	3041	4N35	311	2N5292	312	GE-FET-2	312	RVTMK10-2
298	2SA777R	3041	4N36	311	2N5399	312	GRF-3	312	RVTMK10-B
298	2SA778	3041	4N37	311	2N707	312	HA2001	312	S1211N
298	2SA817	3042	H11A2	311	2N745A	312	HA2010	312	S1212N
298	13-23506-2	3042	IL12	311	28C165	312	HEP-F0015	312	S1213N
298	86-610-2	3042	IL5	311	28C285	312	HEP-F0021	312	S1214N
298	121-1036	3042	MCT2	311	28C285A	312	HEP801	312	S1215N
298	260P16802	3042	MCT2E	311	28C319	312	HEP802	312	S1216N
298	1044-6888	3042	TIL111	311	28C396	312	HP2000411B	312	S1221N
298	2057A2-446	3042	TIL114	311	28C440	312	HP200191A	312	S1222N
298	171556	3043	TIL116	311	28C441	312	HP200191A-0	312	S1223N
298	1417303	3045	H11C1	311	28C628	312	HP200191AO	312	S1224N
298	1417303-1	3046	H11C1	311	28C651	312	HP200191B0	312	S1225N
298	1417303-2	3046	H11C2	311	28C652	312	HP200191BO	312	S1226N
298	43029485	3046	H11C3	311	28C654	312	HP2003010-0	312	S1231N
298	4100905254	3046	H11C4	311	28C731	312	HP2003010-O	312	S1232N
299	C1017	3046	H11C5	311	28C844	312	HP2004011B	312	S1233N
299	C1018	3046	H11C6	311	28C852	312	HP2004110O	312	S1234N
299	DMCE101800	3046	MC82	311	28C991	312	HK-00049	312	S1235N
299	GE-236	3046	MCS2400	311	28C992	312	HK-00330	312	S1236N
299	HX-50106	306	C1760	311	13-0064	312	IC743046	312	S1241N
299	IP20-0131	306	C1760-3	311	19-020-072	312	IP20-0010	312	S1242N
299	KTRE017	306	DBCE176000S	311	12193115	312	IP20-0011	312	SE5819
299	MPS9411	306	EA15X380	311	576-0004-008	312	IP20-0012	312	SE5-0996
299	MPS9411A1	306	E08-0184	311	56803	312	IP20-0035	312	SE85819
299	MPS9411T	306	GE-276	311	36848	312	IP20-0078	312	SFB145
299	Q-17115C	306	01-031760	311	36849	312	IP20-0305	312	SK19
299	01-031101B	306	01-031846	311	36850	312	IT108	312	SK3116
299	28C1017	306	28C1760	311	56997	312	J308	312	SFP024
299	28C1018	306	28C1760-2	311	37431	312	JE1033B	312	S8-3704
299	28C1018B	306	28C1760-3	311	37432	312	JP-1033	312	S83534-4
299	09-302102	306	8-763-113-00	311	37433	312	JP1033	312	S83586
299	09-302135	307	A623	311	37894	312	JP1033B	312	S83672
299	13-23526-6	307	A623-0	311	38063	312	JP1033G	312	S83704
299	52-020-173-0	307	A623A	311	38190	312	JP1033S	312	S83735
299	302-679-1	307	A623B	311	40469	312	JNJ61673	312	SU2076
299	302-680	307	A623C	311	40471	312	K10	312	SU2077
299	1000-140	307	A623D	311	61026	312	K17	312	SU2080
299	1087-01	307	A623G	311	61273	312	K22Y(1 GATE)	312	SU2081
299	4802-3268-000	307	A623R	311	111193-001	312	K23(1 GATE)	312	SX-58Y
299	4804-3267-901	307	A623Y	311	760284	312	K23A	312	SX3819
299	5001-004	307	A624	311	1700035	312	K24(1 GATE)	312	TO1-044
299	5001-044	307	A624A	311$	28C547	312	K25(1 GATE)	312	T1-741
299	8000-00003-041	307	A624B	312	A04201	312	K25P(1 GATE)	312	T1208
299	8000-00003-042	307	A624C	312	A066-110	312	K30B(1 GATE)	312	T1814
299	8000-00003-043	307	A624D	312	A068-106	312	K30GR(1 GATE)	312	T1834
300	C1013	307	A624GN	312	A068-107	312	K31C	312	T1859
300	C1013C	307	A624L	312	A11744	312	K31C(1 GATE)	312	T1888
300	C1013D	307	A624LQ	312	A11745	312	K33(1 GATE)	312	TAA330
300	C1014	307	A624R	312	A192	312	K33E	312	TE-500-B
300	C1014B	307	A624Y	312	A194	312	K33E(1 GATE)	312	T8500
300	C1014C	307	28A623	312	A195	312	K33P	312	TI-741
300	C1014CD	307	28A623-0	312	A196	312	K33P(1 GATE)	312	T1741
300	C1014D	307	28A623A	312	A2652-919	312	K33GR	312	TIS-88B
300	C1014D1	307	28A623B	312	A514-040296	312	K34	312	T1814
300	C1243	307	28A623C	312	B-6001	312	K34(1 GATE)	312	T1834
300	C1243-24	307	28A623D	312	BC264C	312	K34(E)	312	T1842
300	C1243C	307	28A623G	312	BF244	312	K34C	312	T1858
300	C1243C1	307	28A623R	312	BF244A	312	K34C(1 GATE)	312	T1859
300	C1243D	307	28A623Y	312	BF244B	312	K34D	312	T1878
300	C1243D1	307	28A624	312	BF244C	312	K34E	312	T1879
300	C1243D2	307	28A624A	312	BF245	312	K35(1 GATE)	312	T1888
300	C1243E	307	28A624B	312	BF245A	312	K49H(1 GATE)	312	TNJ61672(2SK25)
300	GB-273	307	28A624C	312	BF245B	312	K49HK(1 GATE)	312	TNJ61673
300	IP20-0172	307	28A624D	312	BF245C	312	K49I(1 GATE)	312	TNJ61673(2SK24)
300	28C1013	307	28A624E	312	BF246	312	K49M(1 GATE)	312	TR-8027
300	28C1243-24	307	28A624G	312	BF247	312	K55(1 GATE)	312	TR-FET-1
300	28C1243C	307	28A624GN	312	BF256	312	K55D(1 GATE)	312	TR-U1650E
300	28C1243C1	307	28A624L	312	BF256A	312	K55E(1 GATE)	312	TR-U1835B
300	28C1243C2	307	28A624LQ	312	BF256B	312	KA4559	312	TR06011
300	28C1243D	307	28A624R	312	BF256C	312	KE4416	312	TR06014
300	28C1013B	307	28A624Y	312	BF348	312	LB5105	312	TR228735120325
300	28C1013C	307	28A703	312	BF821	312	L85484	312	TR2BK55
300	28C1013D	309K	LM309KC	312	BF821A	312	L85485	312	TR5528
300	28C1014	309K	SO309K	312	BF868	312	MJF10335	312	TU334
300	28C1014B	309K	S07805CK	312	BF868P	312	MJF1033G	312	TV-83
300	28C1014C	311	BF152	312	BFW10	312	MK-1-O	312	TV80
300	28C1014CD	311	BF153	312	BFW11	312	MK-10-E	312	U1177
300	28C1014D	311	BF154	312	BFW12	312	MK10	312	U1178
300	28C1014D1	311	BF158	312	BFW13	312	MK10-2	312	U1181
300	28C1014E	311	BF159	312	BFW54	312	MK5485	312	U1277
300	28C1243	311	BF160	312	BFW55	312	MPF-102	312	U1278
300	28C1243-24	311	BF303	312	BFW56	312	MPF101	312	U1279
300	28C1243C	311	BF350	312	C-36582	312	MPF103	312	U1280
300	28C1243C1	311	BFY63	312	C764	312	MPF104	312	U1285
300	28C1243C2	311	BFY90	312	CDC731	312	MPF105	312	U1286
300	28C1243D	311	BSX20	312	D1101	312	MPF107	312	U1323
300	28C1243D2	311	BSX28	312	D1102	312	MPF112	312	U1324
300	28C1243E	311	BSX32	312	D1301	312	MPS566B	312	U1325
300	09-302046	311	BSX39	312	D1302	312	NE4304	312	U1714
300	09-305049	311	BSX87A	312	D1303	312	NF520	312	U1715
300	031-058-0	311	BSX88A	312	DDCY002002	312	NF522	312	U1837E
300	1009-05	311	C163	312	DDCY006001	312	NF523	312	U1994E
300	4802-00016	311	C285	312	DS-88	312	NF531	312	U3012
300	5001-010	311	C285A	312	D388	312	NF533	312	UC100
300	7398-6120P	311	C319	312	E103	312	NF5485	312	UC110
300	7398-6120P1	311	C375-0	312	E300	312	NF5486	312	UC120
300	8000-00004-087	311	C387	312	EA15X165	312	NKT800112	312	UC155
300	171162-287	311	C387A			312	NKT800113	312	UC200
300	43029486	311	C396			312	NKT800211	312	UC201
300MP	171174-2	311	C423			312	NKT800212	312	UC210
302	C1728	311	C428					312	UC220
302	C1728-3	311	C440						
302	C1728D								
302	GE-275								
302	RE204								

ECG	Industry Standard No.	ECG	Industry Standard No.	ECG	Industry Standard No.	ECG	Industry Standard No.	ECG	Industry Standard No.
312	UC714	312	28K42-CM1	312	0105-0012	312	127214	313	2SC475K
312	UC734	312	28K42CM1	312	106M	312	144859	313	2SC476
312	UC734 E	312	28K43	312	108-0049-08	312	207417	313	2SC540
312	UC756	312	28K49	312	108-0068-12	312	226517	313	2SC605
312	V159	312	28K49B2	312	121-731	312	452077	313	2SC605(Q)
312	V1650E-1	312	28K49P	312	121-756	312	489751-208	313	2SC605K
312	V1650E-4	312	28K49H	312	121-858	312	5160098	313	2SC605M
312	V185	312	28K49H1	312	121-860	312	610164-1	313	2SC605Q
312	V1833E	312	28K49H2	312	123-005	312	618580	313	2SC606
312	V828K49P-1	312	28K49HK	312	132-049	312	760268	313	2SC606A
312	W1P	312	28K49I	312	173-1(SYLVANIA)	312	916082	313	2SC640
312	WP1	312	28K49M	312	182-009-9-001	312	916100	313	2SC640B
312	WP2	312	28K54	312	182-014-9-002	312	970253	313	2SC695
312	ZEN123	312	28K55	312	182-014-9-003	312	985715	313	2SC804DH
312	001-027030	312	28K55C	312	182-015-9-001	313	2006623-47	313	28CP11
312	01-070030	312	28K55D	312	182-021-9-001	313	2327132	313	90-46
312	1W11706	312	28K55DE	312	182-029-9-001	313	2327142	313	90-47
312	2N2968A	312	28K55B	312	182-039-9-001	313	3412004-1912	313	1848
312	2N3066	312	28K55R	312	182-044-9-001	313	5320032	313	1016-85
312	2N3067	312	03A09	312	182-044-9-002	313	5320583	313	1210-17
312	2N3068	312	3830B	312	182-045-9-001	313	5320702	313	1634-17
312	2N3070	312	4-324	312	182-046-9-001	313	5320942	313	2134-17
312	2N3071	312	04-38190-01	312	182-056-9-001	313	5320943	313	129604
312	2N3084	312	006-0004443	312	200-053	313	5321422	313	171319
312	2N3085	312	007-0214-00	312	200-064	313	5321501	313	171319
312	2N3086	312	007-0214-01	312	220-008001	313	7853090-01	313$	C1158
312	2N3087	312	07-07158	312	229-0192-20	313	7910134-01	313$	C1159
312	2N3088	312	07-07159	312	260-10-006	313	7938318	313$	28C1158
312	2N3088A	312	8-723-302-00	312	260P22001	313	8110014	313$	28C1159
312	2N3089	312	8-905-706-901	312	260P22002	313	26010006	315	C1475
312	2N3089A	312	8P111	312	260P22003	313	30400021	315	C1475-1
312	2N3436	312	09-304017	312	364-10048	313	43022134	315	GE-279
312	2N3437	312	09-305014	312	417-140	313	43024225	315	28C1475
312	2N3438	312	09-305021	312	417-169	313	55460014-001	315	28C1475-1
312	2N3452	312	09-305023	312	417-194	313	62608528	315	28C1475-3
312	2N3453	312	09-305031	312	417-211	313	80053501	315	28C1475-4
312	2N3454	312	09-305032	312	417-231	313	C182	315	28C1475A
312	2N3455	312	09-305133	312	417-246	313	C182Q	315	8-729-665-47
312	2N3456	312	09-305135	312	417-252	313	C183	315	8-760-413-10
312	2N3460	312	09-309074	312	417-253	313	C183E	315	09-302202
312	2N3465	312	10-004	312	430-25762	313	C183J	315	90-112
312	2N3466	312	13-22690-1	312	576-0006-003	313	C183K	315	90-177
312	2N3684A	312	13-22692-1	312	588U	313	C183L	315	90-454
312	2N3685	312	13-22692-2	312	654-1(SYLVANIA)	313	C183M	315	921-1010
312	2N3685A	312	13-28654-1	312	690V116H22	313	C183P	315	2322-17
312	2N3686	312	13-28654-2	312	700-04	313	C183Q	315	2340-17
312	2N3686A	312	13-34375-1	312	734EU	313	C183R	315	8000-00030-008
312	2N3687	312	13-34375-2	312	753-4000-024	313	C183W	315	741858
312	2N3819	312	13-34378-1	312	753-4000-025	313	C184	316	A430
312	2N3821	312	13-34378-2	312	753-6000-019	313	C184H	316	A485
312	2N3967	312	13-34378-3	312	753-9000-019	313	C184J	316	C325
312	2N3968	312	13-43112-1	312	800-535-00	313	C184L	316	C325A
312	2N3969A	312	13-44290	312	800-535-01	313	C185	316	C328A
312	2N4117	312	13-44291	312	921-126B	313	C185A	316	C329
312	2N4117A	312	14-2002-01	312	921-203B	313	C185J	316	C329B
312	2N4118	312	14-700-01	312	921-231B	313	C185M	316	C329C
312	2N4119	312	14-700-02	312	991-01-1706	313	C185Q	316	C392
312	2N4119A	312	14-700-03	312	991-01-3055	313	C185R	316	C464
312	2N4222	312	14-700-04	312	991-011576	313	C266	316	C464C
312	2N4222A	312	14-700-05	312	991-011706	313	C429	316	C465
312	2N4223	312	14-700-06	312	1002-01	313	C429J	316	C466
312	2N4338	312	14-710-21	312	1009-127	313	C429X	316	C466H
312	2N4867	312	14-713-31	312	1041-70	313	C430	316	C567
312	2N4867A	312	14-713-32	312	1042-02	313	C430H	316	C568
312	2N4868	312	14-714-13	312	1095-01	313	C430W	316	C611
312	2N4868A	312	16027	312	16128K24E	313	C469	316	C612
312	2N5045	312	19-020-115	312	1859-17	313	C469A	316	C707
312	2N5046	312	020-1110-016	312	1934-17	313	C469F	316	C707H
312	2N5047	312	020-1110-021	312	2000-101	313	C469K	316	GE-213
312	2N5163	312	020-1110-022	312	2000-104	313	C469Q	316	GE-214
312	2N5245	312	020-1112-005	312	2000-105	313	C469R	316	GE-86
312	2N5246	312	020-1112-006	312	2000-107	313	C475	316	8CA3021
312	2N5247	312	020-1112-007	312	2020-04	313	C475K	316	8K3093
312	2N5248	312	020-1112-008	312	2032-56	313	C476	316	2N2708
312	2N5277	312	020-1112-009	312	2056-75	313	C540	316	2N2857
312	2N5484	312	20R3	312	2057A2-445	313	C605	316	2N3478
312	2N5486	312	21A113-002	312	2057B149-12	313	C605(NEC)	316	2N3570
312	2N5555	312	21M196	312	2058-02	313	C605(Q)	316	2N3571
312	2N5955	312	21M224	312	2074-17	313	C605K	316	2N3572
312	2NJ233B	312	21M412	312	2335-17	313	C605L	316	2N3839
312	2BJ11	312	21M534	312	2336-17	313	C605M	316	2N3880
312	2BJ12	312	022-2876-005	312	2450-17	313	C605Q	316	2N3953
312	2BK1033B	312	022-2876-006	312	3511(SEARS)	313	C605TW	316	2N4934
312	2BK104	312	022-2876-009	312	3511(WARDS)	313	C606	316	2N5024
312	2BK104H	312	24MW723	312	3512(RCA)	313	C606(NEC)	316	2N5053
312	2BK13	312	24MW736	312	3512(SEARS)	313	C640	316	2N5054
312	2BK17	312	24MW989	312	3512(WARDS)	313	C640B	316	2N5179
312	2BK17-0	312	24T-026-001	312	3819(RCA)	313	C695	316	2N5650
312	2BK17A	312	33K59	312	4802-00010	313	CP11	316	2N5651
312	2BK17B	312	34-6018-2	312	4811-0000-015	313	EQ8-19	316	2N5652
312	2BK17BL	312	44T-300-105	312	4811-0000-025	313	EQ8-20	316	2N6304
312	2BK17OR	312	44T-300-113	312	5001-047	313	GE-278	316	2N6305
312	2BK17R	312	46-86316-3	312	5096	313	TM313	316	2N6389
312	2BK17Y	312	48-134944	312	5459	313	1A0020	316	28C1424
312	2BK22Y	312	48-137023	312	6005	313	1A0029	316	28C1790
312	2BK23	312	48-137343	312	6013	313	2D038	316	28C251
312	2BK23A	312	48-3003A09	312	8000-00004-080	313	2SC0182	316	28C251A
312	2BK23A540	312	48-43467J01	312	8000-00004-081	313	2SC182Q	316	28C252
312	2BK24	312	48-90252A14	312	8000-00004-P080	313	2SC183	316	28C253
312	2BK24C	312	48-90234A39	312	8000-00004-P081	313	2SC183E	316	28C325
312	2BK24D	312	48-97046A47	312	8000-00005-001	313	2SC183J	316	28C325A
312	2BK24DR	312	48-97046A48	312	8000-00009-178	313	2SC183K	316	28C325C
312	2BK24E	312	48-97177A01	312	8000-00010-017	313	2SC183M	316	28C325B
312	2BK24F	312	48-97177A06	312	8000-00011-054	313	2SC183P	316	28C329A
312	2BK24G	312	48-97177B01	312	8000-00011-055	313	2SC183Q	316	28C329
312	2BK25	312	488134944	312	8000-0004-P080	313	2SC183R	316	28C329B
312	2BK25C	312	488137343	312	8000-00004-P081	313	2SC183W	316	28C329C
312	2BK25D	312	48X97177H01	312	8000-00049-062	313	2SC184	316	28C390
312	2BK25E	312	57A149-12	312	8000-0005-001	313	2SC184H	316	28C392
312	2BK25ET	312	57A150-12	312	8010-173	313	2SC184J	316	28C464
312	2BK25F	312	57A31-1	312	8440-121	313	2SC184L	316	28C464C
312	2BK25G	312	57A31-4	312	8710-161	313	2SC184R	316	28C465
312	2BK304	312	57B149-12	312	8840-161	313	2SC185	316	28C466
312	2BK31	312	57B150-12	312	8910-141	313	2SC185A	316	28C466H
312	2BK31C	312	57B169-12	312	18410-141	313	2SC185J	316	28C567
312	2BK32B	312	57L106-9	312	23606	313	2SC185M	316	28C568
312	2BK33	312	63-13926	312	28810-171	313	2SC185Q	316	28C611
312	2BK33(E)	312	065-001	312	36582	313	2SC185R	316	28C653
312	2BK33D	312	065-002	312	38510-161	313	2SC185V	316	28C707
312	2BK33E	312	81-461255065-6	312	40461	313	2SC266	316	28C707H
312	2BK33F	312	86-477-2	312	40468	313	2SC429	316	09-302061
312	2BK330R	312	86-500-2	312	42396	313	2SC429J	316	13-10521-50
312	2BK35H	312	86-5095-2	312	43296	313	2SC429X	316	46-8677-3
312	2BK34	312	86-5096-2	312	44766	313	2SC430	316	90-49
312	2BK34(E)	312	86-5123-2	312	45810-161	313	2SC430H	316	121-585
312	2BK34B	312	87-0001	312	48009	313	2SC430W	316	121-585B
312	2BK34C	312	90-179	312	48009(1)	313	2SC469	316	121-687
312	2BK34D	312	90-50	312	58810-160	313	2SC469A	316	150-152
312	2BK34E	312	90-59	312	71748-*	313	2SC469F	316	176-044-9-002
312	2BK37	312	90-606	312	88060-141	313	2SC469K	316	260-10-056
312	2BK37N	312	90-607	312	88510-171	313	2SC469Q	316	2000-214
312	2BK37K	312	90-608	312	95133	313	2SC469R		
312	2BK37L	312	90-613	312	95228	313	2SC475		
312	2BK42	312	90-62	312	95231				

ECG	Industry Standard No.	ECG	Industry Standard No.	ECG	Industry Standard No.	ECG	Industry Standard No.	ECG	Industry Standard No.
316	40295	324	D120C	329	576-0004-004	373	EP15X32	375	28D586A
316	40296	324	D120H	329	576-0004-005	373	28C1382	375	28D586D
316	40894	324	D120HA	329	576-0004-006	373	28C1382-Q	375	28D586Y
316	40895	324	D120HB	329	576-0004-007	373	28C1382Y	375	28D587
316	40896	324	D120HC	329	576-0004-009	373	28C1903	375	28D401
316	40897	324	EA15X246	329	576-0004-011	373	28C2028-2	375	28D402
316B	0566	324	PTC123	329	576-0004-012	373	28C2028-B/20	375	28D478
316B	28C566	324	2N1479	329	576-0004-104	373	28C2028/2	375	28D608
318	SD1074	324	2N1480	329	2904-059	373	28C2028B	375	28D94
319	A67-08-760	324	2N1481	329	40080	373	28C2028B/20	375	8-729-316-12
319P	C598(PA-1)	324	28C484	329	40081	373	28C2036	375	46-86535-3
319P	0399	324	28C484BL	329	40081V1	373	28D414	375B	28C2238
319P	C957	324	28C484R	329	40082	373	28D415	375B	28D724
319P	CGB-60	324	28C484Y	329	40082A2	373	13-40344-1	376	A6319403
319P	EA15X7173	324	28C485	329	40581	373	48-90234A36	376	C1569LBQ
319P	EA15X7174	324	28C485BL	331	BD242A	373	48-90343A85	376	C1569LBR
319P	EA15X7178	324	28C485C	331	TIP41	373	146596	376	C1569LBY
319P	EA15X7179	324	28C485Y	331	TIP41A	373	2320843	376	C1569R
319P	SE5055	324	28C486	331	TIP41Y	373#	1212WT	376	C1756A
319P	TG28C2057-C	324	28C486Y	331	2N6099	373#	1212WTA	376	C1756C
319P	TG28C2057-D	324	28C510	331	2N6100	373#	1212WTB	376	C1756K
319P	TG28C927-D-A	324	28C511	331	2N6101	373#	1212WTC	376	01-051514
319P	TG28C927-D-A	324	28C512	331	2N6102	373#	1212WTD	376	28C1446
319P	TG28C927-E-A	324	28C512-0	331	2N6103	373#	28C1212	376	28C1446B
319P	TM319	324	28C512-R	331	2N6486	373#	28C1212A	376	28C1446C
319P	TV-33	324	28C513	331	2N6487	373#	28C1212AA	376	28C1446L
319P	TV-34	324	28C513-0	331	28D525	373#	28C1212AB	376	28C1446LB
319P	TV-35	324	28C513R	331	28D525-0	373#	28C1212ABWT	376	28C1446P
319P	TV-50	324	28D120	331	28D525-0	373#	28C1212AC	376	28C1446PQ
319P	TV37	324	28D120A	331	28D525Y	373#	28C1212ACWT	376	28C1446Q
319P	01-032057	324	28D120B	331	86X0074-001	373#	28C1212AD	376	28C1446R
319P	28C598	324	28D120C	331	417-298	373#	28C1212AWT	376	28C1514
319P	28C399	324	28D120H	332	D45IB	373#	28C1212AWTA	376	28C1514BK
319P	28C957	324	28D120HA	332	TIP42	373#	28C1212AWTB	376	28C1514BVC
319P	13-10321-75	324	28D120HC	332	TIP42A	373#	28C1212AWTC	376	28C1514CVC
319P	13-23160-3	324	28D121	332	TIP42B	373#	28C1212AWTD	376	28C1569
319P	13-23822	324	28D121A	332	2N6488	373#	28C1212B	376	28C1569-0
319P	13-23824-2	324	28D121B	332	2N6489	373#	28C1212C	376	28C1569BK
319P	34-6015-15	324	28D121HA	332	2N6490	373#	28C1212D	376	28C1569K
319P	34-6015-16	324	28D121HB	332	2N6491	373#	28C1212WT	376	28C1569LBQ
319P	34-6015-17	324	40347	332	28A1010V	373#	28C1212WTA	376	28C1569LBR
319P	34-6015-19	324	40367	332	13-36443-1	373#	28C1212WTB	376	28C1569LBY
319P	34-6015-29	324/427	28D222	332	86X0075-001	373#	28C1212WTC	376	28C1569Y
319P	46-86112-3	324/427	28D223	335	SD1076	373#	28C1212WTD	376	28C1749
319P	46-86538-3	326	2A4360	337	2N5689	373#	28C1567	376	28C1755
319P	57A139-4	326	2N5460	338	2N5690	373#	28C1567A	376	28C1755C
319P	57A139-4-4-6	326	2N5461	341	BF341	373#	28C1847	376	28C1756
319P	57A139-4	326	2N5462	345	40955	373#	28C1847Q	376	28C1756B
319P	57B138-4	326	2N5463	346	C571	373#	28C2497Q	376	28C1756C
319P	57B139-4	326	417-241	346	M9591	373#	28C2497R	376	28C1756D
319P	86-185-2	327	BDY57	346	M9631	373#	28D668	376	28C1756E
319P	86X0052-001	327	BDY58	346	M9703	373#	28D669	376	28C1756K
319P	100N3	327	TIP516	346	MRP207	374	A496-0	376	28C1756M
319P	3535	327	28D5038	346	SD1080	374	A496Y	376	28C1756MC
319P	3535(RCA)	327	28C5039	346	2N4427	374	A505	376	28C1757
319P	124757	327	2N5671	346	2N5421	374	A505-0	376	28C1819
319P	133171	327	2N5672	346	2N5687	374	A505-R	376	28C1819M
319P	610181-1	327	2N6338	346	28C1303	374	A505-Y	376	28C2068
319P	1473535-1	327	2N6339	346	28C730	374	BD170	376	28C2068LB
319P	1475680-1	327	2N6340	346	28C821	374	28A496	376	28C2085
319P	2320471	327	2N6341	346	28C822	374	28A496-0	376	28C2242
319P	23114163	327	13-40342-1	346	48-869591	374	28A496-R	376	46-86505-3
319P	2003038208	327	1723-0405	346	48-869631	374	28A496Y	376	48-90420A92
319P	6621004000	327	1723-0410	346	40280	374	28A505	376	2003S151-432
320	GE-288	327	1723-0605	349	2N5589	374	28A505-0	376	260F35S101
320	2N6084	327	1723-0610	349	2N5590	374	28A505-R	376	2321472
321	28D470	327	1723-0805	350	MM1681	374	28A505-Y	376B	06120083
321	28D470A	327	1723-0810	350	P78769	374	28A682	376B	28C1722
321	28D470B	327	1723-1005	350	P78837	374	28A899	376B	28C1722B
322	MPS-U31	327	1723-1010	350	SD1014	374	28B548	376B	28C1723
322	MPSU31	327	1723-1205	350	SD1069	374	28B549	377	D44C1
322	2N3781	327	1723-1210	350	SD1133	374	13-40345-1	377	D44C2
322	28C2074	327	1723-1405	350	SD1143	374	144033	377	D44C3
322	28C2074C	327	1723-1410	350	2N4127	374	2320855	377	D44C4
322	28C2074T	327	1723-1605	350	2N6081	374#	A973	377	D44C5
323	A486(JAPAN)	327	1723-1610	350	40973	374#	A743A	377	D44C6
323	A510	327	40444	351	SD1218	374#	A743AA	377	D44C7
323	A510-0	328	BD111A	351	2N4128	374#	A743AB	377	D44C8
323	A510-R	328	BD112	351	2N5591	374#	A743AC	377	D44C8B
323	A511	328	BD116	351	2N6082	374#	A743AD	377	D44C9
323	A511-0	328	BD118	351	2N6083	374#	A743B	377	D44CBB
323	A511-R	328	BD121	351	40974	374#	A743C	377	D44H10
323	A512	328	BD123	353	MM4020	374#	A743D	377	D44H11
323	A512-0	328	BD145	354	2N6094	374#	28A743	377	D44H7
323	A512-R	328	BDY61	354	MM4021	374#	28A743A	377	2N6098
323	A513	328	BLY12	354	2N6095	374#	28A743AA	377	2N6121
323	A513-0	328	C646	354	2N6096	374#	28A743AB	377	2N6122
323	A513-R	328	C647	356	MM4023	374#	28A743AC	377	2N6123
323	A516	328	C647Q	356	2N6097	374#	28A743AD	378	D44C1
323	A516A	328	C647R	357	2N5711	374#	28A743B	378	D44C2
323	B510	328	C768	358	105604	374#	28A743C	378	D44C3
323	B510S	328	C939	358	817062	374#	28A743D	378	D44C4
323	BC460	328	C939D	359	BLY93	374#	28A794	378	D44C5
323	BC461	328	C939L	359	2N5642	374#	28A886	378	D44C6
323	2N3660	328	D151	359	2N5712	374#	28A886V	378	D44C7
323	2N5679	328	D161	359	2N5713	374#	28A886VR	378	D44C8
323	2N5680	328	D177	359$	28C6690	374#	28B648	378	D44C9
323	28A486	328	D47	362	2N5648	374#	28B649	378	D45H7
323	28A510	328	TIP509	362	2N5698	375	C1448A	378	2N6124
323	28A510-0	328	TIP510	362	2N5914	375	C1448P	378	2N6125
323	28A510-R	328	2N3919	363	2N5645	375	C1448Q	378	2N6126
323	28A511	328	2N4111	363$	28C891	375	C1448R	378	28A671
323	28A511-0	328	2N4113	364	2N5648	375	C1448B	378	28A671A
323	28A511-R	328	2N5614	364$	28C892	375	D152	378	28A671B
323	28A512	328	2N5616	369	C1828	375	D424-0	378	28A671D
323	28A512-0	328	2N5618	369	2N5660	375	Q51078	378	28A671K
323	28A512-R	328	2N5622	369	28C1078	375	V8280189OA/1E	378	28A671KA
323	28A513	328	2N5881	369	28C1828	375	28C1089	378	28A671KB
323	28A513-0	328	2N5882	369	28D546	375	28C1089B	378	28A671KC
323	28A513-R	328	2N6354	369	28D766	375	28C1089C	379	28C6477
323	28A516	328	2N6496	369	28D766Q	375	28C1089D	379	2N6478
323	28A516A	328	28C1866	369	28D766R	375	28C1409	379	28C2573
323	40595	328	28C1869	373	BD139	375	28C1409A	379	57A205-14
323	06100058	328	28C646	373	BF456	375	28C1409AB	379	121-1028-01
324	BC461	328	28C647	373	BF457	375	28C1409B	380	28D586
324	C484R	328	28C647Q	373	C1212	375	28C1409C	380	28D586R
324	C484Y	328	28C647R	373	C1212A	375	28C1410	382	C499
324	C485	328	28C768	373	C1212AA	375	28C1410B	382	D400D
324	C485BL	328	28C959	373	C1212AB	375	28C1448	382	D400E
324	C485C	328	28C939D	373	C1212AC	375	28C1448A	382	D400F
324	C485Y	328	28C939L	373	C1212AD	375	28C1448P	382	28C2236
324	C486	328	28D151	373	C1212AWT	375	28C1448Q	382	28C2236-Q
324	C486BL	328	28D161	373	C1212AWTA	375	28C1448R	382	28C2236Y
324	C486Y	328	28D177	373	C1212AWTB	375	28C1683	382	28D490
324	C512	328	28D575	373	C1212AWTC	375	28C1683LA	382	28D400
324	C512-Q	328	28D47	373	C1212AWTD	375	28C1683P	382	28D400D
324	C512-R	328	23114864	373	C1212B	375	28C1683R	382	28D400E
324	C513	329	13-0079	373	C1212C	375	28C2073	382	28D400F
324	C513-0	329	13-0079A	373	C1212D	375	28C2167	382	28D438
324	C513R	329	576-0004-001	373	C1382	375	28D152	382	28D438E
324	D120			373	C1382-Q	375	28D586		
324	D120A			373	C1382Y				
324	D120B			373	D600				

ECG	Industry Standard No.	ECG	Industry Standard No.	ECG	Industry Standard No.	ECG	Industry Standard No.	ECG	Industry Standard No.
382	2SD666	385	2SC412	390	1971487	4001	4001	451	FB0654B
382	2SD666C	385	2SC807	390	'971503	4002B	GE-4002	451	MPF102
382$	2SC1670	385	2SC807A	390	3464648-1	4002B	4002	451	MPF106
382$	2SC1670J	385	2SD165	390	3464648-2	4006B	CD4006B	451	NPC10B
382$	2SC1670JW	385	2SD166	390	4450040P1	4006B	4006	451	NPC10BA
382$	2SC2235	385	2SD235	390	94818400	4007	CD4007UBE	451	2N5485
383	A5T4026	386	MJ13335	391	TIP34	4007	MC14007	451	2N5485-1
383	A5T4028	387	2N6277	391	TIP34A	4011B	CD4011	451	2SK41
383	A751	388	DTS409	391	TIP34B	4011B	CD4011BE	451	2SK41D
383	A751Q	388	MJ15024	391	TIP34C	4011B	DD8Y089001	451	2SK41E
383	A751QR	388	2SC1585	391	TIP3055	4011B	F4011	451	2SK41B2
383	A752	388	2SC1585F	392	TIP35	4011B	F4011PC	451	2SK41P
383	A752P	388	2SC1585H	392	TIP35A	4011B	GE-4011	451	2SK61
383	A752Q	388	2SC1586	392	TIP35B	4011B	MB84011	451	2SK61GR
383	A752R	388	2SC1870	392	TIP35C	4011B	MB84011-U	451	2SK61Y
383	A7528	388	2SC2256	393	BD246	4011B	MB84011U	451	2SK63
383	BFR80TO5	388$	2SD735	393	BD246A	4011B	MB84011V	451	38K23
383	C917	389	BD160	393	BD246B	4011B	MC14011B	451	4G2
383	TC8101	389	BDX32	393	BD246C	4011B	MC14011BCP	451	5G2
383	TC8103	389	C1086	393	T1P2955	4011B	MC14011CP	451	129424
383	2SA751	389	C1154	393	T1P36A	4011B	MSM4011R8	4518B	CD4518B
383	2SA751P	389	C1170	393	TIP2955	4011B	TC4011BP	4518B	CD4518BE
383	2SA751Q	389	C558	393	TIP36	4011B	TC4011P	4518B	CM4518
383	2SA751QR	389	C999	393	TIP36A	4011B	UPD4011C	4518B	GE-4518
383	2SA751R	389	C999A	393	TIP36B	4011B	40-035-0	452	2N4416
383	2SA7518	389	D200	393	TIP36C	4011B	51810655A18	452	2N4416A
383	2SA752	389	D300	394	TIP51	4011B	307-113-9-001	452	2N4417
383	2SA752P	389	D818	394	TIP52	4011B	307-152-9-012	452	2N5078
383	2SA752Q	389	DTS801	394	TIP53	4011B	905-126	452	2N5397
383	2SA7528	389	MJ12003	394	TIP54	4011B	1147-09	452	2N5592
383	2SA7528	389	SDT402	395	MM4049	4011B	4011	452	13-28654-3
383	2SA773	389	2SB540	395	2N4080	4011B	4011(IC)	452	13-28654-4
383	2SB560	389	2SC1086	396	BP257	4011B	4011-PC	4520B	CD4520B
383	2SB741	389	2SC1086M	396	B8W-68	4011B	4011PC	453	DDCY001002
383$	2SA949	389	2SC1100	396	B8W69	4011B	84011	454	3N159
383$	2SA965	389	2SC1153	396	C1447-0	4011B	84011U	454	3N200
384	BDY78	389	2SC1154	396	C1447LB	4012B	CD4012BE	454	3N201
384	BDY79	389	2SC1617	396	C1447R	4012B	GE-4012	454	3N202
384	BUY38	389	2SC1170	396	C959	4013B	CD4013	454	3N203
384	C873BL	389	2SC1170A	396	D317A	4013B	CD4013AE	454	3N204
384	D284	389	2SC1170B	396	D318A	4013B	CD4013BE	454	3N205
384	MJ2249	389	2SC1174	396	ER8120	4013B	F4013	454	3N206
384	MJ2250	389	2SC1348	396	ER8140	4013B	GE-4013	454	38K40
384	MJ3101	389	2SC1875	396	ER8160	4013B	51-10655A19	454	38K40I
384	SDF6901	389	2SC1875K	396	ER8180	4015B	CD4015BE	454	38K40M
384	SDF6905	389	2SC1875L	396	ER8200	4015B	F4015	454	13-10521-70
384	SE3040	389	2SC1891	396	ER8225	4015B	GE-4015	454	40819
384	SE3041	389	2SC1892	396	ER8250	4016B	CD4016AE	454	40820
384	2N3054	389	2SC1893	396	ER8275	4016B	CD4016BE	454	40821
384	2N3441	389	2SD558	396	ER8301	4016B	F4016	454	40822
384	2N3584	389	2SD643	396	ER8325	4016B	GE-4016	454$	38211
384	2N3766	389	2SD643A	396	MJ421	4016B	HD4016	456	D1180
384	2N3767	389	280937	396	RCA1A15	4016B	MSM4016	456	LDP603
384	2N3878	389	280937A	396	SK3044	4016B	51-90433A13	456	LDP604
384	2N4910	389	280937B	396	001-021140	4017B	CD4017BE	456	U1180
384	2N4911	389	280937YL	396	2N1715	4017B	F4017	456	U322
384	2N4912	389	280999	396	2N1717	4017B	GE-4017	456	2N4220
384	2N5050	389	280999A	396	2N3019	4018B	F4018B	456	2N4221
384	2N5051	389	2SD200	396	2N3440	4019B	CD4019BE	456	2N4221A
384	2N5052	389	2SD200A	396	2N698	4019B	F4019	456	2N4223
384	2N5202	389	2SD246	396	2N699	4019B	GE-4019	457	K37S821
384	2N6077	389	2SD299	396	2SC1062	4020B	CD4020BE	457	094
384	2N6078	389	2SD299B	396	2SC1447	4020B	F4020	457	E101
384	2N6079	389	2SD299Y	396	2SC1447-0	4020B	GE-4020	457	E102
384	2N6261	389	2SD300	396	2SC1447LB	4021B	CD4021BE	457	FB3819
384	2N6372	389	2SD300B	396	2SC1447R	4021B	F4021	457	ITE3066
384	2N6373	389	2SD517	396	2SD318A	4021B	GE-4021	457	ITE3067
384	2N6374	389	2SD575	396	121-911	4022B	CD4022B	457	ITE4339
384	2SC1055	389	2SD627	396	40321	4023B	CD4023BE	457	ITE4340
384	2SC1055H	389	2SD764	396	40321V1	4023B	CD4023C	457	K47
384	2SC2354	389	2SD765	396	40321V2	4023B	GE-4023	457	K47M
384	2SC508	389	13-33181-1	396	40346	4023B	MSM4023	457	K2S684
384	280833	389	13-33181-2	396	40355	4024B	CD4024BE	457	K2S103
384	2SC833BL	389	13-33181-3	396	40412	4024B	GE-4024	457	MPF111
384	2SD284	389	13-45016-1	396/427	2N4063	4024B	15-45184-1	457	NF4302
384	992-003172	389$	2SC1922	396/427	2N4064	4025B	CD4025BE	457	NF4303
384	992-004092	389$	2SC1942	396/427	40346V2	4025B	GE-4025	457	NKTB0111
384	1678	390	BD245	396/427	4041 2V2	4025B	MC14025BE	457	NKTB0215
384	7510	390	BD245A	397	MM4000	4027B	CD4027BE	457	NKTB0216
384	40250	390	BD245B	397	MM4001	4027B	F4027	457	NPC211N
385	C244	390	BD245C	397	MM4002	4027B	GE-4027	457	NPC212N
385	C245	390	ES69(ELCOM)	397	RCA1A16	4030B	GE-4030	457	NPC214N
385	C246	390	ETS-005	397	2N3224	404/175	40250V1	457	NPC215N
385	C407	390	G23-76	397	2N4930	4040B	CD4040BE	457	NPC216N
385	C408	390	GE-255	397	2N4931	4040B	GE-4040	457	T1858
385	C409	390	IP20-0028	397	2N5415	4042B	CD4042BE	457	2N3051
385	C410	390	LN78533	397	2N5416	4042B	GE-4042	457	2N4302
385	C411	390	M9666	397	40327	4043B	GE-4043B	457	2N4303
385	C807	390	MU-26-1C	398	2SA766	4044B	CD4044B	457	2N4304
385	C807A	390	T-23-71	398	2SA766B	4049	CD4009UBE	457	2N5457
385	D165	390	T1P33	398	2SA816	4049	CD4049UBE	457	2N5458
385	D166	390	T1P33A	398	2SA940	4049	GE-4049	457	2N5668
385	D181	390	TA7200	398	2SA969	4049	MC14049CP	457	26S718
385	DTS410	390	TA7201	398	2SB546	4049	SK4049	457	2SK44
385	DTS423	390	TA7202	398	2SB568	4049	4049	457	2SK44C
385	DTS430	390	TIP33	398$	2SA840	4050B	CD4050BE	457	2SK44D
385	MJ13014	390	TIP33A	398$	2SA968	4050B	MSM4050	457	2SK47
385	MJ3430	390	TIP33B	398$	2SB547	4050B	SK4050	457	2SK47M
385	MJ423	390	TIP33C	398$	2SB628	4050B	51-90433A11	457	302
385	SDT-411	390	T2223	399	C1573	4051B	CD4051BE	457	71686-4
385	SDT1611	390	TR1025	399	2SC1573	4051B	GE-4051	459	K19
385	SDT1612	390	TR2327574	399	2SC1811	4051B	MC14051	459	K19(1 GATE)
385	SDT1613	390	2N5034	399	2SC1885	4051B	SK4051	459	K19(GR)
385	SDT1614	390	2N5035	399	2SC1941K	4052B	CD4052BE	459	K19BL(1 GATE)
385	SDT1616	390	2N5036	399	2SC1941L	4052B	GE-4052B	459	K19C
385	SDT1617	390	2N5037	399	2SC1941M	4053B	CD4053B	459	K19QC(1 GATE)
385	SDT1618	390	7-0115-000	399	2SC2228	4053B	4053	459	K19GR
385	SDT423	390	7-0197	399	2SC2228C	4055B	CD4055BE	459	K19QR(1 GATE)
385	SDT431	390	7-0197-00	399	2SC2228D	4055B	GE-4055	459	K19Y
385	DTS430	390	8-1075	399	2SC2228M	4060B	CD4060B	459	K19Y(1 GATE)
385	DTS431	390	09-305138	399	2SC2229	4066B	CD4066B	459	MPF109
385	2N3445	390	13-28396-1	399	2SC2229Y	4069	CD4069UB	459	U1282
385	2N3446	390	14-601-18	399	2SC2230	4069	51-90433A10	459	U1283
385	2N3713	390	14-601-20	399	2SC2230AGR	4075B	CD4075B	459	U1284
385	2N3714	390	14-601-22	399	2SC2230AY	4081B	CD4081B	459	2N3684
385	2N3715	390	14-601-24	399	2SC2230Y	4081B	CD4081BE	459	2N3822
385	2N3716	390	30-005072	399	2SC2271N	411/128	SC4244	459	2N4869
385	2N4071	390	31-0108	399	2SC2482	411/128	86-5109-2	459	2N4869A
385	2N5264	390	48-869660	400	CD4000BE	411/128	121-279	459	2N5459
385	2N5632	390	52-025-004-0	4000	GE-4000	427/128	C960	459	2N5543
385	2N5633	390	52-079-010	4000	SCL4000	427/128	C960B	459	2N5544
385	2N5634	390	711-001	4001B	CD4001	427/128	C960C	459	2N5592
385	2N5758	390	766-'00999	4001B	CD4001AE	427/128	C960S	459	2N5593
385	2N5759	390	1074-117	4001B	CD4001BE	427/128	C960SA	459	2N5594
385	2SC1868	390	1081-7104	4001B	GE-4001	427/128	C960SB	459	2SK19
385	2SC244	390	5036-1	4001B	SCL4001	427/128	C960SC	459	2SK19(GR)
385	2SC245	390	5036-2	4001B	TC4001P	427/128	C960SD	459	2SK19-GR
385	2SC246	390	28396	4001B	51810655A17	427/282	C522	459	2SK19BL
385	2SC407	390	38804	4001B	51810655A17	427/282	C523	459	2SK19BL
385	2SC408	390	61007	4001B	544-3001-103	427/282	C524	459	2SK19B
385	2SC409	390	76251	4001B	905-125	427/324	D223	459	2SK19C
385	2SC410	390	2822'7			451	E211	459	2SK19E
385	2SC411	390	1833404			451	E305	459	2SK19GR
								459	2SK19H
								459	2SK19K
								459	2SK19V

ECG	Industry Standard No.
459	2SK19Y
459	2SK40
459	2SK40-3
459	2SK40A
459	2SK40B
459	2SK40C
459	2SK40D
459	2SK84
459	71686-5
459	71686-6
459	5321502
459$	2N4340
461	2N5452
461	2N5453
461	2N5454
462	M101
466	E203
466$	E232
472	C320
472	2N5913
472	2SC320
472	40972
473	PT3501
473	48-01-003
473	40290
474	2SC1589
475	2N3926
475	2SC572
475	2SC637
475	40281
476	M9610
476	2N3927
476	2SC573
476	2SC638
476	48-869316
476	48-869610
476	40282
486	C741
486	2N5697
486	2SC741
486	2SC994
486	40964
486	40965
487	2SC890
5000A	MZ500-1
5000A	MZ92-2.4
5000A	1N371
5001A	MZ92-2.5
5002A	MZ500-2
5002A	1N465
5002A	1N465A
5002A	1S41
5004A	SK3770
5005A	A-120420
5005A	AZ746
5005A	BZ102-3V4
5005A	HEP-Z0206
5005A	HS2033A
5005A	HS7033
5005A	KS033B
5005A	KS2033B
5005A	LR33H
5005A	MR33C-H
5005A	MZ500-4
5005A	MZ70-3.3A
5005A	MZ70-3.3B
5005A	MZ92-3.3
5005A	PD6002
5005A	ST-41
5005A	1N5226
5005A	1N5226A
5005A	1N5226B
5005A	1N5518
5005A	1N5518A
5005A	1N746
5005A	1S2033
5005A	1S7033
5005A	1S7033A
5005A	1S7033B
5005A	2A44
5005A	31Z6
5005A	31Z6A
5005A	48-82256C26
5005A	93A39-49
5005A	1102(ZENER)
5005A	5508
5006A	AZ747
5006A	BZX55C3V6
5006A	BZX83C3V6
5006A	RD4AL
5006A	1N466
5006A	1N5227
5006A	1N5227A
5006A	1N5519
5006A	1N5519A
5006A	1N703
5006A	1N703A
5006A	1N747
5006A	1S2036
5006A	1S7036
5006A	1S7036A
5006A	1S7036B
5006A	1S753
5006A	2A25(ZENER)
5006A	13-33187-14
5006A	61E43
5006A	650C0
5006A	5509
5006A	9971
5006A	134883461543
5007A	AZ748
5007A	HEP-Z0208
5007A	MZ1004
5007A	MZ500-6
5007A	RD4A
5007A	RD4AM
5007A	1N1927
5007A	1N1954
5007A	1N1981
5007A	1N467
5007A	1N467A
5007A	1N5228
5007A	1N5228A
5007A	1N5228B
5007A	1N5520
5007A	1N5520A
5007A	1N748
5007A	1S7039
5007A	1S7039A
5007A	1S7039B
5007A	34-8057-55
5007A	650C1
5007A	650C2
5007A	5510
5008A	AZ749
5008A	BZX83C4V3
5008A	BZY64
5008A	RD4AN
5008A	TR4S
5008A	1A58
5008A	1N374
5008A	1N375
5008A	1N5229
5008A	1N5229A
5008A	1N5521
5008A	1N5521A
5008A	1N749
5008A	1N749A
5008A	1S2043
5008A	1S47
5008A	1S7043
5008A	1S7043A
5008A	1S7043B
5008A	1S754
5008A	2A18
5008A	2A47
5008A	13-33187-19
5008A	650C5
5008A	650C6
5008A	5511
5008A	9972
5009A	BZY83D4V7
5009A	BZY85D4V7
5009A	BZY88C4V7
5009A	HEP-Z0210
5009A	MZ500-8
5009A	RD4R7EB
5009A	RD5AL
5009A	RD5AM
5009A	1N1928
5009A	1N1955
5009A	1N1982
5009A	1N468
5009A	1N468A
5009A	1N5230
5009A	1N5230A
5009A	1N5522
5009A	1N5522A
5009A	1N5728
5009A	1N705
5009A	1N705A
5009A	1N750
5009A	1N761
5009A	1N761-1
5009A	1S2047
5009A	1S7047
5009A	1S7047A
5009A	1S7047B
5009A	1S755
5009A	35Z6
5009A	35Z6A
5009A	48-82256C07
5009A	651
5009A	651C0
5009A	651C1
5009A	651C2
5009A	651C3
5009A	1104
5009A	5512
5009A	71780
5009A	134882256303
5009A	134882256307
500A	A95-5314
500A	GE-518
500A	MH914
500A	MH920
500A	MH932
500A	MH987A01
500A	MH987A02
500A	MH987A03
500A	TVM778
500A	28-32-0X
500A	32-29778-2
500A	32-29778-3
500A	32-33057-3
500A	32-33057-4
500A	32-33057-5
500A	32-35894-1
500A	32-35894-5
500A	66F-054-3
500A	76-14327-2
500A	76-14327-3
500A	76-14327-4
500A	76-14327-5
500A	76-14327-6
500A	76-14327-7
500A	76-14327-8
500A	93A96-1
500A	93A96-3
500A	93B96-1
500A	93D96-1
500A	93D96-2
500A	93D96-3
500A	212-00102
500A	212-00104
500A	800-000-025
500A	72162-2
5010A	AZ751
5010A	AZ751A
5010A	BZX46C5V1
5010A	BZX71C5V1
5010A	BZX79C5V1
5010A	BZX83C5V1
5010A	BZY57
5010A	BZY65
5010A	BZY83C5V1
5010A	BZY85C5V1
5010A	BZY88C5V1
5010A	HD3003309
5010A	HEP-Z0211
5010A	MZ-5
5010A	MZ500-9
5010A	RD5.1E
5010A	RD5AN
5010A	RT6922
5010A	RVDRD5A1E
5010A	RZ5.1
5010A	RZZ5.1
5010A	TVSEQA01-05T
5010A	WZ-052
5010A	WZ050
5010A	WZ052
5010A	1N4095
5010A	1N4689
5010A	1N5231
5010A	1N5231A
5010A	1N5231B
5010A	1N5523
5010A	1N5523A
5010A	1N5729
5010A	1N751
5010A	1N751A
5010A	1S190
5010A	1S2051
5010A	1S2051A
5010A	1S330
5010A	1S3305
5010A	1S330A
5010A	1S4735
5010A	1S7051
5010A	1S7051A
5010A	1S7051B
5010A	05-480205
5010A	05Z5.1L
5010A	09-306191
5010A	19-080-014
5010A	19-090-014
5010A	48-137337
5010A	48-83461E10
5010A	48-83461E40
5010A	48-90343A55
5010A	61E10
5010A	61E23
5010A	61E40
5010A	651C4
5010A	651C5
5010A	651C6
5010A	651C7
5010A	651C8
5010A	5513
5010A	8000-00049-021
5010A	9973
5010A	134882256315
5010A	134883461510
5010A	134883461523
5010A	134883461540
5011A	AZ752
5011A	BZX71C5V6
5011A	BZY78
5011A	BZY78P
5011A	BZY83D5V6
5011A	BZY85D5V6
5011A	BZY88C5V6
5011A	HEP-Z0212
5011A	MZ500-10
5011A	RD5.1EC
5011A	RD5.6EB
5011A	RD5R6EB
5011A	RD6AL
5011A	RD6AN
5011A	SD32
5011A	VZ-054
5011A	VZ-056
5011A	WZ-054
5011A	WZ-056
5011A	1N1929
5011A	1N1956
5011A	1N1983
5011A	1N5232
5011A	1N5232A
5011A	1N5524
5011A	1N5524A
5011A	1N5730
5011A	1N708
5011A	1N752A
5011A	1N762-1
5011A	1S2056
5011A	1S7056
5011A	1S7056A
5011A	1S7056B
5011A	48-137322
5011A	48-82256C12
5011A	651C9
5011A	652C1
5011A	652C2
5011A	1105
5011A	5514
5011A	134882256312
5012A	BZY88C6V2
5012A	BZZ10
5012A	EQA01-06S
5012A	EQA01-06SB
5012A	HZ6B
5012A	HZ6B2
5012A	HZ6C
5012A	UZ6.2B
5012A	VZ-058
5012A	VZ-061
5012A	ZL6
5012A	1N5233
5012A	1N5233A
5012A	1N5233B
5012A	1N706
5012A	1N706A
5012A	1N762
5012A	1S191
5012A	1S470
5012A	1S756
5012A	019-003928
5012A	46-86559-3
5012A	48-83461E27
5012A	61E27
5012A	103-308A
5012A	652
5012A	652C3
5012A	652C4
5012A	652C5
5012A	652C6
5012A	5330392
5012A	62045280
5012A	134883461527
5013A	A-125278
5013A	AZ753
5013A	BZX71C6V2
5013A	BZY66
5013A	DX-0530
5013A	EQA01-06
5013A	HEP-Z0214
5013A	MZ-6
5013A	MZ500-11
5013A	RD6.2E
5013A	RD6.2EB
5013A	RH-EX0048CEZZ
5013A	UZ6.2
5013A	VZ-063
5013A	WZ-061
5013A	WZ-063
5013A	WZ061
5013A	ZEN500
5013A	1N5234
5013A	1N5234A
5013A	1N5525
5013A	1N5525A
5013A	1N5731
5013A	1N753
5013A	1S2062
5013A	1S7062
5013A	1S7062A
5013A	1S7062B
5013A	02Z6.2A
5013A	46-86457-3
5013A	46-86501-3
5013A	48-155146
5013A	48-82256C01
5013A	652C8
5013A	1074-123
5013A	5515
5013A	134882256301
5014A	.25N6.8
5014A	AZ754
5014A	AZ957
5014A	AZ957A
5014A	BZX71C6V8
5014A	BZY83D6V8
5014A	BZY85D6V8
5014A	BZY88C6V8
5014A	BZZ11
5014A	EQA01-07S
5014A	HEP-Z0215
5014A	HM6.8
5014A	HM6.8A
5014A	HZ-7
5014A	HZ-7A
5014A	HZ-7B
5014A	HZ-7C
5014A	HZ7
5014A	HZ7A
5014A	HZ7B
5014A	MZ500-12
5014A	OAZ244
5014A	PD6010
5014A	RD6.2EC
5014A	RD6.8E
5014A	RD6.8EB
5014A	RD7A
5014A	RD7AM
5014A	RD7AN
5014A	SZ7
5014A	TMD04
5014A	TMD04A
5014A	TR7SA
5014A	VZ-065
5014A	VZ-067
5014A	VZ-069
5014A	WZ-065
5014A	WZ-067
5014A	WZ-069
5014A	Z0C6.8
5014A	Z1008
5014A	Z5C6.8
5014A	Z5D6.8
5014A	ZB6.8
5014A	ZE6.8
5014A	ZG6.8
5014A	ZP6.8
5014A	1N1930
5014A	1N1957
5014A	1N1984
5014A	1N378
5014A	1N469
5014A	1N469A
5014A	1N470
5014A	1N470A
5014A	1N5235
5014A	1N5235A
5014A	1N5526
5014A	1N5526A
5014A	1N5732
5014A	1N707
5014A	1N707A
5014A	1N707A(ZENER)
5014A	1N710
5014A	1N754A
5014A	1N763-1
5014A	1N763-2
5014A	1N957
5014A	1N957A
5014A	1S192
5014A	1S471
5014A	1S7068
5014A	1S7068A
5014A	1S7068B
5014A	1S757
5014A	8-719-168-07
5014A	09-306052
5014A	34-8057-42
5014A	34-8057-49
5014A	39Z6
5014A	39Z6A
5014A	48-137279
5014A	48-82256C02
5014A	48-82256C19
5014A	48-82256C23
5014A	48-82256C47
5014A	48-83461E25
5014A	61E25
5014A	103-279-14A
5014A	0114-0026/4460
5014A	653C1
5014A	653C2
5014A	653C3
5014A	653C4
5014A	1106
5014A	1125-2608
5014A	3755-1
5014A	3755-2
5014A	5516
5014A	9606
5014A	2330631
5014A	2330634
5014A	06200012
5014A	62123672
5014A	134882256302
5014A	134882256319
5014A	134882256323
5014A	134882256347
5014A	134883461525
5015A	.25N7.5
5015A	AZ755
5015A	AZ958
5015A	AZ958A
5015A	BZ-162
5015A	BZX71C7V5
5015A	BZY67
5015A	BZY88C7V5
5015A	BZZ12
5015A	HM7.5
5015A	HM7.5A
5015A	HZ-7GR
5015A	HZ-7SV
5015A	RD8.2E
5015A	1N4693
5015A	1N5236
5015A	1N5236A
5015A	1N5236B
5015A	1N5527
5015A	1N5527A
5015A	1N5733
5015A	1N711
5015A	1N755
5015A	1N755A
5015A	1N958
5015A	1N958A
5015A	1S7075
5015A	1S7075A
5015A	1S7075B
5015A	09-306212
5015A	13-33187-26
5015A	15-123106
5015A	48-82256C44
5015A	653C6
5015A	653C7
5015A	1107
5015A	5517
5015A	9607
5015A	134882256344
5016A	.25N8.2
5016A	AZ756
5016A	AZ959
5016A	AZ959A
5016A	BZX71C8V2
5016A	BZY83D8V2
5016A	BZY85C8V2
5016A	BZY85D8V2
5016A	BZY88C8V2
5016A	BZZ13
5016A	HEP-Z0217
5016A	HM8.2
5016A	HM8.2A
5016A	MZ500-14
5016A	OAZ246
5016A	PD6012
5016A	TMD06
5016A	TMD06A
5016A	TR9SA
5016A	VZ-081
5016A	VZ-083
5016A	WZ-083
5016A	Z0C8.2
5016A	Z0D8.2
5016A	Z1010
5016A	Z1C8.2
5016A	Z1D8.2
5016A	Z5C8.2
5016A	Z5D8.2
5016A	ZB8.2
5016A	ZE8.2
5016A	ZG8.2
5016A	ZP8.2
5016A	1N1931
5016A	1N1958
5016A	1N1985
5016A	1N5237
5016A	1N5237A
5016A	1N5237B
5016A	1N5528
5016A	1N5528A
5016A	1N5734
5016A	1N5853
5016A	1N712
5016A	1N764-1
5016A	1N959
5016A	1N959A
5016A	1S2082
5016A	1S7082
5016A	1S7082A
5016A	1S7082B
5016A	41Z6
5016A	41Z6A
5016A	48-137328
5016A	48-82256C09
5016A	48-82256C16
5016A	48-83461E32
5016A	61E32
5016A	653C9
5016A	1108
5016A	5518
5016A	13030285
5016A	134882256308
5016A	134882256309
5016A	134882256316
5016A	134883461532
5017A	N-756A
5017A	VZ-085
5017A	VZ-088
5017A	WZ-085
5017A	WZ-088
5017A	1N1313
5017A	1N1313A
5017A	1N5238
5017A	1N5238A
5017A	1S758
5017A	48-82256C45
5017A	134882256345
5018A	.25N9.1
5018A	AZ757
5018A	AZ960
5018A	AZ960A
5018A	BZX71C9V2
5018A	BZX72
5018A	BZX72A
5018A	BZX72B
5018A	BZX72C
5018A	BZX83C9V1
5018A	BZY68
5018A	BZY83C9V1
5018A	BZY85C9V1
5018A	BZY88C9V1
5018A	HEP-Z0219
5018A	HM9.1
5018A	HM9.1A
5018A	HZ-9C
5018A	HZ9
5018A	HZ9C2
5018A	KDD0015
5018A	MA1091
5018A	MZ250
5018A	MZ500-15
5018A	RD9.1EB
5018A	RD9.1EB2

ECG	Industry Standard No.
5018A	RD9AM
5018A	RD9AN
5018A	VZ-090
5018A	VZ-092
5018A	VZ-094
5018A	XZ-090
5018A	1N5239
5018A	1N5239A
5018A	1N5529
5018A	1N5529A
5018A	1N5735
5018A	1N713
5018A	1N757
5018A	1N764-3
5018A	1S194
5018A	1S2091(ZENER)
5018A	1S2091A
5018A	1S472
5018A	1S7091
5018A	1S7091A
5018A	1S7091B
5018A	48-82256C18
5018A	48-82256C38
5018A	48-83461E15
5018A	51-02007-12
5018A	61E15
5018A	200X8220-878
5018A	654C9
5018A	1109
5018A	2000-305
5018A	5519
5018A	146260
5018A	62045612
5018A	6615005000
5018A	134882256318
5018A	134882256338
5018A	134883461515
5019A	.25N10
5019A	A42X00041-01
5019A	AZ758
5019A	AZ961
5019A	AZ961A
5019A	BZX71C10
5019A	BZY83D10
5019A	BZY85D10
5019A	BZY88C10
5019A	BZY94C10
5019A	EQA01-10
5019A	HEP-Z0220
5019A	HM10
5019A	HM10A
5019A	HZ-11
5019A	HZ-11A
5019A	HZ11A
5019A	MZ-10
5019A	MZ500-16
5019A	R-7097
5019A	RD10E
5019A	RD10EB
5019A	RD11AL
5019A	RD11AM
5019A	VZ-096
5019A	VZ-098
5019A	VZ-100
5019A	WZ-096
5019A	WZ-098
5019A	Z-1012
5019A	Z1012
5019A	1N1932
5019A	1N1959
5019A	1N1986
5019A	1N5240
5019A	1N5240A
5019A	1N5530
5019A	1N5530A
5019A	1N5736
5019A	1N701
5019A	1N714
5019A	1N714B
5019A	1N758
5019A	1S2100
5019A	1S7100
5019A	1S7100A
5019A	1S7100B
5019A	1S759
5019A	02Z10A
5019A	13-33187-6
5019A	13-33187-7
5019A	48-137393
5019A	48-82256C11
5019A	48-82256C32
5019A	655C9
5019A	1110
5019A	5520
5019A	62258942
5019A	134882256311
5019A	134882256332
501B	APS-AH1
501B	GE-520
501B	PTC602
501B	S88569
501B	TVM108
501B	38A4148-000
501B	38A4148-001
501B	46-86330-3
501B	66X0045-001
501B	212-108
501B	212-109
501B	212-110
502	A04299-201
502	D2M
502	D4G
502	DIJ71961
502	GECR-4
502	HS20
502	HS6/1
502	HVT-22DA
502	MF12/16
502	PTC210
502	SELEN-58
502	TV13-11K60
502	TVM567
502	4-686227-3
502	13S
502	15-108045
502	34-8061-1
502	38-11016
502	46-86227-3
502	46-86279-3
502	46-86291-3
502	48-137081
502	48-137801
502	48S155002
502	53A022-10
502	212-66
502	212-67
502	264P08901
502	4686-227-3
502	1445740-502
502	2330921
502	23115157
502	6611003000
5020A	.25N11
5020A	AZ962
5020A	AZ962A
5020A	BZX71C11
5020A	BZY88C11
5020A	BZY94C11
5020A	EQA01-11S
5020A	HM11
5020A	HM11A
5020A	HZ-11C
5020A	HZ11B
5020A	HZ11C
5020A	HZ11Y
5020A	MZ500-17
5020A	OA126/10
5020A	RD11AN
5020A	RD11E
5020A	RD11EB
5020A	RH-EX0021TAZZ
5020A	VZ-110
5020A	1N5241
5020A	1N5241A
5020A	1N5531A
5020A	1N5737
5020A	1N715
5020A	1N715A
5020A	1N962
5020A	1N962A
5020A	1N962B
5020A	1S2110
5020A	1S2110A
5020A	1S473
5020A	1S7110
5020A	1S7110A
5020A	1S7110B
5020A	48-82256C34
5020A	1111
5020A	5521
5020A	2331161
5020A	134882256334
5021A	.25N12
5021A	A-120077
5021A	AZ759
5021A	AZ963
5021A	AZ963A
5021A	BZX71C12
5021A	BZY69
5021A	BZY83D12
5021A	BZY85D12
5021A	BZY88C12
5021A	BZY94C12
5021A	HEP-Z0222
5021A	HM12
5021A	HM12A
5021A	HZ-12
5021A	HZ-12A
5021A	MZ500-18
5021A	02Z12GR
5021A	RD12E
5021A	RD12EA
5021A	RD12EB
5021A	RD12EB1Z
5021A	RD13AL
5021A	SZ-RD12EB
5021A	SZ-UZ-12B
5021A	TR12SA
5021A	VZ-120
5021A	WZ-120
5021A	Z1014
5021A	Z12.0
5021A	ZEN502
5021A	1N1933
5021A	1N1960
5021A	1N1987
5021A	1N5242
5021A	1N5242A
5021A	1N5532
5021A	1N5532A
5021A	1N5738
5021A	1N716
5021A	1N716(ZENER)
5021A	1N759
5021A	1N963
5021A	1N963A
5021A	1S2119A
5021A	1S2120
5021A	1S7120
5021A	1S7120A
5021A	1S7120B
5021A	02Z12GR
5021A	8-719-112-24
5021A	13-33187-11
5021A	21A037-001
5021A	48-82256C17
5021A	48-82256C25
5021A	103-279-21A
5021A	1112
5021A	5522
5021A	2331154
5021A	134882256317
5021A	134882256325
5022A	.25N13
5022A	A694X1
5022A	A694X1-OA
5022A	AW01-13
5022A	AWOL-13
5022A	AZ964
5022A	AZ964A
5022A	BZX71C13
5022A	BZY83C13V5
5022A	BZY85C13
5022A	BZY88C13
5022A	BZY94C13
5022A	HM13
5022A	HM13A
5022A	HZ-12B
5022A	HZ-12C
5022A	MZ500-19
5022A	RD-13AN
5022A	RD13AM
5022A	RD13E
5022A	RD13EA
5022A	RD13EB
5022A	RT8200
5022A	TR12S
5022A	TR12SB
5022A	VZ-125
5022A	VZ-130
5022A	WZ-125
5022A	WZ-130
5022A	1N1315
5022A	1N1315A
5022A	1N5243
5022A	1N5243A
5022A	1N5533
5022A	1N5533A
5022A	1N5739
5022A	1N717
5022A	1N766
5022A	1N964
5022A	1N964A
5022A	1N964B
5022A	1S2130
5022A	1S2130A
5022A	1S760
5022A	4-2020-06600
5022A	09-306173
5022A	48-82256C48
5022A	48-82256C50
5022A	48-83461E02
5022A	48-83461E31
5022A	61E02
5022A	61E31
5022A	903-73B
5022A	2006-17
5022A	2021-17
5022A	2093A41-82
5022A	5523
5022A	99201-319
5022A	134335
5022A	233148
5022A	530145-130
5022A	985961
5022A	2327074
5022A	5330054
5022A	5330054H
5022A	134882256348
5022A	134882256350
5022A	134883461502
5022A	134883461531
5023A	.25N14
5023A	RD16AL
5023A	RZ14
5023A	RZZ15
5023A	TR12SC
5023A	VZ-135
5023A	VZ-140
5023A	WZ-135
5023A	WZ-140
5023A	1N1961
5023A	1N5244
5023A	1N5244A
5023A	1N5534
5023A	1N5534A
5023A	1S338
5023A	1S474
5023A	48-82256C13
5023A	1113
5023A	134882256313
5024A	.25NN15
5024A	AZ965
5024A	AZ965A
5024A	BZX71C15
5024A	BZX79C15
5024A	BZY83D15
5024A	BZY85D15
5024A	BZY88C15
5024A	BZY94C15
5024A	HEP-Z0225
5024A	HM15
5024A	HM15A
5024A	HZ-15
5024A	MZ500-20
5024A	RD15EA
5024A	RD16AM
5024A	VZ-145
5024A	VZ-150
5024A	WZ-145
5024A	Z1016
5024A	1N1934
5024A	1N1988
5024A	1N5245
5024A	1N5245A
5024A	1N5535
5024A	1N5535A
5024A	1N5740
5024A	1N718
5024A	1N965
5024A	1N965A
5024A	1S2150
5024A	1S7150
5024A	1S7150A
5024A	1S7150B
5024A	13-33187-18
5024A	14-515-23
5024A	48-137330
5024A	48-137394
5024A	48-82256C14
5024A	5330541
5024A	134882256314
5025A	.25N16
5025A	AZ966
5025A	AZ966A
5025A	BZX71C16
5025A	BZY85C16
5025A	BZY88C16
5025A	BZY94C16
5025A	HM16
5025A	HM16A
5025A	MZ500-21
5025A	RD16A
5025A	RH-EX0065CEZZ
5025A	RZ16
5025A	VZ-157
5025A	VZ-162
5025A	WZ-157
5025A	WZ-162
5025A	1N1316
5025A	1N1316A
5025A	1N5246
5025A	1N5246A
5025A	1N5536
5025A	1N5536A
5025A	1N5741
5025A	1N719
5025A	1N767
5025A	1N966
5025A	1N966A
5025A	1S7160A
5025A	1S761
5025A	5525
5026A	.25N17
5026A	BZY83C16V5
5026A	BZY85C16V5
5026A	RD16AN
5026A	1N5247
5026A	1N5247A
5026A	1N5537
5026A	1N5537A
5026A	1116
5026A	5325
5027A	.25N18
5027A	AZ18
5027A	AZ967
5027A	AZ967A
5027A	BZX71C18
5027A	BZY83D18
5027A	BZY85D18
5027A	BZY88C18
5027A	BZY94C18
5027A	HEP-Z0228
5027A	HM18
5027A	HM18A
5027A	PD6020
5027A	RZZ18
5027A	SV4018
5027A	SV4018A
5027A	SZ18
5027A	1N1935
5027A	1N1962
5027A	1N1989
5027A	1N5248
5027A	1N5248A
5027A	1N5538
5027A	1N5538A
5027A	1N5742
5027A	1N720
5027A	1N720B
5027A	1N967
5027A	1N967A
5027A	1N967B
5027A	1S339
5027A	1S475
5027A	34-8057-43
5027A	48-82256C24
5027A	48-83461E18
5027A	48-83461E33
5027A	61E18
5027A	61E33
5027A	1118
5027A	5526
5027A	134882256324
5027A	134883461518
5027A	134883461533
5028A	.25N19
5028A	RD19A
5028A	RD19AL
5028A	RD19AM
5028A	1N1317
5028A	1N1317A
5028A	1N5249
5028A	1N5249A
5028A	1N768
5028A	1S762
5028A	48-83461E24
5028A	61E24
5028A	134883461524
5029A	.25N20
5029A	AZ968
5029A	AZ968A
5029A	BZX71C20
5029A	BZY88C20
5029A	BZY94C20
5029A	HEP-Z0230
5029A	HM20
5029A	HM20A
5029A	RZ20
5029A	SZ20
5029A	1N5250
5029A	1N5250A
5029A	1N5540
5029A	1N5540A
5029A	1N5743
5029A	1N721
5029A	1N968
5029A	1N968A
5029A	1N968B
5029A	1S2200
5029A	13-33187-3
5029A	34-8057-44
5029A	48-83461E22
5029A	61E22
5029A	1120
5029A	5527
5029A	134883461522
503	A04299-202
503	A04299-251
503	D2N
503	EDMF-15B
503	ES57X13
503	ES57X21
503	GECR-5
503	H598
503	H598-10
503	HS10/1
503	HS15/16
503	HS15/1C
503	MF20/1B
503	NTC-16
503	PTC211
503	RE200
503	RF03C06
503	SELEN-52
503	TV13-12K60
503	TV24540
503	TVM531
503	TVM569
503	1-531-028
503	5-30119.1
503	05-370151
503	09-306430
503	13-43382-1
503	13-55010-1
503	15-123104
503	15-123231
503	46-86313-3
503	46-86520-3
503	48-137082
503	48S137082
503	48X90234A05
503	103-239
503	103-239-01
503	264P08902
503	132574
503	132966
503	140995
503	489752-054
503	530116-1
503	530119-5
503	1445470-501
503	1445470-503
503	4202107370
5030A	.25N22
5030A	AZ22
5030A	AZ969
5030A	AZ969A
5030A	BZX71C22
5030A	BZY83D22
5030A	BZY85D22
5030A	BZY88C22
5030A	BZY94C22
5030A	HEP-Z0231
5030A	HM22
5030A	HM22A
5030A	RZ22A
5030A	RZZ22
5030A	1N1936
5030A	1N1963
5030A	1N1990
5030A	1N5251
5030A	1N5251A
5030A	1N5251B
5030A	1N5541
5030A	1N5541A
5030A	1N5744
5030A	1N722
5030A	1N969A
5030A	1S2220
5030A	1S476
5030A	48-137306
5030A	5528
5031A	.25N24
5031A	AZ970
5031A	AZ970A
5031A	BZX71C24
5031A	BZY88C24
5031A	BZY94C24
5031A	HM24
5031A	HM24A
5031A	MZ224A
5031A	1N1318
5031A	1N1318A
5031A	1N5252
5031A	1N5252A
5031A	1N5542
5031A	1N5542A
5031A	1N5745
5031A	1N723
5031A	1N970
5031A	1N970A
5031A	1S2240
5031A	1S763
5031A	48-82256C49
5031A	48-83461E26
5031A	61E26
5031A	5529
5031A	134882256331
5031A	134882256349
5031A	134883461526
5032A	.25N25
5032A	BZY83C24V5
5032A	RZ25
5032A	1N4710
5032A	1N5253
5032A	1N5253A
5032A	1N5543
5032A	1N5543A
5032A	1N769-3
5032A	13-33187-15
5033A	.25N27
5033A	AZ971
5033A	AZ971A
5033A	BZY94C27
5033A	HEP-Z0234
5033A	HM27
5033A	HM27A
5033A	MZ500-26
5033A	RD29AL
5033A	RZZ27
5033A	1N1937
5033A	1N1964
5033A	1N1991
5033A	1N5254
5033A	1N5254A
5033A	1N5746
5033A	1N724
5033A	1S2270
5033A	1S764
5033A	48-137062
5033A	48-82256C20
5033A	5530
5034A	RD29A
5034A	RD29AM
5034A	1N1319
5034A	1N1319A
5034A	1N5255
5034A	1N5255A
5034A	1N5544
5034A	1N5544A
5034A	1S477
5035A	.25N30
5035A	AZ972
5035A	AZ972A
5035A	BZY88C30
5035A	BZY94C30
5035A	HM30
5035A	HM30A
5035A	RD29AN
5035A	RZ30
5035A	1N5256
5035A	1N5256A
5035A	1N5545
5035A	1N5545A
5035A	1N5747
5035A	1N725
5035A	1N972
5035A	1N972A
5035A	1S2300
5035A	5531
5036A	.25N33
5036A	AZ973
5036A	AZ973A
5036A	BZY85C33
5036A	BZY94C33
5036A	HM33
5036A	HM33A
5036A	MZ500-28
5036A	RD35AL

ECG	Industry Standard No.	ECG	Industry Standard No.	ECG	Industry Standard No.	ECG	Industry Standard No.	ECG	Industry Standard No.
5036A	RZ33A	5042A	BZY94C56	505	11386-9	506	BA145	506	RH-DX0104CEZZ
5036A	1N1320	5042A	HM56	505	134777	506	BA148	506	RH-EX0024CEZZ
5036A	1N1320A	5042A	HM56A	505	530119-7	506	BB-10	506	RS1296(SEARS)
5036A	1N1938	5042A	RZ56A	505	530119-9	506	BB-108	506	RSLND0003CEZZ
5036A	1N1965	5042A	1N1941	505	530149-9	506	BB-4	506	RT1686
5036A	1N1992	5042A	1N1942	505	1476930	506	BB2	506	RU-1N
5036A	1N5257	5042A	1N1968	505	2330032	506	BB4	506	S-15H
5036A	1N5257A	5042A	1N1995	505	14769302	506	BB68	506	S-34
5036A	1N5546	5042A	1N5263	505	4202106270	506	BBG8	506	S1-RECT-158
5036A	1N5546A	5042A	1N5263A	505	4202108070	506	BR-10S	506	S10110
5036A	1N5748	5042A	1N5754	505	4202108170	506	BY105	506	S10120
5036A	1N726	5042A	1N732	5050A	-.25N00	506	BY127	506	S12110
5036A	1N973	5042A	1N732B	5050A	HM100	506	BY133	506	S12120
5036A	1N973A	5042A	1N979	5050A	HM100A	506	BY157	506	S12130
5036A	182330	5042A	1N979A	5050A	1N1944	506	BY158	506	S12140
5036A	2A64	5042A	5538	5050A	1N1971	506	BYX22-1200	506	S1801-04
5036A	5532	5043A	1N1323	5050A	1N1998	506	BYX57-500	506	S1D23-13
5037A	-.25N36	5043A	1N1323A	5050A	1N5271	506	BYX57-600	506	S1D25-15
5037A	AZ974	5043A	1N5264	5050A	1N5271A	506	C-2C	506	S1D26
5037A	AZ974A	5043A	1N5264A	5050A	1N738	506	CGJ-1	506	S1D30-13
5037A	BZY94C36	5043A	5Z860	5050A	1N738B	506	D172-F	506	S1D30-15
5037A	HM36	5043A	5Z860A	5050A	1N985	506	D1F	506	S1F
5037A	HM56A	5044A	.25N62	5050A	1N985A	506	D1F	506	S34
5037A	RD35AM	5044A	BZY94C62	5050A	13-33187-2	506	D431-F	506	SB-1
5037A	RD36EB	5044A	HM62	5051A	.25N110	506	D8G	506	SB-2
5037A	RH-EX0053CEZZ	5044A	HM62A	5051A	HM110	506	D8L	506	SB-2B
5037A	1N5258	5044A	MZ500-35	5051A	HM110A	506	D8M	506	SB-2CH
5037A	1N5258A	5044A	1N5265	5051A	1N5272	506	DQ-1N	506	SB-2T
5037A	1N5749	5044A	1N5265A	5051A	1N5272A	506	DQ-1NR	506	SB2C
5037A	1N727	5044A	1N5755	5051A	1N986	506	DG1N	506	SD-1
5037A	1N974	5044A	1N980	5051A	1N986A	506	DG1NR	506	SD-1(BOOST)
5037A	1N974A	5044A	1N980A	5052A	.25N120	506	DIJ72163	506	SD-1-30DA
5037A	18478	5045A	.25N68	5052A	HM120	506	DIJ72167	506	SD-1A4P
5037A	5533	5045A	BZY94C68	5052A	HM120A	506	DIJ72170	506	SD-1HF
5038A	.25N39	5045A	HM68	5052A	1N1945	506	DIJ72171	506	SD-1C
5038A	AZ975	5045A	HM68A	5052A	1N1999	506	DIJ72174	506	SD-1VHF
5038A	AZ975A	5045A	RZ68A	5052A	1N5273	506	DIJ72290	506	SD-22
5038A	BZX46C39	5045A	1N969	5052A	1N5273A	506	DIJ72291	506	SD-ERC26-13
5038A	BZY94C39	5045A	1N996	5052A	1N740	506	DS-113A	506	SD1-11
5038A	HM39	5045A	1N5266	5052A	1N987	506	DS-113B	506	0000000SD1A
5038A	HM39A	5045A	1N5266A	5052A	1N987A	506	DS-15A(SANYO)	506	SELEN-92
5038A	RD35AE	5045A	1N5756	5052A	13-33187-13	506	DS-15B(SANYO)	506	SF1
5038A	RD39A	5045A	1N734	5053A	.25N130	506	DS113B	506	SI-RECT-114
5038A	1N1321	5045A	1N734B	5053A	HM130	506	DS133B	506	SI-RECT-124
5038A	1N1321A	5045A	1N981	5053A	HM130A	506	DS16NE(SONY)	506	SI-RECT-162
5038A	1N1939	5045A	1N981A	5053A	1N1327	506	DT230B	506	SI-RECT-170
5038A	1N1966	5045A	48-82256005	5053A	1N5274	506	ED23(ELCOM)	506	SI-RECT-180
5038A	1N1993	5045A	5540	5053A	1N5274A	506	EP16X11	506	SI-RECT-206
5038A	1N5259	5045A	134882256305	5053A	1N741	506	EP16X24	506	SI-RECT-224
5038A	1N5259A	5046A	.25N75	5053A	1N988	506	EP6X11	506	SI-RECT-36
5038A	1N5750	5046A	BZX79C75	5053A	1N988A	506	EPB22-15	506	SI-RECT-44
5038A	1N728A	5046A	BZY94C75	5053A	1N988B	506	ERB24-06	506	SID1-1
5038A	1N975	5046A	HM75	5054A	.25N140	506	ERB24-02B	506	SK3130
5038A	1N975A	5046A	HM75A	5054A	1N5275	506	ERB24-04A	506	SK3175
5038A	5534	5046A	1N5267	5054A	1N5275A	506	ERB24-04B	506	SM-150-6(FOCUS)
5039A	.25N43	5046A	1N5267A	5055A	.25N150	506	ERB24-06D	506	SR-15
5039A	AZ976	5046A	1N735B	5055A	HM150	506	ERC27-13	506	SR156
5039A	AZ976A	5046A	1N982	5055A	HM150A	506	ES16X16	506	SR1FM-2
5039A	BZY94C43	5046A	1N982A	5055A	1N1946	506	ES16X20	506	SR1HM-12
5039A	HM43	5046A	5541	5055A	1N2000	506	ES16X28	506	SR1HM-16
5039A	HM43A	5047A	.25N82	5055A	1N5276	506	ES16X31	506	SR1HM-4
5039A	1N260	5047A	HM82	5055A	1N5276A	506	ES57X8	506	SR1HM-8
5039A	1N5260A	5047A	HM82A	5055A	1N741B	506	ES57X9	506	SR25
5039A	1N5751	5047A	MZ500-38	5055A	1N742	506	ET57X39	506	T-E1145
5039A	1N729	5047A	RZ82A	5055A	1N742B	506	ET57X40	506	T-E1146
5039A	1N976	5047A	1N1970	5055A	1N989	506	F114E	506	T-E1153
5039A	1N976A	5047A	1N1997	5055A	13-33187-10	506	FDH600	506	TC-0.2
5039A	5535	5047A	1N5268	5056A	HM160	506	FG-2N	506	TC-0.2P11/1
504	EDMP25B	5047A	1N5268A	5056A	HM160A	506	FG2FC	506	TCO.1P
504	EP-5641H-2	5047A	1N983	5056A	1N5277	506	FT-1M	506	TSTD-15
504	GECR-6	5047A	1N983A	5056A	1N5277A	506	FT-1P	506	TV24167
504	H830-16	5047A	13-33187-1	5057A	1N743	506	FT1M	506	TV24169
504	H830-15	5048A	1N1325	5057A	1N990	506	FU10	506	TV24219
504	MP30/1B	5048A	1N5269	5057A	1N990A	506	FU1K	506	TV24617
504	TVM533	5048A	1N5269A	5057A	1N5278	506	GB-511	506	TV24648
504	13-37933-2	5048A	13-33187-5	5057A	1N5278A	506	GRU2A	506	TV34
504	34-8062-4	5049A	.25N91	5058A	HM180	506	H815	506	TVM535
504	46-86292-3	5049A	HM91	5058A	HM180A	506	HD-20008	506	TVM537
504	46-86454-3	5049A	HM91A	5058A	1N947	506	HEP-R0606	506	TVS-0V-02
504	46-86451-3	5049A	MZ500-39	5058A	1N2001	506	HF-17	506	TVS-2B-2C
504	46-86454-3	5049A	1N5270	5058A	1N5279	506	HF-1B	506	TVS-D01NR
504	46-86547-3	5049A	1N5270A	5058A	1N5279A	506	HP-8D-1%	506	TVS-FR1P
504	48-3555047	5049A	1N737	5058A	1N744	506	HPSD-14	506	TVS-FR2P
504	93A95-*	5049A	1N737B	5058A	1N744B	506	HPSD-1A	506	TVS-FT1N
504	93A93-2	5049A	1N984	5058A	1N991	506	HPSD-1B	506	TVS-FT1P
504	103-239-02	5049A	1N984A	5058A	1N991A	506	HPSD-1C	506	TVS-FP1PC
504	103-293-02	5049A	D2Y	5059A	1N5280	506	HPSD-1C(SEARS)	506	TVS-HP-SD1%
504	117-440	505	E13-021-01	5059A	1N5280A	506	HPSD-1Z	506	TVS-SP-1
504	212-95	505	E13-021-03	506	.7872	506	HPSD1	506	TVS10D1
504	264P04206	505	EP57X22	506	.7J12	506	HSFD-1A(SONY)	506	TVSSB-2-H5W
504	129203	505	E857X11	506	A04093-A	506	HSFD1	506	TVSAPB01
504	530119-8	505	GECR-7	506	A04093-X	506	IR10D3L	506	TVSBB2
504	1445470-504	505	H386-9	506	A04093A	506	IR1P	506	TVSERB24-04D
504	2330034	505	H386C1D2Y	506	A04093X	506	IR3D	506	TVSERB24-06
504	2330381	505	H8-20	506	A04230	506	J241183	506	TVSERB24-06A
504	23115215	505	H815/1	506	A04230-A	506	J241215	506	TVSERB24-06B
504	23115938	505	H820-1	506	A04241-A	506	JC-00017	506	TVSFSFPC
504	62747501	505	H820-1A	506	A04294	506	L00.09M1113	506	TVSFR2-005
504	4202107570	505	H820/16S	506	A04294-1	506	LM1932	506	TVSFR2-02
5040A	.25N45	505	H820/1AS	506	A08	506	MR-852	506	TVSFR2-02G
5040A	.25N47	505	H820/1B	506	A14PD1	506	MR801	506	TVSFR2-02Q
5040A	BZY94C47	505	H820/1C	506	A14PD2	506	MR811	506	TVSFR2-04
5040A	HM47	505	H825-1	506	A14PD3	506	MR812	506	TVSFR2-10
5040A	HM47A	505	H825/16	506	A2458	506	MR814	506	TVSFT1M
5040A	RD47A	505	H825/1B	506	A6N1	506	MR816	506	TVSHF-1
5040A	1N1322	505	H825/1BS	506	A6N5	506	N315835L1	506	TVSHFT
5040A	1N1322A	505	H825/1C	506	A6N9	506	NTC--17	506	TVSHFSD-1C
5040A	1N1940	505	H830-1C	506	A7568250	506	OP66	506	TVSHFSD-1%
5040A	1N1967	505	H830/1C	506	A7580910	506	OVO2	506	TVSHFSD1
5040A	1N1994	505	H89/17N	506	A7582000	506	PH108	506	TVSRA-1Z
5040A	1N5261	505	MP25/1B	506	A7N1	506	RO606	506	TVSRA1Z
5040A	1N5261A	505	PTC212	506	A7N5	506	RA-Z	506	TVSS-34
5040A	1N5752	505	PTC2'3	506	A7N9	506	RA1	506	TVSS15
5040A	1N730	505	SELEN-40	506	AD-29B	506	RFJ70149	506	TVSS1D30-15
5040A	1N977	505	TV20-10K80	506	AD-29S	506	RFJ70971	506	TVSS34
5040A	1N977A	505	TV20-S	506	AD29S	506	RFJ70976	506	TVSSB-2
5040A	18479	505	TVM540	506	AD8	506	RH-1B	506	TVSSB2C
5040A	5536	505	TV8-H-339W	506	AF1(DIODE)	506	RH-DX0004TAZZ	506	TVSUP2
5041A	.25N50	505	TV8-H-399W	506	A04093-X	506	RH-DX0017CEZZ	506	TVSUFSD-1
5041A	.25N52	505	2HT-6	506	A04230-A	506	RH-DX0028CEZZ	506	UO6C
5041A	BZY94C51	505	4-2021-07570	506	A04241-A	506	RH-DX0029CEZZ	506	UP10-1C
5041A	HM51	505	4-2021-08070	506	B24-06C	506	RH-DX0039TAZZ	506	UO6C
5041A	HM51A	505	09-3063*0	506	B2620	506	RH-DX0045CEZZ	506	VO3B(HITACHI)
5041A	RD50	505	10K80	506	B3N1	506	RH-DX0051CEZZ	506	VO6C(BOOST)
5041A	1N5262	505	28-31-01	506	B3N5	506	RH-DX0056TAZZ	506	VO6C(HITACHI)
5041A	1N5262A	505	34-8062-*	506	B3N9	506	RH-DX0062CEZZ	506	VO9
5041A	1N5753	505	46-86294-3	506	B6N1	506	RH-DX0063CEZZ	506	VO9-E
5041A	1N731	505	488137''4	506	B6N5	506	RH-DX0065CEZZ	506	VO9A
5041A	1N731B	505	57X'*	506	B6N9	506	RH-DX0065CZZZ	506	VO9B
5041A	1N978	505	103-258	506	B7978850	506	RH-DX0081CEZZ	506	V11J
5041A	1N978A	505	420-2'06-270	506	B7N1	506	RH-DX0090CEZZ	506	V11L
5041A	48-11-010			506	B7N5	506	RH-DX0092CEZZ	506	VP-SD1%
5041A	5537			506	B7N9	506	RH-DX0096CEZZ		
5042A	.25N56			506	BA'33				

ECG	Industry Standard No.
506	VHD1S1834//-1
506	V0-9C
506	V09E
506	VS-SD-1B
506	1.5E12
506	1.5J12
506	1N1409
506	1N2327
506	1N2357
506	1N2503
506	1N2504
506	1N2507
506	1N2508
506	1N2618
506	1N2619
506	1N2776
506	1N2777
506	1N2778
506	1N2779
506	1N2884
506	1N2885
506	1N3233
506	1N3244
506	1N3487
506	1N3724
506	1N3725
506	1N3915
506	1N4146
506	1N4155
506	1N4252
506	1N4826
506	1N487
506	1N487A
506	1N487B
506	1N488
506	1N488A
506	1N488B
506	1N4934
506	1N4935
506	1N4936
506	1N4945
506	1N4947
506	1N4948
506	1N5188
506	1N5189
506	1N5190
506	1N5418
506	1N5419
506	1N543
506	1N5433
506	1N5434
506	1N543A
506	1N549
506	1N5552
506	1N5553
506	1N5554
506	1N5616
506	1N5618
506	1N5620
506	1N5621
506	1N693
506	1N893
506	1N923
506	1NJ70976
506	1R10D3L
506	1R31
506	1R3B
506	1R5TH61
506	1S1226
506	1S1237
506	1S1238
506	1S1349
506	1S1517
506	1S1517A
506	1S1832
506	1S1855
506	1S1882
506	1S1920
506	1S1921A
506	1S1921B
506	1S1921C
506	1S1921D
506	1S1921E
506	1S1921F
506	1S2306
506	1S2307
506	1S2308
506	1S2309
506	1S2711
506	1S2756
506	1S312(HITACHI)
506	1S315(HITACHI)
506	2B-2
506	2SB-2C
506	2SC-2C
506	03-936011
506	4-2020-03-700
506	4-2020-03700
506	4-2020-03800
506	4-2020-06500
506	4-2020-06700
506	4-2020-07800
506	4-2020-07801
506	4-2020-08000
506	4-2020-14400
506	4-2020-14600
506	4-2020-15100
506	4-2021-04570
506	4-2021-04870
506	4-2021-06970
506	4-2021-07970
506	4-2021-10470
506	4-202104570
506	4-202104870
506	05-112404
506	05-936010
506	5B-15H(SHARP)
506	5B-2-H5W(SHARP)
506	8-719-906-15
506	09-306030
506	09-306031
506	09-306141
506	09-306144
506	09-306237
506	09-306260
506	09-306392
506	09-306418
506	09-306423
506	09-306425
506	10D-2(CROWN)
506	10D-2B-4
506	10D2(CROWN)
506	10D2(DAMPER)
506	10D5A
506	10D5D
506	10R11B
506	10R12B
506	10R13B
506	10R14B
506	11/10
506	12G4
506	13-29663-1
506	13-39073-1
506	13-55030-1
506	13-55031-1
506	13-55031-2
506	13-55031-3
506	13-55078-1
506	14-514-03
506	14-514-03/53
506	14-514-07
506	14-514-07/57
506	14-514-14
506	14-514-17/67
506	14-514-18
506	14-514-18/68
506	14-514-20/70
506	14-514-23
506	14-514-23/73
506	14/514-12/62
506	14/514-14/64
506	14/514-19/69
506	15-108013
506	15-108026
506	15-108035
506	15-108040
506	15-108041
506	15-108043
506	15-108044
506	15-108047
506	15-108048
506	15-123102
506	15-123103
506	15-123242
506	21A004-000
506	21A110-005
506	21A110-008
506	21A110-009
506	21A110-013
506	21A110-014
506	21A112-006
506	21A119-068
506	28-22-13
506	28-22-22
506	34-8054-24
506	34-8054-25
506	34-8054-8
506	34-8057-22
506	34-8057-24
506	46-86106-3
506	46-861149-3
506	46-861179-3
506	46-86179-3
506	46-8620-3
506	46-86280-3
506	46-86281-3
506	46-86282-3
506	46-86308-3
506	46-86320-3
506	46-86324-3
506	46-86328-3
506	46-86331-3
506	46-86382-3
506	46-86391-3
506	46-86393-2
506	46-86393-3
506	46-86395-3
506	46-86440-3
506	46-86489-3
506	46-86511-3
506	46-86519-3
506	46-86525-1
506	46-86525-3
506	48-134921
506	48-134939
506	48-134978
506	48-137112
506	48-137533
506	48-137546
506	48-137551
506	48-155108
506	48-155125
506	48-191A05
506	48-191A05A
506	48-355013
506	48-40235G02
506	48-41266G01
506	48-8240C10
506	48-8240C12
506	48-82466H12
506	48-82466H13
506	48-82466H14
506	48-82466H15
506	48-82466H16
506	48-82466H18
506	48-82466H19
506	48-82466H21
506	48-82466H25
506	48-82466H28
506	48-86308-3
506	48-90158A01
506	48-90234A12
506	48-90343A53
506	48B41226G01
506	48C40235G02
506	48S137533
506	48S137546
506	48S137551
506	48S155107
506	48S155108
506	48S191A05
506	48S191A05A
506	48S191A08
506	53A014-1
506	53A015-1
506	53A017-1
506	53A022-5
506	53B010-8
506	53L001-14
506	53L001-5
506	55-001
506	56-15
506	66X0036-001
506	66X0038-001
506	66X0055-001
506	66X36-1
506	73A60-11
506	75R11B
506	75R12B
506	75R13B
506	75R14B
506	86-105-3
506	86-117-3
506	86-118-3
506	87-104-3
506	91A08
506	93A40-1
506	93A60-1
506	93A60-10
506	93A60-11
506	93A60-3
506	93A69-1
506	93A71-1
506	93B52-1
506	93B52-2
506	93B58-1
506	93B60-3
506	93C52-1
506	93C60-3
506	93C69-1
506	094-012
506	1000(ADMIRAL)
506	100R11B
506	100R12B
506	100R13B
506	100R14B
506	103-112
506	103-160
506	103-193
506	103-196
506	103-244
506	103-247
506	103-263
506	121GI241
506	123-018
506	148-3
506	150-012-9-001
506	151-045-9-003
506	200X8130-171
506	212-68
506	212-80
506	229-1301-25
506	229-5100-235
506	264P01701
506	264P02301
506	264P03601
506	264P03603
506	264P03605
506	264P04701
506	264P04705
506	264P06603
506	264P09101
506	264P09501
506	264P13002
506	388B
506	400(QUASAR)
506	601X0049-066
506	601X0375-006
506	690V080H49
506	690V080H50
506	690V080H51
506	690V081H91
506	800-029-30
506	800-282
506	903-68B
506	1010-8116
506	1019-1385
506	1038-1788
506	1050(GE)
506	1051
506	1051(GE)
506	1059-2848
506	1061-6290
506	1061-8387
506	1872-3
506	2093A41-67
506	2093A41-69
506	2093A41-72
506	2093A41-73
506	2093A41-96
506	6171-18
506	12685
506	0015107
506	21008-002
506	34424
506	72137
506	72161-1
506	125529
506	125844
506	125848
506	126527
506	126826(DIODE)
506	126856
506	131148
506	131318
506	131475
506	131476
506	132148
506	132418
506	132501
506	132509
506	132547
506	132548
506	132549
506	133615
506	133616
506	135341
506	135380
506	135734
506	136606
506	137075
506	140503
506	147617-11
506	147617-12
506	161032
506	161033
506	185022
506	194917
506	233062
506	489752-016
506	489752-026
506	489752-096
506	530097-3
506	530113-1001
506	530122-2
506	530123-3
506	530123-4
506	530123-5
506	530123-7
506	530136-T
506	530144
506	530148-1003
506	530148-1004
506	530148-3
506	530148-4
506	530179-1002
506	530184-1002
506	530972-14
506	552007
506	575047
506	0575047H
506	0575049
506	0575049H
506	0575054
506	0575066
506	984260
506	1390655
506	1415721-1
506	1415721-35
506	1445470-502
506	1446149-1
506	1446149-2
506	1471872-3
506	1472460-13
506	1474778-14
506	1474778-21
506	1474778-5
506	1474778-6
506	1476049-2
506	1476161-12
506	1476171-17
506	1476171-18
506	1476171-22
506	1476171-25
506	1476171-26
506	1476171-32
506	1476171-34
506	1476183-2
506	1476183-5
506	1476183-6
506	2311598
506	2330101
506	2330191
506	2330211
506	2330253
506	2330356
506	2331121
506	2331141
506	06200010
506	06200030
506	8519510-2
506	23114074
506	23115094
506	23115131
506	23115145
506	23115146
506	23115192
506	23115298
506	23115338
506	23115377
506	23115912
506	23115913
506	23115914
506	23115960
506	23115981
506	23115999
506	37568200
506	37568250
506	37582000
506	62051302
506	62051310
506	62522712
506	62562625
506	62564520
506	62734728
506	62761318
506	62761326
506	62801328
506	2008100076
506	2008130040
506	2008130171
506	4202003700
506	4202003900
506	4202006700
506	4202007601
506	4202007700
506	4202007801
506	4202014400
506	4202014600
506	4202015100
506	4202020500
506	4202020900
506	4202022300
506	4202023000
506	4202104570
506	4202104870
506	4202106970
506	4202107970
506	4202110270
506	4203008000
506	6611000900
506	6611001000
506	6611001100
506	6611001200
506	6611001500
506	6611002300
506	6613002200
506	13488240310
506	13488240312
506	44202007800
506	134882466812
506	134882466813
506	134882466814
506	134882466815
506	134882466816
506	134882466817
506	134882466818
506	134882466821
506	134882466825
506	134882466827
506	134882466828
5060A	.25N200
5060A	HM200
5060A	HM200A
5060A	1N5281
5060A	1N5281A
5060A	1N745
5060A	1N745B
5060A	1N992
5060A	1N992A
5063A	BZY88C3V3
5065A	MLV4372A
5066A	AZ3.3
5066A	AZ746A
5066A	BZX96C3V3
5066A	BZY85B3V3
5066A	BZY85C3V3
5066A	BZY88C3V9
5066A	EZ3(ELCOM)
5066A	GEZD-3.3
5066A	HEP-Z0401
5066A	KS033A
5066A	KS2033A
5066A	MLV746A
5066A	PD6043
5066A	PS8901
5066A	RE100
5066A	RZ3.3
5066A	RZZ3.3
5066A	1D3.3
5066A	1D3.3A
5066A	1D3.3B
5066A	1M3.3ZS10
5066A	1M3.3ZS5
5066A	1N3821
5066A	1N3821A
5066A	1N466A
5066A	1N4684
5066A	1N4728
5066A	1N4728A
5066A	1N5988B
5066A	1N746A
5066A	1S2033A
5066A	1Z3.3A
5066A	1Z3.3T10.5
5066A	14-0104-6
5066A	019-002691
5066A	41-0906
5066A	48-83461E03
5066A	56-50
5066A	61E03
5066A	523-2003-339
5067A	AZ3.9
5067A	AZ748A
5067A	BZ3.9
5067A	BZY85C3V9
5067A	BZY92C3V9
5067A	GEZD-3.9
5067A	GLA39B
5067A	HEP-Z0403
5067A	KS2039A
5067A	MGLA39A
5067A	PD6004A
5067A	PD6045
5067A	PS8903
5067A	RE102
5067A	RZ3.9
5067A	RZZ3.9
5067A	001-023036
5067A	1D3.9
5067A	1D3.9A
5067A	1D3.9B
5067A	1M3.9ZS10
5067A	1M3.9ZS5
5067A	1N1507
5067A	1N1507A
5067A	1N1518
5067A	1N1518A
5067A	1N1981A
5067A	1N3823
5067A	1N3823A
5067A	1N4686
5067A	1N4730
5067A	1N4730A
5067A	1N748A
5067A	1S2039A
5067A	1S994
5067A	1Z3.9
5067A	1Z3.9A
5067A	05-480204
5067A	09-306380
5067A	13Z6AF
5067A	13Z6F
5067A	48-82256C46
5067A	48-83461E19
5067A	61E19
5067A	93A39-34
5067A	96-5248-09
5067A	96-5248-11
5067A	264P09707
5067A	523-2003-399
5067A	134882256346
5067A	134883461519
5068A	AZ4.3
5068A	AZ749A
5068A	BZ4.3
5068A	BZY92C4V3
5068A	GEZD-4.3
5068A	RE103
5068A	1D4.3
5068A	1D4.3A
5068A	1D4.3B
5068A	1M4.3ZS10
5068A	1M4.3ZS5
5068A	1N3824
5068A	1N3824A
5068A	1N4687
5068A	1N4731
5068A	1N4731A
5068A	1N5229B
5068A	1N5845B
5068A	1S220
5068A	1Z4.3
5068A	1Z4.3A
5068A	1ZC4.3T10.5
5068A	5D-5A-L
5068A	09-306289
5068A	14Z6AF
5068A	14Z6F
5068A	61E13
5068A	523-2003-439
5068A	72165-3
5068A	134883461513
5069A	AZ4.7
5069A	BZ4.7
5069A	BZY56
5069A	BZY83C4V7
5069A	BZY85C4V7
5069A	GEZD-4.7
5069A	GLA47B
5069A	HEP-Z0405
5069A	HZ5
5069A	KS2047A
5069A	MGLA47A
5069A	OA126/5
5069A	OAZ200
5069A	PD6006A
5069A	PD6047
5069A	PS8905
5069A	RE104

ECG	Industry Standard No.
5069A	RZ4.7
5069A	RZZ4.7
5069A	001-023037
5069A	1/4M4.7AZ5
5069A	1D4.7
5069A	1D4.7B
5069A	1M4.7ZS10
5069A	1M4.7ZS5
5069A	1N1484
5069A	1N1508
5069A	1N1508A
5069A	1N1519
5069A	1N1519A
5069A	1N2032
5069A	1N2032A
5069A	1N3825
5069A	1N3825A
5069A	1N4688
5069A	1N4732
5069A	1N4732A
5069A	1N5230B
5069A	1N5846B
5069A	1N674
5069A	1N750A
5069A	1S2047A
5069A	1S994S
5069A	1Z4.7
5069A	1Z4.7A
5069A	13-0164
5069A	15Z6AF
5069A	15Z6F
5069A	24MW1107
5069A	48-97046A14
5069A	86-109-1
5069A	523-2003-479
5069A	2606-303
5069A	4822-130-30284
5069A	137647
507	D3A
507	DS-1U
507	EDS-1
507	H890
507	SB-2(CENTERING)
507	SD-1AHF
507	T-E1029
507	T-E1108
507	TVM546
507	1N4718
507	1N4933
507	1N5415
507	1NJ71126
507	1NJ71186
507	13-33189-1
507	34-8054-16
507	53B010-9
507	86-23-1
507	4800-224
507	115867
507	530106-1
507	530106-1001
507	817120
507	1464778-9
507	1474778-9
507	23115051
507	26010011
5070A	.25T5.8
5070A	.25T5.8A
5070A	AW01-06
5070A	AZ-058
5070A	BZ-058
5070A	EQB01-06
5070A	FV-24
5070A	GEZD-6.0
5070A	HZ-6B
5070A	IP20-0203
5070A	J24262
5070A	MZ206
5070A	MZ306
5070A	MZ306C
5070A	MZ70MA
5070A	MZ92-6.0B
5070A	NZ-206
5070A	PZW
5070A	QB106P
5070A	QD-ZMZ306CE
5070A	RE108
5070A	RH-EX0019CEZZ
5070A	RVDCD0033
5070A	RVDEQA0106S
5070A	RVDMZ-206
5070A	RVDR5R6EB
5070A	RVDRD5R6EB
5070A	SV124
5070A	TR6S
5070A	TVSZB1-6
5070A	Z6
5070A	ZB1-6
5070A	ZQ-6
5070A	1N1779
5070A	1N2033
5070A	1N2033-2
5070A	1N5849B
5070A	1N762-2
5070A	1N816
5070A	1S1956
5070A	1S347
5070A	1S37
5070A	1S480
5070A	05-480306
5070A	09-306073
5070A	09-306208
5070A	09-306383
5070A	46-86460-3
5070A	48-155225
5070A	48S155080
5070A	53K001-4
5070A	57S-06
5070A	61B05
5070A	264D00507
5070A	264P03301
5070A	2093A41-153
5070A	2106-17
5070A	38510-331
5070A	2008230042
5070A	2012230042
5070A	4202020100
5070A	134883461505
5071A	.25T6.8
5071A	.25T6.8A
5071A	.25T7.1
5071A	.25T7.1A
5071A	.7JZ6.8
5071A	.7Z6.8A
5071A	.7Z6.8B
5071A	.7Z6.8C
5071A	.7Z6.8D
5071A	.7ZM6.8A
5071A	.7ZM6.8B
5071A	.7ZM6.8C
5071A	.7ZM6.8D
5071A	A7286201
5071A	AW-01-07
5071A	AW01-7
5071A	AZ-065
5071A	AZ-067
5071A	AZ-069
5071A	AZ6.8
5071A	AZ754A
5071A	AZ957B
5071A	BZ-065
5071A	BZ-067
5071A	BZ-069
5071A	BZ6.8
5071A	BZX29C6V8
5071A	BZX46C6V8
5071A	BZX55C6V8
5071A	BZX79C6V8
5071A	BZX83C6V8
5071A	BZY60
5071A	BZY83C6V8
5071A	BZY85C6V8
5071A	BZY92C6V8
5071A	DS-159
5071A	DS-31
5071A	EA16X123
5071A	EDZ-20
5071A	EQA01-07
5071A	EQA01-07R
5071A	EQA01-07RE
5071A	ERB-07RE
5071A	ETD-HZ7C
5071A	EZ6(ELCOM)
5071A	GEZD-6.8
5071A	HEP-Z0409
5071A	HF-20033
5071A	HW6.8
5071A	HW6.8A
5071A	IP20-0186
5071A	KZ6
5071A	KZ6A
5071A	KZ6ZA
5071A	MZ207
5071A	MA1068
5071A	MZ-207
5071A	MZ-7
5071A	MZ207
5071A	MZ207(ZENER)
5071A	MZ207-02A
5071A	OAZ204
5071A	PD6010A
5071A	PD6051
5071A	PS8909
5071A	QA01-07RE
5071A	QA107RE
5071A	RO7A
5071A	RD-6.8EB
5071A	RD6.8F
5071A	RD6.8FA
5071A	RD6.8FB
5071A	RE110
5071A	RT5215
5071A	RT5471
5071A	RVDRD7AN
5071A	RZ6.8
5071A	RZZ6.8
5071A	SD-49
5071A	SD42
5071A	SD49
5071A	SK3334
5071A	TR7SB
5071A	TV24981
5071A	TVSQA01-07RE
5071A	UZ8706
5071A	UZ8806
5071A	VHERD6R8EE/-1
5071A	WY-157
5071A	WY-157A
5071A	WZ069
5071A	XC70
5071A	XZ-070
5071A	XZ070
5071A	XZ070
5071A	YZ060
5071A	ZOB6.8
5071A	Z1106
5071A	Z1106-C
5071A	Z1B6.8
5071A	Z5B6.8
5071A	Z6.8
5071A	ZB6.8B
5071A	ZF6.8
5071A	ZM6.8B
5071A	ZOD6.8
5071A	ZPD618
5071A	ZZ6.8
5071A	001-0082-00
5071A	01-2303-3
5071A	1/4M6.8Z5
5071A	1M6.8Z
5071A	1M6.8Z10
5071A	1M6.8Z5
5071A	1M6.8ZS10
5071A	1M6.8ZS5
5071A	1N1510
5071A	1N1510A
5071A	1N1521
5071A	1N1521A
5071A	1N1767
5071A	1N1767A
5071A	1N2034
5071A	1N2034-2
5071A	1N2034A
5071A	1N3016
5071A	1N3016A
5071A	1N3016B
5071A	1N3444
5071A	1N3675
5071A	1N3675A
5071A	1N3829
5071A	1N3829A
5071A	1N4158
5071A	1N4158A
5071A	1N4323
5071A	1N4323A
5071A	1N4400
5071A	1N4400A
5071A	1N4628
5071A	1N4692
5071A	1N4736
5071A	1N4736A
5071A	1N5235B
5071A	1N5526B
5071A	1N5559
5071A	1N5559A
5071A	1N5851B
5071A	1N710A
5071A	1N957B
5071A	1S1957
5071A	1S2113A
5071A	1S222
5071A	1S3006
5071A	1S332M
5071A	1S3332
5071A	1S481
5071A	1Z6.8
5071A	1Z6.8A
5071A	02Z6.8A
5071A	3L4-3505-2
5071A	3L4-3506-12
5071A	3L4-3506-2
5071A	4-2020-11300
5071A	05-110106
5071A	05-69669-01
5071A	05Z6.8
5071A	09-306215
5071A	09-306325
5071A	09-306419
5071A	09-307088
5071A	14-515-27
5071A	14-515-30
5071A	19Z6AF
5071A	19Z6F
5071A	24MW742
5071A	34-8057-10
5071A	46-86566-3
5071A	48-127021
5071A	48-82256C37
5071A	48-83461E42
5071A	56-58
5071A	56-6
5071A	56-63
5071A	56-71
5071A	61E42
5071A	63-11147
5071A	63-11659
5071A	63-12110
5071A	86-0510
5071A	86-110-1
5071A	93A39-11
5071A	96-5248-01
5071A	100-218
5071A	100-437
5071A	103-140
5071A	103-140A
5071A	0114-0260
5071A	142-016
5071A	152-042-9-002
5071A	264P17401
5071A	523-2003-689
5071A	523-2503-689
5071A	642-236
5071A	0648R
5071A	800-030-00
5071A	903-166B
5071A	903-180
5071A	903-335
5071A	1011-02
5071A	1063-4145
5071A	1063-6454
5071A	2017-113
5071A	2093A41-70
5071A	2341-17
5071A	4822-130-30132
5071A	99201-312
5071A	126851
5071A	132616
5071A	530073-1034
5071A	530145-689
5071A	530157-689
5071A	740628
5071A	740953
5071A	1471878-3
5071A	2092055-0018
5071A	2092055-0710
5071A	2330302
5071A	2330632
5071A	2330643
5071A	2330791
5071A	3596398
5071A	5330051
5071A	5330312
5071A	8052192
5071A	23115908
5071A	42024925
5071A	62564490
5071A	2008230057
5071A	4202011300
5071A	4202011400
5071A	134882256337
5071A	134883461542
5072A	.25T8.2
5072A	.25T8.2A
5072A	.7JZ8.2
5072A	.7Z8.2A
5072A	.7Z8.2B
5072A	.7Z8.2C
5072A	.7Z8.2D
5072A	.7ZM8.2A
5072A	.7ZM8.2B
5072A	.7ZM8.2C
5072A	.7ZM8.2D
5072A	OA01-08R
5072A	A72-86-700
5072A	AW-01-08
5072A	AW01-08
5072A	AW01-08J
5072A	AZ-077
5072A	AZ-079
5072A	AZ-081
5072A	AZ-083
5072A	AZ756A
5072A	AZ8.2
5072A	AZ959B
5072A	B-1808
5072A	B1-8.2
5072A	BZ-077
5072A	BZ-079
5072A	BZ-080
5072A	BZ-0800
5072A	BZ-081
5072A	BZ-083
5072A	BZ080
5072A	BZ081
5072A	BZ085
5072A	BZ8.2
5072A	BZX29C8V2
5072A	BZX46C8V2
5072A	BZX55C8V2
5072A	BZX79C8V2
5072A	BZX83C8V2
5072A	BZY62
5072A	BZY83C8V2
5072A	BZY92C8V2
5072A	C02
5072A	CI-8.2
5072A	C02
5072A	D1A(ZENER)
5072A	D1U
5072A	D6M
5072A	DDAY008003
5072A	DIA
5072A	DS-149
5072A	DS149
5072A	DX-0729
5072A	DZ-081
5072A	DZ0820
5072A	EDZ-24
5072A	EQA01-08
5072A	EQA01-08R
5072A	EQB-0108
5072A	EQB01-08
5072A	EZ8(ELCOM)
5072A	GE-X11
5072A	GEZD-8.2
5072A	HD3003109
5072A	HEP-Z0411
5072A	HW8.2
5072A	HW8.2A
5072A	KZ-8A
5072A	MA1082
5072A	ME408-02C
5072A	MZ-08
5072A	MZ-208
5072A	MZ-8
5072A	MZ208
5072A	MZ408-02C
5072A	MZ8
5072A	02Z8.2A
5072A	OAZ206
5072A	PD6012A
5072A	PD6053
5072A	PS8911
5072A	QA01-082
5072A	QA01-08R
5072A	QD-ZMZ408CE
5072A	R-1S333Y
5072A	RD-13AL
5072A	RD-2.2E
5072A	RD-8.2E
5072A	RD-8.2EC
5072A	RD8
5072A	RD8.2
5072A	RD8.2C
5072A	RD8.2EA
5072A	RD8.2EB
5072A	RD8.2EC
5072A	RD8.2EK
5072A	RD8.2F
5072A	RD8.2FA
5072A	RD8.2FB
5072A	RE112
5072A	RF-8.2E
5072A	RF8.2E
5072A	RT6105
5072A	RT6105(G.E.)
5072A	RT6923
5072A	RVD1N4738
5072A	RZ8.2
5072A	RZZ8.2
5072A	SD-8(ZENER)
5072A	SK3136
5072A	SVD02Z8.2A
5072A	SZ200-8.2
5072A	TVSYZ-080
5072A	TVSYZ080
5072A	UZ8.2B
5072A	UZ8708
5072A	UZ8808
5072A	WE081
5072A	WM-079
5072A	WM-079A
5072A	WM-079B
5072A	WM-079C
5072A	WM-079D
5072A	WM-081
5072A	WM-081A
5072A	WM-081B
5072A	WM-081C
5072A	WM-081D
5072A	WM-083
5072A	WM-083A
5072A	WM-083B
5072A	WM-083C
5072A	WM-083D
5072A	WM-085
5072A	WM-085A
5072A	WM-085B
5072A	WM-085C
5072A	WM-085D
5072A	WN-079
5072A	WN-079A
5072A	WN-079B
5072A	WN-079C
5072A	WN-079D
5072A	WN-081
5072A	WN-081A
5072A	WN-081B
5072A	WN-081C
5072A	WN-081D
5072A	WN-083
5072A	WN-083A
5072A	WN-083B
5072A	WN-083C
5072A	WN-083D
5072A	WN-085
5072A	WN-085A
5072A	WN-085B
5072A	WN-085C
5072A	WN-085D
5072A	WO-079
5072A	WO-079A
5072A	WO-079B
5072A	WO-079C
5072A	WO-079D
5072A	WO-081
5072A	WO-081A
5072A	WO-081B
5072A	WO-081C
5072A	WO-081D
5072A	WO-083
5072A	WO-083A
5072A	WO-083B
5072A	WO-083C
5072A	WO-083D
5072A	WO-085
5072A	WO-085A
5072A	WO-085B
5072A	WO-085C
5072A	WO-085D
5072A	WP-079
5072A	WP-079A
5072A	WP-079B
5072A	WP-079C
5072A	WP-079D
5072A	WP-081
5072A	WP-081A
5072A	WP-081B
5072A	WP-081C
5072A	WP-081D
5072A	WP-083
5072A	WP-083A
5072A	WP-083B
5072A	WP-083C
5072A	WP-083D
5072A	WP-085
5072A	WP-085A
5072A	WP-085B
5072A	WP-085C
5072A	WP-085D
5072A	WQ-079
5072A	WQ-079A
5072A	WQ-079B
5072A	WQ-079C
5072A	WQ-079D
5072A	WQ-081
5072A	WQ-081A
5072A	WQ-081B
5072A	WQ-081C
5072A	WQ-081D
5072A	WQ-083
5072A	WQ-083A
5072A	WQ-083B
5072A	WQ-083C
5072A	WQ-083D
5072A	WQ-085
5072A	WQ-085A
5072A	WQ-085B
5072A	WQ-085C
5072A	WQ-085D
5072A	WR-079
5072A	WR-079A
5072A	WR-079B
5072A	WR-079C
5072A	WR-079D
5072A	WR-081
5072A	WR-081A
5072A	WR-081B
5072A	WR-081C
5072A	WR-081D
5072A	WR-083
5072A	WR-083A
5072A	WR-083B
5072A	WR-083C
5072A	WR-083D
5072A	WR-085
5072A	WR-085A
5072A	WR-085B
5072A	WR-085C
5072A	WR-085D
5072A	WS-079
5072A	WS-079A
5072A	WS-079B
5072A	WS-079C
5072A	WS-079D
5072A	WS-081
5072A	WS-081A
5072A	WS-081B
5072A	WS-081C
5072A	WS-081D
5072A	WS-083
5072A	WS-083A
5072A	WS-083B
5072A	WS-083C
5072A	WS-083D
5072A	WS-085
5072A	WS-085A
5072A	WS-085B
5072A	WS-085C
5072A	WS-085D
5072A	WT-079
5072A	WT-079A
5072A	WT-079B
5072A	WT-079C
5072A	WT-079D
5072A	WT-081
5072A	WT-081A
5072A	WT-081B
5072A	WT-081C
5072A	WT-081D
5072A	WT-083
5072A	WT-083A
5072A	WT-083B
5072A	WT-083C
5072A	WT-083D
5072A	WT-085
5072A	WT-085A
5072A	WT-085B
5072A	WT-085C
5072A	WT-085D
5072A	WU-079
5072A	WU-079A
5072A	WU-079B
5072A	WU-079C
5072A	WU-079D
5072A	WU-081
5072A	WU-081A
5072A	WU-081B
5072A	WU-081C
5072A	WU-081D
5072A	WU-083
5072A	WU-083A
5072A	WU-083B
5072A	WU-083C
5072A	WU-083D
5072A	WU-085
5072A	WU-085A
5072A	WU-085B
5072A	WU-085C
5072A	WU-085D
5072A	WV-079

ECG	Industry Standard No.	ECG	Industry Standard No.	ECG	Industry Standard No.	ECG	Industry Standard No.	ECG	Industry Standard No.
5072A	WV-079A	5072A	1N1425	5072A	25810-54	5074A	RH-EXO015CEZZ	5074A	05-060110
5072A	WV-079B	5072A	1N1511	5072A	25840-54	5074A	RZZ11	5074A	052I1
5072A	WV-079C	5072A	1N1511A	5072A	28810-65	5074A	SO11-0	5074A	09-306428
5072A	WV-079D	5072A	1N1522	5072A	37510-54	5074A	SK3139	5074A	09-306429
5072A	WV-081	5072A	1N1522A	5072A	88060-54	5074A	SZ-AWO1-11	5074A	13-22319-0
5072A	WV-081A	5072A	1N1769	5072A	88510-54	5074A	SZ-RD11FB	5074A	21A037-020
5072A	WV-081B	5072A	1N1769A	5072A	93018	5074A	SZ11	5074A	21A103-045
5072A	WV-081C	5072A	1N1875	5072A	175043-069	5074A	SZ671-0	5074A	2426AF
5072A	WV-081D	5072A	1N1875A	5072A	216024	5074A	SZ671-W	5074A	2426F
5072A	WV-083	5072A	1N1875B	5072A	245117	5074A	TVS-RD11A	5074A	46-86458-3
5072A	WV-083A	5072A	1N2035A	5072A	741869	5074A	TVSAWO1-11	5074A	46-86462-3
5072A	WV-083B	5072A	1N3018	5072A	741870	5074A	TVSQA01-11SE	5074A	46-86487-3
5072A	WV-083C	5072A	1N3018A	5072A	742922	5074A	TVSZB1-11	5074A	46-86504-3
5072A	WV-083D	5072A	1N3018B	5072A	1221900	5074A	WM-110	5074A	48-155048
5072A	WV-085	5072A	1N3433	5072A	1956197	5074A	WM-110A	5074A	48-155081
5072A	WV-085A	5072A	1N3445	5072A	2002209-11	5074A	WM-110B	5074A	48-355062
5072A	WV-085B	5072A	1N3677	5072A	2092055-0002	5074A	WM-110C	5074A	488155048
5072A	WV-085C	5072A	1N3677A	5072A	2092055-0017	5074A	WM-110D	5074A	488155081
5072A	WV-085D	5072A	1N3677B	5072A	2092055-0027	5074A	WN-110	5074A	488155084
5072A	WW-079A	5072A	1N4160	5072A	2092055-0712	5074A	WN-110A	5074A	93A39-13
5072A	WW-079B	5072A	1N4160A	5072A	2330307	5074A	WN-110B	5074A	93A39-44
5072A	WW-079C	5072A	1N4160B	5072A	7201090	5074A	WN-110C	5074A	93A55-1
5072A	WW-079D	5072A	1N4325	5072A	7202782	5074A	WN-110D	5074A	103-246
5072A	WW-081	5072A	1N4325A	5072A	7287079	5074A	WO-110	5074A	103-279-20
5072A	WW-081A	5072A	1N4325B	5072A	7296476	5074A	WO-110A	5074A	264P03302
5072A	WW-081B	5072A	1N4402	5072A	7297258	5074A	WO-110B	5074A	523-2003-110
5072A	WW-081C	5072A	1N4402A	5072A	7305469	5074A	WO-110D	5074A	1059-9140
5072A	WW-081D	5072A	1N4630	5072A	79\0111-01	5074A	WO110C	5074A	10020
5072A	WW-083	5072A	1N4694	5072A	23115277	5074A	WP-110	5074A	125499
5072A	WW-083A	5072A	1N4738	5072A	23115374	5074A	WP-110A	5074A	126863
5072A	WW-083B	5072A	1N4738A	5072A	26010044	5074A	WP-110B	5074A	127177
5072A	WW-083C	5072A	1N5528B	5072A	73010780	5074A	WP-110C	5074A	489752-091
5072A	WW-083D	5072A	1N5561	5073A	.25T8.7	5074A	WP-110D	5074A	922950
5072A	WW-085	5072A	1N5561A	5073A	.25T8.7A	5074A	WQ-110	5074A	2327078
5072A	WW-085A	5072A	1N5734B	5073A	.25T8.8	5074A	WQ-110A	5074A	2331174
5072A	WW-085B	5072A	1N5853B	5073A	.25T8.8A	5074A	WQ-110B	5074A	23115368
5072A	WW-085C	5072A	1N664	5073A	AZ-085	5074A	WQ-110C	5074A	4202011200
5072A	WW-085D	5072A	1N712A	5073A	AZ-088	5074A	WQ-110D	5074A	4202011500
5072A	WX-079	5072A	1N756	5073A	BZ-085	5074A	WR-110	5074A	4202012501
5072A	WX-079A	5072A	1N756A	5073A	BZ-088	5074A	WR-110A	5074A	4202109070
5072A	WX-079B	5072A	1N959B	5073A	D8U	5074A	WR-110B	5074A	4202110870
5072A	WX-079C	5072A	1R8.2	5073A	DL172172	5074A	WR-110C	5074A	4202116670
5072A	WX-079D	5072A	1R8.2A	5073A	EA16X402	5074A	WR-110D	5075A	.25T15.8
5072A	WX-081	5072A	1R8.2B	5073A	ED491130	5074A	WS-110	5075A	.25T15.8A
5072A	WX-081A	5072A	1S193	5073A	HZ-9	5074A	WS-110A	5075A	.25T16
5072A	WX-081B	5072A	131958	5073A	HZ-9B	5074A	WS-110B	5075A	.25T16A
5072A	WX-081C	5072A	182082A	5073A	H93B	5074A	WS-110C	5075A	.4T16
5072A	WX-081D	5072A	182115A	5073A	MB6356	5074A	WS-110D	5075A	.4T16A
5072A	WX-083	5072A	183008	5073A	RD9AL	5074A	WT-110	5075A	.4T16B
5072A	WX-083A	5072A	18333	5073A	RD9B	5074A	WT-110A	5075A	.7JZ16
5072A	WX-083B	5072A	13333Y	5073A	RE113	5074A	WT-110B	5075A	.7Z16A
5072A	WX-083C	5072A	128.2	5073A	WZ085	5074A	WT-110C	5075A	.7Z16B
5072A	WX-083D	5072A	128.2T10	5073A	YAAD017	5074A	WT-110D	5075A	.7Z16C
5072A	WX-085	5072A	128.2T20	5073A	001-0101-01	5074A	WU-110	5075A	.7Z16D
5072A	WX-085A	5072A	128.2T5	5073A	1N2035-1	5074A	WU-110A	5075A	.72M16A
5072A	WX-085B	5072A	1Z08.2T10	5073A	1N4695	5074A	WU-110B	5075A	.72M16B
5072A	WX-085C	5072A	1Z08.2T20	5073A	1N5294B	5074A	WU-110C	5075A	.72M16C
5072A	WX-085D	5072A	1Z08.2T5	5073A	1N5854B	5074A	WU-110D	5075A	.72M16D
5072A	WY-079	5072A	1ZB825	5073A	1N764	5074A	WV-110	5075A	AWO1-16
5072A	WY-079A	5072A	02Z8.2A	5073A	181717	5074A	WV-110A	5075A	AZ-157
5072A	WY-079B	5072A	3L4-3505-1	5073A	18224	5074A	WV-110B	5075A	AZ-162
5072A	WY-079C	5072A	3L4-3506-7	5073A	2M214	5074A	WV-110C	5075A	AZ9565B
5072A	WY-079D	5072A	4-856	5073A	24MW670	5074A	WV-110D	5075A	BZ-157
5072A	WY-081	5072A	05-04800-02	5073A	46-86194-3	5074A	WW-110	5075A	BZ162
5072A	WY-081A	5072A	05-540082	5073A	48-137577	5074A	WW-110A	5075A	BZX29C16
5072A	WY-081B	5072A	05Z8.2	5073A	754-9030-009	5074A	WW-110B	5075A	BZX46C16
5072A	WY-081C	5072A	8-9V	5073A	129904(RCA)	5074A	WW-110C	5075A	BZX55C16
5072A	WY-081D	5072A	09-306196	5073A	5330322	5074A	WW-110D	5075A	BZX83C16
5072A	WY-083	5072A	09-306238	5074A	.25T11	5074A	WX-110	5075A	BZT92C16
5072A	WY-083A	5072A	09-306242	5074A	.25T11A	5074A	WX-110A	5075A	DDAY009003
5072A	WY-083B	5072A	09-306314	5074A	.4T11	5074A	WX-110B	5075A	DDAY009007
5072A	WY-083C	5072A	09-306351	5074A	.4T11A	5074A	WX-110C	5075A	EQA01-16
5072A	WY-083D	5072A	09-306356	5074A	.4T11B	5074A	WX-110D	5075A	EQB01-16
5072A	WY-085	5072A	09-306378	5074A	.7JZ11	5074A	WZ-110	5075A	EZ16(ELCOM)
5072A	WY-085B	5072A	09-306381	5074A	.7Z11A	5074A	WZ-110A	5075A	GEZD-16
5072A	WY-085C	5072A	10DO8	5074A	.7Z11B	5074A	WZ-115	5075A	HEP-Z0419
5072A	WY-085D	5072A	21A037-012	5074A	.7Z11C	5074A	Z0414	5075A	HW16
5072A	WZ-079A	5072A	21A119-073	5074A	.7Z11D	5074A	ZB-11	5075A	HW16A
5072A	WZ-081	5072A	2126AF	5074A	.72M1	5074A	ZB1-11	5075A	HZ16H
5072A	WZ-081A	5072A	2126P	5074A	.72M11A	5074A	ZB11	5075A	QD-ZBZ162XJ
5072A	WZ-083A	5072A	32-0025	5074A	.72M11B	5074A	ZB11A	5075A	RD16A-N
5072A	WZ-083B	5072A	46-41763001	5074A	.72M11C	5074A	ZB11B	5075A	RD16C-*
5072A	WZ081	5072A	46-86742-3	5074A	A4ZXO0480-01	5074A	ZM11B	5075A	RD16E-M
5072A	XZ-082	5072A	46-86483-3	5074A	AWO1-11	5074A	1M11Z	5075A	RD16E-N
5072A	X2064	5072A	48-03073A06	5074A	AZ-105	5074A	1M11Z10	5075A	RD16EB
5072A	X2084	5072A	48-134912	5074A	AZ-110	5074A	1M11Z5	5075A	RD16F
5072A	YAAD021	5072A	48-137421	5074A	AZ11	5074A	1M11ZS10	5075A	RD16FA
5072A	YEAD010	5072A	48-40458A06	5074A	A2962B	5074A	1M11ZS5	5075A	RD16FB
5072A	YEAD014	5072A	48-40458A064	5074A	BZ-105	5074A	1N1772	5075A	RE122
5072A	YBAD024	5072A	48-41763001	5074A	BZ-110	5074A	1N1772A	5075A	SK3142
5072A	YZ-080	5072A	48B137021	5074A	BZ110	5074A	1N2036-2	5075A	SV139
5072A	Z0217	5072A	51-02006-12	5074A	BZX29C11	5074A	1N3021	5075A	TVSEQA01-068
5072A	Z0411	5072A	56-4885	5074A	BZX46C11	5074A	1N3021A	5075A	TVSZB1-15
5072A	ZOB8.2	5072A	56-54	5074A	BZX55C11	5074A	1N3021B	5075A	UZ8716
5072A	Z101	5072A	63-8825	5074A	BZX79C11	5074A	1N3680	5075A	UZ8816
5072A	Z1108	5072A	63-9783	5074A	BZX83C11	5074A	1N3680A	5075A	WL-157
5072A	Z1108-C	5072A	81-46123044-3	5074A	BZT85C11	5074A	1N4163	5075A	WL-157A
5072A	Z1A8.2A	5072A	86-114-1	5074A	BZT85C11	5074A	1N4163A	5075A	WL-157B
5072A	Z1A8.2B	5072A	86-135-1	5074A	BZY92C11	5074A	1N4328	5075A	WL-157C
5072A	Z1BB.2	5072A	100-219	5074A	D8-189	5074A	1N4328A	5075A	WL-157D
5072A	Z2A8.2	5072A	100-521	5074A	D8189	5074A	1N4405	5075A	WL-162
5072A	Z2A8ZF	5072A	123-019	5074A	EQA01-11	5074A	1N4405A	5075A	WL-162A
5072A	Z4A8.2	5072A	123B-004	5074A	EQA01-11	5074A	1N4433	5075A	WL-162B
5072A	Z5B8.2	5072A	152-042-9-001	5074A	EQA01-11E	5074A	1N4698	5075A	WL-162C
5072A	Z8.2	5072A	152-054-9-001	5074A	EQA01-11Z	5074A	1N4741	5075A	WL-162D
5072A	Z8.2A	5072A	157-009-9-001	5074A	EQB01-11	5074A	1N4741A	5075A	WL-167
5072A	Z8.2B	5072A	201A0723	5074A	EQB01-11V	5074A	1N5241B	5075A	WL-167A
5072A	Z8.2C	5072A	260-10-044	5074A	EQB01-11Z	5074A	1N5531B	5075A	WL-167B
5072A	Z8.2D	5072A	412(ZENER)	5074A	EZ11(ELCOM)	5074A	1N5564	5075A	WL-167C
5072A	ZA8.2	5072A	523-2003-829	5074A	GEZD-11	5074A	1N5564A	5075A	WL-167D
5072A	ZA8.2A	5072A	523-2005-829	5074A	HD30000109-0	5074A	1N5857B	5075A	WM-157
5072A	ZA8.2B	5072A	754-5710-219	5074A	HD30001090	5074A	1N765-2	5075A	WM-157A
5072A	ZB*-8	5072A	754-5750-282	5074A	HD3000113	5074A	1R11	5075A	WM-157B
5072A	ZB8.2B	5072A	754-5850-284	5074A	HD3001009	5074A	1R11A	5075A	WM-157C
5072A	ZF-8.2	5072A	754-9000-082	5074A	HD30010090	5074A	1R11B	5075A	WM-157D
5072A	ZP8.2	5072A	919-001-1214	5074A	HD3001809	5074A	18196	5075A	WM-162
5072A	ZP8.2P	5072A	1042-19	5074A	HEP-Z0414	5074A	18961	5075A	WM-162A
5072A	ZPBA	5072A	1048-9839	5074A	HW1	5074A	18218A	5075A	WM-162B
5072A	ZM8.2A	5072A	1074-122	5074A	HW11A	5074A	18226	5075A	WM-162C
5072A	ZM8.2B	5072A	2064-17	5074A	IWO1-09J	5074A	183011	5075A	WM-162D
5072A	ZM8.2C	5072A	2079-42	5074A	LPM11A	5074A	18336	5075A	WM-167
5072A	ZM8.2D	5072A	2203	5074A	NTC-20	5074A	18483	5075A	WM-167A
5072A	ZPD8.2	5072A	2203-17	5074A	QA01-11SE	5074A	WO1-08J	5075A	WM-167B
5072A	ZSP88.2	5072A	2339-17	5074A	QA1-11M	5074A	1Z11	5075A	WM-167C
5072A	ZZ8.2	5072A	2343-17	5074A	QA111M	5074A	1Z11A	5075A	WM-167D
5072A	001-0152-00	5072A	2451-17	5074A	QA111SE	5074A	1Z11T10	5075A	WN-157
5072A	001-0161-00	5072A	4808-0000-009	5074A	QB11Z	5074A	1Z11T20	5075A	WN-157A
5072A	001-0163-13	5072A	6115(G.E.)	5074A	RD11ZA	5074A	1ZC11Z10	5075A	WN-157B
5072A	1M8.2Z	5072A	8910-54	5074A	RD11EM	5074A	1ZC11T5	5075A	WN-157C
5072A	1M8.2Z10	5072A	0011193	5074A	RD11F	5074A	02Z11A	5075A	WN-157D
5072A	1M8.2Z5	5072A	18410-43	5074A	RD11FA	5074A	3L4-3505-4	5075A	WN-162
5072A	1M8.2ZS10	5072A	18600-54	5074A	RD11FB	5074A	4-2020-11500	5075A	WN-162A
5072A	1M8.2ZS5			5074A	RE116	5074A	4-2020-12300	5075A	WN-162B
				5074A	RE117	5074A	4-2021-10870	5075A	WN-162C

ECG	Industry Standard No.
5077A	1N989A
5077A	1R18
5077A	1R18A
5077A	1R18B
5077A	1S1966
5077A	1S199
5077A	1S2123A
5077A	1S2180A
5077A	1S233
5077A	1S3018
5077A	1S485
5077A	1Z18
5077A	1Z18A
5077A	1Z18T10
5077A	1Z18T20
5077A	1Z18T5
5077A	1ZC18T10
5077A	1ZC18T5
5077A	02Z18A
5077A	4-2020-09200
5077A	4-2020-12000
5077A	05-950177
5077A	09-306355
5077A	14-515-05
5077A	46-86392-3
5077A	48-155182
5077A	48-82256C53
5077A	48-83461E37
5077A	61E37
5077A	62-21574
5077A	103-252
5077A	523-2003-180
5077A	919-003309-1
5077A	71411-3
5077A	97202-222
5077A	99202-222
5077A	130132
5077A	141187
5077A	530073-22
5077A	530073-30
5077A	1477046-10
5077A	2002209-7
5077A	4202009200
5077A	4202012000
5077A	4202012800
5077A	134882256353
5077A	134883461537
5078A	.25T19
5078A	.25T19A
5078A	.7JZ19
5078A	.7Z19A
5078A	.7Z19B
5078A	.7Z19C
5078A	.7Z19D
5078A	.7ZM19A
5078A	.7ZM19B
5078A	.7ZM19C
5078A	.7ZM19D
5078A	AZ-187
5078A	AZ-192
5078A	BZ-187
5078A	BZ-192
5078A	CD3212055
5078A	ED498150
5078A	EQB01-19
5078A	GEZD-19
5078A	RE125
5078A	RH-EX0013CEZZ
5078A	RH-EX013CEZZ
5078A	SV143
5078A	1M19Z10
5078A	1M19Z5
5078A	1N2039
5078A	1N2039-2
5078A	1N2039A
5078A	1N5249B
5078A	1N5539B
5078A	1N5865B
5078A	1N768-2
5078A	1N768A
5078A	1S234
5078A	103-279-28A
5078A	469-199-19
5078A	523-2003-190
5079A	.25T20
5079A	.25T20A
5079A	.4T20
5079A	.4T20A
5079A	.4T20B
5079A	.7JZ20
5079A	.7Z20A
5079A	.7Z20B
5079A	.7Z20C
5079A	.7Z20D
5079A	.7ZM20A
5079A	.7ZM20B
5079A	.7ZM20C
5079A	.7ZM20D
5079A	A054-232
5079A	AW01-20
5079A	AZ-197
5079A	AZ968B
5079A	B66X0040-002
5079A	BZ-197
5079A	BZ-210
5079A	BZ210
5079A	BZX29C20
5079A	BZX46C20
5079A	BZX55C20
5079A	BZX55C22
5079A	BZX79C20
5079A	BZX83C20
5079A	BZY83C20
5079A	BZY85C20
5079A	BZY92C20
5079A	ES16X41
5079A	EZ20(ELCOM)
5079A	GEZD-20
5079A	HEP-Z0421
5079A	HW20
5079A	HW20A
5079A	MZ1000-20
5079A	P3309
5079A	R9381
5079A	RD20E
5079A	RD20F
5079A	RD20FA
5079A	RD20FB
5079A	RE126
5079A	RH-EX0037CEZZ
5079A	SD-31
5079A	001-023032
5079A	1M20Z
5079A	1M20Z10
5079A	1M20Z20
5079A	1M20Z3
5079A	1M20Z5
5079A	1M20ZS10
5079A	1N1778
5079A	1N1779A
5079A	1N2039-3
5079A	1N3027
5079A	1N3027A
5079A	1N3027B
5079A	1N3686
5079A	1N3686A
5079A	1N4169
5079A	1N4169A
5079A	1N4334
5079A	1N4334A
5079A	1N4411
5079A	1N4411A
5079A	1N4639
5079A	1N4707
5079A	1N4747
5079A	1N4747A
5079A	1N5250B
5079A	1N5570
5079A	1N5570A
5079A	1N721A
5079A	1N768-3
5079A	1R20
5079A	1R20A
5079A	1R20B
5079A	1S1967
5079A	1S2200A
5079A	1S235
5079A	1S3020
5079A	1Z20
5079A	1Z20A
5079A	1Z20T10
5079A	1Z20T20
5079A	1Z20T5
5079A	1ZC20T10
5079A	1ZC20T5
5079A	13-14879-6
5079A	13-14879-7
5079A	13-33179-5
5079A	21A037-009
5079A	48-82256C39
5079A	48-82256C57
5079A	56-45
5079A	86-139-1
5079A	86-66-1
5079A	86-73-1
5079A	86-88-1
5079A	93A39-5
5079A	93A39-7
5079A	229-1301-30
5079A	523-2003-200
5079A	919-00-3309-1
5079A	919-003309
5079A	919-015618-1
5079A	43072
5079A	71411-4
5079A	62669462
5079A	62761296
5079A	134882256339
5079A	134882256357
508	GESS-3A3
5080A	.25T22
5080A	.25T22A
5080A	.4T22
5080A	.4T22A
5080A	.4T22B
5080A	.7JZ22
5080A	.7Z22A
5080A	.7Z22B
5080A	.7Z22C
5080A	.7Z22D
5080A	.7ZM22A
5080A	.7ZM22B
5080A	.7ZM22C
5080A	.7ZM22D
5080A	AW01-22
5080A	AW0122
5080A	AZ969B
5080A	BZ-220
5080A	BZ230
5080A	BZX29C22
5080A	BZX46C22
5080A	BZX79C22
5080A	BZX83C22
5080A	BZY83C22
5080A	BZY85C22
5080A	BZY92C22
5080A	D6U
5080A	EA16X38
5080A	ES16X29
5080A	EZ22(ELCOM)
5080A	GEZD-22
5080A	HEP-Z0422
5080A	HW22
5080A	HW22A
5080A	MZ92-24B
5080A	RD22
5080A	RD22E
5080A	RD22F
5080A	RD22FA
5080A	RD22FB
5080A	RE127
5080A	RZ22
5080A	SD31
5080A	TVSQA01-25A
5080A	TVSQA01-25R
5080A	001-023031
5080A	01-2303-0
5080A	01-2303-1
5080A	1M22Z
5080A	1M22Z10
5080A	1M22Z5
5080A	1M22ZS10
5080A	1M22ZS5
5080A	1N1429
5080A	1N1516
5080A	1N1516A
5080A	1N1527
5080A	1N1527A
5080A	1N1880
5080A	1N1880A
5080A	1N3028
5080A	1N3028A
5080A	1N3028B
5080A	1N3438
5080A	1N3450
5080A	1N3687
5080A	1N3687A
5080A	1N4170
5080A	1N4170A
5080A	1N4335
5080A	1N4335A
5080A	1N4412
5080A	1N4412A
5080A	1N4640
5080A	1N4708
5080A	1N4748
5080A	1N4748A
5080A	1N5571
5080A	1N5571A
5080A	1N668
5080A	1N722A
5080A	1N769-1
5080A	1N969
5080A	1N969B
5080A	1R22
5080A	1R22A
5080A	1R22B
5080A	1S1968
5080A	1S2220A
5080A	1S236
5080A	1S3022
5080A	1Z22
5080A	1Z22A
5080A	1Z22T10
5080A	1Z22T20
5080A	1Z22T5
5080A	1ZC22T10
5080A	1ZC22T5
5080A	34-8057-31
5080A	48S137442
5080A	53N001-4
5080A	86-11-1
5080A	86-111-1
5080A	103-212
5080A	523-2003-220
5080A	1091
5080A	143777
5080A	530073-2
5080A	2337063
5080A	5330059
5080A	62638338
5080A	4202108570
5081A	.7JZ24
5081A	.7Z24A
5081A	.7Z24B
5081A	.7Z24D
5081A	.7ZM24A
5081A	.7ZM24B
5081A	.7ZM24C
5081A	.7ZM24D
5081A	A7288504
5081A	AW01-24
5081A	AZ970B
5081A	BZ-230
5081A	BZ-240
5081A	ED511918
5081A	EP16X16
5081A	EQA01-24RR
5081A	GEZD-24
5081A	HEP-Z0423
5081A	HW24
5081A	HW24A
5081A	QA01-25RA
5081A	RD-24A
5081A	RD24A
5081A	RD24AL
5081A	RD24AM
5081A	RD24AN
5081A	RD24F
5081A	RD24FA
5081A	RD24FB
5081A	RE128
5081A	RH-EX0034CEZZ
5081A	TR14001-1
5081A	TVSQA01-25RA
5081A	1M24Z
5081A	1M24Z10
5081A	1M24Z5
5081A	1M24ZS10
5081A	1M24ZS5
5081A	1N1780
5081A	1N2040
5081A	1N2040A
5081A	1N3029B
5081A	1N3688
5081A	1N3688A
5081A	1N4171
5081A	1N4171A
5081A	1N4336
5081A	1N4336A
5081A	1N4413
5081A	1N4641
5081A	1N4749
5081A	1N4749A
5081A	1N5252B
5081A	1S237
5081A	1Z24
5081A	1Z24A
5081A	1Z24T10
5081A	1Z24T20
5081A	1Z24T5
5081A	1ZC24T10
5081A	1ZC24T5
5081A	02Z24A
5081A	05-933945
5081A	8-719-906-24
5081A	46-86529-3
5081A	66X0040-007
5081A	070-049
5081A	86-113-1
5081A	86-91-1
5081A	93A39-1
5081A	93A39-45
5081A	103-105
5081A	103-105-01
5081A	103-248
5081A	103-278
5081A	103-289
5081A	121-289
5081A	130-105
5081A	1040-9373
5081A	72168-2
5081A	99201-325
5081A	137655
5081A	142738
5081A	530073-24
5081A	23115328
5081A	62540311
5082A	.25T25
5082A	.25T25A
5082A	.7JZ25
5082A	.7Z25A
5082A	.7Z25B
5082A	.7Z25C
5082A	.7Z25D
5082A	.7ZM25A
5082A	.7ZM25B
5082A	.7ZM25C
5082A	.7ZM25D
5082A	A04344-026
5082A	BZ-250
5082A	HR200
5082A	J241216
5082A	RE129
5082A	RH-EX0012CEZZ
5082A	1M25Z
5082A	1M25Z10
5082A	1M25Z5
5082A	1N2040-3
5082A	1N2040-4
5082A	1N5253B
5082A	1N5543B
5082A	1N5869B
5082A	1S238
5082A	13-33179-8
5082A	33R
5082A	48-355050
5082A	48-8225EC42
5082A	48-83461E11
5082A	61E11
5082A	86-117-1
5082A	93A39-14
5082A	523-2003-250
5082A	2092055-0024
5082A	134882256342
5082A	134883461511
5083A	.25T28
5083A	.25T28A
5083A	BZ-280
5083A	GEZD-28
5083A	HEP-Z0425
5083A	SV172
5083A	1N4712
5083A	1N5255B
5083A	1N5871B
5083A	1N769-4
5083A	1S487
5083A	46-86447-3
5083A	48-82256C36
5083A	134882256336
5084A	.25T30
5084A	.25T30A
5084A	.4T30A
5084A	.4T30B
5084A	.7JZ30
5084A	.7Z30A
5084A	.7Z30B
5084A	.7Z30C
5084A	.7Z30D
5084A	.7ZM30A
5084A	.7ZM30B
5084A	.7ZM30C
5084A	.7ZM30D
5084A	AW01-30
5084A	AZ972B
5084A	BZ-290
5084A	BZ-300
5084A	BZX29C30
5084A	BZX46C30
5084A	BZX79C30
5084A	BZX83C30
5084A	BZY85C30
5084A	EQA01-30
5084A	EQA01-Z4RA
5084A	EZ30(ELCOM)
5084A	GEZD-30
5084A	HD3000201-0
5084A	HW30
5084A	HW30A
5084A	RD30F
5084A	RD30FA
5084A	1M30Z
5084A	1M30Z10
5084A	1M30Z5
5084A	1M30ZS10
5084A	1M30ZS5
5084A	1N1782
5084A	1N1782A
5084A	1N2387
5084A	1N3031
5084A	1N3031A
5084A	1N3031B
5084A	1N3452
5084A	1N3690
5084A	1N3690A
5084A	1N3690B
5084A	1N4173
5084A	1N4173A
5084A	1N4173B
5084A	1N4338
5084A	1N4338A
5084A	1N4338B
5084A	1N4415
5084A	1N4415A
5084A	1N4643
5084A	1N4713
5084A	1N4751
5084A	1N4751A
5084A	1N5256B
5084A	1N5545B
5084A	1N5574
5084A	1N5574A
5084A	1N5747B
5084A	1N5872B
5084A	1N725A
5084A	1N972B
5084A	1R30
5084A	1R30A
5084A	1R30B
5084A	1S1971
5084A	1S2300A
5084A	1S3030
5084A	1Z30
5084A	1Z30A
5084A	1Z30T10
5084A	1Z30T20
5084A	1Z30T5
5084A	1ZC30T10
5084A	1ZC30T5
5084A	013-768
5084A	56-76
5084A	56-79
5084A	66X0040-004
5084A	523-2003-300
5084A	2327075
5084A	23115990
5084A	41029498
5085A	.25T36
5085A	.25T36A
5085A	.4T36
5085A	.4T36A
5085A	.4T36B
5085A	.7JZ36
5085A	.7Z36A
5085A	.7Z36B
5085A	.7Z36C
5085A	.7Z36D
5085A	.7ZM36A
5085A	.7ZM36B
5085A	.7ZM36C
5085A	.7ZM36D
5085A	AZ974B
5085A	B310(ZENER)
5085A	BZ-340
5085A	BZ-350
5085A	BZX29C36
5085A	BZX46C36
5085A	BZX55C36
5085A	BZX60C36
5085A	BZX79C36
5085A	BZX83C36
5085A	BZY92C36
5085A	EQA01-35
5085A	EZ36(ELCOM)
5085A	GEZD-36
5085A	HEP-Z0427
5085A	HW36
5085A	HW36A
5085A	RD36F
5085A	RD36FA
5085A	RD36FB
5085A	1M36Z
5085A	1M36Z10
5085A	1M36Z5
5085A	1M36ZS10
5085A	1M36ZS5
5085A	1N1784
5085A	1N3033
5085A	1N3033A
5085A	1N3033B
5085A	1N3692
5085A	1N3692A
5085A	1N4175
5085A	1N4175A
5085A	1N4340
5085A	1N4340A
5085A	1N4417
5085A	1N4417A
5085A	1N4645
5085A	1N4715
5085A	1N4753
5085A	1N4753A
5085A	1N5258B
5085A	1N5576
5085A	1N5576A
5085A	1N727A
5085A	1N974B
5085A	1R36
5085A	1R36A
5085A	1R36B
5085A	1R39
5085A	1S242
5085A	1S3036
5085A	1S488
5085A	1Z36
5085A	1Z36A
5085A	4-2020-13300
5085A	13-0017
5085A	86-119-1
5085A	96-5248-12
5085A	103-279-37
5085A	523-2003-360
5085A	133543
5085A	1477046-5
5085A	4202013300
5086A	.25T39
5086A	.25T39A
5086A	.4T39
5086A	.4T39A
5086A	.4T39B
5086A	.7JZ39
5086A	.7Z39A
5086A	.7Z39B
5086A	.7Z39C
5086A	.7Z39D
5086A	.7ZM39A
5086A	.7ZM39B
5086A	.7ZM39C
5086A	.7ZM39D
5086A	AZ975B
5086A	BZX29C39
5086A	BZX55C39
5086A	BZX79C39
5086A	BZX83C39
5086A	EZ39(ELCOM)
5086A	GEZD-39
5086A	HEP-Z0428
5086A	HW39
5086A	HW39A
5086A	RD39F
5086A	RD39FA
5086A	RD39FB
5086A	1M39Z
5086A	1M39Z10
5086A	1M39Z5
5086A	1M39ZS10
5086A	1M39ZS5
5086A	1N1785
5086A	1N1883
5086A	1N1883A
5086A	1N3034
5086A	1N3034A
5086A	1N3034B
5086A	1N3441
5086A	1N3454
5086A	1N3693
5086A	1N3693A
5086A	1N4176
5086A	1N4176A
5086A	1N4341
5086A	1N4341A
5086A	1N4418
5086A	1N4418A
5086A	1N4646
5086A	1N4716
5086A	1N4754
5086A	1N4754A
5086A	1N5259B
5086A	1N5577
5086A	1N5577A

ECG	Industry Standard No.
5086A	1N728
5086A	1N975B
5086A	1R39A
5086A	1R39B
5086A	1S243
5086A	1S3039
5086A	1Z39
5086A	1Z39A
5086A	03-933935
5086A	48-82256C58
5086A	56-55
5086A	56-66
5086A	93A39-35
5086A	103-270
5086A	523-2003-390
5086A	134882256358
5087A	.25T43
5087A	.25T43A
5087A	.4T43
5087A	.4T43A
5087A	.4T43B
5087A	.7JZ43
5087A	.7Z43A
5087A	.7Z43B
5087A	.7Z43C
5087A	.7Z43D
5087A	.7ZM43A
5087A	.7ZM43B
5087A	.7ZM43C
5087A	.7ZM43D
5087A	AZ976B
5087A	BZX29C43
5087A	BZX46C43
5087A	BZX79C43
5087A	BZX83C43
5087A	BZX85C43
5087A	GEZD-43
5087A	HW43
5087A	HW43A
5087A	1M43Z
5087A	1M43Z10
5087A	1M43Z5
5087A	1M43ZS10
5087A	1M43ZS5
5087A	1N1786
5087A	1N1786A
5087A	1N3035
5087A	1N3035A
5087A	1N3035B
5087A	1N3694
5087A	1N3694A
5087A	1N4177
5087A	1N4177A
5087A	1N4177B
5087A	1N4342
5087A	1N4342A
5087A	1N4342B
5087A	1N4419
5087A	1N4419A
5087A	1N4419B
5087A	1N4647
5087A	1N4717
5087A	1N4755
5087A	1N4755A
5087A	1N5260B
5087A	1N5578
5087A	1N5578A
5087A	1N729A
5087A	1N976B
5087A	1R43
5087A	1R43A
5087A	1R43B
5087A	1S244
5087A	1S3043
5087A	1Z43
5087A	1Z43A
5087A	56-72
5087A	523-2003-430
5088A	.25T47
5088A	.25T47A
5088A	.4T47
5088A	.4T47A
5088A	.4T47B
5088A	.7JZ45
5088A	.7JZ47
5088A	.7Z45A
5088A	.7Z45B
5088A	.7Z45C
5088A	.7Z45D
5088A	.7Z47A
5088A	.7Z47B
5088A	.7Z47C
5088A	.7Z47D
5088A	.7ZM47A
5088A	.7ZM47B
5088A	.7ZM47C
5088A	.7ZM47D
5088A	BZX29C47
5088A	BZX79C47
5088A	EZ47(ELCOM)
5088A	GEZD-47
5088A	HEP-Z0430
5088A	HW47
5088A	HW47A
5088A	1M47Z
5088A	1M47Z10
5088A	1M47Z5
5088A	1M47ZS10
5088A	1M47ZS5
5088A	1N1787
5088A	1N1884
5088A	1N1884A
5088A	1N3036
5088A	1N3036A
5088A	1N3036B
5088A	1N3442
5088A	1N3455
5088A	1N3695
5088A	1N3695A
5088A	1N4178
5088A	1N4178A
5088A	1N4343
5088A	1N4343A
5088A	1N4420
5088A	1N4420A
5088A	1N4648
5088A	1N4756
5088A	1N4756A
5088A	1N5261B
5088A	1N5579
5088A	1N5579A
5088A	1N730A
5088A	1N977B
5088A	1R47
5088A	1R47A
5088A	1R47B
5088A	1S246
5088A	1S3047
5088A	1S489
5088A	1Z47
5088A	1Z47A
5088A	145-470
5088A	523-2003-450
5088A	523-2003-470
5089A	.25T51
5089A	.25T51A
5089A	.4T51
5089A	.4T51A
5089A	.4T51B
5089A	.7JZ50
5089A	.7JZ51
5089A	.7Z50A
5089A	.7Z50B
5089A	.7Z50C
5089A	.7Z50D
5089A	.7Z51A
5089A	.7Z51B
5089A	.7Z51C
5089A	.7Z51D
5089A	.7Z52A
5089A	.7Z52B
5089A	.7Z52C
5089A	.7Z52D
5089A	.7ZM50A
5089A	.7ZM50B
5089A	.7ZM50C
5089A	.7ZM50D
5089A	.7ZM51A
5089A	.7ZM51B
5089A	.7ZM51C
5089A	.7ZM51D
5089A	.7ZM52A
5089A	.7ZM52B
5089A	.7ZM52C
5089A	.7ZM52D
5089A	BZX29C51
5089A	BZX79C51
5089A	GEZD-51
5089A	HW51
5089A	HW51A
5089A	RM-25
5089A	RM25
5089A	SK-1W50
5089A	TVSRM25
5089A	TVSRM26V
5089A	1M51Z
5089A	1M51Z10
5089A	1M51Z5
5089A	1M51ZS10
5089A	1M51ZS5
5089A	1N1788
5089A	1N1788A
5089A	1N3037
5089A	1N3037A
5089A	1N3037B
5089A	1N3696
5089A	1N3696A
5089A	1N3696B
5089A	1N4179A
5089A	1N4179B
5089A	1N4321
5089A	1N4344
5089A	1N4344A
5089A	1N4344B
5089A	1N4421
5089A	1N4421A
5089A	1N4757
5089A	1N4757A
5089A	1N5262B
5089A	1N5580
5089A	1N5580A
5089A	1N5580B
5089A	1N731A
5089A	1N978B
5089A	1R51
5089A	1R51A
5089A	1R51B
5089A	1S247
5089A	1S248
5089A	1S3051
5089A	1T263
5089A	1T264
5089A	1Z51
5089A	1Z51A
5089A	523-2003-500
5089A	523-2003-510
5089A	523-2003-520
509	GESS-3AT2
5090A	.25T56
5090A	.25T56A
5090A	.7Z56B
5090A	.7ZM56A
5090A	.7ZM56B
5090A	.7ZM56C
5090A	.7ZM56D
5090A	BZX29C56
5090A	BZX79C56
5090A	EZ56(ELCOM)
5090A	GEZD-56
5090A	HEP-Z0432
5090A	HW56
5090A	HW56A
5090A	LPM56A
5090A	PTC513
5090A	RH-EX0057CEZZ
5090A	RM257
5090A	SK-1W55
5090A	SK3342
5090A	ZD56B
5090A	1M56Z
5090A	1M56Z10
5090A	1M56Z5
5090A	1M56ZS10
5090A	1M56ZS5
5090A	1N1789
5090A	1N1789A
5090A	1N1885
5090A	1N1885A
5090A	1N3038
5090A	1N3038A
5090A	1N3038B
5090A	1N3456
5090A	1N3697
5090A	1N3697A
5090A	1N4180
5090A	1N4180A
5090A	1N4180B
5090A	1N4345
5090A	1N4345A
5090A	1N4345B
5090A	1N4422
5090A	1N4422A
5090A	1N4422B
5090A	1N4758
5090A	1N4758A
5090A	1N5263B
5090A	1N5581
5090A	1N5581A
5090A	1N732A
5090A	1N979B
5090A	1R56
5090A	1R56A
5090A	1R56B
5090A	1S249
5090A	1S3056
5090A	1Z56
5090A	1Z56A
5090A	48-82256C29
5090A	48-83461E09
5090A	61E09
5090A	523-2003-560
5090A	134882256329
5090A	134883461509
5091A	1N5264B
5092A	.25T68
5092A	.25T68A
5092A	.7JZ68
5092A	.7Z68A
5092A	.7Z68B
5092A	.7Z68C
5092A	.7Z68D
5092A	.7ZM68A
5092A	.7ZM68B
5092A	.7ZM68C
5092A	.7ZM68D
5092A	BZX79C68
5092A	EZ68(ELCOM)
5092A	GEZD-68
5092A	HW68
5092A	HW68A
5092A	RM-26
5092A	1M68Z
5092A	1M68Z10
5092A	1M68Z5
5092A	1M68ZS10
5092A	1M68ZS5
5092A	1N1431
5092A	1N1791
5092A	1N1791A
5092A	1N1886
5092A	1N1886A
5092A	1N3040
5092A	1N3040A
5092A	1N3040B
5092A	1N3457
5092A	1N3699
5092A	1N3699A
5092A	1N4182
5092A	1N4182A
5092A	1N4182B
5092A	1N4347
5092A	1N4347A
5092A	1N4347B
5092A	1N4424
5092A	1N4424A
5092A	1N4424B
5092A	1N4760
5092A	1N4760A
5092A	1N5266B
5092A	1N5583
5092A	1N5583A
5092A	1N670
5092A	1N734A
5092A	1N981B
5092A	1R68
5092A	1R68A
5092A	1R68B
5092A	1S251
5092A	1S3068
5092A	1Z68
5092A	1Z68A
5092A	13-0106
5092A	48-155158
5092A	56-68
5092A	523-2003-680
5093A	.25T75
5093A	.25T75A
5093A	.7JZ75
5093A	.7Z75A
5093A	.7Z75B
5093A	.7Z75C
5093A	.7Z75D
5093A	.7ZM75A
5093A	.7ZM75B
5093A	.7ZM75C
5093A	.7ZM75D
5093A	A7272130
5093A	GEZD-75
5093A	HW75
5093A	HW75A
5093A	1M75Z
5093A	1M75Z10
5093A	1M75Z3
5093A	1M75Z5
5093A	1M75ZS10
5093A	1M75ZS5
5093A	1N1792
5093A	1N1792A
5093A	1N3041
5093A	1N3041A
5093A	1N3041B
5093A	1N3700
5093A	1N3700A
5093A	1N4183
5093A	1N4183A
5093A	1N4183B
5093A	1N4348
5093A	1N4348A
5093A	1N4348B
5093A	1N4425
5093A	1N4425A
5093A	1N4425B
5093A	1N4761
5093A	1N4761A
5093A	1N5267B
5093A	1N5584
5093A	1N5584A
5093A	1N5584B
5093A	1N735A
5093A	1R75
5093A	1R75A
5093A	1R75B
5093A	1S252
5093A	1S3075
5093A	1Z72
5093A	1Z75
5093A	1Z75A
5093A	48-83461E04
5093A	61E04
5093A	070-025
5093A	523-2003-750
5093A	134883461504
5094A	.25T87
5094A	.25T87A
5094A	GEZD-87
5094A	1N1325A
5094A	1N5269B
5094A	1N5885B
5094A	530157-870
5095A	.25T91
5095A	.25T91A
5095A	.7JZ91
5095A	.7Z91A
5095A	.7Z91B
5095A	.7Z91C
5095A	.7Z91D
5095A	.7ZM91A
5095A	.7ZM91B
5095A	.7ZM91C
5095A	.7ZM91D
5095A	GEZD-91
5095A	HW91
5095A	HW91A
5095A	1M91Z
5095A	1M91Z10
5095A	1M91Z3
5095A	1M91Z5
5095A	1M91ZS10
5095A	1M91ZS5
5095A	1N1794
5095A	1N1794A
5095A	1N3043
5095A	1N3043A
5095A	1N3043B
5095A	1N3702
5095A	1N3702A
5095A	1N4185
5095A	1N4185A
5095A	1N4185B
5095A	1N4350
5095A	1N4350A
5095A	1N4350B
5095A	1N4427
5095A	1N4427A
5095A	1N4763
5095A	1N4763A
5095A	1N5270B
5095A	1N5586
5095A	1N5586A
5095A	1N5586B
5095A	1N737A
5095A	1R91A
5095A	1R91B
5095A	1S254
5095A	1S3091
5095A	1Z91
5095A	1Z91A
5095A	48-83461E16
5095A	61E16
5095A	523-2003-910
5095A	134883461516
5096A	.25T100
5096A	.25T100A
5096A	.4T100
5096A	.4T100A
5096A	.4T100B
5096A	.4T110B
5096A	.7JZ100
5096A	.7Z100A
5096A	.7Z100B
5096A	.7Z100C
5096A	.7Z100D
5096A	.7ZM100A
5096A	.7ZM100B
5096A	.7ZM100C
5096A	.7ZM100D
5096A	GEZD-100
5096A	HEP-Z0438
5096A	HW100
5096A	HW100A
5096A	RE133
5096A	1M100Z
5096A	1M100Z10
5096A	1M100Z5
5096A	1M100ZS10
5096A	1M100ZS5
5096A	1N1432
5096A	1N1795
5096A	1N1795A
5096A	1N1888
5096A	1N1888A
5096A	1N3044
5096A	1N3044A
5096A	1N3044B
5096A	1N3459
5096A	1N3703
5096A	1N3703A
5096A	1N4186
5096A	1N4186A
5096A	1N4186B
5096A	1N4351
5096A	1N4351A
5096A	1N4351B
5096A	1N4428
5096A	1N4428A
5096A	1N4428B
5096A	1N4764
5096A	1N4764A
5096A	1N5271B
5096A	1N5587
5096A	1N5587A
5096A	1N5587B
5096A	1N671
5096A	1N738A
5096A	1R100
5096A	1R100A
5096A	1R100B
5096A	1S255
5096A	1S3100
5096A	1Z100
5096A	1Z100A
5096A	2D7A
5096A	4-2020-14100
5096A	21A037-016
5096A	46-86467-3
5096A	48-82256C30
5096A	48-83461E06
5096A	48-83461E14
5096A	61E06
5096A	61E14
5097A	.25T120
5097A	.25T120A
5097A	.4T120
5097A	.4T120A
5097A	.4T120B
5097A	.7JZ120
5097A	.7Z120A
5097A	.7Z120B
5097A	.7Z120C
5097A	.7Z120D
5097A	.7ZM120A
5097A	.7ZM120B
5097A	.7ZM120C
5097A	.7ZM120D
5097A	GEZD-120
5097A	HEP-Z0440
5097A	HW120
5097A	HW120A
5097A	RH-EX0054CEZZ
5097A	1M120Z
5097A	1M120Z10
5097A	1M120Z3
5097A	1M120Z5
5097A	1M120ZS10
5097A	1M120ZS5
5097A	1M20ZS5
5097A	1N1797
5097A	1N1889
5097A	1N3046
5097A	1N3046A
5097A	1N3046B
5097A	1N3098
5097A	1N3098A
5097A	1N3460
5097A	1N3705
5097A	1N3705A
5097A	1N4188
5097A	1N4188A
5097A	1N4353
5097A	1N4353A
5097A	1N4430
5097A	1N4430A
5097A	1N4858A
5097A	1N5273B
5097A	1N5589
5097A	1N5589A
5097A	1N987B
5097A	1R120
5097A	1R120A
5097A	1R120B
5097A	1S258
5097A	1S3120
5097A	229-1301-29
5097A	99201-242
5097A	135406
5098A	.25T130
5098A	.25T130A
5098A	.4T130
5098A	.4T130A
5098A	.4T130B
5098A	.7JZ130
5098A	.7Z130A
5098A	.7Z130B
5098A	.7Z130C
5098A	.7Z130D
5098A	.7ZM130A
5098A	.7ZM130B
5098A	.7ZM130C
5098A	.7ZM130D
5098A	GEZD-130
5098A	HEP-Z0254
5098A	1EZ130D5
5098A	1M130Z
5098A	1M130Z10
5098A	1M130Z3
5098A	1M130Z5
5098A	1M130ZS10
5098A	1M130ZS5
5098A	1N1798
5098A	1N1798A
5098A	1N3047
5098A	1N3047A
5098A	1N3047B
5098A	1N3706
5098A	1N3706A
5098A	1N4189
5098A	1N4189A
5098A	1N4189B
5098A	1N4354
5098A	1N4354A
5098A	1N4354B
5098A	1N4431
5098A	1N4431A
5098A	1N4431B
5098A	1N4859
5098A	1N4859A
5098A	1N5274B
5098A	1N5590
5098A	1N5590A
5098A	1N5590B
5098A	1N5890
5098A	1N5890A
5098A	1N741A
5098A	1R130
5098A	1R130A
5098A	1R130B
5098A	1S259
5098A	103-208
5099A	.25T140
5099A	.7JZ140
5099A	.7Z140A
5099A	.7Z140B
5099A	.7Z140C
5099A	.7Z140D
5099A	.7ZM140A
5099A	.7ZM140B
5099A	.7ZM140C
5099A	.7ZM140D
5099A	GEZD-140
5099A	HEP-Z0255
5099A	1EZ140D5
5099A	1M140Z10
5099A	1M140Z5
5099A	1N5275B
5099A	1N5891
5099A	1N5891A
5099A	1N5891B
5099A	1R140
5099A	1R140A

ECG	Industry Standard No.
5099A	1R140B
5099A	1S260
5099A	72165-5
510	GESS-3DB3
5100A	.25T150
5100A	.25T150A
5100A	.4T150
5100A	.4T150A
5100A	.4T150B
5100A	.7JZ150
5100A	.7Z150A
5100A	.7Z150B
5100A	.7Z150C
5100A	.7Z150D
5100A	.7ZM150A
5100A	.7ZM150B
5100A	.7ZM150C
5100A	.7ZM150D
5100A	GEZD-150
5100A	HEP-Z0442
5100A	HW150
5100A	HW150A
5100A	1M150Z
5100A	1M150Z10
5100A	1M150Z3
5100A	1M150Z5
5100A	1M150ZS10
5100A	1M150ZS5
5100A	1N1433
5100A	1N1799
5100A	1N1890
5100A	1N3048
5100A	1N3048A
5100A	1N3048B
5100A	1N3099
5100A	1N3461
5100A	1N3707
5100A	1N3707A
5100A	1N4190
5100A	1N4190A
5100A	1N4355
5100A	1N4355A
5100A	1N4432
5100A	1N4432A
5100A	1N4860
5100A	1N4860A
5100A	1N5591
5100A	1N5591A
5100A	1N672
5100A	1N742A
5100A	1N989B
5100A	1R150
5100A	1R150A
5100A	1R150B
5100A	1S261
5101A	.25T160
5101A	.25T160A
5101A	.4T160
5101A	.4T160A
5101A	.4T160B
5101A	.7JZ160
5101A	.7Z160A
5101A	.7Z160B
5101A	.7Z160C
5101A	.7Z160D
5101A	.7ZM160A
5101A	.7ZM160B
5101A	.7ZM160C
5101A	.7ZM160D
5101A	GEZD-160
5101A	HW160
5101A	HW160A
5101A	1EZ160D5
5101A	1M160Z
5101A	1M160Z10
5101A	1M160Z3
5101A	1M160Z5
5101A	1M160ZS10
5101A	1M160ZS5
5101A	1N1800
5101A	1N1800A
5101A	1N3049
5101A	1N3049A
5101A	1N3049B
5101A	1N3708
5101A	1N3708A
5101A	1N4191
5101A	1N4191A
5101A	1N4356
5101A	1N4356A
5101A	1N4356B
5101A	1N4433
5101A	1N4433A
5101A	1N5277B
5101A	1N5592
5101A	1N5592A
5101A	1N5592B
5101A	1N743A
5101A	1N990B
5101A	1R160
5101A	1R160A
5101A	1R160B
5101A	72165-6
5102A	.25T175
5102A	GEZD-170
5102A	1N5278B
5102A	1N5894B
5103A	.25T180
5103A	.25T180A
5103A	.4T180
5103A	.4T180A
5103A	.4T180B
5103A	.7JZ180
5103A	.7Z180A
5103A	.7Z180B
5103A	.7Z180C
5103A	.7Z180D
5103A	.7ZM180A
5103A	.7ZM180B
5103A	.7ZM180C
5103A	.7ZM180D
5103A	GEZD-180
5103A	HEP-Z0444
5103A	1EZ180D5
5103A	1M180Z
5103A	1M180Z10
5103A	1M180Z3
5103A	1M180Z5
5103A	1M180ZS10
5103A	1M180ZS5
5103A	1N1801
5103A	1N1801A
5103A	1N2001A
5103A	1N3050
5103A	1N3050A
5103A	1N3050B
5103A	1N3100
5103A	1N3100A
5103A	1N3462
5103A	1N3709
5103A	1N3709A
5103A	1N4192
5103A	1N4192A
5103A	1N4192B
5103A	1N4357
5103A	1N4357A
5103A	1N4357B
5103A	1N4434
5103A	1N4434A
5103A	1N4434B
5103A	1N5279B
5103A	1N5593
5103A	1N5593A
5103A	1N5593B
5103A	1N744A
5103A	1N991B
5103A	1R180
5103A	1R180A
5103A	1R180B
5103A	1S3180
5103A	62-21573
5104A	.7JZ200
5104A	GEZD-190
5104A	1EZ190D5
5104A	1N5280B
5104A	1N5896B
5105A	.25T200
5105A	.25T200A
5105A	.4T200
5105A	.4T200A
5105A	.4T200B
5105A	.7Z200A
5105A	.7Z200B
5105A	.7Z200C
5105A	.7Z200D
5105A	.7ZM200A
5105A	.7ZM200B
5105A	.7ZM200C
5105A	.7ZM200D
5105A	GEZD-200
5105A	HEP-Z0445
5105A	HW200
5105A	HW200A
5105A	UZ8120
5105A	UZ8220
5105A	1/4M200Z5
5105A	1EZ200D5
5105A	1M200Z
5105A	1M200Z10
5105A	1M200Z3
5105A	1M200Z5
5105A	1M200ZS10
5105A	1M200ZS5
5105A	1N1802
5105A	1N3051
5105A	1N3051A
5105A	1N3051B
5105A	1N3710
5105A	1N3710A
5105A	1N4193
5105A	1N4193A
5105A	1N4358
5105A	1N4358A
5105A	1N4435
5105A	1N4435A
5105A	1N5281B
5105A	1N5594
5105A	1N5594A
5105A	1N5594B
5105A	1N5897B
5105A	1N6031B
5105A	1N745A
5105A	1N992B
5105A	1R200
5105A	1R200A
5105A	1R200B
5105A	63-5058
511	GESS-2AV2
5111A	GE5ZD-3.3
5111A	1N5008
5111A	1N5008A
5111A	1N5333
5112A	GE5ZD-3.6
5112A	1N5009
5112A	1N5009A
5112A	1N5334
5112A	1N5334A
5113A	GE5ZD-3.9
5113A	1N5010
5113A	1N5010A
5113A	1N5335
5113A	1N5335A
5114A	GE5ZD-4.3
5114A	1N5011
5114A	1N5336
5114A	1N5336A
5115A	GE5ZD-4.7
5115A	1N5012
5115A	1N5012A
5115A	1N5337
5115A	1N5337A
5116A	GE5ZD-5.1
5116A	PS8906
5116A	1D5.1B
5116A	1N5013
5116A	1N5338
5116A	1N5338A
5116A	1ZM5.1T5
5116A	23115921
5117A	GE5ZD-5.6
5117A	1N5014
5117A	1N5339
5117A	1N5339A
5118A	GE5ZD-6.0
5118A	1N5340
5118A	1N5340A
5118A	2VR6.2
5119A	GE5ZD-6.2
5119A	1N4460
5119A	1N5015
5119A	1N5341
5119A	1N5341A
512	GESS-6DW4
5120A	BZX55D6V8
5120A	EZ3-7(ELCOM)
5120A	GE5ZD-6.8
5120A	UZ5706
5120A	UZ5806
5120A	UZ706
5120A	UZ806
5120A	1.5DKZ6.8
5120A	1.5JZ6.8
5120A	1.5Z6.8
5120A	1N3785
5120A	1N3785A
5120A	1N4461
5120A	1N4954
5120A	1N5016
5120A	1N5063
5120A	1N5342
5120A	1N5342A
5120A	3TZ6.8B
5120A	3VR6.8B
5120A	59Z6A
5120A	96-5249-02
5121A	GE5ZD-7.5
5121A	HEP-Z2510
5121A	SZ75A
5121A	001-023042
5121A	1.5DKZ7.5A
5121A	1.5JZ7.5A
5121A	1N3786
5121A	1N3786A
5121A	1N4462
5121A	1N4955
5121A	1N5017
5121A	1N5064
5121A	1N5343
5121A	1N5343A
5121A	3TZ7.5D
5121A	3VR7.5B
5121A	5ZS7.5A
5121A	5ZS7.5B
5121A	103-194
5121A	7717
5122A	BZX55D8V2
5122A	GE5ZD-8.2
5122A	1.5DKZ8.2
5122A	1.5JZ8.2
5122A	1.5Z8.2A
5122A	1N3787
5122A	1N3787A
5122A	1N4463
5122A	1N4956
5122A	1N5018
5122A	1N5065
5122A	1N5344
5122A	1N5344A
5122A	1S4008A
5122A	2VR8.2B
5122A	3TZ8.2B
5122A	1308
5122A	7718
5123A	1N5345
5123A	1N5345A
5123A	2VR8.5B
5123A	3VR8.5B
5123A	5ZS8.7A
5124A	GE5ZD-9.1
5124A	HEP-Z2513
5124A	SZ91A
5124A	ZD9.1
5124A	1.5DKZ9.1A
5124A	1.5JZ9.1A
5124A	1N3788
5124A	1N3788A
5124A	1N4464
5124A	1N4831
5124A	1N4831A
5124A	1N4957
5124A	1N5019
5124A	1N5019A
5124A	1N5066
5124A	1N5346
5124A	1N5346A
5124A	3TZ9.1D
5124A	3VR9.1B
5124A	5ZS9.1A
5124A	62Z6A
5124A	7719
5125A	BZX55D10
5125A	GE5ZD-10
5125A	HEP-Z2514
5125A	1.5DKZ10A
5125A	1N3789
5125A	1N3789A
5125A	1N4465
5125A	1N4832A
5125A	1N4958
5125A	1N5020
5125A	1N5067
5125A	1N5347
5125A	1N5347A
5125A	3TZ10D
5125A	3VR10B
5125A	5ZS10A
5125A	96-5344-01
5125A	523-2004-100
5125A	7720
5126A	BZX70C11
5126A	GE5ZD-11
5126A	1.5DKZ11A
5126A	1.5JZ11A
5126A	1N3790
5126A	1N3790A
5126A	1N4466
5126A	1N4833
5126A	1N4833A
5126A	1N4959
5126A	1N5021
5126A	1N5068
5126A	1N5348
5126A	1N5348A
5126A	1N5348B
5126A	1S4011A
5126A	2VR11A
5126A	3VR11B
5126A	5ZS11B
5126A	1311
5127A	BZX70C12
5127A	EZ3-12(ELCOM)
5127A	GE5ZD-12
5127A	HEP-Z2516
5127A	SZ12C
5127A	ZD12
5127A	ZY12
5127A	1.5DKZ12A
5127A	1.5JZ12A
5127A	1N3791
5127A	1N3791A
5127A	1N4467
5127A	1N4834
5127A	1N4834A
5127A	1N4883
5127A	1N4960
5127A	1N5022
5127A	1N5349
5127A	1N5349A
5127A	1S4012A
5127A	2VR12A
5127A	3TZ12A
5127A	5ZS12B
5127A	919-00-3309-2
5127A	1312
5127A	530073-32
5128A	BZY95C13
5128A	GE5ZD-13
5128A	1.5DKZ13A
5128A	1.5JZ13A
5128A	1N3792
5128A	1N3792A
5128A	1N4468
5128A	1N4835
5128A	1N4835A
5128A	1N4961
5128A	1N5023
5128A	1N5069
5128A	1N5350
5128A	1N5350A
5128A	1S4013A
5128A	2VR13A
5128A	3R13A
5128A	3TZ13D
5128A	3VR13B
5128A	5ZS13A
5128A	66Z6A
5128A	103-256
5129A	EZ3-14(ELCOM)
5129A	GE5ZD-14
5129A	1.5JZ14A
5129A	1.5M14Z
5129A	1.5Z14A
5129A	1N5024
5129A	1N5070
5129A	1N5118
5129A	1N5351
5129A	1N5351A
5129A	2VR14A
5129A	5ZS14A
5129A	48-83461E21
5129A	61E21
5129A	67Z6A
5129A	134883461521
513	D6C
513	EP57X3
513	GE-513
513	H445-2
513	H484
513	R-12C
513	REN513
513	TV-215
513	28-31-00
513	48-137397
513	48-69723A01
513	48-69723A02
513	48-69723AD2-185
513	48-69723B02
513	48D69723A01
513	48D69723A02-185
513	48S137397
513	86-114-3
513	103-215
513	1371
513	23115321
513	23115334
5130A	BZY95C15
5130A	EZ3-15(ELCOM)
5130A	GE5ZD-15
5130A	HEP-Z2519
5130A	MZ5915
5130A	SZ15C
5130A	ZD15
5130A	ZM15B
5130A	ZY15
5130A	001-023041
5130A	1.5DKZ15A
5130A	1.5JZ15A
5130A	1N3793
5130A	1N3793A
5130A	1N4469
5130A	1N4836
5130A	1N4836A
5130A	1N4962
5130A	1N5025
5130A	1N5071
5130A	1N5352
5130A	1N5352A
5130A	1S4015
5130A	7724
5131A	BZY95C16
5131A	EZ3-16(ELCOM)
5131A	GE5ZD-16
5131A	1.5DKZ16A
5131A	1.5JZ16A
5131A	1N3794
5131A	1N3794A
5131A	1N4470
5131A	1N4837
5131A	1N4837A
5131A	1N4963
5131A	1N5026
5131A	1N5072
5131A	1N5353
5131A	1N5353A
5131A	1S4016A
5131A	3R16A
5131A	3VR16B
5131A	5ZS16
5131A	5ZS16A
5131A	5ZS16B
5131A	7725
5132A	GE5ZD-17
5132A	1.5JZ17A
5132A	1.5Z17D
5132A	1N5027
5132A	1N5354
5132A	1N5354A
5132A	3TZ17
5132A	3TZ17A
5132A	3TZ17B
5132A	3TZ17C
5132A	3TZ17D
5132A	5ZS17
5132A	5ZS17A
5132A	5ZS17B
5133A	BZX70C18
5133A	GE5ZD-18
5133A	HEP-Z2522
5133A	MZ5918
5133A	SZ18C
5133A	1.5DKZ18A
5133A	1.5JZ18A
5133A	1N3795
5133A	1N3795A
5133A	1N4471
5133A	1N4838
5133A	1N4838A
5133A	1N4964
5133A	1N5028
5133A	1N5073
5133A	1N5355
5133A	1N5355A
5133A	1S4018
5133A	7726
5133A	530073-1030
5134A	GE5ZD-19
5134A	1.5JZ19A
5134A	1.5M19Z
5134A	1N5029
5134A	1N5356
5134A	1N5356A
5134A	3TZ19
5134A	5ZS19
5134A	5ZS19A
5134A	5ZS19B
5135A	BZY95C20
5135A	GE5ZD-20
5135A	1.5DKZ20
5135A	1.5JZ20
5135A	1.5Z20D
5135A	1N3796
5135A	1N3796A
5135A	1N3950
5135A	1N4472
5135A	1N4839
5135A	1N4839A
5135A	1N4881
5135A	1N4884
5135A	1N4965
5135A	1N5030
5135A	1N5357
5135A	1N5357A
5135A	919-017406-010
5135A	530073-1021
5136A	BZX70C22
5136A	EP16X2
5136A	GE5ZD-22
5136A	HEP-Z2525
5136A	MZ5922
5136A	SZ22C
5136A	1.5DKZ22A
5136A	1.5JZ22A
5136A	1.5Z22D
5136A	1N3797
5136A	1N3797A
5136A	1N4473
5136A	1N4840
5136A	1N4966
5136A	1N5031
5136A	1N5074
5136A	1N5358
5136A	1N5358A
5136A	1S4022
5136A	7728
5136A	530166-1006
5137A	BZX70C24
5137A	GE5ZD-24
5137A	HEP-Z2526
5137A	1.5DKZ24A
5137A	1.5JZ24A
5137A	1.5Z24D
5137A	1N3798
5137A	1N3798A
5137A	1N4474
5137A	1N4841
5137A	1N4841A
5137A	1N4967
5137A	1N5032
5137A	1N5075
5137A	1N5359
5137A	1N5359A
5137A	1S4024A
5137A	48-83461E35
5137A	61E35
5137A	86-95-1
5137A	103-144
5137A	103-144-01
5137A	7729
5137A	530073-1029
5137A	134883461535
5138A	GE5ZD-25
5138A	1.5JZ25
5138A	1.5M25Z
5138A	1N3951
5138A	1N5033
5138A	1N5360
5138A	1N5360A
5138A	1N5360B
5138A	5ZS25
5138A	5ZS25A
5138A	5ZS25B
5138A	5329
5139A	BZX70C27
5139A	GE5ZD-27
5139A	HEP-Z2528
5139A	MZ5927
5139A	SZ27C
5139A	1.5DKZ27A
5139A	1.5JZ27A
5139A	1.5Z27A
5139A	1N3799
5139A	1N3799A
5139A	1N4475
5139A	1N4842
5139A	1N4842A
5139A	1N4842B
5139A	1N4968
5139A	1N5034
5139A	1N5076
5139A	1N5361
5139A	1N5361A
5139A	1S4027
5139A	96-5250-02
5139A	7730
5140A	GE5ZD-28
5140A	1N5362
5140A	1N5362A
5140A	2VR28
5140A	2VR28A
5140A	2VR28B

ECG	Industry Standard No.
5140A	5ZS28
5140A	5ZS28A
5140A	5ZS28B
5141A	BZX70C30
5141A	EZ3-30(ELCOM)
5141A	EZ5-30(ELCOM)
5141A	GE5ZD-30
5141A	HEP-Z2530
5141A	1.5DKZ30
5141A	1N3800
5141A	1N3800A
5141A	1N4476
5141A	1N4843
5141A	1N4843A
5141A	1N4969
5141A	1N5035
5141A	1N5077
5141A	1N5363
5141A	1N5363A
5141A	1S4030A
5141A	61E28
5141A	7731
5142A	BZX70C33
5142A	GE5ZD-33
5142A	HEP-Z2531
5142A	MZ5933
5142A	SX33
5142A	1.5DKZ33B
5142A	1.5JZ33A
5142A	1N3801
5142A	1N3801A
5142A	1N4477
5142A	1N4503
5142A	1N4844
5142A	1N4844A
5142A	1N4844B
5142A	1N4970
5142A	1N5036
5142A	1N5078
5142A	1N5364
5142A	1N5364A
5142A	1S4033A
5142A	5332
5142A	7732
5143A	MZ1000-26
5143A	1N4478
5143A	1N4845
5143A	1N4845A
5143A	1N4882
5143A	1N4971
5143A	1N5037
5143A	1N5079
5144A	1N4479
5144A	1N4846
5144A	1N4846A
5144A	1N4972
5144A	1N5038
5144A	1N5080
5144A	1N5081
5145A	1N4480
5145A	1N4847
5145A	1N4847A
5145A	1N4973
5145A	1N5039
5145A	1N5082
5146A	1N4481
5146A	1N4848
5146A	1N4848A
5146A	1N4974
5146A	1N5040
5146A	1N5040A
5146A	1N5041
5146A	1N5083
5146A	1N5084
5147A	1N4482
5147A	1N4849
5147A	1N4849A
5147A	1N4975
5147A	1N5042
5147A	1N5043
5147A	1N5044
5147A	1N5044A
5147A	1N5085
5147A	1N5086
5148A	1N4483
5148A	1N4850
5148A	1N4850A
5148A	1N4976
5148A	1N5045
5148A	1N5087
5149A	1N5088
5149A	1N5122
5149A	5ZS60B
515	D2201A
515	D2201B
515	D2201D
515	D2201F
515	D2201M
515	D2201N
515	D2601M
515	MR2272
515	SK3043
515	13-43956-1
515	34-8054-17
515	66X0054-001
515	103-312
515	40642
515	40643
515	40644
515	44936
515	126857
515	126858
515	126859
515	135386
515	1476171-12
515	1476171-19
515	1476171-20
515	1476171-24
515	1476171-27
515	1476171-35
515	06200037
5150A	1N4484
5150A	1N4851
5150A	1N4851A
5150A	1N4889
5150A	1N4977
5150A	1N5046
5150A	1N5089
5151A	1N4485
5151A	1N4852
5151A	1N4852A
5151A	1N4978
5151A	1N5047
5151A	1N5090
5152A	1N4486
5152A	1N4853
5152A	1N4853A
5152A	1N4979
5152A	1N5048
5152A	1N5092
5153A	1N4487
5153A	1N4854
5153A	1N4854A
5153A	1N4854B
5153A	1N4980
5153A	1N5049
5153A	1N5093
5153A	1N5094
5155A	1N4096
5155A	1N4488
5155A	1N4855
5155A	1N4855A
5155A	1N4981
5155A	1N5050
5155A	1N5095
5156A	MZ310
5156A	1N4097
5156A	1N4489
5156A	1N4856
5156A	1N4856A
5156A	1N4982
5156A	1N5051
5157A	1N4490
5157A	1N4857
5157A	1N4857A
5157A	1N4857B
5157A	1N4983
5157A	1N5096
5158A	SX120
5158A	1N4491
5158A	1N4858
5158A	1N4984
5158A	1N5097
5159A	1N4492
5159A	1N4985
5159A	1N5098
5160A	1N5099
5160A	1N5126
5161A	1N4098
5161A	1N4493
5161A	1N4986
5162A	1N4494
5162A	1N4987
5162A	1N5100
5162A	3R160B
5162A	103-206
5163A	1N5101
5163A	1N5127
5164A	1N4495
5164A	1N4988
5164A	1N5102
5165A	1N5103
5165A	1N5128
5166A	UZ120
5166A	UZ220
5166A	UZ5120
5166A	UZ5220
5166A	1N4496
5166A	1N4504
5166A	1N4989
5166A	1N5104
5166A	2VR200
5176AK	1N2041
5177A	GZ5.1
5177AK	1N2041-2
5183A	61Z6A
5185A	GZ9.1
5185A	5524
5186A	1N4198
5188A	0AZ230
5188A	ZX12
5188A	919-011340
5189AK	1N2046
519	A-113367(DIODE)
519	A-11790169
519	A-1634 1125
519	A-1901-0050-1
519	A-1901-0196-1
519	A-1901-044-1
519	A-1901-0461-1
519	A-2008-9140
519	A-36617
519	A11159761
519	A488-A0060
519	A5010005
519	A65-P11324-0001
519	A72-49-600
519	A7249601
519	A9218
519	A9228-3
519	B269-3345
519	BA202
519	BA317
519	BAC SHIMI
519	BACSHIMI
519	BACSIM1
519	BAW62
519	BAX-13
519	BAX13
519	BAX91C/TF102
519	BAX95TF600
519	BPS-S-50
519	BRN-SPEC-24-12
519	C23018
519	C256125-011
519	CD-5038
519	CD5002
519	CD6016
519	CD6016-013-689
519	CD6161P1NO13-75
519	CG23018
519	D-215-1
519	D919039-2
519	DASZ158800
519	DDAY047001
519	DDAY047005
519	DDAY048007
519	DDAY048012
519	DDAY048013
519	DDAY048014
519	DDAY069001
519	DS-113
519	DS1005-1X8628
519	DS448
519	DSI-104-2
519	DX-0273
519	DX-0299
519	DX7429
519	EA16X122
519	EA16X135
519	EA16X152
519	EDH6023
519	EMS 72272
519	EMS72258
519	FD-1029-GP
519	FD01880
519	FD06193
519	FD100
519	FD111
519	FD600
519	FD777
519	FDH444
519	FDH999
519	FR-1U
519	G657061
519	G657123
519	GD101
519	GD102
519	GE-514
519	GVL20226
519	GVL20265
519	GVL20327-1
519	HD20001210
519	HF-20063
519	HF-20124
519	HP5082-2800
519	IC743049
519	IP20-0021
519	IP20-0061
519	IP20-0216
519	IP20-0282
519	IR10E6X
519	ITT200
519	ITT73N
519	IX8055-379005N
519	M8489
519	M8489A
519	M8513-0
519	MA150
519	MA150TA
519	MA161
519	MA162
519	MER-65-L11324
519	MP4300158
519	ON120623
519	ON143285
519	ON206068
519	PO4-41-0025-001
519	PO4-42-0011
519	PO4410025-003
519	PO4410042-001
519	PO4410042-002
519	PS30802
519	PS63A205
519	PT4-2268-011
519	PT4-2268-01B
519	PT4-2311-011
519	PT40063
519	Q-1N914
519	Q-23115C
519	R3410-P1
519	RE94
519	RH-DX0048CEZZ
519	RI4010658
519	RKZ12003
519	RKZ120101
519	S-2064-G
519	S074-005-0001
519	SC-1016
519	SE5-0247-C
519	SE5-0456
519	SK3100
519	SM-C-706156
519	SN4448
519	ST/123/CR
519	ST/146/CR
519	ST22546-1
519	ST32012-0037
519	T-10010
519	TA198785-2
519	TA8532787
519	WG851
519	X-19031-A
519	Y56001-21
519	ZA29312
519	1-0002-0001
519	001-0112-00
519	001-026060
519	1-13989
519	1A35
519	1A69425-1
519	1A99812-1001
519	1B2992
519	1N3062
519	1N3600
519	1N4148
519	1N4149
519	1N4153
519	1N4154
519	1N4305
519	1N4446
519	1N4447
519	1N4448
519	1N4449
519	1N4454
519	1N5282
519	1N916
519	1S-2472
519	1S1553
519	1S1588
519	1S1588V
519	1S2075K
519	1S2076-27
519	1S2076-TF1
519	1S2076A
519	1S2076A-07
519	1S2134
519	1S2471
519	1S2472
519	1S2473
519	1S2473-T72
519	1S2473H
519	1S2473HC
519	1S2473K
519	1S2473VE
519	1S954
519	1SS55
519	03-034-042
519	3L4-3002-01
519	04-000653
519	04-000655-1
519	4-2020-13600
519	05-180953
519	05-182076
519	05A06
519	006-0000004
519	006-6400902
519	007-25005-01
519	007-25013-01
519	007-25016-01
519	007-6016-00
519	007-6060-00
519	8-719-923-76
519	09-306135
519	09-306244
519	11-9-156438
519	13-0003
519	13-0003(PACE)
519	13-17596-10
519	13-17596-2
519	13-17596-3
519	13-17596-4
519	13-17596-5
519	13-17596-7
519	13-17596-8
519	13-17596-9
519	13-547604
519	14-514-12
519	14-514-62
519	14-514-64
519	019-003676-334
519	19A115250
519	19A115661P1
519	19Q820
519	21-608-4148-006
519	21A108-003
519	022-0163-00
519	022-2823-005
519	23-001R03A10
519	23-001R03AA10
519	034-032-0
519	35-010-04
519	38-8057-6
519	40-09297
519	43A114346-1
519	43A114346-P1
519	43A114832-3
519	43A175989P1
519	46-86309-3
519	48-03005A06
519	48-155047
519	48-155060
519	48-155077
519	48-8240C07
519	48-8240C11
519	48-8240C14
519	48-8240C16
519	48-8240C17
519	48-8240C18
519	48-82420C03
519	48-82420C04
519	48-90420A03
519	48D82420C03
519	48D82420C04
519	51-08001-11
519	51-47-28
519	56-56
519	65-P11305-0001
519	65-P11324-0001
519	65-P11325-0001
519	66A10319
519	66X0046-001
519	70-943-083-002
519	70-943-083-003
519	71-13/51/60
519	77S1017
519	79E104-2
519	93C64-1
519	93C64-2
519	93C64-3
519	93D60-5
519	93EVHD1N4148//
519	094-007
519	102-412
519	103-141
519	106-007
519	116,666
519	118-02900
519	118-02902
519	123-017(DIODE)
519	123-024
519	144-1
519	150-066-9-001
519	151-015-9-001
519	151-035-9-001
519	151-035-9-004
519	151-051-9-001
519	151-059-9-001
519	151-060-9-001
519	151-064-9-001
519	151-064-9-002
519	151-069-9-001
519	152-0333-00
519	152-141-1
519	165-432-2-48-1
519	201X2010-144
519	201X2010-159
519	212-96
519	241ABNG4148
519	260-10-057
519	264P04501
519	264P04502
519	264P04507
519	269M01201
519	296L002B01
519	311-0126-001
519	349-0002-001
519	349-0002-002
519	353-3024-000
519	353-3289-000
519	353-3338-000
519	353-3627-010
519	353-3627-020
519	353-3663-010
519	353-3687-010
519	353-3687-020
519	380-1000
519	380-1001
519	394-1571-1
519	400-1417-101
519	402-004-02
519	429-10036-0A
519	429-10036-0B
519	429-10054-0A
519	479-0547-001
519	479-1012-001
519	479-1013-001
519	479-1013-002
519	479-1055-001
519	479-1163-001
519	479-1229-001
519	479-1248-001
519	523-0006-002
519	523-0007-001
519	523-0013-201
519	523-006-002
519	523-1000-883
519	523-1500-883
519	540-028-00
519	540-033-00
519	606-6003-101
519	606-6003-102
519	754-9000-473
519	754-9052-473
519	755-845049
519	800-102-001
519	800-102-101-1
519	919-010873-050
519	926C206P1
519	943-086
519	943-087
519	943-087-1
519	943-105-001
519	1010A
519	1042-15
519	1074-118
519	1100-9487
519	1410-169
519	1543-00
519	1703-8662
519	2000-302
519	2000-303
519	2000-317
519	2000-332
519	2005-2981
519	2019-45
519	2065-08
519	2409-17
519	2477-173
519	4805-1241-200
519	4820-0201
519	5001-120
519	5001-128
519	5001-146
519	5084-600-4521-0
519	5084-600-452I-0
519	6129-P1
519	8000-00006-281
519	8000-00042-007
519	10895
519	11305-0001
519	11352-78
519	011950
519	011956
519	12255-235
519	014007-1
519	014007-2
519	014024
519	22164-000
519	41173
519	44002
519	50213-5
519	59840-1
519	71119-2
519	103514
519	104762
519	110667
519	115060-102
519	118335
519	119264-001
519	120068
519	129682-001
519	129708-101
519	132553
519	137057-001
519	139029
519	144581
519	144691
519	146571
519	146576
519	150834
519	181619
519	184798
519	185335
519	188056
519	197464
519	198773-1
519	198775-1
519	198776-1
519	198779-1
519	198785
519	198785-1
519	198785-2
519	198799-4
519	198810-1
519	198813-1
519	202862-518
519	211083
519	280017
519	309481
519	326309-10A
519	338307
519	379005N
519	462580-1
519	497442
519	497616
519	497616-1
519	497616-2
519	504720-1
519	529658
519	530072-1006
519	530072-1018
519	530135-2
519	530144-1001
519	530144-1004
519	562654
519	615154-1
519	618639-1
519	619011
519	619087-1
519	619526-1
519	720608-13
519	720609-1
519	726654
519	760037
519	800743-001
519	801724
519	802008
519	860001-153
519	900545-2

ECG	Industry Standard No.
519	900546-2
519	908705-1
519	908705-2
519	908742-1
519	921150-021
519	922943-1
519	925075-501B
519	925252-1
519	925252-102
519	925252-2
519	925252-4
519	925253-3
519	925297
519	925297-1
519	930293
519	932022-0001
519	932033-1
519	932050-1
519	943502
519	969099-1
519	992157-1
519	1223771
519	1417872-11
519	1471872-11
519	1471872-5
519	1680008-01
519	1780174-1
519	1800009
519	1905490-1
519	1908519
519	1950003
519	2014400
519	2132524-1
519	2182124-1
519	2185494-1
519	2185494-2
519	2330351
519	2330352
519	2331351
519	2546059
519	2621786-1
519	2865141
519	3160000-00-08A
519	3180006
519	3263029-10
519	3311133
519	3464611
519	4010577-0701
519	04410025-001
519	04410025-005
519	04410042-001
519	5330131
519	5330571
519	5330572
519	5330761
519	5573505-1
519	5573505-2
519	06200013
519	7012157
519	7090612
519	8120126
519	8511632-5
519	8513070-1
519	8523209-1
519	8532787-2
519	8532787-3
519	8532787-4
519	8533519-1
519	8549020-1
519	10041264-101
519	10669652
519	10676144
519	11083979
519	11103439
519	11159860
519	11268821
519	11787173
519	12947206
519	12997748
519	14714661
519	14714703
519	15039472
519	15039480
519	16270175
519	16658600
519	17854700
519	18341300
519	24553500
519	24553501
519	25175800
519	34621786-1
519	51631000
519	440009905
519	2012010144
519	4202013600
519	4202107670
519	8001100001
519	8001100003
519	13488240307
519	13488240311
519	13488240314
519	13488240316
519	13488240317
519	13488240318
519	16411992473
519	134882420303
519	134882420304
519$	FD333
5196A	1N1820
52	2320432
5200AK	1N2049C
521	MH987A04
521	212-102
521	212-103
521	212-128
521	212-129
521	212-130
521	212-131
521	212-132
521	212-133
521	212-135
521	212-136
521	212-139
521	212-139-01
521	212-139-02
521	212-140
5212A	1N1831
5212A	1N1831A
5216A	1N4226
522	EP62X41
522	GE-516
522	GE-517
522	MH913
522	MH919
522	MH919D01
522	MH931
522	MH983A02
522	MH983A03
522	MH983A04
522	28-35-01
522	32-33057-2
522	32-33094-1
522	32-33094-2
522	32-33094-3
522	32-33094-4
522	32-33094-5
522	32-33094-6
522	32-35894-2
522	32-35894-3
522	32-35894-4
522	32-35894-6
522	32-35894-7
522	32-39704-1
522	32-39704-2
522	66F112-1
522	66F112-2
522	66F181-1
522	66X0060-001
522	66X0060-002
522	86-106-3
522	93A91-1
522	93A91-2
522	93A91-3
522	93A91-4
522	93B91-1
522	93D91-1
522	93D91-2
522	93D91-3
522	93D91-4
522	93D91-5
522	93D91-6
522	212-139-03
522	212-142
522	72162-1
522	530153-1
522	530165-1
522	530165-2
522	530165-3
522	530165-4
523	MH1203
523	MH1204
523	MH1205
523	MH1221A01
523	REN523
523	14B348-1
523	14B348-2
523	14B348-3
523	14B348-4
523	14B348-5
523	32-39091-1
523	32-39091-2
523	32-39091-4
523	32-39091-5
523	32-39091-6
523	32-39091-9
523	66F159-1
523	66F159-2
523	66F159-3
523	212-141
523	212-141-01
523	3107-108-40501
523	72180
5232A	10T200
5248A	1207
525	A7579500
525	EF200
525	ERB26-20
525	ERB26-20L
525	ERB26-20M
525	ERB26-20MV
525	ERC26-13L
525	GH1F
525	HC-2
525	MB-1D
525	MB-1F
525	RC-2
525	RC-2V
525	RC2
525	RH-DX0077CEZZ
525	SB-2C
525	SB-2CGL
525	SF-1
525	SID30-13
525	SID30-132
525	V11N
525	1TH61
525	8-719-305-15
525	21A110-071
525	46-86494-3
525	46-86560-3
525	48-155152
525	103-261-02
525	103-287
525	103-305
525	201X2100-164
525	138736
525	143594
525	146316
525	2330564
525	2332152
525	23115965
525	2012100119
525	4202022900
5254A	1212
526A	32-39091-3
526A	32-39091-7
526A	32-39091-8
526A	212-141-02
526A	212-141-03
526A	212-141-04
528	212-145
528	212-145-01
528	212-145-02
528	212-146
529	MH1201
529	MH1222A01
529	MH1222A02
529	N2A-1
529	N2A-2
529	86-127-3
529	530165-11
529	530165-13
529	530165-17
529	530165-6
530	GE-540
530	MH1206
530	MH1207
530	MH1209
530	MH1220A01
530	57-90
530	57-98
530	212-143
530	3107-108-40401
5304	W005M
5304	W02
5305	VE68
5305	W06M
5305	2FB600
5305	10DB6P
5306	W08M
5306	2FB800
5307	2FB1000
531	MH915C02
531	138752
531	139001
531	1464607-7
531	1464607-8
531	1826065-3
5312	FWLD100
5312	FWLD50
5312	KBL005
5312	KBPC1005
5312	KBPC6005
5312	KBPC8005
5312	M4B-31-22
5312	S4VB10
5312	5331102
5313	FWLD200
5313	KBL02
5313	KBPC102
5313	KBPC602
5313	KBPC802
5313	S2VB10
5313	VS248
5314	FWLD400
5314	KBL04
5314	KBPC104
5314	KBPC604
5314	KBPC804
5314	VS448
5315	FWLD600
5315	KBPC106
5315	KBPC606
5315	KBPC806
5315	VS648
5316	FWLD800
5316	KBPC108
532	GE-519
532	MH915C01
532	57-83
532	135691
532	137031
532	137646
532	137693
532	138019
532	138421
532	138907
532	1464607-1
532	1464607-2
532	1464607-3
532	1464607-4
532	1464607-5
532	1464607-6
532	1464984-1
532	1464984-2
532	1466860-1
532	1826065-1
5322	KBPC25-005
5322	SCBA05
5322	250JB05L
5324	KBPC25-04
5324	250JB4L
5326	KBPC25-06
533	MH970C02
533	1464607-10
533	1464607-9
5332	VM08
5332	VM18
5332	VM25
5332	VM28
5332	VM48
534	MH943
534	MH985A01
534	530165-10
534	530165-15
534	530165-16
534	530165-5
534	530165-8
535	MH1222A03
535	530165-14
535	530165-18
536A	MH353
536A	MH988A03
537	145208
537	1466860-2
538	MH1030A01
538	93D99-2
538	93D99-3
538	93D99-4
539	MH1030A02
539	15-123060
539	93A99-7
539	93A99-8
539	93D99-5
539	93D99-6
539	93D99-7
539	93D99-8
539	530165-12
539	530165-7
5400	C103Y
5400	EC103Y
5400	ID100
5400	MCR101
5400	MCR102
5400	T1C44
5400	T1C60
5400	TIC44
5400	2N2679
5400	2N2687
5400	2N3005
5400	2N3027
5400	2N3030
5400	2N4144
5400	2N4145
5400	2N5060
5400	2N876
5400	2N877
5401	C103YY
5401	ID101
5401	MCR103
5401	T1C45
5401	TIC45
5401	2N2680
5401	2N2688
5401	2N3006
5401	2N3028
5401	2N3031
5401	2N3259
5401	2N4146
5401	2N5061
5401	2N878
5401	2SF101
5402	EC103A
5402	ID102
5402	MCR104
5402	SD-2N5062
5402	T1C46
5402	TIC46
5402	2N2689
5402	2N3007
5402	2N3029
5402	2N3032
5402	2N4147
5402	2N5062
5402	2N879
5402	2SF102
5402	2SF102A
5402	2SF522
5402	86-110-3
5403	ID103
5403	MCR115
5403	2N4148
5403	2N5063
5403	2N880
5404	C103B
5404	EC103B
5404	EP15X10
5404	EP15X106
5404	ID104
5404	MCR120
5404	RTC0120
5404	T1C47
5404	T1C63
5404	T1C64
5404	TIC47
5404	2N2690
5404	2N3008
5404	2N4149
5404	2N5064
5404	2N881
5405	EC103D
5405	ID105
5405	ID106
5405	IP105
5405	IP106
5405	MCR100-6
5405	RTC0130
5405	RTJ0225
5405	RTJ0230
5405	2SF106
5406	MCR100-8
5408	S0501M
5408	S0503M
5408	S1001M
5408	S1003M
5408	S2001M
5408	S2003M
5408	2N1595
5408	2N1596
5408	2N1597
5408	11T4
5409	S4001M
5409	S4003M
5409	2N1598
5409	2N1599
5411	MCR106-1
5411	MCR107-1
5411	S0303LS3
5411	S0303RS3
5412	MCR106-2
5412	MCR107-2
5412	S0503LS3
5412	SF1A11
5412	SF1A11A
5413	MCR106-3
5413	MCR107-3
5413	SF1B11
5413	SF1B11A
5414	HEP-R1218
5414	MCR106-4
5414	MCR107-4
5414	MCR406-4
5414	MCR407-4
5414	S2003LS3
5414	S2003RS3
5414	SF1D11
5414	SF1D11A
5414	2N6239
5414	86-113-3
5415	2N6240
5416	2N6241
5421	MCR406-1
5421	2N6236
5421	13-33183-1
5422	MCR406-2
5422	2N6237
5422	48-83875D05
5422	48D83875D05
5422	75D05
5422	134883875405
5423	MCR406-3
5423	SCR104(ELCOM)
5423	2N6238
5431	MCR407-1
5431	48-83875D04
5431	48D83875D04
5431	75D04
5431	134883875404
5432	MCR407-2
5433	MCR407-3
5442	2N4441
5444	MCR3000-2
5444	MCR3000-4
5444	2N4442
5446	MCR3000-6
5446	2N4443
5448	MCR3000-8
5448	2N4444
5452	C106Q
5452	C106Q1
5452	C106Q2
5452	C106Q3
5452	C106Q4
5452	C106Y
5452	C106Y1
5452	C106Y2
5452	C106Y3
5452	C106Y4
5452	C107Q
5452	C107Q1
5452	C107Q2
5452	C107Q3
5452	C107Q4
5452	C107Y
5452	C107Y1
5452	C107Y2
5452	C107Y3
5452	C107Y4
5452	IR106Q1
5452	IR106Q2
5452	IR106Q3
5452	IR106Q4
5452	IR106Q41
5452	IR106Y1
5452	IR106Y1-C
5452	IR106Y2
5452	S0303RS2
5452	S106Q
5452	S106Y
5452	S106Y1
5452	S107Y1
5452	S2062Q
5452	S2062Y
5452	TC106Q1
5452	TC106Q2
5452	TC106Q3
5452	TC106Q4
5452	TC106Y1
5452	TC106Y2
5452	TC106Y3
5452	TC106Y4
5452	TIC106Y
5452	TX-145
5452	108Q
5452	108Y
5453	C106F
5453	C106F1
5453	C106F2
5453	C106F3
5453	C106F4
5453	C107F
5453	C107F1
5453	C107F2
5453	C107F3
5453	C107F4
5453	GE-X5
5453	IR106F1
5453	IR106F2
5453	IR106F3
5453	IR106F4
5453	IR106F41
5453	IRC5
5453	S106F
5453	S106F1
5453	S107F
5453	S107F1
5453	S2062F
5453	TC106F1
5453	TC106F2
5453	TC106F3
5453	TC106F4
5453	TIC106F
5453	11T4S
5453	108F
5454	C106A
5454	C106A1
5454	C106A2
5454	C106A3
5454	C106A4
5454	C107A
5454	C107A1
5454	C107A2
5454	C107A3
5454	C7A4
5454	IR106A1
5454	IR106A1-C
5454	IR106A2
5454	IR106A3
5454	IR106A4
5454	IR106A41
5454	RE168
5454	S1003RS2
5454	S106A
5454	S106A1
5454	S107A
5454	S107A1
5454	S2060A
5454	S2061A
5454	S2062A
5454	SFIR3B41
5454	TC106A1
5454	TC106A2
5454	TC106A3
5454	TC106A4
5454	TIC106A
5454	2SF941
5454	3C6100
5454	48-83875D01
5454	48D83875D01
5454	75D01
5454	106A
5454	107A(SCR)
5454	108A
5454	134883875401
5455	C106B
5455	C106B1
5455	C106B2
5455	C106B3
5455	C106B4
5455	C107B
5455	C107B1
5455	C107B2
5455	C107B3
5455	C107B4
5455	CU-12E
5455	GEMR-5
5455	IR106B1
5455	IR106B1-C
5455	IR106B2
5455	IR106B3
5455	IR106B4
5455	IR106B41
5455	IRC20
5455	RE171
5455	S106B
5455	S106B1
5455	S107B
5455	S107B1
5455	S2003RS2
5455	S2060B

ECG	Industry Standard No.
5455	S2061B
5455	S2062B
5455	SF1R3D41
5455	SFIR3D41
5455	SK3597
5455	TA2888
5455	TA2889
5455	TC106B1
5455	TC106B2
5455	TC106B3
5455	TC106B4
5455	TIC106B
5455	1RC20
5455	2SF942
5455	19B200248-P1
5455	19B200248-P2
5455	19B200248-P3
5455	48-8375D02
5455	48-83875D06
5455	48D8375D02
5455	48D83875D06
5455	53-1517
5455	75D02
5455	75D06
5455	106B
5455	107B(SCR)
5455	108B
5455	2321264
5455	134883875402
5455	134883875406
5456	BT100A-300R
5456	C106C
5456	C106C1
5456	C106C2
5456	C106C3
5456	C106C4
5456	C107C
5456	C107C1
5456	C107C2
5456	C107C3
5456	C107C4
5456	IR106C1
5456	IR106C1-C
5456	IR106C2
5456	IR106C3
5456	IR106C4
5456	IR106C41
5456	S106C
5456	S107C
5456	S2060C
5456	S2061C
5456	TC106C1
5456	TC106C2
5456	TC106C3
5456	TC106C4
5456	TIC106C
5456	106C
5456	107C
5456	108C
5457	C106D
5457	C106D1
5457	C107D
5457	S106D
5457	S107D
5457	86X0081-001
5457	86X0081-002
5461	C122F
5461	C122F1
5461	S0306L
5461	S0308L
5461	S0506L
5461	S0508L
5461	S2060F
5461	S2060Q
5461	S2060Y
5461	S2061F
5461	S2061Q
5461	S2061Y
5461	T1C116F
5461	T1C126F
5461	2N6394
5462	C122A1
5462	S1006L
5462	S1008L
5462	S2800A
5462	T1C116A
5462	T1C126A
5462	40867
5463	C122B
5463	C122B1
5463	S2006L
5463	S2008L
5463	S2800B
5463	T1C116B
5463	T1C126B
5463	40868
5465	C122C1
5465	C122D
5465	C122D1
5465	S2800D
5465	S4006L
5465	S4008L
5465	T1C116C
5465	T1C116D
5465	40869
5470	2N1600
5471	2N1601
5472	2N1602
5474	2N1603
5474	2N1604
5474	2N2653
5480	C110
5480	C150
5480	2N4167
5481	2N4168
5482	2N1772
5482	2N1772A
5482	2N4169
5483	C15B
5483	GE-X1
5483	MCR2305-2
5483	MCR2305-4
5483	MCR2315-4
5483	RTN0220
5483	2N1773
5483	2N1773A
5483	2N1774
5483	2N1774A
5483	2N4170
5483	2N4178
5483	90270020
5484	C154H
5484	MCR2315-5
5484	2N1775

ECG	Industry Standard No.
5484	2N1775A
5484	2N1776
5484	2N1776A
5484	2N4171
5484	2N4179
5485	C15D
5485	MCR2315-6
5485	RE175
5485	RTN0230
5485	RTN0240
5485	2N1777
5485	2N1777A
5485	2N4172
5485	2N4180
5486	BT101-500R
5486	2N1778A
5486	2N4173
5487	RTN0250
5487	RTN0260
5487	2N4174
5491	C20A
5491	C20F
5491	C20U
5491	S0306H
5491	S0308H
5491	S0310H
5491	S0506H
5491	S0508H
5491	S0510H
5491	S1006H
5491	S1008H
5491	S1010H
5491	SPS08
5491	SPS18
5492	C20B
5492	C220B
5492	S2006H
5492	S2008H
5492	S2010H
5492	SPS28
5494	C220C
5494	C220D
5494	GEMR-4
5494	S4006H
5494	S4008H
5494	S4010H
5494	SPS48
5496	C20D
5496	S6006H
5496	S6008H
5496	S6010H
5500	C36U
5500	RTU0102
5500	2N1842
5500	2N1842A
5501	C36F
5501	2N1843
5501	2N1843A
5501	2N1843B
5502	C36A
5502	RTU0110
5502	2N1844
5502	2N1844A
5502	2N1844B
5503	C36G
5503	2N1845
5503	2N1845A
5504	C36B
5504	RTU0120
5504	2N1845B
5504	2N1846
5504	2N1846A
5504	2N1846B
5505	2N1847
5505	2N1847A
5505	2N1847B
5506	C36C
5506	2N1848
5506	2N1848A
5506	2N1848B
5507	C36D
5507	RTU0130
5507	RTU0140
5507	S2610D
5507	2N1849
5507	2N1849A
5507	2N1849B
5508	2N1850
5508	2N1850A
551	MR2266
551	MR2273
551	SID30-15
551	TD-13
551	TD-15
551	TD15
551	TVSTD15M
551	1N1443
551	1N1443A
551	13-39072-1
551	13-39072-2
551	2331381
5511	2N3228
5512	2N3525
5513	GE-700
5513	S2710B
5513	S2710D
5513	S2710M
5513	2N4101
5514	C222B
5514	MCR3818-1
5514	MCR3818-2
5514	MCR3818-3
5514	MCR3818-4
5514	PS08
5514	PS18
5514	PS220
5514	PS28
5514	RTS0202
5514	RTS0205
5514	RTS0210
5514	RTS0220
5514	S0306G
5514	S0308G
5514	S0310G
5514	S0315G
5514	S0506G
5514	S0508G
5514	S0510G
5514	S0515G
5514	S1006G
5514	S1008G
5514	S1010G
5514	S1015G
5514	S2006G

ECG	Industry Standard No.
5514	S2008G
5514	S2010G
5514	S2015G
5514	S6200A
5514	S6200B
5514	2N5164
5514	2N5165
5514	40737
5514	40738
5514	40749
5515	MCR2604-4
5515	MCR2604-5
5515	MCR2604-6
5515	MCR3818-5
5515	MCR3818-6
5515	PS320
5515	PS38
5515	PS48
5515	RTS0230
5515	RTS0240
5515	S4006G
5515	S4008G
5515	S4010G
5515	S4015G
5515	S6200D
5515	2N5166
5515	40739
5516	MCR3818-7
5516	MCR3818-8
5516	PS520
5516	PS58
5516	PS68
5516	RTS0250
5516	RTS0260
5516	S6006G
5516	S6008G
5516	S6010G
5516	S6015G
5516	S6200M
5516	2N5167
5516	40740
5517	MCR2835-1
5517	MCR2835-2
5517	MCR2835-3
5517	MCR2835-4
5517	MCR3835-1
5517	MCR3835-2
5517	MCR3835-3
5517	MCR3835-4
5517	RTS0702
5517	RTS0705
5517	RTS0710
5517	RTS0720
5517	S0325G
5517	S0335G
5517	S0525G
5517	S1025G
5517	S2025G
5517	S2035G
5517	2N3870
5517	2N3871
5518	S4025G
5518	S4035G
5518	2N3872
5519	S6025G
5519	S6035G
5519	2N3873
552	A7568300
552	A7978850
552	A7978855
552	BB-6
552	BB10
552	BB6
552	D5G
552	D5H
552	EP57X4
552	EP57X5
552	ERB-24-06A
552	ERB24
552	ERB24-04C
552	ERB24-04D
552	ERB24-06
552	ERB24-06B
552	ERB24-06C
552	ERB2406A
552	ERB28-04
552	ERC24-04
552	FU-1M
552	FU-1MA
552	FU-1N
552	FU-1NA
552	FU-IM
552	FU1N
552	FU1NA
552	MR1337-1
552	MR1337-2
552	MR1337-3
552	MR1337-4
552	MR1337-5
552	RA-1Z
552	RE2
552	RF-1A
552	RH-1
552	RH-1V
552	RH-DX0043TAZZ
552	RH-DX0067TAZZ
552	RH-DX0073CEZZ
552	RH-DX0085TAZZ
552	RH-DX0086TAZZ
552	RH-DX0100CEZZ
552	RH-DX0101CEZZ
552	RH-DX0106CEZZ
552	RH-DX0114CEZZ
552	RH1M
552	RH1Z
552	RM1
552	RM1Z
552	RU1
552	RU2
552	S-15
552	S1R80
552	S5295G
552	S5295J
552	SD-ERB24-04D
552	SIB01-01
552	SIB01-02
552	SIB01-04
552	SIR60
552	TVS10D2
552	TVSB01-02
552	TVSB24-06C
552	TVSB24-06D
552	TVSBB10
552	TVSS1R20

ECG	Industry Standard No.
552	TVSS1R80
552	UF2
552	V09C
552	V09E
552	V09G
552	1N4942
552	1N4944
552	1N4946
552	1N5617
552	1N5619
552	1S1834
552	1S1835
552	1S2244
552	1S2245
552	1S2246
552	1S2277
552	1S2278
552	1S2279
552	1S2316
552	1S2317
552	1S2775FA-1
552	05-860002
552	05-931609
552	05-935201
552	8-719-200-02
552	8-719-320-11
552	8-719-320-31
552	8-719-900-93
552	8-719-901-19
552	8-719-901-93
552	10D-2
552	10D8
552	10E2
552	13-31014-6
552	13-37868-1
552	13-41122-2
552	13-43777-1
552	13-43777-2
552	46-86507-3
552	48-137347
552	48-137348
552	48-155041
552	48-155099
552	48-155107
552	48-155178
552	48-155223
552	48-355048
552	48-90343A54
552	48-90420A97
552	48-90420A98
552	48S137347
552	48S137348
552	48S155056
552	86-104-3
552	86-116-3
552	86-65-1
552	86-67-3
552	93A60-7
552	93C60-7
552	93ERH-DX0155//
552	103-261-04
552	103-284
552	103-284A
552	103-298-05A
552	103-96
552	201X2100-126
552	201X2120-009
552	264P10102
552	264P10201
552	137606
552	140971
552	140972
552	142569
552	146136
552	146137
552	530151-1001
552	2330551
552	2330562
552	06200014
552	2012130234
552	4202110470
5520	C30U
5520	C37U
5520	MCR3918-1
5520	2N681
5520	10RC2
5520	10RC2A
5521	C30F
5521	C37F
5521	S0325H
5521	S0525H
5521	SPS020
5521	2N5168
5521	2N682
5521	10RC5
5521	10RC5A
5522	C30A
5522	C37A
5522	MCR3918-3
5522	NL-C37A
5522	S1025H
5522	SPS120
5522	2N3896
5522	2N683
5522	10RC10
5523	2N684
5523	10RC15
5523	10RC15A
5524	C30B
5524	C37B
5524	MCR3918-4
5524	NL-C37B
5524	S2025H
5524	2N2888
5524	2N3897
5524	2N5169
5524	2N685
5524	10RC20
5524	10RC20A
5525	2N2889
5525	2N686
5525	10RC25
5526	C30C
5526	C37C
5526	MCR3918-5
5526	NL-C37C
5526	SPS320
5526	2N687
5526	10RC30
5526	10RC30A
5527	C230D
5527	C30D
5527	C37D
5527	NL-C37D

ECG	Industry Standard No.
5527	S4025H
5527	SPS420
5527	2N3898
5527	2N688
5527	2SF77
5527	10RC40
5527	10RC40A
5527	21A040-015
5527	40379
5528	C37E
5528	SPS520
5528	2N689
5528	2SF139
5529	C37M
5529	MCR3918-8
5529	RTU0160
5529	S2610M
5529	S6015H
5529	S6025H
5529	SPS620
5529	2N5171
5529	2N690
553	DDAY090001
553	MC301
553	1S1544
553	1S1545
553	1S2186
553	129095
553	1477022-1
5531	S8025H
5531	2N692
5540	C35U
5540	C38U
5540	MCR2935-1
5540	MCR3935-1
5540	NL-C35U
5540	13RC2
5540	16RC2A
5540	18RC2
5540	22RC2
5541	C35F
5541	MCR2935-2
5541	MCR3935-2
5541	NL-C35F
5541	S0335H
5541	S0535H
5541	T400002208
5541	13RC5
5541	16RC5A
5541	18RC5
5541	22RC5
5542	C35A
5542	MCR2935-3
5542	MCR3935-3
5542	NL-C35A
5542	RE170
5542	S1035H
5542	T400012208
5542	13RC10
5542	16RC10A
5542	18RC10
5542	22RC10
5543	C35B
5543	C35G
5543	GE-X4
5543	MCR3935-4
5543	RE173
5543	S2035H
5544	C35C
5544	IR30C
5544	MCR3935-5
5545	C35D
5545	GEMR-3
5545	IR30D
5545	MCR3935-6
5545	RE176
5545	S4035H
5546	C137E
5546	C38E
5546	MCR3935-7
5547	BFW61
5547	C137M
5547	C35M
5547	MCR3935-8
5547	S6035H
5547	2N3899
5548	C137N
5548	C137S
555	BA182
555	MPN3401
555	1S2692
555	13-10321-34
556	32-43737-1
5562	S6420A
5562	2N6168
5566	S6420M
5566	2N6170
558	HF-1
558	HF-1A
558	HF-1C
558	HF-1Z
558	HF1
558	HF1A
558	HF1B
558	HF1C
558	HF1Z
558	S-4C
558	S1R-80
558	U07L
558	UF-1
558	UF-1A
558	UF-1C
558	1S2281
558	1S2319
558	1S2324
558	1S2593
558	13-33172-1
558	13-33172-2
558	264P03604
5600	MAC77-1
5600	2N6068
56004	BS7-02A
56004	BS9-02A
56004	IT28
56004	MAC15-4
56004	Q2015R5
56004	SC141A
56004	SC141A1
56004	SC141B
56004	SC141B2
56004	SC141B3
56004	SC141B4
56004	SC141B5
56004	SC141B6

ECG	Industry Standard No.	ECG	Industry Standard No.	ECG	Industry Standard No.	ECG	Industry Standard No.	ECG	Industry Standard No.
56004	SC143B	5643	2N5757	5800	RB-05	5801	1N5920P	5802	1N3228
56004	SC151B	5645	IT010	5800	RT7848	5801	1P644	5802	1N3239
56004	TIC226B	5645	IT06	5800	S105A	5801	18020	5802	1N3248
56004	TIC236B	5645	IT08	5800	S1M1	5801	18051	5802	1N3748
56004	2N6542	5645	IT110	5800	S1M2	5801	182416	5802	1N3755
56004	2N6542A	5645	IT16	5800	S2A06	5801	1840	5802	1N3757
56004	2N6546	5645	IT210	5800	S2M2	5801	18431	5802	1N3866
56004	2N6546A	5645	IT310	5800	S91A	5801	381E	5802	1N3938
56006	B87-04A	5645	IT36	5800	S92A	5801	7-6006-00	5802	1N3952
56006	MAC15-5	5645	IT38	5800	SB12	5801	1501	5802	1N3981
56006	MAC15-6	5645	IT410	5800	S091	5801	1581	5802	1N4141
56006	Q4015R5	5645	IT46	5800	SI05	5801	20A1	5802	1N4517
56006	SC141C	5645	IT48	5800	SI05A	5801	20C1	5802	1N4721
56006	SC141C2	5645	IT510	5800	SJ053F	5801	30R1	5802	1N4721R
56006	SC141C3	5645	IT56	5800	SLA81A	5801	3031	5802	1N4999
56006	SC141C4	5645	IT58	5800	SLA21B	5801	100-525	5802	1N5005
56006	SC141C5	5645	IT610	5800	SLA21C	5801	0135-1	5802	1N5056
56006	SC141C6	5645	IT66	5800	SLA5197	5801	250R1B	5802	1N5199
56006	SC141D2	5645	IT68	5800	SOD30AL	5801	300R1B	5802	1N5393
56006	SC141D3	5645	T2500B	5800	SOD30BL	5801	398B	5802	1N5402
56006	SC141D4	5645	T2800B	5800	SOD30DL	5801	540-015	5802	1N5624
56006	SC141D6	5645	T2800D	5800	SOD50AL	5801	919-007776	5802	18021
56006	SC149D	5645	T2800C	5800	SOD50BL	5801	919-010829	5802	18032
56006	SC151D	5645	T2800D	5800	SOD50CL	5801	919-011211	5802	181072
56006	TIC226D	5645	T2800M	5800	SOD50DL	5801	919-013036	5802	182410
56006	TIC236D	5645	T2850A	5800	8P-1-15A	5801	2454-17	5802	182417
56006	2N6543	5645	T2850B	5800	SR132-1	5801	5210DC-1	5802	18432
56006	2N6543A	5645	T2850D	5800	SR1422	5801	5923	5802	18921(RECT)
56006	2N6547	5645	1T10	5800	SR1598	5801	40266	5802	382
56006	2N6547A	5645	1T26	5800	SR1643A	5801	61088	5802	382E
56008	B87-05A	5645	1T510	5800	SR1DM-2	5801	71449-1	5802	13-18481-1
56008	B89-05A	5650	T2300A	5800	SR3AM1	5801	720454	5802	13-18481-3
56008	MAC15-7	5650	T2301A	5800	SR475	5801	3460758-2	5802	13-34368-1
56008	MAC15-8	5650	T2302A	5800	SW-05	5801	06200065	5802	1502
56008	Q5015R5	5651	L2001M5	5800	TB5	5802	A-132591	5802	1502D
56008	Q6015R5	5651	L2001M7	5800	T050	5802	A15B	5802	1582
56008	SC141E	5651	T2300B	5800	TI-56	5802	A83C	5802	20A2
56008	SC141E2	5651	T2301B	5800	TV24651	5802	A83C	5802	2002
56008	SC141E3	5651	T2302B	5800	TVS-181893	5802	B56-15	5802	23-0003
56008	SC141E4	5652	L4001M5	5800	TVS-181922	5802	BYX30/150	5802	35-1008
56008	SC141E5	5652	L4001M7	5800	UB205	5802	E3026(ELCOM)	5802	86-5028-3
56008	SC141E6	5652	T2300D	5800	UT2005	5802	HEP-R0092	5802	86-5032-3
56008	SC149E	5652	T2301D	5800	UT3005	5802	HEP162	5802	96-5106-01
56008	SC149M	5652	T2302D	5800	WR-006	5802	HR-200	5802	96-5149-01
56008	SC151E	5673	GE-X12	5800	WR-030	5802	MP549	5802	96-5184-01
56008	SC151M	5673	MAC5573	5800	WR-040	5802	MP551	5802	157-737
56008	2N6544	5673	Q2010H	5800	Z810	5802	NGF3003	5802	150R2B
56008	2N6544A	5673	Q2015H	5800	Z8270	5802	PGR-24	5802	250R2B
56008	2N6548	5673	SC240B	5800	1.5E05	5802	R-1B	5802	300R2B
56008	2N6548A	5673	SC245B	5800	1.5J05	5802	R-5B	5802	399B
5601	MAC77-2	5673	SC250B	5800	1N1644	5802	R-5C	5802	599B
5601	2N6069	5673	TIC222B	5800	1N2013	5802	S-12	5802	399B
5601	SC160B	5673	TIC232B	5800	1N2390	5802	S-12A	5802	475-018
5601	SC160D	5673	TIC242B	5800	1N2399	5802	S1020	5802	501-363-2
56014	T4140B	5673	2N5569	5800	1N2408	5802	S1220	5802	919-001449
56014	T4140D	5673	2N5573	5800	1N2895BA	5802	S229	5802	919-001449-1
56014	T4150B	5675	MAC5574	5800	1N3072	5802	S236	5802	919-001449-3
56014	T4150D	5675	Q4010H	5800	1N3237	5802	S2A20	5802	919-002240-1
5602	MAC77-3	5675	Q4015H	5800	1N3246	5802	S2CN1	5802	919-005045
5602	2N6070	5675	SC240D	5800	1N4159	5802	S2GR2	5802	919-008862
56022	MAC040688	5675	SC245D	5800	1N4719	5802	S3A2	5802	919-010459
56022	Q2040D	5675	SC250D	5800	1N4816	5802	S5M1	5802	919-010623
56022	T6420B	5675	TIC222D	5800	1N4997	5802	S4CN1	5802	919-010829-1
56022	2N6163	5675	TIC232D	5800	1N5171	5802	S4GR2	5802	1412-182
56024	MAC40689	5675	2N5570	5800	1N5197	5802	SA2M1	5802	1901-0045
56024	SC265D2	5675	2N5574	5800	1N5391	5802	SD6(DUAL)	5802	5522-8
56024	T6420D	5676	SC240E	5800	1N645	5802	SD91A	5802	7921
56024	2N6164	5676	SC245E	5800	1S030	5802	SD91B	5802	031450
56026	MAC40690	5676	SC250E	5800	1S2408	5802	SD92	5802	34405
56026	T6420M	5676	TIC222E	5800	1S2415	5802	SD92A	5802	36147
56026	2N6165	5676	TIC232E	5800	1S430	5802	SD928	5802	56591
5603	MAC77-4	5676	TIC242E	5800	2R05	5802	SD205	5802	38052
5603	2N6071	5677	Q6010M	5800	3A15	5802	SI18	5802	39804
5604	MAC77-5	5677	T4110M	5800	3A30	5802	SI1A	5802	40267
5604	2N6072	5677	T4111M	5800	3E05	5802	SIG1/100	5802	46914
5604	13-39678-1	5680	MAC036-1	5800	380S2	5802	SJ203F	5802	61807
5604	13-39678-3	5681	MAC036-2	5800	15-18481-2	5802	SI-5	5802	71449
5604	13-39678-4	5682	MAC036-3	5800	15005	5802	SLA23B	5802	121180
5605	MAC77-6	5683	MAC036-4	5800	15805	5802	SLA23C	5802	206180
5605	2N6073	5683	TIC252B	5800	30805	5802	SLA5199	5802	206180
5616	MAC10-4	5683	MAC036-5	5800	48-137098	5802	SOD100AL	5802	206190
5616	MAC10-6	5685	MAC036-6	5800	399A	5802	SOD100BL	5802	475018
5624	MAC11-4	5685	TIC262D	5800	720453	5802	SOD100CL	5802	506911
5624	2N6154	5685	2N5274	5801	A15A	5802	SOD100DS	5802	530071-1
5626	MAC11-6	5686	MAC036-7	5801	AE3B	5802	SOD200AL	5802	630065
5626	2N6155	5686	SPT525	5801	A83B	5802	SOD200BL	5802	654420
5632	SC146A	5686	TIC252E	5801	B56-33	5802	SOD200DL	5802	720455
5632	SC146A2	5686	TIC262E	5801	HEP-R0091	5802	SOD200DL	5802	801730
5632	SC146A3	5693	Q2025H	5801	S1010	5802	SR-100	5802	7570014
5632	SC146A4	5693	SC260B	5801	S1210	5802	SR-114	5802	8521587-101
5632	SC146A5	5693	SPT040	5801	S1082	5802	SR1762	5803	A15C
5632	SC146A6	5693	SPT140	5801	S2M1	5802	SR210	5803	81030
5633	SC146B	5693	SPT230	5801	S3A1	5802	SR220	5803	81230
5633	SC146B2	5693	SPT240	5801	S91B	5802	SR507	5803	UT211
5633	SC146B3	5693	T6411B	5801	SA1M1	5802	SSIC0810	5803	UT212
5633	SC146B4	5693	TIC272B	5801	S11	5802	SSIC1110	5803	1.5J3
5633	SC146B5	5693	2N5444	5801	SJ103F	5802	SSIC1210	5803	1N1221B
5633	SC146B6	5695	2N6160	5801	SLA22A	5802	TB100	5803	1N1558
5633	T2801B	5695	Q4025H	5801	SLA22B	5802	TB200	5803	1N1565
5633	T2802B	5695	SC260D	5801	SLA22C	5802	TC100	5803	1N1648
5634	SC146C	5695	SPT330	5801	SLA5198	5802	TC200	5803	1N1649
5634	SC146C2	5695	SPT430	5801	SR3AM-3	5802	UB215	5803	1N1910
5634	SC146C3	5695	T6411D	5801	UB210	5802	UB220	5803	1N2017
5634	SC146C4	5695	TIC272D	5801	UT2010	5802	UT12	5803	1N2018
5634	SC146C5	5695	2N5445	5801	UT236	5802	UT13	5803	1N2393
5634	T2802C	5695	2N6161	5801	UT249	5802	UT2020	5803	1N2402
5634	SC141D	5697	BTU0660	5801	UT251	5802	UT234	5803	1N2411
5635	SC146D	5697	Q5025H	5801	UT261	5802	UT242	5803	1N2861A
5635	SC146D2	5697	Q6025H	5801	UT3010	5802	UT252	5803	1N3076
5635	SC146D3	5697	SC265M	5801	ZR11	5802	UT262	5803	1N3077
5635	SC146D4	5697	SPT530	5801	ZR601	5802	UT3020	5803	1N4819
5635	SC146D5	5697	SPT540	5801	ZR271	5802	ZR12	5803	1N5057
5635	SC146D6	5697	SPT630	5801	001-024061	5802	ZR602	5803	1N5173
5635	T2801D	5697	SPT640	5801	1.5B1	5802	Z8272	5803	1N5394
5635	T2802D	5697	T6411M	5801	1.5J1	5802	Z8701	5803	1N5403
5636	Q16	5697	TIC272M	5801	1N1556	5802	Z8702	5803	1645
5636	SC146E	5697	2N5275	5801	1N1553	5802	1.5B2	5803	1646
5636	SC146E3	5697	2N5446	5801	1N1645	5802	1.5J2	5803	182411
5636	SC146E4	5800	A15F	5801	1N2014	5802	1K2	5803	182418
5636	SC146E5	5800	A15U(DIODE)	5801	1N2391	5802	1N1055	5803	1842
5636	SC146E6	5800	AC050(RECTIFIER)	5801	1N2400	5802	1N220B	5803	3A500
5636	T2801E	5800	A83A	5801	1N2409	5802	1N557	5803	3AM4
5636	T2802E	5800	A83A	5801	1N2895A	5802	1N564	5803	383E
5637	SC146M	5800	BA136A	5801	1N3073	5802	1N646	5803	20A3
5637	T2801M	5800	BA178	5801	1N3227	5802	1N647	5803	2003
5637	T2802M	5800	BTM-50	5801	1N3238	5802	1N2015	5803	3083
5641	2N5754	5800	HEP-R0090	5801	1N3247	5802	1N2016	5803	150R3B
5641	L2001M9	5800	HEP161	5801	1N3486	5802	1N2392	5803	200R3B
5641	2N3528	5800	OA10	5801	1N4140	5802	1N2401	5803	250R3B
5641	2N5755			5801	1N4720	5802	1N2410	5803	300R3B
5641	1473619-3			5801	1N4817	5802	1N2860A	5803	A15D
5642	L4001M9			5801	1N4998	5802	1N3074	5804	AC050
5642	2N5756			5801	1N5004	5802	1N3075	5804	AC400
5642	137876			5801	1N5055	5802	1N3082		

ECG	Industry Standard No.	ECG	Industry Standard No.	ECG	Industry Standard No.	ECG	Industry Standard No.	ECG	Industry Standard No.
5804	AE3D	5804	540-008	5808	1N5398	5835	1N1539R	5842	1N2517
5804	A83D	5804	919-002440-2	5808	18058	5835	1N1540R	5842	1N2525
5804	BY206	5804	1901-0028	5808	1S46	5835	1N1582R	5842	1N2530
5804	B304(ELCOM)	5804	1901-0036	5808	3AP8	5835	1N1583R	5842	1N2541
5804	HEP-R0094	5804	1901-0388	5808	3B8	5835	3PR10	5842	1N2552
5804	PT-72130-1	5804	1901-0389	5808	20A8	5835	3PR20	5842	1N256
5804	PTC201	5804	3585	5808	305B	5836	AM51	5842	1N2852
5804	R-4A	5804	35604	5809	AC1000	5836	AM52	5842	1N554
5804	R-81805	5804	38074	5809	S1O100	5836	AM54	5842	1N555
5804	RPJ70147	5804	720456	5809	S1090	5836	BYX38/300	5842	1N613
5804	RV1476	5804	720458	5809	S3A10	5836	1N1117	5842	1N613A
5804	S1040	5804	3430063-1	5809	S110	5836	1N1125	5842	1N614
5804	S1240	5804	3430065-I	5809	SSIC0880	5836	1N1553	5842	1N614A
5804	S13A	5804	8521587-102	5809	SSIC1280	5836	1N1584	5842	1S1355
5804	S14A	5804	62020911	5809	UT347	5836	1N2028	5842	18415
5804	S2A30	5805	A15E	5809	UT363	5836	1N2292	5842	3C60
5804	S2A40	5805	BY226	5809	UT364	5836	1N2292A	5842	3P60
5804	S3A3	5805	S1050	5809	1N2502	5836	1N2514	5842	3JC12
5804	S3A4	5805	UT214	5809	1N2506	5836	1S661	5842	60P1
5804	S3CN1	5805	UT237	5809	1N2774	5836	3P30	5842	6085
5804	S3GR2	5805	UT245	5809	1N2775	5837	1N1038	5842	0245
5804	S4M1	5805	UT265	5809	1N2866	5837	1N1044	5843	305M
5804	S500B	5805	1N1223B	5809	1N2868	5837	1N1050	5843	BYX38/900R
5804	S500C	5805	1N1560	5809	1N3107	5837	1N1541R	5843	GB-5009
5804	S5CN1	5805	1N1567	5809	1N3232	5837	1N1584R	5843	1N1128R
5804	S5M1	5805	1N1652	5809	1N3243	5837	1S661R	5843	1N1544R
5804	SA3M1	5805	1N1764A	5809	1N3252	5837	3PR30	5843	1N1587R
5804	SD101	5805	1N2395	5809	1N3565	5838	AM41	5843	1N555R
5804	SD93	5805	1N2404	5809	1N3752	5838	AM42	5843	3C60R
5804	SD93A	5805	1N2413	5809	1N3761	5838	AM44	5843	3FR60
5804	SD94	5805	1N2863A	5809	1N3869	5838	BAV205	5846	BAV208
5804	SD94A	5805	1N3229	5809	1N3957	5838	GB-5004	5846	BAV209
5804	SE-0.5A	5805	1N4821	5809	1N4725	5838	1N1039	5846	BAV218
5804	SE05A	5805	1N5175	5809	1N5005	5838	1N1045	5846	BAV219
5804	SG005	5805	1N5396	5809	3B10	5838	1N1051	5846	BAV228
5804	SG305	5805	1S44	5809	20A10	5838	1N1118	5846	BAV229
5804	SJ405P	5805	3A500	5810	30310	5838	1N1125A	5846	BAV309
5804	SLA1103	5805	20A5	5814	13-34901-1	5838	1N1126	5846	BAV318
5804	SLA24A	5806	3082	5818	D2412A	5838	1N126A	5846	BAV319
5804	SLA24B	5806	3085	5818	D2412B	5838	1N1448	5846	BAV328
5804	SLA24C	5806	A15M	5818	D2412F	5838	1N1541	5846	BAV329
5804	SLA25A	5806	AC500	5820	D2412C	5838	1N1542	5846	BY209
5804	SLA25B	5806	AC600	5820	D2412D	5838	1N1554	5846	BY219
5804	SLA25C	5806	BY207	5820$	BYX61-400	5838	1N1585	5846	BY228
5804	SLA5200	5806	BY217	5822	BYX62-600	5838	1N2029	5846	BY229
5804	SR1378-2	5806	S1060	5822	BYX66-600	5838	1N2293	5846	BY308
5804	SR3AM6	5806	S3A5	5822	D2412M	5838	1N2293A	5846	BY309
5804	SR3AM8	5806	S3A6	5830	N5I2006	5838	1N2515	5846	BY318
5804	SSIC1120	5806	SI5	5830	AM1	5838	1N255	5846	BY319
5804	SSIC0820	5806	SLA5201	5830	AM2	5838	1N2850	5846	BY328
5804	SSIC1220	5806	SSIC1240	5830	AM4	5838	1N332	5846	BY329
5804	TS4	5806	UT2060	5830	AM5	5838	1N333	5846	PR9000
5804	UR225	5806	UT215	5830	1N1537	5838	1N335	5846	PR9001
5804	UT2040	5806	UT238	5830	1N1581	5838	1N341	5846	1N1235
5804	UT213	5806	UT247	5830	1N2026	5838	1N342	5846	1N1235A
5804	UT235	5806	UT257	5830	1N2216	5838	1N343	5846	1N1235B
5804	UT244	5806	UT267	5830	1N2217	5838	1N344	5846	1N1236B
5804	UT254	5806	1N1222A	5830	1N2266	5838	1N552	5846	1N2222
5804	UT264	5806	1N1223	5830	1N2267	5838	1N553	5846	1N2222A
5804	UT3040	5806	1N1223A	5830	1N2289	5838	1N611	5846	1N2223
5804	WRR1955	5806	1N1224B	5830	1N2289A	5838	1N611A	5846	1N2223A
5804	WRR1956	5806	1N1568	5830	1N2290A	5838	1N612	5846	1N2231
5804	X-23305-3	5806	1N1653	5830	1N2348	5838	1N612A	5846	1N2532
5804	X-23305-4	5806	1N2375	5830	1N2534	5838	S012	5846	1N2542
5804	ZR13	5806	1N2396	5830	1N2535	5839	GE-5005	5846	1N2543
5804	ZR14	5806	1N2405	5830	1N607	5839	1N1585R	5846	1N2553
5804	ZR604	5806	1N2414	5830	1N607A	5839	3PR40	5846	1N2554
5804	ZS274	5806	1N2864A	5830	305A	5840	AM51	5846	1N562
5804	1.5B4	5806	1N3081	5831	1N1046	5840	AM52	5846	1S1356
5804	1.5V4	5806	1N3084	5831	1N1537R	5840	AM54	5846	1S1629
5804	1N1222	5806	1N3106	5831	1N1581R	5840	BAV206	5846	18417
5804	1N1559	5806	1N3230	5831	3PR5	5840	BAV216	5846	3070
5804	1N1566	5806	1N3241	5832	AM1	5840	BAV226	5846	3080
5804	1N1650	5806	1N3250	5832	AM12	5840	BAV306	5846	3P80
5804	1N1651	5806	1N5658	5832	1N1115	5840	BAV316	5846	3LC12
5804	1N1763A	5806	1N3750	5832	1N1551	5840	BAV326	5846	305F
5804	1N2019	5806	1N3759	5832	1N1582	5840	BY216	5846	305B
5804	1N2020	5806	1N3868	5832	1N2349	5840	BY306	5848	BY2001
5804	1N2394	5806	1N3940	5832	1N2512	5840	BY316	5848	BY2002
5804	1N2403	5806	1N3983	5832	1S146	5840	BY326	5848	BY2101
5804	1N2412	5806	1N4723	5832	3P10	5840	1N1119	5848	BY2102
5804	1N2862A	5806	1N4822	5832	305B	5840	1N1127	5848	BY2201
5804	1N3078	5806	1N5001	5832	1N035	5840	1N1233	5848	BY2202
5804	1N3079	5806	1N5007	5834	AM21	5840	1N1233A	5848	BY3001
5804	1N3083	5806	1N5176	5834	AM22	5840	1N1233B	5848	BY3002
5804	1N3240	5806	1N5201	5834	AM24	5840	1N1555	5848	BY3101
5804	1N3249	5806	1N5206	5834	NA21	5840	1N1586	5848	BY3102
5804	1N3697	5806	1N5597	5834	1N1036	5840	1N2030	5848	BY3201
5804	1N3749	5806	1S1890	5834	1N1037	5840	1N2218	5848	BY3202
5804	1N3756	5806	1S45	5834	1N1041	5840	1N2268	5848	BYX38/900
5804	1N3758	5806	3AP6	5834	1N1042	5840	1N2269	5848	PR9002
5804	1N3867	5806	3B6	5834	1N1043	5840	1N2529	5848	PR9003
5804	1N3959	5806	1586	5834	1N1047	5840	1N2540	5848	S1000
5804	1N3982	5806	20A6	5834	1N1048	5840	1N2551	5848	1N1444
5804	1N4089	5806	3086	5834	1N1049	5840	1N2851	5848	1N1444A
5804	1N4442	5808	A15N	5834	1N1116	5840	3C50	5848	1N2224
5804	1N4722	5808	AC800	5834	1N1124	5840	3P50	5848	1N2224A
5804	1N4820	5808	A83F	5834	1N1124A	5840	5P50D	5848	1N2225
5804	1N5000	5808	BY208	5834	1N1538	5840	50P1	5848	1N2225A
5804	1N5006	5808	S1070	5834	1N1539	5840	505K	5848	1N2272
5804	1N5058	5808	S3A8	5834	1N1540	5841	1N1127R	5848	1S565
5804	1N5174	5808	SD-7	5834	1N1552	5841	1N1543R	5848	1S1557
5804	1N5200	5808	SG805	5834	1N1583	5841	1N1586R	5848	1S1630
5804	1N5395	5808	SR18	5834	1N2027	5841	1N1554R	5848	18419
5804	1N5404	5808	UT258	5834	1N2550	5841	3C50R	5848	3C100
5804	1P647	5808	UT268	5834	1N2513	5841	3PR50	5848	3090
5804	1S023	5808	UT361	5834	1N253	5842	AM64	5848	3P100
5804	1S034	5808	UT362	5834	1N254	5842	BAV207	5848	3NC12
5804	1S1073	5808	1N2374	5834	1N336	5842	BAV217	5848	305V
5804	1S2550	5808	1N2397	5834	1N337	5842	BAV227	5848	5052
5804	1S2379	5808	1N2398	5834	1N338	5842	BAV307	5849	GB-5013
5804	1S2412	5808	1N2406	5834	1N339	5842	BAV317	5849	PR9002R
5804	1S2419	5808	1N2407	5834	1N340	5842	BAV327	5849	PR9003R
5804	1S43	5808	1N2415	5834	1N345	5842	BY227	5849	1N563R
5804	1S434	5808	1N2416	5834	1N346	5842	BY307	5849	3C100R
5804	1S922(RECT)	5808	1N2501	5834	1N347	5842	BY317	5849	3090R
5804	1WS4	5808	1N2505	5834	1N348	5842	BY327	5849	3PR100
5804	3-1477	5808	1N2772	5834	1N349	5842	GE-500B	5850	1N1341
5804	3S43	5808	1N2773	5834	1N550	5842	1N120	5850	1N1341A
5804	3SM4	5808	1N2865	5834	1N551	5842	1N127A	5850	1N1341B
5804	15C4	5808	1N2867	5834	1N608	5842	1N128	5850	1N2491
5804	1S84	5808	1N3080	5834	1N608A	5842	1N128A	5850	407A
5804	20A4	5808	1N3251	5834	1N609	5842	1N1234	5851	1N1341AR
5804	200A	5808	1N3242	5834	1N609A	5842	1N1234A	5851	1N1341BR
5804	3085	5808	1N3251	5834	1N610	5842	1N1234B	5851	1N1612R
5804	3084	5808	1N3751	5834	1N610A	5842	1N1543	5851	6FR5
5804	48-137327	5808	1N3760	5834	1S1352	5842	1N1544	5851	6FR5-D
5804	48-137340	5808	1N4514	5834	1S1660	5842	1N1587	5851	6FR5A
5804	152-0047	5808	1N4585	5834	3P20	5842	1N2031	5851	407RA
5804	152-0047-00	5808	1N4586	5834	3P20D-C	5842	1N2219	5852	AYY10-120
5804	200RB4	5808	1N4724	5834	2085	5842	1N2220		
5804	211-58	5808	1N5002	5834	305C	5842	1N2221		
5804	250R4B	5808	1N5052	5834	305D	5842	1N2270		
		5808	1N5053	5835	1N1538R	5842	1N2271		

ECG	Industry Standard No.	ECG	Industry Standard No.	ECG	Industry Standard No.	ECG	Industry Standard No.	ECG	Industry Standard No.
5852	1N342	5862	8518383-10	5874	19B200011-P3	5882	8600	5890	304V
5852	1N342A	5863	GE-5025	5874	19B200011-P4	5882	86AN12	5890	304Z
5852	1N342B	5863	1N348AR	5874	20I3	5882	86AN6	5890	367V
5852	1N342BR	5863	1N348BR	5874	20J2	5882	SJ603	5890	367Z
5852	1N613	5863	1N348R	5874	20J3P	5882	SJ603E	5890	404V
5852	1N2492	5863	407RM	5874	20N2	5882	SJ603K	5890	404Z
5852	1N3569	5866	1N3649	5874	20N3	5882	SJ604	5890	408V
5852	10H3P	5866	1N3987	5874	20P2	5882	SJ604E	5890	408Z
5852	366B	5866	1N3988	5874	20P3	5882	SJ604K	5890	437V
5852	407B	5866	6A700	5874	30J2	5882	SL500	5891	BYX39-1000R
5852	530091-1	5866	6A800	5874	40H3	5882	SL600	5891	BYX40-1000R
5853	1N342R	5868	BY4001	5874	40P3	5882	TM56	5891	BYX42-900R
5853	1N342AR	5868	BY4002	5874	252-0014	5882	TR1125	5891	GE-5045
5853	1N615R	5868	BY4101	5874	478C	5882	TR1126	5891	MR1130R
5853	6FR10	5868	BY4102	5874	40110	5882	ZR206	5891	PR9010R
5853	6FR10-D	5868	BY4201	5875	GE-5033	5882	1N1206A	5891	PR9011R
5853	6FR10A	5868	BY4202	5875	MR1122R	5882	1N1616	5891	1N3672AR
5853	6FR10B	5868	BY7001	5875	1N1201AR	5882	1N2258	5891	1N3672R
5853	407RB	5868	BY7002	5875	1N1201BR	5882	1N2258A	5891	1N3673AR
5854	GE-5016	5868	BY7101	5875	1N1201R	5882	1N2259	5891	1N3673R
5854	1N343	5868	BY7102	5875	1N1202AR	5882	1N2259A	5891	1N4014R
5854	1N343A	5868	BY7201	5875	1N1202BR	5882	1N4508	5891	1N4015R
5854	1N343B	5868	BY7202	5875	1N1202R	5882	62P60	5891	12C100R
5854	1N344	5868	GE-502B	5875	12FR10B	5882	46R2	5891	12C90R
5854	1N344A	5868	PR9006	5875	12FR15	5882	66R2	5891	12FR100
5854	1N344B	5868	PR9007	5875	12FR15A	5882	404M	5891	12FR100A
5854	1N614	5868	1N2242	5875	12FR15B	5882	408M	5891	12FR100B
5854	1N2493	5868	1N2242A	5875	12FR20	5882	227675	5891	12FR90
5854	1N3570	5868	1N2243	5875	12FR20A	5882	227724	5891	12FR90A
5854	20H3P	5868	1N2243A	5875	12FR20B	5882	227801	5891	12FR90B
5854	407C	5868	1N2559	5875	408RA	5882	229042	5891	12NF11
5854	407D	5868	1N2560	5875	408RC	5882	230773	5891	337V
5855	GE-5017	5868	1N2563	5876	MR1123	5882	232519	5891	337Z
5855	1N343AR	5868	1N2564	5876	1N1203A	5882	235157	5891	367RV
5855	1N343BR	5868	1N2574	5876	1N1623	5882	256729	5891	367RZ
5855	1N343R	5868	1N2575	5876	1N2252	5882	262648	5892	1N3615
5855	1N344AR	5868	1N3650	5876	1N2252A	5882	263807	5892	16P5
5855	1N344BR	5868	1N3919	5876	1N2253A	5882	270779	5892	1N3616
5855	1N344R	5868	1N3989	5876	404F	5882	300532	5894	16P10
5855	1N614R	5868	1N3990	5876	408F	5882	300735	5896	TR150
5855	6FR15	5868	1N4459	5876	40111	5883	GE-5041	5896	1N3617
5855	6FR15A	5868	181627	5877	1N1203AR	5883	MR1126R	5896	16P15
5855	6FR15B	5868	6A1000	5877	1N1203BR	5883	1N1206AR	5896	16P20
5855	6FR20	5868	6A900	5877	1N1203R	5883	1N1206BR	5896	324-0118
5855	6FR20-D	5868	6C100	5877	12FR30	5883	1N1206R	5896	3680
5855	6FR20A	5868	6090	5877	12FR30A	5883	12FR60	5896	368D
5855	6FR20B	5868	6P100	5877	12FR30B	5883	12FR60A	5896	409C
5855	20H3PN	5868	6P100-D	5877	408RF	5883	12FR60B	5896	409D
5855	407RC	5868	6P100A	5877	408RH	5883	66R2B	5898	16P50
5855	407RD	5868	6P100B	5877	40111R	5883	408RM	59	2057A2-277
5856	1N345	5868	6P90	5878	BYX42/300	5883	40114R	5900	1N3619
5856	1N345A	5868	6P90A	5878	BYX48/300	5883	6142190	5900	1N3620
5856	1N345B	5868	6P90B	5878	GE-5056	5886	S417	5900	5M40
5856	1N2252A	5868	6NC12	5878	HEP-R0134	5886	S57	5900	8AM40
5856	1N2494	5868	50R2B	5878	MR1124	5886	S58	5900	16P40
5856	1N3571	5868	341V	5878	SL3	5886	S800	5900	40Q4
5856	30H3P	5868	341Z	5878	SL400	5886	SL800	5900	368H
5856	407F	5868	366V	5878	TR1123	5886	SL800X	5900	409H
5857	1N345AR	5868	366Z	5878	TR1124	5886	TR1128	5902	16250
5857	1N345BR	5868	407V	5878	1N1204A	5886	1N2260A	5902	16P50
5857	1N345R	5868	407Z	5878	1N1615	5886	1N4012	5904	1N3622
5857	6FR30	5868	441V	5878	1N1624	5886	1N4013	5904	5M60
5857	6FR30A	5868	441Z	5878	1N2254	5886	18427	5904	16P60
5857	6FR30B	5868	446V	5878	1N2254A	5886	12A700	5904	368M
5857	407RF	5868	446Z	5878	1N2255	5886	12A800	5904	409M
5858	GE-5020	5869	BYX48-900R	5878	1N2255A	5886	12P80B	5908	1N3623
5858	1N346	5869	GE-5029	5878	1N4507	5886	40115	5910	BYX38-900
5858	1N346A	5869	PR9006R	5878	4043P	5890	BY5001	5910	1N3624
5858	1N346B	5869	PR9007R	5878	404C	5890	BY5002	5910	8AM100
5858	1N2253	5869	1N3989R	5878	404D	5890	BY5101	5910	16P100
5858	1N2253A	5869	1N3990R	5878	404H	5890	BY5102	5910	368V
5858	1N2254	5869	6C100R	5878	408C	5890	BY5201	5910	368Z
5858	1N2254A	5869	6090R	5878	408D	5890	BY5202	5910	409V
5858	1N2255	5869	6PR100	5878	408H	5890	BY8001	5910	409Z
5858	1N2255A	5869	6PR100A	5878	40112	5890	BY8002	5911	BYX38-900R
5858	1N2495	5869	6PR90	5879	GE-5037	5890	BY8101	5911	1N3624R
5858	1N3572	5869	6PR90A	5879	MR1124R	5890	BY8102	5911	368RV
5858	6GC12	5869	6PR90B	5879	1N1204AR	5890	BY8201	5911	368RZ
5858	40H3P	5869	346V	5879	1N1204BR	5890	BY8202	5911	409RV
5858	407H	5869	346Z	5879	1N1204R	5890	BYX39-1000	5911	409RZ
5859	GE-5021	5869	366RV	5879	12FR40	5890	BYX40-1000	5916	BYX46-200
5859	1N346AR	5869	366RZ	5879	12FR40A	5890	BYX42-900	5916	GE-5064
5859	1N346BR	5869	407RV	5879	12FR40B	5890	BYX49-900	5916	S2020
5859	1N346R	5869	407RZ	5879	408RD	5890	BYX48/900	5916	TR152
5859	6FR40	5870	HEP-R0130	5879	40112R	5890	GE-5044	5917	GE-5065
5859	6FR40-D	5870	MR1120	5880	BAV806	5890	HEP-R0138	5920	BYX46-400
5859	6FR40A	5870	SL120	5880	BAV816	5890	MR1130	5920	BYX66-400
5859	6FR40B	5870	TR1120	5880	BAV826	5890	PR9011	5920	GE-5068
5859	40H3PN	5870	1N1199A	5880	BY806	5890	S10AN12	5920	S2125
5859	407RH	5870	1N2246	5880	BY816	5890	S10AN6	5920	S2130
5860	BAV706	5870	1N2246A	5880	BY826	5890	S59	5920	S2135
5860	BAV716	5870	40108	5880	MR1123R	5890	S9AN12	5920	S2140
5860	BAV726	5871	MR1120R	5880	MR1125	5890	S9AN6	5920	40Q3
5860	BY706	5871	1N1199AR	5880	1N1205A	5890	SJ10003EK	5924	BYX58-600
5860	BY716	5871	1N1199R	5880	1N2256	5890	SL10	5924	BYX46-600
5860	BY726	5871	12FR5	5880	1N2256A	5890	SL1000	5924	S2045
5860	1N347	5871	12FR5A	5880	1N2257	5890	SL1000X	5925	GE-5073
5860	1N347A	5871	12FR5B	5880	1N2257A	5890	TR1130	5928	16P80
5860	1N2256	5871	40108R	5880	12P50	5890	12B100A	5932	BYX66-1000
5860	1N2256A	5872	MR1121	5880	404K	5890	1N2262	5932	GE-5076
5860	1N2257	5872	TR1121	5880	408K	5890	1N2262A	5932	S2090
5860	1N2257A	5872	1N1200A	5880	40113	5890	1N2263	5933	GE-5077
5860	1N2496	5872	1N2248	5881	MR1125R	5890	1N2263A	5940	A40F
5860	1N2516	5872	1N2248A	5881	1N1205AR	5890	1N2585	5940	1N3208
5860	1N3573	5872	046-40209	5881	1N1205BR	5890	1N2586	5940	405A
5860	6P50	5872	304B	5881	1N1205R	5890	1N2596	5942	A40A
5860	50H3P	5872	404B	5881	12FR50	5890	1N2597	5942	P1004
5860	407K	5872	408B	5881	12FR50A	5890	1N2607	5942	1N3209
5861	1N347AR	5872	40109	5881	12FR50B	5890	1N3672	5944	A40B
5861	1N347BR	5873	MR1121R	5881	408RK	5890	1N3672A	5944	BY313
5861	1N347RB	5873	1N1200AR	5881	40113R	5890	1N3673	5944	GE-5048
5861	1N347R	5873	1N1200R	5882	BAV807	5890	1N3924	5944	HEP-R0161
5861	1N616	5873	12FR10	5882	BAV817	5890	1N4014	5944	HEP-R0162
5861	407RK	5873	046-07037	5882	BAV827	5890	1N4015	5944	1N2021
5862	BAV707	5873	232-0007	5882	BY807	5890	1N4458	5944	1N2786
5862	BAV717	5873	408RB	5882	BY817	5890	1N4510	5944	1N3210
5862	BAV727	5873	40109R	5882	BY827	5890	1N4578	5944	5P15
5862	BY707	5873	GE-5032	5882	BYX39/600	5890	18945	5944	5P20
5862	BY717	5873	GEMR-1	5882	BYX42/600	5890	18930	5944	51-10650A01
5862	BY727	5874	HEP-R0131	5882	BYX48/600	5890	11R108	5944	0200
5862	GE-5024	5874	HEP-R0132	5882	FD-1029-JE	5890	12A1000	5944	0201
5862	1N348	5874	MR1122	5882	GE-5040	5890	12A900	5944	0221
5862	1N348A	5874	TM22	5882	HEP-R0136	5890	12C100	5944	405B
5862	1N348B	5874	TR1122	5882	MR1126	5890	12C90	5944	405C
5862	1N2258	5874	1N1202A	5882	R0130	5890	12F100	5944	405D
5862	1N2258A	5874	1N1612	5882	R0131	5890	12F100A	5945	A41B
5862	1N2259	5874	1N1622	5882	R0132	5890	12F90	5945	8AN20
5862	1N2259A	5874	1N2250	5882	R0134	5890	12F90A	5946	A40C
5862	1N2497	5874	1N2250A	5882	R0136	5890	12F90B	5946	1N3211
5862	1N3574	5874	1N4506	5882	S415	5890	70R2S	5948	12GC11
5862	6P60	5874	19B200011-P1	5882	S56			5952	A40M
5862	46R2S	5874	19B200011-P2	5882	S5AN12			5952	12JC11
5862	60H3P			5882	S5AN6			5953	A41X
5862	407M								

ECG	Industry Standard No.
5953	A41M
5966	1N3493
5980	HEP-R0250
5980	83105
5980	83205
5980	001-024070
5980	1N1183
5980	1N1183A
5980	1N1191
5980	1N1191A
5980	1N3304
5980	1N1434
5980	1N2154
5980	1N248
5980	1N248A
5980	1N248B
5980	1S2446
5980	20R2
5980	2S005
5980	35005
5980	35H5
5980	40A50
5980	40HP5
5980	57-18
5980	371A
5980	402A
5980	403A
5980	3105
5980	40208
5980	198765-1
5980	8518382-1
5981	1N1183R
5981	1N1191R
5981	1N2154R
5981	5Q5
5981	2S005R
5981	35005R
5981	40HPR5
5981	335A
5981	336A
5981	435A
5981	436A
5981	40208R
5981	198765-2
5982	83110
5982	83210
5982	TR100
5982	1N1184
5982	1N1184A
5982	1N1192
5982	1N1192A
5982	1N1302
5982	1N1435
5982	1N2155
5982	1N249A
5982	1N249B
5982	1N249C
5982	301
5982	401
5982	10M10
5982	21R18
5982	25C10
5982	35C10
5982	40A100
5982	40HP10
5982	302B
5982	303B
5982	371B
5982	402B
5982	417B
5982	418B
5982	419B
5982	3110
5982	198765-3
5982	7285354
5982	8518382-2
5983	1N1184R
5983	1N1192R
5983	1N2155R
5983	25C10R
5983	35C10R
5983	40HPR10
5983	232-0008
5983	335B
5983	371RB
5983	417RB
5983	419RB
5983	435B
5983	3110R
5983	40209R
5983	198765-4
5983	8518382-9
5986	GE-5096
5986	HEP-R0251
5986	HEP-R0253
5986	RN120
5986	81AN51
5986	81AN40
5986	81BN31
5986	81BN40
5986	83115
5986	83120
5986	83215
5986	83220
5986	ST4-10
5986	ST4-20
5986	TR151
5986	TR200
5986	1N1185
5986	1N1185A
5986	1N1186
5986	1N1186A
5986	1N1193
5986	1N1193A
5986	1N1194
5986	1N1194A
5986	1N3304
5986	1N1436
5986	1N2156
5986	1N248C
5986	N249
5986	1N250A
5986	1N250B
5986	1N4525
5986	1S1654
5986	182447
5986	182448
5986	18936
5986	3C2
5986	4C2
5986	10R20
5986	21R28
5986	22R22
5986	25C15
5986	25C05
5986	31R2
5986	32R2
5986	35C15
5986	35C20
5986	35H10
5986	35H20
5986	40A150
5986	40A200
5986	40HP15
5986	40HP20
5986	96-5246-01
5986	302A
5986	302C
5986	302D
5986	303A
5986	303C
5986	303D
5986	371C
5986	371D
5986	402C
5986	402D
5986	403B
5986	403C
5986	403D
5986	417A
5986	417C
5986	417D
5986	418A
5986	418C
5986	418D
5986	419A
5986	419C
5986	419D
5986	3115
5986	3120
5986	40210
5986	198764-5
5986	198765-5
5986	8518382-3
5987	GE-5097
5987	R3110
5987	R3115
5987	R3120
5987	R3210
5987	R3215
5987	R3220
5987	81AN51R
5987	81AN40R
5987	81BN31R
5987	81BN40R
5987	1N1185R
5987	1N1186AR
5987	1N1186R
5987	1N1193R
5987	1N1194R
5987	1S1654R
5987	25C15R
5987	35C20R
5987	35H10R
5987	35H20
5987	40HPR15
5987	40HPR20
5987	335C
5987	335D
5987	336C
5987	336D
5987	371RC
5987	371RD
5987	417RA
5987	417RC
5987	417RD
5987	419RA
5987	419RC
5987	419RD
5987	435C
5987	435D
5987	436B
5987	436C
5987	436D
5987	3115R
5987	3120R
5987	40210R
5987	198764-6
5987	1662259-6
5987	8518382-10
5988	BYX52-300
5988	83125
5988	83130
5988	83225
5988	83250
5988	TR252
5988	1N1187A
5988	1N1195
5988	1N1195A
5988	1N1306
5988	1N2157
5988	1N2282
5988	1S1655
5988	1S2449
5988	10M30
5988	21R58
5988	25C30
5988	35C30
5988	40A300
5988	40HP30
5988	302E
5988	302F
5988	303E
5988	303F
5988	371F
5988	402F
5988	403F
5988	417F
5988	418F
5988	419F
5988	3125
5988	3130
5988	40211
5988	198765-7
5988	BYX52-300R
5989	1N1187R
5989	1N1195R
5989	1N2157R
5989	40HPR30
5989	335F
5989	336F
5989	371RF
5989	417RF
5989	419RF
5989	435F
5989	3125R
5989	3130R
5989	40211R
5989	198765-8
5990	GE-5100
5990	GEMR-2
5990	HEP-R0254
5990	HEP-R0255
5990	R43HZ
5990	R48Z
5990	83135
5990	83140
5990	83235
5990	83240
5990	GE-5104
5990	ST3-10
5990	ST3-20
5990	ST3-30
5990	ST3-40
5990	ST4-30
5990	ST4-40
5990	TR251
5990	TR300
5990	TR351
5990	TR352
5990	TR400
5990	TR401
5990	TR402
5990	1N1187
5990	1N1188
5990	1N1196
5990	1N1196A
5990	1N1437
5990	1N2158
5990	1N2283
5990	1N250C
5990	1N4526
5990	1S1841
5990	3C4
5990	4C4
5990	10M40
5990	2083
5990	2084
5990	21R48
5990	24R2
5990	25C40
5990	34R2
5990	35C40
5990	35H30
5990	35H40
5990	40A400
5990	40HP40
5990	40R5
5990	4085
5990	302G
5990	302H
5990	303G
5990	303H
5990	371H
5990	402H
5990	403H
5990	417H
5990	418H
5990	419H
5990	3135
5990	3140
5990	3235
5990	40212
5990	198765-9
5991	GB-5101
5991	R3125
5991	R3130
5991	R3135
5991	R3140
5991	R3225
5991	R3230
5991	R3235
5991	R3240
5991	1N1188R
5991	1N1196R
5991	1N2158R
5991	25C30R
5991	25C40R
5991	35C30R
5991	35C40R
5991	35HR30
5991	35HR40
5991	40HPR40
5991	335H
5991	336H
5991	371RH
5991	417RH
5991	419RH
5991	435H
5991	436F
5991	436H
5991	2244-1
5991	3140R
5991	40212R
5991	198765-10
5992	83145
5992	83150
5992	83245
5992	83260
5992	TR502
5992	1N1189
5992	1N1189A
5992	1N1197
5992	1N1197A
5992	1N2159
5992	1N2284
5992	1S2451
5992	10M50
5992	21R58
5992	25C50
5992	40A500
5992	40HP50
5992	302K
5992	303K
5992	371K
5992	402K
5992	403K
5992	417K
5992	418K
5992	419K
5992	3145
5992	3245
5992	40213
5992	198765-11
5992	3680322-3
5993	1N1189R
5993	1N1197R
5993	1N2159R
5993	25C50R
5993	40HPR50
5993	335K
5993	336K
5993	371RK
5993	417RK
5993	419RK
5993	435K
5993	3150R
5993	40213R
5993	198765-12
5994	BYX52-600
5994	BYX56-600
5994	GE-5104
5994	HEP-R0256
5994	HEP-R0257
5994	R-63HZ
5994	R0160
5994	R0161
5994	R0162
5994	R0164
5994	R0250
5994	R0251
5994	R0253
5994	R0254
5994	R0255
5994	R0256
5994	R0257
5994	R6HZ
5994	83160
5994	83250
5994	TR602
5994	1N1190
5994	1N1190A
5994	1N1198A
5994	1N1438
5994	1N2160
5994	1N2285
5994	1N4527
5994	1S1842
5994	3C6
5994	4A1122
5994	4A162
5994	4A2122
5994	4A232
5994	4A262
5994	406
5994	4G1122
5994	4G132
5994	4G162
5994	4G2122
5994	4G232
5994	4G262
5994	6DC11
5994	10M60
5994	20HA3
5994	21R68
5994	25C60
5994	26R2
5994	26R28
5994	36R2
5994	36R28
5994	40A600
5994	40HP60
5994	124-0165
5994	302M
5994	303M
5994	371M
5994	402M
5994	403M
5994	417M
5994	418M
5994	419M
5994	3160
5994	40214
5994	198764-13
5994	198765-13
5994	225265
5994	227676
5994	229088
5994	230756
5994	232520
5994	235299
5994	237509
5994	240077
5994	241420
5994	255903
5994	256122
5994	256728
5994	256730
5994	258884
5994	262310
5994	265115
5994	300312
5994	300733
5994	BYX52-600R
5995	BYX56-600R
5995	GE-5105
5995	83145
5995	83150
5995	83160
5995	83245
5995	83260
5995	1N1190R
5995	1N1198R
5995	1N2160R
5995	25C60R
5995	40HPR60
5995	127A905PO1
5995	128A157PO1
5995	128A157PO2
5995	131A246PO1
5995	131A246PO2
5995	222-1
5995	222-2
5995	335M
5995	336M
5995	371RM
5995	417RM
5995	419RM
5995	435M
5995	436M
5995	436K
5995	1321-4051
5995	1471-4729
5995	40214R
5995	198765-14
5995	1851490
5995	4420022-P1
5995	4420022-P2
5995	24561601-E
5995	24561602-E
5995	24561603
5998	1N2286
5998	12LC11
600	HV-23
600	HV-25(ELGIN)
600	HV26
600	M8513
600	M8513A-R
600	M8513AO
600	M8513R
6002	BYX52-900
6002	BYX56-1000
6002	GE-5108
6002	PR9014
6002	PR9015
6002	PR9025
6002	PR9026
6002	R4101022
6002	R4101040
6002	1N2287
6002	1N3768
6002	1N4529
6002	1S1844
6002	2WMT10
6002	3010
6002	4010
6002	10M100
6002	12NC11
6002	21R108
6002	25C100
6002	25C90
6002	30R25
6002	35C100
6002	35C90
6002	40HP100
6002	40HP90
6002	40R28
6002	302V
6002	302Z
6002	303V
6002	303Z
6002	371V
6002	371Z
6002	402Z
6002	403V
6002	403Z
6002	418V
6002	418Z
6002	419V
6003	BYX52-900R
6003	BYX56-1000R
6003	GE-5109
6003	PR9014R
6003	PR9015R
6003	R4111022
6003	R4111040
6003	1N3767R
6003	1N3768R
6003	13R418
6003	25C100R
6003	25C90R
6003	35C100R
6003	35C90R
6003	40HP100R
6003	40HP90R
6003	335V
6003	335Z
6003	336V
6003	336Z
6003	371NV
6003	371RZ
6003	417RV
6003	417RZ
6003	419NV
6003	436V
6003	436Z
6006	D2540A
6006	D2540B
6006	D2540F
6008	D2540D
601	DX-0150
601	EA16X68
601	EA16X69
601	EA16X73
601	HP-20105
601	HV-80
601	HV0000206
601	HV0000502
601	HV80
601	IP20-0120
601	KB-162
601	KB-16205
601	KB-16205A
601	KB-165
601	KB-265
601	KB-269
601	KB162N
601	KB165
601	KB169
601	KB265S
601	KB265A
601	KB269
601	M8513A
601	MA-26
601	MA-26-1
601	MA26
601	MA26A
601	MV-13
601	MV1
601	MV3
601	RH-VXQ0004TAZZ
601	RVDVD1150L
601	83016R
601	SV-9
601	VD-1122
601	VD-1124
601	VD1120
601	VD1121
601	VD1122
601	VD1123
601	VD1124
601	VD1150M
601	181209
601	181210
601	1S1212
601	1S1212A
601	21A111-002
601	48-555045
601	48M555046
601	265P03301
601	2017-111
601	8000-00011-045
601	129475
601	2003069-5
6010	D2540M

ECG	Industry Standard No.	ECG	Industry Standard No.	ECG	Industry Standard No.	ECG	Industry Standard No.	ECG	Industry Standard No.
6020	PR9034	6054	TR153	612	1S2687D	614	2002207-2	6409	SU44
6020	1N2128	6054	TR203	612	13352	614	5330661	6409	ZEN129
6020	1N2128A	6054	1N4136	612	13352M	614	5330852	6409	2N2647
6020	1N2446	6054	25H15	612	1848	614	17210430	6409	2SH22
6020	1N2458	6054	25H15A	612	18553	614	TVSMPC574J	6410	2D030
6020	2505	6054	25H20A	612	1885	614	8-759-157-40	6410	2N4870
6020	25HB5	6054	70H15	612	1886	615	103-237	6410	2N4871
6020	50005	6054	70H15A	612	022-2823-501	615	5350611	6410	132650
6020	60HP5	6054	70H20	612	48-45323901	6154	1N1399	6410	656064
6022	1N2447	6054	70H20A	612	754-4000-553	6154	1N3288	703A	C555A
6022	1N2459	6055	GB-5129	612	903-47	6154	1N3289	703A	FU5D770331X
6022	60HP10	606	SV03	612	903-47B	6154	1N3290	703A	FU5D770339
6026	PR9036	6060	GE-5132	612	914-000-7-00	6154	1N3291	703A	HC1000109
6026	1N2130	6060	83640	613	117729	6156	1N3292	703A	HC1000109-0
6026	1N2130A	6060	TR253	613	125399	6156	1N3293	703A	HC1000111-0
6026	1N2131	6060	TR353	613	126149	6158	1N3295	703A	IC101-109
6026	1N2131A	6060	TR403	613	129359	6158	18940	703A	IC40
6026	1N2448	6060	1N1401	613	256717	6159	1N3295R	703A	ICP-1
6026	1N2449	6060	1N4137	613	741052	616	BB105B	703A	ICP-1-6826
6026	1N2460	6060	25H40	613	EA16X117	616	BB109	703A	IRF03B
6026	1N2461	6060	70H40	613	GO1-406-A	616	182207	703A	IP20-0174
6026	1N2788	6060	70H40A	613	G222	616	182208	703A	L103I1
6026	1N3142	6061	GE-5133	613	ITT310	616	103-146	703A	LA703E
6026	1N3968	6064	40JH3R	613	ITT310G	616	103-176	703A	LM703L
6026	20JE5	6064	GE-5136	613	MV1634	6354	MR12378B	703A	LM703LM
6026	20M20	6064	1N4138	613	MV2107	6354	1N4045	703A	NJ703N
6026	25G20	6064	25H50	613	MV832	6354	1N4048	703A	NJM-703N
6026	25HB15	6064	25H50A	613	PC136	6355	181643R	703A	NJM703
6026	50020	6064	25H60	613	RB195	6356	A90M	703A	PA7703
6026	60HP15	6064	25H60A	613	SC-15	6400A	BB847(ELCOM)	703A	PAT703E
6026	60HP20	6064	50JH5	613	SC-20	6400A	HEP310	703A	PAT703X
6027	PR9036R	6064	60JH3	613	001-0130-00	6400A	IR2160	703A	QA703E
6027	R3415	6064	70H50	613	1N4792	6400A	M9264	703A	RE300-IC
6027	R3420	6064	70H50A	613	1N5687	6400A	001	703A	SL7059
6027	1N2130R	6064	70H60	613	1N5702	6400A	2D022	703A	SL7283
6027	1N2131R	6065	GE-5137	613	1S3518R	6400A	2D022-211	703A	SL7308
6027	1N2788R	607	SV04	613	18554	6400A	2N1671	703A	SL7531
6027	1N3968R	6072	GE-5140	613	18555	6400A	2N2160	703A	SL7593
6027	25G20R	6072	R4101070	613	18893	6400A	2SB12	703A	SL8020
6027	50020R	6072	836100	613	13-33177-1	6400A	2SB22	703A	T18
6030	1N2450	6072	83690	613	603000002	6400A	4JD5B29	703A	TRO-9005
6030	1N2451	6072	181615	613	BB104	6400A	4JX5B670	703A	TRO-9006
6030	1N2462	6072	25H100	613	BB109G	6400A	14-593-01	703A	1E703E
6030	1N2463	6072	25H90	614	BB104	6400A	14-593-03	703A	01F
6030	60HP30	6072	25H90A	614	BB109	6400A	48-134792	703A	01F-8L8020
6034	PR9038	6072	70H100	614	BE-55	6400A	48-137058	703A	4-007
6034	1N2134	6072	70H100A	614	BJ-155	6400A	48-137165	703A	4-008
6034	1N2134A	6072	70H90	614	BN-55	6400A	48-137282	703A	09-308004
6034	1N2135	6072	70H90A	614	D7N	6400A	48-869206	703A	09-308013
6034	1N2135A	6073	GE-5141	614	DS-55	6400A	488869206	703A	09-308019
6034	1N2452	6073	R4111070	614	D555	6400A	48B869264	703A	13-076002
6034	1N2453	610	BA243	614	DVV004	6400A	107-0021-00	703A	13-1-6
6034	1N2464	610	CF3	614	EX011	6400A	417-183	703A	13-10-6
6034	1N2465	610	EA16X166	614	EA16X127	6400A	417-187	703A	13-11-6
6034	1N2789	610	FV1043	614	EA16X177	6400A	417-181	703A	13-9-6
6034	1N3969	610	HF-20062	614	EDS-0042	6400A	576-0005-001	703A	15-26587-1
6034	20M40	610	MC100	614	FCDO070ANC	6400A	753-9010-021	703A	19-020-079
6034	25GC12	610	MC100A	614	GR-90	6400A	945	703A	020-114-007
6034	25HB35	610	MV2101	614	HD40001060	6400A	2160	703A	020-114-008
6034	25HB40	610	MV2201	614	HD4000909	6400A	4792	703A	202BM
6034	50040	610	PC139	614	HP-20007	6400A	5001	703A	2085AH
6034	60HP35	610	PC140	614	IP20-0151	6400A	70399-1	703A	022A
6034	60HP40	610	PC141	614	MV1638	6400A	740855	703A	46-5002-4
6038	1N2454	610	SMV1172	614	MV2109	6401	D5E-37	703A	46-5002-7
6038	1N2466	610	1N4786	614	MV2111	6401	D5E-44	703A	51-10302A01
6038	60HP50	610	1N4801	614	MV2209	6401	D5E37	703A	51B10302A01
604	HV-15	610	1N5681	614	MV834	6401	D5E44	703A	56A1-1
604	HV15	610	1N5696	614	MV9600	6401	GB-410	703A	56C1-1
604	MA-23B	610	1S2090	614	MV9601	6401	GB-410	703A	56D1-1
604	MA-25A	610	1S2139A	614	NV004	6401	GBE4870	703A	96-5238-01
604	MA23	610	1S2180	614	NV009	6401	M9256	703A	96-5238-02
604	MA23A	610	1S2239	614	PC128	6401	SU110	703A	998022
604	MA23B	610	05-320301	614	PG533	6401	2N2646	703A	998022-1
604	MA25	610	13-22154-500	614	QD-082688DJ	6401	2SB118	703A	221-31
604	MA25A	610	51-06002-00	614	RVDSD113	6401	2SH18	703A	307-001-9-001
604	QVM800B	610	151-072-9-001	614	SC20	6401	2SH18K	703A	442-20
604	1N2326	610	4809-0000-001	614	SD113	6401	2SH18L	703A	442-8
604	523-1000-326	610	760204-0001	614	SD39	6401	2SH18M	703A	880-101-00
6040	PR9040	611	BA120	614	SV0-201	6401	2SH18N	703A	1000-25
6040	1N2138	611	BA121	614	SVC0053	6401	2SH19	703A	2020-1
6040	1N2138A	611	IP20-0204	614	V33	6401	2SH19K	703A	2020-2
6040	1N2139	611	ITT210	614	V933	6401	2SH19L	703A	2020-3
6040	1N2139A	611	MV-201	614	001-0160-00	6401	2SH19M	703A	8000-00012-041
6040	1N2455	611	MV2203	614	001-0176-00	6401	2SH19N	703A	09005
6040	1N2467	611	PC135	614	1N3182	6401	2SH20	703A	09006
6040	1N3970	611	V910	614	1N4794	6401	48-869256	703A	26587
6040	20M60	611	1N4788	614	1N5450A	6401	48B869256	703A	26587-1
6040	25060	611	1N5683	614	1N5470A	6401	86-668-2	703A	036001
6040	25HB60	611	18145	614	1N5689	6401	96-5269-01	703A	55810-167
6040	25C212	611	1S1501	614	1S1503	6401	242-997	703A	612020-1
6040	50060	611	1S1502	614	1S1558	6401	70399-2	703A	612020-2
6040	60HP60	611	1S1923	614	1S1658	6401	654001	703A	612020-3
6042	1N2456	611	1S2139	614	1S1658FA-3	6401	801525	703A	09308004
6042	1N2457	611	1S2236	614	1S2688	6401	801531	703A	62885976
6042	1N2468	611	1S2238	614	1S2689	6402	D13T	704	CA3011
6042	1N2469	611	1S2518	614	1SV50	6402	GB-X17	704	CA3014
6042	PR9043	611	1S3518	614	1SV53	6402	MPU131	704	GEIC-205
6044	PR9044	611	09-306359	614	1SV68	6402	2N6027	704	GEIC-84
6044	R4101060	611	151-062-9-001	614	314-3503-5	6402	2N6028	704	HA1103
6044	20M100	611	754-0102-139	614	314-3503-6	6402	13-33184-1	704	HA1104
6044	250100	611	8003-110B	614	05-182688	6402	121U2AT	704	M5113
6044	25NC12	611	9000-00038-010	614	05-200310	6402	801535	704	R3502
6044	500100	611	168652	614	05-472209	6403	MB84991	704	R3502-1
6044	50090	611	171814	614	05-780251	6403	2N4992	704	R3502-2
6045	PR9043R	611	922214	614	13-14278-2	6403	2N4990	704	SM-A-595819-1
6045	PR9044R	612	BB122	614	46-86289-3	6406	MP220	704	T1A
6045	R4111060	612	BB142	614	48-137487	6406	1N5758	704	TA117
6045	50C100R	612	HP-20005	614	48A42583A01	6406	1N5758A	704	4-082-664-0001
6045	50C90R	612	IP20-0185	614	533001-15	6407	HEP311	704	09-308017
605	HV-46	612	ITT-410	614	56-4835	6407	MP128	704	19A116796-1
605	HV46	612	MV-12	614	66X0050-001	6407	ST-2	704	32-25555-1
605	HV46GR	612	MV1626	614	103-189	6407	1N3301	704	32-25555-2
605	KB-262	612	MV2204	614	103-59	6407	1N5760	704	32-25555-3
605	KB262	612	MV2205	614	103-47-01	6407	1N5761A	704	32-25555-4
605	MA26W	612	QDCTT410XQ	614	151-030-9-004	6408	A72-83-300	704	51-10276A01
605	MA26WA	612	V12	614	153-008-9-001	6408	D30	704	51-10276A01
605	RVDVD1210L	612	V912	614	523-0009-049	6408	D3202Y	704	51810276A01
605	RVDVD1210M	612	1S3554	614	754-9002-687	6408	GE-X13	704	57028
605	RVDVD1211L	612	1N4789	614	903-1413	6408	GT-35	704	66F015
605	RVDVD1212L	612	1N5684	614	903-171B	6408	MP132	704	86X0024-001
605	RVDVD1213	612	1N5699	614	903-67B	6408	RB190	704	86X0027-001
605	SV-02	612	1S1895	614	903-96B	6408	SK3523	704	266F001-01
605	SV-03	612	1S1924	614	903-99B	6408	ST2	704	266P00101
605	SV-04	612	1S2085	614	1048-9938	6408	1N5760A	704	266P00102
605	SV-3A	612	1S2085A	614	2061-45	6408	1N5761	704	276A01
605	VD1210L	612	1S2087	614	5001-196	6408	2S2093	704	417-419
605	VD1212	612	1S2087A	614	115099	6409	D5E-43	704	5113E
605	VD1213	612	1S2267	614	123276	6409	D5E-45	704	8005-3
605	22A001-17	612	1S2268	614	123726	6409	D5E43	704	8007-0
605	48-355046	612	1S2687	614	136164	6409	D5E45	704	8007-1
605	168692			614	171984	6409	GE487I	704	8007-3
605	233061			614	172722	6409	HEP-89002	704	8007-4
6054	GB-5128			614	741689	6409	RS-2029	704	8008-1
6054	TR103			614	741968				
				614	742732				
				614	1223925				

ECG	Industry Standard No.	ECG	Industry Standard No.	ECG	Industry Standard No.	ECG	Industry Standard No.	ECG	Industry Standard No.
704	8008-3	710	46-1347-3	712	61A030-6	715	C6071P	720	09-308071
704	8009-0	710	46-1348-3	712	66P077-1	715	CA3071	720	15-34379-1
704	8009-4	710	46-1356-5	712	73C180475	715	FF274	720	020-114-009
704	80053	710	51-10408A01	712	73C180475-1	715	GEIC-6	720	020-114-009
704	80070	710	51810408A01	712	73C180475-4	715	HEP-C6071P	720	051-0012-00
704	80071	710	998042	712	73C180475-8	715	IC510	720	51-10566A02
704	80073	710	998072	712	73C180475-9	715	LM3071N	720	57A32-22
704	80074	710	133-002	712	73C180475004	715	MC1371P	720	442-9
704	80083	710	515-10408A-01	712	73C180476-5	715	MC1371PQ	720	7292DC
704	80090	710	3633-1	712	73C182186-1	715	N5071	720	729PC
704	80094	710	126871	712	73C182186-2	715	N5071A	720	740-9016-105
704	95298	710	129871	712	73C182186-3	715	PTC719	720	1001-0036/446Q
704	113561	710	1462432-1	712	73C182186-5	715	RE508-IC	720	3531-031-000
704	119609	710	1462454-1	712	082	715	SN76243	720	8840-171
704	730547	710	23119004	712	082-1	715	SN76243N	720	09011
704	1473502-1	710	23119007	712	86X0053-001	715	TVCM-9	720	18600-155
704	1473502-2	711	CA3044	712	86X53-1	715	ULN2127A	720	25810-165
705A	06089	711	CA3044V1	712	998082-1	715	09-308045	720	34379-1
705A	GEIC-3	711	CA344Y1	712	200X2110-269	715	09-308046	720	58840-202
705A	HEP-C6089	711	GEIC-207	712	221-48	715	13-40-6	720	5180028
705A	IC502	711	IC-6	712	266P30102	715	15-37703-1	720	740622
705A	PTC708	711	TA7050M	712	266P30109	715	21A101-016	720	2327422
705A	RE502-IC	711	TVSAN220	712	1010-9940	715	46-5002-13	720	C6056P
705A	SL20721	711	106	712	1165	715	56A5-1	720	E84053
705A	TVCM-1	711	13-27-6	712	2005-1	715	56D5-1	720	EX4055
705A	14-2007-00	711	21A101-002	712	2005-2	715	56L102	720	EX4053
705A	14-2007-02	711	21A101-2	712	2516	715	69B1Z	722	F767PC
705A	14-2007-04	711	46-1340-3	712	2516-1	715	69B2Z	722	GEIC-9
705A	221-36	711	46-1346-3	712	3686-1	715	095	722	HEP-C6056P
705A	221-37	711	46-5002-6	712	7221A	715	095-1	722	IC515
705A	221-39	711	417-202	712	33201-1	715	095A	722	KLH4793
705A	442-10	711	3677-1	712	33201-2	715	998095-1	722	LM1307
706	CA3041	711	24451	712	35059-1	715	221-43	722	LM1307N
706	IC-24(ELCOM)	711	126604	712	80710	715	781DC	722	MA767
706	IC-401	711	126609	712	130122	715	781PC	722	MC1307P
706	IC401	711	1462445-1	712	130751	715	2030-1	722	MC1307PQ
706	R2434-1	711	23119003	712	144026	715	37703-1	722	PC-20024
706	13-28-6	711	23119005	712	612005-1	715	612030-1	722	PC-20050
706	21A101-001	712	AN221	712	612005-2	715	612069-1	722	PTC721
706	21A101-1	712	AN240	712	1462516	715	62674970	722	RE309-IC
706	46-5002-10	712	AN240D	712	1462516-001	715	2002500711	722	SC5740PQ
706	57A29	712	AN240P	712	1462516-1	718	C1-1004	722	SC5743P
706	57A29-1	712	AN240PN	712	1464686-1	718	C6074P	722	SC9426P
706	57A29-2	712	C6063P	712	2560042	718	DM-14	722	SI21864
706	57C29	712	C6083P	712	2560092	718	GEIC-8	722	SN75110N
706	57C29-1	712	CA3065	712	2560201	718	HC1000117-0	722	SN76110N-07
706	57C29-2	712	CA3065/7P	712	5351951	718	HEP-594	722	SN76110
706	80114	712	CA3065E	712	57001009	718	HEP-C6074P	722	SN76110N
706	80287	712	CA3065RC	712	37001011	718	HEP-C6094P	722	T1E
706	80827	712	CA3065RCA	712	62674954	718	IC-9(PHILCO)	722	T1Z
706	1472434-1	712	EP84X2	712	62734469	718	IC511	722	T2C
707	SL21441	712	EP84X6	712	2002110206	718	KLH5489	722	TVCM-10
708	C6062P	712	GEIC-147	712	2002110269	718	LM1304	722	ULN2128A
708	C6082P	712	GEIC-148	712	4206004000	718	LM1304N	722	06B1M
708	CA2111AE	712	GEIC-2	712	4206104970	718	MC1304P	722	06B1Z
708	CA2111AQ	712	HA11107	712	4206105470	718	MC1304PQ	722	006B2M
708	CA2111E	712	HA1124	712	4206105770	718	PTC709	722	06B2Z
708	GEIC-10	712	HA1124D	712	4206105870	718	RE319-IC	722	09-017
708	GEL2111	712	HA1124B	712	6644000100	718	SC5177P	722	10-47674-01
708	GEL2111AL1	712	HA1125	713	CA3072	718	SC5199P	722	13-52-6
708	GEL2111P1	712	HA1125A	713	EP84X3	718	SC9314P	722	13-60-6
708	HEP-C6062P	712	HA1128	713	EP84X9	718	SL22756	722	15-34049
708	HEP-C6082P	712	HA1144	713	GEIC-5	718	SN76104	722	15-34049-1
708	LM2111	712	HA1144	713	GEL2114	718	SN76104N	722	15-34049-3
708	LM2111N	712	HEP-C6063P	713	IC508	718	T1J	722	15-34505
708	MC1357	712	HEP-C6083P	713	MC1329P	718	T1N	722	15-34503-1
708	MC1357P	712	IC507	713	N5072A	718	T2F	722	15-71420-1
708	MC1357PQ	712	LA1363	713	RE506-IC	718	T1J	722	15-71420-1
708	N5111	712	LA1365	713	09-308047	718	TVCM-9	722	21M588
708	PTC701	712	LM3065	713	09-308048	718	ULN2120A	722	51-10382A01
708	RE511-IC	712	LM3065N	713	14-2010-01	718	09-004	722	51-10559A01
708	SN76643N	712	M5143P	713	14-2010-02	718	13-26-6	722	51-10566A01
708	SN76653N	712	MA3065	713	5606-1	718	020-1114-006	722	51-1056PA01
708	T2J	712	MC13558	713	73C180837-1	718	21-B1	722	51-10592A01
708	TA6220	712	MC1358P	713	73C180837-2	718	21B1M	722	51810559A01
708	TVCM-4	712	MC1358PQ	713	73C180837-3	718	21B1Z	722	51810566A01
708	ULN2111A	712	NS-3065	713	73C180837P2	718	21B2Z	722	51810592A01
708	ULN2111N	712	NTC-21	713	096A	718	46-5002-9	722	053-1
708	07B1Z	712	PTC726	713	998096-1	718	51-10382A	722	57A32-10
708	07B2B	712	R2516-1	713	221-46	718	51-10422A	722	998053
708	07B3B	712	R25161	713	221-51	718	51-10422A01	722	998053-1
708	07B3C	712	RE305-IC	713	221-52	718	51-10422A02	722	0205
708	07B3D	712	REN712	713	442-55	718	51-10422A07	722	221-65
708	07B3M	712	RH-1XO043CEZZ	713	1010-9973	718	51-10617A01	722	221-79
708	07B3Z	712	RH-IXOO43CEZZ	713	62756291	718	51810382A	722	221-79-01
708	14-2008-01	712	SC9436P	714	AMD780	718	51810382A01	722	437-26551A
708	15-34048-1	712	SI-MC1358P	714	C6070P	718	51810422A	722	442-16
708	15-34048-2	712	SL21654	714	C3070	718	51810422A01	722	740-2001-307
708	19A11644P1	712	SN76664N	714	CA3070G	718	51810437A01	722	767DC
708	221-61	712	SN76665N	714	GEIC-4	718	51810617A01	722	767PC
708	442-28	712	SN76665N	714	HEP-C6070P	718	63-15345	722	905-38B
708	740-2002-111	712	SN76666	714	IC509	718	181-000200	722	905-46B
708	2007-2	712	SN76666N	714	LM3070N	718	732DC	722	1043-1344
708	2007-3	712	TA5814	714	MC1370P	718	732PC	722	2006-1
708	7402	712	TA7071P	714	MC1370PQ	718	2021-1	722	2006-2
708	7423	712	TA7176AP	714	N5070	718	2021-2	722	3533
708	612007-1	712	TA7176PFA-1	714	N5070B	718	09004	722	6551
708	612007-2	712	TM712	714	PTC715	718	45380	722	09017
708	612007-3	712	TVCM-11	714	RE507-IC	718	183013	722	34049-1
709	A455	712	TVSAN241	714	REN714	718	612021-1	722	34503-1
709	DM-5	712	TVSAN241D	714	SL21122	718	612021-2	722	45381
709	DM11	712	TVSSN76665	714	SN76242	718	7312294	722	61566
709	DM11A	712	ULN2165A	714	SN76242N	718	SN76131N	722	95291
709	DM31	712	ULN2165N	714	TA5649	72	2N3621	722	95294
709	FEL2113	712	01-121365	714	TVCM-8	72	2N3625	722	183014
709	GEIC-11	712	02-091128	714	ULN2124A	72	2N3629	722	612006-1
709	GEL2113	712	02-121365	714	02-561410	720	FUN14LHO26	722	612006-2
709	GEL2113AL1	712	02-343065	714	09-308079	720	GEIC-7	722	760522-0002
709	GEL2113P1	712	02-403065	714	13-42-6	720	HA1115W	722	999053-1
709	IC505	712	4-2060-04000	714	15-37702-1	720	HA1115	722	5351042
709	PTC-703	712	4-2060-04600	714	21A101-017	720	HA1115W	722	43126551
709	RE301-IC	712	05B1Z	714	46-5002-12	720	HEP-595	722	43126551-A
709	SN76642N	712	05B2Z	714	5604-1	720	HEP-C6068P	723	06101P
709	TVCM-5	712	09-308100	714	5604-1	720	LM1305	723	CA3075
709	TVSSN76642N	712	13-29-6	714	561101	720	LM1305N	723	CA3075D
709	ULN2113A	712	13-35059-1	714	094	720	MC1305P	723	CA3075E
709	ULN2113N	712	13-64-6	714	094-1	720	MC1305P-C	723	GEIC-15
709	15-34452	712	15-33201	714	094A	720	PC-20008	723	HEP-C6101P
709	15-34452-1	712	15-33201-1	714	998094-1	720	PC20018	723	LM3075N
709	19A11644S	712	15-33201-2	714	221-42	720	PTC713	723	MC1375P
709	051-0022-00	712	15-35059-1	714	221-87	720	RE320-IC	723	REN723
709	34452-1	712	15-35059-2	714	780DC	720	SC5118P	723	SN76032
709	916112	712	15-43300-1	714	780PC	720	SC5182P	723	SN76675N
709	7311325	712	21A120-008	714	2031-1	720	SC5741P	723	TVCM-16
709	7952980	712	44-135-5	714	37702-1	720	SN76105	723	51-10594A01
709	CA3042	712	46-1361-5	714	612031-1	720	SN76105N	723	51810594A01
709	GEIC-89	712	46-5002-15	714	612070-1	720	TR2327422	723	88-9842P
709	LA1342	712	51-13753A11	714	62674962	720	TR4104-2327421	723	88-9842R
709	R2432-1	712	51-90305A61	714	2002500812	720	TRO-9011	723	88-9842S
710	T18-1	712	51819753A09	714	2002501006	720	TVCM-7	723	221-90
710	TA-7051P	712	51815753A11	715	AMD781	720	ULN2122A	723	2665-2
710	TA7051P	712	56A3-1			720	001-0036		
710	19A116797-1	712	5603-1			720	4-009		
710	042A	712	56D5-1			720	09-011		

ECG	Industry Standard No.	ECG	Industry Standard No.	ECG	Industry Standard No.	ECG	Industry Standard No.	ECG	Industry Standard No.
723	5815	731	612024-1	7400	SK7400	7404	C3004P	7408	1741-0200
723	9692-1	736	GEIC-17	7400	SL16795	7404	DM7404N	7408	1820-0870
723	619692-1	736	REN736	7400	SN7400	7404	FD-1073-BJ	7408	2473-2109
723	3596354	736	221-89	7400	SN7400A	7404	FJH241	7408	7408-6A
723	4082665-0001	736	753TC	7400	SN7400N	7404	FLH211	7408	7408-9A
723	4082665-0003	737	AE-907	7400	T7400B1	7404	GB-7404	7408	7408A
723	4082665-1	737	AE907	7400	TD1401	7404	HD2522	7408	7408N
723	4082665-2	737	AE907-51	7400	TD1401P	7404	HD2522P	7408	740BPC
723	4082665-3	737	DM-41	7400	TD3400A	7404	HD7404	7408	43202
724	CA3028A	737	DM41	7400	TD3400AP	7404	HD7404A	7408	94152
724	CA3028A	737	GEIC-16	7400	TD3400P	7404	HEP-C3004P	7408	138315
724	CA3028AP	737	LM1841N	7400	TL7400N	7404	HL19000	7408	310254
724	CA3028A8	737	SN76669N	7400	U87400A	7404	HL55862	7408	374109-1
724	CA3028B	737	TVCM-18	7400	U87400J	7404	HL56421	7408	5175460
724	CA3028BF	737	ULN2136A	7400	WEP7400	7404	IC-7404	7408	9003398-03
724	CA3028Ba	737	314-9007-01	7400	ZN7400E	7404	IC-84(ELCOM)	7408	9003398-04
724	CA3053	737	314-9007-1	7400	006-0000146	7404	ITT7404N	7408	51330005
724	GEIC-86	737	314-9007-51	7400	007-1695001	7404	J1000-7404	7408	GEIC-31
724	LM3028	737	3149007-51	7400	09-308022	7404	J4-1004	740A	M380
724	LM3028A	737	88-9574	7400	9N00	7404	KS20967-L2	740A	RE321-IC
724	LM3028B	737	2136	7400	9N00DC	7404	LB5006	740A	TVCM-35
724	LM3053	737	2136D	7400	9N00PC	7404	M53204	740A	46-5002-23
724	TA7045	737	2136P	7400	19A116180P1	7404	M53204P	740A	301-576-14
724	TA7045M	737	2136PC	7400	51-10611A11	7404	MB418	7410	A05(I.C.)
724	09-308002	737	9341258	7400	51810611A11	7404	MC7404L	7410	C3010P
724	09-308003	738	A1368	7400	68A9025	7404	MC7404P	7410	FD-1073-BN
724	57832-1	738	C6075P	7400	78A20001OP4	7404	MIC7404J	7410	FJH121
724	133-003	738	CA1398E	7400	225A6946-P000	7404	MIC7404N	7410	GE-7410
724	133-005	738	EP84X12	7400	236-0005	7404	N7404A	7410	HD2507
724	307-008-9-001	738	GEIC-29	7400	301-576-4	7404	N7404P	7410	HD2507P
724	551-1-011-022	738	HEP-C6075P	7400	398-13223-1	7404	PA7001/527	7410	HD7410
724	551-1-011-032	738	IC-297(ELCOM)	7400	435-21026-0A	7404	SK7404	7410	HD7410P
724	1042-11	738	LA1368	7400	443-11	7404	SL16796	7410	HEP-C5010P
724	1820-0306	738	M5190	7400	1065-4861	7404	SN7404N	7410	HL19001
724	8000-0004-P090	738	M5190P	7400	1741-0051	7404	T7404B1	7410	HL56899
724	9694-1	738	MC1398	7400	1820-0054	7404	TD3404A	7410	IC-7410
724	38446-00000	738	MC1398P	7400	7400	7404	TD3404AP	7410	IC-86(ELCOM)
724	38446-00010	738	MC1398PQ	7400	7400-6A	7404	TL7404N	7410	ITT7410N
724	38446-00020	738	MC1938	7400	7400-9A	7404	U87404A	7410	J1000-7410
724	74004-1	738	RE313-IC	7400	7400/9N0Q	7404	U87404J	7410	J4-1010
724	74004-2	738	REN738	7400	7400A	7404	ZN7404E	7410	LB5001
724	243028	738	SN76298N	7400	7400PC	7404	007-1695301	7410	M53210
724	619694-1	738	TA6405	7400	8000-00038-004	7404	9N04DC	7410	M53210P
725	C6055L	738	TVCM-27	7400	10302-04	7404	9N04PC	7410	MB401
725	GEIC-19	738	ULN2298A	7400	11216-1	7404	19A116180P20	7410	MB602
725	HEP-C6055L	738	13-56-6	7400	43200	7404	51-10611A12	7410	MC7410L
725	IC1303P	738	15-39075-1	7400	55001	7404	51810611A12	7410	MC7410P
725	LM1303	738	15-43705-1	7400	138511	7404	68A9028	7410	MIC7410J
725	LM1303N	738	44B1	7400	339300	7404	156-0148-00	7410	MIC7410N
725	MC1303	738	44B1B	7400	339500-2	7404	225A6946-P004	7410	N7410A
725	MC1303L	738	44B1Z	7400	573401-1	7404	256-0007	7410	N7410P
725	MC1303P	738	46-13124-3	7400	558875	7404	398-13224-1	7410	PA7001/520
725	SC2914P	738	51-70177A03	7400	611563	7404	398-13632-1	7410	RLH111
725	SC5116L	738	51D70177A03	7400	760011	7404	435-21028-0A	7410	SK7410
725	SL22108	738	51G10679A03	7400	800024-001	7404	443-18	7410	SL16801
725	SN76151	738	51M70177A03	7400	930347-3	7404	1741-0143	7410	SN7410N
725	TBA231	738	51S10655A03	7400	2610786	7404	1806	7410	T7410B1
725	0831M	738	51810655A03A	7400	3520041-001	7404	1820-0174	7410	T7410D1
725	082M	738	51810655B03	7400	4663001D907	7404	1820-0894	7	
725	020-1114-003	738	71-70177A03	7400	5359031	7404	7404-6A	7410	TD1402P
725	020-1114-005	738	71D70177A03	7400	9003151-03	7404	7404-9A	7410	TD3410A
725	133-004	738	71M70177A03	7400	51520000	7404	7404A	7410	TD3410
725	442-41	738	177A03	7400	51520012	7404	7404PC	7410	TD3410P
725	739DC	738	229-1301-44	7401	GEIC-194	7404	8000-00028-042	7410	TL7410N
725	739PC	738	266P60502	7401	ITT7401N	7404	11202-1	7410	UPB202C
725	2008-1	738	1010-9932	7401	1741-0085	7404	015040/7	7410	UPB202D
725	2008-2	738	1351	7401	7401-6A	7404	43201	7410	UPB7410DC
725	612008-1	738	1352	7401	7401-9A	7404	55003	7410	U87410A
725	612008-2	738	2044-1	7401	55032	7404	138314	7410	U87410J
726	CA3011	738	10655A03	7402	C3002P	7404	373404-1	7410	ZN7410E
726	CA3012	738	10655A03A	7402	DM7402N	7404	508590	7410	006-0000147
726	GEIC-81	738	10655B03	7402	FD-1073-BQ	7404	611565	7410	007-1695901
726	GEIC-83	738	39075-1	7402	FJH221	7404	800587-001	7410	9N10DC
727	A3148	738	70177A03	7402	FLH191	7404	801806	7410	9N10PC
727	CA3048	738	612044-1	7402	GE-7402	7404	930347-13	7410	19A116180P4
727	CA3052	738	4206009200	7402	HD25111	7404	3520048-001	7410	49A00055-001
727	CA3052E	738	4206009?00	7402	HD2511P	7404	4663001A909	7410	68A9050
727	E2495	739	C6072P	7402	HD7402	7404	9001551-02	7410	225A6946-P010
727	GEIC-210	739	GEIC-30	7402	HD7402P	7404	9001551-03	7410	435-21030-0A
727	LA3148	739	HEP-C6072P	7402	HEP-C3002P	7404	51320002	7410	443
727	MBD03	739	MC1324P	7402	HL19004	7406	DM7406N	7410	1741-0234
727	15-34005-1	739	MC1324PQ	7402	IC-7402	7406	FLH481	7410	1820-0068
727	15-34202-1	739	MC1326P	7402	IC-82(ELCOM)	7406	GE-7406	7410	7410
727	51-04488D03	739	MC1326PQ	7402	ITT7402N	7406	HD7406	7410	7410-
727	51R04488D03	739	RE312-IC	7402	J1000-7402	7406	HD7406P	7410	7410-9A
727	34005-1	739	TVCM-21	7402	J4-1002	7406	IC-104(ELCOM)	7410	7410PC
727	34202-1	739	13-57-6	7402	LB5008	7406	ITT7406N	7410	10302-03
727	81336-1	739	13-67-6	7402	M53202	7406	MC7406L	7410	11200-1
727	81336-2	739	22B1B	7402	M53202P	7406	MC7406P	7410	55005
727	813362	739	46-5002-14	7402	MB417	7406	N7406A	7410	373405-1
728	CA3066	739	221-49	7402	MC7402L	7406	N7406P	7410	558877
728	CA3066E	739	1326PC	7402	MC7402P	7406	RH-IX0038PAZZ	7410	611566
728	EP84X7	739	1347	7402	MIC7402J	7406	SK7406	7410	800023-001
728	GEIC-22	739	1348	7402	MIC7402N	7406	SN7406N	7410	3520042-001
728	SN76266	739	612072-1	7402	N7402A	7406	T7406B1	7410	4663001A912
728	86X0055-001	74	2N2632	7402	N7402P	7406	TL7406N	7410	9003091-02
728	86X55-1	748	2N3852	7402	PA7001/525	7406	007-1696901	7410	9003091-03
728	132514	748	2N3853	7402	SK7402	7406	9N06DC	7410	51320003
728	1462559-1	7400	A00	7402	SL16795	7406	9N06PC	74100	D2503
729	CA3067	7400	C3000P	7402	SN7402	7406	68A9032	74100	D2503P
729	CA3067E	7400	DDEYO30001	7402	SN7402N	7406	1607A80	74100	EP84X19
729	GEIC-23	7400	DM7400	7402	TD7402N	7406	7406	74123	FLK121
729	LM3067N	7400	DM7400N	7402	TD3402A	7406	7406PC	74123	GE-74123
729	SN76267	7400	FD-1073-BF	7402	TD3402AP	7406	55036	74123	HD2561
729	ULN2267A	7400	FJH131	7402	TL7402N	7406	373429-1	74123	HD2561P
729	86X0055-001	7400	FLH101	7402	U87402A	7406	800651-001	74123	IC-74123
729	86X56-1	7400	GE-7400	7402	U87402J	7406	50254200	74123	ITT74123N
729	86X56-2	7400	HD7400	7402	XAA104	7408	DM7408N	74123	N74122A
729	274	7400	HD7400P	7402	ZN7402E	7408	FD-1073-BM	74123	N74123B
729	2560	7400	HEP-C3000P	7402	007-1696201	7408	FLH381	74123	N74123P
729	132515	7400	HL18998	7402	9N02DC	7408	GE-7408	74123	RH-IX00A1PAZZ
729	1462560	7400	HL56420	7402	9N02PC	7408	HD2550	74123	SK74123
729	1462560-001	7400	IC-80(ELCOM)	7402	19A116180P3	7408	HD2550P	74123	SN74123N
729	1462560-1	7400	IP20-0205	7402	43A223009	7408	HL55763	74123	TL74123N
73	2N3622	7400	J1000-7400	7402	68A9027	7408	IC-102(ELCOM)	74123	UPB74123C
73	CA3068	7400	ITT7400N	7402	435-21027-0A	7408	IC-7408	74123	9N123DC
731	C6085P	7400	J1000-7400	7402	443-46	7408	ITT7408N	74123	9N123PC
731	CA3120E	7400	J4-1002	7402	1741-0119	7408	KS21282-L1	74123	1462
731	GEIC-13	7400	LB3000	7402	1820-0528	7408	L-612099	74123	74123
731	HEP-C6085P	7400	M53200	7402	7402-6A	7408	MC7408L	74123	74123PC
731	IC-33	7400	M53200P	7402	7402-9A	7408	MC7408P	7413	DM7413N
731	LM3445	7400	MB400	7402	7402PC	7408	N7408A	7413	FLH551
731	TVCM-15	7400	MB601	7402	11207-1	7408	N7408P	7413	HD2545
731	ULN2125A	7400	MC7400L	7402	55002	7408	SL14971	7413	HD2545P
731	14-2054-01	7400	MC7400P	7402	138313	7408	SL16798	7413	IC-103(ELCOM)
731	24B1AH	7400	MIC7400N	7402	558876	7408	SL17869	7413	ITT7413N
731	24B1B	7400	N7400A	7402	611564	7408	SN7408N	7413	MIC7413N
731	24B1Z	7400	N7401A	7402	800080-001	7408	U87408A	7413	N7413A
731	46-5002-33	7400	N7401P	7402	930347-11	7408	U87408AI	7413	N7413P
731	221-45	7400	PA7001/521	7402	2610783	7408	007-1699301	7413	SN7413N
731	221-45-01	7400	REN7400	7402	7012166	7408	9N08DC	7413	TL7413N
731	2024-1			7402	51520001	7408	9N08PC		
731	811177			7404	A03	7408	435-21029-0A		
						7408	443-45		

ECG	Industry Standard No.
7413	UPB7413C
7413	9N13DC
7413	9N13PC
7413	443-44
7413	443-44-2854
7413	601-0100865
7413	3531-021-000
7413	7413PC
7413	55027
74145	DM74145N
74145	FLL111T
74145	GE-74145
74145	HD2555
74145	HD2555P
74145	ITT74145N
74145	MB443
74145	MC74145P
74145	MIC74145J
74145	MIC74145N
74145	N74145B
74145	PA7001/593
74145	SN74145N
74145	TL74145N
74145	US74145A
74145	007-1696801
74145	19-130-004
74145	443-87
74145	1542
74145	74145
74145	74145DC
74145	74145PC
74145	93145DC
74145	93145PC
74150	DM74150N
74150	FLY111
74150	GE-74150
74150	HD2548
74150	HD2548P
74150	HD74150
74150	HD74150P
74150	HL55861
74150	ITT74150N
74150	MC74150P
74150	MIC74150J
74150	MIC74150N
74150	N74150F
74150	N74150N
74150	SK74150
74150	SN74150N
74150	TL74150N
74150	UPB2150D
74150	UPB74150C
74150	1741-1042
74150	74150DC
74150	74150PC
74150	93150DC
74150	93150PC
74154	DM74154N
74154	FLY141
74154	GE-74154
74154	HD2580
74154	HD2580P
74154	ITT74154N
74154	MIC74154J
74154	MIC74154N
74154	N74154F
74154	N74154N
74154	SN74154N
74154	TL74154N
74154	US74154A
74154	74154DC
74154	74154PC
74154	93154DC
74154	93154PC
74192	FLJ241
74192	GE-74192
74192	HD2541
74192	HD2541P
74192	HL56429
74192	ITT74192N
74192	M53392
74192	M53392P
74192	MC74192P
74192	N74192B
74192	N74192F
74192	SK74192
74192	SN74192N
74192	TD34192A
74192	TD34192AP
74192	TL74192N
74192	007-1698301
74192	443-66
74192	9360DC
74192	9360PC
74192	74192DC
74192	74192PC
74192	611731
74193	FLJ251
74193	GE-74193
74193	HD2542
74193	HD2542P
74193	HL56430
74193	ITT74193N
74193	M53393
74193	M53393P
74193	MC74193P
74193	N74193B
74193	N74193F
74193	SK74193
74193	SN74193N
74193	T74193B1
74193	TL74193N
74193	007-1698401
74193	43C216447
74193	43C216447P1
74193	443-162
74193	9366DC
74193	9366PC
74193	11204-1
74193	74193
74193	74193DC
74193	74193PC
74193	138320
74193	611730
74193	800386-001
74193	7012142-03
74196	DM74196N
74196	FLJ381
74196	GE-74196
74196	HD2572
74196	HD2572P
74196	SN74196J
74196	SN74196N
74196	TL74196N
74196	443-628
74196	74196DC
74196	74196PC
74196	93196DC
74196	93196PC
742	GEIC-213
742	612076-2
7420	A06(I.C.)
7420	C3020P
7420	DM7420N
7420	EP84X11
7420	FD-1073-BR
7420	FJH111
7420	FLH121
7420	GE-7420
7420	HD2504
7420	HD2504P
7420	HD7420
7420	HD7420P
7420	HEP-C3020P
7420	HL19003
7420	HL56422
7420	IC-87(ELCOM)
7420	ITT7420N
7420	LB3002
7420	M53220
7420	M53220P
7420	MB402
7420	MB603
7420	MC7420L
7420	MC7420P
7420	MIC7420J
7420	MIC7420N
7420	N7420A
7420	N7420F
7420	PA7001/519
7420	SK7420
7420	SL16800
7420	SN7420J
7420	SN7420N
7420	T7420B1
7420	T7420D1
7420	TD1403
7420	TD1403P
7420	TD3420A
7420	TD3420AP
7420	TD3420P
7420	TL7420N
7420	UPB203D
7420	UPB7420C
7420	US7420A
7420	US7420J
7420	ZN7420E
7420	007-1695101
7420	9N20PC
7420	19A116180P5
7420	49A0006-000
7420	68A9033
7420	225A6946-P020
7420	435-21033-0A
7420	443-2
7420	1741-0325
7420	1820-0069
7420	7420(IC)
7420	7420-6A
7420	7420-9A
7420	7420PC
7420	10302-02
7420	11205-1
7420	55006
7420	138318
7420	373406-1
7420	558878
7420	611567
7420	800020-001
7420	930347-1
7420	930347-10
7420	9004076
7420	9004076-03
7420	9004076-04
7420	51320004
7420	885540026-3
7427	DM7427N
7427	FLH621
7427	GE-7427
7427	HD7427
7427	HD7427P
7427	SK7427
7427	SN7427N
7427	US7427A
7427	9N27DC
7427	9N27PC
7427	443-65
7427	7427PC
743	AE904
743	CA758E
743	DM-44
743	GEIC-214
743	GEIC-32
743	LM1800N
743	MC1311P
743	RE317-IC
743	SN76116N
743	ULN2244A
743	3L4-9004-4
7430	55007
7432	DM7432N
7432	GE-7432
7432	HD7432
7432	HD7432P
7432	ITT7432N
7432	KS21282-L3
7432	L-612107
7432	N7432A
7432	N7432F
7432	SK7432
7432	SN7432N
7432	US7432A
7432	US7432J
7432	9N32DC
7432	9N32PC
7432	7432PC
7432	138381
7432	4511424
744	CA3123E
744	DM-20
744	DM-32
744	DM20
744	DM32
744	GEIC-215
744	GEIC-24
744	IC-607(ELCOM)
744	LM1820
744	LM1820N
744	SK3171
744	SN76635N
744	TVCM-19
744	U6A7720354
744	U6A7720395
744	U9A7720354
744	U9A7720395
744	UA720
744	UA720DC
744	UA720PC
744	ULN2137A
744	ULN2137N
744	15-36995-1
744	88-9302
744	88-9302R
744	88-9302S
744	221-107
744	720DC
744	720PC
744	720SDC
744	720SPC
744	889302
744	7935181
744	8898302
744	9341899
7441	C3041P
7441	FJL101
7441	GE-7441
7441	HD2518
7441	HD2518P
7441	HEP-C3041P
7441	IC-89(ELCOM)
7441	M53241
7441	M53241P
7441	MC7441AL
7441	MIC7441AJ
7441	MIC7441AN
7441	N7441B
7441	N7441F
7441	T7441AB1
7441	TD3441A
7441	TD3441AP
7441	US7441A
7441	ZN7441AE
7441	007-1697801
7441	376-0099
7441	1741-1190
7441	7441-6A
7441	7441-9A
7441	7441DC
7441	7441PC
7441	9315DC
7441	9315PC
7447	FLL121T
7447	FLL121V
7447	GE-7447
7447	HD2532
7447	HD2532P
7447	IC-101(ELCOM)
7447	ITT7447AN
7447	J1000-7447
7447	J4-1047
7447	M53247
7447	M53247P
7447	MC7447L
7447	MC7447P
7447	N7447B
7447	N7447F
7447	SK7447
7447	SN7447AN
7447	TD3447A
7447	TD3447AP
7447	TL7447AN
7447	TVCM-503
7447	US7447A
7447	443-36
7447	7447
7447	7447BDC
7447	7447BPC
7447	7447DC
7447	7447PC
7447	9357B
7447	9357BDC
7447	9357BPC
7448	FLH551
7448	GE-7448
7448	ITT7448N
7448	M53248
7448	M53248A
7448	M53248P
7448	MC7448L
7448	MC7448P
7448	N7448B
7448	SN7448N
7448	TL7448N
7448	US7448A
7448	7448DC
7448	7448PC
7448	9358
7448	9358DC
7448	9358PC
745	GEIC-216
745	IC-200(ELCOM)
745	MC1306P
7451	A12(I.C.)
7451	DM7451N
7451	FD-1073-BW
7451	FJH161
7451	FLH161
7451	GE-7451
7451	HD2505
7451	HD2505P
7451	HD7451
7451	HD7451P
7451	IC-91(ELCOM)
7451	ITT7451N
7451	MC7451L
7451	MC7451P
7451	MIC7451J
7451	MIC7451N
7451	N7451A
7451	N7451F
7451	PA7001/523
7451	SK7451
7451	SN7451N
7451	T7451B1
7451	T7451D1
7451	TD1419
7451	TD1419P
7451	TD3451A
7451	TD3451AP
7451	TD3451P
7451	TL7451N
7451	UPB207D
7451	UPB7451C
7451	US7451A
7451	US7451J
7451	ZN7451E
7451	9N51DC
7451	9N51PC
7451	435-21034-0A
7451	1741-0564
7451	1820-0063
7451	7451-6A
7451	7451-9A
7451	7451PC
7451	10302-05
7451	373715-1
7451	930347-12
7451	3520045-001
7451	51320016
746	C6059P
746	GEIC-217
746	HEP-C6059P
746	MC1350
746	MC1350P
746	SN76600
746	SN76600P
746	09-010
746	46-5002-17
746	051-0021-00
746	2906-005
746	8000-0058-006
746	09010
746	76600P
746	916111
747	C6079P
747	GEIC-218
747	HEP-C6079P
747	IC-18(PHILCO)
747	IC-202(ELCOM)
747	M5169P
747	MC1330P
747	RE310-IC
747	SC9430P
747	SN76530P
747	TDA1330P
747	15-39060-1
747	21A101-015
747	46-5002-18
747	229-1301-42
747	266P10201
747	266P10202
747	442-56
747	39060-1
747	45385
747	23119014
7473	B0075660
7473	C3073P
7473	DM7473N
7473	EA33X8385
7473	FJJ121
7473	FLJ121
7473	GE-7473
7473	HD2515
7473	HD2515P
7473	HD7473AP
7473	HD7473P
7473	HEP-C3073P
7473	HL19002
7473	IC-95(ELCOM)
7473	ITT7473N
7473	M53273
7473	M53273P
7473	M7641
7473	MC7473L
7473	MC7473P
7473	MIC7473J
7473	MIC7473N
7473	N7473
7473	N7473A
7473	N7473F
7473	PA7001/531
7473	SK7473
7473	SL16806
7473	SL17242
7473	SN7473
7473	SN7473N
7473	T7473B1
7473	TD1409
7473	TD1409P
7473	TD3473A
7473	TD3473AP
7473	TL7473N
7473	US7473A
7473	US7473J
7473	XAA107
7473	ZN7473E
7473	9N73DC
7473	9N73PC
7473	19A116180P15
7473	43A223025
7473	49A0002-000
7473	236-0009
7473	435-23006-0A
7473	443-5
7473	477-0412-004
7473	1030-25
7473	1820-0075
7473	7473-6A
7473	7473-9A
7473	7473PC
7473	43205
7473	138403
7473	558881
7473	930347-7
7473	3520043-001
7473	9004093-03
7473	2002010072
7474	A15
7474	DM7474N
7474	F7474PC
7474	FJJ131
7474	FLJ141
7474	GE-7474
7474	HD2510
7474	HD2510P
7474	HD7474
7474	HD7474P
7474	HEPC7474P
7474	HL18999
7474	HL56425
7474	IC-7474
7474	IC-97(ELCOM)
7474	IP20-0206
7474	IP20-0316
7474	M53274
7474	M53274P
7474	M5374
7474	M5374P
7474	MB420
7474	MIC7474J
7474	MIC7474N
7474	N7474A
7474	N7474F
7474	PA7001/529
7474	REN7474
7474	SK7474
7474	SL16807
7474	SN7474
7474	SN7474N
7474	T7474B1
7474	TD3474A
7474	TD3474AP
7474	TD3474P
7474	TL7474N
7474	TVCM-502
7474	US7474A
7474	US7474J
7474	WEP7474
7474	YEAM53274P
7474	ZN7474E
7474	007-1699801
7474	9N74
7474	9N74DC
7474	9N74PC
7474	19A116180P16
7474	43A223026P1
7474	49A0012-000
7474	075-045037
7474	90-39
7474	90-67
7474	435-23007-0A
7474	443-6
7474	1820-0077
7474	7474
7474	7474-6A
7474	7474-9A
7474	7474/9N74
7474	7474PC
7474	8000-00038-007
7474	11213-1
7474	55011
7474	373409-1
7474	558882
7474	611571
7474	800400-001
7474	881916
7474	286
7474	3520046-001
7474	4663001A905
7474	9003152
7474	9003152-01
7475	B01
7475	C30
7475	DM7475N
7475	FJJ181
7475	FLJ151
747	
7475	HD2517
7475	HD2517P
7475	HD7475
7475	HD7475P
7475	HEP-C3075P
7475	HL
7475	IC-96(ELCOM)
7475	ITT747
7475	J4-1075
7475	M53275
7475	M53275
7475	MC7475L
7475	MC7475P
7475	MIC7475J
7475	MIC7475N
7475	N74
7475	SK7475
7475	SM63
7475	SM73(I.C.)
7475	SN7475N
7475	T7475B1
7475	TD3475A
7475	TD
7475	TL7475N
7475	US7475A
7475	US7475J
7475	ZN7475E
7475	40-065-19-027
7475	49A0000
7475	51-
7475	51S10611A16
7475	68A9041
7475	443-13
7475	1741-0747
7475	1820-0301
7475	7475-6A
7475	7475-9A
7475	7475DC
7475	7475PC
7475	9375DC
7475	9375PC
7475	373713-1
7475	611065
7475	800382-001
7475	7011203-02
7	
7475	36188000
7476	B02
7476	DM7476N
7476	FJJ191
7476	FLJ131
7476	GE-7476
7476	HD2516
7476	HD2516P
7476	HL19010
7476	IC-74
7476	IC-99(ELCOM)
7476	ITT7476N
7476	J1000-7476
7476	J4-1076
7476	M53276
7476	M53276P
7476	MC7476L
7476	MC7476P
7476	MIC7476J
7476	MIC7476N
7476	N7476B
7476	N7476F
7476	SK7476
7476	SL16808
7476	SN7476N
7476	T7476B1
7476	TL7476N

ECG	Industry Standard No.	ECG	Industry Standard No.	ECG	Industry Standard No.	ECG	Industry Standard No.	ECG	Industry Standard No.
7476	US7476A	749	62736283	74LS123	SK74LS123	783	MC1364P	790	MC1328P
7476	XAA108	749	2002120012	74LS138	74LS138PC	783	MC1364PQ	790	MC1328PQ
7476	ZN7476H	749	4206105170	74LS151	SK74LS151	783	RE503-IC	790	SN76246
7476	SN76DC	7490	A17(I.C.)	74LS151	SN74LS151N	783	SI-MC1364P	790	SN76246N
7476	SN76RC	7490	C5800P	74LS157	SK74LS157	783	SN76564N	790	TA5912
7476	43A22302B	7490	DDEY029001	74LS157	SN74LS157N	783	T2062	790	TVCM-2
7476	68A9042	7490	DM7490	74LS174	SK74LS174	783	ULN2264A	790	ULN2114A
7476	443-16	7490	DM7490N	74LS174	SN74LS174N	783	02-561201	790	ULN2114N
7476	1348A14HO1	7490	FJJ141	74LS175	SK74LS175	783	56A20-1	790	ULN2228A
7476	7476	7490	FLJ161	74LS175	SN74LS175N	783	56D20-1	790	13-41-6
7476	7476-6A	7490	GE-7490	74LS193	SN74LS193N	783	6182E	790	15-37704-1
7476	7476-9A	7490	HD2519	74LS20	SK74LS20	783	730180843-1	790	29B1
7476	7476PC	7490	HD2519P	74LS20	SN74LS20N	783	730180843-2	790	29B17
7476	55012	7490	HD7490A	74LS20	74LS20PC	783	730180843-3	790	29B13
7476	72185	7490	HD7490AP	74LS27	SK74LS27	783	730180843-4	790	29B1Z
7476	373414-1	7490	HEP-C5800P	74LS27	SN74LS27N	783	097-1	790	46-5002-11
7476	611870	7490	HL19015	74LS30	SK74LS30	783	998097	790	56A6-1
7476	760015	7490	IC-98(ELCOM)	74LS30	SN74LS30N	783	998097-1	790	5605-1
7476	9004300-03	7490	ITT7490N	74LS32	SK74LS32	783	442-59	790	5606-1
748	GEIC-115	7490	J1000-7490	74LS32	SN74LS32	783	1010-9965	790	56LI03
748	GEIC-183	7490	J4-1090	74LS367	SK74LS367	783	1097	790	081-1
748	GEIC-26	7490	LB3150	74LS367	SN74LS367N	783	2061-1	790	998081
748	LM1351N	7490	M53290	74LS51	SK74LS51	783	612061-1	790	998081-1
748	MC1351P	7490	M53290P	74LS51	SN74LS51N	783	62736305	790	221-62-1
748	MC1351PQ	7490	MC7490L	74LS73	SK74LS73	784	CA3020	790	229-1301-41
748	RE331-IC	7490	MC7490P	74LS73	SN74LS73N	784	GEIC-236	790	746PC
748	RH-IX0001TAZZ	7490	MIC7490J	74LS74A	SK74LS74	784	HA1302	790	2029-1
748	RH-IX0001TAZZ	7490	MIC7490N	74LS75	SK74LS75	784	KD2115	790	2029-2
748	SC5265P	7490	N7490A	74LS75	SN74LS75N	784	11505	790	37704-1
748	SN76651N	7490	N7490P	74LS85	SK74LS85	784	12505	790	612029-1
748	TA7072P	7490	SK7490	74LS85	SN74LS85N	784	12505RB	790	612029-2
748	TA7073P	7490	SN7490AN	74LS93	SK74LS93	784	38265-00000	790	612029-3
748	108	7490	SN7490N	74LS93	SN74LS93	784	38265-00010	790	62674989
748	51010679A13	7490	T7490B1	74LS93	741593	784	38265-00020	791	CA31213
7480	FLH221	7490	TD3490A	75	2N2877	784	38265-00030	791	GEIC-231
7480	MC7480P	7490	TD3490AP	75	2N2878	784	3006206-00	791	GEIC-33
7485	FLH431	7490	TD7490P	75	2N2878A	784	8508331	791	IC-325(ELCOM)
7485	GB-7485	7490	TL7490N	750	C6061P	784	8508403-1	791	RE324-IC
7485	HD7485	7490	TVCM-505	750	GEIC-219	785	CA3035	791	REN791
7485	HD7485P	7490	US7490A	750	HEP-06061P	785	CA3035V	791	TVCM-34
7485	HL56426	7490	US7490J	750	IC-204(ELCOM)	785	CA3035V1	791	02-561171
7485	MB448	7490	XAA109	750	MC1355P	785	GEIC-226	791	09-308099
7485	N7485A	7490	1-000-099-00	750	MC1355PQ	785	R3528-1	791	15-59209-1
7485	N7485P	7490	19-130-005	750	TRO9007	785	R3528-1(RCA)	791	56A17-1
7485	SK7485	7490	19A116180-24	750	09-007	785	442-4	791	56D17-1
7485	SN7485N	7490	443-7	750	09007	785	1133-14	791	221-69
7485	TL7485N	7490	443-7-16088	750	45386	785	3528-1	791	2002301300
7485	UPB2085H	7490	733W00039	753	680-1	785	3991-303-112	795	C6099P
7485	UPB74850	7490	905-102	755	4050	785	4020	795	GEIC-232
7485	007-149600'	7490	1808	760	06001	785	127166	795	HEP-06099P
7485	7485DC	7490	1820-0055	760	181-000100	785	1473528-1	795	MC1349P
7485	7485PC	7490	7490	77	MM8002	786	AE900	795	15-39061-1
7485	9385DC	7490	7490-6A	77	130-144	786	AE902	797	CA3126EM
7485	9385PC	7490	7490-9A	77	130-150	786	CA3043	797	CA3126EM1
7486	DM7486N	7490	7490DC	77	130-172	786	GEIC-227	797	CA3126Q
7486	PD-1073-CA	7490	7490PC	77	130-174	786	3L4-9002-01	797	GEIC-233
7486	FLH341	7490	9390DC	772A	142251	786	3L4-9002-1	797	R5158-1
7486	GE-7486	7490	9390PC	778A	C6102P	786	115CT	797	RE333-IC
7486	HD2526	7490	16088	778A	CA1458S	786	86313	797	U1B2262A
7486	HD2526P	7490	102005	778A	GEIC-220	786	CA3088E	797	51-13753A19
7486	HD7486	7490	373427-1	778A	HEP-06102P	786	GEIC-228	797	787PC
7486	HD7486P	7490	558883	778A	LM1458N	786	R4437-1	797	1063-5019
7486	HL19014	7490	611572	778A	MC1458	786	R4437-2	797	5158
7486	ITT7486N	7490	760013	778A	MC1458CP'	786	3130-3193-512	797	5158-1
7486	K820967-L3	7490	801808	778A	MC1458P	786	4437-1	797	1386090(RCA)
7486	M53286	7490	7012167-02	778A	MC1458P1	786	4437-2	797	141279
7486	M53286P	7492	C5801P	778A	N5558V	786	4437-3	797	1415742-5
7486	MB449	7492	DM7492N	778A	SN72558	787	9144-60	797	1465158-1
7486	MC7486L	7492	FJJ152	778A	SN72558P	787	9695-1	797	1465158-2
7486	MC7486P	7492	FLJ171	778A	2G28A102	787	136145	798	C6066P
7486	MIC7486J	7492	GE-7492	778A	1458CP'	787	619693-1	798	CA3125E
7486	MIC7486N	7492	HD2521	778A	1458P1	787	1464437-1	798	GEIC-234
7486	N7486A	7492	HD2521P	778A	34502-1	787	1464437-2	798	HEP-06066P
7486	N7486P	7492	HD7492A	778A	142648	787	1464437-3	798	SC5204P
7486	SK7486	7492	HD7492AP	779A	06081P	788	CA3089E	798	T1K
7486	SN7486N	7492	HEP-C5801P	779A	GEIC-221	788	DM-51	798	TA6404
7486	T7486B'	7492	IC-100(ELCOM)	779A	MC1344P	788	GEIC-229	798	51-10425A01
7486	T7486D1	7492	ITT7492N	780	C6060G	788	HA1137	798	51-10655B01
7486	TL7486N	7492	J1000-7492	780	CA3064/5A	788	HA1137P	798	51-70177A01
7486	UPB2086D	7492	J4-1092	780	CA50641	788	HA1137W	798	51D70177A01
7486	US7486A	7492	K820969-L3	780	FUST7064393	788	IC-324(ELCOM)	798	51M70177A01
7486	US7486J	7492	M53292	780	GEIC-20	788	MC1389P	798	51M70177B01
7486	9N86DC	7492	M53292P	780	GEIC-222	788	R4438-1	798	51810425A01
7486	9N86PC	7492	MC7492L	780	HEP-06060G	788	R4438-1(RCA)	798	51810655A01
7486	19A116180-18	7492	MC7492P	780	HEP-06060P	788	R4438-2	798	51810655B01-5
7486	19A116180P18	7492	MIC7492J	780	LM3064	788	R4438-2(RCA)	798	177A01
7486	40-065-19-029	7492	MIC7492N	780	LM3064H	788	RE332-IC	798	177B01
7486	435-21035-0A	7492	N7492A	780	MC1364G	788	REN788	798	10655A01
7486	1741-0804	7492	N7492P	780	R3677-1	788	T2T	798	10655A01-3
7486	7486PC	7492	SL6809	780	R3677-2	788	T2T-2	798	70177A01
7486	359486	7492	SN7492AN	780	R3677-3	788	TDA1200	798	EA33X8364
7486	373410-1	7492	TD3492A	780	3730-6	788	TRO9018	799	GEIC-34
7486	611066	7492	TD3492AP	780	46-5002-16	788	TVCM-39	799	HC1000417
7486	611844	7492	TD7492P	780	097	788	07B2Z	799	IC-521(ELCOM)
7486	900'349-02	7492	TL7492N	780	097A	788	09-018	799	MC1312P
7486	900'349-03	7492	US7492A	780	3064T	788	51-10658A01	799	MXC-1312
7486	51320001B	7492	US7492J	780	3677-2	788	51-10658A02	799	MXC1312A
749	AN238	7492	19A116180-27	780	3677-3	788	51-10658A03	799	MC1312P
749	AN2388	7492	436-10010-0A	780	130130	788	51-10658A01	799	SC5747
749	C6076P	7492	7492-6A	780	137245	788	88-20372	799	21M506
749	CA1352E	7492	7492-9A	780	1462577	788	221-108	799	46-500230
749	GEIC-97	7492	7492DC	780	1463677-1	788	810	799	86X0064-001
749	HA1139	7492	7492PC	780	1463677-2	788	2077-1	801	C6096P
749	HEP-06076P	7492	9392PC	780	1463677-3	788	3531-030-000	801	CA1510E
749	M5183	7492	373712-1	781	CA5076	788	4438-1	801	ETI-22
749	M5183-8098	7492	611573	781	KLH4794	788	4438-1(RCA)	801	G09-028-A
749	M5183P	7492	1000100	781	SL23649	788	4438-2	801	GEIC-155
749	M1352P	7492	1000100-000	781	SN76676L	788	4438-2(RCA)	801	GEIC-35
749	MC1352PQ	7492	9003445-03	781	T2M	788	4438-3	801	HA1156
749	RE304-IC	74C00	HD1-74C00	781	19-130-001	788	4438-3(RCA)	801	HA1156-6C
749	RH-IX0004CEZZ	74C00	MM74C00N	781	51-10637A01	788	09018	801	HA1156W
749	RH-IX0004CEZZ	74C02	HD1-74C02	781	51810637A01	788	9144-61	801	HC1000401Q
749	SC9431P	74C08	HD1-74C08	781	2802-1	788	10658A01	801	HEP-06096P
749	SI-MC1352P	74C192	MM74C192	781	2802-2	788	1840-0148	801	LM1310
749	SN76650	74C74	HD1-74C74	781	3076HC	788	29810-179	801	LM1310N
749	SN76650N	74C76	HD1-74C76	781	3076PC	788	38510-169	801	MC1310
749	TA7074P	74C90	HD1-74C90	781	4794(KLH)	788	45391	801	MC1310A
749	TA7074PGL	74LS00	DM74LS00N	781	5757	788	45810-167	801	MC1310P
749	09-508090	74LS00	SK74LS00	781	80755	788	136146	801	RE334-IC
749	21A120-001	74LS00	SN74LS00N	781	4082802-0001	788	612077-1	801	REN801
749	46-1365-3	74LS00	74LS00N	781	4082802-0002	788	1464435-1	801	SN76115
749	46-5002-28	74LS02	SK74LS02	781	4082802-1	788	1464438-2	801	SN76115N
749	51-70177A05	74LS02	SN74LS02N	781	4082802-2	788	1464438-3	801	TRO9027
749	51M70177A05	74LS04	SK74LS04	783	06100P	788	3596810	801	TR2567171
749	51810655A05	74LS04	SN74LS04N	783	CA3064E	788	4154011370	801	ULN2210A
749	71-70177A05	74LS04	74LS04PC	783	EP84X4	790	CA3090AQ	801	14-2012-01
749	730180838-1	74LS08	SK74LS08	783	EP84X5	790	AMD746	801	15-40183-1
749	730180838-3	74LS08	SN74LS08N	783	GEIC-21	790	C6057P	801	051-0050-00
749	177A05	74LS10	SK74LS10	783	GEIC-225	790	EX62-X	801	051-0090-00
749	266P10101	74LS10	SN74LS10N	783	HEP-06100P	790	GEIC-18	801	051-04164J01
749	266P10103	74LS10	74LS10PC	783	LM3064N	790	GEIC-230	801	51-04164J01
749	70177A05			783	MC1364	790	GEL3072P1	801	51-42908J01
749	23119019					790	HEP-06057P	801	56-4834
749	40306400					790	LM746P	801	57A32-32

ECG	Industry Standard No.	ECG	Industry Standard No.	ECG	Industry Standard No.	ECG	Industry Standard No.	ECG	Industry Standard No.
801	613001-10	85	C372-R	85	C715EV	85	2SC1634	85	2SC458LGB
801	75B12	85	C372-Y	85	C715P	85	2SC1675	85	2SC458LGBM
801	87-0234	85	C372-Z	85	C715XL	85	2SC1675L	85	2SC458LGC
801	221-91	85	C3722GR	85	C752	85	2SC1675M	85	2SC458LGD
801	221-91-01	85	C372H	85	C752G	85	2SC1685	85	2SC458LGG
801	280-0002	85	C372Y	85	C828	85	2SC1685P	85	2SC458LGS
801	442-46	85	C373	85	C828-0	85	2SC1685Q	85	2SC458M
801	2057A32-33	85	C373BL	85	C828-0P	85	2SC1685R	85	2SC458P
801	2075-1	85	C373G	85	C828A	85	2SC1685S	85	2SC458RGS
801	003515	85	C373GR	85	C828AP	85	2SC1740	85	2SC458V
801	5652-HA1156	85	C373W	85	C828AQ	85	2SC1740L	85	2SC458VC
801	5723P	85	C376	85	C828AR	85	2SC1740P	85	2SC459
801	5724P	85	C377	85	C828AS	85	2SC1740Q	85	2SC459B
801	26810-157	85	C379	85	C828E	85	2SC1740R	85	2SC459C
801	29810-180	85	C380	85	C828F	85	2SC1740S	85	2SC459D
801	38510-170	85	C380-0	85	C828FR	85	2SC1741	85	2SC460(A)
801	45810-168	85	C380-0	85	C828H	85	2SC1766	85	2SC460(B)
801	612075-1	85	C380A	85	C828K	85	2SC1766C	85	2SC460A
801	916113	85	C380A(0)	85	C828LR	85	2SC1815	85	2SC460C
801	4154011560	85	C380A-0(TV)	85	C828N	85	2SC1815GR	85	2SC460D
801	4157017102	85	C380A-R	85	C828P	85	2SC1815Y	85	2SC460G
801	4159761157	85	C380A-R(TV)	85	C828Q	85	2SC1815YW	85	2SC460GB
801	5147013102	85	C380AO	85	C828QRS	85	2SC1853	85	2SC460H
802	GEIC-235	85	C380ATV	85	C828R	85	2SC1854	85	2SC460K
802	MC1314P	85	C380AY	85	C828S	85	2SC1854C	85	2SC461A
802	TR09033	85	C380D	85	C828T	85	2SC1854S	85	2SC461B
803	GEIC-237	85	C380R	85	C828W	85	2SC1959Y	85	2SC461C
803	MC1315P	85	C380Y	85	C828Y	85	2SC1973	85	2SC461K
803	TR09034	85	C383	85	C829	85	2SC2000	85	2SC529
804	C6090	85	C383G	85	C829A	85	2SC2000L	85	2SC529A
804	GEIC-27	85	C383T	85	C829B	85	2SC2001	85	2SC621
804	GEL277	85	C383W	85	C829BC	85	2SC2001L	85	2SC631
804	HEP-C6090	85	C383Y	85	C829C	85	2SC2021	85	2SC631A
804	PA277	85	C388	85	C829D	85	2SC2021Q	85	2SC632
804	RTN1001	85	C388A	85	C829R	85	2SC2021R	85	2SC632A
804	SN76177	85	C388ATV	85	C829X	85	2SC2021S	85	2SC633
804	SN76177ND	85	C454	85	C829Y	85	2SC2076	85	2SC633-7
804	09-308078	85	C454(A)	85	C839	85	2SC2076B	85	2SC633G
804	13-100000	85	C454A	85	C839(H)	85	2SC2076C	85	2SC633H
804	13-1000000	85	C454B	85	C839(J)	85	2SC2076CD	85	2SC634
804	13-59-6	85	C454C	85	C839(M)	85	2SC2076D	85	2SC634A
804	13-61-6	85	C454D	85	C839A	85	2SC2310	85	2SC641
804	45B17	85	C454L	85	C839B	85	2SC363	85	2SC641B
804	45B1AH	85	C454LA	85	C839C	85	2SC366	85	2SC641C
804	45B1D	85	C458	85	C839D	85	2SC366G	85	2SC641K
804	45B12	85	C458(C)	85	C839F	85	2SC366R	85	2SC709
804	69-2401	85	C458A	85	C839H	85	2SC366Y	85	2SC709B
804	69-2403	85	C458AD	85	C839J	85	2SC367	85	2SC709C
804	69-3116	85	C458B	85	C839L	85	2SC367G	85	2SC709CD
804	88-9304	85	C458BC	85	C839M	85	2SC367R	85	2SC709D
804	96-5774-01	85	C458BL	85	C839N	85	2SC367Y	85	2SC711
804	2045-1	85	C458C	85	C839S	85	2SC372	85	2SC711(E)
804	2277P	85	C458CLG	85	C870	85	2SC372(0)	85	2SC711AE
804	3535-110-50008	85	C458CM	85	C870BL	85	2SC372-1	85	2SC711AF
804	3535-110-50009	85	C458D	85	C870E	85	2SC372-2	85	2SC711D
804	7204A	85	C458G	85	C870F	85	2SC372-0	85	2SC711E
804	45387	85	C458GLB	85	C871	85	2SC372-R	85	2SC711F
804	45395	85	C458K	85	C871BL	85	2SC372-Y	85	2SC711H
804	50009	85	C458KA	85	C871D	85	2SC372-Z	85	2SC715
804	95293	85	C458KB	85	C871E	85	2SC372/4454C	85	2SC715A
804	183044	85	C458KC	85	C871F	85	2SC372BL	85	2SC715B
804	612045-1	85	C458KD	85	C871G	85	2SC372G	85	2SC715C
804	1440977	85	C458L	85	C912	85	2SC372GR	85	2SC715D
804	2402277	85	C458LB	85	C944	85	2SC372H	85	2SC715E
804	2412275	85	C458LG	85	C944A	85	2SC372X	85	2SC715BJ
804	3535110050009	85	C458LG(B)	85	C944B	85	2SC372Y	85	2SC715EV
804	3535110500009	85	C458LGA	85	C944C	85	2SC373	85	2SC715P
805	DM-19	85	C458LGB	85	C944D	85	2SC373BL	85	2SC715XL
806	DM-35	85	C458LGBM	85	C944K	85	2SC373G	85	2SC752
806	DM35	85	C458LGC	85	C945	85	2SC373GR	85	2SC752G
807	GEIC-240	85	C458LGD	85	C945(R)	85	2SC373W	85	2SC752G-0
81	EP15X29	85	C458LGG	85	C945-0	85	2SC376	85	2SC752G-R
81	2N2641	85	C458LGS	85	C945A	85	2SC377	85	2SC752G-Y
81	2N2644	85	C458M	85	C945AP	85	2SC377A	85	2SC752R
81	2N2722	85	C458P	85	C945AQ	85	2SC379	85	2SC752T
81	2N2913	85	C458RGS	85	C945B	85	2SC380	85	2SC828
81	2N2917	85	C458TOK	85	C945C	85	2SC380-0	85	2SC828-0
81	2N2974	85	C458VC	85	C945D	85	2SC380-Y	85	2SC828-0P
81$	2N2639	85	C459	85	C945E	85	2SC380A	85	2SC828A
81$	2N2640	85	C459B	85	C945F	85	2SC380A-0(O)	85	2SC828AP
81$	2N2642	85	C459D	85	C945G	85	2SC380A-0(TV)	85	2SC828AQ
81$	2N2645	85	C460	85	C945H	85	2SC380A-0	85	2SC828AR
81$	2N2720	85	C460(A)	85	C945K	85	2SC380A-0(TV)	85	2SC828AS
81$	2N2721	85	C460(B)	85	C945L	85	2SC380A-R	85	2SC828B
814	56D8-1	85	C460A	85	C945M	85	2SC380A-R(TV)	85	2SC828C
815	CA1391E	85	C460C	85	C945O	85	2SC380AO	85	2SC828E
815	GEIC-244	85	C460D	85	C945P	85	2SC380ATV	85	2SC828F
815	MC1391P	85	C460G	85	C945Q	85	2SC380AY	85	2SC828FR
815	SN76591P	85	C460GB	85	C945QL	85	2SC380D	85	2SC828H
815	14-2014-01	85	C460H	85	C945QP	85	2SC380R	85	2SC828HR
815	56A23-1	85	C460K	85	C945R	85	2SC380Y	85	2SC828K
815	86X0084-001	85	C460L	85	C945S	85	2SC383	85	2SC828LR
815	221-141	85	C461	85	C945T	85	2SC383G	85	2SC828LS
815	612082-2	85	C461A	85	C945TR	85	2SC383T	85	2SC828N
82	2N2804	85	C461B	85	C945W	85	2SC383W	85	2SC828P
82	2N2807	85	C461C	85	C945X	85	2SC383Y	85	2SC828PQ
82$	2N2802	85	C461E	85	D227	85	2SC388	85	2SC828Q
82$	2N2803	85	C461L	85	D227A	85	2SC388A	85	2SC828QRS
82$	2N2805	85	C529	85	D227B	85	2SC388ATV	85	2SC828R
82$	2N2806	85	C529A	85	D227C	85	2SC454	85	2SC828RS
834	CA239AE	85	C621	85	D227D	85	2SC454(A)	85	2SC828RST
834	CA239AG	85	C631	85	D227E	85	2SC454B	85	2SC828S
834	CA239E	85	C631A	85	D227F	85	2SC454C	85	2SC828W
834	CA239G	85	C632	85	D227L	85	2SC454D	85	2SC828Y
834	CA339AE	85	C632A	85	D227R	85	2SC454L	85	2SC829A
834	CA339AG	85	C633	85	D227S	85	2SC454LA	85	2SC829B
834	CA339E	85	C633-7	85	D227W	85	2SC458	85	2SC829B/4454C
834	CA339G	85	C633A	85	D327	85	2SC458(C)	85	2SC829BC
834	LM239AD	85	C633G	85	D327A	85	2SC458AD	85	2SC829BY
834	LM2901N	85	C633H	85	D327B	85	2SC458B	85	2SC829C
834	LM339AD	85	C634	85	D327C	85	2SC458BC	85	2SC829D
834	LM339AN	85	C634A	85	D327D	85	2SC458BL	85	2SC829F
834	LM339N	85	C641	85	D327E	85	2SC458C	85	2SC829R
836	CA1394E	85	C641B	85	D327F	85	2SC458CLG	85	2SC829X
85	C1359	85	C709	85	G05-036B	85	2SC458CM	85	2SC829Y
85	C1359A	85	C709B	85	G05-036C	85	2SC458D	85	2SC839
85	C1359B	85	C709C	85	G05-413B	85	2SC458GLB	85	2SC839(H)
85	C1359C	85	C709CD	85	G05-413C	85	2SC458K	85	2SC839(J)
85	C1633	85	C709D	85	G05-413D	85	2SC458KA	85	2SC839(L)
85	C1675	85	C711	85	ST1402D	85	2SC458KB	85	2SC839(M)
85	C1675K	85	C711(E)	85	ST1702M	85	2SC458KC	85	2SC839A
85	C1675L	85	C711A	85	ST1702N	85	2SC458KD	85	2SC839B
85	C1675M	85	C711AE	85	2SC1310	85	2SC458L	85	2SC839C
85	C1685	85	C711D	85	2SC1311	85	2SC458LB	85	2SC839D
85	C1685P	85	C711E	85	2SC1359	85	2SC458LC	85	2SC839E
85	C1685Q	85	C711F	85	2SC1359A	85	2SC458LG	85	2SC839F
85	C1973	85	C715	85	2SC1359EC	85	2SC458LG(B)	85	2SC839H
85	C363	85	C715A	85	2SC1359C	85	2SC458LGA	85	2SC839J
85	C366	85	C715B	85	2SC1359Q			85	2SC839JI
85	C367	85	C715C	85	2SC1652				
85	C372	85	C715D	85	2SC1633				
85	C372-1	85	C715E						
85	C372-2	85	C715EJ						
85	C372-0								

ECG	Industry Standard No.	ECG	Industry Standard No.	ECG	Industry Standard No.	ECG	Industry Standard No.	ECG	Industry Standard No.
85	2SC839K	87	D16	87	2SD73A	909	156-0015-00	923	723DC
85	2SC839L	87	D17	87	2SD73B	909	221-30	923	AMU6A7723393
85	2SC839M	87	D18	87	2SD73C	909	349-212-003	923	FU6A7723393
85	2SC839N	87	D180	87	2SD73E	909	351-7140-010	923	FU9A7723393
85	2SC839R	87	D180A	87	2SD74	909	477-0376-001	923	G39050782
85	2SC839S	87	D180B	87	2SD74A	909	586-024	923	GEIC-260
85	2SC839Y	87	D180C	87	2SD74A	909	709HC	923	HA17723
85	2SC870	87	D180M	87	2SD74B	909	9300613-P1	923	IC-20(PHILCO)
85	2SC870BL	87	D188	87	2SD74C	909	1081K94-7	923	IC-53(ELCOM)
85	2SC870D	87	D188A	87	2SD74D	909	1820-0058	923	IC-531(ELCOM)
85	2SC870E	87	D188B	87	2SD80	909	1820-0248	923	IO53(ELCOM)
85	2SC870P	87	D188C	87	2SD81	909	2036-68	923	ITT723(D.I.P.)
85	2SC871	87	D201	87	2SD82	909	2036-72	923	LM723C
85	2SC871BL	87	D201(0)	87	2SD83	909	2056-93	923	LM723D
85	2SC871D	87	D201M	87	2SD84	909	55986-1	923	LM723CN
85	2SC871E	87	D201Y	87	2SD88	909	130221	923	MC1723CL
85	2SC871P	87	D202	87	2SD88A	909	198410-1	923	N5723A
85	2SC871Q	87	D203	88	A626	909	505342	923	P8-801
85	2SC872	87	D26A	88	A627	909	517999	923	RB6-P8-801
85	2SC925	87	D26B	88	A658	909	710398-28	923	SG723CD
85	2SC944	87	D26C	88	A756	909	932030-1	923	SG723CN
85	2SC944K	87	D50	88	A756A	909	989709	923	SK3165
85	2SC945(R)	87	D51	88	A756B	909	1604609-2	923	SL21385
85	2SC945-0	87	D53	88	A756C	909	1872425-1	923	SL22310
85	2SC945-R	87	D73	88	A757	909	2391773	923	SL22935
85	2SC945A	87	D73A	88	A757A	909	5113642	923	SL23325
85	2SC945A/D	87	D73B	88	A757B	909	7528157P1	923	TDBO723A
85	2SC945AK	87	D73C	88	A757C	909	11242096	923	Z10003
85	2SC945AP	87	D73E	88	A837	909	16765333-001	923	007-1669901
85	2SC945AQ	87	D74	88	B506	909	30677435-001	923	13-5020-6
85	2SC945AR	87	D74A	88	B506A	909	84667800	923	19-11025-00
85	2SC945B	87	D74B	88	B541	909	91011500	923	44A393611
85	2SC945C	87	D74C	88	B550	909D	FU6A7709393	923	44A393611-001
85	2SC945CK	87	D74D	88	2SA1007	909D	GEIC-250	923	46-136284-P2
85	2SC945D	87	D74E	88	2SA626	909D	KS-20971-L1	923	46-5002-20
85	2SC945E	87	D80	88	2SA626L	909D	LM709CN	923	156-0071-00
85	2SC945F	87	D81	88	2SA627	909D	MC1709CP2	923	224HACAO723
85	2SC945G	87	D82	88	2SA656	909D	MC1709P2	923	398-13222-1
85	2SC945K	87	D83	88	2SA657	909D	N5709A	923	723DC
85	2SC945L	87	D84	88	2SA658	909D	PA7000/591	923	723FC
85	2SC945LP	87	D88	88	2SA756	909D	PA7001/502	923	1677-1149
85	2SC945LPQ	87	D88A	88	2SA756B	909D	RC709DP	923	142349
85	2SC945LQ	87	2SC1030	88	2SA756C	909D	SK3552	923	157800
85	2SC945M	87	2SC1030A	88	2SA757	909D	SN72709N	923	157800-2760
85	2SC9450	87	2SC1030B	88	2SA757A	909D	15-37833-1	923	950419
85	2SC945P	87	2SC1030C	88	2SA757B	909D	15-37833-2	923	1802677-1
85	2SC945PO	87	2SC1030D	88	2SA757C	909D	43A212040-2	923	2899414
85	2SC945PQ	87	2SC1111	88	2SA808	909D	179-46447-03	923	3008340
85	2SC945Q	87	2SC1112	88	2SA837	909D	442-7	923	04040751-001
85	2SC945QL	87	2SC1403	88	2SB506	909D	709DC	923	7012109-01
85	2SC945QP	87	2SC1618	88	2SB506A	909D	709PC	928	LM358AH
85	2SC945QR	87	2SC1618B	88	2SB506B	909D	37833-2	928	LM358H
85	2SC945R	87	2SC1619	88	2SB506C	909D	41180	928	LM358R
85	2SC945S	87	2SC1777	88	2SB506D	909D	658583	928	LM2904N
85	2SC945T	87	2SC2337	88	2SB541	909D	04040501-1	928M	LM358AN
85	2SC945TK	87	2SC493	88	2SB550	91	2SA1019	928M	LM358AQ
85	2SC945TP	87	2SC493-BL	88	2SB653	91	2SA893	928M	LM358N
85	2SC945TR	87	2SC493-R	88	2SB654	91	2SA893AE	928M	LM358P
85	2SC945X	87	2SC493-R	89	2SB905	910	LM710CH	94	A6773802
85	2SC945Y	87	2SC519	90	2SC1890	910	MC1710CG	94	D78411
85	2SD227	87	2SC519A	90	2SC1890A	910	N5710T	94	MJ1800
85	2SD227A	87	2SC520	90	2SC1890AD	910	RC710T	94	MJ3028
85	2SD227B	87	2SC520A	90	2SC1890AE	910	SG710CT	94	MJ410
85	2SD227C	87	2SC521	90	2SC1890E	910	SN72710L	94	MJ411
85	2SD227D	87	2SC521A	90	2SC1890F	910	710BC	94	RCA411
85	2SD227E	87	2SC665	900	CA3000	910D	LM710CN	94	SDT410
85	2SD227P	87	2SC665H	900	GEIC-245	910D	N5710A	94	SDT411
85	2SD227L	87	2SC665HA	901	CA3001	910D	SG710CQ	94	SPC430
85	2SD227R	87	2SC665HB	901	GEIC-180	910D	SN72710N	94	2N5239
85	2SD227S	87	2SC675	901	HA1110	911	LM711CH	94	2N5240
85	2SD227V	87	2SC677	901	TA7031M	911	N5711K	94	2N5858
85	2SD227W	87	2SC736	903	CA3010	911	RC711T	94	2N5859
85	2SD227X	87	TA7031M	903	CA3010A	911	SG711CT	94	2SC1050
85	2SD327	87	2SD124	903	CA3015	911D	LM711CN	94	2SC1050D
85	2SD327A	87	2SD124A	903	CA3015A	911D	N5711A	94	2SC1050E
85	2SD327B	87	2SD124AH	903	GEIC-288	911D	SG711CQ	94	2SC1050P
85	2SD327C	87	2SD124AHA	903	HA1301	911D	SN72711N	94	2SC1106
85	2SD327D	87	2SD124AHB	904	CA3018	912	CA3045	94	2SC1106A
85	2SD327E	87	2SD125	904	CA3018A	912	CA3045P	94	2SC1106C
85	2SD327F	87	2SD125A	904	GEIC-289	912	CA3046	94	2SC1106K
85	2SD327Y	87	2SD125AH	904	1463681-1	912	CA3086	94	2SC1106L
85	2SD467	87	2SD125AHA	905	CA3019	912	GEIC-172	94	2SC1106M
85	2SD467B	87	2SD126	905	GEIC-290	912	HA1127	94	2SC1106Q
85	2SD467C	87	2SD126A	906	CA3026	912	LM3045D	94	2SC1114
85	2SD592	87	2SD126AH	906	CA3049	912	LM3046N	94	2SC1195
85	2SD592B	87	2SD126AHA	906	CA3049T	912	LM3086N	94	2SC1829
85	700-154	87	2SD126AHB	906	GEIC-246	912	MC3346P	94	2SD198
85	700-325	87	2SD126HA	906	1477-5352	912	544-2002-008	94	2SD198A
85	2000-268	87	2SD126HB	907	CA3059	912	2010-5409	94	2SD198AP
86	2SC1629A	87	2SD15	907	GEIC-247	912	4082626-0001	94	2SD198AR
86	2SC1629M	87	2SD16	908	CA3029	913	CA3047	94	2SD198AR
86	2SC1768	87	2SD17	908	CA3030	913	CA3047A	94	2SD198H
86	48-155140	87	2SD18	908	CA3030A	914	CA3058	94	2SD198HQ
86S	2SC1629	87	2SD180	908	CA3037	914	CA3059	94	2SD198HR
87	C1030	87	2SD180A	908	CA3037A	914	CA3079	94	2SD198Q
87	C1030A	87	2SD180B	908	CA3038	914	GEIC-256	94	2SD198R
87	C1030B	87	2SD180C	908	CA3038A	915	FU587717593	94	2SD198S
87	C1030C	87	2SD180D	908	GEIC-248	915	G39050381	94	2SD198V
87	C1030D	87	2SD180K	909	CD541	915	IC35(ELCOM)	94	2SD632
87	C1111	87	2SD180L	909	FU587709393	915	SL21384	94	2SD632Q
87	C493	87	2SD180M	909	G39007681	915	SL21577	94	8-729-345-42
87	C493-BL	87	2SD188	909	HEP-C6105P	915	82M432B2	94	57A196-10
87	C493-R	87	2SD188A	909	HL24630	915	156-0151-00	94	57B196-10
87	C493-Y	87	2SD188B	909	LM709C	915	179-46447-21	94	57B256-10
87	C519	87	2SD188C	909	MC1709CG	915	715HC	94	2003145404
87	C519A	87	2SD188L	909	MC1709CP1	915	1820-0476	941	A1820-0203-1
87	C520A	87	2SD188M	909	N5709T	915	157564	941	A1826-0007-1
87	C521	87	2SD201	909	PA7001/501	915	290279B-2	941	AD741
87	C521A	87	2SD201(0)	909	PA7709	915	9101600	941	AD741C
87	C665	87	2SD201M	909	PA7709C	915	50254400	941	AD741H
87	C665H	87	2SD201MY	909	RC709T	916	CA3081	941	AMU587741393
87	C665HA	87	2SD201O	909	SG709CT	916	GEIC-257	941	CA3741
87	C665HB	87	2SD201Q	909	SLO7OCQ	916	M2032-330	941	CA3741CB
87	C675	87	2SD201Y	909	SLO7055	916	51-84320A32	941	CA3741CS
87	C736	87	2SD202	909	SLO8066	916	51B84320A32	941	CA3741CT
87	D124	87	2SD203	909	SL20927	916	1858-0023	941	CA3741S
87	D124AH	87	2SD217	909	SL21923	917	CA3054	941	CA3741T
87	D124AHA	87	2SD218	909	SL23524	917	GEIC-258	941	CA3747CT
87	D125	87	2SD26	909	SL23482	917	51-10541AO1	941	CA741CT
87	D125A	87	2SD26A	909	SN72709L	922	LM311H	941	FU587741393
87	D125AH	87	2SD26B	909	TAA521	922M	LM311N	941	GEIC-263
87	D125AHA	87	2SD26C	909	TAA521/709	923	AMU6A7723393	941	HA17741M
87	D125AHB	87	2SD371	909	13-53-6	923	LM723	941	HL24510
87	D126	87	2SD388	909	17-12096-1	923	LM723CH	941	HL24593
87	D126A	87	2SD428	909	19A116549P1	923	MC1723CG	941	IC-40(ELCOM)
87	D126AH	87	2SD50	909	32-807072-1	923	MC1723G	941	IC317(ELCOM)
87	D126AHA	87	2SD51	909	43A21040-1	923	N5723L	941	I040(ELCOM)
87	D126AHB	87	2SD51A	909	46-15629P1	923	SG723CT	941	LM741C
87	D126H	87	2SD53	909	51-25789H	923	TBA281	941	LM741CH
87	D126HA	87	2SD53A	909	126-40	923	TBA281/723	941	MC1741CG
87	D126HB	87	2SD673	909	133P80057	923	TBA281/723C	941	ON587840
87	D15	87	2SD674					941	N5741T
		87	2SD73					941	PA7001/503

ECG	Industry Standard No.	ECG	Industry Standard No.	ECG	Industry Standard No.	ECG	Industry Standard No.	ECG	Industry Standard No.
941	PA7741	947D	SG747CN	HIDIV-12	33-1390-3				
941	PA7741C	947D	SN72747J	HIDIV-12	230199-1				
941	RC741T	947D	SN72747N	HIDIV-2	D5404				
941	SC51750	947D	SS8747CP	HIDIV-2	PP-3				
941	SC741T	948	A40D	HIDIV-2	GE-FR9				
941	SG741CT	949	GEIC-25	HIDIV-2	13-10-4				
941	SK3514	949	749	HIDIV-2	13-12-4				
941	SL20929	949	749DHC	HIDIV-2	13-18-4				
941	SL21673	949	138681	HIDIV-2	13-19-4				
941	SL23486	950	LM78L12ACZ	HIDIV-2	13-20-4				
941	SMC750123-1	955M	CA555CE	HIDIV-3	13-9-4				
941	SN2741L	955M	GEIC-269	HIDIV-3	GEFR-10				
941	SS8741CJ	955M	IC-520(ELCOM)	HIDIV-3	63-9897				
941	TBA221	955M	J1000-NE555	HIDIV-3	63-9898				
941	TBA221/741C	955M	J4-1555	HIDIV-3	63-9898-02				
941	TBA222	955M	M51841P	HIDIV-3	800-616				
941	TBA222/741	955M	MC1455P1						
941	ULN2741D	955M	NE555JG						
941	19-09234	955M	NE555V						
941	19A116297F3	955M	SK3564						
941	19A116297F3-10	955M	SN72555JP						
941	19A116297F3-9	955M	SN72555P						
941	44A332168-001	955M	555						
941	44A332168-002	960	GEIC-190						
941	44A332169-002	960	LM340T-5.0						
941	68A7672P1	960	LM341P-5.0						
941	77C710891-2	960	TA78005P						
941	131A8471	961	LM320T-5.0						
941	156-0049-00	961	SN72905						
941	179-46447-08	962	LM340T-6.0						
941	351-1029-020	962	LM341P-6.0						
941	477-0542-001	963	LM320T-6.0						
941	741HC	963	SN72906						
941	1081K94-6	964	GEIC-191						
941	1081K94-9	964	LM340T-8.0						
941	1479-0273	964	LM341P-8.0						
941	2014-6684	966	LM340T-12						
941	68994-1	966	LM341P-12						
941	92138-01	966	LM342P-12						
941	153270	966	2361461						
941	183532	967	LM320T-12						
941	188660-01	967	MC7912CT						
941	193207	967	SN72912						
941	301915-1	968	LM340T-15						
941	508511	968	LM341P-15						
941	618483-1	969	LM320T-15						
941	618984-1	969	SN72915						
941	717399-4	97	MJ3029						
941	717399-49	97	2N6249						
941	1802520-001	97	2N6250						
941	2710002	97	2N6251						
941	2797658-616030A	972	LM340T-24						
941	3007680-00	972	LM341P-24						
941	11728100	973	LM1496H						
941	18458117	973	N5596K						
941	39126500	973	SG1496T						
941	46156741P1	973D	LM1496J						
941	50251300	973D	LM1496N						
941D	AD741CN	975	GEIC-173						
941D	AMU6A7741393	975	LM301AN						
941D	FU6A7741393	975	LM301AP						
941D	FU9A7741393	975	LM748CN						
941D	GEIC-264	975	ML748CS						
941D	HC1000217-0	975	SG301AM						
941D	MC1741CP2	975	SN72301AP						
941D	N5741A	975	SN72748P						
941D	SG741CD	975	SS8301AP						
941D	SG741CN	976	LM307N						
941D	SN72741N	976	LM307P						
941D	TBA221A/741C	976	ML3078						
941D	741DC	976	SG307M						
941M	A1820-0219-1	976	SG307N						
941M	C6052P	977	LM78L05						
941M	CA741CS	977	LM78L05ACH						
941M	FU9T7741393	977	LM78L05ACZ						
941M	GEIC-265	977	LM78L05CH						
941M	HEP-C6052P	977	LM78L05CZ						
941M	IC-295(ELCOM)	977	MC78L05						
941M	KS-21177	977	MC78L05ACP						
941M	LM741CN	977	MC78L05C						
941M	MC1741CP1	977	MC78L05CP						
941M	N5741V	977	NJM78L05A						
941M	RC741DN	977	REN977						
941M	RC741DP	977	02-781050						
941M	SG741CM	977	78L05						
941M	SL22745	978	LM556CN						
941M	SL23059	978	NE556A						
941M	SL23252	978	SG556CN						
941M	SL23496	981	LM78L08ACZ						
941M	SN72471P	981	LM78L08CH						
941M	SN72741P	981	LM78L08CZ						
941M	STX49007	981	MC78L08ACP						
941M	T3B	987	LM224AD						
941M	TA7504P	987	LM224D						
941M	007-1669602	987	LM2902N						
941M	13-5018-6	987	LM324AN						
941M	19-10298-00	987	LM324J						
941M	020-1114-015	987	LM324N						
941M	31-1012	987	LM565CN						
941M	34-194	989	NE565A						
941M	44A417779-001	990	LM377						
941M	51-10715A01	990	LM377N						
941M	51810715A01	992	CA3401E						
941M	133P80104	992	LM2900N						
941M	156-0067-00	992	LM3301N						
941M	156-0067-06	992	LM3401N						
941M	398-13227	992	LM3900N						
941M	551-008-00	HIDIV-1	D5403						
941M	733W00021	HIDIV-1	PP-2						
941M	741TC	HIDIV-1	GE-FR8						
941M	1820-0216-1	HIDIV-1	63-9893						
941M	1820-0217-1	HIDIV-1	63-9894						
941M	137875(I.C.)	HIDIV-1	63-9895						
941M	144178	HIDIV-1	63-9896						
941M	150580	HIDIV-1	6606-1						
941M	150580-2285	HIDIV-1	6606-10						
941M	615246-1	HIDIV-1	6606-2						
941M	615268-101	HIDIV-1	6606-3						
941M	660388-02	HIDIV-1	6606-4						
941M	2392152	HIDIV-1	6606-5						
941M	2610043-03	HIDIV-1	6606-6						
941M	14500016-002	HIDIV-1	6606-7						
941M	261004303	HIDIV-1	6606-8						
941M	2002800034	HIDIV-1	6606-9						
947	AMU5B7747393	HIDIV-1	7500871-1						
947	CA747CT	HIDIV-1	7500871-2						
947	GEIC-268	HIDIV-1	7500871-3						
947	LM747C	HIDIV-1	7500871-4						
947	LM747CH	HIDIV-1	7500871-5						
947	SG747CT	HIDIV-1	7500871-6						
947	747HC	HIDIV-1	7500871-7						
947	CA747CE	HIDIV-1	7500871-8						
947D	CA747CP	HIDIV-1	7500871-9						
947D	ML747CP	HIDIV-12	33-1390-1						
		HIDIV-12	33-1390-2						

SK	Industry Standard No.	SK	Industry Standard No.	SK	Industry Standard No.	SK	Industry Standard No.	SK	Industry Standard No.
3003	A12	3003	2N63	3003	9L-4A4-7	3004	A25-1006	3004	B120
3003	A1466-1	3003	2N631	3003	9L-4A5	3004	A49	3004	B135
3003	A1466-19	3003	2N64	3003	9L-4A5L	3004	A52	3004	B135B
3003	A149L-4	3003	2N65	3003	9L-4A6	3004	A53	3004	B135C
3003	A149L-49	3003	2N81	3003	9L-4A6-1	3004	A58	3004	B135E
3003	A66-1-70	3003	2N82	3003	9L-4A7	3004	A615-1010	3004	B36
3003	A66-1-705	3003	2846	3003	9L-4A7-1	3004	A615-1011	3004	B136A
3003	A66-1A9G	3003	2847	3003	9L-4A8	3004	A64	3004	B136B
3003	A9L-4-70	3003	2SA12	3003	9L-4A82	3004	A65	3004	B136C
3003	A9L-4-705	3003	2SA12A	3003	9L-4A9	3004	A66	3004	B136U
3003	A9L-4A9G	3003	2SA12B	3003	9L-4A9G	3004	A66-2-70	3004	B153
3003	ACY-22	3003	2SA12C	3003	12-1-274	3004	A66-2-705	3004	B154
3003	B156	3003	2SA12D	3003	21-36	3004	A66-2A9G	3004	B155
3003	B156A	3003	2SA12H	3003	31-0017	3004	A66-3-70	3004	B155A
3003	B156AA	3003	2SA12V	3003	34-6016-50	3004	A66-3-705	3004	B155B
3003	B156AB	3003	2SA203B	3003	57D1-26	3004	A66-3A9G	3004	B161
3003	B156AC	3003	2SB110	3003	57D1-27	3004	A8P-2-70	3004	B162
3003	B156B	3003	2SB111	3003	57D1-42	3004	A8P-2-705	3004	B163
3003	B156C	3003	2SB114	3003	66-1	3004	A8P-2A9G	3004	B164
3003	B156D	3003	2SB115	3003	066-1-12	3004	A909-1011	3004	B165
3003	B156F	3003	2SB156	3003	066-1-12-7	3004	A909-1012	3004	B166
3003	EA2134	3003	2SB156A	3003	66-1-70	3004	A909-1013	3004	B168
3003	L5021	3003	2SB156AA	3003	66-1-70-12	3004	AC-107	3004	B170
3003	OO-318	3003	2SB156AC	3003	66-1-70-12-7	3004	AC-113	3004	B171
3003	OO-70	3003	2SB156B	3003	66-10	3004	AC-113A	3004	B171A
3003	OO-72	3003	2SB156BK	3003	66-11	3004	AC-114	3004	B171B
3003	OO-73	3003	2SB156C	3003	66-12	3004	AC-116	3004	B172
3003	OO-74	3003	2SB156D	3003	66-13	3004	AC-117	3004	B172A
3003	OO-75	3003	2SB156E	3003	66-14	3004	AC-117A	3004	B172AP
3003	OO58	3003	2SB156F	3003	66-15	3004	AC-117P	3004	B172B
3003	OO60	3003	2SB156G	3003	66-16	3004	AC-122	3004	B172C
3003	OO72	3003	2SB156GN	3003	66-17	3004	AC-123	3004	B172D
3003	R83310	3003	2SB156H	3003	66-18	3004	AC-125	3004	B172F
3003	R85406	3003	2SB156J	3003	66-19	3004	AC-126	3004	B172H
3003	RT3229	3003	2SB156K	3003	66-1A	3004	AC-128	3004	B172P
3003	SF8871	3003	2SB156L	3003	66-1AO	3004	AC-132	3004	B172R
3003	T-00014	3003	2SB156M	3003	66-1AOR	3004	AC-150	3004	B173
3003	TO003	3003	2SB156P	3003	66-1A19	3004	AC-151	3004	B173A
3003	TO004	3003	2SB156R	3003	66-1A2	3004	AC-152	3004	B173B
3003	TO005	3003	2SB156Y	3003	66-1A21	3004	AC-154	3004	B173C
3003	TO012	3003	2SB168	3003	66-1A3	3004	AC-155	3004	B173L
3003	TO015	3003	2SB169	3003	66-1A3P	3004	AC-156	3004	B174
3003	TO033	3003	2SB172	3003	66-1A4	3004	AC-161	3004	B175
3003	T1013	3003	2SB172A	3003	66-1A4-7	3004	AC-162	3004	B175A
3003	T1740	3003	2SB172A-P	3003	66-1A5	3004	AC-165	3004	B175B
3003	TA-1697	3003	2SB172AL	3003	66-1A5L	3004	AC-166	3004	B175C
3003	TA-1706	3003	2SB172AL	3003	66-1A6	3004	AC-167	3004	B175E
3003	TO103	3003	2SB172B	3003	66-1A6-3	3004	AC-168	3004	B176
3003	TO104	3003	2SB172C	3003	66-1A7	3004	AC-169	3004	B176-0
3003	TRO2063-1	3003	2SB172D	3003	66-1A7-1	3004	AC105	3004	B176-P
3003	TRO4	3003	2SB172E	3003	66-1A8	3004	AC106	3004	B176-PR
3003	TRO4G	3003	2SB172F	3003	66-1A82	3004	AC151	3004	B176M
3003	WEP250	3003	2SB172FN	3003	66-1A9	3004	AC151R	3004	B176PRC
3003	ZA105604	3003	2SB172G	3003	66-1A9G	3004	AC191	3004	B176R
3003	1-21-274	3003	2SB172GN	3003	86-25-2	3004	AC192	3004	B178
3003	1A0055	3003	2SB172H	3003	86-29-2	3004	ACY-32	3004	B178-0
3003	1A0056	3003	2SB172J	3003	86-39-2	3004	ACY27	3004	B178-8
3003	002-011900	3003	2SB172K	3003	86-45-2	3004	ACY28	3004	B178A
3003	2N1009	3003	2SB172L	3003	86-49-2	3004	ACY29	3004	B178C
3003	2N106	3003	2SB172M	3003	86-61-2	3004	ACY30	3004	B178D
3003	2N107	3003	2SB172OR	3003	86-72-2	3004	ACY31	3004	B178M
3003	2N108	3003	2SB172R	3003	86-74-2	3004	ACY34	3004	B178N
3003	2N1097	3003	2SB172X	3003	089-222	3004	ACY35	3004	B178T
3003	2N1098	3003	2SB172Y	3003	121-148	3004	ACY36	3004	B178U
3003	2N1144	3003	2SB176	3003	121-163	3004	ACY41	3004	B178V
3003	2N1145	3003	2SB176A	3003	121-192	3004	ACY41-1	3004	B178X
3003	2N1265	3003	2SB176B	3003	121-274	3004	ACY44	3004	B178Y
3003	2N130	3003	2SB176C	3003	121-61	3004	ACY44-1	3004	B180
3003	2N131	3003	2SB176D	3003	121-64	3004	AE-50	3004	B180A
3003	2N131A	3003	2SB176E	3003	149-4	3004	AFB-11A-1008	3004	B181
3003	2N132	3003	2SB176F	3003	149L-4-12	3004	AFB-11H-1008	3004	B181A
3003	2N132A	3003	2SB176G	3003	149L-4-12-8	3004	AFB-11H-1010	3004	B183
3003	2N133	3003	2SB176GN	3003	189L-4	3004	AR-102	3004	B184
3003	2N133A	3003	2SB176H	3003	189L-4-12	3004	AR-103	3004	B186
3003	2N1352	3003	2SB176J	3003	189L-4-127	3004	AR-104	3004	B186-1
3003	2N1353	3003	2SB176K	3003	189L-4L	3004	AS33868	3004	B186-K
3003	2N138	3003	2SB176L	3003	189L-4L8	3004	ASY48	3004	B186A
3003	2N138A	3003	2SB176M	3003	650-108	3004	ASY48-IV	3004	B186AG
3003	2N138B	3003	2SB176O	3003	690V052H23	3004	ASY48-V	3004	B186B
3003	2N1427	3003	2SB176OR	3003	1466-1	3004	ASY48-VI	3004	B186BT
3003	2N175	3003	2SB176P	3003	1466-1-12	3004	ASY49	3004	B186G
3003	2N207	3003	2SB176PL	3003	1466-1-12-8	3004	ASY50	3004	B186L
3003	2N207A	3003	2SB176PR	3003	1866-1	3004	ASY52	3004	B186Y
3003	2N207B	3003	2SB176PRC	3003	1866-1-12	3004	ASY54	3004	B187
3003	2N223	3003	2SB176R	3003	1866-1-127	3004	ASY55	3004	B187AA
3003	2N265	3003	2SB176R(1)	3003	1866-1L	3004	ASY56	3004	B187B
3003	2N272	3003	2SB176RG	3003	1866-1L8	3004	ASY70	3004	B187C
3003	2N281	3003	2SB176X	3003	2057B2-23	3004	ASY70-IV	3004	B187D
3003	2N283	3003	2SB176Y	3003	3112	3004	ASY70-VI	3004	B187G
3003	2N302	3003	2SB34	3003	12127-4	3004	ASY90	3004	B187R
3003	2N303	3003	2SB39	3003	44616-1	3004	ASY91	3004	B187B
3003	2N319	3003	2SB43	3003	46776-2	3004	AT-50	3004	B188
3003	2N320	3003	2SB46	3003	47737-2	3004	AT-6A	3004	B200
3003	2N321	3003	2SB47	3003	49139-2	3004	A210H	3004	B200A
3003	2N322	3003	2SB50	3003	81153-9	3004	A210M	3004	B201
3003	2N324	3003	2SB51	3003	95203	3004	A210N	3004	B202
3003	2N36	3003	2SB52	3003	116203	3004	A220H	3004	B218
3003	2N37	3003	2SB76	3003	116206	3004	A220M	3004	B219
3003	2N38	3003	2SB89	3003	218502	3004	A220N	3004	B220
3003	2N38A	3003	2SB92	3003	223371	3004	A230H	3004	B220A
3003	2N402	3003	03-156B	3003	223484	3004	A230M	3004	B221
3003	2N403	3003	03-57-301	3003	223485	3004	A230N	3004	B221A
3003	2N405	3003	4-66-1A7-1	3003	225593	3004	A8874	3004	B222
3003	2N407	3003	4-9L-4A7-1	3003	226287	3004	ATAP1	3004	B223
3003	2N408	3003	9L-4	3003	551051	3004	ATAP2	3004	B225
3003	2N41	3003	09L-4-12	3003	0573034	3004	ATOP	3004	B226
3003	2N422	3003	09L-4-12-7	3003	057314ZH	3004	B-1058	3004	B227
3003	2N43	3003	9L-4-70	3003	801520	3004	B-22-3	3004	B241
3003	2N466	3003	9L-4-70-12	3003	802389-2	3004	B-22-4	3004	B257
3003	2N467	3003	9L-4-70-12-7	3003	815139	3004	B-23	3004	B261
3003	2N508	3003	9L-40	3003	980375	3004	B-23-1	3004	B262
3003	2N522	3003	9L-41	3003	980376	3004	B-23-2	3004	B263
3003	2N523	3003	9L-42	3003	995002	3004	B-24-1	3004	B264
3003	2N535	3003	9L-43	3003	8023892	3004	B-26	3004	B266
3003	2N535A	3003	9L-44	3003	A-514-027662	3004	B-F1A	3004	B266P
3003	2N535B	3003	9L-45	3004	A059-106	3004	B101	3004	B266Q
3003	2N536	3003	9L-46	3004	A059-107	3004	B102	3004	B267
3003	2N569	3003	9L-47	3004	A059-116	3004	B1022	3004	B268
3003	2N570	3003	9L-48	3004	A061-114	3004	B1022-1	3004	B269
3003	2N571	3003	9L-49	3004	A065-106	3004	B1058-1	3004	B270
3003	2N572	3003	9L-4A	3004	A069-105	3004	B106	3004	B270A
3003	2N582	3003	9L-4AO	3004	A12B	3004	B108A	3004	B270B
3003	2N584	3003	9L-4AOR	3004	A12D	3004	B108B	3004	B270C
3003	2N59	3003	9L-4A1	3004	A1466-2	3004	B109	3004	B270D
3003	2N59A	3003	9L-4A19	3004	A1466-29	3004	B111K	3004	B270E
3003	2N59C	3003	9L-4A2	3004	A1466-3	3004	B112	3004	B271
3003	2N60	3003	9L-4A21	3004	A1466-39	3004	B113	3004	B272
3003	2N60A	3003	9L-4A3	3004	A148P-2	3004	B114	3004	B273
3003	2N60B	3003	9L-4A3P	3004	A148P-29	3004	B115	3004	B295
3003	2N60C	3003	9L-4A4	3004	A174	3004	B1154	3004	B299
3003	2N61			3004	A1877EV	3004	B1154-1	3004	B302
3003	2N61A			3004	A25-1004	3004	B116	3004	B3030
3003	2N61B			3004	A25-1005	3004	B117		
3003	2N61C					3004	B117K		

SK	Industry Standard No.	SK	Industry Standard No.	SK	Industry Standard No.	SK	Industry Standard No.	SK	Industry Standard No.	SK	Industry Standard No.
3004	B303A	3004	B60	3004	D031	3004	HB-54	3004	OC-71N	3004	RS-1192
3004	B303B	3004	B601-1009	3004	D038	3004	HB-56	3004	OC-75N	3004	RS-1540
3004	B303C	3004	B60A	3004	D043	3004	HB-77C	3004	OC-77	3004	RS-1541
3004	B303K	3004	B61	3004	D135	3004	HB-85	3004	OC-79	3004	RS-1542
3004	B315	3004	B65	3004	D043	3004	HB186	3004	OC-81DD	3004	RS-1543
3004	B316	3004	B66	3004	DS-14	3004	HB187	3004	OC304-1	3004	RS-1544
3004	B317	3004	B66H	3004	DS-26	3004	HB459	3004	OC305	3004	RS-1546
3004	B32	3004	B67	3004	E-070	3004	HB75	3004	OC305-1	3004	RS-1548
3004	B32-0	3004	B67A	3004	E-158	3004	HB77	3004	OC305-2	3004	RS-1555
3004	B32-1	3004	B71	3004	E-2465	3004	HB77B	3004	OC41	3004	RS-2004
3004	B32-2	3004	B72	3004	E241	3004	HEP-00004	3004	OC41A	3004	RS-2005
3004	B32-4	3004	B73	3004	E24106	3004	HJ43	3004	OC711	3004	RS-2350
3004	B321	3004	B73B	3004	E2453	3004	HM-08014	3004	OC76	3004	RS-2351
3004	B322	3004	B73GR	3004	E2465	3004	HR-2	3004	OC81	3004	RS-2352
3004	B323	3004	B75	3004	E2467	3004	HR-3	3004	OC81D	3004	RS-2353
3004	B326	3004	B75AH	3004	EA0009	3004	HR-30	3004	OF-129	3004	RS-2354
3004	B327	3004	B75B	3004	EA1346	3004	HR-39	3004	PTC109	3004	RS-2355
3004	B328	3004	B75C	3004	EA15X164	3004	HR-7	3004	PTC135	3004	RS-2367
3004	B329	3004	B75F	3004	EA15X19	3004	HR2A	3004	PXB-103	3004	RS-2675
3004	B329K	3004	B75H	3004	EA15X2	3004	HR53	3004	PXB-113	3004	RS-2677
3004	B32N	3004	B75LB	3004	EA15X203	3004	HR7A	3004	PXC-101	3004	RS-2689
3004	B33-4	3004	B76	3004	EA15X207	3004	HS-15	3004	PXC-101AB	3004	RS-2697
3004	B336	3004	B77	3004	EA15X212	3004	HS-22D	3004	Q-16	3004	RS-3275
3004	B330	3004	B77(B)	3004	EA15X23	3004	HS102	3004	Q-6	3004	RS-3276
3004	B33D	3004	B77A	3004	EA15X25	3004	HT200540	3004	Q-7	3004	RS-3282
3004	B33B	3004	B77AA	3004	EA15X28	3004	HT200540A	3004	Q-8	3004	RS-3283
3004	B33F	3004	B77AB	3004	EA15X3	3004	HT200541	3004	QOV60526	3004	RS-3284
3004	B34	3004	B77AC	3004	EA15X36	3004	HT200541A	3004	QOV60528	3004	RS-3285
3004	B345	3004	B77AD	3004	EA15X4	3004	HT200541B-Q	3004	Q1-7C	3004	RS-3286
3004	B346	3004	B77AH	3004	EA15X67	3004	HT200561	3004	Q2-7C	3004	RS-3299
3004	B346K	3004	B77AP	3004	EA15X6840	3004	HT200561A	3004	Q2N2428	3004	RS-3301
3004	B347	3004	B77B	3004	EA15X7	3004	HT200561B	3004	Q2N2613	3004	RS-3308
3004	B348	3004	B77B-11	3004	EA15X8	3004	HT200561C	3004	Q2N406	3004	RS-3310
3004	B348R	3004	B77C	3004	EA15X8442	3004	HT200561C-0	3004	Q35218	3004	RS-3316
3004	B349	3004	B77D	3004	EA2195	3004	HT200751B	3004	Q40263	3004	RS-3316-1
3004	B350	3004	B77H	3004	EA2176	3004	HT200770B	3004	Q82378	3004	RS-3316-2
3004	B37	3004	B77V	3004	EB0001	3004	HT200771	3004	R-120	3004	RS-3318
3004	B370	3004	B77VRED	3004	EB0003	3004	HT200771A	3004	R-152	3004	RS-3904
3004	B370A	3004	B78	3004	EC0102A	3004	HT200771B	3004	R-16	3004	RS-3925
3004	B370AA	3004	B79	3004	EC0158	3004	HT200771C	3004	R-164	3004	RS-406
3004	B370AB	3004	B89	3004	ED55	3004	HT201721A	3004	R-23-1003	3004	RS-5008
3004	B370AC	3004	B89A	3004	ED56	3004	HT201721D	3004	R-23-1004	3004	RS-5206
3004	B370AHA	3004	B89AH	3004	ED57	3004	HT201725A	3004	R-24-1001	3004	RS-5311
3004	B370AHB	3004	B89H	3004	EO-44A	3004	HT201782A	3004	R-24-1002	3004	RS-5406
3004	B370B	3004	B90	3004	EQO-15	3004	HT203701	3004	R-242	3004	RS-5505
3004	B370C	3004	B92	3004	EQG-9	3004	HT203701A	3004	R-245	3004	RS-5506
3004	B370D	3004	B94	3004	ER15X17	3004	HT203701B	3004	R-291	3004	RS-5530
3004	B370P	3004	B95	3004	ER15X18	3004	IP-65	3004	R-28B186	3004	RS-5531
3004	B370V	3004	B97	3004	ES15X31	3004	IRTR85	3004	R-28B187	3004	RS-5532
3004	B371	3004	B98	3004	ES15X32	3004	J241164	3004	R-28B303	3004	RS-5533
3004	B371D	3004	BCM1002-4	3004	ES15X4	3004	J241190	3004	R-530	3004	RS-5534
3004	B376	3004	BCM1002-5	3004	ES15X49	3004	J24869	3004	R-593	3004	RS-5535
3004	B376G	3004	BF1	3004	ES15X50	3004	J24870	3004	R-608A	3004	RS-5536
3004	B377	3004	BF1A	3004	ES15X53	3004	J24934	3004	R-64	3004	RS-5541
3004	B378A	3004	BF2	3004	ES15X55	3004	JP5063	3004	R-66	3004	RS-5542
3004	B379	3004	C10227	3004	ES15X8	3004	JP5064	3004	R-67	3004	RS-5544
3004	B379-2	3004	C10230-3	3004	ES3120	3004	JR5	3004	R-83	3004	RS-5551
3004	B379A	3004	C11021	3004	ES3122	3004	KD2101	3004	R06-1007	3004	RS-5552
3004	B379B	3004	C1458	3004	ES3123	3004	KR-Q1010	3004	R06-1008	3004	RS-5553
3004	B37A	3004	CB161	3004	ES3124	3004	KR-Q1011	3004	R06-1009	3004	RS-5554
3004	B37B	3004	CB248	3004	ES3125	3004	KR-Q1012	3004	R06-1010		
3004	B37C	3004	CB249	3004	ES3126	3004	KV-1	3004	R24-1003		
3004	B37F	3004	CE0360/7839	3004	ET3	3004	KV-4	3004	R24-1004		
3004	B38	3004	CE0562/7839	3004	ET4	3004	L5022A	3004	R60-1004		
3004	B380	3004	CK22	3004	ET5	3004	L5025	3004	R60-1005		
3004	B380A	3004	CK22A	3004	ETTB-28B176	3004	L5025A	3004	R60-1006		
3004	B381	3004	CK22B	3004	ETTB-28B176A	3004	M4315	3004	R65		
3004	B382	3004	CK22C	3004	ETTB-28B176B	3004	M4567	3004	RE3		
3004	B383	3004	CK64	3004	ETTB-28B176R	3004	M4596	3004	RE4		
3004	B383-1	3004	CK64A	3004	ES15X25	3004	M4597	3004	REN102		
3004	B383-2	3004	CK64B	3004	F20-1006	3004	M4597GRN	3004	REN102A		
3004	B386	3004	CK64C	3004	F20-1007	3004	M4597RED	3004	REN158		
3004	B389	3004	CK65	3004	F20-1008	3004	M4607				
3004	B39	3004	CK65A	3004	F20-1009	3004	M4627				
3004	B40	3004	CK65B	3004	F215-1006	3004	M75517-1				
3004	B400	3004	CK65C	3004	F215-1007	3004	M9148				
3004	B400A	3004	CK66	3004	F215-1008	3004	M9198				
3004	B400B	3004	CK66A	3004	F215-1009	3004	MA1700				
3004	B400K	3004	CK66B	3004	FB420	3004	MN-53				
3004	B421	3004	CK66C	3004	FB421	3004	N57B2-15				
3004	B422	3004	CK67	3004	FD-1029-EE	3004	N57B2-25				
3004	B423	3004	CK67A	3004	FPR50-1005	3004	N57B2-3				
3004	B427	3004	CK67B	3004	FPR50-1006	3004	N57B2-6				
3004	B42B	3004	CK67C	3004	G0006	3004	N57B2-7				
3004	B43	3004	CK721	3004	G0007	3004	NAP-TZ-10				
3004	B439	3004	CK722	3004	GC1184	3004	N269				
3004	B439A	3004	CK725	3004	GC250	3004	NKT123				
3004	B43A	3004	CK727	3004	GC343	3004	NKT126				
3004	B44	3004	CK751	3004	GC408	3004	NKT133				
3004	B440	3004	CK754	3004	GC466	3004	NKT222				
3004	B443	3004	CK790	3004	GC520	3004	NKT223				
3004	B443A	3004	CK791	3004	GC551	3004	NKT223A				
3004	B443B	3004	CK793	3004	GC552	3004	NKT224				
3004	B454	3004	CK794	3004	GC578	3004	NKT224A				
3004	B455	3004	CK870	3004	GC579	3004	NKT225				
3004	B459	3004	CK871	3004	GC580	3004	NKT225A				
3004	B459-Q	3004	CK872	3004	GC581	3004	NKT226				
3004	B459A	3004	CK875	3004	GC639	3004	NKT226A				
3004	B459B	3004	CK878	3004	GC680	3004	NKT227				
3004	B459C	3004	CK879	3004	GC682	3004	NKT227A				
3004	B459D	3004	CK882	3004	GC856	3004	NKT228				
3004	B46	3004	CK888	3004	GC864	3004	NKT228A				
3004	B460	3004	CT1009	3004	GE-52	3004	NKT231				
3004	B460A	3004	CTP-2001-1001	3004	GE-X9	3004	NKT231A				
3004	B460B	3004	CTP-2001-1002	3004	GER-A	3004	NKT232				
3004	B47	3004	CTP-2001-1003	3004	GET-113	3004	NKT232A				
3004	B470	3004	CTP-2001-1004	3004	GET-113A	3004	NKT244				
3004	B475	3004	CTP-2001-1009	3004	GET-114	3004	NKT246				
3004	B475A	3004	CTP-2006-1001	3004	GET898	3004	NKT247				
3004	B475B	3004	CTP-2006-1002	3004	GI2	3004	NKT247A				
3004	B475D	3004	CTP-2006-1003	3004	GI4	3004	NKT273				
3004	B475E	3004	CTP1032	3004	GP139	3004	NKT278				
3004	B475F	3004	CTP1033	3004	GP139A	3004	NKT308				
3004	B475G	3004	CTP1034	3004	GP139B	3004	OC-307				
3004	B475P	3004	CTP1035	3004	GT-109	3004	OC-308				
3004	B475Q	3004	CTP1036	3004	GTE-2	3004	OC-330				
3004	B48	3004	CTP2076-1001	3004	HB-00054	3004	OC-34				
3004	B482	3004	CTP2076-1002	3004	HB-00056	3004	OC-340				
3004	B486	3004	CTP2076-1003	3004	HB-00156	3004	OC-341				
3004	B49	3004	CTP2076-1004	3004	HB-00172	3004	OC-342				
3004	B496	3004	CTP2076-1005	3004	HB-00173	3004	OC-343				
3004	B497	3004	CTP2076-1006	3004	HB-00175	3004	OC-350				
3004	B498	3004	CTP2076-1007	3004	HB-00176	3004	OC-351				
3004	B5	3004	CTP2076-1008	3004	HB-00178	3004	OC-360				
3004	B52	3004	CTP2076-1009	3004	HB-00186	3004	OC-362				
3004	B53	3004	CTP2076-1010	3004	HB-00187	3004	OC-363				
3004	B54	3004	CTP2076-1011	3004	HB-156	3004	OC-364				
3004	B54B	3004	CTP2076-1012	3004	HB-172	3004	OC-38				
3004	B54F	3004	CXL1012A	3004	HB-173	3004	OC-602				
3004	B54Y	3004	CXL158	3004	HB-175	3004	OC-604				
3004	B56	3004	D008	3004	HB-32	3004	OC-65				
3004	B56A	3004	D018	3004	HB-33	3004	OC-66				
3004	B56C	3004	D021			3004	OC-71				
						3004	OC-71A				

SK	Industry Standard No.	SK	Industry Standard No.	SK	Industry Standard No.	SK	Industry Standard No.	SK	Industry Standard No.
3004	RS-5555	3004	ST-302	3004	TR310225	3004	1-522210111	3004	2N205
3004	RS-5556	3004	ST-304	3004	TR310235	3004	1-522211200	3004	2N206
3004	RS-5557	3004	ST-332	3004	TR310252	3004	1-522211328	3004	2N207BLU
3004	RS-5558	3004	SYL-107	3004	TR320A	3004	1-522216500	3004	2N215
3004	RS-5602	3004	SYL-108	3004	TR320AN	3004	1-6207190405	3004	2N217
3004	RS-5704	3004	SYL-1583	3004	TR321A	3004	1-801-005-23	3004	2N220
3004	RS-57042	3004	SYL-1668	3004	TR323	3004	1-801-006-12	3004	2N224
3004	RS-57062	3004	SYL107A	3004	TR323A	3004	1-801-006-14	3004	2N225
3004	RS-5708	3004	SYL108A	3004	TR323AN	3004	1-801-308	3004	2N226
3004	RS-5708-2	3004	SYL1583A	3004	TR38117	3004	1-801-308-24	3004	2N227
3004	RS-5709	3004	SYL1655	3004	TR508	3004	002-005100	3004	2N2271
3004	RS-5711	3004	SYL1655A	3004	TR508A	3004	002-006600	3004	2N23
3004	RS-5717	3004	SYL1665	3004	TR508AN	3004	002-006800	3004	2N237
3004	RS-5717-3	3004	SYL1668A	3004	TR5826	3004	002-006900	3004	2N2374
3004	RS-5717-6	3004	SYL1668BA	3004	TR650	3004	002-007300	3004	2N2375
3004	RS-5720	3004	SYL2248A	3004	TR650A	3004	002-008400	3004	2N2376
3004	RS-5731	3004	SYL2249A	3004	TR653	3004	002-011000	3004	2N238
3004	RS-5733	3004	SYL2300A	3004	TR653A	3004	002-11800	3004	2N24
3004	RS-5734	3004	SYL3613	3004	TR6826	3004	002-11900	3004	2N241
3004	RS-5736	3004	SYL3613A	3004	TR71	3004	2D013	3004	2N241A
3004	RS-5737	3004	T-126	3004	TR72	3004	2D013-109	3004	2N2428
3004	RS-5742	3004	T-129	3004	TR721	3004	2D013-13	3004	2N2429
3004	RS-5743	3004	T-130	3004	TR721A	3004	2D013-160	3004	2N2447
3004	RS-5743-2	3004	T-23	3004	TR722	3004	2D013-54	3004	2N2448
3004	RS-5743-3	3004	T-3321	3004	TR722A	3004	2D016	3004	2N2449
3004	RS-5744	3004	T-3322	3004	TR763	3004	2D036	3004	2N2450
3004	RS-5744-3	3004	T-39	3004	TR763A	3004	2D039	3004	2N2468
3004	RS-5749	3004	T-52150	3004	TR81	3004	201027	3004	2N2469
3004	RS-5852	3004	T-52150Z	3004	TRA-32	3004	20270	3004	2N25
3004	RS-5854	3004	T-52151	3004	TRA-33	3004	20271	3004	2N2564
3004	RS-686	3004	T-52151Z	3004	T8-1	3004	2G302	3004	2N2565
3004	RS-687	3004	T-95	3004	T8-13	3004	2G306	3004	2N26
3004	RS3717	3004	T-HU60U	3004	T8-14	3004	2G308	3004	2N2613
3004	RS880	3004	T-Q5025	3004	T8-15	3004	2G321	3004	2N2614
3004	RS5533	3004	T-Q5026	3004	T8-3	3004	2G322	3004	2N266
3004	RS5534	3004	T-Q5027	3004	TV-61	3004	2G323	3004	2N270
3004	RS5536	3004	T0038	3004	TV-61A	3004	2G324	3004	2N270A
3004	RS5541	3004	T0039	3004	TV24115	3004	2G374A	3004	2N273
3004	RS5543	3004	T0040	3004	TV24154	3004	2G381	3004	2N279
3004	RS5545	3004	T0041	3004	TV24156	3004	2G381A	3004	2N280
3004	RS5717	3004	T1001	3004	TV24189	3004	2G384	3004	2N282
3004	RS5731	3004	T10010	3004	TV2428	3004	2G385	3004	2N284
3004	RS5732	3004	T1005	3004	TV2429	3004	2G386	3004	2N284A
3004	RS5737	3004	T1006	3004	TV24370	3004	2G387	3004	2N291
3004	RS5743-1	3004	T1008	3004	TVS-28A71B	3004	2G508	3004	2N2953
3004	RS5745	3004	T1042	3004	TVS-28B171	3004	2G509	3004	2N315A
3004	RS5746	3004	T1043	3004	TVS-28B172	3004	2N1008	3004	2N316A
3004	RS5747	3004	T1046	3004	TVS-28B172F	3004	2N1008A	3004	2N32
3004	RS5748	3004	T1047	3004	TVS-28B176	3004	2N1008B	3004	2N323
3004	RS5750	3004	T13000	3004	TVS28B171A	3004	2N104	3004	2N32A
3004	RS5751	3004	T1310	3004	TVS28B171B	3004	2N1044	3004	2N34
3004	RS5766	3004	T1352	3004	TVS28B324	3004	2N1045	3004	2N3427
3004	RS5767	3004	T1546	3004	V10/15A	3004	2N105	3004	2N3428
3004	R86840	3004	T1573	3004	V10/15A18	3004	2N1056	3004	2N359
3004	R86843	3004	T1574	3004	V10/30A	3004	2N1057	3004	2N360
3004	R86843A	3004	T1577	3004	V10/30A18	3004	2N109	3004	2N361
3004	R86846	3004	T1583	3004	V10/50A	3004	2N109/5	3004	2N362
3004	R86846A	3004	T1903	3004	V10/50A18	3004	2N1124	3004	2N362B
3004	R87568	3004	T3321	3004	VFT-2745H	3004	2N1125	3004	2N363
3004	RT-4625	3004	T3323	3004	VM-30244	3004	2N1126	3004	2N367
3004	R2185	3004	T50339A	3004	VS-28B171	3004	2N1127	3004	2N368
3004	RT2329	3004	T52159	3004	VS-28B172FN	3004	2N1129	3004	2N369
3004	RT2330	3004	T59247	3004	VS-28B176	3004	2N113	3004	2N381
3004	RT2331	3004	T59249	3004	VS-28B178	3004	2N1130	3004	2N382
3004	RT3097	3004	T61	3004	VS-28B178A	3004	2N1175	3004	2N385
3004	RT3098	3004	T72	3004	VS-28B324	3004	2N1175A	3004	2N39
3004	RT3230	3004	T74	3004	W3	3004	2N1189	3004	2N40
3004	RT3251	3004	T82	3004	WC19863	3004	2N1190	3004	2N404A
3004	RT3363	3004	T84	3004	WC19864	3004	2N1191	3004	2N406
3004	RT3364	3004	TA-1575	3004	WEP253	3004	2N1192	3004	2N42
3004	RT3365	3004	TA-1730	3004	WEP254	3004	2N1193	3004	2N43A
3004	RT3449	3004	TC3123041557	3004	WEP631	3004	2N1194	3004	2N44
3004	RT3468	3004	TF-30	3004	WEP632	3004	2N1265A	3004	2N44A
3004	RT3566	3004	TF-65	3004	WTV15MG	3004	2N1273	3004	2N460
3004	RT3568	3004	TF-66	3004	WTV20VMG	3004	2N1273BLU	3004	2N461
3004	RT5468	3004	TF75	3004	WTV30VMG	3004	2N1273GRN	3004	2N462
3004	RT5637	3004	TF77	3004	WTV30VHG	3004	2N1273ORN	3004	2N464
3004	RT6205	3004	TF80/302	3004	WTV30VLG	3004	2N1273RED	3004	2N465
3004	RT6604	3004	TIA-01	3004	WTV30VMG	3004	2N1273YEL	3004	2N47
3004	RT6734	3004	TIA04	3004	WTVA36	3004	2N1287	3004	2N48
3004	RT6735	3004	TK-23C	3004	WTVB5A	3004	2N1287A	3004	2N483-6M
3004	RT6736	3004	TK-40C	3004	WTVBA6A	3004	2N130A	3004	2N483B
3004	RT6990	3004	TK-41C	3004	X1C1644	3004	2N1348	3004	2N48A
3004	RT7407	3004	TK-42C	3004	X45C-HO6	3004	2N1370	3004	2N49
3004	RT8602	3004	TK-45C	3004	XB1	3004	2N1371	3004	2N50
3004	RV1475	3004	TK40	3004	XB102	3004	2N1372	3004	2N508A
3004	S-1348	3004	TK40A	3004	XB103	3004	2N1373	3004	2N51
3004	S-1349	3004	TK40C	3004	XB104	3004	2N1374	3004	2N52
3004	S-1639	3004	TK40CA	3004	XB112	3004	2N1375	3004	2N524
3004	S-95201	3004	TK41	3004	XB113	3004	2N1376	3004	2N524A
3004	S-95204	3004	TK41A	3004	XB114	3004	2N1377	3004	2N525
3004	0822.3504-040	3004	TK42	3004	XB1A	3004	2N1378	3004	2N525A
3004	0822.3504-060	3004	TK42A	3004	XB2	3004	2N1379	3004	2N526
3004	0822.3505-910	3004	TK45C	3004	XB2A	3004	2N1380	3004	2N526A
3004	S4248	3004	TM102A	3004	XB3	3004	2N1381	3004	2N527
3004	S99201	3004	TNJ-60070	3004	XB3A	3004	2N1382	3004	2N527A
3004	S99203	3004	TNJ-60074	3004	XB3B	3004	2N1383	3004	2N53
3004	SC-12	3004	TNJ-60079	3004	XB3BN	3004	2N1392	3004	2N54
3004	SC-66	3004	TNJ-60282	3004	XB3C	3004	2N1413	3004	2N55
3004	SC-68	3004	TNJ-60283	3004	XB3C-1	3004	2N1414	3004	2N56
3004	SC-69	3004	TNJ-60728	3004	XB4-1	3004	2N1415	3004	2N563
3004	SC-73	3004	TNJ61221	3004	XC101	3004	2N1432	3004	2N564
3004	SC63	3004	TNJ61222	3004	XC121	3004	2N1446	3004	2N565
3004	SC66	3004	TNJ61282	3004	XC171	3004	2N1447	3004	2N566
3004	SC68	3004	TNJ72283	3004	XG8	3004	2N1448	3004	2N568
3004	SC69	3004	TNJ72289	3004	XJ13	3004	2N1449	3004	2N573
3004	SC73	3004	TR-320	3004	XJ13-1	3004	2N1451	3004	2N573BRN
3004	SE-5-0819	3004	TR-320A	3004	Y363	3004	2N1452	3004	2N573GRN
3004	SPT-306	3004	TR-321	3004	YV1	3004	2N151	3004	2N573RED
3004	SPT-322	3004	TR-321A	3004	YV2	3004	2N159	3004	2N591
3004	SPT-323	3004	TR-323A	3004	ZTR-B54	3004	2N1705	3004	2N59B
3004	SPT-327	3004	TR-482A	3004	ZTR-B56	3004	2N1706	3004	2N59D
3004	SPT-337	3004	TR-508A	3004	001-012011	3004	2N1707	3004	2N60Y
3004	SPT-552	3004	TR-045	3004	001-012060	3004	2N180	3004	2N60R
3004	SPT221	3004	TR06	3004	01-1201-0	3004	2N181	3004	2N60R
3004	SPT221A	3004	TR14	3004	1-21-106	3004	2N186	3004	2N610
3004	SPT222	3004	TR14C	3004	1-21-107	3004	2N186A	3004	2N611
3004	SPT222A	3004	TR15	3004	1-21-120	3004	2N187	3004	2N613
3004	SPT351	3004	TR15C	3004	1-21-148	3004	2N187A	3004	2N62
3004	SMB447610	3004	TR2083-75	3004	1-21-164	3004	2N188	3004	2N632
3004	SMB454549	3004	TR310011	3004	1-21-184	3004	2N188A	3004	2N633
3004	SO-25	3004	TR310012	3004	1-21-191	3004	2N189	3004	2N650
3004	SO-88	3004	TR310017	3004	1-21-192	3004	2N190	3004	2N650A
3004	SS0001	3004	TR310018	3004	1-21-225	3004	2N191	3004	2N651
3004	SS0001A	3004	TR3100227	3004	1-21-226	3004	2N192	3004	2N651A
3004	SS0002	3004	TR310026	3004	1-21-227	3004	2N1924	3004	2N652
3004	SS0002A	3004	TR310075	3004	1-21-232	3004	2N1925	3004	2N652A
3004	SS0003A	3004	TR310107	3004	1-21-246	3004	2N1926	3004	2N653
3004	SS0004	3004	TR310125	3004	1-21-266	3004	2N195	3004	2N654
3004	SS0004A	3004	TR310136	3004	1-21-267	3004	2N196	3004	2N655
3004	SS0005A	3004	TR310149	3004	1-21-272	3004	2N197	3004	2N655GRN
3004	ST-122	3004	TR310153	3004	1-21-95	3004	2N198	3004	2N655RED
3004	ST-123	3004	TR310159	3004	1-21-96	3004	2N199		
3004	ST-301	3004	TR310164	3004	1-52221011	3004	2N204		

SK	Industry Standard No.	SK	Industry Standard No.	SK	Industry Standard No.	SK	Industry Standard No.	SK	Industry Standard No.
3004	2N680	3004	28A520R	3004	2SB119H	3004	2SB134	3004	2SB157J
3004	2N71	3004	28A52R	3004	2SB111H	3004	2SB134A	3004	2SB157K
3004	2N73	3004	28A52Y	3004	2SB111J	3004	2SB134B	3004	2SB157L
3004	2N74	3004	2SA53	3004	2SB111K	3004	2SB134C	3004	2SB157M
3004	2N76	3004	2SA53A	3004	2SB111L	3004	2SB134D	3004	2SB157OR
3004	2N77	3004	2SA53B	3004	2SB111M	3004	2SB134E	3004	2SB157R
3004	2N79	3004	2SA53C	3004	2SB111OR	3004	2SB134F	3004	2SB157X
3004	2N799	3004	2SA53D	3004	2SB111R	3004	2SB134G	3004	2SB157Y
3004	2N80	3004	2SA53B	3004	2SB111X	3004	2SB134GN	3004	2SB158A
3004	2N826	3004	2SA53F	3004	2SB111Y	3004	2SB134H	3004	2SB158B
3004	2N96	3004	2SA53G	3004	2SB112A	3004	2SB134J	3004	2SB158C
3004	2NJ9A	3004	2SA53L	3004	2SB112B	3004	2SB134K	3004	2SB158D
3004	2NJ9D	3004	2SA53M	3004	2SB112C	3004	2SB134L	3004	2SB158E
3004	2N8121	3004	2SA530R	3004	2SB112D	3004	2SB134M	3004	2SB158F
3004	2N832	3004	2SA53R	3004	2SB112F	3004	2SB134OR	3004	2SB158G
3004	2B187	3004	2SA53Y	3004	2SB112G	3004	2SB134R	3004	2SB158GN
3004	2814	3004	2SB-3783	3004	2SB112H	3004	2SB134X	3004	2SB158H
3004	2815	3004	2SB-3812	3004	2SB112J	3004	2SB134Y	3004	2SB158J
3004	2815A	3004	2SB-3813	3004	2SB112K	3004	2SB135	3004	2SB158K
3004	28163	3004	2SB-P1A	3004	2SB112L	3004	2SB135A	3004	2SB158L
3004	28179	3004	2SB100	3004	2SB112M	3004	2SB135B	3004	2SB158M
3004	28189	3004	2SB100A	3004	2SB1120R	3004	2SB135C	3004	2SB158OR
3004	2822	3004	2SB100B	3004	2SB112R	3004	2SB135D	3004	2SB158R
3004	2824	3004	2SB100C	3004	2SB112X	3004	2SB135E	3004	2SB158X
3004	2832	3004	2SB100D	3004	2SB112Y	3004	2SB135F	3004	2SB158Y
3004	2833	3004	2SB100B	3004	2SB113	3004	2SB135G	3004	2SB159A
3004	2834	3004	2SB100F	3004	2SB113A	3004	2SB135GN	3004	2SB159B
3004	2837	3004	2SB100G	3004	2SB113B	3004	2SB135H	3004	2SB159C
3004	2838	3004	2SB100GN	3004	2SB113C	3004	2SB135J	3004	2SB159D
3004	2839	3004	2SB100H	3004	2SB113D	3004	2SB135K	3004	2SB159E
3004	2840	3004	2SB100J	3004	2SB113E	3004	2SB135L	3004	2SB159F
3004	2843	3004	2SB100K	3004	2SB113F	3004	2SB135M	3004	2SB159G
3004	2844	3004	2SB100L	3004	2SB113G	3004	2SB135OR	3004	2SB159GN
3004	2854	3004	2SB100M	3004	2SB113GN	3004	2SB135R	3004	2SB159H
3004	2856	3004	2SB100OR	3004	2SB113H	3004	2SB135X	3004	2SB159J
3004	28A100D	3004	2SB100R	3004	2SB113J	3004	2SB135Y	3004	2SB159K
3004	28A100E	3004	2SB100X	3004	2SB113K	3004	2SB136	3004	2SB159M
3004	28A100F	3004	2SB100Y	3004	2SB113L	3004	2SB136A	3004	2SB159N
3004	28A100G	3004	2SB101	3004	2SB113M	3004	2SB136B	3004	2SB159OR
3004	28A100H	3004	2SB101A	3004	2SB113R	3004	2SB136C	3004	2SB159R
3004	28A100J	3004	2SB101B	3004	2SB113X	3004	2SB136D	3004	2SB159X
3004	28A100K	3004	2SB101C	3004	2SB113Y	3004	2SB136E	3004	2SB159Y
3004	28A100M	3004	2SB101D	3004	2SB114A	3004	2SB136F	3004	2SB160A
3004	28A100OR	3004	2SB101E	3004	2SB114B	3004	2SB136G	3004	2SB160B
3004	28A100R	3004	2SB101F	3004	2SB114C	3004	2SB136GN	3004	2SB160C
3004	28A100Y	3004	2SB101G	3004	2SB114D	3004	2SB136H	3004	2SB160D
3004	28A204A	3004	2SB101GN	3004	2SB114E	3004	2SB136J	3004	2SB160E
3004	28A204B	3004	2SB101H	3004	2SB114F	3004	2SB136K	3004	2SB160F
3004	28A204C	3004	2SB101J	3004	2SB114G	3004	2SB136L	3004	2SB160GN
3004	28A204D	3004	2SB101K	3004	2SB114GN	3004	2SB136M	3004	2SB160H
3004	28A204B	3004	2SB101L	3004	2SB114H	3004	2SB136OR	3004	2SB160J
3004	28A204F	3004	2SB101M	3004	2SB114J	3004	2SB136U	3004	2SB160K
3004	28A204G	3004	2SB101OR	3004	2SB114K	3004	2SB150	3004	2SB160L
3004	28A204H	3004	2SB101R	3004	2SB114L	3004	2SB150A	3004	2SB160M
3004	28A204J	3004	2SB101X	3004	2SB114M	3004	2SB150B	3004	2SB160OR
3004	28A204K	3004	2SB101T	3004	2SB1140R	3004	2SB150C	3004	2SB160X
3004	28A204L	3004	2SB102	3004	2SB114R	3004	2SB150D	3004	2SB160Y
3004	28A204M	3004	2SB102A	3004	2SB114X	3004	2SB150E	3004	2SB161
3004	28A204OR	3004	2SB102B	3004	2SB114Y	3004	2SB150F	3004	2SB161A
3004	28A204R	3004	2SB102C	3004	2SB115A	3004	2SB150G	3004	2SB161B
3004	28A204X	3004	2SB102D	3004	2SB115B	3004	2SB150GN	3004	2SB161C
3004	28A204Y	3004	2SB102E	3004	2SB115C	3004	2SB150H	3004	2SB161D
3004	28A205A	3004	2SB102F	3004	2SB115D	3004	2SB150J	3004	2SB161E
3004	28A205B	3004	2SB102G	3004	2SB115E	3004	2SB150K	3004	2SB161F
3004	28A205C	3004	2SB102GN	3004	2SB115F	3004	2SB150L	3004	2SB161G
3004	28A205D	3004	2SB102H	3004	2SB115G	3004	2SB150M	3004	2SB161GN
3004	28A205E	3004	2SB102J	3004	2SB115GN	3004	2SB150OR	3004	2SB161H
3004	28A205F	3004	2SB102K	3004	2SB115H	3004	2SB150R	3004	2SB161J
3004	28A205G	3004	2SB102L	3004	2SB115J	3004	2SB150X	3004	2SB161K
3004	28A205GN	3004	2SB102M	3004	2SB115K	3004	2SB150Y	3004	2SB161L
3004	28A205H	3004	2SB102OR	3004	2SB115L	3004	2SB153	3004	2SB161M
3004	28A205J	3004	2SB102R	3004	2SB115M	3004	2SB153A	3004	2SB161OR
3004	28A205K	3004	2SB102X	3004	2SB1150R	3004	2SB153B	3004	2SB161R
3004	28A205L	3004	2SB102Y	3004	2SB115R	3004	2SB153C	3004	2SB161X
3004	28A205M	3004	2SB103	3004	2SB115X	3004	2SB153D	3004	2SB161Y
3004	28A205OR	3004	2SB103A	3004	2SB115Y	3004	2SB153E	3004	2SB162
3004	28A205R	3004	2SB103C	3004	2SB116	3004	2SB153F	3004	2SB162A
3004	28A205X	3004	2SB103D	3004	2SB116A	3004	2SB153G	3004	2SB162B
3004	28A205Y	3004	2SB103E	3004	2SB116B	3004	2SB153GN	3004	2SB162C
3004	28A206A	3004	2SB103F	3004	2SB116C	3004	2SB153H	3004	2SB162D
3004	28A206B	3004	2SB103G	3004	2SB116D	3004	2SB153J	3004	2SB162E
3004	28A206C	3004	2SB103GN	3004	2SB116E	3004	2SB153K	3004	2SB162F
3004	28A206D	3004	2SB103H	3004	2SB116F	3004	2SB153L	3004	2SB162G
3004	28A206E	3004	2SB103J	3004	2SB116G	3004	2SB153M	3004	2SB162GN
3004	28A206F	3004	2SB103K	3004	2SB116GN	3004	2SB153OR	3004	2SB162H
3004	28A206G	3004	2SB103L	3004	2SB116H	3004	2SB153R	3004	2SB162J
3004	28A206GN	3004	2SB103M	3004	2SB116J	3004	2SB153X	3004	2SB162K
3004	28A206H	3004	2SB1030R	3004	2SB116K	3004	2SB153Y	3004	2SB162L
3004	28A206J	3004	2SB103R	3004	2SB116L	3004	2SB154	3004	2SB162M
3004	28A206K	3004	2SB103X	3004	2SB116M	3004	2SB154A	3004	2SB162OR
3004	28A206L	3004	2SB103Y	3004	2SB116OR	3004	2SB154B	3004	2SB162R
3004	28A206M	3004	2SB104	3004	2SB116R	3004	2SB154C	3004	2SB162T
3004	28A206OR	3004	2SB104A	3004	2SB116X	3004	2SB154D	3004	2SB162Y
3004	28A206R	3004	2SB104B	3004	2SB116Y	3004	2SB154E	3004	2SB163
3004	28A206X	3004	2SB104C	3004	2SB117	3004	2SB154F	3004	2SB163A
3004	28A206Y	3004	2SB104D	3004	2SB117A	3004	2SB154G	3004	2SB163B
3004	28A250A	3004	2SB104E	3004	2SB117B	3004	2SB154GN	3004	2SB163C
3004	28A250B	3004	2SB104F	3004	2SB117C	3004	2SB154H	3004	2SB163D
3004	28A250C	3004	2SB104G	3004	2SB117D	3004	2SB154J	3004	2SB163E
3004	28A250D	3004	2SB104GN	3004	2SB117E	3004	2SB154K	3004	2SB163F
3004	28A250B	3004	2SB104H	3004	2SB117F	3004	2SB154L	3004	2SB163G
3004	28A250F	3004	2SB104J	3004	2SB117G	3004	2SB154M	3004	2SB163GN
3004	28A250G	3004	2SB104K	3004	2SB117GN	3004	2SB154OR	3004	2SB163H
3004	28A250GN	3004	2SB104L	3004	2SB117H	3004	2SB154R	3004	2SB163J
3004	28A250H	3004	2SB104M	3004	2SB117J	3004	2SB154X	3004	2SB163K
3004	28A250J	3004	2SB104OR	3004	2SB117K	3004	2SB155A	3004	2SB163L
3004	28A250L	3004	2SB104R	3004	2SB117L	3004	2SB155B	3004	2SB163OR
3004	28A250M	3004	2SB104X	3004	2SB117M	3004	2SB155C	3004	2SB163R
3004	28A250Y	3004	2SB104Y	3004	2SB117OR	3004	2SB155D	3004	2SB163X
3004	28A303	3004	2SB110A	3004	2SB117R	3004	2SB155E	3004	2SB163Y
3004	28A49	3004	2SB110B	3004	2SB117X	3004	2SB155F	3004	2SB164A
3004	28A49A	3004	2SB110C	3004	2SB117Y	3004	2SB155G	3004	2SB164B
3004	28A49B	3004	2SB110D	3004	2SB120	3004	2SB155GN	3004	2SB164C
3004	28A49C	3004	2SB110E	3004	2SB120A	3004	2SB155H	3004	2SB164D
3004	28A49D	3004	2SB110F	3004	2SB120B	3004	2SB155J	3004	2SB164E
3004	28A49B	3004	2SB110G	3004	2SB120C	3004	2SB155K	3004	2SB164F
3004	28A49F	3004	2SB110GN	3004	2SB120D	3004	2SB155L	3004	2SB164G
3004	28A49G	3004	2SB110H	3004	2SB120E	3004	2SB155M	3004	2SB164GN
3004	28A49L	3004	2SB110J	3004	2SB120F	3004	2SB155OR	3004	2SB164H
3004	28A49M	3004	2SB110K	3004	2SB120G	3004	2SB155R	3004	2SB164J
3004	28A49OR	3004	2SB110L	3004	2SB120GN	3004	2SB155X	3004	2SB164L
3004	28A49R	3004	2SB110M	3004	2SB120H	3004	2SB155Y	3004	2SB164M
3004	28A49Y	3004	2SB110OR	3004	2SB120J	3004	2SB156/7825B	3004	2SB164OR
3004	28A52	3004	2SB110X	3004	2SB120K	3004	2SB157A	3004	2SB164R
3004	28A52A	3004	2SB110Y	3004	2SB120L	3004	2SB157B	3004	2SB164X
3004	28A52B	3004	2SB111A	3004	2SB120M	3004	2SB157C	3004	2SB164Y
3004	28A52C	3004	2SB111B	3004	2SB1200R	3004	2SB157D	3004	2SB165
3004	28A52D	3004	2SB111C	3004	2SB120R	3004	2SB157E	3004	2SB165A
3004	28A52E	3004	2SB111D	3004	2SB120X	3004	2SB157F	3004	2SB165B
3004	28A52F	3004	2SB111E	3004	2SB120Y	3004	2SB157G	3004	2SB165C
3004	28A52G	3004	2SB111F	3004	2SB13	3004	2SB157GN	3004	2SB165D
3004	28A52L	3004	2SB111G			3004	2SB157H		
3004	28A52M								

SK	Industry Standard No.
3004	2SB165E
3004	2SB165F
3004	2SB165G
3004	2SB165GN
3004	2SB165H
3004	2SB165J
3004	2SB165K
3004	2SB165M
3004	2SB165OR
3004	2SB165R
3004	2SB165X
3004	2SB165Y
3004	2SB166
3004	2SB166A
3004	2SB166B
3004	2SB166C
3004	2SB166D
3004	2SB166E
3004	2SB166F
3004	2SB166G
3004	2SB166GN
3004	2SB166H
3004	2SB166J
3004	2SB166K
3004	2SB166L
3004	2SB166M
3004	2SB166OR
3004	2SB166R
3004	2SB166X
3004	2SB166Y
3004	2SB167A
3004	2SB167B
3004	2SB167BK
3004	2SB167C
3004	2SB167D
3004	2SB167E
3004	2SB167F
3004	2SB167G
3004	2SB167GN
3004	2SB167H
3004	2SB167J
3004	2SB167K
3004	2SB167L
3004	2SB167M
3004	2SB167OR
3004	2SB167R
3004	2SB167X
3004	2SB167Y
3004	2SB168A
3004	2SB168B
3004	2SB168C
3004	2SB168D
3004	2SB168E
3004	2SB168F
3004	2SB168G
3004	2SB168GN
3004	2SB168H
3004	2SB168J
3004	2SB168K
3004	2SB168L
3004	2SB168M
3004	2SB168OR
3004	2SB168R
3004	2SB168X
3004	2SB168Y
3004	2SB169A
3004	2SB169B
3004	2SB169C
3004	2SB169D
3004	2SB169E
3004	2SB169F
3004	2SB169G
3004	2SB169GN
3004	2SB169H
3004	2SB169J
3004	2SB169K
3004	2SB169L
3004	2SB169M
3004	2SB169OR
3004	2SB169R
3004	2SB169X
3004	2SB169Y
3004	2SB170
3004	2SB170A
3004	2SB170B
3004	2SB170C
3004	2SB170D
3004	2SB170E
3004	2SB170F
3004	2SB170G
3004	2SB170GN
3004	2SB170H
3004	2SB170J
3004	2SB170K
3004	2SB170L
3004	2SB170M
3004	2SB170R
3004	2SB170X
3004	2SB170Y
3004	2SB171
3004	2SB171A
3004	2SB171B
3004	2SB171C
3004	2SB171D
3004	2SB171E
3004	2SB171F
3004	2SB171G
3004	2SB171GN
3004	2SB171H
3004	2SB171J
3004	2SB171K
3004	2SB171L
3004	2SB171M
3004	2SB171OR
3004	2SB171X
3004	2SB171Y
3004	2SB173
3004	2SB173A
3004	2SB173B
3004	2SB173BL
3004	2SB173C
3004	2SB173CL
3004	2SB173D
3004	2SB173E
3004	2SB173F
3004	2SB173G
3004	2SB173GN
3004	2SB173H
3004	2SB173J
3004	2SB173K
3004	2SB173L
3004	2SB173M
3004	2SB173OR
3004	2SB173R
3004	2SB173X
3004	2SB173Y
3004	2SB174
3004	2SB174A
3004	2SB174B
3004	2SB174C
3004	2SB174D
3004	2SB174E
3004	2SB174F
3004	2SB174G
3004	2SB174GN
3004	2SB174H
3004	2SB174K
3004	2SB174L
3004	2SB174M
3004	2SB174OR
3004	2SB174R
3004	2SB174X
3004	2SB174Y
3004	2SB175
3004	2SB175A
3004	2SB175B
3004	2SB175B-1
3004	2SB175BL
3004	2SB175C
3004	2SB175CL
3004	2SB175D
3004	2SB175E
3004	2SB175F
3004	2SB175G
3004	2SB175GN
3004	2SB175H
3004	2SB175L
3004	2SB175M
3004	2SB175OR
3004	2SB175R
3004	2SB175X
3004	2SB175Y
3004	2SB177
3004	2SB177A
3004	2SB177B
3004	2SB177C
3004	2SB177D
3004	2SB177E
3004	2SB177F
3004	2SB177G
3004	2SB177GN
3004	2SB177H
3004	2SB177K
3004	2SB177L
3004	2SB177M
3004	2SB177OR
3004	2SB177R
3004	2SB177X
3004	2SB177Y
3004	2SB178
3004	2SB1780
3004	2SB1780A
3004	2SB1780B
3004	2SB1780C
3004	2SB1780D
3004	2SB1780E
3004	2SB1780F
3004	2SB1780G
3004	2SB1780GN
3004	2SB1780H
3004	2SB1780J
3004	2SB1780K
3004	2SB1780L
3004	2SB1780M
3004	2SB1780R
3004	2SB1780X
3004	2SB1780Y
3004	2SB178A
3004	2SB178B
3004	2SB178C
3004	2SB178D
3004	2SB178E
3004	2SB178F
3004	2SB178G
3004	2SB178GN
3004	2SB178H
3004	2SB178J
3004	2SB178K
3004	2SB178L
3004	2SB178M
3004	2SB178N
3004	2SB178OR
3004	2SB178R
3004	2SB178S
3004	2SB178T
3004	2SB178TC
3004	2SB178TS
3004	2SB178U
3004	2SB178V
3004	2SB178Y
3004	2SB183
3004	2SB183A
3004	2SB183B
3004	2SB183C
3004	2SB183D
3004	2SB183E
3004	2SB183F
3004	2SB183G
3004	2SB183GN
3004	2SB183H
3004	2SB183J
3004	2SB183K
3004	2SB183L
3004	2SB183M
3004	2SB183OR
3004	2SB183R
3004	2SB183X
3004	2SB183Y
3004	2SB184
3004	2SB184A
3004	2SB184B
3004	2SB184C
3004	2SB184D
3004	2SB184E
3004	2SB184F
3004	2SB184G
3004	2SB184GN
3004	2SB184H
3004	2SB184J
3004	2SB184K
3004	2SB184L
3004	2SB184M
3004	2SB184OR
3004	2SB184R
3004	2SB184X
3004	2SB184Y
3004	2SB185
3004	00002SB185-0
3004	002SB18500
3004	2SB185A
3004	002SB185AA
3004	2SB185F
3004	2SB185P
3004	000002SB186
3004	0002SB186-0
3004	2SB186-7
3004	0002SB1860
3004	2SB186A
3004	2SB186AG
3004	2SB186B
3004	2SB186BY
3004	2SB186C
3004	2SB186D
3004	2SB186E
3004	2SB186F
3004	2SB186G
3004	2SB186GN
3004	2SB186H
3004	2SB186J
3004	2SB186K
3004	2SB186L
3004	2SB186M
3004	2SB186OR
3004	2SB186R
3004	2SB186X
3004	2SB186Y
3004	00002SB187
3004	2SB187A
3004	2SB187AA
3004	2SB187B
3004	2SB187BK
3004	2SB187C
3004	2SB187D
3004	2SB187E
3004	2SB187F
3004	2SB187G
3004	2SB187GN
3004	2SB187H
3004	2SB187K
3004	2SB187L
3004	2SB187M
3004	2SB187OR
3004	2SB187R
3004	2SB187S
3004	2SB187X
3004	2SB187Y
3004	2SB188
3004	2SB188A
3004	2SB188B
3004	2SB188C
3004	2SB188D
3004	2SB188E
3004	2SB188F
3004	2SB188G
3004	2SB188GN
3004	2SB188H
3004	2SB188J
3004	2SB188K
3004	2SB188L
3004	2SB188M
3004	2SB188OR
3004	2SB188R
3004	2SB188X
3004	2SB188Y
3004	2SB189
3004	2SB189A
3004	2SB189B
3004	2SB189C
3004	2SB189D
3004	2SB189E
3004	2SB189F
3004	2SB189G
3004	2SB189GN
3004	2SB189H
3004	2SB189J
3004	2SB189K
3004	2SB189L
3004	2SB189M
3004	2SB189OR
3004	2SB189R
3004	2SB189X
3004	2SB189Y
3004	2SB199
3004	2SB199A
3004	2SB199B
3004	2SB199C
3004	2SB199D
3004	2SB199E
3004	2SB199F
3004	2SB199G
3004	2SB199GN
3004	2SB199H
3004	2SB199J
3004	2SB199L
3004	2SB199M
3004	2SB199OR
3004	2SB199R
3004	2SB199X
3004	2SB199Y
3004	2SB200
3004	2SB200A
3004	2SB200B
3004	2SB200C
3004	2SB200D
3004	2SB200E
3004	2SB200F
3004	2SB200G
3004	2SB200GN
3004	2SB200H
3004	2SB200J
3004	2SB200K
3004	2SB200L
3004	2SB200M
3004	2SB200OR
3004	2SB200R
3004	2SB200X
3004	2SB200Y
3004	2SB202
3004	2SB202A
3004	2SB202B
3004	2SB202C
3004	2SB202D
3004	2SB202E
3004	2SB202F
3004	2SB202G
3004	2SB202GN
3004	2SB202H
3004	2SB202J
3004	2SB202L
3004	2SB202M
3004	2SB202OR
3004	2SB202R
3004	2SB202X
3004	2SB202Y
3004	2SB203A
3004	2SB203B
3004	2SB203C
3004	2SB203D
3004	2SB203E
3004	2SB203F
3004	2SB203G
3004	2SB203GN
3004	2SB203H
3004	2SB203J
3004	2SB203K
3004	2SB203L
3004	2SB203M
3004	2SB203OR
3004	2SB203R
3004	2SB203X
3004	2SB203Y
3004	2SB218
3004	2SB218A
3004	2SB218B
3004	2SB218C
3004	2SB218D
3004	2SB218E
3004	2SB218F
3004	2SB218G
3004	2SB218GN
3004	2SB218H
3004	2SB218J
3004	2SB218K
3004	2SB218L
3004	2SB218M
3004	2SB218OR
3004	2SB218R
3004	2SB218X
3004	2SB218Y
3004	2SB219
3004	2SB219A
3004	2SB219B
3004	2SB219E
3004	2SB219F
3004	2SB219G
3004	2SB219H
3004	2SB219J
3004	2SB219K
3004	2SB219L
3004	2SB219R
3004	2SB219X
3004	2SB219Y
3004	2SB22
3004	2SB220
3004	2SB220A
3004	2SB220B
3004	2SB220C
3004	2SB220D
3004	2SB220E
3004	2SB220F
3004	2SB220GN
3004	2SB220H
3004	2SB220J
3004	2SB220K
3004	2SB220L
3004	2SB220M
3004	2SB220R
3004	2SB220X
3004	2SB220Y
3004	2SB221
3004	2SB221A
3004	2SB221B
3004	2SB221C
3004	2SB221D
3004	2SB221E
3004	2SB221F
3004	2SB221G
3004	2SB221GN
3004	2SB221H
3004	2SB221J
3004	2SB221K
3004	2SB221L
3004	2SB221M
3004	2SB221OR
3004	2SB221R
3004	2SB221X
3004	2SB221Y
3004	2SB222
3004	2SB222A
3004	2SB222B
3004	2SB222C
3004	2SB222D
3004	2SB222E
3004	2SB222F
3004	2SB222G
3004	2SB222GN
3004	2SB222H
3004	2SB222J
3004	2SB222K
3004	2SB222L
3004	2SB222M
3004	2SB222OR
3004	2SB222R
3004	2SB222X
3004	2SB222Y
3004	2SB223
3004	2SB223A
3004	2SB223B
3004	2SB223C
3004	2SB223D
3004	2SB223E
3004	2SB223F
3004	2SB223G
3004	2SB223Y
3004	2SB224
3004	2SB224A
3004	2SB224B
3004	2SB224C
3004	2SB224D
3004	2SB224E
3004	2SB224F
3004	2SB224G
3004	2SB224GN
3004	2SB224H
3004	2SB224K
3004	2SB224L
3004	2SB224M
3004	2SB224OR
3004	2SB224R
3004	2SB224X
3004	2SB224Y
3004	2SB225
3004	2SB225A
3004	2SB225B
3004	2SB225C
3004	2SB225D
3004	2SB225E
3004	2SB225F
3004	2SB225G
3004	2SB225GN
3004	2SB225H
3004	2SB225K
3004	2SB225L
3004	2SB225M
3004	2SB225OR
3004	2SB225R
3004	2SB225X
3004	2SB225Y
3004	2SB226
3004	2SB226A
3004	2SB226B
3004	2SB226C
3004	2SB226D
3004	2SB226E
3004	2SB226F
3004	2SB226G
3004	2SB226GN
3004	2SB226J
3004	2SB226K
3004	2SB226L
3004	2SB226M
3004	2SB226OR
3004	2SB226R
3004	2SB226X
3004	2SB226Y
3004	2SB227
3004	2SB227A
3004	2SB227B
3004	2SB227C
3004	2SB227D
3004	2SB227F
3004	2SB227G
3004	2SB227GN
3004	2SB227H
3004	2SB227J
3004	2SB227K
3004	2SB227L
3004	2SB227M
3004	2SB227OR
3004	2SB227R
3004	2SB227X
3004	2SB227Y
3004	2SB23
3004	2SB23A
3004	2SB23B
3004	2SB23C
3004	2SB23D
3004	2SB23E
3004	2SB23F
3004	2SB23G
3004	2SB23GN
3004	2SB23H
3004	2SB23J
3004	2SB23K
3004	2SB23L
3004	2SB23M
3004	2SB23OR
3004	2SB23R
3004	2SB23X
3004	2SB23Y
3004	2SB24
3004	2SB241
3004	2SB241A
3004	2SB241B
3004	2SB241C
3004	2SB241D
3004	2SB241E
3004	2SB241F
3004	2SB241G
3004	2SB241GN
3004	2SB241H
3004	2SB241J
3004	2SB241K
3004	2SB241L
3004	2SB241M
3004	2SB241OR
3004	2SB241R
3004	2SB241X
3004	2SB241Y
3004	2SB24A
3004	2SB24B
3004	2SB24C
3004	2SB24D
3004	2SB24E
3004	2SB24F
3004	2SB24G
3004	2SB24GN
3004	2SB24H
3004	2SB24J
3004	2SB24K
3004	2SB24L
3004	2SB24M
3004	2SB24OR
3004	2SB24R
3004	2SB24X
3004	2SB24Y
3004	2SB257
3004	2SB257A
3004	2SB257B
3004	2SB257C
3004	2SB257D
3004	2SB257E
3004	2SB257F
3004	2SB257G
3004	2SB257GN
3004	2SB257H
3004	2SB257K
3004	2SB257L
3004	2SB257M
3004	2SB257OR
3004	2SB257R
3004	2SB257X
3004	2SB257Y
3004	2SB261
3004	2SB261A
3004	2SB261B
3004	2SB261C
3004	2SB261D
3004	2SB261E
3004	2SB261F
3004	2SB261G
3004	2SB261GN
3004	2SB261H
3004	2SB261J
3004	2SB261K
3004	2SB261L
3004	2SB261M
3004	2SB261OR
3004	2SB261R
3004	2SB261X

SK	Industry Standard No.	SK	Industry Standard No.	SK	Industry Standard No.	SK	Industry Standard No.	SK	Industry Standard No.
3004	28B261Y	3004	28B269G	3004	28B299R	3004	28B326Y	3004	28B347E
3004	28B262	3004	28B269GN	3004	28B299X	3004	28B327	3004	28B347F
3004	28B262A	3004	28B269J	3004	28B299Y	3004	28B327A	3004	28B347G
3004	28B262B	3004	28B269K	3004	28B302	3004	28B327B	3004	28B347GN
3004	28B262C	3004	28B269L	3004	28B302A	3004	28B327C	3004	28B347H
3004	28B262D	3004	28B269M	3004	28B302B	3004	28B327D	3004	28B347J
3004	28B262E	3004	28B269OR	3004	28B302C	3004	28B327E	3004	28B347K
3004	28B262F	3004	28B269R	3004	28B302D	3004	28B327F	3004	28B347L
3004	28B262G	3004	28B269X	3004	28B302E	3004	28B327G	3004	28B347M
3004	28B262GN	3004	28B269Y	3004	28B302F	3004	28B327GN	3004	28B347OR
3004	28B262H	3004	28B270	3004	28B302G	3004	28B327H	3004	28B347R
3004	28B262J	3004	28B270A	3004	28B302H	3004	28B327J	3004	28B347X
3004	28B262K	3004	28B270B	3004	28B302J	3004	28B327K	3004	28B348
3004	28B262L	3004	28B270C	3004	28B302K	3004	28B327L	3004	28B348A
3004	28B262M	3004	28B270D	3004	28B302L	3004	28B327M	3004	28B348B
3004	28B262OR	3004	28B270E	3004	28B302M	3004	28B327OR	3004	28B348C
3004	28B262R	3004	28B272	3004	28B302OR	3004	28B327R	3004	28B348D
3004	28B262X	3004	28B272A	3004	28B302R	3004	28B327X	3004	28B348E
3004	28B262Y	3004	28B272B	3004	28B302X	3004	28B327Y	3004	28B348F
3004	28B263	3004	28B272C	3004	28B302Y	3004	28B328	3004	28B348G
3004	28B263A	3004	28B272D	3004	000028B303	3004	28B328A	3004	28B348GN
3004	28B263B	3004	28B272F	3004	00028B303-0	3004	28B328B	3004	28B348H
3004	28B263C	3004	28B272G	3004	00028B3030	3004	28B328D	3004	28B348K
3004	28B263D	3004	28B272GN	3004	28B303A	3004	28B328E	3004	28B348L
3004	28B263E	3004	28B272J	3004	28B303B	3004	28B328F	3004	28B348M
3004	28B263F	3004	28B272K	3004	28B303BK	3004	28B328G	3004	28B348OR
3004	28B263G	3004	28B272L	3004	28B303C	3004	28B328GN	3004	28B348Q
3004	28B263GN	3004	28B272M	3004	28B303D	3004	28B328H	3004	28B348R
3004	28B263H	3004	28B272OR	3004	28B303E	3004	28B328J	3004	28B348Y
3004	28B263J	3004	28B272R	3004	28B303F	3004	28B328K	3004	28B34A
3004	28B263K	3004	28B272X	3004	28B303G	3004	28B328L	3004	28B34B
3004	28B263L	3004	28B272Y	3004	28B303GN	3004	28B328M	3004	28B34C
3004	28B263M	3004	28B273	3004	28B303H	3004	28B328OR	3004	28B34D
3004	28B263OR	3004	28B273A	3004	28B303J	3004	28B328R	3004	28B34E
3004	28B263R	3004	28B273B	3004	28B303K	3004	28B328X	3004	28B34F
3004	28B263X	3004	28B273C	3004	28B303L	3004	28B328Y	3004	28B34G
3004	28B263Y	3004	28B273D	3004	28B303M	3004	28B329	3004	28B34GN
3004	28B264	3004	28B273F	3004	28B3030R	3004	28B329A	3004	28B34H
3004	28B264A	3004	28B273G	3004	28B303R	3004	28B329B	3004	28B34K
3004	28B264B	3004	28B273GN	3004	28B303X	3004	28B329C	3004	28B34L
3004	28B264C	3004	28B273H	3004	28B303Y	3004	28B329D	3004	28B34M
3004	28B264D	3004	28B273J	3004	28B314	3004	28B329E	3004	28B340R
3004	28B264E	3004	28B273K	3004	28B314A	3004	28B329F	3004	28B34R
3004	28B264F	3004	28B273L	3004	28B314B	3004	28B329G	3004	28B34X
3004	28B264G	3004	28B273M	3004	28B314C	3004	28B329GN	3004	28B34Y
3004	28B264GN	3004	28B273OR	3004	28B314D	3004	28B329H	3004	28B350
3004	28B264H	3004	28B273R	3004	28B314E	3004	28B329J	3004	28B350A
3004	28B264J	3004	28B273X	3004	28B314F	3004	28B329K	3004	28B350C
3004	28B264K	3004	28B273Y	3004	28B314GN	3004	28B329L	3004	28B350D
3004	28B264L	3004	28B290A	3004	28B314H	3004	28B329M	3004	28B350F
3004	28B264M	3004	28B290B	3004	28B314J	3004	28B329OR	3004	28B350G
3004	28B264OR	3004	28B290C	3004	28B314K	3004	28B329R	3004	28B350GN
3004	28B264R	3004	28B290D	3004	28B314L	3004	28B329X	3004	28B350H
3004	28B264X	3004	28B290E	3004	28B314M	3004	28B329Y	3004	28B350J
3004	28B264Y	3004	28B290F	3004	28B314OR	3004	28B32A	3004	28B350K
3004	28B265	3004	28B290G	3004	28B314R	3004	28B32B	3004	28B350L
3004	28B265A	3004	28B290GN	3004	28B314X	3004	28B32C	3004	28B350M
3004	28B265B	3004	28B290H	3004	28B314Y	3004	28B32D	3004	28B350OR
3004	28B265C	3004	28B290J	3004	28B315	3004	28B32E	3004	28B350R
3004	28B265D	3004	28B290K	3004	28B315A	3004	28B32F	3004	28B350X
3004	28B265E	3004	28B290L	3004	28B315B	3004	28B32G	3004	28B350Y
3004	28B265F	3004	28B290M	3004	28B315C	3004	28B32GN	3004	28B364A
3004	28B265G	3004	28B290OR	3004	28B315D	3004	28B32H	3004	28B364C
3004	28B265GN	3004	28B290R	3004	28B315E	3004	28B32J	3004	28B364D
3004	28B265H	3004	28B290X	3004	28B315F	3004	28B32K	3004	28B364F
3004	28B265J	3004	28B290Y	3004	28B315G	3004	28B32M	3004	28B364G
3004	28B265K	3004	28B291A	3004	28B315GN	3004	28B320R	3004	28B364GN
3004	28B265L	3004	28B291B	3004	28B315H	3004	28B32R	3004	28B364H
3004	28B265M	3004	28B291C	3004	28B315J	3004	28B32X	3004	28B364J
3004	28B265OR	3004	28B291D	3004	28B315L	3004	28B32Y	3004	28B364K
3004	28B265X	3004	28B291E	3004	28B315M	3004	28B33	3004	28B364M
3004	28B265Y	3004	28B291F	3004	28B315OR	3004	28B33(5)	3004	28B364R
3004	28B266	3004	28B291G	3004	28B315R	3004	28B33-4	3004	28B364Y
3004	28B266A	3004	28B291GN	3004	28B315X	3004	28B33A	3004	28B365C
3004	28B266B	3004	28B291H	3004	28B315Y	3004	28B33B	3004	28B365D
3004	28B266C	3004	28B291J	3004	28B316	3004	28B33BK	3004	28B365E
3004	28B266D	3004	28B291K	3004	28B316A	3004	28B33C	3004	28B365F
3004	28B266E	3004	28B291L	3004	28B316B	3004	28B33D	3004	28B365G
3004	28B266F	3004	28B291M	3004	28B316C	3004	28B33E	3004	28B365GN
3004	28B266G	3004	28B291OR	3004	28B316D	3004	28B33F	3004	28B365H
3004	28B266GN	3004	28B291R	3004	28B316E	3004	28B33G	3004	28B365J
3004	28B266H	3004	28B291X	3004	28B316F	3004	28B33GN	3004	28B365K
3004	28B266J	3004	28B291Y	3004	28B316G	3004	28B33H	3004	28B365L
3004	28B266K	3004	28B292A	3004	28B316GN	3004	28B33J	3004	28B365M
3004	28B266L	3004	28B292C	3004	28B316H	3004	28B33K	3004	28B365OR
3004	28B266M	3004	28B292D	3004	28B316OR	3004	28B33L	3004	28B365Y
3004	28B266OR	3004	28B292E	3004	28B316R	3004	28B33M	3004	28B377
3004	28B266P	3004	28B292G	3004	28B316X	3004	28B330R	3004	28B370
3004	28B266Q	3004	28B292GN	3004	28B316Y	3004	28B33R	3004	28B370A
3004	28B266R	3004	28B292H	3004	28B32	3004	28B33X	3004	28B370B
3004	28B266X	3004	28B292J	3004	28B32-0	3004	28B33Y	3004	28B370D
3004	28B266Y	3004	28B292K	3004	28B32-1	3004	28B345	3004	28B370F
3004	28B267	3004	28B292L	3004	28B32-2	3004	28B345A	3004	28B370P
3004	28B267A	3004	28B292M	3004	28B324(E)	3004	28B345B	3004	28B370G
3004	28B267B	3004	28B292OR	3004	28B240	3004	28B345C	3004	28B370GN
3004	28B267C	3004	28B292R	3004	28B240A	3004	28B345D	3004	28B370H
3004	28B267D	3004	28B292X	3004	28B240B	3004	28B345E	3004	28B370J
3004	28B267E	3004	28B293	3004	28B240C	3004	28B345F	3004	28B370K
3004	28B267F	3004	28B293A	3004	28B240D	3004	28B345G	3004	28B370L
3004	28B267G	3004	28B293B	3004	28B240E	3004	28B345H	3004	28B370M
3004	28B267GN	3004	28B293C	3004	28B240F	3004	28B345J	3004	28B370OR
3004	28B267H	3004	28B293D	3004	28B240G	3004	28B345K	3004	28B370R
3004	28B267J	3004	28B293E	3004	28B240GN	3004	28B345L	3004	28B370X
3004	28B267K	3004	28B293F	3004	28B240H	3004	28B345M	3004	28B370Y
3004	28B267L	3004	28B293G	3004	28B240J	3004	28B345OR	3004	28B371
3004	28B267M	3004	28B293GN	3004	28B240K	3004	28B345R	3004	28B371A
3004	28B267OR	3004	28B293H	3004	28B240L	3004	28B345X	3004	28B371C
3004	28B267R	3004	28B293J	3004	28B240M	3004	28B345Y	3004	28B371D
3004	28B267X	3004	28B293K	3004	28B240OR	3004	28B346	3004	28B371E
3004	28B267Y	3004	28B293L	3004	28B240R	3004	28B346A	3004	28B371F
3004	28B268	3004	28B293M	3004	28B240X	3004	28B346C	3004	28B371G
3004	28B268A	3004	28B293OR	3004	28B240Y	3004	28B346D	3004	28B371GN
3004	28B268B	3004	28B293R	3004	28B24C	3004	28B346E	3004	28B371H
3004	28B268C	3004	28B293X	3004	28B24GN	3004	28B346F	3004	28B371J
3004	28B268D	3004	28B293Y	3004	28B24M	3004	28B346G	3004	28B371K
3004	28B268E	3004	28B299	3004	28B240R	3004	28B346GN	3004	28B371L
3004	28B268F	3004	28B299A	3004	28B24X	3004	28B346H	3004	28B371M
3004	28B268G	3004	28B299B	3004	28B24Y	3004	28B346J	3004	28B371OR
3004	28B268GN	3004	28B299C	3004	28B326	3004	28B346K	3004	28B371R
3004	28B268H	3004	28B299D	3004	28B326A	3004	28B346L	3004	28B371X
3004	28B268J	3004	28B299F	3004	28B326B	3004	28B346M	3004	28B371Y
3004	28B268K	3004	28B299G	3004	28B326C	3004	28B346OR	3004	28B376
3004	28B268L	3004	28B299GN	3004	28B326D	3004	28B346Q	3004	28B376A
3004	28B268M	3004	28B299H	3004	28B326E	3004	28B346R		
3004	28B268OR	3004	28B299J	3004	28B326F	3004	28B346X		
3004	28B268R	3004	28B299L	3004	28B326G	3004	28B346Y		
3004	28B268X	3004	28B299M	3004	28B326GN	3004	28B347		
3004	28B268Y	3004	28B299OR	3004	28B326H	3004	28B347A		
3004	28B269			3004	28B326J	3004	28B347B		
3004	28B269A			3004	28B326K	3004	28B347D		
3004	28B269C			3004	28B326L				
3004	28B269D			3004	28B326M				
3004	28B269E			3004	28B326OR				
3004	28B269F			3004	28B326R				
				3004	28B326X				

SK	Industry Standard No.	SK	Industry Standard No.	SK	Industry Standard No.	SK	Industry Standard No.	SK	Industry Standard No.
3004	2SB376B	3004	2SB381C	3004	2SB38K	3004	2SB401K	3004	2SB417X
3004	2SB376C	3004	2SB381D	3004	2SB38L	3004	2SB401L	3004	2SB417Y
3004	2SB376D	3004	2SB381E	3004	2SB38M	3004	2SB401M	3004	2SB422
3004	2SB376E	3004	2SB381F	3004	2SB380R	3004	2SB401OR	3004	2SB422A
3004	2SB376F	3004	2SB381GN	3004	2SB38R	3004	2SB401R	3004	2SB422B
3004	2SB376G	3004	2SB381H	3004	2SB38X	3004	2SB401X	3004	2SB422C
3004	2SB376GN	3004	2SB381J	3004	2SB38Y	3004	2SB401Y	3004	2SB422D
3004	2SB376J	3004	2SB381K	3004	2SB392A	3004	2SB402A	3004	2SB422P
3004	2SB376K	3004	2SB381L	3004	2SB392B	3004	2SB402B	3004	2SB422G
3004	2SB376L	3004	2SB381M	3004	2SB392C	3004	2SB402D	3004	2SB422GN
3004	2SB376M	3004	2SB381OR	3004	2SB392D	3004	2SB402G	3004	2SB422H
3004	2SB376OR	3004	2SB381R	3004	2SB392E	3004	2SB402GN	3004	2SB422J
3004	2SB376R	3004	2SB381X	3004	2SB392F	3004	2SB402H	3004	2SB422K
3004	2SB376Y	3004	2SB381Y	3004	2SB392G	3004	2SB402J	3004	2SB422L
3004	2SB377	3004	2SB382	3004	2SB392GN	3004	2SB402K	3004	2SB422M
3004	2SB377B	3004	2SB382A	3004	2SB392H	3004	2SB402L	3004	2SB422OR
3004	2SB378	3004	2SB382B	3004	2SB392J	3004	2SB402M	3004	2SB422R
3004	2SB3783A	3004	2SB382C	3004	2SB392K	3004	2SB402OR	3004	2SB422X
3004	2SB3783B	3004	2SB382E	3004	2SB392L	3004	2SB402R	3004	2SB422Y
3004	2SB3783C	3004	2SB382F	3004	2SB3920R	3004	2SB4020R	3004	2SB423
3004	2SB3783D	3004	2SB382G	3004	2SB392R	3004	2SB402R	3004	2SB423A
3004	2SB3783E	3004	2SB382GN	3004	2SB392X	3004	2SB402X	3004	2SB423B
3004	2SB3783F	3004	2SB382H	3004	2SB392Y	3004	2SB402Y	3004	2SB423C
3004	2SB3783G	3004	2SB382J	3004	2SB393A	3004	2SB403A	3004	2SB423D
3004	2SB3783GN	3004	2SB382K	3004	2SB393B	3004	2SB403B	3004	2SB423E
3004	2SB3783H	3004	2SB382L	3004	2SB393C	3004	2SB403C	3004	2SB423F
3004	2SB3783J	3004	2SB3820R	3004	2SB393D	3004	2SB403D	3004	2SB423GN
3004	2SB3783K	3004	2SB382R	3004	2SB393E	3004	2SB403E	3004	2SB423H
3004	2SB3783L	3004	2SB382X	3004	2SB393F	3004	2SB403F	3004	2SB423J
3004	2SB3783M	3004	2SB382Y	3004	2SB393G	3004	2SB403G	3004	2SB423K
3004	2SB37830R	3004	2SB383	3004	2SB393GN	3004	2SB403GN	3004	2SB423L
3004	2SB3783R	3004	2SB383-1	3004	2SB393H	3004	2SB403H	3004	2SB423M
3004	2SB3783X	3004	2SB383-2	3004	2SB393J	3004	2SB403J	3004	2SB4230R
3004	2SB3783Y	3004	2SB383A	3004	2SB393K	3004	2SB403K	3004	2SB423R
3004	2SB378A	3004	2SB383B	3004	2SB393L	3004	2SB403L	3004	2SB423X
3004	2SB378B	3004	2SB383C	3004	2SB393M	3004	2SB403M	3004	2SB423Y
3004	2SB378C	3004	2SB383D	3004	2SB3930R	3004	2SB4030R	3004	2SB439
3004	2SB378D	3004	2SB383E	3004	2SB393R	3004	2SB403R	3004	2SB439A
3004	2SB378E	3004	2SB383F	3004	2SB393X	3004	2SB403X	3004	2SB439B
3004	2SB378F	3004	2SB383G	3004	2SB393Y	3004	2SB403Y	3004	2SB439C
3004	2SB378GN	3004	2SB383GN	3004	2SB394A	3004	2SB405-1	3004	2SB439D
3004	2SB378J	3004	2SB383H	3004	2SB394B	3004	2SB405-R	3004	2SB439E
3004	2SB378K	3004	2SB383GN	3004	2SB394C	3004	2SB405AG	3004	2SB439F
3004	2SB378L	3004	2SB383H	3004	2SB394D	3004	2SB405DK	3004	2SB439G
3004	2SB378M	3004	2SB383J	3004	2SB394E	3004	2SB405EK	3004	2SB439GN
3004	2SB3780R	3004	2SB383K	3004	2SB394F	3004	2SB405F	3004	2SB439H
3004	2SB378R	3004	2SB383L	3004	2SB394GN	3004	2SB405GN	3004	2SB439J
3004	2SB378Y	3004	2SB383M	3004	2SB394H	3004	2SB4050N	3004	2SB439L
3004	2SB379	3004	2SB3830R	3004	2SB394K	3004	2SB405J	3004	2SB439M
3004	2SB379-2	3004	2SB383R	3004	2SB394L	3004	2SB405L	3004	2SB4390R
3004	2SB379A	3004	2SB383X	3004	2SB394M	3004	2SB405M	3004	2SB439X
3004	2SB379B	3004	2SB383Y	3004	2SB3940R	3004	2SB4050R	3004	2SB439Y
3004	2SB379D	3004	2SB384	3004	2SB394R	3004	2SB405X	3004	2SB43A
3004	2SB379D	3004	2SB384A	3004	2SB394R	3004	2SB405Y	3004	2SB43B
3004	2SB379F	3004	2SB384B	3004	2SB394X	3004	2SB408	3004	2SB43C
3004	2SB379G	3004	2SB384C	3004	2SB394Y	3004	2SB408A	3004	2SB43E
3004	2SB379GN	3004	2SB384D	3004	2SB395A	3004	2SB408C	3004	2SB43F
3004	2SB379GN	3004	2SB384E	3004	2SB395B	3004	2SB408D	3004	2SB43G
3004	2SB379H	3004	2SB384F	3004	2SB395C	3004	2SB408E	3004	2SB43GN
3004	2SB379J	3004	2SB384GN	3004	2SB395D	3004	2SB408F	3004	2SB43H
3004	2SB379K	3004	2SB384H	3004	2SB395E	3004	2SB408G	3004	2SB43J
3004	2SB379L	3004	2SB384J	3004	2SB395F	3004	2SB408GN	3004	2SB43K
3004	2SB379M	3004	2SB384K	3004	2SB395G	3004	2SB408H	3004	2SB43L
3004	2SB3790R	3004	2SB384K	3004	2SB395GN	3004	2SB408J	3004	2SB43M
3004	2SB379R	3004	2SB384M	3004	2SB395H	3004	2SB408K	3004	2SB430R
3004	2SB379Y	3004	2SB3840R	3004	2SB395J	3004	2SB408L	3004	2SB43X
3004	2SB37A	3004	2SB384R	3004	2SB395K	3004	2SB408M	3004	2SB43Y
3004	2SB37B	3004	2SB384X	3004	2SB395M	3004	2SB4080R	3004	2SB44
3004	2SB37C	3004	2SB384Y	3004	2SB3950R	3004	2SB408R	3004	2SB440A
3004	2SB37D	3004	2SB385	3004	2SB395R	3004	2SB408X	3004	2SB440B
3004	2SB37E	3004	2SB385A	3004	2SB395X	3004	2SB408Y	3004	2SB440C
3004	2SB37F	3004	2SB385B	3004	2SB395Y	3004	2SB40A	3004	2SB440E
3004	2SB37G	3004	2SB385C	3004	2SB396	3004	2SB40B	3004	2SB440F
3004	2SB37GN	3004	2SB385D	3004	2SB396A	3004	2SB40C	3004	2SB440G
3004	2SB37H	3004	2SB385E	3004	2SB396B	3004	2SB40D	3004	2SB440GN
3004	2SB37J	3004	2SB385F	3004	2SB396C	3004	2SB40E	3004	2SB440H
3004	2SB37K	3004	2SB385G	3004	2SB396D	3004	2SB40F	3004	2SB440K
3004	2SB37L	3004	2SB385GN	3004	2SB396E	3004	2SB40G	3004	2SB440L
3004	2SB37M	3004	2SB385H	3004	2SB396F	3004	2SB40GN	3004	2SB440M
3004	2SB370R	3004	2SB385J	3004	2SB396G	3004	2SB40H	3004	2SB4400R
3004	2SB37R	3004	2SB385K	3004	2SB396GN	3004	2SB40J	3004	2SB440X
3004	2SB37X	3004	2SB385L	3004	2SB396H	3004	2SB40K	3004	2SB440Y
3004	2SB37Y	3004	2SB385M	3004	2SB396J	3004	2SB40L	3004	2SB443
3004	2SB38	3004	2SB3850R	3004	2SB396K	3004	2SB40M	3004	2SB443A
3004	2SB380	3004	2SB385R	3004	2SB396L	3004	2SB40R	3004	2SB443B
3004	2SB380A	3004	2SB385Y	3004	2SB396M	3004	2SB400R	3004	2SB443C
3004	2SB380B	3004	2SB387	3004	2SB3960R	3004	2SB40R	3004	2SB443D
3004	2SB380C	3004	2SB387A	3004	2SB396R	3004	2SB40X	3004	2SB443F
3004	2SB380D	3004	2SB387B	3004	2SB396X	3004	2SB40Y	3004	2SB443G
3004	2SB380E	3004	2SB387D	3004	2SB396Y	3004	2SB415C	3004	2SB443GN
3004	2SB380GN	3004	2SB387J	3004	2SB39A	3004	2SB415D	3004	2SB443H
3004	2SB380J	3004	2SB387P	3004	2SB39B	3004	2SB415E	3004	2SB443K
3004	2SB380K	3004	2SB387P	3004	2SB39D	3004	2SB415F	3004	2SB443L
3004	2SB380L	3004	2SB387GN	3004	2SB39E	3004	2SB415G	3004	2SB443M
3004	2SB3800R	3004	2SB387H	3004	2SB39F	3004	2SB415GN	3004	2SB4430R
3004	2SB380R	3004	2SB387J	3004	2SB39G	3004	2SB415H	3004	2SB443X
3004	2SB380Y	3004	2SB387K	3004	2SB39GN	3004	2SB415J	3004	2SB443Y
3004	2SB381	3004	2SB387K	3004	2SB39H	3004	2SB415K	3004	2SB443X
3004	2SB3812A	3004	2SB387M	3004	2SB39J	3004	2SB415L	3004	2SB443L
3004	2SB3812B	3004	2SB3870R	3004	2SB39K	3004	2SB415R	3004	2SB443M
3004	2SB3812C	3004	2SB387X	3004	2SB39L	3004	2SB416A	3004	2SB4430R
3004	2SB3812F	3004	2SB387Y	3004	2SB39R	3004	2SB416B	3004	2SB443X
3004	2SB3812G	3004	2SB389	3004	2SB39OR	3004	2SB416C	3004	2SB443Y
3004	2SB3812GN	3004	2SB389A	3004	2SB39X	3004	2SB416D	3004	2SB4440D
3004	2SB3812H	3004	2SB389B	3004	2SB39Y	3004	2SB416E	3004	2SB444C
3004	2SB3812J	3004	2SB389BK	3004	2SB40	3004	2SB416F	3004	2SB444D
3004	2SB3812K	3004	2SB389C	3004	2SB400	3004	2SB416G	3004	2SB444E
3004	2SB3812L	3004	2SB389D	3004	2SB400A	3004	2SB416H	3004	2SB444F
3004	2SB3812M	3004	2SB389E	3004	2SB400BK	3004	2SB416J	3004	2SB444G
3004	2SB38120R	3004	2SB389F	3004	2SB400C	3004	2SB416K	3004	2SB444GN
3004	2SB3812R	3004	2SB389G	3004	2SB400D	3004	2SB416L	3004	2SB444H
3004	2SB3812X	3004	2SB389H	3004	2SB400E	3004	2SB416M	3004	2SB444K
3004	2SB3812Y	3004	2SB389J	3004	2SB400F	3004	2SB4160R	3004	2SB444L
3004	2SB3813A	3004	2SB389H	3004	2SB400G	3004	2SB416R	3004	2SB444M
3004	2SB3813B	3004	2SB389J	3004	2SB400GN	3004	2SB416X	3004	2SB4440R
3004	2SB3813C	3004	2SB389K	3004	2SB400H	3004	2SB416Y	3004	2SB444R
3004	2SB3813D	3004	2SB389L	3004	2SB400J	3004	2SB417A	3004	2SB444X
3004	2SB3813E	3004	2SB389M	3004	2SB400K	3004	2SB417B	3004	2SB444Y
3004	2SB3813F	3004	2SB3890R	3004	2SB400L	3004	2SB417C	3004	2SB44A
3004	2SB3813G	3004	2SB389R	3004	2SB400M	3004	2SB417D	3004	2SB44B
3004	2SB3813GN	3004	2SB389X	3004	2SB4000R	3004	2SB417F	3004	2SB44C
3004	2SB3813H	3004	2SB38A	3004	2SB400X	3004	2SB417G	3004	2SB44D
3004	2SB3813J	3004	2SB38B	3004	2SB400Y	3004	2SB417GN	3004	2SB44E
3004	2SB3813K	3004	2SB38C	3004	2SB401A	3004	2SB417H	3004	2SB44F
3004	2SB3813L	3004	2SB38D	3004	2SB401C	3004	2SB417H	3004	2SB44G
3004	2SB3813M	3004	2SB38E	3004	2SB401D	3004	2SB417K	3004	2SB44GN
3004	2SB38130R	3004	2SB38F	3004	2SB401E	3004	2SB417K	3004	2SB44H
3004	2SB3813X	3004	2SB38F	3004	2SB401F	3004	2SB417L	3004	2SB44J
3004	2SB3813Y	3004	2SB38GN	3004	2SB401G	3004	2SB417M	3004	2SB44K
3004	2SB3581A	3004	2SB38GN	3004	2SB401GN	3004	2SB4170R	3004	2SB44L
3004	2SB3581B	3004	2SB38J	3004	2SB401H	3004	2SB417R	3004	2SB44M
								3004	2SB4400R

SK	Industry Standard No.	SK	Industry Standard No.	SK	Industry Standard No.	SK	Industry Standard No.	SK	Industry Standard No.
3004	2SB44R	3004	2SB482D	3004	2SB50C	3004	2SB56G	3004	2SB68B
3004	2SB44X	3004	2SB482E	3004	2SB50D	3004	2SB56GN	3004	2SB68C
3004	2SB44Y	3004	2SB482F	3004	2SB50E	3004	2SB56H	3004	2SB68D
3004	2SB457B	3004	2SB482G	3004	2SB50F	3004	2SB56J	3004	2SB68E
3004	2SB457C	3004	2SB482GN	3004	2SB50G	3004	2SB56K	3004	2SB68F
3004	2SB457D	3004	2SB482H	3004	2SB50GN	3004	2SB56L	3004	2SB68G
3004	2SB457E	3004	2SB482J	3004	2SB50H	3004	2SB56OR	3004	2SB68GN
3004	2SB457F	3004	2SB482K	3004	2SB50J	3004	2SB56R	3004	2SB68H
3004	2SB457G	3004	2SB482L	3004	2SB50K	3004	2SB56X	3004	2SB68J
3004	2SB457GN	3004	2SB482M	3004	2SB50L	3004	2SB56Y	3004	2SB68K
3004	2SB457H	3004	2SB482X	3004	2SB50M	3004	2SB57	3004	2SB68L
3004	2SB457J	3004	2SB482Y	3004	2SB50OR	3004	2SB57A	3004	2SB68M
3004	2SB457K	3004	2SB486	3004	2SB50R	3004	2SB57B	3004	2SB68OR
3004	2SB457L	3004	2SB486A	3004	2SB50X	3004	2SB57C	3004	2SB68R
3004	2SB457M	3004	2SB486B	3004	2SB50Y	3004	2SB57D	3004	2SB68X
3004	2SB457R	3004	2SB486C	3004	2SB516C	3004	2SB57E	3004	2SB68Y
3004	2SB457X	3004	2SB486D	3004	2SB516CD	3004	2SB57F	3004	2SB71
3004	2SB457Y	3004	2SB486E	3004	2SB516D	3004	2SB57G	3004	2SB71A
3004	2SB459	3004	2SB486F	3004	2SB516P	3004	2SB57GN	3004	2SB71B
3004	2SB459A	3004	2SB486G	3004	2SB51A	3004	2SB57H	3004	2SB71C
3004	2SB459B	3004	2SB486GN	3004	2SB51B	3004	2SB57J	3004	2SB71D
3004	2SB459C	3004	2SB486H	3004	2SB51C	3004	2SB57K	3004	2SB71E
3004	2SB459C-2	3004	2SB486J	3004	2SB51D	3004	2SB57L	3004	2SB71F
3004	2SB459D	3004	2SB486K	3004	2SB51E	3004	2SB57M	3004	2SB71G
3004	2SB459E	3004	2SB486L	3004	2SB51F	3004	2SB57OR	3004	2SB71GN
3004	2SB459F	3004	2SB486M	3004	2SB51G	3004	2SB57R	3004	2SB71H
3004	2SB459G	3004	2SB486OR	3004	2SB51GN	3004	2SB57X	3004	2SB71J
3004	2SB459GN	3004	2SB486R	3004	2SB51H	3004	2SB57Y	3004	2SB71K
3004	2SB459H	3004	2SB486X	3004	2SB51J	3004	2SB59	3004	2SB71L
3004	2SB459J	3004	2SB486Y	3004	2SB51K	3004	2SB59A	3004	2SB71M
3004	2SB459K	3004	2SB48A	3004	2SB51L	3004	2SB59B	3004	2SB71OR
3004	2SB459L	3004	2SB48B	3004	2SB51M	3004	2SB59C	3004	2SB71R
3004	2SB459M	3004	2SB48C	3004	2SB51OR	3004	2SB59D	3004	2SB71X
3004	2SB459OR	3004	2SB48D	3004	2SB51R	3004	2SB59E	3004	2SB71Y
3004	2SB459R	3004	2SB48E	3004	2SB51X	3004	2SB59F	3004	2SB72
3004	2SB459X	3004	2SB48F	3004	2SB51Y	3004	2SB59G	3004	2SB72A
3004	2SB459Y	3004	2SB48GN	3004	2SB52A	3004	2SB59GN	3004	2SB72B
3004	2SB460	3004	2SB48H	3004	2SB52B	3004	2SB59J	3004	2SB72C
3004	2SB460A	3004	2SB48J	3004	2SB52C	3004	2SB59K	3004	2SB72D
3004	2SB460B	3004	2SB48K	3004	2SB52D	3004	2SB59L	3004	2SB72E
3004	2SB461A	3004	2SB48M	3004	2SB52E	3004	2SB59M	3004	2SB72F
3004	2SB461B	3004	2SB48OR	3004	2SB52F	3004	2SB59OR	3004	2SB72G
3004	2SB461C	3004	2SB48R	3004	2SB52G	3004	2SB59R	3004	2SB72GN
3004	2SB461D	3004	2SB48X	3004	2SB52GN	3004	2SB59X	3004	2SB72H
3004	2SB461E	3004	2SB48Y	3004	2SB52H	3004	2SB59Y	3004	2SB72J
3004	2SB461F	3004	2SB49	3004	2SB52J	3004	2SB60	3004	2SB72K
3004	2SB461G	3004	2SB495E	3004	2SB52K	3004	2SB60A	3004	2SB72L
3004	2SB461H	3004	2SB495F	3004	2SB52L	3004	2SB60B	3004	2SB72M
3004	2SB461J	3004	2SB495G	3004	2SB52M	3004	2SB60C	3004	2SB72OR
3004	2SB461K	3004	2SB495GN	3004	2SB52OR	3004	2SB60D	3004	2SB72R
3004	2SB461L	3004	2SB495H	3004	2SB52R	3004	2SB60E	3004	2SB72X
3004	2SB461M	3004	2SB495J	3004	2SB52X	3004	2SB60F	3004	2SB72Y
3004	2SB461OR	3004	2SB495K	3004	2SB52Y	3004	2SB60G	3004	2SB73
3004	2SB461R	3004	2SB495L	3004	2SB53	3004	2SB60GN	3004	2SB73A
3004	2SB461X	3004	2SB495M	3004	2SB53A	3004	2SB60H	3004	2SB73A-1
3004	2SB461Y	3004	2SB495OR	3004	2SB53B	3004	2SB60J	3004	2SB73B
3004	2SB46A	3004	2SB495R	3004	2SB53C	3004	2SB60K	3004	2SB73C
3004	2SB46B	3004	2SB495X	3004	2SB53D	3004	2SB60L	3004	2SB73D
3004	2SB46C	3004	2SB495Y	3004	2SB53E	3004	2SB60M	3004	2SB73E
3004	2SB46D	3004	2SB496	3004	2SB53F	3004	2SB60OR	3004	2SB73F
3004	2SB46E	3004	2SB496A	3004	2SB53G	3004	2SB60R	3004	2SB73GN
3004	2SB46F	3004	2SB496B	3004	2SB53H	3004	2SB60X	3004	2SB73H
3004	2SB46G	3004	2SB496C	3004	2SB53J	3004	2SB60Y	3004	2SB73J
3004	2SB46GN	3004	2SB496D	3004	2SB53K	3004	2SB61A	3004	2SB73K
3004	2SB46H	3004	2SB496E	3004	2SB53L	3004	2SB61B	3004	2SB73L
3004	2SB46J	3004	2SB496F	3004	2SB53M	3004	2SB61C	3004	2SB73M
3004	2SB46K	3004	2SB496G	3004	2SB53OR	3004	2SB61D	3004	2SB73OR
3004	2SB46L	3004	2SB496GN	3004	2SB53R	3004	2SB61E	3004	2SB73R
3004	2SB46M	3004	2SB496H	3004	2SB53X	3004	2SB61F	3004	2SB73X
3004	2SB460R	3004	2SB496J	3004	2SB53Y	3004	2SB61G	3004	2SB73Y
3004	2SB46R	3004	2SB496K	3004	2SB54	3004	2SB61GN	3004	2SB74A
3004	2SB46X	3004	2SB496L	3004	2SB54A	3004	2SB61H	3004	2SB74B
3004	2SB46Y	3004	2SB496M	3004	2SB54B	3004	2SB61J	3004	2SB74C
3004	2SB470	3004	2SB496OR	3004	2SB54BA	3004	2SB61K	3004	2SB74D
3004	2SB470A	3004	2SB496R	3004	2SB54C	3004	2SB61L	3004	2SB74E
3004	2SB470B	3004	2SB496X	3004	2SB54D	3004	2SB61M	3004	2SB74F
3004	2SB470C	3004	2SB496Y	3004	2SB54E	3004	2SB61OR	3004	2SB74G
3004	2SB470D	3004	2SB497	3004	2SB54F	3004	2SB61R	3004	2SB74GN
3004	2SB470E	3004	2SB497A	3004	2SB54G	3004	2SB61X	3004	2SB74H
3004	2SB470F	3004	2SB497B	3004	2SB54GN	3004	2SB61Y	3004	2SB74J
3004	2SB470G	3004	2SB497C	3004	2SB54H	3004	2SB65	3004	2SB74K
3004	2SB470GN	3004	2SB497D	3004	2SB54J	3004	2SB65A	3004	2SB74L
3004	2SB470H	3004	2SB497E	3004	2SB54K	3004	2SB65B	3004	2SB74M
3004	2SB470J	3004	2SB497F	3004	2SB54L	3004	2SB65C	3004	2SB74OR
3004	2SB470K	3004	2SB497GN	3004	2SB54L1	3004	2SB65D	3004	2SB74X
3004	2SB470L	3004	2SB497H	3004	2SB54M	3004	2SB65F	3004	2SB74X
3004	2SB470M	3004	2SB497J	3004	2SB54OR	3004	2SB65G	3004	2SB74Y
3004	2SB470OR	3004	2SB497K	3004	2SB54R	3004	2SB65GN	3004	2SB75
3004	2SB470R	3004	2SB497L	3004	2SB54Y	3004	2SB65H	3004	2SB75A
3004	2SB470X	3004	2SB497M	3004	2SB55	3004	2SB65J	3004	2SB75AH
3004	2SB470Y	3004	2SB497OR	3004	2SB55A	3004	2SB65K	3004	2SB75B
3004	2SB475	3004	2SB497R	3004	2SB55B	3004	2SB65L	3004	2SB75C
3004	2SB475A	3004	2SB497X	3004	2SB55C	3004	2SB65M	3004	2SB75C-4
3004	2SB475B	3004	2SB497Y	3004	2SB55D	3004	2SB65OR	3004	2SB75D
3004	2SB475C	3004	2SB498	3004	2SB55E	3004	2SB65R	3004	2SB75E
3004	2SB475D	3004	2SB498A	3004	2SB55F	3004	2SB65X	3004	2SB75GN
3004	2SB475E	3004	2SB498B	3004	2SB55G	3004	2SB65Y	3004	2SB75H
3004	2SB475F	3004	2SB498C	3004	2SB55GN	3004	2SB66A	3004	2SB75J
3004	2SB475G	3004	2SB498D	3004	2SB55H	3004	2SB66B	3004	2SB75L
3004	2SB475GN	3004	2SB498E	3004	2SB55J	3004	2SB66C	3004	2SB75M
3004	2SB475H	3004	2SB498F	3004	2SB55K	3004	2SB66D	3004	2SB75OR
3004	2SB475J	3004	2SB498G	3004	2SB55L	3004	2SB66F	3004	2SB75R
3004	2SB475K	3004	2SB498GN	3004	2SB55M	3004	2SB66G	3004	2SB75Y
3004	2SB475M	3004	2SB498H	3004	2SB55OR	3004	2SB66GN	3004	2SB76A
3004	2SB475OR	3004	2SB498J	3004	2SB55R	3004	2SB66J	3004	2SB76B
3004	2SB475P	3004	2SB498K	3004	2SB55X	3004	2SB66K	3004	2SB76C
3004	2SB475PL	3004	2SB498L	3004	2SB55Y	3004	2SB66L	3004	2SB76D
3004	2SB475Q	3004	2SB498M	3004	2SB56	3004	2SB66M	3004	2SB76E
3004	2SB475R	3004	2SB498OR	3004	2SB565	3004	2SB66OR	3004	2SB76F
3004	2SB475Y	3004	2SB498R	3004	2SB565A	3004	2SB66R	3004	2SB76G
3004	2SB47A	3004	2SB498X	3004	2SB565B	3004	2SB66X	3004	2SB76GN
3004	2SB47B	3004	2SB498Y	3004	2SB565C	3004	2SB66Y	3004	2SB76H
3004	2SB47C	3004	2SB49A	3004	2SB565D	3004	2SB67	3004	2SB76J
3004	2SB47D	3004	2SB49C	3004	2SB565E	3004	2SB67A	3004	2SB76K
3004	2SB47E	3004	2SB49D	3004	2SB565F	3004	2SB67B	3004	2SB76L
3004	2SB47F	3004	2SB49E	3004	2SB565G	3004	2SB67C	3004	2SB76M
3004	2SB47G	3004	2SB49F	3004	2SB565GN	3004	2SB67D	3004	2SB76OR
3004	2SB47GN	3004	2SB49G	3004	2SB565H	3004	2SB67E	3004	2SB76R
3004	2SB47H	3004	2SB49GN	3004	2SB565J	3004	2SB67F	3004	2SB76X
3004	2SB47J	3004	2SB49H	3004	2SB565K	3004	2SB67G	3004	2SB76Y
3004	2SB47K	3004	2SB49J	3004	2SB565L	3004	2SB67GN	3004	2SB77
3004	2SB47L	3004	2SB49K	3004	2SB565M	3004	2SB67H	3004	2SB77(B)
3004	2SB47M	3004	2SB49L	3004	2SB565OR	3004	2SB67J	3004	2SB77A
3004	2SB47OR	3004	2SB49M	3004	2SB565R	3004	2SB67K	3004	2SB77A/P
3004	2SB47R	3004	2SB49OR	3004	2SB565X	3004	2SB67L	3004	2SB77AB
3004	2SB47X	3004	2SB49R	3004	2SB565Y	3004	2SB67M	3004	2SB77AH
3004	2SB47Y	3004	2SB49X	3004	2SB56A	3004	2SB67OR	3004	2SB77B
3004	2SB48	3004	2SB49Y	3004	2SB56B	3004	2SB67R	3004	2SB77B-11
3004	2SB482	3004	2SB50A	3004	2SB56C	3004	2SB67X	3004	2SB77C
3004	2SB482A	3004	2SB50B	3004	2SB56CK	3004	2SB67Y	3004	2SB77D
3004	2SB482B			3004	2SB56D	3004	2SB67	3004	2SB77E
3004	2SB482C			3004	2SB56E	3004	2SB68		
				3004	2SB56F	3004	2SB68A		

SK	Industry Standard No.
3004	2SB77F
3004	2SB77G
3004	2SB77GN
3004	2SB77H
3004	2SB77K
3004	2SB77L
3004	2SB77M
3004	2SB77R
3004	2SB77V
3004	2SB77X
3004	2SB77Y
3004	2SB78
3004	2SB78A
3004	2SB78B
3004	2SB78C
3004	2SB78D
3004	2SB78E
3004	2SB78F
3004	2SB78G
3004	2SB78GN
3004	2SB78H
3004	2SB78J
3004	2SB78K
3004	2SB78L
3004	2SB78M
3004	2SB78OR
3004	2SB78R
3004	2SB78X
3004	2SB78Y
3004	2SB79
3004	2SB79A
3004	2SB79B
3004	2SB79C
3004	2SB79D
3004	2SB79E
3004	2SB79F
3004	2SB79G
3004	2SB79GN
3004	2SB79H
3004	2SB79J
3004	2SB79K
3004	2SB79L
3004	2SB79M
3004	2SB79OR
3004	2SB79R
3004	2SB79X
3004	2SB79Y
3004	2SB85A
3004	2SB85B
3004	2SB85C
3004	2SB85D
3004	2SB85E
3004	2SB85F
3004	2SB85G
3004	2SB85GN
3004	2SB85H
3004	2SB85J
3004	2SB85K
3004	2SB85L
3004	2SB85M
3004	2SB85OR
3004	2SB85R
3004	2SB85X
3004	2SB85Y
3004	2SB87A
3004	2SB87B
3004	2SB87C
3004	2SB87D
3004	2SB87E
3004	2SB87F
3004	2SB87G
3004	2SB87GN
3004	2SB87H
3004	2SB87J
3004	2SB87K
3004	2SB87L
3004	2SB87M
3004	2SB87OR
3004	2SB87R
3004	2SB87X
3004	2SB87Y
3004	2SB89A
3004	2SB89C
3004	2SB89D
3004	2SB89E
3004	2SB89F
3004	2SB89G
3004	2SB89GN
3004	2SB89J
3004	2SB89K
3004	2SB89L
3004	2SB89M
3004	2SB89OR
3004	2SB89R
3004	2SB89X
3004	2SB89Y
3004	2SB90
3004	2SB90A
3004	2SB90B
3004	2SB90C
3004	2SB90D
3004	2SB90E
3004	2SB90F
3004	2SB90G
3004	2SB90GN
3004	2SB90H
3004	2SB90J
3004	2SB90K
3004	2SB90L
3004	2SB90M
3004	2SB90OR
3004	2SB90R
3004	2SB90X
3004	2SB90Y
3004	2SB91
3004	2SB91A
3004	2SB91B
3004	2SB91C
3004	2SB91D
3004	2SB91E
3004	2SB91G
3004	2SB91GN
3004	2SB91H
3004	2SB91J
3004	2SB91K
3004	2SB91L
3004	2SB91M
3004	2SB91OR
3004	2SB91R
3004	2SB91X
3004	2SB91Y
3004	2SB92A
3004	2SB92B
3004	2SB92C
3004	2SB92D
3004	2SB92E
3004	2SB92F
3004	2SB92G
3004	2SB92GN
3004	2SB92H
3004	2SB92J
3004	2SB92K
3004	2SB92L
3004	2SB92M
3004	2SB92X
3004	2SB92Y
3004	2SB94
3004	2SB94A
3004	2SB94B
3004	2SB94C
3004	2SB94D
3004	2SB94E
3004	2SB94F
3004	2SB94G
3004	2SB94GN
3004	2SB94H
3004	2SB94J
3004	2SB94K
3004	2SB94M
3004	2SB94OR
3004	2SB94R
3004	2SB94X
3004	2SB94Y
3004	2SB95
3004	2SB95A
3004	2SB95B
3004	2SB95C
3004	2SB95D
3004	2SB95E
3004	2SB95F
3004	2SB95G
3004	2SB95GN
3004	2SB95H
3004	2SB95J
3004	2SB95L
3004	2SB95M
3004	2SB95OR
3004	2SB95R
3004	2SB95X
3004	2SB95Y
3004	2SB97
3004	2SB97A
3004	2SB97B
3004	2SB97C
3004	2SB97D
3004	2SB97F
3004	2SB97G
3004	2SB97GN
3004	2SB97H
3004	2SB97K
3004	2SB97L
3004	2SB97M
3004	2SB97OR
3004	2SB97R
3004	2SB97X
3004	2SB97Y
3004	2SB98
3004	2SB98A
3004	2SB98B
3004	2SB98C
3004	2SB98D
3004	2SB98E
3004	2SB98F
3004	2SB98G
3004	2SB98GN
3004	2SB98H
3004	2SB98J
3004	2SB98K
3004	2SB98L
3004	2SB98M
3004	2SB98OR
3004	2SB98R
3004	2SB98X
3004	2SB98Y
3004	2SB99
3004	2SB99A
3004	2SB99B
3004	2SB99C
3004	2SB99D
3004	2SB99E
3004	2SB99F
3004	2SB99G
3004	2SB99GN
3004	2SB99H
3004	2SB99J
3004	2SB99K
3004	2SB99L
3004	2SB99M
3004	2SB99OR
3004	2SB99R
3004	2SB99X
3004	2SB99Y
3004	2SBF1
3004	2SBF1A
3004	2SBF2
3004	2SBF2A
3004	2SBM77
3004	2T2001
3004	2T230
3004	2T231
3004	2T3
3004	2TN45A
3004	2TN56
3004	2TN95
3004	2TN95A
3004	2V362
3004	2V363
3004	03-0020-0
3004	03-0022
3004	03-0023-0
3004	03-57-302
3004	03-57-304
3004	003-H03
3004	04-57-303
3004	4-66-2A7-1
3004	4-66-3A7-1
3004	4-686163-3
3004	4-686195-3
3004	4-686196-3
3004	4-68681-2
3004	4-68681-3
3004	4-8P-2A7-1
3004	4JD1A73
3004	4JX1A520
3004	4JX1A520B
3004	4JX1A520C
3004	4JX10850
3004	6-0000155A
3004	6-13
3004	6-53
3004	6-53A
3004	6-63
3004	6-63A
3004	6A12989
3004	6A12990
3004	6D122
3004	6D122R
3004	07-07119
3004	07-1075-01
3004	7-59-029/3477
3004	7-59-060/3477
3004	07-6015-16
3004	8-0062
3004	8-0205400
3004	8-0205600
3004	8-0236400
3004	8-0236430
3004	8-0243900
3004	8-619-030-007
3004	8-619-030-008
3004	8-619-030-009
3004	8-619-030-014
3004	8-697-020-567
3004	8-697-020-568
3004	8-697-020-569
3004	8-905-605-016
3004	8-905-605-030
3004	8-905-605-032
3004	8-905-605-050
3004	8-905-605-051
3004	8-905-605-075
3004	8-905-605-090
3004	8-905-605-091
3004	8-905-605-230
3004	8-905-605-232
3004	8-905-605-234
3004	8-905-605-292
3004	8-905-605-305
3004	8-905-606-750
3004	8-905-606-800
3004	8-905-606-815
3004	8-905-606-817
3004	8-905-606-885
3004	8-905-613-010
3004	8-905-613-640
3004	8-905-613-710
3004	8-905-613-955
3004	8-905-615-156
3004	08P-2
3004	08P-2-12-7
3004	8P-2-70
3004	8P-2-70-12
3004	8P-2-70-12-7
3004	8P-20
3004	8P-21
3004	8P-22
3004	8P-23
3004	8P-24
3004	8P-25
3004	8P-26
3004	8P-27
3004	8P-28
3004	8P-29
3004	8P-2A
3004	8P-2A0
3004	8P-2A0R
3004	8P-2A1
3004	8P-2A19
3004	8P-2A2
3004	8P-2A21
3004	8P-2A3
3004	8P-2A3P
3004	8P-2A4
3004	8P-2A4-7
3004	8P-2A5
3004	8P-2A5L
3004	8P-2A6
3004	8P-2A6-2
3004	8P-2A7
3004	8P-2A7-1
3004	8P-2A8
3004	8P-2A82
3004	8P-2A9
3004	8P-2A9G
3004	09-301001
3004	09-301002
3004	09-301002-6
3004	09-301003
3004	09-301004
3004	09-301005
3004	09-301006
3004	09-301007
3004	09-301008-18
3004	09-301009
3004	09-301016
3004	09-301020
3004	09-301023
3004	09-301025
3004	09-301025-6
3004	09-301026
3004	09-301032
3004	09-301036
3004	09-301048
3004	09-30104B
3004	09-301056
3004	09-301072
3004	09-301074
3004	09-30126
3004	09-30313
3004	9-511410100
3004	9-511410200
3004	9-511410900
3004	9-511413500
3004	9-5201
3004	9-5217
3004	9-5218
3004	9-5224-1
3004	9-9104
3004	9-9201
3004	9-9202
3004	9-9203
3004	10-2SB54
3004	10-2SB56
3004	12-1-106
3004	12-1-107
3004	12-1-120
3004	12-1-148
3004	12-1-164
3004	12-1-184
3004	12-1-191
3004	12-1-227
3004	12-1-232
3004	12-1-246
3004	12-1-266
3004	12-1-267
3004	12-1-272
3004	12-1-95
3004	12-1-96
3004	12A9275
3004	12A9275-1
3004	13-087005
3004	13-14085-10
3004	13-14085-18
3004	13-14085-23
3004	13-14085-35
3004	13-14085-60
3004	13-14085-71
3004	13-14085-9
3004	13-14888-3
3004	13-15805-1
3004	13-15836-1
3004	13-18304-1
3004	13-23785-1
3004	13-35792-1
3004	13-67599-3
3004	13-67599-9
3004	13-94096-2
3004	14-577-10
3004	14-602-05
3004	14-602-05A
3004	14MW69
3004	15-2221011
3004	15-22210111
3004	15-22211200
3004	15-22211328
3004	15-22216500
3004	16-207190405
3004	19-020-034
3004	19-020-035
3004	19-020-036
3004	19A115077-P1
3004	19A115077-P2
3004	19A115281-P1
3004	19A115674-P2
3004	19AR14-1
3004	19AR14-2
3004	19AR16-1
3004	19AR16-2
3004	19AR19-1
3004	19AR19-2
3004	19AR25
3004	19AR26
3004	19AR27
3004	19AR32
3004	19AR7-1
3004	19AR7-2
3004	19B2000132-P1
3004	19B2000132-P2
3004	19B2000132-P3
3004	19B2000132-P4
3004	19B200054-P1
3004	19B200061-P2
3004	19B200061-P3
3004	19B200061-P4
3004	19B200063-P1
3004	19B200210-P1
3004	19B200210-P2
3004	19B200210-P3
3004	19C300073-P1
3004	19C300073-P2
3004	19C300073-P3
3004	19C300073-P4
3004	19C300073-P5
3004	19C300073-P6
3004	19C300074-P2
3004	19C300128-P1
3004	19C300128-P2
3004	19C300128-P3
3004	19C300128-P6
3004	19C300128-P7
3004	19C300128-P8
3004	19C300138-P4
3004	20A0007
3004	20A0009
3004	20A0015
3004	20C72
3004	21-34
3004	21-37
3004	21A005-000
3004	21A015-001
3004	21A015-006
3004	21A038-000
3004	21A039-000
3004	21A040-005
3004	21A040-021
3004	21A040-022
3004	21A040-057
3004	21A040-058
3004	21A040-079
3004	21A053-000
3004	21A054-000
3004	21A055-000
3004	21A074-000
3004	22-002007
3004	022-3504-040
3004	022-3504-060
3004	022-3505-910
3004	23-5014
3004	23-5017
3004	24MW107
3004	24MW1083
3004	24MW1084
3004	24MW1115
3004	24MW115
3004	24MW116
3004	24MW132
3004	24MW15
3004	24MW16
3004	24MW178
3004	24MW179
3004	24MW185
3004	24MW263
3004	24MW28
3004	24MW29
3004	24MW370
3004	24MW384
3004	24MW43
3004	24MW598
3004	24MW599
3004	24MW60
3004	24MW601
3004	24MW614
3004	24MW615
3004	24MW69
3004	24MW70
3004	24MW777
3004	24MW78
3004	24MW780
3004	24MW781
3004	24MW782
3004	24MW789
3004	24MW824
3004	24MW83
3004	24MW84
3004	24MW853
3004	24MW856
3004	24MW857
3004	24MW892
3004	24MW893
3004	025-10031
3004	25B378
3004	25T1
3004	026-100005
3004	026-100012
3004	026-100018
3004	26MW613
3004	26T1
3004	27T403
3004	27T404
3004	27T405
3004	29V0038H03
3004	29V12H01
3004	31-0008
3004	31-0018
3004	31-0025
3004	31-0026
3004	31-0053
3004	31-0070
3004	31-0107
3004	31-0153
3004	31-0172
3004	31-0188
3004	31-0247
3004	31-025
3004	31-22005400
3004	33-1000-00
3004	33-1001-00
3004	33-1009-01
3004	33-1019-00
3004	33-1020-00
3004	34-6000-15
3004	34-6000-28
3004	34-6000-29
3004	34-6000-31
3004	34-6000-32
3004	34-6000-33
3004	34-6000-34
3004	34-6000-4
3004	34-6000-5
3004	34-6000-6
3004	34-6000-7
3004	34-6000-8
3004	34-6001-10
3004	34-6001-11
3004	34-6001-13
3004	34-6001-14
3004	34-6001-16
3004	34-6001-17
3004	34-6001-18
3004	34-6001-19
3004	34-6001-20
3004	34-6001-21
3004	34-6001-22
3004	34-6001-23
3004	34-6001-26
3004	34-6001-29
3004	34-6001-30
3004	34-6001-31
3004	34-6001-33
3004	34-6001-42
3004	34-6001-47
3004	34-6001-7
3004	34-6001-72
3004	34-6001-76
3004	34-6001-8
3004	34-6001-9
3004	34-6008
3004	34-6009
3004	34-6015-44A
3004	34-6016-11
3004	35P1
3004	35P2
3004	35P2C
3004	35T1
3004	36J003-1
3004	36P1
3004	36P1C
3004	36P1F
3004	36P2F
3004	36P3
3004	36P3A
3004	36P3C
3004	36P4
3004	36P4C
3004	36P5
3004	36P5C
3004	36P7
3004	36P7C
3004	36P7T
3004	36P8
3004	36P8C
3004	38T1
3004	39A9
3004	39T1
3004	40-601
3004	40P1
3004	40P2
3004	42-17143
3004	42-19671
3004	42-19863
3004	42-19863A
3004	42-19864
3004	42-19864A
3004	42-20222
3004	42-21406
3004	42-22534
3004	42-27376
3004	42X230
3004	42X233
3004	42X308
3004	42X309
3004	42X311
3004	43P2
3004	43P4
3004	43P4C
3004	43P6
3004	43P6A
3004	43P7
3004	43P7A
3004	43P7G

SK	Industry Standard No.
3004	45X1A502C
3004	46-163-3
3004	46-8610-3
3004	46-8611
3004	46-8611-3
3004	46-86195-3
3004	46-86373-3
3004	46-86430-3
3004	46-8664-3
3004	46-8665-3
3004	46-8666-3
3004	46-8679-1
3004	46-8679-2
3004	46-8679-3
3004	46-8680-1
3004	46-8680-2
3004	46-8680-3
3004	46-8681-1
3004	46-8681-2
3004	48-10073A02
3004	48-10074A02
3004	48-124219
3004	48-124286
3004	48-124297
3004	48-124303
3004	48-124304
3004	48-124306
3004	48-124309
3004	48-124318
3004	48-124319
3004	48-124353
3004	48-124354
3004	48-124355
3004	48-134573
3004	48-134621
3004	48-134632
3004	48-35P1
3004	48-36P1
3004	48-36P3
3004	48-39P3
3004	48-43P3
3004	48-43P4
3004	48-63029A18
3004	48-63029A19
3004	48-63029A91
3004	48-63029A93
3004	48-63029A94
3004	48-869249
3004	48-869253
3004	48-869282
3004	48-869475
3004	48-869475A
3004	48A124315
3004	48C125237
3004	48K134458
3004	48K35P1
3004	48K36P1
3004	48K36P3
3004	48K43P3
3004	48K43P4
3004	48P1
3004	48R134407
3004	48R134573
3004	48R134621
3004	48R134632
3004	48R869148
3004	48R869249
3004	48R869253
3004	48R869282
3004	48R869475
3004	48R869475A
3004	48S134407
3004	48S134408
3004	48S134909
3004	48X97046A53
3004	48X97046A54
3004	48X97046A55
3004	48X97162A06
3004	48X97162A07
3004	48X97238A05
3004	50B173-S
3004	50B175A
3004	50B175B
3004	50B423
3004	50B54
3004	50BU75-C
3004	51D170
3004	52004
3004	53P157
3004	55-1016
3004	56-8091
3004	56-8091A
3004	56-8091B
3004	56-8091C
3004	56-8091D
3004	56-8092
3004	56-8092A
3004	56-8092B
3004	57A2-15
3004	57A2-24
3004	57A2-4
3004	57B6
3004	57C6-1
3004	57C6-22
3004	57C6-25
3004	57D1-105
3004	57D1-11
3004	57D1-111
3004	57D1-121
3004	57D1-43
3004	57D1-44
3004	57D1-70
3004	57D1127
3004	57D1143
3004	57D126
3004	57D143
3004	57D156
3004	57D169
3004	57D170
3004	57D188
3004	57D189
3004	57D3-6
3004	57D68
3004	57P1
3004	61B004-1
3004	61B005-1
3004	61B006-1
3004	61B009-1
3004	61B015-1
3004	61B017-1
3004	61B019-1
3004	61B020-1
3004	61B021-1
3004	61B022-2
3004	61B022-3
3004	61B023-1
3004	61C002-1
3004	61C003-1
3004	61J004-1
3004	63P3
3004	66-2
3004	066-2-12
3004	066-2-12-7
3004	66-2-70
3004	66-2-70-12
3004	66-2-70-12-7
3004	66-20
3004	66-21
3004	66-22
3004	66-23
3004	66-24
3004	66-25
3004	66-26
3004	66-27
3004	66-28
3004	66-29
3004	66-2A
3004	66-2A0
3004	66-2A0R
3004	66-2A1
3004	66-2A19
3004	66-2A2
3004	66-2A21
3004	66-2A3
3004	66-2A3P
3004	66-2A4
3004	66-2A4-7
3004	66-2A5
3004	66-2A5L
3004	66-2A6
3004	66-2A6-4
3004	66-2A7
3004	66-2A7-1
3004	66-2A8
3004	66-2A82
3004	66-2A9
3004	66-2A9G
3004	66-3
3004	066-3-12
3004	066-3-12-7
3004	66-3-70
3004	66-3-70-12
3004	66-3-70-12-7
3004	66-30
3004	66-33
3004	66-34
3004	66-35
3004	66-36
3004	66-37
3004	66-38
3004	66-3A
3004	66-3A0
3004	66-3A0R
3004	66-3A1
3004	66-3A19
3004	66-3A2
3004	66-3A21
3004	66-3A3
3004	66-3A3P
3004	66-3A4
3004	66-3A4-7
3004	66-3A5
3004	66-3A5L
3004	66-3A6
3004	66-3A6C
3004	66-3A7
3004	66-3A7-1
3004	66-3A8
3004	66-3A82
3004	66-3A9
3004	66-3A9G
3004	66-6023
3004	66-6023-00
3004	66-6024-00
3004	66-6025-00
3004	66-6026-00
3004	66-6027-00
3004	66-6028-00
3004	66-6033
3004	79P1
3004	80-205400
3004	80-205600
3004	80-2226314
3004	80-236400
3004	80-236430
3004	80-243900
3004	80P1
3004	81-46125002-9
3004	81-46125003-7
3004	81-46125004-5
3004	81-46125009-4
3004	81-46125010-2
3004	81-46125011-0
3004	81-46125029-2
3004	86-114-2
3004	86-126-2
3004	86-128-2
3004	86-129-2
3004	86-130-2
3004	86-156-2
3004	86-156-2A
3004	86-159-2
3004	86-16-2
3004	86-169-2
3004	86-30-2
3004	86-303-2
3004	86-304-2
3004	86-305-2
3004	86-419-2
3004	86-421-2
3004	86-476-2
3004	86-497-2
3004	86-50-2
3004	86-5000-2
3004	86-5006-2
3004	86-5067-2
3004	86-77-2
3004	86-80-2
3004	86-82-2
3004	86-83-2
3004	86-84-2
3004	86-93-2
3004	86-98-2
3004	86A318
3004	86X0017-001
3004	86X0018-001
3004	86X0037-002
3004	87-0018
3004	87-0019
3004	87-0020
3004	87-0021
3004	93A9
3004	95-11410100
3004	95-11410200
3004	95-11410900
3004	95-11413500
3004	95-201
3004	95-203
3004	95-204
3004	95-208
3004	95-209
3004	95-212
3004	95-213
3004	95-214
3004	95-217
3004	95-218
3004	95-222-1
3004	95-224-1
3004	96-5032-01
3004	96-5033-01
3004	96-5033-02
3004	96-5033-03
3004	96-5033-04
3004	96-5085-01
3004	96-5085-02
3004	96-5098-01
3004	96-5101-01
3004	96-5102-01
3004	96XZ6053-09N
3004	96XZ6053-27N
3004	96XZ6053/51N
3004	96XZ801/50N
3004	97P1
3004	99AT6
3004	99B5
3004	99S002
3004	99S004A
3004	99S005
3004	99S010
3004	99S010A
3004	99S011
3004	99S011A
3004	100B63
3004	0102-0371
3004	0000103
3004	0000104
3004	110-01563-00
3004	112-003
3004	112-004
3004	112-034923
3004	112-200525
3004	116-091
3004	116-201
3004	116-203
3004	116-206
3004	116-685
3004	116-686
3004	116-757
3004	116-997
3004	120-00-19
3004	120-001195
3004	120-002013
3004	120-002014
3004	120-002521
3004	120-002748
3004	120-004727
3004	120-004729
3004	121-1032
3004	121-1034
3004	121-106
3004	121-107
3004	121-164
3004	121-184
3004	121-19
3004	121-190
3004	121-191
3004	121-193
3004	121-226
3004	121-227
3004	121-246
3004	121-266
3004	121-267
3004	121-27
3004	121-272
3004	121-291
3004	121-305
3004	121-306
3004	121-307
3004	121-309
3004	121-311
3004	121-314
3004	121-319
3004	121-320
3004	121-327
3004	121-328
3004	121-34
3004	121-347
3004	121-348
3004	121-372
3004	121-374
3004	121-375
3004	121-395
3004	121-396
3004	121-399
3004	121-408
3004	121-409
3004	121-437
3004	121-46
3004	121-47
3004	121-490
3004	121-52
3004	121-543
3004	121-544
3004	121-632
3004	121-633
3004	121-640
3004	121-68
3004	121-69
3004	121-734
3004	121-95
3004	121-96
3004	129-17
3004	129-18
3004	129-32
3004	129-8
3004	129-8-1
3004	129-8-1A
3004	129-8-2
3004	130-40095
3004	130-40236
3004	130-40352
3004	0131-000100
3004	0131-000101
3004	0131-000102
3004	0131-000563
3004	0131-001419
3004	0131-001426
3004	0131-002656
3004	132-001
3004	132-010
3004	132-090
3004	132-90
3004	148P-2
3004	148P-2-12
3004	148P-2-12-8
3004	152-221011
3004	154A3675-105
3004	154A3676-205
3004	154A3679
3004	154A8681
3004	175-006-9-001
3004	175-007-9-001
3004	175-008-9-001
3004	188P-2
3004	188P-2-12
3004	188P-2-127
3004	188P-2L
3004	188P-2L8
3004	205A2-210
3004	231-0009
3004	264P06301
3004	296-60-9
3004	296-62-9
3004	297V003H03
3004	297V003H09
3004	297V003M01
3004	297V003M07
3004	297V004H03
3004	297V004H06
3004	297V004H10
3004	297V004H14
3004	297V004H15
3004	297V004H16
3004	297V004M01
3004	297V025H02
3004	297V025H04
3004	297V025H05
3004	297V025H15
3004	297V027H01
3004	297V032H01
3004	297V033H01
3004	297V037B02
3004	297V037H01
3004	297V037H02
3004	297V040H01
3004	297V040H08
3004	297V040H10
3004	297V040H11
3004	297V040H12
3004	297V040H13
3004	297V040H16
3004	297V042C01
3004	297V042C02
3004	297V042C03
3004	297V042C04
3004	297V042H01
3004	297V042H02
3004	297V043H01
3004	297V050C02
3004	297V050H03
3004	297V051C03
3004	297V051C04
3004	297V051H01
3004	297V051H02
3004	297V051H03
3004	297V051H04
3004	297V052C01
3004	297V052H01
3004	297V052H02
3004	297V052H04
3004	297V053C01
3004	297V053H01
3004	297V076B01
3004	297V076C01
3004	297V081C01
3004	301
3004	310-188
3004	324-0029
3004	324-0133
3004	324-0142
3004	324-0143
3004	324-0144
3004	324-0146
3004	324-144
3004	325-0025-329
3004	325-0025-330
3004	325-0025-331
3004	325-0028-79
3004	325-0028-80
3004	325-0028-81
3004	325-0028-83
3004	325-0030-315
3004	325-0030-317
3004	325-0030-318
3004	325-0030-319
3004	325-0047-516
3004	325-0054-310
3004	325-0054-311
3004	325-0670
3004	325-0670-1
3004	325-1370-18
3004	325-1376-56
3004	325-1376-57
3004	325-1376-58
3004	325-1378-20
3004	325-1378-22
3004	325-1442-8
3004	352
3004	353
3004	353-9312-001
3004	394-3074-2
3004	394-3074-5
3004	394-3097-1
3004	394-3097-2
3004	410-012-0150
3004	410-013-0240
3004	417-48
3004	421-11
3004	421-12
3004	421-13
3004	421-14
3004	421-15
3004	421-19
3004	421-7143
3004	421-8109
3004	421-9671
3004	421-9862A
3004	421-9863
3004	421-9863A
3004	421-9864
3004	421-9864A
3004	422-0222
3004	422-1406
3004	422-2535
3004	473B6-2
3004	473B6-2A
3004	473B6-4
3004	473B6-5
3004	473B6-7
3004	573-529
3004	576-0001-003
3004	576-0001-009
3004	576-0001-014
3004	576-0002-004
3004	576-0002-012
3004	576-001
3004	576-005
3004	602-040
3004	605-030
3004	606-020
3004	610-035
3004	610-035-1
3004	610-036
3004	610-036-1
3004	610-036-2
3004	610-036-3
3004	610-036-4
3004	610-036-5
3004	610-036-7
3004	610-036-8
3004	610-040
3004	610-040-1
3004	610-040-2
3004	610-043
3004	610-079
3004	610-080
3004	614X10
3004	614X5
3004	614X6
3004	614X7
3004	614X9
3004	617-50
3004	617-52
3004	650-105
3004	650-106
3004	650-107
3004	660-227
3004	660B
3004	675-154
3004	675-155
3004	675-156
3004	690V034H30
3004	690V034H31
3004	690V034H39
3004	690V040H61
3004	690V040H62
3004	690V047H58
3004	690V047H59
3004	690V047H60
3004	690V052H24
3004	690V054H21
3004	690V056H33
3004	690V056H34
3004	690V056H90
3004	690V057H27
3004	690V057H28
3004	690V059H20
3004	690V059H52
3004	690V059H55
3004	690V061H98
3004	690V061H99
3004	690V063H16
3004	690V063H17
3004	690V063H50
3004	690V063H51
3004	690V066H46
3004	690V066H47
3004	690V068H30
3004	690V068H31
3004	690V077H37
3004	690V080H44
3004	690V084H61
3004	690V084H63
3004	690V086H39
3004	690V088H52
3004	690V089H90
3004	690V094H18
3004	690V094H19
3004	690V094H20
3004	690V097H59
3004	690V104H53
3004	690V104H54
3004	720-35019
3004	720-35019A
3004	750-137
3004	750-138
3004	750-139
3004	750-140
3004	792-286
3004	792-287
3004	792-288
3004	792-289
3004	792-290
3004	800-502-00
3004	802-05400
3004	802-05600
3004	802-22631U
3004	802-36400
3004	802-36430
3004	802-43900
3004	808-304
3004	808-305
3004	808-306
3004	808-307
3004	808-308
3004	808-310
3004	808-311
3004	815-181D
3004	901-000-6-51
3004	916-31001-1
3004	916-31012-6
3004	921-100B
3004	921-147B
3004	921-14B
3004	921-150B
3004	921-153B
3004	921-217B
3004	921-222B
3004	921-223B
3004	921-238B

SK	Industry Standard No.	SK	Industry Standard No.	SK	Industry Standard No.	SK	Industry Standard No.	SK	Industry Standard No.
3004	921-27	3004	2057B2-86	3004	18611	3004	81506-9A	3004	223486
3004	921-273B	3004	2093A38-23	3004	18731	3004	81506-9B	3004	224696
3004	921-274B	3004	2106-119	3004	18737	3004	81506-9C	3004	224857
3004	921-27A	3004	2106-121	3004	23114-061	3004	81507-0	3004	226924
3004	921-318B	3004	2106-122	3004	25114-101	3004	81507-0A	3004	230253
3004	921-319B	3004	2106-123	3004	25114-102	3004	81507-0B	3004	230259
3004	921-35	3004	2112-17	3004	25114-103	3004	81507-0C	3004	230524
3004	921-35A	3004	2402-453	3004	25114-104	3004	81507-0D	3004	230525
3004	921-35B	3004	2402-455	3004	27125-150	3004	81507-4	3004	231140-21
3004	921-36B	3004	2402-457	3004	27125-170	3004	81510-4	3004	234076
3004	921-38	3004	2405-453	3004	27125-330	3004	81511-4	3004	234630
3004	921-38A	3004	2408-326	3004	27125-340	3004	81511-8	3004	235194
3004	921-38B	3004	2490	3004	27125-350	3004	81512-0	3004	236709
3004	921-39	3004	2490A	3004	27125-480	3004	81512-0A	3004	238417
3004	921-39A	3004	2495-014	3004	27125-540	3004	81512-0B	3004	238418
3004	921-39B	3004	2495-080	3004	30204	3004	81512-0C	3004	242221
3004	921-40	3004	2495-588	3004	30206	3004	81512-0D	3004	243939
3004	921-40A	3004	2497-473	3004	30207	3004	81512-0E	3004	255728
3004	921-40B	3004	2497-496	3004	30208-1	3004	81513-6	3004	257340
3004	921-45	3004	2606-286	3004	30208-2	3004	81515-8	3004	257473
3004	921-45A	3004	2606-287	3004	30244	3004	81516-0	3004	262113
3004	921-45B	3004	2606-291	3004	34098	3004	81516-0A	3004	266686
3004	921-51B	3004	2612	3004	34493P	3004	81516-0B	3004	269374
3004	921-52B	3004	2704-384	3004	35045	3004	81516-0C	3004	292240-1
3004	921-67	3004	2704-385	3004	35454	3004	81516-0D	3004	300538
3004	921-67A	3004	2704-386	3004	35590	3004	81516-0E	3004	300540
3004	921-67B	3004	2781	3004	35628	3004	81516-0F	3004	300541
3004	921-68	3004	3004-856	3004	35792	3004	81516-0G	3004	309421
3004	921-68A	3004	3517	3004	35820	3004	81516-0H	3004	310017
3004	921-68B	3004	3518	3004	35952	3004	81516-0I	3004	310159
3004	930X10	3004	3546	3004	35953	3004	81516-0J	3004	310201
3004	930X7	3004	3550	3004	35954	3004	86452	3004	346016-1
3004	930X8	3004	3564	3004	35955	3004	86832	3004	346016-11
3004	930X9	3004	3586	3004	36557	3004	86842	3004	373003
3004	951-1	3004	3598-2	3004	36558	3004	087003	3004	373117
3004	964-17142	3004	3600	3004	37549	3004	94008	3004	373119
3004	964-19863	3004	3637	3004	37550	3004	94039	3004	489751-045
3004	964-19864	3004	3907	3004	37551	3004	95201	3004	489751-109
3004	965〓1	3004	3970QL	3004	38176	3004	95204	3004	489751-114
3004	972X10	3004	4041-000-10160	3004	38177	3004	95208	3004	510007
3004	972X11	3004	4041-000-20120	3004	40034-1	3004	95209	3004	537200
3004	972X12	3004	4041-000-30180	3004	40034-2	3004	95212	3004	0563012H
3004	972X9	3004	0004201	3004	40034-3	3004	95213	3004	573001
3004	987〓1	3004	0004202	3004	40034VM	3004	95214	3004	0573001-14
3004	988〓1	3004	4313	3004	40035-1	3004	95217	3004	0573001H
3004	989〓1	3004	4315	3004	40035-2	3004	95218	3004	0573002
3004	990〓1	3004	4348	3004	40035-3	3004	95219	3004	0573003
3004	991-011222	3004	4349	3004	40036-1	3004	95222-1	3004	0573003H
3004	991-011223	3004	4450	3004	40036-2	3004	95224-1	3004	0573004
3004	99121	3004	4462	3004	40036-3	3004	95224-3	3004	0573005
3004	992〓1	3004	44660RN	3004	40038-1	3004	99204	3004	0573005-14
3004	1001	3004	4468BRN	3004	40038-2	3004	99205	3004	0573005H
3004	1002(SQUELCH)	3004	4469RED	3004	40038-3	3004	99217	3004	0573011
3004	1002-05	3004	44710RN	3004	40253	3004	99218	3004	0573012
3004	1003	3004	4471YEL	3004	40263	3004	100693	3004	0573012H
3004	1005-17	3004	44720RN	3004	40329	3004	101973	3004	0573018H
3004	1007-17	3004	4474YEL	3004	40359	3004	101974	3004	0573022
3004	1009	3004	44750RN	3004	40395	3004	103562	3004	0573022H
3004	1009(SEARS)	3004	4476BLU	3004	40490	3004	104444	3004	0573023
3004	1013-16	3004	4477PUR	3004	44067-2	3004	107274	3004	0573023H
3004	1014	3004	4510	3004	47394-2	3004	111957	3004	0573024-14
3004	1021-17	3004	4553BLU	3004	49341	3004	111959	3004	0573025
3004	1023-17	3004	4553BRN	3004	61008-8	3004	112297	3004	573029
3004	1032	3004	4553ORN	3004	64071-1	3004	116201	3004	0573036
3004	1033-5	3004	4553RED	3004	66008-2	3004	116205	3004	0573036H
3004	1033-7	3004	4553YEL	3004	67193-85	3004	116286	3004	0573056
3004	1052-17	3004	4562	3004	68504-63	3004	116685	3004	573105
3004	1060-17	3004	4563	3004	68895-13	3004	116686	3004	573114
3004	1102-17A	3004	4564	3004	69107-45	3004	116757	3004	0573114H
3004	1178-3453	3004	4565	3004	71193-2	3004	116997	3004	0573117
3004	1192	3004	4596	3004	72117	3004	116998	3004	0573117-14
3004	1241A	3004	4607	3004	72784-22	3004	117208	3004	573118
3004	1248	3004	4627	3004	72784-23	3004	117209	3004	573119
3004	1316-17	3004	4686-163-3	3004	73100-9	3004	117210	3004	573125
3004	1320	3004	4686-196-3	3004	75960CE	3004	117727	3004	0573131
3004	1329	3004	4686-81-2	3004	77052-3	3004	117728	3004	0573142
3004	1330	3004	4686-81-3	3004	77053-2	3004	120075	3004	0573152
3004	1347-17	3004	4907-976	3004	77272-0	3004	120143	3004	573153
3004	1360	3004	5612-370	3004	77272-1	3004	120144	3004	0573153H
3004	1362-17	3004	5612-75C	3004	77272-5	3004	120546	3004	573184
3004	1362-17A	3004	5612-77C	3004	77272-9	3004	120909-24.4	3004	0573187
3004	1364-17	3004	5766-25	3004	77273-2	3004	122243	3004	573328
3004	1413-160	3004	6440	3004	77273-3	3004	122244	3004	0573422
3004	1413-175	3004	6452	3004	77273-6	3004	122901	3004	0573422H
3004	1459	3004	7215-0	3004	77273-7	3004	123379	3004	0573429
3004	1466-2	3004	8000-00001-068	3004	080003	3004	123791	3004	0573742
3004	1466-2-12	3004	8000-00003-058	3004	080004	3004	123806	3004	576001
3004	1466-2-12-8	3004	8000-00003-039	3004	080043	3004	124626	3004	576005
3004	1466-3	3004	8000-00004-088	3004	080047	3004	126093	3004	650030
3004	1466-3-12	3004	8000-00004-P088	3004	080052	3004	126093-1	3004	610035
3004	1466-3-12-8	3004	8070-4	3004	80818VM	3004	126093-2	3004	610035-2
3004	1476-17	3004	8072-4	3004	081001	3004	126093-4	3004	610036
3004	1476-17-6	3004	8073-4	3004	081018	3004	126187	3004	610036-4
3004	1777-17	3004	8999-202	3004	081019	3004	126276	3004	610036-5
3004	1866-2	3004	9012HE	3004	081038	3004	126945	3004	610036-6
3004	1866-2-12	3004	9012HF	3004	081046	3004	127114	3004	610036-7
3004	1866-2-127	3004	9390	3004	081047	3004	127589	3004	610036-8
3004	1866-2L	3004	9391	3004	081048	3004	127962	3004	610040
3004	1866-2L8	3004	9400-8	3004	081049	3004	128543	3004	610040-1
3004	1866-3	3004	9400-9	3004	081050	3004	129286	3004	610040-2
3004	1866-3-12	3004	9401-7	3004	081056	3004	129802	3004	610043
3004	1866-3-127	3004	9403-7	3004	81404-4A	3004	130200-00	3004	610043-1
3004	1866-3L	3004	9403-9	3004	81500-3	3004	130200-02	3004	610043-2
3004	1866-3L8	3004	10036	3004	81501-5	3004	130400-95	3004	610043-3
3004	1919-17	3004	10037	3004	81502-2	3004	130402-36	3004	610043-6
3004	1919-17A	3004	12110-6	3004	81502-2A	3004	130403-52	3004	610043-7
3004	1954-17	3004	12110-7	3004	81502-2B	3004	137093	3004	610079
3004	1960-17	3004	12112-0	3004	81502-3B	3004	161705	3004	610079-1
3004	2012	3004	12114-8	3004	81502-4	3004	165976	3004	610080
3004	2042-17	3004	12116-4	3004	81502-4A	3004	166882	3004	610080-1
3004	2057A100-24	3004	12118-4	3004	81502-4B	3004	166883	3004	610088
3004	2057A100-8	3004	12119-1	3004	81502-9	3004	167998	3004	610088-1
3004	2057A2-148	3004	12119-2	3004	81502-9A	3004	167999	3004	610088-2
3004	2057A2-165	3004	12122-5	3004	81502-9B	3004	168907	3004	610089-3
3004	2057A2-206	3004	12122-6	3004	81503-0	3004	168954	3004	650196
3004	2057A2-241	3004	12127-7	3004	81503-0A	3004	169359	3004	650859-1
3004	2057A2-288	3004	12123-2	3004	81503-0B	3004	169360	3004	650859-2
3004	2057A2-329	3004	12124-6	3004	81503-1	3004	169361	3004	650859-3
3004	2057A2-088	3004	12126-6	3004	81503-1A	3004	169773	3004	651012
3004	2057B100-7	3004	12126-7	3004	81503-1B	3004	171017	3004	651236
3004	2057B2-135	3004	12127-2	3004	81503-4	3004	171049	3004	660059
3004	2057B2-142	3004	12195	3004	81503-4A	3004	171162-074	3004	660060
3004	2057B2-29	3004	12196	3004	81503-4B	3004	171162-075	3004	660082
3004	2057B2-32	3004	16958	3004	81503-4C	3004	171162-076	3004	660082
3004	2057B2-34	3004	17047-1	3004	81503-8A	3004	171162-080	3004	731009
3004	2057B2-4	3004	17143	3004	81503-8B	3004	171522	3004	740417
3004	2057B2-43	3004	18109	3004	81503-8C	3004	171916	3004	770525
3004	2057B2-44	3004	18530	3004	81505-8	3004	171917	3004	770524
3004	2057B2-49	3004	18601	3004	81505-8A	3004	175006-186	3004	772721
3004	2057B2-52			3004	81505-8B	3004	218503	3004	802032-2
3004	2057B2-72			3004	81505-8C	3004	222509	3004	802032-4
3004	2057B2-78			3004	81505-8X	3004	223124	3004	802033-3
3004	2057B2-83			3004	81506-9	3004	223366		
						3004	223483		

SK	Industry Standard No.	SK	Industry Standard No.	SK	Industry Standard No.	SK	Industry Standard No.	SK	Industry Standard No.
3004	802054-0	3004	2000625-32	3004	95114101-00	3005	E826	3005	RS-1550
3004	802056-0	3004	2000625-33	3004	95114102-00	3005	ET1	3005	RS-2683
3004	802189-7	3004	2000625-34	3004	95114109-00	3005	ET2	3005	RS-2684
3004	802189-8	3004	2000646-108	3004	95114135-00	3005	GA52829	3005	RS-2685
3004	802263-0	3004	2000604-9	3004	120001195	3005	GA53149	3005	RS-2686
3004	802263-1	3004	2001653-23	3004	120002013	3005	GA53242	3005	RS-2687
3004	802415-2	3004	2001653-24	3004	120002014	3005	GC1302	3005	RS-2688
3004	802439-0	3004	2001809-48	3004	120002521	3005	GC181	3005	RS-2690
3004	802560	3004	2001809-48A	3004	120002748	3005	GC182	3005	RS-2691
3004	810044A	3004	2001809-48B	3004	120004493	3005	GC360	3005	RS-2692
3004	815003	3004	2002151-19	3004	120004494	3005	GC460	3005	RS-2694
3004	815015	3004	2002153-71	3004	120004495	3005	GC461	3005	RS-2695
3004	815022	3004	2002153-78	3004	152221011	3005	GC462	3005	RS-2696
3004	815022A	3004	2002210-110	3004	1522210111	3005	GC521	3005	RS-3277
3004	815022B	3004	2002211-24	3004	1522211200	3005	GC532	3005	RS-3278
3004	815023	3004	2002211-25	3004	1522211221	3005	GC60	3005	RS-3279
3004	815023A	3004	2003073-14	3004	1522211328	3005	GC61	3005	RS-3288
3004	815023B	3004	2003073-15	3004	1522216500	3005	GET880	3005	RS-3309
3004	815024	3004	2003073-8	3004	2004305600	3005	GET881	3005	RS-3867
3004	815024A	3004	2004358-123	3004	2004305618	3005	GET882	3005	RS-3868
3004	815024B	3004	2004358-168	3004	3122005400	3005	GET887	3005	RS-3907
3004	815029	3004	2006226-14	3004	4100104753	3005	GET888	3005	RS-3913
3004	815029A	3004	2006431-49	3004	4102100220	3005	GET889	3005	RS-3929
3004	815029B	3004	2006441-113	3004	9511410100	3005	GET890	3005	RS-5104
3004	815029C	3004	2008235-109	3004	9511410200	3005	GET891	3005	RS-5105
3004	815030	3004	2047102	3004	9511410900	3005	GET892	3005	RS-5106
3004	815030A	3004	2076945-0701	3004	9511413500	3005	GET895	3005	RS-5401
3004	815030B	3004	2090924-0008	3004	16200190186	3005	GET896	3005	RS-5504
3004	815031	3004	2090924-8	3004	16201190022	3005	GET897	3005	RS-5511
3004	815031A	3004	2091244-0018	3004	16201190186	3005	G11	3005	RS-5540
3004	815031B	3004	2091578-0702	3004	16208190187	3005	GT-759R	3005	RS3281
3004	815034	3004	2091578-1	3005	A101Z	3005	GT-760R	3005	RS3287
3004	815034A	3004	2092693-1	3005	A1243	3005	GT-761R	3005	RS3914
3004	815034C	3004	2092693-8	3005	A142	3005	GT153	3005	RS3915
3004	815038B	3004	2320261	3005	A1474-3	3005	GT1604	3005	RS5302
3004	815038C	3004	2320302	3005	A1474-39	3005	GT1605	3005	RS5303
3004	815058	3004	2320302H	3005	A1488C	3005	GT1606	3005	RS5402
3004	815058A	3004	2320422-1	3005	A148809	3005	GT1607	3005	RS5403
3004	815058B	3004	2320423	3005	A169	3005	GT269	3005	R86824
3004	815058C	3004	2495014	3005	A172	3005	GT2694	3005	SC43
3004	815058X	3004	2495080	3005	A172A	3005	GT348	3005	SC44
3004	815069	3004	2495388	3005	A173	3005	GT762R	3005	SC46
3004	815069A	3004	2495388-1	3005	A173B	3005	GT764	3005	SPT-307
3004	815069B	3004	2495568-2	3005	A182	3005	GT83	3005	SPT-319
3004	815069C	3004	2497473	3005	A202	3005	GT832	3005	SPT223
3004	815070	3004	2497473-1	3005	A202A	3005	GT87	3005	SPT226
3004	815070A	3004	2855296-01	3005	A202C	3005	GT88	3005	SPT227
3004	815070B	3004	3004856	3005	A36	3005	HA15	3005	SPT228
3004	815070C	3004	4036558-P2	3005	A44	3005	HF3H	3005	SPT229
3004	815070D	3004	4037145-P1	3005	A474	3005	HF3M	3005	SPT251
3004	815074	3004	4037145-P2	3005	A55	3005	HF6H	3005	SPT252
3004	815104	3004	4037804-P1	3005	A74-3-70	3005	HF6M	3005	SPT253
3004	815114	3004	4907976	3005	A74-3-705	3005	HJ228	3005	SM-217
3004	815118	3004	5073004	3005	A74-3A9G	3005	HJ41	3005	ST-28B
3004	815120	3004	5320101	3005	A88C-70	3005	HJ606	3005	ST-28C
3004	815120A	3004	5320295	3005	A88C-705	3005	HJ600	3005	ST-37C
3004	815120B	3004	5320296	3005	A88C19G	3005	HM-00049	3005	ST-37D
3004	815120C	3004	5320485	3005	AA1	3005	HR-4	3005	SYL-105
3004	815120D	3004	5955748	3005	ACY40	3005	HR-5	3005	SYL-106
3004	815136	3004	7210027-003	3005	AF-101	3005	HR-8	3005	SYL-160
3004	815158	3004	7278422	3005	AF127	3005	HR-9	3005	SYL-160B
3004	815160	3004	7278423	3005	AF128	3005	HR4A	3005	SYL-224B
3004	815160A	3004	7279941	3005	AF223	3005	HR6	3005	SYL-2249
3004	815160B	3004	7281510	3005	ALZ10	3005	HR8A	3005	SYL-2250
3004	815160C	3004	7284751	3005	A8Y-26	3005	HR9A	3005	SYL1588
3004	815160D	3004	7294135	3005	A8Y-27	3005	HT1001510	3005	SYL1690
3004	815160E	3004	7852775-01	3005	A8Y76	3005	JR15	3005	SYL1697
3004	815160H	3004	7852776-01	3005	A8Y77	3005	M4364	3005	SYL1717
3004	815177	3004	7852778-01	3005	A8Y80	3005	M4365	3005	SYL2120
3004	815178	3004	7852779-01	3005	AT-15	3005	MA286	3005	SYL2247
3004	815179	3004	7853351-01	3005	AT-4	3005	MA287	3005	T-109
3004	815181	3004	7853352-01	3005	B103	3005	MA288	3005	T-116
3004	815181-B	3004	7853354-01	3005	B104	3005	NKT105	3005	T-152148
3004	815181A	3004	7853356-01	3005	B142B	3005	NKT106	3005	T-2439
3004	815181B	3004	7855296-01	3005	B142C	3005	NKT128	3005	T-2440
3004	815181C	3004	7855329-01	3005	B290	3005	NKT129	3005	T-2441
3004	815181D	3004	7910071-01	3005	B291	3005	NKT132	3005	T-46
3004	815195	3004	7910588-01	3005	B292	3005	NKT141	3005	T-47
3004	815196	3004	7910589-01	3005	B292A	3005	NKT142	3005	T-48
3004	815228A	3004	7910801-01	3005	B314	3005	NKT143	3005	T-52148Z
3004	910050-2	3004	8020322	3005	B378	3005	NKT144	3005	T-52149
3004	910062-1	3004	8020324	3005	B387	3005	NKT162	3005	T-52149Z
3004	910070-6	3004	8020333	3005	B392	3005	NKT163	3005	T-78
3004	910094-4	3004	8020334	3005	B393	3005	NKT163/25	3005	T-Q5020
3004	980144	3004	8020540	3005	B394	3005	NKT16325	3005	TO-101
3004	980148	3004	8020560	3005	B395	3005	NKT164	3005	TO-102
3004	980149	3004	8021897	3005	B396	3005	NKT164/25	3005	T1003
3004	980153	3004	8021898	3005	B401	3005	NKT16425	3005	T1011
3004	980508	3004	8022630	3005	B402	3005	NKT202	3005	T1012
3004	980510	3004	8022631	3005	B403	3005	NKT206	3005	T11618
3004	980511	3004	8112023	3005	B408	3005	NKT207	3005	T1232
3004	980836	3004	8112027	3005	B416	3005	NKT221	3005	T1233
3004	980877	3004	8112071	3005	B417	3005	NKT222281	3005	T1289
3004	980960	3004	8112090	3005	BA6	3005	NKT222282	3005	T1291
3004	980961	3004	8112146	3005	BA6A	3005	NKT243	3005	T1298
3004	981148	3004	8112161	3005	BE6	3005	NKT261	3005	T1299
3004	981672	3004	8112162	3005	BE6A	3005	NKT262	3005	T1305
3004	981673	3004	8945077-12	3005	CK13	3005	NKT263	3005	T1306
3004	981674	3004	9100502	3005	CK13A	3005	NKT264	3005	T1312
3004	981675	3004	9100621	3005	CK14	3005	NKT42	3005	T1322
3004	982151	3004	9100706	3005	CK14A	3005	NKT43	3005	T1326
3004	982152	3004	9100944	3005	CK16	3005	NKT62	3005	T1327
3004	982244	3004	09301020	3005	CK16A	3005	NKT63	3005	T1328
3004	982283	3004	11220018/7611	3005	CK17	3005	NKT64	3005	T1334
3004	982284	3004	12090924-4	3005	CK17A	3005	NKT72	3005	T1346
3004	982285	3004	13020000	3005	CK25	3005	NKT73	3005	T1454
3004	982375	3004	13020002	3005	CK25A	3005	NKT74	3005	T1459
3004	982531	3004	13040095	3005	CK26	3005	OC-130	3005	T1474
3004	982532	3004	13040236	3005	CK26A	3005	OC-140	3005	T1510
3004	983405	3004	13040352	3005	CK27	3005	OC-410	3005	T152148
3004	983406	3004	23114021	3005	CK27A	3005	OC-44	3005	T1877
3004	983407	3004	23114061	3005	CK661	3005	OC-45	3005	T1902
3004	983408	3004	23114211	3005	CK662	3005	OC-46	3005	T1904
3004	983409	3004	25114101	3005	CK759	3005	OC-47	3005	T1961
3004	984221	3004	25114102	3005	CK760	3005	PADT23	3005	T2038
3004	984228	3004	25114103	3005	CK761	3005	PT-530	3005	T2039
3004	985468A	3004	25114104	3005	CK766	3005	PT-530A	3005	T2040
3004	985469A	3004	27125330	3005	CK768	3005	PTC102	3005	T2091
3004	985470A	3004	27125340	3005	CK776	3005	Q-1	3005	T2122
3004	986302	3004	27125350	3005	CK776A	3005	Q-1A	3005	T2159
3004	986766	3004	31220054-00	3005	D019	3005	Q-4	3005	T2172
3004	995003	3004	62041528	3005	D078	3005	R-119	3005	T2173
3004	1471101-15	3004	62041536	3005	E2412	3005	R-163	3005	T2256
3004	1473378-1	3004	62045930	3005	E2451	3005	R-186	3005	T2257
3004	1473598-1	3004	62052570	3005	ED52	3005	R-227	3005	T2258
3004	1473598-2	3004	62054536	3005	EG53	3005	R-244	3005	T2259
3004	1956016	3004	80203350	3005	ED54B	3005	R-424	3005	T2260
3004	1960643	3004	80205400	3005	EK159	3005	R-425	3005	T2261
3004	1961897	3004	80205400	3005	EO105	3005	R-488	3005	T3005
3004	2000287-28	3004	80222631U	3005	EO65	3005	R-506	3005	T50944
3004	2000625-31	3004	80236400	3005	EO66	3005	R258	3005	T52147
		3004	80243900	3005	EO67	3005	R31	3005	T52147Z
		3004	92005600	3005	EO68	3005	REN100	3005	TA-1575B
				3005	E825	3005	RS-1539	3005	TA-1655B

SK	Industry Standard No.	SK	Industry Standard No.	SK	Industry Standard No.	SK	Industry Standard No.	SK	Industry Standard No.
3005	TA-1704	3005	2N1176A	3005	2N578	3005	28A135GN	3005	28A148K
3005	TA-1763	3005	2N1176B	3005	2N579	3005	28A135H	3005	28A148L
3005	TA-1763A	3005	2N1185	3005	2N580	3005	28A135J	3005	28A148M
3005	TA-1778	3005	2N1186	3005	2N581	3005	28A135K	3005	28A148OR
3005	TA-1782	3005	2N1187	3005	2N583	3005	28A135L	3005	28A148R
3005	TA-1783	3005	2N1188	3005	2N586	3005	28A135M	3005	28A148X
3005	TG28A201	3005	2N123	3005	2N592	3005	28A1350R	3005	28A148Y
3005	TG28A201C	3005	2N1266	3005	2N593	3005	28A135R	3005	28A149
3005	TI-563	3005	2N1280	3005	2N597	3005	28A135X	3005	28A149A
3005	TIA03	3005	2N1281	3005	2N598	3005	28A135Y	3005	28A149B
3005	TIA05	3005	2N1282	3005	2N599	3005	28A137	3005	28A149C
3005	TIXA-03	3005	2N1284	3005	2N600	3005	28A137A	3005	28A149D
3005	TIXA-04	3005	2N1316	3005	2N614	3005	28A137B	3005	28A149E
3005	TIXA-05	3005	2N1317	3005	2N615	3005	28A137C	3005	28A149F
3005	TIXA01	3005	2N1318	3005	2N616	3005	28A137D	3005	28A149G
3005	TIXA02	3005	2N1319	3005	2N617	3005	28A137E	3005	28A149GN
3005	TNJ7227B	3005	2N1343	3005	2N674	3005	28A137F	3005	28A149H
3005	T0101	3005	2N1344	3005	2N801	3005	28A137G	3005	28A149J
3005	T0102	3005	2N1345	3005	2N802	3005	28A1370N	3005	28A149L
3005	TR-C44	3005	2N1346	3005	2N803	3005	28A137H	3005	28A149OR
3005	TR-C44A	3005	2N1347	3005	2N804	3005	28A137J	3005	28A149R
3005	TR-C45A	3005	2N1349	3005	2N805	3005	28A137K	3005	28A149X
3005	TR310015	3005	2N135	3005	2N806	3005	28A137L	3005	28A149Y
3005	TR310161	3005	2N1350	3005	2N807	3005	28A137M	3005	28A14A
3005	T8-601	3005	2N1351	3005	2N808	3005	28A1370R	3005	28A14B
3005	T8-602	3005	2N1354	3005	2N809	3005	28A137R	3005	28A14C
3005	T8672A	3005	2N1355	3005	2N810	3005	28A137X	3005	28A14D
3005	T8672B	3005	2N1356	3005	2N811	3005	28A137Y	3005	28A14E
3005	T867SA	3005	2N1357	3005	2N812	3005	28A13A	3005	28A14G
3005	T8673B	3005	2N136	3005	2N813	3005	28A13B	3005	28A14L
3005	TV24152	3005	2N1361A	3005	2N814	3005	28A13C	3005	28A14M
3005	TV2434	3005	2N137	3005	2N83	3005	28A13D	3005	28A14OR
3005	TV4152	3005	2N139	3005	2N993	3005	28A13G	3005	28A14R
3005	TV8-28A171	3005	2N1393	3005	2NJ5A	3005	28A13L	3005	28A14X
3005	V10/18	3005	2N140	3005	2NJ6	3005	28A13M	3005	28A14Y
3005	V10/28	3005	2N1469	3005	2NJ8A	3005	28A130R	3005	28A15
3005	V10/28J	3005	2N1470	3005	28111	3005	28A13R	3005	28A15-6
3005	V6/2R	3005	2N1471	3005	2812	3005	28A13X	3005	28A151
3005	V6/2RC	3005	2N1570	3005	2813	3005	28A13Y	3005	28A151A
3005	V6/4R	3005	2N1581	3005	28155	3005	28A14	3005	28A151B
3005	V6/4RC	3005	2N1583	3005	28159	3005	28A142	3005	28A151C
3005	V6/8R	3005	2N1584	3005	28160	3005	28A142A	3005	28A151D
3005	VPQ-2745F	3005	2N1664	3005	28167	3005	28A142B	3005	28A151E
3005	VP8-2745J	3005	2N1670	3005	28174	3005	28A142C	3005	28A151F
3005	VFY-2745E	3005	2N1673	3005	28178	3005	28A142D	3005	28A151G
3005	VSP2745	3005	2N1683	3005	2825	3005	28A142E	3005	28A151GN
3005	W1	3005	2N1684	3005	2830	3005	28A142F	3005	28A151H
3005	WTV12MC	3005	2N1729	3005	2831	3005	28A142G	3005	28A151J
3005	WTV20MC	3005	2N1731	3005	2832	3005	28A1420N	3005	28A151K
3005	WTV6MC	3005	2N1743	3005	28420N	3005	28A142H	3005	28A151L
3005	WTVAT6A	3005	2N1744	3005	2845	3005	28A142J	3005	28A151M
3005	WTVB5	3005	2N1754	3005	2849	3005	28A142K	3005	28A151OR
3005	WTVBA6	3005	2N1782	3005	2851	3005	28A142L	3005	28A151R
3005	WTVBB6	3005	2N1784	3005	2852	3005	28A142M	3005	28A151X
3005	WTVBMC	3005	2N1853	3005	2853	3005	28A1420R	3005	28A151Y
3005	X78	3005	2N1940	3005	2860	3005	28A142R	3005	28A152
3005	XA101	3005	2N1954	3005	2891	3005	28A142Y	3005	28A152A
3005	XA102	3005	2N1955	3005	28A103	3005	28A144	3005	28A152B
3005	XA103	3005	2N1956	3005	28A103A	3005	28A144A	3005	28A152C
3005	XA104	3005	2N1957	3005	28A103B	3005	28A144B	3005	28A152D
3005	XA111	3005	2N1997	3005	28A103C	3005	28A144C	3005	28A152E
3005	XA112	3005	2N1998	3005	28A103CA	3005	28A144D	3005	28A152F
3005	XB10	3005	2N1999	3005	28A103CAK	3005	28A144E	3005	28A152G
3005	XB9	3005	2N2171	3005	28A103CG	3005	28A144F	3005	28A152GN
3005	001-01202-1	3005	2N2172	3005	28A103D	3005	28A144G	3005	28A152H
3005	001-01203-1	3005	2N218	3005	28A103DA	3005	28A144GN	3005	28A152J
3005	1-21-100	3005	2N219	3005	28A103E	3005	28A144H	3005	28A152K
3005	1-21-102	3005	2N2209	3005	28A103F	3005	28A144J	3005	28A152L
3005	1-21-104	3005	2N249	3005	28A103G	3005	28A144K	3005	28A152M
3005	1-21-105	3005	2N252	3005	28A103K	3005	28A144L	3005	28A152OR
3005	1-21-128	3005	2N267	3005	28A103L	3005	28A144M	3005	28A152R
3005	1-21-161	3005	2N269	3005	28A103M	3005	28A1440R	3005	28A152X
3005	1-21-162	3005	2N27	3005	28A103OR	3005	28A144R	3005	28A152Y
3005	1-21-179	3005	2N271	3005	28A103R	3005	28A144X	3005	28A15A
3005	1-21-180	3005	2N271A	3005	28A103Y	3005	28A144Y	3005	28A15BK
3005	1-21-186	3005	2N30	3005	2813	3005	28A145	3005	28A15C
3005	1-21-234	3005	2N31	3005	28A130	3005	28A145A	3005	28A15D
3005	1-21-235	3005	2N311	3005	28A130A	3005	28A145B	3005	28A15E
3005	1-21-236	3005	2N315	3005	28A130B	3005	28A145C	3005	28A15F
3005	1-21-240	3005	2N315B	3005	28A130C	3005	28A145D	3005	28A15G
3005	1-21-241	3005	2N316	3005	28A130D	3005	28A145E	3005	28A15H
3005	1-21-254	3005	2N317	3005	28A130F	3005	28A145G	3005	28A15K
3005	1-21-273	3005	2N317A	3005	28A130G	3005	28A145GN	3005	28A15L
3005	1-21-275	3005	2N327	3005	28A130GN	3005	28A145K	3005	28A15M
3005	1-21-289	3005	2N394	3005	28A130H	3005	28A145M	3005	28A15OR
3005	1-21-73	3005	2N394A	3005	28A130J	3005	28A145OR	3005	28A15R
3005	1-21-74	3005	2N395	3005	28A130K	3005	28A145R	3005	28A15U
3005	1-21-75	3005	2N396	3005	28A130L	3005	28A145X	3005	28A15V
3005	1-21-76	3005	2N396A	3005	28A130M	3005	28A145Y	3005	28A15Y
3005	1-21-78	3005	2N397	3005	28A130R	3005	28A146	3005	28A16
3005	1-21-83	3005	2N409	3005	28A130X	3005	28A146A	3005	28A160
3005	1-21-91	3005	2N410	3005	28A130Y	3005	28A146B	3005	28A160A
3005	1-21-92	3005	2N411	3005	28A131	3005	28A146C	3005	28A160B
3005	1-21-93	3005	2N412	3005	28A131A	3005	28A146D	3005	28A160C
3005	1N765-1	3005	2N413	3005	28A131B	3005	28A146E	3005	28A160D
3005	2B	3005	2N413A	3005	28A131C	3005	28A146F	3005	28A160F
3005	2C	3005	2N414	3005	28A131D	3005	28A146G	3005	28A160GN
3005	2D	3005	2N414A	3005	28A131E	3005	28A146GN	3005	28A160H
3005	201024	3005	2N414B	3005	28A131F	3005	28A146H	3005	28A160J
3005	201025	3005	2N414C	3005	28A131G	3005	28A146K	3005	28A160K
3005	201026	3005	2N415	3005	28A131GN	3005	28A146L	3005	28A160L
3005	20138	3005	2N415A	3005	28A131H	3005	28A146M	3005	28A160M
3005	20139	3005	2N416	3005	28A131J	3005	28A146OR	3005	28A160OR
3005	20140	3005	2N417	3005	28A131K	3005	28A146X	3005	28A160R
3005	20319	3005	2N425	3005	28A131L	3005	28A146Y	3005	28A160X
3005	20320	3005	2N426	3005	28A131M	3005	28A147	3005	28A160Y
3005	20345	3005	2N427	3005	28A1310R	3005	28A147A	3005	28A167
3005	20371	3005	2N428	3005	28A131R	3005	28A147B	3005	28A167B
3005	20371A	3005	2N428A	3005	28A131X	3005	28A147C	3005	28A167C
3005	20374	3005	2N450	3005	28A131Y	3005	28A147D	3005	28A167D
3005	20383	3005	2N481	3005	28A132	3005	28A147E	3005	28A167E
3005	20394	3005	2N482	3005	28A132A	3005	28A147F	3005	28A167F
3005	20396	3005	2N483	3005	28A132B	3005	28A147G	3005	28A167G
3005	20397	3005	2N484	3005	28A132C	3005	28A147H	3005	28A167GN
3005	20524	3005	2N485	3005	28A132D	3005	28A147J	3005	28A167H
3005	20525	3005	2N486	3005	28A132F	3005	28A147K	3005	28A167J
3005	20526	3005	2N486B	3005	28A132G	3005	28A147L	3005	28A167K
3005	20527	3005	2N503	3005	28A132H	3005	28A147M	3005	28A167L
3005	20577	3005	2N505	3005	28A132J	3005	28A147OR	3005	28A167M
3005	20603	3005	2N518	3005	28A132K	3005	28A147R	3005	28A167OR
3005	20604	3005	2N519	3005	28A132L	3005	28A147X	3005	28A167R
3005	2N1107	3005	2N519A	3005	28A132M	3005	28A147Y	3005	28A167X
3005	2N1113	3005	2N520	3005	28A1320R	3005	28A148	3005	28A167Y
3005	2N1115	3005	2N520A	3005	28A132X	3005	28A148A	3005	28A168
3005	2N1115A	3005	2N521	3005	28A132Y	3005	28A148B	3005	28A168A
3005	2N111A	3005	2N521A	3005	28A135	3005	28A148C	3005	28A168B
3005	2N112	3005	2N522A	3005	28A135A	3005	28A148D	3005	28A168C
3005	2N1122	3005	2N525A	3005	28A135B	3005	28A148E	3005	28A168D
3005	2N1122A	3005	2N529	3005	28A135C	3005	28A148F	3005	28A168E
3005	2N112A	3005	2N530	3005	28A135D	3005	28A148G		
3005	2N114	3005	2N531	3005	28A135E	3005	28A148GN		
3005	2N1171	3005	2N532	3005	28A135F	3005	28A148J		
3005	2N1176	3005	2N533	3005	28A135G				

SK	Industry Standard No.	SK	Industry Standard No.	SK	Industry Standard No.	SK	Industry Standard No.	SK	Industry Standard No.
3005	28A168F	3005	28A17B	3005	28A19B	3005	28A211B	3005	28A277X
3005	28A168G	3005	28A17C	3005	28A198A	3005	28A211C	3005	28A277Y
3005	28A168GN	3005	28A17D	3005	28A198B	3005	28A211D	3005	28A278
3005	28A168H	3005	28A17E	3005	28A198C	3005	28A211E	3005	28A278A
3005	28A168J	3005	28A17F	3005	28A198D	3005	28A211F	3005	28A278B
3005	28A168K	3005	28A17G	3005	28A198E	3005	28A211G	3005	28A278C
3005	28A168L	3005	28A17H	3005	28A198F	3005	28A211GN	3005	28A278D
3005	28A168M	3005	28A17M	3005	28A198G	3005	28A211H	3005	28A278E
3005	28A168OR	3005	28A170R	3005	28A198GN	3005	28A211J	3005	28A278F
3005	28A168R	3005	28A17R	3005	28A198H	3005	28A211K	3005	28A278G
3005	28A168X	3005	28A17X	3005	28A198J	3005	28A211L	3005	28A278GN
3005	28A168Y	3005	28A17Y	3005	28A198K	3005	28A211M	3005	28A278H
3005	28A169	3005	28A18	3005	28A198L	3005	28A211OR	3005	28A278J
3005	28A169A	3005	28A180	3005	28A198M	3005	28A211R	3005	28A278K
3005	28A169B	3005	28A180A	3005	28A198OR	3005	28A211X	3005	28A278L
3005	28A169C	3005	28A180B	3005	28A198R	3005	28A211Y	3005	28A278M
3005	28A169D	3005	28A180C	3005	28A198X	3005	28A212	3005	28A278OR
3005	28A169E	3005	28A180D	3005	28A198Y	3005	28A212A	3005	28A278R
3005	28A169F	3005	28A180E	3005	28A201	3005	28A212B	3005	28A278X
3005	28A169G	3005	28A180F	3005	28A201-0	3005	28A212C	3005	28A278Y
3005	28A169GN	3005	28A180G	3005	28A201-0	3005	28A212D	3005	28A279
3005	28A169H	3005	28A180GN	3005	28A201B	3005	28A212E	3005	28A279A
3005	28A169J	3005	28A180H	3005	28A201CL	3005	28A212F	3005	28A279B
3005	28A169K	3005	28A180J	3005	28A201D	3005	28A212G	3005	28A279C
3005	28A169L	3005	28A180K	3005	28A201E	3005	28A212GN	3005	28A279D
3005	28A169M	3005	28A180L	3005	28A201F	3005	28A212H	3005	28A279E
3005	28A169OR	3005	28A180M	3005	28A201G	3005	28A212J	3005	28A279F
3005	28A169R	3005	28A180OR	3005	28A201GN	3005	28A212K	3005	28A279G
3005	28A169X	3005	28A180R	3005	28A201H	3005	28A212L	3005	28A279GN
3005	28A169Y	3005	28A180X	3005	28A201J	3005	28A212M	3005	28A279H
3005	28A16A	3005	28A180Y	3005	28A201K	3005	28A212OR	3005	28A279J
3005	28A16B	3005	28A181	3005	28A201L	3005	28A212R	3005	28A279K
3005	28A16C	3005	28A181A	3005	28A201M	3005	28A212X	3005	28A279L
3005	28A16D	3005	28A181B	3005	28A201R	3005	28A212Y	3005	28A279OR
3005	28A16E	3005	28A181C	3005	28A201OR	3005	28A217	3005	28A279R
3005	28A16F	3005	28A181D	3005	28A201R	3005	28A217A	3005	28A279X
3005	28A16L	3005	28A181E	3005	28A201TV	3005	28A217B	3005	28A279Y
3005	28A16M	3005	28A181F	3005	28A201TVO	3005	28A217C	3005	28A28
3005	28A16OR	3005	28A181G	3005	28A201X	3005	28A217D	3005	28A282
3005	28A16R	3005	28A181GN	3005	28A201Y	3005	28A217E	3005	28A282A
3005	28A16X	3005	28A181H	3005	28A202	3005	28A217F	3005	28A282B
3005	28A16Y	3005	28A181J	3005	28A202A	3005	28A217G	3005	28A282C
3005	28A17	3005	28A181K	3005	28A202AP	3005	28A217GN	3005	28A282D
3005	28A170	3005	28A181L	3005	28A202B	3005	28A217H	3005	28A282E
3005	28A170A	3005	28A181M	3005	28A202C	3005	28A217J	3005	28A282F
3005	28A170B	3005	28A181OR	3005	28A202D	3005	28A217K	3005	28A282G
3005	28A170C	3005	28A181R	3005	28A202D-4	3005	28A217L	3005	28A282GN
3005	28A170D	3005	28A181X	3005	28A202E	3005	28A217M	3005	28A282H
3005	28A170E	3005	28A181Y	3005	28A202F	3005	28A217OR	3005	28A282J
3005	28A170F	3005	28A182	3005	28A202G	3005	28A217R	3005	28A282K
3005	28A170G	3005	28A182A	3005	28A202GN	3005	28A217X	3005	28A282L
3005	28A170GN	3005	28A182B	3005	28A202H	3005	28A217Y	3005	28A282M
3005	28A170H	3005	28A182C	3005	28A202J	3005	28A248	3005	28A2820R
3005	28A170J	3005	28A182D	3005	28A202K	3005	28A248A	3005	28A282R
3005	28A170K	3005	28A182F	3005	28A202L	3005	28A248B	3005	28A282X
3005	28A170L	3005	28A182G	3005	28A202M	3005	28A248C	3005	28A282Y
3005	28A170M	3005	28A182GN	3005	28A2020R	3005	28A248D	3005	28A283
3005	28A170OR	3005	28A182H	3005	28A202R	3005	28A248E	3005	28A283A
3005	28A170R	3005	28A182J	3005	28A202Y	3005	28A248F	3005	28A283B
3005	28A170X	3005	28A182K	3005	28A203AA	3005	28A248G	3005	28A283C
3005	28A170Y	3005	28A182L	3005	28A207	3005	28A248GN	3005	28A283D
3005	28A171	3005	28A182M	3005	28A207A	3005	28A248H	3005	28A283B
3005	28A171A	3005	28A1820R	3005	28A207B	3005	28A248J	3005	28A283F
3005	28A171B	3005	28A182R	3005	28A207C	3005	28A248K	3005	28A283G
3005	28A171C	3005	28A182X	3005	28A207D	3005	28A248L	3005	28A283GN
3005	28A171D	3005	28A182Y	3005	28A207E	3005	28A248M	3005	28A283H
3005	28A171F	3005	28A183	3005	28A207F	3005	28A248OR	3005	28A283J
3005	28A171G	3005	28A183A	3005	28A207G	3005	28A248R	3005	28A283K
3005	28A171GN	3005	28A183B	3005	28A207GN	3005	28A248X	3005	28A283L
3005	28A171H	3005	28A183C	3005	28A207H	3005	28A248Y	3005	28A283M
3005	28A171J	3005	28A183D	3005	28A207J	3005	28A254	3005	28A2830R
3005	28A171K	3005	28A183E	3005	28A207K	3005	28A254A	3005	28A283R
3005	28A171L	3005	28A183F	3005	28A207L	3005	28A254C	3005	28A283X
3005	28A171M	3005	28A183G	3005	28A207M	3005	28A254D	3005	28A283Y
3005	28A171OR	3005	28A183GN	3005	28A207OR	3005	28A254E	3005	28A284
3005	28A171R	3005	28A183H	3005	28A207R	3005	28A254F	3005	28A284A
3005	28A171X	3005	28A183J	3005	28A207X	3005	28A254G	3005	28A284B
3005	28A171Y	3005	28A183K	3005	28A207Y	3005	28A254H	3005	28A284C
3005	28A172	3005	28A183L	3005	28A208	3005	28A254J	3005	28A284D
3005	28A172A	3005	28A183M	3005	28A208A	3005	28A254K	3005	28A284E
3005	28A172B	3005	28A1830R	3005	28A208B	3005	28A254L	3005	28A284F
3005	28A172C	3005	28A183R	3005	28A208C	3005	28A254M	3005	28A284G
3005	28A172D	3005	28A183X	3005	28A208D	3005	28A254OR	3005	28A284GN
3005	28A172E	3005	28A183Y	3005	28A208E	3005	28A254R	3005	28A284J
3005	28A172F	3005	28A188	3005	28A208F	3005	28A254X	3005	28A284K
3005	28A172G	3005	28A188A	3005	28A208G	3005	28A254Y	3005	28A284L
3005	28A172GN	3005	28A188B	3005	28A208GN	3005	28A255	3005	28A284M
3005	28A172H	3005	28A188C	3005	28A208H	3005	28A255A	3005	28A284OR
3005	28A172J	3005	28A188D	3005	28A208J	3005	28A255B	3005	28A284R
3005	28A172K	3005	28A188E	3005	28A208K	3005	28A255C	3005	28A284X
3005	28A172L	3005	28A188F	3005	28A208L	3005	28A255D	3005	28A284Y
3005	28A172M	3005	28A188G	3005	28A208M	3005	28A255E	3005	28A28A
3005	28A172OR	3005	28A188GN	3005	28A208OR	3005	28A255F	3005	28A28B
3005	28A172R	3005	28A188H	3005	28A208R	3005	28A255G	3005	28A28C
3005	28A172X	3005	28A188J	3005	28A208X	3005	28A255GN	3005	28A28D
3005	28A172Y	3005	28A188K	3005	28A208Y	3005	28A255H	3005	28A28E
3005	28A173	3005	28A188L	3005	28A209	3005	28A255J	3005	28A28F
3005	28A173A	3005	28A188M	3005	28A209A	3005	28A255K	3005	28A28G
3005	28A173B	3005	28A188OR	3005	28A209B	3005	28A255L	3005	28A28H
3005	28A173C	3005	28A188R	3005	28A209C	3005	28A255M	3005	28A28L
3005	28A173D	3005	28A188X	3005	28A209D	3005	28A255OR	3005	28A28M
3005	28A173E	3005	28A188Y	3005	28A209F	3005	28A255X	3005	28A280R
3005	28A173F	3005	28A189	3005	28A209G	3005	28A255Y	3005	28A28X
3005	28A173G	3005	28A189A	3005	28A209GN	3005	28A26	3005	28A28Y
3005	28A173GN	3005	28A189B	3005	28A209H	3005	28A26A	3005	28A296
3005	28A173H	3005	28A189C	3005	28A209J	3005	28A26B	3005	28A296A
3005	28A173J	3005	28A189D	3005	28A209K	3005	28A26C	3005	28A296B
3005	28A173K	3005	28A189E	3005	28A209L	3005	28A26D	3005	28A296C
3005	28A173L	3005	28A189F	3005	28A209M	3005	28A26E	3005	28A296D
3005	28A173M	3005	28A189G	3005	28A209OR	3005	28A26F	3005	28A296E
3005	28A1730R	3005	28A189GN	3005	28A209R	3005	28A26G	3005	28A296F
3005	28A173R	3005	28A189H	3005	28A209X	3005	28A26L	3005	28A296G
3005	28A173X	3005	28A189J	3005	28A209Y	3005	28A26M	3005	28A296GN
3005	28A173Y	3005	28A189K	3005	28A210	3005	28A26OR	3005	28A296H
3005	28A174	3005	28A189L	3005	28A210A	3005	28A26R	3005	28A296J
3005	28A174A	3005	28A189M	3005	28A210B	3005	28A26X	3005	28A296K
3005	28A174B	3005	28A1890R	3005	28A210C	3005	28A26Y	3005	28A296L
3005	28A174C	3005	28A189R	3005	28A210D	3005	28A277A	3005	28A296M
3005	28A174D	3005	28A189X	3005	28A210E	3005	28A277B	3005	28A296OR
3005	28A174E	3005	28A189Y	3005	28A210F	3005	28A277C	3005	28A296R
3005	28A174F	3005	28A18A	3005	28A210G	3005	28A277D	3005	28A296X
3005	28A174G	3005	28A18B	3005	28A210GN	3005	28A277E	3005	28A296Y
3005	28A174GN	3005	28A18C	3005	28A210H	3005	28A277F	3005	28A297A
3005	28A174H	3005	28A18D	3005	28A210J	3005	28A277G	3005	28A297B
3005	28A174J	3005	28A18E	3005	28A210K	3005	28A277H	3005	28A297C
3005	28A174K	3005	28A18F	3005	28A210L	3005	28A277J	3005	28A297D
3005	28A174L	3005	28A18G	3005	28A210M	3005	28A277K	3005	28A297E
3005	28A174M	3005	28A18L	3005	28A210OR	3005	28A277L	3005	28A297F
3005	28A174OR	3005	28A18M	3005	28A210R	3005	28A277M	3005	28A297G
3005	28A174R	3005	28A18OR	3005	28A210X	3005	28A277OR	3005	28A297H
3005	28A174X	3005	28A18R	3005	28A210Y			3005	28A297J
3005	28A174Y	3005	28A18X	3005	28A211			3005	28A297K
3005	28A17A	3005	28A18Y	3005	28A211A			3005	28A297L
								3005	28A297M
								3005	28A297OR
								3005	28A297R

SK	Industry Standard No.	SK	Industry Standard No.	SK	Industry Standard No.	SK	Industry Standard No.	SK	Industry Standard No.	SK	Industry Standard No.
3005	28A297X	3005	28A39L	3005	28A50A	3005	2V633	3005	48-134471	3005	48-134472
3005	28A297Y	3005	28A39M	3005	28A50B	3005	4-74-3A7-1	3005	48-134472	3005	74-3
3005	28A30	3005	28A390R	3005	28A50C	3005	4-88C17-1	3005	48-134473	3005	74-3-70
3005	28A305	3005	28A39R	3005	28A50D	3005	4JD1A17	3005	48-134474	3005	74-3-70-12
3005	28A305A	3005	28A39X	3005	28A50E	3005	4JX1A520D	3005	48-134475	3005	74-3-70-12-7
3005	28A305B	3005	28A39Y	3005	28A50F	3005	4JX1A520E	3005	48-134476	3005	74-30
3005	28A305C	3005	28A40	3005	28A50G	3005	4JX1C850A	3005	48-134477	3005	74-31
3005	28A305D	3005	28A406	3005	28A50L	3005	4JX2A60	3005	48-134494	3005	74-32
3005	28A305E	3005	28A406A	3005	28A50M	3005	09-300005	3005	48-134495	3005	74-33
3005	28A305F	3005	28A406B	3005	28A50R	3005	09-300011	3005	48-134496	3005	74-34
3005	28A305G	3005	28A406C	3005	28A50X	3005	09-300017	3005	48-134499	3005	74-35
3005	28A305GN	3005	28A406D	3005	28A50Y	3005	9-5120A	3005	48-134500	3005	74-36
3005	28A305H	3005	28A406E	3005	28A51	3005	011-E01	3005	48-134507	3005	74-37
3005	28A305J	3005	28A406F	3005	28A51A	3005	12-1-100	3005	48-134507	3005	74-38
3005	28A305K	3005	28A406G	3005	28A51B	3005	12-1-102	3005	48-134509	3005	74-39
3005	28A305L	3005	28A406GN	3005	28A51C	3005	12-1-103	3005	48-134510	3005	74-3A
3005	28A305M	3005	28A406H	3005	28A51D	3005	12-1-104	3005	48-134512	3005	74-3A0
3005	28A305OR	3005	28A406J	3005	28A51F	3005	12-1-105	3005	48-134535	3005	74-3A1
3005	28A305R	3005	28A406K	3005	28A51G	3005	12-1-128	3005	48-134538	3005	74-3A19
3005	28A305X	3005	28A406L	3005	28A51L	3005	12-1-161	3005	48-134539	3005	74-3A2
3005	28A305Y	3005	28A406M	3005	28A51M	3005	12-1-162	3005	48-134540	3005	74-3A21
3005	28A30A	3005	28A406OR	3005	28A51OR	3005	12-1-179	3005	48-134541	3005	74-3A3
3005	28A30B	3005	28A406X	3005	28A51R	3005	12-1-180	3005	48-134542	3005	74-3A3P
3005	28A30C	3005	28A406Y	3005	28A51X	3005	12-1-186	3005	48-134543	3005	74-3A4
3005	28A30D	3005	28A407	3005	28A51Y	3005	12-1-234	3005	48-134544	3005	74-3A4-7
3005	28A30E	3005	28A407A	3005	28A54	3005	12-1-235	3005	48-134553	3005	74-3A5
3005	28A30P	3005	28A407B	3005	28A54A	3005	12-1-236	3005	48-134554	3005	74-3A5L
3005	28A30G	3005	28A407C	3005	28A54B	3005	12-1-240	3005	48-134555	3005	74-3A6
3005	28A30L	3005	28A407D	3005	28A54C	3005	12-1-241	3005	48-134556	3005	74-3A6-3
3005	28A30M	3005	28A407E	3005	28A54D	3005	12-1-254	3005	48-134557	3005	74-3A7
3005	28A30X	3005	28A407F	3005	28A54E	3005	12-1-273	3005	48-134558	3005	74-3A7-1
3005	28A30Y	3005	28A407G	3005	28A54F	3005	12-1-275	3005	48-134559	3005	74-3A8
3005	28A31	3005	28A407GN	3005	28A54G	3005	12-1-289	3005	48-134562	3005	74-3A82
3005	28A31A	3005	28A407H	3005	28A54L	3005	12-1-73	3005	48-134563	3005	74-3A9
3005	28A31B	3005	28A407J	3005	28A54M	3005	12-1-74	3005	48-134564	3005	74-3A9G
3005	28A31C	3005	28A407K	3005	28A54OR	3005	12-1-75	3005	48-134565	3005	77
3005	28A31D	3005	28A407L	3005	28A54R	3005	12-1-76	3005	48-134567	3005	78BLK
3005	28A31E	3005	28A407M	3005	28A54X	3005	12-1-78	3005	48-134603	3005	78GRN
3005	28A31F	3005	28A407OR	3005	28A54Y	3005	12-1-83	3005	48-134604	3005	78RED
3005	28A31G	3005	28A407R	3005	28A55	3005	12-1-91	3005	48-134610	3005	78YEL
3005	28A31L	3005	28A407X	3005	28A55A	3005	12-1-92	3005	48-134625	3005	86-133-2
3005	28A31M	3005	28A407Y	3005	28A55B	3005	12-1-93	3005	48-134626	3005	86-21-2
3005	28A31OR	3005	28A40A	3005	28A55C	3005	012-E02	3005	48-134631	3005	86-22-2
3005	28A31X	3005	28A40B	3005	28A55D	3005	12A6240	3005	48-134636		
3005	28A31Y	3005	28A40C	3005	28A55E	3005	13-18944-2	3005	48-134637		
3005	28A32	3005	28A40D	3005	28A55F	3005	13-50484-1	3005	48-134641		
3005	28A32A	3005	28A40E	3005	28A55G	3005	13-50486-1	3005	48-134655		
3005	28A32B	3005	28A40F	3005	28A55L	3005	13-50631-1	3005	48-134656		
3005	28A32C	3005	28A40G	3005	28A55M	3005	13-50944-1	3005	48-134657		
3005	28A32E	3005	28A40L	3005	28A55OR	3005	019-003324	3005	48-134956		
3005	28A32F	3005	28A40M	3005	28A55R	3005	019-003342	3005	48-137199		
3005	28A32L	3005	28A40R	3005	28A55X	3005	019-003343	3005	48-63029A92		
3005	28A32M	3005	28A40X	3005	28A55Y	3005	19-020-033	3005	48-63075A76		
3005	28A32X	3005	28A40Y	3005	28A61A	3005	19A115301-P1	3005	48-63077A03		
3005	28A32Y	3005	28A41	3005	28A61C	3005	19A115301-P2	3005	48-63078A59		
3005	28A330	3005	28A414	3005	28A61D	3005	19B2000129-P1	3005	48-63078A60		
3005	28A330A	3005	28A414A	3005	28A61E	3005	24MW11	3005	48-63078A61		
3005	28A330B	3005	28A414B	3005	28A61G	3005	24MW77	3005	48-63078A64		
3005	28A330C	3005	28A414C	3005	28A61K	3005	27T401	3005	48-63082A15		
3005	28A330D	3005	28A414D	3005	28A61L	3005	27T402	3005	48X869001		
3005	28A330E	3005	28A414E	3005	28A61OR	3005	31-0253	3005	488134860		
3005	28A330F	3005	28A414F	3005	28A61R	3005	34-6000-18	3005	488134861		
3005	28A330G	3005	28A414G	3005	28A61X	3005	34-6000-19	3005	488134862		
3005	28A330GN	3005	28A414GN	3005	28A61Y	3005	34-6000-3	3005	5782-23		
3005	28A330H	3005	28A414H	3005	28A65	3005	34-6001-43	3005	5706-6A		
3005	28A330J	3005	28A414J	3005	28A65A	3005	34-6001-44	3005	5706-6B		
3005	28A330K	3005	28A414K	3005	28A65B	3005	34-6015-34	3005	5706-6C		
3005	28A330L	3005	28A414L	3005	28A65C	3005	42-19682	3005	57D1-80		
3005	28A330M	3005	28A414M	3005	28A65D	3005	43A111449	3005	57D1-8A		
3005	28A330X	3005	28A414OR	3005	28A65F	3005	43P1	3005	57D1E0		
3005	28A330Y	3005	28A414R	3005	28A65G	3005	43P3	3005	57D1E4		
3005	28A35	3005	28A414X	3005	28A65K	3005	45X2	3005	61B026-1		
3005	28A35A	3005	28A415	3005	28A65L	3005	46-86123-3	3005	62-18415		
3005	28A35B	3005	28A415A	3005	28A65M	3005	46-8636-3	3005	62-18416		
3005	28A35C	3005	28A415B	3005	28A65OR	3005	48-124307	3005	62-18417		
3005	28A35D	3005	28A415C	3005	28A65R	3005	48-124308	3005	62-18423		
3005	28A35E	3005	28A415D	3005	28A65X	3005	48-124314	3005	62-18424		
3005	28A35F	3005	28A415E	3005	28A65Y	3005	48-124315	3005	63-10200		
3005	28A35G	3005	28A415F	3005	28A66	3005	48-124327	3005	63-18416		
3005	28A35L	3005	28A415G	3005	28A66A	3005	48-124328	3005	63-25946		
3005	28A35M	3005	28A415H	3005	28A66B	3005	48-124343	3005	63-26849		
3005	28A35X	3005	28A415J	3005	28A66C	3005	48-124344	3005	63-27278		
3005	28A35Y	3005	28A415K	3005	28A66D	3005	48-124345	3005	63-27279		
3005	28A36	3005	28A415L	3005	28A66E	3005	48-124357	3005	63-27280		
3005	28A36A	3005	28A415M	3005	28A66F	3005	48-124358	3005	63-27567		
3005	28A36B	3005	28A415OR	3005	28A66G	3005	48-124359	3005	63-29662		
3005	28A36C	3005	28A415R	3005	28A66K	3005	48-124370	3005	63-29663		
3005	28A36D	3005	28A415X	3005	28A66L	3005	48-124371	3005	63-29664		
3005	28A36F	3005	28A415Y	3005	28A66M	3005	48-124378	3005	63-29863		
3005	28A36G	3005	28A41A	3005	28A66OR	3005	48-124379	3005	63-7547		
3005	28A36L	3005	28A41B	3005	28A66R	3005	48-124380	3005	64II		
3005	28A36M	3005	28A41C	3005	28A66X	3005	48-124398	3005	070-020		
3005	28A36OR	3005	28A41D	3005	28A66Y	3005	48-124443	3005	74-3		
3005	28A36R	3005	28A41E	3005	28A75A	3005	48-124445	3005	74-3-70		
3005	28A36X	3005	28A41F	3005	28A75B	3005	48-124446	3005	74-3-70-12		
3005	28A36Y	3005	28A41G	3005	28A75C	3005	48-125229	3005	74-3-70-12-7		
3005	28A37	3005	28A41L	3005	28A75D	3005	48-125230	3005	74-30		
3005	28A37A	3005	28A410R	3005	28A75E	3005	48-125231	3005	74-31		
3005	28A37B	3005	28A41R	3005	28A75F	3005	48-125232	3005	74-32		
3005	28A37C	3005	28A41X	3005	28A75G	3005	48-125237	3005	74-33		
3005	28A37D	3005	28A41Y	3005	28A75H	3005	48-125238	3005	74-34		
3005	28A37E	3005	28A44	3005	28A75K	3005	48-125239	3005	74-35		
3005	28A37F	3005	28A44A	3005	28A75L	3005	48-125240	3005	74-36		
3005	28A37G	3005	28A44B	3005	28A75M	3005	48-125242	3005	74-37		
3005	28A37L	3005	28A44C	3005	28A75OR	3005	48-125296	3005	74-38		
3005	28A37M	3005	28A44D	3005	28A75R	3005	48-128303	3005	74-39		
3005	28A37OR	3005	28A44F	3005	28A75X	3005	48-134415	3005	74-3A		
3005	28A37X	3005	28A44G	3005	28A75Y	3005	48-134417	3005	74-3A0		
3005	28A37Y	3005	28A44L	3005	28A759N	3005	48-134418	3005	74-3A1		
3005	28A38	3005	28A44M	3005	28B142B	3005	48-134419	3005	74-3A19		
3005	28A38A	3005	28A44X	3005	28B142C	3005	48-134420	3005	74-3A2		
3005	28A38B	3005	28A44Y	3005	28B292A	3005	48-134421	3005	74-3A21		
3005	28A38C	3005	28A474	3005	0028C203	3005	48-134422	3005	74-3A3		
3005	28A38D	3005	28A474A	3005	0028C203A	3005	48-134423	3005	74-3A3P		
3005	28A38E	3005	28A474B	3005	0028C203AA	3005	48-134424	3005	74-3A4		
3005	28A38F	3005	28A474C	3005	28H203	3005	48-134425	3005	74-3A4-7		
3005	28A38G	3005	28A474D	3005	2TN15	3005	48-134426	3005	74-3A5		
3005	28A38L	3005	28A474E	3005	2TN52	3005	48-134427	3005	74-3A5L		
3005	28A38M	3005	28A474G	3005	2TN48	3005	48-134428	3005	74-3A6		
3005	28A380R	3005	28A474GN	3005	2TN49	3005	48-134429	3005	74-3A6-3		
3005	28A38R	3005	28A474H	3005	2TN52	3005	48-134432	3005	74-3A7		
3005	28A38X	3005	28A474J	3005	2TN53	3005	48-134433	3005	74-3A7-1		
3005	28A38Y	3005	28A474L	3005	2V464	3005	48-134443	3005	74-3A8		
3005	28A39	3005	28A474M	3005	2V465	3005	48-134444	3005	74-3A82		
3005	28A39A	3005	28A474OR	3005	2V466	3005	48-134445	3005	74-3A9		
3005	28A39B	3005	28A474R	3005	2V467	3005	48-134446	3005	74-3A9G		
3005	28A39C	3005	28A474Y	3005	2V482	3005	48-134450	3005	77		
3005	28A39D	3005	28A50	3005	2V483	3005	48-134458	3005	78BLK		
3005	28A39E			3005	2V484	3005	48-134462	3005	78GRN		
3005	28A39F			3005	2V486	3005	48-134468	3005	78RED		
3005	28A39G			3005	2V631	3005	48-134469	3005	78YEL		
3005				3005	2V632	3005	48-134470	3005	86-133-2		

SK	Industry Standard No.	SK	Industry Standard No.	SK	Industry Standard No.	SK	Industry Standard No.	SK	Industry Standard No.
3005	86-27-2	3005	260P2100	3005	38209	3005	980426	3006	AF239B
3005	86-28-2	3005	260P21001	3005	40269	3005	980432	3006	AF240
3005	86-32-2	3005	260P21002	3005	40403	3005	980434	3006	AF256
3005	86-33-2	3005	297V008H01	3005	72784-21	3005	980438	3006	AF279
3005	86-46-2	3005	297V011H02	3005	080274	3005	980459	3006	AFY12
3005	86-47-2	3005	297V012H01	3005	81502-0	3005	1221649	3006	AFY14
3005	86-48-2	3005	297V012H02	3005	81502-OA	3005	2000648-23	3006	AFY15
3005	86-509-2	3005	297V012H03	3005	81502-OB	3005	2001812-65	3006	AFY16
3005	86-59-2	3005	297V012H05	3005	81502-1	3005	2091211-0014	3006	AFY34
3005	86-60-2	3005	297V012H06	3005	81502-1A	3005	2091241-0013	3006	AFY37
3005	86-73-2	3005	297V012H09	3005	81502-1B	3005	2091241-13	3006	AFY39
3005	86-75-2	3005	297V017H01	3005	81502-5A	3005	2091241-13A	3006	AFY40
3005	86-78-2	3005	297V017H02	3005	81502-5B	3005	2091241-14	3006	AFY40K
3005	86-79-2	3005	297V019B01	3005	81502-7	3005	2091241-15	3006	AFY40R
3005	87-0016	3005	297V019B01	3005	81502-7A	3005	2091241-15A	3006	AFY41
3005	88C	3005	297V020H02	3005	81502-7B	3005	2091241-7	3006	AFY42
3005	088C-12	3005	297V020M01	3005	81502-7C	3005	2320154	3006	AFZ11
3005	088C-12-7	3005	297V021H01	3005	81502-8	3005	2320512	3006	AFZ12
3005	88C-70	3005	297V021H02	3005	81502-8A	3005	2320513	3006	APB-11H-1001
3005	88C-70-12-7	3005	297V021H03	3005	81502-8B	3005	2320514	3006	APB-11H-1004
3005	88C0	3005	297V022E01	3005	81502-8C	3005	2320514-1	3006	AR103
3005	88C1	3005	297V026H03	3005	81503-6	3005	2970038H05	3006	AR104
3005	88C10	3005	297V038E01	3005	81503-6A	3005	4036612-P1	3006	ASY30
3005	88C10R	3005	297V038E02	3005	81503-6B	3005	4036612-P2	3006	ASY57
3005	88C11	3005	297V038E03	3005	81503-6C	3005	4036707-P1	3006	ASY58
3005	88C119	3005	297V038E05	3005	81503-7	3005	4036707-P2	3006	ASY59
3005	88C12	3005	297V038E07	3005	81503-7A	3005	4036937-P1	3006	ASY63
3005	88C121	3005	297V038E09	3005	81503-7B	3005	4036937-P2	3006	ASZ20
3005	88C13	3005	297V043H02	3005	81503-7C	3005	4037993-P1	3006	ASZ20N
3005	88C13P	3005	297V044E01	3005	81503-8	3005	4037993-P2	3006	AT-1
3005	88C14	3005	297V054C01	3005	81504-1	3005	4038260-P1	3006	AT-14
3005	88C14-7	3005	297V054C02	3005	81504-1A	3005	4038260-P2	3006	AT-2
3005	88C15	3005	297V054H01	3005	81504-1B	3005	6100724-2	3006	AT-3
3005	88C15L	3005	297V054H02	3005	81504-1C	3005	7276211	3006	AT-8
3005	88C16	3005	297V055C01	3005	81504-3	3005	7278421	3006	AT-9
3005	88C16D	3005	297V055H01	3005	81504-3A	3005	7279379	3006	AT/RP1
3005	88C17	3005	297V065H01	3005	81504-3B	3005	7279782	3006	AT/RP2
3005	88C17-1	3005	297V065H02	3005	81504-3C	3005	7279788	3006	AT/813
3005	88C18	3005	297V065H03	3005	81506-5	3005	7279789	3006	B240
3005	88C18Z	3005	310-189	3005	81506-5A	3005	7279940	3006	B240A
3005	88C19	3005	350	3005	81506-5C	3005	7281307	3006	B601-1006
3005	88C19G	3005	36521	3005	81506-6	3005	7281308	3006	B601-1007
3005	88C2	3005	421-9682	3005	81506-6A	3005	7281309	3006	B601-1008
3005	88C3	3005	52021	3005	81506-6B	3005	7281891	3006	BCM1002-1
3005	88C4	3005	5211T	3005	81506-6C	3005	8516986	3006	C102
3005	88C5	3005	690V010H42	3005	81510-3	3005	8989441-2	3006	C10215-2
3005	88C6	3005	690V047H56	3005	81510-5	3005	27125240	3006	C125
3005	88C7	3005	690V047H57	3005	81511-5	3005	27125360	3006	CXL160
3005	88C8	3005	690V012H96	3005	81511-6	3005	43024833	3006	DO63
3005	88C9	3005	800-505-00	3005	81511-7	3005	43049626	3006	DO79
3005	88C0-70-12	3005	921-65	3005	95103	3005	62087684	3006	DO86
3005	93A9-3	3005	921-65A	3005	95120A	3005	62118946	3006	D174
3005	93A9-4	3005	921-65B	3005	117616	3005	120001192	3006	D8-24
3005	94T1	3005	921-66	3005	117617	3005	A1220	3006	D8-34
3005	95-120A	3005	921-66A	3005	120545	3006	A127	3006	D8-35
3005	99BA6	3005	921-66B	3005	121153	3006	A1377	3006	D8-36
3005	99BB6	3005	930X6	3005	121154	3006	A1465-1	3006	D8-41
3005	101-12	3005	1145	3005	221856	3006	A1465-19	3006	D8-42
3005	112-001	3005	1146	3005	223372	3006	A1488B	3006	D8-51
3005	120-001192	3005	1340	3005	223473	3006	A1488B9	3006	D8-52
3005	121-102	3005	1350	3005	223475	3006	A163	3006	D8-53
3005	121-103	3005	1390	3005	224584	3006	A376	3006	D8-56
3005	121-104	3005	1400	3005	226181	3006	A378	3006	D8-62
3005	121-105	3005	1410	3005	226338	3006	A379	3006	D8-63
3005	121-128	3005	1474-3	3005	227752	3006	A411	3006	D8-64
3005	121-1330	3005	1474-3-12	3005	233507	3006	A42	3006	D8-65
3005	121-1350	3005	1474-3-12-8	3005	233945	3006	A435	3006	E-2462
3005	121-1360	3005	1488C	3005	244007	3006	A525	3006	E2450
3005	121-1390	3005	1488C-12	3005	256126	3006	A525A	3006	EA0002
3005	121-1400	3005	1488C-12-8	3005	257470	3006	A525B	3006	EA0007
3005	121-147	3005	1874-3	3005	266702	3006	A57	3006	EA0053
3005	121-160	3005	1874-3-12	3005	310223	3006	A59	3006	EA1337
3005	121-161	3005	1874-3-127	3005	489751-113	3006	A615-1008	3006	EA1338
3005	121-162	3005	1874-3L	3005	610074-2	3006	A615-1009	3006	EA1339
3005	121-179	3005	1874-3L8	3005	815020	3006	A65-1-70	3006	EA1340
3005	121-180	3005	1888C	3005	815020A	3006	A65-1-705	3006	EA1342
3005	121-186	3005	1888C-12	3005	815020B	3006	A65-1A9G	3006	EA15X11
3005	121-205	3005	1888C-127	3005	815021	3006	A70F	3006	EA15X13
3005	121-206	3005	1888CL8	3005	815021A	3006	A70L	3006	EA15X40
3005	121-207	3005	2057A2-263	3005	815021B	3006	A70MA	3006	EA15X41
3005	121-208	3005	2057A2-60	3005	815025	3006	A84	3006	EA15X43
3005	121-209	3005	2057B2-124	3005	815025A	3006	A86	3006	ECG160
3005	121-210	3005	2057B2-60	3005	815025B	3006	A88B-70	3006	ER15X11
3005	121-211	3005	2093A9-3	3005	815027	3006	A88B-705	3006	ER15X12
3005	121-212	3005	2093A9-4	3005	815027A	3006	A88B19G	3006	ER15X13
3005	121-213	3005	3425	3005	815027B	3006	A89	3006	ER15X24
3005	121-219	3005	3434	3005	815027C	3006	A909-1008	3006	ER15X25
3005	121-220	3005	3435	3005	815028	3006	AF-105A	3006	ER15X26
3005	121-221	3005	3458	3005	815028A	3006	AF-106	3006	FB401
3005	121-222	3005	3540	3005	815028B	3006	AF-109	3006	FB402
3005	121-234	3005	3603	3005	815028C	3006	AF-121	3006	FB403
3005	121-235	3005	3746	3005	815036	3006	AF-166	3006	FB440
3005	121-236	3005	3750	3005	815036A	3006	AF-182	3006	GC1092
3005	121-240	3005	4566	3005	815036B	3006	AF102	3006	GC1093X3
3005	121-254	3005	4567	3005	815036C	3006	AF106	3006	GE-245
3005	121-273	3005	4484	3005	815037	3006	AF110	3006	GE-3
3005	121-275	3005	4486	3005	815037A	3006	AF111	3006	GER-A-D
3005	121-310	3005	12110-0	3005	815037B	3006	AF112	3006	GET-672
3005	121-333	3005	12110-3	3005	815037C	3006	AF113	3006	GET-672A
3005	121-334	3005	12110-4	3005	81503B	3006	AF114N	3006	GET-673
3005	121-335	3005	12110-5	3005	815041	3006	AF115N	3006	HA240
3005	121-397	3005	12112-8	3005	815041A	3006	AF116N	3006	HA353
3005	121-45	3005	12116-1	3005	815041B	3006	AF117N	3006	HA525
3005	121-53	3005	12116-2	3005	815041C	3006	AF118	3006	HEP-2
3005	121-54	3005	12117-9	3005	815043	3006	AF119	3006	HEP-G0002
3005	121-62	3005	12118-0	3005	815043A	3006	AF146	3006	HEP-G0003
3005	121-63	3005	12118-6	3005	815043B	3006	AF147	3006	HF50M
3005	121-65	3005	12123-4	3005	815043C	3006	AF148	3006	HR-40
3005	121-66	3005	12123-5	3005	815055	3006	AF149	3006	HR-43
3005	121-67	3005	12123-6	3005	815056	3006	AF150	3006	HR-45
3005	121-72	3005	12124-0	3005	815057	3006	AF164	3006	HR20
3005	121-73	3005	12124-1	3005	815065	3006	AF165	3006	HR20A
3005	121-74	3005	12125-4	3005	815065A	3006	AF167	3006	HR21
3005	121-75	3005	12127-5	3005	815065B	3006	AF168	3006	HR21A
3005	121-76	3005	12127-5	3005	815065C	3006	AF169	3006	HR22
3005	121-78	3005	12128-9	3005	815066	3006	AF170	3006	HR22A
3005	121-80	3005	12173	3005	815066A	3006	AF171	3006	HR22B
3005	121-81	3005	12174	3005	815066B	3006	AF172	3006	HR24
3005	121-82	3005	12175	3005	815066C	3006	AF178	3006	HR24A
3005	121-83	3005	12176	3005	815101	3006	AF179	3006	HR25
3005	121-84	3005	12178	3005	815103	3006	AF180	3006	HR25A
3005	121-85	3005	12183	3005	815105	3006	AF181	3006	HR26
3005	121-86	3005	12191	3005	815107	3006	AF185	3006	HR26A
3005	121-87	3005	12192	3005	815109	3006	AF186	3006	HR27
3005	121-88	3005	12193	3005	815115	3006	AF186G	3006	HR27A
3005	121-89	3005	18540	3005	815116	3006	AF186W	3006	HR40
3005	121-90	3005	27125-360	3005	815117	3006	AF193	3006	HR42
3005	121-91	3005	34099	3005	815308A	3006	AF200	3006	HR43
3005	121-92	3005	34119	3005	825065	3006	AF201	3006	HR44
3005	121-93	3005	34219	3005	922896	3006	AF201C	3006	HR45
3005	121-94	3005	34220	3005	980136	3006	AF202	3006	HR46
3005	122-229	3005	34221	3005	980316	3006	AF202L	3006	HT102341
3005	235	3005	36816			3006	AF2028	3006	HT102341A
3005	260P1300					3006	AF239		

SK	Industry Standard No.	SK	Industry Standard No.	SK	Industry Standard No.	SK	Industry Standard No.	SK	Industry Standard No.
3006	HT102341B	3006	TI403	3006	2SA117J	3006	2SA134K	3006	2SA156OR
3006	HT102541C	3006	TIXM-201	3006	2SA117K	3006	2SA134L	3006	2SA156R
3006	HT102551	3006	TIXM-203	3006	2SA117L	3006	2SA134M	3006	2SA156X
3006	HT102551A	3006	TIXM-205	3006	2SA117M	3006	2SA134OR	3006	2SA156Y
3006	HT103501	3006	TIXM-206	3006	2SA117OR	3006	2SA134R	3006	2SA157A
3006	HT200541C	3006	TIXM202	3006	2SA117R	3006	2SA134X	3006	2SA157B
3006	JP5062	3006	TIXM204	3006	2SA117X	3006	2SA134Y	3006	2SA157C
3006	JR100	3006	TK1228-001	3006	2SA117Y	3006	2SA136A	3006	2SA157D
3006	JR200	3006	TM160	3006	2SA118A	3006	2SA136B	3006	2SA157E
3006	JR30	3006	TNJ-60067	3006	2SA118B	3006	2SA136C	3006	2SA157F
3006	MPR1097	3006	TNJ-60068	3006	2SA118C	3006	2SA136D	3006	2SA157G
3006	N5782-11	3006	TNJ-60069	3006	2SA118D	3006	2SA136E	3006	2SA157GN
3006	N5782-13	3006	TNJ-60071	3006	2SA118E	3006	2SA136F	3006	2SA157H
3006	N5782-14	3006	TNJ-60073	3006	2SA118F	3006	2SA136G	3006	2SA157J
3006	N5782-22	3006	TNJ-60077	3006	2SA118G	3006	2SA136GN	3006	2SA157K
3006	NKT121	3006	TNJ-60279	3006	2SA118GN	3006	2SA136H	3006	2SA157L
3006	NKT122	3006	TNJ-60280	3006	2SA118H	3006	2SA136J	3006	2SA157M
3006	NKT124	3006	TNJ-60281	3006	2SA118J	3006	2SA136K	3006	2SA157OR
3006	NKT125	3006	TR-8001	3006	2SA118K	3006	2SA136L	3006	2SA157R
3006	NKT252	3006	TR-8002	3006	2SA118L	3006	2SA136M	3006	2SA157X
3006	OC-169	3006	TR-8003	3006	2SA118M	3006	2SA136OR	3006	2SA157Y
3006	OC-615	3006	TR11C	3006	2SA118OR	3006	2SA136R	3006	2SA159A
3006	OC169R	3006	TR17	3006	2SA118R	3006	2SA136X	3006	2SA159B
3006	OC170V	3006	TR18	3006	2SA118X	3006	2SA136Y	3006	2SA159C
3006	OC170R	3006	TR18C	3006	2SA118T	3006	2SA139A	3006	2SA159D
3006	OC171R	3006	TR1R26	3006	2SA121A	3006	2SA139B	3006	2SA159E
3006	OC171W	3006	TR2R26	3006	2SA121B	3006	2SA139C	3006	2SA159F
3006	ON174	3006	TR310068	3006	2SA121C	3006	2SA139D	3006	2SA159GN
3006	PADT20	3006	TR310123	3006	2SA121D	3006	2SA139E	3006	2SA159H
3006	PADT21	3006	TR310124	3006	2SA121E	3006	2SA139F	3006	2SA159J
3006	PADT22	3006	TR310139	3006	2SA121F	3006	2SA139G	3006	2SA159K
3006	PADT24	3006	TR310147	3006	2SA121G	3006	2SA139GN	3006	2SA159L
3006	PADT25	3006	TR310150	3006	2SA121GN	3006	2SA139J	3006	2SA159M
3006	PADT26	3006	TR5R26	3006	2SA121H	3006	2SA139K	3006	2SA159OR
3006	PADT27	3006	TR4R26	3006	2SA121J	3006	2SA139L	3006	2SA159R
3006	PADT28	3006	TR8001	3006	2SA121K	3006	2SA139M	3006	2SA159X
3006	PADT30	3006	TR8002	3006	2SA121L	3006	2SA139OR	3006	2SA159Y
3006	PADT31	3006	TR8003	3006	2SA121M	3006	2SA139R	3006	2SA161A
3006	PADT51	3006	TV24158	3006	2SA121R	3006	2SA139X	3006	2SA161B
3006	PQ27	3006	TV24172	3006	2SA121X	3006	2SA139Y	3006	2SA161C
3006	PQ30	3006	TV24351	3006	2SA121Y	3006	2SA141A	3006	2SA161D
3006	PT2A	3006	TV2455	3006	2SA122A	3006	2SA141B	3006	2SA161E
3006	PT28	3006	TV2479	3006	2SA122B	3006	2SA141C	3006	2SA161F
3006	PTC107	3006	TVS-28A103	3006	2SA122C	3006	2SA141D	3006	2SA161G
3006	Q5044	3006	V205	3006	2SA122D	3006	2SA141E	3006	2SA161GN
3006	R9531	3006	VS-28A385L	3006	2SA122E	3006	2SA141F	3006	2SA161H
3006	R9532	3006	VS-28A71	3006	2SA122F	3006	2SA141G	3006	2SA161J
3006	R9601	3006	VS-28A71B	3006	2SA122G	3006	2SA141GN	3006	2SA161K
3006	R9602	3006	VS-28A71B8	3006	2SA122GN	3006	2SA141H	3006	2SA161L
3006	RE27	3006	VS-28C585L	3006	2SA122K	3006	2SA141K	3006	2SA161M
3006	REN126	3006	V12	3006	2SA122L	3006	2SA141L	3006	2SA161OR
3006	REN160	3006	WBF3	3006	2SA122M	3006	2SA141M	3006	2SA161R
3006	RS-101	3006	WBF635	3006	2SA122OR	3006	2SA141OR	3006	2SA161X
3006	RS-103	3006	XA131	3006	2SA122R	3006	2SA141R	3006	2SA161Y
3006	RS-2002	3006	XB9	3006	2SA122X	3006	2SA141X	3006	2SA161Z
3006	RS-3892	3006	XG1	3006	2SA122Y	3006	2SA141Y	3006	2SA162A
3006	RS-3898	3006	XG10	3006	2SA123A	3006	2SA143A	3006	2SA162B
3006	RS-3900	3006	XG11	3006	2SA123B	3006	2SA143B	3006	2SA162C
3006	RS-3902	3006	XG12	3006	2SA123C	3006	2SA143C	3006	2SA162D
3006	RS-3903	3006	XG2	3006	2SA123D	3006	2SA143D	3006	2SA162E
3006	RS-3911	3006	XG24	3006	2SA123E	3006	2SA143E	3006	2SA162GN
3006	RS-5208	3006	XG3	3006	2SA123F	3006	2SA143F	3006	2SA162H
3006	RS3905	3006	XG5	3006	2SA123G	3006	2SA143G	3006	2SA162J
3006	RS3912	3006	XJ72	3006	2SA123GN	3006	2SA143GN	3006	2SA162K
3006	RS3986	3006	XJ75	3006	2SA123H	3006	2SA143H	3006	2SA162L
3006	RS3999	3006	1-21-135	3006	2SA123J	3006	2SA143J	3006	2SA162M
3006	RT6988	3006	1-21-137	3006	2SA123K	3006	2SA143K	3006	2SA162OR
3006	S-1640	3006	1-21-139	3006	2SA123L	3006	2SA143L	3006	2SA162R
3006	S01	3006	1-21-157	3006	2SA123M	3006	2SA143M	3006	2SA162X
3006	S02	3006	1-21-228	3006	2SA123OR	3006	2SA1430R	3006	2SA162Y
3006	S03	3006	1-21-229	3006	2SA123R	3006	2SA143R	3006	2SA163A
3006	SC-71	3006	1-21-230	3006	2SA123X	3006	2SA143X	3006	2SA163B
3006	SC-72	3006	1-21-231	3006	2SA123Y	3006	2SA143Y	3006	2SA163C
3006	SC-74	3006	1-21-233	3006	2SA124A	3006	2SA153A	3006	2SA163D
3006	SC-78	3006	1-522210131	3006	2SA124C	3006	2SA153B	3006	2SA163E
3006	SC-79	3006	1-522210300	3006	2SA124D	3006	2SA153C	3006	2SA163F
3006	SC-80	3006	1-522211021	3006	2SA124E	3006	2SA153D	3006	2SA163G
3006	SPT-163	3006	1-522211921	3006	2SA124F	3006	2SA153E	3006	2SA163GN
3006	SPT-358	3006	1-522214400	3006	2SA124G	3006	2SA153F	3006	2SA163H
3006	T-99	3006	1-522214411	3006	2SA124GN	3006	2SA153G	3006	2SA163J
3006	T-Q5021	3006	1-522214435	3006	2SA124H	3006	2SA153GN	3006	2SA163K
3006	T-Q5022	3006	1-522214821	3006	2SA124J	3006	2SA153H	3006	2SA163L
3006	T-Q5034	3006	1-522214831	3006	2SA124K	3006	2SA153J	3006	2SA163M
3006	T-Q5035	3006	1-522217400	3006	2SA124L	3006	2SA153K	3006	2SA163OR
3006	T-Q5038	3006	002-007200	3006	2SA124M	3006	2SA153L	3006	2SA163R
3006	T1052	3006	2N1023	3006	2SA124OR	3006	2SA153M	3006	2SA163X
3006	T1224	3006	2N1066	3006	2SA124R	3006	2SA1530R	3006	2SA163Y
3006	T1225	3006	2N1285	3006	2SA124X	3006	2SA153R	3006	2SA164A
3006	T1250	3006	2N1301	3006	2SA124Y	3006	2SA153X	3006	2SA164B
3006	T1387	3006	2N1749	3006	2SA125A	3006	2SA153Y	3006	2SA164C
3006	T1389	3006	2N1788	3006	2SA125B	3006	2SA154A	3006	2SA164E
3006	T1390	3006	2N1789	3006	2SA125C	3006	2SA154B	3006	2SA164F
3006	T1391	3006	2N2084	3006	2SA125D	3006	2SA154C	3006	2SA164GN
3006	T1400	3006	2N2190	3006	2SA125E	3006	2SA154D	3006	2SA164H
3006	T1401	3006	2N2191	3006	2SA125F	3006	2SA154E	3006	2SA164J
3006	T1403	3006	2N2238	3006	2SA125G	3006	2SA154F	3006	2SA164K
3006	T1657	3006	2N2363	3006	2SA125GN	3006	2SA154G	3006	2SA164L
3006	T1690	3006	2N2495	3006	2SA125H	3006	2SA154GN	3006	2SA164M
3006	T1691	3006	2N2496	3006	2SA125J	3006	2SA154H	3006	2SA164OR
3006	T1692	3006	2N2654	3006	2SA125K	3006	2SA154J	3006	2SA164R
3006	T1737	3006	2N2996	3006	2SA125L	3006	2SA154K	3006	2SA164X
3006	T1738	3006	2N300	3006	2SA125M	3006	2SA154L	3006	2SA164Y
3006	T1814	3006	2N3074	3006	2SA125OR	3006	2SA154M	3006	2SA165A
3006	T2015	3006	2N3588	3006	2SA125R	3006	2SA154OR	3006	2SA165B
3006	T2016	3006	2N987	3006	2SA125X	3006	2SA154R	3006	2SA165C
3006	T2017	3006	28175	3006	2SA125Y	3006	2SA154X	3006	2SA165D
3006	T2019	3006	2SA-4551	3006	2SA133A	3006	2SA154Y	3006	2SA165E
3006	T2021	3006	2SA-4561	3006	2SA133B	3006	2SA155A	3006	2SA165F
3006	T2022	3006	2SA116A	3006	2SA133C	3006	2SA155B	3006	2SA165G
3006	T2379	3006	2SA116B	3006	2SA133D	3006	2SA155C	3006	2SA165GN
3006	T2584	3006	2SA116C	3006	2SA133E	3006	2SA155D	3006	2SA165H
3006	T253	3006	2SA116D	3006	2SA133F	3006	2SA155F	3006	2SA165J
3006	T280	3006	2SA116E	3006	2SA133G	3006	2SA155G	3006	2SA165L
3006	T367	3006	2SA116F	3006	2SA133GN	3006	2SA155GN	3006	2SA165M
3006	T368	3006	2SA116G	3006	2SA133H	3006	2SA155H	3006	2SA165OR
3006	T373	3006	2SA116GN	3006	2SA133J	3006	2SA155J	3006	2SA165R
3006	T374	3006	2SA116H	3006	2SA133K	3006	2SA155K	3006	2SA165X
3006	T449	3006	2SA116J	3006	2SA133L	3006	2SA155L	3006	2SA165Y
3006	TA-1846	3006	2SA116K	3006	2SA133M	3006	2SA155M	3006	2SA166A
3006	TA-1847	3006	2SA116L	3006	2SA133OR	3006	2SA155OR	3006	2SA166B
3006	TA-1860	3006	2SA116M	3006	2SA133R	3006	2SA155R	3006	2SA166C
3006	TA-1861	3006	2SA116OR	3006	2SA133X	3006	2SA155X	3006	2SA166D
3006	TA1829	3006	2SA116R	3006	2SA133Y	3006	2SA155Y	3006	2SA166E
3006	TA1860	3006	2SA116X	3006	2SA134A	3006	2SA156A	3006	2SA166F
3006	TA1990	3006	2SA116Y	3006	2SA134B	3006	2SA156B	3006	2SA166G
3006	TI-338	3006	2SA117A	3006	2SA134C	3006	2SA156C	3006	2SA166H
3006	TI-387	3006	2SA117B	3006	2SA134D	3006	2SA156D	3006	2SA166J
3006	TI-388	3006	2SA117C	3006	2SA134E	3006	2SA156E	3006	2SA166K
3006	TI-400	3006	2SA117D	3006	2SA134F	3006	2SA156F		
3006	TI-401	3006	2SA117E	3006	2SA134G	3006	2SA156G		
3006	TI-403	3006	2SA117F	3006	2SA134H	3006	2SA156GN		
3006	TI400	3006	2SA117G	3006	2SA134H	3006	2SA156H		
3006	TI401	3006	2SA117GN	3006	2SA134G	3006	2SA156K		
3006	TI402	3006	2SA117H	3006	2SA134J	3006	2SA156L		
								3006	2SA156M

SK	Industry Standard No.	SK	Industry Standard No.	SK	Industry Standard No.	SK	Industry Standard No.	SK	Industry Standard No.
3006	2SA166L	3006	2SA223J	3006	2SA239L	3006	2SA260H	3006	2SA280F
3006	2SA166M	3006	2SA223K	3006	2SA239M	3006	2SA260J	3006	2SA280G
3006	2SA166OR	3006	2SA223L	3006	2SA230OR	3006	2SA260K	3006	2SA280GN
3006	2SA166R	3006	2SA223M	3006	2SA239R	3006	2SA260L	3006	2SA280H
3006	2SA166X	3006	2SA223OR	3006	2SA239X	3006	2SA260M	3006	2SA280J
3006	2SA166Y	3006	2SA223R	3006	2SA239Y	3006	2SA260R	3006	2SA280K
3006	2SA175A	3006	2SA223X	3006	2SA240	3006	2SA260X	3006	2SA280L
3006	2SA175B	3006	2SA223Y	3006	2SA240A	3006	2SA260Y	3006	2SA280M
3006	2SA175C	3006	2SA225A	3006	2SA240B2	3006	2SA261	3006	2SA280OR
3006	2SA175D	3006	2SA225B	3006	2SA240C	3006	2SA261A	3006	2SA280R
3006	2SA175E	3006	2SA225C	3006	2SA240D	3006	2SA261B	3006	2SA280X
3006	2SA175F	3006	2SA225D	3006	2SA240E	3006	2SA261C	3006	2SA280Y
3006	2SA175G	3006	2SA225F	3006	2SA240F	3006	2SA261E	3006	2SA281A
3006	2SA175GN	3006	2SA225G	3006	2SA240G	3006	2SA261F	3006	2SA281B
3006	2SA175H	3006	2SA225GN	3006	2SA240GN	3006	2SA261G	3006	2SA281C
3006	2SA175J	3006	2SA225H	3006	2SA240H	3006	2SA261GN	3006	2SA281D
3006	2SA175K	3006	2SA225J	3006	2SA240J	3006	2SA261H	3006	2SA281E
3006	2SA175L	3006	2SA225K	3006	2SA240K	3006	2SA261K	3006	2SA281F
3006	2SA175M	3006	2SA225L	3006	2SA240L	3006	2SA261L	3006	2SA281G
3006	2SA175OR	3006	2SA225M	3006	2SA240M	3006	2SA261M	3006	2SA281GN
3006	2SA175R	3006	2SA225OR	3006	2SA240OR	3006	2SA261OR	3006	2SA281H
3006	2SA175X	3006	2SA225R	3006	2SA240X	3006	2SA261R	3006	2SA281J
3006	2SA175Y	3006	2SA225X	3006	2SA240Y	3006	2SA261X	3006	2SA281K
3006	2SA197A	3006	2SA225Y	3006	2SA241	3006	2SA261Y	3006	2SA281L
3006	2SA197B	3006	2SA227	3006	2SA241A	3006	2SA262	3006	2SA281M
3006	2SA197C	3006	2SA227A	3006	2SA241B	3006	2SA262A	3006	2SA281OR
3006	2SA197D	3006	2SA227B	3006	2SA241C	3006	2SA262B	3006	2SA281R
3006	2SA197R	3006	2SA227C	3006	2SA241D	3006	2SA262C	3006	2SA281X
3006	2SA197F	3006	2SA227D	3006	2SA241E	3006	2SA262D	3006	2SA281Y
3006	2SA197G	3006	2SA227E	3006	2SA241F	3006	2SA262E	3006	2SA289A
3006	2SA197GN	3006	2SA227F	3006	2SA241G	3006	2SA262F	3006	2SA289B
3006	2SA197H	3006	2SA227G	3006	2SA241GN	3006	2SA262G	3006	2SA289C
3006	2SA197J	3006	2SA227GN	3006	2SA241H	3006	2SA262GN	3006	2SA289D
3006	2SA197K	3006	2SA227H	3006	2SA241J	3006	2SA262H	3006	2SA289E
3006	2SA197L	3006	2SA227J	3006	2SA241K	3006	2SA262J	3006	2SA289F
3006	2SA197M	3006	2SA227K	3006	2SA241L	3006	2SA262K	3006	2SA289G
3006	2SA197OR	3006	2SA227L	3006	2SA241M	3006	2SA262L	3006	2SA289GN
3006	2SA197R	3006	2SA227M	3006	2SA241OR	3006	2SA262M	3006	2SA289H
3006	2SA197X	3006	2SA227OR	3006	2SA241R	3006	2SA262OR	3006	2SA289J
3006	2SA197Y	3006	2SA227R	3006	2SA241X	3006	2SA262R	3006	2SA289K
3006	2SA213A	3006	2SA227X	3006	2SA241Y	3006	2SA262X	3006	2SA289L
3006	2SA213B	3006	2SA227Y	3006	2SA246A	3006	2SA262Y	3006	2SA289M
3006	2SA213C	3006	2SA229	3006	2SA246B	3006	2SA263	3006	2SA289OR
3006	2SA213D	3006	2SA229A	3006	2SA246C	3006	2SA263A	3006	2SA289R
3006	2SA213E	3006	2SA229B	3006	2SA246D	3006	2SA263B	3006	2SA289X
3006	2SA213F	3006	2SA229C	3006	2SA246E	3006	2SA263C	3006	2SA289Y
3006	2SA213G	3006	2SA229D	3006	2SA246F	3006	2SA263D	3006	2SA290A
3006	2SA213GN	3006	2SA229E	3006	2SA246G	3006	2SA263E	3006	2SA290B
3006	2SA213H	3006	2SA229F	3006	2SA246GN	3006	2SA263F	3006	2SA290C
3006	2SA213J	3006	2SA229G	3006	2SA246H	3006	2SA263G	3006	2SA290D
3006	2SA213K	3006	2SA229GN	3006	2SA246J	3006	2SA263GN	3006	2SA290E
3006	2SA213L	3006	2SA229H	3006	2SA246K	3006	2SA263H	3006	2SA290F
3006	2SA213M	3006	2SA229J	3006	2SA246L	3006	2SA263J	3006	2SA290G
3006	2SA213OR	3006	2SA229K	3006	2SA246M	3006	2SA263K	3006	2SA290GN
3006	2SA213R	3006	2SA229L	3006	2SA246OR	3006	2SA263L	3006	2SA290H
3006	2SA213X	3006	2SA229OR	3006	2SA246R	3006	2SA263M	3006	2SA290J
3006	2SA213Y	3006	2SA229X	3006	2SA246X	3006	2SA263OR	3006	2SA290K
3006	2SA214A	3006	2SA229Y	3006	2SA246Y	3006	2SA263R	3006	2SA290L
3006	2SA214B	3006	2SA230	3006	2SA247A	3006	2SA263X	3006	2SA290M
3006	2SA214C	3006	2SA230A	3006	2SA247B	3006	2SA263Y	3006	2SA290R
3006	2SA214D	3006	2SA230B	3006	2SA247C	3006	2SA264	3006	2SA290X
3006	2SA214E	3006	2SA230C	3006	2SA247D	3006	2SA264A	3006	2SA291A
3006	2SA214F	3006	2SA230D	3006	2SA247E	3006	2SA264B	3006	2SA291B
3006	2SA214G	3006	2SA230E	3006	2SA247F	3006	2SA264C	3006	2SA291D
3006	2SA214GN	3006	2SA230F	3006	2SA247GN	3006	2SA264D	3006	2SA291E
3006	2SA214H	3006	2SA230G	3006	2SA247H	3006	2SA264E	3006	2SA291F
3006	2SA214J	3006	2SA230GN	3006	2SA247J	3006	2SA264F	3006	2SA291G
3006	2SA214K	3006	2SA230H	3006	2SA247K	3006	2SA264G	3006	2SA291GN
3006	2SA214L	3006	2SA230J	3006	2SA247L	3006	2SA264GN	3006	2SA291H
3006	2SA214M	3006	2SA230K	3006	2SA247M	3006	2SA264H	3006	2SA291J
3006	2SA214OR	3006	2SA230L	3006	2SA247OR	3006	2SA264K	3006	2SA291K
3006	2SA214R	3006	2SA230M	3006	2SA247R	3006	2SA264L	3006	2SA291L
3006	2SA214X	3006	2SA230R	3006	2SA247X	3006	2SA264M	3006	2SA291M
3006	2SA214Y	3006	2SA230X	3006	2SA247Y	3006	2SA264R	3006	2SA291OR
3006	2SA215A	3006	2SA230Y	3006	2SA251A	3006	2SA264X	3006	2SA291R
3006	2SA215B	3006	2SA234A	3006	2SA251B	3006	2SA264Y	3006	2SA291X
3006	2SA215C	3006	2SA234C	3006	2SA251C	3006	2SA265	3006	2SA291Y
3006	2SA215D	3006	2SA234D	3006	2SA251D	3006	2SA265A	3006	2SA292A
3006	2SA215E	3006	2SA234E	3006	2SA251E	3006	2SA265B	3006	2SA292B
3006	2SA215F	3006	2SA234G	3006	2SA251F	3006	2SA265C	3006	2SA292C
3006	2SA215G	3006	2SA234GN	3006	2SA251G	3006	2SA265D	3006	2SA292D
3006	2SA215GN	3006	2SA234H	3006	2SA251H	3006	2SA265E	3006	2SA292E
3006	2SA215H	3006	2SA234J	3006	2SA251J	3006	2SA265F	3006	2SA292F
3006	2SA215J	3006	2SA234K	3006	2SA251K	3006	2SA265G	3006	2SA292G
3006	2SA215K	3006	2SA234L	3006	2SA251L	3006	2SA265GN	3006	2SA292GN
3006	2SA215L	3006	2SA234M	3006	2SA251M	3006	2SA265H	3006	2SA292H
3006	2SA215M	3006	2SA234OR	3006	2SA251OR	3006	2SA265J	3006	2SA292K
3006	2SA215OR	3006	2SA234R	3006	2SA251R	3006	2SA265K	3006	2SA292L
3006	2SA215R	3006	2SA234X	3006	2SA251X	3006	2SA265L	3006	2SA292M
3006	2SA215X	3006	2SA234Y	3006	2SA251Y	3006	2SA265M	3006	2SA292OR
3006	2SA215Y	3006	2SA235	3006	2SA252A	3006	2SA265OR	3006	2SA292R
3006	2SA216A	3006	2SA235A	3006	2SA252B	3006	2SA265R	3006	2SA292X
3006	2SA216B	3006	2SA235B	3006	2SA252C	3006	2SA265X	3006	2SA292Y
3006	2SA216C	3006	2SA235C	3006	2SA252D	3006	2SA265Y	3006	2SA293A
3006	2SA216D	3006	2SA235D	3006	2SA252E	3006	2SA267A	3006	2SA293B
3006	2SA216E	3006	2SA235E	3006	2SA252F	3006	2SA267B	3006	2SA293C
3006	2SA216F	3006	2SA235F	3006	2SA252G	3006	2SA267C	3006	2SA293D
3006	2SA216G	3006	2SA235G	3006	2SA252H	3006	2SA267D	3006	2SA293E
3006	2SA216GN	3006	2SA235GN	3006	2SA252J	3006	2SA267E	3006	2SA293G
3006	2SA216H	3006	2SA235H	3006	2SA252K	3006	2SA267F	3006	2SA293GN
3006	2SA216J	3006	2SA235K	3006	2SA252L	3006	2SA267G	3006	2SA293H
3006	2SA216K	3006	2SA235L	3006	2SA252OR	3006	2SA267GN	3006	2SA293J
3006	2SA216L	3006	2SA235M	3006	2SA252R	3006	2SA267H	3006	2SA293K
3006	2SA216M	3006	2SA235OR	3006	2SA252X	3006	2SA267J	3006	2SA293L
3006	2SA216OR	3006	2SA235R	3006	2SA252Y	3006	2SA267K	3006	2SA293M
3006	2SA216R	3006	2SA235X	3006	2SA253A	3006	2SA267L	3006	2SA293OR
3006	2SA216X	3006	2SA235Y	3006	2SA253B	3006	2SA267M	3006	2SA293R
3006	2SA216Y	3006	2SA238A	3006	2SA253C	3006	2SA267OR	3006	2SA293X
3006	2SA219A	3006	2SA238B	3006	2SA253D	3006	2SA267R	3006	2SA293Y
3006	2SA219B	3006	2SA238C	3006	2SA253E	3006	2SA267X	3006	2SA294A
3006	2SA219C	3006	2SA238D	3006	2SA253F	3006	2SA267Y	3006	2SA294B
3006	2SA219D	3006	2SA238E	3006	2SA253G	3006	2SA268A	3006	2SA294C
3006	2SA219E	3006	2SA238F	3006	2SA253GN	3006	2SA268B	3006	2SA294D
3006	2SA219F	3006	2SA238G	3006	2SA253H	3006	2SA268C	3006	2SA294F
3006	2SA219G	3006	2SA238GN	3006	2SA253J	3006	2SA268D	3006	2SA294GN
3006	2SA219GN	3006	2SA238H	3006	2SA253K	3006	2SA268E	3006	2SA294H
3006	2SA219H	3006	2SA238J	3006	2SA253L	3006	2SA268F	3006	2SA294J
3006	2SA219J	3006	2SA238L	3006	2SA253M	3006	2SA268G	3006	2SA294K
3006	2SA219K	3006	2SA239	3006	2SA2530R	3006	2SA268GN	3006	2SA294L
3006	2SA219L	3006	2SA239A	3006	2SA253R	3006	2SA268H	3006	2SA294M
3006	2SA219M	3006	2SA239B	3006	2SA253X	3006	2SA268J	3006	2SA294OR
3006	2SA219OR	3006	2SA239C	3006	2SA253Y	3006	2SA268K	3006	2SA294R
3006	2SA219R	3006	2SA239D	3006	2SA260	3006	2SA268L	3006	2SA294X
3006	2SA219X	3006	2SA239E	3006	2SA260A	3006	2SA268M	3006	2SA294Y
3006	2SA219Y	3006	2SA239G	3006	2SA260B	3006	2SA268OR	3006	2SA295A
3006	2SA223A	3006	2SA239GN	3006	2SA260D	3006	2SA268R	3006	2SA295B
3006	2SA223B	3006	2SA239H	3006	2SA260E	3006	2SA268X	3006	2SA295D
3006	2SA223C	3006	2SA239J	3006	2SA260F	3006	2SA268Y	3006	2SA295E
3006	2SA223F	3006	2SA239K	3006	2SA260G	3006	2SA280A	3006	2SA295F
3006	2SA223G				2SA260GN	3006	2SA280B		
3006	2SA223GN					3006	2SA280C		
3006	2SA223H					3006	2SA280D		
						3006	2SA280E		

SK	Industry Standard No.
3006	2SA295G
3006	2SA295GN
3006	2SA295H
3006	2SA295J
3006	2SA295K
3006	2SA295L
3006	2SA295OR
3006	2SA295R
3006	2SA295Y
3006	2SA301A
3006	2SA301B
3006	2SA301C
3006	2SA301D
3006	2SA301E
3006	2SA301F
3006	2SA301G
3006	2SA301GN
3006	2SA301H
3006	2SA301J
3006	2SA301K
3006	2SA301L
3006	2SA301M
3006	2SA301R
3006	2SA301X
3006	2SA301Y
3006	2SA308A
3006	2SA308B
3006	2SA308C
3006	2SA308D
3006	2SA308E
3006	2SA308F
3006	2SA308G
3006	2SA308GN
3006	2SA308H
3006	2SA308J
3006	2SA308K
3006	2SA308L
3006	2SA308M
3006	2SA308OR
3006	2SA308R
3006	2SA308X
3006	2SA308Y
3006	2SA309A
3006	2SA309B
3006	2SA309C
3006	2SA309D
3006	2SA309E
3006	2SA309F
3006	2SA309G
3006	2SA309GN
3006	2SA309H
3006	2SA309J
3006	2SA309K
3006	2SA309L
3006	2SA309M
3006	2SA309OR
3006	2SA309R
3006	2SA309X
3006	2SA309Y
3006	2SA310A
3006	2SA310B
3006	2SA310C
3006	2SA310D
3006	2SA310E
3006	2SA310F
3006	2SA310G
3006	2SA310GN
3006	2SA310H
3006	2SA310J
3006	2SA310K
3006	2SA310L
3006	2SA310M
3006	2SA310OR
3006	2SA310R
3006	2SA310X
3006	2SA310Y
3006	2SA321-1
3006	2SA321A
3006	2SA321B
3006	2SA321C
3006	2SA321D
3006	2SA321E
3006	2SA321F
3006	2SA321G
3006	2SA321GN
3006	2SA321H
3006	2SA321J
3006	2SA321K
3006	2SA321L
3006	2SA321M
3006	2SA321OR
3006	2SA321R
3006	2SA321X
3006	2SA321Y
3006	2SA324A
3006	2SA324B
3006	2SA324C
3006	2SA324D
3006	2SA324E
3006	2SA324G
3006	2SA324GN
3006	2SA324H
3006	2SA324K
3006	2SA324L
3006	2SA324M
3006	2SA324OR
3006	2SA324R
3006	2SA324X
3006	2SA324Y
3006	2SA325A
3006	2SA325B
3006	2SA325D
3006	2SA325E
3006	2SA325G
3006	2SA325GN
3006	2SA325H
3006	2SA325J
3006	2SA325K
3006	2SA325L
3006	2SA325M
3006	2SA325OR
3006	2SA325R
3006	2SA325X
3006	2SA325Y
3006	2SA326A
3006	2SA326B
3006	2SA326C
3006	2SA326D
3006	2SA326E
3006	2SA326F
3006	2SA326G
3006	2SA326GN
3006	2SA326H
3006	2SA326J
3006	2SA326K
3006	2SA326L
3006	2SA326M
3006	2SA326OR
3006	2SA326R
3006	2SA326X
3006	2SA326Y
3006	2SA329C
3006	2SA329D
3006	2SA329E
3006	2SA329F
3006	2SA329G
3006	2SA329GN
3006	2SA329H
3006	2SA329J
3006	2SA329K
3006	2SA329L
3006	2SA329M
3006	2SA329OR
3006	2SA329R
3006	2SA329X
3006	2SA329Y
3006	2SA331A
3006	2SA331B
3006	2SA331C
3006	2SA331D
3006	2SA331E
3006	2SA331F
3006	2SA331G
3006	2SA331GN
3006	2SA331H
3006	2SA331J
3006	2SA331K
3006	2SA331L
3006	2SA331M
3006	2SA331OR
3006	2SA331R
3006	2SA331X
3006	2SA331Y
3006	2SA335A
3006	2SA335B
3006	2SA335C
3006	2SA335D
3006	2SA335E
3006	2SA335F
3006	2SA335G
3006	2SA335GN
3006	2SA335H
3006	2SA335J
3006	2SA335K
3006	2SA335L
3006	2SA335M
3006	2SA335OR
3006	2SA335R
3006	2SA335X
3006	2SA335Y
3006	2SA337A
3006	2SA337B
3006	2SA337C
3006	2SA337D
3006	2SA337E
3006	2SA337F
3006	2SA337G
3006	2SA337GN
3006	2SA337H
3006	2SA337J
3006	2SA337K
3006	2SA337L
3006	2SA337M
3006	2SA337OR
3006	2SA337R
3006	2SA337X
3006	2SA337Y
3006	2SA3410A
3006	2SA342A
3006	2SA342B
3006	2SA342C
3006	2SA342D
3006	2SA342E
3006	2SA342F
3006	2SA342G
3006	2SA342GN
3006	2SA342H
3006	2SA342J
3006	2SA342K
3006	2SA342L
3006	2SA342M
3006	2SA342OR
3006	2SA342R
3006	2SA342X
3006	2SA342Y
3006	2SA343
3006	2SA343A
3006	2SA343B
3006	2SA343C
3006	2SA343D
3006	2SA343E
3006	2SA343F
3006	2SA343G
3006	2SA343H
3006	2SA343J
3006	2SA343K
3006	2SA343L
3006	2SA343M
3006	2SA343OR
3006	2SA343R
3006	2SA343X
3006	2SA343Y
3006	2SA344A
3006	2SA344B
3006	2SA344C
3006	2SA344D
3006	2SA344E
3006	2SA344F
3006	2SA344G
3006	2SA344GN
3006	2SA344H
3006	2SA344J
3006	2SA344K
3006	2SA344L
3006	2SA344M
3006	2SA344OR
3006	2SA344R
3006	2SA344X
3006	2SA344Y
3006	2SA348
3006	2SA348A
3006	2SA348B
3006	2SA348C
3006	2SA348D
3006	2SA348E
3006	2SA348F
3006	2SA348G
3006	2SA348GN
3006	2SA348H
3006	2SA348J
3006	2SA348K
3006	2SA348L
3006	2SA348M
3006	2SA348OR
3006	2SA348R
3006	2SA348X
3006	2SA348Y
3006	2SA355B
3006	2SA355C
3006	2SA355D
3006	2SA355E
3006	2SA355F
3006	2SA355G
3006	2SA355H
3006	2SA355J
3006	2SA355K
3006	2SA355L
3006	2SA355M
3006	2SA355OR
3006	2SA355R
3006	2SA355X
3006	2SA355Y
3006	2SA358-3
3006	2SA358A
3006	2SA358B
3006	2SA358C
3006	2SA358D
3006	2SA358E
3006	2SA358F
3006	2SA358G
3006	2SA358GN
3006	2SA358H
3006	2SA358J
3006	2SA358K
3006	2SA358L
3006	2SA358M
3006	2SA358OR
3006	2SA358R
3006	2SA358X
3006	2SA358Y
3006	2SA359A
3006	2SA359B
3006	2SA359C
3006	2SA359D
3006	2SA359E
3006	2SA359F
3006	2SA359G
3006	2SA359GN
3006	2SA359H
3006	2SA359J
3006	2SA359K
3006	2SA359L
3006	2SA359M
3006	2SA359OR
3006	2SA359R
3006	2SA359X
3006	2SA359Y
3006	2SA360A
3006	2SA360B
3006	2SA360C
3006	2SA360E
3006	2SA360F
3006	2SA360GN
3006	2SA360J
3006	2SA360L
3006	2SA360M
3006	2SA360OR
3006	2SA360R
3006	2SA360X
3006	2SA360Y
3006	2SA364A
3006	2SA364B
3006	2SA364C
3006	2SA364D
3006	2SA364E
3006	2SA364F
3006	2SA364G
3006	2SA364GN
3006	2SA364H
3006	2SA364J
3006	2SA364K
3006	2SA364L
3006	2SA364M
3006	2SA364OR
3006	2SA364R
3006	2SA364X
3006	2SA364Y
3006	2SA365A
3006	2SA365B
3006	2SA365C
3006	2SA365D
3006	2SA365E
3006	2SA365F
3006	2SA365G
3006	2SA365GN
3006	2SA365H
3006	2SA365J
3006	2SA365K
3006	2SA365L
3006	2SA365M
3006	2SA365OR
3006	2SA365R
3006	2SA365X
3006	2SA365Y
3006	2SA366A
3006	2SA366B
3006	2SA366C
3006	2SA366D
3006	2SA366E
3006	2SA366F
3006	2SA366G
3006	2SA366GN
3006	2SA366H
3006	2SA366J
3006	2SA366K
3006	2SA366L
3006	2SA366M
3006	2SA366OR
3006	2SA366R
3006	2SA366X
3006	2SA366Y
3006	2SA367A
3006	2SA367B
3006	2SA367C
3006	2SA367D
3006	2SA367E
3006	2SA367F
3006	2SA367G
3006	2SA367GN
3006	2SA367H
3006	2SA367J
3006	2SA367K
3006	2SA367L
3006	2SA367M
3006	2SA367OR
3006	2SA367R
3006	2SA367X
3006	2SA367Y
3006	2SA376A
3006	2SA376B
3006	2SA376C
3006	2SA376D
3006	2SA376E
3006	2SA376F
3006	2SA376GN
3006	2SA376H
3006	2SA376J
3006	2SA376K
3006	2SA376L
3006	2SA376OR
3006	2SA376R
3006	2SA376X
3006	2SA376Y
3006	2SA377
3006	2SA377A
3006	2SA377B
3006	2SA377C
3006	2SA377D
3006	2SA377E
3006	2SA377F
3006	2SA377G
3006	2SA377GN
3006	2SA377H
3006	2SA377J
3006	2SA377K
3006	2SA377L
3006	2SA377M
3006	2SA377OR
3006	2SA377R
3006	2SA377X
3006	2SA377Y
3006	2SA378
3006	2SA379
3006	2SA391A
3006	2SA391B
3006	2SA391C
3006	2SA391D
3006	2SA391E
3006	2SA391F
3006	2SA391G
3006	2SA391GN
3006	2SA391H
3006	2SA391J
3006	2SA391K
3006	2SA391L
3006	2SA391M
3006	2SA391OR
3006	2SA391R
3006	2SA391X
3006	2SA391Y
3006	2SA392A
3006	2SA392B
3006	2SA392C
3006	2SA392D
3006	2SA392E
3006	2SA392F
3006	2SA392G
3006	2SA392GN
3006	2SA392H
3006	2SA392J
3006	2SA392K
3006	2SA392L
3006	2SA392M
3006	2SA392OR
3006	2SA392R
3006	2SA392X
3006	2SA392Y
3006	2SA394A
3006	2SA394B
3006	2SA394C
3006	2SA394D
3006	2SA394E
3006	2SA394F
3006	2SA394G
3006	2SA394GN
3006	2SA394H
3006	2SA394J
3006	2SA394K
3006	2SA394L
3006	2SA394M
3006	2SA394OR
3006	2SA394R
3006	2SA394X
3006	2SA394Y
3006	2SA395A
3006	2SA395B
3006	2SA395C
3006	2SA395D
3006	2SA395E
3006	2SA395F
3006	2SA395G
3006	2SA395GN
3006	2SA395H
3006	2SA395J
3006	2SA395K
3006	2SA395L
3006	2SA395M
3006	2SA395OR
3006	2SA395R
3006	2SA395X
3006	2SA395Y
3006	2SA398A
3006	2SA398B
3006	2SA398C
3006	2SA398D
3006	2SA398E
3006	2SA398F
3006	2SA398G
3006	2SA398GN
3006	2SA398H
3006	2SA398J
3006	2SA398K
3006	2SA398L
3006	2SA398M
3006	2SA398OR
3006	2SA398X
3006	2SA398Y
3006	2SA399A
3006	2SA399B
3006	2SA399C
3006	2SA399D
3006	2SA399E
3006	2SA399F
3006	2SA399G
3006	2SA399GN
3006	2SA399H
3006	2SA399J
3006	2SA399K
3006	2SA399L
3006	2SA399M
3006	2SA399OR
3006	2SA399X
3006	2SA399Y
3006	2SA403
3006	2SA403A
3006	2SA403B
3006	2SA403C
3006	2SA403D
3006	2SA403E
3006	2SA403F
3006	2SA403G
3006	2SA403GN
3006	2SA403H
3006	2SA403J
3006	2SA403K
3006	2SA403L
3006	2SA403M
3006	2SA403OR
3006	2SA403R
3006	2SA403X
3006	2SA403Y
3006	2SA404
3006	2SA404A
3006	2SA404B
3006	2SA404C
3006	2SA404D
3006	2SA404E
3006	2SA404F
3006	2SA404G
3006	2SA404GN
3006	2SA404H
3006	2SA404K
3006	2SA404L
3006	2SA404M
3006	2SA404OR
3006	2SA404R
3006	2SA404X
3006	2SA404Y
3006	2SA412A
3006	2SA412B
3006	2SA412C
3006	2SA412D
3006	2SA412E
3006	2SA412F
3006	2SA412G
3006	2SA412GN
3006	2SA412H
3006	2SA412J
3006	2SA412K
3006	2SA412L
3006	2SA412M
3006	2SA412OR
3006	2SA412R
3006	2SA412X
3006	2SA412Y
3006	2SA420
3006	2SA420A
3006	2SA420B
3006	2SA420C
3006	2SA420D
3006	2SA420E
3006	2SA420F
3006	2SA420G
3006	2SA420H
3006	2SA420J
3006	2SA420K
3006	2SA420L
3006	2SA420M
3006	2SA420R
3006	2SA420X
3006	2SA420Y
3006	2SA426GN
3006	2SA427A
3006	2SA427B
3006	2SA427C
3006	2SA427D
3006	2SA427E
3006	2SA427F
3006	2SA427G
3006	2SA427GN
3006	2SA427H
3006	2SA427J
3006	2SA427K
3006	2SA427L
3006	2SA427M
3006	2SA427OR
3006	2SA427R
3006	2SA427X
3006	2SA427Y
3006	2SA428A
3006	2SA428B
3006	2SA428C
3006	2SA428D
3006	2SA428E
3006	2SA428F
3006	2SA428G
3006	2SA428GN
3006	2SA428H
3006	2SA428J
3006	2SA428K
3006	2SA428L
3006	2SA428M
3006	2SA428OR
3006	2SA428R
3006	2SA428X
3006	2SA428Y
3006	2SA432
3006	2SA432A
3006	2SA432B
3006	2SA432C
3006	2SA432D
3006	2SA432E
3006	2SA432F
3006	2SA432G
3006	2SA432GN
3006	2SA432H
3006	2SA432K
3006	2SA432L
3006	2SA432M
3006	2SA432OR
3006	2SA432R
3006	2SA432X
3006	2SA432Y
3006	2SA433
3006	2SA433A
3006	2SA433B
3006	2SA433C
3006	2SA433D
3006	2SA433E
3006	2SA433F

SK	Industry Standard No.
3006	2SA433G
3006	2SA433GN
3006	2SA433H
3006	2SA433K
3006	2SA433L
3006	2SA433M
3006	2SA433OR
3006	2SA433R
3006	2SA433X
3006	2SA433Y
3006	2SA434
3006	2SA434A
3006	2SA434B
3006	2SA434C
3006	2SA434D
3006	2SA434E
3006	2SA434F
3006	2SA434G
3006	2SA434GN
3006	2SA434H
3006	2SA434J
3006	2SA434K
3006	2SA434L
3006	2SA434M
3006	2SA434OR
3006	2SA434R
3006	2SA434X
3006	2SA434Y
3006	2SA435
3006	2SA435A
3006	2SA435B
3006	2SA435C
3006	2SA435D
3006	2SA435E
3006	2SA435F
3006	2SA435G
3006	2SA435GN
3006	2SA435H
3006	2SA435J
3006	2SA435L
3006	2SA435M
3006	2SA435OR
3006	2SA435R
3006	2SA435Y
3006	2SA436A
3006	2SA436B
3006	2SA436C
3006	2SA436D
3006	2SA436E
3006	2SA436F
3006	2SA436G
3006	2SA436GN
3006	2SA436H
3006	2SA436J
3006	2SA436K
3006	2SA437A
3006	2SA437B
3006	2SA437C
3006	2SA437D
3006	2SA437E
3006	2SA437F
3006	2SA437G
3006	2SA437GN
3006	2SA437H
3006	2SA437J
3006	2SA437K
3006	2SA437L
3006	2SA437M
3006	2SA437OR
3006	2SA437R
3006	2SA437X
3006	2SA437Y
3006	2SA438A
3006	2SA438B
3006	2SA438C
3006	2SA438D
3006	2SA438E
3006	2SA438F
3006	2SA438G
3006	2SA438GN
3006	2SA438H
3006	2SA438J
3006	2SA438K
3006	2SA438L
3006	2SA438M
3006	2SA438OR
3006	2SA438R
3006	2SA438X
3006	2SA438Y
3006	2SA440
3006	2SA440AL
3006	2SA440B
3006	2SA440C
3006	2SA440D
3006	2SA440E
3006	2SA440F
3006	2SA440G
3006	2SA440GN
3006	2SA440H
3006	2SA440J
3006	2SA440K
3006	2SA440L
3006	2SA440M
3006	2SA440OR
3006	2SA440R
3006	2SA440X
3006	2SA440Y
3006	2SA446A
3006	2SA446B
3006	2SA446C
3006	2SA446D
3006	2SA446E
3006	2SA446F
3006	2SA446G
3006	2SA446GN
3006	2SA446H
3006	2SA446J
3006	2SA446K
3006	2SA446L
3006	2SA446M
3006	2SA446OR
3006	2SA446R
3006	2SA446X
3006	2SA446Y
3006	2SA447A
3006	2SA447B
3006	2SA447C
3006	2SA447D
3006	2SA447E
3006	2SA447F
3006	2SA447G
3006	2SA447GN
3006	2SA447H
3006	2SA447J
3006	2SA447K
3006	2SA447L
3006	2SA447M
3006	2SA447OR
3006	2SA447R
3006	2SA447X
3006	2SA447Y
3006	2SA453A
3006	2SA453B
3006	2SA453C
3006	2SA453D
3006	2SA453E
3006	2SA453F
3006	2SA453G
3006	2SA453GN
3006	2SA453H
3006	2SA453J
3006	2SA453K
3006	2SA453L
3006	2SA453M
3006	2SA453OR
3006	2SA453R
3006	2SA453X
3006	2SA453Y
3006	2SA454A
3006	2SA454B
3006	2SA454C
3006	2SA454D
3006	2SA454E
3006	2SA454F
3006	2SA454G
3006	2SA454GN
3006	2SA454H
3006	2SA454J
3006	2SA454K
3006	2SA454L
3006	2SA454OR
3006	2SA454R
3006	2SA454X
3006	2SA454Y
3006	2SA455A
3006	2SA455B
3006	2SA455C
3006	2SA455D
3006	2SA455E
3006	2SA455F
3006	2SA455G
3006	2SA455GN
3006	2SA455H
3006	2SA455J
3006	2SA455K
3006	2SA455L
3006	2SA455M
3006	2SA455OR
3006	2SA455R
3006	2SA455X
3006	2SA455Y
3006	2SA456A
3006	2SA456B
3006	2SA456C
3006	2SA456D
3006	2SA456E
3006	2SA456F
3006	2SA456G
3006	2SA456GN
3006	2SA456H
3006	2SA456J
3006	2SA456K
3006	2SA456L
3006	2SA456M
3006	2SA456OR
3006	2SA456R
3006	2SA456X
3006	2SA456Y
3006	2SA457A
3006	2SA457B
3006	2SA457C
3006	2SA457D
3006	2SA457E
3006	2SA457F
3006	2SA457G
3006	2SA457GN
3006	2SA457H
3006	2SA457J
3006	2SA457K
3006	2SA457L
3006	2SA457M
3006	2SA457OR
3006	2SA457R
3006	2SA457X
3006	2SA457Y
3006	2SA466A
3006	2SA466B
3006	2SA466C
3006	2SA466D
3006	2SA466E
3006	2SA466F
3006	2SA466G
3006	2SA466GN
3006	2SA466H
3006	2SA466J
3006	2SA466K
3006	2SA466L
3006	2SA466M
3006	2SA466OR
3006	2SA466R
3006	2SA466X
3006	2SA466Y
3006	2SA477A
3006	2SA477B
3006	2SA477C
3006	2SA477D
3006	2SA477E
3006	2SA477F
3006	2SA477G
3006	2SA477GN
3006	2SA477H
3006	2SA477J
3006	2SA477K
3006	2SA477L
3006	2SA477M
3006	2SA477OR
3006	2SA477R
3006	2SA477X
3006	2SA477Y
3006	2SA478A
3006	2SA478B
3006	2SA478C
3006	2SA478D
3006	2SA478E
3006	2SA478F
3006	2SA478G
3006	2SA478GN
3006	2SA478H
3006	2SA478J
3006	2SA478K
3006	2SA478L
3006	2SA478M
3006	2SA478OR
3006	2SA478R
3006	2SA478X
3006	2SA478Y
3006	2SA479A
3006	2SA479B
3006	2SA479C
3006	2SA479D
3006	2SA479E
3006	2SA479F
3006	2SA479G
3006	2SA479GN
3006	2SA479H
3006	2SA479J
3006	2SA479K
3006	2SA479L
3006	2SA479M
3006	2SA479OR
3006	2SA479R
3006	2SA479X
3006	2SA479Y
3006	2SA507
3006	2SA507A
3006	2SA507B
3006	2SA507C
3006	2SA507D
3006	2SA507E
3006	2SA507F
3006	2SA507G
3006	2SA507GN
3006	2SA507H
3006	2SA507J
3006	2SA507K
3006	2SA507L
3006	2SA507OR
3006	2SA507R
3006	2SA507X
3006	2SA507Y
3006	2SA525A
3006	2SA525B
3006	2SA525C
3006	2SA525D
3006	2SA525E
3006	2SA525F
3006	2SA525GN
3006	2SA525H
3006	2SA525J
3006	2SA525K
3006	2SA525L
3006	2SA525M
3006	2SA525OR
3006	2SA525X
3006	2SA58A
3006	2SA58B
3006	2SA58C
3006	2SA58D
3006	2SA58E
3006	2SA58F
3006	2SA58G
3006	2SA58H
3006	2SA58J
3006	2SA58K
3006	2SA58L
3006	2SA58M
3006	2SA58OR
3006	2SA58R
3006	2SA58X
3006	2SA58Y
3006	2SA60A
3006	2SA60B
3006	2SA60C
3006	2SA60D
3006	2SA60E
3006	2SA60F
3006	2SA60G
3006	2SA60H
3006	2SA60K
3006	2SA60L
3006	2SA60M
3006	2SA60R
3006	2SA60X
3006	2SA60Y
3006	2SA67A
3006	2SA67B
3006	2SA67C
3006	2SA67D
3006	2SA67E
3006	2SA67F
3006	2SA67G
3006	2SA67H
3006	2SA67K
3006	2SA67L
3006	2SA67OR
3006	2SA67R
3006	2SA67X
3006	2SA67Y
3006	2SA69A
3006	2SA69B
3006	2SA69C
3006	2SA69D
3006	2SA69E
3006	2SA69F
3006	2SA69G
3006	2SA69H
3006	2SA69K
3006	2SA69L
3006	2SA69M
3006	2SA69OR
3006	2SA69R
3006	2SA69X
3006	2SA69Y
3006	2SA74A
3006	2SA74B
3006	2SA74C
3006	2SA74D
3006	2SA74E
3006	2SA74F
3006	2SA74G
3006	2SA74H
3006	2SA74K
3006	2SA74L
3006	2SA74M
3006	2SA74OR
3006	2SA74R
3006	2SA74X
3006	2SA74Y
3006	2SA76A
3006	2SA76B
3006	2SA76C
3006	2SA76D
3006	2SA76E
3006	2SA76F
3006	2SA76G
3006	2SA76H
3006	2SA76K
3006	2SA76L
3006	2SA76OR
3006	2SA76R
3006	2SA76X
3006	2SA76Y
3006	2SA77E
3006	2SA77F
3006	2SA77G
3006	2SA77H
3006	2SA77K
3006	2SA77L
3006	2SA77M
3006	2SA77OR
3006	2SA77R
3006	2SA77X
3006	2SA77Y
3006	2SA87A
3006	2SA87B
3006	2SA87C
3006	2SA87D
3006	2SA87E
3006	2SA87F
3006	2SA87G
3006	2SA87H
3006	2SA87K
3006	2SA87L
3006	2SA87M
3006	2SA87OR
3006	2SA87R
3006	2SA87X
3006	2SA87Y
3006	2SA90
3006	2SAU03H
3006	2SC101XL
3006	2T201
3006	2T203
3006	03-57-200
3006	4-432
3006	4-65-1A7-1
3006	4-88B17-1
3006	6-69
3006	6-69X
3006	07-3080-06
3006	07-4233-19
3006	07-4235-13
3006	07-4235-73
3006	8-0050400
3006	8-905-605-320
3006	8-905-606-001
3006	8-905-606-003
3006	8-905-606-007
3006	8-905-606-008
3006	8-905-606-010
3006	8-905-606-016
3006	8-905-606-051
3006	8-905-606-075
3006	8-905-606-077
3006	8-905-606-090
3006	8-905-606-105
3006	8-905-606-106
3006	8-905-606-120
3006	8-905-606-142
3006	8-905-606-152
3006	8-905-606-154
3006	8-905-606-155
3006	8-905-606-158
3006	8-905-606-165
3006	8-905-606-168
3006	8-905-606-180
3006	8-905-606-211
3006	8-905-606-225
3006	8-905-606-241
3006	8-905-606-255
3006	8-905-606-256
3006	8-905-606-349
3006	8-905-606-351
3006	8-905-606-352
3006	8-905-606-360
3006	8-905-606-375
3006	8-905-606-390
3006	8-905-606-391
3006	8-905-606-392
3006	8-905-606-405
3006	8-905-606-419
3006	8-905-606-420
3006	8-905-606-423
3006	8-905-706-790
3006	8D
3006	8E
3006	8F
3006	09-300021
3006	09-300024
3006	09-300028
3006	09-304011
3006	9-5108
3006	9-5110
3006	9-5111
3006	9-5116
3006	9-5117
3006	9-5118
3006	9-5119
3006	9-5120
3006	9-5121
3006	9-5122
3006	9-5123
3006	9-5124
3006	9-9105
3006	9-9106
3006	9-9107
3006	9-9108
3006	9-9120
3006	9-9121
3006	12-1-135
3006	12-1-137
3006	12-1-139
3006	12-1-157
3006	12-1-228
3006	12-1-229
3006	12-1-230
3006	12-1-231
3006	12-1-233
3006	13-14085-28
3006	13-14085-30
3006	13-18946-1
3006	13-18947-1
3006	13-18948-1
3006	13-18948-2
3006	14-600-01
3006	14-600-04
3006	14-600-10
3006	14-600-16
3006	15-22210131
3006	15-22210300
3006	15-22211021
3006	15-22211921
3006	15-22214400
3006	15-22214411
3006	15-22214435
3006	15-22214821
3006	15-22214831
3006	15-22217400
3006	19-020-031
3006	19-020-032
3006	19A115140-P1
3006	19A115140-P2
3006	19A115192-P1
3006	19A115192-P2
3006	19A115628-P1
3006	19A115635-1
3006	19A115636-P1
3006	19A115665-P1
3006	19A115665-P2
3006	19AR13-1
3006	19AR13-2
3006	19AR13-3
3006	19AR13-4
3006	19AR18
3006	19AR24
3006	19B2000130-P1
3006	19B2000130-P2
3006	19C300216-P1
3006	020-1110-006
3006	21A048-000
3006	21A049-000
3006	21A050-000
3006	21A050-001
3006	23-PT284-122
3006	23-PT284-123
3006	24MW44
3006	24MW55
3006	24MW59
3006	31-0001
3006	31-0002
3006	31-0003
3006	31-0141
3006	31-0150
3006	31-0168
3006	31-0180
3006	31-0241
3006	31-0241-1
3006	31-21004900
3006	31-21007744
3006	31-21024033
3006	31-21024044
3006	31-21047111
3006	31-2104733
3006	31-21050611
3006	31-21050622
3006	34-119
3006	34-220
3006	34-221
3006	34-298
3006	34-6
3006	34-6000-10
3006	34-6000-11
3006	34-6000-12
3006	34-6000-13
3006	34-6000-14
3006	34-6000-17
3006	34-6000-20
3006	34-6000-25
3006	34-6000-26
3006	34-6000-58
3006	34-6000-59
3006	34-6000-60
3006	34-6000-61
3006	34-6000-62
3006	34-6000-63
3006	34-6000-67
3006	34-6000-68
3006	34-6000-76
3006	34-6000-77
3006	34-6000-78
3006	34-6000-79
3006	34-6000-80
3006	34-6000-81
3006	34-6000-82
3006	34-6000-9
3006	42-19792
3006	42-22778
3006	42-22779
3006	42-22780
3006	42-22781
3006	42-22784
3006	42-23965P
3006	46-86300-3
3006	48-97271A06
3006	488134956
3006	48X97046A16
3006	57B2-11
3006	57B2-13
3006	57B2-14
3006	57B2-22
3006	57C5-2
3006	57D1-107
3006	57D1-114
3006	57D1-96
3006	58B2-14
3006	065-1-12
3006	065-1-12-7
3006	65-1-70
3006	65-1-70-12
3006	65-1-70-12-7
3006	65-10
3006	65-11
3006	65-12
3006	65-13
3006	65-15
3006	65-16
3006	65-17
3006	65-18
3006	65-19
3006	65-1A
3006	65-1A0
3006	65-1A0R
3006	65-1A1
3006	65-1A19
3006	65-1A2
3006	65-1A21
3006	65-1A3
3006	65-1A3P
3006	65-1A4
3006	65-1A4-7

SK	Industry Standard No.
3006	65-1A5
3006	65-1A5L
3006	65-1A6
3006	65-1A6-5
3006	65-1A7
3006	65-1A7-1
3006	65-1A8
3006	65-1A82
3006	65-1A9
3006	65-1A9G
3006	070-001
3006	80-050400
3006	86-100-2
3006	86-102-2
3006	86-112-2
3006	86-117-2
3006	86-135-2
3006	86-136-2
3006	86-149-2
3006	86-150-2
3006	86-162-2
3006	86-163-2
3006	86-164-2
3006	86-179-2
3006	86-180-2
3006	86-181-2
3006	86-253-2
3006	86-254-2
3006	86-278-2
3006	86-279-2
3006	86-296-2
3006	86-363-2
3006	86-367-2
3006	86-368-2
3006	86-373-2
3006	86-374-2
3006	86-376-2
3006	86-449-9
3006	86-88-2
3006	86-89-2
3006	86-90-2
3006	86-91-2
3006	86-92-2
3006	088B
3006	088B-12
3006	088B-12-7
3006	88B-70
3006	88B-70-12
3006	88B-70-12-7
3006	88B0
3006	88B1
3006	88B10
3006	88B10R
3006	88B11
3006	88B119
3006	88B12
3006	88B121
3006	88B13
3006	88B13P
3006	88B14
3006	88B14-7
3006	88B15
3006	88B15L
3006	88B16
3006	88B16E
3006	88B17
3006	88B17-1
3006	88B18
3006	88B182
3006	88B19
3006	88B19G
3006	88B2
3006	88B3
3006	88B4
3006	88B5
3006	88B6
3006	88B7
3006	88B8
3006	88B9
3006	95-108
3006	95-110
3006	95-116
3006	95-117
3006	95-118
3006	95-119
3006	95-120
3006	95-121
3006	95-122
3006	95-123
3006	95-124
3006	96-5062-01
3006	96-5095-01
3006	96-5138-01
3006	96-5139-01
3006	96-5140-01
3006	96XZ801/37N
3006	998006
3006	998007
3006	108-1
3006	108-2
3006	108-3
3006	108-4
3006	115-227
3006	115-228
3006	115-229
3006	116-202
3006	116-683
3006	116-684
3006	120-001190
3006	120-002214
3006	121-132
3006	121-135
3006	121-137
3006	121-139
3006	121-157
3006	121-304
3006	121-349
3006	121-350
3006	121-351
3006	121-352
3006	121-353
3006	121-356
3006	121-358
3006	121-359
3006	121-384
3006	121-385
3006	121-411
3006	121-412
3006	121-413
3006	121-414
3006	121-415
3006	121-426
3006	121-427
3006	121-428

SK	Industry Standard No.
3006	121-429
3006	121-432
3006	121-538
3006	121-539
3006	121-540
3006	121-697
3006	121-698
3006	0131-000418
3006	0131-000419
3006	0131-000498
3006	0131-000802
3006	0131-000859
3006	0131-000862
3006	0131-000863
3006	0131-001182
3006	0131-001314
3006	0131-001332
3006	0131-001433
3006	0131-001434
3006	0131-001435
3006	0131-001436
3006	0131-001697
3006	0131-003029
3006	132-019
3006	132-020
3006	132-027
3006	157T1
3006	162T1
3006	207A1
3006	297V024H03
3006	297V064B01
3006	324-0095
3006	324-0187
3006	324-132
3006	325-0028-85
3006	353-9301-002
3006	421-9792
3006	422-2778
3006	422-2779
3006	422-2780
3006	573-518
3006	576-003-009
3006	601-040
3006	602-075
3006	603-020
3006	603-030
3006	603-040
3006	604-030
3006	604-080
3006	609-020
3006	610-050
3006	610-051
3006	610-053
3006	610-055
3006	617-54
3006	617-57
3006	690V040H56
3006	690V057H25
3006	690V119H95
3006	690V119H96
3006	800-504-00
3006	800-50400
3006	921-1A
3006	921-1B
3006	921-2A
3006	921-2B
3006	921-3A
3006	921-3B
3006	921-4A
3006	955-1
3006	955-2
3006	955-3
3006	972X1
3006	972X2
3006	972X3
3006	972X4
3006	972X5
3006	1119-56
3006	1465-1
3006	1465-1-12
3006	1465-1-12-8
3006	1488B
3006	1488B-12
3006	1488B-12-8
3006	1865-1
3006	1865-1-127
3006	1865-1L
3006	1888B
3006	1888B-127
3006	1888BL
3006	1888BL8
3006	2015
3006	2020
3006	2021
3006	2057A2-149
3006	2057A2-166
3006	2057A2-231
3006	2057A2-232
3006	2057A2-252
3006	2057B2-149
3006	2057B2-157
3006	2057B2-158
3006	2057B2-159
3006	2057B2-89
3006	2057B2-90
3006	2057B2-93
3006	2057B2A2-118
3006	2106-120
3006	2215-17
3006	2495-078
3006	2495-079
3006	2495-082
3006	2495-376
3006	2495-377
3006	2495-378
3006	2606-295
3006	2700
3006	6990
3006	9510-7
3006	12113-5
3006	12113-7
3006	12113-9
3006	12115-7
3006	12122-8
3006	12122-9
3006	12123-0
3006	12123-1
3006	12123-3
3006	18541
3006	30238
3006	30239
3006	30240
3006	68504-62
3006	69107-43

SK	Industry Standard No.
3006	77271-8
3006	080026
3006	080027
3006	080028
3006	080061
3006	080244
3006	080245
3006	080266
3006	080267
3006	080269
3006	95107
3006	95108
3006	95110
3006	95121
3006	95124
3006	99105
3006	99106
3006	99107
3006	99108
3006	111954
3006	111955
3006	111956
3006	115227
3006	115228
3006	115229
3006	116202
3006	116207
3006	116208
3006	116209
3006	116683
3006	116684
3006	117725
3006	117726
3006	117866
3006	119013
3006	124625
3006	129347
3006	171016
3006	171039
3006	175006-181
3006	175006-182
3006	175006-183
3006	175006-184
3006	175006-185
3006	224586
3006	224587
3006	225600
3006	229133
3006	232676
3006	232680
3006	234015
3006	234631
3006	235200
3006	265771
3006	310132
3006	310162
3006	310204
3006	0510079
3006	0510079H
3006	0573335
3006	573336
3006	573371
3006	573398
3006	573405
3006	573406
3006	0573428
3006	601040
3006	602075
3006	603020
3006	603030
3006	603040
3006	604030
3006	604080
3006	609020
3006	610050
3006	610050-1
3006	610050-2
3006	610050-3
3006	610051
3006	610051-1
3006	610051-2
3006	610051-4
3006	610053
3006	610053-1
3006	610053-2
3006	610055
3006	610055-1
3006	610055-2
3006	610055-3
3006	660084
3006	660085
3006	785897-01
3006	815193
3006	815197
3006	815234
3006	980140
3006	980142
3006	980146
3006	980435
3006	980441
3006	980505
3006	980506
3006	980507
3006	980509
3006	980514A
3006	980545A
3006	980636A
3006	981143
3006	981144
3006	981145
3006	981146
3006	982322
3006	982374
3006	982497
3006	985446A
3006	2000646-104
3006	2002153-58
3006	2002153-59
3006	2002153-60
3006	2002153-76
3006	2003073-11
3006	2003073-12
3006	2003073-13
3006	2091247-005
3006	2092417-0707
3006	2092417-0708
3006	2092417-0709
3006	2092417-0710
3006	2092417-1
3006	2092417-2
3006	2092417-3
3006	2092417-6
3006	2092418-0710
3006	2092418-1

SK	Industry Standard No.
3006	2092418-2
3006	2092418-5
3006	2092418-6
3006	2092418-7
3006	2092693-9
3006	2495078
3006	2495079
3006	2495082
3006	2495376
3006	2495377
3006	4036715-P1
3006	4036715-P2
3006	4036923-P1
3006	4036923-P2
3006	4036962-P1
3006	4036962-P2
3006	4036963-P1
3006	4036963-P2
3006	4036965-P1
3006	4036965-P2
3006	4037410-P1
3006	4037410-P2
3006	4037764-P1
3006	4037764-P2
3006	4038359-P1
3006	4038359-P2
3006	4038406-P1
3006	4038406-P2
3006	7284513
3006	7292308
3006	7852900-01
3006	31210049-00
3006	31210077-44
3006	31210240-33
3006	31210240-44
3006	31210471-11
3006	31210506-11
3006	31210506-22
3006	43022055
3006	43025620
3006	62034076
3006	62046406
3006	62087641
3006	62593733
3006	80050400
3006	80146620
3006	80146630
3006	120001190
3006	120002214
3006	120002513
3006	120002515
3006	120002518
3006	120002520
3006	120002656
3006	120004492
3006	312104733
3006	1522211921
3006	1522214400
3006	1522214831
3006	1522217400
3006	3121004900
3006	3121007744
3006	3121024033
3006	3121024044
3006	3121047111
3006	3121050611
3006	3121050622
3006/160	A059-104
3006/160	A061-107
3006/160	A061-110
3006/160	A061-111
3006/160	A069-121
3006/160	A101AA
3006/160	A101AY
3006/160	A101B
3006/160	A101BA
3006/160	A101BB
3006/160	A101BC
3006/160	A101BX
3006/160	A101C
3006/160	A101CA
3006/160	A101CV
3006/160	A101CX
3006/160	A101E
3006/160	A101QA
3006/160	A101Y
3006/160	A103A
3006/160	A103B
3006/160	A103C
3006/160	A103CG
3006/160	A103DA
3006/160	A104D
3006/160	A104P
3006/160	A105
3006/160	A109
3006/160	A113
3006/160	A117
3006/160	A118
3006/160	A124
3006/160	A125
3006/160	A126
3006/160	A131
3006/160	A134
3006/160	A1383
3006/160	A14-1001
3006/160	A14-1002
3006/160	A14-1003
3006/160	A15-1001
3006/160	A15-1002
3006/160	A15-1003
3006/160	A188
3006/160	A189
3006/160	A234B
3006/160	A234C
3006/160	A235A
3006/160	A235C
3006/160	A38
3006/160	A39
3006/160	A43
3006/160	A45
3006/160	A45-1
3006/160	A45-2
3006/160	A45-3
3006/160	A51
3006/160	A56
3006/160	A67
3006/160	A69
3006/160	A72
3006/160	A72BLU
3006/160	A72BRN
3006/160	A72ORN
3006/160	A72WHT
3006/160	A75
3006/160	A76

SK	Industry Standard No.
3006/160	A77A
3006/160	A77B
3006/160	A77C
3006/160	A77D
3006/160	A78
3006/160	A79
3006/160	A80
3006/160	A82
3006/160	A83
3006/160	A87
3006/160	OA90
3006/160	A909-1009
3006/160	A909-1010
3006/160	A94
3006/160	AC164
3006/160	AF108
3006/160	AF114
3006/160	AF115
3006/160	AF116
3006/160	AF117
3006/160	AF120
3006/160	AF124
3006/160	AF125
3006/160	AF126
3006/160	AF129
3006/160	AF130
3006/160	AF131
3006/160	AF132
3006/160	AF133
3006/160	AF138
3006/160	AF144
3006/160	AFY10
3006/160	AFY11
3006/160	AFY18
3006/160	AG134
3006/160	APB-11H-1007
3006/160	ASZ10
3006/160	ASZ11
3006/160	ASZ21
3006/160	ASZ30
3006/160	ATRF1
3006/160	ATRF2
3006/160	ATS13
3006/160	B235A
3006/160	B384
3006/160	B385
3006/160	B444
3006/160	B444A
3006/160	B444B
3006/160	B74
3006/160	CK4
3006/160	CK4A
3006/160	CK762
3006/160	CK766A
3006/160	DAT1A
3006/160	DAT2
3006/160	EA15X133
3006/160	EA15X140
3006/160	EA15X141
3006/160	EA2133
3006/160	EA2491
3006/160	EA2497
3006/160	EA2498
3006/160	ES14
3006/160	ES3110
3006/160	ES3111
3006/160	ES3112
3006/160	ES3113
3006/160	ES3114
3006/160	ES3115
3006/160	ES3116
3006/160	ES41
3006/160	ESA213
3006/160	ESA233
3006/160	ET12
3006/160	G0008
3006/160	GET871
3006/160	GET872
3006/160	GET875
3006/160	GET885
3006/160	GT46
3006/160	GT47
3006/160	GT5117
3006/160	GT5148
3006/160	GT5149
3006/160	GT5151
3006/160	GT5153
3006/160	GT66
3006/160	GTJ33141
3006/160	GTJ33229
3006/160	GTJ33230
3006/160	HA-269
3006/160	HA-350
3006/160	HA-350A
3006/160	HA-353
3006/160	HA-354
3006/160	HA471
3006/160	HEP-G0001
3006/160	HEP-G0008
3006/160	HEP-G0010
3006/160	HT101011X
3006/160	HT103501A
3006/160	HT103531C
3006/160	HT103541B
3006/160	HT200541B
3006/160	M4363
3006/160	M4366
3006/160	M4367
3006/160	M4368
3006/160	M4454
3006/160	M4456
3006/160	M4457
3006/160	M4501
3006/160	M4509
3006/160	M4586
3006/160	M4589
3006/160	M4603
3006/160	M4604
3006/160	M4632
3006/160	MAS20
3006/160	MAS21
3006/160	MAS23
3006/160	N-EA15X133
3006/160	NKT103
3006/160	NKT127
3006/160	NKT131
3006/160	NKT151
3006/160	NKT152
3006/160	NKT153/25
3006/160	NKT15325
3006/160	NKT154/25
3006/160	NKT249
3006/160	NKT252
3006/160	NKT253

SK	Industry Standard No.
3006/160	NKT254
3006/160	NKT255
3006/160	NKT265
3006/160	NKT270
3006/160	NKT675
3006/160	NKT676
3006/160	NKT677
3006/160	OC-170
3006/160	OC-171
3006/160	OC130
3006/160	OC331
3006/160	OC42
3006/160	OC43
3006/160	P1L
3006/160	R581
3006/160	R9600
3006/160	RE15
3006/160	RS-2003
3006/160	RS5753-2
3006/160	RS6821
3006/160	RS6822
3006/160	RT5063
3006/160	RT5466
3006/160	S-55TB
3006/160	SFT357P
3006/160	T1388
3006/160	T1524BRN
3006/160	T1654
3006/160	TIX-M14
3006/160	TIX-M15
3006/160	TIX-M16
3006/160	TNJ71248
3006/160	TNJ72287
3006/160	TR02012
3006/160	TRA-22
3006/160	TVS-2SA385
3006/160	TVS-2SA385A
3006/160	V10/1SJ
3006/160	V15/20R
3006/160	VFL-2744K
3006/160	VFW-27450
3006/160	VS-2SA358
3006/160	VS-2SA378
3006/160	VS-2SA379
3006/160	ZEN311
3006/160	ZEN312
3006/160	2A
3006/160	2B
3006/160	2G
3006/160	2G344
3006/160	2G601
3006/160	2G602
3006/160	2MC
3006/160	2N1003
3006/160	2N1108RED
3006/160	2N1111A
3006/160	2N1111B
3006/160	2N1111RED
3006/160	2N111B
3006/160	2N111M1
3006/160	2N112M1
3006/160	2N1180
3006/160	2N1204
3006/160	2N1204A
3006/160	2N1213
3006/160	2N1214
3006/160	2N1215
3006/160	2N1216
3006/160	2N1225
3006/160	2N123A
3006/160	2N1264
3006/160	2N1395
3006/160	2N1396
3006/160	2N1397
3006/160	2N1398
3006/160	2N1399
3006/160	2N1400
3006/160	2N1401
3006/160	2N1401A
3006/160	2N1402
3006/160	2N140M2
3006/160	2N1499
3006/160	2N1499A
3006/160	2N1499B
3006/160	2N1515
3006/160	2N1516
3006/160	2N1517
3006/160	2N1726
3006/160	2N1727
3006/160	2N1728
3006/160	2N1742
3006/160	2N1745
3006/160	2N1746
3006/160	2N1748
3006/160	2N1748A
3006/160	2N1785
3006/160	2N1786
3006/160	2N1787
3006/160	2N2089
3006/160	2N2090
3006/160	2N2091
3006/160	2N2188
3006/160	2N2189
3006/160	2N2207
3006/160	2N2208
3006/160	2N231-YEL-RED
3006/160	2N231BLU
3006/160	2N231YEL
3006/160	2N2451
3006/160	2N247/33
3006/160	2N2487
3006/160	2N2488
3006/160	2N2489
3006/160	2N2494
3006/160	2N299
3006/160	2N3075
3006/160	2N33
3006/160	2N344
3006/160	2N345
3006/160	2N346
3006/160	2N499
3006/160	2N499A
3006/160	2N500BLU
3006/160	2N500RED
3006/160	2N500WHT
3006/160	2N501
3006/160	2N501/18
3006/160	2N501A
3006/160	2N502
3006/160	2N502A
3006/160	2N502B
3006/160	2N588
3006/160	2N588A
3006/160	2N591-6M
3006/160	2N604A
3006/160	2N695
3006/160	2N72
3006/160	2N741
3006/160	2N75
3006/160	2N85
3006/160	2N86
3006/160	2N87
3006/160	2N88
3006/160	2N89
3006/160	2N90
3006/160	2N982
3006/160	2N983
3006/160	2N984
3006/160	2N991
3006/160	2N992
3006/160	2NJ52
3006/160	2NJ53
3006/160	2S141
3006/160	2S142
3006/160	2S143
3006/160	2S145
3006/160	2S146
3006/160	2S176
3006/160	2S96
3006/160	2S97
3006/160	2S98
3006/160	2SA1010Y
3006/160	2SA116
3006/160	2SA117
3006/160	2SA118
3006/160	2SA121
3006/160	2SA122
3006/160	2SA123
3006/160	2SA124
3006/160	2SA125
3006/160	2SA133
3006/160	2SA134
3006/160	2SA136
3006/160	2SA139
3006/160	2SA141B
3006/160	2SA141C
3006/160	2SA143
3006/160	2SA153
3006/160	2SA154
3006/160	2SA155
3006/160	2SA156
3006/160	2SA157
3006/160	2SA159
3006/160	2SA161
3006/160	2SA162
3006/160	2SA163
3006/160	2SA164
3006/160	2SA165
3006/160	2SA166
3006/160	2SA175
3006/160	2SA197
3006/160	2SA213
3006/160	2SA214
3006/160	2SA215
3006/160	2SA216
3006/160	2SA219
3006/160	2SA223
3006/160	2SA225
3006/160	2SA238
3006/160	2SA246
3006/160	2SA246Y
3006/160	2SA247
3006/160	2SA251
3006/160	2SA252
3006/160	2SA253
3006/160	2SA267
3006/160	2SA268
3006/160	2SA280
3006/160	2SA281
3006/160	2SA288A
3006/160	2SA289
3006/160	2SA290
3006/160	2SA291
3006/160	2SA292
3006/160	2SA293
3006/160	2SA295
3006/160	2SA301
3006/160	2SA308
3006/160	2SA309
3006/160	2SA310
3006/160	2SA321
3006/160	2SA324
3006/160	2SA324F
3006/160	2SA325
3006/160	2SA326
3006/160	2SA329
3006/160	2SA335
3006/160	2SA337
3006/160	2SA344
3006/160	2SA355
3006/160	2SA355A
3006/160	2SA359
3006/160	2SA364
3006/160	2SA365
3006/160	2SA366
3006/160	2SA372
3006/160	2SA376
3006/160	2SA391
3006/160	2SA392
3006/160	2SA393
3006/160	2SA393A
3006/160	2SA394
3006/160	2SA395
3006/160	2SA398
3006/160	2SA399
3006/160	2SA412
3006/160	2SA427
3006/160	2SA428
3006/160	2SA436
3006/160	2SA437
3006/160	2SA438
3006/160	2SA446
3006/160	2SA447
3006/160	2SA45
3006/160	2SA45-1
3006/160	2SA45-2
3006/160	2SA45-3
3006/160	2SA453
3006/160	2SA454
3006/160	2SA455
3006/160	2SA456
3006/160	2SA457
3006/160	2SA466
3006/160	2SA466-2
3006/160	2SA466-3
3006/160	2SA466BLK
3006/160	2SA466BLU
3006/160	2SA466YEL
3006/160	2SA477
3006/160	2SA478
3006/160	2SA479
3006/160	2SA67
3006/160	2SA69
3006/160	2SA72
3006/160	2SA72BLU
3006/160	2SA72BLU-BLU
3006/160	2SA72BRN
3006/160	2SA72ORN
3006/160	2SA72WHT
3006/160	2SA76
3006/160	2SA77A
3006/160	2SA77B
3006/160	2SA77C
3006/160	2SA78
3006/160	2SA80
3006/160	2SA87
3006/160	2SB25B
3006/160	2SB444A
3006/160	2SB444B
3006/160	6-50
3006/160	6-62X
3006/160	6-70
3006/160	6-71
3006/160	6-72
3006/160	6-89
3006/160	07-2012-04
3006/160	07-3012-04
3006/160	07-3350-57
3006/160	8-0060
3006/160	09-300006
3006/160	09-300027
3006/160	09-300029
3006/160	09-300079
3006/160	9-9101
3006/160	9-9102
3006/160	9-9103
3006/160	13-14085-31
3006/160	13-14085-32
3006/160	13-14085-33
3006/160	13-14886-1
3006/160	13-14887-1
3006/160	13-14889-1
3006/160	13-18951-1
3006/160	13-18951-2
3006/160	14-582-01
3006/160	19A115087-P1
3006/160	19A115098
3006/160	19A115098-P1
3006/160	19A115099-P1
3006/160	21M006
3006/160	21M007
3006/160	022-5311-770
3006/160	022-5311-780
3006/160	24MW27
3006/160	24MW271
3006/160	24MW303
3006/160	24MW74
3006/160	24MW816
3006/160	27T412
3006/160	29V008M01
3006/160	29V011H01
3006/160	29V012H01
3006/160	031
3006/160	31-0041
3006/160	31-0042
3006/160	31-0108
3006/160	31-0123
3006/160	31-0124
3006/160	31-0132
3006/160	31-0184
3006/160	31-0191
3006/160	31-0228
3006/160	34-8001-43
3006/160	44-13
3006/160	48-124255
3006/160	48-124256
3006/160	48-124305
3006/160	48-124310
3006/160	48-124311
3006/160	48-124346
3006/160	48-124349
3006/160	48-124350
3006/160	48-124360
3006/160	48-124364
3006/160	48-124365
3006/160	48-124366
3006/160	48-124367
3006/160	48-125228
3006/160	48-125278
3006/160	48-128096
3006/160	48-134101
3006/160	48-134404
3006/160	48-134405
3006/160	48-134406
3006/160	48-134414
3006/160	48-134434
3006/160	48-134454
3006/160	48-134456
3006/160	48-134457
3006/160	48-134479
3006/160	48-134480
3006/160	48-134481
3006/160	48-134508
3006/160	48-134514
3006/160	48-134536
3006/160	48-134545
3006/160	48-134547
3006/160	48-134561
3006/160	48-134591
3006/160	48-134600
3006/160	48-134601
3006/160	48-134602
3006/160	48-134635
3006/160	48-134681
3006/160	48-134682
3006/160	48-134683
3006/160	48-134697
3006/160	48-134711
3006/160	48-134795
3006/160	48-134797
3006/160	48-134798
3006/160	48-134860
3006/160	48-63029A90
3006/160	48-63075A74
3006/160	48-63078A65
3006/160	48-63082A24
3006/160	48-97046A08
3006/160	48-97046A09
3006/160	48-97238A01
3006/160	48-97238A02
3006/160	48-97271A04
3006/160	48X97238A01
3006/160	48X97238A02
3006/160	48X97238A03
3006/160	50A103
3006/160	50A103K
3006/160	50P2
3006/160	50P3
3006/160	51P2
3006/160	51P4
3006/160	55P2
3006/160	55P3
3006/160	57C16
3006/160	57C5-1
3006/160	57C5-3
3006/160	57C5-4
3006/160	57C6-16
3006/160	57D168
3006/160	57D187
3006/160	57D5-1
3006/160	57D5-2
3006/160	57D5-4
3006/160	61B003-1
3006/160	61P1
3006/160	61P10
3006/160	65-1
3006/160	70.00.730
3006/160	76
3006/160	86-119-2
3006/160	86-132-2
3006/160	86-280-2
3006/160	86-449-2
3006/160	86X0013-001
3006/160	86X0014-001
3006/160	089-220
3006/160	95-111
3006/160	96XZ6051-28N
3006/160	96XZ6051-35N
3006/160	96XZ6051-36N
3006/160	96XZ6053-10N
3006/160	96XZ6053-24N
3006/160	0101-0034
3006/160	0101-0222
3006/160	104-17
3006/160	104-19
3006/160	104-21
3006/160	111-6935
3006/160	112-002
3006/160	115-275
3006/160	116-072
3006/160	116-756
3006/160	120-00190
3006/160	120-00213
3006/160	120-002513
3006/160	120-002515
3006/160	120-002518
3006/160	120-002520
3006/160	120-002656
3006/160	120-004048
3006/160	120-004492
3006/160	121-113
3006/160	121-187
3006/160	121-228
3006/160	121-229
3006/160	121-230
3006/160	121-231
3006/160	121-233
3006/160	121-261
3006/160	121-262
3006/160	121-263
3006/160	121-292
3006/160	121-294
3006/160	121-295
3006/160	121-296
3006/160	121-297
3006/160	121-298
3006/160	121-299
3006/160	121-360
3006/160	121-554
3006/160	121-555
3006/160	121-714
3006/160	129-11
3006/160	145-T1B
3006/160	260P13001
3006/160	297V042H03
3006/160	297V042H04
3006/160	297V045H01
3006/160	297V045H02
3006/160	297V063C01
3006/160	297V070C01
3006/160	297V077C01
3006/160	310-068
3006/160	310-123
3006/160	310-124
3006/160	310-139
3006/160	310-191
3006/160	310-68
3006/160	324-0098
3006/160	324-0099
3006/160	324-0106
3006/160	324-0121
3006/160	324-0129
3006/160	324-0136
3006/160	324-0138
3006/160	325-0036-536
3006/160	325-0036-565
3006/160	421-20
3006/160	421-20B
3006/160	421-21
3006/160	421-21B
3006/160	421-22
3006/160	421-22B
3006/160	455-1
3006/160	576-0003-008
3006/160	617-56
3006/160	690V040H57
3006/160	690V040H58
3006/160	690V040H59
3006/160	690V040H60
3006/160	690V057H59
3006/160	690V057H62
3006/160	690V066H44
3006/160	690V066H45
3006/160	690V066H49
3006/160	690V073H59
3006/160	690V077H34
3006/160	690V077H36
3006/160	690V080H37
3006/160	690V080H39
3006/160	690V080H40
3006/160	690V085H42
3006/160	921-4B
3006/160	964-16598
3006/160	1951-17
3006/160	2015-00
3006/160	2020-00
3006/160	2021-00
3006/160	2057A2-37
3006/160	2057A2-65
3006/160	2057A2-66
3006/160	2057A2-80
3006/160	2057B2-141
3006/160	2057B2-35
3006/160	2057B2-65
3006/160	2057B2-66
3006/160	2057B2-67
3006/160	2057B2-68
3006/160	2057B2-70
3006/160	2057B2-71
3006/160	2057B2-77
3006/160	2057B2-81
3006/160	2057B2-88
3006/160	2791
3006/160	2797
3006/160	2799
3006/160	3961
3006/160	4364
3006/160	4454
3006/160	4456
3006/160	4457
3006/160	4501
3006/160	4509
3006/160	4545
3006/160	4589
3006/160	4595
3006/160	4603
3006/160	4604
3006/160	4605
3006/160	4621
3006/160	4677
3006/160	6154
3006/160	6155
3006/160	6162
3006/160	18493
3006/160	30221
3006/160	30222
3006/160	30223
3006/160	30230
3006/160	30293
3006/160	36559
3006/160	36560
3006/160	36563
3006/160	69107-42
3006/160	69107-44
3006/160	80071
3006/160	80114
3006/160	080225
3006/160	080277
3006/160	95116
3006/160	95118
3006/160	95119
3006/160	95120
3006/160	95122
3006/160	95123
3006/160	99120
3006/160	116756
3006/160	122725
3006/160	123511
3006/160	126185
3006/160	128938
3006/160	165667
3006/160	249588
3006/160	310221
3006/160	310224
3006/160	573330
3006/160	573366
3006/160	0573427
3006/160	0573471
3006/160	0573518
3006/160	604112
3006/160	740946
3006/160	740947
3006/160	740948
3006/160	980626
3006/160	985611
3006/160	2091241-5A
3006/160	2320011
3006/160	2495379
3006/160	2495488-1
3006/160	2495488-2
3006/160	7279779
3006/160	7852899-01
3006/160	7910783-01
3006/160	62049774
3006/160	62123648
3006/160	1522210131
3006/160	1522210300
3006/160	1522211021
3006/160	1522214411
3006/160	1522214435
3006/160	1522214821
3006/160	16000190201
30060160	48-124312
3007	A-1384
3007	A059-100
3007	A059-101
3007	A059-102
3007	A059-103
3007	A1-3
3007	A101
3007	A101X
3007	A102
3007	A102A
3007	A102AA
3007	A102AB
3007	A102BA
3007	A102BN
3007	A102CA
3007	A102TV
3007	A103
3007	A103CA
3007	A104
3007	A122
3007	A1465A
3007	A1465A9
3007	A14A8-1
3007	A14A8-19
3007	A203
3007	A203B
3007	A221
3007	A222
3007	A234
3007	A292
3007	A294
3007	A331
3007	A341
3007	A341-0B
3007	A354
3007	A354A
3007	A354B

SK	Industry Standard No.
3007	A358
3007	A360
3007	A367
3007	A368
3007	A440A
3007	A469
3007	A470
3007	A471
3007	A472
3007	A517
3007	A518
3007	A58
3007	A60
3007	A61
3007	A65A-70
3007	A65A-705
3007	A65A19G
3007	A74
3007	A77
3007	OA8-1
3007	OA8-1-12
3007	OA8-1-12-7
3007	A8-1-70
3007	A8-1-70-12
3007	A8-1-70-12-7
3007	A8-10
3007	A8-11
3007	A8-12
3007	A8-13
3007	A8-14
3007	A8-15
3007	A8-16
3007	A8-17
3007	A8-18
3007	A8-19
3007	A8-1A
3007	A8-1A0
3007	A8-1A0R
3007	A8-1A1
3007	A8-1A19
3007	A8-1A2
3007	A8-1A21
3007	A8-1A3
3007	A8-1A3P
3007	A8-1A4
3007	A8-1A4-7
3007	A8-1A5
3007	A8-1A5L
3007	A8-1A6
3007	A8-1A6-4
3007	A8-1A7
3007	A8-1A7-1
3007	A8-1A8
3007	A8-1A82
3007	A8-1A9
3007	A8-1A9G
3007	A85
3007	A92
3007	A93
3007	AA8-1-70
3007	AA8-1-705
3007	AA8-1A9G
3007	CB156
3007	CB244
3007	CB254
3007	D026
3007	D073
3007	DS-22
3007	FS2299
3007	HA-353C
3007	HEP-G0009
3007	HF12H
3007	HF12M
3007	HF20H
3007	HF20M
3007	JR10
3007	JR30X
3007	L2091241-2
3007	M4545
3007	M75516-2
3007	M75516-2P
3007	M75516-2R
3007	M755162-P
3007	M755162-R
3007	MC101
3007	R425
3007	RE69
3007	REN69
3007	S-80T
3007	S-874TB
3007	OS22.3511-770
3007	OS22.3511-780
3007	OS22.3511-790
3007	OS22.3516-380
3007	SO-65A
3007	S065
3007	ST-103
3007	T1033
3007	T1831
3007	TA-1628
3007	TA-1658
3007	TA-1659
3007	TA-1660
3007	TA-1662
3007	TA-1731
3007	TA-1757
3007	TA-1796
3007	TA-1797
3007	TA-1798
3007	TA-1828
3007	TA-1829
3007	TA1828
3007	TR07
3007	TR07C
3007	TR310069
3007	TR310193
3007	TR310232
3007	TR760
3007	TR761
3007	TR762
3007	VS-2SA103
3007	WEP637
3007	XA121
3007	XA122
3007	XA123
3007	XA124
3007	XA126
3007	ZEN313
3007	ZEN314
3007	1-21-150
3007	1-21-190
3007	1-21-256
3007	1-21-258
3007	1-21-259
3007	1-21-260
3007	1-522210921
3007	1-522216600
3007	002-007100
3007	002-007400
3007	2N1109
3007	2N1111
3007	2N1177
3007	2N1178
3007	2N1179
3007	2N1224
3007	2N1524
3007	2N1524-1
3007	2N1524-2
3007	2N1526
3007	2N1638
3007	2N1750
3007	2N2092
3007	2N247
3007	2N248
3007	2N2671
3007	2N2672
3007	2N274
3007	2N308
3007	2N309
3007	2N310
3007	2N370
3007	2N370/33
3007	2N370A
3007	2N371
3007	2N371/33
3007	2N372
3007	2N372/33
3007	2N373
3007	2N374
3007	2N384
3007	2N384/33
3007	2N642
3007	2N643
3007	2N644
3007	2N645
3007	2N990
3007	2NJ50
3007	2NJ51
3007	2S110
3007	2SA101
3007	2SA101A
3007	2SA101AA
3007	2SA101AY
3007	2SA101B
3007	2SA101BA
3007	2SA101BB
3007	2SA101BC
3007	2SA101BX
3007	2SA101C
3007	2SA101CA
3007	2SA101CV
3007	2SA101CX
3007	2SA101D
3007	2SA101E
3007	2SA101F
3007	2SA101G
3007	2SA101H
3007	2SA101K
3007	2SA101L
3007	2SA101M
3007	2SA101OR
3007	2SA101R
3007	2SA101X
3007	2SA101XBX
3007	2SA101Y
3007	2SA101YA
3007	2SA101Z
3007	2SA102
3007	2SA102(BA)
3007	2SA102A
3007	2SA102AA
3007	2SA102AB
3007	2SA102B
3007	2SA102BA
3007	2SA102BA-2
3007	2SA102BN
3007	2SA102C
3007	2SA102CA
3007	2SA102CA-1
3007	2SA102D
3007	2SA102E
3007	2SA102F
3007	2SA102G
3007	2SA102H
3007	2SA102K
3007	2SA102L
3007	2SA102M
3007	2SA102OR
3007	2SA102TV-2
3007	2SA102Y
3007	2SA104
3007	2SA104A
3007	2SA104B
3007	2SA104C
3007	2SA104D
3007	2SA104E
3007	2SA104F
3007	2SA104G
3007	2SA104H
3007	2SA104K
3007	2SA104L
3007	2SA104M
3007	2SA104OR
3007	2SA104P
3007	2SA104R
3007	2SA104X
3007	2SA104Y
3007	2SA106
3007	2SA106B
3007	2SA106C
3007	2SA106D
3007	2SA106E
3007	2SA106F
3007	2SA106G
3007	2SA106H
3007	2SA106K
3007	2SA106L
3007	2SA106M
3007	2SA106OR
3007	2SA106R
3007	2SA106X
3007	2SA106Y
3007	2SA107
3007	2SA107A
3007	2SA107B
3007	2SA107C
3007	2SA107D
3007	2SA107E
3007	2SA107F
3007	2SA107G
3007	2SA107H
3007	2SA107K
3007	2SA107L
3007	2SA107M
3007	2SA107OR
3007	2SA107R
3007	2SA107X
3007	2SA107Y
3007	2SA108
3007	2SA108A
3007	2SA108B
3007	2SA108C
3007	2SA108D
3007	2SA108E
3007	2SA108F
3007	2SA108G
3007	2SA108H
3007	2SA108K
3007	2SA108L
3007	2SA108M
3007	2SA108OR
3007	2SA108R
3007	2SA108X
3007	2SA108Y
3007	2SA109
3007	2SA109A
3007	2SA109B
3007	2SA109C
3007	2SA109D
3007	2SA109E
3007	2SA109F
3007	2SA109G
3007	2SA109K
3007	2SA109L
3007	2SA109M
3007	2SA109OR
3007	2SA109R
3007	2SA109X
3007	2SA109Y
3007	2SA110
3007	2SA110A
3007	2SA110B
3007	2SA110C
3007	2SA110D
3007	2SA110E
3007	2SA110F
3007	2SA110G
3007	2SA110K
3007	2SA110L
3007	2SA110M
3007	2SA110OR
3007	2SA110R
3007	2SA110X
3007	2SA110Y
3007	2SA111
3007	2SA111A
3007	2SA111B
3007	2SA111C
3007	2SA111D
3007	2SA111E
3007	2SA111F
3007	2SA111G
3007	2SA111K
3007	2SA111L
3007	2SA111M
3007	2SA111OR
3007	2SA111R
3007	2SA111X
3007	2SA111Y
3007	2SA112
3007	2SA112A
3007	2SA112B
3007	2SA112C
3007	2SA112D
3007	2SA112E
3007	2SA112F
3007	2SA112G
3007	2SA112GN
3007	2SA112H
3007	2SA112K
3007	2SA112L
3007	2SA112M
3007	2SA112OR
3007	2SA112R
3007	2SA112X
3007	2SA112Y
3007	2SA113
3007	2SA113A
3007	2SA113B
3007	2SA113C
3007	2SA113D
3007	2SA113E
3007	2SA113F
3007	2SA113G
3007	2SA113GN
3007	2SA113H
3007	2SA113J
3007	2SA113L
3007	2SA113M
3007	2SA113R
3007	2SA113X
3007	2SA113Y
3007	2SA114
3007	2SA114A
3007	2SA114B
3007	2SA114C
3007	2SA114D
3007	2SA114E
3007	2SA114F
3007	2SA114G
3007	2SA114H
3007	2SA114K
3007	2SA114L
3007	2SA114M
3007	2SA114OR
3007	2SA114R
3007	2SA114X
3007	2SA114Y
3007	2SA115
3007	2SA115A
3007	2SA115B
3007	2SA115C
3007	2SA115D
3007	2SA115E
3007	2SA115F
3007	2SA115G
3007	2SA115GN
3007	2SA115H
3007	2SA115J
3007	2SA115K
3007	2SA115L
3007	2SA115M
3007	2SA115OR
3007	2SA115R
3007	2SA115X
3007	2SA115Y
3007	2SA141
3007	2SA203
3007	2SA221
3007	2SA222
3007	2SA233
3007	2SA233A
3007	2SA233B
3007	2SA233C
3007	2SA233D
3007	2SA233E
3007	2SA233F
3007	2SA233G
3007	2SA233GN
3007	2SA233H
3007	2SA233J
3007	2SA233K
3007	2SA233L
3007	2SA233M
3007	2SA233OR
3007	2SA233R
3007	2SA233X
3007	2SA233Y
3007	2SA234
3007	2SA236
3007	2SA236A
3007	2SA236B
3007	2SA236C
3007	2SA236D
3007	2SA236E
3007	2SA236F
3007	2SA236G
3007	2SA236GN
3007	2SA236H
3007	2SA236J
3007	2SA236K
3007	2SA236L
3007	2SA236M
3007	2SA236OR
3007	2SA236R
3007	2SA236X
3007	2SA236Y
3007	2SA257
3007	2SA257A
3007	2SA257B
3007	2SA257C
3007	2SA257D
3007	2SA257E
3007	2SA257F
3007	2SA257G
3007	2SA257GN
3007	2SA257H
3007	2SA257J
3007	2SA257K
3007	2SA257L
3007	2SA257OR
3007	2SA257R
3007	2SA257X
3007	2SA257Y
3007	2SA258
3007	2SA258A
3007	2SA258B
3007	2SA258C
3007	2SA258D
3007	2SA258E
3007	2SA258F
3007	2SA258G
3007	2SA258GN
3007	2SA258H
3007	2SA258J
3007	2SA258K
3007	2SA258L
3007	2SA258M
3007	2SA258OR
3007	2SA258R
3007	2SA258X
3007	2SA258Y
3007	2SA270
3007	2SA270A
3007	2SA270B
3007	2SA270C
3007	2SA270D
3007	2SA270E
3007	2SA270F
3007	2SA270G
3007	2SA270GN
3007	2SA270H
3007	2SA270J
3007	2SA270K
3007	2SA270L
3007	2SA270M
3007	2SA270R
3007	2SA270X
3007	2SA270Y
3007	2SA294
3007	2SA331
3007	2SA341
3007	2SA341-OB
3007	2SA341OB
3007	2SA351
3007	2SA351A
3007	2SA351A-2
3007	2SA351B
3007	2SA351C
3007	2SA351D
3007	2SA351E
3007	2SA351F
3007	2SA351G
3007	2SA351GN
3007	2SA351K
3007	2SA351L
3007	2SA351M
3007	2SA351OR
3007	2SA351R
3007	2SA351X
3007	2SA351Y
3007	2SA354
3007	2SA354A
3007	2SA354B
3007	2SA354BK
3007	2SA354C
3007	2SA354D
3007	2SA354E
3007	2SA354F
3007	2SA354G
3007	2SA354GN
3007	2SA354H
3007	2SA354J
3007	2SA354K
3007	2SA354L
3007	2SA354M
3007	2SA354OR
3007	2SA354R
3007	2SA354Y
3007	2SA358
3007	2SA360
3007	2SA360G
3007	2SA360H
3007	2SA360K
3007	2SA361
3007	2SA361A
3007	2SA361B
3007	2SA361C
3007	2SA361D
3007	2SA361E
3007	2SA361F
3007	2SA361G
3007	2SA361GN
3007	2SA361H
3007	2SA361J
3007	2SA361K
3007	2SA361L
3007	2SA361M
3007	2SA361OR
3007	2SA361R
3007	2SA361X
3007	2SA361Y
3007	2SA367
3007	2SA368
3007	2SA368A
3007	2SA368B
3007	2SA368C
3007	2SA368D
3007	2SA368E
3007	2SA368F
3007	2SA368G
3007	2SA368GN
3007	2SA368H
3007	2SA368J
3007	2SA368K
3007	2SA368L
3007	2SA368M
3007	2SA368OR
3007	2SA368R
3007	2SA368X
3007	2SA368Y
3007	2SA43
3007	2SA440A
3007	2SA469
3007	2SA469A
3007	2SA469B
3007	2SA469C
3007	2SA469D
3007	2SA469E
3007	2SA469F
3007	2SA469G
3007	2SA469GN
3007	2SA469H
3007	2SA469J
3007	2SA469K
3007	2SA469L
3007	2SA469M
3007	2SA469OR
3007	2SA469R
3007	2SA469Y
3007	2SA470
3007	2SA470A
3007	2SA470B
3007	2SA470C
3007	2SA470D
3007	2SA470E
3007	2SA470F
3007	2SA470G
3007	2SA470GN
3007	2SA470H
3007	2SA470J
3007	2SA470K
3007	2SA470L
3007	2SA470M
3007	2SA470OR
3007	2SA470R
3007	2SA470Y
3007	2SA471
3007	2SA471-1
3007	2SA471-2
3007	2SA471-3
3007	2SA471A
3007	2SA471B
3007	2SA471C
3007	2SA471D
3007	2SA471E
3007	2SA471F
3007	2SA471G
3007	2SA471GN
3007	2SA471H
3007	2SA471J
3007	2SA471K
3007	2SA471L
3007	2SA471M
3007	2SA471OR
3007	2SA471R
3007	2SA471Y
3007	2SA472
3007	2SA472A
3007	2SA472B
3007	2SA472C
3007	2SA472D
3007	2SA472E
3007	2SA472F
3007	2SA472G
3007	2SA472GN
3007	2SA472H
3007	2SA472J
3007	2SA472K
3007	2SA472L
3007	2SA472M
3007	2SA472OR
3007	2SA472R
3007	2SA472Y
3007	2SA475
3007	2SA517
3007	2SA518
3007	2SA518A
3007	2SA518B
3007	2SA518C
3007	2SA518D
3007	2SA518E
3007	2SA518F
3007	2SA518G
3007	2SA518GN
3007	2SA518H
3007	2SA518J
3007	2SA518K
3007	2SA518L
3007	2SA518M

SK	Industry Standard No.	SK	Industry Standard No.	SK	Industry Standard No.	SK	Industry Standard No.	SK	Industry Standard No.
3007	28A518OR	3007	20MC	3007	1040059	3008	R60-1001	3008	1-21-243
3007	28A518R	3007	022-3511-770	3007	111117	3008	R60-1002	3008	1-21-244
3007	28A518Y	3007	022-3511-780	3007	111118	3008	R60-1003	3008	1-21-257
3007	28A57	3007	022-3511-790	3007	111313	3008	R714	3008	002-006300
3007	28A58	3007	022-3516-380	3007	117618	3008	R715	3008	2N1065
3007	28A59	3007	24MW368	3007	117658	3008	R8-1554	3008	2N1108
3007	28A59A	3007	24MW61	3007	125330	3008	R8-3322	3008	2N1110
3007	28A59B	3007	40D1547	3007	223369	3008	R8-3323	3008	2N1226
3007	28A59C	3007	051-0062	3007	223474	3008	R8-3324	3008	2N128
3007	28A59D	3007	051-0063	3007	225311	3008	R8-3862	3008	2N129
3007	28A59F	3007	5705-10	3007	225594	3008	R8-3863	3008	2N1313
3007	28A59G	3007	5705-9	3007	310030	3008	R8-3866	3008	2N1384
3007	28A59L	3007	57D1-88	3007	853864-0	3008	R8-5107	3008	2N1425
3007	28A59M	3007	57D1-89	3007	985445	3008	R8-5108	3008	2N1524
3007	28A590R	3007	57D130	3007	985446	3008	R8-5109	3008	2N1525
3007	28A59R	3007	57D131	3007	1061854-2	3008	R8-5201	3008	2N1527
3007	28A59X	3007	57D132	3007	2000648-21	3008	R8-5205	3008	2N1631
3007	28A59Y	3007	57D186	3007	2000648-22	3008	R8-5207	3008	2N1632
3007	28A60	3007	65A	3007	2002151-18	3008	R8-5209	3008	2N1633
3007	28A61	3007	065A-12	3007	2076393	3008	R8-5301	3008	2N1634
3007	28A73	3007	065A-12-7	3007	2076403	3008	R8-5305	3008	2N1635
3007	28A73A	3007	65A-70	3007	2076403-0703	3008	R8-5306	3008	2N1636
3007	28A73B	3007	65A-70-12	3007	2092693-2	3008	R8-5312	3008	2N1637
3007	28A73C	3007	65A-70-12-7	3007	2092693-3	3008	R8-5313	3008	2N1638
3007	28A73D	3007	65A0	3007	2092693-4	3008	R8-5752	3008	2N1639
3007	28A73E	3007	65A1	3007	2320161	3008	R8-5753	3008	2N1678
3007	28A73F	3007	65A10	3007	7279781	3008	R8-5754	3008	2N1864
3007	28A73G	3007	65A10R	3007	8538640	3008	R8-5755	3008	2N1865
3007	28A73H	3007	65A119	3007	62044888	3008	R8-5756	3008	2N1866
3007	28A73K	3007	65A12	3007	62087668	3008	R8-5757	3008	2N1867
3007	28A73L	3007	65A121	3007	120002213	3008	R8-5758	3008	2N2093
3007	28A730M	3007	65A13	3007	120002216	3008	R8-5759	3008	2N231
3007	28A730R	3007	65A13P	3007	1522210921	3008	R8-5760	3008	2N232
3007	28A73R	3007	65A14	3007	1522216600	3008	R8-5761	3008	2N240
3007	28A73X	3007	65A14-7	3008	A059-105	3008	R8-5762	3008	2N331
3007	28A73Y	3007	65A15	3008	A1465B	3008	R8-5802	3008	2N393
3007	28A74	3007	65A15L	3008	A1465B9	3008	R8-5818	3008	2N500
3007	28A75	3007	65A16	3008	A25-1001	3008	R8-684	3008	2N504
3007	28A77	3007	65A16-3	3008	A25-1002	3008	R8-685	3008	2N544
3007	28A83	3007	65A17	3008	A25-1003	3008	R81554	3008	2N544/33
3007	28A83A	3007	65A17-1	3008	A350	3008	R85109	3008	2N602
3007	28A83B	3007	65A18	3008	A350A	3008	R85207	3008	2N602A
3007	28A83D	3007	65A182	3008	A352	3008	R85314	3008	2N603
3007	28A83E	3007	65A19	3008	A353	3008	RT-61014	3008	2N603A
3007	28A83F	3007	65A19G	3008	A360	3008	RT3361	3008	2N604
3007	28A83G	3007	65A2	3008	A385	3008	RT3362	3008	2N605
3007	28A83H	3007	65A3	3008	A468	3008	RT3466	3008	2N606
3007	28A83K	3007	65A4	3008	A65B-70	3008	RT3467	3008	2N607
3007	28A83L	3007	65A5	3008	A65B-705	3008	RT5467	3008	2N608
3007	28A83M	3007	65A6	3008	A65B19G	3008	S-70T	3008	2N624
3007	28A830R	3007	65A7	3008	A70	3008	S-87TB	3008	2N640
3007	28A83X	3007	65A8	3008	A71	3008	S-88TB	3008	2N641
3007	28A84	3007	65A9	3008	A71AB	3008	S-95101	3008	2N694
3007	28A84A	3007	81-46125001-1	3008	A71AC	3008	S-95102	3008	2N725
3007	28A84B	3007	86-151-2	3008	A71B8	3008	S-95103	3008	2N994
3007	28A84C	3007	86-18-2	3008	ACR810-101	3008	899101	3008	28109
3007	28A84D	3007	86-36-2	3008	ACR810-102	3008	899102	3008	28112
3007	28A84E	3007	86-37-2	3008	ACR810-103	3008	899103	3008	2835
3007	28A84F	3007	86-38-2	3008	ACR83-1001	3008	899104	3008	2836
3007	28A84G	3007	112-000267	3008	ACR83-1002	3008	SB-100	3008	2858
3007	28A84H	3007	116-207	3008	ACR83-1003	3008	SPT-315	3008	2892
3007	28A84K	3007	116-208	3008	AP105	3008	SPT-316	3008	2892A
3007	28A84L	3007	116-209	3008	AP134	3008	SPT-317	3008	2893
3007	28A84M	3007	120-002213	3008	AP135	3008	SPT-320	3008	2893A
3007	28A840R	3007	120-002216	3008	AP36	3008	SPT-357	3008	28A105
3007	28A84X	3007	121-256	3008	AP137	3008	SPT308	3008	28A105A
3007	28A84Y	3007	121-258	3008	AP138/20	3008	T-1363	3008	28A105B
3007	28A85	3007	121-259	3008	AP166	3008	T-1364	3008	28A105C
3007	28A85L	3007	121-260	3008	A01	3008	T-1460	3008	28A105D
3007	28A92	3007	121-290	3008	AR1-05	3008	T-6028	3008	28A105E
3007	28A92A	3007	121-44	3008	ASY-24	3008	T-6029	3008	28A105H
3007	28A92B	3007	121-49	3008	AT-5	3008	T-6030	3008	28A105K
3007	28A92C	3007	154T1	3008	C10291	3008	T-6052	3008	28A105M
3007	28A92D	3007	297V036H02	3008	CB157	3008	T1028	3008	28A1050R
3007	28A92E	3007	324-0130	3008	CB158	3008	T1034	3008	28A105R
3007	28A92F	3007	324-0137	3008	CK28	3008	T1038	3008	28A105X
3007	28A92G	3007	324-0145	3008	CK28A	3008	T116	3008	28A105Y
3007	28A92H	3007	353-9001-002	3008	D020	3008	T1166	3008	28A106A
3007	28A92K	3007	576-0003-009	3008	DS-25	3008	T1314	3008	28A128
3007	28A92L	3007	576-0003-013	3008	E-065	3008	T1524	3008	28A128A
3007	28A92M	3007	576-0003-014	3008	E-066	3008	T1548	3008	28A128B
3007	28A920R	3007	576-0003-015	3008	E-067	3008	T1618	3008	28A128C
3007	28A92R	3007	576-0003-024	3008	E-068	3008	T1788	3008	28A128D
3007	28A92X	3007	576-0003-025	3008	ER15X14	3008	T2020	3008	28A128E
3007	28A92Y	3007	576-2000-990	3008	ER15X15	3008	T2024	3008	28A128F
3007	28A93	3007	576-2000-993	3008	ER15X16	3008	T2025	3008	28A128G
3007	28A93A	3007	690V056B27	3008	E815X61	3008	T2026	3008	28A128GN
3007	28A93B	3007	690V056H29	3008	E815X63	3008	T2322	3008	28A128H
3007	28A93C	3007	690V056H30	3008	ES3	3008	T2323	3008	28A128J
3007	28A93F	3007	690V066H89	3008	GC11148	3008	T2324	3008	28A128K
3007	28A93G	3007	1465A	3008	GC282	3008	T281	3008	28A128L
3007	28A93H	3007	1465A-12	3008	GC283	3008	T282	3008	28A128M
3007	28A93L	3007	1465A-12-8	3008	GC284	3008	T348	3008	28A1280R
3007	28A93M	3007	1865A	3008	GET-692	3008	T50818	3008	28A128R
3007	28A930R	3007	1865A-12	3008	GET-873A	3008	T811	3008	28A128Y
3007	28A93R	3007	1865A-127	3008	GET-883	3008	TA-1650A	3008	28A129
3007	28A93X	3007	1865AL8	3008	HA-0010Z	3008	TA-1755	3008	28A129A
3007	28A93Y	3007	2904-016	3008	HA-354B	3008	TA-1756	3008	28A129B
3007	2T204	3007	2904-029	3008	HA-49	3008	TK1228-1001	3008	28A129C
3007	2T204A	3007	3008(CB)	3008	HA-52	3008	TNJ-60362	3008	28A129D
3007	2T205	3007	3009	3008	HA-53	3008	TNJ-60363	3008	28A129E
3007	2T205A	3007	3009(MIXER)	3008	HA12	3008	TNJ-60364	3008	28A129F
3007	03-57-002	3007	3010(IF)	3008	HA201	3008	TNJ-60608	3008	28A129GN
3007	3MC	3007	3012	3008	HA330	3008	TR310019	3008	28A129H
3007	4-65A17-1	3007	3024	3008	HP12N	3008	TR310065	3008	28A129J
3007	6MC	3007	3025	3008	HR50	3008	TR310155	3008	28A129K
3007	09-300012	3007	3551A	3008	HR51	3008	TR310156	3008	28A129M
3007	09-300015	3007	3613	3008	HR52	3008	TR310157	3008	28A1290R
3007	09-300016	3007	06008	3008	HT101021A	3008	TR310158	3008	28A129R
3007	12-1-150	3007	9403-3	3008	L2091241-3	3008	TR310251	3008	28A129X
3007	12-1-190	3007	9403-6	3008	M4605	3008	TR310255	3008	28A129Y
3007	12-1-256	3007	9510-1	3008	M4621	3008	TRA-10R	3008	28A150
3007	12-1-258	3007	9510-2	3008	MA393	3008	TRA-11R	3008	28A150A
3007	12-1-259	3007	12115-0	3008	MA393A	3008	TRA-12R	3008	28A150B
3007	12-1-260	3007	12119-0	3008	MA393B	3008	TRA-2	3008	28A150C
3007	12MC	3007	12125-6	3008	MA393C	3008	TRA-22A	3008	28A150D
3007	14A8-1	3007	12125-8	3008	MA393G	3008	TRA-22B	3008	28A150F
3007	14A8-1-12	3007	12125-9	3008	MA393R	3008	TRA-23	3008	28A150G
3007	15-22210921	3007	12126-0	3008	MA822	3008	TRA-23A	3008	28A150GN
3007	15-22216600	3007	30215	3008	MDB35	3008	TRA-23B	3008	28A150H
3007	18A8-1	3007	30216	3008	N57B2-17	3008	TRA-24	3008	28A150J
3007	18A8-1-12	3007	30231	3008	N57B2-18	3008	TRA-24A	3008	28A150K
3007	18A8-1-127	3007	35002	3008	N57B2-19	3008	TRA-24B	3008	28A150L
3007	18A8-1L	3007	080224	3008	N57B2-23	3008	TS620	3008	28A150M
3007	18A8-1LB	3007	080228	3008	NA-1114-1002	3008	TS621	3008	28A1500R
3007	019-003515	3007	080258	3008	NA1022-1001	3008	VF55K	3008	28A150R
3007	019-003778	3007	94035	3008	00-390	3008	V3-28A385	3008	28A150X
3007	019-003778	3007	94036	3008	00-612	3008	XJ71	3008	28A176
		3007	95101	3008	00-613	3008	ZJ72	3008	28A176A
		3007	99121	3008	00-614	3008	ZJ73	3008	28A176B
		3007	100678	3008	Q2N1526	3008	1-21-138		
		3007	101078	3008	Q40359	3008	1-21-189		
		3007	101087	3008	R-28A222	3008	1-21-242		
		3007	104009	3008	R-539				

SK	Industry Standard No.	SK	Industry Standard No.	SK	Industry Standard No.	SK	Industry Standard No.	SK	Industry Standard No.
3008	2SA176C	3008	2SA224F	3008	2SA272Y	3008	2SA29R	3008	2SA322F
3008	2SA176D	3008	2SA224G	3008	2SA273	3008	2SA29X	3008	2SA322G
3008	2SA176E	3008	2SA224GN	3008	2SA273A	3008	2SA29Y	3008	2SA322GN
3008	2SA176F	3008	2SA224H	3008	2SA273B	3008	2SA307	3008	2SA322H
3008	2SA176G	3008	2SA224J	3008	2SA273C	3008	2SA307A	3008	2SA322L
3008	2SA176GN	3008	2SA224K	3008	2SA273D	3008	2SA307B	3008	2SA322M
3008	2SA176H	3008	2SA224L	3008	2SA273E	3008	2SA307C	3008	2SA322OR
3008	2SA176K	3008	2SA224M	3008	2SA273F	3008	2SA307D	3008	2SA322X
3008	2SA176M	3008	2SA224OR	3008	2SA273G	3008	2SA307E	3008	2SA322Y
3008	2SA176L	3008	2SA224R	3008	2SA273GN	3008	2SA307F	3008	2SA323
3008	2SA176N	3008	2SA224X	3008	2SA273H	3008	2SA307G	3008	2SA323A
3008	2SA176OR	3008	2SA224Y	3008	2SA273J	3008	2SA307GN	3008	2SA323B
3008	2SA176R	3008	2SA237	3008	2SA273K	3008	2SA307J	3008	2SA323C
3008	2SA176X	3008	2SA237A	3008	2SA273L	3008	2SA307K	3008	2SA323D
3008	2SA176Y	3008	2SA237B	3008	2SA273M	3008	2SA307L	3008	2SA323E
3008	2SA19	3008	2SA237C	3008	2SA273OR	3008	2SA307M	3008	2SA323F
3008	2SA19A	3008	2SA237D	3008	2SA273X	3008	2SA307OR	3008	2SA323G
3008	2SA19B	3008	2SA237E	3008	2SA273Y	3008	2SA307R	3008	2SA323GN
3008	2SA19C	3008	2SA237F	3008	2SA274A	3008	2SA307X	3008	2SA323J
3008	2SA19D	3008	2SA237G	3008	2SA274B	3008	2SA307Y	3008	2SA323K
3008	2SA19E	3008	2SA237GN	3008	2SA274C	3008	2SA311	3008	2SA323L
3008	2SA19F	3008	2SA237H	3008	2SA274D	3008	2SA311A	3008	2SA323M
3008	2SA19G	3008	2SA237L	3008	2SA274E	3008	2SA311B	3008	2SA323R
3008	2SA19L	3008	2SA237M	3008	2SA274F	3008	2SA311C	3008	2SA323X
3008	2SA19M	3008	2SA237OR	3008	2SA274G	3008	2SA311D	3008	2SA323Y
3008	2SA19OR	3008	2SA237R	3008	2SA274GN	3008	2SA311E	3008	2SA329A
3008	2SA19R	3008	2SA237X	3008	2SA274H	3008	2SA311F	3008	2SA329B
3008	2SA19Y	3008	2SA237Y	3008	2SA274J	3008	2SA311G	3008	2SA332
3008	2SA20	3008	2SA249	3008	2SA274K	3008	2SA311GN	3008	2SA332A
3008	2SA201A	3008	2SA256	3008	2SA274L	3008	2SA311H	3008	2SA332B
3008	2SA20A	3008	2SA256A	3008	2SA274M	3008	2SA311J	3008	2SA332C
3008	2SA20B	3008	2SA256B	3008	2SA274OR	3008	2SA311K	3008	2SA332D
3008	2SA20C	3008	2SA256C	3008	2SA274R	3008	2SA311L	3008	2SA332E
3008	2SA20D	3008	2SA256D	3008	2SA274X	3008	2SA311M	3008	2SA332F
3008	2SA20E	3008	2SA256E	3008	2SA274Y	3008	2SA311OR	3008	2SA332G
3008	2SA20F	3008	2SA256F	3008	2SA275	3008	2SA311R	3008	2SA332GN
3008	2SA20G	3008	2SA256G	3008	2SA275A	3008	2SA311X	3008	2SA332H
3008	2SA20L	3008	2SA256GN	3008	2SA275B	3008	2SA311Y	3008	2SA332J
3008	2SA20M	3008	2SA256H	3008	2SA275C	3008	2SA312	3008	2SA332K
3008	2SA200R	3008	2SA256J	3008	2SA275D	3008	2SA312A	3008	2SA332L
3008	2SA20R	3008	2SA256K	3008	2SA275E	3008	2SA312B	3008	2SA332M
3008	2SA20X	3008	2SA256L	3008	2SA275F	3008	2SA312D	3008	2SA332OR
3008	2SA20Y	3008	2SA256M	3008	2SA275G	3008	2SA312E	3008	2SA332X
3008	2SA21	3008	2SA256OR	3008	2SA275GN	3008	2SA312G	3008	2SA332Y
3008	2SA218	3008	2SA256X	3008	2SA275H	3008	2SA312GN	3008	2SA338
3008	2SA218A	3008	2SA256Y	3008	2SA275J	3008	2SA312H	3008	2SA338A
3008	2SA218B	3008	2SA259	3008	2SA275L	3008	2SA312J	3008	2SA338B
3008	2SA218C	3008	2SA259B	3008	2SA275M	3008	2SA312K	3008	2SA338C
3008	2SA218D	3008	2SA259C	3008	2SA275OR	3008	2SA312L	3008	2SA338D
3008	2SA218E	3008	2SA259D	3008	2SA275R	3008	2SA312M	3008	2SA338E
3008	2SA218F	3008	2SA259E	3008	2SA275X	3008	2SA312OR	3008	2SA338F
3008	2SA218G	3008	2SA259F	3008	2SA275Y	3008	2SA312R	3008	2SA338G
3008	2SA218GN	3008	2SA259G	3008	2SA285	3008	2SA312X	3008	2SA338GN
3008	2SA218H	3008	2SA259GN	3008	2SA285B	3008	2SA312Y	3008	2SA338H
3008	2SA218J	3008	2SA259H	3008	2SA285C	3008	2SA313	3008	2SA338J
3008	2SA218K	3008	2SA259J	3008	2SA285E	3008	2SA313A	3008	2SA338K
3008	2SA218L	3008	2SA259L	3008	2SA285F	3008	2SA313B	3008	2SA338L
3008	2SA218M	3008	2SA259R	3008	2SA285G	3008	2SA313C	3008	2SA338M
3008	2SA218OR	3008	2SA259OR	3008	2SA285GN	3008	2SA313D	3008	2SA338OR
3008	2SA218R	3008	2SA259X	3008	2SA285H	3008	2SA313E	3008	2SA338R
3008	2SA218X	3008	2SA259Y	3008	2SA285J	3008	2SA313F	3008	2SA338X
3008	2SA218Y	3008	2SA266	3008	2SA285L	3008	2SA313G	3008	2SA338Y
3008	2SA21A	3008	2SA266A	3008	2SA285M	3008	2SA313GN	3008	2SA339
3008	2SA21B	3008	2SA266B	3008	2SA285OR	3008	2SA313H	3008	2SA339A
3008	2SA21C	3008	2SA266C	3008	2SA285R	3008	2SA313J	3008	2SA339B
3008	2SA21D	3008	2SA266D	3008	2SA285X	3008	2SA313K	3008	2SA339C
3008	2SA21E	3008	2SA266E	3008	2SA285Y	3008	2SA313L	3008	2SA339D
3008	2SA21F	3008	2SA266F	3008	2SA286	3008	2SA313M	3008	2SA339E
3008	2SA21G	3008	2SA266GN	3008	2SA286A	3008	2SA313OR	3008	2SA339F
3008	2SA21L	3008	2SA266H	3008	2SA286B	3008	2SA313R	3008	2SA339G
3008	2SA21M	3008	2SA266J	3008	2SA286C	3008	2SA313X	3008	2SA339GN
3008	2SA21OR	3008	2SA266K	3008	2SA286D	3008	2SA313Y	3008	2SA339H
3008	2SA21R	3008	2SA266L	3008	2SA286F	3008	2SA314	3008	2SA339K
3008	2SA21X	3008	2SA266M	3008	2SA286G	3008	2SA314A	3008	2SA339L
3008	2SA21Y	3008	2SA266OR	3008	2SA286GN	3008	2SA314B	3008	2SA339M
3008	2SA220	3008	2SA266R	3008	2SA286H	3008	2SA314C	3008	2SA339OR
3008	2SA220A	3008	2SA266X	3008	2SA286M	3008	2SA314D	3008	2SA339R
3008	2SA220B	3008	2SA266Y	3008	2SA286OR	3008	2SA314E	3008	2SA339X
3008	2SA220C	3008	2SA269	3008	2SA286R	3008	2SA314F	3008	2SA339Y
3008	2SA220D	3008	2SA269A	3008	2SA286X	3008	2SA314G	3008	2SA341A
3008	2SA220E	3008	2SA269B	3008	2SA286Y	3008	2SA314GN	3008	2SA341B
3008	2SA220F	3008	2SA269C	3008	2SA287	3008	2SA314H	3008	2SA341C
3008	2SA220G	3008	2SA269D	3008	2SA287A	3008	2SA314J	3008	2SA341D
3008	2SA220GN	3008	2SA269F	3008	2SA287B	3008	2SA314K	3008	2SA341E
3008	2SA220H	3008	2SA269GN	3008	2SA287C	3008	2SA314L	3008	2SA341F
3008	2SA220J	3008	2SA269H	3008	2SA287D	3008	2SA314M	3008	2SA341G
3008	2SA220K	3008	2SA269J	3008	2SA287E	3008	2SA314OR	3008	2SA341H
3008	2SA220L	3008	2SA269K	3008	2SA287F	3008	2SA314R	3008	2SA341J
3008	2SA220M	3008	2SA269L	3008	2SA287G	3008	2SA314X	3008	2SA341K
3008	2SA220OR	3008	2SA269M	3008	2SA287GN	3008	2SA314Y	3008	2SA341L
3008	2SA220R	3008	2SA269OR	3008	2SA287H	3008	2SA315	3008	2SA341M
3008	2SA220X	3008	2SA269X	3008	2SA287J	3008	2SA315A	3008	2SA341OR
3008	2SA220Y	3008	2SA269Y	3008	2SA287K	3008	2SA315B	3008	2SA341R
3008	2SA221A	3008	2SA271	3008	2SA287L	3008	2SA315C	3008	2SA341X
3008	2SA221B	3008	2SA271A	3008	2SA287M	3008	2SA315D	3008	2SA341Y
3008	2SA221C	3008	2SA271B	3008	2SA287OR	3008	2SA315E	3008	2SA342
3008	2SA221D	3008	2SA271C	3008	2SA287X	3008	2SA315F	3008	2SA350
3008	2SA221E	3008	2SA271E	3008	2SA287Y	3008	2SA315G	3008	2SA350A
3008	2SA221F	3008	2SA271F	3008	2SA288	3008	2SA315GN	3008	2SA350B
3008	2SA221G	3008	2SA271G	3008	2SA288B	3008	2SA315H	3008	2SA350BK
3008	2SA221GN	3008	2SA271GN	3008	2SA288C	3008	2SA315J	3008	2SA350C
3008	2SA221H	3008	2SA271H	3008	2SA288D	3008	2SA315K	3008	2SA350D
3008	2SA221J	3008	2SA271J	3008	2SA288E	3008	2SA315L	3008	2SA350E
3008	2SA221K	3008	2SA271K	3008	2SA288F	3008	2SA315M	3008	2SA350F
3008	2SA221L	3008	2SA271L	3008	2SA288G	3008	2SA315OR	3008	2SA350G
3008	2SA221M	3008	2SA271M	3008	2SA288GN	3008	2SA315R	3008	2SA350GN
3008	2SA221OR	3008	2SA271OR	3008	2SA288J	3008	2SA315X	3008	2SA350H
3008	2SA221R	3008	2SA271R	3008	2SA288K	3008	2SA315Y	3008	2SA350J
3008	2SA221X	3008	2SA271X	3008	2SA288L	3008	2SA316	3008	2SA350K
3008	2SA221Y	3008	2SA271Y	3008	2SA288M	3008	2SA316A	3008	2SA350L
3008	2SA222A	3008	2SA272	3008	2SA288OR	3008	2SA316B	3008	2SA350OR
3008	2SA222B	3008	2SA272A	3008	2SA288X	3008	2SA316C	3008	2SA350R
3008	2SA222C	3008	2SA272B	3008	2SA288Y	3008	2SA316D	3008	2SA350T
3008	2SA222D	3008	2SA272D	3008	2SA29	3008	2SA316E	3008	2SA350TY
3008	2SA222E	3008	2SA272E	3008	2SA298	3008	2SA316F	3008	2SA350Y
3008	2SA222F	3008	2SA272F	3008	2SA298A	3008	2SA316G	3008	2SA352
3008	2SA222G	3008	2SA272G	3008	2SA298B	3008	2SA316GN	3008	2SA352A
3008	2SA222GN	3008	2SA272GN	3008	2SA298C	3008	2SA316H	3008	2SA352C
3008	2SA222H	3008	2SA272H	3008	2SA298G	3008	2SA316J	3008	2SA352D
3008	2SA222J	3008	2SA272J	3008	2SA298OR	3008	2SA316K	3008	2SA352F
3008	2SA222K	3008	2SA272K	3008	2SA298X	3008	2SA316L	3008	2SA352G
3008	2SA222L	3008	2SA272M	3008	2SA298Y	3008	2SA316M	3008	2SA352GN
3008	2SA222M	3008	2SA272OR	3008	2SA29A	3008	2SA316OR	3008	2SA352H
3008	2SA222OR	3008	2SA272R	3008	2SA29B	3008	2SA316R	3008	2SA352J
3008	2SA222R	3008	2SA272X	3008	2SA29C	3008	2SA316X	3008	2SA352K
3008	2SA222X			3008	2SA29D	3008	2SA316Y	3008	2SA352L
3008	2SA222Y			3008	2SA29E	3008	2SA322	3008	2SA352M
3008	2SA224			3008	2SA29F	3008	2SA322A		
3008	2SA224A			3008	2SA29G	3008	2SA322B		
3008	2SA224B			3008	2SA29L	3008	2SA322C		
3008	2SA224C			3008	2SA29M	3008	2SA322D		
3008	2SA224D					3008	2SA322E		
3008	2SA224E								

SK	Industry Standard No.	SK	Industry Standard No.	SK	Industry Standard No.	SK	Industry Standard No.	SK	Industry Standard No.
3008	2BA352OR	3008	2BA383H	3008	2BA70H	3008	31-0004	3008	86-348-2
3008	2BA352R	3008	2BA383J	3008	2BA70K	3008	31-0006	3008	86-86-2
3008	2BA352Y	3008	2BA383K	3008	2BA70L	3008	31-0015	3008	86-87-2
3008	2BA353	3008	2BA383L	3008	2BA70MA	3008	31-0016	3008	86-99-2
3008	2BA353A	3008	2BA383M	3008	2BA70OR	3008	31-0065	3008	863001-001
3008	2BA353AL	3008	2BA383OR	3008	2BA70R	3008	31-0134	3008	87-0015
3008	2BA353B	3008	2BA383X	3008	2BA70Y	3008	31-0155	3008	87-0017
3008	2BA353C	3008	2BA383Y	3008	2BA71	3008	31-0139	3008	0000101
3008	2BA353CL	3008	2BA384	3008	2BA71A	3008	31-0170	3008	0102
3008	2BA353D	3008	2BA384A	3008	2BA71AB	3008	31-0178	3008	121-138
3008	2BA353B	3008	2BA384B	3008	2BA71AC	3008	31-0190	3008	121-145
3008	2BA353F	3008	2BA384C	3008	2BA71B	3008	31-0217	3008	121-146
3008	2BA353G	3008	2BA384D	3008	2BA71BS	3008	34-6000-83	3008	121-150
3008	2BA353GN	3008	2BA384E	3008	2BA71C	3008	34-6000-84	3008	121-153
3008	2BA353H	3008	2BA384F	3008	2BA71D	3008	34-6000-85	3008	121-154
3008	2BA353J	3008	2BA384G	3008	2BA71E	3008	34-6016-28	3008	121-170
3008	2BA353K	3008	2BA384GN	3008	2BA71F	3008	34-6016-29	3008	121-189
3008	2BA353L	3008	2BA384L	3008	2BA71G	3008	37T1	3008	121-242
3008	2BA353M	3008	2BA384M	3008	2BA71H	3008	42-21403	3008	121-243
3008	2BA353OR	3008	2BA384OR	3008	2BA71K	3008	44P1	3008	121-244
3008	2BA353R	3008	2BA384R	3008	2BA71L	3008	46-8613-3	3008	121-257
3008	2BA353Y	3008	2BA384X	3008	2BA71M	3008	46-8625-1	3008	121-312
3008	2BA356	3008	2BA384Y	3008	2BA71OR	3008	46-8625-3	3008	121-313
3008	2BA356A	3008	2BA385	3008	2BA71R	3008	46-8625-4	3008	121-381
3008	2BA356B	3008	2BA385A	3008	2BA71Y	3008	46-8625-5	3008	121-48
3008	2BA356C	3008	2BA385B	3008	2BA71XA	3008	46-8625-6	3008	121-493
3008	2BA356D	3008	2BA385C	3008	2BA80A	3008	46-86256-1	3008	154A3676
3008	2BA356E	3008	2BA385D	3008	2BA80B	3008	46-86256-2	3008	154A3677
3008	2BA356F	3008	2BA385E	3008	2BA80D	3008	46-86256-3	3008	154A5679-5110
3008	2BA356G	3008	2BA385F	3008	2BA80E	3008	48-123522	3008	229-510-227
3008	2BA356GN	3008	2BA385G	3008	2BA80F	3008	48-123536	3008	297V011801
3008	2BA356H	3008	2BA385GN	3008	2BA80G	3008	48-124316	3008	297V012H08
3008	2BA356J	3008	2BA385H	3008	2BA80H	3008	48-124347	3008	297V012H10
3008	2BA356K	3008	2BA385J	3008	2BA80K	3008	48-124348	3008	297V012H11
3008	2BA356L	3008	2BA385K	3008	2BA80L	3008	48-124351	3008	297V012H14
3008	2BA356M	3008	2BA385L	3008	2BA80M	3008	48-124352	3008	297V012H15
3008	2BA356OR	3008	2BA385M	3008	2BA800R	3008	48-124377	3008	297V026801
3008	2BA356R	3008	2BA385OR	3008	2BA80R	3008	48-128093	3008	297V034801
3008	2BA357	3008	2BA385R	3008	2BA80X	3008	48-128094	3008	297V035801
3008	2BA357A	3008	2BA385Y	3008	2BA80Y	3008	48-128095	3008	297V038801
3008	2BA357B	3008	2BA400	3008	2BA81	3008	48-134372	3008	297V038H06
3008	2BA357C	3008	2BA400A	3008	2BA82	3008	48-134537	3008	297V038H10
3008	2BA357D	3008	2BA400B	3008	2BA82A	3008	48-134605	3008	297V038H11
3008	2BA357E	3008	2BA400C	3008	2BA82B	3008	48-134680	3008	297V038H12
3008	2BA357F	3008	2BA400D	3008	2BA82C	3008	48-134684	3008	297V065001
3008	2BA357G	3008	2BA400E	3008	2BA82D	3008	48-134796	3008	297V065002
3008	2BA357GN	3008	2BA400F	3008	2BA82E	3008	48-134859	3008	297V065003
3008	2BA357H	3008	2BA400G	3008	2BA82F	3008	48-134861	3008	310-190
3008	2BA357J	3008	2BA400GN	3008	2BA82G	3008	48-134862	3008	324-0027
3008	2BA357K	3008	2BA400H	3008	2BA82H	3008	48-134880	3008	324-0028
3008	2BA357L	3008	2BA400J	3008	2BA82K	3008	48-644676	3008	324-0131
3008	2BA357M	3008	2BA400K	3008	2BA82M	3008	48-644677	3008	325-0036-564
3008	2BA357OR	3008	2BA400L	3008	2BA820R	3008	48-645867	3008	325-1375-10
3008	2BA357R	3008	2BA400M	3008	2BA82R	3008	48-97162A03	3008	325-1375-11
3008	2BA357X	3008	2BA400R	3008	2BA82X	3008	48-97258A03	3008	325-1375-12
3008	2BA357Y	3008	2BA400X	3008	2BA82Y	3008	48-97271A05	3008	325-1376-53
3008	2BA369	3008	2BA400Y	3008	2BA83C	3008	48K134494	3008	325-1376-54
3008	2BA369A	3008	2BA468	3008	2BA94	3008	48K134495	3008	353-9001-001
3008	2BA369C	3008	2BA468A	3008	2BA94A	3008	48K134496	3008	353-9001-003
3008	2BA369D	3008	2BA468B	3008	2BA94C	3008	48K134601	3008	353-9002-002
3008	2BA369E	3008	2BA468C	3008	2BA94D	3008	48K134796	3008	417-68
3008	2BA369F	3008	2BA468D	3008	2BA94E	3008	48K134797	3008	421-16
3008	2BA369G	3008	2BA468E	3008	2BA94F	3008	48I134797	3008	421-17
3008	2BA369GN	3008	2BA468F	3008	2BA94G	3008	50A102	3008	421-26
3008	2BA369H	3008	2BA468G	3008	2BA94H	3008	50A52	3008	421-6
3008	2BA369J	3008	2BA468GN	3008	2BA94K	3008	051-0079	3008	421-7
3008	2BA369K	3008	2BA468H	3008	2BA94L	3008	56P2	3008	421-8
3008	2BA369L	3008	2BA468J	3008	2BA94M	3008	56P4	3008	422-1403
3008	2BA369M	3008	2BA468K	3008	2BA940R	3008	57A2-19	3008	576-0003-01Q
3008	2BA369R	3008	2BA468L	3008	2BA94R	3008	57B2-02	3008	601-054
3008	2BA369X	3008	2BA468M	3008	2BA94X	3008	57B2-17	3008	610-052
3008	2BA369Y	3008	2BA468OR	3008	2BA94Y	3008	57B2-18	3008	610-056
3008	2BA380	3008	2BA468R	3008	03-57-001	3008	57B2-19	3008	610-074
3008	2BA380A	3008	2BA468T	3008	03-57-101	3008	57B2-6	3008	612-60039
3008	2BA380C	3008	2BA476	3008	03-57-201	3008	5705-5	3008	612-60039A
3008	2BA380D	3008	2BA476A	3008	4-2073	3008	57D1-81	3008	614X1
3008	2BA380E	3008	2BA476B	3008	4-276	3008	57D1-85	3008	614X2
3008	2BA380F	3008	2BA476C	3008	4-279	3008	57D1-86	3008	614X3
3008	2BA380G	3008	2BA476D	3008	4-280	3008	57D1-87	3008	614X4
3008	2BA380H	3008	2BA476E	3008	4-65B17-1	3008	57D1-97	3008	690V034H29
3008	2BA380J	3008	2BA476F	3008	4-686256-1	3008	61-260039	3008	690V047H54
3008	2BA380K	3008	2BA476G	3008	4-686256-2	3008	61-260039A	3008	690V047H55
3008	2BA380L	3008	2BA476GN	3008	4-686256-3	3008	65B	3008	690V056H31
3008	2BA380M	3008	2BA476H	3008	6-1260039	3008	065B-12-7	3008	690V056H32
3008	2BA380OR	3008	2BA476J	3008	6-1260039A	3008	65B-70	3008	690V056H99
3008	2BA380R	3008	2BA476K	3008	6-60	3008	65B-70-12	3008	690V063H14
3008	2BA380Y	3008	2BA476L	3008	6-60A	3008	65B-70-12-7	3008	690V063H15
3008	2BA381	3008	2BA476M	3008	6-60B	3008	65B10	3008	690V068H29
3008	2BA381A	3008	2BA476OR	3008	6-60C	3008	65B10R	3008	690V073H85
3008	2BA381B	3008	2BA476R	3008	6-60D	3008	65B11	3008	690V077H35
3008	2BA381E	3008	2BA476X	3008	6-60E	3008	65B119	3008	690V110H94
3008	2BA381F	3008	2BA476Y	3008	6-60F	3008	65B12	3008	690V119H97
3008	2BA381G	3008	2BA57A	3008	6-60P	3008	65B121	3008	800-50500
3008	2BA381GN	3008	2BA57B	3008	6-61	3008	65B13	3008	801-04900
3008	2BA381H	3008	2BA57C	3008	6-61A	3008	65B13P	3008	801-05200
3008	2BA381K	3008	2BA57D	3008	6-61B	3008	65B14	3008	801-05500
3008	2BA381L	3008	2BA57E	3008	6-61D	3008	65B14-7	3008	916-31019-3
3008	2BA381M	3008	2BA57F	3008	6-61E	3008	65B15	3008	916-31019-3B
3008	2BA381OR	3008	2BA57G	3008	6-61F	3008	65B15L	3008	921-10B
3008	2BA381R	3008	2BA57L	3008	6-61P	3008	65B16	3008	921-11B
3008	2BA381X	3008	2BA57M	3008	6-61T	3008	65B16-2	3008	921-9B
3008	2BA381Y	3008	2BA57OR	3008	6-62	3008	65B17	3008	972X6
3008	2BA382	3008	2BA57R	3008	6-62A	3008	65B17-1	3008	972X7
3008	2BA382A	3008	2BA57X	3008	6-62B	3008	65B18	3008	972X8
3008	2BA382B	3008	2BA57Y	3008	6-62C	3008	65B182	3008	1033--1
3008	2BA382C	3008	2BA64	3008	6-62D	3008	65B19	3008	1033-2
3008	2BA382D	3008	2BA64A	3008	6-62E	3008	65B19G	3008	1033-3
3008	2BA382E	3008	2BA64C	3008	6-62P	3008	65B3	3008	1033-4
3008	2BA382F	3008	2BA64D	3008	6A11301	3008	65B4	3008	1241-719
3008	2BA382G	3008	2BA64E	3008	07-3015-05	3008	65B5	3008	1465B
3008	2BA382GN	3008	2BA64F	3008	8-0050500	3008	65B6	3008	1465B-12
3008	2BA382H	3008	2BA64G	3008	8-0104900	3008	65B7	3008	1465B-12-8
3008	2BA382J	3008	2BA64GN	3008	8-0105200	3008	65B8	3008	1865-1-12
3008	2BA382K	3008	2BA64H	3008	8-0105300	3008	65B9	3008	1865B
3008	2BA382L	3008	2BA64J	3008	09-300002	3008	80-050500	3008	1865B-12
3008	2BA382M	3008	2BA64K	3008	12-1-138	3008	80-104900	3008	1865B-127
3008	2BA382OR	3008	2BA64L	3008	12-1-189	3008	80-105200	3008	1865BL
3008	2BA382R	3008	2BA64M	3008	12-1-242	3008	80-105300	3008	1865BL8
3008	2BA382X	3008	2BA64OR	3008	12-1-243	3008	86-101-2	3008	2057B2-104
3008	2BA382Y	3008	2BA64R	3008	12-1-244	3008	86-108-2	3008	2057B2-105
3008	2BA383	3008	2BA64X	3008	12-1-257	3008	86-109-2	3008	2057B2-377
3008	2BA383A	3008	2BA64Y	3008	14-600-19	3008	86-111-2	3008	2057B2-41
3008	2BA383C	3008	2BA70	3008	020-00023	3008	86-115-2	3008	2057B2-42
3008	2BA383D	3008	2BA70-0B	3008	21-32	3008	86-116-2	3008	2057B2-48
3008	2BA383F	3008	2BA70A	3008	21-33	3008	86-131-2	3008	2057B2-50
3008	2BA383GN	3008	2BA70B	3008	21A040-031	3008	86-20-2	3008	2057B2-51
		3008	2BA70C	3008	24MW111	3008	86-281-2	3008	2057B2-79
		3008	2BA70D	3008	24MW157	3008	86-282-2	3008	2057B2-80
		3008	2BA70P	3008	24MW205	3008	86-283-2	3008	2478
		3008	2BA70Q	3008	24MW34	3008	86-295-2	3008	2478A
				3008	24MW352	3008	86-311-2	3008	2478B
				3008	24MW353	3008	86-312-2	3008	2487B
				3008	24MW441			3008	2488
				3008	24MW991			3008	2488A

SK	Industry Standard No.	SK	Industry Standard No.	SK	Industry Standard No.	SK	Industry Standard No.	SK	Industry Standard No.
3008	2489	3008	980958	3009	AD-157	3009	B425	3009	K04774
3008	2489A	3008	980959	3009	AD-159	3009	B426	3009	K4-520
3008	2495-012	3008	982150	3009	AD104	3009	B426BL	3009	K4-521
3008	2495-013	3008	982267	3009	AD105	3009	B426R	3009	K04774
3008	2495-200	3008	982289	3009	AD130	3009	B426Y	3009	KT1017
3008	2606-292	3008	983236	3009	AD131	3009	B445	3009	L-417-29BLK
3008	3010	3008	983271	3009	AD138	3009	B446	3009	L-417-29GRN
3008	003460	3008	983272	3009	AD139	3009	B449	3009	L-417-29WHT
3008	3665	3008	985445A	3009	AD149-01	3009	B449P	3009	L-417-60
3008	3666	3008	1471104-5	3009	AD149-02	3009	B449P	3009	M4531
3008	3679	3008	2000646-106	3009	AD149B	3009	B471	3009	M4463
3008	3851	3008	2001653-20	3009	AD149C	3009	B471-2	3009	M4570
3008	4363	3008	2001653-21	3009	ADY22	3009	B471A	3009	M4582
3008	4365	3008	2002151-18A	3009	ADY24	3009	B471B	3009	M4582BRN
3008	4632	3008	2091241-0005	3009	ADY28	3009	B472	3009	M4583
3008	4686-256-1	3008	2091241-0014	3009	AR-10	3009	B472A	3009	M4583RED
3008	4686-256-2	3008	2091241-0015	3009	AR-11	3009	B472B	3009	M4584
3008	4686-256-3	3008	2091241-0719	3009	AR-12	3009	B62	3009	M4584GRN
3008	9403-8	3008	2091241-1	3009	AR-13	3009	B64	3009	M4608
3008	9510-3	3008	2091241-2	3009	AR-14	3009	B69	3009	M4619
3008	12113-8	3008	2091241-3	3009	AR-4	3009	B80	3009	M4619RED
3008	12118-9	3008	2495012	3009	AR-5	3009	B81	3009	M4620
3008	12124-2	3008	2495013	3009	AR-6	3009	B82	3009	M4620GRN
3008	12124-3	3008	2495200	3009	AR-7	3009	B83	3009	M4649
3008	12124-4	3008	2498837	3009	AR-8	3009	B84	3009	M4722BLU
3008	12125-7	3008	7279780	3009	AR-9	3009	BC1073	3009	M4722GRN
3008	30213	3008	7287940	3009	AR8P404R	3009	BC1073A	3009	M4722PUR
3008	30214	3008	8010490	3009	AS215	3009	BC1073A	3009	M4722RED
3008	30217	3008	8010520	3009	AS215	3009	BC1274	3009	M4722YEL
3008	30218	3008	8010530	3009	AS216	3009	BC1274A	3009	M4730
3008	34100	3008	8014711	3009	AS217	3009	BC1274B	3009	M4766
3008	34118	3008	8014712	3009	AS218	3009	BDY62	3009	M4767
3008	34342	3008	43024834	3009	ATC-TR-14	3009	BF5	3009	M4888
3008	34389	3008	61260039	3009	ATC-TR-5	3009	C337B	3009	M4888B
3008	35070	3008	61260039A	3009	ATC-TR-6	3009	050BA042	3009	M7031
3008	35168	3008	62087025	3009	AU1D2	3009	CDT1309	3009	M84
3008	35169	3008	80050500	3009	AUY-21	3009	CDT1310	3009	M84B
3008	35170	3008	80104900	3009	AUY10	3009	CDT1311	3009	MA4670
3008	35677	3008	80105200	3009	AUY21A	3009	CDT1349	3009	MN194
3008	35678	3008	80105300	3009	AUZ22A	3009	CDT1349A	3009	MN22
3008	35815	3009	2295100227	3009	AY105	3009	CDT1350	3009	MN23
3008	35816	3009	A059-108	3009	B-1914	3009	CDT1350A	3009	MN29BLK
3008	35817	3009	A059-115	3009	B10064	3009	CQT1075	3009	MN29GRN
3008	35818	3009	A1124C	3009	B10069	3009	CQT1076	3009	MN29PUR
3008	35824	3009	A12163	3009	B1017	3009	CQT1077	3009	MN29WHT
3008	40004	3009	A12178	3009	B107A	3009	CQT1110A	3009	M46
3008	40006	3009	A14-586-01	3009	B10912	3009	CQT1111	3009	MN73BLK
3008	40262	3009	A1414A	3009	B10913	3009	CQT1111A	3009	MN73WHT
3008	48937-2	3009	A1414A9	3009	B1151	3009	CQT940A	3009	MN76
3008	48937-2	3009	A144A-1	3009	B1151A	3009	CQT940B	3009	MP1014
3008	61102-0	3009	A144A-19	3009	B151B	3009	CQT940BA	3009	MP1509-2
3008	080001	3009	A1465C	3009	B1152	3009	CRT1544	3009	MP1509-3
3008	080072	3009	A146509	3009	B1152A	3009	CRT1545	3009	MP2060
3008	080206	3009	A1484A	3009	B1152B	3009	CRT1552	3009	MP2060-1
3008	080236	3009	A1484A9	3009	B1181	3009	CRT1553	3009	MP2737A
3008	080275	3009	A14A-70	3009	B119	3009	CRT3602A	3009	MP2138A
3008	080275	3009	A14A-705	3009	B119A	3009	CT1122	3009	MP2139A
3008	080276	3009	A15927	3009	B122	3009	CT1124	3009	MP2142A
3008	81506-4	3009	A2090056-1	3009	B123	3009	CT1124A	3009	MP2143A
3008	81506-4A	3009	A2090056-27	3009	B123A	3009	CT1124B	3009	MP2444A
3008	81506-4B	3009	A2090056-5	3009	B124	3009	CTP1106	3009	MP3611
3008	81506-4C	3009	A2091859-0025	3009	B126	3009	CTP1119	3009	MP3612
3008	81506-7A	3009	A2091859-0720	3009	B126A	3009	CTP1265	3009	MP3613
3008	81506-7B	3009	A2091859-10	3009	B127	3009	CTP1266	3009	MP3614
3008	81506-7C	3009	A2091859-11	3009	B1274	3009	CTP1306	3009	MP3615
3008	81506-8	3009	A218012D8	3009	B1274A	3009	CTP1307	3009	MP3617
3008	81506-8A	3009	A28B240A	3009	B1274B	3009	CTP1508	3009	MP825
3008	81506-8B	3009	A28B242A	3009	B127A	3009	CTP1511	3009	N57B4-2
3008	81506-8C	3009	A28B248A	3009	B128A	3009	CTP1512	3009	N57B4-4
3008	94038	3009	A30302	3009	B129	3009	CTP1513	3009	NA2-ZZ-8
3008	95102	3009	A34-6001-1	3009	B131A	3009	CTP1730	3009	NKT-401
3008	95111	3009	A34-6002-17	3009	B132	3009	D081	3009	NKT-402
3008	99101	3009	A34715	3009	B134	3009	D8-503	3009	NKT-403
3008	99102	3009	A35084	3009	B1368A	3009	D8-515	3009	NKT-404
3008	99103	3009	A35201	3009	B1368B	3009	D8-520	3009	NKT-405
3008	100089	3009	A35260	3009	B1368C	3009	DTG1011	3009	NKT-415
3008	112296	3009	A36896	3009	B1368D	3009	DTG1040	3009	NKT-416
3008	115275	3009	A417-62	3009	B137	3009	DU6	3009	NKT-451
3008	116072	3009	A4247	3009	B138	3009	EA1341	3009	NKT-452
3008	117724	3009	A4347	3009	B140	3009	EA15X10	3009	NKT-453
3008	117824	3009	A48-134727	3009	B141	3009	EA15X12	3009	NKT-454
3008	119526	3009	A48-134731	3009	B142	3009	EA15X15	3009	NKT-501
3008	125790	3009	A48-134907	3009	B143	3009	EA15X173	3009	NKT-503
3008	125972	3009	A48-137102	3009	B143P	3009	EA15X26	3009	NKT-504
3008	129589	3009	A48-137213	3009	B144	3009	EA15X33	3009	NKT450
3008	162002-040	3009	A48-137215	3009	B144P	3009	EA15X35	3009	OC-16
3008	162002-041	3009	A48-137216	3009	B145	3009	EA15X38	3009	OC-22
3008	162002-042	3009	A48-137217	3009	B146	3009	EA15X53	3009	OC-23
3008	162002-40	3009	A48-137218	3009	B147	3009	EA15X88	3009	OC-24
3008	162002-41	3009	A48-137219	3009	B149	3009	EA1700	3009	OC-25
3008	166908	3009	A48-137220	3009	B177	3009	EGG-6	3009	OC-28
3008	166909	3009	A4A-1-70	3009	B1904	3009	EQG-8	3009	OC-29
3008	171162-169	3009	A4A-1-705	3009	B215	3009	ER15X10	3009	OC-30A
3008	198010-1	3009	A4A-1A9G	3009	B216	3009	ES313	3009	OC-35
3008	223487	3009	A5253	3009	B216A	3009	ES15X17	3009	OC-36
3008	225594A	3009	A57B124-10	3009	B217	3009	ES15X43	3009	OC19
3008	231706	3009	A57J5-1	3009	B217A	3009	ES15X45	3009	OC20
3008	252681	3009	A57M3-7	3009	B217G	3009	ES15X51	3009	OC27
3008	261586	3009	A57M3-8	3009	B217U	3009	ES15X78	3009	OC30
3008	310157	3009	A62-18427	3009	B224	3009	ES503	3009	P-31898
3008	310158	3009	A63-18427	3009	B228	3009	ES7	3009	P-5-30
3008	489751-108	3009	A65C-70	3009	B229	3009	ES9	3009	P1A
3008	573303	3009	A65C-705	3009	B230	3009	ET15X17	3009	P1E-1
3008	573329	3009	A65G190	3009	B239	3009	ET15X4	3009	P1K
3008	573402	3009	A660097	3009	B239A	3009	ET15X43	3009	P1KBLK
3008	601054	3009	A690V081H97	3009	B246	3009	ET15X5	3009	P1KBLU
3008	610052	3009	A7279039	3009	B247	3009	ET6	3009	P1KBRN
3008	610052-1	3009	A7279049	3009	B248	3009	F67E	3009	P1KGRN
3008	610056	3009	A7285774	3009	B248A	3009	G19	3009	P1KORN
3008	610056-1	3009	A7285778	3009	B249	3009	GC4087	3009	P1KRED
3008	610056-2	3009	A7289047	3009	B249A	3009	GC4125	3009	P1KYEL
3008	610056-3	3009	A7290594	3009	B250A	3009	GC4156	3009	P1R
3008	610056-4	3009	A7291252	3009	B254	3009	GC4251	3009	P1T
3008	610074	3009	A7297043	3009	B255	3009	GC4267-2	3009	P2D
3008	611020	3009	A7297092	3009	B256	3009	GC641	3009	P2DBLU
3008	741050	3009	A7297093	3009	B26	3009	GE-3	3009	P2DBRN
3008	772768	3009	A76-11770	3009	B282	3009	GET-572	3009	P2DGRN
3008	815064	3009	A84A-70	3009	B283	3009	GPT3008/40	3009	P2DORN
3008	815064A	3009	A84A-705	3009	B284	3009	GP1493	3009	P2DRED
3008	815064B	3009	A84A19G	3009	B285	3009	GP1494	3009	P2DYEL
3008	815064C	3009	A87404F	3009	B295	3009	GP1622	3009	P2R
3008	815067A	3009	A94004	3009	B28B241	3009	GP1882	3009	P3EBLK
3008	815067B	3009	A964-17887	3009	B28B244	3009	GPT-16	3009	P75534
3008	815067C	3009	A97A83	3009	B337	3009	GTX-2001	3009	P75534-1
3008	815068	3009	A992-00-1192	3009	B338	3009	HA-1202	3009	P8879
3008	815068A	3009	AA28B240A	3009	B355	3009	HP19	3009	PA-10889-1
3008	815068B	3009	ACY16	3009	B356	3009	HP20	3009	PA-10889-2
3008	815068C	3009	AD-140	3009	B391	3009	HR101A	3009	PA-10890
3008	980372	3009	AD-148	3009	B407	3009	HR102	3009	PA-10890-1
3008	980373	3009	AD-149	3009	B407-0	3009	HR102C	3009	PAD750
3008	980374	3009	AD-150	3009	B407TV	3009	HR103	3009	PAR-12
3008	980833	3009	AD-152	3009	B413	3009	HR103A	3009	PB110
3008	980834	3009	AD-156	3009	B414	3009	JR40	3009	POWER12
3008	980835			3009	B424	3009	JR40		

SK	Industry Standard No.	SK	Industry Standard No.	SK	Industry Standard No.	SK	Industry Standard No.	SK	Industry Standard No.
3009	POWER25	3009	SP1596RED	3009	V30/10DP	3009	2N1529	3009	2N257B
3009	POWER99	3009	SP1603-3	3009	V30/20DP	3009	2N1529A	3009	2N257G
3009	PQ51	3009	SP1651	3009	V30/30DP	3009	2N1530	3009	2N257W
3009	PS-1	3009	SP176	3009	V60/10DP	3009	2N1530A	3009	2N2612
3009	PT-12	3009	SP1938	3009	V60/10P	3009	2N1531	3009	2N268
3009	PT-150	3009	SP1950	3009	V60/20DP	3009	2N1531A	3009	2N268A
3009	PT-155	3009	SP2045	3009	V60/30DP	3009	2N1532	3009	2N2836
3009	PT-176	3009	SP2046	3009	V60/30P	3009	2N1532A	3009	2N285
3009	PT-234	3009	SP2048	3009	VFU-274513	3009	2N1533	3009	2N285A
3009	PT-235	3009	SP2072	3009	VFP-2746C	3009	2N1534	3009	2N285B
3009	PT-235A	3009	SP2076	3009	VFP-6537C	3009	2N1534A	3009	2N2869
3009	PT-236	3009	SP2155	3009	VFU-2746B	3009	2N1535	3009	2N2869/2N301
3009	PT-236A	3009	SP2234	3009	VM-30203	3009	2N1535A	3009	2N2870
3009	PT-236B	3009	SP230	3009	W5	3009	2N1535B	3009	2N297
3009	PT-242	3009	SP2358	3009	W9	3009	2N1536	3009	2N297A
3009	PT-25	3009	SP2361	3009	WTV199PWR	3009	2N1536A	3009	2N301
3009	PT-255	3009	SP2361BLU	3009	WTV25PWR	3009	2N1537	3009	2N301A
3009	PT-256	3009	SP2361BRN	3009	X1005	3009	2N1537A	3009	2N301B
3009	PT-285	3009	SP2361GRN	3009	XB14	3009	2N1538	3009	2N301G
3009	PT-285A	3009	SP2361ORN	3009	XB5	3009	2N1538A	3009	2N301W
3009	PT-3	3009	SP2361RED	3009	XB7	3009	2N1539	3009	2N307
3009	PT-301	3009	SP2361YEL	3009	XC141	3009	2N1539A	3009	2N307A
3009	PT-301A	3009	SP334	3009	XC142	3009	2N1544	3009	2N307B
3009	PT-307	3009	SP441D	3009	XC155	3009	2N1544A	3009	2N3132
3009	PT-307A	3009	SP4418	3009	XN12A	3009	2N1549A	3009	2N3212
3009	PT-3A	3009	SP47	3009	XN12B	3009	2N155	3009	2N3213
3009	PT-40	3009	SP485B	3009	XN12C	3009	2N1553	3009	2N3214
3009	PT-50	3009	SP485BLK	3009	XN12E	3009	2N1553A	3009	2N3215
3009	PT-554	3009	SP485BLU	3009	Z20	3009	2N1557	3009	2N325
3009	PT-555	3009	SP485BRN	3009	ZEN325	3009	2N1557A	3009	2N350
3009	PT-6	3009	SP485W	3009	001-01204-0	3009	2N156	3009	2N350A
3009	PT236C	3009	SP485WHT	3009	001-012040	3009	2N157	3009	2N351
3009	PT30	3009	SP486WHT	3009	001-01205-0	3009	2N157A	3009	2N351A
3009	PT0105	3009	SP819R	3009	001-012050	3009	2N158	3009	2N352
3009	Q60074	3009	SP875	3009	001-012051	3009	2N158A	3009	2N353
3009	R3515	3009	SP891BLU	3009	1-21-270	3009	2N1653	3009	2N3611
3009	R516	3009	SP891GRN	3009	1-21-271	3009	2N1666	3009	2N3612
3009	R7167	3009	SP891R	3009	002-007000	3009	2N1667	3009	2N3613
3009	R7293	3009	SP891WHT	3009	002-008100	3009	2N1668	3009	2N3614
3009	R7620	3009	SZ-235	3009	002-010100	3009	2N1669	3009	2N3615
3009	R8313	3009	SYL109	3009	002-017000	3009	2N1755	3009	2N3616
3009	R8659	3009	T-101	3009	002-9700	3009	2N1756	3009	2N3617
3009	R8-105	3009	T-127	3009	2AD140	3009	2N1757	3009	2N3618
3009	R8-2006	3009	T-Q5036	3009	2D004	3009	2N1758	3009	2N375
3009	R8-3858-1	3009	T1040	3009	2D004-9	3009	2N1759	3009	2N376
3009	R8-5613	3009	T1041	3009	2D015	3009	2N176	3009	2N376A
3009	R8-5835	3009	T1167	3009	2D016-45	3009	2N176-1YEL	3009	2N378
3009	R8-5855	3009	T1366	3009	2D016-54	3009	2N176-3PUR	3009	2N379
3009	R83358-1	3009	T1366A	3009	2D021	3009	2N176-4PUR	3009	2N380
3009	R83359-1	3009	T1367	3009	2D021-11	3009	2N176-5WHT	3009	2N386
3009	R83858	3009	T1367A	3009	2D021-56	3009	2N176-6WHT	3009	2N387
3009	R83858-1	3009	T1368	3009	2D021-8	3009	2N1760	3009	2N392
3009	R83959	3009	T1368A	3009	2D023	3009	2N1761	3009	2N399
3009	R85612	3009	T1369	3009	2G220	3009	2N1762	3009	2N400
3009	R85613	3009	T1370	3009	2G223	3009	2N176A	3009	2N401
3009	R85614	3009	T1370A	3009	2G225	3009	2N176G	3009	2N418
3009	R85616	3009	T142	3009	2N1007	3009	2N176W	3009	2N419
3009	8-39T	3009	T1601	3009	2N1011	3009	2N178	3009	2N420
3009	8-40T	3009	T1602	3009	2N1014	3009	2N179	3009	2N420A
3009	8-40TB	3009	TA-1614	3009	2N1020	3009	2N1971	3009	2N456
3009	8-41T	3009	TA-1682	3009	2N1021	3009	2N2061	3009	2N456A
3009	8-42T	3009	TA-1682A	3009	2N1021A	3009	2N2061A	3009	2N456B
3009	8-43T	3009	TA-1705	3009	2N1022	3009	2N2062	3009	2N458
3009	8-46T	3009	TA-1765	3009	2N1022A	3009	2N2062A	3009	2N458A
3009	8-48T	3009	TA-1766	3009	2N1029	3009	2N2063	3009	2N458B
3009	8-49T	3009	TA-1773	3009	2N1029A	3009	2N2063A	3009	2N459A
3009	8-58TB	3009	TA-1794	3009	2N1029B	3009	2N2064	3009	2N511
3009	8-95253	3009	TA-1881	3009	2N1029C	3009	2N2064A	3009	2N511A
3009	8-95253-1	3009	TA-1890	3009	2N1030	3009	2N2065	3009	2N511B
3009	0822.3640-050	3009	TA-1891	3009	2N1030A	3009	2N2065A	3009	2N512
3009	BC-70	3009	TA-2	3009	2N1030B	3009	2N2066	3009	2N512A
3009	BC70	3009	TA2301	3009	2N1030C	3009	2N2067	3009	2N512B
3009	SE-40022	3009	TF-80/30	3009	2N1031	3009	2N2067B	3009	2N513
3009	SFT190	3009	TFT78/30Z	3009	2N1031A	3009	2N2067D	3009	2N513A
3009	SFT191	3009	TFT8/60	3009	2N1031B	3009	2N2067W	3009	2N513B
3009	SFT192	3009	TI-266A	3009	2N1031C	3009	2N2069	3009	2N514
3009	SFT212	3009	TI-269	3009	2N1032	3009	2N2070	3009	2N514A
3009	SFT214	3009	TI-366	3009	2N1032A	3009	2N2137	3009	2N514B
3009	SFT238	3009	TI-366A	3009	2N1032B	3009	2N2137A	3009	2N538
3009	SFT240	3009	TI-367	3009	2N1032C	3009	2N2138	3009	2N538A
3009	SFT250	3009	TI-367A	3009	2N1033	3009	2N2138A	3009	2N539
3009	SP-1108	3009	TI-368	3009	2N1040	3009	2N2139	3009	2N539A
3009	SP-148-3	3009	TI-368A	3009	2N1041	3009	2N2139A	3009	2N540
3009	SP-1482-5	3009	TI-369	3009	2N1046	3009	2N2140	3009	2N540A
3009	SP-1483	3009	TI-369A	3009	2N1046B	3009	2N2140A	3009	2N553
3009	SP-1556-2	3009	TI-370	3009	2N1120	3009	2N2141	3009	2N554
3009	SP-1603	3009	TI-370A	3009	2N1136	3009	2N2141A	3009	2N555
3009	SP-1603-1	3009	TI3012	3009	2N1136A	3009	2N2142	3009	2N561
3009	SP-1603-2	3009	TI3027	3009	2N1136B	3009	2N2142A	3009	2N589
3009	SP-404T	3009	TI3028	3009	2N1137	3009	2N2143	3009	2N618
3009	SP-441	3009	TNJ60454	3009	2N1137A	3009	2N2143A	3009	2N627
3009	SP-485	3009	TNJ72318	3009	2N1137B	3009	2N2144	3009	2N628
3009	SP-486	3009	TO-012	3009	2N1138	3009	2N2144A	3009	2N629
3009	SP-486W	3009	TO-015	3009	2N1138A	3009	2N2145	3009	2N637
3009	SP-634	3009	TQ5036	3009	2N1138B	3009	2N2145A	3009	2N637A
3009	SP-649	3009	TQ5064	3009	2N1146	3009	2N2146	3009	2N637B
3009	SP-649-1	3009	TR-01	3009	2N1146A	3009	2N2146A	3009	2N638
3009	SP-834	3009	TR-01B	3009	2N1146B	3009	2N2212	3009	2N638A
3009	SP-880	3009	TR-01C	3009	2N1146C	3009	2N2288	3009	2N638B
3009	SP-880-1	3009	TR-01E	3009	2N1147	3009	2N2289	3009	2N639
3009	SP-880-3	3009	TR-8006	3009	2N1147A	3009	2N2290	3009	2N639A
3009	SP-891	3009	TRO1C	3009	2N1147B	3009	2N2291	3009	2N639B
3009	SP-891B	3009	TRO2	3009	2N1147C	3009	2N2292	3009	2N665
3009	SP-891W	3009	TRO2C	3009	2N1159	3009	2N2293	3009	2N669
3009	SP1013B	3009	TR16C	3009	2N1160	3009	2N2294	3009	2N677
3009	SP1323	3009	TR333	3009	2N1162	3009	2N2295	3009	2N677A
3009	SP1403	3009	TR56	3009	2N1162A	3009	2N2296	3009	2N677B
3009	SP1481	3009	TRA-7R	3009	2N1163	3009	2N230	3009	2N677C
3009	SP1481-1	3009	TRA-7RM	3009	2N1163A	3009	2N234	3009	2N678
3009	SP1481-2	3009	TRA-8R	3009	2N1164	3009	2N234A	3009	2N678A
3009	SP1481-3	3009	TS-173	3009	2N1164A	3009	2N235	3009	2N678B
3009	SP1481-4	3009	TS-610	3009	2N1165	3009	2N2357	3009	2N678C
3009	SP1481-5	3009	TS-612	3009	2N1165A	3009	2N235A	3009	2826
3009	SP1482	3009	TS-613	3009	2N1166	3009	2N235B	3009	2826A
3009	SP1482-2	3009	TS-614	3009	2N1166A	3009	2N236	3009	2841
3009	SP1482-3	3009	TS1657	3009	2N1172	3009	2N236A	3009	2841A
3009	SP1482-4	3009	TS176	3009	2N1245	3009	2N236B	3009	2842
3009	SP1482-6	3009	TS612	3009	2N1246	3009	2N242	3009	28A298D
3009	SP1482-7	3009	TS614	3009	2N1291	3009	2N2423	3009	28A298E
3009	SP1483-1	3009	TT-1083	3009	2N1292	3009	2N2446	3009	28A298F
3009	SP1483-2	3009	TV24214	3009	2N1293	3009	2N250	3009	28A298GN
3009	SP1483-3	3009	TV24337	3009	2N1295	3009	2N250A	3009	28A298H
3009	SP1550-3	3009	TV28B126F	3009	2N1297	3009	2N251	3009	28A298J
3009	SP1556	3009	TV8-28B126	3009	2N1314	3009	2N251A	3009	28A298K
3009	SP1556-1	3009	TV8-28B126F	3009	2N1314R	3009	2N255	3009	28A298L
3009	SP1556-3	3009	TV828B449	3009	2N1359	3009	2N255A	3009	28A298M
3009	SP1563-2	3009	TX103-1	3009	2N1360	3009	2N256	3009	28A298R
3009	SP1595BLK	3009	UTRA-7RM	3009	2N1361	3009	2N256A	3009	28A384H
3009	SP1595BLU	3009	V145	3009	2N1362	3009	2N257	3009	28A384J
3009	SP1595GRN	3009	V15/10DP	3009	2N1363	3009	2N257A	3009	28B107
3009	SP1595RED	3009	V15/20DP	3009	2N1364			3009	28B107A
3009	SP1596BLK	3009	V15/30DP	3009	2N1365			3009	28B119
3009	SP1596BLU			3009	2N1419			3009	28B119A
3009	SP1596GRN								

SK	Industry Standard No.	SK	Industry Standard No.	SK	Industry Standard No.	SK	Industry Standard No.	SK	Industry Standard No.
3009	28B122	3009	28B407TV	3009	8P060	3009	48-125267	3009	65015L
3009	28B123	3009	28B407TV-2	3009	8P860	3009	48-125332	3009	65016
3009	28B123A	3009	28B407Y	3009	09-301010	3009	48-129934	3009	65016-4
3009	28B124	3009	28B41	3009	09-301052	3009	48-129935	3009	65017
3009	28B126	3009	28B42	3009	9-5250	3009	48-129936	3009	65017-1
3009	28B126A	3009	28B426	3009	9-5251	3009	48-129937	3009	65018
3009	28B126B	3009	28B426R	3009	11-0399	3009	48-134302	3009	65C182
3009	28B126C	3009	28B426Y	3009	11-0400	3009	48-134447	3009	65019
3009	28B126CB	3009	28B449	3009	12-1-270	3009	48-134448	3009	65019G
3009	28B126F	3009	28B449A	3009	12-1-271	3009	48-134449	3009	65C2
3009	28B126G	3009	28B449B	3009	12M2	3009	48-134463	3009	65C3
3009	28B126H	3009	28B449C	3009	13-14735	3009	48-134487	3009	65C4
3009	28B126V	3009	28B449B	3009	13-14735A	3009	48-134488	3009	65C5
3009	28B127	3009	28B449F	3009	13-18034	3009	48-134493	3009	65C6
3009	28B127A	3009	28B449L	3009	13-18034-1	3009	48-134519	3009	65C7
3009	28B127M	3009	28B449M	3009	13-18034A	3009	48-134560	3009	65C8
3009	28B128A	3009	28B466	3009	13-22739-1	3009	48-134570	3009	65C9
3009	28B129	3009	28B467	3009	13-22741	3009	48-134574	3009	67P1
3009	28B131	3009	28B471	3009	13-22741-1	3009	48-134575	3009	67P3
3009	28B131A	3009	28B471-2	3009	14-573-10	3009	48-134592	3009	68P1
3009	28B132	3009	28B471A	3009	14-574-10	3009	48-134611	3009	68P1B
3009	28B137	3009	28B471B	3009	14-578-10	3009	48-134612	3009	76-11770
3009	28B138	3009	28B471D	3009	14-586-01	3009	48-134613	3009	77-271491-1
3009	28B140	3009	28B471B	3009	14-589-01	3009	48-134634	3009	78-5050
3009	28B141	3009	28B471F	3009	14-590-01	3009	48-134638	3009	80-050700
3009	28B142	3009	28B471GN	3009	14-601-01	3009	48-134639	3009	83-1056
3009	28B143	3009	28B471H	3009	14-601-03	3009	48-134644	3009	84
3009	28B143P	3009	28B471J	3009	14-601-04	3009	48-134645	3009	84A
3009	28B144	3009	28B471K	3009	14-601-05	3009	48-134646	3009	084A-12
3009	28B144P	3009	28B471L	3009	14-601-06	3009	48-134647	3009	084A-12-7
3009	28B145	3009	28B471M	3009	14-601-07	3009	48-134649	3009	84A-70
3009	28B146	3009	28B471R	3009	14-601-08	3009	48-134670	3009	84A-70-12
3009	28B147	3009	28B471X	3009	14-601-09	3009	48-134672	3009	84A-70-12-7
3009	28B149	3009	28B471Y	3009	14-601-11	3009	48-134696	3009	84A0
3009	28B151	3009	28B472A	3009	14-604-08	3009	48-134722	3009	84A1
3009	28B152	3009	28B472B	3009	14A	3009	48-134723	3009	84A10
3009	28B152A	3009	28B62	3009	014A-12	3009	48-134727	3009	84A10R
3009	28B19	3009	28B64	3009	014A-12-7	3009	48-134730	3009	84A12
3009	28B20	3009	28B69	3009	14A0	3009	48-134731	3009	84A121
3009	28B21	3009	28B80	3009	14A1	3009	48-134738	3009	84A13
3009	28B215	3009	28B81	3009	14A10	3009	48-134746	3009	84A13P
3009	28B217	3009	28B82	3009	14A10R	3009	48-134750	3009	84A14
3009	28B217A	3009	28B83	3009	14A11	3009	48-134751	3009	84A14-7
3009	28B217G	3009	28B91P	3009	14A12	3009	48-134757	3009	84A15
3009	28B217U	3009	28BP5	3009	14A13	3009	48-134758	3009	84A15L
3009	28B228	3009	28PT212	3009	14A13P	3009	48-134763	3009	84A16
3009	28B229	3009	2T3011	3009	14A14	3009	48-134764	3009	84A16B
3009	28B230	3009	2T3021	3009	14A14-7	3009	48-134766	3009	84A17
3009	28B239	3009	2T3022	3009	14A15	3009	48-134767	3009	84A17-1
3009	28B239A	3009	2T3030	3009	14A15L	3009	48-134907	3009	84A18
3009	28B240	3009	2T3031	3009	14A16	3009	48-134930	3009	84A182
3009	28B240A	3009	2T3032	3009	14A16-5	3009	48-134947	3009	84A19
3009	28B242	3009	2T3033	3009	14A17	3009	48-134974	3009	84A19G
3009	28B242A	3009	2T3041	3009	14A17-1	3009	48-134977	3009	84A2
3009	28B242B	3009	2T3042	3009	14A18	3009	48-137031	3009	84A3
3009	28B242C	3009	2T3043	3009	14A19	3009	48-137078	3009	84A4
3009	28B242D	3009	03-57-501	3009	14A19G	3009	48-137102	3009	84A5
3009	28B242E	3009	4-14A17-1	3009	14A2	3009	48-137118	3009	84A6
3009	28B242F	3009	4-435	3009	14A3	3009	48-137119	3009	84A7
3009	28B242G	3009	4-4A-1A7-1	3009	14A4	3009	48-137120	3009	84A8
3009	28B242GN	3009	4-65C17-1	3009	14A5	3009	48-137122	3009	84A9
3009	28B242H	3009	4-686213-3	3009	14A6	3009	48-137123	3009	84AA1
3009	28B242J	3009	004-8000	3009	14A7	3009	48-137124	3009	84AA19
3009	28B242K	3009	4A-1	3009	14A8	3009	48-137215	3009	84B
3009	28B242L	3009	04A-1-12-7	3009	14A9	3009	48-137216	3009	86-120-2
3009	28B242M	3009	4A-1-70	3009	17A4422-1	3009	48-137217	3009	86-127-2
3009	28B242OR	3009	4A-1-70-12	3009	19A115101	3009	48-137218	3009	86-141-2
3009	28B242R	3009	4A-1-70-12-7	3009	19A115101-P1	3009	48-137219	3009	86-142-2
3009	28B242X	3009	4A-10	3009	19A115267P1	3009	48-137220	3009	86-146-2
3009	28B242Y	3009	4A-11	3009	19A115341P1	3009	48-3P1	3009	86-147-2
3009	28B246	3009	4A-12	3009	19A115361-P1	3009	48-57B2	3009	86-173-2
3009	28B247	3009	4A-13	3009	19A115385-P1	3009	48-57B42	3009	86-173-9
3009	28B248	3009	4A-14	3009	19A115561	3009	48K125230	3009	86-19-2
3009	28B248A	3009	4A-15	3009	19AR31	3009	48K59P1	3009	86-230-2
3009	28B249	3009	4A-16	3009	19C300113-P1	3009	48K57B2	3009	86-231-2
3009	28B249A	3009	4A-17	3009	20A0017	3009	48R134722	3009	86-232-2
3009	28B25	3009	4A-18	3009	20A0041	3009	48B134695	3009	86-235-2
3009	28B250	3009	4A-19	3009	20A0042	3009	48B134746	3009	86-248-2
3009	28B250A	3009	4A-1A	3009	20A0074	3009	48B134747	3009	86-313-2
3009	28B252	3009	4A-1AO	3009	21-28	3009	48B134759	3009	86-317-2
3009	28B253A	3009	4A-1AOR	3009	21A097-000	3009	48B134761	3009	86-319-2
3009	28B254	3009	4A-1A1	3009	022-3640-050	3009	48B134766	3009	86-353-2
3009	28B255	3009	4A-1A19	3009	23-5009	3009	48B134767	3009	86-354-2
3009	28B26	3009	4A-1A2	3009	23B-210-025	3009	48B134947	3009	86-5043-2
3009	28B26A	3009	4A-1A21	3009	026-100003	3009	48B134974	3009	86-5058-2
3009	28B27	3009	4A-1A3	3009	026-100028	3009	48B137031	3009	86-5088-2
3009	28B28	3009	4A-1A3P	3009	27T406	3009	49P1C	3009	86-5943-2
3009	28B282	3009	4A-1A4	3009	31-0192	3009	57B124-10	3009	86-62-2
3009	28B283	3009	4A-1A4-7	3009	31-0240	3009	57B4-1	3009	86-63-2
3009	28B284	3009	4A-1A5	3009	33-1004-00	3009	57B4-2	3009	86-8-2
3009	28B285	3009	4A-1A5L	3009	34-6001-1	3009	57B4-4	3009	86X0009-001
3009	28B29	3009	4A-1A6	3009	34-6002-10	3009	5706-12	3009	86X0015-001
3009	28B295	3009	4A-1A6-4	3009	34-6002-11	3009	5706-2	3009	86X2
3009	28B30	3009	4A-1A7	3009	34-6002-13	3009	5706-23	3009	094-013
3009	28B31	3009	4A-1A7-1	3009	34-6002-14	3009	5706-3	3009	95-11414000
3009	28B337	3009	4A-1A82	3009	34-6002-17	3009	5709-2	3009	95-250
3009	28B337A	3009	4A-1A9	3009	34-6002-18	3009	57D1-119	3009	95-251
3009	28B337B	3009	4A-1A9G	3009	34-6002-18A	3009	57D4-1	3009	96-5026-01
3009	28B337BK	3009	4JX8P404	3009	34-6002-19	3009	57D4-2	3009	96-5045-01
3009	28B337C	3009	4JX8P404	3009	34-6002-2	3009	57D6-12	3009	96-5064-01
3009	28B337D	3009	4JX8P409	3009	34-6002-20	3009	57D9-1	3009	96-5081-01
3009	28B337E	3009	6-88	3009	34-6002-22	3009	57D9-2	3009	96-5086-02
3009	28B337F	3009	8-0050700	3009	34-6002-22	3009	5TJ5-1	3009	96-5100-01
3009	28B337G	3009	8-905-605-624	3009	34-6002-3	3009	57M3-7	3009	96-5100-03
3009	28B337GN	3009	8-905-605-635	3009	34-6002-34	3009	57M3-8	3009	96-5125-01
3009	28B337H	3009	8-905-605-636	3009	34-6002-4	3009	61-782	3009	96-5143-01
3009	28B337J	3009	8-905-605-637	3009	34-6002-5	3009	610005-1	3009	96-5143-02
3009	28B337LB	3009	8-905-606-720	3009	34-6002-6	3009	62-18427	3009	96-51430-02
3009	28B337M	3009	8-905-613-215	3009	34-6002-7	3009	63-18427	3009	96-5148-01
3009	28B337R	3009	8-905-613-250	3009	34-6002-8	3009	63-29451	3009	96-5155-01
3009	28B337X	3009	8A12991	3009	34-6002-9	3009	63-29459	3009	96-5192-01
3009	28B337Y	3009	8L201	3009	38P1	3009	63-8706	3009	96XZ801/06N
3009	28B338	3009	8L201B	3009	38P1C	3009	65C	3009	96XZ801/10N
3009	28B338H	3009	8L201O	3009	39P1	3009	0650-12	3009	96XZ801/34X
3009	28B355	3009	8L201R	3009	39P1C	3009	0650-12-7	3009	97A83
3009	28B361	3009	8L201V	3009	42-16599	3009	650-70	3009	998001
3009	28B375	3009	8L301V	3009	42-22834	3009	650-70-12	3009	998014
3009	28B391	3009	8L404	3009	42-23968P	3009	650-70-12-7	3009	998014A
3009	28B392M	3009	8P40	3009	46-86136-3	3009	6500	3009	998015
3009	28B407	3009	8P404	3009	46-8615-3	3009	65010	3009	106P1
3009	28B407A	3009	8P404B	3009	46-8617-3	3009	65010R	3009	1061AG
3009	28B407B	3009	8P404F	3009	46-86213-3	3009	65011	3009	112-524
3009	28B407C	3009	8P404M	3009	46-8634-3	3009	65019	3009	115-063
3009	28B407D	3009	8P404M-1	3009	46-8638-3	3009	65012	3009	115-268
3009	28B407E	3009	8P404N	3009	47P1	3009	65121	3009	115-269
3009	28B407G	3009	8P404ORN	3009	48-124204	3009	65013	3009	120-004887
3009	28B407GN	3009	8P404R	3009	48-124246	3009	65013P	3009	121-1124
3009	28B407H	3009	8P404T	3009	48-124247	3009	65014	3009	121-270
3009	28B407J	3009	8P404V	3009	48-124285	3009	65014-7	3009	121-271
3009	28B407K	3009	8P415C	3009	48-124302	3009	65015	3009	121-308
3009	28B407M	3009	8P416C	3009	48-124332			3009	121-371
3009	28B407R			3009	48-124356			3009	121-389
				3009	48-125204			3009	121-398
				3009	48-125208				

SK	Industry Standard No.	SK	Industry Standard No.	SK	Industry Standard No.	SK	Industry Standard No.	SK	Industry Standard No.
3009	121-418	3009	2057A2-302	3009	190425A	3009	2091859-0025	3010	T-81
3009	121-830	3009	2057B124-10	3009	214396	3009	2091859-0712	3010	T-Q5039
3009	121-Z9006	3009	2057B2-133	3009	216986	3009	2091859-0714	3010	T-Q5050
3009	122-1625	3009	2347-17	3009	217892	3009	2091859-0716	3010	TK490
3009	124-1	3009	2780	3009	218012DB	3009	2091859-0718	3010	TNJ72284
3009	129-13	3009	2780-3	3009	219301	3009	2091859-0720	3010	TRO1065
3009	129-5	3009	2780-4	3009	219361	3009	2091859-0723	3010	TR310236
3009	129-6	3009	2780-5	3009	219440	3009	2091859-10	3010	TV24143
3009	129-7	3009	2901-010	3009	219940	3009	2091859-11	3010	TV8AP801
3009	129-9	3009	2904-008	3009	221602	3009	2091859-4	3010	W4
3009	0131-000192	3009	2904-014	3009	221605	3009	2091859-6	3010	WC19862
3009	0131-000336	3009	3512	3009	221940	3009	2091859-8	3010	WEF641A
3009	0131-000337	3009	3577	3009	221941	3009	2091859-9	3010	XG33
3009	0131-001425	3009	4247	3009	222915	3009	2320541	3010	X826
3009	144A-1	3009	4331	3009	223565	3009	2900014	3010	2N1059
3009	144A-1-12	3009	4347	3009	223490	3009	3460553-4	3010	2N1101
3009	144A-1-12-8	3009	4463	3009	223576	3009	4036598-P1	3010	2N1102
3009	146-T1	3009	4582BRN	3009	223810	3009	4036733-P1	3010	2N1173
3009	147-E1	3009	4583RED	3009	224503	3009	4037607-P1	3010	2N1173W
3009	154X3680	3009	4584GRN	3009	225595	3009	4082501-0001A	3010	2N1174
3009	171-001-9-001	3009	4608	3009	225596	3009	4082501-001	3010	2N1251
3009	171-015-9-001	3009	4619RED	3009	225925	3009	7274653	3010	2N1431
3009	173A3936	3009	4620GRN	3009	225927	3009	7279039	3010	2N148B
3009	173A3965	3009	4649	3009	226634	3009	7279049	3010	2N214
3009	173A4419-2	3009	4686-213-3	3009	226999	3009	7286665	3010	2N214A
3009	173A4419	3009	4722BLU	3009	227566	3009	7285774	3010	2N228
3009	173A4419-1	3009	4722GRN	3009	227804	3009	7285778	3010	2N229
3009	173A4419-2	3009	4720ORN	3009	228229	3009	7287110	3010	2N2430
3009	173A4419-3	3009	4722PUR	3009	228230	3009	7289041	3010	2N2707
3009	173A4419-8	3009	4722RED	3009	228558	3009	7289047	3010	2N2930
3009	173A4419-9	3009	4722YEL	3009	228559	3009	7290594	3010	2N306
3009	173A4420	3009	4727	3009	231140-11	3009	7291252	3010	2N306A
3009	173A4420-1	3009	4730	3009	231140-33	3009	7294796	3010	2N55
3009	173A4420-5	3009	4801-1100-011	3009	231672	3009	7297043	3010	2N4105
3009	173A4421-1	3009	4888A	3009	231797	3009	7297092	3010	2N507
3009	173A4422-1	3009	4888B	3009	232194	3009	7297093	3010	2N516
3009	173A4436	3009	8883-2	3009	232674	3009	7298079	3010	2N649
3009	173A4469	3009	8999-115	3009	232675	3009	7570003-01	3010	28C36
3009	184A-1	3009	9005-0	3009	234077	3009	20918596-2	3010	28D-P1
3009	184A-1-12	3009	9405-2	3009	234178	3009	23114011	3010	28D-P1A
3009	184A-1L	3009	9404-0	3009	234566	3009	23114033	3010	28D11
3009	184A-1L8	3009	12127-0	3009	235312	3009	23311006	3010	28D170BC
3009	231-0006-03	3009	12127-1	3009	236935	3009	62051509	3010	28D25
3009	231-0015	3009	12163	3009	237452	3009	62084200	3010	28D352
3009	295V041R04	3009	014382	3009	242183	3009	62084936	3010	28D352D
3009	297V040R15	3009	15024	3009	242838	3009	80050700	3010	28D352E
3009	297V041R01	3009	15027	3009	256068	3009	95114140-00	3010	28D352F
3009	297V041R02	3009	15354-3	3009	256071	3009	120004887	3010	28D37
3009	297V041R03	3009	15927	3009	257341	3009	951141000	3010	28D37A
3009	297V041R04	3009	16599	3009	257403	3009	310720490070	3010	28D37B
3009	297V041R07	3009	16959	3009	257536	3010	A122-1962	3010	28D37C
3009	297V062001	3009	17887	3009	258990	3010	A13-86416-1	3010	28D61
3009	297V062005	3009	17945	3009	261970	3010	A1465-29	3010	28D62
3009	310-192	3009	23311-006	3009	262114	3010	A14665-2	3010	28D72
3009	353-9201-001	3009	25661-020	3009	262570	3010	A16A1	3010	28D72-2C
3009	417-141	3009	30203	3009	263956	3010	A16A2	3010	28D72-3C
3009	417-20	3009	30211	3009	270744	3010	A2T682	3010	28D72-4C
3009	417-216	3009	30246	3009	270745	3010	A42X210	3010	28D72-6
3009	417-32	3009	30246A	3009	270746	3010	A48-869254	3010	28D72A
3009	417-44	3009	30302	3009	270780	3010	A48-869283	3010	28D72B
3009	417-45	3009	34022	3009	270785	3010	A48-869476	3010	28D72C
3009	417-46	3009	34298	3009	275612	3010	A48-869476A	3010	28D72D
3009	417-62	3009	34315	3009	309412	3010	A44JX2A822	3010	28D72E
3009	417-90	3009	34425	3009	0573040	3010	A514-023553	3010	28D72F
3009	417-99	3009	34715	3009	0573166	3010	A5705	3010	28D72GA
3009	421-24	3009	35044	3009	0573205	3010	A65-2-70	3010	28D72H
3009	421-25	3009	35084	3009	602032	3010	A65-2-705	3010	28D72J
3009	421-6599	3009	35201	3009	603031	3010	A65-2A9G	3010	28D72K
3009	473A31	3009	35231	3009	610039	3010	AC-127	3010	28D72L
3009	572-0040-051	3009	35260	3009	610039-1	3010	AC-172	3010	28D72M
3009	576-0002-002	3009	35349	3009	610049-1	3010	AC127	3010	28D72R
3009	576-0040-051	3009	35728	3009	610067	3010	AC127-132	3010	28D72X
3009	602-032	3009	35951	3009	610067-1	3010	AC176	3010	28D72Y
3009	610-039	3009	36303	3009	610067-2	3010	AC187	3010	28D96
3009	610-039-1	3009	36304	3009	610067-D	3010	AC194	3010	04-00072-01
3009	610-067	3009	36304-4	3009	610068	3010	AC194X	3010	4-65-2A7-1
3009	610-068	3009	36312	3009	610068-1	3010	A8Y61	3010	6A12992
3009	642-176	3009	36395	3009	610106	3010	C277C	3010	07-07167
3009	642-272	3009	36477	3009	610111-5	3010	D083	3010	8-0050300
3009	690L270R02	3009	36555	3009	610111-5	3010	D085	3010	8-0050600
3009	690L297R02	3009	36896	3009	610111-7	3010	D8-44	3010	8-0052800
3009	690V081R09	3009	36910	3009	610152-1	3010	B-2466	3010	8-723-650
3009	690V081R97	3009	40022	3009	617871-1	3010	B2466	3010	8-905-605-105
3009	800-507-00	3009	40050	3009	617539-1	3010	EQH-1	3010	8-905-605-108
3009	800-507O0	3009	40051	3009	652085	3010	E815X48	3010	8-905-605-109
3009	800-518-00	3009	40254	3009	660077	3010	E815X71	3010	8-905-605-110
3009	884-250-001	3009	40462	3009	660097	3010	E815X72	3010	8-905-605-111
3009	884-250-010	3009	40623	3009	660103	3010	E815X74	3010	8-905-605-112
3009	924-17945	3009	43074	3009	801518	3010	G18	3010	8-905-605-113
3009	964-17887	3009	50447-4	3009	801519	3010	GT336	3010	8-905-605-365
3009	991-01-1216	3009	057040	3009	815067	3010	HD-00072	3010	8-905-605-384
3009	992-00-1192	3009	61010-6	3009	815137	3010	HT400721	3010	8-905-605-390
3009	992-00-8870	3009	66010-3	3009	815138	3010	HT400721A	3010	8-905-613-015
3009	992-001192	3009	70434	3009	980134	3010	HT400721B	3010	8-905-613-062
3009	992-01-1216	3009	71448-1	3009	980155	3010	HT400721C	3010	09-03506
3009	992-01-1218	3009	71488-4	3009	980155	3010	HT400721D	3010	09-303013
3009	992-011218	3009	71488-5	3009	980437	3010	HT400721B	3010	09-303030
3009	1024-17	3009	71488-6	3009	983056	3010	HT400700B	3010	09-307075
3009	1124	3009	72856-63	3009	983874	3010	IE-850	3010	9-5222-2
3009	1124A	3009	080048	3009	983975	3010	IP20-0008	3010	13-18654-1
3009	1124B	3009	080050	3009	984261	3010	IP20-0076	3010	13-27050-1
3009	1124C	3009	81506-7	3009	984431	3010	IP20-0160	3010	19AR6-1
3009	1413-168	3009	81513-7	3009	985036	3010	J241185	3010	19AR6-2
3009	1413-172	3009	88832	3009	985431	3010	J24868	3010	19AR6-3
3009	1414A	3009	90050	3009	985443	3010	N5782-8	3010	21A015-005
3009	1414A-12	3009	94004	3009	985447	3010	NA30	3010	21A040-061
3009	1414A-12-8	3009	94024	3009	985449	3010	NKT701	3010	2T810
3009	1465C	3009	94025	3009	985453	3010	NKT703	3010	42-19862
3009	1465C-12	3009	94026	3009	985459	3010	NKT713	3010	45N1
3009	1465C-12-8	3009	94032	3009	986080	3010	NKT717	3010	45N2
3009	1484A	3009	94034	3009	988413	3010	NKT751	3010	46-86211-3
3009	1484A-12	3009	94040	3009	988468	3010	NKT752	3010	46-8671-3
3009	1484A-12-8	3009	95250	3009	989387	3010	NKT781	3010	48R134665
3009	1559-17	3009	95251	3009	995001	3010	PTC134	3010	48R869254
3009	1559-17A	3009	95253	3009	995014	3010	Q0V60527	3010	48R869283
3009	1814A	3009	95250	3009	995015	3010	Q0V60537	3010	48R869476
3009	1814A-12	3009	110515	3009	1471036-20	3010	Q2N4105	3010	48R869476A
3009	1814A-127	3009	115063	3009	1472446-1	3010	Q5030	3010	56-8100
3009	1814AL	3009	115268	3009	1961480	3010	Q5050	3010	56-8100A
3009	1859-14	3009	115269	3009	1961835	3010	R-28D187	3010	56-8100B
3009	1859-16	3009	116093	3009	1965017	3010	RB5	3010	56-8100C
3009	1865C	3009	121243	3009	2000646-113	3010	RB6	3010	56-8100D
3009	1865C-12	3009	122792	3009	2006607-59	3010	RB-1524	3010	5782-8
3009	1865C-127	3009	123792	3009	2090056-1	3010	RB8407	3010	5706-21
3009	1865CL	3009	162002-033	3009	2090056-27	3010	RB8441	3010	5706-5
3009	1865CL8	3009	162002-095	3009	2090056-5	3010	RB8445	3010	065-2
3009	1884A	3009	170307-1	3009	2091858-0712	3010	RT3096	3010	065-2-12
3009	1884A-12	3009	170376	3009	2091858-11	3010	RT7400	3010	65-2-70
3009	1884A-127	3009	170376-1	3009	2091859-0008	3010	RT7944	3010	65-2-70-12
3009	1884AL	3009	170479-1	3009	2091859-0011	3010	S-95202	3010	65-2-70-12-7
3009	1884AL8	3009	170666-1			3010	SC-56	3010	65-20
3009	1888B-12	3009	175043-023			3010	SC56	3010	65-21
3009	2002	3009	175043-81			3010	SE-7001	3010	65-22
3009	2007-01	3009	190425			3010	SYL-152	3010	65-23

SK	Industry Standard No.	SK	Industry Standard No.	SK	Industry Standard No.	SK	Industry Standard No.	SK	Industry Standard No.
3010	65-24	3010	5320306	3011	2SD77GN	3011A	2SD33X	3012	2SB236
3010	65-25	3010	5320361	3011	2SD77H	3011A	2SD33Y	3012	2SB237
3010	65-26	3010	5320475	3011	2SD77J	3011A	2SD43	3012	2SB237-12A
3010	65-27	3010	8512001-2	3011	2SD77L	3011A	2SD43A	3012	2SB237-12B
3010	65-28	3010	13040096	3011	2SD77M	3011A	2SD44	3012	2SB258A
3010	65-29	3010	23114045	3011	2SD77OR	3011A	2SD63	3012	2SB258B
3010	65-2A	3010	27125310	3011	2SD77P	3011A	2SD64	3012	2SB258C
3010	65-2AO	3010	62087609	3011	2SD77R	3011A	2SD65	3012	2SB258D
3010	65-2A1	3010	80050300	3011	2SD77X	3011A	2SD65-1	3012	2SB258E
3010	65-2A19	3010	80050600	3011	2SD77Y	3011A	2SD65A	3012	2SB258F
3010	65-2A2	3010	80052800	3011	2SDP1A	3011A	2SD65B	3012	2SB258G
3010	65-2A21	3011	A48-124220	3011	2SE4002	3011A	2SD65C	3012	2SB258GN
3010	65-2A3	3011	A48-125236	3011	2T54	3011A	2SD65D	3012	2SB258H
3010	65-2A3P	3011	A65-4-705	3011	2T55	3011A	2SD65E	3012	2SB258J
3010	65-2A4	3011	AC181	3011	2T551	3011A	2SD65F	3012	2SB258L
3010	65-2A4-7	3011	A07	3011	2T56	3011A	2SD65G	3012	2SB258M
3010	65-2A5	3011	ASY28	3011	2T57	3011A	2SD65GN	3012	2SB258GR
3010	65-2A5L	3011	ASY29	3011	2T58	3011A	2SD65H	3012	2SB258R
3010	65-2A6	3011	ASY53	3011	2T67	3011A	2SD65J	3012	2SB258X
3010	65-2A6-1	3011	C50A	3011	2T78	3011A	2SD65K	3012	2SB258Y
3010	65-2A7	3011	C60	3011	2T78R	3011A	2SD65L	3012	2SB551
3010	65-2A7-1	3011	D1P	3011	3N22	3011A	2SD65M	3012	3S50
3010	65-2A8	3011	D8-1	3011	3N23	3011A	2SD65OR	3012	3N51
3010	65-2A82	3011	ET8	3011	3N23A	3011A	2SD65R	3012	3N52
3010	65-2A9	3011	ET9	3011	3N23B	3011A	2SD65Y	3012	4-77A17-1
3010	65-2A9G	3011	GC1056	3011	3N23C	3011A	2SD66	3012	4-92-1A7-1
3010	80-050300	3011	GE-5	3011	3N29	3011A	2SD75	3012	5AA5-1
3010	80-050600	3011	GE-6	3011	3N30	3011A	2SD75A	3012	19A115094-P1
3010	80-052800	3011	GE-7	3011	3N31	3011A	2SD75B	3012	19A115487-P1
3010	86-13-2	3011	GT-1200	3011	3N36	3011A	2SD75C	3012	19A115540-P1
3010	86-14-2	3011	GT2884	3011	3N37	3011A	2SD75E	3012	27T407
3010	86-24-2	3011	GT2886	3011	4-65-4A7-1	3011A	394-3102-1	3012	48-134622
3010	86-25-2	3011	GT2888	3011	4JX1A813	3011A	38199	3012	53N49
3010	86-35-2	3011	HC-00730	3011	27T408	3011A	38200	3012	63N50
3010	86-58-2	3011	HD-187	3011	030-034-Q	3012	A1477A	3012	73N51
3010	86-6-2	3011	MHT2008	3011	65-4AO	3012	A1477A9	3012	O77A
3010	86-76-2	3011	00139	3011	65-4AOR	3012	A1492-1	3012	O77A-12
3010	95-202	3011	00141	3011	65-4A6-2	3012	A1492-19	3012	O77A-12-7
3010	95-222-2	3011	PTC108	3011	86-5003-2	3012	A77A-70	3012	77A-70
3010	95-224-2	3011	R-63	3011	86-5004-2	3012	A77A-705	3012	77A-70-12
3010	96-5205-01	3011	RB2	3011	86-5007-2	3012	A77A19G	3012	77A-70-12-7
3010	978ZU	3011	REN101	3011	86-5008-2	3012	A92-1-70	3012	77AO
3010	116-687	3011	R81513	3011	86-5011-2	3012	A92-1-705	3012	77A1
3010	121-641	3011	R81530	3011	86-5012-2	3012	A92-1A9G	3012	77A10
3010	121-762CL	3011	R81532	3011	86-5013-2	3012	AA5	3012	77A10R
3010	130-40096	3011	R81534	3011	86-5015-2	3012	ADZ11	3012	77A119
3010	130-40347	3011	R82356	3011	86-5016-2	3012	ADZ12	3012	77A12
3010	174-002-9-001	3011	R83306	3011	86-5017-2	3012	B236	3012	77A121
3010	421-9862	3011	SYL1279	3011	86-5018-2	3012	B237	3012	77A13P
3010	576-0002-013	3011	SYL1312	3011	86-5026-2	3012	B237-12A	3012	77A14
3010	601-065	3011	SYL1313	3011	86-5029-2	3012	B237-12B	3012	77A14-7
3010	614XB	3011	SYL1326	3011	86-5047-2	3012	B351	3012	77A15
3010	660-228	3011	SYL1327	3011	86-5048-2	3012	B92-1A21	3012	77A15L
3010	690V102H39	3011	SYL1380	3011	86-5060-2	3012	CB1P4	3012	77A16
3010	800-50300	3011	SYL1408	3011	86-5061-2	3012	CTP1509	3012	77A16-1
3010	800-506-00	3011	SYL1454	3011	86-5062-2	3012	DS-501	3012	77A17
3010	800-50600	3011	SYL1537	3011	86-5080-2	3012	DS970	3012	77A17-1
3010	800-528	3011	SYL1591	3011	86-5087-2	3012	DU47	3012	77A18
3010	800-528-00	3011	SYL1617	3011	86-81-2	3012	DU7	3012	77A182
3010	808-509	3011	SYL1750	3011	998A7	3012	EA15X6	3012	77A19
3010	921-5A	3011	T-Q5031	3011	121-141Q	3012	E810	3012	77A19G
3010	921-5B	3011	T-Q5032	3011	121-6	3012	ES21	3012	77A2
3010	921-6A	3011	TA-1759	3011	121-60	3012	ES501	3012	77A3
3010	921-6B	3011	TA-1767	3011	127-7	3012	ET7	3012	77A4
3010	1349-17	3011	TA-1771	3011	129-30	3012	GE-4	3012	77A5
3010	1371-17	3011	TA-1772	3011	297V002M04	3012	HEP-06004	3012	77A6
3010	1465-2	3011	TQ5039	3011	297V002M05	3012	IG-100	3012	77A7
3010	1465-2-12	3011	TR10	3011	110263	3012	JC100	3012	77A8
3010	1465-2-12-8	3011	TR10C	3011	224820	3012	JEM1	3012	77A9
3010	1515	3011	WEP641	3011	258993	3012	JEM2	3012	83N52
3010	1865-2	3011	2N1010	3011	650860	3012	JEM3	3012	86-512-2
3010	1865-2-12	3011	2N1017	3011	13040089	3012	JEM4	3012	092-1
3010	1865-2L	3011	2N1018	3011A	A2092418	3012	JEM5	3012	092-1-12
3010	1865-2L8	3011	2N1058	3011A	A8Y72	3012	M4553PUR	3012	92-1-70-12
3010	1906-17	3011	2N1102/5	3011A	C179	3012	M4622	3012	92-1-70-12-7
3010	1958-17	3011	2N126	3011A	C406	3012	MHT180	3012	92-10
3010	2057A2-167	3011	2N1302	3011A	D195	3012	MHT1801Q	3012	92-11
3010	5001-512	3011	2N148C/D	3011A	GE-X8	3012	MHT1807	3012	92-12
3010	5614-77C	3011	2N1747	3011A	RT-119	3012	MHT1808	3012	92-13
3010	8000-00004-086	3011	2N1781	3011A	2N102	3012	MHT1809	3012	92-14
3010	8000-00032-027	3011	2N193	3011A	2SC179	3012	MHT181	3012	92-15
3010	27125-310	3011	2N194	3011A	2SC179A	3012	MHT230	3012	92-16
3010	030812-1	3011	2N194A	3011A	2SC179B	3012	MHT2305	3012	92-17
3010	37279	3011	2N1993	3011A	2SC179C	3012	MP503	3012	92-18
3010	37552	3011	2N211	3011A	2SC179D	3012	MP503A	3012	92-19
3010	40037-1	3011	2N212	3011A	2SC179E	3012	MP507	3012	92-1A
3010	40037-2	3011	2N216	3011A	2SC179F	3012	MP507A	3012	92-1AO
3010	40037-3	3011	2N233	3011A	2SC179G	3012	POWER40	3012	92-1AOR
3010	40037VM	3011	2N233A	3011A	2SC179GN	3012	POWER500	3012	92-1A1
3010	40038VM	3011	2N439	3011A	2SC179H	3012	POWER60	3012	92-1A19
3010	46590-2	3011	2N439A	3011A	2SC179J	3012	POWER80	3012	92-1A2
3010	46591-2	3011	2N440	3011A	2SC179K	3012	PT-501	3012	92-1A21
3010	46592-2	3011	2N440A	3011A	2SC179L	3012	PTC106	3012	92-1A3
3010	46593-2	3011	2N448	3011A	2SC179M	3012	QP8-6623N	3012	92-1A3P
3010	47645-2	3011	2N449	3011A	2SC179OR	3012	RE8	3012	92-1A4
3010	80817YM	3011	2N515	3011A	2SC179R	3012	RS-106	3012	92-1A4-7
3010	81507-5	3011	2N517	3011A	2SC179X	3012	SFT-265	3012	92-1A5
3010	81507-6	3011	2N815	3011A	2SC179Y	3012	SFT-266	3012	92-1A5L
3010	95224-2	3011	2N816	3011A	2SC406	3012	SFT-267	3012	92-1A6
3010	95224-4	3011	2N817	3011A	2SC406A	3012	SFT264	3012	92-1A6-1
3010	104080	3011	2N818	3011A	2SC406B	3012	ST-106	3012	92-1A7
3010	114095	3011	2N819	3011A	2SC406C	3012	ST-107	3012	92-1A7-1
3010	111958	3011	2N820	3011A	2SC406D	3012	ST-109	3012	92-1A8
3010	130400-96	3011	2N825	3011A	2SC406E	3012	ST-110	3012	92-1A82
3010	130403-47	3011	2N94	3011A	2SC406F	3012	ST-111	3012	92-1A9
3010	168953	3011	2N94A	3011A	2SC406G	3012	ST-112	3012	92-1A9G
3010	170783-3	3011	2S011	3011A	2SC406GN	3012	TA-3	3012	119-0016
3010	171005	3011	2S050A	3011A	2SC406H	3012	TRO3	3012	122-1028
3010	223482	3011	2S050B	3011A	2SC406J	3012	TRO3C	3012	122-1028A
3010	225300	3011	2S050C	3011A	2SC406K	3012	TRO34	3012	148-134622
3010	226797	3011	2S050E	3011A	2SC406L	3012	TS-609	3012	417-177
3010	23140-45	3011	2S050F	3011A	2SC406M	3012	1G100	3012	1417-177
3010	243837	3011	2S050G	3011A	2SC406OR	3012	2N1099	3012	1477A
3010	250400	3011	2S050GN	3011A	2SC406X	3012	2N1100	3012	1477A-12
3010	269367	3011	2S050H	3011A	2SC406Y	3012	2N1202	3012	1477A-12-8
3010	0573037	3011	2S050J	3011A	2SD162	3012	2N1203	3012	1492-1
3010	0573037H	3011	2S050K	3011A	2SD167	3012	2N1261	3012	1492-1-12
3010	0573139	3011	2S050L	3011A	2SD195	3012	2N1262	3012	1492-1-12-8
3010	601065	3011	2S050M	3011A	2SD195A	3012	2N1263	3012	1877A
3010	610151-5	3011	2S050OR	3011A	2SD33	3012	2N1433	3012	1877A-12
3010	815075	3011	2S050R	3011A	2SD33A	3012	2N1434	3012	1877A-127
3010	815076	3011	2S050X	3011A	2SD33B	3012	2N1435	3012	1877AL
3010	815232	3011	2S050Y	3011A	2SD33C	3012	2N1523	3012	1877AL8
3010	1817017	3011	2S060	3011A	2SD33D	3012	2N2210	3012	1892-1
3010	2003073-16	3011	2SD77	3011A	2SD33E	3012	2N2266	3012	1892-1-12
3010	2010952-14	3011	2SD77-A	3011A	2SD33F	3012	2N2267	3012	1892-1-127
3010	2091260-1	3011	2SD77A	3011A	2SD33G	3012	2N2268	3012	1892-1L
3010	2091260-2	3011	2SD77B	3011A	2SD33GN	3012	2N2269	3012	1892-1L8
3010	2091260-3	3011	2SD77C	3011A	2SD33H	3012	2N2728	3012	2009
3010	2320331	3011	2SD77D	3011A	2SD33J	3012	2N2730	3012	2010
3010	4038256-P1	3011	2SD77E	3011A	2SD33K	3012	2N2731	3012	4573
3010	4038256-P2	3011	2SD77F	3011A	2SD33L	3012	2N2732	3012	4597
		3011	2SD77G	3011A	2SD33OR	3012	2N2793	3012	4597GRN

SK	Industry Standard No.
3012	4622
3012	6818
3012	7172-54
3012	14573
3012	217119
3012	217230
3012	226789
3012	229045
3012	233305
3012	233307
3012	233508
3012	234078
3012	243215
3012	243815
3012	253704
3012	256480
3012	257242
3012	257243
3012	257534
3012	261401
3012	261488
3012	262309
3012	660144
3012	660144A
3012	980052
3012	980052A
3012	980462
3012	980462A
3012	980463
3012	980463A
3012	981969
3012	981969A
3012	983945
3012	985432
3012	988414
3012	988977
3012	989171
3012	989615
3012	989692
3012	989693
3012	1221028
3012	4036831-P1
3012	4036832-P1
3012	5983945A
3012	5985432
3012	5988414
3012	5988977
3012	5989171
3012	5989615
3012	5989692
3012	5989693
3012	7276605
3012	7279005
3012	7279007
3012	7279011
3012	7279017
3012	7279025
3012	7279027
3012	7279033
3012	7279073
3012	7279076
3012	7279293
3012	7279298
3012	7279566
3012	7279793
3012	7280281
3012	7282315
3012	7288072
3012	7288073
3012	7288076
3012	7288079
3012	7299720
3012	57276605
3012	57279005
3012	57279007
3012	57279009
3012	57279011
3012	57279017
3012	57279025
3012	57279027
3012	57279033
3012	57279073
3012	57279076
3012	57279293
3012	57279298
3012	57279566
3012	57279793
3012	57280281
3012	57282315
3012	57288072
3012	57288073
3012	57288076
3012	57288079
3012	57299720
3012/105	ADY26
3012/105	B235
3012/105	B238-12A
3012/105	B238-12B
3012/105	B238-12C
3012/105	B258
3012/105	B259
3012/105	B260
3012/105	B331
3012/105	B332
3012/105	B333
3012/105	B352
3012/105	B353
3012/105	B430
3012/105	B433
3012/105	GE-240
3012/105	L5022
3012/105	MP525
3012/105	2N1157
3012/105	2N1157A
3012/105	2N1358
3012/105	2N1358A
3012/105	2N1358M
3012/105	2N1501
3012/105	2N1502
3012/105	2N1518
3012/105	2N1519
3012/105	2N1520
3012/105	2N1521
3012/105	2N1522
3012/105	2N173
3012/105	2N174
3012/105	2N174A
3012/105	2N1970
3012/105	2N1980
3012/105	2N1981
3012/105	2N1982
3012/105	2N2075
3012/105	2N2075A
3012/105	2N2076
3012/105	2N2076A
3012/105	2N2077
3012/105	2N2077A
3012/105	2N2078
3012/105	2N2078A
3012/105	2N2079
3012/105	2N2079A
3012/105	2N2080
3012/105	2N2080A
3012/105	2N2081
3012/105	2N2081A
3012/105	2N2082
3012/105	2N2082A
3012/105	2N2152A
3012/105	2N2153
3012/105	2N2153A
3012/105	2N2154A
3012/105	2N2155
3012/105	2N2155A
3012/105	2N2156A
3012/105	2N2157
3012/105	2N2157A
3012/105	2N2158
3012/105	2N2158A
3012/105	2N2159
3012/105	2N2159A
3012/105	2N2490
3012/105	2N2491
3012/105	2N2492
3012/105	2N2493
3012/105	2N277
3012/105	2N278
3012/105	2N290
3012/105	2N3311
3012/105	2N3312
3012/105	2N3313
3012/105	2N3314
3012/105	2N3315
3012/105	2N441
3012/105	2N442
3012/105	2N443
3012/105	2N459
3012/105	2N575
3012/105	2N575A
3012/105	2SB235
3012/105	2SB235A
3012/105	2SB238-12A
3012/105	2SB238-12B
3012/105	2SB238-12C
3012/105	2SB258
3012/105	2SB259
3012/105	2SB260
3012/105	2SB331
3012/105	2SB331H
3012/105	2SB332
3012/105	2SB332H
3012/105	2SB333
3012/105	2SB352
3012/105	2SB352D
3012/105	2SB353
3012/105	13-16592-1
3012/105	1413-159
3013	A13-22741-2
3013	A1477B
3013	A1477B9
3013	A168P1
3013	A48-10075A01
3013	A48-10075A02
3013	A48-10075A03
3013	A48-10075A04
3013	A48-10075A05
3013	A48-10075A06
3013	A48-10075A07
3013	A48-10075A08
3013	A48-10103A01
3013	A48-10103A02
3013	A48-10103A03
3013	A48-10103A04
3013	A48-10103A05
3013	A48-10103A06
3013	A48-10103A07
3013	A48-10103A08
3013	A48-10103A09
3013	A48-10103A10
3013	A48-10103A11
3013	A48-64978A10
3013	A48-64978A11
3013	A48-64978A24
3013	A642-271
3013	A660031
3013	A815203-5
3013	A86X0030-100
3013	AD149
3013	B337B
3013	GE-13MP
3013	W9MP
3013	4-77B17-1
3013	7-2
3013	9-5250-1
3013	9-5257
3013	13-14604-1A
3013	13-14604-1B
3013	13-14604-1C
3013	13-14604-1D
3013	13-14604-1E
3013	13-14777-1A
3013	13-14777-1B
3013	13-14777-1C
3013	13-14777-1D
3013	13-14778-1A
3013	13-14778-1B
3013	13-14778-1C
3013	13-14778-1D
3013	13-22741-2
3013	026-100004
3013	026-100020
3013	48-10075A01
3013	48-10075A02
3013	48-10075A03
3013	48-10075A04
3013	48-10075A05
3013	48-10075A06
3013	48-10075A07
3013	48-10075A08
3013	48-10103A01
3013	48-10103A02
3013	48-10103A03
3013	48-10103A04
3013	48-10103A05
3013	48-10103A06
3013	48-10103A07
3013	48-10103A08
3013	48-10103A09
3013	48-10103A10
3013	48-10103A11
3013	48-64978A10
3013	48-64978A11
3013	48-64978A24
3013	077B
3013	077B-12
3013	077B-12-7
3013	77B-70
3013	77B-70-12
3013	77B-70-12-7
3013	77B0
3013	77B1
3013	77B10
3013	77B10R
3013	77B11
3013	77B119
3013	77B12
3013	77B121
3013	77B13
3013	77B13P
3013	77B14
3013	77B14-7
3013	77B15
3013	77B15L
3013	77B16
3013	77B16-2
3013	77B17
3013	77B17-1
3013	77B18
3013	77B182
3013	77B19
3013	77B19G
3013	77B2
3013	77B3
3013	77B4
3013	77B5
3013	77B6
3013	77B7
3013	77B8
3013	77B9
3013	86X0030-100
3013	95-250-1
3013	95-257
3013	96B-3A65
3013	115-281
3013	115-282
3013	115-283
3013	115-284
3013	144N1
3013	168P1
3013	0418
3013	642-271
3013	800-253
3013	964-16599
3013	1477B
3013	1477B-12
3013	1477B-12-8
3013	1877B
3013	1877B-12
3013	1877B-127
3013	1877BL
3013	1877BL8
3013	67085-0
3013	95250-1
3013	170668-1
3013	170850-1
3013	660031
3013	670850
3013	670850-1
3013	815203-5
3013	62054212
3013	310720440170
3013	310720490190
3014	A1477C
3014	A1477C9
3014	A77C-70
3014	A77C-705
3014	A77C19G
3014	AD103
3014	AD132
3014	AD140
3014	ADY23
3014	AU101
3014	AUY19
3014	AUY20
3014	B-1511
3014	B10162
3014	B10163
3014	B10474
3014	B10475
3014	B1085
3014	B134A
3014	B134C
3014	B1368
3014	B1368E
3014	B1368F
3014	B179
3014	B410
3014	B411
3014	CDT1320
3014	CDT1321
3014	CM2550
3014	CQT1110
3014	CQT1112
3014	CRT1602
3014	CTP1133
3014	CTP1135
3014	CTP1504
3014	DT1040
3014	DT401
3014	ES18
3014	G6013
3014	HEP-G6000
3014	HEP-G6003
3014	HEP-G6013
3014	M4888A
3014	MP1509-1
3014	NKT452-S1
3014	P8890
3014	P8890A
3014	POWER299
3014	RS3959-1
3014	SFT213
3014	SFT239
3014	SP1271
3014	SP2493
3014	SP2541
3014	SP441G
3014	SP744
3014	SP891
3014	TA2672
3014	TI3029
3014	TR35144
3014	TR35144A
3014	TS173
3014	TS610
3014	TS613
3014	V60/20DP
3014	VFG2745B
3014	VFP2746C
3014	VFP6537C
3014	VFU2746B
3014	WTV12PWR
3014	WTV40PWR
3014	WTV6PWR
3014	WTV99PWR
3014	XC156
3014	2AD149
3014	2G210
3014	2G222
3014	2G224
3014	2G240
3014	2N1182
3014	2N1227
3014	2N1227-3
3014	2N1227-4
3014	2N1227-4R
3014	2N1227A
3014	2N1294
3014	2N1296
3014	2N1430
3014	2N1652
3014	2N1905
3014	2N2066A
3014	2N2147
3014	2N2148
3014	2N3125
3014	2N4241
3014	2N633B
3014	2SA416
3014	2SA416A
3014	2SA416B
3014	2SA416C
3014	2SA416D
3014	2SA416E
3014	2SA416F
3014	2SA416G
3014	2SA416GN
3014	2SA416H
3014	2SA416J
3014	2SA416K
3014	2SA416L
3014	2SA416M
3014	2SA4160R
3014	2SA416R
3014	2SA416X
3014	2SA416Y
3014	2SB132A
3014	2SB318
3014	2SB362
3014	2SB410
3014	2SB411
3014	2SB413
3014	2SB413A
3014	2SB413B
3014	2SB413C
3014	2SB413D
3014	2SB413E
3014	2SB413F
3014	2SB413G
3014	2SB413GN
3014	2SB413H
3014	2SB413J
3014	2SB413K
3014	2SB413L
3014	2SB413M
3014	2SB4130R
3014	2SB413R
3014	2SB413X
3014	2SB413Y
3014	2SB414
3014	2SB414B
3014	2SB414C
3014	2SB414D
3014	2SB414E
3014	2SB414F
3014	2SB414G
3014	2SB414GN
3014	2SB414H
3014	2SB414J
3014	2SB414K
3014	2SB414L
3014	2SB414M
3014	2SB4140R
3014	2SB414R
3014	2SB414X
3014	2SB414Y
3014	2SB424
3014	2SB445
3014	2SB445A
3014	2SB445B
3014	2SB445C
3014	2SB445D
3014	2SB445E
3014	2SB445F
3014	2SB445G
3014	2SB445GN
3014	2SB445H
3014	2SB445J
3014	2SB445K
3014	2SB445L
3014	2SB445M
3014	2SB4450R
3014	2SB445R
3014	2SB445X
3014	2SB445Y
3014	2SB446
3014	2SB446A
3014	2SB446B
3014	2SB446C
3014	2SB446D
3014	2SB446E
3014	2SB446F
3014	2SB446G
3014	2SB446GN
3014	2SB446H
3014	2SB446J
3014	2SB446K
3014	2SB446L
3014	2SB446M
3014	2SB4460R
3014	2SB446R
3014	2SB446X
3014	2SB446Y
3014	2SB84
3014	2SB84A
3014	2SB84B
3014	2SB84C
3014	2SB84D
3014	2SB84E
3014	2SB84F
3014	2SB84G
3014	2SB84GN
3014	2SB84H
3014	2SB84J
3014	2SB84K
3014	2SB84L
3014	2SB84M
3014	2SB840R
3014	2SB84R
3014	2SB84X
3014	2SB84Y
3014	4-77C17-1
3014	8H303
3014	13-14735-1
3014	13-14777-1
3014	21A064-000
3014	23-5042
3014	48-125288
3014	48-134651
3014	48-134888
3014	48-134938
3014	53P153
3014	57D0-23
3014	77C
3014	077C-12
3014	077C-12-7
3014	77C-70
3014	77C-70-12
3014	77C-70-12-7
3014	77C0
3014	77C1
3014	77C10
3014	77C10R
3014	77C11
3014	77C119
3014	77C12
3014	77C121
3014	77C13
3014	77C13P
3014	77C14
3014	77C14-7
3014	77C15
3014	77C15L
3014	77C16
3014	77C16-3
3014	77C17
3014	77C17-1
3014	77C18
3014	77C182
3014	77C19
3014	77C19G
3014	77C2
3014	77C3
3014	77C4
3014	77C5
3014	77C6
3014	77C7
3014	77C8
3014	77C9
3014	81-27126130-7
3014	81-27126130-7A
3014	81-27126130-7B
3014	86-5039-2
3014	86X0030-001
3014	99S013
3014	99S013A
3014	112-202147
3014	121-1134
3014	121-363
3014	129-10
3014	130-104
3014	146T1
3014	147T1
3014	231-0011
3014	297V041H05
3014	297V041H06
3014	417-60
3014	992-008-890
3014	992-008890
3014	1413-178
3014	1477C
3014	1477C-12
3014	1477C-12-8
3014	1877C
3014	1877C-12
3014	1877C-127
3014	1877CL
3014	1877CL8
3014	2243
3014	2577
3014	35144
3014	35885A
3014	35885B
3014	38094
3014	40421
3014	40612
3014	40626
3014	50477-4
3014	51650
3014	66009-5
3014	80416C
3014	115281
3014	115282
3014	115283
3014	115284
3014	119724
3014	125703
3014	125761
3014	162002-062
3014	162002-062A
3014	170407-1
3014	171162-082
3014	171162-086
3014	224873
3014	230523
3014	233509
3014	660095
3014	1473515-1
3014	1960584
3014	1961479
3014	1966079
3014	2006514-59
3014	2091859-0713
3014	2091859-0715
3014	2091859-0717
3014	2091859-2
3015	A146B-3
3015	A146B-39
3015	A6B-3-70
3015	A6B-3-705
3015	A6B-3A9G

SK	Industry Standard No.	SK	Industry Standard No.	SK	Industry Standard No.	SK	Industry Standard No.	SK	Industry Standard No.
3015	4-6B-3A7-1	3016	AS15	3016	DI-52S	3016	M680C	3016	RF34383
3015	6B-3	3016	A82	3016	DI-7	3016	M68A	3016	RFJ-30704
3015	06B-3-12	3016	A83	3016	DI645	3016	M68B	3016	RFJ-31218
3015	06B-3-12-7	3016	A84	3016	DI646	3016	M69	3016	RFJ-31362
3015	6B-3-70	3016	A85	3016	DIJ72168	3016	M690	3016	RFJ-31363
3015	6B-3-70-12	3016	B-31	3016	DR200	3016	M690B	3016	RFJ-33292
3015	6B-30	3016	B1B	3016	DR300	3016	M690C	3016	RFJ-60366
3015	6B-31	3016	B1C1	3016	DR400	3016	M700C	3016	RFJ70971
3015	6B-32	3016	B200C40	3016	DR435	3016	M702C	3016	RFJ70974
3015	6B-33	3016	B250C100	3016	DR5101	3016	M70C	3016	RFL-30596
3015	6B-34	3016	B250C100TD	3016	DR8102	3016	M9206	3016	RPM-33160
3015	6B-35	3016	B250C125	3016	DS-0065	3016	MA-110	3016	RPS-33118
3015	6B-36	3016	B250C125K4	3016	E-075L	3016	MA2	3016	RPV60500
3015	6B-37	3016	B250C125N2	3016	E1	3016	MA203	3016	RG1004
3015	6B-38	3016	B250C125X4	3016	E102	3016	MA215	3016	RH-DX0068TA2Z
3015	6B-39	3016	B250C150	3016	E1410	3016	MA242	3016	RH-DX0081CEZZ
3015	6B-3A	3016	B250C150K4	3016	E146	3016	MC010	3016	RL44
3015	6B-3A0	3016	B250C75	3016	E140350	3016	MH680	3016	RLP10
3015	6B-3A1	3016	B250C75K4	3016	E4	3016	MM2	3016	RS-1264
3015	6B-3A19	3016	B250C75K45	3016	EA16X8	3016	MP-01	3016	RS-3570
3015	6B-3A21	3016	B250C7K4S	3016	EA57X1	3016	MPZ215	3016	RS-3727
3015	6B-3A3	3016	B205	3016	EA57X10	3016	MR-1	3016	RS-6344
3015	6B-3A3P	3016	B30C250	3016	EA57X11	3016	MR1237FB	3016	RS-6461
3015	6B-3A4	3016	B300350-1	3016	EA57X3	3016	MR1237FL	3016	RS-6471
3015	6B-3A4-7	3016	B300600	3016	EA57X8	3016	MR1237SB	3016	RS220AF
3015	6B-3A5	3016	B30C600CB	3016	ED-4	3016	MR1237SL	3016	RS230AF
3015	6B-3A5L	3016	B350C600	3016	ED-6	3016	MR1247FB	3016	RS6705
3015	6B-3A6	3016	B4B9	3016	ED1804	3016	MR1247PL	3016	RT1840
3015	6B-3A7	3016	B405	3016	ED1892	3016	MR1247SB	3016	S-10
3015	6B-3A7-1	3016	B409	3016	ED2106	3016	MR1247SL	3016	S-17
3015	6B-3A8	3016	B5C1	3016	ED2108	3016	MR1267	3016	S-17A
3015	6B-3A8Z	3016	BA-100	3016	ED2109	3016	MR2273	3016	S-262
3015	6B-3A9	3016	BA-104	3016	ED2110	3016	MRB-20C	3016	S-3MX
3015	6B-3A99	3016	BA105	3016	ED224550	3016	MRV-20C	3016	S1-1
3015	13-14604-1	3016	BA127	3016	ED2842	3016	NA-22	3016	S101
3015	13-14778-1	3016	BAY17	3016	ED2843	3016	NA-25	3016	S102
3015	13-18198-1	3016	BAY164	3016	ED2844	3016	NA-33	3016	S103
3015	32-16599	3016	BB-6	3016	ED2915	3016	NA-36	3016	S104
3015	42-21443	3016	BR42	3016	ED2916	3016	NA-46	3016	S13
3015	146B-3	3016	BR44	3016	ED2917	3016	NA13	3016	S16B
3015	146B-3-12	3016	BR52	3016	ED2918	3016	NA21	3016	S19
3015	146B-3-12-8	3016	BY101	3016	ED2919	3016	NA32	3016	S1A06
3015	186B-3	3016	BY112	3016	ED3001A	3016	NA35	3016	S1A60
3015	186B-3-12	3016	BY113	3016	ED3002A	3016	NA42	3016	S1AR2
3015	186B-3-127	3016	BY122	3016	ED3003B	3016	NA45	3016	S1SM-150-02
3015	186B-3L	3016	BY123	3016	ED5	3016	NL-10	3016	S1SW-05-02
3015	186B-3L8	3016	BY153	3016	ED8-0004	3016	NL15	3016	S201
3015	422-1443	3016	BYX26-60	3016	EM502	3016	NL20	3016	S202
3015	560004	3016	BYX36/150	3016	ER102D	3016	NL25	3016	S203
3015	815203-3	3016	BYX36/300	3016	ER103E	3016	NL30	3016	S204
3016	A-04049-B	3016	BYY-31	3016	ER21	3016	OA180	3016	S20ND400
3016	A-04091-A	3016	BYY-32	3016	ER22	3016	OA200	3016	S20NH400
3016	A-04092	3016	BYY-35	3016	ER31	3016	OA202	3016	S217
3016	A-04092-B	3016	BYY33	3016	ER41	3016	OA210	3016	S221
3016	A-04093	3016	BYY34	3016	ER57X2	3016	OA214	3016	S222
3016	A-04093A	3016	BYY89	3016	ER57X3	3016	OY-5061	3016	S223
3016	A-04212-A	3016	C10110	3016	ER57X4	3016	OY-5062	3016	S224
3016	A-04212-B	3016	C1181C1E1C	3016	ER82406A	3016	OY101	3016	S235
3016	A-04242	3016	C1B	3016	ERD300	3016	OY5063	3016	S238
3016	A-10105	3016	C1H	3016	ERD400	3016	OY5064	3016	S240
3016	A-10113	3016	CA102DA	3016	ET200	3016	P105	3016	S253
3016	A-10118	3016	CC102BA	3016	ET55-25	3016	P1D5	3016	S254
3016	A-1946	3016	CC102DA	3016	ET57X25	3016	P205	3016	S26
3016	A-95-5281	3016	CD1122	3016	ET57X29	3016	P2D5	3016	S2A10
3016	A-95-5289	3016	CD1123	3016	ET57X30	3016	P3D5	3016	S2030
3016	A04731	3016	CD1124	3016	ET57X35	3016	P405	3016	S2040
3016	A132-1	3016	CD1125	3016	F05	3016	P505	3016	S2040A
3016	A13AA2	3016	CD1126	3016	F1	3016	PA-069	3016	S2B20
3016	A13B2	3016	CD1127	3016	F10148	3016	PA-10556	3016	S2B60-1
3016	A13P2	3016	C500	3016	F4	3016	PA-3	3016	S31
3016	A1B1	3016	C8502	3016	FD3	3016	PA-320	3016	S32
3016	A1B9	3016	CER68	3016	FR1P	3016	PA-320A	3016	S33
3016	A105	3016	CER680	3016	FST2	3016	PA-320B	3016	S34
3016	A109	3016	CER680A	3016	FST3	3016	PAO71	3016	S3A06
3016	A2B1	3016	CER680B	3016	FT1N	3016	PA305	3016	S3AR1
3016	A209	3016	CER68A	3016	FU-1MA	3016	PA310	3016	S428
3016	A2E5	3016	CER68B	3016	FU-1NA	3016	PA315	3016	S44
3016	A3B1	3016	CER69	3016	FW500	3016	PA315A	3016	S46
3016	A3B3	3016	CER690	3016	FW4500	3016	PA325A	3016	S4A06
3016	A3B5	3016	CER690B	3016	G100D	3016	PA325B	3016	S4AR1
3016	A3B9	3016	CER70	3016	G100G	3016	PA330A	3016	S4FN300
3016	A3C5	3016	CER700A	3016	G2	3016	PA330B	3016	S81
3016	A3D5	3016	CER700C	3016	G657	3016	PA340A	3016	S82
3016	A4B1	3016	CER70A	3016	G659	3016	PA340B	3016	S83
3016	A514-023626	3016	CF102DA	3016	G701	3016	PE-401	3016	S91
3016	A514-025607	3016	CH119D	3016	GD12E	3016	PE401N	3016	S91-A
3016	A514-028072	3016	CL1	3016	GD400	3016	PE402	3016	S91H
3016	A514-028073	3016	CL2	3016	GD72B/5	3016	PE403	3016	S93A
3016	A514-033903	3016	CL3	3016	GJ4M	3016	PE502	3016	S94
3016	A5B2	3016	C0D1532	3016	H-881	3016	PH-108	3016	SB01
3016	A5B5	3016	C0D1533	3016	HA200	3016	PH1021	3016	SB302
3016	A5B9	3016	C0D1551	3016	HAR20	3016	PH25022/1	3016	SB315
3016	A5C1	3016	C0D15524	3016	HB2	3016	PS-025	3016	SB332
3016	A5C5	3016	C0D1553	3016	HC80	3016	PS-035	3016	SB333
3016	A5D2	3016	C0D15534	3016	HP-08W05	3016	PS-040	3016	SB393
3016	A5D5	3016	C0D15544	3016	HP-1B	3016	PS-120	3016	SC305
3016	A7C	3016	C0D1612	3016	HP-1Z	3016	PS010	3016	SC4116
3016	A7D	3016	C0D1613	3016	HPBD1	3016	PS015	3016	SCA1
3016	A7G	3016	CP102BA	3016	HPBD12	3016	PS125	3016	SD-1L
3016	AA100	3016	CP102DA	3016	HGR2	3016	PS130	3016	SD-2
3016	AA200	3016	CR1035	3016	INJ61227	3016	PS140	3016	SD-201
3016	AAT-22	3016	CT200	3016	IP20-0025	3016	PS2247	3016	SD07
3016	AG100D	3016	CT300	3016	JB16C4	3016	PS2412	3016	SD12
3016	AJ-30	3016	CX-0037	3016	JCN1	3016	PS2413	3016	SD1CUP
3016	AJ10	3016	CX-0039	3016	JCN4	3016	PS415	3016	SD1HP
3016	AJ15	3016	CX-0040	3016	JCV-2	3016	PS420	3016	SD1X
3016	AJ25	3016	CX-0048	3016	JCV-3	3016	PS425	3016	SD1Z
3016	AJ35	3016	CXO037	3016	K1B5	3016	PS430	3016	SD45
3016	AJ40	3016	CX9001	3016	K105	3016	PS435	3016	SD51
3016	AM-010	3016	D004	3016	K205	3016	PS440	3016	SD80
3016	AM-020	3016	D1448	3016	K4B5	3016	PT-510	3016	SD94AB
3016	AM-025	3016	D15C	3016	K505	3016	PT-515	3016	SD95
3016	AM-030	3016	D18	3016	K00BC11/10	3016	PT-550	3016	SD95A
3016	AM-035	3016	D1H	3016	K00BC11/8	3016	PV-8	3016	SE-05
3016	AM-22	3016	D25	3016	K00BC21/5	3016	Q1B	3016	SE-05X
3016	AM-33	3016	D25A	3016	KD2103	3016	Q1H	3016	SE058
3016	AMO10	3016	D25B	3016	KD2104	3016	Q4B	3016	SE30B26A
3016	AMO20	3016	D25C	3016	KLH4567	3016	R-106379	3016	SFR135
3016	AM13	3016	D28	3016	KSKE400C200	3016	R-113321	3016	SFR151
3016	AM21	3016	D2H	3016	KSKE400C500	3016	R-113392	3016	SFR152
3016	AM23	3016	D3U	3016	LA300	3016	R-1A	3016	SFR153
3016	AM24	3016	D45C	3016	LL-2	3016	R859-2	3016	SFR251
3016	AM32	3016	D4M	3016	LRR-200	3016	RA132BA	3016	SFR252
3016	AM410	3016	D6623A	3016	LRR-300	3016	RD-26235-1	3016	SG-1198
3016	AM42	3016	D6624A	3016	LRR-400	3016	RD-29799P	3016	SG105
3016	AM420	3016	D6625A	3016	LRR-50	3016	RD-31903P	3016	SGR100
3016	AM425	3016	DA000	3016	LRR-500	3016	RD-5472	3016	SH4D1
3016	AM43	3016	DD-003	3016	M-0027	3016	RF-3160	3016	SH4D2
3016	AM430	3016	DD-006	3016	M12	3016	RF-32101-8	3016	SI4D1
3016	AM435	3016	DD176C	3016	M124J779-1	3016	RF-32101R	3016	SI100B
3016	AM440	3016	DD2321	3016	M1H	3016	RF-6235-1	3016	SI91G
3016	AN-1	3016	DH4R2	3016	M4736	3016	RF29799P	3016	SIL-200
3016	AS14	3016	DI-428	3016	M680A	3016	RF32412-3	3016	SIR-80
				3016	M680B			3016	SJ101F

SK	Industry Standard No.	SK	Industry Standard No.	SK	Industry Standard No.	SK	Industry Standard No.	SK	Industry Standard No.
3016	SJ102F	3016	T-H13557	3016	1C2	3016	1N440	3016	18588
3016	SJ201F	3016	T-H86105	3016	1D2	3016	1N440B	3016	18605
3016	SJ202F	3016	TC-136	3016	1E1	3016	1N441	3016	1871
3016	SJ301F	3016	TPR-120	3016	1E2	3016	1N441B	3016	1881
3016	SJ302F	3016	TG-11	3016	1E3	3016	1N442	3016	1890
3016	SL-2	3016	TG-12	3016	1E4	3016	1N442B	3016	1892
3016	SL-3	3016	TG-21	3016	1E5	3016	1N443	3016	18442
3016	SL833	3016	TG-22	3016	1EA10A	3016	1N443B	3016	1T2011
3016	SL833A	3016	TG-31	3016	1F2	3016	1N444	3016	1T2012
3016	SL91	3016	TG-32	3016	1HY50	3016	1N449	3016	1T2013
3016	SIA1100	3016	TG-41	3016	1N1028	3016	1N456A	3016	1T22
3016	SIA1101	3016	TG-42	3016	1N1029	3016	1N457	3016	1T504
3016	SIA1102	3016	TG-51	3016	1N1030	3016	1N457A	3016	1TB06
3016	SIA1487	3016	TG-52	3016	1N1031	3016	1N458	3016	1V3074A20
3016	SIA1488	3016	TG20A	3016	1N1032	3016	1N458A	3016	1V3074A21
3016	SIA1489	3016	TI-55	3016	1N1033	3016	1N459	3016	1VA10
3016	SIA1692	3016	TJ-5A	3016	1N1041	3016	1N459A	3016	2A4113
3016	SIA1693	3016	TJ10A	3016	1N1042	3016	1N460	3016	2B4
3016	SIA1694	3016	TJ15A	3016	1N1043	3016	1N460A	3016	2G13
3016	SIA1695	3016	TJ25A	3016	1N1047	3016	1N461	3016	2G8
3016	SIA2610	3016	TJ30A	3016	1N1048	3016	1N461A	3016	2G805
3016	SIA2611	3016	TJ35A	3016	1N1049	3016	1N462	3016	2S-16E
3016	SIA2612	3016	TJ40A	3016	1N1052	3016	1N462A	3016	2T502
3016	SIA3193	3016	TK-10	3016	1N1053	3016	1N463	3016	2T503
3016	SIA440B	3016	TK-30	3016	1N1056	3016	1N463A	3016	2T504
3016	SIA441	3016	TK20	3016	1N1057	3016	1N464	3016	2W4A
3016	SIA441B	3016	TKF10	3016	1N1081	3016	1N464A	3016	2X94116
3016	SIA442	3016	TKF20	3016	1N1082	3016	1N482	3016	03-931609
3016	SIA442B	3016	TL11	3016	1N1083	3016	1N482A	3016	03-931971
3016	SIA537	3016	TL12	3016	1N1084	3016	1N482B	3016	3A152
3016	SIA538	3016	TL21	3016	1N1095	3016	1N483	3016	3A200
3016	SIA601	3016	TM-33	3016	1N1100	3016	1N483A	3016	3A252
3016	SIA602A	3016	TM-43	3016	1N1101	3016	1N483B	3016	3A81
3016	SIA603	3016	TP101	3016	1N1102	3016	1N484	3016	3A82
3016	SIA603A	3016	TP201	3016	1N1103	3016	1N484A	3016	3B81
3016	SM-10	3016	TR-2880	3016	1N1104	3016	1N484B	3016	3B82
3016	SM-150B	3016	TS-2A	3016	1N1169	3016	1N486	3016	3C81
3016	SM11	3016	TSB-245	3016	1N1169A	3016	1N486A	3016	3C82
3016	SM120	3016	TV-24104	3016	1N1218	3016	1N530	3016	3D81
3016	SM130	3016	TV-24125	3016	1N1218A	3016	1N531	3016	3D82
3016	SM200	3016	TV-2496	3016	1N1219	3016	1N533	3016	3E81
3016	SM230	3016	TV24136	3016	1N1219A	3016	1N534	3016	3E82
3016	SM240	3016	TV24155	3016	1N1251	3016	1N536	3016	3F81
3016	SM30	3016	TV24191	3016	1N1252	3016	1N538	3016	3F82
3016	SM483	3016	TV24200	3016	1N1253	3016	1N539	3016	3G152
3016	SM486	3016	TV24224	3016	1N1254	3016	1N540	3016	3G252
3016	SM487	3016	TVM-M204B	3016	1N1255	3016	1N550	3016	3G8
3016	SM488	3016	TVM-M19D22/1	3016	1N1256	3016	1N599	3016	3G81
3016	SM510	3016	TVS-D8-1K	3016	1N1415	3016	1N599A	3016	3M810
3016	SM645	3016	TVS-P002P11/2	3016	1N1439	3016	1N600	3016	3M820
3016	SM646	3016	TVS10DC4	3016	1N1440	3016	1N600A	3016	3M830
3016	SM710	3016	TVS10DC4R	3016	1N1441	3016	1N601	3016	3M840
3016	SM720	3016	TVSERB2A-06A	3016	1N1442	3016	1N601A	3016	3M85
3016	SM730	3016	TVSERB2A-06B	3016	1N1486	3016	1N602	3016	3M850
3016	SP-1	3016	TVSE8A06	3016	1N1487	3016	1N602A	3016	38B629
3016	SPN-01	3016	TVSUFSD1F	3016	1N1488	3016	1N603	3016	3T501
3016	SR-101-1	3016	TW10	3016	1N1489	3016	1N603A	3016	3T503
3016	SR-101-2	3016	TX1N3190	3016	1N1490	3016	1N604	3016	3T504
3016	SR-112	3016	TX1N645	3016	1N1491	3016	1N604A	3016	004-002700
3016	SR-120	3016	U-2400-03	3016	1N2070	3016	1N605	3016	004-003300
3016	SR-120-1	3016	U-633	3016	1N2072	3016	1N605A	3016	004-003400
3016	SR-14	3016	U213	3016	1N2073	3016	1N645A	3016	004-003500
3016	SR-22	3016	U214	3016	1N2074	3016	1N658	3016	004-003600
3016	SR-23	3016	UT-16	3016	1N2075	3016	1N663	3016	004-004000
3016	SR-24	3016	UT17	3016	1N2076	3016	1N673	3016	004-004100
3016	SR-27	3016	UT18	3016	1N2077	3016	1N819	3016	4-2020-07300
3016	SR-28	3016	UT22	3016	1N2078	3016	1N846	3016	4-2020-07601
3016	SR-3	3016	UT221	3016	1N2079	3016	1N847	3016	4-2020-07700
3016	SR-30	3016	UT224	3016	1N2080	3016	1N848	3016	4-2020-08001
3016	SR-401	3016	V-270D1	3016	1N2081	3016	1N849	3016	4-3540012
3016	SR-846-2	3016	V-442	3016	1N2082	3016	1N850	3016	4-686149-3
3016	SR-889	3016	VO6BX4	3016	1N2083	3016	1N851	3016	4AJ4DX520
3016	SR-9005	3016	VO9-E	3016	1N2084	3016	1N857	3016	4JA10DX3
3016	SR1024	3016	V11J	3016	1N2085	3016	1N858	3016	4JA10DX32
3016	SR145	3016	V3074A20	3016	1N2088	3016	1N859	3016	4JA10BX3
3016	SR151	3016	V3074A21	3016	1N2090	3016	1N860	3016	4JA211A
3016	SR1668	3016	VB-400	3016	1N2091	3016	1N861	3016	4JA2FX355
3016	SR1693	3016	VB100	3016	1N2092	3016	1N862	3016	4JA2X355
3016	SR1731-5	3016	VB300	3016	1N2093	3016	1N868	3016	4JA4DX520
3016	SR1766	3016	VB500	3016	1N2094	3016	1N869	3016	4T501
3016	SR1984	3016	VFA-2745C	3016	1N2095	3016	1N870	3016	4T502
3016	SR1K-8	3016	VS-1	3016	1N2103	3016	1N871	3016	4T503
3016	SRIT	3016	VS-102	3016	1N2104	3016	1N872	3016	4T504
3016	SR200B	3016	VS-FR-1	3016	1N2105	3016	1N873	3016	5-30086.1
3016	SR2121	3016	VS-FR-1P	3016	1N2106	3016	1N879	3016	5-30088.1
3016	SR27	3016	VS-PH9D522/1	3016	1N2107	3016	1N880	3016	5-30094.1
3016	SR2A-1	3016	VS-BD-1B	3016	1N2108	3016	1N883	3016	5-30095.1
3016	SR2A-2	3016	VS-T002P11/2	3016	1N316	3016	1N884	3016	5-30098.1
3016	SR2A-4	3016	VSG-20024	3016	1N316A	3016	1NJ61676	3016	5-30099.1
3016	SR2A-8	3016	WC-14020	3016	1N317	3016	1NJ70980	3016	5-30099.4
3016	SR390-2	3016	WC-14027	3016	1N317A	3016	1NJ71126	3016	5-30106.1
3016	SR3943	3016	WR-200	3016	1N318	3016	1NJ71186	3016	5-30109.1
3016	SR500	3016	WR300	3016	1N318A	3016	1R30	3016	5-30113.1
3016	SR500B	3016	WR400	3016	1N319	3016	1R5H	3016	5-30120.1
3016	SR6324	3016	WRE-981	3016	1N319A	3016	18100	3016	5A1
3016	SR6385	3016	XS-10	3016	1N320	3016	18101	3016	5A2
3016	SR6560	3016	XS-31	3016	1N320A	3016	18102	3016	5A3
3016	SR6617	3016	XS17	3016	1N323	3016	18103	3016	5A4
3016	SR76	3016	XS17A	3016	1N323A	3016	18106A	3016	5A4D
3016	SS455	3016	XU604	3016	1N324	3016	18110	3016	5A4D-C
3016	ST-2040P	3016	YSG-V47-1-3	3016	1N324A	3016	18110A	3016	5D1
3016	ST12	3016	ZJ252B	3016	1N325	3016	18111	3016	5E1
3016	SV-1238E	3016	ZR-1025	3016	1N3257	3016	18112	3016	5E2
3016	SW-05-01	3016	ZR-1031	3016	1N3258	3016	18113	3016	5E4
3016	SW-05V	3016	ZR-1035	3016	1N325A	3016	18113A	3016	5GD
3016	SW05A	3016	ZR-1076	3016	1N326	3016	181224	3016	5GPH
3016	SW05B	3016	ZR-500	3016	1N326A	3016	18123	3016	5GJ/FR1N
3016	SW05BB	3016	ZR-590A	3016	1N327	3016	18124	3016	5H4D1
3016	SX-642	3016	ZR-61	3016	1N327A	3016	18125	3016	5H750M
3016	SX623	3016	ZR-63	3016	1N3298	3016	18146	3016	5J-F1
3016	SX631	3016	ZR62	3016	1N350	3016	18147	3016	5MA2
3016	T-0150	3016	ZS-10B	3016	1N351	3016	18148	3016	5MF1
3016	T-065	3016	ZS-20A	3016	1N352	3016	18149	3016	5M810
3016	T-075	3016	ZS-20B	3016	1N353	3016	18150	3016	5M820
3016	T-100	3016	ZS-23	3016	1N354	3016	18151A	3016	5M830
3016	T-13	3016	ZS-25	3016	1N359	3016	18151S	3016	5M840
3016	T-13G	3016	ZS-30A	3016	1N359A	3016	181621	3016	5M85
3016	T-14	3016	ZS-30B	3016	1N360	3016	181622	3016	5M850
3016	T-140	3016	ZS-31A	3016	1N360A	3016	181623	3016	5N1
3016	T-200	3016	ZS-31B	3016	1N361	3016	181624	3016	6-59010
3016	T-220	3016	ZS-50	3016	1N361A	3016	181691	3016	6GA175D
3016	T-260	3016	ZS-71	3016	1N362	3016	18180/5GB	3016	6RW62HY
3016	T-300	3016	ZS-73	3016	1N362A	3016	18181	3016	08-0821
3016	T-400	3016	ZS10A	3016	1N363	3016	18188?	3016	8-22
3016	T-450	3016	ZS72	3016	1N363A	3016	1820	3016	8-25
3016	T-50	3016	001-024050	3016	1N3669	3016	182310	3016	8-38
3016	T-500	3016	001-024051	3016	1N3895	3016	18309	3016	8-710-222-21
3016	T-550	3016	001-024052	3016	1N400	3016	18313	3016	8-719-205-10
3016	T-650	3016	1-531-105-11	3016	1N4009	3016	1835	3016	8G015
3016	T-E1029	3016	1-531-106-17	3016	1N400B	3016	18457	3016	8C015RE
3016	T-E1064	3016	1-534-105-13	3016	1N4043	3016	18557	3016	09-306311
3016	T-E1078	3016	1-8259							10-12
3016	T-E1078A	3016	1B05J20							10-42
3016	T-E1089	3016	1B10J20							

SK	Industry Standard No.	SK	Industry Standard No.	SK	Industry Standard No.	SK	Industry Standard No.	SK	Industry Standard No.
3016	10-7	3016	46AX19	3016	86-4-1	3016	304B	3016	2784
3016	10A590B	3016	46AX30	3016	86-40-3	3016	307A	3016	2786
3016	10AG2	3016	46AX5	3016	86-42-3	3016	307B	3016	3074A20
3016	10AL2	3016	46AX7	3016	86-46-3	3016	307C	3016	3074A21
3016	10A8	3016	46BD11	3016	86-49-3	3016	307D	3016	3700-153
3016	10AT2	3016	46BR7	3016	86-50-3	3016	307F	3016	3843
3016	10B	3016	48-01	3016	86-5009-3	3016	307H	3016	4101-685
3016	10C1	3016	48-1005	3016	86-5010-3	3016	307K	3016	004567
3016	10C2	3016	48-137205	3016	86-5011-3	3016	310	3016	4686-149-3
3016	10D3	3016	48-137291	3016	86-5015-3	3016	310-4	3016	5416
3016	10H	3016	48-137316	3016	86-5024-3	3016	0311	3016	7211-9
3016	10K	3016	48-40235G01	3016	86-5027-3	3016	0312	3016	7212-3
3016	10M	3016	48-67120A01	3016	86-5029-3	3016	0320	3016	7212-3A
3016	10N1	3016	48-67120A02	3016	86-51-3	3016	320A	3016	7213-7
3016	10R2B	3016	48-67120A03	3016	86-54-3	3016	320B	3016	7568
3016	11-085013	3016	48-67120A04	3016	86-57-3	3016	320C	3016	8300-8
3016	12C2	3016	48-67120A05	3016	86-58-3	3016	320D	3016	8500-206
3016	12J2	3016	48-67120A09	3016	86-59-3	3016	320F	3016	8864-1
3016	13-085039	3016	48A41508A01	3016	86-62-3	3016	320H	3016	9100-1
3016	13-1002-1	3016	48A41508A02	3016	86-63-3	3016	320K	3016	9300-5
3016	13-1032-1-3	3016	48B66629A03	3016	86-7-1	3016	0321	3016	9300-6
3016	13-14094-16	3016	48C66037A03	3016	86-89-1	3016	0322	3016	9300-7
3016	13-14627-1	3016	48C66037A04	3016	86-9-3	3016	324-0117	3016	9301-1
3016	13-16104-8	3016	48C66037A05	3016	88-125	3016	324-0119	3016	9302-3
3016	13-16637-3	3016	48C66037A10	3016	88-831	3016	324-0120	3016	9348-3
3016	13-17557-1	3016	48C66037A12	3016	88-832	3016	324-0135	3016	9650-001
3016	13-17825-1	3016	48C67926A01	3016	88-833	3016	324-0147	3016	11746
3016	13-18458-1	3016	48D66037A03	3016	91A06	3016	324-0162	3016	12602
3016	13-39073-1	3016	48D66037A04	3016	93A12-1	3016	325-0047-517	3016	12720
3016	13J2	3016	48D66037A05	3016	93A97-2	3016	325-0054-312	3016	12736
3016	13P1	3016	48D66037A08	3016	93B1-18	3016	325-0135-B	3016	12768
3016	14J2	3016	48D66656A02	3016	93B1-20	3016	325-0141-23	3016	12786
3016	015-002	3016	48D67120A01	3016	93B1-21	3016	325-1441-10	3016	12788
3016	015-006	3016	48D67120A06	3016	93B1-6	3016	325-1441-11	3016	12803
3016	15R1	3016	48D67120A07	3016	93B12-1	3016	325-1446-29	3016	12837
3016	16-2	3016	48K64169	3016	93B12-2	3016	366B	3016	013339
3016	16C4	3016	48K646954	3016	93B12-3	3016	369-2	3016	13782
3016	17-10	3016	48K647829	3016	93B24-2	3016	369-3	3016	14027
3016	019-002935	3016	48K746831	3016	93B24-3	3016	380H61	3016	14588-9
3016	019-003870-013	3016	48K752497	3016	93B27-2	3016	385Q	3016	17443
3016	019-003870-020	3016	48M355025	3016	93B27-3	3016	386BW	3016	021154
3016	19-040-002	3016	48R10001-A01	3016	93B30-3	3016	386CX	3016	025056
3016	19-040-004	3016	48R10062A02	3016	93B41-4	3016	386CY	3016	27123-050
3016	20A8	3016	48R134671	3016	93B42-2	3016	386DW	3016	27123-070
3016	20B8	3016	48810062A05	3016	93B45-1	3016	386DY	3016	27123-100
3016	20H	3016	48810062A05A	3016	93B47-1	3016	421-4027	3016	27123-120
3016	20K	3016	488134736	3016	93B48-2	3016	421-7443	3016	34174
3016	20M	3016	488155037	3016	93B58-1	3016	421-7445A	3016	34394
3016	20N1	3016	488155040	3016	93C1-20	3016	421-9865	3016	34894
3016	21-810	3016	488155041	3016	93C1-21	3016	422-1362	3016	35287A
3016	21-810-2	3016	488155063	3016	93C16-2	3016	422-1866	3016	352877A
3016	21A008-001	3016	488155079	3016	93C24-2	3016	422-2540	3016	35500
3016	21A008-002	3016	488155099	3016	93C30-1	3016	435-40012	3016	36549
3016	21A008-003	3016	488191A02	3016	93C30-3	3016	454-A2534-1	3016	36554
3016	21A020-001	3016	488191A06	3016	93C42-2	3016	471-010	3016	37126-1
3016	21A103-012	3016	49-3112	3016	93C02-1	3016	474-025	3016	41616
3016	21A6	3016	50A	3016	93C53-2	3016	517-0021	3016	44465-3
3016	21A7	3016	50B5	3016	94-1066-1	3016	517-0025	3016	44465-4
3016	21B-14	3016	50D2	3016	96-5022-01	3016	517-0031	3016	44465-5
3016	21M283	3016	50D4	3016	96-5022-1	3016	517-0033	3016	44465-6
3016	21M302	3016	50E05	3016	96-5023-01	3016	530-111-1	3016	46140-2
3016	23-0003	3016	50E1	3016	96-5117-01	3016	532-341	3016	47126
3016	23B8C101	3016	50E2	3016	100A	3016	532-341A	3016	47126-1
3016	24-198	3016	50E3	3016	100R2B	3016	540-008	3016	47126-1A
3016	24DP1	3016	50E4	3016	101B6	3016	540-014	3016	47126-2
3016	027-000306	3016	051-0003	3016	102D	3016	575-028	3016	47126-3A
3016	027-000312	3016	53-5051-2	3016	103-245	3016	575-042	3016	47126-4
3016	28-21-01	3016	53A022-4	3016	103-29	3016	575-048	3016	47126-4A
3016	28-22-01	3016	56-5	3016	103Q3125	3016	575-050	3016	47126-6
3016	28-22-01	3016	57-0006	3016	103Q3125A	3016	580-029	3016	50745-06
3016	28-22-04	3016	57-27	3016	108	3016	604B	3016	50745-07
3016	28-22-07	3016	57-42	3016	108A4	3016	606-113	3016	059210
3016	28-22-21	3016	57-42A	3016	0110-0209	3016	607-113	3016	59840
3016	28-22-11	3016	60C	3016	110-629	3016	608-030	3016	59844-3
3016	28-22-15	3016	61-7728	3016	110-635	3016	608-101	3016	61011-2
3016	28-25-01	3016	62J2	3016	110-636	3016	608-113	3016	66682-42
3016	28-254566-1	3016	63J2	3016	110-672	3016	609-030	3016	68177
3016	28-26-01	3016	64J2	3016	110-684	3016	609-113	3016	68177A
3016	28-26-01	3016	065	3016	111-1	3016	610C	3016	68504-78
3016	30A8	3016	65P117	3016	112-500-0-501	3016	612	3016	69115
3016	30B5	3016	65P124-2	3016	12-601-0-102	3016	624-0009	3016	69213-78
3016	30K	3016	65P155	3016	13A7739	3016	624-0010	3016	70064-7
3016	32-0045	3016	65P155	3016	120-001300	3016	690V051H33	3016	70066-3
3016	33-0006	3016	65P206	3016	124J490	3016	690V058H22	3016	70066-4
3016	34-8048-2	3016	65P284	3016	126-4	3016	690V059H52	3016	70372-2
3016	34-8054-1	3016	65P297	3016	145A9254	3016	690V041H08	3016	71588-1
3016	34-8054-2	3016	66-6030-00	3016	145A9786	3016	690V069H39	3016	71588-2
3016	34-8054-23	3016	66-6031-00	3016	150	3016	0700	3016	71588-5
3016	34-8054-27	3016	66X0023-001	3016	151-011-9-011	3016	700-137	3016	71588-6
3016	34-8054-9	3016	66X0023-002	3016	154A3992	3016	0727-50	3016	72041
3016	34-8055-2	3016	66X0023-004	3016	162J2	3016	753-2000-535	3016	72058A
3016	34-8055-3	3016	66X0023-005	3016	173A3981	3016	800-006-00	3016	72060
3016	40A8	3016	66X0023-007	3016	173A3981-1	3016	822-4	3016	72097
3016	40B5	3016	66X0023-008	3016	173A4393	3016	822-5	3016	72101
3016	40D6665A03	3016	66X0023-1	3016	196-654	3016	919-001172	3016	72103
3016	40H	3016	66X0033-000	3016	0200	3016	919-001172-1	3016	72113
3016	40J2	3016	66X0033-001	3016	212-18	3016	919-004326	3016	72113A
3016	40K	3016	66X0037-001	3016	212-23	3016	919-004799	3016	72119
3016	40KR	3016	66X0055-001	3016	212-25	3016	919-007109	3016	72123
3016	40M	3016	66X0053-001	3016	212-27	3016	919-007394	3016	72123A
3016	40N1	3016	66X21	3016	212-33	3016	919-007776	3016	74200-8
3016	40T3P	3016	66X23	3016	212-35	3016	919-011212	3016	74200-9
3016	41J2	3016	66X24	3016	212-36	3016	936-10	3016	75230-9
3016	42-14027	3016	66X25	3016	212-37	3016	964-17443	3016	77190-7
3016	42-17443	3016	66X26	3016	212-39	3016	964-19865	3016	080040
3016	42-17443A	3016	70P40	3016	212-40	3016	977-18B	3016	81513-8
3016	42-19865	3016	70B40	3016	212-41	3016	977-19B	3016	81514-2
3016	42-21866	3016	70T40	3016	212-42	3016	1002-0219	3016	83008
3016	42-22540	3016	72Z	3016	212-46	3016	1018-6963	3016	91001
3016	42-23350A	3016	75	3016	212-47	3016	1043-1534	3016	93005
3016	42A11	3016	75D1	3016	212-50	3016	1065-9928	3016	93006
3016	42A23	3016	75D2	3016	212-51	3016	1138	3016	93007
3016	42B16	3016	75E1	3016	212-58	3016	1139-17	3016	93011
3016	42B2	3016	75E2	3016	212-62	3016	1263A	3016	93012
3016	42X244	3016	75E3	3016	212-65	3016	1412-170	3016	93023
3016	42X244B	3016	75E5	3016	212-77	3016	2060-024	3016	100520
3016	42X245	3016	078-0016	3016	212-79	3016	2093A12-1	3016	100624
3016	42X245B	3016	078-1696	3016	215-51	3016	2093A41-89	3016	100403
3016	42X25	3016	078-2400	3016	215-58	3016	2093B38-14	3016	103318
3016	42X32	3016	78-254566-1	3016	220-003001	3016	2093B4-6	3016	104081
3016	43-540012	3016	78-254566-4	3016	232-1011	3016	2093B41-10	3016	104213
3016	45A2F355	3016	78-271143-1	3016	264I03607	3016	2093B41-12	3016	106379
3016	46-46-3	3016	83-43-3	3016	264P09001	3016	2093B41-6	3016	110043
3016	46-86149-3	3016	85-5	3016	264P10102	3016	2093B41-9	3016	110351
3016	46-86339-3	3016	86-0016-01	3016	264P10103	3016	2102-074	3016	110388
3016	46-86402-3	3016	86-0516	3016	291-04	3016	2400-23	3016	110496
3016	46-86416-3	3016	86-1-3	3016	295V027001	3016	2400-462	3016	110629
3016	46-86486-3	3016	86-21-1	3016	295V027001-1	3016	2405-462	3016	110636
3016	46-86497-3	3016	86-22-3	3016	295V028C02	3016	2405-463	3016	110873
3016	46-8661-3	3016	86-23-3	3016	295V029001	3016	2762	3016	111086
3016	46AR6	3016	86-26-1	3016	296V002H08	3016	2763	3016	111516
3016	46AR8	3016	86-28-3	3016	296V006H03	3016	2766	3016	111642
3016	46AX19	3016	86-30-3	3016	0300	3016	2777	3016	111776
3016	46AX17	3016	86-54-3	3016	302	3016	2779	3016	111820

SK	Industry Standard No.
3016	112017
3016	112018
3016	112528
3016	112826
3016	113039
3016	113392
3016	114013
3016	115039
3016	115559
3016	115599
3016	116052
3016	117145
3016	117145A
3016	118244
3016	118825
3016	120544
3016	122129
3016	124812
3016	127396
3016	130338-01
3016	130389-00
3016	130607
3016	131245
3016	161021
3016	161029
3016	161037
3016	214105
3016	215669
3016	216449
3016	216817
3016	218612
3016	219935
3016	221128
3016	222611
3016	223215
3016	223216
3016	223323
3016	223358
3016	223462
3016	223467
3016	223724
3016	224159
3016	224597
3016	224774
3016	225200
3016	225592
3016	226058
3016	226182
3016	226237
3016	226546
3016	226788
3016	227348
3016	227565
3016	227720
3016	229522
3016	230218
3016	231339
3016	231665
3016	231669
3016	231923
3016	232203
3016	233561
3016	233597
3016	234552
3016	234565
3016	234611
3016	234761
3016	235313
3016	235543
3016	235546
3016	237227
3016	237453
3016	237929
3016	239221
3016	240456
3016	240594
3016	240603
3016	242029
3016	242226
3016	243364
3016	258882
3016	259368
3016	259878
3016	260429
3016	261463
3016	261596
3016	261898
3016	261975
3016	262112
3016	262872
3016	265164
3016	265235
3016	265634
3016	276097
3016	300233
3016	300315
3016	300524
3016	300550
3016	301586
3016	330003
3016	330018
3016	330019
3016	348048-2
3016	348053-3
3016	348054-1
3016	348054-10
3016	348054-2
3016	348054-5
3016	348054-9
3016	348055-2
3016	348055-3
3016	464010
3016	464070
3016	489752-035
3016	530057-1
3016	530063-11
3016	530063-12
3016	530063-6
3016	530063-7
3016	530071-1
3016	530072-11
3016	530082-2
3016	530086-1
3016	530088-1
3016	530094-1
3016	530095-1
3016	530098-1
3016	530098-1001
3016	530099-1
3016	530099-3
3016	530099-4
3016	530106-1
3016	530109-1
3016	530123-4
3016	551026
3016	0552006
3016	552007
3016	0552010
3016	575028
3016	575042
3016	0575050
3016	580029
3016	606113
3016	607113
3016	608030
3016	608101
3016	608113
3016	609030
3016	609113
3016	610112
3016	612112
3016	613020
3016	613130
3016	614010
3016	700647
3016	700663
3016	700664
3016	740951
3016	742008
3016	742009
3016	752309
3016	801711
3016	801714
3016	801715
3016	801716
3016	815142
3016	817066
3016	817068
3016	817104
3016	817109
3016	817111
3016	817112
3016	817114
3016	817117
3016	817121
3016	817122
3016	817128
3016	817130
3016	817133
3016	817135
3016	817138
3016	817140
3016	817141
3016	817143
3016	817148
3016	817166
3016	817172
3016	852158-7.1
3016	921608
3016	924605-3
3016	924801-5
3016	924805-5
3016	924805-8
3016	972571-4
3016	972571-7
3016	973962-1
3016	980143
3016	980964
3016	981739
3016	992171
3016	1476171-29
3016	1956486
3016	1961843
3016	1962594
3016	2000625-36
3016	2000626-32
3016	2000646-115
3016	2000646-119
3016	2000646-120
3016	2002332-57
3016	2003073-67
3016	2013019-117
3016	2060041
3016	2330561
3016	2330612
3016	4101685
3016	5955749
3016	7027023
3016	7043216
3016	7100628
3016	7100630
3016	7570014-02
3016	7570215-21
3016	8521587-1
3016	9246053
3016	9248015
3016	9248055
3016	9248058
3016	13033801
3016	13038900
3016	19080002
3016	23115300
3016	23115337
3016	23115338
3016	23115947
3016	27123050
3016	27123070
3016	27123100
3016	27123120
3016	41025773
3016	41025849
3016	41027063
3016	43540012
3016	62761334
3016	120001300
3016	2008130084
3016	4139104002
3016	4202006700
3016	4202007600
3016	4202007802
3016	4202008500
3016	4202018500
3016	4202110470
3017B	.5E05
3017B	.5E1
3017B	.5E2
3017B	.5E3
3017B	.5E4
3017B	.5E5
3017B	.5E6
3017B	.5J05
3017B	.5J1
3017B	.5J2
3017B	.5J3
3017B	.5J4
3017B	.5J5
3017B	.5J6
3017B	A14A
3017B	B294
3017B	BB-109
3017B	BCF5
3017B	BF26235-5
3017B	COD11556
3017B	COD15564
3017B	D1172S
3017B	DI56
3017B	DS16A
3017B	ECG117
3017B	ECR600-2
3017B	EER600-2
3017B	GER4005
3017B	HF-SD-1A
3017B	HR-2A
3017B	NA62
3017B	NA63
3017B	NA65
3017B	NA66
3017B	PS030
3017B	PS110
3017B	PS410
3017B	PT6D22/1
3017B	RE50
3017B	REN117
3017B	RFS61436
3017B	RT-215
3017B	RT214
3017B	OS22.3905-001
3017B	SD1101
3017B	SD1102
3017B	SD1103
3017B	SD1104
3017B	SR1A-12
3017B	T30155
3017B	TL2
3017B	TL22
3017B	TL31
3017B	TL42
3017B	TL51
3017B	TL61
3017B	V148
3017B	V171
3017B	001-0081-00
3017B	1-531-105-13
3017B	1EA20A
3017B	1EA30A
3017B	1EA40A
3017B	1EA50A
3017B	1EA60A
3017B	1N193
3017B	1N368
3017B	1R9I
3017B	1S1694
3017B	1S311
3017B	1S963
3017B	4-686116-3
3017B	4-686147-3
3017B	4-686148-3
3017B	4-686150-3
3017B	4-686151-3
3017B	4-686177-3
3017B	4-68687-3
3017B	5-30088.2
3017B	5-30111.1
3017B	5-30122.1
3017B	09-306042
3017B	09-306042(RECT)
3017B	09-306366
3017B	09-306431
3017B	14-0072-1
3017B	14-0072-2
3017B	14-0072-3
3017B	15-108024
3017B	24MW175
3017B	24MW196
3017B	40-0502
3017B	46-86116-3
3017B	46-86150-3
3017B	46-86151-3
3017B	46-86807-3
3017B	48-10001-A03
3017B	48M355016
3017B	48M355023
3017B	48X90234A02
3017B	53J002-2
3017B	86-67-8
3017B	113-998
3017B	115-867
3017B	126-7
3017B	200-6582-22
3017B	215-760E
3017B	264P03606
3017B	2093A3D-2
3017B	2093A4-2
3017B	2093A6-1
3017B	2093A6-2
3017B	2411-17
3017B	3069
3017B	4686-116-3
3017B	4686-147-3
3017B	4686-148-3
3017B	4686-150-3
3017B	4686-151-3
3017B	4686-177-3
3017B	4686-87-3
3017B	7213-0
3017B	9302-2
3017B	9302-2A
3017B	93027
3017B	219245
3017B	223489
3017B	223720
3017B	225267
3017B	227015
3017B	265236
3017B	266583
3017B	348054-11
3017B	348054-14
3017B	348054-6
3017B	402003173
3017B	2295100232
3017B	2295100234
3017B	2295100235
3018	A054-148
3018	A054-170
3018	A066-111
3018	A1G
3018	A1P/4922
3018	A1R/4924
3018	A1U
3018	A1Z
3018	A2479
3018	A2M
3018	A2M-1
3018	A2N
3018	A2N-1
3018	A2Y
3018	A67-37-940
3018	A6708850
3018	AR200
3018	AR209
3018	AR210
3018	AR211
3018	AR213
3018	AR213V
3018	AR218
3018	AR218RO
3018	AR219
3018	AR219YY
3018	AR220
3018	AR220GY
3018	AR222
3018	AR222BY
3018	AR224
3018	B1H
3018	B2Z
3018	BC456
3018	BF173
3018	BF194
3018	BF194B
3018	BF195
3018	BF195D
3018	BF200
3018	BF254
3018	BFY-47
3018	BFY-48
3018	BFY10
3018	BFY11
3018	BFY49
3018	C122
3018	C1390A
3018	C1674L
3018	C206
3018	C2485078-1
3018	C2485079-1
3018	C398(FA-1)
3018	C399
3018	C56
3018	C605Q
3018	C645
3018	C683
3018	C683A
3018	C683B
3018	C683V
3018	C738
3018	C738D
3018	C739
3018	C739C
3018	C903D
3018	C920
3018	C920E
3018	C922
3018	C9426
3018	CCS2008F015
3018	CDC-8001
3018	CDC12030B
3018	CDC5000
3018	CDC5000-1B
3018	CS-1330
3018	CS-1386E
3018	CS-6225G
3018	CS-9016D
3018	CS-9018
3018	CS1244X
3018	CS1330
3018	CS1330A
3018	CS1361E
3018	CS1361F
3018	CS1585H
3018	CS2006F
3018	CS2008
3018	CS6225E
3018	CS9016
3018	CS9018D
3018	CS9021G-I
3018	CT1012
3018	CT1013
3018	D006
3018	D009
3018	D072
3018	D159
3018	D160
3018	DBCZ039404
3018	DBCZ083905
3018	DDBY209003
3018	DS-71
3018	DS-72
3018	DS-73
3018	DS-78
3018	DS-81
3018	DS78
3018	DS81
3018	E13-004-00
3018	E9625
3018	EA1123
3018	EA15X130
3018	EA15X134
3018	EA15X135
3018	EA15X189
3018	EA15X239
3018	EA15X240
3018	EA15X350
3018	EA15X351
3018	EA15X370
3018	EA15X4064
3018	EA15X7120
3018	EA15X7140
3018	EA15X7173
3018	EA15X7174
3018	EA15X7178
3018	EA15X7179
3018	EA15X7215
3018	EA15X7233
3018	EA15X7234
3018	EA15X7235
3018	EA15X7236
3018	EA15X7244
3018	EA15X7264
3018	EA15X8589
3018	EA2131
3018	EA2132
3018	EA2429
3018	EA2494
3018	EA2496
3018	EA2812
3018	EA3713
3018	EA3763
3018	ED-1502C
3018	ED1502A
3018	ED1502B
3018	ED592K
3018	ED592M
3018	EL434
3018	EQF-0009
3018	EQS-0018
3018	EQS-0100
3018	EQS-0196
3018	EQS-0198
3018	EQS-139
3018	EQS-18
3018	EQS-19
3018	ES10186
3018	ES10188
3018	ES15046
3018	ES15047
3018	ES15X102
3018	ES15X127
3018	ES15X56
3018	ES15X57
3018	ES15X80
3018	ES15X81
3018	ES15X96
3018	ES15X97
3018	EW163
3018	EW164
3018	F73216
3018	G04-041B
3018	G05-003-A
3018	G05-003-B
3018	G05-050-C
3018	G05-063-R
3018	G05-065-A
3018	G9423
3018	G9625
3018	H9423
3018	H9618
3018	H9623
3018	HC-00380
3018	HC-00394
3018	HC-00535
3018	HC-00711
3018	HC-00784
3018	HC-00829
3018	HC-00929
3018	HC-00930
3018	HC-01047
3018	HC-01359
3018	HC-537
3018	HC-561
3018	HC-772
3018	HC381
3018	HC460
3018	HC461
3018	HC784
3018	HF309301E
3018	HR-59
3018	HR-60
3018	HR63
3018	HR76
3018	HR77
3018	HR78
3018	HR79
3018	HR80
3018	HR81
3018	HR84
3018	HR87
3018	HS-40014
3018	HS-40020
3018	HS-40054
3018	HS-40055
3018	HT303711A
3018	HT303711B
3018	HT303711B-0
3018	HT303711C
3018	HT303801
3018	HT303801A
3018	HT303801B
3018	HT303801B-0
3018	HT303801C
3018	HT306451A
3018	HT307720B
3018	HT307721C
3018	HT307721D
3018	HT308291A
3018	HT308291A-0
3018	HT308291B
3018	HT308291B-0
3018	HT308291B0
3018	HT309291E
3018	HT309301C
3018	HT309301D
3018	HT309301E
3018	HT309301F
3018	HX-50110
3018	HX-50113
3018	I9631
3018	IC743040
3018	IC743041
3018	IE460B
3018	IP20-0029
3018	IP20-0034
3018	IP20-0037
3018	IP20-0038
3018	IP20-0039
3018	IP20-0110
3018	IRTR71
3018	J107
3018	J241177
3018	J241188
3018	J241189
3018	J24561
3018	J24562
3018	J24563
3018	J24701
3018	J24812
3018	JSP7005
3018	JSP7006
3018	KLH4792
3018	KTR0710C
3018	KTR0839C
3018	KTR1687C
3018	LM-1129
3018	LM-1130
3018	LM-1132
3018	LM-1133
3018	LM-1147
3018	LM-1148
3018	LM-1155
3018	LM1110B
3018	LM1120B
3018	LM1120C

SK	Industry Standard No.	SK	Industry Standard No.	SK	Industry Standard No.	SK	Industry Standard No.	SK	Industry Standard No.
3018	LZ1016H	3018	SK-31024-3	3018	TV17	3018	28C1026H	3018	28C148G
3018	LZ1016I	3018	SKA4074	3018	TV17A	3018	28C1026J	3018	28C148GN
3018	LZB1016	3018	SKA9013	3018	TV18	3018	28C1026K	3018	28C148H
3018	LZT1016	3018	SM-4304-8	3018	TV22	3018	28C1026L	3018	28C148J
3018	M140-1	3018	SP4168	3018	TV24102	3018	28C1026M	3018	28C148K
3018	M140Q-1	3018	SPS-1551	3018	TV24160	3018	28C1026OR	3018	28C148L
3018	M401	3018	SPS-1473	3018	TV24161	3018	28C1026X	3018	28C148M
3018	M4757	3018	SPS-1476	3018	TV24203	3018	28C1032	3018	28C148R
3018	M4904	3018	SPS-952-2	3018	TV24204	3018	00028C1032A	3018	28C148X
3018	M546	3018	SP81351	3018	TV24209	3018	00028C1032B	3018	28C148Y
3018	M612	3018	SP81352	3018	TV24210	3018	28C1032RL	3018	28C159A
3018	M613	3018	SP81353	3018	TV24380	3018	00028C1032C	3018	28C159B
3018	M614	3018	SP82110	3018	TV24383	3018	28C1032D	3018	28C159C
3018	M9032	3018	SP82111	3018	TV32	3018	28C1032E	3018	28C159E
3018	M924	3018	SP8220	3018	TV33	3018	28C1032F	3018	28C159F
3018	MC9427	3018	SP82224	3018	TV35	3018	28C1032G	3018	28C159GN
3018	MB113B	3018	SP82265	3018	TV36	3018	28C1032GN	3018	28C159H
3018	MB9001	3018	SPS2265-2	3018	TV37	3018	28C1032H	3018	28C159J
3018	MP83563	3018	SP82320	3018	TV38	3018	28C1032J	3018	28C159K
3018	MP83693	3018	SP83370	3018	TV39	3018	28C1032K	3018	28C159M
3018	MP87513	3018	SP840	3018	TV40	3018	28C1032M	3018	28C159OR
3018	MP89423G	3018	SP84143	3018	TV43	3018	28C1032OR	3018	28C159R
3018	MP89423H	3018	SP84167	3018	TV44	3018	28C1032X	3018	28C159X
3018	MP89423I	3018	SP84168	3018	TV46	3018	28C1032Y	3018	28C159Y
3018	MP89426A.B	3018	SX-3825	3018	TV56	3018	28C1070	3018	28C160A
3018	MP89427	3018	SX3825	3018	TV8-28C446	3018	28C1123	3018	28C160B
3018	MP89427B.C	3018	SX3826	3018	TV8-280563	3018	28C1123A	3018	28C160C
3018	MP89604	3018	SX408	3018	TV8-280563A	3018	28C1123B	3018	28C160D
3018	MP89604P	3018	T-483	3018	TV8-280683	3018	28C1123C	3018	28C160E
3018	MP89623E.G	3018	T-484	3018	TV8-280683V	3018	28C1123D	3018	28C160P
3018	MP89625	3018	T-486	3018	TV8-280762	3018	28C1123E	3018	28C160G
3018	MP89625C	3018	T-Q5049	3018	TV8280538	3018	28C1123F	3018	28C160GN
3018	MP89625D	3018	T-Q5071	3018	TV8280920-0Q	3018	28C1123GN	3018	28C160H
3018	MP89625B	3018	T-Q5078	3018	U1585E	3018	28C1123H	3018	28C160K
3018	MP89625P	3018	T-Q5079	3018	V8-28C208	3018	28C1123J	3018	28C160L
3018	MP89625G	3018	T-Q5086	3018	V8-280563	3018	28C1123K	3018	28C160M
3018	MP89632T	3018	T-Q5106	3018	V8-280645A	3018	28C1123L	3018	28C160X
3018	M83694	3018	T1408	3018	V8-28C762	3018	28C1123M	3018	28C160Y
3018	N-EA15X130	3018	T1909	3018	V828C1855//-1	3018	28C1123OR	3018	28C17B
3018	N-EA15X134	3018	T1818	3018	V8280394-Y-1	3018	28C1123R	3018	28C17C
3018	N-EA15X135	3018	T1894	3018	V8280460B-1	3018	28C1123X	3018	28C17D
3018	N4T	3018	T1898	3018	W23	3018	28C1123Y	3018	28C17F
3018	NC89018D	3018	T1828	3018	WEP720	3018	28C1126	3018	28C17G
3018	NJ100A	3018	T1X-M14	3018	ZEN108	3018	28C1126A	3018	28C170N
3018	NL100B	3018	T1X-M15	3018	ZJ40	3018	28C1126B	3018	28C17H
3018	NR-431AG	3018	T2634	3018	280555B	3018	28C1126F	3018	28C17J
3018	NR-431A8	3018	T3536	3018	01-030394	3018	28C1126GN	3018	28C17K
3018	NR421	3018	T3539	3018	01-030682	3018	28C1126J	3018	28C17L
3018	NR421DG	3018	T9423	3018	1-801-003	3018	28C1126K	3018	28C17M
3018	NR461AF	3018	T9631	3018	1-801-309	3018	28C1126L	3018	28C170R
3018	NTC-4	3018	TG28C1293	3018	1B55A	3018	28C1126M	3018	28C17R
3018	PM195	3018	TG28C1293A	3018	1B55A/7825B	3018	28C1126OR	3018	28C17X
3018	PM195A	3018	TG28C927-C-A	3018	1B55B	3018	28C1126R	3018	28C17Y
3018	PTC126	3018	TG28C927-D-A	3018	2AG	3018	28C1126X	3018	28C185
3018	Q-00284R	3018	TG28C927-E-A	3018	2AH	3018	28C1126Y	3018	28C185M
3018	Q-00284R-3	3018	TI398A	3018	2N2615	3018	28C1182B	3018	28C185Q
3018	Q-00284R-3	3018	TNJ-60604	3018	2N3287	3018	28C1182C	3018	28C185R
3018	Q35259	3018	TNJ-60605	3018	2N3288	3018	28C1182D	3018	28C185V
3018	Q8E1001	3018	TNJ61217	3018	2N3563	3018	28C1320A	3018	28C195A
3018	Q8E5020	3018	TNJ61218	3018	2N3563-1	3018	28C1320B	3018	28C195B
3018	QT-00460CBB	3018	TNJ70478	3018	2N3564	3018	28C1320D	3018	28C195C
3018	QT-00659XDA	3018	TNJ70478-1	3018	2N3665	3018	28C1320E	3018	28C195D
3018	R-28C535	3018	TNJ70480	3018	2N3826	3018	28C1320P	3018	28C195B
3018	R-28C772	3018	TNJ70484	3018	2N3827	3018	28C1320Q	3018	28C195F
3018	R89	3018	TNJ71498	3018	2N3855	3018	28C1320GN	3018	28C195G
3018	RB-7102	3018	TNJ71937	3018	2N4255	3018	28C1320H	3018	28C195GN
3018	RB-7104	3018	TNJ71963	3018	2N4435	3018	28C1320J	3018	28C195H
3018	RB-7106	3018	TNJ71964	3018	2N5130	3018	28C1320K	3018	28C195K
3018	RB-7107	3018	TNJ71965	3018	2N5133	3018	28C1320L	3018	28C195L
3018	RB-7108	3018	TNJ72277	3018	2N5180	3018	28C1320N	3018	28C195M
3018	RB-7109	3018	TNJ72278	3018	2N5830A	3018	28C1320R	3018	28C195OR
3018	RB-7110	3018	TNT-839	3018	28C-4033	3018	28C1320X	3018	28C195R
3018	R8T726	3018	TNT-840	3018	28C-F11	3018	28C1320Y	3018	28C195X
3018	R87112	3018	TNT-841	3018	28C-F11A	3018	28C134A	3018	28C195Y
3018	R87114	3018	TNT-843	3018	28C-F11B	3018	28C134C	3018	28C196A
3018	R87143	3018	TQ1	3018	28C-F11C	3018	28C134D	3018	28C196B
3018	R87173	3018	TQ3	3018	28C-F11D	3018	28C134E	3018	28C196C
3018	R8T222	3018	TQ5	3018	28C-F11E	3018	28C134F	3018	28C196D
3018	R87233	3018	TQ6	3018	28C-F11F	3018	28C134G	3018	28C196E
3018	R8T238	3018	TQ7	3018	28C-F11G	3018	28C134GN	3018	28C196F
3018	R8T523	3018	TQ8	3018	28C-F11GN	3018	28C134H	3018	28C196G
3018	RT2915	3018	TQ9	3018	28C-F11H	3018	28C134J	3018	28C196GN
3018	RT5464	3018	TR-1R35	3018	28C-F11J	3018	28C134K	3018	28C196H
3018	RT5465	3018	TR-21	3018	28C-F11K	3018	28C134L	3018	28C196J
3018	R6201	3018	TR-24	3018	28C-F11L	3018	28C134M	3018	28C196K
3018	R6204	3018	TR-2R35	3018	28C-F11M	3018	28C134OR	3018	28C196L
3018	R6601	3018	TR-33	3018	28C-F11OR	3018	28C134R	3018	28C196M
3018	R6602	3018	TR-3R35	3018	28C-F11R	3018	28C134X	3018	28C196OR
3018	R6732	3018	TR-4R35	3018	28C-F11X	3018	28C134Y	3018	28C196X
3018	R6733	3018	TR-8004	3018	28C-F11Y	3018	28C135A	3018	28C196Y
3018	RT6991	3018	TR-8004-5	3018	28C-F14	3018	28C135B	3018	28C197A
3018	RT7704	3018	TR-8028	3018	28C-F14A	3018	28C135C	3018	28C197B
3018	RT8330	3018	TR-8029	3018	28C-F14B	3018	28C135D	3018	28C197C
3018	RT8668	3018	TR-8030	3018	28C-F14C	3018	28C135E	3018	28C197D
3018	RT8669	3018	TR-8031	3018	28C-F14D	3018	28C135F	3018	28C197E
3018	RV1473	3018	TR-8032	3018	28C-F14E	3018	28C135G	3018	28C197F
3018	RV2354	3018	TR-8034	3018	28C-F14F	3018	28C135GN	3018	28C197G
3018	S-1079	3018	TR01026	3018	28C-F14G	3018	28C135H	3018	28C197GN
3018	S-1286	3018	TR05073507	3018	28C-F14GN	3018	28C135J	3018	28C197H
3018	S-95125	3018	TR112	3018	28C-F14H	3018	28C135L	3018	28C197J
3018	S-95125A	3018	TR2083-71	3018	28C-F14J	3018	28C135M	3018	28C197K
3018	S-95126	3018	TR21C	3018	28C-F14K	3018	28C135OR	3018	28C197L
3018	S-95126A	3018	TR228735045311	3018	28C-F14L	3018	28C135R	3018	28C197M
3018	0822.3640-080	3018	TR228735046011	3018	28C-F14M	3018	28C135X	3018	28C197OR
3018	S25805	3018	TR228735048617	3018	28C-F14OR	3018	28C135Y	3018	28C197R
3018	S326690	3018	TR228735048618	3018	28C-F14R	3018	28C137A	3018	28C197X
3018	340204	3018	TR22C	3018	28C-F14X	3018	28C137B	3018	28C197Y
3018	85021	3018	TR28C1342	3018	28C-F14Y	3018	28C137C	3018	28C1990B
3018	SC1227F	3018	TR28C371	3018	28C1023	3018	28C137D	3018	28C206
3018	SC1227G	3018	TR28C372	3018	28C1023A	3018	28C137F	3018	28C206A
3018	SB-1419	3018	TR28C384	3018	28C1023B	3018	28C137G	3018	28C206B
3018	SB-3001	3018	TR28C535	3018	28C1023C	3018	28C137GN	3018	28C206C
3018	SB-3005	3018	TR310230	3018	28C1023D	3018	28C137H	3018	28C206D
3018	SB-5002	3018	TR310244	3018	28C1023E	3018	28C137J	3018	28C206E
3018	SB-5003	3018	TR310250	3018	28C1023F	3018	28C137K	3018	28C206F
3018	SB-5006	3018	TR36	3018	28C1023G	3018	28C137L	3018	28C206G
3018	SB-5020	3018	TR38	3018	28C1023GN	3018	28C137M	3018	28C206GN
3018	SB-5021	3018	TR5320326	3018	28C1023H	3018	28C137OR	3018	28C206H
3018	SB-5050	3018	TRPLC711	3018	28C1023J	3018	28C137R	3018	28C206J
3018	SB1001	3018	TSC614	3018	28C1023K	3018	28C137X	3018	28C206K
3018	SB3001	3018	TT-204	3018	28C1023L	3018	28C137Y	3018	28C206L
3018	SB3002	3018	TT-204A	3018	28C1023M	3018	28C137I	3018	28C206M
3018	SB5001	3018	TT-204AB	3018	28C1023OR	3018	28C1390WI	3018	28C206OR
3018	SB5002	3018	TT-204C	3018	28C1023R	3018	28C1417VP	3018	28C206R
3018	SB5004	3018	TV-15	3018	28C1023X	3018	28C1424	3018	28C206X
3018	SB5020	3018	TV-20	3018	28C1023Y	3018	28C148A	3018	28C206Y
3018	SB5021	3018	TV-32	3018	28C1026	3018	28C148B	3018	28C287A
3018	SB5023	3018	TV116	3018	28C1026A	3018	28C148D	3018	28C324B
3018	SB5024	3018	TV15A	3018	28C1026C	3018	28C148E	3018	28C324C
3018	SB5030	3018	TV15C	3018	28C1026D	3018	28C148F		
3018	SB5050	3018	TV16	3018	28C1026F				
3018	SB5055			3018	28C1026G				
3018	SB5056			3018	28C1026GN				
3018	SB6006								

SK	Industry Standard No.	SK	Industry Standard No.	SK	Industry Standard No.	SK	Industry Standard No.	SK	Industry Standard No.
3018	2SC324D	3018	2SC399C	3018	2SC466A	3018	2SC561H	3018	2SC657H
3018	2SC324E	3018	2SC399D	3018	2SC466B	3018	2SC561J	3018	2SC657X
3018	2SC324F	3018	2SC399E	3018	2SC466C	3018	2SC561K	3018	2SC657Y
3018	2SC324G	3018	2SC399F	3018	2SC466D	3018	2SC561L	3018	2SC662A
3018	2SC324GN	3018	2SC399G	3018	2SC466E	3018	2SC561M	3018	2SC674
3018	2SC324J	3018	2SC399GN	3018	2SC466F	3018	2SC561N	3018	2SC674V
3018	2SC324K	3018	2SC399H	3018	2SC466G	3018	2SC561OR	3018	2SC683
3018	2SC324L	3018	2SC399J	3018	2SC466GN	3018	2SC561R	3018	2SC683A
3018	2SC324M	3018	2SC399K	3018	2SC466J	3018	2SC561X	3018	2SC683B
3018	2SC324OR	3018	2SC399L	3018	2SC466K	3018	2SC561Y	3018	2SC683C
3018	2SC324R	3018	2SC399M	3018	2SC466L	3018	2SC562	3018	2SC683D
3018	2SC324X	3018	2SC399OR	3018	2SC466M	3018	2SC562-0	3018	2SC683E
3018	2SC324Y	3018	2SC399R	3018	2SC466OR	3018	2SC562A	3018	2SC683F
3018	2SC335	3018	2SC399X	3018	2SC466R	3018	2SC562C	3018	2SC683G
3018	2SC337	3018	2SC399Y	3018	2SC466X	3018	2SC562C	3018	2SC683GN
3018	2SC337B	3018	2SC4033A	3018	2SC466Y	3018	2SC562F	3018	2SC683H
3018	2SC349R	3018	2SC4033B	3018	2SC469	3018	2SC562G	3018	2SC683J
3018	2SC361	3018	2SC4033C	3018	2SC469A	3018	2SC562GN	3018	2SC683K
3018	2SC361A	3018	2SC4033D	3018	2SC469B	3018	2SC562H	3018	2SC683L
3018	2SC361B	3018	2SC4033F	3018	2SC469C	3018	2SC562J	3018	2SC683M
3018	2SC361C	3018	2SC4033G	3018	2SC469D	3018	2SC562K	3018	2SC683OR
3018	2SC361D	3018	2SC4033GN	3018	2SC469E	3018	2SC562L	3018	2SC683R
3018	2SC361E	3018	2SC4033H	3018	2SC469G	3018	2SC562M	3018	2SC683V
3018	2SC361F	3018	2SC4033J	3018	2SC469GN	3018	2SC562OR	3018	2SC683X
3018	2SC361G	3018	2SC4033K	3018	2SC469H	3018	2SC562R	3018	2SC683Y
3018	2SC361GN	3018	2SC4033L	3018	2SC469J	3018	2SC562X	3018	2SC688
3018	2SC361H	3018	2SC4033M	3018	2SC469K	3018	2SC562Y	3018	2SC688A
3018	2SC361J	3018	2SC4033OR	3018	2SC469L	3018	2SC563	3018	2SC688B
3018	2SC361K	3018	2SC4033R	3018	2SC469M	3018	2SC563A	3018	2SC688C
3018	2SC361L	3018	2SC4033X	3018	2SC469OR	3018	2SC563C	3018	2SC688D
3018	2SC361M	3018	2SC4033Y	3018	2SC469Q	3018	2SC563D	3018	2SC688E
3018	2SC361OR	3018	2SC404A	3018	2SC469R	3018	2SC563E	3018	2SC688F
3018	2SC361R	3018	2SC404B	3018	2SC469X	3018	2SC563F	3018	2SC688G
3018	2SC361X	3018	2SC404C	3018	2SC469Y	3018	2SC563G	3018	2SC688GN
3018	2SC361Y	3018	2SC404D	3018	2SC472	3018	2SC563GN	3018	2SC688H
3018	2SC370-0	3018	2SC404E	3018	2SC472A	3018	2SC563H	3018	2SC688J
3018	2SC370A	3018	2SC404F	3018	2SC472B	3018	2SC563J	3018	2SC688K
3018	2SC370B	3018	2SC404G	3018	2SC472C	3018	2SC563K	3018	2SC688L
3018	2SC370C	3018	2SC404GN	3018	2SC472D	3018	2SC563M	3018	2SC688M
3018	2SC370D	3018	2SC404H	3018	2SC472E	3018	2SC563OR	3018	2SC688OR
3018	2SC370E	3018	2SC404J	3018	2SC472F	3018	2SC563R	3018	2SC688R
3018	2SC370GN	3018	2SC404K	3018	2SC472G	3018	2SC563X	3018	2SC688X
3018	2SC370L	3018	2SC404L	3018	2SC472GN	3018	2SC563Y	3018	2SC688Y
3018	2SC370M	3018	2SC404M	3018	2SC472H	3018	2SC605	3018	2SC70
3018	2SC370OR	3018	2SC404OR	3018	2SC472J	3018	2SC605B	3018	0002SC710B
3018	2SC370R	3018	2SC404R	3018	2SC472K	3018	2SC605C	3018	0002SC710C
3018	2SC370X	3018	2SC404X	3018	2SC472L	3018	2SC605D	3018	2SC736R
3018	2SC370Y	3018	2SC404Y	3018	2SC472M	3018	2SC605E	3018	2SC738F
3018	2SC371D	3018	2SC429A	3018	2SC472OR	3018	2SC605G	3018	2SC739
3018	2SC372OA	3018	2SC429B	3018	2SC472R	3018	2SC605GN	3018	2SC739A
3018	2SC372OB	3018	2SC429C	3018	2SC472X	3018	2SC605H	3018	2SC739B
3018	2SC372OC	3018	2SC429D	3018	2SC472Y	3018	2SC605J	3018	2SC739D
3018	2SC372OD	3018	2SC429E	3018	2SC478	3018	2SC605L	3018	2SC739F
3018	2SC372OF	3018	2SC429F	3018	2SC529B	3018	2SC605OR	3018	2SC739G
3018	2SC372OG	3018	2SC429G	3018	2SC529C	3018	2SC605Q	3018	2SC739GN
3018	2SC372OGN	3018	2SC429GN	3018	2SC529D	3018	2SC605R	3018	2SC739H
3018	2SC372OH	3018	2SC429H	3018	2SC529E	3018	2SC605X	3018	2SC739J
3018	2SC372OJ	3018	2SC429K	3018	2SC529F	3018	2SC606A	3018	2SC739K
3018	2SC372OK	3018	2SC429L	3018	2SC529G	3018	2SC606B	3018	2SC739L
3018	2SC372OL	3018	2SC429M	3018	2SC529GN	3018	2SC606C	3018	2SC739M
3018	2SC372OX	3018	2SC429OR	3018	2SC529H	3018	2SC606D	3018	2SC739M
3018	2SC372OY	3018	2SC429R	3018	2SC529J	3018	2SC606E	3018	2SC739OR
3018	2SC371	3018	2SC429Y	3018	2SC529K	3018	2SC606F	3018	2SC739R
3018	2SC378A	3018	2SC430A	3018	2SC529L	3018	2SC606G	3018	2SC739T
3018	2SC378B	3018	2SC430B	3018	2SC529M	3018	2SC606GN	3018	2SC739Y
3018	2SC378C	3018	2SC430C	3018	2SC529OR	3018	2SC606H	3018	2SC74A
3018	2SC378D	3018	2SC430D	3018	2SC529R	3018	2SC606J	3018	2SC74B
3018	2SC378E	3018	2SC430E	3018	2SC529X	3018	2SC606K	3018	2SC74C
3018	2SC378F	3018	2SC430F	3018	2SC529Y	3018	2SC606L	3018	2SC74D
3018	2SC378G	3018	2SC430G	3018	2SC543	3018	2SC606OR	3018	2SC74E
3018	2SC378GN	3018	2SC430GN	3018	2SC543A	3018	2SC606R	3018	2SC74F
3018	2SC378H	3018	2SC430H	3018	2SC543B	3018	2SC606X	3018	2SC74G
3018	2SC378J	3018	2SC430J	3018	2SC543C	3018	2SC606Y	3018	2SC74GN
3018	2SC378K	3018	2SC430K	3018	2SC543D	3018	2SC629	3018	2SC74H
3018	2SC378L	3018	2SC430L	3018	2SC543E	3018	2SC629A	3018	2SC74J
3018	2SC378M	3018	2SC430M	3018	2SC543G	3018	2SC629B	3018	2SC74K
3018	2SC378OR	3018	2SC430OR	3018	2SC543GN	3018	2SC629C	3018	2SC74L
3018	2SC378R	3018	2SC430R	3018	2SC543H	3018	2SC629D	3018	2SC74M
3018	2SC378X	3018	2SC430X	3018	2SC543J	3018	2SC629E	3018	2SC74P
3018	2SC378T	3018	2SC430Y	3018	2SC543K	3018	2SC629F	3018	2SC74X
3018	2SC380O	3018	2SC455	3018	2SC543L	3018	2SC629GN	3018	2SC74Y
3018	2SC380OA	3018	2SC455A	3018	2SC543M	3018	2SC629H	3018	2SC758OR
3018	2SC380OB	3018	2SC455B	3018	2SC543OR	3018	2SC629J	3018	2SC771
3018	2SC380OD	3018	2SC455C	3018	2SC543R	3018	2SC629K	3018	2SC771A
3018	2SC380OE	3018	2SC455D	3018	2SC543X	3018	2SC629L	3018	2SC771B
3018	2SC380OG	3018	2SC455E	3018	2SC543Y	3018	2SC629M	3018	2SC771BX
3018	2SC380OGN	3018	2SC455F	3018	2SC544A	3018	2SC629OR	3018	2SC771C
3018	2SC380OH	3018	2SC455G	3018	2SC544AG	3018	2SC629X	3018	2SC771D
3018	2SC380OJ	3018	2SC455GN	3018	2SC544B	3018	2SC638C	3018	2SC771E
3018	2SC380OK	3018	2SC455H	3018	2SC544F	3018	2SC645B-1	3018	2SC771F
3018	2SC380OL	3018	2SC455J	3018	2SC544G	3018	2SC645C	3018	2SC771G
3018	2SC380OM	3018	2SC455K	3018	2SC544GN	3018	2SC645D	3018	2SC771GN
3018	2SC380OX	3018	2SC455L	3018	2SC544H	3018	2SC645E	3018	2SC771H
3018	2SC380OY	3018	2SC455M	3018	2SC544J	3018	2SC645F	3018	2SC771J
3018	2SC394O	3018	2SC455OR	3018	2SC544K	3018	2SC645G	3018	2SC771K
3018	2SC394OA	3018	2SC455R	3018	2SC544L	3018	2SC645GN	3018	2SC771L
3018	2SC394OB	3018	2SC455X	3018	2SC544M	3018	2SC645H	3018	2SC771M
3018	2SC394OC	3018	2SC455Y	3018	2SC544OR	3018	2SC645J	3018	2SC771OR
3018	2SC394OD	3018	2SC464	3018	2SC544R	3018	2SC645K	3018	2SC771R
3018	2SC394OG	3018	2SC464A	3018	2SC544X	3018	2SC645L	3018	2SC771X
3018	2SC394OF	3018	2SC464B	3018	2SC544Y	3018	2SC645M	3018	2SC771Y
3018	2SC394OGN	3018	2SC464D	3018	2SC545	3018	2SC645OR	3018	2SC786
3018	2SC394OH	3018	2SC464E	3018	2SC545A	3018	2SC645R	3018	2SC786A
3018	2SC394OJ	3018	2SC464F	3018	2SC545G	3018	2SC645X	3018	2SC786B
3018	2SC394OK	3018	2SC464GN	3018	2SC545C	3018	2SC645Y	3018	2SC786C
3018	2SC394OL	3018	2SC464H	3018	2SC545D	3018	2SC657	3018	2SC786D
3018	2SC394OM	3018	2SC464J	3018	2SC545E	3018	2SC657A	3018	2SC786E
3018	2SC394OR	3018	2SC464K	3018	2SC55A	3018	2SC657B	3018	2SC786F
3018	2SC394OX	3018	2SC464L	3018	2SC55B	3018	2SC657C	3018	2SC786G
3018	2SC394OY	3018	2SC464M	3018	2SC55C	3018	2SC657D	3018	2SC786GN
3018	2SC398	3018	2SC464OR	3018	2SC55D	3018	2SC657E	3018	2SC786H
3018	2SC398(FA-1)	3018	2SC464R	3018	2SC55P	3018	2SC657F	3018	2SC786J
3018	2SC398A	3018	2SC464X	3018	2SC55Q	3018	2SC657G	3018	2SC786K
3018	2SC398B	3018	2SC464Y	3018	2SC55GN	3018	2SC657GN	3018	2SC786L
3018	2SC398C	3018	2SC465	3018	2SC55H	3018	2SC657H	3018	2SC786M
3018	2SC398D	3018	2SC465A	3018	2SC55J	3018	2SC657J	3018	2SC786OR
3018	2SC398E	3018	2SC465B	3018	2SC55K	3018	2SC657K	3018	2SC786R
3018	2SC398F	3018	2SC465C	3018	2SC55L	3018	2SC657L	3018	2SC786X
3018	2SC398G	3018	2SC465D	3018	2SC55M	3018	2SC657M	3018	2SC786Y
3018	2SC398GN	3018	2SC465E	3018	2SC55OR	3018	2SC657OR	3018	2SC828B
3018	2SC398H	3018	2SC465F	3018	2SC55R	3018	2SC657R	3018	2SC837
3018	2SC398J	3018	2SC465G	3018	2SC55X	3018	2SC657C	3018	2SC837A
3018	2SC398K	3018	2SC465GN	3018	2SC55Y	3018	2SC657D	3018	2SC837B
3018	2SC398L	3018	2SC465J	3018	2SC56	3018	2SC657E	3018	2SC837C
3018	2SC398M	3018	2SC465K	3018	2SC561A	3018	2SC657F	3018	2SC837D
3018	2SC398OR	3018	2SC465L	3018	2SC561B	3018	2SC657G	3018	2SC837E
3018	2SC398R	3018	2SC465M	3018	2SC561C	3018	2SC657GN	3018	2SC837F
3018	2SC398X	3018	2SC465OR	3018	2SC561D	3018	2SC657H	3018	2SC837G
3018	2SC398Y	3018	2SC465X	3018	2SC561E	3018	2SC657J	3018	2SC837GN
3018	2SC399	3018	2SC465R	3018	2SC561F	3018	2SC657K	3018	2SC837L
3018	2SC399A	3018	2SC465Y	3018	2SC561G	3018	2SC657M	3018	2SC837H
3018	2SC399B	3018	2SC466	3018	2SC561GN	3018	2SC657OR	3018	2SC837J

SK	Industry Standard No.
3018	2SC837K
3018	2SC837L
3018	2SC837M
3018	2SC837OR
3018	2SC837R
3018	2SC837X
3018	2SC837Y
3018	2SC918
3018	2SC918A
3018	2SC918B
3018	2SC918C
3018	2SC918D
3018	2SC918E
3018	2SC918F
3018	2SC918G
3018	2SC918GN
3018	2SC918H
3018	2SC918J
3018	2SC918K
3018	2SC918L
3018	2SC918M
3018	2SC918OR
3018	2SC918R
3018	2SC918X
3018	2SC918Y
3018	2SC920
3018	2SC920-OQ
3018	2SC920A
3018	2SC920B
3018	2SC920C
3018	2SC920CL
3018	2SC920D
3018	2SC920E
3018	2SC920F
3018	2SC920G
3018	2SC920GN
3018	2SC920H
3018	2SC920L
3018	2SC920M
3018	2SC920OR
3018	2SC920X
3018	2SC920Y
3018	2SC921
3018	2SC921A
3018	2SC921B
3018	2SC921C
3018	2SC921CL
3018	2SC921D
3018	2SC921E
3018	2SC921F
3018	2SC921G
3018	2SC921GN
3018	2SC921H
3018	2SC921J
3018	2SC921K
3018	2SC921L
3018	2SC921M
3018	2SC921OR
3018	2SC921R
3018	2SC921W
3018	2SC921X
3018	2SC921Y
3018	2SC922
3018	2SC922L
3018	2SC922M
3018	2SC971
3018	2SC997
3018	2SC997A
3018	2SC997B
3018	2SC997C
3018	2SC997D
3018	2SC997E
3018	2SC997F
3018	2SC997G
3018	2SC997GN
3018	2SC997H
3018	2SC997J
3018	2SC997K
3018	2SC997L
3018	2SC997M
3018	2SC997OR
3018	2SC997R
3018	2SC997X
3018	2SC997Y
3018	2SCF1A
3018	2SCF1B
3018	2SCF1C
3018	2SCF1D
3018	2SCF1E
3018	2SCF1F
3018	2SCF1G
3018	2SCF1GN
3018	2SCF1H
3018	2SCF1J
3018	2SCF1K
3018	2SCF1L
3018	2SCF1M
3018	2SCF1OR
3018	2SCF1R
3018	2SCF1X
3018	2SCF1Y
3018	2SCF2A
3018	2SCF2B
3018	2SCF2C
3018	2SCF2D
3018	2SCF2E
3018	2SCF2G
3018	2SCF2GN
3018	2SCF2H
3018	2SCF2J
3018	2SCF2K
3018	2SCF2L
3018	2SCF2M
3018	2SCF2R
3018	2SCF2X
3018	2SCF2Y
3018	2SCM39J
3018	2SCM39X
3018	2SCS429
3018	2SCS429A
3018	2SCS429B
3018	2SCS429C
3018	2SCS429D
3018	2SCS429E
3018	2SCS429F
3018	2SCS429G
3018	2SCS429GN
3018	2SCS429H
3018	2SCS429K
3018	2SCS429L
3018	2SCS429M
3018	2SCS429OR
3018	2SCS429R
3018	2SCS429X
3018	2SCS429Y
3018	2SCS430
3018	2SCS430A
3018	2SCS430B
3018	2SCS430C
3018	2SCS430D
3018	2SCS430E
3018	2SCS430F
3018	2SCS430GN
3018	2SCS430J
3018	2SCS430K
3018	2SCS430L
3018	2SCS430M
3018	2SCS430R
3018	2SCS430X
3018	2SCS430Y
3018	2SCS461
3018	2SCS461A
3018	2SCS461B
3018	2SCS461C
3018	2SCS461D
3018	2SCS461E
3018	2SCS461F
3018	2SCS461G
3018	2SCS461GN
3018	2SCS461H
3018	2SCS461J
3018	2SCS461K
3018	2SCS461L
3018	2SCS461M
3018	2SCS461OR
3018	2SCS461R
3018	2SCS461X
3018	2SCS461Y
3018	2SCS469
3018	2SCS469A
3018	2SCS469B
3018	2SCS469C
3018	2SCS469D
3018	2SCS469E
3018	2SCS469F
3018	2SCS469G
3018	2SCS469GN
3018	2SCS469H
3018	2SCS469J
3018	2SCS469K
3018	2SCS469L
3018	2SCS469M
3018	2SCS469R
3018	2SCS469X
3018	2SCS469Y
3018	003-01
3018	03A03
3018	3L4-6007-10
3018	3L4-6007-11
3018	3L4-6007-12
3018	3L4-6007-13
3018	3L4-6007-14
3018	3L4-6007-19
3018	3L4-6007-20
3018	3L4-6007-21
3018	3L4-6007-22
3018	3L4-6007-23
3018	3L4-6007-35
3018	04-00461-02
3018	04-00535-02
3018	04-00535-06
3018	04-01585-08
3018	04-15850-06
3018	04-46000-02
3018	4-850
3018	07-07129
3018	07-07166
3018	7-23
3018	7-59-019/3477
3018	7-59-020/3477
3018	7-59-021/3477
3018	7-59-023/3477
3018	7-8
3018	09-3002006
3018	09-302002
3018	09-302006
3018	09-302016
3018	09-302037
3018	09-302063
3018	09-302073
3018	09-302079
3018	09-302086
3018	09-302095
3018	09-302114
3018	09-302128
3018	09-302138
3018	09-302142
3018	09-302143
3018	09-302151
3018	09-302191
3018	09-302201
3018	09-302206
3018	09-302216
3018	09-302224
3018	09-302241
3018	09-304042
3018	09-304043
3018	09-304048
3018	09-305011
3018	09-305033
3018	09-305041
3018	09-305050
3018	09-305051
3018	09-305093
3018	09-305094
3018	09-309006
3018	09-309065
3018	09-309069
3018	10-2SA49
3018	10-2SC380
3018	12X047
3018	13-0009
3018	13-0010
3018	13-0020
3018	13-0040
3018	13-0063
3018	13-10321-29
3018	13-10321-31
3018	13-10321-32
3018	13-10321-46
3018	13-14085-16
3018	13-14085-17
3018	13-14085-26
3018	13-14085-3
3018	13-14085-4
3018	13-15808-1
3018	13-15835-1
3018	13-23013-2
3018	13-23160-2
3018	13-23160-3
3018	13-23309-5
3018	13-26577-1
3018	13-29947-1
3018	13-31013-1
3018	13-31013-5
3018	13-32362-1
3018	14-602-31
3018	14-602-34
3018	14-602-46
3018	14-602-77
3018	14-602-77B
3018	14-603-02
3018	14-603-02A
3018	14-603-05-2
3018	16-21426
3018	019-003929
3018	19-020-070
3018	020-00024
3018	020-00025
3018	020-00028
3018	21A040-045
3018	21A040-046
3018	21A040-047
3018	21A040-053
3018	21A040-054
3018	21A040-055
3018	21A040-063
3018	21A040-091
3018	21A112-007
3018	21A112-084
3018	21M091
3018	21M093
3018	21M094
3018	21M095
3018	21M099
3018	21M140
3018	21M151
3018	21M152
3018	21M153
3018	21M154
3018	21M161
3018	21M178
3018	21M179
3018	21M188
3018	21M387
3018	21M577
3018	24MW1022
3018	24MW1057
3018	24MW1058
3018	24MW1081
3018	24MW1106
3018	24MW287
3018	24MW361
3018	24MW535
3018	24MW593
3018	24MW595
3018	24MW653
3018	24MW656
3018	24MW673
3018	24MW700
3018	24MW737
3018	24MW738
3018	24MW793
3018	24MW812
3018	24MW813
3018	24MW814
3018	24MW815
3018	24MW863
3018	24MW953
3018	24MW957
3018	24MW990
3018	025-100012
3018	025-100036
3018	025-100037
3018	025-100038
3018	25B-1
3018	25B2
3018	31-0048
3018	31-0049
3018	31-0050
3018	31-0052
3018	31-0054
3018	31-0098
3018	34-6001-12
3018	34-6001-64
3018	34-6001-79
3018	34-6015-16
3018	34-6015-17
3018	34-6015-18
3018	34-6015-19
3018	34-6015-20
3018	34-6015-22
3018	34-6015-25
3018	34-6015-37
3018	34-6015-47
3018	34-6015-48
3018	34-6015-50
3018	34-6015-51
3018	34-6015-52
3018	34-6015-8
3018	34-6015-9
3018	41N
3018	42-21401
3018	42-21402
3018	42-22532
3018	42-22785
3018	42-23960P
3018	42-23962P
3018	42-23963P
3018	42-27372
3018	42-27373
3018	42-28205
3018	43-025764
3018	46-119-3
3018	46-86107-3
3018	46-86108-3
3018	46-86112-3
3018	46-86114-3
3018	46-86117-3
3018	46-86118-3
3018	46-86119-3
3018	46-86120-3
3018	46-86126-3
3018	46-86127-3
3018	46-86131-3
3018	46-86172-3
3018	46-86207-3
3018	46-86209-3
3018	46-86224-3
3018	46-86239-3
3018	46-86240-3
3018	46-86251-3
3018	46-86252-3
3018	46-86265-3
3018	46-86285-3
3018	46-86295-3
3018	46-86299-3
3018	46-86301-3
3018	46-86302-3
3018	46-86352-3
3018	46-86353-3
3018	46-86354-3
3018	46-86357-3
3018	46-86434-3
3018	46-86435-3
3018	46-8647
3018	46-86538-3
3018	48-134190A16
3018	48-134805
3018	48-134825
3018	48-134857
3018	48-137351
3018	48-137491
3018	48-137612
3018	48-155087
3018	48-3003A03
3018	48-41815J02
3018	48-41816J01
3018	48-41816J02
3018	48-43992J01
3018	48M355002
3018	48M355052
3018	48M355053
3018	48P63076A81
3018	48P63076A82
3018	48P63077A31
3018	48P63078A52
3018	48P63078A70
3018	48P63079A97
3018	48P63082A25
3018	48P63082A27
3018	48P65146A61
3018	48P65146A62
3018	48S134773
3018	48S134774
3018	48S134775
3018	48S134776
3018	48S134783
3018	48S134784
3018	48S134785
3018	48S134804
3018	48S134805
3018	48S134807
3018	48S134825
3018	48S134826
3018	48S134827
3018	48S134837
3018	48S134840
3018	48S134844
3018	48S134845
3018	48S134855
3018	48S134857
3018	48S134879
3018	48S134894
3018	48S134904
3018	48S134922
3018	48S134925
3018	48S134926
3018	48S134932
3018	48S134937
3018	48S134948
3018	48S134949
3018	48S134950
3018	48S134961
3018	48S134962
3018	48S134963
3018	48S134964
3018	48S134981
3018	48S137003
3018	48S137110
3018	48S137158
3018	48S137190
3018	48S137191
3018	48S137192
3018	48S137350
3018	48S137351
3018	48S137386
3018	48S137543
3018	48S155087
3018	48X90232A03
3018	48X90232A10
3018	48X90232A17
3018	48X90232A18
3018	48X90232A19
3018	48X97046A17
3018	48X97046A18
3018	48X97046A19
3018	48X97046A20
3018	48X97046A21
3018	48X97177A03
3018	48X97177A13
3018	50C380-0
3018	50C394-0
3018	50C394-R
3018	50C784-R
3018	51N3M
3018	56-4826
3018	56-8086
3018	56-8086A
3018	56-8086B
3018	56-8086C
3018	56-8087
3018	56-8087B
3018	56-8087C
3018	56-8088
3018	56-8088A
3018	56-8088C
3018	57A126-12
3018	57A142-2
3018	57A143-12
3018	57A144-12
3018	57A164-4
3018	57A177-12
3018	57A21-12
3018	57A21-13
3018	57A21-14
3018	57A21-17
3018	57A21-45
3018	57A249-4
3018	57B102-4
3018	57B103-4
3018	57B141-4
3018	57B142-4
3018	57B21-6
3018	57B21-7
3018	57B21-9
3018	69-1810
3018	69-1811
3018	69-1812
3018	69-1813
3018	73N1
3018	81-46125007-8
3018	81-46125012-8
3018	81-46125013-6
3018	81-46125030-0
3018	81-46125032-6
3018	81-46125033-4
3018	86-100002
3018	86-100003
3018	86-100004
3018	86-100006
3018	86-100007
3018	86-185-2
3018	86-204-2
3018	86-389-2
3018	86-442-2
3018	86-465-2
3018	86-515-2
3018	86-526-2
3018	86-593-2
3018	86-593-9
3018	86-594-2
3018	86-596-2
3018	86-597-2
3018	86-619-2
3018	86-620-2
3018	86-621-2
3018	86X0019-001
3018	86X0060-001
3018	86X0061-001
3018	87-0002
3018	90-180
3018	90-45
3018	90-452
3018	90-453
3018	90-457
3018	90-60
3018	90-601
3018	90-602
3018	90-604
3018	90-612
3018	91N1
3018	99S031
3018	99S032
3018	99S055
3018	99S056
3018	99S067-1
3018	100N1
3018	100N1P
3018	102-0394-25
3018	102-0454-02
3018	102-0460-02
3018	102-0461-02
3018	102-0535-02
3018	102-1047-03
3018	102-1342-02
3018	0103-0568
3018	0103-0568/4460
3018	105-001-05
3018	105-00107-09
3018	105-02005-07
3018	105-02006-05
3018	105-02008-01
3018	105-904-85
3018	105-904-86
3018	105-904-87
3018	106-001
3018	106-002
3018	106-351
3018	115-13
3018	121-377
3018	121-378
3018	121-379
3018	121-380
3018	121-453
3018	121-462
3018	121-470
3018	121-471
3018	121-500
3018	121-502
3018	121-505
3018	121-506
3018	121-510
3018	121-521
3018	121-523
3018	121-526
3018	121-580
3018	121-600
3018	121-723
3018	121-732
3018	121-735
3018	121-779
3018	121-849
3018	121-856
3018	121-883
3018	121-885
3018	121-951
3018	121-968
3018	121-974
3018	122-7
3018	128N2
3018	129-27
3018	139N1
3018	139N2
3018	142N1
3018	142N1P
3018	145N1
3018	145N1P
3018	150N1
3018	150N2
3018	150N3
3018	176-026-9-001
3018	176-042-9-001
3018	176-042-9-003
3018	176-056-9-001
3018	176-056-9-003
3018	176-060-9-002
3018	176-072-9-005
3018	176-073-9-001
3018	176-074-9-001
3018	185-002
3018	186-001
3018	229-0180-119
3018	229-0180-32
3018	229-0180-33
3018	229-0190-31
3018	229-0191-29
3018	229-0191-30
3018	229-0260-18
3018	229-5100-31V

SK	Industry Standard No.
3018	229-5100-32V
3018	250-0380
3018	260P1601
3018	260P16101
3018	260P17601
3018	260P70403
3018	260P70501
3018	325-1378-18
3018	325-1378-19
3018	344-6015-8
3018	344-6015-9
3018	366-1
3018	576-0003-029
3018	612-16
3018	613
3018	614-12
3018	634-1
3018	642-229
3018	680-1
3018	690V010H40
3018	690V080H36
3018	690V088H46
3018	690V088H47
3018	690V088H49
3018	690V089H46
3018	690V089H86
3018	690V103H28
3018	690V109H46
3018	690V116H20
3018	690V120H89
3018	0701
3018	715EN
3018	753-2000-007
3018	753-2000-460
3018	753-4000-668
3018	753-4000-929
3018	753-9000-839
3018	753-9001-674
3018	753-9020-784
3018	753-9050-785
3018	772A
3018	772B
3018	772BJ
3018	772BL
3018	772BM
3018	772BN
3018	772BY
3018	772C
3018	772D
3018	772E
3018	772F
3018	772G
3018	800-53600
3018	800-557-00
3018	824-1
3018	902-001-7-18
3018	916-31024-5
3018	916-31025-4
3018	921-1013
3018	921-102
3018	921-102A
3018	921-102B
3018	921-114B
3018	921-115B
3018	921-116B
3018	921-141B
3018	921-142B
3018	921-143B
3018	921-145B
3018	921-176B
3018	921-177B
3018	921-181B
3018	921-210B
3018	921-211B
3018	921-213B
3018	921-226B
3018	921-232B
3018	921-233B
3018	921-235B
3018	921-264B
3018	921-265
3018	921-265B
3018	921-301B
3018	921-312B
3018	921-349B
3018	921-350B
3018	921-351
3018	921-379
3018	921-56B
3018	921-57B
3018	921-58B
3018	921-59B
3018	921-62
3018	921-62A
3018	921-62B
3018	921-63
3018	921-63A
3018	921-63B
3018	921-64
3018	921-64A
3018	921-64B
3018	921-64C
3018	921-84
3018	921-84A
3018	921-84B
3018	921-85
3018	921-85A
3018	921-85B
3018	921-86
3018	921-86A
3018	921-86B
3018	921-92B
3018	929C
3018	930B
3018	930D
3018	930DX
3018	930E
3018	930EX
3018	930X3
3018	947-1
3018	991-003304
3018	1000-135
3018	1002-02
3018	1002-68
3018	1003-01
3018	1010-8066
3018	1013-15
3018	1016-85
3018	1024G
3018	1025G
3018	1048-9904
3018	1049-0060
3018	1063-5381
3018	1077-07
3018	1081-3087
3018	1081-3301
3018	1207
3018	1373-17
3018	1373-17A
3018	1402E
3018	1420-1-1
3018	1455-7-4
3018	1501
3018	1502B
3018	1502D
3018	1931-17
3018	1998-17
3018	1999-17
3018	2000-204
3018	2000-205
3018	2000-214
3018	2022-03
3018	2022-244
3018	2048-17
3018	2057A2-109
3018	2057A2-116
3018	2057A2-117
3018	2057A2-120
3018	2057A2-128
3018	2057A2-157
3018	2057A2-158
3018	2057A2-159
3018	2057A2-207
3018	2057A2-216
3018	2057A2-217
3018	2057A2-218
3018	2057A2-220
3018	2057A2-221
3018	2057A2-237
3018	2057A2-257
3018	2057A2-273
3018	2057A2-274
3018	2057A2-275
3018	2057A2-304
3018	2057A2-305
3018	2057A2-322
3018	2057A2-323
3018	2057A2-325
3018	2057A2-326
3018	2057A2-331
3018	2057A2-356
3018	2057A2-395
3018	2057A2-432
3018	2057A2-454
3018	2057A2-466
3018	2057A2-475
3018	2057A2-477
3018	2057A2-478
3018	2057A2-501
3018	2057A2-502
3018	2057A2-503
3018	2057A2-527
3018	2057A2-539
3018	2057A2-540
3018	2057A2-558
3018	2057A2-559
3018	2057B-113
3018	2057B2-101
3018	2057B2-102
3018	2057B2-116
3018	2057B2-117
3018	2057B2-125
3018	2057B2-138
3018	2057B2-139
3018	2127-17
3018	2197-17
3018	2212-17
3018	2213-17
3018	2224-17
3018	2226-17
3018	2284-17
3018	2291-17
3018	2448-17
3018	2450
3018	2473
3018	2476
3018	2477
3018	2495-520
3018	2495-521
3018	2510-101
3018	2510-102
3018	2634
3018	2904-053
3018	3003
3018	3029
3018	3508
3018	3510
3018	3524
3018	3524-2
3018	3527
3018	3527-1
3018	3568
3018	3571-1
3018	3652
3018	3652-2
3018	3657-1
3018	3657-2
3018	4167
3018	4168
3018	4169
3018	4684-120-3
3018	4801-0000-003
3018	4801-0000-016
3018	4801-0000-035
3018	4825
3018	4857
3018	5001-032
3018	5001-542
3018	5001-543
3018	5613-1342
3018	5613-1342C
3018	5613-460
3018	5613-460A
3018	5613-460B
3018	5613-460C
3018	5613-461
3018	5613-461C
3018	5613-535
3018	5613-535A
3018	5613-535B
3018	5613-535C
3018	7112
3018	7116
3018	7117
3018	7233
3018	7233S
3018	7236
3018	7236S
3018	7261
3018	7264
3018	7638
3018	7639
3018	8000-00004-298
3018	8000-00005-003
3018	8000-00006-005
3018	8000-00009-177
3018	8000-00028-037
3018	8000-00035-003
3018	8000-00041-046
3018	8000-00043-019
3018	8010-174
3018	8010-175
3018	8440-122
3018	8840-162
3018	8910-142
3018	9011F
3018	9011G
3018	9011H
3018	9016G
3018	9018F
3018	9018G
3018	9033G
3018	9410A
3018	9426B
3018	9426C
3018	9623
3018	28810-172
3018	28810-173
3018	38510-162
3018	40234
3018	40235
3018	40236
3018	40237
3018	40244
3018	40245
3018	40478
3018	40479
3018	45810-162
3018	45810-163
3018	55170-1
3018	55810-162
3018	55810-164
3018	58840-192
3018	61009-1
3018	61009-2
3018	61755
3018	000073230
3018	000073231
3018	000073332
3018	082020
3018	88060-142
3018	88510-172
3018	95170-1
3018	101434
3018	116199
3018	118822
3018	119414
3018	119554
3018	119555
3018	119556
3018	119557
3018	119635
3018	122902
3018	122904
3018	124024
3018	124623
3018	124624
3018	125141
3018	125329
3018	126025
3018	126704
3018	129050
3018	129392
3018	129393
3018	129698
3018	130278
3018	130403-04
3018	131545
3018	134142
3018	134144
3018	136168
3018	136240
3018	136430
3018	138946
3018	165392
3018	165931
3018	165932
3018	166906
3018	168567
3018	169194
3018	169196
3018	170388
3018	170753-1
3018	170756-1
3018	170794
3018	171029
3018	171030
3018	171031
3018	171038
3018	171044
3018	171162-026
3018	171162-186
3018	171162-278
3018	171162-279
3018	171553
3018	171983
3018	172761
3018	183015
3018	183016
3018	183018
3018	183019
3018	185236
3018	489751-037
3018	489751-038
3018	489751-039
3018	489751-047
3018	489751-052
3018	489751-127
3018	489751-128
3018	489751-173
3018	514045
3018	515041S
3018	0573474
3018	0573475
3018	0573485
3018	0573486
3018	0573487
3018	0573487H
3018	0573508
3018	0573509
3018	0573509H
3018	0573510
3018	0573510H
3018	0573511H
3018	0573607
3018	610041-2
3018	610073-1
3018	610100-1
3018	610128-2
3018	610139-1
3018	610142-1
3018	610145-1
3018	610150-1
3018	610150-3
3018	610151-1
3018	610151-3
3018	652072
3018	701678-00
3018	740949
3018	740950
3018	741726
3018	760253
3018	815173
3018	815206
3018	965632
3018	971460
3018	972305
3018	972306
3018	972307
3018	972417
3018	972418
3018	972419
3018	972420
3018	983233
3018	983234
3018	983235
3018	983742
3018	984191
3018	984192
3018	984876
3018	984877
3018	984878
3018	986576
3018	986693
3018	986694
3018	1223781
3018	1471122-6
3018	1471122-7
3018	1472634-1
3018	1473524-1
3018	1473524-2
3018	1473527-1
3018	1473532-1
3018	1473546-1
3018	1473546-2
3018	1473571-1
3018	1473576
3018	1473577-1
3018	1473579-1
3018	1473595-1
3018	1473610-1
3018	1473657-1
3018	1473657-2
3018	1473680-1
3018	1473686-1
3018	1700020
3018	1700032
3018	1700033
3018	2003779-22
3018	2003779-23
3018	2003779-24
3018	2003779-25
3018	2006513-39
3018	2006582-101
3018	2006681-94
3018	2092417-0017
3018	2092417-0018
3018	2092417-0019
3018	2092417-0711
3018	2092417-0712
3018	2092417-0713
3018	2092417-0715
3018	2092417-0716
3018	2092418-0022
3018	2092418-0023
3018	2092418-0024
3018	2092418-0716
3018	2092418-0717
3018	2092418-0718
3018	2092418-0719
3018	2092418-0720
3018	2092418-0721
3018	2093308-0704
3018	2093308-0704A
3018	2093308-0705
3018	2093308-0706
3018	2093308-0725
3018	2320031
3018	2320041
3018	2320041H
3018	2320043
3018	2320141
3018	2320471
3018	2320471-1
3018	2497094-1
3018	2497094-2
3018	2498456-2
3018	2498482-2
3018	2498902-1
3018	2498902-2
3018	2498903-1
3018	2498903-2
3018	3539307-001
3018	3539307-002
3018	3596260
3018	3596261
3018	3596440
3018	3597103
3018	3597104
3018	3597114
3018	3597260
3018	3597261
3018	5320051
3018	5320326H
3018	5320851
3018	5320861
3018	06120025
3018	06120026
3018	6208839
3018	7294910
3018	7297980
3018	7914009-01
3018	7914010-01
3018	8031837
3018	8036683
3018	8037722
3018	8249600
3018	23114007
3018	23114017
3018	23114044
3018	23114078
3018	23114157
3018	23114280
3018	23114282
3018	23114966
3018	23126183
3018	26010026
3018	26010051
3018	30200062
3018	43021067
3018	43027618
3018	43029471
3018	43029472
3018	44089001
3018	62105275
3018	62105291
3018	62563273
3018	62563281
3018	62614765
3018	62636696
3018	62695919
3018	62695927
3018	62789247
3018	70260450
3018	80366830
3018	80366840
3018	83078402
3018	83093005
3018	83167403
3018	83167405
3018	83167503
3018	83167505
3018	89960404
3018	93038040
3018	93039440
3018	93082920
3018	2003038208
3018	2003038812
3018	4100208291
3018	4100208292
3018	4100208293
3018	4100210472
3018	4100213591
3018	4104204603
3018	4104204612
3018	4104205352
3018	4108296047
3018	4108296238
3018	4108296255
3018	4108296257
3018	6621005000
3018	16103190930
3018	16104190668
3018	16106190772
3018	16112190772
3018	16114190772
3019	BF-115
3019	BFY-37
3019	BSY-62
3019	BSY-72
3019	BSY-73
3019	BSY-74
3019	BSY-80
3019	BSY-95
3019	C389
3019	C389-0
3019	C389R
3019	EA15X396
3019	ET15X2
3019	GM308
3019	IRTR83
3019	M4756
3019	S-1019
3019	S-1041
3019	S-2617
3019	S-5328E
3019	S-5670-E
3019	S1041-16GN
3019	S2020
3019	S2038
3019	S2617
3019	S3019
3019	S32669
3019	SDS240
3019	SE-1010
3019	SE-1019
3019	SE-1044
3019	SE-3002
3019	SE-3019
3019	SE-5001
3019	SE1010
3019	SE1019
3019	SE3001Y
3019	SL-100
3019	SPS428
3019	SYL-2300
3019	SYL-4131
3019	T-H2SC313
3019	T-H2SC387
3019	T1S-18
3019	T3530
3019	TC-0918
3019	TI-407
3019	TI-408
3019	TI-409
3019	TI-410
3019	TI-417
3019	TI-418
3019	TI-419
3019	TMT-2427
3019	TNJ-60066
3019	TNJ71173
3019	TR1512-80
3019	TV24387
3019	TV24806
3019	VS-2SC206
3019	VS-2SC288A
3019	VS-2SC645
3019	VS-2SC684
3019	W8
3019	XT15X3
3019	ZDT-30
3019	ZDT-31
3019	ZEN104
3019	2N2865
3019	2N3289
3019	2N3290
3019	2N3291
3019	2N3292
3019	2N3293

SK	Industry Standard No.
3019	2N3294
3019	2N3407
3019	2N917
3019	2SC171
3019	2SC174
3019	2SC174A
3019	2SC186
3019	2SC187
3019	2SC266
3019	2SC268
3019	2SC287
3019	2SC288
3019	2SC288AB
3019	2SC29
3019	2SC313
3019	2SC389
3019	2SC389-0
3019	2SC389A
3019	2SC389B
3019	2SC389C
3019	2SC389D
3019	2SC389E
3019	2SC389F
3019	2SC389G
3019	2SC389GN
3019	2SC389H
3019	2SC389J
3019	2SC389K
3019	2SC389L
3019	2SC389M
3019	2SC389R
3019	2SC389Y
3019	2SC463
3019	2SC948
3019	3S004
3019	15-088003
3019	16GN
3019	16L64
3019	19A115342-2
3019	24T-016
3019	24T-016-010
3019	025-100026
3019	46-8629
3019	46-8677-3
3019	47-2
3019	48P65112A65
3019	48P65113A88
3019	48P65123A67
3019	48P65123A95
3019	48P65144A72
3019	48P65174A24
3019	48S134756
3019	48X134902
3019	57A21-18
3019	90-46
3019	229-0240-25
3019	260P09201
3019	260P10601
3019	991-002232
3019	991-002873
3019	991-013057
3019	991-013068
3019	1420-2-2
3019	3530
3019	3530-2
3019	3617
3019	026237
3019	36203
3019	113398
3019	114267
3019	129604
3019	134417
3019	170906
3019	170906-1
3019	171009
3019	231140-36
3019	231140-37
3019	231140-43
3019	489751-049
3019	489751-058
3019	651955-3
3019	1408649-1
3019	1473529-1
3019	1473617-1
3019	23114037
3020	CDC12108
3020	CDQ10035
3020	CDQ10036
3020	DS-67W
3020	ECG123
3020	EQS-64
3020	HC-00644
3020	HR48
3020	HT800191H
3020	RE12
3020	REN123
3020	T-Q5063
3020	TK1228-1008
3020	TK1228-1009
3020	TM123
3020	TVS2SC968
3020	WEP53
3020	002-006500
3020	2SA828A
3020	2SC1211
3020	2SC1211E
3020	2SCNJ107
3020	0018
3020	022-006500
3020	026-100017
3020	48S134841
3020	48S134918
3020	69-1814
3020	297V059H01
3020	690V103H29
3020	690V103H32
3020	8000-0004-P085
3020	15835-1
3020	20738
3020	567312
3020	5700045452
3020/123	A0-54-195
3020/123	CCS6229H
3020/123	CDC8000-1
3020/123	CDC8002
3020/123	CDC8002-1
3020/123	CDC9002-18
3020/123	CJ-5208
3020/123	CS-2004C
3020/123	CS9022LE
3020/123	D232
3020/123	DDBY410002
3020/123	EA15X152
3020/123	EA15X241
3020/123	EA15X245
3020/123	EA15X349
3020/123	EA15X356
3020/123	EA15X371
3020/123	EA15X8502
3020/123	EA1873
3020/123	FCS9013HG
3020/123	G05-413B
3020/123	G05-413C
3020/123	G05-413D
3020/123	HT304971B
3020/123	IP20-0214
3020/123	M4853
3020/123	MJE9411T
3020/123	Q-00469
3020/123	QOV60529
3020/123	QOV60530
3020/123	RS-7409
3020/123	RS-7411
3020/123	RS-7412
3020/123	RS-7413
3020/123	RS-7504
3020/123	RS-7606
3020/123	RS-7607
3020/123	RS-7609
3020/123	RS-7610
3020/123	RS-7611
3020/123	RS-7612
3020/123	RS-7613
3020/123	RS-7614
3020/123	RS-7622
3020/123	RS-7623
3020/123	RS7518
3020/123	RS7519
3020/123	RT5208
3020/123	S-1221A
3020/123	S-1331W
3020/123	SE-6002
3020/123	SJ-570
3020/123	SM-4508-B
3020/123	SM-5564
3020/123	SM-5643
3020/123	SM-7815
3020/123	SM-7836
3020/123	SPS-4075
3020/123	ST-1242
3020/123	ST-1243
3020/123	ST-1244
3020/123	ST-1290
3020/123	SYL-1182
3020/123	T-1416
3020/123	T-256
3020/123	T139
3020/123	T1810B
3020/123	T35A-5
3020/123	T59235A
3020/123	TAC-047
3020/123	TI-415
3020/123	TI-416
3020/123	TI-492
3020/123	TI-493
3020/123	TI-494
3020/123	TI-495
3020/123	TMT-1543
3020/123	TNJ-60076
3020/123	TR-8014
3020/123	TR-8025
3020/123	TR010602-1
3020/123	TR22
3020/123	TR2SC373
3020/123	TR2SC735
3020/123	TRA-34
3020/123	TRA-36
3020/123	TRA-4
3020/123	TRA-4A
3020/123	TRA-4B
3020/123	TRA-9R
3020/123	TT-1097
3020/123	TV21
3020/123	TV28
3020/123	TV42
3020/123	TV57
3020/123	TV57A
3020/123	TV58
3020/123	TV58A
3020/123	TV59
3020/123	TV59A
3020/123	TV60
3020/123	TV60A
3020/123	TVS-CS1255H
3020/123	TVS2SC828
3020/123	TVS2SC828P
3020/123	TVS2SC828R
3020/123	TX-100-1
3020/123	TX-100-2
3020/123	TX-101-12
3020/123	TX-107-1
3020/123	TX-107-10
3020/123	TX-107-12
3020/123	TX-107-13
3020/123	TX-107-16
3020/123	TX-107-3
3020/123	TX-107-4
3020/123	TX-107-5
3020/123	TX-107-6
3020/123	TX-108-1
3020/123	TX-112-1
3020/123	TX-119-1
3020/123	VM-30209
3020/123	VM-30241
3020/123	VM-30242
3020/123	VS-2SC-458
3020/123	VS-2SC538
3020/123	001-021010
3020/123	001-021040
3020/123	001-021050
3020/123	001-021060
3020/123	001-021090
3020/123	001-021133
3020/123	2N1150
3020/123	2N296
3020/123	2N5201
3020/123	2S014
3020/123	2SC1641R
3020/123	2SC182
3020/123	2SC853KLM
3020/123	5-70004503
3020/123	09-302226
3020/123	13-23323-6
3020/123	13-29392-2
3020/123	21A040-082
3020/123	21M182
3020/123	025-100017
3020/123	43-023212
3020/123	46-86183-3
3020/123	46-86381
3020/123	46-86405-3
3020/123	051-0049
3020/123	75W-005
3020/123	86-444-2
3020/123	86-567-2
3020/123	121-768
3020/123	121-862
3020/123	176-074-9-004
3020/123	260P31303
3020/123	297V061C01
3020/123	753-2000-107
3020/123	902-000-2-04
3020/123	1048-9912
3020/123	2000-209
3020/123	26810-155
3020/123	040001
3020/123	95242-1
3020/123	1473608-002
3020/123	5321261
3020/123	5321291
3020/123	23114053
3020/123	23114054
3020/123	23114095
3020/123	23114216
3020/123	43021415
3020/123	43024879
3020/123	43026284
3020/123	48134842
3020/123	88125010-9
3020/123	93073260
3020/123	93073350
3020/123	93073440
3020/123	93097120
3020/123	229005013
3020/123	229005014
3020/123	229005015
3020/123	4109213354
3021	A1N
3021	A2U
3021	A7B
3021	AR-15
3021	AR-18
3021	C1059
3021	C1102
3021	C1102A
3021	C1102B
3021	C1102C
3021	C1105
3021	C1105A
3021	C1105B
3021	C1105C
3021	C514
3021	C515
3021	C515A
3021	C582
3021	C582A
3021	C582C
3021	C635A
3021	C685
3021	C685A
3021	C685B
3021	C685GU
3021	C685P
3021	C685Y
3021	C795
3021	C795A
3021	D11C201B20
3021	D11C203B20
3021	D11C205B20
3021	D11C207B20
3021	D11C210B20
3021	D11C211B20
3021	E2460
3021	EP15X16
3021	EP15X35
3021	ES15X95
3021	FBN-37605
3021	FBN-38982
3021	FT300
3021	GE-12
3021	HC515
3021	HEP-S5011
3021	HR106
3021	HR107
3021	J24566
3021	M4872
3021	M4885
3021	MJ2251
3021	MJ3201
3021	MT3202
3021	Q34450
3021	Q5075CLY
3021	Q5075DLY
3021	Q5075DXY
3021	Q5075ELY
3021	Q5075EXY
3021	Q5075FXY
3021	Q5075ZMY
3021	Q5104Z
3021	Q5113ZLM
3021	Q5113ZMM
3021	R2444
3021	RS-7315
3021	RS-7316
3021	RS-7317
3021	RS-7318
3021	RS7310
3021	RS7311
3021	RS7312
3021	RS7313
3021	RS7315
3021	RS7316
3021	RS7317
3021	RS7318
3021	RS7320
3021	RS7321
3021	RS7327
3021	RS7328
3021	RS7329
3021	RS7330
3021	RS7365
3021	RS7366
3021	RS7367
3021	RS7368
3021	RT7311
3021	S2059
3021	SC-4004
3021	SC-4131
3021	SC-4131-1
3021	SC-4167
3021	SC-727
3021	SC4004
3021	SC727
3021	SC777
3021	SE7006
3021	SE7020
3021	SE7030
3021	SJ1165
3021	SJ1201
3021	SJ1286
3021	SJ805
3021	SJ806
3021	SK5184A
3021	STX0015
3021	T-Q5057
3021	T-Q5075
3021	T-Q5104
3021	TA2509
3021	TA2509A
3021	TG2SD24Y
3021	TNJ72147
3021	TNJ72153
3021	TR-23
3021	TR-23C
3021	TR-8005
3021	TR1605LP
3021	TR23C
3021	TR262-2
3021	TR266-2
3021	TR8005
3021	TRS1005
3021	TRS1005LP
3021	TRS1205
3021	TRS1205LP
3021	TRS1405
3021	TRS1405LP
3021	TRS1605
3021	TRS1805
3021	TRS1805LP
3021	TRS2005
3021	TRS2005LP
3021	TRS2255
3021	TRS2255LP
3021	TRS2505
3021	TRS2505LP
3021	TRS2755
3021	TRS2755LP
3021	TRS3015LP
3021	TRS3016LC
3021	TRS4016LC
3021	TRS5016LC
3021	TRS6016LC
3021	TV113
3021	TV122
3021	TX-100-3
3021	TX-101-11
3021	TX-101-8
3021	TX-102-4
3021	TX-111-1
3021	TX100-5
3021	TX101-11
3021	TX101-8
3021	TX102-4
3021	TX107-13
3021	TX111-1
3021	001-02115-0
3021	001-021150
3021	001-021151
3021	001-02116-0
3021	001-021160
3021	001-021162
3021	1-801-301-13
3021	1-801-301-14
3021	1-801-301-15
3021	1A0038
3021	002-009100
3021	2N2204
3021	2N4298
3021	2N4299
3021	2N6425
3021	2SC101
3021	2SC101A
3021	2SC101B
3021	2SC101C
3021	2SC101D
3021	2SC101E
3021	2SC101F
3021	2SC101G
3021	2SC101GN
3021	2SC101H
3021	2SC101J
3021	2SC101K
3021	2SC101L
3021	2SC101M
3021	2SC1010R
3021	2SC101R
3021	2SC101X
3021	2SC101Y
3021	2SC1059
3021	2SC1105
3021	2SC1105A
3021	2SC1105B
3021	2SC1105C
3021	2SC1105K
3021	2SC1105L
3021	2SC1105M
3021	2SC1235AL
3021	2SC1235AM
3021	2SC1391
3021	2SC1391VL
3021	2SC1456L
3021	2SC1456M
3021	2SC514
3021	2SC515
3021	2SC515A
3021	2SC515AM
3021	2SC515AX
3021	2SC515BK
3021	2SC515Y
3021	2SC582
3021	2SC582A
3021	2SC582B
3021	2SC582BC
3021	2SC582C
3021	2SC685
3021	2SC685A
3021	2SC685B
3021	2SC685BK
3021	2SC685GU
3021	2SC685P
3021	2SC685S
3021	2SC685Y
3021	2SC795
3021	2SC795A
3021	2SD136D
3021	2SD136E
3021	2SD136F
3021	2SD136G
3021	2SD136GN
3021	2SD136J
3021	2SD136K
3021	2SD136L
3021	2SD136M
3021	2SD1360R
3021	2SD136R
3021	2SD136X
3021	2SD136Y
3021	2SD156
3021	2SD156B
3021	2SD156C
3021	2SD156F
3021	2SD190
3021	2SD24
3021	2SD24B
3021	2SD24C
3021	2SD24CK
3021	2SD24D
3021	2SD24E
3021	2SD24F
3021	2SD24K
3021	2SD24KC
3021	2SD24KD
3021	2SD24Y
3021	2SD24YB
3021	2SD24YD
3021	2SD24YE
3021	2SD24YK
3021	2SD24YLC
3021	2SD24YLD
3021	2SD24YLE
3021	2SD24YM
3021	2SD326
3021	3-19
3021	3L4-6006-1
3021	7B1
3021	7B13
3021	7B2
3021	7C1
3021	7C2
3021	7C3
3021	7D1
3021	7D2
3021	7D3
3021	7E1
3021	7E2
3021	7E3
3021	7G1
3021	7G2
3021	7G3
3021	7G4
3021	8-760-343-10
3021	09-302146
3021	09-302156
3021	09-302160
3021	09-302185
3021	09-303028
3021	9-5252
3021	9-5252-1
3021	9-5252-2
3021	9-5252-3
3021	9-5252-4
3021	11C10B1
3021	11C11B1
3021	11C1B1
3021	11C3B1
3021	11C3B3
3021	11C5B1
3021	11C7B1
3021	13-14085-29
3021	13-18282
3021	13-18282-1
3021	13-18359
3021	13-18359-1
3021	13-18359-3
3021	13-18359A
3021	13-23543-1
3021	19A115623-P1
3021	19A115623-P2
3021	19AR35
3021	21-35
3021	21A112-025
3021	21A112-029
3021	34-6002-21
3021	34-6002-26
3021	34-6002-46
3021	34-6002-62
3021	42-18310
3021	46-86384-3
3021	46-86386-3
3021	46-86439-3
3021	48-134920
3021	48-134972
3021	48-137207
3021	48-63026A45
3021	48-90232A07
3021	48X90232A07
3021	57A158-10
3021	57A192-10
3021	57B158-1
3021	57B158-2
3021	57B158-3
3021	57B158-4
3021	57B158-5
3021	57B158-6
3021	57B158-7
3021	57B158-8
3021	57B158-9
3021	57B195-10
3021	57C6-10
3021	57C6-14
3021	57C6-24
3021	57D6-10
3021	059
3021	71N1
3021	71N1T
3021	71N2
3021	71N2T
3021	86-227-2
3021	86-236-2
3021	86-259-2
3021	86-260-2
3021	86-261-2
3021	86-275-2
3021	86-487-2
3021	86-487-3
3021	86-624-9
3021	86X0028-001
3021	95-252
3021	95-252-1
3021	95-252-2

SK	Industry Standard No.
3021	95-252-3
3021	95-252-4
3021	96-5132-01
3021	96-5135-01
3021	96N1
3021	116-075
3021	116-1
3021	117-1
3021	121-315
3021	121-436
3021	121-451
3021	121-582
3021	121-713
3021	132-028
3021	132-033
3021	132-059
3021	132-515
3021	132-516
3021	132-521
3021	132-522
3021	132-523
3021	132-524
3021	132-525
3021	132-526
3021	154A5943
3021	154A5943-1
3021	158-10
3021	169-257
3021	169-284
3021	171-003-9-001
3021	260P09402
3021	260P15100
3021	260P15108
3021	260P16202
3021	260P16208
3021	260P21208
3021	297V060H01
3021	297V060H02
3021	297V060H03
3021	297V071C03
3021	297V071H03
3021	325-1442-9
3021	421-18
3021	421-8310
3021	573-515
3021	610-071
3021	800-122
3021	800-158
3021	800-172
3021	800-203
3021	800-204
3021	800-321
3021	921-156B
3021	921-61B
3021	921-88
3021	921-88A
3021	921-88B
3021	992-00-3172
3021	2057A2-223
3021	2057A2-81
3021	2057A2-84
3021	2057B123-10
3021	2057B2-47
3021	2057B2-58
3021	2057B2-84
3021	2417
3021	2491
3021	2491A
3021	2491B
3021	3520
3021	3520-1
3021	3566
3021	4872
3021	7311
3021	7317
3021	16232
3021	18310
3021	19278
3021	25114-143
3021	30234
3021	30245
3021	30257
3021	36634
3021	37584
3021	37730
3021	40264
3021	40313
3021	40318
3021	40328
3021	40374
3021	40422
3021	40423
3021	40425
3021	40426
3021	40427
3021	40491
3021	40546
3021	40547
3021	95252
3021	95252-1
3021	95252-2
3021	95252-3
3021	95252-4
3021	99252
3021	99252-1
3021	99252-3
3021	116075
3021	118279
3021	118686
3021	123275
3021	123275-14
3021	123375
3021	126138
3021	126188
3021	146286-01
3021	237450
3021	240388
3021	244357
3021	263561
3021	263857
3021	489751-043
3021	0573415
3021	573515
3021	0573515H
3021	610071
3021	610071-1
3021	610071-2
3021	770768-3170756
3021	815166
3021	815166-4
3021	815167-3
3021	815175
3021	815175H
3021	815180-3
3021	815180-4
3021	815180-7
3021	816135
3021	970255
3021	982376
3021	984608
3021	984932
3021	1018734-001
3021	1471117-1
3021	1473520-1
3021	1969113
3021	2004746-87
3021	2320221
3021	2320223
3021	2320228
3021	2320931
3021	2321001
3021	2321403
3021	2327182
3021	23114200
3021	23114268
3021	23114277
3021	23114315
3021	25114143
3021	62372540
3021	62563370
3021/124	C1104
3021/124	C1104A
3021/124	C1104B
3021/124	C1104C
3021/124	C1160K
3021/124	C1304
3021/124	SC-4244
3021/124	T-Q5080
3021/124	TR23
3021/124	TV24211
3021/124	TVS-2SC840
3021/124	TVS-2SC840A
3021/124	001-021140
3021/124	2N4240
3021/124	2N4296
3021/124	2N4297
3021/124	2SC1161
3021/124	2SC1304BK
3021/124	2SC825
3021/124	2SC825A
3021/124	2SC825B
3021/124	2SC825C
3021/124	2SD158
3021/124	2SD158A
3021/124	2SD158B
3021/124	2SD158C
3021/124	2SD158F
3021/124	2SD159
3021/124	2SD159A
3021/124	2SD159B
3021/124	2SD159C
3021/124	2SD159F
3021/124	2SD259
3021/124	09-302188
3021/124	19-020-045
3021/124	57B158-10
3021/124	86-177-2
3021/124	86-257-2
3021/124	96-5267-01
3021/124	260P20101
3021/124	260P24803
3021/124	690V094H17
3021/124	40322
3021/124	40424
3021/124	982523
3021/124	23114084
3021/124	23114266
3022	GEIC-82
3022	TA7028M
3023	GEIC-84
3023	HA1103
3023	HA1104
3023	M5113
3023	R3502
3023	R3502-1
3023	R3502-2
3023	T1A
3023	TA117
3023	4-082-664-0001
3023	09-308017
3023	19A116796-1
3023	32-23555-4
3023	51-10276A01
3023	51-110276A01
3023	51S10276A01
3023	57C28
3023	66F015
3023	86X0024-001
3023	86X0027-001
3023	266P001-01
3023	266P00101
3023	266P00102
3023	276A01
3023	417-119
3023	5113T
3023	8005-3
3023	8007-3
3023	8007-4
3023	8008-3
3023	8009-0
3023	8009-4
3023	80073
3023	95298
3023	118361
3023	119609
3023	730547
3023	1473502-2
3024	AO-54-175
3024	A054-156
3024	A054-160
3024	A054-186
3024	A054-206
3024	A059-110
3024	A1314
3024	A1341
3024	A2471
3024	A3011112
3024	A466
3024	A5B
3024	AQ2
3024	AQ3
3024	AQ5
3024	AR203
3024	AR203R
3024	AT-12
3024	AT-7
3024	AT339
3024	AT380
3024	AT381
3024	AT382
3024	AT383
3024	AT384
3024	AT385
3024	AT386
3024	AT387
3024	AT388
3024	AT440
3024	AT441
3024	AT442
3024	AT443
3024	AT444
3024	AT445
3024	AT446
3024	AT470
3024	AT471
3024	AT472
3024	AT473
3024	AT474
3024	AT475
3024	AT476
3024	AT477
3024	AT478
3024	AT479
3024	ATC-TR-13
3024	ATC-TR-4
3024	ATC-TR-7
3024	B3746
3024	BC103
3024	BC103C
3024	BC117
3024	BC120
3024	BC138
3024	BC140
3024	BC140-10
3024	BC140-16
3024	BC140-6
3024	BC140C
3024	BC140D
3024	BC141
3024	BC141-10
3024	BC141-16
3024	BC141-6
3024	BC142
3024	BC144
3024	BC174
3024	BC211
3024	BC216
3024	BC216A
3024	BC216B
3024	BC254
3024	BC255
3024	BC286
3024	BC288
3024	BC301
3024	BC340-10
3024	BC340-16
3024	BC340-6
3024	BC341-10
3024	BC341-6
3024	BC429
3024	BCW46
3024	BCW47
3024	BCW48
3024	BCW49
3024	BCW95
3024	BCY443
3024	BCY46
3024	BCY47
3024	BCY48
3024	BCY49
3024	BCY65
3024	BCY66
3024	BE71
3024	BFR36
3024	BFS29
3024	BFX17
3024	BFX50
3024	BFX51
3024	BFX52
3024	BFX61
3024	BFX68
3024	BFX68A
3024	BFX69
3024	BFX69A
3024	BFX74
3024	BFX84
3024	BFX85
3024	BFX86
3024	BFX97
3024	BFY13
3024	BFY14
3024	BFY15
3024	BFY17
3024	BFY27
3024	BFY34
3024	BFY40
3024	BFY44
3024	BFY46
3024	BFY51
3024	BFY52
3024	BFY53
3024	BFY55
3024	BFY56
3024	BFY56A
3024	BFY66
3024	BFY67
3024	BFY67A
3024	BFY67C
3024	BFY68
3024	BFY68A
3024	BFY70
3024	BFY99
3024	BLY61
3024	BSV51
3024	BSW10
3024	BSW26
3024	BSW27
3024	BSW28
3024	BSW29
3024	BSW35
3024	BSW65
3024	BSW66
3024	BSX23
3024	BSX33
3024	BSX45
3024	BSX46
3024	BSX60
3024	BSX61
3024	BSX62
3024	BSX62B
3024	BSX62C
3024	BSX62D
3024	BSX63B
3024	BSX63C
3024	BSX70
3024	BSX71
3024	BSX72
3024	BSX95
3024	BSX96
3024	BSY44
3024	BSY45
3024	BSY46
3024	BSY51
3024	BSY52
3024	BSY53
3024	BSY54
3024	BSY55
3024	BSY56
3024	BSY71
3024	BSY77
3024	BSY78
3024	BSY81
3024	BSY82
3024	BSY83
3024	BSY84
3024	BSY85
3024	BSY86
3024	BSY87
3024	BSY88
3024	BSY90
3024	BSY92
3024	C1008
3024	C1072
3024	C1072A
3024	C108
3024	C109
3024	C109A
3024	C113
3024	C114
3024	C115
3024	C117
3024	C121
3024	C1218
3024	C124
3024	C130
3024	C188
3024	C188A
3024	C188AB
3024	C189
3024	C19
3024	C190
3024	C210
3024	C211
3024	C216
3024	C217
3024	C218
3024	C22
3024	C220
3024	C221
3024	C222
3024	C226
3024	C227
3024	C228
3024	C229
3024	C236
3024	C247
3024	C2485076-3
3024	C2485077-2
3024	C249
3024	C268
3024	C268A
3024	C30
3024	C306
3024	C307
3024	C308
3024	C309
3024	C32A
3024	C353
3024	C353A
3024	C36579
3024	C420
3024	C426
3024	C443
3024	C46
3024	C479
3024	C497
3024	C497-O
3024	C497-R
3024	C497-Y
3024	C498
3024	C498-O
3024	C498-R
3024	C498-Y
3024	C503
3024	C503-O
3024	C503-Y
3024	C504
3024	C504-O
3024	C504-Y
3024	C504GR
3024	C51
3024	C516
3024	C560
3024	C61
3024	C708
3024	C708A
3024	C708AA
3024	C708AB
3024	C708AC
3024	C708B
3024	C708C
3024	C797
3024	C816
3024	C816K
3024	C826
3024	C827
3024	C875
3024	C875-1
3024	C875-1C
3024	C875-1D
3024	C875-1E
3024	C875-1F
3024	C875-2
3024	C875-2C
3024	C875-2D
3024	C875-2E
3024	C875-2F
3024	C875-3
3024	C875-3C
3024	C875-3D
3024	C875-3E
3024	C875-3F
3024	C875D
3024	C875E
3024	C875F
3024	C876
3024	C876C
3024	C876D
3024	C876E
3024	C876TV
3024	C876TVE
3024	C876TVEF
3024	C934
3024	C959A
3024	C959B
3024	C959C
3024	C959D
3024	C959M
3024	C959S
3024	C959SA
3024	C959SB
3024	C959SC
3024	C959SD
3024	C97
3024	C972
3024	C972C
3024	C972D
3024	C972E
3024	C97A
3024	CCS6168G
3024	CDC5008
3024	CDC5028A
3024	CDQ10011
3024	CDQ10012
3024	CDQ10014
3024	CDQ10033
3024	CDQ10044
3024	CDQ10048
3024	CDQ10051
3024	CDQ10052
3024	CDQ10053
3024	CDQ10057
3024	CDQ10058
3024	CDZ15000
3024	CE4003D
3024	CI3704
3024	CI3705
3024	CI3706
3024	CJ5210
3024	CJ5213
3024	CJ5214
3024	CJ5215
3024	CM0770
3024	CN2484
3024	CS1129E
3024	CS1225H
3024	CS1225HF
3024	CS1229K
3024	CS1248
3024	CS1248I
3024	CS1248T
3024	CS1255H
3024	CS1295H
3024	CS1305
3024	CS1453G
3024	CS1462I
3024	CS1464H
3024	CS1591LE
3024	CS1609F
3024	CS1613
3024	CS1711
3024	CS1893
3024	CS1990
3024	CS2484
3024	CS3704
3024	CS3705
3024	CS3706
3024	CS5449
3024	CS5450
3024	CS5451
3024	CS9103B
3024	CS9103C
3024	CS956
3024	CXL128
3024	D11C10B1
3024	D11C11B1
3024	D11C1B1
3024	D11C3B1
3024	D11C5B1
3024	D11C7B1
3024	D204L
3024	D7A30
3024	D7A31
3024	D7A32
3024	DT1110
3024	DT1111
3024	DT1112
3024	DT1120
3024	DT1121
3024	DT1122
3024	DT1311
3024	DT1321
3024	DT1510
3024	DT1511
3024	DT1512
3024	DT1520
3024	DT1521
3024	DT1522
3024	DT1621
3024	E2441
3024	E2449
3024	EA0081
3024	EA0090
3024	EA15X102
3024	EA15X144
3024	EA15X7635
3024	EA1684
3024	ECG128
3024	EL214
3024	EP15X33
3024	EQS131
3024	ES15X93
3024	ET15X36
3024	ET15X8
3024	ETS-070
3024	EU15X27
3024	EU15X34
3024	F3560
3024	F3561
3024	F3589
3024	F4709
3024	FCS1229
3024	FD-1029-MM
3024	FS1331
3024	FS2042
3024	FS27233
3024	FT001
3024	FT0019H
3024	FT002
3024	FT003

SK	Industry Standard No.	SK	Industry Standard No.	SK	Industry Standard No.	SK	Industry Standard No.	SK	Industry Standard No.
3024	PT004A	3024	RS8113	3024	TR-8021	3024	2SC1072M	3024	2SC118G
3024	PT004A	3024	RT-114	3024	TR-8023	3024	2SC1072OR	3024	2SC118GN
3024	PT027	3024	RT-188	3024	TR-8024	3024	2SC1072R	3024	2SC118H
3024	GB-243	3024	RT141	3024	TR1001	3024	2SC1072X	3024	2SC118J
3024	GVL20077	3024	RT154	3024	TR1003	3024	2SC1072Y	3024	2SC118L
3024	HC-00268	3024	RT188	3024	TR1005	3024	2SC108	3024	2SC118M
3024	HC-01209	3024	RT482	3024	TR25	3024	2SC108A	3024	2SC118R
3024	HR13A	3024	RT483	3024	TR28C482	3024	2SC108B	3024	2SC118RR
3024	HR28	3024	RT484	3024	TV-26	3024	2SC108C	3024	2SC118X
3024	HR29	3024	RT5151	3024	TV-28	3024	2SC108D	3024	2SC118Y
3024	HR67	3024	RT5152	3024	TV-41	3024	2SC108E	3024	2SC119
3024	HT104861	3024	RT5401	3024	TV-45	3024	2SC108F	3024	2SC119A
3024	HT104861A	3024	RT5402	3024	TV-59	3024	2SC108G	3024	2SC119B
3024	HT104861B	3024	RT5403	3024	TV-67	3024	2SC108GN	3024	2SC119D
3024	HT304861	3024	RT5404	3024	TV23	3024	2SC108H	3024	2SC119F
3024	HT304861A	3024	RT7945	3024	TV41	3024	2SC108J	3024	2SC119G
3024	HT304971C	3024	S1368	3024	TV45	3024	2SC108K	3024	2SC119GN
3024	HT304971A	3024	S1421	3024	TV51	3024	2SC108L	3024	2SC119H
3024	HT307341	3024	S1514	3024	TV52	3024	2SC108M	3024	2SC119K
3024	HT307341A	3024	S1516	3024	TV53	3024	2SC108OR	3024	2SC119L
3024	HT309680B	3024	S1517	3024	TVS-280582	3024	2SC108R	3024	2SC119M
3024	HT8001210	3024	S1523	3024	TVS-280582A	3024	2SC108X	3024	2SC119OR
3024	HT8001310	3024	S1525	3024	TVS28C1255	3024	2SC108Y	3024	2SC119R
3024	IR-TR53	3024	S15660	3024	TVS28C1255HP	3024	2SC109	3024	2SC119X
3024	IRTR87	3024	S1642	3024	TVS280696	3024	2SC109A	3024	2SC119Y
3024	KD2218	3024	S1683	3024	TVS28D968	3024	2SC109B	3024	2SC1206
3024	LDA404	3024	S1689	3024	TX-100-4	3024	2SC109C	3024	2SC120A
3024	LDA405	3024	S1762	3024	TX101-9	3024	2SC109D	3024	2SC120B
3024	LDA406	3024	S1773	3024	V120RH	3024	2SC109F	3024	2SC120C
3024	LDA408	3024	S1777	3024	V139	3024	2SC109G	3024	2SC120D
3024	LDS200	3024	S17900	3024	V146	3024	2SC109GN	3024	2SC120F
3024	LDS201	3024	S18000	3024	V154	3024	2SC109J	3024	2SC120G
3024	M4689	3024	S1864	3024	V166	3024	2SC109K	3024	2SC120GN
3024	M9138	3024	S1874	3024	V172	3024	2SC109L	3024	2SC120H
3024	M9228	3024	S19386	3024	V177	3024	2SC109OR	3024	2SC120J
3024	ME1075	3024	S2104	3024	VS-CS1255H	3024	2SC109R	3024	2SC120K
3024	MB8001	3024	S2118	3024	VS-CS1255HP	3024	2SC109X	3024	2SC120L
3024	MB8002	3024	S21549	3024	VS28C1213AC/1A	3024	2SC109Y	3024	2SC120M
3024	MB8003	3024	S2209	3024	XS30	3024	2SC112	3024	2SC120OR
3024	MH9414	3024	S2369	3024	ZT1479	3024	2SC112A	3024	2SC120R
3024	MHT2414	3024	S2371	3024	ZT1481	3024	2SC112B	3024	2SC120X
3024	MHT2418	3024	S2400	3024	ZT1613	3024	2SC112C	3024	2SC120Y
3024	MHT4401	3024	S2400B	3024	ZT1700	3024	2SC112D	3024	2SC121
3024	MHT4411	3024	S2401	3024	ZT190	3024	2SC112E	3024	2SC121A
3024	MHT4412	3024	S2401A	3024	ZT191	3024	2SC112F	3024	2SC121B
3024	MHT4413	3024	S2401B	3024	ZT192	3024	2SC112G	3024	2SC121C
3024	MHT4451	3024	S2401C	3024	ZT193	3024	2SC112H	3024	2SC121D
3024	MHT4483	3024	S2402	3024	ZT3512	3024	2SC112J	3024	2SC121E
3024	MHT4511	3024	S2402A	3024	ZT3866	3024	2SC112K	3024	2SC121F
3024	MHT4512	3024	S2402B	3024	ZT66A	3024	2SC112L	3024	2SC121G
3024	MHT4513	3024	S2402C	3024	ZT66B	3024	2SC112OR	3024	2SC121GN
3024	MHT7401	3024	S2427	3024	ZT66C	3024	2SC112R	3024	2SC121H
3024	MHT7411	3024	S24598	3024	ZT90	3024	2SC112X	3024	2SC121J
3024	MHT7412	3024	S24614	3024	ZT93	3024	2SC112Y	3024	2SC121L
3024	MHT7414	3024	S24616	3024	ZT94	3024	2SC113	3024	2SC121M
3024	MHT7417	3024	S2487	3024	001-021110	3024	2SC113A	3024	2SC121OR
3024	MHT9001	3024	S2526	3024	001-021111	3024	2SC113B	3024	2SC121R
3024	MHT9002	3024	S2648	3024	001-02119-0	3024	2SC113C	3024	2SC121X
3024	MHT9004	3024	S27233	3024	001-021190	3024	2SC113D	3024	2SC121Y
3024	MHT9005	3024	S2794	3024	001-021290	3024	2SC113G	3024	2SC122
3024	MM181OA	3024	S2992	3024	01-2110	3024	2SC113GN	3024	2SC122A
3024	MM1943	3024	S409F	3024	01-2111	3024	2SC113H	3024	2SC122B
3024	MM2266	3024	S6804	3024	01-2114	3024	2SC113J	3024	2SC122C
3024	MM486	3024	SC1229E	3024	1A0066	3024	2SC113K	3024	2SC122D
3024	MM487	3024	SDD1220	3024	1A348R	3024	2SC113L	3024	2SC122E
3024	MM488	3024	SDD420	3024	1A34R	3024	2SC113M	3024	2SC122G
3024	MM511	3024	SB-8001	3024	1N3112	3024	2SC113OR	3024	2SC122GN
3024	MM512	3024	SB6020A	3024	002-010600	3024	2SC113R	3024	2SC122H
3024	MM513	3024	SB6021	3024	002-012500	3024	2SC113X	3024	2SC122J
3024	MP89410A	3024	SB6021A	3024	2N1092	3024	2SC113Y	3024	2SC122K
3024	MR39933	3024	SB6022	3024	2N1472	3024	2SC114	3024	2SC122L
3024	MS2991	3024	SB6023	3024	2N1613	3024	2SC114A	3024	2SC122M
3024	MST-10	3024	SB7005	3024	2N1711	3024	2SC114B	3024	2SC122OR
3024	MT1613	3024	SB7015	3024	2N1764	3024	2SC114C	3024	2SC122R
3024	MT1711	3024	SB8001	3024	2N2017	3024	2SC114D	3024	2SC122X
3024	NPC115	3024	SB8002	3024	2N2102	3024	2SC114E	3024	2SC122Y
3024	NPC187	3024	SB8010	3024	2N2106	3024	2SC114F	3024	2SC123
3024	NPC189	3024	SB8012	3024	2N2107	3024	2SC114G	3024	2SC123A
3024	NS2100	3024	SB8041	3024	2N2108	3024	2SC114GN	3024	2SC123B
3024	NS2101	3024	SB8042	3024	2N2195A	3024	2SC114J	3024	2SC123C
3024	NS9400	3024	SB8510	3024	2N2594	3024	2SC114K	3024	2SC123D
3024	NS9420	3024	SB8520	3024	2N2895	3024	2SC114L	3024	2SC123E
3024	NS9500	3024	SB8521	3024	2N2897	3024	2SC114M	3024	2SC123F
3024	NS9540	3024	SPT443	3024	2N2J324	3024	2SC114OR	3024	2SC123G
3024	NS9728	3024	SPT443A	3024	2N3053	3024	2SC114R	3024	2SC123GN
3024	NS9729	3024	SPT445	3024	2N343A	3024	2SC114X	3024	2SC123H
3024	NS9730	3024	SQ5013	3024	2N3469	3024	2SC115	3024	2SC123J
3024	NS9731	3024	SS5997B	3024	2N3568	3024	2SC115A	3024	2SC123K
3024	0A-10	3024	SM6251	3024	2N3569	3024	2SC115B	3024	2SC123L
3024	00V60529	3024	SN166	3024	2N4237	3024	2SC115C	3024	2SC123M
3024	ORP-2	3024	SN167	3024	2N4238	3024	2SC115D	3024	2SC123OR
3024	P633024	3024	SPS-0122	3024	2N4239	3024	2SC115E	3024	2SC123R
3024	P633024G	3024	SPS-4077	3024	2N4943	3024	2SC115F	3024	2SC123Y
3024	PEP2001	3024	SPS0122	3024	2N4944	3024	2SC115G	3024	2SC124
3024	PET9021	3024	SPS4038	3024	2N4946	3024	2SC115GN	3024	2SC124A
3024	PET9022	3024	SPS4300	3024	2N4960	3024	2SC115H	3024	2SC124C
3024	PG33024G	3024	SPS4309	3024	2N4961	3024	2SC115J	3024	2SC124D
3024	PL1083	3024	SPS6124	3024	2N4962	3024	2SC115K	3024	2SC124F
3024	PL1084	3024	ST-LM2682	3024	2N4963	3024	2SC115L	3024	2SC124G
3024	PRT-104-4	3024	ST6573	3024	2N4964	3024	2SC115M	3024	2SC124GN
3024	PT3502	3024	ST6574	3024	2N5188	3024	2SC115OR	3024	2SC124H
3024	PT612	3024	STL4280	3024	2N5211	3024	2SC115R	3024	2SC124J
3024	PT850	3024	T-339	3024	2N5450	3024	2SC115X	3024	2SC124K
3024	PT850A	3024	T-Q5081	3024	2N5451	3024	2SC115Y	3024	2SC124L
3024	PT896	3024	T-Q5099	3024	2N5820	3024	2SC117	3024	2SC124M
3024	Q-01169	3024	T13015	3024	2N6178	3024	2SC117B	3024	2SC124OR
3024	Q-01169A	3024	T1340A3H	3024	2N656	3024	2SC117C	3024	2SC124R
3024	Q-01169C	3024	T164213	3024	2N656A	3024	2SC117D	3024	2SC124X
3024	QA-10	3024	T1706	3024	2N657	3024	2SC117F	3024	2SC124Y
3024	QA-8	3024	T1706A	3024	2N697	3024	2SC117G	3024	2SC130
3024	QQV60529	3024	T1706B	3024	2N697A	3024	2SC117GN	3024	2SC130A
3024	R3608-1	3024	T1706C	3024	2N699A	3024	2SC117H	3024	2SC130B
3024	R3608-2	3024	T1811B	3024	2N699B	3024	2SC117J	3024	2SC130C
3024	R5048	3024	T1811G	3024	2S017	3024	2SC117L	3024	2SC130D
3024	R7613	3024	T23-94	3024	2S018	3024	2SC117M	3024	2SC130GN
3024	R8915	3024	T247	3024	2S019	3024	2SC117R	3024	2SC130H
3024	RC2270	3024	T452	3024	2S020	3024	2SC117X	3024	2SC130J
3024	RCA1A01	3024	TA6200	3024	2S742	3024	2SC117Y	3024	2SC130K
3024	RCA1A06	3024	TI3015	3024	2S742A	3024	2SC118	3024	2SC130M
3024	RCA1A02	3024	TI64213	3024	2S745	3024	2SC118A	3024	2SC130OR
3024	RCA1A17	3024	TI811G	3024	2SC1072	3024	2SC118B	3024	2SC130R
3024	RCA1A18	3024	TI892M	3024	2SC1072B	3024	2SC118C	3024	2SC130X
3024	RE17	3024	TM128	3024	2SC1072C	3024	2SC118D	3024	2SC130Y
3024	REN128	3024	TNJ70482	3024	2SC1072D	3024	2SC118E	3024	2SC1386H
3024	RS-2014	3024	TNJ71035	3024	2SC1072E	3024	2SC118F	3024	2SC1399E
3024	RS7678	3024	TNJ71234	3024	2SC1072F			3024	2SC140
3024	RS8101	3024	TNJ72288	3024	2SC1072G			3024	2SC140A
3024	RS8103	3024	TQ5055	3024	2SC1072GN				
3024	RS8105	3024	TR-25	3024	2SC1072H				
3024	RS8107	3024	TR-280482	3024	2SC1072J				
3024	RS8109	3024	TR-4R31	3024	2SC1072K				
3024	RS8111	3024	TR-5R31	3024	2SC1072L				
		3024	TR-7R31						

SK	Industry Standard No.
3024	2SC140C
3024	2SC140D
3024	2SC140E
3024	2SC140F
3024	2SC140G
3024	2SC140GN
3024	2SC140H
3024	2SC140J
3024	2SC140K
3024	2SC140L
3024	2SC140M
3024	2SC140OR
3024	2SC140R
3024	2SC140X
3024	2SC140Y
3024	2SC152R
3024	2SC163A
3024	2SC163B
3024	2SC163C
3024	2SC163D
3024	2SC163E
3024	2SC163F
3024	2SC163G
3024	2SC163GN
3024	2SC163H
3024	2SC163J
3024	2SC163K
3024	2SC163L
3024	2SC163M
3024	2SC163OR
3024	2SC163R
3024	2SC163X
3024	2SC163Y
3024	2SC188
3024	2SC188A
3024	2SC188AB
3024	2SC188B
3024	2SC188C
3024	2SC188D
3024	2SC188E
3024	2SC188F
3024	2SC188G
3024	2SC188GN
3024	2SC188J
3024	2SC188K
3024	2SC188L
3024	2SC188M
3024	2SC188OR
3024	2SC188R
3024	2SC188X
3024	2SC188Y
3024	2SC189
3024	2SC189A
3024	2SC189B
3024	2SC189C
3024	2SC189D
3024	2SC189E
3024	2SC189G
3024	2SC189GN
3024	2SC189H
3024	2SC189J
3024	2SC189K
3024	2SC189L
3024	2SC189M
3024	2SC189OR
3024	2SC189R
3024	2SC189X
3024	2SC189Y
3024	2SC190
3024	2SC190A
3024	2SC190B
3024	2SC190C
3024	2SC190D
3024	2SC190E
3024	2SC190F
3024	2SC190GN
3024	2SC190H
3024	2SC190J
3024	2SC190K
3024	2SC190L
3024	2SC190M
3024	2SC190R
3024	2SC190X
3024	2SC190Y
3024	2SC19A
3024	2SC19B
3024	2SC19C
3024	2SC19D
3024	2SC19E
3024	2SC19F
3024	2SC19G
3024	2SC19GN
3024	2SC19H
3024	2SC19J
3024	2SC19K
3024	2SC19L
3024	2SC19M
3024	2SC19OR
3024	2SC19R
3024	2SC19X
3024	2SC19Y
3024	2SC20
3024	2SC208A
3024	2SC208B
3024	2SC208C
3024	2SC208D
3024	2SC208E
3024	2SC208F
3024	2SC208G
3024	2SC208GN
3024	2SC208H
3024	2SC208J
3024	2SC208K
3024	2SC208L
3024	2SC208M
3024	2SC208OR
3024	2SC208R
3024	2SC208X
3024	2SC208Y
3024	2SC20A
3024	2SC20B
3024	2SC20C
3024	2SC20D
3024	2SC20E
3024	2SC20F
3024	2SC20G
3024	2SC20GN
3024	2SC20H
3024	2SC20J
3024	2SC20K
3024	2SC20L
3024	2SC20M
3024	2SC20R
3024	2SC20X

SK	Industry Standard No.
3024	2SC20Y
3024	2SC210
3024	2SC210A
3024	2SC210B
3024	2SC210C
3024	2SC210D
3024	2SC210E
3024	2SC210F
3024	2SC210G
3024	2SC210GN
3024	2SC210H
3024	2SC210J
3024	2SC210K
3024	2SC210L
3024	2SC210M
3024	2SC210R
3024	2SC210X
3024	2SC210Y
3024	2SC211
3024	2SC211A
3024	2SC211B
3024	2SC211C
3024	2SC211D
3024	2SC211F
3024	2SC211G
3024	2SC211GN
3024	2SC211H
3024	2SC211J
3024	2SC211K
3024	2SC211L
3024	2SC211M
3024	2SC211OR
3024	2SC211R
3024	2SC211X
3024	2SC211Y
3024	2SC212A
3024	2SC212B
3024	2SC212C
3024	2SC212D
3024	2SC212E
3024	2SC212F
3024	2SC212G
3024	2SC212GN
3024	2SC212H
3024	2SC212J
3024	2SC212K
3024	2SC212L
3024	2SC212M
3024	2SC212OR
3024	2SC212R
3024	2SC212X
3024	2SC212Y
3024	2SC213A
3024	2SC213B
3024	2SC213C
3024	2SC213D
3024	2SC213E
3024	2SC213F
3024	2SC213G
3024	2SC213GN
3024	2SC213H
3024	2SC213J
3024	2SC213K
3024	2SC213L
3024	2SC213M
3024	2SC213OR
3024	2SC213R
3024	2SC213X
3024	2SC213Y
3024	2SC214A
3024	2SC214B
3024	2SC214C
3024	2SC214D
3024	2SC214E
3024	2SC214F
3024	2SC214G
3024	2SC214GN
3024	2SC214H
3024	2SC214J
3024	2SC214K
3024	2SC214L
3024	2SC214M
3024	2SC214OR
3024	2SC214X
3024	2SC214Y
3024	2SC215A
3024	2SC215B
3024	2SC215C
3024	2SC215D
3024	2SC215E
3024	2SC215F
3024	2SC215G
3024	2SC215GN
3024	2SC215H
3024	2SC215J
3024	2SC215K
3024	2SC215L
3024	2SC215M
3024	2SC215OR
3024	2SC215R
3024	2SC215X
3024	2SC215Y
3024	2SC216
3024	2SC216A
3024	2SC216B
3024	2SC216C
3024	2SC216D
3024	2SC216E
3024	2SC216F
3024	2SC216G
3024	2SC216GN
3024	2SC216H
3024	2SC216J
3024	2SC216K
3024	2SC216L
3024	2SC216M
3024	2SC216OR
3024	2SC216R
3024	2SC216X
3024	2SC216Y
3024	2SC217
3024	2SC217A
3024	2SC217B
3024	2SC217C
3024	2SC217D
3024	2SC217E
3024	2SC217F
3024	2SC217G
3024	2SC217GN
3024	2SC217H
3024	2SC217J
3024	2SC217K
3024	2SC217L
3024	2SC217M
3024	2SC217OR

SK	Industry Standard No.
3024	2SC217R
3024	2SC217X
3024	2SC217Y
3024	2SC218C
3024	2SC218D
3024	2SC218E
3024	2SC218F
3024	2SC218G
3024	2SC218GN
3024	2SC218H
3024	2SC218J
3024	2SC218L
3024	2SC218M
3024	2SC218OR
3024	2SC218R
3024	2SC218X
3024	2SC218Y
3024	2SC22
3024	2SC220
3024	2SC220A
3024	2SC220B
3024	2SC220C
3024	2SC220D
3024	2SC220E
3024	2SC220F
3024	2SC220G
3024	2SC220GN
3024	2SC220H
3024	2SC220K
3024	2SC220L
3024	2SC220M
3024	2SC220X
3024	2SC220Y
3024	2SC221
3024	2SC221A
3024	2SC221B
3024	2SC221C
3024	2SC221D
3024	2SC221E
3024	2SC221F
3024	2SC221G
3024	2SC221GN
3024	2SC221H
3024	2SC221J
3024	2SC221K
3024	2SC221L
3024	2SC221M
3024	2SC221OR
3024	2SC221R
3024	2SC221X
3024	2SC221Y
3024	2SC222
3024	2SC222A
3024	2SC222B
3024	2SC222C
3024	2SC222D
3024	2SC222E
3024	2SC222F
3024	2SC222G
3024	2SC222GN
3024	2SC222H
3024	2SC222J
3024	2SC222L
3024	2SC222M
3024	2SC222OR
3024	2SC222R
3024	2SC222X
3024	2SC222Y
3024	2SC223A
3024	2SC223B
3024	2SC223C
3024	2SC223D
3024	2SC223E
3024	2SC223F
3024	2SC223G
3024	2SC223GN
3024	2SC223H
3024	2SC223J
3024	2SC223L
3024	2SC223M
3024	2SC223OR
3024	2SC223R
3024	2SC223X
3024	2SC223Y
3024	2SC224A
3024	2SC224B
3024	2SC224C
3024	2SC224D
3024	2SC224E
3024	2SC224F
3024	2SC224G
3024	2SC224GN
3024	2SC224H
3024	2SC224J
3024	2SC224K
3024	2SC224L
3024	2SC224M
3024	2SC224OR
3024	2SC224R
3024	2SC224X
3024	2SC224Y
3024	2SC225
3024	2SC225A
3024	2SC225B
3024	2SC225C
3024	2SC225D
3024	2SC225E
3024	2SC225F
3024	2SC225G
3024	2SC225GN
3024	2SC225H
3024	2SC225J
3024	2SC225L
3024	2SC225M
3024	2SC225OR
3024	2SC225R
3024	2SC225X
3024	2SC225Y
3024	2SC226
3024	2SC226A
3024	2SC226B
3024	2SC226C
3024	2SC226D
3024	2SC226E
3024	2SC226F
3024	2SC226G
3024	2SC226GN
3024	2SC226H
3024	2SC226J
3024	2SC226K
3024	2SC226L
3024	2SC226M
3024	2SC226OR
3024	2SC226R

SK	Industry Standard No.
3024	2SC226X
3024	2SC226Y
3024	2SC227
3024	2SC227A
3024	2SC227B
3024	2SC227C
3024	2SC227D
3024	2SC227E
3024	2SC227F
3024	2SC227G
3024	2SC227GN
3024	2SC227H
3024	2SC227J
3024	2SC227K
3024	2SC227L
3024	2SC227M
3024	2SC227OR
3024	2SC227R
3024	2SC227X
3024	2SC227Y
3024	2SC228
3024	2SC228A
3024	2SC228B
3024	2SC228C
3024	2SC228D
3024	2SC228E
3024	2SC228F
3024	2SC228G
3024	2SC228GN
3024	2SC228H
3024	2SC228J
3024	2SC228K
3024	2SC228L
3024	2SC228M
3024	2SC228OR
3024	2SC228R
3024	2SC228X
3024	2SC228Y
3024	2SC229A
3024	2SC229B
3024	2SC229C
3024	2SC229D
3024	2SC229E
3024	2SC229F
3024	2SC229G
3024	2SC229GN
3024	2SC229H
3024	2SC229J
3024	2SC229K
3024	2SC229L
3024	2SC229M
3024	2SC229OR
3024	2SC229R
3024	2SC229X
3024	2SC229Y
3024	2SC22A
3024	2SC22B
3024	2SC22D
3024	2SC22E
3024	2SC22F
3024	2SC22G
3024	2SC22GN
3024	2SC22H
3024	2SC22J
3024	2SC22K
3024	2SC22L
3024	2SC22M
3024	2SC22OR
3024	2SC22R
3024	2SC22X
3024	2SC22Y
3024	2SC231A
3024	2SC231B
3024	2SC231C
3024	2SC231D
3024	2SC231E
3024	2SC231F
3024	2SC231G
3024	2SC231GN
3024	2SC231H
3024	2SC231J
3024	2SC231K
3024	2SC231L
3024	2SC231M
3024	2SC231OR
3024	2SC231R
3024	2SC231X
3024	2SC231Y
3024	2SC232A
3024	2SC232B
3024	2SC232C
3024	2SC232D
3024	2SC232E
3024	2SC232F
3024	2SC232G
3024	2SC232GN
3024	2SC232H
3024	2SC232J
3024	2SC232K
3024	2SC232L
3024	2SC232M
3024	2SC232OR
3024	2SC232R
3024	2SC232X
3024	2SC232Y
3024	2SC233A
3024	2SC233B
3024	2SC233C
3024	2SC233D
3024	2SC233F
3024	2SC233G
3024	2SC233GN
3024	2SC233H
3024	2SC233J
3024	2SC233L
3024	2SC233M
3024	2SC233OR
3024	2SC233R
3024	2SC233X
3024	2SC233Y
3024	2SC234
3024	2SC234A
3024	2SC234B
3024	2SC234C
3024	2SC234D
3024	2SC234E
3024	2SC234F
3024	2SC234G
3024	2SC234GN
3024	2SC234H
3024	2SC234J
3024	2SC234K
3024	2SC234L
3024	2SC234M

SK	Industry Standard No.
3024	2SC234OR
3024	2SC234R
3024	2SC234X
3024	2SC234Y
3024	2SC236
3024	2SC236A
3024	2SC236B
3024	2SC236C
3024	2SC236D
3024	2SC236E
3024	2SC236F
3024	2SC236G
3024	2SC236GN
3024	2SC236H
3024	2SC236J
3024	2SC236K
3024	2SC236L
3024	2SC236M
3024	2SC236OR
3024	2SC236R
3024	2SC236X
3024	2SC236Y
3024	2SC238A
3024	2SC238B
3024	2SC238C
3024	2SC238D
3024	2SC238E
3024	2SC238F
3024	2SC238G
3024	2SC238GN
3024	2SC238H
3024	2SC238J
3024	2SC238K
3024	2SC238L
3024	2SC238M
3024	2SC238OR
3024	2SC238R
3024	2SC238X
3024	2SC238Y
3024	2SC23A
3024	2SC23B
3024	2SC23D
3024	2SC23E
3024	2SC23F
3024	2SC23G
3024	2SC23GN
3024	2SC23H
3024	2SC23J
3024	2SC23K
3024	2SC23L
3024	2SC23M
3024	2SC23OR
3024	2SC23R
3024	2SC23X
3024	2SC23Y
3024	2SC247
3024	2SC247A
3024	2SC247B
3024	2SC247C
3024	2SC247D
3024	2SC247E
3024	2SC247F
3024	2SC247G
3024	2SC247GN
3024	2SC247H
3024	2SC247J
3024	2SC247L
3024	2SC247M
3024	2SC247OR
3024	2SC247R
3024	2SC247X
3024	2SC247Y
3024	2SC248A
3024	2SC248B
3024	2SC248C
3024	2SC248D
3024	2SC248E
3024	2SC248F
3024	2SC248G
3024	2SC248J
3024	2SC248K
3024	2SC248L
3024	2SC248M
3024	2SC248OR
3024	2SC248R
3024	2SC248X
3024	2SC248Y
3024	2SC249
3024	2SC249A
3024	2SC249B
3024	2SC249C
3024	2SC249D
3024	2SC249E
3024	2SC249F
3024	2SC249G
3024	2SC249GN
3024	2SC249H
3024	2SC249J
3024	2SC249K
3024	2SC249L
3024	2SC249M
3024	2SC249OR
3024	2SC249R
3024	2SC249X
3024	2SC249Y
3024	2SC24A
3024	2SC24B
3024	2SC24C
3024	2SC24D
3024	2SC24E
3024	2SC24F
3024	2SC24G
3024	2SC24GN
3024	2SC24H
3024	2SC24J
3024	2SC24K
3024	2SC24L
3024	2SC24M
3024	2SC24OR
3024	2SC24R
3024	2SC24X
3024	2SC24Y
3024	2SC268A
3024	2SC27A
3024	2SC27B
3024	2SC27D
3024	2SC27E
3024	2SC27F
3024	2SC27G
3024	2SC27GN
3024	2SC27H
3024	2SC27J
3024	2SC27K
3024	2SC27L

SK	Industry Standard No.	SK	Industry Standard No.	SK	Industry Standard No.	SK	Industry Standard No.	SK	Industry Standard No.
3024	2SC27M	3024	2SC3OE	3024	2SC443	3024	2SC502M	3024	2SC590D
3024	2SC27OR	3024	2SC3OF	3024	2SC443A	3024	2SC502OR	3024	2SC590E
3024	2SC27R	3024	2SC3OG	3024	2SC443B	3024	2SC502R	3024	2SC590F
3024	2SC27X	3024	2SC3OGN	3024	2SC443C	3024	2SC502X	3024	2SC590G
3024	2SC27Y	3024	2SC3OH	3024	2SC443D	3024	2SC502Y	3024	2SC590GN
3024	2SC291	3024	2SC3OJ	3024	2SC443E	3024	2SC503	3024	2SC590H
3024	2SC291A	3024	2SC3OL	3024	2SC443F	3024	2SC503-O	3024	2SC590J
3024	2SC291B	3024	2SC3OM	3024	2SC443G	3024	2SC503-Y	3024	2SC590K
3024	2SC291C	3024	2SC3OR	3024	2SC443GN	3024	2SC503A	3024	2SC590L
3024	2SC291D	3024	2SC3OX	3024	2SC443H	3024	2SC503B	3024	2SC590M
3024	2SC291E	3024	2SC3OY	3024	2SC443J	3024	2SC503C	3024	2SC590OR
3024	2SC291F	3024	2SC31	3024	2SC443K	3024	2SC503D	3024	2SC590X
3024	2SC291G	3024	2SC31OA	3024	2SC443L	3024	2SC503E	3024	2SC594A
3024	2SC291GN	3024	2SC31OF	3024	2SC443M	3024	2SC503F	3024	2SC594C
3024	2SC291H	3024	2SC31OG	3024	2SC443OR	3024	2SC503G	3024	2SC594D
3024	2SC291J	3024	2SC31OGN	3024	2SC443R	3024	2SC503GN	3024	2SC594E
3024	2SC291K	3024	2SC31OH	3024	2SC443X	3024	2SC503R	3024	2SC594F
3024	2SC291L	3024	2SC31OJ	3024	2SC443Y	3024	2SC503H	3024	2SC594G
3024	2SC291M	3024	2SC31OK	3024	2SC46	3024	2SC503J	3024	2SC594GN
3024	2SC291OR	3024	2SC31OL	3024	2SC46A	3024	2SC503K	3024	2SC594H
3024	2SC291R	3024	2SC31OM	3024	2SC46B	3024	2SC503L	3024	2SC594J
3024	2SC291X	3024	2SC31OOR	3024	2SC46C	3024	2SC503M	3024	2SC594K
3024	2SC291Y	3024	2SC31OR	3024	2SC46E	3024	2SC503OR	3024	2SC594L
3024	2SC292A	3024	2SC31OX	3024	2SC46F	3024	2SC503X	3024	2SC594M
3024	2SC292B	3024	2SC31OY	3024	2SC46G	3024	2SC504	3024	2SC594OR
3024	2SC292C	3024	2SC31A	3024	2SC46GN	3024	2SC504-O	3024	2SC594R
3024	2SC292D	3024	2SC31B	3024	2SC46H	3024	2SC504-Y	3024	2SC594X
3024	2SC292F	3024	2SC31C	3024	2SC46J	3024	2SC504A	3024	2SC594Y
3024	2SC292G	3024	2SC31D	3024	2SC46K	3024	2SC504B	3024	2SC59A
3024	2SC292GN	3024	2SC31E	3024	2SC46L	3024	2SC504C	3024	2SC59B
3024	2SC292H	3024	2SC31F	3024	2SC46M	3024	2SC504D	3024	2SC59C
3024	2SC292J	3024	2SC31G	3024	2SC46OR	3024	2SC504E	3024	2SC59D
3024	2SC292K	3024	2SC31GN	3024	2SC46X	3024	2SC504F	3024	2SC59E
3024	2SC292L	3024	2SC31H	3024	2SC46Y	3024	2SC504G	3024	2SC59F
3024	2SC292M	3024	2SC31J	3024	2SC479	3024	2SC504GN	3024	2SC59G
3024	2SC292OR	3024	2SC31K	3024	2SC479A	3024	2SC504GR	3024	2SC59GN
3024	2SC292R	3024	2SC31L	3024	2SC479B	3024	2SC504H	3024	2SC59H
3024	2SC292X	3024	2SC31M	3024	2SC479C	3024	2SC504J	3024	2SC59J
3024	2SC292Y	3024	2SC31OR	3024	2SC479D	3024	2SC504K	3024	2SC59K
3024	2SC293A	3024	2SC31R	3024	2SC479F	3024	2SC504L	3024	2SC59L
3024	2SC293B	3024	2SC31X	3024	2SC479G	3024	2SC504M	3024	2SC59M
3024	2SC293C	3024	2SC31Y	3024	2SC479GN	3024	2SC504OR	3024	2SC59OR
3024	2SC293D	3024	2SC32	3024	2SC479H	3024	2SC504R	3024	2SC59R
3024	2SC293E	3024	2SC32A	3024	2SC479J	3024	2SC504X	3024	2SC59Y
3024	2SC293F	3024	2SC32B	3024	2SC479K	3024	2SC51	3024	2SC61
3024	2SC293G	3024	2SC32C	3024	2SC479L	3024	2SC512-O	3024	2SC610
3024	2SC293GN	3024	2SC32D	3024	2SC479M	3024	2SC512A	3024	2SC610A
3024	2SC293H	3024	2SC32F	3024	2SC479OR	3024	2SC512B	3024	2SC610B
3024	2SC293J	3024	2SC32G	3024	2SC479R	3024	2SC512C	3024	2SC610C
3024	2SC293K	3024	2SC32GN	3024	2SC479X	3024	2SC512D	3024	2SC610D
3024	2SC293L	3024	2SC32H	3024	2SC479Y	3024	2SC512E	3024	2SC610E
3024	2SC293M	3024	2SC32J	3024	2SC497	3024	2SC512G	3024	2SC610F
3024	2SC293OR	3024	2SC32K	3024	2SC497A	3024	2SC512GN	3024	2SC610G
3024	2SC293R	3024	2SC32L	3024	2SC497B	3024	2SC512H	3024	2SC610GN
3024	2SC293X	3024	2SC32M	3024	2SC497C	3024	2SC512J	3024	2SC610H
3024	2SC293Y	3024	2SC32OR	3024	2SC497D	3024	2SC512K	3024	2SC610J
3024	2SC30	3024	2SC32R	3024	2SC497E	3024	2SC512M	3024	2SC610K
3024	2SC306	3024	2SC32X	3024	2SC497F	3024	2SC512OR	3024	2SC610L
3024	2SC306A	3024	2SC32Y	3024	2SC497GN	3024	2SC512X	3024	2SC610M
3024	2SC306B	3024	2SC352B	3024	2SC497H	3024	2SC512Y	3024	2SC610OR
3024	2SC306C	3024	2SC352C	3024	2SC497J	3024	2SC513A	3024	2SC610R
3024	2SC306D	3024	2SC352D	3024	2SC497L	3024	2SC513B	3024	2SC610X
3024	2SC306E	3024	2SC352E	3024	2SC497M	3024	2SC513C	3024	2SC610Y
3024	2SC306F	3024	2SC352F	3024	2SC497OR	3024	2SC513D	3024	2SC614A
3024	2SC306G	3024	2SC352G	3024	2SC497R	3024	2SC513E	3024	2SC614B
3024	2SC306GN	3024	2SC352GN	3024	2SC497X	3024	2SC513F	3024	2SC614GN
3024	2SC306H	3024	2SC352H	3024	2SC497Y	3024	2SC513G	3024	2SC614H
3024	2SC306J	3024	2SC352J	3024	2SC498	3024	2SC513GN	3024	2SC614J
3024	2SC306K	3024	2SC352K	3024	2SC498A	3024	2SC513H	3024	2SC614K
3024	2SC306L	3024	2SC352L	3024	2SC498B	3024	2SC513J	3024	2SC614L
3024	2SC306M	3024	2SC352M	3024	2SC498C	3024	2SC513K	3024	2SC614M
3024	2SC306OR	3024	2SC352OR	3024	2SC498D	3024	2SC513L	3024	2SC614OR
3024	2SC306R	3024	2SC352R	3024	2SC498F	3024	2SC513M	3024	2SC614R
3024	2SC306X	3024	2SC352X	3024	2SC498G	3024	2SC513OR	3024	2SC614X
3024	2SC306Y	3024	2SC352Y	3024	2SC498GN	3024	2SC513X	3024	2SC614Y
3024	2SC307	3024	2SC353	3024	2SC498H	3024	2SC513Y	3024	2SC61A
3024	2SC307A	3024	2SC353A	3024	2SC498J	3024	2SC516	3024	2SC61B
3024	2SC307B	3024	2SC353B	3024	2SC498K	3024	2SC516A	3024	2SC61C
3024	2SC307C	3024	2SC353C	3024	2SC498L	3024	2SC516B	3024	2SC61D
3024	2SC307D	3024	2SC353D	3024	2SC498M	3024	2SC516C	3024	2SC61E
3024	2SC307E	3024	2SC353E	3024	2SC498OR	3024	2SC516D	3024	2SC61F
3024	2SC307F	3024	2SC353F	3024	2SC498X	3024	2SC516E	3024	2SC61G
3024	2SC307G	3024	2SC353G	3024	2SC498Y	3024	2SC516F	3024	2SC61GN
3024	2SC307GN	3024	2SC353GN	3024	2SC49A	3024	2SC516G	3024	2SC61H
3024	2SC307H	3024	2SC353H	3024	2SC49B	3024	2SC516GN	3024	2SC61J
3024	2SC307J	3024	2SC353J	3024	2SC49C	3024	2SC516H	3024	2SC61K
3024	2SC307K	3024	2SC353K	3024	2SC49D	3024	2SC516J	3024	2SC61M
3024	2SC307L	3024	2SC353L	3024	2SC49F	3024	2SC516K	3024	2SC61OR
3024	2SC307M	3024	2SC353M	3024	2SC49G	3024	2SC516L	3024	2SC61R
3024	2SC307OR	3024	2SC353OR	3024	2SC49H	3024	2SC516M	3024	2SC61Y
3024	2SC307R	3024	2SC353R	3024	2SC49J	3024	2SC516OR	3024	2SC69
3024	2SC307X	3024	2SC353X	3024	2SC49K	3024	2SC516R	3024	2SC696B
3024	2SC307Y	3024	2SC353Y	3024	2SC49L	3024	2SC516X	3024	2SC696H
3024	2SC308	3024	2SC376A	3024	2SC49M	3024	2SC516Y	3024	2SC69A
3024	2SC308A	3024	2SC376B	3024	2SC49OR	3024	2SC51A	3024	2SC69B
3024	2SC308B	3024	2SC376C	3024	2SC49X	3024	2SC51B	3024	2SC69C
3024	2SC308C	3024	2SC376D	3024	2SC501	3024	2SC51C	3024	2SC69D
3024	2SC308D	3024	2SC376E	3024	2SC501A	3024	2SC51D	3024	2SC69E
3024	2SC308E	3024	2SC376F	3024	2SC501B	3024	2SC51E	3024	2SC69F
3024	2SC308F	3024	2SC376G	3024	2SC501C	3024	2SC51F	3024	2SC69G
3024	2SC308G	3024	2SC376GN	3024	2SC501D	3024	2SC51G	3024	2SC69GN
3024	2SC308GN	3024	2SC376H	3024	2SC501E	3024	2SC51H	3024	2SC69H
3024	2SC308H	3024	2SC376J	3024	2SC501F	3024	2SC51J	3024	2SC69J
3024	2SC308J	3024	2SC376K	3024	2SC501G	3024	2SC51K	3024	2SC69K
3024	2SC308K	3024	2SC376L	3024	2SC501GN	3024	2SC51L	3024	2SC69L
3024	2SC308L	3024	2SC376M	3024	2SC501H	3024	2SC51M	3024	2SC69M
3024	2SC308M	3024	2SC376OR	3024	2SC501J	3024	2SC51OX	3024	2SC69OR
3024	2SC308OR	3024	2SC376X	3024	2SC501K	3024	2SC51R	3024	2SC69R
3024	2SC308R	3024	2SC376Y	3024	2SC501L	3024	2SC51X	3024	2SC69X
3024	2SC308X	3024	2SC423A	3024	2SC501M	3024	2SC51Y	3024	2SC69Y
3024	2SC308Y	3024	2SC423G	3024	2SC501OR	3024	2SC564AB	3024	2SC708
3024	2SC309	3024	2SC423GN	3024	2SC501R	3024	2SC564C	3024	2SC708A
3024	2SC309A	3024	2SC423H	3024	2SC501X	3024	2SC564D	3024	2SC708AB
3024	2SC309B	3024	2SC423J	3024	2SC502A	3024	2SC564E	3024	2SC708AC
3024	2SC309C	3024	2SC423K	3024	2SC502B	3024	2SC564F	3024	2SC708AH
3024	2SC309D	3024	2SC423L	3024	2SC502C	3024	2SC564G	3024	2SC708B
3024	2SC309E	3024	2SC423M	3024	2SC502D	3024	2SC564GN	3024	2SC708C
3024	2SC309F	3024	2SC423OR	3024	2SC502E	3024	2SC564H	3024	2SC708D
3024	2SC309G	3024	2SC423R	3024	2SC502F	3024	2SC564J	3024	2SC708E
3024	2SC309GN	3024	2SC423X	3024	2SC502G	3024	2SC564K	3024	2SC708F
3024	2SC309H	3024	2SC425A	3024	2SC502GN	3024	2SC564L	3024	2SC708GN
3024	2SC309J	3024	2SC425G	3024	2SC502H	3024	2SC564M	3024	2SC708H
3024	2SC309K	3024	2SC425GN	3024	2SC502J	3024	2SC564OR	3024	2SC708L
3024	2SC309L	3024	2SC425H	3024	2SC502K	3024	2SC564PL	3024	2SC708M
3024	2SC309M	3024	2SC425K	3024	2SC502L	3024	2SC564QC	3024	2SC708OR
3024	2SC309OR	3024	2SC425L			3024	2SC564X	3024	2SC708R
3024	2SC309R	3024	2SC425M			3024	2SC564Y	3024	2SC708X
3024	2SC309X	3024	2SC425OR			3024	2SC59	3024	2SC708Y
3024	2SC309Y	3024	2SC425R			3024	2SC590A	3024	2SC727C
3024	2SC30A	3024	2SC425X			3024	2SC590B	3024	2SC727D
3024	2SC30B	3024	2SC425Y			3024	2SC590C		
3024	2SC30C								
3024	2SC30D								

SK	Industry Standard No.
3024	2SC727E
3024	2SC727F
3024	2SC727G
3024	2SC727GN
3024	2SC727H
3024	2SC727J
3024	2SC727K
3024	2SC727L
3024	2SC727M
3024	2SC727R
3024	2SC727X
3024	2SC727Y
3024	2SC744A
3024	2SC797
3024	2SC797A
3024	2SC797B
3024	2SC797C
3024	2SC797D
3024	2SC797E
3024	2SC797F
3024	2SC797G
3024	2SC797GN
3024	2SC797H
3024	2SC797J
3024	2SC797K
3024	2SC797L
3024	2SC797M
3024	2SC797OR
3024	2SC797R
3024	2SC797X
3024	2SC797Y
3024	2SC798A
3024	2SC798B
3024	2SC798C
3024	2SC798D
3024	2SC798E
3024	2SC798G
3024	2SC798GN
3024	2SC798H
3024	2SC798J
3024	2SC798K
3024	2SC798L
3024	2SC798M
3024	2SC798OR
3024	2SC798R
3024	2SC798X
3024	2SC798Y
3024	2SC816
3024	2SC816A
3024	2SC816B
3024	2SC816C
3024	2SC816D
3024	2SC816E
3024	2SC816F
3024	2SC816G
3024	2SC816GN
3024	2SC816H
3024	2SC816K
3024	2SC816L
3024	2SC816M
3024	2SC816OR
3024	2SC816R
3024	2SC816X
3024	2SC816Y
3024	2SC826
3024	2SC826A
3024	2SC826B
3024	2SC826C
3024	2SC826D
3024	2SC826E
3024	2SC826F
3024	2SC826G
3024	2SC826H
3024	2SC826J
3024	2SC826K
3024	2SC826L
3024	2SC826M
3024	2SC826OR
3024	2SC826R
3024	2SC826X
3024	2SC826Y
3024	2SC827
3024	2SC827A
3024	2SC827B
3024	2SC827C
3024	2SC827D
3024	2SC827E
3024	2SC827F
3024	2SC827G
3024	2SC827GN
3024	2SC827H
3024	2SC827J
3024	2SC827K
3024	2SC827M
3024	2SC827OR
3024	2SC827R
3024	2SC827X
3024	2SC827Y
3024	2SC875
3024	2SC875-1
3024	2SC875-1C
3024	2SC875-1D
3024	2SC875-1E
3024	2SC875-2
3024	2SC875-2C
3024	2SC875-2D
3024	2SC875-2E
3024	2SC875-3
3024	2SC875-3C
3024	2SC875-3D
3024	2SC875-3E
3024	2SC875B
3024	2SC875C
3024	2SC875D
3024	2SC875DL
3024	2SC875E
3024	2SC875EL
3024	2SC875F
3024	2SC875G
3024	2SC875GN
3024	2SC875J
3024	2SC875K
3024	2SC875L
3024	2SC875M
3024	2SC875OR
3024	2SC875X
3024	2SC875Y
3024	2SC876
3024	2SC876A
3024	2SC876B
3024	2SC876C
3024	2SC876D
3024	2SC876E
3024	2SC876F
3024	2SC876G
3024	2SC876GN
3024	2SC876H
3024	2SC876J
3024	2SC876K
3024	2SC876L
3024	2SC876M
3024	2SC876OR
3024	2SC876R
3024	2SC876TVE
3024	2SC876TVF
3024	2SC876X
3024	2SC876Y
3024	2SC934A
3024	2SC934B
3024	2SC934GN
3024	2SC934H
3024	2SC934J
3024	2SC934K
3024	2SC934L
3024	2SC934M
3024	2SC934OR
3024	2SC934R
3024	2SC934X
3024	2SC934Y
3024	2SC97
3024	2SC972
3024	2SC972A
3024	2SC972B
3024	2SC972C
3024	2SC972D
3024	2SC972E
3024	2SC972F
3024	2SC972G
3024	2SC972GN
3024	2SC972H
3024	2SC972J
3024	2SC972K
3024	2SC972L
3024	2SC972M
3024	2SC972OR
3024	2SC972R
3024	2SC972X
3024	2SC972Y
3024	2SC97A
3024	2SC97B
3024	2SC97C
3024	2SC97D
3024	2SC97E
3024	2SC97F
3024	2SC97G
3024	2SC97GN
3024	2SC97H
3024	2SC97J
3024	2SC97K
3024	2SC97L
3024	2SC97M
3024	2SC97OR
3024	2SC97R
3024	2SC97X
3024	2SC97Y
3024	2SCM98F
3024	2SD121C
3024	2SD121D
3024	2SD121E
3024	2SD121F
3024	2SD121G
3024	2SD121GN
3024	2SD121J
3024	2SD121K
3024	2SD121L
3024	2SD121M
3024	2SD121OR
3024	2SD121R
3024	2SD121X
3024	2SD121Y
3024	2SD142X
3024	2SD172M
3024	2SD182
3024	2SD182A
3024	2SD182B
3024	2SD182C
3024	2SD182D
3024	2SD182E
3024	2SD182F
3024	2SD182G
3024	2SD182GN
3024	2SD182H
3024	2SD182J
3024	2SD182K
3024	2SD182L
3024	2SD182OR
3024	2SD182R
3024	2SD182X
3024	2SD182Y
3024	2SD183
3024	2SD183B
3024	2SD183E
3024	2SD183GN
3024	2SD183K
3024	2SD183L
3024	2SD183OR
3024	2SD183R
3024	2SD204
3024	2SD204A
3024	2SD204B
3024	2SD204BL
3024	2SD204C
3024	2SD204D
3024	2SD204E
3024	2SD204F
3024	2SD204G
3024	2SD204GA
3024	2SD204GN
3024	2SD204H
3024	2SD204J
3024	2SD204K
3024	2SD204L
3024	2SD204M
3024	2SD204R
3024	2SD204Y
3024	2SD205A
3024	2SD205B
3024	2SD205C
3024	2SD205D
3024	2SD205E
3024	2SD205F
3024	2SD205G
3024	2SD205GN
3024	2SD205H
3024	2SD205J
3024	2SD205L
3024	2SD205M
3024	2SD205OR
3024	2SD205R
3024	2SD205X
3024	2SD205Y
3024	2SD219
3024	2SD219A
3024	2SD219B
3024	2SD219C
3024	2SD219D
3024	2SD219E
3024	2SD219G
3024	2SD219GN
3024	2SD219H
3024	2SD219J
3024	2SD219K
3024	2SD219L
3024	2SD219M
3024	2SD219OR
3024	2SD219R
3024	2SD219X
3024	2SD219Y
3024	2SD220
3024	2SD220A
3024	2SD220B
3024	2SD220C
3024	2SD220D
3024	2SD220E
3024	2SD220G
3024	2SD220GN
3024	2SD220H
3024	2SD220J
3024	2SD220K
3024	2SD220L
3024	2SD220M
3024	2SD220OR
3024	2SD220X
3024	2SD220Y
3024	2SD221
3024	2SD222A
3024	2SD222B
3024	2SD222C
3024	2SD222D
3024	2SD222E
3024	2SD222F
3024	2SD222G
3024	2SD222GN
3024	2SD222H
3024	2SD222J
3024	2SD222K
3024	2SD222L
3024	2SD222M
3024	2SD222OR
3024	2SD222R
3024	2SD222X
3024	2SD222Y
3024	2SDU84GN
3024	3L4-6007-4
3024	4-3023223
3024	4-3023844
3024	4-397
3024	4-398
3024	4-686182-3
3024	7-0051-00
3024	8-0053300
3024	8-905-705-410
3024	8Q-3-13
3024	09-302171
3024	09-305076
3024	9-5220
3024	9-5226-004
3024	9-5226-4
3024	11C1536
3024	13-105698-1
3024	13-14085-15A
3024	13-14085-91
3024	13-14605-1
3024	13-15833-1
3024	13-23840-1
3024	13-26666
3024	13-26666-1
3024	13-27443-1
3024	13-28392-1
3024	13-28392-3
3024	13-28394-1
3024	13-28394-2
3024	13-28394-3
3024	13-28471-1
3024	13-34371-1
3024	13-55018-4
3024	13-55064-1
3024	14-602-24
3024	14-602-43
3024	14-602-49
3024	14-602-59
3024	14-603-01
3024	14-603-07
3024	17-443
3024	19-020-075
3024	19-3934-643
3024	19A115300-1
3024	19A115300-2
3024	19A115304-2
3024	19C300115-P1
3024	020-1110-008
3024	020-1111-018
3024	20-1680-174
3024	20A0076
3024	21A015-018
3024	21A015-019
3024	21A015-026
3024	21A105-004
3024	21A105-006
3024	21M185
3024	21M448
3024	23-5039
3024	23B114053
3024	23B114054
3024	24MW663
3024	24MW674
3024	24MW714
3024	026-100013
3024	31-0083
3024	34-6001-51
3024	34-6001-83
3024	34-6015-23
3024	34-6015-30
3024	34-6016-22
3024	34-6016-27
3024	34-6016-30
3024	34-6016-33
3024	41B581014
3024	42-19642
3024	42-21233
3024	43-023223
3024	43-023844
3024	46-86233-3
3024	46-86371-3
3024	46-86381-3
3024	46-86426-3
3024	46-86427-3
3024	48-01-007
3024	48-134689
3024	48-134838
3024	48-134941
3024	48-134953
3024	48-134982
3024	48-137005
3024	48-137041
3024	48-137088
3024	48-137307
3024	48-137988
3024	48-859248
3024	48-869138
3024	48-869170
3024	48-869228
3024	48-869464
3024	48-90232A08
3024	48-97127A04
3024	48K869228
3024	48M355037
3024	48R859428
3024	48R869138
3024	48R869170
3024	48R869464
3024	48S134765
3024	48S134899
3024	48S134941
3024	48S134942
3024	48S134953
3024	48S134992
3024	48S137041
3024	53P169
3024	55-1084
3024	55-642
3024	57A104-8-6
3024	57B109-9
3024	57B113-9
3024	57B128-9
3024	57B167-9
3024	57B170-9
3024	57B2-57
3024	57C109-9
3024	57C12-4
3024	57C14-1
3024	57C14-2
3024	57C14-3
3024	57M1-33
3024	57M1-34
3024	57M2-11
3024	57M2-14
3024	57M2-18
3024	57M2-7
3024	57M3-3
3024	61-747
3024	61-813
3024	61C004-1
3024	63-11989
3024	63-12989
3024	63-12990
3024	63-13214
3024	63-13215
3024	72N1B
3024	76-14090-1
3024	76N1B
3024	76N2B
3024	76N3B
3024	77-271490
3024	77-271490-1
3024	77N2B
3024	79F114-1
3024	79F114-3
3024	80-053300
3024	81T2
3024	86-207-2
3024	86-208-2
3024	86-210-2
3024	86-211-2
3024	86-234-2
3024	86-266-2
3024	86-267-2
3024	86-273-2
3024	86-330-2
3024	86-336-2
3024	86-393-2
3024	86-428-2
3024	86-428-9
3024	86-440-2
3024	86-441-2
3024	86-452-2
3024	86-463-2
3024	86-463-3
3024	86-5073-2
3024	86-5093-2
3024	86-510-2
3024	94N1V
3024	95-220
3024	95-226-004
3024	95-226-4
3024	96-5107-01
3024	96-5107-02
3024	96-5170-01
3024	96-5180-01
3024	96-5180-02
3024	96-5203-01
3024	96-5204-01
3024	96-5244-01
3024	96-5252-01
3024	96-5256-01
3024	96XZ6053/38N
3024	998074
3024	0103-0014R
3024	0103-0014S
3024	0103-0051
3024	0103-0503
3024	0103-0503S
3024	0103-0607
3024	0103-0616
3024	0104-0013
3024	104H01
3024	107N1
3024	112-203053
3024	112-361
3024	112-525
3024	119-0077
3024	121-639CL
3024	121-676
3024	121-722
3024	121-737CL
3024	121-766
3024	121-773CL
3024	121-844
3024	125A137
3024	0131-000561
3024	0131-001429
3024	0131-001430
3024	131-005807
3024	132-021B
3024	132-022
3024	132-038
3024	132-066
3024	132N1
3024	173A4490-5
3024	176-004
3024	226-4
3024	231-0013
3024	248-38104-1
3024	260P10003
3024	260P10003A
3024	260P10005
3024	260P12401
3024	260P17002
3024	270-950-037-02
3024	297V049H06
3024	297V061C05
3024	297V062C06
3024	297V072C05
3024	297V074C12
3024	297V082B01
3024	297V083C04
3024	324-019
3024	324-6011
3024	324-6013
3024	325-1513-46
3024	342-1
3024	344-6011-2
3024	344-6013-1B
3024	345-2
3024	349-1
3024	349-2
3024	352-043-010
3024	352-0479-010
3024	352-0816-010
3024	353-9301-001
3024	360-1
3024	392-1
3024	394-3127-1
3024	394-3127-2
3024	394-3127-3
3024	411-237
3024	417-114
3024	417-133
3024	417-136
3024	417-137
3024	417-155
3024	417-224
3024	417-233
3024	417-237
3024	417-49
3024	417-59
3024	417-87
3024	417-88
3024	417-89
3024	422-1233
3024	422-1404
3024	422-2158
3024	430-23223
3024	430-23844
3024	448A662
3024	454A104
3024	491A948
3024	555-4
3024	565-1
3024	573-532
3024	577R819H01
3024	608-2
3024	615-1
3024	622-1
3024	625-1
3024	639
3024	639CL
3024	660-125
3024	660-145
3024	686-0012
3024	686-0112
3024	686-0130
3024	690V075H62
3024	690V086H90
3024	703-2
3024	703B
3024	729-3
3024	737
3024	753-8510-470
3024	773CL
3024	800-512-00
3024	800-513-00
3024	800-515-00
3024	800-521-02
3024	800-521-03
3024	800-522-03
3024	800-53300
3024	914F298-1
3024	921-188B
3024	921-236B
3024	921-337B
3024	921-60B
3024	964-22158
3024	991-00-2888
3024	991-01-1305
3024	1008
3024	1010-7993
3024	1010-8041
3024	1040-7
3024	1099-0950
3024	1157
3024	1282
3024	1414-157
3024	1414-186
3024	1414-189
3024	1933-17
3024	2057A100-17
3024	2057A2-151
3024	2057A2-230
3024	2057A2-295
3024	2057B100-17
3024	2057B107-8
3024	2057B109-9
3024	2057B113-9
3024	2057B126-12
3024	2057B129-9
3024	2057B144-12
3024	2057B153-9
3024	2057B156-9
3024	2093A38-24

SK	Industry Standard No.
3024	2498-163
3024	2904-032
3024	3034
3024	3202-51-01
3024	3202-5H01
3024	3532-1
3024	3555-3
3024	3591
3024	3608-2
3024	3615
3024	3622-2
3024	4473-M-12
3024	4473-M3
3024	4473-N
3024	4483
3024	4686-182-3
3024	4689
3024	4822-130-40356
3024	5613-1213D
3024	6151
3024	6651-486
3024	8102-209
3024	8304
3024	8554-9
3024	8868-6
3024	8880-3
3024	8883-4
3024	9367-1
3024	9405-0
3024	9405-1
3024	16082
3024	018077
3024	20810-93
3024	27125-380
3024	030011
3024	030011-1
3024	030011-2
3024	36579
3024	37464
3024	37649
3024	37694
3024	37800
3024	38270
3024	38716
3024	40309
3024	40309V1
3024	40309V2
3024	40311
3024	40314
3024	40315
3024	40315V1
3024	40317
3024	40320
3024	40323
3024	40326
3024	40360
3024	40361
3024	40366
3024	40407
3024	40409
3024	40457
3024	40501
3024	40539
3024	40611
3024	40616
3024	40635
3024	41053
3024	43021-198
3024	43117
3024	50137-2
3024	51194
3024	51194-01
3024	51194-02
3024	51194-03
3024	61828
3024	66007-4
3024	000073110
3024	88803
3024	88834
3024	94041
3024	94050
3024	94051
3024	94066
3024	94070
3024	95220
3024	110669
3024	111278
3024	112360
3024	112361
3024	119822
3024	126721
3024	126725
3024	127376
3024	127845
3024	129051
3024	132175
3024	132327
3024	132328
3024	136424
3024	136696
3024	137241
3024	140501
3024	141767
3024	141783
3024	143042
3024	165735
3024	165736
3024	186342A
3024	221158
3024	230214
3024	231375
3024	232841
3024	233735
3024	233944
3024	234024
3024	236287
3024	239612
3024	276160
3024	276413
3024	276415
3024	301606
3024	346015-23
3024	346015-30
3024	346016-17
3024	346016-27
3024	489751-129
3024	489751-144
3024	0573527
3024	573532
3024	602122
3024	605113
3024	608122
3024	610213-1
3024	660070
3024	660074
3024	702407-00
3024	800073-6
3024	800073-7
3024	916034
3024	984196
3024	984229
3024	988993
3024	1417342-1
3024	1417345-2
3024	1417381-2
3024	1471120-14
3024	1471123-3
3024	1471123-4
3024	1471123-5
3024	1473555-3
3024	1473565-001
3024	1473566-1
3024	1473608-1
3024	1473608-2
3024	1473608-3
3024	1473615-1
3024	1473622-2
3024	1473625-1
3024	1815154-9
3024	1815156
3024	1815157
3024	1815159
3024	1960083-1
3024	1967799
3024	1967799-1
3024	1967801
3024	2006334-155
3024	2006623-148
3024	2097013-0702
3024	2320233
3024	2320646-1
3024	2320647-1
3024	2327022
3024	2327292
3024	2327293
3024	2327332
3024	2327403
3024	2485077-2
3024	2485077-3
3024	5320612
3024	06120028
3024	7902310
3024	8000736
3024	8000737
3024	8524457
3024	11166535
3024	19901403
3024	22130009
3024	23114058
3024	23114195
3024	23114232
3024	23200596-1
3024	43020731
3024	43021198
3024	43023223
3024	62054298
3024	62208945
3024	62530915
3024	62638265
3024	62691298
3024	80421980
3024	181515417
3024	16105191229
3024	16156197229
3024/128	A107
3024/128	A175
3024/128	A2416
3024/128	A311
3024/128	A3K
3024/128	A485
3024/128	A490
3024/128	BC174A
3024/128	BC174B
3024/128	BC232A
3024/128	BC232B
3024/128	BC313
3024/128	BCW94K
3024/128	BCW95K
3024/128	BFW33
3024/128	BFX94
3024/128	BFX95
3024/128	BFX96
3024/128	BFY12
3024/128	BFY16
3024/128	BFY19
3024/128	BFY47
3024/128	BFY72
3024/128	BSX22
3024/128	BSX63
3024/128	BSY25
3024/128	BSY91
3024/128	C120
3024/128	C1220E
3024/128	C213
3024/128	C214
3024/128	C215
3024/128	C224
3024/128	C225
3024/128	C231
3024/128	C248
3024/128	C27
3024/128	C292
3024/128	C31
3024/128	C310
3024/128	C376
3024/128	C484R
3024/128	C484Y
3024/128	C49
3024/128	C512
3024/128	C512-O
3024/128	C512-R
3024/128	C513
3024/128	C513-O
3024/128	C513R
3024/128	C59
3024/128	C590
3024/128	C594
3024/128	C628
3024/128	C69
3024/128	C776
3024/128	C776(Y)
3024/128	C776Y
3024/128	C798
3024/128	C803
3024/128	C959
3024/128	CDQ10045
3024/128	CDQ10046
3024/128	CK419
3024/128	CK422
3024/128	CK474
3024/128	CK475
3024/128	CP408
3024/128	CP409
3024/128	CS-2143
3024/128	CS1255HF
3024/128	EA15X249
3024/128	EA15X360
3024/128	EA15X7592
3024/128	EA15X7639
3024/128	EQS-131
3024/128	EQS-78
3024/128	FCS-9013F
3024/128	FCS-9013G
3024/128	FCS9013
3024/128	FCS9013F
3024/128	FCS9013G
3024/128	FCS9013H
3024/128	FCS9013HH
3024/128	FCS9014C
3024/128	G05-055-D
3024/128	G05-413A
3024/128	G05-415-B
3024/128	GE-45
3024/128	HD-00471
3024/128	HEP-S5026
3024/128	HR13
3024/128	HS-40016
3024/128	HT312131CQ
3024/128	IP20-0165
3024/128	IP20-0211
3024/128	LM1501H
3024/128	MH1501
3024/128	MPS-U01
3024/128	MPS3642
3024/128	MPSU01A
3024/128	PET4001
3024/128	PTC125
3024/128	RV2068
3024/128	SE6020
3024/128	SPS5450
3024/128	T-342
3024/128	T1007
3024/128	TA2710
3024/128	TR-8022
3024/128	TR01062-1
3024/128	TR23A
3024/128	TV-49
3024/128	TV26
3024/128	VS2SD467-C/-1
3024/128	ZT210
3024/128	ZT211
3024/128	1A0027
3024/128	1A0048
3024/128	2N1252
3024/128	2N1253
3024/128	2N1479
3024/128	2N1480
3024/128	2N1481
3024/128	2N1482
3024/128	2N1893
3024/128	2N2405
3024/128	2N2846
3024/128	2N2848
3024/128	2N2849
3024/128	2N2850
3024/128	2N2851
3024/128	2N2852
3024/128	2N2853
3024/128	2N2854
3024/128	2N2855
3024/128	2N2856
3024/128	2N3262
3024/128	2N3506
3024/128	2N3507
3024/128	2N3567
3024/128	2N3877
3024/128	2N3948
3024/128	2N5108
3024/128	2N5129
3024/128	2N5233
3024/128	2N5234
3024/128	2N5235
3024/128	2N5249
3024/128	2N5249A
3024/128	2N5309
3024/128	2N5310
3024/128	2N5311
3024/128	2N5320HS
3024/128	2N5421
3024/128	2N5449
3024/128	2N5682
3024/128	2N5784
3024/128	2N5785
3024/128	2N5881
3024/128	2N5882
3024/128	2N699
3024/128	2S103
3024/128	2S104
3024/128	2S741
3024/128	2S744
3024/128	2SC120
3024/128	2SC1220E
3024/128	2SC1385H
3024/128	2SC1627Y
3024/128	2SC189F
3024/128	2SC2060
3024/128	2SC2060Q
3024/128	2SC208
3024/128	2SC213
3024/128	2SC214
3024/128	2SC215
3024/128	2SC223
3024/128	2SC224
3024/128	2SC229
3024/128	2SC23
3024/128	2SC231
3024/128	2SC232
3024/128	2SC233
3024/128	2SC238
3024/128	2SC24
3024/128	2SC248
3024/128	2SC27
3024/128	2SC292
3024/128	2SC298-4
3024/128	2SC310
3024/128	2SC310B
3024/128	2SC310C
3024/128	2SC310D
3024/128	2SC310E
3024/128	2SC352
3024/128	2SC352A
3024/128	2SC376
3024/128	2SC456D
3024/128	2SC484
3024/128	2SC484R
3024/128	2SC484Y
3024/128	2SC510
3024/128	2SC512
3024/128	2SC513
3024/128	2SC513-O
3024/128	2SC564A
3024/128	2SC564P
3024/128	2SC564Q
3024/128	2SC564R
3024/128	2SC590
3024/128	2SC590Y
3024/128	2SC594
3024/128	2SC614C
3024/128	2SC614F
3024/128	2SC614G
3024/128	2SC615
3024/128	2SC615A
3024/128	2SC615B
3024/128	2SC615C
3024/128	2SC615D
3024/128	2SC615E
3024/128	2SC615G
3024/128	2SC65
3024/128	2SC727
3024/128	2SC727A
3024/128	2SC727B
3024/128	2SC776
3024/128	2SC776(Y)
3024/128	2SC776Y
3024/128	2SC798
3024/128	2SC803
3024/128	2SC960
3024/128	2SC960A
3024/128	2SC960B
3024/128	2SC960D
3024/128	2SC960S
3024/128	2SD120
3024/128	2SD120A
3024/128	2SD120B
3024/128	2SD120C
3024/128	2SD120H
3024/128	2SD121
3024/128	2SD121A
3024/128	2SD121B
3024/128	2SD121H
3024/128	2SD205
3024/128	2SD219F
3024/128	2SD220F
3024/128	2SD222
3024/128	3L4-6007-41
3024/128	6A12988
3024/128	09-301071
3024/128	09-302119
3024/128	09-309031
3024/128	13-27432-1
3024/128	13-32630-3
3024/128	13-39114-1
3024/128	13-39114-2
3024/128	14-602-55
3024/128	14-602-74
3024/128	17-458
3024/128	019-003692
3024/128	23-5037
3024/128	24M125
3024/128	24MW1152
3024/128	34-6015-28
3024/128	46-86182-3
3024/128	46-86589-3
3024/128	46-86591-3
3024/128	48-137505
3024/128	48-21598B01
3024/128	48S134988
3024/128	50-40105-08
3024/128	57A167-9
3024/128	57A219-14
3024/128	57A241-14
3024/128	57A280-14
3024/128	86-284-2
3024/128	86X0071-001
3024/128	90-31
3024/128	90-66
3024/128	96-5165-01
3024/128	99S091-1
3024/128	121-708
3024/128	121-710
3024/128	121-843
3024/128	12103115
3024/128	0124
3024/128	130
3024/128	130-144
3024/128	130-150
3024/128	130-172
3024/128	130-174
3024/128	130-245
3024/128	167-9
3024/128	306-1
3024/128	344-6012-1
3024/128	417-109
3024/128	576-0001-004
3024/128	884-250-011
3024/128	921-230B
3024/128	991-013544
3024/128	1004
3024/128	1018
3024/128	1039-0441
3024/128	1067
3024/128	1071-4913
3024/128	1804-17
3024/128	1866-17
3024/128	2017-110
3024/128	2057A2-404
3024/128	3565
3024/128	4822-130-40314
3024/128	5613-1209
3024/128	5613-1209C
3024/128	8000-00049-061
3024/128	35001
3024/128	36803
3024/128	36848
3024/128	36849
3024/128	36850
3024/128	36997
3024/128	37431
3024/128	37432
3024/128	37433
3024/128	37894
3024/128	38063
3024/128	38190
3024/128	38281
3024/128	40084
3024/128	40314V1
3024/128	40314V2
3024/128	40315V2
3024/128	40317V1
3024/128	40317V2
3024/128	40320V1
3024/128	40320V2
3024/128	40323V1
3024/128	40323V2
3024/128	40326V2
3024/128	40360V1
3024/128	40360V2
3024/128	40361V2
3024/128	40408
3024/128	43115
3024/128	44763
3024/128	71447-1
3024/128	71447-3
3024/128	88686
3024/128	95241-3
3024/128	111193-001
3024/128	112525
3024/128	123139
3024/128	128940
3024/128	131243
3024/128	138763
3024/128	165737
3024/128	167569
3024/128	171162-113
3024/128	171162-188
3024/128	248017
3024/128	742729
3024/128	916033
3024/128	2320647
3024/128	8114057
3024/128	22114210
3024/128	23114314
3024/128	43025538
3024/128	43025539
3024/128	43029484
3024/128	62389486
3024/128	62638222
3024/128	62734639
3024/128	62766255
3024/128	4109213174
3025	A054-109
3025	A1016
3025	A116084
3025	A116284
3025	A119983
3025	A126724
3025	A1473549-1
3025	A1473616-1
3025	A170
3025	A2057B110-9
3025	A2057B112-9
3025	A2057B114-9
3025	A2057B115-9
3025	A2057B116-9
3025	A2057B121-9
3025	A2057B122-9
3025	A2057B145-12
3025	A2057B163-12
3025	A2482
3025	A297074C11
3025	A297V073C01
3025	A297V073C02
3025	A297V082B03
3025	A30278
3025	A3523
3025	A353-9008-001
3025	A3533
3025	A3533-1
3025	A3616-1
3025	A36577
3025	A4037764-2
3025	A40410
3025	A417-138
3025	A417-170
3025	A417-234
3025	A417-43
3025	A4478
3025	A495
3025	A4950
3025	A527
3025	A528
3025	A532
3025	A532A
3025	A532B
3025	A532C
3025	A532D
3025	A532E
3025	A532F
3025	A537
3025	A537A
3025	A537AA
3025	A537AB
3025	A537AC
3025	A537AH
3025	A537B
3025	A537C
3025	A537H
3025	A546
3025	A546A
3025	A546B
3025	A546E
3025	A546H
3025	A551
3025	A551C
3025	A551D
3025	A551E
3025	A552
3025	A560
3025	A571
3025	A594
3025	A594-O
3025	A594-R
3025	A594-Y
3025	A595C
3025	A604
3025	A606
3025	A606S
3025	A608-C
3025	A608-E
3025	A608A
3025	A608B
3025	A608C
3025	A608D
3025	A608E
3025	A608F
3025	A608G
3025	A628
3025	A628A
3025	A628D
3025	A628E

SK	Industry Standard No.
3025	A628F
3025	A800-511-00
3025	A800-516-00
3025	A8015613
3025	A815185
3025	A815185E
3025	A83P2B
3025	A880-250-107
3025	A94063
3025	AR304
3025	AR308
3025	AT2848
3025	AT394
3025	AT395
3025	AT396
3025	AT397
3025	AT398
3025	AT460
3025	AT461
3025	AT462
3025	AT463
3025	AT464
3025	AT465
3025	AT466
3025	AT467
3025	AT468
3025	AT480
3025	AT481
3025	AT482
3025	AT483
3025	AT484
3025	AT485
3025	B5493957-4
3025	B5493957-5
3025	B5493957-6
3025	BC139
3025	BC143
3025	BC160
3025	BC160-10
3025	BC160-16
3025	BC160-6
3025	BC161
3025	BC161-10
3025	BC161-16
3025	BC161-6
3025	BC287
3025	BC303
3025	BC360-10
3025	BC360-16
3025	BC360-6
3025	BC361-10
3025	BC361-6
3025	BC404A
3025	BC404VI
3025	BC477
3025	BC477A
3025	BCY17
3025	BCY18
3025	BCY19
3025	BCY21
3025	BCY22
3025	BCY23
3025	BCY24
3025	BCY25
3025	BCY26
3025	BCY27
3025	BCY28
3025	BCY30
3025	BCY31
3025	BCY32
3025	BCY33
3025	BCY34
3025	BCY67
3025	BCY96
3025	BCY96B
3025	BCY97
3025	BCY97B
3025	BDY70
3025	BFW91
3025	BFX38
3025	BFX39
3025	BFX40
3025	BFX41
3025	BFX74A
3025	BFX87
3025	BFX88
3025	BSV15
3025	BSV16
3025	BSW23
3025	BSX40
3025	BSX41
3025	BTX-097
3025	C118
3025	C119
3025	C201
3025	C36577
3025	C610
3025	CDC-9000-1B
3025	CDC9002
3025	CE4002D
3025	CJ5209
3025	CS-6228G
3025	CS1237
3025	CS1369
3025	CS2142
3025	CS9102B
3025	CS9128
3025	CXL129
3025	E2498
3025	EA0087
3025	ECG129
3025	EL264
3025	EP15X4
3025	ES51
3025	ET15X33
3025	ETS-069
3025	ETS-071
3025	F2041
3025	FCS1795D
3025	GE-244
3025	GET2904
3025	GET2905
3025	HA7530
3025	HA7531
3025	HA7630
3025	HA7631
3025	HA7632
3025	HEP-S3012
3025	HEP-S5023
3025	HS40032
3025	HT104971A0
3025	IRTR88
3025	J241015
3025	K1181
3025	KD2120
3025	LDA450
3025	M4478
3025	M828
3025	MH9460A
3025	MM4005
3025	MM4006
3025	MM4019
3025	MPS9460A
3025	MR3934
3025	MSJ7505
3025	PTC141
3025	QA-11
3025	QA-17
3025	QA-9
3025	R2270-76963
3025	RCA1A02
3025	RCA1A05
3025	RCA1A08
3025	RCA1A19
3025	RE18
3025	REN129
3025	REN159
3025	RS8100
3025	RS8102
3025	RS8104
3025	RS8106
3025	RS8108
3025	RS8110
3025	RS8112
3025	RT5230
3025	RV1472
3025	RV2069
3025	S-437
3025	S1430
3025	S1431
3025	S1477
3025	S1520
3025	S1698
3025	S1863
3025	S1983
3025	S2117
3025	S2368
3025	S2370
3025	S2398C
3025	S24597
3025	S24612
3025	S24612A
3025	S24615
3025	S2771
3025	S2991
3025	S2993
3025	S2994
3025	S2995
3025	S3012
3025	S33886A
3025	SDT3321
3025	SDT3322
3025	SDT3501
3025	SDT3502
3025	SDT3503
3025	SE8540
3025	SE8541
3025	SE8542
3025	SL3101
3025	SM3987
3025	SOS0121
3025	SPS-29
3025	SPS-4076
3025	SPS-4078
3025	SPS0121
3025	SPS2226
3025	SPS4010
3025	SPS4301
3025	SPS4310
3025	SPS4312
3025	SPS4462
3025	SPS4492
3025	SPS4497
3025	ST72039
3025	ST72040
3025	STC5610
3025	STC5611
3025	STC5612
3025	STX0011
3025	SX61
3025	T-482
3025	T1275
3025	T1276
3025	T1808
3025	T1808A
3025	T1808B
3025	T1808C
3025	T1808D
3025	T1808E
3025	T246
3025	T459
3025	TIS93M
3025	TM129
3025	TR-28
3025	TR-8020
3025	TR1002
3025	TR1004
3025	TR22A
3025	TR28
3025	TV29
3025	TVS2SA546
3025	TVS2SC1256
3025	TVS2SC1256HG
3025	V180
3025	VS-CS1256HG
3025	W15
3025	WEP242
3025	ZT286
3025	002-010300-6
3025	002-010700
3025	002-012600
3025	002-9800-12
3025	2N1197
3025	2N1228
3025	2N1229
3025	2N1230
3025	2N1231
3025	2N1232
3025	2N1233
3025	2N1234
3025	2N1254
3025	2N1255
3025	2N1258
3025	2N1259
3025	2N1275
3025	2N1429
3025	2N2904
3025	2N2904A
3025	2N2905
3025	2N2905A
3025	2N3120
3025	2N3133
3025	2N3134
3025	2N327B
3025	2N328B
3025	2N329
3025	2N329A
3025	2N329B
3025	2N330
3025	2N330A
3025	2N3344
3025	2N3345
3025	2N3494
3025	2N3502
3025	2N3503
3025	2N3645
3025	2N3671
3025	2N3677
3025	2N3719
3025	2N3720
3025	2N3762
3025	2N3763
3025	2N3764
3025	2N3776
3025	2N3780
3025	2N3867
3025	2N3868
3025	2N4026
3025	2N4027
3025	2N4028
3025	2N4029
3025	2N4030
3025	2N4031
3025	2N4032
3025	2N4033
3025	2N4036
3025	2N4314
3025	2N4404
3025	2N4405
3025	2N4406
3025	2N4407
3025	2N4412
3025	2N4412A
3025	2N4414
3025	2N4414A
3025	2N4928
3025	2N5022
3025	2N5023
3025	2N5111
3025	2N5160
3025	2N5782
3025	2N5821
3025	2N6180
3025	2S021
3025	2S022
3025	2S023
3025	2S301
3025	2S3010
3025	2S302
3025	2S302A
3025	2S303
3025	2S304
3025	2S305
3025	2SA497
3025	2SA497R
3025	2SA497Y
3025	2SA498
3025	2SA498Y
3025	2SA501
3025	2SA502A
3025	2SA502B
3025	2SA502C
3025	2SA502D
3025	2SA502E
3025	2SA502F
3025	2SA502G
3025	2SA502GN
3025	2SA502H
3025	2SA502J
3025	2SA502K
3025	2SA502L
3025	2SA502M
3025	2SA502OR
3025	2SA502R
3025	2SA502X
3025	2SA502Y
3025	2SA503
3025	2SA503-O
3025	2SA503-R
3025	2SA503-Y
3025	2SA504
3025	2SA504-O
3025	2SA504-R
3025	2SA504-Y
3025	2SA510-RD
3025	2SA510A
3025	2SA510B
3025	2SA510C
3025	2SA510D
3025	2SA510E
3025	2SA510F
3025	2SA510G
3025	2SA510GN
3025	2SA510H
3025	2SA510J
3025	2SA510K
3025	2SA510L
3025	2SA510M
3025	2SA510R
3025	2SA510X
3025	2SA510Y
3025	2SA512-RD
3025	2SA512A
3025	2SA512B
3025	2SA512C
3025	2SA512D
3025	2SA512E
3025	2SA512F
3025	2SA512G
3025	2SA512GN
3025	2SA512H
3025	2SA512J
3025	2SA512K
3025	2SA512M
3025	2SA512R
3025	2SA512X
3025	2SA512Y
3025	2SA527
3025	2SA528
3025	2SA532
3025	2SA532A
3025	2SA532B
3025	2SA532C
3025	2SA532D
3025	2SA532E
3025	2SA532F
3025	2SA537
3025	2SA537A
3025	2SA537AH
3025	2SA537B
3025	2SA537C
3025	2SA537H
3025	2SA543
3025	2SA546
3025	2SA546A
3025	2SA546B
3025	2SA546E
3025	2SA546H
3025	2SA548G
3025	2SA548GN
3025	2SA548OR
3025	2SA548R
3025	2SA548Y
3025	2SA551
3025	2SA551C
3025	2SA551D
3025	2SA551E
3025	2SA552
3025	2SA560
3025	2SA560A
3025	2SA594
3025	2SA594-R
3025	2SA594-Y
3025	2SA604
3025	2SA606
3025	2SA607E
3025	2SA607F
3025	2SA607G
3025	2SA607GN
3025	2SA607H
3025	2SA607J
3025	2SA607OR
3025	2SA607R
3025	2SA607X
3025	2SA607Y
3025	2SA612
3025	2SA742
3025	2SA742H
3025	2SC512L
3025	3L4-6010-4
3025	3L4-6010-8
3025	4-30203845
3025	4-3023222
3025	4-686170-3
3025	4-686229-3
3025	4-686230-3
3025	4-686235-3
3025	4-686238-3
3025	7-14A
3025	8-0051600
3025	8-0052700
3025	8-2410300
3025	8-905-706-545
3025	8-905-713-810
3025	09-305075
3025	9-5226-003
3025	9-5226-1
3025	9-5226-3
3025	13-14085-92
3025	13-28394-4
3025	13-28394-4A
3025	13-28394-5
3025	13-28394-5A
3025	13-28394-6
3025	13-32631
3025	13-34004-1A
3025	14-602-28
3025	14-602-28A
3025	14-602-60
3025	14-602-79
3025	14-602-79A
3025	14-607-29A
3025	19A115180-2
3025	020-1110-009
3025	020-1111-017
3025	20A0075
3025	24MW727
3025	026-100019
3025	29A4
3025	32E64
3025	34-6001-86
3025	34-6016-23
3025	34-6016-23A
3025	34-6016-32A
3025	039
3025	41-0500
3025	41-0500A
3025	42-19643
3025	42-20739
3025	42-21232
3025	43-0203845
3025	43-023222
3025	46-86403-3
3025	46-86425-3
3025	46-86574-3
3025	48-134478
3025	48-134478A
3025	48-134951
3025	48-134951A
3025	48-65177A77
3025	48-65177A77A
3025	48-869426
3025	48-90232A06
3025	48-90232A06A
3025	48-90232A09
3025	48-90232A09A
3025	48-90232A12A
3025	48-90232A15
3025	48-90232A15A
3025	48R869426
3025	48S134910
3025	48S134943
3025	48S137045
3025	48X97177A14
3025	53P170
3025	55-643
3025	57A108-6-8
3025	57B110-9
3025	57B112-9
3025	57B112-9A
3025	57B114-9
3025	57B114-9A
3025	57B115-9
3025	57B115-9A
3025	57B116-9
3025	57B116-9A
3025	57B148-1
3025	57B148-10
3025	57B148-11
3025	57B148-12
3025	57B148-12A
3025	57B148-2
3025	57B148-3
3025	57B148-4
3025	57B148-5
3025	57B148-6
3025	57B148-7
3025	57B148-8
3025	57B148-9
3025	57B163-12
3025	57B163-12A
3025	57B168-9
3025	57B171-9
3025	57C110-9
3025	57C122-9
3025	57C148-12
3025	57C148-12A
3025	57C23-1
3025	57C23-2
3025	57C23-3
3025	57C6-15
3025	57C6-26
3025	57C6-26A
3025	57C6-31
3025	57C6-31A
3025	57C6-33
3025	57C6-33A
3025	57M1-13
3025	57M1-13A
3025	57M1-21
3025	57M1-21A
3025	57M1-22
3025	57M1-25
3025	57M1-25A
3025	57M2-10
3025	57M2-10A
3025	57M2-15
3025	57M2-15A
3025	000062
3025	79F114-2A
3025	79F114-4
3025	79F114-4A
3025	80-051600
3025	80-052700
3025	82-410300
3025	83P2B
3025	86-329-2
3025	86-334-2
3025	86-431-2
3025	86-431-9
3025	86-5064-2
3025	86-5064-2A
3025	86-5082-2
3025	86-5082-2A
3025	95-226-003
3025	95-226-1
3025	95-226-3
3025	96-5176-01
3025	99P2
3025	99P2B
3025	0101-439
3025	106KBA
3025	112-2
3025	112-2A
3025	121-1007
3025	121-765
3025	121-774CL
3025	121-952
3025	0131-001427
3025	0131-001428
3025	131-005808
3025	132-007
3025	132-032
3025	132-039
3025	134-1P
3025	134P2
3025	158P1M
3025	174-25566-01
3025	177-001
3025	177-001-9-001
3025	183P1
3025	226-3
3025	297V073C01
3025	297V073C02
3025	297V074C11
3025	297V080C01
3025	297V082B02
3025	297V082B03
3025	297V083C03
3025	303-2
3025	353-9008-001
3025	353-9304-004
3025	380-0171-000
3025	386-7184P1
3025	393-1
3025	417-138
3025	417-170
3025	417-234
3025	417-43
3025	422-0739
3025	422-1232
3025	422-1405
3025	422-2008
3025	430-203845
3025	430-23222
3025	433M852
3025	610-083
3025	631-1
3025	690V086H89
3025	690V110H55
3025	774
3025	774CL
3025	800-284
3025	800-511-00
3025	800-516-00
3025	800-51600
3025	800-52700
3025	824-10300
3025	880-250-107
3025	881-250-107
3025	921-270B
3025	921-315B
3025	921-47B
3025	964-20739
3025	1184G
3025	1285
3025	1414-185
3025	1414-187
3025	1889-17
3025	2057A2-150

SK	Industry Standard No.
3025	2057B110-9
3025	2057B112-9
3025	2057B114-9
3025	2057B115-9
3025	2057B116-9
3025	2057B121-9
3025	2057B122-9
3025	2057B145-12
3025	2057B163-12
3025	2448
3025	2482
3025	2502
3025	3523
3025	3533-1
3025	3590
3025	3616
3025	3616-1
3025	4367-001
3025	4478
3025	4686-170-3
3025	4686-229-3
3025	4686-230-3
3025	4686-235-3
3025	4686-238-3
3025	8000-004-P089
3025	8303
3025	8400-1
3025	8400-1A
3025	8400-1B
3025	9405-2
3025	09800-12
3025	10036-001
3025	10300-12
3025	018069
3025	20739
3025	23114-050
3025	23114-051
3025	30278
3025	36577
3025	40319
3025	40362
3025	40406
3025	40410
3025	40537
3025	40538
3025	40634
3025	41052
3025	43107
3025	43116
3025	61009-9
3025	61011-0
3025	84001
3025	84001A
3025	84001B
3025	94063
3025	94064
3025	116090
3025	116284
3025	119983
3025	124616
3025	125142
3025	126724
3025	130404-29
3025	131242-12
3025	136423
3025	147112-7
3025	233969
3025	236433
3025	240402
3025	241052
3025	242422
3025	242460
3025	242958
3025	262638
3025	297074C11
3025	564671
3025	0573542
3025	0573559
3025	610099
3025	610099-1
3025	610099-2
3025	610110
3025	610129-1
3025	815185
3025	815185B
3025	970762-6
3025	1471112-12
3025	1471112-8-9
3025	1471112-9
3025	1473516-1
3025	1827322
3025	2004746-116
3025	2004746-117
3025	2006436-37
3025	2320242
3025	2320243
3025	2327282
3025	2327283
3025	5320111
3025	06100030
3025	7026019
3025	7570032-01
3025	8015613
3025	13040429
3025	43023222
3025	43024880
3025	43026285
3025	62208953
3025	62278566
3025	62393696
3025	62593660
3025	62638257
3025	80051600
3025	80052700
3025	82410300
3025	430203845
3025/129	A054-223
3025/129	A066-113
3025/129	A066-118
3025/129	A068-109
3025/129	A171
3025/129	A2SA564FR
3025/129	A495-0
3025/129	A495Y
3025/129	A497
3025/129	A530
3025/129	A530H
3025/129	A539
3025/129	A539L
3025/129	A539S
3025/129	A544
3025/129	A548
3025/129	A550
3025/129	A550A
3025/129	A550AQ
3025/129	A550R
3025/129	A561
3025/129	A561-O
3025/129	A561-R
3025/129	A561-Y
3025/129	A561GR
3025/129	A562-R
3025/129	A562-Y
3025/129	A565
3025/129	A565A
3025/129	A565B
3025/129	A565C
3025/129	A568
3025/129	A569
3025/129	A569J
3025/129	A592Y
3025/129	A597
3025/129	A5T2907
3025/129	A5T3644
3025/129	A5T3645
3025/129	A5T3905
3025/129	A5T3906
3025/129	A5T4125
3025/129	A5T4126
3025/129	A603
3025/129	A608-F
3025/129	A610
3025/129	A610B
3025/129	A611
3025/129	A611-4B
3025/129	A637
3025/129	A640A
3025/129	A640B
3025/129	A640C
3025/129	A640D
3025/129	A640E
3025/129	A640M
3025/129	A643W
3025/129	A683
3025/129	A718
3025/129	A719
3025/129	A719Q
3025/129	A719S
3025/129	A723
3025/129	A723A
3025/129	A723B
3025/129	A723C
3025/129	A723D
3025/129	A723E
3025/129	A723F
3025/129	A731
3025/129	A733
3025/129	A733A
3025/129	A733B
3025/129	A733C
3025/129	A733D
3025/129	A733E
3025/129	A733F
3025/129	A733H
3025/129	A733I
3025/129	A733P
3025/129	A733Q
3025/129	A733R
3025/129	A758
3025/129	A758B
3025/129	A759A
3025/129	A759B
3025/129	AT333
3025/129	AT391
3025/129	AT392
3025/129	AT393
3025/129	B266A-1
3025/129	B266B-1
3025/129	BC116
3025/129	BC116A
3025/129	BC157B
3025/129	BC231A
3025/129	BC261
3025/129	BC261B
3025/129	BC315
3025/129	BC381
3025/129	BC405
3025/129	BC405A
3025/129	BC405B
3025/129	BC406B
3025/129	BC415A
3025/129	BC415B
3025/129	BC416A
3025/129	BC416B
3025/129	BC417
3025/129	BC418
3025/129	BC419
3025/129	BC478
3025/129	BC478A
3025/129	BC479
3025/129	BC479B
3025/129	BC512
3025/129	BC512A
3025/129	BC513
3025/129	BC513A
3025/129	BC514A
3025/129	BCW62
3025/129	BCW62A
3025/129	BCW63
3025/129	BCW63A
3025/129	BCW64A
3025/129	BCW69
3025/129	BCW69R
3025/129	BCW79-10
3025/129	BCW79-16
3025/129	BCW80-10
3025/129	BCW80-16
3025/129	BCW96
3025/129	BCW96K
3025/129	BCW97
3025/129	BCW97K
3025/129	BF-832
3025/129	BFS37
3025/129	BFS37A
3025/129	BFS40
3025/129	BFS40A
3025/129	BFS41
3025/129	BFS69
3025/129	BFV82
3025/129	BFV82A
3025/129	BFV82B
3025/129	BFV82C
3025/129	BFV86
3025/129	BFV86A
3025/129	BFV86B
3025/129	BFV86C
3025/129	BFW90
3025/129	BFX29
3025/129	BFX30
3025/129	BFX65
3025/129	BFY18
3025/129	BR-832
3025/129	BSV43A
3025/129	BSV44A
3025/129	BSV45A
3025/129	BSV47A
3025/129	BSV48A
3025/129	BSV49A
3025/129	BSV96
3025/129	BSV97
3025/129	BSV98
3025/129	BT832
3025/129	C202
3025/129	C402
3025/129	C564P
3025/129	C564Q
3025/129	CDC-9000-1D
3025/129	CDC496
3025/129	CDC9000-1
3025/129	CDC9000-1B
3025/129	CDC9002-1C
3025/129	CS1170F
3025/129	CS1228
3025/129	CS1251E
3025/129	CS9129
3025/129	CS9129B
3025/129	D29A4
3025/129	EQR-0016
3025/129	ES15X128
3025/129	EW183
3025/129	F3574
3025/129	FCS-9012F
3025/129	FCS-9012G
3025/129	FCS1170F
3025/129	FCS9012
3025/129	FCS9012HG
3025/129	G03-017-B
3025/129	GC1145
3025/129	GE-21
3025/129	HS-40031
3025/129	HS-40050
3025/129	HT104951B-0
3025/129	HT104951C
3025/129	HT105611
3025/129	HT105611A
3025/129	HT105611B
3025/129	HT105641C
3025/129	HT105642B
3025/129	HT106731B
3025/129	HT600011F
3025/129	HT6000210
3025/129	LM1502H
3025/129	M4815
3025/129	M4943
3025/129	M4989
3025/129	MA113
3025/129	ME0404
3025/129	MH1502
3025/129	MPS-U51
3025/129	MPS3638
3025/129	MPS3638A
3025/129	MPS4355
3025/129	MPS6524
3025/129	MPSA56
3025/129	MPSU52
3025/129	MPSU55
3025/129	P1H
3025/129	P1J
3025/129	P1W
3025/129	P2E
3025/129	P2W
3025/129	P5B
3025/129	PL1031
3025/129	PL1033
3025/129	PL1034
3025/129	PL1101
3025/129	PL1102
3025/129	PL1103
3025/129	PL1104
3025/129	Q-00984R
3025/129	Q5102P
3025/129	Q5102Q
3025/129	Q5102R
3025/129	Q5121Q
3025/129	RE26
3025/129	RS-110
3025/129	RS-2021
3025/129	RS-2022
3025/129	RS-2023
3025/129	RS-2024
3025/129	RV2353
3025/129	S3655
3025/129	SJE514
3025/129	SPS-952
3025/129	ST8014
3025/129	SX3702
3025/129	T1S38
3025/129	T276
3025/129	T3570
3025/129	TIS37
3025/129	TR-19
3025/129	TR20
3025/129	TV-44
3025/129	TVS2SA543
3025/129	TVS2SA564
3025/129	TVS2SC5640
3025/129	TVS2SC564R
3025/129	V410A
3025/129	XA-495
3025/129	XA494
3025/129	XA495(C)
3025/129	XA495AC
3025/129	XA495C
3025/129	XS19
3025/129	2N1256
3025/129	2N1257
3025/129	2N2175
3025/129	2N2177
3025/129	2N2968
3025/129	2N2970
3025/129	2N3765
3025/129	2N3781
3025/129	2SA516
3025/129	2SA516A
3025/129	2SA548
3025/129	2SA571
3025/129	2SA575K
3025/129	2SA575L
3025/129	2SA597
3025/129	2SA607
3025/129	2SA607A
3025/129	2SA607B
3025/129	2SA607C
3025/129	2SA607D
3025/129	2SA607K
3025/129	2SA607L
3025/129	2SA607M
3025/129	2SM610B
3025/129	09-302161
3025/129	09-305058
3025/129	09-309030
3025/129	13-0006
3025/129	13-0043
3025/129	13-0044
3025/129	13-006
3025/129	13-14085-13
3025/129	13-22582-1
3025/129	13-28393-1A
3025/129	13-28393-2
3025/129	13-28393-2A
3025/129	13-32631-3
3025/129	13-33178-1
3025/129	13-34367-1
3025/129	13-34369-1
3025/129	14-602-75
3025/129	19A115531-P1
3025/129	32-20739
3025/129	42-21405
3025/129	42-22810
3025/129	46-86429-3
3025/129	48S134989
3025/129	48S137032
3025/129	48S137127
3025/129	48S137173
3025/129	48S155122
3025/129	055
3025/129	57A168-9
3025/129	57A220-14
3025/129	57A240-14
3025/129	57B132-10
3025/129	57B175-9
3025/129	74P1
3025/129	86-345-2
3025/129	86-396-2
3025/129	86X0072-001
3025/129	96-5258-01
3025/129	99P1
3025/129	99P1M
3025/129	99S092-1
3025/129	0101-0439
3025/129	121-707
3025/129	121-709
3025/129	126P1
3025/129	0131-001420
3025/129	132-029
3025/129	134P1
3025/129	157
3025/129	168-9
3025/129	177-018-9-001
3025/129	177-020-9-001
3025/129	853-0300-643
3025/129	991-013058
3025/129	1414-173
3025/129	1935-17
3025/129	2000-218
3025/129	2057B2-150
3025/129	2904-038
3025/129	3513
3025/129	3533
3025/129	5611-695
3025/129	5611-695C
3025/129	8000-00004-P089
3025/129	38271
3025/129	40319V1
3025/129	40319V2
3025/129	40362V1
3025/129	40362V2
3025/129	94052
3025/129	95226-1
3025/129	95226-3
3025/129	116084
3025/129	116091
3025/129	131242
3025/129	132830
3025/129	172336
3025/129	255821HS
3025/129	489751-032
3025/129	610134-1
3025/129	988992
3025/129	1473570-1
3025/129	1473590-1
3025/129	2003073-0702
3025/129	2321521
3025/129	7570013-01
3025/129	23114050
3025/129	23114051
3025/129	23114081
3025/129	23114124
3025/129	23114136
3025/129	23114137
3025/129	23114293
3025/129	23114313
3025/129	62389478
3025/129	62711205
3025/129	91056140
3025/129	16009090545
3026	A054-224
3026	A068-114
3026	A1Y
3026	AR18
3026	ES15X98
3026	EW183B
3026	FBN-35469
3026	FBN-36486
3026	GE-23
3026	HEP-S5019
3026	HT304911
3026	HT304911A
3026	HT304911B
3026	IP20-0212
3026	IRTR-57
3026	IRTR57
3026	PTC112
3026	Q-00769
3026	Q-00769A
3026	Q-00769B
3026	Q-00769C
3026	Q-01269
3026	Q-01269B
3026	Q-01269C
3026	RCA40250
3026	S21520
3026	SD141
3026	TA2402
3026	TA2402A
3026	TR1591
3026	TR1593
3026	TR2327852
3026	TR26-1
3026	TV109
3026	TV112
3026	TVS2SD226
3026	TVS2SD226-O
3026	TVS2SD226A
3026	TVS2SD226B
3026	TVS2SD226C
3026	TVS2SD226D
3026	TVS2SD226P
3026	WEP703
3026	XI-549
3026	XT-549A
3026	2N3054
3026	2N6260
3026	2N6261
3026	2SC1024
3026	2SC1024-E
3026	2SC1024A
3026	2SC1024B
3026	2SC1024C
3026	2SC1024D
3026	2SC1024E
3026	2SC1024F
3026	2SC1024G
3026	2SC1024L
3026	2SC1024Y
3026	2SC4116
3026	2SC4116A
3026	2SC4116B
3026	2SC4116C
3026	2SC4116D
3026	2SC4116E
3026	2SC4116F
3026	2SC4116G
3026	2SC4116GN
3026	2SC4116H
3026	2SC4116J
3026	2SC4116K
3026	2SC4116L
3026	2SC4116M
3026	2SC4116OR
3026	2SC4116R
3026	2SC4116X
3026	2SC4116Y
3026	2SC489
3026	2SC489A
3026	2SC489B
3026	2SC489C
3026	2SC489D
3026	2SC489E
3026	2SC489F
3026	2SC489G
3026	2SC489GN
3026	2SC489H
3026	2SC489J
3026	2SC489K
3026	2SC489L
3026	2SC489M
3026	2SC489OR
3026	2SC489R
3026	2SC489X
3026	2SC489Y
3026	2SD103
3026	2SD103-O
3026	2SD103-Y
3026	2SD129
3026	2SD143
3026	2SD146
3026	2SD146UK
3026	2SD152
3026	2SD152A
3026	2SD152B
3026	2SD152C
3026	2SD152D
3026	2SD152E
3026	2SD152F
3026	2SD152G
3026	2SD152GN
3026	2SD152H
3026	2SD152J
3026	2SD152K
3026	2SD152L
3026	2SD152M
3026	2SD152OR
3026	2SD152R
3026	2SD152X
3026	2SD152Y
3026	2SD155
3026	2SD155L
3026	2SD236
3026	2SD236A
3026	2SD236B
3026	2SD236C
3026	2SD236D
3026	2SD236E
3026	2SD236F
3026	2SD236G
3026	2SD236GN
3026	2SD236H
3026	2SD236J
3026	2SD236K
3026	2SD236L
3026	2SD236M
3026	2SD236R
3026	2SD236X
3026	2SD236Y
3026	2SD254
3026	2SD255
3026	2SD28
3026	2SD28A
3026	2SD28B
3026	2SD28C
3026	2SD28D
3026	2SD28E
3026	2SD28F
3026	2SD28G
3026	2SD28GN
3026	2SD28H
3026	2SD28J
3026	2SD28K
3026	2SD28L
3026	2SD28M
3026	2SD28OR
3026	2SD28R
3026	2SD28X
3026	2SD28Y
3026	2SD29
3026	2SD291
3026	2SD291-O
3026	2SD291B
3026	2SD291BL

SK	Industry Standard No.	SK	Industry Standard No.	SK	Industry Standard No.	SK	Industry Standard No.	SK	Industry Standard No.
3026	2SD291C	3026	559-1516-001	3027	AMF104	3027	S2741	3027	2SC242D
3026	2SD291D	3026	638BJ	3027	AMF105	3027	S305A	3027	2SC242E
3026	2SD291E	3026	686-021-0-0	3027	AMF115	3027	893SEE133	3027	2SC242F
3026	2SD291G	3026	992-003172	3027	AMF116	3027	893SR165	3027	2SC242G
3026	2SD291H	3026	992-004092	3027	AMF117	3027	SDT9201	3027	2SC242GN
3026	2SD291J	3026	1040-02	3027	AMF117A	3027	SDT9205	3027	2SC242H
3026	2SD291K	3026	1042-02	3027	AMF118	3027	SDT9206	3027	2SC242J
3026	2SD291L	3026	1043-7309	3027	AMF118A	3027	SDT9210	3027	2SC242K
3026	2SD291M	3026	1123-3335	3027	AMF119	3027	SDT9308	3027	2SC242L
3026	2SD291OR	3026	1414-180	3027	AMF119A	3027	SE-3033	3027	2SC242M
3026	2SD292	3026	1534-8931	3027	AMF120	3027	SE3033	3027	2SC242OR
3026	2SD292-0	3026	1714-0402	3027	AMF120A	3027	SE3035	3027	2SC242R
3026	2SD292A	3026	1714-0405	3027	AMF210	3027	SE3036	3027	2SC242Y
3026	2SD292B	3026	1714-0602	3027	AMF210A	3027	SE9002	3027	2SC244A
3026	2SD292BL	3026	1714-0605	3027	AT-10	3027	SE9080	3027	2SC244B
3026	2SD292C	3026	1714-0802	3027	AT-1856	3027	SJ1106	3027	2SC244C
3026	2SD292D	3026	1714-0805	3027	AT3260	3027	SJ1470	3027	2SC244E
3026	2SD292E	3026	1714-1002	3027	ATC-TR-15	3027	SJ2000	3027	2SC244F
3026	2SD292F	3026	1714-1005	3027	B170000	3027	SJ2008	3027	2SC244G
3026	2SD292G	3026	1936-17	3027	B170001	3027	SJ3464	3027	2SC244GN
3026	2SD292H	3026	2110N-134	3027	B170002	3027	SJ3604	3027	2SC244J
3026	2SD292J	3026	2163-17	3027	B170003	3027	SJ3678	3027	2SC244K
3026	2SD292K	3026	2180-155	3027	B170004	3027	SJ619	3027	2SC244L
3026	2SD292L	3026	3152-159	3027	B170005	3027	SJ619-1	3027	2SC244M
3026	2SD292M	3026	4686-130-3	3027	B170006	3027	SJ820	3027	2SC244OR
3026	2SD292OR	3026	4686-226-3	3027	B170007	3027	SJ8701	3027	2SC244R
3026	2SD29A	3026	5320-003	3027	B170009	3027	SJ9110	3027	2SC244T
3026	2SD29B	3026	7312	3027	B170010	3027	STC-1035	3027	2SC493A
3026	2SD29E	3026	8000-00005-008	3027	B170011	3027	STC-1035A	3027	2SC493B
3026	2SD29F	3026	9404-9	3027	B170012	3027	STC-1036	3027	2SC493C
3026	2SD29GN	3026	9409-4	3027	B170013	3027	STC-1036A	3027	2SC493E
3026	2SD29J	3026	12044-0021	3027	B170014	3027	STC1080	3027	2SC493F
3026	2SD29K	3026	50256	3027	B170015	3027	STC1081	3027	2SC493G
3026	2SD29L	3026	40250	3027	B170016	3027	STC1082	3027	2SC493GN
3026	2SD29M	3026	40250V1	3027	B170018	3027	STC1083	3027	2SC493H
3026	2SD290R	3026	40310	3027	B170019	3027	STC1084	3027	2SC493J
3026	2SD29R	3026	40310V1	3027	B170020	3027	STC4253	3027	2SC493K
3026	2SD29X	3026	40312	3027	B170021	3027	STX0014	3027	2SC493L
3026	2SD29Y	3026	40312V1	3027	B170022	3027	STX0027	3027	2SC493M
3026	2SD315	3026	40316	3027	B170024	3027	STX0032	3027	2SC493OR
3026	2SD315D	3026	40324	3027	B170025	3027	T-Q5105	3027	2SC493R
3026	2SD315E	3026	40372	3027	B5020	3027	TA2577A	3027	2SC493X
3026	2SD49	3026	40910	3027	BD111	3027	TA7199	3027	2SC493Y
3026	2SD49A	3026	42342	3027	BD130	3027	TM130	3027	2SC494BL
3026	2SD49B	3026	70019-1	3027	BD142	3027	TNJ72148	3027	2SC664
3026	2SD49C	3026	70019-5	3027	BDX10	3027	TR1007	3027	2SC664B
3026	2SD49D	3026	94049	3027	BDY17	3027	TR1077	3027	2SC664C
3026	2SD49E	3026	94094	3027	BDY38	3027	TR1490	3027	2SC736A
3026	2SD49F	3026	165738	3027	BDY39	3027	TR1491	3027	2SC736B
3026	2SD49G	3026	167958	3027	BDY55	3027	TR1492	3027	2SC736C
3026	2SD49GN	3026	175007-277	3027	BLY10	3027	TR1493	3027	2SC736D
3026	2SD49J	3026	198038-3	3027	BLY11	3027	TR26	3027	2SC736E
3026	2SD49K	3026	198038-4	3027	BN7133	3027	TR26C	3027	2SC736F
3026	2SD49L	3026	230084	3027	BN7214	3027	TR271 TR26	3027	2SC736G
3026	2SD49M	3026	262116	3027	BUY10	3027	W16	3027	2SC736GN
3026	2SD490R	3026	489751-174	3027	BUY11	3027	WEP704	3027	2SC736H
3026	2SD49R	3026	489751-175	3027	C21	3027	XI-548	3027	2SC736J
3026	2SD49X	3026	532003	3027	C494	3027	XT-548A	3027	2SC736K
3026	2SD49Y	3026	700191	3027	C664B	3027	ZT1487	3027	2SC736L
3026	2SD56	3026	700195	3027	C664C	3027	ZT1488	3027	2SC736M
3026	2SD56A	3026	1851516	3027	C765	3027	ZT1489	3027	2SC736OR
3026	2SD56B	3026	1851518	3027	C793	3027	ZT1490	3027	2SC736X
3026	2SD56C	3026	1971296	3027	C793BL	3027	ZT1702	3027	2SC736Y
3026	2SD56D	3026	2320084	3027	C793R	3027	001-021270	3027	2SC765
3026	2SD56E	3026	2220222	3027	C793T	3027	001-040201	3027	2SC765A
3026	2SD56F	3026	3130057	3027	C794R	3027	2D010	3027	2SC765B
3026	2SD56G	3026	3152159	3027	CXL130	3027	2N1069	3027	2SC765C
3026	2SD56GN	3026	3468841-1	3027	DS-509	3027	2N1070	3027	2SC765D
3026	2SD56H	3026	4080838-0001	3027	DS-514	3027	2N2305	3027	2SC765E
3026	2SD56J	3026	4080838-0002	3027	DS514	3027	2N3055	3027	2SC765G
3026	2SD56K	3026	4080838-2	3027	DT4011	3027	2N3232	3027	2SC765GN
3026	2SD56L	3026	4080866-1	3027	DT4110	3027	2N3233	3027	2SC765H
3026	2SD56M	3026	4080866-2	3027	DT4111	3027	2N3235	3027	2SC765J
3026	2SD56OR	3026	5320003	3027	DT4120	3027	2N3448	3027	2SC765K
3026	2SD56R	3026	7570022-01	3027	DT4121	3027	2N4395	3027	2SC765L
3026	2SD56X	3026	10644417	3027	EA15Z10Q	3027	2N4396	3027	2SC765M
3026	2SD56Y	3026	12363800	3027	EC961	3027	2N6253	3027	2SC765OR
3026	2SD90	3026	15077183	3027	ECG130	3027	2N6569	3027	2SC765R
3026	2SD90A	3026	23114220	3027	ETB-003	3027	2S033	3027	2SC765X
3026	2SD90B	3026	23114221	3027	F82003-1	3027	2S034	3027	2SC768A
3026	2SD90C	3026	62734612	3027	GB-14	3027	2SC21	3027	2SC768B
3026	2SD90D	3026	94029220	3027	HBP-87004	3027	2SC21A	3027	2SC768C
3026	2SD90E	3027	A054-154	3027	HT30494	3027	2SC21B	3027	2SC768D
3026	2SD90F	3027	A112363	3027	HT30494X	3027	2SC21C	3027	2SC768E
3026	2SD90GN	3027	A13-0032	3027	HT401191	3027	2SC21E	3027	2SC768F
3026	2SD90J	3027	A13-17918-1	3027	HT401191A	3027	2SC21F	3027	2SC768G
3026	2SD90K	3027	A13-23594-1	3027	HT401191B	3027	2SC21G	3027	2SC768GN
3026	2SD90L	3027	A13-33188-2	3027	IR-TR59	3027	2SC21GN	3027	2SC768H
3026	2SD90M	3027	A14-601-10	3027	K4-525	3027	2SC21H	3027	2SC768J
3026	2SD90Y	3027	A14-601-12	3027	M4715	3027	2SC21J	3027	2SC768K
3026	2SD91	3027	A14-601-13	3027	M4882	3027	2SC21K	3027	2SC768L
3026	2SD91A	3027	A18	3027	MHT7601	3027	2SC21L	3027	2SC768M
3026	2SD91B	3027	A18-4	3027	MHT7602	3027	2SC21M	3027	2SC768OR
3026	2SD91C	3027	A241B	3027	MHT7603	3027	2SC21OR	3027	2SC768R
3026	2SD91E	3027	A2B	3027	MHT7077	3027	2SC21R	3027	2SC768X
3026	2SD91F	3027	A2B-2	3027	MHT7608	3027	2SC21X	3027	2SC768Y
3026	2SD91GN	3027	A2EBRN	3027	MHT7609	3027	2SC21Y	3027	2SC793
3026	2SD91J	3027	A2EBRN-1	3027	MJ2801	3027	2SC240A	3027	2SC793A
3026	2SD91K	3027	A28	3027	MJ2802	3027	2SC240B	3027	2SC793B
3026	2SD91M	3027	A28-3	3027	MJ480	3027	2SC240C	3027	2SC793C
3026	2SD910R	3027	A3L4-6001-01	3027	MJ481	3027	2SC240D	3027	2SC793D
3026	2SD91R	3027	A3TE120	3027	P2271	3027	2SC240E	3027	2SC793E
3026	2SD91Y	3027	A3TE230	3027	P3139	3027	2SC240F	3027	2SC793F
3026	4-686130-3	3027	A3TE240	3027	PN350	3027	2SC240G	3027	2SC793G
3026	4-686226-3	3027	A3TX003	3027	PP3000	3027	2SC240GN	3027	2SC793GN
3026	007-0074	3027	A3TX004	3027	PP3003	3027	2SC240H	3027	2SC793H
3026	13-0014	3027	A3U	3027	PTC0119	3027	2SC240J	3027	2SC793J
3026	19-00-3485	3027	A3U-4	3027	PTC140	3027	2SC240K	3027	2SC793K
3026	019-003485	3027	A43023843	3027	QP-11	3027	2SC240L	3027	2SC793L
3026	19A115527	3027	A4J	3027	QP-12	3027	2SC240M	3027	2SC793M
3026	23-5051	3027	A4ARED-1	3027	QP-8	3027	2SC240OR	3027	2SC793OR
3026	31-0010	3027	A48	3027	QP-8-P	3027	2SC240R	3027	2SC793R
3026	34-6002-43	3027	A48-1	3027	R227075497	3027	2SC240X	3027	2SC793X
3026	42-20960	3027	A4Z	3027	R2982	3027	2SC240Y	3027	2SC793Y
3026	44A351A637-001	3027	A522	3027	R4369	3027	2SC241A	3027	2SC794
3026	46-861130-3	3027	A522-3	3027	RC1700	3027	2SC241B	3027	2SC794A
3026	46-86349-3	3027	A6LBLK	3027	RC8242	3027	2SC241C	3027	2SC794B
3026	46-86350-3	3027	A6LBLK-1	3027	RB19	3027	2SC241D	3027	2SC794C
3026	46-86496-3	3027	A6LBRN	3027	REN130	3027	2SC241E	3027	2SC794D
3026	48-90343A56	3027	A6LBRN-1	3027	S-305	3027	2SC241F	3027	2SC794E
3026	488134936	3027	A6LRED	3027	S-305A	3027	2SC241G	3027	2SC794F
3026	051-0151	3027	A6LRED-1	3027	S-356	3027	2SC241H	3027	2SC794G
3026	119-0068	3027	A6N	3027	S1691	3027	2SC241K	3027	2SC794GN
3026	121-588	3027	A6N-6	3027	S1692	3027	2SC241L	3027	2SC794H
3026	125B410	3027	A7-13	3027	S1865	3027	2SC241M	3027	2SC794J
3026	129-23	3027	A7-13	3027	S1905	3027	2SC241OR	3027	2SC794K
3026	200-018	3027	A7M	3027	S1905A	3027	2SC241R	3027	2SC794L
3026	231-0008	3027	A80052402	3027	S1907	3027	2SC241X	3027	2SC794M
3026	260P*6009	3027	A80414120	3027	S1977634	3027	2SC241Y	3027	2SC794OR
3026	314-6006	3027	A80414130	3027	S2003-1	3027	2SC242A	3027	2SC794R
3026	352-0581-020	3027	AEX-8230B	3027	S2241	3027	2SC242B	3027	2SC794RA
3026	352-0581-030	3027	AEX9846	3027	S2392	3027	2SC242C		
3026	422-0960	3027	AM3235	3027	S2403B				
				3027	S2403C				

SK	Industry Standard No.	SK	Industry Standard No.	SK	Industry Standard No.	SK	Industry Standard No.	SK	Industry Standard No.
3027	2SC794X	3027	2SD164X	3027	2SD211F	3027	2SD70H	3027	44A390244-001
3027	2SC794Y	3027	2SD164Y	3027	2SD211G	3027	2SD70J	3027	44A417716-001
3027	2SC851	3027	2SD16A	3027	2SD211GN	3027	2SD70K	3027	46-86388-3
3027	2SC851A	3027	2SD16B	3027	2SD211H	3027	2SD70L	3027	48-134701
3027	2SC851B	3027	2SD16C	3027	2SD211J	3027	2SD70M	3027	48-134701A
3027	2SC851C	3027	2SD16D	3027	2SD211K	3027	2SD70R	3027	48-134715
3027	2SC851D	3027	2SD16E	3027	2SD211L	3027	2SD70X	3027	48-134715A
3027	2SC851E	3027	2SD16F	3027	2SD211M	3027	2SD70Y	3027	48-137008
3027	2SC851F	3027	2SD16G	3027	2SD211OR	3027	2SD80A	3027	48-137008A
3027	2SC851G	3027	2SD16GN	3027	2SD211R	3027	2SD80B	3027	48-137027
3027	2SC851GN	3027	2SD16H	3027	2SD211X	3027	2SD80C	3027	48-137027A
3027	2SC851H	3027	2SD16J	3027	2SD211Y	3027	2SD80D	3027	48-137036A
3027	2SC851K	3027	2SD16K	3027	2SD212	3027	2SD80E	3027	48-137053
3027	2SC851L	3027	2SD16L	3027	2SD212A	3027	2SD80F	3027	48-137053A
3027	2SC851M	3027	2SD16M	3027	2SD212B	3027	2SD80GN	3027	48-137079
3027	2SC851OR	3027	2SD16OR	3027	2SD212C	3027	2SD80G	3027	48-137079A
3027	2SC851R	3027	2SD16R	3027	2SD212D	3027	2SD80H	3027	48-137175
3027	2SC851X	3027	2SD16X	3027	2SD212E	3027	2SD80J	3027	48-137175A
3027	2SC851Y	3027	2SD16Y	3027	2SD212F	3027	2SD80K	3027	48-137180
3027	2SD12	3027	2SD172	3027	2SD212G	3027	2SD80L	3027	48-137180A
3027	2SD124A	3027	2SD172A	3027	2SD212GN	3027	2SD80R	3027	48-137251
3027	2SD124C	3027	2SD172C	3027	2SD212H	3027	2SD80X	3027	48-15
3027	2SD124E	3027	2SD172D	3027	2SD212J	3027	2SD80Y	3027	488155053
3027	2SD124F	3027	2SD172E	3027	2SD212K	3027	2SD81A	3027	488155067
3027	2SD124G	3027	2SD172F	3027	2SD212L	3027	2SD81B	3027	50C401
3027	2SD124GN	3027	2SD172G	3027	2SD212M	3027	2SD81C	3027	57A256-10
3027	2SD124H	3027	2SD172GN	3027	2SD212OR	3027	2SD81E	3027	57B175-9A
3027	2SD124J	3027	2SD172H	3027	2SD212R	3027	2SD81F	3027	57M3-1
3027	2SD124K	3027	2SD172J	3027	2SD212X	3027	2SD81G	3027	57M3-1A
3027	2SD124L	3027	2SD172K	3027	2SD212Y	3027	2SD81GN	3027	57M3-4
3027	2SD124M	3027	2SD172L	3027	2SD26D	3027	2SD81H	3027	57M3-4A
3027	2SD124OR	3027	2SD172OR	3027	2SD26E	3027	2SD81J	3027	57M3-6
3027	2SD124R	3027	2SD172R	3027	2SD26F	3027	2SD81K	3027	57M3-6A
3027	2SD124X	3027	2SD172X	3027	2SD26G	3027	2SD81L	3027	63-11991
3027	2SD124Y	3027	2SD172Y	3027	2SD26GN	3027	2SD81M	3027	63-11991A
3027	2SD125B	3027	2SD173	3027	2SD26H	3027	2SD81OR	3027	63-8707
3027	2SD125C	3027	2SD173A	3027	2SD26J	3027	2SD81R	3027	63-8707A
3027	2SD125E	3027	2SD173C	3027	2SD26K	3027	2SD81X	3027	070
3027	2SD125F	3027	2SD173D	3027	2SD26L	3027	2SD81Y	3027	80-052402
3027	2SD125G	3027	2SD173E	3027	2SD26M	3027	2SD82A	3027	80-414120
3027	2SD125GN	3027	2SD173F	3027	2SD26OR	3027	2SD82B	3027	80-414130
3027	2SD125H	3027	2SD173G	3027	2SD26R	3027	2SD82C	3027	085
3027	2SD125J	3027	2SD173GN	3027	2SD26X	3027	2SD82D	3027	86-5084-2
3027	2SD125K	3027	2SD173H	3027	2SD26Y	3027	2SD82E	3027	86-5084-2A
3027	2SD125L	3027	2SD173J	3027	2SD41	3027	2SD82F	3027	86-5101-2
3027	2SD125M	3027	2SD173K	3027	2SD41A	3027	2SD82G	3027	86-5101-2A
3027	2SD125OR	3027	2SD173M	3027	2SD41B	3027	2SD82GN	3027	86-665-2
3027	2SD125R	3027	2SD173OR	3027	2SD41C	3027	2SD82H	3027	86-665-9
3027	2SD125X	3027	2SD173X	3027	2SD41D	3027	2SD82J	3027	96-5162-03
3027	2SD125Y	3027	2SD173Y	3027	2SD41E	3027	2SD82K	3027	96-5162-04
3027	2SD146A	3027	2SD174	3027	2SD41F	3027	2SD82L	3027	96-5164-02
3027	2SD146B	3027	2SD174F	3027	2SD41G	3027	2SD82M	3027	96-5164-03
3027	2SD146C	3027	2SD175	3027	2SD41GN	3027	2SD820R	3027	96-5201-01
3027	2SD146D	3027	2SD175A	3027	2SD41H	3027	2SD82R	3027	96-5207-01
3027	2SD146E	3027	2SD175B	3027	2SD41J	3027	2SD82X	3027	100T2
3027	2SD146F	3027	2SD175C	3027	2SD41K	3027	2SD82Y	3027	100T2A
3027	2SD146G	3027	2SD175D	3027	2SD41L	3027	2SDP2A	3027	100X2
3027	2SD146GN	3027	2SD175E	3027	2SD41M	3027	2SDP2B	3027	100X6
3027	2SD146H	3027	2SD175F	3027	2SD41OR	3027	2SDP2C	3027	100X6A
3027	2SD146J	3027	2SD175P	3027	2SD41R	3027	2SDP2D	3027	104T2
3027	2SD146K	3027	2SD175G	3027	2SD41X	3027	2SDP2F	3027	104T2A
3027	2SD146L	3027	2SD175GN	3027	2SD41Y	3027	2SDP2G	3027	111N4
3027	2SD146OR	3027	2SD175H	3027	2SD50A	3027	2SDP2GN	3027	111N4A
3027	2SD146R	3027	2SD175J	3027	2SD50B	3027	2SDP2H	3027	112-203055
3027	2SD146X	3027	2SD175K	3027	2SD50D	3027	2SDP2J	3027	112-363
3027	2SD146Y	3027	2SD175L	3027	2SD50E	3027	2SDP2K	3027	121-726
3027	2SD147	3027	2SD175M	3027	2SD50F	3027	2SDP2L	3027	121-726A
3027	2SD147A	3027	2SD175OR	3027	2SD50G	3027	2SDP2M	3027	123B-001
3027	2SD147B	3027	2SD175R	3027	2SD50GN	3027	2SDP20R	3027	0131-001597
3027	2SD147C	3027	2SD175X	3027	2SD50H	3027	2SDP2R	3027	0131-002068
3027	2SD147D	3027	2SD175Y	3027	2SD50J	3027	2SDP2X	3027	0132
3027	2SD147F	3027	2SD176	3027	2SD50K	3027	2SDP2Y	3027	132-070
3027	2SD147GN	3027	2SD176A	3027	2SD50L	3027	2SDU47X	3027	132-085
3027	2SD147H	3027	2SD176B	3027	2SD50M	3027	3L4-6001-01	3027	132-541
3027	2SD147J	3027	2SD176C	3027	2SD50R	3027	3TE140	3027	140N2
3027	2SD147K	3027	2SD176D	3027	2SD50X	3027	3TE250	3027	156-043
3027	2SD147L	3027	2SD176E	3027	2SD50Y	3027	3TE240	3027	156-043A
3027	2SD147M	3027	2SD176G	3027	2SD51B	3027	3TX005	3027	161N4
3027	2SD147OR	3027	2SD176GN	3027	2SD51C	3027	3TX004	3027	172-003-9-001
3027	2SD147R	3027	2SD176H	3027	2SD51D	3027	4-3023843	3027	172-003-9-001A
3027	2SD147X	3027	2SD176J	3027	2SD51E	3027	7-12	3027	173A4491-2
3027	2SD147Y	3027	2SD176K	3027	2SD51F	3027	7-13	3027	173A4491-2A
3027	2SD15A	3027	2SD176L	3027	2SD51G	3027	007-7450301	3027	173A4491-4
3027	2SD15B	3027	2SD176M	3027	2SD51GN	3027	8-0052402	3027	176-040-9-001
3027	2SD15C	3027	2SD176OR	3027	2SD51H	3027	8-0414120	3027	180T2
3027	2SD15D	3027	2SD176X	3027	2SD51J	3027	8-0414130	3027	180T2A
3027	2SD15E	3027	2SD176Y	3027	2SD51K	3027	8-905-706-555	3027	180T2B
3027	2SD15F	3027	2SD17A	3027	2SD51L	3027	8-905-706-557	3027	180T2C
3027	2SD15G	3027	2SD17C	3027	2SD51M	3027	8-905-713-101	3027	181T2A
3027	2SD15GN	3027	2SD17D	3027	2SD51OR	3027	8-905-713-556	3027	181T2C
3027	2SD15H	3027	2SD17E	3027	2SD51R	3027	09-302122	3027	211-40140-18
3027	2SD15J	3027	2SD17F	3027	2SD51X	3027	10-13002-003	3027	231-0004
3027	2SD15K	3027	2SD17GN	3027	2SD51Y	3027	10-13002-004	3027	260P24008
3027	2SD15L	3027	2SD17H	3027	2SD53B	3027	13-0032	3027	297V061001A
3027	2SD15M	3027	2SD17J	3027	2SD53C	3027	13-17918-1	3027	297V061002A
3027	2SD15OR	3027	2SD17K	3027	2SD53E	3027	13-23594-1	3027	378-44
3027	2SD15R	3027	2SD17L	3027	2SD53F	3027	13-33188-2	3027	378-44A
3027	2SD15X	3027	2SD17M	3027	2SD53G	3027	14-40325A	3027	386-7183P1
3027	2SD15Y	3027	2SD17R	3027	2SD53GN	3027	14-40363A	3027	386-7183P1A
3027	2SD163	3027	2SD17X	3027	2SD53H	3027	14-40369A	3027	394-3135A
3027	2SD163A	3027	2SD17Y	3027	2SD53J	3027	14-40421A	3027	417-139A
3027	2SD163B	3027	2SD89B	3027	2SD53K	3027	14-40464A	3027	417-212
3027	2SD163C	3027	2SD89C	3027	2SD53L	3027	14-40465A	3027	417-212A
3027	2SD163D	3027	2SD89D	3027	2SD53M	3027	14-40466A	3027	417-215A
3027	2SD163E	3027	2SD89E	3027	2SD53OR	3027	14-40471	3027	417-273
3027	2SD163F	3027	2SD89F	3027	2SD53R	3027	14-40934-1	3027	422-0961
3027	2SD163G	3027	2SD89GN	3027	2SD53X	3027	14-601-10	3027	430-53843
3027	2SD163GN	3027	2SD89H	3027	2SD53Y	3027	14-601-13	3027	472-0946-001
3027	2SD163H	3027	2SD89J	3027	2SD68	3027	14-601-15	3027	571-844
3027	2SD163J	3027	2SD89K	3027	2SD68A	3027	14-601-15A	3027	571-844A
3027	2SD163K	3027	2SD89L	3027	2SD68B	3027	14-601-16	3027	686-0243
3027	2SD163L	3027	2SD89M	3027	2SD68C	3027	14-601-16A	3027	686-0243-0
3027	2SD163M	3027	2SD89OR	3027	2SD68D	3027	19A126813	3027	686-143
3027	2SD163OR	3027	2SD89X	3027	2SD68E	3027	19A126813A	3027	686-143A
3027	2SD163R	3027	2SD89Y	3027	2SD68F	3027	20-1111-002	3027	700-113
3027	2SD163X	3027	2SD201A	3027	2SD68G	3027	20-1111-003	3027	800-510-00
3027	2SD163Y	3027	2SD201B	3027	2SD68GN	3027	020-1111-008	3027	800-510-01
3027	2SD164	3027	2SD201C	3027	2SD68H	3027	23-5035	3027	800-524-02
3027	2SD164A	3027	2SD201F	3027	2SD68J	3027	23-5038	3027	800-524-02A
3027	2SD164B	3027	2SD201G	3027	2SD68L	3027	23-5041	3027	800-524-03
3027	2SD164C	3027	2SD201GN	3027	2SD68M	3027	025	3027	800-524-03A
3027	2SD164E	3027	2SD201H	3027	2SD68OR	3027	34-1000	3027	800-52402
3027	2SD164F	3027	2SD201J	3027	2SD68R	3027	34-1000A	3027	800-524-04A
3027	2SD164G	3027	2SD201K	3027	2SD68X	3027	34-1028	3027	804-14120
3027	2SD164GN	3027	2SD201L	3027	2SD68Y	3027	34-1028A	3027	804-14130
3027	2SD164H	3027	2SD201X	3027	2SD70A	3027	34-6002-32	3027	921-1011
3027	2SD164J	3027	2SD211A	3027	2SD70B	3027	34-6002-32A	3027	991-01-3063
3027	2SD164K	3027	2SD211B	3027	2SD70C	3027	41-0318	3027	992-00-3139
3027	2SD164L	3027	2SD211C	3027	2SD70D	3027	41-0318A	3027	992-00-3139A
3027	2SD164M	3027	2SD211D	3027	2SD70E	3027	42-20961	3027	992-002271
3027	2SD164OR	3027	2SD211D	3027	2SD70F	3027	42-20961A	3027	992-00271A
3027	2SD164R	3027	2SD211E	3027	2SD70G	3027	43-023843	3027	992-003139
					2SD70GN			3027	992-004091

SK	Industry Standard No.
3027	992-017169
3027	1045-7844
3027	1045-7851
3027	1064-6032
3027	1069-7032
3027	1080-5364
3027	1806-17
3027	1806-17A
3027	1854-0563
3027	2000-221
3027	2015-1
3027	2015-1A
3027	2015-2
3027	2015-2A
3027	2015-3
3027	2015-3A
3027	3146-977
3027	3152-170
3027	3665-2
3027	3683-1
3027	3714H1
3027	3714H1A
3027	4701
3027	4715
3027	4715A
3027	4802-00005
3027	4802-00005A
3027	4882
3027	7214A
3027	7514
3027	8000-00006-006
3027	8000-00006-190
3027	12536
3027	12536A
3027	16001
3027	16176
3027	16201
3027	16230
3027	16234
3027	16240
3027	16292
3027	23114-070
3027	023762
3027	28474
3027	30276
3027	36892
3027	37888
3027	38272
3027	38474
3027	38494
3027	38897
3027	39921
3027	40251
3027	40325
3027	40363
3027	40464
3027	40465
3027	40633
3027	40934-1
3027	43060
3027	43114
3027	43168
3027	52329
3027	52360
3027	60085
3027	60130
3027	70008-0
3027	70008-3
3027	94065
3027	94065A
3027	130014
3027	138194
3027	198034-1
3027	198034-3
3027	198034-4
3027	198039-0507
3027	198039-6
3027	198039-7
3027	231378
3027	232359
3027	232359A
3027	241657
3027	267791
3027	318835
3027	0573562
3027	610111-2
3027	610111-4
3027	610140-1
3027	610161-2
3027	700080
3027	700080A
3027	700083
3027	700083A
3027	793356-1
3027	801537
3027	815246-1
3027	984259
3027	984259A
3027	1417320-1
3027	1417356-2
3027	1471135-1
3027	1473665-2
3027	2003073-91
3027	2003073-91A
3027	2057199-0701
3027	2057199-070BA
3027	2057199-701
3027	2057323-0501
3027	2327052
3027	2327053
3027	2327172
3027	3146977
3027	3146977A
3027	3152170
3027	3152170A
3027	3463101-1
3027	4080320-0501
3027	4080320-050B
3027	4080866-000A
3027	4080866-8006
3027	4082886-001
3027	7026018
3027	7306982
3027	7309160
3027	7937586
3027	9340311
3027	10713642
3027	20278600
3027	23114070
3027	23114070A
3027	23114100
3027	23114108
3027	30200101
3027	43023843

SK	Industry Standard No.
3027	43024216
3027	51126850
3027	62737425
3027	74140226-001
3027	74140226-003
3027	80052402
3027/130	B170008
3027/130	B170017
3027/130	B170023
3027/130	B170026
3027/130	BD112
3027/130	BD113
3027/130	BD116
3027/130	BD118
3027/130	BD145
3027/130	BDY23
3027/130	BLY12
3027/130	BLY15
3027/130	BLY47
3027/130	BLY48
3027/130	C244
3027/130	C493
3027/130	C49Y
3027/130	C520A
3027/130	C521
3027/130	C521A
3027/130	C736
3027/130	C768
3027/130	D81
3027/130	DT1610
3027/130	IP20-0028
3027/130	MJ2840
3027/130	MJ2841
3027/130	SDT9207
3027/130	SE3032
3027/130	TNJ72320
3027/130	TR1009
3027/130	TR2327574
3027/130	2N3445
3027/130	2N3446
3027/130	2N3447
3027/130	2N4913
3027/130	2N4914
3027/130	2N4915
3027/130	2N5036
3027/130	2N5067
3027/130	2N5068
3027/130	2N5869
3027/130	2N5870
3027/130	2SC1080
3027/130	2SC1080R
3027/130	2SC1080Y
3027/130	2SC1618
3027/130	2SC1618B
3027/130	2SC1619
3027/130	2SC1777
3027/130	2SC244
3027/130	2SC493
3027/130	2SC49Y
3027/130	2SC521
3027/130	2SC521A
3027/130	2SC736
3027/130	2SC768
3027/130	2SD124
3027/130	2SD15
3027/130	2SD16
3027/130	2SD17
3027/130	2SD189R
3027/130	2SD201
3027/130	2SD201M
3027/130	2SD201Y
3027/130	2SD26
3027/130	2SD26A
3027/130	2SD26B
3027/130	2SD26C
3027/130	2SD371
3027/130	2SD428
3027/130	2SD50
3027/130	2SD51
3027/130	2SD51A
3027/130	2SD53
3027/130	2SD53A
3027/130	2SD70
3027/130	2SD80
3027/130	2SD81
3027/130	2SD82
3027/130	09-305138
3027/130	13-41738-1
3027/130	14-601-17
3027/130	14-601-23
3027/130	21A112-124
3027/130	52-025-004-0
3027/130	57A196-10
3027/130	57B213-11
3027/130	0135-1
3027/130	176-049-9-002
3027/130	181T2B
3027/130	394-3135
3027/130	800-537-02
3027/130	800-537-03
3027/130	1074-117
3027/130	1081-7104
3027/130	1414-179
3027/130	2057A100-16
3027/130	2057B100-16
3027/130	5380-73
3027/130	16088
3027/130	16114
3027/130	36534
3027/130	36535
3027/130	40636
3027/130	55810-166
3027/130	60154
3027/130	132776
3027/130	138195
3027/130	282217
3027/130	1971487
3027/130	1971503
3027/130	4082886-0002
3027/130	62573741
3027/U30	96-5117-91
3028	GE-24MP
3028	218-26
3029	CXL130MP
3029	DF-2
3029	ECG130MP
3029	GE-15MP
3029	W16MP
3029	2SC494A
3029	2SC494B
3029	2SC494C
3029	2SC494D
3029	2SC494E
3029	2SC494F

SK	Industry Standard No.
3029	2SC494G
3029	2SC494GN
3029	2SC494H
3029	2SC494J
3029	2SC494K
3029	2SC494L
3029	2SC494M
3029	2SC494OR
3029	2SC494R
3029	2SC494X
3029	2SC494Y
3029	8-0053702
3029	80-053702
3029	132-542
3029	161N2
3029	800-53702
3029	171174-1
3029	80053702
3031A	A04
3031A	A04331-021
3031A	A04331-043
3031A	A10164
3031A	A10165
3031A	A10C
3031A	A114C
3031A	A1302
3031A	A1D1
3031A	A1D9
3031A	A1E1
3031A	A1E5
3031A	A1E9
3031A	A2460
3031A	A2481
3031A	A2D5
3031A	A2D9
3031A	A300
3031A	A3D5
3031A	A3D9
3031A	A3E1
3031A	A3E3
3031A	A3E5
3031A	A3E9
3031A	A400
3031A	A4212-A
3031A	A4D1
3031A	A4D5
3031A	A4E1
3031A	A909-1019
3031A	AA300
3031A	ACR810-108
3031A	AG100G
3031A	AM-6-5
3031A	AM-G-22
3031A	AM-G-5
3031A	AM-G-5A
3031A	AM-G-5C
3031A	AM040
3031A	AN-G-5B
3031A	AR19
3031A	AR20
3031A	AW-01-12
3031A	B1D1
3031A	B1D5
3031A	B1E1
3031A	B3D5
3031A	B3D9
3031A	B3E1
3031A	B3E5
3031A	B4D5
3031A	B4D9
3031A	B4E1
3031A	B5D1
3031A	B60C300
3031A	BYX22/400
3031A	BYX36/600
3031A	C248507
3031A	CA102FA
3031A	CA250
3031A	CB250
3031A	CC102HA
3031A	CD1115
3031A	CD1116
3031A	CD1151
3031A	CDR-2
3031A	CE0398/7839
3031A	CER700B
3031A	CER70B
3031A	CER70C
3031A	CL4
3031A	C049
3031A	CP102FA
3031A	CP102HA
3031A	CT3003
3031A	CY40
3031A	D-05
3031A	D028
3031A	D101167
3031A	D10168
3031A	D1J70544
3031A	D2J
3031A	D3A
3031A	D6462
3031A	DD177C
3031A	DD2066
3031A	DD236
3031A	DE14
3031A	DG1PR
3031A	DH-001
3031A	DIJ71959
3031A	DR1PR
3031A	DR4
3031A	DR5
3031A	DR670
3031A	DR671
3031A	DR695
3031A	DR698
3031A	DR699
3031A	DR826
3031A	DR81
3031A	DS-16A
3031A	DS-19
3031A	DS-1M
3031A	DS-2N
3031A	DS2N
3031A	E0788-C
3031A	E10171
3031A	E1415
3031A	EA040
3031A	EA1448
3031A	EA16X2
3031A	EA16X33
3031A	EA16X34
3031A	EA2140
3031A	EA2499

SK	Industry Standard No.
3031A	EA2501
3031A	ED3003A
3031A	ED3003B
3031A	ED3004A
3031A	ED3004B
3031A	EM1021
3031A	EM503
3031A	EM504
3031A	EP16X4
3031A	ER103D
3031A	ERB22-15
3031A	ERV-02F2150
3031A	ES57X5
3031A	ES57X8
3031A	ES57X9
3031A	ET400
3031A	F14B
3031A	F14D
3031A	FA4
3031A	FG-2NA
3031A	FR10
3031A	FR1MD
3031A	FU1U
3031A	G00-502A
3031A	G4
3031A	GE42-7
3031A	GER4004
3031A	GI-P100-D
3031A	H400
3031A	H616
3031A	H617
3031A	HA400
3031A	HD20-003-01
3031A	HD2000
3031A	HD2000-301
3031A	HD2000301-0
3031A	HD2000307
3031A	HD2000308
3031A	HF-20052
3031A	HFSD12
3031A	HGR4
3031A	HR-05A
3031A	HR5
3031A	HS3103
3031A	HS3104
3031A	HT8000710
3031A	HV46GR
3031A	IR20
3031A	J241271
3031A	J24570
3031A	JB-BB1A
3031A	JC-SG005
3031A	K2D5
3031A	K2E5
3031A	K4-557
3031A	K4D5
3031A	K4E5
3031A	K5D5
3031A	K5E5
3031A	KC2DP12/1
3031A	KE-262
3031A	KS-05X
3031A	LA600
3031A	M604HT
3031A	M700A
3031A	M700B
3031A	M70A
3031A	M70B
3031A	MB-01
3031A	MC030
3031A	MC030A
3031A	MC030B
3031A	MC040
3031A	MC040A
3031A	MC1523
3031A	MC1524
3031A	MD137
3031A	MD138
3031A	MH70
3031A	MP225
3031A	MP300
3031A	MP400
3031A	MP651
3031A	MR1337-4
3031A	MR2272
3031A	MS35H
3031A	MS36H
3031A	MS5H
3031A	MSR-V5
3031A	MT44
3031A	OA131
3031A	OA132
3031A	OA81
3031A	OG-30L125
3031A	P-10115
3031A	P21316
3031A	P21317
3031A	P7C5
3031A	PA070
3031A	PA330
3031A	PA340
3031A	PD110
3031A	PD111
3031A	PD135
3031A	PE504
3031A	PS4560
3031A	PS5300
3031A	PS5301
3031A	PS5302
3031A	PS627
3031A	PS628
3031A	PS629
3031A	PS632
3031A	PS633
3031A	PS636
3031A	PS637
3031A	Q59
3031A	Q60
3031A	R-1
3031A	R10DC
3031A	RE504
3031A	RFC61197
3031A	RFJ70432
3031A	RFJ70703
3031A	RFJ70931
3031A	RG100G
3031A	RS6471
3031A	RT-2669
3031A	RT3443
3031A	RT4069
3031A	RT5070
3031A	RT5472
3031A	RT5909
3031A	RT5911

SK	Industry Standard No.
3031A	RT6729
3031A	RT6791
3031A	RT7634
3031A	RT7849
3031A	RT7850
3031A	RT8839
3031A	RVD08C22/1A
3031A	RVD10DC1
3031A	RVD2DP
3031A	RVDSD-1Y
3031A	S-05-005
3031A	S-05-01
3031A	S-050
3031A	S-0501
3031A	S1-B01-02
3031A	S10149
3031A	S1A
3031A	S1B01-0226
3031A	S1B02-CR1
3031A	S1BD1-02
3031A	S36
3031A	S750
3031A	S750C
3031A	S93H
3031A	SB-2T
3031A	SB-3F01
3031A	SCA3
3031A	SCA4
3031A	SCE4
3031A	SD-1A4F
3031A	SD-470
3031A	SD104
3031A	SD105
3031A	SD202
3031A	SD470
3031A	SD500
3031A	SD500C
3031A	SDS-113
3031A	SE05
3031A	SFR154
3031A	SFR164
3031A	SFR253
3031A	SH-1A
3031A	SH4D3
3031A	SH4D4
3031A	SI-RECT-144
3031A	SI-RECT-174
3031A	SI-RECT-222
3031A	SI-RECT-226
3031A	SI-RECT-59
3031A	SIB0201CR
3031A	SID01E
3031A	SID02E
3031A	SL-030T
3031A	SL93
3031A	SLA1490
3031A	SLA14AB
3031A	SLA14C
3031A	SLA15AB
3031A	SM40
3031A	SM513
3031A	SM740
3031A	SP1U-2
3031A	SR-13H
3031A	SR-1849-1
3031A	SR-2
3031A	SR-4
3031A	SR-5
3031A	SR1694
3031A	SR1695
3031A	SR40
3031A	SRIDM-1
3031A	SRK-2
3031A	SRLFM-1
3031A	SS0010
3031A	SS0009
3031A	ST14
3031A	STB576
3031A	SW-05-005
3031A	SW0501
3031A	SX644
3031A	SX645
3031A	T-E1102A
3031A	T-E1108
3031A	T-E1124
3031A	T10144
3031A	T21602
3031A	T21649
3031A	T42692-001
3031A	TA400
3031A	TA7803
3031A	TH803
3031A	TI58
3031A	TI59
3031A	TI60
3031A	TIR03
3031A	TIR04
3031A	TL32
3031A	TL41
3031A	TR330027
3031A	TUS-185D
3031A	TV24167
3031A	TV24586
3031A	TV34232
3031A	TVM56
3031A	TVS-1S1850
3031A	TVS-1S1906
3031A	TVS-DG-1N-R
3031A	TVS-FR-2PC
3031A	TVS-FR1-PC
3031A	TVS-FR2P
3031A	TVS-FU1N
3031A	TVS-HF-SD1Z
3031A	TVS-UPSD-1
3031A	TVSFR1P
3031A	TVSFR2PC
3031A	TVS0A71
3031A	UFSD-1
3031A	UR125
3031A	USFD-1
3031A	USFD-1A
3031A	UT225
3031A	UT226
3031A	UT227
3031A	UT25
3031A	UT26
3031A	V10158
3031A	VAMV-1
3031A	WD013
3031A	WD014
3031A	WD015
3031A	WO-6A
3031A	XS16A
3031A	YAAD004

SK	Industry Standard No.
3031A	YR-011
3031A	YSG-V47-7-51-1
3031A	Z330611
3031A	ZS-24
3031A	ZS-33A
3031A	ZS-33B
3031A	ZS-34A
3031A	ZS-34B
3031A	ZS-53
3031A	ZS103
3031A	ZS104
3031A	ZS123
3031A	ZS124
3031A	ZS173
3031A	ZS174
3031A	ZS174B
3031A	ZS22
3031A	ZS34
3031A	ZS94
3031A	001-0072-00
3031A	001-0077-00
3031A	1-530-012-11
3031A	1-6501190016
3031A	1A12214
3031A	1A12407
3031A	1A15790
3031A	1A50
3031A	1B05J40
3031A	1ET3
3031A	1ET4
3031A	1GZ61
3031A	1HY40
3031A	1N1037
3031A	1N1039
3031A	1N1044
3031A	1N1045
3031A	1N1050
3031A	1N1255A
3031A	1N2070A
3031A	1N2115
3031A	1N2116
3031A	1N315
3031A	1N486B
SK	1N507
3031A	1N5212
3031A	1N5216
3031A	1N522
3031A	1N532
3031A	1N553
3031A	1R3J
3031A	1R9H
3031A	1S054
3031A	1S114
3031A	1S1664
3031A	1S1851
3031A	1S1851R
3031A	1S193
3031A	1S1943
3031A	1S204
3031A	1S2461
3031A	1S399
3031A	1S9413
3031A	1T503
3031A	03-0018-0
3031A	03-0018-6
3031A	3A254
3031A	3G254
3031A	3GA
3031A	3L4-3001-5
3031A	3L4-3001-8
3031A	4-2020-07600
3031A	4-202R101
3031A	4-686105-3
3031A	4-686189-3
3031A	4-686199-3
3031A	4-686212-3
3031A	4-68689-3
3031A	4D4
3031A	4G8
3031A	4GA
3031A	5GF
3031A	8-0001300
3031A	8-0001400
3031A	8D4
3031A	09-306100
3031A	09-306115
3031A	09-306157
3031A	09-306172
3031A	09-306263
3031A	09-306264
3031A	09-306285
3031A	09-306323
3031A	10-102005
3031A	10B-4
3031A	10C
3031A	10C3
3031A	10C4
3031A	10C4D
3031A	10D-5
3031A	10D-V
3031A	10D2L
3031A	10DC
3031A	10DC05N
3031A	10DC05R
3031A	10DC5
3031A	11-0429
3031A	11-0769
3031A	11-0771
3031A	11-085024
3031A	12-085040
3031A	12-100008
3031A	12-102001
3031A	13-085024
3031A	13-085027
3031A	13-087027
3031A	13-14094-12
3031A	13-14094-17
3031A	13-14627-4
3031A	13-15465-1
3031A	13-29165-1
3031A	13-55029-1
3031A	15-085043
3031A	15-100004
3031A	15-108011
3031A	15-108049
3031A	16-501190016
3031A	018-00006
3031A	018-00007
3031A	018-00009
3031A	19-085022
3031A	21A007-000
3031A	21A103-018
3031A	21A103-050
3031A	21M433
3031A	21M469
3031A	022D
3031A	24MW1108
3031A	24MW1113
3031A	24MW227
3031A	24MW246
3031A	24MW269
3031A	24MW602
3031A	24MW605
3031A	24MW779
3031A	24MW829
3031A	24MW862
3031A	24MW864
3031A	24MW871
3031A	24MW924
3031A	24MW974
3031A	24MW975
3031A	025-100029
3031A	30H3P
3031A	32-0042
3031A	32-0047
3031A	32-0048
3031A	32-0059
3031A	33-0026
3031A	33G59122
3031A	34-8054-15
3031A	34-8057-8
3031A	40C
3031A	42-21408
3031A	42-23350
3031A	42-28202
3031A	44-530
3031A	46-8601-3
3031A	46-86105-3
3031A	46-86146-3
3031A	46-86199-3
3031A	46-86212-3
3031A	46-86303-3
3031A	46-86365-3
3031A	46-86383-3
3031A	46-86420-3
3031A	46-86463-3
3031A	46AR10
3031A	46BR9
3031A	48-10062A02
3031A	48-137347
3031A	48-137348
3031A	48-191A02
3031A	48-191A08
3031A	48-67120A10
3031A	48-86148
3031A	48S10062A01
3031A	48S134959
3031A	48S137133
3031A	48X90234A01
3031A	48X97168A06
3031A	48X97172A01
3031A	51-03007-06
3031A	53A022-5
3031A	53B003-1
3031A	53B010-3
3031A	53B012-1
3031A	53C005-3
3031A	53K001-9
3031A	53N001-2
3031A	53N004-7
3031A	53N004-8
3031A	53N004-9
3031A	62A02
3031A	65-085013
3031A	66-2246
3031A	66X0023-003
3031A	66X0038-001
3031A	70-270050
3031A	75R4B
3031A	80-001300
3031A	80-001400
3031A	81-46123004-7
3031A	81-46123005
3031A	84A20
3031A	86-0006
3031A	86-27-1
3031A	86-32-1
3031A	86-65-9
3031A	86-72-3
3031A	089-252
3031A	93A42-7
3031A	93A60-2
3031A	93A60-3
3031A	93A60-8
3031A	93A67-1
3031A	100R3B
3031A	100R4B
3031A	103-191
3031A	103-261
3031A	103-261-0
3031A	108B6
3031A	109B6
3031A	110B6
3031A	0112
3031A	120-003148
3031A	124-0178
3031A	130-30189
3031A	130-30256
3031A	164J2
3031A	174-1
3031A	174-2
3031A	174-3
3031A	212-61
3031A	212-94
3031A	229-5100-233
3031A	264P00601
3031A	264P04801
3031A	325-4610-100
3031A	359F
3031A	359H
3031A	384D
3031A	385F
3031A	385H
3031A	386AW
3031A	386FW
3031A	386FX
3031A	0402
3031A	412
3031A	422-1408
3031A	539J2F
3031A	540-010
3031A	540J2F
3031A	617-46
3031A	617-53
3031A	624
3031A	660-230
3031A	690V053H57
3031A	690V086H92
3031A	690V109H44
3031A	750-141
3031A	792-238
3031A	800-014-00
3031A	800-01400
3031A	800-016-00
3031A	851-0372-130
3031A	854-0372-020
3031A	903-169B
3031A	903Y00212
3031A	916-33003-2
3031A	936-20
3031A	964-21866
3031A	977-2B
3031A	977-3B
3031A	1001-11
3031A	1034
3031A	1040
3031A	1044
3031A	1849
3031A	1849R
3031A	1850
3031A	1850R
3031A	1950-17
3031A	1982-17
3031A	2061B45-35
3031A	2093A38-13
3031A	2093A41-104
3031A	2093A41-42
3031A	2093A41-49
3031A	2093A41-51
3031A	2093A41-54
3031A	2093A41-66
3031A	2093A41-86
3031A	2093A42-7
3031A	2093B41-28
3031A	2110N-42
3031A	2211B
3031A	2252
3031A	2302
3031A	2402
3031A	2495-489
3031A	2498-513
3031A	4051-300-10150
3031A	4686-105-3
3031A	4686-189-3
3031A	4686-199-3
3031A	4686-212-3
3031A	4686-89-3
3031A	5203N1
3031A	5632-W06A
3031A	5632-W06B
3031A	5641-MV11
3031A	7213
3031A	7214-6
3031A	7215-2
3031A	8000-00005-022
3031A	8000-00005-152
3031A	8000-0004-P062
3031A	8000-00049-010
3031A	8000-004-P061
3031A	8000-004-P064
3031A	8000-004-P067
3031A	8000-004-P068
3031A	8800-201
3031A	8800-206
3031A	8999-205
3031A	12844
3031A	16681
3031A	17002
3031A	36503
3031A	41001-4
3031A	41020-618
3031A	41023-224
3031A	41023-225
3031A	42221
3031A	47126-003
3031A	50745-02
3031A	50745-03
3031A	50745-04
3031A	50745-05
3031A	50745-08
3031A	58840-116
3031A	000072020
3031A	77190-8
3031A	77271-4
3031A	100617
3031A	118873
3031A	123702
3031A	124098
3031A	125458
3031A	125529
3031A	125549
3031A	126856
3031A	127176
3031A	127379
3031A	127993
3031A	129213
3031A	130052
3031A	135284
3031A	135571
3031A	139328
3031A	141872-6
3031A	166593
3031A	166922
3031A	167413
3031A	168339
3031A	169116
3031A	169565
3031A	169765
3031A	170133
3031A	170970-1
3031A	171561
3031A	172721
3031A	241295
3031A	270642
3031A	348057-8
3031A	489752-022
3031A	489752-025
3031A	489752-043
3031A	489752-075
3031A	489752-096
3031A	489752-097
3031A	530051-2
3031A	530072-1009
3031A	530084-4
3031A	530088-1004
3031A	530088-4
3031A	530111-1001
3031A	530116-1003
3031A	530972-14
3031A	651038
3031A	801723
3031A	817068P
3031A	817164
3031A	922092
3031A	922094
3031A	922183
3031A	922567
3031A	973935-20
3031A	973936-18
3031A	973936-20
3031A	973936-21
3031A	981953
3031A	982253
3031A	982377
3031A	986414
3031A	988049
3031A	1471872-5
3031A	1472460-13
3031A	1472460-14
3031A	1472460-16
3031A	1474778-10
3031A	1474778-7
3031A	1474788-10
3031A	1476690-2
3031A	1967813
3031A	2002336-115
3031A	2003069-1
3031A	2006422-133
3031A	2006441-91
3031A	2006582-23
3031A	2010957-49
3031A	2057062-0702
3031A	2330611
3031A	5330101H
3031A	5340022H
3031A	7027038
3031A	7570011-02
3031A	7570014-05
3031A	7570215-34
3031A	7852264-01
3031A	7853358-01
3031A	7855284-01
3031A	7910805-01
3031A	7910872-01
3031A	7910873-01
3031A	7914011-01
3031A	13030189
3031A	13030256
3031A	22115183
3031A	41020618
3031A	41023224
3031A	41023225
3031A	62522674
3031A	62522739
3031A	62749558
3031A	0063050118
3031A	70270050
3031A	70270390
3031A	70270720
3031A	80001300
3031A	80001400
3031A	87219410
3031A	87500720
3031A	96421866
3031A	2295100233
3031A	4202108270
3032A	A04093-A
3032A	A04212-A
3032A	A059-114
3032A	A10N
3032A	A13M2
3032A	A1H1
3032A	A1H5
3032A	A1H9
3032A	A1K1
3032A	A1K5
3032A	A1K9
3032A	A2461
3032A	A2462
3032A	A2H1
3032A	A2H4
3032A	A2H5
3032A	A2H9
3032A	A2K1
3032A	A2K4
3032A	A2K5
3032A	A2K9
3032A	A3H1
3032A	A3H3
3032A	A3H5
3032A	A3H9
3032A	A3K1
3032A	A3K3
3032A	A3K5
3032A	A3K9
3032A	A4H1
3032A	A4H5
3032A	A4H9
3032A	A4K1
3032A	A4K5
3032A	A4K9
3032A	A5H1
3032A	A5H2
3032A	A5H5
3032A	A5H9
3032A	A5K1
3032A	A5K2
3032A	A5K5
3032A	A5K9
3032A	A759500
3032A	A800
3032A	AA800
3032A	ATC-SR-3
3032A	B1H1
3032A	B1H5
3032A	B1H9
3032A	B1K1
3032A	B1K5
3032A	B1K9
3032A	B2K1
3032A	B2K5
3032A	B2K9
3032A	B3H1
3032A	B3H5
3032A	B3H9
3032A	B3K1
3032A	B3K5
3032A	B3K9
3032A	B4H1
3032A	B4H5
3032A	B4H9
3032A	B4K1
3032A	B4K5
3032A	B4K9
3032A	B5H1
3032A	B5H5
3032A	B5H9
3032A	B5K1
3032A	B5K5
3032A	B5K9
3032A	BA145
3032A	BAY15
3032A	BAY16
3032A	BB2
3032A	BY100
3032A	BY100S
3032A	BY103
3032A	BY108
3032A	BY109
3032A	BY118
3032A	BY119
3032A	BY129
3032A	BYX10
3032A	BYX55009
3032A	BYY91
3032A	C0410
3032A	CA102PA
3032A	CA102RA
3032A	CC102PA
3032A	CC102RA
3032A	CE508
3032A	CER72
3032A	CER720
3032A	CER720A
3032A	CER720B
3032A	CER720C
3032A	CER72A
3032A	CER72B
3032A	CER72C
3032A	CER72D
3032A	CER72F
3032A	CP102PA
3032A	CP102RA
3032A	CL7
3032A	CL8
3032A	CODI537
3032A	CODI538
3032A	CODI617
3032A	CP102
3032A	CP102PA
3032A	CP102RA
3032A	CX-0054
3032A	CY80
3032A	D800
3032A	D85C
3032A	D88
3032A	D8HZ
3032A	DA058
3032A	DA2068
3032A	DD058
3032A	DD2068
3032A	DD268
3032A	DE-201
3032A	DI650
3032A	DR700
3032A	DR800
3032A	DRS107
3032A	DS-16B
3032A	E1M3
3032A	E1N3
3032A	EA080
3032A	ED2911
3032A	ED2923
3032A	ED3007
3032A	ED3007A
3032A	ED3007B
3032A	EDJ-363
3032A	EM408
3032A	EM507
3032A	EM508
3032A	ER186
3032A	ER308
3032A	ER801
3032A	ER81
3032A	ERC01-06
3032A	ERD700
3032A	ERD800
3032A	ES16X25
3032A	ES57X12
3032A	ESKE125C500
3032A	ETD-V06C
3032A	EW169
3032A	F8
3032A	FA8
3032A	FR2-02C
3032A	FR2-06
3032A	FT10
3032A	FU-1N
3032A	G100K
3032A	G8
3032A	GER4006
3032A	H109
3032A	H625
3032A	HA800
3032A	HS3108
3032A	HSFD-1A
3032A	IR2E
3032A	JCN7
3032A	JCV7
3032A	K1H5
3032A	K1K5
3032A	K2H5
3032A	K2K5
3032A	K3H5
3032A	K3K5
3032A	K4H5
3032A	K4K5
3032A	K5H5
3032A	K5K5
3032A	KSKE12C500
3032A	M702
3032A	M72
3032A	M720
3032A	M720A
3032A	M720B
3032A	M720C
3032A	M72A
3032A	M72B
3032A	M72C
3032A	M72D
3032A	M82
3032A	M8HZ
3032A	M91A06
3032A	MC070
3032A	MC070A
3032A	MC080
3032A	MC080A
3032A	MC1527
3032A	MC1528
3032A	MM7

SK	Industry Standard No.	SK	Industry Standard No.	SK	Industry Standard No.	SK	Industry Standard No.	SK	Industry Standard No.
3032A	MM8	3032A	TW80	3032A	488137348	3032A	530088-1003	3033	DP100
3032A	MR2266	3032A	U05E	3032A	488191A07	3032A	530088-3	3033	DR1000
3032A	MR801	3032A	V03C	3032A	50D8	3032A	530136-2	3033	DR900
3032A	MT84	3032A	V06G	3032A	50E7	3032A	530162-1	3033	DR82
3032A	NA74	3032A	V09	3032A	50E8	3032A	530162-1001	3033	DU1000
3032A	NA75	3032A	V09A	3032A	53A010-1	3032A	817179A	3033	DU800
3032A	NA76	3032A	V09B	3032A	53B010-5	3032A	972571-5	3033	E10
3032A	NA84	3032A	Z1A103-018	3032A	53J001-1	3032A	981955	3033	E8
3032A	NA85	3032A	ZS108	3032A	53J002-1	3032A	981959	3033	E9
3032A	NA86	3032A	ZS78	3032A	56-15	3032A	984794	3033	ED2847
3032A	OA211	3032A	ZS78A	3032A	67-1000-00	3032A	984882	3033	ED2848
3032A	OT5067	3032A	ZS78B	3032A	67J2	3032A	1303256	3033	ED2849
3032A	P10115A	3032A	1D8	3032A	67J2A	3032A	1474872-1	3033	ED2910
3032A	P10156A	3032A	1E7	3032A	75D8	3032A	2001786-142	3033	ED2912
3032A	P580	3032A	1E8	3032A	75E7	3032A	2006582-22	3033	ED2913
3032A	P6RP8	3032A	1ET7	3032A	75E8	3032A	2330251	3033	ED2922
3032A	PA380	3032A	1ET8	3032A	75R7B	3032A	2330251H	3033	ED2924
3032A	PD114	3032A	1F8	3032A	75R8B	3032A	2330253	3033	ED3008
3032A	PD115	3032A	1HY100	3032A	80A8	3032A	5330102H	3033	ED3008A
3032A	PD913	3032A	1HY80	3032A	80H	3032A	62047895	3033	ED3008B
3032A	PD914	3032A	1N1226A	3032A	86-67-2	3032A	62562625	3033	ED3009
3032A	PE408	3032A	1N1258	3032A	86-67-9	3032A	62749067	3033	ED3009A
3032A	PE508	3032A	1N1259	3032A	86-75-3	3032A	95115115-00	3033	ED3009B
3032A	PH9D522/1	3032A	1N3196	3032A	86-85-3	3032A	4202003173	3033	ED3010
3032A	P82416	3032A	1N3280	3032A	93A69-1	3032A	4202006800	3033	ED3010A
3032A	PT580	3032A	1N4006	3032A	93A78-1	3032A	4202104570	3033	ED3010B
3032A	PTC203	3032A	1N4250	3032A	93A79-6	3032A	9511511500	3033	EF100
3032A	R080	3032A	1N517	3032A	95-11511500	3033	.5B10	3033	EM410
3032A	R8	3032A	1N5214	3032A	100B7B	3033	.5E7	3033	EM510
3032A	RA-2	3032A	1N5218	3032A	100R8B	3033	.5E8	3033	EP1000
3032A	RO080	3032A	1N597	3032A	103-203	3033	.5J7	3033	EP800
3032A	RPA70597	3032A	1N853	3032A	103-216	3033	.5J8	3033	ER1001
3032A	RPA70600	3032A	1N854	3032A	103-228	3033	.7B10	3033	ER107D
3032A	RH-DI0003TAZZ	3032A	1N875	3032A	103-284	3033	.7E7	3033	ER108D
3032A	RM-2AV	3032A	1S05B	3032A	103-51	3033	.7E8	3033	ER187
3032A	RM2A	3032A	1S106	3032A	108B-E2	3033	.7J10	3033	ER310
3032A	S-05/01	3032A	1S1061	3032A	124-0028	3033	.7J7	3033	ERD1000
3032A	S100	3032A	1S1065	3032A	167J2	3033	.7J8	3033	ERD900
3032A	S1C	3032A	1S107	3032A	200-6582	3033	A059-118	3033	F110
3032A	S1R20	3032A	1S117	3032A	212-80	3033	A068-104	3033	G10
3032A	S20	3032A	1S1668	3032A	212-95	3033	A14F	3033	G100M
3032A	S208	3032A	1S1668F	3032A	264-701508	3033	A1M1	3033	GSR1
3032A	S257	3032A	1S315	3032A	264P01508	3033	A1M5	3033	H1000
3032A	S258	3032A	1S401	3032A	264P03001	3033	A1M9	3033	H621
3032A	S28	3032A	1S848	3032A	264P04501	3033	A2M1	3033	H626
3032A	S2E60	3032A	1896	3032A	264P04702	3033	A2M4	3033	H800
3032A	S7AR1	3032A	1897	3032A	264P04703	3033	A2M5	3033	HA1000
3032A	S8AR1	3032A	1T507	3032A	264P04901	3033	A2M9	3033	H83110
3032A	SA-2B	3032A	1T508	3032A	264P06601	3033	A2MA	3033	H801
3032A	SC8	3032A	2-64701508	3032A	290V034C01	3033	A3M1	3033	J10
3032A	SC8A	3032A	2HR3J	3032A	3208	3033	A3M3	3033	K1M5
3032A	SCA8	3032A	2T507	3032A	359P	3033	A3M5	3033	K2M5
3032A	SCE8	3032A	2T508	3032A	359S	3033	A3M9	3033	K3M5
3032A	SD-1REF	3032A	2W6A	3032A	385P	3033	A4M1	3033	K4M5
3032A	SD-6AUF	3032A	2W7A	3032A	385S	3033	A4M5	3033	K5M5
3032A	SD2B	3032A	3A258	3032A	420-2003-173	3033	A4M9	3033	M102
3032A	SD8	3032A	3T507	3032A	420-2104-570	3033	A5M1	3033	M702B
3032A	SD98	3032A	3T508	3032A	523-0013-201	3033	A5M2	3033	M73
3032A	SD98A	3032A	4-2020-03200	3032A	601X0048-066	3033	A5M5	3033	M730
3032A	SE05C	3032A	4-2020-03900	3032A	601X0152-066	3033	A5M9	3033	M730A
3032A	SPR258	3032A	4-2020-05200	3032A	601X0227-066	3033	A725EH2AB1	3033	M730B
3032A	SPR268	3032A	4-2020-06700	3032A	690V08OH50	3033	AA1000	3033	M730C
3032A	SH1B	3032A	4-2020-08500	3032A	690V08OH54	3033	AD4007	3033	M73A
3032A	SH4DB	3032A	4-686106-3	3032A	754-2000-001	3033	AH1005	3033	M73B
3032A	SI-RECT-044	3032A	4-686139-3	3032A	903-17	3033	AH1010	3033	M73C
3032A	SI-RECT-102	3032A	4-686179-3	3032A	903-17A	3033	AH1015	3033	MC090
3032A	SI-RECT-102A	3032A	4-68697-3	3032A	903-17C	3033	AHB05	3033	MC090A
3032A	SIA18AB	3032A	4T507	3032A	903-36	3033	AHB10	3033	MC100
3032A	SIA18C	3032A	4T508	3032A	903-36A	3033	AHB15	3033	MC100A
3032A	SIA2616	3032A	5-30082.3	3032A	903-56B	3033	B1M1	3033	MC1529
3032A	SIA3196	3032A	5-30082.4	3032A	903-36C	3033	B1M5	3033	MM10
3032A	SIA560	3032A	5-30088.3	3032A	903-36D	3033	B1M9	3033	MM9
3032A	SM1500	3032A	5A8	3032A	903-36F	3033	B2M1	3033	MR990
3032A	SM170	3032A	5A8D	3032A	1016-79	3033	B2M5	3033	NA104
3032A	SM180	3032A	5GJ	3032A	1061-8361	3033	B2M9	3033	NA105
3032A	SM270	3032A	5MA8	3032A	1070	3033	B3M1	3033	NS81021
3032A	SM280	3032A	807	3032A	1070-0623	3033	B3M5	3033	P1000
3032A	SM517	3032A	8GA	3032A	1080	3033	B3M9	3033	P5100
3032A	SM518	3032A	09-306050	3032A	1084	3033	B4M1	3033	P6RP10
3032A	SM70	3032A	09-306063	3032A	2093A41-65	3033	B4M5	3033	PA300
3032A	SM71	3032A	09-306083	3032A	2093A41-77	3033	B4M9	3033	PD116
3032A	SM73	3032A	09-306088	3032A	2093A52-1	3033	B5M1	3033	PD915
3032A	SM770	3032A	09-306176	3032A	2093A69-1	3033	B5M5	3033	PD916
3032A	SM780	3032A	09-306192	3032A	2093A78-1	3033	B5M9	3033	PE410
3032A	SM80	3032A	9-511511500	3032A	2093A79-6	3033	BAY23	3033	PE510
3032A	SM81	3032A	10A98	3032A	2802	3033	BAY90	3033	PH109
3032A	SM83	3032A	1008	3032A	4686-106-3	3033	BTB1000	3033	PS1140
3032A	SR-1K-2A	3032A	10D-7K	3032A	4686-139-3	3033	BTM1000	3033	PS2417
3032A	SR-1K-2B	3032A	10D7	3032A	4686-179-3	3033	BXY10	3033	R1000
3032A	SR-1K-2C	3032A	10D8	3032A	4686-97-3	3033	BY1001	3033	R5
3032A	SR-1K-2D	3032A	10DC-2	3032A	04970	3033	BY1002	3033	R5B
3032A	SR-9001	3032A	10R7B	3032A	5001-089	3033	BY104	3033	R5C
3032A	SR1PM-8	3032A	10R8B	3032A	7215-1	3033	BY1101	3033	R6
3032A	ST18	3032A	13-16104-9	3032A	23115-042	3033	BY1102	3033	S0010G
3032A	SW05C	3032A	13-17174-1	3032A	23115-072	3033	BY1201	3033	S10AR1
3032A	SW1C	3032A	13-17174-2	3032A	42021	3033	BY1202	3033	S1D
3032A	T-E1157	3032A	13-17174-5	3032A	44006	3033	BY128	3033	S210
3032A	T-E1171	3032A	13-26614-1	3032A	47126-8	3033	BXY22/800	3033	S233
3032A	T600	3032A	13-33376-1	3032A	47126-10	3033	BYX36	3033	S24
3032A	T600X	3032A	15-100002	3032A	47126-9	3033	BYX37	3033	S260
3032A	TA7804	3032A	15-108005	3032A	53008-4	3033	C1.0E02	3033	S2E100
3032A	TA7805	3032A	15-108006	3032A	055228	3033	CA102VA	3033	S61
3032A	TA800	3032A	15-108020	3032A	055228H	3033	CA152VA	3033	S62
3032A	TH800	3032A	15-108021	3032A	61436	3033	CB100	3033	S9AR1
3032A	TH808	3032A	15-108022	3032A	70205-8A	3033	CC102VA	3033	SC10
3032A	TIR07	3032A	18J2	3032A	72172-1	3033	CC152VA	3033	SC10A
3032A	TIR08	3032A	18J2F	3032A	90203-6	3033	CE510	3033	SCA10
3032A	TK800	3032A	21A037-003	3032A	120471	3033	CER73	3033	SCE10
3032A	TKF80	3032A	26-4701508	3032A	126885	3033	CER730	3033	SD-1B
3032A	TM86	3032A	31-194	3032A	127992	3033	CER730A	3033	SD1C
3032A	T88	3032A	32-0061	3032A	131502	3033	CER730B	3033	SD2
3032A	TV24266	3032A	34-8054-14	3032A	131950	3033	CER730C	3033	SD7
3032A	TV24298	3032A	35-1003	3032A	167034	3033	CER73A	3033	SD800
3032A	TV24617	3032A	42-051	3032A	169363	3033	CER73B	3033	SD910
3032A	TV8	3032A	46-86139-3	3032A	222867	3033	CER73C	3033	SD910A
3032A	TVM55	3032A	46-86147-3	3032A	223757	3033	CER73D	3033	SE05D
3032A	TVM511	3032A	46-86307-3	3032A	226922	3033	CER73F	3033	SE1730
3032A	TVM550	3032A	46-86321-3	3032A	234553	3033	CP102VA	3033	SH1C
3032A	TVS-DS1K	3032A	46-86326-5	3032A	280217	3033	C0D1617	3033	SI1000E
3032A	TVS1P20	3032A	46-86551-5	3032A	348054-7	3033	C0D1618	3033	SI1001K
3032A	TVS1S1906	3032A	46-86555-5	3032A	348057-9	3033	CP102VA	3033	SID00K
3032A	TVSC0410	3032A	46-86562-3	3032A	489752-013	3033	CP103	3033	SL608
3032A	TVSDS-2K	3032A	46-8697-3	3032A	489752-015	3033	CP152VA	3033	SL610
3032A	TVSPR2-005	3032A	46AR15	3032A	489752-028	3033	CY100	3033	SL708
3032A	TVSPR2-04	3032A	46AR50	3032A	489752-051	3033	D1000	3033	SL710
3032A	TVSPR2-10	3032A	46AX13	3032A	489752-066	3033	D108	3033	SLA001
3032A	TVSHFSD-1C	3032A	46AX55	3032A	489752-092	3033	D1K	3033	SIA19AB
3032A	TVSMR-1M	3032A	46AX56	3032A	489752-094	3033	D4L	3033	SIA19C
3032A	TVSS15	3032A	46BR27	3032A	530072-1014	3033	DER1	3033	SIA2617
3032A	TVSS1P20	3032A	48-191406	3032A	530082-1002	3033	DI-650	3033	SIA4561
3032A	TVSS3-2	3032A	48-191407	3032A	530082-1004	3033	DICR1	3033	SM100
3032A	TVSS4C	3032A	48-355048	3032A	530082-3	3033	DID	3033	SM101
3032A	TVSSA-2B	3032A	488134939	3032A	530082-4	3033	DIK	3033	SM150D
3032A	TVSSA2B	3032A	488137347	3032A	530088-1002				

SK	Industry Standard No.	SK	Industry Standard No.	SK	Industry Standard No.	SK	Industry Standard No.	SK	Industry Standard No.
3033	SM200	3033	5AD10D	3033A	P14J	3035	B318	3035	13-16607-1
3033	SM300	3033	5AD8	3033A	GE-509	3035	B319	3035	13-17607-1
3033	SM520	3033	5ADBDC	3033A	GER4007	3035	B320	3035	13-17607B
3033	SM800	3033	5B8	3033A	QRT-210	3035	B342	3035	13-55009-1
3033	SR-35	3033	5G8	3033A	203	3035	B343	3035	13-55009-2
3033	SR-9007	3033	5MA10	3033A	9203-8	3035	B34N	3035	19A115056-P1
3033	SR150	3033	7D210	3033A	99203-3	3035	B357	3035	31-0175
3033	SW05D	3033	7D210A	3033A	9203-5	3035	B358	3035	31-0196
3033	SW1D	3033	8-905-305-400	3033A	99203-6	3035	B359	3035	34-6002-30
3033	T1000	3033	8-905-405-105	3033A	99203-7	3035	B360	3035	34-6002-31
3033	TIRO9	3033	8-905-405-160	3033A	99203-9	3035	B361	3035	34-6002-49
3033	TIR10	3033	8-905-405-170	3033A	125105	3035	B362	3035	46-86518-3
3033	TK1000	3033	8D10	3033A	132149	3035	B375-2B	3035	96-5231-01
3033	TKP100	3033	8D8	3033A	141489	3035	B375-5B	3035	116-086
3033	TVS-FR10	3033	10AG10	3033A	147015	3035	B375A-2B	3035	116-087
3033	TVS-FT-1 N	3033	10AT10	3034	E815X52	3035	B375A-5B	3035	116-088
3033	TVS-FT-1 P	3033	10AT8	3034	HEP-06007	3035	B375A-NB	3035	116-089
3033	TVS-FT10	3033	10C10	3034	P1G	3035	B375ZV	3035	121-370
3033	TVS-SD-1 B	3033	10D10	3034	T-Q5028	3035	B390	3035	324-0128
3033	TVS-SP-1	3033	10G4	3034	T370	3035	B425Y	3035	2057B2-136
3033	TVSBB10	3033	10GA	3034	TA2083	3035	B432	3035	40439
3033	TW100	3033	10R10B	3034	TA2188	3035	B447	3035	40440
3033	UT345	3033	10R9B	3034	TNJ-60075	3035	B464	3035	115068
3033	1010	3033	19-100001	3034	TNJ-60080	3035	B465	3035	116086
3033	108	3033	19A115024-P6	3034	TR16	3035	B468	3035	135744
3033	1D10	3033	21A103-013	3034	TV24162	3035	B468A	3035	171004
3033	1E10	3033	21A103-015	3034	TV24163	3035	B468C	3035	256288
3033	1EF10	3033	22-004004	3034	TV24341	3035	B468D	3035	243168
3033	198	3033	025-10029	3034	TVS-28B126V	3035	E815X54	3036	0573212
3033	1LE11	3033	25-7	3034	VS-28B126	3035	E815X77	3036	0573212H
3033	1N1096	3033	25D10	3034	2N1073	3035	E815X26	3036	1473648-1
3033	1N1105	3033	25K10	3034	2N1073A	3035	E815X40	3036	1962323
3033	1N1225	3033	34-8050-10	3034	2N1073B	3035	GC4094	3036	2320833
3033	1N1225A	3033	35-1000	3034	2N3730	3035	HEP-06008	3036	A08-1050115
3033	1N1226	3033	35-1029	3034	2N3732	3035	HEP235	3036	B0301-049
3033	1N1226B	3033	46AR15	3034	28B216	3035	P1F	3036	BDX41
3033	1N1260	3033	46AR16	3034	28B216A	3035	P1Y	3036	BDY-10
3033	1N1261	3033	46AR18	3034	28B339	3035	P4F	3036	BDY11
3033	1N1407	3033	46AR52	3034	28B339H	3035	Q5030	3036	D8-519
3033	1N1408	3033	46AR59	3034	28B340	3035	R8132	3036	EP15X29
3033	1N1443	3033	46AX12	3034	28B340H	3035	8B5-0399	3036	G181-725-001
3033	1N1443A	3033	46AX14	3034	28B425	3035	8P59	3036	HEP-87002
3033	1N1492	3033	46AX59	3034	28B425Y	3035	8P62	3036	IRT859
3033	1N1730	3033	46AX70	3034	28B472	3035	TA1928A	3036	P6500A
3033	1N1730A	3033	46AX82	3034	13-15806-1	3035	TV111	3036	PTC116
3033	1N321A	3033	46AX85	3034	13-16604-1	3035	TV114	3036	S200
3033	1N322	3033	46BD101	3034	13-17608-1	3035	TV24142	3036	ST101
3033	1N322A	3033	46BD27	3034	13-17608-2	3035	TV24468	3036	STC4252
3033	1N328	3033	46BD50	3034	13-17608B	3035	TV24540	3036	STC4254
3033	1N3281	3033	46BD32	3034	13-17608C	3035	TV24568	3036	STC4255
3033	1N3282	3033	46BD33	3034	13-17609-1	3035	TVS-28B448	3036	T841
3033	1N328A	3033	46BD34	3034	13-21606-1	3035	2N4346	3036	T842
3033	1N329	3033	46BD58	3034	324-0116	3035	28B128	3036	T843
3033	1N329A	3033	46BD39	3034	608-112	3035	28B128B	3036	T844
3033	1N364	3033	46BD52	3034	690V080H42	3035	28B128C	3036	TA7201
3033	1N364A	3033	46BR18	3034	690V080H43	3035	28B128D	3036	TR-8018
3033	1N365	3033	46BR21	3034	205782-134	3035	28B128E	3036	WEP247
3033	1N365A	3033	46BR62	3034	23114-097	3035	28B128P	3036	001-021280
3033	1N5929	3033	46BR63	3034	36537	3035	28B128G	3036	283667
3033	1N4011	3033	46BR64	3034	116089	3035	28B128H	3036	2N3771
3033	1N4251	3033	46BR68	3034	119723	3035	28B128V	3036	2N3772
3033	1N445	3033	48-137290	3034	121244	3035	28B130	3036	2N3863
3033	1N445B	3033	50D10	3034	171003	3035	28B231	3036	2N3864
3033	1N509	3033	50R10	3034	231140-04	3035	28B232	3036	2N4130
3033	1N510	3033	60DE10	3034	231140-09	3035	28B234	3036	2N5034
3033	1N518	3033	63-11762	3034	231140-26	3035	28B234N	3036	2N6257
3033	1N525	3033	070-004	3034	240403	3035	28B234V	3036	28CM93D
3033	1N526	3033	070-005	3034	275845	3035	28B251	3036	28D118
3033	1N547	3033	070-006	3034	489751-115	3035	28B251A	3036	28D118A
3033	1N560	3033	070-007	3034	0573199	3035	28B252A	3036	28D118BL
3033	1N561	3033	070-008	3034	0573199H	3035	28B253	3036	28D118C
3033	1N563	3033	070-009	3034	608112	3035	28B274	3036	28D118D
3033	1N598	3033	070-010	3034	802425-0	3035	28B274A	3036	28D118R
3033	1N606	3033	070-013	3034	980150	3035	28B274C	3036	28D118Y
3033	1N855	3033	070-014	3034	1471125-3	3035	28B274D	3036	28D119
3033	1N856	3033	070-015	3034	2000646-109	3035	28B274E	3036	28D119A
3033	1N864	3033	070-016	3034	2000646-110	3035	28B274F	3036	28D119B
3033	1N865	3033	070-017	3034	2320201	3035	28B274G	3036	28D119BL
3033	1N866	3033	070-028	3034	8024250	3035	28B274H	3036	28D119G
3033	1N867	3033	070-030	3034	23114004	3035	28B274J	3036	28D119J
3033	1N876	3033	070-032	3034	23114009	3035	28B274K	3036	28D119R
3033	1N877	3033	070-033	3034	23114006	3035	28B274L	3036	28D119Y
3033	1N878	3033	75D10	3034	23114097	3035	28B274M	3036	28D151
3033	1N886	3033	75R10	3034	62051450	3035	28B274OR	3036	28D151A
3033	1N887	3033	75R10B	3035	AL100	3035	28B274R	3036	28D151C
3033	1N888	3033	75R9B	3035	AL102	3035	28B274X	3036	28D151E
3033	1N889	3033	93A71-1	3035	AL103	3035	28B274Y	3036	28D151F
3033	1NE11	3033	93B65-1	3035	AL210	3035	28B275	3036	28D151G
3033	181066	3033	930118-1	3035	AI200	3035	28B276	3036	28D1510N
3033	18108	3033	930118-2	3035	AU103	3035	28B296	3036	28D151H
3033	18109	3033	96-5178-01	3035	AU104	3035	28B300	3036	28D151J
3033	181223	3033	100D10	3035	AU105	3035	28B301	3036	28D151K
3033	181225	3033	100R10B	3035	AU106	3035	28B309	3036	28D151L
3033	181233	3033	100R9B	3035	AU107	3035	28B310	3036	28D151M
3033	181234	3033	120-004061	3035	AU108	3035	28B311	3036	28D1510R
3033	181255A	3033	137-684	3035	AU110	3035	28B312	3036	28D151R
3033	181347	3033	137-828	3035	AU111	3035	28B313	3036	28D151Y
3033	181348	3033	264P03605	3035	AU112	3035	28B319	3036	3TE120
3033	18138	3033	264P04303	3035	AU113	3035	28B320	3036	4-3023190
3033	181695	3033	0307	3035	AUY28	3035	28B342	3036	13-28532-1
3033	18397	3033	0317	3035	AUY38	3035	28B343	3036	24MW978
3033	18398	3033	320P	3035	B102000	3035	28B357	3036	43-023190
3033	18402	3033	320Z	3035	B102001	3035	28B358	3036	43A167885-P2
3033	18686	3033	0327	3035	B102002	3035	28B359	3036	44A391593-001
3033	18687	3033	359V	3035	B102003	3035	28B360	3036	48F232796
3033	18850	3033	359Z	3035	B103000	3035	28B361	3036	938E165
3033	1898	3033	384K	3035	B103001	3035	28B375	3036	96-5315-01
3033	1899	3033	384M	3035	B103002	3035	28B375-2B	3036	1160P475
3033	18058	3033	384P	3035	B103004	3035	28B375-5B	3036	119-0075
3033	18058	3033	3848	3035	B126P	3035	28B375A-2B	3036	152N2C
3033	18509	3033	384V	3035	B128	3035	28B375A-5B	3036	161N4C
3033	1T510	3033	547J2P	3035	B128V	3035	28B375A-NB	3036	172-024-9-005
3033	2HR3M	3033	1090	3035	B231	3035	28B375ZV	3036	417-139
3033	2K02	3033	1104	3035	B232	3035	28B390	3036	430-23190
3033	28J8A	3033	2095A71-1	3035	B234	3035	28B432	3036	1414-188
3033	28509	3033	2330-191	3035	B234N	3035	28B447	3036	1431-7184
3033	2T510	3033	7510	3035	B251	3035	28B464	3036	1561-0808
3033	2W9A	3033	7702-1A	3035	B251A	3035	28B465	3036	1561-0810
3033	3A1510	3033	7704-1	3035	B252	3035	28B468	3036	1561-1010
3033	3A158	3033	7706-1	3035	B252A	3035	28B468A	3036	1723-0405
3033	3A2510	3033	7708-1	3035	B253	3035	28B468B	3036	1723-0410
3033	3G1510	3033	7710-1	3035	B253A	3035	28B468C	3036	1723-0605
3033	3G158	3033	7711-1	3035	B275	3035	28B468D	3036	1723-0610
3033	3G2510	3033	7712-1	3035	B276	3035	28V341V	3036	1723-0805
3033	30258	3033	7713-1	3035	B296	3035	8-729-447-53	3036	1723-0810
3033	3T510	3033	59844-2	3035	B300	3035	8-905-605-775	3036	1723-1005
3033	3T510	3033	239429	3035	B301	3035	8-905-605-908	3036	1723-1010
3033	4-2020-08000	3033	242141	3035	B309	3035	8-905-613-232	3036	1723-1205
3033	4T509	3033	275891	3035	B310	3035	8-905-613-295	3036	1723-1210
3033	4T510	3033	530106-1001	3035	B311	3035	09-301073	3036	1723-1405
3033	5A10	3033	530113-2	3035	B312	3035	09-301079	3036	1723-1410
3033	5A10D-C	3033	984795	3035	B313			3036	1723-1605
3033	5AD10	3033	2330191					3036	1723-1610
3033	5AD10C	3033A	P14H						

SK	Industry Standard No.	SK	Industry Standard No.	SK	Industry Standard No	SK	Industry Standard No.	SK	Industry Standard No.
3036	1723-1805	3039	2SC103K	3039	2SC186F	3039	2SC272GN	3039	2SC316K
3036	1723-1810	3039	2SC103L	3039	2SC186G	3039	2SC272H	3039	2SC316L
3036	1763-0415	3039	2SC103M	3039	2SC186GN	3039	2SC272J	3039	2SC316M
3036	1763-0420	3039	2SC103X	3039	2SC186H	3039	2SC272K	3039	2SC316OR
3036	1763-0425	3039	2SC103Y	3039	2SC186J	3039	2SC272L	3039	2SC316R
3036	1763-0615	3039	2SC117A	3039	2SC186K	3039	2SC272M	3039	2SC316X
3036	1763-0620	3039	2SC117B	3039	2SC186L	3039	2SC272OR	3039	2SC316Y
3036	1763-0625	3039	2SC117C	3039	2SC186M	3039	2SC272R	3039	2SC356A
3036	1763-0815	3039	2SC117D	3039	2SC186OR	3039	2SC272X	3039	2SC356B
3036	1763-0825	3039	2SC117E	3039	2SC186R	3039	2SC272Y	3039	2SC356D
3036	1763-1015	3039	2SC117F	3039	2SC186X	3039	2SC282A	3039	2SC356E
3036	1763-1020	3039	2SC117G	3039	2SC186Y	3039	2SC282B	3039	2SC356F
3036	1763-1025	3039	2SC117GN	3039	2SC187A	3039	2SC282C	3039	2SC356G
3036	1763-1215	3039	2SC117J	3039	2SC187B	3039	2SC282D	3039	2SC356GN
3036	1763-1220	3039	2SC117K	3039	2SC187C	3039	2SC282E	3039	2SC356H
3036	1763-1225	3039	2SC117L	3039	2SC187D	3039	2SC282F	3039	2SC356J
3036	1763-1415	3039	2SC117M	3039	2SC187E	3039	2SC282G	3039	2SC356K
3036	1763-1420	3039	2SC117OR	3039	2SC187F	3039	2SC282GN	3039	2SC356L
3036	1763-1425	3039	2SC117R	3039	2SC187G	3039	2SC282J	3039	2SC356OR
3036	1854-0245	3039	2SC117X	3039	2SC187H	3039	2SC282K	3039	2SC356R
3036	2057A100-26	3039	2SC117Y	3039	2SC187J	3039	2SC282L	3039	2SC356X
3036	2071	3039	2SC127A	3039	2SC187K	3039	2SC282M	3039	2SC356Y
3036	2842-875	3039	2SC127B	3039	2SC187L	3039	2SC282OR	3039	2SC360A
3036	3002	3039	2SC127C	3039	2SC187M	3039	2SC282R	3039	2SC360C
3036	5001-508	3039	2SC127D	3039	2SC187OR	3039	2SC282X	3039	2SC360E
3036	10003	3039	2SC127E	3039	2SC187R	3039	2SC282Y	3039	2SC360F
3036	16267	3039	2SC127F	3039	2SC187X	3039	2SC286A	3039	2SC360G
3036	037085	3039	2SC127G	3039	2SC199A	3039	2SC286B	3039	2SC360GN
3036	37967	3039	2SC127GN	3039	2SC199C	3039	2SC286C	3039	2SC360H
3036	38268	3039	2SC127H	3039	2SC199D	3039	2SC286D	3039	2SC360J
3036	38473	3039	2SC127J	3039	2SC199E	3039	2SC286E	3039	2SC360K
3036	38491	3039	2SC127K	3039	2SC199F	3039	2SC286F	3039	2SC360L
3036	38965	3039	2SC127L	3039	2SC199G	3039	2SC286G	3039	2SC360M
3036	39455	3039	2SC127M	3039	2SC199GN	3039	2SC286GN	3039	2SC360OR
3036	39466	3039	2SC127OR	3039	2SC199H	3039	2SC286H	3039	2SC360R
3036	39616	3039	2SC127R	3039	2SC199J	3039	2SC286J	3039	2SC360X
3036	39635	3039	2SC127X	3039	2SC199K	3039	2SC286K	3039	2SC360Y
3036	39751	3039	2SC127Y	3039	2SC199L	3039	2SC286L	3039	2SC363A
3036	40411	3039	2SC155A	3039	2SC199M	3039	2SC286M	3039	2SC363B
3036	60350	3039	2SC155B	3039	2SC199OR	3039	2SC286OR	3039	2SC363C
3036	60885	3039	2SC155C	3039	2SC199R	3039	2SC286R	3039	2SC363D
3036	61456	3039	2SC155D	3039	2SC199X	3039	2SC286X	3039	2SC363E
3036	61868	3039	2SC155E	3039	2SC199Y	3039	2SC287B	3039	2SC363F
3036	62287	3039	2SC155F	3039	2SC263A	3039	2SC287C	3039	2SC363G
3036	98484-001	3039	2SC155GN	3039	2SC263B	3039	2SC287D	3039	2SC363GN
3036	133550	3039	2SC155H	3039	2SC263C	3039	2SC287E	3039	2SC363H
3036	166918	3039	2SC155J	3039	2SC263D	3039	2SC287F	3039	2SC363J
3036	167542	3039	2SC155K	3039	2SC263E	3039	2SC287G	3039	2SC363K
3036	171053-1	3039	2SC155L	3039	2SC263F	3039	2SC287GN	3039	2SC363L
3036	236854	3039	2SC155M	3039	2SC263G	3039	2SC287H	3039	2SC363OR
3036	239713	3039	2SC155OR	3039	2SC263GN	3039	2SC287J	3039	2SC363R
3036	267878	3039	2SC155R	3039	2SC263H	3039	2SC287K	3039	2SC363X
3036	291509	3039	2SC155X	3039	2SC263J	3039	2SC287L	3039	2SC363Y
3036	610161-4	3039	2SC155Y	3039	2SC263K	3039	2SC287M	3039	2SC37A
3036	1851517	3039	2SC156A	3039	2SC263L	3039	2SC287OR	3039	2SC37B
3036	4832800	3039	2SC156B	3039	2SC263M	3039	2SC287R	3039	2SC37C
3036	9000630	3039	2SC156C	3039	2SC263OR	3039	2SC287X	3039	2SC37D
3036	11194628	3039	2SC156D	3039	2SC263R	3039	2SC287Y	3039	2SC37E
3036	43023190	3039	2SC156E	3039	2SC263X	3039	2SC288B	3039	2SC37F
3037	WEP247MP	3039	2SC156F	3039	2SC266A	3039	2SC288C	3039	2SC37GN
3038	A4B	3039	2SC156G	3039	2SC266B	3039	2SC288D	3039	2SC37GN
3038	EA15X373	3039	2SC156GN	3039	2SC266C	3039	2SC288E	3039	2SC37H
3038	HT313592B	3039	2SC156H	3039	2SC266D	3039	2SC288F	3039	2SC37J
3038	KTR0815C	3039	2SC156J	3039	2SC266E	3039	2SC288G	3039	2SC37K
3038	QT-C1359XAN	3039	2SC156K	3039	2SC266F	3039	2SC288GN	3039	2SC37L
3038	QT-C1687XAN	3039	2SC156L	3039	2SC266G	3039	2SC288H	3039	2SC37M
3038	RQ-4448	3039	2SC156M	3039	2SC266GN	3039	2SC288J	3039	2SC37OR
3038	8-B1250108	3039	2SC156OR	3039	2SC266H	3039	2SC288K	3039	2SC37R
3038	09-302020	3039	2SC156R	3039	2SC266J	3039	2SC288L	3039	2SC37X
3038	87-0235	3039	2SC156X	3039	2SC266K	3039	2SC288M	3039	2SC37Y
3038	87-0235-C	3039	2SC156Y	3039	2SC266L	3039	2SC288OR	3039	2SC386A
3038	87-0235B	3039	2SC171A	3039	2SC266M	3039	2SC288R	3039	2SC386B
3038	88-1250108	3039	2SC171B	3039	2SC266OR	3039	2SC288X	3039	2SC386C
3038	881-250108	3039	2SC171C	3039	2SC266R	3039	2SC288Y	3039	2SC386D
3038	2057A2-219	3039	2SC171D	3039	2SC266X	3039	2SC289A	3039	2SC386E
3038	3519-2	3039	2SC171E	3039	2SC266Y	3039	2SC289B	3039	2SC386F
3038	5001-541	3039	2SC171F	3039	2SC268B	3039	2SC289C	3039	2SC386G
3038	5001-544	3039	2SC171G	3039	2SC268C	3039	2SC289D	3039	2SC386GN
3038	8868-7	3039	2SC171GN	3039	2SC268D	3039	2SC289E	3039	2SC386H
3038	27125-460	3039	2SC171H	3039	2SC268E	3039	2SC289F	3039	2SC386J
3038	27125-470	3039	2SC171J	3039	2SC268F	3039	2SC289G	3039	2SC386K
3038	133218	3039	2SC171K	3039	2SC268G	3039	2SC289GN	3039	2SC386L
3038	171162-004	3039	2SC171L	3039	2SC268GN	3039	2SC289H	3039	2SC386M
3038	1223782	3039	2SC171M	3039	2SC268H	3039	2SC289J	3039	2SC386OR
3038	5320074	3039	2SC171OR	3039	2SC268J	3039	2SC289K	3039	2SC386R
3039	A430	3039	2SC171R	3039	2SC268K	3039	2SC289L	3039	2SC386X
3039	C1360	3039	2SC171X	3039	2SC268L	3039	2SC289M	3039	2SC386Y
3039	C287A	3039	2SC171Y	3039	2SC268M	3039	2SC289OR	3039	2SC390
3039	C288	3039	2SC172B	3039	2SC268OR	3039	2SC289R	3039	2SC390A
3039	C288A	3039	2SC172C	3039	2SC268R	3039	2SC289X	3039	2SC391
3039	C36578	3039	2SC172D	3039	2SC268X	3039	2SC289Y	3039	2SC391A
3039	C645A	3039	2SC172E	3039	2SC268Y	3039	2SC29A	3039	2SC391B
3039	C645B	3039	2SC172F	3039	2SC269B	3039	2SC29B	3039	2SC391C
3039	C645C	3039	2SC172G	3039	2SC269C	3039	2SC29C	3039	2SC391E
3039	C656	3039	2SC172GN	3039	2SC269D	3039	2SC29D	3039	2SC391F
3039	C707	3039	2SC172H	3039	2SC269E	3039	2SC29E	3039	2SC391G
3039	C707H	3039	2SC172J	3039	2SC269F	3039	2SC29F	3039	2SC391GN
3039	C738C	3039	2SC172K	3039	2SC269G	3039	2SC29G	3039	2SC391H
3039	C800	3039	2SC172L	3039	2SC269GN	3039	2SC29GN	3039	2SC391J
3039	C920Q	3039	2SC172OR	3039	2SC269H	3039	2SC29H	3039	2SC391K
3039	C920R	3039	2SC172R	3039	2SC269J	3039	2SC29J	3039	2SC391L
3039	C921	3039	2SC172Y	3039	2SC269K	3039	2SC29K	3039	2SC391M
3039	C921C1	3039	2SC174B	3039	2SC269L	3039	2SC29L	3039	2SC391OR
3039	C921L	3039	2SC174C	3039	2SC269M	3039	2SC29M	3039	2SC391R
3039	C921M	3039	2SC174D	3039	2SC269OR	3039	2SC290A	3039	2SC391X
3039	GE-214	3039	2SC174E	3039	2SC269R	3039	2SC29R	3039	2SC391Y
3039	HT303801A0	3039	2SC174F	3039	2SC269X	3039	2SC29X	3039	2SC392A
3039	HT303801C0	3039	2SC174G	3039	2SC269Y	3039	2SC29Y	3039	2SC392B
3039	IRTR80	3039	2SC174GN	3039	2SC271A	3039	2SC313A	3039	2SC392C
3039	SP838	3039	2SC174H	3039	2SC271B	3039	2SC313B	3039	2SC392D
3039	TA7303	3039	2SC174J	3039	2SC271C	3039	2SC313D	3039	2SC392E
3039	TRO1073	3039	2SC174K	3039	2SC271D	3039	2SC313E	3039	2SC392F
3039	TRO1074	3039	2SC174L	3039	2SC271E	3039	2SC313G	3039	2SC392GN
3039	2N2857	3039	2SC174M	3039	2SC271H	3039	2SC313GN	3039	2SC392H
3039	2N3478	3039	2SC174R	3039	2SC271P	3039	2SC313K	3039	2SC392K
3039	2N5024	3039	2SC174X	3039	2SC271Q	3039	2SC313J	3039	2SC392L
3039	2N5179	3039	2SC174Y	3039	2SC271GN	3039	2SC313K	3039	2SC392M
3039	2N5652	3039	2SC185B	3039	2SC271H	3039	2SC313L	3039	2SC392OR
3039	2N6304	3039	2SC185C	3039	2SC271J	3039	2SC313M	3039	2SC392R
3039	2N6305	3039	2SC185E	3039	2SC271K	3039	2SC313OR	3039	2SC392X
3039	2N6389	3039	2SC185F	3039	2SC271L	3039	2SC313R	3039	2SC392Y
3039	2SC1026-R	3039	2SC185G	3039	2SC271M	3039	2SC313X	3039	2SC392M
3039	2SC1026BL	3039	2SC185GN	3039	2SC271OR	3039	2SC313Y	3039	2SC392OR
3039	2SC1026GR	3039	2SC185H	3039	2SC271R	3039	2SC316A	3039	2SC392R
3039	2SC1033	3039	2SC185K	3039	2SC271X	3039	2SC316B	3039	2SC392X
3039	2SC103C	3039	2SC185L	3039	2SC271Y	3039	2SC316C	3039	2SC392Y
3039	2SC103D	3039	2SC185X	3039	2SC272A	3039	2SC316D	3039	2SC397A
3039	2SC103F	3039	2SC185Y	3039	2SC272B	3039	2SC316E	3039	2SC397B
3039	2SC103G	3039	2SC186A	3039	2SC272C	3039	2SC316F	3039	2SC397C
3039	2SC103GN	3039	2SC186B	3039	2SC272D	3039	2SC316G	3039	2SC397D
3039	2SC103H	3039	2SC186C	3039	2SC272E	3039	2SC316GN	3039	2SC397E
3039	2SC103J	3039	2SC186E	3039	2SC272F	3039	2SC316H	3039	2SC397F
						3039	2SC316J	3039	2SC397G

SK	Industry Standard No.	SK	Industry Standard No.	SK	Industry Standard No.	SK	Industry Standard No.	SK	Industry Standard No.
3039	2SC397GN	3039	2SC612GN	3039	2SC662K	3039	2SC918LF	3039/316	A1P-1A
3039	2SC397H	3039	2SC612H	3039	2SC662L	3039	2SC918XL	3039/316	A1P/4923
3039	2SC397J	3039	2SC612J	3039	2SC662M	3039	2SC947A	3039/316	A1P/4923-1
3039	2SC397K	3039	2SC612K	3039	2SC662OR	3039	2SC947B	3039/316	A1R
3039	2SC397L	3039	2SC612L	3039	2SC662R	3039	2SC947C	3039/316	A1R-1
3039	2SC397M	3039	2SC612M	3039	2SC662X	3039	2SC947D	3039/316	A1R-2
3039	2SC397OR	3039	2SC612OR	3039	2SC662Y	3039	2SC947F	3039/316	A1R/4925
3039	2SC397R	3039	2SC612R	3039	2SC705A	3039	.2SC947P	3039/316	A1R/4926
3039	2SC397X	3039	2SC612X	3039	2SC705G	3039	2SC947G	3039/316	A2057B2-115
3039	2SC397Y	3039	2SC612Y	3039	2SC705J	3039	2SC947GN	3039/316	A210
3039	2SC39B	3039	2SC613A	3039	2SC705K	3039	2SC947H	3039/316	A2746
3039	2SC39C	3039	2SC613B	3039	2SC705L	3039	2SC947J	3039/316	A2R-2
3039	2SC39D	3039	2SC613C	3039	2SC705M	3039	2SC947K	3039/316	A2P
3039	2SC39E	3039	2SC613D	3039	2SC705OR	3039	2SC947L	3039/316	A417-154
3039	2SC39F	3039	2SC613E	3039	2SC705R	3039	2SC947M	3039/316	A417-19
3039	2SC39GN	3039	2SC613F	3039	2SC705X	3039	2SC947OR	3039/316	A417-190
3039	2SC39H	3039	2SC613G	3039	2SC705Y	3039	2SC947R	3039/316	A417-205
3039	2SC39J	3039	2SC613GN	3039	2SC707	3039	2SC947X	3039/316	A429-0981-12
3039	2SC39K	3039	2SC613H	3039	2SC707A	3039	2SC947Y	3039/316	A451
3039	2SC39L	3039	2SC613J	3039	2SC707B	3039	2SC948A	3039/316	A455
3039	2SC39M	3039	2SC613K	3039	2SC707C	3039	2SC948B	3039/316	A46-86101-3
3039	2SC39OR	3039	2SC613L	3039	2SC707D	3039	2SC948D	3039/316	A46-86109-3
3039	2SC39R	3039	2SC613M	3039	2SC707F	3039	2SC948F	3039/316	A46-86110-3
3039	2SC39X	3039	2SC613OR	3039	2SC707G	3039	2SC948G	3039/316	A46-86133-3
3039	2SC39Y	3039	2SC613R	3039	2SC707GN	3039	2SC948GN	3039/316	A46-86301-3
3039	2SC405A	3039	2SC613X	3039	2SC707H	3039	2SC948H	3039/316	A46-86302-3
3039	2SC405B	3039	2SC613Y	3039	2SC707J	3039	2SC948J	3039/316	A46-181-19
3039	2SC405D	3039	2SC629-31	3039	2SC707K	3039	2SC948K	3039/316	A467
3039	2SC405E	3039	2SC629-41	3039	2SC707L	3039	2SC948L	3039/316	A48-63076A81
3039	2SC405F	3039	2SC63A	3039	2SC707M	3039	2SC948M	3039/316	A481
3039	2SC405G	3039	2SC63B	3039	2SC707OR	3039	2SC948OR	3039/316	A483
3039	2SC405GN	3039	2SC63D	3039	2SC707R	3039	2SC948R	3039/316	A486
3039	2SC405H	3039	2SC63E	3039	2SC707X	3039	2SC948X	3039/316	A489
3039	2SC405J	3039	2SC63F	3039	2SC707Y	3039	2SC948Y	3039/316	A492
3039	2SC405K	3039	2SC63G	3039	2SC748	3039	2SC957A	3039/316	A4E
3039	2SC405L	3039	2SC63GN	3039	2SC748A	3039	2SC957AL	3039/316	A4Y-1
3039	2SC405M	3039	2SC63H	3039	2SC748C	3039	2SC957B	3039/316	A640
3039	2SC405OR	3039	2SC63J	3039	2SC748D	3039	2SC957C	3039/316	A640L
3039	2SC405R	3039	2SC63K	3039	2SC748E	3039	2SC957D	3039/316	A643L
3039	2SC405X	3039	2SC63L	3039	2SC748F	3039	2SC957E	3039/316	A643B
3039	2SC405Y	3039	2SC63M	3039	2SC748P	3039	2SC957F	3039/316	A667TRED
3039	2SC40A	3039	2SC63OR	3039	2SC748G	3039	2SC957G	3039/316	A67-08-76Q
3039	2SC40B	3039	2SC63R	3039	2SC748GN	3039	2SC957GN	3039/316	A6V
3039	2SC40C	3039	2SC63X	3039	2SC748H	3039	2SC957H	3039/316	AT310
3039	2SC40D	3039	2SC641A	3039	2SC748J	3039	2SC957J	3039/316	AT311
3039	2SC40B	3039	2SC641B	3039	2SC748K	3039	2SC957K	3039/316	AT312
3039	2SC40F	3039	2SC641D	3039	2SC748L	3039	2SC957L	3039/316	AT313
3039	2SC40G	3039	2SC641F	3039	2SC748M	3039	2SC957M	3039/316	AT314
3039	2SC40GN	3039	2SC641GN	3039	2SC748OR	3039	2SC957OR	3039/316	AT315
3039	2SC40H	3039	2SC641H	3039	2SC748R	3039	2SC957R	3039/316	AT316
3039	2SC40J	3039	2SC641J	3039	2SC748X	3039	2SC957X	3039/316	AT318
3039	2SC40K	3039	2SC641L	3039	2SC748Y	3039	2SC957XL	3039/316	AT319
3039	2SC40L	3039	2SC641M	3039	2SC761A	3039	2SC957Y	3039/316	AT321
3039	2SC40OR	3039	2SC641OR	3039	2SC761B	3039	2SC98A	3039/316	AT322
3039	2SC40X	3039	2SC641R	3039	2SC761C	3039	2SC98B	3039/316	AT323
3039	2SC40Y	3039	2SC641X	3039	2SC761D	3039	2SC98C	3039/316	AT324
3039	2SC463A	3039	2SC641Y	3039	2SC761E	3039	2SC98D	3039/316	AT325
3039	2SC463B	3039	2SC649C	3039	2SC761F	3039	2SC98E	3039/316	AT326
3039	2SC463C	3039	2SC649D	3039	2SC761G	3039	2SC98F	3039/316	AT327
3039	2SC463D	3039	2SC649E	3039	2SC761GN	3039	2SC98G	3039/316	AT328
3039	2SC463F	3039	2SC649F	3039	2SC761H	3039	2SC98GN	3039/316	AT330
3039	2SC463G	3039	2SC649G	3039	2SC761J	3039	2SC98H	3039/316	AT338
3039	2SC463GN	3039	2SC649GN	3039	2SC761K	3039	2SC98J	3039/316	A2Y
3039	2SC463J	3039	2SC649H	3039	2SC761L	3039	2SC98L	3039/316	BC127
3039	2SC463K	3039	2SC649J	3039	2SC761M	3039	2SC98M	3039/316	BC128
3039	2SC463L	3039	2SC649K	3039	2SC761OR	3039	2SC98OR	3039/316	BC155
3039	2SC463M	3039	2SC649L	3039	2SC761R	3039	2SC98R	3039/316	BC156
3039	2SC463OR	3039	2SC649M	3039	2SC761X	3039	2SC98X	3039/316	BC188
3039	2SC463R	3039	2SC649OR	3039	2SC762B	3039	2SC98Y	3039/316	BC189
3039	2SC463X	3039	2SC649R	3039	2SC762C	3039	2SC99A	3039/316	BC295
3039	2SC463Y	3039	2SC649X	3039	2SC762D	3039	2SC99B	3039/316	BC442
3039	2SC477A	3039	2SC649Y	3039	2SC762E	3039	2SC99C	3039/316	BC510C
3039	2SC477B	3039	2SC656	3039	2SC762F	3039	2SC99D	3039/316	BCM1002-2
3039	2SC477C	3039	2SC656A	3039	2SC762G	3039	2SC99B	3039/316	BP121
3039	2SC477D	3039	2SC656B	3039	2SC762GN	3039	2SC99F	3039/316	BP123
3039	2SC477E	3039	2SC656C	3039	2SC762H	3039	2SC99G	3039/316	BP125
3039	2SC477F	3039	2SC656D	3039	2SC762J	3039	2SC99GN	3039/316	BP127
3039	2SC477G	3039	2SC656E	3039	2SC762K	3039	2SC99J	3039/316	BP195C
3039	2SC477GN	3039	2SC656F	3039	2SC762L	3039	2SC99K	3039/316	BP196
3039	2SC477H	3039	2SC656GN	3039	2SC762M	3039	2SC99L	3039/316	BP197
3039	2SC477J	3039	2SC656H	3039	2SC762X	3039	2SC99M	3039/316	BP198
3039	2SC477K	3039	2SC656J	3039	2SC787A	3039	2SC99OR	3039/316	BP199
3039	2SC477L	3039	2SC656K	3039	2SC787B	3039	2SC99R	3039/316	BP216
3039	2SC477M	3039	2SC656L	3039	2SC787C	3039	2SC99X	3039/316	BP217
3039	2SC477OR	3039	2SC656M	3039	2SC787D	3039	2SC99Y	3039/316	BP218
3039	2SC477R	3039	2SC656OR	3039	2SC787E	3039	2SD562A	3039/316	BP219
3039	2SC477X	3039	2SC656R	3039	2SC787F	3039	2SD562B	3039/316	BP220
3039	2SC477Y	3039	2SC656X	3039	2SC787GN	3039	2SD562C	3039/316	BP229
3039	2SC567	3039	2SC658G	3039	2SC787H	3039	2SD562D	3039/316	BP230
3039	2SC567B	3039	2SC658GN	3039	2SC787K	3039	2SD562E	3039/316	BP233-3
3039	2SC567C	3039	2SC658H	3039	2SC787L	3039	2SD562K	3039/316	BP240
3039	2SC567D	3039	2SC658J	3039	2SC787M	3039	2SD562L	3039/316	BP241
3039	2SC567E	3039	2SC658K	3039	2SC787OR	3039	2SD562OR	3039/316	BP310
3039	2SC567F	3039	2SC658L	3039	2SC787R	3039	2SD562R	3039/316	BP311
3039	2SC567G	3039	2SC658M	3039	2SC787X	3039	2SD562X	3039/316	BP314
3039	2SC567GN	3039	2SC658OR	3039	2SC787Y	3039	2SD562Y	3039/316	BP332
3039	2SC567H	3039	2SC658R	3039	2SC79A	3039	7-28	3039/316	BP333
3039	2SC567J	3039	2SC658X	3039	2SC79B	3039	09-302061	3039/316	BP334
3039	2SC567K	3039	2SC658Y	3039	2SC79C	3039	045-1	3039/316	BP335
3039	2SC567L	3039	2SC659A	3039	2SC79E	3039	045-2	3039/316	BPS13E
3039	2SC567M	3039	2SC659B	3039	2SC79F	3039	48-40247901	3039/316	BPS13F
3039	2SC567OR	3039	2SC659C	3039	2SC79GN	3039	121-585	3039/316	BPS13G
3039	2SC567R	3039	2SC659D	3039	2SC79H	3039	121-585B	3039/316	BPS14E
3039	2SC567X	3039	2SC659E	3039	2SC79J	3039	121-687	3039/316	BPS14F
3039	2SC567Y	3039	2SC659F	3039	2SC79K	3039	176-044-9-002	3039/316	BPS14G
3039	2SC605TW	3039	2SC659G	3039	2SC79L	3039	2427	3039/316	BPS15E
3039	2SC611A	3039	2SC659GN	3039	2SC79M	3039	3539	3039/316	BPS15F
3039	2SC611B	3039	2SC659H	3039	2SC790R	3039	40295	3039/316	BPS15G
3039	2SC611D	3039	2SC659J	3039	2SC79R	3039	40894	3039/316	BPS16E
3039	2SC611E	3039	2SC659K	3039	2SC79X	3039	40895	3039/316	BPS16G
3039	2SC611F	3039	2SC659L	3039	2SC79Y	3039	40896	3039/316	BPS18
3039	2SC611G	3039	2SC659M	3039	2SC800	3039	40897	3039/316	BPS18R
3039	2SC611GN	3039	2SC659OR	3039	2SC80A	3039	129572	3039/316	BPS19
3039	2SC611H	3039	2SC659R	3039	2SC80B	3039/316	A060-100	3039/316	BPS19R
3039	2SC611J	3039	2SC659X	3039	2SC80C	3039/316	A061-105	3039/316	BPS20
3039	2SC611K	3039	2SC659Y	3039	2SC80D	3039/316	A061-106	3039/316	BPS20R
3039	2SC611L	3039	2SC662B	3039	2SC80E	3039/316	A061-108	3039/316	BPX32
3039	2SC611M	3039	2SC662C	3039	2SC80F	3039/316	A061-109	3039/316	BFY69
3039	2SC611OR	3039	2SC662D	3039	2SC80G	3039/316	A061-112	3039/316	BFY69A
3039	2SC611R	3039	2SC662E	3039	2SC80GN	3039/316	A069-101	3039/316	BFY69B
3039	2SC611X	3039	2SC662F	3039	2SC80H	3039/316	A069-102	3039/316	B8V53P
3039	2SC611Y	3039	2SC662G	3039	2SC80J	3039/316	A069-103	3039/316	B8V54P
3039	2SC612A	3039	2SC662GN	3039	2SC80K	3039/316	A069-114	3039/316	C1023-0
3039	2SC612B	3039	2SC662H	3039	2SC80L	3039/316	A069-116	3039/316	C1023-Y
3039	2SC612C	3039	2SC662-I	3039	2SC80OR	3039/316	A069-119	3039/316	C1026
3039	2SC612D			3039	2SC80R	3039/316	A121-585	3039/316	C1026Q
3039	2SC612E			3039	2SC80X	3039/316	A121-585B	3039/316	C1026Y
3039	2SC612F			3039	2SC80Y	3039/316	A121-687	3039/316	C1032
3039	2SC612G			3039	2SC918AL	3039/316	A130	3039/316	C1032Y
						3039/316	A1300RN	3039/316	C1047
						3039/316	A1G-1	3039/316	C171
						3039/316	A1P	3039/316	C174
						3039/316	A1P-1	3039/316	C174A
								3039/316	C182Q

SK	Industry Standard No.
3039/316	C185
3039/316	C185A
3039/316	C185M
3039/316	C185Q
3039/316	C185R
3039/316	C186
3039/316	C187
3039/316	C266
3039/316	C282
3039/316	C361
3039/316	C362
3039/316	C463
3039/316	C469
3039/316	C469A
3039/316	C469F
3039/316	C469Q
3039/316	C469R
3039/316	C545
3039/316	C545A
3039/316	C545B
3039/316	C545C
3039/316	C545D
3039/316	C545E
3039/316	C605
3039/316	C605TW
3039/316	C606
3039/316	C629
3039/316	C649
3039/316	C652
3039/316	C657
3039/316	C695
3039/316	C705
3039/316	C705B
3039/316	C705C
3039/316	C705D
3039/316	C705E
3039/316	C705F
3039/316	C761
3039/316	C761Y
3039/316	C761Z
3039/316	C762
3039/316	C922A
3039/316	C922B
3039/316	C922C
3039/316	CS9016E
3039/316	D10B1051
3039/316	D10B1055
3039/316	D10G1051
3039/316	D10G1052
3039/316	D16K1
3039/316	D16K2
3039/316	D16K3
3039/316	D16K4
3039/316	D1Y
3039/316	D26B1
3039/316	D26B2
3039/316	D26C1
3039/316	D26C2
3039/316	D26C3
3039/316	D26E2
3039/316	D26G1
3039/316	D308
3039/316	DDBY261002
3039/316	E1A
3039/316	ED1502
3039/316	ES15X87
3039/316	ES15X88
3039/316	F4706
3039/316	FS1308
3039/316	GM770
3039/316	HC00930
3039/316	HC537
3039/316	HC545
3039/316	HEP-S0010
3039/316	HS-40045
3039/316	HT3037310
3039/316	HT3038013-B
3039/316	HV0000302
3039/316	K2001
3039/316	K2501
3039/316	K2502
3039/316	K2503
3039/316	K2509
3039/316	K4002
3039/316	M91
3039/316	M9481
3039/316	M9482
3039/316	MF1161
3039/316	MF1162
3039/316	MF1163
3039/316	MF1164
3039/316	MM709
3039/316	MPS2369
3039/316	MPS2823
3039/316	MPS2894
3039/316	MPS6528
3039/316	MPS6529
3039/316	MPS6540
3039/316	MPS6569
3039/316	MPS6570
3039/316	MPS834
3039/316	MPS918
3039/316	NPC1075
3039/316	NS1510
3039/316	NS3039
3039/316	NS3040
3039/316	NS3041
3039/316	NS9710
3039/316	PL1024
3039/316	PL1025
3039/316	PL1026
3039/316	PL1051
3039/316	PL1113
3039/316	PM194
3039/316	PT4816
3039/316	PT4830
3039/316	R118
3039/316	R8529
3039/316	RT5204
3039/316	RT6787
3039/316	RT7703
3039/316	RT8333
3039/316	S1122
3039/316	S1126
3039/316	S1308
3039/316	S1897
3039/316	S2131
3039/316	S2132
3039/316	S2133
3039/316	S2134
3039/316	S2135
3039/316	S21648
3039/316	S3020
3039/316	S5327B
3039/316	SATCHZ339
3039/316	SE-5023
3039/316	SE2002
3039/316	SE2397
3039/316	SE3003
3039/316	SE5035
3039/316	SGS87231
3039/316	SK7181
3039/316	SM5796
3039/316	SPS20
3039/316	SPS3003
3039/316	SPS3787
3039/316	SPS3929
3039/316	SPS3948
3039/316	SPS3952
3039/316	SPS3968
3039/316	SPS3971
3039/316	SPS4
3039/316	SPS4002
3039/316	SPS4005
3039/316	SPS4008
3039/316	SPS4016
3039/316	SPS4030
3039/316	SPS4043
3039/316	SPS4050
3039/316	SPS4051
3039/316	SPS4068
3039/316	SPS4080
3039/316	SPS4091
3039/316	SPS4145
3039/316	SPS43-1
3039/316	SPS4610
3039/316	SPS856
3039/316	SPS860
3039/316	SS4042
3039/316	ST10
3039/316	ST1026
3039/316	ST1050
3039/316	ST1051
3039/316	ST13
3039/316	ST1336
3039/316	ST15
3039/316	ST1694
3039/316	ST29
3039/316	ST30
3039/316	ST3030
3039/316	ST3031
3039/316	ST31
3039/316	ST32
3039/316	ST33
3039/316	ST34
3039/316	ST35
3039/316	ST40
3039/316	ST41
3039/316	ST415
3039/316	ST42
3039/316	ST43
3039/316	ST44
3039/316	ST45
3039/316	ST60
3039/316	ST61
3039/316	ST6120
3039/316	ST62
3039/316	ST70
3039/316	ST71
3039/316	ST72
3039/316	ST80
3039/316	ST82
3039/316	ST9
3039/316	ST903
3039/316	ST904
3039/316	ST904A
3039/316	ST905
3039/316	ST910
3039/316	SX3001
3039/316	T1003-521
3039/316	T1202
3039/316	T308
3039/316	T3568
3039/316	TA2503
3039/316	TA7319
3039/316	TC0914
3039/316	TC2369A
3039/316	TC2483
3039/316	TC2484
3039/316	TI3016
3039/316	TI431
3039/316	TIS108
3039/316	TIS18
3039/316	TIS24
3039/316	TIS412
3039/316	TIS84
3039/316	TIS85
3039/316	TIS86
3039/316	TIS87
3039/316	TIS97
3039/316	TIS99
3039/316	TIXS09
3039/316	TIXS10
3039/316	TIXS28
3039/316	TIXS29
3039/316	TIXS30
3039/316	TIXS31
3039/316	TMT696
3039/316	TMT697
3039/316	TMT839
3039/316	TMT840
3039/316	TMT841
3039/316	TMT842
3039/316	TMT843
3039/316	TNJ60447
3039/316	TNJ60448
3039/316	TNJ60449
3039/316	TNJ61729
3039/316	TNJ61730
3039/316	TNJ61731
3039/316	TNJ72150
3039/316	TNJ72275
3039/316	TNJ72368
3039/316	TNJ72701
3039/316	TQ5049
3039/316	TR01042
3039/316	TR24
3039/316	TR310249
3039/316	TV-55
3039/316	TV1000
3039/316	TV2403
3039/316	TV2404
3039/316	TV24148
3039/316	TV24313
3039/316	TV24573
3039/316	TV24574
3039/316	TV24589
3039/316	TV50
3039/316	TV54
3039/316	TVS-2S288A
3039/316	TVS-2SC185A
3039/316	TVS-2SC313
3039/316	TVS-2SC429A
3039/316	TVS-2SC466
3039/316	TVS-TIS18
3039/316	TVS2SC562
3039/316	V118
3039/316	V120PH
3039/316	V129
3039/316	V143
3039/316	V220
3039/316	V221
3039/316	V222
3039/316	V405
3039/316	V435
3039/316	VS-2SC446
3039/316	VS-2SC683
3039/316	W29
3039/316	XC371
3039/316	XS1
3039/316	XS14
3039/316	XS15
3039/316	XS2
3039/316	XS3
3039/316	XS4
3039/316	XS40
3039/316	XS6
3039/316	ZA100962E
3039/316	ZEN105
3039/316	ZEN109
3039/316	1-801-304-15
3039/316	2N162
3039/316	2N162A
3039/316	2N2192
3039/316	2N2192A
3039/316	2N2193
3039/316	2N2196
3039/316	2N2639
3039/316	2N2640
3039/316	2N2641
3039/316	2N2642
3039/316	2N2643
3039/316	2N2644
3039/316	2N3082
3039/316	2N3083
3039/316	2N3137
3039/316	2N337
3039/316	2N3648
3039/316	2N3662
3039/316	2N3682
3039/316	2N3825
3039/316	2N3845
3039/316	2N3845A
3039/316	2N3846
3039/316	2N3855A
3039/316	2N3856
3039/316	2N3856A
3039/316	2N3860
3039/316	2N4254
3039/316	2N5031
3039/316	2N5032
3039/316	2N5222
3039/316	2N780
3039/316	000028606
3039/316	2SC1023-O
3039/316	2SC1023-Y
3039/316	0002SC1026B
3039/316	2SC1026Y
3039/316	2SC1117
3039/316	2SC1117H
3039/316	2SC1636
3039/316	2SC185A
3039/316	2SC185J
3039/316	2SC187R
3039/316	2SC187Y
3039/316	2SC360B
3039/316	2SC641C
3039/316	2SC641K
3039/316	2SC649
3039/316	2SC649A
3039/316	2SC649B
3039/316	2SC694Z
3039/316	2SC705
3039/316	2SC705B
3039/316	2SC705C
3039/316	2SC705D
3039/316	2SC705E
3039/316	2SC705F
3039/316	2SC761
3039/316	2SC761Y
3039/316	2SC761Z
3039/316	2SC762
3039/316	2SC787
3039/316	2SC787A
3039/316	2SC947
3039/316	2SC948E
3039/316	2SC957
3039/316	2T919
3039/316	3N71
3039/316	3N72
3039/316	3N73
3039/316	7-10
3039/316	7-26
3039/316	7-9
3039/316	7A30
3039/316	7A31
3039/316	7A32
3039/316	09-302032
3039/316	09-302036
3039/316	09-305132
3039/316	10-2
3039/316	12-23163-3
3039/316	13-23163-2
3039/316	13-23824-2
3039/316	13-26009-1
3039/316	13-26576-1
3039/316	13-26576-2
3039/316	13-26577-3
3039/316	14-602-63
3039/316	14-603-05
3039/316	14-603-06
3039/316	018B
3039/316	19-020-072
3039/316	21A112-006
3039/316	23-PT275-121
3039/316	24MW594
3039/316	24MW596
3039/316	24MW597
3039/316	24MW805
3039/316	24MW958
3039/316	24T-013-003
3039/316	24T013003
3039/316	24T013005
3039/316	24T016001
3039/316	24T016005
3039/316	25T-002
3039/316	31-0206
3039/316	34-6015-38
3039/316	46-86101-3
3039/316	46-86109-3
3039/316	46-86110-3
3039/316	46-86113-3
3039/316	46-86133-3
3039/316	48-134787
3039/316	48-134789
3039/316	48-134837
3039/316	48-134845
3039/316	48-134904
3039/316	48-134922
3039/316	48-134923
3039/316	48-134924
3039/316	48-134925
3039/316	48-134926
3039/316	48-134932
3039/316	48-134945
3039/316	48-134961
3039/316	48-134962
3039/316	48-134963
3039/316	48-134964
3039/316	48-134965
3039/316	48-134966
3039/316	48-134981
3039/316	48-137071
3039/316	48-137075
3039/316	48-137076
3039/316	48-137077
3039/316	48-137197
3039/316	48-43351A02
3039/316	48-63076A81
3039/316	48-63076A82
3039/316	48-63078A52
3039/316	48-65112A65
3039/316	48-65112A67
3039/316	48-97046A05
3039/316	48-97046A06
3039/316	48-97046A07
3039/316	48-97127A02
3039/316	48-97127A03
3039/316	48-97127A06
3039/316	48-97162A01
3039/316	48-97162A02
3039/316	57A21-16
3039/316	57B139-4
3039/316	57B21-17
3039/316	61B007-2
3039/316	62-19581
3039/316	74
3039/316	86-525-2
3039/316	87-0003
3039/316	90T2
3039/316	100N3
3039/316	100N3P
3039/316	121-480
3039/316	130-185
3039/316	1300RN
3039/316	176-025-9-002
3039/316	176-060-9-001
3039/316	417-154
3039/316	417-19
3039/316	417-190
3039/316	417-205
3039/316	465-181-19
3039/316	576-0003-023
3039/316	642-254
3039/316	642-260
3039/316	690V086H94
3039/316	690V086H95
3039/316	715FB
3039/316	772B1
3039/316	772D1
3039/316	772EH
3039/316	772FE
3039/316	909-27125-160
3039/316	916-31025-5B
3039/316	921-152B
3039/316	921-202B
3039/316	921-204B
3039/316	921-257B
3039/316	921-258B
3039/316	929DX
3039/316	930C
3039/316	991-01-1316
3039/316	1039-0482
3039/316	1075
3039/316	1852-17
3039/316	2057A2-258
3039/316	2057A2-259
3039/316	2057A2-272
3039/316	2057B2-115
3039/316	2225-17
3039/316	3027
3039/316	3881
3039/316	4756
3039/316	4789
3039/316	8000-00041-040
3039/316	16190
3039/316	0023645
3039/316	0023829
3039/316	25671-020
3039/316	25671-021
3039/316	25671-023
3039/316	40294
3039/316	40413
3039/316	40414
3039/316	40469
3039/316	40470
3039/316	40517
3039/316	125138
3039/316	129145
3039/316	129146
3039/316	129509
3039/316	129510
3039/316	129511
3039/316	129512
3039/316	129513
3039/316	129573
3039/316	236039
3039/316	489751-121
3039/316	0573474H
3039/316	610042
3039/316	772738
3039/316	772739
3039/316	2003342-244
3039/316	2006681-95
3039/316	2092693-0724
3039/316	2092693-0725
3039/316	2320141H
3039/316	13038115
3039/316	23114163
3039/316	23114165
3039/316	43022860
3039/316	6621004000
3040	A1S
3040	A35
3040	C1048
3040	C1048B
3040	C1048C
3040	C1048D
3040	C1048E
3040	C1048F
3040	C154
3040	C154B
3040	C154C
3040	C154H
3040	C500
3040	C500R
3040	C500Y
3040	C526
3040	HT8000101
3040	S3002
3040	T-481
3040	TA7292
3040	TA7293
3040	TV-19
3040	TV19
3040	TV24435
3040	TVS-2SC58
3040	TVS-2SC58A
3040	2N5184
3040	2N5185
3040	2SC1048
3040	2SC1048B
3040	2SC1048C
3040	2SC1048D
3040	2SC1048E
3040	2SC1048F
3040	2SC1089B
3040	2SC1089D
3040	2SC154
3040	2SC154A
3040	2SC154B
3040	2SC154C
3040	2SC154H
3040	2SC500
3040	2SC500R
3040	2SC526
3040	2SC58
3040	2SC58A
3040	2SC58B
3040	2SC58D
3040	2SC58E
3040	2SC58F
3040	2SC58G
3040	2SC58GN
3040	2SC58H
3040	2SC58J
3040	2SC58K
3040	2SC58L
3040	2SC58M
3040	2SC58OR
3040	2SC58R
3040	2SC58X
3040	2SC58Y
3040	2SC805
3040	2SC805A
3040	2SC805B
3040	2SC805C
3040	2SC805D
3040	2SC805E
3040	2SC805F
3040	2SC805G
3040	2SC805GN
3040	2SC805H
3040	2SC805J
3040	2SC805L
3040	2SC805OR
3040	2SC805R
3040	2SC805X
3040	2SC805Y
3040	2SC856
3040	2SC856-02
3040	2SC856A
3040	2SC856B
3040	2SC856C
3040	2SC856D
3040	2SC856E
3040	2SC856F
3040	2SC856G
3040	2SC856GN
3040	2SC856J
3040	2SC856K
3040	2SC856L
3040	2SC856M
3040	2SC856OR
3040	2SC856R
3040	2SC856X
3040	2SC856Y
3040	2SC857
3040	2SC857H
3040	2SC857K
3040	4-686145-3
3040	4-686232-3
3040	09-302019
3040	09-302099
3040	34-6001-65
3040	46-86232-3
3040	57A104-1
3040	57A104-2
3040	57A104-3
3040	57A104-4
3040	57A104-5
3040	57A104-6
3040	57A104-7
3040	57A104-8
3040	57A207-8
3040	57A236-11
3040	57A261-10
3040	99S033
3040	121-445
3040	121-743
3040	121-744
3040	121-792
3040	260P10301
3040	260P22101
3040	4686-145-3
3040	4686-232-3
3040	23114-052
3040	40354
3040	40355
3040	40459

SK	Industry Standard No.
3040	116081
3040	231140-28
3040	236282
3040	984227
3040	2320051
3040	2320051H
3040	2320191
3040	2320892
3040	06120001
3040	23114028
3040	23114052
3040	23114125
3040	62051493
3040	62543914
3040	62737395
3040	62737522
3041	A54-3
3041	A5A
3041	A5A-1
3041	A5A-2
3041	A5A-3
3041	A6C
3041	A6C-1
3041	A6C-2
3041	A6C-3
3041	A6D
3041	AR28
3041	AR38
3041	ATC-TR-19
3041	D28A6
3041	DS-66L
3041	EQS-66
3041	EQS-67
3041	G06-717-B
3041	G06-717-C
3041	IP20-0083
3041	M9556
3041	M9582
3041	ME-5
3041	P1V
3041	P1V-2
3041	P1V-3
3041	P1V-4
3041	P2T
3041	P2T-1
3041	SJE255
3041	SJE256
3041	SJE284
3041	SJE288
3041	SJE289
3041	SJE320
3041	SJE401
3041	SJE402
3041	SJE724
3041	SJE743
3041	SJE781
3041	SJE785
3041	STX0026
3041	T1P29
3041	TR01045
3041	TR01056-5
3041	VS-2SC58
3041	VS-2SC58A
3041	VS-2SC58B
3041	VS-2SC58C
3041	VS2SC1983//-1
3041	1C73045
3041	2SC1983
3041	3L4-6012-4
3041	3L4-6012-5
3041	3L4-6012-55
3041	3L4-6012-58
3041	3L4-6012-6
3041	3L4-6012-8
3041	4-686234-3
3041	13-14085-121
3041	24MW1143
3041	48-137080
3041	48-137256
3041	114N2P
3041	0123
3041	153N5C
3041	155N1
3041	331-1
3041	364-6004
3041	449
3041	753-4001-931
3041	0773
3041	1009-05A
3041	3567-2
3041	4491-6
3041	4491-9
3041	4686-234-3
3041	5614-359C
3041	41342
3041	41344
3041	132495
3041	141020
3041	171162-164
3041	610155-1
3041	910952
3041	1417331-1
3041	1471132-002
3041	2006623-88
3041	3596091
3041	3596092
3041	3596449
3041	4080879-0006
3041	22114225
3042	2SF1188
3042	2SF1188A
3042	2SF1188B
3042	2SF1188C
3042	2SF1188D
3042	2SF1188E
3042	2SF1188F
3042	2SF1188G
3042	2SF1188H
3042	2SF1188K
3042	2SF1188L
3042	2SF1188M
3042	2SF1188N
3042	2SF1188R
3042	2SF1188Y
3042	53704D
3042	126899
3042	131346
3042	132325
3042	140763
3042	1415762-1
3042	1473588-8
3042/230	GE-700
3043	A04093-X
3043	A04230-A
3043	A04241-A
3043	DG1NR
3043	ES16X16
3043	ES16X28
3043	ES16X31
3043	FT-1M
3043	FU-1M
3043	HFSD-1A
3043	HFSD-1Z
3043	SD-1AHF
3043	SR1FM-2
3043	T-E1144
3043	T-E1153
3043	TVS5B-2-H5W
3043	4-2020-03700
3043	4-2020-03800
3043	4-2021-04570
3043	4-2021-04870
3043	4-2021-06970
3043	13-33189-1
3043	53A017-1
3043	264P03603
3043	264P04701
3043	601X0375-006
3043	21008-002
3043	147617-11
3043	147617-12
3043	489752-026
3043	530144
3043	530144-1001
3043	973936-12
3043	973936-7
3043	1446149-1
3043	1446149-2
3043	1471872-3
3043	4202104870
3043	4202106970
3044	A1409
3044	A1M
3044	A247
3044	A8V
3044	AT350
3044	AT351
3044	BC100
3044	C1012
3044	C1012A
3044	C1056
3044	C1103
3044	C1103A
3044	C1103B
3044	C1103C
3044	C1103L
3044	C273
3044	C470
3044	C505
3044	C505-0
3044	C505-R
3044	C506
3044	C506-0
3044	C506-R
3044	C507
3044	C507-0
3044	C507-R
3044	C507-Y
3044	C589
3044	C58A
3044	C627
3044	C65
3044	C65B
3044	C65N
3044	C65Y
3044	C65YA
3044	C65YB
3044	C65YTV
3044	C66
3044	C686
3044	C70
3044	C743A
3044	C746A
3044	C788
3044	C805
3044	C818
3044	C857
3044	C857H
3044	C868
3044	C869
3044	C995
3044	CXL154
3044	ECG154
3044	EP15X18
3044	ES15X59
3044	ET15X34
3044	GE-40
3044	HEP-S3033
3044	HEP-S3034
3044	HEP-S3035
3044	HEP-S5025
3044	M4819
3044	M4843
3044	PTC117
3044	RS-2008
3044	RT-110
3044	S1366
3044	S3033
3044	S3034
3044	S3035
3044	S40205
3044	TG2SC65Y
3044	TM154
3044	TV-70
3044	TVS2SC58A
3044	W28
3044	WEP712
3044	2SC1012
3044	2SC1033B
3044	2SC1033C
3044	2SC1033D
3044	2SC1033E
3044	2SC1033F
3044	2SC1033G
3044	2SC1033GN
3044	2SC1033H
3044	2SC1033J
3044	2SC1033K
3044	2SC1033L
3044	2SC1033OR
3044	2SC1033R
3044	2SC1033X
3044	2SC1033Y
3044	2SC1056
3044	2SC1103
3044	2SC1103A
3044	2SC1103L
3044	2SC505
3044	2SC506
3044	2SC506-R
3044	2SC507
3044	2SC589
3044	2SC65A
3044	2SC65B
3044	2SC65C
3044	2SC65D
3044	2SC65E
3044	2SC65F
3044	2SC65G
3044	2SC65GN
3044	2SC65H
3044	2SC65K
3044	2SC65L
3044	2SC65M
3044	2SC65OR
3044	2SC65R
3044	2SC65X
3044	2SC65Y
3044	2SC65YA
3044	2SC788
3044	2SC788B
3044	2SC788C
3044	2SC788D
3044	2SC788E
3044	2SC788F
3044	2SC788G
3044	2SC788GN
3044	2SC788J
3044	2SC788K
3044	2SC788L
3044	2SC788M
3044	2SC788OR
3044	2SC788R
3044	2SC788X
3044	2SC788Y
3044	2SC995
3044	48-134819
3044	48-134843
3044	48-134919
3044	48-137364
3044	48-137415
3044	48M355012
3044	48S134927
3044	57B136-1
3044	57B136-10
3044	57B136-11
3044	57B136-2
3044	57B136-3
3044	57B136-4
3044	57B136-5
3044	57B136-6
3044	57B136-7
3044	57B136-8
3044	57B136-9
3044	57B207-8
3044	57C12-2
3044	86-612-2
3044	86-629-2
3044	121-241
3044	121-473
3044	135N1
3044	144-4
3044	656-4
3044	1010-7951
3044	1076-1559
3044	2057A2-261
3044	2057B2-140
3044	2114-0
3044	3687
3044	4819
3044	4853
3044	6517
3044	014558
3044	126709
3044	126710
3044	0320051
3044	610075-1
3044	610135-1
3044	1449098-1
3044	1473552-1
3044	1473656-4
3044/154	A5T
3044/154	A675315F
3044/154	A675318A
3044/154	BC145
3044/154	BSW69
3044/154	BSY79
3044/154	C407
3044/154	C472Y
3044/154	OC58
3044/154	C727
3044/154	C728
3044/154	C88
3044/154	C88A
3044/154	ES15X107
3044/154	ES15X82
3044/154	FBN-L108
3044/154	GE-235
3044/154	M7476
3044/154	MPSA42
3044/154	SMC-583259
3044/154	TT2SC983-0-A
3044/154	TT2SC983-O-A
3044/154	TT2SC983-R-A
3044/154	TT2SC983-Y-A
3044/154	TV-70B
3044/154	TVS-2SC645C
3044/154	001-021100
3044/154	2N3440
3044/154	2SC876TV
3044/154	4-686252-3
3044/154	4-8134842
3044/154	13-29437-1
3044/154	14-609-01A
3044/154	14-609-02
3044/154	14-609-02A
3044/154	14-609-05
3044/154	46-86344-3
3044/154	46-86346-3
3044/154	48S137364
3044/154	86-5109-2
3044/154	998070-1
3044/154	998077-1
3044/154	417-294
3044/154	481-34842
3044/154	1027G
3044/154	1187
3044/154	1414-184
3044/154	2057A2-468
3044/154	4686-252-3
3044/154	22939
3044/154	40321V1
3044/154	40321V2
3044/154	40367
3044/154	40390
3044/154	116088
3044/154	139044
3044/154	237075
3044/154	346016-63
3044/154	1473553-1
3044/154	1473622-002
3044/154	1473656-1
3044/154	1473656-2
3044/154	1473687-1
3044/154	1968959
3044/154	2320946
3044/154	7311074
3044/154	23114249
3044/154	23114250
3044/154	23114251
3044/154	62734647
3045	A104-8
3045	A12-1
3045	A12-2
3045	A13-15809-1
3045	A13-28432-1
3045	A13-55062-1
3045	A14-602-19
3045	A14-602-36
3045	A14-602-37
3045	A3170757
3045	A3545
3045	A3547
3045	A3552
3045	A3553
3045	A3582
3045	A48-134648
3045	AA133
3045	AA1M
3045	AA2A
3045	AA2K
3045	AA2Z
3045	AA310
3045	AA3M
3045	BD115
3045	GE-256
3045	Q5217
3045	REN154
3045	TV25
3045	2N739A
3045	2SC1012B
3045	2SC1012C
3045	2SC1012D
3045	2SC1012F
3045	2SC1012G
3045	2SC1012GN
3045	2SC1012H
3045	2SC1012J
3045	2SC1012K
3045	2SC1012L
3045	2SC1012M
3045	2SC1012R
3045	2SC1012X
3045	2SC1012Y
3045	2SC470A
3045	2SC470B
3045	2SC470C
3045	2SC470D
3045	2SC470E
3045	2SC470F
3045	2SC470G
3045	2SC470GN
3045	2SC470H
3045	2SC470J
3045	2SC470K
3045	2SC470L
3045	2SC470M
3045	2SC470OR
3045	2SC470R
3045	2SC470X
3045	2SC470Y
3045	2SC65YTV1
3045	2SC686A
3045	2SC686B
3045	2SC686C
3045	2SC686D
3045	2SC686E
3045	2SC686F
3045	2SC686G
3045	2SC686GN
3045	2SC686H
3045	2SC686J
3045	2SC686K
3045	2SC686L
3045	2SC686M
3045	2SC686OR
3045	2SC686R
3045	2SC686X
3045	2SC686Y
3045	2SC728C
3045	2SC728D
3045	2SC728E
3045	2SC728F
3045	2SC728G
3045	2SC728GN
3045	2SC728H
3045	2SC728J
3045	2SC728K
3045	2SC728L
3045	2SC728M
3045	2SC728OR
3045	2SC728R
3045	2SC728X
3045	2SC728Y
3045	2SC743A
3045	2SC746A
3045	2SC868
3045	2SC869
3045	12-2
3045	3545
3045	3552
3045	3553
3045	3582
3045	4686-210-3
3045	25762-010
3045	25762-012
3045	40389
3045	40544
3045	239103
3045/225	A111T2
3045/225	A116081
3045/225	A121-361
3045/225	A12546
3045/225	A126705
3045/225	A127712
3045/225	A2057B104-8
3045/225	A2090
3045/225	A25114130
3045/225	A25762-010
3045/225	A25762-012
3045/225	A2620
3045/225	A2A
3045/225	A2K
3045/225	A2Z
3045/225	A310
3045/225	A3170717
3045/225	A3M
3045/225	A417-115
3045/225	A46-867-3
3045/225	A4648
3045/225	A48-134819
3045/225	A48-134843
3045/225	A48-134853
3045/225	A48-134898
3045/225	A48-134919
3045/225	A48-134927
3045/225	A48-137002
3045/225	A48-137035
3045/225	A4819
3045/225	A4838
3045/225	A4843
3045/225	A4853
3045/225	A573501
3045/225	A57C12-1
3045/225	A57C12-2
3045/225	A57D1-122
3045/225	A57M2-16
3045/225	A57M2-17
3045/225	A610075-1
3045/225	A6181-1
3045/225	A63-18426
3045/225	A7253
3045/225	A86-213-2
3045/225	A86-214-2
3045/225	A86-215-2
3045/225	A86-316-2
3045/225	A8Z
3045/225	B5D
3045/225	BF108
3045/225	BF109
3045/225	BF110
3045/225	BF111
3045/225	BF114
3045/225	BF117
3045/225	BF118
3045/225	BF119
3045/225	BF140
3045/225	BF140A
3045/225	BF140R
3045/225	BF140S
3045/225	BF155R
3045/225	BF155S
3045/225	BF156
3045/225	BF157
3045/225	BF157B
3045/225	BF174
3045/225	BF177
3045/225	BF178
3045/225	BF179
3045/225	BF179A
3045/225	BF179B
3045/225	BF179C
3045/225	BF186
3045/225	BF257
3045/225	BF258
3045/225	BF259
3045/225	BF292A
3045/225	BF292B
3045/225	BF292C
3045/225	BF294
3045/225	BF305
3045/225	BFW37
3045/225	BFW45
3045/225	BFX34
3045/225	BFY43
3045/225	BFY45
3045/225	BFY57
3045/225	BFY65
3045/225	BFY80
3045/225	BN7253
3045/225	BSW32
3045/225	BSW70
3045/225	BSX21
3045/225	CDC744
3045/225	CDQ10013
3045/225	CDQ10015
3045/225	CDQ10034
3045/225	CDQ10037
3045/225	CDQ10047
3045/225	CDQ10049
3045/225	DT1003
3045/225	DT1602
3045/225	DT1603
3045/225	DT1612
3045/225	DT1613
3045/225	FT34C
3045/225	FT34D
3045/225	IRTR78
3045/225	M4839
3045/225	M4927
3045/225	M7002
3045/225	ME1110
3045/225	ME1120
3045/225	MJ420
3045/225	MJ421
3045/225	MM3000
3045/225	MM3002
3045/225	MM3005
3045/225	MM3009
3045/225	MM3100
3045/225	MM3101
3045/225	MM7087
3045/225	MM7088
3045/225	MT1893
3045/225	MT698
3045/225	MT699
3045/225	MT870
3045/225	MT871
3045/225	MT910
3045/225	MT911
3045/225	MT912
3045/225	N1X
3045/225	N2XA
3045/225	NS6212
3045/225	Q7
3045/225	RE23
3045/225	S17862
3045/225	S2986
3045/225	SC843
3045/225	SE7002

SK	Industry Standard No.
3045/225	SE7010
3045/225	SE7016
3045/225	SE7017
3045/225	SFT186
3045/225	SFT187
3045/225	SP8400
3045/225	SP8401
3045/225	SS524
3045/225	SX60M
3045/225	T-Q5082
3045/225	TIS100
3045/225	TIS101
3045/225	TIS102
3045/225	TIS103
3045/225	TR301
3045/225	TRS100
3045/225	TRS101
3045/225	TRS120
3045/225	TRS140
3045/225	TRS160
3045/225	TRS180
3045/225	TRS200
3045/225	TRS225
3045/225	TRS250
3045/225	TRS275
3045/225	TRS301
3045/225	TRS3011
3045/225	TRS3012
3045/225	TV24164
3045/225	TVS-2SC526
3045/225	2N1052
3045/225	2N1053
3045/225	2N1054
3045/225	2N1207
3045/225	2N1493
3045/225	2N1572
3045/225	2N1573
3045/225	2N1574
3045/225	2N2509
3045/225	2N2510
3045/225	2N2618
3045/225	2N3388
3045/225	2N3389
3045/225	2N3700
3045/225	2N3923
3045/225	2N4063
3045/225	2N4068
3045/225	2N4269
3045/225	2N4270
3045/225	2N4410
3045/225	2N4924
3045/225	2N4925
3045/225	2N4926
3045/225	2N4927
3045/225	2N5058
3045/225	2N5059
3045/225	2N5174
3045/225	2N5175
3045/225	2N5176
3045/225	2N738
3045/225	2N739
3045/225	2N740
3045/225	2N740A
3045/225	2S743
3045/225	2S743A
3045/225	2S745A
3045/225	2S746
3045/225	2S746A
3045/225	2SC1012A
3045/225	2SC1012E
3045/225	2SC470
3045/225	2SC500Y
3045/225	2SC506-0
3045/225	2SC522
3045/225	2SC524
3045/225	2SC525
3045/225	2SC64
3045/225	2SC65YTV
3045/225	2SC66
3045/225	2SC686
3045/225	2SC728
3045/225	2SC728A
3045/225	2SC728B
3045/225	2SC780
3045/225	2SC780G
3045/225	2SC818
3045/225	2SC88
3045/225	2SC88A
3045/225	2SC95
3045/225	2SC996
3045/225	09-302130
3045/225	13-15809-1
3045/225	13-28432-1
3045/225	13-55062-1
3045/225	14-602-19
3045/225	14-602-36
3045/225	14-602-37
3045/225	34-6001-82
3045/225	46-86210-3
3045/225	46-867-3
3045/225	48-134648
3045/225	48-134853
3045/225	48-134898
3045/225	48-134927
3045/225	48-137002
3045/225	48-137035
3045/225	48S155006
3045/225	56A7-1
3045/225	57B104-8
3045/225	57C12-1
3045/225	57D1-122
3045/225	57M2-16
3045/225	57M2-17
3045/225	86-213-2
3045/225	86-214-2
3045/225	86-215-2
3045/225	86-316-2
3045/225	99S101-1
3045/225	104-8
3045/225	111T2
3045/225	121-361
3045/225	260P10801
3045/225	417-115
3045/225	417-250
3045/225	690V080H38
3045/225	2057B104-8
3045/225	2620
3045/225	4648
3045/225	4838
3045/225	4843
3045/225	6181-1
3045/225	12546
3045/225	40327V2
3045/225	40346V2
3045/225	40392
3045/225	40412V2
3045/225	126705
3045/225	126722
3045/225	126726
3045/225	127712
3045/225	0573501
3045/225	610148-3
3045/225	652231
3045/225	652321
3045/225	699414-164
3045/225	1473547-1
3045/225	1473567-1
3045/225	1473567-2
3045/225	1473584-1
3045/225	2321101
3045/225	2321111
3045/225	3170717
3045/225	3170757
3045/225	25114130
3045/225	62674997
3046	PT855
3046	TR31
3046	2SCF5R
3046	576-0004-006
3046	576-0004-035
3046	4006
3046	40080
3047	EA15X412
3047	G05-015C
3047	MM1809
3047	MM1809A
3047	PT856
3047	PT888
3047	Q-00869
3047	Q-00869A
3047	Q-00869B
3047	Q-00969
3047	Q-00969A
3047	Q-00969B
3047	Q-17115C
3047	SE-8010
3047	SM7991
3047	2SC116
3047	2SC116T
3047	2SC12
3047	2SC12A
3047	2SC12B
3047	2SC12C
3047	2SC12D
3047	2SC12E
3047	2SC12F
3047	2SC12G
3047	2SC12GN
3047	2SC12H
3047	2SC12J
3047	2SC12K
3047	2SC12L
3047	2SC12M
3047	2SC12OR
3047	2SC12R
3047	2SC12X
3047	2SC12Y
3047	2SC150
3047	2SC150A
3047	2SC150B
3047	2SC150C
3047	2SC150D
3047	2SC150E
3047	2SC150F
3047	2SC150G
3047	2SC150GN
3047	2SC150H
3047	2SC150J
3047	2SC150K
3047	2SC150L
3047	2SC150M
3047	2SC150OR
3047	2SC150R
3047	2SC150T
3047	2SC150Y
3047	2SC163
3047	2SC164A
3047	2SC164B
3047	2SC164C
3047	2SC164D
3047	2SC164E
3047	2SC164F
3047	2SC164G
3047	2SC164GN
3047	2SC164H
3047	2SC164J
3047	2SC164K
3047	2SC164L
3047	2SC164OR
3047	2SC164X
3047	2SC164Y
3047	2SC478-4
3047	2SC478D
3047	2SC614D
3047	2SC696G
3047	2SC774
3047	2SC775
3047	2SCF5A
3047	2SCF5B
3047	2SCF5C
3047	2SCF5D
3047	2SCF5E
3047	2SCF5F
3047	2SCF5G
3047	2SCF5GN
3047	2SCF5H
3047	2SCF5J
3047	2SCF5K
3047	2SCF5L
3047	2SCF5M
3047	2SCF5OR
3047	2SCF5X
3047	2SCF5Y
3047	2SCF6
3047	2SCF6A
3047	2SCF6B
3047	2SCF6C
3047	2SCF6D
3047	2SCF6E
3047	2SCF6F
3047	2SCF6G
3047	2SCF6GN
3047	2SCF6H
3047	2SCF6J
3047	2SCF6K
3047	2SCF6L
3047	2SCF6M
3047	2SCF6OR
3047	2SCF6R
3047	2SCF6X
3047	2SCF6Y
3047	09-302015
3047	09-302055
3047	09-302068
3047	09-302081
3047	09-302135
3047	9TR11
3047	9TR4
3047	9TR6
3047	9TR9
3047	13-0064
3047	19-3349
3047	90-112
3047	120-004884
3047	176-018-9-001
3047	176-029-9-002
3047	176-029-9-003
3047	260Z00401
3047	296-81-9
3047	386-7181P2
3047	576-0004-004
3047	576-0004-008
3047	660-128
3047	1000-139
3047	1001-01
3047	1004-02
3047	1009-03
3047	1009-03-17
3047	1010-14
3047	1010-17
3047	1040-04
3047	1041-76
3047	1482-17
3047	1879-17
3047	1879-17A
3047	1916-17
3047	4001
3047	8000-00003-041
3047	8000-00003-042
3047	8000-00005-009
3047	8000-00005-010
3047	8000-00006-001
3047	9404-2
3047	94042
3047	101435
3047	110699
3047	237028
3047	0573517
3047	0573519
3047	740857
3047	916119
3047	1223785
3047	120004884
3048	2SC106A
3048	2SC106B
3048	2SC106C
3048	2SC106G
3048	2SC106GN
3048	2SC106H
3048	2SC106J
3048	2SC106K
3048	2SC106L
3048	2SC106M
3048	2SC106OR
3048	2SC106R
3048	2SC106X
3048	2SC106Y
3048	2SC111A
3048	2SC111B
3048	2SC111C
3048	2SC111E
3048	2SC111F
3048	2SC111G
3048	2SC111GN
3048	2SC111H
3048	2SC111J
3048	2SC111K
3048	2SC111L
3048	2SC111M
3048	2SC111OR
3048	2SC111R
3048	2SC111X
3048	2SC111Y
3048	2SC147A
3048	2SC147B
3048	2SC147C
3048	2SC147D
3048	2SC147F
3048	2SC147G
3048	2SC147GN
3048	2SC147H
3048	2SC147J
3048	2SC147K
3048	2SC147L
3048	2SC147M
3048	2SC147R
3048	2SC147X
3048	2SC147Y
3048	2SC456B
3048	2SC456C
3048	2SC456E
3048	2SC456F
3048	2SC456G
3048	2SC456GN
3048	2SC456H
3048	2SC456J
3048	2SC456K
3048	2SC456L
3048	2SC456M
3048	2SC456OR
3048	2SC456R
3048	2SC456X
3048	2SC456Y
3048	13-0079
3048	13-0079A
3048	576-0004-001
3048	576-0004-005
3048	576-0004-007
3048	576-0004-009
3048	576-0004-011
3048	576-0004-012
3048	2904-059
3048	40081
3048	40081V1
3048	40082
3048	40082A2
3048	40581
3048/329	A066-116
3048/329	A67-15-280
3048/329	B3465
3048/329	BC412
3048/329	BSX32
3048/329	BSX59
3048/329	C147
3048/329	C163
3048/329	C293
3048/329	C311
3048/329	C456
3048/329	C456-0
3048/329	C456A
3048/329	C456D
3048/329	C478
3048/329	C478(D)
3048/329	C478-4
3048/329	C478D
3048/329	C502
3048/329	C571
3048/329	C580
3048/329	C614
3048/329	C614D
3048/329	C614E
3048/329	C614F
3048/329	C614G
3048/329	C615
3048/329	C615A
3048/329	C615B
3048/329	C615C
3048/329	C615D
3048/329	C615E
3048/329	C615G
3048/329	C774
3048/329	C775
3048/329	CF6
3048/329	D152
3048/329	DS-512
3048/329	EQS-57
3048/329	HEP-S3001
3048/329	HEP-S3013
3048/329	HT309714A-0
3048/329	IRTR64
3048/329	IRTR65
3048/329	MM1810
3048/329	MM8004
3048/329	PT1537
3048/329	PT2677C
3048/329	PT3141C
3048/329	Q-00869C
3048/329	Q-00969C
3048/329	Q-01069
3048/329	Q-01069A
3048/329	Q-01069B
3048/329	Q-01184R
3048/329	Q-01284R
3048/329	S3001
3048/329	S3004
3048/329	S3013
3048/329	TR-65
3048/329	2SC106
3048/329	2SC111D
3048/329	2SC147
3048/329	2SC293
3048/329	2SC307T
3048/329	2SC311
3048/329	2SC456
3048/329	2SC456-0
3048/329	2SC456A
3048/329	2SC478(D)
3048/329	2SC502
3048/329	2SC614
3048/329	2SC614E
3048/329	2SC615F
3048/329	2SC696
3048/329	2SC696(D)
3048/329	2SC696D
3048/329	2SC696E
3048/329	2SC696F
3048/329	2SC696I
3048/329	09-302030
3048/329	09-302035
3048/329	09-302082
3048/329	09-302212
3048/329	09-309062
3048/329	019-003691
3048/329	19-020-038
3048/329	19-020-076
3048/329	19-020-077
3048/329	19-3692
3048/329	022-3640-081
3048/329	022-3640-082
3048/329	90-177
3048/329	90-454
3048/329	142-007
3048/329	260-10-010
3048/329	296-58-9
3048/329	386-7182P1
3048/329	417-128
3048/329	576-0004-013
3048/329	576-0036-213
3048/329	1000-141
3048/329	1001-07
3048/329	1005-02
3048/329	1009-04-17
3048/329	1011-01
3048/329	1096-11
3048/329	1098-15
3048/329	1183-17
3048/329	2027
3048/329	2322-17
3048/329	2340-17
3048/329	2904-037
3048/329	4005(CB)
3048/329	4007
3048/329	4011
3048/329	4802-00007
3048/329	4802-00017
3048/329	7364-6053P1
3048/329	8000-00003-043
3048/329	8000-00030-008
3048/329	8000-0004-P083
3048/329	8000-0004-P084
3048/329	9404-3
3048/329	36213
3048/329	36213A2
3048/329	38789
3048/329	082022
3048/329	082033
3048/329	94043
3048/329	101436
3048/329	111279
3048/329	652092
3048/329	760284
3048/329	1223784
3048/329	1223786
3048/329	2006514-61
3048/329	2006607-61
3048/329	5320501
3048/329	26010010
3048/329	26010058
3048/329	16304198000
3049	A059-111
3049	C312
3049	RE202
3049	SE-3034
3049	SE3034
3049	SM7989
3049	2SC312
3049	019-003935
3049	120-004885
3049	120-004886
3049	172-001
3049	172-001-9-001
3049	1068-17A
3049	2904-030
3049	40446
3049	40582
3049	112362
3049	112527
3049	120004885
3049	120004886
3049/224	C517
3049/224	C517C
3049/224	D223
3049/224	EQS-56
3049/224	GE-219
3049/224	Q-01069C
3049/224	Q-18115C
3049/224	2SC517
3049/224	2SC517C
3049/224	2SCF8
3049/224	2SD223
3049/224	172-014-9-002
3049/224	185-008
3049/224	501-068
3049/224	1041-77
3049/224	4013
3049/224	8000-00011-051
3050	BFS28
3050	BFS28R
3050	MOS3635
3050	REN221
3050	SFE303
3050	TA7149
3050	TA7150
3050	TA7151
3050	ZEN124
3050	3SK32B
3050	3SK32B-6
3050	3SK32E
3050	3SK32E-4
3050	3SK35
3050	3SK35BL
3050	3SK35G
3050	13-10321-53
3050	045
3050	46-86396-3
3050	047
3050	86-540-2
3050	121-782
3050	132-045
3050	132-047
3050	576-0006-221
3050	635-1
3050	3618
3050	3635
3050	6351
3050	40600
3050	40601
3050	40602
3050	40603
3050	40604
3050	135324
3050	135963
3050	1473588-2
3050	1473635-2
3050	2327232
3050/221	AR501
3050/221	AR502
3050/221	DDCY104001
3050/221	DDCY104003
3050/221	DS-105
3050/221	EA15X402
3050/221	EA15X405
3050/221	IP20-0157
3050/221	MFE121
3050/221	MFE130-712
3050/221	MFE131
3050/221	MFP121
3050/221	SFC8999
3050/221	SPF274
3050/221	SPF512
3050/221	TA7189
3050/221	TA7262
3050/221	TA7399
3050/221	TR08004
3050/221	TR2327431
3050/221	W1U
3050/221	YEAN3SK39Q
3050/221	3L4-6503-1
3050/221	3L4-6503-2
3050/221	3SK37
3050/221	3SK40
3050/221	3SK40M
3050/221	3SK49
3050/221	3SK59
3050/221	3SK59BL
3050/221	3SK59GR
3050/221	7-40
3050/221	09-305040
3050/221	13-0118
3050/221	13-28583-1
3050/221	022-2876-008
3050/221	24MW1122
3050/221	041A
3050/221	46-8646
3050/221	48-137488
3050/221	48-137567
3050/221	48-64978A39
3050/221	57A267-4
3050/221	86-464-2
3050/221	86-605-2
3050/221	86-606-2
3050/221	86-625-2
3050/221	86-632-2
3050/221	90-178
3050/221	90-58
3050/221	99S041
3050/221	99S045A
3050/221	99S046
3050/221	121-1024
3050/221	121-783

SK	Industry Standard No.
3050/221	121-784
3050/221	121-785
3050/221	121-786
3050/221	121-787
3050/221	121-953
3050/221	182-038-9-001
3050/221	182-138-9-001
3050/221	203-1
3050/221	229-0192-18
3050/221	576-0006-222
3050/221	1251-1-1
3050/221	2000-102
3050/221	2000-103
3050/221	2022-06
3050/221	2359-17
3050/221	8000-00042-013
3050/221	40673
3050/221	42396
3050/221	95132
3050/221	127980
3050/221	129980
3050/221	134450
3050/221	171206-6
3050/221	610166-1
3050/221	610203-1
3050/221	610203-2
3050/221	610203-3
3050/221	610203-4
3050/221	610203-5
3050/221	610203-6
3050/221	741860
3050/221	986930
3050/221	1408694-1
3050/221	1473558-1
3050/221	1473618-1
3050/221	1473635-1
3050/221	1473651-1
3050/221	2327431
3050/221	3596401
3050/221	3596402
3050/221	4084114-0001
3050/221	4084114-0002
3051	A22B
3051	A22B1
3051	A22D
3051	A22D1
3051	A22M
3051	A514-035596
3051	A6A1
3051	A6A5
3051	A6A9
3051	A6B1
3051	A6B5
3051	A6B9
3051	A6C5
3051	A6C9
3051	A6D1
3051	A6D5
3051	A6D9
3051	A6E1
3051	A6E5
3051	A6E9
3051	A6F1
3051	A6F5
3051	A6F9
3051	A6G1
3051	A6G5
3051	A6G9
3051	A6H1
3051	A6H5
3051	A6H9
3051	A6K1
3051	A6K5
3051	A6K9
3051	A6M1
3051	A6M5
3051	A6M9
3051	A7G1
3051	A7G5
3051	A7G9
3051	A7H1
3051	A7H5
3051	A7H9
3051	A7K1
3051	A7K5
3051	A7K9
3051	A7M1
3051	A7M5
3051	A7M9
3051	AB1000
3051	AB600
3051	AB800
3051	AC200
3051	AE1B
3051	AE1C
3051	AE1D
3051	AE1E
3051	AE1F
3051	AE1G
3051	B172-10
3051	B172-100
3051	B172-20
3051	B172-5
3051	B172-60
3051	B172-70
3051	B172-80
3051	B172-90
3051	B6A1
3051	B6A5
3051	B6A9
3051	B6B1
3051	B6B5
3051	B6B9
3051	B6C5
3051	B6C9
3051	B6G1
3051	B6G5
3051	B6G9
3051	B6H1
3051	B6H5
3051	B6H9
3051	B6K1
3051	B6K5
3051	B6K9
3051	B6M1
3051	B6M5
3051	B6M9
3051	B7G1
3051	B7G5
3051	B7G9
3051	B7H1
3051	B7H5
3051	B7H9
3051	B7K1
3051	B7K5
3051	B7K9
3051	B7M1
3051	B7M5
3051	B7M9
3051	BH4R6
3051	BY172
3051	BY173
3051	BYX38
3051	BYY60
3051	BYY61
3051	CF152VA
3051	DIJ72292
3051	DS-1.5-2
3051	DS1M
3051	ES16X33
3051	ESK-1
3051	GE-512
3051	GM3Y
3051	HD20005100
3051	IP20-0164
3051	MA242RC
3051	MR-1C
3051	MR1030A
3051	MR1031
3051	MR1031A
3051	MR1031B
3051	MR1032A
3051	MR1032B
3051	MR2369
3051	NS2006
3051	NS2007
3051	NS2008
3051	NS3006
3051	NS3007
3051	NS3008
3051	PS2346
3051	PS2347
3051	R0091
3051	R0094
3051	R0096
3051	R0097
3051	R0098
3051	R350
3051	RA132AA
3051	RA132DA
3051	RA132MA
3051	RA132PA
3051	RA132RA
3051	RA132VA
3051	RE92
3051	RFJ71123
3051	RH-DX0091CEZZ
3051	RT210
3051	RVDSD-1Z
3051	RVDSG-5N
3051	RVDSG-5P
3051	RZ-3
3051	S3V20
3051	S6BR2
3051	S7BR2
3051	S8-06
3051	S8-10
3051	S808
3051	S8BR2
3051	S9BR2
3051	SCBR10
3051	SCBR6
3051	SD-1Y
3051	SE15B
3051	SE15C
3051	SE15D
3051	SG-5N
3051	SG-5P
3051	SI-RECT-208
3051	SI8A
3051	SIDA05K
3051	SIDA05N
3051	SIDA05P
3051	SJ1003F
3051	SR-499
3051	SR154
3051	SR2B12
3051	SR2B16
3051	SR2B20
3051	SR3AM-4
3051	SSIB0140
3051	SSIB0180
3051	SSIB0640
3051	SSIB0680
3051	SSIC1740
3051	SSIC1780
3051	SVDAS2HB10
3051	SW-1
3051	SW-1-01
3051	SZA100
3051	TM104
3051	TM105
3051	TM61
3051	TM84
3051	TM85
3051	TVSDS1M
3051	U05B
3051	U05C
3051	001-02406-0
3051	001-02406-1
3051	1-506-319
3051	1.5B05
3051	1.5C05
3051	1AS027
3051	1AS029
3051	1LF11
3051	1N5172
3051	1N5177
3051	1N5178
3051	1S1076
3051	1S1891
3051	1S1892
3051	1WS10
3051	2A100
3051	2A200
3051	2E10
3051	2E6
3051	2E8
3051	3BZ61
3051	3DZ61
3051	3GZ61
3051	5B2
3051	8-719-911-54
3051	8-719-912-54
3051	09-306053
3051	09-306379
3051	012-1025-002
3051	13-17174-4
3051	13-29165-2
3051	022-2823-001
3051	24MW1118
3051	030
3051	30R10
3051	30R8
3051	032
3051	035
3051	46-86558-3
3051	60LA
3051	065-016
3051	070-031
3051	070-035
3051	070-041
3051	96-5193-01
3051	96-5194-01
3051	106-013
3051	150R10B
3051	150R8B
3051	150R9B
3051	200R10B
3051	200R1B
3051	200R7B
3051	200R8B
3051	200R9B
3051	0248
3051	503-T10192
3051	523-0017-001
3051	903-334
3051	919-002440
3051	919-002440-1
3051	919-011339
3051	919-013061
3051	919-013079
3051	2000-348
3051	8000-00011-105
3051	12550
3051	23115-085
3051	131815
3051	132815
3051	229805
3051	235541
3051	237070
3051	237421
3051	239097
3051	240055
3051	240076
3051	243115
3051	489752-029
3051	560020S
3051	1471405-1
3051	1474778-13
3051	1476171-13
3051	1800017
3051	2330332
3051	2330771
3051	41020687
3051	53015201
3051	4202021100
3051/156	A04212A
3051/156	A04212B
3051/156	A10A
3051/156	A10B
3051/156	A10M
3051/156	A14F
3051/156	A7978850
3051/156	AM1
3051/156	AM11
3051/156	AM12
3051/156	AM2
3051/156	AM31
3051/156	AM34
3051/156	AM4
3051/156	AM41
3051/156	AM44
3051/156	AM5
3051/156	BAV205
3051/156	BY105
3051/156	BYY31
3051/156	D1201F
3051/156	D1201N
3051/156	D1201P
3051/156	DRS106
3051/156	DRS108
3051/156	EC100
3051/156	ER102
3051/156	HC71
3051/156	HC710
3051/156	HC72
3051/156	HC720
3051/156	HC73
3051/156	HC730
3051/156	KSKE125C200
3051/156	MA100
3051/156	MH71
3051/156	MH710
3051/156	MH72
3051/156	MH720
3051/156	MH730
3051/156	MP549
3051/156	P100J
3051/156	P150J
3051/156	P800
3051/156	PD100
3051/156	PD60
3051/156	PD80
3051/156	PTC204
3051/156	QDSSR3AMBE
3051/156	R0092
3051/156	S10AR2
3051/156	S16
3051/156	S16A
3051/156	S18
3051/156	S2VB10
3051/156	S3A6
3051/156	S415
3051/156	S417
3051/156	S4VB10
3051/156	S56
3051/156	S57
3051/156	S58
3051/156	S59
3051/156	S600
3051/156	S6AR2
3051/156	S800
3051/156	S8AR2
3051/156	SA2Z
3051/156	SD910S
3051/156	SD96S
3051/156	SD98S
3051/156	SI-RECT-110
3051/156	SI-RECT-112
3051/156	SR3AM
3051/156	SR3AM2
3051/156	TM62
3051/156	TM65
3051/156	TRSR3AM
3051/156	TVSSA-2H
3051/156	V03
3051/156	V03E
3051/156	001-024061
3051/156	1N2026
3051/156	1N2027
3051/156	1N2028
3051/156	1N2029
3051/156	1N2030
3051/156	1N4718
3051/156	1N4942
3051/156	1N4944
3051/156	1N4945
3051/156	1N4946
3051/156	1N4947
3051/156	1N4948
3051/156	1N5190
3051/156	1N5198
3051/156	1N5200
3051/156	1N5415
3051/156	1N5418
3051/156	1N5419
3051/156	1N5552
3051/156	1N5553
3051/156	1N5554
3051/156	1S1660
3051/156	1S1661
3051/156	1S1661R
3051/156	1S1829
3051/156	3F20D-C
3051/156	3GC12
3051/156	09-306424
3051/156	10B10
3051/156	10B8
3051/156	13-34901-1
3051/156	20S5
3051/156	30D2
3051/156	34-8054-10
3051/156	46-86148-3
3051/156	53A001-34
3051/156	86-128-3
3051/156	098GI219
3051/156	305A
3051/156	305B
3051/156	305C
3051/156	305D
3051/156	523-4001-001
3051/156	919-001449
3051/156	919-001449-1
3051/156	919-001449-3
3051/156	919-002240-1
3051/156	919-002440-2
3051/156	919-008862
3051/156	919-010459
3051/156	919-010623
3051/156	919-010829
3051/156	919-010829-1
3051/156	919-011211
3051/156	919-013036
3051/156	1042-17
3051/156	1063-8971
3051/156	2794
3051/156	5923
3051/156	44007
3051/156	131318
3051/156	133266
3051/156	530071-2
3051/156	1474778-14
3051/156	06200065
3051/156	23115913
3051/156	23115966
3051/156	37568250
3051/156	2008100093
3052	A061-116
3052	A065-111
3052	B463
3052	B463BL
3052	B463E
3052	B463R
3052	B463Y
3052	E24107
3052	EA15X139
3052	EA15X154
3052	ES10110
3052	GE-30
3052	HB367
3052	HT204736
3052	IRTR-50
3052	P3R-3
3052	P3T
3052	PTC120
3052	RE68
3052	SP2551
3052	TNJ70483
3052	TR50
3052	WEP642
3052	WEP643
3052	2N4078
3052	2N5897
3052	2SB256
3052	2SB368
3052	2SB368A
3052	2SB368B
3052	2SB368C
3052	2SB368D
3052	2SB368E
3052	2SB368F
3052	2SB368G
3052	2SB368GN
3052	2SB368H
3052	2SB368J
3052	2SB368K
3052	2SB368L
3052	2SB368M
3052	2SB368X
3052	2SB368Y
3052	2SB458
3052	2SB458B
3052	2SB458C
3052	2SB462
3052	2SB463
3052	2SB463A
3052	2SB463B
3052	2SB463BL
3052	2SB463C
3052	2SB463D
3052	2SB463E
3052	2SB463F
3052	2SB463G
3052	2SB463GN
3052	2SB463H
3052	2SB463J
3052	2SB463K
3052	2SB463L
3052	2SB463M
3052	2SB463R
3052	2SB463XL
3052	2SB463Y
3052	4-142
3052	4-48
3052	8A10521
3052	8A11083
3052	8A11721
3052	8A13164
3052	8P-505
3052	8P505
3052	09-301075
3052	21A015-003
3052	22-002001
3052	22-002008
3052	22-002009
3052	24MW994
3052	43-025834
3052	48-137267
3052	48-137268
3052	48-137269
3052	48-137270
3052	48-137271
3052	48-137308
3052	48-97046A31
3052	48S13270
3052	48S137270
3052	48S40172G01
3052	120-003150
3052	171-016-9-001
3052	220-002001
3052	296-61-9
3052	430-25834
3052	1840-17
3052	1945-17
3052	2402-456
3052	7219-3
3052	8000-00003-040
3052	27126-060
3052	123808
3052	167285
3052	171162-083
3052	175043-065
3052	310035
3052	489751-163
3052	0573030
3052	0573030-14
3052	740471
3052	984521
3052	1474778-21
3052	2320092
3052	7855295-01
3052	8112143
3053	AR29
3053	BC178B
3053	MJ4645
3053	PTC127
3053	RCA1A16
3053	TA2819
3053	2N5415
3054	A8E
3054	A8K
3054	B23-82
3054	B5E
3054	BD107
3054	C932
3054	C932E
3054	CXL152
3054	CXL196
3054	DDBY407001
3054	DDBY407004
3054	EA15X7121
3054	ECG196
3054	EP15X11
3054	EP15X14
3054	EP15X30
3054	EP15X43
3054	ES58
3054	GE-241
3054	HR-68
3054	IRTR-55
3054	IRTR55
3054	IRTR92
3054	J241252
3054	RCA1C05
3054	RCA1C10
3054	RCA1C14
3054	RE21
3054	REN196
3054	RT150
3054	RT152
3054	S2D153
3054	STX3326
3054	T23-93
3054	TA-7155
3054	TA7155
3054	TA7362
3054	TA7782
3054	TA7783
3054	TA7784
3054	TA8231
3054	TA8232
3054	TA8233
3054	TIP-24
3054	TIP29XA
3054	TV-112
3054	VS2SD476-C/-1
3054	X735-40C
3054	ZSD235Y
3054	01-040389
3054	01-040476
3054	2N5294
3054	2N5297
3054	2N5496
3054	2N5497
3054	2N6291
3054	2N6292
3054	2N6293
3054	2SC1826
3054	2SC1826P
3054	2SC2317
3054	2SC789A
3054	2SC789B
3054	2SC789C
3054	2SC789D
3054	2SC789E
3054	2SC789F
3054	2SC789G
3054	2SC789GN
3054	2SC789H

SK	Industry Standard No.
3054	2SC789J
3054	2SC789K
3054	2SC789L
3054	2SC789M
3054	2SC7890R
3054	2SC789X
3054	2SC9320R
3054	2SD234A
3054	2SD234B
3054	2SD234C
3054	2SD234D
3054	2SD234E
3054	2SD234F
3054	2SD234G
3054	2SD234GA
3054	2SD234GN
3054	2SD234GR
3054	2SD234H
3054	2SD234J
3054	2SD234K
3054	2SD234L
3054	2SD234M
3054	2SD234N
3054	2SD2340R
3054	2SD234X
3054	2SD234Y
3054	2SD2350
3054	2SD235A
3054	2SD235B
3054	2SD235C
3054	2SD235E
3054	2SD235F
3054	2SD235GN
3054	2SD235H
3054	2SD235J
3054	2SD235L
3054	2SD235M
3054	2SD2350R
3054	2SD235X
3054	2SD314A
3054	2SD314B
3054	2SD314F
3054	2SD314GN
3054	2SD314H
3054	2SD314L
3054	2SD314M
3054	2SD314N
3054	2SD314R
3054	2SD314Y
3054	2SD317A
3054	2SD343A
3054	2SD343B
3054	2SD343C
3054	2SD343D
3054	2SD343H
3054	2SD343J
3054	2SD343K
3054	2SD365A
3054	2SD389A
3054	2SD389AF
3054	2SD389AFP
3054	2SD389AP
3054	2SD389AQ
3054	2SD476
3054	2SD476A
3054	2SD476B
3054	2SD476C
3054	2SD476D
3054	2SD570
3054	2SD876
3054	3L4-6011-1
3054	3L4-6012-56
3054	09-303021
3054	13-33742-1
3054	13-34002-1
3054	13-34838-1
3054	13-43635-1
3054	14-608-02A
3054	14-609-03A
3054	14-909-23
3054	020-1111-005
3054	46-86347-3
3054	46-86459-3
3054	48M355039
3054	48M355040
3054	48M355059
3054	48R869676
3054	48S137323
3054	57A244-14
3054	57B244-14
3054	69-1818
3054	86-568-2
3054	96-5232-01
3054	96-5232-02
3054	99S083-1
3054	99S100
3054	121-770CL
3054	121-772CL
3054	121-853
3054	121-880
3054	121-967
3054	121-976
3054	132-5
3054	153N1
3054	195N1C
3054	260P14202
3054	432-1
3054	632-3
3054	690V110H89
3054	1010-7936
3054	1061-8338
3054	1069
3054	1125
3054	2001
3054	3632-1
3054	3632-3
3054	6099-2
3054	7364-1
3054	8471
3054	16029
3054	16039
3054	16115
3054	16163
3054	40624
3054	40627
3054	40632
3054	60132
3054	60133
3054	60416
3054	131161
3054	131239
3054	134771
3054	135735
3054	140626
3054	140979
3054	239517
3054	309689
3054	610111-8
3054	610195-1
3054	1417364-1
3054	1471132-3
3054	1471132-5
3054	1473612-1
3054	1473621-1
3054	1473623-1
3054	1473632-3
3054	1473640-1
3054	4080187-0506
3054	4080187-0507
3054	4080835-0002
3054	4080866-0006
3054	4080866-0012
3054	4080879-0011
3054	23114923
3054	23114969
3054	62734596
3054/196	A066-114
3054/196	A68-23-560
3054/196	A6D-1
3054/196	A6D-2
3054/196	A6D-3
3054/196	A9M
3054/196	AG6
3054/196	B-1790
3054/196	B-1823
3054/196	B143004
3054/196	B143011
3054/196	B143012
3054/196	B143018
3054/196	B143019
3054/196	B143026
3054/196	B143027
3054/196	B1U148
3054/196	B2J
3054/196	B3547
3054/196	B3548
3054/196	B3550
3054/196	B3551
3054/196	B3577
3054/196	B3578
3054/196	B3580
3054/196	B3584
3054/196	B3585
3054/196	B3586
3054/196	B3588
3054/196	B3589
3054/196	B5002
3054/196	B5021
3054/196	B5022
3054/196	B5031
3054/196	B5032
3054/196	BC430
3054/196	BD106
3054/196	BD106A
3054/196	BD106B
3054/196	BD107A
3054/196	BD107B
3054/196	BD109
3054/196	BD124
3054/196	BD135
3054/196	BD137
3054/196	BD139
3054/196	BD162
3054/196	BD163
3054/196	BD433
3054/196	BDY12
3054/196	BDY13
3054/196	BDY34
3054/196	BR101B
3054/196	BSV60
3054/196	BUY24
3054/196	C325
3054/196	C325E
3054/196	C789
3054/196	C789-0
3054/196	C789-0
3054/196	C789-R
3054/196	C789-Y
3054/196	D18A12
3054/196	D234
3054/196	D234-0
3054/196	D234-0
3054/196	D235
3054/196	D235-0
3054/196	D235-0
3054/196	D235D
3054/196	D235R
3054/196	D235Y
3054/196	D28A12
3054/196	D40D1
3054/196	D44CBB
3054/196	EA15X327
3054/196	EA15X333
3054/196	EA15X8119
3054/196	EA15X8602
3054/196	EP15X19
3054/196	EP15X22
3054/196	EP15X68
3054/196	ES81
3054/196	ETTD-235
3054/196	G03-007C
3054/196	HC-00496
3054/196	HEP-S5003
3054/196	HP57
3054/196	HT311621B
3054/196	HT402352B
3054/196	IP20-0007
3054/196	IP20-0036
3054/196	IRTR76
3054/196	MU9611
3054/196	MU9611T
3054/196	PLE-48
3054/196	PN66
3054/196	PP3250
3054/196	PP3310
3054/196	PP3312
3054/196	PTC110
3054/196	Q-11115C
3054/196	Q-12115C
3054/196	RCA29
3054/196	RCA29A
3054/196	RS-2017
3054/196	RS-2020
3054/196	S12020-04
3054/196	S1D153
3054/196	S5472
3054/196	SD345
3054/196	SDA345
3054/196	SDB345
3054/196	SDK345
3054/196	SDM345
3054/196	SDN345
3054/196	SJE100
3054/196	SJE106
3054/196	SJE113
3054/196	SJE1519
3054/196	SJE1520
3054/196	SJE203
3054/196	SJE211
3054/196	SJE222
3054/196	SJE228
3054/196	SJE229
3054/196	SJE237
3054/196	SJE242
3054/196	SJE244
3054/196	SJE246
3054/196	SJE248
3054/196	SJE253
3054/196	SJE254
3054/196	SJE261
3054/196	SJE262
3054/196	SJE271
3054/196	SJE272
3054/196	SJE274
3054/196	SJE278
3054/196	SJE280
3054/196	SJE305
3054/196	SJE340
3054/196	SJE404
3054/196	SJE405
3054/196	SJE407
3054/196	SJE42
3054/196	SJE513
3054/196	SJE515
3054/196	SJE583
3054/196	SJE721
3054/196	SJE737
3054/196	SJE769
3054/196	SJE784
3054/196	SP8660
3054/196	STX0013
3054/196	T344
3054/196	T611-1
3054/196	T612-1
3054/196	TA2911
3054/196	TA7156
3054/196	TA7311
3054/196	TA7312
3054/196	TA7313
3054/196	TA7314
3054/196	TA7315
3054/196	TA7316
3054/196	TA7318
3054/196	TA7363
3054/196	TIP-14
3054/196	TIP-31
3054/196	TIP-31A
3054/196	TIP-31B
3054/196	TIP29A
3054/196	TIP31
3054/196	TIP31A
3054/196	TNJ70540
3054/196	TNJ71252
3054/196	TR2327203
3054/196	TR2SD330E
3054/196	TR57
3054/196	TV-117
3054/196	TVSSJE5472-1
3054/196	V176
3054/196	VS2SC1173-Y-3
3054/196	01-031173
3054/196	01-040243
3054/196	01-572784
3054/196	01-572791
3054/196	01-572861
3054/196	2N5191
3054/196	2N5192
3054/196	2N5293
3054/196	2N5295
3054/196	2N5296
3054/196	2N5298
3054/196	2N6121
3054/196	2N6122
3054/196	2N6123
3054/196	2N6289
3054/196	2N6290
3054/196	2SC1061
3054/196	2SC1061B
3054/196	2SC1061C
3054/196	2SC1173-0
3054/196	2SC1173-Y
3054/196	2SC1173A
3054/196	2SC1173C
3054/196	2SC1173R
3054/196	2SC1418
3054/196	2SC1418A
3054/196	2SC1418B
3054/196	2SC1418C
3054/196	2SC1418D
3054/196	2SC1568R
3054/196	2SC789
3054/196	2SC789-0
3054/196	2SC789R
3054/196	2SC789Y
3054/196	2SC790
3054/196	2SC790Y
3054/196	2SD154
3054/196	2SD234
3054/196	2SD234-0
3054/196	2SD234-0
3054/196	2SD234R
3054/196	2SD235
3054/196	2SD235-0
3054/196	2SD235-0
3054/196	2SD235D
3054/196	2SD235G
3054/196	2SD235GR
3054/196	2SD235R
3054/196	002SD235RY
3054/196	2SD235Y
3054/196	2SD314
3054/196	2SD314C
3054/196	2SD314D
3054/196	2SD314E
3054/196	2SD317
3054/196	2SD317F
3054/196	2SD317P
3054/196	2SD318Q
3054/196	2SD343
3054/196	2SD365
3054/196	2SD365-0
3054/196	2SD365F
3054/196	2SD365H
3054/196	2SD365P
3054/196	2SD365Q
3054/196	2SD366
3054/196	2SD366-0
3054/196	2SD366P
3054/196	2SD366Q
3054/196	2SD389
3054/196	2SD389AFO
3054/196	2SD389B
3054/196	2SD389BLB
3054/196	2SD389BLB-P
3054/196	2SD389BP
3054/196	2SD389LP
3054/196	2SD389P
3054/196	2SD389Q
3054/196	2SD390
3054/196	2SD390P
3054/196	2SD390Q
3054/196	2SD477
3054/196	2SD526
3054/196	8-905-706-801
3054/196	8-905-713-110
3054/196	09-302083
3054/196	09-302102
3054/196	09-302132
3054/196	09-302164
3054/196	09-302236
3054/196	09-303018
3054/196	09-303022
3054/196	09-303029
3054/196	09-303032
3054/196	09-303033
3054/196	09-30318
3054/196	09-305095
3054/196	09-305140
3054/196	09-306377
3054/196	11-0772
3054/196	13-14085-93
3054/196	13-28222-3
3054/196	13-28336-2
3054/196	13-28469-2
3054/196	13-2P64
3054/196	13-32636-1
3054/196	13-32638-1
3054/196	13-32640-1
3054/196	13-32642-1
3054/196	13-33925-1
3054/196	13-39004-1
3054/196	13-39099-1
3054/196	13-39884-1
3054/196	13-41628-3
3054/196	13-43005-1
3054/196	13-4P63
3054/196	14-608-01
3054/196	14-608-02
3054/196	14-609-03
3054/196	14-609-04
3054/196	14-903-23
3054/196	14-905-23
3054/196	14-908-23
3054/196	15-123065
3054/196	19A115200-P1
3054/196	21A040-052
3054/196	21A112-095
3054/196	21M466
3054/196	24MW778
3054/196	24MW977
3054/196	31-0101
3054/196	33-090
3054/196	34-6002-50
3054/196	34-6002-52
3054/196	34-6002-56
3054/196	34-6015-64
3054/196	34-6016-31
3054/196	34-6016-53
3054/196	42-23459
3054/196	42-28212
3054/196	46-86234-3
3054/196	46-86348-3
3054/196	46-86374-3
3054/196	46-86400-3
3054/196	46-86500-3
3054/196	46-86516-3
3054/196	48-137145
3054/196	48-137146
3054/196	48-137147
3054/196	48-137148
3054/196	48-137437
3054/196	48-137473
3054/196	48-137549
3054/196	48S137309
3054/196	48S137311
3054/196	48S137369
3054/196	48S155013
3054/196	48S155062
3054/196	48S43240G01
3054/196	57A187-12
3054/196	57A205-14
3054/196	57A214-12
3054/196	57A250-14
3054/196	57A278-14
3054/196	57A279-14
3054/196	57A286-10
3054/196	57B205-14
3054/196	57B250-14
3054/196	61-309689
3054/196	62B046-1
3054/196	62B046-2
3054/196	62B046-4
3054/196	73C180829-12
3054/196	81-23860400-3
3054/196	81-23860400A
3054/196	81-23860400B
3054/196	86-344-2
3054/196	86-529-2
3054/196	86-601-2
3054/196	86-613-2
3054/196	86-648-2
3054/196	86-663-2
3054/196	86X0059-001
3054/196	86X0059-002
3054/196	86X0073-001
3054/196	90-600
3054/196	96-5225-01
3054/196	102-1061-01
3054/196	103-0235-85
3054/196	121-719
3054/196	121-804
3054/196	121-808
3054/196	121-822
3054/196	121-874
3054/196	121-887
3054/196	121-927
3054/196	121-966
3054/196	121-966-01
3054/196	121-992
3054/196	131N2
3054/196	132-046
3054/196	132-072
3054/196	132-080
3054/196	132-081
3054/196	132-3
3054/196	133-1
3054/196	133-3
3054/196	142-004
3054/196	142-008
3054/196	149N1
3054/196	149N2004
3054/196	149N4
3054/196	157N3
3054/196	172-010-9-001
3054/196	172-014-9-001
3054/196	172-014-9-003
3054/196	172-014-9-007
3054/196	172-031-9-003
3054/196	173A-4490-7
3054/196	173A-4491-5
3054/196	176-042-9-007
3054/196	176-055-9-004
3054/196	185-007
3054/196	190N1C
3054/196	190N3C
3054/196	195N1
3054/196	211A6380-3
3054/196	211A6381-2
3054/196	260P21106
3054/196	260P37108
3054/196	353-9502-001
3054/196	617-117
3054/196	686-0210
3054/196	753-2001-173
3054/196	753-4001-932
3054/196	0772
3054/196	800-533-00
3054/196	800-546-00
3054/196	866-6
3054/196	921-1009
3054/196	1041-74
3054/196	1043-1286
3054/196	1043-1328
3054/196	1043-7358
3054/196	1045-7828
3054/196	01057-1
3054/196	1080-130
3054/196	1081-3368
3054/196	1087-01
3054/196	1116
3054/196	1968-17
3054/196	1969-17
3054/196	2000-216
3054/196	2011
3054/196	2057A100-49
3054/196	2082-6
3054/196	2085-17
3054/196	2382-17
3054/196	2408-17
3054/196	5001-044
3054/196	5001-053
3054/196	6380-3
3054/196	6381-2
3054/196	7252
3054/196	7413
3054/196	7423
3054/196	8000-00011-050
3054/196	8000-00028-039
3054/196	8000-00028-041
3054/196	16113
3054/196	16164
3054/196	16181
3054/196	16182
3054/196	16207
3054/196	16241
3054/196	30272
3054/196	39458
3054/196	39824
3054/196	40613
3054/196	40618
3054/196	40621
3054/196	40622
3054/196	40629
3054/196	40630
3054/196	40631
3054/196	41500
3054/196	42942
3054/196	43163
3054/196	44699
3054/196	50200-9
3054/196	95258-1
3054/196	95263-1
3054/196	127978
3054/196	128056
3054/196	131075
3054/196	131257
3054/196	131849
3054/196	132445
3054/196	132446
3054/196	132499
3054/196	132573
3054/196	134774
3054/196	136648
3054/196	137065
3054/196	137369
3054/196	138192
3054/196	142691
3054/196	171174-2
3054/196	183034
3054/196	309459
3054/196	610111-6
3054/196	610153-1
3054/196	610153-2
3054/196	610153-3
3054/196	610153-4
3054/196	610153-5
3054/196	610157-3
3054/196	610162-8
3054/196	610195-3
3054/196	657180
3054/196	717101
3054/196	1445829-503
3054/196	1445829-504
3054/196	1471132-2
3054/196	1471132-4
3054/196	1471132-6
3054/196	1471133-1
3054/196	1473567-4
3054/196	1473611-1
3054/196	1473612-11

SK	Industry Standard No.	SK	Industry Standard No.	SK	Industry Standard No.	SK	Industry Standard No.	SK	Industry Standard No.
3054/196	1473629-3	3056	CHA4Z5.1	3057	AZ5.6	3057	WO-056	3057	WX-054D
3054/196	1473632-1	3056	CHMZ5.1	3057	BZ-052	3057	WO-056A	3057	WX-056
3054/196	1473636-1	3056	DX-0987	3057	BZ5.6	3057	WO-056B	3057	WX-056A
3054/196	1473681-1	3056	EA16X136	3057	BZX55C5V6	3057	WO-056C	3057	WX-056B
3054/196	1700036	3056	EP16X25	3057	BZX73C5V6	3057	WO-056D	3057	WX-056C
3054/196	2320482B	3056	EQA01-05S	3057	BZY58	3057	WO-058	3057	WX-056D
3054/196	2320483	3056	FZ5.1T10	3057	BZY83C5V6	3057	WO-058A	3057	WX-058
3054	2320485	3056	FZ5.1T5	3057	BZY85C5V6	3057	WO-058B	3057	WX-058A
3054	2320486	3056	G5.1T10	3057	BZY92C5V6	3057	WO-058C	3057	WX-058B
3054	2320602	3056	G5.1T20	3057	CD6S	3057	WO-058D	3057	WX-058C
3054	2320652	3056	G5.1T5	3057	CHA4Z5.6	3057	WP-054	3057	WX-058D
3054	2320845	3056	GEZD-5.0	3057	CHMZ5.6	3057	WP-054A	3057	WY-054
3054	2321302	3056	GEZD-5.1	3057	CXL5011	3057	WP-054B	3057	WY-054A
3054	2327206	3056	GIA47	3057	D850	3057	WP-054C	3057	WY-054C
3054/196	3438867	3056	GIA47A	3057	DX-0061	3057	WP-054D	3057	WY-054D
3054/196	3596446	3056	HEP-Z0406	3057	DX-1132	3057	WP-056	3057	WY-056
3054/196	3596448	3056	HS2051	3057	ECG136	3057	WP-056A	3057	WY-056A
3054/196	4080187-0502	3056	HS7051	3057	ECG136A	3057	WP-056C	3057	WY-056B
3054/196	4080879-0001	3056	IP20-0086	3057	EDZ-19	3057	WP-056D	3057	WY-056C
3054/196	4080879-0015	3056	K82047B	3057	EQA0105T	3057	WP-058	3057	WY-056D
3054/196	53204228	3056	K82051B	3057	ERA0106R	3057	WP-058A	3057	WY-058
3054/196	5320432	3056	KS34A	3057	EVR4	3057	WP-058B	3057	WY-058A
3054/196	5320433	3056	KS34AF	3057	EVR4A	3057	WP-058C	3057	WY-058B
3054/196	5320492H	3056	KS34B	3057	EVR4B	3057	WP-058D	3057	WY-058C
3054/196	5320642	3056	KS34BF	3057	FZ5.6T10	3057	WQ-054	3057	WY-058D
3054/196	5320671	3056	KS35A	3057	FZ5.6T5	3057	WQ-054A	3057	WZ-054A
3054/196	5321301	3056	KS35AF	3057	G5.6T10	3057	WQ-054C	3057	WZ-056
3054/196	06120030	3056	KS047A	3057	G5.6T20	3057	WQ-054D	3057	WZ-058
3054/196	6212096	3056	KS047B	3057	G5.6T5	3057	WQ-056	3057	WZ060
3054/196	8034903	3056	KS051A	3057	GEZD-5.6	3057	WQ-056A	3057	WZ523
3054/196	23114253	3056	LPM4.7	3057	GIA56	3057	WQ-056C	3057	XZ055
3054/196	23114254	3056	LPM4.7A	3057	GIA56A	3057	WQ-056D	3057	Z-1140
3054/196	26010024	3056	LR47CH	3057	GIA56B	3057	WQ-058	3057	Z-1145
3054/196	26010059	3056	LR51CH	3057	HD30042090	3057	WQ-058A	3057	Z-1150
3054/196	43027213	3056	LR56CH	3057	HEP-Z0407	3057	WQ-058B	3057	Z-1155
3054/196	43027214	3056	M4Z4.7	3057	H87056	3057	WQ-058D	3057	Z-1160
3054/196	43027987	3056	M4Z4.7-20	3057	J24186	3057	WR-054	3057	Z-1165
3054/196	62734671	3056	M4Z4.7A	3057	KS2056A	3057	WR-054A	3057	Z-1170
3054/196	62743182	3056	M4Z5.1	3057	KS2056B	3057	WR-054B	3057	Z-1250
3054/196	62766212	3056	M4Z5.1-20	3057	KS56A	3057	WR-054C	3057	Z-1255
3054/196	2004547806	3056	M4Z5.1A	3057	KS56AF	3057	WR-054D	3057	Z-1260
3054/196	6621007200	3056	MGIA47	3057	KS36B	3057	WR-056	3057	Z-1265
3055	AZ3.6	3056	MGIA47B	3057	KS36BF	3057	WR-056A	3057	Z-1270
3055	AZ747A	3056	MGIA51	3057	KS056A	3057	WR-056C	3057	Z-1540
3055	BZ3.6	3056	MGIA51A	3057	KS056B	3057	WR-056D	3057	Z-1545
3055	BZX85C3V6	3056	MGIA51B	3057	LPM5.6	3057	WR-058	3057	Z-1555
3055	CD31-00002	3056	MR47C-H	3057	LPM5.6B	3057	WR-058A	3057	Z-1560
3055	CHMZ3.6	3056	MR51C-H	3057	LZ5.6	3057	WR-058B	3057	Z-1565
3055	FZ3.6T10	3056	MR51B-H	3057	M4Z5.6	3057	WR-058C	3057	Z-1570
3055	FZ3.6T5	3056	MZ1005	3057	M4Z5.6-20	3057	WR-058D	3057	Z-5140
3055	GEZD-3.6	3056	MZ205	3057	M4Z5.6A	3057	WS-054	3057	Z-5145
3055	GIA39	3056	MZ4624	3057	M758	3057	WS-054A	3057	Z-5150
3055	GIA39A	3056	MZ4625	3057	MC6007	3057	WS-054C	3057	Z-5155
3055	HEP-Z0402	3056	MZ92-4.7A	3057	MC6007A	3057	WS-054D	3057	Z-5160
3055	H82036	3056	MZ92-5.1A	3057	MC6107	3057	WS-056A	3057	Z-5165
3055	H82039	3056	OA2201	3057	MC6107A	3057	WS-056B	3057	Z-5170
3055	H87036	3056	OA2209	3057	MD752	3057	WS-056C	3057	Z-5240
3055	H87039	3056	OAZ240	3057	MD752A	3057	WS-058A	3057	Z-5245
3055	K82039B	3056	OAZ241	3057	MEZ5.6T10	3057	WS-058B	3057	Z-5250
3055	K830A	3056	PD6006	3057	MEZ5.6T5	3057	WS-058D	3057	Z-5255
3055	K830AF	3056	PD6007	3057	MGIA56	3057	WT-054A	3057	Z-5260
3055	K830B	3056	PD6007A	3057	MGIA56B	3057	WT-054C	3057	Z-5265
3055	K830BF	3056	PD6048	3057	MR56C-H	3057	WT-054D	3057	Z-5270
3055	K831A	3056	PR515	3057	MTZ607	3057	WT-056	3057	Z-5540
3055	K832A	3056	PR605	3057	MTZ607A	3057	WT-056A	3057	Z-5545
3055	K832AF	3056	PR804	3057	MZ4626	3057	WT-056C	3057	Z-5550
3055	K832B	3056	PR83017	3057	MZ5A	3057	WT-056D	3057	Z-5555
3055	K832BF	3056	PS6468	3057	MZ92-5.6A	3057	WT-058A	3057	Z-5560
3055	K8033A	3056	QD-ZMZ205XE	3057	OAZ202	3057	WT-058C	3057	Z-5565
3055	K8033B	3056	Q25.1T10	3057	OAZ242	3057	WT-058D	3057	Z-5570
3055	K8036A	3056	Q25.1T5	3057	PD6008	3057	WU-054A	3057	Z0212
3055	K8039A	3056	RD-5A	3057	PD6008A	3057	WU-054C	3057	Z0407
3055	K8039B	3056	RD5B	3057	PD6049	3057	WU-054D	3057	ZOD5.6
3055	LPM3.6	3056	RH-EXOO24TAZZ	3057	PD6049C	3057	WU-056A	3057	Z1104
3055	LPM3.6A	3056	SV122	3057	PS8907	3057	WU-056C	3057	Z1B5.6
3055	LR39CH	3056	SZ200	3057	PTC602	3057	WU-056D	3057	Z1D5.6
3055	M4Z3.3	3056	SZ200-5	3057	QD-ZRD56PAA	3057	WU-058A	3057	Z2A56F
3055	M4Z3.3A	3056	SZ4.7	3057	QRT-236	3057	WU-058B	3057	Z4O5.6
3055	M4Z3.9	3056	SZ5	3057	QZ5.6T10	3057	WU-058C	3057	Z4D5.6
3055	M4Z3.9A	3056	SZ5.1	3057	QZ5.6T5	3057	WU-058D	3057	Z4I5.6
3055	MGIA39	3056	TMD01	3057	RD-6AM	3057	WV-054	3057	Z5.6
3055	MGIA39B	3056	TMD01A	3057	RD5-6E	3057	WV-054A	3057	Z5B5.6
3055	MR36H	3056	TV5RD5A	3057	RD6	3057	WV-054C	3057	Z5O5.6
3055	MR39C-H	3056	WP-052D	3057	RE107	3057	WV-054D	3057	Z5D5.6
3055	MZ4620	3056	001-0163-02	3057	REN136	3057	WV-056A	3057	ZB5.6
3055	MZ500-5	3056	001-02303-0	3057	RY1181	3057	WV-056C	3057	ZB5.6B
3055	MZ92-3.9A	3056	1D5.1A	3057	RZ5.6	3057	WV-058A	3057	ZD5.6A
3055	PD6003	3056	1EZ5.1	3057	RZZ5.6	3057	WV-058B	3057	ZB5.6
3055	PD6003A	3056	1N3826	3057	SV123	3057	WV-058C	3057	ZBO5.6
3055	PD6004	3056	1N3826A	3057	SZ5.6	3057	WV-058D	3057	ZF5.6
3055	PD6044	3056	1N4654	3057	TM136	3057	WW-054	3057	ZG5.6
3055	PTC601	3056	1N4733	3057	TMD02	3057	WW-054A	3057	ZO85.6
3055	Q23.6T10	3056	1N4733A	3057	TMD02A	3057	WW-054C	3057	ZP5.6
3055	RE101	3056	1S320	3057	VHEHZ6B3///1A	3057	WW-054D	3057	ZS5.6
3055	RZ3.6	3056	1Z5.1T10	3057	VR5.6	3057	WW-056	3057	ZS5.6A
3055	RZZ3.6	3056	1ZP5.1T10	3057	VR5.6B	3057	WW-056A	3057	ZS5.6B
3055	SZ3.6	3056	1ZP5.1T20	3057	WBP603	3057	WW-056B	3057	ZT5.6
3055	TIXD747	3056	1ZP5.1T5	3057	WM-054	3057	WW-056C	3057	ZT5.6A
3055	TVSRD4AM	3056	1ZM5.1T10	3057	WM-054A	3057	WW-056D	3057	ZT5.6B
3055	1/4LZ3.6D	3056	1ZM5.1T20	3057	WM-054B	3057	WW-058	3057	ZZ5.6
3055	1/4LZ3.6D5	3056	10-085027	3057	WM-054C	3057	WW-058A	3057	001-0163-04
3055	1D3.6	3056	13-33187-12	3057	WM-054D	3057	WW-058B	3057	1-210
3055	1D3.6A	3056	14-509-02	3057	WM-056	3057	WW-058C	3057	1/4AZ5.6D
3055	1D3.6B	3056	14-515-02	3057	WM-056A	3057	WW-058D	3057	1/4AZ5.6D10
3055	1EZ3.6	3056	1626	3057	WM-056B	3057	WX-054	3057	1/4AZ5.6D5
3055	1N3307	3056	1626A	3057	WM-056C	3057	WX-054A	3057	1/4LZ5.6D10
3055	1N3373	3056	20B409	3057	WM-056D	3057	WX-054B	3057	1/4LZ5.6D5
3055	1N3822	3056	20B410	3057	WM-058	3057	WX-054C	3057	1/4M5.6AZ
3055	1N4650	3056	34-8057-37	3057	WM-058A			3057	1/4M5.6AZ1Q
3055	1N4729	3056	3626	3057	WM-058B			3057	1/4M5.6AZ5
3055	1N4729A	3056	48-134698	3057	WM-058C			3057	1D5.6
3055	123.6	3056	48-134991	3057	WM-058D			3057	1D5.6A
3055	123.6A	3056	56-16	3057	WN-054			3057	1D5.6B
3055	1ZC3.6	3056	56-44	3057	WN-054A			3057	1EZ5.6
3055	1ZM3.6T10	3056	56-59	3057	WN-054B			3057	1N1509A
3055	1ZM3.6T20	3056	264 P02501	3057	WN-054C			3057	1N1520A
3055	1ZM3.6T5	3056	523-2003-519	3057	WN-054D			3057	1N1765A
3055	2Z6P	3056	919-01-3058	3057	WN-056			3057	1N3512
3055	29P1	3056	2093A41-90	3057	WN-056A			3057	1N4655
3055	32Z6	3056	2774	3057	WN-056B			3057	1N4690
3055	32Z6A	3056	4822-130-30264	3057	WN-056C			3057	1N4734
3055	3326	3056	5001-197	3057	WN-056D			3057	1N4734A
3055	3326A	3056	741738	3057	WN-058			3057	1NJ61225
3055	56-70	3056	760298	3057	WN-058A			3057	1S2111
3055	93039-1	3056	986578	3057	WN-058B			3057	1S2111A
3055	1102	3057	7852904-01	3057	WN-058C			3057	1S221
3055	1103	3057	.25T5.6	3057	WN-058D			3057	13692
3056	.75N5.1	3057	.25T5.6B	3057	WO-054			3057	1T5.6
3056	AZ-050	3057	.4T5.6	3057	WO-054A			3057	1T5.6B
3056	AZ5.1	3057	.4T5.6A	3057	WO-054B				
3056	B25.1	3057	.4T5.6B	3057	WO-054C				
3056	BZY92C5V1	3057	.75N5.6	3057	WO-054D				
3056	CD3122	3057	AZ-056						

SK	Industry Standard No.	SK	Industry Standard No.	SK	Industry Standard No.	SK	Industry Standard No.	SK	Industry Standard No.
3057	1TA5.6	3058	PD6009A	3058	WQ-063	3058	Z106.2	3059	A059-117
3057	1TA5.6A	3058	PD6050	3058	WQ-063A	3058	Z1D6.2	3059	A2126B
3057	1Z05.6	3058	P81325	3058	WQ-063B	3058	Z2A62P	3059	AZ-073
3057	1ZP5.6T10	3058	P88908	3058	WQ-063C	3058	Z4XL6.2	3059	AZ-075
3057	1ZP5.6T20	3058	PTC503	3058	WQ-063D	3058	Z4XL6.2B	3059	AZ7.5
3057	1ZP5.6T5	3058	QRT-237	3058	WQ-065	3058	Z5B6.2	3059	A3958B
3057	1ZM5.6T10	3058	QZ6.2T10	3058	WQ-065A	3058	Z5D6.2	3059	B071
3057	1ZN5.6T20	3058	QZ6.2T5	3058	WQ-065B	3058	Z6.2	3059	BZ-071
3057	1ZN5.6T5	3058	RE109	3058	WQ-065C	3058	Z801	3059	CDZ-318-75
3057	0ZZ5.6	3058	REN137	3058	WQ-065D	3058	ZB6.2	3059	D-00369C
3057	0ZZ5.6A	3058	RZ6.2	3058	WR-061	3058	ZB6.2A	3059	EA16X88
3057	3L4-3506-21	3058	RZZ6.2	3058	WR-061A	3058	ZB6.2B	3059	ED2-2
3057	4-436	3058	S1A3	3058	WR-061B	3058	ZD6.2A	3059	EP16X12
3057	6V-200	3058	S3004-1715	3058	WR-061C	3058	ZD6.2B	3059	EP36X12
3057	07-5331-86A	3058	SA821	3058	WR-061D	3058	ZP6.2-4	3059	EQA01-072
3057	07-5331-86A	3058	SA821A	3058	WR-063	3058	Z0B6.2	3059	EQB01-07
3057	07-5331-86C	3058	SA823	3058	WR-063A	3058	Z0D6.2	3059	ES57X7
3057	07-5331-86C	3058	SA823A	3058	WR-063B	3058	ZP6.2	3059	EZD-RD75
3057	07-5331-86D	3058	SA825	3058	WR-063C	3058	ZT6.2	3059	FV21
3057	8-905-421-109	3058	SA825A	3058	WR-063D	3058	ZT6.2A	3059	GEZD-7.5
3057	09-306109	3058	SA827	3058	WR-065	3058	ZT6.2B	3059	HD3000309
3057	10-016	3058	SA827A	3058	WR-065A	3058	Z26.2	3059	HD3000409
3057	012-1023-007	3058	SA829	3058	WR-065B	3058	001-02303-3	3059	HE-Z0049
3057	13-67544-1	3058	SA829A	3058	WR-065C	3058	1/4AZ6.2D5	3059	HEP-Z0410
3057	17Z6	3058	SB821	3058	WR-065D	3058	1/4LZ6.2D	3059	IP20-0027
3057	17Z6A	3058	SB821A	3058	WS-061	3058	1/4LZ6.2D5	3059	MZ1007
3057	21A119-041	3058	SB823	3058	WS-061A	3058	1/4M6.2AZ5	3059	MZ2078
3057	34-8057-32	3058	SB823A	3058	WS-061B	3058	1D6.2	3059	PTC504
3057	37Z6	3058	SB825	3058	WS-061C	3058	1D6.2A	3059	R7030
3057	37Z6A	3058	SB825A	3058	WS-061D	3058	1D6.2B	3059	RD-7.5EB
3057	48-134993	3058	SB827	3058	WS-063	3058	1D6.28A	3059	RDTR5EB
3057	73-31	3058	SB827A	3058	WS-063A	3058	1D6.28B	3059	RB111
3057	93A39-19	3058	SB829	3058	WS-063B	3058	1M753A	3059	RVDBQA0107S
3057	100-10	3058	SB829A	3058	WS-063C	3058	1N1485	3059	RVDRDTR5EB
3057	100-135	3058	SC821	3058	WS-063D	3058	1N1766	3059	SVD02Z8
3057	103-29007	3058	SC821A	3058	WS-065	3058	1N1766A	3059	SZ-7
3057	530-073-31	3058	SC823	3058	WS-065A	3058	1N3496	3059	T17-A
3057	65200	3058	SC823A	3058	WS-065B	3058	1N3497	3059	TR-7B8
3057	903-120B	3058	SC825	3058	WS-065C	3058	1N3498	3059	TR-7SB
3057	903-333	3058	SC825A	3058	WS-065D	3058	1N3499	3059	TVS-RD7A
3057	2362-17	3058	SC827	3058	WT-061	3058	1N3500	3059	Z-1010
3057	5001-160	3058	SC827A	3058	WT-061A	3058	1N3513	3059	001-0099-01
3057	169199	3058	SC829	3058	WT-061B	3058	1N4499	3059	001-0099-02
3057	171842	3058	SC829A	3058	WT-061C	3058	1N4656	3059	1A7.5M
3057	225301	3058	SU0650	3058	WT-061D	3058	1N4735	3059	1A7.5MA
3057	348057-17	3058	SUM6010	3058	WT-063	3058	1N4735A	3059	1C0020
3057	530073-31	3058	SUM6011	3058	WT-063A	3058	1N675	3059	1BZ-5210
3057	530145-1569	3058	SUM6020	3058	WT-063B	3058	1N709	3059	1EZ7.5Z
3057	530145-569	3058	SUM6021	3058	WT-063C	3058	1N709A	3059	1N3017B
3057	530157-569	3058	SV0625	3058	WT-063D	3058	1N709B	3059	1N4737
3057	1800012	3058	SVC650	3058	WT-065	3058	1N821	3059	1N4737A
3057	62522666	3058	SVM601	3058	WT-065A	3058	1N821A	3059	1S2114A
3058	.25Z6.2	3058	SVM6010	3058	WT-065B	3058	1N822	3059	18223
3058	.25Z6.2B	3058	SVM6011	3058	WT-065C	3058	1N823	3059	18332
3058	.4T6.2	3058	SVM602	3058	WT-065D	3058	1N823A-4	3059	1T7-5
3058	.4T6.2A	3058	SVM6020	3058	WU-061	3058	1N824	3059	1T7.5B
3058	.4T6.2B	3058	SVM6021	3058	WU-061A	3058	1N825	3059	1TA7.5
3058	.4T6.2BA	3058	SVM605	3058	WU-061B	3058	1N827	3059	1TA7.5A
3058	.75N6.2	3058	SVM61	3058	WU-061C	3058	1N827A	3059	1Z7.5
3058	AZ-061	3058	SZ-RD6.2EB	3058	WU-061D	3058	1N829	3059	1ZM7.5T10
3058	AZ-063	3058	SZ6.2	3058	WU-063	3058	1N829A	3059	1ZM7.5T20
3058	AZ6.2	3058	SZ6.2A	3058	WU-063A	3058	181715	3059	0Z27.5A
3058	BZX10	3058	TIX0753	3058	WU-063B	3058	182112	3059	3L4-3505-3
3058	BZX5506V2	3058	TM137	3058	WU-063C	3058	182112A	3059	4-2021-05470
3058	BZX79C6V2	3058	TM137A	3058	WU-063D	3058	1Z6.2	3059	05-110108
3058	BZY59	3058	TMD03	3058	WU-065	3058	1Z6.2B	3059	09-306181
3058	BZY8306V2	3058	TMD03A	3058	WU-065A	3058	1ZA6.2	3059	09-306197
3058	BZY8506V2	3058	TRR6	3058	WU-065B	3058	1ZA6.2A	3059	09-306277
3058	BZY88/C6V2	3058	WEP103	3058	WU-065C	3058	1Z6.2T10	3059	13-14879-4
3058	CHAZ6.2	3058	WM-061	3058	WU-065D	3058	1Z6.2T5	3059	14-515-04
3058	CHAZ6.2A	3058	WM-061A	3058	WV-061	3058	1Z06.2	3059	14-515-13
3058	CXL137	3058	WM-061B	3058	WV-061A	3058	1ZM6.2	3059	2026AP
3058	CXL6013	3058	WM-061C	3058	WV-061B	3058	1ZM6.2T10	3059	2026P
3058	D-00469C	3058	WM-061D	3058	WV-061C	3058	1ZM6.2T20	3059	21A103-049
3058	D3H	3058	WM-063	3058	WV-061D	3058	1ZM6.2T5	3059	24MW207
3058	D5W	3058	WM-063A	3058	WV-063	3058	2MWT42	3059	33-0025
3058	D6.2	3058	WM-063B	3058	WV-063A	3058	0ZZ6.2	3059	34-8057-16
3058	DDAY008001	3058	WM-063C	3058	WV-063B	3058	0ZZ62A	3059	34-8057-27
3058	DIA72293	3058	WM-063D	3058	WV-063C	3058	4-2020-07500	3059	38A64C
3058	E21431	3058	WM-065	3058	WV-063D	3058	6.2SR1	3059	44T-300-94
3058	E2486	3058	WM-065A	3058	WV-065	3058	6.2SR1A	3059	46-86225-3
3058	EC0137	3058	WM-065B	3058	WV-065A	3058	6.2SR2	3059	46-86456-3
3058	EC0137A	3058	WM-065C	3058	WV-065B	3058	6.2SR2A	3059	53B015-1
3058	ED6.2EB	3058	WM-065D	3058	WV-065C	3058	6.2SR3	3059	56-51
3058	EVR5	3058	WN-061	3058	WV-065D	3058	6.2SR3A	3059	56-8096
3058	EVR5A	3058	WN-061A	3058	WW-061	3058	6.2SR4	3059	93C39-10
3058	EVR5B	3058	WN-061B	3058	WW-061A	3058	6.2SR4A	3059	96-5124-02
3058	FV-22	3058	WN-061C	3058	WW-061B	3058	09-306055	3059	1009-06
3058	FZ6.2T10	3058	WN-061D	3058	WW-061C	3058	09-306183	3059	1041-67
3058	G01-036A	3058	WN-063	3058	WW-061D	3058	09-306286	3059	2000-325
3058	GEZD-6.2	3058	WN-063A	3058	WW-063	3058	18Z6	3059	2093A41-53
3058	GlA62	3058	WN-063B	3058	WW-063A	3058	18Z6A	3059	2330-021
3058	GlA62A	3058	WN-063C	3058	WW-063B	3058	38Z6	3059	4822-130-30287
3058	GlA62B	3058	WN-063D	3058	WW-063C	3058	38Z6A	3059	05470
3058	GZ6.2	3058	WN-065	3058	WW-063D	3058	46-86394-3	3059	7574
3058	HEP-Z0408	3058	WN-065A	3058	WW-065	3058	48-137170	3059	8000-00005-019
3058	HS2062	3058	WN-065B	3058	WW-065A	3058	48S10641D62	3059	8000-00011-103
3058	J241179	3058	WN-065C	3058	WW-065B	3058	488137170	3059	72135
3058	KD2503	3058	WN-065D	3058	WW-065C	3058	488137387	3059	115123
3058	KS2062A	3058	WO-061A	3058	WW-065D	3058	53B011-2	3059	171162-172
3058	KS2062B	3058	WO-061B	3058	WX-061	3059	53J004-1	3059	239219
3058	KS37A	3058	WO-061C	3058	WX-061A	3059	66X0040-003	3059	281917
3058	KS37AF	3058	WO-061D	3058	WX-061B	3059	121-0041	3059	985011
3058	K3062A	3058	WO-063	3058	WX-061C	3059	123B-005	3059	984252
3058	LR62CH	3058	WO-063A	3058	WX-061D	3059	151-030-9-003	3059	2005779-26
3058	M4Z6.2	3058	WO-063B	3058	WX-063	3059	152-047-9-001	3059	2006607-63
3058	M4Z6.2-20	3058	WO-063C	3058	WX-063A	3059	152-052-9-002	3059	2092055-5
3058	M4Z6.2A	3058	WO-063D	3058	WX-063B	3059	65207	3059	2330021
3058	MC6008	3058	WO-065	3058	WX-063C	3059	5001-131	3059	2330302H
3058	MC6008A	3058	WO-065A	3058	WX-063D	3059	8000-00004-239	3059	23115984
3058	MC6108	3058	WO-065B	3058	WX-065	3059	8000-00005-020	3059	4202105470
3058	MC6108A	3058	WO-065C	3058	WX-065A	3059	72168-3	3060	.25T9.1
3058	MD753	3058	WO-065D	3058	WX-065B	3059	168653	3060	.25T9.1B
3058	MD753A	3058	WP-061	3058	WX-065C	3059	302540	3060	.4Z9.1
3058	MGlA62	3058	WP-061A	3058	WX-065D	3059	530073-5	3060	.4Z9.1A
3058	MGlA62A	3058	WP-061B	3058	WY-061	3059	530166-1004	3060	.7JZ9.1
3058	MGlA62B	3058	WP-061C	3058	WY-061A	3059	2002209-4	3060	.7Z9.1
3058	MR62D-H	3058	WP-061D	3058	WY-061B	3059	4037413-P1	3060	.7Z9.1B
3058	MR62B-H	3058	WP-063	3058	WY-061C	3059	26010029	3060	.7Z9.1C
3058	MR62H	3058	WP-063A	3058	WY-061D	3059	26010046	3060	.7Z9.1D
3058	MTZ608	3058	WP-063B	3058	WY-063	3059	.25T7.5	3060	.7ZM9.1A
3058	MTZ608A	3058	WP-063C	3058	WY-063A	3059	.25T7.5B	3060	.7ZM9.1B
3058	MZ4627	3058	WP-063D	3058	WY-063B	3059	.75N7.5	3060	.7ZM9.1C
3058	MZ6.2	3058	WP-065	3058	WY-063C	3059	.7J27.5	3060	.7ZM9.1D
3058	MZ6.2A	3058	WP-065A	3058	WY-065	3059	.7Z7.5A	3060	A061-119
3058	MZ6.2B	3058	WP-065B	3058	WY-065A	3059	.7Z7.5B	3060	A068-103
3058	MZ6.2T5	3058	WP-065C	3058	WY-065B	3059	.7Z7.5D	3060	AW-01-09
3058	MZ605	3058	WP-065D	3058	WY-065C	3059	.7Z7.5D	3060	AW01-09
3058	MZ610	3058	WQ-061A	3058	WY-065D	3059	.7ZM7.5A	3060	AW01-9
3058	MZ620	3058	WQ-061 B	3058	WZ065	3059	.7ZM7.5B	3060	AW09
3058	MZ640	3058	WQ-061 C	3058	Z0214	3059	A04332-007	3060	AZ-090
3058	MZ92-6.2A	3058	WQ-061 D	3058	Z0408	3059	A04344-007	3060	AZ-092
3058	0AZ203			3058	Z0D6.2			3060	AZ9.1
3058	0AZ270			3058	Z1B6.2			3060	B090
3058	PD6009								

SK	Industry Standard No.	SK	Industry Standard No.	SK	Industry Standard No.	SK	Industry Standard No.	SK	Industry Standard No.
3060	B2090	3060	TMD07	3060	WS-090B	3060	WZ-092	3060	56-19
3060	B2090	3060	TYSRD9AL	3060	WS-090C	3060	WZ-092A	3060	56-46
3060	BX909	3060	UZ8T09	3060	WS-090D	3060	WZ-094	3060	56-62
3060	BXY63	3060	UZ8809	3060	WS-092	3060	WZ-094A	3060	75Z9.1
3060	BZ-090	3060	UZ9.1B	3060	WS-092A	3060	WZO90	3060	75Z9.1B
3060	BZ-090(ZENER)	3060	WD90	3060	WS-092C	3060	WZO92	3060	86-0005
3060	BZ-0900	3060	WEP104	3060	WS-092D	3060	WZ9.1	3060	86-37-1
3060	BZ-0901	3060	W091	3060	WS-094	3060	X092	3060	86-58-1
3060	BZ-094	3060	WM-088	3060	WS-094A	3060	XZ-092	3060	93C39-3
3060	BZ090	3060	WM-088A	3060	WS-094B	3060	XZO90	3060	100-139
3060	BZ090.1Z9	3060	WM-088B	3060	WS-094D	3060	YEABO15	3060	100-14
3060	BZ1-9	3060	WM-088C	3060	WT-088	3060	Z0219	3060	103-272
3060	BZ9.1	3060	WM-088D	3060	WT-088B	3060	Z0412	3060	106-010
3060	BZX46C9V1	3060	WM-090	3060	WT-088C	3060	Z1109	3060	0114-0017
3060	BZX55C9V1	3060	WM-090A	3060	WT-088D	3060	ZB1-9	3060	0114-0090
3060	BZXT1C9V1	3060	WM-090B	3060	WT-090	3060	ZB1-9V	3060	120-004879
3060	BZXT9C9V1	3060	WM-090C	3060	WT-090A	3060	ZB9.1A	3060	123-620
3060	CA-092	3060	WM-090D	3060	WT-090C	3060	ZB9.1B	3060	142-012
3060	CD31-00012	3060	WM-092	3060	WT-090D	3060	ZBI-09	3060	145-118
3060	CD31-12039	3060	WM-092A	3060	WT-092	3060	ZQ9.1A	3060	151-031-9-006
3060	CD31I2039	3060	WM-092B	3060	WT-092B	3060	Z89.1	3060	152-008-9-001
3060	CF-092	3060	WM-092C	3060	WT-092C	3060	ZT9.1	3060	152-012-9-001
3060	CXL139	3060	WM-092D	3060	WT-092D	3060	ZT9.1A	3060	152-051-9-001
3060	CXL5018	3060	WM-094A	3060	WT-094	3060	ZV9.1A	3060	152-082-9-001
3060	CZ-092	3060	WM-094B	3060	WT-094A	3060	ZV9.1B	3060	185-015
3060	CZ-92	3060	WM-094C	3060	WT-094C	3060	ZV9.1	3060	186-015
3060	CZO92	3060	WM-094D	3060	WT-094D	3060	ZX9.1	3060	230-0023
3060	CZO94	3060	WN-088	3060	WU-088	3060	ZZ9.1	3060	690V105H32
3060	D-18	3060	WN-088A	3060	WU-088B	3060	001-0163-15	3060	903-119B
3060	D1-8	3060	WN-088C	3060	WU-088C	3060	1/4A9.1	3060	903-179
3060	D176	3060	WN-088D	3060	WU-088D	3060	1/4A9.1A	3060	903T00228
3060	DAAYO10092	3060	WN-090	3060	WU-090	3060	1/4A9.1B	3060	919-01-1213
3060	DDAYO10002	3060	WN-090A	3060	WU-090A	3060	1A12688	3060	1001-12
3060	DDAY126001	3060	WN-090B	3060	WU-090C	3060	1A9.1M	3060	1042-18
3060	DI-8	3060	WN-090C	3060	WU-090D	3060	1A9.1MA	3060	1074-120
3060	DIJ72165	3060	WN-090D	3060	WU-092	3060	1B9.1Z	3060	1080-10
3060	DX-0087	3060	WN-092	3060	WU-092A	3060	1B9.1Z10	3060	1858A
3060	DX-0728	3060	WN-092A	3060	WU-092C	3060	1N1770	3060	2000-308
3060	DZ10	3060	WN-092B	3060	WU-092D	3060	1N1770A	3060	2041-06
3060	EO771-3	3060	WN-092D	3060	WU-094	3060	1N3019A	3060	2093A41-159
3060	EA16X162	3060	WN-094A	3060	WU-094A	3060	1N3678	3060	2093A41-16
3060	EA16X82	3060	WN-094B	3060	WU-094C	3060	1N3855A	3060	2093A41-186
3060	EA3866	3060	WN-094C	3060	WU-094D	3060	1N4103	3060	2093A41-24
3060	ECG139	3060	WN-094D	3060	WV-088	3060	1N4161A	3060	2093A41-24A
3060	ECG139A	3060	WO-088	3060	WV-088A	3060	1N4161B	3060	2150-17
3060	EDZ-23	3060	WO-088A	3060	WV-088B	3060	1N4326	3060	4653
3060	EQ-098	3060	WO-088B	3060	WV-088D	3060	1N4326A	3060	5001-125
3060	EQA-01098	3060	WO-088C	3060	WV-090	3060	1N4326B	3060	5001-152
3060	EQA01-09	3060	WO-088D	3060	WV-090A	3060	1N4660	3060	5403-MS
3060	EQA01-09R	3060	WO-090	3060	WV-090B	3060	1N4739	3060	6432-3
3060	EQBO1-09	3060	WO-090A	3060	WV-090C	3060	1N4739A	3060	8000-00004-065
3060	EVR9	3060	WO-090B	3060	WV-090D	3060	1R9.1A	3060	8000-00005-021
3060	G01036	3060	WO-090C	3060	WV-092	3060	18130(ZENER)	3060	8000-00006-232
3060	G9.1Z10	3060	WO-090D	3060	WV-092B	3060	181717L	3060	8000-00011-043
3060	G9.1T20	3060	WO-092	3060	WV-092C	3060	181728	3060	8000-00041-018
3060	G9.1T5	3060	WO-092A	3060	WV-092D	3060	182116	3060	8000-00043-021
3060	GEZD-9.1	3060	WO-092B	3060	WV-094	3060	182116A	3060	8000-00043-068
3060	GIA91	3060	WO-092C	3060	WV-094A	3060	0000018334	3060	8710-54
3060	GIA91A	3060	WO-092D	3060	WV-094B	3060	18334N	3060	8840-55
3060	GIA91B	3060	WO-094	3060	WW-088	3060	1T9.1	3060	79408
3060	HD3000401	3060	WO-094A	3060	WW-088A	3060	1T9.1B	3060	112530
3060	HD3001T090	3060	WO-094C	3060	WW-088B	3060	1TA9.1	3060	236266
3060	HEP-Z0412	3060	WO-094D	3060	WW-088C	3060	1TA9.1A	3060	280317
3060	HP-20004	3060	WP-088	3060	WW-088D	3060	1Z9.1	3060	602081
3060	HF-20011	3060	WP-088A	3060	WW-090	3060	1Z9.1A	3060	741739
3060	HF-20065	3060	WP-088B	3060	WW-090B	3060	1Z9.1B	3060	T81717
3060	H82091	3060	WP-088C	3060	WW-090C	3060	1Z9.1D10	3060	922358
3060	H87091	3060	WP-088D	3060	WW-090D	3060	1Z9.1D5	3060	922524
3060	HW9.1	3060	WP-090	3060	WW-092	3060	1Z9.1T5	3060	922603
3060	HW9.1A	3060	WP-090A	3060	WW-092B	3060	1ZC9.1	3060	1223927
3060	HW9.1B	3060	WP-090B	3060	WW-092C	3060	1ZF9.1Z10	3060	2002209-1
3060	IC743047	3060	WP-090C	3060	WW-092D	3060	1ZF9.1T20	3060	2003779-26
3060	IP20-0019	3060	WP-090D	3060	WW-094	3060	1ZM9.1Z10	3060	5330011
3060	KS2091A	3060	WP-092	3060	WW-094A	3060	1ZM9.1T20	3060	70260540
3060	KS2091B	3060	WP-092A	3060	WW-094B	3060	1ZM9.1T5	3060	87500920
3060	KS41A	3060	WP-092C	3060	WW-094C	3060	02Z-9.1A	3060	87600920
3060	KS41AF	3060	WP-092D	3060	WW-094D	3060	02Z102A	3060	.25T10
3060	LPM9.1	3060	WP-094	3060	WX-088	3060	02Z9.1A	3061	.25T10B
3060	LPM9.1A	3060	WP-094A	3060	WX-088B	3060	05-990094	3061	.4T10
3060	LR91CH	3060	WP-094B	3060	WX-088C	3060	05Z9.1L	3061	.4T10A
3060	M4Z9.1	3060	WP-094C	3060	WX-088D	3060	08	3061	.7Z10
3060	M4Z9.1-20	3060	WP-094D	3060	WX-090	3060	08-08125	3061	.7Z10A
3060	M4Z9.1A	3060	WQ-088	3060	WX-090A	3060	8-905-421-118	3061	.7Z10B
3060	M92	3060	WQ-088B	3060	WX-090B	3060	8-905-421-228	3061	.7Z10C
3060	MD757	3060	WQ-088C	3060	WX-090C	3060	8C91	3061	.7Z10D
3060	MD757A	3060	WQ-088D	3060	WX-090D	3060	09-306124	3061	.7ZM10A
3060	MGIA91	3060	WQ-090	3060	WX-092	3060	09-306158	3061	.7ZM10B
3060	MGIA91A	3060	WQ-090A	3060	WX-092B	3060	09-306165	3061	.7ZM10C
3060	MGIA91B	3060	WQ-090B	3060	WX-092C	3060	09-306180	3061	AW-01-1Q
3060	MR91C-H	3060	WQ-090C	3060	WX-092D	3060	09-306194	3061	AZ-098
3060	MR91E-H	3060	WQ-090D	3060	WX-094	3060	09-306232	3061	AZ-100
3060	MTZ613	3060	WQ-092	3060	WX-094B	3060	09-306239	3061	AZ10
3060	MTZ613A	3060	WQ-092A	3060	WX-094C	3060	09-306241	3061	B094
3060	MZ-209	3060	WQ-092C	3060	WY-088	3060	09-306247	3061	BZ-100
3060	MZ92-9.1A	3060	WQ-092D	3060	WY-088B	3060	09-306287	3061	BZX46O10
3060	MZ92-9.1B	3060	WQ-094A	3060	WY-088B	3060	09-306327	3061	BZXT9C10
3060	MZX9.1	3060	WQ-094B	3060	WY-088D	3060	09-306332	3061	BZY85C10
3060	OA2207	3060	WQ-094C	3060	WY-090	3060	09-306351	3061	BZY85C10
3060	OA2212	3060	WQ-094D	3060	WY-090A	3060	09-306375	3061	C4016
3060	OA2247	3060	WR-088	3060	WY-090B	3060	09-306382	3061	CHZ10
3060	OA2272	3060	WR-088A	3060	WY-090C	3060	9D141003-12	3061	CHZ10A
3060	P21344	3060	WR-088B	3060	WY-090D	3060	9D15	3061	CKOO51
3060	PD6013	3060	WR-088C	3060	WY-092	3060	9D15	3061	CXL140
3060	PD6013A	3060	WR-088D	3060	WY-092A	3060	10-013	3061	CXL5019
3060	PD6054	3060	WR-090	3060	WY-092B	3060	13-085028	3061	CZD010
3060	PS8912	3060	WR-090B	3060	WY-092C	3060	13-085042	3061	CZD010-5
3060	PTC505	3060	WR-090D	3060	WY-092D	3060	19A115528-P3	3061	DZ10A
3060	Q-25115C	3060	WR-088	3060	WY-094	3060	19B200379-P1	3061	BR4
3060	QD-ZRD9EXAA	3060	WR-088B	3060	WY-094A	3060	21A103-064	3061	EA16X150
3060	QRT-240	3060	WR-088C	3060	WY-094B	3060	21A108-002	3061	EA16X157
3060	RD-91	3060	WR-088D	3060	WY-094C	3060	21M214	3061	ECG140
3060	RE114	3060	WR-090	3060	WY-094D	3060	21M584	3061	ECG140A
3060	REN139	3060	WR-090A	3060	WY-088	3060	21M585	3061	EQBO1-10
3060	RPJ71480	3060	WR-090B	3060	WY-088B	3060	022-2823-002	3061	G01-012-F
3060	RP-9A	3060	WR-090D	3060	WY-088B	3060	2226AF	3061	GEZD-10
3060	RT-240	3060	WR-092	3060	WY-088C	3060	2226P	3061	GIA100
3060	RT8339	3060	WR-092A	3060	WY-090A	3060	24MW331	3061	GIA100A
3060	RVDIN4739	3060	WR-092B	3060	WY-090B	3060	24MW998	3061	GIA100B
3060	RZ9.1	3060	WR-092C	3060	WY-090C	3060	32-0005	3061	HD30001200
3060	RZZ9.1	3060	WR-092D	3060	WY-090D	3060	33R-09	3061	HEP-Z0413
3060	SC91	3060	WR-094	3060	WY-092	3060	42-Z6A	3061	HN-00061
3060	SD-27	3060	WR-094A	3060	WY-092A	3060	42Z6	3061	HS2100
3060	SD632	3060	WR-094B	3060	WY-092B	3060	46-86421-3	3061	HS7100
3060	SO-632	3060	WR-094C	3060	WY-092C	3060	48-03073A09	3061	IP20-0233
3060	SV9	3060	WR-094D	3060	WY-092	3060	48-134653	3061	J24872
3060	SVDU2Z9.5A	3060	WS-088	3060	WY-092A	3060	48-134957	3061	KS100A
3060	SZ-200-9V	3060	WS-088A	3060	WY-092B	3060	48-41873J02	3061	KS100B
3060	SZ-9	3060	WS-088B	3060	WY-092C	3060	48-44080J05	3061	KS2100A
3060	SZ9	3060	WS-088A	3060	WY-094	3060	5029.1C	3061	KS2100B
3060	SZ9.1	3060	WS-088B	3060	WY-094A	3060	051-0024	3061	KS42A
3060	SZC9	3060	WS-088D	3060	WY-094B	3060	52-053-013-0	3061	
3060	SZT-9	3060	WS-090	3060	WY-094D	3060	53A022-6	3061	
3060	SZT9	3060		3060		3060	53A022-7	3061	
3060	TM139A	3060	WZ-090	3060	WZ-090	3060	53K001-14	3061	KR42RP

SK	Industry Standard No.
3061	LR100CH
3061	LZ10
3061	M10Z
3061	M4Z10
3061	M4Z10-20
3061	M4Z10A
3061	MA1100
3061	MC6014
3061	MC6014A
3061	MGLA100
3061	MGLA100A
3061	MGLA100B
3061	MR100C-H
3061	MR100H
3061	MTZ614
3061	MTZ614A
3061	MZ-210B
3061	MZ10
3061	MZ1000-13
3061	MZ1010
3061	MZ10A
3061	MZ310A
3061	MZ92-10A
3061	02Z-10A
3061	PC1879-004
3061	PD6014
3061	PD6014A
3061	PD6055
3061	PS1511
3061	PS1512
3061	PS1513
3061	PS1514
3061	PS1515
3061	PS1516
3061	PS1517
3061	PS8913
3061	PTC506
3061	QRT-241
3061	QZ10T10
3061	QZ10T5
3061	RA-26
3061	RD-10EB
3061	RD10EB-2
3061	RD10F
3061	RE115
3061	REN140
3061	RT-241
3061	RVD1N4740
3061	RZ10
3061	RZZ10
3061	S6-10
3061	S6-10A
3061	S6-10B
3061	S6-10C
3061	SV133
3061	SZ10
3061	SZL9
3061	TIXD758
3061	TM140
3061	TMD08
3061	TMD08A
3061	TR228736003026
3061	TR330028
3061	TVSRA-26
3061	TVSRD11
3061	TVSRD11A
3061	UZ8710
3061	UZ8810
3061	VHEWZ-100-1F
3061	VHEWZ-100//1F
3061	WEP101
3061	WM-098
3061	WM-098A
3061	WM-098B
3061	WM-098C
3061	WM-098D
3061	WM-100
3061	WM-100A
3061	WM-100B
3061	WM-100C
3061	WM-100D
3061	WN-098
3061	WN-098A
3061	WN-098B
3061	WN-098C
3061	WN-098D
3061	WN-100
3061	WN-100A
3061	WN-100B
3061	WN-100C
3061	WN-100D
3061	WO-098
3061	WO-098A
3061	WO-098B
3061	WO-098C
3061	WO-098D
3061	WO-100A
3061	WO-100B
3061	WO-100C
3061	WO-100D
3061	WP-098
3061	WP-098A
3061	WP-098B
3061	WP-098C
3061	WP-098D
3061	WP-100
3061	WP-100A
3061	WP-100B
3061	WP-100C
3061	WP-100D
3061	WQ-098
3061	WQ-098A
3061	WQ-098B
3061	WQ-098C
3061	WQ-098D
3061	WQ-100
3061	WQ-100A
3061	WQ-100B
3061	WQ-100C
3061	WQ-100D
3061	WR-098
3061	WR-098A
3061	WR-098B
3061	WR-098C
3061	WR-098D
3061	WR-100
3061	WR-100A
3061	WR-100B
3061	WR-100C
3061	WR-100D
3061	WS-098
3061	WS-098A
3061	WS-098B
3061	WS-098C
3061	WS-098D
3061	WS-100
3061	WS-100A
3061	WS-100B
3061	WS-100C
3061	WS-100D
3061	WT-098
3061	WT-098A
3061	WT-098B
3061	WT-098C
3061	WT-098D
3061	WT-100
3061	WT-100A
3061	WT-100B
3061	WT-100C
3061	WT-100D
3061	WU-098
3061	WU-098A
3061	WU-098B
3061	WU-098C
3061	WU-098D
3061	WU-100
3061	WU-100A
3061	WU-100B
3061	WU-100C
3061	WU-100D
3061	WV-098
3061	WV-098A
3061	WV-098B
3061	WV-098C
3061	WV-098D
3061	WV-100
3061	WV-100A
3061	WV-100B
3061	WV-100C
3061	WV-100D
3061	WW-098
3061	WW-098A
3061	WW-098B
3061	WW-098C
3061	WW-098D
3061	WW-100
3061	WW-100A
3061	WW-100B
3061	WW-100C
3061	WW-100D
3061	WX-098
3061	WX-098A
3061	WX-098B
3061	WX-098C
3061	WX-098D
3061	WX-100
3061	WX-100A
3061	WX-100B
3061	WX-100C
3061	WX-100D
3061	WY-098
3061	WY-098A
3061	WY-098B
3061	WY-098C
3061	WY-098D
3061	WY-100
3061	WY-100A
3061	WY-100B
3061	WY-100C
3061	WY-100D
3061	WZ-100
3061	WZ-100A
3061	WZ096
3061	WZ10
3061	WZ533
3061	XZ-096
3061	XZ-100
3061	XZ098
3061	Z0220
3061	Z0413
3061	Z0B10
3061	Z0C10
3061	Z0D10
3061	Z10
3061	Z10K
3061	Z1110
3061	Z1210
3061	Z1B10
3061	Z1C10
3061	Z1D10
3061	Z5B10
3061	Z5C10
3061	Z5D10
3061	ZB10
3061	ZB10A
3061	ZB10B
3061	ZB10X
3061	ZE10
3061	ZF10
3061	ZG10
3061	ZP10
3061	ZZ10
3061	001-0127-00
3061	001-023035
3061	1-20398
3061	1/2Z33T5
3061	1/4A10
3061	1/4A10A
3061	1/4A10B
3061	1A10M
3061	1A10MA
3061	1E10Z
3061	1N3020B
3061	1N3518
3061	1N4104
3061	1N4295
3061	1N4295A
3061	1N4740
3061	1N4740A
3061	1N5856B
3061	1R10
3061	1R10A
3061	1S195
3061	1S2117
3061	1S2117A
3061	1T10B
3061	1TA10
3061	1TA10A
3061	1Z10
3061	1ZCT10
3061	1ZM10T10
3061	1ZM10T20
3061	1ZM10T5
3061	02Z-10A
3061	05Z-10
3061	8-719-937-10
3061	10EB1
3061	10VJ
3061	13DD02F
3061	019-003411
3061	21M493
3061	23Z6AF
3061	23Z6F
3061	24MW1065
3061	24MW1125
3061	24MW743
3061	32-0049
3061	43Z6
3061	43Z6A
3061	44T-300-93
3061	46-86323-3
3061	48-41873J03
3061	56-67
3061	070-011
3061	070-024
3061	86-118-1
3061	96-5116-01
3061	103-Z9010
3061	145-100
3061	157-100
3061	179-4
3061	187-6
3061	523-2001-100
3061	523-2003-100
3061	523-2005-100
3061	523-2503-100
3061	630-077
3061	903-337
3061	1006-5985
3061	1040-09
3061	1065-6775
3061	2040-17
3061	5632-HZ11A
3061	8000-00028-047
3061	8000-00058-004
3061	26810-52
3061	38510-53
3061	42020-737
3061	45810-54
3061	58810-86
3061	58840-117
3061	93030
3061	119594
3061	171560
3061	190404
3061	228007
3061	262111
3061	262546
3061	263424
3061	300732
3061	489752-040
3061	489752-109
3061	530073-3
3061	530145-100
3061	530157-1100
3061	610122
3061	741867
3061	988050
3061	1223773
3061	1800013
3061	2002209-5
3061	2002209-9
3061	4036392-P2
3061	42020414
3061	42020737
3061	2008220125
3061	16602995856
3062	.25T12A
3062	.25T12B
3062	.75N12
3062	.7JZ12
3062	.7Z12A
3062	.7Z12B
3062	.7Z12C
3062	.7ZM12A
3062	.7ZM12B
3062	.7ZM12C
3062	.7ZM12D
3062	A054-151
3062	0A12612
3062	A2474
3062	A36539
3062	A7287500
3062	A7287513
3062	AU2012
3062	AV2012
3062	AV5
3062	AW-01-12C
3062	AW01-12V
3062	AZ-120
3062	AZ12
3062	0AZ213
3062	BZ-12
3062	BZ120
3062	BZX17
3062	BZX46C12
3062	BZX79C12
3062	BZY18
3062	BZY83C12
3062	BZY85C12
3062	BZY94/C12
3062	C4018
3062	CD31-00015
3062	CD31-12022
3062	CD4116
3062	CD4117
3062	CD4118
3062	CD4121
3062	CD4122
3062	CR956
3062	CX-0052
3062	CXL142
3062	CXL5021
3062	CZD012-5
3062	D-00484R
3062	D2G-3
3062	D2G-4
3062	DS-108
3062	DZ12A
3062	E0771-7
3062	E1852
3062	E262
3062	EA1318
3062	EA16X162
3062	EA16X29
3062	EA3719
3062	ECG142
3062	ECG142A
3062	EP16X36
3062	EQA01-12
3062	EQA01-12R
3062	EQA01-12S
3062	EQB01
3062	EQB01-12
3062	EQB01-12A
3062	EQB01-12B
3062	EQB01-12BV
3062	EQB01-12R
3062	EQB01-12Z
3062	EQG01-12A
3062	ES10234
3062	ET16X15
3062	EVR12
3062	EVR12A
3062	EVR12B
3062	FZ12A
3062	FZ12T10
3062	FZ12T5
3062	G01-037-A
3062	G12T10
3062	G12T20
3062	G12T5
3062	GEZD-12
3062	H089
3062	HD3000101-0
3062	HEP-Z0415
3062	HS2120
3062	HS7120
3062	HW12A
3062	HW12B
3062	HZ-212
3062	KS120A
3062	KS120B
3062	KS2120A
3062	KS2120B
3062	KS44A
3062	KS44AF
3062	KS44B
3062	KS44BF
3062	LPM12
3062	LPM12A
3062	LPZT12
3062	LR120CH
3062	M4Z12
3062	M4Z12-20
3062	M4Z12A
3062	MC6016
3062	MC6016A
3062	MC6116
3062	MC6116A
3062	MD759
3062	MD759A
3062	MEZ12T10
3062	MEZ12T5
3062	MTZ616
3062	MTZ616A
3062	MZ-11
3062	MZ-212
3062	MZ1000-15
3062	MZ1012
3062	MZ12
3062	MZ12A
3062	MZ12B
3062	MZ212
3062	MZ500.18
3062	MZ92-12A
3062	N-EA16X29
3062	NGP5010
3062	0A126-12
3062	0A126/12
3062	0AZ213
3062	0AZ273
3062	PD6016
3062	PD6016A
3062	PD6057
3062	PR617
3062	PS10022B
3062	PS10066
3062	PS8915
3062	PTC507
3062	QRT-243
3062	QZ12T10
3062	QZ12T5
3062	R-7103
3062	R2D
3062	RD12EC
3062	RD12FB
3062	RE118
3062	REN142
3062	RH-EX0038CEZZ
3062	RH-EX0047CEZZ
3062	RS1348
3062	RT-243
3062	RZ12
3062	RZZ12
3062	S1R12B
3062	S1R13B
3062	SD53
3062	SFZ716
3062	SI-RECT-228
3062	SI-RECT-230
3062	SV1017
3062	SV135
3062	SV4012
3062	SV4012A
3062	SW01
3062	SZ-EQB01-12A
3062	SZ-RD12FB
3062	SZ12
3062	SZ12.0
3062	SZ671-B
3062	SZ671-G
3062	T-E1068
3062	T-E1077
3062	T-E1140
3062	TM142
3062	TR14002-6
3062	TS-337
3062	TVS-RD13D
3062	TVS-RD13M
3062	TVS-RD13P
3062	TVS1N741H
3062	TVSEQA01-125
3062	TVSEQB01-12
3062	TVSQA01-12S
3062	TVSRD12EBH
3062	TZ12
3062	TZ12A
3062	TZ12B
3062	TZ12C
3062	UZ8712
3062	UZ8812
3062	V160
3062	VR12
3062	VR12A
3062	VR12B
3062	WEP105
3062	WEP1112
3062	WM-120
3062	WM-120A
3062	WM-120B
3062	WM-120C
3062	WM-120D
3062	WN-120
3062	WN-120A
3062	WN-120B
3062	WN-120C
3062	WN-120D
3062	WO-120
3062	WO-120A
3062	WO-120B
3062	WO-120C
3062	WO-120D
3062	WP-120
3062	WP-120A
3062	WP-120B
3062	WP-120C
3062	WP-120D
3062	WQ-120
3062	WQ-120A
3062	WQ-120B
3062	WQ-120C
3062	WQ-120D
3062	WR-120
3062	WR-120A
3062	WR-120B
3062	WR-120C
3062	WR-120D
3062	WS-120
3062	WS-120A
3062	WS-120B
3062	WS-120C
3062	WS-120D
3062	WT-120
3062	WT-120A
3062	WT-120B
3062	WT-120C
3062	WT-120D
3062	WU-120
3062	WU-120A
3062	WU-120C
3062	WU-120D
3062	WV-120
3062	WV-120A
3062	WV-120B
3062	WV-120C
3062	WV-120D
3062	WW-120
3062	WW-120A
3062	WW-120B
3062	WW-120C
3062	WW-120D
3062	WX-120
3062	WX-120A
3062	WX-120B
3062	WX-120C
3062	WX-120D
3062	WY-120
3062	WY-120A
3062	WY-120B
3062	WY-120C
3062	WY-120D
3062	WZ12
3062	WZ535
3062	WZ917
3062	X330302
3062	XZ-122
3062	Z-1112C
3062	Z0-12
3062	Z0-12A
3062	Z0-12B
3062	Z0222
3062	Z0415
3062	Z0B-12
3062	Z0B12
3062	Z0C12
3062	Z0D12
3062	Z1112
3062	Z1112C
3062	Z1212
3062	Z12K
3062	0Z12T10
3062	0Z12T5
3062	Z1B12
3062	Z1C12
3062	Z1D12
3062	Z2A120F
3062	Z4B12
3062	Z4X12
3062	Z4XL12
3062	Z4XL12B
3062	Z5B12
3062	Z5B2
3062	Z5C12
3062	Z5D12
3062	ZA12
3062	ZA12B
3062	ZB1-12
3062	ZB12
3062	ZB12A
3062	ZB12B
3062	ZC-140
3062	ZC012
3062	ZD012
3062	ZD12A
3062	ZD12B
3062	ZE12
3062	ZE12A
3062	ZE12B
3062	ZEC12
3062	ZENER-122
3062	ZF12
3062	ZF12A
3062	ZF12B
3062	ZG12
3062	ZH12
3062	ZH12A
3062	ZH12B
3062	ZJ12
3062	ZJ12A
3062	ZJ12B
3062	Z0-12
3062	Z0-12A
3062	Z0-12B
3062	Z0B12
3062	ZP12
3062	ZP12A
3062	ZP12B
3062	ZQ12
3062	ZQ12A

SK	Industry Standard No.	SK	Industry Standard No.	SK	Industry Standard No.	SK	Industry Standard No.	SK	Industry Standard No.
3062	ZG12B	3062	41-0905	3063	C2D015	3063	WEP607	3063	ZF15B
3062	ZR50B793-1	3062	42-27541	3063	C2D015-5	3063	WL-150	3063	ZG-15
3062	Z812	3062	46-86379-3	3063	D1ZBLU	3063	WL-150A	3063	ZG-15B
3062	Z812A	3062	48-137177	3063	D1ZYEL	3063	WL-150B	3063	ZG15
3062	Z812B	3062	48-137272	3063	D5B	3063	WL-150C	3063	ZG15A
3062	ZZ12	3062	488137000	3063	D215A	3063	WL-150D	3063	ZG15B
3062	ZZ12A	3062	488155105	3063	BA16X124	3063	WM-150	3063	ZH-15
3062	ZZ12B	3062	48X90233A07	3063	ECG145	3063	WM-150A	3063	ZH-15A
3062	ZU12	3062	48X97048A08	3063	ECG145A	3063	WM-150C	3063	ZH-15B
3062	ZU12A	3062	50Z12	3063	EP16X5	3063	WM-150D	3063	ZH15
3062	ZU12B	3062	50Z12A	3063	EQA01-15R	3063	WN-150	3063	ZH15A
3062	ZV12	3062	50Z12B	3063	EQB01-15	3063	WN-150A	3063	ZH15B
3062	ZV12A	3062	50Z12C	3063	EQB01-15Z	3063	WN-150C	3063	ZJ-15
3062	ZV12B	3062	56-4837	3063	EVR15	3063	WN-150D	3063	ZJ-15A
3062	ZT12A	3062	56-51	3063	EVR15A	3063	WO-150	3063	ZJ-15B
3062	ZT12B	3062	56-57	3063	EVR15B	3063	WO-150A	3063	ZJ15
3062	ZE12	3062	63-9942	3063	EZ150	3063	WO-150C	3063	ZJ15A
3062	1/4A12	3062	65-085004	3063	FA8009	3063	WP-150	3063	ZJ15B
3062	1/4A12A	3062	75N12	3063	FA8010	3063	WP-150A	3063	Z0-15
3062	1A12C	3062	75Z12	3063	FA8011	3063	WP-150D	3063	Z0-15A
3062	1A12MA	3062	75Z12A	3063	FA8012	3063	WQ-150	3063	Z0-15AC
3062	1AC12	3062	75Z12B	3063	FZ15A	3063	WQ-150A	3063	Z0-15B
3062	1AC12A	3062	75Z12C	3063	FZ15T10	3063	WQ-150C	3063	Z0-15BY
3062	1AC12B	3062	86-0521	3063	FZ15T5	3063	WQ-150D	3063	Z0B-15
3062	1C12Z	3062	86-31-1	3063	G15T10	3063	WR-150	3063	Z0B15
3062	1C12ZA	3062	86-85-1	3063	G15T20	3063	WR-150A	3063	Z0D-15
3062	1DZ12	3062	93A39-12	3063	GEZD-15	3063	WR-150C	3063	Z0D15
3062	1B12Z	3062	93A39-43	3063	GZ15T5	3063	WS-150	3063	ZP15
3062	1B12Z10	3062	93B39-12	3063	HEP-Z0418	3063	WS-150A	3063	ZP15A
3062	1B12Z5	3062	93B39-13	3063	H82150	3063	WS-150D	3063	ZP15B
3062	1BZ12	3062	93G39-12	3063	H37150	3063	WT-150	3063	ZQ-15
3062	1M12Z	3062	93G39-13	3063	HW15	3063	WT-150A	3063	ZQ-15A
3062	1M12Z10	3062	93G39-2	3063	HW15A	3063	WT-150C	3063	ZQ-15B
3062	1M12ZB10	3062	93G39-6	3063	HW15B	3063	WT-150D	3063	ZQ15
3062	1M12ZB5	3062	96-5091-01	3063	K8150A	3063	WU-150	3063	ZQ15A
3062	1N426	3062	96-5133-01-02	3063	K8150B	3063	WU-150A	3063	ZQ15B
3062	1N513	3062	103-158	3063	K8Z150A	3063	WU-150C	3063	ZQ15X
3062	1N513A	3062	103-279-01	3063	K8Z150B	3063	WU-150D	3063	Z8-15A
3062	1N524	3062	103-Z9003	3063	K846	3063	WW-150	3063	Z8-15B
3062	1N524A	3062	160	3063	K846AF	3063	WW-150A	3063	Z815
3062	1N773	3062	264P03303	3063	K846BF	3063	WW-150C	3063	Z815A
3062	1N773A	3062	264P09301	3063	LPM15	3063	WW-150D	3063	Z815B
3062	1N1877	3062	296L003B01	3063	LPM15A	3063	WX-150	3063	ZT-15
3062	1N1877A	3062	296V017H01	3063	LPZ215	3063	WX-150A	3063	ZT-15A
3062	1N1960A	3062	296V018B01	3063	LR150CH	3063	WX-150C	3063	ZT-15B
3062	1N960B	3062	800-004-00	3063	M15Z	3063	WY-150	3063	ZT15
3062	1N2037-1	3062	800-023-00	3063	M4Z15	3063	WY-150A	3063	ZT15A
3062	1N3022	3062	800-035-00	3063	M4Z15-20	3063	WY-150C	3063	ZT15B
3062	1N3022A	3062	903-97B	3063	M4Z15A	3063	WZ-150	3063	ZU-15
3062	1N3022B	3062	919-00-1445-002	3063	MC6018	3063	WZ-920	3063	ZU-15A
3062	1N3520	3062	919-001445-2	3063	MC6018A	3063	WZ15	3063	ZU-15B
3062	1N3537	3062	1002-4404	3063	MC6118	3063	WZ150	3063	ZU15
3062	1N3681	3062	2093A39-12	3063	MC6118A	3063	WZ538	3063	ZU15A
3062	1N3681A	3062	2093A80-1	3063	MEZ15T10	3063	WZ920	3063	ZU15B
3062	1N3681B	3062	004887	3063	MEZ15T5	3063	XB152	3063	ZV-15
3062	1N4106	3062	5122	3063	MN1500-H	3063	XZ-152	3063	ZV-15A
3062	1N4106A	3062	5222	3063	MTZ618	3063	Z-15K	3063	ZV-15B
3062	1N4164	3062	36539	3063	MTZ618A	3063	Z0225	3063	ZV15
3062	1N4164A	3062	42025-850	3063	MZ1000-17	3063	Z0418	3063	ZV15A
3062	1N4164B	3062	71411-2	3063	MZ15A	3063	Z0B-15	3063	ZV15B
3062	1N4329	3062	120504	3063	MZ15T20	3063	Z0C-15	3063	ZY-15
3062	1N4329A	3062	125126	3063	MZ92-15A	3063	Z0015	3063	ZY-15A
3062	1N4329B	3062	130328	3063	0Z15T10	3063	ZD015	3063	ZY15
3062	1N4406	3062	130380	3063	0Z15T5	3063	Z1114	3063	ZY15A
3062	1N4406A	3062	165358	3063	PD-6018	3063	Z1214	3063	ZY15B
3062	1N4406B	3062	224780	3063	PD-6018A	3063	Z15	3063	ZZ-15
3062	1N4634	3062	31720B	3063	PD-6059	3063	Z15K	3063	ZZ15
3062	1N4663	3062	0330302	3063	PD6018	3063	Z1B-15	3063	Z215
3062	1N4742	3062	530073-1013	3063	PD6018A	3063	Z1B15	3063	001-02303-4
3062	1N4742A	3062	530073-13	3063	PD6059	3063	Z10-15	3063	001-023034
3062	1N4896	3062	530073-14	3063	PR-620	3063	Z1015	3063	1/4A15
3062	1N5858A	3062	530073-6	3063	PR620	3063	Z1D15	3063	1/4A15A
3062	1N665	3062	530118-2	3063	P8-1006B	3063	Z2A-150F	3063	1/4A15B
3062	1N765	3062	530145-120	3063	P8-8917	3063	Z2A150F	3063	1/4M15Z
3062	1R12	3062	530163-120	3063	P810024B	3063	Z4B-15	3063	1/4M15Z10
3062	1R12A	3062	760304	3063	P810068	3063	Z4B15	3063	1/4M15Z5
3062	1R12B	3062	801731	3063	P88917	3063	Z4X-15	3063	1/4Z15D10
3062	1S2119	3062	817155	3063	PT0509	3063	Z4X15	3063	1/4Z15D5
3062	1S227	3062	817208	3063	QRT-245	3063	Z5B-15	3063	1/4Z15T5
3062	1S337	3062	922693	3063	QZ-15T10	3063	Z5C-15	3063	1A15M
3062	1S337-Y	3062	1474777-1	3063	QZ-15T5	3063	Z5D-15	3063	1A15MA
3062	1S337A	3062	2002209-10	3063	QZ15T10	3063	Z5D15	3063	1AC15
3062	1S696	3062	2330241	3063	QZ15T5	3063	ZA15	3063	1AC15A
3062	1T12	3062	2337101	3063	R2E	3063	ZA15A	3063	1AC15B
3062	1T12A	3062	23115919	3063	RD15EB	3063	ZA15B	3063	1C15Z
3062	1T12B	3062	30600140	3063	RD15F	3063	ZB-15A	3063	1C15ZA
3062	1Z12	3062	2012230109	3063	RD15FA	3063	ZB1-15	3063	1D08
3062	1Z12A	3062	2012230312	3063	RD15FB	3063	ZB15	3063	1E15Z
3062	1Z12B	3062	4120501200	3063	RD15B	3063	ZB15A	3063	1E15Z10
3062	1Z12Z	3062	4129501200	3063	RD16H	3063	ZB15B	3063	1E15Z5
3062	1Z12Z10	3063	.25Z15	3063	RE121	3063	ZC-015	3063	1EZ15
3062	1Z12Z20	3063	.25Z15A	3063	REN145	3063	ZC015	3063	1M15Z
3062	1Z12T5	3063	.25Z15B	3063	RT-245	3063	ZD-15	3063	1M15Z10
3062	1ZB12	3063	.4Z15	3063	RT3671	3063	ZD-15B	3063	1M15Z5
3062	1ZB12B	3063	.75N5	3063	RT3671A	3063	ZD15	3063	1N1427
3062	1ZE12	3063	.7Z15	3063	RZ-15AB	3063	ZD15A	3063	1N1514
3062	1ZF12Z10	3063	.7Z14A	3063	RZ15	3063	ZD15B	3063	1N1514A
3062	1ZF12Z20	3063	.7Z14C	3063	RZ15A	3063	ZE-15A	3063	1N1525
3062	1ZF12T5	3063	.7Z15D	3063	SU5	3063	ZE-15B	3063	1N1525A
3062	1ZM12Z10	3063	.7ZM15A	3063	SV-1020A	3063	ZE15	3063	1N1775
3062	1ZM12Z20	3063	.7ZM15B	3063	SV-138A	3063	ZE15A	3063	1N1775A
3062	1ZM12T5	3063	.7ZM15C	3063	SV-4015A	3063	ZE15B	3063	1N1878
3062	1ZT12	3063	.7ZM15D	3063	SV1020	3063	ZEN508	3063	1N1878A
3062	1ZT12B	3063	AV-2015	3063	SV138	3063	ZF-15A	3063	1N2038-1
3062	02Z12A	3063	AV2015	3063	SV4015	3063	ZF-15B	3063	1N3024
3062	003-009200	3063	AW01-15	3063	SV4015A	3063	ZF15	3063	1N3024A
3062	003-009700	3063	AZ-15	3063	SZ-150	3063	ZF15A	3063	1N3024B
3062	003-010000	3063	AZ-150	3063	SZ-15B			3063	1N3522
3062	03-933943	3063	AZ15	3063	SZ-200-15			3063	1N3683
3062	3/4Z12D10	3063	B-6002	3063	SZ-200-15A			3063	1N3683A
3062	3/4Z12D5	3063	BZ-150	3063	SZ15			3063	1N3683B
3062	4-2020-12400	3063	BZ-X19	3063	SZ15.0			3063	1N4109
3062	4W01-13	3063	BZ150	3063	SZ150			3063	1N4166
3062	05-933943	3063	BZX-46C15	3063	TM145			3063	1N4166A
3062	8-905-421-234	3063	BZX-79C15	3063	TR14002-12			3063	1N4166B
3062	8-905-421-319	3063	BZX19	3063	TVS-ZB1-15			3063	1N4331
3062	09-306179	3063	BZX46C15	3063	TVSBA01-07R			3063	1N4331A
3062	09-306391	3063	BZY-19	3063	TVSBB01-15			3063	1N4331B
3062	1205Y	3063	BZY-83C15	3063	TVSQA01-07R			3063	1N4408
3062	13-14879-2	3063	BZY19	3063	TVSQB01-15ZB			3063	1N4408A
3062	13-5532-1	3063	BZY83/C15	3063	TZ-15A			3063	1N4408B
3062	14-509-01	3063	BZY83/D15	3063	TZ-15AB			3063	1N4636
3062	14-515-01	3063	BZY85C15	3063	TZ-15BC			3063	1N4665
3062	14-515-03	3063	BZY85/D15	3063	TZ-15C			3063	1N4702
3062	16Z4	3063	BZY85C015	3063	TZ15			3063	1N4744
3062	18-085002	3063	BZY92/C15	3063	TZ15A			3063	1N4744A
3062	21A037-008	3063	BZY92C015	3063	TZ15B			3063	1N666
3062	25Z6	3063	BZY94/C15	3063	TZ15C			3063	1R15
3062	25Z6A	3063	C4020	3063	UZ-15C			3063	1R15A
3062	34-8057-14	3063	CD31-00017	3063	UZ15			3063	1R15B
3062	34-8057-33	3063	CD31-10365	3063	UZB715			3063	1S137
3062	34-8057-53	3063	CD31-12024	3063	UZ8815			3063	1S2121
3062	36-6343	3063	CXL145	3063	V-1266			3063	1S2121A
				3063	V126			3063	1S230
								3063	1S597

SK	Industry Standard No.
3063	1S697
3063	1T15
3063	1T15B
3063	1T19-15B
3063	1TA15
3063	1TA15A
3063	1Z15
3063	1Z15A
3063	1Z15B
3063	1Z15C
3063	1Z15D
3063	1Z15D10
3063	1Z15D5
3063	1Z15T10
3063	1Z15T20
3063	1ZB15
3063	1ZB15B
3063	1ZC15
3063	1ZC15T10
3063	1ZC15T5
3063	1ZF15T10
3063	1ZF15T20
3063	1ZF15T5
3063	1ZM15T10
3063	1ZM15T20
3063	1ZM15T5
3063	1ZT15
3063	1ZT15B
3063	02Z15A
3063	003-009100
3063	3L4-3506-29
3063	8-905-421-128
3063	8-905-421-239
3063	8-905-421-715
3063	09-306367
3063	13-14879-1
3063	13-14879-3
3063	13-14879-5
3063	13-33179-2
3063	14-515-07
3063	14-515-73
3063	25N15
3063	28Z6
3063	28Z6A
3063	34-8057-12
3063	48S137330
3063	48S155104
3063	50Z15
3063	50Z15A
3063	50Z15B
3063	50Z15C
3063	56-25
3063	62-20643
3063	66X0040-009
3063	070-021
3063	73-15
3063	75N5
3063	75Z15
3063	75Z15A
3063	75Z15B
3063	75Z15C
3063	93A39-24
3063	96-5110-02
3063	96-5110-03
3063	115Z4
3063	137-759
3063	200X8220-531
3063	215Z4
3063	264P04003
3063	523-2003-150
3063	523-2503-150
3063	642-255
3063	690V081H92
3063	1006-1737
3063	1015
3063	2396-17
3063	5124
3063	5224
3063	126852
3063	127382
3063	129946
3063	134444
3063	140973
3063	233150
3063	530073-1017
3063	530073-12
3063	530073-17
3063	530073-9
3063	1473777-2
3063	1474717-2
3063	1474777-002
3063	1474777-2
3063	1477046-1
3063	1477081-501
3063	5300073-15
3063	62166916
3063	62746483
3063	4202012700
3063	4202970330
3064	.25T27
3064	.25T27A
3064	.25T27B
3064	.4T27
3064	.4T27A
3064	.4T27B
3064	.75N27
3064	.7JZ27
3064	.7Z27A
3064	.7Z27C
3064	.7Z27D
3064	.7ZM27A
3064	.7ZM27B
3064	.7ZM27C
3064	.7ZM27D
3064	AB2027
3064	AV2027
3064	AV2027A
3064	AW01-27
3064	AZ27
3064	AZ27A
3064	AZ971B
3064	BZX25
3064	BZX25A
3064	BZX79C27
3064	BZX79C27A
3064	BZY85C27
3064	BZY92C27
3064	C4026
3064	CD31-00023
3064	CD31-10361
3064	CD31-12030
3064	D1W
3064	D2Z-2
3064	DZ27A
3064	EVR27
3064	EVR27A
3064	EVR27B
3064	FZ27A
3064	FZ27T10
3064	FZ27T5
3064	G27T10
3064	G27T20
3064	G27T5
3064	GEZD-27
3064	GZ27
3064	HEP-Z0424
3064	HM27B
3064	HS2270
3064	HS7270
3064	HW27
3064	HW27A
3064	HW27B
3064	LPM27
3064	LPM27A
3064	LPZT27
3064	M27Z
3064	M4Z27
3064	M4Z27-20
3064	M4Z27A
3064	MC6024
3064	MC6024A
3064	MC6124
3064	MC6124A
3064	MEZ27T10
3064	MEZ27T5
3064	MZ27A
3064	MZ92-27A
3064	QZ27T10
3064	QZ27T10A
3064	QZ27T5
3064	QZ27T5A
3064	RZ27
3064	RZ27A
3064	RZ27AC
3064	SV4027
3064	SV4027-6
3064	SV4027A
3064	SV4027A-6
3064	TC27A5A
3064	TC27A5A-5
3064	TVSZB1-27
3064	TVSZB1-29
3064	TX27A
3064	ZA27B
3064	1A27M
3064	1A27MA
3064	1AC27
3064	1AC27A
3064	1AC27B
3064	1C27Z
3064	1C27ZA
3064	1D27Z5
3064	1E27Z
3064	1E27Z10
3064	1EZ27
3064	1M27Z
3064	1M27Z10
3064	1M27Z5
3064	1M27ZS10
3064	1M27ZS5
3064	1N1430
3064	1N1517
3064	1N1517A
3064	1N1528
3064	1N1528A
3064	1N1781
3064	1N1781A
3064	1N1881
3064	1N1881A
3064	1N3030
3064	1N3030A
3064	1N3030B
3064	1N3528
3064	1N3689
3064	1N3689A
3064	1N3689B
3064	1N4172
3064	1N4172A
3064	1N4172B
3064	1N4337
3064	1N4337A
3064	1N4337B
3064	1N4414
3064	1N4414A
3064	1N4414B
3064	1N4642
3064	1N4671
3064	1N4711
3064	1N4750
3064	1N4750A
3064	1N669
3064	1R27
3064	1R27A
3064	1R27B
3064	1S239
3064	1S700
3064	1T27
3064	1T27B
3064	1TA27
3064	1TA27A
3064	1Z27
3064	1Z27A
3064	1Z27B
3064	1Z27C
3064	1Z27D
3064	1Z27D10
3064	1Z27D5
3064	1Z27T10
3064	1Z27T20
3064	1Z27T5
3064	1ZB27
3064	1ZB27B
3064	1ZC27
3064	1ZC27T10
3064	1ZC27T5
3064	1ZF27T10
3064	1ZF27T20
3064	1ZF27T5
3064	1ZT27
3064	1ZT27B
3064	25N27
3064	48-137034
3064	50Z27
3064	50Z27A
3064	50Z27B
3064	50Z27C
3064	56-47
3064	75Z27
3064	75Z27A
3064	75Z27B
3064	75Z27C
3064	1027
3064	1127
3064	5130
3064	5230
3064	99202-226
3064	138174
3064	267272
3064	530073-18
3064	7570021-12
3065	CXL222
3065	DS-102
3065	DS-106
3065	ECG222
3065	EP15X64
3065	GE-FET-4
3065	IP20-0218
3065	KD2130
3065	MEM630
3065	MFE130
3065	MFE3006
3065	MPF121
3065	RE199
3065	REN222
3065	SFD2285
3065	SFE303424
3065	T1731
3065	TA2644
3065	TA7274
3065	TA7374
3065	TA7669
3065	TA8242
3065	TM222
3065	01-080045
3065	03A10
3065	3N187
3065	3SK30
3065	3SK30A
3065	3SK30B
3065	3SK30C
3065	3SK39
3065	3SK39Q
3065	3SK39R
3065	3SK49Q
3065	13-10321-37
3065	033-014-0
3065	48-3003A10
3065	121-1030
3065	121-826
3065	417-240
3065	576-0003-224
3065	576-0006-227
3065	3588-2
3065	6227
3065	40823
3065	141332
3065	258017
3065	610358-1
3065	610358-2
3065	610358-3
3065	760005
3065	1417372-1
3065/222	3N140
3065/222	3N141
3065/222	3SK45
3065/222	3SK45-B-09
3065/222	3SK45B
3065/222	3SK45B09
3065/222	7-117-02
3065/222	86-513-2
3065/222	40819
3065/222	40820
3065/222	40821
3065/222	40822
3065/222	40841
3065/222	2068491-704
3066	A-95-5296
3066	A04735-A
3066	A95-5296
3066	B66X0035-001
3066	ET57X32
3066	ET58X32
3066	EU57X32
3066	FR-1033
3066	GECR-1
3066	HS-7/1
3066	HS-8/1
3066	HS1/7
3066	HS1/9
3066	HS9/1
3066	JT-E1095
3066	K122J176-2
3066	PTC208
3066	R1
3066	RE167
3066	RF32103-1
3066	RF32103R
3066	S913
3066	S926
3066	SR-32
3066	SR9000
3066	T-E1095
3066	1N4103A
3066	1T9.1A
3066	3MA
3066	004-003200
3066	004-03200
3066	5-30132.1
3066	6RS18PH110BEB1
3066	6RS18PH110BHB1
3066	6RS18PH110BMB1
3066	6RS7PH130BCB1
3066	6RS7PH30BCB1
3066	13-16106-1
3066	15-108012
3066	15-108017
3066	19AR30
3066	21A3
3066	23B210679-1
3066	28-23-01
3066	34-8053-2
3066	34-8053-3
3066	34-8053-4
3066	46-86180-3
3066	48-66653A003
3066	48-66653A005
3066	48-66653APT015
3066	48-66653APT1003
3066	53B011-3
3066	61-8969
3066	66X0035-001
3066	86-44-3
3066	93A57-1
3066	93B57-1
3066	110BH1
3066	212-85
3066	212-85B
3066	295V003H01
3066	295V003M01
3066	295V033B01
3066	800-012-00
3066	2093A57-1
3066	23115-022
3066	72110
3066	126860
3066	530096
3066	530132
3066	530132-1
3066	605131
3066	700021-00
3066	817123
3066	983995
3066	1440977-1
3066	62605618
3066/118	BZX14
3066/118	D1G
3066/118	D2Z
3066/118	D3F
3066/118	EVR9A
3066/118	EVR9B
3066/118	HS10/1
3066/118	M4653
3066/118	NGP5007
3066/118	PS10019B
3066/118	PS10063
3066/118	TZ9.1C
3066/118	UZ9.1
3066/118	Z2A91F
3066/118	Z4XL9.1B
3066/118	ZH9.1A
3066/118	ZQ9.1B
3066/118	ZS9.1A
3066/118	ZS9.1B
3066/118	ZT9.1B
3066/118	1EZ9.1
3066/118	1N3019B
3066/118	1N4161
3066/118	1N4403
3066/118	1N4403A
3066/118	1N4403B
3066/118	1R9.1B
3066/118	1Z9.1C
3066/118	1Z9.1T10
3066/118	1Z9.1T20
3066/118	1ZF9.1T5
3066/118	48-137130
3066/118	48-137164
3066/118	50Z9.1
3066/118	50Z9.1A
3066/118	50Z9.1B
3066/118	75Z9.1A
3066/118	75Z9.1C
3066/118	86-35-1
3066/118	2309
3066/118	2795A
3066/118	113397
3066/118	126861
3066/118	530096-1
3066/118	982254
3066/118	1440977
3066/118	1956197
3066/118	1965019
3067	D2M
3067	D4G
3067	DIJ71961
3067	GECR-4
3067	HS1011
3067	HS20
3067	HVT-22DA
3067	MF12/16
3067	PTC210
3067	R1A
3067	SELEN-58
3067	TV13-11K60
3067	TVM567
3067	4-686227-3
3067	13S
3067	15-108045
3067	34-8061-1
3067	38-11016
3067	46-86227-3
3067	46-86279-3
3067	46-86291-3
3067	48-137081
3067	48-137801
3067	48S155002
3067	53A022-10
3067	212-66
3067	212-67
3067	264P08901
3067	4686-227-3
3067	1445740-502
3067	2330921
3067	6611003000
3067/502	HS-20
3067/502	HS15/1
3067/502	HS20/1
3067/502	HS20/1C
3067/502	4-2021-07570
3067/502	2330032
3068	A04299-202
3068	D2N
3068	EDMF-15B
3068	ES57X13
3068	ES57X21
3068	GECR-5
3068	H598
3068	H598-10
3068	HS15/16
3068	HS15/1C
3068	MF20/1B
3068	NTC-16
3068	PTC211
3068	RE200
3068	RF03C06
3068	RF10K35
3068	SELEN-52
3068	TV13-12K60
3068	TVM531
3068	TVM569
3068	1-531-028
3068	5-30119.1
3068	05-370151
3068	13-43382-1
3068	13-55010-1
3068	15-123104
3068	15-123231
3068	46-86313-3
3068	46-86520-3
3068	48-137082
3068	48S137082
3068	48X90234A05
3068	103-239
3068	103-239-01
3068	264P08902
3068	132966
3068	140995
3068	489752-054
3068	530119-5
3068	1445470-501
3068	1445470-503
3068	4202107370
3068/503	HS20-1
3068/503	HS20-1A
3068/503	HS20/1B
3068/503	HS6/1
3068/503	TVM533
3068/503	09-306310
3068/503	129203
3068/503	1445470-502
3070	CA3044
3070	CA3044V1
3070	CA344V1
3070	GEIC-207
3070	IC-6
3070	TA7050M
3070	TVSAN220
3070	13-27-6
3070	21A101-002
3070	21A101-2
3070	46-1340-3
3070	46-1346-3
3070	46-5002-6
3070	417-202
3070	24451
3070	126604
3070	126609
3070	1462445-1
3070	23119003
3070	23119005
3070/711	3677-3
3071	A3148
3071	CA3048
3071	CA3052
3071	CA3052E
3071	E2495
3071	GEIC-210
3071	LA3148
3071	M8D03
3071	15-34005-1
3071	15-34202-1
3071	51-04488D03
3071	51R04488D03
3071	34005-1
3071	34202-1
3071	81336-1
3071	81336-2
3071	813362
3072	AN221
3072	AN240
3072	AN240D
3072	AN240P
3072	AN240PN
3072	C6063P
3072	C6083P
3072	CA3065
3072	CA3065/7P
3072	CA3065E
3072	CA3065FC
3072	CA3065RCA
3072	EP84X2
3072	EP84X6
3072	GEIC-147
3072	GEIC-148
3072	GEIC-2
3072	HA11107
3072	HA1124
3072	HA1124D
3072	HA1125
3072	HA1125A
3072	HA1128
3072	HA1141
3072	HA1154
3072	HEP-C6063P
3072	HEP-C6083P
3072	IC507
3072	LA1363
3072	LA1365
3072	LM3065
3072	LM3065N
3072	M5143P
3072	MA3065
3072	MC1358
3072	MC1358P
3072	MC1358PQ
3072	NS-3065
3072	NTC-21
3072	PTC726
3072	R2516-1
3072	R25161
3072	RE305-IC
3072	REN712
3072	RH-1X0043CEZZ
3072	RH-IX0043CEZZ
3072	SC9436P
3072	SI-MC1358P
3072	SL21654
3072	SN76664N
3072	SN76665
3072	SN76665N
3072	SN76666
3072	SN76666N
3072	TA5814
3072	TA7071P
3072	TA7176AP
3072	TA7176PFA-1
3072	TM712
3072	TVCM-11
3072	TVSAN241
3072	TVSAN241D
3072	TVSSN76665
3072	ULN2165A
3072	ULN2165N
3072	01-121365
3072	02-091128
3072	02-121365
3072	02-343065
3072	02-403065
3072	4-2060-04000
3072	4-2060-04600

SK	Industry Standard No.
3072	05B1Z
3072	05B2Z
3072	09-308100
3072	13-29-6
3072	13-35059-1
3072	13-64-6
3072	15-33201
3072	15-33201-1
3072	15-33201-2
3072	15-35059-1
3072	15-35059-2
3072	15-45300-1
3072	21A120-008
3072	46-13145-3
3072	46-1361-3
3072	46-5002-15
3072	51-13753A11
3072	51-90305A61
3072	51S13753A09
3072	51S13753A11
3072	56A3-1
3072	56C3-1
3072	56D3-1
3072	61A030-6
3072	66F077-1
3072	73C180475
3072	73C180475-1
3072	73C180475-4
3072	73C180475-7
3072	73C180475-8
3072	73C180475-9
3072	73C180475004
3072	73C182186-1
3072	73C182186-2
3072	73C182186-3
3072	73C182186-4
3072	082
3072	082-1
3072	86X0053-001
3072	86X53-1
3072	99S082-1
3072	200X2110-269
3072	221-48
3072	266P30102
3072	266P30109
3072	1010-9940
3072	1165
3072	2005-1
3072	2005-2
3072	2516
3072	2516-1
3072	3686
3072	3686-1
3072	7221A
3072	33201-1
3072	33201-2
3072	35059-1
3072	80710
3072	130122
3072	130751
3072	144026
3072	612005-1
3072	612005-2
3072	1462516
3072	1462516-001
3072	1462516-1
3072	1464686-1
3072	2360042
3072	2360092
3072	2360201
3072	5351351
3072	37001009
3072	37001011
3072	62674954
3072	62734469
3072	4206004000
3072	4206004600
3072	4206104970
3072	4206105470
3072	4206105770
3072	4206105870
3072	6644000100
3072/712	AN241
3072/712	AN241D
3072/712	AN241P
3072/712	AN241PD
3072/712	B0316403
3072/712	EA16X35
3072/712	MPC32C
3072/712	TA7176P
3072/712	02-165143
3072/712	8-759-101-60
3072/712	09-308033
3072/712	51-13753A04
3072/712	51M33801A01
3072/712	56D8-1
3072/712	23119978
3073	CA3066
3073	CA3066E
3073	EP84X7
3073	GEIC-22
3073	SN76266
3073	73C182077
3073	86X0055-001
3073	86X55-1
3073	132314
3073	1462559-1
3073/728	5158-1
3074	CA3067
3074	CA3067E
3074	GEIC-23
3074	LM3067N
3074	SN76267
3074	ULN2267A
3074	86X0056-001
3074	86X56-1
3074	86X56-2
3074	274
3074	2560
3074	132315
3074	1462560
3074	1462560-001
3074	1462560-1
3075	AMD780
3075	C6070P
3075	CA3070
3075	CA3070G
3075	GEIC-4
3075	HEP-C6070P
3075	IC509
3075	LM3070N
3075	MC1370P
3075	MC1370PQ
3075	N5070
3075	PTC715
3075	RE307-IC
3075	REN714
3075	SL21122
3075	SN76242
3075	SN76242N
3075	TA5649
3075	TVCM-8
3075	ULN2124A
3075	02-561410
3075	09-308079
3075	13-42-6
3075	15-37702-1
3075	21A101-017
3075	46-5002-12
3075	56A4-1
3075	56C4-1
3075	56D4-1
3075	56L101
3075	094
3075	094-1
3075	094A
3075	99S094-1
3075	221-42
3075	780DC
3075	780PC
3075	2031-1
3075	37702-1
3075	612031-1
3075	612070-1
3075	62674962
3076	AMD781
3076	C6071P
3076	CA3071
3076	FF274
3076	GEIC-6
3076	HEP-C6071P
3076	IC510
3076	LM3071N
3076	MC1371P
3076	MC1371PQ
3076	N5071
3076	PTC719
3076	RE308-IC
3076	SN76243
3076	SN76243N
3076	TVCM-9
3076	ULN2127A
3076	09-308045
3076	09-308046
3076	13-40-6
3076	15-37703-1
3076	21A101-016
3076	46-5002-13
3076	56A5-1
3076	56D5-1
3076	56L102
3076	69B1Z
3076	69B2Z
3076	095
3076	095-1
3076	095A
3076	99S095-1
3076	221-43
3076	781DC
3076	781PC
3076	2030-1
3076	37703-1
3076	612030-1
3076	612069-1
3076	62674970
3076/715	56C5-1
3077	AMD746
3077	GEIC-18
3077	GEIC-230
3077	GEL3072P1
3077	SN76246
3077	SN76246N
3077	TA5912
3077	TVCM-2
3077	ULN2114A
3077	ULN2114N
3077	612029-3
3077	62674989
3077/790	CA3072
3077/790	EP84X3
3077/790	EP84X9
3077/790	GEIC-5
3077/790	GEL2114
3077/790	HA1124S
3077/790	IC508
3077/790	MC1329P
3077/790	RE306-IC
3077/790	09-308047
3077/790	09-308048
3077/790	14-2010-01
3077/790	73C180837-1
3077/790	73C180837-2
3077/790	73C180837-3
3077/790	73C180837P2
3077/790	096A
3077/790	99S096-1
3077/790	221-46
3077/790	221-51
3077/790	221-52
3077/790	442-33
3077/790	1010-9973
3077/790	62736291
3078	CA3090AQ
3078/789	51-44837J04
3079	C687
3079	SDT9203
3079	SDT9204
3079	SDT9208
3079	SDT9209
3079	T-Q5083
3079	TR68
3079	2SC1079Y
3079	2SC1104
3079	2SC42
3079	2SC42A
3079	2SC687
3079	2SD161
3079	2SD312
3079	13-331182-1
3079	14-601-26
3079	21A112-033
3079	48S155105
3079	121-829
3079	260P19009
3079	260P19909
3079	260P21608
3079	417-286
3079	637-1
3079	1092
3079	3637-1
3079	40369
3079	235997
3079	2320281
3080	A1000
3080	AD-29B
3080	ERC04-10
3080	HFSD1B
3080	MR1M
3080	NTC-18
3080	RE49
3080	RE90
3080	REN90
3080	RM-2C
3080	SD-1C
3080	SI-RECT-136
3080	TA7806
3080	TVS-S8-2C
3080	TVS1P80
3080	TVSS1D30-15
3080	WEP160
3080	1N4007
3080	1NZ61
3080	1S1830
3080	05-112406
3080	10DI
3080	46-86553-3
3080	48-137540
3080	48-155083
3080	48-155198
3080	48M355021
3080	48S155083
3080	48S155085
3080	53-1086
3080	094-011
3080	201X3130-109
3080	1076-1674
3080	136635
3081	A7580111
3081	B2H1
3081	B2H5
3081	B2H9
3081	DS2M
3081	ECG125
3081	ERCC4-06
3081	GE-531
3081	HF-20071
3081	PT9C22/1
3081	PTC205
3081	RA2C
3081	RE51
3081	REN125
3081	RT8665
3081	S1B-0306
3081	S2V
3081	S3G4
3081	SI-10A
3081	SI-1A
3081	SI-2A
3081	SI-3A
3081	SI-4A
3081	SI-5A
3081	SI-6A
3081	SI-7A
3081	SI-8A
3081	SI-RECT-204
3081	SIB-0306
3081	TM125
3081	TVSS3G4
3081	WEP170
3081	1N4818
3081	1N5054
3081	1N5054A
3081	1R5DZ61
3081	1R5GZ61
3081	5A10D
3081	15-108046
3081	21A110-072
3081	48-90343A62
3081	53A001-4
3081	66X0023-009
3081	264P14701
3081	8000-00006-147
3081	144051
3081	530111-1002
3081	765713
3081	2331991
3081	06200005
3081	06200009
3081	20130109
3081	41027991
3081	4202008700
3081/125	A054-229
3081/125	A7579500
3081/125	A7A1
3081/125	A7A5
3081/125	A7A9
3081/125	AB50
3081/125	B7A1
3081/125	B7A5
3081/125	B7A9
3081/125	DS1K
3081/125	E24100
3081/125	E24101
3081/125	FR2-02
3081/125	GE-510
3081/125	HC50
3081/125	JA-KCDP
3081/125	KBF-02
3081/125	KCDP12-1
3081/125	P100A
3081/125	P150A
3081/125	R0080
3081/125	R0081
3081/125	R0082
3081/125	R0086
3081/125	R008A
3081/125	R0090
3081/125	R210
3081/125	R250
3081/125	RH-DX0041CEZZ
3081/125	RH-DX0042CEZZ
3081/125	S2VC
3081/125	SA-2C
3081/125	SD-1Z
3081/125	V09C
3081/125	1-531-027
3081/125	1G2C1
3081/125	1G2Z1
3081/125	1N1217
3081/125	1N1217A
3081/125	1N2524
3081/125	1N2535
3081/125	1N5170
3081/125	1R5BZ61
3081/125	1S2367
3081/125	2A30
3081/125	2E05
3081/125	2EM015
3081/125	8-719-901-93
3081/125	13-43956-1
3081/125	34-8057-11
3081/125	46-86494-3
3081/125	53L001-14
3081/125	3529
3081/125	3529A
3081/125	3529B
3081/125	3529C
3081/125	126855
3081/125	1474778-4
3081/125	1476171-21
3081/125	2012100126
3082	B474
3082	B474-3
3082	B474-4
3082	B474-6D
3082	B474MP
3082	B474S
3082	B474V10
3082	B474V4
3082	B474Y
3082	CXL226
3082	ECG226
3082	ES61(ELCOM)
3082	GE-49
3082	RE83
3082	REN226
3082	2SB474
3082	2SB474-2
3082	2SB474-3
3082	2SB474S
3082	2SB474V10
3082	2SB474Y
3082	8200-204
3083	A764
3083	B512A
3083	B513A
3083	CXL153
3083	CXL197
3083	ECG197
3083	EP15X44
3083	GE-250
3083	GE-69
3083	IRTR56
3083	M7310
3083	P3U
3083	PIV
3083	PIV-1
3083	PIV-2
3083	PIV-3
3083	QP-31
3083	RCA1C06
3083	RCA1C11
3083	RE22
3083	RT151
3083	RT155
3083	TA7742
3083	TA8210
3083	TA8212
3083	TIP3A
3083	WEP246
3083	WEPS3031
3083	XA1199
3083	01-020566
3083	2N6106
3083	2N6107
3083	2N6108
3083	2SA473B
3083	2SA473R
3083	2SA473Y
3083	2SA769
3083	2SB507E
3083	2SB512A
3083	2SB513A
3083	2SC636
3083	13-283361-1
3083	13-34839-1
3083	020-1111-004
3083	46-86581-3
3083	48-137259
3083	57A245-14
3083	57A251-14
3083	57B251-14
3083	69-1819
3083	99S099
3083	195P2C
3083	417-289
3083	1071
3083	4781
3083	16167
3083	16169
3083	16175
3083	62277
3083	140625
3083	147624-1
3083	309690
3083	610195-2
3083	1473624-1
3083	4082873-0001
3083	06100033
3083	23114939
3083	485137370
3083/197	A566
3083/197	A566A
3083/197	A566B
3083/197	A566C
3083/197	A613
3083/197	A614
3083/197	A616
3083/197	B-1695
3083/197	B507
3083/197	B508
3083/197	B509
3083/197	B512
3083/197	B513
3083/197	B514
3083/197	B515
3083/197	BC460
3083/197	BC461
3083/197	BD136
3083/197	BD138
3083/197	BD140
3083/197	BD434
3083/197	BFW87
3083/197	BFW88
3083/197	D41D2
3083/197	D45C1
3083/197	D45C2
3083/197	D45C3
3083/197	D45C7
3083/197	D45C8
3083/197	D45C9
3083/197	EA15X328
3083/197	EA15X334
3083/197	EA15X8118
3083/197	EA15X8130
3083/197	EA15X8601
3083/197	EA15X8605
3083/197	EP15X15
3083/197	EP15X23
3083/197	HA-00496
3083/197	HA-00699
3083/197	HEP-S5007
3083/197	HEP-S5018
3083/197	HF51
3083/197	HF58
3083/197	IRTR77
3083/197	MU9661
3083/197	MU9661T
3083/197	NPS6518
3083/197	P2K
3083/197	P2T-2
3083/197	P2T-3
3083/197	P2T-4
3083/197	P3M
3083/197	P3N
3083/197	P3N-1
3083/197	P3N-2
3083/197	P3N-3
3083/197	P3N-4
3083/197	P3P
3083/197	P3P-1
3083/197	P3P-2
3083/197	P3P-3
3083/197	P3P-4
3083/197	P3P-5
3083/197	P5H
3083/197	P5L
3083/197	P8H
3083/197	PTC111
3083/197	RCA1C04
3083/197	RCA1C13
3083/197	RCA30
3083/197	RCA30A
3083/197	RS-2025
3083/197	S2041
3083/197	SDA445
3083/197	SDB445
3083/197	SDL445
3083/197	SDM445
3083/197	SDN445
3083/197	SJE108
3083/197	SJE111
3083/197	SJE112
3083/197	SJE114
3083/197	SJE1518
3083/197	SJE202
3083/197	SJE210
3083/197	SJE221
3083/197	SJE227
3083/197	SJE231
3083/197	SJE241
3083/197	SJE243
3083/197	SJE245
3083/197	SJE257
3083/197	SJE265
3083/197	SJE267
3083/197	SJE273
3083/197	SJE275
3083/197	SJE276
3083/197	SJE277
3083/197	SJE279
3083/197	SJE283
3083/197	SJE403
3083/197	SJE408
3083/197	SJE584
3083/197	SJE723
3083/197	SJE736
3083/197	SJE768
3083/197	SJE797
3083/197	SJE799
3083/197	SPS1437
3083/197	STX0020
3083/197	T345
3083/197	T396
3083/197	TA7741
3083/197	TIP30
3083/197	TIP32
3083/197	TR2327723
3083/197	TVS2SB546
3083/197	W19
3083/197	01-010473
3083/197	01-572774
3083/197	002-012300
3083/197	2N4388
3083/197	2N4899
3083/197	2N5193
3083/197	2N5194
3083/197	2N5195
3083/197	2N6109
3083/197	2N6124
3083/197	2N6125
3083/197	2SA473
3083/197	2SA489
3083/197	2SA623-Q
3083/197	2SA624
3083/197	2SA624A
3083/197	2SA624B
3083/197	2SA624C
3083/197	2SA624D
3083/197	2SA624E
3083/197	2SA624G
3083/197	2SA624GN
3083/197	2SA624L
3083/197	2SA624LG
3083/197	2SA624R
3083/197	2SA624Y
3083/197	2SA748Q
3083/197	2SA754
3083/197	2SA754A
3083/197	2SA754B
3083/197	2SA754C
3083/197	2SA754D
3083/197	2SA816
3083/197	2SB507
3083/197	2SB508
3083/197	2SB512
3083/197	2SB512P
3083/197	2SB513
3083/197	2SB513P
3083/197	2SB513Q

SK	Industry Standard No.	SK	Industry Standard No.	SK	Industry Standard No.	SK	Industry Standard No.	SK	Industry Standard No.
3083/197	2SB566	3084	3L4-6013-5	3087	D6726	3087	HE-10002	3087	RL232G
3083/197	2SB566A	3084	3L4-6013-55	3087	D8410	3087	HE-10003	3087	RL246
3083/197	2SB566C	3084	3L4-6013-58	3087	DC-15	3087	HE-10040	3087	RL31
3083/197	2SB566D	3084	3L4-6013-6	3087	DDMV-1	3087	HE-1N34	3087	RL32
3083/197	3L4-6013-56	3084	3L4-6013-8	3087	DDMV-2	3087	HE-1N34A	3087	RL32G
3083/197	004-1	3084	8P9253	3087	DG1834	3087	HE-1N60	3087	RL34
3083/197	09-300078	3084	21M465	3087	DGM-2	3087	HE-1N60P	3087	RL34G
3083/197	09-300090	3084	48-137161Q	3087	DGM-3	3087	HE-18188	3087	RL41G
3083/197	09-305134	3084	114P1P	3087	DI-2	3087	HE-18426	3087	RL52
3083/197	11-0773	3084	121-926	3087	DIJ70542	3087	HE-18446	3087	RS2801
3083/197	13-34047-3	3084	121-977	3087	DIJ71776	3087	HEP-R9134	3087	RT1008
3083/197	13-36443-1	3084	0131-005353	3087	DIJ71778	3087	HEP-R9135	3087	RT1106
3083/197	13-39100-1	3084	41501	3087	DIJ72549	3087	H05002	3087	RT1108
3083/197	13-39819-1	3084	171162-195	3087	DK20	3087	H05004	3087	RT1184
3083/197	20A0059	3084	3596100	3087	DK21	3087	H05006	3087	RT2334
3083/197	20A0060	3084	3596101	3087	DR291	3087	H05007	3087	RT2452
3083/197	21A1412-094	3084	4104007380	3087	DR351	3087	H05008	3087	RT2694
3083/197	024	3085	B502	3087	DR352	3087	H05009	3087	RT5072
3083/197	31-0102	3085	B503	3087	DR385	3087	H05078	3087	RT3099
3083/197	33-096	3085	GE-26	3087	DR434	3087	H05079	3087	RT3233
3083/197	34-1003	3085	J241258	3087	DR449	3087	H05085	3087	RT3336
3083/197	34-6002-57	3085	SDT3509	3087	DS-359	3087	H05088	3087	RT3469
3083/197	34-6016-54	3085	SDT3513	3087	D827	3087	H091	3087	RT4293
3083/197	42-28213	3085	SDT3707	3087	D855	3087	HV15	3087	RT4644
3083/197	46-86235-3	3085	SDT3710	3087	DS816685	3087	IC743050	3087	RT61012
3083/197	46-86411-3	3085	2N5954	3087	E0018	3087	IN60-1	3087	RT7636
3083/197	48-137153	3085	2N5955	3087	EA16X1	3087	INJ60284	3087	RT7689
3083/197	48-137154	3085	28A483	3087	EA16X11	3087	IP20-0015	3087	RT8671
3083/197	48-137155	3085	28A566	3087	EA16X140	3087	IR5JA	3087	RVD1K110
3083/197	48-137156	3085	28A566A	3087	EA16X22	3087	IT22	3087	RVD2-1K110
3083/197	48-137157	3085	28A566B	3087	EA16X9	3087	IT23	3087	0822.3901-001
3083/197	48-137512	3085	28A566Q	3087	ED219464	3087	IT23G	3087	0822.3902-001
3083/197	48-137472	3085	28A614	3087	ED4	3087	J242	3087	S054
3083/197	48-137550	3085	28A616	3087	ED60	3087	J243	3087	S06
3083/197	48-90420A06	3085	638-1	3087	EDG-6	3087	J2441	3087	SD-12
3083/197	488137310	3085	3639-1	3087	EP16X21	3087	J320041	3087	SD-46-2
3083/197	488137312	3085	3647-1	3087	ES16X12	3087	J685	3087	SD-630
3083/197	488137370	3085	135351	3087	ES16X2	3087	JT-E1014	3087	SD12B
3083/197	488155066	3085	1473638-1	3087	ES16X5	3087	JT-E1031	3087	SD12B
3083/197	57A188-12	3086	CXL226MP	3087	ES16X7	3087	K115J511-2	3087	SD12M
3083/197	57A206-14	3086	EC0226MP	3087	ET16X1	3087	K3	3087	SD13
3083/197	57A277-14	3086	IRER94MP	3087	ET16X19	3087	K4-550	3087	SD14
3083/197	57B206-14	3086	R-288474	3087	ET16X20	3087	K6	3087	SD15
3083/197	61-309690	3086	RE83MP	3087	ET16X21	3087	K60	3087	SD16
3083/197	730180830-12	3086	REN226MP	3087	ET41X37	3087	K882	3087	SD21A
3083/197	86-507-2	3086	4-48(SEARS)	3087	ETD-1N60	3087	KDD-0013	3087	SD34
3083/197	86-530-2	3086	09-301030	3087	EU16X19	3087	KGE41959	3087	SD56
3083/197	86-602-2	3086	65-80001	3087	EW166	3087	M51	3087	SD701-02
3083/197	121-803	3086	1042-10	3087	EW167	3087	M60	3087	SPT104
3083/197	121-873	3086	5001-049	3087	F136	3087	M8489	3087	SPT108
3083/197	121-886	3086	22881	3087	F20-1010	3087	M95	3087	SIS20
3083/197	121-994	3087	A054-105	3087	F20-1012	3087	MA51A	3087	SQ46
3083/197	121-997	3087	A066-120	3087	F20-1013	3087	MA55	3087	SV30
3083/197	0131-002049	3087	A068-101	3087	F20-1014	3087	MA8	3087	SVDOA79
3083/197	132-024	3087	A069-109	3087	F20303	3087	MA90	3087	SVD20A79
3083/197	132-079	3087	A069-115	3087	F215-1010	3087	MA900	3087	T-B1177
3083/197	134-1	3087	A07	3087	F215-1012	3087	MC2526	3087	T11
3083/197	149P1	3087	A090	3087	F215-1013	3087	MD-60A	3087	T12G
3083/197	149P2003	3087	A15-1007	3087	F215-1014	3087	MD34	3087	T17
3083/197	157P4	3087	A20371	3087	FD1980	3087	MD34A	3087	T18
3083/197	177-023-9-001	3087	A2419	3087	FD222	3087	MD46	3087	T20
3083/197	195P2	3087	A2420	3087	FPR6D-1006	3087	MD60	3087	T20G
3083/197	202-1	3087	A2476	3087	FS19	3087	MN34A	3087	T21
3083/197	202N1	3087	A25-1007	3087	FV-23	3087	MN60	3087	T21238
3083/197	260P34008	3087	A30	3087	FV23	3087	N48	3087	T21271
3083/197	0770	3087	A514-022057	3087	G00-004-A	3087	NC29	3087	T22
3083/197	1043-1294	3087	A514-042791	3087	G00-013-8	3087	NJM-703N	3087	T23
3083/197	1045-7856	3087	A556-142	3087	G1010	3087	NTC-14	3087	T23G
3083/197	02057-1	3087	A615-1012	3087	G1288	3087	NU34	3087	T26G
3083/197	7351	3087	A7001800	3087	G156	3087	0A-91	3087	T8G
3083/197	7420	3087	A909-1015	3087	G157	3087	0A134Q	3087	T9
3083/197	16165	3087	A909-1017	3087	G159	3087	0A150	3087	T9G
3083/197	30ZT1	3087	0A91	3087	G199	3087	0A159	3087	TC311200600
3083/197	95258-2	3087	AA111	3087	G1HA	3087	0A160	3087	TB1014
3083/197	95263-2	3087	AA112	3087	G297	3087	0A161	3087	TB1031
3083/197	128057	3087	AA112P	3087	G409	3087	0A172	3087	TP34
3083/197	132447	3087	AA113	3087	G498	3087	0A174	3087	TP34A
3083/197	132448	3087	AA117	3087	G580	3087	0A47	3087	TRO575002
3083/197	132571	3087	AA118	3087	G5C	3087	0A50	3087	TR12001-4
3083/197	132574	3087	AA131	3087	G5F	3087	0A541	3087	TR228736002003
3083/197	144076	3087	AA132	3087	G5K	3087	0A6	3087	TR320008
3083/197	166919	3087	AA134	3087	G766	3087	0A7	3087	TR320039
3083/197	185035	3087	AA135	3087	G788	3087	0A70	3087	TR320041
3083/197	570030	3087	AA136	3087	G7D	3087	0A71	3087	TR320048
3083/197	610157-4	3087	AA137	3087	G7E	3087	0A71C	3087	TR48
3083/197	610162-7	3087	AA143	3087	G7F	3087	0A72	3087	TVS-0A90
3083/197	610159-4	3087	AA218	3087	G7G	3087	0A73	3087	TVS-0A91
3083/197	610202-1	3087	AAY139	3087	G814	3087	0A73C	3087	TVS-0A95
3083/197	610202-2	3087	AAY15	3087	G815	3087	0A74	3087	TV80A90
3083/197	610227-1	3087	AAY18	3087	G816	3087	0A74A	3087	UP-SD1
3083/197	657181	3087	AAY30	3087	G820	3087	0A81C	3087	V115
3083/197	1445829-501	3087	AAY33	3087	G825	3087	0A85	3087	V117
3083/197	1445829-502	3087	AAY46	3087	G844	3087	0A85C	3087	V135
3083/197	1473628-3	3087	AAZ18	3087	G846	3087	0A9	3087	V50260-16
3083/197	2320884	3087	ACR810-107	3087	G847	3087	0A90	3087	V50A260-36
3083/197	2321281	3087	ACR83-1007	3087	GB-1	3087	0A91	3087	V74
3083/197	2321581	3087	AF83-160-1017	3087	GC5012	3087	0A95	3087	VD12
3083/197	3596451	3087	AR35	3087	GD-25	3087	OP173	3087	VHD1N34A///-1
3083/197	3596452	3087	B28	3087	GD-26	3087	OS816308	3087	VHD1N60-1
3083/197	3596453	3087	B30	3087	GD-30	3087	OS816685	3087	VHD1834///-1
3083/197	3596454	3087	B601-1011	3087	GD1001	3087	P10155	3087	WD1
3083/197	7570030-01	3087	B692213	3087	GD12	3087	PBE3322	3087	X18
3083/197	62393661	3087	C21480	3087	GD13E	3087	P093	3087	YEADO32
3083/197	62734450	3087	C5005	3087	GD1E	3087	PTC207	3087	ZR-1.5
3083/197	62743794	3087	CB163	3087	GD363B	3087	PTC207M	3087	ZEN430
3083/197	2004047510	3087	CD000	3087	GD401	3087	Q49	3087	001-0010-00
3083/197	2004596801	3087	CD0014	3087	GD402	3087	Q50	3087	001-01501-0
3083/197	6664003100	3087	CG12B	3087	GD403	3087	Q51	3087	001-015010
3084	AR-30	3087	CG64H	3087	GD404	3087	QRT-200	3087	001-015011
3084	AR26	3087	CG65H	3087	GD406	3087	QVD1KP114	3087	1-016
3084	AR27	3087	CG66H	3087	GD409	3087	R1106	3087	1-017
3084	AR30	3087	CG74H	3087	GD4E	3087	R1107	3087	1A11306
3084	AR37	3087	CGD462	3087	GD5E	3087	R1109	3087	1A14384
3084	AR44	3087	CGD685	3087	GD663	3087	R1667	3087	100029
3084	B23-79	3087	CK705	3087	GD6E	3087	R1889	3087	100039
3084	P2U	3087	CK706	3087	GD73E/4	3087	R2164	3087	1G02
3084	P2U-2	3087	CK706A	3087	GD73E/3	3087	R2334	3087	1GD2
3084	RV2356	3087	CK706P	3087	GD74E/3	3087	R5096	3087	1GD4
3084	TA7743	3087	CK715	3087	GD74E/4	3087	R5522	3087	1GD6
3084	TA8211	3087	CT461	3087	GD74E/5	3087	R60-1007	3087	1K261
3084	WEP700	3087	CTP-2001-1010	3087	GD8E	3087	R7743	3087	1N10
3084	2M6110	3087	CTP-2006-1004	3087	GE-X66	3087	R7892	3087	1N100
3084	2M6111	3087	CTP461	3087	GE363B	3087	R7893	3087	1N100A
3084	3L4-6011-11	3087	CV425	3087	GED05B850	3087	R8060	3087	1N107
3084	3L4-6011-14	3087	CV442	3087	G00-003-A	3087	R8061	3087	1N109
3084	3L4-6011-2	3087	CXO036	3087	GP2354	3087	R8219	3087	1N1093
3084	3L4-6011-1	3087	CXO041	3087	GPM1NA	3087	R8257	3087	1N111
3084	3L4-6011-52	3087	D-00169C	3087	GPM1NB	3087	R8475	3087	1N112
3084	3L4-6011-53	3087	D-00204R	3087	HC-30	3087	R8887	3087	1N113
3084	3L4-6011-9	3087	D-00269C	3087	HD10-001-01	3087	R8970	3087	1N114
3084	3L4-6013-15	3087	D-00669C	3087	HD1000101	3087	R9590	3087	1N115
3084	3L4-6013-2	3087	D093	3087	HE-0A90	3087	RF1811	3087	1N116
3084	3L4-6013-3	3087	D1-2	3087	HE-10001	3087	RF60034	3087	1N116A
3084	3L4-6013-4	3087	D4R			3087	RFJ60614		

SK	Industry Standard No.	SK	Industry Standard No.	SK	Industry Standard No.	SK	Industry Standard No.	SK	Industry Standard No.
3087	1N117	3087	1N618	3087	1T22AJ	3087	21A103-019	3087	56-2
3087	1N117A	3087	1N62	3087	1T22B	3087	21A103-022	3087	56-20
3087	1N118	3087	1N63	3087	1T22G	3087	21A103-046	3087	56-26
3087	1N118A	3087	1N631	3087	1T231	3087	21A119-005	3087	56-3
3087	1N119	3087	1N632	3087	1T236	3087	21K60	3087	56-4
3087	1N119A	3087	1N636	3087	1T23J	3087	21M289	3087	56-4886
3087	1N120	3087	1N63A	3087	1T23M	3087	21M325	3087	56-8
3087	1N120A	3087	1N64G	3087	1V9002	3087	21M594	3087	56-8101
3087	1N125	3087	1N65	3087	02-1001-1221-3	3087	022-2823-006	3087	57D1-1
3087	1N126	3087	1N65A	3087	2XAA112	3087	022-2823-008	3087	57D1-2
3087	1N126A	3087	1N66	3087	2XAA113	3087	022-3901-001	3087	57D1-54
3087	1N128	3087	1N66A	3087	03-0021-0	3087	022-3902-001	3087	57D1-62
3087	1N134	3087	1N67	3087	003-0050Q	3087	24MW1029	3087	61-59395
3087	1N139	3087	1N67A	3087	03-931051	3087	24MW1030	3087	62-10254
3087	1N140	3087	1N68A	3087	03-931771	3087	24MW199	3087	62-10655
3087	1N142	3087	1N69	3087	4-2020-03500	3087	24MW243	3087	62-12034
3087	1N143	3087	1N695	3087	4-2020-03600	3087	24MW860	3087	62-15318
3087	1N144	3087	1N695A	3087	4-2020-05600	3087	24MW87	3087	62-16769
3087	1N145	3087	1N698	3087	4-2021-05870	3087	24MW967	3087	62-16841
3087	1N148	3087	1N69A	3087	05-00000-00	3087	27F1	3087	62-19846
3087	1N1561	3087	1N70	3087	05-170034	3087	28J1	3087	63-10739
3087	1N1562	3087	1N70A	3087	05-180034	3087	29-505	3087	63-11074
3087	1N191	3087	1N71	3087	05-180188	3087	31-0039	3087	63-11879
3087	1N192	3087	1N72	3087	05-490091	3087	32-0000	3087	63-12158
3087	1N198	3087	1N72G	3087	05-490095	3087	32-0001	3087	63-12607
3087	1N198A	3087	1N73	3087	05-610046	3087	32-0002	3087	63-12645
3087	1N198B	3087	1N74	3087	05-931771	3087	32-0003	3087	63-12754
3087	1N22	3087	1N75A	3087	05A03	3087	32-0004	3087	63-12755
3087	1N265	3087	1N76	3087	7-0005	3087	32-0023	3087	63-12756
3087	1N267	3087	1N76A	3087	7-0006	3087	32-0029	3087	63-12757
3087	1N268	3087	1N76C	3087	07-5134-14	3087	32-0036	3087	63-13080
3087	1N270	3087	1N76G	3087	07-5134-14A	3087	32-18537	3087	63-22724
3087	1N273	3087	1N770	3087	07-5134-14B	3087	34-8002-1	3087	63-25933
3087	1N276	3087	1N771	3087	07-5134-14C	3087	34-8002-2	3087	63-26382
3087	1N277	3087	1N771A	3087	8-619-030-011	3087	34-8002-3	3087	63-28250
3087	1N278	3087	1N771B	3087	8-697-020-571	3087	34-8002-4	3087	63-28888
3087	1N279	3087	1N772A	3087	8-719-026-11	3087	34-8002-5	3087	63-8381
3087	1N2801	3087	1N773	3087	8-905-305-007	3087	34-8002-6	3087	63-8955
3087	1N281	3087	1N773A	3087	8-905-305-020	3087	34-8002-7	3087	63-9523
3087	1N283	3087	1N774	3087	8-905-305-065	3087	34-8022-7	3087	065-0013
3087	1N285	3087	1N774A	3087	8-905-305-318	3087	34-8057-23	3087	65-085010
3087	1N287	3087	1N775	3087	8-905-305-330	3087	34-8057-25	3087	65-085012
3087	1N288	3087	1N776	3087	8-905-305-336	3087	34-8057-26	3087	66X0020-000
3087	1N289	3087	1N777	3087	8-905-305-338	3087	34-8057-28	3087	66X0020-001
3087	1N290	3087	1N781	3087	8-905-305-359	3087	34-8057-30	3087	66X0039-001
3087	1N292	3087	1N805	3087	8-905-305-342	3087	34-8057-6	3087	66X0043-001
3087	1N294	3087	1N81	3087	8-905-305-348	3087	42-22537	3087	66X0047-001
3087	1N295	3087	1N81A	3087	8-905-305-405	3087	42-22539	3087	66X0047-901
3087	1N295X	3087	1N835	3087	8-905-305-555	3087	42-22755	3087	76-14196-1
3087	1N296	3087	1N84	3087	8-905-305-561	3087	42-27378	3087	77-271031-1
3087	1N297	3087	1N86	3087	8-905-305-580	3087	42-27380	3087	77-271032-1
3087	1N297A	3087	1N87GA	3087	8-905-305-635	3087	42-27381	3087	78-271199-1
3087	1N298A	3087	1N88	3087	8-905-313-010	3087	42A14	3087	78-271228-1
3087	1N304	3087	1N89	3087	8-905-313-011	3087	46-8616-3	3087	79F015
3087	1N305	3087	1N90	3087	8-905-313-100	3087	46-86168-3	3087	80-60-1
3087	1N306	3087	1N909	3087	8-905-313-101	3087	46-86200-3	3087	81-2T123150-8
3087	1N307	3087	1N90G	3087	8-905-313-120	3087	46-86253-3	3087	81-46123013-8
3087	1N308	3087	1N90GA	3087	8-905-405-077	3087	46-86266-3	3087	86-10-1
3087	1N309	3087	1N910	3087	8-905-405-838	3087	46-86436-3	3087	86-125-1
3087	1N310	3087	1N911	3087	8A01	3087	46-8646-3	3087	86-146-1
3087	1N3110	3087	1N949	3087	09-306002	3087	46-86484-3	3087	86-5007-3
3087	1N312	3087	1N95	3087	09-306009	3087	46-8688-3	3087	86-64-1
3087	1N3121	3087	1N96	3087	09-306010	3087	48-03005A03	3087	86-74-1
3087	1N3125	3087	1N96A	3087	09-306012	3087	48-134587	3087	87-10-0
3087	1N3146	3087	1N97	3087	09-306020	3087	48-134588	3087	87-10-1
3087	1N3204	3087	1N97A	3087	09-306024	3087	48-137299	3087	090A64-1
3087	1N3287N	3087	1N98	3087	09-306040	3087	48-137495	3087	93.24.401
3087	1N34	3087	1N98A	3087	09-306049	3087	48-155039	3087	93.24.601
3087	1N3465	3087	1N99	3087	09-306051	3087	48-60082A97	3087	93.24.604
3087	1N3466	3087	1N994	3087	09-306091	3087	48-60077A06	3087	93A105-1
3087	1N3469	3087	1N996	3087	09-306093	3087	48-61074B01	3087	93A110-1
3087	1N3470	3087	1NA4	3087	09-306107	3087	48-61767B01	3087	93A27-1
3087	1N3484	3087	1NA4G	3087	09-306108	3087	48-62334A02	3087	93A27-8
3087	1N34A	3087	1NJ33233	3087	9D12	3087	48-63006A56	3087	93A33-1
3087	1N34A-Z	3087	1NJ60284	3087	9D16	3087	48-63029A20	3087	93A38-1
3087	1N34A8	3087	1NJ61224	3087	9PI1100310	3087	48-63077A32	3087	93A64-3
3087	1N349	3087	1NJ61675	3087	10-085001	3087	48-63084A06	3087	93B25-3
3087	1N349A	3087	1NJ71185	3087	10-085005	3087	48-644587	3087	93B27-1
3087	1N34M	3087	1S10078	3087	11-085001	3087	48-644681	3087	93B38-1
3087	1N35	3087	1S12	3087	11-085004	3087	48-647311	3087	93B38-5
3087	1N355	3087	1S127	3087	11-085007	3087	48-647313	3087	93B41-1
3087	1N3564	3087	1S13	3087	11-085015	3087	48-65837A02	3087	93B41-29
3087	1N356	3087	1S1589	3087	11-085022	3087	48-65937A02	3087	93B41-3
3087	1N3567	3087	1S1701	3087	12-085005	3087	48-67020A11	3087	93B38-1
3087	1N3567B	3087	1S17D1	3087	12-085029	3087	48-711052	3087	93C2-6
3087	1N3753	3087	1S186	3087	12-085034	3087	48-739300	3087	93C218
3087	1N38	3087	1S187	3087	12-085035	3087	48-741280	3087	93C27-1
3087	1N38A	3087	0001S188	3087	12-085038	3087	48-86168-3	3087	93C64-1
3087	1N38B	3087	1S188A	3087	12-087003	3087	48-861-3	3087	96-5007-01
3087	1N3991	3087	1S188AM	3087	13-004	3087	48-86200-3	3087	96-5059-01
3087	1N40	3087	1S188FM	3087	13-085012	3087	48-90210A01	3087	96XZ778/21N
3087	1N41	3087	1S188FM1A	3087	13-14094-1	3087	48-90233A01	3087	96XZ778/44N
3087	1N417	3087	1S188FM2	3087	13-14094-15	3087	48-90233A06	3087	100-00340-00
3087	1N418	3087	1S188FMA	3087	13-14094-2	3087	48-97048A02	3087	100-0125
3087	1N419	3087	1S188G	3087	13-14094-5	3087	48-97177A15	3087	100-160
3087	1N43	3087	1S188TV	3087	13-16235-8	3087	48061074B01	3087	102-02
3087	1N435	3087	1S189	3087	13-17204-1	3087	48010346A02	3087	102-207
3087	1N44	3087	1S318	3087	13-23917-1	3087	48M355008	3087	103-114
3087	1N447	3087	1S32	3087	13-55046-1	3087	48M355009	3087	103-22
3087	1N45	3087	1S33	3087	132-1005	3087	48010346A02	3087	103-31
3087	1N452	3087	1S34	3087	14-504-01	3087	48S134587	3087	103-34
3087	1N454	3087	1S34A	3087	14-510-01	3087	48S137299	3087	103-73
3087	1N455	3087	1S34B	3087	14-511-01	3087	48S155039	3087	103-79
3087	1N45A	3087	1S354	3087	14-512-01	3087	48S155078	3087	103-87
3087	1N46	3087	1S355	3087	14-513-01	3087	48X97048A19	3087	105-02
3087	1N47	3087	1S357	3087	14-514-01	3087	48X97168A04	3087	106-008
3087	1N476	3087	1S4260	3087	14-514-06	3087	48X97239A01	3087	0112-0026
3087	1N477	3087	1S426GPM	3087	14-514-08	3087	51IN60P	3087	0112-0037
3087	1N478	3087	1S428	3087	14-514-09	3087	51I8348	3087	0112-0046
3087	1N479	3087	1S441	3087	14-514-12	3087	51I81834	3087	120-004498
3087	1N48	3087	1S442	3087	14-514-21	3087	52-051-017-0	3087	120-004499
3087	1N48A	3087	1S446	3087	14-514-22	3087	53A001-1	3087	120-005299
3087	1N49	3087	1S447	3087	15-085002	3087	53A006-1	3087	121-31
3087	1N497	3087	1S447P	3087	15-085003	3087	53B001-1	3087	123-015
3087	1N498	3087	1S448	3087	15-085009	3087	53B004-1	3087	130-30281
3087	1N50	3087	1S449	3087	019-001918	3087	53B005-2	3087	130-30301
3087	1N500	3087	1S451	3087	019-001980	3087	53C001-1	3087	137-824
3087	1N51	3087	1S452	3087	019-002718	3087	53C006-1	3087	141-003
3087	1N52	3087	1S453	3087	019-00301980	3087	53C006-2	3087	142-011
3087	1N527	3087	1S454	3087	019-005043	3087	53C009-2	3087	150-001-9-007
3087	1N52A	3087	1S467R	3087	19-080-009	3087	53C020-1	3087	150-004-9-001
3087	1N54	3087	1S5454	3087	19-085005	3087	53CD01-1	3087	150-005-9-001
3087	1N54A	3087	1S74	3087	019-301980	3087	53X003-1	3087	186-015
3087	1N54G	3087	1S75	3087	19A115086-P1	3087	53X006-1	3087	229-5100-231
3087	1N54GA	3087	1S76	3087	20-1680-175	3087	53N004-14	3087	264D01001
3087	1N56	3087	1S78	3087	20A70	3087	53N004-5	3087	264P00401
3087	1N56A	3087	1S78B	3087	20A79	3087	53T001-1	3087	264P00801
3087	1N57	3087	1S79	3087	20A9M	3087	53T001-6	3087	264P03802
3087	1N57A	3087	1S80	3087	21A103-010	3087	56-1	3087	294-42-9
3087	1N58	3087	1T188	3087	21A103-017	3087	56-10	3087	296-42-9
3087	1N58A	3087	1T213	3087		3087	56-11	3087	296L002B01
3087	1N616	3087	1T22A	3087		3087		3087	296V001H01
3087	1N617	3087		3087		3087		3087	296V002H02
								3087	296V002H05

SK	Industry Standard No.
3087	296V002H06
3087	296V002H07
3087	296V002M01
3087	296V006H02
3087	296V007H02
3087	296V012H01
3087	296V015B01
3087	296V015H01
3087	296V020B01
3087	296V024B01
3087	296V024B02
3087	324-0014
3087	324-0105
3087	324-0141
3087	324-0160
3087	325-0025-327
3087	325-0028-86
3087	325-0028-87
3087	325-0031-335
3087	325-0036-562
3087	325-1376-60
3087	400A
3087	400D
3087	403-1
3087	464-100-19
3087	464-103-19
3087	464-106-19
3087	464-111-19
3087	464-113-19
3087	503-T21472
3087	510IN60
3087	510IS34
3087	521-145
3087	523-1000-067
3087	523-1000-295
3087	523-1000-326
3087	523-1002-326
3087	523-1500-067
3087	523-2003-001
3087	524-457
3087	525-877
3087	0575-005
3087	617-15
3087	617-17
3087	630-002
3087	630-079
3087	642-028
3087	642-102
3087	642-119
3087	642-132
3087	642-199
3087	650
3087	656-142
3087	690V034H32
3087	690V034H74
3087	690V037H91
3087	690V047H61
3087	690V052H50
3087	690V052H68
3087	690V066H48
3087	690V067H09
3087	690V073H60
3087	690V083H89
3087	690V098H52
3087	690V68H32
3087	690V73H60
3087	742
3087	754-4000-088
3087	792-292
3087	800-003-00
3087	800-005-00
3087	800-020-00
3087	800-022-00
3087	800-039-00
3087	800-517-00
3087	805-1060
3087	808-312
3087	903-00390
3087	903-114B
3087	903-12B
3087	903-18B
3087	903-23
3087	903-23A
3087	903-23B
3087	903-23C
3087	903-23D
3087	903-23E
3087	903-30B
3087	903-45B
3087	903-65B
3087	903-83B
3087	916-32000-7
3087	919-01-0867
3087	919-010867
3087	1002-08
3087	1002-09
3087	1002-17
3087	1012-17
3087	1030-17
3087	1033-1916
3087	1042-12
3087	1042-13
3087	1063-3553
3087	1074-24
3087	1077-2325
3087	1077-2760
3087	1119-17
3087	01122-0073
3087	1410-171
3087	1489-17
3087	1550
3087	1778-17
3087	1947-17
3087	1956-17
3087	2093A25-3
3087	2093A33-1
3087	2093A38-1
3087	2093A38-10
3087	2093A38-27
3087	2093A38-33
3087	2093A38-40
3087	2093A41
3087	2093A41-148
3087	2093A41-150
3087	2093A41-154
3087	2093A41-187
3087	2093A41-2
3087	2093A77-1
3087	2093B41-11
3087	2102-010
3087	2102-025
3087	2102-028
3087	2106-124
3087	2180-41

SK	Industry Standard No.
3087	2279-13
3087	2280-13
3087	2281-13
3087	2282-13
3087	2283-13
3087	2284-13
3087	2290-13
3087	2402-459
3087	2405-458
3087	2408-330
3087	2603-186
3087	2703-389
3087	2704-388
3087	2789
3087	2796
3087	3500
3087	3505
3087	4001-230
3087	4041-200-40100
3087	4354
3087	4801-00628
3087	4801-00629
3087	4822-130-30281
3087	4822-130-30311
3087	4822-130-30312
3087	4828-4
3087	5120A90
3087	8000-00003-044
3087	8000-00003-045
3087	8000-00004-045
3087	8000-00004-060
3087	8000-00005-016
3087	8000-00005-017
3087	8000-00011-060
3087	8000-00012-041
3087	8000-00038-009
3087	8000-0004-042
3087	8000-0004-063
3087	8000-00041-015
3087	9861B-43
3087	11252
3087	0012060
3087	12808
3087	12850
3087	25201-001
3087	25840-55
3087	28287
3087	031033
3087	031040
3087	42020
3087	43959
3087	46287-4
3087	48287
3087	48287-4
3087	50505-01
3087	53092-1
3087	55810-51
3087	55810-52
3087	057001
3087	057001H
3087	059395
3087	59840-1
3087	68504-77
3087	71119-2
3087	71467-1
3087	72080
3087	72129A
3087	085002
3087	085004
3087	085006
3087	085016
3087	95000
3087	95002
3087	95004
3087	95007
3087	95014
3087	95017
3087	95018
3087	100844
3087	104152
3087	105517
3087	110610
3087	111207
3087	111605
3087	112524
3087	112529
3087	115101
3087	116048
3087	122166
3087	126177
3087	129494
3087	134180
3087	135872
3087	143595
3087	161006
3087	161016
3087	161038
3087	162002-39
3087	166273
3087	169501
3087	170373
3087	171162-042
3087	171162-269
3087	190716
3087	195617
3087	202315
3087	245517
3087	489752-003
3087	489752-042
3087	489752-076
3087	489850-004
3087	0517826
3087	0517829
3087	0525002
3087	0526224
3087	530065-1
3087	530065-1002
3087	530065-2
3087	530065-3
3087	530072-1015
3087	530072-7
3087	530092-1001
3087	530092-2
3087	530116-1
3087	0573024
3087	0575001
3087	0575001H
3087	0575002
3087	0575002H
3087	0575005H
3087	575009
3087	0575019
3087	0575067
3087	0575099

SK	Industry Standard No.
3087	576063
3087	0577001
3087	601030
3087	610030
3087	615010
3087	740402
3087	740952
3087	741866
3087	742004
3087	760101-0005
3087	760101-0006
3087	761113
3087	765722
3087	771909
3087	771911
3087	772740
3087	785278-01
3087	817032
3087	817077
3087	817125
3087	817158
3087	817159
3087	922021
3087	922604
3087	972258-6
3087	980514
3087	981150
3087	981153
3087	981207
3087	981522
3087	981676
3087	982065
3087	982822
3087	983239
3087	984200
3087	984226
3087	988994
3087	1223770
3087	1476179-001
3087	1800002
3087	2000757-18
3087	2001786-134
3087	2002151-020
3087	2002151-20
3087	2003069-4
3087	2004357-106
3087	2006431-50
3087	2092055-0001
3087	2092055-0007
3087	2092055-001
3087	2092055-0708
3087	2092055-0713
3087	2092055-1
3087	2092055-7
3087	2495084
3087	2495380
3087	2498530
3087	5330335
3087	5330721
3087	06200001
3087	06200017
3087	7279893
3087	7282358
3087	7570016-02
3087	7570016-03
3087	7851947-01
3087	7852223-01
3087	7852223-01A
3087	7852782-01
3087	7853357-01
3087	7855282-01
3087	8051060
3087	8052446
3087	8121002
3087	13030301
3087	17101881
3087	17101882
3087	20001786-134
3087	20115070
3087	22115181
3087	25115102
3087	27123150
3087	27123240
3087	27123270
3087	41029009
3087	41522875
3087	41624836
3087	43300203
3087	62564539
3087	62761261
3087	80510600
3087	80521881
3087	231150050
3087	1522270100
3087	1522270101
3087	2008000064
3087	4120100090
3087	4120200090
3087	4202003500
3087	6612009000
3087	9511510200
3087	16411190188
3087	16412190410
3088	A054-103
3088	A054-187
3088	A2473
3088	A692X13-4
3088	OA90LF
3088	OA90Z
3088	AA779
3088	B-30P
3088	B-3P
3088	CA90
3088	CB106
3088	CD-0000
3088	CD101
3088	CE0495/7839
3088	CT-2002
3088	CT2002
3088	CT2005
3088	CT2007
3088	CT2008
3088	CX-0045
3088	D-00284R
3088	D1J70542
3088	D1J70543
3088	D286
3088	D2R31
3088	DAAY001002
3088	DANZ006000
3088	DDAY001001
3088	DDAY001002
3088	DDAY001004
3088	DDAY001010

SK	Industry Standard No.
3088	DDAY001022
3088	DG1N60
3088	DIJ61224
3088	DIJ70644
3088	DIJ70645
3088	DIJ70646
3088	DIJ72166
3088	DIJ72294
3088	DS410
3088	DX-0161
3088	DX-0162
3088	DX-0725
3088	EA16X27
3088	EA16X48
3088	EA16X5
3088	EA16X97
3088	EA2137
3088	EA2502
3088	EA2606
3088	EA3127
3088	EA3718
3088	ED-46
3088	ED-60
3088	ED46
3088	EDG-0003
3088	EDG-0006
3088	EDG-3
3088	EP16X3
3088	ES10189
3088	ES10224
3088	ES10225
3088	ES15054
3088	ES16X3
3088	ES16X6
3088	ES36X103
3088	EU16X2
3088	EYV420D1R5JB
3088	G00-003-A
3088	G00-008-A
3088	G00-009-A
3088	G01-803-A
3088	GD-29
3088	GD5004
3088	HD1000105
3088	HD10001050
3088	HD1000301
3088	HD1000302
3088	HE-10024
3088	HE-10025
3088	HE-10044
3088	HE-1024
3088	HE-CD0000
3088	HF-10024
3088	HN-00003
3088	INJ61675
3088	IP20-0016
3088	IP20-0060
3088	IP20-0283
3088	ITT102
3088	ITT301
3088	ITT718
3088	J24567
3088	J24820
3088	NTC-13
3088	OA-90
3088	OA-90G
3088	OA90
3088	OA90GA
3088	OA90M
3088	OA90Z
3088	OA99
3088	PTC206
3088	PTC206M
3088	Q-22115C
3088	QD-G1N60PXT
3088	QD-G1N60XXT
3088	QD-G1S32XXT
3088	QVM800B
3088	R7028
3088	R7029
3088	RT1667
3088	RT2451
3088	RT4880
3088	RT5212
3088	RT5213
3088	RT5379
3088	RT5470
3088	RT6728
3088	RT7851
3088	RV1478
3088	SD-1N60
3088	SD-1N60S
3088	SD020
3088	SS0007
3088	SS0008
3088	SU-31
3088	SVDOA70
3088	T-E1031
3088	TR0575005
3088	TR2083-41
3088	TV241013
3088	V10916-3
3088	YAAD009
3088	YEAD1N60P
3088	ZTR-1N60
3088	001-0081
3088	1-037/2207
3088	1-12689
3088	1A12689
3088	1DG2
3088	1K60
3088	1K60A
3088	1N46A
3088	00001N60
3088	1N60-5
3088	1N60-M3
3088	1N60-T
3088	1N60-Z
3088	1N60/7825B
3088	1N60A
3088	1N60AM
3088	1N60C
3088	1N60D
3088	1N60F
3088	1N60FD1
3088	1N60FM
3088	1N60G
3088	1N60GA
3088	1N60GB
3088	1N60P
3088	1N60S
3088	1N60TV
3088	1N60TVGL
3088	1NJ70973

SK	Industry Standard No.
3088	1S188FM-1
3088	1S426
3088	1S4266FM
3088	00001S446D
3088	1S60P
3088	1T23S
3088	1T261
3088	2-0A90
3088	2MW665
3088	03-160
3088	3L4-2001-1
3088	3L4-2001-1A
3088	3L4-2003-1
3088	3L4-2003-4
3088	3L4-3002-7
3088	4-2020-08600
3088	4-202A16
3088	4-282
3088	4-852
3088	4-853
3088	4-854
3088	4-855
3088	4-857
3088	05-00060-00
3088	05-00060-01
3088	05-00160-01
3088	05-170060
3088	05-490090
3088	05-932510
3088	7-59-001/3477
3088	08-08111
3088	08-08112
3088	8-719-422-21
3088	09-306036
3088	09-306037
3088	09-306061
3088	09-306219
3088	09-306290
3088	09-306331
3088	09-306336
3088	09-306349
3088	09-306370
3088	09-306376
3088	9DI2
3088	10-085014
3088	10-085018
3088	13-085002
3088	13-12001-0
3088	13-12003-0
3088	13-14094-11
3088	13-14094-14
3088	13-14890-1
3088	13-35621-1
3088	13-55166-1
3088	14-514-05
3088	14-514-11
3088	14-514-55
3088	14-514-61
3088	19-085018
3088	020-00030
3088	20A90
3088	20A90M
3088	20A90MLF
3088	20A90Z
3088	21A103-006
3088	21A103-016
3088	21A103-048
3088	21A103-052
3088	21A103-055
3088	21A109-001
3088	21A109-002
3088	21A109-022
3088	21M228
3088	21M288
3088	21M432
3088	21M568
3088	22-1-005
3088	22-1-129
3088	022-2823-003
3088	022-2823-007
3088	24MW1043
3088	24MW1051
3088	24MW665
3088	24MW785
3088	24MW820
3088	025-100027
3088	32-0013
3088	34-8022-6
3088	34-8057-29
3088	34-8057-3
3088	36E004-1
3088	42-23969
3088	42-27543
3088	42-27544
3088	42-28199
3088	46-8619-3
3088	48-155061
3088	48-155114
3088	48-41768G01
3088	48-86289-3
3088	48-86343-3
3088	48-90343A66
3088	48B41768G01
3088	48C65837A02
3088	48K644681
3088	48X97048A05
3088	48X97048A06
3088	48X97168A01
3088	48X97168A03
3088	51-04001-01
3088	52-050-021-0
3088	53A022-2
3088	53B010-1
3088	53K001-5
3088	53N001-7
3088	53N003-1
3088	53N003-2
3088	53N004-11
3088	53N004-6
3088	53T001-4
3088	53T001-5
3088	53V001-2
3088	56-8093
3088	65-14
3088	66X0049-001
3088	66X0049-002
3088	66X0049-100
3088	66X0051-001
3088	69-1820
3088	69-2922
3088	81-46123001-3
3088	81-46123006-2
3088	81-46123015-3
3088	86-0002

SK	Industry Standard No.
3088	86-0008
3088	86-22-1
3088	86-88-3
3088	089-236
3088	089-248
3088	089-293
3088	93A25-1
3088	93A25-2
3088	93A25-3
3088	93A41-2
3088	93A8-1
3088	93A83-1
3088	094-014
3088	94-42-9
3088	96-0008
3088	100-0051
3088	100-00910-07
3088	100-00914-10
3088	100-0124
3088	100-12
3088	100-136
3088	100-180
3088	100-181
3088	100-215
3088	100-436
3088	103-192
3088	103-74
3088	0112-0019
3088	0112-0028
3088	0112-0028-6438
3088	0112-0028/4460
3088	0112-0073
3088	0112-0082
3088	120-004730
3088	123-013
3088	130-40229
3088	150-001-9-005
3088	150-006-9-001
3088	150-013-9-001
3088	150-014-9-001
3088	185-013
3088	200X8000-026
3088	201X2000-118
3088	209-31
3088	230-0006
3088	264D00612
3088	264D00701
3088	264D00801
3088	264D00901
3088	264P01301
3088	264P01305
3088	264Z00701
3088	354-9001-001
3088	354-9101-002
3088	503-T21271
3088	510A90
3088	510ED46
3088	600X0096-066
3088	600X0097-066
3088	601X0150-066
3088	601X0151-066
3088	617-156
3088	642-221
3088	690V059H63
3088	690V092H85
3088	690V103H54
3088	690V110H88
3088	690V119H14
3088	754-1003-030
3088	754-2000-009
3088	754-5900-090
3088	754-9000-460
3088	903-103B
3088	903-108B
3088	903-113B
3088	903-13B
3088	903-167B
3088	903-168B
3088	903-16B
3088	903-171B
3088	903-27
3088	903-27B
3088	903-29B
3088	903-34
3088	903-34A
3088	903-34B
3088	903-34C
3088	903-34D
3088	903-34E
3088	903-37B
3088	903-43B
3088	903-51B
3088	903-54B
3088	903-8B
3088	903-92B
3088	903-9B
3088	914-000-4-00
3088	914-001-7-00
3088	1000-17
3088	1001-10
3088	1006-9292
3088	1016-77
3088	1019-6699
3088	1040-08
3088	1041-65
3088	1042-3
3088	1048-9870
3088	1063-8591
3088	1206-17
3088	1207-17
3088	1977-17
3088	1980-17
3088	2000-301
3088	2093A38-21
3088	2093A38-30
3088	2093A38-31
3088	2093A38-32
3088	2093A38-5
3088	2093A41-14
3088	2093A41-167
3088	2093A41-169
3088	2093A41-29
3088	2093A41-38
3088	2093A41-59
3088	2093A41-92
3088	2093A8-1
3088	2151-17
3088	2196-17
3088	2232-17
3088	2282-17
3088	2510-31
3088	2606-296
3088	3322-6
3088	03571
3088	4041-200-10180
3088	04770
3088	5001-080
3088	5001-134
3088	5001-141
3088	5001-161
3088	5361-1N60P
3088	5631-1N60
3088	5631-1N60P
3088	8000-00004-063
3088	8000-00005-015
3088	8000-00005-018
3088	8000-00005-023
3088	8000-00006-007
3088	8000-00011-042
3088	8000-00041-016
3088	8010-53
3088	8710-53
3088	8840-54
3088	8910-53
3088	10181
3088	18410-42
3088	18600-53
3088	25810-53
3088	25840-53
3088	26810-51
3088	27840-42
3088	28810-64
3088	37510-53
3088	38510-52
3088	45810-53
3088	55166
3088	67590
3088	72013
3088	000072090
3088	72128A
3088	000072160
3088	88060-53
3088	88510-53
3088	100017
3088	102989
3088	112526
3088	117659
3088	117760
3088	119199
3088	120617
3088	129157
3088	129158
3088	129474
3088	139634
3088	144585
3088	165572
3088	168910
3088	169362
3088	216001
3088	216003
3088	226344
3088	489752-001
3088	489752-031
3088	489752-049
3088	489752-125
3088	500009G
3088	0517022
3088	530065-1002A
3088	530065-1003
3088	0537820
3088	0575004
3088	0575005
3088	0575007
3088	740954
3088	742730
3088	771910
3088	972259-8
3088	988997
3088	988998
3088	992143
3088	1223931
3088	1471872-11
3088	2002336-20
3088	2006422-132
3088	2006441-122
3088	2095083
3088	2495083
3088	3596062
3088	5330331
3088	5330332
3088	5330731
3088	5330732
3088	06200002
3088	06200003
3088	7100460
3088	7852438-01
3088	7855282
3088	8121014
3088	13040229
3088	22115182
3088	22115192
3088	30600010
3088	37000918
3088	41027612
3088	41027613
3088	41027992
3088	41029290
3088	41527380
3088	43200103
3088	62009616
3088	87100600
3088	87100605
3088	87204260
3088	87500620
3088	600000060
3088	2008000001
3088	2008000026
3088	2012000092
3088	2012000100
3088	2012000118
3088	4120100600
3088	4120129000
3088	4120200602
3088	4129100600
3088	4129100602
3088	4129200602
3088	4202104770
3088	6612004000
3088	16400690060
3088	16401190188
3088	16415490000
3089	AA015
3089	ET16X14
3089	EU16X14
3089	EU16X4
3089	G01A
3089	HEP-R0700
3089	JT-1601-41
3089	K3E
3089	M8482
3089	MBD101
3089	RE48
3089	SD-404
3089	SD-51
3089	SD82
3089	SD82A
3089	SD82AG
3089	SP82A
3089	TH-1S750
3089	TV24103
3089	TV24103A
3089	TV24103B
3089	TV24103C
3089	TV24103D
3089	TV24159
3089	TV24182
3089	TVS-182G
3089	TVS-1N82G
3089	TVS-1S750
3089	TVS-82G
3089	TVS-S82
3089	TVS-SD82A
3089	1N82A
3089	1N82AG
3089	1N82G
3089	1S1925
3089	1S1926
3089	1S1926K
3089	1S2198
3089	1S750
3089	1SS16
3089	4JB2C6
3089	4JBC12
3089	07-5160-15A
3089	07-5160-15B
3089	07-5160-15C
3089	09-306077
3089	09-306089
3089	09-306209
3089	09-306216
3089	13-31014-4
3089	15-085018
3089	19A115322
3089	19A115322-P1
3089	23-PT275-124
3089	24E-002
3089	24E-022
3089	025-100008
3089	46-861-3
3089	46-8613
3089	46-86134-3
3089	46-8625
3089	46-8643
3089	46-8643-3
3089	47-4
3089	48-65112A73
3089	48-65113A84
3089	48-674297U
3089	48-674970
3089	48-742970
3089	48-90234A66
3089	48P65112A73
3089	57D1-65
3089	62-17232
3089	62-19260
3089	63-250128-2
3089	76-13570-65
3089	76-13848-23
3089	86-12-1
3089	93A43-1
3089	93A43-2
3089	93A59-1
3089	93D112-57
3089	103-49
3089	103-60
3089	103-61
3089	103-65
3089	229-0240-20
3089	264P03401
3089	2090A43-1
3089	2093A43-2
3089	2093A59-1
3089	8000-00012-038
3089	107729
3089	116314
3089	119662
3089	127532
3089	129556
3089	131214
3089	134074
3089	134264
3089	143162
3089	147922-2
3089	489752-020
3089	489752-052
3089	489752-089
3089	489752-090
3089	575037
3089	970047
3089	970759
3089	1442415-2
3089	1442415-3
3089	1471922-1
3089	62034297
3089	62139455
3089/112	ED7
3089/112	07-5160-15
3089/112	13-55333-1
3089/112	15-085037
3089/112	93A77-1
3089/112	103-202
3090	A066-121
3090	A42X00340-01
3090	A48-63078A52
3090	AA114
3090	AA121
3090	AA123
3090	AA139
3090	AA143S
3090	AA144
3090	AAZ10
3090	C60(DIODE)
3090	CB393
3090	CD-0000N
3090	CR0000
3090	CR101/6515
3090	CR102/6515
3090	CXL109
3090	DA90
3090	DI-8(DIODE)
3090	DS-18(DELCO)
3090	DS410(G.E.)
3090	DS410R(G.E.)
3090	DX6873
3090	E21430
3090	ECG109
3090	ED12(ELCOM)
3090	ED4(ELCOM)
3090	ED6(ELCOM)
3090	ED9(ELCOM)
3090	EDG-1
3090	ES16X103
3090	ETD-SD46
3090	FA-1(DIODE)
3090	G00003A
3090	G00009A
3090	G100
3090	G769
3090	G770
3090	GD1P
3090	GD3638-00
3090	GD556
3090	GP-354
3090	GS73E/3
3090	HD10000101
3090	HD10000302
3090	HD1000101-0
3090	HD10001010
3090	HD1468
3090	HD2149
3090	HD2155
3090	HE-10027
3090	HF-20008
3090	J241245
3090	J24911
3090	J24913
3090	J24914
3090	K52
3090	KD27
3090	KR-Q0005
3090	M8489-A
3090	MA3
3090	MC308
3090	MD604A
3090	MN51
3090	R-1S188
3090	R-7051
3090	RE47
3090	REN109
3090	RT5214
3090	RT5908
3090	RT5939
3090	RT6119
3090	RT6179
3090	RT6180
3090	RT6181
3090	RT6182
3090	RT6183
3090	RT6184
3090	RT6189
3090	RT6619
3090	RT7330
3090	RV1479
3090	S-21271
3090	S21271
3090	S3030G
3090	S3577G
3090	S3603G
3090	S3776B
3090	S3838GA
3090	S3885G
3090	S93.24.401
3090	S93.24.601
3090	S93.24.604
3090	SD-150
3090	SD-60
3090	SD46
3090	SDH-2
3090	SFD112
3090	SV-31(DIODE)
3090	SVD20A70
3090	SVD0A70
3090	SVD0A90
3090	T-E1014
3090	TM109
3090	WEP134
3090	001-0000-00
3090	001-0022-00
3090	00001S4460
3090	003-009000
3090	103-23-01
3090	103-23-1
3090	103-Z9001
3090	1512
3090	003016
3090	127017
3090	755722
3090	972258-8
3090	972258-9
3090	984163
3090	984666
3090	1476179-1
3090/109	MV4
3090/109	1S46
3090/109	2AA119
3090/109	13-29867-1
3090/109	103-23
3090/109	800-002-00
3091	OA180
3091	A36508
3091	OA7
3091	A73
3091	OA81
3091	OA9
3091	AA120
3091	AA130
3091	AA138
3091	AA140
3091	AA142
3091	AAY22
3091	AAY27
3091	B-30
3091	CG86H
3091	CGD1029
3091	CGD591
3091	CTP573
3091	CX-0036
3091	CX-0042
3091	DHD800
3091	DK19
3091	DR365
3091	DR427
3091	DR464
3091	ED6
3091	ES16X14
3091	ES16X4
3091	EU16X1
3091	EW168
3091	G158
3091	G198
3091	G200
3091	G700
3091	G702
3091	G789
3091	G790
3091	G821
3091	G822
3091	G823
3091	G824
3091	G845
3091	G868
3091	G869
3091	GD11E
3091	GD405
3091	GD72E/3
3091	GD72E/4
3091	GPM2NA
3091	H087
3091	H091
3091	H316
3091	H614
3091	HG1090
3091	HG5808
3091	NGP3002
3091	OA79
3091	RF33550-1
3091	RL252
3091	RL42
3091	RS1811
3091	SFD107
3091	SX-990
3091	T13
3091	T13G
3091	T14
3091	T14G
3091	T21G
3091	T22G
3091	T24G
3091	T27G
3091	TC3112006000
3091	TG-28
3091	TR320007
3091	TVS-OA70
3091	TVS-OA90
3091	TVS-OA91
3091	V-10916-3
3091	V-210C
3091	VD11
3091	1-425-636
3091	1N103
3091	1N104
3091	1N105
3091	1N132
3091	1N133
3091	1N295A
3091	1N3122
3091	1N3287
3091	1N3287W
3091	1N3467
3091	1N3468
3091	1N3483
3091	1N3592
3091	1N4088
3091	1N569
3091	1N571
3091	1N64
3091	1N64A
3091	1N64B
3091	1N64GA
3091	1N760
3091	1N87
3091	1N87A
3091	1N87G
3091	1N87S
3091	1N87T
3091	1P541
3091	1P542
3091	1S15
3091	1S188FMI
3091	1S188MPX
3091	1S188P
3091	1S82
3091	1T23
3091	1T23G
3091	1T262
3091	003-006700
3091	003-007500
3091	003-009600
3091	11-085008
3091	11-085014
3091	12-085006
3091	13-0004
3091	14-514-72
3091	21A009-000
3091	21A009-002
3091	21A020-005
3091	22-004003
3091	30P1
3091	48-63590A01
3091	48-97168A01
3091	48-97168A03
3091	48-97168A07
3091	48C134587
3091	86-20-1
3091	86-45-1
3091	93B41-2
3091	93C25-3
3091	93C8-1
3091	103-19
3091	103-44
3091	105-03
3091	150-002-9-001
3091	324-0108
3091	916-32003-2
3091	2093A41-50
3091	72128
3091	72129
3091	119919
3091	161001
3091	320007
3091	500001
3091	0525002H
3091	530063-10
3091	530063-4
3091	530072-15
3091	530072-8
3091	530092-1
3091	530105-1
3091	530105-1001
3091	801722

SK	Industry Standard No.	SK	Industry Standard No.	SK	Industry Standard No.	SK	Industry Standard No.	SK	Industry Standard No.
3091	942677-4	3093	WN-125	3094	.7ZM14B	3094	1M14Z	3095	1N4339B
3091	982270	3093	WN-125A	3094	.7ZM14C	3094	1M14Z1Q	3095	1N4416
3091	982271	3093	WN-125B	3094	.7ZM14D	3094	1M14Z5	3095	1N4416A
3091	982275	3093	WN-125C	3094	AZ-140	3094	1N2037-3	3095	1N4416B
3091	982290	3093	WN-125D	3094	AZ7	3094	1N4108	3095	1N4673
3091	985999	3093	WO-125	3094	BZ140	3094	1N4701	3095	1N4714
3091	2000648-26	3093	WO-125A	3094	CXL144	3094	1N766-3	3095	1N4752
3091	2008299-2	3093	WO-125C	3094	CXL5024	3094	1T14	3095	1N4752A
3092	.25T11.5	3093	WO-125D	3094	CZD014	3094	1T14B	3095	1Z33
3092	.25T11.5B	3093	WP-125	3094	CZD014-5	3094	1TA14	3095	1Z33A
3092	.4T11.5	3093	WP-125A	3094	D1T	3094	1TA14A	3095	1Z33B
3092	.4T11.5A	3093	WP-125B	3094	D1ZRED	3094	1ZF14T1Q	3095	18Z41
3092	AV01-07	3093	WP-125C	3094	D4P	3094	1ZF14T2Q	3095	1Z33
3092	AZ-115	3093	WP-125D	3094	EO771-6	3094	1ZF14T5	3095	1TA33
3092	CXL141	3093	WQ-125	3094	ECG144	3094	09-306278	3095	1TA33A
3092	D2G	3093	WQ-125A	3094	EC0144A	3094	27Z6	3095	1Z33A
3092	D2G-1	3093	WQ-125C	3094	EQAO1-14R	3094	27Z6A	3095	1Z33B
3092	D2G-2	3093	WQ-125D	3094	EQAO1-14RD	3094	34M14Z	3095	1Z33C
3092	D3N	3093	WR-125	3094	FZ14T10	3094	34M14Z1Q	3095	1Z33D
3092	D4N	3093	WR-125A	3094	FZ14T5	3094	34M14Z5	3095	1Z33D10
3092	EA16X4	3093	WR-125B	3094	GEZD-14	3094	34Z14Q	3095	1Z33D5
3092	EVR11	3093	WR-125D	3094	HD5001109	3094	34Z14D1Q	3095	1Z33T10
3092	EVR11A	3093	WS-125	3094	HD30011090	3094	34Z14D5	3095	1Z33T20
3092	EVR11B	3093	WS-125B	3094	HEP-Z0417	3094	46-86341-3	3095	1Z33T5
3092	FA8006	3093	WS-125C	3094	M4659	3094	46-86401-3	3095	1ZF33
3092	FA8006A	3093	WS-125D	3094	M2I014	3094	48-134659	3095	1ZF33B
3092	FA8007	3093	WT-125	3094	OA126/14	3094	48-137017	3095	1ZF33T10
3092	FA8008	3093	WT-125A	3094	QZ14T10	3094	48-137048	3095	1ZF33T20
3092	GEZD-11.5	3093	WT-125B	3094	QZ14T5	3094	488137017	3095	1ZF33T5
3092	HD3000109-0	3093	WT-125C	3094	RE120	3094	50Z14	3095	1ZT33B
3092	HW11B	3093	WT-125D	3094	REN144	3094	50Z14A	3095	13-33186-1
3092	LPM11	3093	WU-125	3094	SV1019	3094	50Z14B	3095	48-134663
3092	M4850	3093	WU-125A	3094	SV137	3094	50Z14C	3095	48-134859
3092	M4851	3093	WU-125B	3094	SV4014	3094	75Z14	3095	488137266
3092	ME-110MA	3093	WU-125C	3094	SV4014A	3094	75Z14A	3095	488137272
3092	MZ-11	3093	WU-125D	3094	SZ14	3094	75Z14B	3095	50Z33
3092	SV11021	3093	WV-125	3094	SZ961-R	3094	75Z14C	3095	50Z33A
3092	SVC1125	3093	WV-125A	3094	TVSQA01-14RD	3094	103-Z9012	3095	50Z33B
3092	SVC1150	3093	WV-125C	3094	VR14	3094	4552	3095	50Z33C
3092	SVM1020	3093	WV-125D	3094	VR14A	3094	165740	3095	52-053-005-0
3092	SVM1105	3093	WW-125	3094	VR14B	3094	530073-1023	3095	56C7-1
3092	SVM111	3093	WW-125A	3094	WEP606	3094	530073-23	3095	75Z33
3092	SZ11.0	3093	WW-125B	3094	WL-140	3094	2330305	3095	75Z33A
3092	TVS-ZB1-11	3093	WW-125C	3094	WL-140A	3095	.25T33	3095	75Z33B
3092	TVS1N4741	3093	WW-125D	3094	WL-140B	3095	.25T33A	3095	75Z33C
3092	TVS1N4741A	3093	WW-130	3094	WL-140D	3095	.25T33B	3095	86-94-1
3092	ZB1-11	3093	WW-130A	3094	WM-140	3095	.4T33	3095	93A39-40
3092	1E11Z	3093	WW-130C	3094	WM-140A	3095	.4T33A	3095	103-256
3092	1E11Z10	3093	WW-130D	3094	WM-140B	3095	.4T33B	3095	152-006-9-001
3092	1N3580	3093	WX-125	3094	WM-140D	3095	.75N33	3095	152-029-9-002
3092	1N3580A	3093	WX-125A	3094	WN-140	3095	.7Z33	3095	264P11009
3092	1N3580B	3093	WX-125C	3094	WN-140A	3095	.7Z33A	3095	1081-3186
3092	1N3581	3093	WX-125D	3094	WN-140B	3095	.7Z33B	3095	1133
3092	1N3581A	3093	WY-125	3094	WN-140D	3095	.7Z33D	3095	4663
3092	1N3581B	3093	WY-125A	3094	WO-140	3095	.72M33A	3095	485B
3092	1N3582A	3093	WY-125B	3094	WO-140A	3095	.72M33B	3095	5132
3092	1N3582B	3093	WY-125C	3094	WO-140C	3095	.72M33D	3095	5232
3092	1N3584	3093	WY-125D	3094	WO-140D	3095	AA10	3095	8000-00006-01Q
3092	1N3584A	3093	WZ-125	3094	WP-140	3095	AA10-1	3095	179733A
3092	1N3584B	3093	WZ13B	3094	WP-140A	3095	AN155	3095	0099202-128
3092	1N4299	3093	ZO1-13	3094	WP-140B	3095	AV10	3095	99202-228
3092	1N4299A	3093	1N2046	3094	WP-140D	3095	AV2033	3095	136721
3092	1N4303	3093	1N4896A	3094	WQ-140	3095	AW-01-33	3095	141302
3092	1N4583	3093	1N4897	3094	WQ-140A	3095	AWO1-33	3095	142670
3092	1N4583A	3093	1N4897A	3094	WQ-140C	3095	BZX27	3095	146138
3092	1N4583B	3093	1N4898	3094	WR-140	3095	BZY29033	3095	1478164-1
3092	1N492	3093	1N4898A	3094	WR-140A	3095	BZY92033	3095	2327076
3092	1N492A	3093	1N4899	3094	WR-140C	3095	CD31-00025	3096	.25T55
3092	1N492B	3093	1N4899A	3094	WS-140	3095	CD31-12032	3096	.25T55B
3092	1N493	3093	1N4900A	3094	WS-140A	3095	D2E	3096	.7Z56
3092	1N493A	3093	1N4901	3094	WS-140C	3095	D4H	3096	.7Z56A
3092	1N493B	3093	1N4901A	3094	WS-140D	3095	D4J	3096	.7Z56D
3092	1N941	3093	1N4902	3094	WT-140	3095	DZ33A	3096	AV2055
3092	1N941A	3093	1N4902A	3094	WT-140A	3095	EQAO1-32R	3096	EVR56
3092	1N941B	3093	1N4903	3094	WT-140C	3095	EQBO1-33	3096	EVR56A
3092	1N942	3093	1N4903A	3094	WT-140D	3095	FZ33A	3096	EVR56B
3092	1N945	3093	1N4904	3094	WU-140	3095	GEZD-33	3096	GEZD-55
3092	1N945A	3093	1N4904A	3094	WU-140A	3095	HEP-Z0426	3096	LPM56
3092	1N945B	3093	1N4905	3094	WU-140C	3095	H82330	3096	SV4055
3092	1N946	3093	1N4905A	3094	WV-140	3095	H87330	3096	SV4055A
3092	1N946A	3093	1N4906	3094	WV-140A	3095	HW33	3096	SZ1200
3092	1N946B	3093	1N4906A	3094	WV-140C	3095	HW33A	3096	1A56M
3092	18135	3093	1N4907	3094	WW-140	3095	HW33B	3096	1A56MA
3092	1T11.5	3093	1N4907A	3094	WW-140A	3095	LPM33	3096	1Z55
3092	1T11.5B	3093	1N4908	3094	WW-140C	3095	LPM33A	3096	1Z55B
3092	1TA11.5	3093	1N4908A	3094	WX-140	3095	LPZZ33	3096	1TA55
3092	1TA11.5A	3093	1N4909	3094	WX-140A	3095	M4665	3096	1TA55A
3092	14-615-18	3093	1N4909A	3094	WX-140C	3095	M4858	3096	1Z55
3092	34-8057-56	3093	1N4910	3094	WY-140	3095	M4Z33	3096	1Z55B
3092	48-134850	3093	1N4910A	3094	WY-140A	3095	MC6026	3096	48-134643
3092	48-134851	3093	1N4911	3094	WY-140B	3095	MC6026A	3096	48-134971
3092	48-137188	3093	1N4911A	3094	WY-140C	3095	MC6126	3096	75Z56
3092	50Z11	3093	1N4912	3094	WY-140D	3095	MC6126A	3096	75Z56A
3092	50Z11A	3093	1N4912A	3094	WZ14	3095	MZ33A	3096	75Z56B
3092	50Z11B	3093	1N4913	3094	WZ537	3095	PTC512	3096	75Z56C
3092	50Z11C	3093	1N4913A	3094	WZ919	3095	RD-35A	3096	0086
3092	56-55	3093	1N4914	3094	ZO417	3095	RE130	3097	79949
3092	75Z11	3093	1N4914A	3094	Z4X14	3095	RH-EX0033CEZZ	3097	.25Z62
3092	75Z11A	3093	1N4915	3094	Z4X14A	3095	RH-IX0037CEZZ	3097	.25Z62A
3092	75Z11B	3093	1N4915A	3094	Z4XL14	3095	RS-35	3097	.25Z62B
3092	75Z11C	3093	18136	3094	Z4XL14B	3095	SV4033	3097	.75N62
3092	21Z4	3093	1T12.8	3094	ZB1-14	3095	SV4033A	3097	.7JZ62
3092	297W019B04	3093	1T12.8B	3094	ZEN507	3095	TO33A5A	3097	.7Z62B
3092	311Z4	3093	1T13	3094	ZXL-14	3095	TR-14002-10	3097	.7Z62C
3092	412Z4	3093	1T13A	3094	ZXY14B	3095	TVS8AN155	3097	.7Z62D
3092	4850	3093	1T13B	3094	1/4A12B	3095	1A33M	3097	.7ZM62A
3092	71411-1	3093	1TA12.8	3094	1/4A14	3095	1A33MA	3097	.7ZM62B
3093	.25T12.8	3093	1TA12.8A	3094	1/4A14A	3095	1AO33	3097	.7ZM62C
3093	.25T12.8A	3093	4-2020-06600	3094	1/4A14B	3095	1AO33A	3097	.7ZM62D
3093	.25T12.8B	3093	7-23(SARKES)	3094	1/4Z14D	3095	1AO33B	3097	AV2062
3093	.4T12.8	3093	7-24(SARKES)	3094	1/4Z14D1Q	3095	1C33Z	3097	BZX29062
3093	.4T12.8A	3093	7-25(SARKES)	3094	1/4Z14D5	3095	1C33ZA	3097	BZX94062
3093	AWO1-15	3093	86-61-1	3094	1E14Z	3095	1E33Z	3097	CD3171
3093	BZ-125	3093	93A39-31	3094	1E14Z1Q	3095	1E33Z1Q	3097	GEZD-62
3093	CXL143	3093	264P71402	3094	1E14Z5	3095	1Z33Z5	3097	HEP-Z0433
3093	D1ZV10	3093	600X0101-066			3095	1M33Z	3097	HW62
3093	D5W	3093	4169(PENNCREST)			3095	1M33Z10	3097	HW62A
3093	D2I2	3093	171092-1			3095	1M33ZB5	3097	HW62B
3093	QRT-244	3093	530157-130			3095	1N1882	3097	LPM62
3093	RH-EX0017CEZZ	3093	2331152			3095	1N1882A	3097	M4Z62
3093	T-E1106	3094	.25T14			3095	1N3032	3097	M4Z62A
3093	U213B	3094	.25T14A			3095	1N3032A	3097	PTC514
3093	VZ-125	3094	.25T14B			3095	1N3530	3097	RB132
3093	WEP605	3094	.4T14			3095	1N3691	3097	SV4062
3093	WL-125	3094	.4T14A			3095	1N3691A	3097	SV4062A
3093	WL-125A	3094	.7JZ14			3095	1N3691B	3097	1AC62
3093	WL-125B					3095	1N4174	3097	1AC62A
3093	WL-125D					3095	1N4174A	3097	1AC62B
3093	WM-125					3095	1N4174B	3097	1C62Z
3093	WM-125A					3095	1N4339A		
3093	WM-125B								
3093	WM-125C								
3093	WM-125D								

SK	Industry Standard No.
3097	1C62ZA
3097	1E62Z
3097	1E62Z10
3097	1E62Z5
3097	1M62Z
3097	1M62Z10
3097	1M62Z5
3097	1M62ZS10
3097	1M62ZS5
3097	1N3039
3097	1N3039A
3097	1N3039B
3097	1N3698
3097	1N3698A
3097	1N3698B
3097	1N4181
3097	1N4181A
3097	1N4181B
3097	1N4346
3097	1N4346A
3097	1N4346B
3097	1N4423
3097	1N4423A
3097	1N4423B
3097	1N4759
3097	1N4759A
3097	1N733
3097	1N733A
3097	1R62
3097	1R62A
3097	1R62B
3097	1S250
3097	1T62
3097	1T62B
3097	1TA62
3097	1TA62A
3097	1Z62
3097	1Z62A
3097	1Z62B
3097	1Z62C
3097	1Z62D
3097	1Z62D10
3097	1Z62D5
3097	48-134955
3097	50Z56
3097	50Z56A
3097	50Z56B
3097	50Z62
3097	50Z62A
3097	50Z62B
3097	50Z62C
3097	66X0040-012
3097	75Z62
3097	75Z62A
3097	75Z62B
3097	75Z62C
3097	93A39-25
3097	5139
3097	5239
3097	138107
3097	62746491
3098	.25T82
3098	.25T82A
3098	.25T82B
3098	.75N82
3098	.7JZ82
3098	.7Z82A
3098	.7Z82B
3098	.7Z82C
3098	.7Z82D
3098	.7ZM82A
3098	.7ZM82B
3098	.7ZM82C
3098	.7ZM82D
3098	AV2082
3098	BZX29C82
3098	CD3174
3098	DZ82A
3098	EVR82
3098	EVR82A
3098	EVR82B
3098	FZ82A
3098	GEZD-82
3098	HEP-Z0436
3098	HW82
3098	HW82A
3098	HW82B
3098	LPM82
3098	LPM82A
3098	M4699
3098	M4704
3098	M4Z82
3098	MZ82A
3098	PS6325
3098	PTC515
3098	SV4082
3098	SV4082A
3098	TC82A5B
3098	1A82M
3098	1A82MA
3098	1AC82
3098	1AC82A
3098	1AC82B
3098	1C82Z
3098	1C82ZA
3098	1E82Z
3098	1E82Z10
3098	1E82Z5
3098	1M82Z
3098	1M82Z10
3098	1M82Z5
3098	1M82ZS10
3098	1M82ZS5
3098	1N1793
3098	1N1793A
3098	1N1887
3098	1N1887A
3098	1N3042
3098	1N3042A
3098	1N3042B
3098	1N3701
3098	1N3701A
3098	1N3701B
3098	1N4184
3098	1N4184A
3098	1N4184B
3098	1N4349
3098	1N4349A
3098	1N4349B
3098	1N4426
3098	1N4426A
3098	1N4426B
3098	1N4762
3098	1N4762A

SK	Industry Standard No.
3098	1N736
3098	1N736A
3098	1R82
3098	1R82A
3098	1R82B
3098	1S253
3098	1T82
3098	1T82B
3098	1TA82
3098	1TA82A
3098	1Z82
3098	1Z82A
3098	1Z82B
3098	1Z82C
3098	1Z82D
3098	1Z82D10
3098	1Z82D5
3098	1ZB82
3098	1ZB82B
3098	1ZT82
3098	1ZT82B
3098	8-905-421-315
3098	13-0029
3098	48-134699
3098	48-134704
3098	50Z82
3098	50Z82A
3098	50Z82B
3098	50Z82C
3098	75Z82
3098	75Z82A
3098	75Z82B
3098	75Z82C
3098	187-1
3098	4699
3098	4704
3098	5142
3098	5242
3099	.25T110
3099	.25T110B
3099	.4T110
3099	.4T110A
3099	.75N110
3099	.7Z110A
3099	.7Z110C
3099	.7Z110D
3099	.7ZM110A
3099	.7ZM110B
3099	.7ZM110C
3099	.7ZM110D
3099	AV2110
3099	EVR110
3099	EVR110A
3099	EVR110B
3099	HEP-Z0439
3099	HW110
3099	HW110A
3099	LPM110
3099	LPM110A
3099	M4728
3099	M4Z110
3099	PTC516
3099	TC110A5B
3099	ZH110
3099	1A110M
3099	1A110MA
3099	1AC110
3099	1AC110A
3099	1AC110B
3099	1C110Z
3099	1C110ZA
3099	1E110Z
3099	1E110Z10
3099	1E110Z5
3099	1M110Z
3099	1M110Z10
3099	1M110Z3
3099	1M110Z5
3099	1M110ZS10
3099	1M110ZS5
3099	1N1796
3099	1N1796A
3099	1N3045
3099	1N3045A
3099	1N3045B
3099	1N3704
3099	1N3704A
3099	1N3704B
3099	1N4187
3099	1N4187A
3099	1N4187B
3099	1N4352
3099	1N4352A
3099	1N4352B
3099	1N4429
3099	1N4429A
3099	1N4429B
3099	1N739
3099	1N739A
3099	1R110
3099	1R110A
3099	1R110B
3099	1S257
3099	1Z110
3099	1Z110A
3099	1Z110B
3099	1Z110C
3099	1Z110D
3099	1Z110D10
3099	1Z110D5
3099	4-2020-14000
3099	17-1
3099	48-134728
3099	50Z110
3099	50Z110A
3099	50Z110B
3099	50Z110C
3099	56-18
3099	56-29
3099	56-48
3099	75Z110
3099	75Z110A
3099	75Z110B
3099	75Z110C
3099	187-2
3099	4728
3099	5145
3099	5245
3099	530073-1020
3099	1960642
3099/118	ZV9.1
3100	A066-119
3100	A691M4
3100	A72-49-600

SK	Industry Standard No.
3100	A7249601
3100	BA317
3100	BAW62
3100	BAX-13
3100	BAX13
3100	CXL519
3100	D1J70545
3100	DASZ158800
3100	DDAY047001
3100	DDAY047005
3100	DDAY048007
3100	DDAY048012
3100	DDAY048014
3100	DDAY069001
3100	DI-20
3100	DS448
3100	DX-0255
3100	DX-0273
3100	DX-0299
3100	EA16X135
3100	ECG519
3100	EVD-3
3100	EX16X10
3100	FD111
3100	FD777
3100	FDH444
3100	FDH666
3100	FDN666
3100	GE-514
3100	HF-20061
3100	HF-20063
3100	HF-20124
3100	HSFD-1Z
3100	HV23(DIODE)
3100	IC743049
3100	IP20-0021
3100	IP20-0061
3100	IP20-0216
3100	IP20-0282
3100	IR10E6X
3100	ITT73N
3100	KR-162
3100	LD34R
3100	M8640E
3100	MA150
3100	MA150TA
3100	MA161
3100	MA162
3100	NTC-15
3100	Q-1N914
3100	Q-23115C
3100	R8314
3100	RE94
3100	REN94
3100	RH-DX0048CEZZ
3100	RS-253S
3100	RS-267S
3100	RS-272US
3100	RS-280S
3100	SN4448
3100	TM519
3100	TR2083-44
3100	UO-5E
3100	VHD1S2076-//-1
3100	WEP925
3100	WH1012
3100	001-0112-00
3100	1N215
3100	1N3062
3100	1N3600
3100	1N4148
3100	1N4149
3100	1N4153
3100	1N4154
3100	1N4305
3100	1N4446
3100	1N4447
3100	1N4448
3100	1N4449
3100	1N4454
3100	1N5282
3100	1S-2472
3100	1S1207
3100	1S1553
3100	1S1555V
3100	1S1588
3100	1S1992
3100	1S2075
3100	1S2075K
3100	1S2134
3100	1S2471
3100	1S2472
3100	1S2473
3100	1S2473-T72
3100	1S2473H
3100	1S2473HC
3100	1S2473K
3100	1S2473VE
3100	1S2597
3100	1S3016R
3100	1S954
3100	05-182076
3100	05-330161
3100	05A01
3100	05A05
3100	05A06
3100	09-3060111
3100	09-306135
3100	09-306200
3100	09-306244
3100	9DI1
3100	012-1021-001
3100	13-0003
3100	13-17596-10
3100	13-17596-2
3100	13-17596-3
3100	13-17596-4
3100	13-17596-5
3100	13-17596-7
3100	13-17596-8
3100	13-17596-9
3100	14-514-62
3100	14-514-64
3100	21A108-003
3100	022-2823-005
3100	23-001R03A10
3100	23-001R03AA10
3100	034-032-0
3100	46-86309-3
3100	48-03005A06
3100	48-155047
3100	48-155060
3100	48C134816
3100	48G10346A01

SK	Industry Standard No.
3100	48M355014
3100	48M355035
3100	48S10577A13
3100	48S67120A13
3100	51-08001-11
3100	53B001-9
3100	56-56
3100	66X0046-001
3100	66X0056-001
3100	69-1822
3100	86-574-2
3100	93C27-2
3100	93C27-3
3100	93EVHD1N4148//
3100	094-007
3100	102-412
3100	106-007
3100	123-017(DIODE)
3100	144-1
3100	150-066-9-001
3100	151-015-9-001
3100	151-035-9-001
3100	151-035-9-004
3100	151-051-9-001
3100	151-059-9-001
3100	151-064-9-001
3100	151-064-9-002
3100	151-069-9-001
3100	171-1
3100	201X2010-144
3100	201X2010-159
3100	212-96
3100	264P04501
3100	264P04502
3100	264P04507
3100	523-0006-002
3100	523-0007-001
3100	523-1500-883
3100	754-9000-473
3100	754-9052-473
3100	1010A
3100	1042-15
3100	1074-118
3100	2000-302
3100	2000-303
3100	2000-317
3100	2000-332
3100	2019-45
3100	2409-17
3100	003114
3100	003307
3100	4805-1241-200
3100	5001-120
3100	5001-128
3100	5001-146
3100	8000-00006-281
3100	8000-00042-007
3100	144581
3100	144691
3100	146571
3100	146576
3100	280017
3100	530072-1006
3100	530135-2
3100	530144-1004
3100	760037
3100	923147
3100	1223771
3100	1476690-1
3100	1800009
3100	2006441-132
3100	2330351
3100	2331351
3100	5330131
3100	5330572
3100	06200013
3100	8120126
3100	2012010144
3100	4129400003
3100	4202013600
3100	4202107670
3100	16411992473
3100/519	A01
3100/519	A054-228
3100/519	A066-12
3100/519	A691M5
3100/519	A691M5-2
3100/519	A7246711
3100/519	AG30
3100/519	ALC-1A
3100/519	AV0000105-0
3100/519	B-1599
3100/519	B-1702
3100/519	BA104
3100/519	BA147
3100/519	BA167
3100/519	BA168
3100/519	BA170
3100/519	BA174
3100/519	BA182
3100/519	BA216
3100/519	BA219
3100/519	BA316
3100/519	BAW24
3100/519	BAW24A
3100/519	BAW24B
3100/519	BAW25
3100/519	BAW25A
3100/519	BAW25B
3100/519	BAW45
3100/519	BAW53
3100/519	BAW59
3100/519	BAW59A
3100/519	BAW59B
3100/519	BAW63A
3100/519	BAW63B
3100/519	BAW99
3100/519	BAX-16
3100/519	BAX28
3100/519	BAX30
3100/519	BAX74
3100/519	BAX87
3100/519	BAX87A
3100/519	BAX88TF11
3100/519	BAX89A
3100/519	BAY31
3100/519	BAY36
3100/519	BAY41
3100/519	BAY41A
3100/519	BAY41B
3100/519	BAY52
3100/519	BAY68
3100/519	BSA01
3100/519	BSA02

SK	Industry Standard No.
3100/519	BSA11
3100/519	BZ1021V4
3100/519	C-10-20A
3100/519	C10-22C
3100/519	C1A
3100/519	CD-0014N
3100/519	CD-0021
3100/519	CD-20
3100/519	CD-37
3100/519	CD-84857
3100/519	CD0000NC
3100/519	CD37
3100/519	CD37A
3100/519	CD8457
3100/519	CD860011
3100/519	CD86003
3100/519	CD860037
3100/519	CDG-00
3100/519	CDG-21
3100/519	CDG-22
3100/519	CDG00
3100/519	CDG025
3100/519	CDG24
3100/519	CDG27
3100/519	CDJ-00
3100/519	CE37
3100/519	D1-20
3100/519	D300
3100/519	D353
3100/519	D5V
3100/519	D7Z
3100/519	DAAY003002
3100/519	DAAY004001
3100/519	DC8457
3100/519	DDAY004001
3100/519	DDAY048001
3100/519	DDAY048008
3100/519	DFFY007001
3100/519	DHD8001
3100/519	DI-1
3100/519	DI-3
3100/519	DIC
3100/519	DIJ70486
3100/519	DIJ71711
3100/519	DIJ71895-1
3100/519	DIJ71960
3100/519	DIJ72163
3100/519	DIJ72164
3100/519	DND800
3100/519	DS-117
3100/519	DS-31
3100/519	DS-410
3100/519	DS104
3100/519	DS443
3100/519	DSA150
3100/519	DSS953
3100/519	E1176R
3100/519	E13-017-01
3100/519	EA1405
3100/519	EA1661
3100/519	EA16X101
3100/519	EA16X110
3100/519	EA16X134
3100/519	EA16X146
3100/519	EA16X39
3100/519	EA16X49
3100/519	EA16X60
3100/519	EA16X61
3100/519	EA16X68
3100/519	EA16X69
3100/519	EA16X84
3100/519	EA16X92
3100/519	EA2607
3100/519	EA2608
3100/519	EA3447
3100/519	ED514721
3100/519	ED515790
3100/519	ED516420
3100/519	ED536062
3100/519	ED560913
3100/519	EDS-0001
3100/519	EDS-0014
3100/519	EDS-0024
3100/519	EDS-1
3100/519	EDS-25
3100/519	EP-2798
3100/519	EP16X1
3100/519	EP16X20
3100/519	EP16X23
3100/519	EP16X27
3100/519	EP6X10
3100/519	ES1627
3100/519	ES16X23
3100/519	ES16X30
3100/519	ES16X32
3100/519	ES16X40
3100/519	ETD-SG9150
3100/519	ETD1S2788
3100/519	ETDCDG21
3100/519	FCD0003PC
3100/519	FCD0014NCS
3100/519	FDH400
3100/519	FDH6229
3100/519	FDH900
3100/519	G00-014-A
3100/519	G00-803-A
3100/519	G01-083-A
3100/519	G01-209-B
3100/519	G01-803A
3100/519	G01209
3100/519	G01211
3100/519	G01803
3100/519	G1010A
3100/519	GE6063
3100/519	G01209
3100/519	G01211
3100/519	G01803
3100/519	GP2-345
3100/519	HD2000206
3100/519	HD2001105
3100/519	HD20011050
3100/519	HP-20048
3100/519	HP-20060
3100/519	HP-20064
3100/519	HP-20095
3100/519	HP-DS410
3100/519	HP-MV2
3100/519	HN-00002
3100/519	HV-25
3100/519	HV-27
3100/519	HV00001050
3100/519	HV0000405
3100/519	IC743051
3100/519	INJ61433

SK	Industry Standard No.
3100/519	INJ61677
3100/519	INJ61725
3100/519	IP20-0018
3100/519	IP20-0020
3100/519	IP20-0145
3100/519	IP20-0184
3100/519	IS-446D
3100/519	ITT7215
3100/519	ITT73
3100/519	ITT921
3100/519	J241182
3100/519	J241242
3100/519	J24755
3100/519	JL40A
3100/519	JM-40
3100/519	JM40
3100/519	KB102
3100/519	KD300A
3100/519	KGE46109
3100/519	KLH4763
3100/519	KX-1
3100/519	LM-1159
3100/519	M1-301
3100/519	M150-1
3100/519	M8640
3100/519	MA206
3100/519	MC2326
3100/519	MC5321
3100/519	MDP173
3100/519	ME-4
3100/519	MI-301
3100/519	MI301
3100/519	MPC3500
3100/519	MPN-3401
3100/519	MPS9444
3100/519	MPS9606
3100/519	MPS9606H
3100/519	MPS9606I
3100/519	MPS9644
3100/519	MPS9646
3100/519	MPS9646G
3100/519	MPS9646H
3100/519	MPS9646J
3100/519	MSS1000
3100/519	MV-12
3100/519	MZ-00
3100/519	MZ2360
3100/519	MZ2361
3100/519	09-306113
3100/519	OA205
3100/519	OF156
3100/519	OF162
3100/519	OF66
3100/519	P-6006
3100/519	P1172
3100/519	P1172-1
3100/519	P4326
3100/519	P7394
3100/519	PD137
3100/519	PS700
3100/519	PS720
3100/519	Q-21115C
3100/519	Q-24115C
3100/519	QD-SMA150XN
3100/519	QD-SS1555XT
3100/519	R-7027
3100/519	RA1
3100/519	REJ71253
3100/519	RH-DX0025CEZZ
3100/519	RH-DX0033TAZZ
3100/519	RH-DX0046CEZZ
3100/519	RH-DX0054CEZZ
3100/519	RS1428
3100/519	RT1689
3100/519	RT218
3100/519	RT5217
3100/519	RT5554
3100/519	RT7946
3100/519	RV1226
3100/519	RV2071
3100/519	RVDKB265J2
3100/519	RVDVD1250M
3100/519	S-3016R
3100/519	S-34
3100/519	S04
3100/519	S04-1
3100/519	S04A-1
3100/519	S04B-1
3100/519	S1-RECT-35
3100/519	S1428
3100/519	S180
3100/519	S1D50B851-A
3100/519	S3004-1716
3100/519	S3072C
3100/519	S502
3100/519	S502A
3100/519	S502B
3100/519	S506
3100/519	S506A
3100/519	S506B
3100/519	S509
3100/519	S509A
3100/519	S509B
3100/519	SB-2CGL
3100/519	SB01-02
3100/519	SD-110
3100/519	SD-43
3100/519	SD-5
3100/519	SD100
3100/519	SD600
3100/519	SD630
3100/519	SD974
3100/519	SFD43
3100/519	SFD83
3100/519	SG9150
3100/519	SI-RECT-152
3100/519	SI-RECT-35
3100/519	SP101
3100/519	SR23
3100/519	STB01-02
3100/519	SV-3
3100/519	SV-3B
3100/519	SV-3C
3100/519	SV-8
3100/519	SV-9
3100/519	SVDVD1121
3100/519	SX780
3100/519	T-E1118
3100/519	T-E1119
3100/519	T-E1121
3100/519	T-E1138
3100/519	T12
3100/519	T12-1
3100/519	T12-2
3100/519	T12A
3100/519	T12B
3100/519	T12C
3100/519	T16
3100/519	TD-15-BL
3100/519	TD81515
3100/519	TD960016-1M
3100/519	TF34
3100/519	TF44
3100/519	TF44J
3100/519	TR2083-42
3100/519	TR2337011
3100/519	TSC136
3100/519	TVM-TCO.2P11/2
3100/519	TVS-1S1211
3100/519	TVS-0A81
3100/519	TVS1S2076
3100/519	TVSBAX-13
3100/519	TVSBAX13
3100/519	TVSGP2-354
3100/519	TVSMA26
3100/519	TVSRA1A
3100/519	TZ1153
3100/519	UG1888
3100/519	US1555
3100/519	V1112
3100/519	V50260-10
3100/519	V50260-36
3100/519	VAR
3100/519	VD1127
3100/519	VHD1S1555-R-1
3100/519	VHD1S2076-1
3100/519	VHV1S1209-1
3100/519	WD4
3100/519	WG-1010A
3100/519	WG-1012
3100/519	WG-599
3100/519	WG-713
3100/519	000WG1010
3100/519	WG1010A
3100/519	WG1010B
3100/519	WG1012
3100/519	WG1014A
3100/519	WG1021
3100/519	WG713
3100/519	WG714
3100/519	WS100
3100/519	WS100A
3100/519	WS100B
3100/519	WS100C
3100/519	WS200
3100/519	WS200A
3100/519	WS200B
3100/519	WS200C
3100/519	WS300
3100/519	WS300A
3100/519	WS300B
3100/519	WS300C
3100/519	WSD002C
3100/519	YAAD010
3100/519	YAAD018
3100/519	YD1121
3100/519	ZS142
3100/519	ZS40
3100/519	ZS4F
3100/519	ZTR-W06B
3100/519	1-001/2207
3100/519	1-014/2207
3100/519	001-026010
3100/519	1-16549
3100/519	001-226010
3100/519	1A16549
3100/519	1A16551
3100/519	1N135
3100/519	1N137A
3100/519	1N137B
3100/519	1N138
3100/519	1N138A
3100/519	1N138B
3100/519	1N1448
3100/519	1N194
3100/519	1N194A
3100/519	1N195
3100/519	1N196
3100/519	1N200
3100/519	1N202
3100/519	1N203
3100/519	1N204
3100/519	1N205
3100/519	1N2075K
3100/519	1N210
3100/519	1N211
3100/519	1N212
3100/519	1N213
3100/519	1N216
3100/519	1N217
3100/519	1N218
3100/519	1N251
3100/519	1N251A
3100/519	1N252A
3100/519	1N300
3100/519	1N300A
3100/519	1N300B
3100/519	1N301
3100/519	1N301A
3100/519	1N301B
3100/519	1N303
3100/519	1N303A
3100/519	1N303B
3100/519	1N3123
3100/519	1N3124
3100/519	1N3147
3100/519	1N3197
3100/519	1N3206
3100/519	1N3207
3100/519	1N3223
3100/519	1N331
3100/519	1N3471
3100/519	1N3550
3100/519	1N3575
3100/519	1N3576
3100/519	1N3577
3100/519	1N3593
3100/519	1N3594
3100/519	1N3595
3100/519	1N3598
3100/519	1N3599
3100/519	1N3601
3100/519	1N3602
3100/519	1N3603
3100/519	1N3604
3100/519	1N3605
3100/519	1N3606
3100/519	1N3607
3100/519	1N3608
3100/519	1N3609
3100/519	1N3653
3100/519	1N3654
3100/519	1N3666
3100/519	1N3668
3100/519	1N3722
3100/519	1N380
3100/519	1N381
3100/519	1N382
3100/519	1N383
3100/519	1N386
3100/519	1N3864
3100/519	1N3872
3100/519	1N3873
3100/519	1N389
3100/519	1N390
3100/519	1N391
3100/519	1N392
3100/519	1N393
3100/519	1N394
3100/519	1N3956
3100/519	1N4087
3100/519	1N4147
3100/519	1N414B
3100/519	1N4150
3100/519	1N4151
3100/519	1N4152
3100/519	1N4242
3100/519	1N4243
3100/519	1N4308
3100/519	1N4309
3100/519	1N431
3100/519	1N4311
3100/519	1N4315
3100/519	1N4318
3100/519	1N432
3100/519	1N4322
3100/519	1N432A
3100/519	1N432B
3100/519	1N433
3100/519	1N433A
3100/519	1N433B
3100/519	1N434
3100/519	1N434B
3100/519	1N4363
3100/519	1N4375
3100/519	1N4382
3100/519	1N4392
3100/519	1N4395
3100/519	1N4395A
3100/519	1N444B
3100/519	1N4450
3100/519	1N4453
3100/519	1N4455
3100/519	1N4531
3100/519	1N4532
3100/519	1N4533
3100/519	1N4534
3100/519	1N4536
3100/519	1N4547
3100/519	1N4548
3100/519	1N456
3100/519	1N457M
3100/519	1N458M
3100/519	1N459M
3100/519	1N460B
3100/519	1N4726
3100/519	1N4727
3100/519	1N4827
3100/519	1N483AM
3100/519	1N483BM
3100/519	1N483C
3100/519	1N484C
3100/519	1N4861
3100/519	1N4862
3100/519	1N4863
3100/519	1N4933
3100/519	1N4949
3100/519	1N5194
3100/519	1N5195
3100/519	1N5196
3100/519	1N5208
3100/519	1N5210
3100/519	1N5219
3100/519	1N5220
3100/519	1N5315
3100/519	1N5316
3100/519	1N5317
3100/519	1N5318
3100/519	1N5319
3100/519	1N5412
3100/519	1N5414
3100/519	1N5426
3100/519	1N5605
3100/519	1N5607
3100/519	1N5610
3100/519	1N5711
3100/519	1N5712
3100/519	1N5713
3100/519	1N5719
3100/519	1N622
3100/519	1N625A
3100/519	1N625M
3100/519	1N626A
3100/519	1N626M
3100/519	1N627
3100/519	1N627A
3100/519	1N628
3100/519	1N628A
3100/519	1N629
3100/519	1N629A
3100/519	1N633
3100/519	1N643
3100/519	1N658A
3100/519	1N658M
3100/519	1N659A
3100/519	1N660
3100/519	1N660A
3100/519	1N661
3100/519	1N661A
3100/519	1N662
3100/519	1N662A
3100/519	1N663A
3100/519	1N663M
3100/519	1N690
3100/519	1N696
3100/519	1N697
3100/519	1N778
3100/519	1N779
3100/519	1N788
3100/519	1N789
3100/519	1N789M
3100/519	1N790
3100/519	1N790M
3100/519	1N791
3100/519	1N791M
3100/519	1N792
3100/519	1N792M
3100/519	1N793
3100/519	1N793M
3100/519	1N794
3100/519	1N795
3100/519	1N796
3100/519	1N797
3100/519	1N798
3100/519	1N799
3100/519	1N800
3100/519	1N801
3100/519	1N802
3100/519	1N803
3100/519	1N804
3100/519	1N806
3100/519	1N807
3100/519	1N808
3100/519	1N809
3100/519	1N810
3100/519	1N811
3100/519	1N811M
3100/519	1N812
3100/519	1N812M
3100/519	1N813
3100/519	1N813M
3100/519	1N814
3100/519	1N814M
3100/519	1N815
3100/519	1N815M
3100/519	1N817
3100/519	1N818
3100/519	1N818M
3100/519	1N837
3100/519	1N837A
3100/519	1N838
3100/519	1N839
3100/519	1N840
3100/519	1N840M
3100/519	1N841
3100/519	1N842
3100/519	1N843
3100/519	1N844
3100/519	1N845
3100/519	1N890
3100/519	1N891
3100/519	1N892
3100/519	1N897
3100/519	1N898
3100/519	1N899
3100/519	1N900
3100/519	1N901
3100/519	1N902
3100/519	1N903
3100/519	1N903A
3100/519	1N903AM
3100/519	1N903M
3100/519	1N904
3100/519	1N904A
3100/519	1N904AM
3100/519	1N904M
3100/519	1N905
3100/519	1N905A
3100/519	1N905AM
3100/519	1N905M
3100/519	1N906
3100/519	1N906A
3100/519	1N906AM
3100/519	1N906M
3100/519	1N907
3100/519	1N907A
3100/519	1N907AM
3100/519	1N907M
3100/519	1N908
3100/519	1N908A
3100/519	1N908AM
3100/519	1N908M
3100/519	1N914
3100/519	1N914A
3100/519	1N914B
3100/519	1N914M
3100/519	1N915
3100/519	1N916A
3100/519	1N916B
3100/519	1N917
3100/519	1N920
3100/519	1N921
3100/519	1N922
3100/519	1N924
3100/519	1N925
3100/519	1N926
3100/519	1N927
3100/519	1N928
3100/519	1N930
3100/519	1N931
3100/519	1N932
3100/519	1N933
3100/519	1N934
3100/519	1N948
3100/519	1N993
3100/519	1N995
3100/519	1N997
3100/519	1N999
3100/519	1NJ61433
3100/519	1NJ61677
3100/519	1NJ61725
3100/519	1NJ71224
3100/519	1R10D3K
3100/519	1R2
3100/519	1R4
3100/519	1S1007
3100/519	1S1052
3100/519	1S1053
3100/519	1S1212A
3100/519	1S1213
3100/519	1S1214
3100/519	1S1215
3100/519	1S1216
3100/519	1S1217
3100/519	1S1218
3100/519	1S1219
3100/519	1S1220
3100/519	1S130
3100/519	1S1302
3100/519	1S1303
3100/519	1S131
3100/519	1S132
3100/519	1S1420H
3100/519	1S1473
3100/519	1S1532
3100/519	1S1544
3100/519	1S1545
3100/519	1S155
3100/519	1S155-1
3100/519	1S1555-S
3100/519	1S1585
3100/519	1S1586
3100/519	1S1650
3100/519	1S1651
3100/519	1S1690
3100/519	1S17
3100/519	1S18
3100/519	1S180
3100/519	1S182
3100/519	1S1825
3100/519	1S1993
3100/519	1S1994
3100/519	1S1995
3100/519	1S2074H
3100/519	1S20751C
3100/519	1S2091
3100/519	1S2091(DIODE)
3100/519	1S2092
3100/519	1S2097
3100/519	1S2098
3100/519	1S2144
3100/519	1S2144A
3100/519	1S2186
3100/519	1S2186GR
3100/519	1S2276
3100/519	1S2460
3100/519	1S2473T
3100/519	1S2788
3100/519	1S2788B
3100/519	1S306
3100/519	1S307
3100/519	1S322
3100/519	1S358
3100/519	1S38
3100/519	1S444
3100/519	1S473
3100/519	1S642
3100/519	1S84
3100/519	1S89
3100/519	1S920
3100/519	1S921
3100/519	1S951
3100/519	1S952
3100/519	1S955
3100/519	1S983
3100/519	1S994A
3100/519	1S99A
3100/519	1SS53
3100/519	1T243M
3100/519	1T40
3100/519	1X9179
3100/519	1X9805
3100/519	1X9809
3100/519	2-1K60
3100/519	2D165
3100/519	2S10032
3100/519	03-931641
3100/519	03-931642
3100/519	03-931645
3100/519	3L4-2001-3
3100/519	3L4-2001-4
3100/519	3L4-2003-3
3100/519	3L4-3001-1
3100/519	3L4-3001-7
3100/519	3L4-3002-10
3100/519	3L4-3002-25
3100/519	3L4-3002-31
3100/519	3L4-3002-32
3100/519	4-1724
3100/519	4-1726
3100/519	4-2020-05800
3100/519	4-2020-06100
3100/519	4-2020-06200
3100/519	4-2020-06400
3100/519	4-2020-10100
3100/519	4-2020-15600
3100/519	4-2021-05000
3100/519	4-2021-07470
3100/519	4-2021-07670
3100/519	4-3033
3100/519	4-3034
3100/519	05-02160-01
3100/519	5GB
3100/519	6X97174A01
3100/519	6X97174XA08
3100/519	7-0013
3100/519	08-08117
3100/519	08-08119
3100/519	8-905-405-098
3100/519	8-905-421-300
3100/519	08A159-007
3100/519	09-306060
3100/519	09-306062
3100/519	09-306110
3100/519	09-306111
3100/519	09-306112
3100/519	09-306113
3100/519	09-306134
3100/519	09-306145
3100/519	09-306151
3100/519	09-306154
3100/519	09-306159
3100/519	09-306161
3100/519	09-306163
3100/519	09-306168
3100/519	09-306195
3100/519	09-306198
3100/519	09-306199
3100/519	09-306202
3100/519	09-306206
3100/519	09-306211
3100/519	09-306212
3100/519	09-306220
3100/519	09-306221
3100/519	09-306223
3100/519	09-306231
3100/519	09-306236
3100/519	09-306248
3100/519	09-306266
3100/519	09-306276
3100/519	09-306283
3100/519	09-306288
3100/519	09-306309
3100/519	09-306313
3100/519	09-306326
3100/519	09-306368
3100/519	09-306373
3100/519	09-306390
3100/519	09-306426
3100/519	09-307045

SK	Industry Standard No.
3100/519	09-307055
3100/519	09-307075
3100/519	09-307080
3100/519	09-307081
3100/519	09-307082
3100/519	09-307085
3100/519	09-307089
3100/519	9D11
3100/519	9D14
3100/519	9DI3
3100/519	10-010
3100/519	10-10112
3100/519	012-1020-005
3100/519	012-1024-001
3100/519	12/1N10
3100/519	13-085015
3100/519	13-085022
3100/519	13-085026
3100/519	13-10321-54
3100/519	13-10321-55
3100/519	13-17174-3
3100/519	13-22017-0
3100/519	13-22606-0
3100/519	13-22609-0
3100/519	13-29687-2
3100/519	13-34056-1
3100/519	13-67590-1
3100/519	14-514-02
3100/519	14-514-15
3100/519	14-514-17
3100/519	14-514-20
3100/519	14-514-65
3100/519	14-514-70
3100/519	15-085005
3100/519	15-108025
3100/519	15-108042
3100/519	15-123101
3100/519	16X39
3100/519	19-080-001
3100/519	19-080-008
3100/519	20A11
3100/519	21A008-008
3100/519	21A009
3100/519	21A009-008
3100/519	21A009-009
3100/519	21A103-065
3100/519	21A108-001
3100/519	21A108-004
3100/519	21A110-004
3100/519	21A111-001
3100/519	21M330
3100/519	21M386
3100/519	21M415
3100/519	21M562
3100/519	21M586
3100/519	022-2823-004
3100/519	24-28201
3100/519	24MW1109
3100/519	24MW667
3100/519	24MW744
3100/519	24MW825
3100/519	24MW858
3100/519	24MW861
3100/519	24MW894
3100/519	24MW956
3100/519	32-0022
3100/519	32-0057
3100/519	32-0063
3100/519	34-8057-13
3100/519	34P4
3100/519	36D-32
3100/519	39A69-2
3100/519	42-22538
3100/519	44T-300-96
3100/519	46-61267-3
3100/519	46-61307-3
3100/519	46-80309-3
3100/519	46-861187-3
3100/519	46-86178-3
3100/519	46-86184-3
3100/519	46-86186-3
3100/519	46-86250-3
3100/519	46-86254-3
3100/519	46-86338-3
3100/519	46-86380-3
3100/519	46-86431-3
3100/519	46-86481-3
3100/519	46-8676-3
3100/519	48-03005A01
3100/519	48-03005A05
3100/519	48-134781
3100/519	48-134816
3100/519	48-137385
3100/519	48-137514
3100/519	48-137573
3100/519	48-155159
3100/519	48-40738P01
3100/519	48-43265G01
3100/519	48-67120A11
3100/519	48-97048A01
3100/519	48-V34816
3100/519	48D67120A13
3100/519	48S10577A01
3100/519	48S10577A11
3100/519	48S134816
3100/519	48S137167
3100/519	48S155047
3100/519	48S155060
3100/519	48S155077
3100/519	49S1
3100/519	53A018-1
3100/519	53A020-1
3100/519	53A022-3
3100/519	53A030-1
3100/519	53B001-7
3100/519	53D001-4
3100/519	53D001-8
3100/519	53E003-2
3100/519	53L001-15
3100/519	53N001-5
3100/519	53N004-12
3100/519	53T001-7
3100/519	56-24
3100/519	56-27
3100/519	56-28
3100/519	56-4832
3100/519	59E001-1
3100/519	62-18782
3100/519	62-20223
3100/519	62-20319
3100/519	62-20437
3100/519	62-20597
3100/519	62-26597
3100/519	065-012
3100/519	66X0062-001
3100/519	72-18
3100/519	77A01
3100/519	77A02
3100/519	77A11
3100/519	77A13
3100/519	81-46123023-7
3100/519	81-46123038-5
3100/519	86-0513
3100/519	86-0522
3100/519	86-100013
3100/519	86-100014
3100/519	86-147-1
3100/519	86-41-1
3100/519	86-62-1
3100/519	86-67-3
3100/519	86-70-1
3100/519	86-74-9
3100/519	86-76-1
3100/519	86-77-1
3100/519	86-78-1
3100/519	88-77-1
3100/519	089-241
3100/519	93A60-5
3100/519	93A60-6
3100/519	93A64-3
3100/519	93A69-2
3100/519	93B48-1
3100/519	93B48-3
3100/519	93B48-4
3100/519	93B64-1
3100/519	100-00310-09
3100/519	100-011-20
3100/519	100-01120-09
3100/519	100-11
3100/519	100-120
3100/519	100-125
3100/519	100-130
3100/519	100-137
3100/519	100-161
3100/519	100-184
3100/519	100-216
3100/519	100-435
3100/519	102-339
3100/519	103-145-01
3100/519	103-159
3100/519	103-222
3100/519	103-240
3100/519	103-261-02
3100/519	103-42
3100/519	106-009
3100/519	112-500-0-50
3100/519	120-003147
3100/519	120-004877
3100/519	120A02
3100/519	120A11
3100/519	120A13
3100/519	123-016
3100/519	123-017
3100/519	123-022
3100/519	123-025
3100/519	123B-003
3100/519	130-30265
3100/519	130-30274
3100/519	130-30702
3100/519	142-009
3100/519	142-010
3100/519	150-030-9-002
3100/519	150-040-9-002
3100/519	150-1
3100/519	151-001-9-001
3100/519	151-002-9-001
3100/519	151-011-9-001
3100/519	151-021-9-001
3100/519	151-024-9-001
3100/519	151-029-9-001
3100/519	151-030-9-001
3100/519	151-030-9-002
3100/519	151-030-9-006
3100/519	151-032-9-001
3100/519	151-032-9-002
3100/519	151-034-9-001
3100/519	151-035-9-003
3100/519	151-040-9-001
3100/519	151-040-9-002
3100/519	151-042-9-001
3100/519	151-045-9-002
3100/519	151-049-9-001
3100/519	151-066-9-001
3100/519	151-067-9-001
3100/519	151-072-9-001
3100/519	170-1
3100/519	185-011
3100/519	185-012
3100/519	186-011
3100/519	186-012
3100/519	186-015
3100/519	209-32
3100/519	230-0014
3100/519	260-10-011
3100/519	264D00101
3100/519	264D00209
3100/519	264P03703
3100/519	265D00702
3100/519	265Z00101
3100/519	344-6005-6
3100/519	360-32
3100/519	596-5
3100/519	601X0224-066
3100/519	601X0225-066
3100/519	610-017-706
3100/519	642-126
3100/519	642-275
3100/519	690V088H20
3100/519	690V118H57
3100/519	690V118H58
3100/519	690V119H15
3100/519	754-2000-011
3100/519	754-2009-150
3100/519	754-2509-150
3100/519	754-2720-021
3100/519	754-5750-283
3100/519	800-036-00
3100/519	800-040-00
3100/519	800-042-00
3100/519	863-254B
3100/519	863-567B
3100/519	863-776B
3100/519	903-100B
3100/519	903-112B
3100/519	903-116B
3100/519	903-118B
3100/519	903-121B
3100/519	903-177B
3100/519	903-178
3100/519	903-41B
3100/519	903-42B
3100/519	903-48B
3100/519	903-58B
3100/519	903-72B
3100/519	903-82B
3100/519	903-84B
3100/519	914-000-2-00
3100/519	914-000-6-00
3100/519	914-001-1-00
3100/519	919-00-4326
3100/519	919-00-4799
3100/519	919-00-9929
3100/519	919-01-0873
3100/519	919-01-1215
3100/519	919-01-1307
3100/519	919-01-3072
3100/519	921-342B
3100/519	977-1
3100/519	991-00-1172
3100/519	991-00-1172-1
3100/519	1009-07
3100/519	1009-08
3100/519	1010
3100/519	1010-143
3100/519	1011-0302
3100/519	1018-3259
3100/519	1018-9884
3100/519	1033-0911
3100/519	1033-0983
3100/519	1033-0991
3100/519	1040-07
3100/519	1040-10
3100/519	1040-9332
3100/519	1041-66
3100/519	1042-14
3100/519	1042-16
3100/519	1042-23
3100/519	1043-10
3100/519	1044-8983
3100/519	1045-7802
3100/519	1048-6421
3100/519	1048-9987
3100/519	1062GE
3100/519	1076-1484
3100/519	1077-2341
3100/519	1079-01
3100/519	1079-89
3100/519	1092-16
3100/519	1212
3100/519	1699-17
3100/519	1756-17
3100/519	1846-17
3100/519	1851-17
3100/519	2000-318
3100/519	2010-03
3100/519	2022-07
3100/519	2041-05
3100/519	2061A45-38
3100/519	2065-17
3100/519	2066-17
3100/519	2076
3100/519	2093A38-37
3100/519	2093A41-105
3100/519	2093A41-116
3100/519	2093A41-158
3100/519	2093A41-165
3100/519	2093A41-171
3100/519	2093A41-172
3100/519	2093A41-76
3100/519	2093A41-96
3100/519	2102-029
3100/519	2110N-41
3100/519	2209-17
3100/519	2231-17
3100/519	2285-17
3100/519	2328
3100/519	2328-17
3100/519	3017
3100/519	4014-200-30110
3100/519	4801-00154
3100/519	5001-083
3100/519	5001-107
3100/519	5001-144
3100/519	5001-145
3100/519	5001-156
3100/519	5001-162
3100/519	5001-164
3100/519	5631-MA150
3100/519	06102
3100/519	8000-00003-046
3100/519	8000-00004-061
3100/519	8000-00004-062
3100/519	8000-00004-064
3100/519	8000-00004-066
3100/519	8000-00004-184
3100/519	8000-00006-008
3100/519	8000-00010-109
3100/519	8000-00011-046
3100/519	8000-00028-045
3100/519	8000-00038-008
3100/519	8000-0004-066
3100/519	8000-00041-019
3100/519	8010-52
3100/519	8710-52
3100/519	8840-53
3100/519	8910-52
3100/519	9440
3100/519	9646
3100/519	12101
3100/519	15122
3100/519	18600-52
3100/519	25810-52
3100/519	25840-52
3100/519	27840-43
3100/519	28810-63
3100/519	37510-52
3100/519	38510-330
3100/519	53016R
3100/519	58840-113
3100/519	67055
3100/519	72146
3100/519	72147
3100/519	72152
3100/519	72158
3100/519	72163-1
3100/519	085003
3100/519	87756
3100/519	88060-52
3100/519	88060-55
3100/519	88510-52
3100/519	95003
3100/519	101154
3100/519	123805
3100/519	125397
3100/519	125588
3100/519	125964
3100/519	129375
3100/519	129475
3100/519	130046
3100/519	136162
3100/519	136163
3100/519	136688
3100/519	139065
3100/519	168706
3100/519	169114
3100/519	169117
3100/519	169558
3100/519	0170301
3100/519	170733-1
3100/519	170857
3100/519	171162-196
3100/519	171162-271
3100/519	171840
3100/519	171841
3100/519	171843
3100/519	172201
3100/519	172253
3100/519	172547
3100/519	172722
3100/519	183033
3100/519	193717
3100/519	199596
3100/519	206617
3100/519	240564
3100/519	248817
3100/519	249217
3100/519	249508-3
3100/519	0517133
3100/519	0517828
3100/519	0526232
3100/519	530063-13
3100/519	530072-1002
3100/519	530072-1008
3100/519	530072-1010
3100/519	530072-1011
3100/519	530072-18
3100/519	530092-1002
3100/519	530116-1001
3100/519	530144-1002
3100/519	530144-1003
3100/519	530146-1
3100/519	530146-2
3100/519	530150-1
3100/519	530179-1
3100/519	530181-1
3100/519	530181-1003
3100/519	534001H
3100/519	0576054
3100/519	741100
3100/519	817190
3100/519	860011
3100/519	922433
3100/519	922873
3100/519	986934
3100/519	988995
3100/519	992150
3100/519	1421207-1
3100/519	1470872-6
3100/519	1471072-4
3100/519	1471872-10
3100/519	1471872-12
3100/519	1471872-15
3100/519	1471872-16
3100/519	1471876-6
3100/519	1474778-16
3100/519	1476930
3100/519	2003069-2
3100/519	2003069-5
3100/519	2003069-6
3100/519	2092055-0714
3100/519	2337011
3100/519	3596061
3100/519	5330133
3100/519	5330212H
3100/519	5330261
3100/519	5330334
3100/519	5340022
3100/519	5340051
3100/519	7201212
3100/519	7855283-01
3100/519	7914014-01
3100/519	7932638
3100/519	13030274
3100/519	17200000
3100/519	17200240
3100/519	22692012
3100/519	23115194
3100/519	23115249
3100/519	23115273
3100/519	23115285
3100/519	23115331
3100/519	23115897
3100/519	26010030
3100/519	26010032
3100/519	26010047
3100/519	26010057
3100/519	37246602
3100/519	37246711
3100/519	37275350
3100/519	41622859
3100/519	62522720
3100/519	62697571
3100/519	62741406
3100/519	62761350
3100/519	70270250
3100/519	87227880
3100/519	2008010028
3100/519	2008010094
3100/519	2008010110
3100/519	2008010127
3100/519	2008010131
3100/519	2008100027
3100/519	2012010159
3100/519	2012010165
3100/519	4129100000
3100/519	4202006100
3100/519	4202006200
3100/519	4202007900
3100/519	4202010100
3100/519	4202015600
3100/519	4202018700
3100/519	4202021900
3100/519	4202107470
3100/519	4202109170
3100/519	6611003200
3100/519	6612006000
3100/519	16405990022
3101	CA3041
3101	13-28-6
3101	21A101-001
3101	21A101-1
3101	57A29-2
3101	57C29
3101	57C29-1
3101	57C29-2
3101	1472434-1
3101/706	171179-027
3101/706	171179-036
3101/706	172252
3102	CA3042
3102	GEIC-89
3102	LA1342
3102	R2432-1
3102	T-2062
3102	T1H
3102	TA-7051P
3102	TA7051P
3102	19A116797-1
3102	042A
3102	46-1347-3
3102	46-1348-3
3102	46-1356-3
3102	51-10408A01
3102	51S10408A01
3102	99S042
3102	99S072
3102	133-002
3102	515-10408A-01
3102	126871
3102	129871
3102	1462432-1
3102	1462434-1
3102	23119004
3102	23119007
3102/710	R2434-1
3102/710	2
3102/710	09-308089
3102/710	072
3102/710	39901-0001
3102/710	000074020
3102/710	2360401
3102/710	4206002400
3102/710	4206004400
3103A	RCA1A15
3103A	2N3439
3103A	121-911
3103A	610190-3
3104A	A3Y
3104A	A5F
3104A	A5T-1
3104A	BDY65
3104A	BFR22
3104A	BFY50
3104A	BSW67
3104A	BSW68
3104A	C1157
3104A	C212
3104A	C235-0
3104A	C780AG-0
3104A	C780AG-R
3104A	E13-007-00
3104A	E13-010-00
3104A	E13-011-00
3104A	E13-012-00
3104A	IRTR74
3104A	IRTR79
3104A	J241250
3104A	LM-1154
3104A	LM-1156
3104A	LM-1157
3104A	MJE-200
3104A	Q5163D
3104A	SJE103
3104A	TA7739
3104A	TG2SD386Y-D-A
3104A	TG2SD386Y-E-A
3104A	TV-120
3104A	W14
3104A	ZEN205
3104A	2N2297
3104A	2N3114
3104A	2N3500
3104A	2N3501
3104A	2N3712
3104A	2N6175
3104A	2SC1089
3104A	2SC1089C
3104A	2SC1156
3104A	2SC1157
3104A	2SC305
3104A	2SC305A
3104A	2SC305B
3104A	2SC305C
3104A	2SC305D
3104A	2SC305E
3104A	2SC3
3104A	2SC305G
3104A	
3104A	2SC305H
3104A	2
3104A	2SC305K
3104A	2
3104A	2SC305M
3104A	2SC305OR
3104A	2SC305R
3104A	2SC305X
3104A	2SC305Y
3104A	2SC511
3104A	2SC5120
3104A	2SC5130
3104A	2SC513R
3104A	2SC560
3104A	2SC959
3104A	2SC959A
3104A	2SC959B
3104A	2SC959C
3104A	2
3104A	2SC959M
3104A	09-
3104A	09-304046
3104A	09-304052
3104A	09-304055
3104A	011-00
3104A	012-00
3104A	13-29974-4
3104A	13-33174-2
3104A	13-33176-1
3104A	13-35089-1
3104A	14-901-12
3104A	14-904-12

SK	Industry Standard No.
3104A	19-3935-641
3104A	46-86327-3
3104A	46-86466-3
3104A	46-86506-3
3104A	46-86509-3
3104A	48S137093
3104A	48S155070
3104A	57B211-8
3104A	86-556-2
3104A	86-624-2
3104A	86-628-2
3104A	86-628-9
3104A	86-631-2
3104A	86-672-2
3104A	99S105-1
3104A	121-989
3104A	121-990
3104A	260P22203
3104A	260P22204
3104A	656-2
3104A	1061-8346
3104A	1061-8353
3104A	1061-8668
3104A	1061-8908
3104A	1063-5142
3104A	1063-7916
3104A	1169
3104A	3613-3
3104A	3633-1
3104A	40321
3104A	40327
3104A	40346
3104A	40412
3104A	40885
3104A	126703
3104A	131139
3104A	131140
3104A	133265
3104A	134772
3104A	135716
3104A	610144-101
3104A	610144-2
3104A	610144-3
3104A	1473545-1
3104A	1473555-2
3104A	1473613-4
3104A	1473628-2
3104A	1473633-1
3104A	23114336
3104A	23114344
3104A	23114974
3104A	23114993
3104A	6621001800
3105	CXL166
3105	CXL167
3105	KBP02
3105	MDA920-2
3105	RVD8MB4
3105	SR1B
3105	W02
3105	ZEN434
3105	125993
3106	CXL168
3107	KCO.8C22/1
3107	2FB1000
3108	D2Y
3108	E13-021-01
3108	E13-021-03
3108	EP57X2
3108	ES57X11
3108	GECR-7
3108	H386-9
3108	H386C1D2Y
3108	HS20/16S
3108	HS20/1AS
3108	HS25-1
3108	HS25/16
3108	HS25/1B
3108	HS25/1BS
3108	HS25/1C
3108	HS30-1C
3108	HS30/1C
3108	HS9/17N
3108	MF25/1B
3108	PTC212
3108	PTC213
3108	SELEN-40
3108	TV20-10K80
3108	TV20-S
3108	TVM540
3108	TVS-H-339W
3108	TVS-H-399W
3108	2HT-6
3108	4-2021-08070
3108	10K80
3108	28-31-01
3108	34-8062-1
3108	46-86294-3
3108	48S137114
3108	57X11
3108	103-258
3108	420-2106-270
3108	11386-9
3108	134777
3108	530119-7
3108	530119-9
3108	530149-9
3108	14769302
3108	4202106270
3108	4202108070
3108	4202108170
3108/505	A04299-201
3108/505	A04299-251
3108/505	EDMF25B
3108/505	EP-5641H-2
3108/505	GECR-6
3108/505	HS30-16
3108/505	HS30-1B
3108/505	MF30/1B
3108/505	13-37933-2
3108/505	34-8062-4
3108/505	46-86292-3
3108/505	46-86334-3
3108/505	46-86451-3
3108/505	46-86454-3
3108/505	46-86547-3
3108/505	48-355047
3108/505	73C180476-5
3108/505	93A93-1
3108/505	93A93-2
3108/505	103-239-02
3108/505	103-293-02
3108/505	117-A40
3108/505	530119-8
3108/505	530124-3
3108/505	1445470-504
3108/505	2330034
3108/505	2330381
3108/505	23115215
3108/505	23115938
3108/505	62747501
3108/505	4202107570
3109	A-95-5295
3109	A04710
3109	B66X0036-001
3109	ET57X31
3109	EU57X31
3109	GECR-2
3109	PTC209
3109	R2
3109	SELEN-48
3109	SR-31
3109	SR9001
3109	T-E1107
3109	TVM-537
3109	2MA
3109	004-003100
3109	004-03100
3109	4-2020-06300
3109	4-2020-09400
3109	6RS36PH13BJK1
3109	6RS36PH13BLJ1
3109	6RS6PH13BCJ1
3109	6RS6PH13BJJ1
3109	6RS6PH13BKJ1
3109	6RS6PH13BLJ1
3109	6RS6PH13BMJ1
3109	13-16105-3
3109	21A4
3109	33G59113
3109	34-8056-1
3109	34-8057-15
3109	61-8968
3109	86-45-3
3109	93C40
3109	93C40-1
3109	295V031B
3109	295V034C01
3109	690V039H17
3109	800-011-00
3109	72111
3109	97251-3
3109	113391
3109	126862
3109	817124
3109	972571-2
3109	972571-3
3109	4202009400
3109/119	DIJ72291
3109/119	HFSD-1
3109/119	66X0036-001
3109/119	135341
3109/119	138736
3109/119	161033
3109/119	489752-016
3109/119	489752-073
3109/119	530097-3
3110	A04727
3110	B66X0041-001
3110	E13-013-03
3110	E13-013-04
3110	ET57X26
3110	GECR-3
3110	GI-TVC3
3110	M105064
3110	PTC404
3110	R-3
3110	R3
3110	R8560
3110	RCC7022
3110	RE91
3110	RSLNA0004CEZZ
3110	SELEN-42
3110	SR-37
3110	1TV24237
3110	004-003700
3110	6RS51GX12
3110	10B-4-C4
3110	13-17596-1
3110	15-108007
3110	15-108038
3110	21A005
3110	21A008-016
3110	21A500
3110	21A500-000
3110	34-8058
3110	34-8058-1
3110	34-8058-7
3110	41-001
3110	48-66653A02
3110	48-90068A01
3110	48C66653A02
3110	51GX14
3110	51GX3
3110	66X0041-001
3110	66X41-1
3110	86-37-3
3110	86-55-3
3110	86-56-3
3110	86-56-3C
3110	86-56-3F
3110	93A53-2
3110	93A75-1
3110	93B19-1
3110	93B53-2
3110	93C1-2021
3110	188-70-48
3110	212-63
3110	295V031B02
3110	295V031C02
3110	301-1
3110	324-0015
3110	690V037H92
3110	690V039H54
3110	2090A53-2
3110	2093A75-1
3110	7209A
3110	105064
3110	489752-072
3110	530087-2
3110	530097-2
3110	575995
3110	817149
3110	1470990-1
3110	1887048
3110	62047259
3110	62522704
3110/120	ET57X33
3110/120	KC2G
3110/120	R105064
3110/120	TVC-3
3110/120	6RS56PHL3BKJ1
3110/120	09-306104
3110/120	10B-Y
3110/120	21A006-000
3110/120	93B53-1
3110/120	93C19-1
3110/120	72109
3110/120	105604
3110/120	113321
3110/120	199985
3110/120	282601
3110/120	817127
3110/120	6846503
3111	B2B
3111	BU207
3111	E13-009-00
3111	Q5140ZXQ
3111	Q5140ZXR
3111	S5020
3111	SJ5526
3111	T-Q5084
3111	TNJ72149
3111	TNJ72319
3111	TR67
3111	TV-119
3111	TV-125
3111	TV108
3111	TV118
3111	TV124
3111	2SC1433
3111	2SC1894
3111	2SC665HB
3111	2SD575
3111	2SD69
3111	2SD725
3111	009-00
3111	09-302176
3111	09-304057
3111	13-37870-1
3111	13-37870-2
3111	13-43633-1
3111	34-6002-59
3111	34-6002-63
3111	34-6002-64
3111	46-86390-3
3111	46-86415-3
3111	46-86486-3
3111	57A198-11
3111	86-626-9
3111	86-633-2
3111	99S079-1
3111	99S103-1
3111	99S103-2
3111	99S103-3
3111	203X3189-408
3111	260P08908
3111	260P09701
3111	260P11209
3111	260P33408
3111	690V080H45
3111	1073
3111	204117
3111	489751-119
3111	1473669-1
3111	62737441
3112	A04201
3112	A194
3112	A195
3112	A196
3112	B-6001
3112	BF244
3112	BF245
3112	BF246
3112	BF247
3112	BF348
3112	BFS21
3112	BFS21A
3112	BFW13
3112	C674
3112	C764
3112	C94
3112	C94E
3112	C95
3112	D1101
3112	D1102
3112	D1301
3112	D1302
3112	D1303
3112	E101
3112	E103
3112	E203
3112	E232
3112	EA15X169
3112	EA3278
3112	EQF-3
3112	ET491051
3112	FE-100
3112	FE0654B
3112	FE100
3112	FE100A
3112	FE102
3112	FE102A
3112	FE104A
3112	FE400
3112	FE402
3112	FE402A
3112	FE404A
3112	FF400
3112	G08-005L
3112	GE-FET-1
3112	HA2001
3112	HA2010
3112	HF200301B
3112	HF200301B0
3112	HF200301E
3112	IP20-0010
3112	IT210
3112	ITE3066
3112	ITE3067
3112	ITE4339
3112	ITE4340
3112	JNJ61673
3112	M100
3112	M101
3112	MK5485
3112	MPF101
3112	MPF103
3112	MPF104
3112	MPF109
3112	NE4304
3112	NF4302
3112	NF4303
3112	NF520
3112	NF522
3112	NF523
3112	NF531
3112	NF533
3112	NF5485
3112	NF5486
3112	NKT800112
3112	NKT800113
3112	NKT80111
3112	NKT80211
3112	NKT80212
3112	NKT80213
3112	NKT80214
3112	NKT80215
3112	NKT80216
3112	NPC211N
3112	NPC212N
3112	NPC214N
3112	NPC215N
3112	NPC216N
3112	NPC312N
3112	P1087
3112	QA-20
3112	RE46
3112	REN133
3112	RT175
3112	RT176
3112	S1211N
3112	S1212N
3112	S1213N
3112	S1214N
3112	S1215N
3112	S1216N
3112	S1221N
3112	S1222N
3112	S1223N
3112	S1224N
3112	S1225N
3112	S1226N
3112	S1231N
3112	S1232N
3112	S1233N
3112	S1234N
3112	S1235N
3112	S1236N
3112	SE3819
3112	SFE145
3112	SPF024
3112	SS3586
3112	SU2076
3112	SU2077
3112	SU2081
3112	T1S14
3112	TAA320
3112	TIS14
3112	TIS78
3112	TIS79
3112	TNJ61673
3112	U1177
3112	U1178
3112	U1180
3112	U1181
3112	U1277
3112	U1278
3112	U1279
3112	U1280
3112	U1283
3112	U1284
3112	U1285
3112	U1286
3112	U1322
3112	U1323
3112	U1324
3112	U1325
3112	U1714
3112	U1715
3112	U3012
3112	UC100
3112	UC110
3112	UC120
3112	UC20
3112	UC220
3112	UC756
3112	V159
3112	V1650E-4
3112	V1833E
3112	WEP801
3112	WF1
3112	WK5458
3112	WK5459
3112	1W11706
3112	2D031
3112	2N3066
3112	2N3067
3112	2N3068
3112	2N3070
3112	2N3071
3112	2N3084
3112	2N3085
3112	2N3086
3112	2N3087
3112	2N3088
3112	2N3088A
3112	2N3089
3112	2N3089A
3112	2N3366
3112	2N3367
3112	2N3368
3112	2N3369
3112	2N3370
3112	2N3436
3112	2N3437
3112	2N3438
3112	2N3452
3112	2N3453
3112	2N3454
3112	2N3455
3112	2N3456
3112	2N3457
3112	2N3460
3112	2N3465
3112	2N3466
3112	2N3684
3112	2N3684A
3112	2N3685
3112	2N3685A
3112	2N3686
3112	2N3686A
3112	2N3687
3112	2N3687A
3112	2N3821
3112	2N3822
3112	2N3967
3112	2N3967A
3112	2N3968
3112	2N3968A
3112	2N3969
3112	2N3969A
3112	2N4117
3112	2N4117A
3112	2N4118
3112	2N4119
3112	2N4119A
3112	2N4222
3112	2N4222A
3112	2N4223
3112	2N4302
3112	2N4303
3112	2N4304
3112	2N4338
3112	2N4339
3112	2N4416
3112	2N4416A
3112	2N4417
3112	2N4867
3112	2N4867A
3112	2N4868
3112	2N4868A
3112	2N4869
3112	2N4869A
3112	2N5045
3112	2N5046
3112	2N5047
3112	2N5267
3112	2N5268
3112	2N5269
3112	2N5270
3112	2N5277
3112	2N5358
3112	2N5359
3112	2N5391
3112	2N5452
3112	2N5453
3112	2N5454
3112	2N5458
3112	2N5543
3112	2N5544
3112	2N5556
3112	2N5557
3112	2N5558
3112	2N5561
3112	2N5562
3112	2N5563
3112	2N5648
3112	2N5649
3112	2N5716
3112	2N5717
3112	2N5718
3112	2SJ11
3112	2SJ12
3112	2SK11
3112	2SK11-R
3112	2SK11-Y
3112	2SK12
3112	2SK12-R
3112	2SK12-Y
3112	2SK15-R
3112	2SK15-Y
3112	2SK17-OR
3112	2SK24
3112	2SK24DR
3112	2SK24E
3112	2SK30
3112	2SK30-O
3112	2SK30-O
3112	2SK30-R
3112	2SK30-Y
3112	2SK30A
3112	2SK30A-Y
3112	2SK30AD
3112	2SK30AGR
3112	2SK30AY
3112	2SK30D
3112	2SK30GR
3112	2SK30Y
3112	2SK35
3112	2SK35-O
3112	2SK35-1
3112	2SK35-2
3112	2SK35A
3112	2SK35BL
3112	2SK35C
3112	2SK35GN
3112	2SK35Y
3112	2SK40
3112	2SK40A
3112	2SK40B
3112	2SK40C
3112	2SK40D
3112	2SK43
3112	2SK43A
3112	2SK43B
3112	2SK43C
3112	2SK43D
3112	2SK43E
3112	2SK43F
3112	2SK43GN
3112	2SK43H
3112	2SK43J
3112	2SK43K
3112	2SK43L
3112	2SK43M
3112	2SK43OR
3112	2SK43R
3112	2SK43X
3112	2SK43Y
3112	2SK44C
3112	2SK44D
3112	3SK20H
3112	3SK21H
3112	07-07159
3112	8-905-706-901
3112	09-305021
3112	09-305023
3112	09-305031
3112	09-305133
3112	13-28654-4
3112	14-2002-01
3112	020-1110-016
3112	020-1110-018
3112	020-1110-021
3112	020-1112-005
3112	022-2876-005
3112	46-86336-3
3112	48-134944
3112	57A149-12

SK	Industry Standard No.
3112	57A31-4
3112	86-477-2
3112	86-5095-2
3112	86-5096-2
3112	86-5122-2
3112	90-608
3112	108-0068-12
3112	121-860
3112	123-003
3112	173-1
3112	182-014-9-002
3112	182-014-9-003
3112	182-046-9-001
3112	364-10048
3112	417-169
3112	417-194
3112	417-231
3112	576-0006-003
3112	753-4000-024
3112	921-1019
3112	991-01-1706
3112	991-01-3055
3112	1002-01
3112	1612SK24E
3112	2000-104
3112	2010-17
3112	2057B149-12
3112	2065-55
3112	2450-17
3112	6003
3112	8000-00004-080
3112	8000-00010-017
3112	8440-121
3112	71686-4
3112	95133
3112	95228
3112	95231
3112	226517
3112	452077
3112	618580
3112	5320583
3112	5321501
3112	43024225
3114	A2057013-0004
3114	A642S
3114	A829F
3114	AT11
3114	BCY79A
3114	CS2005
3114	CS9129-B1
3114	ECG290
3114	EQR-1
3114	HT104941C0
3114	HT105621B0
3114	MMT3798
3114	NTC-10
3114	NTC-6
3114	REN290
3114	REN62
3114	RT-121
3114	TD400
3114	TD401
3114	TD402
3114	TD500
3114	TD501
3114	TD502
3114	TD550
3114	TR-1032-1
3114	TV-47
3114	VS2SA562-0/1E
3114	WEP51
3114	WEP715
3114	WEP716
3114	WEP717
3114	WEP911
3114	2N2303/46
3114	2N3911
3114	2N3912
3114	2N4971A
3114	2N4972A
3114	2N4982A
3114	2N5086A
3114	2N5087A
3114	2N5142A
3114	2N5143A
3114	2N5226A
3114	2N5227A
3114	2N5354A
3114	2N5355A
3114	2N5356A
3114	2N5365A
3114	2N5366A
3114	2N5367A
3114	2N5372A
3114	2N5373A
3114	2N5374A
3114	2N5375A
3114	2N5378A
3114	2N5379A
3114	2N5382A
3114	2N5383A
3114	2N5447A
3114	2N5448A
3114	2S3020A
3114	2S3021A
3114	2S3030A
3114	2S3040A
3114	2S3210A
3114	2S321A
3114	2S3220A
3114	2S3221A
3114	2S322AB
3114	2S3230A
3114	2S323A
3114	2S3240A
3114	2S324A
3114	2S326A
3114	2S327A
3114	2SA494-YE
3114	2SA465
3114	2SA467A
3114	2SA467B
3114	2SA467C
3114	2SA467D
3114	2SA467E
3114	2SA467F
3114	2SA467GN
3114	2SA467H
3114	2SA467J
3114	2SA467K
3114	2SA467L
3114	2SA467M
3114	2SA467R
3114	2SA467X
3114	2SA482A
3114	2SA482B
3114	2SA482C
3114	2SA482D
3114	2SA482E
3114	2SA482F
3114	2SA482G
3114	2SA482GN
3114	2SA482H
3114	2SA482J
3114	2SA482K
3114	2SA482L
3114	2SA482M
3114	2SA4820R
3114	2SA482R
3114	2SA482X
3114	2SA482Y
3114	2SA494-GR-1
3114	2SA494-OR
3114	2SA4940
3114	2SA494A
3114	2SA494B
3114	2SA494C
3114	2SA494D
3114	2SA494E
3114	2SA494F
3114	2SA494G
3114	2SA494GN
3114	2SA494GR
3114	2SA494H
3114	2SA494J
3114	2SA494K
3114	2SA494L
3114	2SA494M
3114	2SA4940R
3114	2SA494R
3114	2SA494X
3114	2SA494Y
3114	2SA495-1
3114	2SA495-OR
3114	2SA495-RD
3114	2SA495-YL
3114	2SA495B
3114	2SA495C
3114	2SA495E
3114	2SA495F
3114	2SA495K
3114	2SA495L
3114	2SA495M
3114	2SA4950F
3114	2SA4950R
3114	2SA495W1
3114	2SA509
3114	2SA509-0
3114	2SA509-O
3114	2SA509-OR
3114	2SA509-R
3114	2SA509-RD
3114	2SA509-Y
3114	2SA509-YE
3114	2SA509A
3114	2SA509B
3114	2SA509BL
3114	2SA509C
3114	2SA509D
3114	2SA509E
3114	2SA509F
3114	2SA509G
3114	2SA509GN
3114	2SA509GR
3114	2SA509GR-1
3114	2SA509H
3114	2SA509J
3114	2SA509K
3114	2SA509L
3114	2SA509M
3114	2SA5090R
3114	2SA509R
3114	2SA509T
3114	2SA509X
3114	2SA509Y
3114	2SA510-OR
3114	2SA511-G
3114	2SA511-OR
3114	2SA511-RD
3114	2SA511A
3114	2SA511B
3114	2SA511C
3114	2SA511D
3114	2SA511E
3114	2SA511F
3114	2SA511G
3114	2SA511GN
3114	2SA511H
3114	2SA511J
3114	2SA511K
3114	2SA511L
3114	2SA511M
3114	2SA5110R
3114	2SA511R
3114	2SA511X
3114	2SA511Y
3114	2SA512-OR
3114	2SA512-OR1
3114	2SA512L
3114	2SA513-OR
3114	2SA513-RD
3114	2SA513A
3114	2SA513B
3114	2SA513C
3114	2SA513D
3114	2SA513E
3114	2SA513F
3114	2SA513G
3114	2SA513GN
3114	2SA513H
3114	2SA513J
3114	2SA513K
3114	2SA513L
3114	2SA513M
3114	2SA5130R
3114	2SA513R
3114	2SA513X
3114	2SA513Y
3114	2SA522AL
3114	2SA522B
3114	2SA522C
3114	2SA522D
3114	2SA522E
3114	2SA522F
3114	2SA522G
3114	2SA522GN
3114	2SA522H
3114	2SA522J
3114	2SA522K
3114	2SA522L
3114	2SA522M
3114	2SA5220R
3114	2SA522R
3114	2SA522X
3114	2SA522Y
3114	2SA530H1
3114	2SA530R
3114	2SA539A
3114	2SA539B
3114	2SA539C
3114	2SA539D
3114	2SA539E
3114	2SA539F
3114	2SA539G
3114	2SA539GN
3114	2SA539H
3114	2SA539J
3114	2SA5390R
3114	2SA539R
3114	2SA539X
3114	2SA539Y
3114	2SA542G
3114	2SA544A
3114	2SA544B
3114	2SA544C
3114	2SA544D
3114	2SA544E
3114	2SA544F
3114	2SA544G
3114	2SA544GN
3114	2SA544H
3114	2SA544J
3114	2SA544K
3114	2SA544L
3114	2SA544M
3114	2SA5440R
3114	2SA544R
3114	2SA544X
3114	2SA544Y
3114	2SA561-0
3114	2SA561-OR
3114	2SA561-RD
3114	2SA561-YL
3114	2SA561GRN
3114	2SA562-R
3114	2SA562-RD
3114	2SA562-YE
3114	2SA5620
3114	2SA562D
3114	2SA562E
3114	2SA562GRN
3114	2SA565AB
3114	2SA565BA
3114	2SA565E
3114	2SA565F
3114	2SA565G
3114	2SA565GN
3114	2SA565H
3114	2SA565J
3114	2SA565L
3114	2SA565M
3114	2SA5650R
3114	2SA565R
3114	2SA565X
3114	2SA565Y
3114	2SA567D
3114	2SA567E
3114	2SA567F
3114	2SA567G
3114	2SA567GN
3114	2SA567H
3114	2SA567J
3114	2SA567K
3114	2SA567L
3114	2SA567M
3114	2SA5670R
3114	2SA567X
3114	2SA569A
3114	2SA569B
3114	2SA569C
3114	2SA569D
3114	2SA569E
3114	2SA569F
3114	2SA569G
3114	2SA569H
3114	2SA569K
3114	2SA569L
3114	2SA569M
3114	2SA5690R
3114	2SA569X
3114	2SA570A
3114	2SA570B
3114	2SA570C
3114	2SA570D
3114	2SA570E
3114	2SA570F
3114	2SA570G
3114	2SA570H
3114	2SA570J
3114	2SA570K
3114	2SA570L
3114	2SA570M
3114	2SA5700R
3114	2SA570X
3114	2SA608
3114	2SA608-0
3114	2SA608A
3114	2SA608B
3114	2SA608BL
3114	2SA608C
3114	2SA608D
3114	2SA608E
3114	2SA608F
3114	2SA608G
3114	2SA608K
3114	2SA608L
3114	2SA608M
3114	2SA6080R
3114	2SA608P
3114	2SA608R
3114	2SA628
3114	2SA628A
3114	2SA628AA
3114	2SA628D
3114	2SA628E
3114	2SA628EF
3114	2SA628F
3114	2SA628GN
3114	2SA628H
3114	2SA628J
3114	2SA628K
3114	2SA628M
3114	2SA6280R
3114	2SA628R
3114	2SA628X
3114	2SA628Y
3114	2SA629A
3114	2SA629B
3114	2SA629C
3114	2SA629D
3114	2SA629E
3114	2SA629F
3114	2SA629G
3114	2SA629GN
3114	2SA629H
3114	2SA629J
3114	2SA629K
3114	2SA629L
3114	2SA629M
3114	2SA6290R
3114	2SA629R
3114	2SA629X
3114	2SA629Y
3114	2SA642G
3114	2SA642GN
3114	2SA642H
3114	2SA642J
3114	2SA642K
3114	2SA642M
3114	2SA6420R
3114	2SA642X
3114	2SA673
3114	2SA673E
3114	2SA673F
3114	2SA673G
3114	2SA673GN
3114	2SA673H
3114	2SA673J
3114	2SA673K
3114	2SA673L
3114	2SA673M
3114	2SA6730R
3114	2SA677-0
3114	2SA677A
3114	2SA677B
3114	2SA677C
3114	2SA677D
3114	2SA677E
3114	2SA677F
3114	2SA677G
3114	2SA677GN
3114	2SA677H
3114	2SA677J
3114	2SA677K
3114	2SA677L
3114	2SA677M
3114	2SA6770R
3114	2SA677R
3114	2SA677X
3114	2SA677Y
3114	2SA678A
3114	2SA678B
3114	2SA678C
3114	2SA678D
3114	2SA678F
3114	2SA678G
3114	2SA678GN
3114	2SA678H
3114	2SA678J
3114	2SA678K
3114	2SA678L
3114	2SA678M
3114	2SA6780R
3114	2SA678R
3114	2SA678X
3114	2SA678Y
3114	2SC673B
3114	3N90
3114	3N91
3114	3N92
3114	3N93
3114	3N94
3114	3N95
3114	020-1110-004
3114	48M355007
3114	69-1815
3114	69-1817
3114	666-1
3114	1050B
3114	1147
3114	1254
3114	1294
3114	2022-01
3114	2057A2-480
3114	3540-1
3114	3549
3114	3559
3114	3562
3114	3563
3114	3581
3114	3620
3114	4822-130-40348
3114	9012HG
3114	10508
3114	06100034
3114/290	A-195C
3114/290	A066-113A
3114/290	A066-113AB
3114/290	A066-118A
3114/290	A0666-118
3114/290	A068-109A
3114/290	A1030
3114/290	A112-000172
3114/290	A112-000185
3114/290	A112-000187
3114/290	A116078
3114/290	A118284
3114/290	A119730
3114/290	A120P1
3114/290	A121-1
3114/290	A121-1RED
3114/290	A121-444
3114/290	A121-446
3114/290	A121-495
3114/290	A121-496
3114/290	A121-497
3114/290	A121-497WHT
3114/290	A121-602
3114/290	A121-603
3114/290	A121-679
3114/290	A121-699
3114/290	A121-746
3114/290	A121-774
3114/290	A1214
3114/290	A121467
3114/290	A121659
3114/290	A122GRN
3114/290	A122YEL
3114/290	A124047
3114/290	A124755
3114/290	A12594
3114/290	A126524
3114/290	A126700
3114/290	A126707
3114/290	A126715
3114/290	A126718
3114/290	A126719
3114/290	A12888
3114/290	A129-34
3114/290	A129697
3114/290	A129699
3114/290	A130-149
3114/290	A130-40315
3114/290	A130-40429
3114/290	A130139
3114/290	A1471114-1
3114/290	A1473563-1
3114/290	A1473570-1
3114/290	A1473574-1
3114/290	A1473590-1
3114/290	A1473591-1
3114/290	A1473597-1
3114/290	A1558-17
3114/290	A177A
3114/290	A177AB
3114/290	A178A
3114/290	A178AB
3114/290	A178B
3114/290	A178BA
3114/290	A179A
3114/290	A179AC
3114/290	A179B
3114/290	A179BB
3114/290	A1844-17
3114/290	A1867-17
3114/290	A190429
3114/290	A200-052
3114/290	A20030703-0702
3114/290	A2057013-0701
3114/290	A2057013-0702
3114/290	A2057013-0703
3114/290	A2057A2-198
3114/290	A2057B106-12
3114/290	A2057B108-6
3114/290	A20K
3114/290	A20KA
3114/290	A22008
3114/290	A23114050
3114/290	A23114051
3114/290	A23114550
3114/290	A2448
3114/290	A2498512
3114/290	A2798
3114/290	A297L012C01
3114/290	A297V073C03
3114/290	A297V073C04
3114/290	A30270
3114/290	A30290
3114/290	A3513
3114/290	A3540
3114/290	A3549
3114/290	A3559
3114/290	A3562
3114/290	A3563
3114/290	A3574
3114/290	A3581
3114/290	A4086
3114/290	A4087
3114/290	A4126
3114/290	A41440
3114/290	A417-116
3114/290	A417-132
3114/290	A417-153
3114/290	A417-176
3114/290	A417-182
3114/290	A417-184
3114/290	A417-196
3114/290	A417-200
3114/290	A417-201
3114/290	A417-235
3114/290	A43023845
3114/290	A4310
3114/290	A4442
3114/290	A4745
3114/290	A4802-00004
3114/290	A4815
3114/290	A4822-130-40348
3114/290	A4844
3114/290	A489751-028
3114/290	A489751-031
3114/290	A514-044910
3114/290	A5226-1
3114/290	A5320111
3114/290	A562
3114/290	A562-0
3114/290	A562Y
3114/290	A576-0001-002
3114/290	A576-0001-013
3114/290	A59625-1
3114/290	A59625-10
3114/290	A59625-11
3114/290	A59625-12
3114/290	A59625-2
3114/290	A59625-3
3114/290	A59625-4
3114/290	A59625-5
3114/290	A59625-6
3114/290	A59625-7
3114/290	A59625-8
3114/290	A59625-9
3114/290	A610074-1
3114/290	A610083
3114/290	A610083-1
3114/290	A610083-2
3114/290	A610083-3
3114/290	A610110-1
3114/290	A610120-1
3114/290	A629
3114/290	A642
3114/290	A642A
3114/290	A642B
3114/290	A642C
3114/290	A642D
3114/290	A642E
3114/290	A642F
3114/290	A642L
3114/290	A65-09-220
3114/290	A650232E
3114/290	A650235G
3114/290	A650238A

SK	Industry Standard No.
3114/290	A650923F
3114/290	A650925H
3114/290	A669
3114/290	A672
3114/290	A672A
3114/290	A672B
3114/290	A672C
3114/290	A673
3114/290	A673A
3114/290	A673AA
3114/290	A673AB
3114/290	A673AC
3114/290	A673AD
3114/290	A673AE
3114/290	A673B
3114/290	A673C
3114/290	A673D
3114/290	A677
3114/290	A678
3114/290	A701584-00
3114/290	A701589-00
3114/290	A71687-1
3114/290	A753-4004-248
3114/290	A7570013-01
3114/290	A7576004-01
3114/290	A78331
3114/290	A800-523-01
3114/290	A800-523-02
3114/290	A800-527-00
3114/290	A815199
3114/290	A815199-6
3114/290	A815211
3114/290	A815213
3114/290	A815229
3114/290	A815247
3114/290	A829
3114/290	A829A
3114/290	A829B
3114/290	A829C
3114/290	A829D
3114/290	A829E
3114/290	A833
3114/290	A8405
3114/290	A844
3114/290	A8540
3114/290	A8867
3114/290	A921-70B
3114/290	A94037
3114/290	A95227
3114/290	A95232
3114/290	A970246
3114/290	A970248
3114/290	A970251
3114/290	A970254
3114/290	A984193
3114/290	A991-01-0098
3114/290	A991-01-1225
3114/290	A991-01-1319
3114/290	A991-01-3058
3114/290	AC9082
3114/290	AC9083
3114/290	AC9084
3114/290	AC9085
3114/290	ACDC9000-1
3114/290	AD29A-4
3114/290	AD29A-5
3114/290	AD29A-6
3114/290	AD29A-9
3114/290	AD29E-1
3114/290	AD29E-2
3114/290	AD29E10
3114/290	AD29E4
3114/290	AD29E5
3114/290	AD29E6
3114/290	AD29E7
3114/290	AD29E8
3114/290	AD29E9
3114/290	AD30A1
3114/290	AD30A2
3114/290	AD30A3
3114/290	AD30A4
3114/290	AD30A5
3114/290	AF21490
3114/290	AF3570
3114/290	AF3590
3114/290	AF699
3114/290	AFCS1170F
3114/290	AFS24226
3114/290	AFT0019M
3114/290	AFT052
3114/290	AFT1341
3114/290	AFT1746
3114/290	AT-11
3114/290	AT331
3114/290	AT331A
3114/290	AT332
3114/290	AT332A
3114/290	AT335A
3114/290	AT410
3114/290	AT410-1
3114/290	AT412
3114/290	AT412-1
3114/290	AT413
3114/290	AT413-1
3114/290	AT414
3114/290	AT414-1
3114/290	AT415
3114/290	AT415-1
3114/290	AT416
3114/290	AT416-1
3114/290	AT417
3114/290	AT417-1
3114/290	AT418
3114/290	AT418-1
3114/290	AT419
3114/290	AT419-1
3114/290	AT430
3114/290	AT430-1
3114/290	AT431
3114/290	AT431-1
3114/290	AT432
3114/290	AT432-1
3114/290	AT433
3114/290	AT433-1
3114/290	AT434
3114/290	AT434-1
3114/290	AT435
3114/290	AT435-1
3114/290	AT436
3114/290	AT436-1
3114/290	AT437
3114/290	AT437-1
3114/290	AT438
3114/290	AT438-1
3114/290	AT451
3114/290	AT451-1
3114/290	AT452
3114/290	AT452-1
3114/290	AT453
3114/290	AT453-1
3114/290	AT454
3114/290	AT454-1
3114/290	AT455
3114/290	AT455-1
3114/290	B-1426
3114/290	B1J
3114/290	B1N-1
3114/290	B1N-2
3114/290	B2A
3114/290	B2E
3114/290	B2G
3114/290	B2M-1
3114/290	B2M-2
3114/290	B2M-3
3114/290	B2S
3114/290	B2W
3114/290	B2Y
3114/290	BC126
3114/290	BC126-1
3114/290	BC137
3114/290	BC137-1
3114/290	BC153
3114/290	BC153-1
3114/290	BC154
3114/290	BC154-1
3114/290	BC157
3114/290	BC157-1
3114/290	BC158-1
3114/290	BC158A-1
3114/290	BC158B
3114/290	BC158B-1
3114/290	BC159-1
3114/290	BC159A
3114/290	BC159A-1
3114/290	BC159B
3114/290	BC159B-1
3114/290	BC177-1
3114/290	BC177A
3114/290	BC177A-1
3114/290	BC177B
3114/290	BC177B-1
3114/290	BC177V
3114/290	BC177V-1
3114/290	BC177V1
3114/290	BC177V1-1
3114/290	BC178
3114/290	BC178-1
3114/290	BC178A
3114/290	BC178A-1
3114/290	BC178B-1
3114/290	BC178D
3114/290	BC178D-1
3114/290	BC178V
3114/290	BC178V-1
3114/290	BC178V1
3114/290	BC178V1-1
3114/290	BC179
3114/290	BC179-1
3114/290	BC179A
3114/290	BC179A-1
3114/290	BC179B
3114/290	BC179B-1
3114/290	BC181A
3114/290	BC181A-1
3114/290	BC187
3114/290	BC187-1
3114/290	BC192
3114/290	BC192-1
3114/290	BC196
3114/290	BC196-1
3114/290	BC196A
3114/290	BC196A-1
3114/290	BC196B-1
3114/290	BC196V1
3114/290	BC196V1-1
3114/290	BC200-1
3114/290	BC201
3114/290	BC201-1
3114/290	BC202
3114/290	BC202-1
3114/290	BC203
3114/290	BC203-1
3114/290	BC204
3114/290	BC204-1
3114/290	BC204A
3114/290	BC204A-1
3114/290	BC204B
3114/290	BC204B-1
3114/290	BC204V
3114/290	BC204V-1
3114/290	BC204V1
3114/290	BC204V1-1
3114/290	BC205
3114/290	BC205-1
3114/290	BC205A
3114/290	BC205A-1
3114/290	BC205B
3114/290	BC205B-1
3114/290	BC205V
3114/290	BC205V-1
3114/290	BC205V1
3114/290	BC205V1-1
3114/290	BC206
3114/290	BC206-1
3114/290	BC206B
3114/290	BC206B-1
3114/290	BC212
3114/290	BC212-1
3114/290	BC212K
3114/290	BC212K-1
3114/290	BC212KA
3114/290	BC212KA-1
3114/290	BC212KB
3114/290	BC212KB-1
3114/290	BC212L
3114/290	BC212L-1
3114/290	BC212LA
3114/290	BC212LA-1
3114/290	BC212LB
3114/290	BC212LB-1
3114/290	BC213
3114/290	BC213-1
3114/290	BC213K
3114/290	BC213K-1
3114/290	BC213KA
3114/290	BC213KA-1
3114/290	BC213KB
3114/290	BC213KB-1
3114/290	BC213KC
3114/290	BC213KC-1
3114/290	BC213L
3114/290	BC213L-1
3114/290	BC213LA
3114/290	BC213LA-1
3114/290	BC213LB
3114/290	BC213LB-1
3114/290	BC213LC
3114/290	BC213LC-1
3114/290	BC214
3114/290	BC214-1
3114/290	BC214K
3114/290	BC214K-1
3114/290	BC214KA
3114/290	BC214KA-1
3114/290	BC214KB
3114/290	BC214KB-1
3114/290	BC214KC
3114/290	BC214KC-1
3114/290	BC214L
3114/290	BC214L-1
3114/290	BC214LA
3114/290	BC214LA-1
3114/290	BC214LB
3114/290	BC214LB-1
3114/290	BC214LC
3114/290	BC214LC-1
3114/290	BC221
3114/290	BC221-1
3114/290	BC225-1
3114/290	BC250
3114/290	BC250-1
3114/290	BC250A
3114/290	BC250A-1
3114/290	BC250B
3114/290	BC250B-1
3114/290	BC250C
3114/290	BC250C-1
3114/290	BC251
3114/290	BC251A
3114/290	BC251A-1
3114/290	BC251B
3114/290	BC251B-1
3114/290	BC251C
3114/290	BC251C-1
3114/290	BC252
3114/290	BC252-1
3114/290	BC252A
3114/290	BC252A-1
3114/290	BC252B
3114/290	BC252B-1
3114/290	BC252C
3114/290	BC252C-1
3114/290	BC253
3114/290	BC253-1
3114/290	BC253A
3114/290	BC253A-1
3114/290	BC253B
3114/290	BC253B-1
3114/290	BC253C
3114/290	BC253C-1
3114/290	BC256
3114/290	BC256-1
3114/290	BC256A
3114/290	BC256A-1
3114/290	BC256B
3114/290	BC256B-1
3114/290	BC257-1
3114/290	BC258
3114/290	BC258-1
3114/290	BC259
3114/290	BC259-1
3114/290	BC25BB
3114/290	BC260
3114/290	BC260-1
3114/290	BC260A
3114/290	BC260A-1
3114/290	BC260B
3114/290	BC260B-1
3114/290	BC260C
3114/290	BC260C-1
3114/290	BC261A
3114/290	BC261A-1
3114/290	BC261B-1
3114/290	BC261C
3114/290	BC261C-1
3114/290	BC262
3114/290	BC262-1
3114/290	BC262A
3114/290	BC262A-1
3114/290	BC262B
3114/290	BC262B-1
3114/290	BC262C
3114/290	BC262C-1
3114/290	BC263
3114/290	BC263A
3114/290	BC263A-1
3114/290	BC263B
3114/290	BC263B-1
3114/290	BC263C
3114/290	BC263C-1
3114/290	BC266
3114/290	BC266-1
3114/290	BC266A
3114/290	BC266B
3114/290	BC281
3114/290	BC281-1
3114/290	BC281A
3114/290	BC281A-1
3114/290	BC281B
3114/290	BC281B-1
3114/290	BC281C
3114/290	BC281C-1
3114/290	BC283
3114/290	BC283-1
3114/290	BC291
3114/290	BC291-1
3114/290	BC291A
3114/290	BC291A-1
3114/290	BC291D
3114/290	BC291D-1
3114/290	BC292
3114/290	BC292A
3114/290	BC292A-1
3114/290	BC292D
3114/290	BC292D-1
3114/290	BC307A
3114/290	BC308
3114/290	BC308A
3114/290	BC309
3114/290	BC309A
3114/290	BC325
3114/290	BC325A
3114/290	BC326
3114/290	BC326A
3114/290	BC400
3114/290	BCW35
3114/290	BCW37
3114/290	BCW37A
3114/290	BCW56
3114/290	BCW56A
3114/290	BCW57A
3114/290	BCW58A
3114/290	BCW59A
3114/290	BCY10
3114/290	BCY10A
3114/290	BCY11
3114/290	BCY11A
3114/290	BCY12
3114/290	BCY12A
3114/290	BCY29
3114/290	BCY35
3114/290	BCY35A
3114/290	BCY37
3114/290	BCY37A
3114/290	BCY38
3114/290	BCY38A
3114/290	BCY39
3114/290	BCY39A
3114/290	BCY40
3114/290	BCY40A
3114/290	BCY54
3114/290	BCY54A
3114/290	BCY70
3114/290	BCY70A
3114/290	BCY71
3114/290	BCY71A
3114/290	BCY72
3114/290	BCY72A
3114/290	BCY78
3114/290	BCY78A
3114/290	BCY79
3114/290	BCY90
3114/290	BCY90A
3114/290	BCY90B
3114/290	BCY90B-1
3114/290	BCY91
3114/290	BCY91A
3114/290	BCY91B
3114/290	BCY91B-1
3114/290	BCY92
3114/290	BCY92A
3114/290	BCY92B
3114/290	BCY92B-1
3114/290	BCY93
3114/290	BCY93A
3114/290	BCY93B
3114/290	BCY93B-1
3114/290	BCY94
3114/290	BCY94A
3114/290	BCY94B
3114/290	BCY94B-1
3114/290	BCY95
3114/290	BCY95A
3114/290	BCY95B
3114/290	BCY95B-1
3114/290	BCY98
3114/290	BCY98A
3114/290	BCY98B
3114/290	BCY98B-1
3114/290	BCZ10
3114/290	BCZ10A
3114/290	BCZ10B
3114/290	BCZ10C
3114/290	BCZ11
3114/290	BCZ11A
3114/290	BCZ12
3114/290	BCZ12A
3114/290	BCZ12B
3114/290	BCZ12C
3114/290	BCZ13
3114/290	BCZ13A
3114/290	BCZ14
3114/290	BCZ14A
3114/290	BFV25
3114/290	BFV26
3114/290	BJ1A
3114/290	BJ1B
3114/290	BJ1C
3114/290	BJ2
3114/290	BJ2A
3114/290	BJ2B
3114/290	BJ2C
3114/290	BJ2D
3114/290	BJ3
3114/290	BJ3A
3114/290	BJ3B
3114/290	BJ4
3114/290	BJ4A
3114/290	BJ4B
3114/290	BJ4C
3114/290	BJ4D
3114/290	BJ5
3114/290	BJ56
3114/290	BJ5A
3114/290	BJ5B
3114/290	BMT1991
3114/290	BMT2303
3114/290	BMT2411
3114/290	BMT2412
3114/290	BSV21
3114/290	BSV21A
3114/290	BSW-21A
3114/290	BSW-22
3114/290	BSW-22A
3114/290	BSW-24
3114/290	BSW-44
3114/290	BSW-44A
3114/290	BSW-45
3114/290	BSW-45A
3114/290	BSW-72
3114/290	BSW-73
3114/290	BSW-74
3114/290	BSW-75
3114/290	BSW21
3114/290	BSW21A
3114/290	BSW22
3114/290	BSW22A
3114/290	BSW24
3114/290	BSW44
3114/290	BSW44A
3114/290	BSW45
3114/290	BSW45A
3114/290	BSW72
3114/290	BSW73
3114/290	BSW74
3114/290	BSW75
3114/290	BSX36
3114/290	BSY-40
3114/290	BSY-41
3114/290	BSY40
3114/290	BSY41
3114/290	C9082
3114/290	C9083
3114/290	C9084
3114/290	C9085
3114/290	CCS2005B
3114/290	CCS9015
3114/290	CDC10000-1E
3114/290	CE4005C
3114/290	CK942
3114/290	CS-2005B
3114/290	CS-2005C
3114/290	CS1228E
3114/290	CS1251F
3114/290	CS1294E
3114/290	CS1294H
3114/290	CS1303
3114/290	CS1308
3114/290	CS1312G
3114/290	CS3702
3114/290	CS3703
3114/290	CS3906
3114/290	CS5447
3114/290	CS5448
3114/290	CS7228G
3114/290	CS9012
3114/290	CS9012E
3114/290	CS9012E-F
3114/290	CS9012H
3114/290	CS9012HE
3114/290	CS9012HF
3114/290	CS9012HG
3114/290	CS9012HH
3114/290	CS9015C
3114/290	CS9015C2
3114/290	CS9015D
3114/290	CS9102
3114/290	CS9128-B2
3114/290	CS9129B2
3114/290	D29A5
3114/290	D29A9
3114/290	D29E1
3114/290	D29E10
3114/290	D29E2
3114/290	D29E4
3114/290	D29E5
3114/290	D29E6
3114/290	D29E7
3114/290	D29E8
3114/290	D29E9
3114/290	D30A1
3114/290	D30A4
3114/290	D30A5
3114/290	DBAZ073304
3114/290	DDBY003001
3114/290	DDBY004001
3114/290	DDBY008001
3114/290	DDCY007002
3114/290	DS-82
3114/290	DS-83
3114/290	DS-86
3114/290	E13-001-02
3114/290	E13-001-03
3114/290	E13-001-04
3114/290	E13-006-02
3114/290	EA15X194
3114/290	EA15X242
3114/290	EA15X268
3114/290	EA15X385
3114/290	EA15X395
3114/290	EA3714
3114/290	EN2894A
3114/290	EP15X17
3114/290	EP15X21
3114/290	EP15X26
3114/290	EP15X51
3114/290	EP15X90
3114/290	EQR-0038
3114/290	ES15X9
3114/290	ES15X90
3114/290	ET350335
3114/290	ET539122
3114/290	EWQ202
3114/290	F21490
3114/290	F3570
3114/290	F3590
3114/290	F699
3114/290	FCS8550
3114/290	FCS9012H
3114/290	FCS9012HE
3114/290	FCS9015B
3114/290	FS24226
3114/290	FT0019M
3114/290	FT052
3114/290	FT1341
3114/290	FT1746
3114/290	G03007
3114/290	G03014
3114/290	G13638
3114/290	GE-269
3114/290	GE-48
3114/290	GE-65
3114/290	GET3638
3114/290	GI3638
3114/290	GI3638A
3114/290	GI3644
3114/290	GI3702
3114/290	GM760
3114/290	GME040-1
3114/290	GT1644
3114/290	HA-00495
3114/290	HA00562
3114/290	HEP-S0032
3114/290	HF-47
3114/290	HF47
3114/290	HR-71
3114/290	HR71
3114/290	HS-40027
3114/290	HS-40035
3114/290	HS-40040
3114/290	HS-40053
3114/290	HS-40057
3114/290	HT100
3114/290	HT101
3114/290	HT105611B0
3114/290	HT105611C
3114/290	HT106731B0
3114/290	HT107211T
3114/290	HT305642B
3114/290	HX-50112

SK	Industry Standard No.
3114/290	I9680
3114/290	IC743043
3114/290	IP20-0009
3114/290	IP20-0046
3114/290	IP20-0159
3114/290	IP20-0192
3114/290	IP20-0213
3114/290	IP20-0217
3114/290	J241253
3114/290	J241259
3114/290	J24640
3114/290	J9680
3114/290	J9697
3114/290	JA1050
3114/290	JA1050G
3114/290	JA1050GL
3114/290	K4-505
3114/290	K9682
3114/290	KLH4746
3114/290	KSA495Y
3114/290	LJ152
3114/290	LJ152B
3114/290	LJ152G
3114/290	LM-1149
3114/290	LM-1150
3114/290	LM-1151
3114/290	LM-1153
3114/290	LM1404
3114/290	M4442
3114/290	M446
3114/290	M4590
3114/290	M4745
3114/290	M644
3114/290	M65A
3114/290	M65B
3114/290	M65C
3114/290	M65D
3114/290	M65E
3114/290	M65F
3114/290	M7127
3114/290	M829A
3114/290	M829B
3114/290	M829C
3114/290	M829D
3114/290	M829E
3114/290	M829F
3114/290	M833
3114/290	M9514
3114/290	M9531
3114/290	ME0404-1
3114/290	ME0404-2
3114/290	ME501
3114/290	MF3304
3114/290	MM3726
3114/290	MM3905
3114/290	MM3906
3114/290	MM4048
3114/290	MPS1572
3114/290	MPS3639
3114/290	MPS3640
3114/290	MPS3703
3114/290	MPS5086
3114/290	MPS6562
3114/290	MPS6563
3114/290	MPS9666
3114/290	MPS9680I/J
3114/290	MPS9682
3114/290	MPS9682I
3114/290	MPS9682J
3114/290	MPS9682K
3114/290	MPS9682T
3114/290	MS7505
3114/290	MS9667
3114/290	MS9681
3114/290	MT0404
3114/290	MT0404-1
3114/290	MT0404-2
3114/290	MT0411
3114/290	MT0413
3114/290	MT1131
3114/290	MT1131A
3114/290	MT1132
3114/290	MT1132A
3114/290	MT1132B
3114/290	MT1254
3114/290	MT1255
3114/290	MT1256
3114/290	MT1257
3114/290	MT1258
3114/290	MT1259
3114/290	MT1420
3114/290	MT1991
3114/290	MT2303
3114/290	MT2411
3114/290	MT2412
3114/290	MT726
3114/290	MT869
3114/290	NB121
3114/290	NJ101B
3114/290	NKT20329
3114/290	NKT20329A
3114/290	NKT20339
3114/290	NR-621AU
3114/290	NR601BT
3114/290	NR621AT
3114/290	NR621EU
3114/290	NR631AY
3114/290	NS1000
3114/290	NS1000A
3114/290	NS1001
3114/290	NS1001A
3114/290	NS1672
3114/290	NS1672A
3114/290	NS1673
3114/290	NS1674
3114/290	NS1674A
3114/290	NS1675
3114/290	NS1675A
3114/290	NS1861
3114/290	NS1861A
3114/290	NS1862
3114/290	NS1863
3114/290	NS1863A
3114/290	NS1864
3114/290	NS1864A
3114/290	NS404
3114/290	NS6001
3114/290	NS6062
3114/290	NS6062A
3114/290	NS6063
3114/290	NS6063A
3114/290	NS6064
3114/290	NS6064A
3114/290	NS6065
3114/290	NS6065A
3114/290	NS6211
3114/290	NS6211A
3114/290	NS6241
3114/290	NS661
3114/290	NS662
3114/290	NS663
3114/290	NS664
3114/290	NS665
3114/290	NS666
3114/290	NS667
3114/290	NS668
3114/290	NS732
3114/290	NS732A
3114/290	OC200
3114/290	OC201
3114/290	OC202
3114/290	OC203
3114/290	OC204
3114/290	OC205
3114/290	OC206
3114/290	OC207
3114/290	OC430
3114/290	OC430K
3114/290	OC440
3114/290	OC440K
3114/290	OC443
3114/290	OC443K
3114/290	OC445
3114/290	OC445K
3114/290	OC449
3114/290	OC449K
3114/290	OC450
3114/290	OC460
3114/290	OC460K
3114/290	OC463
3114/290	OC463K
3114/290	OC465
3114/290	OC465K
3114/290	OC466
3114/290	OC466K
3114/290	OC467
3114/290	OC467K
3114/290	OC468
3114/290	OC468K
3114/290	OC469
3114/290	OC469K
3114/290	OC470
3114/290	OC470K
3114/290	OC700
3114/290	OC700A
3114/290	OC700B
3114/290	OC702
3114/290	OC702A
3114/290	OC702B
3114/290	OC704
3114/290	OC740
3114/290	OC740G
3114/290	OC740M
3114/290	OC7400
3114/290	OC742
3114/290	OC742G
3114/290	OC742M
3114/290	OC7420
3114/290	P1C
3114/290	P1D
3114/290	P1N
3114/290	P1N-1
3114/290	P1N-2
3114/290	P1N-3
3114/290	P1P
3114/290	P1P-1
3114/290	P2A
3114/290	P2G
3114/290	P2M-1
3114/290	P2M-2
3114/290	P2M-3
3114/290	P2P
3114/290	P3C
3114/290	P3CA
3114/290	P5C
3114/290	PA1000
3114/290	PA1001
3114/290	PA1001A
3114/290	Q-36
3114/290	Q-36A
3114/290	Q5087Z
3114/290	Q5100A
3114/290	Q5102
3114/290	Q5116C
3114/290	Q5116CA
3114/290	Q5135
3114/290	Q5205
3114/290	Q6521
3114/290	QA-21
3114/290	QA-21A
3114/290	QQC61689
3114/290	QQC61689A
3114/290	QT-A0719XAN
3114/290	QT-A0719XCN
3114/290	QT-A0719XHN
3114/290	QT-A0733XAA
3114/290	R8967
3114/290	R8967A
3114/290	R8969
3114/290	R8969A
3114/290	RE62
3114/290	RS7665
3114/290	RT3065
3114/290	RT3065A
3114/290	RT3071
3114/290	RT3071A
3114/290	RT8670
3114/290	RT8895
3114/290	RV1059
3114/290	RV2260
3114/290	RV2351
3114/290	RV2355
3114/290	RVTS22411
3114/290	S0026
3114/290	S1350
3114/290	S1350A
3114/290	S1367
3114/290	S2091
3114/290	S2128
3114/290	S2129
3114/290	S2130
3114/290	S2525
3114/290	S2645
3114/290	S2645A
3114/290	S2988
3114/290	S2988A
3114/290	S3639
3114/290	S3640
3114/290	S3640A
3114/290	S3655A
3114/290	SA310
3114/290	SA310A
3114/290	SA311
3114/290	SA311A
3114/290	SA312
3114/290	SA312A
3114/290	SA313
3114/290	SA314
3114/290	SA314A
3114/290	SA315
3114/290	SA315A
3114/290	SA316
3114/290	SA316A
3114/290	SA412
3114/290	SA412A
3114/290	SA413
3114/290	SA413A
3114/290	SA414
3114/290	SA414A
3114/290	SA415
3114/290	SA415A
3114/290	SA50
3114/290	SA50A
3114/290	SA51
3114/290	SA52
3114/290	SA52A
3114/290	SA52AC
3114/290	SA52B
3114/290	SA52BC
3114/290	SA53
3114/290	SA53A
3114/290	SA54
3114/290	SA55
3114/290	SA55A
3114/290	SA56
3114/290	SA56A
3114/290	SA70
3114/290	SA70A
3114/290	SF8014
3114/290	SHA7530
3114/290	SHA7531
3114/290	SHA7532
3114/290	SHA7533
3114/290	SHA7534
3114/290	SHA7536
3114/290	SHA7537
3114/290	SHA7538
3114/290	SK1639
3114/290	SK1639D
3114/290	SK1640
3114/290	SK1856
3114/290	SK1856A
3114/290	SK2604
3114/290	SK2604A
3114/290	SK6345
3114/290	SK6346
3114/290	SK6347
3114/290	SK6347A
3114/290	SKA4129
3114/290	SM4547
3114/290	SM4719
3114/290	SNT204
3114/290	SNT204A
3114/290	SPS1097
3114/290	SPS1097A
3114/290	SPS1523
3114/290	SPS1523A
3114/290	SPS2269
3114/290	SPS2272
3114/290	SPS3724
3114/290	SPS3724A
3114/290	SPS3786
3114/290	SPS3786A
3114/290	SPS3924
3114/290	SPS3924A
3114/290	SPS3927
3114/290	SPS3927A
3114/290	SPS3931
3114/290	SPS3931A
3114/290	SPS3987
3114/290	SPS3988
3114/290	SPS3988A
3114/290	SPS3990
3114/290	SPS3990A
3114/290	SPS4007
3114/290	SPS4007A
3114/290	SPS4013
3114/290	SPS4013A
3114/290	SPS4014
3114/290	SPS4014A
3114/290	SPS4018
3114/290	SPS4018A
3114/290	SPS4019
3114/290	SPS4019A
3114/290	SPS4025
3114/290	SPS4025A
3114/290	SPS4026
3114/290	SPS4026A
3114/290	SPS4027
3114/290	SPS4027A
3114/290	SPS4028
3114/290	SPS4028A
3114/290	SPS4031
3114/290	SPS4031A
3114/290	SPS4054
3114/290	SPS4054A
3114/290	SPS4056
3114/290	SPS4056A
3114/290	SPS4064
3114/290	SPS4064A
3114/290	SPS4072
3114/290	SPS4072A
3114/290	SPS4073
3114/290	SPS4073A
3114/290	SPS4076
3114/290	SPS4076A
3114/290	SPS4078
3114/290	SPS4078A
3114/290	SPS4082
3114/290	SPS4082A
3114/290	SPS4086
3114/290	SPS4086A
3114/290	SPS4087
3114/290	SPS4087A
3114/290	SPS4090
3114/290	SPS4090A
3114/290	SPS42
3114/290	SPS42A
3114/290	SPS4302
3114/290	SPS4314
3114/290	SPS4314A
3114/290	SPS4348
3114/290	SPS4348A
3114/290	SPS4354
3114/290	SPS4354A
3114/290	SPS4355
3114/290	SPS4365
3114/290	SPS4365A
3114/290	SPS4452
3114/290	SPS4452A
3114/290	SPS4458
3114/290	SPS4458A
3114/290	SPS4460
3114/290	SPS4460A
3114/290	SPS4473
3114/290	SPS4473A
3114/290	SPS4480
3114/290	SPS4480A
3114/290	SPS4489
3114/290	SPS4489A
3114/290	SPS5007
3114/290	SPS5007-1
3114/290	SPS5007-1A
3114/290	SPS5007-2
3114/290	SPS5007-2A
3114/290	SPS5007A
3114/290	SPS514
3114/290	SPS514A
3114/290	SPS5451
3114/290	SPS6109
3114/290	SPS6109A
3114/290	SS1606
3114/290	SS1606A
3114/290	SS1906
3114/290	SS1906A
3114/290	SS2503
3114/290	SS2503A
3114/290	SSA43
3114/290	SSA43A
3114/290	SSA43A-1
3114/290	SSA46
3114/290	SSA46A
3114/290	SSA48
3114/290	SSA48A
3114/290	ST8033
3114/290	ST8033A
3114/290	ST8034
3114/290	ST8034A
3114/290	ST8035
3114/290	ST8035A
3114/290	ST8036
3114/290	ST8036A
3114/290	ST8065
3114/290	ST8065A
3114/290	ST8500
3114/290	ST8500A
3114/290	ST8509
3114/290	ST8509A
3114/290	SX3702A
3114/290	SX61M
3114/290	SX61MA
3114/290	T-Q5077
3114/290	T-Q5087
3114/290	T1-503
3114/290	T1-503A
3114/290	T1-743
3114/290	T1-743A
3114/290	T1-744
3114/290	T1-744A
3114/290	T1-752
3114/290	T1-752A
3114/290	T1-906
3114/290	T1S03
3114/290	T1S03A
3114/290	T1S04
3114/290	T1S04A
3114/290	T1S37A
3114/290	T1S38A
3114/290	T1S53
3114/290	T1S53A
3114/290	T1S54
3114/290	T1S5A
3114/290	T1S61A
3114/290	T1S91
3114/290	T1S91A
3114/290	T1S93
3114/290	T1S93A
3114/290	T3570A
3114/290	T9681
3114/290	TG2SA608
3114/290	TG2SA608C
3114/290	TI-503
3114/290	TI-743
3114/290	TI-744
3114/290	TI-752
3114/290	TI-906
3114/290	TIS-03
3114/290	TIS04
3114/290	TIS38
3114/290	TIS53
3114/290	TIS54
3114/290	TIS61
3114/290	TIS61A
3114/290	TIS61B
3114/290	TIS61C
3114/290	TIS61D
3114/290	TIS61E
3114/290	TIS61M
3114/290	TIS91
3114/290	TIS93
3114/290	TM290
3114/290	TN061690
3114/290	TN061703
3114/290	TN2SA733-Q
3114/290	TN2SA733-R
3114/290	TNJ70481
3114/290	TNJ71773
3114/290	TNJ71774
3114/290	TNJ72154
3114/290	TR-1030-1
3114/290	TR-1030-2
3114/290	TR-1032-2
3114/290	TR-19A
3114/290	TR-20
3114/290	TR-20A
3114/290	TR-30
3114/290	TR-30A
3114/290	TR-4R38
3114/290	TR-6R35
3114/290	TR-6R35A
3114/290	TR-8026
3114/290	TR0055
3114/290	TR01053-1
3114/290	TR02051-1
3114/290	TR02062-1
3114/290	TR02062-6
3114/290	TR02063-8
3114/290	TR1000
3114/290	TR1000A
3114/290	TR1030
3114/290	TR1030A
3114/290	TR1032
3114/290	TR1032A
3114/290	TR1034
3114/290	TR1034A
3114/290	TR2327743
3114/290	TR30
3114/290	TR8007A
3114/290	TR8026A
3114/290	TR8055
3114/290	TV-47A
3114/290	TV-87
3114/290	TV-93
3114/290	TV24214A
3114/290	TV24363
3114/290	TV24363A
3114/290	TV24495A
3114/290	TVS-2SA564
3114/290	TVS-2SA564A
3114/290	TVS-2SA564P
3114/290	TVS-CS1303
3114/290	TVS-CS1303A
3114/290	TVS2SA564-O
3114/290	TVS2SA564P
3114/290	TVS2SA564Q
3114/290	V152
3114/290	V152A
3114/290	V162
3114/290	V162A
3114/290	V410
3114/290	VS2SA495-O/1E
3114/290	VS2SA495-Y/1E
3114/290	VS2SA562-Y/1E
3114/290	VS2SA673-B/1E
3114/290	VS2SA673-C/1A
3114/290	VS2SA673-C/1E
3114/290	VS2SA844-D/-1
3114/290	VS2SB561-C/-1
3114/290	W21
3114/290	XA-495C
3114/290	XA495D
3114/290	Y410
3114/290	ZEN107
3114/290	ZT131
3114/290	ZT131A
3114/290	ZT152
3114/290	ZT152A
3114/290	ZT153
3114/290	ZT153A
3114/290	ZT154
3114/290	ZT154A
3114/290	ZT180
3114/290	ZT180A
3114/290	ZT181
3114/290	ZT181A
3114/290	ZT182
3114/290	ZT182A
3114/290	ZT183
3114/290	ZT183A
3114/290	ZT184
3114/290	ZT184A
3114/290	ZT187
3114/290	ZT187A
3114/290	ZT280
3114/290	ZT280A
3114/290	ZT281
3114/290	ZT281A
3114/290	ZT282
3114/290	ZT282A
3114/290	ZT283
3114/290	ZT283A
3114/290	ZT284
3114/290	ZT284A
3114/290	ZT287
3114/290	ZT287A
3114/290	ZTX530
3114/290	ZTX530A
3114/290	ZTX530B
3114/290	ZTX530C
3114/290	ZTX530D
3114/290	ZTX531
3114/290	ZTX531A
3114/290	ZTX531B
3114/290	01-010495
3114/290	01-010564
3114/290	01-010628
3114/290	01-010673
3114/290	01-010719
3114/290	01-010844
3114/290	01-572811
3114/290	1V68611A47
3114/290	1V68611A47A
3114/290	1W11700
3114/290	1W11700A
3114/290	1W11711
3114/290	1W9728
3114/290	1W9728A
3114/290	1W9782
3114/290	1W9782A
3114/290	1W9810
3114/290	1W9810A
3114/290	1W9810S
3114/290	1W9810SA
3114/290	2D027
3114/290	2N1034
3114/290	2N1035
3114/290	2N1036
3114/290	2N1037
3114/290	2N1428
3114/290	2N1606
3114/290	2N1623
3114/290	2N2104
3114/290	2N2105
3114/290	2N327A
3114/290	2N328
3114/290	2N328A
3114/290	2N3798A
3114/290	2N3799A
3114/290	2N4142
3114/290	2N495
3114/290	2N496
3114/290	2N5221
3114/290	02P1BC
3114/290	2SA1015-O
3114/290	2SA1015Y
3114/290	2SA1029
3114/290	2SA1029C
3114/290	2SA495
3114/290	2SA495-O
3114/290	2SA495-G

SK	Industry Standard No.
3114/290	2SA495-O
3114/290	2SA495-R
3114/290	2SA495-Y
3114/290	2SA495A
3114/290	2SA495D
3114/290	2SA495G
3114/290	2SA495G-GR
3114/290	2SA495G-O
3114/290	2SA495G-R
3114/290	2SA495G-Y
3114/290	2SA495H
3114/290	2SA495J
3114/290	2SA495R
3114/290	2SA495W
3114/290	2SA495X
3114/290	2SA495Y
3114/290	2SA561
3114/290	2SA561-O
3114/290	2SA561-R
3114/290	2SA561-Y
3114/290	2SA561GR
3114/290	2SA561R
3114/290	2SA561Y
3114/290	2SA562
3114/290	2SA562-O
3114/290	2SA562-O
3114/290	2SA562-Y
3114/290	2SA562G
3114/290	2SA562GR
3114/290	2SA562R
3114/290	2SA562Y
3114/290	2SA567A
3114/290	2SA567C
3114/290	2SA569
3114/290	2SA628G
3114/290	2SA629
3114/290	2SA642
3114/290	2SA642A
3114/290	2SA642B
3114/290	2SA642C
3114/290	2SA642D
3114/290	2SA642E
3114/290	2SA642F
3114/290	2SA642L
3114/290	2SA642R
3114/290	2SA642S
3114/290	2SA642V
3114/290	2SA642W
3114/290	2SA642Y
3114/290	2SA659
3114/290	2SA659A
3114/290	2SA659B
3114/290	2SA659C
3114/290	2SA659D
3114/290	2SA659E
3114/290	2SA659F
3114/290	2SA659G
3114/290	2SA659L
3114/290	2SA659P
3114/290	2SA659R
3114/290	2SA659Y
3114/290	2SA672
3114/290	2SA672A
3114/290	2SA672B
3114/290	2SA672C
3114/290	2SA673A
3114/290	2SA673AB
3114/290	2SA673AC
3114/290	2SA673AD
3114/290	2SA673AS
3114/290	2SA673B
3114/290	2SA673C
3114/290	2SA673D
3114/290	2SA677
3114/290	2SA678
3114/290	2SA678(C)
3114/290	2SA678E
3114/290	2SA719
3114/290	2SA719P
3114/290	2SA719Q
3114/290	2SA719R
3114/290	2SA719RS
3114/290	2SA719S
3114/290	2SA720
3114/290	2SA720A
3114/290	2SA720P
3114/290	2SA720Q
3114/290	2SA720R
3114/290	2SA720S
3114/290	2SA723
3114/290	2SA723A
3114/290	2SA723B
3114/290	2SA723C
3114/290	2SA723D
3114/290	2SA723E
3114/290	2SA723F
3114/290	2SA723R
3114/290	2SA733
3114/290	2SA733A
3114/290	2SA733AP
3114/290	2SA733H
3114/290	2SA733I
3114/290	2SA733IQ
3114/290	2SA733P
3114/290	2SA733Q
3114/290	2SA733QP
3114/290	2SA733R
3114/290	2SA741H
3114/290	2SA825
3114/290	2SA826
3114/290	2SA826P
3114/290	2SA826Q
3114/290	2SA826RY
3114/290	2SA836
3114/290	2SA836D
3114/290	2SA836E
3114/290	2SA841
3114/290	2SA844
3114/290	2SA844C
3114/290	2SA844D
3114/290	2SA844E
3114/290	2SA952
3114/290	2SB741
3114/290	2SC564T
3114/290	07-07113
3114/290	09-300036
3114/290	09-300037A
3114/290	09-300043
3114/290	09-300064
3114/290	09-300070
3114/290	09-300076
3114/290	09-300080
3114/290	09-300081
3114/290	09-305024

SK	Industry Standard No.
3114/290	13-0006A
3114/290	13-0043A
3114/290	13-0044A
3114/290	13-0061
3114/290	13-0061A
3114/290	13-16570-1
3114/290	13-16570-1A
3114/290	13-16570-2
3114/290	13-16570-2A
3114/290	13-22582-1A
3114/290	13-23325-5
3114/290	14-602-11
3114/290	14-602-11A
3114/290	14-602-54
3114/290	15-088002
3114/290	15-088002A
3114/290	020-1110-005
3114/290	020-1110-027
3114/290	21A040-049
3114/290	21A040-050
3114/290	21A040-059
3114/290	21A112-002
3114/290	21A112-003
3114/290	21A118-032
3114/290	022-2876-004
3114/290	030-007-0
3114/290	31-0055
3114/290	34-6016-64
3114/290	36-1
3114/290	42-22810A
3114/290	42-27536
3114/290	46-86170-3
3114/290	46-86406-3
3114/290	46-86517-3
3114/290	46-86546-3
3114/290	48-134830
3114/290	48-134830A
3114/290	48-134831
3114/290	48-134831A
3114/290	48-134832
3114/290	48-134832A
3114/290	48-155065
3114/290	48-155095
3114/290	48-155156
3114/290	48-90343A59
3114/290	48-90343A68
3114/290	48S155035
3114/290	48S155045
3114/290	48S155095
3114/290	57A148-12A
3114/290	57A281-14
3114/290	57B197-12
3114/290	57B201-14
3114/290	65B1
3114/290	86-100009
3114/290	86-298-2
3114/290	86-298-20
3114/290	86-423-2
3114/290	86-5079-2
3114/290	86-5104-2
3114/290	86-570-2
3114/290	86-610-9
3114/290	86-669-2
3114/290	86X0066-001
3114/290	86X0066-003
3114/290	87-423-2
3114/290	93P1AA
3114/290	98P1
3114/290	98P10
3114/290	99S073
3114/290	100-0495-15
3114/290	100-0673-04
3114/290	121-1016
3114/290	121-417
3114/290	121-495
3114/290	121-608
3114/290	121-765-01
3114/290	121-774
3114/290	121-838
3114/290	121-845
3114/290	121-875
3114/290	121-879
3114/290	0124A
3114/290	147P2
3114/290	177-006-9-002
3114/290	177-007-9-001
3114/290	177-019-9-003
3114/290	177-025-9-002
3114/290	200X4082-614
3114/290	200X4101-500
3114/290	260P16504
3114/290	260P26201
3114/290	294
3114/290	353-9317-001
3114/290	829A
3114/290	0831
3114/290	921-1016
3114/290	921-182B
3114/290	921-308B
3114/290	972-659E-0
3114/290	991-011319
3114/290	991-012328
3114/290	1010-7738
3114/290	1016
3114/290	1039-0060
3114/290	1040-9068
3114/290	1043-7374
3114/290	1044-0295
3114/290	1061-8312
3114/290	1061-8807
3114/290	1061-9068
3114/290	1063-5423
3114/290	1063-5431
3114/290	1063-5449
3114/290	1063-6926
3114/290	1080G
3114/290	1214
3114/290	1414-158
3114/290	2000-201
3114/290	2000-202
3114/290	2020-01
3114/290	2057A2-529
3114/290	2057A2-561
3114/290	2904-045
3114/290	3574-1
3114/290	4745
3114/290	004746
3114/290	5001-540
3114/290	8000-00003-037
3114/290	12594
3114/290	30270
3114/290	71687-1
3114/290	084001C
3114/290	95239-1

SK	Industry Standard No.
3114/290	121659
3114/290	124047
3114/290	126700
3114/290	126718
3114/290	126719
3114/290	129697
3114/290	129699
3114/290	130215
3114/290	137155
3114/290	256317
3114/290	268717
3114/290	514067S
3114/290	514068S
3114/290	514072S
3114/290	610083-4
3114/290	610099-5
3114/290	610134-6
3114/290	741729
3114/290	986931
3114/290	1223914
3114/290	1224031
3114/290	1471112-10
3114/290	1473540-1
3114/290	1473562-1
3114/290	1473563-1
3114/290	1473574-1
3114/290	1473591-1
3114/290	1473620-1
3114/290	1473627-1
3114/290	1473666-1
3114/290	1473683-1
3114/290	2320162
3114/290	2320632
3114/290	2320637
3114/290	2320681
3114/290	2321351
3114/290	2327262
3114/290	2327387
3114/290	2487340
3114/290	2487341
3114/290	2498512
3114/290	3596063
3114/290	3596118
3114/290	3596340
3114/290	3596341
3114/290	3650238A
3114/290	5320042H
3114/290	5320043H
3114/290	5320592
3114/290	5320593
3114/290	5321184
3114/290	5321252
3114/290	5321253
3114/290	06100035
3114/290	06100053
3114/290	7576004-01
3114/290	8111230
3114/290	8115009
3114/290	12901503
3114/290	20030703-0702
3114/290	23114138
3114/290	23114321
3114/290	23114550
3114/290	26010016
3114/290	26010027
3114/290	43021168
3114/290	43023845
3114/290	48137195A
3114/290	62256893
3114/290	62389702
3114/290	62438096
3114/290	62563265
3114/290	62563311
3114/290	62590661
3114/290	62608498
3114/290	62734663
3114/290	62737433
3114/290	62737468
3114/290	62752742
3114/290	62766247
3114/290	632049518
3114/290	2004056215
3114/290	2004060858
3114/290	2004067350
3114/290	2004082614
3114/290	2004085415
3114/290	2004356122
3114/290	2004700116
3114/290	4100006285
3114/290	6623001900
3114/290	6623002300
3114/I90	P2Y
3115	A6317800
3115	BU105
3115	BU108
3115	BU115
3115	BU204
3115	C1046
3115	DTS702
3115	DTS704
3115	DTS802
3115	DTS804
3115	EP15X28
3115	ES15X94
3115	GE-237
3115	GE-38
3115	IRTR93
3115	MJ8400
3115	Q5140XP
3115	Q5140Z
3115	Q5140ZXP
3115	Q514Z
3115	RE32
3115	ST-BU208
3115	TG2SC1046N-A
3115	TG2SC1295
3115	TG2SC12950
3115	2SC1005
3115	2SC1005A
3115	2SC1046
3115	2SC1046K
3115	2SC1046N
3115	2SC1295
3115	2SC1309
3115	2SD577
3115	2SD649
3115	09-302187
3115	09-304056
3115	14-601-27
3115	21A112-036
3115	21A112-103
3115	46-86389-3
3115	46-86477-3
3115	48-90343A52

SK	Industry Standard No.
3115	48S137341
3115	48S155001
3115	48S155076
3115	57A186-11
3115	57A186-12
3115	57A213-11
3115	57B198-11
3115	57B199-11
3115	86-563-2
3115	86-563-9
3115	86-564-3
3115	86-564-9
3115	121-759
3115	121-759X
3115	121-831
3115	121-985
3115	260P08901
3115	260P33108
3115	260P39008
3115	335-1
3115	370-1
3115	1010-8025
3115	1062-7511
3115	1190
3115	3649-1
3115	140976
3115	140977
3115	1417335-1
3115	1417366-1
3115	1417370-1
3115	1473649-1
3115	23114343
3115	62741449
3115	2003189109
3115/165	A6317900
3115/165	A6738701
3115/165	A6771373
3115/165	BU208
3115/165	C1045
3115/165	C1045B
3115/165	C1045C
3115/165	C1045D
3115/165	C1045E
3115/165	C1045R
3115/165	C1170
3115/165	C1171
3115/165	C999
3115/165	C999A
3115/165	DTS013
3115/165	DTS701
3115/165	DTS801
3115/165	Q5119D
3115/165	SJ5525
3115/165	2SC1045B
3115/165	2SC1045C
3115/165	2SC1045D
3115/165	2SC1045E
3115/165	2SC1045R
3115/165	2SC1100
3115/165	2SC1101
3115/165	2SC1101A
3115/165	2SC1101B
3115/165	2SC1101C
3115/165	2SC1101D
3115/165	2SC1101E
3115/165	2SC1101F
3115/165	2SC1101L
3115/165	2SC1153
3115/165	2SC1170
3115/165	2SC1367
3115/165	2SC1891
3115/165	2SC999
3115/165	2SC999A
3115/165	2SD200
3115/165	2SD200A
3115/165	2SD246
3115/165	8-729-345-42
3115/165	48S155090
3115/165	121-758
3115/165	121-758X
3115/165	137718
3115/165	610217-5
3115/165	610245-1
3115/165	2003064314
3115/165	6621002400
3116	BF245A
3116	BF245B
3116	BF245C
3116	BF256A
3116	BF256B
3116	BF256C
3116	DDCY006001
3116	DS-88
3116	E300
3116	E305
3116	EA15X394
3116	EA15X400
3116	EA15X401
3116	FE3819
3116	GE-FET-3
3116	HEP-F0015
3116	HEP-F0021
3116	HF20019IA0
3116	HF20019IB0
3116	HK-00049
3116	IP20-0035
3116	IP20-0078
3116	IRTRFE100
3116	IT108
3116	ITE4416
3116	J308
3116	K10
3116	K19(GR)
3116	K23A
3116	K34(E)
3116	K34E
3116	K42
3116	K47
3116	K49
3116	KE3684
3116	KE4416
3116	KE5103
3116	KE5105
3116	LDF603
3116	LDF604
3116	LS5485
3116	MJF10335
3116	MJF1033G
3116	MK-10-E
3116	MK102
3116	MPF-102
3116	MPF108
3116	NF500
3116	NF501

SK	Industry Standard No.
3116	NF506
3116	NF550
3116	NPC108
3116	NPC108A
3116	QA-18
3116	QKT0033XBE
3116	RE45
3116	RT8331
3116	S1243N
3116	SFB6183
3116	T1834
3116	TE500
3116	TIS-88
3116	TR-8027
3116	TR2083-70
3116	TV80
3116	U1837E
3116	U1994E
3116	UC155
3116	UC201
3116	UC714
3116	VS2SK49F-1
3116	WEP802
3116	ZEN123
3116	2N3823
3116	2N4224
3116	2N5078A
3116	2N5103A
3116	2N5104A
3116	2N5105A
3116	2N5163
3116	2N5163A
3116	2N5278A
3116	2N5360A
3116	2N5361
3116	2N5362
3116	2N5363
3116	2N5364
3116	2N5392
3116	2N5393
3116	2N5394
3116	2N5395
3116	2N5396
3116	2N5397
3116	2N5398
3116	2N5457
3116	2N5459
3116	2N5484
3116	2N5485
3116	2N5486
3116	2N5592
3116	2N5593
3116	2N5594
3116	2N5668
3116	2N5669
3116	2N5670
3116	2N5949
3116	2N5950
3116	2N5951
3116	2N5952
3116	2N5953
3116	2NJ233B
3116	2SK1033B
3116	2SK22Y
3116	2SK54
3116	2SK83
3116	2SK84
3116	3SK22
3116	3SK22-Y
3116	3SK22GR
3116	3SK22Y
3116	3SK23
3116	3SK34
3116	04-38190-01
3116	4G2
3116	5G2
3116	09-305135
3116	13-28654-5
3116	13-34375-2
3116	13-34378-2
3116	13-34378-3
3116	16-17
3116	020-1112-002
3116	020-1112-008
3116	21A110-003
3116	022-2876-009
3116	24MW652
3116	33K59
3116	34-6018-2
3116	48S137070
3116	57A150-12
3116	57B149-12
3116	57B150-12
3116	90-179
3116	90-606
3116	90-607
3116	90-613
3116	99S045
3116	0105-0012
3116	107M
3116	108-0049-08
3116	182-029-9-001
3116	182-039-9-001
3116	182-044-9-001
3116	182-044-9-002
3116	182-045-9-001
3116	200-053
3116	1009-01
3116	1042-01
3116	1095-01
3116	1859-17
3116	2000-101
3116	2000-105
3116	2074-17
3116	2336-17
3116	3819
3116	4811-0000-025
3116	5001-046
3116	5001-047
3116	8000-00011-053
3116	8000-00049-062
3116	23606
3116	45810-161
3116	71686-5
3116	71686-6
3116	129424
3116	916100
3116	5320942
3116	5321422
3117	A244
3117	A2464
3117	A2465
3117	A248
3117	A482

SK	Industry Standard No.
3117	AT340
3117	AT341
3117	AT342
3117	AT343
3117	AT344
3117	AT345
3117	AT346
3117	BF155
3117	BF161
3117	BF166
3117	BF167
3117	BF168
3117	BF169
3117	BF175
3117	BF180
3117	BF181
3117	BF182
3117	BF183
3117	BF184
3117	BF185
3117	BF206
3117	BF207
3117	BF208
3117	BF209
3117	BF212
3117	BF213
3117	BF214
3117	BF215
3117	BF222
3117	BF226
3117	BF232
3117	BF251
3117	BF252
3117	BF260
3117	BF261
3117	BF270
3117	BF271
3117	BF273
3117	BF273C
3117	BF273D
3117	BF274
3117	BF274B
3117	BF274C
3117	BF287
3117	BF288
3117	BF290
3117	BF302
3117	BF303
3117	BF304
3117	BF306
3117	BFX31
3117	BFX60
3117	BFX62
3117	BFX73
3117	BFX77
3117	BFY79
3117	C1222
3117	C682
3117	C682A
3117	C682B
3117	CCS6225F
3117	CCS6226G
3117	CCS6227F
3117	CCS9017
3117	CCS9017G925
3117	CCS9018H924
3117	CS1014H
3117	CS1120F
3117	CS1284G
3117	CS1284H
3117	CS2715
3117	CS2716
3117	CS3662
3117	CS3663
3117	CS3707
3117	CS3708
3117	CS3709
3117	CS3710
3117	CS3711
3117	CS929
3117	CS930
3117	D4C28
3117	D4C29
3117	D4C30
3117	D4C31
3117	D4D20
3117	D4D21
3117	D4D22
3117	ES15X104
3117	ES15X105
3117	ES15X106
3117	ES15X123
3117	ES15X65
3117	ES15X66
3117	ES15X67
3117	ES15X79
3117	F3535
3117	FT118
3117	GM0380
3117	NKT35219
3117	S15650
3117	S15657
3117	S15658
3117	S15659
3117	SE1002-1
3117	SE1002-2
3117	SE3005
3117	SE4020
3117	SE4021
3117	SE4022
3117	SE5022
3117	SE5051
3117	SE5052
3117	T-Q5055
3117	T3535
3117	T576-1
3117	TG2SC927
3117	TG2SC927A
3117	TG2SC927C
3117	TH-H2SC313
3117	TNJ72151
3117	TV-33
3117	TV-34
3117	TV-50
3117	TV-54
3117	TV24382
3117	TV24437
3117	TV24571
3117	TV81
3117	2N3338
3117	2N3339
3117	2N4434
3117	2N4435
3117	2N5650
3117	2N5651
3117	2N917A
3117	2N918
3117	2SC1035
3117	2SC1035A
3117	2SC1035B
3117	2SC1035C
3117	2SC1035D
3117	2SC1035E
3117	2SC1035F
3117	2SC1035GN
3117	2SC1035H
3117	2SC1035J
3117	2SC1035L
3117	2SC1035M
3117	2SC1035OR
3117	2SC1035R
3117	2SC1035X
3117	2SC1035Y
3117	2SC1036
3117	2SC1036A
3117	2SC1036B
3117	2SC1036C
3117	2SC1036D
3117	2SC1036E
3117	2SC1036F
3117	2SC1036G
3117	2SC1036GN
3117	2SC1036H
3117	2SC1036J
3117	2SC1036K
3117	2SC1036L
3117	2SC1036M
3117	2SC1036OR
3117	2SC1036R
3117	2SC1036X
3117	2SC1036Y
3117	2SC313C
3117	2SC313H
3117	2SC348
3117	2SC348A
3117	2SC348B
3117	2SC348C
3117	2SC348D
3117	2SC348E
3117	2SC348F
3117	2SC348G
3117	2SC348GN
3117	2SC348H
3117	2SC348J
3117	2SC348K
3117	2SC348L
3117	2SC348M
3117	2SC348OR
3117	2SC348X
3117	2SC348Y
3117	2SC390B
3117	2SC390C
3117	2SC390D
3117	2SC390E
3117	2SC390F
3117	2SC390G
3117	2SC390GN
3117	2SC390H
3117	2SC390J
3117	2SC390K
3117	2SC390L
3117	2SC390M
3117	2SC390OR
3117	2SC390R
3117	2SC390X
3117	2SC390Y
3117	2SC463H
3117	2SC466H
3117	2SC663
3117	2SC663A
3117	2SC663B
3117	2SC663C
3117	2SC663D
3117	2SC663E
3117	2SC663F
3117	2SC663G
3117	2SC663GN
3117	2SC663H
3117	2SC663J
3117	2SC663K
3117	2SC663L
3117	2SC663M
3117	2SC663OR
3117	2SC663R
3117	2SC663X
3117	2SC663Y
3117	2SC674B
3117	2SC674C
3117	2SC674CK
3117	2SC674CL
3117	2SC674D
3117	2SC674F
3117	2SC682
3117	2SC682A
3117	2SC682B
3117	2SC682C
3117	2SC682D
3117	2SC682E
3117	2SC682F
3117	2SC682G
3117	2SC682GN
3117	2SC682H
3117	2SC682J
3117	2SC682K
3117	2SC682L
3117	2SC682M
3117	2SC682OR
3117	2SC682R
3117	2SC682X
3117	2SC682Y
3117	2SC860
3117	2SC860A
3117	2SC860B
3117	2SC860C
3117	2SC860D
3117	2SC860E
3117	2SC860F
3117	2SC860G
3117	2SC860GN
3117	2SC860H
3117	2SC860J
3117	2SC860K
3117	2SC860L
3117	2SC860M
3117	2SC860X
3117	2SC860Y
3117	2SC863
3117	2SC863A
3117	2SC863B
3117	2SC863C
3117	2SC863D
3117	2SC863E
3117	2SC863F
3117	2SC863G
3117	2SC863GN
3117	2SC863H
3117	2SC863J
3117	2SC863K
3117	2SC863L
3117	2SC863M
3117	2SC863OR
3117	2SC863R
3117	2SC863X
3117	2SC863Y
3117	2SC928
3117	2SC928A
3117	2SC928B
3117	2SC928C
3117	2SC928D
3117	2SC928E
3117	2SC928F
3117	2SC928G
3117	2SC928GN
3117	2SC928H
3117	2SC928J
3117	2SC928K
3117	2SC928L
3117	2SC928M
3117	2SC928OR
3117	2SC928R
3117	2SC928X
3117	2SC928Y
3117	09-302072
3117	09-302152
3117	09-305006
3117	10-1
3117	13-26577-2
3117	14-653-21
3117	21A112-010
3117	34-6015-15
3117	34-6015-27
3117	34-6015-29
3117	34-6015-31
3117	34-6015-62
3117	46-86112-2
3117	46-8672-3
3117	48-134756
3117	48-65146A61
3117	48-65146A62
3117	48-65173A78
3117	48-90232A01
3117	48-90232A17
3117	48-90232A18
3117	48-97177A02
3117	48-97177A03
3117	51A180-4
3117	57A139-1
3117	57A139-2
3117	57A139-3
3117	57A139-4
3117	57A141-4
3117	57C142-4
3117	86-186-2
3117	86-205-2
3117	86-262-2
3117	86-263-2
3117	86-289-2
3117	86-290-2
3117	86X0052-001
3117	102-4
3117	121-460
3117	121-461
3117	121-504
3117	121-692
3117	121-704
3117	121-754
3117	121-760
3117	121-761
3117	121-924
3117	130-152
3117	181N1
3117	181N1D
3117	181N2
3117	181N2D
3117	229-0204-6
3117	260P24901
3117	297V074C01
3117	576-0003-026
3117	600X0093-086
3117	600X0094-086
3117	916-31025-4B
3117	921-55B
3117	1952-17
3117	2057A2-195
3117	3476
3117	3579
3117	3680
3117	40238
3117	40239
3117	40240
3117	119824
3117	119825
3117	133171
3117	137383
3117	171162-118
3117	610181-1
3117	610181-2
3117	1473586
3117	2092417-0714
3117	62539348
3117	62748273
3118	A160
3118	A161
3118	A162
3118	A3T2894
3118	A3T2906
3118	A3T2906A
3118	A3T2907
3118	A514-040296
3118	BC158
3118	BC158A
3118	BC159
3118	BC196B
3118	BC200
3118	BC225
3118	BC257
3118	BC559
3118	BCW29
3118	BCW29R
3118	BCW30
3118	BCW30R
3118	BCW57
3118	BCW58
3118	BCW59
3118	BCW61
3118	BCW61A
3118	BCW61B
3118	BCW61C
3118	BCW61D
3118	BCZ13B
3118	BCZ13C
3118	BCZ13D
3118	BCZ13E
3118	BCZ13F
3118	BCZ13G
3118	BCZ13H
3118	BCZ14B
3118	BCZ14C
3118	BCZ14D
3118	BCZ14E
3118	BCZ14F
3118	BF340
3118	BF340A
3118	BF340B
3118	BF340C
3118	BF340D
3118	BF341
3118	BF341A
3118	BF341B
3118	BF341C
3118	BF341D
3118	BF342
3118	BF342A
3118	BF342B
3118	BF342C
3118	BF342D
3118	BF516
3118	BFS14A
3118	BFS14B
3118	BFS14C
3118	BFS14D
3118	BFS16
3118	BFS16A
3118	BFS16B
3118	BFS16C
3118	BFS16D
3118	BFS16F
3118	BFS26
3118	BFS26A
3118	BFS26B
3118	BFS26C
3118	BFS26D
3118	BFS26E
3118	BFS26F
3118	BFS26G
3118	BFS31
3118	BFS32
3118	BFS32P
3118	BFS33P
3118	BFS34P
3118	BFV20
3118	BFV21
3118	BFV22
3118	BFV29
3118	BFV30
3118	BFV33
3118	BI-82
3118	BJ10
3118	BJ11
3118	BJ11A
3118	BJ11B
3118	BJ12
3118	BJ12B
3118	BJ12C
3118	BJ13
3118	BJ13A
3118	BJ13B
3118	BJ14
3118	BJ14A
3118	BJ15
3118	BJ160
3118	BJ161
3118	BJ161A
3118	BJ161B
3118	BJ161C
3118	BJ6
3118	BJ6A
3118	BJ6B
3118	BJ6C
3118	BJ6D
3118	BJ7
3118	BJ7A
3118	BJ7B
3118	BJ7C
3118	BJ7D
3118	BJ8
3118	BJ8A
3118	BJ8B
3118	BJ8C
3118	BJ8D
3118	BJ9
3118	BJ9A
3118	BJ9B
3118	BJ9C
3118	BJ9D
3118	BR-82
3118	BSV55A
3118	BSV55AP
3118	BSV55P
3118	CXL106
3118	D29A6
3118	D30A2
3118	D30A3
3118	DS-68
3118	DS68
3118	EA15X69
3118	EA15X70
3118	EA15X71
3118	EN2907
3118	EN3504
3118	EN722
3118	F103P
3118	FCS9015C
3118	FS24954
3118	G03-407-Y
3118	GE-22
3118	GET3638A
3118	GME0404
3118	GME0404-1
3118	GME0404-2
3118	IRTR52
3118	MMCM2907
3118	MMT3905
3118	MMT71
3118	MPS404
3118	MPS404A
3118	MPS6134
3118	MPS6534M
3118	MPS6580
3118	MPSH81
3118	MT0412
3118	MT0463
3118	P3Y
3118	P4B
3118	P5D
3118	PTC103
3118	PTC131
3118	RE53
3118	RE63
3118	REN106
3118	REN63
3118	RT-115
3118	RT115
3118	TD2905
3118	TM106
3118	TP4257
3118	TP4258
3118	TR02020-2
3118	TR54
3118	TVS2SA607
3118	TVS2SA609
3118	V655
3118	W7
3118	WEP52
3118	ZEN101
3118	2N1640
3118	2N1641
3118	2N1642
3118	2N2802
3118	2N2803
3118	2N2804
3118	2N2805
3118	2N2806
3118	2N2807
3118	2N3209
3118	2N3245
3118	2N4058
3118	2N4059
3118	2N4060
3118	2N4061
3118	2N4062
3118	2N4080
3118	2N4122
3118	2N4313
3118	2N4389
3118	2N4421
3118	2N4453
3118	2N4872
3118	2N5055
3118	2N5140
3118	2N5141
3118	2N5208
3118	2SA480A
3118	2SA480B
3118	2SA480C
3118	2SA480D
3118	2SA480E
3118	2SA480F
3118	2SA480G
3118	2SA480GN
3118	2SA480H
3118	2SA480J
3118	2SA480K
3118	2SA480L
3118	2SA480M
3118	2SA480OR
3118	2SA480R
3118	2SA480X
3118	2SA480Y
3118	2SA4930
3118	2SA4950R
3118	2SA499A
3118	2SA499B
3118	2SA499C
3118	2SA499D
3118	2SA499E
3118	2SA499F
3118	2SA499G
3118	2SA499GN
3118	2SA499H
3118	2SA499J
3118	2SA499K
3118	2SA499L
3118	2SA499M
3118	2SA499OR
3118	2SA499R
3118	2SA499X
3118	2SA499Y
3118	2SA500A
3118	2SA500B
3118	2SA500C
3118	2SA500D
3118	2SA500E
3118	2SA500F
3118	2SA500G
3118	2SA500GN
3118	2SA500H
3118	2SA500J
3118	2SA500K
3118	2SA500L
3118	2SA500M
3118	2SA500OR
3118	2SA500R
3118	2SA500X
3118	2SA500Y
3118	2SA525
3118	2SA525R
3118	2SA525Y
3118	2SA530A
3118	2SA530B
3118	2SA530C
3118	2SA530D
3118	2SA530E
3118	2SA530F
3118	2SA530G
3118	2SA530GN
3118	2SA530GR
3118	2SA530J
3118	2SA530K
3118	2SA530L
3118	2SA530M
3118	2SA530X
3118	2SA530Y
3118	00002SA550
3118	2SA550A
3118	2SA550A(Q)
3118	2SA550A(R)
3118	2SA550AB
3118	2SA550AQ
3118	2SA550B

SK	Industry Standard No.
3118	2SA550BL
3118	2SA550C
3118	2SA550D
3118	2SA550P
3118	2SA550Q
3118	2SA550R
3118	2SA550S
3118	2SA550Y
3118	2SA5670
3118	2SA5670R
3118	2SA567B
3118	2SA567GR
3118	2SA567R
3118	2SA567Y
3118	2SA568
3118	2SA5680
3118	2SA5680R
3118	2SA568A
3118	2SA568B
3118	2SA568C
3118	2SA568D
3118	2SA568E
3118	2SA568F
3118	2SA568G
3118	2SA568GN
3118	2SA568H
3118	2SA568J
3118	2SA568K
3118	2SA568L
3118	2SA568M
3118	2SA5680R
3118	2SA568R
3118	2SA568X
3118	2SA568Y
3118	2SA5690
3118	2SA5690R
3118	2SA569GN
3118	2SA569R
3118	2SA569Y
3118	2SA570GN
3118	2SA570R
3118	2SA570Y
3118	2SA603A
3118	2SA603B
3118	2SA603C
3118	2SA603D
3118	2SA603E
3118	2SA603F
3118	2SA603G
3118	2SA603GN
3118	2SA603H
3118	2SA603J
3118	2SA603K
3118	2SA603L
3118	2SA603M
3118	2SA6030R
3118	2SA603R
3118	2SA603X
3118	2SA603Y
3118	2SA628B
3118	2SA628C
3118	2SA628L
3118	2SA640
3118	2SA704
3118	2SA704A
3118	2SA704B
3118	2SA704C
3118	2SA704D
3118	2SA704E
3118	2SA704F
3118	2SA704G
3118	2SA704GN
3118	2SA704H
3118	2SA704J
3118	2SA704K
3118	2SA704L
3118	2SA704M
3118	2SA7040R
3118	2SA704R
3118	2SA704X
3118	2SA704Y
3118	2SA736
3118	03A04
3118	6-49
3118	13-34367-3
3118	13-40083-1
3118	19AR21
3118	36T1
3118	40C2PW8V1SP
3118	48-3003A04
3118	48S134831
3118	48S137314
3118	48S137321
3118	50-40204-10
3118	52-010-109-0
3118	57B135-12
3118	57D1-52
3118	76-1
3118	86-608-2
3118	0131-001438
3118	132-031
3118	134P1AA
3118	246P1
3118	260P36001
3118	364-1
3118	576-0003-012
3118	597-1
3118	991-011576
3118	991-011706
3118	1013
3118	1186
3118	1341
3118	2381-17
3118	3549-1
3118	8000-00049-056
3118	130139
3118	131647
3118	141711
3118	610134-4
3118	610246-1
3118	1473549-1
3118	1473549-2
3118	1473592-1
3118	1473597-1
3118	1473597-2
3118	1473616-1
3118	43027619
3118	62279664
3118	62539291
3118	62539313
3118	81073304
3118	910678870
3118	2004049514
3119/113	A40-6704
3119/113	A40-6722
3119/113	A4201
3119/113	A42946
3119/113	A42946B
3119/113	A86-9-1
3119/113	A95-5280
3119/113	A95-5297
3119/113	ALC1
3119/113	ALC1A
3119/113	B-46-110
3119/113	B522-893
3119/113	B527-062
3119/113	C05-03C
3119/113	C08P1
3119/113	C10
3119/113	C10-02A
3119/113	C10-13B
3119/113	C10-18B
3119/113	C10-1B
3119/113	C10-31A
3119/113	C10-38C
3119/113	C10-47B
3119/113	CM
3119/113	D1B
3119/113	D1C
3119/113	D4
3119/113	D51
3119/113	DD04
3119/113	E1176ALC1A
3119/113	EP16X6
3119/113	ES57X6
3119/113	F209
3119/113	FSA1177
3119/113	FSA1178
3119/113	G5019
3119/113	H615
3119/113	HD2000710
3119/113	K112
3119/113	K112C
3119/113	K115J510-1
3119/113	K115J510-2
3119/113	K117J460-1
3119/113	K117J460-2
3119/113	K122
3119/113	K122C
3119/113	K1615
3119/113	K8533137
3119/113	M109474
3119/113	M8534992
3119/113	P15
3119/113	PTC407
3119/113	R109328
3119/113	R109474
3119/113	R3057
3119/113	R3314
3119/113	R3533
3119/113	R4409
3119/113	R4666
3119/113	R8474
3119/113	RE87
3119/113	REJ701148-1
3119/113	REJ70148
3119/113	REJ70148A
3119/113	RER023
3119/113	RF33426-7
3119/113	RF5464
3119/113	RF5464-1P
3119/113	RF5465
3119/113	RF5465-1P
3119/113	RFJ60313
3119/113	RFJ70148
3119/113	RSLNB0001CEZZ
3119/113	RVDC08P1
3119/113	SD4
3119/113	SDD-C10
3119/113	SDD4
3119/113	SELEN-26
3119/113	SELEN-38
3119/113	SLEN-26
3119/113	SR-0004
3119/113	SR-13
3119/113	SR20
3119/113	SR29
3119/113	SR6
3119/113	SR9002
3119/113	SRR13
3119/113	SVD1S1850
3119/113	T-E1086A
3119/113	T-E1176
3119/113	TQ5020
3119/113	TV24226
3119/113	TVM-526
3119/113	TVM554
3119/113	TVM554A
3119/113	TVS-K112C
3119/113	1B2C1
3119/113	1NJ70972
3119/113	1S1579
3119/113	2SC1625YLBGL1
3119/113	003-002200
3119/113	4-2020-05000
3119/113	4-2020-05400
3119/113	6G4
3119/113	6GC1
3119/113	6GC1BY1
3119/113	6GX1BY1
3119/113	7K705M
3119/113	7VM705M
3119/113	09-306101
3119/113	09-306193
3119/113	9LR2
3119/113	9LR2-1
3119/113	9LR2-24
3119/113	9LR2-3
3119/113	9LR2-S
3119/113	9LR2-S1
3119/113	9LR21
3119/113	12/3-04
3119/113	13-31014-2
3119/113	13-85943-1
3119/113	13-85943-2
3119/113	13-85943-3
3119/113	14-501-01
3119/113	14-501-02
3119/113	15-085038
3119/113	15-085039
3119/113	15-085047
3119/113	15-108001
3119/113	15-108002
3119/113	15-108009
3119/113	19AR29
3119/113	19AR29-1
3119/113	21A002
3119/113	21A002-000
3119/113	21A2
3119/113	27-226
3119/113	027-300226
3119/113	27-C226
3119/113	32-0062
3119/113	33G59019
3119/113	34-8034-7
3119/113	34-8037
3119/113	34-8037-1
3119/113	34-8037-2
3119/113	34-8037-3
3119/113	34-8037-4
3119/113	34-9037-1
3119/113	46-86220-3
3119/113	46-86332-3
3119/113	48-90235A01
3119/113	48S134916
3119/113	48S134917
3119/113	53B007-1
3119/113	53B010-2
3119/113	62-16712
3119/113	62-18337
3119/113	62-19734
3119/113	66X0024-000
3119/113	66X0025-000
3119/113	66X0025-000-001
3119/113	66X0025-001
3119/113	66X218
3119/113	66X25-0
3119/113	66XZ18
3119/113	86-18-1
3119/113	86-18-1A
3119/113	86-9-1
3119/113	86-97-1
3119/113	93A5-10
3119/113	93A5-9
3119/113	93B5-1
3119/113	93B5-10
3119/113	93B5-3
3119/113	93B5-3-6
3119/113	93B5-3-8
3119/113	93B5-3-9
3119/113	93B5-5
3119/113	93B5-6
3119/113	93B5-8
3119/113	93B5-9
3119/113	93C5-10
3119/113	93C5-5
3119/113	93C5-6
3119/113	93C5-7
3119/113	93C5-8
3119/113	93C5-9
3119/113	93C59
3119/113	93K2-1
3119/113	96X2778/27N
3119/113	103-20
3119/113	103-32
3119/113	103-43
3119/113	165
3119/113	227-200001
3119/113	264P00501
3119/113	264P00506
3119/113	269V004-H01
3119/113	296V004H01
3119/113	420-2005-000
3119/113	977-14B
3119/113	1615C
3119/113	2093A5-10
3119/113	2113
3119/113	10031
3119/113	12871
3119/113	53093-1
3119/113	72053
3119/113	72148
3119/113	100471
3119/113	100581
3119/113	103872
3119/113	107474
3119/113	109328
3119/113	489752-044
3119/113	489765-005
3119/113	530093-1
3119/113	530127-5
3119/113	607101
3119/113	611132
3119/113	616010
3119/113	633977
3119/113	661010
3119/113	700055-00
3119/113	744002
3119/113	744006
3119/113	817074
3119/113	817126
3119/113	982361
3119/113	1107832-10
3119/113	1107832-11
3119/113	1107832-7
3119/113	1107832-8
3119/113	1107832-9
3119/113	2004107-40
3119/113	2006512-79
3119/113	2006512-80
3119/113	62522682
3119/113	2003162538
3119/113	4202005000
3120	A86-4-1
3120	CXL114
3120	D5
3120	DD05
3120	ECG114
3120	EU16X8
3120	K118J966-1
3120	K122D
3120	K1616
3120	P16
3120	RE88
3120	REN114
3120	RF5794
3120	SDD5
3120	SR10
3120	SR15
3120	TVM526
3120	TVMTC00921-3
3120	TVS-TC009M21/3
3120	WEP165A
3120	Y100
3120	6GD1
3120	9LR2-2
3120	012-1022-002
3120	14-503-01
3120	14-503-02
3120	14-503-03
3120	14-503-04
3120	14-503-08
3120	14-504-04
3120	19AR4-1
3120	21/3
3120	21/3.92
3120	86-3-1
3120	1616C
3120	2093A5-2
3120	489752-017
3120	530045-2
3120	530045-3
3120	530045-4
3120	1045494-1
3120/114	GE-6GD1
3120/114	PTC406
3120/114	SD5
3120/114	SR-15
3120/114	T-E1145
3120/114	93A5-2
3120/114	93B5-4
3120/114	817062
3121	C08P1R
3121	D6
3121	D7
3121	DD06
3121	FSA1169
3121	FSA1202
3121	K118J966-2
3121	K118J966-4
3121	K118J9663
3121	K1617
3121	P17
3121	RE89
3121	SDD6
3121	6GX1
3121	9LR2-4
3121	12/1N
3121	48S137164
3121	86-0007
3121	166
3121	1617C
3121	5203RNI
3121	107268
3121	107628
3121	1107832-6
3121	2001786-139
3121/115	D3G
3121/115	HD2000610
3121/115	SD6
3121/115	0110-0011
3122	A066-143
3122	A322
3122	A4F
3122	A67-33-540
3122	A670722D
3122	A673351K
3122	A673354K
3122	A673355H
3122	A673355K
3122	A675419H
3122	A757
3122	ALS-8922
3122	AR303
3122	BC-1072
3122	BC-1082
3122	BC-1086
3122	BC-1096
3122	BC-1690
3122	BC170
3122	BC170A
3122	BC170B
3122	BC170C
3122	BC173C
3122	BC209/7825B
3122	BC239B
3122	BC548
3122	BCY51I
3122	BD153
3122	BF255
3122	BFS36
3122	BFS36A
3122	BFS36B
3122	BFS36C
3122	BFS38
3122	BFS38A
3122	BFS42
3122	BFS42A
3122	BFS42B
3122	BFS42C
3122	BFS43
3122	BFS43A
3122	BFS43B
3122	BFS43C
3122	BFY37I
3122	BFY39-1
3122	BFY39I
3122	BR67
3122	C100
3122	C100-OY
3122	C1000
3122	C1000-BL
3122	C1000-GR
3122	C1000-Y
3122	C1000Y
3122	C103
3122	C1033
3122	C1033A
3122	C104
3122	C104A
3122	C105
3122	C110
3122	C111
3122	C111E
3122	C1175
3122	C1175C
3122	C1175D
3122	C1175E
3122	C1175F
3122	C1211
3122	C1213
3122	C1213A
3122	C1213AA
3122	C1213AB
3122	C1213AC
3122	C1213AD
3122	C1213B
3122	C1213C
3122	C1213D
3122	C131
3122	C132
3122	C133
3122	C134
3122	C134B
3122	C135
3122	C1359
3122	C1359B
3122	C1359C
3122	C136
3122	C1372Y
3122	C138A
3122	C157
3122	C158
3122	C159
3122	C160
3122	C166
3122	C167
3122	C16A
3122	C18
3122	C182
3122	C191
3122	C203
3122	C204
3122	C205
3122	C218A
3122	C230
3122	C232
3122	C233
3122	C234
3122	C237
3122	C239
3122	C267
3122	C267A
3122	C28
3122	C281
3122	C283
3122	C29
3122	C299
3122	C302
3122	C37
3122	C371
3122	C371(0)
3122	C371-0
3122	C371-O
3122	C371-R
3122	C371-R-1
3122	C3710
3122	C371B
3122	C371G
3122	C371R
3122	C440
3122	C441
3122	C442
3122	C45
3122	C459
3122	C459B
3122	C459D
3122	C460
3122	C460A
3122	C460B
3122	C460C
3122	C460D
3122	C460GB
3122	C460H
3122	C460K
3122	C460L
3122	C461
3122	C461A
3122	C461B
3122	C461C
3122	C461E
3122	C461L
3122	C47
3122	C476
3122	C48
3122	C529
3122	C53
3122	C536
3122	C536A
3122	C536AG
3122	C536B
3122	C536C
3122	C536D
3122	C536DK
3122	C536E
3122	C536ED
3122	C536EH
3122	C536EJ
3122	C536EN
3122	C536ER
3122	C536ET
3122	C536EZ
3122	C536F
3122	C536F1
3122	C536F2
3122	C536FC
3122	C536FS
3122	C536FS6
3122	C536FZ
3122	C536G
3122	C536GF
3122	C536GK
3122	C536GT
3122	C536GV
3122	C536GY
3122	C536H
3122	C536W
3122	C537-01
3122	C537A
3122	C537B
3122	C537C
3122	C537D
3122	C537D2
3122	C537E
3122	C537EF
3122	C537EH
3122	C537EJ
3122	C537EK
3122	C537F
3122	C537F1
3122	C537F2
3122	C537FC
3122	C537FV
3122	C537G
3122	C537GF
3122	C537GI
3122	C537H
3122	C537HT
3122	C537W
3122	C54
3122	C55
3122	C602E
3122	C63
3122	C64
3122	C650
3122	C650B
3122	C651
3122	C67
3122	C68
3122	C702

SK	Industry Standard No.	SK	Industry Standard No.	SK	Industry Standard No.	SK	Industry Standard No.	SK	Industry Standard No.
3122	C735	3122	EA15X325	3122	R4057	3122	2SC1000-Y	3122	2SC133GN
3122	C735(FA-3)	3122	EA15X336	3122	R8889	3122	2SC1000A	3122	2SC133H
3122	C735(O)	3122	EA15X408	3122	R8900	3122	2SC1000B	3122	2SC133J
3122	C735-O	3122	EA15X7118	3122	RE10	3122	2SC1000BL	3122	2SC133K
3122	C735-o	3122	EA15X7643	3122	RE67	3122	2SC1000C	3122	2SC133L
3122	C735-R	3122	EA15X8529	3122	REN108	3122	2SC1000D	3122	2SC133M
3122	C735Y	3122	EA2738	3122	REN67	3122	2SC1000E	3122	2SC1330R
3122	C763	3122	EA3211	3122	RH120	3122	2SC1000F	3122	2SC133X
3122	C763(C)	3122	EA5990	3122	RS-805U8	3122	2SC1000G	3122	2SC133Y
3122	C763B	3122	EN2219	3122	R3128	3122	2SC1000GN	3122	2SC1359
3122	C763C	3122	EP15X20	3122	R8136	3122	2SC1000GR	3122	2SC1359(A)
3122	C763D	3122	EQ8-0165	3122	RT6737	3122	2SC1000H	3122	2SC1359(B)
3122	C773E	3122	EQ8-0932	3122	RT6989	3122	2SC1000J	3122	2SC1359(C,B)
3122	C828	3122	ES15052	3122	RT8337	3122	2SC1000K	3122	2SC1359A
3122	C828Y	3122	ES15226	3122	RT8666	3122	2SC1000L	3122	2SC1359B
3122	C829	3122	ES46	3122	RT8865	3122	2SC1000M	3122	2SC1359C
3122	C829A	3122	ET579462	3122	S0002	3122	2SC1000X	3122	2SC1372Y
3122	C829B	3122	ET517263	3122	S1016	3122	2SC1000Y	3122	2SC138B
3122	C829BC	3122	ETT-CDC-12000	3122	S1240	3122	2SC1010	3122	2SC138C
3122	C829C	3122	ETTC-930D	3122	S1241	3122	2SC1010A	3122	2SC138D
3122	C829D	3122	FCS8050C	3122	S133-1	3122	2SC1010B	3122	2SC138E
3122	C829R	3122	FS2043	3122	S169N	3122	2SC1010C	3122	2SC138F
3122	C829X	3122	FT008	3122	S2034	3122	2SC1010D	3122	2SC138G
3122	C829Y	3122	FT008A	3122	S2045	3122	2SC1010E	3122	2SC138GN
3122	C838	3122	FT023	3122	S2090	3122	2SC1010G	3122	2SC138H
3122	C838(H)	3122	FT024	3122	S2125	3122	2SC1010GN	3122	2SC138J
3122	C838(J)	3122	FT025	3122	S2140	3122	2SC1010H	3122	2SC138L
3122	C838(K)	3122	FT026	3122	SE-1001	3122	2SC1010K	3122	2SC138M
3122	C838(M)	3122	FT053	3122	SE1001-1	3122	2SC1010L	3122	2SC1380R
3122	C838A	3122	GO5-037-A	3122	SE1001-2	3122	2SC1010M	3122	2SC138R
3122	C838B	3122	GO5-055-C	3122	SP1714	3122	2SC1010X	3122	2SC138X
3122	C838C	3122	GO5-055-E	3122	SK3434A	3122	2SC1010Y	3122	2SC139A
3122	C838D	3122	GO5415	3122	SP8-41	3122	2SC104B	3122	2SC139B
3122	C838E	3122	GE-10	3122	SP82271	3122	2SC104C	3122	2SC139C
3122	C838F	3122	GE-268	3122	SP83937	3122	2SC104D	3122	2SC139E
3122	C838H	3122	GET2483	3122	SP8394O	3122	2SC104E	3122	2SC139F
3122	C838J	3122	GET929	3122	SX3709	3122	2SC104F	3122	2SC139G
3122	C838K	3122	GET930	3122	SX3711	3122	2SC104G	3122	2SC139GN
3122	C838L	3122	G13706	3122	SX55	3122	2SC104GN	3122	2SC139H
3122	C838M	3122	GO5415	3122	T1341A3K	3122	2SC104H	3122	2SC139J
3122	C838R	3122	G89014	3122	T1855	3122	2SC104J	3122	2SC139K
3122	C853KLM	3122	G89014I	3122	T650	3122	2SC104K	3122	2SC139M
3122	C853L	3122	G89014J	3122	T76	3122	2SC104L	3122	2SC1390R
3122	C859GK	3122	G89014K	3122	T9016H	3122	2SC104M	3122	2SC139R
3122	C87	3122	G89018F	3122	TE3606	3122	2SC104OR	3122	2SC139X
3122	C907H	3122	G89023H	3122	TE3606A	3122	2SC104X	3122	2SC139Y
3122	C907HA	3122	G89023I	3122	TE5309	3122	2SC104Y	3122	2SC1537-O
3122	C933BB	3122	G89023J	3122	TE5310	3122	2SC110A	3122	2SC1537B
3122	C933O	3122	G89023K	3122	TE5311	3122	2SC110B	3122	2SC1537S
3122	C933D	3122	HC-00573	3122	TG28C1175	3122	2SC110C	3122	2SC1538
3122	C933B	3122	HC-00732	3122	TG28C1175C	3122	2SC110D	3122	2SC15A
3122	C933F	3122	HC-00735	3122	TG28C536	3122	2SC110F	3122	2SC15B
3122	C933FP	3122	HC-01317	3122	TG28C536-D-A	3122	2SC110G	3122	2SC15C
3122	C933FPC	3122	HC-01318	3122	TG28C536-D-B	3122	2SC110GN	3122	2SC15D
3122	C933FPD	3122	HC-01335	3122	TG28C536-E-A	3122	2SC110H	3122	2SC15F
3122	C933FPF	3122	HD-00261	3122	TG28C536-E-B	3122	2SC110J	3122	2SC15G
3122	C933FPP	3122	HR65	3122	TG28C536-F	3122	2SC110K	3122	2SC15GN
3122	C933FPG	3122	HR66	3122	TG28C536-F-A	3122	2SC110L	3122	2SC15N
3122	C933G	3122	HT304601BO	3122	TG28C536C	3122	2SC110M	3122	2SC15H
3122	CD0000	3122	HT308281D	3122	TG28C536E	3122	2SC110OR	3122	2SC15J
3122	CDB000-1	3122	HT308291C	3122	TI-412	3122	2SC110X	3122	2SC15K
3122	CDC-13000-1	3122	HT313171R	3122	TI-414	3122	2SC110Y	3122	2SC15L
3122	CDC-13000-1D	3122	HT313172A	3122	TI-714	3122	2SC1175C	3122	2SC15M
3122	CDC8011B	3122	HT313172T	3122	TI-714A	3122	2SC1175D	3122	2SC15OR
3122	CP5	3122	IP20-0041	3122	TI-751	3122	2SC1175E	3122	2SC15R
3122	C12711	3122	IP20-0122	3122	TI-806G	3122	2SC1175F	3122	2SC15X
3122	C12712	3122	IP20-0179	3122	TI-907	3122	2SC1213	3122	2SC15Y
3122	C12713	3122	IRTR51	3122	TI-908	3122	2SC1213A	3122	2SC1641
3122	C12714	3122	J241255	3122	TI850	3122	2SC1213AB	3122	2SC1641Q
3122	C12923	3122	J241256	3122	TI898	3122	2SC1213AC	3122	2SC1641R
3122	C12924	3122	J24875	3122	TNJ7O479	3122	2SC1213B	3122	2SC166A
3122	C12925	3122	J24932	3122	TNJ71037	3122	2SC1213BC	3122	2SC166B
3122	C12926	3122	JA1350	3122	TR-8036	3122	2SC1213C	3122	2SC166C
3122	C13390	3122	JA1350B	3122	TR-8042	3122	2SC1213D	3122	2SC166D
3122	C13391	3122	JA1350W	3122	TR-8O149C	3122	2SC1213E	3122	2SC166E
3122	C13391A	3122	JA7010	3122	TR01015	3122	2SC1213F	3122	2SC166F
3122	C13392	3122	KGE41054	3122	TR01040	3122	2SC1213G	3122	2SC166G
3122	C13393	3122	KGE41055	3122	TR2083-72	3122	2SC1213GN	3122	2SC166GN
3122	C13394	3122	KGE46538	3122	TR2083-73	3122	2SC1213H	3122	2SC166H
3122	C13395	3122	LM1138	3122	TR2327293	3122	2SC1213J	3122	2SC166J
3122	C13396	3122	LM1138E/F	3122	TR2327333	3122	2SC1213K	3122	2SC166K
3122	C13397	3122	LM1138G/F	3122	TR2327363	3122	2SC1213L	3122	2SC166L
3122	C13398	3122	LM1138H/I	3122	TR33	3122	2SC1213M	3122	2SC166M
3122	C13402	3122	M7006	3122	TR53	3122	2SC1213OR	3122	2SC166OR
3122	C13403	3122	M7014	3122	T89013	3122	2SC1213Y	3122	2SC166R
3122	C13404	3122	M774	3122	TV-5T	3122	2SC220-003	3122	2SC166X
3122	C13405	3122	M7740RN	3122	TVB-C81255HP	3122	2SC222	3122	2SC166Y
3122	C13414	3122	M779	3122	TV8280828Q	3122	2SC222E	3122	2SC167A
3122	C13415	3122	M780	3122	VS-28C324H	3122	2SC222U	3122	2SC167B
3122	C13416	3122	M786	3122	V8280375-1R-1	3122	2SC131A	3122	2SC167C
3122	C13417	3122	M787	3122	V8280373-//1E	3122	2SC131B	3122	2SC167D
3122	C13900	3122	MP1014-2	3122	V8280373G-1	3122	2SC131C	3122	2SC167E
3122	C13900A	3122	MPS-706	3122	V8280374-B-1	3122	2SC131F	3122	2SC167F
3122	C13901	3122	MPS3710	3122	V8280732-V1F	3122	2SC131G	3122	2SC167G
3122	C14256	3122	MPS6507	3122	V8280733B-1	3122	2SC131GN	3122	2SC167GN
3122	C14424	3122	MPS6511-8	3122	V8280784-R1F	3122	2SC131H	3122	2SC167H
3122	C14425	3122	MPS9416	3122	W10	3122	2SC131J	3122	2SC167J
3122	CJ-5206	3122	MPS9416A	3122	WEP458	3122	2SC131L	3122	2SC167K
3122	CJ-5207	3122	MPS9625H	3122	WEP460	3122	2SC131M	3122	2SC167L
3122	C8-1255P	3122	MPSA05	3122	WEP505	3122	2SC131N	3122	2SC167M
3122	C8-2007H	3122	MPSA16	3122	WEP723	3122	2SC131OR	3122	2SC167OR
3122	C8-6168H	3122	MPSA17	3122	WEP724	3122	2SC131R	3122	2SC167X
3122	C8-9011	3122	MPX9623	3122	WEP728	3122	2SC131T	3122	2SC167Y
3122	C81238F	3122	NB211	3122	WEP729	3122	2SC131X	3122	2SC16B
3122	C88050	3122	NB211BI	3122	WEP735	3122	2SC132A	3122	2SC16C
3122	C89016G	3122	NKT10339	3122	XBAN282941	3122	2SC132B	3122	2SC16D
3122	C89017P	3122	NKT10419	3122	ZEN117	3122	2SC132C	3122	2SC16E
3122	C89017G	3122	NKT10439	3122	ZEN118	3122	2SC132D	3122	2SC16F
3122	C89417	3122	NKT10519	3122	ZEN126	3122	2SC132E	3122	2SC16G
3122	D33D24	3122	NKT12329	3122	ZE66	3122	2SC132F	3122	2SC16GN
3122	D33D29	3122	NKT12429	3122	001-02011	3122	2SC132GN	3122	2SC16H
3122	D33D30	3122	NKT13329	3122	01-030536	3122	2SC132H	3122	2SC16J
3122	D4D24	3122	NKT13429	3122	01-030734	3122	2SC132J	3122	2SC16K
3122	DBCZO37300	3122	NTC-11	3122	01-030755	3122	2SC132K	3122	2SC16L
3122	DBCZO37S000	3122	PT2040A	3122	01-030765	3122	2SC132L	3122	2SC16M
3122	DBCZO7S04	3122	PT3500	3122	01-030828	3122	2SC132M	3122	2SC16OR
3122	DBCZ094504	3122	PTC101	3122	01-030829	3122	2SC1320R	3122	2SC16R
3122	DDBY216002	3122	Q-00469C	3122	01-031166	3122	2SC132R	3122	2SC16X
3122	DDBY219001	3122	Q-00569C	3122	01-031213	3122	2SC132X	3122	2SC16Y
3122	DDBY224001	3122	Q-0115C	3122	01-031318	3122	2SC132Y	3122	2SC1739
3122	DDBY2S9001	3122	Q-08115C	3122	2N2625	3122	2SC133C	3122	2SC1740
3122	DDBY262001	3122	Q-09115C	3122	2N3510	3122	2SC133D	3122	2SC1740P
3122	DDBY270001	3122	Q-13115C	3122	2N3880	3122	2SC133E	3122	2SC1740Q
3122	DDBY301001	3122	Q5124	3122	2N3973	3122	2SC133F	3122	2SC1740QH
3122	D8-66	3122	Q8054	3122	2N3974	3122	2SC133G	3122	2SC1740QJ
3122	D8-76	3122	QT-C0372XAT	3122	2N4140			3122	2SC1740R
3122	D866	3122	QT-C0735XBT	3122	2N4141			3122	2SC1740RH
3122	D876	3122	QT-C0828XAN	3122	2N5220			3122	2SC1740S
3122	EA15X161	3122	QT-C0828XDN	3122	2N783			3122	2SC1766C
3122	EA15X190	3122	QT-C0829XAN	3122	2N784			3122	2SC1854
3122	EA15X256	3122	QT-C0829XBN	3122	2SC1000			3122	2SC1854C
3122	EA15X258	3122	QT-C131BXDN	3122	2SC1000-BL			3122	2SC1854S
3122	EA15X259	3122	QT-C131BXDN	3122	2SC1000-GR			3122	2SC18A
3122	EA15X267							3122	2SC18B

SK	Industry Standard No.
3122	2SC18C
3122	2SC18D
3122	2SC18E
3122	2SC18F
3122	2SC18G
3122	2SC18GN
3122	2SC18H
3122	2SC18J
3122	2SC18K
3122	2SC18L
3122	2SC18M
3122	2SC18OR
3122	2SC18R
3122	2SC18X
3122	2SC191A
3122	2SC191B
3122	2SC191C
3122	2SC191D
3122	2SC191E
3122	2SC191F
3122	2SC191G
3122	2SC191GN
3122	2SC191H
3122	2SC191J
3122	2SC191K
3122	2SC191L
3122	2SC191M
3122	2SC191OR
3122	2SC191R
3122	2SC191X
3122	2SC191Y
3122	2SC192A
3122	2SC192B
3122	2SC192C
3122	2SC192D
3122	2SC192E
3122	2SC192F
3122	2SC192G
3122	2SC192GN
3122	2SC192H
3122	2SC192J
3122	2SC192K
3122	2SC192L
3122	2SC192M
3122	2SC192OR
3122	2SC192R
3122	2SC192X
3122	2SC192Y
3122	2SC193A
3122	2SC193B
3122	2SC193C
3122	2SC193D
3122	2SC193E
3122	2SC193F
3122	2SC193G
3122	2SC193GN
3122	2SC193H
3122	2SC193J
3122	2SC193K
3122	2SC193L
3122	2SC193M
3122	2SC193OR
3122	2SC193R
3122	2SC193X
3122	2SC194A
3122	2SC194B
3122	2SC194C
3122	2SC194D
3122	2SC194E
3122	2SC194F
3122	2SC194G
3122	2SC194GN
3122	2SC194H
3122	2SC194J
3122	2SC194K
3122	2SC194L
3122	2SC194M
3122	2SC194OR
3122	2SC194X
3122	2SC194Y
3122	2SC200A
3122	2SC200B
3122	2SC200C
3122	2SC200D
3122	2SC200E
3122	2SC200F
3122	2SC200G
3122	2SC200GN
3122	2SC200H
3122	2SC200J
3122	2SC200K
3122	2SC200L
3122	2SC200M
3122	2SC200R
3122	2SC200X
3122	2SC200Y
3122	2SC201A
3122	2SC201B
3122	2SC201C
3122	2SC201D
3122	2SC201E
3122	2SC201F
3122	2SC201G
3122	2SC201GN
3122	2SC201H
3122	2SC201J
3122	2SC201K
3122	2SC201L
3122	2SC201M
3122	2SC201R
3122	2SC201X
3122	2SC201Y
3122	2SC202A
3122	2SC202B
3122	2SC202C
3122	2SC202D
3122	2SC202E
3122	2SC202F
3122	2SC202G
3122	2SC202GN
3122	2SC202H
3122	2SC202J
3122	2SC202K
3122	2SC202L
3122	2SC202M
3122	2SC202OR
3122	2SC202R
3122	2SC202X
3122	2SC202Y
3122	2SC203B
3122	2SC203C
3122	2SC203D
3122	2SC203E
3122	2SC203F
3122	2SC203G
3122	2SC203GN
3122	2SC203H
3122	2SC203J
3122	2SC203K
3122	2SC203L
3122	2SC203M
3122	2SC203OR
3122	2SC203R
3122	2SC203X
3122	2SC203Y
3122	2SC204A
3122	2SC204B
3122	2SC204C
3122	2SC204D
3122	2SC204E
3122	2SC204F
3122	2SC204G
3122	2SC204GN
3122	2SC204H
3122	2SC204J
3122	2SC204K
3122	2SC204L
3122	2SC204M
3122	2SC204OR
3122	2SC204R
3122	2SC204X
3122	2SC204Y
3122	2SC205A
3122	2SC205B
3122	2SC205C
3122	2SC205D
3122	2SC205F
3122	2SC205G
3122	2SC205GN
3122	2SC205H
3122	2SC205J
3122	2SC205K
3122	2SC205L
3122	2SC205M
3122	2SC205OR
3122	2SC205R
3122	2SC205X
3122	2SC205Y
3122	2SC2076
3122	2SC2076C
3122	2SC2076CB
3122	2SC2076CD
3122	2SC2076D
3122	2SC230A
3122	2SC230B
3122	2SC230C
3122	2SC230D
3122	2SC230E
3122	2SC230F
3122	2SC230G
3122	2SC230GN
3122	2SC230H
3122	2SC230J
3122	2SC230K
3122	2SC230L
3122	2SC230M
3122	2SC230OR
3122	2SC230R
3122	2SC230X
3122	2SC230Y
3122	2SC237A
3122	2SC237B
3122	2SC237C
3122	2SC237D
3122	2SC237E
3122	2SC237F
3122	2SC237G
3122	2SC237GN
3122	2SC237H
3122	2SC237J
3122	2SC237K
3122	2SC237L
3122	2SC237M
3122	2SC237OR
3122	2SC237R
3122	2SC237X
3122	2SC237Y
3122	2SC239A
3122	2SC239B
3122	2SC239C
3122	2SC239D
3122	2SC239E
3122	2SC239F
3122	2SC239G
3122	2SC239GN
3122	2SC239H
3122	2SC239J
3122	2SC239K
3122	2SC239L
3122	2SC239M
3122	2SC239OR
3122	2SC239R
3122	2SC239X
3122	2SC239Y
3122	2SC267B
3122	2SC267C
3122	2SC267D
3122	2SC267E
3122	2SC267F
3122	2SC267G
3122	2SC267GN
3122	2SC267H
3122	2SC267J
3122	2SC267K
3122	2SC267L
3122	2SC267M
3122	2SC267OR
3122	2SC267R
3122	2SC267X
3122	2SC267Y
3122	2SC26A
3122	2SC26B
3122	2SC26C
3122	2SC26D
3122	2SC26E
3122	2SC26F
3122	2SC26G
3122	2SC26GN
3122	2SC26H
3122	2SC26J
3122	2SC26K
3122	2SC26L
3122	2SC26M
3122	2SC26OR
3122	2SC26R
3122	2SC26X
3122	2SC26Y
3122	2SC283A
3122	2SC283B
3122	2SC283C
3122	2SC283D
3122	2SC283E
3122	2SC283F
3122	2SC283G
3122	2SC283GN
3122	2SC283H
3122	2SC283J
3122	2SC283K
3122	2SC283L
3122	2SC283M
3122	2SC283OR
3122	2SC283R
3122	2SC283X
3122	2SC283Y
3122	2SC284A
3122	2SC284B
3122	2SC284C
3122	2SC284D
3122	2SC284E
3122	2SC284F
3122	2SC284G
3122	2SC284GN
3122	2SC284J
3122	2SC284K
3122	2SC284L
3122	2SC284M
3122	2SC284OR
3122	2SC284R
3122	2SC284X
3122	2SC284Y
3122	2SC285B
3122	2SC285C
3122	2SC285D
3122	2SC285E
3122	2SC285F
3122	2SC285G
3122	2SC285GN
3122	2SC285H
3122	2SC285J
3122	2SC285K
3122	2SC285L
3122	2SC285M
3122	2SC285OR
3122	2SC285R
3122	2SC285X
3122	2SC285Y
3122	2SC301A
3122	2SC301B
3122	2SC301C
3122	2SC301D
3122	2SC301E
3122	2SC301F
3122	2SC301G
3122	2SC301GN
3122	2SC301H
3122	2SC301J
3122	2SC301K
3122	2SC301L
3122	2SC301M
3122	2SC301OR
3122	2SC301R
3122	2SC301X
3122	2SC301Y
3122	2SC302A
3122	2SC302B
3122	2SC302C
3122	2SC302D
3122	2SC302E
3122	2SC302F
3122	2SC302G
3122	2SC302GN
3122	2SC302H
3122	2SC302J
3122	2SC302K
3122	2SC302L
3122	2SC302M
3122	2SC302OR
3122	2SC302R
3122	2SC302X
3122	2SC302Y
3122	2SC318B
3122	2SC318C
3122	2SC318D
3122	2SC318E
3122	2SC318F
3122	2SC318G
3122	2SC318GN
3122	2SC318H
3122	2SC318J
3122	2SC318K
3122	2SC318L
3122	2SC318M
3122	2SC318OR
3122	2SC318R
3122	2SC318X
3122	2SC318Y
3122	2SC319
3122	2SC319A
3122	2SC319B
3122	2SC319C
3122	2SC319D
3122	2SC319E
3122	2SC319F
3122	2SC319G
3122	2SC319GN
3122	2SC319H
3122	2SC319J
3122	2SC319K
3122	2SC319L
3122	2SC319M
3122	2SC319OR
3122	2SC319R
3122	2SC319X
3122	2SC319Y
3122	2SC320A
3122	2SC320B
3122	2SC320C
3122	2SC320D
3122	2SC320E
3122	2SC320F
3122	2SC320G
3122	2SC320GN
3122	2SC320H
3122	2SC320J
3122	2SC320K
3122	2SC320L
3122	2SC320M
3122	2SC320OR
3122	2SC320R
3122	2SC320X
3122	2SC320Y
3122	2SC321A
3122	2SC321B
3122	2SC321C
3122	2SC321D
3122	2SC321E
3122	2SC321F
3122	2SC321G
3122	2SC321GN
3122	2SC321J
3122	2SC321K
3122	2SC321L
3122	2SC321M
3122	2SC321OR
3122	2SC321R
3122	2SC321X
3122	2SC321Y
3122	2SC323A
3122	2SC323B
3122	2SC323C
3122	2SC323D
3122	2SC323E
3122	2SC323F
3122	2SC323G
3122	2SC323GN
3122	2SC323H
3122	2SC323J
3122	2SC323K
3122	2SC323L
3122	2SC323M
3122	2SC323OR
3122	2SC323R
3122	2SC323X
3122	2SC323Y
3122	2SC33
3122	2SC33A
3122	2SC33B
3122	2SC33C
3122	2SC33D
3122	2SC33E
3122	2SC33F
3122	2SC33G
3122	2SC33GN
3122	2SC33H
3122	2SC33J
3122	2SC33K
3122	2SC33L
3122	2SC33M
3122	2SC33OR
3122	2SC33R
3122	2SC33X
3122	2SC33Y
3122	2SC350A
3122	2SC350B
3122	2SC350C
3122	2SC350D
3122	2SC350E
3122	2SC350F
3122	2SC350G
3122	2SC350GN
3122	2SC350J
3122	2SC350K
3122	2SC350L
3122	2SC350M
3122	2SC350OR
3122	2SC350R
3122	2SC350X
3122	2SC350Y
3122	2SC366G
3122	2SC366R
3122	2SC366Y
3122	2SC368
3122	2SC368A
3122	2SC368B
3122	2SC368C
3122	2SC368D
3122	2SC368E
3122	2SC368F
3122	2SC368G
3122	2SC368GN
3122	2SC368H
3122	2SC368J
3122	2SC368K
3122	2SC368L
3122	2SC368M
3122	2SC368OR
3122	2SC368R
3122	2SC368X
3122	2SC368Y
3122	2SC371
3122	2SC371(O)
3122	2SC371-O
3122	2SC371-O
3122	2SC371-ORG-G
3122	2SC371-R
3122	2SC371-R-1
3122	2SC371-RED-G
3122	2SC371-T
3122	2SC371A
3122	2SC371B
3122	2SC371C
3122	2SC371D
3122	2SC371E
3122	2SC371F
3122	2SC371G
3122	2SC371GN
3122	2SC371H
3122	2SC371J
3122	2SC371K
3122	2SC371L
3122	2SC371M
3122	2SC371O
3122	2SC371OR
3122	2SC371R
3122	2SC371T
3122	2SC371X
3122	2SC371Y
3122	2SC3720
3122	2SC379A
3122	2SC379B
3122	2SC379C
3122	2SC379D
3122	2SC379E
3122	2SC379F
3122	2SC379G
3122	2SC379GN
3122	2SC379H
3122	2SC379J
3122	2SC379K
3122	2SC379L
3122	2SC379M
3122	2SC379OR
3122	2SC379R
3122	2SC379X
3122	2SC379Y
3122	2SC381
3122	2SC381-O
3122	2SC381-O
3122	2SC381A
3122	2SC381B
3122	2SC381BN
3122	2SC381BN-1
3122	2SC381C
3122	2SC381D
3122	2SC381E
3122	2SC381F
3122	2SC381G
3122	2SC381GN
3122	2SC381H
3122	2SC381J
3122	2SC381K
3122	2SC381L
3122	2SC381M
3122	2SC381OR
3122	2SC381R
3122	2SC381RL
3122	2SC381X
3122	2SC381Y
3122	2SC395B
3122	2SC395C
3122	2SC395D
3122	2SC395E
3122	2SC395F
3122	2SC395G
3122	2SC395GN
3122	2SC395H
3122	2SC395J
3122	2SC395K
3122	2SC395L
3122	2SC395M
3122	2SC395OR
3122	2SC395X
3122	2SC395Y
3122	2SC396A
3122	2SC396B
3122	2SC396C
3122	2SC396D
3122	2SC396E
3122	2SC396F
3122	2SC396G
3122	2SC396GN
3122	2SC396H
3122	2SC396J
3122	2SC396K
3122	2SC396L
3122	2SC396M
3122	2SC396OR
3122	2SC396R
3122	2SC396X
3122	2SC396Y
3122	2SC400A
3122	2SC400B
3122	2SC400C
3122	2SC400D
3122	2SC400E
3122	2SC400F
3122	2SC400G
3122	2SC400GN
3122	2SC400H
3122	2SC400J
3122	2SC400K
3122	2SC400L
3122	2SC400M
3122	2SC400OR
3122	2SC400R
3122	2SC400X
3122	2SC400Y
3122	2SC459
3122	2SC459A
3122	2SC459C
3122	2SC459D
3122	2SC459E
3122	2SC459F
3122	2SC459G
3122	2SC459GN
3122	2SC459H
3122	2SC459J
3122	2SC459K
3122	2SC459L
3122	2SC459M
3122	2SC459OR
3122	2SC459R
3122	2SC459X
3122	2SC459Y
3122	00002SC460
3122	2SC460(A)
3122	2SC460(B)
3122	2SC460-5
3122	2SC460-B
3122	2SC460-C
3122	2SC460A
3122	2SC460B
3122	00002SC460C
3122	2SC460D
3122	2SC460E
3122	2SC460F
3122	2SC460G
3122	2SC460GN
3122	2SC460H
3122	2SC460J
3122	2SC460K
3122	2SC460L
3122	2SC460M
3122	2SC460OR
3122	2SC460R
3122	2SC460X
3122	2SC460Y
3122	00002SC461
3122	2SC461-A
3122	2SC461-B
3122	2SC461A
3122	2SC461AL
3122	2SC461B
3122	2SC461BK
3122	2SC461BL
3122	2SC461C
3122	2SC461EP
3122	2SC461L
3122	2SC476A
3122	2SC476B
3122	2SC476C
3122	2SC476D
3122	2SC476E
3122	2SC476F
3122	2SC476G
3122	2SC476GN
3122	2SC476H
3122	2SC476J
3122	2SC476K
3122	2SC476L

SK	Industry Standard No.	SK	Industry Standard No.	SK	Industry Standard No.	SK	Industry Standard No.	SK	Industry Standard No.
3122	28C476M	3122	28C540Y	3122	28C62C	3122	28C708B	3122	28C796C
3122	28C476OR	3122	28C54A	3122	28C62D	3122	28C713	3122	28C796D
3122	28C476R	3122	28C54B	3122	28C62E	3122	28C713A	3122	28C796E
3122	28C476X	3122	28C54C	3122	28C62F	3122	28C713B	3122	28C796F
3122	28C476Y	3122	28C54D	3122	28C62G	3122	28C713C	3122	28C796G
3122	28C47A	3122	28C54E	3122	28C62GN	3122	28C713D	3122	28C796GN
3122	28C47B	3122	28C54F	3122	28C62H	3122	28C713E	3122	28C796H
3122	28C47C	3122	28C54G	3122	28C62J	3122	28C713F	3122	28C796J
3122	28C47D	3122	28C54GN	3122	28C62L	3122	28C713G	3122	28C796K
3122	28C47E	3122	28C54H	3122	28C62M	3122	28C713GN	3122	28C796L
3122	28C47G	3122	28C54J	3122	28C62OR	3122	28C713H	3122	28C796OR
3122	28C47GN	3122	28C54K	3122	28C62R	3122	28C713J	3122	28C796R
3122	28C47H	3122	28C54L	3122	28C62X	3122	28C713K	3122	28C796X
3122	28C47J	3122	28C54M	3122	28C62Y	3122	28C713L	3122	28C796Y
3122	28C47K	3122	28C540R	3122	28C640A	3122	28C713M	3122	28C828
3122	28C47L	3122	28C54X	3122	28C640C	3122	28C713OR	3122	0002BC829
3122	28C47M	3122	28C54Y	3122	28C640D	3122	28C713R	3122	28C829A
3122	28C47OR	3122	28C587B	3122	28C640B	3122	28C713X	3122	28C829AK
3122	28C47R	3122	28C587C	3122	28C640F	3122	28C713Y	3122	28C829B
3122	28C47X	3122	28C587D	3122	28C640G	3122	28C714	3122	28C829B/4454C
3122	28C47Y	3122	28C587E	3122	28C640GN	3122	28C714A	3122	28C829BC
3122	28C494	3122	28C587F	3122	28C640J	3122	28C714B	3122	28C829BJ
3122	28C52A	3122	28C587G	3122	28C640K	3122	28C714C	3122	28C829BK
3122	28C52B	3122	28C587GN	3122	28C640L	3122	28C714D	3122	28C829BY
3122	28C52C	3122	28C587H	3122	28C640M	3122	28C714E	3122	28C829C
3122	28C52D	3122	28C587J	3122	28C640R	3122	28C714F	3122	28C829CL
3122	28C52E	3122	28C587K	3122	28C640G	3122	28C714G	3122	28C829D
3122	28C52F	3122	28C587L	3122	28C640X	3122	28C714GN	3122	28C829E
3122	28C52G	3122	28C587M	3122	28C650	3122	28C714H	3122	28C829G
3122	28C52GN	3122	28C587OR	3122	28C650B	3122	28C714J	3122	28C829GN
3122	28C52H	3122	28C587R	3122	28C654A	3122	28C714K	3122	28C829H
3122	28C52J	3122	28C587X	3122	28C654B	3122	28C714L	3122	28C829K
3122	28C52K	3122	28C587Y	3122	28C654C	3122	28C714M	3122	28C829L
3122	28C52L	3122	28C588A	3122	28C654D	3122	28C714OR	3122	28C829M
3122	28C52M	3122	28C588B	3122	28C654E	3122	28C714R	3122	28C829OR
3122	28C52OR	3122	28C588C	3122	28C654F	3122	28C714X	3122	28C829R
3122	28C52X	3122	28C588D	3122	28C654G	3122	28C714Y	3122	28C829X
3122	28C52Y	3122	28C588E	3122	28C654GN	3122	28C716	3122	28C829Y
3122	28C531	3122	28C588F	3122	28C654H	3122	28C716A	3122	28C835
3122	0002BC531F	3122	28C588G	3122	28C654J	3122	28C716B	3122	28C838(H)
3122	28C56	3122	28C588GN	3122	28C654K	3122	28C716C	3122	28C838(J)
3122	28C56A	3122	28C588H	3122	28C654L	3122	28C716D	3122	28C838(K)
3122	28C56AG	3122	28C588J	3122	28C654M	3122	28C716E	3122	28C838(L)
3122	28C56B	3122	28C588K	3122	28C654OR	3122	28C716G	3122	28C838(M)
3122	28C56C	3122	28C588L	3122	28C654R	3122	28C716J	3122	28C838-2
3122	28C56D	3122	28C588M	3122	28C654X	3122	28C716K	3122	28C838A
3122	28C56DK	3122	28C588OR	3122	28C654Y	3122	28C716L	3122	28C838B
3122	28C56E	3122	28C588R	3122	28C655A	3122	28C716M	3122	28C838C
3122	28C56EH	3122	28C588X	3122	28C655B	3122	28C716R	3122	28C838D
3122	28C56EJ	3122	28C588Y	3122	28C655C	3122	28C716Y	3122	28C838E
3122	28C56EN	3122	28C595A	3122	28C655D	3122	28C735	3122	28C838P
3122	28C56ER	3122	28C595B	3122	28C655E	3122	28C735(O)	3122	28C838M
3122	28C56ER	3122	28C595C	3122	28C655F	3122	28C735-O	3122	28C838K
3122	28C56ET	3122	28C595D	3122	28C655G	3122	28C735-GRN	3122	28C838L
3122	28C56F	3122	28C595E	3122	28C655GN	3122	28C735-O	3122	28C838M
3122	28C56F1	3122	28C595F	3122	28C655H	3122	28C735-ORG	3122	28C847A
3122	28C56F2	3122	28C595G	3122	28C655J	3122	28C735-RED	3122	28C847B
3122	28C56PC	3122	28C595H	3122	28C655L	3122	28C735-YEL	3122	28C847C
3122	28C56P86	3122	28C595J	3122	28C655OR	3122	28C735A	3122	28C847D
3122	28C56PZ	3122	28C595L	3122	28C655R	3122	28C735B	3122	28C847E
3122	28C56G	3122	28C595M	3122	28C655X	3122	28C735C	3122	28C847F
3122	28C56GK	3122	28C595OR	3122	28C655Y	3122	28C735D	3122	28C847GN
3122	28C56GL	3122	28C595R	3122	28C67A	3122	28C735E	3122	28C847H
3122	28C56GP	3122	28C595X	3122	28C67B	3122	28C735F	3122	28C847J
3122	28C56GT	3122	28C595Y	3122	28C67C	3122	28C735G	3122	28C847K
3122	28C56GY	3122	28C596A	3122	28C67D	3122	28C735GN	3122	28C847L
3122	28C56GY	3122	28C596B	3122	28C67E	3122	28C735GR	3122	28C847OR
3122	28C56GZ	3122	28C596C	3122	28C67F	3122	28C735H	3122	28C847R
3122	28C56H	3122	28C596D	3122	28C67G	3122	28C735J	3122	28C847X
3122	28C56K	3122	28C596E	3122	28C67GN	3122	28C735K	3122	28C848A
3122	28C56W	3122	28C596F	3122	28C67H	3122	28C735L	3122	28C848B
3122	28C56X	3122	28C596G	3122	28C67J	3122	28C735M	3122	28C848C
3122	28C57	3122	28C596GN	3122	28C67K	3122	28C735O	3122	28C848D
3122	28C57(F)	3122	28C596H	3122	28C67L	3122	28C735OR	3122	28C848E
3122	28C57(G)	3122	28C596J	3122	28C67M	3122	28C735R	3122	28C848GN
3122	28C57-01	3122	28C596K	3122	28C67OR	3122	28C735X	3122	28C848H
3122	28C57-EV	3122	28C596L	3122	28C67R	3122	28C735Y	3122	28C848J
3122	28C57BK	3122	28C596M	3122	28C67X	3122	28C738	3122	28C848K
3122	28C57C	3122	28C596OR	3122	28C67Y	3122	28C738A	3122	28C848M
3122	28C57D	3122	28C596R	3122	28C689	3122	28C738B	3122	28C848OR
3122	28C57D2	3122	28C596X	3122	28C689A	3122	28C738C	3122	28C848R
3122	28C57E	3122	28C596Y	3122	28C689B	3122	28C738D	3122	28C848X
3122	28C57EP	3122	28C619A	3122	28C689C	3122	28C738E	3122	28C849A
3122	28C57EH	3122	28C619B	3122	28C689D	3122	28C739C	3122	28C849C
3122	28C57EJ	3122	28C619C	3122	28C689E	3122	28C741H	3122	28C849D
3122	28C57EK	3122	28C619D	3122	28C689F	3122	28C741J	3122	28C849F
3122	28C57P-C7	3122	28C619E	3122	28C689GN	3122	28C741K	3122	28C849G
3122	28C57P1	3122	28C619F	3122	28C689H	3122	28C741L	3122	28C849GN
3122	28C57P2	3122	28C619G	3122	28C689J	3122	28C741M	3122	28C849H
3122	28C57PC	3122	28C619H	3122	28C689K	3122	28C741OR	3122	28C849J
3122	28C57PJ	3122	28C619J	3122	28C689L	3122	28C741R	3122	28C849K
3122	28C57PK	3122	28C619K	3122	28C689M	3122	28C741X	3122	28C849L
3122	28C57PV	3122	28C619L	3122	28C689OR	3122	28C741Y	3122	28C849M
3122	28C57G	3122	28C619M	3122	28C689R	3122	28C752	3122	28C849OR
3122	28C57G1	3122	28C619OR	3122	28C689X	3122	28C752A	3122	28C849R
3122	28C57G2	3122	28C619R	3122	28C689Y	3122	28C752B	3122	28C849X
3122	28C57GP	3122	28C619X	3122	28C68A	3122	28C752C	3122	28C849Y
3122	28C57GPL	3122	28C619Y	3122	28C68B	3122	28C752D	3122	28C850A
3122	28C57GI	3122	28C621A	3122	28C68C	3122	28C752E	3122	28C850B
3122	28C57H	3122	28C621B	3122	28C68D	3122	28C752P	3122	28C850C
3122	28C57HT	3122	28C621C	3122	28C68E	3122	28C752G	3122	28C850D
3122	28C57W	3122	28C621D	3122	28C68F	3122	28C752GN	3122	28C850E
3122	28C57WF	3122	28C621E	3122	28C68GN	3122	28C752H	3122	28C850F
3122	28C53A	3122	28C621F	3122	28C68H	3122	28C752J	3122	28C850G
3122	28C53B	3122	28C621G	3122	28C68J	3122	28C752K	3122	28C850GN
3122	28C53C	3122	28C621GN	3122	28C68K	3122	28C752L	3122	28C850H
3122	28C53D	3122	28C621H	3122	28C68L	3122	28C752M	3122	28C850J
3122	28C53E	3122	28C621J	3122	28C68M	3122	28C752OR	3122	28C850K
3122	28C53F	3122	28C621K	3122	28C68OR	3122	28C752R	3122	28C850L
3122	28C53G	3122	28C621L	3122	28C68R	3122	28C752X	3122	28C850M
3122	28C53GN	3122	28C621M	3122	28C68X	3122	28C752Y	3122	28C850OR
3122	28C53H	3122	28C621OR	3122	28C694F	3122	28C763	3122	28C850X
3122	28C53J	3122	28C621R	3122	28C694G	3122	28C763(C)	3122	28C850Y
3122	28C53L	3122	28C621X	3122	28C701B	3122	28C763A	3122	28C853
3122	28C53OR	3122	28C621Y	3122	28C709	3122	28C763B	3122	28C853A
3122	28C53R	3122	28C622A	3122	28C709A	3122	28C763C	3122	28C853B
3122	28C53X	3122	28C622B	3122	28C709B	3122	28C763D	3122	28C853C
3122	28C53Y	3122	28C622C	3122	28C709CD	3122	28C763E	3122	28C853D
3122	28C540A	3122	28C622D	3122	28C709D	3122	28C763F	3122	28C853E
3122	28C540B	3122	28C622E	3122	28C709E	3122	28C763G	3122	28C853F
3122	28C540C	3122	28C622F	3122	28C709F	3122	28C763GN	3122	28C853G
3122	28C540D	3122	28C622GN	3122	28C709GN	3122	28C763H	3122	28C853GN
3122	28C540E	3122	28C622H	3122	28C709H	3122	28C763J	3122	28C853H
3122	28C540F	3122	28C622J	3122	28C709J	3122	28C763K	3122	28C853J
3122	28C540G	3122	28C622K	3122	28C709L	3122	28C763L	3122	28C853K
3122	28C540GN	3122	28C622L	3122	28C709M	3122	28C763M	3122	28C853L
3122	28C540H	3122	28C622M	3122	28C709OR	3122	28C763OR	3122	28C853M
3122	28C540J	3122	28C622OR	3122	28C709X	3122	28C763X	3122	28C853OR
3122	28C540K	3122	28C622R	3122	28C709Y	3122	28C763Y	3122	28C853R
3122	28C540L	3122	28C622X	3122		3122	28C796A	3122	28C853X
3122	28C540M	3122	28C622Y	3122		3122	28C796B	3122	28C853Y
3122	28C540OR	3122	28C62A	3122		3122		3122	28C853Z
3122	28C540R	3122	28C62B	3122		3122		3122	28C853GN
3122	28C540X	3122		3122		3122		3122	

SK	Industry Standard No.
3122	2SC853H
3122	2SC853J
3122	2SC853K
3122	2SC853L
3122	2SC853M
3122	2SC853OR
3122	2SC853R
3122	2SC853X
3122	2SC853Y
3122	2SC87A
3122	2SC87B
3122	2SC87C
3122	2SC87D
3122	2SC87E
3122	2SC87F
3122	2SC87G
3122	2SC87GN
3122	2SC87H
3122	2SC87J
3122	2SC87K
3122	2SC87L
3122	2SC87M
3122	2SC87OR
3122	2SC87R
3122	2SC87X
3122	2SC87Y
3122	2SC894A
3122	2SC894B
3122	2SC894C
3122	2SC894D
3122	2SC894E
3122	2SC894F
3122	2SC894G
3122	2SC894GN
3122	2SC894H
3122	2SC894J
3122	2SC894K
3122	2SC894L
3122	2SC894M
3122	2SC894OR
3122	2SC894R
3122	2SC894X
3122	2SC894Y
3122	2SC899A
3122	2SC899B
3122	2SC899C
3122	2SC899D
3122	2SC899E
3122	2SC899F
3122	2SC899G
3122	2SC899GN
3122	2SC899H
3122	2SC899J
3122	2SC899L
3122	2SC899M
3122	2SC899OR
3122	2SC899R
3122	2SC899X
3122	2SC899Y
3122	2SC907
3122	2SC907A
3122	2SC907AH
3122	2SC907B
3122	2SC907C
3122	2SC907D
3122	2SC907E
3122	2SC907F
3122	2SC907G
3122	2SC907GN
3122	2SC907H
3122	2SC907HA
3122	2SC907J
3122	2SC907K
3122	2SC907L
3122	2SC907M
3122	2SC907OR
3122	2SC907R
3122	2SC907X
3122	2SC907Y
3122	2SC920Q
3122	2SC920R
3122	2SC933
3122	2SC933A
3122	2SC933B
3122	2SC933BB
3122	2SC933C
3122	2SC933D
3122	2SC933E
3122	2SC933F
3122	2SC933FP
3122	2SC933FPC
3122	2SC933FPD
3122	2SC933FPE
3122	2SC933FPF
3122	2SC933FPG
3122	2SC933G
3122	2SC933GN
3122	2SC933H
3122	2SC933J
3122	2SC933K
3122	2SC933L
3122	2SC933M
3122	2SC933OR
3122	2SC933R
3122	2SC933X
3122	2SC933Y
3122	2SC943D
3122	2SC943E
3122	2SC943F
3122	2SC943G
3122	2SC943GN
3122	2SC943H
3122	2SC943J
3122	2SC943K
3122	2SC943L
3122	2SC943M
3122	2SC943OR
3122	2SC943R
3122	2SC943X
3122	2SC943Y
3122	2SC966A
3122	2SC966B
3122	2SC966C
3122	2SC966D
3122	2SC966E
3122	2SC966F
3122	2SC966G
3122	2SC966GN
3122	2SC966H
3122	2SC966J
3122	2SC966K
3122	2SC966L
3122	2SC966M
3122	2SC966OR
3122	2SC966R
3122	2SC966X
3122	2SC966Y
3122	2SC967A
3122	2SC967B
3122	2SC967C
3122	2SC967D
3122	2SC967E
3122	2SC967F
3122	2SC967G
3122	2SC967GN
3122	2SC967H
3122	2SC967J
3122	2SC967K
3122	2SC967L
3122	2SC967M
3122	2SC967OR
3122	2SC967R
3122	2SC967X
3122	2SC967Y
3122	2SC98K
3122	2SC991A
3122	2SC991B
3122	2SC991C
3122	2SC991D
3122	2SC991E
3122	2SC991F
3122	2SC991G
3122	2SC991GN
3122	2SC991H
3122	2SC991J
3122	2SC991K
3122	2SC991L
3122	2SC991M
3122	2SC991OR
3122	2SC991R
3122	2SC991X
3122	2SC991Y
3122	2SC992A
3122	2SC992B
3122	2SC992C
3122	2SC992D
3122	2SC992E
3122	2SC992F
3122	2SC992G
3122	2SC992GN
3122	2SC992H
3122	2SC992J
3122	2SC992K
3122	2SC992L
3122	2SC992M
3122	2SC992OR
3122	2SC992R
3122	2SC992X
3122	2SC992Y
3122	2SCF11
3122	2SCF5
3122	2SCNJ100
3122	2SD228
3122	2SD228A
3122	2SD228B
3122	2SD228C
3122	2SD228D
3122	2SD228E
3122	2SD228F
3122	2SD228G
3122	2SD228GN
3122	2SD228H
3122	2SD228J
3122	2SD228K
3122	2SD228L
3122	2SD228M
3122	2SD228OR
3122	2SD228R
3122	2SD228X
3122	2SD228Y
3122	2SD327L
3122	2SD327OR
3122	2SD327R
3122	2SD327Y
3122	2SD336
3122	2SD636-0
3122	3L4-6010-3
3122	4-46
3122	4-47
3122	4JX16A667/O
3122	4JX16A667/R
3122	4JX16A667/Y
3122	4JX16A668/G
3122	4JX16A668/O
3122	4JX16A668/Y
3122	4JX16B670/B
3122	4JX16B670/R
3122	4JX16B670/Y
3122	6-2708
3122	006-6450032
3122	6A10228
3122	07-07124
3122	007-74655-02
3122	007-74655-06
3122	007-74661-01
3122	007-7466101
3122	8-905-706-730
3122	09-002012
3122	09-302003
3122	09-302004
3122	09-302053
3122	09-302075
3122	09-302090
3122	09-302092
3122	09-302093
3122	09-302127
3122	09-302139
3122	09-302173
3122	09-302207
3122	09-302222
3122	09-302244
3122	09-305092
3122	09-305126
3122	09-305148
3122	09-305152
3122	09-309072
3122	9-5216
3122	9TR1
3122	9TR10
3122	9TR3
3122	10-002
3122	10-008
3122	13-0062
3122	13-0065
3122	13-14085-95
3122	13-14085-96
3122	13-14085-97
3122	13-17-6
3122	13-22581
3122	13-34381-1
3122	13-34381-2
3122	13-35226-1
3122	13-35550-1
3122	13-35807-2
3122	14-575-10
3122	14-602-25
3122	14-602-45
3122	14-661-21
3122	14-801-23
3122	014-862
3122	016B12
3122	016B810
3122	016B812
3122	16E1330
3122	16U1
3122	16X1
3122	19-020-071
3122	020-1110-010
3122	020-1110-011
3122	020-1110-012
3122	020-1110-013
3122	020-1110-014
3122	020-1110-017
3122	020-1112-004
3122	21-1
3122	21A040-051
3122	21A112-019
3122	21M160
3122	21M455
3122	21M541
3122	22-001001
3122	24MW1060
3122	24MW1161
3122	24MW660
3122	24MW677
3122	34-6000-65
3122	34-6001-34
3122	38
3122	42-27533
3122	42-27534
3122	42-27535
3122	42-27537
3122	43A128340-3
3122	43A128342-1
3122	43A128342-2
3122	44T-300-103
3122	44T-300-104
3122	46-8257-3
3122	46-86262-3
3122	46-86485-3
3122	46-86540-3
3122	48-134802
3122	48-134803
3122	48-134912
3122	48-134918
3122	48-134997
3122	48-137015
3122	48-40170G01
3122	48-40246G01
3122	48-44885G01
3122	48-44886G01
3122	48-65146A63
3122	48-90232A12
3122	48-90343A73
3122	48-90343A80
3122	48-97238A04
3122	48S137108
3122	48S137109
3122	48S137174
3122	48S155046
3122	48S155117
3122	48S40606J02
3122	48S40607J01
3122	48S44886G01
3122	56-4829
3122	57A265-4
3122	57A268-9
3122	57A283-11
3122	57B2-79
3122	63-10188
3122	63-11825
3122	63-12944
3122	065-006
3122	69-1816
3122	70.01.704
3122	74N1
3122	81-27125530-9
3122	81-27125530-9A
3122	81-27125530-9B
3122	86-306-2
3122	86-550-2
3122	86-607-2
3122	86-609-2
3122	86-650-2
3122	86-661-2
3122	86-675-2
3122	87-0023-T
3122	90-181
3122	90-455
3122	90-603
3122	95-216
3122	95-221
3122	95-225
3122	95-296
3122	96-5161-01
3122	102-0373-00
3122	102-0732-28
3122	102-0735-25
3122	102-0828-17
3122	102-1166-25
3122	0103-0060A
3122	103-0227-18
3122	0103-0531/4460
3122	105-001-04
3122	105-001-07
3122	105-085-33
3122	105-28196-07
3122	106-003
3122	107-8
3122	107BRN
3122	109-1
3122	123-007
3122	129WHT
3122	130-40883
3122	130-40901
3122	135ORN
3122	136RED
3122	142-002
3122	142-003
3122	142-005
3122	147N1
3122	148N3
3122	161-001I
3122	161-011J
3122	161-016H
3122	161-016K
3122	176-016-9-001
3122	176-025-9-001
3122	176-029-9-001
3122	176-037-9-001
3122	176-037-9-003
3122	176-037-9-004
3122	176-047-9-003
3122	176-056-9-005
3122	176-060-9-004
3122	176-075-9-003
3122	185-005
3122	186-003
3122	186-004
3122	186-005
3122	200X3174-006
3122	200X3174-014
3122	200X3174-021
3122	229-1200-36
3122	250-1213
3122	250-1312
3122	260P28107
3122	296-86
3122	296-98-9
3122	297L007H03
3122	353-9318-001
3122	353-9318-002
3122	366-2
3122	404-2
3122	417-107
3122	433-1
3122	465-106-19
3122	512RED
3122	536-2
3122	000546-1
3122	567-0003-011
3122	576-0004-010
3122	595-1
3122	600-224-605
3122	600-229-201
3122	614-2
3122	617-87
3122	000653
3122	0704
3122	753-1372-100
3122	753-2000-100
3122	753-9000-922
3122	853-0301-317
3122	902-003-6-06
3122	909-27125-140
3122	921-111B
3122	921-133B
3122	921-154B
3122	964-2073B
3122	964-27986
3122	991-011313
3122	991-011314
3122	1006-48
3122	1007-3153
3122	1035-80
3122	1041-71
3122	1042-03
3122	1042-7
3122	1045-2951
3122	1045-3082
3122	1049-0100
3122	1049-1744
3122	1063-8963
3122	1074-03
3122	1074-115
3122	1080-01
3122	1080-03
3122	1080-07
3122	1087-02
3122	1210-17
3122	1483
3122	1634-17
3122	1687-17
3122	1710
3122	1751-17
3122	1835-17
3122	1925-17
3122	2000-206
3122	2000-210
3122	2022-05
3122	2041-01
3122	2041-02
3122	2057A2-249
3122	2057A2-260
3122	2057A2-289
3122	2057A2-370
3122	2057A2-391
3122	2057A2-405
3122	2057A2-487
3122	2057A2-507
3122	2057A2-508
3122	2057A2-510
3122	2057A2-518
3122	2057A2-530
3122	2057B119-2
3122	2057B120-12
3122	2057B125-9
3122	2063-17
3122	2064
3122	2065-07
3122	2110N-132
3122	2110N-133
3122	2134-17
3122	2181-17
3122	2208-17
3122	2214-17
3122	2246-17
3122	2302-17
3122	2362-1
3122	2446
3122	2449-17
3122	2904-033
3122	2925
3122	3506
3122	3519
3122	3519-1
3122	3525
3122	3538
3122	3541
3122	3543
3122	3544
3122	3548
3122	3551
3122	3554
3122	3558
3122	3560
3122	3560-2
3122	3561-1
3122	3565-1
3122	3569
3122	4002(PACE)
3122	4010
3122	5001-037
3122	5001-074
3122	5001-505
3122	5001-539
3122	5001-545
3122	5613-871F
3122	7001
3122	7398-6119P
3122	7429
3122	7430
3122	7431
3122	7432
3122	7433
3122	8000-00003-034
3122	8000-00004-085
3122	8000-00004-241
3122	8000-00004-242
3122	8000-00004-P079
3122	8000-00004-P082
3122	8000-00005-004
3122	8000-00009-280
3122	8000-00012-039
3122	8000-00028-038
3122	8000-00028-206
3122	8000-00049-055
3122	8000-00049-057
3122	8440-123
3122	8440-124
3122	9502
3122	9618
3122	9696H
3122	18410-144
3122	25840-163
3122	030010
3122	030010-1
3122	030531-1
3122	35212
3122	37510-161
3122	38510-164
3122	40242
3122	40243
3122	40458
3122	45810-164
3122	053492
3122	71226-5
3122	95221
3122	119721
3122	119823
3122	125143
3122	126699
3122	126708
3122	126717
3122	126720
3122	127263
3122	129029
3122	129049
3122	129425
3122	137875
3122	145258
3122	150117
3122	168660
3122	171162-095
3122	171162-285
3122	202617
3122	221857
3122	228417
3122	231017
3122	256217
3122	256817
3122	256917
3122	282317
3122	0320031
3122	00444028-010
3122	489751-164
3122	515039S
3122	610094-3
3122	610107-2
3122	702415-00
3122	740886
3122	741737
3122	741855
3122	741861
3122	742728
3122	760142
3122	760249
3122	916049
3122	916052
3122	985087
3122	992052
3122	992066
3122	995030
3122	1223912
3122	1223913
3122	1223920
3122	1473519-1
3122	1473538-1
3122	1473544-1
3122	1473551-1
3122	1473560-002
3122	1473560-2
3122	1473561-1
3122	1473565-1
3122	1473569-1
3122	1473614-1
3122	1473614-3
3122	1700035
3122	2003229-25
3122	2006227-51
3122	2092609-0025
3122	2320042
3122	2320643
3122	2320644
3122	2320646
3122	2320664
3122	2487424
3122	3596116
3122	3596117
3122	3596338
3122	3596339
3122	4036887-P8
3122	4036924-P1
3122	4036924-P2
3122	4037586-P1
3122	4037586-P2
3122	4037800-P1
3122	4037800-P2
3122	004440028-002
3122	04440028-006

SK	Industry Standard No.	SK	Industry Standard No.	SK	Industry Standard No.	SK	Industry Standard No.	SK	Industry Standard No.
3122	5320613	3123	HJ62	3123	R67	3123	ST301	3123	2SB108G
3122	5320622	3123	HJX2	3123	R83	3123	ST302	3123	2SB108GN
3122	06120008	3123	HR4	3123	R98	3123	ST303	3123	2SB108H
3122	06120063	3123	HR8	3123	R81539	3123	ST304	3123	2SB108J
3122	7576015-01	3123	HR9	3123	R81540	3123	ST332	3123	2SB108K
3122	8113323	3123	H8170	3123	R81543	3123	ST370	3123	2SB108L
3122	13073525	3123	H817D	3123	R81544	3123	ST37C	3123	2SB108M
3122	23114015	3123	H822D	3123	R81546	3123	ST37D	3123	2SB108OR
3122	23114297	3123	H829D	3123	R81548	3123	SY61583	3123	2SB108X
3122	26010021	3123	HV12	3123	R81550	3123	SYL105	3123	2SB108Y
3122	43029483	3123	HV16	3123	R81555	3123	SYL106	3123	2SB238
3122	43040313	3123	HV17	3123	R82350	3123	SYL107	3123	2SB294
3122	44079004	3123	IRTR82	3123	R82351	3123	SYL108	3123	2SB294A
3122	62589494	3123	JR05	3123	R82352	3123	SYL160	3123	2SB294B
3122	632073518	3123	KD2121	3123	R82353	3123	SYL1608	3123	2SB294C
3122	2003073517	3123	KV4	3123	R82354	3123	SYL1668	3123	2SB294D
3122	2003082823	3123	L5108	3123	R82355	3123	T0031	3123	2SB294F
3122	2003082837	3123	L5121	3123	R82367	3123	T0051	3123	2SB294G
3122	2003174006	3123	L5122	3123	R82675	3123	T1010	3123	2SB294GN
3122	2003174014	3123	M108	3123	R82677	3123	T1023	3123	2SB294H
3122	4100217172	3123	M4313	3123	R82679	3123	T129	3123	2SB294J
3122	4100907633	3123	M4398	3123	R82680	3123	T130	3123	2SB294K
3122	4104204615	3123	M4466ORN	3123	R82683	3123	T1300	3123	2SB294L
3122	4203970101	3123	M4469RED	3123	R82684	3123	T39	3123	2SB294M
3122	6621001000	3123	M4470ORN	3123	R82685	3123	T45	3123	2SB294OR
3122	6621003500	3123	M4471YEL	3123	R82686	3123	T46	3123	2SB294X
3122	6621008100	3123	M4472GRN	3123	R82687	3123	T47	3123	2SB304
3122	6621009100	3123	M4474YEL	3123	R82688	3123	T48	3123	2SB304A
3123	A061-115	3123	M4475ORN	3123	R82689	3123	TP66	3123	2SB35
3123	A1378	3123	M4476BLU	3123	R82694	3123	TIA02	3123	2SB35A
3123	A1384	3123	M4510	3123	R82695	3123	TIA35	3123	2SB35B
3123	A14-1004	3123	M4553BLU	3123	R82696	3123	TIXA03	3123	2SB35C
3123	A14-1005	3123	M4553BRN	3123	R82697	3123	TIXA04	3123	2SB35D
3123	A14-1006	3123	M4553ORN	3123	R83211	3123	TIXA05	3123	2SB35E
3123	A14-1007	3123	M4553GRN	3123	R83275	3123	TNJ60611	3123	2SB35F
3123	A14-1008	3123	M4553RED	3123	R83276	3123	TR320	3123	2SB35GN
3123	A14-1009	3123	M4553YEL	3123	R83277	3123	TR321	3123	2SB35H
3123	A14-1010	3123	M4563	3123	R83278	3123	TR332	3123	2SB35J
3123	A15-1004	3123	M4564	3123	R83279	3123	TR34	3123	2SB35K
3123	A15-1005	3123	M4565	3123	R83280	3123	TR363	3123	2SB35L
3123	AC122-GRN	3123	M4573	3123	R83282	3123	TR45	3123	2SB35M
3123	AC122-RED	3123	M8640A	3123	R83283	3123	TR031	3123	2SB35OR
3123	AC122-YEL	3123	MA112	3123	R83285	3123	TS1266	3123	2SB35X
3123	AC135	3123	MA115	3123	R83286	3123	TS1541	3123	2SB35Y
3123	AC161	3123	MA116	3123	R83288	3123	TS162	3123	2SB66
3123	AC170	3123	MA117	3123	R83299	3123	TS163	3123	2SB86
3123	AC182	3123	MA23	3123	R83301	3123	TS627A	3123	2SB427
3123	ACY23	3123	MA23A	3123	R83309	3123	TS627B	3123	2SB428
3123	ACY32	3123	MA890	3123	R83897	3123	TS630	3123	2SB477C
3123	ACY58	3123	MA891	3123	R83898	3123	TS67BB	3123	2SB66H
3123	AF188	3123	MA892	3123	R83906	3123	TS765	3123	2SB9AH
3123	AR-105	3123	MA893	3123	R8406	3123	WTV15VMG	3123	2SB9H
3123	A8477	3123	MA894	3123	R85104	3123	WTV20VH6	3123	28C492
3123	ASY14-1	3123	MA895	3123	R85106	3123	WTV3MC	3123	2T521
3123	ASY14-2	3123	MA896	3123	R85108	3123	WTV36	3123	2T593
3123	ASY14-3	3123	MA897	3123	R85208	3123	WTVBB6A	3123	3N21
3123	ASY26	3123	MA898	3123	R85312	3123	X-78	3123	4-1546
3123	ASY31	3123	MA899	3123	R85313	3123	X42	3123	6-60X
3123	ASY32	3123	MA901	3123	R85401	3123	2AC128	3123	6-61X
3123	ASY51	3123	MA902	3123	R85502	3123	2G108	3123	6A1O624
3123	ASY67	3123	MA903	3123	R85503	3123	2G109	3123	6A11665
3123	ASY81	3123	MA904	3123	R85504	3123	2G201	3123	6A11668
3123	AT100H	3123	MD836	3123	R85505	3123	2G202	3123	6A12516
3123	AT100M	3123	MM2550	3123	R85506	3123	2G395	3123	6D122V
3123	AT100N	3123	MM2552	3123	R85507	3123	2G605	3123	6L122
3123	AT6	3123	MM2554	3123	R85530	3123	2N109BLU	3123	09-301015
3123	AT6A	3123	MN53BLU	3123	R85532	3123	2N109GRN	3123	2OC71
3123	B-26-1	3123	MN53GRN	3123	R85535	3123	2N1174W	3123	20V-HG
3123	B-5	3123	MN53RED	3123	R85542	3123	2N183	3123	21A040-035
3123	B461	3123	NKT212	3123	R85544	3123	2N184	3123	22-002006
3123	B51	3123	NKT224J	3123	R85563	3123	2N1315	3123	24MW799
3123	CA2D2	3123	NKT225J	3123	R85564	3123	2N1403	3123	30V-HG
3123	DC-10	3123	NKT226J	3123	R85565	3123	2N1637/33	3123	33-1021-00
3123	D814	3123	NKT24	3123	R85566	3123	2N1639/33	3123	45X1A520C
3123	D816	3123	NKT242	3123	R85567	3123	2N174RED	3123	46-8660-3
3123	D822	3123	NKT247J	3123	R85568	3123	2N1853/18	3123	48-124158
3123	E2482	3123	NKT274	3123	R85607	3123	2N185BLU	3123	48-124159
3123	EA1081	3123	NKT32	3123	R85608	3123	2N1960/46	3123	48-124175
3123	ER15X22	3123	NKT4	3123	R85610	3123	2N1961	3123	48-124276
3123	ER15X23	3123	NKT5	3123	R85704	3123	2N1961/46	3123	48-124279
3123	ES19	3123	NKT618J	3123	R85715	3123	2N2406	3123	48-124300
3123	ES23	3123	OC103	3123	R85715-1	3123	2N2621	3123	48-124322
3123	ES3121	3123	OC304	3123	R85733	3123	2N2622	3123	48-124373
3123	FBN-2N1183	3123	OC307	3123	R85734	3123	2N2623	3123	48-12443
3123	FBN-0254759	3123	OC307-1	3123	R85736	3123	2N2624	3123	48-125271
3123	FPR40-1004	3123	OC307-2	3123	R85740-1	3123	2N2625	3123	48-125276
3123	FPR40-1005	3123	OC307-3	3123	R85743	3123	2N2627	3123	48-134407
3123	GC1186	3123	OC308	3123	R85749	3123	2N2628	3123	48-134408
3123	GC1187	3123	OC309	3123	R85765	3123	2N2629	3123	48-134482
3123	GC1257	3123	OC309-1	3123	R85768	3123	2N2630	3123	48-134483
3123	GC4144	3123	OC309-2	3123	R85852	3123	2N2672BLK	3123	48-134653
3123	GC464	3123	OC309-3	3123	R85854	3123	2N2672GRN	3123	48-63078A62
3123	GC5000	3123	OC330	3123	R8593	3123	2N2928	3123	48-63078A68
3123	GC650A	3123	OC34	3123	R8684	3123	2N331M	3123	48-63084A03
3123	GC640	3123	OC340	3123	R8685	3123	2N400	3123	48-63084A04
3123	GET5116	3123	OC350	3123	R8687	3123	2N427A	3123	48-63084A05
3123	GET5117	3123	OC38	3123	RSKK36	3123	2N534	3123	48-644678
3123	GF20	3123	OC45	3123	RT4624	3123	2N782	3123	48-644679
3123	GF21	3123	OC46	3123	S025	3123	2N800	3123	48-97046A03
3123	GF32	3123	OC53	3123	S95203	3123	2N846B	3123	48-97046A10
3123	GT-269	3123	OC54	3123	S95214	3123	2N977	3123	48-97046A34
3123	GT-34HV	3123	OC55	3123	SA128-1	3123	2NJ5D	3123	48-97162A18
3123	GT-348	3123	OC56	3123	SA197	3123	2SA231	3123	48-97162A19
3123	GT-762R	3123	OC59	3123	SA197-1	3123	2SA504	3123	48-97221A03
3123	GT109R	3123	OC65	3123	SA197-2	3123	2SA375	3123	48-97271A01
3123	GT11	3123	OC66	3123	SA197-3	3123	2SA408	3123	48-97271A02
3123	GT12	3123	OC70	3123	SA204	3123	2SA409	3123	48-97271A03
3123	GT1223	3123	OC71	3123	SA205BLU	3123	2SA42	3123	48K56P1
3123	GT13	3123	OC71N	3123	SA205BRN	3123	2SA79	3123	520189
3123	GT132	3123	OC74	3123	SA205ORN	3123	2SA86	3123	57BZ-24
3123	GT20R	3123	OC77M	3123	SA205GRN	3123	2SB105	3123	61B0015-1
3123	GT5116	3123	PA10889-1	3123	SA205RED	3123	2SB105A	3123	61B016-1
3123	GT751	3123	PADT55	3123	SA205WHT	3123	2SB105B	3123	61B018-1
3123	GT766	3123	PT0139	3123	SA205YEL	3123	2SB105C	3123	6581
3123	GT766A	3123	R120	3123	SA240	3123	2SB105D	3123	73A01
3123	GT81R	3123	R152	3123	SP.T184	3123	2SB105E	3123	73A02
3123	GTE1	3123	R164	3123	SP.T237	3123	2SB105F	3123	74A01
3123	HA-201	3123	R242	3123	SP.T251	3123	2SB105G	3123	74A02
3123	HA-330	3123	R255	3123	SP.T252	3123	2SB105GN	3123	74A03
3123	HB-186	3123	R290	3123	SP.T253	3123	2SB105H	3123	86-5001-2
3123	HB-187	3123	R291	3123	SP.T306	3123	2SB105K	3123	86-5042-2
3123	HBP-G6011	3123	R324	3123	SP.T318	3123	2SB105L	3123	86-5063-2
3123	HF50	3123	R35	3123	SP.T321	3123	2SB105M	3123	86-5091-2
3123	HJ15	3123	R36	3123	SP.T337	3123	2SB105OR	3123	86-95-2
3123	HJ17	3123	R424-1	3123	SP.T351	3123	2SB105R	3123	98-1
3123	HJ17D	3123	R428	3123	SP.T352	3123	2SB105X	3123	98-2
3123	HJ22	3123	R497	3123	SP.T353	3123	2SB105Y	3123	98-3
3123	HJ226	3123	R506	3123	SPT337	3123	2SB108	3123	98-4
3123	HJ23	3123	R515	3123	SM217	3123	2SB108A	3123	98-5
3123	HJ230	3123	R52	3123	ST122	3123	2SB108B		
3123	HJ315	3123	R593	3123	ST123	3123	2SB108C		
3123	HJ37	3123	R64	3123	ST28B	3123	2SB108D		
3123	HJ54	3123	R66	3123	ST28C	3123	2SB108E		
3123	HJ60A					3123	2SB108F		

SK	Industry Standard No.
3123	98-6
3123	98-7
3123	101-15
3123	101A
3123	101B
3123	101M
3123	120-004728
3123	121-10
3123	121-11
3123	121-12
3123	121-14
3123	121-300
3123	121-301
3123	121-368
3123	121-552
3123	121-553
3123	121-636
3123	121-79
3123	121-9
3123	130-40456
3123	201A
3123	201B
3123	297V057H01
3123	297V057H02
3123	322T1
3123	323T1
3123	324T1
3123	325T1
3123	326T1
3123	417-79
3123	421-11B
3123	421-12B
3123	421-14B
3123	505ES105
3123	509ES025P
3123	512ES040P
3123	513ES045P
3123	617-70
3123	830
3123	921-37B
3123	941T1
3123	991-011221
3123	1140-17
3123	1277-17
3123	2057A2-147
3123	2402-455
3123	2405-454
3123	2405-455
3123	2405-456
3123	2405-457
3123	2408-328
3123	2603-180
3123	2603-182
3123	2603-183
3123	2703-385
3123	2703-386
3123	3504
3123	4398
3123	4467
3123	4473
3123	4477V10
3123	4553V10
3123	8500-202
3123	8500-203
3123	10038
3123	10039
3123	020156
3123	30202
3123	35218
3123	72003
3123	72004
3123	72006
3123	116996
3123	123877
3123	245568-2
3123	983012
3123	983238
3123	1221648
3123	2001809-47
3123	2006334-31
3123	2243255-1
3123	2495388-2
3123	2495567-2
3123	2495567-3
3123	2497496
3123	5958539
3123	7279281
3123	8526849-1
3123	20912578-1
3123	80285400
3124	A-128
3124	A749
3124	B-66
3124	C1908E
3124	C458B-D
3124	C458BLG
3124	C712
3124	C712A
3124	C712C
3124	C712D
3124	C712E
3124	C712W
3124	CS-1361G
3124	CS-9013
3124	CS1166D-G
3124	CS9013E-F
3124	CS9014B-C
3124	DS-46
3124	DS-75
3124	ECG289
3124	EQS-10
3124	EQS-11
3124	EQS-62
3124	ES15048
3124	FS27604
3124	G05-035D
3124	H-1567
3124	HR11A
3124	HR11B
3124	HR14
3124	HR14A
3124	HR15
3124	HR15A
3124	HR16
3124	HR16A
3124	HR17
3124	HR17A
3124	HR18
3124	HR18A
3124	HR19
3124	HR19A
3124	HR37
3124	HT306441B0
3124	HT3073310
3124	KSC815-0
3124	MPS-2716
3124	MPS6572
3124	MPS9630H.I
3124	PTC139
3124	Q2
3124	REN289
3124	TR01062-7
3124	TR2083-74
3124	WEP1945
3124	WEP536
3124	WEP910
3124	01-031685
3124	2N4086
3124	2N4104
3124	2SC-4012
3124	2SC1209
3124	2SC1209D
3124	2SC1209E
3124	2SC1317A
3124	2SC1317BC
3124	2SC1317C
3124	2SC1317E
3124	2SC1317G
3124	2SC1317GR
3124	2SC1317L
3124	2SC13170R
3124	2SC1317V
3124	2SC1317Y
3124	2SC1318B
3124	2SC1318C
3124	2SC1318E
3124	2SC1318F
3124	2SC1318G
3124	2SC1318GN
3124	2SC1318H
3124	2SC1318J
3124	2SC1318K
3124	2SC1318L
3124	2SC1318M
3124	2SC1318Y
3124	2SC1342D
3124	2SC1342E
3124	2SC1342F
3124	2SC1342G
3124	2SC1342GN
3124	2SC1342H
3124	2SC1342J
3124	2SC1342K
3124	2SC1342L
3124	2SC1342M
3124	2SC13420R
3124	2SC1342R
3124	2SC1342X
3124	2SC1342Y
3124	2SC1364B
3124	2SC1364C
3124	2SC1364D
3124	2SC1364E
3124	2SC1364H
3124	2SC1364K
3124	2SC1364L
3124	2SC1364M
3124	2SC13640R
3124	2SC1364R
3124	2SC1364X
3124	2SC1364Y
3124	2SC157A
3124	2SC157B
3124	2SC157C
3124	2SC157D
3124	2SC157E
3124	2SC157F
3124	2SC157G
3124	2SC157GN
3124	2SC157J
3124	2SC157K
3124	2SC157L
3124	2SC157M
3124	2SC1570R
3124	2SC157R
3124	2SC157Y
3124	2SC158A
3124	2SC158B
3124	2SC158C
3124	2SC158D
3124	2SC158E
3124	2SC158F
3124	2SC158G
3124	2SC158GN
3124	2SC158H
3124	2SC158J
3124	2SC158K
3124	2SC158L
3124	2SC158M
3124	2SC1580R
3124	2SC158R
3124	2SC158X
3124	2SC158Y
3124	2SC182A
3124	2SC182B
3124	2SC182C
3124	2SC182D
3124	2SC182E
3124	2SC182F
3124	2SC182G
3124	2SC182GN
3124	2SC182H
3124	2SC182J
3124	2SC182K
3124	2SC182L
3124	2SC182M
3124	2SC182R
3124	2SC182X
3124	2SC182Y
3124	2SC1908
3124	2SC1908E
3124	2SC2120-O
3124	2SC218B
3124	2SC2210
3124	2SC281BL
3124	2SC281E
3124	2SC281F
3124	2SC281G
3124	2SC281GN
3124	2SC281J
3124	2SC281K
3124	2SC281L
3124	2SC281M
3124	2SC2810R
3124	2SC281R
3124	2SC281X
3124	2SC281Y
3124	2SC28A
3124	2SC28B
3124	2SC28C
3124	2SC28D
3124	2SC28E
3124	2SC28F
3124	2SC28G
3124	2SC28GN
3124	2SC28H
3124	2SC28J
3124	2SC28K
3124	2SC28L
3124	2SC28M
3124	2SC280R
3124	2SC28R
3124	2SC28X
3124	2SC28Y
3124	2SC318AB
3124	2SC362A
3124	2SC362B
3124	2SC362C
3124	2SC362D
3124	2SC362E
3124	2SC362F
3124	2SC362G
3124	2SC362GN
3124	2SC362H
3124	2SC362J
3124	2SC362K
3124	2SC362L
3124	2SC362M
3124	2SC3620R
3124	2SC362R
3124	2SC362X
3124	2SC362Y
3124	2SC383A
3124	2SC383B
3124	2SC383C
3124	2SC383D
3124	2SC383E
3124	2SC383GN
3124	2SC383H
3124	2SC383J
3124	2SC383K
3124	2SC383L
3124	2SC383M
3124	2SC3830R
3124	2SC383R
3124	2SC383X
3124	2SC394(0)
3124	2SC394(0)
3124	2SC394-Y
3124	2SC394A
3124	2SC394AP
3124	2SC394B
3124	2SC394C
3124	2SC394D
3124	2SC394E
3124	2SC394F
3124	2SC394G
3124	2SC394GN
3124	2SC394H
3124	2SC394J
3124	2SC394K
3124	2SC394L
3124	2SC394M
3124	2SC3940
3124	2SC3940R
3124	2SC4012A
3124	2SC4012B
3124	2SC4012C
3124	2SC4012D
3124	2SC4012E
3124	2SC4012F
3124	2SC4012G
3124	2SC4012GN
3124	2SC4012H
3124	2SC4012J
3124	2SC4012L
3124	2SC4012M
3124	2SC4012X
3124	2SC4012Y
3124	2SC401B
3124	2SC401C
3124	2SC401D
3124	2SC401E
3124	2SC401F
3124	2SC401G
3124	2SC401GN
3124	2SC401H
3124	2SC401J
3124	2SC401K
3124	2SC401L
3124	2SC401M
3124	2SC4010R
3124	2SC401R
3124	2SC401Y
3124	2SC402A
3124	2SC402C
3124	2SC402D
3124	2SC402E
3124	2SC402F
3124	2SC402G
3124	2SC402GN
3124	2SC402H
3124	2SC402J
3124	2SC402K
3124	2SC402L
3124	2SC402M
3124	2SC4020R
3124	2SC402R
3124	2SC402X
3124	2SC402Y
3124	2SC403AL
3124	2SC403B
3124	2SC403CG
3124	2SC403D
3124	2SC403E
3124	2SC403F
3124	2SC403G
3124	2SC403GN
3124	2SC403H
3124	2SC403J
3124	2SC403K
3124	2SC403L
3124	2SC403M
3124	2SC4030R
3124	2SC403R
3124	2SC403X
3124	2SC403Y
3124	2SC441A
3124	2SC441B
3124	2SC441C
3124	2SC441D
3124	2SC441E
3124	2SC441F
3124	2SC441G
3124	2SC441GN
3124	2SC441H
3124	2SC441J
3124	2SC441K
3124	2SC441L
3124	2SC441M
3124	2SC4410R
3124	2SC441R
3124	2SC441X
3124	2SC441Y
3124	2SC442A
3124	2SC442B
3124	2SC442C
3124	2SC442D
3124	2SC442E
3124	2SC442F
3124	2SC442G
3124	2SC442GN
3124	2SC442H
3124	2SC442J
3124	2SC442K
3124	2SC442L
3124	2SC442M
3124	2SC4420R
3124	2SC442R
3124	2SC442X
3124	2SC442Y
3124	2SC454(B)
3124	2SC454-3
3124	2SC454-5
3124	2SC454B-6
3124	2SC454BL
3124	2SC454E
3124	2SC454F
3124	2SC454G
3124	2SC454GN
3124	2SC454H
3124	2SC454J
3124	2SC454K
3124	2SC454M
3124	2SC4540R
3124	2SC454R
3124	2SC454X
3124	2SC454Y
3124	2SC458-4
3124	2SC458-5
3124	2SC458B-D
3124	2SC458BK
3124	2SC458LGC-6
3124	2SC458LGR
3124	2SC458R
3124	2SC475A
3124	2SC475B
3124	2SC475C
3124	2SC475D
3124	2SC475E
3124	2SC475F
3124	2SC475G
3124	2SC475GN
3124	2SC475H
3124	2SC475J
3124	2SC475L
3124	2SC475M
3124	2SC4750R
3124	2SC475R
3124	2SC475X
3124	2SC475Y
3124	2SC48A
3124	2SC48B
3124	2SC48D
3124	2SC48E
3124	2SC48F
3124	2SC48G
3124	2SC48GN
3124	2SC48H
3124	2SC48J
3124	2SC48K
3124	2SC48L
3124	2SC48M
3124	2SC480R
3124	2SC48R
3124	2SC48X
3124	2SC48Y
3124	2SC5370
3124	2SC5370A
3124	2SC5370B
3124	2SC5370C
3124	2SC5370D
3124	2SC5370E
3124	2SC5370F
3124	2SC5370G
3124	2SC5370GN
3124	2SC5370H
3124	2SC5370J
3124	2SC5370K
3124	2SC5370L
3124	2SC5370M
3124	2SC5370R
3124	2SC5370X
3124	2SC5370Y
3124	2SC56A
3124	2SC56B
3124	2SC56C
3124	2SC56D
3124	2SC56E
3124	2SC56F
3124	2SC56G
3124	2SC56GN
3124	2SC56H
3124	2SC56J
3124	2SC56K
3124	2SC56L
3124	2SC56M
3124	2SC560R
3124	2SC56R
3124	2SC56X
3124	2SC56Y
3124	2SC620
3124	2SC620A
3124	2SC620B
3124	2SC620C
3124	2SC620CD
3124	2SC620D
3124	2SC620DE
3124	2SC620E
3124	2SC620F
3124	2SC620G
3124	2SC620GN
3124	2SC620H
3124	2SC620J
3124	2SC620K
3124	2SC620L
3124	2SC620M
3124	2SC6200R
3124	2SC620R
3124	2SC620X
3124	2SC620Y
3124	2SC631B
3124	2SC631C
3124	2SC631D
3124	2SC631E
3124	2SC631F
3124	2SC631G
3124	2SC631GN
3124	2SC631H
3124	2SC631J
3124	2SC631K
3124	2SC631L
3124	2SC631M
3124	2SC6310R
3124	2SC631R
3124	2SC631X
3124	2SC631Y
3124	2SC632(1)
3124	2SC632B
3124	2SC632C
3124	2SC632D
3124	2SC632E
3124	2SC632F
3124	2SC632G
3124	2SC632GN
3124	2SC632H
3124	2SC632J
3124	2SC632K
3124	2SC632L
3124	2SC632M
3124	2SC6320R
3124	2SC632R
3124	2SC632X
3124	2SC632Y
3124	2SC633B
3124	2SC633C
3124	2SC633D
3124	2SC633E
3124	2SC633F
3124	2SC633GN
3124	2SC633J
3124	2SC633K
3124	2SC633L
3124	2SC633M
3124	2SC6330R
3124	2SC633R
3124	2SC633X
3124	2SC633Y
3124	2SC634(2)
3124	2SC634AK
3124	2SC634AL
3124	2SC634AXL
3124	2SC634B
3124	2SC634C
3124	2SC634D
3124	2SC634E
3124	2SC634F
3124	2SC634G
3124	2SC634GN
3124	2SC634H
3124	2SC634J
3124	2SC634K
3124	2SC634L
3124	2SC634M
3124	2SC6340R
3124	2SC634R
3124	2SC634X
3124	2SC634Y
3124	2SC648A
3124	2SC648B
3124	2SC648C
3124	2SC648D
3124	2SC648E
3124	2SC648F
3124	2SC648G
3124	2SC648J
3124	2SC648K
3124	2SC648L
3124	2SC648M
3124	2SC6480R
3124	2SC648R
3124	2SC648X
3124	2SC648Y
3124	2SC650C
3124	2SC650D
3124	2SC650E
3124	2SC650F
3124	2SC650G
3124	2SC650GN
3124	2SC650H
3124	2SC650J
3124	2SC650K
3124	2SC650L
3124	2SC650M
3124	2SC650R
3124	2SC650X
3124	2SC650Y
3124	2SC693A
3124	2SC693B
3124	2SC693C
3124	2SC693D
3124	2SC693GN
3124	2SC693J
3124	2SC693K
3124	2SC693L
3124	2SC693M
3124	2SC6930R
3124	2SC693R
3124	2SC693X
3124	2SC693Y
3124	2SC694A
3124	2SC694B
3124	2SC694C
3124	2SC694D
3124	2SC694GN
3124	2SC694H
3124	2SC694J
3124	2SC694K
3124	2SC694L
3124	2SC694M
3124	2SC6940R
3124	2SC694R
3124	2SC694X
3124	2SC694Y
3124	2SC701A
3124	2SC701C
3124	2SC701D
3124	2SC701E
3124	2SC701F
3124	2SC701G

SK	Industry Standard No.
3124	2SC701GN
3124	2SC701H
3124	2SC701J
3124	2SC701K
3124	2SC701L
3124	2SC701M
3124	2SC701OR
3124	2SC701R
3124	2SC701X
3124	2SC701Y
3124	2SC702
3124	2SC702A
3124	2SC702B
3124	2SC702C
3124	2SC702D
3124	2SC702E
3124	2SC702F
3124	2SC702G
3124	2SC702GN
3124	2SC702H
3124	2SC702J
3124	2SC702K
3124	2SC702L
3124	2SC702M
3124	2SC702R
3124	2SC702X
3124	2SC702Y
3124	2SC712
3124	2SC712-CD
3124	2SC712A
3124	2SC712B
3124	2SC712C
3124	2SC712CD
3124	2SC712D
3124	2SC712E
3124	2SC712F
3124	2SC712G
3124	2SC712GN
3124	2SC712H
3124	2SC712J
3124	2SC712K
3124	2SC712L
3124	2SC712M
3124	2SC712OR
3124	2SC712R
3124	2SC712W
3124	2SC712Y
3124	2SC715G
3124	2SC715J
3124	2SC715K
3124	2SC715L
3124	2SC715M
3124	2SC715R
3124	2SC715Y
3124	2SC719
3124	2SC81
3124	2SC814A
3124	2SC814B
3124	2SC814C
3124	2SC814D
3124	2SC814E
3124	2SC814F
3124	2SC814G
3124	2SC814GN
3124	2SC814H
3124	2SC814J
3124	2SC814K
3124	2SC814M
3124	2SC814OR
3124	2SC814R
3124	2SC814X
3124	2SC814Y
3124	2SC815-1
3124	2SC815BK
3124	2SC815D
3124	2SC815E
3124	2SC815G
3124	2SC815GN
3124	2SC815H
3124	2SC815J
3124	2SC815K,L
3124	2SC815LJ
3124	2SC815OR
3124	2SC815R
3124	2SC815X
3124	2SC815Y
3124	2SC81A
3124	2SC81B
3124	2SC81C
3124	2SC81D
3124	2SC81E
3124	2SC81F
3124	2SC81G
3124	2SC81GN
3124	2SC81H
3124	2SC81J
3124	2SC81K
3124	2SC81L
3124	2SC81M
3124	2SC81OR
3124	2SC81R
3124	2SC839G
3124	2SC839GN
3124	2SC839OR
3124	2SC839X
3124	2SC858A
3124	2SC858B
3124	2SC858C
3124	2SC858D
3124	2SC858GA
3124	2SC858GN
3124	2SC858H
3124	2SC858J
3124	2SC858K
3124	2SC858L
3124	2SC858M
3124	2SC858OR
3124	2SC858R
3124	2SC858X
3124	2SC858Y
3124	2SC859A
3124	2SC859B
3124	2SC859D
3124	2SC859GL
3124	2SC859GM
3124	2SC859GN
3124	2SC859H
3124	2SC859J
3124	2SC859K
3124	2SC859L
3124	2SC859M
3124	2SC859OR
3124	2SC859R
3124	2SC859X

SK	Industry Standard No.
3124	2SC859Y
3124	2SC870H
3124	2SC871A
3124	2SC871AM
3124	2SC871B
3124	2SC871C
3124	2SC871GN
3124	2SC871H
3124	2SC871J
3124	2SC871K
3124	2SC871L
3124	2SC871M
3124	2SC871OR
3124	2SC871R
3124	2SC871X
3124	2SC871Y
3124	2SC923G
3124	2SC923GN
3124	2SC923H
3124	2SC923K
3124	2SC923L
3124	2SC923M
3124	2SC923OR
3124	2SC923R
3124	2SC923X
3124	2SC923Y
3124	2SC924A
3124	2SC924B
3124	2SC924C
3124	2SC924D
3124	2SC924G
3124	2SC924GN
3124	2SC924H
3124	2SC924J
3124	2SC924K
3124	2SC924L
3124	2SC924OR
3124	2SC924R
3124	2SC924X
3124	2SC924Y
3124	2SC925A
3124	2SC925B
3124	2SC925C
3124	2SC925D
3124	2SC925E
3124	2SC925F
3124	2SC925G
3124	2SC925GN
3124	2SC925H
3124	2SC925J
3124	2SC925K
3124	2SC925L
3124	2SC925M
3124	2SC925OR
3124	2SC925R
3124	2SC925X
3124	2SC925Y
3124	2SC938D
3124	2SC938E
3124	2SC938F
3124	2SC938G
3124	2SC938GN
3124	2SC938H
3124	2SC938J
3124	2SC938K
3124	2SC938L
3124	2SC938M
3124	2SC938OR
3124	2SC938R
3124	2SC938X
3124	2SC938Y
3124	2SC945GN
3124	2SC945H
3124	2SC945J
3124	2SC945OR
3124	2SC945PJ
3124	2SC945RA
3124	2SCS183A
3124	2SCS183B
3124	2SCS183C
3124	2SCS183D
3124	2SCS183F
3124	2SCS183G
3124	2SCS183GN
3124	2SCS183H
3124	2SCS183J
3124	2SCS183K
3124	2SCS183L
3124	2SCS183M
3124	2SCS183OR
3124	2SCS183R
3124	2SCS183Y
3124	2SCS184
3124	2SCS184A
3124	2SCS184B
3124	2SCS184C
3124	2SCS184D
3124	2SCS184F
3124	2SCS184G
3124	2SCS184GN
3124	2SCS184H
3124	2SCS184K
3124	2SCS184L
3124	2SCS184M
3124	2SCS184OR
3124	2SCS184R
3124	2SCS184X
3124	2SCS184Y
3124	2SD1347
3124	2SD134A
3124	2SD134C
3124	2SD134D
3124	2SD134F
3124	2SD134G
3124	2SD134GN
3124	2SD134H
3124	2SD134J
3124	2SD134K
3124	2SD134L
3124	2SD134M
3124	2SD134OR
3124	2SD134R
3124	2SD134X
3124	2SD134Y
3124	2SD227G
3124	2SD227GN
3124	2SD227H
3124	2SD227J
3124	2SD227K
3124	2SD227LF
3124	2SD227M
3124	2SD227OR
3124	2SD227Y
3124	2SDU780R

SK	Industry Standard No.
3124	7-59-024/3477
3124	13-14085-122
3124	34-6015-43
3124	42-27374
3124	42-27375
3124	48P63005A72
3124	48R869248
3124	48S134405
3124	48S134666
3124	48S134718
3124	48S134719
3124	48S134720
3124	48S134732
3124	48S134733
3124	48S134733A
3124	48S134734
3124	48S134734A
3124	48S134737
3124	48S134768
3124	48S134809
3124	48S134810
3124	48S134811
3124	48S134823
3124	48S134832
3124	48S134838
3124	48S134842
3124	48S134846
3124	48S134853
3124	48S134854
3124	48S134898
3124	48S134905
3124	48S134906
3124	48S134912
3124	48S134935
3124	48S134952
3124	48S137014
3124	48S137111
3124	48S137206
3124	48S40171G01
3124	48X90232A05
3124	48X97046A22
3124	48X97046A23
3124	48X97177A12
3124	58-1
3124	70N1M
3124	86-157-2
3124	86X0036-001
3124	90-56
3124	141-430
3124	142N3P
3124	260P0770
3124	430
3124	433
3124	750A858-319
3124	767
3124	1006
3124	1202
3124	2443
3124	2447
3124	2475
3124	2546
3124	3507
3124	3532
3124	3555
3124	3556
3124	3572
3124	8000-0004-P086
3124	8102-207
3124	8102-208
3124	15810-1
3124	28810-174
3124	000073370
3124	171026
3124	171040
3124	607030
3124	2003239-65
3124	43024878
3124	4100206443
3124	4104213421
3124	4109208280
3124/289	A-1379
3124/289	A-1380
3124/289	A-1567
3124/289	A-158B
3124/289	A-158C
3124/289	A-168
3124/289	A104A
3124/289	A104B
3124/289	A104Y
3124/289	A115
3124/289	A1A
3124/289	A2480
3124/289	AC175
3124/289	AQ4
3124/289	AQ6
3124/289	B-1338
3124/289	B-1910
3124/289	BC-108
3124/289	BC148B
3124/289	BC173
3124/289	BC173A
3124/289	BC337
3124/289	BC338
3124/289	BC414B
3124/289	BC547C
3124/289	BE-66
3124/289	BFY-22
3124/289	BFY-23
3124/289	BFY-23A
3124/289	BFY-24
3124/289	BFY-29
3124/289	BFY-30
3124/289	BFY-39
3124/289	BG-66
3124/289	BG-94
3124/289	BN-66
3124/289	BQ-94
3124/289	BR-66
3124/289	BS-94
3124/289	BSY10
3124/289	BSY11
3124/289	BT-94
3124/289	C1214
3124/289	C1214A
3124/289	C1214B
3124/289	C1214C
3124/289	C1318
3124/289	C1318Q
3124/289	C1342
3124/289	C1342A
3124/289	C1342B
3124/289	C1342C
3124/289	C1675
3124/289	C1675L

SK	Industry Standard No.
3124/289	C1675M
3124/289	C1685
3124/289	C1685P
3124/289	C1685Q
3124/289	C1685S
3124/289	C383
3124/289	C383G
3124/289	C383Y
3124/289	C394
3124/289	C394-O
3124/289	C394-O
3124/289	C394GR
3124/289	C394R
3124/289	C394W
3124/289	C394Y
3124/289	C401(JAPAN)
3124/289	C454
3124/289	C454(A)
3124/289	C454A
3124/289	C454B
3124/289	C454C
3124/289	C454D
3124/289	C454L
3124/289	C454LA
3124/289	C458
3124/289	C458(C)
3124/289	C458A
3124/289	C458B
3124/289	C458BC
3124/289	C458BL
3124/289	C458C
3124/289	C458CLG
3124/289	C458CM
3124/289	C458D
3124/289	C458G
3124/289	C458GLB
3124/289	C458L
3124/289	C458LB
3124/289	C458LG
3124/289	C458LG(B)
3124/289	C458LGA
3124/289	C458LGB
3124/289	C458LGBM
3124/289	C458LGC
3124/289	C458LGD
3124/289	C458LGO
3124/289	C458LGS
3124/289	C458P
3124/289	C458RGS
3124/289	C633
3124/289	C633-7
3124/289	C633A
3124/289	C633G
3124/289	C633H
3124/289	C634
3124/289	C634A
3124/289	C693
3124/289	C693(JAPAN)
3124/289	C693A
3124/289	C693B
3124/289	C693C
3124/289	C693D
3124/289	C693E
3124/289	C693EB
3124/289	C693ET
3124/289	C693F
3124/289	C693FC
3124/289	C693FL
3124/289	C693FU
3124/289	C693G
3124/289	C693GL
3124/289	C693GS
3124/289	C693GU
3124/289	C693GZ
3124/289	C693H
3124/289	C815
3124/289	C815(M)
3124/289	C815A
3124/289	C815B
3124/289	C815C
3124/289	C815F
3124/289	C815K
3124/289	C815L
3124/289	C815M
3124/289	C815S
3124/289	C815SA
3124/289	C815SC
3124/289	C871
3124/289	C871BL
3124/289	C871D
3124/289	C871E
3124/289	C871F
3124/289	C871G
3124/289	C923
3124/289	C923A
3124/289	C923B
3124/289	C923C
3124/289	C923D
3124/289	C923E
3124/289	C923F
3124/289	C941
3124/289	C941-O
3124/289	C941-R
3124/289	C941-Y
3124/289	C945
3124/289	C945(R)
3124/289	C945A
3124/289	C945AP
3124/289	C945AQ
3124/289	C945B
3124/289	C945C
3124/289	C945D
3124/289	C945E
3124/289	C945F
3124/289	C945G
3124/289	C945H
3124/289	C945K
3124/289	C945L
3124/289	C945M
3124/289	C945P
3124/289	C945Q
3124/289	C945QL
3124/289	C945QP
3124/289	C945R
3124/289	C945S
3124/289	C945T
3124/289	C945TQ
3124/289	C945TR
3124/289	C945X
3124/289	CCS2004D303
3124/289	CS-9125B
3124/289	CS1453F
3124/289	CS2001
3124/289	CS2004

SK	Industry Standard No.
3124/289	CS9013
3124/289	CS9013A
3124/289	CS9013B
3124/289	CS9013C
3124/289	CS9013D
3124/289	CS9013E
3124/289	CS9013F
3124/289	CS9013G
3124/289	CS9013H
3124/289	CS9013HE
3124/289	CS9013HF
3124/289	CS9013HG
3124/289	CS9013HH
3124/289	CS9126
3124/289	D16G6
3124/289	D9634
3124/289	DBCZ073304
3124/289	DDBY224003
3124/289	DDBY224004
3124/289	DDBY224006
3124/289	DDBY233001
3124/289	DDBY259002
3124/289	DDBY273001
3124/289	DDBY283001
3124/289	DDBY287001
3124/289	DS-47
3124/289	DS-94
3124/289	EA1549
3124/289	EA15X237
3124/289	EA15X264
3124/289	EA15X353
3124/289	EA15X354
3124/289	EA15X374
3124/289	EA15X378
3124/289	EA15X404
3124/289	EA15X5
3124/289	EA15X7245
3124/289	EA15X7583
3124/289	EA1698
3124/289	EA2771
3124/289	EA4025
3124/289	EL403
3124/289	EP16X7
3124/289	EQS-0061
3124/289	EQS-20
3124/289	ES10231
3124/289	ES15049
3124/289	ET234854
3124/289	ET329218
3124/289	ET398711
3124/289	ETTC-945
3124/289	EW165
3124/289	EWS-78
3124/289	F121-433804
3124/289	G05-035-E
3124/289	G05-035E
3124/289	G05-036D
3124/289	G05-066A
3124/289	G05-706-D
3124/289	G05-706-E
3124/289	GE-210
3124/289	GE-47
3124/289	GV5760
3124/289	GV6063
3124/289	H9696
3124/289	HC-00536
3124/289	HC-00900
3124/289	HC-00923
3124/289	HC-00945
3124/289	HC-56
3124/289	HR-15
3124/289	HR-47
3124/289	HR11
3124/289	HR36
3124/289	HR47
3124/289	HT306442A
3124/289	HT306442B
3124/289	HT30733100
3124/289	HT309002A0
3124/289	HT309451R0
3124/289	HT316751M0
3124/289	HT36441B
3124/289	HX-50063
3124/289	HX-50072
3124/289	HX-50105
3124/289	HX-50107
3124/289	IP20-0032
3124/289	IP20-0040
3124/289	IP20-0191
3124/289	IRTR86
3124/289	KD2102
3124/289	KGE46146
3124/289	KSC945Y
3124/289	LM1540C
3124/289	M4838
3124/289	MPS3694
3124/289	MPS9433J
3124/289	MPS9616J
3124/289	MPS9630
3124/289	MPS9630H
3124/289	MPS9630I
3124/289	MPS9630J
3124/289	MPS9630K
3124/289	MPS9630T
3124/289	MPS9633
3124/289	MPS9633D
3124/289	MPS9696F
3124/289	MPS9696H
3124/289	MS7506H
3124/289	MSG7506
3124/289	MSK5405
3124/289	N3563
3124/289	NB013
3124/289	NJ102C
3124/289	NR-071AU
3124/289	NTC-7
3124/289	ON271
3124/289	ON274
3124/289	PTC123
3124/289	Q-00469A
3124/289	Q-00469B
3124/289	Q1
3124/289	Q5053D
3124/289	Q5053E
3124/289	Q5053F
3124/289	Q5053G
3124/289	Q51210
3124/289	Q5121R
3124/289	Q5180
3124/289	Q5183
3124/289	Q5183P
3124/289	QOV60529
3124/289	QQC61210
3124/289	QSC509

SK	Industry Standard No.
3124/289	QT-C0900XBA
3124/289	QT-C0900XCA
3124/289	QT-C0945ACA
3124/289	QT-C0945AGA
3124/289	QT-C1318XAN
3124/289	R-2SC858
3124/289	RE13
3124/289	RS-279US
3124/289	RS-5851
3124/289	RS-5853
3124/289	RS-5856
3124/289	RS-5857
3124/289	RS-7103
3124/289	RS-7105
3124/289	RT-929H
3124/289	RT2309
3124/289	RT4760
3124/289	RT5435
3124/289	RT5905
3124/289	RT5906
3124/289	RT5907
3124/289	RT6921
3124/289	RT7559
3124/289	RT8047
3124/289	RV2070
3124/289	RV2248
3124/289	S-1061
3124/289	S-1065
3124/289	S-1066
3124/289	S-1068
3124/289	S-1128
3124/289	S-1143
3124/289	S-1221
3124/289	S-1245
3124/289	S-1363
3124/289	S-1364
3124/289	S-1403
3124/289	S-1559
3124/289	S2002
3124/289	S21
3124/289	S27604
3124/289	S4002
3124/289	SC-4044
3124/289	SC-65
3124/289	SC-832
3124/289	SE-1331
3124/289	SE-2001
3124/289	SE-4001
3124/289	SE-4002
3124/289	SE-6001
3124/289	SE5025
3124/289	ST-2SC383W
3124/289	T-Q5093
3124/289	T1340A31
3124/289	TIS60
3124/289	TIS60A
3124/289	TIS60B
3124/289	TIS60C
3124/289	TIS60D
3124/289	TIS60E
3124/289	TIS60M
3124/289	TM289
3124/289	TN2SC945-Q
3124/289	TN2SC945-R
3124/289	TNJ70691
3124/289	TNJ71034
3124/289	TNJ71271
3124/289	TNJ71277
3124/289	TR0573486
3124/289	TR0573491
3124/289	TR2320063
3124/289	TR2327443
3124/289	TR2327444
3124/289	TR2S1570LH
3124/289	TSC767
3124/289	TV-7
3124/289	VS2SC1675M-1
3124/289	VS2SC454-B/1E
3124/289	VS2SC454-C/1E
3124/289	VS2SC454-C/3A
3124/289	VS2SC458
3124/289	VS2SC458-C/1E
3124/289	VS2SC945LP-1
3124/289	001-021210
3124/289	001-021232
3124/289	01-030454
3124/289	01-030458
3124/289	01-030509
3124/289	01-030900
3124/289	01-030945
3124/289	01-031364
3124/289	01-031675
3124/289	1-035/2207
3124/289	01-572814
3124/289	01-680815
3124/289	1A0045
3124/289	002-012000
3124/289	2N1104
3124/289	2N1149
3124/289	2N1151
3124/289	2N1152
3124/289	2N1153
3124/289	2N117
3124/289	2N118
3124/289	2N118A
3124/289	2N119
3124/289	2N120
3124/289	2N1201
3124/289	2N1205
3124/289	2N1247
3124/289	2N1248
3124/289	2N1249
3124/289	2N1267
3124/289	2N1276
3124/289	2N1277
3124/289	2N1278
3124/289	2N1279
3124/289	2N1386
3124/289	2N1387
3124/289	2N1388
3124/289	2N1389
3124/289	2N148
3124/289	2N148A
3124/289	2N1586
3124/289	2N1587
3124/289	2N1589
3124/289	2N1590
3124/289	2N1592
3124/289	2N1593
3124/289	2N160
3124/289	2N160A
3124/289	2N161
3124/289	2N161A
3124/289	2N163
3124/289	2N163A
3124/289	2N1682
3124/289	2N1763
3124/289	2N2198
3124/289	2N2218
3124/289	2N2218A
3124/289	2N2234
3124/289	2N2235
3124/289	2N2330
3124/289	2N2389
3124/289	2N2390
3124/289	2N2396
3124/289	2N2397
3124/289	2N3077
3124/289	2N3078
3124/289	2N3117
3124/289	2N3246
3124/289	2N3268
3124/289	2N332
3124/289	2N332A
3124/289	2N333
3124/289	2N333A
3124/289	2N334
3124/289	2N334A
3124/289	2N335
3124/289	2N335A
3124/289	2N335B
3124/289	2N336
3124/289	2N336A
3124/289	2N338
3124/289	2N338A
3124/289	2N3402
3124/289	2N3403
3124/289	2N3404
3124/289	2N3405
3124/289	2N3721
3124/289	2N3900
3124/289	2N3901
3124/289	2N431
3124/289	2N432
3124/289	2N433
3124/289	2N4425
3124/289	2N470
3124/289	2N471A
3124/289	2N473
3124/289	2N476
3124/289	2N477
3124/289	2N478
3124/289	2N479
3124/289	2N479A
3124/289	2N480
3124/289	2N5126
3124/289	2N5132
3124/289	2N5135
3124/289	2N5136
3124/289	2N5172
3124/289	2N5183
3124/289	2N542A
3124/289	2N546
3124/289	2N547
3124/289	2N548
3124/289	2N549
3124/289	2N550
3124/289	2N551
3124/289	2N552
3124/289	2N696
3124/289	2N701
3124/289	2N707A
3124/289	2N715
3124/289	2N728
3124/289	2N729
3124/289	2N742
3124/289	2N742A
3124/289	2N756A
3124/289	2N757A
3124/289	2N758A
3124/289	2N758B
3124/289	2N759A
3124/289	2N759B
3124/289	2N760A
3124/289	2N760B
3124/289	2N770
3124/289	2N771
3124/289	2N772
3124/289	2N773
3124/289	2N774
3124/289	2N775
3124/289	2N776
3124/289	2N777
3124/289	2N778
3124/289	2N789
3124/289	2N790
3124/289	2N791
3124/289	2N792
3124/289	2N793
3124/289	2N866
3124/289	2N867
3124/289	2S001
3124/289	2S002
3124/289	2S003
3124/289	2S004
3124/289	2S005
3124/289	2S701
3124/289	2S702
3124/289	2S703
3124/289	2S711
3124/289	2S712
3124/289	2SA203P
3124/289	2SC1204
3124/289	2SC1204B
3124/289	2SC1204C
3124/289	2SC1204D
3124/289	002SC1209C
3124/289	2SC1214
3124/289	2SC1214A
3124/289	2SC1214B
3124/289	2SC1214C
3124/289	2SC1214D
3124/289	2SC1317
3124/289	2SC1317B
3124/289	2SC1317D
3124/289	2SC1317P
3124/289	2SC1317Q
3124/289	2SC1317R
3124/289	2SC1317S
3124/289	2SC1318
3124/289	2SC1318A
3124/289	2SC1318P
3124/289	2SC1318Q
3124/289	2SC1318R
3124/289	2SC1318S
3124/289	2SC1318X
3124/289	2SC1342
3124/289	2SC1342A
3124/289	2SC1342B
3124/289	2SC1342C
3124/289	2SC1361
3124/289	2SC1362
3124/289	2SC1363
3124/289	2SC1364
3124/289	2SC1364-6
3124/289	2SC1364A
3124/289	2SC157
3124/289	2SC1570
3124/289	2SC1570LH
3124/289	2SC1571
3124/289	2SC1571G
3124/289	2SC158
3124/289	2SC1632
3124/289	2SC1633
3124/289	2SC1634
3124/289	2SC1675
3124/289	2SC1675K
3124/289	2SC1675L
3124/289	2SC1675M
3124/289	2SC1682
3124/289	2SC1682V
3124/289	2SC1684
3124/289	2SC1684BL
3124/289	2SC1684P
3124/289	2SC1684Q
3124/289	2SC1684R
3124/289	2SC1684S
3124/289	2SC1684T
3124/289	2SC1685
3124/289	2SC1685P
3124/289	2SC1685Q
3124/289	2SC1685R
3124/289	2SC1685S
3124/289	2SC1741
3124/289	2SC1788
3124/289	2SC1788R
3124/289	2SC182Q
3124/289	2SC1959Y
3124/289	2SC2000
3124/289	2SC2000L
3124/289	2SC2012
3124/289	2SC2120
3124/289	2SC2120Y
3124/289	2SC2308
3124/289	2SC28
3124/289	2SC299
3124/289	2SC362
3124/289	2SC367G
3124/289	2SC367R
3124/289	2SC367Y
3124/289	2SC383
3124/289	2SC383F
3124/289	2SC383G
3124/289	2SC383W
3124/289	2SC383Y
3124/289	2SC394
3124/289	2SC394-O
3124/289	2SC394-O
3124/289	2SC394GR
3124/289	2SC394R
3124/289	2SC394W
3124/289	2SC394X
3124/289	2SC394Y
3124/289	2SC401
3124/289	2SC401A
3124/289	2SC403
3124/289	2SC403(C)
3124/289	2SC403A
3124/289	2SC403C
3124/289	2SC441
3124/289	2SC442
3124/289	2SC454
3124/289	2SC454(A)
3124/289	2SC454A
3124/289	2SC454B
3124/289	2SC454C
3124/289	2SC454D
3124/289	2SC454L
3124/289	2SC454LA
3124/289	2SC458
3124/289	2SC458(C)
3124/289	2SC458A
3124/289	2SC458AD
3124/289	2SC458B
3124/289	2SC458BC
3124/289	2SC458BL
3124/289	2SC458C
3124/289	2SC458CLG
3124/289	2SC458CM
3124/289	2SC458D
3124/289	2SC458G
3124/289	2SC458GLB
3124/289	2SC458K
3124/289	2SC458L
3124/289	2SC458LB
3124/289	2SC458LC
3124/289	2SC458LG
3124/289	2SC458LG(B)
3124/289	2SC458LGB
3124/289	2SC458LGC
3124/289	2SC458LGD
3124/289	2SC458LGO
3124/289	2SC458M
3124/289	2SC458P
3124/289	2SC48
3124/289	2SC509
3124/289	2SC509G
3124/289	2SC509Y
3124/289	2SC537F
3124/289	2SC631
3124/289	2SC631A
3124/289	2SC632
3124/289	2SC632A
3124/289	2SC633
3124/289	2SC633-7
3124/289	2SC633A
3124/289	2SC633G
3124/289	2SC633H
3124/289	2SC634
3124/289	2SC634A
3124/289	2SC648
3124/289	2SC648H
3124/289	2SC650A
3124/289	2SC693
3124/289	2SC693E
3124/289	2SC693EB
3124/289	2SC693ET
3124/289	2SC693F
3124/289	2SC693FC
3124/289	2SC693FL
3124/289	2SC693FU
3124/289	2SC693G
3124/289	2SC693GL
3124/289	2SC693GS
3124/289	2SC693GU
3124/289	2SC693GZ
3124/289	2SC693H
3124/289	2SC694
3124/289	2SC694E
3124/289	2SC696A
3124/289	2SC715
3124/289	2SC715A
3124/289	2SC715B
3124/289	2SC715C
3124/289	2SC715D
3124/289	2SC715E
3124/289	2SC715EJ
3124/289	2SC715EV
3124/289	2SC715F
3124/289	2SC715XL
3124/289	2SC814
3124/289	2SC815
3124/289	2SC815(M)
3124/289	2SC815A
3124/289	2SC815B
3124/289	2SC815C
3124/289	2SC815F
3124/289	2SC815K
3124/289	2SC815L
3124/289	2SC815M
3124/289	2SC815S
3124/289	2SC815SA
3124/289	2SC815SC
3124/289	2SC833BL
3124/289	2SC839
3124/289	2SC839(H)
3124/289	2SC839(J)
3124/289	2SC839(L)
3124/289	2SC839(M)
3124/289	2SC839A
3124/289	2SC839B
3124/289	2SC839C
3124/289	2SC839D
3124/289	2SC839E
3124/289	2SC839F
3124/289	2SC839H
3124/289	2SC839J
3124/289	2SC839JI
3124/289	2SC839K
3124/289	2SC839L
3124/289	2SC839M
3124/289	2SC839N
3124/289	2SC839R
3124/289	2SC839S
3124/289	2SC839Y
3124/289	2SC859
3124/289	2SC859E
3124/289	2SC859F
3124/289	2SC859FG
3124/289	2SC859G
3124/289	2SC859GK
3124/289	2SC870
3124/289	00002SC870A
3124/289	00002SC870B
3124/289	2SC870BL
3124/289	2SC870C
3124/289	2SC870D
3124/289	2SC870E
3124/289	2SC870F
3124/289	2SC871
3124/289	2SC871BL
3124/289	2SC871D
3124/289	2SC871E
3124/289	2SC871F
3124/289	2SC871G
3124/289	2SC904
3124/289	2SC923
3124/289	2SC923A
3124/289	2SC923B
3124/289	2SC923C
3124/289	2SC923D
3124/289	2SC923E
3124/289	2SC923F
3124/289	2SC941
3124/289	2SC941-O
3124/289	2SC941-R
3124/289	2SC941-Y
3124/289	2SC941R
3124/289	2SC941Y
3124/289	2SC944
3124/289	2SC944K
3124/289	00002SC945
3124/289	2SC945(R)
3124/289	2SC945A
3124/289	2SC945AK
3124/289	2SC945AP
3124/289	2SC945AQ
3124/289	2SC945AR
3124/289	2SC945B
3124/289	2SC945C
3124/289	2SC945CK
3124/289	2SC945D
3124/289	2SC945E
3124/289	2SC945F
3124/289	2SC945G
3124/289	2SC945K
3124/289	2SC945L
3124/289	2SC945LP
3124/289	2SC945LQ
3124/289	2SC945M
3124/289	2SC945O
3124/289	2SC945P
3124/289	2SC945Q
3124/289	2SC945QL
3124/289	2SC945QP
3124/289	2SC945R
3124/289	2SC945S
3124/289	2SC945T
3124/289	2SC945TK
3124/289	2SC945TP
3124/289	2SC945TQ
3124/289	2SC945TR
3124/289	2SC945X
3124/289	2SC945Y
3124/289	2SC968
3124/289	0002SC968P
3124/289	2SCS183E
3124/289	2SCS184E
3124/289	2SD134
3124/289	2SD227
3124/289	2SD227A
3124/289	2SD227B
3124/289	2SD227C
3124/289	2SD227D
3124/289	2SD227E
3124/289	2SD227F
3124/289	2SD227L
3124/289	2SD227R
3124/289	2SD227S
3124/289	2SD227V
3124/289	2SD227W
3124/289	2SD227X
3124/289	2SD571K
3124/289	2SD591R
3124/289	2SG536GN
3124/289	2T402
3124/289	03A02
3124/289	4-12-1A7-1
3124/289	4-1544
3124/289	4-851
3124/289	5-70005452
3124/289	5-70005503
3124/289	5-7000901504
3124/289	07-1458-85
3124/289	8-0024-3
3124/289	8-726-357-10
3124/289	8-729-663-47
3124/289	8-729-665-47
3124/289	09-302014
3124/289	09-302040
3124/289	09-302044
3124/289	09-302085
3124/289	09-302107
3124/289	09-302125
3124/289	09-302194
3124/289	09-302225
3124/289	09-305048
3124/289	09-305052
3124/289	09-305123
3124/289	09-309070
3124/289	10-003
3124/289	12CLN
3124/289	13-14085-41
3124/289	13-14085-49
3124/289	13-14085-54
3124/289	13-15804-1
3124/289	13-15808-2
3124/289	13-18363
3124/289	13-18363-1A
3124/289	13-18365-1
3124/289	13-18927-1
3124/289	13-18927-1A
3124/289	13-
3124/289	13-23324-6
3124/289	13-23327-4
3124/289	13-23338-3
3124/289	13-2
3124/289	13-23825-1
3124/289	13-28
3124/289	14-515-15
3124/289	14-60
3124/289	14-602-29
3124/289	14-602-30
3124/289	14-602-46A
3124/289	16-147191229
3124/289	16-1
3124/289	21-1L
3124/289	21A040-020
3124/289	21A040-066
3124/289	21A040
3124/289	21A112-045
3124/289	21A112-046
3124/289	21A11
3124/289	21M124
3124/289	21M137
3124/289	21M138
3124/289	21M170
3124/289	21M174
3124/289	21M408
3124/289	21M446
3124/289	21M550
3124/289	21M603
3124/289	22-1
3124/289	24MW1025
3124/289	24MW372
3124/289	24MW826
3124/289	24MW964
3124/289	24MW965
3124/289	25CS58LGBM
3124/289	27T409
3124/289	27T411
3124/289	31-0012
3124/289	31-0013
3124/289	31-0099
3124/289	31-0100
3124/289	31-0187
3124/289	31-0239
3124/289	34
3124/289	34-6001-78
3124/289	34-
3124/289	34-6001-85
3124/289	3
3124/289	34-6015-24
3124/289	42-17444
3124/289	42-18111
3124/289	42
3124/289	42-19670
3124/2	
3124/289	42-21407
3124/289	42-22809
3124/289	42-22811
3124/289	42-22812
3124/289	42-22847
3124/289	42-23348
3124/289	42-23349
3124/289	42-23964
3124/289	42-23964P
3124/289	42-23966
3124/289	42-23966P
3124/289	42-27277
3124/289	42
3124/289	43-023221
3124/289	43-025766
3124/289	43N
3124/289	43N6
3124/289	4
3124/289	44T-300-111
3124/289	44T-300-112
3124/28	
3124/289	46-86173-3
3124/289	46-86536-3
3124/289	46
3124/289	46-8682-3
3124/289	48-
3124/289	48-134823
3124/289	48-155046
3124/289	48-
3124/289	48-155154
3124/289	48-3003A02
3124/289	48-

SK	Industry Standard No.
3124/289	48-90343A83
3124/289	48D67120A11
3124/289	48S134919
3124/289	48S137015
3124/289	48S40170G01
3124/289	48X97046A52
3124/289	500371
3124/289	500372
3124/289	500373
3124/289	53P151
3124/289	53P158
3124/289	53P159
3124/289	53P161
3124/289	53P162
3124/289	53P163
3124/289	53P165
3124/289	53P166
3124/289	56-4827
3124/289	56-8196
3124/289	56-8197
3124/289	57C6-17
3124/289	57C6-19
3124/289	57C6-27
3124/289	57C6-29
3124/289	57C6-30
3124/289	57C6-32
3124/289	57C6-7
3124/289	57D6-19
3124/289	61C001-1
3124/289	70N1
3124/289	70N2
3124/289	75N1
3124/289	86-123-2
3124/289	86-161-2
3124/289	86-170-2
3124/289	86-291-2
3124/289	86-308-2
3124/289	86-5065-2
3124/289	86-5099-2
3124/289	86-521-2
3124/289	86-630-2
3124/289	86-649-2
3124/289	86X0007-204
3124/289	86X0012-001
3124/289	86X0022-001
3124/289	86X0025-001
3124/289	86X0034-001
3124/289	86X0062-001
3124/289	86X0090-001
3124/289	86X7-6
3124/289	86X8-3
3124/289	87-0005
3124/289	87-0006
3124/289	87-0009
3124/289	87-0013
3124/289	87-0212-1
3124/289	87-0227
3124/289	87-0238
3124/289	87-0239-A
3124/289	87-218-U
3124/289	90-140
3124/289	90-32
3124/289	90-33
3124/289	90-458
3124/289	90-459
3124/289	90-605
3124/289	90-614
3124/289	90-70
3124/289	90-71
3124/289	96-5190-01
3124/289	99S012
3124/289	99S012A
3124/289	99S020
3124/289	102-0945-16
3124/289	102-0945-17
3124/289	102-0945-38
3124/289	102-0945-39
3124/289	102-1675-11
3124/289	102-1675-12
3124/289	0103-0088/4460
3124/289	0103-0492
3124/289	0103-93
3124/289	107N2
3124/289	112-203391
3124/289	113-118
3124/289	115-1
3124/289	115-4
3124/289	121-1004
3124/289	121-1017
3124/289	121-276
3124/289	121-277
3124/289	121-278
3124/289	121-279
3124/289	121-422
3124/289	121-431
3124/289	121-434
3124/289	121-581
3124/289	121-610
3124/289	121-766-01
3124/289	121-767
3124/289	121-878
3124/289	123-005
3124/289	129-14
3124/289	129-15
3124/289	0140-5
3124/289	148N1
3124/289	176-031-9-002
3124/289	176-042-9-004
3124/289	176-042-9-006
3124/289	176-047-9-001
3124/289	176-048-9-002
3124/289	176-060-9-003
3124/289	176-062-9-001
3124/289	176-065-9-001
3124/289	185-010
3124/289	200-016
3124/289	0221
3124/289	260P17502
3124/289	260P17701
3124/289	260P17704
3124/289	260P19503
3124/289	297V049H01
3124/289	297V049H03
3124/289	297V049H04
3124/289	297V059H02
3124/289	297V059H03
3124/289	297V061C02
3124/289	297V061C03
3124/289	297V061C04
3124/289	297V061C06
3124/289	297V061C07
3124/289	297V074C02
3124/289	297V074C03
3124/289	297V074C04
3124/289	297V074C10
3124/289	310-187
3124/289	324-0151
3124/289	324-0152
3124/289	324-0154
3124/289	325-0031-304
3124/289	325-0031-305
3124/289	325-0500-12
3124/289	325-0500-13
3124/289	325-1446-26
3124/289	325-1446-27
3124/289	325-1446-28
3124/289	325-1513-29
3124/289	325-1513-30
3124/289	325-1771-16
3124/289	344-6001-1
3124/289	344-6001-2
3124/289	344-6011-1
3124/289	344-6013-4
3124/289	417-127
3124/289	433CL
3124/289	511-515
3124/289	511-519
3124/289	515
3124/289	516
3124/289	517-518
3124/289	576-0001-005
3124/289	576-0036-847
3124/289	600-207-801
3124/289	602X0008-002
3124/289	602X0018-002
3124/289	605
3124/289	642-268
3124/289	660-134
3124/289	690V080H41
3124/289	690V110H34
3124/289	690V110H36
3124/289	693GV
3124/289	700-155
3124/289	700A858-319
3124/289	750A858-328
3124/289	753-1644-100
3124/289	753-4000-010
3124/289	753-6000-002
3124/289	753-8400-230
3124/289	753-8500-380
3124/289	753-9001-675
3124/289	800-250-102
3124/289	800-50100
3124/289	800-522-01
3124/289	800-525-03
3124/289	800-53001
3124/289	853-0300-644
3124/289	902-000-8-04
3124/289	902-002-3-06
3124/289	902-003-0-12
3124/289	902-003-3-17
3124/289	903Y002150
3124/289	904-95B
3124/289	919-013044
3124/289	921-1017
3124/289	921-1022
3124/289	921-205B
3124/289	921-207
3124/289	921-207B
3124/289	921-208B
3124/289	921-209B
3124/289	921-26B
3124/289	921-281B
3124/289	921-7
3124/289	929CA
3124/289	929CU
3124/289	991-002298
3124/289	991-002356
3124/289	991-008393
3124/289	991-011305
3124/289	1003-6754
3124/289	1004-0780
3124/289	1007-3088
3124/289	1010-8082
3124/289	1010-8090
3124/289	1030-21
3124/289	1039-0458
3124/289	1040-01
3124/289	1043-1229
3124/289	1048-9920
3124/289	1049-0092
3124/289	1050-21
3124/289	1080-21
3124/289	1081-9464
3124/289	1096-12
3124/289	1368C/D
3124/289	1374-17
3124/289	1374-17A
3124/289	1524
3124/289	2000-203
3124/289	2000-213
3124/289	2035-5100-53660
3124/289	2035-5100-69372
3124/289	2044-17
3124/289	2057A100-53
3124/289	2057A2-122
3124/289	2057A2-156
3124/289	2057A2-192
3124/289	2057A2-212
3124/289	2057A2-251
3124/289	2057A2-262
3124/289	2057A2-284
3124/289	2057A2-285
3124/289	2057A2-290
3124/289	2057A2-333
3124/289	2057A2-334
3124/289	2057A2-352
3124/289	2057A2-428
3124/289	2057A2-560
3124/289	2057A2-87
3124/289	2057B2-192
3124/289	2057B2-28
3124/289	2057B2-57
3124/289	2057B2-59
3124/289	2057B2-63
3124/289	2057B2-73
3124/289	2065-03
3124/289	2065-04
3124/289	2065-54
3124/289	2121-17
3124/289	2132-17
3124/289	2195-17
3124/289	2204-17
3124/289	2207-17
3124/289	2338-17
3124/289	2495
3124/289	2584
3124/289	2904-034
3124/289	2904-035
3124/289	3018
3124/289	3391
3124/289	3391A
3124/289	3514
3124/289	4002
3124/289	4464
3124/289	4465
3124/289	4470
3124/289	4470M-32
3124/289	4714
3124/289	5001-021
3124/289	5001-038
3124/289	5001-043
3124/289	5001-511
3124/289	7991
3124/289	8000-00005-005
3124/289	8000-00006-280
3124/289	8000-00011-049
3124/289	8000-00041-041
3124/289	8000-00041-042
3124/289	8000-00049-053
3124/289	8000-00049-054
3124/289	8200-202
3124/289	8200-203
3124/289	8440-126
3124/289	9617K
3124/289	15809-1
3124/289	17144
3124/289	17444
3124/289	18410-145
3124/289	20810-91
3124/289	20810-92
3124/289	25840-162
3124/289	27127-550
3124/289	30210
3124/289	30219
3124/289	30224
3124/289	30226
3124/289	30248
3124/289	30254
3124/289	30259
3124/289	33563
3124/289	37585
3124/289	38178
3124/289	38510-350
3124/289	40233
3124/289	45810-166
3124/289	51213-02
3124/289	51442
3124/289	51442-01
3124/289	51442-02
3124/289	51442-03
3124/289	55810-161
3124/289	55810-163
3124/289	000073070
3124/289	000073090
3124/289	000073120
3124/289	000073130
3124/289	000073290
3124/289	000073310
3124/289	000073351
3124/289	000073361
3124/289	000073390
3124/289	000073391
3124/289	86812
3124/289	86822
3124/289	88060-144
3124/289	88060-145
3124/289	88510-174
3124/289	95171-1
3124/289	95171-2
3124/289	95171-3
3124/289	95171-4
3124/289	95216
3124/289	95225
3124/289	95226-2
3124/289	99109-1
3124/289	99109-2
3124/289	116087
3124/289	116588
3124/289	121664
3124/289	122520
3124/289	123941
3124/289	126716
3124/289	134989
3124/289	144044
3124/289	145838
3124/289	146512
3124/289	146570
3124/289	168651
3124/289	169505
3124/289	170967
3124/289	171162-163
3124/289	171162-180
3124/289	171557
3124/289	171676
3124/289	171678
3124/289	213217
3124/289	215074
3124/289	215075
3124/289	224506
3124/289	236285
3124/289	236286
3124/289	346016-16
3124/289	506902
3124/289	0573480H
3124/289	0573491
3124/289	0573492
3124/289	609112
3124/289	610094
3124/289	610148-1
3124/289	610226-1
3124/289	651955-1
3124/289	651955-2
3124/289	740306
3124/289	740887
3124/289	741114
3124/289	741856
3124/289	741857
3124/289	801527
3124/289	916055
3124/289	992129
3124/289	1471115-2
3124/289	1473536-2
3124/289	1473560-1
3124/289	1473572-3
3124/289	1473601-001
3124/289	2002153-77
3124/289	2002621-2
3124/289	2003073-10
3124/289	2003073-9
3124/289	2003168-135
3124/289	2003168-136
3124/289	2320471H
3124/289	2320591
3124/289	2320595
3124/289	2320596
3124/289	2320598
3124/289	2320631
3124/289	2485076-2
3124/289	2485076-3
3124/289	2499950
3124/289	05320064
3124/289	5320326
3124/289	5320813
3124/289	5340001
3124/289	06120005
3124/289	8113327
3124/289	13094517
3124/289	23114349
3124/289	26010023
3124/289	43024859
3124/289	43024873
3124/289	43025059
3124/289	62103272
3124/289	62593725
3124/289	62596279
3124/289	62604727
3124/289	83038004
3124/289	93063470
3124/289	93064440
3124/289	93078420
3124/289	632039418
3124/289	2003189025
3124/289	4100208283
3124/289	4104206442
3124/289	4104208282
3124/289	4104208283
3124/289	6624003200
3124/289	16108190536
3124/289	16109190536
3125	.5E12
3125	.5J12
3125	.7E12
3125	.7J12
3125	A04093A
3125	A04093X
3125	A04294
3125	A04294-1
3125	A7568250
3125	A7568300
3125	A7580910
3125	A7978855
3125	CXL506
3125	D172-F
3125	D431-F
3125	DG-1N
3125	DG-1NR
3125	DG1N
3125	DS16NE
3125	EP57X5
3125	ET57X39
3125	ET57X40
3125	FU10
3125	FU1K
3125	H815
3125	HFSD-1B
3125	LM1932
3125	MB-1D
3125	MR-852
3125	MR811
3125	MR812
3125	MR814
3125	MR816
3125	NTC-17
3125	S1-RECT-158
3125	SD-1VHF
3125	SD-ERC26-13
3125	SF-1
3125	SI-RECT-36
3125	SI-RECT-44
3125	SID30-15
3125	SISD-1
3125	SISD-1HF
3125	TV24217
3125	TV24219
3125	TV24648
3125	TVM535
3125	TVM537
3125	TVS-2B-2C
3125	TVS-DG1NR
3125	TVS-FR1P
3125	TVS-FT1N
3125	TVS-FT1P
3125	TVS-FT1PC
3125	WEP186
3125	1N4937
3125	1N5616
3125	1N5617
3125	1N5618
3125	1N5620
3125	1N5621
3125	1N693
3125	1R5TH61
3125	1S1226
3125	1S1920
3125	1S1921A
3125	1S1921B
3125	1S1921C
3125	1S1921D
3125	1S1921E
3125	1S1921F
3125	09-306031
3125	15-085008
3125	15-108008
3125	21A110-006
3125	21A110-013
3125	46-86284-3
3125	48-191A05
3125	53B010-8
3125	66X0036-002
3125	100C
3125	264P03601
3125	264P10201
3125	601X0049-066
3125	690V080H49
3125	690V081H91
3125	800-029-30
3125	1019-1385
3125	0015107
3125	34424
3125	530123-3
3125	530123-5
3125	530123-7
3125	0575066
3125	984260
3125	23115912
3125	62051302
3125	4202017600
3125	4202020500
3125	4202023000
3125	4202023200
3125	6611000900
3125	6611001000
3125	6611001200
3125	6611002300
3125	6613002200
3126	BB104
3126	CF3
3126	E0011
3126	EA16X117
3126	EA16X127
3126	EA16X166
3126	EDS-0042
3126	EP16X13
3126	FB1043
3126	FCD0070ANC
3126	G01-406-A
3126	GE-90
3126	HD4000109
3126	HD4000909
3126	HF-20005
3126	HF-20007
3126	ITT210
3126	MPS5668
3126	MV2203
3126	MV2205
3126	QDCTT410XQ
3126	RE195
3126	RT6731
3126	RV1017
3126	SD39
3126	SMV1172
3126	001-0130-00
3126	1N3182
3126	1S1352
3126	1S1503
3126	1S1533
3126	1S1558
3126	1S553
3126	1S554
3126	1S85
3126	1S853
3126	1S85V
3126	1S86
3126	1SV53
3126	1T240
3126	1T243
3126	3I4-3503-5
3126	004-2
3126	09-306039
3126	09-306210
3126	09-306359
3126	12-085009
3126	13-14278-2
3126	13-33177-1
3126	24MW1067
3126	34-8057-39
3126	48-137487
3126	48A42383A01
3126	53B001-3
3126	56-4835
3126	66X0050-001
3126	86-84-1
3126	93A41-5
3126	100-01110-01
3126	103-189
3126	103-39
3126	151-030-9-004
3126	151-062-9-001
3126	153-008-9-001
3126	754-0102-139
3126	754-9002-687
3126	903-110B
3126	903-141B
3126	903-47B
3126	903-67B
3126	903-96B
3126	903-99B
3126	1006-5977
3126	1048-9938
3126	2000-309
3126	2000-327
3126	5001-196
3126	115099
3126	117729
3126	123276
3126	123726
3126	125399
3126	126149
3126	129359
3126	136164
3126	168652
3126	170370
3126	171814
3126	171984
3126	741052
3126	741868
3126	742732
3126	760204-0001
3126	1223925
3126	17210430
3126	26010031
3129	CA3011
3129	CA3012
3129	GEIC-81
3129	GEIC-83
3130	A04233
3130	A7582000
3130	AD-29S
3130	AD29S
3130	B2620
3130	BB-10S
3130	BB6S
3130	CDG-24
3130	D1D
3130	D8G
3130	D8L
3130	D8M
3130	DIJ72167
3130	DIJ72170
3130	DIJ72171
3130	DIJ72174
3130	DIJ72290
3130	DS-113A
3130	DS-113B
3130	DS113B
3130	DS133B
3130	EP6X11
3130	EPB22-15

SK	Industry Standard No.
3130	ERC27-13
3130	EU16X20
3130	EV16X20
3130	FG2N
3130	FG2PC
3130	0000FR202
3130	HFSD-14
3130	HFSD-1C
3130	RA-Z
3130	RH-DX0017CEZZ
3130	RH-DX0029CEZZ
3130	RH-DX0045CEZZ
3130	RH-DX0051CEZZ
3130	RH-DX0056TAZZ
3130	RH-DX0065CEZZ
3130	RH-DX0096CEZZ
3130	RSLND0003CEZZ
3130	S-15H
3130	S1D23-13
3130	S1D30-13
3130	S1D30-15
3130	S1P
3130	SB-1
3130	SB-2CH
3130	SD-14
3130	SD1-11
3130	SD22
3130	SELEN-92
3130	SI-RECT-114
3130	SI-RECT-162
3130	SI-RECT-206
3130	SI-RECT-224
3130	SIB01-04
3130	SR1FM-12
3130	SR1HM-8
3130	SR24
3130	SR25
3130	TC-0.2P11/1
3130	TSTD-15
3130	TV24941
3130	TVSFG2PC
3130	TVSFR2-02
3130	TVSFR2-02C
3130	TVSFR2-02G
3130	TVSFT1M
3130	TVSHF-1
3130	TVSHF1
3130	TVSHFSD-1Z
3130	TVSHFSD1
3130	TVSS-34
3130	TVSS34
3130	TVSSB2C
3130	TVSTD15M
3130	TVSUFSD-1
3130	03-936011
3130	4-2020-06500
3130	4-2020-07800
3130	4-2020-07801
3130	4-2020-14400
3130	4-2020-14600
3130	4-2020-15100
3130	09-306148
3130	10D4L
3130	48M355013
3130	48S155061
3130	66X36-1
3130	86-117-3
3130	86-118-3
3130	87-104-3
3130	93A40-1
3130	93A52-7
3130	93A60-1
3130	93A60-11
3130	121GI241
3130	150-012-9-001
3130	151-045-9-003
3130	264P03604
3130	264P04705
3130	264P06603
3130	264P09101
3130	264P09501
3130	1038-1788
3130	1051
3130	2093A41-67
3130	2093A41-72
3130	2093A41-73
3130	72137
3130	72161-1
3130	131475
3130	530113-1001
3130	530122-2
3130	530144-1
3130	530148-1004
3130	530148-3
3130	530148-4
3130	1415721-1
3130	1476171-11
3130	1476171-18
3130	1476171-22
3130	1476171-35
3130	2311598
3130	2331121
3130	23115192
3130	23115298
3130	23115377
3130	23115960
3130	23115981
3130	37568200
3130	37582000
3130	62522712
3130	62734728
3130	62761318
3130	62761326
3130	4202007601
3130	4202007700
3130	4202007801
3130	4202015100
3130	4202020900
3130	4202107970
3130	4203008000
3130	44202007800
3131A	C1828
3131A	UPT221
3131A	2N4233A
3131A	2N5660
3131A	2SC1078
3131A	2SC1828
3131A	2SD546
3131A	2SD766
3131A	2SD766Q
3131A	2SD766R
3132	A640S
3132	BFS18CA
3132	BFS27E
3132	BFS27F
3132	BFS27G
3132	BFW30
3132	BFW63
3132	C1035
3132	C1035C
3132	C1035D
3132	C1035E
3132	C1036
3132	C1182B
3132	C1182C
3132	C1182D
3132	C313C
3132	C313H
3132	C348
3132	C375
3132	C375-0
3132	C375-Y
3132	C382
3132	C382-GR
3132	C382-GY
3132	C382BK
3132	C382BL
3132	C382BN
3132	C382BR
3132	C382G
3132	C382R
3132	C382V
3132	C384
3132	C463H
3132	C464
3132	C465
3132	C466
3132	C466H
3132	C563
3132	C563A
3132	C567
3132	C663
3132	C674B
3132	C674C
3132	C674D
3132	C674E
3132	C674F
3132	C674G
3132	C717
3132	C717B
3132	C717BK
3132	C717BLK
3132	C717C
3132	C717E
3132	C772B
3132	C772BG
3132	C772BH
3132	C772BV
3132	C772BX
3132	C772BY
3132	C772C
3132	C772C1
3132	C772C2
3132	C772CK
3132	C772CL
3132	C772CS
3132	C772CU
3132	C772CV
3132	C772CX
3132	C772D
3132	C772DJ
3132	C772DU
3132	C772DV
3132	C772DX
3132	C772DY
3132	C772E
3132	C772F
3132	C772K
3132	C772KB
3132	C772KC
3132	C772KD
3132	C772KD1
3132	C772KD2
3132	C772R
3132	C772RB-D
3132	C772RD
3132	C787
3132	C837K
3132	C837L
3132	C860
3132	C860C
3132	C860D
3132	C860E
3132	C863
3132	C864
3132	C918
3132	C928
3132	C928B
3132	C928C
3132	C928D
3132	C928E
3132	C929-0
3132	C929DE
3132	C929DU
3132	C947
3132	C948
3132	C957
3132	C997
3132	CS1014G
3132	CS1120C
3132	CS1120D
3132	CS1120E
3132	CS1120H
3132	CS1460E
3132	CS1460H
3132	CS1461X
3132	CS1462F
3132	CS1589E
3132	CS1589F
3132	CS1589S
3132	CS1594E
3132	CS1596E
3132	CS9017H
3132	DDBY295002
3132	ES15X122
3132	G05-037-D
3132	G05-037B
3132	GE-327
3132	HEP-S0008
3132	HS40021
3132	HS40022
3132	HS40023
3132	HS40024
3132	HS40025
3132	HT8001610
3132	HT8001710
3132	IRTR70
3132	KSC1674R
3132	MRF502
3132	NKT16229
3132	NR461AA
3132	PL1021
3132	PL1022
3132	PL1023
3132	PL1066
3132	PL1067
3132	PL1111
3132	PL1112
3132	PTG132
3132	Q5078Z
3132	RS-2011
3132	RV1068
3132	SDS420
3132	SE3019
3132	SE5003
3132	TG2SC1293-A-A
3132	TG2SC1293-B-A
3132	TG2SC1293-C-A
3132	TG2SC1293-D-A
3132	TNJ72276
3132	TV-35
3132	TV-36
3132	TV15
3132	TV24436
3132	TVS-2SC948
3132	WEP1717
3132	WEP709
3132	WEP719
3132	WEP784
3132	XS36
3132	XS37
3132	XS38
3132	XS39
3132	01-031293
3132	01-031360
3132	01-571804
3132	2SC1047
3132	2SC1047A
3132	2SC1047B
3132	2SC1047BCD
3132	2SC1047C
3132	2SC1047D
3132	2SC1047R
3132	2SC1293
3132	2SC1293A
3132	2SC1293B
3132	2SC1360
3132	2SC1674
3132	2SC1674K
3132	2SC1674L
3132	2SC1674M
3132	2SC1856
3132	2SC1923
3132	2SC1923BN
3132	2SC375
3132	2SC375-0
3132	2SC375-Y
3132	2SC375A
3132	2SC375B
3132	2SC375C
3132	2SC375D
3132	2SC375E
3132	2SC375F
3132	2SC375G
3132	2SC375GN
3132	2SC375H
3132	2SC375J
3132	2SC375K
3132	2SC375L
3132	2SC375M
3132	2SC3750R
3132	2SC375R
3132	2SC375X
3132	2SC382
3132	2SC382-V
3132	2SC382A
3132	2SC382B
3132	2SC382BK
3132	2SC382BL
3132	2SC382BN
3132	2SC382BR
3132	2SC382C
3132	2SC382D
3132	2SC382E
3132	2SC382F
3132	2SC382G
3132	2SC382GN
3132	2SC382GR
3132	2SC382H
3132	2SC382J
3132	2SC382K
3132	2SC382L
3132	2SC382M
3132	2SC3820R
3132	2SC382R
3132	2SC382V
3132	2SC382W
3132	2SC382X
3132	2SC382Y
3132	2SC384
3132	2SC384-0
3132	2SC384A
3132	2SC384B
3132	2SC384C
3132	2SC384D
3132	2SC384E
3132	2SC384F
3132	2SC384G
3132	2SC384GN
3132	2SC384H
3132	2SC384J
3132	2SC384K
3132	2SC384L
3132	2SC384M
3132	2SC3840R
3132	2SC384R
3132	2SC384X
3132	2SC384Y
3132	2SC388
3132	2SC388A
3132	2SC388ATV
3132	2SC388B
3132	2SC388C
3132	2SC388D
3132	2SC388E
3132	2SC388F
3132	2SC388G
3132	2SC388GN
3132	2SC388H
3132	2SC388J
3132	2SC388K
3132	2SC388L
3132	2SC388M
3132	2SC3880R
3132	2SC388R
3132	2SC388X
3132	2SC388Y
3132	2SC674E
3132	2SC717
3132	2SC717B
3132	2SC717BK
3132	2SC717BLK
3132	2SC717C
3132	2SC717E
3132	2SC717F
3132	2SC717H
3132	2SC717K
3132	2SC717L
3132	2SC717M
3132	00002SC772
3132	2SC772A
3132	2SC772B
3132	2SC772BG
3132	2SC772BH
3132	2SC772BX
3132	2SC772BY
3132	0002SC772C
3132	2SC772C1
3132	2SC772C2
3132	2SC772CK
3132	2SC772CL
3132	2SC772CS
3132	2SC772CU
3132	2SC772CV
3132	2SC772CX
3132	2SC772D
3132	2SC772DJ
3132	2SC772DU
3132	2SC772DX
3132	2SC772DY
3132	2SC772E
3132	2SC772F
3132	2SC772G
3132	2SC772GN
3132	2SC772H
3132	2SC772J
3132	2SC772K
3132	2SC772KB
3132	2SC772KD
3132	2SC772KD1
3132	2SC772KD2
3132	2SC772L
3132	2SC772M
3132	2SC7720R
3132	2SC772R
3132	2SC772RB-D
3132	2SC772RD
3132	2SC772X
3132	2SC772Y
3132	2SC773
3132	2SC773A
3132	2SC773B
3132	2SC773C
3132	2SC773D
3132	2SC773E
3132	2SC773F
3132	2SC773J
3132	2SC773K
3132	2SC773L
3132	2SC773M
3132	2SC773R
3132	2SC773Y
3132	2SC864
3132	2SC929
3132	2SC929A
3132	2SC929B
3132	2SC929C
3132	2SC929C1
3132	2SC929D
3132	2SC929D1
3132	2SC929DE
3132	2SC929DP
3132	2SC929DU
3132	2SC929DV
3132	2SC929E
3132	2SC929ED
3132	2SC929F
3132	2SC929FK
3132	2SC929G
3132	2SC929GN
3132	2SC929H
3132	2SC929J
3132	2SC929K
3132	2SC929L
3132	2SC929M
3132	2SC9290R
3132	2SC929R
3132	2SC929X
3132	8-905-605-644
3132	8-905-706-044
3132	8-905-706-055
3132	8-905-706-060
3132	8-905-706-070
3132	8-905-706-071
3132	8-905-706-075
3132	8-905-706-080
3132	8-905-706-101
3132	8-905-706-110
3132	09-302129
3132	14-654-21
3132	23-PT274-120
3132	23-PT274-123
3132	23-PT283-122
3132	23-PT283-124
3132	24T021
3132	34-6000-72
3132	41N2M
3132	46-86531-3
3132	48-134904A1G
3132	48-90343A75
3132	57A138-4
3132	57A142-4
3132	57A160-8
3132	57B249-4
3132	057B474H
3132	87-0002-1
3132	96-5131-01
3132	96-5163-01
3132	96-5174-01
3132	96-5175-01
3132	96-5198-01
3132	96-5199-01
3132	96-5235-01
3132	96-5236-01
3132	96-5259-01
3132	96-5260-01
3132	99S044
3132	121-283
3132	121-383
3132	121-823
3132	121-824
3132	130-240
3132	132-008
3132	132-009
3132	132-076
3132	132-082
3132	132-087
3132	132-185
3132	180N1
3132	180N1P
3132	186N1
3132	260P11102
3132	260P17702
3132	576-1
3132	1076-1377
3132	1801
3132	1845-17
3132	2057A2-526
3132	3537
3132	3576
3132	38787
3132	131543
3132	131544
3132	137388
3132	610180-1
3132	2495520
3132	2495521
3132	2495522-1
3132	2495523-1
3132	3939307-002
3132	06120015
3132	23114164
3132	62506350
3132	62748281
3133	C1101
3133	C1101A
3133	C1101B
3133	C1101C
3133	C1101D
3133	C1101E
3133	C1151
3133	C1184
3133	C1184A
3133	C1184B
3133	C1184C
3133	C1184D
3133	C1184E
3133	C642
3133	C936
3133	IRTR68
3133	PTC130
3133	RE31
3133	TNJ72146
3133	2SC1004
3133	2SC1151
3133	2SC1151A
3133	2SC1171
3133	2SC642
3133	2SC936
3133	2SC936A
3133	2SC936BK
3133	09-302157
3133	21A112-023
3133	121-821
3133	123N1
3133	260P09501
3133	260P19108
3133	260P24308
3133	610123-1
3133/164	A6L
3133/164	A8U
3133/164	C1034
3133/164	TV121
3133/164	2SD353
3133/164	48S137344
3134	C6089
3134	GEIC-3
3134	HEP-C6089
3134	IC502
3134	PTC708
3134	RE302-IC
3134	SL20721
3134	TVCM-1
3134	14-2007-00
3134	14-2007-02
3134	14-2007-04
3134	221-36
3134	221-37
3134	221-39
3134	442-10
3135	DM-31
3135	DM11
3135	DM31
3135	IC505
3135	PTC-703
3135	RE301-IC
3135	SN76642N
3135	TVCM-5
3135	TVSSN76642N
3135	15-34452
3135	15-34452-1
3135	19A116445
3135	051-0022-00
3135	34452-1
3135	916112
3135	7311325
3135	7932980
3135/709	C6062P
3135/709	C6082P
3135/709	CA2111AE
3135/709	CA2111AQ
3135/709	CA2111E
3135/709	GEIC-10
3135/709	GEL2111
3135/709	GEL2111AL1
3135/709	GEL2111F1
3135/709	HEP-C6062P
3135/709	HEP-C6082P
3135/709	LM2111
3135/709	LM2111N
3135/709	MC1357
3135/709	MC1357P
3135/709	MC1357PQ
3135/709	N5111
3135/709	PTC701
3135/709	RE311-IC
3135/709	SN76643N
3135/709	TA6220
3135/709	TVCM-4
3135/709	ULN2111A

SK	Industry Standard No.
3135/709	ULN2111N
3135/709	07B1Z
3135/709	07B2B
3135/709	07B3B
3135/709	07B5C
3135/709	07B5D
3135/709	07B5M
3135/709	07B5Z
3135/709	14-2008-01
3135/709	15-34048-1
3135/709	15-34048-2
3135/709	19A11644P1
3135/709	221-34
3135/709	442-28
3135/709	740-2002-111
3135/709	2007-2
3135/709	2007-3
3135/709	612007-1
3135/709	612007-2
3135/709	612007-3
3136	.7Z8.2
3136	.7Z8.2A
3136	.7Z8.2B
3136	.7Z8.2C
3136	.7Z8.2D
3136	.7ZM8.2A
3136	.7ZM8.2B
3136	.7ZM8.2C
3136	.7ZM8.2D
3136	A7Z-86-7QQ
3136	AZ-081
3136	AZ-083
3136	AZ8.2
3136	AZ959B
3136	B-180B
3136	B1-8.2
3136	BZ-080
3136	BZ081
3136	BZ8.2
3136	BZX29C8V2
3136	BZX46C8V2
3136	BZX55C8V2
3136	BZX85C8V2
3136	BZY62
3136	BZY85C8V2
3136	BZY92C8V2
3136	CO2
3136	D6M
3136	DB-149
3136	DZ-081
3136	DZ0820
3136	EC95072
3136	EC95072A
3136	ED8-24
3136	EQA01-08
3136	EQB-0108
3136	EQB01-08
3136	GB-X11
3136	GEZD-8.2
3136	HBR-Z0411
3136	KZ-8A
3136	MA1082
3136	MZ-08
3136	MZ-208
3136	MZ8
3136	OA2206
3136	PD6053
3136	PS8911
3136	QA01-082
3136	QRT-257
3136	RD-2.2M
3136	RD8
3136	RD8.2C
3136	RD8.2EB
3136	RD8.2BC
3136	RD8.2F
3136	RE112
3136	REN5072
3136	RF6.2E
3136	RZ8.2
3136	SZ200-8.2
3136	TM5072
3136	TVSYZ08Q
3136	UZ8708
3136	UZ8808
3136	WE081
3136	WM-079
3136	WM-079A
3136	WM-079B
3136	WM-079C
3136	WM-079D
3136	WM-081
3136	WM-081A
3136	WM-081C
3136	WM-081D
3136	WM-083
3136	WM-083A
3136	WM-083C
3136	WM-083D
3136	WM-085
3136	WM-085A
3136	WM-085C
3136	WM-085D
3136	WN-079
3136	WN-079A
3136	WN-079B
3136	WN-079C
3136	WN-079D
3136	WN-081
3136	WN-081A
3136	WN-081B
3136	WN-081C
3136	WN-081D
3136	WN-083
3136	WN-083A
3136	WN-083B
3136	WN-083C
3136	WN-085
3136	WN-085A
3136	WN-085B
3136	WN-085C
3136	WN-085D
3136	WO-079
3136	WO-079A
3136	WO-079B
3136	WO-079C
3136	WO-079D
3136	WO-081
3136	WO-081A
3136	WO-081B
3136	WO-081C
3136	WO-083
3136	WO-083A
3136	WO-083C
3136	WO-083D
3136	WO-085
3136	WO-085A
3136	WO-085C
3136	WP-079
3136	WP-079A
3136	WP-079B
3136	WP-079C
3136	WP-079D
3136	WP-081
3136	WP-081A
3136	WP-081C
3136	WP-081D
3136	WP-083
3136	WP-083A
3136	WP-083C
3136	WP-083D
3136	WP-085
3136	WP-085A
3136	WP-085C
3136	WP-085D
3136	WQ-079
3136	WQ-079A
3136	WQ-079C
3136	WQ-079D
3136	WQ-081
3136	WQ-081A
3136	WQ-081C
3136	WQ-081D
3136	WQ-083
3136	WQ-083A
3136	WQ-083C
3136	WQ-083D
3136	WQ-085
3136	WQ-085A
3136	WQ-085C
3136	WQ-085D
3136	WR-079
3136	WR-079A
3136	WR-079C
3136	WR-079D
3136	WR-081
3136	WR-081A
3136	WR-081C
3136	WR-081D
3136	WR-083
3136	WR-083A
3136	WR-083C
3136	WR-083D
3136	WR-085
3136	WR-085A
3136	WR-085C
3136	WS-079
3136	WS-079A
3136	WS-079C
3136	WS-081
3136	WS-081A
3136	WS-081C
3136	WS-081D
3136	WS-083
3136	WS-083A
3136	WS-083B
3136	WS-083C
3136	WS-083D
3136	WS-085
3136	WS-085A
3136	WS-085C
3136	WS-085D
3136	WT-079
3136	WT-079A
3136	WT-079B
3136	WT-079C
3136	WT-081
3136	WT-081A
3136	WT-081C
3136	WT-081D
3136	WT-083
3136	WT-083A
3136	WT-083B
3136	WT-083C
3136	WT-083D
3136	WT-085
3136	WT-085A
3136	WT-085B
3136	WT-085C
3136	WT-085D
3136	WU-079
3136	WU-079A
3136	WU-079B
3136	WU-079C
3136	WU-081
3136	WU-081A
3136	WU-081C
3136	WU-083
3136	WU-083A
3136	WU-083B
3136	WU-083C
3136	WU-083D
3136	WU-085
3136	WU-085A
3136	WU-085B
3136	WU-085C
3136	WU-085D
3136	WV-079
3136	WV-079A
3136	WV-079B
3136	WV-079C
3136	WV-081
3136	WV-081A
3136	WV-081B
3136	WV-081C
3136	WV-083
3136	WV-083A
3136	WV-083B
3136	WV-083C
3136	WV-083D
3136	WV-085
3136	WV-085A
3136	WV-085B
3136	WV-085C
3136	WW-079
3136	WW-079A
3136	WW-079B
3136	WW-079C
3136	WW-081
3136	WW-081A
3136	WW-081B
3136	WW-081C
3136	WW-083
3136	WW-083A
3136	WW-083B
3136	WW-083C
3136	WW-083D
3136	WW-085
3136	WW-085A
3136	WW-085B
3136	WW-085C
3136	WW-085D
3136	WX-079
3136	WX-079A
3136	WX-079B
3136	WX-079C
3136	WX-079D
3136	WX-081
3136	WX-081A
3136	WX-081B
3136	WX-081C
3136	WX-083
3136	WX-083A
3136	WX-083C
3136	WX-083D
3136	WX-085
3136	WX-085A
3136	WX-085C
3136	WX-085D
3136	WY-079
3136	WY-079A
3136	WY-079B
3136	WY-079C
3136	WY-081
3136	WY-081A
3136	WY-081B
3136	WY-081C
3136	WY-083
3136	WY-085
3136	WY-085A
3136	WY-085B
3136	WY-085C
3136	WY-085D
3136	WZ081
3136	XZ-082
3136	YAAD021
3136	YEAD010
3136	YEAD014
3136	YEAD024
3136	YZ-080
3136	Z0217
3136	Z0411
3136	Z101
3136	Z110B
3136	Z1A8.2A
3136	Z1A8.2B
3136	Z2AB2F
3136	Z4AB.2
3136	Z8.2A
3136	Z8.2C
3136	Z8.2D
3136	ZA8.2
3136	ZA8.2A
3136	ZA8.2B
3136	ZF8.2
3136	ZF8.2F
3136	ZP8A
3136	ZPD8.2
3136	ZZ8.2
3136	001-0152-00
3136	001-0161-00
3136	001-0165-13
3136	1N1425
3136	1N1511
3136	1N1511A
3136	1N1522
3136	1N1522A
3136	1N1769
3136	1N1769A
3136	1N1875
3136	1N1875A
3136	1N3018
3136	1N3018A
3136	1N3018B
3136	1N3516
3136	1N3677
3136	1N3677A
3136	1N3677B
3136	1N4101
3136	1N4160
3136	1N4160A
3136	1N4160B
3136	1N4325
3136	1N4325A
3136	1N4325B
3136	1N4402A
3136	1N4630
3136	1N4659
3136	1N4694
3136	1N4758
3136	1N4758A
3136	1N5561
3136	1N5561A
3136	1N664
3136	1N756
3136	1N756A
3136	1R8.2
3136	1R8.2A
3136	1R8.2B
3136	1S2115
3136	1S2115A
3136	1S2190
3136	1S2191
3136	1S2192
3136	1S3008
3136	1S333
3136	1S333Y
3136	1Z8.2
3136	1Z8.2T1Q
3136	1Z8.2T2Q
3136	1Z8.2T5
3136	1Z08.2T1Q
3136	1Z08.2T5
3136	1ZD825
3136	02Z-8.2A
3136	02Z8.2A
3136	3L4-3506-7
3136	05-04800-02
3136	05-540082
3136	0528.2
3136	6DM
3136	8-9V
3136	09-306196
3136	09-306258
3136	09-306378
3136	09-306381
3136	10D08
3136	21Z6AF
3136	21Z6F
3136	46-86442-3
3136	46-86483-3
3136	48-03073A06
3136	48-137421
3136	48-40458A06
3136	48-41765C1
3136	86-114-1
3136	100-219
3136	123-019
3136	123B-004
3136	152-042-9-001
3136	152-054-9-001
3136	754-5710-219
3136	754-5750-282
3136	754-9000-082
3136	1042-19
3136	2064-17
3136	2343-17
3136	5001-130
3136	25810-54
3136	25840-54
3136	28810-65
3136	88060-54
3137	741869
3137	23115277
3137	23115374
3137	26010044
3137	A3Z2484
3137	A543
3137	BC226
3137	BC510
3137	BC584
3137	BCW51
3137	BCW73-16
3137	BCW74-16
3137	BCW77-16
3137	BCW78-16
3137	BCW82
3137	BCW82A
3137	BCW82B
3137	BCW83A
3137	BCW83B
3137	BCW84
3137	BCW90K
3137	BCW91K
3137	BF829P
3137	BF830
3137	BF830P
3137	BF860
3137	BFX53
3137	BS825
3137	BSY62B
3137	C1330
3137	C1330A
3137	C1330B
3137	C1330D
3137	C1330D
3137	C1330L
3137	C1346
3137	C1346R
3137	C1547
3137	C1407
3137	C814
3137	C853
3137	C853A
3137	C853B
3137	C853C
3137	C881
3137	C881K
3137	C881L
3137	C971
3137	CDC-8000-1D
3137	CGE-63
3137	CS2023
3137	CXL192
3137	EA15X4531
3137	EA15X7519
3137	GE-88
3137	HEP-80002
3137	HEP-80003
3137	HS5810
3137	HS5812
3137	HS5814
3137	HS5818
3137	HS5820
3137	HS5822
3137	HS6014
3137	HS6016
3137	HT404001E
3137	NR-141ES
3137	PET1001
3137	RE196
3137	REN192
3137	2SC1330
3137	2SC1330L
3137	2SC1330R
3137	2SC1346
3137	2SC1347
3137	2SC1347A
3137	2SC1347B
3137	2SC1347F
3137	2SC1347L
3137	2SC1347R
3137	2SC1347X
3137	2SC1347Y
3137	2SC1406
3137	2SC1406Q
3137	2SC1407
3137	2SC4583LG
3137	2SC881
3137	2SC881A
3137	2SC881B
3137	2SC881C
3137	2SC881D
3137	2SC881F
3137	2SC881G
3137	2SC881GN
3137	2SC881H
3137	2SC881K
3137	2SC881L
3137	2SC881M
3137	2SC881OR
3137	2SC881R
3137	2SC881X
3137	2SC881Y
3137	000028D261
3137	2SD261-0
3137	2SD261A
3137	2SD261B
3137	2SD261C
3137	2SD261D
3137	2SD261E
3137	2SD261F
3137	2SD261G
3137	2SD261GN
3137	2SD261H
3137	2SD261J
3137	2SD261K
3137	2SD261L
3137	2SD261M
3137	2SD261OR
3137	2SD261P
3137	2SD261Q
3137	2SD261R
3137	2SD261S
3137	2SD261U
3137	2SD261V
3137	2SD261W
3137	2SD261X
3137	2SD261Y
3137	4-1792
3137	09-302116
3137	09-303019
3137	09-303519
3137	13-28592-2
3137	21M192
3137	21M193
3137	21M606
3137	488155123
3137	065-008
3137	121-703
3137	8000-00012-040
3137	45810-165
3137	000073300
3137	000073301
3137	000073302
3137	000073303
3137	62727128
3138	A3Z2907A
3138	A545
3138	A545GRN
3138	A545KLM
3138	A545L
3138	A547A
3138	A643
3138	A643A
3138	A643B
3138	A643C
3138	A643D
3138	A643E
3138	A643F
3138	A643R
3138	A643V
3138	A750
3138	BC257A
3138	BC257B
3138	BC258A
3138	BC258B
3138	BC259A
3138	BC259B
3138	BC370
3138	BC406
3138	BC512B
3138	BC513B
3138	BC514
3138	BC514B
3138	BCW52
3138	BCW61BA
3138	BCW61BB
3138	BCW62B
3138	BCW63B
3138	BCW64
3138	BCW64B
3138	BCW70
3138	BCW70R
3138	BCW75-10
3138	BCW75-16
3138	BCW76-10
3138	BCW76-16
3138	BCW92
3138	BCW92K
3138	BCW93
3138	BCW93K
3138	BCX71BG
3138	BCX71BH
3138	BSF44
3138	BSF45
3138	BSV43B
3138	BSV44B
3138	BSV45B
3138	BSV47B
3138	BSV48B
3138	BSV49B

SK	Industry Standard No.	SK	Industry Standard No.	SK	Industry Standard No.	SK	Industry Standard No.	SK	Industry Standard No.
3138	CGE-67	3139	WW-110B	3142	WV-162D	3151	1M24Z10	3156	86-515-2
3138	CS2082	3139	WW-110C	3142	WW-162	3151	1M24Z5	3156	86-541-2
3138	CXL193	3139	WW-110D	3142	WW-162A	3151	1M24Z810	3156	86-549-2
3138	EP15X13	3139	WY-110	3142	WW-162B	3151	1M24Z85	3156	86X0089-001
3138	G3-89	3139	WY-110A	3142	WW-162C	3151	1N5029B	3156	113N1AG
3138	HA-00643	3139	WY-110B	3142	WW-162D	3151	1N4749	3156	113N2
3138	H85811	3139	WY-110C	3142	WX-162	3151	1N4749A	3156	121-752
3138	H85813	3139	WY-110D	3142	WX-162A	3151	1Z24	3156	132-052A
3138	H85815	3139	WZ-110	3142	WX-162B	3151	1Z24A	3156	229-1301-27
3138	H85817	3139	WZ-110A	3142	WX-162C	3151	1Z24T10	3156	417-161
3138	H85819	3139	Z0414	3142	WX-162D	3151	1Z24Z20	3156	417-222
3138	H85821	3139	ZB11A	3142	WY-162	3151	1Z24Z5	3156	576-0007-001
3138	H85823	3139	ZB11B	3142	WY-162A	3151	1ZC24T10	3156	921-01129
3138	H86011	3139	1N3021B	3142	WY-162B	3151	1ZC24T5	3156	921-403
3138	H86013	3139	1N4741	3142	WY-162C	3151	05-933945	3156	924-2209
3138	H86015	3139	1N4741A	3142	WY-162D	3151	8-719-906-24	3156	964-22009
3138	H86017	3139	1S2118	3142	1ZC16Z10	3151	46-86529-3	3156	991-01-0461
3138	QT-A0733XON	3139	1Z11	3142	130762	3151	66X0040-007	3156	991-01-3543
3138	RE197	3139	05-060110	3143	530073-1015	3151	070-049	3156	991-015663
3138	REN193	3139	09-306428	3143	CA3060	3151	86-113-1	3156	2122-5
3138	001-02117-2	3139	09-306429	3144	C6101P	3151	86-91-1	3156	3367
3138	28A545	3139	48-155048	3144	CA3075	3151	93A39-45	3156	22009
3138	28A545A	3139	48-155081	3144	CA3075D	3151	103-105	3156	72054-1
3138	28A545B	3139	48M355062	3144	CA3075B	3151	103-105-01	3156	95296
3138	28A545C	3139	48B155048	3144	GEIC-15	3151	103-278	3156	130040
3138	28A545D	3139	48B155081	3144	HEP-C6101P	3151	103-289	3156	139618
3138	28A545E	3139	48B155084	3144	LM3075N	3151	121-289	3156	530113-1
3138	28A545F	3139	93A39-44	3144	REN723	3151	130-105	3156	610113-1
3138	28A545G	3139	1059-9140	3144	SPC6052	3151	1040-9373	3156	610113-2
3138	28A5450N	3139	922950	3144	TVCM-16	3151	721G8-2	3156	760012
3138	28A545H	3139	925233	3144	51-10594A01	3151	99201-325	3156	801533
3138	28A545J	3139	4202012501	3144	51810594A01	3151	142738	3156	1471136-3
3138	28A545K	3139	4202110870	3144	88-9842F	3151	530073-24	3156	1473604-1
3138	28A545KLM	3139	4202116670	3144	88-9842R	3151	23115328	3156	4082671-0002
3138	28A545L	3139	6615000200	3144	221-90	3151	6240311	3156	62256885
3138	28A545M	3140	AE900	3144	2665-2	3152	GB-720	3157	C555A
3138	28A5450R	3140	AE902	3144	5815	3152	PC-20006	3157	FU5D770331X
3138	28A545R	3140	CA3043	3144	9692-1	3152	STK-011	3157	FU5D770339
3138	28A545X	3140	GEIC-227	3144	619692-1	3152	24MW997	3157	HC1000109
3138	28A545Y	3140	3L4-9002-01	3144	3596354	3152	16740135809	3157	HO1000109-0
3138	28A643	3140	3L4-9002-1	3144	4082665-0001	3153	GB-721	3157	HO1000109-0
3138	28A643A	3140	115C7	3144	4082665-0003	3153	PC-20005	3157	HC10001I1-0
3138	28A643B	3140	86313	3144	4082665-1	3153	STK-015	3157	IC101-109
3138	28A643C	3141	C60690	3144	4082665-2	3153	415200050	3157	ICP-1
3138	28A643D	3141	CA3064	3144	4082665-3	3153	GE-722	3157	ICP-1-6826
3138	28A643E	3141	CA3064/5A	3144/723	2008-1	3154	STK-020	3157	IP20-0174
3138	28A643F	3141	CA3064T	3144/723	2008-2	3154	STK-020B	3157	L103T1
3138	28A643V	3141	FU5B7064393	3144/723	612008-1	3155	STK-020D	3157	LA703E
3138	28A730	3141	GEIC-20	3144/723	612008-2	3155	STK-020B	3157	LM703L
3138	21A112-004	3141	GEIC-222	3145	AZ-177	3155	STK-020B	3157	LM703LM
3138	21M027	3141	HEP-06069G	3145	BZ-177	3155	STK-020F	3157	N4703N
3138	21M459	3141	HEP-06069P	3146	CA3088E	3155	STK-020K	3157	NJM703
3138	1072	3141	ITT3064	3146	GEIC-228	3155	STK-020L	3157	PA7703
3138	1072G	3141	LM3064	3146	R4437-1	3156	A9C	3157	PA7703E
3139	.7Z11A	3141	LM3064H	3146	R4437-2	3156	BA1003	3157	PA7703X
3139	.7Z11B	3141	R3677-1	3146	RT-246	3156	C1280	3157	QA703B
3139	.7Z11C	3141	R3677-2	3146	3130-3193-512	3156	C1280A	3157	RE300-IC
3139	.7Z11D	3141	R3677-3	3146	4437-1	3156	C1280A8	3157	SL7059
3139	.7ZM1	3141	13-30-6	3146	4437-2	3156	C12808	3157	SL7283
3139	.7ZM11A	3141	46-5002-16	3146	4437-3	3156	C1472K	3157	SL7308
3139	.7ZM11B	3141	097	3146	9144-60	3156	C982	3157	SL7331
3139	.7ZM11C	3141	097A	3146	9693-1	3156	CS-3024	3157	SL7593
3139	AWO1-11	3141	442-5	3146	136145	3156	D16P2	3157	SL8020
3139	AZ-110	3141	3064T	3146	619695-1	3156	D26C4	3157	T1B
3139	AZ11	3141	3677-2	3146	1464437-1	3156	D26C5	3157	TRO-9005
3139	BZ110	3141	130150	3146	1464437-2	3156	BA15X247	3157	TRO-9006
3139	ECG5074	3141	137245	3146	1464437-3	3156	BA15X266	3157	1E703E
3139	ECG5074A	3141	1462577	3146/787	7935181	3156	EL401	3157	01F
3139	EQA01-11	3141	1463677-1	3147	CA3089E	3156	GE-64	3157	01F-8L8020
3139	GEZD-11	3141	1463677-2	3147	DR-51	3156	GET5305	3157	4-007
3139	HEP-Z0414	3141	1463677-3	3147	GEIC-229	3156	GET5306	3157	4-008
3139	NTC-20	3141/780	ES717375	3147	MC1389P	3156	GET5307	3157	09-308004
3139	QAO1-11SE	3141/780	3677-1	3147	R4438-1	3156	GET5308	3157	09-308013
3139	QRM-242	3142	.7JZ16	3147	R4438-2	3156	GET5308A	3157	09-308019
3139	RD11F	3142	AZ-162	3147	REN788	3156	HEP-89100	3157	13-076002
3139	RD11FB	3142	ED2-11	3147	TRO9018	3156	IRTR69	3157	13-1-6
3139	RZ116	3142	EQA01-16	3147	TVCM-39	3156	M433	3157	13-10-6
3139	REN5074	3142	RD16E	3147	51-10658A01	3156	MA10	3157	13-11-6
3139	RT-242	3142	RH-EXOO65CEZZ	3147	221-108	3156	MP8-A12	3157	13-9-6
3139	SZ-AW01-11	3142	WL-162D	3147	811	3156	MS7504	3157	13-26587-1
3139	SZ-RD11FB	3142	WM-162	3147	4438-1	3156	PEP95	3157	19-020-079
3139	TVSQA01-11SE	3142	WM-162A	3147	4438-2	3156	PTC153	3157	020-114-007
3139	WM-110	3142	WM-162B	3147	4438-3	3156	RE33	3157	020-114-008
3139	WM-110A	3142	WM-162C	3147	4154011370	3156	89100	3157	20B2M
3139	WM-110B	3142	WM-162	3148	BZ-260	3156	TE2713	3157	2083AH
3139	WM-110C	3142	WN-162A	3149	CA3121E	3156	TE2714	3157	022A
3139	WM-110D	3142	WN-162B	3149	GEIC-231	3156	TN061688	3157	46-5002-4
3139	WN-110	3142	WN-162C	3149	GEIC-233	3156	TV62	3157	46-5002-7
3139	WN-110A	3142	WN-162D	3149	IC-325(ELCOM)	3156	2N2723	3157	51-10302A01
3139	WN-110B	3142	WO-162	3149	RE324-IC	3156	2N2724	3157	51810302A01
3139	WN-110C	3142	WO-162A	3149	REN791	3156	2N5305	3157	56A1-1
3139	WN-110D	3142	WO-162C	3149	TVCM-34	3156	2N5306	3157	56C1-1
3139	WO-110	3142	WO-162D	3149	02-561171	3156	2N5307	3157	56D1-1
3139	WO-110A	3142	WP-162	3149	09-308099	3156	2N5308	3157	96-5238-01
3139	WO-110B	3142	WP-162A	3149	15-39209-1	3156	2N5525	3157	96-5258-02
3139	WO-110D	3142	WP-162B	3149	56A17-1	3156	2N997	3157	998022
3139	WP-110	3142	WP-162D	3149	56D17-1	3156	2N998	3157	998022-1
3139	WP-110A	3142	WQ-162	3149	221-69	3156	2N999	3157	221-31
3139	WP-110B	3142	WQ-162A	3149/791	LM1845	3156	2SC1280	3157	221-32
3139	WP-110C	3142	WQ-162C	3150	.7JZ30	3156	2SC1280A	3157	307-001-9-001
3139	WP-110D	3142	WQ-162D	3151	.7JZ24	3156	2SC1280A8	3157	442-80
3139	WQ-110	3142	WR-162	3151	.7Z24A	3156	2SC1280B	3157	442-8
3139	WQ-110A	3142	WR-162A	3151	.7Z24B	3156	2SC1472K	3157	880-101-00
3139	WQ-110B	3142	WR-162B	3151	.7Z24D	3156	280982	3157	1000-25
3139	WQ-110D	3142	WR-162D	3151	.7ZM24A	3156	8A1002	3157	2020-1
3139	WR-110	3142	WR-167	3151	.7ZM24B	3156	8A1003	3157	2020-2
3139	WR-110A	3142	WR-167A	3151	.7ZM24C	3156	09-309029	3157	2020-3
3139	WR-110B	3142	WR-167C	3151	.7ZM24D	3156	11-0423	3157	09005
3139	WR-110D	3142	WR-167D	3151	A72B8504	3156	11-0775	3157	09006
3139	WS-110	3142	WS-162	3151	AW01-24	3156	13-14085-14	3157	26587
3139	WS-110A	3142	WS-162A	3151	AZ970B	3156	13-29775-1	3157	26587-1
3139	WS-110B	3142	WS-162C	3151	BZ-230	3156	13-33175-1	3157	036001
3139	WS-110D	3142	WS-162D	3151	BZ-240	3156	14-0104-5	3157	55810-167
3139	WT-110	3142	WT-162	3151	ED51191B	3156	14-2000-01	3157	612020-1
3139	WT-110A	3142	WT-162A	3151	EP16X16	3156	14-2000-02	3157	612020-2
3139	WT-110B	3142	WT-162B	3151	EQA01-24RR	3156	14-2000-03	3157	612020-3
3139	WT-110D	3142	WT-162C	3151	GEZD-24	3156	14-2000-04	3157	09308004
3139	WU-110	3142	WU-162	3151	HEP-Z0423	3156	14-2000-05	3157	62383976
3139	WU-110A	3142	WU-162A	3151	HW24	3156	14-2000-23	3158	CA3126EM
3139	WU-110B	3142	WU-162B	3151	HW24A	3156	14-2053-23	3158	CA3126EM1
3139	WV-110	3142	WU-162C	3151	RD-24A	3156	16B670-GRN	3158	CA3126Q
3139	WV-110A	3142	WV-162	3151	RD24A	3156	16P2881	3158	GEIC-233
3139	WV-110B	3142	WV-162B	3151	RD24AL	3156	16P3367	3158	R5158-1
3139	WV-110D			3151	RD24AM	3156	23-5044	3158	RE333-IC
3139	WW-110A			3151	RD24AN	3156	34-6017-3	3158	ULN2262A
				3151	RD24F	3156	34-601703	3158	51-13753A19
				3151	RD24FA	3156	42-22009	3158	787PC
				3151	RD24FB	3156	42-24263	3158	1063-5019
				3151	RE128	3156	48-137392	3158	5158
				3151	TB14001-1	3156	052	3158	138699(RCA)
				3151	TVSQA01-25RA	3156	052A	3158	141279
				3151	1M24Z	3156	53-1173	3158	141574-5
						3156	55-641	3158	1465158-1
						3156	66F039-1	3158	1465158-2
						3156	66F039-2		

SK	Industry Standard No.
3159	C1-1004
3159	C6074P
3159	DM-14
3159	GEIC-8
3159	HC1000117-0
3159	HEP-594
3159	HEP-C6074P
3159	HEP-C6094P
3159	IC-9(PHILCO)
3159	IC511
3159	KLH5489
3159	LM1304
3159	LM1304N
3159	MC1304P
3159	MC1304PQ
3159	PTC709
3159	RE319-IC
3159	SC5177P
3159	SC5199P
3159	SC9314P
3159	SL22756
3159	SN76104
3159	SN76104N
3159	T1J
3159	T1N
3159	T2F
3159	TIJ
3159	ULN2120A
3159	09-004
3159	13-26-6
3159	020-1114-006
3159	21-B1
3159	21B1M
3159	21B1Z
3159	21B2Z
3159	46-5002-9
3159	51-10382A
3159	51-10422A
3159	51-10422A01
3159	51-10422A02
3159	51-10437A01
3159	51-10617A01
3159	51S10382A
3159	51S10382A01
3159	51S10422A
3159	51S10422A01
3159	51S10437A01
3159	51S10617A01
3159	63-15345
3159	181-000200
3159	732DC
3159	732PC
3159	2021-1
3159	2021-2
3159	09004
3159	45380
3159	183013
3159	612021-1
3159	612021-2
3159	7312294
3159/718	T1E
3159/718	T2J
3159/718	15-71420-1
3160	C6096P
3160	CA1310E
3160	ETI-22
3160	GEIC-155
3160	GEIC-35
3160	HA1156
3160	HA1156-6C
3160	HA1156W
3160	HC10004010
3160	HEP-C6096P
3160	LM1310
3160	LM1310N
3160	MC1310
3160	MC1310A
3160	MC1310P
3160	RE334-IC
3160	REN801
3160	SN76115
3160	SN76115N
3160	TR09027
3160	TR2367171
3160	ULN2210A
3160	14-2012-01
3160	15-40183-1
3160	051-0050-00
3160	051-0050-01
3160	51-40464P01
3160	51-42908J01
3160	56-4834
3160	57A32-32
3160	61K001-10
3160	75B1Z
3160	87-0234
3160	221-91
3160	221-91-01
3160	280-0002
3160	442-46
3160	2057A32-33
3160	2075-1
3160	003515
3160	5652-HA1156
3160	5723P
3160	5724P
3160	26810-157
3160	29810-180
3160	38510-170
3160	45810-168
3160	612075-1
3160	916113
3160	4154011560
3160	4157013102
3160	4159761157
3160	5147013102
3160/801	TA7157P
3161	C6056P
3161	ES4053
3161	EX4035
3161	EX4053
3161	F767PC
3161	GEIC-9
3161	HEP-C6056P
3161	IC513
3161	KLH4793
3161	LM1307
3161	LM1307N
3161	MA767
3161	MC1307P
3161	PC-20024
3161	PC-20030
3161	RE309-IC
3161	SC5740PQ
3161	SC5743P
3161	SC9426P
3161	SL21864
3161	SN75110N
3161	SN7610N-07
3161	SN76110
3161	SN76110N
3161	T1Z
3161	T2C
3161	TVCM-10
3161	ULN2128A
3161	06B1M
3161	06B1Z
3161	006B2M
3161	06B2Z
3161	09-017
3161	10-47674-01
3161	13-52-6
3161	13-60-6
3161	15-34049
3161	15-34049-1
3161	15-34049-3
3161	15-34503
3161	15-34503-1
3161	15-34503-2
3161	15-71420-1
3161	21M588
3161	51-10382A01
3161	51-10559A01
3161	51-10566A01
3161	51-1056FA01
3161	51-10592A01
3161	51S10559A01
3161	51S10566A01
3161	51S1056FA01
3161	51S1056GA01
3161	51S10592A01
3161	053-1
3161	57A32-10
3161	99S053
3161	99S053-1
3161	0205
3161	221-65
3161	221-79
3161	221-79-01
3161	431-26551A
3161	442-16
3161	740-2001-307
3161	767DC
3161	767PC
3161	905-38B
3161	905-46B
3161	1043-1344
3161	2006-1
3161	2006-2
3161	6551
3161	34049-1
3161	34503-1
3161	45381
3161	61566
3161	95291
3161	95294
3161	183014
3161	612006-1
3161	612006-2
3161	760522-0002
3161	995053-1
3161	5351042
3161	43126551
3161	43126551-A
3162	C6055L
3162	GEIC-19
3162	HEP-C6055L
3162	IC1303P
3162	LM1303
3162	LM1303N
3162	MC1303
3162	MC1303L
3162	MC1303P
3162	SC2914P
3162	SC5116L
3162	SL22108
3162	SN76131
3162	TBA231
3162	08B1M
3162	08B2M
3162	020-1114-003
3162	020-1114-005
3162	133-004
3162	442-41
3162	739DC
3162	739PC
3162/725	SN76131N
3163	GEIC-17
3163	REN736
3163	221-89
3163	753TC
3163/736	DM-19
3163/736	LM3053
3164	LM723
3164	MC1723CG
3164	MC1723G
3164	N5723L
3164	SG723CT
3164	TBA281
3164	TBA281/723
3164	TBA281/723C
3164	723HC
3164/923	LM723C
3164/923	LM723CN
3165	ECG923D
3165	FU9A7723393
3165	GEIC-260
3165	IC-20
3165	LM723CD
3165	LM723CH
3165	MC1723CL
3165	N5723A
3165	SG723CD
3165	46-5002-20
3165	221-Z9020
3165	723DC
3165	723PC
3166	GEIC-25
3166	749
3166	749DHC
3166	138681
3167	A1368
3167	C6075P
3167	CA1398E
3167	EP84X12
3167	GEIC-29
3167	HEP-C6075P
3167	IC-297(ELCOM)
3167	LA1368
3167	M5190
3167	M5190P
3167	MC1398
3167	MC1398P
3167	MC1398PQ
3167	MC1938
3167	RE313-IC
3167	REN738
3167	SN76298N
3167	TA6405
3167	TVCM-27
3167	ULN2298A
3167	13-56-6
3167	15-39075-1
3167	15-43703-1
3167	44B1
3167	44B1B
3167	44B1Z
3167	46-13124-3
3167	51-70177A03
3167	51D70177A03
3167	51G10679A03
3167	51M70177A03
3167	51S10655A03
3167	51S10655A03A
3167	51S10655B03
3167	71-70177A03
3167	71D70177A03
3167	71M70177A03
3167	177A03
3167	229-1301-44
3167	266P60502
3167	1010-9932
3167	1351
3167	1352
3167	2044-1
3167	10655A03
3167	10655A03A
3167	10655B03
3167	39075-1
3167	70177A03
3167	612044-1
3167	4206009200
3167	4206009700
3168	AN238
3168	C6076P
3168	CA1352E
3168	GEIC-97
3168	HEP-C6076P
3168	M5183
3168	M5183-8098
3168	M5183P
3168	MC1352P
3168	MC1352PQ
3168	RE304-IC
3168	RH-1X0004CEZZ
3168	RH-IX0004CEZZ
3168	SC9431P
3168	SI-MC1352P
3168	TA7074P
3168	TA7074PGL
3168	09-308090
3168	21A120-001
3168	46-1365-3
3168	46-5002-28
3168	51-70177A05
3168	51M70177A05
3168	51S10655A05
3168	71-70177A05
3168	73C180838-1
3168	73C180838-3
3168	177A05
3168	266P10101
3168	266P10103
3168	70177A05
3168	23119019
3168	40306400
3168	4206105170
3168/749	EP84X10
3168/749	GEIC-175
3168/749	MC1375P
3168/749	TVSMPC595C
3169	CA3076
3169	GEIC-223
3169	KLH4794
3169	SL23649
3169	SN76676L
3169	T2M
3169	05-111011
3169	19-130-001
3169	51-10637A01
3169	51S10637A01
3169	2802-1
3169	2802-2
3169	3076HC
3169	3076PC
3169	4794(KLH)
3169	5757
3169	80755
3169	4082802-0001
3169	4082802-0002
3169	4082802-1
3169	4082802-2
3170	C6085P
3170	CA3120E
3170	GEIC-13
3170	IC-33
3170	TVCM-15
3170	ULN2125A
3170	14-2054-01
3170	24B1AH
3170	24B1B
3170	24B1Z
3170	46-5002-33
3170	221-45
3170	221-45-01
3170	2024-1
3170	81177
3170	612024-1
3170/731	001-026060
3170/731	23-PT274-125
3170/731	23-PT283-125
3170/731	2403
3170/731	0518926
3170/731	2002402
3171	CA3123E
3171	DM-32
3171	DM32
3171	ECG744
3171	GEIC-215
3171	GEIC-24
3171	PTC734
3171	RE362-IC
3171	REN744
3171	TM744
3171	UA720
3171	ULN2137A
3171	15-36995-1
3171	88-9302
3171	221-107
3171/744	AN238S
3171/744	HA1139
3171/744	SN76650
3171/744	SN76650N
3171/744	TVSUPC595C
3171/744	51-90305A04
3171/744	62736283
3172	AE904
3172	CA758E
3172	DM-44
3172	GEIC-214
3172	GEIC-32
3172	LM1800N
3172	MC1311P
3172	RE317-IC
3172	SN76116N
3172	ULN2244A
3172	3L4-9004-4
3172/743	HEP-C6085P
3173	A663
3173	BDX18
3173	CXL219
3173	ECG219
3173	HEP-248
3173	HEP-705
3173	HEP-S7003
3173	MJ2955
3173	NKT4055
3173	RE82
3173	REN219
3173	REN61
3173	SJ2001
3173	SJ3520
3173	SJ3679
3173	SJ821
3173	TR2327841
3173	TR29
3173	2N3790
3173	2N3791
3173	2N6246
3173	2N6247
3173	2N6469
3173	2N6594
3173	2SA663
3173	34-1029
3173	0137
3173	760021
3173/219	A648
3173/219	A648A
3173/219	A648B
3173/219	A648C
3173/219	RE61
3173/219	2N4904
3173/219	2N4905
3173/219	2N5871
3173/219	2N5879
3173/219	2SA657
3173/219	2SA658
3173/219	2SA680
3173/219	2SA680R
3173/219	2SA680Y
3173/219	2SA756
3173/219	2SA756A
3173/219	2SA756B
3173/219	2SA756C
3173/219	2SA808
3173/219	2SB531
3173/219	2SB558
3173/219	119722
3173/219	1473514-1
3175	A7246602
3175	AM620
3175	AM620A
3175	AM626
3175	AM632
3175	AM632A
3175	BAW12TF22
3175	BAW13TF23
3175	BAW16
3175	BAW17
3175	BAW18
3175	BAY18
3175	BAY19
3175	BAY20
3175	CXL177
3175	D-00184R
3175	D5G
3175	D5H
3175	DDAY090001
3175	DS-442
3175	DS-97
3175	DS410R
3175	DS97
3175	DX-0270
3175	DX-0543
3175	EP16X10
3175	EP16X22
3175	ES16X27
3175	EU16X11
3175	HEP-R0604
3175	HEP-R0606
3175	N5406
3175	PTC214
3175	RE52
3175	REN177
3175	SD-1S1555
3175	SR1K
3175	VHD1S1553//-1
3175	VHD1S1555//1A
3175	001-0095-00
3175	001-0125-00
3175	001-0151-00
3175	001-0151-01
3175	001-026030
3175	01-1501
3175	01-2405-0
3175	01-2601
3175	01-2601-0
3175	01-2603-0
3175	1N5433
3175	1N5434
3175	1S1210
3175	00001S1555
3175	1S1587
3175	1S2076
3175	1S2076A
3175	1S2144Z
3175	000001S990
3175	1S990A
3175	1S990S
3175	1S990S(DIODE)
3175	04-8054-3
3175	04-8054-7
3175	05-180953
3175	05-181555
3175	05-931642
3175	05-931645
3175	05-936470
3175	8-719-815-55
3175	8-905-406-020
3175	09-306170
3175	09-306171
3175	09-307039
3175	11-0430
3175	11-0781
3175	12-087004
3175	019-002964
3175	19B200249-P1
3175	19B200249-P2
3175	19P1
3175	20A13
3175	22-1-044
3175	24E-006
3175	35-1014
3175	42-24387
3175	046-0134
3175	46-86428-3
3175	48-155077
3175	48-67120A13
3175	48-6720A02
3175	48-90420A03
3175	48-90420A04
3175	051-0020
3175	070-022
3175	070-047
3175	86-104-3
3175	86-105-3
3175	93A64-2
3175	93A64-5
3175	93A64-7
3175	93C60-5
3175	93C60-9
3175	96-5254-01
3175	103-142-01
3175	149-142-01
3175	269M01201
3175	0500
3175	903-311
3175	903-332
3175	2000-125
3175	2304
3175	2405
3175	3008
3175	004763
3175	6171-28
3175	8020-202
3175	000072130
3175	109474
3175	125528
3175	126321
3175	128474
3175	133390
3175	135380
3175	135386
3175	138173
3175	138196
3175	142569
3175	530170-1
3175	530179-1001
3175	530181-1001
3175	741864
3175	1223930
3175	1471872-006
3175	1471872-17
3175	1471872-18
3175	1471872-4
3175	1471872-7
3175	1471872-8
3175	1476171-19
3175	1476171-28
3175	1476171-31
3175	1476171-32
3175	600000150
3176	MM1681
3176	PT8769
3176	PT8837
3176	SD1014
3176	SD1069
3176	2N4127
3176	2N6081
3176	40973
3176/350	SD1218
3176/350	2N5590
3177	2N4128
3177	2N5591
3177	40974
3177/351	GE-288
3177/351	40955
3178A	B1C
3178A	KSD415R
3178A	2N6551
3178A	2904-057
3178A	610228-3
3179A	P4V
3179A	TA7556
3179A	TA7557
3179A	TR2327393
3180	D44E1
3180	D44E2
3180	D44E3
3180	2N6386
3180	2N6387
3180	2N6388
3180	2N6530
3180/263	MJE1100
3180/263	MJE1101
3180/263	MJE1102
3180/263	MJE1103
3180/263	MJE2100
3180/263	MJE2101
3180/263	MJE2103
3180/263	2N6037
3180/263	2SD678
3180/263	2SD679
3181A	D45E1
3181A	D45E2
3181A	D45E3
3181A	2N6666
3181A	2N6667
3181A	2N6668
3181A	2SD699

SK	Industry Standard No.
3182	MJ1000
3182	MJ1001
3182	MJ4000
3182	RCA1000
3182	RCA1001
3182	2N6055
3182	2N6056
3182	2N6492
3182	2N6493
3182	2N6494
3182	2SD692
3182/243	MJ3000
3182/243	MJ3001
3182/243	2N6294
3182/243	2N6295
3182/243	2N6300
3182/243	2N6301
3182/243	2N6383
3182/243	2N6384
3182/243	2N6385
3182/243	2SD582
3182/243	2SD582A
3182/243	46-86551-3
3183A	MJ2500
3183A	MJ4010
3183A	MJ4011
3183A	RCA8350
3183A	RCA8350A
3183A	RCA8350B
3183A	2N6053
3183A	2N6054
3183A	2N6297
3183A	2N6648
3183A	2N6649
3183A	2N6650
3184	CA810M
3184	GEIC-278
3184	TBA810AS
3184	TBA810DS
3184	88-0550
3184	544-2006-001
3184	3130-3248-801
3184	6001
3184	3597049
3187	3N211
3187	3N213
3188A	AR22(PHILCO)
3188A	B1A
3188A	BD148
3188A	BD461
3188A	BD463
3188A	D44H4
3188A	D44H5
3188A	D44H8
3188A	ECG241
3188A	ES66(ELCOM)
3188A	HEP-S5001
3188A	HEP-S5004
3188A	LM1157
3188A	MJE182
3188A	MJE205
3188A	MJE2801
3188A	MJE3055
3188A	RS-2019
3188A	RT8336
3188A	SJE220
3188A	SJE5018
3188A	SJE5019
3188A	SJE5020
3188A	SJE516
3188A	TR-1037-1
3188A	TR-1037-2
3188A	TRO-1057-1
3188A	TRO-1057-3
3188A	TRO-1057-4
3188A	13-0097
3188A	13-28222-2
3188A	13-28222-4
3188A	13-33188-1
3188A	13-40340-1
3188A	3656-1
3188A	3656-2
3188A	3656-4
3188A	4802-3274-200
3188A	43118
3188A	43119
3188A	43120
3188A	140506
3188A	04450037-002
3189A	AR23(PHILCO)
3189A	BD200
3189A	BD464
3189A	CXL183
3189A	D45H10
3189A	D45H2
3189A	ECG183
3189A	ECG242
3189A	ES67(ELCOM)
3189A	GE-56
3189A	HEP-S5002
3189A	HEP-S5005
3189A	HEP-S5008
3189A	HEP-S5009
3189A	HEP-S5010
3189A	MJE104
3189A	MJE105
3189A	MJE2901
3189A	P3N-5
3189A	P3UA
3189A	P4T
3189A	P4V-2
3189A	RE39
3189A	REN183
3189A	REN197
3189A	RS-2027
3189A	RT8335
3189A	SJE517
3189A	TR-1036-1
3189A	TR-1036-2
3189A	TR1036-2
3189A	TR1038-6
3189A	2N5986
3189A	2SB699
3189A	13-28222-1
3190	A5A-4
3190	A5A-5
3190	A5G
3190	A5Y
3190	A6C-2-BLACK
3190	A6C-3-WHITE
3190	A6C-GREEN
3190	A6W
3190	BD131
3190	BD149

SK	Industry Standard No.
3190	BD155
3190	CXL184
3190	EA15X244
3190	ECG184
3190	GE-57
3190	HEP-S5000
3190	HT104961C
3190	HT304961B
3190	HT304961C
3190	HT313681B0
3190	M844
3190	M852
3190	MJE3521
3190	MJE482
3190	MJE483
3190	MJE488
3190	MJE520
3190	MJE521
3190	MJE720
3190	MJE721
3190	MOTMJE521
3190	QP-14
3190	RE40
3190	REN184
3190	S-85
3190	S-86
3190	S8660121-808
3190	SJE133
3190	SJE527
3190	SJE634
3190	SJE669
3190	T-344
3190	TM184
3190	TR76
3190	ZEN202
3190	ZEN210
3190	002-012200
3190	2N4921
3190	2N4922
3190	2N4923
3190	2N5190
3190	2SC2456
3190	2SC2568K
3190	2SC2568L
3190	2SC2568M
3190	314-6005-2
3190	314-6005-3
3190	13-0049
3190	13-14085-126
3190	13-36444-3
3190	14-608-03
3190	14-608-04
3190	14-906-13
3190	14-907-13
3190	14-910-13
3190	14-911-13
3190	48-137037
3190	48-137095
3190	48-137200
3190	48-137277
3190	48-137323
3190	48-41784J03
3190	080
3190	081
3190	86-5100-2
3190	121-Z9002
3190	149N2
3190	162N2
3190	190N3
3190	576-0002-011
3190	642-261
3190	800-533-01
3190	921-01131
3190	936 NPN
3190	2312-17
3190	4822-130-40537
3190	12539
3190	41178
3190	95262-1
3190	134280
3190	171175-1(NPN)
3190	570031
3190	916114
3190	1471140-1
3190	7910073-01
3190	23114132
3190	43024218
3190	2003121216
3190/184	MJE3520
3190/184	S5000
3190/184	SPS5000
3190/184	2SC1162
3190/184	2SC1162MP
3190/184	2SC1568
3190/184	09-302136
3190/184	13-40344-1
3190/184	42-27539
3190/184	46-86505-3
3190/184	48-90234A36
3190/184	86X0042-002
3190/184	121-772X
3190/184	413
3190/184	4218
3190/184	2320843
3191	A8C
3191	BD132
3191	CXL185
3191	EA15X243
3191	ECG185
3191	GE-58
3191	HEP-S5006
3191	HT104961B
3191	MJE3370
3191	MJE3371
3191	MJE370
3191	MJE492
3191	MJE493
3191	MJE711
3191	P3S
3191	P4S
3191	P4U
3191	P4W-1
3191	P4W-2
3191	QP-13
3191	RE41
3191	REN185
3191	SE9570
3191	SE9571
3191	SE9572
3191	SE9573
3191	SJ285
3191	SJE633
3191	T-396
3191	ZEN203

SK	Industry Standard No.
3191	ZEN155
3191	002-012100
3191	2N4918
3191	2N4919
3191	2SA4964
3191	2SA866
3191	2SB632K
3191	2SB744
3191	11-0770
3191	13-28336-1
3191	13-36445-3
3191	48-134987
3191	48-41785J03
3191	079
3191	149P3
3191	162P1
3191	417-145
3191	417-225
3191	624-1
3191	642-266
3191	644-1
3191	690V116H25
3191	3682
3191	41179
3191	95262-2
3191	138193
3191	171033-2(PNP)
3191	171175-1(PNP)
3191	657179
3191	1417359-1
3191	1471141-1
3191	1473682-1
3191	23114131
3191	23114133
3191	43024219
3191	2004074304
3191	310720490060
3191/185	JSP6009
3191/185	P6009
3191/185	SPS5006
3191/185	2N4920
3191/185	2SA505-0
3191/185	13-40345-1
3191/185	42-27538
3191/185	48-90420A02
3191/185	86X0042-001
3191/185	2320855
3192	B143000
3192	B143001
3192	B143003
3192	B143009
3192	B143024
3192	B143025
3192	B3531
3192	B3533
3192	B3537
3192	B3538
3192	B3540
3192	B3541
3192	B3542
3192	B3570
3192	B3576
3192	B3610
3192	B3611
3192	B3612
3192	B3613
3192	B5001
3192	BFS51
3192	BFS59
3192	BLY20
3192	BLY37
3192	BLY38
3192	BLY53
3192	BLY62
3192	BLY78
3192	BLY91
3192	BR100B
3192	CXL186
3192	D42C1
3192	D42C2
3192	D42C3
3192	D42C4
3192	D42C5
3192	D42C6
3192	D42C7
3192	D42C8
3192	D44C88
3192	EA15X160
3192	EA15X248
3192	EA2488
3192	EA3674
3192	ECG186
3192	EP100
3192	EP15X25
3192	ES15227
3192	ET495371
3192	G06-714C
3192	G06714C
3192	GE-28
3192	GEMR-6
3192	HC-01096
3192	HC-01098
3192	HC-01226
3192	HR-69
3192	HT304961C-0
3192	KGE41061
3192	PA8900
3192	PPR1006
3192	PPR1008
3192	PT2620
3192	PT2640
3192	PT2660
3192	PT3503
3192	PT4690
3192	PT600
3192	PT601
3192	PT6618
3192	PT6669
3192	PT6696
3192	RE42
3192	REN186
3192	REN186A
3192	S801
3192	SD1023
3192	SDT3326
3192	SDT4455
3192	SDT4483
3192	SDT4551
3192	SDT4553
3192	SDT4583
3192	SDT4611
3192	SDT4612
3192	SDT4614
3192	SDT4615

SK	Industry Standard No.
3192	SDT5011
3192	SDT5511
3192	SDT5901
3192	SDT5902
3192	SDT5906
3192	SDT6101
3192	SDT6102
3192	SDT6104
3192	SDT6105
3192	SDT6106
3192	SDT9001
3192	SDT9002
3192	SDT9003
3192	SDT9005
3192	SDT9006
3192	SDT9007
3192	SDT9008
3192	STC1800
3192	STC1862
3192	STD9007
3192	STT4451
3192	TIS82
3192	TNJ70450
3192	TNJ72775
3192	TR-1037-3
3192	TRAPLC1013
3192	TV-73
3192	TV-75
3192	TV-82
3192	VX3375
3192	XB401
3192	YAANZ
3192	YAANZ2S1096
3192	ZT3375
3192	ZT600
3192	2SC932
3192	2SC932A
3192	2SC932B
3192	2SC932BK
3192	2SC932C
3192	2SC932D
3192	2SC932E
3192	2SC932F
3192	2SC932G
3192	2SC932GN
3192	2SC932H
3192	2SC932J
3192	2SC932K
3192	2SC932L
3192	2SC932M
3192	2SC932R
3192	2SC932X
3192	2SC932Y
3192	09-302113
3192	13-32634-1
3192	13-34046-1
3192	13-34046-2
3192	13-34046-3
3192	13-34046-4
3192	13-39046-3
3192	13-39098-1
3192	13-43790-1
3192	018-00005
3192	19-020-101
3192	19A115300-P1
3192	19A115300-P2
3192	19A115300-P3
3192	21A112-049
3192	21M180
3192	21M181
3192	21M183
3192	21M184
3192	21M286
3192	21M367
3192	21M369
3192	21M556
3192	31-0066
3192	34-6002-41
3192	46-86360-3
3192	48-155097
3192	48-40382J01
3192	48S40382J01
3192	53-1362
3192	53-1967
3192	0103-0419
3192	0103-0419A
3192	0103-512M
3192	0140-7
3192	172-006-9-001
3192	176-024-9-004
3192	260P21102
3192	260P21308
3192	260P28701
3192	361-1
3192	576-002-001
3192	623-1
3192	753-0101-226
3192	753-9000-096
3192	903Y002152
3192	921-163B
3192	1097(GE)
3192	1116(GE)
3192	1125(GE)
3192	2132
3192	2245-17
3192	2510-105
3192	2853-1
3192	2853-3
3192	2854-1
3192	2854-3
3192	2855-1
3192	2855-3
3192	2856-1
3192	2856-3
3192	8000-00004-185
3192	8000-0004-P185
3192	8710-170
3192	8840-169
3192	20810-94
3192	38448
3192	58810-169
3192	58840-200
3192	60407
3192	60413
3192	61242
3192	62950
3192	000073280
3192	000073320
3192	000073380
3192	88060-146
3192	88510-176
3192	133573
3192	741115
3192	760275

SK	Industry Standard No.
3192	916046
3192	986932
3192	1473632-2
3192	2006436-35
3192	2006436-36
3192	2327152
3192	2327153
3192	2327203
3192	4037647-P1
3192	4037647-P2
3192	43027571
3192	16102190931
3192	16343190142
3192/186	A514-047830
3192/186	BFS50
3192/186	BFX96A
3192/186	D40D2
3192/186	D40D3
3192/186	D40D4
3192/186	D40D5
3192/186	D40D7
3192/186	D40D8
3192/186	EP15X32
3192/186	EP15X34
3192/186	GE-247
3192/186	KSC1096-0
3192/186	KSC1096-Y
3192/186	MJE180
3192/186	MJE181
3192/186	SDT3421
3192/186	SDT3425
3192/186	SDT3426
3192/186	SDT4301
3192/186	SDT4302
3192/186	SDT4304
3192/186	SDT4305
3192/186	SDT4307
3192/186	SDT4308
3192/186	SDT4310
3192/186	SDT4311
3192/186	SDT5001
3192/186	SDT5006
3192/186	SDT5501
3192/186	SDT5506
3192/186	SDT7401
3192/186	SDT7411
3192/186	SJE5402
3192/186	T1P31
3192/186	13-34046-5
3192/186	13-39074-1
3192/186	13-41628-5
3192/186	90-38
3192/186	229-1301-34
3192/186	700-110
3192/186	2017-108
3192/186	7398-6120P
3192/186	7398-6120P1
3192/186	7513
3192/186	8000-00004-P185
3192/186	135739
3192/186	651956
3193	AR25
3193	AR25G
3193	CXL187
3193	D4301
3193	D4302
3193	D4303
3193	D4304
3193	D4305
3193	D4306
3193	D4308
3193	ECG187
3193	EP101A
3193	EP15X24
3193	ES15051
3193	G03703C
3193	GE-29
3193	HA-00634
3193	HA-00636
3193	HB-00564
3193	HR-70
3193	HR104961C
3193	MJE170
3193	MJE171
3193	RE43
3193	REN187
3193	SDT3325
3193	SDT3775
3193	SDT3776
3193	SDT3778
3193	SDT9004
3193	TNJ72774
3193	TR-1036-3
3193	TV-74
3193	2SA636(4)K
3193	2SA636(4)L
3193	09-300067
3193	09-300068
3193	09-302170
3193	13-14085-86
3193	13-32635-1
3193	13-34047-1
3193	13-34047-2
3193	13-34373-2
3193	13-39047-3
3193	13-43791-1
3193	018-00004
3193	19-020-102
3193	21A112-048
3193	21M025
3193	21M026
3193	21M028
3193	21M345
3193	21M395
3193	21M443
3193	34-6002-42
3193	34-6016-65
3193	48-40383J01
3193	48S40383J01
3193	0101-0448
3193	0101-0448A
3193	0101-0466
3193	753-0100-699
3193	853-0300-634
3193	903Y002151
3193	61239
3193	000071120
3193	760276
3193	986933
3193	5320632
3193	22114254
3193/187	GE-248
3193/187	KSA634-Y
3193/187	SDT3505

SK	Industry Standard No.
3193/187	SDT3506
3193/187	SDT3550
3193/187	SDT3552
3193/187	SDT3553
3193/187	2SA623
3193/187	2SA623A
3193/187	2SA623B
3193/187	2SA623C
3193/187	2SA623D
3193/187	2SA623G
3193/187	2SA623R
3193/187	2SA623Y
3193/187	13-23510-4
3193/187	13-34047-4
3193/187	2244-1
3193/187	000071130
3194	B1Y
3194	B2P
3194	C779-R
3194	C779-Y
3194	C783
3194	FD-1029-PB
3194	GE-267
3194	2SC1450
3194	2SC1450S
3194	2SC680
3194	2SC680A
3194	2SC680B
3194	2SC680C
3194	2SC680G
3194	2SC680GN
3194	2SC680H
3194	2SC680J
3194	2SC680K
3194	2SC680L
3194	2SC680M
3194	2SC680R
3194	2SC680X
3194	2SC779
3194	2SC779R
3194	2SC783
3194	48-137528
3194	121-993
3195	GE-277
3195	2N3866
3195	2SC731
3195	2SC731R
3195	61026
3195	61273
3195/311	CS1256H
3195/311	CS1256HG
3195/311	GE-320
3195/311	TVS-CS1256HG
3195/311	2N2476
3195/311	2N2477
3195/311	2N3299
3195/311	2N3300
3195/311	2N4432
3195/311	2N4432A
3195/311	2N4437
3195/311	2SC994
3195/311	36913
3195/311	40578
3197	C1237
3197	C1237E
3197	C1909
3197	C1975
3197	DDBY005001
3197	DDBY230001
3197	DDBY257001
3197	DDBY289001
3197	EA15X363
3197	EA15X372
3197	EA15X414
3197	EQS-0160
3197	FT527
3197	GE-216
3197	GE-322
3197	IC743045
3197	IP20-0135
3197	IP20-0155
3197	MDS1678
3197	MJE-200E
3197	NCB35
3197	NCBW35
3197	PTC186
3197	QS1306
3197	QT-C1306XZA
3197	RE203
3197	01-031909
3197	2SC1237
3197	2SC1237E
3197	2SC1909
3197	2SC1909K
3197	2SC1975
3197	2SC2029
3197	2SC2029-1
3197	2SC2029-B/10
3197	2SC2029/3
3197	2SC2029B
3197	2SC2029C
3197	2SC2393D
3197	2SC597
3197	2SD325A
3197	2SD325B
3197	2SD325F
3197	2SD325G
3197	2SD325GN
3197	2SD325H
3197	2SD325J
3197	2SD325K
3197	2SD325L
3197	2SD325M
3197	2SD325OR
3197	2SD325R
3197	2SD325X
3197	2SD325Y
3197	04S06
3197	09-305136
3197	44T-300-109
3197	48-40004S06
3197	48-40764P01
3197	48-42885P01
3197	90-450
3197	90-609
3197	102-1678-00
3197	106-006
3197	123-009
3197	172-024-9-002
3197	172-028-9-001
3197	176-024-9-002
3197	176-074-9-003
3197	576-0004-105
3197	1042-08
3197	1042-8
3197	1212-4
3197	2000-208
3197	2000-271
3197	2022-04
3197	2041-04
3197	2065-06
3197	2399-17
3197	5001-050
3197	5001-514
3197	8000-00038-001
3197	8000-00043-065
3197	8000-00049-060
3197	239917
3197	279717
3197	741732
3197	741859
3197	742970
3197	916118
3197	5321321
3197	630000003
3197	632197406
3197/235	C1173-GR
3197/235	C1173-R
3197/235	C1173-Y
3197/235	C1173C
3197/235	C1173R
3197/235	C1173Y
3197/235	C1307
3197/235	C1377
3197/235	C1678
3197/235	C1679
3197/235	C1816
3197/235	DDBY228001
3197/235	DDBY231002
3197/235	DDBY278002
3197/235	EQS-0159
3197/235	EQS-141
3197/235	MA6101
3197/235	MA6102
3197/235	MJE200
3197/235	VS2SC1237-1
3197/235	2N3554
3197/235	2SC1173XO
3197/235	2SC130L
3197/235	2SC1377
3197/235	2SC1398
3197/235	2SC1398P
3197/235	2SC1398Q
3197/235	2SC1419
3197/235	2SC1419A
3197/235	2SC1419B
3197/235	2SC1419C
3197/235	2SC1419D
3197/235	2SC1678
3197/235	2SC1678E
3197/235	2SC1679
3197/235	2SC1816
3197/235	2SC1957
3197/235	2SC1964
3197/235	2SC2075
3197/235	2SC2078
3197/235	2SC2092
3197/235	2SC2166
3197/235	2SD288
3197/235	2SD288K
3197/235	2SD288L
3197/235	2SD289
3197/235	2SD325
3197/235	2SD325C
3197/235	2SD325D
3197/235	2SD325E
3197/235	09-302192
3197/235	09-302193
3197/235	09-305137
3197/235	10-009
3197/235	21A118-049
3197/235	022-2876-012
3197/235	44T-300-108
3197/235	52-020-173-0
3197/235	065-210
3197/235	90-610
3197/235	123-008
3197/235	172-013-9-001
3197/235	172-024-9-003
3197/235	172-029-9-001
3197/235	176-024-9-003
3197/235	176-073-9-002
3197/235	176-073-9-012
3197/235	1039-0433
3197/235	1042-09
3197/235	1678
3197/235	2000-212
3197/235	2437-17
3197/235	5001-071
3197/235	8000-00032-028
3197/235	26010053
3197/235	632122617
3198	AD152
3198	AD156
3198	AD162
3198	AD164
3198	AD169
3198	B367
3198	B367A
3198	B367B
3198	B367C
3198	B367H
3198	B448
3198	B473
3198	B473D
3198	B473F
3198	B481D
3198	B481E
3198	CXL131
3198	ECG131
3198	ETTB-367B
3198	G04-704-A
3198	G6016
3198	GE-44
3198	HEP-G6016
3198	HT2046710
3198	OC30A
3198	P3T-1
3198	P3T-2
3198	RE20
3198	REN131
3198	TM131
3198	2SB367
3198	2SB367-4
3198	2SB367-5
3198	2SB367A
3198	2SB367AL
3198	2SB367B
3198	2SB367BL
3198	2SB367C
3198	2SB367D
3198	2SB367H
3198	2SB367J
3198	2SB367K
3198	2SB367M
3198	2SB367OR
3198	2SB448
3198	2SB473
3198	2SB473A
3198	2SB473B
3198	2SB473C
3198	2SB473D
3198	2SB473E
3198	2SB473F
3198	2SB473G
3198	2SB473H
3198	2SB481
3198	8-619-030-015
3198	8-905-613-266
3198	8A10625
3198	8P202
3198	46-86125-3
3198	48-97046A15
3198	48-97046A1S
3198	57M3-9P
3198	61-1906
3198	98P1P
3198	0004203
3198	25658-120
3198	25658-121
3198	000071090
3198	171217-1
3198/131	B368
3198/131	B368A
3198/131	B368B
3198/131	B368H
3198/131	B462
3198/131	B466
3198/131	B467
3198/131	B481
3199	A6G
3199	A8J
3199	B2M
3199	ECG188
3199	GE-83
3199	HEP-S3026
3199	IRTR72
3199	MPSU01
3199	MPSU02
3199	MPSU05
3199	P218-1
3199	RE75
3199	REN188
3199	13-33180-1
3199	13-37526-1
3199	46-86583-3
3199	48-137149
3199	48-137319
3199	48-97177A10
3199	48S137169
3199	48S137572
3199	66F058-2
3199	81-27126100-0
3199	86-422-3
3199	86-660-2
3199	2904-058
3199	3628-3
3199	7316-1
3199	7553
3199	40311A
3199	40311B
3199	40361V3
3199	40385V1
3199	40385V2
3199	40437V1
3199	40437V2
3199	137066
3199	139266
3199	610228-1
3199	1417316-1
3199/188	GE-252
3199/188	HEP-S3024
3199/188	1B-201
3199/188	2SC1429
3199/188	2SC1429-1
3199/188	2SC1429-2
3199/188	13-34003-1
3199/188	13-34372-1
3199/188	86X0073-002
3199/188	921-1020
3199/188	1081-3350
3200	CXL189
3200	ECG189
3200	GE-84
3200	HEP-S3030
3200	HEP-S3031
3200	HEP-S3032
3200	IRTR73
3200	MPS-U56
3200	MPSU51
3200	MPSU51A
3200	P218-2
3200	P2V
3200	RE76
3200	REN189
3200	TM189
3200	TR02053-7
3200	2SA706
3200	2SA861
3200	2SA962
3200	13-34004-1
3200	13-34373-1
3200	13-37527-1
3200	46-86584-3
3200	48-137320
3200	48-97177A11
3200	48S137160
3200	48S137168
3200	86-423-3
3200	86-659-2
3200	121-980
3200	121-980-01
3200	0126(WARDS)
3200	3629-3
3200	132488
3200	1417317-1
3200	1471134-1
3200	1473629-1
3200/189	BC327
3200/189	BC328
3200/189	GE-253
3200/189	HEP-S3028
3200/189	TIP30B
3200/189	921-1021
3201	A278
3201	A279
3201	BFS99
3201	BSS19
3201	BSS20
3201	D40N1
3201	D40N3
3201	D40N5
3201	EP15X27
3201	EP15X61
3201	ES89
3201	GE-27
3201	PT2523
3201	PT2524
3201	PT2525
3201	RCP111A
3201	RCP111B
3201	RCP111C
3201	RCP111D
3201	RCP113A
3201	RCP113B
3201	RCP113C
3201	RCP113D
3201	RCP115
3201	RCP115A
3201	RCP115B
3201	RCP117
3201	RCP117B
3201	RE73
3201	RT1116
3201	RT1893
3201	SF.T187
3201	TQ-5063
3201	TRS3014
3201	01-572088
3201	007-00
3201	8-729-322-78
3201	13-33174-1
3201	46-81187-3
3201	57A193-11
3201	57A208-8
3201	57A312-11
3201	66F074-1
3201	66F074-2
3201	66F074-4
3201	86X0065-001
3201	121-1037
3201	121-868
3201	121-868-01
3201	121-868-02
3201	132-003
3201	144N2
3201	417-245
3201	141295
3201	142671
3201	146826
3201	610144-1
3201	801541
3201	1417352-5
3201/171	BF292
3201/171	BF297
3201/171	BF456
3201/171	BF457
3201/171	BF458
3201/171	BFR18
3201/171	BFR19
3201/171	BFR20
3201/171	BFR57
3201/171	BFR58
3201/171	BFR59
3201/171	2SC2068
3201/171	2SC2068LB
3201/171	2SC2278
3201/171	13-27974-1
3201/171	46-86567-3
3201/171	121-777-01
3201/171	7362-1
3201/171	140259
3201/171	142251
3201/171	610190-4
3201/171	1417362-1
3201/171	1473628-1
3201/171	2003206800
3202	D40D10
3202	D40D11
3202	ECG210
3202	HEP-S3025
3202	HR-73
3202	HR85
3202	RE80
3202	REN210
3202	TM210
3202	09-3-2123
3202	13-33185-1
3202	86-673-2
3202	3107-204-90180
3202/210	C10963ZM
3202/210	C10964ZL
3202/210	D120H
3202/210	D120HA
3202/210	D120HB
3202/210	D120HC
3202/210	D121H
3202/210	D121HA
3202/210	D121HB
3202/210	D328S
3202/210	GE-236
3202/210	2SC10963ZM
3202/210	2SC10964ZL
3202/210	2SD120HA
3202/210	2SD120HB
3202/210	2SD120HC
3202/210	2SD121HA
3202/210	2SD121HB
3202/210	4ZL
3202/210	065-007
3202/210	121-1006
3202/210	172-024-9-004
3202/210	2168-17
3202/210	610190-1
3203	CXL211
3203	D41D1
3203	D41D10
3203	D41D11
3203	D41D4
3203	D41D5
3203	D41D7
3203	D41D8
3203	ECG211
3203	HEP-S3029
3203	HR-74
3203	HR86
3203	RE81
3203	REN211
3203	SPS4092
3203	13299
3203	610361-5
3203/211	A504GR
3203/211	A623
3203/211	A623-0
3203/211	A623A
3203/211	A623B
3203/211	A623C
3203/211	A623D
3203/211	A623G
3203/211	A623R
3203/211	A623Y
3203/211	A624
3203/211	A624A
3203/211	A624B
3203/211	A624C
3203/211	A624D
3203/211	A624GN
3203/211	A624L
3203/211	A624LG
3203/211	A624R
3203/211	A624Y
3203/211	A708
3203/211	A708A
3203/211	A708B
3203/211	A708C
3203/211	A717
3203/211	B510S
3203/211	BCW79-25
3203/211	BCW80-25
3203/211	BCX10
3203/211	P2U-1
3203/211	2N6554
3203/211	2SA503GR
3203/211	2SA504GR
3203/211	2SA703
3203/211	2SA708
3203/211	2SA708A
3203/211	2SA708B
3203/211	2SA708C
3203/211	2SA717
3203/211	2SB510S
3203/211	260P26301
3205/982	221-87
3206	CA1394E
3210	CA3137E
3212	CA3134EM
3212	CA3134QM
3215	C6100P
3215	CA3064E
3215	EP84X4
3215	EP84X5
3215	GEIC-21
3215	GEIC-225
3215	HEP-C6100P
3215	MC1364
3215	MC1364P
3215	MC1364PQ
3215	RE303-IC
3215	SI-MC1364P
3215	SN76564N
3215	T2062
3215	ULN2264A
3215	02-561201
3215	56A20-1
3215	56D20-1
3215	61B2Z
3215	73C180843-1
3215	73C180843-2
3215	73C180843-3
3215	73C180843-4
3215	097-1
3215	99S097
3215	99S097-1
3215	442-59
3215	1010-9965
3215	1097
3215	2061-1
3215	612061-1
3215	62736305
3215/783	MC1364G
3216	C6066P
3216	CA3125E
3216	GEIC-234
3216	HEP-C6066P
3216	SC5204P
3216	T1K
3216	TA6404
3216	51-10425A01
3216	51-10655B01
3216	51-70177A01
3216	51D70177A01
3216	51M70177A01
3216	51M70177B01
3216	51S10425A01
3216	51S10655A01
3216	51S10655A01-3
3216	51S10655B01
3216	177A01
3216	177B01
3216	10655A01
3216	10655A01-3
3216	70177A01
3217	RCA1B04
3217	RCA1B05
3218	TA2800
3218	2N5109
3218	2SC548
3218/278	2SC1365
3218/278	2SC730
3218/278	16065
3219	A678760A
3219	A678971C
3219	C1447-0
3219	C1447LB
3219	C1447R
3219	C1505
3219	C1505K
3219	C1505L
3219	C1505LA
3219	C1505M
3219	C1506
3219	C1507A
3219	C1507K
3219	C1507L
3219	C1507LM
3219	C1507M
3219	C1569-0
3219	C1569LBQ
3219	C1569LBR
3219	C1569LBY

SK	Industry Standard No.
3219	C1569R
3219	C1756A
3219	C1756C
3219	C1756D
3219	C1756K
3219	D44R1
3219	D44R2
3219	GE-325
3219	KSC1507
3219	LM2701
3219	MPSU04
3219	ST-2SC1514
3219	TIP47
3219	TIP48
3219	TVS2SC1505
3219	01-031507
3219	01-031514
3219	01-031756
3219	2SC1446
3219	2SC1446B
3219	2SC1446C
3219	2SC1446L
3219	2SC1446LB
3219	2SC1446P
3219	2SC1446PQ
3219	2SC1446Q
3219	2SC1446R
3219	2SC1447
3219	2SC1447-O
3219	2SC1447R
3219	2SC1505
3219	2SC1505K
3219	2SC1505L
3219	2SC1505LA
3219	2SC1505M
3219	2SC1506
3219	2SC1507
3219	2SC1507A
3219	2SC1507H
3219	2SC1507K
3219	2SC1507L
3219	2SC1507LM
3219	2SC1507M
3219	2SC1514
3219	2SC1514BK
3219	2SC1514BVC
3219	2SC1514CVC
3219	2SC1519
3219	2SC1569
3219	2SC1569-O
3219	2SC1569BK
3219	2SC1569K
3219	2SC1569LBQ
3219	2SC1569LBR
3219	2SC1569LBY
3219	2SC1569R
3219	2SC1569Y
3219	2SC1625
3219	2SC1722
3219	2SC1722S
3219	2SC1723
3219	2SC1749
3219	2SC1755
3219	2SC1755A
3219	2SC1755C
3219	2SC1756
3219	2SC1756A
3219	2SC1756B
3219	2SC1756C
3219	2SC1756D
3219	2SC1756E
3219	2SC1756M
3219	2SC1756MC
3219	2SC1757
3219	2SC1819
3219	2SC1819M
3219	2SC1905
3219	2SC1929
3219	2SC1929Q
3219	2SC1929R
3219	2SC2085
3219	2SC2231Y
3219	2SC2242
3219	2SD669
3219	8-729-372-31
3219	48-155059
3219	48-155110
3219	48-155130
3219	48-155153
3219	48-155213
3219	121-1028-01
3219	200X3151-432
3219	200X3172-208
3219	200X3206-800
3219	06120083
3219	2003172216
3219	2003172305
322	21A118-008
3220	A678971D
3220	BWS66
3220	BWS67
3220	C1520
3220	C1520-1A
3220	C1520-3A
3220	C1520K
3220	C1520KL
3220	C1520L
3220	C1520M
3220	C1521L
3220	C1521LM
3220	D44R4
3220	EP15X50
3220	GE-32
3220	J241233
3220	KSC1520
3220	M9320
3220	M9465
3220	MJE2360
3220	NTC-9
3220	Q5138
3220	Q5138K
3220	Q5138L
3220	Q5138M
3220	Q5160
3220	Q5160Y
3220	RE191
3220	SC441
3220	SM6814
3220	ST84027
3220	ST84028
3220	ST84029
3220	TG2SC1756
3220	TIP49
3220	TIP50
3220	TN2SC1507
3220	TN2SC1507-K-A
3220	TN2SC1507-L-A
3220	TN2SC1507-M-A
3220	TN2SC1520-1
3220	TN2SC1520-K-1A
3220	TN2SC1520-M-1A
3220	TN2SC1520-M-3A
3220	TRS1204
3220	TRS1404
3220	TRS140MP
3220	TRS1604
3220	TRS160MP
3220	TRS1804
3220	TRS180MP
3220	TRS2004
3220	TRS2006
3220	TRS200MP
3220	TRS2254
3220	TRS225MP
3220	TRS2504
3220	TRS250MP
3220	TRS2754
3220	TRS275MP
3220	TRS2804S
3220	TRS2805S
3220	TRS3006
3220	TRS3015
3220	TRS301LC
3220	TRS301MP
3220	TRS3204S
3220	TRS3205S
3220	TRS3254
3220	TRS3255
3220	TRS3742
3220	TRS4926
3220	TRS4927
3220	TV2SC1505
3220	TV2SC1507
3220	001-02114-0
3220	2SC1520
3220	2SC1520-1A
3220	2SC1520-3A
3220	2SC1520I
3220	2SC1520K
3220	2SC1520KL
3220	2SC1520L
3220	2SC1520M
3220	14-602-76
3220	21A112-071
3220	21A112-098
3220	41-0609
3220	46-86469-3
3220	46-86482-3
3220	46-86508-3
3220	46-86556-3
3220	48-355005
3220	48-355038
3220	48-355044
3220	48-869265
3220	48-869286
3220	48-869320
3220	48-869465
3220	48-90343A61
3220	48S155034
3220	48S155059
3220	48S155110
3220	417-232
3220	462-1059
3220	1070-0631
3220	91371
3220	198075-1
3220	618960-1
3220	619010-1
3220	1417362-4
3220	23114982
3220/198	A6319400
3220/198	A6319403
3220/198	C1521K
3220/198	2SC1447LB
3220/198	2SC1683
3220/198	2SC1683LA
3220/198	2SC1683P
3220/198	2SC1683R
3220/198	2SC1756K
3220/198	2SC2637
3220/198	86X0076-001
3220/198	260P35101
3220/198	417-283-13271
3221	RCA8203A
3223	BA402
3223	DDEY004001
3223	EA33X8500
3223	GEIC-92
3223	HC10002050
3223	TA7061
3223	TA7061(AP)
3223	TA7061AP
3223	TA7061P
3223	09-308041
3223	09-308059
3223	90-36
3223	90-73
3223	307-007-9-001
3223	307-020-9-001
3223	307-029-1-001
3223	588-40-201
3223	1021-25
3223	8710-171
3223	8840-170
3223	8910-147
3223	18600-154
3223	25810-164
3223	37510-164
3223	58810-170
3223	58840-201
3223	88060-147
3223	88510-177
3224	GEIC-93
3224	IP20-0014
3224	TA7062P
3224	90-35
3224	90-72
3224	307-007-9-002
3224	307-009-9-002
3224	26010035
3225	BA401
3225	GEIC-91
3225	Q-01369C
3225	TA7060
3225	TA7060P
3225	TA7060P-10
3225	TA7060PR
3225	TA7060PRW
3225	TA7060PW
3225	90-37(IC)
3225	90-74
3225	8000-00005-014
3225	916105
3226	AN342
3226	GEIC-67
3227	AN288
3227	GEIC-64
3228	AN220
3228	GEIC-143
3228	GEIC-51
3228	GEIC-88
3228	HA1108
3229	M5115P
3230	GEIC-287
3230	TA7203
3230	TA7203P
3231	B0319200
3231	EA33X8389
3231	EA33X8396
3231	ECG1155
3231	EICM-0060
3231	ETI-23
3231	GEIC-179
3231	IP20-0161
3231	KIA7205AP
3231	KIA7205P
3231	QQ-M07205AT
3231	QQM07205AT
3231	REN1155
3231	TA7205
3231	TA7205A
3231	TA7205AP
3231	TA7205P
3231	TM1155
3231	WEP949
3231	02-257205
3231	44T-100-120
3231	051-0055-02
3231	051-0055-03
3231	740-9007-205
3231	740-9607-205
3231	5002-031
3231	26810-159
3231	45810-170
3231	0207205
3231	741687
3231	742364
3231	916110
3231	1223910
3232	A5TA
3232	B1E-1
3232	B1G
3232	ES96
3232	GE-224
3232	GE-249
3232	HEP-S3021
3232	HEP-S3022
3232	MPS-U10
3232	MPSU10
3232	MPSU11
3232	R3613-3(RCA)
3232	RE77
3232	ST-2N6558
3232	TV-25
3232	TV49
3232	14-609-01
3232	14-900-12
3232	48-137239
3232	48-137476
3232	60P22204
3232	121-743-01
3232	687-1
3232	3613-2(GE)
3232	40256
3232	40439V1
3232	40439V2
3232	139017
3232	610131-2
3232	1473613-2
3232	1473613-3
3232/191	EP15X10
3232/191	GE-217
3232/191	HEP-S3019
3232/191	SJE1032
3232/191	TR1033-3
3232/191	2N6557
3232/191	2N6558
3232/191	13-45018-1
3232/191	48S137476
3232/191	57A295-8
3232/191	57B193-11
3232/191	86-664-2
3232/191	121-755
3232/191	121-775
3232/191	190N1
3232/191	137648
3232/191	142690
3232/191	146143
3232/191	1473679-1
3233	LM1496H
3233	N5596K
3233	SG1496T
3234	C6059P
3234	GEIC-217
3234	HEP-C6059P
3234	MC1350
3234	MC1350P
3234	SN76600
3234	SN76600P
3234	09-010
3234	46-5002-17
3234	051-0021-00
3234	2906-005
3234	8000-0058-006
3234	09010
3234	76600P
3234	916111
3235	C6072P
3235	GEIC-30
3235	HEP-C6072P
3235	MC1324P
3235	MC1324PQ
3235	MC1326P
3235	MC1326PQ
3235	RE312-IC
3235	TVCM-21
3235	13-57-6
3235	13-67-6
3235	22B1B
3235	46-5002-14
3235	221-49
3235	1326PC
3235	1347
3235	1348
3235	612072-1
3235/739	GEIC-52
3235/739	ULN2224A
3236	GEIC-115
3236	GEIC-183
3236	GEIC-26
3236	LM1351N
3236	MC1351P
3236	MC1351PQ
3236	RE331-IC
3236	RH-1X0001TAZZ
3236	RH-IX0001TAZZ
3236	SC5245P
3236	SN76651N
3236	TA7072P
3236	TA7073P
3236	51G10679A13
3237	C6099P
3237	GEIC-232
3237	HEP-C6099P
3237	MC1349P
3237	15-39061-1
3238	EA33X8364
3238	GEIC-34
3238	HC1000417
3238	IC-521(ELCOM)
3238	MC1312P
3238	MXC-1312
3238	MXC1312A
3238	MXC1312P
3238	SC5747
3238	21M506
3238	46-500230
3238	86X0064-001
3238/799	TVS-2SC901
3239	A67-76-200
3239	ECG236
3239	ESQ-141
3239	GE-332
3239	IP20-0154
3239	PL-172-013-9001
3239	QT-C1307XZA
3239	RE201
3239	REN236
3239	STT2405
3239	STT2406
3239	STT4483
3239	STT9001
3239	STT9002
3239	STT9004
3239	STT9005
3239	TM236
3239	2SC1307
3239	2SC1307-1
3239	2SC1848V
3239	2SC1969
3239	2SC1969B
3239	2SC1974
3239	2SC2043
3239	2SC2184
3239	2SD359E
3239	31-091-3
3239	176-044-9-001
3239/236	EA15X381
3239/236	GE-329
3239/236	01-031306
3239/236	2SC1306
3239/236	2SC1848
3239/236	2SC1848Q
3239/236	2SC1848R
3239/236	2SC2020
3239/236	2SD256
3239/236	2SD330
3239/236	2SD330E
3239/236	2SD359
3239/236	2SD359C
3239/236	2SD359D
3239/236	2SD762
3239/236	022-2876-011
3239/236	31-069-0
3239/236	172-038-9-002
3239/236	8000-00041-044
3239/236	1223918
3239/236	632000002
3241	ES107
3241	GE-258
3241	HEP-S9120
3241	MPS-A65
3241	MPS-A66
3241/232	DS-60
3241/232	13-33175-2
3241/232	964-22008
3243	GEIC-196
3243	TM1160
3244	BC125A
3244	BC182LA
3244	BC183LA
3244	BC383
3244	BC384
3244	BC682
3244	BP412
3244	BPS31P
3244	BSS38
3244	C1279
3244	C1279S
3244	C499
3244	C499R
3244	C499Y
3244	C926A
3244	CS4006
3244	CS4007
3244	EN870
3244	EN871
3244	GS9019H
3244	PBC184
3244	Q5210
3244	TE2484
3244	ZTX302
3244	ZTX303
3244	ZTX331
3244	01-030983
3244	2SC1279
3244	2SC1279S
3244	2SC1416A
3244	2SC1811
3244	2SC1885
3244	2SC2229
3244	2SC2229M
3244	2SC2229Y
3244	2SC2371
3244	2SC2482
3244	2SC499
3244	2SC499-R(FA-1)
3244	2SC499-RY
3244	2SC499-Y(FA-1)
3244	2SC499R
3244	2SC499Y
3244	2SC903
3244	2SC914
3244	2SC915
3244	2SC926
3244	2SC926A
3244	2SC926B
3244	2SC926C
3244	2SC926D
3244	2SC926E
3244	2SC926F
3244	2SC926G
3244	2SC926GN
3244	2SC926H
3244	2SC926J
3244	2SC926K
3244	2SC926L
3244	2SC926M
3244	2SC926OR
3244	2SC926R
3244	2SC926X
3244	2SC926Y
3244	8-729-309-06
3244	46-86572-3
3244	46-86630-3
3244	57A194-11
3244	161-014H
3244	200X3222-907
3244	200X3223-025
3244	200X3224-007
3244	260P35301
3245	A-1854-0023-1
3245	A-1854-0088-1
3245	A-1854-0284-1
3245	A1238
3245	A138
3245	A139
3245	A1460
3245	A2370773
3245	A3J
3245	A644L
3245	A644S
3245	A667-GRN
3245	A667-ORG
3245	A667-RED
3245	A667-YEL
3245	A668-GRN
3245	A668-ORG
3245	A668-YEL
3245	A669-GRN
3245	A669-YEL
3245	A67-07-244
3245	A67-33-340
3245	A748C
3245	A749C
3245	A8R
3245	A9-175
3245	BC155C
3245	BC156C
3245	BC182K
3245	BC183K
3245	BC183KA
3245	BC184KC
3245	BC20C
3245	BC384C
3245	BC385A
3245	BC385B
3245	BC386A
3245	BC386B
3245	BC408B
3245	BC408C
3245	BC520
3245	BC520B
3245	BC520C
3245	BC521
3245	BC521C
3245	BC521D
3245	BC522
3245	BC522C
3245	BC522D
3245	BC522E
3245	BC523
3245	BC523B
3245	BC523C
3245	BCW87
3245	BCW98A
3245	BCW98B
3245	BCW98C
3245	BCW98D
3245	BFR25
3245	BFV37
3245	BFV38
3245	BFV62
3245	BFV89
3245	BFV89A
3245	BFY39-2
3245	BFY39-3
3245	BTX-068
3245	C1006
3245	C1006A
3245	C1006B
3245	C1006C
3245	C1010
3245	C1010A
3245	C1010B
3245	C1010C
3245	C1204D
3245	C1222A
3245	C1222B
3245	C1222C
3245	C1222D
3245	C1313Y
3245	C1537
3245	C1537B
3245	C1537S
3245	C1538
3245	C1538A
3245	C1538S
3245	C1538SA
3245	C1648
3245	C1681BL
3245	C368
3245	C368BL
3245	C368GR
3245	C368V
3245	C372(0)
3245	C372-O
3245	C644
3245	C644C
3245	C644F/494

SK	Industry Standard No.
3245	C644FR
3245	C644FS
3245	C644H
3245	C644HR
3245	C644P
3245	C644PJ
3245	C644Q
3245	C644R
3245	C644RST
3245	C644S
3245	C644S/494
3245	C644T
3245	C694
3245	C694E
3245	C694F
3245	C694G
3245	C694Z
3245	C7076
3245	C732
3245	C732BL
3245	C732GR
3245	C732S
3245	C732V
3245	C732Y
3245	C733
3245	C733-0
3245	C733-0
3245	C733BL
3245	C733GR
3245	C733R
3245	C733V
3245	C733Y
3245	C859
3245	C859E
3245	C859F
3245	C859FG
3245	C859G
3245	CGE-62
3245	CXL199
3245	D26E-5
3245	D26E-7
3245	DDBY299001
3245	EA15X288
3245	ECG199
3245	EP15X3
3245	ET379262
3245	ET380834
3245	ET398777
3245	ET517994
3245	G05-012-G
3245	G05035D
3245	GE-212
3245	GE-62
3245	HC-01000
3245	HE-00930
3245	HEP-S0023
3245	HEP-S0024
3245	HR-14
3245	HR-75
3245	HT308281H
3245	HT310002A
3245	HT313272B
3245	LM1117
3245	LM1117C
3245	LS-0031-AR-218
3245	LS-0095-AR-213
3245	M9197
3245	M9269
3245	M9293
3245	M9329
3245	M9338
3245	M9384
3245	M9416
3245	M9447
3245	M9474
3245	M9486
3245	M9547
3245	M9594
3245	MMT70
3245	MPSA10-BLU
3245	MPSA10-GRN
3245	MPSA10-RED
3245	MPSA10-WHT
3245	MPSA10-YEL
3245	MPSA20-BLU
3245	MPSA20-GRN
3245	MPSA20-RED
3245	MPSA20-WHT
3245	MPSA20-YEL
3245	NR-421AS
3245	P1901-48
3245	PA8260
3245	PA8543
3245	PA9004
3245	PA9005
3245	PBC107
3245	PBC107A
3245	PBC107B
3245	PBC108
3245	PBC108A
3245	PBC108B
3245	PBC108C
3245	PBC109
3245	PBC109B
3245	PBC109C
3245	PS6010-1
3245	Q5121
3245	Q51210
3245	QRT-104
3245	RE192
3245	RE4001
3245	RE4002
3245	RE4010
3245	RE64
3245	REN199
3245	REN64
3245	RLB-17
3245	RT-104
3245	RT-105
3245	RT-107
3245	RT-108
3245	RT-112
3245	S001465
3245	S006793
3245	S006927
3245	S007764
3245	SE5-0938
3245	SE5-0938-55
3245	SE5-0938-56
3245	SE5-0938-57
3245	SPS4814
3245	ST.082.114.016
3245	ST2T003
3245	ST33026
3245	ST7100
3245	T01-047
3245	TE2711
3245	TE2712
3245	TE2921
3245	TE2922
3245	TE2923
3245	TE2924
3245	TE2925
3245	TE2926
3245	TE3390
3245	TE3391
3245	TE3391A
3245	TE3392
3245	TE3393
3245	TE3394
3245	TE3395
3245	TE3396
3245	TE3397
3245	TE3398
3245	TE3843
3245	TE3844
3245	TE3845
3245	TE3854
3245	TE3854A
3245	TE3855
3245	TE3855A
3245	TE3859
3245	TE3860
3245	TE3900
3245	TE3900A
3245	TE3901
3245	TE4256
3245	TE5088
3245	TE5089
3245	TE5249
3245	TM199
3245	TP4067-410
3245	TP4067-411
3245	TR-1993
3245	TR-8040
3245	TR19A
3245	TR4010
3245	TSC-499
3245	VS2SC1681G/1E
3245	VS9-0003-913
3245	X19001-D
3245	ZEN116
3245	ZT930
3245	01-030373
3245	002-104-000
3245	002-9501
3245	002-9502
3245	002-9502-12
3245	2N4087
3245	2N4087A
3245	2SC1006
3245	2SC1006A
3245	2SC1006C
3245	2SC1222A
3245	2SC1222B
3245	2SC1222C
3245	2SC1222D
3245	2SC1327S
3245	2SC1537
3245	2SC1538A
3245	2SC1538S
3245	2SC1538SA
3245	2SC1647
3245	2SC1647Q
3245	2SC1647RY
3245	2SC1648
3245	2SC1648E
3245	2SC1648S
3245	2SC1648SH
3245	2SC1815-0
3245	2SC370-G
3245	2SC370-T
3245	2SC372-0
3245	2SC3721GR
3245	2SC372A
3245	2SC372B
3245	2SC372C
3245	2SC372D
3245	2SC372E
3245	2SC372F
3245	2SC372GN
3245	2SC372J
3245	2SC372K
3245	2SC372L
3245	2SC372M
3245	2SC3720
3245	2SC3720R
3245	2SC372R
3245	2SC373-14
3245	2SC373A
3245	2SC373AL
3245	2SC373B
3245	2SC373C
3245	2SC373D
3245	2SC373E
3245	2SC373F
3245	2SC373GN
3245	2SC373H
3245	2SC373J
3245	2SC373K
3245	2SC373L
3245	2SC373M
3245	2SC3730R
3245	2SC373R
3245	2SC373X
3245	2SC373Y
3245	2SC380-R
3245	2SC380A-0
3245	2SC380AR
3245	2SC380B
3245	2SC380B-Y
3245	2SC380C
3245	2SC380C-Y
3245	2SC380D-Y
3245	2SC380E
3245	2SC380E-Y
3245	2SC380F
3245	2SC380F-Y
3245	2SC380G
3245	2SC380GN
3245	2SC380H
3245	2SC380J
3245	2SC380K
3245	2SC380L
3245	2SC380M
3245	2SC3800
3245	2SC3800R
3245	2SC380X
3245	2SC528
3245	2SC644
3245	2SC644C
3245	2SC644F
3245	2SC644FR
3245	2SC644FS
3245	2SC644G
3245	2SC644H
3245	2SC644HR
3245	2SC644HS
3245	2SC644P
3245	2SC644PJ
3245	2SC644Q
3245	2SC644R
3245	2SC644S
3245	2SC644T
3245	2SC732
3245	2SC732A
3245	2SC732B
3245	2SC732BL
3245	2SC732BL-1
3245	2SC732C
3245	2SC732D
3245	2SC732E
3245	2SC732F
3245	2SC732G
3245	2SC732GN
3245	2SC732GR
3245	2SC732GR/4454C
3245	2SC732H
3245	2SC732J
3245	2SC732L
3245	2SC732M
3245	2SC7320R
3245	2SC732R
3245	2SC732S
3245	2SC732V
3245	2SC732X
3245	2SC732Y
3245	2SC733
3245	2SC733-0
3245	2SC733-0
3245	2SC733-ORG
3245	2SC733-YEL
3245	2SC733A
3245	2SC733B
3245	2SC733BL
3245	2SC733BLK
3245	2SC733C
3245	2SC733D
3245	2SC733E
3245	2SC733F
3245	2SC733G
3245	2SC733GN
3245	2SC733GR
3245	2SC733H
3245	2SC733J
3245	2SC733K
3245	2SC733L
3245	2SC733M
3245	2SC7330R
3245	2SC733Q
3245	2SC733R
3245	2SC733S-BL
3245	2SC733V
3245	2SC733X
3245	2SC733Y
3245	2SC734(R)
3245	2SC734-GRN
3245	2SC734-ORG
3245	2SC734-RED
3245	2SC734-YEL
3245	2SC734A
3245	2SC734B
3245	2SC734C
3245	2SC734D
3245	2SC734E
3245	2SC734F
3245	2SC734G
3245	2SC734GN
3245	2SC734H
3245	2SC734J
3245	2SC734K
3245	2SC734K/GR
3245	2SC734L
3245	2SC734M
3245	2SC7340R
3245	2SC734R
3245	2SC734X
3245	00002SC858
3245	2SC858E
3245	2SC858F
3245	2SC858FG
3245	2SC858G
3245	2SD599
3245	3N74
3245	3N75
3245	3N76
3245	3N77
3245	3N78
3245	3N79
3245	4-1791
3245	09-302097
3245	09-309059
3245	11B551-2
3245	11B551-3
3245	11B552-2
3245	11B552-3
3245	11B554-2
3245	11B554-3
3245	11B555-2
3245	11B555-3
3245	11C551-2
3245	11C551-3
3245	11C553-2
3245	11C553-3
3245	16A667-GRN
3245	16A667-ORG
3245	16A667-RED
3245	16A667-YEL
3245	16A668-GRN
3245	16A668-ORG
3245	16A668-YEL
3245	16A669-GRN
3245	16A669-YEL
3245	16L24
3245	16L25
3245	16L45
3245	16L65
3245	019-003675-205
3245	019-004094
3245	21A404-066
3245	21M087
3245	21M096
3245	24-001326
3245	24-001354
3245	27A10489-101-11
3245	31-027-0
3245	34-1009
3245	34-1019
3245	42-9029-31B
3245	42-9029-31L
3245	42-9029-31P
3245	42-9029-31Q
3245	42-9029-60D
3245	43B168613-1
3245	48-134813
3245	48-137325
3245	48-137390
3245	48-355004
3245	48-40170-G01
3245	48-40247G02
3245	48-40606J02
3245	48-40607J01
3245	48-45N2
3245	48-869269
3245	48-869293
3245	48-869338
3245	48-869384
3245	48-869409
3245	48-869416
3245	48-869447
3245	48-869474
3245	48-869486
3245	48-869497
3245	48-869547
3245	48-869594
3245	48-90343A79
3245	48K869269
3245	48K869293
3245	48K869409
3245	48K869447
3245	48K869474
3245	48K869486
3245	48R869338
3245	48R869384
3245	48R869416
3245	48R869547
3245	48R869594
3245	57B203-14
3245	61-309686
3245	66-P11112-0001
3245	66F026-1
3245	67A9060
3245	68A7366-1
3245	74-01-772
3245	86A327
3245	87-0230
3245	87-0230-1
3245	87-0231
3245	91T6
3245	92T6
3245	96-5346-01
3245	127-115
3245	130-40216
3245	0131-004792
3245	0131-005347
3245	0131-005348
3245	0131-005349
3245	0131-005350
3245	147-7016-01
3245	151-0341-00
3245	151-0456-00
3245	176-006-9-002
3245	195
3245	207A17
3245	215-37567
3245	250-0373
3245	250-0711
3245	296-55-9
3245	297L007C
3245	297L007C03
3245	297L015C01
3245	311D916P01
3245	353-9306-006
3245	353-9306-007
3245	400-1569-101
3245	417-108-13163
3245	417-126-12903
3245	417-226-13163
3245	462-1038-01
3245	462-1066-01
3245	462-2004
3245	536F(JVC)
3245	536GT
3245	599P3430
3245	686-257-0
3245	700A-858-328
3245	772-101-00
3245	858
3245	859GK
3245	921-133
3245	921-196
3245	921-206
3245	921-239B
3245	921-240B
3245	929(0)
3245	929D(JVC)
3245	930(0V)
3245	930DU
3245	943-742-002
3245	1472-8349
3245	1479-7963
3245	1515(NPN)
3245	1854-SRK1-1
3245	2000-245
3245	2020-05
3245	2057A2-373
3245	2057A2-385
3245	2057A2-436
3245	2057A2-542
3245	2057A2-543
3245	2057B-59
3245	2101
3245	2271
3245	2510-103
3245	3107-204-90150
3245	4039-00
3245	4039-01
3245	4824-0014
3245	4824-0014-02
3245	5613-871
3245	8000-00009-089
3245	8020-204
3245	8394
3245	8910-144
3245	9033-2
3245	9033-3
3245	9033-4
3245	9033-5
3245	9033GREEN
3245	9033RED
3245	9033WHITE
3245	9330-688-30112
3245	09500
3245	09501
3245	16237
3245	19500-253
3245	20103
3245	22810-174
3245	26810-153
3245	38478
3245	41175
3245	45337-A
3245	51429
3245	58810-164
3245	58810-167
3245	000073080
3245	000073350
3245	000073360
3245	000073373
3245	000073374
3245	90934-35
3245	99240-269
3245	100292
3245	136282
3245	138001
3245	138019-001
3245	138789-4
3245	138789-5
3245	144034
3245	147115-7
3245	147115-8
3245	147122P7
3245	150714-1
3245	162002-085
3245	168716
3245	171162-100
3245	171162-204
3245	171162-247
3245	171162-288
3245	181012
3245	200200
3245	200200-700
3245	268003
3245	281001-53
3245	281001-83
3245	323934
3245	510584
3245	530130-1
3245	531841-002
3245	532775
3245	610079-2
3245	610128-1
3245	610128-5
3245	610128-6
3245	618165-1
3245	618181-1
3245	740438
3245	740439
3245	740440
3245	740442
3245	803696
3245	1417308-1
3245	1417308-2
3245	1443024-1
3245	1471115-11
3245	1471115-3
3245	1471115-4
3245	1471120-11
3245	1471122
3245	1472475-1
3245	1473614-2
3245	1612738-1
3245	2498665-2
3245	2498665-3
3245	3755862
3245	4000921
3245	4906093
3245	7576015-02
3245	7576015-03
3245	7576015-04
3245	7576015-05
3245	7852452-01
3245	8722248-2
3245	10183001
3245	10545506
3245	11706998-2
3245	20088308
3245	23114217
3245	23114258
3245	50210700-00
3245	50210700-10
3245	62638249
3245	80052101
3245	82073206
3245	83073306
3245	436004601
3245	436006401
3245/199	A670720K
3245/199	A670729B
3245/199	AT16
3245/199	AT17
3245/199	BC121
3245/199	BC122
3245/199	BC123
3245/199	BC173B
3245/199	BC182KA
3245/199	BC182KB
3245/199	BC183KB
3245/199	BC183KC
3245/199	BC184K
3245/199	BC184KB
3245/199	BC239A
3245/199	BC239C
3245/199	BC382B
3245/199	BC382C
3245/199	BC383B
3245/199	BC383C
3245/199	BC384B
3245/199	BC413B
3245/199	BC413C
3245/199	BC414C
3245/199	BFY41
3245/199	BSW-68
3245/199	C1204
3245/199	C1204B
3245/199	C1204C
3245/199	C1633
3245/199	C335
3245/199	C372
3245/199	C372-1

SK	Industry Standard No.
3245/199	C372-2
3245/199	C372-0
3245/199	C372-R
3245/199	C372-Y
3245/199	C372-Z
3245/199	C372GR
3245/199	C372H
3245/199	C372Y
3245/199	C373
3245/199	C373BL
3245/199	C373G
3245/199	C373GR
3245/199	C373W
3245/199	C380
3245/199	C380-0
3245/199	C380A
3245/199	C380A-R
3245/199	C380A0
3245/199	C380AY
3245/199	C380R
3245/199	C380Y
3245/199	C648
3245/199	C648H
3245/199	C734
3245/199	C734-0
3245/199	C734-0
3245/199	C734-R
3245/199	C734-Y
3245/199	C734GR
3245/199	C734Y
3245/199	C833BL
3245/199	C907
3245/199	C907A
3245/199	C907AC
3245/199	C907AD
3245/199	C907AH
3245/199	C907C
3245/199	C907D
3245/199	CK420
3245/199	CS4003
3245/199	CS4061
3245/199	DS13
3245/199	EN1613
3245/199	EN1711
3245/199	EN956
3245/199	MPS6564
3245/199	MPSA09
3245/199	MPSA10
3245/199	MPSK20
3245/199	MPSK21
3245/199	MPSK22
3245/199	NPC069
3245/199	NPC069-98
3245/199	S0024
3245/199	TE2715
3245/199	TE2716
3245/199	TE3707
3245/199	TE3708
3245/199	TE3709
3245/199	TE3710
3245/199	TE3711
3245/199	TE3859A
3245/199	VS2SC372-Y/1E
3245/199	ZEN121
3245/199	ZTX107
3245/199	ZTX108
3245/199	ZTX109
3245/199	01-030733
3245/199	2N2925
3245/199	2SC1815
3245/199	2SC1815Y
3245/199	2SC372
3245/199	2SC372(0)
3245/199	2SC372-1
3245/199	2SC372-2
3245/199	2SC372-0
3245/199	2SC372-R
3245/199	2SC372-Y
3245/199	2SC372-Z
3245/199	2SC372/4454C
3245/199	2SC372BL
3245/199	2SC372G
3245/199	2SC372GR
3245/199	2SC372H
3245/199	2SC372X
3245/199	2SC372Y
3245/199	2SC373
3245/199	2SC373BL
3245/199	2SC373G
3245/199	2SC373GR
3245/199	2SC373W
3245/199	2SC380
3245/199	2SC380-0
3245/199	2SC380A
3245/199	2SC380A-R
3245/199	2SC380ATV
3245/199	2SC380AY
3245/199	2SC380D
3245/199	2SC380R
3245/199	2SC380Y
3245/199	2SC734
3245/199	2SC734-0
3245/199	2SC734-0
3245/199	2SC734-R
3245/199	2SC734GR
3245/199	2SC7340
3245/199	2SC734Y
3245/199	2SC907AC
3245/199	2SC907AD
3245/199	2SC930D
3245/199	2SC930E
3245/199	022-2876-003
3245/199	31-002-0
3245/199	48R869641
3245/199	700-156
3245/199	921-404
3245/199	921-408
3245/199	977-64197
3245/199	1473586-2
3245/199	23114126
3245/199	23114941
3245/199	23114975
3245/199	35050411
3245/199	35050412
3245/199	35062712
3245/199	632037218
3245/199	1760579001
3245/199	2003038092
3246	C1187
3246	C1687
3246	C784
3246	C784-0
3246	C784-0
3246	C784BN
3246	C784R
3246	C784Y
3246	C785BN
3246	C785D
3246	C785E
3246	C785R
3246	ECG229
3246	EL75
3246	HEP-S0020
3246	HEP-S0033
3246	HX-50091
3246	KSC1187R
3246	MPS9426
3246	MPS9426A
3246	MPS9426B
3246	MPS9426BC
3246	MPS9426C
3246	REN229
3246	TM229
3246	01-031686
3246	01-031688
3246	2SC1128
3246	2SC1128(M)
3246	2SC1128(S)
3246	2SC1128A
3246	2SC1128B
3246	2SC1128BL
3246	2SC1128C
3246	2SC1128D
3246	2SC1128G
3246	2SC1128H
3246	2SC1128M
3246	2SC1128R
3246	2SC1128S
3246	2SC1128Y
3246	2SC1129
3246	2SC1129(M)
3246	2SC1129(R)
3246	2SC1129-0
3246	2SC1129A
3246	2SC1129B
3246	2SC1129BL
3246	2SC1129C
3246	2SC1129G
3246	2SC1129M
3246	2SC1129R
3246	2SC1129Y
3246	2SC1187
3246	2SC1393
3246	2SC1393K
3246	2SC1393M
3246	2SC1394
3246	2SC1686
3246	2SC1687
3246	2SC1727
3246	2SC1855
3246	2SC2009
3246	2SC2057
3246	2SC2057D
3246	2SC378-YEL
3246	2SC784
3246	2SC784(BN)
3246	2SC784-0
3246	2SC784-6
3246	2SC784-BN
3246	2SC784-0
3246	2SC784-R
3246	2SC784A
3246	2SC784B
3246	2SC784BN
3246	2SC784BN-1
3246	2SC784C
3246	2SC784D
3246	2SC784E
3246	2SC784F
3246	2SC784G
3246	2SC784GN
3246	2SC784H
3246	2SC784J
3246	2SC784K
3246	2SC784L
3246	2SC784M
3246	2SC7840
3246	2SC7840R
3246	2SC784R
3246	2SC784RA
3246	2SC784X
3246	2SC784Y
3246	2SC785
3246	2SC785-0
3246	2SC785A
3246	2SC785B
3246	2SC785BL
3246	2SC785BN
3246	2SC785BR
3246	2SC785C
3246	2SC785D
3246	2SC785E
3246	2SC785F
3246	2SC785G
3246	2SC785GN
3246	2SC785GR
3246	2SC785H
3246	2SC785K
3246	2SC785L
3246	2SC785M
3246	2SC785R
3246	2SC785RA
3246	2SC785V
3246	2SC785X
3246	2SC785Y
3246	2SC8600R
3246	2SC860R
3246	2SC927F
3246	2SC927G
3246	2SC927GN
3246	2SC927H
3246	2SC927J
3246	2SC927K
3246	2SC927L
3246	2SC927M
3246	2SC9270R
3246	2SC927R
3246	2SC927X
3246	2SC927XL
3246	2SC927Y
3246	003-00
3246	13-23824-1
3246	44T-300-110
3246	46-86512-3
3246	57A180-4
3246	121-503
3246	121-507
3246	121-508
3246	121-509
3246	121-954
3246	249-1L
3246	249N1
3246	753-1303-801
3246	3676-1
3246	000073140
3246	125144
3246	137338
3246	610249-1
3246	5321901
3246	06120073
3246/229	C785
3246/229	C927
3246/229	C927B
3246/229	C927C
3246/229	C927CJ
3246/229	C927CU
3246/229	C927CW
3246/229	C927D
3246/229	C927E
3246/229	ED1502D
3246/229	EP15X54
3246/229	FCS9016
3246/229	FCS9016D
3246/229	FCS9016E
3246/229	FCS9016F
3246/229	FCS9016G
3246/229	GE-61
3246/229	JSP7001
3246/229	JSP7001B
3246/229	S0020
3246/229	TG2SC2057C
3246/229	01-031855
3246/229	01-032057
3246/229	01-571794
3246/229	2SC1688
3246/229	2SC250
3246/229	2SC927
3246/229	2SC927A
3246/229	2SC927B
3246/229	2SC927C
3246/229	2SC927CJ
3246/229	2SC927CT
3246/229	2SC927CU
3246/229	2SC927D
3246/229	2SC927E
3246/229	13-23824-3
3246/229	48-155088
3246/229	57A179-4
3246/229	121-522
3246/229	121-524
3246/229	676-1
3246/229	753-5751-359
3246/229	3535
3246/229	124757
3246/229	1473535-1
3246/229	1473676-1
3246/229	2321511
3246/229	2003185506
3247	A-1853-0050-1
3247	A-1853-0066-1
3247	A-1853-0077-1
3247	A-1853-0086-1
3247	A-1853-0098-1
3247	A564-0
3247	A564AS
3247	A564AT
3247	A564S
3247	A564T
3247	A565Q
3247	A578
3247	A578A
3247	A578B
3247	A578C
3247	A579
3247	A579A
3247	A579B
3247	A579C
3247	A721S
3247	A721T
3247	A721U
3247	A722
3247	A722T
3247	A722U
3247	A725F
3247	A725G
3247	A725H
3247	A726
3247	A726F
3247	A726G
3247	A726H
3247	BC158C
3247	BC159C
3247	BC178C
3247	BC179C
3247	BC206C
3247	BC258C
3247	BC320A
3247	BC320C
3247	BC321
3247	BC321A
3247	BC321B
3247	BC321C
3247	BC322
3247	BC322B
3247	BC322C
3247	BC526A
3247	BC526B
3247	BC526C
3247	BCW59B
3247	BCW61BC
3247	BCW61BD
3247	BCX71BJ
3247	BCX71BK
3247	BTX-084
3247	C7
3247	C8
3247	D29F1
3247	D29F2
3247	D29F3
3247	D29F4
3247	DW7655
3247	ECG234
3247	FX3964
3247	HEP-S0006
3247	M7
3247	M9412
3247	PM1120
3247	RE193
3247	REN234
3247	S026094
3247	SE5-0909
3247	TM234
3247	TR2SA763
3247	2SA564-0
3247	2SA564ABQ-1
3247	2SA564AS
3247	2SA564AT
3247	2SA564FQ-1
3247	2SA564FR-1
3247	2SA5640R
3247	2SA564S
3247	2SA564T
3247	2SA572
3247	2SA578
3247	2SA578A
3247	2SA578B
3247	2SA578C
3247	2SA579
3247	2SA579A
3247	2SA579B
3247	2SA579C
3247	2SA666S
3247	2SA702
3247	2SA721
3247	2SA721R
3247	2SA721S
3247	2SA721T
3247	2SA721U
3247	2SA722
3247	2SA722S
3247	2SA722T
3247	2SA722U
3247	2SA725
3247	2SA725F
3247	2SA725G
3247	2SA725H
3247	2SA726
3247	2SA726F
3247	2SA726G
3247	2SA726H
3247	2SA763
3247	2SA763-W
3247	2SA763-WL-3
3247	2SA763-WL-4
3247	2SA763-WL-5
3247	2SA763-WL-6
3247	2SA763-WN
3247	2SA763-WN-3
3247	2SA763-WN-4
3247	2SA763-WN-5
3247	2SA763-WN-6
3247	2SA763-Y
3247	2SA763-YL
3247	2SA763-YL-3
3247	2SA763-YL-4
3247	2SA763-YL-5
3247	2SA763-YL-6
3247	2SA763-YN
3247	2SA763-YN-3
3247	2SA763-YN-4
3247	2SA763-YN-5
3247	2SA763-YN-6
3247	2SA786
3247	2SB774R
3247	2SB774S
3247	11-691501
3247	11-691502
3247	14-803-32
3247	34-1007
3247	34-1008
3247	40-09437
3247	40-09952
3247	40-11253
3247	42-9029-40W
3247	42-9029-60B
3247	42-9029-70Q
3247	48-355006
3247	48-869412
3247	48-869467
3247	48-90234A38
3247	48-90234A99
3247	70-943-773-001
3247	207A14
3247	352-0773-010
3247	352-0773-020
3247	352-0773-030
3247	417-153-13431
3247	417-196-13262
3247	461-2001
3247	1038-9922
3247	1063-8435
3247	57009
3247	75561-18
3247	75561-21
3247	75596-1
3247	75596-2
3247	75596-3
3247	75596-4
3247	138049-001
3247	138049-004
3247	144030
3247	144858
3247	146902
3247	610083-2
3247	610134-2
3247	610134-5
3247	50211400-00
3247	50211400-01
3247	50211410-10
3247	50211410-11
3247	632049514
3247/234	A564-0
3247/234	A564A
3247/234	A564ABQ
3247/234	A564A0
3247/234	A564AP
3247/234	A564AQ
3247/234	A564F
3247/234	A564FQ
3247/234	A564FR
3247/234	A564J
3247/234	A564P
3247/234	A564POR
3247/234	A564Q
3247/234	A564QHD
3247/234	A564QR
3247/234	A564R
3247/234	A572
3247/234	A572-1
3247/234	BC206A
3247/234	BC307C
3247/234	BC308B
3247/234	BC308C
3247/234	BC320
3247/234	BC478B
3247/234	BCW58B
3247/234	ED1802M
3247/234	HEP-S0031
3247/234	MPSK72
3247/234	TP3638
3247/234	TP3638A
3247/234	TPS6516
3247/234	TPS6517
3247/234	TPS6518
3247/234	TPS6519
3247/234	TPS6522
3247/234	TPS6523
3247/234	VS2SA854-Q/1E
3247/234	ZEN122
3247/234	001-021170
3247/234	001-021171
3247/234	001-021172
3247/234	001-021173
3247/234	2SA564
3247/234	2SA564-0
3247/234	2SA564A
3247/234	2SA564ABQ
3247/234	2SA564A0
3247/234	2SA564AP
3247/234	2SA564AQ
3247/234	2SA564AR
3247/234	2SA564D
3247/234	2SA564F
3247/234	2SA564FQ
3247/234	2SA564FR
3247/234	2SA564G
3247/234	2SA564H
3247/234	2SA564J
3247/234	2SA564P
3247/234	2SA564POR
3247/234	2SA564Q
3247/234	2SA564QHD
3247/234	2SA564QR
3247/234	2SA564R
3247/234	2SA564Y
3247/234	2SA666Q
3247/234	2SA666R
3247/234	2SA854
3247/234	2SA854Q
3247/234	2SB774
3247/234	48-155045
3247/234	200X4085-415
3247/234	4053
3248	C1096A
3248	C1096B
3248	C1096C
3248	C1096D
3248	C1096K
3248	C1096L
3248	C1096M
3248	C1096W
3248	DBBY003001
3248	DDBY227001
3248	KTR1096C
3248	YAAN2SC1096
3248	YAAN2SC1096K
3248	YAAN2SC1096L
3248	YAAN2SC1096M
3248	YAAN2SC1096N
3248	YAANL2SC1096
3248	YAANL2SC1096K
3248	YAANL2SC1096L
3248	YAANL2SC1096M
3248	YAANL2SC1096N
3248	YAANZ2SC1096
3248	YAANZ2SC1096K
3248	YAANZ2SC1096L
3248	YAANZ2SC1096M
3248	2SC1096
3248	2SC1096-3ZM
3248	2SC1096-4ZL
3248	2SC1096A
3248	2SC1096B
3248	2SC1096C
3248	2SC1096D
3248	2SC1096E
3248	2SC1096K
3248	2SC1096L
3248	2SC1096M
3248	2SC1096N
3248	2SC1096W
3248	5321235
3249	GEIC-135
3249	RE327-IC
3249	TA70639
3249	09-308076
3249	61A001-12
3249	740-9000-566
3249	1077-2408
3249	003522
3249	26810-158
3249	37510-166
3249	916084
3250	C1475
3250	C1475-1
3250	GE-279
3250	2SC1313
3250	2SC1313B
3250	2SC1313F
3250	2SC1313G
3250	2SC1313H
3250	2SC1313Y
3250	2SC1313YE
3250	2SC1313YG
3250	2SC1313YH
3250	2SC1475
3250	2SC1475-1
3250	2SC1475-3
3250	2SC1475-4
3250	2SC1475A
3250	2SC1475D
3250	2SC369-BLU-G
3250	2SC369-GRN-G
3250	2SC369A
3250	2SC369B
3250	2SC369C
3250	2SC369D
3250	2SC369E
3250	2SC369F
3250	2SC369G/BL
3250	2SC369G/GR
3250	2SC369H
3250	2SC369J
3250	2SC369K
3250	2SC369L
3250	2SC369M
3250	2SC3690R
3250	2SC369R
3250	2SC369Y
3250	8-760-413-10

SK	Industry Standard No.
3250	09-302202
3250	921-1010
3250	741858
3250/315	C1313F
3250/315	C1313G
3250/315	C1313H
3250/315	C1313YF
3250/315	C1313YG
3250/315	C1313YH
3250/315	C369BL
3250/315	C369G
3250/315	C369G-BL
3250/315	C369G-GR
3250/315	C369GR
3250/315	EP15X5
3250/315	SE4010
3250/315	2N5088
3250/315	2SC369
3250/315	2SC369BL
3250/315	2SC369G
3250/315	2SC369GR
3250/315	2SC369V
3250/315	2SC900(L)
3250/315	2SD438
3250/315	2SD438E
3250/315	121-1014
3250/315	144035
3250/315	144040
3251	C1760
3251	C1760-3
3251	DBCZ176003
3251	EA15X380
3251	EQS-0184
3251	GE-276
3251	01-031760
3251	01-031846
3251	2SC1760
3251	2SC1760-2
3251	2SC1760-3
3251	2SC1761
3251	8-763-113-00
3252	C1728
3252	C1728-3
3252	C1728D
3252	GE-275
3252	RE204
3252	2SC1728
3252	2SC1728-3
3252	2SC1728D
3252	09-302219
3252	8000-00043-064
3252	16304190031
3252/302	MPS-U31
3252/302	2SC2074
3252/302	2SC2074C
3252/302	2SC2074Y
3253	A671656K
3253	DDBY256001
3253	GE-270
3253	NCB14
3253	WEP701
3253	01-031957
3253	2SC2314
3253	2SC2314D
3253	2SC495
3253	2SC495-0
3253	2SD612E
3253	10-006
3253	44T-300-107
3253	48-40734P01
3253	48-42884P01
3253	106-005
3253	176-029-9-004
3253	2000-270
3253	8888-00005-013
3253	5321311
3253/295	C1449
3253/295	NCBV14
3253/295	2SC1382
3253/295	2SC1382-Q
3253/295	2SC1382Y
3253/295	2SC1449
3253/295	2SC1449M
3253/295	2SC1846
3253/295	2SC1846B
3253/295	2SC1846P
3253/295	2SC1846Q
3253/295	2SC1846R
3253/295	2SC1846S
3253/295	2SC2028
3253/295	2SC2028-B/20
3253/295	2SC2028/2
3253/295	2SC2036
3253/295	2SC2091
3253/295	2SC2497Q
3253/295	2SC2497R
3253/295	2SD612
3253/295	2SD612K
3253/295	48-90343A85
3253/295	144856
3253/295	632000001
3254	CA3035
3254	CA3035V
3254	CA3035V1
3254	GEIC-226
3254	R3528-1
3254	442-4
3254	1133-14
3254	3528-1
3254	3991-303-112
3254	4020
3254	127166
3254	1473528-1
3255	CA1391E
3255	GEIC-244
3255	MC1391P
3255	SN76591P
3255	14-2014-01
3255	56A23-1
3255	86X0084-001
3255	221-141
3255	612082-2
3256	2N6288
3257	MJ3701
3257	SDT3575
3257	SDT3701
3257	SDT3703
3257	SDT3706
3257	SDT3709
3257	SDT3712
3257	SDT3715
3257	SDT3716
3257	SDT3720
3257	SDT3721
3257	SDT3725
3257	SDT3726
3257	SDT3729
3257	SDT3733
3257	2N4387
3257	2N4898
3257	2N5956
3257	2N6312
3258	2N3202
3258	2N3208
3258	2N3774
3258	2N3778
3258	2N3782
3258	2N4234
3258	2N5110
3258	2N5783
3259	2N3203
3259	2N3775
3259	2N3779
3259	2N4235
3260	2N3773
3260	2N5632
3260	2N5634
3260	2SC1584
3260	16598
3260	43104
3260	67586
3261	CXL175
3261	ECG175
3261	GE-246
3261	REN175
3261	SJ1172
3261	SJ2095
3261	SJ3680
3261	SJ811
3261	TM175
3261	WEP241
3261/175	MJ2252
3261/175	MJ3202
3261/175	MJ400
3261/175	RE14
3261/175	2N3585
3261/175	2N3739
3261/175	2SC1304
3261/175	2SC508
3261/175	2SD157
3261/175	2SD157A
3261/175	2SD157B
3261/175	2SD157C
3262	QRT-101
3262	2N4037
3263	2N2270
3265	C151
3265	RE70
3265	REN70
3265	2SC151
3265	2SC151A
3265	2SC151B
3265	2SC151C
3265	2SC151D
3265	2SC151E
3265	2SC151G
3265	2SC151GN
3265	2SC151H
3265	2SC151J
3265	2SC151K
3265	2SC151L
3265	2SC151M
3265	2SC151OR
3265	2SC151R
3265	2SC151Y
3265	2SC152
3265	2SC152A
3265	2SC152B
3265	2SC152C
3265	2SC152D
3265	2SC152E
3265	2SC152F
3265	2SC152G
3265	2SC152GN
3265	2SC152H
3265	2SC152J
3265	2SC152K
3265	2SC152L
3265	2SC152M
3265	2SC152OR
3265	2SC152Y
3268	GE-72
3268	2N5239
3268	2SD93
3269	MJ410
3269	2SC1768
3269	2SD388
3271	C1168
3271	C490
3271	C491
3271	C491BL
3271	C491R
3271	C491Y
3271	2N3583
3271	2N3584
3271	2N3738
3271	2SC1168
3271	2SC1168X
3271	2SC490
3271	2SC490-RED
3271	2SC490-YEL
3271	2SC491
3271	2SC491A
3271	2SC491B
3271	2SC491BL
3271	2SC491C
3271	2SC491D
3271	2SC491E
3271	2SC491F
3271	2SC491G
3271	2SC491GN
3271	2SC491H
3271	2SC491J
3271	2SC491K
3271	2SC491L
3271	2SC491M
3271	2SC491OR
3271	2SC491R
3271	2SC491X
3271	2SC491Y
3271	36771100
3272	2N4231
3273	D44C1
3274	A670
3274	A670A
3274	A670B
3274	A670C
3274	B434
3274	B434-0
3274	B434-R
3274	B434-Y
3274	B435
3274	B435-0
3274	B435-R
3274	B435-Y
3274	B511
3274	B511C
3274	B511D
3274	D45H4
3274	D45H5
3274	ECG153
3274	M9348
3274	MJE1290
3274	REN153
3274	SD1445
3274	SDJ445
3274	SDK445
3274	SJ1284
3274	TM153
3274	2SA490
3274	2SA490-0
3274	2SA490A
3274	2SA490B
3274	2SA490C
3274	2SA490D
3274	2SA490E
3274	2SA490F
3274	2SA490G
3274	2SA490GN
3274	2SA490H
3274	2SA490J
3274	2SA490K
3274	2SA490L
3274	2SA490M
3274	2SA490OR
3274	2SA490R
3274	2SA490Y
3274	2SA490YA
3274	2SA670
3274	2SA670A
3274	2SA670B
3274	2SA670C
3274	2SA755
3274	2SA755A
3274	2SA755B
3274	2SA755C
3274	2SA768
3274	2SB434
3274	2SB434-0
3274	2SB435
3274	2SB435-0
3274	2SB435R
3274	002SB435RY
3274	2SB435Y
3274	2SB511
3274	2SB511C
3274	2SB511D
3274	2SB511E
3274	146081
3274	2004049081
3274/153	D45C4
3274/153	D4505
3274/153	D4506
3274/153	MJE2490
3274/153	2N6132
3274/153	2N6133
3274/153	2SA671
3274/153	2SA671A
3274/153	2SA671B
3274/153	2SA671C
3274/153	2SA671K
3274/153	2SA671KA
3274/153	2SA671KB
3274/153	2SA671KC
3274/153	13-41629-4
3274/153	06100007
3275	A-1854-0358-1
3275	A-1854-0365-1
3275	A-1854-0474-1
3275	A-1854-0533
3275	A133
3275	A86-565-2
3275	A8N
3275	BC532
3275	BC533
3275	ES97
3275	HEP-S0001
3275	HEP-S0005
3275	RE78
3275	SE5-0745
3275	TR32
3275	2D020
3275	2D020-173
3275	2D020-174
3275	2N1154
3275	2N1155
3275	2N1156
3275	2N4945
3275	2N719
3275	2N719A
3275	13-35807-1
3275	13-39851-1
3275	020-1110-038
3275	21A112-074
3275	34-6016-41
3275	34-6016-56
3275	034A
3275	48-137238
3275	48-137278
3275	48-137332
3275	48-137386
3275	57B181-12
3275	68A7702P1
3275	73B140585-21
3275	73B140585-22
3275	73B140585-23
3275	73B140585-24
3275	73B140585-25
3275	73B140585-26
3275	73B140585-27
3275	86-557-2
3275	86-566-2
3275	86-611-2
3275	86-615-2
3275	86X0049-001
3275	99S034
3275	99S040
3275	99S060-1
3275	99S061-1
3275	590-591731
3275	590-591811
3275	139269
3275	140624
3275/194	BCY85
3275/194	BCY86
3275/194	GE-220
3275/194	MPS6560
3275/194	MPS8098
3275/194	MPS8099
3275/194	MPSL01
3275/194	2N5550
3275/194	2SC212
3275/194	2SC2235
3275/194	2SD666
3275/194	2SD666C
3275/194	86-400-2
3275/194	120073
3275/194	139268
3275/194	143796
3275/194	143797
3275/194	1417318-1
3276	GEIC-216
3276	IC-200(ELCOM)
3276	MC1306P
3277	GEIC-235
3277	MC1314P
3277	TR09033
3278	GEIC-237
3278	MC1315P
3278	TR09034
3279	C6079P
3279	GEIC-218
3279	HEP-C6079P
3279	IC-18(PHILCO)
3279	IC-202(ELCOM)
3279	ITT1330
3279	M5169P
3279	MC1330P
3279	RE310-IC
3279	SC9430P
3279	SN76530P
3279	TDA1330P
3279	15-39060-1
3279	21A101-015
3279	46-5002-18
3279	229-1301-42
3279	266P10201
3279	266P10202
3279	442-56
3279	39060-1
3279	45385
3279	23119014
3279/747	21A120-002
3280	C6061P
3280	GEIC-219
3280	HEP-C6061P
3280	IC-204(ELCOM)
3280	MC1355P
3280	MC1355PQ
3280	TR09007
3280	09-007
3280	09007
3280	45386
3281	EICM-14
3281	G09-007-A
3281	G09-007-B
3281	G09-029-C
3281	G09-029-D
3281	G09007
3281	G09007B
3281	GEIC-94
3281	MPC566HB
3281	MPC566HC
3281	MPC566HD
3281	RE329-IC
3281	TA-7063P
3281	TA7063
3281	TA7063P
3281	TA7063P-B
3281	TA7063P-C
3281	TA7063P-D
3281	TA7063P-0
3281	09-308036
3281	09-308043
3281	051-0011-00
3281	051-0011-00-04
3281	051-0011-01
3281	051-0011-02
3281	051-0011-03
3281	051-0011-04
3281	57A32-19
3281	266P30706
3281	740583
3281	916067
3282	GEIC-182
3282	TA7204P
3283	GEIC-95
3283	23119011
3283	23119022
3284	GEIC-41
3284	GEIC-98
3284	HA1140
3284	LA1353
3284	RE328-IC
3284	REN1080
3284	TA7075
3284	TA7075P
3284	46-13101-3
3284	4206002600
3284	4206003900
3285	GEIC-101
3285	RH-1X0024CEZZ
3285	TA7102P
3285	TA7102P(FA-1)
3285	TA7102P(FA-2)
3285	23119016
3285	23119028
3286	B0313600
3286	GEIC-109
3286	TA7148P
3286	46-1394-3
3286	23119995
3286	6644001700
3287	B0313700
3287	GEIC-110
3287	TA7149P
3287	46-1397-3
3287	23119994
3287	6644001800
3288	A1200
3288	A1201
3288	A1201B
3288	A1201C
3288	A1201T
3288	EA3282
3288	EA33X8367
3288	G09-015-B
3288	GEIC-43
3288	LA1200
3288	LA1201
3288	000LA1201B
3288	LA1201C
3288	LA1201C-W
3288	LA1201T
3288	LA1201W
3288	LA1202
3288	RE335-IC
3288	RT7399
3288	8-759-812-01
3288	09-308063
3288	14LQ007
3288	19-076001
3288	24MW1028
3288	41C-407
3288	46-1343-3
3288	307-005-9-001
3288	740-8160-190
3288	905-30B
3288	917-1201-0
3288	1012
3288	1077-2382
3288	003501
3288	61443
3288	7910112
3288	7910112-01
3288	7910122-01
3289	GEIC-71
3289	LA1366N
3289	4-2060-07500
3289	4206007500
3290	GEIC-72
3290	LA1367
3290	4-2060-05200
3290	4206005200
3291	GE-723
3291	STK-025
3291	STK-025G
3291	STK-032
3293	C1417C
3293	C351
3293	C385A
3293	CXL107
3293	DDBY269001
3293	DS-781
3293	EA15X7231
3293	ECG107
3293	GE-11
3293	NJ202B
3293	REN107
3293	TM107
3293	WEP535
3293	2SC1215
3293	2SC1215C
3293	2SC1215D
3293	2SC1215E
3293	2SC1215F
3293	2SC1215G
3293	2SC1215GN
3293	2SC1215H
3293	2SC1215J
3293	2SC1215K
3293	2SC1215L
3293	2SC1215M
3293	2SC1215OR
3293	2SC1215R
3293	2SC1215Y
3293	2SC1395
3293	2SC1417
3293	2SC1417D
3293	2SC1417F
3293	2SC1417G
3293	2SC1417H
3293	2SC1417U
3293	2SC1417V
3293	2SC1417VW
3293	2SC1417W
3293	2SC1730
3293	2SC1730L
3293	2SC1789
3293	2SC1906
3293	2SC351
3293	2SC351A
3293	2SC351B
3293	2SC351C
3293	2SC351D
3293	2SC351E
3293	2SC351F
3293	2SC351G
3293	2SC351GN
3293	2SC351H
3293	2SC351J
3293	2SC351K
3293	2SC351L
3293	2SC351M
3293	2SC351OR
3293	2SC351R
3293	2SC351Y
3293	2SC385
3293	2SC385A
3293	2SC385B
3293	2SC385C
3293	2SC385D
3293	2SC385E
3293	2SC385F
3293	2SC385G
3293	2SC385GN
3293	2SC385H
3293	2SC385J
3293	2SC385K
3293	2SC385L
3293	2SC385M
3293	2SC385OR
3293	2SC385R
3293	2SC385Y
3293	2SC387
3293	2SC387A
3293	2SC387G
3293	00002SC535
3293	2SC535(B)
3293	2SC535A
3293	2SC535AL
3293	2SC535B
3293	2SC535C
3293	2SC535D
3293	2SC535E
3293	2SC535F
3293	2SC535G
3293	2SC535GN
3293	2SC535H
3293	2SC535J

SK	Industry Standard No.
3293	2SC535K
3293	2SC535L
3293	2SC535M
3293	2SC535OR
3293	2SC535R
3293	2SC535X
3293	2SC535Y
3293	2SC668
3293	2SC668A
3293	2SC668B
3293	2SC668C
3293	2SC668C1
3293	2SC668CD
3293	2SC668D0
3293	2SC668D1
3293	2SC668DV
3293	2SC668DX
3293	2SC668DZ
3293	2SC668E
3293	2SC668E1
3293	2SC668E2
3293	2SC668EP
3293	2SC668EV
3293	2SC668F
3293	2SC668G
3293	2SC668GN
3293	2SC668H
3293	2SC668K
3293	2SC668L
3293	2SC668M
3293	2SC668OR
3293	2SC668R
3293	2SC668X
3293	2SC668Y
3293	2SC684
3293	2SC684A
3293	2SC684B
3293	2SC684BK
3293	2SC684C
3293	2SC684D
3293	2SC684E
3293	2SC684F
3293	2SC684G
3293	2SC684GN
3293	2SC684H
3293	2SC684J
3293	2SC684K
3293	2SC684L
3293	2SC684M
3293	2SC684R
3293	2SC684X
3293	2SC684Y
3293	1100-9446
3293	5321431
3293	06120009
3293	2003190604
3293/107	C387A(FA-3)
3293/107	GE-86
3293/107	01-031906
3293/107	121-612
3293/107	200X3190-604
3293/107	4792
3295	AN253
3295	GEIC-59
3297	ECG280
3297	GE-262
3297	RCA1B06
3297	REN280
3297	2N1487
3297	2N1488
3297	2N1489
3297	2N4347
3297	2N4348
3297	2SC1079
3297	2SC1079R
3297	2SC1402
3297	2SC897C
3297	2SC898
3297	2SC898C
3297	2SD125AHB
3297	2SD180F
3297	2SD180GN
3297	2SD180H
3297	2SD180J
3297	2SD180R
3297	2SD180X
3297	2SD180Y
3297	2SD188D
3297	2SD188E
3297	2SD188F
3297	2SD188G
3297	2SD188GN
3297	2SD188H
3297	2SD188J
3297	2SD188OR
3297	2SD188R
3297	2SD188X
3297	2SD188Y
3297	2SD322
3297	2SD322A
3297	2SD322B
3297	2SD322C
3297	2SD323
3297	2SD323A
3297	2SD323B
3297	2SD323C
3297	2SD425
3297	2SD427
3297	96-5310-01
3297/280	2SC1111
3297/280	2SC665
3297/280	2SC665H
3297/280	2SC665HA
3297/280	2SD124A
3297/280	2SD124AH
3297/280	2SD124AHA
3297/280	2SD124AHB
3297/280	2SD125
3297/280	2SD125A
3297/280	2SD125AH
3297/280	2SD125AHA
3297/280	2SD126
3297/280	2SD126A
3297/280	2SD126AH
3297/280	2SD126AHA
3297/280	2SD126AHB
3297/280	2SD126H
3297/280	2SD126HA
3297/280	2SD126HB
3297/280	2SD180
3297/280	2SD180A
3297/280	2SD180B
3297/280	2SD180C
3297/280	2SD180D

SK	Industry Standard No.
3297/280	2SD180K
3297/280	2SD180L
3297/280	2SD180M
3297/280	2SD188
3297/280	2SD188A
3297/280	2SD188B
3297/280	2SD188C
3297/280	2SD188L
3297/280	2SD188M
3297/280	2SD217
3297/280	2SD673
3297/280	2SD674
3298	C1017
3298	C1018
3298	DBCZ101800
3298	IP20-0131
3298	KTR1017
3298	MPS9411
3298	01-031018
3298	2SC1017
3298	2SC1018
3298	2SC1018B
3298	13-23326-6
3298	302-679-1
3298	302-680
3298	1000-140
3298	4802-3268-000
3298	4804-3267-901
3298	5001-004
3298/299	576-0004-104
3299	A066-117
3299	C-F8
3299	C777
3299	C778
3299	C778B
3299	C799
3299	C799K
3299	CF8
3299	EQS-60
3299	EQS-86
3299	IP20-0005
3299	Q-01384R
3299	RE198
3299	S20446
3299	SC609
3299	2SC-F8
3299	2SC777
3299	2SC778
3299	2SC778B
3299	2SC778D
3299	2SC799
3299	2SC799K
3299	2SC866
3299	07-07141
3299	07-07164
3299	09-302050
3299	09-302051
3299	09-302056
3299	09-302117
3299	09-302169
3299	09-305091
3299	9TR5
3299	19-020-078
3299	90-110
3299	90-175
3299	172-007-9-001
3299	172-008-9-001
3299	172-009-9-001
3299	1001-08
3299	1009-04
3299	1040-05
3299	1068-17
3299	2062-17
3299	2062-17A
3299	2167-17
3299	2418-17
3299	4802-00008
3299	4802-00019
3299	8000-00004-083
3299	8000-00004-084
3299	8000-00004-P083
3299	8000-00004-P084
3299	8000-00005-011
3299	8000-00006-002
3299	8000-00006-230
3299	8000-00011-052
3299	082025
3299	082029
3299	0573541
3299	2006607-62
3299	5320151
3299	5320511
3299	26010022
3299/237	C697A
3299/237	C697F
3299/237	2SC1239
3299/237	2SC697
3299/237	2SC697A
3299/237	2SC697B
3299/237	2SC697D
3299/237	2SC697E
3299/237	2SC697F
3299/237	2SC697H
3299/237	2SC697J
3300/517	MH915001
3300/517	MH915002
3300/517	135691
3300/517	137646
3300/517	137693
3300/517	138019
3300/517	138752
3300/517	139001
3300/517	1464607-7
3300/517	1464607-8
3300/517	1826065-3
3301/531	GE-519
3301/531	57-83
3301/531	137031
3301/531	138421
3301/531	1464607-1
3301/531	1464607-2
3301/531	1464607-3
3301/531	1464607-4
3301/531	1464607-5
3301/531	1464607-6
3301/531	1464984-1
3301/531	1464984-2
3301/531	1826065-1
3302	MH970002
3302	1464607-10
3302	1464607-9
3302/533	145208
3302/533	1466860-1
3302/533	1466860-2

SK	Industry Standard No.
3303	EP62X41
3303	GE-516
3303	GE-517
3303	MH913
3303	MH919
3303	MH919D01
3303	MH931
3303	MH983A02
3303	MH983A03
3303	MH983A04
3303	28-35-01
3303	32-33057-2
3303	32-33094-1
3303	32-33094-3
3303	32-33094-4
3303	32-33094-6
3303	32-35894-2
3303	32-35894-3
3303	32-35894-4
3303	32-35894-6
3303	32-35894-7
3303	32-39704-1
3303	32-39704-2
3303	66F112-1
3303	66F112-2
3303	66F181-1
3303	66X0060-001
3303	66X0060-002
3303	93A91-1
3303	93A91-2
3303	93A91-3
3303	93A91-4
3303	93B91-1
3303	93D91-1
3303	93D91-2
3303	93D91-4
3303	93D91-5
3303	93D91-6
3303	212-139-03
3303	212-142
3303	72162-1
3303	530153-1
3303	530165-1
3303	530165-2
3303	530165-3
3303	530165-4
3303/522	32-29778-3
3303/522	72162-2
3304	A95-5314
3304	GE-518
3304	MH914
3304	MH920
3304	MH932
3304	MH987A01
3304	MH987A02
3304	MH987A03
3304	TVM778
3304	28-32-0X
3304	32-29778-2
3304	32-33057-3
3304	32-33057-4
3304	32-33057-5
3304	32-35894-1
3304	32-35894-5
3304	66F-054-3
3304	76-14327-2
3304	76-14327-3
3304	76-14327-4
3304	76-14327-5
3304	76-14327-6
3304	76-14327-7
3304	76-14327-8
3304	93A96-1
3304	93A96-3
3304	93B96-1
3304	93D96-1
3304	93D96-2
3304	93D96-3
3304	212-00102
3304	212-00104
3304	800-000-025
3305	MH943
3305	MH985A01
3305	530165-15
3305	530165-16
3305	530165-5
3306	MH1203
3306	MH1204
3306	MH1205
3306	MH1221A01
3306	REN523
3306	14B348-1
3306	14B348-2
3306	14B348-3
3306	14B348-4
3306	14B348-5
3306	32-39091-1
3306	32-39091-2
3306	32-39091-4
3306	32-39091-5
3306	32-39091-6
3306	32-39091-9
3306	66F159-1
3306	66F159-2
3306	66F159-3
3306	212-141
3306	212-141-01
3306	3107-108-40501
3306/523	32-39091-3
3306/523	32-39091-7
3306/523	32-39091-8
3306/523	212-141-02
3306/523	212-141-03
3306/523	212-141-04
3306/523	212-145
3306/523	212-145-01
3306/523	212-145-02
3306/523	212-146
3307	MH1201
3307	MH1222A01
3307	MH1222A02
3307	N2A
3307	N2A-1
3307	N2A-2
3307	86-127-3
3307	530165-11
3307	530165-13
3307	530165-17
3307	530165-6
3307/529	MH1222A03
3307/529	32-43737-1
3307/529	530165-14
3307/529	530165-18
3308	GE-540

SK	Industry Standard No.
3308	MH1206
3308	MH1207
3308	MH1209
3308	MH1220A01
3308	57-90
3308	57-98
3308	212-143
3308	3107-108-40401
3309	MH1030A02
3309	15-123060
3309	93A99-7
3309	93A99-8
3309	93D99-5
3309	93D99-6
3309	93D99-7
3309	93D99-8
3309	530165-12
3309	530165-7
3309/539	530165-10
3309/539	530165-8
3310	MH1030A01
3310	93D99-2
3310	93D99-3
3310	93D99-4
3311	A-042313
3311	A-1.5-01
3311	A02
3311	A054-230
3311	A065-101
3311	A069-112
3311	A10169
3311	A200
3311	A23
3311	A3A1
3311	A3A3
3311	A3A5
3311	A3A9
3311	A4A5
3311	A4A9
3311	A50
3311	A5A5
3311	A692T5
3311	A692T5-0
3311	A692X16-0
3311	A7102001
3311	A75-68-500
3311	A7568500
3311	A7572100
3311	A7572200
3311	A909-1018
3311	AAZ18D
3311	AC150
3311	AC30
3311	ACR83-1008
3311	AD10
3311	AD100
3311	AD150
3311	AD200
3311	AD50
3311	AE10
3311	AE100
3311	AE150
3311	AE1A
3311	AE200
3311	AE30
3311	AE50
3311	AM-G-10
3311	AM300
3311	AM300A
3311	AM301
3311	AM301A
3311	AM302A
3311	AM303
3311	AM303A
3311	AM304
3311	AM304A
3311	AM305A
3311	AM306
3311	AM306A
3311	AM307
3311	AM307A
3311	AM308A
3311	AS11
3311	B-1501U
3311	B-1881U
3311	B-1882U
3311	B01-02
3311	B1A1
3311	B1A5
3311	B1A9
3311	B2A1
3311	B2A5
3311	B36564
3311	B3A1
3311	B4A1
3311	B4A5
3311	B601-1012
3311	BA108
3311	BA114
3311	BA119
3311	BA128
3311	BA129
3311	BA130
3311	BA148
3311	BA153
3311	BAY44
3311	BAY86
3311	BAY87
3311	BB2A184
3311	BBIA
3311	BR51400-1
3311	BR51401-2
3311	BV25
3311	BY126
3311	BY141
3311	C10159
3311	CA10
3311	CA100
3311	CA150
3311	CA200
3311	CA50
3311	CB10
3311	CB150
3311	CB200
3311	CB50
3311	CD-2
3311	CD-2N
3311	CD1111
3311	CD1112
3311	CD1113
3311	CD1114
3311	CD1117
3311	CD1121
3311	CD1141

SK	Industry Standard No.
3311	CD1142
3311	CD1143
3311	CD1147
3311	CD1148
3311	CD1149
3311	CD13332
3311	CD13333
3311	CD13334
3311	CD13335
3311	CD13336
3311	CD13337
3311	CD13338
3311	CD13339
3311	CD37A2
3311	CDG005
3311	CER67
3311	CER670A
3311	CER67C
3311	CL010
3311	CL025
3311	CS16B
3311	CX-0047
3311	D-00384R
3311	D05
3311	D1201A
3311	D1201B
3311	D16U3
3311	D16U4
3311	D220
3311	D3Z
3311	D4R26
3311	DD2320
3311	DDAY002001
3311	DDAY042001
3311	DDAY103001
3311	DDBY002001
3311	DDE-201
3311	DH-002
3311	DHD805
3311	DHD806
3311	DIJ70488
3311	DIJ71958
3311	DIS1S
3311	DR1
3311	DR1100
3311	DR2
3311	DR3
3311	DR668
3311	DR669
3311	DR848
3311	DR863
3311	DS-130
3311	DS-130B
3311	DS-130C
3311	DS-130E
3311	DS-130YB
3311	DS-16E
3311	DS-16NE
3311	DS-1N
3311	DS-1U
3311	DS-2M
3311	DS-38
3311	DS-79
3311	DS-IM
3311	DS130ND
3311	DS130YE
3311	DS2K
3311	DS38
3311	DX-0445
3311	DX-0475
3311	DX-0721
3311	DX-1039
3311	E03155-001
3311	E03155-002
3311	E10172
3311	E13-20-00
3311	E21
3311	EA005
3311	EA15X14
3311	EA1672
3311	EA16X149
3311	EA16X21
3311	EA16X30
3311	EA16X55
3311	EA16X71
3311	EA2741
3311	EA3827
3311	EA3989
3311	EA57X14
3311	EC401
3311	EC402
3311	ED1502E
3311	ED2914
3311	ED3000
3311	ED3000A
3311	ED494583
3311	ED511097
3311	EDS-0002
3311	EDS-0017
3311	EDS-17
3311	EM401
3311	EP57X12
3311	ER2
3311	ERB11-01
3311	ERB12-02
3311	ES10233
3311	ES15056
3311	ES16X13
3311	F-14A
3311	F10124
3311	F10180
3311	F14A
3311	F14C
3311	OF164
3311	FD333
3311	FD3389
3311	FG-2N
3311	FM211N
3311	FPR50-1011
3311	FR-1MD
3311	FR1
3311	FR1U
3311	FR2P
3311	FRH-101
3311	FRH101
3311	G00-534-A
3311	G00-535-B
3311	G00535A
3311	G00-543-A
3311	G00-551-A
3311	G00535
3311	G01
3311	G02
3311	G1242

SK	Industry Standard No.	SK	Industry Standard No.	SK	Industry Standard No.	SK	Industry Standard No.	SK	Industry Standard No.
3311	G222	3311	MA350	3311	QRT-212	3311	SM-1K	3311	ZS101
3311	G296	3311	MA351	3311	QRT-213	3311	SM150-01	3311	ZS102
3311	G2A	3311	MC020	3311	R-7026	3311	SM150-11	3311	ZS120
3311	GER4001	3311	MC020A	3311	R-81347	3311	SM205	3311	ZS121
3311	GER4002	3311	MC021	3311	R122C	3311	SM4	3311	ZS122
3311	GER4003	3311	MC021A	3311	R154B	3311	SM5	3311	ZS30
3311	GI-1M385	3311	MC022	3311	R3/2H	3311	SM505	3311	ZS31
3311	GI-300D	3311	MC022A	3311	R6/2H	3311	SM705	3311	ZS32
3311	GIO8B	3311	MC023	3311	R66-8504	3311	SP1K-1	3311	ZS51
3311	GI3992-17	3311	MC023A	3311	R8/2H	3311	SR-05K-2	3311	ZS7
3311	GI411	3311	MC1520	3311	RA-1B	3311	SR-1	3311	ZS9
3311	GI420	3311	MC1521	3311	RA-1E	3311	SR-131-1	3311	ZS90
3311	GPO8B	3311	MC1522	3311	RA1B	3311	SR-1668	3311	ZS91
3311	GP230	3311	MC170	3311	RA1Z	3311	SR-1K-2	3311	ZS92
3311	GP250	3311	MC19	3311	RPJ70487	3311	SR1104	3311	ZTR-W06C
3311	GSM482	3311	MC456	3311	RPJ70643	3311	SR-31-1	3311	ZW2
3311	GSM483	3311	MCV	3311	RPJ72360	3311	SI152	3311	1A10425
3311	GSM51	3311	MD04	3311	RPJ72787	3311	SR1731-3	3311	1A16550
3311	GSM52	3311	MD134	3311	RG100B	3311	SR1D1M	3311	1B261
3311	GSM53	3311	MD135	3311	RG100D	3311	SR1DM	3311	1005
3311	GSM54	3311	MD136	3311	RGP-10D	3311	SR1DM-1	3311	1D261
3311	H50	3311	MH67	3311	RH-DX0039TAZZ	3311	SR1PM-1	3311	1D221
3311	H618	3311	MMO	3311	RH-DX0055TAZZ	3311	SR1PM-4	3311	1D261
3311	H619	3311	MP-5115	3311	RH-DX0059TAZZ	3311	SR1HM-2	3311	1E05
3311	H620	3311	MP100	3311	RH-DX0066TAZZ	3311	SR1HM-4	3311	1ET02
3311	H899	3311	MP5113	3311	RH-DX0073TAZZ	3311	SR1K-1	3311	1F14A
3311	HA50	3311	MP9602	3311	RH-DX0083CEZZ	3311	SR1K-1K	3311	1FM2
3311	HB3	3311	MP89601	3311	RH-DY0008EZZ	3311	SR1K-4	3311	1GA
3311	HC-39	3311	MP89602	3311	RM1Z	3311	SR1K1	3311	1N1046
3311	HC-68	3311	MQ3/2	3311	RM1ZM	3311	SR1K2	3311	1N2069A
3311	HCV	3311	MQ6/2	3311	RQ-409B	3311	SRK1	3311	1N4001
3311	HD2000110	3311	MQ8/2	3311	RQ-4232	3311	SS321	3311	1N4002
3311	HD20001100	3311	MR-150-01	3311	RT6322	3311	SS322	3311	1N4003
3311	HD2000301	3311	MR1C	3311	RT6332	3311	SS324	3311	1N4003GP
3311	HD20003010	3311	MR2261	3311	RT8231	3311	SS334	3311	1N4402
3311	HD2000413	3311	MR9600	3311	RT8840	3311	SS337	3311	1N485
3311	HD2000703	3311	MR9601	3311	RT8841	3311	STY-3	3311	1N485A
3311	HD2000903	3311	MR9602	3311	RV-2289	3311	SV01A	3311	1N485B
3311	HD2001310	3311	MS11H	3311	RV06	3311	SVD10D-1	3311	1N485C
3311	HP-20047	3311	MS12H	3311	RV06/7825B	3311	SVDVD1223	3311	1N503
3311	HP-20067	3311	MS13H	3311	RV1189	3311	SW-05-02	3311	1N511
3311	HP-20083	3311	MS14H	3311	RV1424	3311	SX641	3311	1N519
3311	HP-20084	3311	MS1H	3311	RV2072	3311	SX643	3311	1N5211
3311	HP20066	3311	MS2H	3311	RV2250	3311	T-E1133	3311	1N5215
3311	HGR1	3311	MS3H	3311	RV2289	3311	T-E1155	3311	1N5320
3311	HN-0000B	3311	MS4H	3311	RVD10D1	3311	Z10175	3311	1N625
3311	HN-00018	3311	MS0	3311	RVD10D1LP	3311	T21312	3311	1N626
3311	HN-00032	3311	MBR-500	3311	RVD13854	3311	T21333	3311	1N645
3311	HP-5A	3311	MBR500	3311	RVD4B265J2	3311	T3/2	3311	1N645A
3311	HR-5B	3311	MT021	3311	RVDD5-410	3311	T30155-001	3311	1N645B
3311	HR5A	3311	MT021A	3311	RVDKB16205	3311	T30155-1	3311	1N659
3311	HR5A8E	3311	MVA-05A	3311	RVDBR3AM2J	3311	T8/2	3311	1N681
3311	HS1001	3311	MY-1	3311	RVDVD1212L	3311	TA7802	3311	1N91
3311	HS1002	3311	N-02	3311	S-1.5-0	3311	TA7996	3311	1N919
3311	HS1003	3311	N-EA16X30	3311	S-2	3311	TC-0.2	3311	1R2A
3311	HS1007	3311	NL5	3311	S-5277B	3311	TF20	3311	1R2D
3311	HS1008	3311	NPC0010	3311	S-58	3311	TP21	3311	1R2E
3311	HS1009	3311	0101	3311	S0501	3311	TP22	3311	1R5B
3311	HS1010	3311	0A127	3311	S1-RECT-102	3311	TP23	3311	1R5U
3311	HS1012	3311	0A128	3311	S1-RECT-154	3311	TI-53	3311	1R91
3311	HS1020	3311	0A129	3311	S1.5	3311	TI54	3311	1R9J
3311	HV0000705	3311	0A130	3311	S1D901-02	3311	TI57	3311	1R9U
3311	I538	3311	M129	3311	S129	3311	TK5	3311	1R0R
3311	IP20-0054	3311	08-16308	3311	S14	3311	TL1	3311	1S1004
3311	IP20-0120	3311	08S-16308	3311	S1801-02	3311	TMD41	3311	1S119
3311	IP20-0167	3311	08S36885	3311	S1B-01-02	3311	TMD42	3311	1S1554
3311	IR24	3311	P10156	3311	00000S1B01	3311	TMD45	3311	1S1666
3311	ITT350	3311	P1A5	3311	S1B01-01	3311	TR-77	3311	0000181849
3311	ITT992	3311	P185	3311	S1B02-03C	3311	TR1N4002	3311	1S185
3311	J101183	3311	P20	3311	S1B02-06CE	3311	TR2327031	3311	1S1850R
3311	J20437	3311	P2A5	3311	S1B02-06CRE	3311	TR2327041	3311	1S1881AM
3311	J241100	3311	P2B5	3311	S1B02-C	3311	TRC159	3311	1S1885
3311	J241102	3311	P3/2H	3311	S1B02-CR	3311	TV24278	3311	1S1885-3
3311	J241142	3311	P3A5	3311	S1B0201CR	3311	TV24282	3311	1S1886
3311	J241183	3311	P3B5	3311	S1B1	3311	TV24554	3311	1S1941
3311	J241260	3311	P4A5	3311	S1D51C052-19	3311	TV24582	3311	1S1942
3311	J24871	3311	P4B5	3311	S1D51C169-1	3311	TVM546	3311	1S2371
3311	J24877	3311	P5A5	3311	S22	3311	TVS1S1950	3311	1S268AA
3311	J24935	3311	P6/2H	3311	S220	3311	TVSB01-02	3311	1S268T
3311	J24939	3311	P6A5	3311	S22A	3311	TVSB01-2	3311	1S2775PA-1
3311	J24940	3311	P6B5	3311	S2VB	3311	TVSB2	3311	1S3016
3311	JC-00012	3311	P7A5	3311	S2VC10	3311	TVSB2M	3311	1S849
3311	JC-00014	3311	P8/2H	3311	S35	3311	TVSERB24-04D	3311	1S849R
3311	JC-00017	3311	PA305A	3311	S4001	3311	TVSERB24-06	3311	1S885
3311	JC-00033	3311	PD101	3311	S40A	3311	TVSFR2-06	3311	1S940
3311	JC-00035	3311	PD102	3311	S43	3311	TVSPT1P	3311	1S941
3311	JC-00044	3311	PD103	3311	S431	3311	TVSHFSD-1A	3311	1S801-02
3311	JC-00045	3311	PD104	3311	S47	3311	TVSJL41A	3311	1SD1212
3311	JC-00047	3311	PD105	3311	S48	3311	TVSMR1C	3311	1SD2
3311	JC-00051	3311	PD106	3311	S49	3311	TVSS1R20	3311	1SR1K
3311	JC-00055	3311	PD107	3311	S5089-A	3311	TVSSB-2	3311	1S854
3311	JC-00059	3311	PD107A	3311	S58	3311	TVSSVO2	3311	1T27B
3311	JC-0016B	3311	PD108	3311	S58R	3311	TW3	3311	2-5BR
3311	JO-K805	3311	PD122	3311	S72	3311	TWV	3311	2D-02
3311	J000049	3311	PD125	3311	S73	3311	UT11	3311	2GA
3311	JD-00040	3311	PD129	3311	S75	3311	UT15	3311	2NSM-1
3311	JD-8D1D	3311	PD130	3311	SB-3	3311	UT21	3311	2SB1K
3311	K1A5	3311	PD131	3311	SC05	3311	V-06B	3311	03-3016
3311	K2A5	3311	PD132	3311	SCA05	3311	VO-6	3311	4-1807
3311	K4A5	3311	PD133	3311	SCP5E	3311	VO-6B	3311	4-2020-03173
3311	KC01.3C	3311	PD134	3311	SD-02	3311	VO-6C	3311	4-2021-1027Q
3311	KC01.3G	3311	PD910	3311	SD-15	3311	VO-6C-401	3311	4-686184-3
3311	KC01.3G12/1X2	3311	PH204	3311	SD-16D	3311	VO5-C	3311	4-686186-3
3311	KC01.3G22/1A	3311	PS005	3311	SD-1LA	3311	VO6-C	3311	4-686201-3
3311	KC02DP12/1N	3311	PS105	3311	SD-1X	3311	VO6B	3311	05-03016-01
3311	KC02DP22/1	3311	PS2207	3311	SD040	3311	VO6C	3311	04-04001-02
3311	KDDO032	3311	PS2208	3311	SD05	3311	VO9U	3311	05-540001
3311	K0E41007	3311	PS2209	3311	SD102	3311	VHD1S1885//-1	3311	05-540112
3311	K8-05	3311	PS2411	3311	SD18	3311	VO6-C	3311	05A07
3311	L3/2H	3311	PS4559	3311	000000SD1Y	3311	W03B	3311	5B-1
3311	L6/2H	3311	PS4725	3311	SDR25	3311	WC120	3311	50-D
3311	L8/2H	3311	PS603	3311	SELEN-70	3311	WD001	3311	5H
3311	LM-1158	3311	PS604	3311	SELEN-701	3311	WD002	3311	05V-50
3311	LM-1160	3311	PS605	3311	SI4D05	3311	WD003	3311	8-719-200-02
3311	LM-1862	3311	PS609	3311	SI-RECT-178	3311	WD004	3311	8-719-900-63
3311	LP1H	3311	PS610	3311	SI-RECT-25	3311	WD005	3311	8-719-901-02
3311	LP2H	3311	PS611	3311	SI-RECT-37	3311	WD006	3311	8-719-908-03
3311	LP3H	3311	PS615	3311	SI-RECT-69	3311	WD007	3311	09-306034
3311	LP4H	3311	PS616	3311	SI-RECT-74	3311	WD008	3311	09-306046
3311	M-31	3311	PS617	3311	SI-RECT-75	3311	WD009	3311	09-306059
3311	M172A	3311	PS621	3311	SI-RECT-77	3311	WD010	3311	09-306119
3311	M3016	3311	PS622	3311	S1B-01-022	3311	WD011	3311	09-306177
3311	M4B-31-22	3311	PS623	3311	S1B01	3311	WD012	3311	09-306213
3311	M67	3311	Q3/2	3311	S1B01-01	3311	WEP156	3311	09-306214
3311	M670	3311	Q52	3311	S1B01-02	3311	YAAD019	3311	09-306249
3311	M670A	3311	Q53	3311	S1B02-CR	3311	YAAD020	3311	09-306250
3311	M67A	3311	Q54	3311	SJ051E	3311	YAAD022	3311	09-306254
3311	M67B	3311	Q55	3311	SJ052B	3311	YBAD009	3311	09-306255
3311	M67C	3311	Q56	3311	SK1K-2	3311	YRAD030	3311	09-306312
3311	M8599	3311	Q57	3311	SLA11AB	3311	ZS-21	3311	09-306350
3311	M9312	3311	Q58	3311	SLA536	3311	ZS-32A	3311	09-306365
3311	M9314	3311	Q6/2	3311	SLA599A	3311	ZS-32B	3311	09-306389
3311	M9317	3311	Q8/2	3311	SM-1-005	3311	ZS-52	3311	09-306394
3311	M9319	3311	QD-SS1885XT	3311	SM-1-47	3311	ZS100		
		3311	QD-SSR1KX4P						

SK	Industry Standard No.
3311	09-306421
3311	09-306422
3311	10-012
3311	10-085006
3311	10-085009
3311	10-085010
3311	10-085030
3311	10B-2
3311	10D
3311	10D-05
3311	10D-06
3311	10D05
3311	10D1
3311	10D3G
3311	10D5B
3311	10D5E
3311	10DC-2C
3311	10DC-2J
3311	10DC0B
3311	10DC0H
3311	10DC2F
3311	10DRV
3311	10DZ
3311	10E-1
3311	10E-2
3311	10E-4D
3311	10E-7L
3311	10E1
3311	10E2
3311	10I10
3311	10J2
3311	10J2F
3311	10T200
3311	11-085003
3311	11-102001
3311	11-120007
3311	11/1
3311	12-100003
3311	12B-2
3311	12P2
3311	13-14094-24
3311	13-14094-39
3311	13-14094-42
3311	13-14094-54
3311	13-14094-6
3311	13-14261-1
3311	13-14261-3
3311	13-22463-1
3311	13-23160-5
3311	13-39860-1
3311	13-41122-1
3311	13-41122-4
3311	13-41123-2
3311	13-43766-1
3311	13P2
3311	14P2
3311	15-085041
3311	15-100003
3311	15-108037
3311	15-123105
3311	15P2
3311	16C-4
3311	16P2
3311	17-410
3311	17P2
3311	18P2
3311	19-040-003
3311	19P2
3311	20-22-08
3311	21A001-000
3311	21A040-44
3311	21A103-007
3311	21A103-021
3311	21A103-044
3311	21A103-058
3311	21A103-070
3311	21A103-104
3311	21A110-012
3311	21A119-030
3311	21M248
3311	21M315
3311	21M317
3311	21M416
3311	21M417
3311	21M434
3311	21M435
3311	21M436
3311	21M487
3311	21M519
3311	21M545
3311	22-1-075
3311	23J2
3311	24J2
3311	24MW1066
3311	24MW1092
3311	24MW1123
3311	24MW1124
3311	24MW1144
3311	24MW1146
3311	24MW1162
3311	24MW241
3311	24MW267
3311	24MW268
3311	24MW619
3311	24MW768
3311	24MW851
3311	24MW867
3311	24MW995
3311	025-100016
3311	025-100035
3311	25J2
3311	26J2
3311	27J2
3311	28-22-21
3311	28J2
3311	30-8054-7
3311	32-0037
3311	32-0038
3311	32-0046
3311	32-0050
3311	33-0002
3311	33-0024
3311	34-8054-7
3311	34-8057-18
3311	42-22835
3311	42-23975
3311	42-27202
3311	42-27278
3311	42-27463
3311	42-28058
3311	44T-300-95
3311	46-16261-3
3311	46-86184
3311	46-86187-3
3311	46-86261-3
3311	46-86270-3
3311	46-86433-3
3311	46-86437-3
3311	46-86438-3
3311	46-86503-3
3311	46-8687-3
3311	46-8689-3
3311	46AR2
3311	46AR3
3311	46AX2
3311	46AX3
3311	46BX2
3311	46BX3
3311	48-03005A07
3311	48-10062A01
3311	48-134959
3311	48-134990
3311	48-137212
3311	48-155041
3311	48-155099
3311	48-155126
3311	48-155178
3311	48-155193
3311	48-191A04
3311	48-40739P01
3311	48-41508A01
3311	48-60022A98
3311	48-66629A02
3311	48-66629A05
3311	48-90343A93
3311	48-97048A18
3311	48-97168A02
3311	48-97168A11
3311	48B43265G01
3311	48C40235G01
3311	48M355045
3311	48S134958
3311	48S134990
3311	48S155100
3311	48S155106
3311	48S155126
3311	48S191A04
3311	48S40235G02
3311	48X97048A07
3311	48X97048A10
3311	48X97168A10
3311	48X97168A16
3311	48X97222A02
3311	49-1042
3311	52A011-1
3311	52SD-1
3311	53A001-12
3311	53A001-3
3311	53A001-35
3311	53A001-9
3311	53A011-1
3311	53B001-5
3311	53B003-2
3311	53B020-2
3311	53C003-1
3311	53C007-1
3311	53C009-1
3311	53D001-7
3311	53E010-1
3311	53F001-1
3311	53F002-1
3311	53K001-6
3311	53K001-7
3311	53K001-8
3311	53N001-3
3311	53N002-2
3311	53T001-3
3311	53Y001-1
3311	56-8097
3311	56-8198
3311	56-8199
3311	66X0044-001
3311	66X0048-001
3311	69-1823
3311	69-2246
3311	72-15
3311	81-27123100-3
3311	81-27123300-3
3311	81-46123011-2
3311	81-46123012-2
3311	81-46123014-6
3311	81-46123018-7
3311	81-46123022-9
3311	81-46123034-4
3311	81-46123035-1
3311	86-0511
3311	86-111-3
3311	86-23-1
3311	86-52-1
3311	86-80-1
3311	86-80-3
3311	86-89-3
3311	91A04
3311	91A11
3311	93A104-1
3311	93A60-7
3311	93A97-3
3311	93B42-7
3311	93C42-7
3311	93ERH-DX0155//
3311	0100
3311	100-10121-05
3311	100-10132-00
3311	100-13
3311	100-138
3311	100-217
3311	100-438
3311	100-520
3311	100-527
3311	102B6
3311	103-141
3311	103-145
3311	103-261-01
3311	103-261-04
3311	103-76
3311	103B6
3311	104B6
3311	105B6
3311	106B6
3311	107B6
3311	0110-0141
3311	0110-0141/4460
3311	0111
3311	117-134-11
3311	130A17268
3311	130YE
3311	0131-0026
3311	0131-0044
3311	135-3
3311	135P4
3311	140-013
3311	0140-8
3311	0140-9
3311	146D1
3311	151-030-9-005
3311	151-040-7-003
3311	151-1
3311	162-005A
3311	180-1
3311	200X9120-224
3311	201X3120-255
3311	212-254
3311	212-7
3311	212-71
3311	212-94B
3311	0234
3311	0244
3311	264D00505
3311	264D01112
3311	264P00602
3311	264P04402
3311	264P05001
3311	264P05002
3311	264P08002
3311	264P08801
3311	264P10105
3311	264P13002
3311	264Z00201
3311	325-0031-338
3311	325-0076-315
3311	325-0081-110
3311	354-9110-001
3311	359A
3311	0400
3311	0401
3311	410
3311	0411
3311	429-0989-68
3311	430-31
3311	0450
3311	0460
3311	523-0501-002
3311	523-1500-881
3311	536J2F
3311	0557-010
3311	600
3311	601
3311	602X0019-000
3311	608
3311	610
3311	610-001-103
3311	614
3311	616
3311	617-62
3311	618
3311	620
3311	622
3311	690V094H35
3311	690V114H36
3311	700A858-285
3311	700A858-286
3311	710
3311	750A858-285
3311	751-9001-124
3311	754-2000-005
3311	754-5000-021
3311	754-5750-284
3311	754-9000-953
3311	800-018-00
3311	800-024-00
3311	808-206
3311	903-00394
3311	903-104B
3311	903-105B
3311	903-14B
3311	903-156B
3311	903-15B
3311	903-164B
3311	903-17B
3311	903-28B
3311	903-303
3311	903-49B
3311	903-69B
3311	903-79B
3311	977-13B
3311	977-19
3311	977-20B
3311	977-218
3311	977-22B
3311	977-23B
3311	977-25B
3311	977-27B
3311	977-28B
3311	1000-129
3311	1002-6219
3311	1003-11
3311	1007
3311	1007-0951
3311	1010-9494
3311	1022-5548
3311	1024
3311	1038-1697
3311	1043-7382
3311	1045-0534
3311	1048-9946
3311	1048-9995
3311	1049-3435
3311	1061-8379
3311	1061-8916
3311	1061-8924
3311	1077-2366
3311	1077-3836
3311	1562
3311	1713-17
3311	1854-17
3311	1855-17
3311	1885-17
3311	1970-16
3311	2000-304
3311	2000-320
3311	2045-17
3311	2052
3311	2093A38-28
3311	2093A38-4
3311	2093A41-103
3311	2093A41-108
3311	2093A41-110
3311	2093A41-126
3311	2093A41-131
3311	2093A41-139
3311	2093A41-151
3311	2093A41-156
3311	2093A41-173
3311	2093A41-180
3311	2093A41-185
3311	2093A41-189
3311	2093A41-197
3311	2093A41-43
3311	2093A41-6
3311	2093A41-62
3311	2093A41-75
3311	2093A41-81
3311	2093B41-8
3311	2116-17
3311	2148-17
3311	2152
3311	2199-17
3311	2200-17
3311	2206
3311	2206-17
3311	2233-17
3311	2283-17
3311	2330-201
3311	2330-252
3311	2402-461
3311	2402-462
3311	2402-463
3311	2405-459
3311	2510-32
3311	2522
3311	2523
3311	2606-299
3311	2704-389
3311	003015
3311	3023
3311	4686-184-3
3311	4686-186-3
3311	4686-201-3
3311	4806-0000-004
3311	5001-163
3311	5051-300-10150
3311	5205NI
3311	5380-21
3311	5601-MS
3311	5632-W03B
3311	7962
3311	8102-206
3311	8710-55
3311	8840-56
3311	10909
3311	25115-115
3311	25202-002
3311	25810-51
3311	25840-51
3311	27113-100
3311	27840-41
3311	28810-61
3311	35306
3311	35333
3311	40024VM
3311	44001
3311	44002
3311	44003
3311	45810-51
3311	47126-3
3311	47126-7
3311	50745-01
3311	055210H
3311	58840-111
3311	71779
3311	71794
3311	000072050
3311	72176
3311	77271-3
3311	88641
3311	104325
3311	104615
3311	119596
3311	121468
3311	123804
3311	125127
3311	126176
3311	127695
3311	127835
3311	129334
3311	129348
3311	130045
3311	133543
3311	135281
3311	144050
3311	144052
3311	144860
3311	145839
3311	166920
3311	167543
3311	167544
3311	169590
3311	170297
3311	170750
3311	171162-252
3311	171305
3311	171416
3311	171657
3311	172551
3311	202463
3311	225410
3311	228560
3311	233011
3311	235382
3311	248917
3311	252817
3311	259315
3311	348054-15
3311	348058-2
3311	489752-108
3311	501010
3311	0510006
3311	0517132
3311	0517750
3311	0517750H
3311	530072-1019
3311	530072-5
3311	530072-6
3311	530135-1
3311	530135-1003
3311	530135-3
3311	530171-1
3311	530171-1001
3311	530171-3
3311	530180-1001
3311	0537640
3311	0552006H
3311	0552010H
3311	740183
3311	740289
3311	740570
3311	741865
3311	760202-0003
3311	771907
3311	772713
3311	817088
3311	817179
3311	817180
3311	922311
3311	922860
3311	972216
3311	982526
3311	983413
3311	984254
3311	986935
3311	988051
3311	1223772
3311	1223928
3311	1223929
3311	1471872-001
3311	1471872-004
3311	1471872-008
3311	1471872-1
3311	1471872-13
3311	1472171-32
3311	1474778-2
3311	1474778-3
3311	1474778-5
3311	1474778-8
3311	1477046-5
3311	1966808
3311	2003073-68
3311	2004358-142
3311	2006582-21
3311	2008292-89
3311	2008299-3
3311	2008302-41
3311	2330201
3311	2330254
3311	2330352
3311	2330553
3311	2330721
3311	2331142
3311	2337071
3311	2486836
3311	2498513
3311	5330041(HR-5A)
3311	5330041H
3311	5330042
3311	5330101
3311	5330336
3311	5330341
3311	5330371
3311	5330381
3311	5330431
3311	5331102
3311	5340021
3311	7027005
3311	7027031
3311	7070692
3311	7200953
3311	7570011-01
3311	7570024-02
3311	7852439-01
3311	8120037
3311	8120133
3311	9739636-20
3311	13030192
3311	13030313
3311	13050221
3311	17210010
3311	17240010
3311	22115405
3311	23115185
3311	23115199
3311	23115263
3311	23115914
3311	25115115
3311	27113100
3311	30600040
3311	32510001
3311	41013566
3311	41021007
3311	41029499
3311	41520419
3311	43600101
3311	62744162
3311	87400020
3311	90270060
3311	91190032
3311	602000002
3311	1523270105
3311	2009120101
3311	2009120187
3311	4120910010
3311	4139000005
3311	4202003200
3311	4202022300
3311	4202023400
3311	4202110270
3311	6611001700
3311	6611001800
3311	6611007300
3311	16501090016
3311	16501190005
3311	16501190016
3311	16505090005
3311	16516390010
3311	16629291210
3311	933000611112
3312	D1201D
3312	EP57X4
3312	ESA-06
3312	ESA-10C
3312	ESA-10N
3312	GI3002
3312	QRT-214
3312	RA-1
3312	RH-DX0038CEZZ
3312	RH-DX0063CEZZ
3312	RH-DX0081TAZZ
3312	RH1B
3312	RM-1V
3312	S1-RECT-155
3312	V06E
3312	1N4004
3312	05-000104
3312	05-174004
3312	05-931609
3312	05-931971
3312	05-935201
3312	10D4
3312	48-155063
3312	48-155235
3312	48-155236

SK	Industry Standard No.
3312	48-90343A64
3312	48-90343A91
3312	48-90343A92
3312	48-90420A79
3312	48-90420A80
3312	93A52-1
3312	93A60-9
3312	93A97-1
3312	2021-05
3312	44004
3312	2330256
3312	6200039
3312	2013120208
3313	A7509201
3313	A7547053
3313	CXL116
3313	CXL117
3313	00000DS131
3313	ECG116
3313	FR2
3313	GE-1N5061
3313	M4B31-13
3313	NTC-19
3313	PTC202
3313	QRT-215
3313	REN116
3313	0000000S15
3313	TM116
3313	W06A
3313	W06B
3313	W06C
3313	WEP158
3313	WEP158A
3313	001-0153-00
3313	1JH11F
3313	1N4005
3313	1S1888
3313	004-003900
3313	05-750010
3313	05-931601
3313	8-719-901-13
3313	10D-6
3313	15BD11
3313	48-155136
3313	48S155054
3313	93A60-14
3313	103-254
3313	103-254-01
3313	212-76-02
3313	1045-0518
3313	1059-7961
3313	003007
3313	4403
3313	99203-005
3313	0099203-007
3313	126320
3313	132814
3313	137652
3313	138175
3313	530171-1003
3313	1471872-6
3313	2008100130
3313	4202021000
3313	4202023100
3313/116	A7568700
3313/116	D1201M
3313/116	RH-DX0043TAZZ
3313/116	RH-DX0114CEZZ
3313/116	SIR60
3313/116	V06-E
3313/116	05-190061
3313/116	15-123243
3313/116	48-90420A97
3313/116	212-76
3313/116	44005
3313/116	132148
3313/116	530151-1001
3313/116	6611001100
3314	57-58
3314	57-59
3314	06200037
3315	CXL507
3315	34-8054-16
3316	ERB24-02B
3316	QRT-218
3317	D2201D
3317	ERB24-04A
3317	ERB24-04B
3317	ERB24-04C
3317	ERB24-04D
3317	SD-ERB24-04D
3317	44936
3318	D2201M
3318	ERB24-06
3318	ERB24-06A
3318	ERB24-06B
3318	ERB24-06C
3318	ERB24-06D
3318	8-719-901-92
3318	103-284A
3318	103-298-05A
3318	201X2100-126
3318	44937
3319	BB105B
3319	1S2692
3319	1S2692AB
3319	103-176
3322	BA244
3322	129095
3322	1477022-1
3323	FV1043
3323	HF-20062
3323	MV1620
3323	MV2101
3323	MV2201
3323	PC139
3323	PC140
3323	PC141
3323	1N4786
3323	1N4801
3323	1N5681
3323	1N5696
3323	1S2090
3323	1S2139A
3323	1S2239
3323	13-22154-500
3323	51-06002-00
3323	4809-0000-001
3323/610	MC301
3324	BA121
3324	IP20-0204
3324	MV-201
3324	PC135
3324	V910
3324	1N4788
3324	1N5683
3324	1S145
3324	1S1501
3324	1S1502
3324	1S1923
3324	1S2139
3324	1S2236
3324	1S2336
3324	1S351S
3324	8000-00038-010
3324	922214
3324/611	1S2208
3324/611	103-146
3325	BB122
3325	BB142
3325	IP20-0185
3325	ITT-410
3325	MV1626
3325	MV2104
3325	MV2105
3325	V12
3325	V912
3325	1N3554
3325	1N4789
3325	1N5684
3325	1N5699
3325	1S1895
3325	1S1924
3325	1S2085
3325	1S2085A
3325	1S2087
3325	1S2087A
3325	1S2267
3325	1S2268
3325	1S2687
3325	1S2687D
3325	1S352
3325	1S352M
3325	1S48
3325	022-2823-501
3325	48-45323G01
3325	754-4000-553
3325	903-47
3325	914-000-7-00
3325	256717
3325/612	1S351M
3325/612	05-200310
3325/612	741689
3326	MV1634
3326	MV2107
3326	MV832
3326	PC136
3326	1N4792
3326	1N5687
3326	1N5702
3326	1S351R
3326	1S555
3326	1S893
3327	BB109G
3327	BB-55
3327	BJ-155
3327	BR-55
3327	D7N
3327	DS-55
3327	DVV004
3327	EA16X177
3327	HD40001060
3327	IP20-0151
3327	MV1638
3327	MV2109
3327	MV2111
3327	MV2209
3327	MV834
3327	MV9600
3327	MV9601
3327	NV004
3327	NV009
3327	PC128
3327	PG333
3327	QD-CS2688DJ
3327	RVDSD113
3327	SD113
3327	SVC-201
3327	SVC0053
3327	V33
3327	V933
3327	001-0160-00
3327	001-0176-00
3327	1N4794
3327	1N5450A
3327	1N5470A
3327	1N5689
3327	1S1658
3327	1S1658PA-3
3327	1S2688
3327	1S2689
3327	1SV50
3327	1SV68
3327	3L4-3503-6
3327	05-182688
3327	05-472209
3327	05-780251
3327	46-86289-3
3327	53K001-15
3327	103-47-01
3327	523-0009-049
3327	2061-45
3327	2002207-2
3327	5330661
3327	5330852
3327/614	BB109
3327/614	ITT310
3327/614	ITT310G
3327/614	SC-15
3327/614	SC-20
3327/614	SVDSC20
3327/614	1S2207
3327/614	1S2638
3327/614	603000002
3328	GEIC-31
3328	LM380
3328	RE321-IC
3328	TVCM-35
3328	46-5002-23
3328	301-576-14
3330	AZ3.3
3330	BZX96C3V3
3330	BZY85B3V3
3330	EZ3(ELCOM)
3330	GEZD-3.3
3330	HEP-Z0401
3330	PS8901
3330	RE100
3330	RZZ3.3
3330	1D3.3
3330	1D3.3A
3330	1D3.3B
3330	1M3.3ZS10
3330	1M3.3ZS5
3330	1N3506
3330	1N3821
3330	1N3821A
3330	1N4649
3330	1N4684
3330	1N4728
3330	1N4728A
3330	1N5988B
3330	1Z3.3A
3330	1Z3.3T10.5
3330	14-0104-6
3330	019-002691
3330	41-0906
3330	48-83461E03
3330	56-50
3330	61E03
3330	523-2003-339
3331	AZ3.9
3331	AZ748A
3331	BZY92C3V9
3331	GEZD-3.9
3331	HEP-Z0403
3331	RE102
3331	RZZ3.9
3331	1D3.9
3331	1D3.9A
3331	1D3.9B
3331	1M3.9ZS10
3331	1M3.9ZS5
3331	1N1507
3331	1N1507A
3331	1N1518
3331	1N1518A
3331	1N3508
3331	1N3823
3331	1N3823A
3331	1N4651
3331	1N4686
3331	1N4730
3331	1N4730A
3331	1S994
3331	1Z3.9
3331	1Z3.9A
3331	05-480204
3331	09-306380
3331	13Z6AF
3331	13Z6F
3331	48-82256C46
3331	48-83461E19
3331	61E19
3331	93A39-34
3331	96-5248-09
3331	96-5248-11
3331	264P09707
3331	523-2003-399
3331	134882256346
3331	134883461519
3332	AZ4.3
3332	BZ4.3
3332	BZY92C4V3
3332	GEZD-4.3
3332	RE103
3332	1D4.3
3332	1D4.3A
3332	1D4.3B
3332	1M4.3ZS10
3332	1M4.3ZS5
3332	1N3509
3332	1N3824
3332	1N3824A
3332	1N4652
3332	1N4687
3332	1N4731
3332	1N4731A
3332	1N5845B
3332	1S220
3332	1Z4.3
3332	1Z4.3A
3332	1ZC4.3T10.5
3332	5D-5A-L
3332	14Z6AF
3332	14Z6F
3332	61E13
3332	523-2003-439
3332	72165-3
3332	134883461513
3333	GEZD-4.7
3333	HEP-Z0405
3333	HZ5
3333	RE104
3333	1D4.7
3333	1D4.7B
3333	1M4.7ZS10
3333	1M4.7ZS5
3333	1N1519
3333	1N1519A
3333	1N3510
3333	1N3825
3333	1N3825A
3333	1N4653
3333	1N4688
3333	1N4732
3333	1N4732A
3333	1N674
3333	15Z6AF
3333	15Z6F
3333	24MW1107
3333	523-2003-479
3334	.25T6.8
3334	.25T6.8A
3334	.25T7.1
3334	.25T7.1A
3334	.4T6.8
3334	.4T6.8B
3334	.7JZ6.8
3334	.7Z6.8A
3334	.7Z6.8B
3334	.7Z6.8C
3334	.7Z6.8D
3334	.7ZM6.8A
3334	.7ZM6.8B
3334	.7ZM6.8C
3334	.7ZM6.8D
3334	A7286201
3334	AW01-07
3334	AW01-7
3334	AZ-065
3334	AZ-067
3334	AZ-069
3334	AZ6.8
3334	BZ-065
3334	BZ-067
3334	BZ-069
3334	BZ6.8
3334	BZX29C6V8
3334	BZX79C6V8
3334	BZX83C6V8
3334	BZY60
3334	BZY83C6V8
3334	BZY85C6V8
3334	BZY92C6V8
3334	CXL5014
3334	CXL5071
3334	EA16X123
3334	ECG5071
3334	ECG5071A
3334	EQA01-07
3334	EQA01-07RE
3334	ETD-HZ7C
3334	GEZD-6.8
3334	HEP-Z0409
3334	HW6.8
3334	HW6.8A
3334	IP20-0186
3334	KZ6
3334	KZ6A
3334	KZ6ZA
3334	MZ-207
3334	MZ207(ZENER)
3334	MZ207-02A
3334	PD6051
3334	PS8909
3334	QRT-238
3334	RE110
3334	REN5071
3334	RT5471
3334	RVDRD7AN
3334	RZ6.8
3334	RZZ6.8
3334	TR7SB
3334	TV24981
3334	TVSQA01-07RE
3334	UZ8706
3334	UZ8806
3334	VHERD6R8EE/-1
3334	XZ-070
3334	XZ070
3334	YZ060
3334	ZOB6.8
3334	Z1106
3334	Z1106-C
3334	Z1B6.8
3334	Z5B6.8
3334	Z6.8
3334	ZF6.8
3334	ZPD618
3334	ZZ6.8
3334	001-0082-00
3334	01-2303-3
3334	1M6.8Z
3334	1M6.8Z10
3334	1M6.8Z5
3334	1M6.8ZS10
3334	1M6.8ZS5
3334	1N1510
3334	1N1510A
3334	1N1521
3334	1N1521A
3334	1N1767
3334	1N2034
3334	1N2034-2
3334	1N2034A
3334	1N3016
3334	1N3016A
3334	1N3016B
3334	1N3444
3334	1N3514
3334	1N3675
3334	1N3675A
3334	1N3829
3334	1N3829A
3334	1N4099
3334	1N4158
3334	1N4158A
3334	1N4323
3334	1N4323A
3334	1N4400
3334	1N4400A
3334	1N4628
3334	1N4657
3334	1N4692
3334	1N4736
3334	1N4736A
3334	1N5559
3334	1N5559A
3334	1N5851B
3334	1S2113A
3334	1S222
3334	1S3006
3334	1S332M
3334	1S3332
3334	1S481
3334	1Z6.8
3334	1Z6.8A
3334	3L4-3506-12
3334	4-2020-11300
3334	05-110106
3334	05-110107
3334	05-69669-01
3334	09-306215
3334	09-306325
3334	09-307088
3334	14-515-27
3334	14-515-30
3334	19Z6AF
3334	19Z6F
3334	24MW742
3334	34-8057-10
3334	48-127021
3334	48-82256C37
3334	53A001-5
3334	56-58
3334	56-6
3334	56-63
3334	56-71
3334	61E42
3334	63-12110
3334	86-110-1
3334	93A39-11
3334	96-5248-01
3334	103-140
3334	142-016
3334	264P17401
3334	0276.8A
3334	523-2003-689
3334	523-2503-689
3334	0648R
3334	800-030-00
3334	903-166B
3334	903-180
3334	903-335
3334	1011-02
3334	1063-4145
3334	1063-6454
3334	2093A41-70
3334	2341-17
3334	4822-130-30132
3334	2092055-0710
3334	2330302
3334	2330632
3334	2330643
3334	2330791
3334	3596398
3334	23115908
3334	4202024300
3334	134882256337
3334	134883461542
3335	.25T20
3335	.25T20A
3335	.4T20
3335	.4T20A
3335	.4T20B
3335	.7JZ20
3335	.7Z20A
3335	.7Z20B
3335	.7Z20C
3335	.7Z20D
3335	.7ZM20A
3335	.7ZM20B
3335	.7ZM20C
3335	.7ZM20D
3335	A054-232
3335	AW01-20
3335	AZ-197
3335	B66X0040-002
3335	BZ-197
3335	BZ-210
3335	BZ210
3335	BZX29C20
3335	BZX46C20
3335	BZX55C20
3335	BZX79C20
3335	BZX83C20
3335	BZY83C20
3335	BZY85C20
3335	BZY92C20
3335	EZ20(ELCOM)
3335	GEZD-20
3335	HEP-Z0421
3335	HW20
3335	HW20A
3335	P3309
3335	R9381
3335	RD20E
3335	RD20F
3335	RD20FA
3335	RD20FB
3335	RE126
3335	RH-EX0037CEZZ
3335	SD-31
3335	001-023032
3335	1M20Z
3335	1M20Z10
3335	1M20Z20
3335	1M20Z3
3335	1M20Z5
3335	1M20ZS10
3335	1N1778
3335	1N2039-3
3335	1N3027
3335	1N3027A
3335	1N3027B
3335	1N3525
3335	1N3686
3335	1N3686A
3335	1N4169
3335	1N4169A
3335	1N4334
3335	1N4334A
3335	1N4411
3335	1N4411A
3335	1N4639
3335	1N4668
3335	1N4707
3335	1N4747
3335	1N4747A
3335	1N5570
3335	1N5570A
3335	1N768-3
3335	1R20
3335	1R20A
3335	1R20B
3335	1S1967
3335	1S235
3335	1S3020
3335	1Z20
3335	1Z20A
3335	1Z20T10
3335	1Z20T20
3335	1Z20T5
3335	1ZC20T10
3335	1ZC20T5
3335	13-14879-6
3335	13-14879-7
3335	13-33179-5
3335	21A037-009
3335	48-82256C39
3335	48-82256C57
3335	56-45
3335	86-139-1
3335	86-66-1
3335	86-73-1
3335	86-88-1
3335	93A39-5
3335	93A39-7
3335	229-1301-30
3335	523-2003-200
3335	919-00-3309-1
3335	919-003309
3335	919-015618-1
3335	43072
3335	71411-4
3335	62669462
3335	62761296
3335	134882256339
3335	134882256357

SK	Industry Standard No.
3336	.25T22
3336	.25T22A
3336	.4T22
3336	.4T22A
3336	.4T22B
3336	.7JZ22
3336	.7Z22A
3336	.7Z22B
3336	.7Z22C
3336	.7Z22D
3336	.7ZM22A
3336	.7ZM22B
3336	.7ZM22C
3336	.7ZM22D
3336	AW01-22
3336	AW0122
3336	BZ-220
3336	BZ230
3336	BZX29C22
3336	BZX46C22
3336	BZX79C22
3336	BZX83C22
3336	BZY83C22
3336	BZY85C22
3336	BZY92C22
3336	D6U
3336	EA16X38
3336	ES16X29
3336	EZ22(ELCOM)
3336	GEZD-22
3336	HEP-Z0422
3336	HW22
3336	HW22A
3336	MZ92-24B
3336	RD22
3336	RD22E
3336	RD22F
3336	RD22FA
3336	RD22FB
3336	RE127
3336	RZ22
3336	SD31
3336	TVSQA01-25A
3336	001-023031
3336	01-2303-0
3336	01-2303-1
3336	1M22Z
3336	1M22Z10
3336	1M22Z5
3336	1M22ZS10
3336	1M22ZS5
3336	1N1429
3336	1N1516
3336	1N1516A
3336	1N1527
3336	1N1527A
3336	1N1880
3336	1N1880A
3336	1N3028
3336	1N3028A
3336	1N3028B
3336	1N3438
3336	1N3450
3336	1N3526
3336	1N3687
3336	1N3687A
3336	1N4170
3336	1N4170A
3336	1N4335
3336	1N4335A
3336	1N4412
3336	1N4412A
3336	1N4640
3336	1N4669
3336	1N4708
3336	1N4748
3336	1N4748A
3336	1N5571
3336	1N5571A
3336	1N668
3336	1R22
3336	1R22A
3336	1R22B
3336	1S1968
3336	1S236
3336	1S3022
3336	1Z22
3336	1Z22A
3336	1Z22T10
3336	1Z22T20
3336	1Z22T5
3336	1ZC22T10
3336	1ZC22T5
3336	34-8057-31
3336	48S137442
3336	53N001-4
3336	86-11-1
3336	86-111-1
3336	523-2003-220
3336	1091
3336	143777
3336	2337063
3336	5330059
3336	62638338
3336	4202108570
3337	.25T36
3337	.25T36A
3337	.4T36
3337	.4T36A
3337	.4T36B
3337	.7JZ36
3337	.7Z36A
3337	.7Z36B
3337	.7Z36C
3337	.7Z36D
3337	.7ZM36A
3337	.7ZM36B
3337	.7ZM36C
3337	.7ZM36D
3337	B310(ZENER)
3337	BZ-340
3337	BZ-350
3337	BZX29C36
3337	BZX46C36
3337	BZX55C36
3337	BZX60C36
3337	BZX79C36
3337	BZX83C36
3337	BZY92C36
3337	EQA01-35
3337	EZ36(ELCOM)
3337	GEZD-36
3337	HEP-Z0427
3337	HW36
3337	HW36A
3337	RD36F
3337	RD36FA
3337	RD36FB
3337	1M36Z
3337	1M36Z10
3337	1M36Z5
3337	1M36ZS10
3337	1M36ZS5
3337	1N1784
3337	1N3033
3337	1N3033A
3337	1N3033B
3337	1N3531
3337	1N3692
3337	1N3692A
3337	1N4175
3337	1N4175A
3337	1N4340
3337	1N4340A
3337	1N4417
3337	1N4417A
3337	1N4645
3337	1N4674
3337	1N4715
3337	1N4753
3337	1N4753A
3337	1N5576
3337	1N5576A
3337	1R36
3337	1R36A
3337	1R36B
3337	1S242
3337	1S3036
3337	1S488
3337	1Z36
3337	1Z36A
3337	4-2020-13300
3337	13-0017
3337	86-119-1
3337	96-5248-12
3337	523-2003-360
3337	4202013300
3338	.25T39
3338	.25T39A
3338	.4T39
3338	.4T39A
3338	.4T39B
3338	.7JZ39
3338	.7Z39A
3338	.7Z39B
3338	.7Z39C
3338	.7Z39D
3338	.7ZM39A
3338	.7ZM39B
3338	.7ZM39C
3338	.7ZM39D
3338	BZX29C39
3338	BZX79C39
3338	BZX83C39
3338	EZ39(ELCOM)
3338	GEZD-39
3338	HEP-Z0428
3338	HW39
3338	HW39A
3338	RD39F
3338	RD39FA
3338	RD39FB
3338	1M39Z
3338	1M39Z10
3338	1M39Z5
3338	1M39ZS10
3338	1M39ZS5
3338	1N1785
3338	1N1883
3338	1N1883A
3338	1N3034
3338	1N3034A
3338	1N3034B
3338	1N3441
3338	1N3454
3338	1N3532
3338	1N3693
3338	1N3693A
3338	1N4176
3338	1N4176A
3338	1N4341
3338	1N4341A
3338	1N4418
3338	1N4418A
3338	1N4646
3338	1N4675
3338	1N4716
3338	1N4754
3338	1N4754A
3338	1N5577
3338	1N5577A
3338	1R39A
3338	1R39B
3338	1S243
3338	1S3039
3338	1Z39
3338	1Z39A
3338	03-933935
3338	48-82256C58
3338	56-55
3338	56-66
3338	93A39-35
3338	103-270
3338	523-2003-390
3338	134882256358
3339	.25T43
3339	.25T43A
3339	.4T43
3339	.4T43A
3339	.4T43B
3339	.7JZ43
3339	.7Z43A
3339	.7Z43B
3339	.7Z43C
3339	.7Z43D
3339	.7ZM43A
3339	.7ZM43B
3339	.7ZM43C
3339	.7ZM43D
3339	BZX29C43
3339	BZX46C43
3339	BZX79C43
3339	BZX83C43
3339	BZX85C43
3339	GEZD-43
3339	HW43
3339	HW43A
3339	1M43Z
3339	1M43Z10
3339	1M43Z5
3339	1M43ZS10
3339	1M43ZS5
3339	1N1786
3339	1N1786A
3339	1N3035
3339	1N3035A
3339	1N3035B
3339	1N3533
3339	1N3694
3339	1N3694A
3339	1N4177
3339	1N4177A
3339	1N4177B
3339	1N4342
3339	1N4342A
3339	1N4342B
3339	1N4419
3339	1N4419A
3339	1N4419B
3339	1N4647
3339	1N4676
3339	1N4717
3339	1N4755
3339	1N4755A
3339	1N5578
3339	1N5578A
3339	1R43
3339	1R43A
3339	1R43B
3339	1S244
3339	1S3043
3339	1Z43
3339	1Z43A
3339	523-2003-430
3340	.25T47
3340	.25T47A
3340	.4T47
3340	.4T47A
3340	.4T47B
3340	.7JZ45
3340	.7JZ47
3340	.7Z45A
3340	.7Z45B
3340	.7Z45C
3340	.7Z45D
3340	.7Z47A
3340	.7Z47B
3340	.7Z47C
3340	.7Z47D
3340	.7ZM47A
3340	.7ZM47B
3340	.7ZM47C
3340	.7ZM47D
3340	BZX29C47
3340	BZX79C47
3340	EZ47(ELCOM)
3340	GEZD-47
3340	HEP-Z0430
3340	HW47
3340	HW47A
3340	1M47Z
3340	1M47Z10
3340	1M47Z5
3340	1M47ZS10
3340	1M47ZS5
3340	1N1787
3340	1N1884
3340	1N1884A
3340	1N3036
3340	1N3036A
3340	1N3036B
3340	1N3442
3340	1N3455
3340	1N3534
3340	1N3695
3340	1N3695A
3340	1N4178
3340	1N4178A
3340	1N4343
3340	1N4343A
3340	1N4420
3340	1N4420A
3340	1N4648
3340	1N4677
3340	1N4756
3340	1N4756A
3340	1N5579
3340	1N5579A
3340	1R47
3340	1R47A
3340	1R47B
3340	1S246
3340	1S3047
3340	1S489
3340	1Z47
3340	1Z47A
3340	145-470
3340	523-2003-450
3340	523-2003-470
3341	.25T51
3341	.25T51A
3341	.4T51
3341	.4T51A
3341	.4T51B
3341	.7JZ50
3341	.7JZ51
3341	.7Z50A
3341	.7Z50B
3341	.7Z50C
3341	.7Z50D
3341	.7Z51A
3341	.7Z51B
3341	.7Z51C
3341	.7Z51D
3341	.7Z52A
3341	.7Z52B
3341	.7Z52C
3341	.7Z52D
3341	.7ZM50A
3341	.7ZM50B
3341	.7ZM50C
3341	.7ZM50D
3341	.7ZM51A
3341	.7ZM51B
3341	.7ZM51C
3341	.7ZM51D
3341	.7ZM52A
3341	.7ZM52B
3341	.7ZM52C
3341	.7ZM52D
3341	BZX29C51
3341	BZX79C51
3341	GEZD-51
3341	HW51
3341	HW51A
3341	RM-25
3341	RM25
3341	SK-1W50
3341	TVSRM25
3341	1M51Z
3341	1M51Z10
3341	1M51Z5
3341	1M51ZS10
3341	1M51ZS5
3341	1N1788
3341	1N1788A
3341	1N3037
3341	1N3037A
3341	1N3037B
3341	1N3696
3341	1N3696A
3341	1N3696B
3341	1N4179A
3341	1N4179B
3341	1N4321
3341	1N4344
3341	1N4344A
3341	1N4344B
3341	1N4421
3341	1N4421A
3341	1N4757
3341	1N4757A
3341	1N5580
3341	1N5580A
3341	1N5580B
3341	1R51
3341	1R51A
3341	1R51B
3341	1S247
3341	1S248
3341	1S3051
3341	1Z51
3341	1Z51A
3341	523-2003-500
3341	523-2003-510
3341	523-2003-520
3341	139069
3342	.25T56
3342	.25T56A
3342	.7ZM56A
3342	.7ZM56B
3342	.7ZM56C
3342	.7ZM56D
3342	BZX29C56
3342	BZX79C56
3342	EZ56(ELCOM)
3342	GEZD-56
3342	HEP-Z0432
3342	HW56
3342	HW56A
3342	SK-1W55
3342	1M56Z
3342	1M56Z10
3342	1M56Z5
3342	1M56ZS10
3342	1M56ZS5
3342	1N1789
3342	1N1789A
3342	1N1885
3342	1N1885A
3342	1N3038
3342	1N3038A
3342	1N3038B
3342	1N3456
3342	1N3697
3342	1N3697A
3342	1N4180
3342	1N4180A
3342	1N4345
3342	1N4345A
3342	1N4345B
3342	1N4422
3342	1N4422A
3342	1N4422B
3342	1N4758
3342	1N4758A
3342	1N5581
3342	1N5581A
3342	1R56
3342	1R56A
3342	1R56B
3342	1S249
3342	1S3056
3342	1Z56
3342	1Z56A
3342	48-82256C29
3342	48-83461E09
3342	61E09
3342	523-2003-560
3342	134882256329
3342	134883461509
3343	.25T68
3343	.25T68A
3343	.7JZ68
3343	.7Z68A
3343	.7Z68B
3343	.7Z68C
3343	.7Z68D
3343	.7ZM68A
3343	.7ZM68B
3343	.7ZM68C
3343	.7ZM68D
3343	BZX79C68
3343	EZ68(ELCOM)
3343	GEZD-68
3343	HW68
3343	HW68A
3343	1M68Z
3343	1M68Z10
3343	1M68Z3
3343	1M68Z5
3343	1M68ZS10
3343	1M68ZS5
3343	1N1431
3343	1N1791
3343	1N1791A
3343	1N1886
3343	1N1886A
3343	1N3040
3343	1N3040A
3343	1N3040B
3343	1N3457
3343	1N3699
3343	1N3699A
3343	1N4182
3343	1N4182A
3343	1N4182B
3343	1N4347
3343	1N4347A
3343	1N4347B
3343	1N4424
3343	1N4424A
3343	1N4424B
3343	1N4760
3343	1N4760A
3343	1N5583
3343	1N5583A
3343	1N670
3343	1R68
3343	1R68A
3343	1R68B
3343	1S251
3343	1S3068
3343	1Z68
3343	1Z68A
3343	13-0106
3343	56-68
3343	523-2003-680
3344	.7JZ75
3344	.7Z75A
3344	.7Z75B
3344	.7Z75C
3344	.7Z75D
3344	.7ZM75A
3344	.7ZM75B
3344	.7ZM75C
3344	.7ZM75D
3344	A7272130
3344	GEZD-75
3344	HW75
3344	HW75A
3344	1M75Z
3344	1M75Z10
3344	1M75Z3
3344	1M75Z5
3344	1M75ZS10
3344	1M75ZS5
3344	1N1792
3344	1N1792A
3344	1N3041
3344	1N3041A
3344	1N3041B
3344	1N3700
3344	1N3700A
3344	1N4183
3344	1N4183A
3344	1N4183B
3344	1N4348
3344	1N4348A
3344	1N4348B
3344	1N4425
3344	1N4425A
3344	1N4425B
3344	1N4761
3344	1N4761A
3344	1N5584
3344	1N5584A
3344	1N5584B
3344	1R75
3344	1R75A
3344	1R75B
3344	1S252
3344	1S3075
3344	1Z72
3344	1Z75
3344	1Z75A
3344	48-83461E04
3344	61E04
3344	070-025
3344	523-2003-750
3344	134883461504
3345	.25T91
3345	.25T91A
3345	.7JZ91
3345	.7Z91A
3345	.7Z91B
3345	.7Z91C
3345	.7Z91D
3345	.7ZM91A
3345	.7ZM91B
3345	.7ZM91C
3345	.7ZM91D
3345	GEZD-91
3345	HW91
3345	HW91A
3345	1M91Z
3345	1M91Z10
3345	1M91Z3
3345	1M91Z5
3345	1M91ZS10
3345	1M91ZS5
3345	1N1794
3345	1N1794A
3345	1N3043
3345	1N3043A
3345	1N3043B
3345	1N3702
3345	1N3702A
3345	1N4185
3345	1N4185A
3345	1N4185B
3345	1N4350
3345	1N4350A
3345	1N4350B
3345	1N4427
3345	1N4427A
3345	1N4763
3345	1N4763A
3345	1N5586
3345	1N5586A
3345	1N5586B
3345	1R91A
3345	1R91B
3345	1S254
3345	1S3091
3345	1Z91
3345	1Z91A
3345	48-83461E16
3345	61E16
3345	523-2003-910
3345	134883461516
3346	.25T100
3346	.25T100A
3346	.4T100
3346	.4T100A
3346	.4T100B
3346	.4T110B
3346	.7JZ100
3346	.7Z100A
3346	.7Z100B

SK	Industry Standard No.
3346	.7Z100C
3346	.7Z100D
3346	.7ZM100A
3346	.7ZM100B
3346	.7ZM100C
3346	.7ZM100D
3346	GEZD-100
3346	HEP-Z0438
3346	HW100
3346	HW100A
3346	RE133
3346	1M100Z
3346	1M100Z10
3346	1M100Z5
3346	1M100ZS10
3346	1M100ZS5
3346	1N1432
3346	1N1795
3346	1N1795A
3346	1N1888
3346	1N1888A
3346	1N3044
3346	1N3044A
3346	1N3044B
3346	1N3459
3346	1N3703
3346	1N3703A
3346	1N4186
3346	1N4186A
3346	1N4186B
3346	1N4351
3346	1N4351A
3346	1N4351B
3346	1N4428
3346	1N4428A
3346	1N4428B
3346	1N4764
3346	1N4764A
3346	1N5587
3346	1N5587A
3346	1N5587B
3346	1N671
3346	1R100
3346	1R100A
3346	1R100B
3346	1S255
3346	1S3100
3346	1Z100
3346	1Z100A
3346	2D7A
3346	4-2020-14100
3346	21A037-016
3346	46-86467-3
3346	48-82256C30
3346	48-83461E06
3346	48-83461E14
3346	61E06
3346	61E14
3347	.25T120
3347	.25T120A
3347	.4T120
3347	.4T120A
3347	.4T120B
3347	.7JZ120
3347	.7Z120A
3347	.7Z120B
3347	.7Z120C
3347	.7Z120D
3347	.7ZM120A
3347	.7ZM120B
3347	.7ZM120C
3347	.7ZM120D
3347	GEZD-120
3347	HEP-Z0440
3347	HW120
3347	HW120A
3347	RH-EX0054CEZZ
3347	1M120Z
3347	1M120Z10
3347	1M120Z3
3347	1M120Z5
3347	1M120ZS10
3347	1M120ZS5
3347	1N1797
3347	1N1889
3347	1N3046
3347	1N3046A
3347	1N3046B
3347	1N3098
3347	1N3098A
3347	1N3460
3347	1N3705
3347	1N3705A
3347	1N4188
3347	1N4188A
3347	1N4353
3347	1N4353A
3347	1N4430
3347	1N4430A
3347	1N4858A
3347	1N5589
3347	1N5589A
3347	1R120
3347	1R120A
3347	1R120B
3347	1S258
3347	1S3120
3347	229-1301-29
3348	.25T130
3348	.25T130A
3348	.4T130
3348	.4T130A
3348	.4T130B
3348	.7JZ130
3348	.7Z130A
3348	.7Z130B
3348	.7Z130C
3348	.7Z130D
3348	.7ZM130A
3348	.7ZM130B
3348	.7ZM130C
3348	.7ZM130D
3348	GEZD-130
3348	1EZ130D5
3348	1M130Z
3348	1M130Z10
3348	1M130Z3
3348	1M130Z5
3348	1M130ZS10
3348	1M130ZS5
3348	1N1798
3348	1N1798A
3348	1N3047
3348	1N3047A
3348	1N3047B
3348	1N3706
3348	1N3706A
3348	1N4189
3348	1N4189A
3348	1N4189B
3348	1N4354
3348	1N4354A
3348	1N4354B
3348	1N4431
3348	1N4431A
3348	1N4431B
3348	1N4859
3348	1N4859A
3348	1N5590
3348	1N5590A
3348	1N5590B
3348	1N5890
3348	1N5890A
3348	1R130
3348	1R130A
3348	1R130B
3348	1S259
3349	.25T140
3349	.7JZ140
3349	.7Z140A
3349	.7Z140B
3349	.7Z140C
3349	.7Z140D
3349	.7ZM140A
3349	.7ZM140B
3349	.7ZM140C
3349	.7ZM140D
3349	GEZD-140
3349	1EZ140D5
3349	1M140Z10
3349	1M140Z5
3349	1N5891
3349	1N5891A
3349	1N5891B
3349	1R140
3349	1R140A
3349	1R140B
3349	1S260
3349	72165-5
3350	.25T150
3350	.25T150A
3350	.4T150
3350	.4T150A
3350	.4T150B
3350	.7JZ150
3350	.7Z150A
3350	.7Z150B
3350	.7Z150C
3350	.7Z150D
3350	.7ZM150A
3350	.7ZM150B
3350	.7ZM150C
3350	.7ZM150D
3350	GEZD-150
3350	HEP-Z0442
3350	HW150
3350	HW150A
3350	1M150Z
3350	1M150Z10
3350	1M150Z3
3350	1M150Z5
3350	1M150ZS10
3350	1M150ZS5
3350	1N1433
3350	1N1799
3350	1N1890
3350	1N3048
3350	1N3048A
3350	1N3048B
3350	1N3099
3350	1N3461
3350	1N3707
3350	1N3707A
3350	1N4190
3350	1N4190A
3350	1N4355
3350	1N4355A
3350	1N4432
3350	1N4432A
3350	1N4860
3350	1N4860A
3350	1N5591
3350	1N5591A
3350	1N672
3350	1R150
3350	1R150A
3350	1R150B
3351	.25T160
3351	.25T160A
3351	.4T160
3351	.4T160A
3351	.4T160B
3351	.7JZ160
3351	.7Z160A
3351	.7Z160B
3351	.7Z160C
3351	.7Z160D
3351	.7ZM160A
3351	.7ZM160B
3351	.7ZM160C
3351	.7ZM160D
3351	GEZD-160
3351	HW160
3351	HW160A
3351	1EZ160D5
3351	1M160Z
3351	1M160Z10
3351	1M160Z3
3351	1M160Z5
3351	1M160ZS10
3351	1M160ZS5
3351	1N1800
3351	1N1800A
3351	1N3049
3351	1N3049A
3351	1N3049B
3351	1N3708
3351	1N3708A
3351	1N4191
3351	1N4191A
3351	1N4356
3351	1N4356A
3351	1N4356B
3351	1N4433
3351	1N4433A
3351	1N5592
3351	1N5592A
3351	1N5592B
3351	1R160
3351	1R160A
3351	1R160B
3351	72165-6
3352	.25T175
3352	GEZD-170
3352	1N5894B
3353	.25T180
3353	.25T180A
3353	.4T180
3353	.4T180A
3353	.4T180B
3353	.7JZ180
3353	.7Z180A
3353	.7Z180B
3353	.7Z180C
3353	.7Z180D
3353	.7ZM180A
3353	.7ZM180B
3353	.7ZM180C
3353	.7ZM180D
3353	GEZD-180
3353	HEP-Z0444
3353	1EZ180D5
3353	1M180Z
3353	1M180Z10
3353	1M180Z3
3353	1M180Z5
3353	1M180ZS10
3353	1M180ZS5
3353	1N1801
3353	1N1801A
3353	1N3050
3353	1N3050A
3353	1N3050B
3353	1N3100
3353	1N3100A
3353	1N3462
3353	1N3709
3353	1N3709A
3353	1N4192
3353	1N4192A
3353	1N4192B
3353	1N4357
3353	1N4357A
3353	1N4357B
3353	1N4434
3353	1N4434A
3353	1N4434B
3353	1N5593
3353	1N5593A
3353	1N5593B
3353	1R180
3353	1R180A
3353	1R180B
3353	1S3180
3353	62-21573
3354	.7JZ200
3354	GEZD-190
3354	1EZ190D5
3354	1N5896B
3355	.25T200
3355	.25T200A
3355	.4T200
3355	.4T200A
3355	.4T200B
3355	.7Z200A
3355	.7Z200B
3355	.7Z200C
3355	.7Z200D
3355	.7ZM200A
3355	.7ZM200B
3355	.7ZM200C
3355	.7ZM200D
3355	GEZD-200
3355	HEP-Z0445
3355	HW200
3355	HW200A
3355	UZ8120
3355	UZ8220
3355	1M200Z
3355	1M200Z10
3355	1M200Z3
3355	1M200Z5
3355	1M200ZS10
3355	1M200ZS5
3355	1N1802
3355	1N3051
3355	1N3051A
3355	1N3051B
3355	1N3710
3355	1N3710A
3355	1N4193
3355	1N4193A
3355	1N4358
3355	1N4358A
3355	1N4435
3355	1N4435A
3355	1N5594
3355	1N5594A
3355	1N5594B
3355	1N5897B
3355	1N6031B
3355	1R200
3355	1R200A
3355	1R200B
3355	63-5058
3356	C710
3356	C710(B)
3356	C710(D)
3356	C710-1
3356	C710-2
3356	C710-4
3356	C710B
3356	C710BC
3356	C710C
3356	C710D
3356	C710E
3356	C930
3356	C930B
3356	C930BV
3356	C930CK
3356	C930D
3356	C930DH
3356	C930DT
3356	C930DZ
3356	C930E
3356	C930EV
3356	EA15X364
3356	IP20-0002
3356	QT-C0710XAE
3356	QT-C0710XBE
3356	01-030710
3356	01-030930
3356	2SC710
3356	2SC710(B)
3356	2SC710(C)
3356	2SC710(D)
3356	2SC710-1
3356	2SC710-2
3356	2SC710-4
3356	2SC710-OR
3356	2SC710AL
3356	2SC710BC
3356	2SC710D
3356	2SC710E
3356	2SC710F
3356	2SC710G
3356	2SC710GN
3356	2SC710H
3356	2SC710K
3356	2SC710L
3356	2SC710M
3356	2SC710X
3356	2SC710XL
3356	2SC710Y
3356	2SC930
3356	2SC930A
3356	2SC930B
3356	2SC930BB
3356	2SC930BK
3356	2SC930BV
3356	2SC930C
3356	2SC930CK
3356	2SC930CL
3356	2SC930CS
3356	2SC930DB
3356	2SC930DC
3356	2SC930DH
3356	2SC930DS
3356	2SC930DT
3356	2SC930DT-2
3356	2SC930DX
3356	2SC930DZ
3356	2SC930EP
3356	2SC930ET
3356	2SC930EV
3356	2SC930EX
3356	2SC930F
3356	2SC930G
3356	2SC930GN
3356	2SC930H
3356	2SC930J
3356	2SC930K
3356	2SC930L
3356	2SC930M
3356	2SC930OR
3356	2SC930R
3356	2SC930X
3356	2SC930Y
3356	8-729-671-14
3356	8-729-803-04
3356	022-2876-007
3356	2000-268
3356	1223911
3356	4100900102
3356	4100900103
3356	4100900104
3357	C1095(6)
3357	C1226A
3357	C1226AP
3357	C1226AQ
3357	C1226C
3357	C1226F
3357	C1226P
3357	C1226Q
3357	DDBY003002
3357	DDBY227004
3357	ECG186A
3357	EQS-89
3357	ET453611
3357	G05-416-C
3357	PT-029
3357	2SC1095
3357	2SC1095(6)
3357	2SC1095L
3357	2SC1095M
3357	2SC1098
3357	2SC1098(4)K
3357	2SC1098(4)L
3357	2SC1098A
3357	2SC1098C
3357	2SC1098D
3357	2SC1098L
3357	2SC1098M
3357	2SC1226
3357	2SC1226-0
3357	2SC1226A
3357	2SC1226AC
3357	2SC1226AP
3357	2SC1226AQ
3357	2SC1226AR
3357	2SC1226B
3357	2SC1226C
3357	2SC1226CF
3357	2SC1226F
3357	2SC1226P
3357	2SC1226Q
3357	2SC1226QR
3357	2SC1226R
3357	2SC1226SC
3357	2SC1336
3357	2SC931
3357	2SC931C
3357	2SC931D
3357	2SC931E
3357	04-11620-01
3357	07-07165
3357	09-302080
3357	09-302121
3357	09-302123
3357	09-302126
3357	09-309071
3357	48-155042
3357	48-155074
3357	48-355039
3357	48-41784J04
3357	48-90343A76
3357	48S155042
3357	48S155074
3357	90-111
3357	90-176
3357	90-451
3357	90-75
3357	106-004
3357	123-011
3357	123-011A
3357	123B-002
3357	142-006
3357	172-011-9-001
3357	172-038-9-003
3357	176-042-9-005
3357	185-001
3357	186-008
3357	186-009
3357	260P22801
3357	853-0301-096
3357	1074-116
3357	1080-06
3357	2041-03
3357	2081-17
3357	2160-17
3357	2202-17
3357	2334-17
3357	5001-064
3357	5001-075
3357	8000-00041-043
3357	8910-146
3357	18410-147
3357	38510-168
3357	000073381
3358	A3301
3358	G09-009-A
3358	GEIC-38
3358	LA3301
3358	PC-20007
3358	RE326-IC
3358	REN1006
3358	RVILA3301
3358	09-308034
3358	09-308064
3358	21M532
3358	87-0004
3358	740-5903-301
3358	740-9003-301
3358	1077-2390
3358	2091-49
3358	000074010
3358	171179-051
3358	172272
3358	741098
3358	916072
3358	16233010
3358	90200090
3359	A679
3359	B555
3359	B555-0
3359	ECG281
3359	GE-263
3359	2SA679
3359	2SA753
3359	2SA753A
3359	2SA753B
3359	2SA753C
3359	2SB555
3359	2SB555-0
3359	2SB556
3359	2SB557
3359/281	A626
3359/281	A627
3359/281	2SA626
3359/281	2SA626L
3359/281	2SA627
3359/281	2SA744
3359/281	2SB653
3359/281	2SB654
3360	ECG280MP
3360	GE-262MP
3361	GEIC-160
3361	HA1202
3362	HA1211
3363	GEIC-169
3363	HA1319
3364	HA1406
3364	HA1406-2
3364	HA1406-3
3364	HA1406-4
3364	051-0036-01
3364	051-0036-02
3364	051-0036-03
3365	A1364N
3365	B0306000
3365	B0306004
3365	GEIC-149
3365	GEIC-36
3365	GEIC-96
3365	HA1126D
3365	LA1364N
3365	PTC754
3365	RE325-IC
3365	REN1004
3365	RH-1X0020CEZZ
3365	RH-IX0020CEZZ
3365	TA7070P
3365	TA7070PFA-1
3365	TA7070PGL
3365	46-1357-3
3365	46-1369-3
3365	200X2100-022
3365	905-105
3365	5351361
3366	GEIC-159
3366	HA1201
3368	GEIC-162
3368	HA1306
3368	000HA1306U
3368	HA1306W
3369	GEIC-167
3369	HA1314
3370	GEIC-168
3370	HA1316
3371	GEIC-170
3371	HA1322
3371	HA1322C
3371	051-0036-00
3371	740781
3372	170964
3375	AE-907
3375	AE907
3375	AE907-51
3375	DM-41
3375	DM41
3375	GEIC-16
3375	LM1841N
3375	SN76669N
3375	TVCM-18
3375	ULN2136A
3375	3L4-9007-01
3375	3L4-9007-1
3375	3L4-9007-51
3375	3L49007-51
3375	88-9574

SK	Industry Standard No.
3375	2136
3375	2136D
3375	2136P
3375	2136PC
3375	9341238
3375/737	DM11A
3375/737	FEL2113
3375/737	GEIC-11
3375/737	GEL2113
3375/737	GEL2113AL1
3375/737	GEL2113F1
3375/737	ULN2113A
3375/737	ULN2113N
3376	GEIC-37
3376	LA4031P
3377	GE5ZD-3.3
3377	1N5008
3377	1N5008A
3377	1N5333
3378	GE5ZD-3.6
3378	1N5009
3378	1N5009A
3378	1N5334
3378	1N5334A
3379	GE5ZD-3.9
3379	1N5010
3379	1N5010A
3379	1N5335
3379	1N5335A
3380	GE5ZD-4.3
3380	1N5011
3380	1N5336
3380	1N5336A
3381	GE5ZD-4.7
3381	1N5012
3381	1N5012A
3381	1N5337
3381	1N5337A
3382	GE5ZD-5.1
3382	1N5013
3382	1N5338
3382	1N5338A
3383	GE5ZD-5.6
3383	1N5014
3383	1N5339
3383	1N5339A
3384	GE5ZD-6.0
3384	1N5340
3384	1N5340A
3385	GE5ZD-6.2
3385	1N4460
3385	1N5015
3385	1N5341
3385	1N5341A
3386	BZX55D6V8
3386	EZ3-7(ELCOM)
3386	GE5ZD-6.8
3386	UZ5706
3386	UZ5806
3386	UZ706
3386	UZ806
3386	1.5DKZ6.8
3386	1.5JZ6.8
3386	1.5Z6.8
3386	1N3785
3386	1N3785A
3386	1N4461
3386	1N5016
3386	1N5063
3386	1N5342
3386	1N5342A
3386	3TZ6.8B
3386	3VR6.8B
3386	59Z6A
3386	96-5249-02
3387	GE5ZD-7.5
3387	HEP-Z2510
3387	SZ75A
3387	1.5DKZ7.5A
3387	1.5JZ7.5A
3387	1N3786
3387	1N3786A
3387	1N4462
3387	1N4955
3387	1N5017
3387	1N5064
3387	1N5343
3387	1N5343A
3387	3TZ7.5D
3387	3VR7.5B
3387	5ZS7.5A
3387	5ZS7.5B
3387	7717
3388	BZX55D8V2
3388	GE5ZD-8.2
3388	1.5DKZ8.2
3388	1.5JZ8.2
3388	1.5Z8.2A
3388	1N3787
3388	1N3787A
3388	1N4463
3388	1N4956
3388	1N5018
3388	1N5065
3388	1N5344
3388	1N5344A
3388	1S4008A
3388	2VR8.2B
3388	3TZ8.2B
3388	1308
3388	7718
3389	1N5345
3389	1N5345A
3389	2VR8.5B
3389	3VR8.5B
3389	5ZS8.7A
3390	GE5ZD-9.1
3390	HEP-Z2513
3390	SZ91A
3390	1.5DKZ9.1A
3390	1.5JZ9.1A
3390	1N3788
3390	1N3788A
3390	1N4464
3390	1N4957
3390	1N5019
3390	1N5019A
3390	1N5066
3390	1N5346
3390	1N5346A
3390	3TZ9.1D
3390	3VR9.1B
3390	5ZS9.1A
3390	62Z6A
3390	7719
3391	BZX55D10
3391	GE5ZD-10
3391	HEP-Z2514
3391	1.5DKZ10A
3391	1N3789
3391	1N3789A
3391	1N4465
3391	1N4958
3391	1N5020
3391	1N5067
3391	1N5347
3391	1N5347A
3391	3TZ10D
3391	3VR10B
3391	5ZS10A
3391	96-5344-01
3391	523-2004-100
3391	7720
3392	BZX70C11
3392	GE5ZD-11
3392	1.5DKZ11A
3392	1.5JZ11A
3392	1N3790
3392	1N3790A
3392	1N4466
3392	1N4959
3392	1N5021
3392	1N5068
3392	1N5348
3392	1N5348A
3392	1N5348B
3392	1S4011A
3392	2VR11A
3392	3VR11B
3392	5ZS11B
3392	1311
3393	BZX70C12
3393	EZ3-12(ELCOM)
3393	GE5ZD-12
3393	HEP-Z2516
3393	SZ12C
3393	1.5DKZ12A
3393	1.5JZ12A
3393	1N3791
3393	1N3791A
3393	1N4467
3393	1N4883
3393	1N4960
3393	1N5022
3393	1N5349
3393	1N5349A
3393	1S4012A
3393	2VR12A
3393	3TZ12A
3393	5ZS12B
3393	919-00-3309-2
3393	1312
3393	530073-32
3394	BZY95C13
3394	GE5ZD-13
3394	1.5DKZ13A
3394	1.5JZ13A
3394	1N3792
3394	1N3792A
3394	1N4468
3394	1N4961
3394	1N5023
3394	1N5069
3394	1N5350
3394	1N5350A
3394	1S4013A
3394	2VR13A
3394	3R13A
3394	3TZ13D
3394	3VR13B
3394	5ZS13A
3394	66Z6A
3395	EZ3-14(ELCOM)
3395	GE5ZD-14
3395	1.5JZ14A
3395	1.5M14Z
3395	1.5Z14A
3395	1N5024
3395	1N5070
3395	1N5118
3395	1N5351
3395	1N5351A
3395	2VR14A
3395	5ZS14A
3395	48-83461E21
3395	61E21
3395	67Z6A
3395	134883461521
3396	BZY95C15
3396	EZ3-15(ELCOM)
3396	GE5ZD-15
3396	HEP-Z2519
3396	MZ5915
3396	SZ15C
3396	001-023041
3396	1.5DKZ15A
3396	1.5JZ15A
3396	1N3793
3396	1N3793A
3396	1N4469
3396	1N4962
3396	1N5025
3396	1N5071
3396	1N5352
3396	1N5352A
3396	1S4015
3396	7724
3397	BZY95C16
3397	EZ3-16(ELCOM)
3397	GE5ZD-16
3397	1.5DKZ16A
3397	1.5JZ16A
3397	1N3794
3397	1N3794A
3397	1N4470
3397	1N4963
3397	1N5026
3397	1N5072
3397	1N5353
3397	1N5353A
3397	1S4016A
3397	3R16A
3397	3VR16B
3397	5ZS16
3397	5ZS16A
3397	5ZS16B
3397	7725
3398	GE5ZD-17
3398	1.5JZ17A
3398	1.5Z17D
3398	1N5027
3398	1N5354
3398	1N5354A
3398	3TZ17
3398	3TZ17A
3398	3TZ17B
3398	3TZ17C
3398	3TZ17D
3398	5ZS17
3398	5ZS17A
3398	5ZS17B
3399	BZX70C18
3399	GE5ZD-18
3399	HEP-Z2522
3399	MZ5918
3399	1.5DKZ18A
3399	1.5JZ18A
3399	1N3795
3399	1N3795A
3399	1N4471
3399	1N4964
3399	1N5028
3399	1N5073
3399	1N5355
3399	1N5355A
3399	1S4018
3399	7726
3399	530073-1030
3400	GE5ZD-19
3400	1.5JZ19A
3400	1.5M19Z
3400	1N5029
3400	1N5356
3400	1N5356A
3400	3TZ19
3400	5ZS19
3400	5ZS19A
3400	5ZS19B
3401	BZY95C20
3401	GE5ZD-20
3401	1.5DKZ20
3401	1.5JZ20
3401	1.5Z20D
3401	1N3796
3401	1N3796A
3401	1N4472
3401	1N4881
3401	1N4884
3401	1N4965
3401	1N5030
3401	1N5357
3401	1N5357A
3401	530073-1021
3402	BZX70C22
3402	EP16X2
3402	HEP-Z2525
3402	MZ5922
3402	SZ22C
3402	1.5DKZ22A
3402	1.5JZ22A
3402	1.5Z22D
3402	1N3797
3402	1N3797A
3402	1N4473
3402	1N4966
3402	1N5031
3402	1N5074
3402	1N5358
3402	1N5358A
3402	1S4022
3402	7728
3402	530166-1006
3403	BZX70C24
3403	HEP-Z2526
3403	1.5DKZ24A
3403	1.5JZ24A
3403	1.5Z24D
3403	1N3798
3403	1N3798A
3403	1N4474
3403	1N4967
3403	1N5032
3403	1N5359
3403	1N5359A
3403	1S4024A
3403	48-83461E35
3403	61E35
3403	86-95-1
3403	7729
3403	530073-1029
3403	134883461535
3404	GE5ZD-25
3404	1.5JZ25
3404	1.5M25Z
3404	1N5033
3404	1N5360B
3404	5ZS25
3404	5ZS25A
3404	5ZS25B
3404	5329
3405	BZX70C27
3405	GE5ZD-27
3405	HEP-Z2528
3405	MZ5927
3405	SZ27C
3405	1.5DKZ27A
3405	1.5JZ27A
3405	1.5Z27A
3405	1N3799
3405	1N3799A
3405	1N4475
3405	1N4968
3405	1N5034
3405	1N5076
3405	1N5361
3405	1N5361A
3405	1S4027
3405	96-5250-02
3405	7730
3406	GE5ZD-28
3406	1N5362
3406	1N5362A
3406	2VR28
3406	2VR28A
3406	2VR28B
3406	5ZS28
3406	5ZS28A
3406	5ZS28B
3407	BZX70C30
3407	EZ3-30(ELCOM)
3407	EZ5-30(ELCOM)
3407	GE5ZD-30
3407	HEP-Z2530
3407	1.5DKZ30
3407	1N3800
3407	1N3800A
3407	1N4476
3407	1N4969
3407	1N5035
3407	1N5077
3407	1N5363
3407	1N5363A
3407	1S4030A
3407	61E28
3407	7731
3408	BZX70C33
3408	GE5ZD-33
3408	HEP-Z2531
3408	MZ5933
3408	SX33
3408	1.5DKZ33B
3408	1.5JZ33A
3408	1N3801
3408	1N3801A
3408	1N4477
3408	1N4503
3408	1N4970
3408	1N5036
3408	1N5078
3408	1N5364
3408	1N5364A
3408	1S4033A
3408	5332
3408	7732
3409	MZ1000-26
3409	1N4478
3409	1N4882
3409	1N4971
3409	1N5037
3409	1N5079
3410	1N4479
3410	1N4972
3410	1N5038
3410	1N5080
3410	1N5081
3411	1N4480
3411	1N4973
3411	1N5039
3411	1N5082
3412	1N4481
3412	1N4974
3412	1N5040
3412	1N5040A
3412	1N5041
3412	1N5083
3412	1N5084
3413	1N4482
3413	1N4975
3413	1N5042
3413	1N5043
3413	1N5044
3413	1N5044A
3413	1N5085
3413	1N5086
3414	1N4483
3414	1N4976
3414	1N5045
3414	1N5087
3415	1N5088
3415	1N5122
3416	1N4484
3416	1N4889
3416	1N4977
3416	1N5046
3416	1N5089
3417	1N4485
3417	1N4978
3417	1N5047
3417	1N5090
3418	1N4486
3418	1N4979
3418	1N5048
3418	1N5092
3419	1N4487
3419	1N4980
3419	1N5049
3419	1N5093
3421	1N4096
3421	1N4488
3421	1N4981
3421	1N5050
3421	1N5095
3422	1N4097
3422	1N4489
3422	1N4982
3422	1N5051
3423	1N4490
3423	1N4983
3423	1N5096
3424	SX120
3424	1N4491
3424	1N4984
3424	1N5097
3425	1N4492
3425	1N4985
3425	1N5098
3425	1N5099
3426	1N5126
3427	1N4098
3427	1N4493
3427	1N4986
3428	1N4494
3428	1N4987
3428	1N5100
3428	3R160B
3428	103-206
3429	1N5101
3429	1N5127
3430	1N4495
3430	1N4988
3430	1N5102
3431	1N5103
3431	1N5128
3432	UZ120
3432	UZ220
3432	UZ5120
3432	UZ5220
3432	1N4496
3432	1N4504
3432	1N4989
3432	1N5104
3432	2VR200
3433	A5T5059
3433	GE-222
3433	HX-50108
3433	01-031921
3433	2N5551
3433	8-729-213-01
3433	13-37869-1
3433	46-86345-3
3433	46-86647-3
3433	48-134846
3433	48-155051
3433	48-355012
3433	54E
3433	86-651-2
3433	200X3192-101
3433	1573PQ
3433	95226-4
3433	143041
3433	5321214
3433/287	C1573
3433/287	C780
3433/287	C780AG
3433/287	C780G
3433/287	JE9014C
3433/287	VS2SC1921//1B
3433/287	01-571941
3433/287	01-572831
3433/287	2SC1573
3433/287	2SC1670
3433/287	2SC1670J
3433/287	2SC1670JW
3433/287	2SC2271N
3433/287	2SC780AG
3433/287	2SC780AG-0
3433/287	2SC780AG-R
3433/287	2SC780AG-Y
3433/287	2321472
3434	A549A
3434	A549AH
3434	A638
3434	A778AK
3434	A778K
3434	BF299
3434	GE-223
3434	GS9022H
3434	GS9022I
3434	HX-50109
3434	2SA778AK
3434	2SA835
3434	8-762-020-00
3434	48-355058
3434	161-015G
3434/288	A6532841
3434/288	A778A
3434/288	A840
3434/288	BF298
3434/288	DDBY104002
3434/288	GS9022F
3434/288	GS9022G
3434/288	JE9015B
3434/288	MPS9411A1
3434/288	2SA778A
3434/288	2SA840
3434/288	13-23506-2
3434/288	13-39115-2
3436	GE-724
3436	STK-056
3437	BD250C
3437	CXL180
3437	ECG180
3437	GE-74
3437	HEP-S7001
3437	HF23
3437	MJ4502
3437	RE74
3437	REN180
3437	SJ3637
3437	SJE264
3437	SJE764
3437	2N4398
3437	2N5883
3437	2N5884
3437	2N6329
3437	2N6330
3437	2N6331
3437/180	BD246
3437/180	BD246A
3437/180	BD246B
3437/180	BD246C
3437/180	SDT9202
3437/180	2N3792
3437/180	2N4906
3437/180	2N5880
3438	A1C
3438	A8M
3438	A9K
3438	A9P
3438	B176001
3438	B176002
3438	B176003
3438	BD144
3438	BD160
3438	BDX12
3438	BDY54
3438	BLY49
3438	BLY50
3438	BU106
3438	BU107
3438	BU110
3438	BU120
3438	BUY20
3438	BUY21
3438	BUY21A
3438	BUY22
3438	BUY35
3438	BUY44
3438	C1185
3438	C1185A
3438	C1185B
3438	C1185C
3438	C1185K
3438	C1185L
3438	C1185M
3438	C240
3438	C241
3438	C242
3438	C243
3438	C245
3438	C246
3438	C270
3438	C42
3438	C42A
3438	C43
3438	C44
3438	C519
3438	C519A
3438	C586
3438	C889

SK	Industry Standard No.	SK	Industry Standard No.	SK	Industry Standard No.	SK	Industry Standard No.	SK	Industry Standard No.
3438	D110	3438	14-601-14	3440	2N6473	3444	A12-1A90	3444	A9E
3438	D110-R	3438	21A112-031	3440	2N6474	3444	A128	3444	A9F
3438	D110-Y	3438	21A112-107	3440	2SD358	3444	A128A	3444	A9G
3438	D111	3438	21A118-124	3440	2SD381	3444	A137	3444	A9H
3438	D111-R	3438	34-6002-61	3440	2SD382	3444	A137(NPN)	3444	A9J
3438	D111-Y	3438	46-86455-3	3440	121-1008	3444	A1379	3444	A98
3438	D165	3438	46-86475-3	3440	121-970	3444	A1380	3444	A9T
3438	D166	3438	48-134900	3440	121-987	3444	A1381	3444	A9V
3438	D177	3438	48-137326	3440/291	3S15X125	3444	A1412-1	3444	A9W
3438	D181	3438	48-86475-3	3440/291	2N5051	3444	A153	3444	A9Y
3438	D189	3438	48-8692790	3440/291	2SC1409	3444	A156	3444	ALD-35
3438	D189A	3438	48-869337	3440/291	2SC1409A	3444	A1567	3444	AN
3438	D202	3438	48-869408	3440/291	2SC1409AB	3444	A1567-1	3444	AR-107
3438	D203	3438	60P19009	3440/291	2SC1409B	3444	A157	3444	AR-108
3438	D284	3438	63-12273	3440/291	2SC1409C	3444	A157A	3444	AR-200
3438	D285	3438	63-13216	3440/291	2SC1410	3444	A157B	3444	AR-201
3438	D334	3438	64N1	3440/291	2SC1410B	3444	A157C	3444	AR-202
3438	D334A	3438	96-5284-01	3440/291	2SD525	3444	A158	3444	AR107
3438	D334R	3438	132-025	3440/291	2SD525-0	3444	A158A	3444	AR108
3438	D45	3438	184T2C	3440/291	2SD525Y	3444	A158B	3444	AR204
3438	D46	3438	260P21901	3440/291	2SD880	3444	A158C	3444	AR205
3438	D47	3438	260P21908	3440/291	46-86533-3	3444	A159	3444	AR206
3438	D59	3438	269P19009	3440/291	121-970-02	3444	A159A	3444	AR208
3438	D60	3438	334-377	3440/291	121-987-02	3444	A159B	3444	AR306
3438	D67	3438	417-101	3441	RCA30B	3444	A159C	3444	AR306(BLUE)
3438	D67B	3438	417-158	3441	2N3741	3444	A168	3444	AR306(ORANGE)
3438	D67C	3438	417-204	3441	2N6475	3444	A1B	3444	AT329
3438	D67D	3438	574-844	3441	2N6476	3444	A1P	3444	AT335
3438	D67E	3438	800-550-00	3441	2SA775	3444	A1H	3444	AT336
3438	D83	3438	2057B157-9	3441	2SA775C	3444	A1J	3444	AT337
3438	D84	3438	2057B175-9	3441	2SB528	3444	A1L	3444	AT347
3438	D88A	3438	126900	3441	2SB537	3444	A1T	3444	AT348
3438	DT4303	3438	135352	3441	2SB537L	3444	A1T-1	3444	AT349
3438	DT4304	3438	240404	3441	2SB596-0	3444	A1V	3444	AT370
3438	DT4305	3438	417214	3441	57B188-12	3444	A1VE	3444	AT400
3438	DT4306	3438	610064-1	3441	86-614-2	3444	A1W	3444	AT401
3438	DT80710	3438	1473564-1	3441	121-1009	3444	A2019ZC	3444	AT402
3438	DT8103	3438	1473637-1	3441	121-969	3444	A20372	3444	AT403
3438	DT8104	3438	2320271	3441	121-988	3444	A2410	3444	AT404
3438	DT83705	3438	2320273	3441	06100008	3444	A24100	3444	AT405
3438	DT83705A	3438	7293818	3441/292	2SB596	3444	A2411	3444	AT406
3438	DT83705B	3438	7293819	3441/292	121-969-02	3444	A2412	3444	AT407
3438	DT8401	3438	7301664	3441/292	121-988-02	3444	A2413	3444	AT420
3438	DT8410	3438	7301665	3442	17045	3444	A2434	3444	AT421
3438	DT8411	3438	7301666	3443	D60	3444	A246	3444	AT422
3438	DT8413	3438	7302699	3443	EP57X3	3444	A246(AMC)	3444	AT423
3438	DT8423	3438	7303304	3443	GB-513	3444	A2466	3444	AT424
3438	DT8423M	3438	7313063	3443	H445-2	3444	A2468	3444	AT425
3438	DT8430	3438	7913605	3443	H484	3444	A2469	3444	AT426
3438	DT8431	3438	23114208	3443	R-12C	3444	A2470	3444	AT427
3438	DT8431M	3438	23114317	3443	REN513	3444	A248(AMC)	3444	AT490
3438	E13-008-00	3438	6621002300	3443	TV-215	3444	A249	3444	AT491
3438	G4N1	3438	A1D	3443	28-31-00	3444	A2499	3444	AT492
3438	HHF-85020	3438	A1DJ	3443	48-137397	3444	A2B	3444	AT493
3438	M4900	3438	A1DJ	3443	48-69723A01	3444	A2BRN	3444	AT494
3438	M9408	3439	A2417	3443	48-69723A02	3444	A2PGRN	3444	AT495
3438	MJ1800	3439	A3H-1	3443	48-69723AD2-185	3444	A2J	3444	AWH-24
3438	MJ3010	3439	A6A	3443	48-69723B02	3444	A301	3444	B-1421
3438	MJ3026	3439	A6M	3443	48D69723A01	3444	A306	3444	B-1433
3438	MJ3027	3439	A6Z	3443	48D69723A02-185	3444	A307	3444	B-1666
3438	MJ3029	3439	A6ZH	3443	481137397	3444	A323	3444	B-169
3438	MJE423	3439	A8T	3443	86-114-3	3444	A344	3444	B-1842
3438	N0400	3439	A9R	3443	103-215	3444	A345	3444	B-1872
3438	PTC118	3439	B176000	3443	1371	3444	A346	3444	B-75583-1
3438	Q5119	3439	B176004	3443	23115321	3444	A3E	3444	B-75583-2
3438	2N5633	3439	B176005	3443	23115334	3444	A3F	3444	B-75589-13
3438	2N5758	3439	B176006	3444	A-11095924	3444	A3G	3444	B-75589-1
3438	2N5759	3439	B176007	3444	A-11237336	3444	A3N	3444	B-75589-3
3438	2N5760	3439	B176009	3444	A-1141 6062	3444	A3R	3444	B12-1-A-21
3438	2N6306	3439	B176010	3444	A-120278	3444	A3T	3444	B12-1A21
3438	2SC035	3439	B176011	3444	A-125332	3444	A3T2221	3444	B13357B
3438	2SC036	3439	B176013	3444	A-1854-0003-1	3444	A3T2221A	3444	B169
3438	2SC1006B	3439	B176014	3444	A-1854-0019-1	3444	A3T2222	3444	B169(JAPAN)
3438	2SC1114	3439	B176015	3444	A-1854-0025-1	3444	A3T2222A	3444	B1X
3438	2SC1185	3439	B176024	3444	A-1854-0027-1	3444	A3T5011	3444	B1P7201
3438	2SC1185A	3439	B176025	3444	A-1854-0071-1	3444	A3T929	3444	B1W
3438	2SC1185B	3439	B176026	3444	A-1854-0094-1	3444	A3T930	3444	B2D
3438	2SC1185C	3439	B176027	3444	A-1854-0099-1	3444	A3W	3444	BA67
3438	2SC1185K	3439	B176028	3444	A-1854-0201-1	3444	A3Z	3444	BA71
3438	2SC1185L	3439	B176029	3444	A-1854-0215-1	3444	A415	3444	BACSH2M1
3438	2SC1185M	3439	B1P	3444	A-1854-0241-1	3444	A43021415	3444	BACSH2M2
3438	2SC240	3439	B1P-1	3444	A-1854-0246-1	3444	A454	3444	BACSH2M3
3438	2SC241	3439	BU109	3444	A-1854-0251-1	3444	A481A002B	3444	BACT2F
3438	2SC242	3439	C41?TV	3444	A-1854-0255-1	3444	A481A0031	3444	BB71
3438	2SC243	3439	ES15X126	3444	A-1854-0354-1	3444	A494(JAPAN)	3444	BC-107
3438	2SC245	3439	ES84	3444	A-1854-0408-1	3444	A4A	3444	BC-107A
3438	2SC246	3439	IRTR67	3444	A-1854-0434-1	3444	A4M	3444	BC-121
3438	2SC407	3439	M4901	3444	A-1854-0471-1	3444	A4N	3444	BC-122
3438	2SC43	3439	M4995	3444	A-1854-0492-1	3444	A4P	3444	BC-123
3438	2SC44	3439	M7511	3444	A-1854-0541-1	3444	A4R	3444	BC-148A
3438	2SC519	3439	MJ3011	3444	A-1854-0554-1	3444	A4U	3444	BC-148B
3438	2SC519A	3439	PTC129	3444	A-567A	3444	A4Y	3444	BC-148C
3438	2SC586	3439	RE30	3444	A.184/5	3444	A4Y-2	3444	BC-167-B
3438	2SC792	3439	T-Q5019	3444	A054-108	3444	A567	3444	BC-71
3438	2SC889	3439	2SC1027	3444	A054-114	3444	A593	3444	BC107
3438	2SD110	3439	2SC41	3444	A054-115	3444	A5C	3444	BC107A
3438	2SD110-R	3439	2SC41?TV	3444	A054-155	3444	A5H	3444	BC107B
3438	2SD110-Y	3439	09-302158	3444	A054-175	3444	A5K	3444	BC108
3438	2SD111	3439	09-302177	3444	A054-195	3444	A5L	3444	BC108A
3438	2SD111-R	3439	29V069002	3444	A054-221	3444	A5M	3444	BC108B
3438	2SD111-Y	3439	34-6002-58	3444	A054-222	3444	A5N	3444	BC109
3438	2SD165	3439	48-134901	3444	A054-225	3444	A5P	3444	BC109B
3438	2SD166	3439	48-134995	3444	A054-233	3444	A5R	3444	BC110
3438	2SD177	3439	48-137134	3444	A054-234	3444	A5S	3444	BC113
3438	2SD18	3439	48-137179	3444	A059-109	3444	A5T2222	3444	BC114
3438	2SD181	3439	48-137203	3444	A06-1-12	3444	A5T5903	3444	BC114TR
3438	2SD189	3439	48-8692730	3444	A065-102	3444	A5T5904	3444	BC115
3438	2SD189A	3439	48137524	3444	A065-103	3444	A5U	3444	BC118
3438	2SD202	3439	63N1	3444	A065-104	3444	A5W	3444	BC125
3438	2SD203	3439	86-221-2	3444	A065-108	3444	A641(NPN)	3444	BC125B
3438	2SD283	3439	86-222-2	3444	A065-109	3444	A649L	3444	BC129
3438	2SD284	3439	86-224-2	3444	A065-110	3444	A649S	3444	BC130
3438	2SD285	3439	86-225-2	3444	A065-113	3444	A6HD	3444	BC131
3438	2SD320	3439	121-452	3444	A066-109	3444	A6J	3444	BC132
3438	2SD334	3439	121-996	3444	A068-112	3444	A6K	3444	BC134
3438	2SD334A	3439	260P24408	3444	A068-108	3444	A6R	3444	BC135
3438	2SD334R	3439	417-239	3444	A068-113	3444	A6S	3444	BC136
3438	2SD45	3439	417-248	3444	A069-102/103	3444	A748B	3444	BC147
3438	2SD46	3439	800-550-10	3444	A069-104	3444	A749B	3444	BC147B
3438	2SD47	3439	1074	3444	A069-104/106	3444	A7A	3444	BC147A
3438	2SD59	3439	3634-1	3444	A069-106	3444	A7B(TRANSISTOR)	3444	BC147B
3438	2SD60	3439	80249-910787	3444	A069-120	3444	A7R	3444	BC148
3438	2SD67B	3439	137352	3444	A069-122	3444	A7S	3444	BC148A
3438	2SD67C	3439	0573526	3444	A106(JAPAN)	3444	A7T	3444	BC148C
3438	2SD67D	3439	610063-1	3444	A108	3444	A88	3444	BC149
3438	2SD67E	3439	1473601	3444	A108A	3444	A88(JAPAN)	3444	BC149A
3438	2SD83	3439	3458854	3444	A108B	3444	A8B	3444	BC149B
3438	2SD84	3439	7304149	3444	A10005-010-A	3444	A8G	3444	BC149Q
3438	2SD88	3439	62734604	3444	A10005-011-A	3444	A8L	3444	BC149C
3438	2SD88A	3439	2003055813	3444	A10005-015-D	3444	A937	3444	BC150
3438	09-302159	3439/163	2SC1154	3444	A111	3444	A937-1	3444	BC151
3438	13-33182-1	3440	RCA1003	3444	A12-1-70	3444	A937-3	3444	BC152
3438	13-33182-2	3440	RCA1C12	3444	A12-1-705	3444	A9B	3444	BC155B
		3440	RCA29B						
		3440	RCA29C						

SK	Industry Standard No.	SK	Industry Standard No.	SK	Industry Standard No.	SK	Industry Standard No.	SK	Industry Standard No.
3444	BC156B	3444	BFV83B	3444	BSY62A	3444	C538AQ	3444	CF2
3444	BC168B	3444	BFV83C	3444	BSY63	3444	C538P	3444	CG1
3444	BC69	3444	BFV85	3444	BSY73	3444	C538Q	3444	CJ5201
3444	BC169A	3444	BFV85A	3444	BSY74	3444	C538R	3444	CJ5202
3444	BC169B	3444	BFV85B	3444	BSY75	3444	C538S	3444	CJ5203
3444	BC169CL	3444	BFV85C	3444	BSY76	3444	C538T	3444	CJ5206
3444	BC171	3444	BFV87	3444	BSY80	3444	C539	3444	CJ5207
3444	BC171A	3444	BFV88	3444	BSY89	3444	C539K	3444	CJ5211
3444	BC171B	3444	BFV88A	3444	BSY93	3444	C539L	3444	CJ5212
3444	BC172	3444	BFV88B	3444	BSY95	3444	C539R	3444	C8-9014
3444	BC172A	3444	BFV88C	3444	BSY95A	3444	C5398	3444	C8-9014B
3444	BC172B	3444	BFW32	3444	BT67	3444	C587	3444	C8-9014D
3444	BC172C	3444	BFW46	3444	BT71	3444	C587A	3444	C81166
3444	BC175	3444	BFW59	3444	BTX-094	3444	C588	3444	C81166D
3444	BC180	3444	BFW60	3444	BTX-095	3444	C593	3444	C81166G
3444	BC180B	3444	BFW68	3444	BTX-096	3444	C596	3444	C81166P
3444	BC182	3444	BFX92	3444	BTX-2367B	3444	C619	3444	C81166G
3444	BC182L	3444	BFX93	3444	BTX068	3444	C619B	3444	C81166H/P
3444	BC183	3444	BFX95A	3444	BU67	3444	C619C	3444	C81166P
3444	BC184	3444	BFY	3444	BU71	3444	C619D	3444	C81168G
3444	BC184B	3444	BFY22	3444	BUC 97704-2	3444	C622	3444	C81168H
3444	BC185	3444	BFY23	3444	BV67	3444	C655	3444	C81229
3444	BC197	3444	BFY23A	3444	BV71	3444	C66-P11111-0001	3444	C81229A
3444	BC197A	3444	BFY24	3444	BW67	3444	C6862400	3444	C81229B
3444	BC197B	3444	BFY25	3444	BW71	3444	C689	3444	C81229C
3444	BC198	3444	BFY26	3444	BX67	3444	C689H	3444	C81229D
3444	BC199	3444	BFY28	3444	BX71	3444	C694D	3444	C81229E
3444	BC207	3444	BFY29	3444	BY67	3444	C713	3444	C81229F
3444	BC207A	3444	BFY30	3444	BY71	3444	C714	3444	C81229G
3444	BC207B	3444	BFY33	3444	B267	3444	C716G	3444	C81229H
3444	BC207BL	3444	BFY371	3444	B271	3444	C7350RN	3444	C81235C
3444	BC208	3444	BFY39	3444	C00 68602300	3444	C773	3444	C81235E
3444	BC208A	3444	BFY39/1	3444	C1007	3444	C773C	3444	C81235G
3444	BC208AL	3444	BFY39/2	3444	C10279-1	3444	C773D	3444	C81235D
3444	BC208B	3444	BFY39/3	3444	C10279-3	3444	C796	3444	C81235H
3444	BC208BL	3444	BFY391	3444	C103(JAPAN)	3444	C847	3444	C81238P
3444	BC208C	3444	BFY73	3444	C1071	3444	C848	3444	C81245F
3444	BC208CL	3444	BFY74	3444	C1244	3444	C849	3444	C81245G
3444	BC209	3444	BFY75	3444	C12711	3444	C850	3444	C81245H
3444	BC209A	3444	BFY76	3444	C13901	3444	C896	3444	C81245I
3444	BC209B	3444	BFY77	3444	C1390I	3444	C899	3444	C81245T
3444	BC209BL	3444	BQ71	3444	C1390J	3444	C899K	3444	C81250B
3444	BC209C	3444	BT71	3444	C1390K	3444	C925	3444	C81257
3444	BC209CL	3444	BT171	3444	C1390V	3444	C934C	3444	C81258
3444	BC210	3444	BN7517	3444	C1390W	3444	C934D	3444	C81259
3444	BC220	3444	BN7518	3444	C1390WH	3444	C934E	3444	C81283A
3444	BC222	3444	BP67	3444	C1390WI	3444	C934F	3444	C81288
3444	BC223A	3444	B267	3444	C1390WX	3444	C934G	3444	C81289
3444	BC223B	3444	B8475	3444	C1390WY	3444	C934P	3444	C81295E
3444	BC233A	3444	B867	3444	C1390X	3444	C943	3444	C81295G
3444	BC238A	3444	BS810	3444	C1390XJ	3444	C943A	3444	C81340I
3444	BC238B	3444	BS821	3444	C1390XK	3444	C943B	3444	C81340I
3444	BC238C	3444	BSV35A	3444	C1390YM	3444	C943C	3444	C81344
3444	BC267	3444	BSV40	3444	C15-1	3444	C9604	3444	C81345
3444	BC268	3444	BSV41	3444	C15-2	3444	C966	3444	C81348
3444	BC269	3444	BSV53	3444	C15-3	3444	C967	3444	C81349
3444	BC270	3444	BSV54	3444	C1639	3444	CA-9011H	3444	C81353
3444	BC280	3444	BSV59	3444	C170	3444	CAM-12	3444	C81361G
3444	BC280A	3444	BSV88	3444	C1739	3444	CB246	3444	C81362
3444	BC280B	3444	BSV89	3444	C1045295DY1	3444	CC1168P	3444	C81363
3444	BC280C	3444	BSV90	3444	C238	3444	C81235G	3444	C81368
3444	BC282	3444	BSV91	3444	C2538-11	3444	CC82004	3444	C81368A
3444	BC284	3444	BSW11	3444	C281A	3444	CC84004	3444	C81368B
3444	BC284A	3444	BSW12	3444	C281B	3444	CC86168	3444	C81368C
3444	BC284B	3444	BSW19	3444	C281C	3444	CC86168P	3444	C81368D
3444	BC289	3444	BSW33	3444	C281C-EP	3444	CC89018E	3444	C81370
3444	BC289A	3444	BSW34	3444	C281D	3444	CD0014NA	3444	C81371
3444	BC289B	3444	BSW39	3444	C281EP	3444	CD0014NG	3444	C81372
3444	BC290	3444	BSW41	3444	C281H	3444	CD0015N	3444	C81383
3444	BC290B	3444	BSW42	3444	C281HA	3444	CD0021	3444	C81420
3444	BC290C	3444	BSW42A	3444	C281HB	3444	CD12000	3444	C81453E
3444	BC408	3444	BSW43	3444	C281HC	3444	CD38	3444	C81585
3444	BC409	3444	BSW43A	3444	C282H	3444	CD446	3444	C81585E/P
3444	BC507A	3444	BSX51	3444	C282HA	3444	CD6019	3444	C81585G
3444	BC507B	3444	BSX52	3444	C282HB	3444	CD6150	3444	C81625
3444	BC508A	3444	BSX53	3444	C282HC	3444	CD6157	3444	C81665
3444	BC508B	3444	BSX82	3444	C284	3444	CD6375	3444	C82004D
3444	BC508C	3444	BSX83	3444	C284H	3444	CD8000	3444	C82006
3444	BC509B	3444	BSX84	3444	C284HA	3444	CD9525	3444	C82219
3444	BC509C	3444	BSX85	3444	C284HB	3444	CDC12000-1C	3444	C82221
3444	BC510B	3444	BSX88	3444	C300	3444	CDC12018C	3444	C82222
3444	BC583	3444	BSX89	3444	C301	3444	CDC1201BC	3444	C82369
3444	BC71	3444	BSX92	3444	C302(JAPAN)	3444	CDC12077P	3444	C82481
3444	BCW34	3444	BSX19	3444	C315	3444	CDC13000-1	3444	C82711
3444	BCW36	3444	BSX24	3444	C317	3444	CDC13000-1B	3444	C82712
3444	BCW60A	3444	BSX25	3444	C317-0	3444	CDC13000-1B	3444	C82713
3444	BCW60AA	3444	BSX30	3444	C317C	3444	CDC13000-1C	3444	C82714
3444	BCW83	3444	BSX38	3444	C318(JAPAN)	3444	CDC13000-1D	3444	C82922
3444	BCY13	3444	BSX44	3444	C318A(JAPAN)	3444	CDC13016A	3444	C82923
3444	BCY15	3444	BSX48	3444	C321	3444	CDC13500-1	3444	C82924
3444	BCY16	3444	BSX49	3444	C321H	3444	CDC2010	3444	C82925
3444	BCY36	3444	BSX51	3444	C321HA	3444	CDC2010C	3444	C83390
3444	BCY42	3444	BSX51A	3444	C321HB	3444	CDC2010D	3444	C83391
3444	BCY43	3444	BSX52	3444	C321HC	3444	CDC25100-6	3444	C83391A
3444	BCY50	3444	BSX52A	3444	C323	3444	CDC2510C	3444	C83392
3444	BCY501	3444	BSX53	3444	C324	3444	CDC25100-G	3444	C83393
3444	BCY51	3444	BSX54	3444	C324A	3444	CDC2510G	3444	C83394
3444	BCY511	3444	BSX66	3444	C324HA	3444	CDC4023A130	3444	C83395
3444	BCY56	3444	BSX67	3444	C350H	3444	CDC430	3444	C83396
3444	BCY57	3444	BSX68	3444	C352	3444	CDC4306813	3444	C83397
3444	BCY58B	3444	BSX69	3444	C352(JAPAN)	3444	CDC745(ZENITH)	3444	C83398
3444	BCY58C	3444	BSX75	3444	C352A	3444	CDC746	3444	C83402
3444	BCY58D	3444	BSX76	3444	C356	3444	CD08000	3444	C83403
3444	BCY59	3444	BSX77	3444	C360	3444	CD08000-1B	3444	C83404
3444	BCY59A	3444	BSX78	3444	C360D	3444	CD08001	3444	C83405
3444	BCY59B	3444	BSX79	3444	C36580	3444	CD08021	3444	C83414
3444	BCY59C	3444	BSX80	3444	C39-207	3444	CD08054	3444	C83415
3444	BCY59D	3444	BSX81	3444	C395	3444	CD086X7-5	3444	C83416
3444	BCY69	3444	BSX87	3444	C395A	3444	CDQ10001	3444	C83417
3444	BCY84A	3444	BSX89	3444	C395R	3444	CDQ10002	3444	C8360
3444	BP-214	3444	BSX90	3444	C400	3444	CDQ10003	3444	C83605
3444	BP-215	3444	BSX91	3444	C400-Q	3444	CDQ10004	3444	C83606
3444	BP-226	3444	BSX94A	3444	C400-R	3444	CDQ10005	3444	C83607
3444	BP115	3444	BSX97	3444	C400-Y	3444	CDQ10006	3444	C83843
3444	BP183A	3444	BSY165	3444	C402(JAPAN)	3444	CDQ10007	3444	C83844
3444	BP189	3444	BSY168	3444	C423B	3444	CDQ10008	3444	C83845
3444	BP224J	3444	BSY17	3444	C423C	3444	CDQ10009	3444	C83854
3444	BP225J	3444	BSY18	3444	C423D	3444	CDQ10010	3444	C83854A
3444	BP248	3444	BSY19	3444	C423E	3444	CDQ10016	3444	C83855
3444	BP250	3444	BSY20	3444	C423F	3444	CDQ10017	3444	C83855A
3444	BP291	3444	BSY21	3444	C424(JAPAN)	3444	CDQ10018	3444	C83859
3444	BP291A	3444	BSY26	3444	C424D	3444	CDQ10019	3444	C83859A
3444	BP293	3444	BSY27	3444	C425B	3444	CDQ10020	3444	C83960
3444	BP293A	3444	BSY28	3444	C425C	3444	CDQ10021	3444	C83900
3444	BP293D	3444	BSY29	3444	C425D	3444	CDQ10022	3444	C83900A
3444	BP321B	3444	BSY34	3444	C425E	3444	CDQ10023	3444	C83901
3444	BP321C	3444	BSY38	3444	C425F	3444	CDQ10024	3444	C83903
3444	BP321D	3444	BSY39	3444	C444	3444	CDQ10025	3444	C83904
3444	BP321E	3444	BSY48	3444	C450	3444	CDQ10026	3444	C83424
3444	BP321F	3444	BSY49	3444	C468	3444	CDQ10027	3444	C84425
3444	BP71	3444	BSY58	3444	C468A	3444	CDQ10028	3444	C85088
3444	BPR11	3444	BSY59	3444	C538	3444	CDQ10032	3444	C85369
3444	BPR26	3444	BSY61	3444	C538A	3444	CE4001E		
		3444	BSY62			3444	CE4003E		

SK	Industry Standard No.	SK	Industry Standard No.	SK	Industry Standard No.	SK	Industry Standard No.	SK	Industry Standard No.
3444	CS6168F	3444	EA15X52	3444	F2448	3444	HR-11	3444	LM1415-6
3444	C86229F	3444	EA15X56	3444	F2584	3444	HR-11A	3444	LM1415-7
3444	C86229G	3444	EA15X58	3444	F302-1	3444	HR-11B	3444	LM1540
3444	C8696	3444	EA15X59	3444	F302-1532	3444	HR-13	3444	LM1566P
3444	C8697	3444	EA15X63	3444	F302-2	3444	HR-13A	3444	LM1818
3444	C8706	3444	EA15X68	3444	F302-2532	3444	HR-14A	3444	LRQ849
3444	C8718	3444	EA15X7112	3444	F306-001	3444	HR-15A	3444	L8-0085-01
3444	C8718A	3444	EA15X7115	3444	F306-022	3444	HR-16	3444	L83705
3444	C8720A	3444	EA15X7119	3444	F3519	3444	HR-16A	3444	M-1002-2
3444	C87229G	3444	EA15X7175	3444	F3532	3444	HR-17	3444	M140-3
3444	C8901/	3444	EA15X7176	3444	F3569	3444	HR-17A	3444	M24
3444	C89011/3490	3444	EA15X72	3444	F3571	3444	HR-18	3444	M24A
3444	C89011	3444	EA15X7232	3444	F566	3444	HR-18A	3444	M24B
3444	C89011D	3444	EA15X73	3444	F572-1	3444	HR-19	3444	M25
3444	C89011B	3444	EA15X75	3444	F587	3444	HR-19A	3444	M25A
3444	C89011F	3444	EA15X7514	3444	F75116	3444	HR-32	3444	M25B
3444	C89011G	3444	EA15X7517	3444	F9600	3444	HR-36	3444	M25B2
3444	C89011G/3490	3444	EA15X7586	3444	F9623	3444	HR-37	3444	M31001
3444	C89011H	3444	EA15X7589	3444	FB6853	3444	HR-38	3444	M3519
3444	C89011I	3444	EA15X76	3444	FBC257	3444	HR-48	3444	M4464
3444	C89014	3444	EA15X7638	3444	FC8116B7813	3444	HS-1168	3444	M4465
3444	C89014/3490	3444	EA15X77	3444	FC8116B8	3444	HS-1229	3444	M447
3444	C89014A	3444	EA15X83	3444	FC8116B8704	3444	HS-40037	3444	M4594
3444	C89014B	3444	EA15X84	3444	FC81229F	3444	HS-40044	3444	M4624
3444	C89014C	3444	EA15X85	3444	FC81229G	3444	HS-40046	3444	M4650
3444	C89014D/3490	3444	EA15X8511	3444	FD-1029-LL	3444	HT303620B	3444	M4705
3444	C89014D	3444	EA15X86	3444	FD-1029-NG	3444	HT3036210	3444	M4714
3444	C89014G	3444	EA15X89	3444	FD-1029-PP	3444	HT303711B0	3444	M4732
3444	C89101B	3444	EA15X9	3444	FMPBA20	3444	HT303721-0	3444	M4734
3444	C89104	3444	EA15X91	3444	FPRA0-1001	3444	HT303721	3444	M4737
3444	C89125B	3444	EA15X96	3444	FPR50-1001	3444	HT3037210	3444	M4739
3444	C8925M	3444	EA15X98	3444	FPR50-1002	3444	HT3037210-0	3444	M4765
3444	C89600-4	3444	EA1628	3444	F81221	3444	HT303721D	3444	M4768
3444	C89600-5	3444	EA1629	3444	F81974	3444	HT303730	3444	M4821
3444	CTP-2001-1007	3444	EA1630	3444	F856999	3444	HT303730A	3444	M4834
3444	CTP-2001-1008	3444	EA1638	3444	FT005	3444	HT3073100	3444	M4840
3444	CXL123A	3444	EA1695	3444	FT006	3444	HT304531	3444	M4841
3444	D048	3444	EA1696	3444	G005-036C	3444	HT304531A	3444	M4842
3444	D053	3444	EA1697	3444	G005-036E	3444	HT304531C	3444	M4842A
3444	D16E7	3444	EA1703	3444	G05-010-A	3444	HT304540B0	3444	M4842C
3444	D16E9	3444	EA1716	3444	G05-011-A	3444	HT304580	3444	M4844
3444	D16BC18	3444	EA1718	3444	G05-015-D	3444	HT304580A	3444	M4852
3444	D1A	3444	EA1735	3444	G05-034-D	3444	HT304580B	3444	M4854
3444	D1B38	3444	EA1872	3444	G05-035-D	3444	HT304580C0	3444	M4898
3444	D294	3444	EA2271	3444	G05-035-C	3444	HT304580K	3444	M4906
3444	D2B38	3444	EA2489	3444	G05-036-E	3444	HT304580Y0	3444	M4926
3444	D328	3444	EA2490	3444	G05-064-A	3444	HT304580Z	3444	M4933
3444	D33D28	3444	EA2739	3444	G05015C	3444	HT304581	3444	M4935
3444	D342	3444	EA2740	3444	G05035E	3444	HT304581A	3444	M4937
3444	D372BL	3444	EA2770	3444	G05036	3444	HT304581B	3444	M54
3444	D4D25	3444	EA3149	3444	G05036B	3444	HT304581B-0	3444	M54A
3444	D4D26	3444	EA4112	3444	G05036C	3444	HT304581C	3444	M54B
3444	D912	3444	ECG123A	3444	G05036D	3444	HT304861B	3444	M54BLK
3444	D917254-2	3444	ED-1402	3444	G05036E	3444	HT305361G	3444	M54BLU
3444	D921881-1	3444	ED1402A	3444	G05037B	3444	HT305361B	3444	M54BRN
3444	D928121	3444	ED1402B	3444	G05059	3444	HT305371B	3444	M54C
3444	DBCZ083906	3444	ED150Z	3444	G395967	3444	HT306441	3444	M54D
3444	DBCZ136406	3444	ED1702L	3444	G395967-2	3444	HT306441A	3444	M54E
3444	DDBY222002	3444	EDC-Q10-1	3444	G9600	3444	HT306441B	3444	M54GRN
3444	DN	3444	EDS-100	3444	G9623	3444	HT306441B-0	3444	M54ORN
3444	D8-67	3444	EL232	3444	G9696	3444	HT306441C	3444	M54RED
3444	D81B	3444	EMS-73500	3444	GC1144	3444	HT307321A	3444	M54WHT
3444	D844	3444	EN2222	3444	GE-20	3444	HT307321B-0	3444	M54YEL
3444	D845	3444	EN3009	3444	GE-3265	3444	HT307322A	3444	M671
3444	D846	3444	EN3013	3444	GE-X16A1938	3444	HT307331B	3444	M7003
3444	D847	3444	EN3014	3444	GET2221	3444	HT307331C	3444	M7015
3444	D867	3444	EN697	3444	GET2222	3444	HT307341B	3444	M7033
3444	D867W	3444	EN706	3444	GET2369	3444	HT307341C-0	3444	M7108
3444	D877	3444	EN708	3444	GET3013	3444	HT308281B	3444	M7108/A5N
3444	DT161	3444	EN930	3444	GET3014	3444	HT308281C	3444	M7109
3444	DW6034/M	3444	EP15X47	3444	GET3646	3444	HT308281G	3444	M7109/A5P
3444	E13-000-03	3444	EP15X49	3444	GET706	3444	HT308281O	3444	M7171
3444	E13-000-04	3444	EP15X7	3444	GET708	3444	HT308282A	3444	M773
3444	E13-002-03	3444	EQ8-100	3444	GET914	3444	HT308282A-0	3444	M773RED
3444	E13-003-00	3444	EQ8-13	3444	G110	3444	HT309842A-0	3444	M775
3444	E13-005-00	3444	EQ8-22	3444	G12711	3444	HT400	3444	M775BRN
3444	E13-005-01	3444	EQ8-5	3444	G12712	3444	HT401	3444	M776
3444	E13-005-02	3444	EQ8-61	3444	G12713	3444	HT800011F	3444	M776GRN
3444	E210	3444	EQ8-9	3444	G12714	3444	HT800011G	3444	M779BLU
3444	E212	3444	ES810222	3444	G12715	3444	HT800011H	3444	M780WHT
3444	E24103	3444	ES810223	3444	G12716	3444	HT800011K	3444	M783
3444	E2430	3444	ES810232	3444	G12921	3444	HT8001B10	3444	M783RED
3444	E2431	3444	ES15050	3444	G12922	3444	HV25	3444	M784
3444	E2436	3444	ES15X1	3444	G12923	3444	HX50002	3444	M7840RN
3444	E2444	3444	ES15X11	3444	G12924	3444	HT3045801C	3444	M785
3444	E2452	3444	ES15X14	3444	G13641	3444	I9A115728-2	3444	M785YEL
3444	E2454	3444	ES15X16	3444	G13643	3444	IC745042	3444	M787BLU
3444	E2455	3444	ES15X20	3444	G13704	3444	IP20-0001	3444	M791
3444	E2459	3444	ES15X23	3444	G13705	3444	IP20-0003	3444	M8105
3444	E2461	3444	ES15X24	3444	G13707	3444	IP20-0006	3444	M818
3444	E2497	3444	ES15X37	3444	G13708	3444	IRTR62	3444	M818WHT
3444	E2499	3444	ES15X42	3444	G13709	3444	IRTR63	3444	M822
3444	EA0092	3444	ES15X58	3444	G13710	3444	J139A	3444	M8221
3444	EA1080	3444	ES15X62	3444	G13711	3444	J241054	3444	M822A
3444	EA1128	3444	ES15X64	3444	GME1001	3444	J241099	3444	M822A-BLU
3444	EA1129	3444	ES15X68	3444	GME1002	3444	J241230	3444	M822B
3444	EA1135	3444	ES15X7	3444	GME2001	3444	J241251	3444	M823
3444	EA1145	3444	ES15X70	3444	GME2002	3444	J24458	3444	M823B
3444	EA1344	3444	ES15X76	3444	GME4001	3444	J24564	3444	M823WHT
3444	EA1345	3444	ES15X83	3444	GME4002	3444	J24565	3444	M827
3444	EA1406	3444	ES15X84	3444	GMB4003	3444	J24624	3444	M827BRN
3444	EA1407	3444	ES15X85	3444	GMB6003	3444	J24625	3444	M828GRN
3444	EA1408	3444	ET15X10	3444	G05036	3444	J24641	3444	M847BLK
3444	EA1451	3444	ET15X11	3444	H102	3444	J24658	3444	M9095
3444	EA1452	3444	ET15X13	3444	HC-00537	3444	J24752	3444	M9159
3444	EA1499	3444	ET15X14	3444	HC-00693	3444	J24753	3444	M91A
3444	EA1564	3444	ET15X15	3444	HC-00828	3444	J24817	3444	M91B
3444	EA1578	3444	ET15X16	3444	HC-00838	3444	J24855	3444	M91BGRN
3444	EA1581	3444	ET15X20	3444	HC-00871	3444	J24874	3444	M91C
3444	EA15X1	3444	ET15X24	3444	HC-00921	3444	J24978	3444	M91CM624
3444	EA15X101	3444	ET15X27	3444	HC-00924	3444	J24906	3444	M91D
3444	EA15X103	3444	ET15X37	3444	HC-01820	3444	J24907	3444	M91E
3444	EA15X111	3444	ET15X41	3444	HC371	3444	J24909	3444	M91F
3444	EA15X112	3444	ET15X42	3444	HC372	3444	J24916	3444	M91FM624
3444	EA15X136	3444	ET15X45	3444	HC373	3444	J310249	3444	M9226
3444	EA15X137	3444	ET234843	3444	HC458	3444	J310250	3444	M9248
3444	EA15X142	3444	ET238894	3444	HC539	3444	JA-H	3444	M9525
3444	EA15X143	3444	ET368021	3444	HC561	3444	JA-L	3444	M9532
3444	EA15X153	3444	ET412626	3444	HCL-29	3444	JA1200	3444	M9563
3444	EA15X157	3444	ET8-0081	3444	HCL-6066	3444	JE9011	3444	M9568
3444	EA15X162	3444	ETTC-458LG	3444	HD-00227	3444	JLM-20	3444	M9570
3444	EA15X163	3444	ETTC-CD12000	3444	HEP-30004	3444	JN271	3444	MA9426
3444	EA15X167	3444	ETTC-CD13000	3444	HEP-80011	3444	JT-1601-40	3444	MAQ7786
3444	EA15X168	3444	ETTC-CD8000	3444	HEP-80022	3444	K4-506	3444	ME-1
3444	EA15X18	3444	ETX18	3444	HEP-80030	3444	KB8339	3444	ME-2
3444	EA15X20	3444	EW165V	3444	HP2	3444	KLH1422	3444	ME-3
3444	EA15X22	3444	EW181	3444	HP3	3444	KLH704	3444	MB1001
3444	EA15X24	3444	EW182	3444	HP4	3444	KP06682	3444	MB1002
3444	EA15X272	3444	EYZP-632	3444	HP5	3444	KR-Q1013	3444	MB2001
3444	EA15X31	3444	EYZP-791	3444	HP6	3444	LM1090E	3444	MB2002
3444	EA15X330	3444	F15840	3444	HP7	3444	LM1090F	3444	MB213
3444	EA15X331	3444	F15840-1	3444	HP8	3444	LM1090G	3444	MB213A
3444	EA15X337	3444	F2443	3444	HKT-158	3444	LM1117D	3444	MB216
3444	EA15X44	3444		3444	HKT-161	3444	LM1403	3444	MB217
3444	EA15X45								

SK	Industry Standard No.	SK	Industry Standard No.	SK	Industry Standard No.	SK	Industry Standard No.	SK	Industry Standard No.
3444	M84001	3444	NS480	3444	QA-12	3444	R87529	3444	S1510
3444	M84002	3444	N86114	3444	QA-13	3444	R87530	3444	S1512
3444	M84003	3444	N86115	3444	QA-14	3444	R87542	3444	S1526
3444	M84003C	3444	N86207	3444	QA-15	3444	R87543	3444	S1527
3444	M84101	3444	N86210	3444	QA-16	3444	R87544	3444	S1529
3444	M84102	3444	N87262	3444	QA-19	3444	R87555	3444	S1530
3444	M84103	3444	N8731	3444	QG0254	3444	R87606	3444	S1533
3444	M84104	3444	N8731A	3444	QOV60530	3444	R87607	3444	S1559
3444	M86001	3444	N8733	3444	QSC380	3444	R87609	3444	S1649
3444	M86002	3444	N8733A	3444	R-280537	3444	R87610	3444	S1568
3444	M86003	3444	N8734	3444	R5283	3444	R87611	3444	S1570
3444	M8900	3444	N8734A	3444	R340	3444	R87612	3444	S1619
3444	M8900A	3444	N8949	3444	R582	3444	R87613	3444	S1620
3444	M8901	3444	P/N10000020	3444	RT163	3444	R87614	3444	S1629
3444	M8901A	3444	P04-44-0028	3444	RT165	3444	R87620	3444	S1697
3444	MBP-25	3444	P04-45-0014-P2	3444	RT249	3444	R87621	3444	S1761
3444	MU9623	3444	P04-45-0014-P5	3444	RT343	3444	R87622	3444	S1761A
3444	MH9623	3444	P04440028-001	3444	RT359	3444	R87623	3444	S1761B
3444	MH9630	3444	P04440028-009	3444	RT360	3444	R87624	3444	S1761C
3444	MJ9623	3444	P04440028-004	3444	RT561	3444	R87625	3444	S1764
3444	MJ9630	3444	P04440028-8	3444	RT582	3444	R87626	3444	S1765
3444	MM1755	3444	P04440032-001	3444	RT887	3444	R87627	3444	S1766
3444	MM1756	3444	P15153	3444	RT953	3444	R87628	3444	S1768
3444	MM1757	3444	P1901-50	3444	R8066	3444	R87634	3444	S1770
3444	MM1758	3444	P480A0028	3444	R8067	3444	R87635	3444	S1772
3444	MPS2926BRN	3444	P480A0029	3444	R8068	3444	R87636	3444	S1784
3444	MPS2926GRN	3444	P5152	3444	R8069	3444	R87637	3444	S1785
3444	MPS2926ORN	3444	P5153	3444	R8070	3444	R87638	3444	S1835
3444	MPS2926RED	3444	P633567	3444	R8115	3444	R87639	3444	S1871
3444	MPS2926YEL	3444	P64447	3444	R8116	3444	R87640	3444	S1891
3444	MP83992	3444	P8393	3444	R8117	3444	R87641	3444	S1891A
3444	MP86351	3444	P8394	3444	R8118	3444	R87642	3444	S1891B
3444	MP86413	3444	P9623	3444	R8119	3444	R87643	3444	S1955
3444	MP86552	3444	PA7001/0001	3444	R8120	3444	R87814	3444	S1993
3444	MP86553	3444	PA9006	3444	R8223	3444	R88442	3444	S2043
3444	MP86554	3444	PEP2	3444	R8224	3444	R88503	3444	S2044
3444	MP86556	3444	PEP5	3444	R8225	3444	R886057332	3444	S2121
3444	MP865611	3444	PEP6	3444	R8243	3444	RT100	3444	S2122
3444	MP86590	3444	PEP7	3444	R8244	3444	RT114	3444	S2123
3444	MP89185	3444	PEP8	3444	R8259	3444	RT2016	3444	S2124
3444	MP89423	3444	PEP9	3444	R8260	3444	RT2332	3444	S2171
3444	MP89433	3444	PET1002	3444	R8261	3444	RT2914	3444	S2172
3444	MP89434J	3444	PET2001	3444	R8305	3444	RT3063	3444	S2225
3444	MP89434K	3444	PET2002	3444	R8312	3444	RT3064	3444	S22543
3444	MP89600	3444	PET3704	3444	R8528	3444	RT3228	3444	S2377
3444	MP89600-5	3444	PET3705	3444	R8530	3444	RT3565	3444	S24591
3444	MP89600F	3444	PET3706	3444	R8543	3444	RT3567	3444	S24596
3444	MP89600Q	3444	PET4002	3444	R8551	3444	RT5202	3444	S2581
3444	MP89600Q/H	3444	PET6001	3444	R8552	3444	RT5206	3444	S2582
3444	MP89600H	3444	PET6002	3444	R8553	3444	RT5207	3444	S2590
3444	MP89604D	3444	PET8000	3444	R8554	3444	RT5551	3444	S2593
3444	MP89604E	3444	PET8001	3444	R8555	3444	RT69221	3444	S2635
3444	MP89604I	3444	PET8002	3444	R8556	3444	RT697M	3444	S2636
3444	MP89604R	3444	PET8003	3444	R8557	3444	RT7322	3444	S2935
3444	MP89611-5	3444	PET8004	3444	R8620	3444	RT7325	3444	S2944
3444	MP89616	3444	PET9002	3444	R8646	3444	RT7326	3444	S2945
3444	MP89618	3444	PL1052	3444	R8647	3444	RT7327	3444	S2984
3444	MP89618H	3444	PL1054	3444	R8648	3444	RT7511	3444	S2985
3444	MP89618I	3444	PM1121	3444	R8658	3444	RT7514	3444	S2989
3444	MP89618J	3444	PRT-101	3444	R8914	3444	RT7515	3444	S2996
3444	MP89623C	3444	PRT-104	3444	R8916	3444	RT7517	3444	S2997
3444	MP89623E	3444	PRT-104-1	3444	R8963	3444	RT7518	3444	S2998
3444	MP89623G	3444	PRT-104-2	3444	R8964	3444	RT7528	3444	S2999
3444	MP89623G/H	3444	PRT-104-3	3444	R8965	3444	RT7557	3444	S34540
3444	MP89623H/I	3444	P8209800	3444	R8966	3444	RT7588	3444	S36999
3444	MP89623I/J	3444	PT1558	3444	R8968	3444	RT7845	3444	0856G
3444	MP89626	3444	PT1559	3444	R9004	3444	RT7943	3444	S6801
3444	MP89631	3444	PT1610	3444	R9005	3444	RT8195	3444	S9631
3444	MP89631I	3444	PT1835	3444	R9006	3444	RT8197	3444	SC1001
3444	MP89631J	3444	PT1836	3444	R9025	3444	RT8198	3444	SC1010
3444	MP89631K	3444	PT1837	3444	R9071	3444	RT8201	3444	SC1168G
3444	MP89631T	3444	PT2760	3444	R9384	3444	RT8332	3444	SC1168H
3444	MP89632	3444	PT2896	3444	R9385	3444	RT929H	3444	SC12290
3444	MP89632H	3444	PT3141	3444	R9483	3444	RV1471	3444	S0350
3444	MP89632I	3444	PT3141A	3444	REN1425A	3444	RV1474	3444	SC4010
3444	MP89632J	3444	PT3151A	3444	RRV504	3444	RV2249	3444	SC4044
3444	MP89632K	3444	PT3151A	3444	RR8068	3444	RVTC81381	3444	SC832
3444	MP89700D	3444	PT3151B	3444	RR8914	3444	RVTC81383	3444	SC842
3444	MP89700B	3444	PT3151C	3444	R8-107	3444	RVTS22410	3444	SD109
3444	MP89700P	3444	PT4-7158	3444	R8-108	3444	RYN121105	3444	SD3000
3444	MPX9623H/I	3444	PT4-7158-012	3444	R8-2009	3444	RYN121105-3	3444	SDD421
3444	MFX9650I	3444	PT4-7158-013	3444	R8-2013	3444	RYN121105-4	3444	SDD821
3444	MQ1	3444	PT4-7158-01A	3444	R8-2016	3444	S001466	3444	SE-0566
3444	MQ2	3444	PT4-7158-021	3444	R81049	3444	S0015	3444	SE1331
3444	MR3932	3444	PT4-7158-022	3444	R81059	3444	S0022	3444	SE2401
3444	MR9604	3444	PT4-7158-023	3444	R815048	3444	S0025	3444	SE2402
3444	MS22B	3444	PT4-7158-02A	3444	R85851	3444	S022010	3444	S3646
3444	M87502R	3444	PT4800	3444	R85853	3444	S022011	3444	SE4001
3444	M87503R	3444	PT627	3444	R85856	3444	S022442B	3444	SE4002
3444	M8RT503	3444	PT703	3444	R85897	3444	S024987	3444	SE4172
3444	M2104	3444	PT720	3444	RS7103	3444	S025232	3444	SE5-0128
3444	MT4101	3444	PT851	3444	RS7105	3444	S025289	3444	SE5-0253
3444	MT4102	3444	PT886	3444	RS7111	3444	S031A	3444	SE5-0274
3444	MT4102A	3444	PT887	3444	RS7121	3444	S037	3444	SE5-0367
3444	MT4103	3444	PT897	3444	RS7127	3444	S0704	3444	SE5-0567
3444	MT6001	3444	PT898	3444	RS7129	3444	S1061	3444	SE5-0608
3444	MT6002	3444	PTC115	3444	RS7132	3444	S1065	3444	SE5-0848
3444	MT6003	3444	PTC136	3444	RS7133	3444	S1066	3444	SE5-0854
3444	MT696	3444	Q-00269	3444	RS7136	3444	S1068	3444	SE5-0855
3444	MT697	3444	Q-00269A	3444	RS7160	3444	S1069	3444	SE5-0887
3444	MT706	3444	Q-00269B	3444	RS7223	3444	S1074	3444	SE5-0888
3444	MT706A	3444	Q-00269C	3444	RS7224	3444	S1074R	3444	SE5-0958-54
3444	MT706B	3444	Q-00369	3444	RS7226	3444	S1128	3444	SE5030A
3444	MT707	3444	Q-00369A	3444	RS7232	3444	S1143	3444	SE5030B
3444	MT708	3444	Q-00369B	3444	RS7234	3444	S12-1-A-3P	3444	SE5151
3444	MT9001	3444	Q-00369C	3444	RS7235	3444	S1221	3444	SE6010
3444	MT9002	3444	Q-00484R	3444	RS7236	3444	S1221A	3444	SE8040
3444	N-EA15X136	3444	Q-00569	3444	RS7241	3444	S1226	3444	SF1001
3444	N-EA15X137	3444	Q-00569A	3444	RS7242	3444	S1242	3444	SF1713
3444	N-EA15X138	3444	Q-00569B	3444	RS7405	3444	S1243	3444	SF1726
3444	N201AX	3444	Q-00669	3444	RS7406	3444	S1245	3444	SF1730
3444	ON47204-1	3444	Q-00669A	3444	RS7407	3444	S1272	3444	SFT713
3444	NJ100B	3444	Q-00669C	3444	RS7408	3444	S1307	3444	SFT714
3444	NPC737	3444	Q-00684R	3444	RS7409	3444	S1309	3444	SH1064
3444	NR041	3444	Q-02115C	3444	RS7410	3444	S1331	3444	S5570
3444	NR041E	3444	Q-03115C	3444	RS7411	3444	S1331N	3444	SK1640A
3444	NR071AU	3444	Q-04115C	3444	RS7412	3444	S1331W	3444	SK1641
3444	NR091BT	3444	Q-05115C	3444	RS7413	3444	S1363	3444	SK5801
3444	NR201AY	3444	Q-07115C	3444	RS7415	3444	S1364	3444	SK5915
3444	NR261AB	3444	Q-10115C	3444	RS7421	3444	S1369	3444	SK8215
3444	NR271AY	3444	Q-14115C	3444	RS7504	3444	S1373	3444	SK8251
3444	NR461	3444	Q-15115C	3444	RS7510	3444	S1374	3444	SKA1080
3444	NR461EH	3444	Q-16115C	3444	RS7513	3444	S1403	3444	SKA1117
3444	N81500	3444	Q-35	3444	RS7513-15	3444	S1405	3444	SKA1395
3444	N81972	3444	Q-RF-2	3444	RS7514	3444	S1419	3444	SKA4141
3444	N81973	3444	Q-8E1001	3444	RS7515	3444	S1420	3444	SK88359
3444	N81974	3444	Q3/6515	3444	RS7516	3444	S1429-3	3444	SL5009
3444	N81975	3444	Q35242	3444	RS7517	3444	S1432	3444	SL7990
3444	N8475	3444	Q5053	3444	RS7517-19	3444	S1443	3444	SM-A-726655
3444	N8476	3444	Q5123E	3444	RS7521	3444	S1453	3444	SM-A-726664
3444	N8477	3444	Q5123F	3444	RS7525	3444	S1476	3444	SM-B-610342
3444	N8478	3444	Q5182	3444	RS7526	3444	S1487	3444	SM-B-686767
3444	N8479			3444	RS7527	3444	S1502	3444	SM-C-583256
				3444	RS7528			3444	SM2700

SK	Industry Standard No.	SK	Industry Standard No.	SK	Industry Standard No.	SK	Industry Standard No.	SK	Industry Standard No.
3444	SM2701	3444	SS3694	3444	T9011A1G	3444	TNJ61219	3444	TVS-28C828A
3444	SM3104	3444	ST-MP89433	3444	T9011AZ	3444	TNJ61220	3444	TVS-28C828Q
3444	SM3117A	3444	ST-MP89700D	3444	TA-6	3444	TNJ70479-1	3444	TVS-828A
3444	SM3505	3444	ST-MP89700E	3444	TAO047	3444	TNJ70537	3444	TV3A1200
3444	SM3986	3444	ST-MP89700F	3444	TC3123036722	3444	TNJ70539	3444	TX100-1
3444	SM4508-B	3444	ST.082.112.005	3444	TC3123036900	3444	TNJ70637	3444	TX100-2
3444	SM5379	3444	ST.082.114.015	3444	TC3123037111	3444	TNJ70638	3444	TX100-3
3444	SM5564	3444	ST/217/Q	3444	TC3123037222	3444	TNJ70639	3444	TX101-12
3444	SM5643	3444	STO1	3444	TC3123037412	3444	TNJ70640	3444	TX102-1
3444	SM576-1	3444	STO2	3444	TD101	3444	TNJ71036	3444	TX102-2
3444	SM576-2	3444	STO3	3444	TD102	3444	TNJ72280	3444	TX107-1
3444	SM5981	3444	STO4	3444	TD201	3444	TNJ72281	3444	TX107-10
3444	SM6773	3444	STO5	3444	TD202	3444	TNJ72783	3444	TX107-12
3444	SM716	3444	STO6	3444	TH420	3444	TNJ72784	3444	TX107-16
3444	SMT545	3444	ST1242	3444	TE2369	3444	TN842	3444	TX107-3
3444	SM7815	3444	ST1243	3444	TE3414	3444	TO-033	3444	TX107-4
3444	SM7836	3444	ST1244	3444	TE3415	3444	TO-038	3444	TX107-5
3444	SM8112	3444	ST1290	3444	TE3605	3444	TO-039	3444	TX107-6
3444	SM8113	3444	ST150	3444	TE3605A	3444	TO-040	3444	TX108-1
3444	SM8978	3444	ST1506	3444	TE3607	3444	TO1-101	3444	TX112-1
3444	SM9008	3444	ST151	3444	TE3704	3444	TO1-104	3444	UPI2222
3444	SM9135	3444	ST152	3444	TE3705	3444	TO1-105	3444	UPI4046-46
3444	SM9253	3444	ST153	3444	TE3903	3444	TP4123	3444	V119
3444	SPC40	3444	ST154	3444	TE3904	3444	TP4124	3444	V169
3444	SPC42	3444	ST155	3444	TE3906	3444	TP86514	3444	V297
3444	SPC50	3444	ST156	3444	TE4123	3444	TP86515	3444	VM30209
3444	SPC51	3444	ST157	3444	TE4124	3444	TP86520	3444	VM30241
3444	SPC52	3444	ST160	3444	TE4424	3444	TP86521	3444	VM30242
3444	SP81045	3444	ST1607	3444	TE4951	3444	TQ4	3444	V828B324
3444	SP81475	3444	ST161	3444	TE4952	3444	TQ5052	3444	V828C206
3444	SP82225	3444	ST162	3444	TE4953	3444	TQ5053	3444	V828C208
3444	SP82270	3444	ST163	3444	TE4954	3444	TQ5054	3444	V828C288A
3444	SP83015	3444	ST175	3444	TE5309A	3444	TQ5060	3444	V828C324H
3444	SP83735	3444	ST176	3444	TE5311A	3444	TR-1033-1	3444	V828C538
3444	SP83751	3444	ST177	3444	TE5368	3444	TR-1033-2	3444	V828C645A
3444	SP83900	3444	ST178	3444	TE5369	3444	TR-1347	3444	V89-0005-913
3444	SP83907	3444	ST180	3444	TE5370	3444	TR-1R33	3444	V89-0006-913
3444	SP83908	3444	ST181	3444	TE5371	3444	TR-21-6	3444	W20
3444	SP83909	3444	ST182	3444	TE5376	3444	TR-21C	3444	W24
3444	SP83923	3444	ST250	3444	TE5377	3444	TR-22	3444	WBP710
3444	SP83925	3444	ST251	3444	TE5449	3444	TR-22C	3444	WBP736
3444	SP83926	3444	ST25A	3444	TE5450	3444	TR-28C367	3444	WBP828
3444	SP83930	3444	ST25C	3444	TE5451	3444	TR-28C373	3444	WBP829
3444	SP83936	3444	ST403	3444	TE697	3444	TR-28C735	3444	WRR1952
3444	SP83938	3444	ST50	3444	TE-78	3444	TR-3R38	3444	WRR1953
3444	SP83951	3444	ST501	3444	TH28C536	3444	TR-4R33	3444	WRR1954
3444	SP83957C	3444	ST502	3444	TH28C693	3444	TR-5R33	3444	X16A1938
3444	SP83967	3444	ST503	3444	TH28C715	3444	TR-5R35	3444	X16A545-7
3444	SP83972	3444	ST504	3444	TI1A6	3444	TR-5R38	3444	X1683960
3444	SP83973	3444	ST506Q	3444	TI24A	3444	TR-6R33	3444	X1683960
3444	SP83999	3444	ST51	3444	TI24B	3444	TR-7R35	3444	X16N1485
3444	SP84003	3444	ST53	3444	TI411	3444	TR-8R35	3444	X19001-A
3444	SP84004	3444	ST54	3444	TI415	3444	TR-9100-18	3444	X6584-C
3444	SP84006	3444	ST55	3444	TI416	3444	TR-BC147B	3444	X735-41
3444	SP84009	3444	ST56	3444	TI417	3444	TR-BRC149C	3444	XA-1071
3444	SP84017	3444	ST57	3444	TI418	3444	TR-RR38	3444	XA-1139
3444	SP84020	3444	ST58	3444	TI419	3444	TR-TR38	3444	XC372
3444	SP84029	3444	ST59	3444	TI420	3444	TRO1037	3444	XC373
3444	SP84032	3444	ST63	3444	TI422	3444	TR1011	3444	XC374
3444	SP84034	3444	ST64	3444	TI430	3444	TR1031	3444	XEJO40017
3444	SP84037	3444	ST6511	3444	TI432	3444	TR1033	3444	XO30
3444	SP84039	3444	ST6512	3444	TI480	3444	TR1993-2	3444	XN-400-318-P1
3444	SP84040	3444	SYL1182	3444	TI482	3444	TR21	3444	X821
3444	SP84041	3444	SYL3460	3444	TI483	3444	TR302	3444	X822
3444	SP84042	3444	T-H28C536	3444	TI484	3444	TR310231	3444	Y49001-21
3444	SP84044	3444	T-H28C693	3444	TI485	3444	TR310243	3444	Y56601-04
3444	SP84045	3444	T-H28C715	3444	TI492	3444	TR310245	3444	Y56601-08
3444	SP84049	3444	T-Q5053	3444	TI493	3444	TR4010-2	3444	Y56601-45
3444	SP84052	3444	T-Q5053C	3444	TI494	3444	TR601	3444	Y56601-49
3444	SP84053	3444	T-Q5073	3444	TI495	3444	TR8004-4	3444	Y56601-51
3444	SP84055	3444	TO1-013	3444	TI496	3444	TR8010	3444	Y56601-73
3444	SP84059	3444	TO1-014	3444	TI54A	3444	TR8014	3444	Y56601-75
3444	SP84060	3444	TO1-101	3444	TI54B	3444	TR8021	3444	Y56601-80
3444	SP84061	3444	TO1-105	3444	TI54C	3444	TR8025	3444	Y56601-86
3444	SP84062	3444	T1-1A6	3444	TI54B	3444	TR8028	3444	Y56601-86-AD
3444	SP84063	3444	T1004671	3444	TI714	3444	TR8029	3444	Y56601-93
3444	SP84066	3444	T1008-834	3444	TI751	3444	TR8030	3444	24MW333
3444	SP84067	3444	T1008834	3444	TI802B	3444	TR8031	3444	ZDT
3444	SP84069	3444	T1340A3I	3444	TI803B	3444	TR8034	3444	ZEN100
3444	SP84074	3444	T1340A3J	3444	TI810B	3444	TR8035	3444	ZEN102
3444	SP84075	3444	T1340A3K	3444	TI904	3444	TR8039	3444	ZEN103
3444	SP84077	3444	T1413	3444	TI907	3444	TR8040	3444	ZEN110
3444	SP84081	3444	T1414	3444	TI908	3444	TR8043	3444	ZEN111
3444	SP84083	3444	T1415	3444	TIA06	3444	TR9100	3444	ZEN112
3444	SP84084	3444	T1416	3444	TIA102	3444	TRA34	3444	ZEN113
3444	SP84085	3444	T1417	3444	TI8113	3444	TRA36	3444	ZEN114
3444	SP84088	3444	T143	3444	TI8114	3444	TRA4	3444	ZEN115
3444	SP84089	3444	T1495	3444	TI822	3444	TRA4A	3444	ZEN119
3444	SP84095	3444	T157	3444	TI823	3444	TRA4B	3444	ZEN120
3444	SP841	3444	T158	3444	TI844	3444	TRA9R	3444	ZT-110
3444	SP84169	3444	T1642B	3444	TI845	3444	TRAPLC711	3444	ZT-62
3444	SP84199	3444	T170	3444	TI846	3444	TRAPLC871	3444	ZT-82
3444	SP84303	3444	T171	3444	TI847	3444	TRAPLC871A	3444	ZT111
3444	SP84313	3444	T1746	3444	TI848	3444	TRBC147B	3444	ZT112
3444	SP84345	3444	T1746A	3444	TI849	3444	TS2221	3444	ZT113
3444	SP84347	3444	T1746B	3444	TI851	3444	TS2222	3444	ZT114
3444	SP84356	3444	T1746C	3444	TI852	3444	TSC499	3444	ZT116
3444	SP84359	3444	T1748	3444	TI855	3444	TSC695	3444	ZT117
3444	SP84360	3444	T1748A	3444	TI871	3444	TST705899A	3444	ZT118
3444	SP84363	3444	T1748B	3444	TI872	3444	TT1097	3444	ZT119
3444	SP84367	3444	T1748C	3444	TI890-2	3444	TV-17	3444	ZT1420
3444	SP84368	3444	T1802	3444	TI892	3444	TV-18	3444	ZT1708
3444	SP84382	3444	T1802A	3444	TI892-BLU	3444	TV-21	3444	ZT20
3444	SP84446	3444	T1802B	3444	TI892-GRN	3444	TV-23	3444	ZT20-1
3444	SP84450	3444	T1804	3444	TI892-GRY	3444	TV-40	3444	ZT20-12
3444	SP84451	3444	T1805	3444	TI892-YEL	3444	TV-46	3444	ZT20-55
3444	SP84453	3444	T185	3444	TI894	3444	TV-51	3444	ZT202
3444	SP84455	3444	T235A013-2	3444	TIX712	3444	TV-52	3444	ZT203
3444	SP84456	3444	T237	3444	TIX812	3444	TV-53	3444	ZT204
3444	SP84457	3444	T2446	3444	TIX813	3444	TV-56	3444	ZT20A
3444	SP84459	3444	T255	3444	TK1228-1010	3444	TV-58	3444	ZT20B
3444	SP84472	3444	T256	3444	TK1228-1011	3444	TV-6	3444	ZT20C
3444	SP84476	3444	T277	3444	TK1228-1012	3444	TV-60	3444	ZT21
3444	SP84478	3444	T291	3444	TM123A	3444	TV-65	3444	ZT21-1
3444	SP84493	3444	T327	3444	TM2613	3444	TV-66	3444	ZT21-12
3444	SP84494	3444	T327-2	3444	TM2711	3444	TV-68	3444	ZT21-55
3444	SP84920	3444	T328	3444	TMT1543	3444	TV-71	3444	ZT21A
3444	SP84942	3444	T339	3444	TN237	3444	TV-84	3444	ZT21B
3444	SP85006-1	3444	T342	3444	TN53	3444	TV-92	3444	ZT21C
3444	SP85006-2	3444	T3565	3444	TN55	3444	TV241077	3444	ZT22
3444	SP85457	3444	T3601	3444	TN56	3444	TV241078	3444	ZT22-1
3444	SP86111	3444	T386	3444	TN59	3444	TV24215	3444	ZT22-12
3444	SP86112	3444	T399	3444	TN60	3444	TV24216	3444	ZT22-55
3444	SP86113	3444	T417	3444	TN61	3444	TV24281	3444	ZT2205
3444	SP86571	3444	T457-16	3444	TN62	3444	TV24372	3444	ZT2206
3444	SP87652	3444	T458-16	3444	TN63	3444	TV24453	3444	ZT22A
3444	SP8817	3444	T460	3444	TN64	3444	TV24454	3444	ZT22B
3444	SP8817N	3444	T461-16	3444	TN80	3444	TV24458	3444	ZT22C
3444	SP8868	3444	T472	3444	TN061689	3444	TV24576	3444	ZT23
3444	SP75844	3444	T615A002	3444	TN061702	3444	TV24655	3444	ZT23-1
3444	SS1-145128	3444	T615A006-1	3444	TNJ-60606	3444	TV28C208	3444	ZT23-12
3444	SS2308	3444	T9011A1C	3444	TNJ-60607	3444	TVS-28C538	3444	ZT23A
3444	SS2504			3444	TNJ60076	3444	TVS-28C538A	3444	ZT23B
						3444	TVS-28C828	3444	ZT23C
								3444	ZT24

SK	Industry Standard No.	SK	Industry Standard No.	SK	Industry Standard No.	SK	Industry Standard No.	SK	Industry Standard No.
3444	ZT24-11	3444	002-03	3444	2N3242	3444	2SC15-1	3444	2SC374H
3444	ZT24-12	3444	02-1078-01	3444	2N3242A	3444	2SC15-2	3444	2SC374J
3444	ZT24-55	3444	002-12000	3444	2N3247	3444	2SC15-3	3444	2SC374K
3444	ZT2476	3444	2-8454-031	3444	2N3261	3444	2SC159	3444	2SC374L
3444	ZT2477	3444	2C8900	3444	2N3301	3444	2SC16	3444	2SC374M
3444	ZT24A	3444	2D002-168	3444	2N3302	3444	2SC160	3444	2SC374R
3444	ZT24B	3444	2D002-169	3444	2N3340	3444	2SC1639	3444	2SC374OR
3444	ZT24C	3444	2N1006	3444	2N3395-WHT	3444	2SC166	3444	2SC374R
3444	ZT323.55	3444	2N1082	3444	2N3396-WHT	3444	2SC167	3444	2SC374V
3444	ZT40	3444	2N1103	3444	2N3397-WHT	3444	2SC16A	3444	2SC374X
3444	ZT402	3444	2N1139	3444	2N3398-BLU	3444	2SC170	3444	2SC374Y
3444	ZT403	3444	2N1140	3444	2N3398-WHT	3444	2SC170A	3444	2S38
3444	ZT404	3444	2N1199	3444	2N3415	3444	2SC170B	3444	2SC395
3444	ZT406	3444	2N1199A	3444	2N3462	3444	2SC170C	3444	2SC395A
3444	ZT41	3444	2N1200	3444	2N3463	3444	2SC170D	3444	2SC395R
3444	ZT42	3444	2N1417	3444	2N3565	3444	2SC170E	3444	2SC3927
3444	ZT43	3444	2N1418	3444	2N3566	3444	2SC170F	3444	2SC400
3444	ZT44	3444	2N1528	3444	2N3641	3444	2SC170G	3444	2SC400-0
3444	ZT50	3444	2N1644A	3444	2N3642	3444	2SC170GN	3444	2SC400-R
3444	ZT60	3444	2N1663	3444	2N3643	3444	2SC170H	3444	2SC400-Y
3444	ZT60-1	3444	2N1674	3444	2N3646	3444	2SC170J	3444	2SC423
3444	ZT60-12	3444	2N1704	3444	2N3693	3444	2SC170K	3444	2SC423C
3444	ZT60-55	3444	2N1708	3444	2N3694	3444	2SC170L	3444	2SC423D
3444	ZT60A	3444	2N1708A	3444	2N4013	3444	2SC170M	3444	2SC423E
3444	ZT60B	3444	2N1840	3444	2N4014	3444	2SC170X	3444	2SC423P
3444	ZT60C	3444	2N1944	3444	2N4074	3444	2SC170Y	3444	2SC424
3444	ZT61	3444	2N1945	3444	2N4227	3444	2SC18	3444	2SC424D
3444	ZT61-1	3444	2N1946	3444	2N4436	3444	2SC191	3444	2SC425
3444	ZT61-12	3444	2N1947	3444	2N4450	3444	2SC192	3444	2SC425B
3444	ZT61-55	3444	2N1948	3444	2N472	3444	2SC193	3444	2SC425C
3444	ZT61A	3444	2N1949	3444	2N472A	3444	2SC194	3444	2SC425D
3444	ZT61B	3444	2N1950	3444	2N474	3444	2SC195	3444	2SC425F
3444	ZT61C	3444	2N1951	3444	2N474A	3444	2SC196	3444	2SC433
3444	ZT62-1	3444	2N1952	3444	2N475	3444	2SC197	3444	2S45
3444	ZT62-12	3444	2N1962	3444	2N475A	3444	2SC200	3444	2SC468
3444	ZT62-15	3444	2N1963	3444	2N480A	3444	2SC201	3444	2SC468A
3444	ZT62C	3444	2N1964	3444	2N4966	3444	2SC202	3444	2SC474
3444	ZT63	3444	2N1965	3444	2N4968	3444	2SC204	3444	2S052
3444	ZT63-1	3444	2N1992	3444	2N4969	3444	2SC205	3444	2S053
3444	ZT63-12	3444	2N2096A	3444	2N4970	3444	2SC237	3444	2SC538
3444	ZT63-55	3444	2N2097A	3444	2N5081	3444	2SC239	3444	2SC538AQ
3444	ZT63A	3444	2N2205	3444	2N5082	3444	2SC26	3444	2SC538AQ
3444	ZT63B	3444	2N2206	3444	2N5107	3444	2SC267	3444	2SC538P
3444	ZT63C	3444	2N2220	3444	2N5127	3444	2SC267A	3444	2SC538Q
3444	ZT64	3444	2N2220A	3444	2N5128	3444	2SC281A	3444	2SC538R
3444	ZT64-1	3444	2N2221	3444	2N5134	3444	2SC281B	3444	2SC538S
3444	ZT64-12	3444	2N2221A	3444	2N5137	3444	2SC281C	3444	2SC538T
3444	ZT64-5	3444	2N2222	3444	2N5186	3444	2SC281C-EP	3444	2SC539
3444	ZT64-55	3444	2N2222A	3444	2N5187	3444	2SC281D	3444	2SC539K
3444	ZT64A	3444	2N2236	3444	2N541	3444	2SC281EP	3444	2SC539L
3444	ZT64B	3444	2N2237	3444	2N543	3444	2SC281H	3444	2SC539R
3444	ZT64C	3444	2N2240	3444	2N543A	3444	2SC281HA	3444	2SC539S
3444	ZT68	3444	2N2241	3444	2N708	3444	2SC281HB	3444	2SC539T
3444	ZT696	3444	2N2242	3444	2N708A	3444	2SC281HC	3444	2S054
3444	ZT697	3444	2N2244	3444	2N745	3444	2SC282	3444	2S055
3444	ZT706	3444	2N2245	3444	2N746	3444	2SC282H	3444	2SC587
3444	ZT706A	3444	2N2246	3444	2N747	3444	2SC282HA	3444	2SC587A
3444	ZT708	3444	2N2247	3444	2N748	3444	2SC282HB	3444	2SC588
3444	ZT80	3444	2N2248	3444	2N749	3444	2SC282HC	3444	2SC593
3444	ZT81	3444	2N2249	3444	2N751	3444	2SC283	3444	2SC595
3444	ZT83	3444	2N2250	3444	2N753	3444	2SC284	3444	2SC596
3444	ZT84	3444	2N2253	3444	2N754	3444	2SC284H	3444	2SC619
3444	ZT87	3444	2N2254	3444	2N784A	3444	2SC284HA	3444	2SC619B
3444	ZT89	3444	2N2255	3444	2N834A	3444	2SC284HB	3444	2SC619C
3444	001-00	3444	2N2256	3444	2N835	3444	2SC284HC	3444	2SC619D
3444	1-0006-0021	3444	2N2257	3444	2N839	3444	2SC284A	3444	2SC622
3444	1-0006-0022	3444	2N2272	3444	2N840	3444	2SC300	3444	2SC622
3444	1-001-003-15	3444	2N2309	3444	2N842	3444	2SC300A	3444	2SC626
3444	001-02020	3444	2N2314	3444	2N844	3444	2SC300B	3444	2SC645
3444	001-02101-0	3444	2N2315	3444	2N915	3444	2SC300C	3444	2SC645A
3444	001-02101-1	3444	2N2318	3444	2N919	3444	2SC300D	3444	2SC645B
3444	001-02102-0	3444	2N2319	3444	2N920	3444	2SC300E	3444	2SC655
3444	001-02103-0	3444	2N2320	3444	2N921	3444	2SC300F	3444	2S67
3444	001-02104-0	3444	2N2531	3444	2N922	3444	2SC300G	3444	2S68
3444	001-02105-0	3444	2N2349	3444	2N929	3444	2SC300GN	3444	2SC737
3444	001-02106-0	3444	2N2368	3444	2N929A	3444	2SC300H	3444	2SC737Y
3444	001-02107-0	3444	2N2368A	3444	2N930	3444	2SC300J	3444	2SC796
3444	001-02108-0	3444	2N2369	3444	2N930A	3444	2SC300K	3444	2SC847
3444	001-02109-0	3444	2N2387	3444	2N947	3444	2SC300L	3444	2SC848
3444	001-02110-0	3444	2N2388	3444	2N957	3444	2SC300M	3444	2SC849
3444	001-02111-1	3444	2N2413	3444	2S101	3444	2SC300R	3444	2SC850
3444	001-02113-2	3444	2N2417	3444	2S102	3444	2SC300OR	3444	2SC87
3444	001-02113-3	3444	2N2427	3444	2S131	3444	2SC300X	3444	2SC896
3444	001-02113-4	3444	2N2432	3444	2S501	3444	2SC300Y	3444	2SC899
3444	001-02113-5	3444	2N2481	3444	2S502	3444	2SC301	3444	2SC899K
3444	001-02121-0	3444	2N2483	3444	2S503	3444	2SC302	3444	2SC906
3444	01-031175	3444	2N2484	3444	2S731	3444	2SC315	3444	2SC906F
3444	1-042/2207	3444	2N2501	3444	2S732	3444	2SC317	3444	2SC917B
3444	1-044/2207	3444	2N2523	3444	2S733	3444	2SC317A	3444	2SC917B
3444	1-21-276	3444	2N2524	3444	2S741A	3444	2SC317B	3444	2SC917C
3444	1-21-277	3444	2N2529	3444	2S744A	3444	2SC317C	3444	2SC917D
3444	1-21-278	3444	2N2530	3444	2S95A	3444	2SC317D	3444	2SC917E
3444	1-21-279	3444	2N2531	3444	2SC100	3444	2SC317E	3444	2SC917F
3444	01-2101	3444	2N2532	3444	2SC1007	3444	2SC317F	3444	2SC917G
3444	01-2101-0	3444	2N2533	3444	2SC103	3444	2SC317G	3444	2SC917GN
3444	001-21011	3444	2N2534	3444	2SC103A	3444	2SC317GN	3444	2SC917H
3444	01-2102	3444	2N2539	3444	2SC104	3444	2SC317H	3444	2SC917J
3444	01-2104	3444	2N2569	3444	2SC104A	3444	2SC317J	3444	2SC917L
3444	01-2105	3444	2N2570	3444	2SC105	3444	2SC317K	3444	2SC917M
3444	01-2106	3444	2N2571	3444	2SC1071	3444	2SC317L	3444	2SC917OR
3444	01-2107	3444	2N2572	3444	2SC110	3444	2SC317M	3444	2SC917R
3444	01-2108	3444	2N2586	3444	2SC111	3444	2SC317OR	3444	2SC917X
3444	01-2109	3444	2N2645	3444	2SC1244	3444	2SC317R	3444	2SC917Y
3444	1-522223720	3444	2N2656	3444	2SC131	3444	2SC317X	3444	2SC934
3444	01-571811	3444	2N2673	3444	2SC132	3444	2SC317Y	3444	2SC934C
3444	1-6147191229	3444	2N2675	3444	2SC133	3444	2SC318	3444	2SC934D
3444	-6171191368	3444	2N2676	3444	2SC134	3444	2SC318A	3444	2SC934E
3444	1-801-004	3444	2N2677	3444	2SC134B	3444	2SC321	3444	2SC934F
3444	1-801-004-17	3444	2N2678	3444	2SC135	3444	2SC321H	3444	2SC934G
3444	1-801-314-1	3444	2N2692	3444	2SC136	3444	2SC321HA	3444	2SC934F
3444	1-801-314-15	3444	2N2693	3444	2SC137	3444	2SC321HB	3444	2SC943
3444	1-801-314-16	3444	2N2719	3444	2SC138	3444	2SC321HC	3444	2SC943A
3444	01-9011-52221-3	3444	2N2831	3444	2SC1380	3444	2SC323	3444	2SC943B
3444	01-9013-72221-3	3444	2N2845	3444	2SC1380A	3444	2SC324	3444	2SC943C
3444	01-9014-22221-3	3444	2N2847	3444	2SC1380A-BL	3444	2SC324A	3444	2SC966
3444	1A0034	3444	2N2926-6	3444	2SC1380A-GR	3444	2SC324H	3444	2SC967
3444	1A0035	3444	2N2926G	3444	2SC138A	3444	2SC324HA	3444	2SC984D
3444	1A0043	3444	2N2926GRN	3444	2SC1388	3444	2SC350	3444	2SC984E
3444	1A0063	3444	2N2926ORN	3444	2SC139	3444	2SC350H	3444	2SC984F
3444	1A0079	3444	2N2938	3444	2SC1390	3444	2SC356	3444	2SC984GN
3444	1A0080	3444	2N2951	3444	2SC1901	3444	2SC360	3444	2SC984J
3444	1A0081	3444	2N2952	3444	2SC1390J	3444	2SC360D	3444	2SC984L
3444	1A4757-1	3444	2N2958	3444	2SC1390K	3444	2SC37	3444	2SC984M
3444	1J1	3444	2N2959	3444	2SC1390L	3444	2SC3724	3444	2SC984OR
3444	1U585F	3444	2N2960	3444	2SC1390V	3444	2SC374A	3444	2SC984R
3444	1U585F/7825B	3444	2N2961	3444	2SC1390W	3444	2SC374B	3444	2SC984Y
3444	1W9723	3444	2N3009	3444	2SC1390WH	3444	2SC374BLK	3444	2SCP-2
3444	1W9787	3444	2N3011	3444	2SC1390WX	3444	2SC374C	3444	2SCF2
3444	002-009500	3444	2N3013	3444	2SC1390X	3444	2SC374D	3444	2T172
3444	002-009501	3444	2N3014	3444	2SC1390XJ	3444	2SC374E	3444	2T2O2
3444	002-009502	3444	2N3115	3444	2SC1390XK	3444	2SC374F	3444	2T2708
3444	002-009900	3444	2N3116	3444	2SC1390Y	3444	2SC374G	3444	2T2785
3444	002-010400	3444	2N3210	3444	2SC1390YM	3444	2SC374GN	3444	2T2857
3444	002-010800	3444	2N3211	3444	2SC15				

SK	Industry Standard No.
3444	2T40
3444	2T403
3444	2T404
3444	2T41
3444	2T42
3444	2T43
3444	2T44
3444	2T918
3444	3-0033
3444	03-1585/G
3444	03A05
3444	03A11
3444	3E-2
3444	3E-3
3444	3L4-6007-02
3444	3L4-6007-03
3444	3L4-6007-04
3444	3L4-6007-08
3444	3L4-6007-09
3444	3L4-6007-1
3444	3L4-6007-2
3444	3L4-6007-3
3444	3L4-6007-37
3444	3L4-6007-38
3444	3L4-6010-03
3444	3L4-6010-6
3444	3L4-6015-01
3444	3N35
3444	004-00
3444	04-00460-03
3444	04-01585-06
3444	04-01585-07
3444	04-02090-02
3444	4-1545
3444	4-1790
3444	4-3023212
3444	4-3023221
3444	4-3025766
3444	4-5145
3444	4-686132-3
3444	4-686143-3
3444	4-686144-3
3444	4-686173-3
3444	4-686183-3
3444	4-686231-3
3444	4-686257-3
3444	4-68682-3
3444	4D20
3444	4D21
3444	4D22
3444	4D24
3444	4D25
3444	4D26
3444	4JX16A567
3444	4JX16A667
3444	4JX16A6670
3444	4JX16A667G
3444	4JX16A667R
3444	4JX16A667Y
3444	4JX16A668
3444	4JX16A6680
3444	4JX16A668G
3444	4JX16A668Y
3444	4JX16A669
3444	4JX16A669G
3444	4JX16A669Y
3444	4JX16A670
3444	4JX16A670G
3444	4JX16B670G
3444	4JX16B670R
3444	4JX16B670Y
3444	4JX16E3860
3444	4JX16E3890
3444	4JX16E3960
3444	4JX1E596
3444	4JX7A972
3444	005-02
3444	5-8
3444	006-0000134
3444	6-6450036
3444	6-90
3444	6-9029-15D
3444	6-9029-15E
3444	6-93
3444	6A10227
3444	6A10422
3444	6A10423
3444	6A10520
3444	6A10851
3444	6A10855
3444	6A11180
3444	6A12681
3444	6A12682
3444	6A12683
3444	6A12725
3444	6A12788
3444	6A12789
3444	6A16399
3444	6E4850/56-0001
3444	6X97047A01
3444	7-0015
3444	07-07125
3444	07-07139
3444	07-07156
3444	7-1(SARKES)
3444	7-15(SARKES)
3444	7-16(SARKES)
3444	7-17
3444	7-18(SARKES)
3444	7-19(SARKES)
3444	7-2(SARKES)
3444	7-20(SARKES)
3444	7-3(SARKES)
3444	7-4(SARKES)
3444	7-5
3444	7-5(SARKES)
3444	7-59-0243477
3444	7-59-068
3444	7-6(SARKES)
3444	7-7(SARKES)
3444	7-8(SARKES)
3444	7A30(SHERWOOD)
3444	7A31(SHERWOOD)
3444	8-00243
3444	8-0050100
3444	8-0051500
3444	8-005202
3444	8-0052102
3444	8-0052302
3444	8-0052600
3444	8-0053001
3444	8-0053400
3444	8-0318250
3444	8-0337390
3444	8-0383940
3444	8-0389910
3444	8-0389930
3444	8-0421980
3444	8-2409501
3444	8-4(BENDIX)
3444	8-697-020-570
3444	8-724-733-30
3444	8-81250109
3444	8-902-0706-071
3444	8-905-014-017
3444	8-905-705-112
3444	8-905-705-403
3444	8-905-705-405
3444	8-905-706-104
3444	8-905-706-201
3444	8-905-706-202
3444	8-905-706-203
3444	8-905-706-206
3444	8-905-706-208
3444	8-905-706-211
3444	8-905-706-215
3444	8-905-706-235
3444	8-905-706-236
3444	8-905-706-238
3444	8-905-706-239
3444	8-905-706-240
3444	8-905-706-242
3444	8-905-706-244
3444	8-905-706-245
3444	8-905-706-246
3444	8-905-706-250
3444	8-905-706-257
3444	8-905-706-260
3444	8-905-706-263
3444	8-905-706-336
3444	8-905-706-606
3444	8-905-707-254
3444	8-905-707-265
3444	8-905-707-313
3444	8A12789
3444	09-302007
3444	09-302012
3444	09-302033
3444	09-302034
3444	09-302039
3444	09-302045
3444	09-302045-12
3444	09-302054
3444	09-302058
3444	09-302062
3444	09-302074
3444	09-302078
3444	09-302101
3444	09-302106
3444	09-302118
3444	09-302124
3444	09-302131
3444	09-302140
3444	09-302148
3444	09-302153
3444	09-302165
3444	09-302172
3444	09-302175
3444	09-302189
3444	09-302215
3444	09-303025
3444	09-304044
3444	09-304045
3444	09-304058
3444	09-305034
3444	09-305062
3444	09-305063
3444	09-305064
3444	09-305065
3444	09-305066
3444	09-305067
3444	09-305068
3444	09-305077
3444	09-305139
3444	09-309012
3444	09-309023
3444	09-309049
3444	09-309050
3444	09-309060
3444	09-309064
3444	09-309076
3444	9-5221
3444	9-5225
3444	9-5226-2
3444	9-5227
3444	9-5296
3444	9-9109-1
3444	9-9109-2
3444	9GR2
3444	09N1
3444	9TR2
3444	9TR21001-02
3444	9TR7
3444	9TRZ1001-02
3444	10-080010
3444	11-085010
3444	012-1-12
3444	012-1-12-7
3444	12-1-276
3444	12-1-277
3444	12-1-278
3444	12-1-279
3444	12-1-70
3444	12-1-70-12
3444	12-1-70-12-7
3444	12-10
3444	12-101001
3444	12-11
3444	12-12
3444	12-13
3444	12-14
3444	12-15
3444	12-16
3444	12-17
3444	12-18
3444	12-19
3444	12-1A
3444	12-1A0
3444	12-1A0R
3444	12-1A1
3444	12-1A19
3444	12-1A2
3444	12-1A21
3444	12-1A3
3444	12-1A3P
3444	12-1A4
3444	12-1A4-7
3444	12-1A4-7B
3444	12-1A5
3444	12-1A5L
3444	12-1A6
3444	12-1A6A
3444	12-1A7
3444	12-1A7-1
3444	12-1A8
3444	12-1A82
3444	12-1A9
3444	12-1A9G
3444	12-4
3444	012E
3444	13-0021
3444	13-0022
3444	13-0024
3444	13-0041
3444	13-0048
3444	13-0058
3444	13-0321-10
3444	13-0321-11
3444	13-0321-12
3444	13-0321-5
3444	13-0321-6
3444	13-0321-7
3444	13-0321-8
3444	13-0321-81
3444	13-0321-9
3444	13-14085-15
3444	13-14085-34
3444	13-14085-50
3444	13-14085-6
3444	13-14085-7
3444	13-14085-83
3444	13-14085-84
3444	13-14085-85
3444	13-14085-89
3444	13-14606-1
3444	13-15840-1
3444	13-15840-2
3444	13-15865-1
3444	13-16769-1
3444	13-18087-1
3444	13-18087-2
3444	13-18158-1
3444	13-18363-1
3444	13-18364-1
3444	13-23160-4
3444	13-23916-1
3444	13-27404-2
3444	13-27433-1
3444	13-29432-1
3444	13-33350-1
3444	13-33595-1
3444	13-33595-2
3444	13-33595-3
3444	13-55061-1
3444	13-55061-2
3444	13-55066-1
3444	13-55066-2
3444	13-55067-1
3444	13-55068-1
3444	13-67583-6
3444	13-67585-4
3444	13-67585-5
3444	13-67585-5/3464
3444	13-67586-3
3444	13-68617-1
3444	14 806 12
3444	14-0104-7
3444	14-1
3444	14-2
3444	14-3
3444	14-583-01
3444	14-601-28
3444	14-602-01
3444	14-602-02
3444	14-602-03
3444	14-602-12
3444	14-602-13
3444	14-602-14
3444	14-602-16
3444	14-602-17
3444	14-602-22
3444	14-602-23
3444	14-602-26
3444	14-602-35
3444	14-602-48
3444	14-602-50
3444	14-602-55A
3444	14-602-61
3444	14-602-62
3444	14-602-69
3444	14-602-78
3444	14-602-80
3444	14-602-81
3444	14-602-87
3444	14-602-89
3444	14-603-03
3444	14-603-04
3444	14-603-10
3444	14-603-11
3444	14-651-12
3444	14-655-13
3444	14-656-21
3444	14-659-12
3444	14-660-12
3444	014-680
3444	014-686
3444	014-698
3444	014-784
3444	14-800-32
3444	14-802-12
3444	14-805-12
3444	14-806-12
3444	14-806-23
3444	14-809-23
3444	14-809-32
3444	14-851-32
3444	14-853-23
3444	14-854-12
3444	14-858-12
3444	14-862-23
3444	14-862-32
3444	14-864-12
3444	14-865-12
3444	14-866-32
3444	15-01999
3444	15-03014-00
3444	15-03100
3444	15-05302
3444	15-05369
3444	15-05650
3444	15-082019
3444	15-09338
3444	15-09980
3444	15-1
3444	15-2
3444	15-22223720
3444	15-875-075-003
3444	16A1(FLEETWOOD)
3444	16A1938
3444	16A2(FLEETWOOD)
3444	16A545-7
3444	16G2
3444	16L42
3444	16L43
3444	16L44
3444	16L62
3444	16L63
3444	16X2
3444	17-451
3444	17-457
3444	017E824
3444	018-00003
3444	18-148A
3444	019-00009
3444	019-00010
3444	019-003675-196
3444	019-003675-203
3444	019-003675-207
3444	019-003675-246
3444	019-003932
3444	019-003934
3444	019-004111
3444	019-004428-002
3444	019-005006
3444	019-005021
3444	19-020-043
3444	19-020-043A
3444	19-020-058
3444	19-020-067
3444	19-020-073
3444	19-020-074
3444	19-020071
3444	19-2-02616
3444	19A115061-P1
3444	19A115061-P2
3444	19A115102-P1
3444	19A115108-P1
3444	19A115108-P2
3444	19A115123-2
3444	19A115123-P1
3444	19A115123-P2
3444	19A115142-P1
3444	19A115142-P2
3444	19A115157-1
3444	19A115167-2
3444	19A115245-P1
3444	19A115245-P2
3444	19A115253-P1
3444	19A115253-P2
3444	19A115315-P1
3444	19A115315-P2
3444	19A115328-1
3444	19A115330
3444	19A115342-P1
3444	19A115342-P2
3444	19A115359-P1
3444	19A115359-P2
3444	19A115362-P1
3444	19A115362-P2
3444	19A115410-P1
3444	19A115410-P2
3444	19A115552P1
3444	19A115552P2
3444	19A115591P1
3444	19A115591P2
3444	19A115720-1
3444	19A115720-2
3444	19A115728-1
3444	19A115728-2
3444	19A115786
3444	19A115786A
3444	19A115910P1
3444	19A115944P1
3444	19A115944P2
3444	19A116631P1
3444	19A116755P1
3444	19A116774-P1
3444	19A116865
3444	19A129207P1
3444	19AR20
3444	19C300114-P1
3444	19C300114-P2
3444	19C300114P1
3444	19C300114P2
3444	19C300114P3
3444	020-1112-001
3444	20A0053
3444	20A0073
3444	20A10849
3444	021-0121-00
3444	21A015-013
3444	21A015-020
3444	21A015-027
3444	21A040-032
3444	21A040-033
3444	21A040-033A
3444	21A040-034
3444	21A040-037
3444	21A040-056
3444	21A040-077
3444	21A040-078
3444	21A040-092
3444	21A040-37
3444	21A112-013
3444	21A112-015
3444	21A112-017
3444	21A112-018
3444	21A112-020
3444	21A112-050
3444	21A112-062
3444	21A112-063
3444	21A112-085
3444	21A112-088
3444	21A112-089
3444	21A112-090
3444	21A112-091
3444	21A112-092
3444	21A112-101
3444	21A112-104
3444	21M084
3444	21M085
3444	21M086
3444	21M122
3444	21M123
3444	21M125
3444	21M139
3444	21M146
3444	21M149
3444	21M150
3444	21M186
3444	21M200
3444	21M205
3444	21M366
3444	21M488
3444	21M520
3444	21M563
3444	21M578
3444	21M579
3444	21M60
3444	21M605
3444	22-001006
3444	22-001007
3444	23
3444	23-5033
3444	23-5052
3444	23-PT274-121
3444	23B114044
3444	23E001-1
3444	24-0003714-1
3444	24-000451
3444	24-000457
3444	24-000653-1
3444	24-001327-1
3444	24-002
3444	24-602-25
3444	24A
3444	24B
3444	24B1
3444	24MW1023
3444	24MW1024
3444	24MW1059
3444	24MW1068
3444	24MW1069
3444	24MW1089
3444	24MW1096
3444	24MW1120
3444	24MW1141
3444	24MW1147
3444	24MW119
3444	24MW333
3444	24MW454
3444	24MW458
3444	24MW460
3444	24MW461
3444	24MW609
3444	24MW655
3444	24MW658
3444	24MW659
3444	24MW676
3444	24MW740
3444	24MW760
3444	24MW773
3444	24MW774
3444	24MW775
3444	24MW776
3444	24MW790
3444	24MW795
3444	24MW796
3444	24MW797
3444	24MW801
3444	24MW807
3444	24MW808
3444	24MW809
3444	24MW817
3444	24MW818
3444	24MW823
3444	24MW854
3444	24MW855
3444	24MW874
3444	24MW899
3444	24MW954
3444	24MW961
3444	24MW988
3444	24MW992
3444	25-0060-4
3444	025-100018
3444	025-100030
3444	025-100040
3444	025-10030
3444	25A1273-001
3444	25A2
3444	025B-YEL
3444	026-100026
3444	31-0007
3444	31-0009
3444	31-0068
3444	31-0069
3444	31-0080
3444	31-0081
3444	31-0082
3444	31-0084
3444	31-0085
3444	31-0104
3444	31-0106
3444	31-0115
3444	31-0116
3444	31-0177
3444	31-058
3444	31-1
3444	31-16
3444	031A
3444	32-13843-2
3444	32-20738
3444	33-00706A
3444	33-070
3444	33-0706
3444	33-071
3444	33-084
3444	34-1010
3444	34-34-6015-43
3444	34-6000-64
3444	34-6000-69
3444	34-6000-70
3444	34-6000-71
3444	34-6001-48
3444	34-6001-49
3444	34-6001-5
3444	34-6001-52
3444	34-6001-53
3444	34-6001-54
3444	34-6001-55
3444	34-6001-56
3444	34-6001-57
3444	34-6001-58
3444	34-6001-60
3444	34-6001-61
3444	34-6001-62
3444	34-6001-63
3444	34-6001-69
3444	34-6001-70

SK	Industry Standard No.
3444	34-6001-73
3444	34-6001-74
3444	34-6001-77
3444	34-6007-1
3444	34-6007-2
3444	34-6007-3
3444	34-6015-1
3444	34-6015-10
3444	34-6015-11
3444	34-6015-13
3444	34-6015-14
3444	34-6015-2
3444	34-6015-21
3444	34-6015-3
3444	34-6015-4
3444	34-6015-41
3444	34-6015-42A
3444	34-6015-43A
3444	34-6015-44
3444	34-6015-5
3444	34-6015-54
3444	34-6015-6
3444	34-6015-60
3444	34-6015-63
3444	34-6015-7
3444	34-6015-80
3444	34-6016-14
3444	34-6016-16
3444	34-6016-18
3444	34-6016-19
3444	34-6016-2
3444	34-6016-24
3444	34-6016-25
3444	34-6016-26
3444	34-6016-3
3444	34-6016-4
3444	34-6016-49
3444	34-6016-49A
3444	34-6016-63
3444	34-6016-7
3444	34-6016-8
3444	35-39306001
3444	35-39306002
3444	037
3444	041
3444	41-0499
3444	042
3444	42-19840
3444	42-21234
3444	42-22158
3444	42-22533
3444	42-22786
3444	42-22787
3444	42-28207
3444	42-9029-31M
3444	42-9029-31R
3444	42-9029-40C
3444	42-9029-40L
3444	42-9029-40Y
3444	42-9029-60A
3444	42-9029-60C
3444	42-9029-70C
3444	42-9029-70F
3444	42-9029-70P
3444	43B140883-1
3444	43B168610
3444	43C168567
3444	044-9667-02
3444	44A333463-001
3444	44A390247
3444	44A390249
3444	44A395994-001
3444	44B311097
3444	45AN4AA
3444	45N2M
3444	45N3
3444	45N4
3444	45N4M
3444	46-86121-3
3444	46-86122-3
3444	46-86143-3
3444	46-86144-3
3444	46-86145-3
3444	46-86152-3
3444	46-86169-3
3444	46-86171-3
3444	46-86192-3
3444	46-86228-3
3444	46-86231-3
3444	46-8624-3
3444	46-86247-2
3444	46-86247-3
3444	46-86257-3
3444	46-86268-3
3444	46-86274-3
3444	46-86310-3
3444	46-86375-3
3444	46-86378-3
3444	46-86404-3
3444	46-86407-3
3444	46-86408-3
3444	46-86409-3
3444	46-86419-3
3444	46-8682-2
3444	46-8695-3
3444	48-123802
3444	48-123803
3444	48-134173
3444	48-134464
3444	48-134465
3444	48-134654
3444	48-134664
3444	48-134665
3444	48-134666
3444	48-134667
3444	48-134668
3444	48-134669
3444	48-134673
3444	48-134674
3444	48-134675
3444	48-134690
3444	48-134691
3444	48-134703
3444	48-134705
3444	48-134714
3444	48-134718
3444	48-134720
3444	48-134721
3444	48-134726
3444	48-134732
3444	48-134733
3444	48-134733A
3444	48-134734
3444	48-134734A
3444	48-134737
3444	48-134739
3444	48-134765
3444	48-134768
3444	48-134775
3444	48-134776
3444	48-134782
3444	48-134785
3444	48-134791
3444	48-134801
3444	48-134804
3444	48-134807
3444	48-134808
3444	48-134809
3444	48-13481
3444	48-134811
3444	48-134817
3444	48-134822
3444	48-134824
3444	48-134839
3444	48-134840
3444	48-134841
3444	48-134842
3444	48-134844
3444	48-134847
3444	48-134848
3444	48-134852
3444	48-134854
3444	48-134889
3444	48-134894
3444	48-134895
3444	48-134896
3444	48-134897
3444	48-134899
3444	48-134903
3444	48-134905
3444	48-134906
3444	48-134928
3444	48-134929
3444	48-134933
3444	48-134933E
3444	48-134935
3444	48-134942
3444	48-134952
3444	48-134970
3444	48-134980
3444	48-134988
3444	48-134992
3444	48-134994
3444	48-134996
3444	48-136665
3444	48-137003
3444	48-137007
3444	48-137010
3444	48-137013
3444	48-137014
3444	48-137019
3444	48-137022
3444	48-137043
3444	48-137044
3444	48-137047
3444	48-137056
3444	48-137057
3444	48-137072
3444	48-137073
3444	48-137083
3444	48-137089
3444	48-137096
3444	48-137101
3444	48-137106
3444	48-137107
3444	48-137108
3444	48-137109
3444	48-137110
3444	48-137111
3444	48-137115
3444	48-137137
3444	48-137138
3444	48-137139
3444	48-137171D
3444	48-137172
3444	48-137174
3444	48-137192
3444	48-137206
3444	48-137257
3444	48-137260
3444	48-137265
3444	48-137336
3444	48-137350
3444	48-137353
3444	48-137354
3444	48-137373
3444	48-137374
3444	48-137377
3444	48-137378
3444	48-137384
3444	48-137398
3444	48-137399
3444	48-137498
3444	48-137500
3444	48-137509
3444	48-137530
3444	48-137543
3444	48-137855
3444	48-137998
3444	48-3003A05
3444	48-3003A11
3444	48-3003A12
3444	48-355002
3444	48-355052
3444	48-40171G01
3444	48-40246G02
3444	48-40606J01
3444	48-43354A81
3444	48-43354A82
3444	48-44885G02
3444	48-60022A13
3444	48-63005A66
3444	48-63005A72
3444	48-63026A47
3444	48-63026A48
3444	48-63076A52
3444	48-63076A83
3444	48-63077A10
3444	48-63077A30
3444	48-63077A31
3444	48-63078A70
3444	48-63078A71
3444	48-63079A97
3444	48-63082A25
3444	48-63082A26
3444	48-63082A27
3444	48-63082A45
3444	48-63082A71
3444	48-65123A94
3444	48-65147A72
3444	48-86376-3
3444	48-869226-0
3444	48-869248
3444	48-869312
3444	48-869325
3444	48-869329
3444	48-869444
3444	48-869525
3444	48-869563
3444	48-869568
3444	48-869570
3444	48-869767
3444	48-90172A01
3444	48-90232A05
3444	48-90232A11
3444	48-90232A13
3444	48-97046A22
3444	48-97046A23
3444	48-97046A24
3444	48-97046A28
3444	48-97046A42
3444	48-97046A43
3444	48-97046A46
3444	48-97046A50
3444	48-97046A52
3444	48-971-A95
3444	48-97127A012
3444	48-97127A013
3444	48-97127A018
3444	48-97127A12
3444	48-97127A13
3444	48-97127A18
3444	48-97127A19
3444	48-97127A24
3444	48-97127A29
3444	48-97127A33
3444	48-97162A04
3444	48-97162A05
3444	48-97162A09
3444	48-97162A12
3444	48-97162A15
3444	48-97162A33
3444	48-97177A09
3444	48-97177A12
3444	48-97177A13
3444	48-971A05
3444	48-P02597A
3444	48B407624P1
3444	48P63078A71
3444	48R869312
3444	48R869325
3444	48R869329
3444	48R869444
3444	48R869525
3444	48R869563
3444	48R869568
3444	48R869570
3444	48R869767
3444	48S134903
3444	48S134933
3444	48S134997
3444	48S137107
3444	48S137115
3444	48S137171
3444	48S137172
3444	48S137300
3444	48S137315
3444	48S137498
3444	48S137530
3444	48S40247G02
3444	48S44885G01
3444	48S44885G02
3444	48X97046A60
3444	48X97046A61
3444	48X97046A62
3444	48X97162A05
3444	48X97162A21
3444	48X97238A04
3444	50C374
3444	50C538
3444	50C644
3444	50C828
3444	50C838
3444	50CJ139
3444	051-0046
3444	051-0047
3444	051-0155
3444	51-47-23
3444	51-47-24
3444	52-020-108-0
3444	53-1110
3444	54-1
3444	54A
3444	54B
3444	54BLU
3444	54C
3444	54D
3444	54F
3444	54GRN
3444	54WHT
3444	54YEL
3444	55-1026
3444	55-1034
3444	55-1082
3444	56-35
3444	56-8089
3444	56-8089A
3444	56-8089C
3444	56-8090
3444	56-8090A
3444	56-8090C
3444	56A22-1
3444	56B22-1
3444	057
3444	57-0004503
3444	57-0005452
3444	57-0005503
3444	57-000901504
3444	57-01491-B
3444	57A1-123
3444	57A1-124
3444	57A1-51
3444	57A1-75
3444	57A105-12
3444	57A11-1
3444	57A117-9
3444	57A118-12
3444	57A119-12
3444	57A120-12
3444	57A121-9
3444	57A125-9
3444	57A135-12
3444	57A136-12
3444	57A140-12
3444	57A15-1
3444	57A15-2
3444	57A15-3
3444	57A15-4
3444	57A152-1
3444	57A152-10
3444	57A152-11
3444	57A152-2
3444	57A152-3
3444	57A152-4
3444	57A152-5
3444	57A152-6
3444	57A152-7
3444	57A152-8
3444	57A152-9
3444	57A153-1
3444	57A153-2
3444	57A153-3
3444	57A153-4
3444	57A153-5
3444	57A153-6
3444	57A153-7
3444	57A153-8
3444	57A153-9
3444	57A156-9
3444	57A16-1
3444	57A166-12
3444	57A181-12
3444	57A184-12
3444	57A191-12
3444	57A199-4
3444	57A2-101
3444	57A2-102
3444	57A2-103
3444	57A2-113
3444	57A2-116
3444	57A2-126
3444	57A2-153
3444	57A2-192
3444	57A2-27
3444	57A2-28
3444	57A2-59
3444	57A2-62
3444	57A2-63
3444	57A2-64
3444	57A2-73
3444	57A2-85
3444	57A2-87
3444	57A2-97
3444	57A200-12
3444	57A201-13
3444	57A202-13
3444	57A203-14
3444	57A204-14
3444	57A21-8
3444	57A24-1
3444	57A24-2
3444	57A24-3
3444	57A24-4
3444	57A252-1
3444	57A253-14
3444	57A27-1
3444	57A282-12
3444	57A6-11
3444	57A6-4
3444	57A6-9
3444	57A7-10
3444	57A7-15
3444	57A7-17
3444	57A7-18
3444	57A7-20
3444	57A7-9
3444	57B105-12
3444	57B107-8
3444	57B117-9
3444	57B118-12
3444	57B119-12
3444	57B120-12
3444	57B125-9
3444	57B126-12
3444	57B129-9
3444	57B136-12
3444	57B140-12
3444	57B143-12
3444	57B144-12
3444	57B146-12
3444	57B152-1
3444	57B152-10
3444	57B152-11
3444	57B152-2
3444	57B152-3
3444	57B152-4
3444	57B152-5
3444	57B152-6
3444	57B152-7
3444	57B152-8
3444	57B152-9
3444	57B153-1
3444	57B153-2
3444	57B153-3
3444	57B153-4
3444	57B153-5
3444	57B153-6
3444	57B153-7
3444	57B153-8
3444	57B153-9
3444	57B156-9
3444	57B182-12
3444	57B184-12
3444	57B191-12
3444	57B194-11
3444	57B2-101
3444	57B2-102
3444	57B2-103
3444	57B2-113
3444	57B2-116
3444	57B2-126
3444	57B2-153
3444	57B2-192
3444	57B2-27
3444	57B2-28
3444	57B2-59
3444	57B2-62
3444	57B2-63
3444	57B2-64
3444	57B2-73
3444	57B2-85
3444	57B2-87
3444	57B2-97
3444	57B200-12
3444	57B202-13
3444	57B253-14
3444	57C11-1
3444	57C121-9
3444	57C15-1
3444	57C15-2
3444	57C15-3
3444	57C15-4
3444	57C156-9
3444	57C16-1
3444	57C24-1
3444	57C24-2
3444	57C24-3
3444	57C24-4
3444	57C27-1
3444	57C6-11
3444	57C6-4
3444	57C6-9
3444	57C7-10
3444	57C7-15
3444	57C7-17
3444	57C7-18
3444	57C7-20
3444	57C7-9
3444	57D1-123
3444	57D1-124
3444	57D1-51
3444	57D1-75
3444	57D136-12
3444	57D14-1
3444	57D14-2
3444	57D14-3
3444	57D6-4
3444	57L2-2
3444	57L3-1
3444	57L3-4
3444	57M1-14
3444	57M1-15
3444	57M1-19
3444	57M1-20
3444	57M1-23
3444	57M1-24
3444	57M1-26
3444	57M1-27
3444	57M1-28
3444	57M1-29
3444	57M1-30
3444	57M1-31
3444	57M1-32
3444	58-1(TRUETONE)
3444	61-1400
3444	61-1401
3444	61-1402
3444	61-1403
3444	61-1404
3444	61-1763
3444	61-746
3444	61-751
3444	61-754
3444	61-755
3444	61-814
3444	61-815
3444	61J001-1
3444	61J002-1
3444	61J003-1
3444	62-16905
3444	62-17550
3444	62-18425
3444	62-18641
3444	62-18642
3444	62-18643
3444	62-18828
3444	62-19280
3444	62-19516
3444	62-19548
3444	62-19837
3444	62-19838
3444	62-20155
3444	62-20240
3444	62-20241
3444	62-20242
3444	62-20243
3444	62-20360
3444	62-21496
3444	62-22038
3444	62-22039
3444	62-22250
3444	62-22251
3444	62-7567
3444	62A11868
3444	63-10377
3444	63-10708
3444	63-10725
3444	63-10732
3444	63-10733
3444	63-10734
3444	63-10735
3444	63-10736
3444	63-10737
3444	63-10860
3444	63-11025
3444	63-11143
3444	63-11289
3444	63-11468
3444	63-11469
3444	63-11470
3444	63-11471
3444	63-11472
3444	63-11660
3444	63-11757
3444	63-11758
3444	63-11759
3444	63-11831
3444	63-11832
3444	63-11833
3444	63-11916
3444	63-11934
3444	63-11935
3444	63-11937
3444	63-12003
3444	63-12004
3444	63-12062
3444	63-12272
3444	63-12605
3444	63-12608
3444	63-12609
3444	63-12641
3444	63-12642
3444	63-12696
3444	63-12697
3444	63-12706
3444	63-12707
3444	63-12750
3444	63-12751
3444	63-12752
3444	63-12753

SK	Industry Standard No.
3444	63-12874
3444	63-12875
3444	63-12877
3444	63-12878
3444	63-12879
3444	63-12933
3444	63-12940
3444	63-12941
3444	63-12942
3444	63-12943
3444	63-12946
3444	63-12948
3444	63-12949
3444	63-12950
3444	63-12951
3444	63-12952
3444	63-12953
3444	63-13419
3444	63-13438
3444	63-13440
3444	63-13441
3444	63-13864
3444	63-13927
3444	63-14032
3444	63-14051
3444	63-14052
3444	63-14057
3444	63-18643
3444	63-19280
3444	63-19282
3444	63-29461
3444	63-7421
3444	63-7567
3444	63-7670
3444	63-8555
3444	63-8701
3444	63-8702
3444	63-9337
3444	63-9338
3444	63-9339
3444	63-9341
3444	63-9516
3444	63-9518
3444	63-9829
3444	63-9830
3444	63-9831
3444	63-9832
3444	63-9833
3444	63-9847
3444	065-004
3444	66-127119
3444	66-F29-1
3444	66A00008A
3444	66A00010A
3444	66F027-1
3444	66F028-1
3444	66F029-1
3444	66F057-1
3444	66F057-2
3444	66M
3444	68A7380-1
3444	68A7715P1
3444	069
3444	70-943-722-001
3444	70-943-754-002
3444	70-943-762-001
3444	70-943-772-002
3444	70N4
3444	71-126268
3444	71N1B
3444	72N2B
3444	73B-140-003-5
3444	73C182081-31
3444	73N1B
3444	75N5AA
3444	76N1
3444	76N1M
3444	76N2
3444	76N2369-000
3444	76N2369-001
3444	77-271453-1
3444	77-271819-1
3444	77-271967-1
3444	77-273001-2
3444	77N2
3444	77N4
3444	77N5
3444	77N6
3444	78N1
3444	78N2B
3444	80-050100
3444	80-051500
3444	80-052102
3444	80-052202
3444	80-052302
3444	80-052600
3444	80-053001
3444	80-053400
3444	80-308-2
3444	80-318250
3444	80-337390
3444	80-383940
3444	80-389910
3444	80-389930
3444	80-421980
3444	81-27125140-7
3444	81-27125140-7A
3444	81-27125140-7B
3444	81-27125160-5
3444	81-27125160-5A
3444	81-27125160-5B
3444	81-27125270-2
3444	81-27125270-2A
3444	81-27125270-2B
3444	81-27125300-7
3444	81-46125016-9
3444	81-46125019-3
3444	81-46125026-8
3444	81-46125027-6
3444	82-409501
3444	85C04
3444	86-0007-004
3444	86-0022-001
3444	86-0029-001
3444	86-0031-001
3444	86-100005
3444	86-100008
3444	86-110-2
3444	86-139-2
3444	86-1392
3444	86-143-2
3444	86-144-2
3444	86-155-2
3444	86-158-2
3444	86-166-2
3444	86-171-2
3444	86-175-2
3444	86-182-2
3444	86-188-2
3444	86-189-2
3444	86-190-2
3444	86-191-2
3444	86-192-2
3444	86-193-2
3444	86-194-2
3444	86-195-2
3444	86-196-2
3444	86-197-2
3444	86-198-2
3444	86-199-2
3444	86-201-2
3444	86-202-2
3444	86-237-2
3444	86-238-2
3444	86-247-2
3444	86-250-2
3444	86-255-2
3444	86-256-2
3444	86-264-2
3444	86-265-2
3444	86-277-2
3444	86-291-9
3444	86-293-2
3444	86-309-2
3444	86-310-2
3444	86-323-2
3444	86-324-2
3444	86-327-2
3444	86-328-2
3444	86-339-2
3444	86-339-9
3444	86-342-2
3444	86-359-2
3444	86-362-2
3444	86-365-2
3444	86-379-2
3444	86-390-2
3444	86-391-2
3444	86-399-1
3444	86-399-2
3444	86-399-9
3444	86-403-2
3444	86-420-2
3444	86-445-2
3444	86-457-2
3444	86-458-2
3444	86-460-2
3444	86-461-2
3444	86-462-2
3444	86-472-2
3444	86-481-1
3444	86-481-2
3444	86-483-2
3444	86-483-3
3444	86-484-2
3444	86-485-2
3444	86-486-2
3444	86-493-2
3444	86-494-2
3444	86-495-2
3444	86-496-2
3444	86-502-2
3444	86-5040-2
3444	86-5041-2
3444	86-5044-2
3444	86-5045-2
3444	86-5046-2
3444	86-5049-2
3444	86-5050-2
3444	86-5051-2
3444	86-5055-2
3444	86-5056-2
3444	86-5081-2
3444	86-5097-2
3444	86-5103-2
3444	86-5110-2
3444	86-5111-2
3444	86-5114-2
3444	86-5117-2
3444	86-514-2
3444	86-515-2(SEARS)
3444	86-520-2
3444	86-534-2
3444	86-539-2
3444	86-548-2
3444	86-551-2
3444	86-554-2
3444	86-559-2
3444	86-560-2
3444	86-561-2
3444	86-565-2
3444	86-573-2
3444	86-595-2
3444	86-598-2
3444	86-599-2
3444	86-646-2
3444	86-655-2
3444	86-702-2
3444	86A334
3444	86A336
3444	86A350
3444	86X0006-001
3444	86X0007-001
3444	86X0007-104
3444	86X0008-001
3444	86X0029-001
3444	86X0031-001
3444	86X0031-002
3444	86X0031-003
3444	86X0032-001
3444	86X0035-001
3444	86X0040-00
3444	86X0040-001
3444	86X0045-001
3444	86X0048-001
3444	86X0050-001
3444	86X0051-001
3444	86X0054-001
3444	86X0058-001
3444	86X0058-002
3444	86X0058-003
3444	86X006-001
3444	86X0063-001
3444	86X007-004
3444	86X007-034
3444	86X0079-001
3444	86X34-1
3444	86X6
3444	86X6-1
3444	86X6-4-518
3444	86X7-2
3444	86X7-3
3444	86X7-4
3444	86X8-1
3444	86X8-2
3444	86X8-4
3444	87-0014
3444	88-1250109
3444	089-223
3444	089-226
3444	90-2213-00-18
3444	90-30
3444	90-48
3444	90-57
3444	90-61
3444	90-65
3444	90-69
3444	91C
3444	91D
3444	91E
3444	93A39-15
3444	94N1
3444	94N1B
3444	94N1R
3444	94N2
3444	95-226-2
3444	95-227
3444	96-5080-02
3444	96-5115-01
3444	96-5115-02
3444	96-5115-03
3444	96-5115-04
3444	96-5115-05
3444	96-5152-01
3444	96-5152-03
3444	96-5153-01
3444	96-5153-03
3444	96-5177-01
3444	96-5187-01
3444	96-5213-01
3444	96-5220-01(NPN)
3444	96-5221-01
3444	96-5228-01
3444	96-5229-01
3444	96-5237-01
3444	96-5255-01
3444	96-5257-01
3444	96-5281-01
3444	96-5290-01
3444	96-5314-01
3444	96XZ6052/52N
3444	96XZ6053/11N
3444	96XZ6053/35N
3444	96XZ6053/36N
3444	96XZ801/14N
3444	99-109-1
3444	99-109-2
3444	99S012E
3444	99S025
3444	99S025A
3444	99S033A
3444	99S035
3444	99S036
3444	99S038
3444	99S085
3444	0101-0491
3444	0101-0540
3444	0103-0014
3444	0103-0014/4460
3444	0103-0088
3444	0103-0088H
3444	0103-0088R
3444	0103-0088S
3444	0103-0473
3444	0103-0482
3444	0103-0491
3444	0103-0491/4460
3444	0103-0504
3444	0103-0540
3444	0103-94
3444	105(ADMIRAL)
3444	105-003-06
3444	105-003-09
3444	105-006-08
3444	105-060-09
3444	105-06007-05
3444	105-08243-05
3444	105-085-54
3444	105-12
3444	112-1A82
3444	112-523
3444	115-225
3444	115-875
3444	116-074
3444	116-078
3444	116-085
3444	116-092
3444	116-588
3444	116-875
3444	119-0054
3444	119-0056
3444	120-004480
3444	120-004482
3444	120-004483
3444	120-004880
3444	120-004882
3444	120-004883
3444	120-1
3444	120-2
3444	120-3
3444	120-7
3444	120-8
3444	120-8A
3444	120BLU
3444	121(SEARS)
3444	121-364
3444	121-365
3444	121-366
3444	121-367
3444	121-369
3444	121-404
3444	121-423
3444	121-448
3444	121-450
3444	121-5065
3444	121-629
3444	121-660
3444	121-662
3444	121-678
3444	121-701
3444	121-706
3444	121-711
3444	121-730
3444	121-737
3444	121-751
3444	121-764
3444	121-767CL
3444	121-768CL
3444	121-773
3444	121-836
3444	121-837
3444	121-889
3444	121-Z9000
3444	121-Z9000A
3444	121G3019
3444	121G3020
3444	122-1
3444	122-2
3444	122-6
3444	123-004
3444	123-010
3444	124N16
3444	0125
3444	125B132
3444	0126
3444	126-12
3444	127
3444	128WHT
3444	129-16
3444	129-21
3444	129-33(PILOT)
3444	129BRN
3444	130-40214
3444	130-40215
3444	130-40294
3444	130-40311
3444	130-40312
3444	130-40313
3444	130-40317
3444	130-40318
3444	130-40357
3444	130-40896
3444	130-40922
3444	0131
3444	0131-000473
3444	0131-000704
3444	0131-001417
3444	0131-001418
3444	0131-001421
3444	0131-001422
3444	0131-001423
3444	0131-001424
3444	0131-001464
3444	0131-001864
3444	0131-004323
3444	132-002
3444	132-004
3444	132-005
3444	132-011
3444	132-017
3444	132-018
3444	132-021
3444	132-023
3444	132-026
3444	132-030
3444	132-041
3444	132-042
3444	132-050
3444	132-051
3444	132-054
3444	132-055
3444	132-057
3444	132-062
3444	132-063
3444	132-069
3444	132-075
3444	132-077
3444	132-501
3444	132-502
3444	132-503
3444	132-504
3444	132-539
3444	132-540
3444	0134
3444	135C44322-542
3444	138-4
3444	140-0007
3444	142-001
3444	142-3
3444	142-4
3444	142N3T
3444	142N4
3444	142N5
3444	146N3
3444	146N5
3444	148N2
3444	148N212
3444	151-0103-00
3444	151-0127-00
3444	151-0190-00
3444	151-0223-00
3444	151-0224-00
3444	151-0302-00
3444	151-0424-00
3444	151N2
3444	151N4
3444	151N5
3444	154A5941
3444	154A5946
3444	156
3444	156WHT
3444	158(SEARS)
3444	165A4383
3444	167N1
3444	167N2
3444	168N1
3444	173A4057
3444	173A4399
3444	173A4416
3444	173A4470-11
3444	173A4470-13
3444	173A4470-32
3444	176-008-9-001
3444	176-014-9-001
3444	176-017-9-001
3444	176-024-9-001
3444	176-042-9-002
3444	176-047-9-002
3444	176-054-9-001
3444	185-009
3444	186-002
3444	186-007
3444	200-057
3444	200-058
3444	200-846
3444	200-862
3444	200-863
3444	207A10
3444	207A29
3444	207A31
3444	207A35
3444	209-846
3444	209-862
3444	209-863
3444	211AVPF3415
3444	211AVTE4275
3444	212-695
3444	218-22
3444	218-23
3444	218-24
3444	218-25
3444	220-001001
3444	220-001002
3444	226-1(SYLVANIA)
3444	229-0050-13
3444	229-0050-14
3444	229-0050-15
3444	229-0180-123
3444	229-0190-90
3444	247-257
3444	247-629
3444	247AS-C1249-001
3444	250-0712
3444	260-10-020
3444	260P07701
3444	260P07702
3444	260P07703
3444	260P07704
3444	260P07705
3444	260P07707
3444	260P08801
3444	260P08801A
3444	260P09902
3444	260P11302
3444	260P11303
3444	260P11304
3444	260P11305
3444	260P11502
3444	260P11503
3444	260P11504
3444	260P11505
3444	260P12001
3444	260P12002
3444	260P14101
3444	260P14102
3444	260P14103
3444	260P14105
3444	260P141103
3444	260P17104
3444	260P17105
3444	260P17106
3444	260P17501
3444	260P17503
3444	260P19101
3444	260P19103
3444	260P19501
3444	260P4002
3444	260Z00402
3444	270-950-030
3444	281
3444	284HC
3444	296-50-9
3444	296-51-9
3444	296-56-9
3444	296-59-9
3444	296-77-9
3444	297L006H01
3444	297L006H02
3444	297L007C02
3444	297L007H01
3444	297L007H02
3444	297L013B01
3444	297V049H05
3444	297V061H01
3444	297V061H02
3444	297V061H03
3444	297V072C06
3444	297V074C06
3444	297V074C07
3444	297V074C08
3444	297V083C02
3444	297V085C01
3444	297V085C02
3444	297V085C03
3444	297V085C04
3444	297V086C02
3444	297V086C03
3444	309-327-926
3444	314-6007-1
3444	314-6007-2
3444	314-6007-3
3444	317-8504-001
3444	319C
3444	324-6005-5
3444	325-0031-303
3444	325-0031-310
3444	325-0042-351
3444	325-0076-306
3444	325-0076-307
3444	325-0076-308
3444	325-0081-100
3444	325-0081-101
3444	325-0574-30
3444	325-0574-31
3444	325-1370-19
3444	325-1370-20
3444	325-1771-15
3444	344-6000-2
3444	344-6000-4
3444	344-6000-5
3444	344-6000-5A
3444	344-6002-3
3444	344-6005-1
3444	344-6005-2
3444	344-6005-5
3444	344-6017-2
3444	344-6017-3
3444	344-6017-5
3444	352-0195-000
3444	352-0197-000
3444	352-0316-00
3444	352-0318-00
3444	352-0318-001
3444	352-0319-000
3444	352-0322-010
3444	352-0349-000
3444	352-0365-000
3444	352-0400-00
3444	352-0400-010
3444	352-0400-030

SK	Industry Standard No.	SK	Industry Standard No.	SK	Industry Standard No.	SK	Industry Standard No.	SK	Industry Standard No.
3444	352-0433-00	3444	536D9	3444	750M63-119	3444	921-269B	3444	1203(GE)
3444	352-0477-00	3444	536F	3444	750M63-120	3444	921-26A	3444	1203-169
3444	352-0506-000	3444	536F8	3444	750M63-146	3444	921-272B	3444	1205(03)
3444	352-0519-00	3444	536FU	3444	753-1828-001	3444	921-275R	3444	1210-17B
3444	352-0546-00	3444	536G(WARDS)	3444	753-2000-003	3444	921-27B	3444	1227-17
3444	352-0569-00	3444	537D	3444	753-2000-004	3444	921-28	3444	1228-17
3444	352-0569-010	3444	537R	3444	753-2000-008	3444	921-28A	3444	1236-3776
3444	352-0569-020	3444	537PV	3444	753-2000-009	3444	921-28B	3444	1272
3444	352-0579-00	3444	537PY	3444	753-2000-011	3444	921-28BLU	3444	1315
3444	352-0579-010	3444	550-026-00	3444	753-2000-711	3444	921-305B	3444	1316
3444	352-0579-020	3444	560-2	3444	753-2000-735	3444	921-314B	3444	1374-17AC
3444	352-0596-010	3444	565-074	3444	753-2000-870	3444	921-345B	3444	1412-1
3444	352-0596-020	3444	570-004503	3444	753-2000-871	3444	921-369	3444	1412-1-12
3444	352-0596-030	3444	570-005503	3444	753-2100-001	3444	921-43B	3444	1412-1-12-8
3444	352-0629-010	3444	572-683	3444	753-2100-008	3444	921-46	3444	1415
3444	352-0661-010	3444	573-469	3444	753-4000-011	3444	921-46A	3444	001422
3444	352-0661-020	3444	573-479	3444	753-4000-101	3444	921-46BK	3444	1424
3444	352-0661-010	3444	573-480	3444	753-4000-537	3444	921-47	3444	1431 8349
3444	352-0675-010	3444	573-481	3444	767(ZENITH)	3444	921-47A	3444	1463
3444	352-0675-020	3444	576-0001-008	3444	767CL	3444	921-47BL	3444	1465
3444	352-0675-030	3444	576-0001-012	3444	772-110	3444	921-48B	3444	1471-4778
3444	352-0675-040	3444	576-0001-018	3444	773(ZENITH)	3444	921-49B	3444	1482
3444	352-0675-050	3444	576-0002-006	3444	776-151	3444	921-50B	3444	1493-17
3444	352-0680-010	3444	576-0003-011	3444	776-183	3444	921-73B	3444	1540
3444	352-0680-020	3444	576-0036-916	3444	776-2(PHILCO)	3444	921-77B	3444	1567
3444	352-0713-030	3444	576-0036-917	3444	776GRN	3444	921-8	3444	1567-0
3444	352-0809	3444	576-0036-920	3444	780WHT	3444	921-93B	3444	1567-2
3444	352-7500-010	3444	576-0036-921	3444	785YEL	3444	921-99B	3444	1705-4834
3444	352-7500-450	3444	586-2	3444	791	3444	9260193-1	3444	1711
3444	352-8000-010	3444	590-593031	3444	800-001-034	3444	9260193-2	3444	1711-17
3444	352-8000-020	3444	595-1(SYLVANIA)	3444	800-101-101-1	3444	9260193-P1M4165	3444	1711MC
3444	352-8000-030	3444	595-2	3444	800-101-102-1	3444	930X1	3444	1712-17
3444	352-8000-040	3444	595-2(SYLVANIA)	3444	800-501-00	3444	930X2	3444	1723-17
3444	352-9036-00	3444	600-188-1-13	3444	800-501-01	3444	935-1	3444	1751B036
3444	352-9079-00	3444	600-188-1-20	3444	800-501-02	3444	964-17444	3444	1799-17
3444	352-9103-000	3444	600-188-1-23	3444	800-501-03	3444	964-20738	3444	1800-17
3444	353-9306-001	3444	600-501-80	3444	800-501-04	3444	964-24584	3444	1812-1
3444	353-9306-002	3444	600X0091-086	3444	800-501-11	3444	991-00-1219	3444	1812-1-12
3444	353-9306-003	3444	601-0100793	3444	800-501-22	3444	991-00-2356/K	3444	1812-1-127
3444	353-9306-004	3444	601-1	3444	800-508-00	3444	991-00-2873	3444	1812-1L
3444	353-9306-005	3444	601-1(RCA)	3444	800-509-00	3444	991-00-3144	3444	1812-1L8
3444	353-9310-001	3444	601-2	3444	800-514-00	3444	991-00-3304	3444	1841-17
3444	353-9314-001	3444	601X0149-086	3444	800-51500	3444	991-00-8393	3444	1854-0003
3444	353-9315-001	3444	602X0018-000	3444	800-521-01	3444	991-00-8393A	3444	1854-0005
3444	353-9319-001	3444	604	3444	800-52102	3444	991-00-8393M	3444	1854-0033
3444	353-9319-002	3444	604(SEARS)	3444	800-522-02	3444	991-00-8394	3444	1854-0353
3444	354-3127-1	3444	606-9601-101	3444	800-522-04	3444	991-00-8394A	3444	1854-0432
3444	355D9	3444	606-9602-101	3444	800-52202	3444	991-00-8394AH	3444	1882-17
3444	365-1	3444	607-030	3444	800-52902	3444	991-00-8395	3444	1883-17
3444	386-1102-P1	3444	609-112	3444	800-526-00	3444	991-01-1219	3444	1884-17
3444	386-1102-P2	3444	610-045-3	3444	800-52600	3444	991-01-1220	3444	1893-17
3444	386-1102-P3	3444	610-045-4	3444	800-529-00	3444	991-01-1306	3444	1915-17
3444	386-7178P1	3444	610-070	3444	800-530-00	3444	991-01-1312	3444	1929-17
3444	386-7185P1	3444	610-070-1	3444	800-530-01	3444	991-01-1318	3444	1932-17
3444	394-3003-1	3444	610-070-2	3444	800-534-00	3444	991-01-1705	3444	1961-17
3444	394-3003-3	3444	610-070-3	3444	800-534-01	3444	991-01-3044	3444	1966-17
3444	394-3003-7	3444	610-076	3444	800-53400	3444	991-01-3056	3444	1984-17
3444	394-3009-9	3444	610-076-1	3444	800-538-00	3444	991-01-3057	3444	1999
3444	400-1371-101	3444	610-076-2	3444	800-544-00	3444	991-01-3068	3444	2001-17
3444	400-2023-101	3444	610-077	3444	800-544-10	3444	991-01-3544	3444	2003-17
3444	400-2025-201	3444	610-077-1	3444	800-544-30	3444	991-01-3683	3444	2004-03
3444	417-105	3444	610-077-2	3444	800-548-00	3444	991-01-3740	3444	2004-04
3444	417-106	3444	610-077-3	3444	801B	3444	991-011219	3444	2004-14
3444	417-108	3444	610-077-4	3444	803-18250	3444	991-011220	3444	2006
3444	417-109-13163	3444	610-077-5	3444	803-37390	3444	991-011306	3444	2008-17
3444	417-110	3444	610-078	3444	803-83940	3444	991-011312	3444	2017-115
3444	417-110-13163	3444	610-078-1	3444	803-89910	3444	991-011318	3444	2018-01
3444	417-114-13163	3444	612-145L	3444	803-89930	3444	991-015587	3444	2026
3444	417-118	3444	614-3	3444	804	3444	992-00-2298	3444	2026-00
3444	417-129	3444	617-10	3444	804-21980	3444	992-00-3144	3444	2057A10-64
3444	417-134	3444	617-161	3444	818WHT	3444	992-01-3738	3444	2057A2-103
3444	417-134-13271	3444	617-53	3444	822	3444	998-0061114	3444	2057A2-113
3444	417-155	3444	617-63	3444	822A	3444	998-0200816	3444	2057A2-121
3444	417-155-13163	3444	617-64	3444	822ABLU	3444	1000-136	3444	2057A2-131
3444	417-171-13163	3444	617-67	3444	822B	3444	1000-137	3444	2057A2-145
3444	417-172	3444	617-68	3444	823B	3444	1001(JULIETTE)	3444	2057A2-146
3444	417-172-13271	3444	617-71	3444	823WHT	3444	1001-02	3444	2057A2-152
3444	417-185	3444	626	3444	824-09501	3444	1001-03	3444	2057A2-153
3444	417-192	3444	626-1	3444	827BRN	3444	1001-04	3444	2057A2-154
3444	417-197	3444	630-076	3444	828GRN	3444	1001-05	3444	2057A2-155
3444	417-213	3444	638	3444	828S	3444	1001-06	3444	2057A2-184
3444	417-217	3444	639(ZENITH)	3444	834-6066	3444	1002-03	3444	2057A2-208
3444	417-226-1	3444	642-174	3444	847BLK	3444	1002-04	3444	2057A2-209
3444	417-228	3444	642-242	3444	853-0300-632	3444	1002(28C537)	3444	2057A2-215
3444	417-229	3444	642-246	3444	853-0300-900	3444	1004-03	3444	2057A2-222
3444	417-233-13163	3444	642-319	3444	853-0300-923	3444	1005	3444	2057A2-225
3444	417-244	3444	660-126	3444	853-0373-110	3444	1005(28C537)	3444	2057A2-226
3444	417-67	3444	660-131	3444	858DB	3444	1005-03	3444	2057A2-264
3444	417-69	3444	660-220	3444	880-250-102	3444	1008-02	3444	2057A2-276
3444	417-77	3444	660-221	3444	880-250-108	3444	1009-02	3444	2057A2-278
3444	417-801-12903	3444	660-222	3444	880-250-109	3444	1009-02-16	3444	2057A2-279
3444	417-91	3444	660-225	3444	881-250-102	3444	1009-17	3444	2057A2-280
3444	417-92	3444	690V0103H27	3444	881-250-108	3444	1019-3852	3444	2057A2-281
3444	417-93	3444	690V04TH97	3444	881-250-109	3444	1020-17	3444	2057A2-294
3444	417-93-12903	3444	690V084H62	3444	903-3	3444	1023G	3444	2057A2-296
3444	417-94	3444	690V086H51	3444	903-3G	3444	1023G(GE)	3444	2057A2-297
3444	421-7444	3444	690V086H88	3444	903Y002149	3444	1024G(GE)	3444	2057A2-300
3444	421-8111	3444	690V088H50	3444	904-95	3444	1026G	3444	2057A2-303
3444	421-9644	3444	690V088H51	3444	904-95A	3444	1026G(GE)	3444	2057A2-305
3444	421-9670	3444	690V088H89	3444	904-96B	3444	1028G	3444	2057A2-316
3444	421-9840	3444	690V092H52	3444	912-1A6A	3444	1028G(GE)	3444	2057A2-324
3444	422-0738	3444	690V092H54	3444	916-31024-3	3444	1029G	3444	2057A2-332
3444	422-1234	3444	690V092H81	3444	916-31025-5	3444	1029G(GE)	3444	2057A2-341
3444	422-1407	3444	690V092H84	3444	916-31026-8B	3444	1034-17	3444	2057A2-374
3444	422-2533	3444	690V092H96	3444	921-01122	3444	1038-1	3444	2057A2-387
3444	422-2534	3444	690V092H97	3444	921-01123	3444	1038-1-10	3444	2057A2-390
3444	430(ZENITH)	3444	690V094H21	3444	921-109B	3444	1038-10	3444	2057A2-396
3444	430-10034	3444	690V097H62	3444	921-117B	3444	1038-15	3444	2057A2-398
3444	430-10034-06	3444	690V098H48	3444	921-120B	3444	1038-15CL	3444	2057A2-399
3444	430-10053-0	3444	690V098H49	3444	921-123B	3444	1038-18	3444	2057A2-401
3444	430-10053-0A	3444	690V098H50	3444	921-124B	3444	1038-18CL	3444	2057A2-412
3444	430-1044-0A	3444	690V099H79	3444	921-125B	3444	1038-21	3444	2057A2-433
3444	430-23212	3444	690V102H71	3444	921-128B	3444	1038-23	3444	2057A2-434
3444	430-25221	3444	690V103H31	3444	921-159B	3444	1038-23CL	3444	2057A2-449
3444	430-25766	3444	690V103H33	3444	921-161B	3444	1038-6	3444	2057A2-452
3444	430-86	3444	690V114H30	3444	921-189B	3444	1038-6CL	3444	2057A2-463
3444	430-87	3444	690V114H33	3444	921-191B	3444	1038-8	3444	2057A2-464
3444	430CL	3444	690V116H21	3444	921-195B	3444	1039-01	3444	2057A2-479
3444	447(ZENITH)	3444	693BP	3444	921-200B	3444	1040-03	3444	2057A2-511
3444	450-1167-2	3444	693P8	3444	921-20BK	3444	1040-155	3444	2057A2-62
3444	450-1261	3444	693G	3444	921-214B	3444	1040-2	3444	2057A2-64
3444	462-0119	3444	693GT	3444	921-215B	3444	1041-72	3444	2057B-85
3444	462-1000	3444	694D	3444	921-225B	3444	1041-73	3444	2057B101-4
3444	462-1009-01	3444	694E	3444	921-228B	3444	1041-75	3444	2057B102-4
3444	462-1061	3444	699	3444	921-229B	3444	1042-04	3444	2057B103-4
3444	462-1063	3444	700A858-318	3444	921-22BG	3444	1042-07	3444	2057B117-9
3444	462-2002	3444	700A858-328	3444	921-234B	3444	1043-07	3444	2057B118-12
3444	472-0491-001	3444	737(ZENITH)	3444	921-237B	3444	1080-20	3444	2057B141-4
3444	472-1198-001	3444	739H01	3444	921-23BK	3444	1080-6396	3444	2057B143-12
3444	515ORN	3444	748(ZENITH)	3444	921-25B	3444	1081-3319	3444	2057B145-12
3444	519-1(RCA)	3444	750C858-123	3444	921-252B	3444	1081-3475	3444	2057B146-12
3444	524WHT	3444	750C858-124	3444	921-255B	3444	1113-03	3444	2057B151-6
3444	536-2(RCA)	3444	750C858-125	3444	921-26	3444	1119-8132	3444	2057B152-12
3444	536D	3444	750D858-212	3444	921-268B	3444	1123-60	3444	2057B2-113

SK	Industry Standard No.
3444	2057B2-121
3444	2057B2-122
3444	2057B2-123
3444	2057B2-130
3444	2057B2-152
3444	2057B2-153
3444	2057B2-154
3444	2057B2-155
3444	2057B2-38
3444	2057B2-62
3444	2057B2-69
3444	2057B2-97
3444	2063-17-12
3444	2064(CROWN)
3444	2132(GE)
3444	2158-1541
3444	2180-153
3444	2180-154
3444	2263
3444	2270
3444	2275-17
3444	2290-17
3444	2320-17
3444	2321-17
3444	2337-17
3444	2443(RCA)
3444	2446(RCA)
3444	2447(RCA)
3444	2475(RCA)
3444	2495(RCA)
3444	2495-166-2
3444	2495-522-4
3444	2495-529
3444	2496-125-2
3444	2510-104
3444	2546(RCA)
3444	2603-184
3444	2787
3444	2854
3444	2904-054
3444	3005(SEARS)
3444	3005-861
3444	3006(SEARS)
3444	3007(SEARS)
3444	03008-1
3444	3011
3444	3026
3444	3107-204-9000
3444	3107-204-90010
3444	3107-204-90020
3444	3505(RCA)
3444	3506(RCA)
3444	3507(WARDS)
3444	3508(RCA)
3444	3509(SEARS)
3444	3509(WARDS)
3444	3510(RCA)
3444	3513(RCA)
3444	3514(WARDS)
3444	3519(RCA)
3444	3521(SEARS)
3444	3523(SEARS)
3444	3525(RCA)
3444	3526(RCA)
3444	3532(RCA)
3444	3538(RCA)
3444	3541(RCA)
3444	3543(RCA)
3444	3544-1
3444	3546(RCA)
3444	3546-1(RCA)
3444	3546-2(RCA)
3444	3548(RCA)
3444	3551(RCA)
3444	3551A(RCA)
3444	3554(RCA)
3444	3555(RCA)
3444	3556-1
3444	3558(RCA)
3444	3560(RCA)
3444	3560-2(RCA)
3444	3561
3444	3561(RCA)
3444	3561-1(RCA)
3444	3569(RCA)
3444	3571(RCA)
3444	3571R
3444	3572(RCA)
3444	3577(RCA)
3444	3577-1
3444	3586(RCA)
3444	3588
3444	3589
3444	3601(RCA)
3444	3601-1
3444	3614-1
3444	3614-3
3444	3625(RCA)
3444	3626
3444	3631-1
3444	3634.0011
3444	3706
3444	3867
3444	3999
3444	4002(PENNCREST)
3444	4003E
3444	4021
3444	4022
3444	4046(SEARS)
3444	4057
3444	4066
3444	4085
3444	4309(AIRLINE)
3444	4473-12
3444	4473-4
3444	4473-5
3444	4473-5X
3444	4473-9
3444	4686-132-3
3444	4686-143-3
3444	4686-144-3
3444	4686-173-3
3444	4686-183-3
3444	4686-231-3
3444	4686-257-3
3444	4686-82-3
3444	4732
3444	4733
3444	4734
3444	4737
3444	4765
3444	4768
3444	4802-00003
3444	4802-00006
3444	4802-00009
3444	4802-00012
3444	4802-00015
3444	4822-130-40184
3444	4822-130-40343
3444	4822-130-40354
3444	4822-130-40361
3444	4822-130-40454
3444	4839
3444	4840
3444	4841
3444	4842
3444	4852
3444	4854
3444	4856-0101
3444	4856-0107
3444	4856-0109
3444	4856-0110
3444	5001-002
3444	5001-014
3444	5001-020
3444	5001-069
3444	5001-072
3444	5226-2
3444	5380-71
3444	5380-72
3444	5613-1335
3444	5613-1335D
3444	5613-4581C
3444	5613-558C
3444	5613-711
3444	5613-711E
3444	5613-870
3444	5613-870F
3444	5721
3444	6136
3444	06246-00
3444	6343-1
3444	6367
3444	6367-1
3444	6854K90-074
3444	6954K90-074
3444	7005G(LOWREY)
3444	7113
3444	7122-5
3444	7129
3444	7171
3444	7172
3444	7176
3444	7306
3444	7306-1
3444	7306-4
3444	7306-5
3444	7318
3444	7318-2
3444	7340-2
3444	7398-6117P1
3444	7398-6118P1
3444	7398-6119P1
3444	7506
3444	7515
3444	7516
3444	7517
3444	7518
3444	7585
3444	7586
3444	7586(GE)
3444	7587
3444	7587(GE)
3444	7588
3444	7588(GE)
3444	7589
3444	7590
3444	7590(GE)
3444	7591(GE)
3444	7637
3444	7641
3444	7675
3444	7676
3444	7992
3444	8000-00003-033
3444	8000-00004-079
3444	8000-00004-082
3444	8000-00004-243
3444	8000-00004-85
3444	8000-00005-002
3444	8000-00005-007
3444	8000-00005-055
3444	8000-00006-003
3444	8000-00009-174
3444	8000-00011-004
3444	8000-00011-048
3444	8000-00029-006
3444	8000-00029-007
3444	8000-00030-007
3444	8000-00032-025
3444	8000-0004-P079
3444	8000-0004-P082
3444	8000-0004-P079
3444	8000-0004-P082
3444	8000-0004-P085
3444	8000-0005-007
3444	8000-0005-009
3444	8003-114
3444	8010-176
3444	8020-205
3444	8074-4
3444	8075-4
3444	8210-1203
3444	8281-1
3444	8302
3444	8504
3444	8509
3444	8600
3444	8602
3444	8614 007 0
3444	8710-163
3444	8710-164
3444	8710-165
3444	8710-166
3444	8710-167
3444	8710-168
3444	8800-202
3444	8800-203
3444	8800-204
3444	8840-163
3444	8840-164
3444	8840-165
3444	8840-166
3444	8840-167
3444	8840-168
3444	8868-8
3444	8886-2
3444	8910-143
3444	8910-145
3444	9013H
3444	9013HF
3444	9013HG
3444	9013HH
3444	9014D
3444	9033
3444	9033(SYLVANIA)
3444	9033-1
3444	9033BROWN
3444	9033G(SYLVANIA)
3444	09502-8
3444	9600
3444	9600-5
3444	9920-6-2
3444	9920-7-2
3444	10226/2
3444	10416-009
3444	11252-3
3444	11522-5
3444	11587-5
3444	11607-4
3444	11607-8
3444	11608-5
3444	11609-2
3444	11658-8
3444	11687-5
3444	012013-1
3444	12112-C
3444	12112-D
3444	12112-E
3444	12112-F
3444	12127-6
3444	12127-7
3444	12127-8
3444	12127-9
3444	12593
3444	14303
3444	15840-1
3444	15841-1
3444	17412-5
3444	18410-143
3444	18410-146
3444	18509
3444	18555
3444	18600-153
3444	19645
3444	22158
3444	22635-002
3444	22635-003
3444	22810-173
3444	23114-046
3444	23114-053
3444	23114-054
3444	23114-082
3444	23114-095
3444	23115-057
3444	23115-058
3444	23316
3444	0023828
3444	25114-116
3444	25114-130
3444	25114-161
3444	25810-161
3444	25810-162
3444	25810-163
3444	25840-161
3444	26810-154
3444	27125-080
3444	27125-140
3444	27125-160
3444	27125-270
3444	27125-300
3444	27125-370
3444	27125-500
3444	27125-530
3444	27840-161
3444	27840-162
3444	30227
3444	30228
3444	30229
3444	30235
3444	30241
3444	30242
3444	30243
3444	30253
3444	30268
3444	30269
3444	30289
3444	030512-1
3444	030512-2
3444	030515
3444	030515-4
3444	030527
3444	030536
3444	030537
3444	030537-1
3444	030537-2
3444	030538
3444	030542
3444	030542-1
3444	030543
3444	030543-1
3444	030543-2
3444	31001
3444	31003
3444	31009
3444	36580
3444	36917
3444	36920
3444	36921
3444	37510-162
3444	37510-163
3444	37884
3444	38283
3444	38510-163
3444	38510-166
3444	38510-167
3444	38788
3444	39096
3444	40217
3444	40218
3444	40219
3444	40220
3444	40221
3444	40222
3444	40283
3444	40397
3444	40398
3444	40399
3444	40400
3444	40405
3444	40456(RCA)
3444	40473
3444	40474
3444	40477
3444	40500
3444	40577
3444	40637
3444	41051
3444	41176
3444	42464
3444	43021-017
3444	43044
3444	43045
3444	43139
3444	45184
3444	48004-07
3444	48004-08
3444	50202-1
3444	51213
3444	51213-01
3444	51213-03
3444	51213-2
3444	51428-01
3444	51429-02
3444	51429-03
3444	51429-3
3444	51441
3444	51441-01
3444	51441-02
3444	51441-03
3444	51545
3444	51547
3444	53200-22
3444	53200-23
3444	53200-51
3444	53400-01
3444	55606
3444	58810-162
3444	58810-163
3444	58810-165
3444	58810-166
3444	58810-168
3444	58840-193
3444	58840-194
3444	58840-195
3444	58840-196
3444	58840-197
3444	58840-198
3444	58840-199
3444	60395
3444	61009-4
3444	61009-4-1
3444	61010-7-2
3444	61011-3-2
3444	61013-2-1
3444	61049
3444	68617
3444	70023-0-00
3444	70023-1-00
3444	70260-14
3444	70260-15
3444	70260-16
3444	70260-20
3444	70511
3444	71226-10
3444	71819-1
3444	72114
3444	72115
3444	72116
3444	72151
3444	72204
3444	72206
3444	72207
3444	72874-52
3444	72963-14
3444	000073100
3444	000073333
3444	74651-02
3444	75561-16
3444	75561-28
3444	75561-3
3444	75561-33
3444	75614-1
3444	75700-04
3444	75700-04-01
3444	75700-05
3444	75700-05-01
3444	75700-05-02
3444	75700-05-03
3444	75700-08
3444	75700-08-02
3444	75700-09-01
3444	75700-09-21
3444	76236
3444	80540
3444	80544
3444	80545
3444	80813VM
3444	80814VM
3444	80815VM
3444	80816VM
3444	81170-6
3444	81513-3
3444	082006
3444	082019
3444	85549
3444	86287
3444	87757
3444	88060-143
3444	88510-173
3444	88510-175
3444	88687
3444	88688
3444	88862
3444	90209-172
3444	90209-182
3444	90326-001
3444	90429
3444	91605
3444	94027
3444	94047
3444	94048
3444	95223
3444	96457-1
3444	99206-1
3444	99206-2
3444	99207-2
3444	100092
3444	100093
3444	101119
3444	101185
3444	102002
3444	103521
3444	104389
3444	110697
3444	111303
3444	112356
3444	112357
3444	112358
3444	112359
3444	112520
3444	112521
3444	112522
3444	112523
3444	113348
3444	113438
3444	113524
3444	115167
3444	115225
3444	115720
3444	115728
3444	115875
3444	116074
3444	116076
3444	116077
3444	116078
3444	116085
3444	116092
3444	116875
3444	118200
3444	118713
3444	119232-001
3444	119258-001
3444	119636
3444	119725
3444	119726
3444	119982
3444	120074
3444	120085
3444	120481
3444	120482
3444	120483
3444	121655
3444	121658
3444	121660
3444	121662
3444	121663
3444	122074
3444	122664
3444	122665
3444	123807
3444	124557
3444	124759
3444	125135
3444	125139
3444	125140
3444	125389-14
3444	126150
3444	126156
3444	126331
3444	126525
3444	126526
3444	126702
3444	126706
3444	126711
3444	126713
3444	126714
3444	129147
3444	129949
3444	130403-13
3444	130403-17
3444	130403-18
3444	130537
3444	131240
3444	131243-12
3444	133743
3444	137614
3444	138378
3444	138789-1
3444	138789-2
3444	138789-3
3444	140622
3444	140858-12
3444	146144-2
3444	146153-1
3444	147115
3444	147115-5
3444	147115-6
3444	147357-1-1
3444	147357-7-1
3444	147363-1
3444	147513
3444	147555-1
3444	150741
3444	150763
3444	150768
3444	161918-28
3444	165668
3444	165827
3444	165828
3444	167688
3444	169197
3444	169679
3444	169680
3444	170967-1
3444	170968-1
3444	171003(TOSHIBA)
3444	171009(TOSHIBA)
3444	171026(TOSHIBA)
3444	171027
3444	171030(TOSHIBA)
3444	171040(SEARS)
3444	171040(TOSHIBA)
3444	171044(TOSHIBA)
3444	171046
3444	171162-005
3444	171162-006
3444	171162-008
3444	171162-009
3444	171162-119
3444	171162-132
3444	171162-143
3444	171162-161
3444	171162-162
3444	171162-190
3444	171162-191
3444	171162-202
3444	171162-286
3444	175007-275
3444	175007-276
3444	175027-021
3444	175043-058
3444	175043-059
3444	175043-060
3444	181023
3444	181214
3444	181504-2
3444	183017
3444	183030
3444	183031

SK	Industry Standard No.
3444	187218
3444	190426
3444	190428
3444	190715
3444	196023-1
3444	196023-2
3444	198003-1
3444	198003-2
3444	198007-3
3444	198013-P1
3444	198023-1
3444	198023-3
3444	198023-4
3444	198023-5
3444	198030
3444	198030-2
3444	198030-3
3444	198030-4
3444	198030-6
3444	198030-7
3444	198031-1
3444	198031-2
3444	198042-2
3444	198042-3
3444	198051-1
3444	198051-2
3444	198051-3
3444	198051-4
3444	198067-1
3444	198581-1
3444	198581-2
3444	198581-3
3444	200064-6-107
3444	200076
3444	200251-5377
3444	202609-0713
3444	202862-947
3444	202907-047P1
3444	202914-417
3444	202915-627
3444	202922-237
3444	204210-002
3444	204969
3444	210074
3444	215081
3444	216445-2
3444	221600
3444	221897
3444	221918
3444	222131
3444	229017
3444	231140-15
3444	231374
3444	232678
3444	234612
3444	234758
3444	234763
3444	235192
3444	235205
3444	235206
3444	237025
3444	237223
3444	238368
3444	239970
3444	240401
3444	242758
3444	242759
3444	244817
3444	262066
3444	265240
3444	266685
3444	267898
3444	267899
3444	268044L
3444	270819
3444	275131
3444	276331
3444	290458LGD
3444	299371-1
3444	300113
3444	301591
3444	302342
3444	304581B
3444	308449
3444	309442
3444	320529
3444	321517
3444	321573
3444	325079
3444	330803
3444	333241
3444	334724-1
3444	335288-4
3444	335774
3444	346015-24
3444	346016-14
3444	346016-18
3444	346016-19
3444	346016-25
3444	346016-26
3444	00352080
3444	379101K
3444	388060
3444	405457
3444	433836
3444	00444028-014
3444	489751-025
3444	489751-026
3444	489751-029
3444	489751-030
3444	489751-040
3444	489751-041
3444	489751-107
3444	489751-122
3444	489751-125
3444	489751-166
3444	489751-172
3444	514023
3444	515043S
3444	515045S
3444	533802
3444	552308
3444	570000-5452
3444	570000-5503
3444	570004-503
3444	570005-452
3444	570005-503
3444	570009-01-504
3444	572683
3444	0573066
3444	0573202
3444	0573418
3444	0573430
3444	0573460
3444	573467
3444	0573468
3444	0573469
3444	0573469H
3444	0573479H
3444	0573481H
3444	0573490
3444	0573523
3444	0573529
3444	0573556
3444	0573981
3444	581034A
3444	581054
3444	581055
3444	600080-413-001
3444	600080-413-002
3444	600098-413-001
3444	601122
3444	602113(SHARP)
3444	602909-2A
3444	603122
3444	604122
3444	610045-3
3444	610045-4
3444	610045-5
3444	610070
3444	610070-1
3444	610070-2
3444	610070-3
3444	610070-4
3444	610076
3444	610076-1
3444	610076-2
3444	610077
3444	610077-1
3444	610077-2
3444	610077-3
3444	610077-4
3444	610077-5
3444	610077-6
3444	610078
3444	610078-1
3444	610078-2
3444	610094-1
3444	610094-2
3444	610132
3444	610132-1
3444	610142-2
3444	610142-3
3444	610142-4
3444	610142-5
3444	610142-7
3444	610143-3
3444	610146-3
3444	610146-5
3444	610147-1
3444	610148-2
3444	610148-2A
3444	610150-2
3444	610151-2
3444	610151-4
3444	610165-1
3444	610165-2
3444	610167-1
3444	610167-2
3444	610168-1
3444	610168-2
3444	610232-1
3444	611428
3444	615093-2
3444	615179-1
3444	615179-2
3444	618072
3444	618126-1
3444	618217-2
3444	618810-2
3444	619006
3444	619006-7
3444	651955
3444	651995-1
3444	651995-2
3444	651995-3
3444	656204
3444	700047-47
3444	700047-49
3444	700181
3444	700230-00
3444	700231-00
3444	702884
3444	720236
3444	720240
3444	740437
3444	757008-02
3444	760236
3444	760239
3444	760251
3444	785278-101
3444	800132-001
3444	803182-5
3444	803369-6
3444	803372-0
3444	803373-0
3444	803733-0
3444	803733-3
3444	803735-3
3444	810000-373
3444	815133
3444	815134
3444	815171
3444	815171D
3444	815174
3444	815174L
3444	815182
3444	815183
3444	815184
3444	815184E
3444	815186
3444	815186C
3444	815186L
3444	815190
3444	815191
3444	815198
3444	815201
3444	815202
3444	815210
3444	815212
3444	815227
3444	815233
3444	815237
3444	845050
3444	848082
3444	883802
3444	911743-1
3444	916009
3444	916028
3444	916030
3444	916031
3444	916050
3444	916059
3444	916091
3444	928103-1
3444	930236
3444	941295-2
3444	941295-3
3444	943720-001
3444	954330-2
3444	959492-2
3444	961544-1
3444	965000
3444	970247
3444	970250
3444	970252
3444	970659
3444	970660
3444	970661
3444	970662
3444	970916
3444	970916-6
3444	972155
3444	972156
3444	972214
3444	972215
3444	980147
3444	980440
3444	982231
3444	982510
3444	982511
3444	982512
3444	983097
3444	983743
3444	984197
3444	984198
3444	984222
3444	984224
3444	984286
3444	984590
3444	984591
3444	984593
3444	984686
3444	984687
3444	984745
3444	984854
3444	984879
3444	985098
3444	985099
3444	985100
3444	985101
3444	985102
3444	985543
3444	986542
3444	986636
3444	987010
3444	988003
3444	988991
3444	995016
3444	995017
3444	995870-1
3444	995870-3
3444	996746
3444	1022612
3444	1127859
3444	1221962
3444	1222123
3444	1222133
3444	1222424
3444	1261915-383
3444	1320135
3444	1320135A
3444	1320135BC
3444	1320135C
3444	1417302-1
3444	1417306-2
3444	1417306-4
3444	1417306-5
3444	1417312-1
3444	1417312-2
3444	1417340-2
3444	1471113-2
3444	1471113-3
3444	1471115-1
3444	1471115-12
3444	1471120-15
3444	1471120-7
3444	1471120-8
3444	1471120-8-9
3444	1472495-1
3444	1473500-1
3444	1473505-1
3444	1473536-001
3444	1473536-1
3444	1473539-1
3444	1473546-3
3444	1473548-1
3444	1473550-1
3444	1473554-1
3444	1473555-1
3444	1473556-1
3444	1473557-1
3444	1473572-1
3444	1473582-1
3444	1473589-1
3444	1473601-1
3444	1473601-2
3444	1473622-1
3444	1473626-1
3444	1473631-1
3444	1476188-1
3444	1501883
3444	1522237-20
3444	1563295-101
3444	1596408
3444	1617510-1
3444	1690019-01
3444	1700008
3444	1700019
3444	1780145-1
3444	1780145-2
3444	1780145-2-001
3444	1780724-1
3444	1780738-1
3444	1815041
3444	1815042
3444	1815043
3444	1815054
3444	1815154
3444	1817005
3444	1817007
3444	1817108
3444	1840399-1
3444	1846282-1
3444	1851515
3444	1950039
3444	1960023
3444	1960177-2
3444	1968958
3444	2000646-103
3444	2000646-107
3444	2006514-60
3444	2006582-25
3444	2006607-60
3444	2006613-77
3444	2006623-145
3444	2006681-96
3444	2010088-49
3444	2010499-52
3444	2041614
3444	2092417-0719
3444	2092417-0720
3444	2092417-0721
3444	2092417-0724
3444	2092417-0725
3444	2092417-17
3444	2092417-18
3444	2092417-19
3444	2092605-0705
3444	2092608-22
3444	2092609-0022
3444	2092609-0023
3444	2092609-0024
3444	2092609-0026
3444	2092609-0027
3444	2092609-0028
3444	2092609-0705
3444	2092609-0706
3444	2092609-0707
3444	2092609-0713
3444	2092609-0715
3444	2092609-0718
3444	2092609-0720
3444	2092609-0721
3444	2092609-1
3444	2092609-2
3444	2092609-3
3444	2092609-5
3444	2093308-0701
3444	2093308-0702
3444	2093308-0703
3444	2093308-0708
3444	2320022
3444	2320111
3444	2320123
3444	2320413
3444	2320441
3444	2320591-1
3444	2320696
3444	2320696-1
3444	2326953
3444	2327023
3444	2327122
3444	2327363
3444	2360924-5601
3444	2469749
3444	2469755
3444	2479692
3444	2479836
3444	2485078-1
3444	2485078-2
3444	2485078-3
3444	2485079-1
3444	2485079-2
3444	2485079-3
3444	2495166-2
3444	2495166-4
3444	2495166-8
3444	2495166-9
3444	2495521-1
3444	2495522-4
3444	2496125-2
3444	2498457-2
3444	2498904-3
3444	2498904-4
3444	2498904-6
3444	2505209
3444	2520063
3444	2530733
3444	2621567-1
3444	2621764
3444	2640843-1
3444	2712080
3444	3005861
3444	3068305-2
3444	3201104-10
3444	3403787
3444	3457107-1
3444	3457632-5
3444	04440028-001
3444	04440028-003
3444	004440028-007
3444	004440028-008
3444	004440028-010
3444	04440028-013
3444	004440028-014
3444	04440052-001
3444	04440052-002
3444	04450002-001
3444	04450002-004
3444	04450002-005
3444	4813466
3444	4906071
3444	4906072
3444	4906073
3444	5294477-1
3444	5294477-2
3444	5320023H
3444	5320024
3444	5320026
3444	5320064H
3444	05320074H
3444	5320241
3444	5320372
3444	5320372H
3444	5320373
3444	6212839
3444	6218945
3444	6984590
3444	6984600
3444	6993650
3444	7002453
3444	7011507
3444	7011507-00
3444	7026014
3444	7026015
3444	7026016
3444	7026020
3444	7071021
3444	7071031
3444	7121105-01
3444	7284137
3444	7286858
3444	7287452
3444	7295197
3444	7296314
3444	7296811
3444	7297053
3444	7297054
3444	7303120
3444	7304380
3444	7314584
3444	7570004
3444	7570004-01
3444	7570005
3444	7570005-01
3444	7570005-02
3444	7570005-03
3444	7570008
3444	7570008-01
3444	7570008-02
3444	7570009-01
3444	7570009-21
3444	7840540-1
3444	7851324-01
3444	7851325
3444	7851326
3444	7851327
3444	7851379-01
3444	7851380-01
3444	7851949-01
3444	7851950-01
3444	7851952-01
3444	7851953-01
3444	7852454
3444	7852454-01
3444	7852455-01
3444	7852459-01
3444	7852781-01
3444	7853092-01
3444	7853094-01
3444	7853463-01
3444	7853464-01
3444	7853465-01
3444	7855291-01
3444	7855292-01
3444	7855293-01
3444	7855294-01
3444	7910584-01
3444	7910585-01
3444	7910586-01
3444	7910587-01
3444	7910804-01
3444	7936256
3444	7936331
3444	8031839
3444	8033690
3444	8033696
3444	8033720
3444	8033730
3444	8033944
3444	8037330
3444	8037332
3444	8037333
3444	8037353
3444	8113034
3444	8113051
3444	8113052
3444	8113060
3444	8113134
3444	8114031
3444	8522468-1
3444	9001630
3444	9002159
3444	9008964-01
3444	9176494
3444	10000020
3444	10015595
3444	10022104-101
3444	10106058
3444	10180722
3444	10545502
3444	10644433
3444	10849792
3444	10896074
3444	11198132
3444	11220046/7825
3444	11802400
3444	11802500
3444	12965471
3444	12994851
3444	13035807-1
3444	13037215
3444	13040216
3444	13040313
3444	13040317
3444	13040318
3444	13040357
3444	13083908
3444	14714760
3444	14714786
3444	15038433
3444	15039456
3444	15039464
3444	16270092
3444	16520001-1
3444	16797300
3444	16797301
3444	18179900
3444	20025153-77
3444	20030703-0701
3444	20052600
3444	23114046
3444	23114118
3444	23114119
3444	23114155
3444	23114212
3444	23114214
3444	23114255
3444	23114275
3444	23114276
3444	23114296
3444	24501000
3444	24553600
3444	24562000
3444	24562001
3444	24562101
3444	24562200
3444	25114116
3444	25114121

SK	Industry Standard No.
3444	26004001
3444	26010020
3444	27125080
3444	27125090
3444	27125140
3444	27125160
3444	27125250
3444	27125270
3444	27125300
3444	27125370
3444	27125380
3444	27125460
3444	27125470
3444	32600025-01-08A
3444	35393060-01
3444	35393060-02
3444	35393060-03
3444	36171100
3444	43021017
3444	43021083
3444	43022861
3444	43023212
3444	43023221
3444	43023844
3444	43024972
3444	43025055
3444	43025056
3444	43025972
3444	43027379
3444	43027620
3444	44011001
3444	44090004
3444	50210104
3444	50210300-00
3444	50210300-01
3444	50210300-11
3444	50210310-10
3444	50210310-11
3444	50210510
3444	50210800-00
3444	50210800-01
3444	50210800-02
3444	50210800-11
3444	51003059
3444	51003092
3444	51122245
3444	51581300
3444	55440048-001
3444	56301500
3444	57000901-504
3444	59700278
3444	62084960
3444	62236302
3444	62379995
3444	62389699
3444	62506334
3444	62506377
3444	62537140
3444	62539283
3444	62563362
3444	62593687
3444	62593707
3444	62618663
3444	62638214
3444	62638230
3444	62638281
3444	62638303
3444	62638311
3444	62652381
3444	62675004
3444	62675012
3444	62691263
3444	62711914
3444	62733942
3444	62737409
3444	62737476
3444	62737484
3444	62737492
3444	62766220
3444	62785225
3444	62793163
3444	75857322
3444	80050100
3444	80051500
3444	80052102
3444	80052202
3444	80052302
3444	80052600
3444	80053001
3444	80053400
3444	80318250
3444	82409501
3444	83037204
3444	83073205
3444	83073206
3444	83073304
3444	83073305
3444	83073504
3444	83094502
3444	86401000
3444	89942601
3444	89962306
3444	89962307
3444	89962308
3444	89962404
3444	89963008
3444	89963009
3444	93037230
3444	93037240
3444	93063240
3444	93063270
3444	93063280
3444	93064450
3444	93073540
3444	93082830
3444	93082840
3444	93094502
3444	93938040
3444	94824101
3444	120004883
3444	436005001
3444	485134922
3444	485134923
3444	485134924
3444	485134925
3444	485134926
3444	570004503
3444	570005452
3444	570005503
3444	881250108
3444	881250109
3444	1522223720
3444	1611819064
3444	1760089001
3444	3539306001
3444	3539306002
3444	3539306003
3444	4104206440
3444	4109208284
3444	4360021001
3444	6621003100
3444	6621003200
3444	6621003400
3444	8001200001
3444	16102190693
3444	16102190930
3444	16104191168
3444	16105190536
3444	16106190537
3444	16109209536
3444	16116190634
3444	16147191229
3444	16171190693
3444	16171190858
3444	16171191368
3444	16172190693
3444	16172190858
3444	16307190632
3444	16377190632
3444	57000901504
3444	134800869525
3444	310720490000
3444	310720490010
3444	310720490020
3444	310720490100
3444	310720490150
3444	404100900160
3445	8000-00058-009
3448	A066-110
3448	A068-106
3448	A068-107
3448	A11744
3448	A11745
3448	A192
3448	A5T3821
3448	BF244A
3448	BF244B
3448	BF244C
3448	BF256
3448	BFS68
3448	BFS68P
3448	BFW10
3448	BFW11
3448	BFW12
3448	BFW54
3448	BFW55
3448	BFW56
3448	BFW61
3448	CDC731
3448	DDCY002002
3448	DS88
3448	EA15X165
3448	EA15X192
3448	EA15X193
3448	EQF-4
3448	F1462
3448	F1463
3448	FT34Y
3448	GRF-3
3448	HF200191A
3448	HF200191A-0
3448	HF200191B-0
3448	HF200411B
3448	HF200411C0
3448	IC743046
3448	IP20-0011
3448	IP20-0012
3448	JE1033B
3448	JF1033
3448	JF1033B
3448	JF1033G
3448	K17
3448	K19
3448	K19GC
3448	K19GR
3448	K19Y
3448	K30Y
3448	K31C
3448	K33E
3448	K33F
3448	K33GR
3448	K34
3448	K34C
3448	K34D
3448	K47M
3448	K49F
3448	K49H
3448	K49I
3448	KA4559
3448	LS5484
3448	MK10
3448	MK10-2
3448	MPF102
3448	MPF106
3448	MPF107
3448	MPF111
3448	NR7916
3448	NS316
3448	PL1091
3448	PL1092
3448	PL1093
3448	PL1094
3448	PTC151
3448	PTC152
3448	PTC161
3448	Q-00169
3448	Q-00169A
3448	Q-00169B
3448	Q-00169C
3448	Q-00184R
3448	QRF-3
3448	QRG-3
3448	RS-2028
3448	RS7916
3448	RT8667
3448	RVTMK10-2
3448	RVTMK10-E
3448	S1241N
3448	S1242N
3448	SES3819
3448	SK19
3448	SS3534-4
3448	SS3704
3448	SX3819
3448	T1-741
3448	T1208
3448	T1S58
3448	T1S59
3448	T1S88
3448	TE-500-E
3448	TI741
3448	TIS34
3448	TIS42
3448	TIS58
3448	TIS59
3448	TIS88
3448	TNJ61672(2SK25)
3448	TR-FET-1
3448	TR-U1650E
3448	TR-U1835E
3448	TR06011
3448	TR06014
3448	TR228735120325
3448	TR5528
3448	TUS34
3448	TV-83
3448	U1282
3448	UC210
3448	UC734
3448	UC734E
3448	V183
3448	W1P
3448	WF2
3448	2N3819
3448	2SK13
3448	2SK17
3448	2SK17-0
3448	2SK170
3448	2SK170R
3448	2SK17A
3448	2SK17B
3448	2SK17BL
3448	2SK17R
3448	2SK17Y
3448	2SK23
3448	2SK23A
3448	2SK25
3448	2SK25C
3448	2SK25D
3448	2SK25E
3448	2SK25ET
3448	2SK25F
3448	2SK25G
3448	2SK304
3448	2SK31
3448	2SK31C
3448	2SK34
3448	2SK34(E)
3448	2SK34A
3448	2SK34B
3448	2SK34C
3448	2SK34D
3448	2SK34E
3448	2SK37
3448	2SK37H
3448	2SK37K
3448	2SK37L
3448	2SK68
3448	2SK68-L
3448	2SK68A
3448	2SK68AL
3448	2SK68AM
3448	2SK68L
3448	2SK68M
3448	2SK68Q
3448	2SK68Y
3448	03A09
3448	3S30B
3448	3SK44
3448	07-07158
3448	09-304017
3448	09-305014
3448	09-305032
3448	09-309074
3448	10-004
3448	13-22690-1
3448	13-22692-1
3448	13-22692-2
3448	13-28654-1
3448	13-28654-2
3448	13-28654-3
3448	13-34375-1
3448	13-34378-1
3448	14-700-03
3448	14-700-04
3448	14-700-05
3448	14-700-06
3448	14-710-21
3448	14-713-31
3448	14-713-32
3448	14-714-13
3448	19-020-115
3448	020-1112-006
3448	21A113-002
3448	21M412
3448	21M534
3448	44T-300-105
3448	44T-300-113
3448	48-3003A09
3448	63-13926
3448	065-001
3448	065-002
3448	90-55
3448	90-62
3448	106M
3448	121-731
3448	121-756
3448	121-858
3448	132-049
3448	173-1(SYLVANIA)
3448	182-009-9-001
3448	182-015-9-001
3448	182-021-9-001
3448	654-1(SYLVANIA)
3448	921-231B
3448	1009-127
3448	1041-70
3448	1934-17
3448	2032-36
3448	2335-17
3448	3511(SEARS)
3448	3511(WARDS)
3448	3512(RCA)
3448	3512(SEARS)
3448	3512(WARDS)
3448	3819(RCA)
3448	5096
3448	6013
3448	8000-0004-P080
3448	8000-0004-P081
3448	8840-161
3448	18410-141
3448	28810-171
3448	36582
3448	88060-141
3448	127214
3448	207417
3448	489751-208
3448	516009S
3448	760268
3448	916082
3448	2327132
3448	5320032
3448	5321502
3448	7853090-01
3448	30400021
3448	55460014-001
3448	80053501
3449	ECG297
3449	GE-271
3449	REN297
3449	TM297
3449	2SC1166-R
3449	2SC1509
3449	2SC1509P
3449	2SC1509Q
3449	2SC1509R
3449	2SC1509S
3449	121-1035
3449	1417318-2
3449/297	MPS9696
3449/297	MPS9696G
3449/297	2SC1166
3449/297	2SC1166-0
3449/297	2SC1166-O
3449/297	2SC1166D
3449/297	2SC1166GR
3449/297	2SC11660
3449/297	2SC1166Y
3449/297	2SC1627
3449/297	2SC1627-0
3449/297	2SC1627AY
3449/297	2SC2021
3449/297	2SC2021Q
3449/297	2SC2021S
3449/297	2SD467
3449/297	2SD467B
3449/297	2SD467C
3449/297	121-1020
344U	2SB536
3450	A707V
3450	A777
3450	A777R
3450	BC238-16
3450	FD-1029-JB
3450	GE-272
3450	MPS9466AT
3450	MPS9467T
3450	MPS9468
3450	MPS9468S
3450	MPS9468T
3450	MPS9476AT
3450	2SA707
3450	2SA707V
3450	2SA777
3450	2SA777P
3450	2SA777Q
3450	2SA777R
3450	2SA777S
3450	2SA817
3450	2SB561
3450	2SB561B
3450	2SB561C
3450	86-610-2
3450	121-1036
3450	1044-6888
3450	2057A2-446
3450	171556
3450	1417303
3450	43029485
3450/298	A661
3450/298	A661GR
3450/298	A661R
3450/298	A661Y
3450/298	A720
3450/298	A720P
3450/298	A720Q
3450/298	A720R
3450/298	A720S
3450/298	A778
3450/298	2SA661
3450/298	2SA661-0
3450/298	2SA661GR
3450/298	2SA661R
3450/298	2SA661Y
3450/298	2SB726S
3450/298	121-1021
3451	GEIC-240
3452	A-1854-0485
3452	A-1854-JBD1
3452	A054-157
3452	A054-158
3452	A054-159
3452	A054-163
3452	A054-164
3452	A054-470
3452	A068-111
3452	A068-112
3452	A1109
3452	A1170
3452	A121-480
3452	A124623
3452	A124624
3452	A125329
3452	A129509
3452	A129510
3452	A129511
3452	A129512
3452	A129513
3452	A129571
3452	A129572
3452	A129573
3452	A129574
3452	A14-602-63
3452	A14-603-05
3452	A14-603-06
3452	A1462
3452	A1518
3452	A1519
3452	A1520
3452	A1521
3452	A176-025-9-002
3452	A19-020-072
3452	A1E
3452	A1G-1A
3452	A1K
3452	A1P-5
3452	A1R-1/4925
3452	A1R-1A
3452	A1R-2/4926
3452	A1R-2A
3452	A1R-5
3452	A1R/4925A
3452	A1R/4926A
3452	A2006681-95
3452	A2092693-0724
3452	A2092693-0725
3452	A24
3452	A245
3452	A2498
3452	A24MW594
3452	A24MW595
3452	A24MW596
3452	A24MW597
3452	A24T-016-016
3452	A2C
3452	A2D
3452	A2F
3452	A2G
3452	A2H
3452	A2L
3452	A2N-2A
3452	A2P-5
3452	A2T
3452	A2T919
3452	A2V
3452	A2W
3452	A31-0206
3452	A32-2809
3452	A3A
3452	A3D
3452	A3H
3452	A3N71
3452	A3N72
3452	A3N73
3452	A3P
3452	A3R
3452	A3SB
3452	A418
3452	A419
3452	A420
3452	A427
3452	A4789
3452	A48-134789
3452	A48-134837
3452	A48-134845
3452	A48-134902
3452	A48-134904
3452	A48-134922
3452	A48-134923
3452	A48-134924
3452	A48-134925
3452	A48-134926
3452	A48-134945
3452	A48-134961
3452	A48-134962
3452	A48-134963
3452	A48-134964
3452	A48-134965
3452	A48-134966
3452	A48-134981
3452	A48-137071
3452	A48-137075
3452	A48-137076
3452	A48-137077
3452	A48-137197
3452	A48-40247G01
3452	A48-43351A02
3452	A48-63076A82
3452	A48-97046A05
3452	A48-97046A06
3452	A48-97046A07
3452	A48-97127A06
3452	A48-97127A12
3452	A48-97127A18
3452	A480
3452	A4851
3452	A4E-5
3452	A4G
3452	A4T
3452	A4Y-1A
3452	A57A144-12
3452	A57A145-12
3452	A5J
3452	A62-19581
3452	A642-254
3452	A642-260
3452	A642-268
3452	A6B
3452	A6E
3452	A6F
3452	A6P
3452	A6T
3452	A6U
3452	A6V-5
3452	A772738
3452	A772739
3452	A7A30
3452	A7A31
3452	A7A32
3452	A7N
3452	A7P
3452	A7U
3452	A7V
3452	A7W
3452	A8S
3452	A909-27125-160
3452	A90T2
3452	A916-31025-5B
3452	A921-59B
3452	A921-62B
3452	A921-63B
3452	A921-64B
3452	A991-01-1316
3452	A9A
3452	A9D
3452	AR200W
3452	AR201
3452	AR201Y
3452	AR202
3452	AR202G
3452	AR212
3452	AR221
3452	AR313
3452	AT520
3452	AX91770
3452	BC111
3452	BC112
3452	BC155A
3452	BC156A

SK	Industry Standard No.	SK	Industry Standard No.	SK	Industry Standard No.	SK	Industry Standard No.	SK	Industry Standard No.
3452	BC194	3452	C81252B	3452	FCS1227G	3452	M9575	3452	R87225
3452	BC194B	3452	C81252C	3452	FC81227G81O	3452	ME3001	3452	R87227
3452	BC195	3452	C1292	3452	FC89018D	3452	ME5001	3452	R87228
3452	BC195CD	3452	C81340D	3452	FC89018E	3452	ME8201	3452	R87229
3452	BC6500	3452	C81340E	3452	FC89018F	3452	ME9021	3452	R87230
3452	BCW31	3452	C81340F	3452	FC89018H	3452	MHM1001	3452	R87231
3452	BCW31R	3452	C81340G	3452	FC890188H	3452	MHM1101	3452	R87237
3452	BCW32	3452	C81340H	3452	FC89066	3452	MM1367/28C684	3452	R87333
3452	BCW32R	3452	C81350	3452	FK2369A	3452	MM1382	3452	R87334
3452	BCW71	3452	C81351	3452	FK2484	3452	MM1945	3452	R87520
3452	BCW71R	3452	C81508G	3452	FK3014	3452	MP83536	3452	R87522
3452	BCW72	3452	C81509E	3452	FK3299	3452	MP86511	3452	R87524
3452	BCW72R	3452	C81509F	3452	FK3300	3452	MS7001T	3452	R87532
3452	BCY87	3452	C81519E	3452	FK914	3452	MS7501T	3452	R87533
3452	BCY88	3452	C81555	3452	FK918	3452	MS7502T	3452	R13069
3452	BCY89	3452	C81661	3452	FM1613	3452	MS8R7502	3452	R13070
3452	BD71	3452	C81834	3452	FM1711	3452	MS87501	3452	R13095
3452	BF162	3452	C82006G	3452	FM2369	3452	MS87502	3452	R13225
3452	BF163	3452	C82008G	3452	FM2369	3452	MS87501	3452	R13226
3452	BF164	3452	C82008H	3452	FM2846	3452	MT101	3452	R13227
3452	BF165	3452	C8429J	3452	FM3014	3452	MT107	3452	R13232
3452	BF175A	3452	C8430H	3452	FM708	3452	MT743	3452	R15061
3452	BF176	3452	C86225F	3452	FM709	3452	MT744	3452	R15200
3452	BF187	3452	C86226F	3452	FMT20A	3452	MT753	3452	R15201
3452	BF188	3452	C86227E	3452	FM870	3452	N-EA15X131	3452	R15205
3452	BF194A	3452	C86227F	3452	FM871	3452	NPC173	3452	R15900
3452	BF223	3452	C89001	3452	FM910	3452	NPC188	3452	R15902
3452	BF233-4	3452	C89018	3452	FM911	3452	NS1356	3452	R15903
3452	BF235	3452	C89018/3490	3452	FM914	3452	NS3300	3452	R15904
3452	BF236	3452	C89018E	3452	FPR40-1003	3452	NS381	3452	R16157
3452	BF262	3452	C89018F	3452	FPR50-1003	3452	NS382	3452	R16158
3452	BF263	3452	C89018F/3490	3452	FPR50-1004	3452	NS6112	3452	R16159
3452	BF264	3452	C89018G	3452	FS32666	3452	NS6113	3452	R16160
3452	BF357	3452	C89124-C2	3452	FS326690	3452	NS7261	3452	R16203
3452	BF817	3452	C89125-B1	3452	FS35529	3452	NS7267	3452	R16660
3452	BF817R	3452	C8918	3452	FS3683	3452	NTC-5	3452	R77320
3452	BFV83	3452	CXL108	3452	FSE1001	3452	P346	3452	R77321
3452	BFV83A	3452	D058	3452	FSE3001	3452	PBC182	3452	R77323
3452	BFV85D	3452	D069	3452	FSE5002	3452	PEP1001	3452	R77324
3452	BFV85E	3452	D087	3452	FSP-1	3452	PET-101-1	3452	RV1467
3452	BFV85F	3452	D088	3452	FSP-164	3452	PET1075	3452	RV1468
3452	BFV85G	3452	D1666	3452	FSP-165	3452	PET3001	3452	RV1469
3452	BFW64	3452	D24A3394	3452	FSP-166	3452	PET8201	3452	RV1470
3452	BFX18	3452	D562	3452	FSP-166-1	3452	PET8250	3452	RVT280645
3452	BFX19	3452	E629	3452	FSP-215	3452	PET8251	3452	RVT281384
3452	BFX20	3452	EA0013	3452	FSP-242-1	3452	PET8300	3452	S-1037
3452	BFX21	3452	EA0091	3452	FSP-270-1	3452	PL1053	3452	S-1058
3452	BFX45	3452	EA0093	3452	FSP-289-1	3452	PL1055	3452	S-1059
3452	BFY78	3452	EA0094	3452	FSP-42	3452	PL1061	3452	S-1062
3452	BFY90B	3452	EA0095	3452	FSP-42-1	3452	PL1062	3452	S-1078
3452	BSV35	3452	EA1343	3452	FT1315	3452	PL1063	3452	S-1153
3452	BSV35B	3452	EA1562	3452	FT1324B	3452	PL1064	3452	S-1227
3452	BSV35C	3452	EA1563	3452	FT1324C	3452	PL1065	3452	S-1276
3452	BSV35D	3452	EA15X113	3452	FT709	3452	PL1081	3452	S-1296
3452	BSV52	3452	EA15X131	3452	FV2369A	3452	PL1082	3452	S-1316
3452	BSV52R	3452	EA15X132	3452	FV2484	3452	PMT1767	3452	S-1317
3452	BSX12	3452	EA15X48	3452	FV3014	3452	Q-00484R-1	3452	S-1318
3452	BSX26	3452	EA15X49	3452	FV3299	3452	Q-00584R-3	3452	S-1360
3452	BSX27	3452	EA15X50	3452	FV3300	3452	QSS3001	3452	S-1361
3452	BSX35	3452	EA15X51	3452	FV914	3452	R-280545	3452	S-1562
3452	BSX92	3452	EA15X54	3452	FV918	3452	R-280668	3452	S0016
3452	BSX93	3452	EA15X55	3452	G05-004A	3452	R06-1001	3452	S0021
3452	BSY22	3452	EA15X177	3452	GME3001	3452	R06-1002	3452	S1009
3452	BSY23	3452	EA15X7228	3452	GME3002	3452	R06-1003	3452	S1019
3452	BE950	3452	EA15X7243	3452	GME9001	3452	R06-1004	3452	S1044
3452	C1023	3452	EA15X7263	3452	GME9002	3452	R06-1005	3452	S1076
3452	C111B	3452	EA15X7587	3452	GME9021	3452	R06-1006	3452	S1078
3452	C185V	3452	EA15X7722	3452	GME9022	3452	R2473	3452	S1079
3452	C2475078-3	3452	EA15X8608	3452	H1V	3452	R2476	3452	S1142
3452	C263	3452	EA15X8610	3452	H442	3452	R2477	3452	S1313
3452	C269	3452	EA15X94	3452	H9625	3452	R62194	3452	S1408
3452	C271	3452	EA1733	3452	HC-00668	3452	R81001	3452	S1409
3452	C272	3452	EA1793	3452	HC-00772	3452	R81002	3452	S1636
3452	C289	3452	EA2493	3452	HC-01830	3452	RR8070	3452	S1674
3452	C3123	3452	EA2495	3452	HC668	3452	RR8116	3452	S1674A
3452	039A	3452	EA2600	3452	HC772	3452	RR8118	3452	S1682
3452	C40	3452	EA2601	3452	HEP-80016	3452	RR8119	3452	S2159
3452	C5359	3452	EA2602	3452	HEP-80021	3452	RS8989	3452	TI862
3452	C561	3452	EA2603	3452	HR58	3452	RS-109	3452	TI863
3452	C613	3452	EA2604	3452	HR59	3452	RS-2015	3452	TI864
3452	C662	3452	EA2605	3452	HR60	3452	RS-7113	3452	TM108
3452	C688	3452	EA3406	3452	HS-1225	3452	RS-7114	3452	WEP56
3452	C740	3452	ECG108	3452	HS-1226	3452	RS-7115	3452	Z12475
3452	C748	3452	EN718A	3452	HS-1227	3452	RS-7202	3452	Z12857
3452	C79	3452	EN916	3452	HS-40017	3452	RS-7512	3452	Z13269A
3452	C828	3452	EN918	3452	HS-40047	3452	RST101	3452	Z13600
3452	CC8-2006D	3452	EP15X55	3452	HT3037010	3452	RST116	3452	Z1709
3452	CC89016D	3452	EQ8-21	3452	HT3037011AO	3452	RST117	3452	Z1917
3452	CC89016E	3452	ES10187	3452	HT3037201	3452	RST118	3452	Z1918
3452	CC89016F	3452	ES15X10	3452	HT3037201A	3452	RST119	3452	Z1X320
3452	CDC12112C	3452	ES15X18	3452	HT3037201B	3452	RST120	3452	Z1X321
3452	CDC12112D	3452	ES15X19	3452	HT3037B20A	3452	RST122	3452	1-041/2207
3452	CDC12112F	3452	ES15X2	3452	HT303941	3452	RST123	3452	1-801-003-12
3452	CDC5038A	3452	ES15X3	3452	HT303941A	3452	RST124	3452	1-801-003-13
3452	CDC5071A	3452	ES15X30	3452	HT303941B	3452	RST125	3452	1-801-003-13
3452	CDC5075B	3452	ES15X6	3452	HT304540A0	3452	RST128	3452	1-801-003-15
3452	CE4010D	3452	ES15X60	3452	HT304601C0	3452	RST135	3452	1-801-305-13
3452	CE4010E	3452	ES15X69	3452	HT304611B	3452	RST138	3452	1-801-306
3452	CP1	3452	ES3266	3452	HT3053510C	3452	RST139	3452	1-801-306-13
3452	CIL-531	3452	ET15X18	3452	HT306451	3452	RST140	3452	1-801-306-14
3452	CIL-532	3452	ET15X19	3452	HT306451B	3452	RST141	3452	1-801-306-15
3452	CS-1244X	3452	ET15X30	3452	HT306962A-0	3452	RST142	3452	01-9016-42221-3
3452	CS-1359	3452	EU15X1	3452	J108	3452	RST144	3452	01-9018-42221-3
3452	CS-1361E	3452	EU15X2	3452	J187	3452	RST145	3452	002-009600
3452	CS-1386H	3452	EU15X3	3452	J24596	3452	RST161	3452	002-009601
3452	CS-2007G	3452	EU15X6	3452	J24635	3452	RST162	3452	002-9601
3452	CS-2008P	3452	EW162	3452	J24636	3452	RST163	3452	002-9601-12
3452	CS-3001B	3452	F20-1001	3452	J24637	3452	RST164	3452	2N1005
3452	CS-461B	3452	F20-1002	3452	J24813	3452	RST165	3452	2N1060
3452	CS-627Q	3452	F20-1003	3452	J24814	3452	RST166	3452	2N2032
3452	CS1014	3452	F20-1004	3452	J24852	3452	RST167	3452	2N2197
3452	CS1014D	3452	F20-1005	3452	J24863	3452	RST168	3452	2N2616
3452	CS1014E	3452	F215-1001	3452	J24904	3452	RST169	3452	2N2711
3452	CS1014F	3452	F215-1002	3452	J24905	3452	RST170	3452	2N2784
3452	CS1018	3452	F215-1003	3452	J24921	3452	RST174	3452	2N2784/52
3452	CS1168B	3452	F215-1004	3452	J24923	3452	RST175	3452	2N3010
3452	CS1225D	3452	F215-1005	3452	J24933	3452	RST176	3452	2N3035
3452	CS1225E	3452	F2427	3452	K4-510	3452	RST177	3452	2N3562
3452	CS1225F	3452	F2450	3452	KD2119	3452	RST201	3452	2N3564
3452	CS1226	3452	F2633	3452	LM110A	3452	RST209	3452	2N4996
3452	CS1226E	3452	F2634	3452	LM1123H	3452	RST210	3452	2N4997
3452	CS1226F	3452	F2636	3452	LM1138G/H	3452	RST211	3452	2N709/52
3452	CS1226G	3452	F5530	3452	LT1016	3452	RST212	3452	2N709A
3452	CS1226H	3452	F9625	3452	LT1016D	3452	RST214	3452	2N717A
3452	CS1227D	3452	F968	3452	LT1016E	3452	RST215	3452	2N849
3452	CS1227E	3452	FC5006	3452	M024	3452	RST216	3452	2N850
3452	CS1227F	3452	FC81168E	3452	M4733	3452	RST217	3452	2N851
3452	CS1227G	3452	FC81168B641	3452	M4837	3452	RST218	3452	2N852
3452	CS1238	3452	FC81227E	3452	M4845	3452	RST219	3452	2N914/51
3452	CS1238G	3452	FC81227E	3452	M4855	3452	RST220	3452	2S006
3452	CS1238H	3452	FC81227EB814	3452	M4857	3452	RST221	3452	2SC0148
3452	CS1238I	3452	FC81227F	3452	M75547-1			3452	2SC263
3452	CS1243E	3452	FC81227F743	3452	M75547-2			3452	2SC269
								3452	2SC271
								3452	2SC272

SK	Industry Standard No.
3452	2SC286
3452	2SC289
3452	2SC39
3452	2SC39A
3452	2SC40
3452	2SC5359
3452	2SC561
3452	2SC613
3452	2SC63
3452	2SC662
3452	0002SC668D
3452	2SC79
3452	2SCF1
3452	2SCS184J
3452	2SCS429J
3452	2SCS430H
3452	2SD562
3452	2SE629
3452	2SQ371
3452	03-460C
3452	03-461B
3452	03-535A
3452	3L4-6007-51
3452	4-0485
3452	4-3022861
3452	4-3025763
3452	4-3025764
3452	4-3025765
3452	4-3025767
3452	4-399
3452	4-400
3452	4-433
3452	4-434
3452	4-443
3452	4-684120-3
3452	4-685285-3
3452	4-686107-3
3452	4-686108-3
3452	4-686112-3
3452	4-686114-3
3452	4-686118-3
3452	4-686119-3
3452	4-686120-3
3452	4-686126-3
3452	4-686127-3
3452	4-686131-3
3452	4-686140-3
3452	4-686169-3
3452	4-686171-3
3452	4-686172-3
3452	4-686207-3
3452	4-686208-3
3452	4-686209-3
3452	4-686224-3
3452	4-686228-3
3452	4-686244-3
3452	4-686251-3
3452	4-68695-3
3452	6A12677
3452	6A12679
3452	7-4
3452	7-44
3452	7-59-0193477
3452	7-59-0203477
3452	7-59-0213477
3452	7-59-0223477
3452	7-59-0233477
3452	7-6
3452	7-7
3452	8-0024-1
3452	8-0024-2
3452	8-0053600
3452	8-0338030
3452	8-0338040
3452	8-0339430
3452	8-0339440
3452	8-0383840
3452	8-0383930
3452	09-302005
3452	09-302009
3452	09-302010
3452	09-302017
3452	09-302060
3452	09-302115
3452	09-302141
3452	09-302149
3452	09-302162
3452	09-302190
3452	09-305069
3452	09-305070
3452	09-305071
3452	09-305072
3452	09-305074
3452	09-309007
3452	09-309013
3452	09-309024
3452	09-309027
3452	09-309028
3452	09-309032
3452	09-309073
3452	09-32124
3452	9-5125
3452	9-5126
3452	9-5127
3452	9-5128
3452	9-5129
3452	9-5130
3452	9-5131
3452	9-5223
3452	10-080009
3452	10-2SC80
3452	10-2SC94
3452	10B1051
3452	10B1055
3452	10B551
3452	10B553
3452	10B555
3452	10B556
3452	10C573
3452	10C574
3452	10G1051
3452	10G1052
3452	10H1051
3452	10H1053
3452	10H551
3452	10H553
3452	11B1052
3452	11B1055
3452	11B551
3452	11B552
3452	11B554
3452	11B555
3452	11C1051
3452	11C1053
3452	11C1057
3452	11C551
3452	11C553
3452	11C557
3452	13-0321-14
3452	13-0321-15
3452	13-0321-16
3452	13-0321-17
3452	13-0321-21
3452	13-1032-5
3452	13-10321-1
3452	13-10321-10
3452	13-10321-11
3452	13-10321-12
3452	13-10321-14
3452	13-10321-15
3452	13-10321-16
3452	13-10321-17
3452	13-10321-2
3452	13-10321-20
3452	13-10321-21
3452	13-10321-26
3452	13-10321-30
3452	13-10321-43
3452	13-10321-5
3452	13-10321-51
3452	13-10321-6
3452	13-10321-7
3452	13-10321-77
3452	13-10321-8
3452	13-10321-9
3452	13-14085-1
3452	13-14085-2
3452	13-14085-24
3452	13-14085-27
3452	13-14085-74
3452	13-14085-75
3452	13-14085-76
3452	13-14085-77
3452	13-15810-1
3452	13-15841-1
3452	13-16744-1
3452	13-18949-1
3452	13-18950-1
3452	13-28584
3452	13-28584-1
3452	13-31013-4
3452	13-32366-2
3452	13-34045-1
3452	13-34045-2
3452	13-35550
3452	13-55020-1
3452	13-55063-1
3452	13-55065-1
3452	13-67583-5
3452	14-32430
3452	14-602-41
3452	14-603-12
3452	14-609-49A
3452	015
3452	15-088004
3452	15-08800U
3452	16-736
3452	16J1
3452	16J2
3452	16K1
3452	16K2
3452	16K3
3452	16L2
3452	16L22
3452	16L23
3452	16L3
3452	16L5
3452	19-020-037
3452	19-020-052
3452	19-020-44
3452	19-19420
3452	19A115249-1
3452	19A115342-1
3452	19A115440-1
3452	19A115440-2
3452	19A115441-1
3452	19A115666-1
3452	19A115925-1
3452	19A123160-1
3452	19A123160-2
3452	020-00026
3452	020-00027
3452	20-00229-001
3452	20-00444-001
3452	20-1
3452	21A015-004
3452	21A015-014
3452	21A040-003
3452	21A040-004
3452	21A040-007
3452	21A040-010
3452	21A040-016
3452	21A040-017
3452	21A040-019
3452	21A050-004
3452	21A105-001
3452	21A112-086
3452	21A112-087
3452	21M476
3452	21M481
3452	22-001002
3452	22-001003
3452	22-001004
3452	22-001005
3452	022-3640-080
3452	022.3640-080
3452	23-PT275-122
3452	24-3564
3452	24A1
3452	24MW1038
3452	24MW1082
3452	24MW654
3452	24MW657
3452	24MW675
3452	24MW724
3452	24MW725
3452	24MW739
3452	24MW827
3452	24MW852
3452	24MW865
3452	24T-002
3452	24T-011-008
3452	24T-013-005
3452	24T-016-001
3452	24T-016-005
3452	24T-016-013
3452	24T-016-015
3452	24T-016-016
3452	025-100003
3452	025-100004
3452	025-100009
3452	025-100013
3452	025-100014
3452	25A
3452	25A1262-005
3452	25A1281-001
3452	25AM624
3452	25B1
3452	25C206
3452	25R
3452	31-0051
3452	31-0097
3452	31-0103
3452	31-0242
3452	31-0243
3452	33H50
3452	34-6001-3
3452	34-6001-6
3452	34-6015-49
3452	34-6016-17
3452	34E31
3452	34H31
3452	41N1
3452	41N2
3452	41N2A
3452	41N2AA
3452	41N2B
3452	41N3
3452	42-19683
3452	42-28203
3452	42-28204
3452	42-28206
3452	43-022861
3452	43-025763
3452	43-025765
3452	43-025767
3452	045-1(SYLVANIA)
3452	045-2(SYLVANIA)
3452	45NP
3452	46-84120-3
3452	46-85285-3
3452	46-86140-3
3452	46-86208-3
3452	46-86244-3
3452	46-86269-3
3452	46-86314-3A
3452	46-86397-3
3452	46-864-3
3452	48-01-004
3452	48-01-010
3452	48-124804
3452	48-124805
3452	48-124808
3452	48-13470
3452	48-134706
3452	48-134709
3452	48-134713
3452	48-134717
3452	48-134719
3452	48-134724
3452	48-134725
3452	48-134772
3452	48-134773
3452	48-134774
3452	48-134777
3452	48-134779
3452	48-134780
3452	48-134783
3452	48-134784
3452	48-134786
3452	48-134800
3452	48-134806
3452	48-134814
3452	48-134818
3452	48-134820
3452	48-134821
3452	48-134826
3452	48-134827
3452	48-134828
3452	48-134855
3452	48-134879
3452	48-134891
3452	48-134892
3452	48-134893
3452	48-134902
3452	48-134908
3452	48-134937
3452	48-134946
3452	48-134948
3452	48-134949
3452	48-134950
3452	48-134960
3452	48-134979
3452	48-134983
3452	48-134985
3452	48-137004
3452	48-137006
3452	48-137033
3452	48-137055
3452	48-137104
3452	48-137105
3452	48-137126
3452	48-137136
3452	48-137140
3452	48-137144
3452	48-137158
3452	48-137166
3452	48-137190
3452	48-137191
3452	48-137194
3452	48-137196
3452	48-137339
3452	48-137352
3452	48-137355
3452	48-137371
3452	48-137372
3452	48-137375
3452	48-137376
3452	48-137388
3452	48-137483
3452	48-43351A03
3452	48-43351A04
3452	48-43351A05
3452	48-63026A46
3452	48-63077A29
3452	48-65113A88
3452	48-65118A64
3452	48-65123A67
3452	48-65123A95
3452	48-65144A72
3452	48-90232A03
3452	48-90232A04
3452	48-90232A10
3452	48-90232A19
3452	48-97046A04
3452	48-97046A18
3452	48-97046A51
3452	48-97177A04
3452	48-97177A07
3452	48-97177A08
3452	48-971A04
3452	48-97762A02
3452	48-K869575
3452	48P63082A45
3452	48P63082A71
3452	48P65146A63
3452	48P65173A78
3452	48S134902
3452	48S134946
3452	48S134960
3452	48S134970
3452	48S134979
3452	48S137006
3452	48X97046A51
3452	48X97162A01
3452	48X97162A02
3452	48X97162A04
3452	48X97162A09
3452	48X97162A10
3452	49-1
3452	50C1047
3452	50C784
3452	50C829
3452	50C829B
3452	50C829C
3452	51
3452	54BLK
3452	54BRN
3452	54ORN
3452	54RED
3452	56-234
3452	57A10-1
3452	57A10-2
3452	57A101-4
3452	57A107-1
3452	57A107-2
3452	57A107-3
3452	57A107-4
3452	57A107-5
3452	57A107-6
3452	57A107-8
3452	57A10A-8-6
3452	57A134-12
3452	57A139-4-6
3452	57A141-1
3452	57A141-2
3452	57A141-3
3452	57A142-1
3452	57A142-3
3452	57A143-1
3452	57A143-10
3452	57A143-11
3452	57A143-2
3452	57A143-3
3452	57A143-4
3452	57A143-5
3452	57A143-7
3452	57A143-8
3452	57A143-9
3452	57A146-12
3452	57A151-6
3452	57A152-12
3452	57A160-1
3452	57A160-2
3452	57A160-3
3452	57A160-4
3452	57A20-1
3452	57A21-1
3452	57A21-10
3452	57A21-15
3452	57A21-2
3452	57A21-3
3452	57A21-4
3452	57A21-5
3452	57A21-6
3452	57A21-7
3452	57A21-9
3452	57A27-2
3452	57A5-6
3452	57A5-7
3452	57A5-8
3452	57A7-1
3452	57A7-2
3452	57A7-3
3452	57A7-4
3452	57A7-5
3452	57A7-6
3452	57A7-7
3452	57B101-4
3452	57B134-12
3452	57B141-1
3452	57B141-2
3452	57B141-3
3452	57B142-1
3452	57B142-2
3452	57B142-3
3452	57B143-1
3452	57B143-10
3452	57B143-11
3452	57B143-2
3452	57B143-3
3452	57B143-4
3452	57B143-5
3452	57B143-6
3452	57B143-7
3452	57B143-8
3452	57B143-9
3452	57B151-6
3452	57B152-12
3452	57B160-1
3452	57B160-2
3452	57B160-3
3452	57B160-4
3452	57B160-5
3452	57B160-6
3452	57B160-7
3452	57B160-8
3452	57B166-12
3452	57B21
3452	57B21-12
3452	57B21-13
3452	57B21-14
3452	57B21-15
3452	57B21-16
3452	57B21-18
3452	57B21-2
3452	57B21-3
3452	57B21-4
3452	57B21-5
3452	57C10-1
3452	57C10-2
3452	57C20-1
3452	57C5-6
3452	57C5-7
3452	57C5-8
3452	57C7-1
3452	57C7-2
3452	57C7-3
3452	57C7-4
3452	57C7-5
3452	57C7-6
3452	57C7-7
3452	57D107-8
3452	57D24-1
3452	57D24-2
3452	57D24-3
3452	66X0007-104
3452	69N1
3452	72N1
3452	72N2
3452	76-13570-39
3452	76-13570-59
3452	76-13866-17
3452	76-13866-18
3452	76-13866-19
3452	76-13866-20
3452	76-13866-59
3452	76-13866-62
3452	78C01
3452	78C02
3452	80-053600
3452	80-338030
3452	80-338040
3452	80-339430
3452	80-339440
3452	80-383840
3452	80-383930
3452	81-46125006-0
3452	86-138-2
3452	86-243-2
3452	86-244-2
3452	86-245-2
3452	86-386-2
3452	86-416-2
3452	86-417-2
3452	86-467-2
3452	86-488-2
3452	86-490-2
3452	86-491-2
3452	86X0007-004
3452	86X0038-001
3452	86X0043-001
3452	86X6029-001
3452	86X7-6013
3452	87-0027
3452	089-214
3452	089-215
3452	089-216
3452	91A
3452	91B
3452	91BGRN
3452	91F
3452	92N1
3452	92N1B
3452	95-125
3452	95-126
3452	95-127
3452	95-128
3452	95-129
3452	95-130
3452	95-131
3452	95-223
3452	96-056-234
3452	96-138-2
3452	96N927
3452	96N932
3452	96NPT
3452	99S016
3452	99S016-1
3452	99S017
3452	99S018
3452	99S018A
3452	99S019
3452	99S019A
3452	99S019B
3452	99S037
3452	99S090-1
3452	100W1
3452	0101-0060A
3452	0101-0531
3452	101-2(ADMIRAL)
3452	101-3(ADMIRAL)
3452	101-4(ADMIRAL)
3452	0103-0060
3452	0103-0060B
3452	0103-0191
3452	0103-0389
3452	0103-0521
3452	0103-0521B
3452	0103-0531
3452	0103-389
3452	103-4
3452	0103-9531/4460
3452	105-001-08
3452	105-00106-00
3452	105-00108-07
3452	105-005-12
3452	105-02004-09
3452	105-06004-00
3452	105-24191-04
3452	112-520
3452	112-521
3452	112-522
3452	113-398
3452	113-938
3452	114-118
3452	114-267
3452	116-073
3452	116-079
3452	116-080
3452	116-082
3452	116-083
3452	116-198
3452	116-199
3452	116-200
3452	118-1
3452	118-2
3452	118-3
3452	118-4
3452	120-004496

SK	Industry Standard No.
3452	120-004497
3452	120-004723
3452	120-004724
3452	120-004725
3452	120-004881
3452	120-005291
3452	120-005292
3452	120-005293
3452	120-005294
3452	120-005295
3452	120-005296
3452	120-005297
3452	120-005298
3452	121-303
3452	121-316
3452	121-317
3452	121-318
3452	121-318L
3452	121-321
3452	121-345
3452	121-472
3452	121-481
3452	121-482
3452	121-483
3452	121-520
3452	121-546B
3452	121-547
3452	121-551
3452	121-560
3452	121-612-16
3452	121-613
3452	121-613-16
3452	121-614
3452	121-614-9
3452	121-616
3452	121-630
3452	121-637
3452	121-638
3452	121-638B
3452	121-742
3452	121-753
3452	121-819
3452	121-827
3452	121-834
3452	121-835
3452	121-841
3452	121-846
3452	121-848
3452	121-851
3452	121-857
3452	121-869
3452	121-884
3452	121-898
3452	121-899
3452	121-900
3452	121-925
3452	121-932
3452	122-A484
3452	128N4
3452	130-138
3452	130-40304
3452	130-40362
3452	130-40421
3452	130-40459
3452	132-015
3452	139-4
3452	0142
3452	142N6
3452	151-0138
3452	151M11
3452	151N1
3452	151N11
3452	151N116
3452	161T2
3452	162T2
3452	173A04490-1
3452	173A04490-2
3452	176-003
3452	176-003-9-001
3452	176-004-9-001
3452	176-005
3452	176-005-9-001
3452	176-006
3452	176-006-9-001
3452	176-007
3452	176-007-9-001
3452	189
3452	200-007
3452	200-010
3452	200-015
3452	200-055
3452	200-056
3452	201-254323-12
3452	201-254323-13
3452	201-254343-12
3452	207A9
3452	217-1
3452	220-001011
3452	220-001012
3452	223
3452	229-0151-3
3452	229-0180-124
3452	229-0180-149
3452	229-0180-34
3452	229-0185-2
3452	229-0185-3
3452	229-0190-29
3452	229-0192-19
3452	229-0204-23
3452	229-0204-4
3452	229-0210-14
3452	229-0214-40
3452	229-0220-19
3452	229-0220-9
3452	229-0248-45
3452	229-0250-10
3452	229-5100-15U
3452	229-5100-15V
3452	229-5100-224
3452	229-5100-225
3452	229-5100-226
3452	229-5100-228
3452	229-5100-33V
3452	232N2
3452	247-016-013
3452	260P07901
3452	260P08001
3452	260P08401
3452	260P10403
3452	260P10501
3452	260P10502
3452	260P1060
3452	260P10602
3452	260P11101

SK	Industry Standard No.
3452	260P11101A
3452	260P16301
3452	260P16302
3452	260P17201
3452	260P17602
3452	260P17603
3452	260P70502
3452	260Z00109
3452	260Z00209
3452	260Z00309
3452	290V02H69
3452	297V070H49
3452	297V072C01
3452	297V072C03
3452	297V072C04
3452	297V074C09
3452	297V078C01
3452	297V078C02
3452	324-0149
3452	324-0150
3452	324-1
3452	325-0028-84
3452	330-1304-8
3452	344-6000-3
3452	344-6000-3A
3452	344-6015-10
3452	344-6015-11
3452	344-6015-7
3452	344-6015-7A
3452	344-6017-6
3452	386-7118P1
3452	386-7188P1
3452	396-7178P1
3452	417-124
3452	417-125
3452	417-83
3452	421-9683
3452	422-1401
3452	422-1402
3452	422-2532
3452	429-0986-12
3452	430-22861
3452	430-25763
3452	430-25764
3452	430-25765
3452	430-25767
3452	499-1
3452	501ES001M
3452	515-521
3452	537PS
3452	546
3452	551
3452	573-472
3452	573-474
3452	573-474A
3452	573-475
3452	573-491
3452	573-494
3452	573-495
3452	573-507
3452	573-509
3452	576-0001-006
3452	576-0003-001
3452	576-0003-002
3452	576-0003-003
3452	576-0003-004
3452	576-0003-005
3452	576-0003-006
3452	576-0003-007
3452	576-0003-018
3452	576-0003-020
3452	576-0003-021
3452	576-0003-027
3452	576-0003-028
3452	576-0006-011
3452	576-0036-918
3452	576-0036-919
3452	600X0092-086
3452	601-113
3452	602-113
3452	602-61
3452	603-113
3452	604-113
3452	605-113
3452	610-041
3452	610-041-1
3452	610-041-2
3452	610-041-3
3452	610-042
3452	610-042-1
3452	610-045
3452	610-045-1
3452	610-045-2
3452	610-069
3452	610-069-1
3452	610-072
3452	610-072-1
3452	610-072-2
3452	610-073
3452	610-073-1
3452	612-16A
3452	613-72
3452	653-202
3452	660-127
3452	668CS
3452	690V010H41
3452	690V028H28
3452	690V028H48
3452	690V028H69
3452	690V028H89
3452	690V02H69
3452	690V049H81
3452	690V060H58
3452	690V060H59
3452	690V070H49
3452	690V070H98
3452	690V075H68
3452	690V081H07
3452	690V084H94
3452	690V084H95
3452	690V084H96
3452	690V086H52
3452	690V086H87
3452	690V086H96
3452	690V088H44
3452	690V088H45
3452	690V088H48
3452	690V103H23
3452	690V103H24
3452	690V103H25
3452	690V103H26
3452	690V103H27
3452	690V110H30
3452	690V110H31

SK	Industry Standard No.
3452	690V110H32
3452	690V110H33
3452	690V114H29
3452	690V114H31
3452	690V116H19
3452	690V118H59
3452	0703
3452	750D858-213
3452	753-0101-047
3452	753-2000-710
3452	773RED
3452	774ORN
3452	775BRN
3452	779BLU
3452	783RED
3452	784ORN
3452	786
3452	787BLU
3452	800-536-00
3452	803-38030
3452	803-38040
3452	803-39430
3452	803-39440
3452	803-83840
3452	803-83930
3452	916-31024-5B
3452	921-119B
3452	921-129B
3452	921-158B
3452	921-170B
3452	921-171B
3452	921-172B
3452	921-173B
3452	921-174B
3452	921-20
3452	921-20A
3452	921-20B
3452	921-21
3452	921-212B
3452	921-21A
3452	921-21B
3452	921-21BK
3452	921-22
3452	921-22A
3452	921-22B
3452	921-23
3452	921-23A
3452	921-23B
3452	921-266B
3452	921-267B
3452	921-30
3452	921-30A
3452	921-30B
3452	921-31
3452	921-313B
3452	921-31A
3452	921-31B
3452	921-32
3452	921-325B
3452	921-32A
3452	921-32B
3452	921-33
3452	921-334B
3452	921-335B
3452	921-336B
3452	921-338B
3452	921-33B
3452	921-34
3452	921-34A
3452	921-34B
3452	921-72B
3452	921-97B
3452	921-98B
3452	930X4
3452	930X5
3452	1002-02A
3452	1002-04-1
3452	1004-17
3452	1011-11(R.F.)
3452	1016-83
3452	1016-84
3452	1106-97
3452	1123-55
3452	1123-56
3452	1123-57
3452	1123-58
3452	1123-59
3452	1229H
3452	1284
3452	1373-17AL
3452	1634-17-14A
3452	1761-17
3452	1792-17
3452	1880-17
3452	1881-17
3452	1890-17
3452	1923-17
3452	1923-17-1
3452	1931-17A
3452	1983-17
3452	2028
3452	2028-00
3452	2032-33
3452	2032-34
3452	2057A-120
3452	2057A-429
3452	2057A2-119
3452	2057A2-127
3452	2057A2-163
3452	2057A2-179
3452	2057A2-201
3452	2057A2-224
3452	2057A2-309
3452	2057A2-310
3452	2057A2-311
3452	2057A2-313
3452	2057A2-314
3452	2057A2-342
3452	2057A2-386
3452	2057A2-392
3452	2057A2-393
3452	2057A2-394
3452	2057A2-402
3452	2057A2-448
3452	2057A2-465
3452	2057A2-504
3452	2057A2-505
3452	2057A2-509
3452	2057B100-12
3452	2057B2-103
3452	2057B2-108
3452	2057B2-109
3452	2057B2-110
3452	2057B2-111

SK	Industry Standard No.
3452	2057B2-112
3452	2057B2-119
3452	2057B2-120
3452	2057B2-127
3452	2057B2-128
3452	2057B2-14
3452	2057B2-160
3452	2057B2-161
3452	2057B2-162
3452	2057B2-64
3452	2057B2-85
3452	2057B2-87
3452	2093A2-289
3452	2180-151
3452	2180-152
3452	2445
3452	2495-166-1
3452	2495-166-4
3452	2495-166-8
3452	2495-166-9
3452	2495-522-1
3452	2495-523-1
3452	2498-507-2
3452	2498-507-3
3452	2498-508-2
3452	2498-508-3
3452	2498-903-2
3452	2498-903-3
3452	2606-294
3452	2634-1
3452	2636
3452	2900-007
3452	3001
3452	3020
3452	3021
3452	3028
3452	3107-204-90080
3452	3227-E
3452	3370
3452	3509
3452	3511
3452	3524-1
3452	3539-307-001
3452	3539-307-002
3452	3572-3
3452	4473-1
3452	4473-11
3452	4473-2
3452	4473-3
3452	4473-6
3452	4473-7
3452	4473-8
3452	4490-1
3452	4587
3452	4685-285-3
3452	4686-107-3
3452	4686-108-3
3452	4686-112-3
3452	4686-114-3
3452	4686-118-3
3452	4686-119-3
3452	4686-120-3
3452	4686-126-3
3452	4686-127-3
3452	4686-131-3
3452	4686-140-3
3452	4686-169-3
3452	4686-171-3
3452	4686-172-3
3452	4686-207-3
3452	4686-208-3
3452	4686-209-3
3452	4686-224-3
3452	4686-228-3
3452	4686-244-3
3452	4686-251-3
3452	4686-95-3
3452	4706
3452	4709
3452	4802-00002
3452	4820
3452	4821
3452	4826
3452	4837
3452	4845
3452	4855
3452	5001-510
3452	5093
3452	5313-461B
3452	5613-46B
3452	6158
3452	6158-3
3452	6185-3
3452	7115
3452	7118
3452	7122
3452	7123
3452	7124
3452	7125
3452	7126
3452	7127
3452	7128
3452	7131
3452	7132
3452	7133
3452	7134
3452	7173
3452	7174
3452	7175
3452	7177
3452	7178
3452	7214
3452	7215
3452	7216
3452	7217
3452	7218
3452	7219
3452	7220
3452	7221
3452	7232
3452	7234
3452	7235
3452	7237
3452	7238
3452	7262
3452	7425
3452	7426
3452	7427
3452	7428
3452	7593-2
3452	7642
3452	8606
3452	8607
3452	8609

SK	Industry Standard No.
3452	8611
3452	8710-162
3452	9300
3452	9300A
3452	9300B
3452	9300Z
3452	9314
3452	9513
3452	9600C
3452	9600F
3452	9600G
3452	9600H
3452	9601
3452	9601-12
3452	9604F
3452	9623F
3452	9623G
3452	9623H
3452	9625F
3452	9625H
3452	9630C
3452	11252-0
3452	11252-1
3452	11252-2
3452	11339-8
3452	11393-8
3452	11426-7
3452	11607-3
3452	11607-9
3452	11608-0
3452	11608-2
3452	11608-3
3452	11619-8
3452	11619-9
3452	11620-0
3452	11620-1
3452	16194
3452	19420
3452	23114-056
3452	23114-057
3452	23114-060
3452	23114-078
3452	23114-104
3452	23125-037
3452	25114-121
3452	30292
3452	35004
3452	35449
3452	36212
3452	36212V1
3452	36578
3452	36581
3452	36847
3452	36918
3452	36919
3452	37383
3452	37384
3452	37694A
3452	37694B
3452	38207
3452	38208
3452	38246A
3452	38511
3452	38511A
3452	38785
3452	38786
3452	38920
3452	38921
3452	39331
3452	39730
3452	39731
3452	39789
3452	40480
3452	40482
3452	41689
3452	41694
3452	043001
3452	43022-860
3452	50957-03
3452	57000-5452
3452	60048
3452	61009-1-1
3452	61009-1-2
3452	61009-2-1
3452	61009-6
3452	61009-6-1
3452	61010-0
3452	61010-0-1
3452	61010-7-1
3452	61013-9-1
3452	61015-0-1
3452	61133
3452	61661
3452	61663
3452	62449
3452	67802
3452	70167-8-00
3452	70231
3452	70260-11
3452	70260-12
3452	70260-13
3452	72949-10
3452	72951-95
3452	72951-96
3452	72979-80
3452	75616-6
3452	75810-17
3452	79855
3452	79856
3452	080006
3452	080021
3452	080022
3452	080023
3452	080041
3452	080042
3452	080059
3452	080060
3452	82716
3452	94044
3452	95125
3452	95126
3452	95127
3452	95128
3452	95129
3452	95130
3452	95131
3452	111943
3452	112355
3452	113938
3452	114143-1
3452	114525
3452	115440
3452	115910
3452	115925

SK	Industry Standard No.
3452	116073
3452	116079
3452	116080
3452	116082
3452	116083
3452	116198
3452	116200
3452	117823
3452	122517
3452	122518
3452	123160
3452	123429
3452	123430
3452	123431
3452	124263
3452	124412
3452	125137
3452	125263
3452	125264
3452	125392
3452	125475-14
3452	125944
3452	125994-14
3452	126023
3452	126024
3452	126670
3452	126698
3452	127693
3452	127792
3452	127793
3452	127794
3452	129144
3452	129392-14
3452	129393-14
3452	129394-14
3452	130403-62
3452	130404-21
3452	130404-59
3452	131221
3452	131648
3452	131844
3452	134263
3452	134419
3452	147245-0-1
3452	147356-9-1
3452	147357-2-1
3452	147357-9-1
3452	148751-147
3452	156931
3452	162002-090
3452	165995
3452	166272
3452	168657
3452	168658
3452	168659
3452	169195
3452	171028
3452	171032
3452	171033
3452	171045
3452	171048
3452	171052
3452	171054
3452	171090-1
3452	171139-1
3452	171140-1
3452	171141-1
3452	171162-027
3452	171162-128
3452	171162-129
3452	171162-130
3452	171162-131
3452	171206-1
3452	171206-2
3452	171206-4
3452	171206-5
3452	171207-1
3452	171207-2
3452	175006-187
3452	175043-062
3452	175043-063
3452	175043-064
3452	175043-100
3452	175043-107
3452	181003-7
3452	181003-8
3452	181003-9
3452	181503-6
3452	181503-7
3452	181503-9
3452	181504-1
3452	181504-7
3452	181506-7
3452	200064-6-103
3452	200064-6-105
3452	209417-0714
3452	227000
3452	229392
3452	231140-01
3452	231140-07
3452	231140-23
3452	231140-31
3452	231140-34
3452	231140-44
3452	232840
3452	236251
3452	236706
3452	236907
3452	237020
3452	237021
3452	237024
3452	237026
3452	237785
3452	237840
3452	241249
3452	241778
3452	241960
3452	242590
3452	242960
3452	243318
3452	243645
3452	245078-3
3452	257540
3452	260565
3452	265074
3452	265241
3452	267797
3452	304900
3452	346015-15
3452	346015-16
3452	346015-17
3452	346015-18
3452	346015-19
3452	346015-20
3452	346015-21
3452	346015-22
3452	346015-25
3452	346015-37
3452	489751-027
3452	489751-131
3452	489751-137
3452	489751-143
3452	489751-145
3452	489751-147
3452	489751-148
3452	489751-162
3452	489751-165
3452	489751-167
3452	489751-168
3452	489751-169
3452	489751-171
3452	489751-206
3452	573101
3452	573472
3452	573494
3452	0573495
3452	0573506
3452	0573506H
3452	0573507H
3452	0573511
3452	0573570
3452	601113
3452	602113
3452	603113
3452	604113
3452	610041
3452	610041-1
3452	610041-3
3452	610042-1
3452	610045
3452	610045-1
3452	610045-2
3452	610046-7
3452	610069
3452	610069-1
3452	610072
3452	610072-1
3452	610072-2
3452	610073
3452	610091
3452	610091-1
3452	610091-2
3452	610092
3452	610092-1
3452	610092-2
3452	610096
3452	610096-1
3452	610100
3452	610100-3
3452	610107-1
3452	610128-4
3452	610142-6
3452	610150
3452	610174-1
3452	613112
3452	815164
3452	815165
3452	815170
3452	815172
3452	815172A
3452	815173A
3452	815173C
3452	815173F
3452	815209
3452	824960-0
3452	910799
3452	916029
3452	916060
3452	916069
3452	964634
3452	964713
3452	965074
3452	965633
3452	965634
3452	970046
3452	970046-1
3452	970046-2
3452	970046A
3452	970244
3452	970245
3452	970249
3452	970309
3452	970309-1
3452	970309-12
3452	970309-2
3452	970309-3
3452	970309-4
3452	970309-5
3452	970310
3452	970310-1
3452	970310-12
3452	970310-2
3452	970310-3
3452	970310-4
3452	970310-5
3452	970332
3452	970332-12
3452	970911
3452	980138
3452	980139
3452	982268
3452	982269
3452	982321
3452	982815
3452	982816
3452	982817
3452	982818
3452	982819
3452	983095
3452	983096
3452	984156
3452	984158
3452	984159
3452	984194
3452	984195
3452	984577
3452	984743
3452	984744
3452	984851
3452	984852
3452	984853
3452	984875
3452	985096
3452	985097
3452	985215
3452	985442A
3452	985443A
3452	985444A
3452	986634
3452	986635
3452	988000
3452	988001
3452	988002
3452	988985
3452	988986
3452	988987
3452	988988
3452	988989
3452	994634
3452	1222463
3452	1408615-1
3452	1408640-1
3452	1471115-13
3452	1471115-14
3452	1472450-1
3452	1473530-1
3452	1473530-2
3452	1473533-1
3452	1473537-1
3452	1473568-1
3452	1473603-1
3452	1473604-3
3452	1473606-1
3452	1473652-1
3452	1810037
3452	1810038
3452	1810039
3452	1815036
3452	1815037
3452	1815039
3452	1815045
3452	1815047
3452	1815067
3452	1815068
3452	1817004
3452	1817005-3
3452	1817006-3
3452	1817008
3452	1817045
3452	1819045
3452	2000646-105
3452	2000757-80
3452	2000804-7
3452	2000804-8
3452	2002332-53
3452	2002332-54
3452	2002332-55
3452	2002332-56
3452	2002620-18
3452	2002620-19
3452	2003342-109
3452	2004746-114
3452	2004746-115
3452	2006431-44
3452	2006513-19
3452	2091859-0711
3452	2092418-0715
3452	2092418-0724
3452	2093308-070
3452	2093308-0700
3452	2093308-0705A
3452	2093308-0706A
3452	2093308-1
3452	2093308-2
3452	2093308-3
3452	2320062
3452	2320063
3452	2320073
3452	2495166-1
3452	2498507-1
3452	2498507-2
3452	2498507-3
3452	2596071
3452	3181972
3452	3596067
3452	3596068
3452	3596069
3452	3596070
3452	3596071
3452	3596072
3452	5320328
3452	6212922
3452	7026011
3452	7026012
3452	7026013
3452	7295195
3452	7295196
3452	7910780-01
3452	8031825
3452	8031836
3452	8033803
3452	8033804
3452	8033943
3452	8037723
3452	8556188
3452	13040304
3452	13040362
3452	13040421
3452	13040459
3452	22901513
3452	22902046
3452	23114001
3452	23114023
3452	23114031
3452	23114034
3452	23114036
3452	23114043
3452	23114056
3452	23114057
3452	23114060
3452	23114082
3452	23114109
3452	23114127
3452	23114171
3452	23114172
3452	23114180
3452	23114181
3452	23114238
3452	23124037
3452	23126620
3452	25114161
3452	26010056
3452	27125210
3452	30200091
3452	43027614
3452	43027615
3452	43027616
3452	43027617
3452	44007301
3452	44008401
3452	55440011-001
3452	55440023-001
3452	62048158
3452	62104678
3452	62506296
3452	62506318
3452	62537124
3452	62539305
3452	62539321
3452	62541555
3452	62543892
3452	62543906
3452	62563303
3452	62563346
3452	62563354
3452	62565122
3452	62605465
3452	62691271
3452	62695595
3452	62695943
3452	62713747
3452	62766239
3452	62789263
3452	80053600
3452	80338030
3452	80338040
3452	80339430
3452	80339440
3452	120004496
3452	120004497
3452	120004880
3452	120004881
3452	120004882
3452	226021014
3452	229018032
3452	229018033
3452	229018034
3452	229020423
3452	229021014
3452	229025010
3452	229510015V
3452	229510031V
3452	229510032V
3452	229510033V
3452	450010201
3452	2290180119
3452	2295100224
3452	2295100225
3452	2295100226
3452	2295100228
3452	16100190668
3452	16102190929
3452	16103190668
3452	16104191225
3452	16104191226
3452	310720490080
3452/108	A321
3452/108	A417
3452/108	BF152
3452/108	BF153
3452/108	BF154
3452/108	BF158
3452/108	BF159
3452/108	BF160
3452/108	BF224
3452/108	BF225
3452/108	BF233-2
3452/108	BF233-5
3452/108	BF234
3452/108	BF237
3452/108	BF238
3452/108	BF594
3452/108	BF595
3452/108	BFS62
3452/108	BFX43
3452/108	BFX44
3452/108	BFY37
3452/108	BFY90
3452/108	BSS26
3452/108	BSX28
3452/108	BSX87A
3452/108	BSX88
3452/108	BSX88A
3452/108	BSY70
3452/108	BSY72
3452/108	C1158
3452/108	C1205
3452/108	C1205A
3452/108	C1205B
3452/108	C1205C
3452/108	C127
3452/108	C155
3452/108	C156
3452/108	C17
3452/108	C172
3452/108	C172A
3452/108	C17A
3452/108	C199
3452/108	C296
3452/108	C313
3452/108	C316
3452/108	C318
3452/108	C318A
3452/108	C370
3452/108	C370F
3452/108	C370G
3452/108	C370H
3452/108	C370J
3452/108	C370K
3452/108	C386
3452/108	C392
3452/108	C405
3452/108	C424
3452/108	C429
3452/108	C429J
3452/108	C429X
3452/108	C430
3452/108	C430H
3452/108	C430W
3452/108	C477
3452/108	C544
3452/108	C544C
3452/108	C544D
3452/108	C544E
3452/108	C612
3452/108	C641B
3452/108	C658
3452/108	C658A
3452/108	C659
3452/108	C74
3452/108	C80
3452/108	C912
3452/108	C924
3452/108	C924E
3452/108	C924F
3452/108	C924M
3452/108	C98
3452/108	C99
3452/108	CE4001C
3452/108	CK476
3452/108	CK477
3452/108	CS-460B
3452/108	CS-9016
3452/108	CS9016/3490
3452/108	CS9016F
3452/108	EA15X7141
3452/108	EN3011
3452/108	EN744
3452/108	EN914
3452/108	FCS9011E
3452/108	FCS9011H
3452/108	FCS9014
3452/108	FCS9014B
3452/108	FCS9014D
3452/108	FV2747C
3452/108	HS-40019
3452/108	ME3002
3452/108	ME3011
3452/108	ME9022
3452/108	MM1803
3452/108	MM1941
3452/108	MPS6539
3452/108	MPS6541
3452/108	MPS6542
3452/108	MPS6543
3452/108	MPS6546
3452/108	MPS6547
3452/108	MPS6548
3452/108	MPS706A
3452/108	Q5083C
3452/108	Q5120P
3452/108	Q5120Q
3452/108	Q5120R
3452/108	RS-7511
3452/108	RS7108
3452/108	RT-930H
3452/108	RT5901
3452/108	RT6202
3452/108	RT8193
3452/108	RT8527
3452/108	S1286
3452/108	ZT2368
3452/108	ZT2369
3452/108	ZT2369A
3452/108	ZT2708
3452/108	ZT2938
3452/108	2N1505
3452/108	2N1506
3452/108	2N1506A
3452/108	2N1507
3452/108	2N1989
3452/108	2N2193A
3452/108	2N2193B
3452/108	2N2195
3452/108	2N2195B
3452/108	2N263
3452/108	2N264
3452/108	2N2651
3452/108	2N2708
3452/108	2N2710
3452/108	2N2712
3452/108	2N2715
3452/108	2N2716
3452/108	2N2729
3452/108	2N2883
3452/108	2N2884
3452/108	2N2921
3452/108	2N2922
3452/108	2N3227
3452/108	2N337A
3452/108	2N3493
3452/108	2N3508
3452/108	2N3509
3452/108	2N3511
3452/108	2N3544
3452/108	2N3570
3452/108	2N3571
3452/108	2N3572
3452/108	2N3600
3452/108	2N3633
3452/108	2N3688
3452/108	2N3689
3452/108	2N3690
3452/108	2N3691
3452/108	2N3692
3452/108	2N3832
3452/108	2N3854
3452/108	2N3854A
3452/108	2N3933
3452/108	2N3983
3452/108	2N3984
3452/108	2N3985
3452/108	2N4072
3452/108	2N4073
3452/108	2N4134
3452/108	2N4135
3452/108	2N4251
3452/108	2N4274
3452/108	2N4275
3452/108	2N4418
3452/108	2N4419
3452/108	2N4449
3452/108	2N4873
3452/108	2N4934
3452/108	2N4935
3452/108	2N4936
3452/108	2N5053
3452/108	2N5054
3452/108	2N5131
3452/108	2N5200
3452/108	2N5272
3452/108	2N5292
3452/108	2N5399
3452/108	2N619
3452/108	2N620
3452/108	2N621
3452/108	2N702
3452/108	2N703
3452/108	2N706
3452/108	2N706A
3452/108	2N706B
3452/108	2N706C
3452/108	2N707
3452/108	2N717
3452/108	2N718
3452/108	2N718A
3452/108	2N743
3452/108	2N743A
3452/108	2N744
3452/108	2N752

SK	Industry Standard No.
3452/108	2N756
3452/108	2N757
3452/108	2N758
3452/108	2N759
3452/108	2N760
3452/108	2N761
3452/108	2N762
3452/108	2N834
3452/108	2N841
3452/108	2N843
3452/108	2N913
3452/108	2N914
3452/108	2N914A
3452/108	2N916
3452/108	2N916A
3452/108	2N988
3452/108	2N989
3452/108	2S512
3452/108	2SC1158
3452/108	2SC1205
3452/108	2SC1205A
3452/108	2SC1205B
3452/108	2SC1205C
3452/108	2SC127
3452/108	2SC155
3452/108	2SC156
3452/108	2SC17
3452/108	2SC172
3452/108	2SC172A
3452/108	2SC17A
3452/108	2SC199
3452/108	2SC296
3452/108	2SC316
3452/108	2SC370
3452/108	2SC370F
3452/108	2SC370G
3452/108	2SC370H
3452/108	2SC370J
3452/108	2SC370K
3452/108	2SC386
3452/108	2SC386A-O(TV)
3452/108	2SC392
3452/108	2SC397
3452/108	2SC405
3452/108	2SC429
3452/108	2SC429J
3452/108	2SC429X
3452/108	2SC430
3452/108	2SC430H
3452/108	2SC430W
3452/108	2SC469F
3452/108	2SC477
3452/108	2SC544
3452/108	2SC544C
3452/108	2SC544D
3452/108	2SC544E
3452/108	2SC606
3452/108	2SC611
3452/108	2SC612
3452/108	2SC641
3452/108	2SC641B
3452/108	2SC658
3452/108	2SC658A
3452/108	2SC659
3452/108	2SC74
3452/108	2SC74
3452/108	2SC80
3452/108	2SC8380
3452/108	2SC924
3452/108	2SC924E
3452/108	2SC924F
3452/108	2SC924M
3452/108	2SC98
3452/108	2SC99
3452/108	13-10321-41
3452/108	13-23822
3452/108	13-32366-1
3452/108	19-020-044
3452/108	19-020-048
3452/108	025-100015
3452/108	46-86264-3
3452/108	46-86376-3
3452/108	054
3452/108	57A138-4-6
3452/108	57B2-1
3452/108	57C7-8
3452/108	90-49
3452/108	121-498
3452/108	121-546
3452/108	121-825
3452/108	121-855
3452/108	200-011
3452/108	921-106B
3452/108	9011E
3452/108	9016D
3452/108	9016E
3452/108	9016F
3452/108	09018
3452/108	9018D
3452/108	38246
3452/108	40351
3452/108	40352
3452/108	40472
3452/108	40475
3452/108	40481
3452/108	95170-2
3452/108	116119
3452/108	125994
3452/108	125995
3452/108	129394
3452/108	129571
3452/108	129574
3452/108	129897
3452/108	129979
3452/108	131848
3452/108	134442
3452/108	136165
3452/108	136239
3452/108	137127
3452/108	171034
3452/108	171915
3452/108	0573486H
3452/108	0573507
3452/108	610232-2
3453	GEIC-213
3453	612076-2
3454	C6057P
3454	EX62-X
3454	HEP-C6057P
3454	LM746N
3454	MC1328P
3454	MC1328PQ
3454	ULN2228A
3454	13-41-6
3454	15-37704-1
3454	29B1
3454	29B17
3454	29B1B
3454	29B1Z
3454	46-5002-11
3454	56A6-1
3454	56D6-1
3454	56L103
3454	081-1
3454	99S081
3454	99S081-1
3454	221-62-1
3454	229-1301-41
3454	746PC
3454	2029-1
3454	2029-2
3454	37704-1
3454	612029-1
3454	612029-2
3454/790	5606-1
3455	C6090
3455	GEIC-27
3455	GEL277
3455	HEP-C6090
3455	PA277
3455	RTN1001
3455	SN76177
3455	SN76177ND
3455	09-308078
3455	13-100000
3455	13-1000000
3455	13-59-6
3455	13-61-6
3455	45B17
3455	45B1AH
3455	45B1D
3455	45B1Z
3455	69-2401
3455	69-2403
3455	69-3116
3455	88-9304
3455	96-5374-01
3455	2045-1
3455	2277P
3455	3535-110-50008
3455	3535-110-50009
3455	7204A
3455	45387
3455	45395
3455	50009
3455	95293
3455	183044
3455	612045-1
3455	2402277
3455	2412275
3455	353511050009
3455/804	LM377
3455/804	LM377N
3457	AN203
3457	AN203AA
3457	AN203BA
3457	AN203BB
3457	AN203C
3457	GEIC-45
3457	IC-554(ELCOM)
3457	09-308011
3457	70270730
3457	4150002031
3458	AN211
3458	AN211A
3458	AN211AB
3458	AN211B
3458	GEIC-48
3458	IC-552(ELCOM)
3458	REN1056
3458	1018-25
3458	741853
3459	AN214
3459	AN214P
3459	AN214PQR
3459	AN214Q
3459	AN214R
3459	G09-013-A
3459	GEIC-49
3459	RE338-IC
3459	REN1058
3459	014-556
3459	88-18920
3459	8000-00032-030
3459	18600-156
3459	741854
3459	5350231
3451	46-86314-3
3460	AN217
3460	AN217AA
3460	AN217AB
3460	AN217BA
3460	AN217BB
3460	AN217CA
3460	AN217CB
3460	AN217P
3460	AN217PBB
3460	GEIC-50
3460	PC-20069
3461	DDEY064001
3461	GEIC-140
3461	MB3202
3461	MPC577H
3461	MX-3389
3461	NJM2201
3461	RE341-M
3461	REN1082
3461	09-308052
3461	61A001-10
3461	307-112-9-007
3461	2056-04
3461	28810-175
3461	741852
3461	916085
3462	LM78L05
3462	LM78L05ACZ
3462	LM78L05CZ
3462	MC78L05
3462	MC78L05ACP
3462	MC78L05C
3462	MC78L05CP
3462	NJM78L05A
3462	REN977
3462	02-781050
3462	78L05
3463	DX-0150
3463	EA16X73
3463	HV-80
3463	HV0000206
3463	HV0000502
3463	HV80
3463	KB-162C5
3463	KB-162C5A
3463	KB162N
3463	KB165
3463	KB169
3463	KB265
3463	KB265A
3463	KB269
3463	MA26
3463	MA26A
3463	MV1
3463	MV3
3463	RVDVD1150L
3463	VD1120
3463	VD1121
3463	VD1122
3463	VD1123
3463	VD1124
3463	VD1150M
3463	1S1209
3463	21A111-002
3463	48-355045
3463	48M355046
3463	265P03301
3463	2017-111
3463/601	M8513-0
3463/601	MV11
3463/601	SVDMA26
3463/601	SVDMA26-1
3464	C1013
3464	C1013C
3464	C1013D
3464	C1014
3464	C1014B
3464	C1014C
3464	C1014CD
3464	C1014D
3464	C1014D1
3464	C1243
3464	C1243-24
3464	C1243C
3464	C1243C1
3464	C1243C2
3464	C1243D
3464	C1243D1
3464	C1243D2
3464	C1243E
3464	GE-273
3464	IP20-0172
3464	PTC193
3464	2SC1013
3464	2SC1013B
3464	2SC1013C
3464	2SC1013D
3464	2SC1014
3464	2SC1014B
3464	2SC1014C
3464	2SC1014CD
3464	2SC1014D
3464	2SC1014D1
3464	2SC1014E
3464	2SC1243
3464	2SC1243-24
3464	2SC1243C
3464	2SC1243C1
3464	2SC1243C2
3464	2SC1243D
3464	2SC1243D1
3464	2SC1243D2
3464	2SC1243E
3464	09-302046
3464	09-305049
3464	031-058-0
3464	1009-05
3464	4802-00016
3464	5001-010
3464	8000-00004-087
3464	171162-287
3464	43029486
3465	C6102P
3465	GEIC-220
3465	HEP-C6102P
3465	LM1458N
3465	MC1458
3465	MC1458CP1
3465	MC1458P
3465	MC1458P1
3465	N5558V
3465	SN72558
3465	SN72558P
3465	1458CP1
3465	1458P1
3465	34502-1
3465	142648
3466	BC558
3466	BC558B
3466	CE4012D
3466	CXL159
3466	ECG159
3466	ED1602C
3466	EP15X53
3466	EP15X60
3466	GE-82
3466	HEP-S0026
3466	HX-50094
3466	HX-50176
3466	MPS-A55
3466	MPS-A56
3466	MPS3702
3466	MPS6518
3466	MPS6519
3466	MPS6522
3466	MPS6523
3466	MPS6533
3466	MPS6534
3466	MPS6535
3466	MPS9680
3466	MPS9680H/I
3466	MPS9680I
3466	MPS9680J
3466	MPS9680T
3466	MPS9681
3466	MPS9681I
3466	MPS9681J
3466	MPS9681K
3466	MPS9681T
3466	MPS9750D
3466	MPS9750F
3466	MPS9750G
3466	MPSA55
3466	MPSA70
3466	P2S
3466	Q-2N5226
3466	0S133.0008T
3466	SPS2274
3466	SPS5008
3466	ST-MPS9682J
3466	ST-MPS9750D
3466	ST1602D
3466	TM159
3466	001-02
3466	001-022010
3466	001-03
3466	1-034/2207
3466	001-04
3466	1-043/2207
3466	001-533-00
3466	01-571591
3466	01-571751
3466	01-572588
3466	1B3096-1
3466	2D017-165
3466	2D017-166
3466	2D017-167
3466	2D017-169
3466	2N1025
3466	2N1026
3466	2N1026A
3466	2N1027
3466	2N1028
3466	2N1118
3466	2N1118A
3466	2N1119
3466	2N1131A
3466	2N1132
3466	2N1132/46
3466	2N1132A
3466	2N1132B
3466	2N1135
3466	2N1135A
3466	2N1219
3466	2N1220
3466	2N1221
3466	2N1222
3466	2N1223
3466	2N1439
3466	2N1440
3466	2N1441
3466	2N1442
3466	2N1443
3466	2N1474
3466	2N1474A
3466	2N1475
3466	2N1607
3466	2N1608
3466	2N1643
3466	2N1676
3466	2N1677
3466	2N1919
3466	2N1920
3466	2N1921
3466	2N1991
3466	2N2002
3466	2N2003
3466	2N2004
3466	2N2005
3466	2N2006
3466	2N2007
3466	2N2121
3466	2N2162
3466	2N2163
3466	2N2164
3466	2N2165
3466	2N2166
3466	2N2167
3466	2N2181
3466	2N2182
3466	2N2183
3466	2N2184
3466	2N2185
3466	2N2186
3466	2N2187
3466	2N2274
3466	2N2275
3466	2N2276
3466	2N2277
3466	2N2278
3466	2N2279
3466	2N2280
3466	2N2281
3466	2N2303
3466	2N2332
3466	2N2333
3466	2N2334
3466	2N2335
3466	2N2336
3466	2N2337
3466	2N2370
3466	2N2371
3466	2N2372
3466	2N2373
3466	2N2377
3466	2N2378
3466	2N2393
3466	2N2394
3466	2N2411
3466	2N2412
3466	2N2424
3466	2N2425
3466	2N2474
3466	2N2595
3466	2N2596
3466	2N2597
3466	2N2601
3466	2N2602
3466	2N2603
3466	2N2604
3466	2N2605
3466	2N2605A
3466	2N2695
3466	2N2696
3466	2N2800/46
3466	2N2801
3466	2N2837
3466	2N2838
3466	2N2861
3466	2N2862
3466	2N2894
3466	2N2894A
3466	2N2906
3466	2N2906A
3466	2N2907
3466	2N2907A
3466	2N2944
3466	2N2944A
3466	2N2945
3466	2N2945A
3466	2N2946
3466	2N2946A
3466	2N3058
3466	2N3059
3466	2N3060
3466	2N3062
3466	2N3072
3466	2N3073
3466	2N3121
3466	2N3135
3466	2N3136
3466	2N3217
3466	2N3218
3466	2N3219
3466	2N3248
3466	2N3249
3466	2N3250
3466	2N3250A
3466	2N3251
3466	2N3251A
3466	2N3305
3466	2N3306
3466	2N3307
3466	2N3308
3466	2N3317
3466	2N3318
3466	2N3319
3466	2N3341
3466	2N3346
3466	2N3451
3466	2N3464
3466	2N3485
3466	2N3485A
3466	2N3486
3466	2N3486A
3466	2N3496
3466	2N3504
3466	2N3505
3466	2N3527
3466	2N3545
3466	2N3546
3466	2N3547
3466	2N3548
3466	2N3549
3466	2N3550
3466	2N3579
3466	2N3580
3466	2N3581
3466	2N3582
3466	2N3638
3466	2N3638A
3466	2N3639
3466	2N3640
3466	2N3644
3466	2N3672
3466	2N3673
3466	2N3702
3466	2N3703
3466	2N3798
3466	2N3799
3466	2N3829
3466	2N3840
3466	2N3842
3466	2N3857
3466	2N3905
3466	2N3906
3466	2N3914
3466	2N3930
3466	2N3931
3466	2N3962
3466	2N3963
3466	2N3964
3466	2N3965
3466	2N3977
3466	2N3978
3466	2N3979
3466	2N4125
3466	2N4207
3466	2N4208
3466	2N4209
3466	2N4228
3466	2N4248
3466	2N4249
3466	2N4250
3466	2N4250A
3466	2N4257
3466	2N4257A
3466	2N4258
3466	2N4258A
3466	2N4284
3466	2N4285
3466	2N4288
3466	2N4289
3466	2N4290
3466	2N4291
3466	2N4354
3466	2N4355
3466	2N4356
3466	2N4359
3466	2N4402
3466	2N4403
3466	2N4411
3466	2N4413
3466	2N4413A
3466	2N4415
3466	2N4415A
3466	2N4452
3466	2N4917
3466	2N4965
3466	2N4971
3466	2N4972
3466	2N4982
3466	2N5086
3466	2N5087
3466	2N5142
3466	2N5143
3466	2N5226
3466	2N5227
3466	2N5356
3466	2N5365
3466	2N5366
3466	2N5367
3466	2N5372
3466	2N5373
3466	2N5374
3466	2N5375
3466	2N5378
3466	2N5379
3466	2N5382
3466	2N5383
3466	2N5447
3466	2N5448

SK	Industry Standard No.
3466	2N5855
3466	2N6003
3466	2N6076
3466	2N721
3466	2N721A
3466	2N722
3466	2N722A
3466	2N858
3466	2N859
3466	2N860
3466	2N861
3466	2N862
3466	2N863
3466	2N864
3466	2N923
3466	2N924
3466	2N925
3466	2N926
3466	2N927
3466	2N928
3466	2N935
3466	2N936
3466	2N937
3466	2N938
3466	2N939
3466	2N940
3466	2N941
3466	2N943
3466	2N944
3466	2N945
3466	2N946
3466	02P1B
3466	2S3020
3466	2S3021
3466	2S3030
3466	2S3040
3466	2S306
3466	2S321
3466	2S3210
3466	2S322
3466	2S3220
3466	2S3221
3466	2S322A
3466	2S323
3466	2S3230
3466	2S324
3466	2S3240
3466	2S326
3466	2S327
3466	2S673C
3466	2SA-NJ101
3466	2SA467G-0
3466	2SA4670R
3466	2SA467Y
3466	2SA480
3466	2SA482
3466	2SA4950
3466	2SA522
3466	2SA522A
3466	2SA530
3466	2SA530H
3466	2SA542
3466	2SA542A
3466	2SA542B
3466	2SA542C
3466	2SA542D
3466	2SA542E
3466	2SA542F
3466	2SA542GN
3466	2SA542H
3466	2SA542J
3466	2SA542L
3466	2SA542M
3466	2SA5420R
3466	2SA542R
3466	2SA542X
3466	2SA542Y
3466	2SA565
3466	2SA565A
3466	2SA565B
3466	2SA565C
3466	2SA565D
3466	2SA565K
3466	2SA567
3466	2SA592Y
3466	2SA603
3466	00002SA609
3466	2SA609A
3466	2SA609B
3466	2SA609C
3466	2SA609D
3466	2SA609E
3466	2SA609F
3466	2SA609G
3466	2SA617K
3466	2SA618K
3466	2SA6661QRS
3466	2SA690D
3466	2SA701
3466	2SA701F
3466	2SA701FJ
3466	2SA701FO
3466	2SA718
3466	2SA7200
3466	2SA945Y
3466	2SANJ101
3466	3L4-6017-01
3466	04-440032-002
3466	04-440032-008
3466	4JX29A529
3466	4JX29A826
3466	006-0000135
3466	006-02
3466	6-31
3466	6-31A
3466	6-38
3466	6-38A
3466	7-0014
3466	007-74004-01
3466	007-74008-01
3466	8-905-706-247
3466	8-905-706-251
3466	8-905-706-253
3466	8-905-706-254
3466	8-905-706-255
3466	8-905-706-256
3466	8-905-706-280
3466	8-905-706-286
3466	8-905-706-287
3466	8-905-706-288
3466	8-905-706-289
3466	8-905-706-290
3466	8-905-713-058
3466	8C200
3466	8C201
3466	8C202
3466	8C203
3466	8C204
3466	8C205
3466	8C206
3466	8C207
3466	8C430
3466	8C430K
3466	8C440
3466	8C440K
3466	8C443
3466	8C443K
3466	8C445
3466	8C445K
3466	8C449
3466	8C449K
3466	8C450
3466	8C460
3466	8C460K
3466	8C463
3466	8C463K
3466	8C465
3466	8C465K
3466	8C466
3466	8C466K
3466	8C467
3466	8C467K
3466	8C468
3466	8C468K
3466	8C469
3466	8C469K
3466	8C470
3466	8C470K
3466	8C700
3466	8C700A
3466	8C700B
3466	8C702
3466	8C702A
3466	8C702B
3466	8C704
3466	8C740
3466	8C7400
3466	8C740G
3466	8C740M
3466	8C742
3466	8C7420
3466	8C742G
3466	8C742M
3466	8Q-3-11
3466	8Q-3-14
3466	09-300037
3466	09-300059
3466	09-300061
3466	09-300062
3466	09-300063
3466	09-300074
3466	09-300077
3466	09-300307
3466	09-30063
3466	09-304012
3466	09-304047
3466	09-304049
3466	09-304050
3466	09-304051
3466	09-305073
3466	09-309038
3466	09-309042
3466	10P1
3466	10P1A
3466	11-691504
3466	13-23826-1
3466	13-23826-1A
3466	13-23826-2
3466	13-23826-2A
3466	13-23826-3
3466	13-23826-3A
3466	13-26386-1
3466	13-26386-1A
3466	13-26386-2
3466	13-26386-2A
3466	13-26386-3
3466	13-28391-1
3466	13-28391-1A
3466	13-28391-2
3466	13-28391-2A
3466	13-29776-1
3466	13-29776-1A
3466	13-29776-2
3466	13-29776-3
3466	13-31013-1/2
3466	13-32364-1
3466	13-36386-1
3466	13-39970-1
3466	13-55069-1
3466	13-55069-1A
3466	14-602-20
3466	14-602-20A
3466	14-602-32
3466	14-602-32A
3466	14-602-42
3466	14-602-42A
3466	14-602-44
3466	14-602-44A
3466	14-602-47
3466	14-602-47A
3466	14-602-54A
3466	14-602-56
3466	14-602-56A
3466	14-602-58
3466	14-602-580
3466	14-602-58A
3466	14-602-600
3466	14-602-68
3466	14-602-88
3466	14-602-90
3466	014-611
3466	014-652
3466	014-652C
3466	014-772
3466	14-803-12
3466	14-804-12
3466	14-808-12
3466	14-855-32
3466	14-856-23
3466	14-857-12
3466	14-857-79
3466	14-863-23
3466	14-864-23
3466	14-867-32
3466	15-01742
3466	15-01913-00
3466	15-02762-00
3466	15-02762-1
3466	15-02762-2
3466	15-02979
3466	15-03099
3466	15-03409-0
3466	15-03409-02
3466	15-03409-1
3466	15-09090-01
3466	15-3
3466	15-30
3466	15-4
3466	15-40
3466	15-5
3466	15-50
3466	15-875-075-001
3466	15Z3
3466	17-459
3466	17-459A
3466	018-00001
3466	018-00002
3466	019-003675-231
3466	019-003675-232
3466	019-003675-234
3466	019-003931
3466	019-004558
3466	019-005010
3466	019-005179
3466	19-1
3466	19-10
3466	19-2
3466	19-20
3466	19-3
3466	19-30
3466	19A115178-P1
3466	19A115178-P2
3466	19A115458-P1
3466	19A115458-P2
3466	19A115653-P1
3466	19A115653-P2
3466	19A115654-P1
3466	19A115654-P2
3466	19A115688-P1
3466	19A115688-P2
3466	19A115706-1
3466	19A115706-2
3466	19A115706-P1
3466	19A115706-P2
3466	19A115768-1
3466	19A115768-2
3466	19A115768-3
3466	19A115768-P1
3466	19A115768-P2
3466	19A115779P1
3466	19A115852P1
3466	19A116223P1
3466	19A116408-1
3466	020-1110-004C
3466	21A015-008
3466	21A015-008A
3466	21A015-009
3466	21A015-009A
3466	21A015-011
3466	21A015-011A
3466	21A015-012
3466	21A015-012A
3466	21A015-025
3466	21A112-001
3466	21A112-047
3466	21A112-065
3466	21A112-075
3466	21A112-093
3466	21A112-100
3466	21M022
3466	21M355
3466	21M581
3466	22-001010
3466	23-1
3466	23-10
3466	23-2
3466	23-20
3466	23-3
3466	23-30
3466	23-LLB
3466	23-PT274-122
3466	24-AWH
3466	24MW1031
3466	24MW1049
3466	24MW1061
3466	24MW661
3466	24MW976
3466	24T-011-011
3466	25-000453
3466	25-000462
3466	25-MEF
3466	27A10533
3466	29-HCL
3466	33-016
3466	33-086
3466	34-1013
3466	34-1022
3466	34-143-12
3466	34-3015-28
3466	34-6001-15
3466	34-6015-26
3466	34-6015-42
3466	34-6016-15
3466	34-6016-15A
3466	34-6016-32
3466	34-6016-47
3466	34-6016-60
3466	34P1AA
3466	35(RCA)
3466	35-ALD
3466	0036-001
3466	42-22008
3466	42-22008A
3466	42-23541
3466	42-23541A
3466	42-28208
3466	42-28211
3466	42-9029-40X
3466	42-9029-60Q
3466	42-9029-70D
3466	42-9029-70E
3466	43A145291-1
3466	43A145291-2
3466	43A167207P1
3466	43A167207P2
3466	43A168064-1
3466	43A168064P1
3466	43A176002
3466	43B168450-1
3466	43B168495-1
3466	43B168566-P1
3466	44A333464
3466	44A333464-1
3466	44A390248-001
3466	44A390256-001
3466	44A390261
3466	44A397905
3466	44A417031-001
3466	44A418041-001
3466	44B238203-1
3466	44B238246
3466	46-86229-3
3466	46-86229-3A
3466	46-86230-3
3466	46-86238-3
3466	46-86238-3A
3466	46-86283-3
3466	46-86293-3
3466	46-86377-3
3466	46-86399-3
3466	46-86412-3
3466	46-86424-3
3466	48-03-002
3466	48-134525
3466	48-134525A
3466	48-134702
3466	48-134702A
3466	48-134745
3466	48-134745A
3466	48-134815
3466	48-134815A
3466	48-134829
3466	48-134829A
3466	48-134833
3466	48-134833A
3466	48-134865
3466	48-134865A
3466	48-134866
3466	48-134866A
3466	48-134867
3466	48-134867A
3466	48-134868
3466	48-134868A
3466	48-134869
3466	48-134869A
3466	48-134870
3466	48-134870A
3466	48-134871
3466	48-134871A
3466	48-134909
3466	48-134909A
3466	48-134910
3466	48-134910A
3466	48-134910F
3466	48-134911
3466	48-134913
3466	48-134913A
3466	48-134914
3466	48-134914A
3466	48-134915
3466	48-134915A
3466	48-134940
3466	48-134940A
3466	48-134943
3466	48-134943A
3466	48-134967
3466	48-134967A
3466	48-134973
3466	48-134973A
3466	48-134975
3466	48-134975A
3466	48-134989
3466	48-134989A
3466	48-137020
3466	48-137020A
3466	48-137032
3466	48-137032A
3466	48-137045
3466	48-137045A
3466	48-137046
3466	48-137061
3466	48-137067
3466	48-137067A
3466	48-137068
3466	48-137068A
3466	48-137069
3466	48-137069A
3466	48-137090
3466	48-137090A
3466	48-137127
3466	48-137127A
3466	48-137173
3466	48-137173A
3466	48-137176
3466	48-137176A
3466	48-137195
3466	48-137318
3466	48-137324
3466	48-137366
3466	48-137379
3466	48-137380
3466	48-137381
3466	48-137382
3466	48-137383
3466	48-137391
3466	48-137502
3466	48-137504
3466	48-3003A06
3466	48-355007
3466	48-40118B01
3466	48-40118B01A
3466	48-42098B01
3466	48-42098B01A
3466	48-43238G01
3466	48-64978A40
3466	48-64978A40A
3466	48-64978A41
3466	48-64978A41A
3466	48-869334
3466	48-869413
3466	48-869526
3466	48-869571
3466	48-869649
3466	48-90165A01
3466	48-90165A01A
3466	48-97046A26
3466	48-97046A27
3466	48-97177A14
3466	48-97177A14A
3466	48R869334
3466	48R869413
3466	48R869526
3466	48R869571
3466	48R869649
3466	48S134815
3466	051-0107
3466	51-47-21
3466	53-1516
3466	55-1083
3466	55-1083A
3466	55-1085
3466	55-1085A
3466	55-152579
3466	056
3466	56-8098
3466	56-8098A
3466	56-8098B
3466	56-8098C
3466	57A1-76A
3466	57A106-12
3466	57A122-9A
3466	57A130-9A
3466	57A133-12
3466	57A137-12
3466	57A137-12A
3466	57A145-12
3466	57A145-12A
3466	57A147-12
3466	57A147-12A
3466	57A148-12
3466	57A15-5
3466	57A15-50
3466	57A157
3466	57A157-9
3466	57A157-90
3466	57A157-9A
3466	57A159-12
3466	57A159-12A
3466	57A174-8
3466	57A175-12
3466	57A178-12
3466	57A185-12
3466	57A189-8
3466	57A19
3466	57A19-1
3466	57A19-10
3466	57A19-1A
3466	57A19-2
3466	57A19-20
3466	57A19-3
3466	57A19-30
3466	57A197-12
3466	57A2-70
3466	57A2-70A
3466	57A2-71
3466	57A2-71A
3466	57A201-14
3466	57A216-12
3466	57A235-12
3466	57A258-8
3466	57A305-12
3466	57B106-12
3466	57B108-6
3466	57B108-6A
3466	57B122-9
3466	57B122-9A
3466	57B130-9
3466	57B130-9A
3466	57B133-12
3466	57B137-12
3466	57B137-12A
3466	57B145-12
3466	57B145-12A
3466	57B147-12
3466	57B147-12A
3466	57B157-9
3466	57B157-9A
3466	57B159-12
3466	57B159-12A
3466	57B175-12
3466	57B178-12
3466	57B185-12
3466	57B189-8
3466	57B2-70
3466	57B2-70A
3466	57B2-71
3466	57B2-71A
3466	57B216-12
3466	57B235-12
3466	57B258-8
3466	57C15-5
3466	57C15-50
3466	57C157-9
3466	57C157-90
3466	57C19-1
3466	57C19-1A
3466	57D1-76
3466	57D1-76A
3466	57D19
3466	57D19-1
3466	57D19-10
3466	57D19-2
3466	57D19-20
3466	57D19-3
3466	57D19-30
3466	62-19452
3466	62-20154
3466	62-20154A
3466	62-20244
3466	62-20244A
3466	62A11871
3466	63-12154
3466	63-12154A
3466	63-12156
3466	63-12156A
3466	63-12157
3466	63-12157A
3466	63-13322
3466	63-13322A
3466	65C1
3466	65D
3466	65D1
3466	65E
3466	65E1
3466	65F
3466	65F1
3466	68-110-02
3466	68A7370-1
3466	68A7370-P3
3466	68A7382-P1
3466	68A7734P1
3466	68A8318-P1
3466	73B-140-005-1
3466	73C180831-1
3466	73C180831-2
3466	81-46125071-4
3466	082.115.015
3466	83P1
3466	83P1A
3466	83P1B

SK	Industry Standard No.
3466	83P1BC
3466	83P1M
3466	83P1MC
3466	83P2
3466	83P2A
3466	83P2AA
3466	83P2AA1
3466	83P2M
3466	83P2M1
3466	83P2N
3466	83P3
3466	83P3A
3466	83P3AA
3466	83P3AA1
3466	83P3B
3466	83P3B1
3466	83P3M
3466	83P3M1
3466	83P4
3466	86-10009
3466	86-178-2
3466	86-178-20
3466	86-183-2
3466	86-183-20
3466	86-216-2
3466	86-217-2
3466	86-217-20
3466	86-218-2
3466	86-218-20
3466	86-219-2
3466	86-233-2
3466	86-246-2
3466	86-246-20
3466	86-251-2
3466	86-251-20
3466	86-276-2
3466	86-276-20
3466	86-286-2
3466	86-286-20
3466	86-294-2
3466	86-294-20
3466	86-340-2
3466	86-340-20
3466	86-406-2
3466	86-407-2
3466	86-459-2
3466	86-475-2
3466	86-482-2
3466	86-501-2
3466	86-527-2
3466	86-528-2
3466	86-533-2
3466	86-547-2
3466	86-552-2
3466	86-600-2
3466	86-622-2
3466	86A335
3466	86X0016-001
3466	86X0016-001A
3466	86X0036-001A
3466	86X0041-001
3466	86X0041-001A
3466	86X0044-001
3466	86X0044-001A
3466	86X0046-001
3466	86X0047-001
3466	86X46
3466	86X47
3466	96-5215-01
3466	96-5282-01
3466	96-5365-01
3466	99P10
3466	99P5
3466	99S039
3466	99S039A
3466	99S084-1
3466	101P1
3466	101P10
3466	102P1
3466	102P10
3466	103P935
3466	103P935A
3466	103PA
3466	104-170
3466	104-17NRCAE
3466	105
3466	106-12
3466	106-120
3466	108-6
3466	108-60
3466	110P1
3466	110P1AA
3466	110P1M
3466	112-000172
3466	112-000185
3466	112-000187
3466	112-10
3466	112-7
3466	112-8
3466	119-0055
3466	120-006604
3466	120P1
3466	120P1M
3466	121-1
3466	121-1019
3466	121-441
3466	121-444
3466	121-446
3466	121-496
3466	121-497
3466	121-602
3466	121-679
3466	121-699
3466	121-861
3466	121-865
3466	121-973
3466	121-978
3466	121-986
3466	123-006
3466	125B133
3466	125P1
3466	125P116
3466	125P1M
3466	129-20
3466	129-34
3466	130-149
3466	130-40315
3466	130-40429
3466	0131-000335
3466	0131-001328
3466	0131-001329
3466	0131-001439
3466	0131-005351
3466	0131-4328
3466	132-056
3466	132-074
3466	0133
3466	134P1A
3466	134P1M
3466	134P4
3466	134P4AA
3466	134P4M
3466	147-7009-01
3466	151-0087-00
3466	151-0124-00
3466	151-0188-00
3466	151-0221-00
3466	151-0221-02
3466	151-0325-00
3466	151-0458-00
3466	151-0459-00
3466	158P2
3466	158P2M
3466	173A4483-1
3466	173A4483-2
3466	177-006-9-001
3466	177-012-9-001
3466	200-052
3466	207V073C04
3466	209-1
3466	209P1
3466	211ATPE3391
3466	212-699
3466	223P1
3466	260-10-016
3466	260P08201
3466	260P11403
3466	260P15201
3466	260P15202
3466	260P15203
3466	260P16502
3466	297L012C01
3466	297L013B02
3466	297V073C03
3466	297V073C04
3466	297V083C01
3466	297V086C01
3466	311D589-P2
3466	344-6017-1
3466	352-0219-000
3466	352-0551-010
3466	352-0551-021
3466	352-0610-030
3466	352-0610-040
3466	352-0636-010
3466	352-0636-020
3466	352-0754-020
3466	352-0778-010
3466	352-0848-020
3466	352-0959-010
3466	352-0959-020
3466	352-0959-030
3466	353-9304-001
3466	386-1
3466	394-3145
3466	00415
3466	417-116
3466	417-116-13165
3466	417-132
3466	417-153
3466	417-168
3466	417-176
3466	417-182
3466	417-184
3466	417-196
3466	417-200
3466	417-201
3466	417-234-13165
3466	417-235
3466	417-235-13262
3466	417-242-8181
3466	417-260-50127
3466	430-20013-0B
3466	430-20018-0A
3466	430-20021
3466	430-20023-0A
3466	430-20026-0
3466	436-404-002
3466	461-1006
3466	461-1055-01
3466	549-2
3466	550-027-00
3466	559-1
3466	570-1
3466	574
3466	576-0001-002
3466	576-0001-013
3466	576-0002-008
3466	576-0003-017
3466	576-0003-019
3466	580R304H01
3466	600X0095-086
3466	601X0417-086
3466	602-56
3466	602-60
3466	620-1
3466	627-1
3466	635
3466	669
3466	686-0325-0
3466	686-2700
3466	690V086H86
3466	690V116H23
3466	690V118H60
3466	690V118H61
3466	700-133
3466	753-2000-101
3466	753-4004-248
3466	800-001-031-1
3466	800-101-108-1
3466	800-523-01
3466	800-523-02
3466	800-525-04
3466	800-527-00
3466	826-1
3466	829
3466	829B
3466	829C
3466	829D
3466	829E
3466	829F
3466	833
3466	921-160B
3466	921-197B
3466	921-254B
3466	921-296B
3466	921-29B
3466	921-332B
3466	921-348B
3466	921-405
3466	921-70
3466	921-70A
3466	921-70B
3466	943-721-001
3466	958-023
3466	991-01-0098
3466	991-01-0462
3466	991-01-1225
3466	991-01-1319
3466	991-01-2328
3466	991-01-3058
3466	991-01-3599
3466	991-011225
3466	1005M19
3466	1042-06
3466	1079-85
3466	1081-4000
3466	1081-4010
3466	1084-9784
3466	1089-6199
3466	1112-8
3466	1125-2582
3466	1236-3750
3466	1314
3466	1414-176
3466	1479-8029
3466	1553-17
3466	1582
3466	1679-7391
3466	1844-17
3466	1850-17
3466	1853-0001-1
3466	1853-0081
3466	1853-0089
3466	1867-17
3466	1940-17
3466	1979-808-10
3466	2017-107
3466	2043-17
3466	2057A100-51
3466	2057A2-182
3466	2057A2-183
3466	2057A2-198
3466	2057A2-298
3466	2057A2-307
3466	2057A2-343
3466	2057A2-353
3466	2057A2-359
3466	2057A2-397
3466	2057A2-400
3466	2057A2-403
3466	2057A2-406
3466	2057A2-430
3466	2057A2-457
3466	2057A2-489
3466	2057B106-12
3466	2057B108-6
3466	2057B147-12
3466	2057B159-12
3466	2158-1558
3466	2220-17
3466	2269
3466	2272
3466	2300.036.096
3466	2798
3466	3019
3466	3549-2
3466	3559-1
3466	3570
3466	3570-1
3466	3570P
3466	3574
3466	3597
3466	3597-1
3466	3597-2
3466	3620-1
3466	3627
3466	3627-1
3466	3631
3466	3634.2011
3466	4086
3466	4087
3466	4151-01
3466	4442
3466	4802-00004
3466	4815
3466	4822-130-40315
3466	4844
3466	4856-0106
3466	5001-048
3466	5001-066
3466	5001-509
3466	5226-1
3466	5611-628
3466	5611-628F
3466	5611-673
3466	5611-673D
3466	5701
3466	6201
3466	6854K90-062
3466	7303-1
3466	7303-2
3466	7363-1
3466	7503
3466	8000-00006-004
3466	8000-0004-P089
3466	8301
3466	8405
3466	8540
3466	8601
3466	8710-169
3466	8867
3466	9015C
3466	9330-767-60112
3466	9330-908-10112
3466	9652H
3466	09800
3466	10300
3466	12888
3466	13162
3466	19680
3466	20011
3466	22008
3466	22595-000
3466	22605-005
3466	26810-152
3466	29076-023
3466	30290
3466	31005
3466	33509-1
3466	41440
3466	43127
3466	45122
3466	45337-C
3466	50203-12
3466	50203-8
3466	59625-10
3466	60719-1
3466	63282
3466	000071150
3466	000071151
3466	71818-1
3466	75561-2
3466	75561-31
3466	75617-1
3466	75617-2
3466	77561-27
3466	78331
3466	83272
3466	87758
3466	87759
3466	90330-001
3466	90432
3466	94037
3466	95227
3466	95227-1
3466	95232
3466	95240-1
3466	96458-1
3466	101497
3466	102001
3466	102260
3466	102263
3466	113182
3466	115517-001
3466	118284
3466	119228-001
3466	119730
3466	121467
3466	123971
3466	123991
3466	124755
3466	126524
3466	126707
3466	126715
3466	131241
3466	132176
3466	132285
3466	132498
3466	133182
3466	133253
3466	135286
3466	137340
3466	138376
3466	139455
3466	140290
3466	140371
3466	140372
3466	140623
3466	141018
3466	141343
3466	141345
3466	141738P63-1
3466	142838
3466	142839
3466	143791
3466	143802
3466	143803
3466	143806
3466	143807
3466	145776
3466	147549-1
3466	147549-2
3466	147663
3466	150742
3466	150753
3466	150758
3466	150762
3466	150771
3466	167690
3466	181015
3466	181030
3466	181034
3466	183032
3466	187217
3466	188180
3466	190429
3466	198024
3466	198036-1
3466	198050
3466	200067
3466	200220
3466	200433
3466	202909-577
3466	202909-587
3466	202911-737
3466	203364
3466	205032
3466	205048
3466	205049
3466	205367
3466	210076
3466	232631
3466	267838
3466	309684
3466	320280
3466	321165
3466	324144
3466	333060-1029
3466	337342
3466	401003-0010
3466	436119-002
3466	489751-028
3466	489751-031
3466	489751-042
3466	489751-097
3466	489751-124
3466	489751-130
3466	489751-146
3466	543995
3466	552503
3466	610074-1
3466	610083
3466	610083-3
3466	610093-1
3466	610099-6
3466	610110-1
3466	610110-2
3466	610120-1
3466	610125-1
3466	610147-2
3466	610158-2
3466	610209-1
3466	610223-1
3466	615180-1
3466	615180-2
3466	615180-3
3466	615180-4
3466	650060
3466	698941-1
3466	701584-00
3466	701589-00
3466	721272
3466	760269
3466	801540
3466	815199
3466	815199-6
3466	815211
3466	815213
3466	815229
3466	815236
3466	815247
3466	838105
3466	891008
3466	900552-17
3466	908864-2
3466	916051
3466	916062
3466	928408-101
3466	932040
3466	932107-1
3466	960106-3
3466	970246
3466	970248
3466	970251
3466	970254
3466	970663
3466	970762
3466	971059
3466	984193
3466	988990
3466	1417330-3
3466	1417330-4
3466	1417347-1
3466	1417363-1
3466	1471112-7
3466	1471112-8
3466	1471114-1
3466	1472501-1
3466	1473501-1
3466	1473523-1
3466	1473559-001
3466	1473559-1
3466	1473570-2
3466	1473581-1
3466	1473599-1
3466	1503097-0
3466	1616226-1
3466	1617032
3466	1700001
3466	1700034
3466	1780142
3466	1780522-1
3466	1780522-2
3466	1780522-2-001
3466	1861223-1
3466	1945294
3466	1950052
3466	1950056-1
3466	1969281
3466	2056606-0701
3466	2057013-0004
3466	2057013-0007
3466	2057013-0008
3466	2057013-0012
3466	2057013-0701
3466	2057013-0702
3466	2057013-0703
3466	04440032-002
3466	04440032-003
3466	04440032-004
3466	04440032-005
3466	04440032-006
3466	04440032-007
3466	04440032-008
3466	04440032-009
3466	04450016-001
3466	04450016-002
3466	23114300
3466	23114301
3466	23114302
3466	6623001100
3466	6623002000
3466	6623002100
3466	6623002200
3466	6624002000
3466/159	A675
3466/159	HEP-75
3466/159	TT2SA495-O
3466/159	TT2SA495-Y
3466/159	001
3466/159	001-022020
3466/159	2N1131
3466/159	2N1917
3466/159	2N1918
3466/159	2N2598
3466/159	2N2599
3466/159	2N2599A
3466/159	2N2600
3466/159	2N2600A
3466/159	2N2800
3466/159	2N2801/46
3466/159	2N3841
3466/159	2N5354
3466/159	2N5355
3466/159	2N864A
3466/159	2SA467
3466/159	2SA467G
3466/159	2SA467G-R
3466/159	2SA467G-Y
3466/159	2SA493
3466/159	2SA493GR
3466/159	2SA493Y
3466/159	2SA494
3466/159	2SA494-GR
3466/159	2SA494-O
3466/159	2SA494-Y
3466/159	2SA499
3466/159	2SA499-O
3466/159	2SA499-R
3466/159	2SA499-Y
3466/159	2SA500
3466/159	2SA500-Y
3466/159	2SA502
3466/159	2SA510
3466/159	2SA510-O
3466/159	2SA510-R
3466/159	2SA511
3466/159	2SA511-O
3466/159	2SA511-R

SK	Industry Standard No.	SK	Industry Standard No.	SK	Industry Standard No.	SK	Industry Standard No.	SK	Industry Standard No.
3466/159	28A512	3472	916063	3488	23119990	3501	4A232	3512	FBN-CP34634
3466/159	28A512-0	3472	916092	3488	6644001100	3501	4A262	3512	FBN-L109
3466/159	28A512-R	3473	GEIC-138	3488	B0313300	3501	406	3512	FBN-L113
3466/159	28A513	3473	MPC575C2	3489	GEIC-106	3501	4G1122	3512	FBN-L148
3466/159	28A513-Q	3473	001-0091	3489	REN1134	3501	4G132	3512	HEP-83023
3466/159	28A513-R	3473	36-0083	3489	TAT145P	3501	4G162	3512	HEP-85014
3466/159	28A539	3473	575C2	3489	46-1395-3	3501	4G2122	3512	RCA1A03
3466/159	28A539K	3473	1001-0091/4460	3489	1061-9161	3501	4G232	3512	SDT5502
3466/159	28A539L	3473	90200100	3489	23119989	3501	4G262	3512	SDT5503
3466/159	28A539M	3475	GEIC-123	3489	6644001400	3501	6DC11	3512	SDT5507
3466/159	28A539S	3475	GEIC-133	3490	B0313400	3501	1OM60	3512	SDT5508
3466/159	28A544	3475	IC-142	3490	GEIC-107	3501	2OHA3	3512	SDT5512
3466/159	28A569J	3475	M51191P	3490	REN1135	3501	2N186B	3512	2N1700
3466/159	28A570	3475	MPC562C	3490	TAT146P	3501	25C60	3512	2N5119
3466/159	28A610	3475	RH-1X0025CEZZ	3490	46-1398-3	3501	26R2	3512	2N5320
3466/159	28A610B	3476	21A101-018	3490	1061-9856	3501	26R28	3512	28C235
3466/159	28A611	3476	GEIC-104	3490	23119988	3501	36R2	3512	28C235A
3466/159	28A611-4E	3476	HC1000505	3490	6644001500	3501	36R28	3512	28C235B
3466/159	28A640A	3476	TAT122AP	3492	GEIC-79	3501	4OA600	3512	28C235C
3466/159	28A640B	3476	TAT122AP-D	3492	LA4032P	3501	4OHF60	3512	28C235D
3466/159	28A640C	3476	TAT122AR	3493	AN360	3501	124-0165	3512	28C235E
3466/159	28A640D	3476	TAT122P-B	3493	GEIC-295	3501	302M	3512	28C235F
3466/159	28A640E	3476	15-14471-1	3493	87-0246	3501	303M	3512	28C235G
3466/159	28A640L	3477	G09-008-B	3494	AN210	3501	371M	3512	28C235H
3466/159	28A640M	3477	G09-008-C	3494	AN210A	3501	402M	3512	28C235J
3466/159	28A640S	3477	G09-017-B	3494	AN210B	3501	403M	3512	28C235K
3466/159	28A675	3477	G09-017-C	3494	AN210C	3501	417M	3512	28C235L
3466/159	28A675A	3477	G09-017-D	3494	AN210D	3501	419M	3512	28C235M
3466/159	28A675B	3477	GEIC-103	3494	GEIC-47	3501	316O	3512	28C235OR
3466/159	28A675C	3477	GEIC-171	3495	AN260	3501	40209	3512	28C235X
3466/159	03A06	3477	TA7120	3495	GEIC-60	3501	40214	3512	28C235Y
3466/159	13-39115-1	3477	TA7120B	3496	AN277	3501	198764-13	3512	28A682
3466/159	19-020-114	3477	TA7120P	3496	AN277AB	3501	198765-13	3512	28C482-GR
3466/159	21A112-102	3477	051-0020-00	3496	AN277B	3501	225265	3512	28C482GR
3466/159	23-5045	3477	051-0020-02	3496	AN277BA	3501	227676	3512	28C482X
3466/159	46-86514-3	3477	051-0035-01	3496	GEIC-63	3501	229088	3512	86-674-2
3466/159	48-155035	3477	051-0035-02	3497	AN362	3501	230756	3512	362A10
3466/159	57B121-9	3477	051-0035-04	3498	GEIC-136	3501	255220	3512	40594
3466/159	86-005135-2	3477	588-40-203	3498	MPC571C	3501	235299	3512	94062
3466/159	86-555-2	3477	740-2007-120	3498	RVIUPC22C	3501	237509	3512	132500
3466/159	0110	3477	740-9037-120	3499	A4030P	3501	240077	3512	223753
3466/159	121-603	3477	25810-166	3499	GEIC-77	3501	241420	3512	241302
3466/159	121-746	3477	25840-166	3499	LA4030P	3501	255903	3512	572001
3466/159	121-777	3477	58810-172	3499	46-13131-3	3501	256122	3512	574003
3466/159	0201	3477	58840-203	3500	AM62	3501	256728	3512	660100
3466/159	260P16503	3477	171179-045	3500	BAV807	3501	256730	3512	1971489
3466/159	202P16603	3477	02071120	3500	BAV817	3501	258884	3512	7311350
3466/159	921-292B	3477	742363	3500	BAV827	3501	262310	3513	A484
3466/159	991-012686	3477	5350251	3500	BY807	3501	265115	3513	G03-404-B
3466/159	991-015614	3477	B0313800	3500	BY817	3501	300312	3513	G03-404-C
3466/159	2057A2-277	3478	GEIC-111	3500	BY827	3501	300733	3513	HEP-85027
3466/159	8000-00004-089	3478	REN1130	3500	BYX39/600	3502	S2710B	3513	HEP-85013
3466/159	033589	3478	TA7150P	3500	BYX42/600	3502	S2710D	3513	Q5209
3466/159	59625-11	3478	46-1393-3	3500	BYX48/600	3502	S2710M	3513	2N5322
3466/159	59625-12	3478	1061-9666	3500	FD-1029-JE	3502	2N4101	3513	2B8684
3466/159	59625-2	3478	23119993	3500	GE-5040	3503	S1003M	3513	40595
3466/159	59625-3	3479	GB-81	3500	HEP-R0136	3503	S2003M	3513	62584
3466/159	59625-4	3479	MP8-A05	3500	MA50	3503	S2600M	3513	94068
3466/159	59625-5	3479	MP8-A06	3500	MR1126	3503	S2610M	3513	559557
3466/159	59625-6	3479	MPSA06	3500	R0130	3503	S2620M	3514	AD741
3466/159	59625-7	3479	MPSA20	3500	R0131	3503	84003M	3514	AD741C
3466/159	59625-8	3479	RB66	3500	R0132	3503	13-33183-1	3514	AD741CH
3466/159	59625-9	3479	2N1564	3500	R0134	3503	40379	3514	AM741HC
3466/159	123940	3479	2N1565	3500	R0136	3504	M23C	3514	CA3741
3466/159	130536	3479	2N1566	3500	85AN12	3504	2N1850	3514	CA3741CT
3466/159	146142	3479	2N1566A	3500	85AN6	3504	2N1850A	3514	CA3741T
3466/159	147603	3479	2N1613/46	3500	86AN12	3504	2N5170	3514	CA741CT
3466/159	170128	3479	2N1711/46	3500	86AR6	3504	MC83935-8	3514	EC0941
3466/159	171555	3479	2N2087	3500	SJ605	3505	S6035H	3514	FU5B7741393
3466/159	181619	3479	2N2317	3500	SJ603E	3505	2N3899	3514	GEIC-263
3466/159	610083-1	3479	2N2380	3500	SJ603K	3506	L4001M9	3514	HA1T741M
3466/159	1417303-1	3479	2N2380A	3500	SJ604	3506	2N5756	3514	HC1000217
3467	B2L	3479	2N2479	3500	SJ604B	3507	T4700D	3514	LM741
3467	BDY24	3479	2N2514	3500	SJ604X	3507	Q4010H	3514	LM741C
3467	BDY25	3479	2N2515	3500	SL500	3508	2N5570	3514	LM741CH
3467	BDY26	3479	2N734	3500	SL600	3508	2N5574	3514	LM741H
3467	BDY27	3479	2N734A	3500	TM56	3509	Q4025H	3514	N5141T
3467	BDY28	3479	2N735	3500	TR1125	3509	SPT330	3514	MC1741CG
3467	BUY69B	3479	2N735A	3500	TR1126	3509	SPT430	3514	N5741T
3467	BUY69C	3479	2N736	3500	1N1206A	3509	T6411D	3514	PA7741
3467	C1477	3479	2N736A	3500	1N1616	3509	TIC207D	3514	PA7741C
3467	CG8-75	3479	2N736B	3500	1N2258	3509	2N5445	3514	RC741T
3467	D321	3479	2N956	3500	1N2258A	3510	2B6161	3514	SC5175G
3467	D351	3480	LA1369	3500	1N2259	3510	FBN-36220	3514	SG741CT
3467	D383	3481	GEIC-69	3500	1N2259A	3510	FBN-36485	3514	SN72741L
3467	MJ413	3481	LA1111	3500	12P60	3510	FBN-36603	3514	SS8741CJ
3467	MJ431	3481	LA1111P	3500	46R2	3510	FBN-36972	3514	TBA221
3467	MJ9000	3481	09-308050	3500	66R2	3510	FBN-36973	3514	TBA221/741C
3467	RCA1B09	3481	1111P	3500	404M	3510	GE-19	3514	TBA222
3467	2N5241	3482	GEIC-178	3500	408M	3510	P04450040-002	3514	TBA222/741
3467	2N6510	3482	TA7055P	3500	227675	3510	P5034	3514	UA741C
3467	2N6511	3483	AN366	3500	227724	3510	P5149	3514	ULN2741D
3467	2N6512	3483	87-0233	3500	227801	3510	S353	3514/941	IC40
3467	2N6513	3484	1156	3500	229042	3510	2N1490	3514/941	MC1741CP2
3467	2N6514	3484	003536	3500	230773	3510	2N1702	3514/941	MC1741CM
3467	2N6544	3485	EA33X8372	3500	232519	3510	2SC520	3515	HF-1C
3467	2SC2122	3485	GEIC-128	3500	235157	3510	2SC520A	3515	R0606
3467	2SD321	3485	K24154	3500	256729	3510	6-137	3515	S-15
3467	2SD383	3485	MPC554	3500	262648	3510	33-052	3515	UP-1
3467	15-123100	3485	MPC554C	3500	263807	3510	27126-100	3515	UP-1C
3467	48-137524	3485	PC554	3500	270779	3510	37334	3515	182463
3467	488155118	3485	09-308038	3500	300532	3510	37475	3516	BAV208
3467	938-5	3485	21M485	3500	500T735	3510	38138	3516	BAV209
3467	40852	3485	36-0041	3501	BYX52-600	3510	38166	3516	BAV218
3467	40853	3485	57A132-29	3501	BYX56-600	3510	198034-2	3516	BAV219
3467	610122-2	3485	57A32-29	3501	GE-5104	3510	983055	3516	BAV228
3467	2321241	3485	61A001-11	3501	HEP-R0256	3510	1968977	3516	BAV229
3467/283	86-564-2	3485	588-40-202	3501	HEP-R0257	3510	3130058	3516	BAV308
3467/283	BDX23	3485	740-9000-554	3501	R-63HZ	3510	3130091	3516	BAV309
3467/283	C675	3485	2057A32-26	3501	R0160	3510	3463100-1	3516	BAV318
3467/283	C807	3485	8710-172	3501	R0162	3511	FBN-38022	3516	BAV319
3467/283	C807A	3485	8910-148	3501	R0164	3511	LN75116	3516	BAV328
3467/283	MJ423	3485	25840-165	3501	R0250	3511	2N6254	3516	BAV329
3467/283	2N5240	3485	28810-176	3501	R0251	3511	2N6371	3516	BY209
3467/283	28C675	3485	58810-171	3501	R0253	3511	6-138	3516	BY219
3468	GEIC-159	3485	000074030	3501	R0254	3511	48P217241	3516	BY228
3468	61K001-13	3485	88510-178	3501	R0255	3511	61-309449	3516	BY229
3468	87-0217	3485	916070	3501	R0256	3511	3772-1	3516	BY308
3470	GEIC-121	3486	GEIC-137	3501	R0257	3511	3772-2	3516	BY309
3471	GEIC-118	3487	TA7130P	3501	R6HZ	3511	37476	3516	BY318
3471	KD6311	3487	TA7130PB	3501	83160	3511	198064-1	3516	BY319
3471	RE537-1C	3487	TA7130PC	3501	83250	3511	1967784	3516	BY328
3471	SAJ72157	3487	02-257130	3501	TR602	3511	4450023-007	3516	BY329
3471	TV8-MPC23C	3487	51844789J02	3501	1N1190	3512	C235	3516	PR9000
3471	TV8-UPC23C	3488	A0311400	3501	1N1190A	3512	C235-0	3516	PR9001
3471	TV8UPC23C	3488	B0311400	3501	1N1198A	3512	C482	3516	1N1235
3471	09-308009	3488	B0311422	3501	1N143B	3512	C482-GR	3516	1N1235A
3471	21A101-005	3488	GEIC-105	3501	1N2160	3512	C482-O	3516	1N1235B
3471	37002001	3488	REN1128	3501	1N2285	3512	C482-Y	3516	1N1236
3472	GEIC-130	3488	TA7124P	3501	1N4527	3512	C482GR	3516	1N1236B
3472	RE542-M	3488	46-1396-3	3501	1S1842	3512	C482X	3516	1N2222
3472	09-308096	3488	1061-9153	3501	3C6	3512	C482Y		
		3488	23119981	3501	4A162	3512	EA15X25Q		

SK	Industry Standard No.	SK	Industry Standard No.	SK	Industry Standard No.	SK	Industry Standard No.	SK	Industry Standard No.
3516	1N2222A	3525	38446-00000	3538	C840A	3551	MC1709CG	3561	TV828C647R
3516	1N2223	3525	38446-00010	3538	C840AC	3551	SN72709L	3561	2N5038
3516	1N2223A	3525	38446-00020	3538	FBN-36488	3551	709BC	3561	2N5039
3516	1N2531	3525	74004-1	3538	HEP-35012	3552	EC0941M	3561	2N6354
3516	1N2532	3525	74004-2	3538	HT4031151E	3552	GEIC-265	3561	26647
3516	1N2542	3525	243028	3538	PN26	3552	HEP-C6052P	3561	28C647Q
3516	1N2543	3525	619694-1	3538	RX34	3552	RC741DN	3561	28C647R
3516	1N2553	3526	CA747CT	3538	YAAN28D141	3552	RC741DP	3562	CT91
3516	1N2554	3526	CA747T	3538	2N3441	3552	SN72741P	3562	C850
3516	1N562	3526	GEIC-268	3538	28767C	3552	741TC	3562	C830A
3516	181356	3526	LM747C	3538	2N6263	3553	C6052P	3562	C830B
3516	181629	3526	LM747CH	3538	2N6264	3553	CA3741CB	3562	C830C
3516	18417	3526	SQ747CT	3538	28C1055	3553	CA3741B	3562	D92D
3516	3070	3526	747HC	3538	28C1055H	3553	CA741CB	3562	283878
3516	3080	3526/947	CA3747CT	3538	28C840	3553	FU9T7741393	3562	2N5879
3516	3780	3526/947	09-308067	3538	28C840A	3553	LM710CN	3562	2N6500
3516	3LC12	3526/947	051-0011-00-05	3538	28C840B	3553	MC1710CG	3562	28C791
3516	305P	3526/947	741HC	3538	28C840C	3553	MC1741CP1	3562	28C791A
3516	305B	3526/947	740502	3538	28C840D	3553	SN72710L	3562	28C791B
3517	GE-5041	3528	2N5416	3538	28C840E	3553	276-010	3562	28C791C
3517	MR1126R	3528/397	2N5679	3538	28C840F	3553	710HC	3562	28C791D
3517	1N1206AR	3528/397	2N5680	3538	28C840Q	3556	CA747CE	3562	28C791P
3517	1N1206BR	3529	2N5189	3538	28C840GN	3556	CA747CF	3562	28C791G
3517	1N1206R	3529	2N5262	3538	28C840H	3556	MJ747CP	3562	28C791GN
3517	12FR60	3529	28C1303	3538	28C840J	3556	SO747CN	3562	28C791H
3517	12FR60A	3529	28C566	3538	28C840K	3556	SN72747J	3562	28C791J
3517	12FR60B	3529	28C566A	3538	28C840L	3556	SN72747N	3562	28C791K
3517	408RM	3529	28C566B	3538	28C840M	3556	S88747CP	3562	28C791M
3517	40114R	3529	28C566D	3538	28C840P	3557	S2060D	3562	28C791OR
3518	BYX56-600R	3529	28C566E	3538	28C840R	3557	SP1R3D41	3562	28C791R
3518	BYX56-600R	3529	28C566F	3538	28C840X	3557	S2060D	3562	28C791X
3518	GE-5105	3529	28C566G	3538	28C840Y	3558	C122C1	3562	28C791Y
3518	R3145	3529	28C566GN	3538	28D102	3558	C122D	3562	28C830A
3518	R3150	3529	28C566J	3538	28D102-0	3558	S2800D	3562	28C830B
3518	R3160	3529	28C566K	3538	28D102-Y	3559	B2H	3562	28C830C
3518	R3245	3529	28C566L	3538	28D102A	3559	RE29	3562	28C830D
3518	R3250	3529	28C566M	3538	28D102B	3559	SDT430	3562	28C830F
3518	R3260	3529	28C566R	3538	28D102C	3559	ST-28D1106	3562	28C830GN
3518	1N1190R	3529	28C566RN	3538	28D102D	3559	STA9564	3562	28C830H
3518	1N1198R	3529	28C566X	3538	28D102E	3559	28C1051C	3562	28C830J
3518	1N2160R	3529	28C566Y	3538	28D102F	3559	46-86522-3	3562	28C830L
3518	25060R	3530	2N1483	3538	28D102G	3559	488137548	3562	28C830M
3518	40HPR60	3530	2N1484	3538	28D102GN	3559	121-449	3562	28C830R
3518	12TA905PO1	3530	2N1485	3538	28D102H	3559	200X5110-607	3562	28C830Y
3518	128A157PO1	3530	2N1486	3538	28D102J	3559	1061-6282	3562	28D130
3518	128A157PO2	3530	2N1701	3538	28D102K	3559	1063-8369	3562	28D130-YEL
3518	131A246PO1	3530	28D184	3538	28D102L	3559	40854	3562	28D130A
3518	131A246PO2	3530	28D184A	3538	28D102R	3559/162	DT8403	3562	28D130B
3518	222-1	3530	28D184B	3538	28D102OR	3559/162	DT8409	3562	28D130BL
3518	222-2	3530	28D184C	3538	28D102X	3559/162	GE-73	3562	28D130D
3518	335M	3530	28D184D	3538	28D102Y	3559/162	MJ3430	3562	28D130F
3518	336M	3530	28D184J	3538	28D656	3559/162	SDT-411	3562	28D130G
3518	371RM	3530	28D184K	3538	19A111527-P1	3559/162	SDT-413	3562	28D130GN
3518	417RM	3530	28D184L	3538	22-001008	3559/162	SDT-423	3562	28D130H
3518	419RM	3530	28D184M	3538	22-001009	3559/162	SDT1611	3562	28D130J
3518	435M	3530	28D184OR	3538	41-0909	3559/162	SDT1612	3562	28D130K
3518	436K	3530	28D184Y	3538	48013001A01	3559/162	SDT1613	3562	28D130Y
3518	436M	3530	36508	3538	488155042	3559/162	SDT1614	3562	28D142
3518	1521-4051	3530	40368	3538	488137535	3559/162	SDT1616	3562	28D142A
3518	1471-4729	3533	SC146D	3538	488155005	3559/162	SDT1617	3562	28D142B
3518	40214R	3533	SC146D2	3538	488155044	3559/162	SDT1618	3562	28D142C
3518	198765-14	3533	SC146D3	3538	86-412-2	3559/162	SDT402	3562	28D142D
3518	1851490	3533	SC146D4	3538	96-5191-01	3559/162	SPC430	3562	28D142E
3518	4420022-P1	3533	SC146D5	3538	132-065	3559/162	ST8402	3562	28D142G
3518	4420022-P2	3533	SC146D6	3538	322-1	3559/162	ST8409	3562	28D142GN
3518	24561601-E	3533	T2801D	3538	1065-9944	3559/162	ST8430	3562	28D142H
3518	24561602-E	3533	T2802D	3538	2295	3559/162	2N5804	3562	28D142J
3518	24561603	3533	TIC246D	3538	119650	3559/162	2N5805	3562	28D142K
3519	2N5757	3534	BD278	3538	139270	3559/162	2N6249	3562	28D142L
3520	Q6010H	3534	2N6098	3538	1417322-1	3559/162	2N6250	3562	28D142M
3520	Q6015H	3534	2N6099	3538	23114038	3559/162	2N6251	3562	28D142OR
3520	T4110M	3534	2N6100	3538	23114961	3559/162	28C1050D	3562	28D142R
3520	T4110M	3534	2N6101	3539	23029	3559/162	28C1050E	3562	28D142Y
3521	SPT525	3534	50200-8	3539	CA3030	3559/162	28C1050F	3562	28D150
3521	TIC252E	3535	A580-040215	3539	CA3030A	3559/162	28C1051	3562	28D226A
3521	TIC262E	3535	A580-040315	3539	CA3037	3559/162	28C1051B	3562	28D226AP
3522	BTU0660	3535	A580-080315	3539	CA3037A	3559/162	28C1051LB	3562	28D226B
3522	Q5025H	3535	A580-080515	3539	CA3038	3559/162	28C1051LC	3562	28D226BP
3522	Q6025E	3535	BDX13	3539	CA3038A	3559/162	28C1051LD	3562	28D226C
3522	SPT530	3535	BUY43	3539	GEIC-248	3559/162	28C1051LE	3562	28D226F
3522	SPT540	3535	BUY46	3540	CA3010	3559/162	28C1051LF	3562	28D226G
3522	SPT630	3535	CXL181	3540	CA3010A	3559/162	28C1106	3562	28D226GN
3522	SPT640	3535	ECG181	3540	CA3015	3559/162	28C1106A	3562	28D226H
3522	T6411M	3535	GE-75	3540	CA3015A	3559/162	28C1106K	3562	28D226J
3522	TIC272E	3535	HEP-87000	3540	GEIC-288	3559/162	28C1106L	3562	28D226K
3522	TIC272M	3535	MJ802	3540	HA1301	3559/162	28C1106M	3562	28D226L
3522	2N5275	3535	PP3001	3541	CA3056A	3559/162	28C1106Q	3562	28D226M
3522	2N5446	3535	PP3004	3541	CA3058	3559/162	28D198	3562	28D226OR
3522	2N6162	3535	PP3007	3541	CA3059	3559/162	28D198A	3562	28D226P
3523	CXL6408	3535	RE57	3541	CA3079	3559/162	28D198AP	3562	28D226Q
3523	REN6408	3535	REN181	3541	GEIC-256	3559/162	28D198AQ	3562	28D226R
3523	1N3301	3535	2N5885	3542	CA3018	3559/162	28D198AR	3562	28D226X
3524	CA3020	3535	2N5886	3542	CA3018A	3559/162	28D198HQ	3562	28D226Y
3524	CA3020A	3535	28D113	3542	GEIC-289	3559/162	28D198HR	3562	28D239
3524	GEIC-236	3535	28D114	3542	1463681-1	3559/162	28D198S	3562	28D239F
3524	HA1302	3535/181	BD181	3543	CA3045	3559/162	28D198R	3562	28D92
3524	KD2115	3535/181	BD182	3543	CA3045F	3559/162	28D198V	3562	28D92D
3524	11505	3535/181	BD183	3543	CA3046	3559/162	28D632	3562	40364
3524	12503	3535/181	C1079	3543	CA3086	3559/162	28D632P	3563	C646
3524	12503RB	3535/181	C1080	3543	GEIC-172	3559/162	28D632Q	3563	REN223
3524	38265-00000	3535/181	2N6259	3543	HA1127	3559/162	28D748	3563	2N5874
3524	38265-00010	3535/181	28C897	3543	LM3045D	3559/162	40850	3563	28C1629
3524	38265-00020	3535/181	28C897A	3543	LM3046N	3560	MJ411	3563	28C1629A
3524	38265-00030	3535/181	28C897B	3543	LM3086N	3560	RCA411	3563	28C1629AQ
3524	300620600-00	3535/181	28C898A	3543	MC3346P	3560	SDT410	3563	28C1629M
3524	8508331	3535/181	28C898B	3543	544-2002-008	3560	SDT411	3563	28C2199
3524	8508403-1	3535/181	139229	3543	2010-5409	3560	SDT413	3563	28C646
3525	CA3028	3535/181	1473669-2	3544	4082626-0001	3560	SDT423	3564	CA555CE
3525	CA3028A	3536	40347V1	3544	CA3054	3560	SDT431	3564	EC0955N
3525	CA3028AF	3537	40346V1	3544	GEIC-258	3560	SPC431	3564	GEIC-269
3525	CA3028AB	3538	B1Z	3545	CA3039	3560	ST8431	3564	MC1455P1
3525	CA3028B	3538	B5C	3545	GEIC-247	3560	28C1454	3564	NE555JG
3525	CA3028BF	3538	BDY71	3545	GEIC-290	3560	28C935	3564	NE555V
3525	CA3028BB	3538	BDY72	3547	CA3000	3560	28D604	3564	SN72555P
3525	CA3053	3538	BDY78	3547	GEIC-245	3560	41506	3565	TM955M
3525	GEIC-86	3538	BUY38	3548	CA3026	3560	2003145404	3565	AMLM201
3525	LM3028	3538	C1024	3548	CA3049	3561	BDY57	3565	AMLM301
3525	LM3028A	3538	C1024-D2	3548	CA3049T	3561	BDY58	3565	AMLM301A
3525	TA7045	3538	C1024B	3548	GEIC-246	3561	TV828C647	3565	CA201AT
3525	TA7045M	3538	C1024C	3548	1477-3352	3561	TV828C647-O	3565	CA201T
3525	09-308002	3538	C1024D	3549	CA3001	3561	TV828C647-P	3565	CA301AT
3525	09-308003	3538	C1024E	3549	GEIC-180	3561	TV828C647A	3565	LM201AH
3525	57A32-1	3538	C1160	3549	HA1110	3561	TV828C647B	3565	LM201H
3525	133-003	3538	C1161	3549	TA7031M	3561	TV828C647C		
3525	133-005	3538	C254	3550	CA3081	3561	TV828C647D		
3525	307-008-9-001	3538	C487	3550	GEIC-257	3561	TV828C647E		
3525	351-1011-022	3538	C488	3550	M2032-330	3561	TV828C647Q		
3525	351-1011-032	3538	C489	3550	51-842320A32				
3525	1042-11	3538	C840	3550	51884320A32				
3525	1820-0306			3550	1858-0023				
3525	8000-0004-P090			3551	CA14588				
3525	9694-1								

SK	Industry Standard No.
3565	LM301AH
3565	LM301AL
3565	LM748CH
3565	ML201AT
3565	ML301T
3565	RC748T
3565	SG201AT
3565	SG201T
3565	SG301AT
3565	SN72301AL
3565	SSS201AJ
3565	SSS301AJ
3565	748BE
3565	748CE
3567	LM311H
3569	CA239AE
3569	CA239AG
3569	CA239E
3569	CA239G
3569	CA339AE
3569	CA339AG
3569	CA339E
3569	CA339G
3569	LM239AD
3569	LM2901N
3569	LM339AD
3569	LM339AN
3569	LM339N
3570	MCR406-1
3570	MCR406-2
3570	MCR406-3
3570	MCR406-4
3570	S1003RS2
3570	S2003RS2
3570	S2060A
3570	S2060B
3570	2N6236
3570	2N6237
3570	2N6238
3571	S2060E
3571	S2060M
3571	S2062D
3572	C122B
3572	C122B1
3572	S2008L
3572	S2800B
3572	40868
3574	T1C116A
3574	T1C116B
3574	T1C116F
3574	T1C126A
3574	T1C126B
3574	T1C126F
3574	2N6394
3575	S6000C
3575	S6100C
3576	S6000E
3576	S6100E
3577	S0301M
3577	S0501M
3577	S0503M
3577	S1001M
3577	S2001M
3577	S2600B
3577	S2620B
3577	2N1595
3577	2N1596
3577	2N1597
3577	2N1600
3577	2N1601
3577	2N1602
3578	S2600D
3578	S2610D
3578	S2620D
3578	TA7554
3578	TA7555
3578	2N1598
3578	2N1599
3578	2N1603
3578	2N1604
3578	2N2653
3579	2N1845B
3579	2N1846
3579	2N1846A
3579	2N1846B
3580	RTU0130
3580	RTU0140
3580	2N1849
3580	2N1849A
3580	2N1849B
3581	C37G
3581	MCR3935-4
3581	RE173
3582	GEMR-3
3582	IR30D
3582	MCR2604-1
3582	MCR2604-2
3582	MCR2604-3
3582	MCR3935-6
3582	RE176
3584	GE-5024
3585	A40M
3586	D2412A
3586	D2412B
3586	D2412F
3587	D2412C
3587	D2412D
3587	D2412M
3588	D2540A
3588	D2540B
3588	D2540F
3589	C220B
3589	D2540D
3589	D2540M
3589	RTU0102
3589	RTU0110
3589	RTU0120
3589	S1008H
3590	CD541
3590	FU5B7709393
3590	FU6A7709393
3590	G39007681
3590	GEIC-250
3590	HL24630
3590	KS-20971-L1
3590	LM709C
3590	LM709CH
3590	LM709CN
3590	MC1709CP1
3590	MC1709CP2
3590	MC1709P2
3590	N5709A
3590	N5709T
3590	PA7000/591
3590	PA7001/501
3590	PA7001/502
3590	PA7709
3590	PA7709C
3590	RC709DP
3590	RC709T
3590	SG709CT
3590	SL07040
3590	SL07055
3590	SL08066
3590	SL20927
3590	SL21923
3590	SL23324
3590	SL23482
3590	TAA521
3590	TAA521/709
3590	13-53-6
3590	15-37833-1
3590	15-37833-2
3590	17-12096-1
3590	19A116549P1
3590	32-807072-1
3590	43A212040-1
3590	43A212040-2
3590	46-136279P1
3590	51-25789H
3590	126-40
3590	133P80057
3590	156-0015-00
3590	179-46447-03
3590	221-30
3590	349-212-003
3590	351-7140-010
3590	442-7
3590	477-0376-001
3590	586-024
3590	709DC
3590	709PC
3590	930C613-P1
3590	1081K94-7
3590	1820-0058
3590	1820-0248
3590	2036-68
3590	2036-72
3590	2036-93
3590	37833-2
3590	41180
3590	55986-1
3590	130221
3590	198410-1
3590	505342
3590	517999
3590	658583
3590	710398-28
3590	932030-1
3590	989709
3590	1604609-2
3590	1872425-1
3590	2391773
3590	04040501-1
3590	5113642
3590	7528157P1
3590	11242096
3590	16763333-001
3590	30677435-001
3590	84667800
3590	91011500
3591	GEIC-190
3591	LM340T-5.0
3591	LM341P-5.0
3591	TA78005P
3591/960	SN72905
3592	LM340T-12
3592	LM341P-12
3592	LM342P-12
3592	2361461
3592/966	SN72912
3593	LM340T-15
3593	LM341P-15
3593/968	SN72915
3595	LM565CN
3595	NE565A
3596	LM307N
3596	LM307P
3596	ML307S
3596	SG307M
3596	SG307N
3597	C106B
3597	C106B1
3597	C106B2
3597	C106B3
3597	C106B4
3597	C107B
3597	C107B1
3597	C107B2
3597	C107B3
3597	C107B4
3597	CU-12E
3597	CXL5452
3597	CXL5453
3597	CXL5454
3597	CXL5455
3597	ECG5455
3597	GEMR-5
3597	IR106B1
3597	IR106B1-C
3597	IR106B2
3597	IR106B3
3597	IR106B4
3597	IR106B41
3597	IRC20
3597	RE171
3597	REN5454
3597	REN5455
3597	S106B
3597	S106B1
3597	S107B
3597	S107B1
3597	S2061B
3597	S2062B
3597	SFIR3D41
3597	TA2888
3597	TA2889
3597	TC106B1
3597	TC106B2
3597	TC106B3
3597	TC106B4
3597	TIC106B
3597	1RC20
3597	2SF942
3597	19B200248-P1
3597	19B200248-P3
3597	48-8375D02
3597	48D8375D02
3597	53-1517
3597	2321264
3597	134883875402
3597	134883875406
3598	C106D
3598	C106D1
3598	C107D
3598	CXL5456
3598	ECG5456
3598	S106D
3598	S107D
3598	86X0081-001
3599	GE-5020
3599	1N1346
3599	1N1346A
3599	1N1346B
3600	GE-5016
3600	1N1344
3600	1N1344A
3600	1N1344B
3601	1N1341
3601	1N1341A
3601	1N1341B
3602	GE-5036
3602	HEP-R0134
3602	1N1204A
3603	GE-5032
3603	GEMR-1
3603	HEP-R0131
3603	HEP-R0132
3603	1N1202A
3604	HEP-R0130
3604	1N1199A
3606	GE-5048
3607	405A
3608	GE-5100
3608	GEMR-2
3608	HEP-R0254
3608	HEP-R0255
3608	R43HZ
3608	R4HZ
3608	S3135
3608	S3140
3608	S3235
3608	S3240
3608	ST3-10
3608	ST3-20
3608	ST3-30
3608	ST3-40
3608	ST4-30
3608	ST4-40
3608	TR251
3608	TR300
3608	TR351
3608	TR352
3608	TR400
3608	TR401
3608	TR402
3608	1N1187
3608	1N1188
3608	1N1188A
3608	1N1196
3608	1N1196A
3608	1N1437
3608	1N2158
3608	1N2283
3608	1N4526
3608	1S1841
3608	3C4
3608	4C4
3608	10M40
3608	20S3
3608	20S4
3608	21R4S
3608	24R2
3608	25C40
3608	34R2
3608	35C40
3608	35H30
3608	35H40
3608	40A400
3608	40HP40
3608	40R3
3608	40S3
3608	302G
3608	302H
3608	303G
3608	303H
3608	371H
3608	402H
3608	403H
3608	417H
3608	419H
3608	3135
3608	3140
3608	3235
3608	40212
3608	198765-9
3609	GE-5096
3609	HEP-R0251
3609	HEP-R0253
3609	RN1120
3609	S1AN31
3609	S1AN40
3609	S1BN31
3609	S1BN40
3609	S3115
3609	S3120
3609	S3215
3609	S3220
3609	TR200
3609	1N1185
3609	1N1185A
3609	1N1186
3609	1N1186A
3609	1N1193
3609	1N1193A
3609	1N1194
3609	1N1194A
3609	1N1304
3609	1N1436
3609	1N249
3609	1N250A
3609	1N250B
3609	1N4525
3609	1S1654
3609	1S2447
3609	1S2448
3609	1S936
3609	3C2
3609	4C2
3609	10M20
3609	21R2S
3609	22R22
3609	25C15
3609	25C20
3609	31R2
3609	32R2
3609	35C15
3609	35C20
3609	35H10
3609	35H20
3609	40A150
3609	40A200
3609	40HF15
3609	40HF20
3609	96-5246-01
3609	302C
3609	302D
3609	303C
3609	303D
3609	371C
3609	371D
3609	402C
3609	402D
3609	403B
3609	403C
3609	403D
3609	417C
3609	417D
3609	419C
3609	419D
3609	3115
3609	3120
3609	40210
3609	198764-5
3609	198765-5
3610	HEP-R0250
3610	S3105
3610	S3205
3610	001-024070
3610	1N1183
3610	1N1183A
3610	1N1191
3610	1N1191A
3610	1N1301
3610	1N1434
3610	1N2154
3610	1N248
3610	1N248A
3610	1N248B
3610	1S2446
3610	20R2
3610	25C05
3610	35C05
3610	35H5
3610	40A50
3610	40HF5
3610	57-18
3610	371A
3610	402A
3610	403A
3610	3105
3610	40208
3610	198765-1
3610	8518382-1
3611	T4700B
3613	C222B
3613	MCR3818-1
3613	MCR3818-2
3613	MCR3818-3
3613	MCR3818-4
3613	PS020
3613	PS08
3613	PS18
3613	PS220
3613	PS28
3613	RTS0202
3613	RTS0205
3613	RTS0210
3613	RTS0220
3613	S0306G
3613	S0308G
3613	S0310G
3613	S0315G
3613	S0506G
3613	S0508G
3613	S0510G
3613	S0515G
3613	S1006G
3613	S1008G
3613	S1010G
3613	S1015G
3613	S2006G
3613	S2008G
3613	S2010G
3613	S2015G
3613	S6200A
3613	S6200B
3613	2N5164
3613	2N5165
3613	40749
3614	S6220B
3615	MCR2835-1
3615	MCR2835-2
3615	MCR2835-3
3615	MCR2835-4
3615	MCR3835-1
3615	MCR3835-2
3615	MCR3835-3
3615	MCR3835-4
3615	PS035
3615	PS135
3615	RTS0702
3615	RTS0705
3615	RTS0710
3615	RTS0720
3615	S0325G
3615	S0335G
3615	S0525G
3615	S1025G
3615	S1035G
3615	S2025G
3615	S2035G
3615	2N3870
3615	2N3871
3619	C1030
3619	C1030A
3619	C1030B
3619	C1030C
3619	C1030D
3619	C939
3619	C939D
3619	C939L
3619	2SC1030
3619	2SC1030A
3619	2SC1030B
3619	2SC1030C
3619	2SC1030D
3619	2SC1030E
3619	2SC1030F
3619	2SC1030G
3619	2SC1030J
3619	2SC1030K
3619	2SC1030L
3619	2SC1030M
3619	2SC1343
3619	2SC1343A
3619	2SC1343B
3619	2SC1343BL
3619	2SC1343C
3619	2SC1343D
3619	2SC1343E
3619	2SC1343F
3619	2SC1343G
3619	2SC1343GN
3619	2SC1343H
3619	2SC1343HA
3619	2SC1343HB
3619	2SC1343J
3619	2SC1343K
3619	2SC1343L
3619	2SC1343M
3619	2SC13430
3619	2SC13430R
3619	2SC1343R
3619	2SC1343X
3619	2SC1343Y
3619	2SC939
3619	2SC939D
3619	2SC939L
3619	2SD218
3620	QT-D0313XAC
3620	2N6102
3620	2N6103
3620	2SD313
3620	2SD313A
3620	2SD313B
3620	2SD313C
3620	2SD313D
3620	2SD313E
3620	2SD313F
3620	2SD313G
3620	2SD313GN
3620	2SD313H
3620	2SD313L
3620	2SD313M
3620	2SD313N
3620	2SD313R
3620	2SD313Y
3621	BDX50
3621	BDX51
3621	BDX60
3621	BDX61
3621	BDY18
3621	BDY19
3621	BDY20
3621	BDY55
3621	BDY56
3621	BDY73
3621	BDY77
3621	RCA1B01
3621	2N3442
3621	2N5630
3621	2N5878
3621	2N6262
3621	13-36440-1
3622	2N5202
3623	2N6212
3623	2N6213
3623	2N6214
3624	2N5344
3624	2N6211
3624	2SA766
3624	2SA766S
3624	2SA969
3625	SJ2009
3625	SJ652
3625	SJ822
3625/218	2N6467
3625/218	2N6468
3626	C1025
3626	2SC1025
3626	2SC1025CTV
3626	2SC1025D
3626	2SC1025E
3626	2SC1025MT
3626	46-86476-3
3627	EC103B
3627	EP15X106
3627	ID104
3627	MCR120
3627	RTC0120
3627	T1C47
3627	T1C63
3627	T1C64
3627	TIC47
3627	2N2690
3627	2N3008
3627	2N4149
3627	2N5064
3627	2N881
3628	D13T1
3628	GE-X17
3628	MPU131
3628	2N6027
3628	13-33184-1
3628	121U2AT
3629	LM309KC
3629	SG309K
3629	SG7805CK
3630	GEIC-191
3630	LM340T-8.0
3630	LM341P-8.0
3632	MAC11-4
3632	2N6154
3633	MAC11-6
3633	2N6155
3634	2N4442
3635	MCR3000-6
3635	2N4443
3636	MCR107-8
3636	MCR3000-8
3636	2N4444
3638	EC103A
3638	ID102
3638	MCR104
3638	SD-2N5062
3638	TIC46
3638	2N2689
3638	2N3007
3638	2N3029

SK	Industry Standard No.	SK	Industry Standard No.	SK	Industry Standard No.	SK	Industry Standard No.	SK	Industry Standard No.
3638	2N3032	3642	48-134752	3648/168	RT7402	3685	C122F1	3709	001-0020-0Q
3638	2N4147	3642	48-134753	3649	A-0205	3685	S0306L	3709	1N541
3638	2SC6062	3642	48-134760	3649	AM-0-11	3685	S0308L	3709	1N542
3638	2N879	3642	48-134761	3649	GEBR-1000	3685	S0506L	3709	1N542MP
3638	2SP102	3642	48-134788	3649	18DB10A	3685	S0508L	3709	1N60MP
3638	2SP102A	3642	48-137121	3649	522-958	3685	S2060Y	3709	18128PM
3638	2SP522	3642	48-137125	3652	Q2025H	3685	S2060Q	3709	18186PM
3638	86-110-3	3642	48-869205	3652	SPT040	3685	S2061P	3709	1850
3641	LM501AP	3642	48-869427	3652	SPT140	3685	S2061Q	3709	1860
3641	LM748CN	3642	57A22-1	3652	SPT230	3685	S2061Y	3709	2A119
3641	ML748CS	3642	57A22-2	3652	SPT240	3686	C122A1	3709	003-004200
3641	SG301AP	3642	57C22-1	3652	T6411B	3686	S1006L	3709	09-306019
3641	SN72301AP	3642	57C22-2	3652	TIC272B	3686	S1008L	3709	09-306057
3641	SSS301AP	3642	62-18429	3652	2N5444	3686	40867	3709	09-306229
3642	AD133	3642	86-370-2	3652	2N6160	3687	S4006L	3709	09-306335
3642	AD142	3642	998057	3653	2N3872	3687	S4008L	3709	09-30636
3642	AJ103	3642	112-7292955	3654	2N3873	3687	40869	3709	12-085041
3642	AL113	3642	121-406	3655	B6420A	3688	CA3401E	3709	13-085029
3642	AT1138	3642	121-419	3655	2N6168	3688	ECG992	3709	13-14094-8
3642	AT1138A	3642	324-0115	3657	S6420M	3688	LM2900N	3709	14-514-13/63
3642	AT1138B	3642	417-113	3657	2N6170	3688	LM3301N	3709	15-20A70
3642	AT1833	3642	417-120	3658	B37-02A	3688	LM3401N	3709	15-20A90
3642	AT1834	3642	417-142	3658	B89-02A	3688	LM3900N	3709	020-00011
3642	AUY29	3642	417-29	3658	IT28	3689	ECG978	3709	020-00012
3642	AUY37	3642	417-42	3658	MAC15-4	3689	LM556CN	3709	24MW122
3642	B113000	3642	992-00-8890	3658	Q2015R5	3689	NE556A	3709	24MW244
3642	B204	3642	992-00-8890L	3658	SC141A1	3689	SN556CN	3709	32-0007
3642	B205	3642	4640	3658	SC141B	3691	LM358AH	3709	32-18539
3642	B206	3642	4640P	3658	SC141B2	3691	LM358H	3709	42-19681
3642	B339	3642	4702	3658	SC141B3	3691	LM358L	3709	42-21362
3642	B339H	3642	4729	3658	SC141B4	3692	LM2904N	3709	42-23972
3642	B340	3642	94010	3658	SC141B5	3692	LM358AN	3709	48-134954
3642	B340H	3642	127828	3658	SC141B6	3692	LM358JG	3709	48-355008
3642	B483	3642	405965-30A	3658	TIC226B	3692	LM358N	3709	48-60154A01
3642	B484	3642	612020	3658	TIC236B	3692	LM358P	3709	48-97048A05
3642	DTG110A	3642	1471124-5	3658	2N6342	3698	GE-5097	3709	53A001-2
3642	DTG110B	3642	1944748	3658	2N6342A	3698	R3110	3709	53N003-3
3642	DTG1200	3642	1960652	3658	2N6345	3698	R3115	3709	53N004-10
3642	DTG2000	3642	1962326	3658	2N6346A	3698	R3120	3709	56-8095
3642	DTG2000A	3642	7279069	3659	B37-04A	3698	R3210	3709	065-014
3642	DTG2100	3642	7287107	3659	MAC15-5	3698	R3215	3709	86-15-1
3642	DTG2100A	3642	7287112	3659	MAC15-6	3698	R3220	3709	86-48-1
3642	DTG2200	3642	7287117	3659	Q4015R5	3698	S1AN31R	3709	103-102
3642	DTG400M	3642	7289097	3659	SC141C	3698	S1AN40R	3709	103-90
3642	DTG600	3642	7292689	3659	SC141C2	3698	S1BN40R	3709	120-001301
3642	DTG601	3642	7297347	3659	SC141C3	3698	1N1185R	3709	150-015-9-001
3642	DTG602	3642	7299780	3659	SC141C4	3698	1N1186R	3709	324-0107
3642	DTG603	3642	7301660	3659	SC141C5	3698	1N1193R	3709	1007-1124
3642	DTG603M	3642	7301661	3659	SC141C6	3698	1N1194R	3709	1010-8173
3642	GB-239	3642/179	AD138/50	3659	SC141D2	3698	1S1654R	3709	1048-9888
3642	GB-76	3642/179	AD143	3659	SC141D3	3698	25C15R	3709	2093A38-35
3642	HEP-G6001	3642/179	AD143B	3659	SC141D4	3698	35C20R	3709	5101N60
3642	HEP-G6009	3642/179	B485	3659	SC141D6	3698	35H10R	3709	5631-20A90
3642	HEP-G6016	3642/179	BDY61	3659	SC149D	3698	35HR20	3709	01119
3642	HEP-G6018	3642/179	DTG110	3659	SC151D	3698	40HFR15	3709	58840-114
3642	HEP626	3642/179	SP-1484	3659	TIC226D	3698	40HFR2Q	3709	085005
3642	HP-19D	3642/179	2N2287	3659	TIC236D	3698	335C	3709	085026
3642	HR-19E	3642/179	2SB485	3659	2N6343	3698	335D	3709	161015
3642	M4701	3642/179	86-370-2YEL	3659	2N6343A	3698	336C	3709	167572
3642	M4702	3642/179	231-0004-03	3659	2N6347	3698	336D	3709	17T162-270
3642	M9181	3643	ECG987	3659	2N6347A	3698	371RC	3709	489752-036
3642	MP110B	3643	LM224AD	3659	B37-05A	3698	371RD	3709	5115348
3642	MP2300A	3643	LM224D	3660	B89-05A	3698	417RA	3709	53006S-14
3642	MP600	3643	LM2902N	3660	MAC15-7	3698	417RC	3709	530072-1001
3642	MP601	3643	LM324AN	3660	MAC15-8	3698	417RD	3709	0575019H
3642	MP602	3643	LM324AJ	3660	Q5015R5	3698	419RA	3709	817160
3642	MP603	3643	LM324N	3660	Q6015R5	3698	419RC	3709	817177
3642	P2Z	3644	SN72748P	3660	SC141E	3698	419RD	3709	984881
3642	QP-1	3645	CA748CT	3660	SC141E2	3698	435C	3709	1471822-11
3642	QP-10	3645	SN72748L	3660	SC141E3	3698	435D	3709	716060
3642	QP-1A	3647	B30C250KP	3660	SC141E4	3698	436B	3709	62154543
3642	QP-2	3647	B30050KP	3660	SC141E5	3698	436C	3709	62381620
3642	QP-3	3647	BD0A	3660	SC141E6	3698	436D	3709	71773660
3642	QP-4	3647	BB-1	3660	SC149M	3698	3115R	3709	4202005600
3642	QP-5	3647	EC03090-002	3660	SC151M	3698	3120R	3710	C1170A
3642	QP-6	3647	ECG167	3660	2N6344	3698	40210R	3710	C1325
3642	QP-7	3647	FW200	3660	2N6344A	3698	198764-6	3710	CGE-79
3642	R836	3647	FWB3001	3660	2N6348	3698	1662258-6	3710	EP15312
3642	SP1029	3647	FWB3002	3660	2N6348A	3698	8518382-10	3710	GE-259
3642	SP1650	3647	MDA920-4	3661	T6420B	3700	GEIC-76	3710	J241227
3642	SP1817	3647	QRT-230	3661	2N6163	3700	LA3350	3710	Q5111ZK
3642	SP2077	3647	RE255	3662	T6420D	3700	051-0088-00	3710	Q5161Z
3642	SP2708	3647	REN166	3663	T6420M	3700	003516	3710	Q5207
3642	SP838	3647	RT-250	3663	2N6165	3703	M5134P	3710	Q5207Z
3642	SP838-1	3647	893.20.709	3664	ECG5600	3704	GEIC-150	3710	2SC1172
3642	TI-3030	3647	893.20.714	3664	ECG5601	3704	HA11228	3710	2SC1172A
3642	TI-3031	3647	SD-W04	3664	ECG5602	3704	HA1144	3710	2SC1172B
3642	TQ5030	3647	SEN2A2	3664	MAC77-3	3704	611001-3	3710	2SC1308
3642	TR1036	3647	SR50253-2	3664	2S6070	3704	2360151	3710	2SC1308K
3642	TR1038	3647	TM166	3665	ECG5603	3706	AN271	3710	2SC1308R
3642	TR35	3647	TM167	3665	MAC77-4	3706	AN271B	3710	2SC1325
3642	2Q226	3647	ZEN433	3665	2N6071	3706	GEIC-61	3710	2SC1325A
3642	2Q227	3647	1-531-024	3666	ECG5604	3706	14LN034	3710	2SC1325AK
3642	2Q228	3647	1B08T20	3666	ECG5605	3707	HA1151	3710	2SC1325AL
3642	2N1651	3647	1B2	3666	MAC77-6	3707	TRHA1151	3710	2SC1325L
3642	2N1751	3647	10DB2A	3666	2N6073	3707	415401151Q	3710	2SC1895
3642	2N1906	3647	18DB2A	3668	LM311N	3708	HA1339	3710	2SC1896
3642	2N1907	3647	25P-3H1P	3669	LM340T-6.0	3708	HA1339A	3710	2SD348
3642	2N1907A	3647	35B11	3669	LM341P-6.0	3708	HA1339A	3710	314-6020-01
3642	2N1908	3647	35BL611	3669/962	SN72906	3709	A054-226	3710	8-729-372-52
3642	2N1908A	3647	63-11148	3670	LM340T-24	3709	A069-118	3710	09-302238
3642	2N2285	3647	63-12287	3670	LM341P-24	3709	A909-1016	3710	09-302245
3642	2N2286	3647	325-0081-109	3671	LM320T-5.0	3709	0A95	3710	13-43463-1
3642	2N2636	3647	690V080E53	3672	LM320T-6.0	3709	APB-160-1020	3710	46-8641-3
3642	2N2637	3647	2093A41-112	3673	LM320T-12	3709	CXL110	3710	46-86492-3
3642	2N2638	3647	4822-130-30261	3673	MC79120T	3709	D010	3710	46-86557-3
3642	2N2691	3647	126849	3674	LM320T-15	3709	DIJ70543	3710	48-155058
3642	2N2691A	3647	199919	3675/971	51-10541A01	3709	E21135	3710	48-355043
3642	2N630	3647	4202104170	3676	H68	3709	E2484	3710	48815505B
3642	2SB483	3647/167	KC2DP221C	3676	W06M	3709	ECG110	3710	48815514O
3642	2SB484	3647/167	8IRB-10	3676	2FB600	3709	E816K21	3710	57A263-11
3642	004-001	3647/167	TVM-PP6D22/1	3676	10DB6P	3709	E816X70	3710	57B1B6-11
3642	8F1555	3647/167	W005M	3677	W08M	3709	G00004A	3710	57B263-11
3642	13-18642-1	3647/167	46-8621-3	3677	2FB800	3709	HP-2008B	3710	86-626-2
3642	13-18642-2	3648	BY179	3678	KBF06	3709	J241	3710	86-633-9
3642	13-18642-2D	3648	ECG168	3678	5B4	3709	JP575005	3710	121-1003
3642	13-18642-2E	3648	FB-200	3678	530002-1	3709	JP575995	3710	189N1
3642	13-18642-2F	3648	FB200	3678	530021-1	3709	N-BA16X27	3710	189N1G
3642	13-18642-3	3648	FWB3003	3678	48106	3709	R886	3710	1039-1290
3642	13-18642-3A	3648	FWB3004	3678	71783	3709	REN110	3710	1061-6274
3642	13-18642-3B	3648	GMO578	3678	129241	3709	RT573B	3710	1074T
3642	14-601-02	3648	KBF04	3678/169	10DB6A	3709	RT5912	3710	1081-3343
3642	30-004-001	3648	MDA920-6	3678/169	10DB6A-C	3709	RT753B	3710	62547
3642	34-6000-33	3648	PTC401	3679	KBPC25-005	3709	S1384	3710	145648
3642	48-134640	3648	SEN2A4	3679	250JB05L	3709	SD46(4)	3710	146823
3642	48-134692	3648	1B4	3680	SCBA05	3709	S046	3710	610189-1
3642	48-134695	3648	10DB4A	3681	KBPC25-04	3709	STYL128	3710	610189-2
3642	48-134729	3648	13-33190-1	3681	250JB4L	3709	T-B1105	3710	23114323
3642	48-134740	3648	18DB4A	3682	KBPC25-06	3709	T4590	3710	23114945
3642	48-134741	3648	18DB4A-C	3683	2N3228	3709	TM110	3710	23114962
3642	48-134742	3648	4822-130-50228	3684	2N3525	3709	TR228736002004	3710	2003117044
3642	48-134743	3648/168	FW400	3685	C122F	3709	TV24122	3710/238	BDX32
3642	48-134748					3709	XD2A	3710/238	BU205
3642	48-134749							3710/238	C1086

SK	Industry Standard No.
3710/238	C642A
3710/238	EP15X126
3710/238	EP15X45
3710/238	GE-37
3710/238	MJ12003
3710/238	NTC-12
3710/238	2SC1004A
3710/238	2SC1086
3710/238	2SC1086M
3710/238	2SC1167
3710/238	2SC1170A
3710/238	2SC1170B
3710/238	2SC1174
3710/238	2SC1358
3710/238	2SC1358A
3710/238	2SC1358K
3710/238	2SC1358K1
3710/238	2SC1358K2
3710/238	2SC1358K3
3710/238	2SC1358P
3710/238	2SC1358Q
3710/238	2SC1358R
3710/238	2SC1413
3710/238	2SC1413A
3710/238	2SC1875
3710/238	2SC1875K
3710/238	2SC1875L
3710/238	2SC1892
3710/238	2SC1893
3710/238	2SC1922
3710/238	2SC1942
3710/238	2SC642A
3710/238	2SC643
3710/238	2SC643A
3710/238	2SC937
3710/238	2SC937A
3710/238	2SC937B
3710/238	2SC937YL
3710/238	2SD299
3710/238	2SD299B
3710/238	2SD299V
3710/238	2SD300
3710/238	2SD300B
3710/238	2SD350
3710/238	2SD350A
3710/238	2SD350Q
3710/238	2SD350T
3710/238	2SD380
3710/238	2SD517
3710/238	2SD627
3710/238	2SD764
3710/238	2SD765
3710/238	2SD903
3710/238	8-729-118-76
3710/238	8-729-341-34
3710/238	13-33181-1
3710/238	13-33181-2
3710/238	13-33181-3
3710/238	13-43463-2
3710/238	13-45016-1
3710/238	15-123230
3710/238	48-155140
3710/238	48-155189
3710/238	48-155224
3710/238	121-1029
3710/238	67544
3710/238	137607
3710/238	142689
3710/238	145671
3710/238	610186-1
3710/238	610194-1
3710/238	610194-3
3710/238	610216-2
3710/238	610216-3
3710/238	610233-1
3710/238	610242-1
3710/238	1473647-1
3710/238	2320291
3710/238	2320299
3710/238	2320961
3710/238	2003189304
3711	GEIC-99
3711	IC-545(ELCOM)
3711	TA7076P
3711	TA7076P(FA-1)
3711	TA7076P(FA-6)
3711	TA7076P(FA-7)
3711	46-1370-3
3711	23119013
3711	23119023
3711	23119030
3711	23119031
3711	23119032
3711	23119033
3711	40306604
3711	4206105370
3712	307-107-9-003
3715	EA15X8524
3715	GE-221
3715	MPS6516
3715	MPS6517
3715	MPS8598
3715	MPS8599
3715	OS133.1003T
3715	1-0006-0023
3715	1-002
3715	1-003
3715	001-02201-0
3715	1W11702
3715	1W11702A
3715	1W8537
3715	1W9148
3715	1W9640
3715	1W9640A
3715	002-009800
3715	002-010300
3715	002-010300A
3715	002-010500
3715	002-010500A
3715	002-010900
3715	002-010900A
3715	002-012800
3715	002-012800A
3715	002-9800-A
3715	2N1196
3715	2N2299
3715	2N2709
3715	2N2927
3715	2N2927/46
3715	2N3012
3715	2N3081
3715	2N3401
3715	2N4121
3715	2N4126
3715	2N4143
3715	2N4889
3715	2N4916
3715	2N5138
3715	2N5138A
3715	2N5139
3715	2N5139A
3715	2N5400
3715	2N5401
3715	2N942
3715	2S307
3715	2S307A
3715	2SA637
3715	2SA778
3715	2SA778K
3715	2SA912
3715	2SA912Q
3715	2SA912R
3715	2SA912S
3715	2SA965
3715	2SC673C
3715	121-933
3716	A2332
3716	A244(AMC)
3716	A3T918
3716	AIE
3716	AR111
3716	AR213(VIOLET)
3716	AR218(ORANGE)
3716	AR218(RED)
3716	BF344
3716	BF345
3716	BFX89
3716	BFY87
3716	BFY87A
3716	C1070
3716	C313(JAPAN)
3716	CS1284B
3716	CS1284F
3716	CXL161
3716	ECG161
3716	ES54(ELCOM)
3716	ES73(ELCOM)
3716	ES86(ELCOM)
3716	EW212
3716	F501
3716	F501(ZENITH)
3716	F501-16
3716	F502
3716	F502(ZENITH)
3716	F523
3716	FS1682
3716	FT45
3716	GE-39
3716	GI3793
3716	HEP709
3716	HEP719
3716	M-128J509-1
3716	M-128J511-3
3716	M75545-1
3716	M9266
3716	M9450
3716	NPC167
3716	NS345
3716	NS406
3716	P/N14-603-02
3716	PTC132
3716	RCA40245
3716	RCA40246
3716	RE2001
3716	RE2002
3716	RE28
3716	RE5001
3716	RE5002
3716	REN161
3716	RF200
3716	S130-138
3716	S130-251
3716	SAB1044
3716	SAB3469
3716	SE5-0249
3716	SE5032
3716	SP4436
3716	ST5641
3716	T1202(GE)
3716	T1X-M16
3716	T381
3716	T381(SEARS)
3716	TA2554
3716	TA2555
3716	TM161
3716	TNJ1173
3716	TR-171
3716	TV20
3716	TV24399
3716	002-011400
3716	002-011500
3716	2N3337
3716	2N5181
3716	2N5182
3716	2N5230
3716	2SC1044
3716	2SC1254
3716	2SC1547
3716	7-10(SARKES)
3716	7-11(SARKES)
3716	7-21(SARKES)
3716	7-22(SARKES)
3716	7-9(SARKES)
3716	13-10321-76
3716	13-10321-78
3716	13-1032176
3716	14-603-08
3716	14-603-09
3716	14-603-13
3716	14-850-12
3716	14-851-12
3716	019-005157
3716	020-1112-003
3716	24T-011-001
3716	24T011-012
3716	48-01-027
3716	48-65112A68
3716	48-65174A24
3716	48-869266
3716	48-869450
3716	48-869481
3716	48-97046A25
3716	48-97046A45
3716	48-97162A30
3716	48-97162A31
3716	48-97162A32
3716	57A102-4
3716	57A103-4
3716	57A119-2
3716	57A142-7
3716	57A24
3716	57B119-2
3716	57C164-4
3716	57D24
3716	66F021-1
3716	66F022-1
3716	66F042-1
3716	86-381-2
3716	91N1B
3716	96-5334-01
3716	96XZ6050/25N
3716	101-1(ADMIRAL)
3716	121-501
3716	129N1
3716	141 402
3716	150-1N
3716	170(RCA)
3716	207A27
3716	229-0190-30
3716	229-1301-22
3716	229-5100-32
3716	352-0658-010
3716	352-0658-020
3716	352-0658-030
3716	352-0658-040
3716	352-0658-050
3716	355D6
3716	355D8
3716	386-7243-P001
3716	417-243
3716	417-258
3716	417-262
3716	536-1(RCA)
3716	576-1(RCA)
3716	617-65
3716	657-31
3716	680-1(RCA)
3716	822-1(SYLVANIA)
3716	824-1(SYLVANIA)
3716	1004(G.E.)
3716	1006(G.E.)
3716	1009(G.E.)
3716	1843-17
3716	1854-0231
3716	2020-02
3716	2057A2-180
3716	2057A2-181
3716	2057A2-185
3716	2057A2-187
3716	2057A2-193
3716	2057A2-196
3716	2057A2-197
3716	2057A2-202
3716	2057A2-204
3716	3476(RCA)
3716	3508(SEARS)
3716	3516(RCA)
3716	3518(RCA)
3716	3579(RCA)
3716	3680-1
3716	4167(AIRLINE)
3716	4167(PENNCREST)
3716	4168(PENNCREST)
3716	4168(WARDS)
3716	4169(WARDS)
3716	4822-130-40304
3716	8000-00032-026
3716	29076-005
3716	29076-006
3716	40246
3716	61558
3716	141402
3716	147115-9
3716	147353-0-1
3716	190714
3716	489752-095
3716	540204
3716	1471115-10
3716	1472633
3716	1473543-1
3716	2316183
3716	3459332-1
3716	3463609-2
3716	3468068-1
3716	3468068-2
3716	3468068-3
3716	3468068-4
3716	4028839
3716	7851650-01
3716	7851651-01
3716	7910108-01
3716	7939165
3716	62565160
3716	62593717
3716	62771240
3716	62789255
3716	450010701
3716/161	C674CV
3716/161	TIXM14
3716/161	TVS2SC684
3716/161	2SC1790
3716/161	2SC251
3716/161	2SC251A
3716/161	2SC252
3716/161	2SC253
3716/161	2SC568
3716/161	2SC653
3716/161	2SC674CV
3716/161	2SC674CZ
3716/161	2SC674G
3716/161	13-10321-79
3716/161	921-1014
3716/161	974-1(SYLVANIA)
3716/161	3535(RCA)
3716/161	40296
3717	A14A10G
3717	A416
3717	A65C-19G
3717	A8P-404-ORN
3717	A8P404-ORN
3717	AA4
3717	AC148
3717	AD131-III
3717	AD131-IV
3717	AD131-V
3717	AD13850
3717	AD143R
3717	AD145
3717	AD149-IV
3717	AD149-V
3717	AD150-IV
3717	AD150-V
3717	AD153
3717	AD159
3717	ADY27
3717	AF280
3717	AL101
3717	AR10
3717	AR11
3717	AR12
3717	AR13
3717	AR14
3717	AR8
3717	AR9
3717	AUY31
3717	AUY33
3717	B126V
3717	B14A-1-21
3717	B25
3717	B25B
3717	B337BK
3717	B337HA
3717	B337HB
3717	B338H
3717	B4A-1-A-21
3717	B65C-1-21
3717	B77B-1-21
3717	B84A-1-21
3717	CQT1129
3717	CST1743
3717	CST1744
3717	CST1745
3717	CST1746
3717	CTP1111
3717	CTP1124
3717	CTP1136
3717	CTP1137
3717	CTP1500
3717	CTP1503
3717	CTP1514
3717	CTP1550
3717	CTP1551
3717	CTP1728
3717	CTP1729
3717	CTP1731
3717	CTP1732
3717	CTP1733
3717	CTP1735
3717	CTP1736
3717	CTP1739
3717	CTP3500
3717	CTP3503
3717	CTP3504
3717	CTP3508
3717	CXL121
3717	DS515
3717	DS520
3717	DT41
3717	DT6110
3717	EA1082
3717	ES21(ELCOM)
3717	FD-1029-ET
3717	FD-1029ET
3717	G04-701-A
3717	GC4045
3717	GC4062
3717	GC4097
3717	GC4111
3717	GC691
3717	GC692
3717	GE-16
3717	GET572
3717	GM428
3717	GP1432
3717	GP420
3717	GTX2001
3717	H103A
3717	HEP-G6005
3717	HEP232
3717	HEP623
3717	HEP624
3717	HEP628
3717	HR-101A
3717	HR105
3717	HR105A
3717	HR105B
3717	I472446-I
3717	M4606
3717	M4722
3717	M501
3717	M9141
3717	M9142
3717	M9202
3717	M9237
3717	M9241
3717	M9255
3717	M9263
3717	M9342
3717	M9436
3717	M9550
3717	MN24
3717	MN25
3717	MN26
3717	MN29
3717	MN32
3717	MN48
3717	MN49
3717	MN73
3717	MP2062
3717	NK14A119
3717	NK4A-1A19
3717	NK65C119
3717	NK77C119
3717	NK84AA19
3717	NKT401
3717	NKT402
3717	NKT403
3717	NKT404
3717	NKT405
3717	NKT406
3717	NKT415
3717	NKT416
3717	NKT454
3717	NKT501
3717	NKT503
3717	NKT504
3717	P31898
3717	P4M
3717	P6480001
3717	P75534-4
3717	P75534-5
3717	P8890L
3717	PA10890-1
3717	PAR12
3717	PC3004
3717	PIK
3717	PS1
3717	PT06
3717	PT12
3717	PT155
3717	PT176
3717	PT234
3717	PT235
3717	PT236B
3717	PT242
3717	PT25
3717	PT255
3717	PT256
3717	PT285
3717	PT285A
3717	PT301
3717	PT301A
3717	PT307
3717	PT307A
3717	PT366B
3717	PT3A
3717	PT40
3717	PT50
3717	PT501
3717	PT555
3717	PT6
3717	PT0-6
3717	R102-41
3717	R2446
3717	R2446-1
3717	R265A
3717	R3512-1
3717	R3515(RCA)
3717	RE11
3717	RE11MP
3717	REN121
3717	RS5835
3717	S-1556-2
3717	S1556-2
3717	S65C-1-3P
3717	S67809
3717	S77C-1-3P
3717	S84A-1-3P
3717	S95253
3717	S95253-1
3717	SDT-3048
3717	SE40022
3717	SP1013A
3717	SP1118
3717	SP148-3
3717	SP1482-5
3717	SP1483
3717	SP1556-2
3717	SP1556-4
3717	SP1603
3717	SP1603-1
3717	SP1603-2
3717	SP1657
3717	SP1801
3717	SP1844
3717	SP2094
3717	SP2341
3717	SP2395
3717	SP404
3717	SP404T
3717	SP441
3717	SP485
3717	SP486W
3717	SP634
3717	SP649
3717	SP649-1
3717	SP834
3717	SP880-1
3717	SP880-3
3717	SP891B
3717	SP891W
3717	SPS91
3717	ST235
3717	T-235
3717	T13029
3717	TA1614
3717	TA1682
3717	TA1682A
3717	TA1705
3717	TA1765
3717	TA1766
3717	TA1773
3717	TA1794
3717	TA1881
3717	TA1890
3717	TA1891
3717	TP80/30
3717	TP80/30Z
3717	TI266A
3717	TI269
3717	TI366
3717	TI366A
3717	TI367
3717	TI367A
3717	TI368
3717	TI368A
3717	TI369
3717	TI369A
3717	TM121
3717	TQ-5064
3717	TR-178(OLSON)
3717	TR-43B
3717	TR-5
3717	TR35524
3717	TR8006
3717	TRA7R
3717	TRA8R
3717	TS-1657
3717	TS-176
3717	TT1083
3717	TV24678
3717	TVS-2SB449F
3717	TVS25126F
3717	TVS2SB126
3717	VFU65326B
3717	VM30203
3717	VS2SB126
3717	VS2SB126F
3717	VS2SB126V
3717	WEP232
3717	XB-5
3717	XB-7
3717	ZEN326
3717	ZEN330
3717	ZEN331
3717	002-008800
3717	002-009701
3717	2D001
3717	2N1218
3717	2N176-1BLU

SK	Industry Standard No.
3717	2N176BLK
3717	2N176BLU
3717	2N176GRN
3717	2N176M
3717	2N176PUR
3717	2N176RED
3717	2N176WHT
3717	2N176YEL
3717	2SB337HA
3717	2SB337HB
3717	2SB338HA
3717	2SB338HB
3717	2SB426BL
3717	2SB449P
3717	4-88A17-1
3717	4A-1-A-7B
3717	4A-1A0R
3717	04A1
3717	04A1-12
3717	6-158
3717	06P1C
3717	7-1(STANDEL)
3717	8-905-613-210
3717	9-51141400
3717	14-579-10
3717	14A1-A82
3717	14A14-7B
3717	18AA-1-82
3717	19A115184-P1
3717	19A115268
3717	19A115376
3717	19A115561-1
3717	42-23968
3717	44A-1A5
3717	48-134582
3717	48-134583
3717	48-134606
3717	48-134744
3717	48-137978
3717	48-869141
3717	48-869142
3717	48-869182
3717	48-869202
3717	48-869237
3717	48-869241
3717	48-869255
3717	48-869342
3717	48-869436
3717	48-869550
3717	48K134583
3717	48K134584
3717	48K869342
3717	48NSP1035
3717	48R134582
3717	48R134606
3717	48R869141
3717	48R869142
3717	48R869202
3717	48R869205
3717	48R869241
3717	48R869255
3717	48R869436
3717	48R869550
3717	57A1-119
3717	57A4-1
3717	57A4-4
3717	57A6-12
3717	57A6-2
3717	57A6-23
3717	57A6-3
3717	57A6-8
3717	57A9-2
3717	57DG-32
3717	62-18428
3717	63-8590
3717	65C14-7B
3717	67P1C
3717	67P2C
3717	67P3C
3717	77-270877-2
3717	77-270878-2
3717	77C14-7B
3717	84A14-7B
3717	85-370-2 BLU
3717	86-5057-2
3717	86-5083-2
3717	86-5089-2
3717	86-5090-2
3717	86-5113-2
3717	86-5125-2
3717	94A-1A6-4
3717	96-5378-01
3717	106P1T
3717	114A-1-82
3717	121-171
3717	121-382
3717	121-793
3717	125-402
3717	125-403
3717	131-000562
3717	165C-182
3717	173A4419-10
3717	173A4419-4
3717	173A4419-5
3717	173A4419-6
3717	173A4419-7
3717	188-826
3717	199-POWER
3717	231-006B
3717	247-624
3717	414A-15
3717	417-30
3717	465C-15
3717	484A15
3717	576-0002-005
3717	576-0040-254
3717	610-067-1
3717	610-067-2
3717	610-067-3
3717	610-068-1
3717	614A-1-5L
3717	665C-1-5L
3717	684A-1-5L
3717	914A-1-6-5
3717	964-17945
3717	965C-16-4
3717	984A-1-6B
3717	991-01-0099
3717	992-08890
3717	1105-15
3717	1561-17
3717	1814AL-8
3717	2446-1(RCA)
3717	3107-204-90070
3717	3515(RCA)
3717	3618-1
3717	4082-501-0001
3717	4570
3717	4722
3717	11252-4
3717	11506-3
3717	11526-8
3717	11526-9
3717	27126-090
3717	30215(RCA)
3717	30216(RCA)
3717	33989-2069
3717	34526
3717	36687
3717	36800-2
3717	36800-3
3717	36800-4
3717	36800-6
3717	36800-7
3717	36971
3717	39893
3717	40051-2
3717	43046
3717	59990-1
3717	61010-6-1
3717	62177
3717	71448
3717	71448-2
3717	71448-3
3717	71448-4
3717	71448-5
3717	71448-6
3717	71448-7
3717	75700-03-01
3717	094013
3717	145134-526
3717	147351-5-1
3717	171004(SEARS)
3717	194474-8
3717	196058-4
3717	196148-0
3717	196183-5
3717	196501-7
3717	196607-9
3717	209185-962
3717	230208
3717	322968-140
3717	650970
3717	651202
3717	652086
3717	655319
3717	660094
3717	702885
3717	702885-00
3717	801522
3717	815246-2
3717	983795
3717	1221615
3717	1407205-1
3717	1407206-1
3717	1471036-14
3717	1471102-41
3717	1472446
3717	1473512-1
3717	2091859-16
3717	2091859-25
3717	2091959-16
3717	3130006
3717	3130109
3717	3460553-2
3717	3462221-1
3717	3462306-1
3717	4037594-P1
3717	4999774
3717	5493158-1
3717	5496663-P1
3717	5496939-P1
3717	5496939-P2
3717	6480001
3717	6480004
3717	7285776
3717	7290593
3717	7292690
3717	7292955
3717	7910072-01
3717	23111006
3717	23311066
3717	43022577
3717	62081579
3717	62084103
3717	62087633
3717	62371566
3717	62736976
3717	131000562
3717	134804290-101
3717	1348042901
3717	4803101109-02
3717	134804290101
3717/121	C20
3717/121	HEP-G6015
3717/121	PT3
3717/121	QP1
3717/121	QP1A
3717/121	QP2
3717/121	QP6
3717/121	QP7
3717/121	R3528-1(RCA)
3717/121	T101
3717/121	2N1540
3717/121	2N1540A
3717/121	2N1541
3717/121	2N1541A
3717/121	2N1542
3717/121	2N1542A
3717/121	2N1543
3717/121	2N1543A
3717/121	2N1545
3717/121	2N1545A
3717/121	2N1546
3717/121	2N1546A
3717/121	2N1547
3717/121	2N1547A
3717/121	2N1548
3717/121	2N1548A
3717/121	2N1550
3717/121	2N1550A
3717/121	2N1551
3717/121	2N1551A
3717/121	2N1552
3717/121	2N1552A
3717/121	2N1554
3717/121	2N1554A
3717/121	2N1555
3717/121	2N1555A
3717/121	2N1556
3717/121	2N1556A
3717/121	2N1558
3717/121	2N1558A
3717/121	2N1559
3717/121	2N1559A
3717/121	2N1560
3717/121	2N1560A
3717/121	2N2832
3717/121	8-905-613-265
3717/121	8-905-613-277
3717/121	8-905-613-282
3717/121	8-905-613-283
3717/121	8-905-613-284
3717/121	5253
3717/121	9925-2
3717/121	801523
3717/121	801538
3718	A13-14604-1A
3718	A13-14604-1B
3718	A13-14604-1C
3718	A13-14604-1D
3718	A13-14604-1E
3718	A13-14777-1
3718	A13-14777-1A
3718	A13-14777-1B
3718	A13-14777-1C
3718	A13-14777-1D
3718	A13-14778-1A
3718	A13-14778-1B
3718	A13-14778-1C
3718	A13-14778-1D
3718	A77B-70
3718	A77B-705
3718	A77B19G
3718	B6B-3-A-21
3718	B6B-3A21
3718	B77C-1-21
3718	CXL121MP
3718	ECG121MP
3718	NK6B-3A19
3718	NK77B119
3718	REN121MP
3718	RS5855
3718	S77B-1-3P
3718	TM121MP
3718	6B-3A4-7B
3718	16B-3A82
3718	46B-3A5
3718	57A3-12
3718	57B3-12
3718	66B-3A5L
3718	77B14-7B
3718	177B-1-82
3718	177C-1-82
3718	477B15
3718	477C15
3718	677C-1-5L
3718	977B1-6-2
3718	977C1-6-3
3718	2780(AIRLINE)
3718	11528-1
3718	11528-2
3718	11528-3
3718	11528-4
3718	36359-4
3718	67085-0-1
3718	7299771
3718	7299803
3718	43020418
3719	AR5
3719	AR6
3719	AR7
3719	AUY21
3719	AUY22
3719	B19
3719	B20
3719	B21
3719	B250
3719	B26(JAPAN)
3719	B26A
3719	B27
3719	B28(JAPAN)
3719	B29
3719	B30(JAPAN)
3719	B31(TRANSISTOR)
3719	CST1739
3719	CST1740
3719	CST1741
3719	CST1742
3719	CTP1104
3719	CTP1108
3719	CTP1109
3719	CTP1117
3719	CXL104
3719	DS503
3719	ECG104
3719	ES13(ELCOM)
3719	ES18(ELCOM)
3719	ES503(ELCOM)
3719	HEP200
3719	HEP230
3719	HJ35
3719	HR101
3719	LS52
3719	M4727
3719	M4974
3719	M4974/P1R
3719	P2C
3719	P3EBLU
3719	P3EGRN
3719	P3ERED
3719	P4D
3719	P4L
3719	P4N
3719	P75534-2
3719	P75534-3
3719	PA10890
3719	PT150
3719	PT235A
3719	PT236
3719	PT236A
3719	PT554
3719	PTC114
3719	RE7
3719	REN104
3719	SP1108
3719	SP1137
3719	SP1600
3719	SP1619
3719	SP1927
3719	SP2247
3719	SP2431
3719	SP26
3719	SP486
3719	SP880
3719	TF78
3719	TI370
3719	TI370A
3719	TR-02
3719	TR-16
3719	TR-16C
3719	TR-172(OLSON)
3719	WEP230
3719	WEP624
3719	WEP628
3719	XN12F
3719	001-01205-1
3719	2N457
3719	2N457A
3719	2N457B
3719	48-137025
3719	48-137026
3719	48-137329
3719	48-869087B
3719	48-869099B
3719	57A124-10
3719	111P5C
3719	111P7C
3719	207A20
3719	207A20A
3719	642-206
3719	642-264
3719	642-316
3719	800-329
3719	1008-17
3719	3107-204-90140
3719	3107-204-90190
3719	4822-130-40233
3719	25661-022
3719	60770
3719	175027-022
3719	1221625
3719	7851322-01
3719	62742585
3719/104	C16
3719/104	C23
3719/104	C24
3719/104	C26
3719/104	MP2061
3719/104	0087
3720	CXL104MP
3720	ECG104MP
3720	RE7MP
3720	TM104MP
3720	WEP230MP
3720	WEP628MP
3720	7-2(STANDEL)
3720	32-16591
3720	48-134747
3720	57A3-10
3720	57A3-11
3720	57A3-7
3720	57A3-8
3720	57A3-9
3720	57B3-10
3720	57B3-11
3720	57B3-7
3720	57B3-8
3720	57B3-9
3720	800-196
3720	95257
3720	4082501-0001
3721	A160(JAPAN)
3721	A168(JAPAN)
3721	A168A
3721	A170(JAPAN)
3721	A171(JAPAN)
3721	A204
3721	A205
3721	A206
3721	A207
3721	A208(JAPAN)
3721	A210(JAPAN)
3721	A211(JAPAN)
3721	A212
3721	A217
3721	A248(JAPAN)
3721	A26
3721	A277(JAPAN)
3721	A278(JAPAN)
3721	A279(JAPAN)
3721	A283
3721	A284
3721	A305
3721	A311(JAPAN)
3721	A414
3721	A415(JAPAN)
3721	A74-3-3A9G
3721	AP138/290
3721	ASY27
3721	B74-3-A-21
3721	B74-3A21
3721	B88C-1-21
3721	CXL100
3721	DS-28(DELCO)
3721	DS21
3721	DS23
3721	DS53
3721	ECG100
3721	GC1159
3721	GC31
3721	GC32
3721	GC33
3721	GC34
3721	GC35
3721	GT761
3721	GT761R
3721	GT762
3721	GTE2
3721	HA-12
3721	HA00052
3721	HA00053
3721	HA202
3721	HA49
3721	NK74-3A19
3721	NK88C119
3721	R119
3721	R163
3721	R186
3721	R227
3721	R244
3721	R488
3721	RS2690
3721	RS2691
3721	RS2692
3721	RS3892
3721	RS5105
3721	RS5511
3721	RS5540
3721	RS5743.3
3721	S88C-1-3P
3721	SF.T172
3721	SF.T173
3721	SF.T174
3721	SFT237
3721	SFT288
3721	SYL2250
3721	T-1877
3721	T-2038
3721	T-2039
3721	T-2040
3721	T-2091
3721	T0101
3721	T0102
3721	T033Z3
3721	T1251
3721	T52148Z
3721	T52149
3721	T52149Z
3721	TA1704
3721	TA1763
3721	TA1763A
3721	TA1778
3721	TA1782
3721	TA1783
3721	TI-364
3721	TIA05A
3721	TIX895
3721	TM100
3721	TNJ60608
3721	TO-101
3721	TO-102
3721	TR-05
3721	TR11
3721	TR321(HFGH1)
3721	TR482A
3721	TRC44
3721	TRC44A
3721	TRC45
3721	TRC45A
3721	TS669A
3721	TS669B
3721	TS669D
3721	TS669E
3721	TS669F
3721	UPI1347
3721	VL/8RJ
3721	2N1309
3721	2N2966
3721	2SA322K
3721	6A12678
3721	19A115208
3721	19A115208-P1
3721	19A115208-P2
3721	21A051-000
3721	48-124389
3721	48-56P1
3721	51D189
3721	57A6-6A
3721	57A6-6B
3721	57A6-6C
3721	63-11585
3721	74-3A0R
3721	88C-70-12
3721	88C14-7B
3721	121-354
3721	174-3A82
3721	474-3A5
3721	488C15
3721	674-3A5L
3721	688C-1-5L
3721	2904-038H05
3721	37833
3721	67193-82
3721	454549
3721	573356
3721	2001653-22
3721	2091241-10
3721	2091241-11
3721	2091241-12
3721	2091241-4
3721	2091241-6
3721	2091241-8
3721	62752319
3721	62752327
3721	62771364
3721/100	A146
3721/100	A147
3721/100	A148
3721/100	A149
3721/100	A198
3721/100	A209
3721/100	A282
3721/100	A31
3721/100	A312
3721/100	A330
3721/100	A332
3721/100	A40
3721/100	GE-2
3721/100	GT2693
3721/100	GT2695
3721/100	HA-15
3721/100	NKT12
3721/100	SFT319
3721/100	T131
3721/100	2N1303
3721/100	2N1305
3721/100	2N1500
3721/100	974-3A6-3
3722	A397
3722	A42X00286-01
3722	B1058
3722	B199
3722	B224(JAPAN)
3722	B335
3722	B51(JAPAN)
3722	CXL102
3722	DE-3181
3722	DRC-81252
3722	ECG102
3722	EE100
3722	FBN-CP2293
3722	FD-1029-BG
3722	FD1029EE
3722	HEP252
3722	HEP629
3722	HEP630
3722	HEP631
3722	HEP632
3722	HEP633

SK	Industry Standard No.
3722	HS23D
3722	HS290
3722	M-P3D
3722	M4327
3722	M4482
3722	M4483
3722	M9250
3722	NS121
3722	NS32
3722	P6460006
3722	P6460037
3722	PBX103
3722	PBX113
3722	PXB103
3722	PXB113
3722	PXC101
3722	PXC101A
3722	PXC101AB
3722	QN2613
3722	R46
3722	RS5825
3722	S413796
3722	SF.T222
3722	SF.T227
3722	SMB447610A
3722	SMB621960
3722	SYL1430
3722	SYL1535
3722	SYL1583
3722	SYL2248
3722	SYL2249
3722	SYL2300
3722	T-2122
3722	T50931B
3722	T51573A
3722	TM102
3722	TNJ70634
3722	TNJ70635
3722	TQSA0-222
3722	TR383(HGFH-2)
3722	TR8007
3722	TRC70
3722	TRC71
3722	TRC72
3722	TS-164
3722	TS13
3722	TS14
3722	TS15
3722	TVS25B172
3722	TVSSA71B
3722	VS2SA385
3722	VS2SB171
3722	VS2SB172
3722	VS2SB176
3722	VS2SB178
3722	VS2SB178A
3722	WRT1114
3722	WTVB6A
3722	ZEN303
3722	ZEN306
3722	001-01206-0
3722	2M1127
3722	2N1404
3722	2N1478
3722	2N200
3722	2N270-5E
3722	2N422A
3722	2N831
3722	2SA396
3722	2SA397
3722	2SB335
3722	2SB34N
3722	3B15
3722	3B15-1
3722	4JX1C1224
3722	6-000105
3722	6-155
3722	6D0000105
3722	12A7239P1
3722	16A787
3722	19-020-015
3722	19C300128-P5
3722	19C300138-P8
3722	48-869148
3722	48-869250
3722	48-97046A53
3722	48-97127A32
3722	48A124327
3722	48K134482
3722	48K134483
3722	48R869250
3722	57A1-11
3722	57A126
3722	57A143
3722	57A169
3722	57A170
3722	57A189
3722	57A6(PNP)
3722	57A6-1
3722	57A6-22
3722	86X00011-001
3722	120-001795
3722	173A3970
3722	324-0139
3722	324-0140
3722	354-3052
3722	421-13C
3722	1001-7663
3722	1113-2875
3722	1166-7821
3722	1180-0182
3722	1321-7724
3722	1321-7732
3722	1344-7321
3722	1510-2718
3722	1850-0040
3722	1850-0040-1
3722	1850-0060
3722	1850-0062
3722	1850-0062-1
3722	1850-0101
3722	1850-0184
3722	1850-0184-1
3722	2061A45-47
3722	2093A41-40
3722	2093A41-41
3722	5052
3722	7239
3722	27910-12153
3722	34871
3722	35086
3722	35950
3722	37677
3722	42324
3722	95255-000
3722	111001
3722	111011
3722	111012
3722	111013
3722	112071
3722	177105
3722	537790
3722	581005
3722	581042
3722	815033
3722	815228A01
3722	982820
3722	984746
3722	985217
3722	992289
3722	3130011
3722	3130025
3722	3130060
3722	3404114-1
3722	3404114-2
3722	5496665-P2
3722	5496665-P3
3722	5496665-P4
3722	5496665-P5
3722	5496665-P6
3722	5496666-P1
3722	5496666-P2
3722	5496666-P3
3722	5496666-P4
3722	5496666-P5
3722	5496666-P6
3722	5496666-P7
3722	5496666-P8
3722	5496667-P1
3722	5496667-P2
3722	5496774-P1
3722	5496774-P2
3722	5496774-P3
3722	5496774-P4
3722	5496774-P6
3722	5496839-P1
3722	6460006
3722	7210036-001
3722	8516861-1
3722	8516986-1
3722	8524440-2
3722	10017663
3722	11132875
3722	11667821
3722	11783453
3722	11800182
3722	11858909
3722	13217724
3722	13217732
3722	13443767
3722	13447321
3722	15102718
3722	54967774-P5
3722	62078502
3722	62088502
3722	62380020
3722	62713755
3722	2791012153
3722/102	ACY17
3722/102	ACY18
3722/102	ACY19
3722/102	AF200U
3722/102	AF201U
3722/102	C32
3722/102	C41
3722/102	GE-1
3722/102	M9092
3722/102	Q11/6515
3722/102	RS5557
3722/102	UPI1353
3722/102	ZT2102
3722/102	2N1307
3723	A3300
3723	EA33X8356
3723	GEIC-42
3723	HC1000403
3723	LA3300
3723	RE336-IC
3723	REN1005
3723	SVI-LA3300
3723	09-308058
3723	051-0038-00
3723	740-5853-300
3723	740-8120-160
3723	754-5853-300
3723	2091-50
3723	740940
3724	LM78L08ACZ
3724	LM78L08CH
3724	LM78L08CZ
3724	MC78L08ACP
3725	HA1177
3727	AN245
3727	AN246
3727	GEIC-301
3727	51-13753A10
3727	51S13753A02
3727	51S13753A10
3727	1052-6408
3728	AN331
3728	GEIC-66
3728	51S13753A08
3729	AN247
3729	AN247P
3729	GEIC-302
3729	M51247
3729	51-13753A18
3729	51S13753A06
3729	1052-6416
3729	1065-2055
3731	TA7159P
3731	09-308094
3731	6207159
3732	MC145109
3732	PLL02
3732	PLL02A
3732	REN1167
3732	TC9100P
3732	TM1167
3732	02-360002
3734	DM-35
3734	DM35
3734/806	9341899
3737	GEIC-75
3737	LA3201
3737	4152032010
3738	GEIC-74
3738	LA3155
3739	LA4420
3742	GEIC-176
3742	RH-IX0001PAZZ
3742	TVSMPC596C2
3742	TVSUPC596C2
3743	AN236
3743	GEIC-281
3743	IC-535(ELCOM)
3743	51-90305A20
3743	144011
3744	HA1366
3745	DDEY123001
3745	HC10001200
3745	51-42211P01
3745	003526
3745	916106
3746	ECG326
3746	REN85
3746	2N4360
3746	2N5460
3746	2N5461
3746	2N5462
3746	2N5463
3746	417-241
3746/326	EL131
3747	A3L
3747	BD157
3747	BD158
3747	BF459
3747	C1501Q
3747	C1501R
3747	E185B121712
3747	ES39(ELCOM)
3747	GE-232
3747	HEP244
3747	M4998
3747	MJE340
3747	MJE341
3747	MJE344
3747	MJE3440
3747	MJE9742
3747	RE24
3747	SJE290
3747	SJE3754
3747	ST-MJE9742
3747	T287
3747	2SC1501
3747	2SC1501Q
3747	2SC1501R
3747	13-35257-1
3747	13-35257-2
3747	51-4
3747	57A127
3747	66F020-1
3747	66F020-2
3747	84C01
3747	84G01
3747	86-165-2
3747	089-4(SYLVANIA)
3747	297V084C01
3747	974-2(SYLVANIA)
3747	974-3
3747	974-4(SYLVANIA)
3747	3567-2(RCA)
3747	39510
3747	06120053
3747/157	C1382
3747/157	C1382-0
3747/157	C1382Y
3747/157	2SC1550
3747/157	2SC1566
3747/157	2SC2258
3747/157	48-155070
3747/157	48-355054
3747/157	48-90420A92
3748	BDY60
3748	BFS22A
3748	BFT33
3748	BFT34
3748	BFT42
3748	BFT43
3748	BFW25
3748	BFW26
3748	BSS14
3748	BSS15
3748	BSS16
3748	BSS42
3748	BSS43
3748	C756
3748	C756-1
3748	C756-1-1
3748	C756-1-2
3748	C756-1-3
3748	C756-1-4
3748	C756-2
3748	C756-2-1
3748	C756-2-2
3748	C756-2-3
3748	C756-2-4
3748	C756-2-5
3748	C756-3
3748	C756-3-1
3748	C756-3-2
3748	C756-3-3
3748	C756-3-4
3748	D121
3748	D121A
3748	D121B
3748	D78
3748	D78A
3748	D78B
3748	D78C
3748	D78D
3748	DBBY001003
3748	M113HA
3748	SDT3422
3748	SDT3423
3748	SDT3424
3748	SDT3427
3748	SDT3428
3748	SDT7402
3748	SDT7412
3748	SDT7414
3748	SDT7415
3748	UPT611
3748	2N3831
3748	2N4001
3748	2N4877
3748	2N5846
3748	2SC486A
3748	2SC486B
3748	2SC486C
3748	2SC486D
3748	2SC486E
3748	2SC486F
3748	2SC486G
3748	2SC486GN
3748	2SC486H
3748	2SC486J
3748	2SC486K
3748	2SC486L
3748	2SC486M
3748	2SC4860R
3748	2SC486R
3748	2SC486X
3748	2SC756
3748	2SC756-1
3748	2SC756-1-1
3748	2SC756-1-2
3748	2SC756-1-3
3748	2SC756-1-4
3748	2SC756-2
3748	2SC756-2-1
3748	2SC756-2-2
3748	2SC756-2-3
3748	2SC756-2-4
3748	2SC756-2-5
3748	2SC756-3
3748	2SC756-3-1
3748	2SC756-3-2
3748	2SC756-3-3
3748	2SC756-3-4
3748	2SC756-4
3748	2SC756-4-1
3748	2SC756-4-2
3748	2SC756-4-3
3748	2SC756-4-4
3748	2SD78
3748	2SD78B
3748	09-302166
3748	52-010-151-0
3748	186-010
3748	1080-05
3748	5001-068
3748/282	BFR77
3748/282	BFR78
3748/282	BFS22
3748/282	BFS23
3748/282	BFT39
3748/282	BFT40
3748/282	BFT41
3748/282	BSS44
3748/282	C485
3748/282	C485BL
3748/282	C485C
3748/282	C485Y
3748/282	C486
3748/282	C486BL
3748/282	C486Y
3748/282	D120
3748/282	D120A
3748/282	D120B
3748/282	D120C
3748/282	2N3019
3748/282	2N3830
3748/282	2N5681
3748/282	2SC484BL
3748/282	2SC485
3748/282	2SC485BL
3748/282	2SC485C
3748/282	2SC485Y
3748/282	2SC486
3748/282	2SC486Y
3749	.25T8.7
3749	.25T8.7A
3749	.25T8.8
3749	.25T8.8A
3749	AZ-085
3749	AZ-088
3749	BZ-085
3749	BZ-088
3749	CXL5073
3749	D8U
3749	DIJ72172
3749	ECG5073
3749	ECG5073A
3749	ED491130
3749	GEZD-8.7
3749	MB6356
3749	RD9B
3749	RE113
3749	REN5073
3749	TM5073
3749	YAAD017
3749	001-0101-01
3749	1N1530
3749	1N1530A
3749	1N2035-12
3749	1N2790
3749	1N4102
3749	1N4297
3749	1N4297A
3749	1N4298
3749	1N4298A
3749	1N4302
3749	1N4302A
3749	1N4695
3749	1N5854B
3749	1N764
3749	1S1717
3749	1S224
3749	2M214
3749	24MW670
3749	48-137577
3749	754-9030-009
3749	5330322
3750	.25T13
3750	.25T13A
3750	.4T13
3750	.4T13A
3750	.4T13B
3750	.7JZ13
3750	.7Z13A
3750	.7Z13B
3750	.7Z13C
3750	.7Z13D
3750	.7ZM13A
3750	.7ZM13B
3750	.7ZM13C
3750	.7ZM13D
3750	A054-231
3750	A066-122
3750	AW01(RCA)
3750	AZ-125
3750	AZ-130
3750	AZ13
3750	BZ-130
3750	BZ130
3750	BZX29C13
3750	BZX46C13
3750	BZX55C13
3750	BZX79C13
3750	BZY85C13V5
3750	DIS
3750	DZ13
3750	EA16X81
3750	ECG143
3750	ECG143A
3750	EQA01-13
3750	EZ13(ELCOM)
3750	GEZD-13
3750	HEP-Z0416
3750	HW13
3750	HW13A
3750	HW13B
3750	LPM13
3750	LPM13A
3750	MA1130
3750	PTC508
3750	RD13F
3750	RD13FA
3750	RD13FB
3750	RD13M
3750	RE119
3750	REN143
3750	RZ13
3750	SD-33
3750	SZ-200-13
3750	SZ13
3750	SZ13.0
3750	SZ961-V
3750	TVSRD13AL
3750	UZ8713
3750	UZ8813
3750	VR13
3750	VR13A
3750	VR13B
3750	WL-130
3750	WL-130A
3750	WL-130B
3750	WL-130C
3750	WL-130D
3750	WM-130
3750	WM-130A
3750	WM-130B
3750	WM-130C
3750	WM-130D
3750	WN-130
3750	WN-130A
3750	WN-130B
3750	WN-130C
3750	WN-130D
3750	WO-130A
3750	WO-130B
3750	WO-130C
3750	WO-130D
3750	WP-130
3750	WP-130A
3750	WP-130B
3750	WP-130C
3750	WP-130D
3750	WQ-130
3750	WQ-130A
3750	WQ-130B
3750	WQ-130C
3750	WQ-130D
3750	WR-130
3750	WR-130A
3750	WR-130B
3750	WR-130C
3750	WR-130D
3750	WS-130
3750	WS-130A
3750	WS-130B
3750	WS-130C
3750	WS-130D
3750	WT-130
3750	WT-130A
3750	WT-130B
3750	WT-130C
3750	WT-130D
3750	WU-130
3750	WU-130A
3750	WU-130B
3750	WU-130C
3750	WU-130D
3750	WV-130
3750	WV-130A
3750	WV-130B
3750	WV-130C
3750	WV-130D
3750	WX-130
3750	WX-130A
3750	WX-130B
3750	WX-130C
3750	WX-130D
3750	WY-130
3750	WY-130A
3750	WY-130B
3750	WY-130C
3750	WY-130D
3750	Z0416
3750	Z4X13
3750	ZA13
3750	ZA13A
3750	ZA13B
3750	ZB1-13
3750	ZB13
3750	ZB13B
3750	ZEN506
3750	ZF13
3750	ZH13
3750	ZH13A
3750	ZH13B
3750	1A13M
3750	1A13MA
3750	1M13ZS10
3750	1M13ZS5
3750	1N1774
3750	1N1774A
3750	1N1783
3750	1N1A13MA
3750	1N2037
3750	1N2037-2
3750	1N227
3750	1N227A
3750	1N3023
3750	1N3023A
3750	1N3023B
3750	1N3521
3750	1N3682
3750	1N3682A

SK	Industry Standard No.
3750	1N3682B
3750	1N4107
3750	1N4165
3750	1N4165A
3750	1N4165B
3750	1N4330
3750	1N4330A
3750	1N4330B
3750	1N4407
3750	1N4407A
3750	1N4635
3750	1N4664
3750	1N4700
3750	1N4743
3750	1N4743A
3750	1N5533B
3750	1N5566
3750	1N5566A
3750	1N5859B
3750	1N766-2
3750	1R13
3750	1R13A
3750	1R13B
3750	1S228
3750	1S3013
3750	1Z13
3750	1Z13A
3750	1Z13T10
3750	1Z13T20
3750	1Z13T5
3750	02Z13A
3750	05Z13
3750	09-306127
3750	21A037-018
3750	26Z6AF
3750	26Z6F
3750	42-19917
3750	46-86296-3
3750	48-137000
3750	48-137209
3750	56-32
3750	75Z13
3750	75Z13A
3750	75Z13B
3750	75Z13C
3750	86-67-1
3750	93A102-2
3750	100-286
3750	103-301-17A
3750	523-2003-130
3750	601X0226-066
3750	601X0402-038
3750	1550-17
3750	2093A41-113
3750	161022
3750	166985
3750	2002331-46
3750	13030401
3750	42023425
3751	.25T15.8
3751	.25T15.8A
3751	.25T16
3751	.25T16A
3751	.4T16
3751	.4T16A
3751	.4T16B
3751	.7Z16A
3751	.7Z16B
3751	.7Z16C
3751	.7Z16D
3751	.7ZM16A
3751	.7ZM16B
3751	.7ZM16C
3751	.7ZM16D
3751	AW01-16
3751	BZ162
3751	BZX29C16
3751	BZX46C16
3751	BZX55C16
3751	BZX83C16
3751	BZY92C16
3751	CXL5075
3751	DDAY009003
3751	ECG5075
3751	ECG5075A
3751	EQB01-16
3751	EZ16(ELCOM)
3751	GEZD-16
3751	HEP-Z0419
3751	HW16
3751	HW16A
3751	HZ16H
3751	QD-ZBZ162XJ
3751	QRT-246
3751	RD16E-M
3751	RD16EB
3751	RD16F
3751	RD16FA
3751	RD16FB
3751	RE122
3751	REN5075
3751	SV139
3751	TM5075
3751	TVSZB1-15
3751	UZ8716
3751	UZ8816
3751	WL-157
3751	WL-157A
3751	WL-157B
3751	WL-157C
3751	WL-157D
3751	WL-162
3751	WL-162A
3751	WL-162B
3751	WL-167
3751	WL-167A
3751	WL-167B
3751	WL-167C
3751	WL-167D
3751	WM-157
3751	WM-157A
3751	WM-157B
3751	WM-157C
3751	WM-157D
3751	WM-167
3751	WM-167A
3751	WM-167B
3751	WM-167C
3751	WM-167D
3751	WN-157
3751	WN-157A
3751	WN-157B
3751	WN-157C
3751	WN-157D
3751	WN-167
3751	WN-167A
3751	WN-167B
3751	WN-167C
3751	WN-167D
3751	WO-157
3751	WO-157A
3751	WO-157B
3751	WO-157C
3751	WO-157D
3751	WO-167
3751	WO-167A
3751	WO-167B
3751	WO-167C
3751	WO-167D
3751	WP-157
3751	WP-157A
3751	WP-157B
3751	WP-157C
3751	WP-157D
3751	WP-167
3751	WP-167A
3751	WP-167B
3751	WP-167C
3751	WP-167D
3751	WQ-157
3751	WQ-157A
3751	WQ-157B
3751	WQ-157C
3751	WQ-157D
3751	WQ-167
3751	WQ-167A
3751	WQ-167B
3751	WQ-167C
3751	WQ-167D
3751	WR-157
3751	WR-157A
3751	WR-157B
3751	WR-157C
3751	WR-157D
3751	WS-157
3751	WS-157A
3751	WS-157B
3751	WS-157C
3751	WS-157D
3751	WS-167
3751	WS-167A
3751	WS-167B
3751	WS-167C
3751	WS-167D
3751	WT-157
3751	WT-157A
3751	WT-157B
3751	WT-157C
3751	WT-157D
3751	WT-167
3751	WT-167A
3751	WT-167B
3751	WT-167C
3751	WT-167D
3751	WU-157
3751	WU-157A
3751	WU-157B
3751	WU-157C
3751	WU-157D
3751	WU-167
3751	WU-167A
3751	WU-167B
3751	WU-167C
3751	WU-167D
3751	WV-157
3751	WV-157A
3751	WV-157B
3751	WV-157C
3751	WV-157D
3751	WV-167
3751	WV-167A
3751	WV-167B
3751	WV-167C
3751	WV-167D
3751	WW-157
3751	WW-157A
3751	WW-157B
3751	WW-157C
3751	WW-157D
3751	WW-167
3751	WW-167A
3751	WW-167B
3751	WW-167C
3751	WW-167D
3751	WX-157
3751	WX-157A
3751	WX-157B
3751	WX-157C
3751	WX-157D
3751	WX-167
3751	WX-167A
3751	WX-167B
3751	WX-167C
3751	WX-167D
3751	WY-157B
3751	WY-157C
3751	WY-157D
3751	WY-167
3751	WY-167A
3751	WY-167B
3751	WY-167C
3751	WY-167D
3751	Z0419
3751	ZB1-16
3751	ZB1-16V
3751	ZB16
3751	ZB16A
3751	ZB16B
3751	ZF16
3751	001-023038
3751	1M16Z
3751	1M16Z5
3751	1M16ZS10
3751	1M16ZS5
3751	1MZ10
3751	1N1776
3751	1N1776A
3751	1N2038
3751	1N2038-2
3751	1N3025
3751	1N3025A
3751	1N3025B
3751	1N3523
3751	1N3684
3751	1N3684A
3751	1N4110
3751	1N4167
3751	1N4167A
3751	1N4167B
3751	1N4332
3751	1N4332A
3751	1N4332B
3751	1N4409
3751	1N4409A
3751	1N4637
3751	1N4666
3751	1N4703
3751	1N4745
3751	1N4745A
3751	1N5246B
3751	1N5536B
3751	1N5568
3751	1N5568A
3751	1N5741B
3751	1N719A
3751	1N966B
3751	1R16
3751	1R16A
3751	1R16B
3751	1S1965
3751	1S2122
3751	1S2122A
3751	1S2160A
3751	1S231
3751	1S3016(ZENER)
3751	1Z16
3751	1Z16A
3751	1Z16T10
3751	1Z16T20
3751	1Z16T5
3751	1ZC16T5
3751	02Z16A
3751	3L4-3506-40
3751	05-933950
3751	05Z16
3751	09-306243
3751	09-306268
3751	09-306324
3751	048(ZENER)
3751	48-83461E01
3751	48-83461E07
3751	56-36
3751	61E01
3751	61E07
3751	070-048
3751	73-17
3751	86-93-1
3751	93A39-50
3751	103-231
3751	103-Z9013
3751	123B-006
3751	142-015
3751	152-057-9-001
3751	173-15
3751	264P10906
3751	523-2003-160
3751	800-028-00
3751	2000-312
3751	2000-326
3751	2120-17
3751	5001-143
3751	99202-221
3751	130761
3751	530073-15
3751	530073-26
3751	2327077
3751	2337065
3751	42027092
3751	134883461501
3751	134883461507
3752	.25T18
3752	.25T18A
3752	.4T18
3752	.4T18A
3752	.4T18B
3752	.7JZ18
3752	.7Z18A
3752	.7Z18B
3752	.7Z18C
3752	.7Z18D
3752	.7ZM18A
3752	.7ZM18B
3752	.7ZM18C
3752	.7ZM18D
3752	AW01-18
3752	AZ-182
3752	AZ967B
3752	BZ-182
3752	BZX29C18
3752	BZX55C18
3752	BZX83C18
3752	BZY83C18
3752	BZY85C18
3752	BZY92C18
3752	EQA01-18
3752	EQB01-18
3752	EZ18(ELCOM)
3752	GEZD-18
3752	HEP-Z0420
3752	HW18
3752	HW18A
3752	PD6020A
3752	PD6061
3752	PS8919
3752	PTC510
3752	QB01-18
3752	RD18F
3752	RD18FA
3752	RD18FB
3752	RE124
3752	RH-EX0022CEZZ
3752	RZ18
3752	SD-11
3752	SV142
3752	1M18Z
3752	1M18Z10
3752	1M18Z5
3752	1M18ZS10
3752	1M18ZS5
3752	1N1428
3752	1N1515
3752	1N1515A
3752	1N1526
3752	1N1526A
3752	1N1777
3752	1N1777A
3752	1N1879
3752	1N1879A
3752	1N3026
3752	1N3026A
3752	1N3026B
3752	1N3437
3752	1N3449
3752	1N3524
3752	1N3685
3752	1N3685A
3752	1N3685B
3752	1N4168
3752	1N4168A
3752	1N4168B
3752	1N4333
3752	1N4333A
3752	1N4333B
3752	1N4410
3752	1N4410A
3752	1N4638
3752	1N4667
3752	1N4705
3752	1N4746
3752	1N4746A
3752	1N5538B
3752	1N5569
3752	1N5569A
3752	1N5742B
3752	1N667
3752	1N768-1
3752	1R18
3752	1R18A
3752	1R18B
3752	1S1966
3752	1S199
3752	1S2123A
3752	1S2180A
3752	1S233
3752	1S3018
3752	1S485
3752	1Z18
3752	1Z18A
3752	1Z18T10
3752	1Z18T20
3752	1Z18T5
3752	1ZC18T10
3752	1ZC18T5
3752	02Z18A
3752	4-2020-09200
3752	4-2020-12000
3752	05-950177
3752	09-306355
3752	14-515-05
3752	48-155182
3752	48-82256C53
3752	48-83461E37
3752	61E37
3752	62-21574
3752	103-252
3752	523-2003-180
3752	919-003309-1
3752	71411-3
3752	97202-222
3752	99202-222
3752	99202-222
3752	130132
3752	141187
3752	530073-22
3752	530073-30
3752	1477046-10
3752	2002209-7
3752	4202009200
3752	4202012000
3752	4202012800
3752	134882256353
3752	134883461537
3753	.25T25
3753	.25T25A
3753	.7JZ25
3753	.7Z25A
3753	.7Z25B
3753	.7Z25C
3753	.7Z25D
3753	.7ZM25A
3753	.7ZM25B
3753	.7ZM25C
3753	.7ZM25D
3753	A04344-026
3753	BZ-250
3753	J241216
3753	RE129
3753	1M25Z
3753	1M25Z10
3753	1M25Z5
3753	1N2040-3
3753	1N5543B
3753	1N5869B
3753	1S238
3753	13-33179-8
3753	33R
3753	48-355050
3753	48-82256C42
3753	48-83461E11
3753	61E11
3753	86-117-1
3753	93A39-14
3753	523-2003-250
3753	2092055-0024
3753	134882256342
3753	134883461511
3754	.25T28
3754	.25T28A
3754	BZ-280
3754	GEZD-28
3754	HEP-Z0425
3754	SV172
3754	1N4712
3754	1N5871B
3754	1N769-4
3754	1S487
3754	46-86447-3
3754	48-82256C36
3754	134882256336
3755	.25T30
3755	.25T30A
3755	.4T30A
3755	.4T30B
3755	.7Z30A
3755	.7Z30B
3755	.7Z30C
3755	.7Z30D
3755	.7ZM30A
3755	.7ZM30B
3755	.7ZM30C
3755	.7ZM30D
3755	AW01-30
3755	AZ972B
3755	BZ-290
3755	BZ-300
3755	BZX29C30
3755	BZX46C30
3755	BZX79C30
3755	BZX83C30
3755	BZY85C30
3755	EQA01-30
3755	EQA01-Z4RA
3755	EZ30(ELCOM)
3755	GEZD-30
3755	HD3000201-0
3755	HW30
3755	HW30A
3755	RD30F
3755	RD30FA
3755	1M30Z
3755	1M30Z10
3755	1M30Z5
3755	1M30ZS10
3755	1M30ZS5
3755	1N1782
3755	1N1782A
3755	1N2387
3755	1N3031
3755	1N3031A
3755	1N3031B
3755	1N3452
3755	1N3529
3755	1N3690
3755	1N3690A
3755	1N3690B
3755	1N4173
3755	1N4173A
3755	1N4173B
3755	1N4338
3755	1N4338A
3755	1N4338B
3755	1N4415
3755	1N4415A
3755	1N4643
3755	1N4672
3755	1N4713
3755	1N4751
3755	1N4751A
3755	1N5574
3755	1N5574A
3755	1N5747B
3755	1N5872B
3755	1N725A
3755	1R30
3755	1R30A
3755	1R30B
3755	1S1971
3755	1S2300A
3755	1S3030
3755	1Z30
3755	1Z30A
3755	1Z30T10
3755	1Z30T20
3755	1Z30T5
3755	1ZC30T10
3755	1ZC30T5
3755	013-768
3755	56-76
3755	56-79
3755	66X0040-004
3755	523-2003-300
3755	2327075
3755	23115990
3755	41029498
3756	GESS-3A3
3757	GESS-3AT2
3758	GESS-3DB3
3759	GESS-2AV2
3760	GESS-6DW4
3761/514	3D83
3762	GEIC-70
3762	LA1222
3764	AD166
3764	AD167
3764	B10142
3764	B10142A
3764	B10142B
3764	B10143
3764	B10143A
3764	B10143B
3764	B1178
3764	B233
3764	B274(JAPAN)
3764	B341
3764	B341H
3764	B341V
3764	DTG1010
3764	DTG1110
3764	ES42(ELCOM)
3764	GE-25
3764	HEP234
3764	H0300
3764	M4459
3764	M4623
3764	M4652
3764	M7342
3764	M7342/P4F
3764	M9090
3764	MN61
3764	MN63
3764	MN64
3764	MP1612
3764	MP1612A
3764	MP1612B
3764	MP1613
3764	MP3730
3764	MP3731
3764	P3H
3764	P3J
3764	P4H
3764	PTC122
3764	R-2001
3764	R2001
3764	R2003
3764	R2460-9
3764	R2494-1
3764	R2500-1
3764	R2964
3764	R3514-1
3764	RE16
3764	SE-5-0399
3764	SE50399
3764	SF.T212
3764	SF.T213
3764	SF.T214
3764	SF.T238
3764	SF.T239
3764	SF.T240
3764	SF.T250
3764	SP1742

SK	Industry Standard No.	SK	Industry Standard No.	SK	Industry Standard No.	SK	Industry Standard No.	SK	Industry Standard No.
3764	TNJ6080	3771	HS7033	3776	BZX8305V1	3779	1N5731	3781	1N3413
3764	TQ5028	3771	KS0035B	3776	BZX57	3779	1N753	3781	1N4693
3764	TR-183(OLSON)	3771	KS2035B	3776	BZY65	3779	182062	3781	1N5236
3764	TV-114	3771	LR33H	3776	BZY83C5V1	3779	182062B	3781	1N5236A
3764	TV106	3771	MR330-H	3776	BZY85C5V1	3779	187062	3781	1N5236B
3764	TV28B126	3771	MZ500-4	3776	BZY88C5V1	3779	187062A	3781	1N5527
3764	TV28B448	3771	PD6002	3776	HEP-20211	3779	187062B	3781	1N5527A
3764	TV28258448	3771	ST-41	3776	RD5AN	3779	DZE6.2A	3781	1N5733
3764	002-012700-12	3771	1N5226	3776	RT6922	3779	46-86457-3	3781	1N711
3764	2N2527	3771	1N5226A	3776	RVDRD5A1E	3779	46-86501-3	3781	1N755
3764	2N2528	3771	1N5226B	3776	WZ-052	3779	65208	3781	1N958
3764	2N3731	3771	1N5518	3776	1N4689	3779	5515	3781	1N958A
3764	28B333	3771	1N5518A	3776	1N5231	3779	601100002	3781	187075
3764	28B341	3771	1N746	3776	1N5231A	3779	134882256301	3781	187075A
3764	28B341H	3771	182033	3776	1N5231B	3779	.25N6.8	3781	187075B
3764	28B341S	3771	187033	3776	1N5523	3780	AZ754	3781	13-33187-26
3764	28B341V	3771	187033B	3776	1N5523A	3780	AZ957	3781	48-82256044
3764	3-20	3771	3126	3776	1N5729	3780	AZ957A	3781	65306
3764	13-21606-1	3771	3126A	3776	1N751	3780	B2811	3781	65307
3764	34-6002-24	3771	48-82256C26	3776	1N751A	3780	ECG5014A	3781	1107
3764	48-134459	3771	1102(ZENER)	3776	18190	3780	HEP-Z0215	3781	5517
3764	48-134623	3771	5508	3776	182051	3780	HM6.8	3781	9607
3764	48-134652	3772	AZ747	3776	182051A	3780	HM6.8A	3781	134882256344
3764	48-134934	3772	BZX55C3V6	3776	183305	3780	HZ-7	3782	.25N8.2
3764	48-137001	3772	BZX83C3V6	3776	184735	3780	HZ-7A	3782	AZ754
3764	48-137178	3772	RD4AL	3776	187051	3780	HZ-7B	3782	AZ959
3764	48-137234	3772	1N466	3776	187051A	3780	HZ-7C	3782	AZ959A
3764	48-137235	3772	1N5227	3776	187051B	3780	PD6010	3782	BZX7108V2
3764	48-137342	3772	1N5227A	3776	09-306191	3780	RD6.8B	3782	BZY83D8V2
3764	48-137342-P47	3772	1N5519	3776	19-080-014	3780	RD6.8BB	3782	BZY85C8V2
3764	48-137367	3772	1N5519A	3776	19-090-014	3780	RD7AM	3782	BZY85D8V2
3764	48-869090	3772	1N703	3776	48-137337	3780	RD7AN	3782	BZY88C8V2
3764	488869090	3772	1N703A	3776	48-83461E10	3780	S27	3782	B2813
3764	73B140-004	3772	1N747	3776	48-83461E40	3780	TMDO4	3782	ECG5016A
3764	86-292-2	3772	182036	3776	61B10	3780	TMDO4A	3782	HEP-20217
3764	116-068	3772	187036	3776	61B23	3780	TR78A	3782	HM8.2
3764	417-112	3772	187036A	3776	61B40	3780	VZ-065	3782	HM8.2A
3764	205TA2-265	3772	187036B	3776	65104	3780	VZ-067	3782	PD6012
3764	2496(RCA)	3772	18753	3776	65105	3780	VZ-069	3782	TMDO6
3764	2500(RCA)	3772	2A25(ZENER)	3776	65106	3780	WZ-065	3782	TMDO6A
3764	3514(RCA)	3772	61843	3776	65107	3780	WZ-067	3782	VZ-079
3764	3583(RCA)	3772	65000	3776	65108	3780	WZ-069	3782	VZ-081
3764	3648(RCA)	3772	5509	3776	5513	3780	Z1008	3782	VZ-083
3764	4459	3772	134883461543	3776	9973	3780	Z5D6.8	3782	WZ-079
3764	4623	3772	AZ748	3776	134882256315	3780	ZB6.8	3782	WZ-083
3764	4652	3772	HEP-Z0208	3776	134883461510	3780	ZB6.8	3782	Z1010
3764	11606-8	3772	MZ1004	3776	134883461525	3780	ZG6.8	3782	Z108.2
3764	11608-6	3773	RD4A	3776	134883461540	3780	1/2Z6.8T5	3782	Z1D8.2
3764	11608-7	3773	RD4AM	3777	AZ752	3780	1/4Z6.8AZ	3782	Z508.2
3764	11608-8	3773	1N1927	3777	BZX7105V6	3780	1/4M6.8AZ10	3782	Z5D8.2
3764	11608-9	3773	1N1954	3777	BZX78	3780	1/4M6.8AZ5	3782	ZB8.2
3764	17607-1	3773	1N1981	3777	BZX78P	3780	1/4M6.8Z	3782	ZB8.2
3764	21606-1	3773	1N467	3777	BZY83D5V6	3780	1/4M6.8Z10	3782	ZG8.2
3764	61218	3773	1N467A	3777	BZY85D5V6	3780	1N1930	3782	ZG6.2
3764	147351-4-1	3773	1N5228	3777	BZY88C5V6	3780	1N1957	3782	1/2Z8.2T5
3764	200064-6-109	3773	1N5228A	3777	ECG5011A	3780	1N1984	3782	1N1931
3764	200064-6-110	3773	1N5228B	3777	HEP-20212	3780	1N2765	3782	1N1958
3764	322968-141	3773	1N5520	3777	RD5.6EB	3780	1N2765A	3782	1N1985
3764	1472494-1	3773	1N5520A	3777	RD6AL	3780	1N3400	3782	1N3199
3764	1472500-1	3773	1N748	3777	SD32	3780	1N3412	3782	1N3401
3764	2314009	3773	187039	3777	1/2Z5.6T5	3780	1N378	3782	1N3414
3764	7292684	3773	187039A	3777	1/4Z5.6T5	3780	1N469	3782	1N430
3764	7297348	3773	187039B	3777	1N1929	3780	1N469A	3782	1N430A
3764	62141042	3773	34-8057-55	3777	1N1929B	3780	1N470	3782	1N5237
3764/127	2N2526	3773	65001	3777	1N956	3780	1N470A	3782	1N5237A
3764/127	8505214-1	3773	65002	3777	1N983	3780	1N470B	3782	1N5237B
3765	A270	3773	5510	3777	1N983B	3780	1N475	3782	1N5528
3765	B3468	3774	AZ749	3777	1N5232	3780	1N475A	3782	1N5528A
3765	C481X	3774	BZX83C4V3	3777	1N5232A	3780	1N5235	3782	1N5734
3765	F81978	3774	BZY64	3777	1N5524	3780	1N5235A	3782	1N711
3765	GR102	3774	RD4AN	3777	1N5524A	3780	1N5526	3782	1N764-1
3765	IC743038	3774	TR43	3777	1N5730	3780	1N5526A	3782	1N959
3765	IC743039	3774	1A58	3777	1N708	3780	1N707	3782	1N959A
3765	IP20-0004	3774	1N374	3777	1N752A	3780	1N707A	3782	182082
3765	IP20-0048	3774	1N375	3777	1N762-1	3780	1N710	3782	182082B
3765	MRF8004	3774	1N5229	3777	182056	3780	1N826	3782	187082
3765	P231961	3774	1N5521	3777	182056B	3780	1N826A	3782	187082A
3765	P2857	3774	1N5521A	3777	187056	3780	1N828	3782	187082B
3765	P2C2677C	3774	1N749	3777	187056A	3780	1N828A	3782	4126
3765	R879	3774	1N749A	3777	187056B	3780	1N957	3782	4126A
3765	R879A	3774	182043	3777	48-137322	3780	1N957A	3782	48-137328
3765	S83935	3774	1847	3777	48-82256C12	3780	18192	3782	48-82256009
3765	T18109	3774	187043	3777	65109	3780	18471	3782	48-82256016
3765	T13888	3774	18754	3777	1105	3780	187068	3782	61B32
3765	TR28C671	3774	2A18	3777	134882256312	3780	187068A	3782	65309
3765	2N2540A	3774	2A47	3778	BZX88C6V2	3780	187068B	3782	1108
3765	2N2949	3774	13-33187-19	3778	BZZ10	3780	18757	3782	5518
3765	2N3426	3774	65005	3778	CXL5012	3780	09-306052	3782	13030285
3765	28C1556	3774	65006	3778	ECG5012A	3780	34-8057-42	3782	134882256308
3765	28C481	3774	5511	3778	ZL6	3780	34-8057-49	3782	134882256309
3765	28C481X	3774	9972	3778	ZM6B	3780	3926	3782	134882256316
3765	2806	3775	BZX85D4V7	3778	1N474	3780	3926A	3782	134883461532
3765	28C781	3775	BZY88C4V7	3778	1N474A	3780	48-137279	3783	CXL5017
3765	13-0028	3775	HEP-Z0210	3778	1N474B	3780	48-82256002	3783	ECG5017A
3765	13-0035	3775	RD4R7EB	3778	1N5233	3780	48-82256019	3783	N-756A
3765	48-869209	3775	RD5AL	3778	1N5233A	3780	48-82256023	3783	VZ-085
3765	48-869491	3775	1N1928	3778	1N5233B	3780	48-82256047	3783	VZ-088
3765	48-869519	3775	1N1955	3778	1N706	3780	48-83461E25	3783	WZ-088
3765	488869209	3775	1N1982	3778	1N706A	3780	61B25	3783	ZM8.7B
3765	48R869491	3775	1N468	3778	1N762	3780	103-279-14A	3783	1N1313
3765	48R869519	3775	1N468A	3778	18191	3780	0114-0026/4460	3783	1N1313A
3765	051-0156	3775	1N5230	3778	18470	3780	65301	3783	1N225
3765	051-0157	3775	1N5230A	3778	18756	3780	65302	3783	1N225A
3765	120-001798	3775	1N5728	3778	019-003928	3780	65303	3783	1N3148
3765	120-003151	3775	1N705	3778	48-83461E27	3780	65304	3783	1N5238
3765	576-0036-212	3775	1N705A	3778	61B27	3780	1106	3783	1N5238A
3765	576-0036-913	3775	1N750	3778	652	3780	3755-1	3783	182067
3765	1100-75	3775	1N761	3778	65203	3780	3755-2	3783	182068
3765	1112-75	3775	1N761-1	3778	65204	3780	5516	3783	182069
3765	1487-17	3775	182047	3778	65205	3780	9606	3783	18212
3765	2027-00	3775	187047	3778	65206	3780	06200012	3783	18213
3765	3868	3775	187047A	3778	134883461527	3780	62123672	3783	18214
3765	4004	3775	187047B	3779	AZ753	3780	134882256302	3783	18758
3765	4004(SEARS)	3775	18755	3779	BZX7106V2	3780	134882256319	3783	48-82256045
3765	8000-00011-151	3775	3526	3779	BZY66	3780	134882256323	3783	134882256345
3765	38122	3775	3526A	3779	ECG5013A	3780	134882256347	3784	.25N9.1
3765	115304	3775	48-82256C07	3779	HEP-Z0214	3780	134883461525	3784	AZ757
3765	198005-1	3775	651	3779	RD6.2B	3781	.25N7.5	3784	AZ960
3765	321264-2	3775	65100	3779	RD6.2EB	3781	AZ755	3784	AZ960A
3765	1700037	3775	65101	3779	UZ6.2	3781	AZ958	3784	BZX7109V2
3765	1700038	3775	65102	3779	VZ-063	3781	AZ958A	3784	BZX72
3766	MZ92-1	3775	65103	3779	WZ-063	3781	BZX7107V5	3784	BZX72A
3766	MZ92-2.4	3775	5512	3779	WZ061	3781	BZY67	3784	BZX72B
3766	13871	3775	71780	3779	ZEN500	3781	BZY88C7V5	3784	BZX72C
3766	MZ92-2.5	3775	134882256303	3779	1/2Z6.2T5	3781	BZZ12	3784	BZX8309V1
3767	M2500-2	3775	134882256307	3779	1/4Z6.2T5	3781	HM7.5	3784	BZY68
3768	1N465	3776	AZ751	3779	1462	3781	HM7.5A	3784	BZY8309V1
3768	1N465A	3776	AZ751A	3779	1N411			3784	BZY85C9V1
3768	1341	3776	BZX4605V1	3779	1N5553			3784	BZY88C9V1
3771	A-120420	3776	BZX7105V1	3779	1N5234			3784	ECG5018A
3771	AZ746	3776	BZX7905V1	3779	1N5234A			3784	HEP-Z0219
3771	BZX102-3V4			3779	1N5525			3784	HM9.1
3771	HEP-Z0206			3779	1N5525A			3784	HM9.1A
3771	HS2033A								

SK	Industry Standard No.
3784	HZ-9C
3784	KDD0015
3784	MA1091
3784	RD9.1EB
3784	RD9AM
3784	RD9AN
3784	VZ-090
3784	VZ-092
3784	VZ-094
3784	Z1C9.1
3784	Z1D9.1
3784	1/2Z9.1T5
3784	1/4M9.1Z
3784	1/4M9.1Z10
3784	1N4770
3784	1N4770A
3784	1N5239
3784	1N5239A
3784	1N5529
3784	1N5529A
3784	1N5735
3784	1N713
3784	1N757
3784	1N764-3
3784	1S194
3784	1S472
3784	1S7091
3784	1S7091A
3784	1S7091B
3784	48-83461E15
3784	51-02007-12
3784	61E15
3784	200X8220-878
3784	201X2220-118
3784	264P02503
3784	654C9
3784	1109
3784	5519
3784	129904
3784	62045612
3784	134882256318
3784	134882256338
3784	134883461515
3785	.25N10
3785	A42X00041-01
3785	AZ758
3785	AZ961
3785	AZ961A
3785	BZX71C10
3785	BZY83D10
3785	BZY85D10
3785	BZY88C10
3785	BZY94C10
3785	ECG5019A
3785	HEP-Z0220
3785	HM10
3785	HM10A
3785	HZ-11
3785	HZ-11A
3785	RD11AL
3785	RD11AM
3785	VZ-096
3785	VZ-098
3785	VZ-100
3785	WZ-096
3785	WZ-098
3785	Z-1012
3785	ZM10A
3785	1/2Z10T5
3785	1N1932
3785	1N1932B
3785	1N1959
3785	1N1986
3785	1N1986B
3785	1N226
3785	1N226A
3785	1N3402
3785	1N3415
3785	1N5240
3785	1N5240A
3785	1N5530
3785	1N5530A
3785	1N5736
3785	1N701
3785	1N714
3785	1N714B
3785	1N758
3785	1S2100
3785	1S7100
3785	1S7100A
3785	1S7100B
3785	1S759
3785	13-33187-7
3785	48-137393
3785	48-82256C11
3785	48-82256C32
3785	655C9
3785	1110
3785	5520
3785	62258942
3785	134882256311
3785	134882256332
3786	.25N11
3786	AZ962
3786	AZ962A
3786	BZX71C11
3786	BZY88C11
3786	BZY94C11
3786	ECG5020A
3786	HM11
3786	HM11A
3786	RD11AN
3786	RD11E
3786	VZ-105
3786	VZ-110
3786	WZ-105
3786	1/2Z11T5
3786	1N1314
3786	1N1314A
3786	1N5241
3786	1N5241A
3786	1N5737
3786	1N715
3786	1N962
3786	1N962A
3786	1N962B
3786	1S2110
3786	1S2110A
3786	1S7110
3786	1S7110A
3786	1S7110B
3786	48-82256C34
3786	1111
3786	5521
3786	134882256334
3787	.25N12
3787	A-120077
3787	AZ759
3787	AZ963
3787	AZ963A
3787	BZX71C12
3787	BZY69
3787	BZY83D12
3787	BZY85D12
3787	BZY88C12
3787	BZY94C12
3787	ECG5021A
3787	HEP-Z0222
3787	HM12
3787	HM12A
3787	HZ-12
3787	HZ-12A
3787	RD12E
3787	RD12EA
3787	RD12EB
3787	RD12EB1Z
3787	SZ-RD12EB
3787	SZ-UZ-12B
3787	TR12SA
3787	VZ-120
3787	WZ-120
3787	Z1014
3787	ZEN502
3787	1N1736
3787	1N1736A
3787	1N1933
3787	1N1987
3787	1N3403
3787	1N3416
3787	1N5242
3787	1N5242A
3787	1N5738
3787	1N716
3787	1N716(ZENER)
3787	1N759
3787	1N963
3787	1N963A
3787	1S2120
3787	1S7120
3787	1S7120A
3787	1S7120B
3787	8-719-112-24
3787	21A037-001
3787	48-82256C17
3787	48-82256C25
3787	103-279-21A
3787	1112
3787	5522
3787	134882256317
3787	134882256325
3788	.25N13
3788	A694X1
3788	A694X1-0A
3788	AW01-13
3788	AZ964
3788	AZ964A
3788	BZX71C13
3788	BZY83C13V5
3788	BZY85C13
3788	BZY88C13
3788	BZY94C13
3788	CXL5022
3788	ECG5022A
3788	HM13
3788	HM13A
3788	RD13AM
3788	RD13ANP
3788	RD13E
3788	RD13EA
3788	RD13EB
3788	RT8200
3788	TM5022
3788	VZ-130
3788	WZ-130
3788	1/2Z13T5
3788	1N1315
3788	1N1315A
3788	1N5243
3788	1N5243A
3788	1N5533
3788	1N5533A
3788	1N5739
3788	1N717
3788	1N766
3788	1N964
3788	1N964A
3788	1N964B
3788	1S2130
3788	1S2130A
3788	1S760
3788	09-306173
3788	48-82256C48
3788	48-82256C50
3788	48-83461E02
3788	48-83461E31
3788	61E02
3788	61E31
3788	903-73B
3788	2006-17
3788	2021-17
3788	2093A41-82
3788	5523
3788	99201-319
3788	134335
3788	233148
3788	530145-130
3788	985961
3788	2327074
3788	5330054
3788	5330054H
3788	134882256348
3788	134882256350
3788	134883461502
3788	134883461531
3789	.25N14
3789	CXL5023
3789	ECG5023A
3789	RZ14
3789	TR12SC
3789	VZ-135
3789	VZ-140
3789	WZ-135
3789	WZ-140
3789	1N1961
3789	1N2766
3789	1N2766A
3789	1N5244
3789	1N5244A
3789	1N5534
3789	1N5534A
3789	1S338
3789	1S474
3789	48-82256C13
3789	1113
3789	134882256313
3790	AZ965
3790	AZ965A
3790	BZX71C15
3790	BZX79C15
3790	BZY83D15
3790	BZY85D15
3790	BZY88C15
3790	BZY94C15
3790	ECG5024A
3790	HEP-Z0225
3790	HM15
3790	HM15A
3790	RD15EA
3790	VZ-145
3790	VZ-150
3790	WZ-145
3790	1/2Z15T5
3790	1N1934
3790	1N1934B
3790	1N1988
3790	1N1988B
3790	1N3404
3790	1N3417
3790	1N5245
3790	1N5245A
3790	1N5535
3790	1N5535A
3790	1N5740
3790	1N718
3790	1N965
3790	1N965A
3790	1S2150
3790	1S7150
3790	1S7150A
3790	1S7150B
3790	14-515-23
3790	48-137330
3790	48-137394
3790	48-82256C14
3790	5330541
3790	134882256314
3791	.25N16
3791	AZ966
3791	AZ966A
3791	BZX71C16
3791	BZY88C16
3791	BZY94C16
3791	CXL5025
3791	ECG5025A
3791	HM16
3791	HM16A
3791	RD16A
3791	RD16B
3791	RZ16
3791	VZ-157
3791	VZ-162
3791	WZ-157
3791	WZ-162
3791	1/2Z16T5
3791	1N1316
3791	1N1316A
3791	1N228
3791	1N228A
3791	1N5246
3791	1N5246A
3791	1N5536
3791	1N5536A
3791	1N5741
3791	1N719
3791	1N767
3791	1N966
3791	1N966A
3791	1S7160A
3791	1S761
3791	5525
3792	.25N17
3792	BZY83C16V5
3792	BZY85C16V5
3792	RD16AN
3792	1N5247
3792	1N5247A
3792	1N5537
3792	1N5537A
3792	5325
3793	.25N18
3793	AZ18
3793	BZX71C18
3793	BZY83D18
3793	BZY85D18
3793	BZY88C18
3793	BZY94C18
3793	HEP-Z0228
3793	HM18
3793	HM18A
3793	PD6020
3793	RZZ18
3793	SV4018
3793	SV4018A
3793	SZ18
3793	1N1935
3793	1N1962
3793	1N1989
3793	1N3418
3793	1N5248
3793	1N5248A
3793	1N5538
3793	1N5538A
3793	1N5742
3793	1N720
3793	1N720B
3793	1N967
3793	1N967A
3793	1N967B
3793	1S339
3793	1S475
3793	34-8057-43
3793	48-82256C24
3793	48-83461E18
3793	48-83461E33
3793	61E18
3793	61E33
3793	1118
3793	5526
3793	134882256324
3793	134883461518
3793	134883461533
3794	.25N19
3794	RD19A
3794	RD19AL
3794	RD19AM
3794	1N1317
3794	1N1317A
3794	1N5249
3794	1N5249A
3794	1N768
3794	1S762
3794	48-83461E24
3794	61E24
3794	134883461524
3795	.25N20
3795	AZ968
3795	AZ968A
3795	BZX71C20
3795	BZY88C20
3795	BZY94C20
3795	HEP-Z0230
3795	HM20
3795	HM20A
3795	RZ20
3795	SZ20
3795	1N5250
3795	1N5250A
3795	1N5540
3795	1N5540A
3795	1N5743
3795	1N721
3795	1N968
3795	1N968A
3795	1N968B
3795	1S2200
3795	34-8057-44
3795	48-83461E22
3795	61E22
3795	1120
3795	5527
3795	134883461522
3796	.25N22
3796	AZ22
3796	AZ969
3796	AZ969A
3796	BZX71C22
3796	BZY83D22
3796	BZY85D22
3796	BZY88C22
3796	BZY94C22
3796	HEP-Z0231
3796	HM22
3796	HM22A
3796	RZZ22
3796	1N1936
3796	1N1963
3796	1N1990
3796	1N3419
3796	1N5251
3796	1N5251A
3796	1N5251B
3796	1N5541
3796	1N5541A
3796	1N5744
3796	1N722
3796	1N969A
3796	1S2220
3796	1S476
3796	48-137306
3796	5528
3797	.25N24
3797	AZ970
3797	AZ970A
3797	BZX71C24
3797	BZY88C24
3797	BZY94C24
3797	HM24
3797	HM24A
3797	MZ224A
3797	1N1318
3797	1N1318A
3797	1N5252
3797	1N5252A
3797	1N5542
3797	1N5542A
3797	1N5745
3797	1N723
3797	1N970
3797	1N970A
3797	1S2240
3797	1S763
3797	48-82256C49
3797	48-83461E26
3797	61E26
3797	5529
3797	134882256331
3797	134882256349
3797	134883461526
3798	.25N25
3798	BZY83C24V5
3798	RZ25
3798	1N4710
3798	1N5253
3798	1N5253A
3798	1N5543
3798	1N5543A
3798	1N769-3
3799	.25N27
3799	AZ971
3799	AZ971A
3799	BZY94C27
3799	HEP-Z0234
3799	HM27
3799	HM27A
3799	RD29AL
3799	RZZ27
3799	1N1937
3799	1N1964
3799	1N1991
3799	1N3420
3799	1N5254
3799	1N5254A
3799	1N5746
3799	1N724
3799	1S2270
3799	1S764
3799	48-137062
3799	48-82256C20
3799	5530
3800	RD29A
3800	RD29AM
3800	1N1319
3800	1N1319A
3800	1N5255
3800	1N5255A
3800	1N5544
3800	1N5544A
3800	1S477
3801	.25N30
3801	BZY88C30
3801	BZY94C30
3801	HM30
3801	HM30A
3801	RD29AN
3801	RZ30
3801	1N3421
3801	1N5256
3801	1N5256A
3801	1N5545
3801	1N5545A
3801	1N5747
3801	1N972
3801	1N972A
3801	1S2300
3801	5531
3802	.25N33
3802	AZ973
3802	AZ973A
3802	BZY85C33
3802	BZY94C33
3802	HM33
3802	HM33A
3802	RD35AL
3802	RZ33A
3802	1N1320
3802	1N1320A
3802	1N1938
3802	1N1965
3802	1N1992
3802	1N3422
3802	1N5257
3802	1N5257A
3802	1N5546
3802	1N5546A
3802	1N5748
3802	1N726
3802	1N973
3802	1N973A
3802	1S2330
3802	2A64
3802	5532
3803	.25N36
3803	AZ974
3803	AZ974A
3803	BZY94C36
3803	HM36
3803	HM36A
3803	RD36EB
3803	RH-EX0053CEZZ
3803	1N5258
3803	1N5258A
3803	1N5749
3803	1N727
3803	1N974
3803	1N974A
3803	1S478
3803	5533
3804	.25N39
3804	AZ975
3804	AZ975A
3804	BZX46C39
3804	BZY94C39
3804	HM39
3804	HM39A
3804	RZ39A
3804	1N1321
3804	1N1321A
3804	1N1939
3804	1N1966
3804	1N1993
3804	1N3423
3804	1N5259
3804	1N5259A
3804	1N5750
3804	1N728A
3804	1N975
3804	1N975A
3804	5534
3805	.25N43
3805	AZ976
3805	AZ976A
3805	BZY94C43
3805	HM43
3805	HM43A
3805	1N5260
3805	1N5260A
3805	1N5751
3805	1N729
3805	1N976
3805	1N976A
3805	5535
3806	.25N45
3806	.25N47
3806	BZY94C47
3806	HM47
3806	HM47A
3806	RZ47A
3806	1N1322
3806	1N1322A
3806	1N1940
3806	1N1967
3806	1N1994
3806	1N3424
3806	1N5261
3806	1N5261A
3806	1N5752
3806	1N730
3806	1N977
3806	1N977A
3806	1S479
3806	5536
3807	.25N50
3807	.25N52
3807	BZY94C51
3807	HM51
3807	HM51A
3807	RZ50
3807	1N5262
3807	1N5262A
3807	1N5753
3807	1N731
3807	1N731B
3807	1N978
3807	1N978A
3807	48-11-010
3807	5537
3808	.25N56
3808	BZY94C56
3808	HM56
3808	HM56A

SK	Industry Standard No.
3808	RZ56A
3808	1N1941
3808	1N1942
3808	1N1968
3808	1N1995
3808	1N3425
3808	1N5263
3808	1N5263A
3808	1N5754
3808	1N732
3808	1N732B
3808	1N979
3808	1N979A
3808	5538
3809	1N1323
3809	1N1323A
3809	1N5264
3809	1N5264A
3809	5ZS60
3809	5ZS60A
3810	.25N62
3810	BZY94C62
3810	HM62
3810	HM62A
3810	1N5265
3810	1N5265A
3810	1N5755
3810	1N980
3810	1N980A
3811	.25N68
3811	BZY94C68
3811	HM68
3811	HM68A
3811	RZ68A
3811	1N1969
3811	1N1996
3811	1N3426
3811	1N5266
3811	1N5266A
3811	1N5756
3811	1N734
3811	1N734B
3811	1N981
3811	1N981A
3811	5540
3811	134882256305
3812	.25N75
3812	BZX79C75
3812	BZY94C75
3812	HM75
3812	HM75A
3812	1N5267
3812	1N5267A
3812	1N735
3812	1N735B
3812	1N982
3812	1N982A
3812	5541
3813	.25N82
3813	HM82
3813	HM82A
3813	RZ82A
3813	1/2Z82T5
3813	1N1943
3813	1N1970
3813	1N1997
3813	1N3427
3813	1N5268
3813	1N5268A
3813	1N983
3813	1N983A
3814	1N1325
3814	1N5269
3814	1N5269A
3815	.25N91
3815	HM91
3815	HM91A
3815	1N5270
3815	1N5270A
3815	1N737
3815	1N737B
3815	1N984
3815	1N984A
3816	.25N100
3816	HM100
3816	HM100A
3816	1N1944
3816	1N1971
3816	1N1998
3816	1N3428
3816	1N5271
3816	1N5271A
3816	1N738
3816	1N738B
3816	1N985
3816	1N985A
3817	.25N110
3817	HM110
3817	HM110A
3817	1N5272
3817	1N5272A
3817	1N986
3817	1N986A
3818	.25N120
3818	HM120
3818	HM120A
3818	1N1945
3818	1N1972
3818	1N1999
3818	1N3429
3818	1N5273
3818	1N5273A
3818	1N740
3818	1N987
3818	1N987A
3819	.25N130
3819	HM130
3819	HM130A
3819	1N1327
3819	1N5274
3819	1N5274A
3819	1N741
3819	1N988
3819	1N988A
3819	1N988B
3820	.25N140
3820	1N5275
3820	1N5275A
3821	.25N150
3821	HM150
3821	HM150A
3821	1N1946
3821	1N1973
3821	1N2000
3821	1N3430
3821	1N5276
3821	1N5276A
3821	1N741B
3821	1N742
3821	1N742B
3821	1N989
3821	13-33187-10
3822	HM160
3822	HM160A
3822	1N5277
3822	1N5277A
3822	1N743
3822	1N990
3822	1N990A
3823	1N5278
3823	1N5278A
3824	HM180
3824	HM180A
3824	1N1947
3824	1N1974
3824	1N2001
3824	1N3431
3824	1N5279
3824	1N5279A
3824	1N744
3824	1N744B
3824	1N991
3824	1N991A
3825	1N5280
3825	1N5280A
3826	.25N200
3826	HM200
3826	HM200A
3826	1N5281
3826	1N5281A
3826	1N745
3826	1N745B
3826	1N992
3826	1N992A
3827	BA521
3827	BA521A
3827	ECG1166
3827	GEIC-316
3827	REN1166
3827	TM1166
3829	HA1137
3829	HA1137P
3829	HA1137W
3829	IC-324(ELCOM)
3829	R4438-1(RCA)
3829	R4438-2(RCA)
3829	RE332-IC
3829	T2T
3829	T2T-2
3829	TDA1200
3829	07B2Z
3829	09-018
3829	51-10658A02
3829	51-10658A03
3829	51S10658A01
3829	88-20372
3829	2077-1
3829	3531-030-000
3829	4438-1(RCA)
3829	4438-2(RCA)
3829	4438-3(RCA)
3829	9144-61
3829	10658A01
3829	18410-148
3829	29810-179
3829	38510-169
3829	45391
3829	45810-167
3829	136146
3829	612077-1
3829	1464438-1
3829	1464438-2
3829	1464438-3
3829	3596810
3829/788	G09-028-A
3831	LM78L05ACH
3831	LM78L05CH
3832	TDA1190Z
3834	ECG132
3834	K33(E)
3834	REN132
3834	2N5245A
3834	2N5246A
3834	2N5247A
3834	2N5248A
3834	2SK19-14
3834	2SK19-Y
3834	2SK19A
3834	2SK19B
3834	2SK19BB
3834	2SK39
3834/132	DDCY001002
3834/132	ES15X92
3834/132	JP1033S
3834/132	001-027030
3834/132	2N5245
3834/132	2N5246
3834/132	2N5247
3834/132	2N5248
3834/132	2SK104
3834/132	2SK104H
3834/132	2SK19
3834/132	2SK19(GR)
3834/132	2SK19BL
3834/132	2SK19GC
3834/132	2SK19GE
3834/132	2SK19GR
3834/132	2SK19H
3834/132	2SK19Y
3834/132	2SK32B
3834/132	2SK33
3834/132	2SK33(E)
3834/132	2SK33E
3834/132	2SK33F
3834/132	2SK33GR
3834/132	2SK33H
3834/132	2SK41
3834/132	2SK41D
3834/132	2SK41E
3834/132	2SK41F
3834/132	2SK42
3834/132	2SK42-CM1
3834/132	2SK42CM1
3834/132	2SK47
3834/132	2SK47M
3834/132	2SK49
3834/132	2SK49F
3834/132	2SK49H
3834/132	2SK49H1
3834/132	2SK49HK
3834/132	2SK49I
3834/132	2SK49M
3834/132	2SK55
3834/132	2SK55C
3834/132	2SK55D
3834/132	2SK55DE
3834/132	2SK55E
3834/132	2SK55R
3834/132	2SK61
3834/132	3G2
3834/132	3SK41
3834/132	3SK41L
3834/132	3SK41M
3834/132	8-723-302-00
3834/132	13-43112-1
3834/132	21A040-015
3834/132	21M196
3834/132	21M224
3834/132	022-2876-006
3834/132	24MW723
3834/132	24MW736
3834/132	24MW989
3834/132	46-86316-3
3834/132	48-137343
3834/132	48-90232A14
3834/132	48-90234A39
3834/132	48-97046A47
3834/132	48-97046A48
3834/132	48-97177A01
3834/132	48-97177A06
3834/132	48S134944
3834/132	48S137343
3834/132	57A31-1
3834/132	57B169-12
3834/132	87-0001
3834/132	90-50
3834/132	200-064
3834/132	220-008001
3834/132	229-0192-20
3834/132	260-10-006
3834/132	260P22001
3834/132	260P22002
3834/132	260P22003
3834/132	417-211
3834/132	430-25762
3834/132	588U
3834/132	734EU
3834/132	753-4000-025
3834/132	753-6000-019
3834/132	753-9000-019
3834/132	800-535-00
3834/132	800-535-01
3834/132	921-126B
3834/132	921-203B
3834/132	2057A2-445
3834/132	4802-00010
3834/132	8000-00004-081
3834/132	8000-00004-P080
3834/132	8000-00004-P081
3834/132	8000-00005-001
3834/132	8000-00009-178
3834/132	8000-00011-054
3834/132	8000-00011-055
3834/132	8000-0005-001
3834/132	8010-173
3834/132	8710-161
3834/132	8910-141
3834/132	38510-161
3834/132	43296
3834/132	58810-160
3834/132	88510-171
3834/132	144859
3834/132	610164-1
3834/132	970253
3834/132	985715
3834/132	2006623-47
3834/132	2327142
3834/132	5320702
3834/132	7910134-01
3834/132	26010006
3834/132	43022134
3834/132	62608528
3835	A4L
3835	AC127-01
3835	AC141
3835	AC141B
3835	AC141K
3835	AC179
3835	AC181K
3835	AC185
3835	AC186
3835	AC187K
3835	AC187R
3835	BD-00072
3835	BTX071
3835	C181
3835	CS1759
3835	E24105
3835	EA15X8443
3835	ECG103A
3835	GC1137
3835	GC1185
3835	GC1423
3835	GC148
3835	GC463
3835	GC465
3835	GC467
3835	GC608
3835	GC609
3835	GE-59
3835	GT35
3835	GT364
3835	GT365
3835	GT366
3835	GT903
3835	HA5010
3835	HT400723
3835	HT400723A
3835	HT400723B
3835	HT403523A
3835	K4-501
3835	NC33
3835	NKT773
3835	PBE3020-2
3835	QQV60527
3835	QQV60537
3835	QQV61772
3835	R1531
3835	R1532
3835	R1534
3835	R1537
3835	R1545
3835	R1547
3835	R1549
3835	R1553
3835	R177
3835	R2356
3835	R2359
3835	R2360
3835	R2364
3835	R2375
3835	R3293
3835	R3573-1
3835	R5050
3835	R5054
3835	R5180
3835	R61
3835	R62
3835	R7362
3835	R79
3835	R80
3835	REN103A
3835	RS1533
3835	RS1549
3835	RS3931
3835	RS8420
3835	RS8443
3835	S2042634
3835	SFT377
3835	SP2188
3835	TM103A
3835	TNJ61734
3835	TQ5050
3835	TQ5062
3835	TR310160
3835	TV24983
3835	WEP641B
3835	XG28
3835	XG29
3835	1-TR-112
3835	002-011700
3835	002-11700
3835	2N149
3835	2N149A
3835	2N1672
3835	2N1672A
3835	2N647
3835	2N647/22
3835	2SC180
3835	2SC180A
3835	2SC180B
3835	2SC180C
3835	2SC180D
3835	2SC180E
3835	2SC180F
3835	2SC180G
3835	2SC180GN
3835	2SC180H
3835	2SC180J
3835	2SC180K
3835	2SC180L
3835	2SC180M
3835	2SC180R
3835	2SC180X
3835	2SC180Y
3835	2SC181
3835	2SC181A
3835	2SC181B
3835	2SC181C
3835	2SC181D
3835	2SC181E
3835	2SC181F
3835	2SC181G
3835	2SC181GN
3835	2SC181H
3835	2SC181J
3835	2SC181L
3835	2SC181M
3835	2SC1810R
3835	2SC181R
3835	2SC181X
3835	2SC181Y
3835	2SC277C
3835	2SC34
3835	2SC35
3835	2SD100
3835	2SD100A
3835	2SD104
3835	2SD105
3835	2SD127
3835	2SD127A
3835	2SD128
3835	2SD128A
3835	2SD170
3835	2SD170C
3835	2SD178
3835	2SD178A
3835	2SD178B
3835	2SD178C
3835	2SD178D
3835	2SD178E
3835	2SD178F
3835	2SD178G
3835	2SD178GN
3835	2SD178H
3835	2SD178J
3835	2SD178K
3835	2SD178L
3835	2SD178M
3835	2SD178X
3835	2SD178Y
3835	2SD186
3835	2SD186A
3835	2SD186B
3835	2SD187
3835	2SD187A
3835	2SD187B
3835	2SD187C
3835	2SD187D
3835	2SD187E
3835	2SD187F
3835	2SD187G
3835	2SD187GN
3835	2SD187H
3835	2SD187J
3835	2SD187K
3835	2SD187L
3835	2SD187M
3835	2SD1870R
3835	2SD187R
3835	2SD187X
3835	2SD187Y
3835	2SD191
3835	2SD192
3835	2SD194
3835	2SD30
3835	2SD30-N
3835	2SD30-O
3835	2SD30A
3835	2SD30B
3835	2SD30C
3835	2SD30D
3835	2SD30E
3835	2SD30F
3835	2SD30G
3835	2SD30GN
3835	2SD30H
3835	2SD30J
3835	2SD30K
3835	2SD30L
3835	2SD30M
3835	2SD30P
3835	2SD30R
3835	2SD30X
3835	2SD30Y
3835	2SD31
3835	2SD31D
3835	2SD32
3835	2SD34
3835	2SD35
3835	2SD36
3835	2SD367
3835	2SD367A
3835	2SD367B
3835	2SD367C
3835	2SD367D
3835	2SD367E
3835	2SD367F
3835	2SD367H
3835	2SD367J
3835	2SD367K
3835	2SD367L
3835	2SD367M
3835	2SD3670R
3835	2SD367P
3835	2SD367R
3835	2SD367X
3835	2SD367Y
3835	2SD38
3835	2SD38Y
3835	2SD75AH
3835	4JX2A616
3835	8Q-3-04
3835	09-303006
3835	09-303012
3835	09-303023
3835	9-5202
3835	9-5224-2
3835	13-14279-1
3835	14-602-21
3835	14-602-52
3835	022-2876-002
3835	34-6001-84
3835	42-21404
3835	45N2A
3835	46-86115-3
3835	55-1027
3835	55-1032
3835	57A1-3
3835	57A1-4
3835	57A1-5
3835	57A1-6
3835	57A1-78
3835	57A6-20
3835	57A6-21
3835	57A6-5
3835	57C6-20
3835	57D1-3
3835	57D1-4
3835	57D1-5
3835	57D1-6
3835	57D1-78
3835	57M1-16
3835	57M1-18
3835	57M2-1
3835	57M2-2
3835	57M2-6
3835	57M2-9
3835	62-13259
3835	63-10383
3835	63-7549
3835	63-7565
3835	63-8473
3835	63-8705
3835	63-9340
3835	65-2A4-7B
3835	86-301-2
3835	86-5-2
3835	86-5005-2
3835	86-5034-2
3835	86-5086-2
3835	114N4U
3835	121-237
3835	121-238
3835	121-247
3835	121-248
3835	121-557
3835	121-558
3835	121-59
3835	121-8
3835	126N1
3835	126N2
3835	130-40314
3835	247-256
3835	297L001H01
3835	297L001H02
3835	297L001M01
3835	297V002H03
3835	297V002H04
3835	297V002H05
3835	324-0122
3835	417-121
3835	642-277
3835	690V067H35
3835	690V081H96
3835	800-537-01
3835	921-46B
3835	921-54B
3835	921-71
3835	921-71A
3835	921-71B
3835	921-7B
3835	1040-80
3835	1104-95
3835	2057B100-4
3835	4822-130-40096
3835	8000-00011-086
3835	8000-00030-009
3835	11668-7
3835	40037

SK	Industry Standard No.
3835	72191
3835	94023
3835	94029
3835	110958
3835	116204
3835	116687
3835	610126-2
3835	815218-4
3835	985735
3835	985735A
3835	1473573-1
3835	2002153-83
3835	5320295H
3835	5320305
3835	7269847
3835	7277066
3835	7851319-01
3835	7851321-01
3835	7851467-01
3835	62119578
3835	2004503018
3836	C1116
3836	D287
3836	ECG284
3836	GE-265
3836	KSD9701
3836	KSD9701A
3836	KSD9704
3836	KSD9707
3836	SJ2519
3836	2N5629
3836	2N5631
3836	2N6302
3836	2N6359
3836	2N6360
3836	2SC1116
3836	2SC1116-0
3836	2SD287
3836	2SD424
3836	2SD675
3836	2SD676
3836	13-37708-1
3836	96-5370-01
3836/284	A6773802
3836/284	BD245B
3836/284	BD245C
3836/284	BDY76
3836/284	C1115
3836/284	C1667
3836/284	D379
3836/284	D379P
3836/284	D379Q
3836/284	D379R
3836/284	D379S
3836/284	2SC1112
3836/284	2SC1115
3836/284	2SC1195
3836/284	2SC1403
3836/284	2SC1585
3836/284	2SC1585F
3836/284	2SC1585H
3836/284	2SC1667
3836/284	2SC1829
3836/284	2SC1869
3836/284	2SC2256
3836/284	2SC2337
3836/284	2SC901
3836/284	2SC901A
3836/284	2SD213
3836/284	2SD214
3836/284	2SD316
3836/284	2SD379
3836/284	2SD379P
3836/284	2SD379Q
3836/284	2SD379R
3836/284	2SD379S
3836/284	2SD665
3836/284	2SD73
3836/284	2SD73A
3836/284	2SD73B
3836/284	2SD73C
3836/284	2SD73E
3836/284	2SD74
3836/284	2SD74A
3836/284	2SD74B
3836/284	2SD74C
3836/284	2SD74D
3836/284	2SD74E
3838	MLV4372A
3839	AD157
3839	AD161
3839	AD165
3839	BDY15
3839	BDY15A
3839	BDY15B
3839	BDY15C
3839	BDY16A
3839	BDY16B
3839	GE-43
3839	IRTR91
3839	PT4
3839	2N4077
3839	8Q-3-23
3839	13-26377-2
3839	21A015-021
3839	57M3-10N
3839	57M3-11
3839	57M3-9N
3839	94N2P
3839	4822-130-40212
3839/155	BDY16
3839/155	RE54
3840	ECG131MP
3840	GE-31MP
3840	J24366
3840	N-EA15X139
3840	RE20MP
3840	REN131MP
3840	W17MP
3840	4-1848
3840	24MW618
3840	33-1002-00
3840	48X97238A06
3840	86-0033-007
3840	96X26054/45X
3840	465-166-19
3840	675-206
3840	753-2000-463
3840	4800-223
3840	081042
3840	171162-090
3840	175043-081
3840	215071
3840	0573031
3840	740443
3841	A6532941
3841	A684
3841	B544
3841	GE-334
3841	PTC142
3841	2SA641BL
3841	2SA641G
3841	2SA641GR
3841	2SA641K
3841	2SA641OR
3841	2SA641R
3841	2SA641Y
3841	2SA666B
3841	2SA666BL
3841	2SA666C
3841	2SA666D
3841	2SA666E
3841	2SA666Y
3841	2SA683
3841	2SA683P
3841	2SA683Q
3841	2SA683R
3841	2SA683S
3841	2SA684
3841	2SA684Q
3841	2SA684R
3841	2SA705A
3841	2SA705B
3841	2SA705C
3841	2SA705D
3841	2SA705E
3841	2SA705F
3841	2SA705G
3841	2SA705GN
3841	2SA705H
3841	2SA705J
3841	2SA705K
3841	2SA705L
3841	2SA705M
3841	2SA705OR
3841	2SA705R
3841	2SA705Y
3841	2SA950-0
3841	2SA950Y
3841	2SA966
3841	2SB544
3841	2SB544E
3841	2SB544F
3841	2SB562
3841	2SB562B
3841	2SB562C
3841	2SB564
3841	2SB564L
3841	2SB598
3841	2SB598F
3841	2SB621
3841	2SB621R
3841	8-729-468-43
3841	8-760-514-10
3841	48-90420A05
3841	56-4831
3841	87-0237-R
3841	87-0237-S
3841	121-Z9005
3841	2032-35
3841	95241-1
3841	145840
3841	2321891
3841/294	A641
3841/294	A641(PNP)
3841/294	A641A
3841/294	A641B
3841/294	A641C
3841/294	A641D
3841/294	A641L
3841/294	A641M
3841/294	A641S
3841/294	A6532921
3841/294	A666
3841/294	A666A
3841/294	A666H
3841/294	A666HR
3841/294	A666R
3841/294	A666S
3841/294	A705
3841/294	A772B1
3841/294	A772D1
3841/294	A772EH
3841/294	A772FE
3841/294	2SA641
3841/294	2SA641A
3841/294	2SA641B
3841/294	2SA641C
3841/294	2SA641D
3841/294	2SA641L
3841/294	2SA641M
3841/294	2SA666
3841/294	2SA666A
3841/294	2SA666H
3841/294	2SA666HR
3841/294	2SA705
3841/294	2SA773
3841/294	2SA950
3841/294	2SA950-0
3841/294	2SB544P1
3841/294	2SB544P1D
3841/294	2SB544P1E
3841/294	2SB544P1F
3841/294	4100905254
3842	2SC1247
3842	2SC2086
3843	BB-4
3843	CGJ-1
3843	EP16X15
3843	ERC26-13L
3843	F114E
3843	HF-1
3843	HF-1A
3843	MB-1F
3843	RE55
3843	REN506
3843	RH-1V
3843	RH-DX0028CEZZ
3843	RH-DX0062CEZZ
3843	RH-DX0073CEZZ
3843	RH-DX0085TAZZ
3843	RH-DX0086TAZZ
3843	RH-DX0101CEZZ
3843	RH-DX0104CEZZ
3843	RH1
3843	RM1
3843	S5295G
3843	S5295J
3843	SB-2C
3843	SB2C
3843	SID30-13
3843	TD-13
3843	TVSRA-1Z
3843	TVSRA1Z
3843	U06C
3843	UF01
3843	V09E
3843	VHD1S1834//-1
3843	1.5E12
3843	1.5J12
3843	1N1409
3843	1N2327
3843	1N2357
3843	1N2503
3843	1N2504
3843	1N2507
3843	1N2508
3843	1N2618
3843	1N2619
3843	1N2776
3843	1N2777
3843	1N2778
3843	1N2779
3843	1N2884
3843	1N2885
3843	1N3233
3843	1N3244
3843	1N3487
3843	1N3724
3843	1N3725
3843	1N3729
3843	1N3731
3843	1N3915
3843	1N4146
3843	1N4155
3843	1N4252
3843	1N4826
3843	1N487
3843	1N487A
3843	1N487B
3843	1N488
3843	1N488A
3843	1N488B
3843	1N4934
3843	1N4935
3843	1N4936
3843	1N543
3843	1N543A
3843	1N549
3843	1N893
3843	1N923
3843	1NJ70976
3843	1R10D3L
3843	1R31
3843	1R3B
3843	1S1237
3843	1S1238
3843	1S1349
3843	1S1517
3843	1S1517A
3843	1S1832
3843	1S1834
3843	1S1835
3843	1S1855
3843	1S1882
3843	1S2306
3843	1S2307
3843	1S2308
3843	1S2309
3843	1S2711
3843	1S2756
3843	4-2021-07970
3843	4-2021-10470
3843	4-202104570
3843	4-202104870
3843	05-110046
3843	05-112404
3843	05-800015
3843	05-936010
3843	09-306030
3843	09-306141
3843	09-306144
3843	09-306237
3843	09-306260
3843	09-306392
3843	09-306418
3843	09-306423
3843	09-306425
3843	10D5A
3843	10D5D
3843	10R11B
3843	10R12B
3843	10R13B
3843	10R14B
3843	11/10
3843	12G4
3843	13-29663-1
3843	13-37868-1
3843	13-55030-1
3843	13-55031-1
3843	13-55031-2
3843	13-55031-3
3843	13-55078-1
3843	14-514-03
3843	14-514-03/53
3843	14-514-07
3843	14-514-07/57
3843	14-514-14
3843	14-514-17/67
3843	14-514-18
3843	14-514-18/68
3843	14-514-20/70
3843	14-514-23
3843	14-514-23/73
3843	14-514-66
3843	14-514-74
3843	15-108013
3843	15-108026
3843	15-108035
3843	15-108040
3843	15-108041
3843	15-108043
3843	15-108044
3843	15-108047
3843	15-108048
3843	15-123102
3843	15-123103
3843	15-123242
3843	21A004-000
3843	21A110-005
3843	21A110-008
3843	21A110-009
3843	21A110-014
3843	21A119-068
3843	28-22-13
3843	28-22-22
3843	34-8054-24
3843	34-8054-25
3843	34-8054-8
3843	34-8057-22
3843	34-8057-24
3843	46-86106-3
3843	46-861149-3
3843	46-861179-3
3843	46-86179-3
3843	46-8620-3
3843	46-86280-3
3843	46-86281-3
3843	46-86282-3
3843	46-86308-3
3843	46-86320-3
3843	46-86324-3
3843	46-86328-3
3843	46-86331-3
3843	46-86382-3
3843	46-86391-3
3843	46-86393-2
3843	46-86393-3
3843	46-86395-3
3843	46-86440-3
3843	46-86489-3
3843	46-86511-3
3843	46-86519-3
3843	46-86525-1
3843	46-86525-3
3843	48-134921
3843	48-134939
3843	48-134978
3843	48-137112
3843	48-137533
3843	48-137546
3843	48-137551
3843	48-155107
3843	48-155108
3843	48-155125
3843	48-191A05A
3843	48-355013
3843	48-40235G02
3843	48-41266G01
3843	48-8240C10
3843	48-8240C12
3843	48-82466H12
3843	48-82466H13
3843	48-82466H14
3843	48-82466H15
3843	48-82466H16
3843	48-82466H17
3843	48-82466H18
3843	48-82466H19
3843	48-82466H21
3843	48-82466H25
3843	48-82466H27
3843	48-82466H28
3843	48-86308-3
3843	48-90158A01
3843	48-90234A12
3843	48-90343A53
3843	48-90343A54
3843	48-90420A98
3843	48B41226G01
3843	48C40235G02
3843	48S137533
3843	48S137546
3843	48S137551
3843	48S155056
3843	48S155107
3843	48S155108
3843	48S191A05
3843	48S191A05A
3843	48S191A08
3843	53A014-1
3843	53A015-1
3843	53L001-5
3843	55-001
3843	73A60-11
3843	75R11B
3843	75R12B
3843	75R13B
3843	75R14B
3843	86-116-3
3843	91A08
3843	93A60-10
3843	93B52-1
3843	93B52-2
3843	93B60-3
3843	93C60-3
3843	93C69-1
3843	094-012
3843	100R11B
3843	100R12B
3843	100R13B
3843	100R14B
3843	103-112
3843	103-160
3843	103-193
3843	103-196
3843	103-244
3843	103-247
3843	103-263
3843	123-018
3843	148-3
3843	151-012-9-001
3843	212-68
3843	229-1301-25
3843	388B
3843	690V080H51
3843	800-282
3843	1010-8116
3843	1059-2848
3843	1061-6290
3843	1061-8387
3843	1872-3
3843	2093A41-69
3843	6171-17
3843	6171-18
3843	12685
3843	113998
3843	125844
3843	125848
3843	126527
3843	131148
3843	132418
3843	132501
3843	132509
3843	132547
3843	132548
3843	132549
3843	133615
3843	133616
3843	135734
3843	136605
3843	136606
3843	137075
3843	140503
3843	161032
3843	185022
3843	194917
3843	233062
3843	530148-1003
3843	530179-1002
3843	530184-1002
3843	0552007H
3843	575047
3843	0575047H
3843	0575049
3843	0575049H
3843	0575054
3843	1390655
3843	1415721-35
3843	1474778-6
3843	1476049-2
3843	1476161-12
3843	1476171-17
3843	1476171-25
3843	1476171-26
3843	1476183-2
3843	1476183-5
3843	1476183-6
3843	2330101
3843	2330356
3843	2331381
3843	2332152
3843	06200010
3843	62051310
3843	62564520
3843	62801328
3843	2008100076
3843	2008130040
3843	2008130171
3843	2012120009
3843	2012130234
3843	4202003700
3843	4202003900
3843	13488240310
3843	13488240312
3843	134882466812
3843	134882466813
3843	134882466814
3843	134882466815
3843	134882466816
3843	134882466817
3843	134882466818
3843	134882466821
3843	134882466825
3843	134882466827
3843	134882466828
3844	2SD470
3844	2SD470A
3844	2SD470B
3844/321	GE-260
3845	A127364
3845	A300043-06
3845	ACY39
3845	GP290
3845	M75537-2
3845	M9177
3845	N4967
3845	NC34
3845	P2F
3845	R-2SB492
3845	RE35
3845	RS-5735
3845	RS5788
3845	OS492
3845	SMB620782-1
3845	ST-28A
3845	TNJ61223
3845	002-011800
3845	2N1038
3845	2N2541
3845	2N660
3845	2N661
3845	2N672
3845	2SB492
3845	2SB492B
3845	7-73004-02
3845	7-73004-03
3845	7-73004-04
3845	7-73004-1
3845	09-301031
3845	22E89
3845	23-5034
3845	42-18109
3845	46-86198-3
3845	48-134431
3845	48-134584
3845	48-134585
3845	48-134856
3845	48-137039
3845	48-869177
3845	129-4
3845	2606-288
3845	2703-387
3845	3456
3845	34966
3845	36340
3845	36673
3845	36695
3845	42321
3845	114504
3845	114504A03
3845	127590
3845	171162-025
3845	194086-3
3845	524966
3845	0573185
3845	620782
3845	2227367
3845	8511724-3
3845	8511724-4
3845/176	RS5735
3845/176	SFT-353
3845/176	SFT353
3845/176	ST-370
3845/176	ST28A
3845/176	2N1274
3845/176	2N1274BLU
3845/176	2N1274BRN
3845/176	2N1274GRN
3845/176	2N1274ORN

SK	Industry Standard No.	SK	Industry Standard No.	SR	Industry Standard No.	SK	Industry Standard No.	SK	Industry Standard No.
3845/176	2N1274PUR	3849	2SC1384F	3854	121-895A	3861	AT73R	3861	TR-09
3845/176	2N1274RED	3849	2SC1384G	3854	921-01124	3861	AT75R	3861	TR-09C
3845/176	2N2648	3849	2SC1384GN	3854	1100-9479	3861	AT76R	3861	TR167
3845/176	2N658	3849	2SC1384H	3854	124774	3861	AT77	3861	TR182
3845/176	2N659	3849	2SC1384J	3854	132825	3861	B65-4A21	3861	TR183
3845/176	2N662	3849	2SC1384K	3854	133249	3861	B92-1-A-21	3861	TR184
3846	B539	3849	2SC1384L	3854	133275	3861	CK261	3861	TR193
3846	B539R	3849	2SC1384M	3854	133690	3861	CK262	3861	TR194
3846	B554	3849	2SC1384QR	3854	141331	3861	E4002	3861	TR211
3846	B554-R	3849	2SC1384Q	3854	142686	3861	EO0101	3861	TR212
3846	GB-266	3849	2SC1384R	3854	142711	3861	E85	3861	TR213
3846	SJ2520	3849	2SC1384S	3854	143316	3861	G101079	3861	TR214
3846	2N5876	3849	2SC1384X	3854	143792	3861	G16506	3861	TR216
3846	2N6030	3849	2SC1384Y	3854	143794	3861	GA53270	3861	TR335
3846	2N6031	3849	2SC1474	3854	143804	3861	GC1034	3861	TR336
3846	22B539	3849	2SC1474-3	3854	146484	3861	GC1035	3861	TR337
3846	22B552	3849	2SC1474-4	3856/310	01-031964	3861	GC452	3861	W2
3846	22B554	3849	2SC1474J	3856/310	2000-211	3861	GC453	3861	WTVSA7
3846	22B554-R	3849	2SC1474S	3857	28P1189	3861	GC454	3861	WTVSK7
3846	22B655	3849	2SD1518	3857	28P1189A	3861	GI5	3861	WTVSQ7
3846	22B656	3849	2SD355	3857	28P1189D	3861	GI6	3861	XA701
3846	96-5371-01	3849	2SD355D	3857	28P1189E	3861	GI6506	3861	XA702
3846/285	A746	3849	2SD355E	3857	28P1189F	3861	GI7	3861	XA703
3846/285	A747	3849	2SD400P1	3857	28P1189G	3861	GT1202	3861	XB3315
3846/285	A747A	3849	2SD400P1D	3857	28P1189H	3861	GT1608	3861	001-01101-0
3846/285	B532	3849	2SD400P1E	3857	28P1189K	3861	GT1609	3861	001-011010
3846/285	B532P	3849	2SD400P1F	3857	28P1189L	3861	GT2765	3861	2M78
3846/285	B532Q	3849	2SD468	3857	28P1189M	3861	GT2766	3861	2N100
3846/285	B532R	3849	2SD468A	3857	28P1189R	3861	GT2767	3861	2N1000
3846/285	2SA1007	3849	2SD468AC	3857	28P1189Y	3861	GT2906	3861	2N1012
3846/285	2SA746	3849	2SD468B	3857	131347	3861	GT3150	3861	2N103
3846/285	22B506	3849	2SD468C	3857	140764	3861	GT905	3861	2N1086
3846/285	22B532	3849	2SD468D	3857	1415762-2	3861	GT947	3861	2N1086A
3846/285	22B532P	3849	2SD468E	3857/231	28C1189L	3861	HA5001	3861	2N1087
3846/285	22B532Q	3849	2SD468F	3857/231	126898	3861	HA5002	3861	2N1090
3846/285	22B532R	3849	2SD468G	3857/231	132326	3861	HA5003	3861	2N1091
3846/285	22B532S	3849	2SD468GN	3858	2N6282	3861	HA5005	3861	2N1112
3847	GB-18	3849	2SD468L	3858	2N6283	3861	HA5009	3861	2N1114
3847	2SC2034	3849	2SD468LN	3858	2N6284	3861	HA5011	3861	2N1121
3848	A15M	3849	2SD468Y	3858	2N6355	3861	HA5012	3861	2N124
3848	A050(RECTIFIER)	3849	2SD471	3858	2N6356	3861	HA5014	3861	2N125
3848	AC500	3849	2SD471K	3858	2N6358	3861	HA5020	3861	2N127
3848	AC600	3849	2SD471L	3858/251	MJ3520	3861	HA5021	3861	2N1288
3848	S3A5	3849	2SD471M	3858/251	MJ3521	3861	HA5022	3861	2N1289
3848	SI5	3849	2SD545	3858/251	MJ4033	3861	HA5023	3861	2N1299
3848	SLA5201	3849	2SD545F	3858/251	MJ4034	3861	HA5024	3861	2N1304
3848	SB1C1240	3849	2SD773K	3858/251	MJ4035	3861	HA5025	3861	2N1308
3848	SW05	3849	8-760-335-10	3858/251	TIP641	3861	HA5026	3861	2N1310
3848	UT2060	3849	46-86550-3	3858/251	TIP642	3861	M2N168A	3861	2N1311
3848	UT215	3849	488155121	3859	2N6286	3861	M4700	3861	2N1366
3848	UT238	3849	56-4830	3859	2N6287	3861	M8120	3861	2N1367
3848	UT247	3849	87-0236-Q	3859/252	MJ2501	3861	MHT2002	3861	2N145
3848	UT257	3849	87-0236-R	3859/252	MJ4030	3861	MHT2003	3861	2N146
3848	UT267	3849	87-0236-S	3859/252	MJ4031	3861	MHT2004	3861	2N147
3848	1N1568	3849	102-1384-17	3859/252	MJ4032	3861	MHT2009	3861	2N1473
3848	1N1653	3849	1049-0167	3859/252	MJ901	3861	MHT2010	3861	2N150
3848	1N2373	3849	2057A2-484	3859/252	TIP645	3861	MI814150-18	3861	2N150A
3848	1N2396	3849	8517	3859/252	TIP646	3861	NA20	3861	2N1510
3848	1N2405	3849	2321881	3859/252	TIP647	3861	NK65-4A19	3861	2N1585
3848	1N2414	3849/293	C1973	3859/252	2N6285	3861	NKT734	3861	2N1605
3848	1N2864A	3849/293	EP15X2	3859/252	2N6298	3861	NKT736	3861	2N1605A
3848	1N3081	3849/293	GB-211	3859/252	2N6299	3861	NKT753	3861	2N1622
3848	1N3084	3849/293	2SC1973	3860	D4001	3861	NR05	3861	2N1624
3848	1N3106	3849/293	2SC2001	3860	D4OC4	3861	NR10	3861	2N164
3848	1N3230	3849/293	2SC2236	3860	EA15X332	3861	NR30	3861	2N164A
3848	1N3241	3849/293	2SC2236-Q	3860	13-35324-1	3861	NR5	3861	2N165
3848	1N3250	3849/293	2SC2236Y	3861	A121-1410	3861	NR700	3861	2N166
3848	1N3658	3849/293	2SC722	3861	A121-15	3861	Q-2	3861	2N167
3848	1N3750	3849/293	2SD400D	3861	A121-16	3861	Q-3	3861	2N167A
3848	1N3759	3849/293	2SD400E	3861	A121-17	3861	Q-5	3861	2N168
3848	1N3868	3849/293	2SD400P	3861	A121-21	3861	Q-9	3861	2N168A
3848	1N3940	3849/293	2SD592	3861	A121-50	3861	R-125	3861	2N1685
3848	1N3983	3849/293	2SD592R	3861	A121-762	3861	R-135	3861	2N169
3848	1N4501	3849/293	86-422-2	3861	A127-7	3861	R-136	3861	2N1694
3848	1N4502	3849/293	172763	3861	A129-30	3861	R-137	3861	2N169A
3848	1N4723	3849/293	4100903554	3861	A13-86420-1	3861	R-1533	3861	2N170
3848	1N4822	3850	2N3528	3861	A13-87433-1	3861	R-202	3861	2N172
3848	1N5001	3852	CA2002	3861	A1396	3861	R-203	3861	2N1732
3848	1N5007	3852	ECG1232	3861	A1465-4	3861	R-33	3861	2N1732
3848	1N5176	3852	LM385B	3861	A1465-49	3861	R-54	3861	2N1779
3848	1N5201	3852	TDA2002	3861	A1858	3861	R117	3861	2N1780
3848	1N5206	3852	544-2006-011	3861	A198794-1	3861	R592	3861	2N1783
3848	1N5397	3852	640000003	3861	A2039-2	3861	RS-104	3861	2N184
3848	1N5625	3854	A6H	3861	A2092418-0711	3861	RS-1536	3861	2N1891
3848	1S1890	3854	CB4013E	3861	A3607	3861	RS-1537	3861	2N1994
3848	1S1922	3854	ECG123AP	3861	A3609	3861	RS-1538	3861	2N1995
3848	1845	3854	ECG233	3861	A3T201	3861	RS-1545	3861	2N1996
3848	34P6	3854	EP15X86	3861	A3T202	3861	RS-1547	3861	2N2085
3848	3E6	3854	GE-17	3861	A3T203	3861	RS-1555	3861	2N2426
3848	1S86	3854	HX-50092	3861	A46-8614-3	3861	RS-2001	3861	2N253
3848	20A6	3854	HX-50097	3861	A4700	3861	RS-2359	3861	2N254
3848	3086	3854	HX-50161	3861	A48-124216	3861	RS-2360	3861	2N2699
3849	C1383	3854	KM917F	3861	A48-124217	3861	RS-2364	3861	2N28
3849	C1383P	3854	KM917G	3861	A48-124218	3861	RS-2365	3861	2N29
3849	C1383Q	3854	MPB8001	3861	A48-124221	3861	RS-2366	3861	2N292
3849	C1383R	3854	MPB9410AK	3861	A48-125233	3861	RS-2375	3861	2N292A
3849	C1383R/494	3854	MPB9418	3861	A48-125234	3861	RS-2374	3861	2N293
3849	C1384	3854	MPB9418T	3861	A48-125235	3861	RS-2375	3861	2N312
3849	C1384Q	3854	MPB9634	3861	A48-128239	3861	S028	3861	2N313
3849	C1384R	3854	MPB9634C	3861	A48-134520	3861	SA-7	3861	2N314
3849	C1384S	3854	MPB9634D	3861	A48-134700	3861	SPT-298	3861	2N356
3849	D355	3854	PN2222	3861	A48-134931	3861	SPT259	3861	2N357
3849	D355B	3854	QRT-107	3861	A4JD381	3861	SPT260	3861	2N358
3849	D400P1	3854	REN233	3861	A65-4-70	3861	SPT261	3861	2N377
3849	D400P1D	3854	TM123AP	3861	A65-4A9G	3861	SK-7	3861	2N377A
3849	D400P1E	3854	01-031815	3861	A86-10-2	3861	SN60	3861	2N385
3849	D400P1F	3854	01-349418	3861	A86-444-2	3861	SN80	3861	2N385A
3849	EA15X413	3854	2N3241	3861	A95115	3861	ST-172	3861	2N388
3849	EO9293	3854	2N3241A	3861	A95211	3861	STI-101	3861	2N388A
3849	HT313831X	3854	2SC377FB	3861	A99807	3861	STI-102	3861	2N444
3849	HT313832C	3854	2SC377C	3861	A99SK5	3861	STI-1297	3861	2N445
3849	HT313841R	3854	2SC377D	3861	A99SK7	3861	STI-1310	3861	2N446
3849	K9B41414	3854	2SC377E	3861	AA2	3861	STI-1311	3861	2N447
3849	MPB9466A	3854	2SC377F	3861	AC130	3861	STI-1987	3861	2N556
3849	REN293	3854	2SC377G	3861	AC187/01	3861	STI-2130	3861	2N557
3849	TM293	3854	2SC377GN	3861	AF192	3861	STI-2131	3861	2N558
3849	WEP773	3854	2SC377H	3861	A834-28	3861	STI-2132	3861	2N576
3849	WEP912	3854	2SC377J	3861	ASY-62	3861	STI-2245	3861	2N576A
3849	2SC1383	3854	2SC377K	3861	ASY73	3861	STI-2246	3861	2N587
3849	2SC1383P	3854	2SC377L	3861	ASY74	3861	SYI-4359	3861	2N594
3849	2SC1383Q	3854	2SC377M	3861	ASY75	3861	SYI792	3861	2N595
3849	2SC1383R,RB	3854	2SC377R	3861	ASY86	3861	T59276	3861	2N596
3849	2SC1383RB	3854	2SC377OR	3861	ASY87	3861	T59277	3861	2N634
3849	2SC1383S	3854	2SC377X	3861	ASY88	3861	TP70	3861	2N634A
3849	2SC1383X	3854	2SC377Y	3861	ASY89	3861	TP71	3861	2N635
3849	2SC1384	3854	13-27432-2	3861	AT52	3861	TP72	3861	2N635A
3849	2SC1384A	3854	13-29033-1	3861	AT521	3861	TIX896	3861	2N636
3849	2SC1384B	3854	13-29033-2	3861	AT53	3861	TK33C	3861	2N636A
3849	2SC1384C	3854	13-29033-3	3861	AT551	3861	TNJ61671	3861	2N679
3849	2SC1384D	3854	13-29033-5	3861	AT71	3861	T74274	3861	2N78
3849	2SC1384E	3854	13-29033-5	3861	AT72	3861	TQ5031	3861	2N78A
		3854	13-43773-1			3861	TQ5032	3861	2N821
		3854	48-137171			3861	TR-08	3861	2N822
		3854	121-895			3861	TR-08C		

SK	Industry Standard No.	SK	Industry Standard No.	SK	Industry Standard No.	SK	Industry Standard No.	SK	Industry Standard No.
3861	2N823	3861	86-11-2	3861	8524402-1	3862	48-869093	3867	2SA1019
3861	2N824	3861	86-12-2	3861	8524402-4	3862	48-869254	3867	2SA685
3861	2N955	3861	86-26-2	3861	8975103-2	3862	48-869283	3867	2SA949
3861	2N955A	3861	86-31-2	3861	62050217	3862	48-869476	3874	M51521L
3861	2N97A	3861	86-4-2	3861	62087617	3862	48-869476A	3876	BA301
3861	2N98	3861	86-44-2	3861/101	A37	3862	48R869092	3876	REN1135
3861	2N98A	3861	95-112	3861/101	D3-8	3862	48R869093	3876	09-308062
3861	2N99	3861	95-113	3861/101	HA5016	3862	57A5	3880	10-001
3861	2SC129	3861	95-114	3861/101	SPT-184	3862	57A6-6	3888	EA33X8363
3861	2SC129A	3861	97N2	3861/101	SQ-7	3862	57B2-5	3888	GEIC-80
3861	2SC129B	3861	99K7	3861/101	2N182	3862	57C5	3888	LA4100
3861	2SC129C	3861	99SO7	3861/101	2N183	3862	57C6-6	3888	LA4101
3861	2SC129D	3861	99SK5	3861/101	2N356A	3862	089-233	3889	LA4102
3861	2SC129E	3861	99SK7	3861/101	2N357A	3862	99L6	3890	GEIC-286
3861	2SC129P	3861	99SQ7	3861/101	2N358A	3862	122-1962	3890	LA4050P
3861	2SC129GN	3861	120-004888	3861/101	2N438	3862	200A	3890	LA4051P
3861	2SC129H	3861	121-100	3861/101	2N438A	3862	202A	3892	LM1496J
3861	2SC129J	3861	121-15	3861/101	2N444A	3862	251M1	3892	LM1496N
3861	2SC129K	3861	121-16	3861/101	2N445A	3862	324-0134	3893	A-1854-0464
3861	2SC129L	3861	121-17	3861/101	2N446A	3862	650-109	3893	A272
3861	2SC129M	3861	121-21	3861/101	2N447A	3862	1033-6	3893	A273
3861	2SC129QR	3861	121-22	3861/101	2N447B	3862	2057B2-46	3893	A4I7032
3861	2SC129R	3861	121-24	3861/101	2N585	3862	38175	3893	A5A-1B
3861	2SC129X	3861	121-25	3861/101	2N97	3862	48385-2	3893	A9V
3861	2SC129Y	3861	121-26	3861/101	121+7	3862	49058-2	3893	AR-17
3861	2SC13	3861	121-302	3861/101	610124-1	3862	95202	3893	B131
3861	2SC14	3861	121-50	3862	EC0103	3862	95222-2	3893	B5000
3861	2SC173	3861	121-51	3862	ET10	3862	4036754-P1	3893	BD220
3861	2SC175	3861	121-70	3862	ET11	3862	4036754-P2	3893	BD221
3861	2SC175B	3861	121-71	3862	GB-8	3862	5492639-P1	3893	BD222
3861	2SC176	3861	121-762	3862	GIB	3862	5492639-P2	3893	BD231A
3861	2SC177	3861	124-N16	3862	ISD-162	3862	62728574	3893	BD239
3861	2SC178	3861	124N1	3862	KD2124	3862/103	A514-032815	3893	BD239A
3861	2SC71	3861	130-40089	3862	M9093	3862/103	AC-157	3893	BD241
3861	2SC72	3861	151-0040	3862	RR20	3862/103	TR08	3893	BD243
3861	2SC73	3861	151-0040-00	3862	REN103	3862/103	20339A	3893	BD271
3861	2SC75	3861	151-025B	3862	SA354B	3862/103	2N1198	3893	BD435
3861	2SC76	3861	165-4A82	3862	SYL103	3862/103	2N1217	3893	BD437
3861	2SC77	3861	465-4A5	3862	SYL104	3862/103	2N1306	3893	BD439
3861	2SC77B	3861	665-4A5L	3862	SYL297	3862/103	2N213A	3893	BD533
3861	2SC77C	3861	965-4A6-2	3862	SYL1329	3862/103	2N646	3893	BD535
3861	2SC78	3861	1034-43	3862	SYL1396	3862/103	2SC128	3893	BDX74
3861	2SC89	3861	1102-63	3862	SYL1524	3862/103	2SD101	3893	BDX75
3861	2SC90	3861	1344-3767	3862	SYL1536	3862/103	2SD19	3893	BLY21
3861	2SC91	3861	1396	3862	SYL1538	3862/103	2SD20	3893	BLY36
3861	2SD P1	3861	1465-4	3862	SYL1539	3862/103	2SD21	3893	BLY65
3861	2T513	3861	1465-4-12	3862	SYL1547	3862/103	2SD22	3893	BLY88
3861	2T52	3861	1465-4-12-8	3862	SYL2134	3862/103	2SD23	3893	BLY89
3861	2T520	3861	1858	3862	SYL2135	3862/103	13-85962-1	3893	BRO5296
3861	2T521	3861	1865-4	3862	SYL2136	3862/103	27125-490	3893	C1060
3861	2T524	3861	1865-4-127	3862	SYL2650	3863	HV-23	3893	C1060A
3861	2T53	3861	1865-4L	3862	SYL4315	3863	HV-23BL	3893	C1060B
3861	2T650	3861	1865-4L8	3862	TR09	3863	M8513	3893	C1060BM
3861	2T71	3861	2039-2	3862	TR338	3863	M8513A-R	3893	C1060C
3861	2T72	3861	3607	3862	TV27	3863	M8513R	3893	C1060D
3861	2T73	3861	3609	3862	WTVL6	3863/600	M8513A	3893	C1061A
3861	2T75R	3861	4700	3862	XB4	3863/600	13-43250-1	3893	C1061B
3861	2T74	3861	43992-2	3862	XNC101	3864	HV-46	3893	C1061C
3861	2T75	3861	45495-2	3862	2-36	3864	HV-46GR	3893	C1061F
3861	2T75R	3861	46490-2	3862	20339	3864	HV46	3893	C1061TB
3861	2T76	3861	46631-2	3862	2N1095	3864	KB262	3893	C36583
3861	2T76R	3861	46774-1	3862	2N1096	3864	MA26W	3893	CII73Y
3861	2T77	3861	46775-2	3862	2N1169	3864	MA26WA	3893	D141
3861	2T77R	3861	49138-2	3862	2N1170	3864	RVDVD1210L	3893	D141H01
3861	2T82	3861	81502-6A	3862	2N1312	3864	RVDVD1210M	3893	D141H9Z
3861	3T201	3861	81502-6B	3862	2N1391	3864	RVDVD1211L	3893	D27O1
3861	3T202	3861	81502-6C	3862	2N1858	3864	RVDVD1213	3893	D27O2
3861	3T203	3861	81502-6D	3862	2N213	3864	SV-02	3893	D27O3
3861	4JD3B1	3861	91021	3862	2N2354	3864	SV-03	3893	D27C4
3861	4JX2A801	3861	95112	3862	2N2482	3864	SV-04	3893	D28A1
3861	6-89X	3861	95113	3862	2N364	3864	SV-3A	3893	D28A10
3861	9-5112	3861	95114	3862	2N365	3864	VD1210L	3893	D28A13
3861	9-5113	3861	95115	3862	2N366	3864	VD1212	3893	D28A2
3861	9-5114	3861	95117	3862	2N625	3864	VD1213	3893	D28A3
3861	12AA2	3861	95211	3862	2N648	3864	22A001-17	3893	D28A4
3861	13-86420-1	3861	101678	3862	2N797	3864	48-355046	3893	D28A5
3861	13-87433-1	3861	103443	3862	2SC128A	3864/605	MV-13	3893	D28A7
3861	19A115103-P1	3861	122061	3862	2SC128B	3864/605	SV03	3893	D28D1
3861	19A115201-P1	3861	122112	3862	2SC128C	3865	2SC1124	3893	D28D10
3861	19A115201-P2	3861	198794-1	3862	2SC128D	3865	2SC1124E	3893	D28D2
3861	19A115546-P1	3861	219016	3862	2SC128E	3865	2SC1127B	3893	D28D3
3861	19A115546-P2	3861	221601	3862	2SC128F	3865	2SC1127H	3893	D28D4
3861	19A115673-P1	3861	221924	3862	2SC128G	3865	2SC1127JR	3893	D28D5
3861	19A115673-P2	3861	223367	3862	2SC128GN	3865	2SC1663	3893	D28D7
3861	19B200065-P1	3861	223368	3862	2SC128H	3865	2SC1665H	3893	D9-513
3861	19B200065-P2	3861	223370	3862	2SC128J	3865	2SC1962	3893	E2496
3861	46-8614-3	3861	223684	3862	2SC128K	3865	8-765-170-01	3893	EA15X99
3861	48-124216	3861	226441	3862	2SC128L	3866	C828AP	3893	EA37I6
3861	48-124217	3861	230209	3862	2SC128M	3866	C828AQ	3893	EA4085
3861	48-124218	3861	230256	3862	2SC128OR	3866	C828AR	3893	EC0152
3861	48-124220	3861	232949	3862	2SC128R	3866	C828BP	3893	EP3053
3861	48-124221	3861	236265	3862	2SC128X	3866	C828FR	3893	EQ8140
3861	48-125233	3861	256127	3862	2SC128Y	3866	C828LR	3893	EB15I286
3861	48-125234	3861	257385	3862	2T51	3866	C828LS	3893	G05705
3861	48-125235	3861	260468	3862	2T522	3866	C828P	3893	G05705
3861	48-125236	3861	270781	3862	2T53	3866	C828Q	3893	HC495
3861	48-128239	3861	300486	3862	2T552	3866	C828R	3893	HT308301BQ
3861	48-134520	3861	300536	3862	2T61	3866	C828R/494	3893	J241241
3861	48-134700	3861	300542	3862	2T62	3866	C828S	3893	L8-0066
3861	48-134931	3861	300774	3862	2T63	3866	C828T	3893	M75545-1
3861	65-4	3861	581024	3862	2T64	3866	T028C228B	3893	M9576
3861	65-4-70	3861	723000-18	3862	2T64R	3866	TN28C1941	3893	M9661
3861	65-4-70-12	3861	723001-19	3862	2T65	3866	2SC1941K	3893	MHT5906
3861	65-4-70-12-7	3861	815026	3862	2T65R	3866	2SC1941L	3893	P/FTY/117
3861	65-40	3861	815026A	3862	2T66	3866	2SC1941M	3893	P4I148
3861	65-41	3861	815026B	3862	2T66R	3866	2SC2228	3893	PT2635
3861	65-42	3861	815026C	3862	2T681	3866	2SC2228C	3893	PT5693
3861	65-43	3861	815026D	3862	2T682	3866	2SC2228D	3893	PT665
3861	65-44	3861	2092418-0711	3862	2T69	3866	2SC2228M	3893	QT-D0325XAC
3861	65-45	3861	4036749-P1	3862	2T84	3866	2SC2230	3893	R3611-1
3861	65-46	3861	4036749-P2	3862	2T85	3866	2SC2230AGR	3893	R3681-1
3861	65-47	3861	4037289-P1	3862	2T85A	3866	2SC2230AY	3893	R612-1
3861	65-48	3861	4037289-P2	3862	2T86	3866	2SC2230Y	3893	R621-1
3861	65-49	3861	4037839-P1	3862	2T89	3866	2SC828A	3893	R652-2
3861	65-4A	3861	4037839-P2	3862	4JX1B850	3866	2SC828A0	3893	REN152
3861	65-4A1	3861	4038264-P1	3862	4JX2B16	3866	2SC828AP	3893	S2042
3861	65-4A19	3861	4038264-P2	3862	4JX2B25	3866	2SC828AQ	3893	SB0319
3861	65-4A2	3861	5492653-P1	3862	4JX2A601	3866	2SC828AR	3893	SC4303
3861	65-4A21	3861	5492653-P2	3862	4JX2A816	3866	2SC828A8	3893	SC4303-1
3861	65-4A3	3861	5492655-P1	3862	4JX2A822	3866	2SC828FR	3893	SC4303-2
3861	65-4A3P	3861	5492655-P2	3862	6A12993	3866	2SC828HR	3893	SC4303B
3861	65-4A4	3861	5492655-P3	3862	13-86416-1	3866	2SC828LR	3893	SD1345
3861	65-4A4-4	3861	5492655-P4	3862	16A1	3866	2SC828LS	3893	SDJ345
3861	65-4A4-7B	3861	5492655-P5	3862	16A2	3866	2SC828P	3893	SDPS102
3861	65-4A5	3861	5492655-P6	3862	019-003317	3866	2SC828PQ	3893	SDT5907
3861	65-4A5L	3861	5492659-P1	3862	019-003318	3866	2SC828Q	3893	SDT6001
3861	65-4A6	3861	8510744-1	3862	019-003319	3866	2SC828Q-6	3893	SDT6011
3861	65-4A7	3861	8521502-1	3862	19A115129-2	3866	2SC828R	3893	SDT6013
3861	65-4A7-1	3861	8521502-2	3862	19A115129-P1	3866	2SC828R-1	3893	SDT6103
3861	65-4A8	3861	8521502-4	3862	24MW130	3866	2SC828S	3893	SDT7511
3861	65-4A82			3862	42X210	3866	2SC828T		
3861	65-4A9			3862	42X310	3866	1039-0961		
3861	86-10-2			3862	48-869092	3867	2SA1017		

SK	Industry Standard No.
3893	SDT7512
3893	SDT7514
3893	SDT7515
3893	SDT9009
3893	SE5-0963
3893	SJE783
3893	SP8416
3893	SPS1436
3893	STC1300
3893	STC1850
3893	STC1860
3893	T01-030
3893	T1486
3893	T1487
3893	TI486
3893	TI487
3893	TM152
3893	V409
3893	W18
3893	XB404
3893	XB408
3893	XB476
3893	ZT1483
3893	ZT1484
3893	ZT1485
3893	ZT1486
3893	ZT1701
3893	ZT2876
3893	1A0058
3893	002-012400
3893	2SC1060
3893	2SC1060A
3893	2SC1060B
3893	2SC1060BM
3893	2SC1060C
3893	2SC1060D
3893	0002SC1061A
3893	2SC1061BT
3893	2SC1061D
3893	2SC1061T
3893	2SC1061TB
3893	2SD141
3893	2SD141A
3893	2SD141B
3893	2SD141C
3893	2SD141E
3893	2SD141H
3893	2SD141H01
3893	2SD141H9Z
3893	2SD141J
3893	2SD360
3893	2SD360C
3893	2SD360D
3893	2SD360E
3893	2SD804HP
3893	3L4-6005-1
3893	3L4-6005-5
3893	3L4-6005-55
3893	3L4-6012-02
3893	3L4-6012-2
3893	3L4-6012-3
3893	4-464
3893	6-0005193
3893	6-5193
3893	007-0112
3893	7-0112-00
3893	007-0112-03
3893	7-0112-04
3893	7-0112-05
3893	007-112-04
3893	8-1074
3893	8P345
3893	8P73BLU
3893	8P73GRN
3893	8P73YEL
3893	09-303031
3893	09-304140
3893	9TR8
3893	10-26-123-313
3893	012-103002
3893	13-14085-88
3893	13-2P64-1
3893	13-39004-2
3893	13-4P63-1
3893	14-0104-1
3893	14-609-00
3893	14-609-06
3893	14-609-08
3893	19-020-066
3893	19A116118-1
3893	19A116118-2
3893	19A116118P1
3893	19A116118P2
3893	21A118-029
3893	24MW662
3893	34-1002
3893	44A395986-001
3893	44A417032-001
3893	46-86317-3
3893	46-86335-3
3893	48-13309X3
3893	48-137396
3893	48-137506
3893	48-355040
3893	48-355059
3893	48-43240G01
3893	48-869576
3893	48-869661
3893	48-97046A30
3893	48-97046A38
3893	48S43241G01
3893	48S44883G01
3893	57B131-10
3893	62B046-3
3893	73C18028-11
3893	73C180828-12
3893	73C180829-11
3893	77-272913-1
3893	77-273715-1
3893	77-273738-1
3893	86-5102-2
3893	86-5107-2
3893	86-5108-2
3893	86-544-2
3893	96-5348-01
3893	96-5357-01
3893	998100-1
3893	111N6C
3893	121-770
3893	129-33
3893	0131-005352
3893	131N2G
3893	132-4
3893	149N2002D
3893	149N2D
3893	149P1D
3893	153N1C
3893	153N2C
3893	153N4C
3893	153N6
3893	173A4490-7
3893	173A4491-5
3893	173A4491-8
3893	195N1D
3893	200-076
3893	207A30
3893	207A33
3893	209-30
3893	211A6382-2
3893	216-001-001
3893	229-1301-35
3893	250-0359
3893	260P12701
3893	260P28401
3893	353-9203-001
3893	417-175-12993
3893	474A410BEP2
3893	474A410BW-2
3893	690L-021H25
3893	750A858-448
3893	753-9010-235
3893	770-045
3893	834-250-011
3893	995-01-6131
3893	1000-138
3893	1000-142
3893	1016-81
3893	1098-14
3893	2017-109
3893	2036-59
3893	2057B2-151
3893	2210-17
3893	2854-2
3893	2855-2
3893	2856-2
3893	3107-204-90182
3893	4490-7
3893	4491-5
3893	4491-8
3893	6382-2
3893	8020-206
3893	8800-205
3893	12020-02
3893	16166
3893	16305
3893	16334
3893	16335
3893	16336
3893	23754
3893	30294
3893	38733
3893	39302
3893	39750
3893	39767
3893	39948
3893	39981
3893	41504
3893	44208
3893	50200-1B
3893	50200-24
3893	50308-0100
3893	60175
3893	60216
3893	60408
3893	60679
3893	60719
3893	60835
3893	60838
3893	60886
3893	60987
3893	61102
3893	61193
3893	61285
3893	61286
3893	61418
3893	61636
3893	61772
3893	61875
3893	61958
3893	61981
3893	62156
3893	62571
3893	62681
3893	79992
3893	95261-1
3893	99252-4
3893	116118
3893	116118-2
3893	132697
3893	137527
3893	138379
3893	146139
3893	153107
3893	167691
3893	171162-265
3893	171162-291
3893	172463
3893	172643
3893	242102
3893	309441
3893	489751-033
3893	489751-044
3893	610149-2
3893	610153-6
3893	610162-4
3893	702886
3893	740856
3893	996817
3893	1417358-1
3893	1473611
3893	1473612
3893	1950160
3893	2320083
3893	2320482
3893	2320651
3893	2875493
3893	3438095
3893	3596447
3893	3755171
3893	4080866-0013
3893	4080866-003
3893	4080866-006
3893	4080866-009
3893	4080866-4
3893	4550106-001
3893	4550106-003
3893	06120018
3893	06120021
3893	6902021H25
3893	7026024
3893	7570031-01
3893	7855298-01
3893	8113102
3893	9001324
3893	9001756
3893	9341510
3893	19020065
3893	50221800
3893	50221800-01
3893	62393688
3893	80053300
3893	430233843
3893/152	BLY35
3893/152	D44C2
3893/152	D44C3
3893/152	D44C4
3893/152	D44C5
3893/152	D44C6
3893/152	D44C7
3893/152	D44C8
3893/152	D44C8B
3893/152	D44C9
3893/152	GE-X18
3893/152	TIP29
3893/152	TV115
3893/152	1A0046
3893/152	1A0059
3893/152	2N5490
3893/152	2N5491
3893/152	2N5492
3893/152	2N5493
3893/152	2N5494
3893/152	2N5495
3893/152	2SC325
3893/152	2SC325A
3893/152	2SC325C
3893/152	2SC325E
3893/152	8-729-316-12
3893/152	13-41628-2
3893/152	19-020-050
3893/152	99S075
3893/152	211A6381-1
3893/152	296
3893/152	514-047830
3893/152	2320432
3894	2N6077
3894	2N6078
3894	2N6079
3894	2N6372
3894	2N6373
3894	2N6374
3895	BD121
3895	BD123
3895	TIP509
3895	TIP510
3895	2N3919
3895	2N4111
3895	2N4113
3895	2N5614
3895	2N5616
3895	2N5618
3895	2N5622
3895	2N6496
3895	2SD375
3895/325	23114864
3895/328	TIP516
3896	RCA120
3896	RCA121
3896	TIP120
3896	TIP121
3896/261	2N6043
3896/261	2N6044
3897	RCA125
3897	RCA126
3897	TIP125
3897	TIP126
3897/262	2N6040
3897/262	2N6041
3899	C1312F
3899	C1312G
3899	C1312H
3899	C1312Y
3899	C1312YF
3899	C1312YG
3899	C1312YH
3899	C1335A
3899	C1335B
3899	C1335C
3899	C1335D
3899	C1335E
3899	C1335F
3899	C711
3899	C711(E)
3899	C711A
3899	C711AE
3899	C711D
3899	C711E
3899	C711F
3899	C711FG
3899	C711G
3899	C900
3899	C900(L)
3899	C900A
3899	C900B
3899	C900C
3899	C900D
3899	C900E
3899	C900F
3899	C900L
3899	C900M
3899	C900U
3899	EA15X386
3899	2SC1312
3899	2SC1312A
3899	2SC1312C
3899	2SC1312D
3899	2SC1312E
3899	2SC1312F
3899	2SC1312G
3899	2SC1312H
3899	2SC1312R
3899	2SC1312Y
3899	2SC1312YF
3899	2SC1312YG
3899	2SC1312YH
3899	2SC1327
3899	2SC1327FS
3899	2SC1327T
3899	2SC1327TU
3899	2SC1327U
3899	2SC1328
3899	2SC1328T
3899	2SC1328U
3899	2SC1335
3899	2SC1335A
3899	2SC1335B
3899	2SC1335C
3899	2SC1335D
3899	2SC1335E
3899	2SC1335F
3899	2SC1344
3899	2SC1344C
3899	2SC1344D
3899	2SC1344E
3899	2SC1344F
3899	2SC1345
3899	2SC1345C
3899	2SC1345D
3899	2SC1345E
3899	2SC1345F
3899	2SC711
3899	2SC711(E)
3899	2SC711A
3899	2SC711AE
3899	2SC711AF
3899	2SC711AG
3899	2SC711B
3899	2SC711C
3899	2SC711D
3899	2SC711E
3899	2SC711F
3899	2SC711FG
3899	2SC711G
3899	2SC711H
3899	2SC900
3899	2SC900(F)
3899	2SC900A
3899	2SC900B
3899	2SC900C
3899	2SC900D
3899	2SC900E
3899	2SC900F
3899	2SC900L
3899	2SC900M
3899	2SC900U
3899	2SC900VE
3899	4100900116
3900	MH353
3900	MH988A03
3911	2SD636
3911	2SD636P
3911	2SD636Q
3911	2SD636R
3911	2SD639
3911	144039
3911	144582
3912	2SB641
3912	2SB641S
3912	2SB642
3912	2SB642Q
3912	2SB642R
3912	2SB643
3912	2SB643Q
3912	2SB643R
3912	2SB643S
3912	144031
3913	2SC1664A
3917	CA810Q
3917	TBA810S
3920	AN262
3920	51-90305A21
3920	61C001-4
3920	144012
3925	EF200
3925	ERB26-20
3925	ERB26-20L
3925	ERB26-20M
3925	ERB26-20MV
3925	GH1F
3925	HC-2
3925	RC-2
3925	RC-2V
3925	RC2
3925	RH-DX0077CEZZ
3925	1S2227
3925	1TH61
3925	21A110-071
3925	46-86560-3
3925	48-155152
3925	103-305
3925	201X2100-164
3925	143594
3925	146316
3925	23115965
3925	2012100119
3925	4202022900
3925/525	RH-DX0100CEZZ
3925/525	SM20
3925/525	4202017200
3929	C1448P
3929	C1448Q
3929	C1448R
3929	C1448S
3929	2N6477
3929	2N6478
3929	2SC1448
3929	2SC1448A
3929	2SC1448P
3929	2SC1448Q
3929	2SC1448R
3929	2SC1448S
3929	2SC2073
3929	2SC2167
3929	2SC2238
3929	2SD386
3929	2SD386A
3929	2SD386D
3929	2SD386Y
3929	2SD387
3929	2SD401
3929	2SD402
3929	2SD478
3929	2SD608
3929	2SD724
3929	2SD758
3930	2SA814
3930	2SA815
3930	2SA899
3930	2SA940
3930	2SA968
3930	2SB546
3930	2SB547
3930	2SB568
3930	2SB628
3931	2SC1681
3931	2SC1681BL
3931	2SC1681GR
3931	2SC1681V
3931	2SC1775
3931	2SC1775E
3931	2SC1775F
3931	2SC1890
3931	2SC1890A
3931	2SC1890AD
3931	2SC1890AE
3931	2SC1890E
3931	2SC1890F
3932	2SA893
3932	2SA893AE
3932	06100058
3933	2SD586
3933	2SD586R
3935	TIP140
3935	TIP141
3935	TIP142
3936	TIP145
3936	TIP146
3936	TIP147
3937	SC146A
3937	SC146A2
3937	SC146A3
3937	SC146A4
3937	SC146A5
3937	SC146A6
3938	SC146B
3938	SC146B2
3938	SC146B3
3938	SC146B4
3938	SC146B5
3938	SC146B6
3938	T2801B
3938	T2802B
3938	TIC246B
3939	SC146E
3939	SC146E3
3939	SC146E4
3939	SC146E5
3939	SC146E6
3939	T2801E
3939	T2802E
3942	GE-X1
3942	MCR2305-2
3942	MCR2305-4
3942	MCR2315-4
3942	RTN0220
3942	2N1773
3942	2N1773A
3942	2N1774
3942	2N1774A
3942	2N4170
3942	2N4178
3942	90270020
3943	MCR2315-6
3943	RTN0230
3943	RTN0240
3943	2N1777
3943	2N1777A
3943	2N4172
3943	2N4180
3944	RTN0250
3944	RTN0260
3944	2N4174
3945	2N5671
3945	2N5672
3945	2N6338
3945	2N6339
3945	2N6340
3945	2N6341
3945/327	2N3713
3945/327	2N3714
3945/327	2N3715
3945/327	2N3716
3945/327	2N4071
3946	MJ13014
3946	2SC1868
3946/385	2SC1866
3946/385	2SC1870
3947	MJ15024
3948	2N6057
3948	2N6058
3948	2N6059
3948	2N6357
3948	2SD803
3949	2N6050
3949	2N6051
3949	2N6052
3950	C103Y
3950	EC103Y
3950	ID100
3950	MCR101
3950	MCR102
3950	T1C44
3950	T1C60
3950	TIC44
3950	2N2679
3950	2N2687
3950	2N3005
3950	2N3027
3950	2N3030
3950	2N4144
3950	2N4145
3950	2N5060
3950	2N876
3950	2N877
3951	EC103D
3951	ID105
3951	ID106
3951	IP105
3951	IP106
3951	MCR100-6
3951	RTC0130
3951	RTJ0225
3951	RTJ0230
3951	2SF106
3952	MCR100-8
3953	ECG5411
3953	ECG5412
3953	MCR106-2
3953	MCR107-2
3953	S0503LS3
3953	SF1A11
3953	SF1A11A
3954	ECG5413
3954	ECG5414
3954	HEP-R1218
3954	MCR106-4
3954	MCR107-4
3954	S2003LS3
3954	S2003RS3
3954	SF1D11

SK	Industry Standard No.
3954	SF1D11A
3954	2N6239
3954	86-113-3
3955	ECG5415
3955	MCR106-6
3955	2N6240
3956	MCR106-8
3956	2N6241
3957	GE-238
3958	ECG390
3958	TIP33
3958	TIP33A
3958	TIP33B
3958	TIP33C
3959	TIP34
3959	TIP34A
3959	TIP34B
3959	TIP34C
3960	ECG392
3960	TIP35
3960	TIP35A
3960	TIP35B
3960	TIP35C
3961	TIP2955
3961	TIP36
3961	TIP36A
3961	TIP36B
3961	TIP36C
3968	AN239
3968	AN239Q
3968	AN239QA
3968	AN239QB
3968	GEIC-300
3969	GEIC-279
3969	TBA800
3969	56A42-1
3975	ITT652
3975	L201
3975	9665PC
3977	2N4220
3977	2N4221
3977	2N4221A
3977/456	2N4340
3977/456	2SK17GR
3978	MJE6043
3978	MJE6044
3978	MJE6045
3979	MJE6040
3979	MJE6041
3979	MJE6042
3983	TIP51
3983	TIP52
3983	TIP53
3983	TIP54
3984	A-1853-0009-1
3984	A-1853-0010-1
3984	A-1853-0034-1
3984	BC205L
3984	BF315
3984	BF316
3984	BF339
3984	BFX12
3984	BFX13
3984	BFX48
3984	BSX29
3984	CS1124G
3984	DS96
3984	ECG106
3984	FK2894
3984	FM2894
3984	FV2894
3984	I50865
3984	K071687
3984	SA495
3984	SA495A
3984	SA496
3984	SA496A
3984	SA496B
3984	SA537
3984	SA538
3984	SA539
3984	SA540
3984	SAC40
3984	SAC40A
3984	SAC40B
3984	SAC42
3984	SAC42A
3984	SAC42B
3984	SL200
3984	SL201
3984	SP12271
3984	ST2-2517
3984	T40
3984	TW135
3984	002-0105-00
3984	2N1024
3984	2N1238
3984	2N1239
3984	2N1240
3984	2N1241
3984	2N1242
3984	2N2176
3984	2N2395
3984	2N2969
3984	2N2971
3984	2N3304
3984	2N3342
3984	2N3576
3984	2N3915
3984	2N4034
3984	2N4035
3984	2N4451
3984	2N495/18
3984	2N5228
3984	2N5352
3984	2N726
3984	2N727
3984	2N865
3984	2N865A
3984	2N869
3984	2N869A
3984	2N978
3984	2N995
3984	2N995A
3984	2N996
3984	2SA524
3984	3N112
3984	3N113
3984	14-807-12
3984	14-861-12
3984	15-008-1
3984	21A062-000
3984	25-001328
3984	34-1011
3984	48-137066
3984	48-97127A015
3984	48-97127A15
3984	57A1-52
3984	57A108-1
3984	57A108-2
3984	57A108-3
3984	57A108-4
3984	57A108-5
3984	57A108-6
3984	57A108-7
3984	57A108-8
3984	62-22529
3984	121-615
3984	136P1
3984	151-0417-00
3984	154A5947-7732
3984	210BWTF4121
3984	297L012C-01
3984	352-0950-010
3984	352-0950-020
3984	417-102
3984	549-1
3984	921-103B
3984	921-112B
3984	921-333B
3984	921-333P
3984	1853-0069
3984	2057A2-200
3984	2057A2-203
3984	4484-1
3984	4484-2
3984	4485-1
3984	4590
3984	4822-130-40369
3984	4822-130-40477
3984	4822-130-40508
3984	4822-130-40614
3984	6855K90
3984	7340
3984	95229
3984	97680
3984	105468
3984	115270-101
3984	124634
3984	131262
3984	135347
3984	150865
3984	160196
3984	405192
3984	531298-001
3984	547684
3984	610102-1
3984	610136-1
3984	986015
3984	2621811
3984	36171000
3984	50211300-00
3984	50211300-01
3984	50211600-00
3984	50211600-01
3984	50211600-02
3984	50211600-10
3984	50211600-12
3984	50211610-10
3984	50211610-12
3984/106	A3C
3984/106	BC205C
3984/106	N-EA15X132
3984/106	TM22
3984/106	2N2178
3984/106	13-40083-2
3984/106	57A108-6A
3984/106	77-271038-1
3984/106	4485
3984/106	136766
3985	FWLD100
3985	FWLD50
3985	KBL005
3985	KBPC1005
3985	KBPC6005
3985	KBPC8005
3986	FWLD200
3986	KBL02
3986	KBPC102
3986	KBPC602
3986	KBPC802
3986	VS248
3987	FWLD400
3987	KBL04
3987	KBPC104
3987	KBPC604
3987	KBPC804
3987	VS448
3988	FWLD600
3988	KBPC106
3988	KBPC606
3988	KBPC806
3988	VS648
3989	FWLD800
3989	KBPC108
3990	A054-142
3990	DDCY103001
3990	ECG220
3990	HEP-F2005
3990	M75561-10RK
3990	W1E
3990	2SK45B
3990	3N128
3990	3N142
3990	3N143
3990	3N152
3990	3SK14
3990	3SK29
3990	3SK33
3990	13-33173-1
3990	13-37900-1
3990	48-13707
3990	48-137070
3990	123-002
3990	132-048
3990	417-206
3990	417-207
3990	921-157B
3990	40467A
3990	40468A
3990	40559A
3990	60793
3990	5320031
3990	5518017
3991	3N159
3991	3N200
3991	3N201
3991	3N202
3991	3N203
3991	3N204
3991	3N205
3991	3N206
3991/454	3N201A
3992	SC160B
3993	SC160D
3995	GE-35
3995	IR-TR61
3995	2SC1152
3995	2SC1152F
3995	2SC1152G
3996	MJE800
3996	MJE801
3996	MJE802
3996	MJE803
3996	2N6039
3997	BD262A
3997	MJE700
3997	MJE701
3997	MJE702
3997	MJE703
3997	2N6034
3997	2N6035
3998	ECG506
3998	EP16X11
3998	EP16X24
3998	GE-511
3998	GH3F
3998	GRU2A
3998	PTC216
3998	TM506
3998	V30N
3998	WEP172
3998	8-719-903-09
3998	200X8130-171
3998	1476171-34
3998	06200030
3998/506	RH1M
3998/506	TD-15
3998/506	05-860002
3998/506	8-719-305-15
3998/506	8-719-320-11
3998/506	8-719-320-31
3998/506	8-719-900-93
3998/506	8-719-901-19
3998/506	13-33172-1
3998/506	13-33172-2
3998/506	13-41122-2
3998/506	46-86507-3
3998/506	48-155223
3998/506	57C27-2
3998/506	103-287
3998/506	201X2120-009
3998/506	140971
3998/506	140972
3998/506	146136
3998/506	146137
3998/506	1476171-33
3998/506	2330551
3999	GE-305
3999	RE93P
3999	010
3999	010-6742
3999	010-6744
3999	57-52
3999	57-56
3999	63-19173
3999	120818
3999	135320
3999	135932
3999	1471908-1
4000	CD4000BE
4000	GE-4000
4000	MM4000
4000	SCL4000
4001	CD4001AE
4001	CD4001BE
4001	CM4001
4001	ECG4001B
4001	F4001
4001	GE-4001
4001	HBF4001
4001	MC14001
4001	MEM4001
4001	MM4001
4001	N4001
4001	RS4001
4001	SCL4001
4001	SW4001
4001	TC4001P
4001	TP4001
4001	51S10655A17
4001	905-125
4002	GE-4002
4002	MM4002
4006	CD4006B
4007	CD4007UBE
4007	MC14007
4009	CD4009UBE
4009	MM4009
4011	CD4011BE
4011	CM4011
4011	ECG4011B
4011	F4011
4011	GE-4011
4011	HBF4011
4011	HD4011
4011	MC14011
4011	MC14011B
4011	MC14011BCP
4011	MEM4011
4011	MM4011
4011	MSM4011RS
4011	N4011
4011	RS4011
4011	SCL4011
4011	SW4011
4011	TC4011BP
4011	TC4011P
4011	TP4011
4011	UPD4011C
4011	40-035-0
4011	51S10655A18
4011	905-126
4012	CD4012BE
4012	CM4012
4012	ECG4012B
4012	F4012
4012	GE-4012
4012	HBF4012
4012	HD4012
4012	MC14012
4012	MM4012
4012	N4012
4012	SCL4012
4012	SW4012
4012	TP4012
4013	CD4013AE
4013	CD4013BE
4013	CM4013
4013	ECG4013B
4013	F4013
4013	GE-4013
4013	HBF4013
4013	HD4013
4013	MC14013
4013	MM4013
4013	N4013
4013	RS4013
4013	SCL4013
4013	SW4013
4013	TP4013
4015	CD4015BE
4015	F4015
4015	GE-4015
4016	CD4016AE
4016	CD4016BE
4016	F4016
4016	GE-4016
4016	HD4016
4016	MSM4016
4016	51-90433A13
4017	CD4017BE
4017	CM4017
4017	ECG4017B
4017	F4017
4017	GE-4017
4017	HBF4017
4017	HD4017
4017	MC14017
4017	MM4017
4017	RS4017
4017	SCL4017
4017	SW4017
4017	TP4017
4018	F4018B
4019	CD4019BE
4019	F4019
4019	GE-4019
40192	MM74C192
4020	CD4020BE
4020	ECG4020B
4020	F4020
4020	GE-4020
4020	RS4020
4021	CD4021BE
4021	ECG4021B
4021	F4021
4021	GE-4021
4022	CD4022B
4023	CD4023BE
4023	CD4023C
4023	CM4023
4023	ECG4023B
4023	F4023
4023	GE-4023
4023	HBF4023
4023	HD4023
4023	MC14023
4023	MM4023
4023	MSM4023
4023	N4023
4023	SCL4023
4023	SW4023
4023	TP4023
4024	CD4024BE
4024	GE-4024
4024	15-45184-1
4025	CD4025BE
4025	GE-4025
4025	MC14025B
4027	CD4027BE
4027	CM4027
4027	ECG4027B
4027	F4027
4027	GE-4027
4027	HBF4027
4027	HD4027
4027	MC14027
4027	MM4027
4027	N4027
4027	RS4027
4027	SCL4027
4027	SW4027
4027	TP4027
4028	ECG4028B
4030	GE-4030
4040	CD4040BE
4040	GE-4040
4042	CD4042BE
4042	GE-4042
4043	CD4043B
4044	CD4044B
4046	ECG980
4049	CD4049UBE
4049	CM4049
4049	ECG4049B
4049	F4049
4049	GE-4049
4049	HD4049
4049	MC14049
4049	MEM4049
4049	MM4049
4049	N4049
4049	RS4049
4049	SCL4049
4049	SW4049
4049	TP4049
4050	CD4050BE
4050	CM4050
4050	ECG4050B
4050	F4050
4050	GE-4050
4050	HD4050
4050	MC14050
4050	MEM4050
4050	MM4050
4050	MSM4050
4050	N4050
4050	RS4050
4050	SCL4050
4050	SW4050
4050	TP4050
4050	51-90433A10
4050	51-90433A11
4051	CD4051BE
4051	CM4051
4051	ECG4051B
4051	F4051
4051	GE-4051
4051	MC14051
4051	MEM4051
4051	MM4051
4051	SCL4051
4051	TP4051
4052	CD4052BE
4052	GE-4052
4053	CD4053B
4055	CD4055BE
4055	GE-4055
4060	CD4060B
4066	CD4066B
4066	ECG4066B
4069	CD4069UB
4075	CD4075B
4081	CD4081B
4081	CD4081BE
4511	ECG4511B
4518	CD4518B
4518	CD4518BE
4518	CM4518
4518	ECG4518B
4518	F4518
4518	GE-4518
4520	CD4520B
6554	1N4045
6554	1N4048
6615	MCR2604-4
6615	MCR2604-5
6615	MCR2604-6
6615	MCR3818-5
6615	MCR3818-6
6615	PS320
6615	PS38
6615	PS48
6615	RTS0230
6615	RTS0240
6615	S4006G
6615	S4008G
6615	S4010G
6615	S4015G
6615	S6200D
6615	2N5166
6616	MCR3818-7
6616	MCR3818-8
6616	PS520
6616	PS58
6616	PS68
6616	RTS0250
6616	RTS0260
6616	S6006G
6616	S6008G
6616	S6010G
6616	S6015G
6616	S6200M
6616	2N5167
6621	C30F
6621	C37F
6621	S0325H
6621	S0525H
6621	SPS020
6621	2N5168
6621	2N682
6621	10RC5
6621	10RC5A
6622	C30A
6622	C37A
6622	NL-C37A
6622	S1025H
6622	SPS120
6622	2N683
6622	10RC10
6624	C30B
6624	C37B
6624	MCR3918-4
6624	NL-C37B
6624	2N2888
6624	2N685
6624	10RC20
6624	10RC20A
6627	C30D
6627	C37D
6627	NL-C37D
6627	S4025H
6627	SPS420
6627	2N688
6627	2SF77
6627	10RC40
6627	10RC40A
6629	C37M
6629	MCR3918-8
6629	RTU0160
6629	S6015H
6629	S6025H
6629	SPS620
6629	2N5171
6629	2N690
6631	2N692
6642	MCR2935-3
6642	MCR3935-3
6642	NL-C35A
6642	RE170
6642	22RC10
6648	C137N
6648	C137S
6752	C106Q
6752	C106Q1
6752	C106Q2
6752	C106Q3
6752	C106Q4
6752	C106Y
6752	C106Y1
6752	C106Y2
6752	C106Y3
6752	C106Y4
6752	C107Q
6752	C107Q1
6752	C107Q2
6752	C107Q3
6752	C107Q4
6752	C107Y
6752	C107Y1
6752	C107Y2
6752	C107Y3
6752	C107Y4
6752	ECG5452
6752	IR106Q1
6752	IR106Q2
6752	IR106Q3
6752	IR106Q4
6752	IR106Y1

SK	Industry Standard No.
6752	IR106Y1-C
6752	IR106Y2
6752	S0303RS2
6752	S106Y
6752	S106Y1
6752	S107Y1
6752	S2062Q
6752	S2062Y
6752	TC106Q1
6752	TC106Q2
6752	TC106Q3
6752	TC106Q4
6752	TC106Y1
6752	TC106Y2
6752	TC106Y3
6752	TC106Y4
6752	TIC106Y
6752	TX-145
6752	108Q
6752	108Y
6753	C106F1
6753	C106F2
6753	C106F3
6753	C106F4
6753	C107F1
6753	C107F2
6753	C107F3
6753	C107F4
6753	ECG5453
6753	GE-X5
6753	IR106F1
6753	IR106F2
6753	IR106F3
6753	IR106F4
6753	IR106F41
6753	IRC5
6753	S106F
6753	S106F1
6753	S107F
6753	S107F1
6753	S2062F
6753	TC106F1
6753	TC106F2
6753	TC106F3
6753	TC106F4
6753	TIC106F
6753	11T4S
6754	C106A1
6754	C106A2
6754	C106A3
6754	C106A4
6754	C107A
6754	C107A1
6754	C107A2
6754	C107A3
6754	C7A4
6754	ECG5454
6754	IR106A1
6754	IR106A1-C
6754	IR106A2
6754	IR106A3
6754	IR106A4
6754	IR106A41
6754	RE168
6754	S106A1
6754	S107A
6754	S107A1
6754	S2061A
6754	S2062A
6754	SFIR3B41
6754	TC106A1
6754	TC106A2
6754	TC106A3
6754	TC106A4
6754	TIC106A
6754	2SF941
6754	3C6100
6754	48-83875D01
6754	48D83875D01
6754	75D01
6754	134883875401
6791	C20A
6791	C20F
6791	C20U
6791	S0306H
6791	S0308H
6791	S0310H
6791	S0506H
6791	S0508H
6791	S0510H
6791	S1006H
6791	S1010H
6791	SPS08
6791	SPS18
6792	C20B
6792	S2006H
6792	S2008H
6792	S2010H
6792	SPS28
6794	S4006H
6794	S4008H
6794	S4010H
6796	S6006H
6796	S6008H
6796	S6010H
7042	AM64
7042	BAV207
7042	BAV217
7042	BAV227
7042	BAV307
7042	BAV317
7042	BAV327
7042	BY227
7042	BY307
7042	BY317
7042	BY327
7042	GE-5008
7042	1N1120
7042	1N1128
7042	1N1234
7042	1N1234A
7042	1N1234B
7042	1N1544
7042	1N1587
7042	1N2031
7042	1N2220
7042	1N2221
7042	1N2270
7042	1N2271
7042	1N2517
7042	1N2523
7042	1N2530
7042	1N2541
7042	1N2552
7042	1N256
7042	1N2852
7042	1N555
7042	1N614
7042	1N614A
7042	1S1355
7042	1S415
7042	3C60
7042	3JC12
7042	60F1
7042	60S5
7042	305M
7048	BY2001
7048	BY2002
7048	BY2101
7048	BY2102
7048	BY2201
7048	BY2202
7048	BY3001
7048	BY3002
7048	BY3101
7048	BY3102
7048	BY3201
7048	BYX38/900
7048	PR9002
7048	PR9003
7048	S1000
7048	1N1444
7048	1N1444A
7048	1N1444B
7048	1N2372
7048	1S1357
7048	1S1630
7048	1S419
7048	3C100
7048	3C90
7048	3NC12
7048	305V
7048	305Z
7049	GE-5013
7049	PR9002R
7049	PR9003R
7049	1N563R
7049	3C100R
7049	3C90R
7049	3FR100
7068	BY4001
7068	BY4002
7068	BY4101
7068	BY4102
7068	BY4201
7068	BY4202
7068	BY7001
7068	BY7002
7068	BY7101
7068	BY7102
7068	BY7201
7068	BY7202
7068	GE-5028
7068	PR9006
7068	PR9007
7068	1N2242
7068	1N2242A
7068	1N2243
7068	1N2243A
7068	1N2559
7068	1N2560
7068	1N2563
7068	1N2564
7068	1N2574
7068	1N2575
7068	1N3650
7068	1N3919
7068	1N3989
7068	1N3990
7068	1N4459
7068	1S1627
7068	6A1000
7068	6A900
7068	6C100
7068	6C90
7068	6F100
7068	6F100-D
7068	6F100A
7068	6F100B
7068	6F90
7068	6F90A
7068	6F90B
7068	6NC12
7068	50R2S
7068	341V
7068	341Z
7068	366V
7068	366Z
7068	407V
7068	407Z
7068	441V
7068	441Z
7068	446V
7068	446Z
7069	BYX48-900R
7069	GE-5029
7069	PR9006R
7069	PR9007R
7069	1N3989R
7069	1N3990R
7069	6C100R
7069	6C90R
7069	6FR100
7069	6FR100A
7069	6FR100B
7069	6FR90
7069	6FR90A
7069	6FR90B
7069	346V
7069	346Z
7069	366RV
7069	366RZ
7069	407RV
7069	407RZ
7090	BY5001
7090	BY5002
7090	BY5101
7090	BY5102
7090	BY5201
7090	BY5202
7090	BY8001
7090	BY8002
7090	BY8101
7090	BY8102
7090	BY8201
7090	BY8202
7090	BYX40-1000
7090	BYX42-900
7090	BYX42/900
7090	BYX48-900
7090	GE-5044
7090	HEP-R0138
7090	MR1130
7090	PR9011
7090	S10AN12
7090	S10AN6
7090	S9AN12
7090	S9AN6
7090	SJ10003EK
7090	SL10
7090	SL1000
7090	SL1000X
7090	TR1130
7090	1EB100A
7090	1N2262
7090	1N2262A
7090	1N2263
7090	1N2263A
7090	1N2585
7090	1N2586
7090	1N2596
7090	1N2597
7090	1N2607
7090	1N2608
7090	1N3672
7090	1N3672A
7090	1N3673
7090	1N3673A
7090	1N3924
7090	1N4014
7090	1N4015
7090	1N4458
7090	1N4510
7090	1S1578
7090	1S545
7090	1S930
7090	11R10S
7090	12A1000
7090	12A900
7090	12C100
7090	12C90
7090	12F100
7090	12F100A
7090	12F100B
7090	12F90
7090	12F90A
7090	12F90B
7090	70R2S
7090	304V
7090	304Z
7090	367V
7090	367Z
7090	404V
7090	404Z
7090	408V
7090	408Z
7090	437V
7091	BYX40-1000R
7091	BYX42-900R
7091	GE-5045
7091	MR1130R
7091	PR9010R
7091	PR9011R
7091	1N3672AR
7091	1N3672R
7091	1N3673AR
7091	1N3673R
7091	1N4014R
7091	1N4015R
7091	12C100R
7091	12C90R
7091	12FR100
7091	12FR100A
7091	12FR100B
7091	12FR90
7091	12FR90A
7091	12FR90B
7091	12NF11
7091	337V
7091	337Z
7091	367RV
7091	367RZ
7096	TR150
7096	1N3617
7096	16F20
7096	324-0118
7096	368C
7096	368D
7096	409C
7096	409D
7100	1N3619
7100	1N3620
7100	8AN40
7100	16F40
7100	40Q4
7100	368H
7100	409H
7104	1N3622
7104	5M60
7104	16F60
7104	368M
7104	409M
7110	BYX38-900
7110	1N3624
7110	8AN100
7110	16F100
7110	368V
7110	368Z
7110	409V
7110	409Z
7111	BYX38-900R
7111	1N3624R
7111	368RV
7111	368RZ
7111	409RV
7111	409RZ
7202	BYX52-900
7202	BYX56-1000
7202	GE-5108
7202	PR9014
7202	PR9015
7202	PR9025
7202	PR9026
7202	R4101022
7202	R4101040
7202	1N2287
7202	1N3768
7202	1N4529
7202	1S1844
7202	2WMT10
7202	3C10
7202	4C10
7202	10M100
7202	21R10S
7202	25C100
7202	25C90
7202	30R25
7202	35C100
7202	35C90
7202	40HF100
7202	40HF90
7202	40R2S
7202	302V
7202	302Z
7202	303V
7202	303Z
7202	371V
7202	371Z
7202	402V
7202	402Z
7202	403V
7202	403Z
7202	418V
7202	418Z
7202	419V
7203	BYX52-900R
7203	BYX56-1000R
7203	GE-5109
7203	PR9014R
7203	PR9015R
7203	R4111022
7203	R4111040
7203	1N3767R
7203	1N3768R
7203	1S1418
7203	25C100R
7203	25C90R
7203	35C100R
7203	35C90R
7203	40HFR100
7203	40HFR90
7203	335V
7203	335Z
7203	336V
7203	336Z
7203	371RV
7203	371RZ
7203	417RV
7203	417RZ
7203	419RV
7203	436V
7203	436Z
7220	PR9034
7220	1N2128
7220	1N2128A
7220	1N2446
7220	1N2458
7220	25G5
7220	25HB5
7220	50C05
7220	60HF5
7226	PR9036
7226	1N2130
7226	1N2130A
7226	1N2131
7226	1N2131A
7226	1N2448
7226	1N2460
7226	1N2461
7226	1N2788
7226	1N3142
7226	1N3968
7226	20JH3
7226	20M20
7226	25G20
7226	25HB15
7226	50C20
7226	60HF15
7226	60HF20
7227	PR9036R
7227	R3415
7227	R3420
7227	1N2130R
7227	1N2131R
7227	1N2788R
7227	1N3968R
7227	25G20R
7227	50C20R
7234	PR9038
7234	1N2134
7234	1N2134A
7234	1N2135
7234	1N2135A
7234	1N2452
7234	1N2453
7234	1N2464
7234	1N2465
7234	1N2789
7234	1N3969
7234	20M40
7234	25HB35
7234	25HB40
7234	50C40
7234	60HF35
7234	60HF40
7240	PR9040
7240	1N2138
7240	1N2138A
7240	1N2139
7240	1N2139A
7240	1N2455
7240	1N2467
7240	1N3970
7240	20M60
7240	25G60
7240	25HB60
7240	50C60
7240	60HF60
7244	PR9043
7244	PR9044
7244	R4101060
7244	20M100
7244	25G100
7244	50C100
7244	50C90
7245	PR9043R
7245	PR9044R
7245	R4111060
7245	50C100R
7245	50C90R
7254	GE-5128
7254	1N4136
7254	25H15
7254	25H15A
7254	25H20A
7254	70H15
7254	70H15A
7254	70H20
7254	70H20A
7260	GE-5132
7260	1N1401
7260	1N4137
7260	25H40
7260	70H40
7260	70H40A
7261	GE-5133
7261	40JH3R
7264	GE-5136
7264	1N4138
7264	25H50
7264	25H50A
7264	25H60
7264	25H60A
7264	50JH3
7264	60JH3
7264	70H50
7264	70H50A
7264	70H60
7272	GE-5140
7272	R4101070
7272	S36100
7272	S3690
7272	1S1615
7272	25H100
7272	25H100A
7272	25H90
7272	25H90A
7272	70H100
7272	70H100A
7272	70H90
7272	70H90A
7273	GE-5141
7273	R4111070
7354	1N3288
7354	1N3289
7354	1N3290
7354	1N3291
7356	1N3292
7356	1N3293
7358	1N3295
7359	1N3295R
7400	DDEY030001
7400	DM7400
7400	DM7400N
7400	ECG7400
7400	FD-1073-BF
7400	FJH131
7400	FLH101
7400	GE-7400
7400	HC10004110
7400	HD2503
7400	HD2503P
7400	HD7400
7400	HD7400P
7400	HEP-C3000P
7400	HL18998
7400	IC-7400
7400	ITT7400N
7400	J1000-7400
7400	J4-1000
7400	KS20967-L1
7400	LB3000
7400	M53200
7400	M53200P
7400	MB400
7400	MB601
7400	MC7400L
7400	MC7400P
7400	MIC7400J
7400	MIC7400N
7400	N7400A
7400	N7400F
7400	PA7001/521
7400	REN7400
7400	SL16793
7400	SN7400
7400	SN7400A
7400	SN7400N
7400	T7400B1
7400	TD1401
7400	TD1401P
7400	TD3400A
7400	TD3400AP
7400	TD3400P
7400	TL7400N
7400	US7400A
7400	US7400J
7400	006-0000146
7400	007-1695001
7400	9N00
7400	9N00DC
7400	9N00PC
7400	19A116180P1
7400	51S10611A11
7400	68A9025
7400	78A200010P4
7400	225A6946-P000
7400	236-0005
7400	301-576-4
7400	398-13223-1
7400	435-21026-0A
7400	443-1
7400	1065-4861
7400	1741-0051
7400	1820-0054
7400	7400
7400	7400-6A
7400	7400-9A
7400	7400/9N00
7400	7400PC
7400	8000-00038-004
7400	10302-04
7400	11216-1
7400	55001
7400	138311
7400	339300
7400	339300-2
7400	373401-1
7400	558875
7400	611563
7400	760011
7400	800024-001
7400	930347-3
7400	2610786
7400	3520041-001
7400	4663001D907
7400	5359031
7400	9003151
7400	51320000
7400	51320012

SK	Industry Standard No.
7401	GEIC-194
7401	ITT7401N
7401	N7401A
7401	N7401F
7401	1741-0085
7401	7401-6A
7401	7401-9A
7401	55032
7402	DM7402N
7402	ECG7402
7402	FD-1073-BG
7402	FJH221
7402	FLH191
7402	GE-7402
7402	HD2511
7402	HD2511P
7402	HD7402
7402	HD7402P
7402	HEP-C3002P
7402	HL19004
7402	IC-7402
7402	ITT7402N
7402	J1000-7402
7402	J4-1002
7402	LB3008
7402	M53202
7402	M53202P
7402	MB417
7402	MC7402L
7402	MC7402P
7402	MIC7402J
7402	MIC7402N
7402	N7402A
7402	N7402F
7402	PA7001/525
7402	SL16795
7402	SN7402N
7402	T7402B1
7402	TD3402A
7402	TD3402AP
7402	TL7402N
7402	US7402A
7402	US7402J
7402	XAA104
7402	ZN7402E
7402	007-1696201
7402	9N02
7402	9N02DC
7402	9N02PC
7402	43A223009
7402	68A9027
7402	435-21027-0A
7402	443-46
7402	1741-0119
7402	1820-0328
7402	7402-6A
7402	7402-9A
7402	7402PC
7402	11207-1
7402	55002
7402	138313
7402	558876
7402	611564
7402	800080-001
7402	930347-11
7402	2610783
7402	7012166
7402	51320001
7404	A03
7404	DM7404N
7404	ECG7404
7404	FD-1073-BJ
7404	FJH241
7404	FLH211
7404	GE-7404
7404	HD2522
7404	HD2522P
7404	HEP-C3004P
7404	HL19000
7404	IC-7404
7404	ITT7404N
7404	J4-1004
7404	KS20967-L2
7404	LB3006
7404	M53204
7404	M53204P
7404	MB418
7404	MC7404L
7404	MC7404P
7404	MIC7404J
7404	MIC7404N
7404	N7404A
7404	N7404F
7404	PA7001/527
7404	SL16796
7404	SN7404N
7404	T7404B1
7404	TD3404A
7404	TD3404AP
7404	TL7404N
7404	US7404A
7404	US7404J
7404	ZN7404E
7404	007-1695301
7404	9N04
7404	9N04DC
7404	9N04PC
7404	19A116180P20
7404	51-10611A12
7404	51S10611A12
7404	68A9028
7404	225A6946-P004
7404	236-0007
7404	398-13224-1
7404	398-13632-1
7404	435-21028-0A
7404	443-18
7404	1741-0143
7404	1806
7404	1820-0174
7404	1820-0894
7404	7404
7404	7404-6A
7404	7404-9A
7404	7404PC
7404	8000-00028-042
7404	9016
7404	11202-1
7404	015040/7
7404	55003
7404	138314
7404	373404-1
7404	508590
7404	611565
7404	800387-001
7404	801806
7404	930347-13
7404	3520048-001
7404	9001551-02
7404	9001551-03
7404	51320002
7405	09017
7406	DM7406N
7406	ECG7406
7406	FLH481
7406	GE-7406
7406	HD7406
7406	HD7406P
7406	ITT7406N
7406	MC7406L
7406	MC7406P
7406	N7406A
7406	N7406F
7406	SN7406N
7406	T7406B1
7406	TL7406N
7406	007-1696901
7406	9N06
7406	9N06DC
7406	9N06PC
7406	68A9032
7406	1607A80
7406	7406PC
7406	55036
7406	373429-1
7406	800651-001
7406	50254200
7408	DM7408N
7408	ECG7408
7408	FD-1073-BM
7408	FLH381
7408	GE-7408
7408	HD2550
7408	HD2550P
7408	IC-7408
7408	ITT7408N
7408	KS21282-L1
7408	L-612099
7408	MC7408L
7408	MC7408P
7408	N7408A
7408	N7408F
7408	SL14971
7408	SL16798
7408	SL17869
7408	SN7408N
7408	US7408A
7408	US7408J
7408	007-1699301
7408	9N08
7408	9N08DC
7408	9N08PC
7408	435-21029-0A
7408	443-45
7408	1741-0200
7408	1820-0870
7408	7408-6A
7408	7408-9A
7408	7408N
7408	7408PC
7408	94152
7408	138315
7408	310254
7408	374109-1
7408	5175460
7408	9003398-03
7408	9003398-04
7408	51330005
7410	DM7410N
7410	ECG7410
7410	FD-1073-BN
7410	FJH121
7410	GE-7410
7410	HD2507
7410	HD2507P
7410	HD7410
7410	HD7410P
7410	HEP-C3010P
7410	HL19001
7410	IC-7410
7410	ITT7410N
7410	J1000-7410
7410	J4-1010
7410	LB3001
7410	M53210
7410	M53210P
7410	MB401
7410	MB602
7410	MC7410L
7410	MC7410P
7410	MIC7410J
7410	MIC7410N
7410	N7410A
7410	N7410F
7410	PA7001/520
7410	SL16801
7410	SN7410N
7410	T7410B1
7410	TD1402
7410	TD1402P
7410	TD3410A
7410	TD3410AP
7410	TD3410P
7410	TL7410N
7410	UPB202C
7410	UPB202D
7410	UPB7410C
7410	US7410A
7410	US7410J
7410	ZN7410E
7410	006-0000147
7410	007-1695901
7410	9N10
7410	9N10DC
7410	9N10PC
7410	19A116180P4
7410	49A0005-000
7410	68A9030
7410	225A6946-P010
7410	435-21030-0A
7410	1741-0234
7410	1820-0068
7410	7410-6A
7410	7410-9A
7410	7410N
7410	7410PC
7410	10302-03
7410	11200-1
7410	55005
7410	373405-1
7410	558877
7410	611566
7410	800023-001
7410	3520042-001
7410	4663001A912
7410	9003091-02
7410	9003091-03
7410	51320003
74122	N74122A
74123	ECG74123
74123	FLK121
74123	GE-74123
74123	HD2561
74123	HD2561P
74123	IC-74123
74123	ITT74123N
74123	N74123B
74123	N74123F
74123	REN74123
74123	RH-IX0041PAZZ
74123	SN74123N
74123	TL74123N
74123	UPB74123C
74123	9N123
74123	9N123DC
74123	9N123PC
74123	74123
74123	74123PC
7413	DM7413N
7413	ECG7413
7413	FLH351
7413	GE-7413
7413	HD2545
7413	HD2545P
7413	IC-103(ELCOM)
7413	ITT7413N
7413	MIC7413N
7413	N7413A
7413	N7413F
7413	RS7413(IC)
7413	SN7413N
7413	TL7413N
7413	UPB7413C
7413	9N13
7413	9N13DC
7413	9N13PC
7413	443-44
7413	443-44-2854
7413	601-0100865
7413	3531-021-000
7413	7413PC
7413	55027
74145	DM74145N
74145	ECG74145
74145	FLL111T
74145	GE-74145
74145	HD2555
74145	HD2555P
74145	ITT74145N
74145	MB443
74145	MC74145P
74145	MIC74145J
74145	MIC74145N
74145	N74145B
74145	PA7001/593
74145	RS74145
74145	SN74145N
74145	TL74145N
74145	US74145A
74145	WEP74145
74145	007-1696801
74145	19-130-004
74145	276-1828
74145	443-87
74145	1542
74145	74145DC
74145	74145PC
74145	93145DC
74145	93145PC
74150	DM74150N
74150	ECG74150
74150	FLY111
74150	GE-74150
74150	HD2548
74150	HD2548P
74150	HD74150
74150	HD74150P
74150	ITT74150N
74150	MC74150P
74150	MIC74150J
74150	MIC74150N
74150	N74150F
74150	N74150N
74150	SN74150N
74150	TL74150N
74150	UPB2150D
74150	UPB74150C
74150	74150DC
74150	74150PC
74150	93150DC
74150	93150PC
74154	ECG74154
74154	FLY141
74154	GE-74154
74154	HD2580
74154	HD2580P
74154	MIC74154J
74154	MIC74154N
74154	N74154F
74154	RS74154
74154	SN74154N
74154	TL74154N
74154	74154DC
74154	74154PC
74154	93154DC
74154	93154PC
74192	DM74192N
74192	ECG74192
74192	FLJ241
74192	GE-74192
74192	HD2541
74192	HD2541P
74192	ITT74192N
74192	MC74192P
74192	N74192B
74192	N74192F
74192	SN74192N
74192	TD34192A
74192	TD34192AP
74192	TL74192N
74192	007-1698301
74192	9360DC
74192	9360PC
74192	74192DC
74192	74192PC
74193	DM74193N
74193	ECG74193
74193	FLJ251
74193	GE-74193
74193	HD2542
74193	HD2542P
74193	ITT74193N
74193	M53393
74193	M53393P
74193	MC74193P
74193	N74193B
74193	N74193F
74193	SN74193N
74193	T74193B1
74193	TL74193N
74193	007-1698401
74193	43C216447
74193	43C216447P1
74193	443-162
74193	9366DC
74193	9366PC
74193	11204-1
74193	74193DC
74193	74193PC
74193	138320
74193	611730
74193	800386-001
74193	7012142-03
74196	DM74196N
74196	ECG74196
74196	FLJ381
74196	GE-74196
74196	HD2572
74196	HD2572P
74196	RS74196
74196	SN74196J
74196	SN74196N
74196	TL74196N
74196	443-628
74196	74196DC
74196	74196PC
74196	93196DC
74196	93196PC
7420	DM7420N
7420	ECG7420
7420	FD-1073-BR
7420	FJH111
7420	FLH121
7420	GE-7420
7420	HD2504
7420	HD2504P
7420	HD7420
7420	HD7420P
7420	HEP-C3020P
7420	HL19003
7420	ITT7420N
7420	LB3002
7420	M53220
7420	M53220P
7420	MB402
7420	MB603
7420	MC7420L
7420	MC7420P
7420	MIC7420J
7420	MIC7420N
7420	N7420A
7420	N7420F
7420	PA7001/519
7420	SL16800
7420	SN7420N
7420	T7420B1
7420	TD1403
7420	TD1403P
7420	TD3420A
7420	TD3420AP
7420	TL7420N
7420	UPB203D
7420	UPB7420C
7420	US7420A
7420	US7420J
7420	ZN7420E
7420	007-1695101
7420	9N20
7420	9N20DC
7420	9N20PC
7420	19A116180P5
7420	68A9033
7420	225A6946-P020
7420	435-21033-0A
7420	443-2
7420	1741-0325
7420	1820-0069
7420	7420-6A
7420	7420-9A
7420	7420PC
7420	10302-02
7420	11205-1
7420	55006
7420	373406-1
7420	558878
7420	611567
7420	800020-001
7420	930347-1
7420	930347-10
7420	9004076
7420	9004076-03
7420	9004076-04
7420	51320004
7420	885540026-3
7427	DM7427N
7427	ECG7427
7427	FLH621
7427	GE-7427
7427	HD7427
7427	HD7427P
7427	SN7427N
7427	US7427A
7427	9N27
7427	9N27DC
7427	9N27PC
7427	443-65
7427	7427PC
7430	55007
7432	DM7432N
7432	ECG7432
7432	FLH631
7432	GE-7432
7432	HD7432
7432	HD7432P
7432	ITT7432N
7432	KS21282-L3
7432	L-612107
7432	N7432A
7432	N7432F
7432	SN7432N
7432	US7432A
7432	US7432J
7432	9N32
7432	9N32DC
7432	9N32PC
7432	7432PC
7432	138381
7441	C3041P
7441	ECG7441
7441	FJL101
7441	GE-7441
7441	HD2518
7441	HD2518P
7441	HEP-C3041P
7441	M53241
7441	M53241P
7441	MC7441AL
7441	MIC7441AJ
7441	MIC7441AN
7441	N7441B
7441	N7441F
7441	T7441AB1
7441	TD3441A
7441	TD3441AP
7441	US7441A
7441	ZN7441AE
7441	007-1697801
7441	376-0099
7441	1741-1190
7441	7441-6A
7441	7441-9A
7441	7441DC
7441	7441PC
7441	9315DC
7441	9315PC
7447	ECG7447
7447	FLL121T
7447	FLL121V
7447	GE-7447
7447	HD2532
7447	HD2532P
7447	ITT7447AN
7447	J1000-7447
7447	J4-1047
7447	M53247
7447	M53247P
7447	MC7447L
7447	MC7447P
7447	N7447B
7447	N7447F
7447	SN7447AN
7447	TD3447A
7447	TD3447AP
7447	TL7447AN
7447	US7447A
7447	443-36
7447	7447BDC
7447	7447BPC
7447	7447DC
7447	7447PC
7447	9357B
7447	9357BDC
7447	9357BPC
7448	ECG7448
7448	FLH551
7448	GE-7448
7448	ITT7448N
7448	M53248
7448	M53248A
7448	M53248P
7448	MC7448L
7448	MC7448P
7448	N7448B
7448	SN7448N
7448	TL7448N
7448	US7448A
7448	7448DC
7448	7448PC
7448	9358
7448	9358DC
7448	9358PC
7451	DM7451N
7451	ECG7451
7451	FD-1073-BW
7451	FJH161
7451	FLH161
7451	GE-7451
7451	HD2505
7451	HD2505P
7451	HD7451
7451	HD7451P
7451	ITT7451N
7451	MC7451L
7451	MC7451P
7451	MIC7451J
7451	MIC7451N
7451	N7451A
7451	N7451F
7451	PA7001/523
7451	SN7451N
7451	T7451B1
7451	TD3451A
7451	TD3451AP
7451	TD3451P
7451	TL7451N
7451	UPB207D
7451	UPB7451C
7451	US7451A
7451	US7451J
7451	9N51
7451	9N51DC
7451	9N51PC
7451	435-21034-0A
7451	1741-0564
7451	1820-0063
7451	7451-6A
7451	7451-9A
7451	7451PC
7451	10302-05
7451	373715-1
7451	930347-12
7451	3520045-001
7451	51320016
7473	DM7473N
7473	ECG7473
7473	FJJ121
7473	FLJ121
7473	GE-7473
7473	HD2515
7473	HD2515P

SK	Industry Standard No.	SK	Industry Standard No.	SK	Industry Standard No.	SK	Industry Standard No.	SK	Industry Standard No.
7473	HEP-C3073P	7475	51-10611A16	7490	J1000-7490	9000	ERB28-04	9004	UT249
7473	HL19002	7475	51810611A16	7490	J4-1090	9000	ERB28-04D	9004	UT251
7473	ITT7473N	7475	68A9041	7490	M53290	9000	ERC24-04	9004	UT261
7473	M53273	7475	443-13	7490	M53290P	9000	RU1	9004	UT3010
7473	M53273P	7475	1741-0747	7490	MC7490L	9000	RU2	9004	ZR11
7473	MC7473L	7475	1820-0301	7490	MC7490P	9000	81B80	9004	ZR601
7473	MC7473P	7475	7475-6A	7490	MIC7490J	9000	TV810D2	9004	ZS271
7473	MIC7473J	7475	7475-9A	7490	MIC7490N	9000	TV8B24-06c	9004	1.5E1
7473	MIC7473N	7475	7475DC	7490	N7490A	9000	TV8B24-06d	9004	1.5J1
7473	N7473A	7475	7475PC	7490	N7490P	9000	13-43777-1	9004	1N1556
7473	N7473P	7475	9375DC	7490	SN7490AN	9000	13-43777-2	9004	1N1563
7473	PA7001/531	7475	9375PC	7490	T7490B1	9000	137606	9004	1N1645
7473	REN7473	7475	373713-1	7490	TD3490A	9000	06200014	9004	1N2014
7473	SL16806	7475	611065	7490	TD3490AP	9000/552	SID30-132	9004	1N2391
7473	SL17242	7475	800382-001	7490	TD3490P	9000/552	SM1-02	9004	1N2400
7473	SN7473N	7475	7011203-02	7490	TL7490N	9000/552	TV810D	9004	1N2409
7473	T7473B1	7475	7011203-03	7490	US7490A	9000/552	TV810D2	9004	1N2859A
7473	TD1409	7476	36188000	7490	US7490J	9000/552	V11N	9004	1N3073
7473	TD1409P	7476	B02	7490	WEP7490	9000/552	10D2	9004	1N3227
7473	TD3473A	7476	DM74176N	7490	XAA109	9000/552	13-39072-1	9004	1N3238
7473	TD3473AP	7476	ECG7476	7490	19-130-005	9000/552	13-39072-2	9004	1N3247
7473	TL7473N	7476	FJJ191	7490	19A116180-24	9000/552	138172	9004	1N3486
7473	US7473A	7476	FLJ131	7490	443-7	9002	DS132A	9004	1N4140
7473	US7473J	7476	GE-7476	7490	443-7-16088	9002	DS132B	9004	1N4720
7473	XAA107	7476	HD2516 .	7490	733W00039	9002	DS18	9004	1N4817
7473	ZN7473E	7476	HD2516P	7490	905-102	9002	00001S1849R	9004	1N4998
9N73	9N73	7476	HL19010	7490	1808	9003	100C1R	9004	1N5004
9N73	9N73DC	7476	IC-7476	7490	1820-0055	9003	A15F	9004	1N5055
9N73	9N73PC	7476	J1000-7476	7490	7490-6A	9003	A15U(DIODE)	9004	1N5392GP
7473	19A116180P15	7476	J4-1076	7490	7490-9A	9003	AE3A	9004	12644
7473	43A223025	7476	M53276	7490	7490DC	9003	A83A	9004	18031
7473	49A0002-000	7476	M53276P	7490	7490PC	9003	BA136A	9004	182416
7473	236-0009	7476	MC7476L	7490	9590DC	9003	BA178	9004	1840
7473	435-23006-0A	7476	MC7476P	7490	9590PC	9003	BTM-50	9004	18431
7473	443-5	7476	MIC7476J	7490	102005	9003	ECG5800	9004	381E
7473	477-0412-004	7476	MIC7476N	7490	373427-1	9003	HEP-R0090	9004	7-6006-00
7473	1030-25	7476	N7476B	7490	558883	9003	R8-05	9004	1501
7473	1820-0075	7476	SL16808	7490	611572	9003	RT7848	9004	1581
7473	7473-6A	7476	SN7476N	7490	760013	9003	S105A	9004	20A1
7473	7473-9A	7476	T7476B1	7490	801808	9003	S1M1	9004	20C1
7473	7473PC	7476	TL7476N	7490	7012167-02	9003	S1M2	9004	30R1
7473	138403	7476	US7476A	7492	C38017	9003	S2M2	9004	30B1
7473	558881	7476	XAA108	7492	DM7492N	9003	S92A	9004	100-525
7473	930347-7	7476	ZN7476E	7492	ECG7492	9003	SB1Z	9004	250R1B
7473	3520043-001	9N76	9N76	7492	FJJ152	9003	SD91	9004	300R1B
7474	DM7474N	9N76	9N76DC	7492	FLJ171	9003	S1O5	9004	398B
7474	ECG7474	9N76	9N76PC	7492	GE-7492	9003	S1O5A	9004	244-17
7474	FJJ131	7476	43A223028	7492	HD2521	9003	SJ053F	9004	5210DO-1
7474	FLJ141	7476	68A9042	7492	HD2521P	9003	SLA21A	9004	40266
7474	GE-7474	7476	443-16	7492	HD7492A	9003	SLA21B	9004	720454
7474	HD2510	7476	1348A14HO1	7492	HD7492AP	9003	SLA21C	9005	A-132591
7474	HD2510P	7476	7476	7492	HEP-C53801P	9003	SLA5197	9005	A15B
7474	HD7474	7476	7476-6A	7492	ITT7492N	9003	SOD30AL	9005	AE5C
7474	HD7474P	7476	7476-9A	7492	J1000-7492	9003	SOD30BL	9005	A83C
7474	HL18999	7476	7476PC	7492	J4-1092	9003	SOD30DL	9005	B56-15
7474	IC-7474	7476	55012	7492	M53292P	9003	SOD50AL	9005	BTX30/150
7474	M53274	7476	72185	7492	MC7492L	9003	SOD50BL	9005	E302(ELCOM)
7474	M53274	7476	373414-1	7492	MC7492P	9003	SOD50CL	9005	ECG5002
7474	M53274P	7476	611870	7492	MC7492P	9003	SOD50DL	9005	HEP-R0092
7474	MB420	7476	760015	7492	MIC7492J	9003	SP-1.5A	9005	HR-200
7474	MIC7474J	7476	9004300-03	7492	MIC7492N	9003	SR132-1	9005	MR501
7474	MIC7474N	7480	PLH221	7492	N7492A	9003	SR422	9005	NU75003
7474	N7474A	7480	MC7480P	7492	N7492P	9003	SR1598	9005	PGR-24
7474	N7474P	7485	DM7485N	7492	SL16809	9003	SR1643A	9005	PT-520
7474	PA7001/529	7485	ECG7485	7492	SN7492AN	9003	SR1DM-2	9005	R-1B
7474	REN7474	7485	FLH431	7492	TD3492A	9003	SR3AM1	9005	R-5B
7474	SL16807	7485	GE-7485	7492	TD3492AP	9003	SR475	9005	R-5C
7474	SN7474N	7485	HD7485	7492	TD7492R	9003	SW-05	9005	S1020
7474	T7474B1	7485	HD7485P	7492	US7492A	9003	TB5	9005	S1220
7474	TD3474A	7485	MB448	7492	US7492J	9003	TC50	9005	S229
7474	TD3474AP	7485	N7485A	7492	19A116180-27	9003	TV24651	9005	S236
7474	TD3474P	7485	N7485P	7492	436-1-0010-0A	9003	TV8-181893	9005	S2420
7474	TL7474N	7485	SN7485N	7492	7492	9003	TV8-181922	9005	S2GR1
7474	US7474A	7485	TL7485N	7492	7492-6A	9003	UR205	9005	S2GR2
7474	US7474J	7485	UPB2085D	7492	7492-9A	9003	UT2005	9005	S3A2
7474	ZN7474E	7485	UPB7485C	7492	7492DC	9003	UT3005	9005	S3M1
7474	007-1699801	7485	007-1696001	7492	7492PC	9003	WR-006	9005	S4GR2
7474	9N74	7485	7485DC	7492	9592DC	9003	ZR10	9005	SA2M1
7474	9N74DC	7485	7485PC	7492	9592PC	9003	ZS270	9005	SD91A
7474	9N74PC	7485	9385DC	7492	373712-1	9003	1.5B05	9005	SD918
7474	19A116180P16	7486	DM7486N	7492	611573	9003	1.5J05	9005	SD92
7474	43A223026P1	7486	ECG7486	7492	1000100	9003	1N1644	9005	SD92A
7474	49A0012-000	7486	FD-1073-CA	7492	9003445-03	9003	1N2013	9005	SD92B
7474	435-23007-0A	7486	FLH341	74LS00	ECG74LS00	9003	1N2390	9005	SG205
7474	443-6	7486	GE-7486	74LS00	SN74LS00N	9003	1N2399	9005	SH1S
7474	1820-0077	7486	HD2526	74LS00	74LS00N	9003	1N2408	9005	S11A
7474	7474-6A	7486	HD7486	74LS02	ECG74LS02	9003	1N2858A	9005	S101/100
7474	7474-9A	7486	HD7486P	74LS02	SN74LS02N	9003	1N3072	9005	SJ203P
7474	7474PC	7486	HL19014	74LS04	ECG74LS04	9003	1N3237	9005	SL-5
7474	8000-00038-007	7486	ITT7486N	74LS04	SN74LS04N	9003	1N3246	9005	SLA23B
7474	11213-1	7486	K820967-L3	74LS04	74LS04PC	9003	1N4139	9005	SLA23C
7474	55011	7486	M53286	74LS08	ECG74LS08	9003	1N4719	9005	SLA5199
7474	373409-1	7486	M53286P	74LS08	SN74LS08N	9003	1N4816	9005	SOD100AL
7474	558882	7486	MB449	74LS10	ECG74LS10	9003	1N4997	9005	SOD100BL
7474	611571	7486	MC7486L	74LS10	SN74LS10N	9003	1N5171	9005	SOD100CL
7474	800400-001	7486	MC7486P	74LS10	74LS10PC	9003	1N5197	9005	SOD100DB
7474	881916	7486	MIC7486J	74LS123	ECG74LS123	9003	1N5391	9005	SOD200AL
7474	2868536-1	7486	MIC7486N	74LS123	SN74LS123N	9003	1P643	9005	SOD200BL
7474	3520046-001	7486	N7486A	74LS138	74LS138PC	9003	18030	9005	SOD200CL
7474	4663001A905	7486	N7486P	74LS151	ECG74LS151	9003	1S2408	9005	SOD200DL
7474	9003152	7486	SN7486N	74LS151	SN74LS151N	9003	1S2415	9005	SR-100
7474	9003152-01	7486	T7486B1	74LS157	ECG74LS157	9003	18430	9005	SR-114
7475	B01	7486	TL7486N	74LS157	SN74LS157N	9003	2P05	9005	SR1762
7475	DM7475N	7486	UPB2086D	74LS161A	ECG74LS161A	9003	3A15	9005	SR210
7475	ECG7475	7486	US7486A	74LS174	ECG74LS174	9003	3430	9005	SR220
7475	FJJ181	7486	US7486J	74LS174	SN74LS174N	9003	3B05	9005	SR507
7475	FLJ151	7486	9N86	74LS175	ECG74LS175	9003	3S05E	9005	SSIC10810
7475	GE-7475	7486	9N86DC	74LS175	SN74LS175B	9003	15005	9005	SSIC1110
7475	HD2517	7486	9N86PC	74LS193	ECG74LS193	9003	15805	9005	SSIC1210
7475	HD2517P	7486	19A116180-18	74LS193	SN74LS193N	9003	30805	9005	TB100
7475	HD7475	7486	19A116180P18	74LS20	ECG74LS20	9003	48-137098	9005	TB200
7475	HD7475P	7486	40-065-19-029	74LS20	SN74LS20N	9003	599A	9005	TC100
7475	HEP-C3075P	7486	435-21035-0A	74LS20	74LS20PC	9003	720453	9005	TC200
7475	HL19012	7486	1741-0804	74LS27	ECG74LS27	9004	A15A	9005	UR215
7475	ITT7475N	7486	7486PC	74LS27	SN74LS27N	9004	AE3B	9005	UR220
7475	J4-1075	7486	339486	74LS30	ECG74LS30	9004	A83B	9005	UT12
7475	M53275	7486	373410-1	74LS30	SN74LS30N	9004	B56-33	9005	UT15
7475	M53275P	7486	611066	74LS32	ECG74LS32	9004	ECG5801	9005	UT220
7475	MC7475L	7486	611844	74LS32	SN74LS32N	9004	HEP-R0091	9005	UT234
7475	MC7475P	7486	9001349-02	74LS367	ECG74LS367	9004	S1010	9005	UT242
7475	MIC7475J	7486	9001349-03	74LS367	SN74LS367N	9004	S1210	9005	UT252
7475	MIC7475N	7486	51320018	74LS51	ECG74LS51	9004	S1GR2	9005	UT262
7475	N7475B	7490	DDEY029001	74LS51	SN74LS51N	9004	S2M1	9005	UT3020
7475	SM63	7490	DM7490	74LS73	ECG74LS73	9004	SA1M1	9005	ZR12
7475	SN7475N	7490	DM7490N	74LS73	SN74LS73N	9004	SI1	9005	ZR602
7475	T7475B1	7490	ECG7490	74LS74	ECG74LS74A	9004	SJ103P	9005	ZS272
7475	TD3475A	7490	FJJ141	74LS75	ECG74LS75	9004	SLA22A	9005	ZS701
7475	TD3475AP	7490	FLJ161	74LS85	ECG74LS85	9004	SLA22B	9005	ZS702
7475	TL7475N	7490	GE-7490	74LS85	SN74LS85N	9004	SLA22C	9005	1.5E2
7475	US7475A	7490	HD2519	74LS93	ECG74LS93	9004	SLA5198	9005	1.5J2
7475	US7475J	7490	HD2519P	74LS93	SN74LS93N	9004	SR3AM-3	9005	1K2
7475	ZN7475E	7490	HEP-C3800P	74LS93	74LS93	9004	UR210	9005	1N1055
7475	40-065-19-027	7490	HL19015	8293	1100-9453	9004	UT2010	9005	1N1220B
7475	49A0000	7490	ITT7490N	9000	BB10	9004	UT256	9005	1N1557

SK	Industry Standard No.
9005	1N1564
9005	1N1646
9005	1N1647
9005	1N2015
9005	1N2016
9005	1N2392
9005	1N2401
9005	1N2410
9005	1N2860A
9005	1N3074
9005	1N3075
9005	1N3082
9005	1N3228
9005	1N3239
9005	1N3248
9005	1N3656
9005	1N3748
9005	1N3755
9005	1N3757
9005	1N3866
9005	1N3938
9005	1N3952
9005	1N3981
9005	1N4141
9005	1N4517
9005	1N4721
9005	1N4999
9005	1N5005
9005	1N5056
9005	1N5199
9005	1N5393
9005	1N5402
9005	1S021
9005	1S032
9005	1S1072
9005	1S2410
9005	1S2417
9005	1S432
9005	3S2E
9005	13-18481-1
9005	13-18481-3
9005	13-34368-1
9005	15C2
9005	15C2D
9005	15S2
9005	20C2
9005	35-1008
9005	86-5028-3
9005	86-5032-3
9005	96-5106-01
9005	96-5149-01
9005	96-5184-01
9005	137-737
9005	150R2B
9005	250R2B
9005	300R2B
9005	399B
9005	399C
9005	399D
9005	475-018
9005	501-363-2
9005	919-005045
9005	1412-182
9005	1901-0045
9005	5522-8
9005	7921
9005	031450
9005	34405
9005	36147
9005	36591
9005	38052
9005	39804
9005	40267
9005	46914
9005	61807
9005	71449
9005	121180
9005	206180
9005	206185
9005	206190
9005	475018
9005	506911
9005	630063
9005	654420
9005	720455
9005	801730
9005	7570014
9005	8521587-101
9006	A15C
9006	ECG5803
9006	S1030
9006	S1230
9006	UT211
9006	UT212
9006	1.5J3
9006	1N1221B
9006	1N1558
9006	1N1565
9006	1N1648
9006	1N1649
9006	1N1910
9006	1N2017
9006	1N2018
9006	1N2393
9006	1N2402
9006	1N2411
9006	1N2861A
9006	1N3076
9006	1N3077
9006	1N4819
9006	1N5057
9006	1N5173
9006	1N5394
9006	1N5403
9006	1P645
9006	1P646
9006	1S2411
9006	1S2418
9006	1S42
9006	3A300
9006	3AF4
9006	3S3E
9006	20A3
9006	20C3
9006	30S3
9006	150R3B
9006	200R3B
9006	250R3B
9006	300R3B
9007	A15D
9007	AC300
9007	AC400
9007	AE3D
9007	AS3D
9007	ECG5804
9007	HEP-R0094
9007	PT-525
9007	PT-540
9007	PT-72130-1
9007	PTC201
9007	R-4A
9007	R-S1805
9007	RFJ70147
9007	RV1476
9007	S1040
9007	S13A
9007	S14A
9007	S2A30
9007	S2A40
9007	S3CN1
9007	S3GR2
9007	S4M1
9007	S500B
9007	S500C
9007	S5CN1
9007	S5M1
9007	SA3M1
9007	SD101
9007	SD93
9007	SD93A
9007	SD94
9007	SD94A
9007	SE05A
9007	SG305
9007	SJ403F
9007	SLA1103
9007	SLA24A
9007	SLA24B
9007	SLA24C
9007	SLA25A
9007	SLA25B
9007	SLA25C
9007	SLA5200
9007	SR1378-2
9007	SR3AM6
9007	SSI1120
9007	SSIC0820
9007	SSIC1220
9007	TS4
9007	UR225
9007	UT2040
9007	UT213
9007	UT235
9007	UT244
9007	UT254
9007	UT264
9007	UT3040
9007	WRR1955
9007	WRR1956
9007	ZR13
9007	ZR14
9007	ZR604
9007	ZS274
9007	1.5E4
9007	1.5J4
9007	1N1222
9007	1N1559
9007	1N1566
9007	1N1650
9007	1N1651
9007	1N1763A
9007	1N2019
9007	1N2020
9007	1N2394
9007	1N2403
9007	1N2412
9007	1N2862A
9007	1N3078
9007	1N3079
9007	1N3083
9007	1N3240
9007	1N3249
9007	1N3657
9007	1N3749
9007	1N3756
9007	1N3758
9007	1N3867
9007	1N3939
9007	1N3982
9007	1N4089
9007	1N4142
9007	1N4722
9007	1N4820
9007	1N5000
9007	1N5006
9007	1N5058
9007	1N5174
9007	1N5395
9007	1N5404
9007	1P647
9007	1S023
9007	1S034
9007	1S1073
9007	1S2350
9007	1S2379
9007	1S2412
9007	1S2419
9007	1S43
9007	1S434
9007	1WS4
9007	3-1477
9007	3SM4
9007	15C4
9007	15S4
9007	20A4
9007	20C4
9007	30R3
9007	48-137327
9007	48-137340
9007	150R4B
9007	152-0047
9007	152-0047-00
9007	211-58
9007	250R4B
9007	1901-0028
9007	1901-0036
9007	1901-0388
9007	1901-0389
9007	35604
9007	38074
9007	720456
9007	720458
9007	3430063-1
9007	8521587-102
9007	62020911
9008	A15E
9008	S1050
9008	UT214
9008	UT237
9008	UT245
9008	UT265
9008	1N1223B
9008	1N1560
9008	1N1567
9008	1N1652
9008	1N1764A
9008	1N2395
9008	1N2404
9008	1N2413
9008	1N2863A
9008	1N3229
9008	1N4821
9008	1N5175
9008	1N5396
9008	1S44
9008	3A500
9008	20A5
9008	30S2
9008	30S5
9009	A15N
9009	AC800
9009	AS3F
9009	S1070
9009	S3A8
9009	SD-7
9009	UT258
9009	UT268
9009	UT361
9009	UT362
9009	1N2374
9009	1N2397
9009	1N2398
9009	1N2406
9009	1N2407
9009	1N2415
9009	1N2416
9009	1N2501
9009	1N2505
9009	1N2772
9009	1N2773
9009	1N2865
9009	1N2867
9009	1N3080
9009	1N3231
9009	1N3242
9009	1N3251
9009	1N3751
9009	1N3760
9009	1N4514
9009	1N4585
9009	1N4586
9009	1N4724
9009	1N5002
9009	1N5052
9009	1N5053
9009	1N5398
9009	1S038
9009	3AF8
9009	3E8
9009	20A8
9009	30S8
9010	AC1000
9010	S10100
9010	S1090
9010	S3A10
9010	SI10
9010	SSIC0880
9010	SSIC1280
9010	UT347
9010	UT363
9010	UT364
9010	1N2502
9010	1N2506
9010	1N2774
9010	1N2775
9010	1N2866
9010	1N2868
9010	1N3107
9010	1N3232
9010	1N3243
9010	1N3252
9010	1N3563
9010	1N3723
9010	1N3752
9010	1N3761
9010	1N3869
9010	1N3957
9010	1N4725
9010	1N5003
9010	3E10
9010	20A10
9010	30S10
9014	FUN14LH026
9014	GEIC-7
9014	HA11115W
9014	HA1115
9014	HA1115W
9014	HEP-595
9014	HEP-C6068P
9014	LM1305
9014	LM1305N
9014	MC1305P
9014	MC1305P-C
9014	PC-20008
9014	PC20018
9014	PTC713
9014	RE320-IC
9014	SC5118P
9014	SC5182P
9014	SC5741P
9014	SN76105
9014	SN76105N
9014	TR2327422
9014	TR4104-2327421
9014	TR0-9011
9014	TVCM-7
9014	ULN2122A
9014	001-0036
9014	4-009
9014	09-011
9014	09-308071
9014	15-34379-1
9014	020-1114-009
9014	020-114-009
9014	051-0012-00
9014	51-10566A02
9014	57A32-22
9014	442-9
9014	729DC
9014	729PC
9014	740-9016-105
9014	1001-0036/4460
9014	3531-031-000
9014	8840-171
9014	18600-155
9014	25810-165
9014	34379-1
9014	58840-202
9014	518022S
9014	740622
9014	2327422
9014/720	MC1307PQ
9021	.25T5.8
9021	.25T5.8A
9021	AW01-06
9021	AZ-058
9021	ECG5070
9021	ECG5070A
9021	EQB01-06
9021	FV-24
9021	GEZD-6.0
9021	HZ-6B
9021	IP20-0203
9021	J24262
9021	MZ-206
9021	MZ306
9021	MZ306C
9021	MZ70MA
9021	MZ92-6.0B
9021	PZW
9021	QB106P
9021	RE108
9021	REN5070
9021	RH-EX0019CEZZ
9021	RVDR5R6EB
9021	RVDRD5R6EB
9021	SV124
9021	TM5070
9021	TR6S
9021	Z6
9021	ZB1-6
9021	1N1779
9021	1N377
9021	1N5849B
9021	1N762-2
9021	1N816
9021	1S1956
9021	1S347
9021	1S480
9021	05-480306
9021	09-306073
9021	09-306208
9021	09-306383
9021	46-86460-3
9021	48-155225
9021	48-83461E05
9021	48S155080
9021	53K001-4
9021	57S-06
9021	103-Z9008
9021	264D00507
9021	264P03301
9021	2093A41-153
9021	2106-17
9021	38510-331
9021	2008230042
9021	2012230042
9021	4202020100
9021	134883461505
9022	.25T17
9022	.25T17A
9022	.7JZ17
9022	.7Z17A
9022	.7Z17B
9022	.7Z17C
9022	.7Z17D
9022	.7ZM17A
9022	.7ZM17B
9022	.7ZM17C
9022	.7ZM17D
9022	AZ-167
9022	AZ-172
9022	BZ-167
9022	BZ-172
9022	D7L
9022	EQA01-17R
9022	GEZD-17
9022	RD16EC
9022	RE123
9022	SV141
9022	1EZ17D5
9022	1M17Z
9022	1M17Z10
9022	1M17Z5
9022	1N2038-3
9022	1N4704
9022	1N5537B
9022	1N5863B
9022	1N767-3
9022	1S232
9022	21A037-017
9022	34-8057-4
9022	523-2003-170
9023	.25T19
9023	.25T19A
9023	.7JZ19
9023	.7Z19A
9023	.7Z19B
9023	.7Z19C
9023	.7Z19D
9023	.7ZM19A
9023	.7ZM19B
9023	.7ZM19C
9023	.7ZM19D
9023	AZ-187
9023	AZ-192
9023	BZ-187
9023	BZ-192
9023	CD3212055
9023	ED498150
9023	EQB01-19
9023	GEZD-19
9023	RE125
9023	RH-EX0013CEZZ
9023	RH-EX013CEZZ
9023	SV143
9023	1M19Z10
9023	1M19Z5
9023	1N2039
9023	1N2039-2
9023	1N5539B
9023	1N5865B
9023	1N768-2
9023	1N768A
9023	1S234
9023	103-279-28A
9023	469-199-19
9023	523-2003-190
9024	.25T87
9024	.25T87A
9024	GEZD-87
9024	1N1325A
9024	1N5885B
9024	530157-870
9038	MRF207
9038	SD1080
9038	2N4427
9038/346	39705
9038/346	3457633-1
9038/346	3457633-2
9039	MJ13335
9040	2N6277
9041	2SD414
9041	2SD415
9041	146596
9041/373	2SC1162A
9041/373	2SC1162B
9041/373	2SC1162C
9041/373	2SC1162CP
9041/373	2SC1162D
9041/373	2SC1162WB
9041/373	2SC1162WT
9041/373	2SC1162WTA
9041/373	2SC1162WTB
9041/373	2SC1162WTC
9041/373	2SC1162WTD
9041/373	2SC1212
9041/373	2SC1212A
9041/373	2SC1212AA
9041/373	2SC1212AB
9041/373	2SC1212AC
9041/373	2SC1212AD
9041/373	2SC1212AWT
9041/373	2SC1212B
9041/373	2SC1212C
9041/373	2SC1212D
9041/373	2SC1212WT
9041/373	2SC1212WTA
9041/373	2SC1368
9041/373	2SC1368B
9041/373	2SC1368C
9041/373	2SC1368D
9041/373	2SC1567
9041/373	2SC1567A
9041/373	2SC1847
9041/373	2SC1847Q
9041/373	2SC495-R
9041/373	2SC495-Y
9041/373	2SC495T
9041/373	2SC496
9041/373	2SC496-O
9041/373	2SC496-R
9041/373	2SC496-Y
9041/373	2SD668
9042	2SA496
9042	2SA496-O
9042	2SA496-R
9042	2SA496-Y
9042	2SA496R
9042	2SA496Y
9042	2SA505
9042	2SA505-R
9042	2SA505-Y
9042	2SB548
9042	2SB549
9042	144033
9042/374	2SA715
9042/374	2SA715B
9042/374	2SA715C
9042/374	2SA715D
9042/374	2SA715WT
9042/374	2SA715WTB
9042/374	2SA715WTC
9042/374	2SA738
9042/374	2SA738B
9042/374	2SA738C
9042/374	2SA738D
9042/374	2SA743
9042/374	2SA743A
9042/374	2SA743AB
9042/374	2SA743AC
9042/374	2SA743B
9042/374	2SA743C
9042/374	2SA743D
9042/374	2SA794
9042/374	2SA886
9042/374	2SA886V
9042/374	2SB648
9042/374	2SB649
9067	SG323
9075	A04210-A
9075	A04231-A
9075	A04284-A
9075	A04716
9075	B30C1000
9075	BD-3A184
9075	BD3A-1B4
9075	BR-1
9075	BY164
9075	ECG166
9075	FW50
9075	JB00036
9075	KBP005
9075	KC08C2219
9075	KC2AP221B
9075	KC2D22/1
9075	M604
9075	MDA100
9075	MDA101
9075	MDA920-1
9075	PD1011
9075	PH25C221
9075	PH9-221
9075	PH9D5221
9075	PH9D522M
9075	PH9DS221
9075	PT6D22-1
9075	R204B
9075	RVD12B-1
9075	RVD2DP22/1B
9075	RVD2DP22/1C
9075	RVDD1245
9075	RVDR154B
9075	S1B0201B
9075	S1RB
9075	S1RB10
9075	S2PB
9075	SELEN-44
9075	SEN2A1
9075	SIB50B794-1

GE	Industry Standard No.	GE	Industry Standard No.	GE	Industry Standard No.	GE	Industry Standard No.	GE	Industry Standard No.
9075	SVD123281P-M	9091	13-29867-2						
9075	SVD81RB10	9091	103-131						
9075	T-E1042	9091	103-142						
9075	T10195	9091	103-295-02						
9075	TI-365A	9091	200X8010-165						
9075	TVM529	9091	903-00393						
9075	TVSW04	9091	119597						
9075	TVSW04M	9091	139706						
9075	W-005	9091	143837						
9075	WR011	9091	1471872-2						
9075	WR030	9091	412090001Q						
9075	WR040	9091/177	182462						
9075	ZA150	9092	ITT654						
9075	1B05J05	9092	L202						
9075	1B08T05	9092	9666PC						
9075	1B1	9093	ITT656						
9075	1N1054	9093	L203						
9075	23B-8851	9093	9667PC						
9075	3BB-20B01	9094	9668PC						
9075	4-2021-04170	9098	D2201A						
9075	7-0003	9098	D2201B						
9075	10B2-B1W	9098	D2201P						
9075	10DB1	9098	D2201N						
9075	11-102-001	9098	D2601M						
9075	11-102003	9098	34-8054-17						
9075	11-108002	9098	66X0054-001						
9075	1202P-114	9098	103-312						
9075	13-67539-1	9098	40642						
9075	13-67539-1/3464	9098	40643						
9075	160-4P	9098	40644						
9075	1604B1P	9098	126857						
9075	24MW192	9098	126858						
9075	48-97305A03	9098	126859						
9075	53B018-1	9098	1476171-12						
9075	53E011-1	9098	1476171-20						
9075	86-68-3	9098	1476171-24						
9075	86-68-3A	9098	1476171-27						
9075	130-30261	9098/515	D2600EF						
9075	325-0042-311	9098/515	RH-DX0106CEZZ						
9075	642-216	9098/515	44938						
9075	690V081H40	9098/515	131476						
9075	977-24B	X7X	23114074						
9075	977-4B								
9075	977-6B								
9075	992-531-01								
9075	04170								
9075	4822-130-30414								
9075	72174-1								
9075	0112945								
9075	125787								
9075	126413								
9075	489752-038								
9075	530120-1								
9075	530152-1								
9075	0551029								
9075	0551029H								
9075	771908								
9075	772714								
9075	984690								
9075	2330011								
9075	2330011H								
9075	2330361								
9075	7851329-01								
9075	7852577-01								
9075	7853099-01								
9075/166	FW100								
9075/166	09-307084								
9075/166	24MW669								
9075/166	46-67120A13								
9075/166	151-029-9-003								
9076	G05-406-C								
9076	28A634								
9076	28A634A								
9076	28A634C								
9076	28A634D								
9076	28A634K								
9076	28A634L								
9076	28A634M								
9076	28A636								
9076	28A636A								
9076	28A636B								
9076	28A636C								
9076	28A636D								
9076	28A636K								
9076	28A636L								
9076	28A636M								
9076	28A699								
9076	28A699-0								
9076	28A699A0								
9076	28A699AP								
9076	28A699AQ								
9076	28A699AR								
9076	28A699P								
9076	28A699Q								
9076	28A699R								
9076	04-07150-01								
9076	09-300073								
9076	48-41785J04								
9076	488155116								
9076	000071131								
9082A	MPT20								
9082A	1N5758								
9082A	1N5758A								
9083A	HEP311								
9083A	MPT28								
9083A	1N5760								
9083A	1N5761A								
9084A	A72-83-300								
9084A	1N5760A								
9084A	1N5761								
9086/518	K122176-1								
9086/518	RF32102R								
9086/518	212-48								
9091	A7275400								
9091	B1121R								
9091	ECG177								
9091	GE-300								
9091	KGE46465								
9091	TM177								
9091	1N3063								
9091	1N3064								
9091	1N3065								
9091	1N3066								
9091	1N3067								
9091	1N3068								
9091	1N3069								
9091	1N3069M								
9091	1N3070								
9091	1N3071								
9091	1S881								
9091	05-110442								
9091	05-180053								
9091	05-330150								

GE	Industry Standard No.
A14A	A14A
A15A	A15A
A15F	A15F
A40M	A40M
BD1	BD1
BR-1000	ECG168
BR-1000	GEBR-1000
BR-1000	KBL02
BR-206	CXL166
BR-206	CXL167
BR-206	ECG166
BR-206	ECG167
BR-206	S4VB10
BR-206	SD-W04
BR-600	A-0205
BR-600	AM-G-11
BR-600	BD-3A184
BR-600	BD3A-1B4
BR-600	BR-1
BR-600	FB-200
BR-600	FB200
BR-600	FW300
BR-600	HEP177
BR-600	HEPR0801
BR-600	HEPR0802
BR-600	HEPR0803
BR-600	HEPR0804
BR-600	KCO.8C22/1
BR-600	KC2AP22/1B
BR-600	KC2DP22/1B
BR-600	MA-110
BR-600	PA9D522/1
BR-600	PD1011
BR-600	PH9D522/1
BR-600	PT6D22/1
BR-600	RVD12B-1
BR-600	RVD2DP22/1C
BR-600	RVD8MB4
BR-600	RVDD124B
BR-600	RVDR154B
BR-600	S1B0201B
BR-600	S1RB
BR-600	SK3105
BR-600	SK3106
BR-600	SVD12B2B1P-M
BR-600	SVDS1RB10
BR-600	T10195
BR-600	TC0.09M22/1
BR-600	TVM529
BR-600	TVSW04
BR-600	TVSW04M
C103A	EC103A
C103B	C103B
C103B	EC103B
C103Y	C103Y
C103Y	EC103Y
C103YY	C103YY
C106A	ECG5412
C106A	ECG5413
C106A	ECG5422
C106A	ECG5423
C106A	ECG5432
C106A	ECG5433
C106A1	S106A
C106B	ECG5414
C106B1	S106B
C106B2	C106B2
C106C	C106C
C106C	ECG5456
C106C1	S106C
C106D	ECG5415
C106D1	MCR106-6
C106D1	S106D
C106D1	S2060D
C106E1	S2060E
C106F	ECG5421
C106F	ECG5431
C106F1	S106F
C106M1	MCR106-8
C106M1	S2060M
C106Y	ECG5411
C107D1	S2062D
C107M1	MCR107-8
C122A	S1008L
C122B	S2600B
C122B	S2620B
C122D	S2600D
C122D	S2620D
C122D	S4008L
C122M	S2600M
C122M	S2620M
C126C	S6000C
C126E	S6000E
C137E	C137E
C137M	C137M
C137N	C137N
C15B	C15B
C15D	C15D
C203D	EC103D
C228A	S1035H
C228D	S4035H
C228F	S0535H
C228M	S6035H
C228U	S0335H
C229A	S1035G
C229D	S4035G
C229M	S6035G
C229U	S0335G
C230A	S1025H
C230B	S2025H
C230B2	S6220B
C230D	C230D
C230D	S4025H
C230F	S0525H
C230M	S6025H
C230U	S0325H
C232A	S1025G
C232A	S6200A
C232B	S2025G
C232D	S4025G
C232F	S0525G
C232M	S6025G
C232U	S0325G
C35M	C35M
C38E	C38E
C611B	C611B
C6A	S1001M
C6B	S2001M
C6D	S4001M
C6F	S0301M
C6F	S0501M
CR-1	R1
CR-1	TV6.5
CR-2	A2458
CR-2	1S2329
CR-2	1SN1835
CR-3	SR-9005
D40C1	D40C1
D40C4	D40C4
D40C5	D40C5
D40C7	2N2723
D40C7	2N2724
D40E1	2N5711
D40E7	SDT5502
D40E7	SDT5507
D40E7	SDT5512
D40E7	2N3830
D40E7	2N4047
D40K2	D40K2
D40K2	MPSU45
D41K2	MPSU95
D44C10	2N6178
D44E3	TIP111
D44H4	2N6102
D44H4	2N6103
D44H7	2N6098
D44H7	2N6099
D44R1	2N5344
D45C10	2N6180
D45C8	D45C8
D45E3	TIP116
DZ-8.2	EDZ-24
FET-1	A066-110
FET-1	A068-106
FET-1	A068-107
FET-1	A11744
FET-1	A11745
FET-1	A192
FET-1	A514-040296
FET-1	A5T3821
FET-1	B-6001
FET-1	BF244A
FET-1	BF244B
FET-1	BF244C
FET-1	BF256
FET-1	BFS68
FET-1	BFS68P
FET-1	BFW10
FET-1	BFW11
FET-1	BFW12
FET-1	BFW54
FET-1	BFW55
FET-1	BFW56
FET-1	BFW61
FET-1	CDC731
FET-1	CXL132
FET-1	CXL133
FET-1	D1101
FET-1	D1102
FET-1	D1177
FET-1	D1178
FET-1	D1180
FET-1	D1181
FET-1	D1420
FET-1	D1421
FET-1	DDCY002002
FET-1	DS-88
FET-1	DS88
FET-1	E103
FET-1	EA15X165
FET-1	EA15X169
FET-1	EA15X192
FET-1	EA15X193
FET-1	EA15X446
FET-1	EA3278
FET-1	ECG132
FET-1	EF2(ELCOM)
FET-1	EF3
FET-1	EL131
FET-1	EQF-3
FET-1	EQF-4
FET-1	ES15X92
FET-1	ET491051
FET-1	F1462
FET-1	F1463
FET-1	FT34Y
FET-1	G08-005L
FET-1	G08005L
FET-1	GE-FET-1
FET-1	G08-005L
FET-1	GRF-3
FET-1	HEP801
FET-1	HEP802
FET-1	HEPF0015
FET-1	HF2000411B
FET-1	HF200191A
FET-1	HF200191A-0
FET-1	HF200191A/
FET-1	HF200191B-0
FET-1	HF200191B0
FET-1	HF200301B
FET-1	HF200301B0
FET-1	HF200301B0
FET-1	HF200301C-0
FET-1	HF200301E
FET-1	HF200411B
FET-1	HK-00330
FET-1	IC743046
FET-1	IP20-0011
FET-1	IP20-0012
FET-1	IRTRFE100
FET-1	JE1033B
FET-1	JF1033S
FET-1	K10
FET-1	K11(1-GATE)
FET-1	K11-0(1-GATE)
FET-1	K11-R(1-GATE)
FET-1	K11-Y(1-GATE)
FET-1	K12(1-GATE)
FET-1	K12-3(1-GATE)
FET-1	K12-GR(1-GATE)
FET-1	K12-R(1-GATE)
FET-1	K120Y(1-GATE)
FET-1	K13(1-GATE)
FET-1	K15-0(1-GATE)
FET-1	K15-GR-(1-GATE)
FET-1	K15-R(1-GATE)
FET-1	K15-Y(1-GATE)
FET-1	K17
FET-1	K17(1-GATE)
FET-1	K170(1-GATE)
FET-1	K170R(1-GATE)
FET-1	K17A(1-GATE)
FET-1	K17B(1-GATE)
FET-1	K17BL(1-GATE)
FET-1	K17GR(1-GATE)
FET-1	K17R(1-GATE)
FET-1	K17Y(1-GATE)
FET-1	K19
FET-1	K19(1-GATE)
FET-1	K19(GR)
FET-1	K19BL
FET-1	K19BL(1-GATE)
FET-1	K19GC
FET-1	K19GC(1-GATE)
FET-1	K19GE(1-GATE)
FET-1	K19GR
FET-1	K19GR(1-GATE)
FET-1	K19Y
FET-1	K19Y(1-GATE)
FET-1	K22(1-GATE)
FET-1	K22-Y(1-GATE)
FET-1	K22Y(1-GATE)
FET-1	K25C(1-GATE)
FET-1	K25D(1-GATE)
FET-1	K25E(1-GATE)
FET-1	K25ET(1-GATE)
FET-1	K25F(1-GATE)
FET-1	K25G(1-GATE)
FET-1	K30-0(1-GATE)
FET-1	K30A(1-GATE)
FET-1	K30AD(1-GATE)
FET-1	K30AGR(1-GATE)
FET-1	K30B(1-GATE)
FET-1	K30C(1-GATE)
FET-1	K30D(1-GATE)
FET-1	K30GR(1-GATE)
FET-1	K30R(1-GATE)
FET-1	K30Y
FET-1	K30Y(1-GATE)
FET-1	K31(1-GATE)
FET-1	K31C
FET-1	K31C(1-GATE)
FET-1	K33(1-GATE)
FET-1	K33(E)
FET-1	K33E
FET-1	K33E(1-GATE)
FET-1	K33F
FET-1	K33F(1-GATE)
FET-1	K33GR
FET-1	K33GR(1-GATE)
FET-1	K34
FET-1	K34(1-GATE)
FET-1	K34(E)
FET-1	K34B(1-GATE)
FET-1	K34C
FET-1	K34C(1-GATE)
FET-1	K34D
FET-1	K34D(1-GATE)
FET-1	K34E
FET-1	K37(1-GATE)
FET-1	K37H(1-GATE)
FET-1	K37L(1-GATE)
FET-1	K42
FET-1	K47
FET-1	K47M
FET-1	K47M(1-GATE)
FET-1	K49
FET-1	K49F
FET-1	K49F(1-GATE)
FET-1	K49H
FET-1	K49H(1-GATE)
FET-1	K49HK(1-GATE)
FET-1	K49I
FET-1	K49I(1-GATE)
FET-1	K49M(1-GATE)
FET-1	KA4559
FET-1	LS5484
FET-1	LS5485
FET-1	MJF10335
FET-1	MK-10
FET-1	MK-10-2
FET-1	MK-10-E
FET-1	MK10
FET-1	MK10-2
FET-1	MK5485
FET-1	MPF-102
FET-1	MPF101
FET-1	MPF102
FET-1	MPF105
FET-1	MPF106
FET-1	MPF107
FET-1	MPF108
FET-1	MPF109
FET-1	MPF111
FET-1	MPF112
FET-1	MPS5668
FET-1	NR7916
FET-1	NS316
FET-1	P1087
FET-1	PL1091
FET-1	PL1092
FET-1	PL1093
FET-1	PL1094
FET-1	PTC151
FET-1	Q-00169
FET-1	Q-00169A
FET-1	Q-00169B
FET-1	Q-00169C
FET-1	Q-00184R
FET-1	QA-18
FET-1	QA-20
FET-1	QRF-3
FET-1	QRF3
FET-1	QRG-3
FET-1	RE46
FET-1	REN132
FET-1	RS-2028
FET-1	RS7916
FET-1	RT-176
FET-1	RT175
FET-1	RT176
FET-1	RT8331
FET-1	RT8667
FET-1	RVTMK10-2
FET-1	RVTMK10-E
FET-1	S1241N
FET-1	S1242N
FET-1	SES3819
FET-1	SFE145
FET-1	SJ1925
FET-1	SK19
FET-1	SK3112
FET-1	SS-3704
FET-1	SS3534-4
FET-1	SS3704
FET-1	SS3735
FET-1	SX3819
FET-1	T1-741
FET-1	T1208
FET-1	T1S34
FET-1	T1S58
FET-1	T1S59
FET-1	T1S88
FET-1	TE-500-E
FET-1	TI-741
FET-1	TI741
FET-1	TIS-88
FET-1	TIS14
FET-1	TIS34
FET-1	TIS42
FET-1	TIS58
FET-1	TIS59
FET-1	TIS88
FET-1	TR-U1650E
FET-1	TR-U1835E
FET-1	TR06011
FET-1	TR06014
FET-1	TR228735120325
FET-1	TR5528
FET-1	TUS34
FET-1	TV-83
FET-1	TV80
FET-1	001-02701-1
FET-1	001-02702-0
FET-1	001-027030
FET-1	01-070030
FET-1	1W1706
FET-1	2D031
FET-1	2N3066
FET-1	2N3067
FET-1	2N3069
FET-1	2N3070
FET-1	2N3071
FET-1	2N3084
FET-1	2N3085
FET-1	2N3086
FET-1	2N3087
FET-1	2N3088
FET-1	2N3089
FET-1	2N3365
FET-1	2N3366
FET-1	2N3367
FET-1	2N3368
FET-1	2N3369
FET-1	2N3370
FET-1	2N3436
FET-1	2N3437
FET-1	2N3438
FET-1	2N3452
FET-1	2N3453
FET-1	2N3454
FET-1	2N3455
FET-1	2N3456
FET-1	2N3457
FET-1	2N3458
FET-1	2N3459
FET-1	2N3460
FET-1	2N3465
FET-1	2N3466
FET-1	2N3608
FET-1	2N3684
FET-1	2N3685
FET-1	2N3686
FET-1	2N3687
FET-1	2N3819
FET-1	2N3822
FET-1	2N3823
FET-1	2N3966
FET-1	2N3967
FET-1	2N3968
FET-1	2N3969
FET-1	2N4117
FET-1	2N4118
FET-1	2N4119
FET-1	2N4220
FET-1	2N4222
FET-1	2N4224
FET-1	2N4302
FET-1	2N4303
FET-1	2N4304
FET-1	2N4338
FET-1	2N4339
FET-1	2N4340
FET-1	2N4341
FET-1	2N4867
FET-1	2N4868
FET-1	2N4869
FET-1	2N5045
FET-1	2N5046
FET-1	2N5047
FET-1	2N5163
FET-1	2N5248
FET-1	2N5267
FET-1	2N5268
FET-1	2N5269
FET-1	2N5270
FET-1	2N5277
FET-1	2N5358
FET-1	2N5359
FET-1	2N5360
FET-1	2N5361
FET-1	2N5362
FET-1	2N5363
FET-1	2N5364
FET-1	2N5391
FET-1	2N5392
FET-1	2N5393
FET-1	2N5394
FET-1	2N5395
FET-1	2N5396
FET-1	2N5452
FET-1	2N5453
FET-1	2N5454
FET-1	2N5457
FET-1	2N5458
FET-1	2N5459
FET-1	2N5543
FET-1	2N5544
FET-1	2N5556
FET-1	2N5557
FET-1	2N5558
FET-1	2N5561
FET-1	2N5562
FET-1	2N5563
FET-1	2N5648
FET-1	2N5649
FET-1	2N5668
FET-1	2N5716
FET-1	2N5717
FET-1	2N5718
FET-2	B9TUI
FET-2	BF245A
FET-2	DDCY001002
FET-2	DDCY006001
FET-2	E300
FET-2	EA15X394
FET-2	EA15X400
FET-2	EA15X401
FET-2	ECG133
FET-2	ECG312
FET-2	EP15X92
FET-2	EQF-0009
FET-2	FE-100
FET-2	FE100
FET-2	GE-FET-2
FET-2	HEPF0010
FET-2	HEPF0021
FET-2	HF200191A0
FET-2	HF200191A0
FET-2	HF200301C-0
FET-2	HF200411C0
FET-2	HK-00049
FET-2	IP20-0010
FET-2	IP20-0035
FET-2	IP20-0078
FET-2	IT108
FET-2	ITE4416
FET-2	J308
FET-2	JF-1033
FET-2	JF1033
FET-2	JF1033B
FET-2	JF1033G
FET-2	K55(1-GATE)
FET-2	K55D(1-GATE)
FET-2	K55E(1-GATE)
FET-2	KE4416
FET-2	KE5105
FET-2	MJF1033G
FET-2	MK102
FET-2	MPF-106
FET-2	MPF103
FET-2	MPF104
FET-2	PTC152
FET-2	QKT-0033XBE
FET-2	QKT0033XBE
FET-2	QT-K0023AAS
FET-2	QT-K0033XBE
FET-2	RE45
FET-2	RT-175
FET-2	RT-8667
FET-2	SK3531
FET-2	TE500
FET-2	TR-FE100
FET-2	TR-U1650E-1
FET-2	TR2083-70
FET-2	TR2SK55
FET-2	TR06011
FET-2	TV83
FET-2	001-02703-0
FET-2	2N4223
FET-2	2N4416
FET-2	2N4417
FET-2	2N5078
FET-2	2N5078A
FET-2	2N5103A
FET-2	2N5104A
FET-2	2N5105A
FET-2	2N5163A
FET-2	2N5245
FET-2	2N5245A
FET-2	2N5246
FET-2	2N5246A
FET-2	2N5247
FET-2	2N5247A
FET-2	2N5248A
FET-2	2N5258
FET-2	2N5278A
FET-2	2N5360A
FET-2	2N5397
FET-2	2N5398
FET-2	2N5484
FET-2	2N5485
FET-2	2N5485-1
FET-2	2N5486
FET-2	2N5555
FET-2	2N5592
FET-2	2N5593
FET-2	2N5594
FET-2	2N5669
FET-2	2N5670
FET-2	2N5949
FET-2	2N5950
FET-2	2N5951
FET-2	2N5952
FET-2	2N5953
FET-2	2NJ233B
FET-3	CXL220
FET-3	CXL221
FET-3	EA15X238
FET-3	ECG220
FET-3	GE-FET-3
FET-3	HEPF2005
FET-3	IP20-0305
FET-3	SK3116
FET-4	AR501
FET-4	AR502
FET-4	CXL222
FET-4	DDCY103001
FET-4	DDCY104001
FET-4	DDCY104003
FET-4	DS-102
FET-4	DS-105
FET-4	DS-106
FET-4	EA15X402
FET-4	EA15X405
FET-4	ECG222
FET-4	EP15X36
FET-4	EP15X40
FET-4	GE-FET-4
FET-4	HEPF2004
FET-4	HEPF2007
FET-4	HEPF2007A
FET-4	IP20-0157

GE	Industry Standard No.
FET-4	IP20-0218
FET-4	MEM564C
FET-4	MEM630
FET-4	MEM680Y
FET-4	MFE-3008
FET-4	MFE121
FET-4	MFE130
FET-4	MFE130-712
FET-4	MFE131
FET-4	MFP121
FET-4	MOS3635
FET-4	MPF121
FET-4	RE199
FET-4	SFC-1616
FET-4	SFC-1617
FET-4	SFC8999
FET-4	SFD2285
FET-4	SFE253
FET-4	SFE303
FET-4	SFE303424
FET-4	SFE427
FET-4	SK3050
FET-4	SK3050/221
FET-4	SK3065
FET-4	SK3065/222
FET-4	SPF274
FET-4	SPF512
FET-4	SPF609
FET-4	T1731
FET-4	T464
FET-4	T513
FET-4	TR08004
FET-4	TR2327431
FET-4	01-080045
H11A1	H11A1
H11A1	IL5
H11A2	H11A2
H11A2	MCT2
H11A2	OPI2152
H11A3	H11A3
H11A3	IL1
H11A3	MCT2E
H11A3	OPI2252
H11A3	TIL114
H11A3	TIL116
H11A4	OPI2151
H11A4	TIL111
H11A5	IL12
H11A5	MCT26
H11B1	CL110
H11B1	H11B1
H11B2	H11B2
H11B2	TIL113
H11C1	H11C1
H11C2	H11C2
H11C3	H11C3
H11C3	MCS2
H11C4	H11C4
H11C5	H11C5
H11C6	H11C6
H11C6	MCS2400
IC-10	CA2111AE
IC-10	CA2111AQ
IC-10	LM2111N
IC-10	SN76653N
IC-11	SN76642N
IC-115	LM1351N
IC-12	LM703LN
IC-13	CA3120E
IC-148	LM3065N
IC-148	SN76666N
IC-15	LM3075N
IC-15	SN76675N
IC-16	LM1841N
IC-172	CA3045
IC-172	CA3045F
IC-172	CA3046
IC-172	CA3086
IC-172	LM3086N
IC-179	BO319200
IC-179	EA33X8389
IC-179	EA33X8396
IC-179	ECG1155
IC-179	EICM-0060
IC-179	ETI-23
IC-179	GEIC-179
IC-179	IP20-0161
IC-179	KIA7205AP
IC-179	KIA7205P
IC-179	QQ-M07205AT
IC-179	QQM07205AT
IC-179	REN1155
IC-179	SK3231
IC-179	TA-7205P
IC-179	TA7205
IC-179	TA7205A
IC-179	TA7205AP
IC-179	TA7205P
IC-179	TM1155
IC-179	TVCM-81
IC-179	02-257205
IC-18	LM746N
IC-180	CA3001
IC-193	GEIC-193
IC-193	M53274P
IC-193	TVCM-502
IC-194	J4-1002
IC-2	SN76665N
IC-205	CA3013
IC-205	CA3014
IC-207	CA3044
IC-207	CA3044V1
IC-21	LM3064N
IC-210	CA3048
IC-210	CA3052
IC-214	CA758E
IC-214	LM1800N
IC-215	CA3123E
IC-215	DM-20
IC-215	DM20
IC-215	DM32
IC-215	GEIC-215
IC-215	IC-607(ELCOM)
IC-215	SK3171
IC-217	SN76600P
IC-218	ITT1330
IC-22	CA3066
IC-220	CA1458S
IC-220	LM1458N
IC-220	N5558V
IC-220	SN72558P
IC-222	CA3064
IC-223	CA3076
IC-225	ITT3064
IC-225	RE313-IC
IC-225	SN76564N
IC-226	CA3035
IC-226	CA3035V1
IC-226	CA3065
IC-227	CA3043
IC-23	CA3067
IC-234	CA3125E
IC-236	CA3020
IC-236	CA3020A
IC-24	DM-32
IC-24	ECG744
IC-24	GEIC-24
IC-24	HA1139A
IC-24	LM1820
IC-24	LM1820A
IC-24	LM1820N
IC-24	PTC734
IC-24	REN744
IC-24	SK3171/744
IC-24	SL23971
IC-24	TM744
IC-24	TVCM-19
IC-244	SN76591P
IC-245	CA3000
IC-246	CA3026
IC-248	CA3029
IC-248	CA3030
IC-248	CA3030A
IC-248	CA3037
IC-248	CA3037A
IC-248	CA3038
IC-248	CA3038A
IC-249	LM709CH
IC-249	N5709T
IC-249	RC709T
IC-249	SK3552
IC-249	SN72709L
IC-250	LM709CN
IC-250	N5709A
IC-250	SN72709N
IC-251	LM710CH
IC-251	N5710T
IC-251	RC710T
IC-251	SG710CT
IC-251	SN72710L
IC-252	LM710CN
IC-252	N5710A
IC-252	SG710CD
IC-252	SG710CN
IC-252	SN72710N
IC-253	LM711CH
IC-253	N5711K
IC-253	RC711T
IC-253	SG711CT
IC-253	SN72711L
IC-254	LM711CN
IC-254	N5711A
IC-254	SG711CD
IC-254	SG711CN
IC-254	SN72711N
IC-255	CA3047
IC-255	CA3047A
IC-255	CA3056A
IC-255	CA3741
IC-255	CA3741CS
IC-255	CA3741S
IC-255	CA3741T
IC-256	CA3058
IC-256	CA3059
IC-256	CA3079
IC-257	SK3551
IC-258	CA3054
IC-259	AMU5R7723393
IC-259	SG723CT
IC-260	AMU6A7723393
IC-260	ECG923D
IC-260	FU6A7723393
IC-260	FU9A7723393
IC-260	G390507S2
IC-260	GEIC-260
IC-260	HA17723
IC-260	IC-20
IC-260	IC-20(PHILCO)
IC-260	IC-53(ELCOM)
IC-260	IC-531(ELCOM)
IC-260	IC53(ELCOM)
IC-260	ITT723(D.I.P.)
IC-260	LM723C
IC-260	LM723CD
IC-260	MC1723CL
IC-260	N5723A
IC-260	PS-801
IC-260	RB6-PS-801
IC-260	SG723CD
IC-260	SG723CN
IC-260	SK3165
IC-260	SK3165/923D
IC-260	SL21385
IC-260	SL22310
IC-260	SL22935
IC-260	SL23325
IC-260	TDB0723A
IC-263	A1820-0203-1
IC-263	A1826-0007-1
IC-263	AD741CH
IC-263	AMU5B7741393
IC-263	CA3741CT
IC-263	CA741CT
IC-263	ECG941
IC-263	FU5B7741393
IC-263	GEIC-263
IC-263	HA17741M
IC-263	HC1000217
IC-263	HL24510
IC-263	HL24593
IC-263	IC-40(ELCOM)
IC-263	IC317(ELCOM)
IC-263	IC40(ELCOM)
IC-263	LM741
IC-263	LM741C
IC-263	LM741CH
IC-263	LM741H
IC-263	M5141T
IC-263	MC1741CG
IC-263	ON187840
IC-263	N5741T
IC-263	PA7001/503
IC-263	PA7741
IC-263	PA7741C
IC-263	RC741T
IC-263	SC5175G
IC-263	SC741T
IC-263	SG741CT
IC-263	SK3514
IC-263	SK3514/941
IC-263	SL20929
IC-263	SL21673
IC-263	SL23486
IC-263	SMC750123-1
IC-263	SN72741L
IC-263	TBA221
IC-263	TBA221/741C
IC-263	TBA222
IC-263	TBA222/741
IC-263	TM941
IC-264	AMU6A7741393
IC-264	LM741CN
IC-264	N5741A
IC-264	SG741CD
IC-264	SG741CN
IC-264	SN72741N
IC-265	A1820-0219-1
IC-265	C6052P
IC-265	CA741CS
IC-265	ECG941M
IC-265	FU9T7741393
IC-265	GEIC-265
IC-265	HEP-C6052P
IC-265	HEPC6052P
IC-265	IC-295(ELCOM)
IC-265	J4-1215
IC-265	KS-21177
IC-265	MC1741CP1
IC-265	N5741V
IC-265	SG741CM
IC-265	SK3552/941M
IC-265	SL22745
IC-265	SL23059
IC-265	SL23252
IC-265	SL23496
IC-265	SN72471P
IC-265	SN72741P
IC-265	STX49007
IC-265	T3B
IC-265	TA7504P
IC-265	TM941M
IC-267	CA747CT
IC-267	LM747CH
IC-268	AMU5B7747393
IC-268	CA747CE
IC-268	CA747T
IC-268	SG747CT
IC-269	ECG955M
IC-269	GEIC-269
IC-269	J4-1555
IC-269	M51841P
IC-269	MC1455P
IC-269	MC1455P1
IC-269	SK3564
IC-269	SK3564/955M
IC-269	TM955M
IC-28	SN76131N
IC-282	QQ-MBA521AX
IC-288	CA3010
IC-288	CA3015
IC-29	CA1398E
IC-316	BA521
IC-316	ECG1166
IC-316	GEIC-316
IC-316	REN1166
IC-316	TM1166
IC-333	TDA1190Z
IC-334	TDA2002
IC-35	CA1310E
IC-35	LM1310N
IC-35	MC1310A
IC-35	SN76115N
IC-4	CA3070
IC-4	LM3070N
IC-4	N5070B
IC-4	SN76242N
IC-4011	TC4011BP
IC-43	CA3041
IC-5	CA3072
IC-5	N5072A
IC-5	SN76246N
IC-6	CA3071
IC-6	LM3071N
IC-6	LM307N
IC-6	N5071A
IC-6	SN76243N
IC-7	SN76105N
IC-8	SN76104N
IC-86	CA3028A
IC-86	CA3028AF
IC-86	CA3028AS
IC-86	CA3028B
IC-86	CA3028BF
IC-86	CA3028BS
IC-86	CA3053
IC-89	CA3042
IC-97	CA1352E
IC-97	SN76650N
IN60	RT6728
L14F1	L14F1
L14G1	TIL81
L14H1	L14H1
LED55B	OP131
LED55B	TIL31
LED56	OP130
LED56	TIL34
MR-1	1N1538
MR-1	1N333
MR-1	1N335
MR-1	1N347
MR-1	1S151
MR-2	A40C
MR-2	A40D
MR-2	1N2156
MR-2	1N2449
MR-3	C20A
MR-3	C20B
MR-3	C20D
MR-3	C20F
MR-3	C20U
MR-3	C30A
MR-3	C30B
MR-3	C30C
MR-3	C30D
MR-3	C30F
MR-3	C30U
MR-3	C35A
MR-3	C35B
MR-3	C35C
MR-3	C35D
MR-3	C35F
MR-3	C35G
MR-3	C35U
MR-3	C36A
MR-3	C36B
MR-3	C36C
MR-3	C36D
MR-3	C36F
MR-3	C36G
MR-3	C36U
MR-3	C37A
MR-3	C37B
MR-3	C37C
MR-3	C37D
MR-3	C37F
MR-3	C37G
MR-3	C37U
MR-3	2N1842
MR-3	2N1842A
MR-3	2N1843
MR-3	2N1843A
MR-3	2N1844
MR-3	2N1844A
MR-3	2N1845
MR-3	2N1845A
MR-3	2N1846
MR-3	2N1846A
MR-3	2N1847
MR-3	2N1847A
MR-3	2N1848
MR-3	2N1848A
MR-3	2N1849
MR-3	2N1849A
MR-3	2N3896
MR-3	2N3897
MR-3	2N3898
MR-3	2N5168
MR-3	2N5169
MR-3	2N5170
MR-3	2N681
MR-3	2N682
MR-3	2N683
MR-3	2N684
MR-3	2N685
MR-3	2N686
MR-3	2N687
MR-3	2N688
MR-4	MCR2604-1
MR-4	MCR2604-2
MR-4	MCR2604-3
MR-4	MCR2604-5
MR-4	2N3228
MR-4	2N3525
MR-4	2N4167
MR-4	2N4168
MR-4	2N4169
MR-4	2N4170
MR-4	2N4171
MR-4	2N4172
MR-4	2N4441
MR-4	2N4443
MR-5	C106A
MR-5	C106B
MR-5	C106F
MR-5	C106Y
MR-5	CXL5452
MR-5	CXL5453
MR-5	CXL5454
MR-5	CXL5455
MR-5	ECG5452
MR-5	ECG5453
MR-5	ECG5454
MR-5	ECG5455
MR-5	GEMR-5
MR-5	IRC20
MR-5	RE-168
MR-5	RE-171
MR-5	RE168
MR-5	RE171
MR-5	SK3597
MR-5	TA2888
MR-5	TA2889
MR-5	2N1595
MR-5	2N1596
MR-5	2N1597
MR-5	2N2322
MR-5	2N2323
MR-5	2N2325
MR-5	2N2326
MR-5	2N3528
MR-5	2N5060
MR-5	2N5061
MR-5	2N5062
MR-5	2N5063
MR-6	GEMR-6
S6012	2N6012
SC141B	ECG5600
SC141B	ECG5601
SC141B	ECG5602
SC141B	ECG5603
SC141B	SC141B
SC141D	ECG5604
SC141D	ECG5605
SC141D	SC141D
SC141M	T2801M
SC143M	T2802M
SC146B	SC146B
SC146D	SC146D
SC146M	SC146M
SC149B	TIC236B
SC149D	TIC236D
SC151B	TIC246B
SC151B2	T4700B
SC151D	SC151D
SC151D	TIC246D
SC151D2	T4700D
SC151M	SC151M
SC240B	SC240B
SC240D	SC240D
SC245B	SC245B
SC245D	SC245D
SC250B	Q2015H
SC250B	SC250B
SC250B4	T4140B
SC250B6	T4150B
SC250D	Q4015H
SC250D	SC250D
SC250D4	T4140D
SC250D6	T4150D
SC250M	Q6015H
SC260B	Q2025H
SC260B	SC260B
SC260D	Q4025H
SC260D	SC260D
SC260E	Q5025H
SC260M	Q6025H
SC265B2	MAC40688
SC265B2	Q2040D
SC265D2	MAC40689
SC265M	SC265M
SC265M2	MAC40690
SP-1519	EP15X29
SP-1520	EP15X30
VR-100	LM78L05ACZ
VR-101	LM341P-5.0
VR-102	LM340T-5.0
VR-106	LM78L08ACZ
VR-107	LM341P-8.0
VR-108	LM340T-8.0
VR-109	LM78L12ACZ
VR-110	LM341P-12
VR-111	LM340T-12
VR-114	LM320T-12
VR-115	LM723CH
X10	GE-X10
X11	CD3112018
X11	DZ-081
X11	RD-8.2#E
X11	SD-8
X11	1ZD8.2
X11	1ZD892
X13	GE-X13
X13	RE190
X17	D13T1
X17	2N6027
X18	GE-X18
X5	GE-X5
X8	GE-X8
X9	GE-X9
ZB-6.2	HZ6C
ZD-10	.25N10
ZD-10	.25T10
ZD-10	.25T10.5
ZD-10	.25T10.5A
ZD-10	.25T10A
ZD-10	.25T10B
ZD-10	.4T10
ZD-10	.4T10A
ZD-10	.4T10B
ZD-10	.7JZ10
ZD-10	.7Z10A
ZD-10	.7Z10B
ZD-10	.7Z10C
ZD-10	.7Z10D
ZD-10	.7ZM10A
ZD-10	.7ZM10B
ZD-10	.7ZM10C
ZD-10	.7ZM10D
ZD-10	A42X00041-01
ZD-10	AW-01-10
ZD-10	AW01-10
ZD-10	AZ-094
ZD-10	AZ-096
ZD-10	AZ-098
ZD-10	AZ-100
ZD-10	AZ10
ZD-10	AZ758
ZD-10	AZ758A
ZD-10	AZ961
ZD-10	AZ961A
ZD-10	AZ961B
ZD-10	B094
ZD-10	B100(ZENER)
ZD-10	B66X0040-005
ZD-10	B094
ZD-10	BZ-096
ZD-10	BZ-098
ZD-10	BZ-100
ZD-10	BZ79C10
ZD-10	BZX29C10
ZD-10	BZX46C10
ZD-10	BZX55C10
ZD-10	BZX71C10
ZD-10	BZX79C10
ZD-10	BZX83C10
ZD-10	BZY83/C10
ZD-10	BZY83/D10
ZD-10	BZY83C10
ZD-10	BZY83D10
ZD-10	BZY85C10
ZD-10	BZY85D10
ZD-10	BZY88C10
ZD-10	BZY92C10
ZD-10	BZY94C10
ZD-10	C4016
ZD-10	CHZ10
ZD-10	CHZ10A
ZD-10	CHZ210A
ZD-10	CX0051
ZD-10	CXL140
ZD-10	CXL5019
ZD-10	CZD010
ZD-10	CZD010-5
ZD-10	DZ10A
ZD-10	E84
ZD-10	EA16X150
ZD-10	EA16X157
ZD-10	ECG140
ZD-10	ECG140A
ZD-10	ECG5019A
ZD-10	EQA01-10S
ZD-10	EQB01-10
ZD-10	EZ10(ELCOM)
ZD-10	G01-012-F
ZD-10	GEZD-10
ZD-10	GLA100
ZD-10	GLA100A
ZD-10	GLA100B
ZD-10	HD30001200
ZD-10	HEP-Z0220
ZD-10	HEP-Z0413
ZD-10	HEP101
ZD-10	HEPZ0413
ZD-10	HM10
ZD-10	HM10A
ZD-10	HM10B
ZD-10	HN-00061
ZD-10	HS2100
ZD-10	HS7100
ZD-10	HW10
ZD-10	HW10A
ZD-10	HZ-11A
ZD-10	HZ11A
ZD-10	HZ11B
ZD-10	IP20-0233
ZD-10	J24872
ZD-10	KD2501
ZD-10	KS100A
ZD-10	KS100B
ZD-10	KS2100A

GE	Industry Standard No.
ZD-10	KS2100B
ZD-10	KS42A
ZD-10	KS42AF
ZD-10	KS42B
ZD-10	KS42BF
ZD-10	KVR10
ZD-10	LMZX-10
ZD-10	LR100CH
ZD-10	LZ10
ZD-10	M10Z
ZD-10	M4Z10
ZD-10	M4Z10-20
ZD-10	M4Z10A
ZD-10	MA1100
ZD-10	MC6014
ZD-10	MC6014A
ZD-10	MGLA100
ZD-10	MGLA100A
ZD-10	MGLA100B
ZD-10	MR100C-H
ZD-10	MR100H
ZD-10	MTZ614
ZD-10	MTZ614A
ZD-10	MZ-10
ZD-10	MZ-210
ZD-10	MZ-210B
ZD-10	MZ10
ZD-10	MZ1000-13
ZD-10	MZ1010
ZD-10	MZ10A
ZD-10	MZ310A
ZD-10	MZ500-16
ZD-10	MZ92-10A
ZD-10	O2Z-10A
ZD-10	OA12610
ZD-10	OZ10T10
ZD-10	OZ10T5
ZD-10	PC1879-004
ZD-10	PD6014
ZD-10	PD6014A
ZD-10	PD6055
ZD-10	PS1511
ZD-10	PS1512
ZD-10	PS1513
ZD-10	PS1514
ZD-10	PS1515
ZD-10	PS1516
ZD-10	PS1517
ZD-10	PS8913
ZD-10	PTC506
ZD-10	QZ10T10
ZD-10	QZ10T5
ZD-10	RA-26
ZD-10	RA26(ZENER)
ZD-10	RD-10EB
ZD-10	RD-11B
ZD-10	RD-9E
ZD-10	RD10E
ZD-10	RD10EA
ZD-10	RD10EB
ZD-10	RD10EB-2
ZD-10	RD10F
ZD-10	RD10FA
ZD-10	RD10FB
ZD-10	RD10H
ZD-10	RD10M
ZD-10	RD11A
ZD-10	RD11AL
ZD-10	RD11AM
ZD-10	RE-115
ZD-10	RE115
ZD-10	REN140
ZD-10	RH-EX0021TAZZ
ZD-10	RT-241
ZD-10	RVD1N4740
ZD-10	RZ10
ZD-10	RZZ10
ZD-10	S6-10
ZD-10	S6-10A
ZD-10	S6-10B
ZD-10	S6-10C
ZD-10	SD27
ZD-10	SK3061
ZD-10	SK3061/140A
ZD-10	SV133
ZD-10	SZ-200-10
ZD-10	SZ10
ZD-10	SZL9
ZD-10	TIXD758
ZD-10	TMD08
ZD-10	TMD08A
ZD-10	TR228736003026
ZD-10	TR330028
ZD-10	TVSRA-26
ZD-10	TVSRD11
ZD-10	TVSRD11A
ZD-10	TVSRM26V
ZD-10	001-0127-00
ZD-10	001-023035
ZD-10	1-20398
ZD-10	1/2Z105T5
ZD-10	1/2Z10T5
ZD-10	1/2Z33T5
ZD-10	1/4A10
ZD-10	1/4A10A
ZD-10	1/4A10B
ZD-10	1A10M
ZD-10	1A10MA
ZD-10	1E10Z
ZD-10	1E10Z10
ZD-10	1M10Z
ZD-10	1M10Z10
ZD-10	1M10Z5
ZD-10	1M10ZS10
ZD-10	1M10ZS5
ZD-10	1N1512
ZD-10	1N1512A
ZD-10	1N1523
ZD-10	1N1523A
ZD-10	1N1744
ZD-10	1N1771
ZD-10	1N1876
ZD-10	1N1876A
ZD-10	1N1932
ZD-10	1N1932A
ZD-10	1N1932B
ZD-10	1N1959
ZD-10	1N1986
ZD-10	1N1986A
ZD-10	1N1986B
ZD-10	1N2036
ZD-10	1N2036A
ZD-10	1N226
ZD-10	1N226A
ZD-10	1N3020
ZD-10	1N3020A
ZD-10	1N3020B
ZD-10	1N3402
ZD-10	1N3415
ZD-10	1N3434
ZD-10	1N3446
ZD-10	1N3518
ZD-10	1N3679
ZD-10	1N3679A
ZD-10	1N3776
ZD-10	1N4095
ZD-10	1N4104
ZD-10	1N4162
ZD-10	1N4162A
ZD-10	1N4295
ZD-10	1N4295A
ZD-10	1N4296
ZD-10	1N4296A
ZD-10	1N4327
ZD-10	1N4327A
ZD-10	1N4404
ZD-10	1N4404A
ZD-10	1N4632
ZD-10	1N4661
ZD-10	1N4697
ZD-10	1N4740
ZD-10	1N4740A
ZD-10	1N4832A
ZD-10	1N5240
ZD-10	1N5240A
ZD-10	1N5240B
ZD-10	1N5348B
ZD-10	1N5530
ZD-10	1N5530A
ZD-10	1N5530B
ZD-10	1N5563
ZD-10	1N5563A
ZD-10	1N5736
ZD-10	1N5856
ZD-10	1N5856A
ZD-10	1N5856B
ZD-10	1N701
ZD-10	1N714
ZD-10	1N714A
ZD-10	1N714B
ZD-10	1N754
ZD-10	1N758
ZD-10	1N758A
ZD-10	1N765-1
ZD-10	1N765A
ZD-10	1N961
ZD-10	1N961A
ZD-10	1N961B
ZD-10	1R10
ZD-10	1R10A
ZD-10	1R10B
ZD-10	1S1165
ZD-10	1S1378
ZD-10	1S1393
ZD-10	1S1393A
ZD-10	1S1409
ZD-10	1S1409R
ZD-10	1S1718/4454C
ZD-10	1S1960
ZD-10	1S2100
ZD-10	1S2100A
ZD-10	1S2117
ZD-10	1S2117A
ZD-10	1S215
ZD-10	1S216
ZD-10	1S225
ZD-10	1S2494
ZD-10	1S2551
ZD-10	1S3010
ZD-10	1S335
ZD-10	1S618
ZD-10	1S695
ZD-10	1S7100
ZD-10	1S7100A
ZD-10	1S7100B
ZD-10	1S759
ZD-10	1S759H
ZD-10	1T10
ZD-10	1T10B
ZD-10	1TA10
ZD-10	1TA10A
ZD-10	1Z10
ZD-10	1Z10A
ZD-10	1Z10T10
ZD-10	1Z10T20
ZD-10	1Z10T5
ZD-10	1ZC10T5
ZD-10	1ZCT10
ZD-10	1ZM10T10
ZD-10	1ZM10T20
ZD-10	1ZM10T5
ZD-10	2S1718/4454C
ZD-10-4	1S195
ZD-10.0	ECG5019
ZD-10.0	HEPZ0220
ZD-10.4	GEZD-10.4
ZD-100	1N1432
ZD-100	1N1795
ZD-100	1N1795A
ZD-100	1N1888
ZD-100	1N1944
ZD-100	1N1971
ZD-100	1N1998
ZD-100	1N3428
ZD-100	1N3459
ZD-100	1N3703
ZD-100	1N3703A
ZD-100	1N4186
ZD-100	1N4186A
ZD-100	1N4186B
ZD-100	1N4351
ZD-100	1N4351A
ZD-100	1N4351B
ZD-100	1N4428
ZD-100	1N4856
ZD-100	1N4856A
ZD-100	1N671
ZD-11	.25N11
ZD-11	AW01-11
ZD-11	AZ11
ZD-11	AZ962
ZD-11	AZ962A
ZD-11	AZ962B
ZD-11	BZ110
ZD-11	BZX46C11
ZD-11	BZX55C11
ZD-11	BZX71C11
ZD-11	BZX79C11
ZD-11	BZX83C11
ZD-11	BZY83C11
ZD-11	BZY85C11
ZD-11	BZY88C11
ZD-11	BZY94C11
ZD-11	CXL141
ZD-11	CXL5020
ZD-11	CXL5074
ZD-11	ECG5020A
ZD-11	ECG5074
ZD-11	ECG5074A
ZD-11	EQA01-11
ZD-11	EQA01-115
ZD-11	EQA01-11SE
ZD-11	EQA01-11Z
ZD-11	EQB01-011Z
ZD-11	GEZD-11
ZD-11	HD3000109-0
ZD-11	HEPZ0414
ZD-11	HM11
ZD-11	HM11A
ZD-11	HZ11
ZD-11	HZ11H
ZD-11	IN962B
ZD-11	MZ-11
ZD-11	MZ11
ZD-11	NTC-20
ZD-11	OA126/10
ZD-11	QA01-11.5E
ZD-11	QA01-11SE
ZD-11	QA1-11M
ZD-11	QA111M
ZD-11	QA111SE
ZD-11	QB01-11ZB
ZD-11	QB111
ZD-11	QB111Z
ZD-11	RD11AN
ZD-11	RD11E
ZD-11	RD11EA
ZD-11	RD11EB
ZD-11	RD11EM
ZD-11	RD11FB
ZD-11	RD11M
ZD-11	RE-116
ZD-11	RE116
ZD-11	RT-242
ZD-11	RZ11
ZD-11	RZZ11
ZD-11	SK3139
ZD-11	SZ-11
ZD-11	SZ-200-11
ZD-11	SZ-AW01-11
ZD-11	SZ-RD11FB
ZD-11	TVS1N4741
ZD-11	TVSQA01-11SE
ZD-11	1/2Z11T5
ZD-11	1N1314
ZD-11	1N1314A
ZD-11	1N3021B
ZD-11	1N3519
ZD-11	1N4105
ZD-11	1N4698
ZD-11	1N4741A
ZD-11	1N5241
ZD-11	1N5241A
ZD-11	1N5241B
ZD-11	1N5531
ZD-11	1N5531A
ZD-11	1N5531B
ZD-11	1N5737
ZD-11	1N5857
ZD-11	1N5857A
ZD-11	1N5857B
ZD-11	1N715
ZD-11	1N715A
ZD-11	1N765-2
ZD-11	1N962
ZD-11	1N962A
ZD-11	1S1176
ZD-11	1S140
ZD-11	1S1738
ZD-11	1S1770
ZD-11	1S196
ZD-11	1S2110
ZD-11	1S2110A
ZD-11	1S2118
ZD-11	1S2118A
ZD-11	1S2552
ZD-11	1S336
ZD-11	1S473
ZD-11	1S7110
ZD-11	1S7110A
ZD-11	1S7110B
ZD-11	1WB-11D
ZD-11.0	.25T11
ZD-11.0	.25T11A
ZD-11.0	.4T11
ZD-11.0	.4T11A
ZD-11.0	.4T11B
ZD-11.0	.7JZ11
ZD-11.0	.7Z11A
ZD-11.0	.7Z11B
ZD-11.0	.7Z11C
ZD-11.0	.7Z11D
ZD-11.0	.7ZM1
ZD-11.0	.7ZM11A
ZD-11.0	.7ZM11B
ZD-11.0	.7ZM11C
ZD-11.0	A42X00480-01
ZD-11.0	AZ-105
ZD-11.0	AZ110
ZD-11.0	BZ-105
ZD-11.0	BZ-110
ZD-11.0	BZX29C11
ZD-11.0	BZY92C11
ZD-11.0	DS189
ZD-11.0	ECG5020
ZD-11.0	EGA01-11
ZD-11.0	EQA01-11S
ZD-11.0	EQB01-11
ZD-11.0	EQB01-11V
ZD-11.0	EQB01-11Z
ZD-11.0	EZ11(ELCOM)
ZD-11.0	HD30000109-0
ZD-11.0	HD30001090
ZD-11.0	HD3000113
ZD-11.0	HD3001009
ZD-11.0	HD30010090
ZD-11.0	HD3001809
ZD-11.0	HEP-Z0414
ZD-11.0	HW11
ZD-11.0	HW11A
ZD-11.0	HZ-11
ZD-11.0	HZ11C
ZD-11.0	HZ11Y
ZD-11.0	IW01-09J
ZD-11.0	RD-11E
ZD-11.0	RD11F
ZD-11.0	RD11FA
ZD-11.0	REN5074
ZD-11.0	SZ671-0
ZD-11.0	SZ671-W
ZD-11.0	TVS-RD11A
ZD-11.0	TVSAW01-11
ZD-11.0	TVSZB1-11
ZD-11.0	1M11Z
ZD-11.0	1M11Z10
ZD-11.0	1M11Z5
ZD-11.0	1M11ZS10
ZD-11.0	1M11ZS5
ZD-11.0	1N1772
ZD-11.0	1N2036-2
ZD-11.0	1N3021
ZD-11.0	1N3021A
ZD-11.0	1N3680
ZD-11.0	1N3680A
ZD-11.0	1N4163
ZD-11.0	1N4163A
ZD-11.0	1N4299
ZD-11.0	1N4299A
ZD-11.0	1N4300
ZD-11.0	1N4300A
ZD-11.0	1N4301
ZD-11.0	1N4301A
ZD-11.0	1N4303
ZD-11.0	1N4303A
ZD-11.0	1N4304
ZD-11.0	1N4304A
ZD-11.0	1N4328
ZD-11.0	1N4328A
ZD-11.0	1N4405
ZD-11.0	1N4405A
ZD-11.0	1N4633
ZD-11.0	1N4662
ZD-11.0	1N4741
ZD-11.0	1N4833
ZD-11.0	1N4833A
ZD-11.0	1N5564
ZD-11.0	1N5564A
ZD-11.0	1N962B
ZD-11.0	1R11
ZD-11.0	1R11A
ZD-11.0	1R11B
ZD-11.0	1S1961
ZD-11.0	1S226
ZD-11.0	1S3011
ZD-11.0	1S483
ZD-11.0	1W01-08J
ZD-11.0	1Z11
ZD-11.0	1Z11A
ZD-11.0	1Z11T10
ZD-11.0	1Z11T20
ZD-11.0	1Z11T5
ZD-11.0	1ZC11T10
ZD-11.0	1ZC11T5
ZD-11.5	HEP604
ZD-11.5	MZ500-17
ZD-11.5	SZ11
ZD-11.5	1N3580
ZD-11.5	1N3580A
ZD-11.5	1N3580B
ZD-11.5	1N3581
ZD-11.5	1N3581A
ZD-11.5	1N3581B
ZD-11.5	1N3582A
ZD-11.5	1N3582B
ZD-11.5	1N3584
ZD-11.5	1N3584A
ZD-11.5	1N3584B
ZD-11.5	1N4583
ZD-11.5	1N4583A
ZD-11.5	1N941
ZD-11.5	1N941A
ZD-11.5	1N941B
ZD-11.5	1N942
ZD-11.5	1N945
ZD-11.5	1N945A
ZD-11.5	1N945B
ZD-11.5	1N946
ZD-11.5	1N946A
ZD-11.5	1N946B
ZD-110	1EZ110D5
ZD-110	1N1796
ZD-110	1N1796A
ZD-110	1N3704
ZD-110	1N3704A
ZD-110	1N3704B
ZD-110	1N4352
ZD-110	1N4352A
ZD-110	1N4352B
ZD-110	1N4429
ZD-110	1N4857
ZD-110	1N4857A
ZD-110	1N4857B
ZD-110	00002SA550
ZD-12	.25N12
ZD-12	.25N13
ZD-12	.25T12
ZD-12	.25T12A
ZD-12	.25T12B
ZD-12	.4T12A
ZD-12	.4T12B
ZD-12	.75N12
ZD-12	.7JZ12
ZD-12	.7Z12A
ZD-12	.7Z12B
ZD-12	.7Z12C
ZD-12	.7Z12D
ZD-12	.7ZM12A
ZD-12	.7ZM12B
ZD-12	.7ZM12C
ZD-12	.7ZM12D
ZD-12	A-120077
ZD-12	A-134166-2
ZD-12	A054-151
ZD-12	OA12612
ZD-12	A2474
ZD-12	A36539
ZD-12	A694X1
ZD-12	A694X1-OA
ZD-12	A7287500
ZD-12	A7287513
ZD-12	AU2012
ZD-12	AV2012
ZD-12	AV5
ZD-12	AW-01-12
ZD-12	AW-01-12C
ZD-12	AW01
ZD-12	AW01-12
ZD-12	AW01-12C
ZD-12	AW01-12V
ZD-12	AW01-13
ZD-12	AWOL-13
ZD-12	AX12
ZD-12	AZ-120
ZD-12	AZ12
ZD-12	AZ13
ZD-12	OAZ213
ZD-12	AZ759
ZD-12	AZ759A
ZD-12	AZ963
ZD-12	AZ963A
ZD-12	AZ963B
ZD-12	AZ964
ZD-12	AZ964A
ZD-12	AZ964B
ZD-12	B66X0040-001
ZD-12	BN7551
ZD-12	BZ-12
ZD-12	BZ-120
ZD-12	BZ120
ZD-12	BZX17
ZD-12	BZX29C12
ZD-12	BZX46C12
ZD-12	BZX46C13
ZD-12	BZX55C12
ZD-12	BZX55C13
ZD-12	BZX71C12
ZD-12	BZX71C13
ZD-12	BZX79C12
ZD-12	BZX79C13
ZD-12	BZX83C12
ZD-12	BZX83C13
ZD-12	BZY18
ZD-12	BZY69
ZD-12	BZY83C12
ZD-12	BZY83C13V5
ZD-12	BZY83D12
ZD-12	BZY85C12
ZD-12	BZY85C13
ZD-12	BZY85D12
ZD-12	BZY88C12
ZD-12	BZY88C13
ZD-12	BZY92C12
ZD-12	BZY94/C12
ZD-12	BZY94C12
ZD-12	BZY94C13
ZD-12	C4018
ZD-12	CD31-00015
ZD-12	CD31-12022
ZD-12	CD4116
ZD-12	CD4117
ZD-12	CD4118
ZD-12	CD4121
ZD-12	CD4122
ZD-12	CR956
ZD-12	CX-0052
ZD-12	CXL142
ZD-12	CXL5021
ZD-12	CZD012-5
ZD-12	D-00484R
ZD-12	D2G-3
ZD-12	D2G-4
ZD-12	D3W
ZD-12	DZ-12
ZD-12	DZ12A
ZD-12	E0771-7
ZD-12	E1852
ZD-12	E262
ZD-12	EA1318
ZD-12	EA16X162
ZD-12	EA16X29
ZD-12	EA16X77
ZD-12	EA2500
ZD-12	EA3719
ZD-12	ECG142
ZD-12	ECG142A
ZD-12	EP16X36
ZD-12	EQA01-12R
ZD-12	EQA01-12S
ZD-12	EQA01-12Z
ZD-12	EQB01
ZD-12	EQB01-02R
ZD-12	EQB01-12
ZD-12	EQB01-12A
ZD-12	EQB01-12B
ZD-12	EQB01-12BV
ZD-12	EQB01-12R
ZD-12	EQB01-12Z
ZD-12	EQG-12A
ZD-12	ES10234
ZD-12	ET16X15
ZD-12	EVR12
ZD-12	EVR12A
ZD-12	EVR12B
ZD-12	EZ12(ELCOM)
ZD-12	FZ1215
ZD-12	FZ12A
ZD-12	FZ12T10
ZD-12	FZ12T5
ZD-12	G01-037-A
ZD-12	G12T10
ZD-12	G12T20
ZD-12	G12T5
ZD-12	GEZD-12
ZD-12	HD3000101-0
ZD-12	HD3002409
ZD-12	HEP-Z0222
ZD-12	HEP-Z0415
ZD-12	HEP105
ZD-12	HP-20041
ZD-12	HM12
ZD-12	HM12A
ZD-12	HM12B
ZD-12	HM13
ZD-12	HM13A
ZD-12	HN-00005
ZD-12	H089
ZD-12	HS2120
ZD-12	HS7120
ZD-12	HW12
ZD-12	HW12A
ZD-12	HW12B
ZD-12	HZ-12
ZD-12	HZ-12A
ZD-12	HZ-12B
ZD-12	HZ-212
ZD-12	IE12Z5
ZD-12	J20438
ZD-12	KD2505
ZD-12	KS120A
ZD-12	KS120B

GE	Industry Standard No.
ZD-12	KS2120A
ZD-12	KS2120B
ZD-12	KS44A
ZD-12	KS44AF
ZD-12	KS44B
ZD-12	KS44BF
ZD-12	LPM12
ZD-12	LPM12A
ZD-12	LPZT12
ZD-12	LR120CH
ZD-12	M12Z
ZD-12	M4Z12
ZD-12	M4Z12-20
ZD-12	M4Z12A
ZD-12	MA1120
ZD-12	MC6016
ZD-12	MC6016A
ZD-12	MC6116
ZD-12	MC6116A
ZD-12	MD759
ZD-12	MD759A
ZD-12	MEZ12T10
ZD-12	MEZ12T5
ZD-12	MR120C-H
ZD-12	MTZ616
ZD-12	MTZ616A
ZD-12	MZ-1000-15
ZD-12	MZ-12
ZD-12	MZ-12B
ZD-12	MZ-212
ZD-12	MZ1000-15
ZD-12	MZ1012
ZD-12	MZ12
ZD-12	MZ12A
ZD-12	MZ12B
ZD-12	MZ212
ZD-12	MZ500-18
ZD-12	MZ500.18
ZD-12	MZ92-12A
ZD-12	N-EA16X29
ZD-12	NGP5010
ZD-12	02Z12A
ZD-12	02Z12GR
ZD-12	OA126-12
ZD-12	OA126/12
ZD-12	OA12612
ZD-12	OAZ213
ZD-12	OAZ230
ZD-12	OAZ273
ZD-12	OZ12T10
ZD-12	OZ12T5
ZD-12	PD6016
ZD-12	PD6016A
ZD-12	PD6057
ZD-12	PR617
ZD-12	PS10022B
ZD-12	PS10066
ZD-12	PS8915
ZD-12	PTC507
ZD-12	QA01-12S
ZD-12	QZ12T10
ZD-12	QZ12T5
ZD-12	R-1348
ZD-12	R-7103
ZD-12	R2D
ZD-12	R7894
ZD-12	R8364
ZD-12	RD-13AN
ZD-12	RD11EC
ZD-12	RD12
ZD-12	RD12E
ZD-12	RD12EA
ZD-12	RD12EB
ZD-12	RD12EC
ZD-12	RD12F
ZD-12	RD12FA
ZD-12	RD12FB
ZD-12	RD12M
ZD-12	RD13AL
ZD-12	RD13AM
ZD-12	RD13EA
ZD-12	RD13EB
ZD-12	RD13K
ZD-12	RE-118
ZD-12	RE118
ZD-12	REN142
ZD-12	RH-EX0015CEZZ
ZD-12	RH-EX0038CEZZ
ZD-12	RH-EX0047CEZZ
ZD-12	RS1348
ZD-12	RT-243
ZD-12	RT8200
ZD-12	RZ12
ZD-12	RZ13
ZD-12	RZZ12
ZD-12	RZZ13
ZD-12	S1R12B
ZD-12	S1R13B
ZD-12	SD-33
ZD-12	SD33
ZD-12	SD53
ZD-12	SFZ716
ZD-12	SI-RECT-228
ZD-12	SI-RECT-230
ZD-12	SK3062
ZD-12	SK3062/142A
ZD-12	SV1017
ZD-12	SV135
ZD-12	SV4012
ZD-12	SV4012A
ZD-12	SW01
ZD-12	SZ-200-12
ZD-12	SZ-EQB01-12A
ZD-12	SZ-RD12EB
ZD-12	SZ-RD12FB
ZD-12	SZ-UZ-12B
ZD-12	SZ12
ZD-12	SZ12.0
ZD-12	SZ671-B
ZD-12	SZ671-G
ZD-12	T-E1068
ZD-12	T-E1077
ZD-12	T-E1106
ZD-12	T-E1140
ZD-12	TE1068
ZD-12	TE1077
ZD-12	TMD10
ZD-12	TMD10A
ZD-12	TR-75
ZD-12	TR-7SB
ZD-12	TR14002-6
ZD-12	TS-337
ZD-12	TVS-RD(M)(P)D
ZD-12	TVS-RD13A
ZD-12	TVS-RD13D
ZD-12	TVS-RD13M
ZD-12	TVS-RD13P
ZD-12	TVS1N741A
ZD-12	TVS1N741H
ZD-12	TVSEQA01-12S
ZD-12	TVSEQB01-12
ZD-12	TVSQA01-12S
ZD-12	TVSRD12EBH
ZD-12	0Z12T10
ZD-12	0Z12T5
ZD-12	001-023037
ZD-12	1/4A12
ZD-12	1/4A12A
ZD-12	1/4A12B
ZD-12	1/4Z12T5
ZD-12	1A12M
ZD-12	1A12MA
ZD-12	1AC12
ZD-12	1AC12A
ZD-12	1AC12B
ZD-12	1B759
ZD-12	1C12Z
ZD-12	1C12ZA
ZD-12	1DZ12
ZD-12	1E12Z
ZD-12	1E12Z10
ZD-12	1E12Z5
ZD-12	1E2Z5
ZD-12	1EZ12
ZD-12	1M12Z
ZD-12	1M12Z10
ZD-12	1M12Z5
ZD-12	1M12ZS10
ZD-12	1M12ZS5
ZD-12	1N1426
ZD-12	1N1513
ZD-12	1N1513A
ZD-12	1N1524
ZD-12	1N1524A
ZD-12	1N1713
ZD-12	1N1736
ZD-12	1N1736A
ZD-12	1N1773
ZD-12	1N1773A
ZD-12	1N1877
ZD-12	1N1877A
ZD-12	1N1933
ZD-12	1N1960
ZD-12	1N1960A
ZD-12	1N1960B
ZD-12	1N1987
ZD-12	1N2037-1
ZD-12	1N2037A
ZD-12	1N2046
ZD-12	1N3022
ZD-12	1N3022A
ZD-12	1N3022B
ZD-12	1N3403
ZD-12	1N3416
ZD-12	1N3435
ZD-12	1N3447
ZD-12	1N3520
ZD-12	1N3537
ZD-12	1N3583
ZD-12	1N3583A
ZD-12	1N3583B
ZD-12	1N3681
ZD-12	1N3681A
ZD-12	1N3681B
ZD-12	1N4106
ZD-12	1N4106A
ZD-12	1N4164
ZD-12	1N4164A
ZD-12	1N4164B
ZD-12	1N4329
ZD-12	1N4329A
ZD-12	1N4329B
ZD-12	1N4406
ZD-12	1N4406A
ZD-12	1N4406B
ZD-12	1N4634
ZD-12	1N4663
ZD-12	1N4699
ZD-12	1N4742
ZD-12	1N4742A
ZD-12	1N4743
ZD-12	1N4743A
ZD-12	1N4834
ZD-12	1N4834A
ZD-12	1N4896
ZD-12	1N4896A
ZD-12	1N5242
ZD-12	1N5242A
ZD-12	1N5242B
ZD-12	1N5532
ZD-12	1N5532A
ZD-12	1N5532B
ZD-12	1N5565
ZD-12	1N5565A
ZD-12	1N5738
ZD-12	1N5858
ZD-12	1N5858A
ZD-12	1N5858B
ZD-12	1N665
ZD-12	1N716
ZD-12	1N716(ZENER)
ZD-12	1N716A
ZD-12	1N716B
ZD-12	1N759
ZD-12	1N759A
ZD-12	1N765
ZD-12	1N766-1
ZD-12	1N942A
ZD-12	1N942B
ZD-12	1N943
ZD-12	1N943A
ZD-12	1N943B
ZD-12	1N944
ZD-12	1N944A
ZD-12	1N944B
ZD-12	1N963
ZD-12	1N963A
ZD-12	1N963B
ZD-12	1R12
ZD-12	1R12A
ZD-12	1R12B
ZD-12	1S141
ZD-12	1S1739
ZD-12	1S1962
ZD-12	1S197
ZD-12	1S2119
ZD-12	1S2119A
ZD-12	1S2120
ZD-12	1S2120A
ZD-12	1S2127
ZD-12	1S2138
ZD-12	1S2196
ZD-12	1S227
ZD-12	1S2496
ZD-12	1S2553
ZD-12	1S3012
ZD-12	1S337
ZD-12	1S337-Y
ZD-12	1S337A
ZD-12	1S337E
ZD-12	1S337Y
ZD-12	1S696
ZD-12	1S7120
ZD-12	1S7120A
ZD-12	1S7120B
ZD-12	1T12
ZD-12	1T12B
ZD-12	1TA12
ZD-12	1TA12A
ZD-12	1Z12
ZD-12	1Z12A
ZD-12	1Z12B
ZD-12	1Z12C
ZD-12	1Z12T10
ZD-12	1Z12T20
ZD-12	1Z12T5
ZD-12	1ZB12
ZD-12	1ZB12B
ZD-12	1ZC12
ZD-12	1ZC12T10
ZD-12	1ZC12T20
ZD-12	1ZC12T5
ZD-12	1ZF12T10
ZD-12	1ZF12T20
ZD-12	1ZF12T5
ZD-12	1ZM12T10
ZD-12	1ZM12T20
ZD-12	1ZM12T5
ZD-12	1ZT12
ZD-12	1ZT12B
ZD-12.0	ECG5021
ZD-12.0	ECG5021A
ZD-12.0	HEPZ0222
ZD-12.0	HEPZ0415
ZD-12.0	HEPZ0416
ZD-120	1N1797
ZD-120	1N1889
ZD-120	1N1945
ZD-120	1N1972
ZD-120	1N1999
ZD-120	1N3429
ZD-120	1N3460
ZD-120	1N3705
ZD-120	1N3705A
ZD-120	1N4353
ZD-120	1N4353A
ZD-120	1N4430
ZD-120	1N4858
ZD-120	1N4858A
ZD-120	1N5889A
ZD-13	.25T12.8
ZD-13	.25T12.8A
ZD-13	.25T12.8B
ZD-13	.25T13
ZD-13	.25T13A
ZD-13	.4T12.8
ZD-13	.4T12.8A
ZD-13	.4T13
ZD-13	.4T13A
ZD-13	.4T13B
ZD-13	.7JZ13
ZD-13	.7Z13A
ZD-13	.7Z13B
ZD-13	.7Z13C
ZD-13	.7Z13D
ZD-13	.7ZM13A
ZD-13	.7ZM13B
ZD-13	.7ZM13C
ZD-13	.7ZM13D
ZD-13	A054-231
ZD-13	A066-122
ZD-13	A066-133
ZD-13	AW01(RCA)
ZD-13	AZ-125
ZD-13	AZ-130
ZD-13	BZ-125
ZD-13	BZ-130
ZD-13	BZ130
ZD-13	BZX29C13
ZD-13	BZY92C13
ZD-13	CDZ318-75
ZD-13	CXL143
ZD-13	D1S
ZD-13	D1ZVIO
ZD-13	DIS
ZD-13	DZ12
ZD-13	DZ13
ZD-13	EA16X81
ZD-13	ECG143
ZD-13	ECG143A
ZD-13	ECG5022
ZD-13	ECG5022A
ZD-13	EQA01-12
ZD-13	EQA01-13
ZD-13	EQA01-13R
ZD-13	EQB01-13
ZD-13	EZ13(ELCOM)
ZD-13	GEZD-13
ZD-13	HB-12B
ZD-13	HEP-Z0416
ZD-13	HEP605
ZD-13	HW13
ZD-13	HW13A
ZD-13	HW13B
ZD-13	HZ12
ZD-13	HZ12H
ZD-13	LPM13
ZD-13	LPM13A
ZD-13	MA1130
ZD-13	MMZ12(06)
ZD-13	PTC508
ZD-13	RD-13A
ZD-13	RD-13AD
ZD-13	RD-13AK
ZD-13	RD-13AK-P
ZD-13	RD-13AKP
ZD-13	RD-13AM
ZD-13	RD-13E
ZD-13	RD-13M
ZD-13	RD13
ZD-13	RD13A
ZD-13	RD13AK
ZD-13	RD13AKP
ZD-13	RD13AN
ZD-13	RD13AN-P
ZD-13	RD13ANP
ZD-13	RD13B
ZD-13	RD13B-Z
ZD-13	RD13E
ZD-13	RD13F
ZD-13	RD13FA
ZD-13	RD13FB
ZD-13	RD13M
ZD-13	RE-119
ZD-13	RE119
ZD-13	REN143
ZD-13	RH-EX0011CEZZ
ZD-13	RH-EX0017CEZZ
ZD-13	RH-EX0019TAZZ
ZD-13	RT-244
ZD-13	SK3093
ZD-13	SK3750
ZD-13	SK3750/143A
ZD-13	SZ-200-13
ZD-13	SZ13
ZD-13	SZ13.0
ZD-13	SZ961-V
ZD-13	SZA-13
ZD-13	TVSRD13AL
ZD-13	1/2Z13T5
ZD-13	1A13M
ZD-13	1A13MA
ZD-13	1M13ZS10
ZD-13	1M13ZS5
ZD-13	1N1315
ZD-13	1N1315A
ZD-13	1N1774
ZD-13	1N1783
ZD-13	1N1A13MA
ZD-13	1N2037
ZD-13	1N2037-2
ZD-13	1N227
ZD-13	1N227A
ZD-13	1N3023
ZD-13	1N3023A
ZD-13	1N3023B
ZD-13	1N3521
ZD-13	1N3682
ZD-13	1N3682A
ZD-13	1N4107
ZD-13	1N4165
ZD-13	1N4165A
ZD-13	1N4330
ZD-13	1N4330A
ZD-13	1N4407
ZD-13	1N4407A
ZD-13	1N4635
ZD-13	1N4664
ZD-13	1N4700
ZD-13	1N4835
ZD-13	1N4835A
ZD-13	1N4897
ZD-13	1N4897A
ZD-13	1N4898
ZD-13	1N4898A
ZD-13	1N4899
ZD-13	1N4899A
ZD-13	1N4900
ZD-13	1N4900A
ZD-13	1N4901
ZD-13	1N4901A
ZD-13	1N4902
ZD-13	1N4902A
ZD-13	1N4903
ZD-13	1N4903A
ZD-13	1N4904
ZD-13	1N4904A
ZD-13	1N4905
ZD-13	1N4905A
ZD-13	1N4906
ZD-13	1N4906A
ZD-13	1N4907
ZD-13	1N4907A
ZD-13	1N4908
ZD-13	1N4908A
ZD-13	1N4909
ZD-13	1N4909A
ZD-13	1N4910
ZD-13	1N4910A
ZD-13	1N4911
ZD-13	1N4911A
ZD-13	1N4912
ZD-13	1N4912A
ZD-13	1N4913
ZD-13	1N4913A
ZD-13	1N4914
ZD-13	1N4914A
ZD-13	1N4915
ZD-13	1N4915A
ZD-13	1N5243
ZD-13	1N5243A
ZD-13	1N5243B
ZD-13	1N5533
ZD-13	1N5533A
ZD-13	1N5533B
ZD-13	1N5566
ZD-13	1N5566A
ZD-13	1N5739
ZD-13	1N5859
ZD-13	1N5859A
ZD-13	1N5859B
ZD-13	1N717
ZD-13	1N717A
ZD-13	1N766
ZD-13	1N766-2
ZD-13	1N766A
ZD-13	1N964
ZD-13	1N964A
ZD-13	1N964B
ZD-13	1R13
ZD-13	1R13A
ZD-13	1R13B
ZD-13	1S1097
ZD-13	1S1166
ZD-13	1S1379
ZD-13	1S1379A
ZD-13	1S1394
ZD-13	1S1394A
ZD-13	1S1771
ZD-13	1S2130
ZD-13	1S2130A
ZD-13	1S228
ZD-13	1S2497
ZD-13	1S2554
ZD-13	1S3013
ZD-13	1S338Q
ZD-13	1S619
ZD-13	1S760
ZD-13	1S760H
ZD-13	1T12.8
ZD-13	1T12.8B
ZD-13	1TA12.8
ZD-13	1TA12.8A
ZD-13	1WB-13D
ZD-13	1Z13
ZD-13	1Z13A
ZD-13	1Z13T10
ZD-13	1Z13T20
ZD-13	1Z13T5
ZD-13	1ZC13T10
ZD-13	1ZC13T5
ZD-130	1EZ130D5
ZD-130	1N1798
ZD-130	1N1798A
ZD-130	1N3706
ZD-130	1N3706A
ZD-130	1N4354
ZD-130	1N4354A
ZD-130	1N4354B
ZD-130	1N4431
ZD-130	1N4859
ZD-130	1N4859A
ZD-130	1N5890
ZD-14	.25N14
ZD-14	.25T14
ZD-14	.25T14A
ZD-14	.25T14B
ZD-14	.4T14
ZD-14	.4T14A
ZD-14	.7JZ14
ZD-14	.7ZM14A
ZD-14	.7ZM14B
ZD-14	.7ZM14C
ZD-14	.7ZM14D
ZD-14	AZ-135
ZD-14	AZ-140
ZD-14	AZ7
ZD-14	BZ-135
ZD-14	BZ-140
ZD-14	BZX55C15
ZD-14	BZY85C13V5
ZD-14	CZD014
ZD-14	CZD014-5
ZD-14	D17
ZD-14	D1T
ZD-14	D1ZRED
ZD-14	D4P
ZD-14	DIT
ZD-14	DS-110
ZD-14	E0771-6
ZD-14	ECG144
ZD-14	ECG144A
ZD-14	ECG5023
ZD-14	ECG5023A
ZD-14	EQB01-14
ZD-14	EZ14(ELCOM)
ZD-14	FZ14T10
ZD-14	FZ14T5
ZD-14	GEZD-14
ZD-14	HD3001109-0
ZD-14	HD3001090
ZD-14	HD3002109
ZD-14	HEP606
ZD-14	M4552
ZD-14	M4659
ZD-14	MZ1014
ZD-14	OA126/14
ZD-14	OA12614
ZD-14	PTC511
ZD-14	QZ14T10
ZD-14	QZ14T5
ZD-14	RD16AL
ZD-14	REN144
ZD-14	RZ14
ZD-14	RZZ15
ZD-14	SK3094
ZD-14	SV1019
ZD-14	SV137
ZD-14	SV4014
ZD-14	SV4014A
ZD-14	SZ-200-14
ZD-14	SZ14
ZD-14	SZ961-R
ZD-14	SZ961-Y
ZD-14	1/4A14A
ZD-14	1/4A14B
ZD-14	1/4Z14D
ZD-14	1/4Z14D10
ZD-14	1/4Z14D5
ZD-14	1E14Z
ZD-14	1E14Z10
ZD-14	1E14Z5
ZD-14	1M14Z
ZD-14	1M14Z10
ZD-14	1M14Z5
ZD-14	1N1961
ZD-14	1N2037-3
ZD-14	1N2766
ZD-14	1N2766A
ZD-14	1N4108
ZD-14	1N4701
ZD-14	1N5244
ZD-14	1N5244A
ZD-14	1N5244B
ZD-14	1N5534
ZD-14	1N5534A
ZD-14	1N5860A
ZD-14	1N766-3
ZD-14	1S1410
ZD-14	1S1410R
ZD-14	1S1963
ZD-14	1S198
ZD-14	1S229
ZD-14	1S2498
ZD-14	1S2555
ZD-14	1S338
ZD-14	1S338U
ZD-14	1S474
ZD-14	1S484
ZD-14	1T14
ZD-14	1T14B
ZD-14	1TA14
ZD-14	1TA14A
ZD-14	1ZF14T10
ZD-14	1ZF14T20
ZD-14	1ZF14T5
ZD-140	1EZ140D5
ZD-140	1N5891
ZD-15	.25N15

GE	Industry Standard No.
ZD-15	.25NN15
ZD-15	.25T15
ZD-15	.25T15A
ZD-15	.25T15B
ZD-15	.4T15
ZD-15	.4T15B
ZD-15	.75N5
ZD-15	.7JZ15
ZD-15	.7Z14A
ZD-15	.7Z14B
ZD-15	.7Z14C
ZD-15	.7Z15D
ZD-15	.7ZM15A
ZD-15	.7ZM15B
ZD-15	.7ZM15C
ZD-15	.7ZM15D
ZD-15	A04166-2
ZD-15	A04166-2
ZD-15	AV-2015
ZD-15	AV2015
ZD-15	AW01-15
ZD-15	AZ-145
ZD-15	AZ-15
ZD-15	AZ-150
ZD-15	AZ15
ZD-15	AZ965
ZD-15	AZ965A
ZD-15	AZ965B
ZD-15	B-6002
ZD-15	BZ-145
ZD-15	BZ-150
ZD-15	BZ-X19
ZD-15	BZ150
ZD-15	BZX-46C15
ZD-15	BZX-71C15
ZD-15	BZX-79C15
ZD-15	BZX19
ZD-15	BZX29C15
ZD-15	BZX71C15
ZD-15	BZX79C15
ZD-15	BZX83C15
ZD-15	BZY-19
ZD-15	BZY-83C15
ZD-15	BZY-83D15
ZD-15	BZY19
ZD-15	BZY83/C15
ZD-15	BZY83/D15
ZD-15	BZY83C15
ZD-15	BZY83C16
ZD-15	BZY83D15
ZD-15	BZY85/C15
ZD-15	BZY85/D15
ZD-15	BZY85C15
ZD-15	BZY85D15
ZD-15	BZY88C15
ZD-15	BZY92/C15
ZD-15	BZY92C15
ZD-15	BZY94/C15
ZD-15	BZY94C15
ZD-15	C4020
ZD-15	CD31-00017
ZD-15	CD31-10365
ZD-15	CD31-12024
ZD-15	CXL145
ZD-15	CXL5024
ZD-15	CZD015
ZD-15	CZD015-5
ZD-15	D1Z
ZD-15	D1ZBLU
ZD-15	D1ZYEL
ZD-15	D5B
ZD-15	DHD805(ZENER)
ZD-15	DZ15A
ZD-15	EA16X124
ZD-15	ECG145
ZD-15	ECG145A
ZD-15	ECG5024A
ZD-15	EP16X5
ZD-15	EQA01-15
ZD-15	EQA01-15R
ZD-15	EQB01-15
ZD-15	EQB01-15Z
ZD-15	EQB01-15ZB
ZD-15	EVR15
ZD-15	EVR15A
ZD-15	EVR15B
ZD-15	EZ15(ELCOM)
ZD-15	EZ150
ZD-15	FA8009
ZD-15	FA8010
ZD-15	FA8011
ZD-15	FA8012
ZD-15	FZ15A
ZD-15	FZ15T10
ZD-15	FZ15T5
ZD-15	G15T10
ZD-15	G15T20
ZD-15	GEZD-15
ZD-15	GT15T5
ZD-15	HD3001109
ZD-15	HEP-Z0225
ZD-15	HEP-Z0418
ZD-15	HEP607
ZD-15	HM15
ZD-15	HM15A
ZD-15	HM15B
ZD-15	HS2150
ZD-15	HS7150
ZD-15	HW15
ZD-15	HW15A
ZD-15	HW15B
ZD-15	HZ15
ZD-15	HZ15H
ZD-15	IP20-0018
ZD-15	KS150A
ZD-15	KS150B
ZD-15	KS2150A
ZD-15	KS2150B
ZD-15	KS46
ZD-15	KS46AF
ZD-15	KS46BF
ZD-15	LPM15
ZD-15	LPM15A
ZD-15	LPZT15
ZD-15	LR150CH
ZD-15	M15Z
ZD-15	M4Z15
ZD-15	M4Z15-20
ZD-15	M4Z15A
ZD-15	MA1150
ZD-15	MC6018
ZD-15	MC6018A
ZD-15	MC6118
ZD-15	MC6118A
ZD-15	MEZ15T10
ZD-15	MEZ15T5
ZD-15	MR150C-H
ZD-15	MTZ618
ZD-15	MTZ618A
ZD-15	MZ1000-17
ZD-15	MZ15A
ZD-15	MZ15T20
ZD-15	MZ500-20
ZD-15	MZ92-15A
ZD-15	OZ15T10
ZD-15	OZ15T5
ZD-15	P38103/507-10
ZD-15	PA8261
ZD-15	PD-6018
ZD-15	PD-6018A
ZD-15	PD-6059
ZD-15	PD6018
ZD-15	PD6018A
ZD-15	PD6059
ZD-15	PR-620
ZD-15	PR620
ZD-15	PS-10068
ZD-15	PS-8917
ZD-15	PS10024B
ZD-15	PS10068
ZD-15	PS8917
ZD-15	PTC509
ZD-15	QA01-07R
ZD-15	QB01-15ZB
ZD-15	QZ-15T10
ZD-15	QZ-15T5
ZD-15	QZ1575
ZD-15	QZ15T10
ZD-15	QZ15T5
ZD-15	R2E
ZD-15	RD-15E
ZD-15	RD-16H
ZD-15	RD-16HA
ZD-15	RD15E
ZD-15	RD15EA
ZD-15	RD15EB
ZD-15	RD15F
ZD-15	RD15FA
ZD-15	RD15FB
ZD-15	RD15M
ZD-15	RD15S
ZD-15	RD16A
ZD-15	RD16AM
ZD-15	RD16H
ZD-15	RE-121
ZD-15	RE121
ZD-15	REN145
ZD-15	RT-245
ZD-15	RT3671
ZD-15	RT3671A
ZD-15	RZ-15AB
ZD-15	RZ15
ZD-15	RZ15A
ZD-15	RZZ156
ZD-15	SK3063
ZD-15	SK3063/145A
ZD-15	SU5
ZD-15	SV-1020A
ZD-15	SV-138A
ZD-15	SV-4015A
ZD-15	SV1020
ZD-15	SV138
ZD-15	SV4015
ZD-15	SV4015A
ZD-15	SZ-150
ZD-15	SZ-15B
ZD-15	SZ-200-15
ZD-15	SZ-200-15A
ZD-15	SZ15
ZD-15	SZ15.0
ZD-15	SZ150
ZD-15	SZ961-B
ZD-15	TR125B
ZD-15	TR12S
ZD-15	TR12SB
ZD-15	TR14002-12
ZD-15	TVS-ZB1-15
ZD-15	TVSEQB01-15
ZD-15	TVSQA01-07R
ZD-15	TVSQA01-15RB
ZD-15	TVSQA01-15S
ZD-15	TVSQB01-15ZB
ZD-15	001-02303-4
ZD-15	001-023034
ZD-15	1/2Z15T5
ZD-15	1/4A15
ZD-15	1/4A15A
ZD-15	1/4A15B
ZD-15	1/4M15Z
ZD-15	1/4M15Z10
ZD-15	1/4M15Z5
ZD-15	1/4Z15D10
ZD-15	1/4Z15D5
ZD-15	1/4Z15T5
ZD-15	1A15M
ZD-15	1A15MA
ZD-15	1AC15
ZD-15	1AC15A
ZD-15	1AC15B
ZD-15	1C15Z
ZD-15	1C15ZA
ZD-15	1D08
ZD-15	1E15Z
ZD-15	1E15Z10
ZD-15	1E15Z5
ZD-15	1EZ15
ZD-15	1M15Z
ZD-15	1M15Z10
ZD-15	1M15Z5
ZD-15	1M15ZS10
ZD-15	1M15ZS5
ZD-15	1N1316
ZD-15	1N1427
ZD-15	1N1514
ZD-15	1N1514A
ZD-15	1N1525
ZD-15	1N1525A
ZD-15	1N1775
ZD-15	1N1775A
ZD-15	1N1878
ZD-15	1N1878A
ZD-15	1N1934
ZD-15	1N1934A
ZD-15	1N1934B
ZD-15	1N1988
ZD-15	1N1988A
ZD-15	1N1988B
ZD-15	1N2038-1
ZD-15	1N2038A
ZD-15	1N228
ZD-15	1N3024
ZD-15	1N3024A
ZD-15	1N3024B
ZD-15	1N3404
ZD-15	1N3417
ZD-15	1N3436
ZD-15	1N3448
ZD-15	1N3522
ZD-15	1N3683
ZD-15	1N3683A
ZD-15	1N3683B
ZD-15	1N4109
ZD-15	1N4166
ZD-15	1N4166A
ZD-15	1N4166B
ZD-15	1N4331
ZD-15	1N4331A
ZD-15	1N4331B
ZD-15	1N4408
ZD-15	1N4408A
ZD-15	1N4408B
ZD-15	1N4636
ZD-15	1N4665
ZD-15	1N4702
ZD-15	1N4744
ZD-15	1N4744A
ZD-15	1N4836
ZD-15	1N4836A
ZD-15	1N5245
ZD-15	1N5245A
ZD-15	1N5245B
ZD-15	1N5535
ZD-15	1N5535A
ZD-15	1N5535B
ZD-15	1N5567
ZD-15	1N5567A
ZD-15	1N5740
ZD-15	1N5861
ZD-15	1N5861A
ZD-15	1N5861B
ZD-15	1N666
ZD-15	1N718
ZD-15	1N718A
ZD-15	1N718B
ZD-15	1N719
ZD-15	1N767
ZD-15	1N767-1
ZD-15	1N965
ZD-15	1N965A
ZD-15	1N965B
ZD-15	1N966
ZD-15	1R15
ZD-15	1R15A
ZD-15	1R15B
ZD-15	1S137
ZD-15	1S142
ZD-15	1S1740
ZD-15	1S1772
ZD-15	1S1964
ZD-15	1S2121
ZD-15	1S2121A
ZD-15	1S2150
ZD-15	1S2150A
ZD-15	1S230
ZD-15	1S301.5
ZD-15	1S3016
ZD-15	1S597
ZD-15	1S697
ZD-15	1S7150
ZD-15	1S7150A
ZD-15	1S7150B
ZD-15	1S990(ZENER)
ZD-15	1S990S
ZD-15	1S990S(ZENER)
ZD-15	1T15
ZD-15	1T15B
ZD-15	1T19-15B
ZD-15	1TA15
ZD-15	1TA15A
ZD-15	1WB-15D
ZD-15	1Z15
ZD-15	1Z15A
ZD-15	1Z15B
ZD-15	1Z15C
ZD-15	1Z15D
ZD-15	1Z15D10
ZD-15	1Z15D5
ZD-15	1Z15T10
ZD-15	1Z15T20
ZD-15	1Z15T5
ZD-15	1ZB15
ZD-15	1ZB15B
ZD-15	1ZC15
ZD-15	1ZC15T10
ZD-15	1ZC15T5
ZD-15	1ZF15T10
ZD-15	1ZF15T20
ZD-15	1ZF15T5
ZD-15	1ZM15T10
ZD-15	1ZM15T20
ZD-15	1ZM15T5
ZD-15	1ZT15
ZD-15	1ZT15B
ZD-15.0	ECG5024
ZD-15.0	HEPZ0225
ZD-15.0	HEPZ0417
ZD-15.0	HEPZ0418
ZD-15.0	HEPZ0419
ZD-150	1N1433
ZD-150	1N1799
ZD-150	1N1890
ZD-150	1N1946
ZD-150	1N1973
ZD-150	1N2000
ZD-150	1N3430
ZD-150	1N3461
ZD-150	1N3707
ZD-150	1N3707A
ZD-150	1N4355
ZD-150	1N4355A
ZD-150	1N4432
ZD-150	1N4860
ZD-150	1N4860A
ZD-150	1N672
ZD-16	.25T15.8
ZD-16	.25T15.8A
ZD-16	.25T16
ZD-16	.25T16A
ZD-16	.4T16
ZD-16	.4T16A
ZD-16	.4T16B
ZD-16	.7JZ16
ZD-16	.7Z16A
ZD-16	.7Z16B
ZD-16	.7Z16C
ZD-16	.7Z16D
ZD-16	.7ZM16A
ZD-16	.7ZM16B
ZD-16	.7ZM16C
ZD-16	.7ZM16D
ZD-16	AW01-16
ZD-16	AZ-157
ZD-16	AZ-162
ZD-16	BZ-157
ZD-16	BZ-162
ZD-16	BZ162
ZD-16	BZX29C16
ZD-16	BZY92C16
ZD-16	DDAY009003
ZD-16	DDAY009007
ZD-16	ECG5025
ZD-16	ECG5025A
ZD-16	ECG5075
ZD-16	ECG5075A
ZD-16	EDZ-11
ZD-16	EQA01-16
ZD-16	EQB01-16
ZD-16	EZ16(ELCOM)
ZD-16	GEZD-16
ZD-16	HEP-Z0419
ZD-16	HS2165
ZD-16	HW16
ZD-16	HW16A
ZD-16	HZ16
ZD-16	HZ16H
ZD-16	MA1160
ZD-16	MZ1016
ZD-16	MZ500-21
ZD-16	QD-ZBZ162XJ
ZD-16	RD16A-N
ZD-16	RD16B
ZD-16	RD16C-Y
ZD-16	RD16E
ZD-16	RD16E-M
ZD-16	RD16E-N
ZD-16	RD16EB
ZD-16	RD16F
ZD-16	RD16FA
ZD-16	RD16FB
ZD-16	RD16M
ZD-16	RE-122
ZD-16	RE122
ZD-16	REN5075
ZD-16	RH-EX0065CEZZ
ZD-16	RT-246
ZD-16	SK3142
ZD-16	SK3751
ZD-16	SK3751/5075A
ZD-16	SV139
ZD-16	SZ-200-16
ZD-16	SZ16
ZD-16	TVSEQA01-06S
ZD-16	TVSZB1-15
ZD-16	001-023038
ZD-16	1M16Z
ZD-16	1M16Z5
ZD-16	1M16ZS10
ZD-16	1M16ZS5
ZD-16	1MZ10
ZD-16	1N1316A
ZD-16	1N1776
ZD-16	1N2038
ZD-16	1N2038-2
ZD-16	1N3025
ZD-16	1N3025A
ZD-16	1N3025B
ZD-16	1N3684
ZD-16	1N3684A
ZD-16	1N4110
ZD-16	1N4167
ZD-16	1N4167A
ZD-16	1N4332
ZD-16	1N4332A
ZD-16	1N4409
ZD-16	1N4409A
ZD-16	1N4637
ZD-16	1N4666
ZD-16	1N4703
ZD-16	1N4745
ZD-16	1N4745A
ZD-16	1N4837
ZD-16	1N4837A
ZD-16	1N5246
ZD-16	1N5246A
ZD-16	1N5246B
ZD-16	1N5536
ZD-16	1N5536A
ZD-16	1N5536B
ZD-16	1N5568
ZD-16	1N5568A
ZD-16	1N5862
ZD-16	1N5862A
ZD-16	1N5862B
ZD-16	1N966A
ZD-16	1N966B
ZD-16	1R16
ZD-16	1R16A
ZD-16	1R16B
ZD-16	1S1167
ZD-16	1S1178
ZD-16	1S1191
ZD-16	1S1380
ZD-16	1S1380A
ZD-16	1S1395
ZD-16	1S1395A
ZD-16	1S1411
ZD-16	1S1411R
ZD-16	1S143
ZD-16	1S1965
ZD-16	1S2122A
ZD-16	1S231
ZD-16	1S2499
ZD-16	1S2556
ZD-16	1S3016(ZENER)
ZD-16	1S620
ZD-16	1Z16
ZD-16	1Z16A
ZD-16	1Z16T10
ZD-16	1Z16T20
ZD-16	1Z16T5
ZD-16	1ZC16T10
ZD-16	1ZC16T5
ZD-160	1EZ160D5
ZD-160	1N1800
ZD-160	1N1800A
ZD-160	1N3708
ZD-160	1N3708A
ZD-160	1N4356
ZD-160	1N4356A
ZD-160	1N4356B
ZD-160	1N4433
ZD-17	1S2122
ZD-18	1N1428
ZD-18	1N1515
ZD-18	1N1515A
ZD-18	1N1526
ZD-18	1N1526A
ZD-18	1N1777
ZD-18	1N1879
ZD-18	1N1935
ZD-18	1N1962
ZD-18	1N1989
ZD-18	1N3418
ZD-18	1N3437
ZD-18	1N3449
ZD-18	1N3524
ZD-18	1N3685
ZD-18	1N3685A
ZD-18	1N4168
ZD-18	1N4168A
ZD-18	1N4333
ZD-18	1N4333A
ZD-18	1N4410
ZD-18	1N4638
ZD-18	1N4667
ZD-18	1N4838
ZD-18	1N4838A
ZD-18	1N667
ZD-180	1EZ180D5
ZD-180	1N1801
ZD-180	1N1801A
ZD-180	1N1947
ZD-180	1N1974
ZD-180	1N2001
ZD-180	1N3431
ZD-180	1N3462
ZD-180	1N3709
ZD-180	1N3709A
ZD-180	1N4357
ZD-180	1N4357A
ZD-180	1N4357B
ZD-180	1N4434
ZD-19	1N1317
ZD-19	1N1317A
ZD-19	1N2039
ZD-2	B100
ZD-20	MZ1000-20
ZD-20	RD-2.2E
ZD-20	1N1778
ZD-20	1N3525
ZD-20	1N3686
ZD-20	1N3686A
ZD-20	1N4169
ZD-20	1N4169A
ZD-20	1N4334
ZD-20	1N4334A
ZD-20	1N4411
ZD-20	1N4639
ZD-20	1N4668
ZD-20	1N4839
ZD-20	1N4839A
ZD-200	1N1802
ZD-200	1N3710
ZD-200	1N3710A
ZD-200	1N4358
ZD-200	1N4358A
ZD-200	1N4435
ZD-22	SD31
ZD-22	1N1429
ZD-22	1N1516
ZD-22	1N1516A
ZD-22	1N1527
ZD-22	1N1527A
ZD-22	1N1880
ZD-22	1N1936
ZD-22	1N1963
ZD-22	1N1990
ZD-22	1N3419
ZD-22	1N3438
ZD-22	1N3450
ZD-22	1N3526
ZD-22	1N3687
ZD-22	1N3687A
ZD-22	1N4170
ZD-22	1N4170A
ZD-22	1N4335
ZD-22	1N4335A
ZD-22	1N4412
ZD-22	1N4640
ZD-22	1N4669
ZD-22	1N4840
ZD-22	1N668
ZD-24	DS-108
ZD-24	1N1318A
ZD-24	1N1780
ZD-24	1N2040
ZD-24	1N3688
ZD-24	1N3688A
ZD-24	1N4171
ZD-24	1N4171A
ZD-24	1N4336
ZD-24	1N4336A
ZD-24	1N4413
ZD-24	1N4641
ZD-24	1N4841
ZD-24	1N4841A
ZD-243	1A348
ZD-243	1A348R
ZD-245	1A8-1A82
ZD-25	HR200
ZD-27	MZ500-26
ZD-27	1N1430
ZD-27	1N1517
ZD-27	1N1517A
ZD-27	1N1528
ZD-27	1N1528A
ZD-27	1N1781
ZD-27	1N1781A
ZD-27	1N1831
ZD-27	1N1831A
ZD-27	1N1881
ZD-27	1N1937
ZD-27	1N1964
ZD-27	1N1991
ZD-27	1N3420
ZD-27	1N3439
ZD-27	1N3451
ZD-27	1N3528
ZD-27	1N3689
ZD-27	1N3689A

GE	Industry Standard No.
ZD-27	1N3689B
ZD-27	1N4172
ZD-27	1N4172A
ZD-27	1N4172B
ZD-27	1N4337
ZD-27	1N4337A
ZD-27	1N4337B
ZD-27	1N4414
ZD-27	1N4642
ZD-27	1N4671
ZD-27	1N4842
ZD-27	1N4842A
ZD-27	1N4842B
ZD-27	1N5746B
ZD-27	1N669
ZD-27	1S2270
ZD-27	1S2270A
ZD-28	1N1319A
ZD-3.3	MLV746A
ZD-3.3	MZ500-4
ZD-3.3	MZ92-3.3
ZD-3.3	1N3506
ZD-3.3	1N4649
ZD-3.6	MZ500-5
ZD-3.6	MZ500-6
ZD-3.6	1N3507
ZD-3.6	1N373
ZD-3.6	1N4650
ZD-3.6	1N466
ZD-3.6	1N466A
ZD-3.9	1N1507
ZD-3.9	1N1507A
ZD-3.9	1N1518
ZD-3.9	1N1518A
ZD-3.9	1N1927
ZD-3.9	1N1954
ZD-3.9	1N1981
ZD-3.9	1N3508
ZD-3.9	1N4651
ZD-3.9	1N467
ZD-3.9	1N467A
ZD-30	1N1782
ZD-30	1N2387
ZD-30	1N3421
ZD-30	1N3452
ZD-30	1N3529
ZD-30	1N3690
ZD-30	1N3690A
ZD-30	1N4173
ZD-30	1N4173A
ZD-30	1N4338
ZD-30	1N4338A
ZD-30	1N4415
ZD-30	1N4643
ZD-30	1N4672
ZD-30	1N4843
ZD-30	1N4843A
ZD-33	DS-189
ZD-33	LPZT33
ZD-33	MZ500-28
ZD-33	1N1320A
ZD-33	1N1783A
ZD-33	1N1882
ZD-33	1N1938
ZD-33	1N1965
ZD-33	1N1992
ZD-33	1N3422
ZD-33	1N3440
ZD-33	1N3453
ZD-33	1N3530
ZD-33	1N3691
ZD-33	1N3691A
ZD-33	1N3691B
ZD-33	1N4174
ZD-33	1N4174A
ZD-33	1N4174B
ZD-33	1N4339
ZD-33	1N4339A
ZD-33	1N4339B
ZD-33	1N4416
ZD-33	1N4644
ZD-33	1N4673
ZD-33	1N4844
ZD-33	1N4844A
ZD-33	1N4844B
ZD-36	MZ1000-26
ZD-36	1N1784
ZD-36	1N3531
ZD-36	1N3692
ZD-36	1N3692A
ZD-36	1N4175
ZD-36	1N4175A
ZD-36	1N4340
ZD-36	1N4340A
ZD-36	1N4417
ZD-36	1N4645
ZD-36	1N4674
ZD-36	1N4845
ZD-36	1N4845A
ZD-36	1N5749
ZD-39	1N1321
ZD-39	1N1321A
ZD-39	1N1785
ZD-39	1N1883
ZD-39	1N1939
ZD-39	1N1966
ZD-39	1N1993
ZD-39	1N3423
ZD-39	1N3441
ZD-39	1N3454
ZD-39	1N3532
ZD-39	1N3693
ZD-39	1N3693A
ZD-39	1N4176
ZD-39	1N4176A
ZD-39	1N4341
ZD-39	1N4341A
ZD-39	1N4418
ZD-39	1N4646
ZD-39	1N4675
ZD-39	1N4846
ZD-39	1N4846A
ZD-39	1N5750
ZD-4.3	1N3509
ZD-4.3	1N374
ZD-4.3	1N4652
ZD-4.7	1N1484
ZD-4.7	1N1508
ZD-4.7	1N1508A
ZD-4.7	1N1519
ZD-4.7	1N1519A
ZD-4.7	1N1928
ZD-4.7	1N1955
ZD-4.7	1N1982
ZD-4.7	1N2032
ZD-4.7	1N3510
ZD-4.7	1N4653
ZD-4.7	1N468
ZD-4.7	1N468A
ZD-4.7	1N5728
ZD-4.7	1N674
ZD-4.7	1S134
ZD-43	1N1786
ZD-43	1N1786A
ZD-43	1N3533
ZD-43	1N3694
ZD-43	1N3694A
ZD-43	1N4177
ZD-43	1N4177A
ZD-43	1N4177B
ZD-43	1N4342
ZD-43	1N4342A
ZD-43	1N4342B
ZD-43	1N4419
ZD-43	1N4647
ZD-43	1N4676
ZD-43	1N4847
ZD-43	1N4847A
ZD-43	1N5751
ZD-44	MZ9
ZD-47	1N1322A
ZD-47	1N1787
ZD-47	1N1884
ZD-47	1N1940
ZD-47	1N1967
ZD-47	1N1994
ZD-47	1N3424
ZD-47	1N3442
ZD-47	1N3455
ZD-47	1N3534
ZD-47	1N3695
ZD-47	1N3695A
ZD-47	1N4178
ZD-47	1N4178A
ZD-47	1N4343
ZD-47	1N4343A
ZD-47	1N4420
ZD-47	1N4648
ZD-47	1N4677
ZD-47	1N4848
ZD-47	1N4848A
ZD-47	1N5752
ZD-5.0	GEZD-6.0
ZD-5.1	BZ-052
ZD-5.1	MZ500-8
ZD-5.1	MZ500-9
ZD-5.1	PTC214
ZD-5.1	SD-6
ZD-5.1	1N1765
ZD-5.1	1N2041
ZD-5.1	1N4654
ZD-5.1	1N5729
ZD-5.6	.25T5.6
ZD-5.6	.25T5.6A
ZD-5.6	.25T5.6B
ZD-5.6	.4T5.6
ZD-5.6	.4T5.6A
ZD-5.6	.4T5.6B
ZD-5.6	.75N5.6
ZD-5.6	A7285900
ZD-5.6	AZ-052
ZD-5.6	AZ-054
ZD-5.6	AZ-056
ZD-5.6	AZ5.6
ZD-5.6	AZ752
ZD-5.6	AZ752A
ZD-5.6	BZ-054
ZD-5.6	BZ-056
ZD-5.6	BZ052
ZD-5.6	BZ5.6
ZD-5.6	BZ6.2
ZD-5.6	BZX29C5V6
ZD-5.6	BZX46C5V6
ZD-5.6	BZX55C5V6
ZD-5.6	BZX71C5V6
ZD-5.6	BZX79C5V6
ZD-5.6	BZX83C5V6
ZD-5.6	BZY58
ZD-5.6	BZY78
ZD-5.6	BZY78P
ZD-5.6	BZY83/C5V6
ZD-5.6	BZY83/D5V6
ZD-5.6	BZY83C5V6
ZD-5.6	BZY83D5V6
ZD-5.6	BZY85/C5V6
ZD-5.6	BZY85/D5V6
ZD-5.6	BZY85C5V6
ZD-5.6	BZY85D5V6
ZD-5.6	BZY88/C5V6
ZD-5.6	BZY88C5V6
ZD-5.6	BZY92/C5V6
ZD-5.6	BZY92C5V6
ZD-5.6	CD0120
ZD-5.6	CD31-00007
ZD-5.6	CD48
ZD-5.6	CHA4Z5.6
ZD-5.6	CHMZ5.6
ZD-5.6	CXL5011
ZD-5.6	D1M
ZD-5.6	DIM
ZD-5.6	DS50
ZD-5.6	DX-0727
ZD-5.6	DX-1132
ZD-5.6	EA16X118
ZD-5.6	EA2608
ZD-5.6	ECG136
ZD-5.6	ECG136A
ZD-5.6	ECG5011
ZD-5.6	ECG5011A
ZD-5.6	EDZ-19
ZD-5.6	EQA01-05T
ZD-5.6	EQA0105T
ZD-5.6	ERA0106R
ZD-5.6	ET16X17
ZD-5.6	EVR4
ZD-5.6	EVR4A
ZD-5.6	EVR4B
ZD-5.6	EZ5R6(ELCOM)
ZD-5.6	FZ5.6T10
ZD-5.6	FZ5.6T5
ZD-5.6	G5.6T10
ZD-5.6	G5.6T20
ZD-5.6	G5.6T5
ZD-5.6	GEZD-5.6
ZD-5.6	GLA56
ZD-5.6	GLA56A
ZD-5.6	GLA56B
ZD-5.6	HEP-Z0212
ZD-5.6	HEP-Z0407
ZD-5.6	HEP603
ZD-5.6	HEPZ0212
ZD-5.6	HEPZ0407
ZD-5.6	HS2056
ZD-5.6	HS7056
ZD-5.6	HZ6
ZD-5.6	HZ6B
ZD-5.6	HZ6B2
ZD-5.6	J241186
ZD-5.6	J24186
ZD-5.6	KS056A
ZD-5.6	KS056B
ZD-5.6	KS2056A
ZD-5.6	KS2056B
ZD-5.6	KS36A
ZD-5.6	KS36AF
ZD-5.6	KS36B
ZD-5.6	KS36BF
ZD-5.6	KS056A
ZD-5.6	KS056B
ZD-5.6	LPM5.6
ZD-5.6	LPM5.6A
ZD-5.6	LZ5.6
ZD-5.6	M4Z5.6
ZD-5.6	M4Z5.6-20
ZD-5.6	M4Z5.6A
ZD-5.6	M758
ZD-5.6	MA1056
ZD-5.6	MC6007
ZD-5.6	MC6007A
ZD-5.6	MC6107
ZD-5.6	MC6107A
ZD-5.6	MD752
ZD-5.6	MD752A
ZD-5.6	MEZ5.6T10
ZD-5.6	MEZ5.6T5
ZD-5.6	MGLA56
ZD-5.6	MGLA56A
ZD-5.6	MGLA56B
ZD-5.6	MR56C-H
ZD-5.6	MTZ607
ZD-5.6	MTZ607A
ZD-5.6	MZ-206
ZD-5.6	MZ-5
ZD-5.6	MZ206
ZD-5.6	MZ4626
ZD-5.6	MZ500-10
ZD-5.6	MZ500.10
ZD-5.6	MZ5A
ZD-5.6	MZ92-5.6A
ZD-5.6	OAZ202
ZD-5.6	OAZ242
ZD-5.6	OZ5.6T10
ZD-5.6	OZ5.6T5
ZD-5.6	PD6008
ZD-5.6	PD6008A
ZD-5.6	PD6049
ZD-5.6	PD6049C
ZD-5.6	PS8907
ZD-5.6	PTC502
ZD-5.6	QD-ZRD56FAA
ZD-5.6	QZ5.6T10
ZD-5.6	QZ5.6T5
ZD-5.6	RD-6A
ZD-5.6	RD-6AM
ZD-5.6	RD5.6B
ZD-5.6	RD5.6EB
ZD-5.6	RD5.6ED
ZD-5.6	RD5.6EK
ZD-5.6	RD5.6F
ZD-5.6	RD5.6FA
ZD-5.6	RD5.6FB
ZD-5.6	RD5.6M
ZD-5.6	RD6
ZD-5.6	RD6A(M)
ZD-5.6	RD6AL
ZD-5.6	RD6AM
ZD-5.6	RE-107
ZD-5.6	RE106
ZD-5.6	RE107
ZD-5.6	REN136
ZD-5.6	RT1306
ZD-5.6	RT1306(G.E.)
ZD-5.6	RV1181
ZD-5.6	RZ5.6
ZD-5.6	RZZ5.6
ZD-5.6	SD-32
ZD-5.6	SD32
ZD-5.6	SFZ708
ZD-5.6	SK3057
ZD-5.6	SK3057/136A
ZD-5.6	SK3342
ZD-5.6	SV123
ZD-5.6	SZ5.6
ZD-5.6	TMD02
ZD-5.6	TMD02A
ZD-5.6	TVSEQA01-05T
ZD-5.6	001-0163-04
ZD-5.6	001-023033
ZD-5.6	1-210
ZD-5.6	1/2Z5.6T5
ZD-5.6	1/2Z6.5T5
ZD-5.6	1/4AZ5.6D
ZD-5.6	1/4AZ5.6D10
ZD-5.6	1/4AZ5.6D5
ZD-5.6	1/4LZ5.6D
ZD-5.6	1/4LZ5.6D10
ZD-5.6	1/4LZ5.6D5
ZD-5.6	1/4M5.6AZ
ZD-5.6	1/4M5.6AZ10
ZD-5.6	1/4M5.6AZ25
ZD-5.6	1/4M5.6AZ5
ZD-5.6	1/4Z5.6T5
ZD-5.6	1D5.6
ZD-5.6	1D5.6A
ZD-5.6	1D5.6B
ZD-5.6	1EZ5.6
ZD-5.6	01K-5.4E
ZD-5.6	01K-5.8E
ZD-5.6	1M5.6ZS10
ZD-5.6	1M5.6ZS5
ZD-5.6	1N1509
ZD-5.6	1N1509A
ZD-5.6	1N1520
ZD-5.6	1N1520A
ZD-5.6	1N1765A
ZD-5.6	1N1820
ZD-5.6	1N1929
ZD-5.6	1N1929A
ZD-5.6	1N1929B
ZD-5.6	1N1956
ZD-5.6	1N1983
ZD-5.6	1N1983A
ZD-5.6	1N1983B
ZD-5.6	1N2033A
ZD-5.6	1N2214
ZD-5.6	1N3399
ZD-5.6	1N3512
ZD-5.6	1N3827
ZD-5.6	1N3827A
ZD-5.6	1N4626
ZD-5.6	1N4655
ZD-5.6	1N4690
ZD-5.6	1N4734
ZD-5.6	1N4734A
ZD-5.6	1N5232
ZD-5.6	1N5232A
ZD-5.6	1N5232B
ZD-5.6	1N5524
ZD-5.6	1N5524A
ZD-5.6	1N5524B
ZD-5.6	1N5730
ZD-5.6	1N5848
ZD-5.6	1N5848A
ZD-5.6	1N5848B
ZD-5.6	1N708
ZD-5.6	1N708A
ZD-5.6	1N752
ZD-5.6	1N752A
ZD-5.6	1N762-1
ZD-5.6	1N762A
ZD-5.6	1NJ61225
ZD-5.6	1S1374
ZD-5.6	1S1374A
ZD-5.6	1S1389
ZD-5.6	1S1389A
ZD-5.6	1S1405
ZD-5.6	1S1405R
ZD-5.6	1S2056
ZD-5.6	1S2056A
ZD-5.6	1S2056B
ZD-5.6	1S2111
ZD-5.6	1S2111A
ZD-5.6	1S211A
ZD-5.6	1S221
ZD-5.6	1S2488
ZD-5.6	1S2544
ZD-5.6	1S2545
ZD-5.6	1S52
ZD-5.6	1S692
ZD-5.6	1S7056
ZD-5.6	1S7056A
ZD-5.6	1S7056B
ZD-5.6	1T5.6
ZD-5.6	1T5.6B
ZD-5.6	1TA5.6
ZD-5.6	1TA5.6A
ZD-5.6	1WB-6A
ZD-5.6	1Z5.6
ZD-5.6	1Z5.6A
ZD-5.6	1ZC5.6
ZD-5.6	1ZC5.6T10.5
ZD-5.6	1ZF5.6T10
ZD-5.6	1ZF5.6T20
ZD-5.6	1ZF5.6T5
ZD-5.6	1ZM5.6T10
ZD-5.6	1ZM5.6T20
ZD-5.6	1ZM5.6T5
ZD-5.8	RD6B
ZD-51	1N1788
ZD-51	1N1788A
ZD-51	1N3696
ZD-51	1N3696A
ZD-51	1N3696B
ZD-51	1N4179A
ZD-51	1N4179B
ZD-51	1N4321
ZD-51	1N4344
ZD-51	1N4344A
ZD-51	1N4344B
ZD-51	1N4421
ZD-51	1N4849
ZD-51	1N4849A
ZD-51	1N5753
ZD-56	1N1789
ZD-56	1N1789A
ZD-56	1N1885
ZD-56	1N1941
ZD-56	1N1942
ZD-56	1N1968
ZD-56	1N1995
ZD-56	1N3425
ZD-56	1N3456
ZD-56	1N3697
ZD-56	1N3697A
ZD-56	1N4180
ZD-56	1N4180A
ZD-56	1N4180B
ZD-56	1N4345
ZD-56	1N4345A
ZD-56	1N4345B
ZD-56	1N4422
ZD-56	1N4850
ZD-56	1N4850A
ZD-6.0	BZX83C6V2
ZD-6.0	BZY88C6V2
ZD-6.0	BZZ10
ZD-6.0	ECG5012
ZD-6.0	ECG5012A
ZD-6.0	ECG5070
ZD-6.0	ECG5070A
ZD-6.0	MZ1006
ZD-6.0	MZ306
ZD-6.0	MZ306C
ZD-6.0	QD-ZMZ306CE
ZD-6.0	QD-ZRD56EAA
ZD-6.0	RD5R6EB
ZD-6.0	RD6A
ZD-6.0	RD6AN
ZD-6.0	RVDRD5R6EB
ZD-6.0	SZA-6
ZD-6.0	1N474
ZD-6.0	1N474A
ZD-6.0	1N474B
ZD-6.0	1N5233
ZD-6.0	1N5233A
ZD-6.0	1N5233B
ZD-6.0	1N706
ZD-6.0	1N706A
ZD-6.0	1N762
ZD-6.0	1N762-2
ZD-6.0	1S1173
ZD-6.0	1S1767
ZD-6.0	1S191
ZD-6.0	1S2112A
ZD-6.0	1S2125
ZD-6.0	1S2128
ZD-6.0	1S2194
ZD-6.0	1S2546
ZD-6.0	1S470
ZD-6.0	1S635
ZD-6.0	1S756
ZD-6.1	AW01-06
ZD-6.2	.25T5.8
ZD-6.2	.25T5.8A
ZD-6.2	.25T6.2
ZD-6.2	.25T6.2A
ZD-6.2	.25T6.2B
ZD-6.2	.4T6.2
ZD-6.2	.4T6.2A
ZD-6.2	.4T6.2B
ZD-6.2	.4T6.8A
ZD-6.2	.75N6.2
ZD-6.2	A-125278
ZD-6.2	AZ-058
ZD-6.2	AZ-061
ZD-6.2	AZ-063
ZD-6.2	AZ6.2
ZD-6.2	AZ753
ZD-6.2	AZ753A
ZD-6.2	BZ-058
ZD-6.2	BZ-061
ZD-6.2	BZ-063
ZD-6.2	BZX10
ZD-6.2	BZX29C6V2
ZD-6.2	BZX46C6V2
ZD-6.2	BZX55C6V2
ZD-6.2	BZX71C6V2
ZD-6.2	BZX79C6V2
ZD-6.2	BZY59
ZD-6.2	BZY66
ZD-6.2	BZY83C6V2
ZD-6.2	BZY85C6V2
ZD-6.2	BZY88/6V2
ZD-6.2	BZY88/C6V2
ZD-6.2	BZY92C6V2
ZD-6.2	C4011
ZD-6.2	CD-0033
ZD-6.2	CD0033
ZD-6.2	CD31-00008
ZD-6.2	CHAZ6.2
ZD-6.2	CHAZ6.2A
ZD-6.2	CHAZ62.A
ZD-6.2	CXL137
ZD-6.2	CXL5013
ZD-6.2	D-00469C
ZD-6.2	D3H
ZD-6.2	D3Y
ZD-6.2	D5W
ZD-6.2	D6.2
ZD-6.2	DDAY008001
ZD-6.2	DIJ72293
ZD-6.2	DS104
ZD-6.2	DS159
ZD-6.2	DX-0530
ZD-6.2	DX1194
ZD-6.2	E21431
ZD-6.2	E2486
ZD-6.2	EA16X80
ZD-6.2	ECG137
ZD-6.2	ECG137A
ZD-6.2	ECG5013
ZD-6.2	ECG5013A
ZD-6.2	ED6.2EB
ZD-6.2	EQA01-067
ZD-6.2	EQA01-06S
ZD-6.2	EQA01-06T
ZD-6.2	EQB01-06
ZD-6.2	EVR5
ZD-6.2	EVR5A
ZD-6.2	EVR5B
ZD-6.2	F16H1
ZD-6.2	FV-22
ZD-6.2	FV-24
ZD-6.2	FV22
ZD-6.2	FZ6.2T10
ZD-6.2	G01-036A
ZD-6.2	G01036A
ZD-6.2	GARE
ZD-6.2	GE6.2
ZD-6.2	GE6.2A
ZD-6.2	GE6.2B
ZD-6.2	GEZD-6.2
ZD-6.2	GLA62
ZD-6.2	GLA62A
ZD-6.2	GLA62B
ZD-6.2	GZ6.2
ZD-6.2	HEP-Z0214
ZD-6.2	HEP-Z0408
ZD-6.2	HEP103
ZD-6.2	HEPZ0214
ZD-6.2	HEPZ0408
ZD-6.2	HS2062
ZD-6.2	HS7062
ZD-6.2	HZ-6B
ZD-6.2	INJ61225
ZD-6.2	IP20-0203
ZD-6.2	J241179
ZD-6.2	J24262
ZD-6.2	J24631
ZD-6.2	KD2503
ZD-6.2	KS062A
ZD-6.2	KS2062A
ZD-6.2	KS2062B
ZD-6.2	KS37A
ZD-6.2	KS37AF
ZD-6.2	KS062A
ZD-6.2	LR62CH
ZD-6.2	M4Z6.2
ZD-6.2	M4Z6.2-20
ZD-6.2	M4Z6.2A
ZD-6.2	MA1062
ZD-6.2	MC6008
ZD-6.2	MC6008A
ZD-6.2	MC6108
ZD-6.2	MC6108A
ZD-6.2	MD753
ZD-6.2	MD753A
ZD-6.2	MGLA62
ZD-6.2	MGLA62A
ZD-6.2	MGLA62B
ZD-6.2	MR62C-H
ZD-6.2	MR62E-H
ZD-6.2	MR62H
ZD-6.2	MTZ608
ZD-6.2	MTZ608A
ZD-6.2	MZ4627
ZD-6.2	MZ500-11
ZD-6.2	MZ6.2
ZD-6.2	MZ6.2A
ZD-6.2	MZ6.2B

GE	Industry Standard No.
ZD-6.2	MZ6.2T5
ZD-6.2	MZ605
ZD-6.2	MZ610
ZD-6.2	MZ620
ZD-6.2	MZ640
ZD-6.2	MZ70MA
ZD-6.2	MZ92-6,0B
ZD-6.2	MZ92-6.0B
ZD-6.2	MZ92-6.2A
ZD-6.2	NGP5002
ZD-6.2	NZ-206
ZD-6.2	OAZ203
ZD-6.2	OAZ210
ZD-6.2	OAZ243
ZD-6.2	OAZ270
ZD-6.2	OZ6.2T10
ZD-6.2	OZ6.2T5
ZD-6.2	PA9267
ZD-6.2	PD6009
ZD-6.2	PD6009A
ZD-6.2	PD6050
ZD-6.2	PS1325
ZD-6.2	PS8908
ZD-6.2	PTC503
ZD-6.2	QZ6.2T10
ZD-6.2	QZ6.2T5
ZD-6.2	R-7093
ZD-6.2	RD-6L
ZD-6.2	RD6.2E
ZD-6.2	RD6.2EB
ZD-6.2	RD6.2F
ZD-6.2	RD6.2FA
ZD-6.2	RD6.2FB
ZD-6.2	RE-109
ZD-6.2	RE109
ZD-6.2	REN137
ZD-6.2	REN5070
ZD-6.2	RH-EX0024CEZZ
ZD-6.2	RH-EX0048CEZZ
ZD-6.2	RT5793
ZD-6.2	RT7539
ZD-6.2	RV6.2
ZD-6.2	RVDCD0033
ZD-6.2	RVDEQA0106S
ZD-6.2	RVDMZ-206
ZD-6.2	RVDR5R6EB
ZD-6.2	RZ6.2
ZD-6.2	RZZ6.2
ZD-6.2	S1A3
ZD-6.2	S3004-1715
ZD-6.2	S3004-1718
ZD-6.2	SA821
ZD-6.2	SA821A
ZD-6.2	SA823
ZD-6.2	SA823A
ZD-6.2	SA825
ZD-6.2	SA825A
ZD-6.2	SA827
ZD-6.2	SA827A
ZD-6.2	SA829
ZD-6.2	SA829A
ZD-6.2	SB821
ZD-6.2	SB821A
ZD-6.2	SB823
ZD-6.2	SB823A
ZD-6.2	SB825
ZD-6.2	SB825A
ZD-6.2	SB827
ZD-6.2	SB827A
ZD-6.2	SB829
ZD-6.2	SB829A
ZD-6.2	SC821
ZD-6.2	SC821A
ZD-6.2	SC823
ZD-6.2	SC823A
ZD-6.2	SC825
ZD-6.2	SC825A
ZD-6.2	SC827
ZD-6.2	SC827A
ZD-6.2	SC829
ZD-6.2	SC829A
ZD-6.2	SI-RECT-140
ZD-6.2	SK3058
ZD-6.2	SK3058/137A
ZD-6.2	SUC650
ZD-6.2	SUM6010
ZD-6.2	SUM6011
ZD-6.2	SUM6020
ZD-6.2	SUM6021
ZD-6.2	SV-02
ZD-6.2	SVC625
ZD-6.2	SVC650
ZD-6.2	SVM601
ZD-6.2	SVM6010
ZD-6.2	SVM6011
ZD-6.2	SVM602
ZD-6.2	SVM6020
ZD-6.2	SVM6021
ZD-6.2	SVM605
ZD-6.2	SVM61
ZD-6.2	SZ-YZ063
ZD-6.2	SZ6.2
ZD-6.2	SZ6.2A
ZD-6.2	TIXD753
ZD-6.2	TMD03
ZD-6.2	TMD03A
ZD-6.2	TP20-0284
ZD-6.2	TR2337123
ZD-6.2	TR6S
ZD-6.2	TRR6
ZD-6.2	TVSZB1-6
ZD-6.2	001-02303-3
ZD-6.2	1/2Z6.2T5
ZD-6.2	1/4AZ6.2D5
ZD-6.2	1/4LZ6.2D
ZD-6.2	1/4LZ6.2D5
ZD-6.2	1/4M6.2AZ5
ZD-6.2	1/4Z6.2T5
ZD-6.2	1A62
ZD-6.2	1D6.2
ZD-6.2	1D6.2A
ZD-6.2	1D6.2B
ZD-6.2	1D6.2SA
ZD-6.2	1D6.2SB
ZD-6.2	01K-6.5E
ZD-6.2	01K6.5E
ZD-6.2	1M6.2ZS10
ZD-6.2	1M6.2ZS5
ZD-6.2	1M753A
ZD-6.2	1N1485
ZD-6.2	1N1735
ZD-6.2	1N1766
ZD-6.2	1N1766A
ZD-6.2	1N1779
ZD-6.2	1N2033
ZD-6.2	1N2033-2
ZD-6.2	1N3411
ZD-6.2	1N3443
ZD-6.2	1N3496
ZD-6.2	1N3497
ZD-6.2	1N3498
ZD-6.2	1N3499
ZD-6.2	1N3500
ZD-6.2	1N3513
ZD-6.2	1N3553
ZD-6.2	1N3828
ZD-6.2	1N3828A
ZD-6.2	1N4010
ZD-6.2	1N429
ZD-6.2	1N4499
ZD-6.2	1N4565
ZD-6.2	1N4565A
ZD-6.2	1N4566
ZD-6.2	1N4566A
ZD-6.2	1N4567
ZD-6.2	1N4567A
ZD-6.2	1N4568
ZD-6.2	1N4568A
ZD-6.2	1N4569
ZD-6.2	1N4569A
ZD-6.2	1N4570
ZD-6.2	1N4570A
ZD-6.2	1N4571
ZD-6.2	1N4571A
ZD-6.2	1N4572
ZD-6.2	1N4572A
ZD-6.2	1N4573
ZD-6.2	1N4573A
ZD-6.2	1N4574
ZD-6.2	1N4574A
ZD-6.2	1N4575
ZD-6.2	1N4575A
ZD-6.2	1N4576
ZD-6.2	1N4576A
ZD-6.2	1N4577
ZD-6.2	1N4577A
ZD-6.2	1N4578
ZD-6.2	1N4578A
ZD-6.2	1N4579
ZD-6.2	1N4579A
ZD-6.2	1N4580
ZD-6.2	1N4580A
ZD-6.2	1N4581
ZD-6.2	1N4581A
ZD-6.2	1N4582
ZD-6.2	1N4582A
ZD-6.2	1N4583B
ZD-6.2	1N4584
ZD-6.2	1N4584A
ZD-6.2	1N4627
ZD-6.2	1N4656
ZD-6.2	1N4691
ZD-6.2	1N4735
ZD-6.2	1N4735A
ZD-6.2	1N5234
ZD-6.2	1N5234A
ZD-6.2	1N5234B
ZD-6.2	1N5525
ZD-6.2	1N5525A
ZD-6.2	1N5525B
ZD-6.2	1N5731
ZD-6.2	1N5849
ZD-6.2	1N5849A
ZD-6.2	1N5849B
ZD-6.2	1N5850
ZD-6.2	1N5850A
ZD-6.2	1N5850B
ZD-6.2	1N675
ZD-6.2	1N709
ZD-6.2	1N709A
ZD-6.2	1N709B
ZD-6.2	1N753
ZD-6.2	1N753A
ZD-6.2	1N816
ZD-6.2	1N821
ZD-6.2	1N821A
ZD-6.2	1N822
ZD-6.2	1N823
ZD-6.2	1N823A
ZD-6.2	1N824
ZD-6.2	1N825
ZD-6.2	1N825A
ZD-6.2	1N827
ZD-6.2	1N827A
ZD-6.2	1N829
ZD-6.2	1N829A
ZD-6.2	1S-331
ZD-6.2	1S1201
ZD-6.2	1S136
ZD-6.2	1S136(ZENER)
ZD-6.2	1S1375
ZD-6.2	1S1375A
ZD-6.2	1S1390
ZD-6.2	1S1390A
ZD-6.2	1S1406
ZD-6.2	1S1406R
ZD-6.2	1S1715
ZD-6.2	1S1735
ZD-6.2	1S1956
ZD-6.2	1S2062
ZD-6.2	1S2062A
ZD-6.2	1S2062B
ZD-6.2	1S2112
ZD-6.2	1S2452
ZD-6.2	1S2453
ZD-6.2	1S2454
ZD-6.2	1S2489
ZD-6.2	1S2547
ZD-6.2	1S2769
ZD-6.2	1S2770
ZD-6.2	1S2771
ZD-6.2	1S2774
ZD-6.2	1S331
ZD-6.2	1S331A
ZD-6.2	1S331AZ
ZD-6.2	1S347
ZD-6.2	1S37
ZD-6.2	1S480
ZD-6.2	1S7062
ZD-6.2	1S7062A
ZD-6.2	1S7062B
ZD-6.2	1T6.2
ZD-6.2	1T6.2B
ZD-6.2	1TA6.2
ZD-6.2	1TA6.2A
ZD-6.2	1TA6.2B
ZD-6.2	1Z6.2
ZD-6.2	1Z6.2A
ZD-6.2	1Z6.2T10
ZD-6.2	1Z6.2T5
ZD-6.2	1ZC6.2
ZD-6.2	1ZM6.2
ZD-6.2	1ZM6.2T10
ZD-6.2	1ZM6.2T20
ZD-6.2	1ZM6.2T5
ZD-6.2	2MW742
ZD-6.6	BZ066
ZD-6.6	GEZD-6.6
ZD-6.6	RD6.2EC
ZD-6.6	1N4611
ZD-6.6	1N4612
ZD-6.6	1N4612A
ZD-6.6	1N4612B
ZD-6.6	1N4612C
ZD-6.6	1N4613
ZD-6.6	1S53
ZD-6.8	.25N6.8
ZD-6.8	.25T6.8
ZD-6.8	.25T6.8A
ZD-6.8	.25T7.1
ZD-6.8	.25T7.1A
ZD-6.8	.4T6.8
ZD-6.8	.4T6.8B
ZD-6.8	.7JZ6.8
ZD-6.8	.7Z6.8A
ZD-6.8	.7Z6.8B
ZD-6.8	.7Z6.8C
ZD-6.8	.7Z6.8D
ZD-6.8	.7ZM6.8A
ZD-6.8	.7ZM6.8B
ZD-6.8	.7ZM6.8C
ZD-6.8	.7ZM6.8D
ZD-6.8	A29035-E
ZD-6.8	A7286201
ZD-6.8	AW01-07
ZD-6.8	AW01-7
ZD-6.8	AZ-065
ZD-6.8	AZ-067
ZD-6.8	AZ-069
ZD-6.8	AZ6.8
ZD-6.8	AZ754
ZD-6.8	AZ754A
ZD-6.8	AZ957
ZD-6.8	AZ957A
ZD-6.8	AZ957B
ZD-6.8	BZ-065
ZD-6.8	BZ-067
ZD-6.8	BZ-069
ZD-6.8	BZ6.8
ZD-6.8	BZX29C6V8
ZD-6.8	BZX46C6V8
ZD-6.8	BZX55C6V8
ZD-6.8	BZX71C6V8
ZD-6.8	BZX79C6V8
ZD-6.8	BZX83C6V8
ZD-6.8	BZY60
ZD-6.8	BZY83C6V8
ZD-6.8	BZY83D6V8
ZD-6.8	BZY85C6V8
ZD-6.8	BZY85D6V8
ZD-6.8	BZY88C6V8
ZD-6.8	BZY92C6V8
ZD-6.8	BZZ11
ZD-6.8	CXL5014
ZD-6.8	CXL5071
ZD-6.8	DS-159
ZD-6.8	EA16X123
ZD-6.8	ECG5014
ZD-6.8	ECG5014A
ZD-6.8	ECG5071
ZD-6.8	ECG5071A
ZD-6.8	EDZ-20
ZD-6.8	EQA01-06
ZD-6.8	EQA01-07
ZD-6.8	EQA01-07R
ZD-6.8	EQA01-07RE
ZD-6.8	EQA01-07S
ZD-6.8	EQA107RE
ZD-6.8	ERB-07RE
ZD-6.8	ERB01-07
ZD-6.8	ETD-HZ7C
ZD-6.8	EZ6(ELCOM)
ZD-6.8	GEZD-6.8
ZD-6.8	HB-7B
ZD-6.8	HEP-Z0215
ZD-6.8	HEP-Z0409
ZD-6.8	HEPZ0215
ZD-6.8	HEPZ0409
ZD-6.8	HF-20033
ZD-6.8	HM6.8
ZD-6.8	HM6.8A
ZD-6.8	HW6.8
ZD-6.8	HW6.8A
ZD-6.8	HZ-7(B)
ZD-6.8	HZ-7B
ZD-6.8	HZ7
ZD-6.8	HZ7B
ZD-6.8	IP20-0186
ZD-6.8	KOS25211-2
ZD-6.8	KZ6
ZD-6.8	KZ6A
ZD-6.8	KZ6ZA
ZD-6.8	M2207
ZD-6.8	MA1068
ZD-6.8	MZ-207
ZD-6.8	MZ-6
ZD-6.8	MZ-7
ZD-6.8	MZ207(ZENER)
ZD-6.8	MZ207-02A
ZD-6.8	MZ207A
ZD-6.8	MZ207B
ZD-6.8	MZ207C
ZD-6.8	MZ307
ZD-6.8	MZ6
ZD-6.8	OAZ204
ZD-6.8	OAZ244
ZD-6.8	OSD-0033
ZD-6.8	PD6010
ZD-6.8	PD6010A
ZD-6.8	PD6051
ZD-6.8	PS8909
ZD-6.8	QA01-07RE
ZD-6.8	QA107RE
ZD-6.8	RO7A
ZD-6.8	RD-6.8EB
ZD-6.8	RD-7AM
ZD-6.8	RD-7E
ZD-6.8	RD6.8E
ZD-6.8	RD6.8E-B1
ZD-6.8	RD6.8EB
ZD-6.8	RD6.8F
ZD-6.8	RD6.8FA
ZD-6.8	RD6.8FB
ZD-6.8	RD7AM
ZD-6.8	RD7AN
ZD-6.8	RD7B
ZD-6.8	RE-110
ZD-6.8	RE110
ZD-6.8	REN5071
ZD-6.8	RT5215
ZD-6.8	RT5471
ZD-6.8	RVDRD7AN
ZD-6.8	RZ6.8
ZD-6.8	RZZ6.8
ZD-6.8	SD-49
ZD-6.8	SD42
ZD-6.8	SD49
ZD-6.8	SK3334
ZD-6.8	SZ-RD6.2EB
ZD-6.8	SZ7
ZD-6.8	TMD04
ZD-6.8	TMD04A
ZD-6.8	TR7SA
ZD-6.8	TR7SB
ZD-6.8	TV24981
ZD-6.8	TVSQA01-07RE
ZD-6.8	001-0082-00
ZD-6.8	001-0099-01
ZD-6.8	1/2Z6.8T5
ZD-6.8	1/4M6.8AZ
ZD-6.8	1/4M6.8AZ10
ZD-6.8	1/4M6.8AZ5
ZD-6.8	1/4M6.8Z
ZD-6.8	1/4M6.8Z10
ZD-6.8	1/4M6.8Z5
ZD-6.8	1C0038
ZD-6.8	1M6.8Z
ZD-6.8	1M6.8Z10
ZD-6.8	1M6.8Z5
ZD-6.8	1M6.8ZS10
ZD-6.8	1M6.8ZS5
ZD-6.8	1N1510
ZD-6.8	1N1510A
ZD-6.8	1N1521
ZD-6.8	1N1521A
ZD-6.8	1N1767
ZD-6.8	1N1930
ZD-6.8	1N1957
ZD-6.8	1N1984
ZD-6.8	1N2034
ZD-6.8	1N2034-2
ZD-6.8	1N2034A
ZD-6.8	1N2765
ZD-6.8	1N2765A
ZD-6.8	1N3016
ZD-6.8	1N3016A
ZD-6.8	1N3016B
ZD-6.8	1N3400
ZD-6.8	1N3412
ZD-6.8	1N3444
ZD-6.8	1N3514
ZD-6.8	1N3675
ZD-6.8	1N3675A
ZD-6.8	1N378
ZD-6.8	1N3829
ZD-6.8	1N3829A
ZD-6.8	1N4099
ZD-6.8	1N4158
ZD-6.8	1N4158A
ZD-6.8	1N4323
ZD-6.8	1N4323A
ZD-6.8	1N4400
ZD-6.8	1N4400A
ZD-6.8	1N4501
ZD-6.8	1N4628
ZD-6.8	1N4657
ZD-6.8	1N469
ZD-6.8	1N4692
ZD-6.8	1N469A
ZD-6.8	1N470
ZD-6.8	1N470A
ZD-6.8	1N470B
ZD-6.8	1N4736
ZD-6.8	1N4736A
ZD-6.8	1N475
ZD-6.8	1N475A
ZD-6.8	1N475B
ZD-6.8	1N5235
ZD-6.8	1N5235A
ZD-6.8	1N5235B
ZD-6.8	1N5526
ZD-6.8	1N5526A
ZD-6.8	1N5526B
ZD-6.8	1N5559
ZD-6.8	1N5559A
ZD-6.8	1N5732
ZD-6.8	1N5851
ZD-6.8	1N5851A
ZD-6.8	1N5851B
ZD-6.8	1N707
ZD-6.8	1N707A
ZD-6.8	1N707A(ZENER)
ZD-6.8	1N710
ZD-6.8	1N710A
ZD-6.8	1N754A
ZD-6.8	1N763-1
ZD-6.8	1N763-2
ZD-6.8	1N826
ZD-6.8	1N826A
ZD-6.8	1N828
ZD-6.8	1N828A
ZD-6.8	1N957
ZD-6.8	1N957A
ZD-6.8	1N957B
ZD-6.8	1S1174
ZD-6.8	1S1202
ZD-6.8	1S1376
ZD-6.8	1S1376A
ZD-6.8	1S192
ZD-6.8	1S1957
ZD-6.8	1S2113
ZD-6.8	1S2113A
ZD-6.8	1S222
ZD-6.8	1S2490
ZD-6.8	1S3006
ZD-6.8	1S332M
ZD-6.8	1S3332
ZD-6.8	1S471
ZD-6.8	1S481
ZD-6.8	1S636
ZD-6.8	1S7068
ZD-6.8	1S7068A
ZD-6.8	1S7068B
ZD-6.8	1S757
ZD-6.8	1WB-7A
ZD-6.8	1Z6.8
ZD-6.8	1Z6.8A
ZD-6.8	1Z68
ZD-6.8	1Z68A
ZD-60	1N1323
ZD-60	1N1323A
ZD-62	MZ500-35
ZD-62	PZW
ZD-62	1N1790
ZD-62	1N3698
ZD-62	1N3698A
ZD-62	1N3698B
ZD-62	1N4181
ZD-62	1N4181A
ZD-62	1N4181B
ZD-62	1N4346
ZD-62	1N4346A
ZD-62	1N4346B
ZD-62	1N4423
ZD-62	1N4851
ZD-62	1N4851A
ZD-63	DOOD
ZD-68	1N1431
ZD-68	1N1791
ZD-68	1N1791A
ZD-68	1N1886
ZD-68	1N1969
ZD-68	1N1996
ZD-68	1N3426
ZD-68	1N3457
ZD-68	1N3699
ZD-68	1N3699A
ZD-68	1N4182
ZD-68	1N4182A
ZD-68	1N4182B
ZD-68	1N4347
ZD-68	1N4347A
ZD-68	1N4347B
ZD-68	1N4424
ZD-68	1N4852
ZD-68	1N4852A
ZD-68	1N670
ZD-7.5	HZ-7
ZD-7.5	HZ-7A
ZD-7.5	HZ7A
ZD-7.5	OSD-0033
ZD-7.5	SZT8
ZD-7.5	1N1768
ZD-7.5	1N2034-3
ZD-7.5	1N3413
ZD-7.5	1N3676
ZD-7.5	1N3676A
ZD-7.5	1N4159
ZD-7.5	1N4159A
ZD-7.5	1N4324
ZD-7.5	1N4324A
ZD-7.5	1N4401
ZD-7.5	1N4629
ZD-7.5	1N4738A
ZD-7.5	1N5017
ZD-7.5	1N5852
ZD-7.5	1S1736
ZD-7.5	1S1768
ZD-7.5	1S332
ZD-7.5	1S617
ZD-75	1N1792
ZD-75	1N1792A
ZD-75	1N3700
ZD-75	1N3700A
ZD-75	1N4183
ZD-75	1N4183A
ZD-75	1N4183B
ZD-75	1N4348
ZD-75	1N4348A
ZD-75	1N4348B
ZD-75	1N4425
ZD-75	1N4853
ZD-75	1N4853A
ZD-8.0	MZ308
ZD-8.2	.25N8.2
ZD-8.2	.25T8.2
ZD-8.2	.25T8.2A
ZD-8.2	.4T8.2
ZD-8.2	.4T8.2A
ZD-8.2	.4T8.2B
ZD-8.2	.7JZ8.2
ZD-8.2	.7Z8.2A
ZD-8.2	.7Z8.2B
ZD-8.2	.7Z8.2C
ZD-8.2	.7Z8.2D
ZD-8.2	.7ZM8.2A
ZD-8.2	.7ZM8.2B
ZD-8.2	.7ZM8.2C
ZD-8.2	.7ZM8.2D
ZD-8.2	OA01-08R
ZD-8.2	A72-86-700
ZD-8.2	AW-01-08
ZD-8.2	AW01-08
ZD-8.2	AW01-08J
ZD-8.2	AZ-077
ZD-8.2	AZ-079
ZD-8.2	AZ-081
ZD-8.2	AZ-083
ZD-8.2	AZ756
ZD-8.2	AZ756A
ZD-8.2	AZ8.2
ZD-8.2	AZ959
ZD-8.2	AZ959A
ZD-8.2	AZ959B
ZD-8.2	B-1808
ZD-8.2	B1-8.2
ZD-8.2	BZ-077
ZD-8.2	BZ-079
ZD-8.2	BZ-0800
ZD-8.2	BZ-081
ZD-8.2	BZ-083
ZD-8.2	BZ080
ZD-8.2	BZ081
ZD-8.2	BZ085
ZD-8.2	BZ8.2
ZD-8.2	BZX29C8V2
ZD-8.2	BZX46C8V2
ZD-8.2	BZX55C8V2
ZD-8.2	BZX71C8V2
ZD-8.2	BZX79C8V2
ZD-8.2	BZX83C8V2
ZD-8.2	BZY62
ZD-8.2	BZY83C8V2
ZD-8.2	BZY83D8V2
ZD-8.2	BZY85C8V2
ZD-8.2	BZY85D8V2
ZD-8.2	BZY88C8V2
ZD-8.2	BZY92C8V2
ZD-8.2	BZZ13
ZD-8.2	C02

GE	Industry Standard No.
ZD-8.2	CI-8.2
ZD-8.2	C02
ZD-8.2	CXL5016
ZD-8.2	CXL5072
ZD-8.2	D1A(ZENER)
ZD-8.2	D1U
ZD-8.2	D6M
ZD-8.2	DDAY008003
ZD-8.2	DIA
ZD-8.2	DS-149
ZD-8.2	DS-49
ZD-8.2	DS149
ZD-8.2	DX-0729
ZD-8.2	DX-081
ZD-8.2	DZ0820
ZD-8.2	ECG5016
ZD-8.2	ECG5016A
ZD-8.2	ECG5072
ZD-8.2	ECG5072A
ZD-8.2	EDZ-14
ZD-8.2	EQA01-08
ZD-8.2	EQA01-08R
ZD-8.2	EQB-0108
ZD-8.2	EQB01-08
ZD-8.2	EZ8(ELCOM)
ZD-8.2	GE-X11
ZD-8.2	GEZD-8.2
ZD-8.2	HD3000409
ZD-8.2	HD3003109
ZD-8.2	HEP-Z0217
ZD-8.2	HEP-Z0411
ZD-8.2	HEPZ0217
ZD-8.2	HEPZ0411
ZD-8.2	HM8.2
ZD-8.2	HM8.2A
ZD-8.2	HW8.2
ZD-8.2	HW8.2A
ZD-8.2	IW01-08J
ZD-8.2	KZ-8A
ZD-8.2	MA1082
ZD-8.2	MB6356
ZD-8.2	ME408-02C
ZD-8.2	MZ-08
ZD-8.2	MZ-208
ZD-8.2	MZ-8
ZD-8.2	MZ1008
ZD-8.2	MZ208
ZD-8.2	MZ408-02C
ZD-8.2	MZ8
ZD-8.2	02Z8.2A
ZD-8.2	05Z8.2U
ZD-8.2	0AZ206
ZD-8.2	0AZ246
ZD-8.2	PD6012
ZD-8.2	PD6012A
ZD-8.2	PD6053
ZD-8.2	PS8911
ZD-8.2	QA01-082
ZD-8.2	QA01-08R
ZD-8.2	QD-ZMZ408CE
ZD-8.2	R-1S333Y
ZD-8.2	RD-13AL
ZD-8.2	RD-8.2A
ZD-8.2	RD-8.2E
ZD-8.2	RD-8.2E(C)
ZD-8.2	RD-8.2EB
ZD-8.2	RD-8.2FB
ZD-8.2	RD8
ZD-8.2	RD8.2
ZD-8.2	RD8.2C
ZD-8.2	RD8.2E
ZD-8.2	RD8.2EA
ZD-8.2	RD8.2EB
ZD-8.2	RD8.2EK
ZD-8.2	RD8.2F
ZD-8.2	RD8.2FA
ZD-8.2	RD8.2FB
ZD-8.2	RD8.2M
ZD-8.2	RE-112
ZD-8.2	RE112
ZD-8.2	REN5072
ZD-8.2	RF-8.2E
ZD-8.2	RF8.2E
ZD-8.2	RT6105
ZD-8.2	RT6105(G.E.)
ZD-8.2	RT6923
ZD-8.2	RVD1N4738
ZD-8.2	RZ8.2
ZD-8.2	RZZ8.2
ZD-8.2	SD-8(ZENER)
ZD-8.2	SK3136
ZD-8.2	SVD02Z8
ZD-8.2	SVD02Z8.2
ZD-8.2	SVD02Z8.2A
ZD-8.2	SZ200-8.2
ZD-8.2	TMD06
ZD-8.2	TMD06A
ZD-8.2	TR9SA
ZD-8.2	TVSYZ-080
ZD-8.2	TVSYZ080
ZD-8.2	001-0161-00
ZD-8.2	1/2Z8.2T5
ZD-8.2	1/2Z82T5
ZD-8.2	1/4M8/27
ZD-8.2	1M8.2Z
ZD-8.2	1M8.2Z10
ZD-8.2	1M8.2Z5
ZD-8.2	1M8.2ZS10
ZD-8.2	1M8.2ZS5
ZD-8.2	1N1425
ZD-8.2	1N1511
ZD-8.2	1N1511A
ZD-8.2	1N1522
ZD-8.2	1N1522A
ZD-8.2	1N1769
ZD-8.2	1N1875
ZD-8.2	1N1875A
ZD-8.2	1N1875B
ZD-8.2	1N1931
ZD-8.2	1N1958
ZD-8.2	1N1985
ZD-8.2	1N2035A
ZD-8.2	1N3018
ZD-8.2	1N3018A
ZD-8.2	1N3018B
ZD-8.2	1N3154
ZD-8.2	1N3154A
ZD-8.2	1N3155
ZD-8.2	1N3155A
ZD-8.2	1N3156
ZD-8.2	1N3156A
ZD-8.2	1N3157
ZD-8.2	1N3157A
ZD-8.2	1N3199

GE	Industry Standard No.
ZD-8.2	1N3401
ZD-8.2	1N3414
ZD-8.2	1N3433
ZD-8.2	1N3445
ZD-8.2	1N3516
ZD-8.2	1N3677
ZD-8.2	1N3677A
ZD-8.2	1N4101
ZD-8.2	1N4160
ZD-8.2	1N4160A
ZD-8.2	1N430
ZD-8.2	1N430A
ZD-8.2	1N430B
ZD-8.2	1N4325
ZD-8.2	1N4325A
ZD-8.2	1N4402
ZD-8.2	1N4402A
ZD-8.2	1N4630
ZD-8.2	1N4659
ZD-8.2	1N4694
ZD-8.2	1N4738
ZD-8.2	1N5237
ZD-8.2	1N5237A
ZD-8.2	1N5237B
ZD-8.2	1N5528
ZD-8.2	1N5528A
ZD-8.2	1N5528B
ZD-8.2	1N5561
ZD-8.2	1N5561A
ZD-8.2	1N5734
ZD-8.2	1N5734B
ZD-8.2	1N5853
ZD-8.2	1N5853A
ZD-8.2	1N5853B
ZD-8.2	1N664
ZD-8.2	1N712
ZD-8.2	1N712A
ZD-8.2	1N756
ZD-8.2	1N756A
ZD-8.2	1N764-1
ZD-8.2	1N959
ZD-8.2	1N959A
ZD-8.2	1N959B
ZD-8.2	1R8.2
ZD-8.2	1R8.2A
ZD-8.2	1R8.2B
ZD-8.2	1S1164
ZD-8.2	1S193
ZD-8.2	1S1958
ZD-8.2	1S2082
ZD-8.2	1S2082A
ZD-8.2	1S2115
ZD-8.2	1S2115A
ZD-8.2	1S2190
ZD-8.2	1S2191
ZD-8.2	1S2192
ZD-8.2	1S3008
ZD-8.2	1S333
ZD-8.2	1S333Y
ZD-8.2	1S7082
ZD-8.2	1S7082A
ZD-8.2	1S7082B
ZD-8.2	1Z8.2
ZD-8.2	1Z8.2A
ZD-8.2	1Z8.2T10
ZD-8.2	1Z8.2T20
ZD-8.2	1Z8.2T5
ZD-8.2	1Z8.3
ZD-8.2	1ZC8.2T10
ZD-8.2	1ZC8.2T20
ZD-8.2	1ZC8.2T5
ZD-8.2	1ZD8-2
ZD-8.2	1ZD8-5
ZD-8.2	1ZD8.2V
ZD-8.2	1ZD825
ZD-8.2	1ZD82S
ZD-8.7	CXL5073
ZD-8.7	EA16X402
ZD-8.7	ECG5073
ZD-8.7	ECG5073A
ZD-8.7	HZ9B
ZD-8.7	RD-8.2EC
ZD-8.7	RD8.2EC
ZD-8.7	RD9B
ZD-8.7	SK3749
ZD-8.7	SK3749/5073A
ZD-8.7	TR209T2
ZD-8.7	001-0163-13
ZD-8.7	1N4775
ZD-8.7	1N4775A
ZD-8.7	1N4776
ZD-8.7	1N4776A
ZD-8.7	1N4777
ZD-8.7	1N4777A
ZD-8.7	1N4778
ZD-8.7	1N4778A
ZD-8.7	1N4779
ZD-8.7	1N4779A
ZD-8.7	1N4780
ZD-8.7	1N4780A
ZD-8.7	1N4781
ZD-8.7	1N4781A
ZD-8.7	1N4782
ZD-8.7	1N4782A
ZD-8.7	1N4783
ZD-8.7	1N4783A
ZD-8.7	1N4784
ZD-8.7	1N4784A
ZD-8.7	1N5238
ZD-8.7	1N5238B
ZD-8.7	1N5854
ZD-8.7	1N5854A
ZD-8.7	1N5854B
ZD-8.7	1S1175
ZD-8.7	1S133
ZD-8.7	1S1377
ZD-8.7	1S1377A
ZD-8.7	1S1392
ZD-8.7	1S1392A
ZD-8.7	1S1408
ZD-8.7	1S1408R
ZD-8.7	1S1737
ZD-8.7	1S1769
ZD-8.7	1S1781
ZD-8.7	1S1782
ZD-8.7	1S1783
ZD-8.7	1S2126
ZD-8.7	1S2129
ZD-8.7	1S2137
ZD-8.7	1S2195
ZD-8.7	1S758H
ZD-8.7	1TWC8H
ZD-8.7	1TWC8L
ZD-8.7	1TWC8M

GE	Industry Standard No.
ZD-8.7	1WB-9D
ZD-82	CD3174
ZD-82	MZ500-38
ZD-82	1,000,111-00
ZD-82	1N1793
ZD-82	1N1793A
ZD-82	1N1887
ZD-82	1N1943
ZD-82	1N1970
ZD-82	1N1997
ZD-82	1N3427
ZD-82	1N3458
ZD-82	1N3701
ZD-82	1N3701A
ZD-82	1N3701B
ZD-82	1N4184
ZD-82	1N4184A
ZD-82	1N4184B
ZD-82	1N4349
ZD-82	1N4349A
ZD-82	1N4349B
ZD-82	1N4426
ZD-82	1N4854
ZD-82	1N4854A
ZD-82	1N4854B
ZD-87	1N1325A
ZD-9.1	.25N9.1
ZD-9.1	.25T8.7
ZD-9.1	.25T8.7A
ZD-9.1	.25T8.8
ZD-9.1	.25T8.8A
ZD-9.1	.25T9.1
ZD-9.1	.25T9.1A
ZD-9.1	.25T9.1B
ZD-9.1	.4T9.1
ZD-9.1	.4T9.1A
ZD-9.1	.4T9.1B
ZD-9.1	.7JZ9.1
ZD-9.1	.7Z9.1A
ZD-9.1	.7Z9.1B
ZD-9.1	.7Z9.1C
ZD-9.1	.7Z9.1D
ZD-9.1	.7ZM9.1A
ZD-9.1	.7ZM9.1B
ZD-9.1	.7ZM9.1C
ZD-9.1	.7ZM9.1D
ZD-9.1	A04234-2
ZD-9.1	A061-119
ZD-9.1	A068-103
ZD-9.1	A04234-2
ZD-9.1	AW-01-07
ZD-9.1	AW-01-09
ZD-9.1	AW01-09
ZD-9.1	AW01-9
ZD-9.1	AW09
ZD-9.1	AZ-085
ZD-9.1	AZ-088
ZD-9.1	AZ-090
ZD-9.1	AZ-092
ZD-9.1	AZ757
ZD-9.1	AZ757A
ZD-9.1	AZ9.1
ZD-9.1	AZ960
ZD-9.1	AZ960A
ZD-9.1	AZ960B
ZD-9.1	B090
ZD-9.1	B2090
ZD-9.1	B66X0040-003
ZD-9.1	BX090
ZD-9.1	BX909
ZD-9.1	BXY63
ZD-9.1	BZ-080
ZD-9.1	BZ-085
ZD-9.1	BZ-088
ZD-9.1	BZ-090
ZD-9.1	BZ-090(ZENER)
ZD-9.1	BZ-0900
ZD-9.1	BZ-0901
ZD-9.1	BZ-092
ZD-9.1	BZ-094
ZD-9.1	BZ090
ZD-9.1	BZ090.1Z9
ZD-9.1	BZ094
ZD-9.1	BZ1-9
ZD-9.1	BZ1-90
ZD-9.1	BZ9.1
ZD-9.1	BZX14
ZD-9.1	BZX29C9V1
ZD-9.1	BZX46C9V1
ZD-9.1	BZX55C9V1
ZD-9.1	BZX71C9V1
ZD-9.1	BZX71C9V2
ZD-9.1	BZX72
ZD-9.1	BZX72A
ZD-9.1	BZX72B
ZD-9.1	BZX72C
ZD-9.1	BZX79C9V1
ZD-9.1	BZX83C9V1
ZD-9.1	BZY63
ZD-9.1	BZY68
ZD-9.1	BZY83C9V1
ZD-9.1	BZY85C9V1
ZD-9.1	BZY88/C9V1
ZD-9.1	BZY88/C9V5
ZD-9.1	BZY88C9V1
ZD-9.1	BZY92C9V1
ZD-9.1	C21383
ZD-9.1	C4015
ZD-9.1	CA-092
ZD-9.1	CD31-00012
ZD-9.1	CD31-12019
ZD-9.1	CD31-12039
ZD-9.1	CD3112039
ZD-9.1	CD3122055
ZD-9.1	CD3212055
ZD-9.1	CDZ-9V
ZD-9.1	CDZ-C9V
ZD-9.1	CF-092
ZD-9.1	CP3212055
ZD-9.1	CXL139
ZD-9.1	CXL5018
ZD-9.1	CZ-092
ZD-9.1	CZ-92
ZD-9.1	CZ092
ZD-9.1	CZ094
ZD-9.1	D-00569C
ZD-9.1	D-18
ZD-9.1	D1-8
ZD-9.1	D176
ZD-9.1	D1G
ZD-9.1	D2Z
ZD-9.1	D3F
ZD-9.1	D8U
ZD-9.1	DAAY010092

GE	Industry Standard No.
ZD-9.1	DDAY010002
ZD-9.1	DDAY010005
ZD-9.1	DDAY126001
ZD-9.1	DI-8
ZD-9.1	DI-8(COURIER)
ZD-9.1	DIJ72165
ZD-9.1	DIJ72172
ZD-9.1	DS-43
ZD-9.1	DS43
ZD-9.1	DS48
ZD-9.1	DS99
ZD-9.1	DX-0087
ZD-9.1	DX-0728
ZD-9.1	DZ10
ZD-9.1	E0771-3
ZD-9.1	EA16X62
ZD-9.1	EA16X74
ZD-9.1	EA16X82
ZD-9.1	EA3866
ZD-9.1	ECG139
ZD-9.1	ECG139A
ZD-9.1	ECG5017
ZD-9.1	ECG5018
ZD-9.1	ECG5018A
ZD-9.1	ED491130
ZD-9.1	EDZ-0045
ZD-9.1	EDZ-23
ZD-9.1	E0771-3
ZD-9.1	EQ-09R
ZD-9.1	EQA-01098
ZD-9.1	EQA-9
ZD-9.1	EQA.9
ZD-9.1	EQA01-01
ZD-9.1	EQA01-09
ZD-9.1	EQA01-09(R)
ZD-9.1	EQA01-09R
ZD-9.1	EQA09R
ZD-9.1	EQB01-09
ZD-9.1	EQB01-90S
ZD-9.1	ETD-RD9.1FB
ZD-9.1	EVR9
ZD-9.1	EVR9A
ZD-9.1	EVR9B
ZD-9.1	EZ9(ELCOM)
ZD-9.1	G01-036-G
ZD-9.1	G01-036-H
ZD-9.1	G01036
ZD-9.1	G01036G
ZD-9.1	G9.1T10
ZD-9.1	G9.1T20
ZD-9.1	G9.1T5
ZD-9.1	GEZD-9.1
ZD-9.1	GLA91
ZD-9.1	GLA91A
ZD-9.1	GLA91B
ZD-9.1	G01036
ZD-9.1	GZ9.1
ZD-9.1	HD3000401
ZD-9.1	HD30017090
ZD-9.1	HEP-Z0219
ZD-9.1	HEP-Z0412
ZD-9.1	HEP104
ZD-9.1	HEPZ0219
ZD-9.1	HEPZ0412
ZD-9.1	HF-1S334
ZD-9.1	HF-1S339
ZD-9.1	HF-20004
ZD-9.1	HF-20011
ZD-9.1	HF-20065
ZD-9.1	HM9.1
ZD-9.1	HM9.1A
ZD-9.1	HM9.1B
ZD-9.1	HN-00012
ZD-9.1	HN-00024
ZD-9.1	HS7091
ZD-9.1	HW9.1
ZD-9.1	HW9.1A
ZD-9.1	HW9.1B
ZD-9.1	HZ-9
ZD-9.1	HZ-9B
ZD-9.1	HZ9
ZD-9.1	HZ9C2
ZD-9.1	HZ9H
ZD-9.1	IC743047
ZD-9.1	IP20-0019
ZD-9.1	J24632
ZD-9.1	KD2504
ZD-9.1	KS2091A
ZD-9.1	KS2091B
ZD-9.1	KS41A
ZD-9.1	KS41AF
ZD-9.1	KS091A
ZD-9.1	LPM9.1
ZD-9.1	LPM9.1A
ZD-9.1	LR91CH
ZD-9.1	M4653
ZD-9.1	M4816
ZD-9.1	M4Z9.1
ZD-9.1	M4Z9.1-20
ZD-9.1	M4Z9.1A
ZD-9.1	M9Z
ZD-9.1	MA1091
ZD-9.1	MD757
ZD-9.1	MD757A
ZD-9.1	ME409-02B
ZD-9.1	MGLA91
ZD-9.1	MGLA91A
ZD-9.1	MGLA91B
ZD-9.1	MR91C-H
ZD-9.1	MR91E-H
ZD-9.1	MTZ613
ZD-9.1	MTZ613A
ZD-9.1	MZ-209
ZD-9.1	MZ-9
ZD-9.1	MZ-92-9.1B
ZD-9.1	MZ090
ZD-9.1	MZ209
ZD-9.1	MZ209A
ZD-9.1	MZ209B
ZD-9.1	MZ209C
ZD-9.1	MZ250
ZD-9.1	MZ309
ZD-9.1	MZ309B
ZD-9.1	MZ310
ZD-9.1	MZ500-15
ZD-9.1	MZ500.15
ZD-9.1	MZ9.1B
ZD-9.1	MZ92-9.1A
ZD-9.1	MZ92-9.1B
ZD-9.1	MZX9.1
ZD-9.1	N-756A
ZD-9.1	NGP5007
ZD-9.1	0AZ207
ZD-9.1	0AZ212

GE	Industry Standard No.
ZD-9.1	0AZ247
ZD-9.1	0AZ272
ZD-9.1	P21344
ZD-9.1	PD6013
ZD-9.1	PD6013A
ZD-9.1	PD6054
ZD-9.1	PS10019B
ZD-9.1	PS10063
ZD-9.1	PS8912
ZD-9.1	PTC505
ZD-9.1	Q-25115C
ZD-9.1	QD-ZMZ409BE
ZD-9.1	QD-ZRD9EXAA
ZD-9.1	R-7097
ZD-9.1	RD-7A
ZD-9.1	RD-9.1E
ZD-9.1	RD-91
ZD-9.1	RD-91E
ZD-9.1	RD-96
ZD-9.1	RD-9A
ZD-9.1	RD-9AL
ZD-9.1	RD-9L
ZD-9.1	RD7A
ZD-9.1	RD9
ZD-9.1	RD9.1E
ZD-9.1	RD9.1EA
ZD-9.1	RD9.1EB
ZD-9.1	RD9.1ED
ZD-9.1	RD9.1EK
ZD-9.1	RD9.1F
ZD-9.1	RD9.1FA
ZD-9.1	RD9.1FB
ZD-9.1	RD9.1M
ZD-9.1	RD9A
ZD-9.1	RD9A(10)
ZD-9.1	RD9A-N
ZD-9.1	RD9AL
ZD-9.1	RD9AM
ZD-9.1	RD9AN
ZD-9.1	RE-114
ZD-9.1	RE114
ZD-9.1	REN139
ZD-9.1	REN5073
ZD-9.1	RFJ71480
ZD-9.1	RP-9A
ZD-9.1	RS1290
ZD-9.1	RT-240
ZD-9.1	RT8339
ZD-9.1	RV2213
ZD-9.1	RVD1N4739
ZD-9.1	RVDMZ209
ZD-9.1	RVDRD11AN
ZD-9.1	RX090
ZD-9.1	RZ9.1
ZD-9.1	RZZ9.1
ZD-9.1	S
ZD-9.1	S0702
ZD-9.1	SA-93794
ZD-9.1	SC91
ZD-9.1	SD-15
ZD-9.1	SD-27
ZD-9.1	SD-632
ZD-9.1	SD-8(PHILCO)
ZD-9.1	SD105
ZD-9.1	SD632
ZD-9.1	SD632(10)
ZD-9.1	SK3060
ZD-9.1	SK3060/139A
ZD-9.1	S0-632
ZD-9.1	SV-9
ZD-9.1	SV-9(ZENER)
ZD-9.1	SVD-1S1717
ZD-9.1	SVD02Z9.5A
ZD-9.1	SVD1S1717
ZD-9.1	SZ-200-8
ZD-9.1	SZ-200-9
ZD-9.1	SZ-200-9V
ZD-9.1	SZ-9
ZD-9.1	SZ9
ZD-9.1	SZ9.1
ZD-9.1	SZC9
ZD-9.1	SZP-9
ZD-9.1	SZT-9
ZD-9.1	SZT9
ZD-9.1	T-750-714
ZD-9.1	T-R9S
ZD-9.1	T21334
ZD-9.1	T21639
ZD-9.1	TMD07
ZD-9.1	TMD07A
ZD-9.1	TR-95
ZD-9.1	TR-95(B)
ZD-9.1	TR-95B
ZD-9.1	TR-9GS
ZD-9.1	TR-9S
ZD-9.1	TR-9SA
ZD-9.1	TR-9SB
ZD-9.1	TR9S
ZD-9.1	TR9SB
ZD-9.1	TVSRD9AL
ZD-9.1	WZ090
ZD-9.1	001-0101-01
ZD-9.1	001-0163-15
ZD-9.1	1-20363
ZD-9.1	1/2Z9.1T5
ZD-9.1	1/4A9.1
ZD-9.1	1/4A9.1A
ZD-9.1	1/4A9.1B
ZD-9.1	1/4M9.1Z
ZD-9.1	1/4M9.1Z10
ZD-9.1	1/4M9.1Z5
ZD-9.1	1A12688
ZD-9.1	1A9.1M
ZD-9.1	1A9.1MA
ZD-9.1	1E9.1Z
ZD-9.1	1E9.1Z10
ZD-9.1	1EZ9.1
ZD-9.1	1M9.1Z
ZD-9.1	1M9.1Z10
ZD-9.1	1M9.1Z5
ZD-9.1	1M9.1ZS10
ZD-9.1	1M9.1ZS5
ZD-9.1	1N1313
ZD-9.1	1N1313A
ZD-9.1	1N1530
ZD-9.1	1N1530A
ZD-9.1	1N1770
ZD-9.1	1N1770A
ZD-9.1	1N2035-1
ZD-9.1	1N2035-12
ZD-9.1	1N2163
ZD-9.1	1N2163A
ZD-9.1	1N2164
ZD-9.1	1N2164A

GE	Industry Standard No.	GE	Industry Standard No.	GE	Industry Standard No.	GE	Industry Standard No.	GE	Industry Standard No.
ZD-9.1	1N2165	ZD-9.1	182068	1	A88C-705	1	GT100	1	RS-2690
ZD-9.1	1N2165A	ZD-9.1	182069	1	A88C19G	1	GT11	1	RS-2691
ZD-9.1	1N2166	ZD-9.1	182091(ZENER)	1	AA1	1	GT12	1	RS-2692
ZD-9.1	1N2166A	ZD-9.1	182091A	1	ACR810-104	1	GT13	1	RS-2694
ZD-9.1	1N2167	ZD-9.1	18116	1	ACR810-105	1	GT153	1	RS-2695
ZD-9.1	1N2167A	ZD-9.1	18116A	1	ACR810-106	1	GT1604	1	RS-2696
ZD-9.1	1N2168	ZD-9.1	18212	1	ACR83-1004	1	GT1605	1	RS-3277
ZD-9.1	1N2168A	ZD-9.1	18213	1	ACR83-1005	1	GT1606	1	RS-3278
ZD-9.1	1N2169	ZD-9.1	18214	1	ACR83-1006	1	GT1607	1	RS-3279
ZD-9.1	1N2169A	ZD-9.1	18217	1	ACY40	1	GT210H	1	RS-3288
ZD-9.1	1N2170	ZD-9.1	18224	1	AF-101	1	GT269	1	RS-3309
ZD-9.1	1N2170A	ZD-9.1	182493	1	AP138/290	1	GT2693	1	RS-3867
ZD-9.1	1N2171	ZD-9.1	182550	1	APZ23	1	GT2694	1	RS-3868
ZD-9.1	1N2171A	ZD-9.1	183009	1	ALZ10	1	GT2695	1	RS-3907
ZD-9.1	1N225	ZD-9.1	0000018334	1	A01	1	GT348	1	RS-3913
ZD-9.1	1N225A	ZD-9.1	18334K	1	ASY-26	1	GT761	1	RS-3914
ZD-9.1	1N2620	ZD-9.1	18334M	1	ASY-27	1	GT761R	1	RS-3915
ZD-9.1	1N2620A	ZD-9.1	18334N	1	ASY26	1	GT762	1	RS-3929
ZD-9.1	1N2620B	ZD-9.1	18472	1	ASY27	1	GT762R	1	RS-5104
ZD-9.1	1N2621	ZD-9.1	18482	1	ASY56N	1	GT764	1	RS-5105
ZD-9.1	1N2621A	ZD-9.1	1855	1	ASY57N	1	GT766	1	RS-5106
ZD-9.1	1N2621B	ZD-9.1	1856	1	ASY58N	1	GT766A	1	RS-5401
ZD-9.1	1N2622	ZD-9.1	187091	1	ASY76	1	GT81H	1	RS-5504
ZD-9.1	1N2622A	ZD-9.1	187091A	1	ASY77	1	GT83	1	RS-5511
ZD-9.1	1N2622B	ZD-9.1	187091B	1	ASY80	1	GT832	1	RS-5540
ZD-9.1	1N2623	ZD-9.1	18757A	1	AT-15	1	GT87	1	RS2690
ZD-9.1	1N2623A	ZD-9.1	18758	1	AT-5	1	GT88	1	RS2691
ZD-9.1	1N2623B	ZD-9.1	1875898	1	B103	1	GTE1	1	RS2692
ZD-9.1	1N2624	ZD-9.1	1T243M	1	B104	1	GTE2	1	RS2696
ZD-9.1	1N2624A	ZD-9.1	1T9.1	1	B290	1	GTV	1	RS3281
ZD-9.1	1N2624B	ZD-9.1	1T9.1A	1	B291	1	HA-00102	1	RS3287
ZD-9.1	1N2790	ZD-9.1	1T9.1B	1	B292	1	HA-12	1	RS3892
ZD-9.1	1N3019	ZD-9.1	1TA9.1	1	B292A	1	HA-52	1	RS3914
ZD-9.1	1N3019A	ZD-9.1	1TA9.1A	1	B314	1	HA-53	1	RS3915
ZD-9.1	1N3019B	ZD-9.1	129.1	1	B387	1	HA00052	1	RS5104
ZD-9.1	1N3108	ZD-9.1	129.1A	1	B392	1	HA00053	1	RS5105
ZD-9.1	1N3148	ZD-9.1	129.1B	1	B393	1	HA15	1	RS5302
ZD-9.1	1N33974	ZD-9.1	129.1C	1	B394	1	HA202	1	RS5303
ZD-9.1	1N3517	ZD-9.1	129.1D15	1	B395	1	HA49	1	RS5402
ZD-9.1	1N3678	ZD-9.1	129.1D5	1	B396	1	HEP2	1	RS5403
ZD-9.1	1N3678A	ZD-9.1	129.1T10	1	B401	1	HEPG0004	1	RS5504
ZD-9.1	1N3855A	ZD-9.1	129.1T20	1	B402	1	HEPG0005	1	RS5511
ZD-9.1	1N4102	ZD-9.1	129.1T5	1	B403	1	HEPG0005P/G	1	RS5540
ZD-9.1	1N4103	ZD-9.1	129.1D10	1	B408	1	HEPG0006	1	RS5743.3
ZD-9.1	1N4161	ZD-9.1	129.1.D10	1	B416	1	HEPG0006P/G	1	S74-3-A-3P
ZD-9.1	1N4161A	ZD-9.1	1ZC9.1	1	B417	1	HEPG0007	1	S88C-1-3P
ZD-9.1	1N4161B	ZD-9.1	1ZC9.1T10	1	B74-3-A-21	1	HEPG0007P/G	1	S043
ZD-9.1	1N4297	ZD-9.1	1ZC9.1T5	1	B74-3A21	1	HJ41	1	S044
ZD-9.1	1N4297A	ZD-9.1	1ZP9.1T10	1	B75A	1	HM-00049	1	S046
ZD-9.1	1N4298	ZD-9.1	1ZP9.1T20	1	B88C-1-21	1	HR4A	1	SF-T171
ZD-9.1	1N4298A	ZD-9.1	1ZP9.1T5	1	BB6	1	HR5	1	SF-T172
ZD-9.1	1N4302	ZD-9.1	1ZM9.1T10	1	BB6A	1	HR8A	1	SF-T173
ZD-9.1	1N4302A	ZD-9.1	1ZM9.1T20	1	C1437	1	HR9A	1	SF-T174
ZD-9.1	1N4326	ZD-9.1	1ZM9.1T5	1	C1438	1	IRTR05	1	SFT-307
ZD-9.1	1N4326A	ZD-9.1	2M214	1	C73	1	IRTR11	1	SFT-319
ZD-9.1	1N4326B	ZD.91	MZ500-39	1	C75	1	J24833	1	SFT223
ZD-9.1	1N4403	ZD.91	1N1794	1	C76	1	K14-0066-4	1	SFT226
ZD-9.1	1N4403A	ZD.91	1N1794A	1	CB-103	1	M4389	1	SFT227
ZD-9.1	1N4403B	ZD.91	1N3702	1	CK14	1	MA100	1	SFT228
ZD-9.1	1N4631	ZD.91	1N3702A	1	CK14A	1	MA286	1	SFT229
ZD-9.1	1N4660	ZD.91	1N4185	1	CK16	1	MA287	1	SFT237
ZD-9.1	1N4695	ZD.91	1N4185A	1	CK16A	1	NK74-3A19	1	SFT251
ZD-9.1	1N4739	ZD.91	1N4185B	1	CK17	1	NK88C119	1	SFT252
ZD-9.1	1N4739A	ZD.91	1N4350	1	CK17A	1	NK812	1	SFT253
ZD-9.1	1N4765	ZD.91	1N4350A	1	CK25	1	NK2128	1	SFT288
ZD-9.1	1N4765A	ZD.91	1N4350B	1	CK25A	1	NK2129	1	SK3005
ZD-9.1	1N4766	ZD.91	1N4427	1	CK26	1	NK2141	1	SM-217
ZD-9.1	1N4766A	ZD.91	1N4855	1	CK26A	1	NK2142	1	ST-28B
ZD-9.1	1N4767	ZD.91	1N4855A	1	CK27	1	NK2143	1	ST-28C
ZD-9.1	1N4767A	ZD.8.7	RE113	1	CK27A	1	NK2144	1	ST-370
ZD-9.1	1N4768	ZD15	181177	1	CK661	1	NK2162	1	ST-37C
ZD-9.1	1N4768A	ZD5.6	DX-0061	1	CK662	1	NK2163	1	ST-37D
ZD-9.1	1N4769	ZD5.6	02Z5.6A	1	CK759	1	NK216325	1	STY-105
ZD-9.1	1N4769A	ZD5.6	RD5.6B	1	CK759A	1	NK2164	1	STY-106
ZD-9.1	1N4770	ZD9.1	1N5855A	1	CK760	1	NK216425	1	STY-160
ZD-9.1	1N4770A	ZJ436C	S6100C	1	CK760A	1	NK2202	1	STY-1608
ZD-9.1	1N4771	ZJ436E	S6100E	1	CK761	1	NK2203	1	STY-2248
ZD-9.1	1N4771A	1	A1243	1	CK768	1	NK2204	1	STY-2249
ZD-9.1	1N4772	1	A146	1	CK776	1	NK2205	1	STY-2250
ZD-9.1	1N4772A	1	A147	1	CK776A	1	NK2206	1	STYL05
ZD-9.1	1N4773	1	A1474-3	1	CXL100	1	NK2207	1	STYL06
ZD-9.1	1N4773A	1	A1474-39	1	D019	1	NK2221	1	STYL1588
ZD-9.1	1N4774	1	A148	1	D078	1	NK222281	1	STYL160
ZD-9.1	1N4774A	1	A1488C	1	DS-22	1	NK222282	1	STYL1608
ZD-9.1	1N4831	1	A1488C9	1	DS-28(DELCO)	1	NK2261	1	STYL1690
ZD-9.1	1N4831A	1	A149	1	DS21	1	NK2262	1	STYL1697
ZD-9.1	1N5238A	1	A160(JAPAN)	1	DS22	1	NK2263	1	STYL1717
ZD-9.1	1N5239	1	A167	1	DS23	1	NK2264	1	STYL2120
ZD-9.1	1N5239A	1	A168(JAPAN)	1	DS53	1	NK242	1	STYL2247
ZD-9.1	1N5239B	1	A168A	1	E2412	1	NK243	1	STYL2250
ZD-9.1	1N5529	1	A169	1	EC0100	1	NK762	1	T-109
ZD-9.1	1N5529A	1	A170(JAPAN)	1	ED52	1	NK763	1	T-116
ZD-9.1	1N5529B	1	A171(JAPAN)	1	ED53	1	NK764	1	T-152148
ZD-9.1	1N5562	1	A172	1	ED54B	1	NK772	1	T-1877
ZD-9.1	1N5562A	1	A172A	1	EK159	1	NK773	1	T-2038
ZD-9.1	1N5735	1	A182	1	EO105	1	NK74	1	T-2039
ZD-9.1	1N5855	1	A198	1	EO65	1	OC-130	1	T-2040
ZD-9.1	1N5855B	1	A204	1	EO66	1	OC-140	1	T-2091
ZD-9.1	1N713	1	A205	1	EO67	1	OC-410	1	T-2439
ZD-9.1	1N713A	1	A206	1	EO68	1	OC-44	1	T-2440
ZD-9.1	1N713B	1	A207	1	ES25	1	OC-45	1	T-2441
ZD-9.1	1N757	1	A208(JAPAN)	1	ES26	1	OC-46	1	T-46
ZD-9.1	1N757A	1	A209	1	FV2747C	1	OC-47	1	T-47
ZD-9.1	1N764	1	A210(JAPAN)	1	GA52829	1	PT-530A	1	T-48
ZD-9.1	1N764-3	1	A211(JAPAN)	1	GA53149	1	PTC102	1	T-52148Z
ZD-9.1	1N935	1	A212	1	GA53242	1	Q-1A	1	T-52149
ZD-9.1	1N935A	1	A217	1	GC1159	1	R-119	1	T-52149Z
ZD-9.1	1N935B	1	A248(JAPAN)	1	GC1302	1	R-163	1	T-78
ZD-9.1	1N936	1	A26	1	GC181	1	R-186	1	T-82
ZD-9.1	1N936A	1	A277(JAPAN)	1	GC182	1	R-227	1	T-Q5020
ZD-9.1	1N936B	1	A278(JAPAN)	1	GC31	1	R-244	1	T0-101
ZD-9.1	1N937	1	A279(JAPAN)	1	GC32	1	R-424	1	T0-102
ZD-9.1	1N937A	1	A282	1	GC33	1	R-425	1	T0101
ZD-9.1	1N937B	1	A283	1	GC34	1	R-488	1	T0102
ZD-9.1	1N938	1	A284	1	GC35	1	R-506	1	T03383
ZD-9.1	1N938A	1	A305	1	GC360	1	R119	1	T1251
ZD-9.1	1N938B	1	A31	1	GC4022	1	R163	1	T1289
ZD-9.1	1N939	1	A311(JAPAN)	1	GC532	1	R186	1	T1291
ZD-9.1	1N939A	1	A312	1	GC60	1	R227	1	T131
ZD-9.1	1N939B	1	A330	1	GC61	1	R244	1	T1322
ZD-9.1	1N960	1	A332	1	GE-1	1	R258	1	T1326
ZD-9.1	1N960A	1	A350A	1	GET880	1	R488	1	T1474
ZD-9.1	1N960B	1	A36	1	GET881	1	R506	1	T1510
ZD-9.1	1R9.1A	1	A40	1	GET882	1	R81	1	T152148
ZD-9.1	1R9.1B	1	A414	1	GET887	1	REN100	1	T1877
ZD-9.1	18130(ZENER)	1	A415(JAPAN)	1	GET888	1	RS-1539	1	T1902
ZD-9.1	18139	1	A44	1	GET889	1	RS-1550	1	T2038
ZD-9.1	181717	1	A55	1	GET890	1	RS-2683	1	T2039
ZD-9.1	181717L	1	A74-3-3A9G	1	GET891	1	RS-2684	1	T2040
ZD-9.1	181718	1	A74-3-70	1	GET892	1	RS-2685	1	T2091
ZD-9.1	18728	1	A74-3-705	1	GET893	1	RS-2686	1	T2122
ZD-9.1	18194	1	A74-3A9G	1	GET896	1	RS-2687	1	T2172
ZD-9.1	181959	1	A88C-70	1	GET897	1	RS-2688		
ZD-9.1	182067			1	GI1				

GE	Industry Standard No.	GE	Industry Standard No.	GE	Industry Standard No.	GE	Industry Standard No.	GE	Industry Standard No.
1	T2173	1	2G394	1	2N521	1N34AS	AA139	1N34AS	2-0A90
1	T2256	1	2G396	1	2N521A	1N34AS	AA140	1N34AS	2A119
1	T2257	1	2G397	1	2N522	1N34AS	AA142	1N34AS	2PD-358
1	T2258	1	2G524	1	2N522A	1N34AS	AA143	1N5624	1N5624
1	T2259	1	2G525	1	2N523	1N34AS	AA143B	1N5625	1N5625
1	T2260	1	2G526	1	2N523A	1N34AS	AA144	1N60	A069-111
1	T2261	1	2G527	1	2N529	1N34AS	AA218	1N60	A069-118
1	T346	1	2G577	1	2N530	1N34AS	AA779	1N60	A30
1	T50944	1	2G603	1	2N531	1N34AS	AAY139	1N60	OA9O2A
1	T52147	1	2G604	1	2N532	1N34AS	AAY15	1N60	O6O
1	T52147Z	1	2G605	1	2N533	1N34AS	AAY18	1N60	CA-90
1	T52148Z	1	2N111	1	2N571	1N34AS	AAY22	1N60	CD101
1	T52149	1	2N111A	1	2N572	1N34AS	AAY27	1N60	CG12-E
1	T52149Z	1	2N112A	1	2N578	1N34AS	AAY30	1N60	CK-706
1	TA-1575B	1	2N113	1	2N579	1N34AS	AAY33	1N60	CT-2002
1	TA-1655B	1	2N114	1	2N580	1N34AS	AAY46	1N60	CT-2005
1	TA-1704	1	2N1171	1	2N581	1N34AS	AAZ10	1N60	CT2008
1	TA-1763	1	2N1176	1	2N582	1N34AS	AAZ18	1N60	CX-0031
1	TA-1763A	1	2N1176A	1	2N586	1N34AS	ACR810-107	1N60	CX-0036
1	TA-1778	1	2N1176B	1	2N592	1N34AS	APS-160-1017	1N60	CX-0042
1	TA-1782	1	2N1185	1	2N593	1N34AS	CD0000	1N60	CX-0045
1	TA-1783	1	2N1186	1	2N597	1N34AS	CX-0033	1N60	D-00284R
1	TA1704	1	2N1187	1	2N598	1N34AS	CX-0047	1N60	D010
1	TA1763	1	2N1188	1	2N599	1N34AS	CXL109	1N60	D1770542
1	TA1763A	1	2N123	1	2N600	1N34AS	CXL110	1N60	D1J70543
1	TA1778	1	2N1264	1	2N614	1N34AS	D-GM-2	1N60	DAAY001002
1	TA1782	1	2N1265/5	1	2N615	1N34AS	D2U	1N60	DDAY001001
1	TA1783	1	2N1280	1	2N616	1N34AS	DG1834	1N60	DDAY001002
1	TI-363	1	2N1281	1	2N617	1N34AS	DGM3	1N60	DDAI001004
1	TI-364	1	2N1284	1	2N674	1N34AS	DLJ70644	1N60	DDAY001010
1	TIA03	1	2N1309	1	2N801	1N34AS	DIJ71778	1N60	DDAY001022
1	TIA05	1	2N1316	1	2N802	1N34AS	D8-32	1N60	DDAY101001
1	TIA05A	1	2N1317	1	2N809	1N34AS	D8-33	1N60	DG1N60
1	TIXB95	1	2N1318	1	2N810	1N34AS	D831	1N60	D8-39
1	TIXA-03	1	2N1343	1	2N811	1N34AS	D832	1N60	D8-0161
1	TIXA-04	1	2N1344	1	2N812	1N34AS	DX-0241	1N60	DX-0162
1	TIXA-05	1	2N1345	1	2N83	1N34AS	E20030	1N60	DX-0725
1	TIXA01	1	2N1346	1	2NJ5A	1N34AS	E295ZZ01	1N60	E1031RXT
1	TIXA02	1	2N1347	1	2NJ6	1N34AS	E50060	1N60	E21135
1	TIXA03	1	2N1349	1	2NJ8A	1N34AS	EA16X140	1N60	EA16X148
1	TIXA04	1	2N135	1	2B111	1N34AS	ECG109	1N60	EA2137
1	TIXA05	1	2N1350	1	2B12	1N34AS	EP16X21	1N60	EA2606
1	TM100	1	2N1351	1	2B13	1N34AS	EU16X19	1N60	EDG-0003
1	TNJ-60610	1	2N1354	1	2B155	1N34AS	GF173	1N60	EDG-0006
1	TNJ-60611	1	2N1355	1	2B159	1N34AS	G00004A	1N60	E81633
1	TNJ-60612	1	2N1356	1	2B160	1N34AS	GD510	1N60	E816X21
1	TNJ60608	1	2N1357	1	2B167	1N34AS	H624	1N60	E836X103
1	T0-101	1	2N136	1	2B174	1N34AS	HD1000101	1N60	EW168
1	T0-102	1	2N1361	1	2B178	1N34AS	HD10001010	1N60	EXY420DIR5JB
1	TQ5020	1	2N137	1	2B25	1N34AS	HEPR9134	1N60	EXY420DIR5JB
1	TR-05	1	2N1393	1	2B30	1N34AS	HEPR9134A	1N60	FA-1
1	TR-06	1	2N1395	1	2B31	1N34AS	HF-20088	1N60	G-1010
1	TR-044	1	2N1469	1	2B45	1N34AS	HV-23BL	1N60	GE129
1	TR-044A	1	2N1470	1	2B49	1N34AS	HV-801	1N60	H087
1	TR-045A	1	2N1471	1	2B51	1N34AS	INJ70980	1N60	H614
1	TR05C	1	2N1570	1	2B52	1N34AS	J241	1N60	HD1000105
1	TR07C	1	2N1581	1	2B53	1N34AS	KGB41959	1N60	HD1000105-O
1	TR104	1	2N1583	1	2B60	1N34AS	K0825201-1	1N60	HB-0A90
1	TR109	1	2N1584	1	2B91	1N34AS	K0825201-2	1N60	HE-10044
1	TR11	1	2N1664	1	2B92	1N34AS	MD-60A	1N60	HB-C00000
1	TR215	1	2N1683	1	2B92A	1N34AS	NTC-14	1N60	HEPR9135
1	TR217	1	2N1684	1	2B93	1N34AS	OA91	1N60	HP-20105
1	TR310015	1	2N1729	1	2B93A	1N34AS	OA95	1N60	HV0000202
1	TR310161	1	2N1731	1	2BA167	1N34AS	PTC207M	1N60	INJ73349
1	TR321(HPGH1)	1	2N1743	1	2SA169	1N34AS	QVD1KP114	1N60	INJ60034
1	TR482	1	2N1744	1	2SA201-N	1N34AS	RE47	1N60	INJ70973
1	TR482A	1	2N1853	1,2	2SA217	1N34AS	RE86	1N60	IP20-0060
1	TR758A	1	2N1940	1	TR05	1N34AS	REN109	1N60	IP20-0283
1	TR759	1	2N1997	1N295	CTP-461	1N34AS	RP35123	1N60	ITT102
1	TR764	1	2N1998	1N295	MA790	1N34AS	RPAJ60172	1N60	ITT301
1	TR792	1	2N2171	1N295	RSWY/16L7Z	1N34AS	RT7538	1N60	ITT718
1	TR801	1	2N2172	1N295	T-700-709	1N34AS	S1811	1N60	J24567
1	TR802	1	2N2209	1N295	T700-709	1N34AS	S2085	1N60	J24628
1	TR844	1	2N249	1N295	1N136	1N34AS	SC20	1N60	J24838
1	TRC44A	1	2N262	1N295	1N137	1N34AS	SD46(4)	1N60	JP575005
1	TRC045	1	2N27	1N295	1N295	1N34AS	SD46R	1N60	M-8513R
1	TRM15	1	2N271	1N295	1N34B2	1N34AS	SPD111	1N60	MA-900
1	TRM21	1	2N271A	1N295	1850	1N34AS	S1811	1N60	MA25(B)
1	T8-601	1	2N273	1N295	1N3292	1N34AS	8182	1N60	MD60A
1	T8-602	1	2N2930	1N3292	1N3292	1N34AS	T-750-713	1N60	MT1889
1	T8615	1	2N2966	1N34A	1875A	1N34AS	T-E1116	1N60	NTC-13
1	T8669A	1	2N30	1N34AS	A02	1N34AS	T-E1177	1N60	0A90
1	T8669B	1	2N302	1N34AS	A054-103	1N34AS	T-G1138	1N60	0A90-FM
1	T8669D	1	2N3075	1N34AS	A054-105	1N34AS	TCM109	1N60	0A90FM
1	T8669E	1	2N308	1N34AS	A054-187	1N34AS	TR0575002	1N60	0A90
1	T8669F	1	2N309	1N34AS	A054-226	1N34AS	TR228736002004	1N60	0A90GA
1	T24152	1	2N31	1N34AS	A066-119	1N34AS	001-0000-00	1N60	0A90LF
1	TV4152	1	2N315	1N34AS	A066-119(GE)	1N34AS	001-015011	1N60	0A90M
1	TV8-28A171	1	2N315A	1N34AS	A066-120	1N34AS	1Q01	1N60	0A90MLF
1	TV8-28B172A	1	2N315B	1N34AS	A066-121	1N34AS	1Q25	1N60	0A99
1	001-01202-1	1	2N317	1N34AS	A068-121	1N34AS	1Q86	1N60	PTC206M
1	001-01203-1	1	2N317A	1N34AS	A069-109	1N34AS	1GD10	1N60	QD-01N60PXT
1	1-21-100	1	2N327	1N34AS	A069-115	1N34AS	1QD5X	1N60	QD-01N60XXT
1	1-21-102	1	2N327A	1N34AS	A07	1N34AS	1K261	1N60	QD-01152XXT
1	1-21-103	1	2N394	1N34AS	A090	1N34AS	1N34	1N60	QVM8008
1	1-21-104	1	2N394A	1N34AS	A15-1007	1N34AS	1N34A	1N60	R-7029
1	1-21-105	1	2N395	1N34AS	A20371	1N34AS	1N34A-Z	1N60	R112524
1	1-21-128	1	2N396	1N34AS	A2419	1N34AS	1N34B	1N60	RT-1689
1	1-21-161	1	2N396A	1N34AS	A2420	1N34AS	1N34GA	1N60	RT1667
1	1-21-162	1	2N397	1N34AS	A2473	1N34AS	1N34M	1N60	RT2451
1	1-21-179	1	2N413	1N34AS	A2476	1N34AS	1N34Z	1N60	RT4880
1	1-21-180	1	2N413A	1N34AS	A25-1007	1N34AS	1N3754	1N60	RT5379
1	1-21-186	1	2N414	1N34AS	A36508	1N34AS	1N4198	1N60	RT5470
1	1-21-234	1	2N414A	1N34AS	A42X00340-01	1N34AS	1N49	1N60	RT5738
1	1-21-235	1	2N414B	1N34AS	A48-63078A52	1N34AS	1N541	1N60	RT5912
1	1-21-236	1	2N414C	1N34AS	A514-022057	1N34AS	1N542	1N60	S3838GA
1	1-21-240	1	2N415	1N34AS	A514-042791	1N34AS	1N60M	1N60	SD-1N60
1	1-21-241	1	2N415A	1N34AS	A556-142	1N34AS	1N636A	1N60	SD-1N60B
1	1-21-254	1	2N416	1N34AS	A615-1012	1N34AS	1N67A	1N60	S0020
1	1-21-273	1	2N417	1N34AS	A69213-4	1N34AS	1811	1N60	SD461
1	1-21-275	1	2N425	1N34AS	A7001800	1N34AS	18129	1N60	SVD0A70
1	1-21-289	1	2N426	1N34AS	A909-1015	1N34AS	1814	1N60	SVD0A90
1	1-21-73	1	2N427	1N34AS	A909-1017	1N34AS	1816	1N60	TE-1014
1	1-21-74	1	2N428	1N34AS	A95	1N34AS	18186FM	1N60	TE-1031
1	1-21-75	1	2N428A	1N34AS	AA111	1N34AS	00018188	1N60	TR0575005
1	1-21-76	1	2N438A	1N34AS	AA112	1N34AS	18188AR	1N60	TR2083-41
1	1-21-78	1	2N439A	1N34AS	AA112P	1N34AS	18188FM-1	1N60	TV241013
1	1-21-83	1	2N440A	1N34AS	AA113	1N34AS	18188FM1	1N60	TV24122
1	1-21-91	1	2N450	1N34AS	AA114	1N34AS	18188FMA	1N60	TV24273
1	1-21-92	1	2N481	1N34AS	AA116	1N34AS	18188FM1A	1N60	1-12689
1	1-21-93	1	2N482	1N34AS	AA117	1N34AS	18188FMA	1N60	1N-22
1	2B	1	2N483	1N34AS	AA118	1N34AS	18188MPX	1N60	1N42
1	2C	1	2N484	1N34AS	AA119	1N34AS	18188TV	1N60	1N542MP
1	2D	1	2N485	1N34AS	AA120	1N34AS	1832	1N60	1N60(TV)(PA-1)
1	201024	1	2N486	1N34AS	AA121	1N34AS	1833	1N60	1N60-1
1	201025	1	2N486B	1N34AS	AA123	1N34AS	1834	1N60	1N60-5
1	201026	1	2N503	1N34AS	AA130	1N34AS	1834A	1N60	1N60-PA1
1	20138	1	2N518	1N34AS	AA131	1N34AS	1834B	1N60	1N60-M3
1	20139	1	2N519	1N34AS	AA132	1N34AS	84426GFM	1N60	1N60-F
1	20140	1	2N519A	1N34AS	AA134	1N34AS	1858	1N60	1N60-S
1	20319	1	2N520	1N34AS	AA135	1N34AS	18589	1N60	1N60-T
1	20320	1	2N5201	1N34AS	AA136	1N34AS	1877	1N60	1N60-Z
1	20383	1	2N520A	1N34AS	AA137	1N34AS	189908		
					AA138	1N34AS	1T261		

GE	Industry Standard No.
1N60	1N60/5490
1N60	1N60/7825B
1N60	1N60A
1N60	1N60AM
1N60	1N60B
1N60	1N60C
1N60	1N60D
1N60	1N60F
1N60	1N60FA1
1N60	1N60FD1
1N60	1N60FM
1N60	1N60FMX
1N60	1N60G
1N60	1N60GA
1N60	1N60GB
1N60	1N60M3
1N60	1N60MP
1N60	1N60P
1N60	1N60R
1N60	1N60S
1N60	1N60SD6Q
1N60	1N60T
1N60	1N60TV
1N60	1N60TV-TOGL
1N60	1N60TVGL
1N60	1N60Z
1N60	1N642
1N60	1N67D
1N60	1N73
1N60	1N995M
1N60	18-188
1N60	1S12
1N60	1S128
1N60	1S188A
1N60	1S188FMI
1N60	1S19
1N60	1S2479
1N60	1S4266
1N60	1S542
1N60	1S60
1N60	1S851
1N60	1S87
1N60	1T22A
1N60	1T238
1N60	2-OA90
1N82A	2AA119
1N82A	M2D101
1N82A	SD-51
1N82A	1N22
1N82A	1N23WP
1N91	IR914
1N91	PTC206
1N91	PTC207
1N91	1N38
1N91	1N38A
1N91	1N38B
2	A174
2	A197
2	A397
2	A42X00286-01
2	ACY-17
2	ACY-18
2	ACY-19
2	ACY-20
2	ACY17
2	ACY18
2	ACY19
2	AF200U
2	AF201U
2	APB-11H-1008
2	B101
2	B102
2	B1022
2	B1058
2	B1154
2	B153
2	B154
2	B161
2	B162
2	B163
2	B164
2	B165
2	B166
2	B199
2	B200
2	B200A
2	B202
2	B203AA
2	B218
2	B219
2	B220
2	B220A
2	B221
2	B221A
2	B222
2	B223
2	B224(JAPAN)
2	B225
2	B226
2	B227
2	B241
2	B257
2	B326
2	B327
2	B328
2	B335
2	B34
2	B547
2	B54N
2	B350
2	B378A
2	B379
2	B379-2
2	B379A
2	B379B
2	B38
2	B380
2	B380A
2	B422
2	B423
2	B470
2	B48
2	B486
2	B497
2	B50
2	B51(JAPAN)
2	B52
2	B53
2	B71
2	B72
2	B85
2	B91
2	B92
2	B95
2	B98
2	CE03607839
2	CK721
2	CK722
2	CK725
2	CK727
2	CK751
2	CK754
2	CK790
2	CK791
2	CK793
2	CK794
2	CK870
2	CK871
2	CK872
2	CK875
2	CK878
2	CK879
2	CK882
2	CK888
2	CM8640E
2	CXL102
2	DE-3181
2	D019
2	D038
2	D043
2	D078
2	DRC-81252
2	DS-53
2	E-158
2	E0105
2	E044A
2	E066
2	E067
2	E068
2	E214B
2	ECG102
2	EE100
2	EO-44A
2	ET3
2	ET4
2	ET5
2	F2480
2	FD-1029-BG
2	FD-1029-EE
2	FD1029EE
2	G0006
2	GC5010
2	GC733B
2	GE-2
2	GET-103
2	GET-113
2	GET-113A
2	GET-114
2	GET898
2	GP20
2	GP21
2	GF32
2	GPT3008/40
2	GI2
2	GI4
2	GT-109
2	GT-269
2	GT-348
2	GT-759R
2	GT-760R
2	GT-761R
2	GT-762R
2	GTE-2
2	HA-201
2	HA-330
2	HB-00178
2	HB-00303
2	HB-00370
2	HB-156
2	HB-54
2	HB-56
2	HB-75B
2	HB-77C
2	HEP252
2	HEP629
2	HEP630
2	HEP631
2	HEP632
2	HEP633
2	HR-61
2	HR2A
2	HR3
2	HR4
2	HR6
2	HR7A
2	HR8
2	HR9
2	H8-15
2	H8-22D
2	H815
2	H8170
2	H825D
2	H8290
2	IP65
2	IRTR04
2	J24834
2	KV-2
2	KV-4D
2	M-75517-1
2	M-P3D
2	M4327
2	M4482
2	M4483
2	M4567
2	M4596
2	M4597
2	M4597GRN
2	M4597RED
2	M4607
2	M4627
2	M8640E(XSTR)
2	M9148
2	M9250
2	MA1700
2	NKT223
2	NKT224
2	NKT225
2	NKT225
2	NKT227
2	NKT228
2	NKT231
2	NKT232
2	NKT247
2	N8121
2	N832
2	OC32
2	OC41
2	OC78
2	P6460006
2	P6460037
2	PBX103
2	PBX113
2	PTC109
2	PXB103
2	PXB113
2	PXC101
2	PXC101A
2	PXC101AB
2	Q-1
2	Q-4
2	QG-0076
2	QN2613
2	Q0V60539
2	R28B492
2	R46
2	REN102
2	RS-2004
2	RS-2005
2	RS-5753-2
2	RS3276
2	RS5825
2	RS6843
2	RS6846
2	0822.3504-040
2	0822.3504-060
2	0822.3505-910
2	S413796
2	SE-5-0819
2	SF.T222
2	SF.T227
2	SF.T318
2	SFT221
2	SFT222
2	SMB447610
2	SMB447610A
2	SMB454549
2	SMB621960
2	SNW-Q-1
2	SNW-Q-4
2	SNW-Q-6
2	SYL107
2	SYL108
2	SYL1430
2	SYL1535
2	SYL1583
2	SYL1655
2	SYL1665
2	SYL1668
2	SYL2248
2	SYL2249
2	SYL2300
2	SYL3613
2	T-2122
2	T-Q5023
2	T-Q5025
2	T-Q5026
2	T50931B
2	T51573A
2	TIA04
2	TK40
2	TK40C
2	TK41
2	TK42
2	TK45C
2	TM102
2	TNJ61221
2	TNJ70634
2	TNJ70635
2	TQ5027
2	TSA0-222
2	TR04
2	TR2083-75
2	TR3100227
2	TR310225
2	TR320
2	TR320A
2	TR321
2	TR321A
2	TR323
2	TR323A
2	TR332
2	TR383(HGFH-2)
2	TR508
2	TR508A
2	TR5R26
2	TR650
2	TR653
2	TR6R26
2	TR721
2	TR722
2	TR763
2	TR8007
2	TRC70
2	TRC71
2	TRC72
2	TS-164
2	TS13
2	TS14
2	TS15
2	TS627A
2	TS627B
2	TS672A
2	TV-61
2	TVS-28172F
2	TVS28A171
2	TVS28A171B
2	TVS28B172A
2	001-01206-0
2	1/4LZ3.6D
2	1/4LZ3.6D5
2	002-00840
2	2G1027
2	2G270
2	2G271
2	2G302
2	2G306
2	2G321
2	2G322
2	2G323
2	2G324
2	2G384
2	2G385
2	2G386
2	2G387
2	2G508
2	2G509
2	2M1127
2	2N1008
2	2N1008A
2	2N1008B
2	2N1009
2	2N1044
2	2N1045
2	2N1056
2	2N1057
2	2N108
2	2N109/5
2	2N1097
2	2N1098
2	2N1124
2	2N1125
2	2N1126
2	2N1128
2	2N1174
2	2N1175
2	2N1175A
2	2N1189
2	2N1190
2	2N1191
2	2N1192
2	2N1193
2	2N1194
2	2N1265
2	2N1265A
2	2N1273
2	2N1273BLU
2	2N1273GRN
2	2N1273ORN
2	2N1273RED
2	2N1273YEL
2	2N1274
2	2N1274BLU
2	2N1274BRN
2	2N1274GRN
2	2N1274ORN
2	2N1274PUR
2	2N1274RED
2	2N1287
2	2N1287A
2	2N130
2	2N130A
2	2N131
2	2N131A
2	2N132
2	2N132A
2	2N133
2	2N1348
2	2N1352
2	2N1353
2	2N1361A
2	2N1370
2	2N1371
2	2N1372
2	2N1373
2	2N1374
2	2N1375
2	2N1376
2	2N1377
2	2N1378
2	2N1379
2	2N1380
2	2N1381
2	2N1382
2	2N1383
2	2N1392
2	2N1404
2	2N1413
2	2N1414
2	2N1432
2	2N1446
2	2N1447
2	2N1448
2	2N1449
2	2N1450
2	2N1451
2	2N1452
2	2N1478
2	2N159
2	2N1705
2	2N1706
2	2N1707
2	2N1854
2	2N187A
2	2N188
2	2N188A
2	2N189
2	2N190
2	2N191
2	2N192
2	2N195
2	2N196
2	2N1961/46
2	2N197
2	2N198
2	2N199
2	2N200
2	2N204
2	2N205
2	2N206
2	2N207
2	2N207A
2	2N207B
2	2N207BLU
2	2N2271
2	2N23
2	2N237
2	2N2374
2	2N2375
2	2N2376
2	2N24
2	2N241
2	2N241A
2	2N2431B
2	2N2468
2	2N2469
2	2N25
2	2N2564
2	2N2565
2	2N26
2	2N266
2	2N270
2	2N270-5E
2	2N270A
2	2N272
2	2N282
2	2N291
2	2N2953
2	2N20374
2	2N303
2	2N310
2	2N316A
2	2N319
2	2N32
2	2N321
2	2N322
2	2N323
2	2N324
2	2N32A
2	2N331M
2	2N342T
2	2N3428
2	2N339
2	2N360
2	2N361
2	2N362
2	2N362B
2	2N363
2	2N367
2	2N368
2	2N369
2	2N381
2	2N382
2	2N383
2	2N39
2	2N40
2	2N402
2	2N403
2	2N404
2	2N404A
2	2N405
2	2N41
2	2N4106
2	2N42
2	2N422
2	2N422A
2	2N43
2	2N43A
2	2N44
2	2N44A
2	2N46
2	2N460
2	2N461
2	2N462
2	2N464
2	2N465
2	2N466
2	2N467
2	2N468
2	2N47
2	2N483-6M
2	2N483B
2	2N487
2	2N505
2	2N508
2	2N508A
2	2N524
2	2N524A
2	2N525
2	2N525A
2	2N526
2	2N526A
2	2N527
2	2N527A
2	2N563
2	2N564
2	2N565
2	2N566
2	2N568
2	2N569
2	2N570
2	2N573
2	2N573BRN
2	2N573ORN
2	2N573RED
2	2N60
2	2N609
2	2N60A
2	2N60B
2	2N60C
2	2N60R
2	2N61
2	2N610
2	2N611
2	2N612
2	2N613
2	2N61A
2	2N61B
2	2N61C
2	2N631
2	2N632
2	2N633
2	2N633B
2	2N65
2	2N650
2	2N650A
2	2N651
2	2N651A
2	2N652
2	2N652A
2	2N654
2	2N655
2	2N655RN
2	2N655RED
2	2N71
2	2N73
2	2N74
2	2N76
2	2N79
2	2N799
2	2N80
2	2N82
2	2N825
2	2N826
2	2N96
2	2NJ5D
2	2NJ9A
2	2NJ9D
2	2N8121
2	2N831
2	2N832
2	2S14
2	2S15
2	2S15A
2	2S163
2	2S1760
2	2S1779
2	2S22
2	2S24
2	2S32
2	2S33
2	2S34
2	2S37
2	2S38
2	2S39
2	2S40
2	2S43
2	2S44
2	2S46

GE	Industry Standard No.
2	2847
2	2854
2	2856
2	28A396
2	28A397
2	28A55
2	28B33(5)
2	28B355
2	28B350
2N1671	2N1671
2N2160	HEP310
2N2160	HEPS9002
2N2160	IR2160
2N2160	2S2160
2N2160	2N4870
2N2160	2N4871
2N2324	2N2324
2N2646	2N2646
2N2647	2N2647
2N2711	2N2711#
2N4992	2N4992
2N5777	2N5777
2N5778	2N5778
2N5779	2N5779
2N5780	2N5780
2N6028	2N6028
2N689	2N689
2N690	2N690
3	AD148
3	B20-001
3	D081
3	DTA1011
3	DTG110(SEARS)
3	P20
3	GC4057
3	GE-3
3	GET-572
3	GTX-2001
3	HR101A
3	HR103A
3	HT204071D
3	IRTR16
3	POWER-12
3	POWER-25
3	POWER-299
3	POWER-99
3	PTC114
3	QG-0074
3	Q00074
3	SP1484
3	TR-3
3	001-01250
3	001-01251
3	001-01252
3	001-01253
3	001-01254
3	001-01255
3	001-01256
3	001-01257
3	001-01258
3	001-01259
3	2.4341.0018
3	2N1039
3	2N1041
3	2N115
3	2N141
3	2N143
3	2N1437
3	2N1438
3	2N158
3	2N158A
3	2N1611
3	2N1612
3	2N2072
3	2N2072
3	2N3126
3	2N3146
3	2N3147
3	2N5618
3	2S416
3	2SA416A
3	2SA416B
3	2SA416C
3	2SA416D
3	2SA416E
3	2SA416F
3	2SA416G
3	2SA416GN
3	2SA416H
3	2SA416J
3	2SA416K
3	2SA416L
3	2SA416M
3	2SA416OR
3	2SA416R
3	2SA416X
3	2SA416Y
3	2SB413
3	2SB413A
3	2SB413B
3	2SB413C
3	2SB413D
3	2SB413E
3	2SB413F
3	2SB413G
3	2SB413GN
3	2SB413H
3	2SB413J
3	2SB413K
3	2SB413L
3	2SB413M
3	2SB413OR
3	2SB413R
3	2SB413X
3	2SB413Y
3	2SB414
3	2SB414B
3	2SB414C
3	2SB414D
3	2SB414E
3	2SB414F
3	2SB414G
3	2SB414GN
3	2SB414H
3	2SB414J
3	2SB414K
3	2SB414L
3	2SB414M
3	2SB414OR
3	2SB414R
3	2SB414X
3	2SB414Y
3	2SB445
3	2SB445A
3	2SB445B
3	2SB445C
3	2SB445D
3	2SB445E
3	2SB445F
3	2SB445G
3	2SB445GN
3	2SB445H
3	2SB445J
3	2SB445K
3	2SB445L
3	2SB445M
3	2SB445OR
3	2SB445R
3	2SB445X
3	2SB445Y
3	2SB446
3	2SB446B
3	2SB446C
3	2SB446D
3	2SB446E
3	2SB446G
3	2SB446GN
3	2SB446H
3	2SB446J
3	2SB446K
3	2SB446L
3	2SB446M
3	2SB446OR
3	2SB446R
3	2SB446X
3	2SB446Y
3	2SB84
3	2SB84A
3	2SB84B
3	2SB84C
3	2SB84D
3	2SB84E
3	2SB84F
3	2SB84G
3	2SB84GN
3	2SB84H
3	2SB84J
3	2SB84K
3	2SB84L
3	2SB84M
3	2SB84OR
3	2SB84X
3	2SB84Y
3	2SC337
4	AA5
4	PT201
4	2SB238
4	2SB238-12A
4	2SB238-12B
4	2SB238-12C
4	2SB258A
4	2SB258B
4	2SB258C
4	2SB258D
4	2SB258E
4	2SB258F
4	2SB258G
4	2SB258GN
4	2SB258H
4	2SB258J
4	2SB258K
4	2SB258L
4	2SB258M
4	2SB258OR
4	2SB258R
4	2SB258X
4	2SB258Y
4	GE-5
5	GT-1200
5	GT1201
5	HC00730
5	TN214
5	2N182#
5	2N183#
5	2N529/N
5	2N530/N
5	2N531/N
5	2N532/N
5	2N533/N
5ZD-10	1N4465
5ZD-10	1N4958
5ZD-10	1N5020
5ZD-11	1N4466
5ZD-11	1N4959
5ZD-11	1N5021
5ZD-12	1N4467
5ZD-12	1N4960
5ZD-12	1N5022
5ZD-13	1N4468
5ZD-13	1N4961
5ZD-13	1N5023
5ZD-13	1N5069
5ZD-14	1N5024
5ZD-14	1N5070
5ZD-14	1N5118
5ZD-15	1N4469
5ZD-15	1N5025
5ZD-16	1N5074
5ZD-16	1N4470
5ZD-16	1N4963
5ZD-16	1N5026
5ZD-16	1N5072
5ZD-17	1N5027
5ZD-18	1N4471
5ZD-18	1N4964
5ZD-18	1N5028
5ZD-18	1N5073
5ZD-19	1N5029
5ZD-20	1N3950
5ZD-20	1N4472
5ZD-20	1N4965
5ZD-20	1N5030
5ZD-22	1N4473
5ZD-22	1N4966
5ZD-22	1N5031
5ZD-22	1N5074
5ZD-24	1N4474
5ZD-24	1N4967
5ZD-24	1N5032
5ZD-24	1N5075
5ZD-25	1N3951
5ZD-25	1N5033
5ZD-27	1N4475
5ZD-27	1N4968
5ZD-27	1N5034
5ZD-27	1N5076
5ZD-30	1N4476
5ZD-30	1N4969
5ZD-30	1N5035
5ZD-33	1N5077
5ZD-33	1N4970
5ZD-33	1N5078
5ZD-6.8	1N4461
5ZD-6.8	1N4954
5ZD-6.8	1N5016
5ZD-7.5	1N4462
5ZD-7.5	1N4955
5ZD-8.2	1N4463
5ZD-8.2	1N4956
5ZD-8.2	1N5018
5ZD-9.1	1N4464
5ZD-9.1	1N4957
6	GE-6
6	IRTR08
6	SY101
6	2N1624
6GC1	A04201
6GC1	B-46-110
6GC1	D8731A
6GC1	RER-023
6GC1	SR-20
6GC1	SR0004
6GD1	A86-4-1
6GD1	B1C045494P1
6GD1	CXL114
6GD1	D5
6GD1	ECG114
6GD1	EU16X8
6GD1	HEPR9003
6GD1	K112D
6GD1	K118J966-1
6GD1	K122D
6GD1	RE88
6GD1	REN114
6GD1	RMVTC000921-3
6GD1	RVTMC000921-3
6GD1	SD5(DUAL)
6GD1	SR14
6GD1	SR15
6GD1	TC-0.09M21/5
6GD1	TC0.09M21/3
6GD1	TM114
6GD1	TVH-526
6GD1	TVMTC000921-3
7	GE-7
7	IRTR10
8	A-567
8	A121-1410
8	A121-15
8	A121-16
8	A121-17
8	A121-21
8	A121-50
8	A121-762
8	A127-7
8	A129-30
8	A13-86420-1
8	A13-87433-1'
8	A1396
8	A1465-4
8	A1465-49
8	A1858
8	A198794-1
8	A2039-2
8	A2092418
8	A2092418-0711
8	A3607
8	A3609
8	A3T201
8	A3T202
8	A3T203
8	A46-8614-3
8	A4700
8	A48-124216
8	A48-124217
8	A48-124218
8	A48-124220
8	A48-124221
8	A48-125233
8	A48-125234
8	A48-125235
8	A48-125236
8	A48-128239
8	A48-134520
8	A48-134700
8	A48-134931
8	A4JD3B1
8	A514-023553
8	A514-032815
8	A65-4-70
8	A65-4-705
8	A65-4A90
8	A86-10-2
8	A86-44-2
8	A95115
8	A95211
8	A99807
8	A998K5
8	A998K7
8	AA2
8	AC130
8	AC157
8	AC187/01
8	AF192
8	A07
8	AS342B
8	ASY-62
8	ASY-72
8	ASY28
8	ASY29
8	ASY53
8	ASY72
8	ASY73
8	ASY74
8	ASY75
8	ASY86
8	ASY87
8	ASY88
8	ASY89
8	AT52
8	AT521
8	AT53
8	AT551
8	AT71
8	AT72
8	AT73R
8	AT75R
8	AT76R
8	AT77
8	B65-4-A-21
8	B65-4&21
8	B92-1-A-21
8	C128
8	C129
8	C13(TRANSISTOR)
8	C14
8	C173
8	C175
8	C176
8	C177
8	C178
8	C36(TRANSISTOR)
8	C50(TRANSISTOR)
8	C50A
8	C60(TRANSISTOR)
8	C71
8	C72
8	C75(JAPAN)
8	C76(JAPAN)
8	C77
8	C77C
8	C78
8	C89
8	C90
8	C91
8	CK261
8	CK262
8	CXL101
8	CXL103
8	D-F1
8	D101
8	D11
8	D161
8	D162
8	D167
8	D19
8	D195
8	D195A
8	D20
8	D21
8	D215
8	D22
8	D23
8	D36
8	D43
8	D43A
8	D44
8	DP1
8	D083
8	D085
8	DS11
8	DS12
8	DS2
8	DS4
8	DS5
8	DS6
8	DS7
8	DS8
8	DS9
8	E2427
8	E2428
8	E2429
8	E4002
8	EC0101
8	ECG103
8	ES5
8	ET10
8	ET11
8	ET8
8	ET9
8	G101079
8	G16
8	G16506
8	G17
8	G18
8	GA53270
8	GC1034
8	GC1035
8	GC1036
8	GC452
8	GC453
8	GC454
8	GE-8
8	GEX8
8	GI5
8	GI6
8	GI6506
8	GI7
8	GI8
8	GT-35
8	GT-903
8	GT-905
8	GT-947
8	GT1200
8	GT1202
8	GT1608
8	GT1609
8	GT229
8	GT2765
8	GT2766
8	GT2767
8	GT2884
8	GT2886
8	GT2888
8	GT2906
8	GT3150
8	GT792
8	GT904
8	GT905
8	GT905R
8	GT947
8	GT948
8	GT948R
8	GT949
8	GT949R
8	HA5001
8	HA5002
8	HA5003
8	HA5005
8	HA5009
8	HA5011
8	HA5012
8	HA5014
8	HA5016
8	HA5020
8	HA5021
8	HA5022
8	HA5023
8	HA5024
8	HA5025
8	HA5026
8	HC-00730
8	HD-187
8	HEP641
8	HTO405190
8	HT345019C
8	IRTR09
8	ISD-162
8	ISD162
8	JR40
8	KD2124
8	M2N168A
8	M4700
8	M8120
8	M9092
8	M9093
8	MHT2002
8	MHT2003
8	MHT2004
8	MHT2008
8	MHT2009
8	MHT2010
8	MIS-14150-18A
8	MIS14150-18
8	NA20
8	NK65-4A19
8	NKT732
8	NKT734
8	NKT736
8	NKT753
8	NR-10
8	NR05
8	NR10
8	NR20
8	NR30
8	NR5
8	NR700
8	OC139
8	OC140
8	OC141
8	PTC108
8	Q-2
8	Q-3
8	Q-5
8	Q-9
8	R-125
8	R-135
8	R-136
8	R-137
8	R-1533
8	R-203
8	R-33
8	R-34
8	R-62
8	R-63
8	R117
8	R12
8	R125
8	R135
8	R136
8	R137
8	R14
8	R1533
8	R202
8	R203
8	R41
8	R592
8	RE-5
8	RE5
8	REN101
8	REN103
8	RS-104
8	RS-1536
8	RS-1537
8	RS-1538
8	RS-1545
8	RS-1547
8	RS-1553
8	RS-2001
8	RS-2359
8	RS-2360
8	RS-2364
8	RS-2365
8	RS-2366
8	RS-2373
8	RS-2374
8	RS-2375
8	RS1513
8	RS1530
8	RS1531
8	RS1532
8	RS1534
8	RS1536
8	RS1537
8	RS1538
8	RS1547
8	RS1553
8	RS2256
8	RS2359
8	RS2360
8	RS2364
8	RS2365
8	RS2366
8	RS2375
8	RT-110
8	RT-119
8	S-1453
8	S028
8	S65-4-A-3P
8	SA-7
8	SA354B
8	SA7
8	SF.T184
8	SFT-298
8	SFT184
8	SFT259
8	SFT260
8	SFT261
8	SFT298
8	SK-7
8	SK3011
8	SK7
8	SN60
8	SN80
8	SNW-Q-2
8	SNW-Q-3
8	SNW-Q-5
8	SQ7
8	ST-172
8	ST172
8	SYL-101
8	SYL-102
8	SYL-1297
8	SYL-1310
8	SYL-1311

GE	Industry Standard No.	GE	Industry Standard No.	GE	Industry Standard No.	GE	Industry Standard No.	GE	Industry Standard No.
8	SYL-1987	8	2N127	8	2N97	11	AT316	11	C381-R
8	SYL-2130	8	2N1288	8	2N97A	11	AT318	11	C381BR
8	SYL-2131	8	2N1289	8	2N98	11	AT319	11	C381R
8	SYL-2132	8	2N1299	8	2N98A	11	AT322	11	C385A
8	SYL-2245	8	2N1302	8	2N99	11	AT323	11	C545
8	SYL-2246	8	2N1304	9	002-009000	11	AT324	11	C545A
8	SYL-4359	8	2N1306	10	A-106	11	AT325	11	C545B
8	SYL101	8	2N1308	10	A-128	11	AT326	11	C545C
8	SYL102	8	2N1310	10	A1086	11	AT327	11	C545D
8	SYL103	8	2N1311	10	A1087	11	AT328	11	C545E
8	SYL104	8	2N1312	10	BC-169-C	11	AT330	11	C56
8	SYL1279	8	2N1366	10	BTX367B	11	AT338	11	C605Q
8	SYL1297	8	2N1367	10	CDC8201	11	AZG	11	C629
8	SYL1310	8	2N1391	10	CM7163	11	AZY	11	C645
8	SYL1311	8	2N145	10	C8-1120I	11	B-75568-2	11	C645A
8	SYL1312	8	2N146	10	C8-1259	11	B-T-1000-139	11	C645B
8	SYL1313	8	2N147	10	C8-1294I	11	B-T1000-139	11	C645C
8	SYL1326	8	2N1473	10	C8-1361G	11	BC121	11	C645G
8	SYL1327	8	2N150	10	C8-6168P	11	BC122	11	C645N
8	SYL1329	8	2N150A	10	C8-9013HE	11	BC123	11	C656
8	SYL1380	8	2N1510	10	C811660-G	11	BC155	11	C657
8	SYL1396	8	2N1585	10	EA1538529	11	BC156	11	C668
8	SYL1408	8	2N1605	10	EA2770(N)	11	BC188	11	C695
8	SYL1454	8	2N1605A	10	EN30	11	BC189	11	C705
8	SYL1524	8	2N1622	10	ET15X54	11	BC442	11	C705B
8	SYL1536	8	2N164	10	GB-10	11	BC510C	11	C705C
8	SYL1537	8	2N164A	10	GE1265	11	BCM1002-2	11	C705D
8	SYL1538	8	2N165	10	GI-2711	11	BE173	11	C705E
8	SYL1539	8	2N166	10	G89014	11	BF121	11	C705F
8	SYL1547	8	2N167	10	G89018P	11	BF125	11	C705TV
8	SYL1591	8	2N167A	10	G89023I	11	BF127	11	C707
8	SYL1617	8	2N168	10	H104	11	BF194	11	C707H
8	SYL1750	8	2N168A	10	HC-00458	11	BF194B	11	C715EV
8	SYL1941	8	2N169	10	HT3037210-0	11	BF195	11	C738
8	SYL1987	8	2N169A	10	HT304508K	11	BF195C	11	C738C
8	SYL2130	8	2N170	10	HT304580C0	11	BF195D	11	C738D
8	SYL2131	8	2N171	10	HT304580YO	11	BF196	11	C739
8	SYL2132	8	2N172	10	HT304581B-0	11	BF197	11	C739C
8	SYL2134	8	2N1730	10	J24846	11	BF198	11	C763
8	SYL2135	8	2N1732	10	K14-0066-12	11	BF200(ZENITH)	11	C763(C)
8	SYL2136	8	2N1779	10	LD8206	11	BF216	11	C772B
8	SYL2245	8	2N1780	10	LM415-6	11	BF217	11	C772CU
8	SYL2246	8	2N1781	10	LM1415-7	11	BF218	11	C772CX
8	SYL2650	8	2N1783	10	M-4721	11	BF219	11	C785
8	SYL4315	8	2N1808	10	O8536G	11	BF220	11	C800
8	SYL4339	8	2N182	10	Q8-0254	11	BF229	11	C835
8	SU792	8	2N183	10	Q80254	11	BF230	11	C837
8	T-Q5031	8	2N184	10	S187B	11	BF240	11	C837P
8	T-Q5039	8	2N185B	10	SM6762	11	BF241	11	C837H
8	T-Q5050	8	2N1891	10	SPS-1475(YT)	11	BF253	11	C837L
8	T59276	8	2N1892	10	T1895	11	BF310	11	C837K
8	T59277	8	2N1993	10	TBB0147B	11	BF311	11	C837WF
8	TA-1759	8	2N1994	10	TC312307222	11	BF329	11	C920
8	TA-1767	8	2N1995	10	TI642B	11	BF332	11	C920E
8	TA-1771	8	2N1996	10	TI894(AFAMP)	11	BF333	11	C921
8	TA-1772	8	2N2085	10	TI894(XSTR)	11	BF333C	11	C921C1
8	TA1620A	8	2N2114A	10	002-008300	11	BF333D	11	C921K
8	TA1620B	8	2N2345	10	2N226Y	11	BF334	11	C921L
8	TA1759	8	2N2354	10	2N3298	11	BF335	11	C921M
8	TA1767	8	2N2426	10	280335	11	BF813E	11	C922
8	TA1771	8	2N2482	11	A-1854-0092-1	11	BF813F	11	C922A
8	TA1772	8	2N253	11	A054-148	11	BF813G	11	C922B
8	TF70	8	2N254	11	A054-170	11	BF814E	11	C922C
8	TF71	8	2N2699	11	A060-100	11	BF814F	11	C922K
8	TF72	8	2N28	11	A061-105	11	BF814G	11	C922L
8	TIX896	8	2N29	11	A061-106	11	BF815B	11	C922M
8	TK330	8	2N292	11	A061-108	11	BF815F	11	C930DE
8	TM101	8	2N292A	11	A061-109	11	BF815G	11	C930EE
8	TM103	8	2N293	11	A061-112	11	BF816E	11	CAC5028A
8	TNJ61671	8	2N312	11	A069-101	11	BF816F	11	CDC12030B
8	TP4274	8	2N313	11	A069-102	11	BF816G	11	CDC5000D
8	TQ5031	8	2N314	11	A069-103	11	BF818	11	C8-6225B
8	TQ5032	8	2N356	11	A069-114	11	BF818R	11	C8-9016D
8	TQ5039	8	2N356A	11	A069-116	11	BF819	11	C81585H
8	TR-08	8	2N357	11	A069-119	11	BF819R	11	C84001
8	TR-08C	8	2N357A	11	A121-585	11	BF820	11	C84195
8	TR-09	8	2N358	11	A121-585B	11	BF820R	11	C84194
8	TR-09C	8	2N358A	11	A121-687	11	BFX32	11	C8901TP
8	TR-10	8	2N364	11	A1G	11	BFY47	11	CT1012
8	TR-10C	8	2N365	11	A1G-1	11	BFY48	11	CT1013
8	TR-159(OLSON)	8	2N366	11	A1M-1	11	BFY49	11	CXJ107
8	TR-160(OLSON)	8	2N377	11	A1P	11	BFY69	11	D006
8	TR08	8	2N377A	11	A1P-1	11	BFY69A	11	D10B1051
8	TR08C	8	2N385	11	A1P-1A	11	BO-71	11	D10B1055
8	TR09	8	2N385A	11	A1P/4922	11	BSV53P	11	D10G1051
8	TR09C	8	2N388	11	A1P/4923	11	BSV54P	11	D10G1052
8	TR10	8	2N388A	11	A1R	11	C1023(JAPAN)	11	D160
8	TR10C	8	2N4105	11	A1R-1	11	C1023-0	11	D1606
8	TR167	8	2N438	11	A1R-2	11	C1023-Y	11	D16K1
8	TR182	8	2N439	11	A1R/4924	11	C1023G	11	D16K2
8	TR183	8	2N440	11	A1R/4925	11	C1026	11	D16K3
8	TR184	8	2N444	11	A1R/4926	11	C1026G	11	D16K4
8	TR193	8	2N444A	11	A20Y7B2-115	11	C1026Y	11	D1Y
8	TR194	8	2N445	11	A2M	11	C1032	11	D26B1
8	TR211	8	2N445A	11	A2M-1	11	C1032G	11	D26B2
8	TR212	8	2N446	11	A2N	11	C1032Y	11	D26C1
8	TR213	8	2N446A	11	A2N-1	11	C1123	11	D26C2
8	TR216	8	2N447	11	A2N-2	11	C1126	11	D26C3
8	TR335	8	2N447A	11	A2P	11	C1159	11	D26E2
8	TR336	8	2N447B	11	A2Y	11	C1360	11	D26G1
8	TR337	8	2N448	11	A3772.01	11	C1390A	11	D087
8	TR338	8	2N449	11	A417-154	11	C1394	11	D088
8	TV27	8	2N556	11	A417-190	11	C1417	11	DS-71
8	TVS28C647A	8	2N557	11	A417-205	11	C1417C	11	DS-72
8	001-01101-0	8	2N558	11	A46-86101-3	11	C1417D	11	DS-73
8	001-011010	8	2N576	11	A46-86109-3	11	C1417D(1)	11	DS-78
8	2-36	8	2N576A	11	A46-86110-3	11	C1417DU	11	DS71
8	20339	8	2N585	11	A46-86133-3	11	C1417F	11	DS72
8	20339A	8	2N587	11	A46-86301-3	11	C1417G	11	DS73
8	2M78	8	2N594	11	A46-86302-3	11	C1417H	11	DS74
8	2N100	8	2N595	11	A484(ZENITH)	11	C1417U	11	DS75
8	2N1000	8	2N596	11	A495	11	C1417V	11	E1A
8	2N1012	8	2N625	11	A496	11	C1417VW	11	EA15X134
8	2N102	8	2N634	11	A4E	11	C1417W	11	EA15X135
8	2N103	8	2N634A	11	A4Y-1	11	C171	11	EA15X4064
8	2N1059-1	8	2N635	11	A6V	11	C174	11	EA15X7125
8	2N1086	8	2N635A	11	A772B1	11	C174A	11	EA15X7140
8	2N1086A	8	2N636	11	A772EH	11	C186	11	EA15X7215
8	2N1087	8	2N636A	11	A772FE	11	C1908E	11	EA15X7231
8	2N1090	8	2N646	11	AR200(GREEN)	11	C266	11	EA15X7264
8	2N1091	8	2N649/22	11	AR219YY	11	C329	11	EA2131
8	2N1095	8	2N679	11	AR220(YELLOW)	11	C329B	11	EA2132
8	2N1096	8	2N78	11	AR220GY	11	C329C	11	EA2494
8	2N1102/5	8	2N78A	11	AR222(BLUE)	11	C551	11	EA2496
8	2N1112	8	2N797	11	AR222(YELLOW)	11	C551(FA)	11	EA2812
8	2N114	8	2N821	11	AR222BY	11	C561	11	EC0107
8	2N112	8	2N822	11	AR224(WHITE)	11	C562	11	EC0108
8	2N112	8	2N823	11	AR224(YELLOW)	11	C375-0	11	EL231
8	2N1121	8	2N824	11	AT310	11	C381	11	EL434
8	2N1169	8	2N955	11	AT311	11	C381-0	11	EN5172
8	2N1170	8	2N955A	11	AT312			11	EP15X37
8	2N1198			11	AT313			11	EP15X38
8	2N1217			11	AT314			11	E08-18
8	2N124			11	AT315			11	E810186
8	2N125								

GE	Industry Standard No.	GE	Industry Standard No.	GE	Industry Standard No.	GE	Industry Standard No.	GE	Industry Standard No.
11	ES10188	11	M4840A	11	SPS860	11	2SC313G	14	A3TEX003
11	ES15046	11	M4904	11	ST11	11	2SC313GN	14	A3TEX004
11	ES15047	11	M9032	11	ST65510	11	2SC313H	14	A3U
11	ES15X102	11	M91	11	T-203	11	2SC313J	14	A3U-4
11	ES15X120	11	MP1161	11	T-Q5049	11	2SC313K	14	A417033
11	ES15X121	11	MP1162	11	T-Q5079	11	2SC313L	14	A43023843
11	ES15X266	11	MP1163	11	T-Q5106	11	2SC313M	14	A4J
11	ES15X67	11	MP1164	11	T3568	11	2SC313OR	14	A4J(RED)
11	ES15X80	11	MM8006	11	T3601(RCA)	11	2SC313R	14	A4JBLK
11	ES15X81	11	MM8007	11	TG28C927A	11	2SC313X	14	A4JRED
11	ES15X82	11	MMT8015	11	TG28C927C	11	2SC313Y	14	A4JRED-1
11	ES15X97	11	MPS2369	11	TI-3016	11	2SC329	14	A48
11	ET15X2	11	MPS2823	11	TIS-1B	11	2SC329B	14	A48-1
11	ET15X3	11	MPS2894	11	TIS108	11	2SC329C	14	A4Z
11	EW163	11	MPS4145	11	TIS412	11	2SC344	14	A515
11	EW164	11	MPS6528	11	TI886	11	2SC344(Y)	14	A522
11	EW165	11	MPS6529	11	TI887	11	2SC344Y	14	A522-3
11	F24T-011-013	11	MPS6531	11	TI897	11	2SC3854	14	A523
11	F24T-011-015	11	MPS6532	11	TI898	11	2SC399PA1	14	A572
11	F24T-016-024	11	MPS6538	11	TI899	11	0000028C460	14	A572-1
11	FI-1025	11	MPS6540	11	TM107	11	0000028C555	14	A6L
11	FS1308	11	MPS6569	11	TM108	11	2SC545	14	A6LBLK
11	FS32669	11	MPS6570	11	TNJ-60605	11	2SC545A	14	A6LBLK-1
11	FSP1	11	MPS834	11	TNJ60069(2SC74)	11	2SC545B	14	A6LBRN
11	FX709	11	MPS918	11	TNJ60447	11	2SC545C	14	A6LBRN-1
11	G04-041B	11	MPS89623C	11	TNJ60448	11	2SC545D	14	A6LRED
11	G040041B	11	MPSH08	11	TNJ60449	11	2SC545E	14	A6LRED-1
11	G05-003-A	11	MPSH09	11	TNJ60604	11	2SC561	14	A6N
11	G05-003-B	11	MPSH24	11	TNJ60605	11	2SC561A	14	A6N-6
11	G05-050-C	11	MR86548	11	TNJ60606	11	2SC561B	14	A7-12
11	GE-1	11	N-EA15X130	11	TNJ60607	11	2SC561C	14	A7-13
11	GM-770	11	N-EA15X134	11	TNJ61217	11	2SC561D	14	A7N
11	GM770	11	N-EA15X135	11	TNJ61218	11	2SC561E	14	A80052402
11	GMB6001	11	NCS9018D	11	TNJ61679	11	2SC561F	14	A80414120
11	GMB6002	11	NJ100A	11	TNJ6172(2SC722)	11	2SC561G	14	A80414130
11	G05-004A	11	NJ202B	11	TNJ61730	11	2SC561GN	14	A8P
11	HC-00780	11	NL100B	11	TNJ61731	11	2SC561H	14	A8U
11	HC-00784	11	NR421	11	TNJ71629	11	2SC561I	14	A8W
11	HC-00829	11	NR421DG	11	TNJ71937	11	2SC561J	14	AEX79846
11	HC-00920	11	NR461AP	11	TNJ71963	11	2SC561K	14	AKX9846
11	HC-01047	11	N31510	11	TNJ72277	11	2SC561L	14	AM3235
11	HC-01359	11	N33039	11	TNJ72279	11	2SC561M	14	AMP-121
11	HC580	11	N33040	11	TR-8010	11	2SC561OR	14	AMP104
11	HC581	11	N33041	11	TR-8043	11	2SC561R	14	AMP105
11	HC594	11	PE5031	11	TR228735046011	11	2SC561X	14	AMP115
11	HC454	11	PL10024	11	TV24610	11	2SC561Y	14	AMP116
11	HC460	11	PL1025	11	TV24380	11	2SC604	14	AMP117
11	HC461	11	PL1026	11	TV24382	11	2SC656	14	AMP117A
11	HC535	11	PL1113	11	TV24383	11	2SC657	14	AMP118
11	HC535A	11	PM194	11	TV24385	11	00028C668D	14	AMP118A
11	HC535B	11	PM195	11	TV24698	11	2SC684-OR	14	AMP119
11	HC537	11	PT4816	11	TV24806	11	2SC695	14	AMP119A
11	HC545	11	PT4830	11	TV57	11	2SC707	14	AMP120
11	HC784	11	Q-00584	12	TV58	12	C685	14	AMP120A
11	HEP721	11	Q-01115C	12	TV60	12	C685A	14	AMP201
11	HEP731	11	Q1/6515	12	TV8-2S288A	12	MJ2251	14	AMP210
11	HEP752	11	Q2/6515	12	TV8-2SC183P	12	MJ2252	14	AMP210A
11	HEP734	11	Q301	12	TV8-2SC183Q	12	MJ3201	14	AMP2919-2
11	HR-58	11	Q4/6515	12	TV8-2SC429A	12	MJ3202	14	AR15
11	HR-59	11	Q6/6515	12	TV8-2SC469A	12	MJ400	14	AR15-L8-0026
11	HR59	11	R-280535	12	TV8-2SC0644	12	2N2204	14	AT-10
11	HR76	11	R-28C772	12	TV8-2SC0645	12	2N3440	14	AT-1856
11	HR79	11	R-28C858	12	TV8-2SC0645A	12	2N3585	14	AT1856
11	HR80	11	R118	12	TV8-2SC0645B	12	2N3738	14	AT3260
11	HT303711A	11	R8529	12	TV8-2SC0645C	12	2N3759	14	ATC-TR-15
11	HT303711B	11	RE3001	12	TVS2S0829B	12	2N4240	14	B-12822-2
11	HT303711C	11	RE3002	12	01-031360	12	2N4296	14	B133550
11	HT3037310	11	RE9	12	1-801-304-15	12	2N4297	14	B133577
11	HT303801	11	REN107	12	1-TR-048	12	2N4298	14	B133684
11	HT303801S-B	11	REN108	12	12535B	12	2N4299	14	B133685
11	HT303801A	11	RS-7201	12	2N1586	12	2SC101	14	B170000
11	HT303801AO	11	RS-7212	12	2N1587	12	2SC101A	14	B170000-ORG
11	HT303801B	11	RS7143	12	2N1589	12	2SC101B	14	B170000-RED
11	HT303801B-O	11	R87222	12	2N1590	12	2SC101C	14	B170000BLK
11	HT303801BO	11	R87523	12	2N1592	12	2SC101D	14	B170000BRN
11	HT303801C	11	R89510	12	2N1593	12	2SC101E	14	B170001
11	HT303801CO	11	R89511	12	2N162	12	2SC101F	14	B170001-BLK
11	HT306451A	11	R89512	12	2N162A	12	2SC101G	14	B170001-BRN
11	HT307720B	11	R112	12	2N2901	12	2SC101GN	14	B170001-RED
11	HT307721C	11	RT5464	12	2N3082	12	2SC101H	14	B170001BLK
11	HT307721D	11	RT5465	12	2N3083	12	2SC101J	14	B170001BRN
11	HT308291A	11	RT6204	12	2N3137	12	2SC101K	14	B170002
11	HT308291A-O	11	RT6601	12	2N337	12	2SC101L	14	B170002-ORG
11	HT308291B	11	RT6602	12	2N3633/46	12	2SC101M	14	B170002-RED
11	HT308291B-O	11	RT6732	12	2N3633/TNT	12	2SC101OR	14	B170003
11	HT308291C	11	RTT7703	12	2N3662	12	2SC101R	14	B170004
11	HT309291B	11	RTT704	12	2N3663	12	2SC101X	14	B170005
11	HT309501D	11	RT8668	12	2N3721	12	2SC101XL	14	B170006
11	HV0000302	11	RT8669	12	2N3825	12	2SC101Y	14	B170007
11	HX50003	11	S-1019	12	2N3826	12	2SC1059	14	B170009
11	J107	11	S-1019(UHP)	12	2N3845	12	2SC1078	14	B170010
11	J241177	11	S-1315	12	2N3845A	12	2SC1102	14	B170011
11	J241188	11	S-2617	12	2N3846	12	2SC1104	14	B170012
11	J241189	11	S1010	12	2N3855	12	2SC1105	14	B170013
11	J24561	11	S1041	12	2N3855A	12	2SC1304	14	B170014
11	J24562	11	S1041-16GN	12	2N3856A	12	2SC1391	14	B170015
11	J24563	11	S1122	12	2N3860	12	2SC2082	14	B170016
11	J24701	11	S1126	12	2N4254	12	2SC0685	14	B170018
11	J24903	11	S1308	12	2N4255	13MP	CXL104MP	14	B170019
11	K121J688-1	11	S1897	14	2N4435	13MP	GE-13MP	14	B170020
11	K2001	11	S2131	14	2N476A	14	A-11166527	14	B170021
11	K2119	11	S2132	14	2N5031	14	A-120327	14	B170022
11	K2120	11	S2133	14	2N5032	14	A-140605	14	B170024
11	K2121	11	S2134	14	2N5132	14	A-18	14	B170025
11	K2122	11	S21648	14	2N709/TNT	14	A-1854-0291-1	14	B17307
11	K2123	11	S25941	14	2N709/TNT	14	A-1854-0294-1	14	B5020
11	K2124	11	S2617(UHP)	14	2N780	14	A-6-67703	14	B66X0040-006
11	K2125	11	S40545	14	2N917/51	14	A-6-67703-A-7	14	BD111
11	K2126	11	SAC-1843	14	000028606	14	A054-154	14	BD112
11	K2127	11	SDD820	14	2882	14	A08-105018	14	BD113
11	K2501	11	2882	14	2SC-313	14	A11236J	14	BD116
11	K2502	11	SE002(1)	14	2SC1023	14	A13-0032	14	BD118
11	K2503	11	SE2020	14	2SC10230	14	A13-17918-1	14	BD121
11	K2509	11	S82397	14	2SC1026	14	A13-23594-1	14	BD130
11	K2601C	11	SKA-4075	14	00028C1026A	14	A13-33188-2	14	BD142
11	K2602C	11	SKA1416	14	00028C1026B	14	A14-601-10	14	BD145
11	K2603C	11	SM-A-595830-12	14	00028C1026C	14	A14-601-12	14	BD181
11	K2604	11	SPS-1473RT	14	2SC1032	14	A14-601-13	14	BD182
11	K2604C	11	SPS-2265	14	2SC1159	14	A18	14	BD183BLK
11	K2615	11	SPS-2266	14	2SC1395	14	A18-4	14	BD245
11	K2616	11	SPS-856	14	2SC1585P	14	A2418	14	BD245A
11	K4002	11	SPS-860	14	2SC1585P,H	14	A2E	14	BDX40
11	KLH4792	11	SP82167	14	2SC1585H	14	A2E-2	14	BDY17
11	LM1110B	11	S822224	14	2SC1919C	14	A2EBLK	14	BDY23
11	LM1120B	11	SP82425	14	2SC2884	14	A2EBRN	14	BDY38
11	LM11016	11	S83787	14	2SC313	14	A2EBRN-1	14	BDY39
11	LM-1252510-1	11	SP84143	14	2SC313-OR	14	A2B	14	BDY53
11	M140-1	11	SP8416S	14	2SC313A	14	A2B-3	14	BLY10
11	M1400-1	11	SP8429	14	2SC313B	14	A3902441	14	BLY11
11	M4757	11	SP843-1	14	2SC313D	14	A3L4-6001	14	BLY12
11	M4789	11	SP85569	14	2SC313D	14	A3L4-6001-01	14	BLY15
11	M4825	11	SP86682	14	2SC313E	14	A3TE120	14	BLY47
11		11	SP8816	14	2SC313F	14	A3TE230	14	BLY48
11		11	SP8856	14		14	A3TE240	14	

GE	Industry Standard No.	GE	Industry Standard No.	GE	Industry Standard No.	GE	Industry Standard No.	GE	Industry Standard No.
14	BN7133	14	MHT7607	14	T-Q5105	16	B142B	16	PAD250
14	BN7214	14	MHT7608	14	T1P3055	16	B142C	16	PQ31
14	BRC-116	14	MHT7609	14	TA2577A	16	B143	16	PT235A
14	BUY10	14	MJ2801	14	TA7068	16	B143P	16	PT235B
14	BUY11	14	MJ2802	14	TA7069	16	B144	16	PT236A
14	C21	14	MJ2840	14	TNJ72148	16	B144P	16	PT336B
14	C244	14	MJ2841	14	TQ-PD-3055	16	B145	16	PT554
14	C49T	14	MJ480	14	TR-1000-7	16	B146	16	PTC105
14	C520A	14	MJ481	14	TR-1039-4	16	B147	16	R3515
14	C521	14	MJE2940	14	TR-28	16	B149	16	RT167
14	C521A	14	N-121122	14	TR-59	16	B19	16	RY620
14	C647Q	14	N-52329	14	TR1007	16	B20	16	R8313
14	C647R	14	N121122	14	TR1039-4	16	B21	16	RE11
14	C664	14	P-10954-1	14	TR1039-4	16	B228	16	RE11MP
14	C664B	14	P-10954-2	14	TR1077	16	B230	16	RE7
14	C664C	14	P-11901-1	14	TR1490	16	B233	16	REN104
14	C736	14	P10619-1	14	TR1491	16	B246	16	RS-1055
14	C765	14	P2271	14	TR1492	16	B249	16	RS-2006
14	C768	14	P3139	14	TR1493	16	B249A	16	RT-124
14	C793	14	P50200-11	14	TR26	16	B250	16	RT-127
14	C793R	14	P5034	14	TR26C	16	B250A	16	RT4762MHP25
14	C793Y	14	P5149	14	TR271TR26	16	B255	16	SF-T191
14	C794R	14	PP-AR15	14	TR59	16	B256	16	SP1013B
14	C851	14	PMC-QPO010	14	TS-1193-736	16	B26(JAPAN)	16	SP1108
14	CD461	14	PMC-QPO012	14	TV8280647	16	B26A	16	SP1137
14	CD461-014-614	14	PN350	14	1-002	16	B27	16	SP1323
14	CII-225-Q	14	PP3000	14	1-003	16	B28(JAPAN)	16	SP1403
14	CP400	14	PP3003	14	001-021270	16	B29	16	SP1600
14	CP401	14	PP3006	14	1B1-0356-00-A	16	B2SB241	16	SP1619
14	CP404	14	PT1941	14	2-0-3055	16	B2SB244	16	SP1651
14	CP405	14	PTC119	14	2CD1988	16	B30(JAPAN)	16	SP176
14	CP406	14	PTC140	14	2D010	16	B31(TRANSISTOR)	16	SP1927
14	CP407	14	QP-11	14	2G3055	16	B355	16	SP2234
14	CP408	14	QP-12	14	2N1069	16	B356	16	SP2247
14	CXL130	14	QP-8	14	2N1070	16	B41	16	SP230
14	D12	14	QPO01200A	14	2N1072	16	B413	16	SP235B
14	D146	14	QP8	14	2N1423	16	B414	16	SP2431
14	D147	14	R135--1	14	2N1490	16	B42	16	SP26
14	D15	14	R2270-75497	14	2N1702	16	B62	16	SP441D
14	D151	14	R2270-78399	14	2N1703	16	B83	16	SP441B
14	D16	14	R22707-8399	14	2N2305	16	BC1073	16	SP47
14	D163	14	R227075497	14	2N2383	16	BC1073A	16	SP486
14	D164	14	R227077499	14	2N2384	16	BC1274	16	SP880
14	D175	14	R2982	14	2N3055	16	BC1274A	16	SP891
14	D180M	14	R4369	14	2N3055-1	16	BC1274B	16	SP891-B
14	D188	14	RC-1700	14	2N3055-10	16	CS21739	16	T-Q5028
14	D188A	14	RC1700	14	2N3055-2	16	CS21740	16	T-Q5036
14	D188B	14	RC8242	14	2N3055-4	16	CS21741	16	T1040
14	D188C	14	RE19	14	2N3055-5	16	CS21742	16	T1041
14	D201(0)	14	REN130	14	2N3055-9	16	CTP1104	16	T1366
14	D201M	14	RT-131	14	2N3055S	16	CTP1108	16	T1366A
14	D201Y	14	RT-154	14	2N3226	16	CTP1109	16	T1367
14	D211	14	S-305	14	2N3232	16	CTP1117	16	T1367A
14	D212	14	S-305-PD	14	2N3233	16	CTP1119	16	T136B
14	D241H	14	S-305A	14	2N3234	16	CTP1740	16	T1368A
14	D26A	14	S-356	14	2N3445	16	CXL104	16	T1369
14	D26B	14	S1685	14	2N3446	16	D080	16	T1369A
14	D26C	14	S1691	14	2N3447	16	D8503	16	T1370
14	D3005YN	14	S1692	14	2N3448	16	DT401	16	T1370A
14	D41	14	S1865	14	2N3667	16	DTG-110	16	T139
14	D53	14	S1905	14	2N3713	16	DTG110(WARDS)	16	TP78
14	D69	14	S1905A	14	2N3714	16	DU6	16	TP78/30
14	D80	14	S1907	14	2N3715	16	ECG104	16	T1370
14	D81	14	S1977634	14	2N3865	16	EQ-8	16	T1370A
14	D82	14	S2003--1	14	2N3864	16	ES13	16	TM104MP
14	D82A	14	S2241	14	2N4130	16	ES13(ELCOM)	16	TR-01MP
14	DD-79D107-1	14	S2392	14	2N4395	16	ES15X45	16	TR-02
14	DP-2	14	S2403B	14	2N4396	16	ES15X51	16	TR-16
14	DS-509	14	S2403C	14	2N4913	16	ES15X78	16	TR-16C
14	DS-514	14	S2471	14	2N4914	16	ES18(ELCOM)	16	TR-172(OLSON)
14	DS-519	14	S2741	14	2N4915	16	ES503(ELCOM)	16	TR01
14	DS509	14	S305	14	2N5614	16	ES7	16	TR01MP
14	DS519	14	S305A	14	2N5622	16	ES9	16	TS-610
14	DT4011	14	S305D	14	2N6253	16	ES9(ELCOM)	16	TS-612
14	DT4110	14	S353	14	2N6371	16	F-67-E	16	TS-613
14	DT4111	14	S356	14	28033	16	GE-16	16	TS-614
14	DT4120	14	S37771	14	28034	16	G04-701-A	16	TVS2SB449(P)
14	DT4121	14	S9938B133	14	2SC1030	16	GP216	16	001-01204-0
14	EA151100	14	S9938B165	14	2SC1030-0R	16	HEP200	16	001-012040
14	EA1740	14	SB5	14	2SC1030A	16	HEP230	16	001-01205-0
14	EC961	14	SB6	14	2SC1030B	16	HEP626	16	001-01205-1
14	ECG130	14	SB7	14	2SC1030B2C	16	HEPG6000	16	001-012050
14	ES816(ELCOM)	14	SCD-T320	14	2SC1030C	16	HEPG6000P/G	16	001-012051
14	ES31(ELCOM)	14	SDT9201	14	2SC1030D	16	HEPG6003	16	002-007000
14	ETB-003	14	SDT9205	14	2SC1030E	16	HEPG6003P/G	16	002-008100
14	FBN-36220	14	SDT9206	14	2SC1030F	16	HEPG6005P/G	16	002-009700
14	FBN-36485	14	SDT9210	14	2SC1030G	16	HEPG6013	16	2D004-9
14	FBN-36603	14	SDT9261	14	2SC1030H	16	HEPG6013P/G	16	2G223
14	FS2003-1	14	SE-3033	14	2SC1030J	16	HJ35	16	2N1007
14	G23-45	14	SE3033	14	2SC1030K	16	HR101	16	2N1040
14	G23-67	14	SE3035	14	2SC1030L	16	IRTRO1	16	2N1136C
14	GE-14	14	SE3036	14	2SC1030M	16	JP40	16	2N1138B
14	GRASS-R2982	14	SE9002	14	2SC10300R	16	K04774	16	2N1146
14	HEP247	14	SE9080	14	2SC1030R	16	K4-520	16	2N1146A
14	HEP704	14	SE8632	14	2SC1030X	16	K4-521	16	2N1146B
14	HEP97002	14	SE8881	14	2SC1030Y	16	K04774	16	2N1146C
14	HP22	14	SJ1106	14	2SC1114	16	L852	16	2N1147
14	HST-9201	14	SJ1470	14	2SC1152	16	M4463	16	2N1147A
14	HST-9205	14	SJ2000	14	2SC1618	16	M4570	16	2N1147B
14	HST-9206	14	SJ2008	14	2SC1618B	16	M4619	16	2N1147C
14	HST-9210	14	SJ3464	14	2SC1777	16	M4620	16	2N1162
14	HT30494	14	SJ3604	14	2SC520	16	M4649	16	2N1162A
14	HT304941X	14	SJ3678	14	2SC520A	16	M4727	16	2N1164
14	HT30494X	14	SJ619	14	2SC521	16	M4887A	16	2N1164A
14	HT401191	14	SJ619-1	14	2SC521A	16	M4974	16	2N1166
14	HT401191A	14	SJ820	14	2SC7	16	M4974/P1R	16	2N1172
14	HT401191B	14	SJ8701	15	SE35(ELCOM)	16	M54	16	2N1227
14	IR-TR59	14	SJ9110	15	S9938E140	16	MH76	16	2N1227-3
14	IRTR-61	14	SK3027	15MP	CX130MP	16	O016	16	2N1227-4
14	IRTR61	15	S9938B140	15MP	ECG130MP	16	O022	16	2N1227-4A
14	KO71964-001	15	SK3510	15MP	GE-15MP	16	O023	16	2N1227-4R
14	K4-525	15MP	SP4231	15MP	HEP704X2	16	O024	16	2N1227A
14	KB-1007	15MP	SPD-80062	15MP	IRTR61X2	16	O025	16	2N1245
14	KB-1993B	15MP	SPT3713	16	A0S9-115	16	O026	16	2N1246
14	KBD1051	15MP	STC-1035	16	A2SB240A	16	O027	16	2N1291
14	KBD1055	15MP	STC-1035A	16	A2SB248A	16	O028	16	2N1314
14	KBD1056	14	STC-1036	16	AD139	16	O029	16	2N1314R
14	KBD3055	14	STC-1036A	16	AD149	16	O035	16	2N1419
14	KBD9707	14	STC-1080	16	AR4	16	O036	16	2N1501
14	LN75497	14	STC01035	16	AR5	16	P1R	16	2N1502
14	M1B-12795B	14	STC01035A	16	AR6	16	P2C	16	2N1609
14	M4715	14	STC01036	16	AR7	16	P2R	16	2N1610
14	M4882	14	STC01036A	16	AUY21	16	P3SBLK	16	2N1668
14	M7543-1	14	STC01080	16	AUY22	16	P3SBLU	16	2N1755
14	M75549-2	14	STC01081	16	B10064	16	P3SBRN	16	2N1756
14	M9244	14	STC01082	16	B10069	16	P3SRED	16	2N1757
14	M9259	14	STC01083	16	B1215	16	P4D	16	2N1758
14	M9278	14	STC01084	16	B123	16	P4L	16	2N1759
14	M9302	14	STC04252	16	B123A	16	P4N	16	2N176
14	M9321	14	STC04253	16	B140	16	P75534-2	16	2N176-1
14	M9515	14	STC04254	16	B141	16	P75534-3	16	2N1760
14	MHT7601	14	STC04255	16	B142	16	PA10890	16	2N1761
14	MHT7602	14	STX0014						
14	MHT7603	14	STX0027						
		14	STX0032						

GE	Industry Standard No.	GE	Industry Standard No.	GE	Industry Standard No.	GE	Industry Standard No.	GE	Industry Standard No.
16	2N1762	17	BC198A	17	2SC377D	18	2N3501	18	2SC114G
16	2N176A	17	BC548	17	2SC377E	18	2N3725	18	2SC114GN
16	2N176G	17	BC548VI	17	2SC377F	18	2N5120	18	2SC114H
16	2N176W	17	BCW48A	17	2SC377G	18	2N5300	18	2SC114J
16	2N178	17	BCX58VII	17	2SC377GN	18	2N5784	18	2SC114K
16	2N179	17	BCX58VIII	17	2SC377H	18	2N5859	18	2SC114L
16	2N2061	17	BE-66	17	2SC377J	18	2N5964	18	2SC114M
16	2N2062	17	BF-180	17	2SC377K	18	2N5965	18	2SC114OR
16	2N2062A	17	BG-66	17	2SC377L	18	2N656A	18	2SC114R
16	2N2063	17	C523583	17	2SC377M	18	2N69B	18	2SC114X
16	2N2063A	17	C760	17	2SC377O	18	2N699A	18	2SC114Y
16	2N2064	17	CC059018F	17	2SC377OR	18	2N699B	18	2SC115
16	2N2064A	17	CC8-2006D	17	2SC377RED	18	2N908	18	2SC115-1
16	2N2065	17	CDC-8000-1	17	2SC377X	18	2S644(8)	18	2SC115-2
16	2N2065A	17	CS-1244X	17	2SC377Y	18	2SC-NJ-107	18	2SC115A
16	2N2066	17	CS-1258	17	2SC426	18	2SC1033	18	2SC115B
16	2N2066A	17	CS-1305	17	2SC427	18	2SC1033A	18	2SC115C
16	2N2067	17	CS-1359	17	2SC428	18	2SC1033B	18	2SC115D
16	2N2067B	17	CS-9018D	18	2SC628E	18	2SC1033C	18	2SC115F
16	2N2067Q	17	CB112OI	18	2SC628F	18	2SC1033D	18	2SC115G
16	2N2067W	17	C81330D	18	2SC717	18	2SC1033E	18	2SC115GN
16	2N2069	17	C86225E	18	2SC717(3RD-IF)	18	2SC1033F	18	2SC115H
16	2N2070	17	D11C1P1	18	2SC717(LAST-IF)	18	2SC1033G	18	2SC115J
16	2N2137	17	D11C2D1B2Q	18	2SC717B	18	2SC1033GN	18	2SC115K
16	2N2137A	17	D11C5P1	18	2SC717BK	18	2SC1033H	18	2SC115L
16	2N2138	17	D133	18	2SC717BLK	18	2SC1033J	18	2SC115M
16	2N2138A	17	D32K1	18	2SC717C	18	2SC1033K	18	2SC115OR
16	2N2139	17	D32K2	18	2SC717E	18	2SC1033L	18	2SC115R
16	2N2139A	17	D4D24	18	2SC717F	18	2SC1033OR	18	2SC115X
16	2N2140	17	DS-75	18	2SC717G	18	2SC1033R	18	2SC115Y
16	2N2140A	17	EA15X7117	18	2SC717GN	18	2SC1033X	18	2SC117
16	2N2141	17	EN10	18	2SC717H	18	2SC1033Y	18	2SC117A
16	2N2141A	17	EQ8-62	18	2SC717K	18	2SC1072	18	2SC117B
16	2N2142	17	F7316	18	2SC717L	18	2SC1072A	18	2SC117C
16	2N2142A	17	FK3484	18	2SC717M	18	2SC1072B	18	2SC117D
16	2N2143	17	FK3494	18	2SC717X	18	2SC1072C	18	2SC117F
16	2N2143A	17	F08104	18	AM	18	2SC1072D	18	2SC117GN
16	2N2144	17	FS-1133	18	AR306BLUE	18	2SC1072F	18	2SC117H
16	2N2144A	17	FT40	18	AR306ORANGE	18	2SC1072G	18	2SC117J
16	2N2145	17	GE-17	18	BC173,B	18	2SC1072GN	18	2SC117K
16	2N2145A	17	GI-2712	18	BC341-06	18	2SC1072H	18	2SC117L
16	2N2146	17	GMO-380	18	CDC587	18	2SC1072J	18	2SC117M
16	2N2146A	17	G05-050-C	18	CDC745	18	2SC1072K	18	2SC117R
16	2N2147	17	HC-00535	18	CF-1295H	18	2SC1072L	18	2SC117X
16	2N2282	17	HC-373	18	CJ-5210	18	2SC1072M	18	2SC117Y
16	2N230	17	HC-535	18	CS-1372	18	2SC1072OR	18	2SC118
16	2N242	17	HC-561	18	CS-9013HG	18	2SC1072R	18	2SC118A
16	2N250	17	HC-772	18	CS1229N	18	2SC1072X	18	2SC118B
16	2N250A	17	HC01830	18	CS9013E-F	18	2SC1072Y	18	2SC118C
16	2N255	17	HT303711A-0	18	CXL108	18	2SC108	18	2SC118D
16	2N255A	17	HT303711B-0	18	E-185B121712	18	2SC1080-R	18	2SC118E
16	2N256	17	HT305551C0	18	EA-15X85517	18	2SC108A	18	2SC118F
16	2N256A	17	HT306451H	18	EA0081	18	2SC108A-0	18	2SC118G
16	2N257	17	HT306962A-Q	18	EC0128	18	2SC108A-R	18	2SC118H
16	2N257A	17	IRTR22	18	ERS100	18	2SC108B	18	2SC118J
16	2N257B	17	IRTR24	18	ERS120	18	2SC108C	18	2SC118L
16	2N257G	17	IRTR51	18	EX744-X	18	2SC108D	18	2SC118M
16	2N257W	17	J24842	18	F3561	18	2SC108E	18	2SC118OR
16	2N2612	17	KB8416	18	F69916	18	2SC108F	18	2SC118R
16	2N285	17	KR8417	18	G05-013C	18	2SC108GN	18	2SC118X
16	2N285A	17	M4970	18	G05-413D	18	2SC108H	18	2SC118Y
16	2N285B	17	MPS-H32	18	GE-18	18	2SC108J	18	2SC119A
16	2N290	17	MPS82926-BRN	18	G05-034-D	18	2SC108K	18	2SC119B
16	2N296	17	MPS82926-ORG	18	G05-036D	18	2SC108L	18	2SC119C
16	2N307	17	MPS82926-RED	18	G05-055-D	18	2SC108M	18	2SC119D
16	2N325	17	MPS82926-YEL	18	G05-413A	18	2SC108OR	18	2SC119E
16	2N375	17	MPSH02	18	HC-00644	18	2SC108R	18	2SC119F
16	2N376	17	MPSH32	18	HEP711	18	2SC108X	18	2SC119G
16	2N376A	17	M83694	18	HEP85026	18	2SC108Y	18	2SC119GN
16	2N378	17	ON271	18	HT304971A0	18	2SC109	18	2SC119H
16	2N380	17	PE5015	18	HT306441A0	18	2SC109A	18	2SC119J
16	2N386	17	PE5010	18	HT306441B-0	18	2SC109A-0	18	2SC119K
16	2N387	17	PE5013	18	HT306441B0	18	2SC109A-R	18	2SC119L
16	2N392	17	RE10	18	HT306441C-0	18	2SC109A-Y	18	2SC119M
16	2N399	17	RS-7127	18	HT307341C-0	18	2SC109B	18	2SC119OR
16	2N400	17	RT8330	18	HT308281O	18	2SC109C	18	2SC119X
16	2N401	17	S1060	18	HT309841B0	18	2SC109D	18	2SC119Y
16	2N419	17	SC108	18	HT309842A-0	18	2SC109E	18	2SC12
16	2N4244	17	SC108A	18	HT800012F	18	2SC109F	18	2SC120
16	2N4247	17	TIS-125	18	HT800191H	18	2SC109G	18	2SC120A
16	2N456	17	TNJ-60604	18	IRTR53	18	2SC109GN	18	2SC120B
16	2N456A	17	TR-51	18	IRTR87	18	2SC109H	18	2SC120C
16	2N456B	17	TR2083-71	18	K14-0066-13	18	2SC109J	18	2SC120D
16	2N457	17	TR51	18	KA1225	18	2SC109K	18	2SC120F
16	2N457A	17	TR8330	18	M4918	18	2SC109L	18	2SC120G
16	2N457B	17	1A0045	18	M4919	18	2SC109OR	18	2SC120GN
16	2N458	17	2D033	18	MP89633	18	2SC109X	18	2SC120H
16	2N458A	17	2N1150/904	18	Q5099E	18	2SC109Y	18	2SC120J
16	2N511	17	2N1151/904A	18	Q5099F	18	2SC112	18	2SC120K
16	2N511A	17	2N1153/910	18	RE17	18	2SC112A	18	2SC120M
16	2N511B	17	2N3043	18	REN128	18	2SC112B	18	2SC120R
16	2N512	17	2N3044	18	RT-114	18	2SC112E	18	2SC120Y
16	2N512A	17	2N3045	18	S1421	18	2SC112F	18	2SC121A
16	2N512B	17	2N3046	18	S1683	18	2SC112G	18	2SC121B
16	2N513	17	2N3047	18	SE-4020	18	2SC112GN	18	2SC121C
16	2N513A	17	2N3048	18	SK3046	18	2SC112H	18	2SC121D
16	2N513B	17	2N5103	18	T-Q5033	18	2SC112J	18	2SC121E
16	2N539	17	2N708/46	18	T-Q5055	18	2SC112L	18	2SC121F
16	2N539A	17	2N914/46	18	T-Q5032	18	2SC112M	18	2SC121G
16	2N540	17	2N930/TNT	18	T-Q5063	18	2SC112OR	18	2SC121GN
16	2N540A	17	2SB382BN	18	TA12	18	2SC112R	18	2SC121H
16	2N618	17	2SC1394	18	TG28065(Y)	18	2SC112X	18	2SC121J
16	2N657	17	2S018A	18	TQ5081	18	2SC112Y	18	2SC121L
16	2N637A	17	2SC30	18	TR-53	18	2SC113	18	2SC121M
16	2N637B	17	2SC30-0R	18	TR-87	18	2SC113A	18	2SC121OR
16	2N638	17	2SC30A	18	TR53	18	2SC113B	18	2SC121R
16	2N638A	17	2SC30B	18	TV25	18	2SC113C	18	2SC121X
16	2N638B	17	2SC30C	18	02-33379-6	18	2SC113D	18	2SC121Y
16	2N639	17	2SC30D	18	2N1155	18	2SC113E	18	2SC122
16	2N639A	17	2SC30E	18	2N1156	18	2SC113F	18	2SC1220
16	2N639B	17	2SC30F	18	2N1156/953	18	2SC113G	18	2SC1220A
16	2N66	17	2SC30G	18	2N1335	18	2SC113GN	18	2SC1220AP
16	2N67	17	2SC30GN	18	2N1336	18	2SC113H	18	2SC1220AQ
16	2N68	17	2SC30H	18	2N1337	18	2SC113J	18	2SC1220AR
16	2S41	17	2SC30J	18	2N1339	18	2SC113K	18	2SC1220P
16	2S41A	17	2SC30L	18	2N1340	18	2SC113L	18	2SC1220Q
16	2S42	17	2SC30M	18	2N1341	18	2SC113M	18	2SC1220QQ
16	2SB123	17	2SC30R	18	2N1943	18	2SC113OR	18	2SC1220R
16	2SB123A	17	2SC30RA	18	2N2086	18	2SC113R	18	2SC122A
16	2SB141	17	2SC30X	18	2N2351	18	2SC113X	18	2SC122B
16	2SB152	17	2SC30Y	18	2N2478	18	2SC113Y	18	2SC122C
16	2SB21	17	2SC377	18	2N2538	18	2SC114	18	2SC122D
16	2SB249A	17	2SC377-BN	18	2N2788	18	2SC114A	18	2SC122E
16	2SB27	17	2SC377-BRN	18	2N2789	18	2SC114B	18	2SC122F
16	2SB29	17	2SC377-0	18	2N2890	18	2SC114C	18	2SC122G
16	2SB471-0	17	2SC377-0R	18	2N2891	18	2SC114D	18	2SC122GN
16	2SC467	17	2SC377-ORG	18	2N3036	18	2SC114E	18	2SC122H
16(2)	B337B	17	2SC377-R	18	2N3038	18	2SC114F		
16(TWO)	EC9104MP	17	2SC377-RED	18	2N3056				
16,237	TRO2	17	2SC377A	18	2N3057				
17	A455	17	2SC377B	18	2N3118				
17	B-66	17	2SC377BN	18	2N342B				
17	B2Z	17	2SC377BRN	18	2N3499				
		17	2SC377C						

GE	Industry Standard No.	GE	Industry Standard No.	GE	Industry Standard No.	GE	Industry Standard No.	GE	Industry Standard No.
18	2SC122J	18	2SC163E	18	2SC2100R	18	2SC218F	18	2SC227
18	2SC122K	18	2SC163F	18	2SC210R	18	2SC218G	18	2SC227A
18	2SC122L	18	2SC163G	18	2SC210X	18	2SC218GN	18	2SC227B
18	2SC122M	18	2SC163GN	18	2SC210Y	18	2SC218H	18	2SC227C
18	2SC122OR	18	2SC163H	18	2SC211	18	2SC218J	18	2SC227D
18	2SC122R	18	2SC163J	18	2SC2110	18	2SC218L	18	2SC227E
18	2SC122X	18	2SC163K	18	2SC211A	18	2SC218M	18	2SC227G
18	2SC122Y	18	2SC163L	18	2SC211B	18	2SC218OR	18	2SC227GN
18	2SC123	18	2SC163M	18	2SC211C	18	2SC218R	18	2SC227H
18	2SC123A	18	2SC163OR	18	2SC211D	18	2SC218X	18	2SC227J
18	2SC123B	18	2SC163R	18	2SC211F	18	2SC218Y	18	2SC227K
18	2SC123C	18	2SC163X	18	2SC211G	18	2SC22	18	2SC227L
18	2SC123D	18	2SC163Y	18	2SC211GN	18	2SC220	18	2SC227M
18	2SC123E	18	2SC188	18	2SC211H	18	2SC220A	18	2SC2270R
18	2SC123F	18	2SC188A	18	2SC211J	18	2SC220B	18	2SC227X
18	2SC123G	18	2SC188AB	18	2SC211K	18	2SC220C	18	2SC227Y
18	2SC123GN	18	2SC188B	18	2SC211L	18	2SC220D	18	2SC228
18	2SC123H	18	2SC188C	18	2SC211M	18	2SC220E	18	2SC228A
18	2SC123J	18	2SC188D	18	2SC2110R	18	2SC220F	18	2SC228B
18	2SC123K	18	2SC188E	18	2SC211R	18	2SC220G	18	2SC228C
18	2SC123L	18	2SC188F	18	2SC211X	18	2SC220GN	18	2SC228D
18	2SC123M	18	2SC188G	18	2SC211Y	18	2SC220H	18	2SC228E
18	2SC123OR	18	2SC188GN	18	2SC212O	18	2SC220K	18	2SC228G
18	2SC123R	18	2SC188J	18	2SC212A	18	2SC220L	18	2SC228GN
18	2SC123X	18	2SC188K	18	2SC212B	18	2SC220M	18	2SC228H
18	2SC123Y	18	2SC188L	18	2SC212C	18	2SC220OR	18	2SC228J
18	2SC124	18	2SC188M	18	2SC212D	18	2SC220X	18	2SC228K
18	2SC124A	18	2SC188OR	18	2SC212E	18	2SC220Y	18	2SC228L
18	2SC124B	18	2SC188R	18	2SC212F	18	2SC221	18	2SC228M
18	2SC124C	18	2SC188X	18	2SC212G	18	2SC221A	18	2SC228OR
18	2SC124D	18	2SC188Y	18	2SC212GN	18	2SC221B	18	2SC228R
18	2SC124E	18	2SC189	18	2SC212H	18	2SC221C	18	2SC228X
18	2SC124F	18	2SC189A	18	2SC212J	18	2SC221D	18	2SC228Y
18	2SC124G	18	2SC189B	18	2SC212K	18	2SC221E	18	2SC229
18	2SC124GN	18	2SC189C	18	2SC212L	18	2SC221F	18	2SC229A
18	2SC124H	18	2SC189D	18	2SC212M	18	2SC221G	18	2SC229B
18	2SC124J	18	2SC189F	18	2SC2120R	18	2SC221GN	18	2SC229C
18	2SC124K	18	2SC189G	18	2SC212R	18	2SC221H	18	2SC229D
18	2SC124L	18	2SC189GN	18	2SC212X	18	2SC221J	18	2SC229E
18	2SC124M	18	2SC189H	18	2SC212Y	18	2SC221K	18	2SC229F
18	2SC124OR	18	2SC189J	18	2SC213	18	2SC221L	18	2SC229G
18	2SC124R	18	2SC189K	18	2SC213A	18	2SC221M	18	2SC229GN
18	2SC124X	18	2SC189L	18	2SC213B	18	2SC2210R	18	2SC229H
18	2SC124Y	18	2SC189M	18	2SC213C	18	2SC221R	18	2SC229J
18	2SC12A	18	2SC189OR	18	2SC213D	18	2SC221X	18	2SC229K
18	2SC12B	18	2SC189R	18	2SC213E	18	2SC221Y	18	2SC229L
18	2SC12C	18	2SC189X	18	2SC213F	18	2SC222	18	2SC229M
18	2SC12D	18	2SC189Y	18	2SC213G	18	2SC222A	18	2SC229OR
18	2SC12E	18	2SC19	18	2SC213GN	18	2SC222B	18	2SC229R
18	2SC12F	18	2SC190	18	2SC213H	18	2SC222C	18	2SC229X
18	2SC12G	18	2SC190A	18	2SC213J	18	2SC222E	18	2SC229Y
18	2SC12GN	18	2SC190B	18	2SC213K	18	2SC222F	18	2SC22A
18	2SC12H	18	2SC190C	18	2SC213L	18	2SC222G	18	2SC22B
18	2SC12J	18	2SC190D	18	2SC213M	18	2SC222GN	18	2SC22D
18	2SC12K	18	2SC190E	18	2SC2130R	18	2SC222H	18	2SC22F
18	2SC12L	18	2SC190F	18	2SC213R	18	2SC222J	18	2SC22G
18	2SC12M	18	2SC190GN	18	2SC213X	18	2SC222L	18	2SC22GN
18	2SC12OR	18	2SC190H	18	2SC213Y	18	2SC222M	18	2SC22H
18	2SC12R	18	2SC190J	18	2SC214	18	2SC2220R	18	2SC22J
18	2SC12X	18	2SC190K	18	2SC214A	18	2SC222R	18	2SC22K
18	2SC12Y	18	2SC190L	18	2SC214B	18	2SC222X	18	2SC22L
18	2SC130	18	2SC190M	18	2SC214C	18	2SC222Y	18	2SC22M
18	2SC130A	18	2SC190OR	18	2SC214D	18	2SC223	18	2SC22OR
18	2SC130B	18	2SC190R	18	2SC214F	18	2SC223A	18	2SC22R
18	2SC130C	18	2SC190X	18	2SC214G	18	2SC223B	18	2SC22X
18	2SC130D	18	2SC190Y	18	2SC214GN	18	2SC223C	18	2SC22Y
18	2SC130F	18	2SC19A	18	2SC214H	18	2SC223D	18	2SC23
18	2SC130GN	18	2SC19B	18	2SC214J	18	2SC223E	18	2SC231
18	2SC130H	18	2SC19C	18	2SC214K	18	2SC223F	18	2SC231A
18	2SC130J	18	2SC19D	18	2SC214L	18	2SC223G	18	2SC231B
18	2SC130K	18	2SC19E	18	2SC214M	18	2SC223GN	18	2SC231C
18	2SC130L	18	2SC19F	18	2SC2140R	18	2SC223H	18	2SC231D
18	2SC130M	18	2SC19G	18	2SC214X	18	2SC223J	18	2SC231E
18	2SC130OR	18	2SC19H	18	2SC214Y	18	2SC223L	18	2SC231F
18	2SC130R	18	2SC19J	18	2SC215	18	2SC223M	18	2SC231G
18	2SC130X	18	2SC19K	18	2SC215A	18	2SC2230R	18	2SC231GN
18	2SC130Y	18	2SC19L	18	2SC215B	18	2SC223R	18	2SC231H
18	2SC131T	18	2SC19M	18	2SC215C	18	2SC223X	18	2SC231J
18	2SC140	18	2SC190R	18	2SC215E	18	2SC223Y	18	2SC231K
18	2SC140A	18	2SC19R	18	2SC215G	18	2SC224	18	2SC231L
18	2SC140C	18	2SC19X	18	2SC215GN	18	2SC224A	18	2SC231M
18	2SC140D	18	2SC19Y	18	2SC215H	18	2SC224B	18	2SC2310R
18	2SC140E	18	2SC20	18	2SC215J	18	2SC224C	18	2SC231R
18	2SC140F	18	2SC208	18	2SC215K	18	2SC224D	18	2SC231X
18	2SC140G	18	2SC208A	18	2SC215L	18	2SC224E	18	2SC231Y
18	2SC140GN	18	2SC208B	18	2SC215M	18	2SC224F	18	2SC232
18	2SC140H	18	2SC208C	18	2SC2150R	18	2SC224G	18	2SC232A
18	2SC140J	18	2SC208D	18	2SC215R	18	2SC224GN	18	2SC232B
18	2SC140K	18	2SC208E	18	2SC215X	18	2SC224H	18	2SC232C
18	2SC140L	18	2SC208F	18	2SC215Y	18	2SC224J	18	2SC232D
18	2SC140OR	18	2SC208G	18	2SC216	18	2SC224K	18	2SC232F
18	2SC140R	18	2SC208GN	18	2SC216A	18	2SC224L	18	2SC232G
18	2SC140X	18	2SC208H	18	2SC216C	18	2SC224M	18	2SC232GN
18	2SC140Y	18	2SC208J	18	2SC216D	18	2SC2240R	18	2SC232H
18	2SC147	18	2SC208K	18	2SC216E	18	2SC224R	18	2SC232J
18	2SC147A	18	2SC208L	18	2SC216F	18	2SC224X	18	2SC232K
18	2SC147B	18	2SC208M	18	2SC216G	18	2SC224Y	18	2SC232L
18	2SC147C	18	2SC208OR	18	2SC216GN	18	2SC225	18	2SC232M
18	2SC147D	18	2SC208R	18	2SC216H	18	2SC225A	18	2SC2320R
18	2SC147F	18	2SC208X	18	2SC216J	18	2SC225B	18	2SC232R
18	2SC147G	18	2SC208Y	18	2SC216K	18	2SC225C	18	2SC232X
18	2SC147GN	18	2SC20A	18	2SC216L	18	2SC225D	18	2SC232Y
18	2SC147H	18	2SC20B	18	2SC216X	18	2SC225E	18	2SC233
18	2SC147J	18	2SC20C	18	2SC216Y	18	2SC225F	18	2SC233A
18	2SC147K	18	2SC20D	18	2SC217	18	2SC225G	18	2SC233B
18	2SC147L	18	2SC20E	18	2SC217A	18	2SC225GN	18	2SC233C
18	2SC147M	18	2SC20F	18	2SC217B	18	2SC225H	18	2SC233D
18	2SC147R	18	2SC20G	18	2SC217C	18	2SC225J	18	2SC233F
18	2SC147X	18	2SC20GN	18	2SC217D	18	2SC225L	18	2SC233G
18	2SC147Y	18	2SC20H	18	2SC217E	18	2SC225M	18	2SC233GN
18	2SC152	18	2SC20J	18	2SC217F	18	2SC2250R	18	2SC233H
18	2SC152A	18	2SC20K	18	2SC217G	18	2SC225R	18	2SC233L
18	2SC152B	18	2SC20L	18	2SC217GN	18	2SC225X	18	2SC233M
18	2SC152C	18	2SC20M	18	2SC217H	18	2SC225Y	18	2SC233OR
18	2SC152D	18	2SC20OR	18	2SC217J	18	2SC226	18	2SC233X
18	2SC152E	18	2SC20X	18	2SC217K	18	2SC226A	18	2SC233Y
18	2SC152F	18	2SC20Y	18	2SC217L	18	2SC226B	18	2SC235
18	2SC152G	18	2SC210	18	2SC217M	18	2SC226C	18	2SC235-0
18	2SC152GN	18	2SC210A	18	2SC2170R	18	2SC226D	18	2SC235A
18	2SC152H	18	2SC210B	18	2SC217R	18	2SC226E	18	2SC235B
18	2SC152J	18	2SC210C	18	2SC217X	18	2SC226F	18	2SC235C
18	2SC152K	18	2SC210D	18	2SC217Y	18	2SC226G	18	2SC235D
18	2SC152L	18	2SC210E	18	2SC218	18	2SC226GN	18	2SC235F
18	2SC152M	18	2SC210F	18	2SC218A	18	2SC226H	18	2SC235G
18	2SC152OR	18	2SC210G	18	2SC218B	18	2SC226J	18	2SC235GN
18	2SC152R	18	2SC210GN	18	2SC218C	18	2SC226K	18	2SC235H
18	2SC152X	18	2SC210H	18	2SC218D	18	2SC226L		
18	2SC152Y	18	2SC210J	18	2SC218E	18	2SC226M		
18	2SC163A	18	2SC210K			18	2SC2260R		
18	2SC163B	18	2SC210L			18	2SC226R		
18	2SC163C	18	2SC210M			18	2SC226X		
18	2SC163D					18	2SC226Y		

GE	Industry Standard No.
18	2SC235J
18	2SC235K
18	2SC235L
18	2SC235M
18	2SC235OR
18	2SC235R
18	2SC235X
18	2SC235Y
18	2SC236
18	2SC236A
18	2SC236B
18	2SC236C
18	2SC236D
18	2SC236E
18	2SC236P
18	2SC236G
18	2SC236GN
18	2SC236H
18	2SC236J
18	2SC236K
18	2SC236L
18	2SC236M
18	2SC236OR
18	2SC236R
18	2SC236X
18	2SC236Y
18	2SC238
18	2SC238A
18	2SC238B
18	2SC238C
18	2SC238D
18	2SC238E
18	2SC238F
18	2SC238G
18	2SC238GN
18	2SC238H
18	2SC238J
18	2SC238K
18	2SC238L
18	2SC238M
18	2SC238OR
18	2SC238R
18	2SC238X
18	2SC238Y
18	2SC23A
18	2SC23B
18	2SC23D
18	2SC23E
18	2SC23F
18	2SC23G
18	2SC23GN
18	2SC23H
18	2SC23J
18	2SC23K
18	2SC23L
18	2SC23M
18	2SC23OR
18	2SC23R
18	2SC23X
18	2SC23Y
18	2SC24
18	2SC247
18	2SC247A
18	2SC247B
18	2SC247C
18	2SC247D
18	2SC247F
18	2SC247G
18	2SC247GN
18	2SC247H
18	2SC247J
18	2SC247L
18	2SC247M
18	2SC247OR
18	2SC247R
18	2SC247X
18	2SC247Y
18	2SC248
18	2SC248A
18	2SC248B
18	2SC248C
18	2SC248D
18	2SC248E
18	2SC248F
18	2SC248GN
18	2SC248J
18	2SC248K
18	2SC248L
18	2SC248M
18	2SC248OR
18	2SC248R
18	2SC248X
18	2SC248Y
18	2SC249
18	2SC249A
18	2SC249B
18	2SC249C
18	2SC249D
18	2SC249E
18	2SC249F
18	2SC249G
18	2SC249GN
18	2SC249H
18	2SC249J
18	2SC249K
18	2SC249L
18	2SC249M
18	2SC249OR
18	2SC249R
18	2SC249X
18	2SC249Y
18	2SC24A
18	2SC24B
18	2SC24C
18	2SC24D
18	2SC24E
18	2SC24F
18	2SC24G
18	2SC24GN
18	2SC24H
18	2SC24J
18	2SC24K
18	2SC24L
18	2SC24M
18	2SC24OR
18	2SC24R
18	2SC24X
18	2SC24Y
18	2SC27
18	2SC27A
18	2SC27B
18	2SC27D
18	2SC27E
18	2SC27F
18	2SC27G
18	2SC27GN
18	2SC27H
18	2SC27J
18	2SC27K
18	2SC27L
18	2SC27M
18	2SC270R
18	2SC27R
18	2SC27X
18	2SC27Y
18	2SC291
18	2SC291A
18	2SC291B
18	2SC291C
18	2SC291D
18	2SC291E
18	2SC291F
18	2SC291G
18	2SC291GN
18	2SC291H
18	2SC291J
18	2SC291K
18	2SC291L
18	2SC291M
18	2SC291OR
18	2SC291R
18	2SC291X
18	2SC291Y
18	2SC292
18	2SC292A
18	2SC292B
18	2SC292C
18	2SC292D
18	2SC292E
18	2SC292F
18	2SC292G
18	2SC292GN
18	2SC292H
18	2SC292J
18	2SC292K
18	2SC292L
18	2SC292M
18	2SC2920R
18	2SC292R
18	2SC292X
18	2SC292Y
18	2SC293A
18	2SC293B
18	2SC293C
18	2SC293D
18	2SC293E
18	2SC293F
18	2SC293G
18	2SC293GN
18	2SC293H
18	2SC293J
18	2SC293K
18	2SC293L
18	2SC293M
18	2SC2930R
18	2SC293R
18	2SC293X
18	2SC293Y
18	2SC306
18	2SC306A
18	2SC306B
18	2SC306C
18	2SC306D
18	2SC306E
18	2SC306F
18	2SC306G
18	2SC306GN
18	2SC306H
18	2SC306J
18	2SC306K
18	2SC306L
18	2SC306M
18	2SC3060R
18	2SC306R
18	2SC306X
18	2SC306Y
18	2SC307
18	2SC307-OR
18	2SC307A
18	2SC307B
18	2SC307C
18	2SC307D
18	2SC307E
18	2SC307F
18	2SC307G
18	2SC307GN
18	2SC307H
18	2SC307J
18	2SC307K
18	2SC307L
18	2SC307M
18	2SC3070R
18	2SC307R
18	2SC307T
18	2SC307X
18	2SC307Y
18	2SC308
18	2SC308A
18	2SC308B
18	2SC308C
18	2SC308D
18	2SC308E
18	2SC308F
18	2SC308G
18	2SC308GN
18	2SC308H
18	2SC308J
18	2SC308K
18	2SC308L
18	2SC308M
18	2SC3080R
18	2SC308R
18	2SC308X
18	2SC308Y
18	2SC309
18	2SC309A
18	2SC309B
18	2SC309C
18	2SC309D
18	2SC309E
18	2SC309F
18	2SC309G
18	2SC309GN
18	2SC309H
18	2SC309J
18	2SC309K
18	2SC309L
18	2SC309M
18	2SC309OR
18	2SC309R
18	2SC309X
18	2SC309Y
18	2SC31
18	2SC310
18	2SC310A
18	2SC310B
18	2SC310C
18	2SC310D
18	2SC310E
18	2SC310F
18	2SC310G
18	2SC310GN
18	2SC310H
18	2SC310J
18	2SC310K
18	2SC310L
18	2SC310M
18	2SC3100R
18	2SC310R
18	2SC310X
18	2SC310Y
18	2SC31A
18	2SC31B
18	2SC31C
18	2SC31D
18	2SC31E
18	2SC31F
18	2SC31G
18	2SC31GN
18	2SC31H
18	2SC31J
18	2SC31K
18	2SC31L
18	2SC31M
18	2SC31OR
18	2SC31R
18	2SC31X
18	2SC31Y
18	2SC32
18	2SC32A
18	2SC32B
18	2SC32C
18	2SC32D
18	2SC32E
18	2SC32F
18	2SC32G
18	2SC32GN
18	2SC32H
18	2SC32J
18	2SC32K
18	2SC32L
18	2SC32M
18	2SC320R
18	2SC32R
18	2SC32X
18	2SC32Y
18	2SC353
18	2SC353A
18	2SC353AC
18	2SC353B
18	2SC353C
18	2SC353D
18	2SC353E
18	2SC353F
18	2SC353G
18	2SC353GN
18	2SC353H
18	2SC353J
18	2SC353K
18	2SC353L
18	2SC353M
18	2SC3530R
18	2SC353R
18	2SC353X
18	2SC353Y
18	2SC376
18	2SC376A
18	2SC376B
18	2SC376C
18	2SC376D
18	2SC376E
18	2SC376F
18	2SC376G
18	2SC376GN
18	2SC376H
18	2SC376J
18	2SC376K
18	2SC376L
18	2SC376M
18	2SC3760R
18	2SC376R
18	2SC376X
18	2SC376Y
18	2SC423G
18	2SC425
18	2SC425A
18	2SC425B
18	2SC425C
18	2SC425D
18	2SC425E
18	2SC425F
18	2SC425G
18	2SC425GN
18	2SC425H
18	2SC425J
18	2SC425K
18	2SC425L
18	2SC425M
18	2SC425OR
18	2SC425R
18	2SC425X
18	2SC425Y
18	2SC443
18	2SC443A
18	2SC443B
18	2SC443C
18	2SC443D
18	2SC443E
18	2SC443F
18	2SC443G
18	2SC443GN
18	2SC443H
18	2SC443J
18	2SC443K
18	2SC443L
18	2SC443M
18	2SC4430R
18	2SC443R
18	2SC443X
18	2SC443Y
18	2SC46
18	2SC46A
18	2SC46B
18	2SC46C
18	2SC46E
18	2SC46F
18	2SC46G
18	2SC46GN
18	2SC46H
18	2SC46J
18	2SC46K
18	2SC46L
18	2SC46M
18	2SC46OR
18	2SC46X
18	2SC46Y
18	2SC479
18	2SC479A
18	2SC479B
18	2SC479C
18	2SC479D
18	2SC479E
18	2SC479F
18	2SC479G
18	2SC479GN
18	2SC479H
18	2SC479J
18	2SC479K
18	2SC479L
18	2SC479M
18	2SC479OR
18	2SC479R
18	2SC479X
18	2SC479Y
18	2SC486-R
18	2SC486-RED
18	2SC486-Y
18	2SC486-YEL
18	2SC49
18	2SC497
18	2SC497-0
18	2SC497-0
18	2SC497-OR
18	2SC497-ORG
18	2SC497-R
18	2SC497-RED
18	2SC497-Y
18	2SC497A
18	2SC497B
18	2SC497C
18	2SC497D
18	2SC497E
18	2SC497F
18	2SC497G
18	2SC497GN
18	2SC497H
18	2SC497J
18	2SC497K
18	2SC497L
18	2SC497M
18	2SC497OR
18	2SC497R
18	2SC497RED
18	2SC497X
18	2SC497Y
18	2SC49A
18	2SC49B
18	2SC49C
18	2SC49D
18	2SC49E
18	2SC49F
18	2SC49G
18	2SC49GN
18	2SC49H
18	2SC49J
18	2SC49K
18	2SC49L
18	2SC49M
18	2SC49OR
18	2SC49X
18	2SC49Y
18	2SC501
18	2SC501-0
18	2SC501-ORG
18	2SC501-R
18	2SC501-RED
18	2SC501-Y
18	2SC501-YEL
18	2SC501A
18	2SC501B
18	2SC501C
18	2SC501D
18	2SC501E
18	2SC501F
18	2SC501G
18	2SC501GN
18	2SC501H
18	2SC501J
18	2SC501K
18	2SC501L
18	2SC501M
18	2SC501R
18	2SC501X
18	2SC501Y
18	2SC502A
18	2SC502C
18	2SC502D
18	2SC502E
18	2SC502F
18	2SC502G
18	2SC502GN
18	2SC502H
18	2SC502J
18	2SC502K
18	2SC502L
18	2SC502M
18	2SC502OR
18	2SC502R
18	2SC502X
18	2SC502Y
18	2SC503
18	2SC503-GR
18	2SC503-0
18	2SC503-Y
18	2SC503A
18	2SC503B
18	2SC503C
18	2SC503D
18	2SC503E
18	2SC503G
18	2SC503GN
18	2SC503H
18	2SC503J
18	2SC503K
18	2SC503L
18	2SC503M
18	2SC503OR
18	2SC503R
18	2SC503X
18	2SC504
18	2SC504-GR
18	2SC504-0
18	2SC504-Y
18	2SC504A
18	2SC504B
18	2SC504C
18	2SC504D
18	2SC504E
18	2SC504F
18	2SC504G
18	2SC504GN
18	2SC504GR
18	2SC504H
18	2SC504J
18	2SC504K
18	2SC504L
18	2SC504M
18	2SC504OR
18	2SC504R
18	2SC504X
18	2SC51
18	2SC512
18	2SC512-O
18	2SC512-O
18	2SC512-ORG
18	2SC512-R
18	2SC512-RED
18	2SC512A
18	2SC512B
18	2SC512C
18	2SC512D
18	2SC512E
18	2SC512F
18	2SC512G
18	2SC512GN
18	2SC512H
18	2SC512J
18	2SC512K
18	2SC512L
18	2SC512M
18	2SC512O
18	2SC512OR
18	2SC512X
18	2SC513
18	2SC513-O
18	2SC513-O
18	2SC513-ORG
18	2SC513-R
18	2SC513-RED
18	2SC513A
18	2SC513B
18	2SC513C
18	2SC513D
18	2SC513F
18	2SC513G
18	2SC513GN
18	2SC513H
18	2SC513J
18	2SC513K
18	2SC513L
18	2SC513M
18	2SC513O
18	2SC513OR
18	2SC513R
18	2SC513X
18	2SC513Y
18	2SC516A
18	2SC516B
18	2SC516C
18	2SC516D
18	2SC516E
18	2SC516F
18	2SC516G
18	2SC516GN
18	2SC516H
18	2SC516J
18	2SC516K
18	2SC516L
18	2SC516M
18	2SC516OR
18	2SC516R
18	2SC516X
18	2SC516Y
18	2SC51A
18	2SC51B
18	2SC51C
18	2SC51D
18	2SC51E
18	2SC51F
18	2SC51G
18	2SC51GN
18	2SC51H
18	2SC51J
18	2SC51K
18	2SC51L
18	2SC51M
18	2SC51OR
18	2SC51R
18	2SC51X
18	2SC51Y
18	2SC564(Q)
18	2SC564(Q)(R)
18	2SC564A
18	2SC564B
18	2SC564C
18	2SC564D
18	2SC564E
18	2SC564F
18	2SC564G
18	2SC564GN
18	2SC564H
18	2SC564J
18	2SC564K
18	2SC564L
18	2SC564M
18	2SC564OR
18	2SC564P
18	2SC564PL
18	2SC564Q
18	2SC564QC
18	2SC564R
18	2SC564S

GE	Industry Standard No.	GE	Industry Standard No.	GE	Industry Standard No.	GE	Industry Standard No.	GE	Industry Standard No.
18	2SC564X	18	2SC708B	20	A11414257	20	A6HD	20	BC-108
18	2SC564Y	18	2SC708C	20	A116	20	A6J	20	BC-1082
18	2SC59	18	2SC708D	20	A12-1-70	20	A6K	20	BC-1086
18	2SC590	18	2SC708E	20	A12-1-705	20	A6R	20	BC-108B
18	2SC590A	18	2SC708F	20	A12-1A9G	20	A6S	20	BC-109B
18	2SC590B	18	2SC708G	20	A128	20	A747	20	BC-109BP
18	2SC590C	18	2SC708GN	20	A128A	20	A747A	20	BC-114
18	2SC590D	18	2SC708H	20	A137	20	A748	20	BC-121
18	2SC590E	18	2SC708HA	20	A171(NPN)	20	A748B	20	BC-122
18	2SC590F	18	2SC708HB	20	A1379	20	A749B	20	BC-123
18	2SC590G	18	2SC708L	20	A1380	20	A7A	20	BC-148A
18	2SC590GN	18	2SC708M	20	A15N1	20	A7E(TRANSISTOR)	20	BC-148B
18	2SC590H	18	2SC708OR	20	A141	20	A7R	20	BC-148C
18	2SC590J	18	2SC708R	20	A1412-1	20	A7S	20	BC-167-B
18	2SC590K	18	2SC708X	20	A142	20	A7T	20	BC-169D
18	2SC590L	18	2SC708Y	20	A143	20	A7Y	20	BC-169B
18	2SC590M	18	2SC727	20	A1472-19	20	A88	20	BC-169C
18	2SC590OR	18	2SC727A	20	A151	20	A88(JAPAN)	20	BC-71
18	2SC590X	18	2SC727B	20	A152	20	A8B	20	BC107
18	2SC590Y	18	2SC727C	20	A153	20	A8G	20	BC107A
18	2SC594	18	2SC727D	20	A156	20	A8L	20	BC107B
18	2SC594A	18	2SC727E	20	A1567	20	A909-1011	20	BC108
18	2SC594C	18	2SC727F	20	A1567-1	20	A909-1012	20	BC108A
18	2SC594D	18	2SC727G	20	A157	20	A909-1013	20	BC108B
18	2SC594E	18	2SC727GN	20	A157A	20	A937	20	BC108C
18	2SC594F	18	2SC727H	20	A157B	20	A937-1	20	BC109
18	2SC594G	18	2SC727J	20	A157C	20	A937-3	20	BC1096
18	2SC594GN	18	2SC727K	20	A158	20	A9B	20	BC109BP
18	2SC594H	18	2SC727L	20	A158A	20	A9E	20	BC109C
18	2SC594J	18	2SC727M	20	A158B	20	A9F	20	BC110
18	2SC594K	18	2SC727R	20	A158C	20	A9G	20	BC113
18	2SC594L	18	2SC727X	20	A159	20	A9H	20	BC114
18	2SC594M	18	2SC727Y	20	A159A	20	A9J	20	BC114TR
18	2SC594OR	18,243	TR87	20	A159B	20	A9S	20	BC115
18	2SC594W	19	D8514	20	A159C	20	A9T	20	BC118
18	2SC594X	19	GE-19	20	A168	20	A9U	20	BC125
18	2SC594Y	19	HEP705	20	A1B	20	A9W	20	BC125B
18	2SC595	19	HT9000410-0	20	A1F	20	A9Y	20	BC129
18	2SC59A	19	HT9000410-0	20	A1H	20	AC-175A	20	BC130
18	2SC59B	19	IRTR26	20	A1J	20	AC-175B	20	BC131
18	2SC59C	19	MJ2800	20	A1L	20	AC-175P	20	BC132
18	2SC59D	19	Q5083B	20	A1T	20	AFC3527	20	BC134
18	2SC59E	19	Q5110Z	20	A1T-1	20	ALD-3141	20	BC135
18	2SC59F	19	SE3032	20	A1V	20	ALD-35	20	BC136
18	2SC59G	19	2N2948	20	A1VE	20	ALS-8922	20	BC146
18	2SC59GN	19	2N3297	20	A1W	20	AN	20	BC147
18	2SC59H	19	2N3917	20	A2019ZC	20	AQ4	20	BC147B
18	2SC59J	19	2N3918	20	A20372	20	AQ6	20	BC147A
18	2SC59K	19	2N4111	20	A2410	20	AR-107	20	BC147B
18	2SC59L	19	2N4112	20	A24100	20	AR-108	20	BC148
18	2SC59M	19	2N4113	20	A2411	20	AR-200	20	BC148A
18	2SC590R	19	2N4114	20	A2412	20	AR-201	20	BC148B
18	2SC59R	19	2N5730	20	A2413	20	AR-202	20	BC148C
18	2SC59X	19	2N5970	20	A2434	20	AR107	20	BC149
18	2SC59Y	19	2SC198	20	A246	20	AR108	20	BC149A
18	2SC61	19	2SC198H	20	A246(AMC)	20	AR200(W)	20	BC149B
18	2SC610	19	2SC198S	20	A2466	20	AR200WHITE	20	BC149C
18	2SC610A	20	A-11095924	20	A2468	20	AR201(Y)	20	BC149G
18	2SC610B	20	A-11237336	20	A2469	20	AR201YELLOW	20	BC150
18	2SC610C	20	A-1141-6062	20	A2470	20	AR202GREEN	20	BC151
18	2SC610D	20	A-120278	20	A248(AMC)	20	AR204	20	BC152
18	2SC610E	20	A-125332	20	A249	20	AR205	20	BC155B
18	2SC610F	20	A-1379	20	A2499	20	AR206	20	BC156B
18	2SC610G	20	A-1380	20	A25A509-016-101	20	AR208	20	BC167
18	2SC610GN	20	A-156	20	A2B	20	AR306	20	BC167B
18	2SC610H	20	A-1567	20	A2BRN	20	AR306(BLUE)	20	BC168
18	2SC610J	20	A-158B	20	A2FGRN	20	AR306(ORANGE)	20	BC168A
18	2SC610K	20	A-158C	20	A2J	20	AT329	20	BC168B
18	2SC610L	20	A-168	20	A2SC538PQR	20	AT335	20	BC168C
18	2SC610M	20	A-1854-0003-1	20	A301	20	AT336	20	BC169
18	2SC610OR	20	A-1854-0019-1	20	A306	20	AT337	20	BC169A
18	2SC610R	20	A-1854-0025-1	20	A307	20	AT347	20	BC169B
18	2SC610X	20	A-1854-0027-1	20	A323	20	AT348	20	BC169C
18	2SC610Y	20	A-1854-0094-1	20	A324	20	AT349	20	BC169CL
18	2SC614A	20	A-1854-0099-1	20	A344	20	AT370	20	BC171
18	2SC614B	20	A-1854-0201-1	20	A345	20	AT400	20	BC171A
18	2SC614GN	20	A-1854-0215-1	20	A346	20	AT401	20	BC171B
18	2SC614H	20	A-1854-0241-1	20	A3E	20	AT402	20	BC172
18	2SC614J	20	A-1854-0246-1	20	A3F	20	AT403	20	BC172A
18	2SC614K	20	A-1854-0251-1	20	A3G	20	AT404	20	BC172B
18	2SC614L	20	A-1854-0255-1	20	A3N	20	AT405	20	BC172C
18	2SC614M	20	A-1854-0354-1	20	A3S	20	AT406	20	BC175
18	2SC6140R	20	A-1854-0434-1	20	A3T	20	AT407	20	BC180
18	2SC614R	20	A-1854-0471-1	20	A3T2221	20	AT420	20	BC180B
18	2SC614X	20	A-1854-0492-1	20	A3T2221A	20	AT421	20	BC182
18	2SC614Y	20	A-1854-0541-1	20	A3T2222	20	AT422	20	BC182A
18	2SC61A	20	A-1854-0554-1	20	A3T2222A	20	AT423	20	BC182L
18	2SC61B	20	A-415	20	A3T3011	20	AT424	20	BC183
18	2SC61C	20	A-494	20	A3T929	20	AT425	20	BC183A
18	2SC61D	20	A-567A	20	A3T930	20	AT426	20	BC183B
18	2SC61E	20	A.184/5	20	A3W	20	AT427	20	BC183L
18	2SC61F	20	A0-54-195	20	A3Z	20	AT490	20	BC184
18	2SC61G	20	A054-108	20	A415	20	AT491	20	BC184B
18	2SC61GN	20	A054-114	20	A42X00434-01	20	AT492	20	BC184L
18	2SC61H	20	A054-115	20	A43021415	20	AT493	20	BC185
18	2SC61J	20	A054-155	20	A454	20	AT494	20	BC186
18	2SC61K	20	A054-173	20	A472	20	AT495	20	BC197
18	2SC61L	20	A054-175	20	A481A0028	20	AWH-24	20	BC197A
18	2SC61M	20	A054-195	20	A481A0031	20	B-1338	20	BC197B
18	2SC61OR	20	A054-221	20	A494	20	B-1421	20	BC198
18	2SC61Y	20	A054-222	20	A494(JAPAN)	20	B-1433	20	BC199
18	2SC69	20	A054-225	20	A4A	20	B-1666	20	BC207
18	2SC69A	20	A054-233	20	A4M	20	B-169	20	BC207B
18	2SC69B	20	A054-234	20	A4N	20	B-1842	20	BC207BL
18	2SC69C	20	A059-109	20	A4P	20	B-1872	20	BC208
18	2SC69D	20	A06-1-12	20	A4R	20	B-722246-2	20	BC208A
18	2SC69E	20	A065-102	20	A4U	20	B-75583-1	20	BC208AL
18	2SC69F	20	A065-103	20	A4V	20	B-75583-202	20	BC208B
18	2SC69G	20	A065-104	20	A4Y-2	20	B-75589-13	20	BC208BL
18	2SC69GN	20	A065-108	20	A514-033338	20	B-75589-3	20	BC208C
18	2SC69H	20	A065-109	20	A54-96-001	20	B-75608-3	20	BC208CL
18	2SC69K	20	A065-110	20	A54-96-002	20	B12-1-A-21	20	BC209
18	2SC69L	20	A065-113	20	A567	20	B12-1A21	20	BC209A
18	2SC69M	20	A066-109	20	A567A	20	B133578	20	BC209B
18	2SC69OR	20	A066-112	20	A593	20	B1K	20	BC209BL
18	2SC69X	20	A068-108	20	A5C	20	B1N	20	BC209C
18	2SC69Y	20	A068-113	20	A5H	20	B1P7201	20	BC209CL
18	2SC708	20	A069-102/103	20	A5K	20	B1R	20	BC210
18	2SC708(A)	20	A069-104	20	A5L	20	B1W	20	BC220
18	2SC708(B)	20	A069-104/106	20	A5M	20	B2D	20	BC222
18	2SC708(C)	20	A069-106	20	A5N	20	B559R	20	BC223A
18	2SC708-OR	20	A069-120	20	A5P	20	B6P	20	BC223B
18	2SC708A	20	A069-122	20	A5R	20	B8780010	20	BC233A
18	2SC708AA	20	A106	20	A5S	20	BA67	20	BC237
18	2SC708AB	20	A106(JAPAN)	20	A5T2222	20	BA71	20	BC237A
18	2SC708AC	20	A108	20	A5T3903	20	BACSH2M1	20	BC237B
18	2SC708AH	20	A108A	20	A5T3904	20	BACSH2M2	20	BC238
18	2SC708AHA	20	A108B	20	A5T4124	20	BACSH2M3	20	BC238A
18	2SC708AHB	20	A10005-010-A	20	A5U	20	BACT2F	20	BC238B
18	2SC708AHC	20	A10005-011-A	20	A5W	20	BB71	20	BC239
		20	A10005-015-D	20	A64?(NPN)	20	BC-107	20	BC267
		20	A111	20	A649L	20	BC-1072	20	BC268
				20	A649S	20	BC-107A	20	BC269
				20	A6H				

GE	Industry Standard No.	GE	Industry Standard No.	GE	Industry Standard No.	GE	Industry Standard No.	GE	Industry Standard No.
20	BC270	20	BFV53	20	BSX87	20	C18	20	C404
20	BC280	20	BFV54	20	BSX89	20	C191	20	C423
20	BC280A	20	BFV55	20	BSX90	20	C192	20	C423B
20	BC280B	20	BFV83B	20	BSX91	20	C193	20	C423C
20	BC280C	20	BFV83C	20	BSX94A	20	C194	20	C423D
20	BC282	20	BFV85	20	BSX97	20	C1945295DY1	20	C423E
20	BC284	20	BFV85A	20	BSY10	20	C195	20	C423F
20	BC284A	20	BFV85B	20	BSY11	20	C196	20	C424(JAPAN)
20	BC284B	20	BFV85C	20	BSY16B	20	C197	20	C424D
20	BC285	20	BFV87	20	BSY17	20	C201(JAPAN)	20	C425
20	BC289	20	BFV88	20	BSY18	20	C202(JAPAN)	20	C425B
20	BC289A	20	BFV88A	20	BSY19	20	C203	20	C425C
20	BC289B	20	BFV88B	20	BSY20	20	C204	20	C425D
20	BC290	20	BFV88C	20	BSY21	20	C205	20	C425E
20	BC290B	20	BFW32	20	BSY24	20	C218	20	C425F
20	BC290C	20	BFW46	20	BSY26	20	C218A	20	C440
20	BC317	20	BFW59	20	BSY27	20	C230	20	C441
20	BC317A	20	BFW60	20	BSY28	20	C2300.037-096	20	C442
20	BC317B	20	BFW68	20	BSY29	20	C237	20	C444
20	BC318	20	BFX92	20	BSY34	20	C238	20	C45
20	BC318A	20	BFX93	20	BSY38	20	C239	20	C450
20	BC318B	20	BFX95A	20	BSY39	20	C2538-11	20	C454(A)
20	BC318C	20	BFY	20	BSY48	20	C26	20	C454A
20	BC319	20	BFY-22	20	BSY49	20	C267	20	C454C
20	BC319B	20	BFY-23	20	BSY58	20	C267A	20	C454D
20	BC319C	20	BFY-23A	20	BSY59	20	C28	20	C454L
20	BC408	20	BFY-24	20	BSY61	20	C281	20	C454LA
20	BC409	20	BFY-29	20	BSY62	20	C281A	20	C458LG(B)
20	BC507A	20	BFY-30	20	BSY62A	20	C281B	20	C468
20	BC507B	20	BFY-39	20	BSY63	20	C281C	20	C468A
20	BC508A	20	BFY22	20	BSY73	20	C281C-EP	20	C47
20	BC508B	20	BFY23	20	BSY74	20	C281D	20	C475
20	BC508C	20	BFY23A	20	BSY75	20	C281EP	20	C475K
20	BC509B	20	BFY24	20	BSY76	20	C281H	20	C476
20	BC509C	20	BFY25	20	BSY80	20	C281HA	20	C48
20	BC510B	20	BFY26	20	BSY89	20	C281HB	20	C48C
20	BC546	20	BFY28	20	BSY93	20	C281HC	20	C52(TRANSISTOR)
20	BC546A	20	BFY29	20	BSY95	20	C282	20	C529
20	BC546B	20	BFY30	20	BSY95A	20	C282H	20	C529A
20	BC547B	20	BFY33	20	BT67	20	C282HA	20	C53
20	BC583	20	BFY37	20	BT71	20	C282HB	20	C537
20	BC71	20	BFY371	20	BTX-070	20	C282HC	20	C537(P)
20	BCW34	20	BFY39	20	BTX-094	20	C283	20	C537(G)
20	BCW36	20	BFY39/1	20	BTX-095	20	C284	20	C537-01
20	BCW60A	20	BFY39/2	20	BTX-096	20	C284H	20	C537A
20	BCW60AA	20	BFY39/3	20	BTX-2367B	20	C284HA	20	C537B
20	BCW65RA	20	BFY391	20	BTX068	20	C284HB	20	C537C
20	BCW65EB	20	BFY63	20	BU67	20	C285	20	C537D
20	BCW83	20	BFY73	20	BU71	20	C285A	20	C537D2
20	BCW94	20	BFY74	20	BU297704-2	20	C29	20	C537E
20	BCW94A	20	BFY75	20	BV67	20	C300	20	C537EF
20	BCW94B	20	BFY76	20	BV71	20	C301	20	C537EH
20	BCW94C	20	BFY77	20	BW67	20	C302(JAPAN)	20	C537EJ
20	BCW94KA	20	BG71	20	BW71	20	C315	20	C537EK
20	BCW94KB	20	BH71	20	BX67	20	C317	20	C537F
20	BCW94KC	20	BI71	20	BX71	20	C317-0	20	C537F1
20	BCY-50	20	BIP7201	20	BY67	20	C317C	20	C537F2
20	BCY-58	20	BN-66	20	BY71	20	C318(JAPAN)	20	C537FC
20	BCY13	20	BN7517	20	BZ67	20	C318A(JAPAN)	20	C537FV
20	BCY15	20	BN7518	20	BZ71	20	C319	20	C537G
20	BCY15	20	BP67	20	C00-68602300	20	C320	20	C537G1
20	BCY36	20	BQ67	20	C100	20	C321	20	C537GF
20	BCY42	20	BR-66	20	C100-0Y	20	C321H	20	C537GI
20	BCY43	20	BR67	20	C1000-Y	20	C321HA	20	C537H
20	BCY50	20	BS-66	20	C1007	20	C321HB	20	C537HT
20	BCY51	20	BS475	20	C10279-1	20	C321HC	20	C537W
20	BCY56	20	BS67	20	C10279-3	20	C323	20	C538
20	BCY57	20	B89011G	20	C103	20	C324	20	C538A
20	BCY58	20	BS810	20	C103(JAPAN)	20	C324A	20	C538AQ
20	BCY58B	20	BS821	20	C1033	20	C324H	20	C538P
20	BCY58C	20	BSV35A	20	C1033A	20	C324HA	20	C538Q
20	BCY58D	20	BSV40	20	C104	20	C328A	20	C538R
20	BCY59	20	BSV41	20	C104A	20	C33	20	C538S
20	BCY59A	20	BSV53	20	C105	20	C350H	20	C538T
20	BCY59B	20	BSV54	20	C1071	20	C352	20	C539
20	BCY59C	20	BSV59	20	C110	20	C352(JAPAN)	20	C539K
20	BCY59D	20	BSV65FA	20	C111	20	C352A	20	C539L
20	BCY69	20	BSV65FB	20	C111E	20	C356	20	C539R
20	BCY84A	20	BSV84	20	C1128BD	20	C360	20	C539S
20	BF-214	20	BSV86	20	C1175	20	C360D	20	C54
20	BF-215	20	BSV87	20	C1175C	20	C363	20	C540
20	BF-226	20	BSV88	20	C1175D	20	C36580	20	C55
20	BF-255	20	BSV89	20	C1175F	20	C366	20	C566
20	BF115	20	BSV90	20	C122	20	C367	20	C587
20	BF183A	20	BSV91	20	C1244	20	C369	20	C587A
20	BF189	20	BSW11	20	C12711	20	C369BL	20	C588
20	BF224J	20	BSW12	20	C131	20	C369G	20	C593
20	BF225J	20	BSW19	20	C132	20	C369G-BL	20	C595
20	BF248	20	BSW33	20	C133	20	C369G-GR	20	C596
20	BF249	20	BSW34	20	C134	20	C369G-V	20	C602E
20	BF250	20	BSW39	20	C134B	20	C369GBL	20	C619
20	BF255	20	BSW41	20	C135	20	C369GGR	20	C619B
20	BF291	20	BSW42	20	C136	20	C369GR	20	C619C
20	BF291A	20	BSW42A	20	C1361	20	C369V	20	C619D
20	BF291B	20	BSW43	20	C1362	20	C37	20	C62(TRANSISTOR)
20	BF293	20	BSW43A	20	C1363	20	C37(TRANSISTOR)	20	C620
20	BF293A	20	BSW51	20	C1364	20	C371	20	C620C
20	BF293D	20	BSW52	20	C1364A	20	C371(0)	20	C620CD
20	BF521A	20	BSW53	20	C1372Y	20	C371-0	20	C620D
20	BF521B	20	BSW58	20	C138A	20	C371-0	20	C620DE
20	BF521C	20	BSW82	20	C1390I	20	C371-0	20	C620E
20	BF521D	20	BSW83	20	C1390J	20	C371-R	20	C621
20	BF521E	20	BSW84	20	C1390K	20	C371O	20	C622
20	BF521F	20	BSW85	20	C1390V	20	C371B	20	C63
20	BF440	20	BSW88	20	C1390W	20	C371G	20	C631
20	BF441	20	BSW89	20	C1390WH	20	C371R	20	C631A
20	BF596	20	BSW92	20	C1390WI	20	C372	20	C632
20	BF71	20	BSX19	20	C1390WX	20	C372(0)	20	C632A
20	BFR11	20	BSX20	20	C1390WY	20	C372-0	20	C633
20	BFR26	20	BSX24	20	C1390X	20	C372-1	20	C633-7
20	BF251P	20	BSX25	20	C1390XJ	20	C372-2	20	C633A
20	BF356A	20	BSX30	20	C1390XK	20	C372-0	20	C633G
20	BF356B	20	BSX38	20	C1390YM	20	C372-R	20	C633H
20	BF356C	20	BSX39	20	C1393	20	C372-Y	20	C634
20	BF358	20	BSX44	20	C151(TRANSISTOR)	20	C372-Z	20	C634A
20	BF358A	20	BSX48	20	C15-1	20	C372GR	20	C64
20	BF842	20	BSX49	20	C15-2	20	C372H	20	C640
20	BF842A	20	BSX51	20	C15-3	20	C372Y	20	C640B
20	BF842B	20	BSX51A	20	C151	20	C377	20	C650B
20	BF842C	20	BSX52	20	C1542	20	C378	20	C654
20	BF843	20	BSX52A	20	C157	20	C379	20	C655
20	BF843A	20	BSX53	20	C158	20	C38	20	C66-P11111-0001
20	BF843B	20	BSX54	20	C159	20	C38(TRANSISTOR)	20	C67
20	BF843C	20	BSX66	20	C16	20	C39-207	20	C68
20	BFV10	20	BSX67	20	C160	20	C395	20	C6862400
20	BFV11	20	BSX68	20	C1639	20	C395A	20	C689
20	BFV12	20	BSX69	20	C166	20	C395R	20	C689H
20	BFV42	20	BSX75	20	C167	20	C396	20	C694D
20	BFV43	20	BSX76	20	C168T	20	C400	20	C702
20	BFV44	20	BSX77	20	C168B	20	C400-0	20	C709
20	BFV46	20	BSX78	20	C16a	20	C400-R	20	C709B
20	BFV47	20	BSX79			20	C400-Y	20	C709C
20	BFV50	20	BSX80			20	C401(JAPAN)	20	C709CD
20	BFV51	20	BSX81			20	C403(C)	20	C709D
						20	C403C		

GE	Industry Standard No.	GE	Industry Standard No.	GE	Industry Standard No.	GE	Industry Standard No.	GE	Industry Standard No.
20	C710	20	C933PPE	20	CDQ1021	20	CS13401	20	CS90140/3490
20	C710(B)	20	C933PPF	20	CDQ1024	20	CS1344	20	CS9014D
20	C710(D)	20	C933PPG	20	CE0360/7839	20	CS1345	20	CS9014G
20	C710-1	20	C933G	20	CE0361/7839	20	CS1348	20	CS9015
20	C710-2	20	C934C	20	CE0362/7839	20	CS1349	20	CS9015B
20	C710-4	20	C934D	20	CE4001B	20	CS1353	20	CS90161(Q)
20	C710B	20	C934B	20	CE4001B	20	CS13610	20	CS90161D
20	C710BC	20	C934F	20	CE4003E	20	CS1362	20	CS9016EF
20	C710C	20	C934G	20	CE4004C	20	CS1363	20	CS9016FG
20	C710D	20	C934P	20	CE4013E	20	CS1368	20	CS9018EP
20	C710E	20	C938	20	CP-2	20	CS1368A	20	CS9018PP
20	C712	20	C938A	20	CP2	20	CS1368B	20	CS9101B
20	C712A	20	C938B	20	CP5	20	CS1368C	20	CS9125B
20	C712C	20	C938C	20	CG1	20	CS1368D	20	CS9126
20	C712D	20	C941	20	CI2712	20	CS1370	20	CS925M
20	C712E	20	C941-0	20	CI2713	20	CS1371	20	CS9600-4
20	C712W	20	C941-0	20	CI2714	20	CS1372	20	CS9600-5
20	C713	20	C941-R	20	CI2923	20	CS1383	20	CTP-2001-1007
20	C714	20	C941-Y	20	CI2924	20	CS1420	20	CTP-2001-1008
20	C715(JAPAN)	20	C941R	20	CI2925	20	CS1453E	20	CXL123A
20	C715A	20	C943	20	CI2926	20	CS1463A	20	D048
20	C715B	20	C943A	20	CI3390	20	CS1585	20	D053
20	C715C	20	C943B	20	CI3591	20	CS1585E/F	20	D16E7
20	C715D	20	C943C	20	CI3591A	20	CS1585G	20	D16E9
20	C715E	20	C944	20	CI3592	20	CS1625	20	D16EC18
20	C715EJ	20	C944A	20	CI3593	20	CS1665	20	D1A
20	C715F	20	C944B	20	CI3594	20	CS183E	20	D1R38
20	C715XLL	20	C944C	20	CI3595	20	CS184E	20	D227
20	C716	20	C944D	20	CI3596	20	CS2001	20	D227A
20	C716B	20	C944K	20	CI3597	20	CS2001H	20	D227B
20	C716C	20	C960	20	CI3598	20	CS2004	20	D227C
20	C716D	20	C9604	20	CI3402	20	CS2004C	20	D227D
20	C716E	20	C966	20	CI3403	20	CS2004D	20	D227E
20	C716F	20	C967	20	CI3404	20	CS2006	20	D227F
20	C716G	20	C968	20	CI3405	20	CS2007G	20	D227L
20	C720	20	C968P	20	CI3414	20	CS2007H	20	D227R
20	C725	20	C984	20	CI3415	20	CS2008	20	D227S
20	C725-0	20	C984A	20	CI3416	20	CS2008G	20	D227W
20	C7335-BL	20	C984B	20	CI3417	20	CS2218	20	D294
20	C735-0	20	C984C	20	CI3900	20	CS2219	20	D2R38
20	C735-R	20	C991	20	CI3900A	20	CS2221	20	D327
20	C7350RN	20	C992	20	CI3901	20	CS2222	20	D327A
20	C741	20	CA-9011H	20	CI4256	20	CS2369	20	D327B
20	C742	20	CAM-12	20	CI4424	20	CS2481	20	D327C
20	C752	20	CB246	20	CI4425	20	CS2711	20	D327D
20	C752G	20	CC1168P	20	CIL-531	20	CS2712	20	D327E
20	C773	20	CC81235G	20	CIL-532	20	CS2713	20	D327F
20	C773C	20	CC82001H	20	CJ-5206	20	CS2714	20	D328
20	C773D	20	CC82004	20	CJ-5207	20	CS2922	20	D33D21
20	C773E	20	CC84004	20	CJ-5208	20	CS2923	20	D33D22
20	C796	20	CC86168	20	CJ-5211	20	CS2924	20	D33D24
20	C815	20	CC86168P	20	CJ-5212	20	CS2925	20	D33D25
20	C815(M)	20	CC89018E	20	CJ5201	20	CS3001B	20	D33D26
20	C815A	20	CD0014NA	20	CJ5202	20	CS3390	20	D33D27
20	C815B	20	CD0014NG	20	CJ5203	20	CS3391	20	D33D28
20	C815C	20	CD0015N	20	CJ5206	20	CS3391A	20	D342
20	C815F	20	CD0021	20	CJ5207	20	CS3392	20	D3728L
20	C815K	20	CD12000	20	CJ5211	20	CS3393	20	D4D25
20	C815L	20	CD38	20	CJ5212	20	CS3394	20	D4D26
20	C815M	20	CD446	20	CMC334-423	20	CS3395	20	D912
20	C815S	20	CD6019	20	CM07701	20	CS3396	20	D917254-2
20	C815SA	20	CD6150	20	CS-1120C1	20	CS3397	20	D921881-1
20	C815SC	20	CD6157	20	CS-1120C2	20	CS3398	20	D926640-1
20	C825	20	CD6375	20	CS-1120D1	20	CS3402	20	D928121
20	C828AS	20	CD8000	20	CS-1120H	20	CS3403	20	DBCZ073304
20	C829	20	CD8000-1	20	CS-1235F	20	CS3404	20	DBC803906
20	C829A	20	CD9825	20	CS-1238F	20	CS3405	20	DBC8136406
20	C829B	20	CDC-13000-1	20	CS-1238P	20	CS3414	20	DDBT222002
20	C829BC	20	CDC-13000-1D	20	CS-1330	20	CS3415	20	DN
20	C829C	20	CDC-8001	20	CS-1361E	20	CS3416	20	D031
20	C829D	20	CDC12000	20	CS-1361F	20	CS3417	20	DRC-87540
20	C829E	20	CDC12000-1C	20	CS-1386B	20	CS360	20	D8-44
20	C829X	20	CDC12018C	20	CS-1386H	20	CS3605	20	D8-45
20	C829Y	20	CDC1201BC	20	CS-460B	20	CS3606	20	D8-46
20	C836M	20	CDC12077P	20	CS-6168G	20	CS3607	20	D8-47
20	C838A	20	CDC121DB	20	CS-6168H	20	CS3843	20	D8-66
20	C838B	20	CDC13000	20	CS-6225P	20	CS3844	20	D8-66L
20	C838C	20	CDC13000-1	20	CS-6225G	20	CS3845	20	D8-66W
20	C838D	20	CDC13000-18	20	CS-6227H	20	CS3854	20	D8-67
20	C838E	20	CDC13000-1B	20	CS-6227F	20	CS3854A	20	D8-67W
20	C838F	20	CDC13000-1C	20	CS-9011	20	CS3855	20	D8-76
20	C838J	20	CDC13000-1D	20	CS-9011F	20	CS3855A	20	D8-77
20	C838K	20	CDC13000C	20	CS-9011G	20	CS3859	20	D81B
20	C838M	20	CDC13016A	20	CS-9011L	20	CS3859A	20	D844
20	C838R	20	CDC13019B	20	CS-9013	20	CS3860	20	D845
20	C839	20	CDC13500-1	20	CS-9014B	20	CS3900	20	D846
20	C839(H)	20	CDC15018	20	CS-9014D	20	CS3900A	20	D847
20	C839(J)	20	CDC2010	20	CS-9104	20	CS3901	20	D866
20	C839(M)	20	CDC2010C	20	CS-9125B	20	CS3903	20	D867
20	C839A	20	CDC2010D	20	CS1068	20	CS3904	20	D867W
20	C839B	20	CDC25100-6	20	CS1166	20	CS4424	20	D876
20	C839C	20	CDC25100	20	CS1166D	20	CS4425	20	D877
20	C839D	20	CDC25100-Q	20	CS1166E	20	CS5088	20	Df161
20	C839E	20	CDC25100G	20	CS1166F	20	CS5369	20	Df1610
20	C839F	20	CDC4023A130	20	CS1166G	20	CS6168F	20	DU3
20	C839H	20	CDC430	20	CS1166H	20	CS6229P	20	DW-6505
20	C839J	20	CDC4306813	20	CS1166H/F	20	CS6229G	20	DW-7375
20	C839L	20	CDC60132	20	CS1168F	20	CS696	20	DW6034/M
20	C839M	20	CDC745(ZENITH)	20	CS1168G	20	CS697	20	E13-000-03
20	C839N	20	CDC746	20	CS1168H	20	CS706	20	E13-000-04
20	C839S	20	CDC8000	20	CS1226N	20	CS718	20	E13-002-03
20	C844	20	CDC8000-1B	20	CS1229	20	CS718A	20	E13-003-00
20	C847	20	CDC8001	20	CS1229A	20	CS720A	20	E13-003-01
20	C848	20	CDC8011B	20	CS1229B	20	CS7229G	20	E13-005-02
20	C849	20	CDC8021	20	CS1229C	20	CS9011	20	E210
20	C850	20	CDC8054	20	CS1229D	20	CS9011(E)(P)	20	E212
20	C87	20	CDC86X7-5	20	CS1229F	20	CS9011(EF)	20	E24103
20	C870	20	CDQ10001	20	CS1229G	20	CS9011(GH)	20	E2430
20	C870BL	20	CDQ10002	20	CS1229H	20	CS9011/3490	20	E2431
20	C870E	20	CDQ10003	20	CS1235G	20	CS9011I	20	E2436
20	C870F	20	CDQ10004	20	CS1235S	20	CS9011D	20	E2444
20	C871	20	CDQ10005	20	CS1236D	20	CS9011E	20	E2452
20	C871BL	20	CDQ10006	20	CS1236H	20	CS9011F	20	E2454
20	C871D	20	CDQ10007			20	CS9011G	20	E2455
20	C871E	20	CDQ10008			20	CS9011G/3490	20	E2459
2J	#1XY-	20	CDQ10009			20	CS9011H	20	E2461
PJ	#1XYT	PJ	#SNYJJY;	PJ	#MYPQ1-	PJ	#MZJYI:	PJ	ZPK2X
PJ	#12K	PJ	#SNYJJY:	PJ	#MYPQ10	PJ	#MZJYYA	PJ	ZPK22
PJ	#12;	PJ	#SNYJJYX	PJ	#MYPKV-	PJ	#MZJYQ	PJ	Z,JJ2P
PJ	#122	PJ	#SNYJJY1	PJ	#MYPKVT	PJ	#MZJYQ,	PJ	Z,YJ1J
PJ	#122I	PJ	#SNYJJYJ	PJ	#MYPKVP	PJ	#MZJYQw	PJ	Z,YYP1
PJ	#2YQ	PJ	#SNYJJYV	PJ	#MYPKV:	PJ	#MZJYQ¢	PJ	Z,YYP2
PJ	#2PQZ	PJ	#SNYJJPY	PJ	#MYPKVB	PJ	#MZJYQS	PJ	Z,YYQV
PJ	#2PV	PJ	#SNYJJPP	PJ	#MYPVJZ	PJ	#MZJYQPZ	PJ	Z,YYKV
PJ	#2QQ	PJ	#SNYJJPK	PJ	#MYPVX	PJ	#MZJYQP-	PJ	Z,YQKK
PJ	#2QQWW	PJ	#SNYJJPV	PJ	#MYPV1	PJ	#MZJYQPT	PJ	Z,YQKV
PJ	#2QQ¢	PJ	#SNYJJP;	PJ	#MYPV2	PJ	#MZJYQPF	PJ	Z,YKJ;
PJ	#2QQS	PJ	#SNYJJP:	PJ	#MYP1;	PJ	#MZJYK	PJ	Z,YKJX
PJ	#2QQZ	PJ	#SNYJJPX	PJ	#MYP11	PJ	#MZJYK~¢>	PJ	Z,YKJ1
PJ	#2QQ-O	PJ	#SNYJJP1	PJ	#MYP1Z	PJ	#MZJYK)QK2J	PJ	Z,YKVY
PJ	#2QQ-O¢	PJ	#SNYJJQV	PJ	#MYP2VZ	PJ	#MZJYK,	PJ	Z,YKVP
PJ	#2QQ-OS	PJ	#SNYJJQ;	PJ	#MYP2VT	PJ	#MZJYKw	PJ	Z,YK22
		PJ	#SNYJY1	PJ	#MYQKJY	PJ	#MZJYKw3¢	PJ	Z,YV;K
								PJ	Z,YVX1

GE	Industry Standard No.	GE	Industry Standard No.	GE	Industry Standard No.	GE	Industry Standard No.	GE	Industry Standard No.
20	EA15X81	20	EN930	20	PPR53-1001	20	H933	20	HT304580
20	EA15X1	20	EP15X1	20	PPR50-1001	20	H934	20	HT304580A
20	EA15X101	20	EP15X2	20	PPR50-1002	20	H9423	20	HT304580B
20	EA15X103	20	EP15X39	20	FR83693	20	H9618	20	HT304580C0
20	EA15X111	20	EP15X47	20	FS1168E641	20	H9696	20	HT304580K
20	EA15X112	20	EP15X48(NPN)	20	FS1168P813	20	HC-00460	20	HT304580Y0
20	EA15X136	20	EP15X49	20	FS1221	20	HC-00461	20	HT304580Z
20	EA15X137	20	EP15X7	20	FS1974	20	HC-00537	20	HT304581
20	EA15X142	20	EP15X8	20	FS2043	20	HC-00693	20	HT304581A
20	EA15X143	20	EP15X9	20	FS27604	20	HC-00828	20	HT304581B
20	EA15X153	20	EPX2	20	FS36999	20	HC-00838	20	HT304581B-0
20	EA15X157	20	EQ8-100	20	FT005	20	HC-00871	20	HT304581C
20	EA15X162	20	EQ8-13	20	FT006	20	HC-00921	20	HT304601B0
20	EA15X163	20	EQ8-22	20	FT008	20	HC-00924	20	HT304601C0
20	EA15X167	20	EQ8-5	20	FT008A	20	HC-00945	20	HT304861B
20	EA15X168	20	EQ8-61	20	FT023	20	HC-01590	20	HT305361E
20	EA15X18	20	EQ8-64	20	FT024	20	HC-01417	20	HT305361Q
20	EA15X189	20	EQ8-9	20	FT025	20	HC-01820	20	HT305371E
20	EA15X190	20	ES10222	20	FT026	20	HC-56	20	HT306441
20	EA15X20	20	ES10223	20	FT053	20	HC00838	20	HT306441A
20	EA15X213	20	ES10232	20	FT3567	20	HC01820	20	HT306441B
20	EA15X22	20	ES15048	20	FT3569	20	HC371	20	HT306441B-0
20	EA15X237	20	ES15050	20	FT5643	20	HC372	20	HT306441B0
20	EA15X239	20	ES15X1	20	FX2368	20	HC373	20	HT306441C
20	EA15X24	20	ES15X11	20	FZ101	20	HC458	20	HT306441C-0
20	EA15X240	20	ES15X12	20	G005-036C	20	HC539	20	HT307321A
20	EA15X246	20	ES15X14	20	G005-036E	20	HC561	20	HT307321B-0
20	EA15X272	20	ES15X16	20	G05-010-A	20	HCL-29	20	HT307321B0
20	EA15X31	20	ES15X20	20	G05-011-A	20	HCL-6066	20	HT307322A
20	EA15X330	20	ES15X23	20	G05-015-D	20	HD-00227	20	HT307331Q
20	EA15X331	20	ES15X24	20	G05-015C	20	HEP50	20	HT307331B
20	EA15X355	20	ES15X37	20	G05-034-D	20	HEP53	20	HT307331C
20	EA15X361	20	ES15X42	20	G05-035-D	20	HEP54	20	HT307331C0
20	EA15X37	20	ES15X58	20	G05-035E	20	HEP55	20	HT307341B
20	EA15X370	20	ES15X62	20	G05-036-B	20	HEP724	20	HT307341C-0
20	EA15X371	20	ES15X64	20	G05-036-C	20	HEP725	20	HT308281B
20	EA15X379	20	ES15X68	20	G05-036-C,D,E	20	HEP728	20	HT308281C
20	EA15X44	20	ES15X7	20	G05-056-D	20	HEP729	20	HT308281Q
20	EA15X45	20	ES15X70	20	G05-036B	20	HEP735	20	HT308281Q
20	EA15X52	20	ES15X76	20	G05-036C	20	HEP738	20	HT308282A
20	EA15X56	20	ES15X83	20	G05-036D	20	HEP80004	20	HT308282A-0
20	EA15X58	20	ES15X84	20	G05-036E	20	HEP80015	20	HT308282A-0
20	EA15X59	20	ES15X85	20	G05-037-A	20	HP-40	20	HT308291B-0
20	EA15X60	20	ES20(ELCOM)	20	G05-037-B	20	HP2	20	HT308291B0
20	EA15X63	20	ES46	20	G05-064-A	20	HF3	20	HT308291B0
20	EA15X68	20	ES46(ELCOM)	20	G05-413B	20	HP4	20	HT308291D0
20	EA15X7112	20	ES53(ELCOM)	20	G05015C	20	HP5	20	HT309842A-0
20	EA15X7113	20	ES85(ELCOM)	20	G05035E	20	HP50	20	HT382801D
20	EA15X7115	20	ET15X10	20	G05036	20	HP6	20	HT400
20	EA15X7118	20	ET15X11	20	G05036B	20	HP7	20	HT401
20	EA15X7119	20	ET15X12	20	G05036C	20	HP8	20	HT800011P
20	EA15X7120	20	ET15X13	20	G05036D	20	HKT-158	20	HT800011Q
20	EA15X7175	20	ET15X14	20	G05036E	20	HKT-161	20	HT800181Q
20	EA15X7176	20	ET15X15	20	G05037B	20	HR-11	20	HV25
20	EA15X72	20	ET15X16	20	G05059	20	HR-11A	20	HX-50063
20	EA15X7232	20	ET15X20	20	G395967	20	HR-11B	20	HX-50072
20	EA15X7262	20	ET15X24	20	G395967-2	20	HR-13	20	HX-50092
20	EA15X73	20	ET15X27	20	G9600	20	HR-13A	20	HX-50097
20	EA15X75	20	ET15X37	20	G9600(G.E.)	20	HR-14	20	HX-50110
20	EA15X7514	20	ET15X41	20	G9623	20	HR-14A	20	HX-50113
20	EA15X7517	20	ET15X42	20	G9696	20	HR-15A	20	HX-50161
20	EA15X7586	20	ET15X45	20	GC1144	20	HR-16	20	HX50002
20	EA15X7588	20	ET234843	20	GC783	20	HR-16A	20	HT304580010
20	EA15X7589	20	ET238894	20	GC784	20	HR-17	20	I9A115728-2
20	EA15X76	20	ET368021	20	GE-20	20	HR-17A	20	IC743042
20	EA15X7643	20	ET412626	20	GET2221	20	HR-18	20	IR460B
20	EA15X77	20	ETS-068	20	GET2221A	20	HR-18A	20	IR850
20	EA15X8122	20	ETT-CDC-12000	20	GET2222	20	HR-19	20	IP20-0001
20	EA15X83	20	ETTC-458LG	20	GET2222A	20	HR-19A	20	IP20-0002
20	EA15X84	20	ETTC-CD12000	20	GET2369	20	HR-32	20	IP20-0003
20	EA15X85	20	ETTC-CD13000	20	GET3013	20	HR-36	20	IP20-0006
20	EA15X8502	20	ETTC-CD8000	20	GET3014	20	HR-37	20	IP20-0041
20	EA15X8511	20	ETX18	20	GET3646	20	HR-38	20	IP20-0122
20	EA15X8518	20	EW165V	20	GET3904	20	HR-48	20	IRTR-62
20	EA15X86	20	EW181	20	GET706	20	HR-60	20	IRTR62
20	EA15X89	20	EW182	20	GET708	20	HR11A	20	IRTR63
20	EA15X9	20	EX499-X	20	GET914	20	HR11B	20	IRTR86
20	EA15X90	20	EX500-X	20	GI-3704	20	HR13A	20	I22218
20	EA15X91	20	EX695-X	20	GI-3705	20	HR14	20	I22219
20	EA15X96	20	EX748-X	20	GI-3706	20	HR14A	20	I22221
20	EA15X98	20	EX888-X	20	G110	20	HR15A	20	I22222
20	EA1628	20	EYZP-632	20	GI2711	20	HR16	20	J12yA
20	EA1629	20	EYZP-791	20	GI2712	20	HR16A	20	J241054
20	EA1630	20	P-302-1	20	GI2713	20	HR17	20	J241099
20	EA1638	20	P-302-1532	20	GI2714	20	HR17A	20	J241230
20	EA1695	20	P-302-2532	20	GI2715	20	HR18	20	J241251
20	EA1696	20	F121-433804	20	GI2716	20	HR18A	20	J24458
20	EA1697	20	F121-546	20	GI2921	20	HR19	20	J24564
20	EA1703	20	F15840	20	GI2922	20	HR19A	20	J24565
20	EA1716	20	F15840-1	20	GI2923	20	HR32	20	J24624
20	EA1718	20	F222	20	GI2924	20	HR36	20	J24625
20	EA1735	20	F2443	20	GI3641	20	HR37	20	J24641
20	EA1778	20	F2448	20	GI3643	20	HR48	20	J24658
20	EA1872	20	F2584	20	GI3704	20	HR62	20	J24752
20	EA2271	20	F302-1	20	GI3705	20	HR63	20	J24753
20	EA2489	20	F302-1532	20	GI3706	20	HR64	20	J24817
20	EA2490	20	F302-2	20	GI3707	20	HR65	20	J24843
20	EA2739	20	F302-2532	20	GI3708	20	HR66	20	J24855
20	EA2770	20	F506-001	20	GI3709	20	HR84(NPN)	20	J24874
20	EA3149	20	F506-022	20	GI3710	20	HS-1168	20	J24878
20	EA4112	20	F5519	20	GI3711	20	HS-1229	20	J24906
20	ECG123	20	F5532	20	GME1001	20	HS-40017	20	J24907
20	ECG123A	20	F5569	20	GME1002	20	HS-40020	20	J24909
20	ECG123AP	20	F5571	20	GME2001	20	HS-40030	20	J24916
20	ED-1402	20	F566	20	GME2002	20	HS-40037	20	J310249
20	ED1402A	20	F572-1	20	GME4001	20	HS-40044	20	J310250
20	ED1402B	20	F587	20	GME4002	20	HS-40046	20	J9613(Q.E.)
20	ED1402C	20	F75116	20	GME4003	20	HS-40055	20	JA-H
20	ED1402D	20	F9600	20	GME6003	20	HS40046	20	JA-L
20	ED1402E	20	F9623	20	G04-041B	20	HT3036201	20	JA1200
20	ED1502	20	F9623F	20	G05-003-A	20	HT303620B	20	J39011G
20	ED1502P	20	FA-1(SEARS)	20	G05-003-B	20	HT3036210	20	J39011H
20	ED1502R	20	FB6853	20	G05-010-A	20	HT303711A0	20	JLM-20
20	ED150Z	20	FBC237	20	G05-011-A	20	HT303711B-0	20	JN271
20	ED1702L	20	FCS-9018PF	20	G05-015-C	20	HT303711B-0	20	J8P7001
20	ED1704L	20	FCS1168P813	20	G05-056-C,D,E	20	HT303711B0	20	J8P7001B
20	EDC-010-1	20	FCS1168Q	20	G05-036B	20	HT303721-0	20	JT-1601-40
20	EDO-219	20	FCS11680704	20	G05-036C	20	HT303721D	20	K4-506
20	ED8-100	20	FCS1229F	20	G05-036E	20	HT3037210-0	20	KDB102
20	EL232	20	FCS1229G	20	G05-037B	20	HT303721D	20	KD2102
20	EL238	20	FCS9011P	20	G05036	20	HT303730	20	KU841055
20	EL642	20	FCS9014(B)	20	G05059	20	HT303730A	20	KLH1422
20	EMB-73500	20	FCS9016	20	G89014J	20	HT303751Q0	20	KLH704
20	EN2219	20	FCS9016E	20	G89014K	20	HT303801A0	20	KPG6682
20	EN2222	20	FD-1029-JA	20	G89023H	20	HT303801B0	20	KR-Q1013
20	EN2484	20	FD-1029-LL	20	G89023J	20	HT303801C0	20	KTE1B
20	EN3009	20	FD-1029-NG	20	G89023K	20	HT304531	20	KLB-23
20	EN3013	20	FD-1029-PP	20	GV6063	20	HT304531A	20	LM-1129
20	EN3014	20	FD-1029-PT	20	GVL200077	20	HT304531B	20	LM-1130
20	EN40	20	FMPS-A20	20	H-1567	20	HT304531C	20	LM-1132
20	EN697	20	FMPSA20	20	H102	20	HT304540A0	20	LM-1133
20	EN706	20	FN-51-1A	20	H1567	20	HT304540B0	20	LM-1147
20	EN708	20	FPR4Q-1001	20	H931	20	HT304540B0	20	LM-1148

GE	Industry Standard No.	GE	Industry Standard No.	GE	Industry Standard No.	GE	Industry Standard No.	GE	Industry Standard No.
20	LM-1155	20	M9159	20	MPS6576	20	NR-071AU	20	PT4-7158-023
20	LM1090E	20	M91A	20	MPS6590	20	NR-261A8	20	PT4-7158-02A
20	LM1090F	20	M91B	20	MPS9185	20	NR-431AU	20	PT4800
20	LM1090G	20	M91BGRN	20	MPS9423	20	NR-431A8	20	PT627
20	LM1117D	20	M91C	20	MPS9423I	20	NR-461A8	20	PT703
20	LM1405	20	M91CM624	20	MPS9426A	20	NRO41E	20	PT720
20	LM1540	20	M91D	20	MPS9426A.B	20	NR071AU	20	PU051
20	LM1566F	20	M91E	20	MPS9427B.C	20	NH091ST	20	PT886
20	LM1614D	20	M91F	20	MPS9427C	20	NR201AY	20	PT887
20	LM1614M	20	M91FM624	20	MPS9433K	20	NR261A8	20	PT897
20	LM1818	20	M9226	20	MPS9600	20	NR271AY	20	PT898
20	LM2152	20	M924	20	MPS9600(G)	20	NR461	20	PTC101
20	LRQ849	20	M9248	20	MPS9600F	20	NR461EH	20	PTC136
20	LS-0085-01	20	M9282	20	MPS9600G	20	N31500	20	Q-00269
20	LS3705	20	M9475	20	MPS9604I	20	N31972	20	Q-00269A
20	LT1016(E)	20	M9525	20	MPS9604D	20	N31973	20	Q-00269B
20	LT1016I,H	20	M9532	20	MPS9604E	20	N31974	20	Q-00269C
20	LT1016T,H	20	M9563	20	MPS9604F	20	N31975	20	Q-00369
20	M-1002-2	20	M9568	20	MPS9604FG	20	N83904	20	Q-00369A
20	M-1002-17-NC	20	M9570	20	MPS9604I	20	N3475	20	Q-00369B
20	M-75557-1	20	MA9426	20	MPS9604R	20	N3476	20	Q-00369C
20	M-75557-2	20	MAQ7786	20	MPS9600T	20	N3477	20	Q-00484R
20	M-75557-3	20	MC9427	20	MPS9611-5	20	N3478	20	Q-00569
20	M-75557-4	20	ME-1	20	MPS9616	20	N3479	20	Q-00569A
20	M-75557-5	20	ME-2	20	MPS9618	20	N3480	20	Q-00569B
20	M-75557-6	20	ME-3	20	MPS9618(J)	20	N86114	20	Q-00569
20	M-8641	20	ME1001	20	MPS9618H	20	N86115	20	Q-00669A
20	M140-3	20	ME1002	20	MPS9618I	20	N86207	20	Q-00669B
20	M24	20	ME2001	20	MPS9618J	20	N86210	20	Q-00669C
20	M24A	20	ME2002	20	MPS9623	20	N87262	20	Q-00684R
20	M24B	20	ME213	20	MPS9623C(P)	20	N8731	20	Q-0115C
20	M25	20	ME213A	20	MPS9623E	20	N3731A	20	Q-02115C
20	M25A	20	ME216	20	MPS9623E.G	20	N8733	20	Q-03115C
20	M25A2	20	ME217	20	MPS9623F	20	N8733A	20	Q-04115C
20	M25B	20	ME4001	20	MPS9623G	20	N8734	20	Q-05115C
20	M25B2	20	ME4002	20	MPS9623G/H	20	N8734A	20	Q-06115C
20	M31001	20	ME4003	20	MPS9623H	20	N8949	20	Q-07115C
20	M3519	20	ME4003C	20	MPS9623H/I	20	09-30906Q	20	Q-10115C
20	M4464	20	ME4101	20	MPS9623I	20	OC-307	20	Q-14115C
20	M4465	20	ME4102	20	MPS9623I/J	20	OC-308	20	Q-15115C
20	M447	20	ME4103	20	MPS9623J	20	OC-318	20	Q-16115C
20	M4594	20	ME4104	20	MPS9623G	20	OC-330	20	Q-2N5225
20	M4624	20	ME6001	20	MPS9626	20	OC-34	20	Q-35
20	M4630	20	ME6002	20	MPS9626G	20	OC-340	20	Q-RF-2
20	M4705	20	ME6005	20	MPS9626H	20	OC-341	20	Q-SE1001
20	M4706	20	ME900	20	MPS9626I	20	OC-342	20	QOV60529
20	M4714	20	ME9002	20	MPS9630	20	OC-343	20	QOV60537
20	M4732	20	ME900A	20	MPS9630H	20	ON047204-2	20	QOV60538
20	M4734	20	ME901	20	MPS9630H.I	20	ON271	20	Q3/6515
20	M4737	20	ME901A	20	MPS9630I	20	ON274	20	V25242
20	M4739	20	MEF-25	20	MPS9630T	20	ON47204-1	20	Q5053
20	M4765	20	MG9623	20	MPS9631	20	P-8393	20	Q5053C
20	M4768	20	MH9623	20	MPS9631(I)	20	P/N10000020	20	Q5073D
20	M4821	20	MH9630	20	MPS9631I	20	P04-44-0028	20	Q5073E
20	M4834	20	MI9623	20	MPS9631J	20	P04-45-0014-P2	20	Q5073F
20	M484	20	MI9630	20	MPS9631K	20	P04-45-0014-P5	20	Q5123B
20	M4840	20	MJ29411T	20	MPS9631S	20	P04440028-001	20	Q5123F
20	M4841	20	MM1755	20	MPS9631T	20	P04440028-009	20	QA-12
20	M4842	20	MM1756	20	MPS9632	20	P04440028-014	20	QA-13
20	M4842A	20	MM1757	20	MPS9632(I)	20	P04440028-8	20	QA-14
20	M4842C	20	MM1758	20	MPS9632(K)	20	P04440032-001	20	QA-15
20	M4844	20	MM3904	20	MPS9632I	20	P15153	20	QA-16
20	M4852	20	MMT3014	20	MPS9632I	20	P1901-50	20	QA-19
20	M4854	20	MMT3904	20	MPS9632J	20	P480A0028	20	QG0254
20	M4898	20	MMT72	20	MPS9632K	20	P480A0029	20	QOV60529
20	M4906	20	MN54	20	MPS9632S	20	P5152	20	QOV60530
20	M4926	20	MP1014-2	20	MPS9632T	20	P5153	20	Q8054
20	M4933	20	MPM5006	20	MPS9633G	20	P633567	20	Q8C580
20	M4935	20	MPS-2716	20	MPS9634D	20	P64447	20	QSE1001
20	M4937	20	MPS-3563	20	MPS9696P	20	P8393	20	QT-C0829XAN
20	M4941	20	MPS-3705	20	MPS9696H	20	P8594	20	QT-C0829XBN
20	M4952	20	MPS-6571	20	MPS9696I	20	P9623	20	QT-C1687XAN
20	M4953	20	MPS-706	20	MPS9700P	20	PA7001/0001	20	QT-CBC546AA
20	M54	20	MPS-96301	20	MPS9A05	20	PA9006	20	R-280537
20	M54A	20	MPS-A10	20	MPSA20	20	PBC183	20	R3273-P1
20	M54B	20	MPS2711	20	MPSBE239	20	PE5001	20	R3273-P2
20	M54BLK	20	MPS8239	20	MPSB07	20	PEE2	20	R3283
20	M54BLU	20	MPS807	20	MPSB20	20	PEE5	20	R3293(GE)
20	M54BRN	20	MPSB20	20	MPX9623	20	PEE6	20	R34-6016-58
20	M54O	20	MPX9623	20	MPX9623H	20	PEE7	20	R340
20	M54O	20	MPX9623H	20	MPX9623H/I	20	PEE8	20	R4097
20	M54E	20	MPX9623H/I	20	MPX9623I	20	PEE9	20	R582
20	M54GRN	20	MPX9623I	20	MPX96301	20	PEE1002	20	R7163
20	M54ORN	20	MPX96301	20	MQ1	20	PEE2001	20	R7165
20	M54RED	20	MPS2924	20	MQ2	20	PEE2002	20	R7249
20	M54WHT	20	MPS2925	20	MR3932	20	PEE3704	20	R7343
20	M54YEL	20	MPS2926	20	MR9604	20	PEE3705	20	R7359
20	M671	20	MPS2926BRN	20	MS822B	20	PEE3706	20	R7360
20	M7003	20	MPS2926ORN	20	MS7502R	20	PEE4002	20	R7361
20	M7006	20	MPS2926RED	20	MS7503R	20	PEE6001	20	R7582
20	M7014	20	MPS2926YEL	20	MSR7503	20	PEE6002	20	R7887
20	M7015	20	MPS3392	20	MT104	20	PEE8000	20	R7953
20	M7033	20	MPS3393	20	MT4101	20	PEE8001	20	R8066
20	M7108	20	MPS3394	20	MT4102	20	PEE8002	20	R8067
20	M7108/A5N	20	MPS3395	20	MT4102A	20	PEE8003	20	R8068
20	M7109	20	MPS3396	20	MT4103	20	PEE9002	20	R8069
20	M7109/A5P	20	MPS3397	20	MT6001	20	PL1052	20	R8070
20	M7171	20	MPS3398	20	MT6002	20	PL1054	20	R8115
20	M75565-1	20	MPS3643	20	MT6003	20	PL4021	20	R8116
20	M773	20	MPS3646	20	MT696	20	PL4051	20	R8117
20	M773RED	20	MPS3704	20	MT697	20	PL4052	20	R8118
20	M774	20	MPS3705	20	MT706	20	PL4053	20	R8119
20	M774ORN	20	MPS3706	20	MT706A	20	PL4054	20	R8120
20	M775	20	MPS3707	20	MT706B	20	PL4055	20	R8223
20	M775BRN	20	MPS3708	20	MT707	20	PM1121	20	R8224
20	M776	20	MPS3709	20	MT708	20	PRT-101	20	R8225
20	M776ORN	20	MPS3711	20	MT9001	20	PRT-104	20	R8244
20	M779BLU	20	MPS3721	20	MT9002	20	PRT-104-1	20	R8259
20	M780WHT	20	MPS3826	20	N-EA15X136	20	PRT-104-2	20	R8260
20	M783	20	MPS3827	20	N-EA15X137	20	PRT-104-3	20	R8261
20	M783RED	20	MPS3393	20	N-EA15X138	20	PS209800	20	R8305
20	M784	20	MPS3992	20	ON047204-2	20	PT1558	20	R8312
20	M784ORN	20	MPS5172	20	N201AY	20	PT1559	20	R8528
20	M785	20	MPS6301	20	N3563	20	PT1610	20	R8530
20	M785YEL	20	MPS6413	20	ON47204-1	20	PT1835	20	R8543
20	M787BLU	20	MPS6520	20	N4T	20	PT1836	20	R8551
20	M791	20	MPS6521	20	N3811(NPN)	20	PT1837	20	R8552
20	M8105	20	MPS6530	20	NC207AL	20	PT2760	20	R8553
20	M818	20	MPS6544	20	NJ100B	20	PT2896	20	R8554
20	M818WHT	20	MPS6545	20	NJ102C	20	PT3141	20	R8555
20	M822	20	MPS6552	20	NK12-1A19	20	PT3141A	20	R8556
20	M822A	20	MPS6553	20	NKT10339	20	PT3141B	20	R8557
20	M822A-BLU	20	MPS6554	20	NKT10419	20	PT3151A	20	R8620
20	M822B	20	MPS6555	20	NKT10439	20	PT3151B	20	R8646
20	M823	20	MPS6556	20	NKT10519	20	PT3151C	20	R8647
20	M823B	20	MPS6611	20	NKT12529	20	PT3500	20	R8648
20	M823WHT	20	MPS6565	20	NKT12429	20	PT3501	20	R8658
20	M827	20	MPS6566	20	NKT13329	20	PT4-7158	20	R8889
20	M827BRN	20	MPS6567	20	NKT13429	20	PT4-7158-012	20	R8914
20	M828GRN	20	MPS6568	20	NL-102	20	PT4-7158-013	20	R8916
20	M847	20	MPS6571	20	NN9017	20	PT4-7158-01A	20	R8953
20	M847BLK	20	MPS6573	20	NPC737	20	PT4-7158-021	20	R8964
20	M9095	20	MPS6574	20	NPS6514	20	PT4-7158-022		
		20	MPS6575	20	NPS6520				

337 GE-20 thru GE-20

GE	Industry Standard No.	GE	Industry Standard No.	GE	Industry Standard No.	GE	Industry Standard No.	GE	Industry Standard No.
20	R8965	20	RS7639	20	S1453	20	SE5-0274	20	SPS4029
20	R8966	20	RS7640	20	S1475	20	SE5-0567	20	SPS4032
20	R8968	20	RS7641	20	S1476	20	SE5-0608	20	SPS4034
20	R9004	20	RS7642	20	S1487	20	SE5-0848	20	SPS4037
20	R9005	20	RS7643	20	S1502	20	SE5-0854	20	SPS4039
20	R9006	20	RS7814	20	S1510	20	SE5-0855	20	SPS4040
20	R9025	20	RS8442	20	S1512	20	SE5-0887	20	SPS4041
20	R9071	20	RS8503	20	S1526	20	SE5-0888	20	SPS4042
20	R9384	20	RS86057332	20	S1527	20	SE5-0938-54	20	SPS4044
20	R9385	20	RT-100	20	S1529	20	SE5030A	20	SPS4045
20	R9483	20	RT-102	20	S1530	20	SE5030B	20	SPS4049
20	R812	20	RT-929-H	20	S1533	20	SE5151	20	SPS4052
20	R813	20	RT-929H	20	S1559	20	SE6010	20	SPS4053
20	REN123	20	RT-930H	20	S15649	20	SE8040	20	SPS4055
20	REN123A	20	RT100	20	S1568	20	SP1001	20	SPS4059
20	RH120	20	RT114	20	S1570	20	SP1713	20	SPS4060
20	RT7504	20	RT2016	20	S1619	20	SP1714	20	SPS4061
20	RR8068	20	RT2332	20	S1620	20	SP1726	20	SPS4062
20	RR8914	20	RT2914	20	S1629	20	SP1730	20	SPS4063
20	RS-107	20	RT3063	20	S1697	20	SPT713	20	SPS4066
20	RS-108	20	RT3064	20	S169N	20	SPT714	20	SPS4067
20	RS-2009	20	RT3228	20	S1761	20	SQC-7202	20	SPS4069
20	RS-2013	20	RT5565	20	S1761A	20	SH1064	20	SPS4074
20	RS-2016	20	RT5567	20	S1761B	20	SJ-570	20	SPS4075
20	RS-5851	20	RT476	20	S1761C	20	SJ570	20	SPS4077
20	RS-5853	20	RT4760	20	S1764	20	SK1640A	20	SPS4081
20	RS-5856	20	RT4761	20	S1765	20	SK1641	20	SPS4083
20	RS-5857	20	RT5202	20	S1766	20	SK3020	20	SPS4084
20	RS-7103	20	RT5206	20	S1768	20	SK3058	20	SPS4085
20	RS-7105	20	RT5207	20	S1770	20	SK3434A	20	SPS4088
20	RS-7124	20	RT5551	20	S1772	20	SK5801	20	SPS4089
20	RS-7129	20	RT6600MHP25	20	S1784	20	SK5915	20	SPS4095
20	RS-7409	20	RT6733	20	S1785	20	SK8215	20	SPS41
20	RS-7411	20	RT6921	20	S1788	20	SK8251	20	SPS4169
20	RS-7412	20	RT6921MHP25	20	S1835	20	SKA-6256	20	SPS4199
20	RS-7413	20	RT69221	20	S1871	20	SKA-6437	20	SPS4236
20	RS-7504	20	RT697M	20	S1891	20	SKA-8105	20	SPS4303
20	RS-7511	20	RT6989	20	S1891A	20	SKA0030	20	SPS4313
20	RS-7606	20	RT7322	20	S1891B	20	SKA1080	20	SPS4344
20	RS-7607	20	RT7325	20	S1955	20	SKA1117	20	SPS4345
20	RS-7609	20	RT7327	20	S1993	20	SKA1395	20	SPS4347
20	RS-7610	20	RT7511	20	S2034	20	SKA4141	20	SPS4356
20	RS-7611	20	RT7514	20	S2043	20	SKB8339	20	SPS4359
20	RS-7612	20	RT7515	20	S2044	20	SL300	20	SPS4360
20	RS-7613	20	RT7517	20	S2121	20	SL7990	20	SPS4363
20	RS-7614	20	RT7518	20	S2122	20	SM-4508-B	20	SPS4367
20	RS-7622	20	RT7528	20	S2123	20	SM-5564	20	SPS4368
20	RS-7623	20	RT7638	20	S2124	20	SM-5643	20	SPS4362
20	RS1049	20	RT7845	20	S2171	20	SM-716	20	SPS4390
20	RS1059	20	RT7943	20	S2172	20	SM-7815	20	SPS4392
20	RS136	20	RT8195	20	S2225	20	SM-7836	20	SPS4443
20	RS15048	20	RT8197	20	S22543	20	SM-A-726655	20	SPS4446
20	RS2914	20	RT8198	20	S23579	20	SM-A-726664	20	SPS4450
20	RS5851	20	RT8201	20	S2397	20	SM-B-610342	20	SPS4451
20	RS5853	20	RT8332	20	S24591	20	SM-B-686767	20	SPS4453
20	RS5856	20	RT929H	20	S24596	20	SM-C-583256	20	SPS4455
20	RS5857	20	RV1471	20	S2581	20	SM07275	20	SPS4456
20	RS7103	20	RV1474	20	S2582	20	SM07286	20	SPS4457
20	RS7105	20	RV2249	20	S2590	20	SM2700	20	SPS4459
20	RS7108	20	RVTC81381	20	S2593	20	SM2701	20	SPS4472
20	RS7111	20	RVTC81383	20	S2635	20	SM3104	20	SPS4476
20	RS7121	20	RVT822410	20	S2636	20	SM3117A	20	SPS4478
20	RS7127	20	RYN121105	20	S26822	20	SM3505	20	SPS4491
20	RS7129	20	RYN121105-3	20	S2935	20	SM3986	20	SPS4493
20	RS7132	20	RYN121105-4	20	S2944	20	SM4508-B	20	SPS4494
20	RS7133	20	S-1061	20	S29445	20	SM5379	20	SPS4498
20	RS7136	20	S-1065	20	S2984	20	SM5564	20	SPS4920
20	RS7160	20	S-1066	20	S2985	20	SM5643	20	SPS4942
20	RS7223	20	S-1068	20	S2989	20	SM576-1	20	SPS5000
20	RS7224	20	S-1128	20	S2996	20	SM576-2	20	SPS5006
20	RS7226	20	S-1143	20	S2997	20	SM5981	20	SPS5006-1
20	RS7232	20	S-1221	20	S2998	20	SM6773	20	SPS5006-2
20	RS7234	20	S-1221A	20	S2999	20	SM716	20	SPS5457
20	RS7235	20	S-1245	20	S32550	20	SM7545	20	SPS6111
20	RS7236	20	S-1331W	20	S34540	20	SM7815	20	SPS6112
20	RS7238	20	S-1363	20	S36999	20	SM7836	20	SPS6113
20	RS7241	20	S-1364	20	S37182	20	SM8112	20	SPS627
20	RS7242	20	S-1403	20	S37214	20	SM8113	20	SPS6571
20	RS7405	20	S-1512	20	S37423	20	SM8978	20	SPS699
20	RS7406	20	S-1533	20	S58763	20	SM9008	20	SPS7652
20	RS7407	20	S-1559	20	S38787	20	SM9135	20	SPS817
20	RS7408	20	S001466	20	S38854	20	SM9253	20	SPS817N
20	RS7409	20	S0015	20	S38560	20	SPC40	20	SPS868
20	RS7410	20	S0022	20	S6801	20	SPC042	20	SPS907
20	RS7411	20	S0025	20	S70.01.704	20	SPC50	20	SQD-2170
20	RS7412	20	S022010	20	S95202	20	SPC51	20	SR20234
20	RS7413	20	S022011	20	S9631	20	SPC52	20	SR75844
20	RS7415	20	S024428	20	SAW-28C372GR	20	SPS-1475	20	SR1-145128
20	RS7421	20	S024987	20	SAW-28C372Y	20	SPS-1475YT	20	SR2508
20	RS7504	20	S025232	20	SAW-28C945R	20	SPS-4075	20	SS2504
20	RS7510	20	S025289	20	SC-4044	20	SPS-41	20	SS3694
20	RS7513	20	S031A	20	SC-4244	20	SPS-4396	20	SS9328
20	RS7513-15	20	S037	20	SC-65	20	SPS-934	20	ST-1242
20	RS7514	20	S0704	20	SC-852	20	SPS1045	20	ST-1243
20	RS7515	20	S1016	20	SC1001	20	SPS1082	20	ST-1244
20	RS7516	20	S1061	20	SC1010	20	SPS1475	20	ST-1290
20	RS7517	20	S1065	20	SC108B	20	SPS1802	20	ST-LM2152
20	RS7517-19	20	S1066	20	SC109A	20	SPS1817	20	ST.082.112.005
20	RS7518	20	S1068	20	SC1168Q	20	SPS1977	20	ST.082.114.015
20	RS7519	20	S1069	20	SC1168H	20	SPS2104	20	ST/217/Q
20	RS7521	20	S1074	20	SC1229Q	20	SPS2129	20	ST01
20	RS7525	20	S1074(R)	20	SC350	20	SPS2142	20	ST02
20	RS7526	20	S1074R	20	SC4010	20	SPS2164	20	ST03
20	RS7527	20	S1090	20	SC4044	20	SPS2194	20	ST04
20	RS7528	20	S1128	20	SC65	20	SPS2225	20	ST05
20	RS7529	20	S1143	20	SC785	20	SPS2270	20	ST06
20	RS7530	20	S12-1-A-3P	20	SC832	20	SPS2415	20	ST1242
20	RS7542	20	S1221	20	SCD-7322	20	SPS3015	20	ST1243
20	RS7543	20	S1221A	20	SCDT323	20	SPS3735	20	ST1244
20	RS7544	20	S1226	20	SD-109	20	SPS3751	20	ST1290
20	RS7555	20	S1240	20	SD109	20	SPS3900	20	ST1402D
20	RS7606	20	S1241	20	SD3000	20	SPS3907	20	ST1402E
20	RS7607	20	S1242	20	SDD821	20	SPS3908	20	ST150
20	RS7609	20	S1243	20	SDD821	20	SPS3909	20	ST1506
20	RS7610	20	S1245	20	SE-0566	20	SPS3915	20	ST151
20	RS7611	20	S1272	20	SE-1002	20	SPS3923	20	ST152
20	RS7612	20	S1307	20	SE-1331	20	SPS3925	20	ST153
20	RS7613	20	S1309	20	SE-2001	20	SPS3926	20	ST154
20	RS7614	20	S133-1	20	SE-3646	20	SPS3930	20	ST155
20	RS7620	20	S1331	20	SE-4001	20	SPS3936	20	ST156
20	RS7621	20	S1331N	20	SE-4002	20	SPS3938	20	ST157
20	RS7622	20	S1331W	20	SE-4010	20	SPS3940	20	ST160
20	RS7623	20	S1363	20	SE-5006	20	SPS3951	20	ST1607
20	RS7624	20	S1364	20	SE-6001	20	SPS3957C	20	ST161
20	RS7625	20	S1369	20	SE-6002	20	SPS3967	20	ST162
20	RS7626	20	S1373	20	SE1331	20	SPS3972	20	ST163
20	RS7627	20	S1374	20	SE2401	20	SPS3973	20	ST1702M
20	RS7628	20	S1403	20	SE2402	20	SPS3999	20	ST1702N
20	RS7634	20	S1405	20	SE4004	20	SPS4003	20	ST175
20	RS7635	20	S1419	20	SE4010	20	SPS4004	20	ST176
20	RS7636	20	S1420	20	SE4172	20	SPS4006	20	ST177
20	RS7637	20	S1429-3	20	SE5-0128	20	SPS4009	20	ST178
20	RS7638	20	S1432	20	SE5-0253	20	SPS4017	20	ST180
20		20	S1443			20	SPS54020	20	ST181

GE	Industry Standard No.	GE	Industry Standard No.	GE	Industry Standard No.	GE	Industry Standard No.	GE	Industry Standard No.
20	ST182	20	TA6	20	TI896	20	TR8031	20	001-02113-3
20	ST250	20	TA7	20	TIX712	20	TR8034	20	001-02113-4
20	ST251	20	TAC-047	20	TIXS12	20	TR8035	20	001-02113-5
20	ST25A	20	TAC047	20	TIX813	20	TR8038	20	001-021132
20	ST25B	20	TC3123036722	20	TK1228-1008	20	TR8039	20	001-021133
20	ST25C	20	TC3123036900	20	TK1228-1009	20	TR8040	20	001-021134
20	ST403	20	TC3123037111	20	TK1228-1010	20	TR8043	20	001-021135
20	ST50	20	TC3123037222	20	TK1228-1011	20	TR86	20	001-021210
20	ST501	20	TC3123037412	20	TK1228-1012	20	TR9100	20	001-021232
20	ST502	20	TD101	20	TM23	20	TRA-34	20	01-030829
20	ST503	20	TD102	20	TM2613	20	TRA-36	20	01-031687
20	ST504	20	TD201	20	TM2711	20	TRA-4	20	1-042/2207
20	ST5060	20	TD202	20	TMT-1543	20	TRA-4A	20	1-044/2207
20	ST51	20	TE1420	20	TMT1543	20	TRA-4B	20	1-21-276
20	ST53	20	TE2369	20	TN37	20	TRA-9R	20	1-21-277
20	ST54	20	TE3414	20	TN53	20	TRA34	20	1-21-278
20	ST55	20	TE3415	20	TN55	20	TRA36	20	1-21-279
20	ST56	20	TE3605	20	TN59	20	TRA4	20	01-2101
20	ST57	20	TE3605A	20	TN60	20	TRA4A	20	01-2101-0
20	ST58	20	TE3606	20	TN61	20	TRA4B	20	001-21011
20	ST59	20	TE3606A	20	TN62	20	TRA9R	20	01-2102
20	ST63	20	TE3607	20	TN63	20	TRAPLC711	20	01-2104
20	ST64	20	TE3704	20	TN64	20	TRAPLC871	20	01-2105
20	ST6511	20	TE3705	20	TN80	20	TRAPLC871A	20	01-2106
20	ST6512	20	TE3903	20	TNC61689	20	TRBC147B	20	01-2107
20	SX3709	20	TE3904	20	TNC61702	20	TR01026	20	01-2108
20	SX3711	20	TE3906	20	TNJ-60076	20	TR010602-1	20	01-2109
20	SX55	20	TE4123	20	TNJ-60606	20	T82221	20	01-30829
20	STL-1182	20	TE4124	20	TNJ-60607	20	T82222	20	03-34418
20	STL1182	20	TE4224	20	TNJ1036	20	T8C499	20	1-522223720
20	STL152	20	TE4951	20	TNJ60076	20	T8C695	20	01-571811
20	STL3460	20	TE4952	20	TNJ61219	20	TST705899A	20	01-571821
20	T-1416	20	TE4953	20	TNJ61220	20	TT-1097	20	6-1471791229
20	T-255	20	TE4954	20	TNJ70479-1	20	TT1097	20	6-171191368
20	T-256	20	TE5309A	20	TNJ70537	20	TV-17	20	1-801-003
20	T-399	20	TE5311A	20	TNJ70539	20	TV-18	20	1-801-004
20	T-H28C536	20	TE5368	20	TNJ70637	20	TV-21	20	1-801-004-17
20	T-H28C693	20	TE5369	20	TNJ70639	20	TV-23	20	01-801-314
20	T-H28C715	20	TE5370	20	TNJ70638	20	TV-40	20	1-801-314-15
20	T-Q5053	20	TE5371	20	TNJ70640	20	TV-42	20	1-801-314-16
20	T-Q5053C	20	TE5376	20	TNJ71036	20	TV-46	20	01-9011-52221-3
20	T-Q5073	20	TE5377	20	TNJ72280	20	TV-51	20	01-9013-72221-3
20	T01-013	20	TE5449	20	TNJ72281	20	TV-52	20	01-9014-22221-3
20	T01-014	20	TE5450	20	TNJ72783	20	TV-53	20	1A0021
20	T01-101	20	TE5451	20	TNJ72784	20	TV-56	20	1A0022
20	T01-104	20	TE697	20	TNT-843	20	TV-58	20	1A0024
20	T01-105	20	TEH0147	20	TNT842	20	TV-6	20	1A0025
20	T1-1A6	20	TP-78	20	T0-033	20	TV-60	20	1A0029
20	T1004671	20	T028C1175(C)	20	T0-038	20	TV-62	20	1A0032
20	T1008-834	20	T28C536	20	T0-039	20	TV-64	20	1A0033
20	T1008834	20	T28C693	20	T0-040	20	TV-65	20	1A0034
20	T1340A31	20	T28C715	20	T01-101	20	TV-66	20	1A0035
20	T1340A3I	20	TI-412	20	T01-104	20	TV-68	20	1A0037
20	T1340A3J	20	TI-413	20	T01-105	20	TV-84	20	1A0043
20	T1340A3K	20	TI-414	20	TP4123	20	TV-92	20	1A0044
20	T1413	20	TI-422	20	TP4124	20	TV17	20	1A0051
20	T1414	20	TI-423	20	TPB6512	20	TV18B	20	1A0065
20	T1415	20	TI-432	20	TPB6513	20	TV241077	20	1A0067
20	T1416	20	TI-433	20	TPB6514	20	TV241078	20	1A0070
20	T1417	20	TI-714	20	TPB6515	20	TV24215	20	1A0076
20	T143	20	TI-714A	20	TPB6520	20	TV24216	20	1A0077
20	T1495	20	TI-751	20	TPB6521	20	TV24281	20	1A0078
20	T157	20	TI-806G	20	TQ-5052	20	TV24372	20	1A0079
20	T158	20	TI-907	20	TQ-5053	20	TV24453	20	1A0080
20	T642B	20	TI-908	20	TQ-5054	20	TV24454	20	1A0081
20	T170	20	TI1A6	20	TQ-5060	20	TV24458	20	1A0083
20	T171	20	TI24A	20	TQ4	20	TV24576	20	1A0084
20	T1746	20	TI24B	20	TQ5052	20	TV24655	20	1A4757-1
20	T1746A	20	TI411	20	TQ5053	20	TV28C208	20	103576
20	T1746B	20	TI415	20	TQ5054	20	TV36	20	1J1
20	T1746C	20	TI416	20	TQ5060	20	TV37	20	1U585F
20	T1748	20	TI417	20	TR-1033-1	20	TV38	20	1U585F/7825B
20	T1748A	20	TI418	20	TR-1033-2	20	TV39	20	1W8358
20	T1748B	20	TI419	20	TR-1033-3	20	TV40	20	1W9723
20	T1748C	20	TI420	20	TR-347	20	TV43	20	1W9787
20	T1802	20	TI422	20	TR-162(OLSON)	20	TV46	20	002-006500
20	T1802A	20	TI424	20	TR-1R33	20	TV48	20	002-009500
20	T1802B	20	TI430	20	TR-21	20	TV56	20	002-009501
20	T1804	20	TI432	20	TR-21-6	20	TV57A	20	002-009502
20	T1805	20	TI480	20	TR-21C	20	TV58A	20	002-009900
20	T1810	20	TI481	20	TR-22	20	TV59A	20	002-010400
20	T1810B	20	TI482	20	TR-22C	20	TV60A	20	002-01000
20	T185	20	TI483	20	TR-24(PHILCO)	20	TV65	20	002-012000
20	T1909	20	TI484	20	TR-28C367	20	TV71	20	002-03
20	T1855	20	TI485	20	TR-28C373	20	TV92	20	02-1078-01
20	T235A013-2	20	TI492	20	TR-28C735	20	TVS-2C8645A	20	002-12000
20	T237	20	TI493	20	TR-3R38	20	TVS-28C206	20	2-8454-031
20	T2446	20	TI494	20	TR-4R33	20	TVS-28C208A	20	2C8900
20	T255	20	TI495	20	TR-5R33	20	TVS-28C538	20	2D002
20	T256	20	TI496	20	TR-5R35	20	TVS-28C538A	20	2D002-168
20	T277	20	TI54A	20	TR-5R38	20	TVS-28C828	20	2D002-169
20	T291	20	TI54B	20	TR-62	20	TVS-28C828A	20	2D002-170
20	T327	20	TI54C	20	TR-6R33	20	TVS-28C828Q	20	2D002-171
20	T327-2	20	TI54E	20	TR-7R35	20	TV8282C538A	20	2D002-175
20	T328	20	TI714	20	TR-800A	20	TV828C644	20	2D002-41
20	T339	20	TI751	20	TR-8014	20	TV828C645B	20	2D017
20	T342	20	TI802B	20	TR-8025	20	TV828C684	20	2D026
20	T5565	20	TI803B	20	TR-8035	20	TV828C828	20	2D026-274
20	T55A-5	20	TI810B	20	TR-8039	20	TV828C828(Q)	20	2D033
20	T5601	20	TI904	20	TR-8042	20	TV828C828A	20	2D038
20	T586	20	TI907	20	TR-86	20	TV828C828P	20	2JMW961
20	T599	20	TI908	20	TR-8R35	20	TV828C828R	20	2N1006
20	T416-16(SEARS)	20	TIA06	20	TR-9100-18	20	TV828C829(B)	20	2N1051
20	T417	20	TIA102	20	TR-BC147B	20	001-00	20	2N1081
20	T457-16	20	TIS-62	20	TR-BR0149C	20	1-0006-0021	20	2N1082
20	T457-16(SEARS)	20	TIS113	20	TR-RR38	20	1-0006-0022	20	2N1103
20	T458-16	20	TIS114	20	TR-TR38	20	1-001-003-15	20	2N1104
20	T459(SEARS)	20	TI822	20	TR01057	20	001-02011	20	2N1116
20	T460	20	TI823	20	TR01073	20	01-02020	20	2N1117
20	T461-16	20	TI844	20	TR01074	20	001-02101-0	20	2N1139
20	T461-16(SEARS)	20	TI845	20	TR1011	20	001-02101-1	20	2N1140
20	T462	20	TI846	20	TR1031	20	001-021010	20	2N1149
20	T462(SEARS)	20	TI847	20	TR1033	20	001-02102-0	20	2N1150
20	T472	20	TI848	20	TR1993-2	20	001-021020	20	2N1151
20	T472(SEARS)	20	TI849	20	TR2085-72	20	001-02103-0	20	2N1152
20	T483(SEARS)	20	TI851	20	TR2085-73	20	001-021030	20	2N1153
20	T484(SEARS)	20	TI852	20	TR28C3677	20	001-02104-0	20	2N117
20	T485(SEARS)	20	TI855	20	TR28C373	20	001-021040	20	2N118
20	T486(SEARS)	20	TI856	20	TR28C735	20	001-02105-0	20	2N118A
20	T59235A	20	TI857	20	TR302	20	001-021050	20	2N119
20	T615A002	20	TI860	20	TR310231	20	001-02106-0	20	2N199
20	T615A006-1	20	TI871	20	TR310243	20	001-021060	20	2N199A
20	T6565	20	TI872	20	TR310245	20	001-02107-0	20	2N1200
20	T9011A1C	20	TI883	20	TR4010-2	20	001-021070	20	2N1201
20	T9011A1G	20	TI890	20	TR60	20	001-02108-0	20	2N1205
20	T9011AZ	20	TI890-2	20	TR62	20	001-021080	20	2N1247
20	T9011G(CD)	20	TI892	20	TR8004-4	20	001-02109-0	20	2N1248
20	T9011G(EP)	20	TI892-BLU	20	TR8010	20	001-021090	20	2N1249
20	T9011H(EP)	20	TI892-GRN	20	TR8014	20	001-02110-0	20	2N1267
20	T9011I(EP)	20	TI892-GRY	20	TR8021	20	001-02111-0	20	2N1276
20	T9011J(GH)	20	TI892-VIO	20	TR8025	20	001-02111-1	20	2N1277
20	TA-6	20	TI892-YEL	20	TR8028	20	001-021110	20	
20	TA198030-4	20	TI894	20	TR8029	20	001-021111	20	
20	TA2A	20	TI895	20	TR8030	20	001-02113-2	20	

GE	Industry Standard No.	GE	Industry Standard No.	GE	Industry Standard No.	GE	Industry Standard No.	GE	Industry Standard No.
20	2N1278	20	2N2369	20	2N3338A	20	2N5082	20	2S102
20	2N1279	20	2N2369A	20	2N3390	20	2N5088	20	2S1182D
20	2N1338	20	2N2387	20	2N3391	20	2N5089	20	2S131
20	2N1386	20	2N2388	20	2N3391A	20	2N5107	20	2S363
20	2N1387	20	2N2389	20	2N3392	20	2N5127	20	2S501
20	2N1588	20	2N2390	20	2N3393	20	2N5128	20	2S502
20	2N1589	20	2N2396	20	2N3394	20	2N5134	20	2S503
20	2N1417	20	2N2397	20	2N3395	20	2N5137	20	2S645
20	2N1418	20	2N2413	20	2N3396	20	2N5144	20	2S701
20	2N1420	20	2N2417	20	2N3397	20	2N5172	20	2S702
20	2N1420A	20	2N2427	20	2N3398	20	2N5183	20	2S703
20	2N148	20	2N2432	20	2N3409	20	2N5186	20	2S711
20	2N148C	20	2N2466	20	2N3410	20	2N5187	20	2S712
20	2N148C/D	20	2N2472	20	2N3411	20	2N5209	20	2S731
20	2N148D	20	2N2473	20	2N3414	20	2N5210	20	2S732
20	2N1528	20	2N2476	20	2N3415	20	2N5219	20	2S733
20	2N1588	20	2N2477	20	2N3416	20	2N5223	20	2S741A
20	2N1591	20	2N2481	20	2N3417	20	2N5224	20	2S951
20	2N1594	20	2N2483	20	2N3423	20	2N5225	20	2SA828A
20	2N160	20	2N2484	20	2N3424	20	2N5232	20	2SB645E
20	2N160A	20	2N2501	20	2N3462	20	2N5232A	20	2SC-F6
20	2N161	20	2N2520	20	2N3463	20	2N5368	20	2SC1007
20	2N161A	20	2N2521	20	2N3498	20	2N5369	20	2SC105
20	2N163	20	2N2522	20	2N3510	20	2N5370	20	2SC1071
20	2N163A	20	2N2523	20	2N3542	20	2N5371	20	2SC1244
20	2N1644	20	2N2524	20	2N3546	20	2N5380	20	2SC1317
20	2N1644A	20	2N2529	20	2N3565	20	2N5381	20	2SC1324
20	2N1665	20	2N2530	20	2N3566	20	2N541	20	2SC1324(C)
20	2N1674	20	2N2531	20	2N3605	20	2N5417	20	2SC1324C
20	2N1682	20	2N2532	20	2N3605A	20	2N5418	20	2SC136
20	2N1704	20	2N2533	20	2N3606	20	2N5419	20	2SC1361
20	2N1708	20	2N2534	20	2N3606A	20	2N542	20	2SC1363
20	2N1708A	20	2N2539	20	2N3607	20	2N5420	20	2SC16D
20	2N1763	20	2N2540	20	2N3641	20	2N5420A	20	2SC1372Y
20	2N1838	20	2N2569	20	2N3643	20	2N543	20	2SC1390
20	2N1839	20	2N2570	20	2N3646	20	2N543A	20	2SC1390(L,Y)
20	2N1840	20	2N2571	20	2N3647	20	2N5456	20	2SC1390(V)
20	2N1944	20	2N2572	20	2N3678	20	2N546	20	2SC1390(W)
20	2N1945	20	2N2586	20	2N3693	20	2N547	20	2SC1390(X)
20	2N1946	20	2N2610	20	2N3694	20	2N548	20	2SC1390(Y)
20	2N1947	20	2N2617	20	2N3704	20	2N549	20	2SC1390I
20	2N1948	20	2N2618/46	20	2N3705	20	2N550	20	2SC1390I(W)
20	2N1949	20	2N2645	20	2N3706	20	2N551	20	2SC1390IW
20	2N1950	20	2N2656	20	2N3707	20	2N552	20	2SC1390J
20	2N1951	20	2N2673	20	2N3708	20	2N5581	20	2SC1390J(X)
20	2N1952	20	2N2674	20	2N3709	20	2N5582	20	2SC1390JX
20	2N1953	20	2N2675	20	2N3710	20	2N560	20	2SC1390K
20	2N1958	20	2N2676	20	2N3711	20	2N5769	20	2SC1390L
20	2N1959	20	2N2677	20	2N3725	20	2N5810	20	2SC1390V
20	2N1962	20	2N2678	20	2N3736	20	2N5814	20	2SC1390W
20	2N1963	20	2N2692	20	2N3793	20	2N5816	20	2SC1390WH
20	2N1964	20	2N2693	20	2N3794	20	2N5818	20	2SC1390WI
20	2N1965	20	2N2713	20	2N3828	20	2N5830	20	2SC1390WX
20	2N1972	20	2N2714	20	2N3843	20	2N5845	20	2SC1390X
20	2N1973	20	2N2719	20	2N3843A	20	2N5845A	20	2SC1390XJ
20	2N1974	20	2N2720	20	2N3844	20	2N622	20	2SC1390XK
20	2N1975	20	2N2721	20	2N3844A	20	2N696	20	2SC1390Y
20	2N1983	20	2N2722	20	2N3858	20	2N697A	20	2SC1390YM
20	2N1984	20	2N2784/TNT	20	2N3858A	20	2N701	20	2SC1634
20	2N1985	20	2N2787	20	2N3859	20	2N707A	20	2SC1639
20	2N1986	20	2N2790	20	2N3859A	20	2N708	20	2SC1641
20	2N1987	20	2N2791	20	2N3862	20	2N708A	20	2SC1641Q
20	2N1988	20	2N2792	20	2N3869	20	2N715	20	2SC1641R
20	2N1992	20	2N2851	20	2N3880	20	2N716	20	2SC1687
20	2N2038	20	2N2845	20	2N3885	20	2N728	20	2SC1739
20	2N2039	20	2N2847	20	2N3900	20	2N729	20	2SC183
20	2N2040	20	2N2868	20	2N3900A	20	2N730	20	2SC183(P)
20	2N2041	20	2N2910	20	2N3901	20	2N731	20	2SC183(Q)(R)
20	2N2049	20	2N2913	20	2N3903	20	2N742	20	2SC183(R)
20	2N2094	20	2N2914	20	2N3904	20	2N742A	20	2SC183-1
20	2N2094A	20	2N2917	20	2N3924	20	2N745	20	2SC183-OR
20	2N2095	20	2N2918	20	2N3947	20	2N746	20	2SC183A
20	2N2095A	20	2N2923	20	2N3975	20	2N747	20	2SC183AP
20	2N2096	20	2N2924	20	2N3976	20	2N748	20	2SC183B
20	2N2096A	20	2N2925	20	2N4013	20	2N749	20	2SC183BK
20	2N2097A	20	2N2926	20	2N4014	20	2N750	20	2SC183C
20	2N2161	20	2N2926-6	20	2N4063	20	2N751	20	2SC183D
20	2N2194	20	2N2926ORN	20	2N4064	20	2N753	20	2SC183E
20	2N2194A	20	2N2926G	20	2N4074	20	2N754	20	2SC183F
20	2N2194B	20	2N2926GRN	20	2N4104	20	2N756A	20	2SC183G
20	2N2198	20	2N2931	20	2N4123	20	2N757A	20	2SC183GN
20	2N2205	20	2N2932	20	2N4124	20	2N758A	20	2SC183H
20	2N2206	20	2N2933	20	2N4137	20	2N758B	20	2SC183J
20	2N2217	20	2N2934	20	2N4256	20	2N759A	20	2SC183K
20	2N2217/51	20	2N2935	20	2N4259	20	2N759B	20	2SC183L
20	2N2218	20	2N2936	20	2N4264	20	2N760A	20	2SC183M
20	2N2218/51	20	2N2937	20	2N4265	20	2N760B	20	2SC183OR
20	2N2218A	20	2N2938	20	2N4286	20	2N770	20	2SC183P
20	2N2219	20	2N2951	20	2N4287	20	2N771	20	2SC183Q
20	2N2219/51	20	2N2952	20	2N4294	20	2N772	20	2SC183R
20	2N2219A	20	2N2954	20	2N431	20	2N773	20	2SC183S
20	2N2220	20	2N2958	20	2N432	20	2N774	20	2SC183W
20	2N2221	20	2N2959	20	2N433	20	2N775	20	2SC183X
20	2N2221A	20	2N2960	20	2N4400	20	2N776	20	2SC183Y
20	2N2222	20	2N2961	20	2N4401	20	2N777	20	2SC1908
20	2N2222A	20	2N2974	20	2N4420	20	2N778	20	2SC1908E
20	2N2222A	20	2N3009	20	2N4421	20	2N785	20	2SC1908H
20	2N2234	20	2N3011	20	2N4422	20	2N784	20	2SC2012
20	2N2235	20	2N3013	20	2N4424	20	2N784A	20	2SC2021Q
20	2N2236	20	2N3014	20	2N4432	20	2N789	20	2SC2021R
20	2N2237	20	2N3015	20	2N4432A	20	2N790	20	2SC2021S
20	2N2240	20	2N3115	20	2N4436	20	2N791	20	2SC26
20	2N2241	20	2N3116	20	2N4450	20	2N792	20	2SC280AO
20	2N2242	20	2N3122	20	2N4470	20	2N793	20	2SC2818
20	2N2244	20	2N3128	20	2N471A	20	2N834/46	20	2SC328A
20	2N2245	20	2N3129	20	2N472	20	2N834A	20	2SC349
20	2N2246	20	2N3130	20	2N472A	20	2N835	20	2SC349R
20	2N2247	20	2N3210	20	2N473	20	2N835/46	20	2SC369G-BL
20	2N2248	20	2N3241	20	2N474	20	2N839	20	2SC369G-GR
20	2N2249	20	2N3241A	20	2N474A	20	2N840	20	2SC369V
20	2N2250	20	2N3242	20	2N475	20	2N842	20	2SC3710
20	2N2251	20	2N3242A	20	2N475A	20	2N844	20	2SC373
20	2N2252	20	2N3246	20	2N476	20	2N866	20	2SC372Y
20	2N2253			20	2N477	20	2N867		
20	2N2254		??						
20	2N2255	20	2N3261	20	2N479	20	2N915	20	2SC423-Q
20	2N2256	20	2N3268	20	2N479A	20	2N919	20	2SC423A
20	2N2257	20	2N3300	20	2N480	20	2N920	20	2SC423B
20	2N2272	20	2N3301	20	2N480A	20	2N921	20	2SC423C
20	2N2309	20	2N3302	20	2N480B	20	2N922	20	2SC423D
20	2N2310	20	2N3310	20	2N4950	20	2N929	20	2SC423E
20	2N2312	20	2N3326	20	2N4951	20	2N929A	20	2SC423F
20	2N2314	20	2N3332	20	2N4952	20	2N930	20	2SC423GN
20	2N2315	20	2N3332A	20	2N4953	20	2N930A	20	2SC423H
20	2N2318	20	2N3333	20	2N4954	20	2N947	20	2SC423J
20	2N2319	20	2N3333A	20	2N4966	20	2N951	20	2SC423K
20	2N2320	20	2N3334	20	2N4967	20	2N957	20	2SC423L
20	2N2330	20	2N3334A	20	2N4968	20	2S001	20	2SC423M
20	2N2331	20	2N3335	20	2N4969	20	2S002	20	2SC423OR
20	2N2349	20	2N3335A	20	2N4970	20	2S003	20	2SC423R
20	2N2353	20	2N3335B	20	2N4994	20	2S004	20	2SC423X
20	2N2353A	20	2N3336	20	2N4995	20	2S005	20	2SC423Y
20	2N2368	20	2N3336A	20	2N5030	20	2S095A	20	2SC433
20	2N2368A			20	2N5081	20	2S101	20	2SC440

GE	Industry Standard No.	GE	Industry Standard No.	GE	Industry Standard No.	GE	Industry Standard No.	GE	Industry Standard No.
20	2SC45	21	CDC-9002-IC	21	Q5116CA	21	2N5208	25	2SB143
20	2SC455	21	CS-1294P	21	QT-A0719XCN	21	2N5228	25	2SB143P
20	2SC455-OR	21	CS-9013HH	21	RB18	21	2N5352	25	2SB144
20	2SC455A	21	CS1124G	21	RB55	21	2N5763	25	2SB144P
20	2SC455B	21	CS1369D	21	REN106	21	2N5857	25	2SB145
20	2SC455C	21	CS2005	21	RT-115	21	2N626	25	2SB146
20	2SC455D	21	CS2005B	21	RT-126	21	2N723	25	2SB147
20	2SC455E	21	CS9102	21	RT115	21	2N726	25	2SB149
20	2SC455F	21	CS9123C1	21	S2	21	2N727	25	2SB151
20	2SC455G	21	CS9128	21	S3	21	2N864A	25	2SB19
20	2SC455GN	21	CS9129(B)	21	S4	21	2N865	25	2SB20
20	2SC455H	21	CS9129-B1	21	S500	21	2N865A	25	2SB215
20	2SC455J	21	CXL106	21	S501	21	2N869	25	2SB216
20	2SC455K	21	D14	21	SA495	21	2N869A	25	2SB216A
20	2SC455L	21	DS-68	21	SA495A	21	2N978	25	2SB217
20	2SC455M	21	D868	21	SA496	21	2N995	25	2SB217A
20	2SC455OR	21	D886	21	SA496B	21	2N995A	25	2SB217Q
20	2SC455R	21	D896	21	SA537	21	2N996	25	2SB217U
20	2SC455Y	21	EA0088	21	SA538	21	2N5564R	25	2SB228
20	2SC458B	21	EA15X69	21	SA539	21	2SA465	25	2SB229
20	2SC458C	21	EA15X70	21	SA540	21	2SA538-G	25	2SB230
20	2SC462	21	EA15X71	21	SAC40	21	2SA572	25	2SB233
20	2SC468	21	ECG106	21	SAC40A	21	2SA572Y	25	2SB239
20	2SC468A	21	ED1802-0	21	SAC40B	21	2SA594N	25	2SB239A
20	2SC474	21	ED1802N	21	SAC42	21	2SA6111	25	2SB240
20	2SC529	21	ED1802N,M	21	SAC42A	21	2SA6661	25	2SB240A
20	2SC531	21	EP15X60	21	SAC42B	21	2SA6661QRS	25	2SB242
20	002SC531F	21	EP15X90	21	SAC44	21	2SA813	25	2SB242B
20	2SC537	21	EQR-0038	21	SI341P	21	2SC719	25	2SB242C
20	2SC537F	21	EQR-1	21	SI342P	21	2SC719Q	25	2SB242D
20	2SC539	21	EX699-X	21	SI343P	22	BC205VI	25	2SB242E
20	2SC539(L)(K)	21	F105P	21	SI351P	22	BI-82	25	2SB242G
20	2SC539(R)	21	F121-60216	21	SI352P	22	CS-9012HP	25	2SB242GN
20	2SC539K	21	F121-603	21	SI353P	22	CS-901ZHP	25	2SB242H
20	2SC539L	21	FCS-9015C	21	SL200	22	CS2941	25	2SB242J
20	2SC539R	21	FCS9012HH	21	SL201	22	EP25	25	2SB242K
20	2SC539S	21	FCS9015C	21	SM62186	22	EX746-X	25	2SB242L
20	2SC539T	21	FCS9015D	21	SP12271	22	FT5040	25	2SB242M
20	2SC554	21	FI-1019	21	SPS2526	22	GE-22	25	2SB242OR
20	2SC558AC	21	FK2894	21	SPS4401	22	HA7507	25	2SB242R
20	2SC593	21	FM2894	21	ST2-2517	22	HA7536	25	2SB242X
20	2SC620	21	FS-2299	21	T-2357	22	HA7804	25	2SB242Y
20	2SC620(C)	21	FS26382	21	T-HU6OU	22	HA7806	25	2SB246
20	2SC620(D)	21	FT1702	21	T2357	22	HA7808	25	2SB247
20	2SC620-OR	21	FV2894	21	T40	22	HA7810	25	2SB248
20	2SC620A	21	GE-21	21	TGZSA608(C)	22	HT104941B-0	25	2SB248A
20	2SC620B	21	GET2904	21	TI8128	22	HT105611B0	25	2SB249
20	2SC620C	21	GET2905	21	TI8198	22	HT105621B0	25	2SB25
20	2SC620CD	21	GET3905	22	TM-22	22	IRTR20	25	2SB250
20	2SC620D	21	GET5906	21	TM106	22	MA0404	25	2SB250A
20	2SC620DE	21	GMB0404	21	TR-1032-1	22	MA0414	25	2SB255
20	2SC620E	21	GMB0404-1	21	TR01053-1	22	ME502	25	2SB255B
20	2SC620F	21	GMB0404-2	21	TR02051-1	22	MP896800	25	2SB26
20	2SC620G	21	GMB0404-1	21	TR02062-1	22	MP89682(I)	25	2SB26A
20	2SC620GN	21	HA-0054	21	TV-71	22	MPSD55	25	2SB274B
20	2SC620H	21	HA9500	21	TV24848	22	NPS404	25	2SB274C
20	2SC620J	21	HA9501	21	002-0105-00	22	PN72	25	2SB274D
20	2SC620K	21	HA9502	22	2H1254	22	R826	25	2SB274E
20	2SC620L	21	HEPS0012	22	2H1255	22	S1477	25	2SB274F
20	2SC620M	21	HEPS0013	22	2H1256	22	SPS-1539WT	25	2SB274G
20	2SC620OR	21	HEPS0019	22	2H1257	22	SPS1593WT	25	2SB274H
20	2SC620R	21	HF-47	22	2H1258	22	T407185152	25	2SB274J
20	2SC620X	21	HF47	22	2H1259	22	TV-47	25	2SB274K
20	2SC620Y	21	HR-84	21	2N1024	22	2N4006	25	2SB274M
20	2SC626	21	HR84	21	2N1131/51	22	2N4007	25	2SB274OR
20	2SC631	21	HS-40035	21	2N1131A/51	22	2N5229	25	2SB274R
20	2SC631A	21	HS-40055	21	2N1132/51	22	2N5910	25	2SB274V
20	2SC631B	21	HT104941C-0	21	2N1132A/51	22	2SA-NJ-101	25	2SB274X
20	2SC631C	21	HT104941CO	21	2N1132A46	22	2SA550A(Q)	25	2SB274Y
20	2SC631D	21	HT104951A-0	21	2N1238	22	2SA550A(R)	25	2SB28
20	2SC631E	21	HT105621B-0	21	2N1239	22	2SA550A(R,Q,S)	25	2SB282
20	2SC631F	21	I50865	21	2N1240	22	2SA550A(S)	25	2SB283
20	2SC631G	21	IP20-0046	21	2N1241	22	2SA550AB	25	2SB284
20	2SC631GN	21	IRTR-54	21	2N1640	22	2SA550AQ	25	2SB285
20	2SC631H	21	IRTR19	21	2N1641	22	2SA550B	25	2SB295
20	2SC631J	21	IRTR28	21	2N1642	22	2SA550BL	25	2SB30
20	2SC631K	21	IRTR30	22	2N2175	22	2SA550C	25	2SB31
20	2SC631L	21	J24832	22	2N2176	22	2SA550D	25	2SB337
20	2SC631M	21	K071687	22	2N2177	22	2SA550P	25	2SB337A
20	2SC631OR	21	LDA452	22	2N2395	22	2SA550Q	25	2SB337B
20	2SC631R	21	LDS202	22	2N2591	22	2SA550R	25	2SB337BK
20	2SC631X	21	LDS203	22	2N2592	22	2SA550S	25	2SB337C
20	2SC631Y	21	LJ152(O)	22	2N2593	22	2SA550Y	25	2SB337D
20	2SC640B	21	LJ1522	22	2N2800/51	22	2SA560	25	2SB337E
20	2SC654	21	M4910	22	2N2801/51	22	2SA560A	25	2SB337F
20	2SC655	21	M652/PIC	22	2N2802	22	2SA56A	25	2SB337G
20	2SC7108B	21	MA0401	22	2N2803	22	2SA638E,F	25	2SB337GN
20	002SC7110B	21	MA0402	22	2N2804	22	2SA720Q,R,S	25	2SB337H
20	002SC710C	21	MA0404-1	22	2N2805	22	2SC677	25	2SB337HA
20	2SC715-OR	21	MA0404-2	22	2N2806	23	A2SD2260P	25	2SB337HB
20	2SC721	21	MA0411	22	2N2807	23	GE-23	25	2SB337J
20	2SC723	21	MA0413	22	2N2927/51	23	HT308301B0	25	2SB337LB
20	2SC723BL	21	MD1333P	22	2N2968	23	IRTR-57	25	2SB337R
20	2SC725	21	MD3134P	23	2N2969	23	M-75543-1	25	2SB337X
20	2SC725-0	21	ME0401	23	2N2970	23	M4936	25	2SB337Y
20	2SC7335-BL	21	ME0402	23	2N2971	23	MHT5901	25	2SB338
20	2SC735	21	ME0463	23	2N3049	23	MHT5911	25	2SB338H
20	2SC7354	21	ME503	23	2N3050	23	Q5101D	25	2SB338HA
20	002SC7350Y	21	ME511	23	2N3051	23	RCA40250	25	2SB338HB
20	2SC7354	21	ME512	23	2N3209	23	SD141	25	2SB355
20	002SC7350Y	21	MMCM2907	23	2N3224	23	TIP-29	25	2SB356
20	2SC737	21	MMT3905	23	2N3245	23	TV-109	25	2SB373
20	2SC737Y	21	MPS-5640	23	2N3304	23	2N3879	25	2SB391
20,47	TR21	21	MPS-A55	23	2N3342	24MP	HEP703X2	25	2SB407
21	A-1853-0009-1	21	MPS3534	23	2N3576	24MP	IRTR57X2	25	2SB407-0
21	A-1853-0010-1	21	MPS3644	23	2N3838	25	B300	25	2SB407-OR
21	A-1853-0034-1	21	MPS6434	24	2N3915	25	2SB107	25	2SB407A
21	A524402	21	MPS6535M	24	2N4034	25	2SB107A	25	2SB407B
21	A524403	21	MPS6535M	24	2N4055	25	2SB119	25	2SB407BK
21	AT11	21	MPS6580	24	2N4058	25	2SB119A	25	2SB407C
21	BC126A	21	MPS9681J	24	2N4059	25	2SB122	25	2SB407D
21	BC177VI	21	MPS9682T	24	2N4060	25	2SB124	25	2SB407E
21	BC178VI	21	MPSH81	24	2N4061	25	2SB126	25	2SB407F
21	BC204VI	21	MT0463	24	2N4062	25	2SB126A	25	2SB407G
21	BC205C	21	NPS6516	24	2N4080	25	2SB126V	25	2SB407GN
21	BC205L	21	NPS6517	24	2N4122	25	2SB127	25	2SB407H
21	BC212VI	21	N83905	24	2N4142	25	2SB127A	25	2SB407J
21	BC415	21	N83906	24	2N4228	25	2SB128	25	2SB407K
21	BC416	21	OC7400	24	2N4313	25	2SB128A	25	2SB407M
21	BCW29	21	P67	24	2N4389	25	2SB128V	25	2SB407R
21	BCW29R	21	PL4031	24	2N4423	25	2SB129	25	2SB407TV
21	BCW30	21	PL4032	24	2N4451	25	2SB130	25	2SB407TV-2
21	BCW30R	21	PL4033	24	2N4453	25	2SB130A	25	2SB407X
21	BCX78VII	21	PL4034	24	2N4872	25	2SB131	25	2SB407Y
21	BCX79VII	21	PN2904	24	2N4937	25	2SB131A	25	2SB41
21	BF315	21	PN2905	24	2N4938	25	2SB132	25	2SB42
21	BF316	21	PN2906	24	2N4940	25	2SB132A	25	2SB424
21	BF339	21	PN2906A	24	2N4941	25	2SB137	25	2SB426
21	BFX12	21	PN2907	24	2N495	25	2SB138		
21	BFX13	21	PN2907A	24	2N495/18	25	2SB140		
21	BFX48	21	PN70	24	2N496	25	2SB142		
21	BSX29	21	PN71	24	2N5140	25	2SB142B		
21	C302	21	Q510ZQ	24	2N5141	25	2SB142C		
21	C401								

GE	Industry Standard No.	GE	Industry Standard No.	GE	Industry Standard No.	GE	Industry Standard No.	GE	Industry Standard No.
25	2SB426BL	28	BLY91	28	TNJ70450	29	D43C3	30	2SB473H
25	2SB426R	28	BR100B	28	TNJ72775	29	D43C4	31MP	GE-31MP
25	2SB426Y	28	B8V60	28	TR-1037-3	29	D43C5	32	D44R5
25	2SB448	28	CXL186	28	TR-55	29	D43C6	32	D44R6
25	2SB449	28	D2BD8	28	TR55	29	D43C8	32	2N1715
25	2SB449F	28	D42C1	28	TRAPLC1013	29	EC0187	32	2N1717
25	2SB449P	28	D42C2	28	TV-73	29	EP101A	32	2N2243
25	2SB466	28	D42C3	28	TV-75	29	ES15051	32	2N2243A
25	2SB467	28	D42C4	28	TV-82	29	G03703C	32	2N3500
25	2SB471	28	D42C5	28	01-031096	29	GE-29	32	2N4001
25	2SB471-2	28	D42C6	28	1A0048	29	HA-00634	32	2N4238
25	2SB471A	28	D42C7	28	2N1067	29	HA-00636	32	2N4239
25	2SB471B	28	D42C8	28	2N1068	29	HA-505	32	2N5660
25	2SB471D	28	D44C8B	28	2N1710	29	HB-00564	32	2SC1056
25	2SB471E	28	DDBY227004	28	2N3296	29	HP12	32	2SC1156
25	2SB471F	28	EA15X160	28	2N3375	29	HP16	32	2SC1157
25	2SB471G	28	EA15X248	28	2N3418	29	HP51	32	2SC1168
25	2SB471GN	28	EA2488	28	2N3420	29	HR-70	32	2SC1448
25	2SB471H	28	EA3674	28	2N3525A	29	HR104961C	32	2SC1456
25	2SB471J	28	EA4055	28	2N3619	29	IR-TR56	32	2SC2085
25	2SB471K	28	EA4085	28	2N3620	29	IRTR-56	32	2SC305
25	2SB471L	28	ECG186	28	2N3623	29	IRTR56	32	2SC305A
25	2SB471M	28	EP100	28	2N3624	29	MJ8100	32	2SC305B
25	2SB471R	28	EP15X32	28	2N3628	29	MJE-371	32	2SC305C
25	2SB471X	28	EP15R	28	2N3852	29	MPB052	32	2SC305D
25	2SB471Y	28	EP15X34	28	2N3853	29	PTC111	32	2SC305B
25	2SB472	28	EQ8-66	28	2N3919	29	PTC142	32	2SC305F
25	2SB472A	28	EQ8-67	28	2N3925	29	RE43	32	2SC305G
25	2SB472B	28	ES15227	28	2N3926	29	REN187	32	2SC305GN
25	2SB862	28	ET495371	28	2N3961	29	RY2356	32	2SC305H
25	2SB64	28	G06-711-B	28	2N4012	29	SDT3525	32	2SC305J
25	2SB69	28	G06-714C	28	2N4040	29	SDT3505	32	2SC305K
25	2SB80	28	G06-717-B	28	2N4041	29	SDT3506	32	2SC305L
25	2SB81	28	G06714C	28	2N4225	29	SDT3550	32	2SC305M
25	2SB82	28	GE-28	28	2N4226	29	SDT3552	32	2SC305R
25	2SB83	28	G03-007C	28	2N4350	29	SDT3553	32	2SC305X
26	GE-26	28	G06-714C	28	2N4428	29	SDT3775	32	2SC305Y
27	BF899	28	G06-717-B	28	2N4429	29	SDT3776	32	2SC444
27	C764	28	HC-01060	28	2N4430	29	SDT3778	32	2SC470
27	D40N1	28	HC-01096	28	2N4440	29	SDT9004	32	2SC470-3
27	D40N3	28	HC-01098	28	2N4976	29	TA7741	32	2SC470-4
27	RCP111A	28	HC-01226	28	2N5079	29	TNJ72774	32	2SC470-5
27	RCP111B	28	HC-495	28	2N5080	29	TR-1036-3	32	2SC470-6
27	RCP111C	28	HP11	28	2N5422	29	TR-56	32	2SC470A
27	RCP111D	28	HP15	28	2N5423	29	TR56	32	2SC470B
27	RCP113A	28	HR-69	28	2N5481	29	TV-74	32	2SC470C
27	RCP113B	28	HT304961C-0	28	2N5482	29	2N5782	32	2SC470D
27	RCP113C	28	HT304961C-0	28	2N5489	29	2N5783	32	2SC470F
27	RCP113D	28	IRTR-55	28	2N5608	29	2SA634	32	2SC470G
27	RCP115	28	IRTR55	28	2N5610	29	2SA636(4)K	32	2SC470GN
27	RCP115B	28	KGB41061	28	2N5612	29	2SA636(4)L	32	2SC470H
27	RCP117	28	KSC1096-X	28	2N5644	29	2SA636(M)	32	2SC470J
27	RCP117B	28	KSC1096-Y	28	2N5645	29	2SA6361	32	2SC470K
27	2N2437	28	M9134	28	2N5688	29	2SA6361,K	32	2SC470L
27	2N2438	28	MJB-200E	28	2N5697	29	2SA700(B)	32	2SC470M
27	2N2439	28	MM3004	28	2N5698	29	2SA861	32	2SC470R
27	2N2909	28	MM4429	28	2N5699	29	2SB523	32	2SC470X
27	2N3119	28	MM4430	28	2N5703	29	2SB669	32	2SC470Y
27	2N6558	28	MSA8508	28	2N5764	30	AD-162	32	2SC505
27	2SC1089	28	MT1070	28	2N5766	30	GE-30	32	2SC505-0
27	2SC500	28	PA8900	28	2N5767	30	G04-704-A	32	2SC505-R
27	2SC500R	28	PPR1006	28	2N6288	30	IRTR-50	32	2SC506-0
27	2SC500Y	28	PPR1008	28	2N6289	30	PR3-3	32	2SC506-R
27	2SC58	28	PT2620	28	2N6290	30	RB20	32	2SC507
27	2SC58A	28	PT2640	28	2N6291	30	SK3052	32	2SC507-0
27	2SC58B	28	PT2660	28	2N6293	30	2N5887	32	2SC507-R
27	2SC58D	28	PT3503	28	2SC10148	30	2N5888	32	2SC507-Y
27	2SC58E	28	PT4690	28	2SC1096	30	2N5889	32	2SC560
27	2SC58F	28	PT600	28	2SC1096(M)	30	2N5901	32	2SC589
27	2SC58G	28	PT601	28	2SC1096-32M	30	2SB256	32	2SC64
27	2SC58GN	28	PT6618	28	2SC1096-42L	30	2SB367	32	2C66
27	2SC58H	28	PT6669	28	2SC1096-OR	30	2SB367(A)	32	2SC686
27	2SC58J	28	PT6696	28	2SC1096A	30	2SB367(B)P	32	2SC686A
27	2SC58K	28	PTC110	28	2SC1096B	30	2SB367-4	32	2SC686B
27	2SC58L	28	PTC143	28	2SC1096C	30	2SB367-5	32	2SC686C
27	2SC58M	28	QT-D0325XAC	28	2SC1096D	30	2SB367-OR	32	2SC686D
27	2SC58OR	28	R842	28	2SC1096E	30	2SB367A	32	2SC686E
27	2SC58R	28	REN186	28	2SC1096F	30	2SB367AL	32	2SC686F
27	2SC58X	28	RT-152	28	2SC1096G	30	2SB367B	32	2SC686G
27	2SC58Y	28	RT152	28	2SC1096GN	30	2SB367B-2	32	2SC686GN
28	A201	28	S715	28	2SC1096H	30	2SB367BL	32	2SC686H
28	A203(NPN)	28	S801	28	2SC1096J	30	2SB367BP	32	2SC686J
28	A208	28	SD1023	28	2SC1096K	30	2SB367C	32	2SC686K
28	A208(NPN)	28	SDT3326	28	2SC1096L	30	2SB367D	32	2SC686L
28	A210	28	SDT3421	28	2SC1096LM	30	2SB367F	32	2SC686M
28	A211	28	SDT3425	28	2SC1096M	30	2SB367G	32	2SC686R
28	A253	28	SDT3426	28	2SC1096N	30	2SB367H	32	2SC686X
28	A271	28	SDT4301	28	2SC1096Q	30	2SB367J	32	2SC686Y
28	A275	28	SDT4302	28	2SC1096R	30	2SB367K	32	2SC70
28	A514-047830	28	SDT4304	28	2SC1096W	30	2SB367L	32	2SC728
28	AR24(PHILCO)	28	SDT4305	28	2SC1096X	30	2SB367M	32	2SC728A
28	B143000	28	SDT4307	28	2SC1096Y	30	2SB367OR	32	2SC728B
28	B143001	28	SDT4308	28	2SC1098	30	2SB367P	32	2SC728C
28	B143003	28	SDT4310	28	2SC1098(4)K	30	2SB367R	32	2SC728D
28	B143009	28	SDT4311	28	2SC1098(4)L	30	2SB367X	32	2SC728E
28	B143010	28	SDT4455	28	2SC1098(L)	30	2SB367Y	32	2SC728F
28	B143015	28	SDT4483	28	2SC1098(M)	30	2SB368	32	2SC728G
28	B143016	28	SDT4551	28	2SC1098A	30	2SB368-OR	32	2SC728GN
28	B143024	28	SDT4553	28	2SC1098B	30	2SB368A	32	2SC728H
28	B143025	28	SDT4583	28	2SC1098C	30	2SB368B	32	2SC728J
28	B3531	28	SDT4611	28	2SC1098D	30	2SB368C	32	2SC728L
28	B3533	28	SDT4612	28	2SC1098L	30	2SB368D	32	2SC728M
28	B3537	28	SDT4614	28	2SC1098M	30	2SB368E	32	2SC728OR
28	B3538	28	SDT4615	28	2SC11070	30	2SB368F	32	2SC728R
28	B3540	28	SDT5001	28	2SC1848	30	2SB368G	32	2SC728X
28	B3541	28	SDT5006	28	2SC1848Q	30	2SB368GN	32	2SC728Y
28	B3542	28	SDT5011	28	2SC1848R	30	2SB368H	35	DTS401
28	B3570	28	SDT5501	28	2SC1848V	30	2SB368J	35	DTS402
28	B3606	28	SDT5506	28	2SC297	30	2SB368K	35	DTS410
28	B3607	28	SDT5511	28	2SC541	30	2SB368L	35	DTS411
28	B3608	28	SDT5901	28	2SC542	30	2SB368M	35	DTS413
28	B3609	28	SDT5902	28	2SC547	30	2SB368X	35	DTS423
28	B3610	28	SDT5906	28	2SC548	30	2SB368Y	35	DTS430
28	B3611	28	SDT6101	28	2SC549	30	2SB458	35	DTS431
28	B3612	28	SDT6102	28	2SC550	30	2SB458A	35	MJ1600
28	B3613	28	SDT6104	28	2SC572	30	2SB458BC	35	MJ3010
28	B3614	28	SDT6105	28	2SC592	30	2SB458BL	35	MJ3026
28	B3747	28	SDT6106	28	2SC598	30	2SB458C	35	MJ3027
28	B3748	28	SDT7401	28	2SC635	30	2SB462	35	MJ3029
28	B3750	28	SDT7411	28	2SC635A	30	2SB463-0	35	MJ9000
28	B5001	28	SDT7414	28	2SC637	30	2SB463BLU	35	2N5264
28	BD107	28	SDT9001	28	2SC691	30	2SB463BLU-Q	35	2N5633
28	BF950	28	SDT9002	28	2SC692	30	2SB463O-BL	35	2N5758
28	BF951	28	SDT9003	29	AR-29	30	2SB463O-R	35	2N5759
28	BFX55	28	SDT9005	29	AR25	30	2SB463O-Y	35	2N5760
28	BFX96A	28	SDT9006	29	AR26	30	2SB463RED	35	2N6211
28	BLY20	28	SDT9007	29	C636	30	2SB463RED-Q	35	2N6212
28	BLY3	28	SDT9008	29	CXL187	30	2SB463YEL	35	2N6213
28	BLY37	28	SJE-5038	29	D27D1	30	2SB463YEL-Q		
28	BLY38	28	SJE-5402	29	D27D2	30	2SB473		
28	BLY53	28	STC1800	29	D27D3	30	2SB473(H)		
28	BLY62	28	STC1862	29	D27D4	30	2SB473D		
28	BLY78	28	STD9007	29	D43C1	30	2SB473F		
		28	STT4451	29	D43C2				
		28	TIS82						

GE	Industry Standard No.	GE	Industry Standard No.	GE	Industry Standard No.	GE	Industry Standard No.	GE	Industry Standard No.
35	28C1050	39	AR213V	39	C674C	39	GM308	39	SF294B
35	28C1195	39	AR218	39	C674CV	39	GMO380	39	SF295
35	28C1454	39	AR218(ORANGE)	39	C674D	39	HEP709	39	SF295C
35	28C241	39	AR218(RED)	39	C674B	39	HEP719	39	SF295D
35	28C243	39	AR218RO	39	C674F	39	H840021	39	SF310
35	28C245	39	AR219	39	C674G	39	H840022	39	SF314
35	28C246	39	AR220	39	C682	39	H840023	39	SF88T231
35	28C270	39	AR222	39	C682A	39	H840024	39	SK3018
35	28C407	39	AT340	39	C682B	39	H840025	39	SK3117
35	28C42	39	AT341	39	C683	39	H78001610	39	SK3132
35	28C42A	39	AT342	39	C683A	39	H78001710	39	SK8937
35	28C43	39	AT343	39	C683B	39	IRTR-71	39	SKA5541
35	28C44	39	AT344	39	C683V	39	IRTR71	39	SKA4768
35	28C492	39	AT345	39	C717	39	IRTR95	39	SM-4304-8
35	28C519	39	AT346	39	C717B	39	IT918	39	SP4436
35	28C519A	39	B1H	39	C717BK	39	IT929	39	SP84343
35	28C586	39	BP155	39	C717BLK	39	IT930	39	SP8906
35	28C687	39	BP161	39	C717C	39	K2857C	39	SP8919
36	A3H	39	BP166	39	C717R	39	K2857P	39	SS4042
36	C681	39	BP167	39	C761	39	K3683C	39	ST5641
36	C681A	39	BP168	39	C761Y	39	K3683P	39	T-Q5071
36	DT8424	39	BP169	39	C761Z	39	K3880C	39	T-Q5086
36	DT8425	39	BP173	39	C762	39	M-128J509-1	39	T1202(GE)
36	MJ3011	39	BP175	39	C786R	39	M-128J511-3	39	T1X-M14
36	MJ3028	39	BP180	39	C787	39	M401	39	T1X-M15
36	MJ3030	39	BP181	39	C860	39	M4756	39	T1X-M16
36	MJ413	39	BP182	39	C860C	39	M546	39	T381
36	MJ423	39	BP183	39	C860D	39	M612	39	T381(SEARS)
36	MJ431	39	BP184	39	C860E	39	M613	39	T576-1
36	RCA1B04	39	BP185	39	C863	39	M614	39	TA2554
36	RCA1B05	39	BP199	39	C864	39	M75545-1	39	TA2555
36	2N3902	39	BP200	39	C918	39	M9266	39	TB80129
36	2N5239	39	BP206	39	C9270J	39	M9450	39	TG28C2057-C
36	2N5240	39	BP207	39	C928	39	M9481	39	TG28C2057-D
36	2N5241	39	BP208	39	C928B	39	MPSH34	39	TG28C2057C
36	2N5466	39	BP209	39	C928C	39	MRD150	39	TG28C927
36	2N5804	39	BP212	39	C928D	39	MRD450	39	TG280927(C)
36	2N5805	39	BP213	39	C928E	39	MRF502	39	TH-H28C313
36	2N5838	39	BP214	39	C947	39	MT3001	39	TI-407
36	2N5839	39	BP215	39	C948	39	MT3002	39	TI-408
36	2N5840	39	BP222	39	C957	39	MT4104	39	TI-409
36	2N6306	39	BP226	39	C997	39	NKT16229	39	TI-492
36	2N6307	39	BP227	39	CC86225F	39	NKT35219	39	TI-493
36	2N6308	39	BP232	39	CC86226G	39	NFC167	39	TI-494
36	2N6510	39	BP251	39	CC86227F	39	NR461AA	39	TI818
36	2N6511	39	BP252	39	CC89017	39	N8406	39	TIX-M14
36	2N6512	39	BP254B	39	CC89017G925	39	P/N14-603-02	39	TIX-M15
36	2N6513	39	BP255C	39	CC89018H924	39	PL1021	39	TIX-M16
36	2N6514	39	BP255D	39	CS-9014	39	PL1022	39	TN5200
36	28C1099	39	BP260	39	C81014G	39	PL1023	39	TNJ1173
36	28C1154	39	BP261	39	C81014H	39	PL1066	39	TNJ71964
36	28C1348	39	BP270	39	C81120C	39	PL1067	39	TNJ72150
36	28C1477	39	BP271	39	C81120D	39	PL1111	39	TNJ72151
36	28C1617	39	BP273	39	C81120F	39	PL1112	39	TNJ72275
36	28C1829	39	BP273C	39	C81120H	39	PTC126	39	TNJ72368
36	28C408	39	BP273D	39	C81284B	39	PTC132	39	TNJ72701
36	28C409	39	BP274	39	C81284F	39	PTG132	39	TR-171
36	28C410	39	BP274B	39	C81284G	39	Q2	39	TR-95(XSTR)
36	28C411	39	BP274C	39	C81284H	39	Q3	39	TR01042
36	28C412	39	BP279	39	C81330	39	Q35259	39	TR01042
36	28C675	39	BP287	39	C81460E	39	Q5	39	TV-16
36	28C681	39	BP288	39	C81460H	39	Q8E5020	39	TV-24599
37	DT8701	39	BP290	39	C81461J	39	RCA40245	39	TV-36
37	DT8801	39	BP302	39	C81461X	39	RCA40246	39	TV-37
37	28C1004	39	BP304	39	C81462F	39	RE2001	39	TV-38
37	28C1004A	39	BP344	39	C81589E	39	RE2002	39	TV-39
37	28C1101	39	BP345	39	C81589F	39	RE28	39	TV-54
37	28C1151	39	BP362	39	C81589B	39	RE5001	39	TV15
37	28C1184	39	BP363	39	C81594E	39	RE5002	39	TV15A
37	28C1367	39	BP516	39	C81596E	39	REN161	39	TV15B
37	28C642	39	BP811	39	C82715	39	RP200	39	TV15C
38	BU105	39	BP818CA	39	C82716	39	R8-2011	39	TV20
38	BU108	39	BP827B	39	C83662	39	RT-113	39	TV24160
38	BU208	39	BP827F	39	C83663	39	S1286	39	TV24161
38	D56W1	39	BP827G	39	C83707	39	S130-138	39	TV24209
38	D56W2	39	BFV27	39	C83708	39	S130-251	39	TV24399
38	DT8702	39	BFV28	39	C83709	39	S15650	39	TV24436
38	DT8802	39	BFV59	39	C83710	39	S15657	39	TV24437
38	DT8804	39	BFV60	39	C83711	39	S15658	39	TV24571
38	MJ105	39	BFV61	39	C89016	39	S15659	39	TV33
38	MJ8400	39	BFV80	39	C89017G	39	S2002	39	TV35
38	2N5467	39	BFW41	39	C89017H	39	S25261	39	TV50
38	28C1005	39	BFW63	39	C3929	39	S27604	39	TV54
38	28C1046	39	BFX31	39	C3930	39	S35232	39	TV8-28C313
38	28C1086	39	BFX47	39	CXL161	39	S4002	39	TV8-28C466
38	28C1100	39	BFX60	39	D26B-4	39	S40204	39	TV8-28C562
38	28C1153	39	BFX62	39	D26G-1	39	S5020	39	TV8-28C563
38	28C1167	39	BFX73	39	D4028	39	S5021	39	TV8-28C563A
38	28C1170	39	BFX77	39	D4029	39	SAB1044	39	TV8-28C683
38	28C1170A	39	BFX89	39	D4030	39	SAB3469	39	TV8-28C683V
38	28C1171	39	BFY69B	39	D4031	39	SDS240	39	TV8-28C684
38	28C1172	39	BFY79	39	D4D21	39	SDS420	39	TV8-28C762
38	28C1174	39	BFY87	39	D4D22	39	SE-5023	39	TV8-28C948
38	28C1185	39	BFZ29	39	D8-78	39	SE-5024	39	TV828C562
38	28C1295	39	C1035	39	D8-85	39	SE-5025	39	TV828C563
38	28C1309	39	C1035C	39	D878	39	SE1002	39	01-032057
38	28C1325	39	C1035D	39	D881	39	SE1002-1	39	01-571794
38	28C1358	39	C1035E	39	D885	39	SE1002-2	39	002-011400
38	28C1413	39	C1036	39	EC0161	39	SE1010	39	002-011500
38	28C1875	39	C1044	39	EC0313	39	SE2001	39	2AG
38	28C1893	39	C1070	39	ED1502B	39	SE2002	39	2AH
38	28C1922	39	C1128	39	EP200	39	SE3001	39	2N2865
39	A1U	39	C1128(S)	39	ES15X104	39	SE3005	39	2N3287
39	A2532	39	C1129	39	ES15X105	39	SE3019	39	2N3288
39	A244	39	C1129(R)	39	ES15X119	39	SE4020	39	2N3289
39	A244(AMC)	39	C1182B	39	ES15X122	39	SE4020/6-04	39	2N3290
39	A2464	39	C1182C	39	ES15X123	39	SE4021	39	2N3291
39	A2465	39	C1182D	39	ES15X56	39	SE4022	39	2N3292
39	A2479	39	C250	39	ES15X65	39	SE5-0249	39	2N3293
39	A248	39	C251	39	ES15X79	39	SE5-0250	39	2N3294
39	A2480	39	C251A	39	ES15X87	39	SE5001	39	2N3337
39	A2746	39	C252	39	ES15X88	39	SE5002	39	2N3338
39	A32918	39	C253	39	ES15X96	39	SE5003	39	2N3339
39	A417-19	39	C313(JAPAN)	39	ES54(ELCOM)	39	SE5004	39	2N338
39	A429-0081-12	39	C313C	39	ES73(ELCOM)	39	SE5021	39	2N4434
39	A430	39	C313H	39	ES86(ELCOM)	39	SE5022	39	2N4435
39	A451	39	C348	39	EW212	39	SE5023	39	2N5130
39	A465-181-19	39	C3657B	39	F3535	39	SE5024	39	2N5133
39	A4664	39	C389	39	F35574	39	SE5025	39	2N5181
39	A481	39	C389-0	39	F501	39	SE5032	39	2N5182
39	A482	39	C389R	39	F501(ZENITH)	39	SE5050	39	2N5660
39	A483	39	C398(PA-1)	39	F501-16	39	SE5051	39	2N5651
39	A484	39	C399	39	F50116	39	SE5052	39	2N5652
39	A484(ADMIRAL)	39	C464	39	F502	39	SE6002	39	2N917
39	A486	39	C465	39	F502(ZENITH)	39	SF167	39	2N917A
39	A489	39	C466	39	F523	39	SF173	39	2N917B
39	A492	39	C466H	39	FS1682	39	SF194	39	28B683
39	A1R	39	C562	39	FT118	39	SF194B	39	28C1035C
39	AR111	39	C562Y	39	FT45	39	SF195	39	28C1044
39	AR209	39	C563	39	GE-39	39	SF195C	39	28C1070
39	AR210	39	C563A	39	GI3793	39	SF195D	39	28C1180
39	AR211	39	C567			39	SF196	39	28C1182
39	AR213	39	C663			39	SF197	39	28C1182B
39	AR213(VIOLET)	39	C674B			39	SF294	39	28C1182C

GE	Industry Standard No.	GE	Industry Standard No.	GE	Industry Standard No.	GE	Industry Standard No.	GE	Industry Standard No.
39	2SC1182D	40	BP292C	40	M4853	40	2N3701	44	BCM1002-6
39	2SC1254	40	BF294	40	M4927	40	2N3712	44	C463(Y)
39	2SC1321	40	BF305	40	M7002	40	2N3742	44	CXL1321
39	2SC1622	40	BF336	40	M819	40	2N3923	44	CXL131MP
39	2SC2009	40	BF337	40	MB1110	40	2N4068	44	E24107
39	2SC2057	40	BF338	40	MB1120	40	2N4269	44	EA15X139
39	2SC2057-C	40	BF355	40	MIS14150/37	40	2N4270	44	EA15X154
39	2SC2057C	40	BFW37	40	MJ420	40	2N4410	44	ECG131
39	2SC2057D	40	BFW45	40	MJ421	40	2N4924	44	ES10110
39	2SC2057B	40	BFX34	40	MM2260	40	2N4925	44	ES29(ELCOM)
39	2SC2057F	40	BFX98	40	MM3000	40	2N4926	44	ES50(ELCOM)
39	2SC527	40	BFY43	40	MM3002	40	2N4927	44	ES7(ELCOM)
39	2SC546	40	BFY45	40	MM3005	40	2N5058	44	ETTB-367B
39	2SC581	40	BFY57	40	MM3009	40	2N5059	44	G04-704-A
39	2SC653	40	BFY65	40	MM3100	40	2N5174	44	G6016
39	2SC663	40	BFY80	40	MM3101	40	2N5175	44	GC4094
39	2SC667	40	BN1253	40	MM7087	40	2N5176	44	GE-44
39	2SC674	40	BSW32	40	MM7088	40	2N5184	44	HB367
39	2SC674(D)	40	BSW69	40	MT1893	40	2N738	44	HEP642
39	2SC674(P)	40	BSX21	40	MT698	40	2N739	44	HEP643
39	2SC674(G)	40	C1012	40	MT699	40	2N739A	44	HEP6016
39	2SC674-B	40	C1012A	40	MT870	40	2N740	44	HT20436A
39	2SC674-P	40	C1048	40	MT871	40	2N740A	44	HT2046710
39	2SC674B	40	C1048B	40	MT910	40	2S743	44	HT204736
39	2SC674C	40	C1048C	40	MT911	40	2S743A	44	IP20-0052
39	2SC674CK	40	C1048D	40	MT912	40	2S746	44	IR-R80
39	2SC674CL	40	C1048E	40	N1X	40	2S746A	44	NKT451
39	2SC674CV	40	C1048F	40	N2XA	40	2SC1012	44	NKT452
39	2SC674CZ	40	C1056	40	NS20-42	40	2SC1012A	44	NKT452-S1
39	2SC674D	40	C1103	40	NS48004	40	2SC1012C	44	NKT453
39	2SC674E	40	C1103(A)	40	NS6212	40	2SC1012D	44	0C30
39	2SC674P	40	C1103A	40	PET1075A	40	2SC1012E	44	0C30A
39	2SC674G	40	C1103B	40	PRT101	40	2SC1012P	44	0C30B
39	2SC674V	40	C1103C	40	PTC117	40	2SC1012G	44	P3E
40	A11172	40	C1103L	40	R2474-2	40	2SC1012GN	44	P3R
40	A116081	40	C116	40	R823	40	2SC1012H	44	P3R-1
40	A121-361	40	C1116-0	40	REN154	40	2SC1012J	44	P3R-2
40	A12546	40	C273	40	RS-2008	40	2SC1012K	44	P3R-3
40	A126705	40	C407	40	S1366	40	2SC1012L	44	P3R-4
40	A13712	40	C470	40	S1407	40	2SC1012M	44	P3T
40	A13712	40	C472Y	40	S1763	40	2SC1012OR	44	P3T-1
40	A130-0RN	40	C500	40	S17862	40	2SC1012R	44	P3T-2
40	A130-Y10	40	C500R	40	S2986	40	2SC1012X	44	PO3010
40	A1409	40	C500Y	40	S3002	40	2SC1012Y	44	PT32
40	A1A	40	C505	40	S3033	40	2SC1048	44	PTC120
40	A1M	40	C505-0	40	S3034	40	2SC1048B	44	Q-01-064R
40	A1B	40	C505-R	40	S3035	40	2SC1048C	44	RO092
40	A2057B104-8	40	C506	40	S39560	40	2SC1048D	44	RB20MP
40	A2090	40	C506-0	40	S40205	40	2SC1048DC	44	REN131
40	A247	40	C506-R	40	SC843(T.I.)	40	2SC1048E	44	REN131MP
40	A247(AMC)	40	C507	40	SE7001	40	2SC1048F	44	RT4762
40	A25114130	40	C507-0	40	SE7002	40	2SC1048N	44	SC42TA
40	A25762-010	40	C507-R	40	SE7010	40	2SC1090	44	SP2551
40	A25762-012	40	C507-Y	40	SE7016	40	2SC1090D	44	TI-1A6
40	A2620	40	C526	40	SE7017	40	2SC1103	44	TI-7A
40	A2A	40	C058	40	SE7050	40	2SC1103(A)	44	TNJ70483
40	A2K	40	C589	40	SE7055	40	2SC1103A	44	TNJ70541
40	A2Z	40	C058A	40	SE7056	40	2SC1103B	44	TR-184(OLSON)
40	A310	40	C627	40	SE8001	40	2SC1103C	44	TR50
40	A3170717	40	C64(JAPAN)	40	SE8002	40	2SC1103L	44	1R9.1
40	A35	40	C65	40	SE8010	40	2SC1116	44	2N2835
40	A35(JAPAN)	40	C65B	40	SPT186	40	2SC1116-0	44	2N4078
40	A3M	40	C65N	40	SPT187	40	2SC1546	44	2N5893
40	A3MA	40	C65Y	40	SK3044	40	2SC273	44	2N5897
40	A417-115	40	C65YA	40	SK8261	40	2SC627	44	28B458B
40	A46-867-3	40	C65YB	40	SM6727	40	2SC65	44	28B463B
40	A4648	40	C65YTV	40	SP8400	40	2SC65-0	44	28B463Y
40	A48-134819	40	C66	40	SP8401	40	2SC65-OR	44	28B481-OR
40	A48-134843	40	C686	40	SS1912	40	2SC65A	44	28B481A
40	A48-134853	40	C70	40	S9524	40	2SC65B	44	28B481B
40	A48-134898	40	C727	40	S86111	40	2SC65C	44	28B481C
40	A48-134919	40	C728	40	STX0028	40	2SC65D	44	28B481D
40	A48-134927	40	C743A	40	SX60M	40	2SC65E	44	28B481E
40	A48-137002	40	C746A	40	T-481	40	2SC65F	44	28B481F
40	A48-137035	40	C788	40	T-Q5082	40	2SC65G	44	28B481G
40	A481	40	C805	40	T481(SEARS)	40	2SC65GN	44	28B481GN
40	A483B	40	C818	40	TA7292	40	2SC65H	44	28B481H
40	A4843	40	C856	40	TA7293	40	2SC65K	44	28B481J
40	A4853	40	C856-02	40	TBH0143	40	2SC65L	44	28B481K
40	A4H	40	C856C	40	TG28C65	40	2SC65M	44	28B481L
40	A573501	40	C857	40	TG28C65Y	40	2SC65N	44	28B481M
40	A57C012-1	40	C857H	40	TI-722	40	2SC65OR	44	28B481N
40	A57C012-2	40	C868	40	TIS100	40	2SC65R	44	28B481OR
40	A57D1-122	40	C869	40	TIS101	40	2SC65X	44	28B481R
40	A57M2-16	40	C88	40	TIS102	40	2SC65Y	44	28B481X
40	A57M2-17	40	C88A	40	TIS103	40	2SC65Y(B)	44(2)	J24366
40	A610075-1	40	C926	40	TNJ60072	40	2SC65YA	44(2)	N-EA15X139
40	A6181-1	40	C926A	40	TNJ72282	40	2SC65YB	44(2)	28B481
40	A63-18426	40	C95	40	TQ5063	40	2SC65YTV	44(TWO)	ECG131MP
40	A7253	40	C995	40	TR301	40	2SC65YTV1	45	2SC164
40	A86-213-2	40	CDC744	40	TR8100	43	PXB-103	45	2SC164B
40	A86-214-2	40	CDQ10013	40	TR8101	43	PXB-113	45	2SC164C
40	A86-215-2	40	CDQ10015	40	TR8120	44	A061-116	45	2SC164D
40	A86-316-2	40	CDQ10034	40	TR8140	44	A065-111	45	2SC164E
40	A8V	40	CDQ10037	40	TR8160	44	AD152	45	2SC164G
40	A8VA	40	CDQ10044	40	TR8180	44	AD155	45	2SC164GN
40	AM9	40	CDQ10045	40	TR8200	44	AD156	45	2SC164H
40	AT350	40	CDQ10046	40	TR8225	44	AD162	45	2SC164J
40	AT351	40	CDQ10047	40	TR8250	44	AD164	45	2SC164K
40	B5D	40	CDQ10049	40	TR8275	44	AD169	45	2SC164L
40	BC100	40	C81347	40	TR8301	44	AD262	45	2SC164OR
40	BC117	40	C76776	40	TR83011	44	AD263	45	2SC164R
40	BC145	40	CXL154	40	TR83012	44	B130	45	2SC164X
40	BD115	40	DT1003	40	TS2776	44	B130A	45	2SC164Y
40	BF108	40	DT1602	40	TV-19	44	B367	45	2SC315
40	BF109	40	DT1603	40	TV-49	44	B367(A)	45	2SC38
40	BF110	40	DT1612	40	TV-70	44	B367A	45	2SC731
40	BF111	40	DT1613	40	TV24164	44	B367B	45	2SC731R
40	BF114	40	ECG154	40	TV24435	44	B367C	45,243	TR25
40	BF117	40	EP15X107	40	TV24499	44	B367H	46	C502
40	BF118	40	EP15X18	40	TV70	44	B368	46	C614
40	BF119	40	ES15X107	40	TVS-2SC526	44	B368A	46	C615
40	BF140	40	ES15X59	40	TVS-2SC58	44	B368B	46	MM8004
40	BF140A	40	ES15X89	40	TVS-2SC58A	44	B368H	46	Q-00969C
40	BF140B	40	ES32(ELCOM)	40	001-021100	44	B448	46	SK3512
40	BF140B	40	ES15X34	40	01-571941	44	B458	46	TIS109
40	BF155R	40	EX743-X		01-572831	44	B458A	46	TIS110
40	BF155S	40	FM1893	40	2N1052	44	B462	46	TIS111
40	BF156	40	FT34C	40	2N1053	44	B463BL	46	TR63
40	BF157	40	FT34D	40	2N1054	44	B463B	46	TR64
40	BF157B	40	FT3641	40	2N1207	44	B463Y	46	TR65
40	BF174	40	GS-40	40	2N1493	44	B466	46	2N1491
40	BF177	40	H932	40	2N1572	44	B467	46	2N5322
40	BF178	40	HEP706	40	2N1573	44	B473	46	2N5322
40	BF179	40	HEP712	40	2N1574	44	B473D	46	2SC106A
40	BF179A	40	HEP713	40	2N2509	44	B473F	46	2SC106B
40	BF179B	40	HEP85024	40	2N2510	44	B473H	46	2SC106C
40	BF179C	40	HT8000101	40	2N2511	44	B481	46	2SC106G
40	BF186	40	IRTR78	40	2N2618	44	B481D	46	2SC106GN
40	BF257	40	M4648	40	2N3114	44	B481E	46	2SC106H
40	BF258	40	M4819	40	2N3388	44	B63	46	2SC106J
40	BF259	40	M4838	40	2N3389			46	2SC106K
40	BF292A	40	M4839	40	2N3526				
40	BF292B	40	M4843	40	2N3700				

GE	Industry Standard No.	GE	Industry Standard No.	GE	Industry Standard No.	GE	Industry Standard No.	GE	Industry Standard No.
46	2SC106L	47	HT313831X	47	2SC158L	47	2SC369BL	47	2SC498-Y
46	2SC106M	47	HT313832C	47	2SC158M	47	2SC369C	47	2SC498-YEL
46	2SC106OR	47	HT313841R	47	2SC158OR	47	2SC369D	47	2SC498A
46	2SC106R	47	IRTR21	47	2SC158R	47	2SC369E	47	2SC498B
46	2SC106X	47	IRT33	47	2SC158X	47	2SC369F	47	2SC498C
46	2SC106Y	47	KGE41414	47	2SC159Y	47	2SC369G	47	2SC498D
46	2SC111	47	KT600	47	2SC1741	47	2SC369-V	47	2SC498E
46	2SC111A	47	KT600F	47	2SC1788	47	2SC369Q/BL	47	2SC498F
46	2SC111B	47	KT600T	47	2SC1788R	47	2SC369Q/GR	47	2SC498G
46	2SC111C	47	MA8007	47	2SC179	47	2SC369GN	47	2SC498GN
46	2SC111D	47	MPS9616A	47	2SC179A	47	2SC369GR	47	2SC498H
46	2SC111E	47	PN2221	47	2SC179B	47	2SC369H	47	2SC498J
46	2SC111F	47	PN2222	47	2SC179C	47	2SC369J	47	2SC498K
46	2SC111G	47	REN293	47	2SC179D	47	2SC369K	47	2SC498L
46	2SC111GN	47	T9418	47	2SC179E	47	2SC369L	47	2SC498M
46	2SC111H	47	TI8133	47	2SC179F	47	2SC369M	47	2SC498OR
46	2SC111J	47	TI8134	47	2SC179G	47	2SC369OR	47	2SC498R
46	2SC111K	47	TM1613	47	2SC179GN	47	2SC369R	47	2SC498RED
46	2SC111L	47	TM1711	47	2SC179H	47	2SC369X	47	2SC498X
46	2SC111M	47	TR-33	47	2SC179J	47	2SC369Y	47	2SC498Y
46	2SC111OR	47	TRO1062-7	47	2SC179K	47	2SC4012A	47	2SC498YEL
46	2SC111R	47	2N1154	47	2SC179L	47	2SC4012B	47	2SC50
46	2SC111X	47	2N1154/951	47	2SC179M	47	2SC4012C	47	2SC50A
46	2SC111Y	47	2N1155/952	47	2SC179OR	47	2SC4012D	47	2SC50B
46	2SC1239	47	2N1409A	47	2SC179R	47	2SC4012E	47	2SC50C
46	2SC456B	47	2N1410A	47	2SC179Y	47	2SC4012F	47	2SC50E
46	2SC456C	47	2N1958A	47	2SC181	47	2SC4012G	47	2SC50F
46	2SC456E	47	2N1958A/51	47	2SC181A	47	2SC4012GN	47	2SC50G
46	2SC456F	47	2N1959A	47	2SC181B	47	2SC4012H	47	2SC50GN
46	2SC456G	47	2N1959A/51	47	2SC181C	47	2SC4012J	47	2SC50H
46	2SC456H	47	2N2224	47	2SC181D	47	2SC4012L	47	2SC50J
46	2SC456J	47	2N2350	47	2SC181E	47	2SC4012M	47	2SC50K
46	2SC456K	47	2N2350A	47	2SC181F	47	2SC4012X	47	2SC50L
46	2SC456L	47	2N2352	47	2SC181G	47	2SC4012Y	47	2SC50M
46	2SC456M	47	2N2352A	47	2SC181GN	47	2SC402	47	2SC50OR
46	2SC456OR	47	2N2410	47	2SC181H	47	2SC402A	47	2SC50R
46	2SC456R	47	2N2410/51	47	2SC181J	47	2SC402B	47	2SC50X
46	2SC456X	47	2N243	47	2SC181L	47	2SC402C	47	2SC50Y
46	2SC456Y	47	2N244	47	2SC181M	47	2SC402D	47	2SC537O
46	2SC571	47	2N2557	47	2SC181OR	47	2SC402E	47	2SC537OA
47	A5T3192	47	2N2863	47	2SC181R	47	2SC402F	47	2SC537OB
47	A5T3391	47	2N2864	47	2SC181X	47	2SC402G	47	2SC537OC
47	A5T3391A	47	2N2886	47	2SC181Y	47	2SC402GN	47	2SC537OD
47	A5T3210	47	2N2900	47	2SC182	47	2SC402H	47	2SC537OE
47	A5T3222	47	2N3299	47	2SC182(Q)	47	2SC402J	47	2SC537OF
47	A7T3391	47	2N3405	47	2SC182(V)	47	2SC402L	47	2SC537OG
47	A7T3391A	47	2N3426	47	2SC182A	47	2SC402M	47	2SC537OGN
47	A8T3391	47	2N3512	47	2SC182B	47	2SC402R	47	2SC537OH
47	A8T3391A	47	2N3724	47	2SC182C	47	2SC402Y	47	2SC537OJ
47	A8T3704	47	2N3891	47	2SC182D	47	2SC441	47	2SC537OK
47	A8T3705	47	2N4046	47	2SC182E	47	2SC441A	47	2SC537OL
47	A8T3706	47	2N4383	47	2SC182F	47	2SC441B	47	2SC537OM
47	BC337	47	2N4384	47	2SC182G	47	2SC441C	47	2SC537OOR
47	BC337-16	47	2N4385	47	2SC182GN	47	2SC441D	47	2SC537OX
47	BC340-06	47	2N4386	47	2SC182H	47	2SC441E	47	2SC537OY
47	BCW65EC	47	2N5106	47	2SC182J	47	2SC441F	47	2SC559
47	BCW66EF	47	2N5812	47	2SC182K	47	2SC441G	47	2SC56
47	BCW66EQ	47	2N6002	47	2SC182L	47	2SC441GN	47	2SC56A
47	BCW66EH	47	2N6006	47	2SC182M	47	2SC441H	47	2SC56B
47	BCW66EW	47	2N6010	47	2SC182Q	47	2SC441J	47	2SC56C
47	BCW90A	47	2N6964	47	2SC182R	47	2SC441K	47	2SC56D
47	BCW90B	47	2SC-4012	47	2SC182V	47	2SC441L	47	2SC56F
47	BCW90C	47	2SC1063	47	2SC182X	47	2SC441M	47	2SC56G
47	BCW90KA	47	2SC115-3	47	2SC182Y	47	2SC441OR	47	2SC56GN
47	BCW90KB	47	2SC115-43	47	2SC1851	47	2SC441R	47	2SC56H
47	BCW90KC	47	2SC1211	47	2SC2001	47	2SC441X	47	2SC56J
47	BCX58IX	47	2SC1211C	47	2SC2001L	47	2SC442	47	2SC56K
47	BCX58X	47	2SC1211D	47	2SC2060	47	2SC442A	47	2SC56L
47	BCX59IX	47	2SC1211E	47	2SC2060Q	47	2SC442C	47	2SC56M
47	BCX59VII	47	2SC1214	47	2SC2081	47	2SC442D	47	2SC56OR
47	BCX59VIII	47	2SC1214(B)	47	2SC2086	47	2SC442F	47	2SC56R
47	BCX59X	47	2SC1214A	47	2SC28	47	2SC442G	47	2SC56X
47	BCX73-16	47	2SC1214B	47	2SC28A	47	2SC442GN	47	2SC594-O
47	BCX73-25	47	2SC1214C	47	2SC28B	47	2SC442H	47	2SC594-R
47	BCX73-40	47	2SC1214D	47	2SC28C	47	2SC442J	47	2SC594-Y
47	BCX74-16	47	2SC1383(P,Q,R)	47	2SC28D	47	2SC442K	47	2SC632
47	BCX74-25	47	2SC1383(S)	47	2SC28E	47	2SC442L	47	2SC632(1)
47	BCX74-40	47	2SC1383,RS	47	2SC28F	47	2SC442M	47	2SC632-OR
47	BFR41	47	2SC1383P	47	2SC28G	47	2SC442OR	47	2SC632A
47	BFR41T05	47	2SC1383Q	47	2SC28GN	47	2SC442R	47	2SC632B
47	BFS52	47	2SC1383R	47	2SC28H	47	2SC442X	47	2SC632C
47	BPT31	47	2SC1383RS	47	2SC28J	47	2SC442Y	47	2SC632D
47	BPT54	47	2SC1383S	47	2SC28K	47	2SC471	47	2SC632E
47	BFV14	47	2SC1383X	47	2SC28L	47	2SC475	47	2SC632F
47	BFV52	47	2SC1385H	47	2SC28M	47	2SC475A	47	2SC632G
47	BFX97A	47	2SC150	47	2SC28OR	47	2SC475AC	47	2SC632GN
47	BS840	47	2SC150-OR	47	2SC28R	47	2SC475B	47	2SC632H
47	BS841	47	2SC150A	47	2SC28X	47	2SC475C	47	2SC632J
47	BSV85	47	2SC150B	47	2SC28Y	47	2SC475D	47	2SC632K
47	BSW49	47	2SC150C	47	2SC352	47	2SC475E	47	2SC632L
47	C1211	47	2SC150D	47	2SC352-OR	47	2SC475F	47	2SC632M
47	C1211C	47	2SC150E	47	2SC352A	47	2SC475G	47	2SC632OR
47	C1211D	47	2SC150F	47	2SC352AC	47	2SC475GN	47	2SC632R
47	C1211E	47	2SC150GN	47	2SC352B	47	2SC475H	47	2SC632Y
47	C1383	47	2SC150H	47	2SC352C	47	2SC475J	47	2SC633
47	C1383P	47	2SC150J	47	2SC352D	47	2SC475K	47	2SC633-7
47	C1383Q	47	2SC150K	47	2SC352E	47	2SC475L	47	2SC633-OR
47	C1383R	47	2SC150L	47	2SC352F	47	2SC475OR	47	2SC633A
47	C1383S	47	2SC150M	47	2SC352G	47	2SC475R	47	2SC633B
47	C1383X	47	2SC150OR	47	2SC352GN	47	2SC475X	47	2SC633C
47	C1384	47	2SC150R	47	2SC352H	47	2SC475Y	47	2SC633D
47	C1384Q	47	2SC150U	47	2SC352J	47	2SC48	47	2SC633F
47	C1384R	47	2SC150X	47	2SC352K	47	2SC48A	47	2SC633G
47	C1384S	47	2SC150Y	47	2SC352L	47	2SC48B	47	2SC633GN
47	C235	47	2SC151B	47	2SC352M	47	2SC48C	47	2SC633H
47	C235-O	47	2SC157	47	2SC352OR	47	2SC48D	47	2SC633J
47	C428	47	2SC157A	47	2SC352R	47	2SC48E	47	2SC633K
47	C644RN	47	2SC157B	47	2SC352X	47	2SC48F	47	2SC633L
47	CS1909B	47	2SC157C	47	2SC352Y	47	2SC48G	47	2SC633M
47	D33D2J2	47	2SC157D	47	2SC362	47	2SC48GN	47	2SC633OR
47	D400D	47	2SC157E	47	2SC362A	47	2SC48H	47	2SC633X
47	D400E	47	2SC157F	47	2SC362B	47	2SC48J	47	2SC633Y
47	D400F	47	2SC157G	47	2SC362C	47	2SC48K	47	2SC694
47	D400P1D	47	2SC157GN	47	2SC362D	47	2SC48L	47	2SC694A
47	D400P1E	47	2SC157J	47	2SC362F	47	2SC48M	47	2SC694B
47	D400P1B	47	2SC157K	47	2SC362G	47	2SC48OR	47	2SC694C
47	D400P1F	47	2SC157L	47	2SC362GN	47	2SC48R	47	2SC694D
47	EA15X250	47	2SC157M	47	2SC362H	47	2SC48X	47	2SC694E
47	EA15X2522	47	2SC157OR	47	2SC362J	47	2SC48Y	47	2SC694F
47	EA15X397	47	2SC157R	47	2SC362K	47	2SC498	47	2SC694G
47	EA15X413	47	2SC157T	47	2SC362L	47	2SC498-Q	47	2SC694GN
47	EA15X8511	47	2SC158	47	2SC362M	47	2SC498-O	47	2SC694H
47	EA15X8521	47	2SC158A	47	2SC362OR	47	2SC498-OR	47	2SC694J
47	ECG293	47	2SC158B	47	2SC362R	47	2SC498-ORG	47	2SC694K
47	ED1702M	47	2SC158C	47	2SC362X	47	2SC498-R	47	2SC694L
47	FT3642	47	2SC158D	47	2SC362Y	47	2SC498-RED	47	2SC694M
47	FX3724	47	2SC158E	47	2SC367			47	2SC694OR
47	G05-415-B	47	2SC158F	47	2SC369			47	2SC694R
47	G05415	47	2SC158GN	47	2SC369-BLU-Q				
47	GE-47	47	2SC158H	47	2SC369-GRN-Q				
47	G05415	47	2SC158J	47	2SC369A				
47	HD-00471	47	2SC158K	47	2SC369B				
47	HEPS0014								

GE	Industry Standard No.	GE	Industry Standard No.	GE	Industry Standard No.	GE	Industry Standard No.	GE	Industry Standard No.
47	28C694X	49	28B463G	50	28A107K	50	28A115Y	50	28A368C
47	28C694Y	49	28B463GN	50	28A107L	50	28A15RD	50	28A368D
47	28C694Z	49	28B463H	50	28A107M	50	28A233	50	28A368E
47	28C701A	49	28B463J	50	28A107OR	50	28A233A	50	28A368F
47	28C701B	49	28B463K	50	28A107R	50	28A233B	50	28A368G
47	28C701C	49	28B463L	50	28A107X	50	28A233C	50	28A368GN
47	28C701D	49	28B463M	50	28A107Y	50	28A233D	50	28A368H
47	28C701E	49	28B463R	50	28A108	50	28A233E	50	28A368J
47	28C701F	49	28B463X	50	28A108A	50	28A233F	50	28A368K
47	28C701G	49	28B463XL	50	28A108B	50	28A233G	50	28A368M
47	28C701GN	49	28B474	50	28A108C	50	28A233GN	50	28A368OR
47	28C701H	49	28B474-2	50	28A108D	50	28A233H	50	28A368R
47	28C701J	49	28B474-3	50	28A108E	50	28A233J	50	28A368X
47	28C701K	49	28B474-4	50	28A108F	50	28A233K	50	28A380
47	28C701L	49	28B474-6D	50	28A108G	50	28A233L	50	28A380A
47	28C701M	49	28B474-OR	50	28A108H	50	28A233M	50	28A380B
47	28C701OR	49	28B474A	50	28A108K	50	28A233OR	50	28A380C
47	28C701R	49	28B474B	50	28A108L	50	28A233R	50	28A380D
47	28C701X	49	28B474C	50	28A108M	50	28A233X	50	28A380E
47	28C701Y	49	28B474D	50	28A108OR	50	28A233Y	50	28A380F
47	28C702	49	28B474E	50	28A108R	50	28A235H	50	28A380G
47	28C702A	49	28B474F	50	28A108X	50	28A236	50	28A380GN
47	28C702B	49	28B474G	50	28A108Y	50	28A236A	50	28A380H
47	28C702C	49	28B474GN	50	28A109	50	28A236B	50	28A380J
47	28C702D	49	28B474H	50	28A109A	50	28A236C	50	28A380K
47	28C702E	49	28B474J	50	28A109B	50	28A236D	50	28A380L
47	28C702F	49	28B474K	50	28A109C	50	28A236E	50	28A380M
47	28C702G	49	28B474L	50	28A109D	50	28A236F	50	28A380OR
47	28C702GN	49	28B474M	50	28A109E	50	28A236G	50	28A380R
47	28C702H	49	28B474MP	50	28A109F	50	28A236GN	50	28A380X
47	28C702J	49	28B474OR	50	28A109G	50	28A236H	50	28A380Y
47	28C702K	49	28B474R	50	28A109H	50	28A236J	50	28A405
47	28C702L	49	28B474S	50	28A109K	50	28A236K	50	28A405-0
47	28C702M	49	28B474V10	50	28A109L	50	28A236L	50	28A468
47	28C702OR	49	28B474V4	50	28A109M	50	28A236M	50	28A468A
47	28C702R	49	28B474X	50	28A109OR	50	28A236OR	50	28A468B
47	28C702X	49	28B474Y	50	28A109R	50	28A236R	50	28A468C
47	28C702Y	49	28B474YE1	50	28A109X	50	28A236X	50	28A468D
47	28C712	49	28B474YEL	50	28A109Y	50	28A236Y	50	28A468E
47	28C712(D)	49(2)	D116	50	28A110	50	28A24	50	28A468F
47	28C712-CD	49(2)	IRTR94	50	28A110A	50	28A240GREEN	50	28A468G
47	28C712A	49(2)	IRTR94MP	50	28A110B	50	28A25	50	28A468GN
47	28C712B	49(2)	R-28B474	50	28A110C	50	28A257	50	28A468H
47	28C712C	49(2)	SK3086	50	28A110D	50	28A257A	50	28A468J
47	28C712CD	49(TWO)	ECG226MP	50	28A110E	50	28A257B	50	28A468K
47	28C712D	50	AC107N	50	28A110F	50	28A257C	50	28A468L
47	28C712DC	50	AF-106	50	28A110G	50	28A257D	50	28A468M
47	28C712E	50	AF-137	50	28A110K	50	28A257E	50	28A468OR
47	28C712F	50	AF306	50	28A110L	50	28A257F	50	28A468R
47	28C712G	50	AST63N	50	28A110M	50	28A257G	50	28A468X
47	28C712GN	50	DS-51	50	28A110OR	50	28A257GN	50	28A468Y
47	28C712H	50	DS-52	50	28A110R	50	28A257H	50	28A469
47	28C712J	50	DS-65	50	28A110X	50	28A257J	50	28A469A
47	28C712K	50	E-2462	50	28A110Y	50	28A257K	50	28A469B
47	28C712L	50	EA18Y11	50	28A111	50	28A257L	50	28A469C
47	28C712M	50	GE-50	50	28A111A	50	28A257OR	50	28A469D
47	28C712OR	50	GT1	50	28A111B	50	28A257R	50	28A469E
47	28C712R	50	GT2	50	28A111C	50	28A257X	50	28A469F
47	28C712W	50	GT24H	50	28A111D	50	28A257Y	50	28A469G
47	28C712X	50	GT3	50	28A111E	50	28A258	50	28A469GN
47	28C712Y	50	HJ75	50	28A111P	50	28A258A	50	28A469H
47	28C715	50	HR-40	50	28A111G	50	28A258B	50	28A469J
47	28C715A	50	HR-43	50	28A111K	50	28A258C	50	28A469K
47	28C715B	50	HR-45	50	28A111L	50	28A258D	50	28A469L
47	28C715C	50	IRTR12	50	28A111M	50	28A258E	50	28A469M
47	28C715D	50	IT205A	50	28A111OR	50	28A258F	50	28A469OR
47	28C715E	50	RS-3901	50	28A111R	50	28A258G	50	28A469R
47	28C715EJ	50	RS-3926	50	28A111X	50	28A258GN	50	28A469X
47	28C715EV	50	SWT1728	50	28A111Y	50	28A258H	50	28A469Y
47	28C715F	50	SWT3588	50	28A112	50	28A258J	50	28A472
47	28C715G	50	TA1990	50	28A112A	50	28A258K	50	28A472-1
47	28C715GN	50	TG28A201(C)	50	28A112B	50	28A258L	50	28A472-2
47	28C715H	50	TRO-2012	50	28A112C	50	28A258M	50	28A472-3
47	28C715J	50	TR105	50	28A112D	50	28A258OR	50	28A472-4
47	28C715K	50	TR139	50	28A112E	50	28A258R	50	28A472-5
47	28C715L	50	TR218	50	28A112F	50	28A258X	50	28A472-6
47	28C715M	50	TRM13	50	28A112G	50	28A258Y	50	28A472A
47	28C715R	50	TRM14	50	28A112GN	50	28A270	50	28A472B
47	28C715X	50	TRM81	50	28A112H	50	28A270A	50	28A472C
47	28C715XL	50	TRO2012	50	28A112K	50	28A270B	50	28A472D
47	28C715Y	50	2N1625	50	28A112L	50	28A270C	50	28A472E
47,62	TR33	50	2N1872	50	28A112M	50	28A270D	50	28A472F
47MP	ECG293MP	50	2N1873	50	28A112OR	50	28A270E	50	28A472G
47MP	GE-47MP	50	2N1874	50	28A112R	50	28A270F	50	28A472GN
48	A5T3504	50	2N1875	50	28A112X	50	28A270G	50	28A472H
48	A5T4250	50	2N3328	50	28A112Y	50	28A270GN	50	28A472J
48	A5T5087	50	2N529/P	50	28A113	50	28A270H	50	28A472K
48	BC418B	50	2N530/P	50	28A113A	50	28A270J	50	28A472L
48	BC419B	50	2N532/P	50	28A113B	50	28A270K	50	28A472M
48	DDBY003001	50	2N533/P	50	28A113C	50	28A270L	50	28A472OR
48	IRTR31	50	2S175	50	28A113D	50	28A270M	50	28A472R
48	QT-A0733XON	50	28471-1	50	28A113E	50	28A270OR	50	28A472X
48	RT-101	50	28A104	50	28A113F	50	28A270R	50	28A472Y
48	TR54	50	28A104A	50	28A113G	50	28A270X	50	28A475
48	01-030733	50	28A104B	50	28A113GN	50	28A270Y	50	28A59
48	2N6005	50	28A104C	50	28A113H	50	28A351	50	28A59A
48	28A503	50	28A104D	50	28A113J	50	28A351A	50	28A59C
48	28A503-0	50	28A104E	50	28A113L	50	28A351A-2	50	28A59D
48	28A503-R	50	28A104F	50	28A113M	50	28A351C	50	28A59E
48	28A503-Y	50	28A104G	50	28A113X	50	28A351D	50	28A59F
48	28A503GR	50	28A104H	50	28A113Y	50	28A351E	50	28A59G
48	28A701PO	50	28A104J	50	28A114	50	28A351F	50	28A59L
48	28A704	50	28A104K	50	28A114A	50	28A351G	50	28A59M
48	28A705	50	28A104L	50	28A114B	50	28A351GN	50	28A59OR
48	28A773	50	28A104M	50	28A114C	50	28A351H	50	28A59R
48,82	TR20	50	28A104OR	50	28A114D	50	28A351K	50	28A59X
48,82,	TR31	50	28A104R	50	28A114E	50	28A351L	50	28A59Y
49	B474	50	28A104X	50	28A114F	50	28A351M	50	28A70-0B
49	B474-2	50	28A104Y	50	28A114G	50	28A351OR	50	28A73
49	B474-3	50	28A106	50	28A114H	50	28A351R	50	28A73A
49	B474-4	50	28A106A	50	28A114K	50	28A351X	50	28A73B
49	B474-6D	50	28A106B	50	28A114L	50	28A351Y	50	28A73C
49	B474MP	50	28A106C	50	28A114M	50	28A361	50	28A73D
49	B474S	50	28A106D	50	28A114OR	50	28A361A	50	28A73E
49	B474V10	50	28A106E	50	28A114R	50	28A361B	50	28A73F
49	B474V4	50	28A106F	50	28A114X	50	28A361C	50	28A73G
49	B474Y	50	28A106G	50	28A114Y	50	28A361D	50	28A73H
49	CXL226	50	28A106H	50	28A115	50	28A361E	50	28A73K
49	CXL226MP	50	28A106J	50	28A115A	50	28A361F	50	28A73L
49	ECG226	50	28A106K	50	28A115B	50	28A361G	50	28A73M
49	ES61(ELCOM)	50	28A106L	50	28A115C	50	28A361GN	50	28A73OR
49	GE-49	50	28A106M	50	28A115D	50	28A361H	50	28A73R
49	RB33	50	28A106OR	50	28A115E	50	28A361K	50	28A73X
49	RB33MP	50	28A106R	50	28A115F	50	28A361L	50	28A73Y
49	REN226	50	28A106X	50	28A115G	50	28A361M	50	28A83
49	REN226MP	50	28A106Y	50	28A115GN	50	28A361OR	50	28A83A
49	SK3082	50	28A107	50	28A115H	50	28A361R	50	28A83B
49	28B463	50	28A107A	50	28A115J	50	28A361X	50	28A83C
49	28B463A	50	28A107B	50	28A115K	50	28A361Y	50	28A83D
49	28B463BL	50	28A107C	50	28A115L	50	28A368	50	28A83E
49	28B463C	50	28A107D	50	28A115OR	50	28A368A	50	28A83F
49	28B463D	50	28A107E	50	28A115R	50	28A368B	50	28A83G
49	28B463F	50	28A107F	50	28A115X				
		50	28A107G						
		50	28A107H						

GE	Industry Standard No.	GE	Industry Standard No.	GE	Industry Standard No.	GE	Industry Standard No.	GE	Industry Standard No.
50	2SA83H	51	2N289	51	2SA123L	51	2SA135E	51	2SA145C
50	2SA83K	51	2N601	51	2SA123M	51	2SA135F	51	2SA145D
50	2SA83L	51	2N6365	51	2SA1230R	51	2SA135G	51	2SA145E
50	2SA83M	51	2N6365A	51	2SA123R	51	2SA135GN	51	2SA145G
50	2SA830R	51	2N84	51	2SA123X	51	2SA135H	51	2SA145GN
50	2SA83R	51	2N961/46	51	2SA123Y	51	2SA135J	51	2SA145K
50	2SA83X	51	2N962/46	51	2SA124	51	2SA135K	51	2SA145M
50	2SA83Y	51	2B144	51	2SA124A	51	2SA135L	51	2SA1450R
50	2SA84	51	2B148	51	2SA124B	51	2SA135M	51	2SA145R
50	2SA84A	51	2B201	51	2SA124C	51	2SA1350R	51	2SA145X
50	2SA84B	51	2B5T	51	2SA124D	51	2SA135R	51	2SA145Y
50	2SA84C	51	2BA-4551	51	2SA124E	51	2SA135X	51	2SA146
50	2SA84D	51	2BA-4561	51	2SA124F	51	2SA135Y	51	2SA146A
50	2SA84E	51	2SA076P	51	2SA124G	51	2SA136	51	2SA146B
50	2SA84F	51	2SA098R	51	2SA124GN	51	2SA136A	51	2SA146C
50	2SA84G	51	2SA100	51	2SA124H	51	2SA136B	51	2SA146D
50	2SA84H	51	2SA100A	51	2SA124J	51	2SA136C	51	2SA146E
50	2SA84K	51	2SA100B	51	2SA124K	51	2SA136D	51	2SA146F
50	2SA84L	51	2SA100C	51	2SA124L	51	2SA136E	51	2SA146G
50	2SA84M	51	2SA100D	51	2SA124M	51	2SA136F	51	2SA146GN
50	2SA840R	51	2SA100E	51	2SA1240R	51	2SA136GN	51	2SA146H
50	2SA84R	51	2SA100F	51	2SA124R	51	2SA136H	51	2SA146J
50	2SA84X	51	2SA100G	51	2SA124X	51	2SA136J	51	2SA146K
50	2SA84Y	51	2SA100H	51	2SA124Y	51	2SA136K	51	2SA146L
50	2SA92	51	2SA100J	51	2SA125	51	2SA136L	51	2SA146OR
50	2SA92A	51	2SA100K	51	2SA125A	51	2SA136M	51	2SA146X
50	2SA92B	51	2SA100M	51	2SA125B	51	2SA1360R	51	2SA146Y
50	2SA92C	51	2SA1000R	51	2SA125C	51	2SA136R	51	2SA147
50	2SA92D	51	2SA100R	51	2SA125D	51	2SA136X	51	2SA147A
50	2SA92E	51	2SA100Y	51	2SA125E	51	2SA136Y	51	2SA147B
50	2SA92F	51	2SA101-0R	51	2SA125F	51	2SA137	51	2SA147C
50	2SA92G	51	2SA101B	51	2SA125G	51	2SA137A	51	2SA147D
50	2SA92H	51	2SA102-0R	51	2SA125GN	51	2SA137B	51	2SA147E
50	2SA92K	51	2SA116	51	2SA125H	51	2SA137C	51	2SA147F
50	2SA92L	51	2SA116A	51	2SA125J	51	2SA137D	51	2SA147G
50	2SA92M	51	2SA116B	51	2SA125K	51	2SA137E	51	2SA147H
50	2SA920R	51	2SA116C	51	2SA125L	51	2SA137F	51	2SA147J
50	2SA92R	51	2SA116D	51	2SA125M	51	2SA137G	51	2SA147K
50	2SA92X	51	2SA116E	51	2SA1250R	51	2SA137GN	51	2SA147M
50	2SA92Y	51	2SA116F	51	2SA125R	51	2SA137H	51	2SA1470R
50	2SA93	51	2SA116G	51	2SA125X	51	2SA137J	51	2SA147R
50	2SA93A	51	2SA116GN	51	2SA125Y	51	2SA137K	51	2SA147X
50	2SA93B	51	2SA116H	51	2SA13	51	2SA137L	51	2SA147Y
50	2SA93C	51	2SA116J	51	2SA130	51	2SA137M	51	2SA148
50	2SA93E	51	2SA116K	51	2SA130A	51	2SA1370R	51	2SA148A
50	2SA93F	51	2SA116L	51	2SA130B	51	2SA137R	51	2SA148B
50	2SA93G	51	2SA116M	51	2SA130C	51	2SA137X	51	2SA148C
50	2SA93H	51	2SA1160R	51	2SA130D	51	2SA137Y	51	2SA148D
50	2SA93K	51	2SA116R	51	2SA130E	51	2SA139	51	2SA148E
50	2SA93L	51	2SA116X	51	2SA130F	51	2SA139A	51	2SA148F
50	2SA93M	51	2SA116Y	51	2SA130G	51	2SA139B	51	2SA148G
50	2SA930R	51	2SA117	51	2SA130GN	51	2SA139C	51	2SA148GN
50	2SA93R	51	2SA117A	51	2SA130H	51	2SA139D	51	2SA148H
50	2SA93X	51	2SA117B	51	2SA130J	51	2SA139E	51	2SA148J
50	2SA93Y	51	2SA117C	51	2SA130K	51	2SA139F	51	2SA148K
50	2SB179	51	2SA117D	51	2SA130L	51	2SA139G	51	2SA148L
50,245	TR12	51	2SA117E	51	2SA130M	51	2SA139GN	51	2SA148M
50,245	TR17	51	2SA117F	51	2SA1300R	51	2SA139J	51	2SA1480R
51	AC150-GRN	51	2SA117G	51	2SA130X	51	2SA139K	51	2SA148R
51	AC150-YEL	51	2SA117GN	51	2SA130Y	51	2SA139L	51	2SA148X
51	AC160-GRN	51	2SA117H	51	2SA131	51	2SA139M	51	2SA148Y
51	AC160-RED	51	2SA117J	51	2SA131A	51	2SA1390R	51	2SA149
51	AC160-VIO	51	2SA117K	51	2SA131B	51	2SA139R	51	2SA149A
51	AC160-YEL	51	2SA117L	51	2SA131C	51	2SA139X	51	2SA149B
51	AC221	51	2SA117M	51	2SA131D	51	2SA139Y	51	2SA149C
51	AF138/20	51	2SA1170R	51	2SA131E	51	2SA13A	51	2SA149D
51	D063	51	2SA117R	51	2SA131F	51	2SA13B	51	2SA149E
51	D079	51	2SA117X	51	2SA131G	51	2SA13C	51	2SA149F
51	D086	51	2SA117Y	51	2SA131GN	51	2SA13D	51	2SA149G
51	E070	51	2SA118	51	2SA131H	51	2SA13G	51	2SA149GN
51	GE-51	51	2SA118A	51	2SA131J	51	2SA13L	51	2SA149H
51	GE-9	51	2SA118B	51	2SA131K	51	2SA13M	51	2SA149J
51	GET-672	51	2SA118C	51	2SA131L	51	2SA130R	51	2SA149K
51	GET-672A	51	2SA118D	51	2SA131M	51	2SA13R	51	2SA149L
51	GET-673	51	2SA118E	51	2SA1310R	51	2SA13X	51	2SA149M
51	GET-692	51	2SA118F	51	2SA131R	51	2SA13Y	51	2SA1490R
51	GET-873A	51	2SA118G	51	2SA131X	51	2SA14	51	2SA149R
51	GET-883	51	2SA118GN	51	2SA131Y	51	2SA141	51	2SA149X
51	HA-00354	51	2SA118H	51	2SA132	51	2SA141A	51	2SA149Y
51	HA-234B	51	2SA118J	51	2SA132A	51	2SA141C	51	2SA14A
51	HA-235A	51	2SA118K	51	2SA132B	51	2SA141D	51	2SA14B
51	HA-235C	51	2SA118L	51	2SA132C	51	2SA141E	51	2SA14C
51	HA269	51	2SA118M	51	2SA132D	51	2SA141F	51	2SA14D
51	HA350	51	2SA1180R	51	2SA132E	51	2SA141G	51	2SA14E
51	HA353	51	2SA118R	51	2SA132F	51	2SA141GN	51	2SA14G
51	HA354	51	2SA118X	51	2SA132G	51	2SA141K	51	2SA14L
51	HF-35	51	2SA118Y	51	2SA132GN	51	2SA141L	51	2SA14M
51	HR1	51	2SA121	51	2SA132H	51	2SA141M	51	2SA140R
51	HR2	51	2SA121A	51	2SA132J	51	2SA1410R	51	2SA14X
51	HR20	51	2SA121B	51	2SA132K	51	2SA141R	51	2SA14Y
51	HR20A	51	2SA121C	51	2SA132L	51	2SA141X	51	2SA15
51	HR21	51	2SA121D	51	2SA132M	51	2SA141Y	51	2SA15-6
51	HR21A	51	2SA121E	51	2SA1320R	51	2SA143	51	2SA151
51	HR22	51	2SA121F	51	2SA132X	51	2SA143A	51	2SA151A
51	HR22A	51	2SA121G	51	2SA132Y	51	2SA143B	51	2SA151B
51	HR22B	51	2SA121GN	51	2SA133	51	2SA143C	51	2SA151C
51	HR24	51	2SA121H	51	2SA133A	51	2SA143D	51	2SA151D
51	HR24A	51	2SA121J	51	2SA133B	51	2SA143F	51	2SA151E
51	HR25	51	2SA121K	51	2SA133C	51	2SA143G	51	2SA151F
51	HR25A	51	2SA121L	51	2SA133D	51	2SA143GN	51	2SA151G
51	HR26	51	2SA121M	51	2SA133E	51	2SA143H	51	2SA151GN
51	HR26A	51	2SA1210R	51	2SA133F	51	2SA143K	51	2SA151H
51	HR27	51	2SA121R	51	2SA133G	51	2SA143L	51	2SA151J
51	HR27A	51	2SA121X	51	2SA133GN	51	2SA143M	51	2SA151K
51	HR7	51	2SA121Y	51	2SA133H	51	2SA1430R	51	2SA151L
51	IRTR06	51	2SA122A	51	2SA133J	51	2SA143R	51	2SA151M
51	IRTR07	51	2SA122B	51	2SA133K	51	2SA143X	51	2SA1510R
51	IRTR17	51	2SA122C	51	2SA133M	51	2SA143Y	51	2SA151R
51	IRTR18	51	2SA122D	51	2SA1330R	51	2SA144	51	2SA151X
51	K0825642-10	51	2SA122E	51	2SA133R	51	2SA144A	51	2SA151Y
51	M100	51	2SA122F	51	2SA133X	51	2SA144B	51	2SA152
51	NHN160	51	2SA122G	51	2SA133Y	51	2SA144C	51	2SA152A
51	R85317	51	2SA122GN	51	2SA134	51	2SA144D	51	2SA152B
51	T-Q5021	51	2SA122H	51	2SA134A	51	2SA144E	51	2SA152C
51	T-Q5022	51	2SA122K	51	2SA134B	51	2SA144F	51	2SA152D
51	T-Q5034	51	2SA122L	51	2SA134C	51	2SA144G	51	2SA152E
51	T-Q5035	51	2SA122M	51	2SA134D	51	2SA144GN	51	2SA152F
51	T-Q5038	51	2SA1220R	51	2SA134E	51	2SA144H	51	2SA152GN
51	TI-389	51	2SA122R	51	2SA134F	51	2SA144J	51	2SA152H
51	TR06	51	2SA122X	51	2SA134G	51	2SA144K	51	2SA152J
51	TRA24	51	2SA122Y	51	2SA134H	51	2SA144L	51	2SA152K
51	TRA24A	51	2SA123	51	2SA134J	51	2SA144M	51	2SA152L
51	1T205A	51	2SA123A	51	2SA134K	51	2SA1440R	51	2SA152M
51	1T264	51	2SA123B	51	2SA134L	51	2SA144R	51	2SA1520R
51	002-00900	51	2SA123C	51	2SA1340R	51	2SA144X	51	2SA152R
51	2M1305	51	2SA123D	51	2SA134R	51	2SA144Y	51	2SA152X
51	2N123/5	51	2SA123E	51	2SA134X	51	2SA145	51	2SA152Y
51	2N123A5	51	2SA123F	51	2SA134Y	51	2SA145A	51	2SA153A
51	2N1394	51	2SA123G	51	2SA135	51	2SA145B	51	2SA153B
51	2N1494	51	2SA123GN	51	2SA135A				
51	2N2009	51	2SA123H	51	2SA135B				
51	2N2097	51	2SA123J	51	2SA135C				
51	2N2169	51	2SA123K	51	2SA135D				
51	2N231								

GE	Industry Standard No.	GE	Industry Standard No.	GE	Industry Standard No.	GE	Industry Standard No.	GE	Industry Standard No.
51	2SA153C	51	2SA160GN	51	2SA167Y	51	2SA174M	51	2SA189F
51	2SA153D	51	2SA160H	51	2SA168	51	2SA174OR	51	2SA189G
51	2SA153E	51	2SA160J	51	2SA168A	51	2SA174R	51	2SA189GN
51	2SA153F	51	2SA160K	51	2SA168B	51	2SA174X	51	2SA189H
51	2SA153G	51	2SA160L	51	2SA168C	51	2SA174Y	51	2SA189J
51	2SA153GN	51	2SA160M	51	2SA168D	51	2SA175	51	2SA189K
51	2SA153H	51	2SA160OR	51	2SA168E	51	2SA175A	51	2SA189L
51	2SA153J	51	2SA160R	51	2SA168F	51	2SA175B	51	2SA189M
51	2SA153K	51	2SA160X	51	2SA168G	51	2SA175C	51	2SA189OR
51	2SA153L	51	2SA160Y	51	2SA168GN	51	2SA175D	51	2SA189R
51	2SA153M	51	2SA161	51	2SA168H	51	2SA175E	51	2SA189X
51	2SA153OR	51	2SA161A	51	2SA168J	51	2SA175F	51	2SA189Y
51	2SA153R	51	2SA161B	51	2SA168K	51	2SA175G	51	2SA18A
51	2SA153X	51	2SA161C	51	2SA168L	51	2SA175GN	51	2SA18B
51	2SA153Y	51	2SA161D	51	2SA168M	51	2SA175H	51	2SA18C
51	2SA154	51	2SA161E	51	2SA168OR	51	2SA175J	51	2SA18D
51	2SA154A	51	2SA161F	51	2SA168R	51	2SA175K	51	2SA18E
51	2SA154B	51	2SA161G	51	2SA168X	51	2SA175L	51	2SA18F
51	2SA154C	51	2SA161H	51	2SA168Y	51	2SA175M	51	2SA18G
51	2SA154D	51	2SA161J	51	2SA169A	51	2SA175OR	51	2SA18H
51	2SA154E	51	2SA161K	51	2SA169B	51	2SA175R	51	2SA18L
51	2SA154F	51	2SA161L	51	2SA169C	51	2SA175X	51	2SA18M
51	2SA154G	51	2SA161M	51	2SA169D	51	2SA175Y	51	2SA18OR
51	2SA154GN	51	2SA161OR	51	2SA169E	51	2SA178	51	2SA18R
51	2SA154H	51	2SA161R	51	2SA169F	51	2SA17A	51	2SA18X
51	2SA154J	51	2SA161X	51	2SA169G	51	2SA17B	51	2SA18Y
51	2SA154K	51	2SA161Y	51	2SA169GN	51	2SA17C	51	2SA19O
51	2SA154L	51	2SA162	51	2SA169H	51	2SA17D	51	2SA191
51	2SA154M	51	2SA162A	51	2SA169J	51	2SA17F	51	2SA192
51	2SA154OR	51	2SA162B	51	2SA169K	51	2SA17G	51	2SA193
51	2SA154R	51	2SA162C	51	2SA169L	51	2SA17H	51	2SA194
51	2SA154X	51	2SA162D	51	2SA169M	51	2SA17L	51	2SA195
51	2SA154Y	51	2SA162E	51	2SA169OR	51	2SA17OR	51	2SA196
51	2SA155	51	2SA162F	51	2SA169R	51	2SA17R	51	2SA197
51	2SA155A	51	2SA162G	51	2SA169X	51	2SA17X	51	2SA197A
51	2SA155B	51	2SA162GN	51	2SA169Y	51	2SA17Y	51	2SA197B
51	2SA155C	51	2SA162H	51	2SA16A	51	2SA18	51	2SA197C
51	2SA155D	51	2SA162J	51	2SA16B	51	2SA180	51	2SA197D
51	2SA155E	51	2SA162K	51	2SA16C	51	2SA180A	51	2SA197E
51	2SA155F	51	2SA162L	51	2SA16D	51	2SA180B	51	2SA197F
51	2SA155G	51	2SA162M	51	2SA16E	51	2SA180C	51	2SA197G
51	2SA155GN	51	2SA162OR	51	2SA16F	51	2SA180D	51	2SA197GN
51	2SA155H	51	2SA162R	51	2SA16L	51	2SA180E	51	2SA197H
51	2SA155J	51	2SA162X	51	2SA16M	51	2SA180F	51	2SA197J
51	2SA155K	51	2SA162Y	51	2SA16OR	51	2SA180G	51	2SA197K
51	2SA155L	51	2SA163	51	2SA16R	51	2SA180GN	51	2SA197L
51	2SA155M	51	2SA163A	51	2SA16X	51	2SA180H	51	2SA197M
51	2SA155OR	51	2SA163B	51	2SA16Y	51	2SA180J	51	2SA197OR
51	2SA155R	51	2SA163C	51	2SA17	51	2SA180K	51	2SA197R
51	2SA155X	51	2SA163D	51	2SA170	51	2SA180L	51	2SA197X
51	2SA155Y	51	2SA163E	51	2SA170A	51	2SA180M	51	2SA197Y
51	2SA156	51	2SA163F	51	2SA170B	51	2SA180OR	51	2SA198
51	2SA156A	51	2SA163G	51	2SA170C	51	2SA180R	51	2SA198A
51	2SA156B	51	2SA163GN	51	2SA170D	51	2SA180X	51	2SA198B
51	2SA156C	51	2SA163H	51	2SA170E	51	2SA180Y	51	2SA198C
51	2SA156D	51	2SA163J	51	2SA170F	51	2SA181	51	2SA198D
51	2SA156E	51	2SA163K	51	2SA170G	51	2SA181A	51	2SA198E
51	2SA156F	51	2SA163L	51	2SA170GN	51	2SA181B	51	2SA198F
51	2SA156G	51	2SA163M	51	2SA170H	51	2SA181C	51	2SA198G
51	2SA156GN	51	2SA163OR	51	2SA170J	51	2SA181D	51	2SA198GN
51	2SA156H	51	2SA163R	51	2SA170K	51	2SA181E	51	2SA198H
51	2SA156J	51	2SA163X	51	2SA170L	51	2SA181F	51	2SA198J
51	2SA156K	51	2SA163Y	51	2SA170M	51	2SA181GN	51	2SA198K
51	2SA156L	51	2SA164	51	2SA170OR	51	2SA181H	51	2SA198L
51	2SA156M	51	2SA164A	51	2SA170R	51	2SA181J	51	2SA198M
51	2SA156OR	51	2SA164B	51	2SA170X	51	2SA181K	51	2SA198OR
51	2SA156R	51	2SA164C	51	2SA170Y	51	2SA181L	51	2SA198R
51	2SA156X	51	2SA164D	51	2SA171	51	2SA181M	51	2SA198X
51	2SA156Y	51	2SA164E	51	2SA171A	51	2SA181OR	51	2SA198T
51	2SA157	51	2SA164F	51	2SA171B	51	2SA181R	51	2SA199
51	2SA157A	51	2SA164G	51	2SA171C	51	2SA181X	51	2SA201
51	2SA157B	51	2SA164GN	51	2SA171D	51	2SA181Y	51	2SA201-O
51	2SA157C	51	2SA164H	51	2SA171E	51	2SA182	51	2SA201A
51	2SA157D	51	2SA164J	51	2SA171F	51	2SA182A	51	2SA201B
51	2SA157E	51	2SA164K	51	2SA171G	51	2SA182B	51	2SA201CL
51	2SA157F	51	2SA164L	51	2SA171GN	51	2SA182C	51	2SA201D
51	2SA157G	51	2SA164M	51	2SA171H	51	2SA182D	51	2SA201E
51	2SA157GN	51	2SA164OR	51	2SA171J	51	2SA182E	51	2SA201F
51	2SA157H	51	2SA164R	51	2SA171K	51	2SA182F	51	2SA201G
51	2SA157J	51	2SA164X	51	2SA171L	51	2SA182G	51	2SA201GN
51	2SA157K	51	2SA164Y	51	2SA171M	51	2SA182GN	51	2SA201H
51	2SA157L	51	2SA165	51	2SA171OR	51	2SA182H	51	2SA201J
51	2SA157M	51	2SA165A	51	2SA171R	51	2SA182J	51	2SA201K
51	2SA157OR	51	2SA165B	51	2SA171X	51	2SA182K	51	2SA201L
51	2SA157R	51	2SA165C	51	2SA171Y	51	2SA182L	51	2SA201M
51	2SA157X	51	2SA165D	51	2SA172	51	2SA182M	51	2SA201O
51	2SA157Y	51	2SA165E	51	2SA172A	51	2SA182OR	51	2SA201OR
51	2SA159	51	2SA165F	51	2SA172B	51	2SA182R	51	2SA201R
51	2SA159A	51	2SA165G	51	2SA172C	51	2SA182X	51	2SA201RV
51	2SA159B	51	2SA165GN	51	2SA172D	51	2SA182Y	51	2SA201RVO
51	2SA159C	51	2SA165H	51	2SA172F	51	2SA183	51	2SA201X
51	2SA159D	51	2SA165J	51	2SA172G	51	2SA183A	51	2SA201Y
51	2SA159E	51	2SA165K	51	2SA172GN	51	2SA183B	51	2SA204
51	2SA159F	51	2SA165L	51	2SA172H	51	2SA183C	51	2SA204A
51	2SA159G	51	2SA165M	51	2SA172J	51	2SA183D	51	2SA204B
51	2SA159GN	51	2SA165OR	51	2SA172K	51	2SA183E	51	2SA204C
51	2SA159H	51	2SA165R	51	2SA172L	51	2SA183F	51	2SA204D
51	2SA159J	51	2SA165X	51	2SA172M	51	2SA183G	51	2SA204E
51	2SA159K	51	2SA165Y	51	2SA172OR	51	2SA183GN	51	2SA204F
51	2SA159L	51	2SA166	51	2SA172R	51	2SA183H	51	2SA204G
51	2SA159M	51	2SA166A	51	2SA172X	51	2SA183J	51	2SA204GN
51	2SA159OR	51	2SA166C	51	2SA172Y	51	2SA183K	51	2SA204H
51	2SA159R	51	2SA166D	51	2SA173	51	2SA183L	51	2SA204J
51	2SA159X	51	2SA166E	51	2SA173A	51	2SA183M	51	2SA204K
51	2SA159Y	51	2SA166F	51	2SA173B	51	2SA183OR	51	2SA204L
51	2SA15A	51	2SA166G	51	2SA173C	51	2SA183R	51	2SA204M
51	2SA15B	51	2SA166GN	51	2SA173D	51	2SA183X	51	2SA204OR
51	2SA15BK	51	2SA166H	51	2SA173E	51	2SA183Y	51	2SA204R
51	2SA15BL	51	2SA166J	51	2SA173F	51	2SA184	51	2SA204X
51	2SA15BLU	51	2SA166K	51	2SA173G	51	2SA188	51	2SA204Y
51	2SA15C	51	2SA166L	51	2SA173GN	51	2SA188A	51	2SA205
51	2SA15D	51	2SA166M	51	2SA173H	51	2SA188B	51	2SA205A
51	2SA15E	51	2SA166OR	51	2SA173J	51	2SA188C	51	2SA205B
51	2SA15F	51	2SA166R	51	2SA173K	51	2SA188D	51	2SA205C
51	2SA15G	51	2SA166X	51	2SA173L	51	2SA188E	51	2SA205D
51	2SA15H	51	2SA166Y	51	2SA173M	51	2SA188F	51	2SA205E
51	2SA15K	51	2SA167A	51	2SA173OR	51	2SA188G	51	2SA205F
51	2SA15L	51	2SA167B	51	2SA173R	51	2SA188GN	51	2SA205G
51	2SA15M	51	2SA167C	51	2SA173X	51	2SA188H	51	2SA205GN
51	2SA15OR	51	2SA167D	51	2SA173Y	51	2SA188J	51	2SA205H
51	2SA15R	51	2SA167E	51	2SA174	51	2SA188K	51	2SA205J
51	2SA15U	51	2SA167F	51	2SA174A	51	2SA188L	51	2SA205K
51	2SA15V	51	2SA167G	51	2SA174B	51	2SA188M	51	2SA205L
51	2SA15VR	51	2SA167GN	51	2SA174C	51	2SA188OR	51	2SA205M
51	2SA15X	51	2SA167H	51	2SA174D	51	2SA188R	51	2SA205OR
51	2SA15Y	51	2SA167J	51	2SA174E	51	2SA188X	51	2SA205R
51	2SA16	51	2SA167K	51	2SA174F	51	2SA188Y	51	2SA205X
51	2SA160	51	2SA167L	51	2SA174G	51	2SA189	51	2SA205Y
51	2SA160A	51	2SA167M	51	2SA174GN	51	2SA189A	51	2SA206
51	2SA160B	51	2SA167OR	51	2SA174H	51	2SA189B	51	2SA206A
51	2SA160C	51	2SA167R	51	2SA174J	51	2SA189C	51	2SA206B
51	2SA160D	51	2SA167X	51	2SA174K	51	2SA189D	51	2SA206C
51	2SA160E			51	2SA174L	51	2SA189E	51	2SA206D
51	2SA160F							51	2SA206E

GE	Industry Standard No.	GE	Industry Standard No.	GE	Industry Standard No.	GE	Industry Standard No.	GE	Industry Standard No.
51	2SA206F	51	2SA213R	51	2SA227E	51	2SA240X	51	2SA254G
51	2SA206G	51	2SA213X	51	2SA227F	51	2SA240Y	51	2SA254H
51	2SA206GN	51	2SA213Y	51	2SA227G	51	2SA241	51	2SA254J
51	2SA206H	51	2SA214	51	2SA227GN	51	2SA241A	51	2SA254K
51	2SA206J	51	2SA214A	51	2SA227H	51	2SA241B	51	2SA254L
51	2SA206K	51	2SA214B	51	2SA227J	51	2SA241C	51	2SA254M
51	2SA206L	51	2SA214C	51	2SA227K	51	2SA241D	51	2SA254OR
51	2SA206M	51	2SA214D	51	2SA227L	51	2SA241E	51	2SA254R
51	2SA206OR	51	2SA214E	51	2SA227M	51	2SA241F	51	2SA254X
51	2SA206R	51	2SA214F	51	2SA227OR	51	2SA241G	51	2SA254Y
51	2SA206X	51	2SA214G	51	2SA227R	51	2SA241GN	51	2SA255A
51	2SA206Y	51	2SA214GN	51	2SA227X	51	2SA241H	51	2SA255B
51	2SA207	51	2SA214H	51	2SA227Y	51	2SA241J	51	2SA255C
51	2SA207A	51	2SA214J	51	2SA229	51	2SA241K	51	2SA255D
51	2SA207B	51	2SA214K	51	2SA229A	51	2SA241L	51	2SA255E
51	2SA207C	51	2SA214L	51	2SA229B	51	2SA241M	51	2SA255F
51	2SA207D	51	2SA214M	51	2SA229C	51	2SA241OR	51	2SA255G
51	2SA207E	51	2SA214OR	51	2SA229D	51	2SA241R	51	2SA255GN
51	2SA207F	51	2SA214R	51	2SA229E	51	2SA241X	51	2SA255H
51	2SA207G	51	2SA214X	51	2SA229F	51	2SA241Y	51	2SA255J
51	2SA207GN	51	2SA214Y	51	2SA229G	51	2SA246	51	2SA255K
51	2SA207H	51	2SA215	51	2SA229GN	51	2SA246A	51	2SA255L
51	2SA207J	51	2SA215A	51	2SA229H	51	2SA246B	51	2SA255M
51	2SA207K	51	2SA215B	51	2SA229J	51	2SA246C	51	2SA255OR
51	2SA207L	51	2SA215C	51	2SA229K	51	2SA246D	51	2SA255R
51	2SA207M	51	2SA215D	51	2SA229L	51	2SA246E	51	2SA255X
51	2SA207OR	51	2SA215E	51	2SA229OR	51	2SA246F	51	2SA255Y
51	2SA207R	51	2SA215F	51	2SA229R	51	2SA246G	51	2SA26
51	2SA207X	51	2SA215G	51	2SA229X	51	2SA246GN	51	2SA260
51	2SA207Y	51	2SA215GN	51	2SA229Y	51	2SA246H	51	2SA260A
51	2SA208	51	2SA215H	51	2SA23	51	2SA246J	51	2SA260B
51	2SA208A	51	2SA215J	51	2SA230	51	2SA246K	51	2SA260C
51	2SA208B	51	2SA215K	51	2SA230A	51	2SA246L	51	2SA260D
51	2SA208C	51	2SA215L	51	2SA230B	51	2SA246M	51	2SA260E
51	2SA208D	51	2SA215M	51	2SA230C	51	2SA246OR	51	2SA260F
51	2SA208E	51	2SA215OR	51	2SA230D	51	2SA246R	51	2SA260GN
51	2SA208F	51	2SA215R	51	2SA230E	51	2SA246V	51	2SA260H
51	2SA208G	51	2SA215X	51	2SA230F	51	2SA246X	51	2SA260J
51	2SA208GN	51	2SA215Y	51	2SA230G	51	2SA246Y	51	2SA260K
51	2SA208H	51	2SA216	51	2SA230GN	51	2SA247	51	2SA260L
51	2SA208J	51	2SA216A	51	2SA230H	51	2SA247A	51	2SA260M
51	2SA208K	51	2SA216B	51	2SA230J	51	2SA247B	51	2SA260OR
51	2SA208L	51	2SA216C	51	2SA230K	51	2SA247C	51	2SA260R
51	2SA208M	51	2SA216D	51	2SA230L	51	2SA247D	51	2SA260X
51	2SA208OR	51	2SA216E	51	2SA230M	51	2SA247E	51	2SA260Y
51	2SA208R	51	2SA216F	51	2SA230OR	51	2SA247F	51	2SA261
51	2SA208X	51	2SA216G	51	2SA230R	51	2SA247G	51	2SA261A
51	2SA208Y	51	2SA216GN	51	2SA230X	51	2SA247GN	51	2SA261B
51	2SA209	51	2SA216H	51	2SA230Y	51	2SA247H	51	2SA261C
51	2SA209A	51	2SA216J	51	2SA234	51	2SA247J	51	2SA261D
51	2SA209B	51	2SA216K	51	2SA234A	51	2SA247K	51	2SA261E
51	2SA209C	51	2SA216L	51	2SA234B	51	2SA247L	51	2SA261F
51	2SA209D	51	2SA216M	51	2SA234C	51	2SA247M	51	2SA261G
51	2SA209E	51	2SA216OR	51	2SA234D	51	2SA247OR	51	2SA261GN
51	2SA209F	51	2SA216R	51	2SA234E	51	2SA247R	51	2SA261H
51	2SA209G	51	2SA216X	51	2SA234F	51	2SA247Y	51	2SA261K
51	2SA209GN	51	2SA216Y	51	2SA234G	51	2SA248	51	2SA261L
51	2SA209H	51	2SA217A	51	2SA234GN	51	2SA248A	51	2SA261M
51	2SA209J	51	2SA217B	51	2SA234H	51	2SA248B	51	2SA261OR
51	2SA209K	51	2SA217C	51	2SA234J	51	2SA248C	51	2SA261R
51	2SA209L	51	2SA217D	51	2SA234K	51	2SA248D	51	2SA261X
51	2SA209M	51	2SA217E	51	2SA234L	51	2SA248F	51	2SA261Y
51	2SA209OR	51	2SA217F	51	2SA234M	51	2SA248G	51	2SA262
51	2SA209R	51	2SA217G	51	2SA234OR	51	2SA248GN	51	2SA262A
51	2SA209X	51	2SA217GN	51	2SA234R	51	2SA248H	51	2SA262B
51	2SA209Y	51	2SA217H	51	2SA234X	51	2SA248J	51	2SA262C
51	2SA210	51	2SA217J	51	2SA234Y	51	2SA248K	51	2SA262D
51	2SA210A	51	2SA217K	51	2SA235	51	2SA248L	51	2SA262E
51	2SA210B	51	2SA217L	51	2SA235A	51	2SA248M	51	2SA262F
51	2SA210C	51	2SA217M	51	2SA235B	51	2SA248OR	51	2SA262G
51	2SA210D	51	2SA217OR	51	2SA235C	51	2SA248X	51	2SA262GN
51	2SA210E	51	2SA217R	51	2SA235D	51	2SA248Y	51	2SA262H
51	2SA210F	51	2SA217X	51	2SA235E	51	2SA251	51	2SA262J
51	2SA210G	51	2SA217Y	51	2SA235F	51	2SA251A	51	2SA262K
51	2SA210GN	51	2SA219	51	2SA235G	51	2SA251B	51	2SA262L
51	2SA210H	51	2SA219A	51	2SA235GN	51	2SA251C	51	2SA262M
51	2SA210J	51	2SA219B	51	2SA235H	51	2SA251D	51	2SA262OR
51	2SA210K	51	2SA219C	51	2SA235K	51	2SA251E	51	2SA262R
51	2SA210L	51	2SA219D	51	2SA235M	51	2SA251F	51	2SA262X
51	2SA210M	51	2SA219E	51	2SA235OR	51	2SA251G	51	2SA262Y
51	2SA210OR	51	2SA219F	51	2SA235R	51	2SA251H	51	2SA263
51	2SA210R	51	2SA219G	51	2SA235X	51	2SA251J	51	2SA263A
51	2SA210X	51	2SA219GN	51	2SA235Y	51	2SA251K	51	2SA263B
51	2SA210Y	51	2SA219H	51	2SA238	51	2SA251L	51	2SA263C
51	2SA211	51	2SA219J	51	2SA238A	51	2SA251M	51	2SA263D
51	2SA211A	51	2SA219K	51	2SA238B	51	2SA251OR	51	2SA263E
51	2SA211B	51	2SA219L	51	2SA238C	51	2SA251R	51	2SA263F
51	2SA211C	51	2SA219M	51	2SA238D	51	2SA251X	51	2SA263G
51	2SA211D	51	2SA219OR	51	2SA238E	51	2SA251Y	51	2SA263GN
51	2SA211E	51	2SA219R	51	2SA238F	51	2SA252	51	2SA263H
51	2SA211F	51	2SA219X	51	2SA238G	51	2SA252A	51	2SA263J
51	2SA211G	51	2SA219Y	51	2SA238GN	51	2SA252B	51	2SA263K
51	2SA211GN	51	2SA22	51	2SA238H	51	2SA252C	51	2SA263L
51	2SA211H	51	2SA223	51	2SA238J	51	2SA252D	51	2SA263M
51	2SA211J	51	2SA223A	51	2SA238K	51	2SA252E	51	2SA263OR
51	2SA211K	51	2SA223B	51	2SA238L	51	2SA252F	51	2SA263R
51	2SA211L	51	2SA223C	51	2SA239	51	2SA252G	51	2SA263X
51	2SA211M	51	2SA223D	51	2SA239A	51	2SA252H	51	2SA263Y
51	2SA211OR	51	2SA223E	51	2SA239B	51	2SA252J	51	2SA264
51	2SA211R	51	2SA223F	51	2SA239C	51	2SA252K	51	2SA264(1)
51	2SA211X	51	2SA223G	51	2SA239D	51	2SA252L	51	2SA264A
51	2SA211Y	51	2SA223GN	51	2SA239E	51	2SA252M	51	2SA264B
51	2SA212	51	2SA223H	51	2SA239F	51	2SA252OR	51	2SA264C
51	2SA212A	51	2SA223J	51	2SA239G	51	2SA252R	51	2SA264D
51	2SA212B	51	2SA223K	51	2SA239GN	51	2SA252X	51	2SA264E
51	2SA212C	51	2SA223L	51	2SA239GREEN	51	2SA252Y	51	2SA264F
51	2SA212D	51	2SA223M	51	2SA239H	51	2SA253	51	2SA264G
51	2SA212E	51	2SA223OR	51	2SA239J	51	2SA253A	51	2SA264GN
51	2SA212F	51	2SA223R	51	2SA239K	51	2SA253B	51	2SA264H
51	2SA212G	51	2SA223X	51	2SA239L	51	2SA253C	51	2SA264K
51	2SA212GN	51	2SA223Y	51	2SA239M	51	2SA253D	51	2SA264L
51	2SA212H	51	2SA225	51	2SA239OR	51	2SA253E	51	2SA264M
51	2SA212J	51	2SA225A	51	2SA239R	51	2SA253F	51	2SA264R
51	2SA212K	51	2SA225B	51	2SA239RED	51	2SA253G	51	2SA264X
51	2SA212L	51	2SA225C	51	2SA239X	51	2SA253GN	51	2SA264Y
51	2SA212M	51	2SA225D	51	2SA239Y	51	2SA253H	51	2SA265
51	2SA212OR	51	2SA225E	51	2SA240	51	2SA253J	51	2SA265A
51	2SA212R	51	2SA225F	51	2SA240A	51	2SA253K	51	2SA265B
51	2SA212X	51	2SA225G	51	2SA240B	51	2SA253L	51	2SA265C
51	2SA212Y	51	2SA225GN	51	2SA240B2	51	2SA253M	51	2SA265D
51	2SA213A	51	2SA225H	51	2SA240BL	51	2SA253OR	51	2SA265E
51	2SA213B	51	2SA225J	51	2SA240C	51	2SA253R	51	2SA265F
51	2SA213C	51	2SA225K	51	2SA240D	51	2SA253Y	51	2SA265G
51	2SA213D	51	2SA225L	51	2SA240E	51	2SA254	51	2SA265GN
51	2SA213E	51	2SA225M	51	2SA240F	51	2SA254A	51	2SA265H
51	2SA213F	51	2SA225OR	51	2SA240G	51	2SA254C	51	2SA265J
51	2SA213G	51	2SA225R	51	2SA240GN	51	2SA254D	51	2SA265K
51	2SA213GN	51	2SA225X	51	2SA240H	51	2SA254E	51	2SA265L
51	2SA213H	51	2SA225Y	51	2SA240J	51	2SA254F	51	2SA265M
51	2SA213J	51	2SA227	51	2SA240K			51	2SA265OR
51	2SA213K	51	2SA227A	51	2SA240L			51	2SA265R
51	2SA213L	51	2SA227B	51	2SA240M			51	2SA265X
51	2SA213M	51	2SA227C	51	2SA240OR			51	2SA265Y
51	2SA213OR	51	2SA227D	51	2SA240RED				

GE	Industry Standard No.	GE	Industry Standard No.	GE	Industry Standard No.	GE	Industry Standard No.	GE	Industry Standard No.
51	28A267	51	28A281L	51	28A292L	51	28A308F	51	28A326GN
51	28A267A	51	28A281M	51	28A292M	51	28A308G	51	28A326J
51	28A267B	51	28A281OR	51	28A292OR	51	28A308GN	51	28A326K
51	28A267C	51	28A281R	51	28A292R	51	28A308H	51	28A326L
51	28A267D	51	28A281X	51	28A292X	51	28A308J	51	28A326M
51	28A267E	51	28A281Y	51	28A292Y	51	28A308K	51	28A326OR
51	28A267F	51	28A282	51	28A293	51	28A308L	51	28A326R
51	28A267G	51	28A282A	51	28A293A	51	28A308M	51	28A326X
51	28A267GN	51	28A282B	51	28A293B	51	28A308OR	51	28A326Y
51	28A267H	51	28A282C	51	28A293C	51	28A308R	51	28A329
51	28A267J	51	28A282D	51	28A293D	51	28A308X	51	28A329A
51	28A267K	51	28A282E	51	28A293E	51	28A308Y	51	28A329B
51	28A267L	51	28A282F	51	28A293G	51	28A309	51	28A329C
51	28A267M	51	28A282G	51	28A293GN	51	28A309A	51	28A329D
51	28A267OR	51	28A282GN	51	28A293H	51	28A309B	51	28A329E
51	28A267R	51	28A282H	51	28A293J	51	28A309C	51	28A329G
51	28A267X	51	28A282J	51	28A293K	51	28A309D	51	28A329GN
51	28A267Y	51	28A282K	51	28A293L	51	28A309E	51	28A329H
51	28A268	51	28A282L	51	28A293M	51	28A309F	51	28A329J
51	28A268A	51	28A282M	51	28A293OR	51	28A309G	51	28A329K
51	28A268B	51	28A282OR	51	28A293R	51	28A309GN	51	28A329L
51	28A268C	51	28A282R	51	28A293X	51	28A309H	51	28A329M
51	28A268D	51	28A282X	51	28A293Y	51	28A309J	51	28A329OR
51	28A268E	51	28A282Y	51	28A294	51	28A309K	51	28A329R
51	28A268F	51	28A283	51	28A294A	51	28A309L	51	28A329X
51	28A268G	51	28A283A	51	28A294B	51	28A309M	51	28A329Y
51	28A268GN	51	28A283B	51	28A294C	51	28A309OR	51	28A32B
51	28A268H	51	28A283C	51	28A294D	51	28A309R	51	28A32C
51	28A268J	51	28A283D	51	28A294E	51	28A309X	51	28A32F
51	28A268K	51	28A283E	51	28A294F	51	28A309Y	51	28A32G
51	28A268L	51	28A283G	51	28A294GN	51	28A30A	51	28A32L
51	28A268M	51	28A283GN	51	28A294H	51	28A30B	51	28A32M
51	28A268OR	51	28A283H	51	28A294J	51	28A30C	51	28A32OR
51	28A268R	51	28A283J	51	28A294K	51	28A30D	51	28A32X
51	28A268X	51	28A283K	51	28A294L	51	28A30E	51	28A32Y
51	28A268Y	51	28A283L	51	28A294M	51	28A30F	51	28A330
51	28A26A	51	28A283M	51	28A294OR	51	28A30G	51	28A330A
51	28A26B	51	28A283OR	51	28A294R	51	28A30L	51	28A330B
51	28A26C	51	28A283R	51	28A294X	51	28A30M	51	28A330C
51	28A26D	51	28A283Y	51	28A294Y	51	28A30OR	51	28A330D
51	28A26E	51	28A284	51	28A295	51	28A30X	51	28A330E
51	28A26F	51	28A284A	51	28A295A	51	28A30Y	51	28A330F
51	28A26G	51	28A284B	51	28A295B	51	28A31	51	28A330G
51	28A26L	51	28A284C	51	28A295C	51	28A310	51	28A330GN
51	28A26M	51	28A284D	51	28A295D	51	28A310A	51	28A330H
51	28A26OR	51	28A284E	51	28A295E	51	28A310B	51	28A330J
51	28A26R	51	28A284F	51	28A295F	51	28A310C	51	28A330K
51	28A26X	51	28A284G	51	28A295G	51	28A310D	51	28A330L
51	28A26Y	51	28A284GN	51	28A295GN	51	28A310E	51	28A330M
51	28A27	51	28A284J	51	28A295H	51	28A310G	51	28A330OR
51	28A271(2)	51	28A284K	51	28A295J	51	28A310GN	51	28A330X
51	28A271(3)	51	28A284L	51	28A295K	51	28A310H	51	28A330Y
51	28A277	51	28A284OR	51	28A295L	51	28A310J	51	28A331
51	28A277A	51	28A284R	51	28A295OR	51	28A310K	51	28A331A
51	28A277B	51	28A284X	51	28A295R	51	28A310L	51	28A331B
51	28A277C	51	28A284Y	51	28A295Y	51	28A310M	51	28A331C
51	28A277D	51	28A289	51	28A296	51	28A310OR	51	28A331D
51	28A277E	51	28A289A	51	28A296A	51	28A310X	51	28A331F
51	28A277F	51	28A289B	51	28A296B	51	28A310Y	51	28A331G
51	28A277G	51	28A289C	51	28A296C	51	28A31A	51	28A331GN
51	28A277GN	51	28A289D	51	28A296D	51	28A31B	51	28A331H
51	28A277H	51	28A289E	51	28A296E	51	28A31C	51	28A331J
51	28A277J	51	28A289F	51	28A296F	51	28A31D	51	28A331K
51	28A277K	51	28A289G	51	28A296G	51	28A31F	51	28A331L
51	28A277L	51	28A289GN	51	28A296GN	51	28A31G	51	28A331M
51	28A277M	51	28A289H	51	28A296H	51	28A31L	51	28A331OR
51	28A277OR	51	28A289J	51	28A296J	51	28A31M	51	28A331R
51	28A277X	51	28A289L	51	28A296K	51	28A31OR	51	28A331X
51	28A277Y	51	28A289M	51	28A296L	51	28A31X	51	28A331Y
51	28A278	51	28A289OR	51	28A296M	51	28A31Y	51	28A335
51	28A278A	51	28A289R	51	28A296OR	51	28A32	51	28A335A
51	28A278B	51	28A289X	51	28A296X	51	28A321	51	28A335B
51	28A278C	51	28A289Y	51	28A296Y	51	28A321A-1	51	28A335C
51	28A278D	51	28A28A	51	28A297	51	28A321B	51	28A335D
51	28A278E	51	28A28B	51	28A297A	51	28A321C	51	28A335E
51	28A278F	51	28A28C	51	28A297B	51	28A321D	51	28A335F
51	28A278G	51	28A28D	51	28A297C	51	28A321E	51	28A335GN
51	28A278GN	51	28A28E	51	28A297E	51	28A321F	51	28A335H
51	28A278J	51	28A28F	51	28A297F	51	28A321G	51	28A335J
51	28A278K	51	28A28G	51	28A297G	51	28A321GN	51	28A335K
51	28A278L	51	28A28L	51	28A297H	51	28A321H	51	28A335L
51	28A278M	51	28A28M	51	28A297J	51	28A321J	51	28A335M
51	28A278OR	51	28A28OR	51	28A297K	51	28A321K	51	28A335OR
51	28A278R	51	28A28X	51	28A297L	51	28A321L	51	28A335R
51	28A278X	51	28A28Y	51	28A297M	51	28A321M	51	28A335X
51	28A278Y	51	28A290A	51	28A297OR	51	28A321OR	51	28A335Y
51	28A279	51	28A290B	51	28A297R	51	28A321R	51	28A337
51	28A279A	51	28A290C	51	28A297X	51	28A321X	51	28A337A
51	28A279B	51	28A290D	51	28A297Y	51	28A321Y	51	28A337B
51	28A279C	51	28A290E	51	28A30	51	28A324	51	28A337C
51	28A279D	51	28A290F	51	28A301	51	28A324A	51	28A337E
51	28A279E	51	28A290G	51	28A301A	51	28A324B	51	28A337F
51	28A279F	51	28A290GN	51	28A301B	51	28A324C	51	28A337G
51	28A279G	51	28A290H	51	28A301C	51	28A324D	51	28A337GN
51	28A279GN	51	28A290J	51	28A301D	51	28A324F	51	28A337H
51	28A279H	51	28A290K	51	28A301E	51	28A324G	51	28A337J
51	28A279J	51	28A290L	51	28A301F	51	28A324GN	51	28A337K
51	28A279K	51	28A290M	51	28A301G	51	28A324H	51	28A337L
51	28A279L	51	28A290OR	51	28A301GN	51	28A324K	51	28A337M
51	28A279OR	51	28A290R	51	28A301H	51	28A324L	51	28A337OR
51	28A279R	51	28A290X	51	28A301J	51	28A324M	51	28A337R
51	28A279X	51	28A290Y	51	28A301K	51	28A324OR	51	28A337X
51	28A279Y	51	28A291	51	28A301L	51	28A324R	51	28A337Y
51	28A28	51	28A291A	51	28A301M	51	28A324X	51	28A341OA
51	28A280	51	28A291B	51	28A301OR	51	28A324Y	51	28A341OB
51	28A280A	51	28A291C	51	28A301R	51	28A325	51	28A342
51	28A280B	51	28A291D	51	28A301X	51	28A325A	51	28A342A
51	28A280C	51	28A291E	51	28A301Y	51	28A325B	51	28A342B
51	28A280D	51	28A291F	51	28A305	51	28A325D	51	28A342C
51	28A280E	51	28A291G	51	28A305A	51	28A325E	51	28A342D
51	28A280F	51	28A291GN	51	28A305B	51	28A325G	51	28A342E
51	28A280G	51	28A291H	51	28A305C	51	28A325GN	51	28A342F
51	28A280GN	51	28A291J	51	28A305D	51	28A325H	51	28A342G
51	28A280H	51	28A291K	51	28A305E	51	28A325J	51	28A342GN
51	28A280J	51	28A291L	51	28A305F	51	28A325K	51	28A342H
51	28A280K	51	28A291M	51	28A305G	51	28A325L	51	28A342J
51	28A280L	51	28A291OR	51	28A305GN	51	28A325M	51	28A342K
51	28A280M	51	28A291R	51	28A305H	51	28A325OR	51	28A342L
51	28A280OR	51	28A291X	51	28A305J	51	28A325R	51	28A342M
51	28A280R	51	28A291Y	51	28A305K	51	28A325X	51	28A342OR
51	28A280X	51	28A292A	51	28A305L	51	28A325Y	51	28A342R
51	28A280Y	51	28A292B	51	28A305M	51	28A326	51	28A342X
51	28A281	51	28A292C	51	28A305OR	51	28A326A	51	28A342Y
51	28A281A	51	28A292E	51	28A305R	51	28A326B	51	28A343
51	28A281B	51	28A292F	51	28A305X	51	28A326C	51	28A343A
51	28A281C	51	28A292G	51	28A305Y	51	28A326D	51	28A343B
51	28A281D	51	28A292GN	51	28A308	51	28A326E	51	28A343C
51	28A281E	51	28A292H	51	28A308A	51	28A326G		
51	28A281F	51	28A292J	51	28A308B				
51	28A281G	51	28A292GN	51	28A308C				
51	28A281GN	51	28A292H	51	28A308D				
51	28A281H	51	28A292J	51	28A308E				
51	28A281J	51	28A292K						
51	28A281K								

GE	Industry Standard No.	GE	Industry Standard No.	GE	Industry Standard No.	GE	Industry Standard No.	GE	Industry Standard No.
51	2SA343D	51	2SA355D	51	2SA367	51	2SA392D	51	2SA406A
51	2SA343E	51	2SA355E	51	2SA367A	51	2SA392E	51	2SA406B
51	2SA343F	51	2SA355F	51	2SA367B	51	2SA392F	51	2SA406C
51	2SA343G	51	2SA355G	51	2SA367C	51	2SA392G	51	2SA406D
51	2SA343H	51	2SA355H	51	2SA367D	51	2SA392GN	51	2SA406E
51	2SA343J	51	2SA355J	51	2SA367E	51	2SA392H	51	2SA406F
51	2SA343K	51	2SA355K	51	2SA367F	51	2SA392J	51	2SA406G
51	2SA343L	51	2SA355L	51	2SA367G	51	2SA392K	51	2SA406GN
51	2SA343M	51	2SA355M	51	2SA367GN	51	2SA392L	51	2SA406H
51	2SA343OR	51	2SA355OR	51	2SA367H	51	2SA392M	51	2SA406J
51	2SA343R	51	2SA355R	51	2SA367J	51	2SA392OR	51	2SA406L
51	2SA343X	51	2SA355X	51	2SA367K	51	2SA392R	51	2SA406M
51	2SA343Y	51	2SA355Y	51	2SA367L	51	2SA392X	51	2SA406OR
51	2SA344	51	2SA358	51	2SA367M	51	2SA392Y	51	2SA406R
51	2SA344A	51	2SA358-3	51	2SA367OR	51	2SA394	51	2SA406X
51	2SA344C	51	2SA358A	51	2SA367R	51	2SA394A	51	2SA406Y
51	2SA344D	51	2SA358B	51	2SA367X	51	2SA394B	51	2SA407
51	2SA344E	51	2SA358C	51	2SA367Y	51	2SA394C	51	2SA407A
51	2SA344F	51	2SA358D	51	2SA36A	51	2SA394D	51	2SA407B
51	2SA344GN	51	2SA358E	51	2SA36B	51	2SA394E	51	2SA407C
51	2SA344H	51	2SA358F	51	2SA36C	51	2SA394F	51	2SA407D
51	2SA344J	51	2SA358G	51	2SA36D	51	2SA394G	51	2SA407E
51	2SA344K	51	2SA358GN	51	2SA36E	51	2SA394GN	51	2SA407F
51	2SA344L	51	2SA358H	51	2SA36F	51	2SA394H	51	2SA407G
51	2SA344M	51	2SA358J	51	2SA36G	51	2SA394J	51	2SA407GN
51	2SA344OR	51	2SA358K	51	2SA36L	51	2SA394K	51	2SA407H
51	2SA344R	51	2SA358L	51	2SA36M	51	2SA394L	51	2SA407J
51	2SA344X	51	2SA358M	51	2SA36OR	51	2SA394M	51	2SA407K
51	2SA345	51	2SA358OR	51	2SA36R	51	2SA394OR	51	2SA407L
51	2SA348	51	2SA358R	51	2SA36X	51	2SA394R	51	2SA407M
51	2SA348A	51	2SA358X	51	2SA36Y	51	2SA394Y	51	2SA407OR
51	2SA348B	51	2SA358Y	51	2SA37	51	2SA395	51	2SA407R
51	2SA348C	51	2SA359	51	2SA376	51	2SA395A	51	2SA407X
51	2SA348D	51	2SA359A	51	2SA376A	51	2SA395B	51	2SA407Y
51	2SA348E	51	2SA359B	51	2SA376B	51	2SA395D	51	2SA40A
51	2SA348F	51	2SA359C	51	2SA376C	51	2SA395E	51	2SA40C
51	2SA348G	51	2SA359D	51	2SA376D	51	2SA395F	51	2SA40D
51	2SA348GN	51	2SA359E	51	2SA376E	51	2SA395G	51	2SA40E
51	2SA348H	51	2SA359F	51	2SA376F	51	2SA395GN	51	2SA40F
51	2SA348J	51	2SA359G	51	2SA376GN	51	2SA395H	51	2SA40G
51	2SA348K	51	2SA359GN	51	2SA376H	51	2SA395J	51	2SA40L
51	2SA348L	51	2SA359H	51	2SA376J	51	2SA395K	51	2SA40M
51	2SA348M	51	2SA359J	51	2SA376K	51	2SA395L	51	2SA40OR
51	2SA348OR	51	2SA359K	51	2SA376L	51	2SA395M	51	2SA40R
51	2SA348R	51	2SA359L	51	2SA376OR	51	2SA395OR	51	2SA40X
51	2SA348X	51	2SA359M	51	2SA376R	51	2SA395R	51	2SA40Y
51	2SA348Y	51	2SA359OR	51	2SA376X	51	2SA395X	51	2SA41
51	2SA35	51	2SA359R	51	2SA376Y	51	2SA395Y	51	2SA412
51	2SA350	51	2SA359X	51	2SA377	51	2SA398	51	2SA412A
51	2SA350A	51	2SA359Y	51	2SA377A	51	2SA398A	51	2SA412B
51	2SA350AV	51	2SA35A	51	2SA377B	51	2SA398B	51	2SA412C
51	2SA350B	51	2SA35B	51	2SA377C	51	2SA398C	51	2SA412D
51	2SA350BK	51	2SA35C	51	2SA377D	51	2SA398D	51	2SA412E
51	2SA350C	51	2SA35D	51	2SA377E	51	2SA398E	51	2SA412F
51	2SA350D	51	2SA35E	51	2SA377F	51	2SA398F	51	2SA412G
51	2SA350E	51	2SA35F	51	2SA377G	51	2SA398G	51	2SA412GN
51	2SA350F	51	2SA35G	51	2SA377GN	51	2SA398GN	51	2SA412H
51	2SA350G	51	2SA35L	51	2SA377H	51	2SA398H	51	2SA412J
51	2SA350GN	51	2SA35M	51	2SA377J	51	2SA398J	51	2SA412K
51	2SA350H	51	2SA35OR	51	2SA377K	51	2SA398K	51	2SA412L
51	2SA350J	51	2SA35X	51	2SA377L	51	2SA398L	51	2SA412M
51	2SA350K	51	2SA35Y	51	2SA377M	51	2SA398M	51	2SA412OR
51	2SA350L	51	2SA36	51	2SA377OR	51	2SA398OR	51	2SA412R
51	2SA350M	51	2SA360	51	2SA377X	51	2SA398R	51	2SA412X
51	2SA350OR	51	2SA360A	51	2SA377Y	51	2SA398X	51	2SA412Y
51	2SA350T	51	2SA360B	51	2SA37A	51	2SA398Y	51	2SA414
51	2SA350TY	51	2SA360C	51	2SA37B	51	2SA399	51	2SA414A
51	2SA350X	51	2SA360E	51	2SA37C	51	2SA399A	51	2SA414B
51	2SA352	51	2SA360F	51	2SA37D	51	2SA399B	51	2SA414C
51	2SA352A	51	2SA360G	51	2SA37E	51	2SA399C	51	2SA414D
51	2SA352B	51	2SA360GN	51	2SA37F	51	2SA399D	51	2SA414E
51	2SA352C	51	2SA360H	51	2SA37G	51	2SA399E	51	2SA414F
51	2SA352E	51	2SA360J	51	2SA37L	51	2SA399F	51	2SA414G
51	2SA352F	51	2SA360K	51	2SA37M	51	2SA399G	51	2SA414GN
51	2SA352GN	51	2SA360L	51	2SA37OR	51	2SA399GN	51	2SA414H
51	2SA352H	51	2SA360M	51	2SA37R	51	2SA399H	51	2SA414J
51	2SA352J	51	2SA360OR	51	2SA37X	51	2SA399J	51	2SA414K
51	2SA352K	51	2SA360R	51	2SA37Y	51	2SA399K	51	2SA414L
51	2SA352L	51	2SA360X	51	2SA38	51	2SA399L	51	2SA414M
51	2SA352M	51	2SA360Y	51	2SA385	51	2SA399M	51	2SA414OR
51	2SA352OR	51	2SA364	51	2SA385A	51	2SA399OR	51	2SA414R
51	2SA352R	51	2SA364A	51	2SA385B	51	2SA399X	51	2SA414X
51	2SA352X	51	2SA364B	51	2SA385C	51	2SA399Y	51	2SA415
51	2SA352Y	51	2SA364C	51	2SA385D	51	2SA39A	51	2SA415A
51	2SA353	51	2SA364D	51	2SA385E	51	2SA39B	51	2SA415B
51	2SA353-AC	51	2SA364E	51	2SA385F	51	2SA39C	51	2SA415C
51	2SA353A	51	2SA364F	51	2SA385G	51	2SA39D	51	2SA415D
51	2SA353AL	51	2SA364G	51	2SA385GN	51	2SA39E	51	2SA415E
51	2SA353B	51	2SA364GN	51	2SA385H	51	2SA39F	51	2SA415F
51	2SA353C	51	2SA364H	51	2SA385J	51	2SA39G	51	2SA415G
51	2SA353CL	51	2SA364J	51	2SA385K	51	2SA39L	51	2SA415H
51	2SA353D	51	2SA364K	51	2SA385L	51	2SA39M	51	2SA415J
51	2SA353E	51	2SA364L	51	2SA385M	51	2SA39OR	51	2SA415K
51	2SA353F	51	2SA364M	51	2SA385OR	51	2SA39R	51	2SA415L
51	2SA353Q	51	2SA364OR	51	2SA385R	51	2SA39X	51	2SA415M
51	2SA353GN	51	2SA364R	51	2SA385X	51	2SA39Y	51	2SA415OR
51	2SA353H	51	2SA364X	51	2SA385Y	51	2SA40	51	2SA415R
51	2SA353J	51	2SA364Y	51	2SA38A	51	2SA403	51	2SA415X
51	2SA353K	51	2SA365	51	2SA38B	51	2SA403A	51	2SA41A
51	2SA353L	51	2SA365A	51	2SA38C	51	2SA403B	51	2SA41B
51	2SA353OR	51	2SA365B	51	2SA38D	51	2SA403C	51	2SA41C
51	2SA353R	51	2SA365C	51	2SA38E	51	2SA403D	51	2SA41D
51	2SA353X	51	2SA365D	51	2SA38F	51	2SA403E	51	2SA41F
51	2SA353Y	51	2SA365E	51	2SA38G	51	2SA403F	51	2SA41G
51	2SA354	51	2SA365F	51	2SA38L	51	2SA403G	51	2SA41L
51	2SA354-B	51	2SA365G	51	2SA38M	51	2SA403GN	51	2SA41M
51	2SA354A	51	2SA365GN	51	2SA38OR	51	2SA403H	51	2SA41R
51	2SA354B	51	2SA365H	51	2SA38R	51	2SA403J	51	2SA41X
51	2SA354BK	51	2SA365J	51	2SA38X	51	2SA403K	51	2SA41Y
51	2SA354C	51	2SA365K	51	2SA38Y	51	2SA403L	51	2SA420
51	2SA354D	51	2SA365L	51	2SA39	51	2SA403M	51	2SA420A
51	2SA354E	51	2SA365M	51	2SA391	51	2SA403OR	51	2SA420B
51	2SA354F	51	2SA365OR	51	2SA391A	51	2SA403R	51	2SA420C
51	2SA354G	51	2SA365R	51	2SA391B	51	2SA403X	51	2SA420D
51	2SA354GN	51	2SA365X	51	2SA391C	51	2SA403Y	51	2SA420E
51	2SA354H	51	2SA365Y	51	2SA391D	51	2SA404	51	2SA420F
51	2SA354J	51	2SA366	51	2SA391E	51	2SA404A	51	2SA420G
51	2SA354K	51	2SA366A	51	2SA391F	51	2SA404B	51	2SA420H
51	2SA354L	51	2SA366B	51	2SA391GN	51	2SA404C	51	2SA420J
51	2SA354M	51	2SA366C	51	2SA391H	51	2SA404D	51	2SA420K
51	2SA354OR	51	2SA366D	51	2SA391J	51	2SA404E	51	2SA420L
51	2SA354R	51	2SA366E	51	2SA391K	51	2SA404G	51	2SA420M
51	2SA354X	51	2SA366F	51	2SA391L	51	2SA404GN	51	2SA420OR
51	2SA354Y	51	2SA366G	51	2SA391M	51	2SA404H	51	2SA420R
51	2SA355	51	2SA366GN	51	2SA391OR	51	2SA404K	51	2SA420X
51	2SA355A	51	2SA366H	51	2SA391R	51	2SA404L	51	2SA420Y
51	2SA355B	51	2SA366J	51	2SA391X	51	2SA404M	51	2SA426
51	2SA355C	51	2SA366K	51	2SA391Y	51	2SA404OR	51	2SA426GN
51		51	2SA366L	51	2SA392	51	2SA404R	51	2SA427
51		51	2SA366M	51	2SA392A	51	2SA404X	51	
51		51	2SA366OR	51	2SA392B	51	2SA404Y	51	
51		51	2SA366R	51	2SA392C	51	2SA406	51	
51		51	2SA366X	51		51		51	
51		51	2SA366Y	51		51		51	

GE	Industry Standard No.	GE	Industry Standard No.	GE	Industry Standard No.	GE	Industry Standard No.	GE	Industry Standard No.
51	28A427A	51	28A438	51	28A455L	51	28A477K	51	28A525L
51	28A427B	51	28A438A	51	28A455M	51	28A477L	51	28A525M
51	28A427C	51	28A438B	51	28A455OR	51	28A477M	51	28A525OR
51	28A427D	51	28A438C	51	28A455R	51	28A477OR	51	28A525R
51	28A427E	51	28A438E	51	28A455X	51	28A477R	51	28A525X
51	28A427F	51	28A438F	51	28A455Y	51	28A477X	51	28A525Y
51	28A427G	51	28A438G	51	28A456	51	28A477Y	51	28A52A
51	28A427GN	51	28A438GN	51	28A456A	51	28A478	51	28A52B
51	28A427H	51	28A438H	51	28A456B	51	28A478A	51	28A52C
51	28A427J	51	28A438J	51	28A456C	51	28A478B	51	28A52D
51	28A427K	51	28A438K	51	28A456D	51	28A478C	51	28A52E
51	28A427L	51	28A438L	51	28A456E	51	28A478D	51	28A52F
51	28A427M	51	28A438M	51	28A456F	51	28A478E	51	28A52G
51	28A427OR	51	28A438OR	51	28A456G	51	28A478F	51	28A52L
51	28A427R	51	28A438R	51	28A456GN	51	28A478G	51	28A52M
51	28A427X	51	28A438X	51	28A456H	51	28A478GN	51	28A52OR
51	28A427Y	51	28A438Y	51	28A456J	51	28A478H	51	28A52R
51	28A428	51	28A44	51	28A456K	51	28A478J	51	28A52X
51	28A428A	51	28A440	51	28A456L	51	28A478K	51	28A52Y
51	28A428B	51	28A440A	51	28A456M	51	28A478L	51	28A53
51	28A428C	51	28A440AL	51	28A456OR	51	28A478M	51	28A536
51	28A428D	51	28A440B	51	28A456R	51	28A478OR	51	28A53A
51	28A428E	51	28A440C	51	28A456X	51	28A478R	51	28A53B
51	28A428F	51	28A440D	51	28A456Y	51	28A478X	51	28A53C
51	28A428G	51	28A440E	51	28A457	51	28A478Y	51	28A53D
51	28A428GN	51	28A440F	51	28A457A	51	28A479	51	28A53E
51	28A428H	51	28A440G	51	28A457B	51	28A479A	51	28A53F
51	28A428J	51	28A440GN	51	28A457C	51	28A479B	51	28A53G
51	28A428K	51	28A440H	51	28A457D	51	28A479C	51	28A53L
51	28A428L	51	28A440J	51	28A457E	51	28A479D	51	28A53M
51	28A428M	51	28A440K	51	28A457F	51	28A479E	51	28A53OR
51	28A428OR	51	28A440L	51	28A457G	51	28A479F	51	28A53R
51	28A428R	51	28A440M	51	28A457GN	51	28A479G	51	28A53X
51	28A428X	51	28A440OR	51	28A457H	51	28A479GN	51	28A53Y
51	28A428Y	51	28A440R	51	28A457J	51	28A479H	51	28A54
51	28A432	51	28A440X	51	28A457K	51	28A479J	51	28A54A
51	28A432A	51	28A440Y	51	28A457L	51	28A479K	51	28A54B
51	28A432B	51	28A446	51	28A457M	51	28A479L	51	28A54C
51	28A432C	51	28A446A	51	28A457OR	51	28A479M	51	28A54D
51	28A432D	51	28A446B	51	28A457R	51	28A479OR	51	28A54E
51	28A432E	51	28A446C	51	28A457X	51	28A479R	51	28A54G
51	28A432F	51	28A446D	51	28A457Y	51	28A479X	51	28A54L
51	28A432G	51	28A446E	51	28A466	51	28A479Y	51	28A54M
51	28A432GN	51	28A446F	51	28A466-2	51	28A48	51	28A540R
51	28A432H	51	28A446G	51	28A466-3	51	28A49	51	28A54R
51	28A432K	51	28A446GN	51	28A466A	51	28A49A	51	28A54X
51	28A432L	51	28A446H	51	28A466B	51	28A49B	51	28A54Y
51	28A432M	51	28A446J	51	28A466BLK	51	28A49C	51	28A55A
51	28A432OR	51	28A446K	51	28A466BLU	51	28A49D	51	28A55B
51	28A432R	51	28A446L	51	28A466C	51	28A49E	51	28A55C
51	28A432X	51	28A446M	51	28A466D	51	28A49F	51	28A55D
51	28A432Y	51	28A446OR	51	28A466E	51	28A49G	51	28A55E
51	28A433	51	28A446R	51	28A466F	51	28A49L	51	28A55F
51	28A433A	51	28A446X	51	28A466G	51	28A49M	51	28A55G
51	28A433B	51	28A446Y	51	28A466GN	51	28A49OR	51	28A55L
51	28A433C	51	28A447	51	28A466H	51	28A49R	51	28A55M
51	28A433D	51	28A447A	51	28A466J	51	28A49X	51	28A55OR
51	28A433E	51	28A447B	51	28A466K	51	28A49Y	51	28A55R
51	28A433F	51	28A447C	51	28A466L	51	28A50	51	28A55X
51	28A433G	51	28A447D	51	28A466M	51	28A507	51	28A55Y
51	28A433GN	51	28A447E	51	28A466OR	51	28A507A	51	28A58A
51	28A433K	51	28A447F	51	28A466R	51	28A507B	51	28A58B
51	28A433L	51	28A447G	51	28A466X	51	28A507C	51	28A58C
51	28A433M	51	28A447H	51	28A466Y	51	28A507D	51	28A58D
51	28A433OR	51	28A447J	51	28A466YEL	51	28A507E	51	28A58E
51	28A433X	51	28A447K	51	28A470	51	28A507F	51	28A58G
51	28A433Y	51	28A447L	51	28A470A	51	28A507G	51	28A58H
51	28A434	51	28A447M	51	28A470B	51	28A507GN	51	28A58J
51	28A434A	51	28A447OR	51	28A470C	51	28A507H	51	28A58K
51	28A434C	51	28A447R	51	28A470D	51	28A507J	51	28A58L
51	28A434D	51	28A447X	51	28A470E	51	28A507K	51	28A58M
51	28A434E	51	28A447Y	51	28A470F	51	28A507L	51	28A58OR
51	28A434F	51	28A44A	51	28A470G	51	28A507OR	51	28A58R
51	28A434G	51	28A44B	51	28A470GN	51	28A507R	51	28A58X
51	28A434GN	51	28A44C	51	28A470H	51	28A507X	51	28A58Y
51	28A434H	51	28A44D	51	28A470J	51	28A507Y	51	28A60A
51	28A434J	51	28A44E	51	28A470K	51	28A50A	51	28A60B
51	28A434K	51	28A44F	51	28A470L	51	28A50B	51	28A60C
51	28A434L	51	28A44G	51	28A470M	51	28A50C	51	28A60D
51	28A434M	51	28A44L	51	28A470OR	51	28A50D	51	28A60F
51	28A434OR	51	28A44M	51	28A470R	51	28A50E	51	28A60G
51	28A434R	51	28A44OR	51	28A470X	51	28A50F	51	28A60H
51	28A434X	51	28A44R	51	28A470Y	51	28A50G	51	28A60K
51	28A434Y	51	28A44X	51	28A471	51	28A50L	51	28A60M
51	28A435	51	28A44Y	51	28A471-1	51	28A50M	51	28A60OR
51	28A435A	51	28A453	51	28A471-2	51	28A50OR	51	28A60R
51	28A435B	51	28A453A	51	28A471-3	51	28A50R	51	28A60X
51	28A435C	51	28A453B	51	28A471A	51	28A50X	51	28A60Y
51	28A435D	51	28A453C	51	28A471B	51	28A50Y	51	28A61A
51	28A435E	51	28A453D	51	28A471C	51	28A51	51	28A61C
51	28A435F	51	28A453E	51	28A471D	51	28A518	51	28A61D
51	28A435G	51	28A453F	51	28A471E	51	28A518-G	51	28A61G
51	28A435GN	51	28A453G	51	28A471F	51	28A518A	51	28A61K
51	28A435H	51	28A453GN	51	28A471G	51	28A518B	51	28A61L
51	28A435J	51	28A453H	51	28A471GN	51	28A518C	51	28A61OR
51	28A435L	51	28A453J	51	28A471H	51	28A518D	51	28A61R
51	28A435M	51	28A453K	51	28A471J	51	28A518E	51	28A61X
51	28A435OR	51	28A453L	51	28A471K	51	28A518F	51	28A61Y
51	28A435R	51	28A453M	51	28A471L	51	28A518G	51	28A65
51	28A435X	51	28A453OR	51	28A471M	51	28A518GN	51	28A65A
51	28A435Y	51	28A453R	51	28A471OR	51	28A518H	51	28A65B
51	28A436	51	28A453X	51	28A471R	51	28A518J	51	28A65C
51	28A436A	51	28A453Y	51	28A471X	51	28A518K	51	28A65D
51	28A436C	51	28A454	51	28A471Y	51	28A518L	51	28A65E
51	28A436D	51	28A454A	51	28A728B	51	28A518M	51	28A65F
51	28A436E	51	28A454B	51	28A474	51	28A518OR	51	28A65G
51	28A436F	51	28A454C	51	28A474A	51	28A518R	51	28A65L
51	28A436G	51	28A454D	51	28A474B	51	28A518X	51	28A65OR
51	28A436GN	51	28A454E	51	28A474C	51	28A518Y	51	28A65R
51	28A436H	51	28A454F	51	28A474D	51	28A51A	51	28A65Y
51	28A436J	51	28A454G	51	28A474E	51	28A51B	51	28A65X-R
51	28A436K	51	28A454GN	51	28A474F	51	28A51C	51	28A66A
51	28A437	51	28A454H	51	28A474G	51	28A51D	51	28A66B
51	28A437A	51	28A454J	51	28A474GN	51	28A51E	51	28A66C
51	28A437B	51	28A454K	51	28A474H	51	28A51F	51	28A66D
51	28A437C	51	28A454L	51	28A474J	51	28A51G	51	28A66E
51	28A437D	51	28A454OR	51	28A474K	51	28A51L	51	28A66F
51	28A437E	51	28A454R	51	28A474L	51	28A51M	51	28A66G
51	28A437F	51	28A454X	51	28A474M	51	28A51OR	51	28A66K
51	28A437G	51	28A454Y	51	28A474OR	51	28A51R	51	28A66L
51	28A437GN	51	28A455	51	28A474X	51	28A51X	51	28A66M
51	28A437H	51	28A455A	51	28A474Y	51	28A51Y	51	28A66OR
51	28A437J	51	28A455B	51	28A477	51	28A52	51	28A66R
51	28A437K	51	28A455C	51	28A477A	51	28A525	51	28A66X
51	28A437L	51	28A455D	51	28A477B	51	28A525A	51	
51	28A437M	51	28A455E	51	28A477C	51	28A525B	51	
51	28A437OR	51	28A455F	51	28A477D	51	28A525C	51	
51	28A437X	51	28A455G	51	28A477E	51	28A525D	51	
51	28A437Y	51	28A455GN	51	28A477F	51	28A525E	51	
51		51	28A455H	51	28A477G	51	28A525F	51	
51		51	28A455J	51	28A477GN	51	28A525GN	51	
51		51	28A455K	51	28A477H	51	28A525H	51	
51		51		51	28A477J	51	28A525K	51	

GE	Industry Standard No.
51	2SA66Y
51	2SA67A
51	2SA67B
51	2SA67C
51	2SA67D
51	2SA67E
51	2SA67F
51	2SA67G
51	2SA67H
51	2SA67K
51	2SA67L
51	2SA67OR
51	2SA67R
51	2SA67X
51	2SA67Y
51	2SA69A
51	2SA69B
51	2SA69C
51	2SA69D
51	2SA69E
51	2SA69F
51	2SA69G
51	2SA69K
51	2SA69L
51	2SA69M
51	2SA69OR
51	2SA69R
51	2SA69X
51	2SA69Z
51	2SA70
51	2SA70-08
51	2SA70-0B
51	2SA70A
51	2SA70B
51	2SA70C
51	2SA70D
51	2SA70E
51	2SA70G
51	2SA70H
51	2SA70K
51	2SA70L
51	2SA70MA
51	2SA70OA
51	2SA70OR
51	2SA70R
51	2SA70X
51	2SA70Y
51	2SA71
51	2SA71A
51	2SA71AB
51	2SA71AC
51	2SA71B
51	2SA71BS
51	2SA71C
51	2SA71D
51	2SA71E
51	2SA71G
51	2SA71H
51	2SA71K
51	2SA71L
51	2SA71M
51	2SA71OR
51	2SA71R
51	2SA71X
51	2SA71Y
51	2SA71YA
51	2SA74
51	2SA74A
51	2SA74C
51	2SA74D
51	2SA74E
51	2SA74F
51	2SA74H
51	2SA74K
51	2SA74L
51	2SA74M
51	2SA74OR
51	2SA74R
51	2SA74X
51	2SA75
51	2SA75A
51	2SA75B
51	2SA75C
51	2SA75D
51	2SA75E
51	2SA75F
51	2SA75G
51	2SA75H
51	2SA75K
51	2SA75L
51	2SA75M
51	2SA75OR
51	2SA75R
51	2SA75X
51	2SA75Y
51	2SA76
51	2SA76A
51	2SA76B
51	2SA76C
51	2SA76D
51	2SA76E
51	2SA76F
51	2SA76G
51	2SA76H
51	2SA76K
51	2SA76L
51	2SA76OR
51	2SA76R
51	2SA76X
51	2SA76Z
51	2SA77
51	2SA77A
51	2SA77B
51	2SA77C
51	2SA77D
51	2SA77E
51	2SA77F
51	2SA77G
51	2SA77H
51	2SA77K
51	2SA77L
51	2SA77M
51	2SA77OR
51	2SA77R
51	2SA77X
51	2SA77Y
51	2SA87
51	2SA87A
51	2SA87B
51	2SA87C
51	2SA87D
51	2SA87E
51	2SA87F
51	2SA87G
51	2SA87H
51	2SA87K
51	2SA87L
51	2SA87M
51	2SA87OR
51	2SA87R
51	2SA87X
51	2SA87Y
51	2SAJ150N
51	2SAU/3H
51	2SAU03H
52	A-102
52	A059-104
52	A059-105
52	A061-107
52	A061-110
52	A061-111
52	A069-121
52	A1-5
52	A100A
52	A100B
52	A100C
52	A101
52	A101AA
52	A101AY
52	A101B
52	A101BA
52	A101BB
52	A101BC
52	A101BX
52	A101C
52	A101CA
52	A101CV
52	A101CX
52	A101E
52	A101QA
52	A101V
52	A101X
52	A101Z
52	A102
52	A102(JAPAN)
52	A102A
52	A102AA
52	A102AB
52	A102B
52	A102BA
52	A102BN
52	A102CA
52	A102EV
52	A103
52	A103A
52	A103B
52	A103C
52	A103CA
52	A103CAK
52	A103CG
52	A103DA
52	A104
52	A104(JAPAN)
52	A104A
52	A104B
52	A104D
52	A104Y
52	A109
52	A111(JAPAN)
52	A112
52	A113
52	A121
52	A122(JAPAN)
52	A123
52	A124
52	A125
52	A131
52	A132(JAPAN)
52	A133(JAPAN)
52	A134
52	A136
52	A137(JAPAN)
52	A1383
52	A1384
52	A139(JAPAN)
52	A14-1001
52	A14-1002
52	A14-1003
52	A141(JAPAN)
52	A141B
52	A141C
52	A142(JAPAN)
52	A142A
52	A142B
52	A142C
52	A143(JAPAN)
52	A144
52	A144C
52	A145
52	A145A
52	A145C
52	A15-1001
52	A15-1002
52	A15-1003
52	A150
52	A154(PNP)
52	A155(PNP)
52	A156(JAPAN)
52	A157(JAPAN)
52	A164(PNP)
52	A165(PNP)
52	A166
52	A175
52	A176
52	A188
52	A189
52	A19
52	A20
52	A201-0
52	A201A
52	A201B
52	A201B
52	A201EV0
52	A202
52	A202(JAPAN)
52	A202A
52	A202B
52	A202C
52	A202D
52	A203
52	A203A
52	A203AA(PNP)
52	A203B(PNP)
52	A203P(PNP)
52	A21
52	A213
52	A214
52	A216
52	A234B
52	A234C
52	A234A
52	A235C
52	A236
52	A237
52	A238
52	A246(JAPAN)
52	A246V
52	A247(JAPAN)
52	A25-1001
52	A25-1002
52	A25-1003
52	A250
52	A251
52	A252
52	A253(JAPAN)
52	A254
52	A255
52	A256
52	A257
52	A258
52	A259
52	A260
52	A261
52	A262
52	A263
52	A264
52	A265
52	A266
52	A267
52	A268
52	A269
52	A270(JAPAN)
52	A271(JAPAN)
52	A272(JAPAN)
52	A273(JAPAN)
52	A274(PNP)
52	A275(JAPAN)
52	A28
52	A280
52	A281
52	A288
52	A288A
52	A289
52	A29
52	A290
52	A292
52	A293
52	A294
52	A295
52	A30(TRANSISTOR)
52	A301(JAPAN)
52	A308
52	A309
52	A310(JAPAN)
52	A313
52	A314
52	A315
52	A316
52	A325
52	A326
52	A329
52	A329A
52	A329B
52	A331
52	A335
52	A337
52	A338
52	A339
52	A343
52	A344(JAPAN)
52	A34C
52	A350C
52	A350H
52	A350R
52	A350T
52	A350TY
52	A350TY
52	A351
52	A351A
52	A351B
52	A352
52	A352A
52	A352B
52	A353
52	A353A
52	A353C
52	A354
52	A354A
52	A354B
52	A355
52	A355A
52	A356
52	A357
52	A358
52	A359
52	A364
52	A365
52	A366
52	A367
52	A368
52	A369
52	A37
52	A376(JAPAN)
52	A38
52	A380
52	A381
52	A382
52	A383
52	A384
52	A385
52	A385A
52	A385D
52	A39
52	A391
52	A392
52	A393
52	A393A
52	A394
52	A395
52	A398
52	A399
52	A412
52	A428
52	A43
52	A436
52	A437
52	A438
52	A446
52	A447
52	A45
52	A45-1
52	A45-2
52	A45-3
52	A453
52	A454(JAPAN)
52	A456
52	A457
52	A466(JAPAN)
52	A466-2
52	A466-3
52	A466BLK
52	A466BLU
52	A466YEL
52	A468
52	A469
52	A470
52	A471
52	A471-1
52	A471-2
52	A471-3
52	A472(JAPAN)
52	A472A
52	A472B
52	A472C
52	A472D
52	A472E
52	A474
52	A476
52	A477
52	A478
52	A479
52	A49
52	A517
52	A518
52	A52
52	A53
52	A57
52	A58
52	A59
52	A60
52	A65B-70
52	A65B-705
52	A65B19G
52	A67
52	A69
52	A71
52	A71AB
52	A71AC
52	A71BS
52	A71Y
52	A72
52	A72BLU
52	A72BRN
52	A720RN
52	A72WHT
52	A75
52	A75B
52	A76
52	A77
52	A77A
52	A77B
52	A77C
52	A77D
52	A78
52	A78B
52	A78C
52	A78D
52	A80
52	A82
52	A83
52	A85
52	A87
52	A90(TRANSISTOR)
52	A909-1009
52	A909-1010
52	A92
52	A93
52	A94
52	AC164
52	AC169
52	ACB810-101
52	ACB810-102
52	ACB810-103
52	ACB83-1001
52	ACB83-1002
52	ACB83-1003
52	AF105
52	AF108
52	AF120
52	AF129
52	AF130
52	AF131
52	AF132
52	AF133
52	AF134
52	AF135
52	AF136
52	AF137
52	AF138
52	AF144
52	AFY10
52	AFY11
52	AFY18
52	AG134
52	AO4091A
52	AFB-11H-1007
52	ASY30
52	ASY57
52	ASY58
52	ASY59
52	ASY210
52	ASY211
52	AS230
52	ATRP1
52	ATRP2
52	AT813
52	B384
52	B385
52	B444
52	B444A
52	B444B
52	C10291
52	CB103
52	CB157
52	CB158
52	CDC-9000
52	CK28
52	CK28A
52	CK4
52	CK4A
52	CK762
52	CK766
52	CK766A
52	CXL102A
52	CXL126
52	D020
52	D026
52	D134
52	D65
52	D66
52	DAT1A
52	DAT2
52	DS-24
52	DS-25
52	DS14
52	DS24
52	DS25
52	DS36
52	DS51
52	DS52
52	E-065
52	E-066
52	E-067
52	E-068
52	E102
52	E105
52	E2451
52	EA15X133
52	EA15X140
52	EA15X141
52	EA2133
52	EA2491
52	EA2497
52	EA2498
52	ECG102A
52	ECG126
52	ECG160
52	EO70
52	ES14
52	ES15X61
52	ES15X73
52	ES5
52	ES3110
52	ES3111
52	ES3112
52	ES3113
52	ES3114
52	ES3115
52	ES3116
52	ES41
52	ESA213
52	ESA233
52	ET12
52	GO0-502A
52	GO008
52	GO010
52	GA-52829
52	GA-55149
52	GA-53242
52	GC1081
52	GE-52
52	GET871
52	GET872
52	GET873
52	GET873A
52	GET874
52	GET875
52	GET883
52	GET885
52	G13
52	GMO378
52	GM380
52	GT-34HV
52	GT-66
52	GT843
52	GT46
52	GT47
52	GT5116
52	GT5117
52	GT5148
52	GT5149
52	GT5151
52	GT5153
52	GT66
52	GTJ33141
52	GTJ33229
52	GTJ33230
52	HA-101
52	HA-15
52	HA-269
52	HA-350
52	HA-350A
52	HA-353
52	HA-353C
52	HA-354
52	HA-354B
52	HA-49
52	HA102
52	HA103
52	HA104
52	HA201
52	HA235
52	HA330
52	HA342
52	HA471
52	HA52
52	HA53
52	HB-186
52	HB-187
52	HB-32
52	HB-33
52	HB422
52	HEP1
52	HEP635
52	HEP636
52	HEP638
52	HEP639
52	HEP640
52	HEP0001
52	HEP0008
52	HR-30
52	HR15
52	HS22D
52	HS29D
52	HT101011X
52	HT101021A
52	HT103501A
52	HT103531C
52	HT105541B
52	HY000005-0
52	I9A115180-2
52	JR05

GE	Industry Standard No.	GE	Industry Standard No.	GE	Industry Standard No.	GE	Industry Standard No.	GE	Industry Standard No.
52	K417-68	52	RS-3323	52	TM102A	52	2N370/33	52	28A102H
52	LU2N544	52	RS-3324	52	TNJ60362	52	2N370A	52	28A102K
52	M4363	52	RS-3862	52	TNJ60363	52	2N371	52	28A102L
52	M4363BLU	52	RS-3863	52	TNJ60564	52	2N371/33	52	28A102M
52	M4363GRN	52	RS-3866	52	TNJ71248	52	2N372	52	28A102OR
52	M4363RN	52	RS-5107	52	TR-11	52	2N372/33	52	28A102TV
52	M4363WHT	52	RS-5108	52	TR-168(OLSON)	52	2N373	52	28A102TV-2
52	M4364	52	RS-5109	52	TRO2012	52	2N374	52	28A102X
52	M4365	52	RS-5201	52	TRO2063-1	52	2N393	52	28A102Z
52	M4366	52	RS-5205	52	TR760	52	2N409	52	28A103
52	M4367	52	RS-5207	52	TR761	52	2N410	52	28A103(CA)
52	M4368	52	RS-5209	52	TR762	52	2N411	52	28A103A
52	M4388	52	RS-5301	52	TRA-10R	52	2N412	52	28A103B
52	M4454	52	RS-5305	52	TRA-11R	52	2N499	52	28A103C
52	M4456	52	RS-5306	52	TRA-12R	52	2N499A	52	28A103CA
52	M4457	52	RS-5312	52	TRA-2	52	2N500	52	28A103CAK
52	M4501	52	RS-5313	52	TRA-22	52	2N500BLU	52	28A103CB
52	M4509	52	RS-5752	52	TRA-22A	52	2N500RED	52	28A103CG
52	M4545	52	RS-5753	52	TRA-22B	52	2N500WHT	52	28A103D
52	M4545BLU	52	RS-5754	52	TRA-23	52	2N501	52	28A103DA
52	M4545WHT	52	RS-5755	52	TRA-23A	52	2N501/18	52	28A103E
52	M4586	52	RS-5756	52	TRA-23B	52	2N501A	52	28A103F
52	M4589	52	RS-5757	52	TRA-24	52	2N502	52	28A103G
52	M4603	52	RS-5758	52	TRA-24A	52	2N502A	52	28A103GA
52	M4604	52	RS-5759	52	TRA-24B	52	2N502B	52	28A103K
52	M4605	52	RS-5760	52	TRM16	52	2N504	52	28A103L
52	M4605RED	52	RS-5761	52	TRM17	52	2N544	52	28A103M
52	M4621	52	RS-5762	52	TV24157	52	2N544/33	52	28A103OR
52	M4632	52	RS-5802	52	1-21-138	52	2N583	52	28A103R
52	M8116	52	RS-5818	52	1-21-189	52	2N584	52	28A103X
52	M8640E	52	RS-684	52	1-21-242	52	2N588	52	28A103Y
52	MA1705	52	RS-685	52	1-21-243	52	2N588A	52	28A122
52	MA820	52	RS2550	52	1-21-244	52	2N591-6M	52	28A201
52	MA821	52	RS5753	52	1-21-257	52	2N602	52	28A201-0
52	MA822	52	RS5753-2	52	002-006300	52	2N602A	52	28A201-OR
52	MA823	52	RS5754	52	002-011600	52	2N603	52	28A201N
52	MD835	52	RS5755	52	2A	52	2N603A	52	28A201TVO
52	N-EA15X133	52	RS5756	52	2AC128	52	2N604	52	28A202-OR
52	N020	52	RS5757	52	2B	52	2N604A	52	28A203
52	N57B2-17	52	RS5758	52	2G	52	2N605	52	28A203A
52	N57B2-18	52	RS5759	52	2G344	52	2N606	52	28A203B
52	N57B2-19	52	RS5760	52	2G601	52	2N607	52	28A203C
52	N57B2-23	52	RS5761	52	2G602	52	2N608	52	28A203D
52	NA5018-1002	52	RS5762	52	2MC	52	2N623	52	28A203P
52	NA5018-1003	52	RS5802	52	2N1003	52	2N624	52	28A213
52	NA5018-1004	52	RS6821	52	2N1004	52	2N640	52	28A255
52	NA5018-1005	52	RS6822	52	2N1042	52	2N641	52	28A290
52	NA5018-1006	52	RT-61014	52	2N1043	52	2N641REDM/P	52	28A292
52	NA5018-1007	52	RT3361	52	2N1107	52	2N642	52	28A305-RED
52	NA5018-1008	52	RT3362	52	2N1108	52	2N643	52	28A305-YELLOW
52	NA5018-1009	52	RT5063	52	2N1108RED	52	2N644	52	28A313
52	NA5018-1010	52	RT5466	52	2N1109	52	2N645	52	28A357
52	NA5018-1011	52	RT5467	52	2N1110	52	2N680	52	28A393
52	NA5018-1022	52	RT5520	52	2N1111	52	2N694	52	28A393A
52	NA5018-1219	52	RT61014	52	2N1111A	52	2N695	52	28A43
52	NA5018-1220	52	S-55TB	52	2N1111B	52	2N72	52	28A45
52	NK65B119	52	S-70T	52	2N1111M1	52	2N75	52	28A45-1
52	NK6105	52	S-80T	52	2N1111RED	52	2N885	52	28A45-2
52	NK6127	52	S-87TB	52	2N111B	52	2N86	52	28A45-3
52	NK6131	52	S-88TB	52	2N1111M2	52	2N87	52	28A517
52	NK6132	52	S-95101	52	2N112M1	52	2N88	52	28A57
52	NK6151	52	S-95102	52	2N1180	52	2N89	52	28A58
52	NK6152	52	S-95103	52	2N1204	52	2N90	52	28A59
52	NK6115325	52	S1332	52	2N1204A	52	2N933	52	28A60
52	NK6115425	52	0822.3516-380	52	2N1213	52	2N986	52	28A67
52	NK6242	52	S65B-1-3P	52	2N1214	52	2NJ50	52	28A69
52	NK6249	52	S70.00.730	52	2N1216	52	2NJ51	52	28A72
52	NK6253	52	SA102	52	2N1224	52	2NJ52	52	28A72BLU
52	NK6254	52	SAW-28B56	52	2N1225	52	2NJ53	52	28A72BLU-BLU
52	NK6255	52	SB-100	52	2N1226	52	2NJ59D	52	28A72BRN
52	NK6265	52	SB100	52	2N123A	52	2S109	52	28A72ORN
52	NK6270	52	SB200	52	2N128	52	2S110	52	28A72WHT
52	NK6618	52	SB5122	52	2N129	52	2S112	52	28A78
52	NK6675	52	SFT-313	52	2N1313	52	2S141	52	28A78B
52	NK6676	52	SFT-316	52	2N1384	52	2S142	52	28A78C
52	NK6677	52	SFT-317	52	2N139	52	2S143	52	28A78D
52	OC-390	52	SFT-320	52	2N1396	52	2S145	52	28A880
52	OC-612	52	SFT-357	52	2N1397	52	2S146	52	28A888
52	OC-613	52	SFT307	52	2N1398	52	2S176	52	28A889
52	OC-614	52	SFT308	52	2N1399	52	2S357Q	52	28A890
52	OC-975	52	SFT317	52	2N140	52	2S35	52	28A891
52	OC130	52	SFT318	52	2N1400	52	2S36	52	28A90
52	OC331	52	SFT319	52	2N1401	52	2S58	52	28B135A
52	OC341	52	SFT320	52	2N1401A	52	2S96	52	28B1760
52	OC342	52	SFT354	52	2N1402	52	2S97	52	28B176FR
52	OC343	52	SFT357	52	2N140M1	52	2S98	52	28B185-0
52	OC351	52	SFT357P	52	2N140M2	52	2SA101	52	28B348(Q)
52	OC361	52	SK3003	52	2N1425	52	2SA101OV	52	28B385
52	OC362	52	SK3007	52	2N1426	52	2SA101A	52	28B503-R
52	OC363	52	SK3008	52	2N1427	52	2SA101AA	52	28B503A-R
52	OC390	52	SM862	52	2N1499	52	2SA101AY	52	28B516(C)
52	OC40	52	S065A	52	2N1499A	52	2SA101B	52	28B516(D)
52	OC41N	52	SPS871	52	2N1499B	52	2SA101BA	52	28B556-R
52	OC42	52	ST-103	52	2N1500	52	2SA101BB	52	28B77A/P
52	OC42N	52	STX0033	52	2N1526	52	2SA101BC	52	0028C0203AA
52	OC43	52	STX0034	52	2N1631	52	2SA101BX	53	A-514-027662
52	OC43N	52	T-1363	52	2N1632	52	2SA101C	53	A059-106
52	OC44	52	T-1364	52	2N1635	52	2SA101CA	53	A059-107
52	OC44N	52	T-1460	52	2N1636	52	2SA101CV	53	A059-116
52	OC45N	52	T-348	52	2N1637/33	52	2SA101CX	53	A061-114
52	OC46N	52	T-6028	52	2N1662	52	2SA101D	53	A061-115
52	OC47N	52	T-6029	52	2N1699	52	2SA101E	53	A065-105
52	OC50	52	T-6030	52	2N2089	52	2SA101P	53	A065-106
52	OC612	52	T-6031	52	2N2090	52	2SA101Q	53	A065-112
52	OC613	52	T-6032	52	2N2092	52	2SA101QA	53	A069-105
52	OC614	52	T1460	52	2N2180	52	2SA101R	53	A069-107
52	OC615	52	T1524	52	2N2207	52	2SA101V	53	A115(JAPAN)
52	P2-999005	52	T1524BRN	52	2N2208	52	2SA101X	53	A116(JAPAN)
52	Q2N1526	52	T1524BRN/RED	52	2N231-YEL-RED	52	2SA101XBX	53	A12
52	Q9/6515	52	T1654	52	2N231BLU	52	2SA101Y	53	A128(JAPAN)
52	R-539	52	T1654BLU	52	2N231RED	52	2SA101YA	53	A12A
52	R104-5	52	T2322	52	2N231YEL	52	2SA101Z	53	A12B
52	R104-6	52	T2323	52	2N2363	52	2SA102	53	A12C
52	R104-7	52	T2324	52	2N2451	52	2SA102(BA)	53	A12D
52	R104-8	52	T280(SEARS)	52	2N2456	52	2SA102BA	53	A12H
52	R336	52	T281(SEARS)	52	2N247	52	2SA102A	53	A12V
52	R337	52	T449(SEARS)	52	2N247/33	52	2SA102AA	53	A138(JAPAN)
52	R581	52	T50818	52	2N248	52	2SA102AB	53	A14-1004
52	R60-1001	52	T52054	52	2N2487	52	2SA102B	53	A14-1005
52	R684	52	T6028	52	2N2488	52	2SA102BA	53	A14-1006
52	R714	52	T6029	52	2N2489	52	2SA102BA-2	53	A14-1007
52	R715	52	T6030	52	2N2494	52	2SA102BN	53	A14-1008
52	RCA34098	52	T6031	52	2N252	52	2SA102C	53	A14-1009
52	RCA34099	52	T6032	52	2N2588	52	2SA102CA	53	A14-1010
52	RCA34100	52	TA-1650A	52	2N275	52	2SA102CA-1	53	A1466-1
52	RCA44098	52	TA-1755	52	2N2783	52	2SA102D	53	A1466-19
52	RE15	52	TA-1756	52	2N301G	52	2SA102E	53	A1466-2
52	RE4	52	TA-1830+	52	2N318	52	2SA102F	53	A1466-29
52	REN102A	52	TA-5	52	2N33	52	2SA102G	53	A1466-3
52	REN126	52	TG2SA201	52	2N344			53	A1466-39
52	RS-1554	52	TG2SA201-0	52	2N345			53	A148P-2
52	RS-2003	52	TG2SA201-N	52	2N346			53	A148P-29
52	RS-3322	52	TG2SA201C	52	2N370			53	A148P2
		52	TK1228-1001					53	A148P2-29

GE	Industry Standard No.	GE	Industry Standard No.	GE	Industry Standard No.	GE	Industry Standard No.	GE	Industry Standard No.
53	A149L-4	53	AC128	53	ASY13-2	53	B173A	53	B322
53	A149L-49	53	AC128-01	53	ASY14	53	B173B	53	B323
53	A15(TRANSISTOR)	53	AC128/01	53	ASY14-1	53	B173C	53	B324
53	A15-1004	53	AC128KO1	53	ASY14-2	53	B173L	53	B324A
53	A15-1005	53	AC128K	53	ASY14-3	53	B174	53	B324B
53	A15BK	53	AC131	53	ASY24	53	B175	53	B324D
53	A15BL	53	AC131-3Q	53	ASY31	53	B175A	53	B324S
53	A15BLU	53	AC132	53	ASY32	53	B175B	53	B324B-1
53	A15H	53	AC132-01	53	ASY48	53	B175C	53	B324P
53	A15K	53	AC133A	53	ASY48-IV	53	B175E	53	B324G
53	A15R	53	AC134	53	ASY48-V	53	B176	53	B324H
53	A15V	53	AC135	53	ASY48-VI	53	B176-0	53	B324I
53	A15VR	53	AC136	53	ASY49	53	B176-P	53	B324J
53	A15Y	53	AC137	53	ASY50	53	B176-PR	53	B324K
53	A16	53	AC138	53	ASY51	53	B176B	53	B324L
53	A17	53	AC139	53	ASY52	53	B176M	53	B324N
53	A173	53	AC142	53	ASY54	53	B176P	53	B324R
53	A173B	53	AC142K	53	ASY55	53	B176PRC	53	B324S
53	A17H	53	AC150	53	ASY56	53	B176R	53	B324V
53	A181	53	AC150GRN	53	ASY70	53	B177(JAPAN)	53	B329
53	A183	53	AC150YEL	53	ASY70-IV	53	B178(JAPAN)	53	B329K
53	A187TV	53	AC151-IV	53	ASY70-VI	53	B178-0	53	B32N
53	A18U	53	AC151-RIV	53	ASY70IV	53	B178-S	53	B33
53	A203(PNP)	53	AC151-RV	53	ASY70V	53	B178A	53	B33-4
53	A203AA	53	AC151-RVI	53	ASY70VI	53	B178C	53	B336
53	A203B	53	AC151-V	53	ASY71	53	B178D	53	B33C
53	A203P	53	AC151-VI	53	ASY90	53	B178M	53	B33D
53	A231	53	AC151-VII	53	ASY91	53	B178N	53	B33R
53	A232	53	AC151IV	53	AT-50	53	B178T	53	B33F
53	A2414	53	AC151R	53	AT-6A	53	B178U	53	B345
53	A25-1004	53	AC151RIV	53	AT100H	53	B178V	53	B346
53	A25-1005	53	AC151RV	53	AT100M	53	B178X	53	B346K
53	A25-1006	53	AC151RVI	53	AT100N	53	B178Y	53	B346Q
53	A28A666PQR	53	AC151V	53	AT10H	53	B180	53	B348
53	A302	53	AC151VI	53	AT10M	53	B180A	53	B348Q
53	A303	53	AC151VII	53	AT10N	53	B181	53	B348R
53	A304	53	AC152	53	AT20H	53	B181A	53	B349
53	A32	53	AC152IV	53	AT20M	53	B183	53	B364
53	A33	53	AC152V	53	AT20N	53	B184	53	B365
53	A374	53	AC152VI	53	AT30H	53	B185	53	B365B
53	A406	53	AC153	53	AT30N	53	B185(O)	53	B366
53	A407	53	AC153K	53	AT5O	53	B185(O)	53	B37
53	A50(TRANSISTOR)	53	AC154	53	AT6	53	B185AA	53	B370
53	A514-027662	53	AC155	53	AT6A	53	B185F	53	B370A
53	A538	53	AC156	53	AT74	53	B185P	53	B370AA
53	A595	53	AC160	53	AT74S	53	B186	53	B370AB
53	A615-1010	53	AC160A	53	ATAF1	53	B186(O)	53	B370AHA
53	A615-1011	53	AC160B	53	ATAF2	53	B186(SANYO)	53	B370AHB
53	A64	53	AC160GRN	53	ATGP	53	B186-1	53	B370B
53	A65	53	AC160RED	53	AU100N	53	B186-K	53	B370C
53	A66	53	AC160YEL	53	B-105B	53	B186A	53	B370D
53	A66-1-70	53	AC161	53	B-22-3	53	B186AG	53	B370P
53	A66-1-705	53	AC162	53	B-22-4	53	B186B	53	B370PB
53	A66-1A9G	53	AC163	53	B-23	53	B186BY	53	B370V
53	A66-2-70	53	AC165	53	B-23-1	53	B186G	53	B371
53	A66-2-705	53	AC166	53	B-23-2	53	B186H	53	B371D
53	A66-2A9G	53	AC167	53	B-24-1	53	B186L	53	B372
53	A66-3-5A9G	53	AC168	53	B-26	53	B186Y	53	B373
53	A66-3-70	53	AC170	53	B-26-1	53	B187	53	B376
53	A66-3-705	53	AC171	53	B-315-1	53	B187(1)	53	B376Q
53	A66-3A9G	53	AC173	53	B-324	53	B187(SANYO)	53	B377
53	A8P-2-70	53	AC176K	53	B-P1A	53	B187AA	53	B377B
53	A8P-2-705	53	AC178	53	B1022-1	53	B187B	53	B378
53	A8P-2A9G	53	AC180	53	B105	53	B187C	53	B37A
53	A9L-4-70	53	AC180K	53	B105B-1	53	B187D	53	B37B
53	A9L-4-705	53	AC182	53	B106	53	B187G	53	B37C
53	A9L-4A9G	53	AC184	53	B108	53	B187K	53	B37E
53	AC-107	53	AC188	53	B108A	53	B187R	53	B37F
53	AC-113	53	AC188/01	53	B108B	53	B187RED	53	B381
53	AC-113A	53	AC188O1	53	B109	53	B187S	53	B382
53	AC-114	53	AC188K	53	B111	53	B187Y	53	B383
53	AC-116	53	AC191	53	B111K	53	B187YEL	53	B383-1
53	AC-117	53	AC192	53	B112	53	B188	53	B383-2
53	AC-117A	53	AC193	53	B113	53	B189	53	B386
53	AC-117P	53	AC193K	53	B114	53	B201	53	B389
53	AC-121IV	53	ACY-21	53	B115	53	B22	53	B39
53	AC-122	53	ACY-22	53	B154-1	53	B22-3	53	B40
53	AC-123	53	ACY-23	53	B116	53	B22-4	53	B400
53	AC-125	53	ACY-32	53	B117	53	B22A	53	B4004B
53	AC-126	53	ACY17-1	53	B117K	53	B22B	53	B400A
53	AC-128	53	ACY18-1	53	B120	53	B22I	53	B400B
53	AC-132	53	ACY19-1	53	B134(JAPAN)	53	B22R	53	B400K
53	AC-150	53	ACY20	53	B134-D	53	B22Y	53	B405
53	AC-151	53	ACY20-1	53	B134-E	53	B23	53	B405-2C
53	AC-152	53	ACY21	53	B135	53	B23-1	53	B405-3C
53	AC-154	53	ACY21-1	53	B135B	53	B23-2	53	B405-4C
53	AC-155	53	ACY22	53	B135C	53	B238	53	B405A
53	AC-156	53	ACY22-1	53	B135E	53	B238-12A	53	B405B
53	AC-161	53	ACY23	53	B136	53	B238-12B	53	B405C
53	AC-162	53	ACY23-V	53	B136-2	53	B238-12C	53	B405D
53	AC-165	53	ACY23-VI	53	B136-3	53	B24	53	B405E
53	AC-166	53	ACY23VI	53	B136A	53	B24-1	53	B405Q
53	AC-167	53	ACY27	53	B136B	53	B26	53	B405H
53	AC-168	53	ACY28	53	B136C	53	B261	53	B405K
53	AC-169	53	ACY29	53	B136U	53	B262	53	B405R
53	AC-N7B	53	ACY30	53	B155	53	B263	53	B405RE
53	AC105	53	ACY31	53	B155A	53	B264	53	B415
53	AC106	53	ACY32	53	B155B	53	B265	53	B415A
53	AC108	53	ACY32-V	53	B156	53	B266	53	B415B
53	AC109	53	ACY32-VI	53	B156A	53	B266P	53	B421
53	AC110	53	ACY32V	53	B156AA	53	B266Q	53	B427
53	AC113	53	ACY32VI	53	B156AB	53	B267	53	B428
53	AC114	53	ACY33	53	B156AC	53	B268	53	B43
53	AC115	53	ACY33-VI	53	B156B	53	B269	53	B431
53	AC116	53	ACY33-VII	53	B156C	53	B270	53	B439
53	AC117	53	ACY33-VIII	53	B156D	53	B270A	53	B439A
53	AC117A	53	ACY33VI	53	B156P	53	B270B	53	B43A
53	AC117B	53	ACY33VII	53	B157	53	B270C	53	B44
53	AC117P	53	ACY34	53	B158	53	B270D	53	B440
53	AC118	53	ACY35	53	B159	53	B270E	53	B443
53	AC119	53	ACY36	53	B160	53	B271	53	B443A
53	AC120	53	ACY38	53	B167	53	B272	53	B443B
53	AC121	53	ACY41	53	B168	53	B273	53	B450
53	AC121-IV	53	ACY41-1	53	B169	53	B293	53	B450A
53	AC121-V	53	ACY44	53	B169(JAPAN)	53	B299	53	B451
53	AC121-VI	53	ACY44-1	53	B170	53	B302	53	B452
53	AC121-VII	53	ADY-27	53	B171	53	B303-Q	53	B452A
53	AC121IV	53	AE-50	53	B171(JAPAN)	53	B303O	53	B453
53	AC121V	53	AF187	53	B171A	53	B303A	53	B454
53	AC121VI	53	AF188	53	B171B	53	B303B	53	B455
53	AC121VII	53	AFB-11A-1008	53	B172	53	B303C	53	B457
53	AC122	53	APB-11B-1010	53	B172A	53	B303H	53	B457-C
53	AC122-30	53	AR-102	53	B172AP	53	B303K	53	B457A
53	AC122-GRN	53	AR-103	53	B172B	53	B304	53	B459
53	AC122-RED	53	AR-104	53	B172C	53	B304A	53	B459-0
53	AC122-VIO	53	AR-105	53	B172D	53	B315	53	B459A
53	AC122-YEL	53	AR102	53	B172E	53	B316	53	B459B
53	AC122GRN	53	A833067	53	B172F	53	B317	53	B459C
53	AC122RED	53	A833868	53	B172H	53	B32	53	B459D
53	AC122YEL	53	ASY12-1	53	B172N	53	B32-0	53	B46
53	AC123	53	ASY12-2	53	B172P	53	B32-1	53	B460
53	AC124	53	ASY13-1	53	B172R	53	B32-2	53	B460A
53	AC125	53		53	B173	53	B32-4	53	B460B
53	AC126	53		53		53	B321	53	

GE	Industry Standard No.	GE	Industry Standard No.	GE	Industry Standard No.	GE	Industry Standard No.	GE	Industry Standard No.
53	B47	53	CK22C	53	EA2135	53	GT109R	53	HR-8A
53	B475	53	CK64	53	EA2136	53	GT122	53	HR-9
53	B475A	53	CK64A	53	EA2176	53	GT1223	53	HR-9A
53	B475B	53	CK64B	53	EB0001	53	GT123	53	HR30
53	B475D	53	CK64C	53	EB0003	53	GT132	53	HR53
53	B475E	53	CK65	53	ECG158	53	GT14	53	HR61
53	B475F	53	CK65A	53	ED55	53	GT14H	53	HS102
53	B475G	53	CK65B	53	ED56	53	GT1665	53	HS117D
53	B475P	53	CK65C	53	ED57	53	GT18	53	HS5
53	B475Q	53	CK66	53	EK136	53	GT20	53	HT1001510
53	B476	53	CK66A	53	EO44A	53	GT20H	53	HT200540
53	B482	53	CK66B	53	EQ-15	53	GT20R	53	HT200540A
53	B49	53	CK66C	53	EQ-9	53	GT222	53	HT200541
53	B494	53	CK67	53	ER15X17	53	GT2696	53	HT200541A
53	B495	53	CK67A	53	ER15X18	53	GT2883	53	HT200541B
53	B495A	53	CK67B	53	ER15X22	53	GT2885	53	HT200541B-0
53	B495C	53	CK67C	53	ER15X23	53	GT2887	53	HT200561
53	B495D	53	CK891	53	ER15X7	53	GT31	53	HT200561A
53	B495T	53	CK892	53	ER15X9	53	GT32	53	HT200561B
53	B496	53	CP800	53	ES15X100	53	GT33	53	HT200561C
53	B498	53	CP801	53	ES15X31	53	GT34	53	HT200561C-0
53	B5	53	CP802	53	ES15X32	53	GT34HV	53	HT200751B
53	B516C	53	CP803	53	ES15X4	53	GT41	53	HT200770B
53	B516CD	53	CU1	53	ES15X49	53	GT42	53	HT200771
53	B516D	53	CS1758	53	ES15X50	53	GT44	53	HT200771A
53	B516P	53	CT1009	53	ES15X53	53	GT45	53	HT200771B
53	B534	53	CT1017	53	ES15X55	53	GT74	53	HT200771C
53	B534A	53	CTP-2001-1001	53	ES15X63	53	GT75	53	HT201721A
53	B535	53	CTP-2001-1002	53	ES15X72	53	GT751	53	HT201721D
53	B54	53	CTP-2001-1003	53	ES15X75	53	GT758	53	HT201725A
53	B54B	53	CTP-2001-1004	53	ES15X8	53	GT759	53	HT201782A
53	B54B	53	CTP-2001-1009	53	ES15X99	53	GT759R	53	HT201861A
53	B54P	53	CTP-2006-1001	53	ES17(ELCOM)	53	GT760	53	HT201871L
53	B54T	53	CTP-2006-1002	53	ES19	53	GT760R	53	HT203243A
53	B55(TRANSISTOR)	53	CTP-2006-1003	53	ES2(ELCOM)	53	GT763	53	HT203701
53	B56	53	CTP1032	53	ES23	53	GT81	53	HT203701A
53	B560	53	CTP1033	53	ES26(ELCOM)	53	GT81HS	53	HT203701B
53	B56A	53	CTP1034	53	ES3(ELCOM)	53	GT81R	53	HT204051
53	B56B	53	CTP1035	53	ES3120	53	GT82	53	HT204051A
53	B56C	53	CTP1036	53	ES3121	53	GTJ33231	53	HT204051C
53	B57	53	CTP2076-1001	53	ES3122	53	GTJ33232	53	HT204051D
53	B58	53	CTP2076-1002	53	ES3123	53	H10	53	HT204051E
53	B59	53	CTP2076-1003	53	ES3124	53	HA1350	53	HT204053
53	B5A	53	CTP2076-1004	53	ES3125	53	HA1360	53	HT204053A
53	B60	53	CTP2076-1005	53	ES3126	53	HA54	53	HT204053B
53	B601-1009	53	CTP2076-1006	53	ES37(ELCOM)	53	HA56	53	HT204051A
53	B601-1010	53	CTP2076-1007	53	ES4(ELCOM)	53	HAM-1	53	HV000102
53	B60A	53	CTP2076-1008	53	ET15I1	53	HB-00054	53	HV0000405
53	B61	53	CTP2076-1009	53	ET15X1	53	HB-00056	53	HV12
53	B65	53	CTP2076-1010	53	ET15X25	53	HB-00156	53	HV16
53	B66	53	CTP2076-1011	53	ET15X31	53	HB-00171	53	HV17
53	B66-1-A-21	53	CTP2076-1012	53	ET15X32	53	HB-00172	53	HV17B
53	B66-1A21	53	CXL158	53	ETTB-2SB176	53	HB-00173	53	HV19
53	B66-2-A-21	53	D-P1A	53	ETTB-2SB176A	53	HB-00175	53	IF-65
53	B66-2A21	53	D008	53	ETTB-2SB176B	53	HB-00176	53	IRTR14
53	B66-3-A-21	53	D018	53	ETTB-2SB176R	53	HB-00186	53	IRTR84
53	B66-3A21	53	D021	53	ETTB-75LB	53	HB-00187	53	IRTR85
53	B66H	53	D030	53	EX15225	53	HB-00324	53	ISBF1
53	B67	53	D031	53	F20-1006	53	HB-00405	53	J241111(2SB370)
53	B67A	53	D038	53	F20-1007	53	HB-172	53	J241164
53	B73	53	D043	53	F20-1008	53	HB-173	53	J241178
53	B73A	53	D059	53	F20-1009	53	HB-175	53	J241190
53	B73B	53	D101B	53	F215-1006	53	HB-475	53	J24626
53	B73C	53	D105B	53	F215-1007	53	HB-85	53	J24639
53	B73GR	53	D117	53	F215-1008	53	HB156	53	J24869
53	B75	53	D126	53	F215-1009	53	HB156C	53	J24870
53	B75AH	53	D135	53	FB420	53	HB171	53	J24934
53	B75B	53	D156	53	FB421	53	HB172	53	J310159
53	B75C	53	D171	53	FBN-CP2293	53	HB175	53	J310224
53	B75P	53	D180	53	FPR40-1004	53	HB176	53	J310252
53	B75H	53	D193	53	FPR40-1005	53	HB178	53	J5063
53	B75LB	53	D352	53	FPR50-1005	53	HB186	53	J5064
53	B76	53	DC-10	53	FPR50-1006	53	HB187	53	JP5063
53	B77	53	DC-12	53	G0005	53	HB263	53	JP5064
53	B77(B)	53	DC-9	53	G0007	53	HB270	53	JR15
53	B77A	53	D8-14	53	G004	53	HB32	53	JR5
53	B77AA	53	D8-19	53	G04-711-E	53	HB324	53	K14-0066-1
53	B77AB	53	D8-26	53	G04-711-F	53	HB33	53	K4-500
53	B77AC	53	D8-8	53	G04-711-G	53	HB365	53	KD2101
53	B77AD	53	DS16	53	G04-711-H	53	HB415	53	KG81000
53	B77AH	53	DS26	53	G11	53	HB459	53	K0825642-20
53	B77AP	53	DS29	53	G12	53	HB475	53	K0825642-40
53	B77B	53	D83	53	G14	53	HB54	53	K0825643-10
53	B77B-11	53	DU4	53	GC1097	53	HB55	53	K0825643-15
53	B77C	53	DU5	53	GC1134	53	HB56	53	K0825651-20
53	B77D	53	E-044A	53	GC1136	53	HB75	53	K0825657-53
53	B77H	53	E-070	53	GC1143	53	HB75C	53	KR-Q0001
53	B77V	53	E-2465	53	GC1145	53	HB77	53	KR-Q0002
53	B77VRED	53	E132	53	GC1150	53	HB77B	53	KR-Q0004
53	B78	53	E158	53	GC1183	53	HB77C	53	KR-Q1010
53	B79	53	E181	53	GC1184	53	HC-00176	53	KR-Q1011
53	B89	53	E181A	53	GC1186	53	HEP-G004	53	KR-Q1012
53	B89A	53	E181B	53	GC1187	53	HEP250	53	KV-1
53	B89AH	53	E181C	53	GC1257	53	HEP251	53	KV-4
53	B89B	53	E181D	53	GC1422	53	HEP253	53	KV1
53	B8P-2-A-21	53	E241	53	GC250	53	HEP254	53	KV2
53	B8P-2A21	53	E24104	53	GC343	53	HEP280	53	KV4
53	B90	53	E24106	53	GC408	53	HEP281	53	L5021
53	B91	53	E241A	53	GC4144	53	HEP634	53	L5022
53	B94	53	E241B	53	GC464	53	HJ15	53	L5022A
53	B97	53	E2445	53	GC466	53	HJ17	53	L5025
53	B9L-4-A-21	53	E2448	53	GC5000	53	HJ17D	53	L5025A
53	B9L-4A21	53	E2453	53	GC520	53	HJ22	53	M-8641A
53	BA6	53	E2465	53	GC521	53	HJ228	53	M108
53	BA6A	53	E2467	53	GC551	53	HJ22D	53	M4313
53	BCM1002-18	53	E2476	53	GC552	53	HJ23	53	M4315
53	BCM1002-3	53	E2480	53	GC578	53	HJ230	53	M4398
53	BCM1002-4	53	E2481	53	GC579	53	HJ23D	53	M4450
53	BCM1002-5	53	E2482	53	GC580	53	HJ315	53	M4462
53	BCY91B	53	E4	53	GC581	53	HJ43	53	M4466
53	BP1	53	EA0009	53	GC588	53	HJ50	53	M4466ORN
53	BP1A	53	EA1081	53	GC639	53	HJ51	53	M4468
53	BP2	53	EA1346	53	GC640	53	HJ54	53	M4468BRN
53	BTX070	53	EA15X164	53	GC680	53	HJ60	53	M4469
53	BX-324	53	EA15X19	53	GC681	53	HJ606	53	M4469RED
53	C10227	53	EA15X2	53	GC682	53	HJ62	53	M4470
53	C10230-3	53	EA15X203	53	GC856	53	HJ71	53	M4470ORN
53	C11021	53	EA15X207	53	GC864	53	HJ72	53	M4471
53	C172	53	EA15X212	53	GB-53	53	HJ73	53	M4471YEL
53	C175B	53	EA15X23	53	GER-A	53	HJ74	53	M4472
53	CA2D2	53	EA15X25	53	GET0-50P	53	HJX2	53	M4472GRN
53	CB161	53	EA15X257	53	GET103	53	HM-08014	53	M4473
53	CB248	53	EA15X28	53	GET113	53	HR-1	53	M4474
53	CB249	53	EA15X3	53	GET113A	53	HR-2	53	M4474YEL
53	CB0363/7839	53	EA15X326	53	GET114	53	HR-3	53	M4475
53	CB213811	53	EA15X36	53	GET0-50P	53	HR-39	53	M4475GRN
53	CUE-52	53	EA15X4	53	G04-711-E	53	HR-4	53	M4476
53	CUB-53	53	EA15X467	53	G04-711-F	53	HR-4A	53	M4476BLU
53	CJ5204	53	EA15X6840	53	G04-711-G	53	HR-5	53	M4477
53	CK13	53	EA15X7	53	G04-711-H	53	HR-6	53	M4477PUR
53	CK13A	53	EA15X8	53	GP139	53	HR-7	53	M4510
53	CK22	53	EA15X8442	53	GP139A	53	HR-7A	53	M4555
53	CK22A	53	EA15X8444	53	GP139B	53	HR-8	53	M4553
53	CK22B	53	EA2134	53	GT109			53	M4553BLU
								53	M4553BRN

GE	Industry Standard No.	GE	Industry Standard No.	GE	Industry Standard No.	GE	Industry Standard No.	GE	Industry Standard No.
53	M4553GRN	53	NKT133	53	0C66	53	R1274	53	RS-1541
53	M4553ORN	53	NKT153/25	53	0C70	53	R152	53	RS-1542
53	M4553RED	53	NKT154/25	53	0C70N	53	R1540	53	RS-1543
53	M4553YEL	53	NKT163/25	53	0C71	53	R1541	53	RS-1544
53	M4562	53	NKT164/25	53	0C711	53	R1542	53	RS-1546
53	M4563	53	NKT208	53	0C71A	53	R1543	53	RS-1548
53	M4564	53	NKT211	53	0C71N	53	R1544	53	RS-1555
53	M4565	53	NKT212	53	0C72	53	R1546	53	RS-2007
53	M4575	53	NKT213	53	0C73	53	R1548	53	RS-2350
53	M4595	53	NKT214	53	0C74	53	R1555	53	RS-2351
53	M5285	53	NKT215	53	0C74N	53	R16	53	RS-2352
53	M75517-1	53	NKT216	53	0C75	53	R164	53	RS-2353
53	M75517-2	53	NKT217	53	0C75N	53	R23-1003	53	RS-2354
53	M75561-7	53	NKT218	53	0C76	53	R23-1004	53	RS-2355
53	M8062A	53	NKT219	53	0C77	53	R2350	53	RS-2367
53	M8062B	53	NKT222	53	0C77M	53	R2351	53	RS-2675
53	M8062C	53	NKT223A	53	0C79	53	R2352	53	RS-2677
53	M8604	53	NKT224A	53	0C80	53	R2353	53	RS-2689
53	M8604A	53	NKT225A	53	0C81	53	R2355	53	RS-2697
53	M8640	53	NKT226A	53	0C810	53	R2366	53	RS-3275
53	M8640A	53	NKT227A	53	0C81D	53	R2367	53	RS-3276
53	M9002	53	NKT228A	53	0C81DD	53	R2373	53	RS-3282
53	M9198	53	NKT229	53	0C81DN	53	R24-1001	53	RS-3283
53	M9249	53	NKT231A	53	0C81N	53	R24-1002	53	RS-3284
53	MA112	53	NKT232A	53	0C83	53	R24-1003	53	RS-3285
53	MA113	53	NKT24	53	0C83N	53	R24-1004	53	RS-3286
53	MA114	53	NKT244	53	0C84	53	R242	53	RS-3289
53	MA115	53	NKT246	53	0C84N	53	R245	53	RS-3299
53	MA116	53	NKT247A	53	OP-129	53	R2482-1	53	RS-3301
53	MA117	53	NKT25	53	OP129	53	R255	53	RS-3308
53	MA1318	53	NKT271	53	P1L	53	R2675	53	RS-3316
53	MA1702	53	NKT273	53	P1L4956	53	R2677	53	RS-3316-1
53	MA1703	53	NKT273	53	P3B	53	R2689	53	RS-3316-2
53	MA1704	53	NKT274	53	P3D	53	R2749	53	RS-3318
53	MA1706	53	NKT275	53	PA10889-1	53	R2749M	53	RS-3904
53	MA1707	53	NKT275A	53	PA10889-2	53	R289	53	RS-3925
53	MA1708	53	NKT275E	53	PA9156	53	R290	53	RS-406
53	MA206	53	NKT275J	53	PA9157	53	R291	53	RS-5008
53	MA23	53	NKT278	53	PA9158	53	R324	53	RS-5206
53	MA23B	53	NKT281	53	PBE3014-1	53	R3275	53	RS-5311
53	MA240	53	NKT303	53	PBE3014-2	53	R3276	53	RS-5406
53	MA25	53	NKT308	53	PBE3020-1	53	R3280(RCA)	53	RS-5502
53	MA393	53	NKT32	53	PBE3162	53	R3282	53	RS-5505
53	MA393A	53	NKT33	53	PBE3162-1	53	R3284	53	RS-5506
53	MA393B	53	NKT4	53	PBE3162-2	53	R3286	53	RS-5530
53	MA393C	53	NKT5	53	PC1066T	53	R3299	53	RS-5531
53	MA393E	53	NKT52	53	PC1067T	53	R3301	53	RS-5532
53	MA393G	53	NKT53	53	PC1068T	53	R35	53	RS-5533
53	MA393R	53	NKT54	53	PC3002	53	R3578-1	53	RS-5534
53	MA815	53	00-304/1	53	PC3005	53	R3598-2	53	RS-5535
53	MA881	53	00-304/2	53	PC3006	53	R364	53	RS-5536
53	MA882	53	00-304/3	53	PC3007	53	R428	53	RS-5541
53	MA883	53	00-305/1	53	PC3009	53	R4348	53	RS-5542
53	MA884	53	00-305/2	53	PC4900-1	53	R4349	53	RS-5544
53	MA885	53	00-306/1	53	PQ28	53	R5051	53	RS-5551
53	MA886	53	00-306/2	53	PQ29	53	R5052	53	RS-5552
53	MA887	53	00-306/3	53	PT0-139	53	R5053	53	RS-5553
53	MA888	53	00-350	53	PT0139	53	R5055	53	RS-5554
53	MA889	53	00-351	53	PT530	53	R5097	53	RS-5555
53	MA890	53	00-360	53	PT530A	53	R5098	53	RS-5556
53	MA891	53	00-362	53	PTC135	53	R5099	53	RS-5557
53	MA892	53	00-363	53	PT0139	53	R5100	53	RS-5558
53	MA893	53	00-364	53	FXC-101	53	R5101	53	RS-5602
53	MA894	53	00-38	53	FXC-101AB	53	R5182	53	RS-5704
53	MA895	53	00-602	53	Q-16	53	R52	53	RS-57042
53	MA896	53	00-604	53	Q-6	53	R530	53	RS-57062
53	MA897	53	00-65	53	Q-7	53	R537	53	RS-5708
53	MA898	53	00-66	53	Q-8	53	R5523	53	RS-5708-2
53	MA899	53	00-70	53	Q0V60526	53	R5524	53	RS-5709
53	MA900	53	00-71	53	Q0V60528	53	R5525	53	RS-5711
53	MA901	53	00-71A	53	Q1	53	R558(T.I.)	53	RS-5717
53	MA902	53	00-71N	53	Q1-7C	53	R56	53	RS-5717-3
53	MA903	53	00-72	53	Q11/6515	53	R563(T.I.)	53	RS-5717-6
53	MA904	53	00-73	53	Q12/6515	53	R5708	53	RS-5720
53	MA909	53	00-74	53	Q16	53	R579(T.I.)	53	RS-5731
53	MA910	53	00-75	53	Q2-7C	53	R60-1004	53	RS-5733
53	MD501	53	00-75N	53	Q2N2428	53	R60-1005	53	RS-5734
53	MD501B	53	00-77	53	Q2N2613	53	R60-1006	53	RS-5736
53	MM1151	53	00-79	53	Q2N406	53	R608	53	RS-5737
53	MM1152	53	00-81DD	53	Q2N4106	53	R608A	53	RS-5742
53	MM1153	53	00110	53	Q35218	53	R64	53	RS-5743
53	MM1154	53	00120	53	Q4	53	R65	53	RS-5743-1
53	MM1742	53	00122	53	Q40263	53	R6553	53	RS-5743-2
53	MN-53	53	00123	53	Q6	53	R66	53	RS-5743-3
53	MN52	53	0C302	53	Q7	53	R67	53	RS-5744
53	MN53	53	0C303	53	Q7/6515	53	R6922	53	RS-5744-3
53	MN53BLU	53	0C304-1	53	Q8/6515	53	R7048	53	RS-5749
53	MN53GRN	53	0C304-2	53	Q90076	53	R7124	53	RS-5852
53	MN53RED	53	0C304N	53	Q0V60526	53	R7127	53	RS-5854
53	MN60	53	0C305	53	Q0V60528	53	R7164	53	RS-686
53	MP1014-1	53	0C305-1	53	Q0V60538	53	R7166	53	RS-687
53	MP1014-4	53	0C305-2	53	QQC61209	53	R7363	53	RS1192
53	MP1014-5	53	0C306	53	QQC61210	53	R7489	53	RS1540
53	MP1014-6	53	0C306-1	53	QQV60526	53	R7490	53	RS1541
53	N-BA2136	53	0C306-3	53	QQV60528	53	R7491	53	RS1542
53	N57B2-15	53	0C307	53	QQV60538	53	R7612	53	RS1543
53	N57B2-25	53	0C307-1	53	QQV60539	53	R7888	53	RS1544
53	N57B2-3	53	0C307-2	53	QR2378	53	R7889	53	RS1546
53	N57B2-6	53	0C307-3	53	R-120	53	R8121	53	RS1548
53	N57B2-7	53	0C308	53	R-152	53	R83	53	RS1555
53	NA-1114-1004	53	0C309	53	R-16	53	R8310	53	RS2351
53	NA-1114-1005	53	0C309-1	53	R-164	53	R8311	53	RS2352
53	NA-1114-1006	53	0C309-2	53	R-25-1003	53	R868	53	RS2353
53	NA-1114-1007	53	0C309-3	53	R-25-1004	53	R8687	53	RS2354
53	NA-1114-1008	53	0C318	53	R-24-1001	53	R8688	53	RS2355
53	NA-1114-1009	53	0C33	53	R-24-1002	53	R8695	53	RS2367
53	NA-1114-1010	53	0C330	53	R-242	53	R8697	53	RS2373
53	NA-1114-1011	53	0C34	53	R-245	53	R87	53	RS2374
53	NA1022-1007	53	0C340	53	R-291	53	R8706	53	RS2675
53	NA5018-1013	53	0C350	53	R-28B186	53	R8707	53	RS2677
53	NA5018-1014	53	0C360	53	R-28B187	53	R8883	53	RS2689
53	NA5018-1015	53	0C364	53	R-28B303	53	R8884	53	RS2697
53	NA5018-1016	53	0C38	53	R-28B405	53	R8885	53	RS2867
53	NAP-T-2-10	53	0C41A	53	R-530	53	R8886	53	RS3211
53	NAP-TZ-10	53	0C45	53	R-56	53	R8971	53	RS3275
53	NC50	53	0C46	53	R-593	53	R9533	53	RS3280
53	NC32	53	0C47	53	R-608A	53	R9534	53	RS3282
53	NE269	53	0C56	53	R-64	53	R9603	53	RS3283
53	NJ181B	53	0C57	53	R-66	53	R9604	53	RS3284
53	NK66-1A19	53	0C58	53	R-67	53	R98	53	RS3285
53	NK66-2A19	53	0C59	53	R-83	53	RCA34101	53	RS3286
53	NK66-3A19	53	0C60	53	R06-1007	53	RCA34106	53	RS3289
53	NK8P-2A19	53	0C601	53	R06-1008	53	RCA3517	53	RS3293
53	NK9L-4A19	53	0C602	53	R06-1009	53	RCA35953	53	RS3299
53	NKT102	53	0C6028P	53	R06-1010	53	RCA35954	53	RS3301
53	NKT104	53	0C6028Q	53	R100-1	53	RCA3858	53	RS3308
53	NKT105	53	0C603	53	R100-8	53	RCA40395	53	RS3310
53	NKT106	53	0C604	53	R100-9	53	RCA40396P	53	RS3316
53	NKT107	53	0C6048P	53	R101-2	53	RE25	53	RS3316-1
53	NKT108	53	0C65	53	R101-3	53	REN158	53	RS3316-2
53	NKT109			53	R101-4	53	RS-102	53	RS3318
53	NKT211			53	R120	53	RS-1192	53	RS3717
53	NKT123			53	R1273	53	RS-1540		
53	NKT126								

GE	Industry Standard No.	GE	Industry Standard No.	GE	Industry Standard No.	GE	Industry Standard No.	GE	Industry Standard No.
53	R83726	53	RT3564	53	SFT131P	53	T1002A	53	TIA01
53	R83857	53	RT3566	53	SFT143	53	T1003	53	TIX90
53	R83866	53	RT3568	53	SFT144	53	T1004	53	TK-23C
53	R83867	53	RT4624	53	SFT145	53	T1005	53	TK-40C
53	R83880	53	RT4625	53	SFT146	53	T1006	53	TK-41C
53	R83897	53	RT5468	53	SFT151	53	T1007	53	TK-42C
53	R83904	53	RT5521	53	SFT152	53	T1008	53	TK-45C
53	R83913	53	RT5522	53	SFT221A	53	T10085	53	TK1228-1002
53	R83925	53	RT5637	53	SFT222A	53	T1009	53	TK1228-1003
53	R83926	53	RT61015	53	SFT232	53	T101	53	TK1228-1004
53	R8406	53	RT61016	53	SFT241	53	T1010	53	TK1228-1005
53	R85008	53	RT6205	53	SFT242	53	T1013	53	TK1228-1006
53	R85102	53	RT6604	53	SFT243	53	T1023	53	TK1228-1007
53	R85105	53	RT6734	53	SFT306	53	T102A	53	TK23C
53	R85202	53	RT6735	53	SFT321	53	T1036	53	TK40A
53	R85203	53	RT6736	53	SFT322	53	T1037	53	TK40CA
53	R85243-2	53	RT6990	53	SFT323	53	T1042	53	TK41A
53	R85401	53	RT7401	53	SFT327	53	T1043	53	TK49C
53	R85406	53	RT7558	53	SFT337	53	T1046	53	TM591
53	R85502	53	RT7846	53	SFT337B	53	T1047	53	TNJ-60070
53	R85503	53	RT8442	53	SFT337V	53	T1076	53	TNJ-60074
53	R85505	53	RT8602	53	SFT351	53	T108	53	TNJ-60079
53	R85506	53	RT8842	53	SFT352	53	T109	53	TNJ-60282
53	R85507	53	RV1180	53	SFT353	53	T116	53	TNJ-60283
53	R85530	53	RV1475	53	SFT367	53	T11618	53	TNJ-60728
53	R85531	53	S-1348	53	SFT523	53	T1202	53	TNJ60070
53	R85532	53	S-1349	53	SFT526	53	T1203	53	TNJ60074
53	R85533	53	S-1639	53	SK3004	53	T126	53	TNJ60079
53	R85534	53	S-95201	53	SM0043	53	T127	53	TNJ60282
53	R85535	53	S-95204	53	SM8341	53	T129	53	TNJ60283
53	R85536	53	S025	53	SM843	53	T130	53	TNJ60365
53	R85541	53	S065	53	SO-25	53	T1300	53	TNJ60610
53	R85542	53	S065A	53	SO-88	53	T13000	53	TNJ60611
53	R85543	53	S088	53	SO25	53	T1310	53	TNJ60612
53	R85544	53	S1348	53	SO65	53	T1327	53	TNJ60728
53	R85545	53	S1349	53	SO88	53	T1328	53	TNJ61222
53	R85551	53	S1639	53	SR1	53	T1334	53	TNJ61282
53	R85552	53	S1672	53	SS0001	53	T1342	53	TNJ61674
53	R85553	53	S1B02C	53	SS0001A	53	T1346	53	TNJ70688
53	R85554	53	S2041635	53	SS0002	53	T1352	53	TNJ72278
53	R85555	53	S4248	53	SS0002A	53	T1363	53	TNJ72283
53	R85556	53	S66-1-A-3P	53	SS0003	53	T1364	53	TNJ72285
53	R85557	53	S66-2-A-3P	53	SS0003A	53	T1546	53	TNJ72287
53	R85558	53	S66-3-A-3P	53	SS0004	53	T1559	53	TNJ72289
53	R85563	53	S685	53	SS0004A	53	T1573	53	TO-005
53	R85564	53	S686	53	SS0005	53	T1574	53	TO-014
53	R85565	53	S687	53	SS0005A	53	T1577	53	TO-041
53	R85566	53	S8P-2-A-3P	53	ST-122	53	T1583	53	TO-103
53	R85567	53	S95201	53	ST-123	53	T1593	53	TO-104
53	R85568	53	S95203	53	ST-301	53	T1594	53	TO101
53	R85602	53	S95204	53	ST-302	53	T1595	53	TO102
53	R85603	53	S95206	53	ST-304	53	T1596	53	TO103
53	R85605	53	S95207	53	ST-332	53	T1597	53	TO104
53	R85607	53	S95214	53	ST122	53	T1598	53	TQ-5051
53	R85608	53	S95218	53	ST123	53	T1599	53	TQ-5061
53	R85610	53	S99201	53	ST28A	53	T160	53	TQ5023
53	R85704	53	S99203	53	ST28B	53	T1740	53	TQ5025
53	R85704-2	53	S99218	53	ST28C	53	T1903	53	TQ5026
53	R857042	53	S9L-4-A-3P	53	ST301	53	T1904	53	TQ5051
53	R857062	53	SA128	53	ST302	53	T1961	53	TQ5061
53	R85708	53	SA128-1	53	ST303	53	T2159	53	TR-04
53	R85708-2	53	SA15V	53	ST304	53	T2439	53	TR-14
53	R85709	53	SA197	53	ST332	53	T2440	53	TR-14C
53	R85711	53	SA197-1	53	ST370	53	T2441	53	TR-15
53	R85717	53	SA197-2	53	ST37C	53	T2515	53	TR-157(OLSON)
53	R85717-1	53	SA197-3	53	ST37D	53	T2517	53	TR-158(OLSON)
53	R85717-3	53	SA204	53	ST37E	53	T282(SEARS)	53	TR-15C
53	R85717-6	53	SA205BLU	53	ST382	53	T3005	53	TR-169(OLSON)
53	R85720	53	SA205BRN	53	STX0096	53	T3321	53	TR-170(OLSON)
53	R85731	53	SA205ORN	53	STX0099	53	T3322	53	TR-2N2641C
53	R85732	53	SA205ORN	53	STX0104	53	T33Z1	53	TR-320
53	R85733	53	SA205RED	53	STX0105	53	T33Z3	53	TR-320A
53	R85734	53	SA205VIO	53	STX0110	53	T45	53	TR-321
53	R85735	53	SA205WHT	53	STX0114	53	T46	53	TR-321A
53	R85736	53	SA205YEL	53	STX0121	53	T47	53	TR-323A
53	R85737	53	SA240	53	STX0123	53	T48	53	TR-482A
53	R85738	53	SA29	53	STX0224	53	T50	53	TR-508A
53	R85740	53	SA318-2	53	STX0260	53	T50339A	53	TR-5826
53	R85740-1	53	SA318-3	53	STX0263	53	T50631	53	TR-6R26
53	R85742	53	SA33	53	STX0264	53	T50933B	53	TR-84
53	R85743	53	SA33BRN	53	STX0265	53	T52150	53	TR-85
53	R85743-1	53	SA33RED	53	STX0268	53	T52150Z	53	TR-045
53	R85743-2	53	SA529	53	STX0269	53	T52151	53	TR-C70
53	R85743-3	53	SA565	53	0000000V31	53	T52151Z	53	TR-C71
53	R857433	53	SA646	53	SYL-107	53	T52159	53	TR-C72
53	R85744	53	SA681	53	SYL-108	53	T59	53	TRO4C
53	R85744-3	53	SA853	53	SYL-1583	53	T59247	53	TR14
53	R85745	53	SB168	53	SYL-1668	53	T59249	53	TR14C
53	R85746	53	SB169	53	SYL107A	53	T60	53	TR15
53	R85747	53	SC-12	53	SYL108A	53	T61	53	TR15C
53	R85748	53	SC-56	53	SYL1583A	53	T72	53	TR2N2614C
53	R85749	53	SC-63	53	SYL1655A	53	T74	53	TR310011
53	R85750	53	SC-66	53	SYL1665A	53	T76	53	TR310012
53	R85751	53	SC-68	53	SYL1668A	53	T77	53	TR310017
53	R85752	53	SC-69	53	SYL2248A	53	T78	53	TR310018
53	R85765	53	SC-73	53	SYL2249A	53	T814	53	TR310075
53	R85766	53	SC12	53	SYL2300A	53	T815	53	TR310107
53	R85767	53	SC45	53	SYL3613A	53	T82	53	TR310125
53	R85768	53	SC56	53	T-00014	53	T83	53	TR310136
53	R85852	53	SC63	53	T-126	53	T84	53	TR310149
53	R85854	53	SC66	53	T-129	53	T87	53	TR310153
53	R86840	53	SC68	53	T-130	53	T95	53	TR310159
53	R86843A	53	SC69	53	T-131	53	T99	53	TR310164
53	R86846A	53	SC73	53	T-25	53	TA-1575	53	TR310227
53	R87168	53	SF.T124	53	T-3321	53	TA-1697	53	TR310235
53	R87568	53	SF.T130	53	T-3322	53	TA-1706	53	TR310251
53	R88406	53	SF.T131P	53	T-33Z1	53	TA-1730	53	TR310252
53	R88421	53	SF.T221	53	T-3323	53	TA-4	53	TR310255
53	R88424	53	SF.T223	53	T-50	53	TA1575	53	TR320AN
53	R88444	53	SF.T237	53	T-52150	53	TA1575B	53	TR323AN
53	R88446	53	SF.T251	53	T-52150Z	53	TA1655B	53	TR81
53	R81350	53	SF.T252	53	T-52151	53	TA1697	53	TR38117
53	RT-185	53	SF.T253	53	T-52151Z	53	TA1706	53	TR382
53	RT-4625	53	SF.T306	53	T-95	53	TA1730	53	TR383
53	RT-61015	53	SF.T321	53	T-Q5027	53	TC3123041557	53	TR43
53	RT-61016	53	SF.T322	53	T00014	53	TF-30	53	TR44
53	RT121	53	SF.T337	53	T0003	53	TF-65	53	TR45
53	RT186	53	SF.T351	53	T0004	53	TF-66	53	TR508AN
53	RT2230	53	SF.T352	53	T0005	53	TF30	53	TR650A
53	RT2329	53	SFT-306	53	T0012	53	T749	53	TR653A
53	RT2330	53	SFT-322	53	T0014	53	T65	53	TR721A
53	RT2331	53	SFT-323	53	T0015	53	T65/30	53	TR722A
53	RT2709	53	SFT-327	53	T0031	53	T65/M	53	TR763A
53	RT3097	53	SFT-337	53	T0033	53	T65/S/30	53	TR81
53	RT3098	53	SFT-352	53	T0038	53	T65M	53	TR84
53	RT3229	53	SFT-353	53	T0039	53	T66	53	TR85
53	RT3230	53	SFT121	53	T0040	53	T66/30	53	TRA-32
53	RT3231	53	SFT122	53	T0041	53	T66/60	53	TRA-33
53	RT3363	53	SFT123	53	T0051	53	T75	53	TRA32
53	RT3364	53	SFT124	53	T100	53	T77	53	TRA33
53	RT3365	53	SFT125	53	T1000	53	T80/302	53	TS-1
53	RT3449	53	SFT125P	53	T1001	53	TG48		
53	RT3467	53	SFT130	53	T10010	53	THU60U		
53	RT3468	53	SFT131	53	T1002	53	TIA-01		

GE	Industry Standard No.	GE	Industry Standard No.	GE	Industry Standard No.	GE	Industry Standard No.	GE	Industry Standard No.
53	TS-1007	53	2D013-109	53	2N407GRN	53	2SA150P	53	2SA221A
53	TS-1266	53	2D013-13	53	2N407J	53	2SA150G	53	2SA221B
53	TS-13	53	2D013-160	53	2N407RED	53	2SA150GN	53	2SA221C
53	TS-14	53	2D013-54	53	2N407WHT	53	2SA150H	53	2SA221D
53	TS-15	53	2D016	53	2N407YEL	53	2SA150J	53	2SA221E
53	TS-162	53	2D016-45	53	2N408	53	2SA150K	53	2SA221F
53	TS-163	53	2D016-54	53	2N408J	53	2SA150L	53	2SA221G
53	TS-165	53	2D021	53	2N408WHT	53	2SA150M	53	2SA221GN
53	TS-166	53	2D021-11	53	2N407	53	2SA150OR	53	2SA221H
53	TS-1727	53	2D021-56	53	2N427A	53	2SA150R	53	2SA221J
53	TS-1728	53	2D021-8	53	2N435	53	2SA150X	53	2SA221K
53	TS-2	53	2D023	53	2N45	53	2SA150Y	53	2SA221L
53	TS-3	53	2D036	53	2N45A	53	2SA176	53	2SA221M
53	TS-603	53	2D039	53	2N469	53	2SA176A	53	2SA221OR
53	TS-604	53	2D021-56	53	2N506	53	2SA176B	53	2SA221R
53	TS-616	53	2D023	53	2N534	53	2SA176C	53	2SA221X
53	TS-617	53	20303	53	2N535	53	2SA176D	53	2SA221Y
53	TS-618	53	20304	53	2N535A	53	2SA176E	53	2SA222
53	TS-627	53	20308	53	2N535B	53	2SA176F	53	2SA222A
53	TS-629	53	20309	53	2N536	53	2SA176G	53	2SA222B
53	TS-739	53	20371	53	2N591	53	2SA176GN	53	2SA222C
53	TS-739B	53	20371A	53	2N591/5	53	2SA176H	53	2SA222D
53	TS-740	53	20374	53	2N591A	53	2SA176J	53	2SA222E
53	TS-765	53	20374A	53	2N62	53	2SA176K	53	2SA222F
53	TS1007	53	20376	53	2N63	53	2SA176L	53	2SA222G
53	TS1266	53	20377	53	2N64	53	2SA176M	53	2SA222GN
53	TS162	53	20381	53	2N653	53	2SA176OR	53	2SA222H
53	TS163	53	20381A	53	2N670	53	2SA176R	53	2SA222J
53	TS164	53	20395	53	2N671	53	2SA176X	53	2SA222K
53	TS165	53	2M1303	53	2N673	53	2SA176Y	53	2SA222L
53	TS166	53	2N104	53	2N675	53	2SA186	53	2SA222M
53	TS1727	53	2N105	53	2N800	53	2SA187TV	53	2SA222OR
53	TS1728	53	2N106	53	2N803	53	2SA19	53	2SA222R
53	TS1792	53	2N107	53	2N804	53	2SA19A	53	2SA222X
53	TS601	53	2N109	53	2N805	53	2SA19B	53	2SA222Y
53	TS602	53	2N109BLU	53	2N806	53	2SA19C	53	2SA224
53	TS603	53	2N109GRN	53	2N807	53	2SA19D	53	2SA224A
53	TS604	53	2N109M1	53	2N808	53	2SA19E	53	2SA224B
53	TS616	53	2N109M2	53	2N81	53	2SA19F	53	2SA224C
53	TS617	53	2N109WHT	53	2N813	53	2SA19G	53	2SA224D
53	TS618	53	2N109YEL	53	2N814	53	2SA19L	53	2SA224E
53	TS619	53	2N1123	53	2N815	53	2SA19M	53	2SA224F
53	TS620	53	2N1127	53	2N816	53	2SA19OR	53	2SA224G
53	TS621	53	2N1129	53	2N817	53	2SA19R	53	2SA224GN
53	TS627	53	2N1130	53	2N818	53	2SA19X	53	2SA224H
53	TS629	53	2N1266	53	2N819	53	2SA19Y	53	2SA224J
53	TS739	53	2N1303	53	2N820	53	2SA20	53	2SA224K
53	TS739B	53	2N1305	53	2OC72	53	2SA200	53	2SA224M
53	TS740	53	2N1307	53	2ZB187	53	2SA202	53	2SA240R
53	TS765	53	2N153A	53	2S189	53	2SA202A	53	2SA224R
53	TV-61A	53	2N138	53	2S273	53	2SA202AP	53	2SA224X
53	TV24115	53	2N138A	53	2SA007H	53	2SA202B	53	2SA224Y
53	TV24154	53	2N138B	53	2SA081C	53	2SA202C	53	2SA231
53	TV24156	53	2N1416	53	2SA105	53	2SA202D	53	2SA232
53	TV24199	53	2N143B	53	2SA105A	53	2SA202D-4	53	2SA237
53	TV24194	53	2N1614	53	2SA105B	53	2SA202E	53	2SA237A
53	TV2428	53	2N1673	53	2SA105C	53	2SA202F	53	2SA237B
53	TV2429	53	2N1681	53	2SA105D	53	2SA202G	53	2SA237C
53	TV2434	53	2N175	53	2SA105E	53	2SA202GN	53	2SA237D
53	TV24370	53	2N180	53	2SA105G	53	2SA202H	53	2SA237E
53	TV24599	53	2N1800	53	2SA105H	53	2SA202J	53	2SA237F
53	TV24945	53	2N181	53	2SA105K	53	2SA202K	53	2SA237G
53	TV24984	53	2N185	53	2SA105L	53	2SA202L	53	2SA237GN
53	TV47	53	2N185BLU	53	2SA105M	53	2SA202M	53	2SA237H
53	TV61	53	2N186	53	2SA105OR	53	2SA202OR	53	2SA237J
53	TV8-28A385	53	2N186A	53	2SA105R	53	2SA202R	53	2SA237K
53	TV8-28A385A	53	2N187	53	2SA105X	53	2SA202X	53	2SA237L
53	TV8-28A385L	53	2N189#	53	2SA105Y	53	2SA202Y	53	2SA237M
53	TV8-28A7MB	53	2N191	53	2SA12	53	2SA203AA	53	2SA237OR
53	TV8-28B1T1	53	2N1925	53	2SA128	53	2SA20A	53	2SA237R
53	TV8-28B171A	53	2N1926	53	2SA128A	53	2SA20B	53	2SA237X
53	TV8-28B172	53	2N1954	53	2SA128B	53	2SA20C	53	2SA237Y
53	TV8-28B172F	53	2N1955	53	2SA128C	53	2SA20D	53	2SA250
53	TV8-28B176	53	2N1956	53	2SA128D	53	2SA20F	53	2SA250A
53	TV8-28B234	53	2N1957	53	2SA128E	53	2SA20G	53	2SA250B
53	TV8-28B324	53	2N1969	53	2SA128F	53	2SA20L	53	2SA250C
53	TV328B171	53	2N2000	53	2SA128G	53	2SA20M	53	2SA250D
53	TV328B171A	53	2N2001	53	2SA128GN	53	2SA20OR	53	2SA250B
53	TV328B171B	53	2N2	53	2SA128H	53	2SA20R	53	2SA250F
53	TV328B172	53	2N2215	53	2SA128J	53	2SA20X	53	2SA250G
53	TV328B324	53	2N2217	53	2SA128L	53	2SA20Y	53	2SA250GN
53	001-012011	53	2N2173	53	2SA128M	53	2SA21	53	2SA250H
53	001-012021	53	2N2217A	53	2SA128OR	53	2SA218	53	2SA250J
53	001-012031	53	2N2217RED	53	2SA128R	53	2SA218A	53	2SA250L
53	001-012060	53	2N2217WHT	53	2SA128T	53	2SA218B	53	2SA250M
53	01-1201-0	53	2N2217YEL	53	2SA129	53	2SA218C	53	2SA250OR
53	1-21-106	53	2N22	53	2SA129A	53	2SA218D	53	2SA250X
53	1-21-107	53	2N220	53	2SA129B	53	2SA218E	53	2SA250Y
53	1-21-120	53	2N222	53	2SA129C	53	2SA218F	53	2SA256
53	1-21-148	53	2N223	53	2SA129D	53	2SA218G	53	2SA256A
53	1-21-164	53	2N224	53	2SA129F	53	2SA218GN	53	2SA256B
53	1-21-184	53	2N225	53	2SA129GN	53	2SA218H	53	2SA256C
53	1-21-191	53	2N226	53	2SA129H	53	2SA218J	53	2SA256D
53	1-21-192	53	2N227	53	2SA129J	53	2SA218K	53	2SA256E
53	1-21-225	53	2N228A	53	2SA129K	53	2SA218L	53	2SA256F
53	1-21-226	53	2N238	53	2SA129L	53	2SA218M	53	2SA256G
53	1-21-227	53	2N238-ORN	53	2SA129M	53	2SA218OR	53	2SA256GN
53	1-21-232	53	2N258D	53	2SA129OR	53	2SA218R	53	2SA256H
53	1-21-246	53	2N258B	53	2SA129R	53	2SA218X	53	2SA256J
53	1-21-266	53	2N258F	53	2SA129X	53	2SA218Y	53	2SA256K
53	1-21-267	53	2N2428	53	2SA129Y	53	2SA21A	53	2SA256L
53	1-21-272	53	2N2429	53	2SA12A	53	2SA21B	53	2SA256M
53	1-21-274	53	2N2431	53	2SA12B	53	2SA21C	53	2SA256OR
53	1-21-95	53	2N2447	53	2SA12C	53	2SA21D	53	2SA256R
53	1-21-96	53	2N2448	53	2SA12D	53	2SA21E	53	2SA256X
53	1-52221011	53	2N2449	53	2SA12H	53	2SA21F	53	2SA256Y
53	1-522210111	53	2N2450	53	2SA12V	53	2SA21G	53	2SA259
53	1-522211200	53	2N2613	53	2SA138	53	2SA21L	53	2SA259B
53	1-522211328	53	2N2614	53	2SA142	53	2SA21M	53	2SA259C
53	1-522216500	53	2N265	53	2SA142A	53	2SA21OR	53	2SA259D
53	1-6207190405	53	2N2706	53	2SA142B	53	2SA21R	53	2SA259E
53	1-801-005	53	2N2707	53	2SA142C	53	2SA21X	53	2SA259F
53	1-801-005-23	53	2N279	53	2SA142D	53	2SA21Y	53	2SA259G
53	1-801-006	53	2N280	53	2SA142E	53	2SA220	53	2SA259GN
53	1-801-006-12	53	2N281	53	2SA142F	53	2SA220A	53	2SA259H
53	1-801-006-14	53	2N283	53	2SA142GN	53	2SA220B	53	2SA259J
53	1-801-308	53	2N284	53	2SA142H	53	2SA220C	53	2SA259L
53	1-801-308-24	53	2N284A	53	2SA142J	53	2SA220D	53	2SA259M
53	1-801-310	53	2N320	53	2SA142K	53	2SA220E	53	2SA259OR
53	1-TR-111	53	2N34	53	2SA142L	53	2SA220F	53	2SA259R
53	1A0055	53	2N34A	53	2SA142M	53	2SA220G	53	2SA259X
53	1A0056	53	2N36	53	2SA142OR	53	2SA220GN	53	2SA259Y
53	1T495	53	2N37	53	2SA142R	53	2SA220H	53	2SA266
53	002-005100	53	2N38	53	2SA142X	53	2SA220J	53	2SA266A
53	002-006600	53	2N38A	53	2SA142Y	53	2SA220K	53	2SA266B
53	002-006800	53	2N406	53	2SA150	53	2SA220L	53	2SA266C
53	002-006900	53	2N406BLU	53	2SA150A	53	2SA220M	53	2SA266D
53	002-007300	53	2N406BRN	53	2SA150B	53	2SA220OR	53	2SA266E
53	002-008400	53	2N406GRN	53	2SA150C	53	2SA220R	53	2SA266F
53	002-011000	53	2N406GRN-YEL	53	2SA150D	53	2SA220X	53	2SA266G
53	002-011900	53	2N406ORN	53	2SA150B	53	2SA220Y	53	2SA266GN
53	002-11800	53	2N406RED	53	2SA150C	53	2SA221	53	2SA266GREEN
53	002-11900	53	2N407	53	2SA150E	53	2SA221-OR	53	2SA266H
53	2D013	53	2N407BLK						

GE	Industry Standard No.	GE	Industry Standard No.	GE	Industry Standard No.	GE	Industry Standard No.	GE	Industry Standard No.
53	28A266J	53	28A286G	53	28A314C	53	28A339K	53	28A384
53	28A266K	53	28A286GN	53	28A314D	53	28A339L	53	28A384A
53	28A266L	53	28A286H	53	28A314E	53	28A339M	53	28A384B
53	28A266M	53	28A286K	53	28A314F	53	28A339OR	53	28A384C
53	28A266OR	53	28A286M	53	28A314G	53	28A339R	53	28A384D
53	28A266R	53	28A286OR	53	28A314GN	53	28A339X	53	28A384E
53	28A266X	53	28A286R	53	28A314H	53	28A339Y	53	28A384F
53	28A266Y	53	28A286X	53	28A314J	53	28A341	53	28A384G
53	28A269	53	28A286Y	53	28A314K	53	28A341-OA	53	28A384GN
53	28A269A	53	28A287	53	28A314L	53	28A341-OB	53	28A384H
53	28A269B	53	28A287A	53	28A314M	53	28A341A	53	28A384J
53	28A269C	53	28A287B	53	28A314OR	53	28A341B	53	28A384K
53	28A269D	53	28A287C	53	28A314R	53	28A341C	53	28A384L
53	28A269E	53	28A287D	53	28A314X	53	28A341D	53	28A384M
53	28A269F	53	28A287E	53	28A314Y	53	28A341E	53	28A384OR
53	28A269GN	53	28A287F	53	28A315	53	28A341F	53	28A384R
53	28A269H	53	28A287G	53	28A315-RED	53	28A341G	53	28A384X
53	28A269J	53	28A287GN	53	28A315A	53	28A341GN	53	28A384Y
53	28A269K	53	28A287H	53	28A315B	53	28A341H	53	28A400
53	28A269L	53	28A287J	53	28A315C	53	28A341J	53	28A400A
53	28A269M	53	28A287K	53	28A315D	53	28A341K	53	28A400B
53	28A269OR	53	28A287L	53	28A315E	53	28A341L	53	28A400C
53	28A269X	53	28A287M	53	28A315F	53	28A341M	53	28A400D
53	28A269Y	53	28A287OR	53	28A315G	53	28A341OR	53	28A400E
53	28A271	53	28A287X	53	28A315GN	53	28A341R	53	28A400F
53	28A271A	53	28A287Y	53	28A315H	53	28A341X	53	28A400GN
53	28A271B	53	28A288	53	28A315J	53	28A341Y	53	28A400H
53	28A271C	53	28A288A	53	28A315K	53	28A350Y	53	28A400J
53	28A271D	53	28A288B	53	28A315L	53	28A356	53	28A400K
53	28A271E	53	28A288C	53	28A315M	53	28A356A	53	28A400L
53	28A271F	53	28A288D	53	28A315OR	53	28A356B	53	28A400M
53	28A271G	53	28A288E	53	28A315R	53	28A356C	53	28A400OR
53	28A271GN	53	28A288F	53	28A315X	53	28A356D	53	28A400R
53	28A271H	53	28A288G	53	28A315Y	53	28A356E	53	28A400X
53	28A271J	53	28A288GN	53	28A316	53	28A356F	53	28A400Y
53	28A271K	53	28A288J	53	28A316-RED	53	28A356G	53	28A476
53	28A271L	53	28A288K	53	28A316A	53	28A356GN	53	28A476A
53	28A271M	53	28A288L	53	28A316B	53	28A356H	53	28A476B
53	28A271OR	53	28A288M	53	28A316C	53	28A356J	53	28A476C
53	28A271R	53	28A288OR	53	28A316D	53	28A356K	53	28A476D
53	28A271X	53	28A288X	53	28A316E	53	28A356L	53	28A476E
53	28A271Y	53	28A288Y	53	28A316F	53	28A356M	53	28A476F
53	28A272	53	28A29	53	28A316G	53	28A356OR	53	28A476G
53	28A272A	53	28A298	53	28A316GN	53	28A356R	53	28A476GN
53	28A272B	53	28A298A	53	28A316H	53	28A357A	53	28A476H
53	28A272C	53	28A298B	53	28A316J	53	28A357B	53	28A476J
53	28A272D	53	28A298C	53	28A316K	53	28A357C	53	28A476K
53	28A272E	53	28A298D	53	28A316L	53	28A357D	53	28A476L
53	28A272F	53	28A298E	53	28A316M	53	28A357E	53	28A476M
53	28A272G	53	28A298F	53	28A316OR	53	28A357F	53	28A476OR
53	28A272GN	53	28A298G	53	28A316R	53	28A357G	53	28A476R
53	28A272H	53	28A298GN	53	28A316X	53	28A357GN	53	28A476X
53	28A272J	53	28A298H	53	28A316Y	53	28A357H	53	28A476Y
53	28A272K	53	28A298J	53	28A317J	53	28A357J	53	28A538
53	28A272L	53	28A298K	53	28A322	53	28A357K	53	28A57A
53	28A272M	53	28A298L	53	28A322A	53	28A357L	53	28A57B
53	28A272OR	53	28A298M	53	28A322B	53	28A357M	53	28A57C
53	28A272R	53	28A298OR	53	28A322C	53	28A357OR	53	28A57D
53	28A272X	53	28A298X	53	28A322D	53	28A357R	53	28A57E
53	28A272Y	53	28A298Y	53	28A322E	53	28A357X	53	28A57G
53	28A273	53	28A29A	53	28A322F	53	28A357Y	53	28A57H
53	28A273A	53	28A29B	53	28A322G	53	28A369	53	28A57L
53	28A273B	53	28A29C	53	28A322GN	53	28A369A	53	28A570M
53	28A273C	53	28A29D	53	28A322H	53	28A369B	53	28A57R
53	28A273D	53	28A29E	53	28A322K	53	28A369C	53	28A57X
53	28A273E	53	28A29F	53	28A322L	53	28A369D	53	28A57Y
53	28A273F	53	28A29G	53	28A322M	53	28A369E	53	28A64
53	28A273G	53	28A29J	53	28A322OR	53	28A369F	53	28A64A
53	28A273GN	53	28A29M	53	28A322X	53	28A369G	53	28A64B
53	28A273H	53	28A29R	53	28A322Y	53	28A369GN	53	28A64C
53	28A273J	53	28A29X	53	28A323	53	28A369H	53	28A64D
53	28A273K	53	28A29Y	53	28A323A	53	28A369J	53	28A64E
53	28A273L	53	28A302	53	28A323B	53	28A369K	53	28A64F
53	28A273M	53	28A303	53	28A323C	53	28A369L	53	28A64GN
53	28A273OR	53	28A307	53	28A323D	53	28A369M	53	28A64H
53	28A273X	53	28A307A	53	28A323E	53	28A369OR	53	28A64J
53	28A273Y	53	28A307B	53	28A323F	53	28A369R	53	28A64K
53	28A274	53	28A307C	53	28A323G	53	28A369X	53	28A64L
53	28A274A	53	28A307D	53	28A323GN	53	28A369Y	53	28A64M
53	28A274C	53	28A307E	53	28A323J	53	28A370	53	28A64OR
53	28A274D	53	28A307F	53	28A323K	53	28A371	53	28A64X
53	28A274E	53	28A307G	53	28A323L	53	28A374	53	28A80
53	28A274F	53	28A307GN	53	28A323M	53	28A381	53	28A80A
53	28A274G	53	28A307J	53	28A323R	53	28A381A	53	28A80B
53	28A274GN	53	28A307K	53	28A323X	53	28A381C	53	28A80D
53	28A274H	53	28A307L	53	28A323Y	53	28A381D	53	28A80E
53	28A274J	53	28A307M	53	28A33	53	28A381E	53	28A80F
53	28A274K	53	28A307OR	53	28A332	53	28A381F	53	28A80G
53	28A274L	53	28A307R	53	28A332A	53	28A381GN	53	28A80H
53	28A274M	53	28A307X	53	28A332B	53	28A381H	53	28A80K
53	28A274OR	53	28A307Y	53	28A332C	53	28A381K	53	28A80L
53	28A274X	53	28A311A	53	28A332D	53	28A381L	53	28A80OR
53	28A274Y	53	28A311B	53	28A332E	53	28A381M	53	28A80R
53	28A275	53	28A311C	53	28A332F	53	28A381OR	53	28A80X
53	28A275A	53	28A311D	53	28A332G	53	28A381R	53	28A80Y
53	28A275B	53	28A311E	53	28A332GN	53	28A381X	53	28A82
53	28A275C	53	28A311F	53	28A332H	53	28A381Y	53	28A82A
53	28A275D	53	28A311G	53	28A332J	53	28A382	53	28A82C
53	28A275E	53	28A311GN	53	28A332K	53	28A382A	53	28A82D
53	28A275F	53	28A311H	53	28A332L	53	28A382B	53	28A82F
53	28A275G	53	28A311J	53	28A332M	53	28A382C	53	28A82G
53	28A275GN	53	28A311K	53	28A332OR	53	28A382D	53	28A82H
53	28A275H	53	28A311L	53	28A332R	53	28A382E	53	28A82K
53	28A275J	53	28A311M	53	28A332X	53	28A382F	53	28A82M
53	28A275L	53	28A311OR	53	28A332Y	53	28A382G	53	28A82R
53	28A275M	53	28A311R	53	28A338	53	28A382GN	53	28A82X
53	28A275OR	53	28A311X	53	28A338A	53	28A382H	53	28A82Y
53	28A275R	53	28A311Y	53	28A338B	53	28A382J	53	28A85
53	28A275X	53	28A312	53	28A338C	53	28A382K	53	28A94
53	28A275Y	53	28A312A	53	28A338D	53	28A382L	53	28A945
53	28A285	53	28A312B	53	28A338E	53	28A382M	53	28A94A
53	28A285B	53	28A312C	53	28A338F	53	28A382OR	53	28A94C
53	28A285C	53	28A312D	53	28A338G	53	28A382R	53	28A94D
53	28A285E	53	28A312E	53	28A338GN	53	28A382X	53	28A94E
53	28A285F	53	28A312F	53	28A338H	53	28A382Y	53	28A94F
53	28A285G	53	28A312G	53	28A338J	53	28A383	53	28A94H
53	28A285GN	53	28A312GN	53	28A338K	53	28A383A	53	28A94K
53	28A285H	53	28A312H	53	28A338L	53	28A383B	53	28A94L
53	28A285J	53	28A312J	53	28A338M	53	28A383C	53	28A94M
53	28A285L	53	28A312K	53	28A338OR	53	28A383D	53	28A940R
53	28A285M	53	28A312L	53	28A338R	53	28A383E	53	28A94R
53	28A285OR	53	28A312M	53	28A338X	53	28A383F	53	28A94X
53	28A285R	53	28A312OR	53	28A338Y	53	28A383G	53	28A94Y
53	28A285X	53	28A312R	53	28A339	53	28A383GN		
53	28A285Y	53	28A312X	53	28A339A	53	28A383H		
53	28A286	53	28A312Y	53	28A339B	53	28A383K		
53	28A286A	53	28A313-RED	53	28A339C	53	28A383L		
53	28A286B	53	28A313-YELLOW	53	28A339D	53	28A383M		
53	28A286C	53	28A313A	53	28A339F	53	28A383OR		
53	28A286D	53	28A313B	53	28A339G	53	28A383R		
53	28A286E	53	28A314A	53	28A339GN	53	28A383X		
53	28A286F	53	28A314B	53	28A339H	53	28A383Y		
				53	28A339J				

GE	Industry Standard No.	GE	Industry Standard No.	GE	Industry Standard No.	GE	Industry Standard No.	GE	Industry Standard No.
53	2SB-3783	53	2SB112A	53	2SB134F	53	2SB156E	53	2SB163L
53	2SB-3812	53	2SB112B	53	2SB134G	53	2SB156F	53	2SB163M
53	2SB-3813	53	2SB112C	53	2SB134GN	53	2SB156G	53	2SB163OR
53	2SB-F1A	53	2SB112D	53	2SB134H	53	2SB156GN	53	2SB163R
53	2SB100	53	2SB112E	53	2SB134J	53	2SB156H	53	2SB163X
53	2SB100A	53	2SB112F	53	2SB134K	53	2SB156J	53	2SB163Y
53	2SB100B	53	2SB112G	53	2SB134L	53	2SB156K	53	2SB164
53	2SB100C	53	2SB112H	53	2SB134M	53	2SB156L	53	2SB164A
53	2SB100D	53	2SB112J	53	2SB134OR	53	2SB156M	53	2SB164B
53	2SB100E	53	2SB112K	53	2SB134R	53	2SB156OR	53	2SB164C
53	2SB100F	53	2SB112L	53	2SB134X	53	2SB156P	53	2SB164D
53	2SB100G	53	2SB112M	53	2SB134Y	53	2SB156R	53	2SB164E
53	2SB100GN	53	2SB112OR	53	2SB135	53	2SB156X	53	2SB164F
53	2SB100H	53	2SB112R	53	2SB135(C)	53	2SB156Y	53	2SB164GN
53	2SB100J	53	2SB112X	53	2SB135B	53	2SB157	53	2SB164H
53	2SB100K	53	2SB112Y	53	2SB135C	53	2SB157A	53	2SB164J
53	2SB100L	53	2SB113	53	2SB135D	53	2SB157B	53	2SB164L
53	2SB100M	53	2SB113A	53	2SB135E	53	2SB157C	53	2SB164M
53	2SB100OR	53	2SB113B	53	2SB135F	53	2SB157D	53	2SB164OR
53	2SB100R	53	2SB113C	53	2SB135G	53	2SB157E	53	2SB164R
53	2SB100X	53	2SB113D	53	2SB135GN	53	2SB157F	53	2SB164X
53	2SB100Y	53	2SB113E	53	2SB135H	53	2SB157G	53	2SB164Y
53	2SB101	53	2SB113F	53	2SB135J	53	2SB157GN	53	2SB165
53	2SB101A	53	2SB113G	53	2SB135K	53	2SB157H	53	2SB165A
53	2SB101B	53	2SB113GN	53	2SB135L	53	2SB157J	53	2SB165B
53	2SB101C	53	2SB113H	53	2SB135M	53	2SB157K	53	2SB165C
53	2SB101D	53	2SB113J	53	2SB135OR	53	2SB157L	53	2SB165D
53	2SB101E	53	2SB113K	53	2SB135R	53	2SB157OR	53	2SB165E
53	2SB101F	53	2SB113L	53	2SB135X	53	2SB157R	53	2SB165F
53	2SB101G	53	2SB113M	53	2SB135Y	53	2SB157X	53	2SB165G
53	2SB101GN	53	2SB113OR	53	2SB136	53	2SB157Y	53	2SB165GN
53	2SB101H	53	2SB113R	53	2SB136(C)	53	2SB158	53	2SB165H
53	2SB101J	53	2SB113X	53	2SB136-2	53	2SB158A	53	2SB165J
53	2SB101K	53	2SB113Y	53	2SB136-3	53	2SB158B	53	2SB165K
53	2SB101L	53	2SB114	53	2SB136A	53	2SB158C	53	2SB165M
53	2SB101M	53	2SB114A	53	2SB136B	53	2SB158D	53	2SB165OR
53	2SB101OR	53	2SB114B	53	2SB136C	53	2SB158E	53	2SB165R
53	2SB101R	53	2SB114C	53	2SB136D	53	2SB158F	53	2SB165X
53	2SB101X	53	2SB114D	53	2SB136E	53	2SB158G	53	2SB165Y
53	2SB101Y	53	2SB114E	53	2SB136F	53	2SB158GN	53	2SB166
53	2SB102	53	2SB114F	53	2SB136G	53	2SB158H	53	2SB166A
53	2SB102A	53	2SB114G	53	2SB136GN	53	2SB158J	53	2SB166B
53	2SB102B	53	2SB114GN	53	2SB136H	53	2SB158K	53	2SB166C
53	2SB102C	53	2SB114H	53	2SB136J	53	2SB158L	53	2SB166D
53	2SB102D	53	2SB114J	53	2SB136K	53	2SB158M	53	2SB166E
53	2SB102E	53	2SB114K	53	2SB136L	53	2SB158OR	53	2SB166F
53	2SB102F	53	2SB114L	53	2SB136M	53	2SB158R	53	2SB166G
53	2SB102G	53	2SB114M	53	2SB136OR	53	2SB158X	53	2SB166GN
53	2SB102GN	53	2SB114OR	53	2SB136R	53	2SB158Y	53	2SB166H
53	2SB102H	53	2SB114R	53	2SB136U	53	2SB159	53	2SB166J
53	2SB102J	53	2SB114X	53	2SB136X	53	2SB159A	53	2SB166K
53	2SB102K	53	2SB114Y	53	2SB15	53	2SB159B	53	2SB166L
53	2SB102L	53	2SB115	53	2SB150	53	2SB159C	53	2SB166M
53	2SB102M	53	2SB115A	53	2SB150A	53	2SB159D	53	2SB166OR
53	2SB102OR	53	2SB115B	53	2SB150B	53	2SB159E	53	2SB166R
53	2SB102R	53	2SB115C	53	2SB150C	53	2SB159F	53	2SB166X
53	2SB102X	53	2SB115D	53	2SB150D	53	2SB159G	53	2SB167
53	2SB102Y	53	2SB115E	53	2SB150E	53	2SB159GN	53	2SB167A
53	2SB103	53	2SB115F	53	2SB150F	53	2SB159H	53	2SB167B
53	2SB103A	53	2SB115G	53	2SB150G	53	2SB159J	53	2SB167BK
53	2SB103B	53	2SB115GN	53	2SB150GN	53	2SB159K	53	2SB167C
53	2SB103C	53	2SB115H	53	2SB150H	53	2SB159L	53	2SB167D
53	2SB103D	53	2SB115J	53	2SB150J	53	2SB159M	53	2SB167E
53	2SB103E	53	2SB115K	53	2SB150K	53	2SB159OR	53	2SB167F
53	2SB103F	53	2SB115L	53	2SB150L	53	2SB159R	53	2SB167G
53	2SB103G	53	2SB115M	53	2SB150M	53	2SB159X	53	2SB167GN
53	2SB103GN	53	2SB115OR	53	2SB150OR	53	2SB159Y	53	2SB167H
53	2SB103H	53	2SB115X	53	2SB150X	53	2SB160	53	2SB167J
53	2SB103J	53	2SB115Y	53	2SB150Y	53	2SB160A	53	2SB167K
53	2SB103K	53	2SB116	53	2SB153	53	2SB160B	53	2SB167L
53	2SB103L	53	2SB116A	53	2SB153A	53	2SB160C	53	2SB167M
53	2SB103M	53	2SB116B	53	2SB153B	53	2SB160D	53	2SB167OR
53	2SB103OR	53	2SB116C	53	2SB153C	53	2SB160E	53	2SB167R
53	2SB103R	53	2SB116D	53	2SB153D	53	2SB160P	53	2SB167X
53	2SB103X	53	2SB116E	53	2SB153E	53	2SB160G	53	2SB167Y
53	2SB103Y	53	2SB116F	53	2SB153F	53	2SB160GN	53	2SB168
53	2SB104	53	2SB116G	53	2SB153G	53	2SB160H	53	2SB168A
53	2SB104A	53	2SB116GN	53	2SB153GN	53	2SB160J	53	2SB168B
53	2SB104B	53	2SB116H	53	2SB153H	53	2SB160K	53	2SB168C
53	2SB104C	53	2SB116J	53	2SB153J	53	2SB160L	53	2SB168D
53	2SB104D	53	2SB116K	53	2SB153K	53	2SB160M	53	2SB168E
53	2SB104E	53	2SB116L	53	2SB153L	53	2SB160R	53	2SB168F
53	2SB104F	53	2SB116M	53	2SB153M	53	2SB160X	53	2SB168G
53	2SB104G	53	2SB116OR	53	2SB153OR	53	2SB160Y	53	2SB168GN
53	2SB104GN	53	2SB116R	53	2SB153X	53	2SB161	53	2SB168H
53	2SB104H	53	2SB116X	53	2SB153Y	53	2SB161A	53	2SB168J
53	2SB104J	53	2SB116Y	53	2SB154	53	2SB161B	53	2SB168K
53	2SB104K	53	2SB117	53	2SB154A	53	2SB161C	53	2SB168L
53	2SB104L	53	2SB117A	53	2SB154B	53	2SB161D	53	2SB168M
53	2SB104M	53	2SB117B	53	2SB154C	53	2SB161E	53	2SB168OR
53	2SB104OR	53	2SB117C	53	2SB154D	53	2SB161F	53	2SB168R
53	2SB104R	53	2SB117D	53	2SB154E	53	2SB161G	53	2SB168Y
53	2SB104X	53	2SB117E	53	2SB154F	53	2SB161GN	53	2SB169
53	2SB104Y	53	2SB117F	53	2SB154G	53	2SB161H	53	2SB169A
53	2SB106	53	2SB117G	53	2SB154GN	53	2SB161J	53	2SB169B
53	2SB108B	53	2SB117GN	53	2SB154H	53	2SB161K	53	2SB169C
53	2SB109	53	2SB117H	53	2SB154J	53	2SB161L	53	2SB169D
53	2SB110	53	2SB117J	53	2SB154K	53	2SB161M	53	2SB169E
53	2SB110A	53	2SB117K	53	2SB154L	53	2SB161OR	53	2SB169F
53	2SB110B	53	2SB117L	53	2SB154M	53	2SB161X	53	2SB169GN
53	2SB110C	53	2SB117M	53	2SB154OR	53	2SB161Y	53	2SB169H
53	2SB110D	53	2SB117OR	53	2SB154R	53	2SB162	53	2SB169J
53	2SB110E	53	2SB117R	53	2SB154X	53	2SB162A	53	2SB169K
53	2SB110F	53	2SB117X	53	2SB154Y	53	2SB162B	53	2SB169L
53	2SB110G	53	2SB117Y	53	2SB155	53	2SB162C	53	2SB169M
53	2SB110GN	53	2SB118	53	2SB155A	53	2SB162D	53	2SB169OR
53	2SB110H	53	2SB120	53	2SB155B	53	2SB162E	53	2SB169R
53	2SB110J	53	2SB120A	53	2SB155C	53	2SB162F	53	2SB169X
53	2SB110K	53	2SB120B	53	2SB155D	53	2SB162GN	53	2SB169Y
53	2SB110L	53	2SB120C	53	2SB155E	53	2SB162H	53	2SB170
53	2SB110M	53	2SB120D	53	2SB155F	53	2SB162J	53	2SB170A
53	2SB110OR	53	2SB120E	53	2SB155G	53	2SB162K	53	2SB170B
53	2SB110R	53	2SB120F	53	2SB155GN	53	2SB162L	53	2SB170C
53	2SB110X	53	2SB120G	53	2SB155H	53	2SB162M	53	2SB170D
53	2SB110Y	53	2SB120GN	53	2SB155J	53	2SB162OR	53	2SB170E
53	2SB111	53	2SB120H	53	2SB155K	53	2SB162R	53	2SB170F
53	2SB111A	53	2SB120J	53	2SB155L	53	2SB162X	53	2SB170G
53	2SB111B	53	2SB120K	53	2SB155M	53	2SB162Y	53	2SB170GN
53	2SB111C	53	2SB120L	53	2SB155R	53	2SB163	53	2SB170H
53	2SB111D	53	2SB120M	53	2SB155X	53	2SB163A	53	2SB170J
53	2SB111E	53	2SB120OR	53	2SB155Y	53	2SB163B	53	2SB170K
53	2SB111F	53	2SB120R	53	2SB156	53	2SB163C	53	2SB170L
53	2SB111G	53	2SB120X	53	2SB156/7825B	53	2SB163D	53	2SB170M
53	2SB111GN	53	2SB120Y	53	2SB156A	53	2SB163E	53	2SB170OR
53	2SB111H	53	2SB13	53	2SB156AA	53	2SB163F	53	2SB170R
53	2SB111J	53	2SB134	53	2SB156AB	53	2SB163G	53	2SB170X
53	2SB111K	53	2SB134-D	53	2SB156AC	53	2SB163GN	53	2SB701
53	2SB111L	53	2SB134-E	53	2SB156B	53	2SB163H	53	2SB171
53	2SB111M	53	2SB134A	53	2SB156BK	53	2SB163J	53	2SB171A
53	2SB111OR	53	2SB134B	53	2SB156C	53	2SB163K	53	2SB171B
53	2SB111R	53	2SB134C	53	2SB156D				
53	2SB111X	53	2SB134D						
53	2SB111Y	53	2SB134E						
53	2SB112								

GE	Industry Standard No.	GE	Industry Standard No.	GE	Industry Standard No.	GE	Industry Standard No.	GE	Industry Standard No.
53	28B171C	53	28B177K	53	28B186E	53	28B202M	53	28B224X
53	28B171D	53	28B177L	53	28B186F	53	28B202OR	53	28B224Y
53	28B171E	53	28B177M	53	28B186G	53	28B202R	53	28B225
53	28B171F	53	28B177OR	53	28B186GN	53	28B202X	53	28B225A
53	28B171G	53	28B177R	53	28B186H	53	28B202Y	53	28B225B
53	28B171GN	53	28B177X	53	28B186J	53	28B203A	53	28B225C
53	28B171H	53	28B177Y	53	28B186K	53	28B203AA	53	28B225E
53	28B171J	53	28B178	53	28B186L	53	28B203B	53	28B225F
53	28B171K	53	28B178-O	53	28B186M	53	28B203C	53	28B225G
53	28B171L	53	28B178-OR	53	28B186O	53	28B203D	53	28B225GN
53	28B171M	53	28B178-S	53	28B186OR	53	28B203F	53	28B225H
53	28B171OR	53	28B180	53	28B186R	53	28B203GN	53	28B225K
53	28B171X	53	28B180A	53	28B186X	53	28B203H	53	28B225L
53	28B171Y	53	28B180B	53	28B186Y	53	28B203J	53	28B225M
53	28B172	53	28B180C	53	000028B187	53	28B203K	53	28B225OR
53	28B172A	53	28B180D	53	28B187(1)	53	28B203L	53	28B225R
53	28B172A-1	53	28B180E	53	28B187(K)	53	28B203M	53	28B225X
53	28B172A-F	53	28B180F	53	000028B187(RED)	53	28B203OR	53	28B225Y
53	28B172AF	53	28B180G	53	28B187-1	53	28B203X	53	28B226
53	28B172AL	53	28B180GN	53	28B187-OR	53	28B203Y	53	28B226A
53	28B172B	53	28B180H	53	28B187A	53	28B218	53	28B226B
53	28B172C	53	28B180J	53	28B187AA	53	28B218A	53	28B226C
53	28B172D	53	28B180K	53	28B187B	53	28B218B	53	28B226D
53	28B172E	53	28B180L	53	28B187BK	53	28B218C	53	28B226E
53	28B172F	53	28B180M	53	28B187C	53	28B218D	53	28B226F
53	28B172FN	53	28B180OR	53	28B187D	53	28B218E	53	28B226G
53	28B172G	53	28B180X	53	28B187E	53	28B218F	53	28B226GN
53	28B172GN	53	28B180Y	53	28B187F	53	28B218G	53	28B226J
53	28B172H	53	28B178A	53	28B187G	53	28B218P	53	28B226K
53	28B172J	53	28B178B	53	28B187GN	53	28B218G	53	28B226L
53	28B172K	53	28B178C	53	28B187H	53	28B218GN	53	28B226M
53	28B172L	53	28B178D	53	28B187K	53	28B218H	53	28B226OR
53	28B172M	53	28B178E	53	28B187L	53	28B218J	53	28B226R
53	28B172OR	53	28B178F	53	28B187M	53	28B218K	53	28B226X
53	28B172P	53	28B178G	53	28B187OR	53	28B218L	53	28B226Y
53	28B172R	53	28B178GN	53	28B187R	53	28B218M	53	28B227
53	28B172X	53	28B178H	53	28B187RED	53	28B218OR	53	28B227A
53	28B172Y	53	28B178J	53	28B187S	53	28B218R	53	28B227B
53	28B173	53	28B178K	53	28B187TV	53	28B218X	53	28B227C
53	28B173(C)	53	28B178L	53	28B187X	53	28B219	53	28B227D
53	28B173A	53	28B178M	53	28B187Y	53	28B219A	53	28B227G
53	28B173B	53	28B178N	53	28B187YEL	53	28B219B	53	28B227GN
53	28B173BL	53	28B178OR	53	28B188	53	28B219F	53	28B227H
53	28B173C	53	28B178R	53	28B188A	53	28B219G	53	28B227J
53	28B173CL	53	28B178S	53	28B188B	53	28B219J	53	28B227K
53	28B173D	53	28B178T	53	28B188C	53	28B219K	53	28B227L
53	28B173E	53	28B178TC	53	28B188D	53	28B219L	53	28B227M
53	28B173F	53	28B178TS	53	28B188E	53	28B219R	53	28B227OR
53	28B173G	53	28B178V	53	28B188F	53	28B219X	53	28B227X
53	28B173GN	53	28B178X	53	28B188G	53	28B219Y	53	28B227Y
53	28B173H	53	28B178Y	53	28B188GN	53	28B22	53	28B2A
53	28B173J	53	28B180	53	28B188H	53	28B22-0	53	28B2B
53	28B173K	53	28B180A	53	28B188J	53	28B22/09-30100	53	28B2C
53	28B173L	53	28B181	53	28B188K	53	28B220	53	28B2D
53	28B173M	53	28B181A	53	28B188L	53	28B220A	53	28B2F
53	28B173OR	53	28B182	53	28B188M	53	28B220B	53	28B2G
53	28B173R	53	28B183	53	28B188OR	53	28B220C	53	28B2GN
53	28B173X	53	28B183A	53	28B188R	53	28B220D	53	28B2H
53	28B173Y	53	28B183B	53	28B188X	53	28B220F	53	28B2J
53	28B174	53	28B183C	53	28B188Y	53	28B220GN	53	28B2K
53	28B174A	53	28B183E	53	28B189	53	28B220H	53	28B2L
53	28B174B	53	28B183F	53	28B189A	53	28B220J	53	28B2M
53	28B174C	53	28B183G	53	28B189B	53	28B220K	53	28B2P
53	28B174D	53	28B183GN	53	28B189C	53	28B220L	53	28B2R
53	28B174E	53	28B183J	53	28B189D	53	28B220M	53	28B2Y
53	28B174F	53	28B183K	53	28B189E	53	28B220OR	53	28B3
53	28B174G	53	28B183L	53	28B189F	53	28B220X	53	28B3A
53	28B174GN	53	28B183M	53	28B189G	53	28B220Y	53	28B3B
53	28B174H	53	28B183OR	53	28B189GN	53	28B221	53	28B3C
53	28B174K	53	28B183R	53	28B189H	53	28B221A	53	28B3D
53	28B174L	53	28B183X	53	28B189J	53	28B221B	53	28B3F
53	28B174M	53	28B183Y	53	28B189K	53	28B221C	53	28B3G
53	28B174OR	53	28B184	53	28B189L	53	28B221D	53	28B3GN
53	28B174R	53	28B184A	53	28B189M	53	28B221E	53	28B3H
53	28B174X	53	28B184B	53	28B189OR	53	28B221F	53	28B3J
53	28B174Y	53	28B184C	53	28B189R	53	28B221G	53	28B3K
53	28B175	53	28B184D	53	28B189X	53	28B221GN	53	28B3L
53	28B175(A)	53	28B184E	53	28B189Y	53	28B221H	53	28B3M
53	28B175(B)	53	28B184G	53	28B190	53	28B221J	53	28B230R
53	28B175(C)	53	28B184GN	53	28B192	53	28B221K	53	28B23R
53	28B175A	53	28B184H	53	28B193	53	28B221L	53	28B23X
53	28B175B	53	28B184J	53	28B194	53	28B221M	53	28B23Y
53	28B175B-1	53	28B184K	53	28B195	53	28B221OR	53	28B24
53	28B175BL	53	28B184L	53	28B196	53	28B221R	53	28B241
53	28B175C	53	28B184M	53	28B197	53	28B221X	53	28B241A
53	28B175CL	53	28B184OR	53	28B198	53	28B221Y	53	28B241B
53	28B175D	53	28B184R	53	28B199	53	28B222	53	28B241C
53	28B175E	53	28B184X	53	28B199A	53	28B222A	53	28B241D
53	28B175F	53	28B184Y	53	28B199B	53	28B222C	53	28B241E
53	28B175G	53	28B185	53	28B199C	53	28B222D	53	28B241F
53	28B175GN	53	28B185(O)	53	28B199D	53	28B222E	53	28B241GN
53	28B175H	53	28B185(O)	53	28B199E	53	28B222F	53	28B241H
53	28B175L	53	28B185(P)	53	28B199F	53	28B222G	53	28B241J
53	28B175M	53	000028B185-O	53	28B199G	53	28B222GN	53	28B241K
53	28B175OR	53	002SB18500	53	28B199GN	53	28B222H	53	28B241L
53	28B175R	53	28B185A	53	28B199H	53	28B222J	53	28B241M
53	28B175X	53	28B185B	53	28B199J	53	28B222K	53	28B241OR
53	28B175Y	53	28B185C	53	28B199L	53	28B222L	53	28B241R
53	28B176	53	28B185D	53	28B199M	53	28B222M	53	28B241V
53	28B176-O	53	28B185E	53	28B199OR	53	28B222OR	53	28B241X
53	28B176-P	53	28B185F	53	28B199R	53	28B222R	53	28B241Y
53	28B176-PR	53	28B185GN	53	28B199X	53	28B222X	53	28B24A
53	28B176A	53	28B185H	53	28B199Y	53	28B222Y	53	28B24B
53	28B176C	53	28B185I	53	28B200	53	28B223	53	28B24C
53	28B176D	53	28B185J	53	28B200A	53	28B223A	53	28B24D
53	28B176E	53	28B185L	53	28B200B	53	28B223B	53	28B24E
53	28B176F	53	28B185M	53	28B200F	53	28B223C	53	28B24F
53	28B176G	53	28B185OR	53	28B200G	53	28B223E	53	28B24GN
53	28B176GN	53	28B185P	53	28B200GN	53	28B223F	53	28B24H
53	28B176H	53	28B185R	53	28B200H	53	28B223G	53	28B24J
53	28B176J	53	28B185X	53	28B200J	53	28B223Y	53	28B24K
53	28B176K	53	28B185Y	53	28B200K	53	28B224	53	28B24L
53	28B176L	53	000028B186	53	28B200L	53	28B224A	53	28B24M
53	28B176M	53	28B186(O)	53	28B200M	53	28B224B	53	28B24OR
53	28B176OR	53	000028B186-O	53	28B200OR	53	28B224C	53	28B24R
53	28B176P	53	28B186-1	53	28B200OOR	53	28B224D	53	28B24X
53	28B176PL	53	28B186-7	53	28B200OOR	53	28B224E	53	28B24Y
53	28B176PRC	53	28B186-K	53	28B200OX	53	28B224F	53	28B257
53	28B176R	53	28B186-O	53	28B200Y	53	28B224G	53	28B257A
53	28B176R(1)	53	28B186-OR	53	28B201	53	28B224GN	53	28B257B
53	28B176RG	53	000028B186Q	53	28B202	53	28B224H	53	28B257C
53	28B176X	53	28B186A	53	28B202A	53	28B224J	53	28B257D
53	28B176Y	53	28B186AG	53	28B202B	53	28B224L	53	28B257E
53	28B177	53	28B186B	53	28B202C	53	28B224M	53	28B257F
53	28B177A	53	28B186BY	53	28B202D	53	28B224OR	53	28B257Q
53	28B177B	53	28B186C	53	28B202F	53	28B224R		
53	28B177C	53	28B186D	53	28B202G				
53	28B177D			53	28B202GN				
53	28B177E			53	28B202H				
53	28B177F			53	28B202J				
53	28B177G			53	28B202L				
53	28B177GN								
53	28B177H								

GE	Industry Standard No.	GE	Industry Standard No.	GE	Industry Standard No.	GE	Industry Standard No.	GE	Industry Standard No.	GE	Industry Standard No.
53	2SB257GN	53	2SB257X	53	2SB292L	53	2SB321	53	2SB32GN	53	2SB346M
53	2SB257H	53	2SB267Y	53	2SB292M	53	2SB322	53	2SB32H	53	2SB346OR
53	2SB257K	53	2SB268	53	2SB292OR	53	2SB323	53	2SB32J	53	2SB346Q
53	2SB257L	53	2SB268A	53	2SB292R	53	2SB324	53	2SB32K	53	2SB346X
53	2SB257M	53	2SB268B	53	2SB292X	53	2SB324(E)	53	2SB32M	53	2SB347
53	2SB257OR	53	2SB268C	53	2SB292Y	53	2SB324(P)	53	2SB32N	53	2SB347A
53	2SB257R	53	2SB268D	53	2SB293	53	2SB324(I)	53	2SB32OR	53	2SB347B
53	2SB257X	53	2SB268E	53	2SB293A	53	2SB324(L)	53	2SB32R	53	2SB347C
53	2SB257Y	53	2SB268F	53	2SB293B	53	2SB324(N)	53	2SB32X	53	2SB347D
53	2SB261	53	2SB268G	53	2SB293C	53	2SB324-,K	53	2SB32Y	53	2SB347P
53	2SB261A	53	2SB268GN	53	2SB293D	53	2SB324-OR	53	2SB33	53	2SB347GN
53	2SB261B	53	2SB268H	53	2SB293E	53	2SB324/4454C	53	2SB33-4	53	2SB347H
53	2SB261C	53	2SB268J	53	2SB293F	53	2SB240	53	2SB33-5	53	2SB347J
53	2SB261D	53	2SB268K	53	2SB293G	53	2SB240A	53	2SB336	53	2SB347K
53	2SB261F	53	2SB268L	53	2SB293GN	53	2SB240B	53	2SB33A	53	2SB347L
53	2SB261P	53	2SB268M	53	2SB293H	53	2SB240C	53	2SB33B	53	2SB347M
53	2SB261G	53	2SB268OR	53	2SB293J	53	2SB240D	53	2SB33BK	53	2SB347OR
53	2SB261GN	53	2SB268R	53	2SB293K	53	2SB240E	53	2SB33C	53	2SB347R
53	2SB261H	53	2SB268X	53	2SB293L	53	2SB240F	53	2SB33D	53	2SB347X
53	2SB261J	53	2SB268Y	53	2SB293M	53	2SB240G	53	2SB33E	53	2SB347Y
53	2SB261K	53	2SB269	53	2SB293OR	53	2SB240GN	53	2SB33F	53	2SB348
53	2SB261L	53	2SB269A	53	2SB293R	53	2SB240H	53	2SB33G	53	2SB348A
53	2SB261M	53	2SB269C	53	2SB293X	53	2SB240J	53	2SB33GN	53	2SB348C
53	2SB261OR	53	2SB269D	53	2SB293Y	53	2SB240K	53	2SB33H	53	2SB348D
53	2SB261R	53	2SB269E	53	2SB299	53	2SB240L	53	2SB33J	53	2SB348E
53	2SB261X	53	2SB269F	53	2SB299A	53	2SB240M	53	2SB33K	53	2SB348F
53	2SB261Y	53	2SB269G	53	2SB299B	53	2SB240OR	53	2SB33L	53	2SB348G
53	2SB262	53	2SB269GN	53	2SB299C	53	2SB240R	53	2SB33M	53	2SB348GN
53	2SB262A	53	2SB269J	53	2SB299D	53	2SB240X	53	2SB33OR	53	2SB348H
53	2SB262B	53	2SB269K	53	2SB299E	53	2SB240Y	53	2SB33R	53	2SB348J
53	2SB262C	53	2SB269L	53	2SB299F	53	2SB244H	53	2SB33X	53	2SB348K
53	2SB262D	53	2SB269M	53	2SB299G	53	2SB324A	53	2SB33T	53	2SB348L
53	2SB262E	53	2SB269OR	53	2SB299GN	53	2SB324C	53	2SB34	53	2SB348M
53	2SB262F	53	2SB269R	53	2SB299H	53	2SB324D	53	2SB345	53	2SB348OR
53	2SB262G	53	2SB269X	53	2SB299J	53	2SB324E	53	2SB345A	53	2SB348Q
53	2SB262GN	53	2SB269Y	53	2SB299K	53	2SB324E-1	53	2SB345B	53	2SB348R
53	2SB262H	53	2SB270	53	2SB299L	53	2SB324E-L	53	2SB345C	53	2SB348X
53	2SB262J	53	2SB270A	53	2SB299M	53	2SB324F	53	2SB345D	53	2SB348Y
53	2SB262K	53	2SB270B	53	2SB299OR	53	2SB324G	53	2SB345E	53	2SB349
53	2SB262L	53	2SB270C	53	2SB299R	53	2SB324GN	53	2SB345F	53	2SB34A
53	2SB262M	53	2SB270D	53	2SB299X	53	2SB324H	53	2SB345G	53	2SB34B
53	2SB262OR	53	2SB270E	53	2SB299Y	53	2SB324I	53	2SB345GN	53	2SB34C
53	2SB262R	53	2SB271	53	2SB302	53	2SB324J	53	2SB345H	53	2SB34D
53	2SB262X	53	2SB272	53	2SB302A	53	2SB324K	53	2SB345J	53	2SB34E
53	2SB262Y	53	2SB272A	53	2SB302B	53	2SB324L	53	2SB345K	53	2SB34F
53	2SB263	53	2SB272B	53	2SB302C	53	2SB324M	53	2SB345L	53	2SB34G
53	2SB263A	53	2SB272C	53	2SB302D	53	2SB324N	53	2SB345M	53	2SB34GN
53	2SB263B	53	2SB272D	53	2SB302E	53	2SB240OR	53	2SB345OR	53	2SB34H
53	2SB263C	53	2SB272E	53	2SB302F	53	2SB324P	53	2SB345R	53	2SB34J
53	2SB263D	53	2SB272F	53	2SB302G	53	2SB324R	53	2SB345X	53	2SB34K
53	2SB263E	53	2SB272G	53	2SB302GN	53	2SB324B	53	2SB345Y	53	2SB34L
53	2SB263G	53	2SB272GN	53	2SB302H	53	2SB324V	53	2SB346	53	2SB34M
53	2SB263GN	53	2SB272J	53	2SB302J	53	2SB324X	53	2SB346(Q)	53	2SB34N
53	2SB263H	53	2SB272K	53	2SB302K	53	2SB324Y	53	2SB346A	53	2SB34OR
53	2SB263J	53	2SB272L	53	2SB302M	53	2SB326	53	2SB346B	53	2SB34R
53	2SB263K	53	2SB272M	53	2SB302OR	53	2SB326A	53	2SB346C	53	2SB34X
53	2SB263L	53	2SB272OR	53	2SB302R	53	2SB326B	53	2SB346D	53	2SB34Y
53	2SB263M	53	2SB272R	53	2SB302X	53	2SB326C	53	2SB346E	53	2SB350A
53	2SB263OR	53	2SB272X	53	2SB302Y	53	2SB326D	53	2SB346F	53	2SB350B
53	2SB263R	53	2SB272Y	53	00002SB303	53	2SB326E	53	2SB346G	53	2SB350C
53	2SB263X	53	2SB273	53	2SB303(O)	53	2SB326F	53	2SB346GN	53	2SB350D
53	2SB263Y	53	2SB273A	53	0002SB303-0	53	2SB326G	53	2SB346H	53	2SB350E
53	2SB264	53	2SB273B	53	0002SB3030	53	2SB326GN	53	2SB346K	53	2SB350F
53	2SB264A	53	2SB273C	53	2SB303A	53	2SB326H	53	2SB346L	53	2SB350G
53	2SB264B	53	2SB273D	53	2SB303B	53	2SB326J				
53	2SB264C	53	2SB273E	53	2SB303BK	53	2SB326K				
53	2SB264D	53	2SB273F	53	2SB303C	53	2SB326L				
53	2SB264E	53	2SB273G	53	2SB303D	53	2SB326M				
53	2SB264G	53	2SB273GN	53	2SB303E	53	2SB326OR				
53	2SB264GN	53	2SB273H	53	2SB303F	53	2SB326R				
53	2SB264H	53	2SB273J	53	2SB303G	53	2SB326X				
53	2SB264J	53	2SB273K	53	2SB303GN	53	2SB326Y				
53	2SB264K	53	2SB273L	53	2SB303H	53	2SB327				
53	2SB264L	53	2SB273M	53	2SB303J	53	2SB327A				
53	2SB264M	53	2SB273OR	53	2SB303K	53	2SB327B				
53	2SB264OR	53	2SB273R	53	2SB303L	53	2SB327C				
53	2SB264R	53	2SB273X	53	2SB303M	53	2SB327D				
53	2SB264X	53	2SB273Y	53	2SB303N	53	2SB327P				
53	2SB264Y	53	2SB290	53	2SB303OR	53	2SB327G				
53	2SB265	53	2SB290A	53	2SB303R	53	2SB327GN				
53	2SB265A	53	2SB290B	53	2SB303X	53	2SB327H				
53	2SB265B	53	2SB290C	53	2SB303Y	53	2SB327J				
53	2SB265C	53	2SB290D	53	2SB314	53	2SB327K				
53	2SB265D	53	2SB290F	53	2SB314A	53	2SB327L				
53	2SB265E	53	2SB290G	53	2SB314B	53	2SB327M				
53	2SB265F	53	2SB290GN	53	2SB314C	53	2SB327OR				
53	2SB265G	53	2SB290H	53	2SB314E	53	2SB327R				
53	2SB265GN	53	2SB290J	53	2SB314F	53	2SB327X				
53	2SB265H	53	2SB290K	53	2SB314GN	53	2SB327Y				
53	2SB265J	53	2SB290L	53	2SB314H	53	2SB328				
53	2SB265K	53	2SB290M	53	2SB314J	53	2SB328A				
53	2SB265L	53	2SB290OR	53	2SB314K	53	2SB328B				
53	2SB265M	53	2SB290R	53	2SB314L	53	2SB328C				
53	2SB265OR	53	2SB290X	53	2SB314M	53	2SB328D				
53	2SB265X	53	2SB290Y	53	2SB314OR	53	2SB328E				
53	2SB265Y	53	2SB291	53	2SB314R	53	2SB328F				
53	2SB266	53	2SB291-GREEN	53	2SB314X	53	2SB328G				
53	2SB266A	53	2SB291-RED	53	2SB314Y	53	2SB328GN				
53	2SB266B	53	2SB291-YELLOW	53	2SB315	53	2SB328H				
53	2SB266C	53	2SB291A	53	2SB315A	53	2SB328J				
53	2SB266D	53	2SB291B	53	2SB315B	53	2SB328K				
53	2SB266F	53	2SB291C	53	2SB315C	53	2SB328L				
53	2SB266G	53	2SB291D	53	2SB315D	53	2SB328M				
53	2SB266GN	53	2SB291E	53	2SB315E	53	2SB328OR				
53	2SB266H	53	2SB291F	53	2SB315F	53	2SB328R				
53	2SB266J	53	2SB291G	53	2SB315G	53	2SB328X				
53	2SB266K	53	2SB291GN	53	2SB315GN	53	2SB328Y				
53	2SB266L	53	2SB291H	53	2SB315H	53	2SB329				
53	2SB266M	53	2SB291J	53	2SB315J	53	2SB329A				
53	2SB266OR	53	2SB291L	53	2SB315L	53	2SB329B				
53	2SB266P	53	2SB291M	53	2SB315M	53	2SB329C				
53	2SB266Q	53	2SB291OR	53	2SB315OR	53	2SB329D				
53	2SB266X	53	2SB291R	53	2SB315R	53	2SB329E				
53	2SB266Y	53	2SB291X	53	2SB315X	53	2SB329G				
53	2SB267	53	2SB291Y	53	2SB315Y	53	2SB329GN				
53	2SB267A	53	2SB292	53	2SB316	53	2SB329H				
53	2SB267B	53	2SB292-BLUE	53	2SB316A	53	2SB329J				
53	2SB267C	53	2SB292-GREEN	53	2SB316B	53	2SB329K				
53	2SB267D	53	2SB292-ORANGE	53	2SB316C	53	2SB329L				
53	2SB267E	53	2SB292-RED	53	2SB316D	53	2SB329M				
53	2SB267G	53	2SB292-YELLOW	53	2SB316E	53	2SB329OR				
53	2SB267GN	53	2SB292A	53	2SB316F	53	2SB329R				
53	2SB267H	53	2SB292B	53	2SB316G	53	2SB329X				
53	2SB267J	53	2SB292C	53	2SB316GN	53	2SB329Y				
53	2SB267K	53	2SB292D	53	2SB316H	53	2SB32A				
53	2SB267L	53	2SB292E	53	2SB316R	53	2SB32B				
53	2SB267M	53	2SB292F	53	2SB316X	53	2SB32C				
53	2SB267OR	53	2SB292G	53	2SB316Y	53	2SB32D				
53	2SB267R	53	2SB292GN	53	2SB317	53	2SB32E				
		53	2SB292H	53	2SB32	53	2SB32F				
		53	2SB292J	53	2SB32-0	53	2SB32G				
		53	2SB292K	53	2SB32-1						
				53	2SB32-2						
				53	2SB32-4						

GE	Industry Standard No.	GE	Industry Standard No.	GE	Industry Standard No.	GE	Industry Standard No.	GE	Industry Standard No.
53	28B350GN	53	28B378E	53	28B383A	53	28B393J	53	28B403
53	28B350H	53	28B378F	53	28B383B	53	28B393K	53	28B403A
53	28B350J	53	28B378GN	53	28B383C	53	28B393L	53	28B403B
53	28B350K	53	28B378J	53	28B383D	53	28B393M	53	28B403C
53	28B350L	53	28B378K	53	28B383F	53	28B3950R	53	28B403D
53	28B350M	53	28B378L	53	28B383GN	53	28B393R	53	28B403E
53	28B350OR	53	28B378M	53	28B383H	53	28B393X	53	28B403F
53	28B350R	53	28B378OR	53	28B383J	53	28B393Y	53	28B403G
53	28B350X	53	28B378R	53	28B383L	53	28B394	53	28B403GN
53	28B350Y	53	28B378X	53	28B383M	53	28B394A	53	28B403H
53	28B364	53	28B378Y	53	28B3830R	53	28B394B	53	28B403J
53	28B364-OR	53	28B379	53	28B383R	53	28B394C	53	28B403K
53	28B364A	53	28B379-2	53	28B383X	53	28B394D	53	28B403L
53	28B364B	53	28B379A	53	28B383Y	53	28B394E	53	28B403M
53	28B364C	53	28B379B	53	28B384	53	28B394F	53	28B403OR
53	28B364D	53	28B379C	53	28B384A	53	28B394R	53	28B403R
53	28B364E	53	28B379D	53	28B384B	53	28B394G	53	28B403X
53	28B364F	53	28B379E	53	28B384C	53	28B394GN	53	28B403Y
53	28B364G	53	28B379F	53	28B384D	53	28B394H	53	000028B405
53	28B364GN	53	28B379G	53	28B384E	53	28B394K	53	28B405-(K)
53	28B364H	53	28B379H	53	28B384F	53	28B394L	53	28B405-0
53	28B364J	53	28B379J	53	28B384G	53	28B394M	53	28B405-1
53	28B364K	53	28B379K	53	28B384GN	53	28B3940R	53	28B405-2C
53	28B364L	53	28B379L	53	28B384H	53	28B394R	53	28B405-3C
53	28B364M	53	28B379M	53	28B384J	53	28B394X	53	28B405-4C
53	28B364R	53	28B379OR	53	28B384K	53	28B394Y	53	28B405-0
53	28B364X	53	28B379R	53	28B384L	53	28B395	53	28B405-R
53	28B364Y	53	28B379Y	53	28B384M	53	28B395A	53	28B405A
53	28B365	53	28B37A	53	28B3840R	53	28B395B	53	28B405AG
53	28B365A	53	28B37B	53	28B384X	53	28B395C	53	28B405B
53	28B365B	53	28B37C	53	28B384Y	53	28B395D	53	28B405BR
53	28B365C	53	28B37D	53	28B385A	53	28B395E	53	28B405C
53	28B365D	53	28B37E	53	28B385B	53	28B395F	53	28B405CK
53	28B365E	53	28B37F	53	28B385C	53	28B395G	53	28B405DK
53	28B365F	53	28B37G	53	28B385D	53	28B395GN	53	28B405E
53	28B365G	53	28B37GN	53	28B385E	53	28B395H	53	28B405EK
53	28B365GN	53	28B37H	53	28B385F	53	28B395J	53	28B405F
53	28B365H	53	28B37J	53	28B385G	53	28B395K	53	28B405G
53	28B365J	53	28B37K	53	28B385GN	53	28B395M	53	28B405GN
53	28B365K	53	28B37L	53	28B385H	53	28B3950R	53	28B405H
53	28B365L	53	28B37M	53	28B385J	53	28B395X	53	28B405J
53	28B365M	53	28B370R	53	28B385K	53	28B395Y	53	28B405K
53	28B365OR	53	28B37R	53	28B385L	53	28B396	53	28B405L
53	28B365R	53	28B37Y	53	28B385M	53	28B396A	53	28B405M
53	28B365X	53	28B38	53	28B385OR	53	28B396B	53	28B405OR
53	28B365Y	53	28B380A	53	28B385X	53	28B396C	53	28B405P
53	28B37	53	28B380B	53	28B385Y	53	28B396D	53	28B405RE
53	28B370	53	28B380C	53	28B387	53	28B396E	53	28B405X
53	28B370-0	53	28B380D	53	28B387A	53	28B396F	53	28B405Y
53	28B370AA	53	28B380E	53	28B387D	53	28B396G	53	28B408
53	28B370AB	53	28B380F	53	28B387B	53	28B396GN	53	28B408A
53	28B370AC	53	28B380GN	53	28B387F	53	28B396H	53	28B408B
53	28B370AHA	53	28B380J	53	28B387G	53	28B396J	53	28B408C
53	28B370AHB	53	28B380K	53	28B387GN	53	28B396K	53	28B408D
53	28B370B	53	28B380L	53	28B387H	53	28B396L	53	28B408E
53	28B370C	53	28B380OR	53	28B387J	53	28B396M	53	28B408F
53	28B370D	53	28B380X	53	28B387K	53	28B3960R	53	28B408G
53	28B370E	53	28B380Y	53	28B387L	53	28B396R	53	28B408GN
53	28B370P	53	28B381	53	28B387M	53	28B396X	53	28B408H
53	28B370GN	53	28B3812A	53	28B387OR	53	28B396Y	53	28B408J
53	28B370H	53	28B3812B	53	28B387R	53	28B39	53	28B408K
53	28B370J	53	28B3812C	53	28B387X	53	28B39A	53	28B408L
53	28B370K	53	28B3812D	53	28B387Y	53	28B39B	53	28B408M
53	28B370L	53	28B3812G	53	28B389	53	28B39C	53	28B408OR
53	28B370M	53	28B3812GN	53	28B389A	53	28B39D	53	28B408R
53	28B370OR	53	28B3812H	53	28B389B	53	28B39E	53	28B408X
53	28B370P	53	28B3812J	53	28B389BK	53	28B39F	53	28B408Y
53	28B370PB	53	28B3812K	53	28B389C	53	28B39G	53	28B40A
53	28B370R	53	28B3812L	53	28B389D	53	28B39GN	53	28B40B
53	28B370V	53	28B3812OR	53	28B389E	53	28B39H	53	28B40C
53	28B370X	53	28B3812R	53	28B389F	53	28B39J	53	28B40D
53	28B370Y	53	28B3812Y	53	28B389G	53	28B39K	53	28B40E
53	28B371	53	28B3813A	53	28B389GN	53	28B39L	53	28B40F
53	28B371A	53	28B3813B	53	28B389H	53	28B39M	53	28B40G
53	28B371C	53	28B3813C	53	28B389J	53	28B390R	53	28B40GN
53	28B371D	53	28B3813D	53	28B389K	53	28B39R	53	28B40H
53	28B371E	53	28B3813E	53	28B389L	53	28B39X	53	28B40J
53	28B371F	53	28B3813F	53	28B389M	53	28B39Y	53	28B40K
53	28B371G	53	28B3813G	53	28B389OR	53	28B40	53	28B40L
53	28B371GN	53	28B3813GN	53	28B389R	53	28B400	53	28B40M
53	28B371H	53	28B3815H	53	28B389X	53	28B400A	53	28B40OR
53	28B371K	53	28B3813J	53	28B389Y	53	28B400B	53	28B40R
53	28B371L	53	28B3813K	53	28B38A	53	28B400BK	53	28B40X
53	28B371M	53	28B3813L	53	28B38B	53	28B400BL	53	28B40Y
53	28B371OR	53	28B3813M	53	28B38C	53	28B400C	53	28B415
53	28B371R	53	28B3813OR	53	28B38D	53	28B400D	53	28B415-OR
53	28B371X	53	28B3813X	53	28B38E	53	28B400E	53	28B415A
53	28B371Y	53	28B381A	53	28B38F	53	28B400F	53	28B415B
53	28B372	53	28B381B	53	28B38G	53	28B400G	53	28B415C
53	28B376	53	28B381C	53	28B38GN	53	28B400GN	53	28B415D
53	28B376A	53	28B381D	53	28B38H	53	28B400H	53	28B415E
53	28B376B	53	28B381E	53	28B38J	53	28B400J	53	28B415F
53	28B376C	53	28B381F	53	28B38K	53	28B400K	53	28B415G
53	28B376D	53	28B381G	53	28B38L	53	28B400L	53	28B415GN
53	28B376E	53	28B381H	53	28B38M	53	28B400M	53	28B415H
53	28B376F	53	28B381J	53	28B38OR	53	28B400OR	53	28B415J
53	28B376G	53	28B381K	53	28B38R	53	28B400R	53	28B415K
53	28B376GN	53	28B381L	53	28B38X	53	28B400X	53	28B415M
53	28B376J	53	28B381M	53	28B38Y	53	28B400Y	53	28B415R
53	28B376K	53	28B381OR	53	28B39	53	28B401	53	28B415X
53	28B376L	53	28B381R	53	28B392	53	28B401A	53	28B415Y
53	28B376M	53	28B381X	53	28B392A	53	28B401B	53	28B416A
53	28B376OR	53	28B381Y	53	28B392B	53	28B401C	53	28B416B
53	28B376X	53	28B382	53	28B392C	53	28B401D	53	28B416C
53	28B376Y	53	28B382A	53	28B392D	53	28B401E	53	28B416D
53	28B377	53	28B382B	53	28B392E	53	28B401F	53	28B416E
53	28B377B	53	28B382BK	53	28B392F	53	28B401G	53	28B416F
53	28B378	53	28B382C	53	28B392G	53	28B401GN	53	28B416G
53	28B3783A	53	28B382D	53	28B392GN	53	28B401H	53	28B416GN
53	28B3783B	53	28B382E	53	28B392H	53	28B401K	53	28B416H
53	28B3783C	53	28B382G	53	28B392J	53	28B401L	53	28B416J
53	28B3783D	53	28B382GN	53	28B392K	53	28B401M	53	28B416K
53	28B3783E	53	28B382H	53	28B392L	53	28B4010R	53	28B416L
53	28B3783F	53	28B382J	53	28B392M	53	28B401R	53	28B416M
53	28B3783G	53	28B382K	53	28B3920R	53	28B401X	53	28B416OR
53	28B3783GN	53	28B382L	53	28B392R	53	28B401Y	53	28B416R
53	28B3783H	53	28B382OR	53	28B392X	53	28B402	53	28B416X
53	28B3783J	53	28B382R	53	28B392Y	53	28B402A	53	28B416Y
53	28B3783K	53	28B382X	53	28B393	53	28B402B	53	28B417
53	28B3783L	53	28B382Y	53	28B393A	53	28B402D	53	28B417A
53	28B3783M	53	28B383	53	28B393B	53	28B402E	53	28B417B
53	28B3783OR	53	28B382K	53	28B393C	53	28B402F	53	28B417C
53	28B3783R	53	28B382L	53	28B393D	53	28B402G	53	28B417D
53	28B3783X	53	28B3820R	53	28B393E	53	28B402GN	53	28B417E
53	28B3783Y	53	28B382R	53	28B393F	53	28B402H	53	28B417F
53	28B378A	53	28B382Y	53	28B393G	53	28B402J	53	28B417GN
53	28B378B	53	28B383	53	28B393GN	53	28B402L	53	28B417H
53	28B378D	53	28B383-1	53	28B393H	53	28B402M	53	28B417J
		53	28B383-2			53	28B4020R	53	28B417K
						53	28B402R		
						53	28B402Y		

GE	Industry Standard No.	GE	Industry Standard No.	GE	Industry Standard No.	GE	Industry Standard No.	GE	Industry Standard No.
53	2SB417L	53	2SB44B	53	2SB47J	53	2SB498	53	2SB54R
53	2SB417M	53	2SB44C	53	2SB47K	53	2SB498A	53	2SB54X
53	2SB417OR	53	2SB44D	53	2SB47L	53	2SB498B	53	2SB54Y
53	2SB417R	53	2SB44E	53	2SB47M	53	2SB498C	53	2SB55
53	2SB417X	53	2SB44F	53	2SB47OR	53	2SB498D	53	2SB55A
53	2SB417Y	53	2SB44G	53	2SB47R	53	2SB498E	53	2SB55B
53	2SB421	53	2SB44GN	53	2SB47X	53	2SB498F	53	2SB55C
53	2SB422	53	2SB44H	53	2SB47Y	53	2SB498G	53	2SB55D
53	2SB422A	53	2SB44J	53	2SB48	53	2SB498GN	53	2SB55E
53	2SB422B	53	2SB44K	53	2SB482	53	2SB498H	53	2SB55F
53	2SB422C	53	2SB44L	53	2SB482A	53	2SB498J	53	2SB55GN
53	2SB422D	53	2SB44M	53	2SB482B	53	2SB498K	53	2SB55H
53	2SB422E	53	2SB44OR	53	2SB482C	53	2SB498L	53	2SB55J
53	2SB422F	53	2SB44R	53	2SB482D	53	2SB498M	53	2SB55K
53	2SB422G	53	2SB44X	53	2SB482E	53	2SB498OR	53	2SB55M
53	2SB422GN	53	2SB44Y	53	2SB482F	53	2SB498R	53	2SB550N
53	2SB422H	53	2SB450	53	2SB482G	53	2SB498X	53	2SB55R
53	2SB422J	53	2SB450A	53	2SB482GN	53	2SB498Y	53	2SB55X
53	2SB422K	53	2SB451	53	2SB482H	53	2SB49A	53	2SB55Y
53	2SB422L	53	2SB452	53	2SB482J	53	2SB49B	53	2SB56
53	2SB422M	53	2SB452A	53	2SB482K	53	2SB49C	53	2SB560
53	2SB422OR	53	2SB453	53	2SB482L	53	2SB49D	53	2SB565
53	2SB422R	53	2SB454	53	2SB482M	53	2SB49E	53	2SB565A
53	2SB422X	53	2SB455	53	2SB482V	53	2SB49F	53	2SB565B
53	2SB422Y	53	2SB457	53	2SB482X	53	2SB49G	53	2SB565C
53	2SB423	53	2SB457-C	53	2SB482Y	53	2SB49GN	53	2SB565D
53	2SB423A	53	2SB457A	53	2SB486	53	2SB49H	53	2SB565F
53	2SB423B	53	2SB457AC	53	2SB486A	53	2SB49J	53	2SB565GN
53	2SB423C	53	2SB457B	53	2SB486B	53	2SB49K	53	2SB565H
53	2SB423D	53	2SB457C	53	2SB486C	53	2SB49L	53	2SB565J
53	2SB423E	53	2SB457D	53	2SB486D	53	2SB49M	53	2SB565K
53	2SB423F	53	2SB457E	53	2SB486E	53	2SB49OR	53	2SB565L
53	2SB423G	53	2SB457F	53	2SB486F	53	2SB49R	53	2SB565M
53	2SB423GN	53	2SB457G	53	2SB486G	53	2SB49X	53	2SB565OR
53	2SB423H	53	2SB457GN	53	2SB486GN	53	2SB49Y	53	2SB565X
53	2SB423J	53	2SB457H	53	2SB486H	53	2SB50	53	2SB565Y
53	2SB423K	53	2SB457J	53	2SB486J	53	2SB50A	53	2SB56A
53	2SB423L	53	2SB457K	53	2SB486K	53	2SB50B	53	2SB56B
53	2SB423M	53	2SB457L	53	2SB486L	53	2SB50C	53	2SB56C
53	2SB423OR	53	2SB457M	53	2SB486M	53	2SB50D	53	2SB56CK
53	2SB423R	53	2SB457OR	53	2SB486OR	53	2SB50E	53	2SB56D
53	2SB423X	53	2SB457R	53	2SB486R	53	2SB50F	53	2SB56E
53	2SB423Y	53	2SB457X	53	2SB486X	53	2SB50G	53	2SB56F
53	2SB43	53	2SB457Y	53	2SB486Y	53	2SB50GN	53	2SB56G
53	2SB431	53	2SB459	53	2SB48A	53	2SB50H	53	2SB56GN
53	2SB439	53	2SB459-O	53	2SB48B	53	2SB50J	53	2SB56H
53	2SB439A	53	2SB459A	53	2SB48C	53	2SB50K	53	2SB56J
53	2SB439B	53	2SB459B	53	2SB48D	53	2SB50L	53	2SB56K
53	2SB439C	53	2SB459C	53	2SB48E	53	2SB50M	53	2SB56L
53	2SB439D	53	2SB459C-2	53	2SB48F	53	2SB50OR	53	2SB56OR
53	2SB439E	53	2SB459D	53	2SB48G	53	2SB50X	53	2SB56X
53	2SB439F	53	2SB459E	53	2SB48GN	53	2SB50Y	53	2SB56Y
53	2SB439G	53	2SB459F	53	2SB48H	53	2SB51	53	2SB57
53	2SB439GN	53	2SB459G	53	2SB48J	53	2SB516	53	2SB57A
53	2SB439H	53	2SB459GN	53	2SB48K	53	2SB516C	53	2SB57B
53	2SB439J	53	2SB459H	53	2SB48L	53	2SB516CD	53	2SB57C
53	2SB439L	53	2SB459J	53	2SB48M	53	2SB516CD(P)	53	2SB57D
53	2SB439M	53	2SB459K	53	2SB48OR	53	2SB516D	53	2SB57E
53	2SB439OR	53	2SB459L	53	2SB48R	53	2SB516P	53	2SB57F
53	2SB439R	53	2SB459M	53	2SB48X	53	2SB51A	53	2SB57G
53	2SB439X	53	2SB459OR	53	2SB48Y	53	2SB51B	53	2SB57GN
53	2SB439Y	53	2SB459R	53	2SB49	53	2SB51C	53	2SB57H
53	2SB43A	53	2SB459X	53	2SB492A	53	2SB51D	53	2SB57J
53	2SB43B	53	2SB459Y	53	2SB492C	53	2SB51E	53	2SB57K
53	2SB43C	53	2SB46	53	2SB492D	53	2SB51F	53	2SB57L
53	2SB43D	53	2SB460	53	2SB492E	53	2SB51G	53	2SB57M
53	2SB43E	53	2SB46A	53	2SB492F	53	2SB51GN	53	2SB570M
53	2SB43F	53	2SB46B	53	2SB492G	53	2SB51H	53	2SB57R
53	2SB43G	53	2SB46C	53	2SB492GN	53	2SB51J	53	2SB57X
53	2SB43GN	53	2SB46D	53	2SB492H	53	2SB51K	53	2SB57Y
53	2SB43H	53	2SB46E	53	2SB492J	53	2SB51L	53	2SB59
53	2SB43J	53	2SB46F	53	2SB492K	53	2SB51M	53	2SB59A
53	2SB43K	53	2SB46G	53	2SB492L	53	2SB510R	53	2SB59B
53	2SB43L	53	2SB46GN	53	2SB492M	53	2SB51R	53	2SB59C
53	2SB43M	53	2SB46H	53	2SB492OR	53	2SB51X	53	2SB59D
53	2SB43OR	53	2SB46J	53	2SB492R	53	2SB51Y	53	2SB59E
53	2SB43R	53	2SB46K	53	2SB492X	53	2SB52	53	2SB59F
53	2SB43X	53	2SB46L	53	2SB492Y	53	2SB52A	53	2SB59GN
53	2SB43Y	53	2SB46M	53	2SB494	53	2SB52B	53	2SB59J
53	2SB44	53	2SB46OR	53	2SB495	53	2SB52C	53	2SB59K
53	2SB440	53	2SB46R	53	2SB495A	53	2SB52D	53	2SB59L
53	2SB440A	53	2SB46X	53	2SB495B	53	2SB52E	53	2SB59M
53	2SB440B	53	2SB46Y	53	2SB495C	53	2SB52F	53	2SB590R
53	2SB440C	53	2SB47	53	2SB495D	53	2SB52G	53	2SB59R
53	2SB440E	53	2SB470	53	2SB495E	53	2SB52GN	53	2SB59X
53	2SB440F	53	2SB470A	53	2SB495F	53	2SB52H	53	2SB59Y
53	2SB440G	53	2SB470B	53	2SB495G	53	2SB52J	53	2SB60
53	2SB440GN	53	2SB470C	53	2SB495GN	53	2SB52K	53	2SB60A
53	2SB440H	53	2SB470D	53	2SB495H	53	2SB52L	53	2SB60B
53	2SB440K	53	2SB470E	53	2SB495J	53	2SB52M	53	2SB60C
53	2SB440L	53	2SB470F	53	2SB495K	53	2SB520R	53	2SB60D
53	2SB440M	53	2SB470G	53	2SB495L	53	2SB52R	53	2SB60E
53	2SB440OR	53	2SB470GN	53	2SB495M	53	2SB52X	53	2SB60F
53	2SB440R	53	2SB470H	53	2SB495OR	53	2SB52Y	53	2SB60G
53	2SB440X	53	2SB470J	53	2SB495R	53	2SB53	53	2SB60GN
53	2SB440Y	53	2SB470K	53	2SB495T	53	2SB54	53	2SB60J
53	2SB443	53	2SB470L	53	2SB495X	53	2SB54(A)	53	2SB60K
53	2SB443A	53	2SB470M	53	2SB495Y	53	2SB54A	53	2SB60L
53	2SB443B	53	2SB470OR	53	2SB496	53	2SB535	22	2SB60M
53	2SB443C	53	2SB470X	53	2SB496A	53	2SB53A	53	2SB60OR
53	2SB443D	53	2SB470Y	53	2SB496B	53	2SB53B	53	2SB60R
53	2SB443E	53	2SB475	53	2SB496C	53	2SB53C	53	2SB60X
53	2SB443F	53	2SB475A	53	2SB496D	53	2SB53D	53	2SB60Y
53	2SB443G	53	2SB475B	53	2SB496E	53	2SB53E	53	2SB61
53	2SB443GN	53	2SB475C	53	2SB496F	53	2SB53F	53	2SB61A
53	2SB443H	53	2SB475D	53	2SB496G	53	2SB53G	53	2SB61B
53	2SB443J	53	2SB475E	53	2SB496GN	53	2SB53GN	53	2SB61C
53	2SB443K	53	2SB475F	53	2SB496H	53	2SB53H	53	2SB61D
53	2SB443L	53	2SB475G	53	2SB496J	53	2SB53J	53	2SB61E
53	2SB443M	53	2SB475GN	53	2SB496K	53	2SB53K	53	2SB61F
53	2SB443OR	53	2SB475H	53	2SB496L	53	2SB53L	53	2SB61G
53	2SB443R	53	2SB475J	53	2SB496M	53	2SB53M	53	2SB61GN
53	2SB443X	53	2SB475K	53	2SB496OR	53	2SB53OR	53	2SB61H
53	2SB443Y	53	2SB475L	53	2SB496R	53	2SB53R	53	2SB61J
53	2SB444	53	2SB475M	53	2SB496X	53	2SB53X	53	2SB61K
53	2SB444OD	53	2SB475OR	53	2SB496Y	53	2SB53Y	53	2SB61L
53	2SB444A	53	2SB475P	53	2SB497	53	2SB54	53	2SB61M
53	2SB444B	53	2SB475PL	53	2SB497A	53	2SB54A	53	2SB61OR
53	2SB444C	53	2SB475Q	53	2SB497B	53	2SB54B	53	2SB61X
53	2SB444D	53	2SB475R	53	2SB497C	53	2SB54BA	53	2SB61Y
53	2SB444E	53	2SB475X	53	2SB497D	53	2SB54C	53	2SB65
53	2SB444F	53	2SB475Y	53	2SB497E	53	2SB54D	53	2SB65A
53	2SB444G	53	2SB476	53	2SB497F	53	2SB54E	53	2SB65B
53	2SB444GN	53	2SB47A	53	2SB497G	53	2SB54F	53	2SB65C
53	2SB444H	53	2SB47B	53	2SB497GN	53	2SB54G		
53	2SB444K	53	2SB47C	53	2SB497H	53	2SB54GN		
53	2SB444L	53	2SB47D	53	2SB497J	53	2SB54H		
53	2SB444M	53	2SB47E	53	2SB497K	53	2SB54J		
53	2SB444OR	53	2SB47F	53	2SB497L	53	2SB54K		
53	2SB444R	53	2SB47G	53	2SB497M	53	2SB54L		
53	2SB444X	53	2SB47GN	53	2SB497OR	53	2SB54L1		
53	2SB444Y	53	2SB47H	53	2SB497R	53	2SB54M		
53	2SB44A			53	2SB497X	53	2SB540R		
				53	2SB497Y				

GE	Industry Standard No.	GE	Industry Standard No.	GE	Industry Standard No.	GE	Industry Standard No.	GE	Industry Standard No.
53	2SB65D	53	2SB74H	53	2SB87E	53	2SB98	56	2N5954
53	2SB65P	53	2SB74J	53	2SB87F	53	2SB98A	56	2N5955
53	2SB65G	53	2SB74K	53	2SB87G	53	2SB98B	56	2N5956
53	2SB65GN	53	2SB74L	53	2SB87GN	53	2SB98C	57	A4K
53	2SB65H	53	2SB74M	53	2SB87H	53	2SB98D	57	A5A-1
53	2SB65J	53	2SB74OR	53	2SB87J	53	2SB98E	57	A5A-2
53	2SB65K	53	2SB74R	53	2SB87K	53	2SB98F	57	A5A-3
53	2SB65L	53	2SB74X	53	2SB87L	53	2SB98G	57	A5A-4
53	2SB65M	53	2SB74Y	53	2SB87OR	53	2SB98GN	57	A5A-5
53	2SB65OR	53	2SB75	53	2SB87R	53	2SB98H	57	A5G
53	2SB65R	53	2SB75AH	53	2SB87X	53	2SB98J	57	A5Y
53	2SB65X	53	2SB75B	53	2SB87Y	53	2SB98K	57	A6C
53	2SB65Y	53	2SB75C	53	2SB89	53	2SB98M	57	A6C(GRN)
53	2SB66	53	2SB75C-4	53	2SB89A	53	2SB98OR	57	A6C-1
53	2SB66A	53	2SB75D	53	2SB89AH	53	2SB98R	57	A6C-1(RED)
53	2SB66B	53	2SB75E	53	2SB89C	53	2SB98X	57	A6C-1-RED
53	2SB66C	53	2SB75F	53	2SB89D	53	2SB98Y	57	A6C-2
53	2SB66D	53	2SB75G	53	2SB89E	53	2SB99	57	A6C-2(BLK)
53	2SB66E	53	2SB75GN	53	2SB89G	53	2SB99A	57	A6C-2-BLACK
53	2SB66F	53	2SB75H	53	2SB89GN	53	2SB99B	57	A6C-3
53	2SB66G	53	2SB75J	53	2SB89H	53	2SB99C	57	A6C-3(WHT)
53	2SB66H	53	2SB75L	53	2SB89J	53	2SB99D	57	A6C-3-WHITE
53	2SB66J	53	2SB75LB	53	2SB89K	53	2SB99E	57	A6C-4
53	2SB66K	53	2SB75M	53	2SB89L	53	2SB99F	57	A6C-GREEN
53	2SB66L	53	2SB75OR	53	2SB89OR	53	2SB99G	57	A6W
53	2SB66M	53	2SB75R	53	2SB89R	53	2SB99GN	57	A7C(MOTOROLA)
53	2SB66OH	53	2SB75X	53	2SB89X	53	2SB99H	57	A7Z
53	2SB66R	53	2SB75Y	53	2SB89Y	53	2SB99J	57	A8F
53	2SB66X	53	2SB76	53	2SB90	53	2SB99K	57	A8Y
53	2SB66Y	53	2SB76A	53	2SB90A	53	2SB99L	57	A8P
53	2SB67	53	2SB76B	53	2SB90B	53	2SB99M	57	B-1790
53	2SB67A	53	2SB76C	53	2SB90C	53	2SB99OR	57	B1C
53	2SB67B	53	2SB76D	53	2SB90D	53	2SB99R	57	B1C-1
53	2SB67C	53	2SB76E	53	2SB90E	53	2SB99Y	57	B1C-2
53	2SB67D	53	2SB76F	53	2SB90F	53	2SBP1	57	B1D-1
53	2SB67E	53	2SB76G	53	2SB90G	53	2SBP1A	57	B1F
53	2SB67F	53	2SB76GN	53	2SB90GN	53	2SBP2	57	B1U
53	2SB67G	53	2SB76H	53	2SB90H	53	2SBP2A	57	B1U148
53	2SB67GN	53	2SB76J	53	2SB90J	53	2SBM77	57	B2J
53	2SB67H	53	2SB76K	53	2SB90K	54	J241111(2SD170)	57	B2V
53	2SB67J	53	2SB76L	53	2SB90L	55	D44H10	57	BD-131
53	2SB67K	53	2SB76OR	53	2SB90M	55	D44H4	57	BD131
53	2SB67L	53	2SB76R	53	2SB900R	55	D44H5	57	BD135
53	2SB67M	53	2SB76X	53	2SB90R	55	D44H7	57	BD137
53	2SB67OR	53	2SB76Y	53	2SB90X	55	D44H8	57	BD139
53	2SB67R	53	2SB77-OR	53	2SB90Y	55	MJE182	57	BD149
53	2SB67X	53	2SB77A	53	2SB91	55	MJE204	57	BD153
53	2SB67Y	53	2SB77AB	53	2SB91A	55	MJE205	57	BD154
53	2SB68	53	2SB77AC	53	2SB91C	55	MJE2801	57	BD165
53	2SB68A	53	2SB77AD	53	2SB91D	55	MJE3055	57	BD167
53	2SB68B	53	2SB77AH	53	2SB91E	55	T1P33	57	BD175
53	2SB68C	53	2SB77AP	53	2SB91F	55	T1P33A	57	BD177
53	2SB68D	53	2SB77B	53	2SB91G	55	2SC1212	57	BD185
53	2SB68E	53	2SB77B-11	53	2SB91GN	55	2SC1212A	57	BD187
53	2SB68F	53	2SB77C	53	2SB91H	55	2SC1212AA	57	BD189
53	2SB68G	53	2SB77D	53	2SB91J	55	2SC1212AB	57	BD195
53	2SB68GN	53	2SB77E	53	2SB91K	55	2SC1212ABWT	57	BD197
53	2SB68H	53	2SB77F	53	2SB91L	55	2SC1212AC	57	BD205
53	2SB68J	53	2SB77G	53	2SB91M	55	2SC1212ACWT	57	BD207
53	2SB68K	53	2SB77GN	53	2SB910R	55	2SC1212AD	57	BD561
53	2SB68L	53	2SB77H	53	2SB91R	55	2SC1212AWT	57	BD575
53	2SB68M	53	2SB77K	53	2SB91X	55	2SC1212AWTA	57	BD577
53	2SB68OR	53	2SB77L	53	2SB91Y	55	2SC1212AWTB	57	BD585
53	2SB68R	53	2SB77M	53	2SB92	55	2SC1212AWTC	57	BD587
53	2SB68X	53	2SB77P	53	2SB92A	55	2SC1212AWTD	57	BD589
53	2SB68Y	53	2SB77PD	53	2SB92B	55	2SC1212B	57	C1162
53	2SB70	53	2SB77R	53	2SB92C	55	2SC1212C	57	C1162A
53	2SB71	53	2SB77RED	53	2SB92D	55	2SC1212D	57	C1162B
53	2SB71A	53	2SB77V	53	2SB92E	55	2SC1212WT	57	C1162DP
53	2SB71B	53	2SB77VRED	53	2SB92F	55	2SC1212WTA	57	C1162D
53	2SB71C	53	2SB77X	53	2SB92G	55	2SC1212WTB	57	C1162MP
53	2SB71D	53	2SB77Y	53	2SB92GN	55	2SC1212WTC	57	C1162WB
53	2SB71E	53	2SB78	53	2SB92H	55	2SC1212WTD	57	C1162WTB
53	2SB71F	53	2SB78A	53	2SB92J	56	AR23(PHILCO)	57	C1162WTC
53	2SB71G	53	2SB78B	53	2SB92K	56	BD170	57	C1162WTD
53	2SB71GN	53	2SB78C	53	2SB92L	56	BD180	57	C1212
53	2SB71H	53	2SB78D	53	2SB92M	56	BD200	57	C1212A
53	2SB71J	53	2SB78E	53	2SB92X	56	BD462	57	C1212AA
53	2SB71K	53	2SB78F	53	2SB92Y	56	BD464	57	C1212AB
53	2SB71L	53	2SB78G	53	2SB94	56	CXL183	57	C1212AD
53	2SB71M	53	2SB78GN	53	2SB94A	56	D2H	57	C1212AWT
53	2SB71OR	53	2SB78H	53	2SB94B	56	D45H1	57	C1212AWTA
53	2SB71X	53	2SB78J	53	2SB94C	56	D45H10	57	C1212AWTC
53	2SB71Y	53	2SB78K	53	2SB94D	56	D45H2	57	C1212AWTD
53	2SB72	53	2SB78L	53	2SB94E	56	D45H4	57	C1212B
53	2SB72A	53	2SB78M	53	2SB94F	56	D45H5	57	C1212C
53	2SB72B	53	2SB78OR	53	2SB94G	56	D45H7	57	C1212D
53	2SB72C	53	2SB78R	53	2SB94GN	56	D45H8	57	C1212WT
53	2SB72D	53	2SB78X	53	2SB94H	56	ECG183	57	C1212WTA
53	2SB72E	53	2SB78Y	53	2SB94J	56	ES67(ELCOM)	57	C1212WTB
53	2SB72F	53	2SB79	53	2SB94K	56	GE-56	57	C1212WTC
53	2SB72G	53	2SB79A	53	2SB94M	56	HEP-85002	57	C1212WTD
53	2SB72GN	53	2SB79B	53	2SB940R	56	HEP-85005	57	C1368
53	2SB72H	53	2SB79D	53	2SB94R	56	HEP-85008	57	C1368C
53	2SB72J	53	2SB79E	53	2SB94X	56	HEP-85009	57	C1386C
53	2SB72K	53	2SB79G	53	2SB94Y	56	HEP-85010	57	C495
53	2SB72L	53	2SB79GN	53	2SB95	56	HEP85002	57	C495-0
53	2SB72M	53	2SB79H	53	2SB95A	56	MJE-2020	57	C495-R
53	2SB72OR	53	2SB79J	53	2SB95B	56	MJE104	57	C496Y
53	2SB72R	53	2SB79K	53	2SB95C	56	MJE105	57	CS9013(HQ)
53	2SB72X	53	2SB79L	53	2SB95D	56	MJE1290	57	CXL186
53	2SB72Y	53	2SB79M	53	2SB95E	56	MJE1291	57	D18A12
53	2SB73	53	2SB79OR	53	2SB95F	56	MJE2020	57	EA1-380
53	2SB73A	53	2SB79X	53	2SB95G	56	MJE2901	57	EA15X244
53	2SB73A-1	53	2SB79Y	53	2SB95GN	56	MJE2955	57	EA15X269
53	2SB73B	53	2SB85	53	2SB95H	56	MJE712	57	EAI-380
53	2SB73C	53	2SB85A	53	2SB95J	56	P3N-5	57	ECG184
53	2SB73D	53	2SB85B	53	2SB95L	56	P3UA	57	ES113
53	2SB73F	53	2SB85C	53	2SB95M	56	P4B-1	57	ES58
53	2SB73G	53	2SB85D	53	2SB95OR	56	P4B-2	57	ES58(ELCOM)
53	2SB73GN	53	2SB85E	53	2SB95R	56	P4T	57	G03-007C
53	2SB73H	53	2SB85F	53	2SB95X	56	P4V-2	57	G03007C
53	2SB73J	53	2SB85G	53	2SB95Y	56	RE39	57	G06-717-C
53	2SB73K	53	2SB85GN	53	2SB97	56	REN183	57	G06-717-D
53	2SB73L	53	2SB85H	53	2SB97A	56	RS-2027	57	G06-717B
53	2SB73M	53	2SB85J	53	2SB97B	56	SJE517	57	GE-57
53	2SB73OR	53	2SB85K	53	2SB97C	56	T1P36A	57	HC-00496
53	2SB73R	53	2SB85L	53	2SB97D	56	TIP34	57	HEP245
53	2SB73S	53	2SB85M	53	2SB97E	56	TIP34A	57	HEP701
53	2SB73X	53	2SB85OR	53	2SB97F	56	TIP34B	57	HT104961C
53	2SB74	53	2SB85R	53	2SB97G	56	TIP37	57	HT304961B
53	2SB74A	53	2SB85X	53	2SB97GN	56	TIP37F	57	HT304961C
53	2SB74C	53	2SB87	53	2SB97H	56	TR-1036-1	57	HT97002B
53	2SB74D	53	2SB87A	53	2SB97K	56	TR-1036-2	57	HT313681BO
53	2SB74F	53	2SB87C	53	2SB97L	56	TRO-2057-1	57	IRTR76
53	2SB74G			53	2SB97OR	56	TRO-2057-3	57	J241250
53	2SB74GN			53	2SB97R	56	TRO-2057-4	57	MOTMJE521
				53	2SB97X	56	TRO-2058-1	57	M3567-2
				53	2SB97Y	56	TRO-2058-5	57	M844
						56	TR1036-2	57	M852
						56	TR1036-3	57	M9556
						56	TR1037-1		
						56	TR1038-5		
						56	TR1038-6		

GE	Industry Standard No.	GE	Industry Standard No.	GE	Industry Standard No.	GE	Industry Standard No.	GE	Industry Standard No.
57	M9582	57	2SC1162	58	P3N-4	58	2SA743AD	59	D72B
57	M9618	57	2SC1368	58	P3P	58	2SA743B	59	D72C
57	MJE-220	57	2SC1368(B)	58	P3P-1	58	2SA743C	59	D72RE
57	MJE180	57	2SC1368B	58	P3P-2	58	2SA743D	59	D75
57	MJE181	57	2SC1368C	58	P3P-3	58	2SA886	59	D75A
57	MJE200	57	2SC1368D	58	P3P-4	58	2SA886V	59	D75AH
57	MJE200E	57	2SC1568	58	P3P-5	58	2SA886VR	59	D75B
57	MJE205K	57	2SC1568(R)	58	P3S	58	2SB632K	59	D75C
57	MJE220	57	2SC1568R	58	P3V	59	A122-1962	59	D75H
57	MJE221	57	2SC2456	58	P4E	59	A13-86416-1	59	D77(TRANSISTOR)
57	MJE222	57	2SC2497Q	58	P4E-3	59	A1465-29	59	D77A
57	MJE223	57	2SC2497R	58	P4E-4	59	A14665-2	59	D77AH
57	MJE224	57	2SC2568K	58	P4J	59	A16A1	59	D77B
57	MJE225	57	2SC2568L	58	P4S	59	A16A2	59	D77C
57	MJE2480	57	2SC2568M	58	P4U	59	A2T682	59	D77D
57	MJE2481	57	2SC496	58	P4V	59	A42X210	59	D77H
57	MJE2482	58	A496(JAPAN)	58	P4V-1	59	A48-869254	59	D77P
57	MJE2483	58	A496-0	58	P4V-1	59	A48-869283	59	D96
57	MJE2520	58	A496-R	58	P4V-2	59	A48-869476	59	E-2466
57	MJE2521	58	A496-Y	58	P5P	59	A48-869476A	59	E24105
57	MJE2522	58	A496Y	58	P5L	59	A4JX2A822	59	E2447
57	MJE2523	58	A505	58	P5R	59	A4L	59	E2466
57	MJE3054	58	A505-0	58	P5S	59	A57C5	59	EA15X8443
57	MJE3520	58	A505-R	58	P6009	59	A65-2-70	59	EC0103A
57	MJE3521	58	A505-Y	58	PLE37	59	A65-2-705	59	EQR-1
57	MJE482	58	A715	58	QP-13	59	A65-2A99	59	ES15X48
57	MJE483	58	A715A	58	RE41	59	AC-127	59	ES15X71
57	MJE488	58	A715B	58	REN185	59	AC-157	59	ES15X74
57	MJE520	58	A715C	58	RS-2025	59	AC-172	59	ES30(ELCOM)
57	MJE520K	58	A715D	58	S89570	59	AC127	59	E85(ELCOM)
57	MJE521	58	A715WTA	58	S89571	59	AC127-01	59	E86(ELCOM)
57	MJE521K	58	A715WTB	58	S89572	59	AC127-132	59	GC1137
57	MJE720	58	A715WTC	58	S89573	59	AC141	59	GC1185
57	MJE721	58	A715WTD	58	SJ285	59	AC141B	59	GC1423
57	MJE9400	58	A738C	58	SJB108	59	AC141K	59	GC148
57	MJE8191	58	A738D	58	SJE111	59	AC172	59	GC285
57	MOTMJ8521	58	A743	58	SJE112	59	AC175	59	GC286
57	PLE52	58	A743A	58	SJE114	59	AC175A	59	GC463
57	PLE52	58	A743AA	58	SJE1519	59	AC175B	59	GC465
57	QP-14	58	A743AB	58	SJE202	59	AC175P	59	GC467
57	RE40	58	A743AC	58	SJE210	59	AC176	59	GC608
57	REN184	58	A743AD	58	SJE221	59	AC179	59	GC609
57	RS-2017	58	A743B	58	SJE227	59	AC181	59	GE-59
57	RS-2020	58	A743C	58	SJE231	59	AC181K	59	GI1658
57	S-85	58	A743D	58	SJE241	59	AC183	59	GT167
57	S-86	58	A8C	58	SJE243	59	AC185	59	GT2768
57	S10153	58	B-1695	58	SJE245	59	AC186	59	GT336
57	S32903	58	BD-132	58	SJE256	59	AC187	59	GT35
57	S8660	58	BD132	58	SJE257	59	AC187K	59	GT564
57	S8660121-808	58	BD136	58	SJE265	59	AC187R	59	GT565
57	SC4133	58	BD138	58	SJE267	59	AC194	59	GT566
57	SJE-649	58	BD140	58	SJE273	59	AC194K	59	GT903
57	SJE100	58	BD151	58	SJE275	59	ASY61	59	HA5010
57	SJE106	58	BD152	58	SJE276	59	ASY62	59	HD-00072
57	SJE113	58	BD156	58	SJE277	59	B65-2-A-21	59	HT400721
57	SJE133	58	BD166	58	SJE279	59	B65-2A21	59	HT400721A
57	SJE1519	58	BD168	58	SJE283	59	BD-00072	59	HT400721B
57	SJE1520	58	BD176	58	SJE288	59	BTX071	59	HT400721C
57	SJE203	58	BD178	58	SJE403	59	C179	59	HT400721D
57	SJE211	58	BD186	58	SJE408	59	C181	59	HT400721E
57	SJE228	58	BD188	58	SJE5442	59	C277C	59	HT400721F
57	SJE229	58	BD190	58	SJE584	59	C34(TRANSISTOR)	59	HT400723
57	SJE237	58	BD196	58	SJE633	59	C35(TRANSISTOR)	59	HT400723A
57	SJE242	58	BD206	58	SJE725	59	C81759	59	HT400723B
57	SJE244	58	BD208	58	SJE736	59	D083	59	HT400770B
57	SJE246	58	BD562	58	SJE743	59	D100A	59	HT403525A
57	SJE248	58	BD576	58	SJE768	59	D104	59	IE-850
57	SJE253	58	BD578	58	SJE797	59	D105	59	IP20-0008
57	SJE254	58	BD580	58	SJE799	59	D127	59	IP20-0076
57	SJE255	58	BD586	58	SK3191	59	D127A	59	IP20-0160
57	SJE261	58	BD588	58	SFS4237	59	D128	59	J241185
57	SJE262	58	BD590	58	STX0020	59	D128A	59	J24868
57	SJE271	58	CXL185	58	T-345	59	D170	59	K4-501
57	SJE272	58	EA15X243	58	T-396	59	D170A	59	N57B2-8
57	SJE274	58	EA15X270	58	T345	59	D170AA	59	NA30
57	SJE278	58	EA15X8605	58	T396	59	D170AB	59	NO35
57	SJE280	58	ECG185	58	TA7520	59	D170AC	59	NK65-2A19
57	SJE284	58	ECG242	58	TR-182(OLSON)	59	D170BC	59	NKT701
57	SJE289	58	EE114	58	TR-77(XSTER)	59	D170C	59	NKT703
57	SJE505	58	E860(ELCOM)	58	TRO20BC-3	59	D170PB	59	NKT713
57	SJE520	58	E868(ELCOM)	58	TR2327393	59	D178	59	NKT717
57	SJE540	58	G03-406-C	58	TR77	59	D178A	59	NKT751
57	SJE401	58	GE-58	58	002-012100	59	D178Q	59	NKT752
57	SJE402	58	HA-00496	58	2N4918	59	D178T	59	NKT773
57	SJE404	58	HEP246	58	2N4919	59	D186	59	NKT781
57	SJE405	58	HEP700	58	2N4920	59	D186A	59	PBB3020-2
57	SJE407	58	HEPS5006	58	2N5193	59	D186B	59	PTC134
57	SJE527	58	HEPS5007	58	2N5194	59	D187	59	Q0V60527
57	SJE5402	58	HT104961B	58	2N5195	59	D187A	59	Q2N4105
57	SJE5439	58	HT70011100	58	2SA496	59	D187R	59	Q5039
57	SJE5441	58	I-B198	58	2SA496(Q)	59	D187Y	59	Q5050
57	SJE583	58	IRTR77	58	2SA496-0	59	D191	59	Q0V60527
57	SJE634	58	JSP6009	58	2SA496-ORG	59	D192	59	Q0V60537
57	SJE649	58	MOTMJE371	58	2SA496-R	59	D194	59	QQV60527
57	SJE669	58	MJE-521	58	2SA496-RED	59	D30	59	QQV60537
57	SJE721	58	MJE105K	58	2SA496-Y	59	D30-0	59	QQV61772
57	SJE724	58	MJE230	58	2SA496-YEL	59	D30-N	59	R-28D187
57	SJE737	58	MJE231	58	2SA964	59	D31	59	R1530
57	SJE769	58	MJE232	58	2SA4960	59	D31D	59	R1531
57	SJE781	58	MJE233	58	2SA4960RG	59	D32	59	R1532
57	SJE784	58	MJE234	58	2SA496R	59	D33	59	R1534
57	SJE785	58	MJE235	58	2SA496RED	59	D33C	59	R1537
57	SK3190	58	MJE2370	58	2SA496Y	59	D34	59	R1538
57	SP8918	58	MJE2371	58	2SA496YEL	59	D35	59	R1545
57	STX0013	58	MJE2490	58	2SA505	59	D352D	59	R1547
57	STX0026	58	MJE2491	58	2SA505-0	59	D352E	59	R1549
57	T-342	58	MJE3370	58	2SA505-R	59	D352P	59	R1553
57	T-344	58	MJE3371	58	2SA505-RED	59	D367	59	R177
57	T1P31	58	MJE370	58	2SA505-Y	59	D367A	59	R2356
57	T1P31A	58	MJE370K	58	2SA505-YEL	59	D367B	59	R2359
57	T344	58	MJE371	58	2SA5050	59	D367C	59	R2360
57	T611-1(RCA)	58	MJE371K	58	2SA505R	59	D367D	59	R2364
57	T612-1(RCA)	58	MJE740	58	2SA505Y	59	D367E	59	R2365
57	TIP14	58	MJE3741	58	2SA715	59	D367F	59	R2374
57	TIP31B	58	MJE492	58	2SA715A	59	D367P	59	R2375
57	TNJ70540	58	MJE493	58	2SA715B	59	D37	59	R3293
57	TR-76	58	MJE711	58	2SA715C	59	D37A	59	R33
57	TRO1045	58	MJE9450	58	2SA715WB	59	D37B	59	R34
57	TRO1056-5	58	MJG5194	58	2SA715WT	59	D37C	59	R3573-1
57	TRO1057-3	58	MOTMJE371	58	2SA715WTA	59	D38	59	R5050
57	TR2327203	58	P1V	58	2SA715WTB	59	D61	59	R5054
57	TR76	58	P1V-1	58	2SA715WTC	59	D62	59	R5056
57	002-012200	58	P1V-2	58	2SA715WTD	59	D63	59	R5179
57	002-3(SYLVANIA)	58	P1V-3	58	2SA738	59	D64	59	R5180
57	2N4921	58	P2T	58	2SA738B	59	D65-1	59	R62
57	2N4922	58	P2T-1	58	2SA738C	59	D72	59	R63
57	2N4923	58	P2T-2	58	2SA743	59	D72-2C	59	R7362
57	2N5190	58	P2T-3	58	2SA743A	59	D72-3C	59	R79
57	2N5191	58	P2T-4	58	2SA743AA	59	D72-4C	59	R80
57	2N5192	58	P3M	58	2SA743AB	59	D72DD	59	RCA40231
57	2N6037	58	P3N-1	58	2SA743AC	59	D72DE	59	RCA40396N
57	2N6038	58	P3N-2			59	D72A	59	RE6
		58	P3N-3					59	REN103A
								59	RS-1524

GE	Industry Standard No.	GE	Industry Standard No.	GE	Industry Standard No.	GE	Industry Standard No.	GE	Industry Standard No.
59	RS1524	59	2SC129Y	61	A670722D	61	C785D	61	LTH1016
59	RS1533	59	2SC13	61	A670729B	61	C785E	61	LTH101K
59	RS1549	59	2SC14	61	A6708850	61	C785R	61	MA4101
59	RS3931	59	2SC173	61	AR303	61	C786	61	MA4103
59	RS8407	59	2SC175	61	BC456	61	C903D	61	MA4104
59	RS8420	59	2SC175B	61	BF240B	61	C917	61	ME1138
59	RS8441	59	2SC176	61	BF241C	61	C917K	61	M8900I
59	RS8443	59	2SC177	61	BF241D	61	C920Q	61	M89003
59	RS8445	59	2SC178	61	BF364	61	C920R	61	MMT918
59	RT3096	59	2SC180	61	BF365	61	C929	61	MPS-H17
59	RT7400	59	2SC180A	61	BF494	61	C929-0	61	MPS3563
59	RT7944	59	2SC180B	61	BF495	61	C929B	61	MPS3693
59	S-95202	59	2SC180C	61	C1047	61	C929C	61	MPS3694
59	S2042634	59	2SC180D	61	C1047B	61	C929C1	61	MPS6507
59	S65-2-A-3P	59	2SC180E	61	C1047C	61	C929D	61	MPS6511-8
59	SE-7001	59	2SC180F	61	C1047D	61	C929D1	61	MPS6512
59	SF.P377	59	2SC180G	61	C1047E	61	C929DE	61	MPS6513
59	SFT-184	59	2SC180GN	61	C1187	61	C929DP	61	MPS6514
59	SPF377	59	2SC180H	61	C1359	61	C929DU	61	MPS6568A
59	SK3010	59	2SC180J	61	C1359A	61	C929DV	61	MPS7513
59	SP215R	59	2SC180K	61	C1359B	61	C929F	61	MPS9426
59	SP2188	59	2SC180L	61	C1359C	61	C929FK	61	MPS9426B
59	SQ-7	59	2SC180M	61	C1674	61	C930	61	MPS9426BC
59	SYL-103	59	2SC180OR	61	C1674K	61	C930B	61	MPS9426C
59	SYL-104	59	2SC180R	61	C1674L	61	C930BK	61	MPS9427
59	SYL-1329	59	2SC180X	61	C1674M	61	C930BV	61	MPS9427B
59	SYL-1396	59	2SC180Y	61	C1675	61	C930C	61	MPS9604
59	SYL-152	59	2SC277C	61	C1675K	61	C930CL	61	MPS9625
59	SYL-1524	59	2SC34	61	C1675L	61	C930CS	61	MPS9625C
59	SYL-2134	59	2SC35	61	C1675M	61	C930D	61	MPS9625E
59	SYL-2135	59	2SC36	61	C1686	61	C930DB	61	MPS9625F
59	SYL-2156	59	2SC41	61	C380	61	C930DT	61	MPS9625H
59	SYL-4315	59	2SC60	61	C380-0	61	C930DT-2	61	MPS9700D
59	SYL1468	59	2SC71	61	C380-0/4454C	61	C930DX	61	MPS9700E
59	T-81	59	2SC72	61	C380A	61	C930E	61	MPSR10
59	T61	59	2SC73	61	C380A(0)	61	C930EP	61	MPSR11
59	TA-1620A	60	AR219(YY)	61	C380A-0(TV)	61	C930EV	61	MPSR19
59	TA-1620B	60	AR220(GY)	61	C380A-R	61	C930EX	61	MPSR30
59	TM103A	60	AR220GREEN	61	C380A-R(TV)	61	C930P	61	MPSR31
59	TNJ61671(2SD72)	60	AR220YELLOW	61	C380AO	61	C9426	61	MPSR37
59	TNJ61734	60	AR222(BY)	61	C380ATV	61	CB4008B	61	NTC-4
59	TNJ72284	60	AR222BLUE	61	C380AY	61	CE4008C	61	Q-00284R
59	TQ5044	60	AR222YELLOW	61	C380D	61	C82006P	61	QT-C0372XAT
59	TQ5050	60	AR224	61	C380R	61	C89016E	61	QT-C0828XAN
59	TQ5062	60	AR224WHITE	61	C380R/4454C	61	C89016F	61	QT-C0828XDN
59	TRO1065	60	AR224YELLOW	61	C380Y	61	D009	61	QT-C0839XDA
59	TR310160	60	B1H(X8TR)	61	C394	61	D072	61	QT-C0460CBB
59	TR310236	60	C84021	61	C394-0	61	DBCZ039404	61	R9600
59	TV-27	60	C84060	61	C394-0	61	DBCZ083905	61	REN229
59	TV24143	60	DDBY261002	61	C394GR	61	DDBY216002	61	R3-805US
59	TV24983	60	DDBY278001	61	C394R	61	DDBY219001	61	R87114
59	1-801-309	60	EA15X351	61	C394W	61	DDBY262001	61	R87173
59	1-TR-112	60	EA3713	61	C394Y	61	DDBY277002	61	R87233
59	002-011700	60	ED1502E	61	C461	61	DDBY287001	61	RT6737
59	002-11700	60	EP15X106	61	C461A	61	DS-74	61	RV1068
59	2N1010	60	EP15X123	61	C461B	61	DS-81	61	RY2354
59	2N1058	60	GE-60	61	C461C	61	E9625	61	S2438
59	2N1059	60	GI-2715	61	C461E	61	EA15X130	61	S2716
59	2N1101	60	GI-2716	61	C461L	61	EA15X180	61	S2718
59	2N1102	60	GI-2921	61	C535	61	EA15X325	61	S27893
59	2N1173W	60	GI-2922	61	C535A	61	EA15X376	61	SE-1001
59	2N1251	60	GI-2923	61	C535B	61	EA15X393	61	SE0001
59	2N126	60	GI-3605	61	C535C	61	EA15X396	61	SE1001
59	2N1431	60	GI-3606	61	C535G	61	EA15X7233	61	SE1001-1
59	2N148A	60	GI-3607	61	C668-0	61	EA15X7234	61	SE1001-2
59	2N148B	60	HC-668	61	C668A	61	EA15X7235	61	SS4001
59	2N149	60	HC00930	61	C668B	61	EA15X7236	61	SE50
59	2N149A	60	IP20-0037	61	C668B1	61	EA15X7244	61	SP115
59	2N1672	60	IP20-0038	61	C668BC2	61	EA15X8601	61	SP115A
59	2N1672A	60	MPS89634B	61	C668C	61	EO0229	61	SP115B
59	2N193	60	QT-C1047XAN	61	C668C1	61	ED-1502C	61	SP115C
59	2N194	60	QT-C1047XBN	61	C668CD	61	ED1502A	61	SP115D
59	2N194A	60	R76787	61	C668D	61	ED1502C	61	SP115E
59	2N211	60	R78333	61	C668D0	61	ED1502D	61	SKA-4076
59	2N212	60	SE3646	61	C668D1	61	ED592K	61	SKA-4590
59	2N213	60	SF534	61	C668D0	61	EL75	61	SM4304-8
59	2N213A	60	SF334B	61	C668DV	61	EP15X41	61	SPF530
59	2N214	60	SF335	61	C668DX	61	EP15X42	61	SP81846
59	2N216	60	SF335C	61	C668DZ	61	EP15X6	61	SP84331
59	2N228	60	SF335X	61	C668E	61	EQ8-0018	61	SP84436
59	2N229	60	T-3568	61	C668E1	61	EQ8-0100	61	SP8920
59	2N233	60	T1886	61	C668E2	61	EQ8-0198	61	SR20226
59	2N233A	60	TI8-18	61	C668EP	61	EQS-139	61	ST-MPS9700D
59	2N2430	60	TR2083-74	61	C668EV	61	ESIKX122	61	ST-MPS9700E
59	2N306	60	TV-15A	61	C668EX	61	FG89016F	61	ST-MPS9700F
59	2N306A	60	01-031047	61	C668F	61	FX3013	61	SX3826
59	2N35	60	1SC683	61	C722	61	FX3014	61	SX408
59	2N507	60	2N4081	61	C763B	61	FX914	61	T408
59	2N515	60	2N4295	61	C763C	61	FX918	61	TI8-18
59	2N516	60	2N4397	61	C763CD	61	G05-037-D	61	T1818
59	2N517	60	2N471	61	C763D	61	G05-063-R	61	T9423
59	2N567	60	2SC1047	61	C772	61	G05-065-A	61	TG28C1293
59	2N647	60	2SC1047A	61	C772BG	61	G9423	61	TG28C1293(A)
59	2N647/22	60	2SC1047BC	61	C772BH	61	G9625	61	TG28C1293-A-A
59	2N649	60	2SC1047BCD	61	C772BV	61	GE-61	61	TG28C1293-B-A
59	2N649/5	60	2SC1047C	61	C772BX	61	GET3563	61	TG28C1293-C-A
59	2N94	60	2SC1047D	61	C772BY	61	HC-00372	61	TG28C1293-D-A
59	2N94A	60	2SC1047E	61	C772C	61	HC-00394	61	TG28C1293A
59	2SC11	60	2SC1047F	61	C772C1	61	HEPS0008	61	TIS125
59	2SC128	60	2SC1047GN	61	C772C2	61	HEPS0010	61	TR-70
59	2SC128A	60	2SC1047GR	61	C772CK	61	HEPS0017	61	TRO573486
59	2SC128B	60	2SC1047H	61	C772CL	61	HEPS0033	61	TRO573507
59	2SC128C	60	2SC1047J	61	C772CS	61	HR77	61	TR28C1342
59	2SC128D	60	2SC1047K	61	C772CV	61	HR78	61	TR70
59	2SC128E	60	2SC1047L	61	C772D	61	H87	61	TV-48
59	2SC128F	60	2SC1047M	61	C772DU	61	HT304601BO	61	01-030380
59	2SC128G	60	2SC1047N	61	C772DV	61	IC743040	61	01-030394
59	2SC128GN	60	2SC1047X	61	C772DX	61	IC743041	61	01-030682
59	2SC128H	60	2SC1047Y	61	C772DY	61	IP20-0040	61	01-030763
59	2SC128J	60	2SC1129	61	C772E	61	IP20-006	61	01-030930
59	2SC128K	60	2SC1129(M)	61	C772F	61	IP20-0110	61	01-031293
59	2SC128L	60	2SC1129(R)	61	C772K	61	IP20-0179	61	01-031674
59	2SC128M	60	2SC1129-0	61	C772KB	61	IRTR-70	61	01-031686
59	2SC128OR	60	2SC1129A	61	C772KC	61	J24844	61	01-30828
59	2SC128R	60	2SC1129B	61	C772KD	61	K14-0066-6	61	1E535A
59	2SC128X	60	2SC1129BL	61	C772KD1	61	K0825671-20	61	1E535A/7825B
59	2SC128Y	60	2SC1129C	61	C772KD2	61	K0825671-21	61	2N2475
59	2SC129	60	2SC1129G	61	C772R	61	K0825671-23	61	2N2475/46
59	2SC129A	60	2SC1129M	61	C772R(JAPAN)	61	KSC11870	61	2N2475/51
59	2SC129B	60	2SC1129R	61	C772RB-D	61	KSC1187R	61	2N2615
59	2SC129C	60	2SC1129Y	61	C772RD	61	KSC16740	61	2N3563
59	2SC129D	60	2SC602	61	C784	61	KSC1674R	61	2N3563-1
59	2SC129E	60	2SC650F	61	C784-0	61	LM1138	61	2N3564
59	2SC129F	59	A-473	61	C784-0	61	LM1138E	61	2N3827
59	2SC129N	61	A066-111	61	C784O	61	LM1138E/F	61	2N5830A
59	2SC129H	61	A154	61	C784A	61	LM1138F	61	2N704/51
59	2SC129J	61	A155	61	C784BN	61	LM1138G	61	28C-4033
59	2SC129K	61	A164	61	C784R	61	LM1138G/F	61	28C-P11
59	2SC129L	61	A165	61	C784Y	61	LM1138H	61	28C-P11A
59	2SC129M	61	A467	61	C785(O)	61	LM1138H/I	61	28C-P11B
59	2SC129R	61	A67-37-940	61	C785BN	61	LM11381	61	28C-P11C
59	2SC129X	61	A670720K	61		61	LT1016H	61	28C-P11D

GE	Industry Standard No.
61	2SC-P11E
61	2SC-P11F
61	2SC-P11G
61	2SC-P11GN
61	2SC-P11H
61	2SC-P11J
61	2SC-P11K
61	2SC-P11L
61	2SC-P11M
61	2SC-P11OR
61	2SC-P11R
61	2SC-P11X
61	2SC-P11Y
61	2SC-P14
61	2SC-P14A
61	2SC-P14B
61	2SC-P14C
61	2SC-P14D
61	2SC-P14E
61	2SC-P14F
61	2SC-P14G
61	2SC-P14GN
61	2SC-P14H
61	2SC-P14J
61	2SC-P14K
61	2SC-P14L
61	2SC-P14M
61	2SC-P14OR
61	2SC-P14R
61	2SC-P14X
61	2SC-P14Y
61	2SC1009
61	2SC1023(O)
61	2SC1023(O)
61	2SC1023-O
61	2SC1023-Y
61	2SC1023A
61	2SC1023B
61	2SC1023D
61	2SC1023E
61	2SC1023F
61	2SC1023GN
61	2SC1023H
61	2SC1023J
61	2SC1023K
61	2SC1023L
61	2SC1023M
61	2SC1023OR
61	2SC1023R
61	2SC1023X
61	2SC1023Y
61	2SC1026(G)
61	2SC1026-O
61	2SC1026-R
61	2SC1026BL
61	2SC1026D
61	2SC1026E
61	2SC1026G
61	2SC1026GN
61	2SC1026GR
61	2SC1026H
61	2SC1026J
61	2SC1026K
61	2SC1026L
61	2SC1026M
61	2SC1026OR
61	2SC1026R
61	2SC1026X
61	2SC1026Y
61	2SC1032(Y)
61	0002SC1032A
61	0002SC1032B
61	2SC1032BL
61	0002SC1032C
61	2SC1032D
61	2SC1032E
61	2SC1032F
61	2SC1032GN
61	2SC1032G
61	2SC1032H
61	2SC1032J
61	2SC1032L
61	2SC1032M
61	2SC1032OR
61	2SC1032R
61	2SC1032X
61	2SC1032Y
61	2SC1123
61	2SC1123A
61	2SC1123B
61	2SC1123C
61	2SC1123D
61	2SC1123E
61	2SC1123F
61	2SC1123GN
61	2SC1123H
61	2SC1123J
61	2SC1123L
61	2SC1123M
61	2SC1123OR
61	2SC1123R
61	2SC1123X
61	2SC1123Y
61	2SC1126
61	2SC1126A
61	2SC1126B
61	2SC1126E
61	2SC1126F
61	2SC1126GN
61	2SC1126J
61	2SC1126K
61	2SC1126L
61	2SC1126M
61	2SC1126OR
61	2SC1126X
61	2SC1126Y
61	2SC1128
61	2SC1128(3RD-IF)
61	2SC1128(M)
61	2SC1128(S)
61	2SC1128-O
61	2SC1128A
61	2SC1128B
61	2SC1128BL
61	2SC1128C
61	2SC1128D
61	2SC1128G
61	2SC1128H
61	2SC1128M
61	2SC1128R
61	2SC1128S
61	2SC1128S(3RDIF)
61	2SC1128Y
61	2SC1187
61	2SC1215
61	2SC1215C
61	2SC1215D
61	2SC1215E
61	2SC1215F
61	2SC1215G
61	2SC1215GN
61	2SC1215H
61	2SC1215J
61	2SC1215K
61	2SC1215L
61	2SC1215M
61	2SC1215OR
61	2SC1215R
61	2SC1215X
61	2SC1215Y
61	2SC1216
61	2SC1293
61	2SC1293(3RD-IF)
61	2SC1293(A)
61	2SC1293A
61	2SC1293B
61	2SC1293C
61	2SC1293D
61	2SC1320(K)
61	2SC1320A
61	2SC1320B
61	2SC1320C
61	2SC1320D
61	2SC1320E
61	2SC1320F
61	2SC1320G
61	2SC1320GN
61	2SC1320H
61	2SC1320J
61	2SC1320K
61	2SC1320L
61	2SC1320M
61	2SC1320OR
61	2SC1320R
61	2SC1320X
61	2SC1320Y
61	2SC134
61	2SC1342
61	2SC1342(A)
61	2SC1342(B)
61	2SC1342(C)
61	2SC1342-OR
61	2SC1342A
61	2SC1342B
61	2SC1342C
61	2SC1342D
61	2SC1342E
61	2SC1342F
61	2SC1342G
61	2SC1342GN
61	2SC1342J
61	2SC1342K
61	2SC1342L
61	2SC1342M
61	2SC1342OR
61	2SC1342R
61	2SC1342X
61	2SC1342Y
61	2SC134A
61	2SC134B
61	2SC134C
61	2SC134D
61	2SC134E
61	2SC134F
61	2SC134G
61	2SC134H
61	2SC134K
61	2SC134L
61	2SC134M
61	2SC134OR
61	2SC134R
61	2SC134X
61	2SC134Y
61	2SC135
61	2SC135A
61	2SC135C
61	2SC135D
61	2SC135E
61	2SC135F
61	2SC135G
61	2SC135GN
61	2SC135H
61	2SC135J
61	2SC135L
61	2SC135M
61	2SC135OR
61	2SC135R
61	2SC135X
61	2SC135Y
61	2SC137
61	2SC137A
61	2SC137B
61	2SC137C
61	2SC137D
61	2SC137E
61	2SC137F
61	2SC137G
61	2SC137GN
61	2SC137H
61	2SC137J
61	2SC137K
61	2SC137L
61	2SC137M
61	2SC137OR
61	2SC137R
61	2SC137X
61	2SC137Y
61	2SC1417
61	2SC1417(V,G)
61	2SC1417(W)
61	2SC1417D
61	2SC1417D(U)
61	2SC1417DU
61	2SC1417F
61	2SC1417G
61	2SC1417H
61	2SC1417U
61	2SC1417R
61	2SC1417VF
61	2SC1417VW
61	2SC1417W
61	2SC148
61	2SC148A
61	2SC148B
61	2SC148D
61	2SC148E
61	2SC148F
61	2SC148G
61	2SC148GN
61	2SC148H
61	2SC148J
61	2SC148K
61	2SC148L
61	2SC148M
61	2SC148OR
61	2SC148R
61	2SC148X
61	2SC159
61	2SC159A
61	2SC159B
61	2SC159C
61	2SC159D
61	2SC159F
61	2SC159GN
61	2SC159H
61	2SC159J
61	2SC159K
61	2SC159L
61	2SC159M
61	2SC1590R
61	2SC159R
61	2SC159X
61	2SC159Y
61	2SC160
61	2SC160A
61	2SC160B
61	2SC160C
61	2SC160D
61	2SC160E
61	2SC160F
61	2SC160G
61	2SC160GN
61	2SC160H
61	2SC160J
61	2SC160K
61	2SC160L
61	2SC160M
61	2SC160OR
61	2SC160R
61	2SC160X
61	2SC160Y
61	2SC1674
61	2SC1674K
61	2SC1674L
61	2SC1674M
61	2SC1686
61	2SC1686B
61	2SC17
61	2SC17A
61	2SC17B
61	2SC17C
61	2SC17D
61	2SC17E
61	2SC17F
61	2SC17G
61	2SC17GN
61	2SC17H
61	2SC17J
61	2SC17K
61	2SC17L
61	2SC170R
61	2SC17R
61	2SC17X
61	2SC17Y
61	2SC184
61	2SC184(R)
61	2SC184-OR
61	2SC184A
61	2SC184AP
61	2SC184B
61	2SC184BK
61	2SC184C
61	2SC184D
61	2SC184E
61	2SC184G
61	2SC184H
61	2SC184J
61	2SC184K
61	2SC184M
61	2SC184OR
61	2SC184R
61	2SC184X
61	2SC184Y
61	2SC1854
61	2SC1854C
61	2SC1854B
61	2SC1855
61	2SC1923
61	2SC1923A
61	2SC1923BN
61	2SC195
61	2SC195A
61	2SC195B
61	2SC195C
61	2SC195D
61	2SC195E
61	2SC195F
61	2SC195G
61	2SC195GN
61	2SC195J
61	2SC195K
61	2SC195L
61	2SC195M
61	2SC1950R
61	2SC195X
61	2SC195Y
61	2SC196
61	2SC196A
61	2SC196B
61	2SC196C
61	2SC196D
61	2SC196F
61	2SC196G
61	2SC196GN
61	2SC196H
61	2SC196J
61	2SC196K
61	2SC196L
61	2SC196M
61	2SC196OR
61	2SC196R
61	2SC196X
61	2SC196Y
61	2SC197
61	2SC197A
61	2SC197B
61	2SC197C
61	2SC197D
61	2SC197E
61	2SC197F
61	2SC197G
61	2SC197GN
61	2SC197H
61	2SC197J
61	2SC197K
61	2SC197L
61	2SC197M
61	2SC197OR
61	2SC197R
61	2SC197X
61	2SC197Y
61	2SC324
61	2SC324A
61	2SC324B
61	2SC324C
61	2SC324D
61	2SC324E
61	2SC324F
61	2SC324G
61	2SC324GN
61	2SC324H
61	2SC324HA
61	2SC324J
61	2SC324K
61	2SC324L
61	2SC324M
61	2SC324OR
61	2SC324R
61	2SC324X
61	2SC324Y
61	2SC361
61	2SC361A
61	2SC361B
61	2SC361C
61	2SC361D
61	2SC361E
61	2SC361F
61	2SC361G
61	2SC361GN
61	2SC361H
61	2SC361J
61	2SC361K
61	2SC361L
61	2SC361M
61	2SC361OR
61	2SC361R
61	2SC361X
61	2SC361Y
61	2SC370
61	2SC370-O
61	2SC370-O
61	2SC370-Q
61	2SC370-T
61	2SC370A
61	2SC370B
61	2SC370C
61	2SC370D
61	2SC370F
61	2SC370G
61	2SC370GN
61	2SC370H
61	2SC370J
61	2SC370K
61	2SC370L
61	2SC370OR
61	2SC370R
61	2SC370X
61	2SC370Y
61	2SC371
61	2SC371(O)
61	2SC371(Q)
61	2SC371-O
61	2SC371-OR
61	2SC371-ORG-Q
61	2SC371-R
61	2SC371-R-1
61	2SC371-RED-Q
61	2SC371-T
61	2SC371A
61	2SC371B
61	2SC371C
61	2SC371D
61	2SC371E
61	2SC371F
61	2SC371G
61	2SC371G-O
61	2SC371G-R
61	2SC371GN
61	2SC371H
61	2SC371J
61	2SC371K
61	2SC371L
61	2SC371M
61	2SC371O
61	2SC371OR
61	2SC371R
61	2SC371R-1
61	2SC371RED-Q
61	2SC371T
61	2SC371X
61	2SC371Y
61	2SC372
61	2SC372(3RD-IF)
61	2SC372(H)
61	2SC372(O)
61	2SC372(Y)
61	2SC372-O
61	2SC372-1
61	2SC372-2
61	2SC372-O
61	2SC372-OR
61	2SC372-ORG
61	2SC372-ORG-Q
61	2SC372-R
61	2SC372-Y
61	2SC372-YEL-Q
61	2SC372-Z
61	2SC372/4545Q
61	2SC372Q
61	2SC3720A
61	2SC3720B
61	2SC3720C
61	2SC3720D
61	2SC3720P
61	2SC3720G
61	2SC3720GN
61	2SC3720H
61	2SC3720J
61	2SC3720K
61	2SC3720L
61	2SC37200R
61	2SC3720R
61	2SC3720X
61	2SC3720Y
61	2SC3724
61	2SC372A
61	2SC372AR
61	2SC372BL
61	2SC372C
61	2SC372D
61	2SC372E
61	2SC372F
61	2SC372G
61	2SC372G-Q
61	2SC372G-Y
61	2SC372GN
61	2SC372GR
61	2SC372H
61	2SC372J
61	2SC372K
61	2SC372L
61	2SC372O
61	2SC3720R
61	2SC372R
61	2SC372X
61	2SC372Y
61	2SC372YEL
61	2SC372YEL-Q
61	2SC372Z
61	2SC378
61	2SC378-ORG
61	2SC378-RED
61	2SC378-YEL
61	2SC378A
61	2SC378B
61	2SC378C
61	2SC378D
61	2SC378F
61	2SC378G
61	2SC378GN
61	2SC378H
61	2SC378J
61	2SC378K
61	2SC378L
61	2SC378M
61	2SC3780R
61	2SC378R
61	2SC378X
61	2SC378Y
61	2SC380
61	2SC380(R)
61	2SC380-O
61	2SC380-O/4545Q
61	2SC380-BRN
61	2SC380-O
61	2SC380-O/4545Q
61	2SC380-OR
61	2SC380-ORG
61	2SC380-R
61	2SC380-RED
61	2SC380-Y
61	2SC380-YEL
61	2SC380/4545Q
61	2SC3800
61	2SC3800A
61	2SC3800B
61	2SC3800C
61	2SC3800D
61	2SC3800E
61	2SC3800F
61	2SC3800G
61	2SC3800GN
61	2SC3800H
61	2SC3800J
61	2SC3800K
61	2SC3800L
61	2SC3800M
61	2SC38000R
61	2SC3800R
61	2SC3800X
61	2SC3800Y
61	2SC380A
61	2SC380A(D)
61	2SC380A(O)
61	2SC380A(R)
61	2SC380A(Y)
61	2SC380A+O
61	2SC380A-O
61	2SC380A-O(TV)
61	2SC380A-O
61	2SC380A-O(TV)
61	2SC380A-R
61	2SC380A-R(TV)
61	2SC380A-Y
61	2SC380AO
61	2SC380AR
61	2SC380ATV
61	2SC380AY
61	2SC380B
61	2SC380B-Y
61	2SC380BY
61	2SC380C
61	2SC380C-Y
61	2SC380CY
61	2SC380D
61	2SC380D-Y
61	2SC380DY
61	2SC380E
61	2SC380E-Y
61	2SC380EY
61	2SC380F
61	2SC380P-Y
61	2SC380PY
61	2SC380Q

GE	Industry Standard No.	GE	Industry Standard No.	GE	Industry Standard No.	GE	Industry Standard No.	GE	Industry Standard No.
61	2SC380GN	61	2SC394GR	61	2SC430J	61	2SC469X	61	2SC563OR
61	2SC380H	61	2SC394GRN	61	2SC430K	61	2SC469Y	61	2SC563R
61	2SC380J	61	2SC394H	61	2SC430L	61	2SC472	61	2SC563X
61	2SC380K	61	2SC394J	61	2SC430M	61	2SC472A	61	2SC563Y
61	2SC380L	61	2SC394K	61	2SC430OR	61	2SC472B	61	2SC605Q
61	2SC380M	61	2SC394L	61	2SC430R	61	2SC472C	61	2SC629
61	2SC380O	61	2SC394M	61	2SC430X	61	2SC472D	61	2SC629-31
61	2SC380OR	61	2SC394O	61	2SC430Y	61	2SC472E	61	2SC629-41
61	2SC380R	61	2SC394O	61	2SC455X	61	2SC472F	61	2SC629A
61	2SC380R/4454C	61	2SC394R	61	2SC459	61	2SC472G	61	2SC629B
61	2SC380RED	61	2SC394RED	61	2SC459A	61	2SC472GN	61	2SC629C
61	2SC380X	61	2SC394W	61	2SC459B	61	2SC472H	61	2SC629D
61	2SC380Y	61	2SC394X	61	2SC459C	61	2SC472J	61	2SC629E
61	2SC380YEL	61	2SC394Y	61	2SC459D	61	2SC472K	61	2SC629F
61	2SC381	61	2SC394YEL	61	2SC459E	61	2SC472L	61	2SC629G
61	2SC381(BN)	61	2SC398	61	2SC459F	61	2SC472M	61	2SC629GN
61	2SC381-0	61	2SC398(PA-1)	61	2SC459G	61	2SC472OR	61	2SC629H
61	2SC381-BN	61	2SC398B	61	2SC459GN	61	2SC472R	61	2SC629J
61	2SC381-BRN	61	2SC398C	61	2SC459H	61	2SC472X	61	2SC629K
61	2SC381-0	61	2SC398D	61	2SC459J	61	2SC472Y	61	2SC629L
61	2SC381-OR	61	2SC398E	61	2SC459K	61	2SC529A	61	2SC629M
61	2SC381-ORG	61	2SC398F	61	2SC459L	61	2SC529B	61	2SC629OR
61	2SC381-R	61	2SC398PA1	61	2SC459M	61	2SC529C	61	2SC629R
61	2SC381-RED	61	2SC398G	61	2SC459OR	61	2SC529D	61	2SC629X
61	2SC381A	61	2SC398GN	61	2SC459R	61	2SC529E	61	2SC629Y
61	2SC381B	61	2SC398H	61	2SC459X	61	2SC529F	61	2SC645
61	2SC381BN	61	2SC398K	61	2SC459Y	61	2SC529G	61	2SC645-OR
61	2SC381BN-1	61	2SC398L	61	2SC460(A)	61	2SC529GN	61	2SC645A
61	2SC381BRN	61	2SC398M	61	2SC460(B)	61	2SC529H	61	2SC645B
61	2SC381C	61	2SC398OR	61	2SC460-5	61	2SC529J	61	2SC645B-1
61	2SC381D	61	2SC398R	61	2SC460-B	61	2SC529K	61	2SC645C
61	2SC381E	61	2SC398X	61	2SC460-C	61	2SC529L	61	2SC645D
61	2SC381F	61	2SC398Y	61	2SC460-OR	61	2SC529M	61	2SC645E
61	2SC381G	61	2SC399	61	2SC460D	61	2SC529OR	61	2SC645F
61	2SC381GN	61	2SC399A	61	2SC460E	61	2SC529R	61	2SC645G
61	2SC381H	61	2SC399B	61	2SC460F	61	2SC529X	61	2SC645GN
61	2SC381J	61	2SC399C	61	2SC460G	61	2SC529Y	61	2SC645GR
61	2SC381K	61	2SC399D	61	2SC460GB	61	2SC543	61	2SC645H
61	2SC381L	61	2SC399E	61	2SC460GN	61	2SC543OR	61	2SC645J
61	2SC381M	61	2SC399F	61	2SC460H	61	2SC543A	61	2SC645K
61	2SC381O	61	2SC399G	61	2SC460J	61	2SC543B	61	2SC645L
61	2SC381OR	61	2SC399GN	61	2SC460K	61	2SC543C	61	2SC645M
61	2SC381R	61	2SC399H	61	2SC460M	61	2SC543D	61	2SC645OR
61	2SC381RED	61	2SC399J	61	2SC460OR	61	2SC543E	61	2SC645R
61	2SC381RL	61	2SC399K	61	2SC460R	61	2SC543F	61	2SC645V
61	2SC381X	61	2SC399L	61	2SC460X	61	2SC543G	61	2SC645X
61	2SC381Y	61	2SC399M	61	2SC460Y	61	2SC543GN	61	2SC645Y
61	2SC383GN	61	2SC399R	61	2SC461(8P)	61	2SC543H	61	2SC657A
61	2SC383K	61	2SC399OR	61	2SC461-8P	61	2SC543J	61	2SC657B
61	2SC384	61	2SC399R	61	2SC461-A	61	2SC543K	61	2SC657C
61	2SC384(O)	61	2SC399X	61	2SC461-B	61	2SC543L	61	2SC657D
61	2SC384(Y)	61	2SC399Y	61	2SC461A	61	2SC543M	61	2SC657E
61	2SC384-O	61	2SC401	61	2SC461AL	61	2SC543OR	61	2SC657F
61	2SC384-O	61	2SC401A	61	2SC461B	61	2SC543R	61	2SC657G
61	2SC384A	61	2SC401B	61	2SC461BF	61	2SC543S	61	2SC657GN
61	2SC384B	61	2SC401C	61	2SC461BK	61	2SC543X	61	2SC657H
61	2SC384C	61	2SC401D	61	2SC461BL	61	2SC543Y	61	2SC657K
61	2SC384D	61	2SC401E	61	2SC461C	61	2SC544	61	2SC657L
61	2SC384E	61	2SC401F	61	2SC461E	61	2SC544A	61	2SC657M
61	2SC384P	61	2SC401G	61	2SC461EP	61	2SC544AG	61	2SC657OR
61	2SC384G	61	2SC401GN	61	2SC461L	61	2SC544B	61	2SC657R
61	2SC384GN	61	2SC401H	61	2SC464	61	2SC544C	61	2SC657X
61	2SC384H	61	2SC401J	61	2SC464A	61	2SC544D	61	2SC657Y
61	2SC384J	61	2SC401K	61	2SC464B	61	2SC544D(VHP)	61	2SC662A
61	2SC384K	61	2SC401L	61	2SC464C	61	2SC544E	61	2SC682
61	2SC384L	61	2SC401M	61	2SC464D	61	2SC544F	61	2SC682(B)
61	2SC384OR	61	2SC401OR	61	2SC464E	61	2SC544G	61	2SC682-OR
61	2SC384R	61	2SC401R	61	2SC464F	61	2SC544GN	61	2SC682A
61	2SC384X	61	2SC401X	61	2SC464G	61	2SC544H	61	2SC682B
61	2SC384Y	61	2SC401Y	61	2SC464GN	61	2SC544K	61	2SC682C
61	2SC388	61	2SC4033A	61	2SC464H	61	2SC544L	61	2SC682D
61	2SC388-OR	61	2SC4033B	61	2SC464J	61	2SC544M	61	2SC682F
61	2SC388A	61	2SC4033C	61	2SC464K	61	2SC544OR	61	2SC682G
61	2SC388A(3RD-IP)	61	2SC4033D	61	2SC464L	61	2SC544X	61	2SC682GN
61	2SC388ATV	61	2SC4033E	61	2SC464M	61	2SC544Y	61	2SC682H
61	2SC388B	61	2SC4033F	61	2SC464OR	61	2SC55	61	2SC682J
61	2SC388C	61	2SC4033G	61	2SC464R	61	2SC556	61	2SC682K
61	2SC388D	61	2SC4033GN	61	2SC464X	61	2SC55A	61	2SC682L
61	2SC388E	61	2SC4033H	61	2SC464Y	61	2SC55B	61	2SC682M
61	2SC388F	61	2SC4033J	61	2SC465	61	2SC55C	61	2SC682OR
61	2SC388G	61	2SC4033K	61	2SC465A	61	2SC55D	61	2SC682R
61	2SC388GN	61	2SC4033L	61	2SC465B	61	2SC55E	61	2SC682X
61	2SC388H	61	2SC4033OR	61	2SC465C	61	2SC55F	61	2SC682Y
61	2SC388J	61	2SC4033R	61	2SC465D	61	2SC55G	61	2SC683
61	2SC388K	61	2SC4033X	61	2SC465E	61	2SC55GN	61	2SC683(B)
61	2SC388L	61	2SC4033Y	61	2SC465F	61	2SC55H	61	2SC683-OR
61	2SC388M	61	2SC404A	61	2SC465G	61	2SC55J	61	2SC683A
61	2SC388OR	61	2SC404B	61	2SC465GN	61	2SC55K	61	2SC683B
61	2SC388R	61	2SC404C	61	2SC465H	61	2SC55L	61	2SC683C
61	2SC388X	61	2SC404D	61	2SC465J	61	2SC55M	61	2SC683D
61	2SC388Y	61	2SC404E	61	2SC465K	61	2SC55OR	61	2SC683E
61	2SC389-0	61	2SC404F	61	2SC465L	61	2SC55R	61	2SC683F
61	2SC389-OR	61	2SC404G	61	2SC465M	61	2SC55Y	61	2SC683G
61	2SC394	61	2SC404GN	61	2SC465OR	61	2SC562	61	2SC683GN
61	2SC394(O)	61	2SC404H	61	2SC465R	61	2SC562-O	61	2SC683H
61	2SC394(O)	61	2SC404J	61	2SC465X	61	2SC562-OR	61	2SC683J
61	2SC394-O	61	2SC404K	61	2SC465Y	61	2SC562A	61	2SC683K
61	2SC394-GR	61	2SC404L	61	2SC466	61	2SC562B	61	2SC683L
61	2SC394-GRN	61	2SC404M	61	2SC466A	61	2SC562C	61	2SC683M
61	2SC394-O	61	2SC404OR	61	2SC466C	61	2SC562D	61	2SC683OR
61	2SC394-OR	61	2SC404R	61	2SC466D	61	2SC562E	61	2SC683R
61	2SC394-ORG	61	2SC404X	61	2SC466E	61	2SC562F	61	2SC683S
61	2SC394-R	61	2SC404Y	61	2SC466F	61	2SC562G	61	2SC683V
61	2SC394-RED	61	2SC429	61	2SC466G	61	2SC562H	61	2SC683X
61	2SC394-Y	61	2SC429A	61	2SC466GN	61	2SC562J	61	2SC683Y
61	2SC394-YEL	61	2SC429B	61	2SC466H	61	2SC562K	61	2SC688
61	2SC3940A	61	2SC429C	61	2SC466J	61	2SC562L	61	2SC688A
61	2SC3940B	61	2SC429D	61	2SC466K	61	2SC562M	61	2SC688B
61	2SC3940C	61	2SC429E	61	2SC466L	61	2SC562OR	61	2SC688C
61	2SC3940D	61	2SC429F	61	2SC466M	61	2SC562R	61	2SC688D
61	2SC3940E	61	2SC429G	61	2SC466OR	61	2SC562Y	61	2SC688E
61	2SC3940F	61	2SC429GN	61	2SC466R	61	2SC563	61	2SC688F
61	2SC3940G	61	2SC429H	61	2SC466X	61	2SC563(3RDIP)	61	2SC688G
61	2SC3940GN	61	2SC429J	61	2SC466Y	61	2SC563-F	61	2SC688GN
61	2SC3940H	61	2SC429K	61	2SC469	61	2SC563-G	61	2SC688H
61	2SC3940J	61	2SC429L	61	2SC469-OR	61	2SC563-OR	61	2SC688J
61	2SC3940K	61	2SC429M	61	2SC469A	61	2SC563A	61	2SC688K
61	2SC3940L	61	2SC429OR	61	2SC469B	61	2SC563A(3RDIP)	61	2SC688L
61	2SC3940M	61	2SC429R	61	2SC469C	61	2SC563B	61	2SC688M
61	2SC3940OR	61	2SC429X	61	2SC469D	61	2SC563C	61	2SC688OR
61	2SC3940R	61	2SC429Y	61	2SC469E	61	2SC563D	61	2SC688X
61	2SC3940X	61	2SC430	61	2SC469G	61	2SC563E	61	2SC688Y
61	2SC3940Y	61	2SC430A	61	2SC469GN	61	2SC563F	61	2SC722
61	2SC394A	61	2SC430B	61	2SC469H	61	2SC563G	61	2SC739
61	2SC394AP	61	2SC430C	61	2SC469J	61	2SC563GN	61	2SC739A
61	2SC394B	61	2SC430D	61	2SC469K	61	2SC563H	61	2SC739B
61	2SC394C	61	2SC430E	61	2SC469L	61	2SC563J	61	2SC739C
61	2SC394D	61	2SC430F	61	2SC469M	61	2SC563K	61	2SC739D
61	2SC394E	61	2SC430G	61	2SC469OR	61	2SC563L	61	2SC739E
61	2SC394F	61	2SC430GN	61	2SC469Q	61	2SC563M	61	2SC739F
61	2SC394GN	61	2SC430H	61	2SC469R	61		61	

GE	Industry Standard No.	GE	Industry Standard No.	GE	Industry Standard No.	GE	Industry Standard No.	GE	Industry Standard No.
61	2SC759G	62	C1204C	62	C693EB	62	C945R	62	M9338
61	2SC759GN	62	C1204D	62	C693ET	62	C945S	62	M9384
61	2SC759H	62	C1215	62	C693F	62	C945T	62	M9409
61	2SC759K	62	C1222	62	C693FC	62	C945TQ	62	M9416
61	2SC759L	62	C1222A	62	C693FL	62	C945TU	62	M9447
61	2SC759M	62	C1222B	62	C693FU	62	C945X	62	M9474
61	2SC759P	62	C1222C	62	C693G	62	CC82004B	62	M9486
61	2SC759R	62	C1222D	62	C693G(JAPAN)	62	CC82004D303	62	M9547
61	2SC759Y	62	C1312	62	C693GL	62	CD441	62	M9594
61	2SD74	62	C1312F	62	C693GS	62	CD562	62	MP8-A05
61	2SD74-O	62	C1312G	62	C693GU	62	CGE-62	62	MP8-A09
61	2SD74-R	62	C1312H	62	C693GZ	62	C89104	62	MP83710
62	A-1853-0404-1	62	C1312Y	62	C693H	62	CXL199	62	MP86572
62	A-1854-0023-1	62	C1312YF	62	C694	62	D057	62	MP89433
62	A-1854-0088-1	62	C1312YG	62	C694E	62	D232	62	MP89433J
62	A-1854-0284-1	62	C1312YH	62	C694F	62	D308	62	MP89434J
62	A066-143	62	C1313	62	C694G	62	D32P1	62	MP89434K
62	A110	62	C1313F	62	C694Z	62	D32P2	62	MP89630J
62	A110(JAPAN)	62	C1313G	62	C711	62	D32P3	62	MP89630K
62	A1238	62	C1313H	62	C711(E)	62	D32P4	62	MP89633C
62	A138	62	C1313Y	62	C711A	62	DBCZ037300	62	MP89633D
62	A139	62	C1313YF	62	C711AE	62	DBCZ094504	62	MP89634
62	A1460	62	C1313YG	62	C711D	62	DBCZ373000	62	MP89634C
62	A2370773	62	C1313YH	62	C711E	62	DDBY224001	62	MPSA16
62	A280538R	62	C1327	62	C711F	62	DDBY299001	62	MPSA17
62	A3J	62	C1327FS	62	C711FG	62	DDBY4233001	62	MPSD06
62	A3K	62	C1327U	62	C711G	62	D813	62	NB013
62	A4B	62	C1328	62	C715	62	EA15X152	62	NPSA20
62	A4F	62	C1328T	62	C732	62	EA15X161	62	NR-421A8
62	A5T3707	62	C1328U	62	C732BL	62	EA15X245	62	P1901-48
62	A5T3708	62	C1335	62	C732GR	62	EA15X258	62	P69941
62	A5T3709	62	C1335A	62	C732B	62	EA15X259	62	PA8260
62	A5T3710	62	C1335B	62	C732V	62	EA15X264	62	PA8543
62	A5T3711	62	C1335C	62	C732I	62	EA15X288	62	PA9004
62	A642L	62	C1335D	62	C733	62	EA15X352	62	PA9005
62	A642B(NPN)	62	C1335E	62	C733-0	62	EA15X353	62	PBC107
62	A644L	62	C1335F	62	C733-O	62	EA15X354	62	PBC107A
62	A644S	62	C1344	62	C733BL	62	EA15X437	62	PBC107B
62	A645L	62	C1344C	62	C733GR	62	EA15XT245	62	PBC108
62	A645S	62	C1344D	62	C733H	62	EA15X7583	62	PBC108A
62	A667-GRN	62	C1344E	62	C733V	62	EA2429	62	PBC108B
62	A667-ORG	62	C1344F	62	C733Y	62	EA2738	62	PBC108C
62	A667-RED	62	C1345	62	C828	62	EA2771	62	PBC109
62	A667-YEL	62	C1345C	62	C828-0	62	EA3211	62	PBC109B
62	A668-GRN	62	C1345P	62	C828-OP	62	EA3763	62	PBC109C
62	A668-ORG	62	C1416	62	C828A	62	EA4025	62	PBT4003
62	A668-YEL	62	C1416BL	62	C828AP	62	ECG199	62	PIT-37
62	A669-GRN	62	C1537	62	C828AQ	62	ED1702N	62	PS6010-1
62	A669-YEL	62	C1537-3	62	C828AR	62	EDC-TR-11-1	62	PTC121
62	A67-07-244	62	C1537B	62	C828B	62	EP15X3	62	Q-00469
62	A67-33-340	62	C1537B	62	C828F	62	EQ8-10	62	Q-00469A
62	A6754194H	62	C1538	62	C828FR	62	EQ8-11	62	Q-00469B
62	A747B	62	C1538A	62	C828H	62	EQ8-131	62	Q-00469C
62	A748C	62	C1538B	62	C828K	62	EQ8-78	62	Q-00569C
62	A749	62	C1538BA	62	C828LR	62	E810231	62	Q-09115C
62	A749C	62	C1635	62	C828LS	62	E815049	62	Q-09115C
62	A76228	62	C1648	62	C828N	62	E815052	62	Q-13115C
62	A8R	62	C1681BL	62	C828P	62	ET379262	62	Q5053D
62	A9-175	62	C1684	62	C828Q	62	ET379462	62	Q5053E
62	AR213(Y)	62	C1685Q	62	C828QRS	62	ET380834	62	Q5053F
62	AR213VIOLET	62	C355	62	C828R	62	ET398711	62	Q5053V
62	AR218(RO)	62	C368	62	C828R/494	62	ET398777	62	Q5121
62	B-1910	62	C368BL	62	C828S	62	ET517263	62	Q51210
62	BC109B	62	C368GR	62	C828T	62	ET517994	62	Q5121Q
62	BC127	62	C368V	62	C828W	62	EW8-78	62	Q5120R
62	BC128	62	C371-R-1	62	C828Y	62	F079	62	Q5180
62	BC155C	62	C373	62	C833BL	62	FSE4002	62	QRT105
62	BC156C	62	C373BL	62	C838	62	G05-012-G	62	QT-C0900XBA
62	BC170	62	C373G	62	C838(H)	62	G05-035-D,E	62	QT-C0900XBD
62	BC170A	62	C373GR	62	C838(J)	62	G05-035-E	62	QT-C0900XCA
62	BC170B	62	C373W	62	C838(K)	62	G05-035D	62	RB192
62	BC170C	62	C374	62	C838(M)	62	G05035D	62	RE4001
62	BC173	62	C374-BL	62	C838L	62	G212	62	RE4002
62	BC173A	62	C374-V	62	C858	62	GE-62	62	RE4010
62	BC173B	62	C374JA	62	C858E	62	GET2483	62	REN199
62	BC209/7825B	62	C400-GR	62	C858F	62	G05-035-D,E	62	RLB-17
62	BC20C	62	C536	62	C858FG	62	G05-035-D,E	62	RB-279U8
62	BC238C	62	C536A	62	C858G	62	G05-035-E	62	RT2309
62	BC239B	62	C536AG	62	C859	62	HC-00373	62	RT520B
62	BC239C	62	C536B	62	C859B	62	HC-00536	62	RT5435
62	BC282B	62	C536C	62	C859F	62	HC-00711	62	RT7559
62	BC382C	62	C536D	62	C859FG	62	HC-00732	62	RT8047
62	BC383B	62	C536DK	62	C859G	62	HC-00900	62	RT8337
62	BC383C	62	C536E	62	C859GK	62	HC-00923	62	RT8666
62	BC384C	62	C536ED	62	C900	62	HC-00929	62	RT8663
62	BC385B	62	C536EH	62	C900(L)	62	HC-00930	62	RV2070
62	BC386A	62	C536EJ	62	C900A	62	HC-01000	62	RV2248
62	BC386B	62	C536EN	62	C900B	62	HC-01335	62	RVTC81473
62	BC408B	62	C536ER	62	C900C	62	HC-537	62	S001465
62	BC408C	62	C536ET	62	C900D	62	HE-00930	62	S0023
62	BC413B	62	C536EZ	62	C900E	62	HEF726	62	S006793
62	BC413C	62	C536F	62	C900F	62	HEF736	62	S006927
62	BC414B	62	C536F1	62	C900L	62	HEF737	62	S007764
62	BC414C	62	C536P2	62	C900M	62	HP17	62	S2038
62	BC520	62	C536PP	62	C900U	62	HF309301E	62	S24592
62	BC520B	62	C536PS	62	C907	62	HR-14	62	S29956
62	BC520C	62	C536FS6	62	C907A	62	HR-15	62	SE5-0127
62	BC521	62	C536FZ	62	C907AC	62	HR-47	62	SE5-0565
62	BC521C	62	C536G	62	C907AD	62	HR-75	62	SE5-0569
62	BC521D	62	C536GF	62	C907AH	62	HR47	62	SE5-0958
62	BC522	62	C536GK	62	C907C	62	HT306442A	62	SE5-0958-55
62	BC522C	62	C536GT	62	C907D	62	HT306442B	62	SE5-0958-56
62	BC522D	62	C536GV	62	C907H	62	HT307533100	62	SE5-0958-57
62	BC522E	62	C536GY	62	C907HA	62	HT308281D	62	SKA-4802
62	BC523	62	C536H	62	C923	62	HT308281H	62	S019806
62	BC523B	62	C536W	62	C923A	62	HT308282B	62	S025094
62	BC523C	62	C644	62	C923B	62	HT309002A0	62	SFB-1476
62	BF340	62	C644C	62	C923C	62	HT309301C	62	SFB-1539(WT)
62	BF341	62	C644F	62	C923D	62	HT309301E	62	SP82216
62	BF342	62	C644F/494	62	C923F	62	HT309301F	62	SP82217
62	BF343	62	C644FH	62	C930CK	62	HT310002A	62	SP82271
62	BF597	62	C644FS	62	C930DH	62	HT313271T	62	SP84272
62	BFR25	62	C644H	62	C930DZ	62	HT313272H	62	SP84814
62	BPT55	62	C644HR	62	C945	62	HT356441B	62	ST-MP89433
62	BFV69	62	C644P	62	C945(R)	62	HX-50107	62	ST.082.114.016
62	BFY89A	62	C644PU	62	C945-0	62	I220-0034	62	STZ0005
62	BFY39-1	62	C644Q	62	C945A	62	J24812	62	ST33026
62	BFY39-2	62	C644R	62	C945AQ	62	J24875	62	ST7100
62	BFY39-3	62	C644RST	62	C945B	62	J24932	62	T-Q5093
62	BTX-068	62	C644S	62	C945C	62	JA1350	62	TO1-047
62	C1000	62	C644S/494	62	C945D	62	JA1350B	62	T100T(ZENITH)
62	C1000-BL	62	C644T	62	C945E	62	JA1350W	62	T1341A3K
62	C1000-GR	62	C648	62	C945F	62	LDA410	62	TE2711
62	C1000Y	62	C648H	62	C945G	62	LDS207	62	TE2712
62	C1006	62	C650	62	C945H	62	LM1117	62	TE2921
62	C1006A	62	C644H	62	C945K	62	LM1117C	62	TE2922
62	C1006B	62	C693	62	C945L	62	LM540C	62	TE2923
62	C1006C	62	C693(JAPAN)	62	C945M	62	LMT540C	62	TE2924
62	C1010	62	C693A	62	C945O	62	L8-0031-AR-218	62	TE2925
62	C1010A	62	C693B	62	C945P	62	L8-0095-AR-213	62	TE2926
62	C1010B	62	C693C	62	C945Q	62	M9197	62	TE3390
62	C1010C	62	C693D	62	C945QL	62	M9269	62	TE3391
62	C1204	62	C693E	62	C945QP	62	M9293	62	TE3391A
62	C1204B	62	C693E(JAPAN)			62	M9329	62	TE3392

GE	Industry Standard No.	GE	Industry Standard No.	GE	Industry Standard No.	GE	Industry Standard No.	GE	Industry Standard No.
62	TE3393	62	2SC1312H	62	2SC644D	62	2SC732GR/4454C	63	BSY62B
62	TE3394	62	2SC1312J	62	2SC644E	62	2SC732GRB	63	COC13000-1C
62	TE3395	62	2SC1312K	62	2SC644F	62	2SC732GRN	63	C1330
62	TE3396	62	2SC1312L	62	2SC644F(H)(S)	62	2SC732H	63	C1330A
62	TE3397	62	2SC1312M	62	2SC644FH	62	2SC732J	63	C1330C
62	TE3398	62	2SC1312OR	62	2SC644FHS	62	2SC732L	63	C1330D
62	TE3843	62	2SC1312R	62	2SC644FR	62	2SC732M	63	C1330L
62	TE3844	62	2SC1312X	62	2SC644FS	62	2SC732OR	63	C1330R
62	TE3845	62	2SC1312Y	62	2SC644G	62	2SC732R	63	C1346
62	TE3854	62	2SC1312YF	62	2SC644GN	62	2SC732S	63	C1346R
62	TE3854A	62	2SC1312YG	62	2SC644H	62	2SC732V	63	C1346B
62	TE3855	62	2SC1312YH	62	2SC644H(S)	62	2SC732V10	63	C1347
62	TE3855A	62	2SC1313	62	2SC644HR	62	2SC732VIO	63	C1347Q
62	TE3859	62	2SC1313B	62	2SC644HS	62	2SC732X	63	C1347R
62	TE3860	62	2SC1313F	62	2SC644J	62	2SC732Y	63	C1347B
62	TE3900	62	2SC1313G	62	2SC644K	62	2SC733	63	C1383R/494
62	TE3900A	62	2SC1313H	62	2SC644L	62	2SC733(GR)	63	C1407
62	TE3901	62	2SC1313Y	62	2SC644M	62	2SC733-0	63	C1407P
62	TE4256	62	2SC1313YE	62	2SC644OR	62	2SC733-B	63	C1407R
62	TE5088	62	2SC1313YG	62	2SC644P	62	2SC733-BL	63	C1407B
62	TE5089	62	2SC1313YH	62	2SC644PJ	62	2SC733-BLU	63	C261
62	TE5249	62	2SC1328	62	2SC644Q	62	2SC733-G	63	C744
62	TE5309	62	2SC1328(U)	62	2SC644R	62	2SC733-GR	63	C814
62	TE5310	62	2SC1328(U)(T)	62	2SC644-0	62	2SC733-GRN	63	C853
62	TE5311	62	2SC1328U	62	2SC644RST	62	2SC733-0	63	C853A
62	TI-415	62	2SC1328U	62	0002SC644S,R,Q	62	2SC733-OR	63	C853B
62	TI-416	62	2SC1344	62	2SC644T	62	2SC733-ORG	63	C853C
62	TI-418	62	2SC1344(E)	62	2SC644X	62	2SC733-Y	63	C853KLM
62	TI-419	62	2SC1344C	62	2SC644Y	62	2SC733-YEL	63	C853L
62	TI-421	62	2SC1344D	62	2SC648	62	2SC733A	63	C881
62	TI-92	62	2SC1344E	62	2SC648H	62	2SC733B	63	C881A
62	TI54D	62	2SC1344F	62	2SC64Y-RST	62	2SC733BL	63	C881B
62	TI8-94	62	2SC1360	62	2SC650	62	2SC733BLK	63	C881C
62	TI8-97	62	2SC1362	62	2SC650-OR	62	2SC733BLU	63	C881K
62	TNJ1034	62	2SC1393	62	2SC650-Y	62	2SC733C	63	C881L
62	TNJ70691	62	2SC1393K	62	2SC650A	62	2SC733D	63	C971
62	TNJ71034	62	2SC1393M	62	2SC650B	62	2SC733E	63	CDC-8000-1D
62	TNJ71271	62	2SC1399	62	2SC650C	62	2SC733ER	63	CDB000-1D
62	TNJ71277	62	2SC1399E	62	2SC650D	62	2SC733F	63	CDQ10044
62	TNJ71965	62	2SC1416	62	2SC650E	62	2SC733G	63	CGB-63
62	TP4067-410	62	2SC1416A	62	2SC650G	62	2SC733GN	63	CJ5206A
62	TP4067-411	62	2SC1416BL	62	2SC650GN	62	2SC733GRN	63	COC13000-1C
62	TR-1993	62	2SC1632	62	2SC650H	62	2SC733H	63	C8-1352
62	TR-8034	62	2SC1647	62	2SC650J	62	2SC733J	63	C8-0143
62	TR-8040	62	2SC1647Q	62	2SC650K	62	2SC733K	63	C81255M
62	TR-BC149C	62	2SC1647RY	62	2SC650L	62	2SC733L	63	C81256HQ
62	TR01014	62	2SC1648	62	2SC650M	62	2SC733M	63	C82023
62	TR01040	62	2SC1648E	62	2SC650OR	62	2SC733O	63	C99417
62	TR105(SPRAGUE)	62	2SC1648S	62	2SC650R	62	2SC733OR	63	CXLJ92
62	TR19A	62	2SC1648SH	62	2SC650X	62	2SC733Q	63	D1103F1
62	TR2327443	62	2SC1681	62	2SC650Y	62	2SC733R	63	D228
62	TR2327444	62	2SC1681-GR	62	2SC693	62	2SC733S	63	D261
62	TR4010	62	2SC1681BL	62	2SC693-OR	62	2SC733S-BL	63	D261A
62	TR01037	62	2SC1681GR	62	2SC693A	62	2SC733V	63	D261B
62	TRPLC711	62	2SC1681V	62	2SC693B	62	2SC733X	63	D261C
62	TSC-499	62	2SC1682	62	2SC693C	62	2SC733Y	63	D261D
62	TSC767	62	2SC1682V	62	2SC693D	62	2SC733YEL	63	D261E
62	01-030373	62	2SC1684	62	2SC693E	63	A188103	63	D261L
62	01-030711	62	2SC1684BL	62	2SC693EB	63	A372484	63	D261L
62	01-030732	62	2SC1684P	62	2SC693ET	63	A417034	63	D261P
62	01-030784	62	2SC1684Q	62	2SC693F	63	A543	63	D261R
62	01-030900	62	2SC1684R	62	2SC693FC	63	AR-203(R)	63	D261V
62	01-031685	62	2SC1684S	62	2SC693FL	63	AR203(R)	63	D261W
62	01-031815	62	2SC1684T	62	2SC693FP	63	AR205RED	63	D28207
62	1A0013	62	2SC1685	62	2SC693FU	63	AR207	63	D336
62	002-04-000	62	2SC1685-0	62	2SC693G	63	AR218ORANGE	63	D336R
62	002-9501	62	2SC1685P	62	2SC693GL	63	AR218RED	63	D336Y
62	002-9502	62	2SC1685Q	62	2SC693GN	63	AT7	63	D33D21J1
62	002-9502-12	62	2SC1685R	62	2SC693GS	63	BC190A	63	D33D22J1
62	2C8537FC	62	2SC1685S	62	2SC693GU	63	BC226	63	D33D24J1
62	2N4141	62	2SC1685T	62	2SC693GZ	63	BC302-4	63	D33D25J1
62	2N5249A	62	2SC1740	62	2SC693H	63	BC302-5	63	D33D27J1
62	2N5824	62	2SC1740L	62	2SC693J	63	BC302-6	63	D33D28J1
62	2N5825	62	2SC1740P	62	2SC693K	63	BC338-16	63	D33D29
62	2N5826	62	2SC1740Q	62	2SC693L	63	BC377	63	D33D29J1
62	2N5827	62	2SC1740QH	62	2SC693M	63	BC378	63	D33D30
62	2N5827A	62	2SC1740QJ	62	2SC693NP	63	BC508	63	D33D30J1
62	2N5828	62	2SC1740R	62	2SC693OR	63	BC509	63	EA15Z274
62	2N5828A	62	2SC1740RH	62	2SC693R	63	BC510	63	EA15X349
62	2N5852	62	2SC1740S	62	2SC693U	63	BC582	63	EA15X4531
62	2N6112	62	2SC1766	62	2SC693X	63	BC582A	63	EA15X7519
62	2SC-Q23	62	2SC1766C	62	2SC693Y	63	BC582B	63	EA3990
62	2SC-NJ-100	62	2SC1775	62	2SC711	63	BC583A	63	EC8192
62	2SC-NJ100	62	2SC1775E	62	2SC711(D)	63	BC583B	63	FD-1029-NS
62	2SC1000	62	2SC1775F	62	2SC711(E)	63	BC584	63	FS2042
62	2SC1000(GR)	62	2SC1815	62	2SC711(P)	63	BCW44	63	FX4960
62	2SC1000-BL	62	2SC1815-0	62	2SC711,A,F,G	63	BCW51	63	GE-63
62	2SC1000-GR	62	2SC1815GR	62	2SC711-OR	63	BCW73-16	63	HD-00261
62	2SC1000-Y	62	2SC1815Y	62	2SC711A	63	BCW74-16	63	HR-67
62	2SC1000A	62	2SC1815YW	62	2SC711A(E)	63	BCW77-16	63	HR38
62	2SC1000B	62	2SC1853	62	2SC711AE	63	BCW78-16	63	HR67
62	2SC1000BL	62	2SC2021	62	2SC711AF	63	BCW82	63	HR82
62	2SC1000C	62	2SC369-BL	62	2SC711AG	63	BCW82A	63	HR83
62	2SC1000D	62	2SC369-GR	62	2SC711AN	63	BCW82B	63	H85810
62	2SC1000E	62	2SC3721	62	2SC711B	63	BCW83A	63	H85812
62	2SC1000F	62	2SC3721GR	62	2SC711C	63	BCW83B	63	H85814
62	2SC1000G	62	2SC373W	62	2SC711D	63	BCW84	63	H85816
62	2SC1000G-BL	62	2SC403	62	2SC711E	63	BCW90K	63	H85818
62	2SC1000G-GR	62	2SC403(C)	62	2SC711FG	63	BCW91	63	H85820
62	2SC1000GN	62	2SC403(SONY)	62	2SC711G	63	BCW91A	63	H85822
62	2SC1000GR	62	2SC403-OR	62	2SC711GN	63	BCW91B	63	H86010
62	2SC1000H	62	2SC403A	62	2SC711H	63	BCW91K	63	H86012
62	2SC1000J	62	2SC403AL	62	2SC711J	63	BCW91KA	63	H86014
62	2SC1000K	62	2SC403B	62	2SC711L	63	BCW91KB	63	H86016
62	2SC1000L	62	2SC403B(SONY)	62	2SC711M	63	BFR16	63	H7314071Q
62	2SC1000M	62	2SC403C	62	2SC711OR	63	BFR40	63	H73909100
62	2SC1000R	62	2SC403C(SONY)	62	2SC711R	63	BFR40TO5	63	HT404001E
62	2SC1000R	62	2SC403CG	62	2SC711X	63	BF851	63	HX-50103
62	2SC1000X	62	2SC403CD	62	2SC711Y	63	BF829P	63	IRTH-53
62	2SC1000Y	62	2SC403E	62	2SC732	63	BF850	63	M9521
62	2SC1006	62	2SC403F	62	2SC732(BL)	63	BF850P	63	MH1501
62	2SC1006A	62	2SC403G	62	2SC732-B	63	BF839	63	MM2193A
62	2SC1006B	62	2SC403GN	62	2SC732-BL	63	BF859	63	MM2261
62	2SC1006C	62	2SC403H	62	2SC732-BLU	63	BF860	63	MM2270
62	2SC1189L	62	2SC403J	62	2SC732-G	63	BF250	63	MPS-A06
62	2SC1204	62	2SC403K	62	2SC732-GR	63	BF29	63	MPS94188
62	2SC1204B	62	2SC403L	62	2SC732-GRN	63	BFX53	63	MPS9418T
62	2SC1204C	62	2SC403M	62	2SC732-OR	63	BFX59F	63	M8101Q
62	2SC1204D	62	2SC403OR	62	2SC732-V	63	BFX92A	63	NS510
62	2SC1222	62	2SC403R	62	2SC732-V10	63	BFY16	63	NN7000
62	2SC1222A	62	2SC403X	62	2SC732-VIO	63	BLY27	63	NN7001
62	2SC1222B	62	2SC403Y	62	2SC732A	63	BLY28	63	NN7002
62	2SC1222C	62	2SC536	62	2SC732B	63	BLY93	63	NN7005
62	2SC1222D	62	2SC536FZ	62	2SC732BL	63	BS823	63	NN7004
62	2SC1222E	62	2SC644	62	2SC732BL-1	63	BSV69	63	NN7005
62	2SC1222U	62	2SC644(F)	62	2SC732BLU	63	BSX12A	63	NR-141E8
62	2SC1312	62	2SC644(H)	62	2SC732C	63	BSX45-10	63	NR-141ET
62	2SC1312A	62	2SC644(R)	62	2SC732D	63	BSX45-16	63	NS1555
62	2SC1312C	62	2SC644(R,S)	62	2SC732E	63	BSX45-6	63	NS1960
62	2SC1312D	62	2SC644(S)	62	2SC732F	63	BSX46-10	63	NS1960
62	2SC1312E	62	2SC644-OR	62	2SC732G	63	BSX46-16	63	NS950
62	2SC1312F	62	2SC644A	62	2SC732GN	63	BSX46-6		
62	2SC1312G	62	2SC644B	62	2SC732GR				
62	2SC1312GN	62	2SC644C						

GE	Industry Standard No.	GE	Industry Standard No.	GE	Industry Standard No.	GE	Industry Standard No.	GE	Industry Standard No.
63	PA9483	63	2SC1407P	65	A-1853-0050-1	65	BC322B	65	2SA564-R	66	A5A-1B
63	PBC184	63	2SC1407Q	65	A-1853-0066-1	65	BC322C	65	2SA564A	66	A68-23-560
63	PBT1001	63	2SC1407R	65	A-1853-0077-1	65	BC526A	65	2SA564A(P)	66	A9V
63	PT2540	63	2SC1407S	65	A-1853-0096-1	65	BC526B	65	2SA564A(R)	66	AR-17
63	PTC178	63	2SC1407X	65	A-1853-0098-1	65	BC526C	65	2SA564A(S)	66	AR-22(X8TR)
63	RR196	63	2SC1537	65	A-1853-0300-1	65	BCW58B	65	2SA564ABQ	66	AR17(GREY)
63	RE70	63	2SC1537(S)	65	A2311	65	BCW59B	65	2SA564ABQ-1	66	AR17GREY
63	REN192	63	2SC1537-0	65	A564	65	BCW61BC	65	2SA564AG	66	ATC-TR-19
63	REN70	63	2SC1537B	65	A564-0	65	BCW61BD	65	2SA564AK	66	B-1823
63	S-522	63	2SC1537S	65	A564A	65	BCW88	65	2SA564AL	66	B131
63	S2104	63	2SC1538	65	A564ABQ	65	BCX71BJ	65	2SA564AO	66	B143004
63	SB1012	63	2SC1538A	65	A564AP	65	BCX71BK	65	2SA564AP	66	B143011
63	SEC1078	63	2SC1538S	65	A564AQ	65	BCY78	65	2SA564AQ	66	B143012
63	SEC1079	63	2SC1538S(A)	65	A564AT	65	BCY79	65	2SA564AR	66	B143018
63	SEC1477	63	2SC1538SA	65	A564F	65	BF243	65	2SA564AS	66	B143019
63	SEC1479	63	2SC261	65	A564PQ	65	BF65	65	2SA564AT	66	B143026
63	SF.T440	63	2SC303	65	A564PR	65	BFW22	65	2SA564B	66	B143027
63	SF.T443	63	2SC304	65	A564J	65	BFX37	65	2SA564C	66	B1D
63	SF.T443A	63	2SC57	65	A564P	65	BTX-084	65	2SA564D	66	B3547
63	SF.T445	63	2SC699	65	A564POR	65	C532000585	65	2SA564E	66	B3548
63	SF.T714	63	2SC700	65	A564Q	65	C673	65	2SA564F	66	B3550
63	SKA4410	64	2N6059	65	A564QHD	65	C7	65	2SA564FQ	66	B3551
63	SP8402	64	2N6356	65	A564QR	65	CB	65	2SA564FQ-1	66	B3577
63	ST1504			65	A564R	65	CD500	65	2SA564PR	66	B3578
63	ST1505			65	A564S	65	D29P1	65	2SA564PR-1	66	B3580
63	ST402			65	A564T	65	D29P2	65	2SA564G	66	B3584
63	ST4341			65	A565Q	65	D29P3	65	2SA564GN	66	B3585
63	ST5061			65	A578	65	D29P4	65	2SA564H	66	B3586
63	STC1356			65	A578A	65	DDBY003003	65	2SA564J	66	B3588
63	TC8100			65	A578B	65	DW-7655	65	2SA564K	66	B3589
63	TC8102			65	A578C	65	DW-7655-LV00223	65	2SA564L	66	B5000
63	TE3417			65	A579	65	ECG234	65	2SA564M	66	B5002
63	TE3859A			65	A579B	65	ES112	65	2SA564OR	66	B5021
63	TE4425			65	A579C	65	ET350335	65	2SA564P	66	B5022
63	TI-485			65	A5T4058	65	ET2P-808	65	2SA564P.A	66	B5031
63	TN79			65	A5T4059	65	PI-1021	65	2SA564POR	66	B5032
63	TN81			65	A5T4060	65	FS1990	65	2SA564Q	66	B5E
63	TNJ71252			65	A5T4061	65	FX3964	65	2SA564QGD	66	BD106
63	TS2218			65	A5T4062	65	G8-65	65	2SA564QHD	66	BD106A
63	TS2219			65	A641	65	HA-00564	65	2SA564QP	66	BD106B
63	TV21			65	A641(JAPAN)	65	HA-00733	65	2SA564QR	66	BD107A
63	TV26			65	A641(PNP)	65	HEP-80006	65	2SA564R	66	BD107B
63	TV28			65	A641A	65	HEP80006	65	2SA564S	66	BD109
63	TV82SA543			65	A641B	65	HEP80031	65	2SA564T	66	BD109-6
63	1A34			65	A641C	65	HT10564IC	65	2SA564X	66	BD124
63	1A34(R)			65	A641D	65	HT105641D	65	2SA564XL	66	BD162
63	1A34B(R)			65	A641L	65	HT105641H	65	2SA564Y	66	BD163
63	1A34R			65	A641M	65	HT105642B	65	2SA578	66	BD220
63	1A38			65	A641B	65	HT107211T	65	2SA578A	66	BD221
63	1A38(R)			65	A666	65	HT35642B	65	2SA578B	66	BD222
63	1A38R			65	A666A	65	LDA454	65	2SA578C	66	BD231A
63	2N1206			65	A666HR	65	LDA455	65	2SA579	66	BD239
63	2N1253A			65	A666QRS	65	LD8257	65	2SA579A	66	BD239A
63	2N1444			65	A666R	65	M7	65	2SA579B	66	BD241
63	2N1714			65	A666S	65	M9412	65	2SA579C	66	BD243
63	2N1716			65	A721	65	M9467	65	2SA666	66	BD243B
63	2N1962/46			65	A721S	65	MI813674/47	65	2SA666QRS	66	BD271
63	2N1964/46			65	A721T	65	MPS9680H/E	65	2SA666A	66	BD433
63	2N1965/46			65	A721U	65	MPSA70-YEL	65	2SA666B	66	BD435
63	2N2239			65	A722	65	MPSB56	65	2SA666BL	66	BD477
63	2N2279/51			65	A722S	65	MPSK70	65	2SA666C	66	BD439
63	2N2368/51			65	A722T	65	MPSK71	65	2SA666D	66	BD533
63	2N2369/51			65	A722U	65	MPSK72	65	2SA666E	66	BD535
63	2N2403			65	A725	65	NR-621AU	65	2SA666HR	66	BDX74
63	2N2404			65	A725F	65	P/3E27LO10CO1	65	2SA666Q	66	BDX75
63	2N245			65	A725G	65	P12407-1	65	2SA666QRS	66	BDY12
63	2N246			65	A725H	65	PM1120	65	2SA666R	66	BDY13
63	2N2484A			65	A726	65	Q5102	65	2SA666S	66	BDY34
63	2N29260			65	A726F	65	Q5102P	65	2SA666Y	66	BLY21
63	2N2926R			65	A726G	65	Q5102Q	65	2SA721	66	BLY35
63	2N2J374			65	A726H	65	Q5102R	65	2SA721Q	66	BLY36
63	2N3052			65	AB25	65	QRT106	65	2SA721R	66	BLY63
63	2N3053/40053			65	BC153	65	RB193	65	2SA721S	66	BLY79
63	2N3252			65	BC154	65	REN234	65	2SA721U	66	BLY88
63	2N3295			65	BC158C	65	RV2353	65	2SA722	66	BLY89
63	2N3374			65	BC159C	65	S026094	65	2SA722S	66	BLY92
63	2N339			65	BC177	65	S1990	65	2SA722T	66	BR101A
63	2N339A			65	BC177A	65	SC158B	65	2SA722U	66	BRC5296
63	2N3403			65	BC177B	65	SC159B	65	2SA725	66	BUY24
63	2N3404			65	BC178	65	SC25BB	65	2SA725F	66	C1060
63	2N342			65	BC178A	65	SC259B	65	2SA725G	66	C1060A
63	2N343			65	BC178B	65	SE5-0909	65	2SA725H	66	C1060B
63	2N3435			65	BC178C	65	SS29A4	65	2SA726	66	C1060BM
63	2N3435B			65	BC179	65	SS29A5	65	2SA726F	66	C1060C
63	2N3633/52			65	BC179A	65	T309	65	2SA726G	66	C1060D
63	2N3981			65	BC179B	65	TROG020-2	65	2SA726H	66	C1061
63	2N3982			65	BC179C	65	TR106(SPRAGUE)	65	2SA726Y	66	C1061A
63	2N4425			65	BC204	65	TR28A763	65	2SA728	66	C1061B
63	2N5320HS			65	BC204A	65	01-010564	65	2SA741H	66	C1061C
63	2N545			65	BC204B	65	001-021170	65	2SA763	66	C1061D
63	2N5820HS			65	BC206	65	001-021171	65	2SA763-W	66	C1061T
63	2N5822			65	BC206A	65	001-021172	65	2SA763-WL-3	66	C1061T-B
63	2N5856			65	BC206B	65	001-021173	65	2SA763-WL-4	66	C1061TB
63	2N6014			65	BC206C	65	2SA494(Y)	65	2SA763-WL-5	66	C1173
63	2N7006-JAN			65	BC225	65	2SA494-GR	65	2SA763-WL-6	66	C1173-0
63	2N709/46			65	BC258C	65	2SA494-GR-1	65	2SA763-WN	66	C1173-QR
63	2N709/51			65	BC307C	65	2SA494-0	65	2SA763-WN-3	66	C1173-O
63	2N709A/51			65	BC308C	65	2SA494-OR	65	2SA763-WN-5	66	C1173-R
63	2N743/46			65	BC320	65	2SA494-Y	65	2SA763-WN-6	66	C1173-Y
63	2N743/51			65	BC320A	65	2SA494A	65	2SA763-Y	66	C1173G
63	2N744/46			65	BC320C	65	2SA494B	65	2SA763-YL	66	C1173R
63	2N744/51			65	BC321	65	2SA494C	65	2SA763-YL-3	66	C1173X
63	2N763/51			65	BC321A	65	2SA494D	65	2SA763-YL-5	66	C1173XO
63	2N784/51			65	BC321B	65	2SA494E	65	2SA763-YL-6	66	C1173Y
63	2N834/51			65	BC321C	65	2SA494F	65	2SA763-YN	66	C1398
63	2N835/51			65	BC322	65	2SA494G	65	2SA763-YN-3	66	C1398Q
63	2N907					65	2SA494GN	65	2SA763-YN-5	66	C1418
63	2N918/46					65	2SA494GR	65	2SA763-YN-6	66	C1418A
63	1N48					65	2SA494H	65	2SA786	66	C1418B
63	2SC1209					65	2SA494J	65	2SA787	66	C1418C
63	002SC1209C					65	2SA494K	65	2SA825	66	C1418D
63	2SC1209D					65	2SA494L	65	2SA825Q	66	C1419
63	2SC1209E					65	2SA494M	65	2SA825R	66	C1419A
63	2SC1330					65	2SA494O	65	2SA842-BL	66	C1419B
63	2SC1330A					65	2SA494OR	65	2SA842-QR	66	C1419D
63	2SC1330B					65	2SA494R	65	2SA854	66	C1450B
63	2SC1330C					65	2SA494X	65	2SA854Q	66	C154
63	2SC1330D					65	2SA494Y	65	2SB774	66	C154B
63	2SC1330L					65	2SA564	65	2SB774S	66	C154C
63	2SC1330R					65	2SA564(O)	65	A-1854-0420-1	66	C154H
63	2SC1346					65	2SA564(O)	66	A-1854-0464	66	C325
63	2SC1346(R)					65	2SA564(P)	66	A066-114	66	C325A
63	2SC1346Q					65	2SA564(Q)	66	A272	66	C325E
63	2SC1346R					65	2SA564(R)	66	A273	66	C36583
63	2SC1468					65	2SA564(S,T)	66	A276	66	C789
63	2SC1386H					65	2SA564(T)	66	A277	66	C789-0
63	2SC1406					65	2SA564-O	66	A28U2260P	66	C789-O
63	2SC1406(P)					65	2SA564-0-1	66	A417U32	66	C789-R
63	2SC1406Q					65	2SA564-OGD	66	A54-3	66	C789-Y
63	2SC1407					65	2SA564-OR			66	C952
63	2SC1407(Q)					65	2SA564-P			66	C952E
63	2SC1407U					65	2SA564-Q			66	CGE-66
63	2SC1407B									66	C1173Y

GE	Industry Standard No.	GE	Industry Standard No.	GE	Industry Standard No.	GE	Industry Standard No.	GE	Industry Standard No.
66	CXL152	66	HR69	66	1A0058	66	2SC1419	67	A5T4026
66	D141	66	HT308301 BO	66	1A0059	66	2SC1419A	67	A5T4028
66	D141H01	66	HT311621B	66	002-012400	66	2SC1419B	67	A643
66	D141H9Z	66	HT402352B	66	002D235RY	66	2SC1419C	67	A643A
66	D154	66	IP20-0007	66	2N1047	66	2SC1419D	67	A643B
66	D184	66	IP20-0036	66	2N1049	66	2SC1450	67	A643C
66	D234	66	IP20-0083	66	2N1483	66	2SC1450S	67	A643D
66	D234-0	66	IP20-0323	66	2N1484	66	2SC1450OS	67	A643E
66	D234-0	66	J241241	66	2N1485	66	2SC2317	67	A643F
66	D234-R	66	LS-0066	66	2N1486	66	2SC234	67	A643L
66	D234-Y	66	M75543-1	66	2N1701	66	2SC234A	67	A643R
66	D235	66	M9576	66	2N1709	66	2SC234B	67	A643S
66	D235-0	66	M9661	66	2N1718	66	2SC234C	67	A643V
66	D235-0	66	MHT5906	66	2N1720	66	2SC234D	67	A643W
66	D235-R	66	MJE201	66	2N1768	66	2SC234E	67	A707
66	D235-Y	66	MJE202	66	2N1769	66	2SC234F	67	A707V
66	D235D	66	MM1619	66	2N1886	66	2SC234G	67	A751
66	D235G	66	MPB111	66	2N2033	66	2SC234H	67	A751Q
66	D235GR	66	MPB112	66	2N2034	66	2SC234J	67	A751QR
66	D235R	66	MSA7505	66	2N2035	66	2SC234K	67	A751R
66	D235Y	66	MSA8505	66	2N2036	66	2SC234L	67	A751S
66	D27C1	66	P/PTV/117	66	2N2339	66	2SC234M	67	A752
66	D27C2	66	P4J148	66	2N2652	66	2SC234OR	67	A752P
66	D27C3	66	PLE-48	66	2N2657	66	2SC234R	67	A752Q
66	D27C4	66	PN66	66	2N2828	66	2SC234X	67	A752R
66	D288	66	PP3250	66	2N2829	66	2SC234Y	67	A752S
66	D288A	66	PP3310	66	2N2877	66	2SC2350	67	A8T4026
66	D288B	66	PP3312	66	2N2878	66	2SC325	67	A8T4028
66	D288C	66	PT2635	66	2N2947	66	2SC325A	67	A8304
66	D288L	66	PT5693	66	2N3138	66	2SC325C	67	AR304GREEN
66	D289	66	PT665	66	2N3140	66	2SC325E	67	AR304RED
66	D289A	66	Q-11115C	66	2N3142	66	2SC4116	67	AR308
66	D289B	66	Q-12115C	66	2N3144	66	2SC4116A	67	BC161-06
66	D289C	66	QT-DO313XAC	66	2N3229	66	2SC4116B	67	BC212A
66	D28A1	66	R3611-1	66	2N3621	66	2SC4116C	67	BC212B
66	D28A10	66	R3681-1	66	2N3622	66	2SC4116D	67	BC213B
66	D28A12	66	R612-1	66	2N3625	66	2SC4116E	67	BC214B
66	D28A13	66	R621-1	66	2N3626	66	2SC4116F	67	BC224
66	D28A2	66	R632-1	66	2N3627	66	2SC4116G	67	BC231B
66	D28A3	66	R632-2	66	2N3629	66	2SC4116GN	67	BC257A
66	D28A4	66	RCA29	66	2N3630	66	2SC4116H	67	BC257B
66	D28A5	66	RCA29/SDH	66	2N3632	66	2SC4116J	67	BC258A
66	D28A6	66	RCA29A	66	2N3675	66	2SC4116K	67	BC258B
66	D28A7	66	RCA29A/SDH	66	2N3744	66	2SC4116L	67	BC259A
66	D28A9	66	RE21	66	2N3745	66	2SC4116M	67	BC259B
66	D28D1	66	REN152	66	2N3747	66	2SC4116OR	67	BC297
66	D28D10	66	RT-150	66	2N3748	66	2SC4116R	67	BC298
66	D28D2	66	RT150	66	2N3818	66	2SC4116X	67	BC304-4
66	D28D3	66	S-310E	66	2N3927	66	2SC4116Y	67	BC304-5
66	D28D4	66	S12020-04	66	2N4127	66	2SC489	67	BC304-6
66	D28D5	66	S1D153	66	2N4128	66	2SC489-R	67	BC307B
66	D28D7	66	S33530	66	2N4307	66	2SC489-Y	67	BC309C
66	D317	66	S37166	66	2N4308	66	2SC489A	67	BC327-16
66	D317A	66	S59262	66	2N4311	66	2SC489B	67	BC361-06
66	D317F	66	SB0319	66	2N4312	66	2SC489C	67	BC370
66	D317P	66	SC4303	66	2N4877	66	2SC489D	67	BC406
66	D318	66	SC4303-1	66	2N5293	66	2SC489E	67	BC461-4
66	D318A	66	SC4303-2	66	2N5294	66	2SC489F	67	BC461-5
66	D325	66	SC4308	66	2N5295	66	2SC489G	67	BC461-6
66	D325C	66	SCD-T33Q	66	2N5296	66	2SC489GN	67	BC512B
66	D325D	66	SD1345	66	2N5297	66	2SC489H	67	BC513B
66	D325E	66	SDA345	66	2N5298	66	2SC489J	67	BC514
66	D325F	66	SDB345	66	2N5334	66	2SC489K	67	BC514AB
66	D350D	66	SDI345	66	2N5424	66	2SC489L	67	BCW45
66	D343	66	SDJ345	66	2N5483	66	2SC489M	67	BCW52
66	D359	66	SDK345	66	2N5492	66	2SC489OR	67	BCW61A
66	D359C	66	SDL345	66	2N5606	66	2SC489R	67	BCW61B
66	D359C2	66	SDN345	66	2N5637	66	2SC489X	67	BCW61BA
66	D359D	66	SDT3422	66	2N5642	66	2SC489Y	67	BCW61BB
66	D359D1	66	SDT5102	66	2N5646	66	2SC491	67	BCW62B
66	D359D2	66	SDT5907	66	2N5689	66	2SC491-BL	67	BCW63B
66	D359E	66	SDT6001	66	2N5690	66	2SC491-BLU	67	BCW64
66	D360C	66	SDT6011	66	2N5700	66	2SC491-R	67	BCW64B
66	D360E	66	SDT6013	66	2N5701	66	2SC491-RED	67	BCW70
66	D365H	66	SDT6031	66	2N5704	66	2SC491-Y	67	BCW70R
66	D366-0	66	SDT6103	66	2N5705	66	2SC491-YEL	67	BCW75-10
66	D366P	66	SDT7402	66	2N5712	66	2SC491A	67	BCW75-16
66	D366Q	66	SDT7412	66	2N5713	66	2SC491B	67	BCW76-10
66	D382	66	SDT7415	66	2N5765	66	2SC491BL	67	BCW76-16
66	D382LM	66	SDT7511	66	2N5768	66	2SC491C	67	BCW86
66	D389	66	SDT7512	66	2N6121	66	2SC491D	67	BCW92
66	D389-0	66	SDT7514	66	2N6122	66	2SC491E	67	BCW92K
66	D389-0P	66	SDT7515	66	26012	66	2SC491F	67	BCW93
66	D389A	66	SDT9009	66	2SB389-0	66	2SC491G	67	BCW93A
66	D389APP	66	S85-0963	66	2SB435R	66	2SC491GN	67	BCW93B
66	D389B	66	SJB-515	66	2SB435Y	66	2SC491H	67	BCW93K
66	D389BL	66	SJE-515	66	2SC1019C	66	2SC491J	67	BCW93KA
66	D389LB	66	SJE42	66	2SC1060	66	2SC491K	67	BCW93KB
66	D44C2	66	SJE513	66	2SC1060(C,D)	66	2SC491L	67	BCX71BG
66	D44C3	66	SJE515	66	2SC1060A	66	2SC491M	67	BCX71BH
66	D44C4	66	SJE678	66	2SC1060B	66	2SC491OR	67	BCY78A
66	D44C5	66	SJE694	66	2SC1060BL	66	2SC491R	67	BCY78B
66	D44C6	66	SJE783	66	2SC1060BM	66	2SC491X	67	BCY78VII
66	D44C7	66	SP8416	66	2SC1060BY	66	2SC491Y	67	BCY78VIII
66	D44C8	66	SP8660	66	2SC1060C	66	2SC551	67	BCY79A
66	D44C8B	66	SP81436	66	2SC1060D	66	2SC552	67	BCY79B
66	D44C9	66	STC1300	66	2SC1060E	66	2SC553	67	BCY79VII
66	D44C8BB	66	STC1850	66	2SC1060F	66	2SC573	67	BCY79VIII
66	D513	66	STC1860	66	2SC1060G	66	2SC585	67	BFR61
66	D90	66	T01-03Q	66	2SC1060GN	66	2SC599	67	BFR80
66	D91	66	T1486	66	2SC1060H	66	2SC600	67	BFR80T05
66	D91P	66	T1487	66	2SC1060J	66	2SC636	67	BFS34
66	DDBX278002	66	T1229	66	2SC1060K	66	2SC638C	67	BFS44
66	DDBX407001	66	T1229X	66	2SC1060L	66	2SC680	67	BFS45
66	DS-513	66	T611-1	66	2SC1060M	66	2SC680(A)	67	BFS96
66	DS513	66	T612-1	66	2SC1060R	66	2SC680A	67	BFS97
66	E2496	66	TA2911	66	2SC1060X	66	2SC680B	67	BFT70
66	EA15X327	66	TA7137	66	2SC1060Y	66	2SC680C	67	BFW20
66	EA15X333	66	TA7156	66	2SC1061	66	2SC680G	67	BFW31
66	EA15X7121	66	TA7262	66	2SC1061(B)	66	2SC680GN	67	B3V43B
66	EA15X8119	66	TA7363	66	2SC1061(C)	66	2SC680H	67	B3V44B
66	EA15X8602	66	TA7554	66	0002SC1061A	66	2SC680J	67	B3V45B
66	EA15X399	66	TA7555	66	2SC1061B	66	2SC680K	67	B3V47B
66	EA5716	66	TI486	66	2SC1061BM	66	2SC680L	67	B3V48B
66	EC0152	66	TI487	66	2SC1061BT	66	2SC680M	67	B3V49B
66	EP-100	66	TIP-14	66	2SC1061C	66	2SC680R	67	CDC9000-1Q
66	EP-422	66	TIP24	66	2SC1061D	66	2SC680X	67	CDC90001.B
66	EP-797	66	TIP29	66	2SC1061KA	66	2SC690	67	CG8-67
66	EP-801	66	TIP29A	66	2SC1061KB	66	2SC703	67	CS-2142
66	EP-944	66	TIP31	66	2SC1061KC	66	2SC704	67	CS1255HF
66	EP15X22	66	TIP31A	66	2SC1061T	66	A3T2907A	67	CS2082
66	EP15X68	66	TR-28	66	2SC1061T-B	66	A545	67	CS2142
66	EP3055	66	TR28D330E	66	2SC1061TB	66	A545GRN	67	CLL193
66	EQ8-140	66	TV-115	66	2SC107	66	A545K	67	D29E08
66	EQ8140	66	TV-117	66	2SC1398	66	A545KLM	67	D29E08J1
66	ES15X86	66	01-040243	66	2SC1398P	66	A545L	67	D29E09
66	ES80(ELCOM)	66	01-040589	66	2SC1398Q	66	A545LM	67	D29E09J1
66	ETTD-235	66	01-572784	66	2SC1409	66	A547	67	D29E10
66	G05705	66	01-572791	66	2SC1409(B)	66	A547A	67	D29E10J1
66	GE-3229	66	01-572861	66	2SC1418	66	A575K	67	D29E1J1
66	GE-66	66	01-57291	66	2SC1418A	66	A575L	67	D29E2J1
66	G05705			66	2SC1418B	66	A5T3505	67	D29E4J1
66	HC495			66	2SC1418C			67	D29E5J1
66	HF57	66	1A0046	66	2SC1418D			67	D29E8J1

GE	Industry Standard No.	GE	Industry Standard No.	GE	Industry Standard No.	GE	Industry Standard No.	GE	Industry Standard No.
67	D29ZFJ1	67	2SA545M	69	CGB-69	69	2N2875	69	2SB537LM
67	D29EBAJ1	67	2SA545OR	69	CXL153	69	3N021	69	2SB699
67	D29E9	67	2SA545R	69	D43C7	69	3N022	69	2SB699Q
67	D29E9J1	67	2SA545X	69	D45C1	69	3N023	73	DTS403
67	EA15X242	67	2SA547	69	D45C2	69	3N024	73	DTS409
67	EA15X273	67	2SA547A	69	D45C5	69	3N025	73	MJ3430
67	EC0193	67	2SA575K	69	D45C4	69	3N026	73	SDF402
67	EP15X13	67	2SA575L	69	D45C5	69	3N199	73	SDT430
67	EP15X21	67	2SA599	69	D45C6	69	3N200	73	SDT431
67	EP15X51	67	2SA599(Y)	69	D45C7	69	3N205	73	2SC1106
67	EP35	67	2SA599Y	69	D45C9	69	3N206	73	2SC558
67	F2041	67	2SA643	69	EA15X3118	69	2N3740A	74	A626
67	FT3644	67	2SA643(R)	69	EA15X328	69	2N4387	74	A627
67	FX3962	67	2SA643(V,R)	69	EA15X334	69	2N4388	74	A648
67	GB-67	67	2SA643(W)	69	EA15X8118	69	2N5112	74	A648A
67	HA-00643	67	2SA643A	69	EA15X8130	69	2N5161	74	A648B
67	HA7520	67	2SA643B	69	EA3715	69	2N5597	74	A648C
67	HA7521	67	2SA643C	69	EC0153	69	2N6021	74	A658
67	HA7522	67	2SA643D	69	EP-421	69	2N6022	74	A663
67	HA7523	67	2SA643E	69	EP-802	69	2N6023	74	A744
67	HA7524	67	2SA643F	69	EP-943	69	2N6024	74	A746
67	HA7526	67	2SA643R	69	EP15X15	69	2N6026	74	BD246
67	HA7527	67	2SA643V	69	EP15X23	69	2N6124	74	BD246A
67	HA7528	67	2SA643W	69	ES81(ELCOM)	69	2N6125	74	BD246C
67	HA7723	67	2SA707	69	EW183	69	2N6126	74	BD250
67	HA7730	67	2SA707V	69	GB-69	69	2SA473	74	BD250A
67	HA7732	67	2SA715WT(C,B)	69	HA-00699	69	2SA473(O)	74	BD250B
67	HA7734	67	2SA728A	69	HA505	69	2SA473(Q)-QR	74	BD250C
67	HA7735	67	2SA751	69	HP58	69	2SA473-O	74	BDX18
67	HA7756	67	2SA751(P)	69	HR70	69	2SA473-R	74	BDX18N
67	HA7737	67	2SA751P	69	L8-0067	69	2SA473-Y	74	BDX20
67	HR-72	67	2SA751Q	69	M9348	69	2SA473B	74	CXL1B0
67	HR105611C	67	2SA751QR	69	MJB101	69	2SA473GR	74	CXL219
67	HR72	67	2SA751R	69	MJB102	69	2SA473Y	74	D341
67	H85811	67	2SA751S	69	MJB103	69	2SA486	74	D341H
67	H85813	67	2SA752	69	MM4020	69	2SA486-RED	74	EA15X124
67	H85815	67	2SA752P	69	N86518	69	2SA486-YEL	74	EC0180
67	H85817	67	2SA752Q	69	P1Y-4	69	2SA489	74	EC0219
67	H85819	67	2SA752R	69	P2B	69	2SA489-O	74	ES64(ELCOM)
67	H85821	67	2SA752S	69	P2K	69	2SA489-Y	74	ES74(ELCOM)
67	H85823	69	2SC673	69	P4W	69	2SA490	74	ES90
67	H86011	69	2SC673(B)	69	P5H	69	2SA490(POWER)	74	ES90(ELCOM)
67	H86013	69	2SC673B	69	P5U	69	2SA490-O	74	GB-74
67	H86015	69	2SC673C	69	P8B	69	2SA490-Y	74	HEP-248
67	H86017	69	2SC673C2	69	P8H	69	2SA490A	74	HEP-705
67	HX-50104	69	2SC673D	69	RCA30	69	2SA490B	74	HEP-87001
67	IRTR-52	69	A-1853-0233-1	69	RCA30A	69	2SA490C	74	HEP-87003
67	MDA6518	69	A-1853-0234-1	69	RB22	69	2SA490D	74	HEPS7001
67	ME513	69	A-1853-0254-1	69	REN153	69	2SA490E	74	HP23
67	MH1502	69	A473(JAPAN)	69	RT-151	69	2SA490F	74	HP25
67	MM4008	69	A473-QR	69	RT-155	69	2SA490G	74	K9-200033L2
67	MRC3798	69	A473-O	69	RT151	69	2SA490GN	74	M9344
67	MPS-A56	69	A473-R	69	RT155	69	2SA490H	74	M9359
67	MPS4354	69	A473-Y	69	S2041	69	2SA490J	74	MJ2267
67	MPS4355	69	A473Y	69	S2042	69	2SA490K	74	MJ2268
67	NN7500	69	A489(JAPAN)	69	S33529	69	2SA490L	74	MJ2901
67	NN7501	69	A489-O	69	S37165	69	2SA490LBG1	74	MJ2940
67	NN7502	69	A489-R	69	S39261	69	2SA490M	74	MJ2941
67	NN7503	69	A489-Y	69	SCD-T334	69	2SA490OR	74	MJ2955
67	NN7504	69	A490(POWER)	69	SD1445	69	2SA490X	74	MJ450
67	NN7505	69	A670	69	SD445	69	2SA490Y	74	MJ4502
67	NN7511	69	A670A	69	SDA445	69	2SA490YA	74	MJ490
67	NN-671ET	69	A670B	69	SDB445	69	2SA490YLBG11	74	MJ491
67	P4B	69	A670C	69	SDI445	69	2SA641	74	NKT4055
67	PTC131	69	A748Q	69	SDJ445	69	2SA641A	74	P1E
67	PTC177	69	A754	69	SDK445	69	2SA641B	74	P1E-1BLK
67	RE197	69	A754A	69	SDL445	69	2SA641BL	74	P1E-1BLU
67	REN193	69	A754B	69	SDM445	69	2SA641C	74	P1E-1GRN
67	RT-121	69	A754C	69	SDN445	69	2SA641D	74	P1E-1RED
67	SC256B	69	A754D	69	SDT-445	69	2SA641G	74	P1E-2BLK
67	SHA7520	69	A755	69	SDT3509	69	2SA641GR	74	P1E-2BLU
67	SHA7521	69	A755A	69	SDT3510	69	2SA641K	74	P1E-2GRN
67	SHA7522	69	A755B	69	SDT3513	69	2SA641L	74	P1E-2RED
67	SHA7523	69	A755C	69	SDT3514	69	2SA641M	74	P1E-2V10
67	SHA7524	69	A755D	69	SDT3701	69	2SA641O	74	P1E-3BLK
67	SHA7527	69	AR-23(XSTR)	69	SDT3702	69	2SA641OR	74	P1E-3BLU
67	SHA7528	69	AR-25	69	SDT3703	69	2SA641R	74	P1E-3GRN
67	SHA7597	69	AR25(G)	69	SDT3704	69	2SA641Y	74	P1E-3RED
67	SHA7598	69	AR25(GREEN)	69	SDT3706	69	2SA670A	74	P1E-3V10
67	SHA7599	69	AR27(GREEN)	69	SDT3707	69	2SA670B	74	P2J
67	ST61000	69	AR27GREEN	69	SDT3708	69	2SA670C	74	P3W
67	TC8101	69	AR37(GREEN)	69	SDT3709	69	2SA671	74	PTC172
67	TCS103	69	AR37GREEN	69	SDT3710	69	2SA748Q	74	RE-82
67	TE3702	69	AR44	69	SDT3711	69	2SA748R	74	RE74
67	TE3703	69	B132	69	SDT3712	69	2SA754	74	RE82
67	TE5086	69	B434	69	SDT3713	69	2SA754A	74	REN180
67	TE5087	69	B434-O	69	SDT3715	69	2SA754B	74	REN219
67	TE5367	69	B434-R	69	SDT3716	69	2SA754C	74	RT-148
67	TI-891	69	B434-Y	69	SDT3717	69	2SA754D	74	S35486
67	TNJ70481	69	B435	69	SDT3721	69	2SA755	74	SC0421
67	TNJ72152	69	B435-O	69	SDT3722	69	2SA755A	74	SC0428
67	TP36538A	69	B435-R	69	SDT3725	69	2SA755B	74	SC0421
67	TQ53	69	B435-Y	69	SDT3726	69	2SA755C	74	SCD421
67	TQ54	69	B507	69	SDT3727	69	2SA755D	74	SCE421
67	TQ55	69	B508	69	SDT3729	69	2SA766	74	SDT3760
67	TQ57	69	B509	69	SDT3730	74	2SA766	74	SDT3764
67	TQ58	69	B511	69	SDT3733	69	2SB434	74	SDT3765
67	TQ59	69	B511C	69	SDT445	69	2SB434(0)	74	SDT3766
67	TQ59A	69	B511D	69	SDT5112	69	2SB434-O	74	SDT3826
67	TQ60	69	B512	69	SE5-0964	69	2SB434-R	74	SDT3827
67	TQ60A	69	B512A	69	SJ1152	69	2SB434-Y	74	SDT3876
67	TV82C81256HG	69	B513	69	SJ1171	69	2SB435-O	74	SDT3876
67	001-02117-2	69	B513A	69	SJ1284	69	2SB435-O	74	SDT3877
67	2N1132B/51	69	B514	69	SJE514	69	2SB435-R	74	SDT9202
67	2N1132BJ46	69	B515	69	SJE677	69	2SB435-Y	74	SDT9207
67	2N1242	69	B537LM	69	SJE695	69	002SB435RY	74	SDT9701
67	2N1243	69	BD223	69	SPR1437	69	2SB507	74	SDT9704
67	2N3081/46	69	BD224	69	ST2T020	69	2SB507E	74	SDT9707
67	2N3244	69	BD225	69	STC5202	69	2SB508	74	SJ1272
67	2N5821HB	69	BD240	69	STC5203	69	2SB509	74	SJ2001
67	2N5823	69	BD240A	69	STC5205	69	2SB511	74	SJ2023
67	2N5855	69	BD242	69	STC5206	69	2SB511C	74	SJ2024
67	2N6015	69	BD242A	69	STC5303	69	2SB511D	74	SJ3507
67	2SA512-ORG	69	BD242B	69	STC5802	69	2SB511E	74	SJ3520
67	2SA512-RED	69	BD244	69	STC5803	69	2SB512	74	SJ3636
67	2SA545	69	BD244A	69	STC5805	69	2SB512A	74	SJ3677
67	2SA545(K)	69	BD244B	69	STC5806	69	2SB512P	74	SJ3679
67	2SA545(L)	69	BD262A	69	STX0029	69	2SB513	74	SJ821
67	2SA545A	69	BD272	69	STX0030	69	2SB513A	74	SJE264
67	2SA545B	69	BD434	69	TA7556	69	2SB513P	74	SJE764
67	2SA545C	69	BD436	69	TA7557	69	2SB513Q	74	ST29045
67	2SA545D	69	BD438	69	TIP30	69	2SB513R	74	ST29046
67	2SA545E	69	BD440	69	TIP30A	69	2SB514	74	ST29047
67	2SA545F	69	BD534	69	TIP32	69	2SB515	74	T1P2955
67	2SA545G	69	BD536	69	TIP32A	69	2SB537	74	TIP2955
67	2SA545GN	69	BDX27	69	TR-8019			74	TIP544
67	2SA545GRN	69	BDX27-10	69	TR2327723			74	TR-29
67	2SA545H	69	BDX27-6	69	TRB019			74	TR02060-7
67	2SA545J	69	BDX28	69	TV-116			74	TR0259-6
67	2SA545K	69	BDX28-10	69	TVS28A483			74	TR1038-4
67	2SA545KLM	69	BDX28-6	69	01-010473			74	TR2327841
67	2SA545L	69	BDY82	69	01-572774			74	TR29
67	2SA545LM	69	BDY83	69	002-012300				

GE	Industry Standard No.	GE	Industry Standard No.	GE	Industry Standard No.	GE	Industry Standard No.	GE	Industry Standard No.
74	2N3171	75	FBN-38022	75	2N6357	76	REN121	81	C1003
74	2N3172	75	FD4500AL	76	2N6359	76	REN121MP	81	CXL159
74	2N3173	75	G181-725-001	75	2SC1629	76	TM121	81	EA15X356
74	2N3174	75	GE-75	75	2SC1629A	76	TM121MP	81	EA15X408
74	2N3183	75	HEPS7000	75	2SC1629AQ	76(2)	ECG121MP	81	FC8-9013P
74	2N3184	75	HEPS7004	75	2SC1629M	77	SDT9301	81	FC8-9013Q
74	2N3185	75	HP24	75	2SC21	77	SDT9302	81	FC89013
74	2N3186	75	IRTR36	75	2SC2199	77	SDT9304	81	FC89013P
74	2N3195	75	KSD1052	75	2SC21A	77	SDT9305	81	FC89013Q
74	2N3196	75	KSD1057	75	2SC21B	77	SDT9307	81	FC89013H
74	2N3197	75	KSD2203	75	2SC21C	77	SDT9308	81	FT3560
74	2N3198	75	KSD3771	75	2SC21D	77	2N3235	81	GB-81
74	2N3789	75	KSD3772	75	2SC21E	80	2SA304	81	HC-00509
74	2N3790	75	KSD9701	75	2SC21F	80	2SA375	81	HS-40016
74	2N3791	75	KSD9701A	75	2SC21G	80	2SA408	81	IP20-0195
74	2N3792	75	KSD9704	75	2SC21GN	80	2SA409	81	IP20-0214
74	2N5621	75	LM511160	75	2SC21H	80	2SB105	81	MP88098
74	2N5623	75	LN75116	75	2SC21J	80	2SB105A	81	MP88099
74	2N5625	75	M9480	75	2SC21K	80	2SB105B	81	MP89417A
74	2N5738	75	M9628	75	2SC21L	80	2SB105C	81	MP89417A-?
74	2N5741	75	M9639	75	2SC21M	80	2SB105D	81	MPE817
74	2N5742	75	M9715	75	2SC21OR	80	2SB105E	81	FN2369
74	2N5745	75	MJ3771	75	2SC21R	80	2SB105F	81	PN2369A
74	2N5879	75	MJ3772	75	2SC21X	80	2SB105G	81	PN2484
74	2N5880	75	MJ5257	75	2SC21Y	80	2SB105GN	81	Q5182
74	2N5883	75	MJ6257	75	2SC240	80	2SB105H	81	R3N289
74	2N5884	75	MJ802	75	2SC240A	80	2SB105K	81	RT5212
74	2N6025	75	NO282CT	75	2SC240C	80	2SB105L	81	TI-481
74	2N6246	75	P-11810-1	75	2SC240D	80	2SB105M	81	01-031166
74	2N6247	75	P-11901-3	75	2SC240E	80	2SB105OR	81	01-031317
74	2N6329	75	PO445-0034-1	75	2SC240F	80	2SB105R	81	01-031688
74	2N6330	75	PO445-0034-2	75	2SC240G	80	2SB105X	81	2N1564
74	2N6331	75	PO4450034-1	75	2SC240GN	80	2SB105Y	81	2N1565
74	2SA626	75	PO4450034-2	75	2SC240H	80	2SB108	81	2N1566
74	2SA626L	75	PO4450037	75	2SC240J	80	2SB108A	81	2N1566A
74	2SA627	75	P6500A	75	2SC240K	80	2SB108C	81	2N1613/46
74	2SA648	75	PMC-QPO040	75	2SC240L	80	2SB108D	81	2N1711/46
74	2SA648A	75	PP3001	75	2SC240M	80	2SB108E	81	2N1837
74	2SA648B	75	PP3004	75	2SC240OR	80	2SB108F	81	2N2087
74	2SA648C	75	PP3007	75	2SC240R	80	2SB108G	81	2N2317
74	2SA658	75	PT7930	75	2SC240X	80	2SB108GN	81	2N2380
74	2SA663	75	PT7931	75	2SC240Y	80	2SB108H	81	2N2380A
74	2SA744	75	PTC116	75	2SC241A	80	2SB108J	81	2N2479
74	2SA746	75	PTC173	75	2SC241C	80	2SB108K	81	2N2514
74	2SB518	75	R2270-75116	75	2SC241D	80	2SB108L	81	2N2515
74	2SB519	75	RE37	75	2SC241E	80	2SB108M	81	2N3402
75	A-1854-0458	75	REN181	75	2SC241F	80	2SB108OR	81	2N3877A
75	A08-1050115	75	RT-149	75	2SC241G	80	2SB108X	81	2N5961
75	A391593	75	S35487	75	2SC241GN	80	2SB108Y	81	2N734
75	A417014	75	SC0321	75	2SC241H	80	2SB294	81	2N734A
75	A417014	75	SC0328	75	2SC241K	80	2SB294A	81	2N735
75	A580-040215	75	SCD321	75	2SC241L	80	2SB294C	81	2N735A
75	A580-040315	75	SCE321	75	2SC241M	80	2SB294D	81	2N736
75	A580-040515	75	SDT1621	75	2SC241OR	80	2SB294E	81	2N736A
75	A580-080215	75	SDT1622	75	2SC241R	80	2SB294F	81	2N736B
75	A580-080315	75	SDT1623	75	2SC241X	80	2SB294G	81	2N956
75	A580-080515	75	SDT1631	75	2SC241Y	80	2SB294GN	81	2SC1166
75	A5V	75	SDT1632	75	2SC242	80	2SB294H	81	2SC1166-O
75	A9N	75	SDT1633	75	2SC242A	80	2SB294K	81	2SC1166-OR
75	AMP201B	75	SDT9303	75	2SC242B	80	2SB294L	81	2SC1166-R
75	AMP201A	75	SDT9306	75	2SC242C	80	2SB294M	81	2SC1166-X
75	AMP210B	75	SDT9309	75	2SC242D	80	2SB294OR	81	2SC1166-Y
75	B-12822-4	75	SJ2047	75	2SC242E	80	2SB294R	81	2SC1166D
75	BO301-049	75	SJ2064	75	2SC242F	80	2SB294X	81	2SC1166GR
75	B170000-YEL	75	SJ3519	75	2SC242G	80	2SB294Y	81	2SC1166O
75	B170001-YEL	75	SK3036	75	2SC242H	80	2SB304	81	2SC1166R
75	B170002-YEL	75	SK3511	75	2SC242J	80	2SB304A	81	2SC1166Y
75	B170003-BLK	75	SK3561	75	2SC242K	80	2SB35	81	2SC162
75	B170003-BRN	75	SK3563	75	2SC242L	80	2SB35A	81	2SC1688
75	B170003-ORG	75	SPC040411	75	2SC242M	80	2SB35B	81	2SC509
75	B170003-RED	75	SPD-80059	75	2SC242OR	80	2SB35C	81	2SC509(O)
75	B170003-YEL	75	SPD-80060	75	2SC242R	80	2SB35D	81	2SC509(Q)
75	B170004-BLK	75	SPD-80061	75	2SC242X	80	2SB35E	81	2SC509(Y)
75	B170004-BRN	75	ST101	75	2SC242Y	80	2SB35F	81	2SC509-O
75	B170004-ORG	75	STC1094	75	2SC244	80	2SB35G	81	2SC509-Y
75	B170004-RED	75	STC2220	75	2SC244A	80	2SB35GN	81	2SC509O
75	B170004-YEL	75	STC2221	75	2SC244B	80	2SB35H	81	2SC509-Q
75	B170005-BLK	75	STC2224	75	2SC244C	80	2SB35J	81	2SC509Y
75	B170005-BRN	75	STC2225	75	2SC244D	80	2SB35K	82	A-1005-725
75	B170005-ORG	75	STC2228	75	2SC244E	80	2SB35L	82	A-113110
75	B170005-RED	75	STC2229	75	2SC244F	80	2SB35M	82	A-120417
75	B170005-YEL	75	STC2300	75	2SC244G	80	2SB35OR	82	A-120526
75	B170006-BLK	75	STT3500	75	2SC244GN	80	2SB35R	82	A-1853-0016-1
75	B170006-BRN	75	T841	75	2SC244J	80	2SB35X	82	A-1853-0020-1
75	B170006-ORG	75	T842	75	2SC244K	80	2SB366	82	A-1853-0027-1
75	B170006-RED	75	T843	75	2SC244L	80	2SB386	82	A-1853-0036-1
75	B170006-YEL	75	T844	75	2SC244M	80	2SB427	82	A-1853-0039-1
75	B170007-BLK	75	TK30551	75	2SC244OR	80	2SB428	82	A-1853-0049-1
75	B170007-BRN	75	TK30552	75	2SC244R	80	2SB460A	82	A-1853-0058-1
75	B170007-ORG	75	TK30555	75	2SC244X	80	2SB460B	82	A-1853-0062-1
75	B170007-RED	75	TK30556	75	2SC244Y	80	2SB461A	82	A-1853-0065-1
75	B170007-YEL	75	TK30557	80	2SC493	81	2SB461B	82	A-1853-0092-1
75	B170008-BLK	75	TK30560	80	2SC493-BL	81	2SB461C	82	A-1853-0099-1
75	B170008-BRN	75	TK9201	80	2SC493-R	81	2SB461D	82	A-1853-0285-1
75	B170008-ORG	75	TR-176(OLSON)	80	2SC493-Y	81	2SB461F	82	A-1853-0321-1
75	B170008-RED	75	TR-36	80	2SC493A	81	2SB461G	82	A-1950
75	B170008-YEL	75	TR-36MP	80	2SC493B	81	2SB461H	82	A054-223
75	B170017	75	TR-8018	80	2SC493C	81	2SB461J	82	A066-113
75	B170023	75	TRO1060-7	80	2SC493D	81	2SB461K	82	A066-113A
75	B170026	75	TR1009A	80	2SC493F	81	2SB461L	82	A066-113AB
75	B177000	75	TR2327574	80	2SC493G	81	2SB461M	82	A066-118
75	BDH11A	75	TR36	80	2SC493GN	81	2SB461OR	82	A066-118A
75	BD249A	75	TR8018	80	2SC493H	81	2SB461R	82	A068-109
75	BD249B	75	TV8-2SC646	80	2SC493J	81	2SB461X	82	A068-109A
75	BD249C	75	TVS2SC1629A	80	2SC493K	81	2SB461Y	82	A1030
75	BDX13	75	001-021280	80	2SC493L	81	2SB477(C)	82	A112-000172
75	BDX41	75	01-040201	80	2SC493M	81	2SB477C	82	A112-000185
75	BDY-10	75	1A1123100-1	80	2SC493OR	81	A5T3392	82	A112-000187
75	BDY10	75	2N3055-3	80	2SC493R	81	A5T3565	82	A11200482
75	BDY11	75	2N3055-6	81	2SC493X	81	A5T4409	82	A116078
75	BDY57	75	2N3055-7	81	2SC493Y	81	A5T5172	82	A118284
75	BDY58	75	2N3055-8	80	2SC664	81	A5T5209	82	A119730
75	BDY76	75	2N3236	80	2SC736	81	A5T5219	82	A12021
75	BUY43	75	2N3237	81	2SC736A	81	A5T5220	82	A121-1
75	BUY46	75	2N3238	81	2SC736B	81	A5T5223	82	A121-1RED
75	CXL181	75	2N3299	81	2SC736C	81	A5T5225	82	A121-444
75	D113	75	2N3771	81	2SC736D	81	A7T3392	82	A121-446
75	D113-O	75	2N3772	81	2SC736E	81	A7T5172	82	A121-495
75	D113-R	75	2N5301	81	2SC736F	81	A8T3392	82	A121-496
75	D113-Y	75	2N5303	81	2SC736G	81	A8T3707	82	A121-497
75	D114	75	2N5629	81	2SC736GN	81	A8T3708	82	A121-497WH2
75	D114-O	75	2N5632	81	2SC736H	81	A8T3709	82	A121-602
75	D114-R	75	2N5885	81	2SC736J	81	A8T3710	82	A121-603
75	D114-Y	75	2N5886	81	2SC736K	81	A8T3711	82	A121-679
75	D132	75	2N6254	81	2SC736L	81	A8T5172	82	A121-699
75	D319	75	2N6257	81	2SC736M	81	BC431	82	A121-746
75	D55	75	2N6258	81	2SC736OR	81	BCW95A	82	A121-774
75	D55A	75	2N6326	81	2SC736R	81	BCW95B	82	A1214
75	EA15X123	75	2N6327	81	2SC736X	81	BCW95KA	82	A121467
75	ECG181	75	2N6328	81	2SC736Y	81	BCW95KB	82	A121659
75	ES43(ELCOM)	75	2N6355					82	A1220RN
75	FBN-56972							82	A122YEL
75	FBN-56973							82	A124047

GE	Industry Standard No.	GE	Industry Standard No.	GE	Industry Standard No.	GE	Industry Standard No.	GE	Industry Standard No.
82	A124755	82	A495-0	82	A618K	82	AFC81170F	82	BC196
82	A12594	82	A495-R	82	A628	82	AF824226	82	BC196-1
82	A126524	82	A495-Y	82	A628A	82	AFT0019M	82	BC196A
82	A126700	82	A4950	82	A628D	82	AFT052	82	BC196A-1
82	A126707	82	A495A	82	A628E	82	AFT1341	82	BC196B
82	A126715	82	A495D	82	A628F	82	AFT1746	82	BC196B-1
82	A126718	82	A495G-GR	82	A629	82	AHJ101	82	BC196V1
82	A126719	82	A495G-0	82	A637	82	AR304(GREEN)	82	BC196V1-1
82	A128888	82	A495G-R	82	A640	82	AR30S(VIOLET)	82	BC200
82	A129-34	82	A495G-Y	82	A640(JAPAN)	82	AT-11	82	BC200-1
82	A129697	82	A495W	82	A640A	82	AT331	82	BC201
82	A129699	82	A495Y	82	A640B	82	AT331A	82	BC201-1
82	A130-149	82	A499	82	A640C	82	AT332	82	BC202
82	A130-40315	82	A499-0	82	A640D	82	AT332A	82	BC202-1
82	A130-40429	82	A499-R	82	A640E	82	AT333	82	BC203
82	A130159	82	A499-Y	82	A640L	82	AT335A	82	BC203-1
82	A137(2HP)	82	A500(JAPAN)	82	A640M	82	AT410	82	BC204-1
82	A1471114-1	82	A500-0	82	A640S	82	AT410-1	82	BC204A-1
82	A1473563-1	82	A500-R	82	A642	82	AT412	82	BC204B-1
82	A1473570-1	82	A500-Y	82	A642(JAPAN)	82	AT412-1	82	BC204V
82	A1473574-1	82	A502	82	A642A	82	AT413	82	BC204V-1
82	A1473590-1	82	A510	82	A642B	82	AT413-1	82	BC204V1
82	A1473591-1	82	A510-0	82	A642C	82	AT414	82	BC204V1-1
82	A1473597-1	82	A510-R	82	A642D	82	AT414-1	82	BC205
82	A1558-17	82	A511	82	A642E	82	AT415	82	BC205-1
82	A160	82	A511-0	82	A642F	82	AT415-1	82	BC205A
82	A161	82	A511-R	82	A642W	82	AT416	82	BC205A-1
82	A162	82	A512	82	A659	82	AT416-1	82	BC205B
82	A171	82	A512-0	82	A659C	82	AT417	82	BC205B-1
82	A177	82	A512-R	82	A659D	82	AT417-1	82	BC205V
82	A177(A)	82	A513	82	A659E	82	AT418	82	BC205V-1
82	A177A	82	A513-0	82	A659F	82	AT418-1	82	BC205V1-1
82	A177AB	82	A513-R	82	A6661QRS	82	AT419	82	BC206-1
82	A178A	82	A514-044910	82	A669	82	AT419-1	82	BC206B-1
82	A178AB	82	A522(JAPAN)	82	A672A	82	AT430	82	BC212
82	A178B	82	A5226-1	82	A672B	82	AT430-1	82	BC212-1
82	A178BA	82	A522A	82	A672C	82	AT431	82	BC212K
82	A179A	82	A530	82	A675	82	AT431-1	82	BC212K-1
82	A179AC	82	A530H	82	A675A	82	AT432	82	BC212KA-1
82	A179B	82	A5320111	82	A675B	82	AT432-1	82	BC212KB
82	A179BB	82	A539	82	A675C	82	AT433	82	BC212KB-1
82	A1844-17	82	A539L	82	A678	82	AT433-1	82	BC212L
82	A1867-17	82	A539B	82	A678(SONY)	82	AT434	82	BC212L-1
82	A1901-5338	82	A54-96-003	82	A690D	82	AT434-1	82	BC212LA
82	A190429	82	A542	82	A701	82	AT435	82	BC212LA-1
82	A200-052	82	A544	82	A701584-00	82	AT435-1	82	BC212LB
82	A2003O703-0702	82	A550	82	A701589-00	82	AT436	82	BC212LB-1
82	A2057T013-0004	82	A550A	82	A701F	82	AT436-1	82	BC212V1
82	A2057T013-0701	82	A550AQ	82	A701FJ	82	AT437	82	BC213
82	A2057T013-0702	82	A550Q	82	A701F0	82	AT437-1	82	BC213-1
82	A2057T013-0703	82	A550R	82	A71687-1	82	AT438	82	BC213A
82	A2057T12-198	82	A550B	82	A718	82	AT438-1	82	BC213K
82	A2057T106-12	82	A561	82	A719Q	82	AT451	82	BC213K-1
82	A2057T108-6	82	A561-0	82	A723	82	AT451-1	82	BC213KA
82	A20K	82	A561-R	82	A723A	82	AT452	82	BC213KA-1
82	A20KA	82	A561-Y	82	A723B	82	AT452-1	82	BC213KB
82	A2200B	82	A561GR	82	A723C	82	AT453	82	BC213KB-1
82	A23114050	82	A564-0	82	A723D	82	AT453-1	82	BC213KC
82	A23114051	82	A565	82	A723E	82	AT454	82	BC213KC-1
82	A23114550	82	A565A	82	A723F	82	AT454-1	82	BC213L
82	A242B	82	A565B	82	A730	82	AT455	82	BC213L-1
82	A2448	82	A565C	82	A731	82	AT455-1	82	BC213LA
82	A2498512	82	A565D	82	A753-4004-248	82	B-1426	82	BC213LA-1
82	A2798	82	A565K	82	A7570013-01	82	B-6288	82	BC213LB
82	A297L012C01	82	A567(JAPAN)	82	A7576004-01	82	B-75561-31	82	BC213LB-1
82	A297V073C03	82	A568	82	A759A	82	B1J	82	BC213LC
82	A297V073C04	82	A569	82	A759B	82	B1N-1	82	BC213LC-1
82	A281564PR	82	A569J	82	A78331	82	B1N-2	82	BC214
82	A30270	82	A576-0001-002	82	A800-523-01	82	B266A-1	82	BC214-1
82	A30290	82	A576-0001-013	82	A800-523-02	82	B266B-1	82	BC214K
82	A3513	82	A592Y	82	A800-527-00	82	B2A	82	BC214K-1
82	A3540	82	A59625-1	82	A815199	82	B2E	82	BC214KA
82	A3549	82	A59625-10	82	A815199-6	82	B20	82	BC214KA-1
82	A3559	82	A59625-11	82	A815211	82	B2M-1	82	BC214KB
82	A3562	82	A59625-12	82	A815213	82	B2M-2	82	BC214KB-1
82	A3563	82	A59625-2	82	A815229	82	B2M-3	82	BC214KC
82	A3574	82	A59625-3	82	A815247	82	B2B	82	BC214KC-1
82	A3581	82	A59625-4	82	A829	82	B2T	82	BC214L
82	A3T2894	82	A59625-5	82	A829A	82	B54731-30	82	BC214L-1
82	A3T2906	82	A59625-6	82	A829B	82	B561	82	BC214LA
82	A3T2906A	82	A59625-7	82	A829C	82	B561B	82	BC214LA-1
82	A3T2907	82	A59625-8	82	A829D	82	B561C	82	BC214LB
82	A402	82	A59625-9	82	A829E	82	B0-261	82	BC214LB-1
82	A4086	82	A5T2604	82	A829F	82	BC116	82	BC214LC
82	A4087	82	A5T2605	82	A833	82	BC116A	82	BC214LC-1
82	A4126	82	A5T2907	82	A8405	82	BC126	82	BC221
82	A41440	82	A5T3638	82	A8540	82	BC126-1	82	BC221-1
82	A417-116	82	A5T3638A	82	A8667	82	BC137	82	BC225-1
82	A417-132	82	A5T3644	82	A8T3702	82	BC137-1	82	BC231A
82	A417-153	82	A5T3645	82	A8T3703	82	BC153-1	82	BC238-16
82	A417-176	82	A5T3905	82	A8T404	82	BC154-1	82	BC250
82	A417-182	82	A5T3906	82	A8T404A	82	BC157	82	BC250-1
82	A417-184	82	A5T404	82	A8T405B	82	BC157-1	82	BC250A
82	A417-196	82	A5T404A	82	A8T4059	82	BC157A	82	BC250A-1
82	A417-200	82	A5T4125	82	A8T406O	82	BC157B	82	BC250B
82	A417-201	82	A5T4126	82	A8T4061	82	BC158	82	BC250B-1
82	A417-235	82	A5T4248	82	A8T4062	82	BC158-1	82	BC250C
82	A3Q23845	82	A5T4249	82	A921-470B	82	BC158A	82	BC250C-1
82	A4310	82	A5T5086	82	A94037	82	BC158A-1	82	BC251
82	A4442	82	A5T5221	82	A945-0	82	BC158B	82	BC251A
82	A467(JAPAN)	82	A5T5226	82	A95227	82	BC158B-1	82	BC251A-1
82	A467-0	82	A5T5227	82	A95252	82	BC159	82	BC251B
82	A467-0(JAPAN)	82	A603	82	A970246	82	BC159-1	82	BC251B-1
82	A467-Y	82	A608-C	82	A970248	82	BC159A	82	BC251C
82	A467-Y(JAPAN)	82	A608-D	82	A970251	82	BC159A-1	82	BC251C-1
82	A467G	82	A608-E	82	A970254	82	BC159B	82	BC252
82	A467G(JAPAN)	82	A608-F	82	A984193	82	BC159B-1	82	BC252-1
82	A467G-0	82	A608A	82	A991-01-0098	82	BC177-1	82	BC252A
82	A467G-0(JAPAN)	82	A608B	82	A991-01-1225	82	BC177A-1	82	BC252A-1
82	A467G-R	82	A608C	82	A991-01-1319	82	BC177B-1	82	BC252B
82	A467G-R(JAPAN)	82	A608D	82	A991-01-3058	82	BC177V	82	BC252B-1
82	A467G-Y	82	A608E	82	AC9082	82	BC177V-1	82	BC252C
82	A467G-Y(JAPAN)	82	A608F	82	AC9083	82	BC177V1	82	BC252C-1
82	A4745	82	A608G	82	AC9084	82	BC177V1-1	82	BC253
82	A480(JAPAN)	82	A609	82	AC9085	82	BC178-1	82	BC253-1
82	A4802-00004	82	A609A	82	ACDC9000-1	82	BC178A-1	82	BC253A
82	A4815	82	A609B	82	AD29E-1	82	BC178B-1	82	BC253A-1
82	A481 A0030	82	A609C	82	AD29E-2	82	BC178D	82	BC253B
82	A482(JAPAN)	82	A609D	82	AD29E10	82	BC178D-1	82	BC253B-1
82	A4822-130-40348	82	A609E	82	AD29E4	82	BC178V	82	BC253C
82	A4844	82	A609F	82	AD29E5	82	BC178V-1	82	BC253C-1
82	A489751-02B	82	A609G	82	AD29E6	82	BC178V1	82	BC256
82	A489751-031	82	A610	82	AD29E7	82	BC178V1-1	82	BC256-1
82	A490(JAPAN)	82	A610074-1	82	AD29E8	82	BC179-1	82	BC256A
82	A490Y	82	A610083	82	AD29E9	82	BC179A-1	82	BC256A-1
82	A493	82	A610083-1	82	AD30A1	82	BC179B-1	82	BC256B
82	A493-0	82	A610083-2	82	AD30A2	82	BC181	82	BC256B-1
82	A493GR	82	A610083-3	82	AD30A3	82	BC181A	82	BC256C
82	A493Y	82	A610110-1	82	AD30A4	82	BC181A-1	82	BC256C-1
82	A494-GR	82	A610120-1	82	AD30A5	82	BC187	82	BC257
82	A494-0	82	A610B	82	AF21490	82	BC187-1	82	BC257-1
82	A494-Y	82	A611	82	AF3570	82	BC190B	82	BC257VI
82	A495(JAPAN)	82	A611-4E	82	AF3590	82	BC192		
82	A495-G	82	A617K	82	AF699	82	BC192-1		

GE	Industry Standard No.	GE	Industry Standard No.	GE	Industry Standard No.	GE	Industry Standard No.	GE	Industry Standard No.
82	BC258	82	BCW79-16	82	BFS40A	82	BSY-41	82	EWQ202
82	BC258-1	82	BCW80-1Q	82	BFS41	82	BSY40	82	ETZP-546
82	BC258VI	82	BCW80-16	82	BF869	82	BSY41	82	ETZP-623
82	BC259	82	BCW82	82	BFV20	82	BT82	82	F209
82	BC259-1	82	BCW96	82	BFV21	82	BT832	82	F21490
82	BC25BB	82	BCW96A	82	BFV22	82	C00-686-0241	82	F3559
82	BC260	82	BCW96B	82	BFV25	82	C00686-0258-0	82	F3570
82	BC260-1	82	BCW96K	82	BFV26	82	C00686602720	82	F3590
82	BC260A	82	BCW96KA	82	BFV29	82	C51909B	82	F3597
82	BC260A-1	82	BCW96KB	82	BFV30	82	C610	82	F549-1
82	BC260B	82	BCW97	82	BFV33	82	C673C2	82	F699
82	BC260B-1	82	BCW97A	82	BFV34	82	C673D	82	FC8-9012-HH
82	BC260C	82	BCW97B	82	BFV82	82	C686-248-0	82	FC8-9012F
82	BC260C-1	82	BCW97K	82	BFV82A	82	C9082	82	FC8-9012G
82	BC261	82	BCW97KA	82	BFV82B	82	C9083	82	FC81170P
82	BC261A	82	BCW97KB	82	BFV82C	82	C9084	82	FC89012H
82	BC261A-1	82	BCY10	82	BFV86	82	C9085	82	FC89012HE
82	BC261B	82	BCY10A	82	BFV86A	82	CC82005B	82	FC89015B
82	BC261B-1	82	BCY11	82	BFV86B	82	CC86228P	82	FD-1029-JP
82	BC261C	82	BCY11A	82	BFV86C	82	CC89015	82	FI-1007
82	BC261C-1	82	BCY12	82	BFW89	82	CD10000-1B	82	FI-1008
82	BC262	82	BCY12A	82	BFW90	82	CD437	82	F824954
82	BC262-1	82	BCY35	82	BFX29	82	CD445	82	FT0019M
82	BC262A	82	BCY35A	82	BFX30	82	CDC-9000-1D	82	FT052
82	BC262A-1	82	BCY37	82	BFX35	82	CDC10000-1E	82	FT1341
82	BC262B	82	BCY37A	82	BFX65	82	CDC496	82	FT1746
82	BC262B-1	82	BCY38	82	BFY94	82	CDC746(ZENITH)	82	F219H
82	BC262C	82	BCY38A	82	BJ10	82	CDC9000-1	82	F219M
82	BC262C-1	82	BCY39	82	BJ11	82	CDC9000-1D	82	FT3638
82	BC263	82	BCY39A	82	BJ11A	82	CE4005C	82	FT5041
82	BC263A	82	BCY40	82	BJ11B	82	CK942	82	G05-407-Y
82	BC263A-1	82	BCY40A	82	BJ12	82	C8-2005B	82	G03007
82	BC263B	82	BCY54	82	BJ12B	82	C8-2005C	82	G13638
82	BC263B-1	82	BCY54A	82	BJ12C	82	C8-9015D	82	GB-82
82	BC263C	82	BCY70	82	BJ13	82	C811210	82	GET3638
82	BC263C-1	82	BCY70A	82	BJ13A	82	C811170P	82	GET3638A
82	BC266	82	BCY71	82	BJ13B	82	C81221P	82	G13638
82	BC266-1	82	BCY71A	82	BJ14	82	C81228	82	G13638A
82	BC266A	82	BCY72	82	BJ14A	82	C81228B	82	G13644
82	BC266B	82	BCY72A	82	BJ15	82	C81251B	82	G13702
82	BC281	82	BCY77IX	82	BJ160	82	C81251P	82	G13703
82	BC281-1	82	BCY77VII	82	BJ161	82	C81294E	82	GM760
82	BC281A	82	BCY77VIII	82	BJ161A	82	C81294H	82	GME040-1
82	BC281A-1	82	BCY90	82	BJ161B	82	C81298	82	G03007
82	BC281B	82	BCY90A	82	BJ161C	82	C81303	82	G03014
82	BC281B-1	82	BCY90B	82	BJ1A	82	C81308	82	G89015H
82	BC281C	82	BCY90B-1	82	BJ1B	82	C81354	82	G89015I
82	BC281C-1	82	BCY91	82	BJ1C	82	C81627	82	G89015J
82	BC285	82	BCY91A	82	BJ2	82	C83702	82	G71644
82	BC285-1	82	BCY91B-1	82	BJ2A	82	C83703	82	HA-00495
82	BC291	82	BCY92	82	BJ2B	82	C83906	82	HA-00610
82	BC291-1	82	BCY92A	82	BJ2C	82	C85447	82	HA7501
82	BC291A	82	BCY92B	82	BJ2D	82	C85448	82	HA7502
82	BC291A-1	82	BCY93	82	BJ3	82	C86228P	82	HA7506
82	BC291D	82	BCY93A	82	BJ3A	82	C89012	82	HA7510
82	BC291D-1	82	BCY93B	82	BJ3B	82	C89012E	82	HA7533
82	BC292	82	BCY93B-1	82	BJ4	82	C89012E-P	82	HA7534
82	BC292A	82	BCY94A	82	BJ4A	82	C89012FG	82	HA7537
82	BC292A-1	82	BCY94B	82	BJ4B	82	C89012HG	82	HA7538
82	BC292D	82	BCY94B-1	82	BJ4C	82	C89012HH	82	HA7543
82	BC292D-1	82	BCY95	82	BJ4D	82	C89012HH/3490	82	HA7597
82	BC307	82	BCY95A	82	BJ5	82	C89015C	82	HA7598
82	BC307A	82	BCY95B	82	BJ56	82	C89015C2	82	HA7599
82	BC307V1	82	BCY98	82	BJ5A	82	C89015D	82	HA7633
82	BC308	82	BCY98A	82	BJ5B	82	C89015EP	82	HA7815
82	BC308A	82	BCY98B-1	82	BJ6	82	C89020E	82	HEP52
82	BC308B	82	BCZ10	82	BJ6A	82	C89020P	82	HEP57
82	BC308BV1	82	BCZ10A	82	BJ6B	82	C89128-B2	82	HEP708
82	BC309	82	BCZ10B	82	BJ6C	82	C89128C1	82	HEP715
82	BC309A	82	BCZ10C	82	BJ6D	82	C89129	82	HEP716
82	BC309B	82	BCZ11	82	BJ7	82	C89129B	82	HEP717
82	BC315	82	BCZ11A	82	BJ7A	82	C89129B1	82	HEP739
82	BC325	82	BCZ12	82	BJ7B	82	C89129B2	82	HEP80026
82	BC325A	82	BCZ12A	82	BJ7C	82	D-50492-01	82	HR-71
82	BC326	82	BCZ12B	82	BJ7D	82	D29A4	82	HR71
82	BC326A	82	BCZ12C	82	BJ8	82	D29A5	82	HR84(PNP)
82	BC328-16	82	BCZ13	82	BJ8A	82	D29A6	82	HS-40027
82	BC381	82	BCZ13A	82	BJ8B	82	D29E1	82	HS-40031
82	BC400	82	BCZ13B	82	BJ8C	82	D29E2	82	HS-40040
82	BC404	82	BCZ13C	82	BJ8D	82	D29E4	82	HS-40050
82	BC404VI	82	BCZ13D	82	BJ9	82	D29E5	82	HS-90028
82	BC405	82	BCZ13F	82	BJ9A	82	D29E6	82	HT100
82	BC405A	82	BCZ13G	82	BJ9B	82	D29E7	82	HT101
82	BC405B	82	BCZ13H	82	BJ9C	82	D29E8	82	HT104941B-O
82	BC406B	82	BCZ14	82	BJ9D	82	D29F5	82	HT104941C-O
82	BC415A	82	BCZ14A	82	BMT1991	82	D29F6	82	HT104941CO
82	BC415B	82	BCZ14B	82	BMT2303	82	D29F7	82	HT104942A
82	BC416A	82	BCZ14C	82	BMT2411	82	D30A1	82	HT104951A-O
82	BC416B	82	BCZ14D	82	BMT2412	82	D30A2	82	HT104951B
82	BC417	82	BCZ14E	82	BO8875/2	82	D30A3	82	HT104951B-O
82	BC418	82	BCZ14F	82	BR-82	82	D30A4	82	HT104951C
82	BC419	82	BP-832	82	BSV21	82	D30A5	82	HT105611A
82	BC432	82	BF340A	82	BSV21A	82	DED4191	82	HT105611B
82	BC478	82	BF340B	82	BSV43A	82	D8-82	82	HT105611BO
82	BC478A	82	BF340C	82	BSV44A	82	D8-83	82	HT105611C
82	BC478B	82	BF340D	82	BSV45A	82	D882	82	HT105612B
82	BC479	82	BF341A	82	BSV47A	82	D883	82	HT105621B-O
82	BC479B	82	BF341B	82	BSV48A	82	E13-001-02	82	HT105621BO
82	BC512	82	BF341C	82	BSV49A	82	E13-001-03	82	HT106731B
82	BC512A	82	BF341D	82	BSV55A	82	E13-001-04	82	HT600001F
82	BC513	82	BF342A	82	BSV55AP	82	E13-006-02	82	HT600011H
82	BC513A	82	BF342B	82	BSV55P	82	E211	82	HT6000210
82	BC514A	82	BF342C	82	BSV96	82	E213	82	HX-50105
82	BC556	82	BF342D	82	BSV97	82	EA15X185	82	HX-50112
82	BC556A	82	BFS14A	82	BSV98	82	EA15X233	82	I81030
82	BC556VI	82	BFS14B	82	BSW-21A	82	EA3714	82	I9680
82	BC557	82	BFS14C	82	BSW-22	82	ECG159	82	IP20-0009
82	BC557A	82	BFS14D	82	BSW-22A	82	ECG290	82	IRTR52
82	BC557VI	82	BFS16	82	BSW-24	82	ED1602C	82	J241225
82	BC558	82	BFS16A	82	BSW-44	82	ED1602D	82	J241226
82	BC558VI	82	BFS16B	82	BSW-44A	82	E13-006-02	82	J241253
82	BC559	82	BFS16C	82	BSW-45	82	EN1132	82	J241259
82	BCW35	82	BFS16D	82	BSW-45A	82	EN2894	82	J24640
82	BCW37	82	BFS26	82	BSW-72	82	EN2894A	82	J9680
82	BCW37A	82	BFS26A	82	BSW-73	82	EN2907	82	J9697
82	BCW56	82	BFS26B	82	BSW-74	82	EN3250	82	JA1050
82	BCW56A	82	BFS26C	82	BSW-75	82	EN3504	82	JA1050G
82	BCW57	82	BFS26D	82	BSW21	82	EN3905	82	JA1050GL
82	BCW57A	82	BFS26E	82	BSW21A	82	EN3906	82	K4-505
82	BCW58	82	BFS26F	82	BSW22	82	EN3962	82	K9682
82	BCW58A	82	BFS26G	82	BSW22A	82	EN722	82	KLH4746
82	BCW59	82	BFS31	82	BSW24	82	EP15X17	82	LJ-152
82	BCW59A	82	BFS32	82	BSW44	82	EP15X26	82	LJ152
82	BCW61	82	BFS32P	82	BSW44A	82	EP15X48	82	LJ152(0)
82	BCW61C	82	BFS33	82	BSW45	82	EP15X48(PNP)	82	LJ152-0
82	BCW61D	82	BFS33P	82	BSW45A	82	EP15X53	82	LJ152B
82	BCW62	82	BFS34P	82	BSW70	82	EPX15X17	82	LM-1149
82	BCW62A	82	BFS37	82	BSW72	82	ES15X101	82	LM-1150
82	BCW63	82	BFS37A	82	BSW73	82	ES15X128	82	LM-1151
82	BCW63A	82	BFS40	82	BSW74	82	ES15X9	82	LM-2989
82	BCW64A	82		82	BSW75	82	ES15X90	82	LM1153
82	BCW69	82		82	BSX36	82	ES34(ELCOM)	82	LM1404
82	BCW69R	82		82	BSY-40	82	ES65(ELCOM)	82	
82	BCW79-1Q	82		82		82		82	

GE	Industry Standard No.	GE	Industry Standard No.	GE	Industry Standard No.	GE	Industry Standard No.	GE	Industry Standard No.
82	LM1795	82	NKT20329	82	P2T	82	SA52B	82	SP84314
82	LS-0079-01	82	NKT20329A	82	P3C	82	SA52BC	82	SP84314A
82	LS-0079-02	82	NKT20339	82	P3CA	82	SA53	82	SP84330
82	M094-585-46	82	NN650	82	P3Z	82	SA53A	82	SP84338
82	M4442	82	NPS404A	82	P480A0022	82	SA54	82	SP84348
82	M446	82	NR-601AT	82	P480A0023	82	SA55	82	SP84348A
82	M4525	82	NR601BT	82	P480A0027	82	SA55A	82	SP84354
82	M4590	82	NR621AT	82	P4C	82	SA56	82	SP84354A
82	M4745	82	NR621AU	82	P4G	82	SA56A	82	SP84355
82	M4815	82	NR621AY	82	P4K	82	SA70	82	SP84365
82	M4815D	82	NS1000	82	P4P	82	SA70A	82	SP84365A
82	M4943	82	NS1000A	82	P4R	82	SCD-T326	82	SP84397
82	M4989	82	NS1001	82	P4Y	82	SE5-0370	82	SP84452
82	M644	82	NS1001A	82	P5B	82	SE5-0798	82	SP84452A
82	M65A	82	NS1672	82	P5C	82	SE5-0831	82	SP84458
82	M65B	82	NS1672A	82	P5D	82	SE5-0949	82	SP84458A
82	M65C	82	NS1673	82	PA1000	82	SE5-1057	82	SP84460
82	M65D	82	NS1674	82	PA1001	82	SE5-1223	82	SP84460A
82	M65E	82	NS1674A	82	PA1001A	82	SF8014	82	SP84473
82	M65F	82	NS1675	82	PI-10,131	82	SHA7530	82	SP84473A
82	M6931	82	NS1861	82	PIC	82	SHA7531	82	SP84480
82	M7127	82	NS1861A	82	PIT-50	82	SHA7532	82	SP84480A
82	M7127/P28	82	NS1862	82	PIT-79	82	SHA7533	82	SP84489
82	M829A	82	NS1863	82	PIT-81	82	SHA7534	82	SP84489A
82	M829B	82	NS1863A	82	PL1031	82	SHA7536	82	SP847
82	M829C	82	NS1864	82	PL1033	82	SHA7537	82	SP84813
82	M829D	82	NS1864A	82	PL1034	82	SHA7538	82	SP84815
82	M829E	82	NS404	82	PL1101	82	SK1639	82	SP85007
82	M829F	82	NS6001	82	PL1102	82	SK1639D	82	SP85007-1
82	M833	82	NS6062	82	PL1103	82	SK1640	82	SP85007-1A
82	M9334	82	NS6062A	82	PL1104	82	SK1856	82	SP85007-2
82	M9514	82	NS6063	82	PN	82	SK1856A	82	SP85007-2A
82	M9526	82	NS6063A	82	PTC103	82	SK2604	82	SP85007A
82	M9527	82	NS6064	82	Q-00984R	82	SK2604A	82	SP85008
82	M9531	82	NS6064A	82	Q-36	82	SK3118	82	SPS514
82	M9571	82	NS6065	82	Q-36A	82	SK5797	82	SPS514A
82	M9649	82	NS6065A	82	Q0415	82	SK5798	82	SPS6458
82	ME0404	82	NS6211	82	Q5087Z	82	SK6345	82	SPS6109
82	ME0404-1	82	NS6211A	82	QA-21	82	SK6346	82	SPS6109A
82	ME0404-2	82	NS6241	82	QA-21A	82	SK6347	82	SPS668
82	ME501	82	NS661	82	Q0-419	82	SK6347A	82	SPS6953
82	MF3504	82	NS662	82	QQ61689	82	SK7664	82	ST1606
82	MM3726	82	NS663	82	QQ61689A	82	SKA-4061	82	ST1606
82	MM3905	82	NS664	82	Q8316	82	SKA1279	82	SS1906A
82	MM3906	82	NS665	82	R8967	82	SKA4129	82	SS2503
82	MM4048	82	NS666	82	R8967A	82	SKA6250	82	SS2503A
82	MM4052	82	NS667	82	R8969	82	SKWB07006	82	SS6724
82	M999	82	NS668	82	R8969A	82	SL119	82	SSA43
82	MPL1000	82	NS732	82	RE63	82	SM-A-726658	82	SSA43A
82	MPS-3638A	82	NS732A	82	REN159	82	SM-B-523974	82	SSA43A-1
82	MPS-5702	82	00415	82	REN63	82	SM-B-574495	82	SSA46
82	MPS1572	82	0C200	82	RS-110	82	SM1507	82	SSA46A
82	MPS3638	82	0C201	82	RS-2021	82	SM4547	82	SSA48
82	MPS3638A	82	0C202	82	RS-2022	82	SM4574A	82	SSA48A
82	MPS3639	82	0C203	82	RS-2023	82	SM4719	82	ST-MPS9682J
82	MPS3640	82	0C204	82	RS-2024	82	SN-400-319-P1	82	ST-MPS9750D
82	MPS3702	82	0C205	82	RS-7665A	82	SNT204	82	ST.082.115.015
82	MPS3703	82	0C206	82	R87665	82	SNT204A	82	ST129-1
82	MPS404	82	0C207	82	RT-103	82	SP70	82	ST6110
82	MPS404A	82	0C430	82	RT3065	82	SP90	82	ST62180
82	MPS5086	82	0C430K	82	RT3065A	82	SPS-1539	82	ST8014
82	MPS6134	82	0C440	82	RT3071	82	SPS-952	82	ST8033
82	MPS6516	82	0C440K	82	RT3071A	82	SPS1097	82	ST8033A
82	MPS6517	82	0C443	82	RT8895	82	SPS1097A	82	ST8034
82	MPS6518	82	0C443K	82	RV2351	82	SPS12	82	ST8034A
82	MPS6519	82	0C445	82	RVTC81382	82	SPS1523	82	ST8035
82	MPS6522	82	0C445K	82	RVT822411	82	SPS1523A	82	ST8035A
82	MPS6523	82	0C449	82	S-1367A	82	SPS2269	82	ST8036
82	MPS6524	82	0C450	82	S0026	82	SPS2272	82	ST8036A
82	MPS6533	82	0C460	82	S017446	82	SPS2279	82	ST8065
82	MPS6534	82	0C460K	82	S019843	82	SPS2279	82	ST8065A
82	MPS6534M	82	0C463	82	S022012	82	SPS3329	82	ST8190
82	MPS6535	82	0C463K	82	S023735	82	SPS3724	82	ST8500
82	MPS6562	82	0C465	82	S1047	82	SPS3724A	82	ST8500A
82	MPS6563	82	0C465K	82	08133.0008T	82	SPS3786	82	ST8509
82	MPS6598	82	0C466	82	08133.1003T	82	SPS3786A	82	ST8509A
82	MPS6599	82	0C466K	82	S1350	82	SPS3924	82	STM73Q
82	MPS9467A-T	82	0C467	82	S1350A	82	SPS3927	82	SX3702
82	MPS9666	82	0C467K	82	S1367	82	SPS3927A	82	SX3702A
82	MPS9680	82	0C468	82	S18100	82	SPS3931	82	SX61M
82	MPS9680H	82	0C468K	82	S1889	82	SPS3931A	82	SX61MA
82	MPS9680H/I	82	0C469	82	S2091	82	SPS3987	82	ST14275
82	MPS9680I	82	0C469K	82	S2128	82	SPS3988	82	T-246
82	MPS9680I/J	82	0C470	82	S2129	82	SPS3988A	82	T-251
82	MPS9680J	82	0C470K	82	S23130	82	SPS3990	82	T-Q5077
82	MPS9680T	82	0C700	82	S24226	82	SPS3990A	82	T-Q5087
82	MPS9681	82	0C700A	82	S2525	82	SPS4000	82	T01-023
82	MPS9681I	82	0C700B	82	S2645	82	SPS4007	82	T1-503
82	MPS9681K	82	0C702	82	S2645A	82	SPS4007A	82	T1-503A
82	MPS9681T	82	0C702A	82	S2988	82	SPS4013	82	T1-743
82	MPS9682	82	0C702B	82	S2988A	82	SPS4013A	82	T1-743A
82	MPS9682-I	82	0C704	82	S3004	82	SPS4014	82	T1-744
82	MPS9682I	82	0C740	82	S31551	82	SPS4014A	82	T1-744A
82	MPS9682J	82	0C7400	82	S3639	82	SPS4018	82	T1-752
82	MPS9682K	82	0C7409	82	S3640A	82	SPS4018A	82	T1-752A
82	MPS9750D	82	0C740M	82	S3655	82	SPS4019	82	T1-90b
82	MPS9750P	82	0C742	82	S3655A	82	SPS4019A	82	T112
82	MPS9750Q	82	0C7420	82	S39094	82	SPS401K	82	T1803
82	MPS97500P	82	0C742M	82	S4249	82	SPS4025	82	T1803A
82	MPSA55	82	P00347100	82	S520	82	SPS4025A	82	T1804
82	MPSA56	82	P00347101	82	S608	82	SPS4026	82	T1804A
82	MPSA70	82	P04-45-0015-P1	82	SA310	82	SPS4026A	82	T1337A
82	MPX9681J	82	P04-45-0016-P1	82	SA310A	82	SPS4027	82	T1838
82	M37505	82	P04-450016-002	82	SA311	82	SPS4027A	82	T1838A
82	M39667	82	P04450016-004	82	SA311A	82	SPS4028	82	T1853
82	M39681	82	P1000A	82	SA312	82	SPS4028A	82	T1853A
82	M8R63	82	P1901-70	82	SA312A	82	SPS4031	82	T1854
82	M8R631	82	P1B	82	SA313	82	SPS4031A	82	T185A
82	MT0404	82	P1C	82	SA314	82	SPS4054	82	T1861A
82	MT0404-1	82	P1C0	82	SA314A	82	SPS4054A	82	T1891
82	MT0404-2	82	P1D	82	SA315	82	SPS4056	82	T1891A
82	MT0411	82	P1H	82	SA315A	82	SPS4056A	82	T1893
82	MT0412	82	P1J	82	SA316	82	SPS4064	82	T1893A
82	MT0413	82	P1N	82	SA316A	82	SPS4064A	82	T276
82	MT1131	82	P1N-1	82	SA410	82	SPS4072	82	T340
82	MT1131A	82	P1N-2	82	SA410A	82	SPS4072A	82	T3570
82	MT1132	82	P1N-3	82	SA411	82	SPS4073	82	T3570A
82	MT1132A	82	P1P	82	SA411A	82	SPS4073A	82	T39
82	MT1132B	82	P1P-1	82	SA412	82	SPS4076	82	T460(SEARS)
82	MT1254	82	P1W	82	SA412A	82	SPS4076A	82	T475(SEARS)
82	MT1255	82	P2A	82	SA413	82	SPS4078	82	T482(SEARS)
82	MT1256	82	P2E	82	SA413A	82	SPS4078A	82	T597-1
82	MT1257	82	P2G	82	SA414	82	SPS4082	82	T900
82	MT1258	82	P2GE	82	SA414A	82	SPS4082A	82	T9681
82	MT1259	82	P2H	82	SA415	82	SPS4086	82	TA198036-2
82	MT1420	82	P2L	82	SA415A	82	SPS4086A	82	TC1i98
82	MT1991	82	P2M-1	82	SA416	82	SPS4087	82	TC1i99
82	MT2303	82	P2M-2	82	SA416A	82	SPS4087A	82	TC1i99B
82	MT2411	82	P2M-3	82	SA50	82	SPS4090	82	TD2905
82	MT2412	82	P2P	82	SA50A	82	SPS4090A	82	TD400
82	MT726	82	P2S	82	SA51	82	SPS42	82	TD401
82	MT869	82	P2W	82	SA52	82	SPS42A	82	TD402
82	NB121	82		82	SA52A	82	SPS4302	82	TD500
82	NJ101B	82		82	SA52AC	82		82	TD501

GE	Industry Standard No.	GE	Industry Standard No.	GE	Industry Standard No.	GE	Industry Standard No.	GE	Industry Standard No.
82	TD502	82	TT28A495-Y-A	82	2N2163	82	2N3672	82	28306
82	TD950	82	TV-44	82	2N2164	82	2N3673	82	28307
82	TE3905	82	TV-47A	82	2N2165	82	2N3702	82	28307A
82	TE4125	82	TV-57	82	2N2166	82	2N3703	82	28321
82	TE4126	82	TV-87	82	2N2167	82	2N3798	82	28321O
82	TE5365	82	TV-93	82	2N2178	82	2N3799	82	28321OA
82	TE5366	82	TV24214	82	2N2181	82	2N3829	82	28321A
82	TE5378	82	TV24214A	82	2N2182	82	2N3840	82	28322
82	TE5379	82	TV24363	82	2N2183	82	2N3857	82	28322O
82	TE5447	82	TV24363A	82	2N2184	82	2N3905	82	28322OA
82	TE5448	82	TV24495	82	2N2185	82	2N3906	82	28321
82	TG28A608	82	TV24495A	82	2N2186	82	2N3910	82	28322
82	TG28A608-D	82	TV44	82	2N2187	82	2N3911	82	28322AB
82	TG28A608-E	82	TV72	82	2N2274	82	2N3912	82	28323
82	TG28A608C	82	TVS-28A564	82	2N2275	82	2N3913	82	28323O
82	TI-428	82	TVS-28A564A	82	2N2276	82	2N3914	82	28323OA
82	TI-429	82	TVS-28A564P	82	2N2277	82	2N3930	82	28323A
82	TI-503	82	TVS-28C564	82	2N2278	82	2N3931	82	28324
82	TI-743	82	TVS28A564	82	2N2279	82	2N3962	82	28324O
82	TI-744	82	TVS28A564-0	82	2N2280	82	2N3963	82	28324OA
82	TI-752	82	TVS28A564A	82	2N2281	82	2N3964	82	28324A
82	TI-890	82	TVS28A564C	82	2N2299	82	2N3965	82	28326
82	TI-905	82	TVS28A564P	82	2N2303	82	2N3977	82	28326A
82	TI-906	82	TVS28A564PY	82	2N2332	82	2N3978	82	28327
82	TI503	82	TVS28A564Q	82	2N2333	82	2N3979	82	2S673C
82	TI743	82	TVS28A607	82	2N2334	82	2N4008	82	28A-NJ101
82	TI744	82	TVS28A609	82	2N2335	82	2N4121	82	2SA402
82	TI752	82	TVS28C564	82	2N2336	82	2N4125	82	2SA467O-0
82	TI803	82	TVS28C564-3	82	2N2337	82	2N4126	82	2SA502-0
82	TI804	82	TVS28C5640	82	2N2370	82	2N4149	82	2SA502-R
82	TI8104	82	TVS28C564R	82	2N2371	82	2N4207	82	2SA502-Y
82	TI8112	82	1-0006-0023	82	2N2372	82	2N4208	82	2SA550A
82	TI837	82	01-010562	82	2N2373	82	2N4209	82	2SA561
82	TI838	82	001-02	82	2N2377	82	2N4221	82	2SA561(O)
82	TI850	82	001-02010	82	2N2378	82	2N4248	82	2SA561-O
82	TI853	82	001-022010	82	2N2393	82	2N4249	82	2SA561-OR
82	TI854	82	001-022020	82	2N2394	82	2N4250	82	2SA561-ORN
82	TI861	82	001-03	82	2N2411	82	2N4250A	82	2SA561-O
82	TI891	82	1-034/2207	82	2N2412	82	2N4257	82	2SA561-OR
82	TI893	82	001-04	82	2N2424	82	2N4257A	82	2SA561-ORG
82	TI893-BLU	82	1-043/2207	82	2N2425	82	2N4258	82	2SA561-R
82	TI893-GRN	82	001-533-00	82	2N2595	82	2N4258A	82	2SA561-RD
82	TI893-GRY	82	01-572588	82	2N2596	82	2N4288	82	2SA561-RED
82	TI893-VIO	82	1B3096-1	82	2N2597	82	2N4289	82	2SA561-Y
82	TI893-YEL	82	1V68611A47	82	2N2598	82	2N4290	82	2SA561-YEL
82	TIX804	82	1V68611A47A	82	2N2599	82	2N4291	82	2SA561-YL
82	TIX805	82	1W11700	82	2N2600	82	2N4354	82	2SA561OR
82	TIX890	82	1W11700A	82	2N2601	82	2N4355	82	2SA561ORN
82	TIX891	82	1W11702	82	2N2602	82	2N4356	82	2SA561R
82	TM1614	82	1W11702A	82	2N2603	82	2N4359	82	2SA561RD
82	TM1712	82	1W11711	82	2N2604	82	2N4402	82	2SA561RED
82	TM2614	82	1W11712	82	2N2605	82	2N4403	82	2SA561Y
82	TM2712	82	1W8537	82	2N2605A	82	2N4411	82	2SA561YEL
82	TN061690	82	1W9148	82	2N2695	82	2N4413	82	2SA561YL
82	TN061703	82	1W9640	82	2N2696	82	2N4413A	82	2SA565K
82	TNO61690	82	1W9640A	82	2N2709	82	2N4415	82	2SA592Y
82	TNO61703	82	1W9728	82	2N2800	82	2N4415A	82	2SA608
82	TNJ71037	82	1W9728A	82	2N2800/46	82	2N4452	82	2SA608(C)
82	TNJ71773	82	1W9782	82	2N2801	82	2N4916	82	2SA608(D)
82	TNJ71774	82	1W9782A	82	2N2837	82	2N4917	82	2SA608(P)
82	TNJ72154	82	1W9810	82	2N2838	82	2N4965	82	2SA608-0
82	TP3638	82	1W9810A	82	2N2861	82	2N4971	82	2SA608-C
82	TP4125	82	1W98108	82	2N2862	82	2N4972	82	2SA608-D
82	TP4126	82	1W98108A	82	2N2894	82	2N4980	82	2SA608-E
82	TP4257	82	002-009800	82	2N2894A	82	2N4981	82	2SA608-L
82	TP4258	82	002-010300	82	2N2906	82	2N4982	82	2SA608-0
82	TP86516	82	002-010300A	82	2N2906A	82	2N5086	82	2SA608-OR
82	TP86517	82	002-010500	82	2N2907	82	2N5087	82	2SA608A
82	TP86518	82	002-010500A	82	2N2907A	82	2N5138	82	2SA608B
82	TP86519	82	002-010900	82	2N2927	82	2N5139	82	2SA608BL
82	TP86522	82	002-010900A	82	2N2927/46	82	2N5142	82	2SA608C
82	TP86523	82	002-012800	82	2N2944	82	2N5143	82	2SA608D
82	TQ-63	82	002-012800A	82	2N2944A	82	2N5226	82	2SA608D(P)
82	TQ-64	82	002-9800-A	82	2N2945	82	2N5227	82	2SA608E
82	TQ61	82	2.01.03.02	82	2N2945A	82	2N5230	82	2SA608F
82	TQ61A	82	2CY39	82	2N2946	82	2N5231	82	2SA608G
82	TQ62	82	2CY39	82	2N2946A	82	2N5354	82	2SA608GN
82	TQ62A	82	2D017-165	82	2N3012	82	2N5355	82	2SA608GH
82	TR-1000-2	82	2D017-166	82	2N3039	82	2N5356	82	2SA608H
82	TR-1030-1	82	2D017-167	82	2N3040	82	2N5365	82	2SA608K
82	TR-1030-2	82	2D017-169	82	2N3058	82	2N5366	82	2SA608L
82	TR-1032-2	82	2D027	82	2N3059	82	2N5367	82	2SA608M
82	TR-167(OLSON)	82	2N1025	82	2N3060	82	2N5372	82	2SA608OR
82	TR-19	82	2N1026	82	2N3061	82	2N5373	82	2SA608P
82	TR-19A	82	2N1026A	82	2N3062	82	2N5374	82	2SA608R
82	TR-20	82	2N1027	82	2N3072	82	2N5375	82	2SA608X
82	TR-20A	82	2N1028	82	2N3073	82	2N5378	82	2SA608Y
82	TR-30	82	2N1118	82	2N3081	82	2N5379	82	000002SA609
82	TR-30A	82	2N1118A	82	2N3121	82	2N5382	82	2SA609A
82	TR-31	82	2N1119	82	2N3135	82	2N5383	82	2SA609B
82	TR-4R38	82	2N1131	82	2N3136	82	2N5447	82	2SA609C
82	TR-6R35	82	2N1131A	82	2N3217	82	2N5448	82	2SA609D
82	TR-6R35A	82	2N1132	82	2N3218	82	2N5999	82	2SA609E
82	TR-8007	82	2N1132/46	82	2N3219	82	2N6003	82	2SA609F
82	TR-8007(FISHER)	82	2N1132A	82	2N3248	82	2N6067	82	2SA609G
82	TR-8026	82	2N1132B	82	2N3249	82	2N6076	82	2SA609GN
82	TR-8037	82	2N1135	82	2N3250	82	2N721	82	2SA609J
82	TRO055	82	2N1135A	82	2N3250A	82	2N721A	82	2SA609K
82	TRO1053-1	82	2N1196	82	2N3251	82	2N722	82	2SA609L
82	TRO2051-1	82	2N1219	82	2N3251A	82	2N722A	82	2SA609OR
82	TRO2051-3	82	2N1220	82	2N3305	82	2N858	82	2SA609R
82	TRO2051-5	82	2N1221	82	2N3306	82	2N859	82	2SA609Y
82	TRO2051-6	82	2N1222	82	2N3307	82	2N860	82	2SA610
82	TRO2062-1	82	2N1223	82	2N3308	82	2N861	82	2SA610B
82	TRO2062-6	82	2N1428	82	2N3317	82	2N862	82	2SA617K
82	TRO2063-8	82	2N1439	82	2N3318	82	2N863	82	2SA618K
82	TRN000A	82	2N1440	82	2N3319	82	2N864	82	2SA628
82	TR1030	82	2N1441	82	2N3341	82	2N923	82	2SA628(EF)
82	TR1030-1	82	2N1442	82	2N3346	82	2N924	82	2SA628(P)
82	TR1030-2	82	2N1443	82	2N3401	82	2N925	82	2SA628-0
82	TR1030A	82	2N1474	82	2N3451	82	2N926	82	2SA628-OR
82	TR1032	82	2N1474A	82	2N3464	82	2N927	82	2SA628A
82	TR1032-1	82	2N1475	82	2N3485	82	2N928	82	2SA628AA
82	TR1032A	82	2N1607	82	2N3485A	82	2N935	82	2SA628AD
82	TR1034	82	2N1608	82	2N3486	82	2N936	82	2SA628AB
82	TR1034A	82	2N1643	82	2N3486A	82	2N937	82	2SA628B
82	TR19	82	2N1676	82	2N3496	82	2N938	82	2SA628C
82	TR2327743	82	2N1677	82	2N3504	82	2N939	82	2SA628D
82	TR28	82	2N1917	82	2N3505	82	2N940	82	2SA628E,P
82	TR30	82	2N1918	82	2N3527	82	2N941	82	2SA628EF
82	TR52	82	2N1919	82	2N3545	82	2N942	82	2SA628F
82	TR8007A	82	2N1920	82	2N3547	82	2N943	82	2SA628G
82	TR8020	82	2N1921	82	2N3548	82	2N944	82	2SA628GN
82	TR8026A	82	2N1991	82	2N3549	82	2N945	82	2SA628H
82	TR8055	82	2N2002	82	2N3550	82	2N946	82	2SA628J
82	TS-21756640	82	2N2003	82	2N3579	82	02P1B	82	2SA628K
82	TS2904	82	2N2004	82	2N3580	82	2S3020	82	2SA628L
82	TS2905	82	2N2005	82	2N3581	82	2S3020A	82	2SA628M
82	TS2906	82	2N2006	82	2N3582	82	2S3021	82	2SA628OR
82	TS2907	82	2N2007	82	2N3638	82	2S3021A	82	2SA628R
82	TS97-7	82	2N2104	82	2N3638A	82	2S3030	82	2SA628X
82	TT28A495-0-A	82	2N2105	82	2N3639	82	2S3030A	82	
82	TT28A495-0-A	82	2N2121	82	2N3640	82	2S3040	82	
82		82	2N2162	82	2N3644	82	2S3040A	82	

GE	Industry Standard No.	GE	Industry Standard No.	GE	Industry Standard No.	GE	Industry Standard No.	GE	Industry Standard No.
82	2SA628Y	86	A1170	86	A642-268	86	BSV35C	86	CDC12112D
82	2SA640	86	A121-480	86	A667RED	86	BSV35D	86	CDC12112E
82	2SA640(M)	86	A124623	86	A6B	86	BSV52	86	CDC12112F
82	2SA640A	86	A124624	86	A6E	86	BSV52R	86	CDC5000-1B
82	2SA640B	86	A125329	86	A6P	86	B3X12	86	CDO5038A
82	2SA640C	86	A129509	86	A6P	86	B3X26	86	CDO5071A
82	2SA640D	86	A129510	86	A6T	86	B3X27	86	CDO5075B
82	2SA640E	86	A129511	86	A6U	86	B3X28	86	CE4001C
82	2SA640L	86	A129512	86	A6V-5	86	B3X35	86	CE4010D
82	2SA640M	86	A129513	86	A715PB	86	B3X87A	86	CE4010E
82	2SA640S	86	A129571	86	A772T38	86	B3X88	86	CP1
82	2SA642	86	A129572	86	A772739	86	B3X88A	86	CIL511
82	2SA642A	86	A129573	86	A7A30	86	B3X92	86	CIL512
82	2SA642B	86	A129574	86	A7A31	86	B3X93	86	CIL513
82	2SA642C	86	A1300RN	86	A7A32	86	B3Y-62	86	CIL521
82	2SA642D	86	A130V100	86	A7N	86	B3Y-72	86	CIL522
82	2SA642E	86	A14-602-63	86	A7P	86	B3Y-73	86	CIL523
82	2SA642F	86	A14-603-05	86	A7U	86	B3Y-74	86	CIL531
82	2SA642G	86	A14-603-06	86	A7V	86	B3Y-80	86	CIL532
82	2SA642GN	86	A1462	86	A7W	86	B3Y-82	86	CIL533
82	2SA642H	86	A1518	86	A88	86	B3Y22	86	CK476
82	2SA642J	86	A1519	86	A909-27125-160	86	B3Y23	86	CK477
82	2SA642K	86	A1520	86	A90T2	86	B3Y70	86	CS-2004C
82	2SA642L	86	A1521	86	A916-31025-58	86	B3Y72	86	CS-2007G
82	2SA642M	86	A154(NPN)	86	A916-31025-5B	86	BT930	86	CS-2007H
82	2SA642OR	86	A155(NPN)	86	A921-59B	86	C1023	86	CS-2008P
82	2SA642R	86	A164(NPN)	86	A921-62B	86	C111B	86	CS-3001B
82	2SA642S	86	A165(NPN)	86	A921-63B	86	C115B	86	CS-461B
82	2SA642V	86	A176-025-9-002	86	A921-64B	86	C1205	86	CS-6227G
82	2SA642W	86	A19-020-072	86	A991-01-1316	86	C1205A	86	CS-9016
82	2SA642X	86	A1E	86	A9A	86	C1205B	86	CS-9016P
82	2SA642Y	86	A1G-1A	86	A9D	86	C1205C	86	CS-9018
82	2SA659O	86	A1K	86	AR200	86	C127	86	CS-9018E
82	2SA6661QRS	86	A1P-/4923	86	AR200W	86	C130	86	CS-9018F
82	2SA672	86	A1P-5	86	AR201	86	C155	86	CS-9018G
82	2SA672(B)	86	A1P/4923-1	86	AR201(YELLOW)	86	C156	86	CS-9018H
82	2SA672A	86	A1R-1/4925	86	AR201Y	86	C17	86	CS81014
82	2SA672B	86	A1R-1A	86	AR202	86	C170	86	CS81014D
82	2SA672C	86	A1R-2/4926	86	AR202(GREEN)	86	C172A	86	CS81014E
82	2SA675	86	A1R-24926	86	AR202G	86	C17A	86	CS81014F
82	2SA675A	86	A1R-2A	86	AR212	86	C185V	86	CS81018
82	2SA675B	86	A1R-5	86	AR221	86	C199	86	CS1168E
82	2SA675C	86	A1R/4925A	86	AR25(ORANGE)	86	C206	86	CS81225D
82	2SA677-O	86	A1R/4926A	86	AR25(WHITE)	86	C2475078-3	86	CS81225E
82	2SA677-OR	86	A2006681-95	86	A313	86	C2485078-1	86	CS81225F
82	2SA677TEL	86	A2092693-0724	86	A520	86	C2485079-1	86	CS81226E
82	2SA678	86	A2092693-0725	86	AX9177Q	86	C263	86	CS81226F
82	2SA678(C)	86	A24	86	B9426	86	C269	86	CS81226G
82	2SA678(SONY)	86	A245	86	BC111	86	C271	86	CS81227
82	2SA678-O	86	A245(AMC)	86	BC112	86	C272	86	CS81227D
82	2SA678-OR	86	A2498	86	BC155A	86	C289	86	CS81227F
82	2SA678A	86	A24MW594	86	BC156A	86	C296	86	CS81227G
82	2SA678B	86	A24MW595	86	BC194	86	C3123	86	CS1238
82	2SA678C	86	A24MW596	86	BC194B	86	C313	86	CS81238G
82	2SA678D	86	A24MW597	86	BC195	86	C316	86	CS81238H
82	2SA678F	86	A24T-016-016	86	BC195CD	86	C318	86	CS81238I
82	2SA678G	86	A2C	86	BC295	86	C318A	86	CS81243E
82	2SA678GN	86	A2D	86	BC6500	86	C370	86	CS81243H
82	2SA678H	86	A2F	86	BCW31	86	C370F	86	CS81244H
82	2SA678J	86	A2G	86	BCW31R	86	C370G	86	CS81244J
82	2SA678K	86	A2H	86	BCW32	86	C370H	86	CS81244X
82	2SA678L	86	A2L	86	BCW32R	86	C370J	86	CS81252C
82	2SA678M	86	A2N-2A	86	BCW71	86	C370K	86	CS1293
82	2SA678OR	86	A2P-5	86	BCW71R	86	C384	86	CS81330A
82	2SA678R	86	A2T	86	BCW72	86	C384-0	86	CS81330B
82	2SA678X	86	A2T919	86	BCW72R	86	C384Y	86	CS81330C
82	2SA678Y	86	A2V	86	BCY87	86	C386	86	CS81340D
82	2SA685	86	A2W	86	BCY88	86	C386A-0(TV)	86	CS81340F
82	2SA690D	86	A31-0206	86	BCY89	86	C387	86	CS81340G
82	2SA701	86	A32-2809	86	BD71	86	C387A	86	CS81340H
82	2SA701P	86	A321	86	BF-115	86	C387A(PA-3)	86	CS1350
82	2SA701PJ	86	A3A	86	BF152	86	C387G	86	CS1351
82	2SA701PO	86	A3C	86	BF153	86	C39(TRANSISTOR)	86	CS1359
82	2SA718	86	A3D	86	BF154	86	C392	86	CS1360
82	2SA7200	86	A3N71	86	BF158	86	C39A	86	CS81361E
82	2SA723	86	A3N72	86	BF159	86	C40	86	CS81361F
82	2SA723A	86	A3N73	86	BF160	86	C405	86	CS1861E
82	2SA723B	86	A3P	86	BF162	86	C406	86	CS1508Q
82	2SA723C	86	A3R	86	BF163	86	C424	86	CS81509E
82	2SA723D	86	A38B	86	BF164	86	C429	86	CS81509F
82	2SA723E	86	A417	86	BF165	86	C429J	86	CS1518E
82	2SA723F	86	A418	86	BF173A	86	C429X	86	CS1555
82	2SA723R	86	A419	86	BF176	86	C430	86	CS1661
82	2SA730	86	A420	86	BF187	86	C430H	86	CS1834
82	2SA731	86	A427	86	BF188	86	C430W	86	CS184J
82	2SA735	86	A427(JAPAN)	86	BF194A	86	C454	86	CS2006G
82	2SA701PJ	86	A473	86	BF223	86	C477	86	CS2008H
82	2SA841	86	A4789	86	BF224	86	C5359	86	CS2008H552
82	2SA841-GR	86	A48-134789	86	BF225	86	C544	86	CS429J
82	2SA945-O	86	A48-134837	86	BF233-2	86	C544C	86	CS430H
82	2SA945Z	86	A48-134845	86	BF233-3	86	C544D	86	CS461F
82	2SANJ101	86	A48-134902	86	BF233-4	86	C544E	86	CS469F
83	B2M	86	A48-134904	86	BF233-5	86	C561	86	CS6225F
83	GE-83	86	A48-134922	86	BF234	86	C568	86	CS6226F
83	HEP83024	86	A48-134923	86	BF235	86	C611	86	CS6227E
83	MPS-U01	86	A48-134924	86	BF236	86	C612	86	CS6227F
83	2SC1013	86	A48-134925	86	BF237	86	C613	86	CS9001
83	2SC1474	86	A48-134926	86	BF238	86	C63(JAPAN)	86	CS9016/349Q
83	2SC1474-3	86	A48-134945	86	BF254	86	C641	86	CS9016Q
83	2SC1474-4	86	A48-134962	86	BF262	86	C641B	86	CS90016Q/349Q
83	2SC1474J	86	A48-134963	86	BF263	86	C649	86	CS9016H
83	2SC1474S	86	A48-134964	86	BF264	86	C658	86	CS9018
84	GE-84	86	A48-134965	86	BF557	86	C658A	86	CS9018/349Q
84	HEP83027	86	A48-134966	86	BF594	86	C659	86	CS9018D
84	HEP83028	86	A48-134981	86	BF595	86	C662	86	CS9018E
84	MPS-U51	86	A48-137071	86	BF628	86	C684(JAPAN)	86	CS9018H
84	2N3660	86	A48-137075	86	BF817	86	C684A	86	CS9018F/349Q
85	EA15X386	86	A48-137076	86	BF817R	86	C684A(JAPAN)	86	CS9018Q
85	FT107B	86	A48-137077	86	BFS58	86	C684B	86	CS9021Q-1
85	GE-85	86	A48-137197	86	BFS62	86	C684BK	86	CS9124-C2
85	HEP80030	86	A48-40247901	86	BFV83	86	C684F	86	CS9124B1
85	MPSA18	86	A48-433514A02	86	BFV83A	86	C735(PA-3)	86	CS9125-B1
85	TRO1015	86	A48-63076A82	86	BFV85D	86	C74	86	CS918
85	01-031327	86	A48-97046A05	86	BFV85E	86	C740	86	CT1300
85	2SC1327	86	A48-97046A06	86	BFV85F	86	C749	86	CT1500
85	2SC1327F8	86	A48-97046A07	86	BFV85G	86	C79	86	CXL10B
85	2SC1327S	86	A48-97127A06	86	BFW64	86	C80	86	D092
85	2SC1327T	86	A48-97127A12	86	BFX18	86	C828	86	D069
85	2SC1327TU	86	A48-97127A18	86	BFX19	86	C912	86	D087
85	2SC1327V	86	A480	86	BFX20	86	C924	86	D088
85	2SC1327U	86	A485	86	BFX21	86	C924E	86	D1666
85	2SC1787	86	A4851	86	BFX43	86	C924F	86	D243394
86	A-1854-0485	86	A497	86	BFX44	86	C924M	86	D562
86	A-1854-JBD1	86	A4E-5	86	BFX45	86	C9634	86	D9634
86	A054-157	86	A4G	86	BFY-37	86	C98	86	DX1018
86	A054-158	86	A4T	86	BFY-47	86	C99	86	E2434
86	A054-159	86	A4Y-1A	86	BFY-48	86	CCS2006D	86	E2435
86	A054-163	86	A57A144-12	86	BFY78	86	CCS2008F015	86	E629
86	A054-164	86	A57A145-12	86	BFY90	86	CC89016D	86	E842
86	A054-470	86	A5J	86	BFY90B	86	CC89016E		
86	A068-111	86	A62-19581	86	B3826	86	CC89018F		
86	A068-112	86	A642-254	86	BSV35	86	CDC12112C		
86	A1109	86	A642-260	86	BSV35B				

GE	Industry Standard No.	GE	Industry Standard No.	GE	Industry Standard No.	GE	Industry Standard No.	GE	Industry Standard No.
86	B843	86	FM720A	86	L842	86	R62194	86	S-1041
86	B844	86	FM870	86	LM1110A	86	RE1001	86	S-1058
86	EAO013	86	FM871	86	LM1123H	86	RE1002	86	S-1059
86	EAOO91	86	FM910	86	LM1138G/H	86	RE8070	86	S-1060
86	EAOO93	86	FM911	86	LT1016	86	RRB116	86	S-1062
86	EAOO94	86	FM914	86	LT1016D	86	RRB118	86	S-1078
86	EAOO95	86	FPR40-1003	86	LT1016E	86	RRB119	86	S-1079
86	EA1343	86	FPR50-1003	86	LT1016I	86	RR8989	86	S-1155
86	EA1562	86	FPR50-1004	86	LTH1016(G.E.)	86	RS-109	86	S-1227
86	EA1563	86	F83266	86	LTH1016(G.E.)	86	RS-2015	86	S-1276
86	EA15X113	86	F8326690	86	MO12	86	RS-7102	86	S-1286
86	EA15X131	86	F835529	86	MO24	86	RS-7104	86	S-1296
86	EA15X132	86	F83683	86	M4709	86	RS-7106	86	S-1316
86	EA15X148	86	F8E1001	86	M4733	86	RS-7107	86	S-1317
86	EA15X49	86	F8E3001	86	M4820	86	RS-7108	86	S-1318
86	EA15X50	86	F8E5002	86	M4826	86	RS-7109	86	S-1360
86	EA15X51	86	F8P-1	86	M4837	86	RS-7110	86	S-1361
86	EA15X54	86	F8P-164	86	M4845	86	RS-7113	86	S-1362
86	EA15X55	86	F8P-165	86	M4855	86	RS-7114	86	S-1408
86	EA15X7141	86	F8P-166	86	M4857	86	RS-7115	86	S-1409
86	EA15X7177	86	F8P-166-1	86	M75547-1	86	RS-7202	86	S-5328E
86	EA15X7228	86	F8P-215	86	M75547-2	86	RS-7512	86	S-5670-E
86	EA15X7243	86	F8P-242-1	86	M9010	86	RS1726	86	S-95125
86	EA15X7263	86	F8P-270-1	86	M9482	86	RS6523	86	S-95125A
86	EA15X7587	86	F8P-289-1	86	M9575	86	RS7101	86	S-95126
86	EA15X7722	86	F8P-42	86	ME3001	86	RS7102	86	S-95126A
86	EA15X8589	86	F8P-42-1	86	ME3002	86	RS7104	86	S0016
86	EA15X8608	86	FT1315	86	ME3011	86	RS7106	86	S0020
86	EA15X8609	86	FT1324B	86	ME5001	86	RS7107	86	S0021
86	EA15X8610	86	FT1324C	86	ME8201	86	RS7109	86	S1009
86	EA15X94	86	FT709	86	ME9021	86	RS7110	86	S1019
86	EA1733	86	FV2369A	86	ME9022	86	RS7112	86	S1037
86	EA1793	86	FV2484	86	MHM1001	86	RS7113	86	S1044
86	EA2493	86	FV3014	86	MHM1101	86	RS7115	86	S1058
86	EA2495	86	FV3299	86	MI1546	86	RS7116	86	S1059
86	EA2600	86	FV3300	86	MM1367/280684	86	RS7117	86	S1062
86	EA2601	86	FV914	86	MM1382	86	RS7118	86	S1076
86	EA2602	86	FV918	86	MM1387	86	RS7119	86	S1078
86	EA2603	86	G05-004A	86	MM1803	86	RS7120	86	S1079
86	EA2604	86	G05004A	86	MM1941	86	RS7122	86	S1142
86	EA2605	86	GE-86	86	MM1945	86	RS7123	86	S1153
86	EA3406	86	GMO380	86	MM709	86	RS7124	86	S1227
86	ED592M	86	GME3001	86	MMCM918	86	RS7125	86	S1276
86	EN3011	86	GME3002	86	MP35536	86	RS7126	86	S1296
86	EN718A	86	GME9001	86	MP86511	86	RS7128	86	S1313
86	EN744	86	GME9002	86	MP86539	86	RS7135	86	S1316
86	EN914	86	GME9021	86	MP86541	86	RS7138	86	S1317
86	EN916	86	H1V	86	MP86542	86	RS7139	86	S1318
86	EN918	86	H442	86	MP86543	86	RS7140	86	S1360
86	EP15X55	86	H9623	86	MP86546	86	RS7141	86	S1361
86	EQ8-21	86	H9625	86	MP86547	86	RS7142	86	S1362
86	ES810187	86	HC-00668	86	MP86548	86	RS7144	86	S1408
86	ES15X10	86	HC-00772	86	MP8706	86	RS7145	86	S1409
86	ES15X18	86	HC-00839	86	MP8706A	86	RS7161	86	S1636
86	ES15X19	86	HC-01830	86	MP8805	86	RS7162	86	S1674
86	ES15X2	86	HC206	86	MP894231	86	RS7163	86	S1674A
86	ES15X3	86	HC645	86	MP894235F	86	RS7164	86	S1682
86	ES15X30	86	HC668	86	MP894239	86	RS7166	86	S2159
86	ES15X6	86	HC772	86	MP894238H	86	RS7167	86	0822.5640-080
86	ES15X60	86	HC829	86	MP89600G/H	86	RS7169	86	S2224
86	ES15X69	86	HEP56	86	MP89600H	86	RS7170	86	S25805
86	ES3266	86	HEP718	86	MP89601	86	RS7174	86	S2617
86	ET15X18	86	HEP720	86	MRP501	86	RS7175	86	S2719
86	ET15X19	86	HEP722	86	MST01T	86	RS7176	86	S3019
86	ET15X21	86	HEP723	86	MST5001	86	RS7177	86	S3020
86	ET15X23	86	HEP727	86	MST5018	86	RS7201	86	S32417
86	ET15X30	86	HEP733	86	MST5028	86	RS7202	86	S32669
86	ET15X7	86	HEP80016	86	MST5002T	86	RS7209	86	S326690
86	ET15X9	86	HEP80020	86	MST7502	86	RS7210	86	S33990
86	EU15X1	86	HR58	86	MST7501	86	RS7211	86	S35233
86	EU15X2	86	HR59	86	MST7502	86	RS7212	86	S5327E
86	EU15X3	86	HR60	86	MST7501	86	RS7214	86	S5328E
86	EU15X6	86	HS-1225	86	MT100	86	RS7215	86	S5670R
86	EW162	86	HS-1226	86	MT101	86	RS7216	86	S95125
86	F15810	86	HS-1227	86	MT102	86	RS7217	86	S95125A
86	F15835	86	H8-40014	86	MT106	86	RS7218	86	S95126
86	F15841	86	H8-40019	86	MT107	86	RS7220	86	S95126A
86	F20-1001	86	H8-40045	86	MT743	86	RS7221	86	SA8CH2339
86	F20-1002	86	H8-40047	86	MT744	86	RS7225	86	SC12272
86	F20-1003	86	H8A0049	86	MT753	86	RS7227	86	SC1227G
86	F20-1004	86	HT3037010	86	N-EA15X131	86	RS7228	86	SCA3021
86	F20-1005	86	HT3037711A-0	86	N-EA15X132	86	RS7229	86	SE-1019
86	F215-1001	86	HT3037201	86	NPC173	86	RS7230	86	SE-1044
86	F215-1002	86	HT3037201A	86	NPC188	86	RS7231	86	SE-1419
86	F215-1003	86	HT3037201B	86	N81356	86	RS7237	86	SE-3001
86	F215-1004	86	HT303720A	86	N83300	86	RS7333	86	SE-3002
86	F215-1005	86	HT303941	86	N8381	86	RS7334	86	SE-3005
86	F2427	86	HT303941A	86	N8382	86	RS7511	86	SE-3019
86	F2450	86	HT303941B	86	N86112	86	RS7512	86	SE-5001
86	F2633	86	HT304540A0	86	N86113	86	RS7520	86	SE-5002
86	F2634	86	HT304601C0	86	N87261	86	RS7522	86	SE-5003
86	F2636	86	HT304611B	86	N87267	86	RS7524	86	SE-5006-14
86	F3530	86	HT305351B0	86	N89710	86	RS7532	86	SE-5020
86	F4706	86	HT305351C0	86	NTC-5	86	RS7533	86	SE-5021
86	P9625(G.E.)	86	HT306451	86	P346	86	RT2915	86	SE-5050
86	P9625	86	HT306451B	86	PA-10556	86	RT3069	86	S1019
86	P96N	86	HT306962A-0	86	PBC182	86	RT3070	86	S1044
86	PC5006	86	HT306681C	86	PEP1001	86	RT3095	86	S1419
86	PC8116BE	86	HT309291C	86	PET-101-1	86	RT3225	86	S3001R
86	PC8116BE641	86	HT313592B	86	PET1075	86	RT3226	86	S3001Y
86	PC81225E	86	HX50001	86	PET3001	86	RT3227	86	S3002
86	PC81227E	86	IRTR80	86	PET8201	86	RT3232	86	S3003
86	PC81227EB14	86	IRTR83	86	PET8250	86	RT5261	86	S5006
86	PC81227F	86	J108	86	PET8251	86	RT5200	86	S5010
86	PC81227F743	86	J187	86	PET8300	86	RT5201	86	S5015
86	PC81227G	86	J24596	86	PL1051	86	RT5205	86	S5020
86	PC81227G810	86	J24635	86	PL1053	86	RT5900	86	S5029
86	PC89011E	86	J24636	86	PL1055	86	RT5901	86	S5030
86	PC89011H	86	J24637	86	PL1061	86	RT5902	86	S5031
86	PC89014B	86	J24813	86	PL1062	86	RT5903	86	S5035
86	PC89014D	86	J24814	86	PL1063	86	RT5904	86	S5036
86	PC89016D	86	J24852	86	PL1064	86	RT6157	86	S504
86	PC89018D	86	J24863	86	PL1065	86	RT6158	86	S5040
86	PC89018E	86	J24904	86	PL1081	86	RT6159	86	S5056
86	PC89018F	86	J24905	86	PL1082	86	RT6160	86	S521
86	PC89018G	86	J24915	86	PM195A	86	RT6201	86	SGB-9742
86	PC89018H	86	J24921	86	PMT1767	86	RT6202	86	SK-31024-3
86	PC89066	86	J24923	86	Q-00284R-3	86	RT6203	86	SK1320
86	FJ1033	86	J24933	86	Q-00384R-3	86	RT6600	86	SKT181
86	FK2369A	86	J8P7005	86	Q-00484R-1	86	RT6991	86	SKA-4074
86	FK2484	86	J8P7006	86	Q-00584R-3	86	RT7320	86	SKA-5248
86	FK3014	86	K2109	86	Q8E3001	86	RT7321	86	SKA-5886
86	FK3299	86	K2110	86	R-28C545	86	RT7323	86	SKA4074
86	FK5300	86	K2111	86	R-28C668	86	RT7324	86	SKA4075
86	FK914	86	K2112	86	R-8530-1	86	RT8193	86	SKA4076
86	FK918	86	K2113	86	R06-1001	86	RT8527	86	SKA4525
86	FM1613	86	K2114	86	R06-1002	86	RT930H	86	SKA5248
86	FM1711	86	K2115	86	R06-1003	86	RV1467	86	SKA5886
86	FM2368	86	K2116	86	R06-1004	86	RV1468	86	SKA9013
86	FM2369	86	K2117	86	R06-1005	86	RV1469	86	SKA9096
86	FM2846	86	K2118	86	R06-1006	86	RV1470	86	SL-100
86	FM3014	86	K3880P	86	R2473	86	RV828C645	86	S1100
86	FM708	86	K4-510	86	R2476	86	RVTC81384	86	SM5796
86	FM709	86	KD2119	86	R2477	86	S-1037		

GE	Industry Standard No.	GE	Industry Standard No.	GE	Industry Standard No.	GE	Industry Standard No.	GE	Industry Standard No.
86	SP4168	86	T9011 I	86	TR310250	86	2N3933	86	2SC1336
86	SP8-1351	86	T9011J	86	TR38	86	2N3953	86	2SC1336JK
86	SP8-1352	86	T9016F	86	TR5320326	86	2N3983	86	2SC155
86	SP8-1353	86	T9016H	86	TR80	86	2N3984	86	2SC155A
86	SP8-1473	86	TA-7	86	TR8004	86	2N3985	86	2SC155B
86	SP8-2111	86	TA2401	86	TR8042	86	2N4072	86	2SC155C
86	SP8-2320	86	TA2503	86	TR83	86	2N4073	86	2SC155D
86	SP8-4145	86	TA7303	86	TS9013	86	2N4134	86	2SC155B
86	SP8-4423	86	TA7319	86	TSC614	86	2N4135	86	2SC155P
86	SP8-917	86	TC-0918	86	TT-204	86	2N4251	86	2SC155GN
86	SP81351	86	TC0914	86	TT-204A	86	2N4252	86	2SC155H
86	SP81352	86	TC0918	86	TT-204AB	86	2N4253	86	2SC155J
86	SP81353	86	TC2369A	86	TT-204B	86	2N4274	86	2SC155K
86	SP820	86	TC2483	86	TT-204C	86	2N4275	86	2SC155L
86	SP82110	86	TC2484	86	TT204	86	2N4292	86	2SC155M
86	SP82111	86	TE2484	86	TT204A	86	2N4293	86	2SC155OR
86	SP82135	86	TE2715	86	TT204AB	86	2N4418	86	2SC155R
86	SP82220	86	TE2716	86	TT204B	86	2N4419	86	2SC155X
86	SP82265	86	TE3707	86	TT204C	86	2N4449	86	2SC155Y
86	SP82265-2	86	TE3708	86	TV-15	86	2N4873	86	2SC156
86	SP82320	86	TE3709	86	TV-15B	86	2N4934	86	2SC156A
86	SP82664	86	TE3710	86	TV-22	86	2N4935	86	2SC156B
86	SP83003	86	TE3711	86	TV-32	86	2N4936	86	2SC156C
86	SP83370	86	TE706	86	TV-7	86	2N4996	86	2SC156D
86	SP838	86	TI-410	86	TV1000	86	2N4997	86	2SC156E
86	SP83929	86	TI-417	86	TV115	86	2N5024	86	2SC156 P
86	SP83937	86	TI-420	86	TV16	86	2N5053	86	2SC156G
86	SP8394B	86	TI-430	86	TV17A	86	2N5054	86	2SC156GN
86	SP83952	86	TI-431	86	TV18	86	2N5131	86	2SC156H
86	SP83968	86	TI-474	86	TV2403	86	2N5180	86	2SC156J
86	SP83971	86	TI-490	86	TV2404	86	2N5200	86	2SC156K
86	SP84	86	TI-495	86	TV24102	86	2N5272	86	2SC156L
86	SP840	86	TI25A	86	TV2414B	86	2N5292	86	2SC156M
86	SP84002	86	TI25B	86	TV24203	86	2N5399	86	2SC156OR
86	SP84005	86	TI3016	86	TV24204	86	2N619	86	2SC156R
86	SP84008	86	TI407	86	TV24513	86	2N621	86	2SC156X
86	SP84016	86	TI408	86	TV24587	86	2N702	86	2SC156Y
86	SP84030	86	TI409	86	TV24573	86	2N703	86	2SC156
86	SP84043	86	TI410	86	TV24574	86	2N706	86	2SC170
86	SP84050	86	TI431	86	TV24589	86	2N706A	86	2SC170A
86	SP84051	86	TIS105	86	TV24684	86	2N706B	86	2SC170B
86	SP84068	86	TI824	86	TV32	86	2N706C	86	2SC170C
86	SP84079	86	TI862	86	TV55	86	2N707	86	2SC170D
86	SP84080	86	TI863	86	TV8-1818	86	2N709	86	2SC170E
86	SP84091	86	TI864	86	TV8-28C185A	86	2N709/52	86	2SC170F
86	SP84145	86	TI885	86	TV8-28C208	86	2N709A	86	2SC170G
86	SP84167	86	TI898A	86	TV8-28C287A	86	2N717	86	2SC170GN
86	SP8428	86	TIX876	86	TV8-28C446	86	2N717A	86	2SC170H
86	SP84399	86	TIX880	86	TV8-28C605	86	2N718	86	2SC170J
86	SP84423	86	TIX309	86	TV8-28C606	86	2N718A	86	2SC170K
86	SP84610	86	TIX810	86	TV828C288A	86	2N743	86	2SC170L
86	SP86155	86	TIX828	86	TV828C466	86	2N743A	86	2SC170M
86	SP9915	86	TIX829	86	TV828C538	86	2N744	86	2SC170OR
86	SP9917	86	TIX830	86	TV828C645	86	2N752	86	2SC170R
86	SR130-1	86	TIX831	86	TV828C645A	86	2N756	86	2SC170X
86	ST10	86	TMT-2427	86	TV828C645C	86	2N757	86	2SC170Y
86	ST1026	86	TMT2427	86	TV828C683	86	2N758	86	2SC172
86	ST1050	86	TMT696	86	TV828C762	86	2N759	86	2SC171A
86	ST1051	86	TMT697	86	TV828C828Q	86	2N760	86	2SC171B
86	ST12	86	TMT839	86	1-041/2207	86	2N761	86	2SC171C
86	ST13	86	TMT840	86	1-801-003-12	86	2N762	86	2SC171E
86	ST1336	86	TMT841	86	1-801-003-13	86	2N854	86	2SC171F
86	ST14	86	TMT842	86	1-801-003-14	86	2N841	86	2SC171G
86	ST15	86	TMT843	86	1-801-003-15	86	2N849	86	2SC171GN
86	ST1694	86	TNJ-60066	86	1-801-305-13	86	2N850	86	2SC171H
86	ST29	86	TNJ60066	86	1-801-306	86	2N851	86	2SC171J
86	ST30	86	TNJ61729	86	1-801-306-13	86	2N852	86	2SC171K
86	ST3030	86	TNJ70478	86	1-801-306-14	86	2N913	86	2SC171L
86	ST3031	86	TNJ70478-1	86	1-801-306-15	86	2N914	86	2SC171M
86	ST31	86	TNJ70479	86	01-9016-42221-3	86	2N914/51	86	2SC171OR
86	ST32	86	TNJ70480	86	01-9018-62221-3	86	2N914A	86	2SC171R
86	ST33	86	TNJ70484	86	1-TR-046	86	2N916	86	2SC171X
86	ST34	86	TNJ71173	86	002-009600	86	2N916A	86	2SC171Y
86	ST35	86	TNJ71498	86	002-009601	86	2N918	86	2SC172
86	ST40	86	TNT-839	86	002-9601	86	2N918/51	86	2SC172A
86	ST41	86	TNT-840	86	002-9601-12	86	2N988	86	2SC172B
86	ST415	86	TNT-841	86	2N1005	86	2N989	86	2SC172C
86	ST42	86	TNT839	86	2N1060	86	2S006	86	2SC172D
86	ST43	86	TNT840	86	2N1505	86	2S512	86	2SC172E
86	ST44	86	TNT841	86	2N1506	86	2SC103	86	2SC172F
86	ST45	86	TNT843	86	2N1506A	86	2SC103A	86	2SC172GN
86	ST47025	86	TP4275	86	2N1507	86	2SC103B	86	2SC172H
86	ST60	86	TQ1	86	2N1989	86	2SC103C	86	2SC172J
86	ST61	86	TQ2	86	2N2032	86	2SC103D	86	2SC172K
86	ST6120	86	TQ3	86	2N2193A	86	2SC103E	86	2SC172L
86	ST62	86	TQ5	86	2N2193B	86	2SC103F	86	2SC172M
86	ST70	86	TQ5049	86	2N2195	86	2SC103G	86	2SC172OR
86	ST71	86	TQ6	86	2N2195B	86	2SC103GN	86	2SC172R
86	ST72	86	TQ7	86	2N2197	86	2SC103H	86	2SC172Y
86	ST80	86	TQ8	86	2N2616	86	2SC103J	86	2SC174
86	ST82	86	TQ9	86	2N263	86	2SC103K	86	2SC174A
86	ST9	86	TR-016	86	2N264	86	2SC103L	86	2SC174B
86	ST903	86	TR-01B	86	2N2651	86	2SC103M	86	2SC174C
86	ST904	86	TR-163(OLSON)	86	2N2708	86	2SC103N	86	2SC174D
86	ST904A	86	TR-1R5	86	2N2710	86	2SC103X	86	2SC174E
86	ST905	86	TR-1R35	86	2N2711	86	2SC1117	86	2SC174F
86	ST910	86	TR-24	86	2N2712	86	2SC1117A	86	2SC174G
86	STE400	86	TR-2R31	86	2N2712BLUE	86	2SC1117B	86	2SC174GN
86	SX-3825	86	TR-2R33	86	2N2715	86	2SC1117C	86	2SC174H
86	SX3001	86	TR-2R35	86	2N2716	86	2SC1117D	86	2SC174J
86	SX3825	86	TR-28C371	86	2N2729	86	2SC1117E	86	2SC174K
86	SX3827	86	TR-28C372	86	2N2784	86	2SC1117F	86	2SC174L
86	SYL-2300	86	TR-28C384	86	2N2784/52	86	2SC1117G	86	2SC174M
86	SYL-4131	86	TR-3R31	86	2N2883	86	2SC1117GN	86	2SC174X
86	SYL4131	86	TR-3R35	86	2N2884	86	2SC1117H	86	2SC174Y
86	T-483	86	TR-3R03	86	2N2921	86	2SC1117J	86	2SC185
86	T-484	86	TR-4R35	86	2N2922	86	2SC1117K	86	2SC1852
86	T-486	86	TR-80	86	2N3010	86	2SC1117L	86	2SC1856
86	T-H28C313	86	TR-8004-4	86	2N3035	86	2SC1117M	86	2SC185A
86	T-H28C387	86	TR-8004-5	86	2N3227	86	2SC1117R	86	2SC185B
86	T-Q5075	86	TR-8028	86	2N337A	86	2SC1117X	86	2SC185C
86	T1003-521	86	TR-8029	86	2N3407	86	2SC1117Y	86	2SC185E
86	T1003521	86	TR-8030	86	2N3493	86	2SC115B	86	2SC185F
86	T1894	86	TR-8031	86	2N3508	86	2SC127A	86	2SC185G
86	T1898	86	TR-8032	86	2N3509	86	2SC127B	86	2SC185GN
86	T1828	86	TR-8038	86	2N3511	86	2SC127C	86	2SC185H
86	T1XM15	86	TR-83	86	2N3544	86	2SC127D	86	2SC185J
86	T1XM17	86	TR01026	86	2N3562	86	2SC127E	86	2SC185K
86	T2634	86	TR010602-1	86	2N3570	86	2SC127F	86	2SC185L
86	T308	86	TR112	86	2N3571	86	2SC127G	86	2SC185M
86	T5530	86	TR1512-80	86	2N3572	86	2SC127GN	86	2SC185OR
86	T5535	86	TR210	86	2N3600	86	2SC127H	86	2SC185Q
86	T5536	86	TR228735045311	86	2N3653	86	2SC127J	86	2SC185R
86	T5539	86	TR228735048617	86	2N3648	86	2SC127K	86	2SC185X
86	T3568(RCA)	86	TR228735048618	86	2N3683	86	2SC127L	86	2SC185Y
86	T386(SEARS)	86	TR22C	86	2N3688	86	2SC127M	86	2SC186
86	T9011CD	86	TR28C371	86	2N3689	86	2SC127OR	86	2SC186A
86	T9011EF	86	TR28C372	86	2N3690	86	2SC127R	86	2SC186B
86	T9011G	86	TR28C384	86	2N3691	86	2SC127X	86	2SC186C
86	T9011GEF	86	TR28C535	86	2N3692	86	2SC127Y	86	2SC186D
86	T9011GH	86	TR310250	86	2N3832				
86	T9011H	86	TR310244	86	2N3854				
86	T9011HEF	86	TR310249	86	2N3854A				
				86	2N3932				

GE	Industry Standard No.	GE	Industry Standard No.	GE	Industry Standard No.	GE	Industry Standard No.	GE	Industry Standard No.
86	2SC186E	86	2SC253F	86	2SC286K	86	2SC356M	86	2SC385R
86	2SC186F	86	2SC253G	86	2SC286L	86	2SC356OR	86	2SC385X
86	2SC186G	86	2SC253GN	86	2SC286M	86	2SC356R	86	2SC385Y
86	2SC186GN	86	2SC253H	86	2SC286OR	86	2SC356X	86	2SC386
86	2SC186H	86	2SC253J	86	2SC286R	86	2SC356Y	86	2SC386-O
86	2SC186J	86	2SC253K	86	2SC286X	86	2SC360	86	2SC386A-O(TV)
86	2SC186K	86	2SC253L	86	2SC286Y	86	2SC360-OR	86	2SC386AO
86	2SC186L	86	2SC253M	86	2SC287	86	2SC360A	86	2SC386B
86	2SC186M	86	2SC253OR	86	2SC287-OR	86	2SC360B	86	2SC386C
86	2SC186OR	86	2SC253R	86	2SC287A	86	2SC360C	86	2SC386D
86	2SC186R	86	2SC253X	86	2SC287B	86	2SC360D	86	2SC386E
86	2SC186X	86	2SC253Y	86	2SC287C	86	2SC360E	86	2SC386F
86	2SC186Y	86	2SC263	86	2SC287D	86	2SC360F	86	2SC386G
86	2SC187	86	2SC263A	86	2SC287E	86	2SC360G	86	2SC386GN
86	2SC187(I)	86	2SC263B	86	2SC287F	86	2SC360GN	86	2SC386H
86	2SC187-OR	86	2SC263C	86	2SC287G	86	2SC360H	86	2SC386J
86	2SC187A	86	2SC263D	86	2SC287GN	86	2SC360J	86	2SC386K
86	2SC187B	86	2SC263E	86	2SC287H	86	2SC360K	86	2SC386L
86	2SC187C	86	2SC263F	86	2SC287J	86	2SC360L	86	2SC386M
86	2SC187D	86	2SC263G	86	2SC287K	86	2SC360M	86	2SC386OR
86	2SC187E	86	2SC263GN	86	2SC287L	86	2SC360OR	86	2SC386R
86	2SC187F	86	2SC263H	86	2SC287M	86	2SC360R	86	2SC386X
86	2SC187G	86	2SC263J	86	2SC287OR	86	2SC360X	86	2SC386Y
86	2SC187H	86	2SC263K	86	2SC287R	86	2SC360Y	86	2SC39
86	2SC187I	86	2SC263L	86	2SC287X	86	2SC363	86	2SC391
86	2SC187J	86	2SC263M	86	2SC287Y	86	2SC363-OR	86	2SC391A
86	2SC187K	86	2SC263OR	86	2SC288	86	2SC363A	86	2SC391B
86	2SC187L	86	2SC263R	86	2SC288A	86	2SC363B	86	2SC391C
86	2SC187M	86	2SC263X	86	2SC288A1	86	2SC363C	86	2SC391D
86	2SC187OR	86	2SC263Y	86	2SC288AB	86	2SC363D	86	2SC391E
86	2SC187R	86	2SC266	86	2SC288B	86	2SC363F	86	2SC391F
86	2SC187X	86	2SC266A	86	2SC288C	86	2SC363G	86	2SC391G
86	2SC187Y	86	2SC266B	86	2SC288D	86	2SC363GN	86	2SC391GN
86	2SC1906	86	2SC266C	86	2SC288E	86	2SC363H	86	2SC391H
86	2SC199	86	2SC266D	86	2SC288F	86	2SC363J	86	2SC391J
86	2SC1990	86	2SC266E	86	2SC288G	86	2SC363K	86	2SC391K
86	2SC1990B	86	2SC266F	86	2SC288GN	86	2SC363L	86	2SC391L
86	2SC199A	86	2SC266G	86	2SC288H	86	2SC363M	86	2SC391M
86	2SC199B	86	2SC266GN	86	2SC288J	86	2SC363OR	86	2SC391OR
86	2SC199C	86	2SC266H	86	2SC288K	86	2SC363R	86	2SC391R
86	2SC199D	86	2SC266J	86	2SC288L	86	2SC363X	86	2SC391X
86	2SC199E	86	2SC266K	86	2SC288M	86	2SC363Y	86	2SC391Y
86	2SC199F	86	2SC266L	86	2SC288OR	86	2SC37	86	2SC392
86	2SC199G	86	2SC266M	86	2SC288R	86	2SC375-O	86	2SC392A
86	2SC199GN	86	2SC266OR	86	2SC288X	86	2SC375A	86	2SC392B
86	2SC199H	86	2SC266R	86	2SC288Y	86	2SC375B	86	2SC392C
86	2SC199J	86	2SC266X	86	2SC289	86	2SC375C	86	2SC392D
86	2SC199K	86	2SC266Y	86	2SC289A	86	2SC375D	86	2SC392E
86	2SC199L	86	2SC269	86	2SC289B	86	2SC375E	86	2SC392F
86	2SC199M	86	2SC269A	86	2SC289C	86	2SC375F	86	2SC392G
86	2SC1990R	86	2SC269B	86	2SC289D	86	2SC375G	86	2SC392GN
86	2SC199R	86	2SC269C	86	2SC289E	86	2SC375GN	86	2SC392H
86	2SC199X	86	2SC269D	86	2SC289F	86	2SC375H	86	2SC392J
86	2SC199Y	86	2SC269E	86	2SC289G	86	2SC375J	86	2SC392K
86	2SC206	86	2SC269F	86	2SC289GN	86	2SC375K	86	2SC392L
86	2SC206-OR	86	2SC269G	86	2SC289H	86	2SC375L	86	2SC392M
86	2SC206A	86	2SC269GN	86	2SC289J	86	2SC375M	86	2SC392OR
86	2SC206B	86	2SC269H	86	2SC289K	86	2SC375OR	86	2SC392R
86	2SC206C	86	2SC269J	86	2SC289L	86	2SC375R	86	2SC392X
86	2SC206D	86	2SC269K	86	2SC289M	86	2SC375X	86	2SC392Y
86	2SC206E	86	2SC269L	86	2SC2890R	86	2SC37A	86	2SC397
86	2SC206G	86	2SC269M	86	2SC289R	86	2SC37B	86	2SC397A
86	2SC206GN	86	2SC269OR	86	2SC289X	86	2SC37C	86	2SC397B
86	2SC206H	86	2SC269R	86	2SC289Y	86	2SC37D	86	2SC397C
86	2SC206J	86	2SC269X	86	2SC29	86	2SC37E	86	2SC397D
86	2SC206K	86	2SC269Y	86	2SC296	86	2SC37F	86	2SC397E
86	2SC206L	86	2SC271	86	2SC29A	86	2SC37G	86	2SC397F
86	2SC206M	86	2SC271A	86	2SC29B	86	2SC37GN	86	2SC397G
86	2SC206OR	86	2SC271B	86	2SC29C	86	2SC37H	86	2SC397GN
86	2SC206R	86	2SC271C	86	2SC29D	86	2SC37J	86	2SC397H
86	2SC206RED	86	2SC271D	86	2SC29E	86	2SC37K	86	2SC397K
86	2SC206WHITE	86	2SC271E	86	2SC29F	86	2SC37L	86	2SC397L
86	2SC206X	86	2SC271F	86	2SC29G	86	2SC37M	86	2SC397M
86	2SC206Y	86	2SC271G	86	2SC29GN	86	2SC37OR	86	2SC397OR
86	2SC250	86	2SC271GN	86	2SC29H	86	2SC37R	86	2SC397R
86	2SC250A	86	2SC271H	86	2SC29J	86	2SC37X	86	2SC397X
86	2SC250B	86	2SC271J	86	2SC29K	86	2SC37Y	86	2SC397Y
86	2SC250C	86	2SC271K	86	2SC29L	86	2SC382	86	2SC39A
86	2SC250D	86	2SC271L	86	2SC29M	86	2SC382(BL)	86	2SC39B
86	2SC250E	86	2SC271M	86	2SC290R	86	2SC382(BN)	86	2SC39C
86	2SC250F	86	2SC271OR	86	2SC29R	86	2SC382(R)	86	2SC39D
86	2SC250G	86	2SC271R	86	2SC29X	86	2SC382-BK	86	2SC39E
86	2SC250GN	86	2SC271X	86	2SC29Y	86	2SC382-BK(1)	86	2SC39F
86	2SC250H	86	2SC271Y	86	2SC316	86	2SC382-BK(2)	86	2SC39G
86	2SC250J	86	2SC272	86	2SC316A	86	2SC382-G	86	2SC39GN
86	2SC250K	86	2SC272A	86	2SC316B	86	2SC382-GR	86	2SC39H
86	2SC250L	86	2SC272B	86	2SC316C	86	2SC382-GY	86	2SC39J
86	2SC250M	86	2SC272C	86	2SC316D	86	2SC382-OR	86	2SC39K
86	2SC250OR	86	2SC272D	86	2SC316E	86	2SC382-R	86	2SC39L
86	2SC250R	86	2SC272F	86	2SC316F	86	2SC382-V	86	2SC39M
86	2SC250X	86	2SC272G	86	2SC316G	86	2SC382A	86	2SC39OR
86	2SC250Y	86	2SC272GN	86	2SC316GN	86	2SC382B	86	2SC39R
86	2SC251	86	2SC272H	86	2SC316H	86	2SC382BK	86	2SC39X
86	2SC251A	86	2SC272J	86	2SC316J	86	2SC382BK1	86	2SC39Y
86	2SC251B	86	2SC272K	86	2SC316K	86	2SC382BK2	86	2SC40
86	2SC251C	86	2SC272L	86	2SC316L	86	2SC382BL	86	2SC405
86	2SC251D	86	2SC272M	86	2SC316M	86	2SC382BN	86	2SC405A
86	2SC251E	86	2SC272OR	86	2SC316OR	86	2SC382BR	86	2SC405B
86	2SC251F	86	2SC272R	86	2SC316R	86	2SC382C	86	2SC405C
86	2SC251G	86	2SC272X	86	2SC316X	86	2SC382D	86	2SC405E
86	2SC251GN	86	2SC272Y	86	2SC316Y	86	2SC382E	86	2SC405F
86	2SC251H	86	2SC282	86	2SC351	86	2SC382F	86	2SC405G
86	2SC251I	86	2SC282A	86	2SC351(PA)	86	2SC382G	86	2SC405GN
86	2SC251J	86	2SC282B	86	2SC351A	86	2SC382GR	86	2SC405H
86	2SC251K	86	2SC282C	86	2SC351B	86	2SC382GY	86	2SC405J
86	2SC251L	86	2SC282D	86	2SC351C	86	2SC382H	86	2SC405K
86	2SC251M	86	2SC282E	86	2SC351D	86	2SC382J	86	2SC405L
86	2SC251OR	86	2SC282F	86	2SC351F	86	2SC382K	86	2SC405M
86	2SC251R	86	2SC282G	86	2SC351PA1	86	2SC382L	86	2SC405OR
86	2SC251X	86	2SC282GN	86	2SC351G	86	2SC382M	86	2SC405R
86	2SC251Y	86	2SC282H	86	2SC351GN	86	2SC382OR	86	2SC405X
86	2SC252	86	2SC282HA	86	2SC351H	86	2SC382R	86	2SC405Y
86	2SC252A	86	2SC282HB	86	2SC351J	86	2SC382V	86	2SC406
86	2SC252B	86	2SC282HC	86	2SC351K	86	2SC382W	86	2SC406A
86	2SC252C	86	2SC282J	86	2SC351L	86	2SC382W,R	86	2SC406B
86	2SC252D	86	2SC282K	86	2SC351M	86	2SC382X	86	2SC406C
86	2SC252E	86	2SC282L	86	2SC351OR	86	2SC382Y	86	2SC406D
86	2SC252F	86	2SC282M	86	2SC351R	86	2SC385	86	2SC406E
86	2SC252G	86	2SC282OR	86	2SC351X	86	2SC385A	86	2SC406F
86	2SC252GN	86	2SC282R	86	2SC351Y	86	2SC385B	86	2SC406G
86	2SC252H	86	2SC282X	86	2SC356	86	2SC385C	86	2SC406GN
86	2SC252J	86	2SC282Y	86	2SC356A	86	2SC385D	86	2SC406H
86	2SC252K	86	2SC286	86	2SC356B	86	2SC385E	86	2SC406J
86	2SC252L	86	2SC286A	86	2SC356C	86	2SC385F	86	2SC406K
86	2SC252M	86	2SC286B	86	2SC356D	86	2SC385G	86	2SC406L
86	2SC252OR	86	2SC286C	86	2SC356E	86	2SC385GN	86	2SC406M
86	2SC252R	86	2SC286D	86	2SC356F	86	2SC385H	86	2SC406OR
86	2SC252X	86	2SC286E	86	2SC356G	86	2SC385J	86	2SC406R
86	2SC252Y	86	2SC286F	86	2SC356GN	86	2SC385K	86	2SC406X
86	2SC253	86	2SC286G	86	2SC356H	86	2SC385L	86	2SC406Y
86	2SC253A	86	2SC286GN	86	2SC356J	86	2SC385M	86	2SC40A
86	2SC253B	86	2SC286GN	86	2SC356K	86	2SC385OR	86	
86	2SC253C	86	2SC286H	86	2SC356L	86		86	
86	2SC253D	86	2SC286J	86		86		86	
86	2SC253E	86		86		86		86	

GE	Industry Standard No.	GE	Industry Standard No.	GE	Industry Standard No.	GE	Industry Standard No.	GE	Industry Standard No.
86	2SC40B	86	2SC612	86	2SC658M	88	BCY59VII	88	2SC1384G
86	2SC40C	86	2SC612A	86	2SC658OR	88	BCY59VIII	88	2SC1384GN
86	2SC40D	86	2SC612B	86	2SC658R	88	GE-88	88	2SC1384H
86	2SC40F	86	2SC612C	86	2SC658X	88	HEP75	88	2SC1384J
86	2SC40P	86	2SC612D	86	2SC658Y	88	SK3137	88	2SC1384K
86	2SC40Q	86	2SC612E	86	2SC659	88	2SC100	88	2SC1384L
86	2SC40GN	86	2SC612F	86	2SC659A	88	2SC100-OY	88	2SC1384M
86	2SC40H	86	2SC612G	86	2SC659B	88	2SC1010	88	2SC1384OR
86	2SC40J	86	2SC612GN	86	2SC659C	88	2SC1010A	88	2SC1384Q
86	2SC40K	86	2SC612H	86	2SC659D	88	2SC1010B	88	2SC1384Q,R
86	2SC40L	86	2SC612J	86	2SC659E	88	2SC1010C	88	2SC1384R
86	2SC40OR	86	2SC612K	86	2SC659F	88	2SC1010D	88	2SC1384S
86	2SC40R	86	2SC612L	86	2SC659G	88	2SC1010E	88	2SC1384X
86	2SC40X	86	2SC612M	86	2SC659GN	88	2SC1010F	88	2SC1384Y
86	2SC40Y	86	2SC612OR	86	2SC659H	88	2SC1010G	88	2SC138A
86	2SC424	86	2SC612R	86	2SC659J	88	2SC1010GN	88	2SC138B
86	2SC424D	86	2SC612X	86	2SC659K	88	2SC1010H	88	2SC138C
86	2SC460A	86	2SC612Y	86	2SC659L	88	2SC1010J	88	2SC138D
86	2SC460J	86	2SC613	86	2SC659M	88	2SC1010K	88	2SC138E
86	00002SC460C	86	2SC613A	86	2SC659OR	88	2SC1010L	88	2SC138F
86	00002SC461	86	2SC613B	86	2SC659R	88	2SC1010OR	88	2SC138G
86	2SC463	86	2SC613C	86	2SC659X	88	2SC1010X	88	2SC138GN
86	2SC463A	86	2SC613D	86	2SC659Y	88	2SC1010Y	88	2SC138H
86	2SC463B	86	2SC613E	86	2SC662	88	2SC104	88	2SC138J
86	2SC463C	86	2SC613F	86	2SC662B	88	2SC104A	88	2SC138L
86	2SC463D	86	2SC613G	86	2SC662C	88	2SC104B	88	2SC138M
86	2SC463E	86	2SC613GN	86	2SC662D	88	2SC104C	88	2SC138OR
86	2SC463F	86	2SC613H	86	2SC662E	88	2SC104D	88	2SC138R
86	2SC463G	86	2SC613J	86	2SC662F	88	2SC104E	88	2SC138S
86	2SC463GN	86	2SC613K	86	2SC662G	88	2SC104F	88	2SC138X
86	2SC463H	86	2SC613L	86	2SC662GN	88	2SC104G	88	2SC139
86	2SC463J	86	2SC613M	86	2SC662H	88	2SC104GN	88	2SC139A
86	2SC463K	86	2SC613OR	86	2SC662J	88	2SC104H	88	2SC139B
86	2SC463L	86	2SC613R	86	2SC662K	88	2SC104J	88	2SC139C
86	2SC463M	86	2SC613X	86	2SC662L	88	2SC104K	88	2SC139D
86	2SC463OR	86	2SC613Y	86	2SC662M	88	2SC104L	88	2SC139E
86	2SC463R	86	2SC63	86	2SC662OR	88	2SC104M	88	2SC139F
86	2SC463X	86	2SC63A	86	2SC662R	88	2SC104OR	88	2SC139G
86	2SC463Y	86	2SC63B	86	2SC662X	88	2SC104R	88	2SC139GN
86	2SC477	86	2SC63D	86	2SC662Y	88	2SC104X	88	2SC139H
86	2SC477A	86	2SC63E	86	2SC668	88	2SC104Y	88	2SC139J
86	2SC477B	86	2SC63F	86	2SC684	88	2SC110	88	2SC139K
86	2SC477C	86	2SC63G	86	2SC684A	88	2SC110A	88	2SC139M
86	2SC477D	86	2SC63GN	86	2SC684B	88	2SC110B	88	2SC139OR
86	2SC477E	86	2SC63H	86	2SC684BK	88	2SC110C	88	2SC139R
86	2SC477F	86	2SC63J	86	2SC684C	88	2SC110D	88	2SC139X
86	2SC477G	86	2SC63K	86	2SC684E	88	2SC110E	88	2SC139Y
86	2SC477GN	86	2SC63L	86	2SC684F	88	2SC110F	88	2SC15
86	2SC477H	86	2SC63M	86	2SC684G	88	2SC110G	88	2SC15-0
86	2SC477J	86	2SC63OR	86	2SC684GN	88	2SC110GN	88	2SC15-1
86	2SC477K	86	2SC63R	86	2SC684H	88	2SC110H	88	2SC15-2
86	2SC477L	86	2SC63X	86	2SC684J	88	2SC110J	88	2SC15-3
86	2SC477M	86	2SC63Y	86	2SC684K	88	2SC110K	88	2SC151
86	2SC477OR	86	2SC641	86	2SC684L	88	2SC110L	88	2SC151A
86	2SC477R	86	2SC641A	86	2SC684M	88	2SC110M	88	2SC151B
86	2SC477X	86	2SC641B	86	2SC684-R	88	2SC110OR	88	2SC151C
86	2SC477Y	86	2SC641C	86	2SC684X	88	2SC110R	88	2SC151D
86	2SC555(B)	86	2SC641D	86	2SC684Y	88	2SC110X	88	2SC151E
86	2SC55-OR	86	2SC641F	86	2SC705	88	2SC110Y	88	2SC151G
86	2SC5359	86	2SC641G	86	2SC705A	88	2SC131	88	2SC151GN
86	2SC555A	86	2SC641GN	86	2SC705B	88	2SC131A	88	2SC151H
86	2SC555ABC	86	2SC641H	86	2SC705C	88	2SC131B	88	2SC151J
86	2SC555AL	86	2SC641J	86	2SC705D	88	2SC131C	88	2SC151K
86	2SC555B	86	2SC641K	86	2SC705E	88	2SC131D	88	2SC151L
86	2SC555C	86	2SC641L	86	2SC705F	88	2SC131E	88	2SC151M
86	2SC555D	86	2SC641M	86	2SC705G	88	2SC131F	88	2SC151OR
86	2SC555E	86	2SC641OR	86	2SC705GN	88	2SC131G	88	2SC151R
86	2SC555F	86	2SC641R	86	2SC705J	88	2SC131GN	88	2SC151Y
86	2SC555G	86	2SC641X	86	2SC705K	88	2SC131H	88	2SC15A
86	2SC555GN	86	2SC641Y	86	2SC705L	88	2SC131J	88	2SC15B
86	2SC555H	86	2SC648A	86	2SC705M	88	2SC131L	88	2SC15C
86	2SC555J	86	2SC648B	86	2SC705OR	88	2SC131M	88	2SC15D
86	2SC555K	86	2SC648C	86	2SC705R	88	2SC131OR	88	2SC15E
86	2SC555L	86	2SC648D	86	2SC705TV	88	2SC131R	88	2SC15F
86	2SC555M	86	2SC648E	86	2SC705TVV	88	2SC131Y	88	2SC15G
86	2SC555OR	86	2SC648F	86	2SC705TW	88	2SC132	88	2SC15GN
86	2SC555R	86	2SC648G	86	2SC705X	88	2SC132A	88	2SC15H
86	2SC555X	86	2SC648GN	86	2SC705Y	88	2SC132B	88	2SC15J
86	2SC555Y	86	2SC648H	86	2SC707A	88	2SC132C	88	2SC15K
86	2SC567	86	2SC648J	86	2SC707B	88	2SC132D	88	2SC15L
86	2SC567A	86	2SC648K	86	2SC707C	88	2SC132E	88	2SC15M
86	2SC567B	86	2SC648L	86	2SC707D	88	2SC132F	88	2SC15OR
86	2SC567C	86	2SC648M	86	2SC707F	88	2SC132G	88	2SC15R
86	2SC567D	86	2SC648OR	86	2SC707G	88	2SC132GN	88	2SC15X
86	2SC567E	86	2SC648R	86	2SC707GN	88	2SC132H	88	2SC15Y
86	2SC567F	86	2SC648X	86	2SC707H	88	2SC132J	88	2SC16
86	2SC567G	86	2SC649	86	2SC707K	88	2SC132K	88	2SC165
86	2SC567GN	86	2SC649A	86	2SC707L	88	2SC132L	88	2SC166
86	2SC567H	86	2SC649B	86	2SC707M	88	2SC132M	88	2SC166A
86	2SC567J	86	2SC649C	86	2SC707OR	88	2SC132OR	88	2SC166B
86	2SC567K	86	2SC649D	86	2SC707R	88	2SC132R	88	2SC166C
86	2SC567L	86	2SC649E	86	2SC707X	88	2SC132Y	88	2SC166D
86	2SC567M	86	2SC649F	86	2SC707Y	88	2SC133	88	2SC166F
86	2SC567OR	86	2SC649G	86	2SC716	88	2SC133A	88	2SC166G
86	2SC567R	86	2SC649GN	86	2SC716A	88	2SC133C	88	2SC166GN
86	2SC567X	86	2SC649H	86	2SC716B	88	2SC133D	88	2SC166H
86	2SC567Y	86	2SC649J	86	2SC716C	88	2SC133E	88	2SC166J
86	2SC568	86	2SC649K	86	2SC716D	88	2SC133F	88	2SC166K
86	2SC568A	86	2SC649L	86	2SC716E	88	2SC133G	88	2SC166L
86	2SC568B	86	2SC649M	86	2SC716F	88	2SC133GN	88	2SC166M
86	2SC568C	86	2SC649OR	86	2SC716G	88	2SC133H	88	2SC166OR
86	2SC568D	86	2SC649R	86	2SC716GN	88	2SC133J	88	2SC166R
86	2SC568E	86	2SC649X	86	2SC716H	88	2SC133K	88	2SC166X
86	2SC568F	86	2SC649Y	86	2SC716J	88	2SC133L	88	2SC166Y
86	2SC568G	86	2SC656A	86	2SC716K	88	2SC133M	88	2SC167
86	2SC568GN	86	2SC656B	86	2SC716L	88	2SC133OR	88	2SC167A
86	2SC568H	86	2SC656C	86	2SC716M	88	2SC133X	88	2SC167B
86	2SC568J	86	2SC656D	86	2SC716OR	88	2SC133Y	88	2SC167C
86	2SC568K	86	2SC656E	86	2SC716R	88	2SC1347	88	2SC167D
86	2SC568L	86	2SC656F	86	2SC716X	88	2SC1347(Q)	88	2SC167F
86	2SC568M	86	2SC656G	86	2SC716Y	88	2SC1347A	88	2SC167G
86	2SC568OR	86	2SC656GN	86	2SC738	88	2SC1347B	88	2SC167GN
86	2SC568R	86	2SC656H	86	2SC738A	88	2SC1347F	88	2SC167H
86	2SC568X	86	2SC656J	86	2SC738B	88	2SC1347L	88	2SC167K
86	2SC568Y	86	2SC656K	86	2SC738C	88	2SC1347Q	88	2SC167M
86	2SC606	86	2SC656L	86	2SC738D	88	2SC1347R	88	2SC167OR
86	2SC611	86	2SC656M	86	2SC738E	88	2SC1347RQ	88	2SC167X
86	2SC611A	86	2SC656OR	86	2SC738F	88	2SC1347S	88	2SC167Y
86	2SC611B	86	2SC656R	86	2SC738G	88	2SC1347T	88	2SC1682-BL
86	2SC611C	86	2SC656X	86	2SC738GN	88	2SC1347X	88	2SC1682-GR
86	2SC611D	86	2SC656Y	86	2SC738H	88	2SC1347Y	88	2SC16A
86	2SC611E	86	2SC658	86	2SC738J	88	2SC1362	88	2SC16B
86	2SC611F	86	2SC658A	86	2SC738K	88	2SC1364B	88	2SC16C
86	2SC611G	86	2SC658B	86	2SC738L	88	2SC138	88	2SC16D
86	2SC611GN	86	2SC658C	86	2SC738M	88	2SC1384	88	2SC16E
86	2SC611H	86	2SC658D	86	2SC738OR	88	2SC1384(Q)	88	2SC16F
86	2SC611J	86	2SC658E	86	2SC738R	88	2SC1384-OR	88	2SC16G
86	2SC611K	86	2SC658F	86	2SC738X	88	2SC1384A	88	2SC16GN
86	2SC611L	86	2SC658G	86	2SC738Y	88	2SC1384B	88	2SC16H
86	2SC611M	86	2SC658GN	87	HEPS3023	88	2SC1384C	88	2SC16J
86	2SC611OR	86	2SC658H	88	BCY58VII	88	2SC1384D		
86	2SC611R	86	2SC658J	88	BCY58VIII	88	2SC1384E		
86	2SC611X	86	2SC658K			88	2SC1384F		
86	2SC611Y	86	2SC658L						

GE	Industry Standard No.	GE	Industry Standard No.	GE	Industry Standard No.	GE	Industry Standard No.	GE	Industry Standard No.
88	2SC16K	88	2SC202F	88	2SC267M	88	2SC317A	88	2SC350E
88	2SC16L	88	2SC202G	88	2SC267OR	88	2SC317B	88	2SC350F
88	2SC16M	88	2SC202GN	88	2SC267R	88	2SC317C	88	2SC350G
88	2SC16OR	88	2SC202H	88	2SC267X	88	2SC317D	88	2SC350GN
88	2SC16R	88	2SC202J	88	2SC267Y	88	2SC317E	88	2SC350H
88	2SC16X	88	2SC202K	88	2SC26A	88	2SC317F	88	2SC350J
88	2SC16Y	88	2SC202L	88	2SC26B	88	2SC317G	88	2SC350K
88	2SC18	88	2SC202M	88	2SC26C	88	2SC317GN	88	2SC350L
88	2SC18B	88	2SC202OR	88	2SC26D	88	2SC317H	88	2SC350M
88	2SC18C	88	2SC202R	88	2SC26E	88	2SC317J	88	2SC350OR
88	2SC18D	88	2SC202X	88	2SC26F	88	2SC317K	88	2SC350R
88	2SC18E	88	2SC202Y	88	2SC26G	88	2SC317L	88	2SC350X
88	2SC18F	88	0028C203	88	2SC26GN	88	2SC317M	88	2SC350Y
88	2SC18G	88	0028C203A	88	2SC26H	88	2SC317OR	88	2SC366
88	2SC18GN	88	2SC203B	88	2SC26J	88	2SC317R	88	2SC368
88	2SC18H	88	2SC203C	88	2SC26K	88	2SC317X	88	2SC368A
88	2SC18J	88	2SC203D	88	2SC26L	88	2SC317Y	88	2SC368B
88	2SC18K	88	2SC203E	88	2SC26M	88	2SC318	88	2SC368BL
88	2SC18L	88	2SC203F	88	2SC26OR	88	2SC318A	88	2SC368C
88	2SC18M	88	2SC203G	88	2SC26R	88	2SC318AB	88	2SC368D
88	2SC18OR	88	2SC203GN	88	2SC26X	88	2SC318B	88	2SC368E
88	2SC18R	88	2SC203H	88	2SC26Y	88	2SC318C	88	2SC368F
88	2SC18X	88	2SC203J	88	2SC283	88	2SC318D	88	2SC368G
88	2SC191	88	2SC203K	88	2SC283A	88	2SC318E	88	2SC368GN
88	2SC191A	88	2SC203L	88	2SC283B	88	2SC318F	88	2SC368GR
88	2SC191B	88	2SC203M	88	2SC283C	88	2SC318G	88	2SC368H
88	2SC191C	88	2SC203OR	88	2SC283D	88	2SC318GN	88	2SC368J
88	2SC191D	88	2SC203R	88	2SC283E	88	2SC318H	88	2SC368K
88	2SC191E	88	2SC203X	88	2SC283F	88	2SC318J	88	2SC368L
88	2SC191F	88	2SC204	88	2SC283G	88	2SC318K	88	2SC368M
88	2SC191G	88	2SC204A	88	2SC283GN	88	2SC318L	88	2SC368OR
88	2SC191GN	88	2SC204C	88	2SC283H	88	2SC318M	88	2SC368R
88	2SC191H	88	2SC204D	88	2SC283J	88	2SC318OR	88	2SC368V
88	2SC191J	88	2SC204E	88	2SC283K	88	2SC318R	88	2SC368X
88	2SC191K	88	2SC204F	88	2SC283L	88	2SC318X	88	2SC368Y
88	2SC191L	88	2SC204G	88	2SC283M	88	2SC318Y	88	2SC379
88	2SC191M	88	2SC204GN	88	2SC283OR	88	2SC319	88	2SC379A
88	2SC191OR	88	2SC204H	88	2SC283R	88	2SC319A	88	2SC379B
88	2SC191R	88	2SC204J	88	2SC283X	88	2SC319B	88	2SC379C
88	2SC191X	88	2SC204K	88	2SC283Y	88	2SC319C	88	2SC379D
88	2SC191Y	88	2SC204L	88	2SC284	88	2SC319D	88	2SC379E
88	2SC192	88	2SC204M	88	2SC284A	88	2SC319E	88	2SC379F
88	2SC192A	88	2SC204OR	88	2SC284C	88	2SC319F	88	2SC379G
88	2SC192B	88	2SC204R	88	2SC284D	88	2SC319G	88	2SC379GN
88	2SC192C	88	2SC204X	88	2SC284E	88	2SC319GN	88	2SC379H
88	2SC192D	88	2SC204Y	88	2SC284F	88	2SC319H	88	2SC379J
88	2SC192E	88	2SC205	88	2SC284GN	88	2SC319J	88	2SC379K
88	2SC192F	88	2SC205A	88	2SC284H	88	2SC319K	88	2SC379L
88	2SC192G	88	2SC205B	88	2SC284HA	88	2SC319L	88	2SC379M
88	2SC192GN	88	2SC205C	88	2SC284HB	88	2SC319M	88	2SC379OR
88	2SC192H	88	2SC205D	88	2SC284HC	88	2SC319OR	88	2SC379R
88	2SC192J	88	2SC205F	88	2SC284J	88	2SC319R	88	2SC379X
88	2SC192K	88	2SC205G	88	2SC284K	88	2SC319X	88	2SC379Y
88	2SC192L	88	2SC205GN	88	2SC284L	88	2SC319Y	88	2SC395
88	2SC192M	88	2SC205H	88	2SC284M	88	2SC320	88	2SC395A
88	2SC192OR	88	2SC205J	88	2SC284OR	88	2SC320A	88	2SC395A-ORG
88	2SC192R	88	2SC205K	88	2SC284R	88	2SC320B	88	2SC395A-RED
88	2SC192X	88	2SC205L	88	2SC284X	88	2SC320C	88	2SC395A-YEL
88	2SC192Y	88	2SC205M	88	2SC284Y	88	2SC320D	88	2SC395B
88	2SC193	88	2SC205OR	88	2SC285	88	2SC320E	88	2SC395C
88	2SC193A	88	2SC205R	88	2SC285A	88	2SC320F	88	2SC395D
88	2SC193B	88	2SC205X	88	2SC285C	88	2SC320G	88	2SC395E
88	2SC193C	88	2SC230	88	2SC285D	88	2SC320GN	88	2SC395F
88	2SC193D	88	2SC230A	88	2SC285E	88	2SC320H	88	2SC395G
88	2SC193E	88	2SC230B	88	2SC285F	88	2SC320J	88	2SC395GN
88	2SC193F	88	2SC230C	88	2SC285G	88	2SC320K	88	2SC395H
88	2SC193G	88	2SC230D	88	2SC285GN	88	2SC320L	88	2SC395J
88	2SC193GN	88	2SC230E	88	2SC285H	88	2SC320M	88	2SC395K
88	2SC193H	88	2SC230F	88	2SC285J	88	2SC320OR	88	2SC395L
88	2SC193J	88	2SC230G	88	2SC285L	88	2SC320X	88	2SC395M
88	2SC193K	88	2SC230GN	88	2SC285M	88	2SC320Y	88	2SC395OR
88	2SC193L	88	2SC230H	88	2SC285OR	88	2SC321	88	2SC395R
88	2SC193M	88	2SC230J	88	2SC285R	88	2SC321A	88	2SC395X
88	2SC193OR	88	2SC230K	88	2SC285X	88	2SC321B	88	2SC395Y
88	2SC193R	88	2SC230L	88	2SC285Y	88	2SC321C	88	2SC396
88	2SC193X	88	2SC230M	88	2SC300	88	2SC321D	88	2SC396A
88	2SC194	88	2SC230R	88	2SC300A	88	2SC321E	88	2SC396B
88	2SC194A	88	2SC230X	88	2SC300B	88	2SC321F	88	2SC396C
88	2SC194B	88	2SC230Y	88	2SC300C	88	2SC321G	88	2SC396D
88	2SC194C	88	2SC237	88	2SC300D	88	2SC321GN	88	2SC396E
88	2SC194D	88	2SC237A	88	2SC300E	88	2SC321H	88	2SC396F
88	2SC194E	88	2SC237B	88	2SC300F	88	2SC321HA	88	2SC396G
88	2SC194F	88	2SC237C	88	2SC300G	88	2SC321HB	88	2SC396GN
88	2SC194G	88	2SC237D	88	2SC300GN	88	2SC321HC	88	2SC396GR
88	2SC194GN	88	2SC237E	88	2SC300H	88	2SC321J	88	2SC396H
88	2SC194H	88	2SC237F	88	2SC300J	88	2SC321K	88	2SC396J
88	2SC194J	88	2SC237G	88	2SC300K	88	2SC321L	88	2SC396K
88	2SC194K	88	2SC237GN	88	2SC300L	88	2SC321M	88	2SC396L
88	2SC194L	88	2SC237H	88	2SC300M	88	2SC321OR	88	2SC396M
88	2SC194M	88	2SC237J	88	2SC300OR	88	2SC321R	88	2SC396OR
88	2SC194OR	88	2SC237K	88	2SC300R	88	2SC321X	88	2SC396R
88	2SC194X	88	2SC237L	88	2SC300X	88	2SC321Y	88	2SC396X
88	2SC194Y	88	2SC237M	88	2SC300Y	88	2SC323	88	2SC396Y
88	2SC200	88	2SC237OR	88	2SC301	88	2SC323A	88	2SC400
88	2SC200A	88	2SC237R	88	2SC301A	88	2SC323B	88	2SC400-0
88	2SC200B	88	2SC237X	88	2SC301B	88	2SC323C	88	2SC400-OR
88	2SC200C	88	2SC239	88	2SC301C	88	2SC323D	88	2SC400-R
88	2SC200D	88	2SC239A	88	2SC301D	88	2SC323E	88	2SC400-Y
88	2SC200E	88	2SC239B	88	2SC301E	88	2SC323F	88	2SC400A
88	2SC200F	88	2SC239C	88	2SC301F	88	2SC323G	88	2SC400B
88	2SC200G	88	2SC239D	88	2SC301G	88	2SC323GN	88	2SC400C
88	2SC200GN	88	2SC239E	88	2SC301GN	88	2SC323H	88	2SC400D
88	2SC200H	88	2SC239F	88	2SC301H	88	2SC323J	88	2SC400E
88	2SC200J	88	2SC239G	88	2SC301J	88	2SC323K	88	2SC400F
88	2SC200K	88	2SC239GN	88	2SC301K	88	2SC323L	88	2SC400G
88	2SC200L	88	2SC239H	88	2SC301L	88	2SC323M	88	2SC400GN
88	2SC200M	88	2SC239J	88	2SC301M	88	2SC323OR	88	2SC400H
88	2SC200OR	88	2SC239K	88	2SC301OR	88	2SC323R	88	2SC400J
88	2SC200X	88	2SC239L	88	2SC301R	88	2SC323X	88	2SC400K
88	2SC200Y	88	2SC239M	88	2SC301Y	88	2SC323Y	88	2SC400L
88	2SC201	88	2SC239OR	88	2SC302	88	2SC33	88	2SC400M
88	2SC201A	88	2SC239R	88	2SC302A	88	2SC33A	88	2SC400OR
88	2SC201B	88	2SC239X	88	2SC302B	88	2SC33B	88	2SC400R
88	2SC201C	88	2SC239Y	88	2SC302C	88	2SC33C	88	2SC400X
88	2SC201D	88	2SC260	88	2SC302D	88	2SC33D	88	2SC400Y
88	2SC201E	88	2SC262	88	2SC302E	88	2SC33E	88	2SC47
88	2SC201F	88	2SC267	88	2SC302F	88	2SC33F	88	2SC476
88	2SC201G	88	2SC267A	88	2SC302G	88	2SC33G	88	2SC476A
88	2SC201GN	88	2SC267B	88	2SC302GN	88	2SC33GN	88	2SC476B
88	2SC201H	88	2SC267C	88	2SC302H	88	2SC33H	88	2SC476C
88	2SC201J	88	2SC267D	88	2SC302J	88	2SC33J	88	2SC476D
88	2SC201K	88	2SC267E	88	2SC302K	88	2SC33K	88	2SC476E
88	2SC201L	88	2SC267F	88	2SC302L	88	2SC33L	88	2SC476F
88	2SC201M	88	2SC267G	88	2SC302M	88	2SC33M	88	2SC476G
88	2SC201OR	88	2SC267GN	88	2SC302OR	88	2SC33OR	88	2SC476GN
88	2SC201X	88	2SC267H	88	2SC302R	88	2SC33R	88	2SC476H
88	2SC201Y	88	2SC267J	88	2SC302X	88	2SC33X	88	2SC476J
88	2SC202	88	2SC267L	88	2SC302Y	88	2SC33Y	88	2SC476K
88	2SC202A			88	2SC317	88	2SC350	88	2SC476L
88	2SC202B					88	2SC350A	88	2SC476M
88	2SC202C					88	2SC350B	88	2SC476OR
88	2SC202D					88	2SC350C	88	2SC476R
88	2SC202E					88	2SC350D	88	2SC476X
								88	2SC476Y

GE	Industry Standard No.	GE	Industry Standard No.	GE	Industry Standard No.	GE	Industry Standard No.	GE	Industry Standard No.
88	2SC47A	88	2SC588OR	88	2SC654H	89	ECG289	89	2SA500G
88	2SC47B	88	2SC588R	88	2SC654J	89	GE-89	89	2SA500GN
88	2SC47C	88	2SC588X	88	2SC654K	89	HEP76	89	2SA500H
88	2SC47D	88	2SC588Y	88	2SC654L	89	IP20-0211	89	2SA500J
88	2SC47E	88	2SC595A	88	2SC654M	89	RV2260	89	2SA500K
88	2SC47G	88	2SC595B	88	2SC654OR	89	SK3138	89	2SA500L
88	2SC47GN	88	2SC595C	88	2SC654R	89	1N4971A	89	2SA500M
88	2SC47H	88	2SC595D	88	2SC654X	89	1N4972A	89	2SA500OR
88	2SC47J	88	2SC595E	88	2SC654Y	89	1N4982A	89	2SA500R
88	2SC47K	88	2SC595F	88	2SC655A	89	2N5055	89	2SA500X
88	2SC47L	88	2SC595G	88	2SC655B	89	2N5086A	89	2SA500Y
88	2SC47M	88	2SC595GN	88	2SC655C	89	2N5087A	89	2SA502
88	2SC47OR	88	2SC595H	88	2SC655D	89	2N5138A	89	2SA502A
88	2SC47R	88	2SC595J	88	2SC655E	89	2N5139A	89	2SA502B
88	2SC47X	88	2SC595K	88	2SC655F	89	2N5142A	89	2SA502C
88	2SC47Y	88	2SC595L	88	2SC655G	89	2N5143A	89	2SA502D
88	2SC52	88	2SC595M	88	2SC655GN	89	2N5226A	89	2SA502E
88	2SC52A	88	2SC595OR	88	2SC655H	89	2N5227A	89	2SA502F
88	2SC52B	88	2SC595R	88	2SC655J	89	2N5354A	89	2SA502G
88	2SC52C	88	2SC595X	88	2SC655K	89	2N5355A	89	2SA502GN
88	2SC52E	88	2SC595Y	88	2SC655L	89	2N5356A	89	2SA502H
88	2SC52F	88	2SC596	88	2SC655M	89	2N5365A	89	2SA502J
88	2SC52G	88	2SC596A	88	2SC655OR	89	2N5366A	89	2SA502K
88	2SC52GN	88	2SC596B	88	2SC655R	89	2N5367A	89	2SA502L
88	2SC52H	88	2SC596C	88	2SC655X	89	2N5372A	89	2SA502M
88	2SC52J	88	2SC596D	88	2SC655Y	89	2N5373A	89	2SA502OR
88	2SC52K	88	2SC596E	88	2SC67	89	2N5374A	89	2SA502R
88	2SC52L	88	2SC596F	88	2SC67A	89	2N5375A	89	2SA502X
88	2SC52M	88	2SC596G	88	2SC67B	89	2N5378A	89	2SA502Y
88	2SC52OR	88	2SC596GN	88	2SC67C	89	2N5379A	89	2SA510
88	2SC52X	88	2SC596H	88	2SC67D	89	2N5382A	89	2SA510-0
88	2SC52Y	88	2SC596J	88	2SC67E	89	2N5383A	89	2SA510-OR
88	2SC53	88	2SC596K	88	2SC67F	89	2N5447A	89	2SA510-R
88	2SC53A	88	2SC596L	88	2SC67G	89	2N5448A	89	2SA510-RD
88	2SC53B	88	2SC596M	88	2SC67GN	89	2S327A	89	2SA510A
88	2SC53C	88	2SC596OR	88	2SC67H	89	2SA494-YE	89	2SA510B
88	2SC53D	88	2SC596R	88	2SC67J	89	2SA467	89	2SA510C
88	2SC53E	88	2SC596X	88	2SC67K	89	2SA467-O	89	2SA510D
88	2SC53F	88	2SC596Y	88	2SC67L	89	2SA467-Y	89	2SA510E
88	2SC53G	88	2SC619	88	2SC67M	89	2SA467A	89	2SA510F
88	2SC53GN	88	2SC619(B)	88	2SC67OR	89	2SA467B	89	2SA510G
88	2SC53H	88	2SC619(C)	88	2SC67R	89	2SA467C	89	2SA510GN
88	2SC53J	88	2SC619A	88	2SC67X	89	2SA467H	89	2SA510H
88	2SC53L	88	2SC619B	88	2SC67Y	89	2SA467J	89	2SA510J
88	2SC53OR	88	2SC619C	88	2SC68	89	2SA467K	89	2SA510K
88	2SC53R	88	2SC619D	88	2SC689	89	2SA467L	89	2SA510L
88	2SC53X	88	2SC619E	88	2SC689A	89	2SA467M	89	2SA510M
88	2SC53Y	88	2SC619F	88	2SC689B	89	2SA467OR	89	2SA510OR
88	2SC54	88	2SC619G	88	2SC689C	89	2SA467R	89	2SA510R
88	2SC540	88	2SC619GN	88	2SC689D	89	2SA467X	89	2SA510X
88	2SC540A	88	2SC619H	88	2SC689E	89	2SA467Y	89	2SA510Y
88	2SC540B	88	2SC619J	88	2SC689G	89	2SA480	89	2SA511
88	2SC540C	88	2SC619K	88	2SC689GN	89	2SA480-OR	89	2SA511-G
88	2SC540D	88	2SC619L	88	2SC689H	89	2SA480A	89	2SA511-O
88	2SC540E	88	2SC619M	88	2SC689J	89	2SA480B	89	2SA511-OR
88	2SC540F	88	2SC619OR	88	2SC689K	89	2SA480C	89	2SA511-R
88	2SC540G	88	2SC619R	88	2SC689L	89	2SA480D	89	2SA511-RD
88	2SC540GN	88	2SC619X	88	2SC689M	89	2SA480E	89	2SA511A
88	2SC540H	88	2SC619Y	88	2SC689OR	89	2SA480F	89	2SA511B
88	2SC540J	88	2SC62	88	2SC689R	89	2SA480G	89	2SA511C
88	2SC540K	88	2SC621	88	2SC689X	89	2SA480GN	89	2SA511D
88	2SC540L	88	2SC621A	88	2SC689Y	89	2SA480H	89	2SA511E
88	2SC540M	88	2SC621B	88	2SC68A	89	2SA480J	89	2SA511F
88	2SC540OR	88	2SC621C	88	2SC68B	89	2SA480K	89	2SA511G
88	2SC540R	88	2SC621D	88	2SC68C	89	2SA480L	89	2SA511GN
88	2SC540X	88	2SC621E	88	2SC68D	89	2SA480M	89	2SA511H
88	2SC540Y	88	2SC621F	88	2SC68E	89	2SA480OR	89	2SA511J
88	2SC54A	88	2SC621G	88	2SC68F	89	2SA480R	89	2SA511K
88	2SC54B	88	2SC621GN	88	2SC68G	89	2SA480X	89	2SA511L
88	2SC54C	88	2SC621H	88	2SC68GN	89	2SA480Y	89	2SA511M
88	2SC54D	88	2SC621J	88	2SC68H	89	2SA482	89	2SA511OR
88	2SC54E	88	2SC621K	88	2SC68J	89	2SA482A	89	2SA511R
88	2SC54F	88	2SC621L	88	2SC68K	89	2SA482B	89	2SA511X
88	2SC54G	88	2SC621M	88	2SC68L	89	2SA482C	89	2SA511Y
88	2SC54GN	88	2SC621OR	88	2SC68M	89	2SA482D	89	2SA512
88	2SC54H	88	2SC621R	88	2SC68OR	89	2SA482E	89	2SA512-0
88	2SC54J	88	2SC621X	88	2SC68R	89	2SA482F	89	2SA512-OR
88	2SC54K	88	2SC621Y	88	2SC68X	89	2SA482G	89	2SA512-OR1
88	2SC54L	88	2SC622	88	2SC68Y	89	2SA482GN	89	2SA512-R
88	2SC54M	88	2SC622A	88	2SC709	89	2SA482H	89	2SA512-RD
88	2SC54OR	88	2SC622B	88	2SC709(B)(C)	89	2SA482J	89	2SA512A
88	2SC54X	88	2SC622C	88	2SC709(C)	89	2SA482L	89	2SA512B
88	2SC54Y	88	2SC622D	88	2SC709A	89	2SA482M	89	2SA512C
88	2SC566	88	2SC622E	88	2SC709B	89	2SA482OR	89	2SA512D
88	2SC566A	88	2SC622F	88	2SC709C	89	2SA482R	89	2SA512E
88	2SC566B	88	2SC622G	88	2SC709CD	89	2SA482X	89	2SA512F
88	2SC566C	88	2SC622GN	88	2SC709D	89	2SA482Y	89	2SA512G
88	2SC566D	88	2SC622H	88	2SC709E	89	2SA493	89	2SA512GN
88	2SC566E	88	2SC622J	88	2SC709F	89	2SA493-0	89	2SA512H
88	2SC566F	88	2SC622K	88	2SC709G	89	2SA493O	89	2SA512J
88	2SC566G	88	2SC622L	88	2SC709GN	89	2SA493OR	89	2SA512K
88	2SC566GN	88	2SC622M	88	2SC709H	89	2SA493Y	89	2SA512L
88	2SC566H	88	2SC622OR	88	2SC709J	89	2SA494O	89	2SA512M
88	2SC566J	88	2SC622R	88	2SC709L	89	2SA495O	89	2SA512OR
88	2SC566K	88	2SC622X	88	2SC709M	89	2SA495OR	89	2SA512R
88	2SC566L	88	2SC622Y	88	2SC709OR	89	2SA499	89	2SA512X
88	2SC566M	88	2SC62A	88	2SC709X	89	2SA499-0	89	2SA512Y
88	2SC566OR	88	2SC62B	88	2SC709Y	89	2SA499-ORG	89	2SA513
88	2SC566R	88	2SC62C	88	2SC713	89	2SA499-R	89	2SA513-0
88	2SC566X	88	2SC62D	88	2SC713A	89	2SA499-RED	89	2SA513-ORG
88	2SC566Y	88	2SC62E	88	2SC713B	89	2SA499-Y	89	2SA513-R
88	2SC587	88	2SC62F	88	2SC713C	89	2SA499-YEL	89	2SA513-RD
88	2SC587A	88	2SC62G	88	2SC713D	89	2SA499A	89	2SA513-RED
88	2SC587B	88	2SC62GN	88	2SC713E	89	2SA499C	89	2SA513A
88	2SC587C	88	2SC62H	88	2SC713F	89	2SA499D	89	2SA513B
88	2SC587D	88	2SC62J	88	2SC713G	89	2SA499E	89	2SA513C
88	2SC587E	88	2SC62L	88	2SC713GN	89	2SA499F	89	2SA513D
88	2SC587F	88	2SC62M	88	2SC713H	89	2SA499G	89	2SA513E
88	2SC587GN	88	2SC62OR	88	2SC713J	89	2SA499GN	89	2SA513F
88	2SC587H	88	2SC62R	88	2SC713K	89	2SA499H	89	2SA513G
88	2SC587J	88	2SC62X	88	2SC713L	89	2SA499J	89	2SA513GN
88	2SC587K	88	2SC62Y	88	2SC713M	89	2SA499K	89	2SA513H
88	2SC587L	88	2SC640	88	2SC713OR	89	2SA499L	89	2SA513J
88	2SC587M	88	2SC640A	88	2SC713R	89	2SA499M	89	2SA513K
88	2SC587OR	88	2SC640C	88	2SC713X	89	2SA499O	89	2SA513L
88	2SC587R	88	2SC640D	88	2SC713Y	89	2SA499OR	89	2SA513M
88	2SC587X	88	2SC640E	88	2SC714	89	2SA499R	89	2SA513OR
88	2SC587Y	88	2SC640F	88	2SC714A	89	2SA499X	89	2SA513R
88	2SC588	88	2SC640G	88	2SC714B	89	2SA499Y	89	2SA513X
88	2SC588A	88	2SC640GN	88	2SC714C	89	2SA500	89	2SA513Y
88	2SC588B	88	2SC640J	88	2SC714D	89	2SA500-0	89	2SA522
88	2SC588C	88	2SC640K	88	2SC714E	89	2SA500-ORG	89	2SA522A
88	2SC588D	88	2SC640L	88	2SC714F	89	2SA500-R	89	2SA522AL
88	2SC588E	88	2SC640M	88	2SC714G	89	2SA500-RED	89	2SA522B
88	2SC588F	88	2SC640R	88	2SC714GN	89	2SA500-Y	89	2SA522C
88	2SC588G	88	2SC640X	88	2SC714H	89	2SA500-YEL	89	2SA522D
88	2SC588GN	88	2SC654A	88	2SC714J	89	2SA500A	89	2SA522E
88	2SC588H	88	2SC654B	88	2SC714K	89	2SA500B	89	2SA522F
88	2SC588J	88	2SC654C	88	2SC714L	89	2SA500C	89	2SA522G
88	2SC588K	88	2SC654D	88	2SC714M	89	2SA500D	89	2SA522GN
88	2SC588L	88	2SC654E	88	2SC714OR	89	2SA500E	89	2SA522H
88	2SC588M	88	2SC654F	88	2SC714R	89	2SA500F	89	2SA522J
		88	2SC654G	88	2SC714X			89	2SA522K
		88	2SC654GN	88	2SC714Y				

GE	Industry Standard No.	GE	Industry Standard No.	GE	Industry Standard No.	GE	Industry Standard No.	GE	Industry Standard No.
89	28A522L	89	28A5680R	89	28A677R	210	GI-2714	210	2SC1317A
89	28A522M	89	28A568A	89	28A677X	210	GI-3403	210	2SC1317B
89	28A522N	89	28A568B	89	28A677Y	210	GI-3405	210	2SC1317BC
89	28A522R	89	28A568C	89	28A704A	210	GI-3415	210	2SC1317C
89	28A522X	89	28A568D	89	28A704B	210	GI-3416	210	2SC1317D
89	28A522Y	89	28A568E	89	28A704C	210	GI-3417	210	2SC1317E
89	28A530	89	28A568F	89	28A704D	210	GI-3641	210	2SC1317G
89	28A530A	89	28A568GN	89	28A704E	210	GI-3642	210	2SC1317GR
89	28A530B	89	28A568H	89	28A704F	210	GI-3643	210	2SC1317L
89	28A530C	89	28A568J	89	28A704G	210	HC-00735	210	2SC1317OR
89	28A530D	89	28A568K	89	28A704GN	210	HEP80002	210	2SC1317P
89	28A530E	89	28A568L	89	28A704H	210	HEP80011	210	2SC1317Q
89	28A530F	89	28A568M	89	28A704J	210	HEP80025	210	2SC1317R
89	28A530G	89	28A568OR	89	28A704K	210	IP20-0039	210	2SC1317S
89	28A530GN	89	28A568R	89	28A704M	210	IP20-0191	210	2SC1317T
89	28A530GR	89	28A568X	89	28A704OR	210	I2120	210	2SC1317Y
89	28A530H	89	28A568Y	89	28A704OR	210	I2121	210	2SC1318
89	28A530H1	89	28A569	89	28A704R	210	KSC815-0	210	2SC1318(P,R)
89	28A530J	89	28A569OR	89	28A704X	210	KSC815-0	210	2SC1318(Q)
89	28A530K	89	28A569A	89	28A704Y	210	LDA400	210	2SC1318A
89	28A530L	89	28A569B	89	28A705A	210	LDA400MP	210	2SC1318B
89	28A530M	89	28A569C	89	28A705B	210	LDA401	210	2SC1318C
89	28A530R	89	28A569D	89	28A705C	210	LDA401MP	210	2SC1318E
89	28A530R	89	28A569E	89	28A705D	210	LDA402	210	2SC1318F
89	28A530X	89	28A569F	89	28A705E	210	LDS210	210	2SC1318G
89	28A530Y	89	28A569G	89	28A705F	210	LM1135	210	2SC1318GN
89	28A539	89	28A569GN	89	28A705G	210	M4937(3RD-IP)	210	2SC1318H
89	28A539(K)	89	28A569H	89	28A705GN	210	MA6001	210	2SC1318HN
89	28A539(L)	89	28A569J	89	28A705H	210	MA6002	210	2SC1318HH
89	28A539(M)	89	28A569K	89	28A705J	210	MA6003	210	2SC1318J
89	28A539A	89	28A569L	89	28A705K	210	MM3903	210	2SC1318K
89	28A539B	89	28A569M	89	28A705L	210	MMT2222	210	2SC1318L
89	28A539C	89	28A569OR	89	28A705M	210	MMT3903	210	2SC1318M
89	28A539D	89	28A569X	89	28A705OR	210	MPS6515	210	2SC1318PR
89	28A539E	89	28A569Y	89	28A705R	210	MPS8001	210	2SC1318Q
89	28A539F	89	28A570	89	28A705X	210	MPS8097	210	2SC1318QR
89	28A539G	89	28A570A	89	28A705Y	210	MPSD05	210	2SC1318S
89	28A539GN	89	28A570B	89	28A826	210	NPC1075	210	2SC1318S,R
89	28A539H	89	28A570C	89	28A826P	210	NPS6512	210	2SC1318X
89	28A539J	89	28A570D	89	28A826Q	210	NPS6513	210	2SC1318Y
89	28A539K	89	28A570E	89	28A826R	210	N33903	210	2SC1335
89	28A539M	89	28A570F	90	28A826RY	210	PB5025	210	2SC1335(E)
89	28A539N	89	28A570G	90	D201	210	PN107	210	2SC1335-OR
89	28A539OR	89	28A570GN	90	0001849	210	PN108	210	2SC1335A
89	28A539R	90	28A570H	121	1S853	210	PN109	210	2SC1335B
89	28A539S	210	28A570J	121	28C558Q	210	PTC139	210	2SC1335C
89	28A539X	210	28A570K	210	A1U(3RDIP)	210	Q-00584R	210	2SC1335D
89	28A539Y	210	28A570L	210	A1U(LASTIP)	210	Q-00584R	210	2SC1335P
89	28A542	210	28A570M	210	A1Z	210	Q5124	210	2SC1335Q
89	28A542A	210	28A570OR	210	A5T4123	210	QT-C0735XBT	210	2SC1335GN
89	28A542B	210	28A570R	210	A673351K	210	QT-C1318BXDN	210	2SC1335H
89	28A542C	210	28A570X	210	A673354K	210	QT-C1S1BXDN	210	2SC1335J
89	28A542D	210	28A570Y	210	A673355H	210	R3576-1	210	2SC1335K
89	28A542E	210	28A603	210	A673355K	210	R3676-1	210	2SC1335L
89	28A542F	210	28A603A	210	BC167A	210	RE66	210	2SC1335M
89	28A542G	210	28A603B	210	BCW60AB	210	RE67	210	2SC1335OR
89	28A542GN	210	28A603C	210	BCW60AC	210	REN233	210	2SC1335R
89	28A542H	210	28A603D	210	BCW60B	210	REN67	210	2SC1335X
89	28A542J	210	28A603E	210	BCW60C	210	S0002	210	2SC1335Y
89	28A542L	210	28A603F	210	BCX19	210	S0024	210	2SC1364
89	28A542M	210	28A603G	210	BCX19R	210	S31866	210	2SC1364-6
89	28A542OR	210	28A603GN	210	BCX20	210	S33755	210	2SC1364-OR
89	28A542R	210	28A603H	210	BCX20R	210	SJ3629	210	2SC1364A
89	28A542X	210	28A603J	210	BCX70AG	210	SK3124	210	2SC1364C
89	28A542Y	210	28A603K	210	BCX70AH	210	SPS918	210	2SC1364D
89	28A543	210	28A603L	210	BCX70AJ	210	T-Q5078	210	2SC1364E
89	28A544	210	28A603M	210	BF819CB	210	TD100	210	2SC1364K
89	28A544A	210	28A603OR	210	BFV40	210	TD200	210	2SC1364L
89	28A544B	210	28A603R	210	BFV41	210	TD2219	210	2SC1364M
89	28A544C	210	28A603X	210	BFV45	210	TD250	210	2SC1364OR
89	28A544E	210	28A603Y	210	BFV49	210	TG28C1175	210	2SC1364R
89	28A544F	210	28A607E	210	BG-94	210	TG28C1175-	210	2SC1364Y
89	28A544G	210	28A607F	210	BG-94	210	TG28C1175-D	210	2SC141
89	28A544GN	210	28A607G	210	BS-94	210	TG28C1175-E	210	2SC142
89	28A544H	210	28A607GN	210	BSW88A	210	TGB0331	210	2SC143
89	28A544K	210	28A607H	210	BSW88B	210	TNJ72276	210	2SC144
89	28A544L	210	28A607J	210	BSW89A	210	TR0573491	210	2SC144A
89	28A544M	210	28A607K	210	BSW89B	210	TR2520063	210	2SC1633
89	28A544OR	210	28A607L	210	BSX38A	210	TR2327293	210	2SC1969BH
89	28A544R	210	28A607M	210	BSX38B	210	TR2327333	210	2SC2000
89	28A544X	210	28A607OR	210	BSX79A	210	TR2327363	210	2SC2000L
89	28A544Y	210	28A607R	210	BSX79B	210	TR24	210	2SC2076
89	28A548	210	28A607X	210	BSX81A	210	TY-20	210	2SC2076B
89	28A548G	210	28A607Y	210	BSX81B	210	TV81	210	2SC2076C
89	28A548GN	89	28A629	210	C1128(3RD-IP)	210	01-030458	210	2SC2076CB
89	28A548OR	89	28A629A	210	C1293A(LAST-IP)	210	01-030735	210	2SC2076CD
89	28A548R	89	28A629B	210	C1293B	210	01-031175	210	2SC2076D
89	28A548Y	89	28A629C	210	C1293B(3RD-IP)	210	01-031318	210	2SC264
89	28A5620	89	28A629D	210	C1293B(LAST-IP)	210	01-031364	210	2SC265
89	28A565	89	28A629E	210	C388	210	2N3946	210	2SC395A-Q
89	28A565(D,C)	89	28A629F	210	C388A	210	2N3973	210	2SC395A-R
89	28A565A	89	28A629G	210	C388ATV	210	2N3974	210	2SC395A-Y
89	28A565AB	89	28A629GN	210	C460	210	2N4100	210	2SC454
89	28A565B	89	28A629H	210	C460(A)	210	2N4140	210	2SC454(A)
89	28A565BA	89	28A629J	210	C460(B)	210	2N4227	210	2SC454(B)
89	28A565C	89	28A629K	210	C460A	210	2N5027	210	2SC454-3
89	28A565D	89	28A629L	210	C460B	210	2N5028	210	2SC454-5
89	28A565E	89	28A629M	210	C460C	210	2N5066	210	2SC454-OR
89	28A565F	89	28A629OR	210	C460D	210	2N5735	210	2SC454A
89	28A565G	89	28A629R	210	C460G	210	2N5736	210	2SC454b
89	28A565GN	89	28A629X	210	C460GB	210	2N5998	210	2SC454b-6
89	28A565H	89	28A629Y	210	C460H	210	2N6000	210	2SC454bL
89	28A565J	89	28A659	210	C460K	210	2N6004	210	2SC454C
89	28A565L	89	28A659(D)	210	C460L	210	2N6008	210	2SC454D
89	28A565M	89	28A659(E)	210	C563A(3RDIP)	210	2N706/46	210	2SC454E
89	28A565OR	89	28A659A	210	C717(FINAL-IP)	210	2N706/51	210	2SC454F
89	28A565R	89	28A659B	210	CP14	210	2N706A/46	210	2SC454G
89	28A565X	89	28A659D	210	CGE-61	210	2N706A/51	210	2SC454GN
89	28A565Y	89	28A659E	210	DBCZ073504	210	2N706B/46	210	2SC454H
89	28A567	89	28A659F	210	DDBY233001	210	2N706B/51	210	2SC454K
89	28A567O	89	28A659G	210	DDBY273001	210	2N709A46	210	2SC454X
89	28A567OR	89	28A659L	210	DDBY283001	210	2N753/46	210	2SC454L
89	28A567A	89	28A659N	210	DDBY301001	210	2N784A/46	210	2SC454LA
89	28A567B	89	28A659P	210	DS-94	210	2N784A/51	210	2SC454M
89	28A567C	89	28A659R	210	E13-004-00	210	2N929/46	210	2SC454OR
89	28A567D	89	28A659Y	210	EA15X241	210	2N930/46	210	2SC454R
89	28A567E	89	28A677	210	EA15X373	210	2N930A/46	210	2SC454X
89	28A567F	89	28A677-0	210	EA15X378	210	2N930B	210	2SC454Y
89	28A567G	89	28A677A	210	ECG233	210	2SC1175	210	2SC458
89	28A567GN	89	28A677B	210	EL403	210	2SC1175(D,E,F)	210	2SC458(C)
89	28A567GR	89	28A677C	210	EN3903	210	2SC1175C	210	2SC458(C,D)
89	28A567H	89	28A677D	210	EP15X20	210	2SC1175CTV	210	2SC458(D)
89	28A567J	89	28A677E	210	EP15X54	210	2SC1175D	210	2SC458(LG)
89	28A567K	89	28A677F	210	EQ8-0165	210	2SC1175E	210	2SC458-4
89	28A567L	89	28A677G	210	EQ8-0192	210	2SC1175F	210	2SC458-5
89	28A567M	89	28A677GN	210	ES15X106	210	2SC1205	210	2SC458-6
89	28A567OR	89	28A677H	210	ES15X127	210	2SC1205A	210	2SC458-OR
89	28A567R	89	28A677J	210	ES15X22	210	2SC1205B	210	2SC458A
89	28A567X	89	28A677K	210	ETTC-930D	210	2SC1205C		
89	28A567Y	89	28A677L	210	FC88050C	210	2SC1247		
89	28A568	89	28A677M	210	FX3500	210	2SC1317		
89	28A5680	89	28A6770R	210	FX4046	210	2SC1317(P)		
89		89		210	G3-210	210	2SC1317(R)		
89		89		210	GET3903	210	2SC1317(S)		
89		89				210	2SC1317-0R		

GE	Industry Standard No.
210	2SC458AD
210	2SC458AK
210	2SC458B-D
210	2SC458BC
210	2SC458BD
210	2SC458BK
210	2SC458BL
210	2SC458BLG
210	2SC458BM
210	2SC458CL
210	2SC458CLG
210	2SC458CM
210	2SC458D
210	2SC458E
210	2SC458F
210	2SC458G
210	2SC458GN
210	2SC458H
210	2SC458J
210	2SC458K
210	2SC458KA
210	2SC458KB
210	2SC458KC
210	2SC458KD
210	2SC458L
210	2SC458L(C)
210	2SC458L6
210	2SC458LB
210	2SC458LC
210	2SC458LD
210	2SC458LG
210	2SC458LG(B)
210	2SC458LG(C)
210	2SC458LG(D)
210	2SC458LGA
210	2SC458LGB
210	2SC458LGBM
210	2SC458LGC-6
210	2SC458LGD
210	2SC458LGO
210	2SC458LGR
210	2SC458LGS
210	2SC458M
210	2SC458OR
210	2SC458P
210	2SC458R
210	2SC458RGS
210	2SC458TOK
210	2SC458V
210	2SC458VC
210	2SC458X
210	2SC458Y
210	2SC460L
210	2SC735(O)
210	2SC735(FA-3)
210	2SC735(O)
210	2SC735(Y)
210	2SC735-0
210	2SC735-GRN
210	2SC735-O
210	2SC735-OR
210	2SC735-ORG
210	2SC735-ORN
210	2SC735-R
210	2SC735-RED
210	2SC735-Y
210	2SC735-YEL
210	2SC735/4454C
210	2SC735A
210	2SC735B
210	2SC735C
210	2SC735D
210	2SC735E
210	2SC735F
210	2SC735FA3
210	2SC735G
210	2SC735GN
210	2SC735GR
210	2SC735GRN
210	2SC735H
210	2SC735J
210	2SC735K
210	2SC735L
210	2SC735M
210	2SC7350
210	2SC7350R
210	2SC7350RN
210	2SC735R
210	2SC735RED
210	2SC735X
210	2SC735Y
210	2SC735Y/4454C
210	2SC735YEL
211	DDBY209003
211	EA15X336
211	EA15X364
211	EA15X365
211	EA15X367
211	EA15X441
211	EQS-0196
211	FCS-9016F
211	FCS-9016G
211	FCS9016G
211	GE-211
211	GI-2926
211	HS-40054
211	IP20-0029
211	QT-C0710XAE
211	QT-C0710XBE
211	QT-C0710XEE
211	SK3356
211	SPS-952-2
211	01-030710
211	2SC1035
211	2SC1035A
211	2SC1035B
211	2SC1035D
211	2SC1035E
211	2SC1035F
211	2SC1035GN
211	2SC1035H
211	2SC1035J
211	2SC1035L
211	2SC1035M
211	2SC10350R
211	2SC1035R
211	2SC1035X
211	2SC1035Y
211	2SC1036
211	2SC1036A
211	2SC1036B
211	2SC1036C
211	2SC1036D
211	2SC1036E
211	2SC1036F
211	2SC1036G
211	2SC1036GN
211	2SC1036H
211	2SC1036J
211	2SC1036K
211	2SC1036L
211	2SC1036M
211	2SC10360R
211	2SC1036R
211	2SC1036X
211	2SC1036Y
211	2SC348
211	2SC348A
211	2SC348B
211	2SC348C
211	2SC348D
211	2SC348E
211	2SC348F
211	2SC348G
211	2SC348GN
211	2SC348H
211	2SC348J
211	2SC348K
211	2SC348L
211	2SC348M
211	2SC3480R
211	2SC348X
211	2SC348Y
211	2SC389
211	2SC389-O
211	2SC389A
211	2SC389B
211	2SC389C
211	2SC389D
211	2SC389E
211	2SC389F
211	2SC389G
211	2SC389GN
211	2SC389H
211	2SC389J
211	2SC389K
211	2SC389L
211	2SC389M
211	2SC389R
211	2SC389X
211	2SC389Y
211	2SC390
211	2SC390A
211	2SC390B
211	2SC390C
211	2SC390D
211	2SC390E
211	2SC390F
211	2SC390G
211	2SC390GN
211	2SC390H
211	2SC390J
211	2SC390K
211	2SC390L
211	2SC390M
211	2SC3900R
211	2SC390R
211	2SC390X
211	2SC390Y
211	2SC663A
211	2SC663B
211	2SC663C
211	2SC663D
211	2SC663E
211	2SC663F
211	2SC663G
211	2SC663GN
211	2SC663H
211	2SC663J
211	2SC663K
211	2SC663L
211	2SC663M
211	2SC6630R
211	2SC663R
211	2SC663X
211	2SC663Y
211	2SC668(C)
211	2SC668(D)
211	2SC668-O
211	2SC668-OR
211	2SC668A
211	2SC668B
211	2SC668B1
211	2SC668BC2
211	2SC668C
211	2SC668C1
211	2SC668C2
211	2SC668CD
211	2SC668DO
211	2SC668D1
211	2SC668DE
211	2SC668DO
211	2SC668DV
211	2SC668DX
211	2SC668DZ
211	2SC668E
211	2SC668E1
211	2SC668E2
211	2SC668EP
211	2SC668EV
211	2SC668EX
211	2SC668F
211	2SC668G
211	2SC668GN
211	2SC668H
211	2SC668K
211	2SC668L
211	2SC668M
211	2SC6680R
211	2SC668R
211	2SC668X
211	2SC668Y
211	2SC710
211	2SC710(B)
211	2SC710(C)
211	2SC710(D)
211	2SC710-1
211	2SC710-2
211	2SC710-4
211	2SC710-OR
211	2SC710AL
211	2SC710B2
211	2SC710BC
211	2SC710D
211	2SC710DE
211	2SC710E
211	2SC710F
211	2SC710G
211	2SC710GN
211	2SC710H
211	2SC710K
211	2SC710L
211	2SC710M
211	2SC7100R
211	2SC710R
211	2SC710X
211	2SC710XL
211	2SC710Y
212	A642S
212	BC182K
212	BC182KA
212	BC182KB
212	BC183K
212	BC183KA
212	BC183KB
212	BC183KC
212	BC184K
212	BC184KB
212	BC184KC
212	BC239A
212	BC384B
212	BC385A
212	BCW87
212	BCW98A
212	BCW98B
212	BCW98C
212	BCW98D
212	BCY55
212	BFS55A
212	BFV37
212	BFV38
212	BFV62
212	C1002
212	C7076
212	CK420
212	CK421
212	CS4003
212	CS4061
212	D24A3391
212	D24A3391A
212	D24A3392
212	D24A3393
212	D24A3900
212	D24A3900A
212	D26C4
212	D26C5
212	D26E-1
212	D26E-5
212	D26E-7
212	DDBY224003
212	DDBY224004
212	DDBY224006
212	EA15X251
212	EA15X404
212	EA15X7635
212	EA15X7639
212	EA2740
212	EN956
212	EQS-0061
212	ETTC-945
212	FOS100
212	FOS101
212	G05-706-D
212	G05-706-E
212	GE-212
212	GET3562
212	GI-2924
212	GI-2925
212	GI-3391
212	GI-3391A
212	GI-3392
212	GI-3393
212	GI-3394
212	GI-3395
212	GI-3396
212	GI-3397
212	GI-3398
212	GI-3707
212	GI-3708
212	GI-3709
212	GI-3710
212	GI-3711
212	GI-3721
212	GI-3900
212	GI-3900A
212	HEPS0024
212	HST5001
212	HT309451LO
212	HT310001F
212	IT122
212	IT918A
212	ITC918A
212	KGE46146
212	KSC945Y
212	L4
212	L5
212	L6
212	L7
212	LID929
212	LID930
212	MMCM930
212	MMT70
212	MPS-5172
212	MPS6564
212	MPSA09
212	MPSA10
212	MPSA10-BLU
212	MPSA10-GRN
212	MPSA10-RED
212	MPSA10-WHT
212	MPSA10-YEL
212	MPSA20-BLU
212	MPSA20-GRN
212	MPSA20-RED
212	MPSA20-WHT
212	MPSA20-YEL
212	MPSK20
212	MPSK21
212	MPSK22
212	NPC069
212	NPC069-98
212	NTC-11
212	NTC-7
212	PL4061
212	PL4062
212	PN929
212	PN930
212	Q5183
212	Q5183P
212	QT-C0945ACA
212	QT-C0945AGA
212	QT-C1359XAN
212	RE64
212	REN64
212	RQ-4448
212	RT-104
212	RT-105
212	RT-107
212	RT-108
212	RT-112
212	RT7326
212	RT7557
212	SC147A
212	SC147B
212	SC148
212	SC148A
212	SC148B
212	SC148C
212	SC149
212	SC149B
212	SCA3244
212	SPS952
212	TG2SC536
212	TG2SC536(C)
212	TG2SC536(E)
212	TG2SC536-D-A
212	TG2SC536-D-B
212	TG2SC536-E
212	TG2SC536-E-A
212	TG2SC536-E-B
212	TG2SC536-F
212	TG2SC536-F-A
212	TG2SC536C
212	TG2SC536E
212	TN2SC945-Q
212	TN2SC945-R
212	TN2SC945R
212	TR2S1570LH
212	01-030536
212	01-030734
212	01-030945
212	01-031359
212	2N1268
212	2N1269
212	2N1270
212	2N1271
212	2N1272
212	2N1390
212	2N2432A
212	2N2639
212	2N2640
212	2N2641
212	2N2642
212	2N2643
212	2N2644
212	2N2694
212	2N2926-BRN
212	2N2926-GRN
212	2N2926-ORG
212	2N2926-RED
212	2N2926-YEL
212	2N3390-U29
212	2N3391-U29
212	2N3392-U29
212	2N3393-U29
212	2N3394-U29
212	2N3395-WHT
212	2N3395-YEL
212	2N3396-ORG
212	2N3396-WHT
212	2N3396-YEL
212	2N3397-ORG
212	2N3397-RED
212	2N3397-WHT
212	2N3397-YEL
212	2N3398-BLU
212	2N3398-ORG
212	2N3398-RED
212	2N3398-WHT
212	2N3708-BLU
212	2N3708-BRN
212	2N3708-GRN
212	2N3708-ORG
212	2N3708-RED
212	2N3708-VIO
212	2N3708-YEL
212	2N4086
212	2N4087
212	2N4087A
212	2N4138
212	2N5249
212	2N841/46
212	2SC1285
212	2SC1310
212	2SC1311
212	2SC1345
212	2SC1345(E)
212	2SC1345C
212	2SC1345D
212	2SC1345E
212	2SC1345F
212	2SC1359
212	2SC1359(A)
212	2SC1359(B)
212	2SC1359(C)
212	2SC1359(C,B)
212	2SC1359A
212	2SC1359B
212	2SC1359BC
212	2SC1359C
212	2SC1359Q
212	2SC1380
212	2SC1380-BL
212	2SC1380-GR
212	2SC1380A
212	2SC1380A-BL
212	2SC1380A-GR
212	2SC1453
212	2SC1542
212	2SC1570
212	2SC1570LH
212	2SC1571
212	2SC1571G
212	2SC1571L
212	2SC1623
212	2SC281
212	2SC281(B)
212	2SC281-OR
212	2SC281A
212	2SC281B
212	2SC281BL
212	2SC281C
212	2SC281C-EP
212	2SC281D
212	2SC281E
212	2SC281EP
212	2SC281F
212	2SC281G
212	2SC281GN
212	2SC281H
212	2SC281HA
212	2SC281HB
212	2SC281HC
212	2SC281J
212	2SC281K
212	2SC281L
212	2SC281M
212	2SC2810R
212	2SC281R
212	2SC281X
212	2SC281Y
212	2SC368-BL
212	2SC368-GR
212	2SC373(GR)
212	2SC373-14
212	2SC373-G
212	2SC373-O
212	2SC373-OR
212	2SC373A
212	2SC373AL
212	2SC373B
212	2SC373BL
212	2SC373C
212	2SC373D
212	2SC373E
212	2SC373F
212	2SC373G
212	2SC373GN
212	2SC373GR
212	2SC373H
212	2SC373J
212	2SC373K
212	2SC373L
212	2SC373M
212	2SC3730R
212	2SC373R
212	2SC373X
212	2SC373Y
212	2SC374
212	2SC374(BL)
212	2SC374(V)
212	2SC374-BL
212	2SC374-OR
212	2SC374-V
212	2SC374A
212	2SC374B
212	2SC374BLK
212	2SC374C
212	2SC374D
212	2SC374E
212	2SC374F
212	2SC374G
212	2SC374GN
212	2SC374H
212	2SC374J
212	2SC374JA
212	2SC374K
212	2SC374L
212	2SC374M
212	2SC3740R
212	2SC374R
212	2SC374V
212	2SC374X
212	2SC374Y
212	2SC378-O
212	2SC378-R
212	2SC378-Y
212	2SC400-O
212	2SC528
212	2SC536(C)
212	2SC536(D)
212	2SC536(E)
212	2SC536-D
212	2SC536-E
212	2SC536-F
212	2SC536-G
212	2SC536-OR
212	2SC536A
212	2SC536A(3RD-IF)
212	2SC536AG
212	2SC536B
212	2SC536C
212	2SC536D
212	2SC536DK
212	2SC536E
212	2SC536ED
212	2SC536EH
212	2SC536EJ
212	2SC536EN
212	2SC536EP
212	2SC536ER
212	2SC536ET
212	2SC536EZ
212	2SC536F
212	2SC536F1
212	2SC536F2
212	2SC536FC
212	2SC536FP
212	2SC536FS
212	2SC536FS6
212	2SC536G
212	2SC536G-1
212	2SC536G1
212	2SC536G2
212	2SC536GF
212	2SC536GJ
212	2SC536GK
212	2SC536GL
212	2SC536GM
212	2SC536GN
212	2SC536GP
212	2SC536GT
212	2SC536GV
212	2SC536GY
212	2SC536GZ
212	2SC536H
212	2SC536J
212	2SC536K
212	2SC536L
212	2SC536M
212	2SC536NP
212	2SC5360R
212	2SC536Q
212	2SC536R
212	2SC536W
212	2SC536X
212	2SC536XL
212	2SC536Y
212	2SC537(F)
212	2SC537(G)

GE	Industry Standard No.	GE	Industry Standard No.	GE	Industry Standard No.	GE	Industry Standard No.	GE	Industry Standard No.
212	2SC537-01	214	MT9003	218	BFW87	221	2SA495-OR	235	2SC154D
212	2SC537-EV	214	MZ409-02B	218	BFW88	221	2SA495-ORG	235	2SC154E
212	2SC537ALC	214	PC-20066	218	CXJ249	221	2SA495-ORG-Q	235	2SC154F
212	2SC537B	214	SK3019	218	ECG189	221	2SA495-R	235	2SC154G
212	2SC537BK	214	SK3039	218	EP-976	221	2SA495-RD	235	2SC154GN
212	2SC537C	214	2M6219	218	ES81	221	2SA495-RED	235	2SC154H
212	2SC537C7	214	2MN6374	218	ES81(ELCOM)	221	2SA495-RED-G	235	2SC154J
212	2SC537D	214	2N1085	218	GE-218	221	2SA495-Y	235	2SC154K
212	2SC537D1	214	2N1941	218	HEP-83030	221	2SA495-YEL	235	2SC154L
212	2SC537D2	214	2N2489A	218	HEP-83031	221	2SA495-YEL-Q	235	2SC154M
212	2SC537E	214	2N2655	218	HEP-83032	221	2SA495-YL	235	2SC154OR
212	2SC537EF	214	2N50588	218	IRTR73	221	2SA495A	235	2SC154R
212	2SC537EH	214	2N50598	218	M9641	221	2SA495B	235	2SC154X
212	2SC537EJ	214	2N5304	218	MP8-U55	221	2SA495C	235	2SC154Y
212	2SC537EK	214	2N5487-1	218	MP8-U56	221	2SA495D	239	A059-108
212	2SC537EV	214	2N5487-3	218	MP8U51	221	2SA495E	239	A1124C
212	2SC537P-C7	214	2N5772	218	MP8U51A	221	2SA495F	239	A12153
212	2SC537P1	214	2N5772	218	MP8U55	221	2SA495G	239	A12178
212	2SC537P2	214	2N6045	218	MP8U56	221	2SA495G-GR	239	A14-586-01
212	2SC537PC	214	2N6218	218	MU9660	221	2SA495G-0	239	A141AA
212	2SC537PC7	214	2N6220	218	MU9660S	221	2SA495G-R	239	A141AA9
212	2SC537PJ	214	2N6221	218	MU9660T	221	2SA495G-Y	239	A141AA-19
212	2SC537PK	214	2N6222	218	MU9661	221	2SA495GN	239	A145C
212	2SC537PV	214	2N6232	218	MU9661T	221	2SA495GR	239	A146509
212	2SC537G	214	2N6232-4	218	P218-2	221	2SA495H	239	A1477C
212	2SC537G1	214	2N6367	218	P2U	221	2SA495J	239	A1477Q9
212	2SC537G2	214	2N6370	218	P2U-1	221	2SA495K	239	A1484A
212	2SC537GP	214	2N6372	218	P2V	221	2SA495L	239	A1484A9
212	2SC537GFL	214	2N6373	218	P3K	221	2SA495M	239	A14A-70
212	2SC537GI	214	2N915A	218	P3Y	221	2SA495O	239	A14A-705
212	2SC537H	214	2SC1621	218	PTC141	221	2SA495OF	239	A14A10G
212	2SC537HT	214	2SC1778	218	RE76	221	2SA495OR	239	A15927
212	2SC537W	215	D360	218	REN189	221	2SA495R	239	A2090056-1
212	2SC537WP	215	DDBY228001	218	S39509	221	2SA495RD	239	A2090056-27
212	2SC538	215	01-031173	218	SJE687	221	2SA495RED	239	A2090056-5
212	2SC538(P)	215	2SC1226	218	SP84099	221	2SA495RED-G	239	A2091859-0025
212	2SC538(R)	215	2SC1226(A)	218	SP8837	221	2SA495W	239	A2091859-072Q
212	2SC538-Q	215	2SC1226(AP)	218	T423	221	2SA495W1	239	A2091859-10
212	2SC538A	215	2SC1226(P)	218	TIF30B	221	2SA495WI	239	A218012DS
212	2SC538A(Q)	215	2SC1226(R)	218	TRO2053-5	221	2SA495X	239	A30302
212	2SC538A-P	215	2SC1226-0	218	TRO2053-7	221	2SA495Y	239	A34-6001-1
212	2SC538A-R	215	2SC1226A	218	02P1BC	221	2SA495YEL	239	A34-6002-17
212	2SC538AQ	215	2SC1226A(P)	219	SK3047	221	2SA495YEL-G	239	A34715
212	2SC538AR	215	2SC1226A(Q)	220	TV19	221	2SA495YL	239	A35084
212	2SC538AS	215	2SC1226A(QPR)	220	2SC268	221	2SA702	239	A35201
212	2SC538K	215	2SC1226A(R)	220	2SC268-0R	221	2SA811	239	A35260
212	2SC538P	215	2SC1226AC	220	2SC268B	221	2SA812	239	A36896
212	2SC538R	215	2SC1226ACF	220	2SC268C	221	2SB642	239	A416
212	2SC538S	215	2SC1226AP	220	2SC268D	221	2SB642Q	239	A417-62
212	2SC538T	215	2SC1226AQ	220	2SC268E	221	2SB642R	239	A4247
212	2SC734	215	2SC1226ARL	220	2SC268F	222	Q5217	239	A4347
212	2SC734(O)	215	2SC1226B	220	2SC268G	222	REN297	239	A48-134727
212	2SC734(R)	215	2SC1226BL	220	2SC268GN	222,271	EC0297	239	A48-134731
212	2SC734(Y)	215	2SC1226C	220	2SC268H	223	2SA835	239	A48-134907
212	2SC734-O	215	2SC1226CF	220	2SC268J	223	2SA835H	239	A48-137102
212	2SC734-G	215	2SC1226D	220	2SC268K	224	HEP80021	239	A48-137214
212	2SC734-GR	215	2SC1226E	220	2SC268L	226	A5B	239	A48-137215
212	2SC734-GRN	215	2SC1226F	220	2SC268M	226	A6D	239	A48-137216
212	2SC734-0	215	2SC1226G	220	2SC268OR	226	A6D-1	239	A48-137217
212	2SC734-OR	215	2SC1226L	220	2SC268R	226	A6D-2	239	A48-137218
212	2SC734-ORG	215	2SC1226O	220	2SC268Y	226	A6D-3	239	A48-137219
212	2SC734-OY	215	2SC1226OR	220	2SC634	226	A6J	239	A48-137220
212	2SC734-R	215	2SC1226P	220	2SC634(2)	226	A6	239	A48-63076A81
212	2SC734-RED	215	2SC1226Q	220	2SC634-0	226	B1M	239	A4A-1-70
212	2SC734-Y	215	2SC1226QR	220	2SC634-OR	226	BC537	239	A4A-1-705
212	2SC734-YEL	215	2SC1226R	220	2SC634A	226	BC538	239	A4A-1A9G
212	2SC734A	215	2SC1226RL	220	2SC634AAK	226	BC430	239	A5253
212	2SC734B	215	2SC1226RLP	220	2SC634AL	226	ES82(ELCOM)	239	A57B124-10
212	2SC734C	215	2SC1226RLQ	220	2SC634AXL	226	GE-226	239	A57L5-1
212	2SC734D	215	2SC1226RLR	220	2SC634B	226	IRTR72	239	A57M3-7
212	2SC734F	215	2SC1226SC	220	2SC634D	226	M9640	239	A57M5-8
212	2SC734G	215	2SC1226Y	220	2SC634E	226	MFS-U07	239	A62-18427
212	2SC734GR	216	A67-76-200	220	2SC634F	226	MPSU01	239	A63-18427
212	2SC734GRN	216	C1173(RF-PWR)	220	2SC634G	226	MPSU01A	239	A65C-19G
212	2SC734H	216	C1237E	220	2SC634GN	226	MPSU02	239	A65C-70
212	2SC734K	216	C1307	220	2SC634H	226	MPSU05	239	A65C-705
212	2SC734K/GR	216	C1377	220	2SC634K	226	MPSU06	239	A65C619G
212	2SC734L	216	C1816	220	2SC634L	226	MU9610	239	A66009T
212	2SC734M	216	C1964	220	2SC634M	226	MU9610P	239	A690V081H97
212	2SC734O	216	C799	220	2SC634OR	226	MU9610T	239	A7279039
212	2SC734R	216	DDBY231002	220	2SC634R	226	MU9611	239	A7279049
212	2SC734RED	216	DDBY307003	220	2SC634X	226	MU9611G	239	A7285774
212	2SC734X	216	ECG236	220	2SC634Y	226	MU9611T	239	A7285778
212	2SC734Y	216	EQ8-0159	221	A650238A	226	P218-1	239	A7289047
212	2SC737YEL	216	EQ8-141	221	A759	226	S8B1107	239	A7290594
213	DDBY259001	216	ESQ-141	221	BC158VI	226	T422	239	A7291252
213	DDBY259002	216	GE-216	221	BC307VI	226	2SC1663	239	A7297043
213	EA15X350	216	H8T5906	221	BC308VI	226	2SC1663H	239	A7297092
213	G05-064A	216	IP20-0154	221	BC418A	226	2SC2194	239	A7297093
213	HT316751MO	216	PL-172-013-9001	221	BC419A	227	HEP83051	239	A76-1177Q
213	Q50078Z	216	PTC186	221	BCX17	227	HEP83032	239	A77C-70
213	ST-2SC383W	216	QT-C1307XZA	221	BCX17R	227	2SA962	239	A77C-705
213	01-031675	216	RE201	221	GI-3638	228	2N3743	239	A77C19G
213	2SC1675	216	REN236	221	GI-3638A	230	HEP83033	239	A84A-70
213	2SC1675K	216	STT2405	221	HA9048	230	HEP83034	239	A84A-705
213	2SC1675L	216	STT2406	221	HA9049	230	HEP83035	239	A84A19G
213	2SC1675M	216	STT4483	221	HA9054	230	HEP85025	239	A8P404-ORN
213	2SC383	216	STT9001	221	HA9055	234	SK3085	239	A8P404-ORN
213	2SC383(3RD-IP)	216	STT9002	221	HA9079	235	A5T5058	239	A8P404F
213	2SC383(T)	216	STT9004	221	KSA495Y	235	BC594	239	A94004
213	2SC383-OR	216	STT9005	221	MMT71	235	ERB140	239	A964-17887
213	2SC383A	216	2N2876	221	NTC-10	235	ERB160	239	A97A83
213	2SC383B	216	2N3253	221	NTC-6	235	ERB180	239	A992-00-1192
213	2SC383C	216	2N5785	221	Q5135	235	ERS200	239	AA2SB240A
213	2SC383D	216	2SC-1307	221	REN62	235	ERS225	239	AA4
213	2SC383E	216	2SC1307	221	SC159	235	ERS250	239	AC148
213	2SC383F	216	2SC1307-1	221	SC159A	235	ERS275	239	AC316
213	2SC383G	216	2SC1377	221	SC259	235	ERS301	239	AD-140
213	2SC383H	216	2SC1816	221	SC259A	235	ERS325	239	AD-148
213	2SC383J	216	2SC1969	221	TT2SA495-0	235	ETP2008	239	AD-149
213	2SC383L	216	2SC1969H	221	TT2SA495-Y	235	ETP5925	239	AD-150
213	2SC383M	216	2SC1969H	221	01-010495	235	GE-235	239	AD-152
213	2SC383OR	216	2SC2393D	221	2N2303/46	235	MHT4402	239	AD-156
213	2SC383R	216	2SC2394D	221	2N2474	235	RT-110	239	AD-157
213	2SC383T	216	2SC490	221	2N3841	235	SK3040	239	AD-159
213	2SC383W	216	2SC490-BLU	221	2N3842	235	2N1613A	239	AD105
213	2SC383X	216	2SC490-R	221	2N4284	235	2N1613B	239	AD104
213	2SC383Y	216	2SC490-RED	221	2N4285	235	2N2102A	239	AD105
214	B2	216	2SC490-Y	221	2N4314	235	2N2726	239	AD130
214	B3	216	2SC490-YEL	221	2S305	235	2N340	239	AD130-III
214	B4	217	ECG188	221	2SA495	235	2N341	239	AD130-IV
214	BT-94	217	GE-217	221	2SA495(0)	235	2N341A	239	AD130-V
214	D26E-2	217	H7618	221	2SA495(R)	235	2N342A	239	AD131
214	D26E-3	217	REN188	221	2SA495(Y)	235	2N5185	239	AD131-III
214	D26E-6	217	TR72	221	2SA495-0	235	2SC1062	239	AD131-IV
214	EA15X415	217	2SC1124-OR	221	2SA495-1	235	2SC154	239	AD131-V
214	GE-214	218	A486(JAPAN)	221	2SA495-G	235	2SC154(C)	239	AD132
214	KD5000	218	BC327	221	2SA495-GN	235	2SC154-OR	239	AD138
214	MT3011	218	BC328	221	2SA495-O	235	2SC154A	239	AD138/50
		218	BC460	221	2SA495-OF	235	2SC154B	239	AD13850
		218	BC461			235	2SC154C	239	AD140
								239	AD143R
								239	AD149

GE	Industry Standard No.	GE	Industry Standard No.	GE	Industry Standard No.	GE	Industry Standard No.	GE	Industry Standard No.
239	AD149-01	239	B337	239	EA15X35	239	MP3611	239	PT236B
239	AD149-02	239	B337A	239	EA15X38	239	MP3612	239	PT236C
239	AD149-IV	239	B337BK	239	EA15X53	239	MP3613	239	PT242
239	AD149-V	239	B337H	239	EA15X88	239	MP3614	239	PT25
239	AD149B	239	B337HA	239	EA1700	239	MP3615	239	PT255
239	AD149C	239	B337HB	239	EQ0-6	239	MP3617	239	PT256
239	AD150	239	B338	239	ER15X10	239	MP825	239	PT285
239	AD150-IV	239	B338H	239	ES15X17	239	N57B4-2	239	PT285A
239	AD150-V	239	B338HA	239	ES15X43	239	N57B4-4	239	PT3
239	AD153	239	B338HB	239	ES18	239	NAP-TZ-8	239	PT30
239	AD159	239	B391	239	ES21(ELCOM)	239	NK14A19	239	PT301
239	ADY22	239	B407	239	ES503	239	NK4A-1A19	239	PT301A
239	ADY23	239	B407-0	239	ET15X17	239	NK65C119	239	PT307
239	ADY24	239	B407TV	239	ET15X4	239	NK77C119	239	PT307A
239	ADY27	239	B424	239	ET15X43	239	NK84AA19	239	PT366B
239	ADY28	239	B426	239	ET15X5	239	NKT-401	239	PT3A
239	AF280	239	B426BL	239	F67E	239	NKT-402	239	PT40
239	AL101	239	B426R	239	FD-1029-ET	239	NKT-403	239	PT50
239	AR-10	239	B426Y	239	FD-1029ET	239	NKT-404	239	PT501
239	AR-11	239	B445	239	G04-701-A	239	NKT-405	239	PT555
239	AR-12	239	B446	239	G19	239	NKT-415	239	PT6
239	AR-13	239	B449	239	G6013	239	NKT-416	239	PTO-6
239	AR-14	239	B449F	239	GC4045	239	NKT-451	239	QP1
239	AR-4	239	B449P	239	GC4062	239	NKT-452	239	QP1A
239	AR-5	239	B471	239	GC4087	239	NKT-453	239	QP2
239	AR-6	239	B471-2	239	GC4097	239	NKT-454	239	QP6
239	AR-7	239	B471A	239	GC4111	239	NKT-501	239	QP7
239	AR-8	239	B471B	239	GC4125	239	NKT-503	239	R102-41
239	AR-9	239	B472	239	GC4156	239	NKT-504	239	R2446
239	AR10	239	B472A	239	GC4251	239	NKT401	239	R265A
239	AR11	239	B472B	239	GC4267-2	239	NKT402	239	R3512-1
239	AR12	239	B4A-1-A-21	239	GC641	239	NKT403	239	R3515(RCA)
239	AR13	239	B64	239	GC691	239	NKT404	239	R516
239	AR14	239	B65C-1-21	239	GC692	239	NKT405	239	R7253
239	AR8	239	B69	239	GM428	239	NKT406	239	R8659
239	AR8P404R	239	B77B-1-21	239	GP1432	239	NKT415	239	RS-5858-1
239	AR9	239	B80	239	GP1493	239	NKT416	239	RS-5613
239	A8215	239	B81	239	GP1494	239	NKT450	239	RS-5835
239	A8Z15	239	B82	239	GP1882	239	NKT454	239	RS-5855
239	A8216	239	B84	239	GP420	239	NKT501	239	RS3358-1
239	A8Z17	239	B84A-1-21	239	GPT-16	239	NKT503	239	RS3359-1
239	A8Z18	239	BDY62	239	GTX2001	239	NKT504	239	RS3858
239	ATC-TR-14	239	BF5	239	H103A	239	OC-16	239	RS3858-1
239	ATC-TR-5	239	C337B	239	HEP232	239	OC-22	239	RS3959
239	ATC-TR-6	239	C50BA042	239	HEP623	239	OC-23	239	RS3959-1
239	AU102	239	CDT1309	239	HEP624	239	OC-24	239	RS5612
239	AUY-21	239	CDT1310	239	HEP628	239	OC-25	239	RS5613
239	AUY10	239	CDT1311	239	HF19	239	OC-26	239	RS5614
239	AUY19	239	CDT1319	239	HP20	239	OC-28	239	RS5835
239	AUY20	239	CDT1320	239	HR-101A	239	OC-29	239	S-1556-2
239	AUY21A	239	CDT1321	239	HR102	239	OC-30A	239	S-39T
239	AUY22A	239	CDT1349	239	HR102C	239	OC-35	239	S-40T
239	AUY31	239	CDT1349A	239	HR103	239	OC-36	239	S-40TB
239	AUY33	239	CDT1350	239	HR105	239	OC19	239	S-41T
239	AV105	239	CDT1350A	239	HR105A	239	OC20	239	S-42T
239	B-1511	239	CM2550	239	HR105B	239	P-3189B	239	S-43T
239	B-1914	239	CQT1075	239	HT2040710A	239	P-Z-30	239	S-46T
239	B10162	239	CQT1076	239	I47B446-1	239	P1A	239	S-48T
239	B10163	239	CQT1077	239	KT1017	239	P1E-1	239	S-49T
239	B1017	239	CQT1110	239	L-417-29BLK	239	P1G	239	S-58TB
239	B10474	239	CQT1110A	239	L-417-29GRN	239	P1K	239	S-95253
239	B10475	239	CQT1111	239	L-417-29WHT	239	P1KBLK	239	S-95253-1
239	B107	239	CQT1111A	239	L-417-60	239	P1KBLU	239	S14A-1-3P
239	B107A	239	CQT1112	239	M4331	239	P1KBRN	239	S1556-2
239	B10B5	239	CQT1129	239	M4582	239	P1KGRN	239	S1556-2
239	B10912	239	CQT940A	239	M4582BRN	239	P1KORN	239	082J.3640-050
239	B10913	239	CQT940B	239	M4583	239	P1KRED	239	S4A-1-A-3P
239	B1151	239	CQT940BA	239	M4583RED	239	P1KYEL	239	S65C-1-3P
239	B1151A	239	CRT1544	239	M4584	239	P1T	239	S67809
239	B1151B	239	CRT1545	239	M4584GRN	239	P2D	239	S770-1-3P
239	B1152	239	CRT1552	239	M4606	239	P2DBLU	239	S84A-1-3P
239	B1152A	239	CRT1553	239	M4608	239	P2DBRN	239	S95253
239	B1152B	239	CRT1602	239	M4619RED	239	P2DORN	239	S95253-1
239	B1181	239	CRT3602A	239	M4620GRN	239	P2DRED	239	SC-70
239	B119	239	CST1743	239	M4722	239	P2DYEL	239	SC70
239	B119A	239	CST1744	239	M4722BLU	239	P3189B	239	SD5-3048
239	B122	239	CST1745	239	M4722GRN	239	P4M	239	SE-40022
239	B124	239	CST1746	239	M4722PUR	239	P6480001	239	SE40022
239	B126	239	CTT1122	239	M4722RED	239	P75534	239	SFT190
239	B126A	239	CTT1124	239	M4722YEL	239	P75534-1	239	SFT191
239	B126F	239	CTT1124A	239	M4730	239	P75534-4	239	SFT192
239	B126V	239	CTT1124B	239	M4766	239	P75534-5	239	SFT213
239	B127	239	CTP1111	239	M4767	239	P6870	239	SFT214
239	B1274	239	CTP1124	239	M4888	239	P6890	239	SFT238
239	B1274A	239	CTP1133	239	M4888A	239	P6890A	239	SFT239
239	B1274B	239	CTP1135	239	M4888B	239	P6890L	239	SFT240
239	B127A	239	CTP1136	239	M501	239	PA-10889-1	239	SFT250
239	B128	239	CTP1137	239	M7031	239	PA-10889-2	239	SK3009
239	B128A	239	CTP1265	239	M843	239	PA-10890	239	SK3014
239	B128V	239	CTP1266	239	M9141	239	PA-10890-1	239	SP-1108
239	B129	239	CTP1306	239	M9142	239	PA10890-1	239	SP-148-3
239	B131(JAPAN)	239	CTP1307	239	M9202	239	PAR-12	239	SP-1482-5
239	B131A	239	CTP1500	239	M9237	239	PAR12	239	SP-1483
239	B132(JAPAN)	239	CTP1503	239	M9241	239	PB110	239	SP-1484
239	B132A	239	CTP1504	239	M9255	239	PC3004	239	SP-1556-2
239	B134	239	CTP1508	239	M9263	239	PIK	239	SP-1603
239	B134A	239	CTP1511	239	M9342	239	POWER12	239	SP-1603-1
239	B134C	239	CTP1513	239	M9436	239	POWER25	239	SP-1603-2
239	B1368	239	CTP1514	239	M9550	239	POWER299	239	SP-4040T
239	B1368A	239	CTP1550	239	MA4670	239	PS-1	239	SP-441
239	B1368B	239	CTP1551	239	MN194	239	PS1	239	SP-485
239	B1368C	239	CTP1728	239	MN22	239	PT-12	239	SP-486
239	B1368D	239	CTP1729	239	MN23	239	PT-155	239	SP-486W
239	B1368E	239	CTP1730	239	MN24	239	PT-176	239	SP-634
239	B1368F	239	CTP1732	239	MN25	239	PT-254	239	SP-649
239	B137	239	CTP1733	239	MN26	239	PT-235	239	SP-649-1
239	B138	239	CTP1735	239	MN29	239	PT-235A	239	SP-834
239	B14A-1-21	239	CTP1736	239	MN29BLK	239	PT-236	239	SP-880
239	B177	239	CTP1739	239	MN29GRN	239	PT-236A	239	SP-880-1
239	B178	239	CTP3500	239	MN29PUR	239	PT-236B	239	SP-880-3
239	B179	239	CTP3503	239	MN29WHT	239	PT-242	239	SP-891
239	B1904	239	CTP3504	239	MN32	239	PT-25	239	SP-891B
239	B215	239	CTP3508	239	MN46	239	PT-255	239	SP-891W
239	B216	239	CXL121	239	MN48	239	PT-256	239	SP1013A
239	B216A	239	CXL121MP	239	MN49	239	PT-285	239	SP1118
239	B217	239	DS-515	239	MN73	239	PT-285A	239	SP1271
239	B217A	239	DS-520	239	MN73BLK	239	PT-301	239	SP148-3
239	B217U	239	DS515	239	MN73WHT	239	PT-301A	239	SP1481
239	B224	239	DS520	239	MP1014	239	PT-307	239	SP1481-1
239	B229	239	DT1040	239	MP1509-1	239	PT-307A	239	SP1481-2
239	B239	239	DT41	239	MP1509-2	239	PT-3A	239	SP1481-3
239	B239A	239	DT6110	239	MP1509-3	239	PT-40	239	SP1481-4
239	B247	239	DTG1011	239	MP2060	239	PT-50	239	SP1481-5
239	B248	239	DTG1040	239	MP2060-1	239	PT-554	239	SP1482
239	B248A	239	DTG110	239	MP2061	239	PT-555	239	SP1482-2
239	B25	239	EA1082	239	MP2062	239	PT-6	239	SP1482-3
239	B25B	239	EA1341	239	MP2137A	239	PTO6	239	SP1482-4
239	B282	239	EA15X10	239	MP2138A	239	PT12	239	SP1482-5
239	B283	239	EA15X12	239	MP2139A	239	PT155	239	SP1482-6
239	B284	239	EA15X15	239	MP2142A	239	PT176	239	SP1482-7
239	B285	239	EA15X173	239	MP2143A	239	PT234	239	SP1483
239	B295	239	EA15X26	239	MP2144A	239	PT235	239	SP1483-1
239	B2W	239	EA15X33						

GE	Industry Standard No.	GE	Industry Standard No.	GE	Industry Standard No.	GE	Industry Standard No.	GE	Industry Standard No.
239	SP1483-2	239	TI3027	239	2N1530A	239	2N350A	241	BRC-5496
239	SP1483-3	239	TI3028	239	2N1531	239	2N351	241	C1448A
239	SP1550-3	239	TI3029	239	2N1531A	239	2N351A	241	CXL196
239	SP1556	239	TI366	239	2N1532	239	2N352	241	D513C
239	SP1556-1	239	TI366A	239	2N1532A	239	2N353	241	D513D
239	SP1556-2	239	TI367	239	2N1533	239	2N3611	241	D513F
239	SP1556-3	239	TI367A	239	2N1534	239	2N3612	241	DDBY407004
239	SP1556-4	239	TI368	239	2N1534A	239	2N3613	241	EOG196
239	SP1556-2	239	TI368A	239	2N1535	239	2N3614	241	EP-276
239	SP1595BLK	239	TI369	239	2N1535A	239	2N3615	241	EP15X11
239	SP1595BLU	239	TI369A	239	2N1535B	239	2N3616	241	EP15X14
239	SP1595GRN	239	TNJ60454	239	2N1536	239	2N3617	241	EP15X43
239	SP1595RED	239	TNJ72318	239	2N1536A	239	2N3618	241	E8103
239	SP1596BLK	239	TO-012	239	2N1537	239	2N379	241	GB-241
239	SP1596BLU	239	TO-015	239	2N1537A	239	2N418	241	HB-68
239	SP1596GRN	239	TQ-5064	239	2N1538	239	2N420	241	HT403131EQ
239	SP1596RED	239	TQ5036	239	2N1538A	239	2N420A	241	IRTR57
239	SP1603	239	TQ5064	239	2N1539	239	2N4241	241	IRTR92
239	SP1603-1	239	TR-01	239	2N1540	239	2N458B	241	J241252
239	SP1603-2	239	TR-01C	239	2N1540A	239	2N459	241	KLE4745
239	SP1603-3	239	TR-178(OLSON)	239	2N1541	239	2N459A	241	M9676
239	SP1657	239	TR-43B	239	2N1541A	239	2N514	241	M9676/NPN
239	SP1801	239	TR-5	239	2N1542	239	2N514A	241	MJE2021
239	SP1844	239	TR-8006	239	2N1542A	239	2N514B	241	MJE2380
239	SP1938	239	TR01C	239	2N1543	239	2N538	241	MJE2381
239	SP1950	239	TR02C	239	2N1543A	239	2N538A	241	MJE2382
239	SP2045	239	TR16C	239	2N1544	239	2N553	241	MJE2383
239	SP2046	239	TR333	239	2N1544A	239	2N554	241	MJE29
239	SP2048	239	TR35144	239	2N1545	239	2N555	241	MJE29A
239	SP2072	239	TR35144A	239	2N1545A	239	2N561	241	MJE29B
239	SP2076	239	TR35524	239	2N1546	239	2N589	241	MJE31
239	SP2094	239	TR8006	239	2N1546A	239	2N627	241	MJE31A
239	SP2155	239	TRA-7R	239	2N1547	239	2N628	241	MJE31B
239	SP2341	239	TRA-7RM	239	2N1547A	239	2N629	241	MJE41
239	SP2361	239	TRA-8R	239	2N1548	239	2N663	241	MJE41A
239	SP2361BLU	239	TRA7R	239	2N1548A	239	2N665	241	MJE41B
239	SP2361BRN	239	TRA8R	239	2N1549	239	2N669	241	R623-1
239	SP2361GRN	239	TS-1657	239	2N1549A	239	2N677C	241	R640-1
239	SP2361ORN	239	TS-173	239	2N155	239	2N678B	241	RCA1010
239	SP2361RED	239	TS-176	239	2N1550	239	2N678C	241	RCA1014
239	SP2361YEL	239	TS1657	239	2N1550A	239	2826	241	RRR196
239	SP2395	239	TS173	239	2N1551	239	2S26A	241	S2D153
239	SP2493	239	TS176	239	2N1551A	239	2SB337-OR	241	S37162
239	SP2541	239	T8610	239	2N1552	239	2SBF5	241	SK3041
239	SP534	239	T8612	239	2N1552A	239	2SC337B	241	SK3054
239	SP404	239	T8613	239	2N1553	239(2)	A13-14604-1A	241	STX3326
239	SP404T	239	T8614	239	2N1553A	239(2)	A13-14604-1B	241	T23-93
239	SP441	239	TT-1083	239	2N1554	239(2)	A13-14604-1C	241	TA-7155
239	SP441G	239	TT1083	239	2N1554A	239(2)	A13-14604-1D	241	TA7155
239	SP485	239	TV24337	239	2N1555	239(2)	A13-14604-1E	241	TA7562
239	SP485B	239	TV24678	239	2N1555A	239(2)	A13-14777-1	241	TA7782
239	SP485BLK	239	TV28B126F	239	2N1556	239(2)	A13-14777-1A	241	TA7783
239	SP485BLU	239	TVS-28B126	239	2N1556A	239(2)	A13-14777-1B	241	TA7784
239	SP485BRN	239	TVS-28B126F	239	2N1557	239(2)	A13-14777-1C	241	TA8231
239	SP485W	239	TVS-28B449F	239	2N1557A	239(2)	A13-14777-1D	241	TA8232
239	SP485WHT	239	TVS28B126	239	2N1558	239(2)	A13-14778-1A	241	TA8233
239	SP486W	239	TV828B449	239	2N1558A	239(2)	A13-14778-1B	241	TB7251
239	SP486WHT	239	1-21-270	239	2N1559	239(2)	A13-14778-1C	241	TIP-24
239	SP634	239	1-21-271	239	2N1559A	239(2)	A13-14778-1D	241	TIP-31
239	SP649	239	002-008800	239	2N156	239(2)	A13-22741-2	241	TIP-31A
239	SP649-1	239	002-009701	239	2N1560	239(2)	A146B-3	241	TIP-31B
239	SP744	239	002-010100	239	2N1560A	239(2)	A146B-39	241	TIP29XA
239	SP819R	239	002-012700	239	2N157	239(2)	A1477B	241	TIP4
239	SP834	239	002-9700	239	2N157A	239(2)	A1477B9	241	TIP41
239	SP875	239	2AD140	239	2N1652	239(2)	A168P1	241	TIP41A
239	SP880-1	239	2D001	239	2N1653	239(2)	A48-10075A01	241	TIP41B
239	SP880-3	239	2D004	239	2N1666	239(2)	A48-10075A02	241	TV-112
239	SP891B	239	2D015	239	2N1667	239(2)	A48-10075A03	241	TV112
239	SP891BLU	239	2G210	239	2N176-1BLU	239(2)	A48-10075A04	241	TV117
239	SP891G	239	2G220	239	2N176-1WHT	239(2)	A48-10075A05	241	01-040313
239	SP891GRN	239	2G222	239	2N176-1YEL	239(2)	A48-10075A06	241	002-1(SYLVANIA)
239	SP891R	239	2G224	239	2N176-3PUR	239(2)	A48-10075A07	241	002-2(SYLVANIA)
239	SP891W	239	2G225	239	2N176-3PUR	239(2)	A48-10075A08	241	002-4(SYLVANIA)
239	SP891WHT	239	2G240	239	2N176-4PUR	239(2)	A48-10103A01	241	2N5490
239	SP89	239	2N1011	239	2N176-5WHT	239(2)	A48-10103A02	241	2N5491
239	SS1606A	239	2N1014	239	2N176-6WHT	239(2)	A48-10103A03	241	2N5493
239	ST-235	239	2N1020	239	2N176BLK	239(2)	A48-10103A04	241	2N5494
239	ST235	239	2N1021	239	2N176BLU	239(2)	A48-10103A05	241	2N5495
239	STL109	239	2N1021A	239	2N176GRN	239(2)	A48-10103A06	241	2N5496
239	T-101	239	2N1022	239	2N176M	239(2)	A48-10103A07	241	2N5497
239	T-127	239	2N1022A	239	2N176PUR	239(2)	A48-10103A08	241	2N6292
239	T-235	239	2N1029	239	2N176RED	239(2)	A48-10103A09	241	2SC1826
239	T1167	239	2N1029A	239	2N176WHT	239(2)	A48-10103A10	241	2SC1826P
239	T1168	239	2N1029B	239	2N176YEL	239(2)	A48-10103A11	241	2SC1826Q
239	T13029	239	2N1029C	239	2N1905	239(2)	A48-64978A10	241	2SC1826R
239	T142	239	2N1030	239	2N1911	239(2)	A48-64978A11	241	2SC1985
239	T1601	239	2N1030A	239	2N2061A	239(2)	A48-64978A24	241	2SC647
239	T1602	239	2N1030B	239	2N2148	239(2)	A642-271	241	2SC647Q
239	TA-1614	239	2N1030C	239	2N2212	239(2)	A660051	241	2SC647R
239	TA-1682	239	2N1031	239	2N2287	239(2)	A6B-3-70	243	A-1141-5932
239	TA-1682A	239	2N1031A	239	2N2288	239(2)	A6B-3-705	243	A-120018
239	TA-1705	239	2N1031B	239	2N2289	239(2)	A6B-3A9G	243	A-1854-0022-1
239	TA-1765	239	2N1031C	239	2N2291	239(2)	A77B-70	243	A-1854-0087-1
239	TA-1766	239	2N1032	239	2N2292	239(2)	A77B-705	243	A-1854-0090-1
239	TA-1773	239	2N1032A	239	2N2293	239(2)	A77B19G	243	AO-54-175
239	TA-1794	239	2N1032B	239	2N2294	239(2)	A81S203-5	243	A054-156
239	TA-1881	239	2N1032C	239	2N2295	239(2)	A86X0030-100	243	A054-160
239	TA-1890	239	2N1033	239	2N2296	239(2)	B6B-3-A-21	243	A054-186
239	TA-1891	239	2N1046	239	2N254	239(2)	B6B-3A21	243	A054-206
239	TA-2	239	2N1046B	239	2N254A	239(2)	B77C-1-21	243	A059-110
239	TA1614	239	2N1073	239	2N2357	239(2)	NK6B-3A19	243	A107
239	TA1682	239	2N1073A	239	2N2358	239(2)	NK77B119	243	A115
239	TA1682A	239	2N1120	239	2N2358A	239(2)	R35855	243	A1314
239	TA1705	239	2N1136	239	2N236	239(2)	S6B-3-A-3P	243	A1541
239	TA1765	239	2N1136A	239	2N236A	239(2)	S77B-1-3P	243	A2471
239	TA1766	239	2N1137	239	2N236B	239(2)	SK3013	243	A249(AMC)
239	TA1773	239	2N1137A	239	2N2423	239(2)	SK3015	243	A25A305020101
239	TA1794	239	2N1138A	239	2N2446	239(2)	2AD149	243	A3011112
239	TA1881	239	2N1159	239	2N251	241	A5A	243	A311
239	TA1890	239	2N1160	239	2N251A	241	A8B	243	A32-2805-50-1
239	TA1891	239	2N1163	239	2N268	241	A8K	243	A322805-50-1
239	TA2301	239	2N1163A	239	2N268A	241	ALB6494612	243	A466
239	TA2672	239	2N1165A	239	2N2832	241	AR-30	243	A47392R-0
239	TP-80/30	239	2N1166A	239	2N2836	241	AR24(RED)	243	A490
239	TP78/30Z	239	2N1168	239	2N2869	241	AR24RED	243	A5B
239	TP78/60	239	2N1182	239	2N2869/2N301	241	AR28	243	A5B-R11130-0001
239	TP80/30	239	2N1218	239	2N2870	241	AR28(RED)	243	A5B
239	TP80/30Z	239	2N1292	239	2N297	241	AR28RED	243	AMP-2971-4
239	TI-266A	239	2N1293	239	2N297A	241	AR30	243	A42
239	TI-269	239	2N1294	239	2N301	241	AR35	243	A25
239	TI-3029	239	2N1295	239	2N301A	241	AR38	243	A25
239	TI-366	239	2N1296	239	2N301B	241	AR38(RED)	243	AR203
239	TI-366A	239	2N1297	239	2N301W	241	AR38RED	243	AR203(RED)
239	TI-367	239	2N1359	239	2N307	241	B23-82	243	AR203R
239	TI-367A	239	2N1360	239	2N307A	241	BD273	243	AT-12
239	TI-368	239	2N1362	239	2N307B	241	BD275	243	AT-12(PHILCO)
239	TI-368A	239	2N3125	239	2N3125	241	BD278	243	AT-7
239	TI-369	239	2N1532	239	2N3212	241	BD441	243	AT12
239	TI-369A	239	2N1212	239	2N3213	241	BD537	243	AT339
239	TI-370	239	2N1213	239	2N3214	241	BDX70	243	AT380
239	TI-370A	239	2N1214	239	2N3215	241	BDX71	243	AT381
239	TI266A	239	2N1215	239	2N350	241	BDX72	243	AT382
239	TI269	239	2N1529					241	BDX73	243	AT383
239	TI3012	239	2N1529A							243	AT384
		239	2N1530								

GE	Industry Standard No.	GE	Industry Standard No.	GE	Industry Standard No.	GE	Industry Standard No.	GE	Industry Standard No.
243	AT385	243	BPY52	243	C292	243	C972	243	DW-6982
243	AT386	243	BPY53	243	C30	243	C972C	243	DW6195
243	AT387	243	BPY55	243	C306	243	C972D	243	E-01381
243	AT388	243	BPY56	243	C307	243	C972E	243	E-167-228
243	AT440	243	BPY56A	243	C308	243	C97A	243	E-2491B
243	AT441	243	BPY66	243	C309	243	C993	243	E2441
243	AT442	243	BPY67	243	C31	243	C99D	243	E2449
243	AT443	243	BPY67A	243	C310	243	C9FF-10A58Q	243	E318-1
243	AT444	243	BPY67C	243	C32	243	CC86168Q	243	EAO090
243	AT445	243	BPY68	243	C32A	243	CC86229H	243	EA1549
243	AT446	243	BPY68A	243	C352A(JAPAN)	243	CD6153	243	EA15X102
243	AT470	243	BPY70	243	C353	243	CD6153-2	243	EA15X144
243	AT471	243	BPY72	243	C353A	243	CDC120700	243	EA152249
243	AT472	243	BPY99	243	C36579	243	CDC5008	243	EA15X57
243	AT473	243	BLY61	243	C376	243	CDC5028A	243	EA1684
243	AT474	243	BSV51	243	C403B(SONY)	243	CDC8000-1	243	EA1698
243	AT475	243	B8W10	243	C4030(SONY)	243	CDC8000-1C	243	EA1873
243	AT476	243	B8W26	243	C420	243	CDC8000-CM	243	EDC-TR11-4
243	AT477	243	B8W27	243	C426	243	CDC8002	243	EDC-TR11-5
243	AT478	243	B8W28	243	C443	243	CDC8002-1	243	EL214
243	AT479	243	B8W29	243	C459	243	CDC9002-18	243	EM873278
243	ATC-TR-13	243	B8W35	243	C459B	243	CDC9002-1C	243	EM873279
243	ATC-TR-4	243	B8W65	243	C46	243	CDQ10011	243	EN1613
243	ATC-TR-7	243	B8W66	243	C479	243	CDQ10012	243	EN1711
243	B-6340	243	B8X22	243	C49	243	CDQ10014	243	EN3904
243	B274(SYLVANIA)	243	B8X23	243	C497	243	CDQ10033	243	EP15X33
243	B3746	243	B8X33	243	C497-0	243	CDQ10048	243	EP15X5
243	B87U0007	243	B8X45	243	C497-R	243	CDQ10051	243	EP16X7
243	BC-119	243	B8X46	243	C497-Y	243	CDQ10052	243	EQ8131
243	BC-138	243	B8X60	243	C498	243	CDQ10053	243	ES15226
243	BC-140	243	B8X61	243	C498-0	243	CDQ10057	243	ES15X93
243	BC-140A	243	B8X62	243	C498-R	243	CDQ10058	243	ES22(ELCOM)
243	BC-140B	243	B8X62B	243	C498-Y	243	CDZ15000	243	ES15X65
243	BC-140C	243	B8X62D	243	C503	243	C84003D	243	ES56(ELCOM)
243	BC-140D	243	B8X63	243	C503-0	243	C13704	243	ES62(ELCOM)
243	BC-141	243	B8X63B	243	C503-Y	243	C13705	243	ESD918964P
243	BC-142	243	B8X63C	243	C503GR	243	C13706	243	ET15X36
243	BC103	243	B8X70	243	C504	243	CJ5210	243	ET15X8
243	BC103C	243	B8X71	243	C504-0	243	CJ5213	243	ET8-070
243	BC119	243	B8X72	243	C504-Y	243	CJ5214	243	EU15X27
243	BC120	243	B8X95	243	C504GR	243	CJ5215	243	EU15X34
243	BC138	243	B8X96	243	C509(0)	243	CK419	243	EX-141216
243	BC140	243	B8Y25	243	C509G	243	CK422	243	F318-1
243	BC140-10	243	B8Y44	243	C51	243	CK474	243	F3560
243	BC140-16	243	B8Y45	243	C512	243	CK475	243	F3565
243	BC140-6	243	B8Y46	243	C512-0	243	CMO770	243	F3589
243	BC140C	243	B8Y51	243	C512-R	243	CN2484	243	F361
243	BC140D	243	B8Y52	243	C513	243	CP2357	243	F4709
243	BC141	243	B8Y53	243	C513-0	243	CP409	243	F625-1
243	BC141-10	243	B8Y54	243	C513R	243	C81129E	243	FBN-CP34634
243	BC141-16	243	B8Y55	243	C516	243	C81225H	243	FBN-L109
243	BC141-6	243	B8Y56	243	C560	243	C81225HF	243	FBN-L113
243	BC142	243	B8Y71	243	C564	243	C81229K	243	FBN-L115
243	BC144	243	B8Y77	243	C564A	243	C8124B	243	FBN-L148
243	BC173C	243	B8Y78	243	C564P	243	C81248I	243	FBTX070
243	BC174	243	B8Y79	243	C564Q	243	C81248T	243	FC81229
243	BC174A	243	B8Y81	243	C564R	243	C81250F	243	FC89013HG
243	BC174B	243	B8Y82	243	C564S	243	C81255H	243	FC89013HH
243	BC211	243	B8Y83	243	C564T	243	C81295H	243	FC89014
243	BC216	243	B8Y84	243	C59	243	C81305	243	FC89014C
243	BC216A	243	B8Y85	243	C594	243	C81352	243	FD-1029-FY
243	BC216B	243	B8Y86	243	C61	243	C81453F	243	FD-1029-GE
243	BC232A	243	B8Y87	243	C628	243	C81455G	243	FD-1029-GM
243	BC232B	243	B8Y88	243	C69	243	C81462I	243	FD-1029-JN
243	BC254	243	B8Y90	243	C708	243	C81464H	243	FD-1029-MM
243	BC255	243	B8Y91	243	C708A	243	C81591LE	243	FD-1029-PA
243	BC286	243	B8Y92	243	C708AA	243	C81609F	243	F81531
243	BC288	243	C1008	243	C708AB	243	C81613	243	F8E7233
243	BC301	243	C1072	243	C708AC	243	C81664	243	FTOOO
243	BC310	243	C1072A	243	C708AH	243	C81711	243	FTO019H
243	BC313	243	C108	243	C708AHA	243	C81893	243	FTO02
243	BC340-10	243	C109	243	C708AHB	243	C81990	243	FTO03
243	BC340-16	243	C109A	243	C708AHC	243	C82484	243	FTO04
243	BC340-6	243	C112(JAPAN)	243	C708B	243	C83704	243	FTOO4A
243	BC341-10	243	C113	243	C708C	243	C83705	243	FTO27
243	BC341-6	243	C114	243	C731R	243	C83706	243	G05-055-C
243	BC429	243	C115	243	C734	243	C85449	243	G05-055-D
243	BC535	243	C117	243	C734-0	243	C85450	243	G05-413A
243	BC537	243	C1175E	243	C734-0	243	C85451	243	G05413A
243	BC538	243	C118(JAPAN)	243	C734-R	243	C86168G	243	G05413B
243	BC737	243	C119(JAPAN)	243	C734-Y	243	C87229F	243	G23-46
243	BC7586	243	C12(TRANSISTOR)	243	C734GR	243	C89013/349Q	243	G5A7A66-2
243	BCW46	243	C120	243	C734Q	243	C89013E	243	GC1615-1
243	BCW47	243	C121	243	C734Y	243	C89013F	243	GE-243
243	BCW48	243	C1218	243	C797	243	C89013G	243	GET929
243	BCW49	243	C1220E	243	C798	243	C89013H	243	GET930
243	BCW94K	243	C124	243	C803	243	C89013HG/349Q	243	G05-015-D
243	BCW95	243	C1429	243	C816	243	C89022LE	243	HC-01209
243	BCW95K	243	C1429-1	243	C816K	243	C89103B	243	HC-01317
243	BCY443	243	C1429-2	243	C826	243	C89103C	243	HC-01318
243	BCY46	243	C147	243	C827	243	C8956	243	HEP243
243	BCY47	243	C150	243	C875	243	D11C10B1	243	HEP736
243	BCY48	243	C150T	243	C875-1	243	D11C11B1	243	HP11(PHILCO)
243	BCY49	243	C188	243	C875-1C	243	D11C11B1	243	HP9
243	BCY65	243	C188A	243	C875-1D	243	D11C12B1	243	HR12A
243	BCY66	243	C188AB	243	C875-1E	243	D11C03B1	243	HR12B
243	BF71	243	C189	243	C875-1F	243	D11C05B1	243	HR12C
243	BF587	243	C19	243	C875-2	243	D11C07B1	243	HR12D
243	BFR36	243	C190	243	C875-2C	243	D182	243	HR12E
243	BFS29	243	C20	243	C875-2D	243	D183	243	HR12F
243	BFS36	243	C208	243	C875-2E	243	D204	243	HR28
243	BFW33	243	C210	243	C875-2F	243	D204L	243	HR29
243	BFX17	243	C211	243	C875-3	243	D219	243	HR81
243	BFX50	243	C213	243	C875-3C	243	D221	243	HS-40039
243	BFX51	243	C214	243	C875-3D	243	D233	243	HS40026
243	BFX52	243	C215	243	C875-3E	243	D7A30	243	HT104861
243	BFX68	243	C216	243	C875-3F	243	D7A31	243	HT104861A
243	BFX68A	243	C217	243	C875BR	243	D7A32	243	HT104861B
243	BFX69	243	C22	243	C875C	243	D911138-1	243	HT304861
243	BFX69A	243	C220	243	C875D	243	D911138-2	243	HT304861A
243	BFX74	243	C221	243	C875E	243	D911138-3	243	HT30491
243	BFX84	243	C222	243	C875F	243	D911138-4	243	HT304971A
243	BFX85	243	C224	243	C876	243	D911138-5	243	HT304971AO
243	BFX86	243	C225	243	C876C	243	D911138-6	243	HT304971B
243	BFX94	243	C226	243	C876D	243	D911138-7	243	HT306441AO
243	BFX95	243	C227	243	C876E	243	DDBY410001	243	HT307341
243	BFX96	243	C228	243	C876F	243	DN20-00453	243	HT307341A
243	BFX97	243	C229	243	C876TV	243	DS-512	243	HT307342B
243	BFY10	243	C23	243	C876TVD	243	D8512	243	HT307342C
243	BFY12	243	C231	243	C876TVE	243	DT1110	243	HT309680B
243	BFY13	243	C232	243	C876TVEF	243	DT1111	243	HT309711A-O
243	BFY14	243	C233	243	C934	243	DT1120	243	HT309841BO
243	BFY15	243	C234	243	C959	243	DT1121	243	HT313181C
243	BFY17	243	C236	243	C959A	243	DT1122	243	HT800121O
243	BFY19	243	C24	243	C959B	243	DT1311	243	HT8001310
243	BFY27	243	C247	243	C959C	243	DT1321	243	I473608-2
243	BFY34	243	C248	243	C959D	243	DT1510	243	I473679-1
243	BFY40	243	C2485076-3	243	C959M	243	DT1511	243	I6191
243	BFY44	243	C2485077-2	243	C959N	243	DT1512	243	I9631
243	BFY46	243	C249	243	C959S	243	DT1520	243	INTRON-108
243	BFY46	243	C268	243	C959SA	243	DT1521	243	J241255
243	BFY50	243	C268A	243	C959SB	243	DT1522	243	J241256
243	BFY50	243	C27	243	C959SC	243	DT1621	243	K071961-001
243	BFY51			243	C959SD	243	DTN2V6	243	KD2118
				243	C97				

GE	Industry Standard No.	GE	Industry Standard No.	GE	Industry Standard No.	GE	Industry Standard No.	GE	Industry Standard No.
243	KGB41054	243	PT3502	243	S27233	243	TNJ72288	243	2N3665
243	KLH5807	243	PT612	243	S2794	243	TQ-5055	243	2N3666
243	KB20180-L1	243	PT850	243	S2992	243	TQ5055	243	2N3877
243	L532-008-012	243	PT850A	243	S409P	243	TQPD3053	243	2N3948
243	L532000162	243	PT888	243	S6804	243	TR-01E	243	2N4237
243	LDA404	243	PT896	243	SC1229E	243	TR-1000-3	243	2N4427
243	LDA405	243	Q-0-172	243	SC1444191005	243	TR-13	243	2N4943
243	LDA406	243	Q-00869	243	SC1444103053	243	TR-164(OLSON)	243	2N4944
243	LDA408	243	Q-00869A	243	SDD1220	243	TR-25	243	2N4946
243	LDS200	243	Q-00869B	243	SDD420	243	TR-28C482	243	2N4960
243	LDS201	243	Q-00969	243	SE-8001	243	TR-31B	243	2N4961
243	M-75536-1	243	Q-00969A	243	SE-8010	243	TR-4R31	243	2N4962
243	M-75536-2	243	Q-00969B	243	SE5-0452	243	TR-5R31	243	2N4963
243	M1X	243	Q-01169	243	SE5-0958	243	TR-8021	243	2N4964
243	M300-1300A	243	Q-01169A	243	SE6001	243	TR-8023	243	2N497
243	M4689	243	Q-01169B	243	SE6006	243	TR-8024	243	2N497A
243	M550	243	Q-01169C	243	SE6020	243	TR-8036	243	2N498
243	M9158	243	Q-0172	243	SE6020A	243	TR01054-7	243	2N5129
243	M9170	243	QA-10	243	SE6021	243	TR1001	243	2N5135
243	M9184	243	QA-8	243	SE6021A	243	TR1003	243	2N5136
243	M9209	243	QA8	243	SE6022	243	TR1005	243	2N5188
243	M9221	243	QQV60529	243	SE6023	243	TR23	243	2N5211
243	M9228	243	QRF-2	243	SE7005	243	TR28C482	243	2N5233
243	M9380	243	R-3552-1	243	SE7015	243	TR36643	243	2N5234
243	M9491	243	R-3553-1	243	SE8012	243	TR8036	243	2N5235
243	M9519	243	R-3555	243	SE8041	243	TR01054-7	243	2N5309
243	M9562	243	R-3580-1	243	SE8042	243	TSC-722	243	2N5310
243	M9591	243	R123	243	SE8510	243	TSC722	243	2N5311
243	M9631	243	R123-1	243	SE8520	243	TV-26	243	2N5421
243	M9703	243	R123-3	243	SE8521	243	TV-28	243	2N5449
243	MB1075	243	R123-4	243	SPT443	243	TV-41	243	2N5450
243	MB8001	243	R123-5	243	SPT443A	243	TV-43	243	2N5451
243	MB8002	243	R15003	243	SPT445	243	TV-45	243	2N5820
243	MB8003	243	R15003P1	243	S05013	243	TV-59	243	2N5881
243	MH9410A	243	R2270-60106	243	SGC7202	243	TV-67	243	2N5882
243	MHT2414	243	R2270-77873D	243	SJ2032	243	TV-70B	243	2N655A
243	MHT2418	243	R227077873D	243	SK3024	243	TV23	243	2N656
243	MHT4401	243	R3508	243	SKA4616	243	TV41	243	2N697
243	MHT4411	243	R3555-3	243	SL301C	243	TV42	243	2N697
243	MHT4412	243	R3593	243	SL301CE	243	TV45	243	2N697L
243	MHT4413	243	R3608	243	SM2716	243	TV51	243	2N697B
243	MHT4451	243	R3608-1	243	SM3978	243	TV52	243	2N720
243	MHT4483	243	R3608-2	243	SM6251	243	TV53	243	2N720A
243	MHT4511	243	R3679	243	SM7991	243	TV59	243	2N870
243	MHT4512	243	R5048	243	SMB-706009D	243	TV60B0	243	2N871
243	MHT4513	243	R7613	243	SMC-583259	243	TVS-280582	243	2N910
243	MHT7401	243	R8915	243	SMC-620774-1	243	TVS-280582A	243	2N911
243	MHT7411	243	R02270	243	SM166	243	TVS280696	243	2N912
243	MHT7412	243	RCA1A01	243	SN167	243	001-02119-0	243	2N981
243	MHT7414	243	RCA1A06	243	SPD-80123	243	001-021190	243	28014
243	MHT7417	243	RCA1A07	243	SP8-0122	243	001-021221	243	28017
243	MHT9001	243	RCA1A15	243	SP8-4077	243	001-021290	243	28018
243	MHT9002	243	RCA1A17	243	SP80122	243	1-035/2207	243	28019
243	MHT9004	243	RCA1A18	243	SP82130	243	01-2110	243	28020
243	MHT9005	243	R8-2014	243	SP83912	243	01-2111	243	28103
243	MM1809	243	R8132	243	SP83914	243	01-2114	243	28104
243	MM1809A	243	R87672	243	SP84038	243	1-2114-Q	243	28741
243	MM181OA	243	R87678	243	SP84300	243	1A0066	243	28742
243	MM1943	243	R88101	243	SP84309	243	1N3112	243	28742A
243	MM2266	243	R88103	243	SP84311	243	1W8995A	243	28744
243	MM306	243	R88105	243	SP84361	243	002-010600	243	28744A
243	MM486	243	R88107	243	SP84391	243	002-012500	243	28745
243	MM487	243	R88109	243	SP84441	243	2B1A20A22AAB	243	28745A
243	MM488	243	R88111	243	SP84490	243	2N1092	243	28C-NJ107
243	MM511	243	R88113	243	SP84495	243	2N1252	243	28C1008
243	MM512	243	RT-141	243	S849	243	2N1253	243	28C1218
243	MM513	243	RT-188	243	SP85809	243	2N1472	243	28C1220(B)
243	MP49Q06063	243	RT141	243	SP86124	243	2N1479	243	28C1220-003
243	MP83642	243	RT154	243	ST-201	243	2N1480	243	28C1220A(QPR)
243	MP89410A	243	RT188	243	ST-213	243	2N1481	243	28C1220E
243	MP89410AJ	243	RT482	243	ST-254-Q	243	2N1613	243	2801383
243	MP89410AK	243	RT483	243	ST-LM2682	243	2N1613L	243	28C1383B
243	MP89410E	243	RT484	243	ST213	243	2N1613B	243	28C1429
243	MP89616J	243	RT5151	243	ST4201	243	2N1700	243	28C1429-1
243	MP8A06	243	RT5152	243	ST4202	243	2N1711	243	28C1429-2
243	MPX9410H	243	RT5203	243	ST4203	243	2N1711A	243	28C2120-0
243	MR9933	243	RT5204	243	ST4204	243	2N1711L	243	28C2120Y
243	MS2991	243	RT5401	243	ST6573	243	2N1711B	243	28C298-4
243	MS7506H	243	RT5402	243	ST6574	243	2N1764	243	28C486A
243	MS7506J	243	RT5403	243	SYL4280	243	2N1889	243	28C486B
243	MS07506	243	RT5404	243	T-04689	243	2N1890	243	28C486C
243	M8K5405	243	RT5905	243	T-291	243	2N1990	243	28C486D
243	MSP99058-1	243	RT5906	243	T-339	243	2N2017	243	28C486E
243	MST-10	243	RT5907	243	T-Q5081	243	2N2102	243	28C486G
243	MT1613	243	RT699M	243	T-Q5099	243	2N2102L	243	28C486GN
243	MT1711	243	RT7945	243	T01-022	243	2N2028	243	28C486H
243	ONOR0540	243	RV1473	243	T15015	243	2N2106	243	28C486J
243	NCR046	243	RV2068	243	T1340A3H	243	2N2107	243	28C486K
243	NCR047	243	RYN12104	243	T164213	243	2N2108	243	28C486L
243	NJ107	243	S001683	243	T1706	243	2N2192	243	28C486M
243	NPC115	243	S007220	243	T1706A	243	2N2192A	243	28C486QR
243	NPC187	243	S1368	243	T1706B	243	2N2192B	243	28C486R
243	NPC189	243	S1514	243	T1706C	243	2N2193	243	28C486X
243	N82100	243	S1516	243	T1811	243	2N2195A	243	28C564T
243	N82101	243	S1517	243	T1811E	243	2N2270	243	28C628
243	N89400	243	S1523	243	T1811G	243	2N2270L	243	A-1853-0041-1
243	N89420	243	S1525	243	T23-94	243	2N2270B	244	A054-109
243	N89500	243	S15660	243	T247	243	2N2297	244	A1016
243	N89540	243	S1642	243	T336-2	243	2N2443	244	A116084
243	N89728	243	S1644	243	T4205L1	243	2N2594	244	A116284
243	N89729	243	S1671	243	T452	243	2N2846	244	A119985
243	N89730	243	S1689	243	T9631	243	2N2848	244	A126724
243	N89731	243	S1762	243	TA198035-1	243	2N2849	244	A1473549-1
243	0A-10	243	S1773	243	TA6200	243	2N2850	244	A1473616-1
243	00V60529	243	S1777	243	TE1990	243	2N2851	244	A170
243	ORF-2	243	S17900	243	T83416	243	2N2852	244	A2057B110-9
243	P-11748-1	243	S18000	243	TP101-A	243	2N2853	244	A2057B112-9
243	P-11905-1	243	S18200	243	TP101-B	243	2N2854	244	A2057B114-9
243	P04-45-0026-P5	243	S1864	243	TP101-D	243	2N2855	244	A2057B115-9
243	P04450026P5	243	S1874	243	TG28C1175C	243	2N2856	244	A2057B116-9
243	P4069	243	S19386	243	TI-424	243	2N2895	244	A2057B121-9
243	P480A0018	243	S21118	243	TI-425	243	2N2897	244	A2057B122-9
243	P4Z	243	S21549	243	TI-475	243	2N2J324	244	A2057B143-12
243	P633024	243	S2209	243	TI-480	243	2N3019	244	A2057B163-12
243	P633024Q	243	S2369	243	TI-482	243	2N3020	244	A2482
243	P6450026	243	S2371	243	TI-483	243	2N3053	244	A297074C11
243	P6786	243	S2400	243	TI-484	243	2N3077	244	A297V073C001
243	P9962-1	243	S2400A	243	TI-496	243	2N3078	244	A297V073C002
243	P9962-2	243	S2400B	243	TI3015	243	2N3107	244	A297V082B03
243	P9962-4	243	S2401	243	T164213	243	2N3108	244	A29V082B03
243	P9962-5	243	S2401A	243	TI8119	243	2N3109	244	A289
243	PEP2001	243	S2401B	243	TI8107	243	2N3117	244	A284550P
243	PBT4001	243	S2401C	243	TI860A	243	2N3262	244	A28A564F
243	PBT9021	243	S2402	243	TI860B	243	2N3439	244	A30278
243	PBT9022	243	S2402A	243	TI860C	243	2N3439	244	A3523
243	PG33024Q	243	S2402B	243	TI860D	243	2N3443A	244	A353-9008-001
243	PIT-74	243	S2402C	243	TI860E	243	2N3469	244	A3533
243	PL1083	243	S2427	243	TI860M	243	2N3506	244	A3533-1
243	PL1084	243	S24598	243	TI892M	243	2N3507	244	A3616-1
243	PMC-Q8-0320	243	S24614	243	TN421	243	2N3567	244	A36577
243	PMC-Q8-0400	243	S24616	243	TNJ70482	243	2N3568	244	A4037764-2
243	PRT-104-4	243	S2487	243	TNJ71035	243	2N3569	244	A40410
243	PT1544	243	S2526	243	TNJ71234	243	2N3642	244	A417-138
243	PT1545	243	S2648						

GE	Industry Standard No.	GE	Industry Standard No.	GE	Industry Standard No.	GE	Industry Standard No.	GE	Industry Standard No.
244	A417-170	244	BC404V1	244	HEP83012	244	ST-021660	244	2N4037
244	A417-234	244	BC477	244	HP10	244	ST72039	244	2N4234
244	A417-43	244	BC477A	244	HP12(PHILCO)	244	ST72040	244	2N4235
244	A447B	244	BC477V1	244	HS40032	244	STC5610	244	2N4236
244	A497(JAPAN)	244	BC527	244	HT1049T1A-0	244	STC5611	244	2N4314
244	A498(JAPAN)	244	BC528	244	HT1049T1A0	244	STC5612	244	2N4404
244	A498Y	244	BC534	244	IP20-0217	244	STX0011	244	2N4405
244	A501	244	BC727	244	IRTR88	244	SX61	244	2N4406
244	A503	244	BC728	244	J241015	244	T-340	244	2N4407
244	A503-0	244	BCY17	244	J24908	244	T-482	244	2N4412
244	A503-R	244	BCY18	244	K071818-001	244	T1275	244	2N4412A
244	A503-Y	244	BCY19	244	K071962-001	244	T1276	244	2N4414
244	A503GR	244	BCY21	244	K1181	244	T1808	244	2N4414A
244	A504	244	BCY22	244	KD2120	244	T1808A	244	2N4890
244	A504-R	244	BCY23	244	KB1007-0004-00	244	T1808B	244	2N4928
244	A504-Y	244	BCY24	244	KLH5808	244	T1808C	244	2N5022
244	A504GR	244	BCY25	244	LDA450	244	T1808D	244	2N5023
244	A516	244	BCY26	244	LN76963	244	T1808E	244	2N5040
244	A516A	244	BCY27	244	M4478	244	T246	244	2N5041
244	A527	244	BCY28	244	M652PIC	244	T334-2	244	2N5042
244	A528	244	BCY29	244	M75561-17	244	T459	244	2N5110
244	A532	244	BCY30	244	M75561-8	244	T475	244	2N5111
244	A532A	244	BCY31	244	M828	244	T1808E	244	2N5242
244	A532B	244	BCY32	244	M9145	244	TI861A	244	2N5243
244	A532C	244	BCY33	244	M9257	244	TI861B	244	2N5821
244	A532E	244	BCY34	244	M9308	244	TI861D	244	2N5865
244	A532F	244	BCY67	244	M9400	244	TI861E	244	28021
244	A537	244	BCY96	244	M9426	244	TI861M	244	28022
244	A537A	244	BCY96B	244	M9432	244	TI893M	244	28023
244	A537AA	244	BCY97	244	M9435	244	TQ63	244	28301
244	A537AB	244	BCY97B	244	M9520	244	TQ63A	244	28S010
244	A537AC	244	BDY70	244	MH9460A	244	TQ64	244	28302
244	A537AH	244	BPT60	244	MM4005	244	TQ64A	244	28S02A
244	A537B	244	BPT61	244	MM4006	244	TR-04C	244	28303
244	A537C	244	BPT62	244	MM4019	244	TR-165(OLSON)	244	28304
244	A537H	244	BPT79	244	MP89460A	244	TR-8020	244	28A497
244	A546	244	BPT80	244	MP89460H	244	TR-88	244	28A497-0
244	A546A	244	BPT81	244	MR3934	244	TR02054-7	244	28A497-ORG
244	A546B	244	BPW44	244	MS75069	244	TR1000	244	28A497-R
244	A546H	244	BPW91	244	MSJ7505	244	TR1002	244	28A497-RED
244	A548	244	BPX38	244	P1M	244	TR1004	244	28A497-Y
244	A551	244	BPX39	244	PMC-Q8-0280	244	TR1012	244	28A497-YEL
244	A551C	244	BPX40	244	Q5100A	244	TR22	244	28A497R
244	A551D	244	BPX41	244	QA-11	244	TR22A	244	28A497RED
244	A551E	244	BPX74A	244	QA-17	244	TR23A	244	28A497Y
244	A552	244	BPX87	244	QA-9	244	TR8037	244	28A498
244	A560	244	BPX88	244	R2270-76963	244	TR88	244	28A498Y
244	A571	244	BPY18	244	RCA1A02	244	TV-29	244	28A501
244	A594	244	BPY64	244	RCA1A05	244	TV29	244	28A504
244	A594-0	244	BSS17	244	RCA1A08	244	TV828A546	244	28A504-0
244	A594-R	244	BSS18	244	RCA1A16	244	01-010844	244	28A504-R
244	A594-Y	244	BSV15	244	RCA1A19	244	002-010300-6	244	28A504-Y
244	A595C	244	BSV16	244	REN129	244	002-010700	244	28A504GR
244	A604	244	BSV82	244	RS8100	244	002-012600	244	28A516
244	A606	244	BSV83	244	RS8102	244	002-9800-12	244	28A516A
244	A606B	244	BSW23	244	RS8104	244	2A12	244	28A527
244	A6532921	244	BSW40	244	RS8106	244	2N1034	244	28A528
244	A6532941	244	BSX40	244	RS8108	244	2N1035	244	28A532
244	A677	244	BSX41	244	RS8110	244	2N1036	244	28A532A
244	A708	244	BTX-071	244	RS8112	244	2N1037	244	28A532B
244	A708A	244	BTX-097	244	RT5230	244	2N1197	244	28A532C
244	A708B	244	C102	244	RV1472	244	2N1228	244	28A532D
244	A708C	244	C106(PNP)	244	RV2069	244	2N1229	244	28A532E
244	A736	244	C112(PNP)	244	S-437	244	2N1230	244	28A532F
244	A742	244	C118	244	S-437F	244	2N1231	244	28A537
244	A800-511-00	244	C119	244	S1430	244	2N1232	244	28A537A
244	A800-516-00	244	C201	244	S1431	244	2N1233	244	28A537AA
244	A8015613	244	C202	244	S1520	244	2N1234	244	28A537AB
244	A836	244	C301A(PNP)	244	S1698	244	2N1254	244	28A537AH
244	A836D	244	C302(PNP)	244	S1863	244	2N1255	244	28A537B
244	A836E	244	C36577	244	S1983	244	2N1256	244	28A537C
244	A83P2B	244	C402	244	S2117	244	2N1257	244	28A537H
244	A844	244	C41001	244	S2274	244	2N1258	244	28A546
244	A844D	244	C5620	244	S2368	244	2N1259	244	28A546A
244	A844E	244	C9080	244	S2370	244	2N1275	244	28A546E
244	A880-250-107	244	C9081	244	S2398C	244	2N1429	244	28A546H
244	A94063	244	CDC-9000-1B	244	S24594	244	2N1606	244	28A550AR
244	AEX-85715	244	CD09000-1B	244	S24597	244	2N1623	244	28A550AB
244	AMP2970-2	244	CDC9002	244	S24612	244	2N1922	244	28A550BC
244	AR304(RED)	244	CE4002D	244	S24612A	244	2N2904	244	28A551
244	AT2848	244	CJ5209	244	S24615	244	2N2904A	244	28A551C
244	AT391	244	CS-6228G	244	S2771	244	2N2905	244	28A551D
244	AT392	244	CS1237	244	S2991	244	2N2905A	244	28A551E
244	AT393	244	CS1256H	244	S2993	244	2N3120	244	28A552
244	AT394	244	CS1312G	244	S2994	244	2N3133	244	28A571
244	AT395	244	CS1369	244	S2995	244	2N3134	244	28A594
244	AT396	244	CS1465H	244	S3012	244	2N3202	244	28A594-0
244	AT397	244	CS72280	244	S3386	244	2N3203	244	28A594-R
244	AT398	244	CS9012/5490	244	S33886	244	2N3208	244	28A594-Y
244	AT460	244	CS90121	244	S33886A	244	2N327B	244	28A595C
244	AT461	244	CS9012F	244	S504-0	244	2N328	244	28A597
244	AT462	244	CS9012FC	244	SC1294H	244	2N328A	244	28A604
244	AT463	244	CS9012HE	244	SC365	244	2N328B	244	28A606
244	AT464	244	CS9012HF	244	SC843	244	2N329	244	28A606B
244	AT465	244	CS9012HG/3490	244	SC1444204037	244	2N329A	244	28A612
244	AT466	244	CS9021HF	244	SC1444291004	244	2N329B	244	28A708
244	AT467	244	CS9102B	244	SDT3321	244	2N330	244	28A708A
244	AT468	244	CXL129	244	SDT3322	244	2N330A	244	28A708B
244	AT480	244	DDBY008001	244	SDT3324	244	2N3343	244	28A708C
244	AT481	244	DS-86	244	SDT3501	244	2N3344	244	28A734
244	AT482	244	E2498	244	SDT3502	244	2N3345	244	28A736
244	AT483	244	EA0086	244	SDT3503	244	2N3468	244	28A742H
244	AT484	244	EA0087	244	SE8540	244	2N3494	244	28A836
244	AT485	244	EA15X194	244	SE8541	244	2N3502	244	28A836D
244	AT5156	244	ECG129	244	SE8542	244	2N3503	244	28A836E
244	B510	244	EL264	244	SFD-23	244	2N3571	244	28A836F
244	B5108	244	EN2905	244	SJ2031	244	2N3577	244	28A844
244	B5493957-4	244	EN3502	244	SK16510006-2	244	2N3719	244	28A844C
244	B5493957-5	244	EP15X4	244	SK16510006-4	244	2N3720	244	28A844D
244	B5493957-6	244	ES51(ELCOM)	244	SK5025	244	2N3762	244	28A844E
244	BC139	244	ET15X33	244	SKA1079	244	2N3763	244	28A950
244	BC143	244	ET15X38	244	SKA4621	244	2N3764	244	28A950-0
244	BC160	244	ET15X39	244	SL3101	244	2N3765	244	28A950Y
244	BC160-10	244	ET8-069	244	SL3111	244	2N3774	244	28B510
244	BC160-16	244	ET8-071	244	SM2718	244	2N3775	244	28B5108
244	BC160-6	244	F3549	244	SM3987	244	2N3776	244	28B561
244	BC161	244	FC81795D	244	SM6728	244	2N3778	244	28B561B
244	BC161-10	244	FC89012	244	SMC449077	244	2N3780	244	28B561C
244	BC161-16	244	FC89012HG	244	SO80121	244	2N3781	244	28B598
244	BC161-6	244	FD-1029-MB	244	SPS-0121	244	2N3782	244	28B598P
244	BC287	244	FD-1029-ML	244	SPS-29	244	2N3867	245	A-1384
244	BC303	244	FD-1029-RB	244	SPS-4076	244	2N3868	245	A01(TRANSISTOR)
244	BC311	244	G03-404-B	244	SPS-4078	244	2N4026	245	A059-100
244	BC360-10	244	G03-404-C	244	SPS0121	244	2N4027	245	A059-101
244	BC360-16	244	GE-244	244	SP82131	244	2N4028	245	A059-102
244	BC360-6	244	HA7530	244	SP82226	244	2N4029	245	A059-103
244	BC361-10	244	HA7531	244	SPS4010	244	2N4030	245	A105
244	BC361-6	244	HA7630	244	SPS4301	244	2N4031	245	A107(JAPAN)
244	BC396	244	HA7631	244	SPS4310	244	2N4032	245	A108(JAPAN)
244	BC404A	244	HA7632	244	SPS4312	244	2N4033	245	A117
		244	HEP242	244	SPS4462				
		244	HEP51	244	SPS4477				
		244	HEP710	244	SPS4492				
				244	SPS4497				
				244	SPS6125				

GE	Industry Standard No.	GE	Industry Standard No.	GE	Industry Standard No.	GE	Industry Standard No.	GE	Industry Standard No.
245	A118	245	A462	245	AF166	245	DS42	245	GT40
245	A122	245	A463	245	AF167	245	DS56	245	HA-234
245	A1220	245	A464	245	AF168	245	DS62	245	HA-268
245	A126	245	A506	245	AF169	245	DS63	245	HA1040
245	A127	245	A507	245	AF170	245	DS64	245	HA12
245	A13(TRANSISTOR)	245	A508	245	AF171	245	DS65	245	HA2190
245	A130(JAPAN)	245	A51	245	AF172	245	DU1	245	HA2356
245	A135	245	A525	245	AF178	245	DU12	245	HA235A
245	A1377	245	A525B	245	AF179	245	DU2	245	HA240
245	A137B	245	A56	245	AF180	245	E2438	245	HA266
245	A14(TRANSISTOR)	245	A61	245	AF181	245	E2439	245	HA267
245	A1465-1	245	A615-1008	245	AF182	245	E2440	245	HA30
245	A1465-19	245	A615-1009	245	AF185	245	E2450	245	HA3210
245	A1465A	245	A65-1-1A9G	245	AF186	245	E2462	245	HA3480
245	A1465A9	245	A65-1-70	245	AF186G	245	E2474	245	HA3670
245	A1465B	245	A65-1-705	245	AF186W	245	E2475	245	HA4400
245	A1465B9	245	A65-1A9G	245	AF193	245	E2477	245	HA525
245	A1488B	245	A65A-70	245	AF200	245	E2478	245	HA70
245	A1488B9	245	A65A-705	245	AF201	245	E2479	245	HEP-Q0003
245	A14A8-1	245	A65A19G	245	AF201C	245	EA0002	245	HEP3
245	A14A8-19	245	A70	245	AF202	245	EA0007	245	HEPF637
245	A14A8-19G	245	A70F	245	AF202L	245	EA1337	245	HEPG0002
245	A151(JAPAN)	245	A70L	245	AF202S	245	EA1338	245	HEPG0003
245	A152(JAPAN)	245	A70MA	245	AF239	245	EA1340	245	HF12H
245	A163	245	A74	245	AF239S	245	EA1342	245	HF12N
245	0A180	245	A79	245	AF240	245	EA15X11	245	HF20H
245	A215	245	0A8-1	245	AF251	245	EA15X13	245	HF20M
245	A218	245	0A8-1-12	245	AF252	245	EA15X27	245	HF3H
245	A219	245	0A8-1-12-7	245	AF253	245	EA15X29	245	HF3M
245	A221	245	A8-1-70	245	AF256	245	EA15X30	245	HF50H
245	A222	245	A8-1-70-1	245	AF267	245	EA15X40	245	HF50M
245	A223	245	A8-1-70-12	245	AF279	245	EA15X41	245	HF6H
245	A224	245	A8-1-70-12-7	245	AFY12	245	EA15X43	245	HF6M
245	A225	245	A8-1-A-4-7B	245	AFY14	245	EA15X5	245	HJ15D
245	A226	245	A8-10	245	AFY15	245	EA15X66	245	HJ32
245	A227	245	A8-11	245	AFY16	245	ED51	245	HJ34
245	A228	245	A8-12	245	AFY18C	245	ER15X11	245	HJ34A
245	A229	245	A8-13	245	AFY18D	245	ER15X12	245	HJ37
245	A230	245	A8-14	245	AFY18E	245	ER15X13	245	HJ55
245	A233	245	A8-15	245	AFY19	245	ER15X14	245	HJ56
245	A233A	245	A8-16	245	AFY34	245	ER15X15	245	HJ57
245	A233B	245	A8-17	245	AFY37	245	ER15X16	245	HJ60A
245	A233C	245	A8-18	245	AFY39	245	ER15X19	245	HJ60C
245	A234	245	A8-19	245	AFY40	245	ER15X20	245	HJ70
245	A234A	245	A8-1A	245	AFY40K	245	ER15X21	245	HR-20
245	A235	245	A8-1A0	245	AFY40R	245	ER15X24	245	HR-20A
245	A235B	245	A8-1A0R	245	AFY41	245	ER15X25	245	HR-21
245	A239	245	A8-1A1	245	AFY42	245	ER15X26	245	HR-21A
245	A240	245	A8-1A19	245	AFZ11	245	ER15X4	245	HR-22
245	A240A	245	A8-1A2	245	AFZ12	245	ER15X5	245	HR-22A
245	A240B2	245	A8-1A21	245	AL210	245	ER15X6	245	HR-22B
245	A240BL	245	A8-1A3	245	APB-11H-1001	245	ES1(ELCOM)	245	HR-24
245	A241	245	A8-1A3P	245	APB-11H-1004	245	ES11(ELCOM)	245	HR-24A
245	A242	245	A8-1A4	245	AR105	245	ES14(ELCOM)	245	HR-25
245	A243	245	A8-1A4-7	245	AR104	245	ES15(ELCOM)	245	HR-25A
245	A244(JAPAN)	245	A8-1A4-7B	245	AR105	245	ES19(ELCOM)	245	HR-26
245	A245(JAPAN)	245	A8-1A5	245	AS34280	245	ES23(ELCOM)	245	HR-26A
245	A276(JAPAN)	245	A8-1A5L	245	ASY-24	245	ES25(ELCOM)	245	HR-27
245	A285	245	A8-1A6	245	ASY63	245	ES41(ELCOM)	245	HR-27A
245	A286	245	A8-1A6-4	245	ASY67	245	ES6(ELCOM)	245	HR40
245	A287	245	A8-1A7	245	AS220	245	ET1	245	HR40836
245	A296	245	A8-1A7-1	245	AS220N	245	ET15X29	245	HR40837
245	A297	245	A8-1A8	245	AS221	245	ET2	245	HR41
245	A298	245	A8-1A82	245	AT-1	245	F73216	245	HR42
245	A306(JAPAN)	245	A8-1A9	245	AT-14	245	FB401	245	HR43
245	A307(JAPAN)	245	A8-1A9G	245	AT-2	245	FB402	245	HR43835
245	A321(JAPAN)	245	0A81	245	AT-3	245	FB403	245	HR44
245	A322	245	A84	245	AT-4	245	FB440	245	HR44836
245	A322(JAPAN)	245	A86	245	AT-6	245	F82299	245	HR45
245	A323(JAPAN)	245	A88-70	245	AT-8	245	G0002	245	HR45838
245	A324(JAPAN)	245	A88-705	245	AT-9	245	G13	245	HR45910
245	A340	245	A88B-70	245	AT/RF1	245	GC1003	245	HR45913
245	A341	245	A88B-705	245	AT/RF2	245	GC1004	245	HR46
245	A341-OA	245	A88B19G	245	AT/813	245	GC1005	245	HR50
245	A341-OB	245	A89	245	AT13	245	GC1006	245	HR51
245	A341-OB	245	A909-1008	245	AT14	245	GC1007	245	HR52
245	A342	245	AA3	245	AT15	245	GC1092	245	HT102341
245	A342A	245	AA8-1-70	245	AT16	245	GC1093	245	HT102341A
245	A345(JAPAN)	245	AA8-1-705	245	AT17	245	GC1093X3	245	HT102341B
245	A346(JAPAN)	245	AA8-1A9G	245	AT4	245	GC1142	245	HT102341C
245	A347	245	AC107	245	AT5	245	GC1146	245	HT102351
245	A348	245	AC107M	245	B240	245	GC1148	245	HT102351A
245	A349	245	AC129	245	B240A	245	GC1149	245	HT103501
245	A360	245	ACY24	245	B51	245	GC1155	245	HT2002541C
245	A361	245	ACZ	245	B601-1006	245	GC1182	245	I12032
245	A362	245	AF-105A	245	B601-1007	245	GC1573	245	J24620
245	A363	245	AF-109	245	B601-1008	245	GC282	245	J24621
245	A372	245	AF-121	245	B65-1-A-21	245	GC283	245	J24622
245	A373	245	AF-166	245	B65-1A21	245	GC284	245	J24623
245	A375	245	AF-182	245	B65A-1-21	245	GC387	245	J310251
245	A376	245	AF101	245	B65B-1-21	245	GC388	245	J5062
245	A377	245	AF102	245	B74	245	GC389	245	JP5062
245	A378	245	AF105A	245	B88B-1-21	245	GC460	245	JR10
245	A379	245	AF106	245	B88A-1A-21	245	GC461	245	JR100
245	A401	245	AF106A	245	BCM1002-1	245	GC462	245	JR200
245	A403	245	AF107	245	BCY501	245	G0630	245	JR30
245	A404	245	AF109	245	BCY511	245	G0630A	245	JR50X
245	A405	245	AF109R	245	BFY37I	245	G0631	245	K75508-1
245	A408	245	AF110	245	BF5263	245	GE-245	245	L2091241-2
245	A409	245	AF111	245	C10215-2	245	GER-A-D	245	L2091241-3
245	A41	245	AF112	245	C10258	245	GET5116	245	L5108
245	A410	245	AF113	245	C10260	245	GET5117	245	L5121
245	A411	245	AF114	245	C10261	245	GET671	245	L5122
245	A413	245	AF114N	245	C10262	245	GET672	245	TF121
245	A417(JAPAN)	245	AF115	245	C125	245	GET672A	245	M351
245	A419(JAPAN)	245	AF115N	245	CB156	245	GET673	245	M4439
245	A42	245	AF116	245	CB244	245	GET691	245	M4484
245	A420(JAPAN)	245	AF116N	245	CB254	245	GET692	245	M4485
245	A421	245	AF117	245	CGE-50	245	GET693	245	M4486
245	A422	245	AF117C	245	CGB-51	245	GFP44	245	M4504
245	A425	245	AF117N	245	CXL160	245	GFP45	245	M4506
245	A426	245	AF118	245	D063	245	GGB-51	245	M4507
245	A430(JAPAN)	245	AF119	245	D073	245	GMO290	245	M4524
245	A431	245	AF121	245	D079	245	GMO375	245	M4526
245	A431A	245	AF121B	245	D086	245	GMO376	245	M4697
245	A432	245	AF122	245	D149	245	GMO377	245	M4860
245	A432A	245	AF124	245	D172	245	GM290	245	M75516-2
245	A433	245	AF125	245	D173	245	GM290A	245	M75516-2B
245	A434	245	AF126	245	D174	245	GM378	245	M75516-2R
245	A435	245	AF127	245	DS-34	245	GM378A	245	M755162-P
245	A435A	245	AF127/01	245	DS-35	245	GM378RED	245	M755162-R
245	A435B	245	AF12B	245	DS-36	245	GM656A	245	M76
245	A440	245	AF137A	245	DS-37	245	GMB75	245	M77
245	A440A	245	AF139	245	DS-41	245	GMB76	245	M78
245	A448	245	AF142	245	DS-42	245	GMB77	245	M78A
245	A450	245	AF143	245	DS-56	245	GMB78	245	M78B
245	A450H	245	AF146	245	DS-62	245	GMB78A	245	M78BLK
245	A451(JAPAN)	245	AF147	245	DS-63	245	GMB78B	245	M78C
245	A451H	245	AF148	245	DS-64	245	GMO290	245	M78D
245	A452	245	AF149	245	DS34	245	GMO375	245	M78GRN
245	A452H	245	AF150	245	DS35	245	GMO376	245	M78RED
245	A460	245	AF164	245	DS37	245	GMO377	245	M78YEL
245	A461	245	AF165	245	DS38(DELCO)	245	GMO378	245	M8124
245		245		245	DS41	245		245	

GE	Industry Standard No.	GE	Industry Standard No.	GE	Industry Standard No.	GE	Industry Standard No.	GE	Industry Standard No.
245	MA1	245	R516A	245	S95106	245	T2191	245	TIXM107
245	MA23A	245	R539	245	S99101	245	T2364	245	TIXM108
245	MC101	245	R558	245	S99102	245	T2379	245	TIXM11
245	MC103	245	R563	245	S99103	245	T2384	245	TIXM13
245	MD420	245	R564	245	S99104	245	T253	245	TIXM14
245	MD831	245	R565	245	SA8-1-A-3P	245	T253(SEARS)	245	TIXM15
245	MD832	245	R579	245	SB101	245	T278	245	TIXM16
245	MD833	245	R593	245	SB102	245	T2788	245	TIXM17
245	MD833A	245	R593A	245	SB103	245	T279	245	TIXM18
245	MD833C	245	R60-1002	245	SC-71	245	T280	245	TIXM19
245	MD833D	245	R60-1003	245	SC-72	245	T281	245	TIXM201
245	MD834	245	R7885	245	SC-74	245	T282	245	TIXM202
245	MD836	245	R7886	245	SC-78	245	T2878	245	TIXM203
245	MD837	245	R7891	245	SC-79	245	T2896	245	TIXM204
245	MD838	245	R7962	245	SC-80	245	T2945	245	TIXM205
245	MD839	245	R8240	245	SC1007	245	T2946	245	TIXM206
245	MD840	245	R8241	245	SC71	245	T348	245	TIXM207
245	MM1139	245	R8242	245	SC72	245	T367	245	TK1228-001
245	MM1199	245	R8559	245	SC74	245	T368	245	TK41C
245	MM2503	245	R8685	245	SC78	245	T373	245	TK42C
245	MM2550	245	R8686	245	SC79	245	T374	245	TNJ-60067
245	MM2552	245	R8692	245	SC80	245	T449	245	TNJ-60068
245	MM2554	245	R8693	245	SE2400	245	T605B	245	TNJ-60069
245	MM2894	245	R8694	245	SF.T163	245	T811	245	TNJ-60071
245	MM380	245	R8703	245	SF.T316	245	TA-1628	245	TNJ-60073
245	MM5000	245	R8704	245	SF.T317	245	TA-1658	245	TNJ-60077
245	MM5001	245	R8705	245	SF.T319	245	TA-1659	245	TNJ-60279
245	MM5002	245	R8881	245	SF.T320	245	TA-1660	245	TNJ-60280
245	MPS1097	245	R8882	245	SF.T354	245	TA-1662	245	TNJ-60281
245	MP89600-5	245	R9531	245	SF.T357	245	TA-1731	245	TNJ-60362
245	MT102351A	245	R9532	245	SF.T358	245	TA-1757	245	TNJ-60363
245	N-020	245	R9601	245	SPT-163	245	TA-1796	245	TNJ-60364
245	N57B2-11	245	R9602	245	SPT-358	245	TA-1797	245	TNJ-60365
245	N57B2-13	245	RE-27	245	SPT120	245	TA-1798	245	TNJ-60608
245	N57B2-14	245	RE27	245	SPT162	245	TA-1828	245	TNJ60063
245	N57B2-22	245	R8-101	245	SPT163	245	TA-1829	245	TNJ60064
245	NA-1114-1001	245	R8-103	245	SPT171	245	TA-1846	245	TNJ60065
245	NA-1114-1002	245	R8-2002	245	SPT172	245	TA-1847	245	TNJ60067
245	NA1022-1001	245	R8-3892	245	SPT173	245	TA-1860	245	TNJ60068
245	NA5018-1001	245	R8-3898	245	SPT174	245	TA-1861	245	TNJ60069
245	NK1302	245	R8-3900	245	SPT268	245	TA1628	245	TNJ60071
245	NK1404	245	R8-3902	245	SPT315	245	TA1650A	245	TNJ60073
245	NK65-1A19	245	R8-3903	245	SPT316	245	TA1658	245	TNJ60077
245	NK65A119	245	R8-3911	245	SPT358	245	TA1659	245	TNJ60279
245	NK88B119	245	R8-5208	245	SK3006	245	TA1660	245	TNJ60280
245	NKA8-1A19	245	R81539	245	SK3730	245	TA1662	245	TNJ60281
245	NKT121	245	R81550	245	SK3770	245	TA1731	245	TNJ60456
245	NKT122	245	R81554	245	SM1297	245	TA1755	245	TNJ70641
245	NKT124	245	R82679	245	SM1600	245	TA1756	245	TO-003
245	NKT125	245	R82680	245	SM217	245	TA1757	245	TO-004
245	NKT251	245	R82683	245	SM2491	245	TA1796	245	TQ-5034
245	NK2574F	245	R82684	245	SM2491I	245	TA1797	245	TQ5021
245	NK2677F	245	R82685	245	SM3014	245	TA1798	245	TQ5022
245	OC-169	245	R82686	245	SMB454760	245	TA1828	245	TQ5034
245	OC-170	245	R82687	245	SO-1	245	TA1829	245	TQ5035
245	OC-171	245	R82688	245	SO-2	245	TA1830	245	TQ5038
245	OC-615	245	R82694	245	SO-3	245	TA1846	245	TR-07
245	OC169	245	R82695	245	SO-65A	245	TA1847	245	TR-11C
245	OC169R	245	R83277	245	S01	245	TA1860	245	TR-12
245	OC170	245	R83278	245	S02	245	TA1861	245	TR-12C
245	OC170N	245	R83279	245	S03	245	TA2322	245	TR-13(RP)
245	OC170R	245	R83288	245	SP85328	245	TI-358	245	TR-13C
245	OC170V	245	R83309	245	SB155	245	TI-587	245	TR-161(OLSON)
245	OC171	245	R83322	245	ST-125	245	TI-588	245	TR-166(OLSON)
245	OC171R	245	R83323	245	ST07279	245	TI-400	245	TR-17
245	OC171V	245	R83324	245	STX0036	245	TI-401	245	TR-17A
245	OC520	245	R83668	245	STX0085	245	TI-403	245	TR-17C
245	OC400	245	R83862	245	STX0087	245	T3010	245	TR-18
245	OC410	245	R83863	245	STX0089	245	T3011	245	TR-18C
245	OC53	245	R83864	245	STX0090	245	T3338	245	TR-1R26
245	OC54	245	R83868	245	STL2189	245	T3363	245	TR-2R26
245	OC55	245	R83898	245	T-163	245	T3364	245	TR-3R26
245	OC615N	245	R83900	245	T-2028	245	T365	245	TR-4R26
245	ON174	245	R83901	245	T-2029	245	T387	245	TR-7831
245	PA10880	245	R83902	245	T-2030	245	T388	245	TR-8001
245	PA9154	245	R83903	245	T-278	245	T389	245	TR-8002
245	PA9155	245	R83905	245	T-279	245	T390	245	TR-8003
245	PADT20	245	R83906	245	T-99	245	T391	245	TRO6C
245	PADT21	245	R83907	245	T1011	245	T393	245	TRO7
245	PADT22	245	R83911	245	T1012	245	T395	245	TR11C
245	PADT23	245	R83912	245	T1028	245	T396	245	TR12C
245	PADT24	245	R83929	245	T1032	245	T397	245	TR13
245	PADT25	245	R83986	245	T1033	245	T398	245	TR13C
245	PADT26	245	R83995	245	T1034	245	T399	245	TR18
245	PADT27	245	R85101	245	T1038	245	T400	245	TR18C
245	PADT28	245	R85106	245	T1166	245	TI401	245	TR1R26
245	PADT30	245	R85107	245	T1224	245	TI402	245	TR2R26
245	PADT31	245	R85108	245	T1225	245	TI403	245	TR310019
245	PADT35	245	R85109	245	T1232	245	TIM-01	245	TR310021
245	PADT40	245	R85201	245	T1233	245	TIM-10	245	TR310065
245	PADT51	245	R85204	245	T1250	245	TIM-11	245	TR310068
245	PIL/4956	245	R85205	245	T1298	245	TIX-M01	245	TR310123
245	PQ27	245	R85206	245	T1299	245	TIX-M02	245	TR310124
245	PQ30	245	R85207	245	T1305	245	TIX-M03	245	TR310139
245	PT2A	245	R85208	245	T1306	245	TIX-M04	245	TR310147
245	PT28	245	R85209	245	T1314	245	TIX-M05	245	TR310150
245	PT855	245	R85301	245	T1387	245	TIX-M06	245	TR310155
245	PT856	245	R85305	245	T1388	245	TIX-M07	245	TR310156
245	PTC107	245	R85306	245	T1389	245	TIX-M08	245	TR310157
245	Q40359	245	R85311	245	T1390	245	TIX-M101	245	TR310158
245	Q5044	245	R85312	245	T1391	245	TIX-M11	245	TR310193
245	QA-1	245	R85313	245	T1400	245	TIX-M17	245	TR310224
245	R-28A222	245	R85314	245	T1401	245	TIX-M201	245	TR310232
245	R1539	245	R85818	245	T1402	245	TIX-M202	245	TR331
245	R1550	245	R8593	245	T1403	245	TIX-M203	245	TR3R26
245	R1554	245	R8684	245	T1454	245	TIX-M204	245	TR4R26
245	R2683	245	R8685	245	T1459	245	TIX-M205	245	TR8001
245	R2684	245	R8686	245	T1461	245	TIX-M206	245	TR8002
245	R2685	245	R8687	245	T1548	245	TIX-M207	245	TR8003
245	R2686	245	RT3466	245	T1618	245	TIX3016	245	TRA10R
245	R2687	245	RT4525	245	T163	245	TIX3016A	245	TRA11R
245	R2688	245	RT6988	245	T1657	245	TIX3032	245	TRA12R
245	R2694	245	S-1640	245	T1690	245	TIX316	245	TRA22A
245	R2695	245	S-371	245	T1691	245	TIX91	245	TRA22B
245	R2696	245	S-8713TB	245	T1692	245	TIX92	245	TRA23
245	R2697	245	S-8714TB	245	T1737	245	TIXM-201	245	TRA23A
245	R3277	245	S01	245	T1738	245	TIXM-203	245	TRA23B
245	R3278	245	S02	245	T1788	245	TIXM-205	245	TRA24C
245	R3279	245	S03	245	T1814	245	TIXM-206	245	TS-615
245	R3287	245	S1640	245	T1831	245	TIXMO1	245	TS-620
245	R3288	245	0822.3511-770	245	T2015	245	TIXMO2	245	TS-621
245	R3309	245	0822.3511-780	245	T2016	245	TIXMO3	245	TS-627A
245	R338	245	0822.3511-790	245	T2017	245	TIXMO4	245	TS-627B
245	R339	245	S65-1-A-3P	245	T2019	245	TIXMO5	245	TS-630
245	R341	245	S65A-1-3P	245	T2020	245	TIXMO6	245	TS-672A
245	R424	245	S684	245	T2021	245	TIXMO7	245	TS-672B
245	R424-1	245	S70T	245	T2022	245	TIXMO8	245	TS-673A
245	R425	245	S87TB	245	T2024	245	TIXM10	245	TS-673B
245	R497	245	S88B-1-3P	245	T2025	245	TIXM101	245	TS630
245	R5102	245	S88TB	245	T2026	245	TIXM103	245	TS669C
245	R5103	245	S95101	245	T2028	245	TIXM104	245	TS672B
245	R515	245	S95102	245	T2029	245	TIXM105	245	TS673A
245	R515A	245	S95103	245	T2030	245	TIXM106	245	TS673B
245	R516(T.I.)	245	S95104					245	TV24137

GE	Industry Standard No.	GE	Industry Standard No.	GE	Industry Standard No.	GE	Industry Standard No.	GE	Industry Standard No.	GE	Industry Standard No.
245	TV24158	245	2N1726	245	2N3281	245	2SA450	246	D292	246	D297
245	TV24166	245	2N1727	245	2N3282	245	2SA450H	246	D297		
245	TV24172	245	2N1728	245	2N3283	245	2SA451	246	D315		
245	TV24229	245	2N1742	245	2N3284	245	2SA451H	246	D34014094		
245	TV24230	245	2N1745	245	2N3285	245	2SA452	246	D49		
245	TV24239	245	2N1746	245	2N3286	245	2SA452H	246	D56		
245	TV24351	245	2N1747	245	2N331	245	2SA460	246	D57		
245	TV2455	245	2N1748	245	2N3320	245	2SA461	246	D58		
245	TV2479	245	2N1748A	245	2N3321	245	2SA462	246	D70		
245	TV8-2SA103	245	2N1749	245	2N3322	245	2SA463	246	D71		
245	TV825A103	245	2N1750	245	2N3323	245	2SA464	246	D92		
245	1-21-135	245	2N1752	245	2N3324	245	2SA506	246	D92D		
245	1-21-137	245	2N1753	245	2N3325	245	2SA508	246	DT3301		
245	1-21-139	245	2N1754	245	2N3371	245	2SA56	246	DT3302		
245	1-21-150	245	2N1782	245	2N3399	245	2SA61	246	E570022-01		
245	1-21-157	245	2N1784	245	2N3400	245	2SA79	246	EA15X121		
245	1-21-190	245	2N1785	245	2N3412	245	2SA81	246	EC0175		
245	1-21-228	245	2N1786	245	2N3443	245	2SA86	246	EB15X98		
245	1-21-229	245	2N1787	245	2N3449	245	2SA88	246	ES36(ELCOM)		
245	1-21-230	245	2N1788	245	2N3588	245	2SA89	246	ES44(ELCOM)		
245	1-21-231	245	2N1789	245	2N3770	245	2SC125	246	ES45(ELCOM)		
245	1-21-233	245	2N1790	245	2N3783	246	A-120304	246	ET8-017		
245	1-21-256	245	2N1853/18	245	2N3784	246	A-1854-0449-1	246	ETTC-28C490		
245	1-21-258	245	2N1864	245	2N3785	246	A054-224	246	EW1853A		
245	1-21-259	245	2N1865	245	2N384	246	A068-114	246	FBN-35469		
245	1-21-260	245	2N1866	245	2N384/33	246	A14743	246	FBN-35903		
245	1-522210131	245	2N1867	245	2N3883	246	A2415	246	FBN-36486		
245	1-522210300	245	2N1868	245	2N3995	246	A27(RCA)	246	FBN-36488		
245	1-522210921	245	2N1960	245	2N509	246	A2SD226PQ	246	FRB-564		
245	1-522211021	245	2N1960/46	245	2N537	246	A6843401	246	GE-246		
245	1-522211921	245	2N1961	245	2N559	246	AD160	246	HEP241		
245	1-522214400	245	2N1961G	245	2N700	246	AR17(PHILCO)	246	HEP703		
245	1-522214411	245	2N1999	245	2N700/18	246	AR17A	246	HEP85019		
245	1-522214435	245	2N2022	245	2N700A	246	AR17B	246	HR-107(PHILCO)		
245	1-522214821	245	2N2048	245	2N700A/18	246	AR18(PHILCO)	246	HR107H		
245	1-522214831	245	2N2048A	245	2N705	246	B133823	246	HT304911		
245	1-522216600	245	2N2059	245	2N705A	246	B1Z	246	HT304911A		
245	1-522217400	245	2N2083	245	2N710	246	B2B	246	HT304911B		
245	002-007100	245	2N2084	245	2N710A	246	B5C	246	HT401301BQ		
245	002-007200	245	2N2091	245	2N711	246	BDY16	246	HT403151E		
245	002-007400	245	2N2093	245	2N711A	246	BDY71	246	HT403152A		
245	2F	245	2N2098	245	2N711B	246	BDY72	246	HT403152B		
245	2G101	245	2N2099	245	2N725	246	BDY78	246	I51-0141-00		
245	2G102	245	2N2100	245	2N741	246	BDY79	246	I6114		
245	2G103	245	2N2168	245	2N741A	246	BLY15A	246	INTRON-127		
245	2G104	245	2N2170	245	2N768	246	BLY47A	246	IP20-0212		
245	2G106	245	2N218	245	2N769	246	BLY48A	246	IR-TR57		
245	2G108	245	2N2188	245	2N779	246	BLY49A	246	J24123I		
245	2G109	245	2N2189	245	2N779A	246	BW7168	246	J24642		
245	2G110	245	2N219	245	2N779B	246	BUY38	246	M9225		
245	2G201	245	2N2190	245	2N781	246	C1024	246	M9274		
245	2G202	245	2N2191	245	2N782	246	C1024-D2	246	M9301		
245	2G301	245	2N2199	245	2N794	246	C1024B	246	M9309		
245	2G345	245	2N2200	245	2N795	246	C1024C	246	M9316		
245	2G382	245	2N2225	245	2N796	246	C1024D	246	M9393		
245	2G401	245	2N2238	245	2N827	246	C1024E	246	M9610		
245	2G402	245	2N2258	245	2N828	246	C1024F	246	MJ2249		
245	2G403	245	2N2259	245	2N828A	246	C1025	246	MJ2250		
245	2G404	245	2N2273	245	2N829	246	C1025CTV	246	MJ3101		
245	2G413	245	2N232	245	2N837	246	C1160	246	MJ4101		
245	2G414	245	2N2360	245	2N838	246	C1160K	246	MJ4102		
245	2G415	245	2N2361	245	2N846	246	C1160L	246	MJ5202		
245	2G416	245	2N2362	245	2N846A	246	C1161	246	MJ5203		
245	2G417	245	2N2381	245	2N846B	246	C1828	246	MJ5204		
245	2J72	245	2N2382	245	2N934	246	C254	246	P3172		
245	2J73	245	2N2398	245	2N960	246	C487	246	P5148		
245	2K48	245	2N2399	245	2N960/46	246	C488	246	P6022A		
245	2N1017	245	2N240	245	2N961	246	C489	246	P6128		
245	2N1018	245	2N2400	245	2N962	246	C490	246	P6804		
245	2N1023	245	2N2401	245	2N963	246	C491	246	FP-AR18		
245	2N1065	245	2N2402	245	2N964	246	C491BL	246	PN26		
245	2N1066	245	2N2415	245	2N964/46	246	C491R	246	PTC112		
245	2N1093	245	2N2416	245	2N964A	246	C491Y	246	Q-00769		
245	2N1094	245	2N2455	245	2N965	246	C680	246	Q-00769A		
245	2N1115	245	2N2495	245	2N966	246	C680A	246	Q-00769B		
245	2N1115A	245	2N2496	245	2N967	246	C680R	246	Q-00769C		
245	2N1122	245	2N2512	245	2N968	246	C791	246	Q-01269		
245	2N1122A	245	2N2587	245	2N969	246	C830	246	Q-01269B		
245	2N1141	245	2N2621	245	2N970	246	C830A	246	Q-01269C		
245	2N1141A	245	2N2622	245	2N971	246	C830B	246	Q5134Z		
245	2N1142	245	2N2623	245	2N972	246	C830C	246	R2096		
245	2N1142A	245	2N2624	245	2N973	246	C840	246	RE34		
245	2N1143	245	2N2625	245	2N974	246	C840A	246	REN175		
245	2N1143A	245	2N2626	245	2N975	246	C840AC	246	S21520		
245	2N1144	245	2N2627	245	2N977	246	C840H	246	S2321		
245	2N1145	245	2N2628	245	2N979	246	C840PQ	246	S2486		
245	2N1158	245	2N2629	245	2N980	246	C895	246	S2527		
245	2N1158A	245	2N2630	245	2N982	246	C893-1007	246	S306A		
245	2N1177	245	2N2635	245	2N983	246	CXL175	246	SJ54		
245	2N1178	245	2N2654	245	2N984	246	D0250	246	SD16901		
245	2N1179	245	2N267	245	2N985	246	D102	246	SD16905		
245	2N1195	245	2N2671	245	2N987	246	D102-0	246	SE3040		
245	2N1282	245	2N2672	245	2N990	246	D102-R	246	SE3041		
245	2N1285	245	2N2672A	245	2N991	246	D102-Y	246	SE9060		
245	2N1300	245	2N26720BLK	245	2N992	246	D103	246	SE9061		
245	2N1301	245	2N26720RN	245	2N993	246	D103-0	246	SE9062		
245	2N1309A	245	2N269	245	2N994	246	D103-R	246	SE9063		
245	2N1385	245	2N2717	245	2SA126	246	D103-Y	246	SJ1172		
245	2N1403	245	2N274	245	2SA127	246	D129	246	SJ2095		
245	2N1404A	245	2N274BLU	245	2SA226	246	D129-BL	246	SJ3408		
245	2N1405	245	2N274WHT	245	2SA228	246	D129-R	246	SJ3447		
245	2N1406	245	2N276	245	2SA242	246	D129-Y	246	SJ3648		
245	2N1407	245	2N2786	245	2SA243	246	D130	246	SJ3680		
245	2N1408	245	2N2786A	245	2SA244	246	D130-R	246	SJ811		
245	2N1409	245	2N2795	245	2SA245	246	D130-Y	246	SK3026		
245	2N1410	245	2N2796	245	2SA276	246	D130BL	246	SK5538		
245	2N1411	245	2N2797	245	2SA306	246	D142	246	SM-A-618687-1		
245	2N1436	245	2N2798	245	2SA327	246	D142M	246	STC-4401		
245	2N1495	245	2N2799	245	2SA340	246	D143	246	STC4401		
245	2N1500/18	245	2N286	245	2SA345	246	D144	246	STX0010		
245	2N1515	245	2N2860	245	2SA346	246	D145	246	STX0016		
245	2N1516	245	2N2873	245	2SA347	246	D146UK	246	T-Q5080		
245	2N1517	245	2N2928	245	2SA349	246	D150	246	T271		
245	2N1517A	245	2N2929	245	2SA362	246	D155	246	TA2402		
245	2N1524	245	2N2942	245	2SA363	246	D155H	246	TA2402A		
245	2N1524-1	245	2N2943	245	2SA372	246	D155K	246	TG28C1025		
245	2N1524-2	245	2N2955	245	2SA373	246	D155L	246	TG28C1025D		
245	2N1524/33	245	2N2956	245	2SA378	246	D226	246	TIP503		
245	2N1525	245	2N2957	245	2SA379	246	D226-0	246	TNJ60451		
245	2N1526/33	245	2N299	245	2SA401	246	D226A	246	TNJ60453		
245	2N1527	245	2N2996	245	2SA410	246	D226AP	246	TNJ72286		
245	2N1562	245	2N2997	245	2SA411	246	D226B	246	TR-180(OLSON)		
245	2N1633	245	2N2998	245	2SA413	246	D226BP	246	TR-188(OLSON)		
245	2N1634	245	2N2999	245	2SA417	246	D226P	246	TR-57		
245	2N1637/33	245	2N300	245	2SA419	246	D226Q	246	TR1591		
245	2N1638	245	2N3074	245	2SA42	246	D238	246	TR1593		
245	2N1639	245	2N3127	245	2SA421	246	D238F	246	TR2327852		
245	2N1639/33	245	2N3148	245	2SA422	246	D254	246	TR26-1		
245	2N1646	245	2N3153	245	2SA425	246	D255	246	TV109		
245	2N1665	245	2N316	245	2SA430	246	D257	246	TV24211		
245	2N1670	245	2N3216	245	2SA431	246	D29	246	TV24487		
245	2N1676	245	2N3267	245	2SA431A	246	D290	246	TVS-28C840		
245	2N1678	245	2N3279	245	2SA448	246	D290L	246	1A0013(YAMAHA)		
245	2N1713	245	2N3280			246	D291	246	1A0027		

GE	Industry Standard No.	GE	Industry Standard No.	GE	Industry Standard No.	GE	Industry Standard No.	GE	Industry Standard No.
246	1A0027(YAMAHA)	247	2SC1162WT	253	ECG211	262	C898A	263	B506
246	1A0048(YAMAHA)	247	2SC1162WTA	253	EP15X24	262	C898B	263	B506C
246	2N3054	247	2SC1162WTB	253	GB-253	262	C898C	263	B506D
246	2N3483	247	2SC1162WTC	253	HEP-83027	262	D118	263	B541
246	2N3766	247	2SC1162WTD	253	HEP-83028	262	D118A	263	B555
246	2N3767	248	2SA636(L)	253	HEP-83029	262	D118BL	263	B555-0
246	2N3779	248	2SC298	253	HR-74	262	D118R	263	BDY69
246	2N3878	249	A9M	253	HR86	262	D118Y	263	ECG281
246	2N4231	250	A1853-0233-1	253	P2U-2	262	D119	263	FD-1029-LW
246	2N4232	250	A417756	253	REN211	262	D119BL	263	GE-263
246	2N4233	250	AR27	253	RS-2026	262	D119R	263	RE-61
246	2N4910	250	AR37	253	SPS4092	262	D119Y	263	RE61
246	2N4911	250	B23-79	253	2SA645	262	D124A	263	REN61
246	2N4912	250	B502	253	2SA706	262	D124AH	263	SAE-2
246	2N5202	250	B503	253	2SA717	262	D124AHA	263	SK3183
246	2N5598	250	B550	255	AEX-82308	262	D124B	263	2N6226
246	2N5600	250	BD274	255	C36566	262	D125	263	2N6227
246	2N5601	250	BD276	255	C493	262	D125A	263	2N6228
246	2N5602	250	BD442	255	C493-BL	262	D125AH	263	2N6229
246	2N5604	250	BD538	255	C493-R	262	D125AHA	263	2N6231
246	2N6233	250	BRC6109	255	C493-Y	262	D125AHB	263	2SA679
246	2N6260	250	CXL197	255	C494	262	D126A	263	2SA679-R
246	2N6261	250	ECG197	255	C494-BL	262	D126AH	263	2SA679R
246	2N6263	250	EP15X44	255	C494-R	262	D126AHA	263	2SA679Y
246	2N6500	250	ES104	255	C494-Y	262	D126H	263	2SA680
246	2SC1024	250	GE-250	255	C494BL	262	D126HA	263	2SA680-R
246	2SC1024(D)	250	16342	255	C646	262	D126HB	263	2SA680T
246	2SC1024(P)	250	KLH4781	255	C647	262	D322	263	2SA753
246	2SC1024-D2	250	KLH5353	255	CXL223	262	D322A	263	2SA753A
246	2SC1024-E	250	M7310	255	D180A	262	D322B	263	2SA753B
246	2SC1024A	250	M9677	255	D180B	262	D322C	263	2SA753C
246	2SC1024B	250	M9677/PNP	255	D180C	262	D323	263	2SA756
246	2SC1024C	250	M9701	255	D180D	262	D323A	263	2SA756A
246	2SC1024D	250	MJE-32B	255	D68B	262	D323B	263	2SA756B
246	2SC1024E	250	MJE-42	255	D68C	262	D323C	263	2SA756C
246	2SC1024F	250	MJE170	255	D68D	262	D425	263	2SA757
246	2SC1024G	250	MJE2010	255	D68E	262	D4250	263	2SA757A
246	2SC1024L	250	MJE210	255	ECG223	262	D73	263	2SA757B
246	2SC1024T	250	MJE30	255	E869(ELCOM)	262	D73A	263	2SA757C
246	2SC1025	250	MJE30A	255	ETS-005	262	D73B	263	2SA758
246	2SC1025CTV	250	MJE30B	255	EX524-X	262	D73C	263	2SA758A
246	2SC1025D	250	MJE32	255	G23-76	262	D73D	263	2SA758B
246	2SC1025E	250	MJE32A	255	GE-255	262	D73E	263	2SA758C
246	2SC1025J	250	MJE42A	255	IP20-0028	262	D74	263	2SA837
246	2SC1025MT	250	MJE42B	255	LN76533	262	D74A	263	2SB506
246	2SC1085	250	MJE710	255	M9666	262	D74B	263	2SB506A
246	2SC1160	250	P3A	255	MU-26-1C	262	D74C	263	2SB506B
246	2SC1160K	250	P3U	255	P04450040-002	262	D74D	263	2SB506C
246	2SC1160L	250	PIV	255	QP-8-7	262	ECG280	263	2SB506D
246	2SC1160M	250	PIV-1	255	R227078533	262	FD-1029-LU	263	2SB541
246	2SC1664A	250	PIV-2	255	T-23-71	262	GE-262	263	2SB555
246	2SC254	250	PIV-3	255	TA7199	262	HT401193A0	263	2SB555-0
246	2SC487	250	QP-31	255	TA7200	262	RCA1B06	265	BD245B
246	2SC488	250	R264-1	255	TA7201	262	S124AHB	265	BD245C
246	2SC508	250	RCA1006	255	TA7202	262	SAE-1	265	BDX50
247	C1095	250	RCA1C11	255	TR1025	262	2N1487	265	BDX51
247	C1095(6)	250	REN197	255	2N5034	262	2N1488	265	BDX60
247	C1095L	250	SK3083	255	2N5035	262	2N1489	265	BDX61
247	C1095M	250	SK3084	255	2N5036	262	2N4347	265	BDY18
247	C1096	250	STH72251	255	2N5037	262	2N4348	265	BDY19
247	C1096(M)	250	TA7742	255	2SC494	262	2N5067	265	BDY20
247	C1096-3ZM	250	TA7743	255	2SC494-BL	262	2N5068	265	BDY55
247	C10963ZM	250	TA8210	255	2SC494-R	262	2N5069	265	BDY56
247	C10964ZL	250	TA8211	255	2SC494-Y	262	2SC1051	265	BDY73
247	C1096A	250	TA8212	255	2SC494A	262	2SC1051C	265	BDY77
247	C1096B	250	TIP32B	255	2SC494B	262	2SC1051D	265	C1667
247	C1096C	250	TIP3A	255	2SC494BL	262	2SC1051E	265	D287
247	C1096D	250	TIP42	255	2SC494C	262	2SC1051F	265	D379
247	C1096K	250	TIP42A	255	2SC494D	262	2SC1051LC	265	D379P
247	C1096L	250	TIP42B	255	2SC494F	262	2SC1051LD	265	D379Q
247	C1096M	250	TV116	255	2SC494G	262	2SC1051LE	265	D379R
247	C1096N	250	01-020566	255	2SC494GN	262	2SC1051LF	265	D379S
247	C1096W	250	2N6036	255	2SC494H	262	2SC1079	265	D798
247	C1098	250	2N6106	255	2SC494K	262	2SC1079R	265	D424
247	C1098A	250	2N6107	255	2SC494L	262	2SC1079Y	265	D424-0
247	C1098B	250	2N6108	255	2SC494M	262	2SC1080	265	D424-R
247	C1098C	250	2N6109	255	2SC494QR	262	2SC1080R	265	ECG284
247	C1098D	250	2N6110	255	2SC494R	262	2SC1080Y	265	GE-265
247	C1098M	250	2N6111	255	2SC494Y	262	2SC1111	265	MJE302
247	C1107Q	250	2N6132	255	2SC646	262	2SC1115	265	RCA1B01
247	C1162WBP	250	2N6133	255	BD123	262	2SC1322	265	8J2519
247	C1226	250	2N6134	262	BD141	262	2SC1343	265	2N3442
247	C1226-0	250	2SA768	262	BD184	262	2SC1343A	265	2N3716
247	C1226A	250	2SB502	262	BDX10	262	2SC1343B	265	2N3773
247	C1226AC	250	2SB503	262	BDX11	262	2SC1343BL	265	2N5630
247	C1226AO	250	2SB550	262	BDX24	262	2SC1343C	265	2N5634
247	C1226AP	250	2SB596	262	BDX53	262	2SC1343D	265	2N5978
247	C1226AQ	250	2SB596-0	262	BDY74	262	2SC1343E	265	2N6259
247	C1226AR	252	C101RB	262	BDY90(AUDIO)	262	2SC1343F	265	2N6262
247	C1226C	252	C1243-24	262	BDY91(AUDIO)	262	2SC1343G	265	2N6354
247	C1226CF	252	D3288	262	BDY92(AUDIO)	262	2SC1343G-R	265	2N6496
247	C1226F	252	D4OD1	262	C1030	262	2SC1343H	265	2SC1667
247	C1226P	252	D4OD10	262	C1030A	262	2SC1343HA	265MP	SK3029
247	C1226Q	252	D4OD11	262	C1030B	262	2SC1343HB	266	REN280
247	C1226QR	252	D4OD14	262	C1030C	262	2SC1343J	266	SK3173
247	C1226R	252	D4OD2	262	C1030D	262	2SC1343K	268	A066-115
247	C1368D	252	D4OD3	262	C1030P	262	2SC1343L	268	A67-33-540
247	C931	252	D4OD4	262	C1051	262	2SC1343M	268	C1213
247	C931C	252	D4OD5	262	C1051C	262	2SC1343O	268	C1213A
247	C931D	252	D4OD7	262	C1051D	262	2SC1343OR	268	C1213AA
247	C931E	252	D4OD8	262	C1051E	262	2SC1343X	268	C1213AB
247	D44C1	252	ECG210	262	C1051F	262	2SC1343Y	268	C1213AC
247	DBBY003001	252	EP15X25	262	C1051LC	262	2SC1402	268	C1213AD
247	DBBY003002	252	ET392927	262	C1051LD	262	2SC665	268	C1213AK
247	DDBY227001	252	GE-252	262	C1051LE	262	2SC665H	268	C1213AKA
247	ECG186A	252	HR-73	262	C1051LF	262	2SC665HA	268	C1213AKB
247	EQS-89	252	HR85	262	C1079	262	2SC665HB	268	C1213AKC
247	ET453611	252	ME-5	262	C1079R	262MP	ECG280MP	268	C1213AKD
247	GG9-41H-C	252	REN210	262	C1079Y	262MP	GE-262MP	268	C1213B
247	IP20-0230	252	RS-2018	262	C1080	263	A679	268	C1213BC
247	NSE-181	252	TV73	262	C1080R	263	A679R	268	C1213C
247	PT-029	252	TV75	262	C1080Y	263	A680	268	C1213D
247	REN186A	252	2N2853-1	262	C1111	263	A680R	268	C1214
247	2SC1095	252	2N2854-1	262	C1115	263	A680Y	268	C1214A
247	2SC1095(6)	252	2N2855-1	262	C1343	263	A753	268	C1214B
247	2SC1095L	252	2N2856-1	262	C1343A	263	A753A	268	C1214C
247	2SC1095M	253	A645	262	C1343C	263	A753B	268	C1214D
247	2SC10963ZM	253	A717	262	C1343H	263	A753C	268	C1317
247	2SC10964ZL	253	AR26	262	C1343HA	263	A756	268	C1317B
247	2SC10960R	253	AR29	262	C1343HB	263	A756A	268	C1317P
247	2SC11070	253	BCW79-25	262	C1402	263	A756B	268	C1317Q
247	2SC1107Q	253	BCW80-25	262	C665	263	A756C	268	C1317R
247	2SC116C(C)	253	BCX10	262	C665H	263	A757	268	C1317S
247	2SC1162(RF-PWR)	253	CXL211	262	C665HA	263	A757A	268	C1317T
247	2SC1162A	253	D41D1	262	C665HB	263	A757B	268	C1342
247	2SC1162B	253	D41D10	262	C897	263	A757C	268	C1342A
247	2SC1162CP	253	D41D11	262	C897A	263	A758	268	C1342B
247	2SC1162D	253	D41D14	262	C897C	263	A758A	268	C1342C
247	2SC1162MP	253	D41D2	262	C898	263	A758B	268	C454B
247	2SC1162WB	253	D41D4			263	A758C	268	C455
247	2SC1162WBP	253	D41D5			263	A837		
		253	D41D7						
		253	D41D8						

GE	Industry Standard No.	GE	Industry Standard No.	GE	Industry Standard No.	GE	Industry Standard No.	GE	Industry Standard No.
268	C458	269	A650372D	269	2SA673E	278	C605K	300	AE100
268	C458(C)	269	A650923F	269	2SA673F	278	C605L	300	AE150
268	C458A	269	A650925H	269	2SA673G	278	C605M	300	AE200
268	C458AD	269	A673A	269	2SA673GN	278	C605TW	300	A830
268	C458B	269	A673AA	269	2SA673H	278	C606	300	A850
268	C458B-D	269	A673AB	269	2SA673J	278	C606(NEC)	300	AG30
268	C458BC	269	A673AC	269	2SA673K	278	CF11	300	AM300
268	C458BK	269	A673AD	269	2SA673L	278	EQS-19	300	AM300A
268	C458BL	269	A673AE	269	2SA673M	278	EQS-20	300	AM301
268	C458BLG	269	A673AS	269	2SA673OR	278	GE-278	300	AM301A
268	C458C	269	A673ASC	269	2SA673R	278	2SC1779	300	AM302
268	C458CLG	269	A673B	269	2SA673WT	278	2SC1790	300	AM302A
268	C458CM	269	A673C	269	2SA673X	278	2SC278-OR	300	AM303
268	C458D	269	A673C2	269	2SA673Y	278	2SC605	300	AM303A
268	C458G	269	A673D	269	2SA693	278	2SC605(L)	300	AM304
268	C458GLB	269	A673WT	269	2SA693C	278	2SC605(Q)	300	AM304A
268	C458K	269	A719	269	2SA719	278	2SC605-OR	300	AM305
268	C458KA	269	A719P	269	2SA719(Q)	278	2SC605A	300	AM305A
268	C458KB	269	A719R	269	2SA719,R	278	2SC605B	300	AM306
268	C458KC	269	A719RS	269	2SA719K	278	2SC605C	300	AM306A
268	C458KD	269	A719S	269	2SA719P	278	2SC605D	300	AM307
268	C458L	269	DDBY004001	269	2SA719PQR	278	2SC605E	300	AM307A
268	C458LB	269	DDCY007002	269	2SA719Q	278	2SC605F	300	AM308
268	C458LG	269	EA15X395	269	2SA719QR	278	2SC605G	300	AM308A
268	C458LGA	269	FC88550	269	2SA719R	278	2SC605GN	300	AM620
268	C458LGB	269	FC88550C	269	2SA719RS	278	2SC605H	300	AM620A
268	C458LGBM	269	GE-269	269	2SA719S	278	2SC605J	300	AM626
268	C458LGC	269	HA00562	269	2SB641	278	2SC605K	300	AM626A
268	C458LGD	269	HS-40057	269	2SB6418	278	2SC605L	300	AM632
268	C458LGG	269	HT106731BO	269	2SB643	278	2SC605M	300	AM632A
268	C458LGS	269	IC743043	269	2SB643Q	278	2SC605N	300	AV0000105-Q
268	C458M	269	IP20-0192	269	2SB643R	278	2SC6050R	300	B-1599
268	C458P	269	Q5116C	269	2SB6438	278	2SC605R	300	B-1702
268	C458RGS	269	Q5205	269	2SB741	278	2SC605TW	300	B614-007-Q
268	C458TOK	269	QT-A0719AXN	270	C1162C	278	2SC605Y	300	BA114
268	C458VC	269	QT-A0719XHN	270	NCBV14	278	2SC606(VHP)	300	BA147
268	C459D	269	REN290	270	2SC1449	278	2SC606A	300	BA167
268	C464C	269	RT8670	270	2SC1449(CB)	278	2SC606B	300	BA168
268	C468(LGR)	269	RV2355	270	2SC1449CB	278	2SC606C	300	BA174
268	C509	269	01-010673	270	2SC1449M	278	2SC606D	300	BA182
268	C509Y	269	01-010719	270	2SC2028	278	2SC606E	300	BA187
268	C590	269	01-030509	270	2SC2028-2	278	2SC606F	300	BA243
268	C590Y	269	01-572811	270	2SC2028/2	278	2SC606G	300	BA243A
268	C735	269	2N5221	270	2SC2028B	278	2SC606GN	300	BA244
268	C735(O)	269	2SA1029C	270	2SC2028B/20	278	2SC606H	300	BA316
268	C735-0	269	2SA1029C	270	2SC2028B20	278	2SC606J	300	BAW10TF20
268	C735B	269	2SA509	270	2SC2236	278	2SC606K	300	BAW11TF21
268	C735F	269	2SA509(A)	270	2SC2236-0	278	2SC606L	300	BAW12TF22
268	C735PA3	269	2SA509(O)	270	2SC2236-0	278	2SC606M	300	BAW13TF23
268	C735GR	269	2SA509-0	270	2SC2236Y	278	2SC606N	300	BAW16
268	C735H	269	2SA509-0	270	A67-70-960	278	2SC606R	300	BAW17
268	C735J	269	2SA509-OR	271	C1166	278	2SC606X	300	BAW18
268	C735K	269	2SA509-R	271	C1166-0	278	2SC606Y	300	BAW24
268	C735L	269	2SA509-RD	271	C1166D	283	BFX59	300	BAW24A
268	C735T	269	2SA509-Y	271	C1166GD	283	C382	300	BAW24B
268	C939D	269	2SA509-YE	271	C11660	283	C382-BK(1)	300	BAW25
268	D471	269	2SA509A	271	C1166R	283	C382-BK(2)	300	BAW25A
268	D471L	269	2SA509B	271	C1166Y	283	C382-GR	300	BAW25B
268	D71L	269	2SA509BL	271	C1209	283	C382-GY	300	BAW45
268	DBCZO73503	269	2SA509C	271	C1209C	283	C382BK	300	BAW53
268	EA15X256	269	2SA509D	271	C1318	283	C382BL	300	BAW59
268	EA15X267	269	2SA509E	271	C1318C	283	C382BN	300	BAW59A
268	EA15X360	269	2SA509F	271	C1318Q	283	C382BR	300	BAW59B
268	ET234854	269	2SA509G	271	C1318R	283	C382Q	300	BAW63A
268	ET329218	269	2SA509G-0	271	C1318S	283	C382R	300	BAW63B
268	ET352146	269	2SA509G-Y	271	C1509	283	C382V	300	BAW99
268	G05-055-D	269	2SA509GN	271	C1509P	283	C383	300	BAX-16
268	GE-268	269	2SA509GR	271	C1509Q	283	C383G	300	BAX16
268	HT312131C0	269	2SA509GR-1	271	C1509R	283	C383T	300	BAX28
268	HT313171R	269	2SA509H	271	C1509S	283	C383W	300	BAX30
268	HT313172A	269	2SA509J	271	C8B050	283	C383Y	300	BAX74
268	IC743044	269	2SA509K	271	D355C	283	C463	300	BAX87
268	IP20-0231	269	2SA509L	271	D355E	283	C463H	300	BAX87A
268	M9389418	269	2SA509M	271	GE-271	283	C927	300	BAX887P11
268	MP89416	269	2SA509R	271	MP89416	283	C927A	300	BAX89A
268	Q-00784R	269	2SA509Q	271	MPS9416A	283	C927B	300	BAY17
268	S36951	269	2SA509RD	271	MPS9416AT	283	C927C	300	BAY18
268	SK3122	269	2SA509T	271	MPS9416ST	283	C927CW	300	BAY19
268	SL403	269	2SA509V	271	MPS9417AT	283	C927D	300	BAY20
268	01-572814	269	2SA509X	271	MPS9417T	283	C927E	300	BAY31
268	2N5220	269	2SA509Y	271	MP89696	283	CGE-60	300	BAY36
268	2S01213	269	2SA509YE	271	MP89696G	283	EA15X7173	300	BAY41
268	2SC1213(B)	269	2SA562	271	NB211	283	EA15X7174	300	BAY41A
268	2SC1213-OR	269	2SA562(O)	271	NB211E1	283	EA15X7178	300	BAY41B
268	2SC1213A	269	2SA562(Y)	271	Q5160	283	EA15X7179	300	BAY52
268	2SC1213A(C)	269	2SA562-0	271	Q80509	283	ECG319	300	BAY68
268	2SC1213AA	269	2SA562-GR	271	2SC1209(C)	283	GE-283	300	BAY73
268	2SC1213AB	269	2SA562-GRN	271	2SC1509	283	NS45006	300	B8A01
268	2SC1213AC	269	2SA562-0	271	2SC1509P	283	SB5055	300	B8A02
268	2SC1213AD	269	2SA562-OR	271	2SC1509Q	283	TG28C927-C-A	300	B8A11
268	2SC1213AK	269	2SA562-ORG	271	2SC1509R	283	TG28C927-D-A	300	BTX58-100
268	2SC1213AKA	269	2SA562-R	271	2SC1509S	283	TG28C927-E-A	300	BTX58-200
268	2SC1213AKB	269	2SA562-RD	271	2SC1627	283	TV-33	300	BYX58-50
268	2SC1213AKC	269	2SA562-RED	271	2SC1627-0	283	TV-34	300	B21021V4
268	2SC1213AKD	269	2SA562-YE	271	2SC1627Y	283	TV-35	300	C-4401
268	2SC1213B	269	2SA562-YEL	271MP	EC0297MP	283	TV-50	300	C-505
268	2SC1213BC	269	2SA611	272	MP89466A	285	DDBY272001	300	C255110-011
268	2SC1213C	269	2SA611-4.E	272	2SA952	285	2SC1973	300	C6141990
268	2SC1213CD	269	2SA673	272	2N4437	291	2N5160	300	CA90
268	2SC1213D	269	2SA673(B)	278	C182	300	CA90	300	CD-00-9
268	2SC1213E	269	2SA673(D)	278	C182Q	300	A-120125(DIODE)	300	CD-0021
268	2SC1213F	269	2SA673-OR	278	C183	300	A-1901-0025-1	300	CD-20
268	2SC1213G	269	2SA673A	278	C183E	300	A-1901-0033-1	300	CD-37
268	2SC1213GN	269	2SA673A(C)	278	C183J	300	A-1901-0053-1	300	CD-37A
268	2SC1213H	269	2SA673AA	278	C183K	300	A-1901-0096-1	300	CD-84857
268	2SC1213J	269	2SA673AB	278	C183L	300	A-1901-0150-1	300	CD0000NC
268	2SC1213K	269	2SA673AC	278	C183M	300	A-1901-0156-1	300	CD0014
268	2SC1213L	269	2SA673AD	278	C183P	300	A-1901-1067-1	300	CD0014(MORSE)
268	2SC1213M	269	2SA673AE	278	C183Q	300	A01	300	CD1224
268	2SC1213OR	269	2SA673AK	278	C183R	300	A01(MOTOROLA)	300	CD37
268	2SC1213X	269	2SA673AKA	278	C183W	300	A04093-X	300	CD37A
268	2SC1213Y	269	2SA673AKB	278	C184	300	A054-228	300	CD5003
268	2SC1959Y	269	2SA673AKC	278	C184H	300	A066-119(SI)	300	CD8547
268	2SC4580/L6	269	2SA673AS	278	C184J	300	A066-12	300	CD86003
268	2SC4580LB	269	2SA673AS(C)	278	C184L	300	A08	300	CD86037
268	2SC4580LGC	269	2SA673ASC	278	C185	300	A42X00390-01	300	CDC8457
268MP	BCG289MP	269	2SA673B	278	C185A	300	A488-A0001	300	CDG-00
268MP	GE-268MP	269	2SA673C	278	C185J	300	A65-P11311-0001	300	CDG-20
269	A0666-118	269	2SA673C2	278	C185M	300	A66X0043-001	300	CDG-20/494
269	A509	269	2SA673D	278	C185Q	300	A91M4	300	CDG-21
269	A509-0			278	C185R	300	A691M5	300	CDG-22
269	A509-Y			278	C287A	300	A691M5-2	300	CDG-24
269	A509R			278	C288	300	A72-49-600	300	CDG00
269	A509Y			278	C469	300	A7246602	300	CDG025
269	A562			278	C469A	300	A7246711	300	CDG22
269	A562-0			278	C469F	300	A7246727	300	CDG24
269	A562-R			278	C469K	300	A7275400	300	CDG24/3490
269	A562-Y			278	C469Q	300	A73-16-179	300	CDG27
269	A562GR			278	C469R	300	AC50(DIODE)	300	CDJ-00
269	A562R			278	C605	300	AD100(DIODE)	300	CE37
269	A562Y			278	C605(NEC)	300	AD150(DIODE)	300	CGE-500
269	A65-09-220			278	C605(Q)	300	AD200(DIODE)	300	CR353
						300	AD30		
						300	AE10		

GE	Industry Standard No.	GE	Industry Standard No.	GE	Industry Standard No.	GE	Industry Standard No.	GE	Industry Standard No.
300	CX-0055	300	E857X12	300	KB-162	300	RT7946	300	TVSB01-2
300	CXL177	300	ETD-889150	300	KB-16205	300	RV1017	300	TVSBAX-13
300	CXL178MP	300	ETD1187788	300	KB-16205A	300	RV1226	300	TVSBAX13
300	D-00184R	300	ETDCDG21	300	KB-165	300	RV2071	300	TV8UF2-354
300	D-12	300	EU16X11	300	KB-262	300	RVDK265J2	300	TV8MA26
300	D1-20	300	EU16X20	300	KB-265	300	RVDVD1150L	300	TVSRA1A
300	D1J70545	300	EVD-3	300	KB-269	300	RVDVD1210L	300	000WG1010
300	D2-77-1	300	EX142-X	300	KB102	300	RVDVD1210M	300	001
300	D200MP	300	EX16X10	300	KB162N	300	RVDVD121L2L	300	1-001/2207
300	D3356	300	EX16X27	300	KB165	300	RVDVD1213	300	001-0095-00
300	D34003220-001	300	ETY-320D1R2J	300	KB169	300	RVDVD1250M	300	001-0125-00
300	D353	300	EYZP-384	300	KB265	300	8-3016R	300	1-014/2207
300	D355	300	OF162	300	KB265A	300	S04	300	001-0151-00
300	D3A	300	FA111	300	KB269	300	S04-1	300	001-0151-01
300	D5V	300	FCD0003PC	300	KD300A	300	S04A-1	300	001-026030
300	D6462	300	FCD0014NC8	300	KGE46109	300	S04B-1	300	001-026060
300	D6726	300	FD-1029-MC	300	KGE46465	300	S074-007-0001	300	01-1501
300	D7T	300	FD100	300	KLH4763	300	1H-RECT-35	300	1-16549
300	D77(DIODE)	300	FD1708	300	KX-1	300	S11	300	01-2601
300	D7E	300	FD1843	300	LM-1159	300	S1428	300	01-2601-0
300	D7Z	300	FD200	300	LM1159	300	S160	300	01-2603-0
300	DA205	300	FD222	300	M-8489A	300	S180	300	1-DI-007
300	DAAX003002	300	FD6451	300	M-8513A	300	S1D50B51-A	300	1A16549
300	DAAY004001	300	FD6489	300	M1-301	300	S1820	300	1A16551
300	DAAY047005	300	FDH-9	300	M150-1	300	S182	300	1D098-001V-022
300	DAB2158800	300	FDH444	300	M26	300	S2087G	300	1N135
300	DC8457	300	FDH6229	300	MB489	300	S3004-1716	300	1N137A
300	DDAX004001	300	FDH666	300	MB489A	300	S3016	300	1N137B
300	DDAY047005	300	FDH694	300	MB513	300	S3016R	300	1N138
300	DDAY048001	300	FDH900	300	M8513A	300	S3072C	300	1N138A
300	DDAY048008	300	FDM1006	300	M8513A-R	300	084.000654	300	1N138B
300	DDAY048013	300	FDM1007	300	M8513A0	300	S502	300	1N1630
300	DDAY090001	300	FDN600	300	M8513R	300	S502A	300	1N1638
300	DDFY004002	300	FDN666	300	M8640B(C-M)	300	S502B	300	1N1839
300	DPFY007001	300	FG-12377	300	M8640E(DIO)	300	S506	300	1N1840
300	DHD800	300	FR1D	300	MA-25A	300	S506A	300	1N1841
300	DHD8001	300	G00-012-A	300	MA-26	300	S506B	300	1N1843
300	DHD805	300	G00-012A	300	MA150	300	S509	300	1N1844
300	DHD806	300	G00-014-A	300	MA150TA	300	S509A	300	1N1845
300	DI-20	300	G00-803-A	300	MA26	300	S509B	300	1N1846
300	DI-3	300	G01-083-A	300	MA26A	300	S7-8	300	1N1847
300	DIJ70486	300	G01-209-B	300	MA26WA	300	SA-93792	300	1N194
300	DIJ70545	300	G01-209B	300	MB513AR	300	SB01-02	300	1N194A
300	DIJ71273	300	G01-217-A	300	MC-301	300	SC12(DIODE)	300	1N195
300	DIJ71711	300	G01-803A	300	MC2	300	SC1431(GE)	300	1N196
300	DIJ71960	300	G01209	300	MC2326	300	SD-110	300	1N200
300	DIJ72164	300	G01209B	300	MC301	300	SD-1AUP	300	1N202
300	DND800	300	G01803	300	MC5321	300	SD-18T555	300	1N203
300	DRT575	300	G01803A	300	MDP173	300	SD-34	300	1N204
300	DS-117	300	G1010A	300	ME-4	300	SD-43	300	1N205
300	DS-31	300	GD-510	300	MI-301	300	SD-5	300	1N207
300	DS-31(DELCO)	300	GD3638	300	MI301	300	SD-5(PHILCO)	300	1N2075K
300	DS-410	300	GB-500	300	ML2812	300	SD-630	300	1N208
300	DS-410(AMPEX)	300	GE414	300	MPC3500	300	SD-7	300	1N210
300	DS-442	300	GE6063	300	MPN-3401	300	SD100	300	1N211
300	DS-442(SEARS)	300	G01209	300	MPN3401	300	SD110	300	1N212
300	DS-97	300	G01211	300	MP89444	300	SD12(PHILCO)	300	1N213
300	DS1-002-0	300	G01803	300	MP89606	300	SD165	300	1N214
300	DS117	300	GP2-345	300	MP89606(H,I)	300	SD43	300	1N215
300	DS410	300	GP2-345/MA161	300	MP89606I	300	SD500	300	1N216
300	DS410(COURIER)	300	GV5760	300	MP89606H	300	SD600	300	1N217
300	DS410(EMERSON)	300	H623	300	MP89644	300	SD630	300	1N218
300	DS410(PANON)	300	H8513	300	MP89646I1	300	SD701-02	300	1N2473
300	DS410(OLYMPIC)	300	H889	300	MP89646G	300	SD974	300	1N251
300	DS410R	300	HD-2000106	300	MP89646H	300	SE5-0966	300	1N251A
300	DS441	300	HD2000106	300	MP89646I	300	SFD43	300	1N252
300	DS442	300	HD2000206	300	MP89646J	300	SFD83	300	1N252A
300	DS442FM	300	HD2001105	300	MP89646M	300	SG-9150	300	1N300
300	DS443	300	HD2001105Q	300	MS81000	300	SG3182	300	1N300A
300	DS448	300	HD6001	300	MV-12	300	SG3183	300	1N300B
300	DS897(DELCO)	300	HD6125	300	MV-13(DIO)	300	SG3198	300	1N301
300	DSA150	300	HE-00829	300	MV-2	300	SG3432	300	1N301A
300	DS8953	300	HE-10030	300	MV3	300	SG3516	300	1N301B
300	DT230A	300	HE-M8489	300	MV3(DIODE)	300	SG3585	300	1N303
300	DT230P	300	HF-20014	300	MZ-00	300	SG5028	300	1N303A
300	DT230G	300	HF-20034	300	MZ207	300	SG5392	300	1N303B
300	DX-0255	300	HF-20048	300	MZ2360	300	SG5400	300	1N3063
300	DX-0543	300	HF-20060	300	MZ2361	300	SG9150	300	1N3065
300	EO9-306112	300	HF-20061	300	N5406	300	SI-RECT-152	300	1N3066
300	E1121R	300	HF-20064	300	N5406(RCA)	300	SI-RECT-35	300	1N3067
300	E1138R	300	HF-20095	300	NE-446AQ	300	SIB-C1-02	300	1N3068
300	E13-017-01	300	HF-20124	300	O9-306113	300	SID50894	300	1N3069M
300	EA1405	300	HF-DB410	300	O9-306195	300	SID50B851	300	1N3070
300	EA1661	300	HF-MV2	300	OA-95	300	SK3100	300	1N3071
300	EA16X101	300	HF20032	300	OA200	300	SMB-541191	300	1N3123
300	EA16X110	300	HF20060	300	OA202	300	SP1U-2	300	1N3124
300	EA16X122	300	HN-00002	300	OA205	300	SR-34	300	1N3147
300	EA16X134	300	HN-0047	300	OF156	300	STB01-02	300	1N3197
300	EA16X146	300	HT-230	300	OF162	300	STB576	300	1N3206
300	EA16X171	300	HV-23	300	P-6006	300	SV-04	300	1N3207
300	EA16X20	300	HV-230(BL)	300	P1172	300	SV-3	300	1N3223
300	EA16X39	300	HV-25	300	P1172-1	300	SV-3A	300	1N3257
300	EA16X49	300	HV-25(DIODE)	300	P4326	300	SV-3B	300	1N3258
300	EA16X60	300	HV-25(RCA)	300	P7394	300	SV-3C	300	1N331
300	EA16X61	300	HV-27	300	PD137	300	SV-8	300	1N3471
300	EA16X68	300	HV00001050	300	P8700	300	SV9	300	1N350
300	EA16X69	300	HV0000206	300	P8720	300	SVDVD1121	300	1N351
300	EA16X75	300	HV0000502	300	PT4-2287-01	300	SX780	300	1N352
300	EA16X75	300	HV23	300	PTC214M	300	T-E1098	300	1N3550
300	EA16X84	300	HV23(DIODE)	300	Q-21115C	300	T-E1118	300	1N3575
300	EA2607	300	HV25(HITACHI)	300	Q-24115C	300	T-E1119	300	1N3576
300	EA3447	300	HV460R	300	QD-818953XA	300	T-E1121	300	1N3577
300	EC0177	300	HV80	300	QD-SMA150XN	300	T12-1	300	1N3593
300	ED21(ELCOM)	300	I85436	300	QD-SS1555XT	300	T12-2	300	1N3594
300	ED31(ELCOM)	300	I964(6	300	QD-SS853XXA	300	T12A	300	1N3595
300	ED32(ELCOM)	300	IC743051	300	R-7026	300	T12B	300	1N3596
300	ED514721	300	INJ61433	300	R-7027	300	T12C	300	1N3598
300	ED515790	300	INJ61677	300	R-7092	300	T151	300	1N3599
300	ED516420	300	INJ61725	300	R8022	300	T152	300	1N3601
300	ED536062	300	IP20-0021	300	R8023	300	T153	300	1N3602
300	ED560913	300	IP20-0023	300	R8314	300	T154	300	1N3603
300	ED8-0001	300	IP20-0061	300	R81A	300	T155	300	1N3604
300	ED8-0014	300	IP20-0145	300	RD1343	300	T16	300	1N3605
300	ED8-0024	300	IP20-0184	300	RE-52	300	T21600	300	1N3606
300	ED8-1	300	IP20-0216	300	RE52	300	TD-15-BL	300	1N3607
300	ED8-25	300	IP20-0282	300	RBJ7I1253	300	TD81515	300	1N3608
300	EH16X20	300	ITT413	300	RF34661	300	TD81518	300	1N3609
300	EP16X15	300	ITT7215	300	RH-DX0046CEZZ	300	TD96016-1M	300	1N3625
300	EP16X20	300	ITT73	300	RH-DX0054CEZZ	300	TE-1029	300	1N3653
300	EP16X22	300	ITT921	300	RHDX0033TAZZ	300	TF34	300	1N3654
300	EP16X23	300	J241182	300	R8-2530	300	TF44	300	1N3666
300	EP16X27	300	J241213	300	R8-2538	300	TF44J	300	1N3668
300	EP16X30	300	J241234	300	R8-2678	300	T-51	300	1N3722
300	EP16X4	300	J241235	300	R8-272U8	300	TI-UG-1888	300	1N380
300	EP2798	300	J241242	300	R8-2803	300	TI-UG1888	300	1N381
300	EP6X10	300	J24755	300	R81428	300	T151	300	1N382
300	ES15057	300	J24912	300	RT-218	300	TR2083-42	300	1N383
300	ES1627	300	JL-40A	300	RT2061(G.E.)	300	TR2083-44	300	1N386
300	ES16X23	300	JL40A	300	RT218	300	TR237011	300	1N3864
300	ES16X24	300	JM-40	300	RT5217	300	T8C136	300	1N3872
300	ES16X27	300	JM40	300	RT5554	300	TV24554	300	1N3873
300	ES16X30	300	JM401	300	RT5909	300	TVS-181211	300	1N389
300	ES16X32	300	KO1208A	300		300	TV8-0A81	300	1N390
300	ES16X4	300	K119J804-5	300		300	TV8-0A95	300	
300	ES16X40								

GE	Industry Standard No.
300	1N391
300	1N392
300	1N393
300	1N394
300	1N3953
300	1N3954
300	1N3956
300	1N4009
300	1N4043
300	1N4048
300	1N4087
300	1N4147
300	1N414A
300	1N4150
300	1N4151
300	1N4152
300	1N4156
300	1N4157
300	1N4242
300	1N4243
300	1N4306
300	1N4307
300	1N4308
300	1N4309
300	1N431
300	1N4311
300	1N4315
300	1N4318
300	1N432
300	1N4322
300	1N432A
300	1N432B
300	1N433
300	1N433A
300	1N433B
300	1N434
300	1N434A
300	1N434B
300	1N4365
300	1N4375
300	1N4376
300	1N4382
300	1N4392
300	1N4395
300	1N4395A
300	1N4444
300	1N4450
300	1N4451
300	1N4453
300	1N4455
300	1N4531
300	1N4532
300	1N4533
300	1N4534
300	1N4536
300	1N4547
300	1N4548
300	1N456
300	1N456A
300	1N457
300	1N457A
300	1N457M
300	1N458
300	1N458A
300	1N458M
300	1N459
300	1N459A
300	1N459M
300	1N460
300	1N4606
300	1N4607
300	1N4608
300	1N460A
300	1N460B
300	1N461
300	1N4610
300	1N461A
300	1N462
300	1N462A
300	1N463
300	1N463A
300	1N464
300	1N464A
300	1N4726
300	1N4727
300	1N4827
300	1N4828
300	1N4829
300	1N482A
300	1N482B
300	1N482C
300	1N483
300	1N4830
300	1N483A
300	1N483AM
300	1N483BM
300	1N483C
300	1N484
300	1N484A
300	1N484C
300	1N485
300	1N485A
300	1N485B
300	1N485C
300	1N486
300	1N4861
300	1N4862
300	1N4863
300	1N4864
300	1N486A
300	1N486B
300	1N4938
300	1N4949
300	1N4950
300	1N5062(SEARS)
300	1N5194
300	1N5195
300	1N5196
300	1N5208
300	1N5209
300	1N5210
300	1N5219
300	1N5220
300	1N5282
300	1N5315
300	1N5316
300	1N5317
300	1N5318
300	1N5319
300	1N5320
300	1N5426
300	1N5605
300	1N5606
300	1N5607
300	1N5610
300	1N568
300	1N5711
300	1N5712
300	1N5713
300	1N5719
300	1N5767
300	1N619
300	1N622
300	1N625
300	1N625A
300	1N625M
300	1N626
300	1N626A
300	1N626M
300	1N627
300	1N627A
300	1N628
300	1N628A
300	1N629
300	1N629A
300	1N633
300	1N643
300	1N643A
300	1N658
300	1N658A
300	1N658M
300	1N659
300	1N659/A
300	1N659A
300	1N660
300	1N660A
300	1N661
300	1N661A
300	1N662
300	1N662A
300	1N663
300	1N663A
300	1N663M
300	1N690
300	1N696
300	1N697
300	1N7098Q
300	1N760
300	1N778
300	1N779
300	1N788
300	1N789
300	1N789M
300	1N790
300	1N790M
300	1N791
300	1N791M
300	1N792
300	1N792M
300	1N793
300	1N793M
300	1N794
300	1N795
300	1N796
300	1N797
300	1N798
300	1N799
300	1N800
300	1N801
300	1N802
300	1N803
300	1N804
300	1N806
300	1N807
300	1N808
300	1N809
300	1N810
300	1N811
300	1N811M
300	1N812
300	1N812M
300	1N813
300	1N813M
300	1N814
300	1N814M
300	1N815
300	1N815M
300	1N817
300	1N818
300	1N818M
300	1N837
300	1N837A
300	1N838
300	1N839
300	1N840
300	1N840M
300	1N841
300	1N842
300	1N843
300	1N844
300	1N845
300	1N890
300	1N891
300	1N892
300	1N897
300	1N898
300	1N899
300	1N900
300	1N901
300	1N902
300	1N903
300	1N903/A
300	1N903A
300	1N903AM
300	1N903M
300	1N904
300	1N904/A
300	1N904A
300	1N904AM
300	1N904M
300	1N905
300	1N905/A
300	1N905A
300	1N905AM
300	1N905M
300	1N906
300	1N906/A
300	1N906A
300	1N906AM
300	1N906M
300	1N907
300	1N907/A
300	1N907A
300	1N907AM
300	1N907M
300	1N908
300	1N908/A
300	1N908A
300	1N908AM
300	1N908M
300	1N915
300	1N916A
300	1N916B
300	1N917
300	1N919
300	1N920
300	1N921
300	1N922
300	1N924
300	1N925
300	1N926
300	1N927
300	1N928
300	1N930
300	1N931
300	1N932
300	1N933
300	1N934
300	1N948
300	1N993
300	1N995
300	1N997
300	1N999
300	1NJ61433
300	1NJ61677
300	1NJ61725
300	1NJ70980
300	1NJ71224
300	1R0
300	1R0E
300	1R0D3K
300	1R2
300	1R3A
300	1R4
300	1S-1555-Z
300	1S-1555V
300	1S-180
300	1S-2144Z
300	1S-2472
300	1S1001
300	1S1052
300	1S1053
300	1S1155
300	1S11941
300	1S1201
300	1S1210
300	1S1212
300	1S1212A
300	1S1213
300	1S1214
300	1S1215
300	1S1216
300	1S1217
300	1S1218
300	1S1218GR
300	1S1219
300	1S1220
300	1S130
300	1S1302
300	1S1303
300	1S1305
300	1S131
300	1S132
300	1S1420H
300	1S144
300	1S1473
300	1S1514
300	1S1515
300	1S1516
300	1S1532
300	1S1534
300	1S1545
300	1S155
300	1S155-1
300	1S1553
300	1S1554
300	0000I1S1555
300	1S1555-1
300	1S1555-8
300	1S1555-Z
300	1S1555PA1
300	1S1555V
300	1S1555Z
300	1S1580
300	1S1585
300	1S1586
300	1S1587
300	1S1589
300	1S1621
300	1S1621-O
300	1S1621-R
300	1S1621-Y
300	1S1650
300	1S1651
300	1S1690
300	1S180
300	1S180B
300	1S181
300	1S181FA
300	1S182
300	1S1825
300	1S1992
300	1S1993
300	1S1994
300	1S1995
300	1S2074H
300	1S20751C
300	1S2076
300	1S2076A-07
300	1S2091
300	1S2091(DIODE)
300	1S2091-BK
300	1S2091-BL
300	1S2091-W
300	1S2091BK
300	1S2091BL
300	1S2091W
300	1S2092
300	1S2097
300	1S2098
300	1S2144
300	1S2144A
300	1S2144Z
300	1S2186
300	1S2186GR
300	1S2276
300	1S2460
300	1S2461
300	1S2462
300	1S2597
300	1S2788
300	1S2788B
300	1S306
300	1S307
300	1S3076
300	1S322
300	1S444
300	1S555
300	1S560
300	1S642
300	1S693
300	1S84
300	1S89
300	1S90R
300	1S920
300	1S921
300	1S922(DIODE)
300	1S923
300	1S941
300	1S942
300	1S951
300	1S952
300	1S955
300	1S983
300	0000I1S990Q
300	1S990-AM
300	1S990A
300	1S990B(DIODE)
300	1S994A
300	1S99A
300	1SC2367
300	1S853
300	1S881
300	1T243
300	1T40
300	1X9179
300	1X9805
300	1X9809
300	2-1X60
300	2D165
300	2P2150M
300	2810032
300#(2)	SVDMA26-2
300(2)	ALC-1A
300(2)	C-10-20A
300(2)	C10-20A
300(2)	C10-22C
300(2)	C1A
300(2)	CD860011
300(2)	D3G
300(2)	DIJ71895-1
300(2)	E1176R
300(2)	EP16X1
300(2)	G01-803-A
300(2)	I8-446D
300(2)	M8569
300(2)	PTC215
300(2)	PTC406
300(2)	RF32426-7
300(2)	31B02-CR1
300(2)	SD-7(PHILCO)
300(2)	8D7
300(2)	31B02-CR1
300(2)	SR-13
300(2)	TR-C72(DIODE)
300(2)	TVM-530
300(2)	TVM-K-112C
300(2)	TVM554A
300(2)	1N1842
300(2)	1N201
300(2)	1N206
300(2)	1N209
300(2)	1N353
300(2)	1N379
300(2)	1N384
300(2)	1N385
300(2)	1N387
300(2)	1N388
300(2)	1N4092
300(2)	1N4093
300(2)	1N4389
300(2)	1N4951
300(2)	1N4952
300(2)	1N5179
300(2)	1N929
300(2)	1S1579
300(2)	1S1701
327	GE-327
327	2SC1789
332	GE-332
332	2SC2043
334	GE-334
334	SK3114
334	2N4093
334	2SA838
334	2SA838C
335	PTC115
338	2SA838B
400	REN177
503A	8115
504	153B
504	SD-2
504	SR1K1
504-A	A109
504-A	ECR-600-2
504A	.5B05
504A	.5B1
504A	.5B2
504A	.5B3
504A	.5B4
504A	.5B5
504A	.5B6
504A	.5J05
504A	.5J1
504A	.5J2
504A	.5J3
504A	.5J4
504A	.5J5
504A	.5J6
504A	.7B05
504A	.7B1
504A	.7B2
504A	.7B3
504A	.7B4
504A	.7B5
504A	.7B6
504A	.7J05
504A	.7J1
504A	.7J2
504A	.7J3
504A	.7J5
504A	.7J6
504A	A-04
504A	A-04049-B
504A	A-04091-A
504A	A-04092
504A	A-04092-B
504A	A-04093
504A	A-04093A
504A	A-04212-A
504A	A-04212-B
504A	A-04226
504A	A-042313
504A	A-04242
504A	A-04901A
504A	A-1.5-01
504A	A-100(RECT.)
504A	A-10105
504A	A-10113
504A	A-10118
504A	A-1946
504A	A-95-5281
504A	A-95-5289
504A	AO3
504A	AO375
504A	AO377
504A	AO4
504A	A04049
504A	A04049B
504A	A04091A
504A	A04092
504A	A04092A
504A	A04092B
504A	A04093
504A	A04093A
504A	A04210A
504A	A04212-B
504A	A04212A
504A	A04212B
504A	A04230-A
504A	A042313
504A	A04293
504A	A04241-A
504A	A04242
504A	A04331-021
504A	A04331-023
504A	A04351-043
504A	A04350-022
504A	A04731
504A	A04901A
504A	A0491A
504A	A05
504A	A054-150
504A	A054-230
504A	A059-114
504A	A06
504A	A061-118
504A	A065-101
504A	A066-124
504A	A068-100
504A	A068-102
504A	A069-112
504A	A100
504A	A100(RECTIFIER)
504A	A101-A
504A	A101-A(RECT.)
504A	A10105
504A	A10113
504A	A10118
504A	A10142
504A	A10164
504A	A10165
504A	A10169
504A	A10C
504A	A10D
504A	A10E
504A	A10M
504A	A10N
504A	A11
504A	A123-7
504A	A13
504A	A13(RECTIFIER)
504A	A132
504A	A132-1
504A	A13A1
504A	A13A2
504A	A13AA2
504A	A13B1
504A	A13B2
504A	A13D1
504A	A13D2
504A	A13E2
504A	A13F2
504A	A14B
504A	A14C
504A	A14D
504A	A14E
504A	A14E2
504A	A14M
504A	A15-1008
504A	A1946
504A	A1A1
504A	A1A5
504A	A1A9
504A	A1B1
504A	A1B5
504A	A1B9
504A	A1C1
504A	A1C5
504A	A1C9
504A	A1D1
504A	A1D5
504A	A1D9
504A	A1E1
504A	A1E5
504A	A1E9
504A	A1F1
504A	A1F5
504A	A1F9
504A	A101
504A	A105
504A	A200
504A	A23
504A	A2421
504A	A2422
504A	A2460
504A	A2461
504A	A2462
504A	A2481
504A	A2485
504A	A25-1008
504A	A2A1
504A	A2A4
504A	A2A5
504A	A2A9
504A	A2B1
504A	A2B4
504A	A2B5

GE	Industry Standard No.	GE	Industry Standard No.	GE	Industry Standard No.	GE	Industry Standard No.	GE	Industry Standard No.
504A	A2B9	504A	A909-1018	504A	B1F1	504A	BA153	504A	CC102MA
504A	A2C1	504A	A909-1019	504A	B1F5	504A	BAX12	504A	CD-2R
504A	A2C4	504A	A95-5281	504A	B1G1	504A	BAY44	504A	CD-4
504A	A2C5	504A	A95-5289	504A	B1G5	504A	BAY64	504A	CD-8457
504A	A209	504A	AA100	504A	B1G9	504A	BAY86	504A	CD-860037
504A	A2D1	504A	AA200	504A	B1H(DIODE)	504A	BAY87	504A	CD5
504A	A2D4	504A	AA300	504A	B1H9	504A	BB-109	504A	CD1111
504A	A2D5	504A	AA400	504A	B200C40	504A	BB-68	504A	CD1112
504A	A2D9	504A	AA50	504A	B250C100	504A	BB107	504A	CD1113
504A	A2E1	504A	AA500	504A	B250C100TD	504A	BB117	504A	CD1114
504A	A2E4	504A	AA600	504A	B250C125	504A	BB127	504A	CD1115
504A	A2E5	504A	AA1-22	504A	B250C125K4	504A	BB1A	504A	CD1116
504A	A2E9	504A	AA218D	504A	B250C125N2	504A	BB2A184	504A	CD1121
504A	A2F1	504A	ACR810-108	504A	B250C125X4	504A	BBIA	504A	CD1122
504A	A2F4	504A	ACR83-1008	504A	B250C150	504A	BC-207	504A	CD1123
504A	A2F5	504A	AD-1UF	504A	B250C150K4	504A	BC-307	504A	CD1124
504A	A2F9	504A	AD10	504A	B250C75	504A	BCP5	504A	CD1125
504A	A2G1	504A	AD100	504A	B250C75K4	504A	BD-107	504A	CD1126
504A	A2G4	504A	AD200	504A	B250C75K41	504A	BD-107(RECT.)	504A	CD1127
504A	A2G5	504A	AD4001	504A	B250C75K45	504A	BD-127	504A	CD1141
504A	A209	504A	AD50	504A	B250C75K5	504A	BD-1A	504A	CD1142
504A	A300	504A	AE1A	504A	B250C75K4S	504A	BDOA	504A	CD1143
504A	A3A1	504A	AG100D	504A	B294	504A	BD117	504A	CD1147
504A	A3A3	504A	AG100G	504A	B294(RECTIFIER)	504A	BD3A-184	504A	CD1148
504A	A3A5	504A	AG100J	504A	B2A1	504A	BE107	504A	CD1149
504A	A3A9	504A	AJ-30	504A	B2A5	504A	BE117	504A	CD1151
504A	A3B1	504A	AJ10	504A	B2A9	504A	BE127	504A	CD13532
504A	A3B3	504A	AJ15	504A	B2B1	504A	BF26235-5	504A	CD13533
504A	A3B5	504A	AJ20	504A	B2B5	504A	BFY87A	504A	CD13534
504A	A3B9	504A	AJ25	504A	B2B9	504A	BH4R1	504A	CD13535
504A	A3C1	504A	AJ30	504A	B2C1	504A	BR42	504A	CD13536
504A	A3C3	504A	AJ35	504A	B2C5	504A	BR44	504A	CD13537
504A	A3C5	504A	AJ40	504A	B209	504A	BR46	504A	CD13538
504A	A3C9	504A	AJ5	504A	B2D1	504A	BR47	504A	CD13539
504A	A3D1	504A	AJ50	504A	B2D5	504A	BR48	504A	CD17A2
504A	A3D3	504A	AJ60	504A	B2D9	504A	BR51400-1	504A	CD4
504A	A3D5	504A	AM-010	504A	B2E1	504A	BR51401-2	504A	CD0005
504A	A3D9	504A	AM-020	504A	B2E5	504A	BR52	504A	CDK-4
504A	A3E1	504A	AM-030	504A	B2E9	504A	BS-1	504A	CDR-2
504A	A3E3	504A	AM-035	504A	B2F1	504A	BS1	504A	CDR-4
504A	A3E5	504A	AM-060	504A	B2F5	504A	BS2	504A	CDR-16B
504A	A3E9	504A	AM-22	504A	B2F9	504A	BTM50	504A	CE0398/7839
504A	A3F1	504A	AM-33	504A	B2G1	504A	BU029	504A	CE502
504A	A3F3	504A	AM-6-5	504A	B2G5	504A	BV25	504A	CE504
504A	A3F5	504A	AM-G-10	504A	B2G9	504A	BY101	504A	CE506
504A	A3F9	504A	AM-G-22	504A	B3001000	504A	BY102	504A	CE0605O
504A	A3G1	504A	AM-G-5	504A	B300250	504A	BY106	504A	CER-69
504A	A3G3	504A	AM-G22	504A	B300250-1	504A	BY107	504A	CER-71
504A	A3G5	504A	AM005	504A	B300350-1	504A	BY111	504A	CER-71CA
504A	A3G9	504A	AM010	504A	B300500	504A	BY112	504A	CER500
504A	A400	504A	AM020	504A	B300600	504A	BY113	504A	CER500A
504A	A400(RECTIFIER)	504A	AM025	504A	B300600CB	504A	BY114	504A	CER500B
504A	A4212-A	504A	AM030	504A	B31	504A	BY115	504A	CER500C
504A	A422-A	504A	AM035	504A	B31(RECTIFIER)	504A	BY116	504A	CER67
504A	A42X00269-01	504A	AM040	504A	B350C600	504A	BY117	504A	CER670
504A	A42X00374-01	504A	AM050	504A	B36564	504A	BY121	504A	CER670A
504A	A4A1	504A	AM060	504A	B3A1	504A	BY122	504A	CER670B
504A	A4A5	504A	AM405	504A	B3A5	504A	BY123	504A	CER670C
504A	A4A9	504A	AM410	504A	B3A9	504A	BY124	504A	CER67A
504A	A4B1	504A	AM415	504A	B3B1	504A	BY125	504A	CER67B
504A	A4B5	504A	AM420	504A	B3B5	504A	BY126	504A	CER67C
504A	A4B9	504A	AM425	504A	B3B9	504A	BY127	504A	CER68
504A	A4C1	504A	AM430	504A	B3C1	504A	BY130	504A	CER680
504A	A4C5	504A	AM435	504A	B3C5	504A	BY134	504A	CER680A
504A	A4C9	504A	AM440	504A	B3C9	504A	BY135	504A	CER680B
504A	A4D1	504A	AM445	504A	B3D1	504A	BY141	504A	CER680C
504A	A4D5	504A	AM450	504A	B3D5	504A	BY153	504A	CER68A
504A	A4D9	504A	AM460	504A	B3E1	504A	BY164	504A	CER68B
504A	A4E1	504A	AM62	504A	B3E5	504A	BYX22/200	504A	CER68C
504A	A4E5	504A	AM65	504A	B3E9	504A	BYX22/400	504A	CER69
504A	A4E9	504A	AM66	504A	B3F1	504A	BYX22/600	504A	CER690
504A	A4F1	504A	AN-1	504A	B3F5	504A	BYX36-500	504A	CER690A
504A	A4F5	504A	AN-G-5B	504A	B3F9	504A	BYX36-600	504A	CER690B
504A	A4F9	504A	A04092A	504A	B3G1	504A	BYX36/150	504A	CER690C
504A	A4G1	504A	A04092A	504A	B3G5	504A	BYX36/300	504A	CER69A
504A	A4G5	504A	A04092B	504A	B3G9	504A	BYX36/600	504A	CER69B
504A	A4G9	504A	A04093	504A	B4A1	504A	BYX60-200	504A	CER69C
504A	A50	504A	A04093-A	504A	B4A5	504A	BYX60-300	504A	CER6B
504A	A50(RECTIFIER)	504A	A04093A	504A	B4A9	504A	BYX60-400	504A	CER70
504A	A500	504A	A04210A	504A	B4B1	504A	BYX60-50	504A	CER700
504A	A514-023626	504A	A04212-A	504A	B4B5	504A	BYX60-500	504A	CER700A
504A	A514-025607	504A	A04212-B	504A	B4B9	504A	BYX60-600	504A	CER700B
504A	A514-027757	504A	A04233	504A	B4C1	504A	BYY-31	504A	CER700C
504A	A514-028072	504A	A04716	504A	B4C5	504A	BYY-32	504A	CER70A
504A	A514-028073	504A	A04901A	504A	B4C9	504A	BYY-35	504A	CER70B
504A	A514-035903	504A	AQ2(PHILCO)	504A	B4D1	504A	BYY31	504A	CER70C
504A	A514-035596	504A	AQ3(PHILCO)	504A	B4D5	504A	BYY32	504A	CER71
504A	A5A1	504A	AR-22(DIO)	504A	B4D9	504A	BYY33	504A	CER710
504A	A5A2	504A	AR16	504A	B4E1	504A	BYY34	504A	CER710A
504A	A5A9	504A	AR17	504A	B4E5	504A	BYY35	504A	CER710B
504A	A5B1	504A	AR18	504A	B4E9	504A	BYY36	504A	CER710C
504A	A5B2	504A	AR19	504A	B4F1	504A	BYY89	504A	CER71A
504A	A5B5	504A	AR20	504A	B4F5	504A	C10110	504A	CER71B
504A	A5B9	504A	AR21	504A	B4F9	504A	C10159	504A	CER71C
504A	A5C1	504A	AR22	504A	B4G1	504A	C10176	504A	CP102DA
504A	A5C2	504A	AR22(RECTIFIER)	504A	B4G5	504A	C1181C1E1C	504A	CH119D
504A	A5C5	504A	AR7C	504A	B4G9	504A	C1B	504A	CLO10
504A	A5C9	504A	AR882	504A	B5A1	504A	C1H	504A	CLO25
504A	A5D1	504A	AS-14	504A	B5A5	504A	C21382	504A	CLO5
504A	A5D2	504A	AS-15	504A	B5A9	504A	C23H12B	504A	CL1
504A	A5D5	504A	AS-2	504A	B5B1	504A	C248507	504A	CL1.5
504A	A5D9	504A	AS-3	504A	B5B5	504A	C2AJ102	504A	CL2
504A	A5E1	504A	AS-4	504A	B5B9	504A	C83-829	504A	CL3
504A	A5E2	504A	AS-5	504A	B5C1	504A	C83-880	504A	CL4
504A	A5E5	504A	AS11	504A	B5C5	504A	C934(RECTIFIER)	504A	CL5
504A	A5E9	504A	AS14	504A	B5C9	504A	CA10	504A	CL6
504A	A5F1	504A	AS15	504A	B5D1	504A	CA100	504A	CL7
504A	A5F2	504A	AS2	504A	B5D5	504A	CA1020A	504A	CL8
504A	A5F5	504A	AS3	504A	B5D9	504A	CA102BA	504A	CLMO5
504A	A5F9	504A	AS4	504A	B5E1	504A	CA102DA	504A	CLM1
504A	A5G1	504A	AS5	504A	B5E5	504A	CA102PA	504A	CMS470
504A	A5G2	504A	AS6	504A	B5E9	504A	CA102HA	504A	CO49
504A	A5G5	504A	B-1501U	504A	B5F1	504A	CA102MA	504A	COD1-6045
504A	A5G9	504A	B-31	504A	B5F5	504A	CA150	504A	COD1-6046
504A	A600	504A	B01-02	504A	B5F9	504A	CA200	504A	COD1-6047
504A	A600(RECTIFIER)	504A	B0102	504A	B5G1	504A	CA250	504A	COD1-6048
504A	A692T5	504A	B12-02	504A	B5G5	504A	CA50	504A	COD1556
504A	A692T5-0	504A	B1A1	504A	B5G9	504A	CA6664	504A	COD1531
504A	A692X16-0	504A	B1A5	504A	B601-1012	504A	CA6665	504A	COD1532
504A	A7102001	504A	B1A9	504A	BA-100	504A	CA7248	504A	COD1533
504A	A75-68-500	504A	B1B	504A	BA-104	504A	CA8314	504A	COD1534
504A	A7509201	504A	B1B1	504A	BA-142-01	504A	CB10	504A	COD1535
504A	A7547053	504A	B1B5	504A	BA100	504A	CB150	504A	COD1536
504A	A7568250	504A	B1B9	504A	BA104	504A	CB20	504A	COD1551
504A	A7568500	504A	B1C1	504A	BA105	504A	CB200	504A	COD1552
504A	A7568700	504A	B1C5	504A	BA108	504A	CB250	504A	COD15524
504A	A7572100	504A	B1C9	504A	BA119	504A	CB5	504A	COD1553
504A	A7572200	504A	B1D1	504A	BA120	504A	CB50	504A	COD15534
504A	A7C	504A	B1D5	504A	BA127	504A	CC102BA	504A	COD1554
504A	A7D	504A	B1D9	504A	BA128	504A	CC102DA	504A	COD15544
504A	A7E	504A	B1E1	504A	BA129	504A	CC102PA	504A	COD1555
504A	A7G	504A	B1E5	504A	BA130	504A	CC102HA	504A	COD1556
504A		504A	B1E9	504A	BA142-01	504A	CC102KA	504A	COD1564
504A		504A		504A		504A		504A	COD1575

GE	Industry Standard No.
504A	COD16047
504A	COD1611
504A	COD1612
504A	COD1613
504A	COD1614
504A	COD1615
504A	COD1616
504A	CODI115524
504A	CODI11556
504A	CODI11531
504A	CODI15534
504A	CODI15544
504A	CODI15564
504A	CODI6045
504A	COD16047
504A	CP102BA
504A	CP102DA
504A	CP102FA
504A	CP102HA
504A	CP102KA
504A	CP102MA
504A	CR/E
504A	CR1034
504A	CR1035
504A	CR801
504A	CS-16E
504A	CS131D(AXIAL)
504A	CS16E
504A	CT-3003
504A	CT-3005
504A	CT-00
504A	CT200
504A	CT300
504A	CT3003
504A	CT3005
504A	CT600
504A	CTN200
504A	CTP-2001-1011
504A	CTP-2001-1012
504A	CX-0035
504A	CX-0037
504A	CX-0039
504A	CX-0040
504A	CX-0048
504A	CX-0054
504A	CX-9001
504A	CX0037
504A	CX0039
504A	CX0040
504A	CX0047
504A	CX0048
504A	CX0049
504A	CX9001
504A	CXL116
504A	CXL117
504A	CY40
504A	CY50
504A	D-00384R
504A	D-05
504A	D004
504A	D01-100
504A	D028
504A	D05
504A	D1-26
504A	D1-528
504A	D1-7
504A	D100
504A	D101167
504A	D10167
504A	D10168
504A	D11728
504A	D1201F
504A	D1445
504A	D1448
504A	D15A
504A	D15C
504A	D18
504A	D1E
504A	D1H
504A	D1J
504A	D1J70544
504A	D1L
504A	D2-1
504A	D200
504A	D21489
504A	D220
504A	D2201M
504A	D2201N
504A	D220M
504A	D227(DIODE)
504A	D25
504A	D25A
504A	D25B
504A	D25C
504A	D2600EF
504A	D28
504A	D2H(DIODE)
504A	D2J
504A	D2X4
504A	D300
504A	D3436
504A	D3R
504A	D3R38
504A	D3R39
504A	D3U
504A	D3V
504A	D3Z
504A	D400
504A	D400(RECT.)
504A	D45C
504A	D45CZ
504A	D48
504A	D4M
504A	D4R26
504A	D4R39
504A	D50
504A	D500
504A	D5G
504A	D5R35
504A	D5R39
504A	D600
504A	D65C
504A	D6623
504A	D6623A
504A	D6624
504A	D6624A
504A	D6625
504A	D6625A
504A	D68
504A	D6HZ
504A	D81K
504A	DA000
504A	DA001
504A	DA002
504A	DAAY002001
504A	DD-000
504A	DD-003
504A	DD-006
504A	DD-007
504A	DD003
504A	DD006
504A	DD007
504A	DD056
504A	DD175C
504A	DD176C
504A	DD177C
504A	DD2066
504A	DD2320
504A	DD2321
504A	DD236
504A	DD266
504A	DDAY002001
504A	DDAY002002
504A	DDAY102001
504A	DDBY002001
504A	DDE-201
504A	DE-201
504A	DB14
504A	DB14A
504A	DB16
504A	DB16A
504A	DB201
504A	D01PR
504A	DH-001
504A	DH-002
504A	DH14
504A	DH14A
504A	DH16
504A	DH16A
504A	DH4R2
504A	DI-1649
504A	DI-1728
504A	DI-428
504A	DI-46
504A	DI-528
504A	DI-55
504A	DI-56
504A	DI-645
504A	DI-646
504A	DI-647
504A	DI-648
504A	DI-649
504A	DI-7
504A	DI-705
504A	DI-71
504A	DI-728
504A	DI1649
504A	DI428
504A	DI46
504A	DI528
504A	DI56
504A	DIE
504A	DIJ
504A	DIJ70488
504A	DIJ70544
504A	DIJ70695
504A	DIJ71958
504A	DIJ71959
504A	DIJ72168
504A	DIL
504A	DIS-18
504A	DIS818
504A	DR1
504A	DR100
504A	DR1100
504A	DR1PR
504A	DR2
504A	DR200
504A	DR3
504A	DR300
504A	DR4
504A	DR400
504A	DR427
504A	DR435
504A	DR5
504A	DR500
504A	DR5101
504A	DR5102
504A	DR600
504A	DR668
504A	DR669
504A	DR670
504A	DR671
504A	DR695
504A	DR698
504A	DR699
504A	DR826
504A	DR848
504A	DR863
504A	DR81
504A	DRS101
504A	DRS102
504A	DRS104
504A	DS-1
504A	DS-10
504A	D8-0065
504A	D8-13
504A	D8-13(COURIER)
504A	D8-130
504A	D8-130B
504A	D8-130C
504A	D8-130E
504A	D8-130YB
504A	D8-130YE
504A	D8-131
504A	D8-131A
504A	D8-131B
504A	D8-132
504A	D8-132A
504A	D8-132B
504A	D8-13B(SANYO)
504A	D8-13B(SANYO)
504A	D8-14(DIODE)
504A	D8-16A
504A	D8-16B
504A	D8-16C(SANYO)
504A	D8-16C(SANYO)
504A	D8-16D
504A	D8-16D(SANYO)
504A	D8-16E
504A	D8-16E(SANYO)
504A	D8-16ND
504A	D8-16NE
504A	D8-16NY
504A	D8-16YA
504A	D8-17
504A	D8-17-6A
504A	D8-18
504A	D8-1M
504A	D8-1N
504A	D8-1P
504A	D8-1U
504A	D8-363
504A	D8-430
504A	D8-79
504A	D8-79(DELCO)
504A	D8-1M
504A	D80065
504A	DB1
504A	D8130
504A	D8130B
504A	D8130E
504A	D8130ND
504A	D8130Y
504A	D8130YC
504A	D8130YE
504A	D8131B
504A	D8132A
504A	D8132B
504A	D8160(G.E.)
504A	D816A
504A	D816C
504A	D816E
504A	D816N
504A	D816NB
504A	D816NC
504A	D816ND
504A	D816NE
504A	D817(ADMIRAL)
504A	D817N
504A	D818
504A	D818M
504A	D81K
504A	D81K7
504A	D81N
504A	D81P
504A	D82K
504A	D82N
504A	D82N22
504A	D838
504A	D838(CRAIG)
504A	D838(SANYO)
504A	D8430
504A	D858(SANYO)
504A	D879
504A	D899
504A	DU400
504A	DU600
504A	DX-0099
504A	DX-0445
504A	DX-0475
504A	DX-0721
504A	DX-1039
504A	DX-1131
504A	DX520
504A	E-0704W
504A	E-075L
504A	E-41
504A	EO3090-002
504A	EO3155-001
504A	EO3155-002
504A	EO704-W
504A	EO751
504A	EO788-C
504A	EO788C
504A	E1
504A	E1010
504A	E1011
504A	E10116
504A	E10157
504A	E10171
504A	E10172
504A	E1018N
504A	E102(ELCOM)
504A	E106(ELCOM)
504A	E1124
504A	E1146J
504A	E1146R
504A	E1153RB
504A	E1156RD
504A	E1157RNA
504A	E1250200
504A	E13-020-00
504A	E13-20-00
504A	E135
504A	E1410
504A	E1411
504A	E1412
504A	E1413
504A	E1415
504A	E143
504A	E1440
504A	E146
504A	E140350
504A	E150L
504A	E2
504A	E21
504A	E24100
504A	E2505
504A	E3
504A	E3006
504A	E300L
504A	E41
504A	E4676B
504A	E5
504A	E500L
504A	E6
504A	E650L
504A	E7441
504A	E750(ELCOM)
504A	E752(ELCOM)
504A	E756(ELCOM)
504A	EA0015
504A	EA0016
504A	EA0031
504A	EA005
504A	EA010
504A	EA020
504A	EA030
504A	EA040
504A	EA060
504A	EA1072
504A	EA1448
504A	EA15X14
504A	EA1672
504A	EA16X13
504A	EA16X149
504A	EA16X2
504A	EA16X21
504A	EA16X30
504A	EA16X33
504A	EA16X34
504A	EA16X55
504A	EA16X71
504A	EA16X8
504A	EA16X92
504A	EA2140
504A	EA2499
504A	EA2501
504A	EA2741
504A	EA3827
504A	EA3989
504A	EA5711
504A	EA57X1
504A	EA57X10
504A	EA57X11
504A	EA57X14
504A	EA57X3
504A	EA57X8
504A	EA75X1
504A	EC401
504A	EC402
504A	EC0116
504A	EC0117
504A	ECR600-2
504A	ED-4
504A	ED-5
504A	ED-6
504A	ED1804
504A	ED1892
504A	ED2106
504A	ED2107
504A	ED2108
504A	ED2109
504A	ED2110
504A	ED224548
504A	ED224550
504A	ED2842
504A	ED2843
504A	ED2844
504A	ED2845
504A	ED2846
504A	ED2914
504A	ED2915
504A	ED2916
504A	ED2917
504A	ED2918
504A	ED2919
504A	ED2920
504A	ED2921
504A	ED3000
504A	ED3000A
504A	ED3001
504A	ED3001A
504A	ED3001B
504A	ED3002
504A	ED3002A
504A	ED3002B
504A	ED3003
504A	ED3003A
504A	ED3003B
504A	ED3003B
504A	ED3004
504A	ED3004A
504A	ED3004B
504A	ED3005
504A	ED3005A
504A	ED3005B
504A	ED3006
504A	ED3006A
504A	ED3006B
504A	ED329128
504A	ED329130
504A	ED4
504A	ED4594583
504A	ED5
504A	ED511097
504A	ED6
504A	ED7
504A	ED-363
504A	ED8-0002
504A	ED8-0017
504A	ED8-11
504A	ED8-17
504A	ED8-24
504A	ED8-4
504A	EER600-2
504A	EG100
504A	EG100H
504A	EH2011/7+12/1
504A	EM-1171
504A	EM1021
504A	EMIJ2
504A	EM401
504A	EM402
504A	EM403
504A	EM404
504A	EM405
504A	EM406
504A	EM407
504A	EM408
504A	EM410
504A	EM501
504A	EM502
504A	EM503
504A	EM504
504A	EM505
504A	EM506
504A	EO704
504A	EP-1428-2H
504A	EP-5619-2/7628
504A	EP1259
504A	EP1259-2
504A	EP1428-2H
504A	EP16X13
504A	EP200
504A	EP3149
504A	EP400
504A	EP57X1
504A	EP57X12
504A	EP57X4
504A	EP5X5
504A	EP600
504A	ER
504A	ER101
504A	ER102
504A	ER103D
504A	ER103B
504A	ER104D
504A	ER105D
504A	ER106D
504A	ER11
504A	ER12
504A	ER181
504A	ER182
504A	ER183
504A	ER184
504A	ER185
504A	ER1B22-15
504A	ER2
504A	ER201
504A	ER21
504A	ER22
504A	ER301
504A	ER31
504A	ER381
504A	ER401
504A	ER41
504A	ER410
504A	ER42
504A	ER501
504A	ER51
504A	ER510
504A	ER57X2
504A	ER57X3
504A	ER57X4
504A	ER601
504A	ER61
504A	ER62
504A	ERB11-01
504A	ERB12-02
504A	ERB12-11
504A	ERB22-15
504A	ERB24
504A	ERB24(GE)
504A	ERB24-02B
504A	ERB24-06
504A	ERB24-06A
504A	ERB28-04
504A	ERB28-04D
504A	ERC24-04
504A	ERD500
504A	ERD400
504A	ERV-02P2150
504A	ES10233
504A	ES15056
504A	ES16X10
504A	ES16X13
504A	ES16X25
504A	ES16X38
504A	ES16X9
504A	ES47X1
504A	ES51
504A	ES57X1
504A	ES57X2
504A	ES57X4
504A	ES57X5
504A	ESA-06
504A	ESA-10C
504A	ESA-10N
504A	ESA06
504A	ESK-1
504A	ESK1
504A	ESK1/06
504A	ESKE40C500
504A	ET16X16
504A	ET16X16X
504A	ET200
504A	ET400
504A	ET51X25
504A	ET52X25
504A	ET55-25
504A	ET55X23
504A	ET55X29
504A	ET57X25
504A	ET57X29
504A	ET57X30
504A	ET57X33
504A	ET57X35
504A	ET57X38
504A	ET6
504A	ETD-10D1
504A	ETD-10D2
504A	ETD-VO6C
504A	EU16X7
504A	EU57X30
504A	EU57X40
504A	EVM511
504A	EX76-X
504A	ETV-420D1R5JA
504A	F-05
504A	F-14A
504A	F-14C
504A	F05
504A	F1
504A	F10124
504A	F10148
504A	F10180
504A	F14
504A	F14-C
504A	F14A
504A	F14B
504A	F14D
504A	0F160
504A	0F164
504A	F2
504A	F20-1015
504A	F20-1016
504A	F215-1016
504A	F215-1017
504A	F3
504A	F4
504A	F5
504A	F6
504A	FA4
504A	FA6
504A	FD-1029-DP
504A	FD-1029-DG
504A	FD1599
504A	FD3
504A	FD300
504A	FD333
504A	FD3389
504A	FD6
504A	FDH400
504A	FG-2NA
504A	FM1J2
504A	FM211N
504A	F05
504A	FPR50-1011
504A	FR-1

GE	Industry Standard No.	GE	Industry Standard No.	GE	Industry Standard No.	GE	Industry Standard No.	GE	Industry Standard No.
504A	FR-1B	504A	H100	504A	HR10	504A	K1.3G22A	504A	MO027
504A	FR-1B(M)	504A	H10174	504A	HR11	504A	K1A5	504A	M12
504A	FR-1HM	504A	H200	504A	HR13	504A	K1B5	504A	M124J7779-1
504A	FR-1M	504A	H20052	504A	HR5A	504A	K1O5	504A	M14
504A	FR-1MD	504A	H300	504A	HR5A8E	504A	K1D5	504A	M150
504A	FR-1N	504A	H400	504A	HR5B	504A	K1E5	504A	M172A
504A	FR-1P	504A	H475	504A	HS1001	504A	K1F5	504A	M1H
504A	FR-1U	504A	H50	504A	HS1002	504A	K1G5	504A	M22
504A	FR-2	504A	H500	504A	HS1003	504A	K1H5	504A	M2497
504A	FR-202	504A	H585	504A	HS1007	504A	K2O0	504A	M5016
504A	FR-2M	504A	H600	504A	HS1008	504A	K2A5	504A	M41223-2
504A	FR-2P	504A	H616	504A	HS1009	504A	K2B5	504A	M42
504A	FR-U	504A	H617	504A	HS1010	504A	K2C5	504A	M4736
504A	FR1	504A	H618	504A	HS1011	504A	K2D5	504A	M4HZ
504A	FR10	504A	H619	504A	HS1012	504A	K2E5	504A	M500
504A	FR1M	504A	H620	504A	HS1020	504A	K2F5	504A	M500A
504A	FR1MB	504A	H625	504A	HS3103	504A	K2G	504A	M500B
504A	FR1MD	504A	H626	504A	HS3104	504A	K2G5	504A	M500C
504A	FR1P	504A	H7126-3	504A	HSPD-1A	504A	K3B5	504A	M604
504A	FR1U	504A	H781	504A	HSPD-1Z	504A	K3C5	504A	M604HT
504A	FR2	504A	H783	504A	HS8000710	504A	K3D5	504A	M62
504A	FR2(S1B1)	504A	H881	504A	HV-100	504A	K3F5	504A	M67
504A	FR2-02	504A	H890	504A	HV-23(ELGIN)	504A	K3G5	504A	M670
504A	FR2-02(DIO)	504A	H891	504A	HV-26	504A	K4-555	504A	M670A
504A	FR2-020	504A	HA100	504A	HV-26G	504A	K4-557	504A	M670B
504A	FR2-06	504A	HA200	504A	HV-46	504A	K4A5	504A	M670C
504A	FR2-0Z	504A	HA300	504A	HV-46GR	504A	K4B5	504A	M67A
504A	0000FR202	504A	HA400	504A	HV0000105	504A	K4C5	504A	M67B
504A	FR2M	504A	HA50	504A	HV0000105-0	504A	K4D5	504A	M67C
504A	FR2P	504A	HA500	504A	HV0000406	504A	K4E5	504A	M680
504A	FRE-101	504A	HA600	504A	HV0000705	504A	K4F5	504A	M680A
504A	FRH101	504A	HAR10	504A	HV26	504A	K4O5	504A	M680B
504A	0000000FRI	504A	HAR15	504A	HV46	504A	K5A5	504A	M680C
504A	FSP-288-1	504A	HAR20	504A	HV46#OR#(DIODE)	504A	K5B5	504A	M68A
504A	FSB2	504A	HB2	504A	HV46(DIODE)	504A	K5C5	504A	M68B
504A	FST3	504A	HB3	504A	HV46GR	504A	K5D5	504A	M68C
504A	FT-1	504A	HC-30	504A	HV46GR(DIODE)	504A	K5E5	504A	M69
504A	FT-10	504A	HC-39	504A	IC743048	504A	K5F5	504A	M690
504A	FT-1N	504A	HC-68	504A	IM4004	504A	K5G5	504A	M690A
504A	FTI	504A	HC500	504A	INJ61434	504A	K8952799	504A	M690B
504A	FT1(SHARP)	504A	HC67	504A	INJ61726	504A	K8953058-1	504A	M690C
504A	FT14A	504A	HC670	504A	IP20-0022	504A	KB-182	504A	M69A
504A	FT1N	504A	HC68	504A	IP20-0024	504A	KB262	504A	M69B
504A	FU-1MA	504A	HC680	504A	IP20-0025	504A	KB265A(RECT)	504A	M69C
504A	FU-1N	504A	HC700	504A	IP20-0054	504A	KC0-0691L/8	504A	M68HZ
504A	FU-1NA	504A	HC71	504A	IP20-0120	504A	KC0-80P	504A	M70
504A	FU-IM	504A	HC710	504A	IP20-0167	504A	KC0.8	504A	M700
504A	FU1M	504A	HC80	504A	IR10E6J	504A	KC0.8P	504A	M700A
504A	FU1N	504A	HCV	504A	IR1D	504A	KC0.8P11	504A	M700B
504A	FU1U	504A	HD-1	504A	IR20	504A	KC0.8CP11/1	504A	M700C
504A	FW100	504A	HD-2000308	504A	IR2A	504A	KC0.8CP12/1	504A	M701B
504A	FW200	504A	HD-3000301	504A	IR2E	504A	KC06911	504A	M70A
504A	FW400	504A	HD20-003-01	504A	IT10D4K	504A	KC06E11/8	504A	M70B
504A	FW500	504A	HD2000	504A	ITT350	504A	KC08C11/10	504A	M70C
504A	FW600	504A	HD2000-301	504A	ITT402	504A	KC08C11/8	504A	M71
504A	FW600A	504A	HD2000-703	504A	ITT992	504A	KC08C1110	504A	M710
504A	FWL100	504A	HD20000903	504A	J-05	504A	KC08C21/5	504A	M710A
504A	FWL200	504A	HD2000110	504A	J-1	504A	KC08C215	504A	M710B
504A	FWL300	504A	HD2000110-0	504A	J-2	504A	KC08C22/19	504A	M710C
504A	G00-502-A	504A	HD20001100	504A	J-4	504A	KC08C221	504A	M71A
504A	G00-534-A	504A	HD2000301	504A	J-6	504A	KC09911/8	504A	M71B
504A	G00-535-B	504A	HD2000301-0	504A	J100	504A	KC1.3C	504A	M71C
504A	G00-535A	504A	HD20003010	504A	J101183	504A	KC1.3C22/1	504A	M8222
504A	G00-535-A	504A	HD2000305	504A	J2-4570	504A	KC1.3C3X11/1	504A	M8399
504A	G00-536A	504A	HD2000307	504A	J20437	504A	KC1.3C12/1X2	504A	M91A01
504A	G00-543-A	504A	HD2000308	504A	J241100	504A	KC1.3C22/12	504A	M91A02
504A	G00-551-A	504A	HD2000413	504A	J241102	504A	KC1.3C22/1A	504A	M91A03
504A	G00502A	504A	HD2000501	504A	J241142	504A	KC13C22/1	504A	M9206
504A	G00535	504A	HD2000510	504A	J241209	504A	KC13C221	504A	M9235
504A	G00535A	504A	HD2000703	504A	J241210	504A	KC2BP22/1B	504A	M9312
504A	G01	504A	HD20007030	504A	J241211	504A	KC2B22/1	504A	M9314
504A	G01211	504A	HD2000903	504A	J241212	504A	KC2D221	504A	M9317
504A	G0Q-535-B	504A	HD2001310	504A	J241214	504A	KC2DP	504A	M9319
504A	G1	504A	HD2002207	504A	J241232	504A	KC2DP11/1	504A	MA-26-1
504A	G100A	504A	HD2003501	504A	J241260	504A	KC2DP12/1N	504A	MA100
504A	G100B	504A	HD6147	504A	J241271	504A	KC2DP12/2	504A	MA102
504A	G100D	504A	HD6865	504A	J241271	504A	KC2DP121N	504A	MA110
504A	G100G	504A	HE-20011	504A	J24630	504A	KC2DP122	504A	MA2
504A	G100J	504A	HE-8D1	504A	J24645	504A	KC2DP221B	504A	MA203
504A	G10119	504A	HEP154	504A	J24647	504A	KC2911/1	504A	MA211
504A	G1242	504A	HEP156	504A	J24756	504A	KC2911/1&12/1	504A	MA215
504A	G130	504A	HEP157	504A	J24871	504A	KC2911/1+12/1	504A	MA242
504A	G2	504A	HEP158	504A	J24877	504A	KC2912/1	504A	MA242C
504A	G222	504A	HEP161	504A	J24919	504A	KC0-80P	504A	MA350
504A	G296	504A	HEP162	504A	J24920	504A	KD203	504A	MA351
504A	G2A	504A	HEPR0050	504A	J24935	504A	KD2104	504A	MA50
504A	G3	504A	HEPR0051	504A	J24939	504A	KB-262	504A	MB01
504A	G4	504A	HEPR0052	504A	J24940	504A	K0E41007	504A	MB244
504A	G5	504A	HEPR0053	504A	J320020	504A	KLH4567	504A	MB257
504A	G6	504A	HEPR0054	504A	JAM702C	504A	KLH4577	504A	MB258
504A	G657	504A	HP-08W05	504A	JB-00030	504A	K8-05	504A	MB269
504A	G659	504A	HP-1A	504A	JB-00036	504A	K8-05X	504A	MB270
504A	G700	504A	HP-1C	504A	JB-BB1A	504A	K805	504A	MC010
504A	G701	504A	HP-20032	504A	JO-00012	504A	K8KE40C200	504A	MC015
504A	G702	504A	HP-20042	504A	JO-00014	504A	K8KE40C50Q	504A	MC020
504A	GD12	504A	HP-20047	504A	JO-00025	504A	L3/2H	504A	MC020A
504A	GE-1N506	504A	HP-20050	504A	JO-00028	504A	L32H	504A	MC021
504A	GE-1N5061	504A	HP-20052	504A	JO-00032	504A	L6/2H	504A	MC021A
504A	GE-504	504A	HP-20067	504A	JO-00033	504A	L62H	504A	MC022
504A	GE-504A	504A	HP-20083	504A	JO-00035	504A	L8/2H	504A	MC022A
504A	GE6566	504A	HP-2008A	504A	JO-00037	504A	LA300	504A	MC023
504A	GEX36	504A	HP-8D-1A	504A	JO-00044	504A	LA600	504A	MC023A
504A	GEX66	504A	HP-8D-1C	504A	JO-00045	504A	LA800	504A	MC025
504A	GEZJ252A	504A	HP-8D-1Z	504A	JO-00047	504A	LAA300	504A	MC030
504A	GEZJ252B	504A	HP-8D1	504A	JO-00049	504A	LAA600	504A	MC030A
504A	GI-100-1	504A	HP0W05	504A	JO-00051	504A	LAA800	504A	MC030B
504A	GI-1N4385	504A	HP0W066	504A	JO-00055	504A	LC.09M	504A	MC035
504A	GI-300D	504A	HP0W05	504A	JO-00059	504A	LL-2	504A	MC040
504A	GI-P100-D	504A	HP8D1A	504A	JO-0025	504A	LL2	504A	MC040A
504A	GI08B	504A	HP8D1B	504A	JO-1001	504A	LM-1158	504A	MC120
504A	GI3002	504A	HP8G005	504A	JO-D816E	504A	LM-1160	504A	MC1521
504A	GI3992-17	504A	HGR-10	504A	JO-K805	504A	LM-1862	504A	MC1522
504A	GI411	504A	HGR-20	504A	JC-8D-1X	504A	LM1160	504A	MC1523
504A	GI420	504A	HGR-30	504A	JC-8D-1Z	504A	LM1862	504A	MC170
504A	GIP100D	504A	HGR-40	504A	JC-8D1Z	504A	LP1H	504A	MC19
504A	GJ4M	504A	HGR-5	504A	JC-8D005	504A	LP2H	504A	MC456
504A	G00535	504A	HGR-6Q	504A	JCN1	504A	LP3H	504A	MCV
504A	G00-502-A	504A	HGR1	504A	JCN2	504A	LP4H	504A	MD04
504A	GP-1	504A	HGR2	504A	JCN3	504A	LRR-100	504A	MD134
504A	GP-25B	504A	HGR3	504A	JCN4	504A	LRR-200	504A	MD135
504A	GP05A	504A	HGR4	504A	JCN5	504A	LRR-300	504A	MD136
504A	GP08B	504A	HIPI	504A	JCN6	504A	LRR-400	504A	MD137
504A	GP08D	504A	HN-00008	504A	JCV-2	504A	LRR-50	504A	MD138
504A	GP1	504A	HN-00018	504A	JCV-3	504A	LRR100	504A	MDA104
504A	GP230	504A	HN-00029	504A	JCV2	504A	LRR200	504A	MDA920-2
504A	GP250	504A	HN-00032	504A	JCV3	504A	LRR300	504A	MDA920-4
504A	GP25G	504A	HP-5A	504A	JD-00040	504A	LRR400	504A	MDA920-6
504A	GSM482	504A	HP205	504A	JD-BB1A	504A	LRR50	504A	MH500
504A	GSM483	504A	HR-05A	504A	JD-8D1D	504A	LRR500	504A	MH67
504A	GSM51	504A	HR-2A	504A	JD-8D1Z	504A	M-0027	504A	MH670
504A	GSM52	504A	HR-5A	504A	J8D1D	504A	M-204B	504A	MH68
504A	GSM53	504A	HR-5AX2	504A	JT-E1024D	504A	M-31	504A	MH680
504A	GSM54	504A	HR-5B	504A	JT-E1064				

GE	Industry Standard No.	GE	Industry Standard No.	GE	Industry Standard No.	GE	Industry Standard No.	GE	Industry Standard No.
504A	MH70	504A	NPC0050	504A	PA600	504A	PT550	504A	RF32645
504A	MH700	504A	NPC0100	504A	PA7615	504A	PT560	504A	RF33976
504A	MH71	504A	NPC108	504A	PA8645	504A	PT5B	504A	RF34383
504A	MH710	504A	NPC108A	504A	PA9160	504A	PT2T130	504A	RF3472
504A	MI-15R	504A	NTC-19	504A	PC0.2P11/2	504A	PTC201	504A	RF34720
504A	MI-15B	504A	NU398B	504A	PC0211/2	504A	PTC202	504A	RFA70597
504A	MJR1C	504A	0101	504A	PC02P1/2	504A	PTC403	504A	RFA70600
504A	MMO	504A	0234	504A	PC4004	504A	PU6C22	504A	RPC61197
504A	MM2	504A	04-8054-3	504A	PD101	504A	PV-8	504A	RFJ-30704
504A	MM3	504A	04-8054-4	504A	PD102	504A	PV8	504A	RPJ-31218
504A	MM4	504A	04-8054-7	504A	PD103	504A	PY-5	504A	RPJ-31362
504A	MM5	504A	0A127	504A	PD104	504A	Q-20115C	504A	RPJ-31363
504A	MM6	504A	0A128	504A	PD105	504A	Q-26115C	504A	RPJ-33292
504A	MP-01	504A	0A129	504A	PD106	504A	Q18	504A	RJ-60366
504A	MP-5115	504A	0A130	504A	PD107	504A	Q1H	504A	RFJ30704
504A	MP01	504A	0A131	504A	PD107A	504A	Q3/2	504A	RFJ31218
504A	MP100	504A	0A132	504A	PD108	504A	Q32	504A	RPJ31362
504A	MP1003-1	504A	0A180	504A	PD110	504A	Q4B	504A	RPJ31363
504A	MP1003-2	504A	0A210	504A	PD111	504A	Q52	504A	RFJ33292
504A	MP1005-4	504A	0A211	504A	PD122	504A	Q53	504A	RJ60173
504A	MP225	504A	0A214	504A	PD125	504A	Q54	504A	RJ60174
504A	MP300	504A	0A8	504A	PD129	504A	Q55	504A	RJ60286
504A	MP400	504A	0F160	504A	PD130	504A	Q56	504A	RJ60366
504A	MP500(RECT.)	504A	0F164	504A	PD131	504A	Q57	504A	RJ60869
504A	MP5113	504A	09-30L125	504A	PD132	504A	Q58	504A	RJ6134
504A	MP651	504A	08-16308	504A	PD133	504A	Q59	504A	RFJ70147
504A	MP9602	504A	0816308	504A	PD134	504A	Q6/2	504A	RFJ70432
504A	MPX-25	504A	0SS-16308	504A	PD135	504A	Q60	504A	RFJ70487
504A	MPX215	504A	0SS-36885	504A	PD154	504A	Q8/2	504A	RFJ70703
504A	MPX25	504A	08S36503	504A	PD155	504A	Q82	504A	RFJ70931
504A	MQ3/2	504A	08S36685	504A	PD910	504A	QD-SS1885XT	504A	RFJ70970
504A	MQ32	504A	08S36685	504A	PE-401	504A	QD-SSR1KX4P	504A	RFJ70974
504A	MQ6/2	504A	0Y-5061	504A	PE401	504A	QD-SV06CXXB	504A	RFJ70973
504A	MQ62	504A	0Y-5062	504A	PE401N	504A	R-106379	504A	RFJ71122
504A	MQ8/2	504A	0Y101	504A	PE402	504A	R-113321	504A	RFJ72360
504A	MQ82	504A	0Y5061	504A	PE403	504A	R-113392	504A	RFJ72787
504A	MR-1	504A	0Y5062	504A	PE404	504A	R-154B	504A	RPJ20432
504A	MR-150-01	504A	0Y5063	504A	PE405	504A	R-2-02	504A	RFL-30596
504A	MR1237FB	504A	0Y5064	504A	PE406	504A	R-7096	504A	RFL30596
504A	MR1237FL	504A	0Y5065	504A	PE502	504A	R-8-1720	504A	RPM-33160
504A	MR1237SB	504A	0Y5066	504A	PE504	504A	R-81264	504A	RPM33160
504A	MR1237SL	504A	P-10115	504A	PE506	504A	R-81347	504A	RPP-33118
504A	MR1247FB	504A	P100	504A	PH-108	504A	R1035	504A	RPP33118
504A	MR1247FL	504A	P100A	504A	PH1021	504A	R106379	504A	RF861436
504A	MR1247SB	504A	P100B	504A	PH204	504A	R10D1	504A	RPV60500
504A	MR1247SL	504A	P100D	504A	PH208	504A	R10DC	504A	RG1004
504A	MR1267	504A	P100G	504A	PH25C22	504A	R113321	504A	RG100B
504A	MR1C	504A	P100J	504A	PH25C22/1	504A	R113392	504A	RG100D
504A	MR2064	504A	P10115	504A	PH25C22/21	504A	R122C	504A	RG100G
504A	MR2065	504A	P10156	504A	PH404	504A	R1329	504A	RG100J
504A	MR2261	504A	P150A	504A	PH9D822	504A	R154B	504A	RG1127
504A	MR2262	504A	P150B	504A	PN204	504A	R1A	504A	RGP-10D
504A	MR2271	504A	P150D	504A	PS-025	504A	R1B	504A	RH-DX0003SEZZ
504A	MR9600	504A	P150G	504A	PS-035	504A	R1K	504A	RH-DX0003TAZZ
504A	MR9601	504A	P150J	504A	PS-040	504A	R204B	504A	RH-DX0008CEZZ
504A	MR9602	504A	P1A5	504A	PS-060	504A	R2159	504A	RH-DX0025CEZZ
504A	MRB-20C	504A	P1B5	504A	PS-120	504A	R2252	504A	RH-DX0026AGZZ
504A	MRV-20C	504A	P1C5	504A	PS005	504A	R2442	504A	RH-DX0038CEZZ
504A	MS11H	504A	P1D5	504A	PS010	504A	R2460-1	504A	RH-DX0039TAZZ
504A	MS12H	504A	P20	504A	PS015	504A	R2460-4	504A	RH-DX0043TAZZ
504A	MS13H	504A	P200	504A	PS020	504A	R3/2H	504A	RH-DX0055TAZZ
504A	MS14H	504A	P21316	504A	PS025	504A	R3285	504A	RH-DX0056CEZZ
504A	MS1H	504A	P21317	504A	PS030	504A	R4A	504A	RH-DX0059TAZZ
504A	MS2H	504A	P21443	504A	PS035	504A	R5970	504A	RH-DX0063SEZZ
504A	MS35H	504A	P2A5	504A	PS040	504A	R5971	504A	RH-DX0064CEZZ
504A	MS36H	504A	P2B5	504A	PS050	504A	R6/2H	504A	RH-DX0066TAZZ
504A	MS3H	504A	P2C5	504A	PS060	504A	R6048	504A	RH-DX0067TAZZ
504A	MS4H	504A	P2D5	504A	PS105	504A	R6110	504A	RH-DX0068TAZZ
504A	MS5C	504A	P3/2H	504A	PS110	504A	R6422	504A	RH-DX0069TAZZ
504A	MS5H	504A	P32H	504A	PS120	504A	R66-8504	504A	RH-DX0072CEZZ
504A	MS5M	504A	P3A5	504A	PS125	504A	R7162	504A	RH-DX0073TAZZ
504A	MSR-500	504A	P3B5	504A	PS130	504A	R7248	504A	RH-DX0079TAZZ
504A	MSR-V5	504A	P3C5	504A	PS135	504A	R7271	504A	RH-DX0081TAZZ
504A	MSR500	504A	P3D5	504A	PS140	504A	R7682	504A	RH-DX0085TAZZ
504A	MSS-1000	504A	P400	504A	PS150	504A	R7954	504A	RH-DX0092CBZZ
504A	MT021	504A	P4A5	504A	PS160	504A	R8024	504A	RH-DX1005APZZ
504A	MT021A	504A	P4B5	504A	PS82207	504A	R8470	504A	RH-DY0003SEZZ
504A	MT022	504A	P4D5	504A	PS82208	504A	R8473	504A	RH-DX0062CEZZ
504A	MT022A	504A	P5A5	504A	PS82209	504A	R855-2	504A	RH1M
504A	MT14	504A	P5B5	504A	PS82247	504A	R8721	504A	RHDIX0043TAZZ
504A	MT24	504A	P5C5	504A	PS82249	504A	R9470	504A	RLP1G
504A	MT44	504A	P5D5	504A	PS2411	504A	R9597	504A	RM-1V
504A	MT64	504A	P6/2H	504A	PS2412	504A	R9A	504A	RM-26
504A	MV-13	504A	P600	504A	PS2413	504A	RA-1	504A	RM1Z
504A	MV-13(BIAS)	504A	P62H	504A	PS405	504A	RA-1B	504A	RM1ZM
504A	MV-5	504A	P6A5	504A	PS410	504A	RA-1Z	504A	RM1ZV
504A	MV1	504A	P6B5	504A	PS415	504A	RA-1ZC	504A	RQ-4098
504A	MV11	504A	P6C5	504A	PS420	504A	RA132BA	504A	RR824-06
504A	MV3(RECTIFIER)	504A	P6D5	504A	PS425	504A	RA1B	504A	RS-1264
504A	MV4	504A	P7776	504A	PS430	504A	RA1Y	504A	RS-1347
504A	MVA-05A	504A	P7A5	504A	PS435	504A	RA1Z	504A	RS-3570
504A	MVA-05A(DIO)	504A	P7B5	504A	PS440	504A	RA1ZC	504A	RS-3727
504A	MVA-05A(RECT)	504A	P705	504A	PS4559	504A	RCC-7022	504A	RS-6344
504A	MY-1	504A	P7D5	504A	PS4560	504A	RD-26235-1	504A	RS-6461
504A	N-02	504A	P8/2H	504A	PS460	504A	RD-29799P	504A	RS-6471
504A	N-41	504A	P82H	504A	PS84725	504A	RD-3	504A	R810
504A	N-EA16X30	504A	P9459	504A	PS5300	504A	RD-31903P	504A	R81234
504A	NA-22	504A	PA-069	504A	PS5301	504A	RD-3472	504A	R81264
504A	NA-35	504A	PA-3	504A	PS5302	504A	RD250	504A	R81720
504A	NA-36	504A	PA-320	504A	PS603	504A	RD26235-1	504A	R81749
504A	NA-46	504A	PA-320A	504A	PS604	504A	RD29799P	504A	R81805
504A	NA13	504A	PA-320B	504A	PS605	504A	RD31903P	504A	R81823
504A	NA22	504A	PA069	504A	PS610	504A	RD3472	504A	R81832
504A	NA25	504A	PA070	504A	PS611	504A	RD3A-1B4	504A	R819M-12
504A	NA32	504A	PA071	504A	PS615	504A	RD9037	504A	R220AF
504A	NA33	504A	PA10556	504A	PS616	504A	RE-49	504A	R230AF
504A	NA35	504A	PA10887	504A	PS617	504A	RE-50	504A	R33570
504A	NA36	504A	PA200	504A	PS621	504A	RE2	504A	R3727
504A	NA42	504A	PA3	504A	PS622	504A	RE3	504A	R86344
504A	NA45	504A	PA300	504A	PS623	504A	RE49	504A	R86461
504A	NA46	504A	PA305	504A	PS627	504A	RE50	504A	R86471
504A	NA62	504A	PA305A	504A	PS628	504A	REJ70643	504A	R86705
504A	NA63	504A	PA310	504A	PS629	504A	REJ70931	504A	R86430
504A	NA65	504A	PA310A	504A	PS633	504A	REN116	504A	RT-212
504A	NA66	504A	PA315	504A	PS637	504A	REN117	504A	RT-213
504A	NP500	504A	PA315A	504A	PT-3	504A	RE820	504A	RT-214
504A	NP501	504A	PA320	504A	PT-505	504A	RF-3160	504A	RT-215
504A	NP506	504A	PA320A	504A	PT-510	504A	RF-32101-8	504A	RT-2669
504A	NP550	504A	PA320B	504A	PT-515	504A	RF-32101R	504A	RT-3858
504A	NL-10	504A	PA325	504A	PT-520	504A	RF-6235-1	504A	RT-4232
504A	NL10	504A	PA325A	504A	PT-525	504A	RF26231-1	504A	RT1595
504A	NL15	504A	PA325B	504A	PT-530	504A	RF26234-1	504A	RT1689
504A	NL20	504A	PA330	504A	PT-535	504A	RF26235-1	504A	RT1840
504A	NL25	504A	PA330A	504A	PT-540	504A	RF26235-2	504A	RT213
504A	NL30	504A	PA330B	504A	PT-550	504A	RF26235-5	504A	RT214
504A	NL40	504A	PA340	504A	PT-560	504A	RF29799P	504A	RT215
504A	NL5	504A	PA340A	504A	PT-5B	504A	RF3160	504A	RT3443
504A	NL50	504A	PA340B	504A	PT5	504A	RF31903P	504A	RT3585
504A	NL60	504A	PA350	504A	PT505	504A	RF32101-8	504A	RT3658
504A	NN50	504A	PA350A	504A	PT510	504A	RF32101-9	504A	RT3981
504A	NP50A	504A	PA360	504A	PT520	504A	RF32101R	504A	RT4050
504A	NP60A	504A	PA360A	504A	PT525			504A	RT4069
504A	NPC0010	504A	PA400	504A	PT540			504A	RT4232

GE	Industry Standard No.	GE	Industry Standard No.	GE	Industry Standard No.	GE	Industry Standard No.	GE	Industry Standard No.
504A	RT4764	504A	S1C01	504A	SB-309C	504A	SD93A	504A	SIB0102
504A	RT5070	504A	S1CB1	504A	SB-3P01	504A	SD93B	504A	SIB02-03C
504A	RT5216	504A	S1D51C052-19	504A	SB-3N	504A	SD94	504A	SIB02-03CR
504A	RT5385	504A	S1D51C169-1	504A	SB01	504A	SD94A	504A	SIB02-CR
504A	RT5472	504A	S1L200	504A	SB1-01-04	504A	SD94AB	504A	SIB0201CR
504A	RT5911	504A	S1R-80	504A	SB1000	504A	SD94B	504A	SIBOL
504A	RT6322	504A	S1RC20R	504A	SB302	504A	SD95	504A	SIBOL-02
504A	RT6332	504A	S1SD-1	504A	SB315	504A	SD950	504A	SIB1
504A	RT6605	504A	S1SD-1HF	504A	SB332	504A	SD95A	504A	SID01E
504A	RT6729	504A	S1SD-1X	504A	SB333	504A	SD96	504A	SID01L
504A	RT6791	504A	S1SM-150-01	504A	SB393	504A	SD96A	504A	SID02E
504A	RT7634	504A	S1SM-150-02	504A	SBR-260	504A	SD968	504A	SID02L
504A	RT7849	504A	S1SW-05-02	504A	SC-110	504A	SDR-25	504A	SID51C169
504A	RT7850	504A	S200	504A	SC-16	504A	SDR25	504A	SIG1/200
504A	RT8199	504A	S201	504A	SC-4	504A	SDS-113	504A	SIG1/600
504A	RT8231	504A	S202	504A	SC05	504A	SDS113	504A	SIL-200
504A	RT8340	504A	S203	504A	SC05E	504A	SE-05	504A	SIL200
504A	RT8839	504A	S204	504A	SC1	504A	SE-05-01	504A	SIR-80
504A	RT8840	504A	S205	504A	SC110	504A	SE-05-02	504A	SIR-RECT-44
504A	RT8841	504A	S206	504A	SC1414	504A	SE-05-2	504A	SIR20
504A	RU1	504A	S20NH400	504A	SC1431	504A	SE-05X	504A	SIR60
504A	RV-2289	504A	S20NH400	504A	SC16	504A	SE-2	504A	SIRECT-102
504A	RV06	504A	S21	504A	SC1631	504A	SE-5	504A	SIRECT-2
504A	RV06/7825B	504A	S217	504A	SC1631(GE)	504A	SE05	504A	SIRECT-36
504A	RV1189	504A	S21B	504A	SC2	504A	SE05B	504A	SIRECT-48
504A	RV1424	504A	S219	504A	SC23-3	504A	SE05D	504A	SIRECT-59
504A	RV1476	504A	S22	504A	SC23-9	504A	SE05S	504A	SIRECT-92
504A	RV1478	504A	0822.3905-001	504A	SC2C	504A	SE05SS	504A	SISD-1X
504A	RV2072	504A	S220	504A	SC305	504A	SE08-01	504A	SISD-K
504A	RV2220	504A	S221	504A	SC4	504A	SE30B26A	504A	SISN-150-01
504A	RV2250	504A	S222	504A	SC4116	504A	SE46	504A	SISW-05-02
504A	RV2289	504A	S223	504A	SC6	504A	SE6	504A	SISW-0502
504A	RV2327	504A	S224	504A	SCA0	504A	SELEN-70	504A	SJ051E
504A	RVD08G22/1A	504A	S22A	504A	SCA1	504A	SELEN-701	504A	SJ051F
504A	RVD10D1	504A	S23	504A	SCA1103	504A	SFB6183	504A	SJ052E
504A	RVD10DC1	504A	S230	504A	SCA2	504A	SFR135	504A	SJ052F
504A	RVD10DC1R	504A	S232	504A	SCA3	504A	SFR151	504A	SJ102F
504A	RVD10E1	504A	S233	504A	SCA4	504A	SFR152	504A	SJ201F
504A	RVD10E11F	504A	S234	504A	SCA5	504A	SFR153	504A	SJ202F
504A	RVD10rd1LF	504A	S235	504A	SCA6	504A	SFR154	504A	SJ301F
504A	RVD18854	504A	S238	504A	SCB1	504A	SFR155	504A	SJ302F
504A	RVD2DP	504A	S239	504A	SCE2	504A	SFR156	504A	SJ401F
504A	RVD2DP22/18	504A	S23A	504A	SCE3	504A	SFR164	504A	SJ402F
504A	RVD2DP221P	504A	S240	504A	SCE4	504A	SFR251	504A	SJ501F
504A	RVD2P2P2/1B	504A	S241	504A	SCE6	504A	SFR252	504A	SJ601F
504A	RVD4R265J2	504A	S243	504A	SCO5	504A	SFR253	504A	SJ60F
504A	RVDDS-410	504A	S250	504A	SC05E	504A	SFR254	504A	SK-1B
504A	RVDGP05A	504A	S251	504A	SCP5E	504A	SFR255	504A	SK1FM
504A	RVDER162C5	504A	S252	504A	SD-02	504A	SFR256	504A	SK1K-2
504A	RVDBD-1	504A	S253	504A	SD-1-211B	504A	SFR264	504A	SK3016
504A	RVDSD-1U	504A	S254	504A	SD-1-211C	504A	SFR266	504A	SK3017
504A	RVDSD-1Y	504A	S255	504A	SD-101	504A	SG-005	504A	SK3017A
504A	RVDSR3AM2N	504A	S256	504A	SD-13	504A	SG-105	504A	SK3030
504A	RVDVD1211L	504A	S26	504A	SD-16A	504A	SG-1A	504A	SK3031
504A	S-05	504A	S262	504A	SD-16D	504A	SG-119B	504A	SK3031A
504A	S-05-005	504A	S2A06	504A	SD-1B	504A	SG-205	504A	SK3174
504A	S-05-01	504A	S2A10	504A	SD-1A	504A	SG-305	504A	SL-030
504A	S-050	504A	S2A30	504A	SD-1AHP	504A	SG-805	504A	SL-030T
504A	S-0501	504A	S2AR1	504A	SD-1B	504A	SG005	504A	SL-2
504A	S-1.5	504A	S2AR2	504A	SD-1C-4P	504A	SG105	504A	SL-3
504A	S-1.5-0	504A	S2C30	504A	SD-1C-UF	504A	SG1198	504A	SL-4
504A	S-10	504A	S2C40	504A	SD-1CUF	504A	SG205	504A	SL-5
504A	S-102V	504A	S2C40A	504A	SD-1HF	504A	SG305	504A	SL-833
504A	S-15	504A	S2B20	504A	SD-1L	504A	SG325	504A	SL-833A
504A	S-15-10	504A	S2B60	504A	SD-1LA	504A	SG3400	504A	SL030/3490
504A	S-17	504A	S2B60-1	504A	SD-1U	504A	SG505	504A	SL030S,T
504A	S-17A	504A	S2VB	504A	SD-1UP	504A	SG805	504A	SLO308
504A	S-2	504A	S30	504A	SD-1X	504A	SGR100	504A	SLO30T
504A	S-262	504A	S31	504A	SD-1Y	504A	SH-1	504A	SL2
504A	S-5MX	504A	S32	504A	SD-201	504A	SH-1A	504A	SL5
504A	S-500B	504A	S33	504A	SD-470	504A	SH-1B	504A	SL833
504A	S-500C	504A	S34	504A	SD-80	504A	SH-1DE	504A	SL833A
504A	S-5277B	504A	S35	504A	SD-91	504A	SH-1S	504A	SL91
504A	S-5B	504A	S36	504A	SD-91A	504A	SH1	504A	SL92
504A	S-58R	504A	S3A06	504A	SD-91B	504A	SH15	504A	SL93
504A	S-798	504A	S3AR1	504A	SD-92	504A	SH1A	504A	SLA-445
504A	S-810	504A	S3MX	504A	SD-92A	504A	SH1B	504A	SLA1095
504A	S-05	504A	S40	504A	SD-92B	504A	SH4D05	504A	SLA1096
504A	S05	504A	S4001	504A	SD-93	504A	SHA4D1	504A	SLA1100
504A	SOS01	504A	S40A	504A	SD-93A	504A	SHA4D2	504A	SLA1101
504A	S1-1	504A	S42B	504A	SD-94	504A	SHA4D3	504A	SLA1102
504A	S1-B01-02	504A	S43	504A	SD-94A	504A	SHA4D4	504A	SLA1104
504A	S1-RECT-102	504A	S431	504A	SD-94AB	504A	SHA4D6	504A	SLA1105
504A	S1.5	504A	S44	504A	SD-95	504A	SHAD-1	504A	SLA11AB
504A	S1.5-01	504A	S46	504A	SD-95A	504A	SHAD1	504A	SLA11AB
504A	S10	504A	S47	504A	SD-T	504A	SI-REC-73	504A	SLA12AB
504A	S101	504A	S48	504A	SD040	504A	SI-RECT-044	504A	SLA12C
504A	S102	504A	S49	504A	SD05	504A	SI-RECT-100	504A	SLA13AB
504A	S103	504A	S4A06	504A	SD07	504A	SI-RECT-100-102	504A	SLA13C
504A	S104	504A	S4AO6	504A	SD1-1	504A	SI-RECT-102	504A	SLA1487
504A	S105	504A	S4AR1	504A	SD1-211B	504A	SI-RECT-112	504A	SLA1488
504A	S106	504A	S4AR2	504A	SD1-211C	504A	SI-RECT-122	504A	SLA1489
504A	S107	504A	S4AR30	504A	SD1-Z	504A	SI-RECT-126	504A	SLA1490
504A	S108	504A	S4C	504A	SD10	504A	SI-RECT-144	504A	SLA1491
504A	S10A	504A	S4FN300	504A	SD101	504A	SI-RECT-154	504A	SLA1492
504A	S10B01-02	504A	S50B9-A	504A	SD102	504A	SI-RECT-155	504A	SLA14AB
504A	S1243N	504A	S5AR1	504A	SD103	504A	SI-RECT-156	504A	SLA14C
504A	S129	504A	S5AR2	504A	SD104	504A	SI-RECT-2	504A	SLA15AB
504A	S13	504A	S5SR	504A	SD1101	504A	SI-RECT-20	504A	SLA15C
504A	S14	504A	S6005	504A	SD1102	504A	SI-RECT-218	504A	SLA1692
504A	0000000S15	504A	S6AR2	504A	SD1103	504A	SI-RECT-220	504A	SLA1693
504A	S16	504A	S72	504A	SD1104	504A	SI-RECT-222	504A	SLA1694
504A	S1600	504A	S75	504A	SD13	504A	SI-RECT-226	504A	SLA1695
504A	S16A	504A	S77	504A	SD18	504A	SI-RECT-25	504A	SLA1696
504A	S16B	504A	S79	504A	0000000SD1AB	504A	SI-RECT-27	504A	SLA1697
504A	S17	504A	S81	504A	SD1CUF	504A	SI-RECT-33	504A	SLA16AB
504A	S17A	504A	S82	504A	SD1DM-4	504A	SI-RECT-34	504A	SLA16C
504A	S18	504A	S83	504A	SD1HP	504A	SI-RECT-37	504A	SLA17AB
504A	S1801-02	504A	S84	504A	SD1L	504A	SI-RECT-39	504A	SLA17C
504A	S18A	504A	S85	504A	SD1LA	504A	SI-RECT-48	504A	SLA2610
504A	S18B	504A	S86	504A	SD1X	504A	SI-RECT-49	504A	SLA2611
504A	S19	504A	S91	504A	000000SD1Y	504A	SI-RECT-53	504A	SLA2612
504A	S191G	504A	S91-A	504A	SD1Z	504A	SI-RECT-59	504A	SLA2613
504A	S19A	504A	S91-H	504A	SD1ZHF	504A	SI-RECT-69	504A	SLA2614
504A	S1A	504A	S92	504A	SD201	504A	SI-RECT-73	504A	SLA2615
504A	S1A06	504A	S92-A	504A	SD202	504A	SI-RECT-74	504A	SLA3193
504A	S1A060	504A	S92-H	504A	SD23	504A	SI-RECT-75	504A	SLA3194
504A	S1A60	504A	S93	504A	SD2A	504A	SI-RECT-77	504A	SLA3195
504A	S1AR1	504A	S93A	504A	SD2B	504A	SI-RECT-84	504A	SLA440
504A	S1AR2	504A	S93H	504A	SD4	504A	SI-RECT-92	504A	SLA440B
504A	S1B	504A	S94	504A	SD45	504A	SI-RECT-94	504A	SLA441
504A	S1B-01-02	504A	S95	504A	SD470	504A	SI-RECT102	504A	SLA441B
504A	0000081B01	504A	SA-2B	504A	SD5	504A	SI100E	504A	SLA442
504A	S1B01-01	504A	A2	504A	SD500C	504A	SI50E	504A	SLA442B
504A	S1B01-02	504A	SA2B	504A	SD501	504A	SI91G	504A	SLA443
504A	S1B01-0226	504A	SA3B	504A	SD6	504A	SIB-01-02	504A	SLA443B
504A	S1B01-06	504A	SAW-181941	504A	SD600C	504A	SIB-01-022	504A	SLA444
504A	S1B0101CR	504A	SAW-181944	504A	SD80	504A	SIB01-01	504A	SLA444A
504A	S1B0102	504A	SB-03	504A	SD91	504A	SIB01-02	504A	SLA444B
504A	S1B02	504A	SB-1Z	504A	SD91A	504A	SIB01-04	504A	SLA445
504A	S1B02-C	504A	SB-3	504A	SD91B	504A	SIB01-06	504A	SLA445B
504A	S1B0201CR	504A	SB-3-02	504A	SD92	504A	SIB01-06B	504A	SLA536
504A	S1B1	504A	SB-309A	504A	SD92A			504A	SLA537
504A	S1BD1-02			504A	SD93			504A	SLA538

GE	Industry Standard No.	GE	Industry Standard No.	GE	Industry Standard No.	GE	Industry Standard No.	GE	Industry Standard No.
504A	SLA539	504A	SR1104	504A	SR6724	504A	T065	504A	TJ-5A
504A	SLA540	504A	SR112	504A	SR76	504A	T075	504A	TJ10A
504A	SLA547	504A	SR114	504A	SR806-126	504A	T10144	504A	TJ15A
504A	SLA599	504A	SR120	504A	SR846-2	504A	T10175	504A	TJ20A
504A	SLA599A	504A	SR1266	504A	SR846-3	504A	T10185	504A	TJ25A
504A	SLA600	504A	SR131-1	504A	SR851	504A	T10453	504A	TJ30A
504A	SLA600A	504A	SR132-1	504A	SR851-121	504A	T1085	504A	TJ35A
504A	SLA601	504A	SR135-1	504A	SR885	504A	T1450	504A	TJ40A
504A	SLA601A	504A	SR1378-1	504A	SR9005	504A	T156	504A	TJ5A
504A	SLA602	504A	SR1378-3	504A	SRIDM	504A	T159	504A	TJ60A
504A	SLA602A	504A	SR13H	504A	SRIDM-1	504A	T1J6G	504A	TK-10
504A	SLA603	504A	SR144	504A	SRIDM-4	504A	T18D81K	504A	TK-30
504A	SLA603A	504A	SR145	504A	SRIEM	504A	T200	504A	TK-40
504A	SLA604	504A	SR1493	504A	SRIEM-1	504A	T21312	504A	TK-41
504A	SLA604A	504A	SR151	504A	SRIEM-2	504A	T21333	504A	TK10
504A	SLA605	504A	SR152	504A	SRIEM-4	504A	T21507	504A	TK11
504A	SLA605A	504A	SR1549	504A	SRIK-2	504A	T21602	504A	TK20
504A	SLA606	504A	SR16	504A	SRIK-8	504A	T21638	504A	TK21
504A	SLA606A	504A	SR1668	504A	SRK-2	504A	T21649	504A	TK30
504A	SM-1	504A	SR1692	504A	SRK1	504A	T21679	504A	TK400
504A	SM-1-005	504A	SR1693	504A	SRLPM-1	504A	T3/2	504A	TK5
504A	SM-1-47	504A	SR1694	504A	SS-1	504A	T300	504A	TK50
504A	SM-10	504A	SR1695	504A	SS00010	504A	T30155	504A	TK60
504A	SM-150B	504A	SR17	504A	SS00010	504A	T30155-001	504A	TK600
504A	SM-1K	504A	SR1731-1	504A	SS0009	504A	T30155-1	504A	TK61
504A	SM1-02	504A	SR1731-2	504A	SS321	504A	T400	504A	TKP10
504A	SM10	504A	SR1731-3	504A	SS322	504A	T42692-001	504A	TKP20
504A	SM105	504A	SR1731-4	504A	SS324	504A	T42692-1R	504A	TKP40
504A	SM11	504A	SR1731-5	504A	SS334	504A	T450	504A	TKP5
504A	SM110	504A	SR1742	504A	SS337	504A	T4590	504A	TKP60
504A	SM120	504A	SR1766	504A	SS455	504A	T500	504A	TL1
504A	SM130	504A	SR18	504A	SSD974	504A	T550	504A	TL11
504A	SM140	504A	SR1984	504A	ST-12	504A	T600	504A	TL12
504A	SM150	504A	SR1A-1	504A	ST-14	504A	T650	504A	TL2
504A	SM160	504A	SR1A-12	504A	ST-2040P	504A	T8/2	504A	TL21
504A	SM20	504A	SR1A-2	504A	ST16	504A	TA100	504A	TL22
504A	SM205	504A	SR1A-4	504A	ST2040P	504A	TA1062	504A	TL31
504A	SM210	504A	SR1A-8	504A	STBOL-02	504A	TA1063	504A	TL32
504A	SM220	504A	SR1A1	504A	STFCN10	504A	TA1064	504A	TL41
504A	SM230	504A	SR1A12	504A	STV-3	504A	TA200	504A	TL42
504A	SM240	504A	SR1A2	504A	SV-01A	504A	TA300	504A	TL51
504A	SM250	504A	SR1A4	504A	SV-01B	504A	TA400	504A	TL61
504A	SM260	504A	SR1A8	504A	SV-03	504A	TA50	504A	TM-33
504A	SM30	504A	SR1D1M	504A	SV-05	504A	TA500	504A	TM-43
504A	SM31	504A	SR1DM	504A	SV-1238B	504A	TA600	504A	TM116
504A	SM4	504A	SR1DM-1	504A	SV-1258E	504A	TA7802	504A	TM117
504A	SM40	504A	SR1DM-2	504A	SV01A	504A	TA7803	504A	TM33
504A	SM483	504A	SR1DM-4	504A	SV02A	504A	TA7904	504A	TM43
504A	SM486	504A	SR1DM1	504A	SV04	504A	TA7996	504A	TM62
504A	SM487	504A	SR1DMX	504A	SV05	504A	TC-136	504A	TM63
504A	SM488	504A	SR1E	504A	SV1238	504A	TCO.2P11/2	504A	TM65
504A	SM5	504A	SR1EM	504A	SV12388	504A	TC02P112	504A	TM66
504A	SM50	504A	SR1EM-1	504A	SV12388E	504A	TC02P12	504A	TMD41
504A	SM505	504A	SR1EM-2	504A	SV1238E	504A	TCOP11/2	504A	TMD42
504A	SM51	504A	SR1EM-X	504A	SVD1D1-1	504A	TC136	504A	TMD45
504A	SM510	504A	SR1EM1	504A	SVD181850	504A	TC3112319300	504A	TNJ61672(RECT)
504A	SM512	504A	SR1EM2	504A	SVDMA26	504A	TE-1011	504A	TP101
504A	SM513	504A	SR1FM	504A	SVDVD1223	504A	TE-1050	504A	TP201
504A	SM514	504A	SR1FM-1	504A	SW-05	504A	TE1010	504A	TP302
504A	SM515	504A	SR1FM-8	504A	SW-05-005	504A	TE1011	504A	TP402
504A	SM516	504A	SR1FM10	504A	SW-05-02	504A	TE1024C	504A	TP4067-409
504A	SM60	504A	SR1FM12	504A	SW-05V	504A	TE1024D	504A	TR-02E
504A	SM645	504A	SR1FMA	504A	SW-1A	504A	TE1029	504A	TR-2880
504A	SM645	504A	SR1HM-12	504A	SWO-5A	504A	TE1042	504A	TR-77
504A	SM705	504A	SR1HM-16	504A	SW05	504A	TE1050	504A	TR1N4002
504A	SM710	504A	SR1HM-2	504A	SW05-01	504A	TE1078	504A	TR2327031
504A	SM720	504A	SR1HM-4	504A	SW05-02	504A	TE1080	504A	TR2327041
504A	SM730	504A	SR1K	504A	SW0501	504A	TE1088	504A	TR2880
504A	SM740	504A	SR1K-1	504A	SW05A	504A	TE1089	504A	TR2A
504A	SM750	504A	SR1K-1K	504A	SW05B	504A	TE1090	504A	TR320020
504A	SM760	504A	SR1K-2	504A	SW05S	504A	TE1097	504A	TR320022
504A	SN-1	504A	SR1K-2/494	504A	SW05SS	504A	TE1108	504A	TR330027
504A	SN-1Z	504A	SR1K-4	504A	SW05V	504A	TP20	504A	TS-2A
504A	SN0303	504A	SR1K-8	504A	SWO-5A	504A	TP21	504A	TS05
504A	SN1	504A	SR1K-Z	504A	SX-642	504A	TP22	504A	TS1
504A	S05	504A	SR1K/494	504A	SX623	504A	TP23	504A	TS2
504A	S0501	504A	SR1K08	504A	SX631	504A	TPR-120	504A	TS2A
504A	S0D200D	504A	SR1K2	504A	SX633	504A	TPR120	504A	TS4
504A	SP-1	504A	SR1K8	504A	SX641	504A	TG-11	504A	TS6
504A	SP1	504A	SR1T	504A	SX642	504A	TG-12	504A	TSB-1000
504A	SP1K-1	504A	SR1Z	504A	SX643	504A	TG-21	504A	TSB-245
504A	SP1K-2	504A	SR200	504A	SX644	504A	TG-22	504A	TSB245
504A	SPN-01	504A	SR200B	504A	SX645	504A	TG-31	504A	TSO159
504A	SPN01	504A	SR205	504A	T-0150	504A	TG-32	504A	TSV-1000
504A	SR-0005	504A	SR2121	504A	T-065	504A	TG-41	504A	TT66K26
504A	SR-0007	504A	SR22	504A	T-075	504A	TG-42	504A	TUS-185D
504A	SR-0008	504A	SR23	504A	T-100	504A	TG-51	504A	TV-24104
504A	SR-05K-2	504A	SR2301	504A	T-13	504A	TG-52	504A	TV-24125
504A	SR-1	504A	SR2301A	504A	T-13G	504A	TG-61	504A	TV-24191
504A	SR-101-1	504A	SR24	504A	T-14	504A	TG-62	504A	TV-24232
504A	SR-101-2	504A	SR27	504A	T-14G	504A	TG12	504A	TV-24266
504A	SR-112	504A	SR28	504A	T-200	504A	TG20A	504A	TV-2496
504A	SR-128	504A	SR2A-1	504A	T-22G	504A	TG21	504A	TV24104
504A	SR-120-1	504A	SR2A-2	504A	T-26G	504A	TG22	504A	TV241073
504A	SR-130-1	504A	SR2A-4	504A	T-300	504A	TG31	504A	TV241074
504A	SR-131-1	504A	SR2A1	504A	T-400	504A	TG32	504A	TV24125
504A	SR-132-1	504A	SR2A12	504A	T-450	504A	TG41	504A	TV24136
504A	SR-136	504A	SR2A2	504A	T-4590	504A	TG42	504A	TV24155
504A	SR-13E	504A	SR2A4	504A	T-50	504A	TG51	504A	TV24191
504A	SR-14	504A	SR2A8	504A	T-500	504A	TG52	504A	TV24193
504A	SR-150-01	504A	SR3	504A	T-550	504A	TG61	504A	TV24200
504A	SR-150-1	504A	SR30	504A	T-600	504A	TG62	504A	TV24222
504A	SR-1668	504A	SR3010	504A	T-650	504A	TH18557	504A	TV24224
504A	SR-17	504A	SR35	504A	T-B01029D	504A	TH400	504A	TV24232
504A	SR-18	504A	SR3582	504A	T-E1011	504A	TH50	504A	TV24234
504A	SR-1849-1	504A	SR390	504A	T-E1024	504A	TH600	504A	TV24266
504A	SR-1HM-2	504A	SR390-2	504A	T-E1024C	504A	TH801	504A	TV24278
504A	SR-1K	504A	SR3943	504A	T-E1024D	504A	TH802	504A	TV24282
504A	SR-1K-2	504A	SR3AM-8	504A	T-E1029	504A	TH803	504A	TV24283
504A	SR-1K2	504A	SR3AM2	504A	T-E1042	504A	TH804	504A	TV24285
504A	SR-1Z	504A	SR3BM-6	504A	T-E1050	504A	TH805	504A	TV24292
504A	SR-22	504A	SR4	504A	T-E1064	504A	TH806	504A	TV24298
504A	SR-23	504A	SR40	504A	T-E1078	504A	TH93105	504A	TV24582
504A	SR-24	504A	SR401	504A	T-E1078A	504A	TI-53	504A	TV24586
504A	SR-27	504A	SR405	504A	T-E1080	504A	TI-55	504A	TV24651
504A	SR-28	504A	SR5	504A	T-E1088	504A	TI-71	504A	TV24941
504A	SR-3	504A	SR50	504A	T-E1089	504A	TI152	504A	TV2496
504A	SR-30	504A	SR500	504A	T-E1090	504A	TI52	504A	TV24979
504A	SR-390	504A	SR500B	504A	T-E1097	504A	TI53	504A	TV34232
504A	SR-4	504A	SR50253-2	504A	T-E1102	504A	TI54	504A	TV4
504A	SR-401	504A	SR50411-1	504A	T-E1102A	504A	TI55	504A	TVC-3
504A	SR-5	504A	SR50517	504A	T-E1108	504A	TI56	504A	TVD81M
504A	SR-76	504A	SR60	504A	T-E1124	504A	TI57	504A	TVM-511
504A	SR-846-2	504A	SR605	504A	T-E1133	504A	TI58	504A	TVM-EH2C
504A	SR-889	504A	SR6134	504A	T-E1138	504A	TI59	504A	TVM-EH2C11
504A	SR-1K-2	504A	SR6324	504A	T-E1144	504A	TI60	504A	TVM-EH2C11/1
504A	SR05K-2	504A	SR6325	504A	T-E1148	504A	TI71	504A	TVM-EH2C12/1
504A	SR1-2	504A	SR6385	504A	T-E1155	504A	TIR01	504A	TVM-M204B
504A	SR1-82	504A	SR6415	504A	T-E1157	504A	TIR02	504A	TVM-PH9D22/1
504A	SR100	504A	SR6560	504A	T-E1171	504A	TIR03	504A	TVM-PT6D22/1
504A	SR101-1	504A	SR6567	504A	T-E1176	504A	TIR04	504A	TVM-TCO.2P11/2
504A	SR101-2	504A	SR6617	504A	T-H18557	504A	TIR05	504A	TVM35
504A	SR1024	504A	SR6723	504A	T-H86105	504A	TIR06		
504A	SR105			504A	T-H80105				
				504A	T0150				

GE	Industry Standard No.	GE	Industry Standard No.	GE	Industry Standard No.	GE	Industry Standard No.	GE	Industry Standard No.
504A	TVM56	504A	1A10425	504A	1N153	504A	1N2613	504A	1N4001
504A	TVMHS151B	504A	1A10952	504A	1N1556	504A	1N2614	504A	1N4002
504A	TVML00-09M1115	504A	1A11184	504A	1N1557	504A	1N2615	504A	1N4003
504A	TVMMS04B	504A	1A11671	504A	1N1558	504A	1N2638	504A	1N4003GP
504A	TVMPH9DZ2/1	504A	1A12214	504A	1N1559	504A	1N2650	504A	1N4004
504A	TV8-185D	504A	1A12407	504A	1N1560	504A	1N2858	504A	1N4005
504A	TV8-1S1850	504A	1A12690	504A	1N1563	504A	1N2858A	504A	1N400B
504A	TV8-1S1906	504A	1A13219	504A	1N1563A	504A	1N2859	504A	1N4089
504A	TV8-DS-1N-R	504A	1A13719	504A	1N1564	504A	1N2859A	504A	1N4139
504A	TV8-DS-1X	504A	1A13720	504A	1N1564A	504A	1N2860	504A	1N4139A
504A	TV8-DS-1M	504A	1A15790	504A	1N1565	504A	1N2860A	504A	1N4141
504A	TV8-DS1K	504A	1A16550	504A	1N1565A	504A	1N2861	504A	1N4245
504A	TV8-DS1M	504A	1A50	504A	1N1566	504A	1N2861A	504A	1N4246
504A	TV8-DS2K	504A	1BO5J20	504A	1N1566A	504A	1N2862	504A	1N4247
504A	TV8-ET1P	504A	1BO5J40	504A	1N1567	504A	1N2862A	504A	1N4364
504A	TV8-FR-1P	504A	1B0J20	504A	1N1567A	504A	1N2863	504A	1N4365
504A	TV8-FR-2PC	504A	1B2	504A	1N1568	504A	1N2863A	504A	1N4366
504A	TV8-FR1-PC	504A	1C0009	504A	1N1568A	504A	1N2864	504A	1N4367
504A	TV8-FR1MD	504A	1C0017	504A	1N58	504A	1N2864A	504A	1N4368
504A	TV8-FR1PC	504A	1C0025	504A	1N1617	504A	1N3072	504A	1N4369
504A	TV8-FR2M	504A	1C0026	504A	1N1618	504A	1N3073	504A	1N4383
504A	TV8-FT-1P	504A	1C0031	504A	1N1619	504A	1N3074	504A	1N4384
504A	TV8-PU1N	504A	1D261	504A	1N1620	504A	1N3075	504A	1N4385
504A	TV8-HP-SD-12	504A	1D2Z1	504A	1N1644	504A	1N3076	504A	1N440
504A	TV8-HPSD1Z	504A	1DC1	504A	1N1645	504A	1N3077	504A	1N440B
504A	TV8-KC2-LP	504A	1EO5	504A	1N1646	504A	1N3078	504A	1N441
504A	TV8-KC2OP12/1	504A	1E1	504A	1N1647	504A	1N3079	504A	1N441B
504A	TV8-KC2OP12/1	504A	1E2	504A	1N1648	504A	1N3080	504A	1N442
504A	TV8-KC2DP12/2	504A	1E3	504A	1N1649	504A	1N3081	504A	1N442B
504A	TV8-OV-O2	504A	1E4	504A	1N1650	504A	1N3082	504A	1N443
504A	TV8-PCO2P11/2	504A	1E5	504A	1N1651	504A	1N3083	504A	1N443B
504A	TV8-PCD2P11/2	504A	1E6	504A	1N1652	504A	1N3084	504A	1N444
504A	TV8-81B02-03C	504A	1EA10A	504A	1N1653	504A	1N3094	504A	1N444B
504A	TV8-81B02-03CR	504A	1EA20A	504A	1N1692	504A	1N3106	504A	1N445
504A	TV8-SD1A	504A	1EA30A	504A	1N1693	504A	1N315	504A	1N445B
504A	TV8-TC009M11/1Q	504A	1EA40A	504A	1N1694	504A	1N315A	504A	1N448
504A	TV8-UFSD-1	504A	1EA50A	504A	1N1695	504A	1N316	504A	1N450
504A	TV8010D1	504A	1EA60A	504A	1N1696	504A	1N3160	504A	1N451
504A	TV80A71	504A	1ETO2	504A	1N1697	504A	1N316A	504A	1N453
504A	TV810D	504A	1ETO5	504A	1N1701	504A	1N317	504A	1N4586
504A	TV810DB	504A	1ET1	504A	1N1702	504A	1N317A	504A	1N4720
504A	TV81850	504A	1ET2	504A	1N1703	504A	1N318	504A	1N480B
504A	TV81M4002	504A	1ET3	504A	1N1704	504A	1N3183	504A	1N4816
504A	TV81P20	504A	1ET4	504A	1N1705	504A	1N3184	504A	1N4820
504A	TV81S1850	504A	1ET5	504A	1N1706	504A	1N3189	504A	1N4821
504A	TV81S1906	504A	1ET6	504A	1N1707	504A	1N3189A	504A	1N4822
504A	TV81S1922G	504A	1PO5	504A	1N1709	504A	1N319	504A	1N5004
504A	TV8590	504A	1F14A	504A	1N1710	504A	1N3190	504A	1N5005
504A	TVSBO1-02	504A	1F2	504A	1N1711	504A	1N3191	504A	1N5006
504A	TVSBBE	504A	1FM2	504A	1N1712	504A	1N3193	504A	1N5007
504A	TVSD1K	504A	1GA	504A	1N1763	504A	1N3194	504A	1N503
504A	TVSD52K	504A	1HY40	504A	1N1763A	504A	1N3195	504A	1N504
504A	TVSD21NR	504A	1HY50	504A	1N1764	504A	1N319A	504A	1N505
504A	TVSDB-1M	504A	1JH11P	504A	1N1764A	504A	1N3203	504A	1N5055
504A	TVSDB1K	504A	1JZ61	504A	1N1907	504A	1N3203A	504A	1N5056
504A	TVSD82M	504A	1M2086	504A	1N1908	504A	1N321	504A	1N5057
504A	TVSRRB24-04D	504A	1M8513A	504A	1N1909	504A	1N3227	504A	1N5058
504A	TVSEBA06	504A	1MA4	504A	1N1910	504A	1N3228	504A	1N5059
504A	TVSFR1P	504A	1N-4002	504A	1N1911	504A	1N3229	504A	1N506
504A	TVSFR1PC	504A	1N1008	504A	1N1912	504A	1N323	504A	1N5060
504A	TVSFR2-06	504A	1N1028	504A	1N1913	504A	1N3230	504A	1N507
504A	TVSFR2PC	504A	1N1029	504A	1N193	504A	1N3237	504A	1N508
504A	TVSFT1P	504A	1N1030	504A	1N2013	504A	1N3238	504A	1N511
504A	TVSPU1N	504A	1N1031	504A	1N2014	504A	1N3239	504A	1N512
504A	TVSPU1N	504A	1N1032	504A	1N2015	504A	1N323A	504A	1N513
504A	TVSHPSD-1A	504A	1N1033	504A	1N2016	504A	1N324	504A	1N514
504A	TVSHPSD12C	504A	1N1052	504A	1N2017	504A	1N3240	504A	1N515
504A	TVSIS1850	504A	1N1053	504A	1N2018	504A	1N3241	504A	1N516
504A	TVSJL41A	504A	1N1054	504A	1N2019	504A	1N3246	504A	1N519
504A	TVSM1-02	504A	1N1055	504A	1N2020	504A	1N3247	504A	1N519B
504A	TV80A71	504A	1N1081	504A	1N2069	504A	1N3248	504A	1N520
504A	TVSPCD2P11/2	504A	1N1081A	504A	1N2069A	504A	1N3249	504A	1N521
504A	TVSRA-1Z	504A	1N1082	504A	1N2070	504A	1N324A	504A	1N5211
504A	TVSRMP5020	504A	1N1082A	504A	1N2070A	504A	1N325	504A	1N5212
504A	TVSS1P20	504A	1N1083	504A	1N2071	504A	1N3250	504A	1N5213
504A	TVSS1R20	504A	1N1083A	504A	1N2071A	504A	1N3253	504A	1N5215
504A	TVSS1RBD	504A	1N1084	504A	1N2072	504A	1N3254	504A	1N5216
504A	TVSS-2	504A	1N1084A	504A	1N2073	504A	1N3255	504A	1N5217
504A	TVSS54RECT	504A	1N10D-4F	504A	1N2074	504A	1N325A	504A	1N522
504A	TVSS4C	504A	1N1100	504A	1N2075	504A	1N326	504A	1N523
504A	TVSSA-2H	504A	1N1101	504A	1N2076	504A	1N326A	504A	1N524
504A	TVSSA2B	504A	1N1102	504A	1N2077	504A	1N327	504A	1N530
504A	TVSSB-2T	504A	1N1103	504A	1N2078	504A	1N3277	504A	1N531
504A	TVSSD1A	504A	1N1104	504A	1N2079	504A	1N3278	504A	1N532
504A	TVSSVO2	504A	1N1105	504A	1N2080	504A	1N3279	504A	1N533
504A	TVSUPSD1P	504A	1N1122A	504A	1N2081	504A	1N327A	504A	1N534
504A	TVSWF2	504A	1N1169	504A	1N2082	504A	1N3298	504A	1N535
504A	TW10	504A	1N1169A	504A	1N2083	504A	1N3493	504A	1N536
504A	TW20	504A	1N1217	504A	1N2084	504A	1N354	504A	1N537
504A	TW3	504A	1N1217A	504A	1N2085	504A	1N3544	504A	1N538
504A	TW30	504A	1N1217B	504A	1N2086	504A	1N3545	504A	1N539
504A	TW40	504A	1N1218	504A	1N2088	504A	1N3546	504A	1N5393
504A	TW5	504A	1N1218A	504A	1N2089	504A	1N3547	504A	1N540
504A	TW50	504A	1N1218B	504A	1N2090	504A	1N3548	504A	1N547
504A	TW60	504A	1N1219	504A	1N2091	504A	1N3549	504A	1N596
504A	TWY	504A	1N1219A	504A	1N2092	504A	1N359	504A	1N599
504A	TX1N3190	504A	1N1219B	504A	1N2093	504A	1N359A	504A	1N599A
504A	TX1N3191	504A	1N1220	504A	1N2094	504A	1N360	504A	1N600
504A	TX1N645	504A	1N1220A	504A	1N2095	504A	1N360A	504A	1N600A
504A	TX1N647	504A	1N1220B	504A	1N2096	504A	1N361	504A	1N601
504A	1&12/1	504A	1N1221	504A	1N2103	504A	1N3612	504A	1N601A
504A	1+12/1	504A	1N1221A	504A	1N2104	504A	1N3613	504A	1N602
504A	001-0077-00	504A	1N1221B	504A	1N2105	504A	1N3614	504A	1N602A
504A	001-0081-00	504A	1N1222	504A	1N2106	504A	1N361A	504A	1N603
504A	001-02405-0	504A	1N1222A	504A	1N2107	504A	1N362	504A	1N603A
504A	001-02405-1	504A	1N1222B	504A	1N2108	504A	1N362A	504A	1N604
504A	001-02405-2	504A	1N1223	504A	1N2115	504A	1N363	504A	1N604A
504A	001-024050	504A	1N1223A	504A	1N2116	504A	1N3639	504A	1N605
504A	001-024051	504A	1N1223B	504A	1N2181	504A	1N363A	504A	1N605A
504A	001-024052	504A	1N1224	504A	1N2323	504A	1N3640	504A	1N606
504A	001-02601-0	504A	1N1251	504A	1N2373	504A	1N3641	504A	1N606A
504A	001-02603-0	504A	1N1252	504A	1N2395	504A	1N3642	504A	1N645
504A	1-101	504A	1N1253	504A	1N2396	504A	1N3656	504A	1N645A
504A	1-20-001-89Q	504A	1N1254	504A	1N2404	504A	1N3657	504A	1N645B
504A	01-2405-0	504A	1N1255	504A	1N2405	504A	1N3658	504A	1N646
504A	01-2405-1	504A	1N1255A	504A	1N2413	504A	1N3669	504A	1N647
504A	01-2405-2	504A	1N1256	504A	1N2414	504A	1N368	504A	1N648
504A	01-2406-1	504A	1N1257	504A	1N2422	504A	1N3748	504A	1N649
504A	1-530-012-11	504A	1N1337-5	504A	1N2423	504A	1N3749	504A	1N673
504A	1-531-027	504A	1N1406	504A	1N2482	504A	1N3755	504A	1N676
504A	1-531-105	504A	1N1415	504A	1N2483	504A	1N3756	504A	1N677
504A	1-531-105-11	504A	1N1439	504A	1N2484	504A	1N3757	504A	1N678
504A	1-531-105-13	504A	1N1440	504A	1N2485	504A	1N3758	504A	1N679
504A	1-531-10513	504A	1N1441	504A	1N2486	504A	1N3759	504A	1N681
504A	1-531-106	504A	1N1442	504A	1N2487	504A	1N3866	504A	1N682
504A	1-531-106-13	504A	1N1486	504A	1N2488	504A	1N3867	504A	1N683
504A	1-531-106-17	504A	1N1487	504A	1N2489	504A	1N3868	504A	1N684
504A	1-531-5-11	504A	1N1488	504A	1N2609	504A	1N3895	504A	1N685
504A	1-534-105-13	504A	1N1489	504A	1N2610	504A	1N3938	504A	1N686
504A	1-534-106-13	504A	1N1490	504A	1N2611	504A	1N3939	504A	1N687
504A	1-6501190016	504A	1N1491	504A	1N2611	504A	1N3940	504A	1N689
504A	1-8259	504A	1N1492	504A	1N2611	504A	1N3952	504A	1N692
504A	1-8259	504A	1N151	504A	1N2611	504A	1N3952	504A	1N819
504A	1-RE-004	504A	1N152	504A	1N2612	504A	1N400	504A	1N846
504A		504A		504A		504A		504A	1N847

GE	Industry Standard No.	GE	Industry Standard No.	GE	Industry Standard No.	GE	Industry Standard No.	GE	Industry Standard No.
504A	1N848	504A	1S1694	504A	1SR1FM4	509	1N2398	510	T8SR3AM
504A	1N849	504A	1S1697	504A	1SR1K	509	1N2406	510	1G201
504A	1N850	504A	1S1698	504A	1S854	509	1N2407	510	1G2Z1
504A	1N851	504A	1S180/5GB	504A	1T2011	509	1N2415	510	1N4817
504A	1N852	504A	1S183	504A	1T2012	509	1N2416	510	1N4819
504A	1N857	504A	1S1848	504A	1T2013	509	1N2424	510	1N5392
504A	1N858	504A	1S1851	504A	1T2014	509	1N2425	510	1N5392GP
504A	1N859	504A	1S1851R	504A	1T2015	509	1N2501	510	1R5QZ61FA-1
504A	1N860	504A	1S1850R	504A	1T2016	509	1N2502	510,513	1C8
504A	1N862	504A	1S1881AM	504A	1T2C1	509	1N2505	510,513	1N3929
504A	1N863	504A	1S1885	504A	1T2Z1	509	1N2506	510,531	B1H1
504A	1N868	504A	1S1885-3	504A	1T37B	509	1N2617	510,531	B1H5
504A	1N869	504A	1S1886	504A	1T501	509	1N2653	510,531	B1H9
504A	1N870	504A	1S1887	504A	1T502	509	1N2772	510,531	B1K1
504A	1N871	504A	1S1888	504A	1T503	509	1N2773	510,531	B1K5
504A	1N872	504A	1S1906	504A	1T504	509	1N2774	510,531	B1K9
504A	1N873	504A	1S1941	504A	1T505	509	1N2775	510,531	B1M1
504A	1N874	504A	1S19413	504A	1T506	509	1N2865	510,531	B1M5
504A	1N879	504A	1S1942	504A	1TB06	509	1N2866	510,531	B1M9
504A	1N880	504A	1S1943	504A	1TS05	509	1N2867	510,531	B2H1
504A	1N881	504A	1S1950	504A	1V3074A20	509	1N2868	510,531	B2H5
504A	1N881B	504A	1S1Z09	504A	1V3074A21	509	1N2878	510,531	B2H9
504A	1N882	504A	1S204	504A	1VA10	509	1N2879	510,531	B2K1
504A	1N883	504A	1S205	504A	1WM6	509	1N2880	510,531	B2K5
504A	1N884	504A	1S206	504A	2-58R	509	1N2881	510,531	B2K9
504A	1N885	504A	1S207	504A	2D-02	509	1N2882	510,531	B2M1
504A	1N91	504A	1S208	504A	2F4	509	1N2883	510,531	B2M5
504A	1N916	504A	1S208/28J2A	504A	2G13	509	1N3107	510,531	B2M9
504A	1N92	504A	1S209	504A	2G8	509	1N3185	510,531	B3H1
504A	1N93	504A	1S209/28J4A	504A	2G805	509	1N3186	510,531	B3H5
504A	1N93A	504A	1S2230	504A	2GA	509	1N3231	510,531	B3H9
504A	1N947	504A	1S2313	504A	2JB138	509	1N3232	510,531	B3K5
504A	1N998	504A	1S2351	504A	2S-16E	509	1N3242	510,531	B3K9
504A	1NC61684	504A	1S2352	504A	2S1K	509	1N3243	510,531	B3M1
504A	1NJ61676	504A	1S2356	504A	2SB-C731	509	1N3251	510,531	B3M5
504A	1NJ61726	504A	1S2357	504A(2)	B-1881U	509	1N3252	510,531	B3M9
504A	1NJ71126	504A	1S2361	504A(2)	B-1882U	509	1N3256	510,531	B5H1
504A	1NJ71186	504A	1S2362	504A(2)	CD2	509	1N3486	510,531	B5H5
504A	1R01	504A	1S2363	504A(2)	CR501/6515	509	1N3563	510,531	B5H9
504A	1R0P	504A	1S2371	504A(2)	DS17	509	1N3723	510,531	B5K1
504A	1R0H	504A	1S2372	504A(2)	EN169B	509	1N3751	510,531	B5K5
504A	1R1D	504A	1S2373	504A(2)	INJ61227	509	1N3752	510,531	B5K9
504A	1R1K	504A	1S2374	504A(2)	JC-V03C	509	1N3760	510,531	B5M1
504A	1R2A	504A	1S2375	504A(2)	KC2DP12/1	509	1N3761	510,531	B5M5
504A	1R2D	504A	1S2376	504A(2)	M702C	509	1N3869	510,531	B5M9
504A	1R2E	504A	1S2401	504A(2)	S1B02-0	509	1N3957	510,531	BX110
504A	1R3D	504A	1S2402	504A(2)	S1B02-CR	509	1N4248	510,531	BY100
504A	1R3G	504A	1S2404	504A(2)	S1RC20	509	1N4249	510,531	BY1001
504A	1R3J	504A	1S2606	504A(2)	SS8	509	1N4514	510,531	BY1002
504A	1R5A	504A	1S268AA	504A(2)	SNO303	509	1N4585	510,531	BY1008
504A	1R5B	504A	1S268T	504A(2)	TV01S1950	509	1N5052	510,531	BY104
504A	1R5G	504A	1S2775FA-1	504A(2)	1005	509	1N5053	510,531	BY105
504A	1R5H	504A	1S3016(RECT)	504A(2)	1C1	509	1S107R	510,531	BY108
504A	1R9	504A	1S3016R	504A(2)	1C2	509	1S108R	510,531	BY109
504A	1R90	504A	1S309	504A(2)	1C4	509	1S109R	510,531	BY1101
504A	1R9H	504A	1S310	504A(2)	1C6	509	1S38	510,531	BY1102
504A	1R9J	504A	1S311	504A(2)	1D1	509	1S39	510,531	BY118
504A	1R9L	504A	1S312	504A(2)	1D2	509	1S47	510,531	BY119
504A	1R9U	504A	1S313	504A(2)	1D6	509	1896R	510,531	BY1200
504A	1S-1209	504A	1S314	504A(2)	0000151849	509	1897R	510,531	BY1201
504A	1S005	504A	1S315	504A(2)	0000151849R	509	1898R	510,531	BY1202
504A	1S031	504A	1S3195	504A(2)	151850	509	1899R	510,531	BY127/500
504A	1S032	504A	1S358	504A(2)	18849R	509	2P5CN1	510,531	BY127/600
504A	1S100	504A	1S358(S)	504A(2)	2AA113	510	A7579500	510,531	BY127/700
504A	1S1004	504A	1S3588	504A(3)	KC-1.303X11/1	510	A7580011	510,531	BY128
504A	1S100R	504A	1S36	504A(3)	KC0-80B11/H12/1	510	A7580111	510,531	BY129
504A	1S101	504A	1S390	504A(3)	KC08CP11I	510	B30C250KP	510,531	BX110
504A	1S101R	504A	1S395	504A(3)	KC08CP121	510	B30C50KP	510,531	BYX12/400
504A	1S102	504A	1S396	504A(3)	KC2DP11/1412/1	510	B7579500	510,531	BYX13/400
504A	1S102R	504A	1S399	504A(3)	RF32412-3	510	860S0300	510,531	BYX22/800
504A	1S103R	504A	1S40	504A(3)	TVM511	510	CTN100	510,531	BYX60-700
504A	1S104	504A	1S400	504A(4)	AFS-160-1021	510	CTN1000	510,531	BYT37
504A	1S104R	504A	1S41	504A(4)	JB1604	510	CTN300	510,531	BYY91
504A	1S105	504A	1S42	504A(4)	KC-08C221	510	CTN400	510,531	C1.0BO2
504A	1S105R	504A	1S43	504A(4)	KC2DP22/1	510	CTN50	510,531	CA102PA
504A	1S106	504A	1S430	504A(4)	RCCT225	510	CTN500	510,531	CA102RA
504A	1S1061	504A	1S431	504A(4)	310149	510	CTN600	510,531	CA102YA
504A	1S1062	504A	1S432	504A(4)	TVM-PH9D522/11	510	CTN800	510,531	CA152VA
504A	1S1063	504A	1S44	504A(4)	1RLD	510	CTP100	510,531	CC102PA
504A	1S1064	504A	1S45	509	ATS9500	510	CTP1000	510,531	CC102RA
504A	1S106A	504A	1S456	509	ABB14	510	CTP200	510,531	CC102VA
504A	1S106R	504A	1S457	509	AR-23(DIO)	510	CTP300	510,531	CC152VA
504A	1S1072	504A	1S458	509	AR-24	510	CTP400	510,531	CB508
504A	1S1096	504A	1S459	509	B24-06B	510	CTP50	510,531	CB510
504A	1S110	504A	1S54	509	B24-06C	510	CTP500	510,531	CER72
504A	1S110A	504A	1S558	509	B3K1	510	CTP600	510,531	CER720A
504A	1S111	504A	1S559	509	BA133	510	CTP800	510,531	CER720B
504A	1S112	504A	1S588	509	BB-2	510	EC0125	510,531	CER720C
504A	1S113	504A	1S60(RECT)	509	BYX55009	510	ERCC4-06	510,531	CER72A
504A	1S113A	504A	1S61	509	C0410	510	FA-4	510,531	CER72B
504A	1S114	504A	1S62	509	DI650	510	GE-510	510,531	CER72C
504A	1S115	504A	1S63	509	DS15A	510	KBF-02	510,531	CER72D
504A	1S116	504A	1S64	509	EDS-0004	510	M41032A	510,531	CER72F
504A	1S118	504A	1S65	509	ERB-24-06A	510	MP549	510,531	CER73
504A	1S119	504A	1S66	509	ERB24-06A	510	MR-1C	510,531	CER730
504A	1S120	504A	1S68	509	ERB24-06B	510	MR1031	510,531	CER730A
504A	1S121	504A	1S685	509	ERB24-06C	510	PH9D5	510,531	CER730B
504A	1S121(RECT)	504A	1S71	509	ERB24-06D	510	PT9G22/1	510,531	CER730C
504A	1S122	504A	1S81	509	ERC01-06	510	QDSSR3AMBE	510,531	CER73A
504A	1S1221	504A	1S83	509	ERC04-10	510	R170	510,531	CER73B
504A	1S1222	504A	1S844	509	GE-509	510	RA2C	510,531	CER73C
504A	1S1224	504A	1S846	509	HEPRO055	510	RB-51	510,531	CER73D
504A	1S123	504A	1S849	509	HEPRO056	510	RE51	510,531	CER73F
504A	1S1230	504A	1S849,R	509	HSC1	510	REN125	510,531	CP102PA
504A	1S1231	504A	1S85V	509	J-10	510	RH-DX0041CEZZ	510,531	CP102RA
504A	1S1232	504A	1S885	509	J-8	510	RH-DX0042CEZZ	510,531	COD1537
504A	1S124	504A	1S90	509	K2M5	510	RH-DX0065CZZ	510,531	COD1538
504A	1S125	504A	1S91	509	MR-1M	510	RM-2AV	510,531	COD1617
504A	1S126	504A	1S91R	509	MR1M	510	RO2A	510,531	COD1618
504A	1S1341	504A	1S92	509	NTC-18	510	RT-210	510,531	CP102
504A	1S1342	504A	1S92R	509	PTC203	510	RT210	510,531	CP102PA
504A	1S1343	504A	1S93	509	RA-2	510	S2V	510,531	CP102RA
504A	1S1344	504A	1S93/8GJ	509	RE-90	510	S2VC	510,531	CP102VA
504A	1S1346	504A	1S93R	509	RE90	510	S2VC10	510,531	CP103
504A	1S136(RECT)	504A	1S94	509	REN90	510	S2VC10R	510,531	CP152VA
504A	1S147	504A	1S941/4454C	509	RM-2C	510	SA-2C	510,531	CP100
504A	1S1472	504A	1S9413	509	RM2C	510	SA2H	510,531	D1000
504A	1S148	504A	1S943	509	SK3033	510	SB2CH	510,531	D105C
504A	1S149	504A	1S94R	509	SK3033A	510	SELEN-44	510,531	D108
504A	1S150	504A	1S95	509	SK3080	510	SI-05A	510,531	D1201A
504A	1S1622	504A	1S95R	509	SN75	510	SI-10A	510,531	D1201B
504A	1S1623	504A	1S963	509	TVSB24-06C	510	SI-1A	510,531	D1201D
504A	1S1624	504A	1SB01-02	509	TVSBB2	510	SI-2A	510,531	D1201M
504A	1S1625	504A	1SD-2	509	TVSC0410	510	SI-3A	510,531	D1201N
504A	1S1664	504A	1SD1212	509	TV5MR-1M	510	SI-4A	510,531	D1201P
504A	1S1665	504A	1SD2	509	1N1108	510	SI-5A	510,531	D1D
504A	1S1666	504A	1SIZ09	509	1N1224A	510	SI-6A	510,531	D1K
504A	1S1668F	504A	1SL1885	509	1N1224B	510	SI-7A	510,531	D4L
504A	1S1691	504A	1S030	509	1N1914	510	SI-8A	510,531	D800
504A	1S1692	504A	1S031	509	1N1915	510	SK3032	510,531	D85C
504A	1S1693	504A	1S032	509	1N1916	510	SK3081	510,531	D88
		504A	1S036	509	1N2374	510	SR3AM	510,531	D8HZ
		504A	1S054	509	1N2397	510	SR3AM-2	510,531	DA058
						510	SR3AM-3		
						510	SR3AM-4		

GE	Industry Standard No.
510,531	DA2068
510,531	DD058
510,531	DD2068
510,531	DD268
510,531	DER1
510,531	DI-650
510,531	DICR1
510,531	DID
510,531	DIK
510,531	DP100
510,531	DR1000
510,531	DR700
510,531	DR800
510,531	DR900
510,531	DRS107
510,531	DRS108
510,531	DS-16A(SANYO)
510,531	DS-1K
510,531	DS16NA
510,531	DS2M
510,531	DU1000
510,531	DU800
510,531	E10
510,531	E108(ELCOM)
510,531	E1M3
510,531	E1N3
510,531	E758(ELCOM)
510,531	E760(ELCOM)
510,531	E8
510,531	E9
510,531	EA080
510,531	EA100
510,531	EC100
510,531	ED2847
510,531	ED2848
510,531	ED2849
510,531	ED2910
510,531	ED2911
510,531	ED2912
510,531	ED2913
510,531	ED2922
510,531	ED2923
510,531	ED2924
510,531	ED3007
510,531	ED3007A
510,531	ED3007B
510,531	ED3008
510,531	ED3008A
510,531	ED3008B
510,531	ED3009
510,531	ED3009A
510,531	ED3009B
510,531	ED3010
510,531	ED3010A
510,531	ED3010B
510,531	EF100
510,531	EM507
510,531	EM508
510,531	EM510
510,531	EP1000
510,531	EP800
510,531	ER1001
510,531	ER107D
510,531	ER108D
510,531	ER186
510,531	ER187
510,531	ER308
510,531	ER310
510,531	ER801
510,531	ER81
510,531	ERD1000
510,531	ERD700
510,531	ERD800
510,531	ERD900
510,531	ESKE125C500
510,531	F10
510,531	F11034
510,531	F14C
510,531	F14E
510,531	F14F
510,531	F14H
510,531	F14J
510,531	F8
510,531	FA8
510,531	FR-10
510,531	FT10
510,531	G10
510,531	G100K
510,531	G100M
510,531	G8
510,531	GI237
510,531	GM1J2
510,531	GSR1
510,531	H-881
510,531	H1000
510,531	H621
510,531	H800
510,531	HA1000
510,531	HA800
510,531	HC72
510,531	HC720
510,531	HC73
510,531	HC730
510,531	HEP159
510,531	HEP160
510,531	HEP170
510,531	HF-20071
510,531	HS3108
510,531	HS3110
510,531	IP20-0026
510,531	JCN7
510,531	JCV7
510,531	K1K5
510,531	K1M5
510,531	K2H5
510,531	K2K5
510,531	K3H5
510,531	K3K5
510,531	K3M5
510,531	K4H5
510,531	K4K5
510,531	K4M5
510,531	K5H5
510,531	K5K5
510,531	K5M5
510,531	KSKE125C200
510,531	KSKE12C500
510,531	L82H
510,531	M102
510,531	M702
510,531	M702B
510,531	M72
510,531	M720
510,531	M720A
510,531	M720B
510,531	M720C
510,531	M72A
510,531	M72B
510,531	M72C
510,531	M73
510,531	M730
510,531	M730A
510,531	M730B
510,531	M730C
510,531	M73A
510,531	M73B
510,531	M73C
510,531	M82
510,531	M8HZ
510,531	M91A06
510,531	MC070
510,531	MC070A
510,531	MC080
510,531	MC080A
510,531	MC090
510,531	MC090A
510,531	MC100
510,531	MC1527
510,531	MC1528
510,531	MC1529
510,531	MH72
510,531	MH720
510,531	MH730
510,531	MM10
510,531	MM7
510,531	MM8
510,531	MM9
510,531	MR990
510,531	MT84
510,531	NA104
510,531	NA105
510,531	NA74
510,531	NA75
510,531	NA76
510,531	NA84
510,531	NA85
510,531	NA86
510,531	NSS1021
510,531	OY5067
510,531	P1000
510,531	P10115A
510,531	P10156A
510,531	P5100
510,531	P580
510,531	P6RP10
510,531	P6RP8
510,531	P800
510,531	PA380
510,531	PD114
510,531	PD115
510,531	PD116
510,531	PD913
510,531	PD914
510,531	PD915
510,531	PD916
510,531	PE408
510,531	PE410
510,531	PE508
510,531	PE510
510,531	PH109
510,531	PS1140
510,531	PS2416
510,531	PS2417
510,531	PT580
510,531	PTC205
510,531	R080
510,531	R5
510,531	R5B
510,531	R5C
510,531	R6
510,531	R8
510,531	R8/2H
510,531	RC080
510,531	RL005
510,531	RL010
510,531	RL020
510,531	RL040
510,531	RL060
510,531	RT8665
510,531	S-05/01
510,531	S-500
510,531	S010G
510,531	S100
510,531	S10AR1
510,531	S10AR2
510,531	S1201F
510,531	S1B-0306
510,531	S1C
510,531	S1D
510,531	S20
510,531	S208
510,531	S210
510,531	S24
510,531	S257
510,531	S258
510,531	S260
510,531	S28
510,531	S2E100
510,531	S3G4
510,531	S61
510,531	S62
510,531	S63
510,531	S750
510,531	S750C
510,531	S7AR1
510,531	S8AR1
510,531	S8AR2
510,531	S9AR1
510,531	SA-2
510,531	SA2Z
510,531	SB-1000
510,531	SB2C
510,531	SC10
510,531	SC10A
510,531	SC8
510,531	SC8A
510,531	SCA10
510,531	SCA8
510,531	SCBR05F
510,531	SCBR1F
510,531	SCBR6F
510,531	SCE10
510,531	SCE8
510,531	SCT1
510,531	SCT2
510,531	SCT3
510,531	SCT4
510,531	SCT5
510,531	SD-6AUF
510,531	SD1C
510,531	SD2
510,531	SD2C
510,531	SD7(RECT)
510,531	SD8
510,531	SD800
510,531	SD910
510,531	SD910A
510,531	SD910S
510,531	SD98
510,531	SD98A
510,531	SD98S
510,531	SE0
510,531	SE05C
510,531	SE1730
510,531	SF1
510,531	SF1CN1
510,531	SF3CN1
510,531	SF4
510,531	SF4CN1
510,531	SF5
510,531	SF6
510,531	SFR258
510,531	SFR268
510,531	SH1B
510,531	SH1C
510,531	SH4D8
510,531	SI-RECT-102A
510,531	SI-RECT-136
510,531	SI-RECT-204
510,531	SI1000E
510,531	SIB-0306
510,531	SID01K
510,531	SID02K
510,531	SISD-1HF
510,531	SK-218
510,531	SL608
510,531	SL610
510,531	SL708
510,531	SL710
510,531	SLA01
510,531	SLA18AB
510,531	SLA18C
510,531	SLA19AB
510,531	SLA19C
510,531	SLA2616
510,531	SLA2617
510,531	SLA3196
510,531	SLA560
510,531	SLA561
510,531	SM-150
510,531	SM-150-005
510,531	SM-150-02
510,531	SM-150A
510,531	SM100
510,531	SM101
510,531	SM103
510,531	SM150-01
510,531	SM150-02
510,531	SM150-11
510,531	SM150-6
510,531	SM150A
510,531	SM150B
510,531	SM150C
510,531	SM150D
510,531	SM150S
510,531	SM150SS
510,531	SM170
510,531	SM180
510,531	SM200
510,531	SM270
510,531	SM280
510,531	SM300
510,531	SM517
510,531	SM518
510,531	SM520
510,531	SM70
510,531	SM71
510,531	SM73(DIODE)
510,531	SM770
510,531	SM780
510,531	SM80
510,531	SM800
510,531	SM81
510,531	SM83
510,531	SO10G
510,531	SOA
510,531	SR-1K-2A
510,531	SR-1K-2B
510,531	SR-1K-2C
510,531	SR-1K-2D
510,531	SR-2
510,531	SR-35
510,531	SR-9001
510,531	SR-9007
510,531	SR150
510,531	SR1FM20
510,531	SR2A-12
510,531	SR2A-8
510,531	ST18
510,531	SW05C
510,531	SW05D
510,531	SW1C
510,531	SW1D
510,531	SWC
510,531	SWD
510,531	T-1000
510,531	T-1000X
510,531	T1000X
510,531	T800
510,531	T800X
510,531	TA1000
510,531	TA7805
510,531	TA7806
510,531	TA800
510,531	TE1064
510,531	TFR105
510,531	TFR110
510,531	TFR140
510,531	TH1000
510,531	TH800
510,531	TH808
510,531	TH810
510,531	TIR07
510,531	TIR08
510,531	TIR09
510,531	TIR10
510,531	TK1000
510,531	TK800
510,531	TKF100
510,531	TKF80
510,531	TM86
510,531	TR320039
510,531	TR320041
510,531	TR320048
510,531	TS3
510,531	TS8
510,531	TV24221
510,531	TV24942
510,531	TV8
510,531	TVS-FR10
510,531	TVS-FT-1N
510,531	TVS-FT10
510,531	TVS-SB-2C
510,531	TVS-SB-20
510,531	TVS-SD-1B
510,531	TVS-SD1B
510,531	TVS1P80
510,531	TVSBB10
510,531	TVSMR1C
510,531	TVSS3G4
510,531	TVSSA-2B
510,531	TVSSF-1
510,531	TVSSID30-15
510,531	TW100
510,531	TW80
510,531	001-0072-00
510,531	1C10
510,531	1D10
510,531	1D8
510,531	1E10
510,531	1E7
510,531	1E8
510,531	1ET10
510,531	1ET7
510,531	1ET8
510,531	1F8
510,531	1G8
510,531	1HY100
510,531	1HY80
510,531	1N1095
510,531	1N1096
510,531	1N1225A
510,531	1N1226A
510,531	1N1258
510,531	1N1259
510,531	1N1260
510,531	1N1261
510,531	1N1407
510,531	1N1408
510,531	1N1443
510,531	1N1443A
510,531	1N1730
510,531	1N1730A
510,531	1N3196
510,531	1N321A
510,531	1N322
510,531	1N3221
510,531	1N322A
510,531	1N328
510,531	1N3280
510,531	1N3281
510,531	1N3282
510,531	1N328A
510,531	1N329
510,531	1N329A
510,531	1N364
510,531	1N364A
510,531	1N365
510,531	1N365A
510,531	1N4006
510,531	1N4007
510,531	1N4011
510,531	1N4250
510,531	1N4251
510,531	1N4361
510,531	1N4818
510,531	1N5054
510,531	1N5054A
510,531	1N509
510,531	1N510
510,531	1N517
510,531	1N518
510,531	1N5214
510,531	1N5218
510,531	1N525
510,531	1N526
510,531	1N548
510,531	1N560
510,531	1N561
510,531	1N597
510,531	1N598
510,531	1N853
510,531	1N854
510,531	1N855
510,531	1N856
510,531	1N864
510,531	1N865
510,531	1N866
510,531	1N867
510,531	1N875
510,531	1N876
510,531	1N877
510,531	1N878
510,531	1N886
510,531	1N887
510,531	1N888
510,531	1N889
510,531	1NJ27
510,531	1R5GZ61
510,531	1R96
510,531	1S1065
510,531	1S1066
510,531	1S107
510,531	1S108
510,531	1S109
510,531	1S117
510,531	1S1223
510,531	1S1225A
510,531	1S1233
510,531	1S1234
510,531	1S1255A
510,531	1S1345
510,531	1S1347
510,531	1S1348
510,531	1S138
510,531	1S1695
510,531	1S1829
510,531	1S210
510,531	1S210/2SJ6A
510,531	1S211
510,531	1S211/2SJ8A
510,531	1S2311
510,531	1S2312
510,531	1S2314
510,531	1S2315
510,531	1S2353
510,531	1S2354
510,531	1S2364
510,531	1S2365
510,531	1S2403
510,531	1S2405
510,531	1S2406
510,531	1S2608
510,531	1S2610
510,531	1S397
510,531	1S398
510,531	1S401
510,531	1S402
510,531	1S557
510,531	1S686
510,531	1S687
510,531	1S848
510,531	1S850
510,531	1S96
510,531	1S97
510,531	1S98
510,531	1S99
510,531	1S038
510,531	1S058
510,531	1T507
510,531	1T508
510,531	1T509
510,531	1T510
510,531	2-64701508
510,531	2HR3J
510,531	2HR3M
510,531	2K02
510,531	2NSM-1
511	A04093-A
511	A04093A2
511	A7568300
511	A7580910
511	A7582000
511	A7978850
511	A7978855
511	AD-29B
511	AD-29S
511	AD29S
511	AD8
511	B2620
511	B3N1
511	B3N5
511	B3N9
511	B6N1
511	B6N5
511	B6N9
511	B7978850
511	B7N1
511	B7N5
511	B7N9
511	BB-10
511	BB-10S
511	BB-4
511	BB-6
511	BB2
511	BB4
511	BB6
511	BB6S
511	BBGS
511	BR-10S
511	BY103
511	BY133
511	BY157
511	BY158
511	BYX22-1200
511	BYX57-500
511	BYX57-600
511	C-2C
511	CGJ-1
511	CXL506
511	CXL507
511	D172-F
511	D1F
511	D431-F
511	D5H
511	D8G
511	D8L
511	D8M
511	DG-1N
511	DG-1NR
511	DG1N
511	DG1NR
511	DGIN
511	DIJ72163
511	DIJ72167
511	DIJ72170
511	DIJ72171
511	DIJ72174
511	DIJ72290
511	DIJ72291
511	DS-113A
511	DS-113B
511	DS-15A(SANYO)
511	DS-15B(SANYO)
511	DS-2M
511	DS113B
511	DS133B
511	DS16NE(SONY)
511	DT230B
511	E.S57X9
511	E1145R
511	E1145RED
511	E1145RJH
511	ECG506
511	ED23(ELCOM)
511	EP16X10
511	EP16X11
511	EP16X24
511	EP57X5
511	EP6X11
511	EPB22-15
511	ERB24-06
511	ERB24-04A
511	ERB24-04B
511	ERB24-04C
511	ERB24-04D
511	ERC26-131
511	ERC26-13L
511	ERC27-13
511	ES-16X16
511	ES16X16
511	ES16X20
511	ES16X28
511	ES16X31
511	ES57X8
511	ES57X9
511	ET57X39
511	ET57X40
511	EV16X20

GE	Industry Standard No.	GE	Industry Standard No.	GE	Industry Standard No.	GE	Industry Standard No.	GE	Industry Standard No.
511	F1143	511	S5295G	511	1N5190	514	D-215-1	514	1S2473VE
511	FDB600	511	S5295J	511	1N543	514	D919039-2	514	182692
511	FG-2N	511	SB-1	511	1N5433	514	DDAY-048001	514	182692AB
511	FG-2N/10D-4	511	SB-1B	511	1N5434	514	DDAY047001	514	1S953
511	FG-2N2	511	SB-2	511	1N543A	514	DDAY048007	514	1S954
511	FG2B	511	SB-2C(CENTERING)	511	1N543	514	DDAY048012	514	1S955
511	FG2PC	511	SB-2B	511	1N5552	514	DDAY048014	529	N2A
511	FR2-005	511	SB-2C	511	1N5553	514	DDAY069001	530	1TH61
511	FR2-02(RECT)	511	SB-2CGL	511	1N5554	514	DS-113	531	.5E10
511	FR2-02C	511	SB-2CH	511	1N5616	514	DS1005-1X862B	531	.5E7
511	FR2-04	511	SB-2T	511	1N5617	514	DSI-104-2	531	.5E8
511	FR2-10	511	SB2G	511	1N5618	514	DX-0150	531	.5J10
511	FT-1M	511	SD-1	511	1N5619	514	DX-0270	531	.5J7
511	FT-1P	511	SD-1(BOOST)	511	1N5620	514	DX-0273	531	.5J8
511	FT1M	511	SD-1-30DA	511	1N5621	514	DX-0299	531	.7E10
511	FT1P	511	SD-1A4P	511	1N693	514	DX7429	531	.7E7
511	FT1T	511	SD-1B(HP)	511	1N893	514	EA16X135	531	.7E8
511	FTIM	511	SD-1BHF	511	1N923	514	EA16X152	531	.7J10
511	FU-1M	511	SD-1C	511	1NJ70976	514	ECG519	531	.7J7
511	FU10	511	SD-1VHF	511	R1003L	514	ED86023	531	.7J8
511	FU1K	511	SD-12	511	R31	514	EMS72258	531	A-11803B
511	FU1NA	511	SD-22	511	R5B	514	EMS72272	531	A04212-A
511	GE-511	511	SD-ERB24-04D	511	R5TH61	514	FD-1029-QP	531	A059-118
511	GH3P	511	SD-ERC26-13	511	1S1226	514	FD01880	531	A068-104
511	GRU2A	511	SD1-11	511	1S1257	514	FD06193	531	A1000
511	H815	511	000000SD1A	511	1S1258	514	FD111	531	A10P
511	HA-1	511	SD1BHP	511	1S1349	514	FD600	531	A114C
511	HD-2000108	511	SD22	511	1S1517	514	FD777	531	A13M2
511	HD-20008	511	SELEN-92	511	1S1517A	514	G657061	531	A1H1
511	HEPR0600	511	SF-1	511	1S1832	514	G657123	531	A1H5
511	HEPR0602	511	SI-RECT-114	511	1S1834	514	GD101	531	A1H9
511	HEPR0604	511	SI-RECT-124	511	1S1855	514	GD102	531	A1K1
511	HEPR0606	511	SI-RECT-158	511	1S1855	514	GE-514	531	A1K5
511	HEPR3010	511	SI-RECT-162	511	1S1882	514	GVL20226	531	A1K9
511	HF-17	511	SI-RECT-170	511	1S1920	514	GVL20265	531	A1M1
511	HF-1Z	511	SI-RECT-180	511	1S1921A	514	GVL20327-1	531	A1M5
511	HF-1B	511	SI-RECT-182	511	1S1921B	514	HD20001210	531	A1M9
511	HF-1Z	511	SI-RECT-206	511	1S1921C	514	HP-20063	531	A2H1
511	HF-SD1/BB-4	511	SI-RECT-224	511	1S1921D	514	HP5082-2800	531	A2H4
511	HP1	511	SI-RECT-36	511	1S1921E	514	HV-80	531	A2H5
511	HP1A	511	SI-RECT-44	511	1S1921F	514	IC743049	531	A2H9
511	HP1B	511	SIB0-1	511	1S2306	514	IR10B6X	531	A2K1
511	HP1C	511	SIB01	511	1S2307	514	ITT200	531	A2K4
511	HP1Z	511	SID23-13	511	1S2308	514	ITT73N	531	A2K5
511	HPSD-1	511	SID23-15	511	1S2309	514	IX8055-379005N	531	A2K9
511	HPSD-14	511	SID30-13	511	1S2711	514	LD34R	531	A2M1
511	HPSD-1A	511	SID30-132	511	1S2756	514	M8513-0	531	A2M4
511	HPSD-1B	511	SID30-15	511	1S312(HITACHI)	514	MA161	531	A2M5
511	HPSD-1C(SEARS)	511	SISD-1	511	1S315(HITACHI)	514	MA162	531	A3H1
511	HPSD-1Z	511	SK3130	511	2B-2	514	MBR-65-L11324	531	A3H5
511	HPSD1	511	SK3175	511	25B-2C	514	MF4300158	531	A3H9
511	HFSD1Z	511	SM-150-6(FOCUS)	511	25C-2C	514	ON120623	531	A3K3
511	HSPD-1A(SONY)	511	SR1156	512	D81M	514	ON143285	531	A3K5
511	HSPD1	511	SRIFM-12	512	ECG5800	514	ON206068	531	A3K9
511	IR10D3L	511	SRIFM-4	512	ECG5801	514	NTC-15	531	A3M1
511	IR1P	511	SRIHM-8	512	ECG5802	514	PO4-41-0025-001	531	A3M5
511	IR3D	511	SR25	512	ECG5803	514	PO4-42-0011	531	A3M9
511	IR1F	511	SRIFM-12	512	ECG5804	514	PO4410025-003	531	A4212A
511	J241183	511	SRIFM-2	512	GM-3	514	PO4410042-001	531	A4H1
511	J241215	511	SRIHM-4	512	GM-3X	514	PO4410042-002	531	A4H5
511	JC-00017	511	SRIHM-8	512	GM-3Z	514	P830802	531	A4H9
511	LC0-09M1113	511	T-2062	512	HEPRO080	514	P863A205	531	A4K1
511	LM1932	511	T-E1145	512	HEPRO081	514	PT4-2268-011	531	A4K5
511	MB-1D	511	T-E1146	512	HEPRO082	514	PT4-2268-01B	531	A4K9
511	MB-1P	511	T-E1153	512	HEPRO084	514	PT4-2311-011	531	A4M1
511	MR-852	511	TA7051/01	512	HEPRO090	514	PT40063	531	A4M5
511	MR2266	511	TC-0-2P11/1	512	HEPRO091	514	Q-13914	531	A4M9
511	MR2272	511	TC-0.2	512	HEPRO092	514	Q-23115C	531	A5H1
511	MR2273	511	TC0.1P	512	HEPRO094	514	R14010658	531	A5H2
511	MR801	511	TD-13	512	HR-200	514	R3410-P1	531	A5H5
511	MR811	511	TD-15	512	PTC204	514	RF-94	531	A5H9
511	MR812	511	TD15	512	R210	514	RR94	531	A5K1
511	MR814	511	TD15M	512	R250	514	REN94	531	A5K2
511	MR816	511	TSTD-15	512	R250P	514	RH-DX0033TAZZ	531	A5K5
511	N815835L1	511	TV24167	512	S1RB10	514	RH-DX0048CEZZ	531	A5K9
511	NTC-17	511	TV24169	512	SI10A	514	RKZ12003	531	A5M1
511	0F66	511	TV24217	512	SI8A	514	RKZ120101	531	A5M2
511	0V02	511	TV24219	512	SJ101P	514	S-2064-G	531	A5M5
511	PH108	511	TV24617	512	SR1EM-4	514	S074-005-0001	531	A5M9
511	PTC216	511	TV24648	512	SR3AM-6	514	SC-1016	531	A725EH2AB1
511	R0606	511	TV34	512	SW-1-01	514	SE5-0247-C	531	A800
511	R1Z	511	TVM535	512	SW-1-02	514	SE5-0456	531	A1000
511	RA-Z	511	TVM537	512	1-531-024	514	SM-C-706156	531	AA800
511	RA1	511	TV8-OY-02	512	1N4719	514	SN4448	531	AD4002
511	R855	511	TV8-2B-2C	512	1N4719R	514	ST/123/CR	531	AD4003
511	REN506	511	TVSRF-1A	512	1N4720R	514	ST/146/CR	531	AD4004
511	RF-1A	511	TVS81R80	512	1N4721	514	ST22546-1	531	AD4005
511	RP10K55	511	1.5E12	512	1N4721R	514	ST32012-0037	531	AD4006
511	RPJ70971	511	1.5J12	512	1N4722	514	T-10010	531	AD4007
511	RPJ70149	511	1N1409	512	1N4722R	514	TA198785-2	531	AH1005
511	RPJ70976	511	1N2327	512	1N4997	514	TAB532787	531	AH1010
511	RH-1	511	1N2357	512	1N4997R	514	TVS182076	531	AH1015
511	RH-1B	511	1N2503	512	1N4998	514	TVS18954	531	AH14
511	RH-1C	511	1N2504	512	1N4998R	514	1-0002-0001	531	AHB05
511	RH-1V	511	1N2507	512	1N4999	514	001-026010	531	AHB10
511	RH-DX0004TAZZ	511	1N2508	512	1N4999R	514	1-13989	531	AHB15
511	RH-DX0017CEZZ	511	1N2618	512	1N5000	514	001-226010	531	AN-G-5C
511	RH-DX0028CEZZ	511	1N2619	512	1N5000R	514	1435	531	AN-05B
511	RH-DX0029CEZZ	511	1N2776	512	1N5197	514	1A69425-1	531	AR23
511	RH-DX0045CEZZ	511	1N2777	512	1N5199	514	1A99812-1001	531	AR24
511	RH-DX0051CEZZ	511	1N2778	512	1N5200	514	182992	531	B4H1
511	RH-DX0056TAZZ	511	1N2779	512	1NE11	514	1N3062	531	B4H5
511	RH-DX0062CEZZ	511	1N2884	512	1S1071	514	1N3600	531	B4H9
511	RH-DX0065CEZZ	511	1N2885	512	1S1892	514	1N4148	531	B4K1
511	RH-DX0073CEZZ	511	1N3233	512	1S1944	514	1N4149	531	B4K5
511	RH-DX0081CEZZ	511	1N3244	512	1S2409	514	1N4153	531	B4K9
511	RH-DX0090CEZZ	511	1N3487	514	A-1634-1125	514	1N4154	531	B4M1
511	RH-DX0096CEZZ	511	1N3724	514	A-1901-0050-1	514	1N4305	531	B4M5
511	RH1	511	1N3725	514	A-1901-0196-1	514	1N4446	531	B4M9
511	RH1B	511	1N3729	514	A-1901-044-1	514	1N4447	531	BAY15
511	R81296(SEARS)	511	1N3731	514	A-1901-0461-1	514	1N4448	531	BAY16
511	RBLND0003CEZZ	511	1N3915	514	A-2008-9140	514	1N4449	531	BAY23
511	RT1686	511	1N4146	514	A-36617	514	1N4454	531	BAY90
511	RU-IN	511	1N4155	514	A7249601	514	1N914	531	BB10
511	RU2	511	1N4252	514	B269-3345	514	1N914/A/B	531	GE-531
511	RU2V	511	1N4826	514	BA202	514	1N914A	531	TVSDB-2K
511	S-15H	511	1N487	514	BA216	514	1N914B	4001	CD4001
511	S-34	511	1N487A	514	BA219	514	1N914M	4001	CD4001BE
511	S-4C	511	1N487B	514	BA317	514	1S1588	4001	CM4001
511	S10010	511	1N488	514	BAX-13	514	1S1589V	4001	ECG4001
511	S10120	511	1N488A	514	BAX13	514	1S2075	4001	ECG4001B
511	S12110	511	1N488B	514	BAX91C/TP102	514	1S2075K	4001	F4001
511	S12120	511	1N4934	514	BAX95TP600	514	1S2076-27	4001	G9-4001
511	S12130	511	1N4935	514	BPS-8-50	514	1S2076-TP1	4001	HBP4001
511	S12140	511	1N4936	514	BRN-SPEC-24-12	514	1S2076-TPI	4001	HD4001
511	S154	511	1N4937	514	C23018	514	1S2076A	4001	HEPC4001P
511	S1B01-04	511	1N4945	514	C256125-011	514	1S2471	4001	MC14001
511	S1D23-13	511	1N4946	514	CD-5038	514	1S2472	4001	MEM4001
511	S1D23-15	511	1N4947	514	CD5002	514	1S2473	4001	MM4001
511	S1D26	511	1N4948	514	CD6016	514	1S2473-T72	4001	N4001
511	S1D30-13	511	1N5188	514	CD6016-013-689	514	1S2473H	4001	RS4001
511	S1D30-15	511	1N5189	514	CD6161P1N013-75	514	1S2473HC		
511	S1P			514	C023018	514	1S2473K		
511	S1R80			514	CXL519	514	1S2473T		

GE	Industry Standard No.
4001	SCL4001
4001	SW4001
4001	TC4001P
4001	TP4001
4011	CD4011
4011	CD4011BE
4011	CM4011
4011	DDEY089001
4011	ECG4011
4011	ECG4011B
4011	F4011
4011	F4011PC
4011	GE-4011
4011	HBF4011
4011	HD4011
4011	MB84011
4011	MB84011U
4011	MB84011V
4011	MC-14011CP
4011	MC14011
4011	MC14011B
4011	MC14011CP
4011	MEM4011
4011	MM4011
4011	N4011
4011	RS4011
4011	SCL4011
4011	SW4011
4011	TC4011P
4011	TP4011
4012	CD4012BE
4012	CM4012
4012	ECG4012
4012	ECG4012B
4012	F4012
4012	GE-4012
4012	HBF4012
4012	HD4012
4012	MC14012
4012	MM4012
4012	N4012
4012	SCL4012
4012	SW4012
4012	TP4012
4013	CD4013
4013	CD4013AE
4013	CD4013BE
4013	ECG4013
4013	ECG4013B
4013	F4013
4013	GE-4013
4013	HBF4013
4013	HD4013
4013	MC14013
4013	MEM4013
4013	MM4013
4013	N4013
4013	RS4013
4013	SCL4013
4013	SW4013
4013	TP4013
4017	CD4017BE
4017	CM4017
4017	ECG4017
4017	ECG4017B
4017	F4017
4017	GE-4017
4017	HBF4017
4017	HD4017
4017	MC14017
4017	MM4017
4017	RS4017
4017	SCL4017
4017	SW4017
4017	TP4017
4020	CD4020BE
4020	CM4020
4020	ECG4020
4020	ECG4020B
4020	F4020
4020	GE-4020
4020	HBF4020
4020	HD4020
4020	HEPC4020P
4020	MC14020
4020	MM4020(IC)
4020	RS4020
4020	SCL4020
4020	SK4020
4020	SW4020
4020	TP4020
4021	CD4021BE
4021	CM4021
4021	ECG4021
4021	ECG4021B
4021	F4021
4021	GE-4021
4021	HBF4021
4021	HD4021
4021	HEPC4021P
4021	MC14021
4021	MM4021
4021	N4021
4021	SCL4021
4021	SK4021
4021	SW4021
4021	TP4021
4023	CD4023BE
4023	CM4023
4023	ECG4023
4023	ECG4023B
4023	F4023
4023	GE-4023
4023	HBF4023
4023	HD4023
4023	MC14023
4023	MM4023
4023	N4023
4023	SCL4023
4023	SW4023
4023	TP4023
4027	CD4027BE
4027	CM4027
4027	ECG4027
4027	ECG4027B
4027	F4027
4027	GE-4027
4027	HBF4027
4027	HD4027
4027	MC14027
4027	MM4027
4027	N4027
4027	RS4027
4027	SCL4027
4027	SW4027
4027	TP4027
4049	CD4049UBE
4049	CM4049
4049	ECG4049
4049	ECG4049B
4049	F4049
4049	GE-4049
4049	HD4049
4049	MC14049
4049	MC14049B
4049	MC14049CP
4049	MEM4049
4049	MM4049
4049	N4049
4049	RS4049
4049	SCL4049
4049	SK4049
4049	SW4049
4049	TP4049
4050	CD4050BE
4050	CM4050
4050	ECG4050
4050	ECG4050B
4050	F4050
4050	GE-4050
4050	HD4050
4050	HEPC4050P
4050	MC14050
4050	MEM4050
4050	MM4050
4050	N4050
4050	RS4050
4050	SCL4050
4050	SK4050
4050	SW4050
4050	TP4050
4051	CD4051BE
4051	CM4051
4051	ECG4051
4051	ECG4051B
4051	F4051
4051	GE-4051
4051	HEPC4051P
4051	MC14051
4051	MEM4051
4051	MM4051
4051	SCL4051
4051	SK4051
4051	TP4051
4518	CD4518BE
4518	ECG4518
4518	ECG4518B
4518	F4518
4518	GE-4518
4518	HD4518
4518	MC14518
4518	MM4518
4518	RS4518
4518	SCL4518
4518	SK4518
4518	SW4518
4518	TP4518
5004	AM13
5004	AM21
5004	AM22
5004	AM23
5004	AM24
5004	AM3
5004	AM32
5004	AM33
5004	AM42
5004	AM43
5004	1N448
5004	1N1553
5004	1N1554
5004	1N2217
5004	1N2267
5004	1N2289
5004	1N2289A
5004	1N2290A
5004	N255
5004	1N2850
5004	N348
5004	1N550
5005	38146
5005	1N1046
5005	1N1047
5005	1N1048
5005	1N1049
5005	1N1050
5005	1N1051
5008	AM53
5008	AM63
5008	1N1119
5008	1N1120
5008	1N119
5008	1N1543
5008	1N1544
5008	1N2218
5008	1N2219
5008	1N2220
5008	1N2221
5008	1N2268
5008	1N2269
5008	1N2270
5008	1N2271
5008	1N256
5008	1N2851
5008	1N2852
5008	1N554
5008	1N555
5008	1N613
5008	1N613A
5008	1N614
5008	1N614A
5012	1N2222
5012	1N2222A
5012	1N2223
5012	1N2223A
5012	1N2224
5012	1N2224A
5012	1N2225
5012	1N2225A
5012	1N2372
5012	1N562
5012	1N563
5017	1N3064
5032	SL103
5036	G02
5036	SL3
5044	1N4014
5044	1N427
5048	NA21
5064	1N3615
7400	DM7400N
7400	ECG7400
7400	GE-7400
7400	HC1000411Q
7400	HD7400
7400	HD7400P
7400	HEPC7400P
7400	ITT7400N
7400	J4-1000
7400	N7400A
7400	N7400N
7400	REN7400
7400	RS7400
7400	SK7400
7400	SN7400A
7400	SN7400N
7400	SN7400N-10
7402	DM7402N
7402	ECG7402
7402	GE-7402
7402	HEPC7402P
7402	ITT7402N
7402	N7402A
7402	RS7402
7402	SK7402
7402	SN7402
7402	SN7402N
7402	SN7402N-10
7404	DM7404N
7404	ECG7404
7404	GE-7404
7404	HEPC7404P
7404	ITT7404N
7404	J4-1004
7404	N-7404A
7404	N7404A
7404	RS7404
7404	SK7404
7404	SN7404N
7404	SN7404N-10
7406	DM7406N
7406	ECG7406
7406	GE-7406
7406	HEPC7406P
7406	ITT7406N
7406	N7406A
7406	RS7406(IC)
7406	SK7406
7406	SN7406N
7406	SN7406N-10
7408	DM7408N
7408	ECG7408
7408	GE-7408
7408	HEPC7408P
7408	ITT7408N
7408	N-7408A
7408	N7408A
7408	RS7408(IC)
7408	SK7408
7408	SN7408N
7408	SN7408N-10
7410	DM7410N
7410	ECG7410
7410	GE-7410
7410	HEPC7410P
7410	IC-7410
7410	ITT7410N
7410	M53210
7410	M53210P
7410	N7410A
7410	RS7410(IC)
7410	SK7410
7410	SN7410N
7410	SN7410N-10
7413	DM7413N
7413	ECG7413
7413	GE-7413
7413	HEPC7413P
7413	ITT7413N
7413	N7413A
7413	RS7413(IC)
7413	SK7413
7413	SN7413N-10
7420	DM7420N
7420	ECG7420
7420	EP84X11
7420	GE-7420
7420	HEPC7420P
7420	IC7420
7420	ITT7420N
7420	N7420A
7420	RS7420
7420	SK7420
7420	SN7420N
7420	SN7420N-10
7427	DM7427N
7427	ECG7427
7427	GE-7427
7427	HEPC7427P
7427	N7427A
7427	RS7427
7427	SK7427
7427	SN7427
7427	SN7427N
7427	SN7427N-10
7432	DM7432N
7432	ECG7432
7432	GE-7432
7432	HEPC7432P
7432	ITT7432N
7432	N7432A
7432	RS7432
7432	SK7432
7432	SN7432N-10
7441	DM7441N
7441	ECG7441
7441	GE-7441
7441	HEPC7441AP
7441	RS7441
7441	SK7441
7441	SN7441N-10
7447	DM7447N
7447	ECG7447
7447	GE-7447
7447	HEPC7447AP
7447	ITT7447AN
7447	J4-1047
7447	N7447B
7447	N7447P
7447	RS7447
7447	SK7447
7447	SN7447N-10
7448	DM7448N
7448	ECG7448
7448	GE-7448
7448	HEPC7448P
7448	ITT7448N
7448	N7448B
7448	RS7448
7448	SK7448
7448	SN7448N-10
7451	DM7451N
7451	ECG7451
7451	GE-7451
7451	HEPC7451P
7451	ITT7451N
7451	N74151A
7451	N7451A
7451	RS7451
7451	SK7451
7451	SN7451N-10
7473	BO075660
7473	DM7473N
7473	ECG7473
7473	GE-7473
7473	HD7473AP
7473	HD7473P
7473	HEPC7473P
7473	ITT7473N
7473	M53273P
7473	M7641
7473	N7473
7473	N7473A
7473	REN7473
7473	RS7473
7473	SK7473
7473	SN7473
7473	SN7473N
7473	SN7473N-10
7473A	TD54T3AP
7473A	N-7473A
7474	DM7474N
7474	ECG7474
7474	F7474PC
7474	GE-7474
7474	HD7474
7474	HD7474P
7474	HEPC7474P
7474	IP20-0316
7474	N7474A
7474	REN7474
7474	RS7474
7474	SK7474
7474	SN-7474
7474	SN7474
7474	SN7474N-10
7475	DM7475N
7475	ECG7475
7475	GE-7475
7475	HEPC7475P
7475	ITT7475N
7475	J4-1075
7475	N7475B
7475	RS7475
7475	SN7475N-10
7476	DM7476N
7476	ECG7476
7476	GE-7476
7476	HEPC7476P
7476	ITT7476N
7476	J4-1076
7476	N7476B
7476	RS7476
7476	SK7476
7476	SN7476N
7476	SN7476N-10
7485	DM7485N
7485	ECG7485
7485	GE-7485
7485	HEPC7485P
7485	N7485A
7485	N7485B
7485	RS7485
7485	SK7485
7485	SN7485N-10
7486	DM7486N
7486	ECG7486
7486	GE-7486
7486	HEPC7486P
7486	ITT7486N
7486	N7486A
7486	RS7486
7486	SK7486
7486	SN7486N-10
7490	DM7490N
7490	ECG7490
7490	GE-7490
7490	HEPC7490AP
7490	ITT7490N
7490	J4-1090
7490	M53290P
7490	N7490A
7490	RS7490
7490	SK7490
7490	SN7490N
7490	SN7490N-10
7492	ECG7492
7492	GE-7492
7492	HEPC7492AP
7492	ITT7492N
7492	J4-1092
7492	N7492A
7492	RS7492
7492	SK7492
7492	SN7492AN
7492	SN7492N-10
74123	DM74123N
74123	ECG74123
74123	EP84X19
74123	GE-74123
74123	HEPC74123P
74123	IC-74123
74123	ITT74123N
74123	N74123B
74123	REN74123
74123	RS74123
74123	SK74123
74123	SN74123N
74123	SN74123N-10
74145	DM74145N
74145	ECG74145
74145	GE-74145
74145	HEPC74145P
74145	ITT74145N
74145	N74145
74145	N74145B
74145	RS74145
74145	SK74145
74145	SN74145N-10
74150	DM74150N
74150	ECG74150
74150	GE-74150
74150	HEPC74150P
74150	ITT74150N
74150	N74150N
74150	RS74150
74150	SK74150
74150	SN74150N-10
74154	DM74154N
74154	ECG74154
74154	GE-74154
74154	HEPC74154P
74154	ITT74154N
74154	N74154N
74154	RS74154
74154	SK74154
74154	SN74154N-10
74192	DM74192N
74192	ECG74192
74192	GE-74192
74192	HEPC74192P
74192	ITT74192N
74192	N74192B
74192	RS74192
74192	SK74192
74192	SN74192
74192	SN74192N-10
74193	DM74193N
74193	ECG74193
74193	GE-74193
74193	HEPC74193P
74193	ITT74193N
74193	N74193B
74193	RS74193
74193	SK74193
74193	SN74193N-10
74196	DM74196N
74196	ECG74196
74196	GE-74196
74196	HEPC74196P
74196	N74196A
74196	RS74196
74196	SN74196
74196	SN74196N-10

RS 276-	Industry Standard No.	RS 276-	Industry Standard No.	RS 276-	Industry Standard No.	RS 276-	Industry Standard No.
007	A1820-0219-1	010	183532	561	BZY83D5V6	561	GLA62
007	C6052P	010	188660-01	561	BZY83D6V8	561	GLA62A
007	CA741CS	010	193207	561	BZY85/C5V6	561	GLA62B
007	EC0941M	010	301915-1	561	BZY85/D5V6	561	GZ6.2
007	FU9T7741393	010	508511	561	BZY85C5V6	561	HB-7B
007	GEIC-265	010	618483-1	561	BZY85C6V2	561	HEP-Z0212
007	HEP-C6052P	010	618984-1	561	BZY85D5V6	561	HEP-Z0214
007	HEPC6052P	010	717399-4	561	BZY85D6V8	561	HRP-Z0215
007	IC-295(ELCOM)	010	717399-49	561	BZY88/6V2	561	HEP-Z0407
007	J4-1215	010	1802520-001	561	BZY88/C5V6	561	HEP-Z0408
007	K8-2177	010	2710002	561	BZY88/C6V2	561	HEP-Z0409
007	MC1741CP1	010	2797658-616030A	561	BZY88C5V6	561	HEP-Z212
007	N5741V	010	3007680-00	561	BZY88C6V2	561	HEP103
007	SG741CM	017	EC0915	561	BZY92/5V6	561	HEP603
007	SK3552	017	FU5T7715393	561	BZY92C5V6	561	HEPZ0212
007	SK3552/941M	017	G39050381	561	BZY92C6V2	561	HEPZ0212A
007	SL22745	017	IC35(ELCOM)	561	BZY92C6V8	561	HEPZ0214
007	SL23059	017	SL21384	561	BZZ10	561	HEPZ0214A
007	SL23252	017	SL21577	561	BZZ11	561	HRPZ0215
007	SL23496	017	TM915	561	C4011	561	HEPZ0215A
007	SN72471P	017	62M432B2	561	CD-0033	561	HEPZ0407
007	SN72741P	017	156-0151-00	561	CD0033	561	HEPZ0408
007	STX49007	017	179-46447-21	561	CD0120	561	HEPZ0409
007	T3B	017	715HC	561	CD31-00007	561	HF-20033
007	TA7504P	017	1820-0476	561	CD31-00008	561	HM6.8
007	TM941M	017	143041	561	CD48	561	HM6.8A
007	UA741TC	017	157564	561	CHA4Z5.6	561	H82056
007	WEP933	017	2902798-2	561	CHAZ6.2	561	H82062
007	007-1669608	067	EC03060	561	CHAZ6.2A	561	H87056
007	13-5018-6	561	.25N6.8	561	CHAZ62.A	561	H87062
007	19-10298-00	561	.25T5.6	561	CHMZ5.6	561	HW6.8
007	020-1114-015	561	.25T5.6A	561	CXL137	561	HW6.8A
007	31-1012	561	.25T5.6B	561	CXL5011	561	HZ-6B
007	34-194	561	.25T5.8	561	CXL5012	561	HZ-6C
007	44A41779-001	561	.25T5.8A	561	CXL5013	561	HZ-7
007	44A41779-001	561	.25T6.2	561	CXL5014	561	HZ-7(B)
007	51-10715A01	561	.25T6.2A	561	CXL5070	561	HZ-7A
007	51810715A01	561	.25T6.2B	561	CXL5071	561	HZ-7B
007	133P80104	561	.25T6.8	561	CXL6	561	HZ-7C
007	156-0067-00	561	.25T6.8A	561	D-00469Q	561	H26(H)
007	156-0067-06	561	.25T7.1	561	D1M	561	H26B
007	398-13227	561	.25T7.1A	561	D3H	561	H26B2
007	551-008-00	561	.4T5.6	561	D3Y	561	H26C
007	733W00021	561	.4T5.6A	561	D5Y	561	H26L
007	741TC	561	.4T5.6B	561	D6.2	561	H27
007	1820-0216-1	561	.4T6.2	561	DDAY008001	561	H27(H)
007	1820-0217-1	561	.4T6.2A	561	DIJ72293	561	H27A
007	137875(I.C.)	561	.4T6.2B	561	DIM	561	H27B
007	137875(U.C.)	561	.4T6.8	561	D8-159	561	H27L
007	144178	561	.4T6.8A	561	DS104	561	IN61225
007	150580	561	.4T6.8B	561	DS159	561	IP20-0186
007	150580-2285	561	.75N5.6	561	D850	561	IP20-0203
007	615246-1	561	.75N6.2	561	DX-0061	561	J241179
007	615268-101	561	.7JZ6.8	561	DX-0530	561	J241186
007	660388-02	561	.7Z6.8A	561	DX-0727	561	J24186
007	2392152	561	.726.8B	561	DX-1132	561	J24262
007	2610043-03	561	.726.8C	561	DX1194	561	J24631
010	A1820-0203-1	561	.726.8A	561	DZM6.8	561	J4-1619
010	A1826-0007-1	561	.72M6.8B	561	E21431	561	K8-2503
010	AD741	561	.72M6.8C	561	E2486	561	K0825211-2
010	AD741C	561	.72M6.8D	561	EA16X118	561	K8056A
010	AD741CH	561	A-125278	561	EA16X123	561	K8056B
010	AM741HC	561	A29035-E	561	EA16X80	561	K8062A
010	AMU5B7741393	561	A7285900	561	EA2608	561	K82056B
010	CA3741	561	A7286201	561	EC0136A	561	K82062A
010	CA3741CS	561	AW01-06	561	EC0136A	561	K82062B
010	CA3741CT	561	AW01-07	561	EC0137	561	K836A
010	CA3741S	561	AW01-7	561	EC0137A	561	K8364AF
010	CA3741T	561	AZ-052	561	EC05011	561	K836B
010	CA741CT	561	AZ-054	561	EC05011A	561	K836BF
010	EC0941	561	AZ-056	561	EC05012	561	K837A
010	FU5B7741393	561	AZ-058	561	EC05012A	561	K837AF
010	GEIC-263	561	AZ-061	561	EC05013	561	K8056A
010	HA17741M	561	AZ-063	561	EC05013A	561	K8056B
010	HC10000217	561	AZ-065	561	EC05014	561	K8062A
010	HL24510	561	AZ-067	561	EC05014A	561	K26
010	HL24593	561	AZ-069	561	EC05070	561	KZ6A
010	IC-40(ELCOM)	561	AZ5.6	561	EC05070A	561	KZ6iA
010	IC317(ELCOM)	561	AZ6.2	561	EC05071A	561	LPM5.6
010	IC040(ELCOM)	561	AZ6.8	561	ED6.2EB	561	LPM5.6A
010	ITT741(METAL CAN)	561	AZ752	561	EDZ-19	561	LR62CH
010	ITT741(METAL-CAN)	561	AZ752A	561	EDZ-20	561	L25.6
010	LM741	561	AZ753	561	EQA01-05?	561	M2207
010	LM741C	561	AZ753A	561	EQA01-06	561	M425.6
010	LM741CH	561	AZ754	561	EQA01-067	561	M425.6-20
010	LM741H	561	AZ754A	561	EQA01-068	561	M425.6A
010	M5141T	561	AZ957	561	EQA01-068B	561	M426.2
010	MC1741CG	561	AZ957A	561	EQA01-06T	561	M426.2-20
010	ON187840	561	AZ957B	561	EQA01-07	561	M426.2A
010	N5741T	561	BZ-052	561	EQA01-07R	561	M758
010	PA7001/503	561	BZ-054	561	EQA01-07RE	561	MA1056
010	PA7741	561	BZ-056	561	EQA01-07S	561	MA1062
010	PA7741C	561	BZ-058	561	EQA01-05T	561	MA1068
010	RC741T	561	BZ-061	561	EQA107RE	561	MC6007
010	SC5175G	561	BZ-063	561	EQB01-06	561	MC6007A
010	SC741T	561	BZ-065	561	ERA0106R	561	MC6008
010	SG741CT	561	BZ-067	561	ERB-07RE	561	MC6008A
010	SK3514	561	BZ-069	561	ERB01-07	561	MC6107
010	SK3514/941	561	BZ052	561	ET16X17	561	MC6107A
010	SL20929	561	BZ066	561	ETD-HZ7C	561	MC610B
010	SL21673	561	BZ5.6	561	EVR4	561	MC6108A
010	SL23486	561	BZ6.2	561	EVR4A	561	MD752
010	SMC750123-1	561	BZ6.8	561	EVR4B	561	MD752A
010	SN72741L	561	BZX10	561	EVR5	561	MD753
010	TBA221	561	BZX2905V6	561	EVR5A	561	MD753A
010	TBA221/741C	561	BZX2906V2	561	EVR5B	561	MEZ5.6T1Q
010	TBA222	561	BZX2906V6	561	EZ5R6(ELCOM)	561	MGLA56
010	TBA222/741	561	BZX2906V8	561	EZ61(ELCOM)	561	MGLA56A
010	TM941	561	BZX4605V6	561	F16H1	561	MGLA56B
010	19-09234	561	BZX4606V2	561	FV-22	561	MGLA62
010	19A116297P3	561	BZX4606V8	561	FV-24	561	MGLA62A
010	19A116297P3-10	561	BZX5505V6	561	FV22	561	MGLA62B
010	19A116297P3-9	561	BZX5505V8	561	FZ5.6T1Q	561	MR6C-H
010	44A332168-001	561	BZX5506V8	561	FZ5.6T5	561	MR62C-H
010	44A332168-002	561	BZX7105V6	561	FZ6.2T1Q	561	MR62B-H
010	44A332169-002	561	BZX7106V2	561	G01-036A	561	MR62H
010	68A7672P1	561	BZX7106V8	561	G01036A	561	MZ2007
010	77C710891-2	561	BZX7905V6	561	G5.6T1Q	561	MZ2607A
010	131A8471	561	BZX7906V2	561	G5.6T5	561	MZ2608
010	156-0049-00	561	BZX7906V8	561	GARE	561	MZ2608A
010	179-46447-08	561	BZX8305V6	561	GB6.2	561	MZ-206
010	276-007	561	BZX8306V2	561	GE6.2A	561	MZ-5
010	276-010	561	BZX8306V8	561	GE6.2B	561	MZ-6
010	351-029-020	561	BZT58	561	GEZD-5.6	561	MZ-7
010	351-1029-020	561	BZT59	561	GEZD-6.0	561	MZ1000-7
010	477-0242-001	561	BZT60	561	GEZD-6.2	561	MZ1000-8
010	741HC	561	BZY66	561	GEZD-6.6	561	MZ1000-9
010	1081K94-6	561	BZY78	561	GEZD-6.8	561	
010	1081K94-9	561	BZY78P				
010	1479-0273	561	BZY83/C5V6				
010	2014-6684	561	BZY83/D5V6				

RS 276-	Industry Standard No.
561	MZ1006
561	MZ206
561	MZ207(ZENER)
561	MZ207-02A
561	MZ207A
561	MZ207B
561	MZ207C
561	MZ306
561	MZ306C
561	MZ307
561	MZ4626
561	MZ4627
561	MZ500-10
561	MZ500-111
561	MZ500-12
561	MZ500.10
561	MZ5A
561	MZ6
561	MZ6.2
561	MZ6.2A
561	MZ6.2T5
561	MZ605
561	MZ610
561	MZ620
561	MZ640
561	MZ70MA
561	MZ92-5.6A
561	MZ92-6.0B
561	MZ92-6.0B
561	MZ92-6.2A
561	NGP5002
561	NZ-206
561	O2Z5.6A
561	OAZ202
561	OAZ203
561	OAZ204
561	OAZ210
561	OAZ242
561	OAZ243
561	OAZ244
561	OAZ270
561	OSD-0033
561	OZ5.6T10
561	OZ5.6T5
561	OZ6.2T10
561	OZ6.2T5
561	PA9267
561	PD6008
561	PD6008A
561	PD6009
561	PD6009A
561	PD6010
561	PD6010A
561	PD6049
561	PD6049C
561	PD6050
561	PD6051
561	PL-152-047-9-001
561	PL-152-052-9-002
561	PS1325
561	PS8907
561	PS8908
561	PS8909
561	PTC502
561	PTC503
561	P2W
561	QA01-07RE
561	QA107RE
561	QD-2MZ306CE
561	QD-ZRD56EAA
561	QD-ZRD56PAA
561	QRT-236
561	QRT-237
561	QRT-238
561	QZ5.6T10
561	QZ5.6T5
561	QZ6.2T10
561	QZ6.2T5
561	R-7093
561	RD7A
561	RD-6.8EB
561	RD-6A
561	RD-6AM
561	RD-6L
561	RD-7AM
561	RD-7E
561	RD5.6B
561	RD5.6B
561	RD5.6B-B
561	RD5.6EB
561	RD5.6ED
561	RD5.6EK
561	RD5.6P
561	RD5.6PA
561	RD5.6PB
561	RD5.6M
561	RD5R6EB
561	RD6
561	RD6.2E
561	RD6.2EB
561	RD6.2EC
561	RD6.2P
561	RD6.2PA
561	RD6.2PB
561	RD6.8E
561	RD6.8B-B1
561	RD6.8EB
561	RD6.8P
561	RD6.8PA
561	RD6.8PB
561	RD6A
561	RD6A(M)
561	RD6AL
561	RD6AM
561	RD6AN
561	RD6B
561	RD7AM
561	RD7AN
561	RD7B
561	RD7H
561	RE-107
561	RE-109
561	RE-110
561	RE106
561	RE107
561	RE108
561	RE109
561	RE110
561	REN136
561	REN137
561	REN5070
561	REN5071
561	RH-EX0024CEZZ
561	RH-EX0048CEZZ
561	RT-236
561	RT-237
561	RT-238
561	RT1306
561	RT1306(G.E.)
561	RT5215
561	RT5471
561	RT5793
561	RT7539
561	RV1181
561	RV6.2
561	RVDCD0033
561	RVDBQA01068
561	RVDMZ-206
561	RVDMZ206
561	RVDR5R6EB
561	RVDRD5R6EB
561	RVDRD7AN
561	RZ5.6
561	RZ6.2
561	RZ6.8
561	RZZ5.6
561	RZZ6.2
561	RZZ6.8
561	S1A3
561	S3004-1715
561	S3004-1718
561	SA821
561	SA821A
561	SA823
561	SA823A
561	SA825
561	SA825A
561	SA827
561	SA827A
561	SA829
561	SA829A
561	SB821
561	SB821A
561	SB823
561	SB823A
561	SB825
561	SB825A
561	SB827
561	SB827A
561	SB829
561	SB829A
561	SC821
561	SC821A
561	SC823
561	SC823A
561	SC825
561	SC825A
561	SC827
561	SC827A
561	SC829
561	SC829A
561	OSD-0033
561	SD-32
561	SD-49
561	SD32
561	SD42
561	SD49
561	SFZ708
561	SI-RECT-140
561	SI-RECT-140/TR-6
561	SI-RECT-140/TR68
561	SK3057
561	SK3057/136A
561	SK3058
561	SK3058/137A
561	SK3334
561	SK3334/5071A
561	SK3342
561	SK3777
561	SK3777/5011
561	SK3778
561	SK3778/5012
561	SK3779
561	SK3779/5013
561	SK3780
561	SK3780/5014
561	SK3780/5014A
561	SUC650
561	SUM6010
561	SUM6011
561	SUM6020
561	SUM6021
561	SV-02
561	SV123
561	SVC625
561	SVC650
561	SVM6010
561	SVM6011
561	SVM602
561	SVM6020
561	SVM6021
561	SVM605
561	SVM61
561	SZ-RD6.2EB
561	SZ-YZ063
561	SZ5.6
561	SZ6.2
561	SZ6.2A
561	SZ7
561	SZA-6
561	TIXD753
561	TM136
561	TM137
561	TM137A
561	TM5070
561	TM5071
561	TMD02
561	TMD02A
561	TMD03
561	TMD03A
561	TMD04
561	TMD04A
561	TP20-0284
561	TR2337123
561	TR68
561	TR78A
561	TR78B
561	TRR6
561	TV24981
561	TV8EQA01-05T
561	TV8QA01-07BE
561	TV8QA01-07RE
561	TV8ZB1-6
561	TZ5.6
561	U535/7825B(ZENER
561	UZ6.2
561	UZ6.2B
561	UZ6.2B
561	UZ8706
561	UZ8806
561	VHEE126B3///1A
561	VHBRD6R8EE/-1
561	VR5.6
561	VR5.6A
561	VR5.6B
561	VZ-054
561	VZ-056
561	VZ-058
561	VZ-061
561	VZ-063
561	VZ-065
561	VZ-067
561	VZ-069
561	VO-61
561	WEP1008
561	WEP103
561	WEP603
561	WM-054
561	WM-054A
561	WM-054B
561	WM-054C
561	WM-054D
561	WM-056
561	WM-056A
561	WM-056B
561	WM-056C
561	WM-056D
561	WM-058
561	WM-058A
561	WM-058B
561	WM-058C
561	WM-058D
561	WM-061
561	WM-061A
561	WM-061B
561	WM-061D
561	WM-063A
561	WM-063B
561	WM-063C
561	WM-063D
561	WM-065
561	WM-065A
561	WM-065B
561	WM-065C
561	WM-065D
561	WN-054
561	WN-054A
561	WN-054B
561	WN-054C
561	WN-054D
561	WN-056
561	WN-056A
561	WN-056B
561	WN-056C
561	WN-056D
561	WN-058
561	WN-058A
561	WN-058B
561	WN-058C
561	WN-058D
561	WN-061
561	WN-061A
561	WN-061C
561	WN-061D
561	WN-063
561	WN-063A
561	WN-063B
561	WN-063C
561	WN-063D
561	WN-065
561	WN-065A
561	WN-065B
561	WN-065C
561	WN-065D
561	WO-054
561	WO-054A
561	WO-054B
561	WO-054C
561	WO-054D
561	WO-056
561	WO-056A
561	WO-056B
561	WO-056C
561	WO-056D
561	WO-058
561	WO-058A
561	WO-058B
561	WO-058C
561	WO-058D
561	WO-061
561	WO-061A
561	WO-061B
561	WO-061C
561	WO-061D
561	WO-063
561	WO-063A
561	WO-063B
561	WO-063D
561	WO-065
561	WO-065A
561	WO-065B
561	WO-065C
561	WO-065D
561	WO-61
561	WP-052D
561	WP-054
561	WP-054A
561	WP-054B
561	WP-054C
561	WP-054D
561	WP-056
561	WP-056A
561	WP-056B
561	WP-056C
561	WP-056D
561	WP-058
561	WP-058A
561	WP-058B
561	WP-058C
561	WP-058D
561	WP-061
561	WP-061A
561	WP-061B
561	WP-061C
561	WP-061D
561	WP-063
561	WP-063A
561	WP-063B
561	WP-063C
561	WP-063D
561	WP-065
561	WP-065A
561	WP-065B
561	WP-065C
561	WP-065D
561	WQ-054
561	WQ-054A
561	WQ-054B
561	WQ-054D
561	WQ-056
561	WQ-056A
561	WQ-056B
561	WQ-056C
561	WQ-056D
561	WQ-058
561	WQ-058A
561	WQ-058B
561	WQ-058C
561	WQ-058D
561	WQ-061
561	WQ-061A
561	WQ-061B
561	WQ-061C
561	WQ-061D
561	WQ-063
561	WQ-063A
561	WQ-063B
561	WQ-063C
561	WQ-063D
561	WQ-065
561	WQ-065A
561	WQ-065C
561	WQ-065D
561	WR-054
561	WR-054A
561	WR-054B
561	WR-054C
561	WR-054D
561	WR-056
561	WR-056A
561	WR-056B
561	WR-056C
561	WR-056D
561	WR-058
561	WR-058A
561	WR-058B
561	WR-058C
561	WR-058D
561	WR-061
561	WR-061A
561	WR-061B
561	WR-061C
561	WR-061D
561	WR-063
561	WR-063A
561	WR-063B
561	WR-063C
561	WR-063D
561	WR-065
561	WR-065A
561	WR-065B
561	WR-065C
561	WR-065D
561	WS-054
561	WS-054A
561	WS-054B
561	WS-054C
561	WS-054D
561	WS-056
561	WS-056A
561	WS-056B
561	WS-056C
561	WS-058
561	WS-058A
561	WS-058B
561	WS-058C
561	WS-058D
561	WS-061
561	WS-061A
561	WS-061B
561	WS-061C
561	WS-061D
561	WS-063
561	WS-063A
561	WS-063B
561	WS-063C
561	WS-063D
561	WS-065
561	WS-065A
561	WS-065B
561	WS-065C
561	WS-065D
561	WT-054
561	WT-054A
561	WT-054B
561	WT-054C
561	WT-054D
561	WT-056
561	WT-056A
561	WT-056B
561	WT-056C
561	WT-056D
561	WT-058
561	WT-058A
561	WT-058B
561	WT-058C
561	WT-058D
561	WT-061A
561	WT-061B
561	WT-061C
561	WT-061D
561	WT-063
561	WT-063A
561	WT-063B
561	WT-063C
561	WT-063D
561	WT-065
561	WT-065A
561	WT-065B
561	WT-065C
561	WT-065D
561	WU-054
561	WU-054A
561	WU-054B
561	WU-054C
561	WU-054D
561	WU-056

RS 276-	Industry Standard No.	RS 276-	Industry Standard No.	RS 276-	Industry Standard No.	RS 276-	Industry Standard No.
561	WU-056A	561	WY-063	561	Z1D5.6	561	1/4AZ5.6D5
561	WU-056B	561	WY-063A	561	Z1D6.2	561	1/4AZ6.2D5
561	WU-056C	561	WY-063B	561	Z2A56F	561	1/4LZ5.6D
561	WU-056D	561	WY-063C	561	Z2A62F	561	1/4LZ5.6D10
561	WU-058	561	WY-063D	561	Z4B5.6	561	1/4LZ5.6D5
561	WU-058A	561	WY-065	561	Z4D5.6	561	1/4LZ6.2D
561	WU-058B	561	WY-065A	561	Z4D5.6	561	1/4LZ6.2D5
561	WU-058C	561	WY-065B	561	Z4X5.6	561	1/4M5.6AZ
561	WU-058D	561	WY-065C	561	Z4X16.2	561	1/4M5.6AZ10
561	WU-061	561	WY-065D	561	Z4X16.2B	561	1/4M5.6AZ25
561	WU-061A	561	WY-157	561	Z5.6	561	1/4M5.6AZ5
561	WU-061B	561	WY-157A	561	Z5140	561	1/4M6.2AZ5
561	WU-061C	561	WZ-052	561	Z5145	561	1/4M6.8AZ
561	WU-061D	561	WZ-054	561	Z5150	561	1/4M6.8AZ10
561	WU-063	561	WZ-054A	561	Z5155	561	1/4M6.8AZ25
561	WU-063A	561	WZ-056	561	Z5160	561	1/4M6.8Z
561	WU-063B	561	WZ-056A	561	Z5165	561	1/4M6.8Z10
561	WU-063C	561	WZ-058	561	Z5170	561	1/4M6.8Z5
561	WU-063D	561	WZ-058A	561	Z5240	561	1/4Z5.6T5
561	WU-065	561	WZ-061	561	Z5250	561	1/4Z6.2T5
561	WU-065B	561	WZ-061A	561	Z5255	561	1A62
561	WU-065C	561	WZ-063	561	Z5260	561	10003B
561	WU-065D	561	WZ-063A	561	Z5265	561	1D5.6
561	WU065A	561	WZ-065	561	Z5270	561	1D5.6A
561	WV-054	561	WZ-067	561	Z5540	561	1D5.6B
561	WV-054A	561	WZ-069	561	Z5545	561	1D6.2
561	WV-054B	561	WZ.061	561	Z5550	561	1D6.2A
561	WV-054C	561	WZ050	561	Z5555	561	1D6.2B
561	WV-054D	561	WZ052	561	Z5560	561	1D6.28A
561	WV-056	561	WZ060	561	Z5565	561	1D6.28B
561	WV-056A	561	WZ061	561	Z5570	561	1EZ5.6
561	WV-056B	561	WZ065	561	Z5B5.6	561	01K-5.4B
561	WV-056C	561	WZ069	561	Z5B6.2	561	01K-5.8B
561	WV-056D	561	WZ5.6	561	Z5B6.8	561	01K6-5.8
561	WV-058	561	WZ523	561	Z5D5.6	561	01K6.5E
561	WV-058A	561	WZ6.2	561	Z5D6.2	561	1M5.6AZ10
561	WV-058B	561	XC070	561	Z5D6.8	561	1M5.6Z810
561	WV-058C	561	XZ-070	561	Z6	561	1M5.6Z85
561	WV-058D	561	XZ055	561	Z6.2	561	1M6.2AZ10
561	WV-061	561	XZ070	561	Z6.8	561	1M6.2Z810
561	WV-061A	561	XZ070	561	Z801	561	1M6.2Z85
561	WV-061B	561	YEADAW01-06	561	ZB1-6	561	1M6.8AZ10
561	WV-061C	561	YZ058	561	ZB1-7	561	1M6.8Z
561	WV-061D	561	YZ060	561	ZB1-8	561	1M6.8Z10
561	WV-063	561	YZ063	561	ZB205	561	1M6.8Z5
561	WV-063A	561	Z-1006	561	ZB206	561	1M6.8Z810
561	WV-063B	561	Z-1104	561	ZB5.6	561	1M6.8Z85
561	WV-063C	561	Z-1140	561	ZB5.6A	561	1N753A
561	WV-063D	561	Z-1145	561	ZB5.6B	561	1N1485
561	WV-065	561	Z-1150	561	ZB6.2	561	1N1509
561	WV-065A	561	Z-1155	561	ZB6.2A	561	1N1509A
561	WV-065B	561	Z-1160	561	ZB6.2B	561	1N1510
561	WV-065C	561	Z-1165	561	ZB6.8	561	1N1510A
561	WV-065D	561	Z-1170	561	ZB6.8A	561	1N1520
561	WW-054	561	Z-1240	561	ZB6.8B	561	1N1520A
561	WW-054A	561	Z-1245	561	ZBI-13	561	1N1521
561	WW-054B	561	Z-1250	561	ZC105	561	1N1521A
561	WW-054C	561	Z-1255	561	ZC106	561	1N1735
561	WW-054D	561	Z-1260	561	ZD5.6A	561	1N1765
561	WW-056	561	Z-1265	561	ZD5.6B	561	1N1765A
561	WW-056A	561	Z-1270	561	ZD56A	561	1N1766
561	WW-056B	561	Z-1540	561	ZD56B	561	1N1766A
561	WW-056C	561	Z-1545	561	ZD6.2A	561	1N1767
561	WW-056D	561	Z-1550	561	ZD6.2B	561	1N1767A
561	WW-058	561	Z-1555	561	ZE5.6	561	1N1779
561	WW-058A	561	Z-1560	561	ZE6.8	561	1N1779A
561	WW-058B	561	Z-1565	561	ZEC5.6	561	1N1929
561	WW-058C	561	Z-1570	561	ZEN500	561	1N1929A
561	WW-058D	561	Z-5140	561	ZF15.6R	561	1N1929B
561	WW-061	561	Z-5145	561	ZF16.2	561	1N1930
561	WW-061A	561	Z-5150	561	ZF25.6R	561	1N1956
561	WW-061B	561	Z-5155	561	ZF26.2	561	1N1957
561	WW-061C	561	Z-5160	561	ZF26.8	561	1N1983
561	WW-063	561	Z-5165	561	ZP5.6	561	1N1983A
561	WW-063A	561	Z-5170	561	ZP6.2	561	1N1983B
561	WW-063B	561	Z-5240	561	ZP6.8	561	1N1984
561	WW-063C	561	Z-5245	561	ZG5.6	561	1N2033
561	WW-063D	561	Z-5250	561	ZG6.8	561	1N2033-2
561	WW-065	561	Z-5255	561	ZH105	561	1N2033A
561	WW-065A	561	Z-5260	561	ZH106	561	1N2034
561	WW-065B	561	Z-5265	561	ZL6	561	1N2034-2
561	WW-065C	561	Z-5270	561	ZM5.6	561	1N2034-3
561	WW-065D	561	Z-5540	561	ZM5.6A	561	1N2034A
561	WX-054	561	Z-5545	561	ZM5.6B	561	1N2214
561	WX-054A	561	Z-5550	561	ZM6	561	1N2765
561	WX-054B	561	Z-5555	561	ZM6.2	561	1N2765A
561	WX-054C	561	Z-5560	561	ZM6.2A	561	1N3016
561	WX-054D	561	Z-5565	561	ZM6.2B	561	1N3016A
561	WX-056	561	Z-5570	561	ZM6.8	561	1N3016B
561	WX-056A	561	Z0212	561	ZM6.8A	561	1N3399
561	WX-056B	561	Z0214	561	ZM6.8B	561	1N3400
561	WX-056C	561	Z0407	561	ZM6A	561	1N3411
561	WX-056D	561	Z040B	561	ZM6B	561	1N3412
561	WX-058	561	Z0B5.6	561	ZOB5.6	561	1N3443
561	WX-058A	561	Z0B6.2	561	ZOB6.2	561	1N3444
561	WX-058B	561	Z0B6.8	561	ZOD6.2	561	1N3496
561	WX-058C	561	Z0D5.6	561	ZOD6.8	561	1N3497
561	WX-058D	561	Z0D6.2	561	ZP5.6	561	1N3498
561	WX-061	561	Z0D6.2	561	ZP6.2	561	1N3499
561	WX-061A	561	Z1006	561	ZP6.8	561	1N3500
561	WX-061B	561	Z1008	561	ZPD6.2	561	1N3512
561	WX-061C	561	Z1104	561	ZPD618	561	1N3513
561	WX-061D	561	Z1104-C	561	ZQ-6	561	1N3514
561	WX-063	561	Z1106	561	ZR205	561	1N3553
561	WX-063A	561	Z1106-C	561	ZR206	561	1N3675
561	WX-063B	561	Z1140	561	ZS5.6	561	1N3675A
561	WX-063C	561	Z1145	561	ZS5.6A	561	1N377
561	WX-063D	561	Z1150	561	ZS5.6B	561	1N378
561	WX-065	561	Z1155	561	ZT5.6	561	1N3827
561	WX-065A	561	Z1160	561	ZT5.6A	561	1N3827A
561	WX-065B	561	Z1165	561	ZT5.6B	561	1N3828
561	WX-065C	561	Z1170	561	ZT6.2	561	1N3828A
561	WX-065D	561	Z1204	561	ZT6.2A	561	1N3829
561	WY-054	561	Z1206	561	ZT6.2B	561	1N3829A
561	WY-054A	561	Z1240	561	ZZ5.6	561	1N4010
561	WY-054B	561	Z1245	561	ZZ6.2	561	1N4099
561	WY-054C	561	Z1250	561	ZZ6.8	561	1N4158
561	WY-054D	561	Z1255	561	001-0082-00	561	1N4158A
561	WY-056	561	Z1260	561	001-0099-01	561	1N429
561	WY-056A	561	Z1270	561	001-0163-04	561	1N4323
561	WY-056B	561	Z1540	561	001-02303-3	561	1N4323A
561	WY-056C	561	Z1545	561	001-023033	561	1N4400
561	WY-056D	561	Z1550	561	1-210	561	1N4400A
561	WY-058	561	Z1555	561	01-2303-33	561	1N4499
561	WY-058A	561	Z1560	561	1/2Z6.6T5	561	1N4501
561	WY-058B	561	Z1565	561	1/2Z6.2T5	561	1N4565
561	WY-058C	561	Z1570	561	1/2Z6.5T5	561	1N4565A
561	WY-058D	561	Z1B5.6	561	1/2Z6.8T5	561	1N4566
561	WY-061	561	Z1B6.2	561	1/4AZ5.6D	561	1N4566A
561	WY-061A	561	Z1B6.8	561	1/4AZ5.6D10	561	1N4567
561	WY-061B	561	Z1C6.2			561	1N4567A
561	WY-061C					561	1N4568
561	WY-061D					561	1N4568A
						561	1N4569

RS 276-	Industry Standard No.	RS 276-	Industry Standard No.	RS 276-	Industry Standard No.	RS 276-	Industry Standard No.	RS 276-	Industry Standard No.
561	1N4569A	561	1N825A	561	1Z6.8A	561	48-83461E38	561	
561	1N4570	561	1N826	561	1Z68	561	48-83461E42	561	
561	1N4570A	561	1N826A	561	1Z68A	561	48-97305A02	561	
561	1N4571	561	1N827	561	1ZO5.6	561	48-97305A05	561	
561	1N4571A	561	1N827A	561	1ZO5.6T10.5	561	488106741D62	561	
561	1N4572	561	1N828	561	1ZO6.2	561	488137170	561	
561	1N4572A	561	1N828A	561	1ZP5.6T10	561	488137387	561	
561	1N4573	561	1N829	561	1ZP5.6T20	561	488195080	561	
561	1N4573A	561	1N829A	561	1ZP5.6T6	561	488097305A02	561	
561	1N4574	561	1N957	561	1ZM5.6T10	561	53A001-5	561	
561	1N4574A	561	1N957A	561	1ZM5.6T20	561	53B011-2	561	
561	1N4575	561	1N957B	561	1ZM5.6T5	561	53J004-1	561	
561	1N4575A	561	1NJ61225	561	1ZM6.2	561	53K001-4	561	
561	1N4576	561	1S-331	561	1ZM6.2T10	561	56-56	561	
561	1N4576A	561	1S1162	561	1ZM6.2T20	561	56-6	561	
561	1N4577	561	1S1173	561	1ZM6.2T5	561	56-63	561	
561	1N4577A	561	1S1174	561	2MW742	561	56-71	561	
561	1N4578	561	1S1201	561	02Z6.2A	561	578-06	561	
561	1N4578A	561	1S1201(H)	561	02Z5.6	561	61E05	561	
561	1N4579	561	1S1202	561	02Z5.6A	561	61B25	561	
561	1N4579A	561	1S136	561	02Z6.2	561	61E27	561	
561	1N4580	561	1S136(ZENER)	561	02Z6.2A	561	61E36	561	
561	1N4580A	561	1S1374A	561	02Z6.4A	561	61E58	561	
561	1N4581	561	1S1375	561	02Z6.8A	561	61E42	561	
561	1N4581A	561	1S1375A	561	02Z62A	561	63-11147	561	
561	1N4582	561	1S1376	561	3I4-3505-2	561	63-11659	561	
561	1N4582A	561	1S1376A	561	3I4-3506-12	561	63-12110	561	
561	1N4583	561	1S1389	561	3I4-3506-21	561	66X0040-003	561	
561	1N4583A	561	1S1389A	561	4-2020-07500	561	73-31	561	
561	1N4583B	561	1S1390	561	4-2020-11300	561	75N5.6	561	
561	1N4584	561	1S1390A	561	4-436	561	75N6.2	561	
561	1N4584A	561	1S1405	561	4-436(SEARS)	561	86-0510	561	
561	1N4611	561	1S1405R	561	4T6.2	561	86-110-1	561	
561	1N4612	561	1S1406	561	4T6.2A	561	93A39-11	561	
561	1N4612A	561	1S1406R	561	4T6.2B	561	93A39-19	561	
561	1N4612B	561	1S1715	561	05-110107	561	93A39-2	561	
561	1N4612C	561	1S1735	561	05-480306	561	93A39-26	561	
561	1N4613	561	1S1736	561	05-69669-01	561	93039-19	561	
561	1N4626	561	1S1767	561	052-6.2L	561	96-5248-01	561	
561	1N4627	561	1S1768	561	0525.6	561	100-10	561	
561	1N4628	561	1S191	561	0526.2	561	100-135	561	
561	1N4655	561	1S192	561	0526.8	561	100-218	561	
561	1N4656	561	1S1956	561	6.2SR1	561	100-437	561	
561	1N4657	561	1S1957	561	6.2SR1A	561	103-140	561	
561	1N469	561	1S2056	561	6.2SR2	561	103-140A	561	
561	1N4690	561	1S2056A	561	6.2SR2A	561	103-279-11	561	
561	1N4691	561	1S2056B	561	6.2SR3	561	103-279-12	561	
561	1N4692	561	1S2062	561	6.2SR3A	561	103-279-13	561	
561	1N469A	561	1S2062A	561	6.2SR4	561	103-279-14	561	
561	1N470	561	1S2062B	561	6.2SR4A	561	103-279-14A	561	
561	1N470A	561	1S2111	561	6V-200	561	103-Z9007	561	
561	1N470B	561	1S2111A	561	07-5331-86	561	103-Z9008	561	
561	1N4734	561	1S2112	561	07-5331-86A	561	111-4-2020-1130Q	561	
561	1N4734A	561	1S2112A	561	07-5331-86B	561	0114-0026/4460	561	
561	1N4735	561	1S2113	561	07-5331-86C	561	0114-0260	561	
561	1N4735A	561	1S2113A	561	07-5331-86D	561	121-0041	561	
561	1N4736	561	1S211A	561	8-719-168-07	561	123B-005	561	
561	1N4736A	561	1S2128	561	84-905-421-109	561	142-016	561	
561	1N474	561	1S2136	561	09-306052	561	151-030-9-003	561	
561	1N474A	561	1S2194	561	09-306055	561	151-031-9-001	561	
561	1N474B	561	1S221	561	09-306073	561	151-031-9-003	561	
561	1N475	561	1S222	561	09-306109	561	152-019-9-001	561	
561	1N475A	561	1S2452	561	09-306185	561	152-042-9-002	561	
561	1N475B	561	1S2453	561	09-306208	561	152-047-9-001	561	
561	1N5232	561	1S2454	561	09-306215	561	152-047-9-004	561	
561	1N5232A	561	1S2488	561	09-306286	561	152-051-9-002	561	
561	1N5232B	561	1S2489	561	09-306325	561	152-052-9-002	561	
561	1N5233	561	1S2490	561	09-306377	561	152-079-9-002	561	
561	1N5233A	561	1S2544	561	09-306383	561	152-079-9-003	561	
561	1N5233B	561	1S2545	561	09-306419	561	15280711(D)	561	
561	1N5234	561	1S2546	561	09-307088	561	203A6137	561	
561	1N5234A	561	1S2547	561	10-016	561	260-10-046	561	
561	1N5234B	561	1S2769	561	012-1023-007	561	260-10-46	561	
561	1N5235	561	1S2770	561	13-33187-12	561	264D00507	561	
561	1N5235A	561	1S2771	561	13-33187-26	561	264P03501	561	
561	1N5235B	561	1S2772	561	13-67544-1	561	264P04005	561	
561	1N5524	561	1S2773	561	13-67544-1/3464	561	264P11003	561	
561	1N5524A	561	1S2774	561	14-515-27	561	264P17401	561	
561	1N5524B	561	1S3006	561	14-515-30	561	276-561	561	
561	1N5525	561	1S331	561	1726	561	0276.8A	561	
561	1N5525A	561	1S331A	561	1726A	561	429-0958-43	561	
561	1N5525B	561	1S331AZ	561	1726AF	561	464.062.15	561	
561	1N5526	561	1S332	561	1726P	561	523-2003-569	561	
561	1N5526A	561	1S332M	561	1826	561	523-2003-629	561	
561	1N5526B	561	1S332Z	561	1826A	561	523-2003-689	561	
561	1N5559	561	1S347	561	1826AF	561	523-2503-689	561	
561	1N5559A	561	1S37	561	1826P	561	530-073-31	561	
561	1N5730	561	1S470	561	019-003928	561	642-236	561	
561	1N5731	561	1S471	561	19A115528-P1	561	064BR	561	
561	1N5732	561	1S480	561	19A11552B-P1	561	65109	561	
561	1N5848	561	1S481	561	1926AF	561	652	561	
561	1N5848A	561	1S52	561	1926F	561	65200	561	
561	1N5848B	561	1S53	561	21A119-041	561	65201	561	
561	1N5849	561	1S615	561	24MW742	561	65202	561	
561	1N5849A	561	1S635	561	34-8054-6(PHILCO)	561	65203	561	
561	1N5849B	561	1S636	561	34-8057-10	561	65204	561	
561	1N5850	561	1S692	561	34-8057-32	561	65205	561	
561	1N5850A	561	1S7056	561	34-8057-42	561	65206	561	
561	1N5850B	561	1S7056A	561	34-8057-49	561	65207	561	
561	1N5851	561	1S7056B	561	3726	561	65208	561	
561	1N5851A	561	1S7062	561	3726A	561	65301	561	
561	1N5851B	561	1S7062A	561	3826	561	65302	561	
561	1N675	561	1S7062B	561	3826A	561	65303	561	
561	1N706	561	1S7068	561	3926	561	65304	561	
561	1N706A	561	1S7068A	561	3926A	561	0684R	561	
561	1N707	561	1S7068B	561	44T-300-92	561	800-000-0Q	561	
561	1N707A	561	1S756	561	46-86394-3	561	903-120B	561	
561	1N707A(ZENER)	561	1S756(H)	561	46-86394-3(10)	561	903-166B	561	
561	1N708	561	1S757	561	46-86457-3	561	903-180	561	
561	1N708A	561	1S226	561	46-86460-3	561	903-333	561	
561	1N709	561	1S239	561	46-86501-3	561	903-335	561	
561	1N709A	561	1S250	561	46-86559-3	561	919-00-1445	561	
561	1N709B	561	1S251	561	46-86566-3	561	1011-02	561	
561	1N710	561	1S252	561	48-10641D62	561	1063-1145	561	
561	1N710A	561	1S253	561	48-127021	561	1063-6454	561	
561	1N752	561	1T5.6	561	48-134993	561	1074-123	561	
561	1N752A	561	1T5.6B	561	48-137170	561	1105	561	
561	1N753	561	1T6.2	561	48-137210	561	1106	561	
561	1N753A	561	1T6.2B	561	48-137279	561	1125-2608	561	
561	1N754A	561	1TA5.6	561	48-137322	561	1411-137	561	
561	1N762	561	1TA5.6A	561	48-137387	561	2000-322	561	
561	1N762-1	561	1TA6.2	561	48-155146	561	2000-324	561	
561	1N762-2	561	1TA6.2A	561	48-40458A04	561	2000-328	561	
561	1N762A	561	1TA6.2B	561	48-82256001	561	2000-329	561	
561	1N763-1	561	1WB-6A	561	48-82256002	561	2093A41-153	561	
561	1N763-2	561	1WB-7A	561	48-82256012	561	2093A41-70	561	
561	1N816	561	1Z5.6	561	48-82256019	561	2106-17	561	
561	1N821	561	1Z5.6A	561	48-82256023	561	2323-17	561	
561	1N821A	561	1Z6.2	561	48-82256037	561	2341-17	561	
561	1N822	561	1Z6.2A	561	48-82256047	561	2362-17	561	
561	1N823	561	1Z6.2T1Q	561	48-83461B05	561	3701	561	
561	1N823A	561	1Z6.2T5	561	48-83461B25	561	3702	561	
561	1N824	561	1Z6.8	561	48-83461B27	561	3703	561	
561	1N825	561		561	48-83461E36	561	3704	561	

RS 276-	Industry Standard No.	RS 276-	Industry Standard No.	RS 276-	Industry Standard No.	RS 276-	Industry Standard No.
561	3755-1	562	AW-01-10	562	CD31-12039	562	GEZD-10.4
561	3755-2	562	AW01-08	562	CD31-12018	562	GEZD-8.2
561	4082-748-0002	562	AW01-08J	562	CD3112039	562	GEZD-8.7
561	4822-130-30132	562	AW01-09	562	CD3122055	562	GEZD-9.1
561	4822-130-30193	562	AW01-10	562	CD3212055	562	GLA100
561	5001-131	562	AW01-9	562	CD3212055(ZENER)	562	GLA100A
561	5001-160	562	AW01-9/CP3112030	562	CDZ-9V	562	GLA100B
561	5514	562	AW09	562	CDZ-C9V	562	GLA91
561	5515	562	AZ-077	562	CF-092	562	GLA91A
561	5516	562	AZ-081	562	CHZ10	562	GLA91B
561	8000-00004-239	562	AZ-083	562	CHZ10A	562	G01036
561	8000-00005-020	562	AZ-085	562	CHZ210A	562	G29.1
561	8000-00006-146	562	AZ-088	562	CI-8.2	562	H)-00012
561	8000-0005-020	562	AZ-090	562	C02	562	HD3001200
561	8000-00058-003	562	AZ-092	562	CP3212055	562	HD3000401
561	9330-092-90112	562	AZ-094	562	CXO051	562	HD3000409
561	9606	562	AZ-096	562	CXL139	562	HD30017090
561	38510-331	562	AZ-098	562	CXL140	562	HD3003109
561	67544	562	AZ-100	562	CXL5016	562	HEP-Z0217
561	72168-3	562	AZ10	562	CXL5017	562	HEP-Z0219
561	99201-208	562	AZ756	562	CXL5018	562	HEP-Z0220
561	99201-210	562	AZ756A	562	CXL5019	562	HEP-Z0411
561	99201-211	562	AZ757	562	CXL5072	562	HEP-Z0412
561	99201-212	562	AZ757A	562	CXL5073	562	HEP-Z0413
561	99201-312	562	AZ758	562	CZ-092	562	HEP101
561	99210-210	562	AZ758A	562	CZ-92	562	HEP104
561	126851	562	AZ8.2	562	CZ092	562	HEPZ0217
561	129938	562	AZ9.1	562	CZ094	562	HEPZ0217A
561	132616	562	AZ959	562	CZD010	562	HEPZ0219
561	134993	562	AZ959A	562	CZD010-5	562	HEPZ0219A
561	137647	562	AZ959B	562	D-00569C	562	HEPZ0220
561	138974	562	AZ960	562	D-18	562	HEPZ0220A
561	161199	562	AZ960A	562	D1-8	562	HEPZ0411
561	168653	562	AZ960B	562	D176	562	HEPZ0412
561	169199	562	AZ961	562	D1A(ZENER)	562	HEPZ0413
561	171842	562	AZ961A	562	D1G	562	HF-18334
561	225301	562	AZ961B	562	D1U	562	HF-18339
561	225516	562	B-180B	562	D2Z	562	HP-20004
561	302540	562	B090	562	D3F	562	HP-20011
561	348057-17	562	B094	562	D6M	562	HP-20065
561	530073-1031	562	B1-8.2	562	D8U	562	HM10
561	530073-1034	562	B100(ZENER)	562	DAAYO10092	562	HM10A
561	530073-31	562	B2090	562	DDAYO08003	562	HM10B
561	530073-5	562	B66X0040-003	562	DDAYO10002	562	HM8-2
561	530145-1569	562	B66X0040-005	562	DDAYO10005	562	HM8.2A
561	530145-569	562	B094	562	DDAY126001	562	HM9.1
561	530145-689	562	BX090	562	DI-8	562	HM9.1A
561	530157-569	562	BX909	562	DI-8(COURIER)	562	HM9.1B
561	530157-689	562	BXY63	562	DIA	562	HN-00012
561	530166-1004	562	BZ-077	562	DIJ72165	562	HN-00024
561	740628	562	BZ-079	562	DIJ72172	562	HN-00061
561	740953	562	BZ-080	562	D8-149	562	HS2091
561	741116	562	BZ-0800	562	DS-43	562	HS2100
561	1471878-3	562	BZ-081	562	DS49	562	HS7091
561	1471898-3	562	BZ-083	562	D8149	562	HS7100
561	1471898-5	562	BZ-085	562	D843	562	HW10
561	1642606B4	562	BZ-088	562	D848	562	HW10A
561	1642606B6	562	BZ-090	562	D899	562	HW8.2
561	1800012	562	BZ-090(ZENER)	562	DX-0Z87	562	HW8.2A
561	2002209-4	562	BZ-0900	562	DX-0728	562	HW9.1
561	2010967-84	562	BZ-0901	562	DX-0729	562	HW9.1A
561	2092055-0016	562	BZ-092	562	DX-081	562	HW9.1B
561	2092055-0018	562	BZ-094	562	DZ-081	562	HW9-81B
561	2092055-0710	562	BZ-096	562	DZ0820	562	HZ-9
561	2330302	562	BZ-098	562	DZ10	562	HZ-9B
561	2330631	562	BZ-100	562	DZ10A	562	HZ-9C
561	2330632	562	BZ080	562	E0771-3	562	HZ11A
561	2330643	562	BZ081	562	E64	562	HZ11B
561	2330791	562	BZ085	562	EA16X150	562	HZ9
561	3596398	562	BZ090	562	EA16X157	562	HZ9(H)
561 56J	181374	562	BZ090.1Z9	562	EA16X402	562	HZ9B
562	.25N10	562	BZ094	562	EA16X62	562	HZ9C2
562	.25N8.2	562	BZ1-9	562	EA16X64	562	HZ9H
562	.25N9.1	562	BZ1-90	562	EA16X82	562	HZ9L
562	.25T10	562	BZ279C10	562	EA3866	562	IC743047
562	.25T10.5	562	BZ8.2	562	ECG139	562	IP20-0019
562	.25T10.5A	562	BZ9.1	562	ECG139A	562	IP20-0233
562	.25T10A	562	BZX14	562	ECG140A	562	IW01-08J
562	.25T10B	562	BZX29C10	562	ECG5016	562	J24652
562	.25T8.2	562	BZX29C8V2	562	ECG5016A	562	J24672
562	.25T8.2A	562	BZX29C9V1	562	ECG5017	562	J4-1620
562	.25T8.7	562	BZX46C10	562	ECG5018	562	KD2501
562	.25T8.7A	562	BZX46C8V2	562	ECG5018A	562	KD2504
562	.25T8.8	562	BZX46C9V1	562	ECG5019	562	KS100A
562	.25T8.8A	562	BZX55C10	562	ECG5019A	562	KS100B
562	.25T9.1	562	BZX55C8V2	562	ECG5072A	562	KS2091A
562	.25T9.1A	562	BZX55C9V1	562	ECG5073	562	KS2091B
562	.25T9.1B	562	BZX71C10	562	ECG5073A	562	KS2100A
562	.4T10	562	BZX71C8V2	562	ED491130	562	KS2100B
562	.4T10A	562	BZX71C9V1	562	EDZ-0045	562	KS41A
562	.4T10B	562	BZX71C9V2	562	EDZ-14	562	KS41AF
562	.4T8.2	562	BZX72	562	EDZ-23	562	KS42A
562	.4T8.2A	562	BZX72B	562	EDZ-24	562	KS42AF
562	.4T8.2B	562	BZX72C	562	E0771-3	562	KS42B
562	.4T9.1	562	BZX79C10	562	EQ-09R	562	KS42BF
562	.4T9.1A	562	BZX79C8V2	562	EQA-00198	562	KS091A
562	.4T9.1B	562	BZX79C9V1	562	EQA-9	562	KVR10
562	.7J210	562	BZX83C10	562	EQA.9	562	KZ-8A
562	.7J28.2	562	BZX83C8V2	562	EQA01-01	562	LMZX-10
562	.7J29.1	562	BZX83C9V1	562	EQA01-08	562	LPM9.1
562	.7Z10A	562	BZT62	562	EQA01-08R	562	LPM9.1A
562	.7Z10B	562	BZT63	562	EQA01-09	562	LR100CH
562	.7Z10C	562	BZT68	562	EQA01-09(R)	562	LR91CH
562	.7Z10D	562	BZY83/C10	562	EQA01-09R	562	LZ10
562	.7Z8.2A	562	BZY83/D10	562	EQA01-10	562	M102
562	.7Z8.2B	562	BZY83C10	562	EQA01-10S	562	M4653
562	.7Z8.2C	562	BZY83C8V2	562	EQA09R	562	M4816
562	.7Z8.2D	562	BZY83C9V1	562	EQB-0108	562	M4Z10
562	.7Z9.1A	562	BZY83D10	562	EQB01-08	562	M4Z10-20
562	.7Z9.1B	562	BZY83D8V2	562	EQB01-09	562	M4Z10A
562	.7Z9.1C	562	BZY83E10	562	EQB01-91	562	M4Z9.1
562	.7Z9.1D	562	BZY85C8V2	562	EQB01-90S	562	M4Z9.1-2Q
562	.7ZM10A	562	BZY85C9V1	562	EQB01-10	562	M4Z9.1A
562	.7ZM10B	562	BZY85D10	562	ETD-RD9.1FB	562	M9Z
562	.7ZM10C	562	BZY85D8V1	562	EVR9	562	MA1082
562	.7ZM10D	562	BZY88/C9V1	562	EVR9A	562	MA1091
562	.7ZM8.2A	562	BZY88/C9V5	562	EVR9B	562	MA1100
562	.7ZM8.2B	562	BZY88C10	562	EZ10(ELCOM)	562	MB6356
562	.7ZM8.2C	562	BZY88C8V2	562	EZ8(ELCOM)	562	MC6014
562	.7ZM8.2D	562	BZY88C9V1	562	EZ9(ELCOM)	562	MC6014A
562	.7ZM9.1A	562	BZY88C9V5	562	G01-012-F	562	MD757
562	.7ZM9.1B	562	BZY92C8V2	562	G01-036-G	562	MD757A
562	.7ZM9.1C	562	BZY92C9V1	562	G01-036-H	562	ME408-02C
562	.7ZM9.1D	562	BZY94C10	562	G01036	562	ME409-02B
562	0A01-08R	562	BZZ13	562	G01036G	562	MGLA100
562	A04234-2	562	C02	562	G9.1T10	562	MGLA100A
562	A061-119	562	C21383	562	G9.1T20	562	MGLA100B
562	A068-103	562	C4015	562	G9.1T5	562	MGLA91
562	A42X00041-01	562	C4016	562	GE-X11	562	MGLA91A
562	A72-86-700	562	CA-092	562	GEZD-10	562	MGLA91B
562	A04234-2	562	CD31-00012	562	GEZD-10-4	562	MR100C-H
562	AW-01-07	562	CD31-12019	562		562	MR100H
562	AW-01-08					562	MR91C-H
562	AW-01-09					562	MR91E-H

RS 276-	Industry Standard No.	RS 276-	Industry Standard No.	RS 276-	Industry Standard No.	RS 276-	Industry Standard No.	RS 276-	Industry Standard No.
562	MT2613	562	RD11AM	562	T-R98	562	WN-079C	562	
562	MT2613A	562	RD7A	562	T21334	562	WN-079D	562	
562	MTZ614	562	RD8	562	T21639	562	WN-081	562	
562	MTZ614A	562	RD8.2	562	TIXD758	562	WN-081A	562	
562	MZ-08	562	RD8.2C	562	TM139	562	WN-081B	562	
562	MZ-10	562	RD8.2E	562	TM139A	562	WN-081C	562	
562	MZ-208	562	RD8.2EA	562	TM140	562	WN-081D	562	
562	MZ-209	562	RD8.2EB	562	TM5016	562	WN-083	562	
562	MZ-210	562	RD8.2EC	562	TM5072	562	WN-083A	562	
562	MZ-210B	562	RD8.2EK	562	TM5075	562	WN-083B	562	
562	MZ-8	562	RD8.2F	562	TMD06	562	WN-083C	562	
562	MZ-9	562	RD8.2PA	562	TMD06A	562	WN-083D	562	
562	MZ-92-9.1B	562	RD8.2FB	562	TMD07	562	WN-085	562	
562	MZ090	562	RD8.2M	562	TMD07A	562	WN-085A	562	
562	MZ10	562	RD8H	562	TMD08	562	WN-085B	562	
562	MZ1000-11	562	RD9	562	TMD08A	562	WN-085C	562	
562	MZ1000-12	562	RD9.1E	562	TR-95	562	WN-085D	562	
562	MZ1000-13	562	RD9.1EA	562	TR-95(B)	562	WN-088	562	
562	MZ1008	562	RD9.1EB	562	TR-95B	562	WN-088A	562	
562	MZ1009	562	RD9.1EB2	562	TR-908	562	WN-088B	562	
562	MZ1010	562	RD9.1EC	562	TR-98	562	WN-088C	562	
562	MZ10A	562	RD9.1ED	562	TR-98A	562	WN-088D	562	
562	MZ208	562	RD9.1EK	562	TR-98B	562	WN-090	562	
562	MZ209	562	RD9.1F	562	TR20092	562	WN-090A	562	
562	MZ209A	562	RD9.1PA	562	TR228736003026	562	WN-090B	562	
562	MZ209B	562	RD9.1FB	562	TR3300Z8	562	WN-090C	562	
562	MZ209C	562	RD9.1M	562	TR98	562	WN-090D	562	
562	MZ250	562	RD9A	562	TR98A	562	WN-092	562	
562	MZ308	562	RD9A(10)	562	TR98B	562	WN-092A	562	
562	MZ309	562	RD9A-N	562	TV9RA-26	562	WN-092B	562	
562	MZ309B	562	RD9AL	562	TVSRD11	562	WN-092C	562	
562	MZ310	562	RD9AM	562	TVSRD11A	562	WN-092D	562	
562	MZ310A	562	RD9AN	562	TVSRD9AL	562	WN-094	562	
562	MZ408-02C	562	RD9B	562	TVSRMZ06V	562	WN-094A	562	
562	MZ409-02B	562	RE-112	562	TVSYZ-080	562	WN-094B	562	
562	MZ500-14	562	RE-114	562	TVSYZ2080	562	WN-094C	562	
562	MZ500-15	562	RE-115	562	TZ8.2	562	WN-094D	562	
562	MZ500-16	562	RE112	562	TZ9.1	562	WN-098	562	
562	MZ500.15	562	RE113	562	TZ9.1A	562	WN-098A	562	
562	MZ8	562	RE114	562	TZ9.1B	562	WN-098B	562	
562	MZ9.1B	562	RE115	562	TZ9.1E	562	WN-098C	562	
562	MZ92-10A	562	REN139	562	UZ8.2B	562	WN-098D	562	
562	MZ92-9.1A	562	REN140	562	UZ8708	562	WN-100	562	
562	MZ92-9.1B	562	REN5072	562	UZ8709	562	WN-100A	562	
562	MZX9.1	562	REN5073	562	UZ8710	562	WN-100B	562	
562	N-756A	562	RF-8.2E	562	UZ8714	562	WN-100C	562	
562	NGP5007	562	RP8.2E	562	UZ8808	562	WN-100D	562	
562	02Z-10A	562	RP471480	562	UZ8809	562	WO-079	562	
562	02Z8.2A	562	RH-EX0021TAZZ	562	UZ8810	562	WO-079A	562	
562	05Z8.2U	562	RP-9A	562	UZ8814	562	WO-079B	562	
562	0A12610	562	RS1290	562	UZ9.1	562	WO-079C	562	
562	0AZ206	562	RT-240	562	UZ9.1B	562	WO-079D	562	
562	0AZ207	562	RT-241	562	UZ9B	562	WO-081	562	
562	0AZ212	562	RT-257	562	VHEWZ-100-1F	562	WO-081A	562	
562	0AZ246	562	RT6105	562	VHEWZ-100//1F	562	WO-081B	562	
562	0AZ247	562	RT6105(G.E.)	562	VHEWZ-1001F	562	WO-081C	562	
562	0AZ272	562	RT6923	562	VHEKZ-090-1	562	WO-081D	562	
562	0Z10T10	562	RT8339	562	VR-9.1	562	WO-083	562	
562	0Z10T5	562	RV2213	562	VR10	562	WO-083A	562	
562	P21344	562	RVD1N4740	562	VR10A	562	WO-083B	562	
562	PC1879-004	562	RVD1N4738	562	VR10B	562	WO-083C	562	
562	PD6012	562	RVD1N4739	562	VR9.1	562	WO-083D	562	
562	PD6012A	562	RVD1N4740	562	VR9.1A	562	WO-085	562	
562	PD6013	562	RVDMZ209	562	VR9.1B	562	WO-085A	562	
562	PD6013A	562	RVDRD11AN	562	VR9B	562	WO-085B	562	
562	PD6014	562	RX090	562	VZ-079	562	WO-085C	562	
562	PD6014A	562	RZ10	562	VZ-081	562	WO-085D	562	
562	PD6053	562	RZ8.2	562	VZ-083	562	WO-088	562	
562	PD6054	562	RZ9.1	562	VZ-085	562	WO-088A	562	
562	PD6055	562	RZZ10	562	VZ-088	562	WO-088B	562	
562	PL-152-042-9-001	562	RZZ8.2	562	VZ-090	562	WO-088C	562	
562	PL-152-044-9-001	562	RZZ9.1	562	VZ-092	562	WO-088D	562	
562	PL-152-051-9-001	562	S	562	VZ-094	562	WO-090	562	
562	PL-152-054-9-001	562	S0702	562	VZ-096	562	WO-090A	562	
562	PS10019B	562	S6-10	562	VZ-098	562	WO-090B	562	
562	PS10063	562	S6-10A	562	VZ-100	562	WO-090C	562	
562	PS1511	562	S6-10B	562	WD90	562	WO-090D	562	
562	PS1512	562	S6-10C	562	WE081	562	WO-092	562	
562	PS1513	562	SA-93794	562	WEP101	562	WO-092A	562	
562	PS1514	562	SC91	562	WEP1010	562	WO-092B	562	
562	PS1515	562	SD-15	562	WEP104	562	WO-092C	562	
562	PS1516	562	SD-27	562	WG91	562	WO-092D	562	
562	PS1517	562	SD-652	562	WM-079	562	WO-094	562	
562	PS8911	562	SD-8	562	WM-079A	562	WO-094A	562	
562	PS8912	562	SD-8(PHILCO)	562	WM-079B	562	WO-094B	562	
562	PS8913	562	SD-8(ZENER)	562	WM-079C	562	WO-094C	562	
562	PTC505	562	SD105	562	WM-079D	562	WO-094D	562	
562	PTC506	562	SD27	562	WM-081	562	WO-098	562	
562	Q-251115C	562	SD632	562	WM-081A	562	WO-098A	562	
562	QA01-082	562	SD632(10)	562	WM-081B	562	WO-098B	562	
562	QA01-08R	562	SK3060	562	WM-081C	562	WO-098C	562	
562	QD-ZMZ408CE	562	SK3060/139A	562	WM-081D	562	WO-098D	562	
562	QD-ZMZ409BE	562	SK3061	562	WM-083	562	WO-100A	562	
562	QD-ZRD9EXAA	562	SK3061/140A	562	WM-083A	562	WO-100B	562	
562	QRT-240	562	SK3136	562	WM-083B	562	WO-100C	562	
562	QRT-241	562	SK3136/5072A	562	WM-083C	562	WO-100D	562	
562	Q210T10	562	SK3749	562	WM-083D	562	WP-079	562	
562	QZ10T5	562	SK3749/5073A	562	WM-085	562	WP-079A	562	
562	R-13333Y	562	SK3782	562	WM-085A	562	WP-079B	562	
562	R-7097	562	SK3782/5016	562	WM-085B	562	WP-079C	562	
562	RA-26	562	SK3783	562	WM-085C	562	WP-079D	562	
562	RA26(ZENER)	562	SK3783/5017	562	WM-085D	562	WP-081	562	
562	RD-10EB	562	SK3784	562	WM-088	562	WP-081A	562	
562	RD-11B	562	SK3784/5018	562	WM-088A	562	WP-081B	562	
562	RD-13AL	562	SK3784/5018A	562	WM-088B	562	WP-081C	562	
562	RD-2.2E	562	SK3785	562	WM-088C	562	WP-081D	562	
562	RD-7A	562	SK3785/5019	562	WM-088D	562	WP-083	562	
562	RD-8.2#E	562	S0-652	562	WM-090	562	WP-083A	562	
562	RD-8.2A	562	SV-9	562	WM-090A	562	WP-083B	562	
562	RD-8.2E	562	SV-9(ZENER)	562	WM-090B	562	WP-083C	562	
562	RD-8.2E(C)	562	SV133	562	WM-090C	562	WP-083D	562	
562	RD-8.2EB	562	SVD-181717	562	WM-090D	562	WP-085	562	
562	RD-8.2EC	562	SVD0278.2A	562	WM-092	562	WP-085A	562	
562	RD-8.2FB	562	SVD0279.5A	562	WM-092A	562	WP-085B	562	
562	RD-9.1E	562	SVD0Z28	562	WM-092B	562	WP-085C	562	
562	RD-91	562	SVD0Z28.2	562	WM-092C	562	WP-085D	562	
562	RD-91E	562	SVD0Z28.2A	562	WM-092D	562	WP-088	562	
562	RD-96	562	SVD0Z29.5A	562	WM-094A	562	-P-088A	562	
562	RD-9A	562	SVD181717	562	WM-094B	562	-P-088B	562	
562	RD-9AL	562	SZ-200-10	562	WM-094C	562	WP-088C	562	
562	RD-9E	562	SZ-200-8	562	WM-094D	562	WP-088D	562	
562	RD-9L	562	SZ-200-9	562	WM-098	562	WP-090	562	
562	RD10E	562	SZ-200-9V	562	WM-098A	562	WP-090A	562	
562	RD10EA	562	SZ-9	562	WM-098B	562	WP-090B	562	
562	RD10EB	562	SZ10	562	WM-098D	562	WP-090C	562	
562	RD10EB-2	562	SZ200-8.2	562	WM-098D	562	WP-090D	562	
562	RD10F	562	SZ9	562	WM-100	562	WP-092	562	
562	RD10FA	562	SZ9.1	562	WM-100A	562	WP-092A	562	
562	RD10FB	562	SZC9	562	WM-100B	562	WP-092B	562	
562	RD10H	562	SZL9	562	WM-100C	562	WP-092C	562	
562	RD10M	562	SZP-9	562	WM-100D	562	WP-092D	562	
562	RD11A	562	SZT-9	562	WN-079	562	WP-094	562	
562	RD11AL	562	SZT9	562	WN-079A	562	WP-094A	562	
562		562	T-750-714	562	WN-079B	562	WP-094B	562	

RS 276-	Industry Standard No.	RS 276-	Industry Standard No.	RS 276-	Industry Standard No.	RS 276-	Industry Standard No.
562	WP-094C	562	WS-088B	562	WV-081	562	WX-094D
562	WP-094D	562	WS-088C	562	WV-081A	562	WX-098
562	WP-098	562	WS-088D	562	WV-081B	562	WX-098A
562	WP-098A	562	WS-090	562	WV-081C	562	WX-098B
562	WP-098B	562	WS-090A	562	WV-081D	562	WX-098C
562	WP-098C	562	WS-090B	562	WV-083	562	WX-098D
562	WP-098D	562	WS-090C	562	WV-083A	562	WX-100
562	WP-100	562	WS-090D	562	WV-083B	562	WX-100A
562	WP-100A	562	WS-092	562	WV-083C	562	WX-100B
562	WP-100B	562	WS-092A	562	WV-083D	562	WX-100C
562	WP-100C	562	WS-092B	562	WV-085	562	WX-100D
562	WP-100D	562	WS-092C	562	WV-085A	562	WY-079
562	WQ-079	562	WS-092D	562	WV-085B	562	WY-079A
562	WQ-079A	562	WS-094	562	WV-085C	562	WY-079B
562	WQ-079B	562	WS-094A	562	WV-085D	562	WY-079C
562	WQ-079C	562	WS-094B	562	WV-088	562	WY-079D
562	WQ-079D	562	WS-094C	562	WV-088A	562	WY-081
562	WQ-081	562	WS-094D	562	WV-088B	562	WY-081A
562	WQ-081A	562	WS-098	562	WV-088C	562	WY-081B
562	WQ-081B	562	WS-098A	562	WV-088D	562	WY-081C
562	WQ-081C	562	WS-098B	562	WV-090	562	WY-081D
562	WQ-081D	562	WS-098C	562	WV-090A	562	WY-083
562	WQ-083	562	WS-098D	562	WV-090B	562	WY-083A
562	WQ-083A	562	WS-100	562	WV-090C	562	WY-083B
562	WQ-083B	562	WS-100A	562	WV-090D	562	WY-083C
562	WQ-083C	562	WS-100B	562	WV-092	562	WY-083D
562	WQ-083D	562	WS-100C	562	WV-092A	562	WY-085
562	WQ-085	562	WS-100D	562	WV-092B	562	WY-085A
562	WQ-085A	562	WT-079	562	WV-092C	562	WY-085B
562	WQ-085B	562	WT-079A	562	WV-092D	562	WY-085C
562	WQ-085C	562	WT-079B	562	WV-094	562	WY-085D
562	WQ-085D	562	WT-079C	562	WV-094A	562	WY-088
562	WQ-088	562	WT-079D	562	WV-094B	562	WY-088A
562	WQ-088A	562	WT-081	562	WV-094C	562	WY-088B
562	WQ-088B	562	WT-081A	562	WV-094D	562	WY-088C
562	WQ-088C	562	WT-081B	562	WV-098	562	WY-088D
562	WQ-088D	562	WT-081C	562	WV-098A	562	WY-090
562	WQ-090	562	WT-081D	562	WV-098B	562	WY-090A
562	WQ-090A	562	WT-083	562	WV-098C	562	WY-090B
562	WQ-090B	562	WT-083A	562	WV-098D	562	WY-090C
562	WQ-090C	562	WT-083B	562	WV-100	562	WY-090D
562	WQ-090D	562	WT-083C	562	WV-100A	562	WY-092
562	WQ-092	562	WT-083D	562	WV-100B	562	WY-092A
562	WQ-092A	562	WT-085	562	WV-100C	562	WY-092B
562	WQ-092B	562	WT-085A	562	WV-100D	562	WY-092C
562	WQ-092C	562	WT-085B	562	WW-079	562	WY-092D
562	WQ-092D	562	WT-085C	562	WW-079A	562	WY-094
562	WQ-094	562	WT-085D	562	WW-079B	562	WY-094A
562	WQ-094A	562	WT-088	562	WW-079C	562	WY-094B
562	WQ-094B	562	WT-088A	562	WW-079D	562	WY-094C
562	WQ-094C	562	WT-088B	562	WW-081	562	WY-094D
562	WQ-094D	562	WT-088C	562	WW-081A	562	WY-098
562	WQ-098	562	WT-088D	562	WW-081B	562	WY-098A
562	WQ-098A	562	WT-090	562	WW-081C	562	WY-098B
562	WQ-098B	562	WT-090A	562	WW-081D	562	WY-098C
562	WQ-098C	562	WT-090B	562	WW-083	562	WY-098D
562	WQ-098D	562	WT-090C	562	WW-083A	562	WY-100
562	WQ-100	562	WT-090D	562	WW-083C	562	WY-100A
562	WQ-100A	562	WT-092	562	WW-083D	562	WY-100B
562	WQ-100B	562	WT-092A	562	WW-085	562	WY-100C
562	WQ-100C	562	WT-092B	562	WW-085A	562	WY-100D
562	WQ-100D	562	WT-092C	562	WW-085B	562	WZ-079
562	WR-079	562	WT-092D	562	WW-085C	562	WZ-079A
562	WR-079A	562	WT-094	562	WW-085D	562	WZ-081
562	WR-079B	562	WT-094A	562	WW-088	562	WZ-081A
562	WR-079C	562	WT-094B	562	WW-088A	562	WZ-083
562	WR-079D	562	WT-094C	562	WW-088B	562	WZ-083A
562	WR-081	562	WT-094D	562	WW-088C	562	WZ-085
562	WR-081A	562	WT-098	562	WW-088D	562	WZ-085A
562	WR-081B	562	WT-098A	562	WW-090	562	WZ-088
562	WR-081C	562	WT-098B	562	WW-090A	562	WZ-088A
562	WR-081D	562	WT-098C	562	WW-090C	562	WZ-090
562	WR-083	562	WT-098D	562	WW-090D	562	WZ-092
562	WR-083A	562	WT-100	562	WW-092	562	WZ-092A
562	WR-083B	562	WT-100A	562	WW-092A	562	WZ-094
562	WR-083C	562	WT-100B	562	WW-092B	562	WZ-094A
562	WR-083D	562	WT-100C	562	WW-092C	562	WZ-096
562	WR-085	562	WT-100D	562	WW-092D	562	WZ-098A
562	WR-085A	562	WU-079	562	WW-094	562	WZ-100
562	WR-085B	562	WU-079A	562	WW-094A	562	WZ-100A
562	WR-085C	562	WU-079B	562	WW-094B	562	WZ-90
562	WR-085D	562	WU-079C	562	WW-094C	562	WZ081
562	WR-088	562	WU-079D	562	WW-094D	562	WZ085
562	WR-088A	562	WU-081	562	WW-098	562	WZ090
562	WR-088B	562	WU-081A	562	WW-098A	562	WZ090A
562	WR-088C	562	WU-081B	562	WW-098B	562	WZ092
562	WR-088D	562	WU-081C	562	WW-098C	562	WZ094
562	WR-090	562	WU-081D	562	WW-098D	562	WZ096
562	WR-090A	562	WU-083	562	WW-100	562	WZ10
562	WR-090B	562	WU-083A	562	WW-100A	562	WZ100
562	WR-090C	562	WU-083B	562	WW-100B	562	WZ533
562	WR-090D	562	WU-083C	562	WW-100C	562	WZ9.1
562	WR-092	562	WU-083D	562	WW-100D	562	X092
562	WR-092A	562	WU-085	562	WX-079	562	X89.1B
562	WR-092B	562	WU-085A	562	WX-079A	562	XZ-064
562	WR-092C	562	WU-085B	562	WX-079B	562	XZ-082
562	WR-092D	562	WU-085C	562	WX-079C	562	XZ-086
562	WR-094	562	WU-085D	562	WX-079D	562	XZ-090
562	WR-094A	562	WU-088	562	WX-081	562	XZ-092
562	WR-094B	562	WU-088A	562	WX-081A	562	XZ-096
562	WR-094C	562	WU-088B	562	WX-081B	562	XZ-100
562	WR-094D	562	WU-088C	562	WX-081C	562	XZ064
562	WR-098	562	WU-088D	562	WX-081D	562	XZ084
562	WR-098A	562	WU-090	562	WX-083	562	XZ090
562	WR-098B	562	WU-090A	562	WX-083A	562	XZ092
562	WR-098C	562	WU-090B	562	WX-083B	562	XZ098
562	WR-098D	562	WU-090C	562	WX-083C	562	XZ102
562	WR-100	562	WU-090D	562	WX-083D	562	YAAD001
562	WR-100A	562	WU-092	562	WX-085	562	YAAD017
562	WR-100B	562	WU-092A	562	WX-085A	562	YAAD021
562	WR-100C	562	WU-092B	562	WX-085B	562	YBAD010
562	WR-100D	562	WU-092C	562	WX-085C	562	YBAD014
562	WS-079	562	WU-092D	562	WX-085D	562	YBAD015
562	WS-079A	562	WU-094	562	WX-088	562	YBAD024
562	WS-079B	562	WU-094A	562	WX-088A	562	YS-080
562	WS-079C	562	WU-094B	562	WX-088B	562	Z-1010
562	WS-079D	562	WU-094C	562	WX-088C	562	Z-1012
562	WS-081	562	WU-094D	562	WX-088D	562	ZO217
562	WS-081A	562	WU-098	562	WX-090	562	ZO219
562	WS-081B	562	WU-098A	562	WX-090A	562	ZO220
562	WS-081C	562	WU-098B	562	WX-090B	562	ZO411
562	WS-081D	562	WU-098C	562	WX-090C	562	ZO412
562	WS-083	562	WU-098D	562	WX-092	562	ZO413
562	WS-083A	562	WU-100	562	WX-092A	562	ZOB10
562	WS-083B	562	WU-100A	562	WX-092B	562	ZOB8.2
562	WS-083C	562	WU-100B	562	WX-092C	562	ZOB9.1
562	WS-083D	562	WU-100C	562	WX-092D	562	ZOC10
562	WS-085	562	WU-100D	562	WX-094	562	ZOC8.2
562	WS-085A	562	WU2N1307	562	WX-094A	562	ZOC9.1
562	WS-085B	562	WV-079A	562	WX-094B	562	ZOD10
562	WS-085C	562	WV-079B	562	WX-094C	562	ZOD8.2
562	WS-085D	562	WV-079C	562		562	ZOD9.1
562	WS-088	562	WV-079D	562		562	Z10
562	WS-088A						

RS 276-	Industry Standard No.	RS 276-	Industry Standard No.	RS 276-	Industry Standard No.	RS 276-	Industry Standard No.
562	Z101	562	ZV9.1A	562	1N2624B	562	1N4783
562	Z1010	562	ZV9.1B	562	1N2790	562	1N4783A
562	Z1012	562	ZWO-9.1	562	1N3018	562	1N4784
562	Z10K	562	ZW9.1	562	1N3018A	562	1N4784A
562	Z1108	562	ZX9.1	562	1N3018B	562	1N4831
562	Z1108-C	562	ZZ10	562	1N3019	562	1N4831A
562	Z1109	562	ZZ8.2	562	1N3019A	562	1N4832A
562	Z1110	562	ZZ9.1	562	1N3019B	562	1N5237
562	Z1110-C	562	001-0101-01	562	1N3020	562	1N5237A
562	Z1208	562	001-0127-00	562	1N3020A	562	1N5237B
562	Z1209	562	001-0152-00	562	1N3020B	562	1N5238
562	Z1210	562	001-0161-00	562	1N3108	562	1N5238A
562	Z1AB.2A	562	001-0163-13	562	1N3148	562	1N5238B
562	Z1AB.2B	562	001-0163-15	562	1N3154	562	1N5239
562	Z1AB.2	562	001-023035	562	1N3154A	562	1N5239A
562	Z1B10	562	1-20363	562	1N3155	562	1N5239B
562	Z1B9.2	562	1-20398	562	1N3155A	562	1N5240
562	Z1B9.1	562	1/2Z105T5	562	1N3156	562	1N5240A
562	Z1C10	562	1/2Z10T5	562	1N3156A	562	1N5240B
562	Z1C8.2	562	1/2Z3335	562	1N3157	562	1N5348B
562	Z1C9.1	562	1/2Z8.2T5	562	1N3157A	562	1N5528
562	Z1D10	562	1/2Z8.2T5	562	1N3199	562	1N5528A
562	Z1D8.2	562	1/2Z9.1T5	562	1N33974	562	1N5528B
562	Z1D9.1	562	1/4A10	562	1N3401	562	1N5529
562	Z2A82P	562	1/4A10A	562	1N3402	562	1N5529A
562	Z2A91P	562	1/4A10B	562	1N3414	562	1N5529B
562	Z4A8.2	562	1/4A9.1	562	1N3415	562	1N5530
562	Z4B9.1	562	1/4A9.1A	562	1N3433	562	1N5530A
562	Z4X9.1	562	1/4A9.1B	562	1N3434	562	1N5530B
562	Z4XL9.1	562	1/4M8/27	562	1N3445	562	1N5561
562	Z4XL9.1B	562	1/4M9.1Z	562	1N3446	562	1N5561A
562	Z5B10	562	1/4M9.1Z10	562	1N3516	562	1N5562
562	Z5B8.2	562	1/4M9.1Z5	562	1N3517	562	1N5562A
562	Z5B9.1	562	1A10M	562	1N3518	562	1N5563
562	Z5C10	562	1A10MA	562	1N3677	562	1N5563A
562	Z5C8.2	562	1A12688	562	1N3677A	562	1N5734
562	Z5C9.1	562	1A9.1M	562	1N3678	562	1N5734B
562	Z5D10	562	1A9.1MA	562	1N3678A	562	1N5735
562	Z5D8.2	562	1B10Z	562	1N3679	562	1N5736
562	Z5D9.1	562	1B10Z10	562	1N3679A	562	1N5853
562	Z694	562	1B9.1Z	562	1N3776	562	1N5853A
562	Z714	562	1B9.1Z10	562	1N3855A	562	1N5853B
562	Z8.2	562	1EZ9.1	562	1N4014	562	1N5854
562	Z8.2A	562	1M10Z	562	1N4095	562	1N5854A
562	Z8.2B	562	1M10Z10	562	1N4101	562	1N5854B
562	Z8.2C	562	1M10Z5	562	1N4102	562	1N5855
562	Z8.2D	562	1M10Z810	562	1N4103	562	1N5855A
562	ZA8.2	562	1M10Z85	562	1N4103A	562	1N5855B
562	ZA8.2A	562	1M8.2Z	562	1N4104	562	1N5856
562	ZA8.2B	562	1M8.2Z10	562	1N4160	562	1N5856A
562	ZA9.1	562	1M8.2Z5	562	1N4160A	562	1N5856B
562	ZA9.1A	562	1M8.2Z810	562	1N4161	562	1N5889A
562	ZA9.1B	562	1M8.2Z85	562	1N4161A	562	1N664
562	ZB-1-9.5	562	1M9.1Z	562	1N4161B	562	1N701
562	ZB1-09	562	1M9.1Z10	562	1N4162	562	1N712
562	ZB1-10A	562	1M9.1Z5	562	1N4162A	562	1N712A
562	ZB1-9	562	1M9.1Z810	562	1N4295	562	1N713
562	ZB1-9V	562	1M9.1Z85	562	1N4295A	562	1N713A
562	ZB10	562	1N1313	562	1N4296	562	1N713B
562	ZB10A	562	1N1313A	562	1N4296A	562	1N714
562	ZB10B	562	1N1425	562	1N4297	562	1N714A
562	ZB10X	562	1N1511	562	1N4297A	562	1N714B
562	ZB209	562	1N1511A	562	1N4298	562	1N754
562	ZB210	562	1N1512	562	1N4298A	562	1N756
562	ZB8.2	562	1N1512A	562	1N430	562	1N756A
562	ZB8.2A	562	1N1522	562	1N4302	562	1N757
562	ZB8.2B	562	1N1522A	562	1N4302A	562	1N757A
562	ZB9.1	562	1N1523	562	1N430A	562	1N758
562	ZB9.1A	562	1N1523A	562	1N430B	562	1N758A
562	ZB9.1B	562	1N1530	562	1N4325	562	1N764
562	ZBI-09	562	1N1530A	562	1N4325A	562	1N764-1
562	ZC109	562	1N1744	562	1N4326	562	1N764-3
562	ZC110	562	1N1769	562	1N4326A	562	1N765-1
562	ZD10B	562	1N1769A	562	1N4326B	562	1N765A
562	ZD9.1	562	1N1770	562	1N4327	562	1N935
562	ZD9.1A	562	1N1770A	562	1N4327A	562	1N935A
562	ZD9.1B	562	1N1771	562	1N4402	562	1N935B
562	ZE10	562	1N1771A	562	1N4402A	562	1N936
562	ZE8.2	562	1N1875	562	1N4403	562	1N936A
562	ZE9.1	562	1N1875A	562	1N4403A	562	1N936B
562	ZF-8.2	562	1N1875B	562	1N4403B	562	1N937
562	ZP10	562	1N1876	562	1N4404	562	1N937A
562	ZP110	562	1N1876A	562	1N4404A	562	1N937B
562	ZP19.1	562	1N1931	562	1N4630	562	1N938
562	ZP210	562	1N1932	562	1N4631	562	1N938A
562	ZP29.1	562	1N1932A	562	1N4632	562	1N938B
562	ZP8.2	562	1N1932B	562	1N4659	562	1N939
562	ZP8.2P	562	1N1958	562	1N4660	562	1N939A
562	ZP8A	562	1N1959	562	1N4661	562	1N939B
562	ZP9.1	562	1N1985	562	1N4694	562	1N959
562	ZQ10	562	1N1986	562	1N4695	562	1N959A
562	ZQ8.2	562	1N1986A	562	1N4697	562	1N959B
562	ZQ9.1	562	1N1986B	562	1N4738	562	1N960
562	ZH109	562	1N2035-1	562	1N4738A	562	1N960A
562	ZH110	562	1N2035-12	562	1N4739	562	1N960B
562	ZH9.1	562	1N2035A	562	1N4739A	562	1N961
562	ZH9.1A	562	1N2036	562	1N4740	562	1N961A
562	ZH9.1B	562	1N203CA	562	1N4740A	562	1N961B
562	ZM10	562	1N2163	562	1N4765	562	1R10
562	ZM10A	562	1N2163A	562	1N4765A	562	1R10A
562	ZM10B	562	1N2164	562	1N4766	562	1R10B
562	ZM8.2	562	1N2164A	562	1N4766A	562	1R8.2
562	ZM8.2A	562	1N2165	562	1N4767	562	1R8.2B
562	ZM8.2B	562	1N2165A	562	1N4767A	562	1S9.1A
562	ZM8.2C	562	1N2166	562	1N4768	562	1S9.1B
562	ZM8.2D	562	1N2166A	562	1N4768A	562	1S1095
562	ZM8.7	562	1N2167	562	1N4769	562	1S1156
562	ZM8.7A	562	1N2167A	562	1N4769A	562	1S1164
562	ZM8.7B	562	1N2168	562	1N4770	562	1S1165
562	ZM9.1	562	1N2168A	562	1N4770A	562	1S1175
562	ZM9.1A	562	1N2169	562	1N4771	562	1S130(ZENER)
562	ZM9.1B	562	1N2169A	562	1N4771A	562	1S133
562	ZP10	562	1N2170	562	1N4772	562	1S1377
562	ZP8.2	562	1N2170A	562	1N4772A	562	1S1377A
562	ZP9.1	562	1N2171	562	1N4773	562	1S1378
562	ZP9.1A	562	1N2171A	562	1N4773A	562	1S1378A
562	ZP9.1B	562	1N225	562	1N4774	562	1S138(ZENER)
562	ZPD8.2	562	1N225A	562	1N4775	562	1S139
562	ZPD9.1	562	1N226	562	1N4775A	562	1S1392
562	ZQ9.1	562	1N226A	562	1N4776	562	1S1392A
562	ZQ9.1A	562	1N2620	562	1N4776A	562	1S1393
562	ZQ9.1B	562	1N2620A	562	1N4777	562	1S1393A
562	ZR209	562	1N2620B	562	1N4777A	562	1S1408
562	ZR209T1	562	1N2621	562	1N4778	562	1S1408R
562	ZR209T3	562	1N2621A	562	1N4778A	562	1S1409
562	ZR210	562	1N2621B	562	1N4779	562	1S1409R
562	ZS89.1	562	1N2622	562	1N4779A	562	1S1717
562	ZS89.1A	562	1N2622A	562	1N4780	562	1S1717L
562	ZS89.1B	562	1N2622B	562	1N4780A	562	1S1718
562	ZSF08.2	562	1N2623	562	1N4781	562	1S1718/4454C
562	ZT9.1	562	1N2623A	562	1N4781A	562	1S1728
562	ZT9.1A	562	1N2623B	562	1N4782	562	1S1737
562	ZT9.1B	562	1N2624	562	1N4782A	562	1S1769
562	ZV9.1	562	1N2624A			562	1S1781

RS 276-	Industry Standard No.	RS 276-	Industry Standard No.	RS 276-	Industry Standard No.	RS 276-	Industry Standard No.
562	181782	562	1ZD825	562	46-86194-3	562	187-6
562	181783	562	1ZD828	562	46-86284-3	562	200X8220-878
562	181793	562	1ZD892	562	46-86323-3	562	201A0723
562	181794	562	1ZF9.1T10	562	46-86421-3	562	201X2220-118
562	181795	562	1ZF9.1T20	562	46-86442-3	562	230-0025
562	18195B	562	1ZF9.1T5	562	46-86483-3(ZENER)	562	260-10-044
562	181959	562	1ZM10T10	562	48-03073A06	562	264P02503
562	181960	562	1ZM10T20	562	48-03073A09	562	264P09703
562	182067	562	1ZM10T5	562	48-I1-C005	562	276-562
562	182068	562	1ZM9.1T10	562	48-134653	562	339-529-002
562	182069	562	1ZM9.1T20	562	48-134912	562	412(ZENER)
562	182082	562	1ZM9.1T5	562	48-134957	562	464-311-19
562	182082A	562	2M214	562	48-137130	562	465-199-19
562	182091(ZENER)	562	281718/4454C	562	48-137164	562	523-2001-100
562	182091A	562	02Z-10A	562	48-137328	562	523-2003-100
562	182100	562	02Z-8.2A	562	48-137393	562	523-2003-829
562	182100A	562	02Z-9.1A	562	48-137421	562	523-2005-100
562	182115	562	02Z102A	562	48-137577	562	523-2005-829
562	182115A	562	02Z10A	562	48-40458A06	562	523-2005-919
562	182116	562	02Z10A-U	562	48-40458A064	562	523-2003-919
562	182116A	562	02Z8.2A	562	48-41763001	562	630-077
562	182117	562	02Z9.1A	562	48-41763002	562	65509
562	182117A	562	3L4-3505-1	562	48-41763003	562	65409
562	182212	562	3L4-3506-7	562	48-41763C1	562	65509
562	182126	562	4-856	562	48-41873J02	562	690Y105H32
562	182129	562	05-04800-02	562	48-41873J03	562	700-159
562	18213	562	05-540082	562	48-44080J05	562	754-5700-282
562	182157	562	05-990094	562	48-82256C08	562	754-5710-219
562	18214	562	052-10	562	48-82256C09	562	754-5750-282
562	18215	562	05Z10	562	48-82256C11	562	754-5850-284
562	18216	562	05Z8.2	562	48-82256C16	562	754-9000-082
562	18217	562	05Z9.1	562	48-82256C18	562	754-9030-009
562	182190	562	05Z9.1L	562	48-82256C22	562	754-903-090
562	182191	562	6DM	562	48-82256C28	562	903-119B
562	182192	562	08	562	48-82256C32	562	903-179
562	182193	562	08-08125	562	48-82256C38	562	903-337
562	182195	562	8-905-421-118	562	48-82256C40	562	903Y00228
562	182224	562	8-905-421-228	562	48-82256C43	562	919-01-1213
562	182225	562	8-9V	562	48-82256C45	562	919-01-1214
562	182492	562	8.75VZENER	562	48-82256C56	562	919-01-3035
562	182493	562	8091	562	48-83461E15	562	1000-12
562	182494	562	09-306042(ZENER)	562	48-83461E32	562	1000-132
562	182549	562	09-306106	562	48A40458A06	562	1001-12
562	182550	562	09-306124	562	488134851	562	1006-5985
562	182551	562	09-306158	562	488134912	562	1010
562	183008	562	09-306165	562	488134957	562	1040-09
562	183009	562	09-306180	562	488137021	562	1040-09(ZENER)
562	183010	562	09-306194	562	488137021(10)	562	1042-18
562	18304	562	09-306196	562	488137164	562	1042-19
562	18333	562	09-306228	562	488155054	562	1048-9839
562	18333Y	562	09-306232	562	488155054(ZENER)	562	1065-6775
562	0000018334	562	09-306235	562	50Z9.1	562	1074-120
562	18334K	562	09-306238	562	50Z9.1A	562	1074-122
562	18334M	562	09-306239	562	50Z9.1B	562	1080-10
562	18334N	562	09-306241	562	50Z9.1C	562	1108
562	18335	562	09-306242	562	051-0024	562	1109
562	18472	562	09-306247	562	51-02006-12	562	1110
562	18482	562	09-306275	562	51-02007-12	562	1411-136
562	1855	562	09-306277	562	52-053-013-0	562	1794-17
562	18550	562	09-306287	562	53A022-6	562	1794-17(10)
562	18551	562	09-306314	562	53A022-7	562	1858A
562	1856	562	09-306327	562	53K001-14	562	2000-305
562	18617	562	09-306332	562	53K001-18	562	2000-308
562	18618	562	09-306351	562	5524	562	2000-327
562	18694	562	09-306354	562	56-19	562	2023-41
562	18695	562	09-306356	562	56-46	562	2040-17
562	187082	562	09-306375	562	56-4885	562	2041-06
562	187082A	562	09-306378	562	56-54	562	2064-177
562	187082B	562	09-306381	562	56-62	562	2079-42
562	187091	562	09-306382	562	56-67	562	2079-93
562	187091A	562	09-306401	562	61E15	562	2087-46
562	187091B	562	9D141003-12	562	61E32	562	2093A41-159
562	187100	562	9D15	562	62-130047	562	2093A41-16
562	187100A	562	9D15	562	63-8825	562	2093A41-163
562	187100B	562	10-0013	562	63-9785	562	2093A41-186
562	18757A	562	10B62	562	66X0040-005	562	2093A41-24
562	18758	562	10D08	562	68X0040-005	562	2093A41-24A
562	18758(H)	562	10EB1	562	070-011	562	2093B41-24
562	18758QB	562	10V	562	070-024	562	2102-032
562	18758H	562	10VJ	562	7529.1	562	2150-17
562	18759	562	13-0002	562	7529.1A	562	2203
562	18759H	562	13-0029	562	7529.1B	562	2203-17
562	18825	562	13-085028	562	7529.1C	562	2309
562	18827	562	13-085042	562	81-46123044-3	562	2339-17
562	18256-08	562	13-33179-4	562	86-0005	562	2343-17
562	1T10	562	13-33187-6	562	86-114-1	562	2451-17
562	1T10B	562	13-33187-7	562	86-118-1	562	2452-17
562	1T243M	562	13DD02F	562	86-135-1	562	2795A
562	1T9.1	562	14-515-12	562	86-35-1	562	003102
562	1T9.1A	562	14-515-16	562	86-37-1	562	4653
562	1T9.1B	562	14-515-19	562	86-58-1	562	4801-00801
562	1TA10	562	14-515-24	562	93A39-3	562	4808-0000-009
562	1TA10A	562	019-003411	562	93A39-30	562	5001-125
562	1TA9.1	562	19-090-008A	562	93A39-6	562	5001-130
562	1TA9.1A	562	19A115528-P3	562	93B39-1	562	5001-152
562	1TWC8H	562	19B200379-P1	562	93B39-3	562	5403-MS
562	1TWC8L	562	21A037-012	562	93B39-3	562	5518
562	1TWC8M	562	21A040-051(DIO)	562	93C39-3	562	5519
562	WB-9D	562	21A103-064	562	96-5116-01	562	5520
562	1Z10	562	21A103-064	562	100-139	562	5632-HZ11A
562	1Z10A	562	21A108-002	562	100-144	562	5635-HZ11A
562	1Z10T10	562	21M214	562	100-219	562	5635-ZB1-10
562	1Z10T20	562	21M307	562	100-521	562	06115
562	1Z10T5	562	21M493	562	100-522	562	6115(G.E.)
562	1Z8.2	562	21M584	562	100-523	562	6432-3
562	1Z8.2A	562	21M585	562	103-272	562	7554
562	1Z8.2T10	562	21Z6AF	562	103-279-16	562	8000-00004-065
562	1Z8.2T20	562	21Z6F	562	103-279-17	562	8000-00005-021
562	1Z8.2T5	562	022-2823-002	562	103-279-18	562	8000-00006-009
562	1Z8.2	562	2226AF	562	103-279-19	562	8000-00006-232
562	1Z9.1	562	2226F	562	103-29010	562	8000-00011-043
562	1Z9.1A	562	2326AF	562	106-010	562	8000-00011-064
562	1Z9.1B	562	2326F	562	0114-0017	562	8000-00028-047
562	1Z9.1C	562	24MW1065	562	0114-0090	562	8000-00041-P065
562	1Z9.1D10	562	24MW1125	562	120-004879	562	8000-00041-018
562	1Z9.1D15	562	24MW331	562	123-019	562	8000-00043-065
562	1Z9.1D5	562	24MW670	562	123-020	562	8000-00043-068
562	1Z9.1T10	562	24MW743	562	123B-004	562	8000-00045-021
562	1Z9.1T20	562	24MW998	562	142-012	562	8000-00057-011
562	1Z9.1T5	562	32-0005	562	145-100	562	8000-00058-004
562	1Z91.1D10	562	32-0025	562	145-118	562	8710-54
562	1Z91.D10	562	32-0049	562	151-012-9-001	562	8840-55
562	1Z10T5	562	33R-09	562	151-031-9-006	562	8910-54
562	1Z08.2T10	562	34-8057-9	562	152-008-9-001	562	0011193
562	1Z08.2T20	562	34-8057-9(PHILCO)	562	152-012-9-001	562	18410-43
562	1Z08.2T5	562	41Z6	562	152-042-9-001	562	18600-54
562	1Z09.1	562	41Z6A	562	152-051-9-001	562	25810-54
562	1Z09.1T10	562	42-Z6A	562	152-054-9-001	562	25840-54
562	1Z09.1T5	562	42Z6	562	152-082-9-001	562	26810-52
562	1ZCT10	562	4326	562	157-009-9-001	562	28810-65
562	1ZD8-2	562	4326A	562	157-100	562	37510-54
562	1ZD8-5	562	44T-300-93	562	179-4	562	38510-53
562	1ZD8.2	562	46-41763001	562	185-015		
562	1ZD8.2V			562	186-016		

RS 276-	Industry Standard No.	RS 276-	Industry Standard No.	RS 276-	Industry Standard No.	RS 276-	Industry Standard No.
562	42020-737	563	A36539	563	EA3719	563	LPM12
562	45810-54	563	A42X00480-01	563	EC0142	563	LPM12A
562	58810-86	563	A694X1	563	EC0142A	563	LPM13
562	58840-117	563	A694X1-0A	563	EC0143	563	LPM13A
562	000072150	563	A7287500	563	EC0143A	563	LP22T12
562	72180	563	A7287513	563	EC05020	563	LR120CH
562	72190	563	AU2012	563	EC05020A	563	M12Z
562	79408	563	AV2012	563	EC05021	563	M4Z12
562	88060-54	563	AV5	563	EC05021A	563	M4Z12-20
562	88510-54	563	AV-01-12C	563	EC05022	563	M4Z12A
562	93018	563	AW01	563	EC05022A	563	MA1120
562	93030	563	AW01(RCA)	563	EC05074	563	MA1130
562	99201-216	563	AW01-11	563	EC05074A	563	MC6016
562	99201-316	563	AW01-12	563	EP16X36	563	MC6016A
562	112530	563	AW01-12C	563	EQA01-11	563	MC6116
562	119594	563	AW01-12V	563	EQA01-115	563	MC6116A
562	129904	563	AW01-13	563	EQA01-118	563	MD759
562	129904(RCA)	563	AW0L-13	563	EQA01-118E	563	MD759A
562	130044	563	AX12	563	EQA01-11Z	563	MEZ12T10
562	130047	563	AZ-105	563	EQA01-12	563	MEZ12T5
562	146260	563	AZ-110	563	EQA01-12R	563	MMZ12(06)
562	147477-7-1	563	AZ-120	563	EQA01-12S	563	MR120C-H
562	171162-272	563	AZ-125	563	EQA01-12Z	563	MT2616
562	171162-292	563	AZ-130	563	EQA01-13	563	MT2616A
562	171560	563	AZ11	563	EQA01-13R	563	MZ-1000-15
562	175043-069	563	AZ110	563	EQB01	563	MZ-11
562	190404	563	AZ12	563	EQB01-011Z	563	MZ-12
562	216024	563	AZ13	563	EQB01-02R	563	MZ-12B
562	228007	563	0AZ213	563	EQB01-11	563	MZ12B
562	236266	563	AZ759	563	EQB01-11V	563	MZ1000-14
562	245117	563	AZ759A	563	EQB01-11Z	563	MZ1000-15
562	245217	563	AZ962	563	EQB01-12	563	MZ1000-16
562	262111	563	AZ962A	563	EQB01-12A	563	MZ1012
562	262546	563	AZ962B	563	EQB01-12B	563	MZ11
562	263424	563	AZ963	563	EQB01-12BV	563	MZ12
562	280317	563	AZ963A	563	EQB01-12R	563	MZ12A
562	300732	563	AZ963B	563	EQB01-12Z	563	MZ12B
562	489752-040	563	AZ964	563	EQB01-13	563	MZ212
562	489752-109	563	AZ964A	563	EQ-12A	563	MZ212
562	530073-3	563	AZ964B	563	ES10234	563	MZ500-17
562	530145-100	563	B66X0040-001	563	ET16X15	563	MZ500-18
562	530145-1100	563	BN7551	563	EVR12	563	MZ500-19
562	530157-1100	563	BZ-105	563	EVR12A	563	MZ500-.18
562	602081	563	BZ-110	563	EVR12B	563	MZ92-12A
562	610122	563	BZ-12	563	EZ11(ELCOM)	563	N-BA16Z29
562	741739	563	BZ-120	563	EZ12(ELCOM)	563	NGP5010
562	741867	563	BZ-125	563	EZ13(ELCOM)	563	NTC-20
562	741869	563	BZ-130	563	FZ1215	563	02Z12A
562	741870	563	BZ110	563	FZ12A	563	02Z12GR
562	742922	563	BZ120	563	FZ12T10	563	0A126-12
562	801726	563	BZ130	563	FZ12T5	563	0A126/10
562	817197	563	BZX17	563	G01-037-A	563	0A126/12
562	922358	563	BZX29C11	563	G12T10	563	0A12612
562	922524	563	BZX29C12	563	G12T20	563	0AZ213
562	922603	563	BZX29C13	563	G12T5	563	0AZ230
562	982254	563	BZX46C11	563	GEZD-11	563	0AZ273
562	988050	563	BZX46C12	563	GEZD-12	563	0Z12T10
562	1221900	563	BZX46C13	563	GEZD-13	563	0Z12T5
562	1223773	563	BZX55C11	563	H)-00005	563	PD6016
562	1223927	563	BZX55C12	563	H089	563	PD6016A
562	1800013	563	BZX55C13	563	HB-12B	563	PD6057
562	1956197	563	BZX71C11	563	HD3000109-0	563	PR617
562	1965019	563	BZX71C12	563	HD3000101-0	563	PS10022B
562	2002209-1	563	BZX71C13	563	HD3000109-0	563	PS10066
562	2002209-11	563	BZX79C11	563	HD3000109-0	563	PS8915
562	2002209-5	563	BZX79C12	563	HD3000113	563	PT0507
562	2002209-9	563	BZX79C13	563	HD3001009	563	PT0508
562	2003779-26	563	BZX83C11	563	HD3001009	563	QA01-11.5Z
562	2092055-0002	563	BZX83C12	563	HD3001809	563	QA01-11SE
562	2092055-0017	563	BZX83C13	563	HD3002409	563	QA01-12B
562	2092055-0027	563	BZY-19	563	HEP-Z0222	563	QA1-11M
562	2092055-0712	563	BZY18	563	HEP-Z0414	563	QA111M
562	2330307	563	BZY19	563	HEP-Z0415	563	QA111SE
562	2331161	563	BZY69	563	HEP-Z0416	563	QB01-11ZB
562 56K	1RB-2A	563	BZY83C11	563	HEP105	563	QB111
563	.25N11	563	BZY83C12	563	HEP604	563	QB111Z
563	.25N12	563	BZY83C13	563	HEP605	563	QRT-242
563	.25N13	563	BZY85C11	563	HEPZ0222	563	QRT-243
563	.25T11	563	BZY85C12	563	HEPZ0222A	563	QRT-244
563	.25T11A	563	BZY85C13	563	HEPZ0414	563	QZ12T10
563	.25T12	563	BZY85D12	563	HEPZ0415	563	QZ12T5
563	.25T12.8	563	BZY88C11	563	HEPZ0416	563	R-1348
563	.25T12.8A	563	BZY88C12	563	HP-20041	563	R-7103
563	.25T12.8B	563	BZY88C13	563	HM11	563	R2D
563	.25T12A	563	BZY92C11	563	HM11A	563	R7894
563	.25T12B	563	BZY92C12	563	HM12	563	R8364
563	.25T13	563	BZY92C13	563	HM12A	563	RD-11E
563	.25T13A	563	BZY94/C12	563	HM12B	563	RD-13A
563	.4T11	563	BZY94C11	563	HM13	563	RD-13AD
563	.4T11A	563	BZY94C12	563	HM13A	563	RD-13AK
563	.4T11B	563	BZY94C13	563	HN-00005	563	RD-13AK-P
563	.4T12.8	563	C4018	563	H089	563	RD-13AKP
563	.4T12.8A	563	CD31-00015	563	HS2120	563	RD-13AM
563	.4T12A	563	CD31-12022	563	HST120	563	RD-13AN
563	.4T12B	563	CD4116	563	HW11	563	RD-13E
563	.4T13	563	CD4117	563	HW11A	563	RD-13M
563	.4T13A	563	CD4118	563	HW11B	563	RD11AN
563	.4T13B	563	CD4121	563	HW12A	563	RD11E
563	.75N12	563	CD4122	563	HW12B	563	RD11EA
563	.7J211	563	CD2318-75	563	HW13	563	RD11EB
563	.7J212	563	CR956	563	HW13A	563	RD11EC
563	.7J213	563	CX-0052	563	HW13B	563	RD11EM
563	.7Z11A	563	CXL141	563	HZ-11	563	RD11P
563	.7Z11B	563	CXL142	563	HZ-11A	563	RD11PA
563	.7Z11C	563	CXL143	563	HZ-11C	563	RD11PB
563	.7Z11D	563	CXL5020	563	HZ-12	563	RD11M
563	.7Z12A	563	CXL5021	563	HZ-12A	563	RD12
563	.7Z12B	563	CXL5022	563	HZ-12B	563	RD12E
563	.7Z12C	563	CXL5074	563	HZ-12C	563	RD12EA
563	.7Z12D	563	CZD012-5	563	HZ-212	563	RD12EB
563	.7Z13A	563	D-00484R	563	HZ11	563	RD12EB1Z
563	.7Z13B	563	D18	563	HZ11(H)	563	RD12EC
563	.7Z13C	563	D12VIO	563	HZ11C	563	RD12F
563	.7Z13D	563	D2G-3	563	HZ11H	563	RD12PA
563	.7ZM1	563	D2G-4	563	HZ11L	563	RD12PB
563	.7ZM11	563	D3W	563	HZ11Y	563	RD12M
563	.7ZM11A	563	D8-108	563	HZ12	563	RD13
563	.7ZM11B	563	D8-189	563	HZ12(H)	563	RD13A
563	.7ZM11C	563	DS189	563	HZ12H	563	RD13AK
563	.7ZM12A	563	DZ-12	563	HZ12L	563	RD13AKP
563	.7ZM12B	563	DZ12	563	IB1225	563	RD13AL
563	.7ZM12C	563	DZ12A	563	IN962B	563	RD13AM
563	.7ZM12D	563	DZ13	563	IW01-09J	563	RD13AN
563	.7ZM13A	563	E0771-7	563	J20438	563	RD13AN-P
563	.7ZM13B	563	E1852	563	KD2505	563	RD13ANP
563	.7ZM13C	563	E262	563	KS120A	563	RD13B
563	.7ZM13D	563	EA1318	563	KS120B	563	RD13B-Z
563	A-120077	563	EA16X162	563	KS2120A	563	RD13E
563	A-134166-2	563	EA16X29	563	KS2120B	563	RD13EA
563	A054-151	563	EA16X77	563	KS44A	563	RD13EB
563	A054-231	563	EA16X81	563	KS44AF	563	RD13F
563	A066-122	563	EA2500	563	KS44B	563	RD13PA
563	A066-133			563	KS44BF	563	RD13FB
563	0A12612					563	RD13K
563	A2474						

RS 276-	Industry Standard No.	RS 276-	Industry Standard No.	RS 276-	Industry Standard No.	RS 276-	Industry Standard No.
563	RD13M	563	VZ-110	563	WR-130B	563	WZ-105
563	RE-116	563	VZ-120	563	WR-130C	563	WZ-110
563	RE-118	563	VZ-125	563	WR-130D	563	WZ-110A
563	RE-119	563	VZ-130	563	WS-110	563	WZ-115
563	RE116	563	WEP105	563	WS-110A	563	WZ-120
563	RE118	563	WEP1112	563	WS-110B	563	WZ-125
563	RE119	563	WEP605	563	WS-110C	563	WZ-130
563	REN142	563	WL-125	563	WS-110D	563	WZ12
563	REN143	563	WL-125A	563	WS-120	563	WZ12.8
563	REN5074	563	WL-125B	563	WS-120A	563	WZ120
563	RH-EXO011CEZZ	563	WL-125C	563	WS-120B	563	WZ130
563	RH-EXO015CEZZ	563	WL-125D	563	WS-120C	563	WZ13B
563	RH-EXO017CEZZ	563	WL-130	563	WS-120D	563	WZ55
563	RH-EXO019TAZZ	563	WL-130A	563	WS-125	563	WZ917
563	RH-EXO038CEZZ	563	WL-130B	563	WS-125A	563	X330302
563	RH-EXO047CEZZ	563	WL-130C	563	WS-125B	563	XZ-122
563	RS1348	563	WL-130D	563	WS-125C	563	Z-1014
563	RT-242	563	WM-110	563	WS-130	563	Z-1112C
563	RT-243	563	WM-110A	563	WS-130A	563	Z-963B
563	RT-244	563	WM-110B	563	WS-130B	563	ZO-12
563	RT8200	563	WM-110C	563	WS-130C	563	ZO-12A
563	RZ11	563	WM-110D	563	WS-130D	563	ZO-12B
563	R212	563	WM-120	563	WT-110	563	ZO1-13
563	R213	563	WM-120A	563	WT-110A	563	ZO222
563	RZZ11	563	WM-120B	563	WT-110B	563	ZO414
563	RZZ12	563	WM-120C	563	WT-110C	563	ZO415
563	RZZ13	563	WM-120D	563	WT-110D	563	ZO416
563	S1R12B	563	WM-125	563	WT-120	563	ZOB-12
563	S1R13B	563	WM-125A	563	WT-120A	563	ZOB12
563	SD-33	563	WM-125B	563	WT-120B	563	ZOC12
563	SD33	563	WM-125C	563	WT-120C	563	ZOD12
563	SD53	563	WM-125D	563	WT-120D	563	Z1014
563	SPZ716	563	WM-130	563	WT-125	563	Z1112
563	SI-RECT-228	563	WM-130A	563	WT-125A	563	Z1112-C
563	SI-RECT-230	563	WM-130B	563	WT-125B	563	Z1112C
563	SK3062	563	WM-130C	563	WT-125C	563	Z111Z
563	SK3062/142A	563	WM-130D	563	WT-125D	563	Z12.0
563	SK3093	563	WN-110	563	WT-130	563	Z1212
563	SK3139	563	WN-110A	563	WT-130A	563	Z12A
563	SK3139/5074A	563	WN-110B	563	WT-130B	563	Z12K
563	SK3750	563	WN-110C	563	WT-130C	563	OZ12T10
563	SK3750/143A	563	WN-110D	563	WT-130D	563	OZ12T5
563	SK3786	563	WN-120	563	WU-110	563	Z1B12
563	SK3786/5020	563	WN-120A	563	WU-110A	563	Z1C12
563	SK3787	563	WN-120B	563	WU-110B	563	Z1D12
563	SK3787/5021	563	WN-120C	563	WU-110C	563	Z2A120P
563	SK3787/5021A	563	WN-120D	563	WU-110D	563	Z4B12
563	SK3788	563	WN-125	563	WU-120	563	Z4X12
563	SK3788/5022	563	WN-125A	563	WU-120A	563	Z4X13
563	SV1017	563	WN-125B	563	WU-120C	563	Z4XL12
563	SV135	563	WN-125C	563	WU-120D	563	Z4XL12B
563	SV4012	563	WN-125D	563	WU-125	563	Z5B12
563	SV4012A	563	WN-130	563	WU-125A	563	Z5BZ
563	SW01	563	WN-130A	563	WU-125B	563	Z5C12
563	SZ-11	563	WN-130B	563	WU-125C	563	Z5D12
563	SZ-200-11	563	WN-130C	563	WU-125D	563	ZA12
563	SZ-200-12	563	WN-130D	563	WU-130	563	ZA12A
563	SZ-200-13	563	WO-110	563	WU-130A	563	ZA12B
563	SZ-AW01-11	563	WO-110A	563	WU-130B	563	ZA13
563	SZ-BQB01-12A	563	WO-110B	563	WU-130C	563	ZA13A
563	SZ-RD11FB	563	WO-120	563	WU-130D	563	ZA13B
563	SZ-RD12EB	563	WO-120A	563	WU-20B	563	ZB-11
563	SZ-RD12FB	563	WO-120B	563	WV-110	563	ZB-31.12
563	SZ-UZ-12B	563	WO-120C	563	WV-110A	563	ZB1-1
563	S211	563	WO-120D	563	WV-110B	563	ZB1-10
563	S212	563	WO-125	563	WV-110C	563	ZB1-11
563	S212.0	563	WO-125A	563	WV-110D	563	ZB1-12
563	S213	563	WO-125B	563	WV-120	563	ZB1-13
563	S213.0	563	WO-125C	563	WV-120A	563	ZB11
563	SZ671-B	563	WO-125D	563	WV-120B	563	ZB11A
563	SZ671-G	563	WO-130A	563	WV-120C	563	ZB11B
563	SZ671-O	563	WO-130B	563	WV-120D	563	ZB12
563	SZ671-W	563	WO-130C	563	WV-125	563	ZB12A
563	SZ961-V	563	WO-130D	563	WV-125A	563	ZB12B
563	SZA-13	563	WO110C	563	WV-125B	563	ZB13
563	SZT8	563	WO130	563	WV-125C	563	ZB13B
563	T-E1068	563	WP-110	563	WV-125D	563	ZB212
563	T-E1077	563	WP-110A	563	WV-130	563	ZC012
563	T-E1106	563	WP-110B	563	WV-130A	563	ZD012
563	T-E1140	563	WP-110D	563	WV-130B	563	ZD112
563	TE1068	563	WP-120	563	WW-110	563	ZD012
563	TE1077	563	WP-120A	563	WW-110A	563	ZD12
563	TM142	563	WP-120B	563	WW-110B	563	ZD12A
563	TM142A	563	WP-120C	563	WW-110C	563	ZD12B
563	TM143	563	WP-120D	563	WW-110D	563	ZE12
563	TM5022	563	WP-125	563	WW-115	563	ZE12A
563	TM5074	563	WP-125A	563	WW-115A	563	ZE12B
563	TMD10	563	WP-125B	563	WW-115B	563	ZEC12
563	TMD10A	563	WP-125D	563	WW-115C	563	ZEN502
563	TR-75	563	WP-130	563	WW-115D	563	ZEN506
563	TR-78B	563	WP-130A	563	WX-120	563	ZENER-122
563	TR128	563	WP-130B	563	WX-120A	563	ZF112
563	TR128A	563	WP-130C	563	WX-120B	563	ZF113
563	TR128B	563	WP-130D	563	WX-120C	563	ZF12
563	TR14002-6	563	WQ-110	563	WX-120D	563	ZF12A
563	TS-337	563	WQ-110A	563	WX-125	563	ZF12B
563	TVS-RD(M)(P)D	563	WQ-110C	563	WX-125A	563	ZF13
563	TVS-RD-13A	563	WQ-110D	563	WX-125B	563	ZF211
563	TVS-RD11A	563	WQ-120	563	WX-125C	563	ZF212
563	TVS-RD13A	563	WQ-120A	563	WX-125D	563	ZF213
563	TVS-RD13D	563	WQ-120B	563	WX-130	563	ZG12
563	TVS-RD13M	563	WQ-120D	563	WX-130A	563	ZH112
563	TVS-RD13P	563	WQ-125	563	WX-130B	563	ZH12
563	TVS1N4741	563	WQ-125A	563	WX-130C	563	ZH12A
563	TVS1N741A	563	WQ-125B	563	WX-130D	563	ZH12B
563	TVS1N741H	563	WQ-125C	563	WY-110	563	ZH13
563	TVSAW01-11	563	WQ-125D	563	WY-110A	563	ZH13A
563	TVSBQA01-125	563	WQ-130	563	WY-110B	563	ZH13B
563	TVSBQB01-12	563	WQ-130A	563	WY-110C	563	ZJ12
563	TVSQA01-11.5E	563	WQ-130B	563	WY-110D	563	ZJ12A
563	TVSQA01-11SE	563	WQ-130C	563	WY-120	563	ZJ12B
563	TVSQA01-12S	563	WQ-130D	563	WY-120A	563	ZM11
563	TVSRD12EBH	563	WR-110	563	WY-120B	563	ZM11A
563	TVSRD13AL	563	WR-110A	563	WY-120C	563	ZM11B
563	TVSZB1-11	563	WR-110B	563	WY-120D	563	ZM12
563	TZ12	563	WR-110C	563	WY-125	563	ZM12A
563	TZ12A	563	WR-110D	563	WY-125A	563	ZM12B
563	TZ12B	563	WR-120	563	WY-125B	563	ZM13
563	TZ12C	563	WR-120A	563	WY-125C	563	ZM13A
563	UZ-12B	563	WR-120B	563	WY-125D	563	ZM13B
563	UZ13B	563	WR-120C	563	WY-130	563	ZO-12
563	UZ8712	563	WR-120D	563	WY-130A	563	ZO-12A
563	UZ8713	563	WR-125	563	WY-130B	563	ZO-12B
563	UZ8812	563	WR-125A	563	WY-130C	563	ZOB12
563	UZ8813	563	WR-125B	563	WY-130D	563	ZP12
563	V160	563	WR-125C			563	ZP12A
563	VR12	563	WR-125D			563	ZP12B
563	VR12A	563	WR-130			563	ZQ12
563	VR12B	563	WR-130A			563	ZQ12A
563	VR13					563	ZQ12B
563	VR13A					563	ZR212
563	VR13B					563	ZR50B793-1
563	VS-RD11AM					563	ZS12
563	VZ-105					563	ZS12A
						563	ZS12B

RS 276-	Industry Standard No.	RS 276-	Industry Standard No.	RS 276-	Industry Standard No.	RS 276-	Industry Standard No.
563	ZT12	563	1N4329A	563	1N944	563	1ZF12T5
563	ZT12A	563	1N4329B	563	1N944A	563	1ZM12Z10
563	ZT12B	563	1N4330	563	1N944B	563	1ZM12Z20
563	ZU12	563	1N4330A	563	1N945	563	1ZM12T5
563	ZU12A	563	1N4405	563	1N945A	563	1ZT12
563	ZU12B	563	1N4405A	563	1N945B	563	1ZT12B
563	ZV12	563	1N4406	563	1N946	563	02Z11A
563	ZV12A	563	1N4406A	563	1N946A	563	02Z12A
563	ZV12B	563	1N4406B	563	1N946B	563	022Z12GR
563	ZX12	563	1N4407	563	1N962	563	02Z13A
563	ZY12	563	1N4407A	563	1N962A	563	003-009200
563	ZY12A	563	1N4633	563	1N962B	563	003-009700
563	ZY12B	563	1N4634	563	1N963	563	003-010000
563	ZZ12	563	1N4635	563	1N963A	563	03-933943
563	001-023037	563	1N4662	563	1N963B	563	3/4Z12D10
563	1/2Z11T5	563	1N4663	563	1N964	563	3/4Z12D5
563	1/2Z13T5	563	1N4664	563	1N964A	563	3L4-3505-4
563	1/4A12	563	1N4698	563	1N964B	563	4-2020-06600
563	1/4A12A	563	1N4699	563	1R11	563	4-2020-11500
563	1/4A12B	563	1N4700	563	1R11A	563	4-2020-12300
563	1/4Z12T5	563	1N4741	563	1R11B	563	4-2020-12400
563	1A12M	563	1N4741A	563	1R12	563	4-2021-09070
563	1A12MA	563	1N4742	563	1R12A	563	4-2021-10870
563	1A13M	563	1N4742A	563	1R12B	563	4JZ4X539
563	1A13MA	563	1N4743	563	1R13	563	4JX24X539
563	1AC12	563	1N4743A	563	1R13A	563	4JZ4X539
563	1AC12A	563	1N4833	563	1R13B	563	4JZ4XL12
563	1AC12B	563	1N4833A	563	1S1097	563	4W01-13
563	1B759	563	1N4834	563	1S1166	563	05-060110
563	1C12Z	563	1N4834A	563	1S1176	563	05-111011
563	1C12ZA	563	1N4835	563	1S1379	563	05-933943
563	1DZ12	563	1N4835A	563	1S1379A	563	05Z11
563	1E12Z	563	1N4896	563	1S1394	563	05Z12
563	1E12Z10	563	1N4896A	563	1S1394A	563	05Z13
563	1E12Z5	563	1N4897	563	1S141	563	8-719-112-24
563	1E2Z5	563	1N4897A	563	1S1410R	563	8-719-930-12
563	1EZ12	563	1N4898	563	1S1738	563	8-905-421-234
563	1M11Z	563	1N4898A	563	1S1739	563	8-905-421-319
563	1M11Z10	563	1N4899	563	1S1771	563	09-306127
563	1M11Z5	563	1N4899A	563	1S196	563	09-306173
563	1M11Z810	563	1N4900	563	1S1961	563	09-306179
563	1M11Z85	563	1N4900A	563	1S1962	563	09-306391
563	1M12Z	563	1N4901	563	1S197	563	09-306428
563	1M12Z10	563	1N4901A	563	1S2110	563	09-306429
563	1M12Z5	563	1N4902	563	1S2110A	563	1205Y
563	1M12Z810	563	1N4902A	563	1S211B	563	12V
563	1M12Z85	563	1N4903	563	1S2118A	563	13-14879-2
563	1M13Z810	563	1N4903A	563	1S2119A	563	13-22319-0
563	1M13Z85	563	1N4904	563	1S2120	563	13-33179-6
563	1N314	563	1N4905	563	1S2120A	563	13-33187-11
563	1N314A	563	1N4905A	563	1S2127	563	13-55352-1
563	1N315	563	1N4906	563	1S2130	563	14-509-01
563	1N315A	563	1N4906A	563	1S2130A	563	14-515-01
563	1N426	563	1N4907	563	1S2138	563	14-515-03
563	1N513	563	1N4907A	563	1S2196	563	14-515-25
563	1N513A	563	1N4908	563	1S226	563	1624
563	1N524	563	1N4908A	563	1S227	563	18-085002
563	1N524A	563	1N4909	563	1S228	563	21A037-001
563	1N713	563	1N4909A	563	1S2495	563	21A037-003
563	1N736	563	1N4910	563	1S2496	563	21A037-008
563	1N736A	563	1N4910A	563	1S2497	563	21A037-018
563	1N772	563	1N4911	563	1S2552	563	21A037-020
563	1N772A	563	1N4911A	563	1S2553	563	21A103-045
563	1N773	563	1N4912	563	1S2554	563	21A119-008
563	1N773A	563	1N4912A	563	1S3011	563	21M432(REG)
563	1N774	563	1N4913	563	1S3013	563	2426AF
563	1N774A	563	1N4913A	563	1S536	563	2426P
563	1N783	563	1N4914	563	1S337	563	25Z6
563	1N877	563	1N4914A	563	1S337-Y	563	25Z6A
563	1N877A	563	1N4915	563	1S337A	563	25Z6AF
563	1N933	563	1N4915A	563	1S337E	563	25Z6P
563	1N960	563	1N5241	563	1S337T	563	2626AF
563	1N960A	563	1N5241A	563	1S338Q	563	2626P
563	1N960B	563	1N5241B	563	1S473	563	34-8057-14
563	1N987	563	1N5242	563	1S483	563	34-8057-33
563	1N1413MA	563	1N5242A	563	1S619	563	34-8057-41
563	1N2036-2	563	1N5242B	563	1S69	563	34-8057-53
563	1N2037	563	1N5243	563	1S7110	563	34-8057-56
563	1N2037-1	563	1N5243A	563	1S7110A	563	36-6343
563	1N2037-2	563	1N5243B	563	1S7120	563	39(SHARP)
563	1N2037A	563	1N5531	563	1S7120A	563	39-13
563	1N227	563	1N5531A	563	1S7120B	563	41-0905
563	1N227A	563	1N5531B	563	1S759(H)	563	42-19917
563	1N3021	563	1N5532	563	1S760	563	42-27541
563	1N3021A	563	1N5532A	563	1S760(H)	563	46-86296-3
563	1N3021B	563	1N5532B	563	1S760H	563	46-86379-3
563	1N3022	563	1N5533	563	1S240-11	563	46-86458-3
563	1N3022A	563	1N5533A	563	1S240-12	563	46-86462-3
563	1N3022B	563	1N5533B	563	1S240-13	563	46-86487-3
563	1N3023	563	1N5564	563	1T12	563	46-86504-3
563	1N3023A	563	1N5564A	563	1T12.8	563	46-86570-3
563	1N3023B	563	1N5565	563	1T12.8B	563	48-137000
563	1N3403	563	1N5565A	563	1T12A	563	48-137177
563	1N3416	563	1N5566	563	1T12B	563	48-137209
563	1N3435	563	1N5566A	563	1TA12	563	48-137272
563	1N3447	563	1N5737	563	1TA12.8	563	48-155081
563	1N3519	563	1N5738	563	1TA12.8A	563	48-155210
563	1N3520	563	1N5739	563	1TA12A	563	48-32000
563	1N3521	563	1N5857	563	1W01-06J	563	48-355062
563	1N3537	563	1N5857A	563	1WB-11D	563	48-82256017
563	1N3583	563	1N5857B	563	1WB-13D	563	48-82256025
563	1N3583A	563	1N5858	563	1Z11	563	48-82256034
563	1N3583B	563	1N5858A	563	1Z11A	563	48-82256048
563	1N3680	563	1N5858B	563	1Z11Z10	563	48-82256050
563	1N3680A	563	1N5859	563	1Z11Z20	563	48-82256054
563	1N3681	563	1N5859A	563	1Z11T5	563	48-83461B02
563	1N3681A	563	1N5859B	563	1Z12	563	48-83461B31
563	1N3681B	563	1N665	563	1Z12A	563	48-97048A08
563	1N3682	563	1N715	563	1Z12B	563	48-97048A16
563	1N3682A	563	1N716	563	1Z12C	563	48X355062
563	1N4106	563	1N716(ZENER)	563	1Z12Z10	563	48X137000
563	1N4106	563	1N716A	563	1Z12Z20	563	48X155048
563	1N4106A	563	1N716B	563	1Z12T5	563	48X155081
563	1N4107	563	1N717	563	1Z13	563	48X155084
563	1N4163	563	1N717A	563	1Z13A	563	48X155103
563	1N4163A	563	1N759	563	1Z13Z10	563	48X32000
563	1N4164	563	1N759A	563	1Z13Z20	563	48X90253A07
563	1N4164A	563	1N765	563	1Z13T5	563	48X97048A08
563	1N4164B	563	1N765-2	563	1ZB12	563	50Z12
563	1N4165	563	1N766	563	1ZB12B	563	50Z12A
563	1N4165A	563	1N766-1	563	1ZC11Z10	563	50Z12B
563	1N4299	563	1N766-2	563	1ZC11T5	563	50Z12C
563	1N4299A	563	1N766-A	563	1ZC12	563	53A001-10
563	1N4300	563	1N941	563	1ZC12Z10	563	53J004-2
563	1N4300A	563	1N941A	563	1ZC12Z20	563	56-52
563	1N4301	563	1N941B	563	1ZC12T5	563	56-4837
563	1N4301A	563	1N942	563	1ZC13Z10	563	56-51
563	1N4303	563	1N942A	563	1ZC13T5	563	56-57
563	1N4303A	563	1N942B	563	1ZF12Z10	563	61B02
563	1N4304	563	1N943	563	1ZF12Z20	563	61Z31
563	1N4304A	563	1N943A			563	63-9942
563	1N4328	563	1N943B			563	65-085004
563	1N4328A					563	75Z12
563	1N4329					563	75Z12

RS 276-	Industry Standard No.	RS 276-	Industry Standard No.	RS 276-	Industry Standard No.	RS 276-	Industry Standard No.
563	75Z12A	563	166985	564	B2Y-83D15	564	IP20-0018
563	75Z12B	563	224780	564	BZY83/C15	564	K8150A
563	75Z12C	563	233148	564	BZY83/D15	564	K8150A
563	75Z13	563	317208	564	BZY83C15	564	K8150B
563	75Z13A	563	0330302	564	BZY83C16	564	KS2150A
563	75Z13B	563	489752-045	564	BZY83D15	564	KS2150B
563	75Z13C	563	489752-091	564	BZY85/C15	564	K846
563	78-271383-1	563	530073-1013	564	BZY85/D15	564	K846AF
563	86-0521	563	530073-13	564	BZY85C13V5	564	K846BF
563	86-25-1	563	530073-14	564	BZY85C15	564	LPM15
563	86-31-1	563	530073-6	564	BZY85C16	564	LPM15A
563	86-61-1	563	530118-2	564	BZY85D15	564	LP2T15
563	86-65-1	563	530145-120	564	BZY88C15	564	LR150CH
563	86-67-1	563	530145-130	564	BZY88C16	564	M152
563	86-85-1	563	530157-130	564	BZY92/C15	564	M4552
563	86-92-1	563	530163-120	564	BZY92C15	564	M4659
563	93A102-2	563	530192-120	564	BZY92C16	564	M4Z15
563	93A39-12	563	610073-13	564	BZY94/C15	564	M4Z15-20
563	93A39-13	563	760304	564	BZY94C15	564	M4Z15A
563	93A39-31	563	801731	564	BZY94C16	564	MA1150
563	93A39-43	563	817155	564	C4020	564	MA1160
563	93A39-44	563	817208	564	CD31-00017	564	MC6018
563	93A55-1	563	922693	564	CD31-10365	564	MC6018A
563	93A80-1	563	922950	564	CD31-12024	564	MC6118
563	93B39-12	563	923233	564	CXL144	564	MC6118A
563	93B39-13	563	985961	564	CXL145	564	MBZ15T1Q
563	93B39-31	563	1444875-1	564	CXL5023	564	MBZ15T5
563	93C39-12	563	1474777-1	564	CXL5024	564	MR150Q-H
563	93C39-13	563	1477046-4	564	CXL5025	564	MT2618
563	93C39-2	563	2002209-10	564	CXL5075	564	MT2618A
563	93C39-6	563	2002331-46	564	CZD014	564	MZ1000-17
563	94A80-1	563	2327071	564	CZD014-5	564	MZ1000-18
563	96-5091-01	563	2327074	564	CZD015	564	MZ1014
563	96-5133-01-02	563	2327078	564	CZD015-5	564	MZ1016
563	96-5133-02	563	2330241	564	D17	564	MZ14
563	96-5248-04	563	2331152	564	D1T	564	MZ15A
563	100-286	563	2331154	564	D1Z	564	MZ15T2Q
563	103-158	563	2331174	564	D1ZBLU	564	MZ214
563	103-246	563	2337101	564	D1ZRED	564	MZ216
563	103-256	563 56L	183012	564	D1ZYEL	564	MZ314
563	103-279-01	564	.25N14	564	D4P	564	MZ316
563	103-279-20	564	.25N15	564	D5B	564	MZ500-20
563	103-279-21	564	.25N16	564	DDAY009003	564	MZ500-21
563	103-279-21A	564	.25NN15	564	DDAY009007	564	MZ92-15A
563	103-279-22	564	.25T14	564	DHD805(ZENER)	564	OA126/14
563	103-308A	564	.25T14A	564	DIT	564	OA12614
563	103-96	564	.25T14B	564	D8-110	564	OZ15T10
563	103-Z9003	564	.25T15	564	DZ15A	564	OZ15T5
563	111-4-2020-11500	564	.25T15.8	564	BO771-6	564	P38103/507-1Q
563	160	564	.25T15.8A	564	EA16X124	564	PA8261
563	232-0009-31	564	.25T15A	564	EC0144	564	PD-6018
563	264P03302	564	.25T15B	564	EC0144A	564	PD-6018A
563	264P03303	564	.25T16	564	EC0145	564	PD-6059
563	264P09301	564	.25T16A	564	EC0145A	564	PD6018
563	264P10502	564	.4T14	564	EC05023	564	PD6018A
563	264P10502(ZENER)	564	.4T14A	564	EC05023A	564	PD6059
563	264P17402	564	.4T15	564	EC05024	564	PR-620
563	276-563	564	.4T15B	564	EC05024A	564	PR620
563	296L005B01	564	.4T16	564	EC05025	564	P8-14068
563	296V011H03	564	.4T16A	564	EC05025A	564	P8-8917
563	296V017H01	564	.4T16B	564	EC05075	564	PS10024B
563	296V018B01	564	.75N5	564	EC05075A	564	PS10068
563	296V019B01	564	EDZ-11	564	EDZ-11	564	PS8917
563	523-2003-110	564	.7JZ14	564	EZ16X5	564	PTC509
563	523-2003-120	564	.7JZ16	564	EQA01-14	564	PTC511
563	523-2003-130	564	.7Z14A	564	EQA01-14R	564	QA01-07R
563	600X0101-066	564	.7Z14B	564	EQA01-14RD	564	QA01-14RD
563	601X0226-066	564	.7Z14D	564	EQA01-15	564	QB01-15ZB
563	601X0226-0666	564	.7Z15D	564	EQA01-15D	564	QD-2BZ16ZXJ
563	601X0402-038	564	.7Z16A	564	EQA01-16	564	QRT-245
563	690V103H53	564	.7Z16B	564	EQB01-14	564	QRT-246
563	690V116H40	564	.7Z16C	564	EQB01-15	564	QZ-15T1Q
563	800-004-00	564	.7Z16D	564	EQB01-15Z	564	QZ-15T5
563	800-023-00	564	.7ZM14A	564	EQB01-15ZB	564	QZ14T1Q
563	800-035-00	564	.7ZM14B	564	EQB01-16	564	QZ14T5
563	903-73B	564	.7ZM14C	564	EVR15	564	QZ15T5
563	903-97B	564	.7ZM14D	564	EVR15A	564	QZ15T1Q
563	919-00-1445-002	564	.7ZM15A	564	EVR15B	564	QZ15T5
563	919-001445-2	564	.7ZM15B	564	EZ14(ELCOM)	564	R2E
563	919-001445-2	564	.7ZM15C	564	EZ15(ELCOM)	564	RD-15E
563	919-011340	564	.7ZM15D	564	EZ150	564	RD-16H
563	919-013058	564	.7ZM16A	564	EZ16(ELCOM)	564	RD-16HA
563	1002-4404	564	.7ZM16B	564	F-677	564	RD15E
563	1033	564	.7ZM16C	564	FA8009	564	RD15EA
563	1057-914Q	564	.7ZM16D	564	FA8010	564	RD15EB
563	1111	564	A04166-2	564	FA8011	564	RD15F
563	1112	564	A04166-2	564	FA8012	564	RD15FA
563	1411-135	564	AV-2015	564	FZ14T1Q	564	RD15FB
563	1550-17	564	AV2015	564	FZ14T5	564	RD15M
563	1766	564	AW01-15	564	FZ15A	564	RD15S
563	2006-17	564	AW01-16	564	FZ15T1Q	564	RD16A
563	2021-17	564	AZ-135	564	FZ15T5	564	RD16A-N
563	2093A39-12	564	AZ-140	564	G15T1Q	564	RD16AL
563	2093A41-113	564	AZ-145	564	G15T2Q	564	RD16AM
563	2093A41-56	564	AZ-15	564	GEZD-14	564	RD16B
563	2093A41-60	564	AZ-150	564	GEZD-15	564	RD16C-Y
563	2093A41-82	564	AZ-157	564	GEZD-16	564	RD16E
563	2093A80-1	564	AZ-162	564	GT15T5	564	RD16B-M
563	003114	564	A215	564	HD3001109	564	RD16E-N
563	3506	564	AZ7	564	HD3001109-Q	564	RD16EB
563	3520	564	AZ965	564	HD30011090	564	RD16F
563	3520(DIO)	564	AZ965A	564	HD3002109	564	RD16FA
563	3520(WARDS)	564	AZ965B	564	HEP-Z0225	564	RD16FB
563	004887	564	AZ966	564	HEP-Z0418	564	RD16H
563	5122	564	AZ966A	564	HEP-Z0419	564	RD16M
563	5222	564	AZ966B	564	HEP606	564	RE-121
563	5521	564	B-6002	564	HEP607	564	RE-122
563	5522	564	BAX79C16	564	HEPZ0225	564	RB120
563	5523	564	BZ-135	564	HEPZ0225A	564	RE121
563	10020	564	BZ-140	564	HEPZ0417	564	RB122
563	35219	564	BZ-145	564	HEPZ0418	564	REN144
563	36539	564	BZ-150	564	HEPZ0419	564	REN145
563	42025-850	564	BZ-157	564	HM15	564	REN5075
563	59991-1	564	BZ-162	564	HM15A	564	RH-EX0065CBZZ
563	71411-2	564	BZ-X19	564	HM15B	564	RT-245
563	99201-110	564	BZ140	564	HM16	564	RT-246
563	99201-219	564	BZ150	564	HM16A	564	RT3671
563	99201-319	564	BZ162	564	HS2150	564	RT3671A
563	99202-220	564	BZX-46C15	564	HS2156	564	RZ-15AB
563	120504	564	BZX-71C15	564	HST15Q	564	RZ14
563	125126	564	BZX-79C15	564	HW15	564	RZ15
563	125499	564	BZX19	564	HW15A	564	RZ15A
563	126836(REGULATOR)	564	BZX29C15	564	HW15B	564	RZ16
563	129940	564	BZX29C16	564	HW16	564	RZ215
563	130328	564	BZX46C15	564	HW16A	564	RZ2156
563	130380	564	BZX46C16	564	HZ-15	564	SK3063
563	132416	564	BZX55C15	564	HZ15	564	SK3063/145A
563	132865	564	BZX55C16	564	HZ15(H)	564	SK3094
563	134335	564	BZX71C15	564	HZ15E	564	SK3094/144A
563	136634	564	BZX71C16	564	HZ15L	564	SK3142
563	141429	564	BZX79C15	564	HZ16	564	SK3751
563	147704-6-4	564	BZX83C15	564	HZ16(H)	564	SK3751/5075A
563	161022	564	BZX83C16	564	HZ16H	564	SK3789
563	165358	564	BZY-83C15	564	HZ16L	564	SK3789/5023
		564	BZY-83C15			564	SK3790

RS 276-	Industry Standard No.	RS 276-	Industry Standard No.	RS 276-	Industry Standard No.	RS 276-	Industry Standard No.
564	SK3790/5024	564	WN-140C	564	WS-157A	564	WX-162
564	SK3790/5024A	564	WN-140D	564	WS-157B	564	WX-162A
564	SK3791	564	WN-150	564	WS-157C	564	WX-162B
564	SK3791/5025	564	WN-150A	564	WS-157D	564	WX-162C
564	SU5	564	WN-150B	564	WS-162	564	WX-162D
564	SV-1020A	564	WN-150C	564	WS-162A	564	WX-167
564	SV-138A	564	WN-150D	564	WS-162B	564	WX-167A
564	SV-4015A	564	WN-157	564	WS-162C	564	WX-167B
564	SV1019	564	WN-157A	564	WS-162D	564	WX-167C
564	SV1020	564	WN-157B	564	WS-167	564	WX-167D
564	SV137	564	WN-157C	564	WS-167A	564	WY-140
564	SV138	564	WN-157D	564	WS-167B	564	WY-140A
564	SV139	564	WN-162	564	WS-167C	564	WY-140B
564	SV4014	564	WN-162A	564	WS-167D	564	WY-140C
564	SV4014A	564	WN-162B	564	WT-140	564	WY-140D
564	SV4015	564	WN-162C	564	WT-140A	564	WY-150
564	SV4015A	564	WN-162D	564	WT-140B	564	WY-150A
564	SZ-150	564	WN-167	564	WT-140C	564	WY-150B
564	SZ-15B	564	WN-167A	564	WT-140D	564	WY-150C
564	SZ-200-14	564	WN-167B	564	WT-150	564	WY-150D
564	SZ-200-15	564	WN-167C	564	WT-150A	564	WY-157B
564	SZ-200-15A	564	WN-167D	564	WT-150B	564	WY-157C
564	SZ-200-16	564	WO-140	564	WT-150C	564	WY-157D
564	SZ14	564	WO-140A	564	WT-150D	564	WY-162
564	SZ15	564	WO-140B	564	WT-157	564	WY-162A
564	SZ15.0	564	WO-140C	564	WT-157A	564	WY-162B
564	SZ150	564	WO-140D	564	WT-157B	564	WY-162C
564	SZ16	564	WO-150	564	WT-157C	564	WY-162D
564	SZ961-B	564	WO-150A	564	WT-157D	564	WY-167
564	SZ961-R	564	WO-150B	564	WT-162	564	WY-167A
564	SZ961-Y	564	WO-150C	564	WT-162A	564	WY-167B
564	TM144	564	WO-150D	564	WT-162B	564	WY-167C
564	TM145	564	WO-157	564	WT-162C	564	WY-167D
564	TM145A	564	WO-157A	564	WT-162D	564	WZ-135
564	TM5023	564	WO-157B	564	WT-167	564	WZ-140
564	TM5075	564	WO-157C	564	WT-167A	564	WZ-145
564	TR125B	564	WO-157D	564	WT-167B	564	WZ-150
564	TR128C	564	WO-162	564	WT-167C	564	WZ-157
564	TR14002-12	564	WO-162A	564	WT-167D	564	WZ-162
564	TR2337103	564	WO-162B	564	WU-140	564	WZ-538
564	TVS-ZB1-15	564	WO-162C	564	WU-140A	564	WZ-920
564	TVSEA01-07R	564	WO-162D	564	WU-140C	564	W214
564	TVSEQA01-06S	564	WO-167	564	WU-140D	564	W2140
564	TVSEQB01-15	564	WO-167A	564	WU-150	564	W215
564	TVSEQB01-15Z	564	WO-167B	564	WU-150A	564	W2150
564	TVSQA01-07R	564	WO-167C	564	WU-150B	564	W2537
564	TVSQA01-15RB	564	WO-167D	564	WU-150C	564	W2538
564	TVSQA01-15S	564	WP-140	564	WU-150D	564	W2919
564	TVSQB01-15ZB	564	WP-140A	564	WU-157	564	W2920
564	TVSZB1-15	564	WP-140C	564	WU-157A	564	X3152
564	TZ-15A	564	WP-140D	564	WU-157B	564	XZ-152
564	TZ-15AB	564	WP-150	564	WU-157D	564	Z-1016
564	TZ-15BC	564	WP-150A	564	WU-162	564	Z-1016A
564	TZ-15C	564	WP-150B	564	WU-162A	564	Z-1114
564	TZ15	564	WP-150C	564	WU-162B	564	Z-15
564	TZ15A	564	WP-150D	564	WU-162C	564	Z-15K
564	TZ15B	564	WP-157	564	WU-162D	564	Z0225
564	TZ15C	564	WP-157A	564	WU-167	564	Z0417
564	UZ-15C	564	WP-157B	564	WU-167A	564	Z0418
564	UZ15	564	WP-157C	564	WU-167B	564	Z0419
564	UZ8715	564	WP-157D	564	WU-167D	564	Z0B-15
564	UZ8716	564	WP-162	564	WV-140	564	Z0B15
564	UZ8815	564	WP-162A	564	WV-140A	564	Z0C-15
564	UZ8816	564	WP-162B	564	WV-140B	564	Z0C15
564	V-1266	564	WP-162C	564	WV-140C	564	Z0D-15
564	V126	564	WP-162D	564	WV-140D	564	Z0D15
564	VR-14	564	WP-167	564	WV-157	564	Z1016
564	VR14	564	WP-167A	564	WV-157A	564	Z1114
564	VR14A	564	WP-167B	564	WV-157B	564	Z1114-C
564	VR14B	564	WP-167C	564	WV-157C	564	Z1214
564	VR15	564	WP-167D	564	WV-157D	564	Z15
564	VR15A	564	WQ-140	564	WV-162	564	Z15K
564	VR15B	564	WQ-140A	564	WV-162A	564	Z1B-15
564	YZ-135	564	WQ-140B	564	WV-162B	564	Z1B15
564	YZ-140	564	WQ-140C	564	WV-162C	564	Z1C-15
564	YZ-145	564	WQ-140D	564	WV-162D	564	Z1C15
564	YZ-150	564	WQ-150	564	WV-167	564	Z1D-15
564	YZ-157	564	WQ-150A	564	WV-167A	564	Z1D15
564	YZ-162	564	WQ-150B	564	WV-167B	564	Z2A-150F
564	WEP606	564	WQ-150C	564	WV-167C	564	Z2A150F
564	WEP607	564	WQ-150D	564	WV-167D	564	Z4B-15
564	WI-140	564	WQ-157	564	WW-125	564	Z4B15
564	WI-140A	564	WQ-157A	564	WW-125A	564	Z4X-15
564	WI-140B	564	WQ-157C	564	WW-125B	564	Z4X14
564	WI-140C	564	WQ-157D	564	WW-125C	564	Z4X14B
564	WI-140D	564	WQ-162	564	WW-130	564	Z4X15
564	WI-150	564	WQ-162A	564	WW-130A	564	Z4XL14
564	WI-150A	564	WQ-162B	564	WW-130B	564	Z4XL14B
564	WI-150B	564	WQ-162C	564	WW-130C	564	Z5B-15
564	WI-150C	564	WQ-162D	564	WW-130D	564	Z5C-15
564	WI-150D	564	WQ-167	564	WW-140	564	Z5D-15
564	WI-157	564	WQ-167A	564	WW-140A	564	Z5D15
564	WI-157A	564	WQ-167B	564	WW-140B	564	ZA-15B
564	WI-157B	564	WQ-167C	564	WW-140C	564	ZA15
564	WI-157C	564	WQ-167D	564	WW-140D	564	ZA15A
564	WI-157D	564	WR-140	564	WW-150	564	ZA15B
564	WI-162	564	WR-140A	564	WW-150A	564	ZA15V
564	WI-162A	564	WR-140B	564	WW-150B	564	ZA27B
564	WI-162B	564	WR-140C	564	WW-150C	564	ZB-15A
564	WI-162C	564	WR-140D	564	WW-150D	564	ZB1-14
564	WI-162D	564	WR-150	564	WW-157	564	ZB1-15
564	WI-167	564	WR-150A	564	WW-157A	564	ZB1-16
564	WI-167A	564	WR-150B	564	WW-157B	564	ZB1-16V
564	WI-167B	564	WR-150C	564	WW-157C	564	ZB13A
564	WI-167C	564	WR-150D	564	WW-157D	564	ZB15
564	WI-167D	564	WR-157	564	WW-162	564	ZB15A
564	WM-140	564	WR-157A	564	WW-162A	564	ZB15B
564	WM-140A	564	WR-157B	564	WW-162B	564	ZB16
564	WM-140B	564	WR-157C	564	WW-162C	564	ZB16A
564	WM-140C	564	WR-157D	564	WW-162D	564	ZB16B
564	WM-140D	564	WR-162	564	WW-167	564	ZB215
564	WM-150	564	WR-162A	564	WW-167A	564	ZC-015
564	WM-150A	564	WR-162B	564	WW-167B	564	ZC015
564	WM-150B	564	WR-162C	564	WW-167C	564	Z0115
564	WM-150C	564	WR-162D	564	WW-167D	564	ZD-015
564	WM-150D	564	WR-167	564	WX-140	564	ZD-15
564	WM-157	564	WR-167A	564	WX-140A	564	ZD-15A
564	WM-157A	564	WR-167B	564	WX-140B	564	ZD-15B
564	WM-157B	564	WR-167C	564	WX-140C	564	ZD015
564	WM-157C	564	WR-167D	564	WX-140D	564	ZD15
564	WM-157D	564	WS-140	564	WX-150	564	ZD15A
564	WM-162	564	WS-140A	564	WX-150A	564	ZD15B
564	WM-162A	564	WS-140B	564	WX-150B	564	ZE-15
564	WM-162B	564	WS-140C	564	WX-150C	564	ZE-15A
564	WM-162C	564	WS-140D	564	WX-150D	564	ZE-15B
564	WM-162D	564	WS-150	564	WX-157	564	ZE15
564	WM-167	564	WS-150A	564	WX-157A	564	ZE15A
564	WM-167A	564	WS-150B	564	WX-157B	564	ZE15B
564	WM-167B	564	WS-150C	564	WX-157C	564	ZEN507
564	WM-167C	564	WS-150D	564	WX-157D	564	ZEN50B
564	WM-167D	564	WS-157			564	ZF-15
564	WN-140					564	ZF-15A
564	WN-140A					564	ZF-15B
564	WN-140B						

RS 276-	Industry Standard No.	RS 276-	Industry Standard No.	RS 276-	Industry Standard No.	RS 276-	Industry Standard No.
564	ZP-16	564	1M15Z10	564	1N965B	564	13-33187-18
564	ZP115	564	1M15Z5	564	1N966	564	14-515-07
564	ZP116	564	1M15Z810	564	1N966A	564	14-515-09
564	ZP15	564	1M15Z85	564	1N966B	564	14-515-11
564	ZP15A	564	1M16Z	564	1R15	564	14-515-23
564	ZP15B	564	1M16Z5	564	1R15A	564	14-515-73
564	ZP16	564	1M16Z810	564	1R15B	564	21A037-017
564	ZP215	564	1M16Z85	564	1R16	564	25N15
564	ZP216	564	1MZ10	564	1R16A	564	2726
564	ZQ-15	564	1N1316	564	1R16B	564	2726A
564	ZQ-15B	564	1N1316A	564	1S1098	564	2726AF
564	ZQ100119	564	1N1427	564	1S1167	564	2726F
564	ZQ15	564	1N1514	564	1S1177	564	2826
564	ZQ15A	564	1N1514A	564	1S1178	564	2826AF
564	ZQ15B	564	1N1525	564	1S1191	564	2826F
564	ZR-15	564	1N1525A	564	1S137	564	34-8057-12
564	ZR-15A	564	1N1544	564	1S1380	564	34M14Z
564	ZR-15B	564	1N1544A	564	1S1380A	564	34M14Z10
564	ZR115	564	1N1775	564	1S1395	564	34M14Z5
564	ZR15	564	1N1775A	564	1S1395A	564	34Z14D10
564	ZR15A	564	1N1776	564	1S1410	564	34Z14D5
564	ZR15B	564	1N1776A	564	1S1411	564	46-86341-3
564	ZJ-15	564	1N1878	564	1S1411R	564	46-86401-3
564	ZJ-15A	564	1N1878A	564	1S142	564	46-86464-3
564	ZJ-15B	564	1N1934	564	1S143	564	048I(ZENER)
564	ZJ15	564	1N1934A	564	1S1740	564	48-134552
564	ZJ15A	564	1N1934B	564	1S1772	564	48-134659
564	ZJ15B	564	1N1961	564	1S1963	564	48-137017
564	ZM14	564	1N1988	564	1S1964	564	48-137048
564	ZM14A	564	1N1988A	564	1S1965	564	48-137298
564	ZM14B	564	1N1988B	564	1S198	564	48-137330
564	ZM15	564	1N2037-3	564	1S2121	564	48-137365
564	ZM15A	564	1N2038	564	1S2121A	564	48-137394
564	ZM15B	564	1N2038-1	564	1S2122	564	48-82256C10
564	ZM16	564	1N2038-2	564	1S2122A	564	48-82256C13
564	ZM16A	564	1N2038A	564	1S2150	564	48-82256C14
564	ZM16B	564	1N228	564	1S2150A	564	48-82256C059
564	ZO-15	564	1N228A	564	1S2160A	564	48-83461E01
564	ZO-15A	564	1N2766	564	1S229	564	48-83461E07
564	ZO-15AC	564	1N2766A	564	1S230	564	48I57017
564	ZO-15B	564	1N3024	564	1S231	564	48I137330
564	ZO-15BY	564	1N3024A	564	1S2498	564	48I155104
564	ZOB-15	564	1N3024B	564	1S2499	564	50Z14
564	ZOB15	564	1N3025	564	1S2555	564	50Z14A
564	ZOD-15	564	1N3025A	564	1S2556	564	50Z14B
564	ZOD15	564	1N3025B	564	1S301.5	564	50Z14C
564	ZP-15	564	1N3404	564	1S3016	564	50Z15
564	ZP-15A	564	1N3417	564	1S3016(ZENER)	564	50Z15A
564	ZP-15B	564	1N3436	564	1S338	564	50Z15B
564	ZP15	564	1N3448	564	1S338U	564	50Z15C
564	ZP15A	564	1N3522	564	1S427	564	53A001-33
564	ZP15B	564	1N3523	564	1S474	564	56-25
564	ZQ-15	564	1N3683	564	1S484	564	56-36
564	ZQ-15A	564	1N3683B	564	1S597	564	61E01
564	ZQ-15B	564	1N3684	564	1S620	564	61E07
564	ZQ15	564	1N3684A	564	1S697	564	62-130761
564	ZQ15A	564	1N3684B	564	1S7150	564	62-130762
564	ZQ15B	564	1N4108	564	1S7150A	564	62-20643
564	ZQ15X	564	1N4109	564	1S7150B	564	66X0040-009
564	ZR215	564	1N4110	564	1S7160A	564	070-021
564	ZR50B921-3	564	1N4166	564	1S761	564	070-048
564	Z8-15A	564	1N4166A	564	1S761(H)	564	73-15
564	Z8-15B	564	1N4166B	564	1S990(ZENER)	564	73-17
564	Z815	564	1N4167	564	1S9908	564	75N5
564	Z815A	564	1N4167A	564	1S9908(ZENER)	564	75Z14
564	Z815B	564	1N4331	564	1S228	564	75Z14A
564	ZT-15	564	1N4331A	564	1SZ40-15	564	75Z14C
564	ZT-15A	564	1N4331B	564	1SZ40-16	564	75Z15
564	ZT-15B	564	1N4332	564	1T14	564	75Z15A
564	ZT15	564	1N4332A	564	1T14B	564	75Z15B
564	ZT15A	564	1N4408	564	1T15	564	75Z15C
564	ZT15B	564	1N4408A	564	1T15B	564	78-271383-3
564	ZU-15	564	1N4408B	564	1T19-15B	564	78-271383-4
564	ZU-15A	564	1N4409	564	1TA14	564	86-93-1
564	ZU-15B	564	1N4409A	564	1TA14A	564	93A39-24
564	ZU15	564	1N4636	564	1TA15	564	93A39-50
564	ZU15A	564	1N4637	564	1TA15A	564	93039-24
564	ZU15B	564	1N4665	564	1WB-15D	564	96-5110-02
564	ZV-15	564	1N4666	564	1Z15	564	96-5110-03
564	ZV-15A	564	1N4701	564	1Z15A	564	96-5248-06
564	ZV-15B	564	1N4702	564	1Z15B	564	103-136
564	ZV15	564	1N4703	564	1Z15C	564	103-231
564	ZV15A	564	1N4744	564	1Z15D	564	103-279-23
564	ZV15B	564	1N4744A	564	1Z15D10	564	103-279-24
564	ZXL-14	564	1N4745	564	1Z15D5	564	103-279-25
564	ZXY-14	564	1N4745A	564	1Z15Z10	564	103-84
564	ZXY14B	564	1N4836	564	1Z15Z20	564	103-Z9012
564	ZY-15	564	1N4836A	564	1Z15T5	564	103-Z9013
564	ZY-15B	564	1N4837	564	1Z16	564	11524
564	ZY15	564	1N4837A	564	1Z16A	564	123E-006
564	ZY15A	564	1N5244	564	1Z16T10	564	137-759
564	ZY15B	564	1N5244A	564	1Z16T20	564	142-015
564	ZZ-15	564	1N5244B	564	1Z16T5	564	150D
564	ZZ15	564	1N5245	564	1ZB15	564	152-057-9-001
564	001-02303-4	564	1N5245A	564	1ZB15B	564	173-15
564	001-023034	564	1N5245B	564	1ZC15	564	200X8220-531
564	001-02303B	564	1N5246	564	1ZC15T10	564	21524
564	1/2Z15T5	564	1N5246A	564	1ZC15T5	564	264F04003
564	1/2Z16T5	564	1N5246B	564	1ZC16T10	564	264F10308
564	1/4A14	564	1N5534	564	1ZC16T5	564	264F10906
564	1/4A14A	564	1N5534A	564	1ZP14Z10	564	276-564
564	1/4A14B	564	1N5534B	564	1ZP14T20	564	523-2003-140
564	1/4A15	564	1N5535	564	1ZP14T5	564	523-2003-150
564	1/4A15A	564	1N5535A	564	1ZP15T10	564	523-2003-160
564	1/4A15B	564	1N5535B	564	1ZP15T20	564	523-2503-150
564	1/4M15Z	564	1N5536	564	1ZP15T5	564	540-013
564	1/4M15Z10	564	1N5536A	564	1ZM15T10	564	642-255
564	1/4M15Z5	564	1N5536B	564	1ZM15T20	564	690Y081H92
564	1/4Z14D	564	1N5567	564	1ZM15T5	564	800-028-00
564	1/4Z14D10	564	1N5567A	564	1ZT15	564	1000-133
564	1/4Z14D5	564	1N5568	564	1ZT15A	564	1006-1737
564	1/4Z15D10	564	1N5568A	564	02Z15A	564	1015
564	1/4Z15D5	564	1N5740	564	02Z16A	564	1113
564	1/4Z15T5	564	1N5741	564	005-009100	564	2000-312
564	1A15M	564	1N5860	564	314-3506-29	564	2000-326
564	1A15MA	564	1N5860A	564	314-3506-40	564	2120-17
564	1AC15	564	1N5860B	564	4-2020-12700	564	2396-17
564	1AC15A	564	1N5861	564	05Z15	564	4552
564	1AC15B	564	1N5861A	564	05Z16	564	5001-143
564	1C15Z	564	1N5861B	564	08-08120	564	5124
564	1C15ZA	564	1N5862	564	8-905-421-128	564	5224
564	1D08	564	1N5862A	564	8-905-421-239	564	5224(ZENER)
564	1E14Z	564	1N5862B	564	8-905-421-715	564	5524
564	1E14Z10	564	1N666	564	09-306024(ZENER)	564	5525
564	1E14Z5	564	1N718	564	09-306110(ZENER)	564	99201-221
564	1E15Z	564	1N718A	564	09-306243	564	99202-221
564	1E15Z10	564	1N718B	564	09-306268	564	99202-221
564	1E15Z5	564	1N719	564	09-306278	564	126582
564	1EZ15	564	1N719A	564	09-306324	564	126852
564	1M14Z	564	1N766-3	564	09-306367	564	126862
564	1M14Z10	564	1N767	564	13-14879-1	564	127382
564	1M14Z5	564	1N767-1	564	13-14879-3		
564	1M15Z	564	1N767-22	564	13-14879-5		
564		564	1N965	564	13-33179-2		
564		564	1N965A				

RS 276-	Industry Standard No.	RS 276-	Industry Standard No.	RS 276-	Industry Standard No.	RS 276-	Industry Standard No.
564	129946	1011	1SR13-50	1067	EC05421	1067	2N2325
564	130761	1011	1SR14-50	1067	EC05422	1067	2N2326
564	130762	1011	1SR55-50	1067	EC05423	1067	2N3528
564	134444	1011	276-1101	1067	EC05431	1067	2N6236
564	140973	1020	BT100A-300R	1067	EC05432	1067	2N6237
564	165740	1020	BT101-500R	1067	EC05433	1067	2N6238
564	171092-1	1020	C1060	1067	EC05452	1067	2SF101
564	233150	1020	C106C1	1067	EC05453	1067	2SF102
564	499752-041	1020	C106C2	1067	EC05454	1067	2SF102A
564	530073-1015	1020	C106C3	1067	EC05455	1067	2SF957
564	530073-1017	1020	C106C4	1067	GE-X5	1067	2SF522
564	530073-1023	1020	C107C	1067	GEMR-5	1067	2SF941
564	530073-12	1020	C107C1	1067	HEP-R1218	1067	2SF942
564	530073-15	1020	C107C2	1067	HEPR1215	1067	306100
564	530073-17	1020	C107C3	1067	HEPR1216	1067	1174
564	530073-23	1020	C107C4	1067	HEPR1217	1067	11248
564	530073-26	1020	C511C	1067	HEPR1218	1067	13-18924-1
564	530073-9	1020	C511H	1067	IR106A1	1067	13-33183-1
564	1473777-2	1020	CXL5456	1067	IR106A1-C	1067	19B200248-P1
564	1474717-2	1020	EC05415	1067	IR106A2	1067	19B200248-P2
564	1474777-002	1020	EC05456	1067	IR106A3	1067	19B200248-P3
564	1474777-11	1020	IR106C1	1067	IR106A4	1067	48-8375D02
564	1474777-2	1020	IR106C1-C	1067	IR106B1	1067	48-83875D01
564	1477046-1	1020	IR106C2	1067	IR106B1-C	1067	48-83875D04
564	1477080-501	1020	IR106C3	1067	IR106B2	1067	48-83875D05
564	1477081-501	1020	IR106C4	1067	IR106B3	1067	48-83875D06
564	1945295A1	1020	IR106C41	1067	IR106B4	1067	48D8375D02
564	2327077	1020	IRW106C1	1067	IR106B41	1067	48D83875D01
564	2330305	1020	S2060C	1067	IR106P1	1067	48D83875D04
564	2337065	1020	S2061C	1067	IR106P2	1067	48D83875D05
564	3100017	1020	TC106C1	1067	IR106P3	1067	48D83875D06
702	EC0990	1020	TC106C2	1067	IR106P4	1067	53-1517
703	AE920	1020	TC106C3	1067	IR106P41	1067	75D01
703	CA2002	1020	TC106C4	1067	IR106Q1	1067	75D02
703	ECG1232	1020	TIC106C	1067	IR106Q2	1067	75D04
703	LM383	1020	W106C1	1067	IR106Q3	1067	75D05
703	LM383T	1020	2N2327	1067	IR106Q4	1067	75D06
703	NA2002	1020	2SF106	1067	IR106Q41	1067	86-113-3
703	NAM383	1020	106C	1067	IR106Y1	1067	86X0081-001
703	TDA2002	1020	107C	1067	IR106Y1-C	1067	106A
703	1M383T	1020	108C	1067	IR106Y2	1067	106B
703	3L4-9020-1	1050	A72-83-300	1067	IRC20	1067	107A(SCR)
703	544-2006-011	1050	CXL6407	1067	IRC5	1067	107B(SCR)
704	BA521	1050	CXL6408	1067	IRW106A1	1067	108A
704	EC01166	1050	EC06407	1067	IRW106Y1	1067	108B
704	GEIC-316	1050	EC06408	1067	J4-1725	1067	108F
704	QQ-MBA521AX	1050	GE-X13	1067	K4-584	1067	108Q
704	REN1166	1050	HEP311	1067	M25C	1067	108Y
704	TM1166	1050	HEPR2002	1067	MCR106-1	1067	183-1(SYLVANIA)
704	1223910	1050	J4-1730(DIAC)	1067	MCR106-2	1067	276-1079
705	BO319200	1050	K4-586(DIAC)	1067	MCR106-3	1067	2321264
705	EA33X8389	1050	MPT28	1067	MCR106-4	1102	DB202
705	EA33X8396	1050	MPT32	1067	MCR107-1	1102	DR202
705	EC01155	1050	RE190	1067	MCR107-2	1102	DS130D
705	EICM-0060	1050	REN6408	1067	MCR107-3	1102	DS130NE
705	ETI-23	1050	SK3523	1067	MCR107-4	1102	DS131A
705	GEIC-179	1050	SK3523/6407	1067	MCR406-1	1102	DS15C
705	IP20-0161	1050	ST-2	1067	MCR406-2	1102	P14AP
705	KIA7205AP	1050	TM6407	1067	MCR406-3	1102	P14BP
705	KIA7205P	1050	1N5760	1067	MCR406-4	1102	GM-1Z
705	QQ-M07205AT	1050	1N5760A	1067	MCR407-1	1102	HEP155
705	QQM07205AT	1050	1N5761	1067	MCR407-2	1102	MD236
705	REN1155	1050	1N5761A	1067	MCR407-3	1102	MS-1
705	SK3231	1050	182093	1067	MCR407-4	1102	MB-2
705	TA-7205P	1050	276-1050	1067	R1215(HEP)	1102	OS-1D
705	TA7205	1067	C106A	1067	R1217(HEP)	1102	QRT-213
705	TA7205A	1067	C106A1	1067	RB-168	1102	RE-52
705	TA7205AP	1067	C106A2	1067	RB-171	1102	RM2Z
705	TA7205P	1067	C106A3	1067	RB168	1102	RVD1B1LF
705	TM1155	1067	C106A4	1067	RB171	1102	S-05-02
705	TVCM-81	1067	C106B	1067	REN5454	1102	S-1.5-02
705	WEP949	1067	C106B1	1067	REN5455	1102	S1G10
705	02-257205	1067	C106B2	1067	S2060A	1102	S1G20
705	44T-100-120	1067	C106B3	1067	S2060B	1102	S2F10
705	44T-300-102	1067	C106B4	1067	S2061A	1102	S2F20
705	051-0055-02	1067	C106P	1067	S2061B	1102	S2G20
705	051-0055-03	1067	C106P1	1067	S2062A	1102	SG-5T
705	307-107-9-003	1067	C106P2	1067	S2062B	1102	SH-1Z
705	740-9007-205	1067	C106P3	1067	S2062P	1102	SM-1Z
705	740-9607-205	1067	C106P4	1067	S2062Y	1102	SM-1-02
705	5002-031	1067	C106Q	1067	SCR104(ELCOM)	1102	SR1D-2
705	26810-159	1067	C106Q1	1067	SF1A11	1102	SR1D-4
705	45810-170	1067	C106Q2	1067	SF1A11A	1102	SR1EM-4
705	0207205	1067	C106Q3	1067	SF1B11	1102	SV-1R
705	741687	1067	C106Q4	1067	SF1B11A	1102	SW-05-01
705	742364	1067	C106Y	1067	SF1D11A	1102	V06C
705	916110	1067	C106Y1	1067	SFR3D41	1102	1BZ61
1000	EC05604	1067	C106Y2	1067	SFR3D41	1102	1DZ61
1000	EC05605	1067	C106Y3	1067	SK3597	1102	1HC-10P
1000	MAC77-5	1067	C106Y4	1067	SK3597/5455	1102	1HC-10R
1000	MAC77-6	1067	C107A	1067	TA2888	1102	1HC-15P
1000	2N6072	1067	C107A1	1067	TA2889	1102	1HC-15R
1000	2N6073	1067	C107A2	1067	TC106A1	1102	1HC-20P
1000	2N6073A	1067	C107A3	1067	TC106A2	1102	1HC-20R
1000	13-39607-4	1067	C107B	1067	TC106A3	1102	1S1208
1000	13-39678-1	1067	C107B1	1067	TC106A4	1102	1S1295
1000	13-39678-3	1067	C107B2	1067	TC106B1	1102	1S1296
1000	13-39678-4	1067	C107B3	1067	TC106B3	1102	1S1297
1001	EC05600	1067	C107B4	1067	TC106B4	1102	1S1298
1001	EC05601	1067	C107P	1067	TC106P1	1102	1S1299
1001	EC05602	1067	C107P1	1067	TC106P2	1102	1S1317
1001	EC05603	1067	C107P2	1067	TC106P3	1102	1S1318
1001	MAC77-1	1067	C107P3	1067	TC106P4	1102	1S1359A
1001	MAC77-2	1067	C107P4	1067	TC106Q1	1102	1S1365
1001	MAC77-3	1067	C107Q	1067	TC106Q2	1102	1S1365A
1001	MAC77-4	1067	C107Q1	1067	TC106Q3	1102	1S146
1001	2N6068	1067	C107Q2	1067	TC106Q4	1102	1S1486
1001	2N6069	1067	C107Q3	1067	TC106Y1	1102	1S1487
1001	2N6070	1067	C107Q4	1067	TC106Y2	1102	1S1546
1001	2N6071	1067	C107Y	1067	TC106Y3	1102	1S1547
1011	BA164	1067	C107Y1	1067	TC106Y4	1102	1S1599
1011	BA217	1067	C107Y2	1067	TC10LB2	1102	1S1600
1011	BA218	1067	C107Y3	1067	TIC106A	1102	1S1604
1011	J4-1600	1067	C107Y4	1067	TIC106B	1102	1S1678
1011	MD234	1067	C511A	1067	TIC106F	1102	1S1684
1011	MD235	1067	C511B	1067	TIC106Y	1102	1S1685
1011	QRT-212	1067	C511P	1067	TM5414	1102	1S1692A
1011	SD-60P	1067	C511G	1067	TM5452	1102	1S1693A
1011	SD20	1067	C511U	1067	TM5454	1102	1S1694A
1011	SR1D-1	1067	C611A	1067	TM5455	1102	1S1695A
1011	W03A	1067	C611B	1067	TX-145	1102	1S1696
1011	1S1146	1067	C611P	1067	W106A1	1102	1S1696A
1011	1S1314	1067	C611G	1067	W106Y1	1102	1S1723
1011	1S1316	1067	C611U	1067	1RC20	1102	1S1724
1011	1S1597	1067	C7A4	1067	2N1595	1102	1S1724R
1011	1S1598	1067	CU-12E	1067	2N1596	1102	1S1725
1011	1S1690A	1067	CXL5452	1067	2N1597	1102	1S1725R
1011	1S1691A	1067	CXL5453	1067	2N2322	1102	1S181-M
1011	1S1180-M	1067	CXL5454	1067	2N2323	1102	1S182-M
1011	1S2140	1067	CXL5455	1067	2N2324	1102	1S1845
1011	1S2390	1067	EC05411			1102	1S1846
1011	1S2589	1067	EC05412			1102	1S1948
1011	1S310H	1067	EC05413				
1011	1SR12-50	1067	EC05414				

RS 276-	Industry Standard No.	RS 276-	Industry Standard No.	RS 276-	Industry Standard No.	RS 276-	Industry Standard No.
1102	182080	1103	182463	1104	A10D	1104	A4E9
1102	182226	1103	182603	1104	A10E	1104	A4F1
1102	182227	1103	182604	1104	A10M	1104	A4F5
1102	182244	1103	182776	1104	A10N	1104	A4F9
1102	182270	1103	18313H	1104	A11	1104	A4G1
1102	182270A	1103	18314H	1104	A123-7	1104	A4G5
1102	182277	1103	18314N	1104	A13	1104	A4G9
1102	182358	1103	18327	1104	A13(RECTIFIER)	1104	A50
1102	182391	1103	18359	1104	A132	1104	A50(RECTIFIER)
1102	182392	1103	18369	1104	A132-1	1104	A500
1102	182590	1103	18370	1104	A13A2	1104	A514-023626
1102	182591	1103	18375	1104	A13AA2	1104	A514-025607
1102	182602	1103	18380	1104	A13B2	1104	A514-027757
1102	182775	1103	18393	1104	A13C2	1104	A514-028072
1102	182832	1103	18661	1104	A13D2	1104	A514-028073
1102	18311H	1103	18811	1104	A13E2	1104	A514-033903
1102	18311N	1103	18812	1104	A13F2	1104	A514-033903
1102	18312A	1103	18816	1104	A14B	1104	A514-035596
1102	18312H	1103	18817	1104	A14C	1104	A5A1
1102	18326	1103	18843	1104	A14D	1104	A5A2
1102	18367	1103	18844N	1104	A14E	1104	A5A5
1102	18368	1103	18871	1104	A14B2	1104	A5A9
1102	18373	1103	18872	1104	A14F	1104	A5B1
1102	18374	1103	18881	1104	A14M	1104	A5B2
1102	18378	1103	18882	1104	A15-1008	1104	A5B5
1102	18379	1103	18947	1104	A1946	1104	A5B9
1102	1841	1103	18R11-400	1104	A1A1	1104	A5C1
1102	18841	1103	18R12-400	1104	A1A5	1104	A5C2
1102	18842	1103	18R13-400	1104	A1A9	1104	A5C5
1102	1884H	1103	18R14-400	1104	A1B1	1104	A5C9
1102	18946	1103	18R30-400	1104	A1B5	1104	A5D1
1102	18R11-100	1103	18R77-400	1104	A1B9	1104	A5D2
1102	18R11-200	1103	18R78-400	1104	A1C1	1104	A5D5
1102	18R12-100	1103	34-8058-1	1104	A1C5	1104	A5D9
1102	18R12-200	1103	103-284	1104	A1C9	1104	A5E1
1102	18R13-100	1103	212-96	1104	A1D1	1104	A5E2
1102	18R13-200	1103	276-1103	1104	A1D5	1104	A5E5
1102	18R14-100	1104	.5E05	1104	A1D9	1104	A5E9
1102	18R14-200	1104	.5E1	1104	A1E1	1104	A5F1
1102	18R30-200	1104	.5E2	1104	A1E5	1104	A5F2
1102	18R55-100	1104	.5E3	1104	A1E9	1104	A5F9
1102	18R55-200	1104	.5E4	1104	A1F1	1104	A5G1
1102	18R77-200	1104	.5E5	1104	A1F5	1104	A5G2
1102	18R78-200	1104	.5E6	1104	A1F9	1104	A5G5
1102	10D2	1104	.5J05	1104	A1G1	1104	A5G9
1102	103-185	1104	.5J1	1104	A1G5	1104	A600
1102	103-261-04	1104	.5J2	1104	A1G9	1104	A600(RECTIFIER)
1102	276-1102	1104	.5J3	1104	A200	1104	A692T5
1103	DB204	1104	.5J4	1104	A23	1104	A692T5-0
1103	DR204	1104	.5J5	1104	A2421	1104	A692X16-0
1103	D8130C	1104	.5J6	1104	A2422	1104	A7102001
1103	D8130NC	1104	.7E05	1104	A2460	1104	A75-68-500
1103	F14CP	1104	.7E1	1104	A2461	1104	A7509201
1103	GM-1	1104	.7E2	1104	A2462	1104	A7547053
1103	J4-1601	1104	.7E3	1104	A2481	1104	A7568250
1103	M8-4	1104	.7E4	1104	A2485	1104	A7568500
1103	O8-4D	1104	.7E5	1104	A25-1008	1104	A7568700
1103	QRT-214	1104	.7E6	1104	A2A1	1104	A7572100
1103	RM1	1104	.7J05	1104	A2A4	1104	A7572200
1103	RM2	1104	.7J1	1104	A2A5	1104	A7C
1103	S-05-04	1104	.7J2	1104	A2B1	1104	A7D
1103	S-1.5-04	1104	.7J3	1104	A2B4	1104	A7E
1103	S1040	1104	.7J5	1104	A2B5	1104	A7G
1103	S1040Z	1104	.7J6	1104	A2B9	1104	A86-4-1
1103	S2040	1104	A-0205	1104	A2C1	1104	A909-1018
1103	SD-61P	1104	A-04	1104	A2C4	1104	A909-1019
1103	SD-05-A	1104	A-04049-B	1104	A2C5	1104	A95-5281
1103	SK3031A	1104	A-04091-A	1104	A2D1	1104	A95-5289
1103	SK3312	1104	A-04092	1104	A2D4	1104	AA100
1103	SM-150-04	1104	A-04092-B	1104	A2D5	1104	AA200
1103	SM-150-A	1104	A-04093	1104	A2D9	1104	AA300
1103	SR1D-6	1104	A-04093A	1104	A2E1	1104	AA400
1103	SR1D-8	1104	A-04212-A	1104	A2E4	1104	AA50
1103	SR1DM-6	1104	A-04212-B	1104	A2E5	1104	AA500
1103	SR1DM-8	1104	A-04226	1104	A2E9	1104	AA600
1103	SR1EM-6	1104	A-042313	1104	A2F1	1104	AAY-22
1103	SR1EM-8	1104	A-04242	1104	A2F4	1104	AAZ18D
1103	SV-4R	1104	A-04901A	1104	A2F5	1104	ACR810-108
1103	SW-05-A	1104	A-1.5-01	1104	A2F9	1104	ACR83-1008
1103	SW-1-03	1104	A-100(RECT.)	1104	A2G1	1104	AD-1UF
1103	SW-1-04	1104	A-10105	1104	A2G4	1104	AD10
1103	SY-23	1104	A-10113	1104	A2G5	1104	AD100
1103	SY-24	1104	A-10118	1104	A2G9	1104	AD200
1103	UF8D-1	1104	A-1946	1104	A300	1104	AD50
1103	V07E	1104	A-95-5281	1104	A341	1104	AD4001
1103	V08E	1104	A-95-5289	1104	A3A3	1104	AD50
1103	V0Z61	1104	A0375	1104	A3A5	1104	AB1A
1103	1HC-25P	1104	A0377	1104	A3A9	1104	AFS-160-1021
1103	1HC-25R	1104	A04	1104	A3B1	1104	AG100D
1103	1HC-30P	1104	A04049	1104	A3B3	1104	AG100G
1103	1HC-30R	1104	A04049B	1104	A3B5	1104	AG100J
1103	1HC-40P	1104	A04091A	1104	A3C1	1104	AJ-30
1103	1HC-40R	1104	A04092	1104	A3C3	1104	AJ10
1103	1N2614	1104	A04092A	1104	A3C5	1104	AJ15
1103	181012	1104	A04092B	1104	A3C9	1104	AJ20
1103	181021	1104	A04093	1104	A3D1	1104	AJ25
1103	181027	1104	A04210A	1104	A3D3	1104	AJ30
1103	181043	1104	A04212-B	1104	A3D5	1104	AJ35
1103	181230H	1104	A04212A	1104	A3D9	1104	AJ40
1103	181319	1104	A04212B	1104	A3E1	1104	AJ5
1103	181320	1104	A042313	1104	A3E3	1104	AJ50
1103	181360	1104	A04233	1104	A3E5	1104	AJ60
1103	18136OA	1104	A04242	1104	A3E9	1104	AM-010
1103	181366	1104	A04331-021	1104	A3F1	1104	AM-020
1103	181366A	1104	A04331-023	1104	A3F3	1104	AM-025
1103	18136Q	1104	A04331-043	1104	A3F5	1104	AM-030
1103	181548	1104	A04350-022	1104	A3F9	1104	AM-035
1103	181601	1104	A04731	1104	A3G1	1104	AM-060
1103	181602	1104	A04901A	1104	A3G3	1104	AM-22
1103	181680	1104	A0491A	1104	A3G5	1104	AM-33
1103	181686	1104	A05	1104	A3G9	1104	AM-6-5
1103	181726	1104	A054-150	1104	A400	1104	AM-9-10
1103	181726R	1104	A054-230	1104	A400(RECTIFIER)	1104	AM-9-22
1103	181727	1104	A059-114	1104	A4212-A	1104	AM-9-5
1103	181727R	1104	A06	1104	A422-A	1104	AM-G22
1103	181728R	1104	A061-118	1104	A42X00269-01	1104	AM005
1103	18183-M	1104	A065-101	1104	A42X00374-01	1104	AM010
1103	181845A	1104	A066-124	1104	A4A1	1104	AM020
1103	181846A	1104	A068-100	1104	A4A5	1104	AM025
1103	181949	1104	A068-102	1104	A4A9	1104	AM030
1103	182081	1104	A069-112	1104	A4B1	1104	AM035
1103	182124	1104	A100	1104	A4B5	1104	AM040
1103	182228	1104	A100(RECTIFIER)	1104	A4B9	1104	AM050
1103	182238	1104	A101-A	1104	A4C1	1104	AM060
1103	182241	1104	A101-A(RECT.)	1104	A4C5	1104	AM13
1103	182245	1104	A10105	1104	A4C9	1104	AM21
1103	182269	1104	A10113	1104	A4D1	1104	AM22
1103	182271	1104	A10118	1104	A4D5	1104	AM23
1103	182278	1104	A10142	1104	A4D9	1104	AM24
1103	182316	1104	A10164	1104	A4E1	1104	AM3
1103	182329	1104	A10165	1104	A4E5	1104	AM32
1103	182359	1104	A10169			1104	AM33
1103	182393	1104	A101A(RECTIFIER)			1104	AM405
1103	182394	1104	A10C			1104	AM410

RS 276-	Industry Standard No.	RS 276-	Industry Standard No.	RS 276-	Industry Standard No.	RS 276-	Industry Standard No.
1104	AM62	1104	B3D1	1104	BYX22/400	1104	CER680C
1104	AM63	1104	B3D5	1104	BYX22/600	1104	CER68A
1104	AM65	1104	B3D9	1104	BYX36-300	1104	CER68B
1104	AM66	1104	B3E1	1104	BYX36-600	1104	CER68C
1104	AN-1	1104	B3E5	1104	BYX36/150	1104	CER69
1104	AN-0-5B	1104	B3E9	1104	BYX36/300	1104	CER690
1104	A04092	1104	B3F1	1104	BYX36/600	1104	CER690A
1104	A04092A	1104	B3F5	1104	BYX60-100	1104	CER690B
1104	A04092B	1104	B3F9	1104	BYX60-200	1104	CER690C
1104	A04093	1104	B3G1	1104	BYX60-300	1104	CER69A
1104	A04093-A	1104	B305	1104	BYX60-400	1104	CER69B
1104	A04093A	1104	B309	1104	BYX60-50	1104	CER69C
1104	A04210A	1104	B4A1	1104	BYX60-500	1104	CER6B
1104	A04212-A	1104	B4A5	1104	BYX60-600	1104	CER70
1104	A04212-B	1104	B4A9	1104	BYY-31	1104	CER700
1104	A04233	1104	B4B1	1104	BYY-32	1104	CER700A
1104	A04716	1104	B4B5	1104	BYY-35	1104	CER700B
1104	A04901A	1104	B4B9	1104	BYY31	1104	CER700C
1104	AQ2(PHILCO)	1104	B4C1	1104	BYY32	1104	CER70A
1104	AQ3(PHILCO)	1104	B4C5	1104	BYY33	1104	CER70B
1104	AR-22(DIO)	1104	B4C9	1104	BYY34	1104	CER70C
1104	AR16	1104	B4D1	1104	BYY35	1104	CER71
1104	AR18	1104	B4D5	1104	BYY36	1104	CER710
1104	AR19	1104	B4D9	1104	BYY89	1104	CER710A
1104	AR20	1104	B4E1	1104	C10110	1104	CER710B
1104	AR21	1104	B4E5	1104	C10159	1104	CER710C
1104	AR22	1104	B4E9	1104	C10176	1104	CER71A
1104	AR22(RECTIFIER)	1104	B4F1	1104	C1181C1E1C	1104	CER71B
1104	AR7C	1104	B4F5	1104	C1B	1104	CER71C
1104	AR882	1104	B4F9	1104	C1H	1104	CF102DA
1104	AS-14	1104	B4G1	1104	C21382	1104	CH119D
1104	AS-15	1104	B405	1104	C23H12B	1104	CL010
1104	AS-2	1104	B409	1104	C24807	1104	CL025
1104	AS-3	1104	B5A1	1104	C24J102	1104	CL05
1104	AS-4	1104	B5A5	1104	C83-829	1104	CL1
1104	AS-5	1104	B5A9	1104	C83-880	1104	CL1.5
1104	AS11	1104	B5B1	1104	C934(RECTIFIER)	1104	CL2.
1104	AS14	1104	B5B5	1104	CA10	1104	CL3
1104	AS15	1104	B5B9	1104	CA100	1104	CL4
1104	AS2	1104	B5C1	1104	CA1020A	1104	CL5
1104	AS3	1104	B505	1104	CA102BA	1104	CL6
1104	AS4	1104	B509	1104	CA102DA	1104	CL7
1104	AS5	1104	B5D1	1104	CA102PA	1104	CL8
1104	AS6	1104	B5D5	1104	CA102HA	1104	CLM05
1104	AW-01-12	1104	B5D9	1104	CA102MA	1104	CLM1
1104	B-1501U	1104	B5E1	1104	CA150	1104	CMB470
1104	B-1881U	1104	B5E5	1104	CA20	1104	C049
1104	B-1882U	1104	B5E9	1104	CA200	1104	CODI-6045
1104	B-31	1104	B5F1	1104	CA250	1104	CODI-6046
1104	B01-02	1104	B5F5	1104	CA50	1104	CODI-6047
1104	B0102	1104	B5F9	1104	CA6664	1104	CODI-6048
1104	B1045494P1	1104	B5G1	1104	CA6665	1104	CODI1556
1104	B12-02	1104	B505	1104	CA7248	1104	CODI531
1104	B1A1	1104	B509	1104	CA8314	1104	CODI532
1104	B1A5	1104	B601-1012	1104	CB10	1104	CODI533
1104	B1A9	1104	B60C300	1104	CB150	1104	CODI534
1104	B1B	1104	BA-100	1104	CB20	1104	CODI535
1104	B1B1	1104	BA-104	1104	CB200	1104	CODI536
1104	B1B5	1104	BA-142-01	1104	CB250	1104	CODI551
1104	B1B9	1104	BA100	1104	CB5	1104	CODI552
1104	B1C1	1104	BA104	1104	CB50	1104	CODI5524
1104	B1C5	1104	BA105	1104	CC102BA	1104	CODI553
1104	B1C9	1104	BA108	1104	CC102DA	1104	CODI5534
1104	B1D1	1104	BA119	1104	CC102PA	1104	CODI554
1104	B1D5	1104	BA120	1104	CC102HA	1104	CODI5544
1104	B1D9	1104	BA127	1104	CC102KA	1104	CODI555
1104	B1E1	1104	BA128	1104	CC102MA	1104	CODI556
1104	B1E5	1104	BA129	1104	CD-2	1104	CODI564
1104	B1E9	1104	BA130	1104	CD-2N	1104	CODI575
1104	B1F1	1104	BA142-01	1104	CD-4	1104	CODI6047
1104	B1F5	1104	BA153	1104	CD-8457	1104	CODI611
1104	B1F9	1104	BAX12	1104	CD-860037	1104	CODI612
1104	B1G1	1104	BAY44	1104	CD05	1104	CODI613
1104	B1G5	1104	BAY64	1104	CD1111	1104	CODI614
1104	B1G9	1104	BAT86	1104	CD1112	1104	CODI615
1104	B1H(DIODE)	1104	BAT87	1104	CD1113	1104	CODI616
1104	B1P9	1104	BB-109	1104	CD1114	1104	CODI115524
1104	B200C40	1104	BB-68	1104	CD1115	1104	CODI11556
1104	B250C100	1104	BB107	1104	CD1116	1104	CODI1531
1104	B250C100TD	1104	BB117	1104	CD1117	1104	CODI15524
1104	B250C125	1104	BB127	1104	CD1121	1104	CODI15534
1104	B250C125K4	1104	BB1A	1104	CD1122	1104	CODI15544
1104	B250C125N2	1104	BB2A184	1104	CD1123	1104	CODI15564
1104	B250C125X4	1104	BBIA	1104	CD1124	1104	CODI6045
1104	B250C150	1104	BC-207	1104	CD1125	1104	CODI6047
1104	B250C150K4	1104	BC-307	1104	CD1126	1104	CP102BA
1104	B250C75	1104	BCF5	1104	CD1127	1104	CP102DA
1104	B250C75K4	1104	BD-107	1104	CD1141	1104	CP102PA
1104	B250C75K41	1104	BD-107(RECT.)	1104	CD1142	1104	CP102HA
1104	B250C75K45	1104	BD-127	1104	CD1143	1104	CP102KA
1104	B250C75K5	1104	BD-1A	1104	CD1147	1104	CP102MA
1104	B250C7K4S	1104	BDOA	1104	CD1148	1104	CR/E
1104	B294	1104	BD117	1104	CD1149	1104	CR1034
1104	B294(RECTIFIER)	1104	BD3A-184	1104	CD1151	1104	CR1035
1104	B2A1	1104	BE107	1104	CD13332	1104	CR501/6515
1104	B2A5	1104	BE117	1104	CD13333	1104	CR801
1104	B2A9	1104	BE127	1104	CD13334	1104	CS-16E
1104	B2B1	1104	BF26235-5	1104	CD13335	1104	C28131D(AXIAL)
1104	B2B5	1104	BFY87A	1104	CD13336	1104	C816E
1104	B2B9	1104	BH4R1	1104	CD13337	1104	CT-3003
1104	B2C1	1104	BR42	1104	CD13338	1104	CT-3005
1104	B2C5	1104	BR44	1104	CD13339	1104	CT100
1104	B2C9	1104	BR46	1104	CD2	1104	CT200
1104	B2D1	1104	BR47	1104	CD37A2	1104	CT300
1104	B2D5	1104	BR48	1104	CD4	1104	CT3003
1104	B2D9	1104	BR51400-1	1104	CDG005	1104	CT3005
1104	B2E1	1104	BR51401-2	1104	CDK-4	1104	CT600
1104	B2E5	1104	B52	1104	CDR-2	1104	CTN200
1104	B2E9	1104	B31	1104	CDR-4	1104	CTP-2001-1011
1104	B2F1	1104	B32	1104	CDS-16B	1104	CTP-2001-1012
1104	B2F5	1104	BTM50	1104	CE0398/7839	1104	CX-0035
1104	B2F9	1104	BU029	1104	CE502	1104	CX-0037
1104	B2G1	1104	BY25	1104	CE504	1104	CX-0039
1104	B2G5	1104	BY101	1104	CE506	1104	CX-0040
1104	B2G9	1104	BY102	1104	CE6050	1104	CX-0048
1104	B30C250	1104	BY106	1104	CER-69	1104	CX-0054
1104	B30C250-1	1104	BY107	1104	CER-71	1104	CX-9001
1104	B30C350-1	1104	BY111	1104	CER-71CA	1104	CXO037
1104	B30C500	1104	BY112	1104	CER500	1104	CXO039
1104	B30C600	1104	BY113	1104	CER500A	1104	CXO040
1104	B30C600CB	1104	BY114	1104	CER500C	1104	CXO047
1104	B31	1104	BY115	1104	CER67	1104	CXO048
1104	B31(RECTIFIER)	1104	BY116	1104	CER670	1104	CXO049
1104	B35C600	1104	BY117	1104	CER670A	1104	CX9001
1104	B36564	1104	BY121	1104	CER670B	1104	CXL114
1104	B3A1	1104	BY124	1104	CER670C	1104	CXL116
1104	B3A5	1104	BY125	1104	CER67A	1104	CXL117
1104	B3A9	1104	BY126	1104	CER67B	1104	CY40
1104	B3B1	1104	BY130	1104	CER67C	1104	CY50
1104	B3B5	1104	BY134	1104	CER68	1104	CY7B
1104	B3B9	1104	BY135	1104	CER680	1104	D-00384R
1104	B3C1	1104	BY141	1104	CER680A	1104	D004
1104	B3C5	1104	BY153	1104	CER680B	1104	D01-100
1104	B3C9	1104	BYX22/200	1104		1104	D028
1104		1104		1104		1104	D05

RS 276-	Industry Standard No.	RS 276-	Industry Standard No.	RS 276-	Industry Standard No.	RS 276-	Industry Standard No.
1104	D1-26	1104	D152S	1104	DS430	1104	ED2914
1104	D1-52S	1104	D156	1104	DS58(SANYO)	1104	ED2915
1104	D1-7	1104	D1645	1104	DS5BN-6	1104	ED2916
1104	D100	1104	D1646	1104	D879	1104	ED2917
1104	D101167	1104	D17	1104	D889	1104	ED2918
1104	D10167	1104	D172S	1104	DT18(CHAN.MASTER)	1104	ED2919
1104	D10168	1104	D1J	1104	DU400	1104	ED2920
1104	D11172S	1104	D1J70488	1104	DU600	1104	ED2921
1104	D1201F	1104	D1J70544	1104	DX-0099	1104	ED3000
1104	D1445	1104	D1J70695	1104	DX-0445	1104	ED3000A
1104	D1448	1104	D1J71958	1104	DX-0475	1104	ED3000B
1104	D15A	1104	D1J71959	1104	DX-0721	1104	ED3001
1104	D15C	1104	D1J72168	1104	DX-1039	1104	ED3001A
1104	D18	1104	DIL	1104	DX-1131	1104	ED3001B
1104	D191	1104	DIS-18	1104	DX520	1104	ED3002
1104	D1E	1104	DI81S	1104	E-0704W	1104	ED3002A
1104	D1H	1104	DR1	1104	E-075L	1104	ED3002B
1104	D1J	1104	DR100	1104	E-41	1104	ED3003
1104	D1J70544	1104	DR1100	1104	EO3155-001	1104	ED3003A
1104	D1L	1104	DR1PR	1104	EO3155-002	1104	ED3003B
1104	D2-1	1104	DR2	1104	EO704-W	1104	ED3003S
1104	D200	1104	DR200	1104	EO75L	1104	ED3004
1104	D21489	1104	DR206	1104	EO788-C	1104	ED3004A
1104	D220	1104	DR3	1104	EO788C	1104	ED3004B
1104	D2201M	1104	DR300	1104	E1	1104	ED3005
1104	D2201N	1104	DR4	1104	E1010	1104	ED3005A
1104	D220M	1104	DR400	1104	E1011	1104	ED3005B
1104	D227(DIODE)	1104	DR427	1104	E10116	1104	ED3006
1104	D25	1104	DR435	1104	E10157	1104	ED3006A
1104	D25A	1104	DR5	1104	E10171	1104	ED3006B
1104	D25B	1104	DR500	1104	E10172	1104	ED329128
1104	D25C	1104	DR5101	1104	E10188	1104	ED329130
1104	D2600EF	1104	DR5102	1104	E102(ELCOM)	1104	ED4
1104	D28	1104	DR600	1104	E106(ELCOM)	1104	ED494583
1104	D2H(DIODE)	1104	DR668	1104	E1124	1104	ED5
1104	D2J	1104	DR669	1104	E1146J	1104	ED511097
1104	D2X4	1104	DR670	1104	E1146R	1104	ED6
1104	D30	1104	DR671	1104	E1153RB	1104	ED7
1104	D300	1104	DR695	1104	E1156RD	1104	EDJ-363
1104	D3436	1104	DR698	1104	E1157RNA	1104	ED8-0002
1104	D3R	1104	DR699	1104	E125C200	1104	ED8-0017
1104	D3R38	1104	DR826	1104	E13-020-00	1104	ED8-0024
1104	D3R39	1104	DR848	1104	E13-20-00	1104	ED8-11
1104	D3U	1104	DR863	1104	E135	1104	ED8-17
1104	D3V	1104	DRS1	1104	E1410	1104	ED8-24
1104	D3Z	1104	DRS101	1104	E1411	1104	ED8-4
1104	D400	1104	DRS102	1104	E1412	1104	EER600-2
1104	D400(RECT.)	1104	DRS104	1104	E1413	1104	EG100
1104	D45C	1104	DRS106	1104	E1415	1104	EG100H
1104	D45CZ	1104	DS-0065	1104	E143	1104	EH2011/7+12/1
1104	D48	1104	DS-1	1104	E1440	1104	EM-1171
1104	D4M	1104	DS-10	1104	E146	1104	EM1021
1104	D4R26	1104	DS-13	1104	E140350	1104	EM1142
1104	D4R39	1104	DS-13(COURIER)	1104	E150L	1104	EM401
1104	D5	1104	DS-130	1104	E2	1104	EM402
1104	D50	1104	DS-130B	1104	E21	1104	EM403
1104	D500	1104	DS-130C	1104	E24100	1104	EM404
1104	D5G	1104	DS-130E	1104	E2505	1104	EM405
1104	D5R35	1104	DS-130YB	1104	E3	1104	EM406
1104	D5R39	1104	DS-130YE	1104	E3006	1104	EM407
1104	D600	1104	DS-131	1104	E300L	1104	EM408
1104	D65C	1104	DS-131A	1104	E41	1104	EM410
1104	D6623	1104	DS-131B	1104	E4676B	1104	EM501
1104	D6623A	1104	DS-132	1104	E5	1104	EM502
1104	D6624	1104	DS-132A	1104	E500L	1104	EM503
1104	D6624A	1104	DS-132B	1104	E6	1104	EM504
1104	D6625	1104	DS-13A(SANYO)	1104	E650L	1104	EM505
1104	D6625A	1104	DS-13B(SANYO)	1104	E7441	1104	EM506
1104	D68	1104	DS-14(DIODE)	1104	E750(ELCOM)	1104	EO704
1104	D6HZ	1104	DS-16A	1104	E752(ELCOM)	1104	EP-1428-2H
1104	D81K	1104	DS-16B	1104	E756(ELCOM)	1104	EP-5619-2/7628
1104	DA000	1104	DS-16B(SANYO)	1104	EA0015	1104	EP1259
1104	DA001	1104	DS-16B(SYLVANIA)	1104	EA0016	1104	EP-1259-2
1104	DA002	1104	DS-16C(SANYO)	1104	EA0031	1104	EP1428-2H
1104	DAAY002001	1104	DS-16D	1104	EA005	1104	EP16X13
1104	DB206	1104	DS-16D(SANYO)	1104	EA010	1104	EP200
1104	DD-000	1104	DS-16E	1104	EA020	1104	EP3149
1104	DD-003	1104	DS-16E(SANYO)	1104	EA030	1104	EP400
1104	DD-006	1104	DS-16ND	1104	EA040	1104	EP57X1
1104	DD-007	1104	DS-16NE	1104	EA050	1104	EP57X12
1104	DD003	1104	DS-16NY	1104	EA060	1104	EP57X4
1104	DD006	1104	DS-16YA	1104	EA1072	1104	EP5X5
1104	DD007	1104	DS-17	1104	EA1448	1104	EP600
1104	DD056	1104	DS-17-6A	1104	EA15X14	1104	ER1
1104	DD175C	1104	DS-18	1104	EA1672	1104	ER101
1104	DD176C	1104	DS-19(RECTIFIER)	1104	EA16X13	1104	ER102
1104	DD177C	1104	DS-1M	1104	EA16X149	1104	ER102D
1104	DD2066	1104	DS-1N	1104	EA1622	1104	ER103D
1104	DD2320	1104	DS-1P	1104	EA16X21	1104	ER103E
1104	DD2321	1104	DS-1U	1104	EA16X30	1104	ER104D
1104	DD236	1104	DS-3S	1104	EA16X33	1104	ER105D
1104	DD266	1104	DS-430	1104	EA16X34	1104	ER106D
1104	DDAY002001	1104	DS-79	1104	EA16X55	1104	ER11
1104	DDAY002002	1104	DS-79(DELCO)	1104	EA16X71	1104	ER12
1104	DDAY042001	1104	DS-1M	1104	EA16X8	1104	ER181
1104	DDAY103001	1104	DS0065	1104	EA16X92	1104	ER182
1104	DDBY002001	1104	DS1	1104	EA2140	1104	ER183
1104	DE-201	1104	DS130	1104	EA2499	1104	ER184
1104	DE-201	1104	DS130B	1104	EA2501	1104	ER185
1104	DE14	1104	DS130E	1104	EA2741	1104	ER1B22-15
1104	DE14A	1104	DS130NB	1104	EA3827	1104	ER2
1104	DE16	1104	DS130ND	1104	EA3989	1104	ER201
1104	DE16A	1104	DS130Y	1104	EA5711	1104	ER21
1104	DE201	1104	DS130YC	1104	EA57X1	1104	ER22
1104	DG1M	1104	DS130YE	1104	EA57X10	1104	ER301
1104	DG1MR	1104	00000DS131	1104	EA57X11	1104	ER31
1104	DG1N	1104	DS131B	1104	EA57X14	1104	ER381
1104	DG1PR	1104	DS132A	1104	EA57X3	1104	ER401
1104	DH-001	1104	DS132B	1104	EA57X8	1104	ER41
1104	DH-002	1104	DS15E	1104	EA75X1	1104	ER410
1104	DH14	1104	DS160(G.E.)	1104	EC401	1104	ER42
1104	DH14A	1104	DS16A	1104	EC402	1104	ER501
1104	DH16	1104	DS16C	1104	ECG114	1104	ER51
1104	DH16A	1104	DS16E	1104	ECG116	1104	ER510
1104	DH4R2	1104	DS16N	1104	ECG117	1104	ER57X2
1104	DI-1649	1104	DS16NB	1104	ECR-600-2	1104	ER57X3
1104	DI-172S	1104	DS16NC	1104	ECR600-2	1104	ER57X4
1104	DI-428	1104	DS16ND	1104	ED-4	1104	ER601
1104	DI-46	1104	DS16NE	1104	ED-5	1104	ER61
1104	DI-52S	1104	DS17	1104	ED-6	1104	ER62
1104	DI-55	1104	DS17(ADMIRAL)	1104	ED1804	1104	ERB-24-06A
1104	DI-56	1104	DS17N	1104	ED1892	1104	ERB11-01
1104	DI-645	1104	DS18	1104	ED2106	1104	ERB12-02
1104	DI-646	1104	DS18N	1104	ED2107	1104	ERB12-11
1104	DI-647	1104	DS1K	1104	ED2108	1104	ERB22-15
1104	DI-648	1104	DS1K7	1104	ED2109	1104	ERB24
1104	DI-7	1104	DS1N	1104	ED2110	1104	ERB24(GE)
1104	DI-705	1104	DS1P	1104	ED224548	1104	ERB24-02B
1104	DI-71	1104	DS2K	1104	ED224550	1104	ERB24-06
1104	DI-72S	1104	DS2N	1104	ED2842	1104	ERB24-06A
1104	DI1649	1104	DS2N22	1104	ED2843	1104	ERB2406A
1104	DI428	1104	DS3S	1104	ED2844	1104	ERB28-04
1104	DI46	1104	DS3S(CRAIG)	1104	ED2845	1104	ERB28-04D
1104		1104	DS3S(SANYO)	1104	ED2846	1104	ERC24-04

RS 276-	Industry Standard No.	RS 276-	Industry Standard No.	RS 276-	Industry Standard No.	RS 276-	Industry Standard No.
1104	ERD300	1104	FU-1N	1104	HC-30	1104	HV46GR
1104	ERD400	1104	FU-1NA	1104	HC-39	1104	HV46GR(DIODE)
1104	ERV-02P2150	1104	FU-IM	1104	HC-68	1104	I538
1104	ES10233	1104	FU1M	1104	HC500	1104	IC743048
1104	ES15056	1104	FU1N	1104	HC67	1104	IM4004
1104	ES16X10	1104	FU1U	1104	HC670	1104	INJ61227
1104	ES16X13	1104	FU1U	1104	HC68	1104	INJ61434
1104	ES16X25	1104	FW100	1104	HC680	1104	INJ61726
1104	ES16X38	1104	FW200	1104	HC700	1104	IP20-0022
1104	ES16X9	1104	FW400	1104	HC71	1104	IP20-0024
1104	ES47X1	1104	FW500	1104	HC710	1104	IP20-0025
1104	ES51	1104	FW600	1104	HC80	1104	IP20-0054
1104	ES57X1	1104	FW600A	1104	HCL9	1104	IP20-0120
1104	ES57X2	1104	FWL100	1104	HCV	1104	IP20-0167
1104	ES57X4	1104	FWL200	1104	HD-1	1104	IR10B6J
1104	ES57X5	1104	FWL300	1104	HD-2000308	1104	IR1D
1104	ES3A-06	1104	G00-502-A	1104	HD-2000300	1104	IR20
1104	ES3A-10C	1104	G00-502A	1104	HD20-003-01	1104	IR2A
1104	ES3A-10N	1104	G00-534-A	1104	HD2000	1104	IR2E
1104	ES3A06	1104	G00-535-B	1104	HD2000-301	1104	IT10D4K
1104	ESK-1	1104	G00-535-A	1104	HD2000703	1104	ITT350
1104	ESK1	1104	G00-536-A	1104	HD2000903	1104	ITT402
1104	ESK1/06	1104	G00-536A	1104	HD2000110	1104	ITT73
1104	ESKE400500	1104	G00-543-A	1104	HD2000110-0	1104	ITT992
1104	ET16X16	1104	G00-551-A	1104	HD2000110Q	1104	J-05
1104	ET16X16X	1104	G00502A	1104	HD2000350-0	1104	J-1
1104	ET200	1104	G00535	1104	HD2000301Q	1104	J-2
1104	ET400	1104	G00535A	1104	HD2000305	1104	J-4
1104	ET51X25	1104	G01	1104	HD2000307	1104	J-6
1104	ET52X25	1104	G01211	1104	HD2000308	1104	J100
1104	ET55-25	1104	G02	1104	HD2000413	1104	J101183
1104	ET55X25	1104	G0Q-535-B	1104	HD2000501	1104	J2-4570
1104	ET55X29	1104	G1	1104	HD2000510	1104	J20437
1104	ET57X25	1104	G100A	1104	HD2000610	1104	J241100
1104	ET57X29	1104	G100B	1104	HD2000703Q	1104	J241102
1104	ET57X30	1104	G100D	1104	HD2000703Q	1104	J241142
1104	ET57X33	1104	G100G	1104	HD2000903	1104	J241209
1104	ET57X35	1104	G100J	1104	HD2001310	1104	J241210
1104	ET57X38	1104	G10119	1104	HD200207	1104	J241211
1104	ET600	1104	G1242	1104	HD2000301	1104	J241212
1104	ETD-10D1	1104	G130	1104	HD6147	1104	J241214
1104	ETD-10D2	1104	G2	1104	HD6865	1104	J241232
1104	ETD-V06C	1104	G222	1104	HE-20011	1104	J241260
1104	EU16X7	1104	G296	1104	HE-SD1	1104	J24127
1104	EU16X8	1104	G2A	1104	HEP154	1104	J241271
1104	EU57X30	1104	G3	1104	HEP156	1104	J24570
1104	EU57X40	1104	G4	1104	HEP157	1104	J24630
1104	EVM511	1104	G5	1104	HEP158	1104	J24645
1104	EW169B	1104	G6	1104	HEPR0050	1104	J24647
1104	EX76-X	1104	G657	1104	HEPR0051	1104	J24756
1104	EYV-420D1R5JA	1104	G659	1104	HEPR0052	1104	J24871
1104	F-05	1104	G700	1104	HEPR0053	1104	J24877
1104	F-14A	1104	G701	1104	HEPR0054	1104	J24919
1104	F-14C	1104	G702	1104	HEPR9003	1104	J24920
1104	F05	1104	GD12	1104	HP-08W05	1104	J24935
1104	F1	1104	GE-1N506	1104	HP-0032	1104	J24939
1104	F10124	1104	GE-1N5061	1104	HP-20042	1104	J24940
1104	F10148	1104	GE-504	1104	HP-20047	1104	J320020
1104	F10180	1104	GE-504A	1104	HP-20050	1104	JAM702C
1104	F14	1104	GE-X36	1104	HP-20052	1104	JB-00030
1104	F14-C	1104	GB42-7	1104	HP-20067	1104	JB-00056
1104	F14A	1104	GE6366	1104	HP-20083	1104	JB-BB1A
1104	F14B	1104	GEX36	1104	HP-20084	1104	JB16C4
1104	0P160	1104	GEX66	1104	HP-20105	1104	JC-00012
1104	0P164	1104	GEZJ252A	1104	HP-SD-1A	1104	JC-00014
1104	F2	1104	GEZJ252B	1104	HP-SD-1C	1104	JC-00025
1104	F20-1015	1104	GI-100-1	1104	HP-SD-1Z	1104	JC-00028
1104	F20-1016	1104	GI-1N4385	1104	HP-SD1	1104	JC-00032
1104	F215-1016	1104	GI-3008	1104	HPOW05	1104	JC-00033
1104	F215-1017	1104	GI-300D	1104	HP20066	1104	JC-00035
1104	F3	1104	GI-P100-D	1104	HPOW05	1104	JC-00037
1104	F4	1104	GI08B	1104	HPSD1A	1104	JC-00044
1104	F5	1104	GI3002	1104	HPSD1B	1104	JC-00045
1104	F6	1104	GI3992-17	1104	HPSG005	1104	JC-00047
1104	FA4	1104	GI411	1104	HG-R10	1104	JC-00049
1104	FA6	1104	GI420	1104	HGR-10	1104	JC-00051
1104	FD-1029-DF	1104	GIP100D	1104	HGR-20	1104	JC-00055
1104	FD-1029-DG	1104	GJ4M	1104	HGR-30	1104	JC-00059
1104	FD1599	1104	GM-1A	1104	HGR-40	1104	JC-0025
1104	FD3	1104	G00535	1104	HGR-5	1104	JC-10D1
1104	FD300	1104	G00-502-A	1104	HGR-60	1104	JC-DS16E
1104	FD333	1104	GP-1	1104	HGR1	1104	JC-K805
1104	FD3389	1104	GP-25B	1104	HGR2	1104	JC-SD-1X
1104	FD6	1104	GP05A	1104	HGR3	1104	JC-SD-1Z
1104	FDH400	1104	GP08B	1104	HGR4	1104	JC-SD1Z
1104	FG-2NA	1104	GP08D	1104	HIP1	1104	JC-SG005
1104	FM1J2	1104	GP1	1104	HN-00008	1104	JC-V03C
1104	FM211N	1104	GP230	1104	HN-00018	1104	JCN1
1104	F05	1104	GP250	1104	HN-00029	1104	JCN2
1104	FPR50-1011	1104	GP25G	1104	HN-00032	1104	JCN3
1104	FR-1	1104	GSM482	1104	HO-5A	1104	JCN4
1104	FR-1H	1104	GSM483	1104	HP205	1104	JCN5
1104	FR-1H(M)	1104	GSM51	1104	HR-05A	1104	JCN6
1104	FR-1HM	1104	GSM52	1104	HR-2A	1104	JCV-2
1104	FR-1M	1104	GSM53	1104	HR-2A(PENNCREST)	1104	JCV-3
1104	FR-1MD	1104	GSM54	1104	HR-5A	1104	JCV2
1104	FR-1N	1104	H)-00008	1104	HR-5AX2	1104	JCV5
1104	FR-1P	1104	H)-00018	1104	HR-5B	1104	JD-00040
1104	FR-1U	1104	H100	1104	HR10	1104	JD-BB1A
1104	FR-2	1104	H10174	1104	HR11	1104	JD-SD1D
1104	FR-202	1104	H200	1104	HR13	1104	JD-SD1Z
1104	FR-2M	1104	H20052	1104	HR5A	1104	JDSS1D
1104	FR-2P	1104	H300	1104	HR5A5E	1104	JT-E1024D
1104	FR-U	1104	H400	1104	HR5B	1104	JT-E1064
1104	FR1	1104	H475	1104	HS1001	1104	K1.3G22A
1104	FR10	1104	H50	1104	HS1002	1104	K112D
1104	FR1M	1104	H500	1104	HS1003	1104	K118J966-1
1104	FR1MB	1104	H585	1104	HS1007	1104	K122D
1104	FR1MD	1104	H600	1104	HS1008	1104	K1A5
1104	FR1P	1104	H616	1104	HS1009	1104	K1B5
1104	FR1U	1104	H618	1104	HS1010	1104	K1C5
1104	FR2	1104	H619	1104	HS1011	1104	K1D5
1104	FR2(S1B1)	1104	H620	1104	HS1012	1104	K1E5
1104	FR2-02	1104	H625	1104	HS1020	1104	K1F5
1104	FR2-02(DIO)	1104	H3105	1104	HS3105	1104	K1G5
1104	FR2-06	1104	H7126-3	1104	HS3104	1104	K1H5
1104	FR2-2Z	1104	H781	1104	HSPD-1A	1104	K200
1104	FR2-0Z	1104	H783	1104	HSPD-1Z	1104	K2A5
1104	0000FR202	1104	H881	1104	HT800071Q	1104	K2B5
1104	FR2M	1104	H890	1104	HV-100	1104	K2D5
1104	FR2P	1104	H891	1104	HV-23(ELGIN)	1104	K2E5
1104	FRH-101	1104	HA100	1104	HV-26	1104	K2F5
1104	FRH101	1104	HA200	1104	HV-26G	1104	K2G
1104	0000000FRI	1104	HA300	1104	HV-46	1104	K2G5
1104	PSP-288-1	1104	HA400	1104	HV-46GR	1104	K3B5
1104	FST2	1104	HA50	1104	HV0000105	1104	K3C5
1104	FST3	1104	HA500	1104	HV0000105-Q	1104	K3D5
1104	FT-1	1104	HA600	1104	HV0000406	1104	K3F5
1104	FT-10	1104	HAR10	1104	HV0000705	1104	K3G5
1104	FT-1N	1104	HAR20	1104	HV26	1104	K4-555
1104	FT14A	1104	HB2	1104	HV46	1104	K4-557
1104	FT1N	1104	HB3	1104	HV46#OR#(DIODE)	1104	K4A5
1104	FU-1MA	1104		1104	HV46(DIODE)	1104	K4B5

RS 276-	Industry Standard No.	RS 276-	Industry Standard No.	RS 276-	Industry Standard No.	RS 276-	Industry Standard No.
1104	K405	1104	M4B31-13	1104	MM6	1104	0A128
1104	K4D5	1104	M4HZ	1104	MP-01	1104	0A129
1104	K4E5	1104	M500	1104	MP-5115	1104	0A130
1104	K4F5	1104	M500A	1104	MP01	1104	0A131
1104	K4O5	1104	M500B	1104	MP100	1104	0A132
1104	K5A5	1104	M500C	1104	MP1003-1	1104	0A180
1104	K5B5	1104	M604HT	1104	MP1003-2	1104	0A210
1104	K5O5	1104	M62	1104	MP1003-4	1104	0A211
1104	K5D5	1104	M67	1104	MP225	1104	0A214
1104	K5E5	1104	M670	1104	MP300	1104	0A8
1104	K5F5	1104	M670A	1104	MP400	1104	0P160
1104	K5O5	1104	M670B	1104	MP500(RECT.)	1104	0P164
1104	KB532799	1104	M670C	1104	MP5113	1104	0G-30L125
1104	KB533058-1	1104	M67A	1104	MP651	1104	0S-16308
1104	KB-182	1104	M67B	1104	MP9602	1104	08-6D
1104	KB262	1104	M67C	1104	MPX-25	1104	0816308
1104	KB265A(RECT)	1104	M68	1104	MPX215	1104	08S-16308
1104	KC-0691L/8	1104	M680	1104	MPX25	1104	08S-36885
1104	KC-08C221	1104	M680A	1104	MQ3/2	1104	08S36503
1104	KC-1.3O3X11/1	1104	M680B	1104	MQ32	1104	08S36685
1104	KC0-8CP	1104	M680C	1104	MQ6/2	1104	08S36885
1104	KC0-8CP11/1&12/1	1104	M68A	1104	MQ62	1104	0Y-5061
1104	KC0-8CP11/1+12/1	1104	M68B	1104	MQ8/2	1104	0Y-5062
1104	KC0.8	1104	M68C	1104	MR-1`	1104	0Y101
1104	KC0.8CP	1104	M69	1104	MR-150-01	1104	0Y5061
1104	KC0.8CP11	1104	M690	1104	MR1237FB	1104	0Y5062
1104	KC0.8CP11/1	1104	M690A	1104	MR1237FL	1104	0Y5063
1104	KC0.8CP11/1&121Y	1104	M690B	1104	MR1237SB	1104	0Y5064
1104	KC0.8CP11/1+121Y	1104	M690C	1104	MR1237SL	1104	0Y5065
1104	KC0.8CP11/H12/1	1104	M69A	1104	MR1247FB	1104	0Y5066
1104	KC0.8CP12/1	1104	M69B	1104	MR1247FL	1104	P-10115
1104	KC06911	1104	M69C	1104	MR1247SB	1104	P100
1104	KC006E11/8	1104	M6HZ	1104	MR1247SL	1104	P100A
1104	KC08C11/10	1104	M70	1104	MR1267	1104	P100B
1104	KC08C11/8	1104	M700	1104	MR1C	1104	P100D
1104	KC008C1110	1104	M700A	1104	MR2064	1104	P100G
1104	KC008C21/5	1104	M700B	1104	MR2065	1104	P100J
1104	KC008C215	1104	M700C	1104	MR2261	1104	P10115
1104	KC008C22/19	1104	M701B	1104	MR2262	1104	P10156
1104	KC008C221	1104	M702C	1104	MR2271	1104	P150A
1104	KC008CP111	1104	M70A	1104	MR9600	1104	P150B
1104	KC008CP121	1104	M70B	1104	MR9601	1104	P150D
1104	KC00911/8	1104	M70C	1104	MR9602	1104	P150G
1104	KC1.3C	1104	M71	1104	MRB-20C	1104	P150J
1104	KC1.3C22/1	1104	M710	1104	MRV-20C	1104	P145
1104	KC1.3O3X11/1	1104	M710A	1104	MS11H	1104	P1B5
1104	KC1.3G	1104	M710B	1104	MS12H	1104	P1C5
1104	KC1.3G12/1X2	1104	M710C	1104	MS13H	1104	P1D5
1104	KC1.3G22/12	1104	M71A	1104	MS14H	1104	P20
1104	KC1.3G22/1A	1104	M71B	1104	MS1H	1104	P200
1104	KC13C22/1	1104	M71C	1104	MS2H	1104	P21316
1104	KC13C221	1104	M8222	1104	MS35H	1104	P21317
1104	KC2BP22/1B	1104	M8399	1104	MS36H	1104	P21443
1104	KC2D221	1104	M91A01	1104	MS3H	1104	P2A5
1104	KC2DP	1104	M91A02	1104	MS4H	1104	P2B5
1104	KC2DP11/1	1104	M91A03	1104	MS5C	1104	P2C5
1104	KC2DP11/1412/1	1104	M9206	1104	MS5H	1104	P2D5
1104	KC2DP12/1	1104	M9235	1104	MSR-500	1104	P3/2H
1104	KC2DP12/1N	1104	M9312	1104	MSR-95	1104	P32H
1104	KC2DP12/2	1104	M9314	1104	MSR500	1104	P3A5
1104	KC2DP121N	1104	M9317	1104	MS8-1000	1104	P3B5
1104	KC2DP122	1104	M9319	1104	MT021	1104	P3C5
1104	KC2DP22/1	1104	MA-110	1104	MT021A	1104	P3D5
1104	KC2DP221	1104	MA-26-1	1104	MT022	1104	P400
1104	KC2DP221B	1104	MA101	1104	MT022A	1104	P4A5
1104	KC2G11/1	1104	MA102	1104	MT14	1104	P4B5
1104	KC2G11/1&12/1	1104	MA110	1104	MT24	1104	P4C5
1104	KC2G11/1+12/1	1104	MA2	1104	MT44	1104	P4D5
1104	KC2G12/1	1104	MA203	1104	MT64	1104	P5A5
1104	KC0-8CP	1104	MA211	1104	MV-13	1104	P5B5
1104	KC0-8CP11/1 12/1	1104	MA215	1104	MV-13(BIAS)	1104	P5C5
1104	KC0-8CP11/1+12/1	1104	MA242	1104	MV-5	1104	P5D5
1104	KD2103	1104	MA242C	1104	MV1	1104	P6/2H
1104	KD2104	1104	MA350	1104	MV11	1104	P600
1104	KE-262	1104	MA351	1104	MV3(RECTIFIER)	1104	P62H
1104	KGE41007	1104	MA50	1104	MVA-05A	1104	P6A5
1104	KLH4567	1104	MB01	1104	MVA-05A(DIO)	1104	P6B5
1104	KLH4577	1104	MB244	1104	MVA-05A(RECT)	1104	P605
1104	KS-05	1104	MB257	1104	MY-1	1104	P6D5
1104	KS-05X	1104	MB258	1104	N-02	1104	P7776
1104	KS05	1104	MB269	1104	N-41	1104	P7A5
1104	KSKE40C200	1104	MB270	1104	N-EA16X30	1104	P7B5
1104	KSKE40C500	1104	MC010	1104	NA-22	1104	P705
1104	L3/2H	1104	MC015	1104	NA-25	1104	P7D5
1104	L32H	1104	MC020	1104	NA-33	1104	P8/2H
1104	L6/2H	1104	MC020A	1104	NA-36	1104	P82H
1104	L62H	1104	MC021	1104	NA-46	1104	P9459
1104	L8/2H	1104	MC021A	1104	NA13	1104	PA-069
1104	LA300	1104	MC022	1104	NA21	1104	PA-3
1104	LA600	1104	MC022A	1104	NA22	1104	PA-320
1104	LA800	1104	MC023	1104	NA25	1104	PA-320A
1104	LAA300	1104	MC023A	1104	NA32	1104	PA-320B
1104	LAA600	1104	MC025	1104	NA33	1104	PA069
1104	LAA800	1104	MC030	1104	NA35	1104	PA070
1104	LC.09M	1104	MC030A	1104	NA36	1104	PA071
1104	LL-2	1104	MC030B	1104	NA42	1104	PA10556
1104	LL2	1104	MC035	1104	NA45	1104	PA10887
1104	LM-1158	1104	MC040	1104	NA46	1104	PA200
1104	LM-1160	1104	MC040A	1104	NA62	1104	PA3
1104	LM-1862	1104	MC1520	1104	NA63	1104	PA300
1104	LM1160	1104	MC1521	1104	NA65	1104	PA305
1104	LM1862	1104	MC1522	1104	NA66	1104	PA305A
1104	LP1H	1104	MC1523	1104	NF500	1104	PA310
1104	LP2H	1104	MC1524	1104	NF501	1104	PA310A
1104	LP3H	1104	MC170	1104	NF506	1104	PA315
1104	LP4H	1104	MC19	1104	NF550	1104	PA315A
1104	LRR-100	1104	MC456	1104	NL-10	1104	PA320
1104	LRR-200	1104	MCV	1104	NL10	1104	PA320A
1104	LRR-300	1104	MD04	1104	NL15	1104	PA320B
1104	LRR-400	1104	MD134	1104	NL20	1104	PA325
1104	LRR-50	1104	MD135	1104	NL25	1104	PA325A
1104	LRR-500	1104	MD136	1104	NL30	1104	PA325B
1104	LRR100	1104	MD137	1104	NL40	1104	PA330
1104	LRR200	1104	MD138	1104	NL5	1104	PA330A
1104	LRR300	1104	MD139	1104	NL50	1104	PA330B
1104	LRR400	1104	MDA104	1104	NL60	1104	PA340
1104	LRR50	1104	MDA920-2	1104	NN50	1104	PA340A
1104	LRR500	1104	MH500	1104	NP50A	1104	PA340B
1104	M-0027	1104	MH67	1104	NP60A	1104	PA350
1104	M-204B	1104	MH670	1104	NPC0010	1104	PA350A
1104	M-31	1104	MH68	1104	NPC0050	1104	PA360
1104	M0027	1104	MH680	1104	NPC0100	1104	PA360A
1104	M12	1104	MH70	1104	NPC108	1104	PA400
1104	M124J779-1	1104	MH700	1104	NPC108A	1104	PA600
1104	M14	1104	MH71	1104	NTC-19	1104	PA7615
1104	M150	1104	MH710	1104	NU398B	1104	PA8645
1104	M172A	1104	MI-15R	1104	0101	1104	PA9160
1104	M H	1104	MI-15S	1104	0234	1104	PC0.2P11/2
1104	M22	1104	MJR1C	1104	04-8054-3	1104	PC02P1/2
1104	M2497	1104	MMO	1104	04-8054-4	1104	PC4004
1104	M3016	1104	MM2	1104	04-8054-7	1104	PD101
1104	M41223-2	1104	MM3	1104	0A127	1104	PD102
1104	M42	1104	MM4	1104		1104	PD103
1104	M4736	1104	MM5				

RS 276-	Industry Standard No.	RS 276-	Industry Standard No.	RS 276-	Industry Standard No.	RS 276-	Industry Standard No.
1104	PD104	1104	Q1B	1104	RFJ-33292	1104	RT6332
1104	PD105	1104	Q1H	1104	RFJ-60366	1104	RT6605
1104	PD106	1104	Q3/2	1104	RFJ30704	1104	RT6729
1104	PD107	1104	Q52	1104	RFJ31218	1104	RT6791
1104	PD107A	1104	Q4B	1104	RFJ31362	1104	RT7634
1104	PD108	1104	Q52	1104	RFJ31363	1104	RT7849
1104	PD110	1104	Q53	1104	RFJ333292	1104	RT7850
1104	PD111	1104	Q54	1104	RFJ60173	1104	RT8199
1104	PD122	1104	Q55	1104	RFJ60174	1104	RT8231
1104	PD125	1104	Q56	1104	RFJ60286	1104	RT8340
1104	PD129	1104	Q57	1104	RFJ60366	1104	RT8839
1104	PD130	1104	Q58	1104	RFJ60869	1104	RT8840
1104	PD131	1104	Q59	1104	RFJ6134	1104	RT8841
1104	PD132	1104	Q6/2	1104	RFJ70432	1104	RU1
1104	PD133	1104	Q60	1104	RFJ70487	1104	RU1A
1104	PD134	1104	Q62	1104	RFJ70703	1104	RV-2289
1104	PD135	1104	Q8/2	1104	RFJ70931	1104	RVO6
1104	PD154	1104	Q82	1104	RFJ70974	1104	RV06/7825B
1104	PD155	1104	QD-881885XT	1104	RFJ70977	1104	RV1189
1104	PD910	1104	QD-88R1KX4P	1104	RFJ711122	1104	RV1424
1104	PE-401	1104	QD-8V06CXXB	1104	RFJ71122	1104	RV1478
1104	PE401	1104	QRT-215	1104	RFJ72360	1104	RV2072
1104	PE401N	1104	R-106379	1104	RFJ72787	1104	RV2220
1104	PE402	1104	R-113321	1104	RFJ70432	1104	RV2250
1104	PE403	1104	R-113392	1104	RFL-30596	1104	RV2289
1104	PE404	1104	R-154B	1104	RFL30596	1104	RV2327
1104	PE405	1104	R-2-02	1104	RFM-33160	1104	RVDO.8C22/1A
1104	PE406	1104	R-7096	1104	RFP-33118	1104	RVD08C22/1A
1104	PE502	1104	R-8-1720	1104	RFP33118	1104	RVD10D1
1104	PE504	1104	R-81264	1104	RFS61436	1104	RVD10DC1
1104	PE506	1104	R-81347	1104	RFV60500	1104	RVD10DC1R
1104	PH-108	1104	R1	1104	RG100B	1104	RVD10E1
1104	PH1021	1104	R1035	1104	RG100D	1104	RVD10E11P
1104	PH204	1104	R106379	1104	RG100G	1104	RVD10E11LP
1104	PH208	1104	R10D1	1104	RG100J	1104	RVD18B54
1104	PH25C22	1104	R10DC	1104	RGP-10D	1104	RVD2DP
1104	PH25C22/1	1104	R113321	1104	RH-DX0003SEZZ	1104	RVD2DP22/18
1104	PH25C22/21	1104	R113392	1104	RH-DX0005TAZZ	1104	RVD2DP22/1B
1104	PH404	1104	R122C	1104	RH-DX0008CEZZ	1104	RVD2DP221P
1104	PH9D522/1	1104	R1329	1104	RH-DX0025CEZZ	1104	RVD2P22/1B
1104	PH9D822	1104	R1A	1104	RH-DX0026AEZZ	1104	RVD4B265J2
1104	PL-151-030-9-005	1104	R1B	1104	RH-DX0030CEZZ	1104	RVDDS-410
1104	PL-151-040-9-003	1104	R1K	1104	RH-DX0039TAZZ	1104	RVDGPO5A
1104	PL-151-045-9-001	1104	R259	1104	RH-DX0043TAZZ	1104	RVDKB16205
1104	PL-151-045-9-004	1104	R2252	1104	RH-DX0055TAZZ	1104	RVDKBI-1
1104	PN204	1104	R2442	1104	RH-DX0056CEZZ	1104	RVDBD-1U
1104	PS-025	1104	R2460-1	1104	RH-DX0059TAZZ	1104	RVDBD-1Y
1104	PS-035	1104	R2460-4	1104	RH-DX0063CEZZ	1104	RVDSR3AM2N
1104	PS-040	1104	R3/2H	1104	RH-DX0064CEZZ	1104	RVDVD1211L
1104	PS-060	1104	R3285	1104	RH-DX0066TAZZ	1104	RVTM000921-3
1104	PS-120	1104	R4A	1104	RH-DX0067TAZZ	1104	S-05
1104	PS005	1104	R5970	1104	RH-DX0068TAZZ	1104	S-05-005
1104	PS010	1104	R5971	1104	RH-DX0069TAZZ	1104	S-05-01
1104	PS015	1104	R6/2H	1104	RH-DX0072CEZZ	1104	S-05-06
1104	PS020	1104	R6048	1104	RH-DX0073TAZZ	1104	S-050
1104	PS025	1104	R6110	1104	RH-DX0079TAZZ	1104	S-0501
1104	PS030	1104	R6422	1104	RH-DX0081TAZZ	1104	S-1.5
1104	PS035	1104	R66-8504	1104	RH-DX0083CEZZ	1104	S-1.5-0
1104	PS040	1104	R7162	1104	RH-DX0092CEZZ	1104	S-1.5-06
1104	PS050	1104	R7248	1104	RH-DX1005APZZ	1104	S-10
1104	PS060	1104	R7271	1104	RH-IX0003SEZZ	1104	S-102V
1104	PS105	1104	R7682	1104	RH-EX0062CEZZ	1104	S-15
1104	PS110	1104	R7954	1104	RH1M	1104	S-15-10
1104	PS120	1104	R8024	1104	RH1Z	1104	S-17
1104	PS125	1104	R8470	1104	RHDX0043TAZZ	1104	S-17A
1104	PS130	1104	R8473	1104	RLF10	1104	S-2
1104	PS135	1104	R855-2	1104	RM-1V	1104	S-262
1104	PS140	1104	R8721	1104	RM-26	1104	S-3MX
1104	PS150	1104	R9470	1104	RM1A	1104	S-500B
1104	PS160	1104	R9597	1104	RM1Z	1104	S-500C
1104	PS2207	1104	R9A	1104	RM1ZM	1104	S-5277B
1104	PS2208	1104	RA-1	1104	RM12V	1104	S-58
1104	PS2209	1104	RA-1B	1104	RM2A	1104	S-58R
1104	PS2247	1104	RA132BA	1104	RMVTC00921-3	1104	S-798
1104	PS2249	1104	RA1B	1104	RQ-4098	1104	S-810
1104	PS2411	1104	RA1Y	1104	RRB24-06	1104	S-05
1104	PS2412	1104	RA1ZC	1104	RS-1264	1104	S05
1104	PS2413	1104	RA2	1104	RS-1347	1104	S0501
1104	PS2415	1104	RCC-7022	1104	RS-3570	1104	S1-1
1104	PS405	1104	RCC7225	1104	RS-3727	1104	S1-B01-02
1104	PS410	1104	RD-26235-1	1104	RS-6344	1104	S1-RBCT-102
1104	PS415	1104	RD-29799P	1104	RS-6461	1104	S1.5
1104	PS420	1104	RD-3	1104	RS-6471	1104	S1.5-01
1104	PS425	1104	RD-31903P	1104	RS10	1104	S10
1104	PS430	1104	RD-3472	1104	RS1234	1104	S101
1104	PS435	1104	RD250	1104	RS1264	1104	S10149
1104	PS440	1104	RD26235-1	1104	RS1542	1104	S102
1104	PS450	1104	RD29799P	1104	RS1720	1104	S103
1104	PS4559	1104	RD31903P	1104	RS1749	1104	S104
1104	PS4560	1104	RD3472	1104	RS1805	1104	S105
1104	PS460	1104	RD5A-1B4	1104	RS1823	1104	S106
1104	PS4725	1104	RD9037	1104	RS1832	1104	S107
1104	PS5300	1104	RE-49	1104	RS1PM-12	1104	S108
1104	PS5301	1104	RE-50	1104	RS220AF	1104	S10A
1104	PS5302	1104	RE2	1104	RS230AF	1104	S10B01-02
1104	PS605	1104	RE5	1104	RS3570	1104	S115
1104	PS610	1104	RE49	1104	RS3727	1104	S1243N
1104	PS611	1104	RE50	1104	RS6344	1104	S129
1104	PS615	1104	RE504	1104	RS6461	1104	S13
1104	PS616	1104	RE88	1104	RS6471	1104	S14
1104	PS617	1104	REJ70643	1104	RS6705	1104	0000000815
1104	PS621	1104	REJ70931	1104	RS8430	1104	S16
1104	PS622	1104	REN114	1104	RT-212	1104	S1600
1104	PS623	1104	REN116	1104	RT-213	1104	S16A
1104	PS627	1104	REN117	1104	RT-214	1104	S16B
1104	PS628	1104	RET20	1104	RT-215	1104	S17
1104	PS629	1104	RF-3160	1104	RT-2669	1104	S17A
1104	PS633	1104	RF-32101-8	1104	RT-3858	1104	S18
1104	PS637	1104	RF-32101R	1104	RT-4232	1104	S1B01-02
1104	PT-3	1104	RF-6235-1	1104	RT1595	1104	S18A
1104	PT-505	1104	RF26231-1	1104	RT1689	1104	S18B
1104	PT-510	1104	RF26234-1	1104	RT1840	1104	S19
1104	PT-515	1104	RF26235-1	1104	RT213	1104	S191G
1104	PT-525	1104	RF26235-2	1104	RT214	1104	S19A
1104	PT-530	1104	RF26235-5	1104	RT215	1104	S1A
1104	PT-540	1104	RF29799P	1104	RT3443	1104	S1A06
1104	PT-550	1104	RF3160	1104	RT3585	1104	S1A060
1104	PT-560	1104	RF31903P	1104	RT3858	1104	S1A60
1104	PT-5B	1104	RF32101-8	1104	RT3981	1104	S1AR1
1104	PT5	1104	RF32101-9	1104	RT4050	1104	S1AR2
1104	PT510	1104	RF32101R	1104	RT4069	1104	S1B
1104	PT550	1104	RF32412-3	1104	RT4232	1104	S1B-01-02
1104	PT560	1104	RF32645	1104	RT4764	1104	0000S1B01
1104	PT5B	1104	RF33976	1104	RT5070	1104	S1B01-01
1104	PT72130	1104	RF34383	1104	RT5216	1104	S1B01-02
1104	PTC202	1104	RF3472	1104	RT5217	1104	S1B01-0226
1104	PTC403	1104	RF34720	1104	RT5395	1104	S1B01-06
1104	PU6022	1104	RFA70597	1104	RT5472	1104	S1B0101CR
1104	PV-8	1104	RFA70600	1104	RT5911	1104	S1B0102
1104	PV8	1104	RFC61197	1104	RT6322	1104	S1B02
1104	PY-5	1104	RFJ-30704			1104	S1B02-0
1104	Q-20115C	1104	RFJ-31218			1104	S1B02-06CE
1104	Q-26115C	1104	RFJ-31362			1104	S1B02-06CRE
		1104	RFJ-31363			1104	S1B02-C

RS 276-	Industry Standard No.	RS 276-	Industry Standard No.	RS 276-	Industry Standard No.	RS 276-	Industry Standard No.
1104	S1B02-CR	1104	SA3B	1104	SD80	1104	SIBOL
1104	S1B0201CR	1104	SAW-1S1941	1104	SD938	1104	SIBOL-02
1104	S1B1	1104	SAW-1S1944	1104	SD94AB	1104	SIB1
1104	S1BD1-02	1104	SB-03	1104	SD94B	1104	SID01E
1104	0000S1B01	1104	SB-12	1104	SD94S	1104	SID01L
1104	S1CN1	1104	SB-3	1104	SD95	1104	SID02E
1104	S1D51C052-19	1104	SB-3-02	1104	SD950	1104	SID02L
1104	S1D51C169-1	1104	SB-309A	1104	SD95A	1104	SID51C169
1104	S1060Z	1104	SB-309C	1104	SD96	1104	SIG1/400
1104	S1L200	1104	SB-3P01	1104	SD96A	1104	SIG1/600
1104	S1R-80	1104	SB-3N	1104	SD96S	1104	SIL-200
1104	S1RC20	1104	SB01	1104	SDR-25	1104	SIL200
1104	S1RO20R	1104	SB1-01-04	1104	SDR25	1104	SIR-80
1104	S1SD-1	1104	SB1000	1104	SDS-113	1104	SIR-RECT-44
1104	S1SD-1HP	1104	SB302	1104	SD5113	1104	SIR20
1104	S1SD-1X	1104	SB315	1104	SE-0.5B	1104	SIR60
1104	S1SM-150-01	1104	SB332	1104	SE-05	1104	SIRECT-102
1104	S1SM-150-02	1104	SB333	1104	SE-05-01	1104	SIRECT-2
1104	S1SW-05-02	1104	SB393	1104	SE-05-02	1104	SIRECT-36
1104	S200	1104	SBR-260	1104	SE-05-2	1104	SIRECT-48
1104	S201	1104	SC-110	1104	SE-05-B	1104	SIRECT-59
1104	S202	1104	SC-16	1104	SE-05A	1104	SIRECT-92
1104	S203	1104	SC-4	1104	SE-05X	1104	SISD-1X
1104	S204	1104	SC05	1104	SE-2	1104	SISD-K
1104	S205	1104	SC05E	1104	SE-5	1104	SISM-150-01
1104	S206	1104	SC1	1104	SE05	1104	SISW-05-02
1104	S20ND400	1104	SC110	1104	SE05B	1104	SISW-0502
1104	S20NH400	1104	SC1414	1104	SE05D	1104	SJ-570
1104	S21	1104	SC1431	1104	SE08B	1104	SJ051E
1104	S217	1104	SC16	1104	SE08BB	1104	SJ051F
1104	S218	1104	SC1631	1104	SE08-01	1104	SJ052E
1104	S219	1104	SC1631(GE)	1104	SE30B26A	1104	SJ052F
1104	S22	1104	SC2	1104	SE46	1104	SJ101F
1104	0S22.3905-001	1104	SC23-3	1104	SE6	1104	SJ102F
1104	S220	1104	SC23-9	1104	SELEN-70	1104	SJ201F
1104	S221	1104	SC2C	1104	SELEN-701	1104	SJ202F
1104	S222	1104	SC305	1104	SPB6183	1104	SJ301F
1104	S223	1104	SC4	1104	SFR35	1104	SJ302F
1104	S224	1104	SC4116	1104	SFR152	1104	SJ401F
1104	S22A	1104	SC6	1104	SFR153	1104	SJ402F
1104	S23	1104	SCA05	1104	SFR154	1104	SJ501F
1104	S230	1104	SCA1	1104	SFR155	1104	SJ601F
1104	S232	1104	SCA1103	1104	SFR156	1104	SJ60F
1104	S233	1104	SCA2	1104	SFR164	1104	SK-1B
1104	S234	1104	SCA3	1104	SFR251	1104	SK1FM
1104	S235	1104	SCA4	1104	SFR252	1104	SK1K-2
1104	S238	1104	SCA5	1104	SFR253	1104	SK3016
1104	S239	1104	SCA6	1104	SFR254	1104	SK3017
1104	S23A	1104	SCB1	1104	SFR255	1104	SK3017A
1104	S240	1104	SCB2	1104	SFR256	1104	SK3017B
1104	S241	1104	SCB3	1104	SFR264	1104	SK3017B/117
1104	S243	1104	SCE4	1104	SFR266	1104	SK3030
1104	S250	1104	SCE6	1104	SG-005	1104	SK3031
1104	S251	1104	SCO5	1104	SG-105	1104	SK3174
1104	S252	1104	SCO5E	1104	SG-1198	1104	SK3313
1104	S253	1104	SCP5E	1104	SG-205	1104	SK3313/116
1104	S254	1104	SD-02	1104	SG-305	1104	SL-030
1104	S255	1104	SD-1-211B	1104	SG-805	1104	SL-030T
1104	S256	1104	SD-1-211C	1104	SG105	1104	SL-2
1104	S26	1104	SD-101	1104	SG1198	1104	SL-3
1104	S262	1104	SD-13	1104	SG323	1104	SL-4
1104	S2A06	1104	SD-16A	1104	SG3400	1104	SL-833
1104	S2A10	1104	SD-16D	1104	SG505	1104	SL-853A
1104	S2AR1	1104	SD-18	1104	SG805	1104	SLO30/3490
1104	S2AR2	1104	SD-1A	1104	SGR100	1104	SLO505,T
1104	S2030	1104	SD-1AHP	1104	SH-1	1104	SLO308
1104	S2040	1104	SD-1B	1104	SH-1A	1104	SLO30T
1104	S2040A	1104	SD-1C-4P	1104	SH-1B	1104	SL2
1104	S2820	1104	SD-1C-UP	1104	SH-1DE	1104	SL5
1104	S2860	1104	SD-1CUP	1104	SH-1S	1104	SL65
1104	S2860-1	1104	SD-1HP	1104	SH1	1104	SL833
1104	S2960	1104	SD-1L	1104	SH15	1104	SL833A
1104	S2VB	1104	SD-1LA	1104	SH1A	1104	SL91
1104	S30	1104	SD-1U	1104	SH4D05	1104	SL92
1104	S31	1104	SD-1UP	1104	SH4D1	1104	SL93
1104	S32	1104	SD-1X	1104	SH4D2	1104	SLA-445
1104	S33	1104	SD-1Y	1104	SH4D3	1104	SLA1095
1104	S34	1104	SD-2	1104	SH4D4	1104	SLA1096
1104	S35	1104	SD-201	1104	SH4D6	1104	SLA1100
1104	S36	1104	SD-470	1104	SHAD-1	1104	SLA1101
1104	S3A06	1104	SD-80	1104	SHAD1	1104	SLA1102
1104	S3AR1	1104	SD-91	1104	SI-REC-73	1104	SLA1104
1104	S3MX	1104	SD-91A	1104	SI-RECT-044	1104	SLA1105
1104	S40	1104	SD-91B	1104	SI-RECT-100	1104	SLA11AB
1104	S4001	1104	SD-92	1104	SI-RECT-100-102	1104	SLA11C
1104	S40A	1104	SD-92A	1104	SI-RECT-102	1104	SLA12AB
1104	S42B	1104	SD-92B	1104	SI-RECT-110/SB-3F	1104	SLA12C
1104	S43	1104	SD-93	1104	SI-RECT-112	1104	SLA13AB
1104	S431	1104	SD-93A	1104	SI-RECT-122	1104	SLA13C
1104	S44	1104	SD-94	1104	SI-RECT-126	1104	SLA1487
1104	S46	1104	SD-94A	1104	SI-RECT-144	1104	SLA1488
1104	S47	1104	SD-94AB	1104	SI-RECT-154	1104	SLA1489
1104	S48	1104	SD-95	1104	SI-RECT-155	1104	SLA1490
1104	S49	1104	SD-95A	1104	SI-RECT-156	1104	SLA1491
1104	S4A06	1104	SD-Y	1104	SI-RECT-2	1104	SLA1492
1104	S4AR1	1104	SD040	1104	SI-RECT-20	1104	SLA14AB
1104	S4AR2	1104	SD05	1104	SI-RECT-218	1104	SLA14C
1104	S4AR30	1104	SD07	1104	SI-RECT-220	1104	SLA15AB
1104	S4C	1104	SD1	1104	SI-RECT-222	1104	SLA15C
1104	S4PN300	1104	SD1-1	1104	SI-RECT-226	1104	SLA1692
1104	S5089-A	1104	SD1-211B	1104	SI-RECT-27	1104	SLA1693
1104	S5AR1	1104	SD1-211C	1104	SI-RECT-33	1104	SLA1694
1104	S5AR2	1104	SD1-Z	1104	SI-RECT-34	1104	SLA1695
1104	S5B	1104	SD10	1104	SI-RECT-37	1104	SLA1696
1104	S5BR	1104	SD102	1104	SI-RECT-39	1104	SLA1697
1104	S6005	1104	SD103	1104	SI-RECT-48	1104	SLA16AB
1104	S6AR1	1104	SD104	1104	SI-RECT-49	1104	SLA16C
1104	S6AR2	1104	SD1101	1104	SI-RECT-53	1104	SLA17AB
1104	S72	1104	SD1102	1104	SI-RECT-59	1104	SLA17C
1104	S73	1104	SD1103	1104	SI-RECT-69	1104	SLA2610
1104	S75	1104	SD1104	1104	SI-RECT-73	1104	SLA2611
1104	S77	1104	SD13	1104	SI-RECT-74	1104	SLA2612
1104	S79	1104	SD18	1104	SI-RECT-75	1104	SLA2613
1104	S81	1104	0000SD1AB	1104	SI-RECT-77	1104	SLA2614
1104	S82	1104	SD1CUF	1104	SI-RECT-84	1104	SLA2615
1104	S83	1104	SD1DM-4	1104	SI-RECT-92	1104	SLA3193
1104	S84	1104	SD1HP	1104	SI-RECT-94	1104	SLA3194
1104	S85	1104	SD1L	1104	SI-RECT102	1104	SLA3195
1104	S86	1104	SD1LA	1104	SI100E	1104	SLA440
1104	S91	1104	SD1ZHP	1104	SI50E	1104	SLA440B
1104	S91-A	1104	SD201	1104	SI91G	1104	SLA441
1104	S91-H	1104	SD202	1104	SIB-01-02	1104	SLA441B
1104	S92	1104	SD23	1104	SIB-01-022	1104	SLA442
1104	S92-A	1104	SD2A	1104	SIB01-01	1104	SLA442B
1104	S92-H	1104	SD2B	1104	SIB01-02	1104	SLA443
1104	S93	1104	SD4	1104	SIB01-06	1104	SLA443B
1104	S93A	1104	SD45	1104	SIB01-06B	1104	SLA444
1104	S93H	1104	SD470	1104	SIB0102	1104	SLA444A
1104	S94	1104	SD5	1104	SIB02-03C	1104	SLA444B
1104	S95	1104	SD5(DUAL)	1104	SIB02-03CR	1104	SLA445
1104	SA-2B	1104	SD500C	1104	SIB02-CR	1104	SLA445B
1104	SA2	1104	SD501	1104	SIB0201CH	1104	SLA536
1104	SA2B	1104	SD6			1104	SLA537
		1104	SD600C			1104	SLA538

RS 276-	Industry Standard No.	RS 276-	Industry Standard No.	RS 276-	Industry Standard No.	RS 276-	Industry Standard No.
1104	SLA539	1104	SR101-1	1104	SR6324	1104	T-E1144
1104	SLA540	1104	SR101-2	1104	SR6325	1104	T-E1148
1104	SLA547	1104	SR1024	1104	SR6385	1104	T-E1155
1104	SLA599	1104	SR105	1104	SR6415	1104	T-E1157
1104	SLA599A	1104	SR1104	1104	SR6560	1104	T-E1171
1104	SLA600	1104	SR112	1104	SR6567	1104	T-E1176
1104	SLA600A	1104	SR114	1104	SR6617	1104	T-H18557
1104	SLA601	1104	SR120	1104	SR6723	1104	T-HS6105
1104	SLA601A	1104	SR266	1104	SR6724	1104	T-HSG105
1104	SLA602	1104	SR131-1	1104	SR76	1104	T0150
1104	SLA602A	1104	SR135-1	1104	SR806-126	1104	T065
1104	SLA603	1104	SR1378-1	1104	SR846-2	1104	T075
1104	SLA603A	1104	SR1378-3	1104	SR846-3	1104	T10144
1104	SLA604	1104	SR13H	1104	SR851	1104	T10175
1104	SLA604A	1104	SR14	1104	SR851-121	1104	T10185
1104	SLA605	1104	SR144	1104	SR889	1104	T10453
1104	SLA605A	1104	SR145	1104	SR9005	1104	T1085
1104	SLA606	1104	SR1493	1104	SRIDM-1	1104	T1450
1104	SLA606A	1104	SR15	1104	SRIDM-4	1104	T156
1104	SM-1	1104	SR151	1104	SRIEM-1	1104	T159
1104	SM-1-005	1104	SR152	1104	SRIEM-2	1104	T1J6G
1104	SM-1-47	1104	SR1549	1104	SRIEM-4	1104	T18DS1K
1104	SM-10	1104	SR16	1104	SRIK-2	1104	T200
1104	SM-150-06	1104	SR166B	1104	SRIK-8	1104	T21312
1104	SM-150-B	1104	SR1692	1104	SRK-2	1104	T21333
1104	SM-150B	1104	SR1693	1104	SRK1	1104	T21507
1104	SM-1K	1104	SR1694	1104	SRLPM-1	1104	T21602
1104	SM1-02	1104	SR1695	1104	SS-10	1104	T21638
1104	SM10	1104	SR17	1104	SS00010	1104	T21649
1104	SM105	1104	SR1731-1	1104	SS00010	1104	T21679
1104	SM11	1104	SR1731-2	1104	SS0009	1104	T3/2
1104	SM110	1104	SR1731-3	1104	SS321	1104	T300
1104	SM120	1104	SR1731-4	1104	SS322	1104	T30155
1104	SM130	1104	SR1731-5	1104	SS324	1104	T30155-001
1104	SM140	1104	SR1742	1104	SS334	1104	T30155-1
1104	SM160	1104	SR1766	1104	SS337	1104	T400
1104	SM20	1104	SR18	1104	SS455	1104	T42692-001
1104	SM205	1104	SR1984	1104	SSD974	1104	T42692-1R
1104	SM210	1104	SR1A-1	1104	ST-12	1104	T450
1104	SM220	1104	SR1A-12	1104	ST-14	1104	T500
1104	SM230	1104	SR1A-2	1104	ST-2040P	1104	T550
1104	SM240	1104	SR1A-4	1104	ST16	1104	T600
1104	SM250	1104	SR1A-8	1104	ST2040P	1104	T650
1104	SM260	1104	SR1A1	1104	STBOL-02	1104	T8/2
1104	SM30	1104	SR1A12	1104	STPCN10	1104	TA100
1104	SM31	1104	SR1A2	1104	STV-3	1104	TA1062
1104	SM4	1104	SR1A4	1104	SV-01A	1104	TA1063
1104	SM40	1104	SR1A8	1104	SV-01B	1104	TA1064
1104	SM483	1104	SR1C-12	1104	SV-02(RECTIFIER)	1104	TA200
1104	SM486	1104	SR1D1M	1104	SV-03	1104	TA300
1104	SM487	1104	SR1DM	1104	SV-05	1104	TA400
1104	SM488	1104	SR1DM-1	1104	SV-1238B	1104	TA50
1104	SM5	1104	SR1DM-10	1104	SV-1238E	1104	TA500
1104	SM50	1104	SR1DM-4	1104	SV-6R	1104	TA600
1104	SM505	1104	SR1DM1	1104	SV01A	1104	TA7802
1104	SM51	1104	SR1DMX	1104	SV02A	1104	TA7803
1104	SM510	1104	SR1E	1104	SV04	1104	TA7804
1104	SM512	1104	SR1EM	1104	SV05	1104	TA7996
1104	SM513	1104	SR1EM-1	1104	SV1238	1104	TC-0.09M21/5
1104	SM514	1104	SR1EM-10	1104	SV12388	1104	TC-136
1104	SM515	1104	SR1EM-12	1104	SV12388E	1104	TC0.09M21/3
1104	SM516	1104	SR1EM-2	1104	SV1238E	1104	TC0.2P11/2
1104	SM60	1104	SR1EM-X	1104	SVD10D-1	1104	TC02P112
1104	SM645	1104	SR1EM1	1104	SVD181850	1104	TC02P12
1104	SM646	1104	SR1EM2	1104	SVDMA26	1104	TC02P11/2
1104	SM705	1104	SR1PM	1104	SVDVD1223	1104	TC136
1104	SM710	1104	SR1PM-1	1104	SW-05-005	1104	TC3112319300
1104	SM720	1104	SR1PM-8	1104	SW-05-02	1104	TB-1011
1104	SM730	1104	SR1PM10	1104	SW-05-B	1104	TB-1050
1104	SM740	1104	SR1PM12	1104	SW-05V	1104	TB1010
1104	SM750	1104	SR1PM4	1104	SW-1-06	1104	TB1011
1104	SM760	1104	SR1PMA	1104	SW-1A	1104	TE1024
1104	SN-1	1104	SR1HM-12	1104	SWO.5A	1104	TE1024C
1104	SN-12	1104	SR1HM-16	1104	SW05	1104	TE1024D
1104	SN0303	1104	SR1HM-2	1104	SW05-01	1104	TE1029
1104	SN1	1104	SR1HM-4	1104	SW05-02	1104	TE1042
1104	SNO303	1104	SR1K	1104	SW0501	1104	TE1050
1104	S05	1104	SR1K-1	1104	SW05A	1104	TE1078
1104	S0501	1104	SR1K-12	1104	SW05B	1104	TE1080
1104	S0D200D	1104	SR1K-1K	1104	SW05B	1104	TE1088
1104	SP-1	1104	SR1K-2	1104	SW05BB	1104	TE1089
1104	SP1	1104	SR1K-2/494	1104	SW05V	1104	TE1090
1104	SP1K-1	1104	SR1K-4	1104	SW0.5A	1104	TE1097
1104	SP1K-2	1104	SR1K-8	1104	SX-642	1104	TE1108
1104	SP1U-2	1104	SR1K-Z	1104	SX623	1104	TP20
1104	SPN-01	1104	SR1K/494	1104	SX633	1104	TP21
1104	SPN01	1104	SR1K08	1104	SX641	1104	TP22
1104	SR-0005	1104	SR1K1	1104	SX642	1104	TP23
1104	SR-0007	1104	SR1K2	1104	SX643	1104	TPR-120
1104	SR-0008	1104	SR1KB	1104	SX644	1104	TPR120
1104	SR-05K-2	1104	SR1T	1104	SX645	1104	TQ-11
1104	SR-1	1104	SR1Z	1104	T-0150	1104	TQ-21
1104	SR-101-1	1104	SR200	1104	T-065	1104	TQ-22
1104	SR-101-2	1104	SR200B	1104	T-075	1104	TQ-31
1104	SR-112	1104	SR205	1104	T-100	1104	TQ-32
1104	SR-120-1	1104	SR2121	1104	T-13	1104	TQ-41
1104	SR-130-1	1104	SR22	1104	T-130	1104	TQ-42
1104	SR-131-1	1104	SR23	1104	T-14	1104	TQ-51
1104	SR-132-1	1104	SR2301	1104	T-140	1104	TQ-52
1104	SR-136	1104	SR2301A	1104	T-200	1104	TQ-61
1104	SR-13H	1104	SR24	1104	T-220	1104	TQ-62
1104	SR-14	1104	SR27	1104	T-260	1104	TQ12
1104	SR-150-01	1104	SR28	1104	T-300	1104	TQ20A
1104	SR-150-1	1104	SR2A-1	1104	T-400	1104	TQ21
1104	SR-166B	1104	SR2A-2	1104	T-450	1104	TQ22
1104	SR-17	1104	SR2A-4	1104	T-4590	1104	TQ31
1104	SR-18	1104	SR2A1	1104	T-50	1104	TQ32
1104	SR-1849-1	1104	SR2A12	1104	T-500	1104	TQ41
1104	SR-1HM-2	1104	SR2A2	1104	T-550	1104	TQ42
1104	SR-1K	1104	SR2A8	1104	T-600	1104	TQ51
1104	SR-1K-2	1104	SR3	1104	T-650	1104	TQ52
1104	SR-1K2	1104	SR30	1104	T-E01029D	1104	TQ61
1104	SR-1Z	1104	SR3010	1104	T-E1011	1104	TQ62
1104	SR-22	1104	SR35	1104	T-E1024	1104	TH18557
1104	SR-23	1104	SR3582	1104	T-E1024C	1104	TH400
1104	SR-24	1104	SR390	1104	T-E1024D	1104	TH50
1104	SR-27	1104	SR390-2	1104	T-E1029	1104	TH600
1104	SR-28	1104	SR3943	1104	T-E1050	1104	TH801
1104	SR-3	1104	SR3AM-8	1104	T-E1064	1104	TH802
1104	SR-30	1104	SR3BM-6	1104	T-E1078	1104	TH803
1104	SR-390	1104	SR4	1104	T-E1078A	1104	TH804
1104	SR-4	1104	SR40	1104	T-E1080	1104	TH805
1104	SR-401	1104	SR401	1104	T-E1088	1104	TH806
1104	SR-5	1104	SR405	1104	T-E1089	1104	THBG105
1104	SR-76	1104	SR5	1104	T-E1090	1104	TI-53
1104	SR-846-2	1104	SR50	1104	T-E1097	1104	TI-55
1104	SR-889	1104	SR500	1104	T-E1102	1104	TI-71
1104	SR-9005	1104	SR500B	1104	T-E1102A	1104	TI152
1104	SR-1K-2	1104	SR50411-1	1104	T-E1108	1104	TI52
1104	SR05K-2	1104	SR50517	1104	T-E1124	1104	TI53
1104	SR1-2	1104	SR60	1104	T-E1133	1104	TI54
1104	SR1-K2	1104	SR605	1104	T-E1138	1104	TI55
1104	SR100	1104	SR6134			1104	TI56

RS 276-	Industry Standard No.	RS 276-	Industry Standard No.	RS 276-	Industry Standard No.	RS 276-	Industry Standard No.
1104	T157	1104	TVDS1M	1104	U13033801	1104	VO-5E
1104	T158	1104	TVH-526	1104	U13102	1104	VO-6A
1104	T159	1104	TVM-511	1104	U212	1104	VO-6C
1104	T160	1104	TVM-EH2C	1104	U212-25	1104	VO3-C
1104	T171	1104	TVM-EH2C11	1104	U213	1104	VO3G
1104	TIR01	1104	TVM-EH2C11/1	1104	U214	1104	VO6-B
1104	TIR02	1104	TVM-EH2C11/1+12/1	1104	U2400-03	1104	VO6B
1104	TIR03	1104	TVM-EH2C12/1	1104	U633	1104	VO6C
1104	TIR04	1104	TVM-M204B	1104	UF-1	1104	VO9E
1104	TIR05	1104	TVM-PH9D22/1	1104	UF2	1104	VOG
1104	TIR06	1104	TVM-PH9D522/11	1104	UPSD-1A	1104	VS-1
1104	TJ-5A	1104	TVM-TC0.2P11/2	1104	UPSD1P	1104	VS-102
1104	TJ10A	1104	TVM35	1104	U05E	1104	VS-DG1N
1104	TJ15A	1104	TVM511	1104	U06C	1104	VS-DG1NR
1104	TJ20A	1104	TVM550	1104	UR105	1104	VS-FR-1
1104	TJ25A	1104	TVM56	1104	UR110	1104	VS-FR-1P
1104	TJ30A	1104	TVM563	1104	UR115	1104	VS-FR-1U
1104	TJ35A	1104	TVMB8151B	1104	UR120	1104	VS-FR1
1104	TJ40A	1104	TVML00.09M1115	1104	UR125	1104	VS-FR1P
1104	TJ5A	1104	TVML00.09M1115	1104	USFD-1	1104	VS-PT-1N
1104	TJ60A	1104	TVMM204B	1104	USFD-1A	1104	VS-PC02P11/2
1104	TK-10	1104	TVMPH9D22/1	1104	UT-112	1104	VS-PH9D522/1
1104	TK-30	1104	TVMTC000921-3	1104	UT-16	1104	VS-SD-12
1104	TK-40	1104	TVS-10DB	1104	UT-254	1104	VS-TC0-2P11/2
1104	TK-41	1104	TVS-185D	1104	UT-258	1104	VS-TC0.2P11/2
1104	TK10	1104	TVS-181850	1104	UT116	1104	VS-TC02P11/2
1104	TK11	1104	TVS-181906	1104	UT14	1104	VS1
1104	TK20	1104	TVS-DG-1N-R	1104	UT15	1104	VS120
1104	TK21	1104	TVS-DS-1K	1104	UT16	1104	VS202
1104	TK30	1104	TVS-DS-1M	1104	UT17	1104	VS9-0001-911
1104	TK400	1104	TVS-DS81K	1104	UT18	1104	VS9-0002-911
1104	TK5	1104	TVS-DS81M	1104	UT21	1104	VS9-0005-911
1104	TK50	1104	TVS-DS82K	1104	UT22	1104	VS9-0006-911
1104	TK60	1104	TVS-ET1P	1104	UT221	1104	VS9-0007-911
1104	TK600	1104	TVS-FR-1P	1104	UT222	1104	VSFR1
1104	TK61	1104	TVS-FR-1P(FR1P)	1104	UT223	1104	VSFR1P
1104	TKP10	1104	TVS-FR-2M	1104	UT224	1104	VSPT1N
1104	TKP20	1104	TVS-FR-2PC	1104	UT225	1104	VS0-20024
1104	TKP40	1104	TVS-FR1-PC	1104	UT226	1104	VS0A70
1104	TKP5	1104	TVS-FR1MD	1104	UT227	1104	VSSD1B
1104	TKP60	1104	TVS-FR1PC	1104	UT228	1104	VSSD1Z
1104	TL11	1104	TVS-FR2M	1104	UT229	1104	VSTC02P11/2
1104	TL12	1104	TVS-PT-1P	1104	UT23	1104	W-06A
1104	TL2	1104	TVS-FU1N	1104	UT231	1104	W/6A
1104	TL21	1104	TVS-HF-SD-12	1104	UT232	1104	WO-6A
1104	TL22	1104	TVS-HFSD1Z	1104	UT233	1104	WOO5M
1104	TL31	1104	TVS-HS7/1	1104	UT258	1104	WO3B
1104	TL32	1104	TVS-KC2-LP	1104	UT24	1104	WO6
1104	TL41	1104	TVS-KC20P12/1	1104	UT25	1104	WO6A
1104	TL42	1104	TVS-KC20P12/1	1104	UT26	1104	WO6B
1104	TL51	1104	TVS-KC2DP12/2	1104	UT27	1104	WO6C
1104	TL61	1104	TVS-OV-02	1104	UU/J	1104	W4002
1104	TM-33	1104	TVS-PC02P11/2	1104	V-06B	1104	WC-14020
1104	TM-43	1104	TVS-PCD2P11/2	1104	V-06C	1104	WC-14027
1104	TM114	1104	TVS-S1B02-03C	1104	V-270D1	1104	WC120
1104	TM116	1104	TVS-S1B02-03CR	1104	V-442	1104	WC14020
1104	TM117	1104	TVS-SD-1	1104	VO-3C	1104	WC14027
1104	TM33	1104	TVS-SD1A	1104	VO-6	1104	WC19865
1104	TM43	1104	TVS-SPN01	1104	VO-6B	1104	WD001
1104	TM62	1104	TVS-TC0.09M11/10	1104	VO-6C	1104	WD002
1104	TM63	1104	TVS-TC009M11/10	1104	VO-6C-401	1104	WD003
1104	TM65	1104	TVS-UFSD-1	1104	VO3-E	1104	WD004
1104	TM66	1104	TVS010D1	1104	VO3C	1104	WD005
1104	TMD41	1104	TVS0A71	1104	VO3G	1104	WD006
1104	TMD42	1104	TVS10D	1104	VO5E	1104	WD007
1104	TMD45	1104	TVS10DB	1104	VO6	1104	WD008
1104	TMJ61672(RECT)	1104	TVS1850	1104	VO6(DIO)	1104	WD009
1104	TP101	1104	TVS1N4002	1104	VO6(RECT)	1104	WD010
1104	TP201	1104	TVS1P20	1104	VO6-C	1104	WD011
1104	TP302	1104	TVS181850	1104	VO6-E	1104	WD012
1104	TP402	1104	TVS181906	1104	VO6A	1104	WD014
1104	TP4067-409	1104	TVS181922G	1104	VO6B	1104	WD015
1104	TR-02E	1104	TVS181950	1104	VO6BX4	1104	WEP1082
1104	TR-2880	1104	TVS550	1104	VO6E	1104	WEP1083
1104	TR-77	1104	TVSBBE	1104	VO6G	1104	WEP156
1104	TRN4002	1104	TVSD1K	1104	VO7G	1104	WEP15B
1104	TR2327031	1104	TVSD52K	1104	VO8G	1104	WEP165A
1104	TR2327041	1104	TVSD91NR	1104	VO9-E	1104	WO-6A
1104	TR2880	1104	TVSDS-1M	1104	VO9E	1104	WO6A
1104	TR2A	1104	TVSDS81K	1104	VO9G	1104	WO6B
1104	TR320020	1104	TVSDS82M	1104	VOG	1104	WR-013
1104	TR330022	1104	TVSERBB-06B	1104	V10158	1104	WR-200
1104	TR330027	1104	TVSERB24-04D	1104	V11189-1	1104	WR006
1104	TS-2A	1104	TVSESA06	1104	V11J	1104	WR013
1104	T805	1104	TVSFR1P	1104	V11L	1104	WR100
1104	T81	1104	TVSFR1PC	1104	V148	1104	WR200
1104	T82	1104	TVSFR2-06	1104	V15920	1104	WR300
1104	T82A	1104	TVSFR2PC	1104	V15C200/80-V&	1104	WR400
1104	T85	1104	TVSPT1P	1104	V15C200/80-VP	1104	WRE-981
1104	T86	1104	TVSFU1N	1104	V171	1104	WR8981
1104	TSB-1000	1104	TVSHFSD-1A	1104	V17L	1104	WT-16X9
1104	TSB-245	1104	TVSHFSD12C	1104	V210C	1104	X5M6
1104	TSB245	1104	TVSIS1850	1104	V270-D1	1104	XA121
1104	TSC159	1104	TVSJL41A	1104	V3074A20	1104	XS-10
1104	TSV-1000	1104	TVSJL41AM	1104	V3074A21	1104	XS-31
1104	T66X26	1104	TVSM1-02	1104	V442	1104	XS10
1104	TUS-185D	1104	TVS0A71	1104	V66	1104	X816
1104	TV-24104	1104	TVSPCD2P11/2	1104	V6C	1104	X816A
1104	TV-24125	1104	TVSRA-12	1104	V78	1104	X817
1104	TV-24191	1104	TVSRMP5020	1104	V9446-4	1104	X817A
1104	TV-24232	1104	TVSS1P20	1104	VAMV-1	1104	X818
1104	TV-24266	1104	TVSS1R80	1104	VARIST-5	1104	X822(RECT.)
1104	TV-2496	1104	TVSS1R8D	1104	VB-11	1104	X823
1104	TY24104	1104	TVSS3-2	1104	VB-400	1104	X823A
1104	TY241073	1104	TVSS34RECT	1104	VB-600	1104	X831
1104	TY241074	1104	TVSS4C	1104	VB100	1104	X340A
1104	TY24125	1104	TVSSA-2	1104	VB300	1104	XU604
1104	TY24136	1104	TVSSA-2H	1104	VB500	1104	Y0-6A
1104	TY24155	1104	TVSSA2B	1104	VB600	1104	Y100
1104	TY24191	1104	TVSSB-2T	1104	VB600A	1104	YAAD004
1104	TY24193	1104	TVSSD1A	1104	VBH600	1104	YAAD007
1104	TY24200	1104	TVSSV02	1104	VC6E	1104	YAAD019
1104	TY24222	1104	TVSUP2	1104	VD-1122	1104	YAAD020
1104	TY24224	1104	TVSUFSD1P	1104	VD-121C	1104	YAAD022
1104	TY24232	1104	TVSWP2	1104	VD-1222	1104	YBAD009
1104	TY24234	1104	TW10	1104	VD10E11F	1104	YEAD030
1104	TY24266	1104	TW20	1104	VD1120	1104	YO44
1104	TY24278	1104	TW3	1104	VD1122	1104	YR-011
1104	TY24282	1104	TW30	1104	VD1213	1104	YR011
1104	TY24283	1104	TW40	1104	VD1E1////-1	1104	YSG-20024
1104	TY24292	1104	TW5	1104	VD6	1104	YSG-V47-1-3
1104	TY24298	1104	TW50	1104	VPA-2745C	1104	YSG-V47-7-51-1
1104	TY24582	1104	TW60	1104	VPA2745C	1104	YSG-V47-7-51-2
1104	TY24586	1104	TWV	1104	VHD181834	1104	Z330611
1104	TY24803	1104	TX1N3190	1104	VHD181885-1	1104	ZCOM-5683-U
1104	TY24941	1104	TX1N3191	1104	VHD181885//-1	1104	ZCOM3679
1104	TY2496	1104	TX1N645	1104	VHD181887//-1	1104	ZJ252B
1104	TY24979	1104	TX1N647	1104	VHD181887//-1	1104	ZR-1025
1104	TY34232	1104	U	1104	VM-PH11D522/1	1104	ZR-1031
1104	TV4	1104	U-2400-03	1104	VM-PH90522/1	1104	ZR-1035
1104	TV6.5	1104	U-422	1104	VM-TC0.2P11/2	1104	ZR-1076
1104	TVC-3	1104	U-633	1104	VM-TC02P11/2	1104	ZR-10B
1104		1104		1104	VMPH11D522-1	1104	ZR-500

RS 276-	Industry Standard No.	RS 276-	Industry Standard No.	RS 276-	Industry Standard No.	RS 276-	Industry Standard No.	RS 276-	Industry Standard No.
1104	ZR-590	1104	1A12214	1104	1N1488	1104	1N319A	1104	1N319A
1104	ZR-590A	1104	1A12407	1104	1N1489	1104	1N320	1104	1N320
1104	ZR-61	1104	1A12690	1104	1N1490	1104	1N3203	1104	1N3203
1104	ZR-63	1104	1A13219	1104	1N1491	1104	1N3204A	1104	1N3204A
1104	ZR1025	1104	1A13719	1104	1N1492	1104	1N321	1104	1N321
1104	ZR1031	1104	1A13720	1104	1N151	1104	1N3229	1104	1N3229
1104	ZR1035	1104	1A15790	1104	1N152	1104	1N3230	1104	1N3230
1104	ZR1076	1104	1A16550	1104	1N153	1104	1N323A	1104	1N323A
1104	ZR15	1104	1A50	1104	1N1543	1104	1N324	1104	1N324
1104	ZR500	1104	1B05J20	1104	1N1553	1104	1N324A	1104	1N324A
1104	ZR590	1104	1B05J40	1104	1N1554	1104	1N325	1104	1N325
1104	ZR590A	1104	1B10J20	1104	1N1560	1104	1N3250	1104	1N3250
1104	ZR60	1104	100009	1104	1N1563A	1104	1N3255	1104	1N3255
1104	ZR61	1104	100017	1104	1N1564A	1104	1N325A	1104	1N325A
1104	ZR62	1104	100025	1104	1N1565A	1104	1N326	1104	1N326
1104	ZR63	1104	100026	1104	1N1566A	1104	1N326A	1104	1N326A
1104	ZR64	1104	100031	1104	1N1567	1104	1N327	1104	1N327
1104	ZR66	1104	1005	1104	1N1567A	1104	1N3277	1104	1N3277
1104	Z8-10B	1104	1C1	1104	1N1568	1104	1N327B	1104	1N327B
1104	Z8-20A	1104	1C2	1104	1N1568A	1104	1N3279	1104	1N3279
1104	Z8-20B	1104	1C4	1104	1N158	1104	1N327A	1104	1N327A
1104	Z8-21	1104	1D1	1104	1N1652	1104	1N3298	1104	1N3298
1104	Z8-23	1104	1D2	1104	1N1653	1104	1N348	1104	1N348
1104	Z8-24	1104	1D261	1104	1N169	1104	1N3493	1104	1N3493
1104	Z8-25	1104	1D2Z1	1104	1N1696	1104	1N354	1104	1N354
1104	Z8-30A	1104	1D6	1104	1N1697	1104	1N3548	1104	1N3548
1104	Z8-30B	1104	1DC1	1104	1N1706	1104	1N3549	1104	1N3549
1104	Z8-31A	1104	1B05	1104	1N1712	1104	1N359	1104	1N359
1104	Z8-31B	1104	1B1	1104	1N1764A	1104	1N359A	1104	1N359A
1104	Z8-32A	1104	1B2	1104	1N1912	1104	1N360	1104	1N360
1104	Z8-32B	1104	1B3	1104	1N1913	1104	1N360A	1104	1N360A
1104	Z8-33A	1104	1B4	1104	1N193	1104	1N361	1104	1N361
1104	Z8-33B	1104	1B5	1104	1N2069	1104	1N3613	1104	1N3613
1104	Z8-34A	1104	1B6	1104	1N2069A	1104	1N3614	1104	1N3614
1104	Z8-34B	1104	1EA10A	1104	1N2070	1104	1N3615	1104	1N3615
1104	Z8-50	1104	1EA20A	1104	1N2070A	1104	1N361A	1104	1N361A
1104	Z8-52	1104	1EA30A	1104	1N2071	1104	1N362	1104	1N362
1104	Z8-53	1104	1EA40A	1104	1N2071A	1104	1N362A	1104	1N362A
1104	Z8-71	1104	1EA50A	1104	1N2072	1104	1N363	1104	1N363
1104	Z8-73	1104	1EA60A	1104	1N2073	1104	1N363A	1104	1N363A
1104	Z8100	1104	1ETO2	1104	1N2074	1104	1N3641	1104	1N3641
1104	Z8101	1104	1ETO5	1104	1N2075	1104	1N3642	1104	1N3642
1104	Z8102	1104	1ET1	1104	1N2076	1104	1N3658	1104	1N3658
1104	Z8103	1104	1ET2	1104	1N2077	1104	1N3669	1104	1N3669
1104	Z8104	1104	1ET3	1104	1N2078	1104	1N368	1104	1N368
1104	Z8108	1104	1ET4	1104	1N2079	1104	1N3759	1104	1N3759
1104	Z810A	1104	1ET5	1104	1N2080	1104	1N3866	1104	1N3866
1104	Z810B	1104	1ET6	1104	1N2081	1104	1N3895	1104	1N3895
1104	Z8120	1104	1PO5	1104	1N2082	1104	1N3940	1104	1N3940
1104	Z8121	1104	1P14A	1104	1N2083	1104	1N400	1104	1N400
1104	Z8122	1104	1P2	1104	1N2084	1104	1N4001	1104	1N4001
1104	Z8123	1104	1FM2	1104	1N2085	1104	1N4002	1104	1N4002
1104	Z8124	1104	1GA	1104	1N2086	1104	1N4003	1104	1N4003
1104	Z8173	1104	1HC-50P	1104	1N2088	1104	1N4003GP	1104	1N4003GP
1104	Z8174	1104	1HC-50R	1104	1N2089	1104	1N4004	1104	1N4004
1104	Z8174B	1104	1HC-60P	1104	1N2091	1104	1N4005	1104	1N4005
1104	Z8S20A	1104	1HC-60R	1104	1N2092	1104	1N4003B	1104	1N4003B
1104	Z8S20B	1104	1HY40	1104	1N2093	1104	1N4247	1104	1N4247
1104	Z821	1104	1HY50	1104	1N2094	1104	1N4368	1104	1N4368
1104	Z822	1104	1JH11P	1104	1N2095	1104	1N4369	1104	1N4369
1104	Z823	1104	1JZ61	1104	1N2096	1104	1N4385	1104	1N4385
1104	Z824	1104	1M2086	1104	1N2103	1104	1N4440	1104	1N4440
1104	Z825	1104	1MB513A	1104	1N2104	1104	1N4440B	1104	1N4440B
1104	Z830	1104	MA4	1104	1N2105	1104	1N4441	1104	1N4441
1104	Z830A	1104	1N-4002	1104	1N2106	1104	1N4441B	1104	1N4441B
1104	Z830B	1104	1N1008	1104	1N2107	1104	1N4442	1104	1N4442
1104	Z831	1104	1N1028	1104	1N2108	1104	1N4442B	1104	1N4442B
1104	Z831A	1104	1N1029	1104	1N2115	1104	1N4443	1104	1N4443
1104	Z831B	1104	1N1030	1104	1N2116	1104	1N4443B	1104	1N4443B
1104	Z832	1104	1N1031	1104	1N2117	1104	1N4444	1104	1N4444
1104	Z832A	1104	1N1032	1104	1N2181	1104	1N4444B	1104	1N4444B
1104	Z832B	1104	1N1033	1104	1N2217	1104	1N4445	1104	1N4445
1104	Z834	1104	1N1046	1104	1N2218	1104	1N4445B	1104	1N4445B
1104	Z834A	1104	1N1047	1104	1N2219	1104	1N4448	1104	1N4448
1104	Z834B	1104	1N1048	1104	1N2220	1104	1N4450	1104	1N4450
1104	Z850	1104	1N1049	1104	1N2221	1104	1N4451	1104	1N4451
1104	Z851	1104	1N1050	1104	1N2267	1104	1N4537	1104	1N4537
1104	Z852	1104	1N1051	1104	1N2268	1104	1N4586	1104	1N4586
1104	Z853	1104	1N1052	1104	1N2269	1104	1N4580B	1104	1N4580B
1104	Z87	1104	1N1053	1104	1N2270	1104	1N4821	1104	1N4821
1104	Z870	1104	1N1081	1104	1N2271	1104	1N4822	1104	1N4822
1104	Z871	1104	1N1081A	1104	1N2289	1104	1N5007	1104	1N5007
1104	Z872	1104	1N1082	1104	1N2289A	1104	1N5503	1104	1N5503
1104	Z873	1104	1N1082A	1104	1N2290A	1104	1N5504	1104	1N5504
1104	Z874	1104	1N1083	1104	1N2323	1104	1N5505	1104	1N5505
1104	Z874B	1104	1N1083A	1104	1N2373	1104	1N5506	1104	1N5506
1104	Z876	1104	1N1084	1104	1N2395	1104	1N5507	1104	1N5507
1104	Z88	1104	1N1084A	1104	1N2396	1104	1N5508	1104	1N5508
1104	Z890	1104	1N10D-4P	1104	1N23WP	1104	1N5511	1104	1N5511
1104	Z891	1104	1N1100	1104	1N2404	1104	1N5512	1104	1N5512
1104	Z892	1104	1N1101	1104	1N2405	1104	1N5513	1104	1N5513
1104	Z894	1104	1N1102	1104	1N2413	1104	1N5514	1104	1N5514
1104	ZTR-W06B	1104	1N1103	1104	1N2414	1104	1N5515	1104	1N5515
1104	ZTR-W06C	1104	1N1104	1104	1N2424	1104	1N5516	1104	1N5516
1104	ZTR-W06B	1104	1N1105	1104	1N2423	1104	1N5519	1104	1N5519
1104	ZW2	1104	1N1119	1104	1N2484	1104	1N519B	1104	1N5198
1104	1&12/1	1104	1N1120	1104	1N2486	1104	1N520	1104	1N520
1104	1+12/1	1104	1N122A	1104	1N2488	1104	1N5521	1104	1N521
1104	001-0077-00	1104	1N1169	1104	1N2489	1104	1N5211	1104	1N5211
1104	001-0081-00	1104	1N1169A	1104	1N255	1104	1N5212	1104	1N5212
1104	001-0153-00	1104	1N1217	1104	1N256	1104	1N5213	1104	1N5213
1104	001-02405-0	1104	1N1217A	1104	1N2615	1104	1N5215	1104	1N5215
1104	001-02405-1	1104	1N1217B	1104	1N2638	1104	1N5216	1104	1N5216
1104	001-02405-2	1104	1N1218	1104	1N2650	1104	1N5217	1104	1N5217
1104	001-024050	1104	1N1218A	1104	1N2850	1104	1N522	1104	1N522
1104	001-024051	1104	1N1218B	1104	1N2851	1104	1N523	1104	1N523
1104	001-024052	1104	1N1219	1104	1N2863	1104	1N524	1104	1N524
1104	001-02601-0	1104	1N1219A	1104	1N2863A	1104	1N530	1104	1N530
1104	001-02603-0	1104	1N1219B	1104	1N2864	1104	1N531	1104	1N531
1104	1-101	1104	1N1222A	1104	1N2864A	1104	1N532	1104	1N532
1104	1-20-001-890	1104	1N1222B	1104	1N3080	1104	1N533	1104	1N533
1104	01-2405-0	1104	1N1223	1104	1N3081	1104	1N534	1104	1N534
1104	01-2405-1	1104	1N1223A	1104	1N3084	1104	1N535	1104	1N535
1104	01-2405-2	1104	1N1223B	1104	1N3106	1104	1N536	1104	1N536
1104	01-2406-1	1104	1N1224	1104	1N315	1104	1N537	1104	1N537
1104	1-530-012-11	1104	1N122B	1104	1N315A	1104	1N538	1104	1N538
1104	1-531-027	1104	1N1251	1104	1N316	1104	1N539	1104	1N539
1104	1-531-105	1104	1N1252	1104	1N3160	1104	1N5396	1104	1N5396
1104	1-531-105-11	1104	1N1253	1104	1N316A	1104	1N5397	1104	1N5397
1104	1-531-105-13	1104	1N1254	1104	1N317	1104	1N540	1104	1N540
1104	1-531-10513	1104	1N1255	1104	1N317A	1104	1N5412	1104	1N5412
1104	1-531-106	1104	1N1255A	1104	1N318	1104	1N547	1104	1N547
1104	1-531-106-13	1104	1N1256	1104	1N3183	1104	1N550	1104	1N550
1104	1-531-106-17	1104	1N1257	1104	1N3184	1104	1N5554	1104	1N5554
1104	1-531-5-11	1104	1N1337-5	1104	1N3189	1104	1N555	1104	1N555
1104	1-534-105-13	1104	1N1406	1104	1N318A	1104	1N596	1104	1N596
1104	1-534-106-13	1104	1N1415	1104	1N319	1104	1N599	1104	1N599
1104	1-6501190016	1104	1N1439	1104	1N3190	1104	1N599A	1104	1N599A
1104	1-8259	1104	1N1441	1104	1N3191	1104	1N600	1104	1N600
1104	1-HE-004	1104	1N1442	1104	1N3193	1104	1N600A	1104	1N600A
1104	1A10425	1104	1N1486	1104	1N3194	1104	1N601	1104	1N601
1104	1A10952	1104	1N1487	1104	1N3195	1104	1N601A	1104	1N601A
1104	1A11184								
1104	1A11671								

RS 276-	Industry Standard No.	RS 276-	Industry Standard No.	RS 276-	Industry Standard No.	RS 276-	Industry Standard No.
1104	1N602	1104	18112	1104	18309	1104	28R1K
1104	1N602A	1104	18113	1104	18310	1104	2T501
1104	1N603	1104	18113A	1104	18311	1104	2T502
1104	1N603A	1104	18114	1104	18312	1104	2T503
1104	1N604	1104	18115	1104	18313	1104	2T504
1104	1N604A	1104	18116	1104	18314	1104	2T505
1104	1N605	1104	18118	1104	18315	1104	2T506
1104	1N605A	1104	18119	1104	18315H	1104	2W3A
1104	1N606	1104	18120	1104	1831595	1104	2W4A
1104	1N606A	1104	18121	1104	18328	1104	2W5A
1104	1N613	1104	18121(RECT)	1104	18358	1104	2X9A116
1104	1N613A	1104	18122	1104	18358(8)	1104	003-001
1104	1N614	1104	181221	1104	183588	1104	03-0019-0
1104	1N614A	1104	181222	1104	1836	1104	003-009400
1104	1N645	1104	181224	1104	18371	1104	003-009900
1104	1N645A	1104	18123	1104	18372	1104	03-3016
1104	1N645B	1104	181230	1104	18376	1104	03-931601
1104	1N646	1104	181231	1104	18377	1104	03-931609
1104	1N647	1104	181231H	1104	18381	1104	03-931971
1104	1N648	1104	181232	1104	18382	1104	3A152
1104	1N649	1104	181232H	1104	18390	1104	3A154
1104	1N673	1104	181232N	1104	18395	1104	3A156
1104	1N676	1104	18124	1104	18396	1104	3A200
1104	1N677	1104	18125	1104	18399	1104	3A252
1104	1N678	1104	18126	1104	18400	1104	3A254
1104	1N679	1104	18134	1104	18404	1104	3A256
1104	1N681	1104	181341	1104	1844	1104	3A81
1104	1N682	1104	181342	1104	1845	1104	3A82
1104	1N683	1104	181343	1104	18456	1104	3BB-20801
1104	1N684	1104	181344	1104	18457	1104	3B81
1104	1N685	1104	181346	1104	18458	1104	3B82
1104	1N686	1104	18136(RECT)	1104	18459	1104	3C81
1104	1N687	1104	181361	1104	0001849	1104	3C82
1104	1N689	1104	181361A	1104	1854	1104	3D81
1104	1N692	1104	181367	1104	18558	1104	3D82
1104	1N819	1104	181367A	1104	18559	1104	3E-64
1104	1N846	1104	18147	1104	18588	1104	3E-65
1104	1N847	1104	181472	1104	1860(RECT)	1104	3EB1
1104	1N848	1104	18148	1104	1861	1104	3E82
1104	1N849	1104	18149	1104	1862	1104	3FB1
1104	1N850	1104	18150	1104	1863	1104	3F82
1104	1N851	1104	181622	1104	1864	1104	3G152
1104	1N852	1104	181623	1104	1865	1104	3G154
1104	1N857	1104	181624	1104	1866	1104	3G156
1104	1N858	1104	181625	1104	18685	1104	3G252
1104	1N859	1104	181664	1104	1871	1104	3G254
1104	1N860	1104	181665	1104	1881	1104	3G256
1104	1N861	1104	181666	1104	1883	1104	3G8
1104	1N862	1104	181667	1104	18844	1104	3GA
1104	1N863	1104	181668	1104	18845	1104	3G81
1104	1N868	1104	181668F	1104	18846	1104	3G82
1104	1N869	1104	181681	1104	18846N	1104	3HB1
1104	1N870	1104	181687	1104	18849	1104	3H82
1104	1N871	1104	181691	1104	18849,R	1104	3L4-3001-5
1104	1N872	1104	181692	1104	1885V	1104	3L4-3001-8
1104	1N873	1104	181693	1104	18871A	1104	3L4-3002-13
1104	1N874	1104	181694	1104	18872A	1104	3M810
1104	1N879	1104	181697	1104	18881A	1104	3M820
1104	1N880	1104	181698	1104	18882A	1104	3M830
1104	1N881	1104	181729	1104	18885	1104	3M840
1104	1N881B	1104	181729R	1104	1890	1104	3M85
1104	1N882	1104	181730	1104	1891	1104	3M850
1104	1N883	1104	181730R	1104	1891R	1104	38B-B732
1104	1N884	1104	18180/50B	1104	1892	1104	38B629
1104	1N885	1104	18183	1104	1892R	1104	3T501
1104	1N91	1104	181845B	1104	1893	1104	3T502
1104	1N92	1104	181846B	1104	1893/SGJ	1104	3T503
1104	1N93	1104	181848	1104	1893R	1104	3T504
1104	1N93A	1104	0000181849	1104	1894	1104	3T505
1104	1N947	1104	0000181849R	1104	18941/4454C	1104	3T506
1104	1N998	1104	181850	1104	18943	1104	3X11/1
1104	1NC61684	1104	181850R	1104	18948	1104	004-002000
1104	1NJ61676	1104	181851	1104	1894R	1104	004-002700
1104	1NJ61726	1104	181851R	1104	1895	1104	004-002800
1104	1NJ71126	1104	181850R	1104	1895R	1104	004-003000
1104	1NJ71186	1104	181881AM	1104	18963	1104	004-003300
1104	1R01	1104	181885	1104	18B01-02	1104	004-003400
1104	1R0F	1104	181885-3	1104	18D-2	1104	004-003600
1104	1R0H	1104	181886	1104	18D1212	1104	004-003900
1104	1R1D	1104	181887	1104	18D2	1104	004-004000
1104	1R1K	1104	181888	1104	18I209	1104	004-004100
1104	1R2A	1104	181890	1104	18L1885	1104	004-03300
1104	1R2D	1104	181906	1104	18030	1104	004-03500
1104	1R2E	1104	181941	1104	18031	1104	004-03600
1104	1R3D	1104	1819413	1104	18032	1104	004-03700
1104	1R3G	1104	181942	1104	18034	1104	4-1807
1104	1R3J	1104	181943	1104	18036	1104	4-2020-03173
1104	1R5A	1104	181944	1104	18054	1104	4-2020-03200
1104	1R5B	1104	181950	1104	18R11-600	1104	4-2020-05200
1104	1R50	1104	181209	1104	18R12-600	1104	4-2020-06800
1104	1R5H	1104	18204	1104	18R13-600	1104	4-2020-07300
1104	1R9	1104	18205	1104	18R1FM4	1104	4-2020-07600
1104	1R90	1104	18206	1104	18R1K	1104	4-2020-07601
1104	1R9H	1104	18207	1104	18R50-600	1104	4-2020-07700
1104	1R9I	1104	18208	1104	1854	1104	4-2020-08500
1104	1R9J	1104	18208/28J2A	1104	1T2011	1104	4-2020-08900
1104	1R9L	1104	18209	1104	1T2012	1104	4-2020-10500
1104	1R9U	1104	18209/28J4A	1104	1T2013	1104	4-2020-13600
1104	1RLD	1104	18229	1104	1T2014	1104	4-2020-14500
1104	1R0P	1104	182230	1104	1T2015	1104	4-2021-04970
1104	1R0H	1104	182239	1104	1T2016	1104	4-2021-08270
1104	18-1209	1104	182242	1104	1T2C1	1104	4-2021-09370
1104	18005	1104	182246	1104	1T2E1	1104	4-2021-10270
1104	18031	1104	182272	1104	1T37B	1104	4-2021-10470
1104	18054	1104	182279	1104	1T501	1104	4-202104170
1104	8054	1104	182310	1104	1T502	1104	4-202R101
1104	18100	1104	182313	1104	1T503	1104	4-3033
1104	181004	1104	182317	1104	1T504	1104	4-3540012
1104	18100R	1104	182351	1104	1T505	1104	4-50
1104	18101	1104	182352	1104	1T506	1104	4-50(SEARS)
1104	181013	1104	182356	1104	1TB06	1104	4-686116-3
1104	18101R	1104	182357	1104	1T805	1104	4-686147-3
1104	18102	1104	182361	1104	1V3074A20	1104	4-686148-3
1104	181022	1104	182362	1104	1V3074A21	1104	4-686149-3
1104	18102B	1104	182363	1104	1VA10	1104	4-686150-3
1104	18102R	1104	182372	1104	1WM6	1104	4-686151-3
1104	18103	1104	182373	1104	02-3002-2/2221-3	1104	4-686177-3
1104	18103R	1104	182374	1104	2-52R	1104	4-68687-3
1104	18104	1104	182375	1104	2AA113	1104	04-8054-3
1104	181044	1104	182376	1104	2D-02	1104	04-8054-4
1104	18104R	1104	182395	1104	2P4	1104	04-8054-7
1104	18105	1104	182396	1104	2G13	1104	4AJ4DX520
1104	18105R	1104	182401	1104	2G8	1104	4AJ4DX52D
1104	18106	1104	182402	1104	2G805	1104	4D4
1104	181061	1104	182404	1104	2GA	1104	4D6
1104	181062	1104	182605	1104	2J8138	1104	4D8
1104	181063	1104	182606	1104	28-16E	1104	40A
1104	181064	1104	18268AA	1104	281K	1104	4I39104002
1104	18106A	1104	18268T	1104	28B-C731	1104	4JA10DX3
1104	18106R	1104	182757	1104	28D18(RECTIFIER)	1104	4JA10DX32
1104	181096	1104	182775PA-1	1104	28J2A	1104	4JA10EX3
1104	18110	1104	182777	1104	28J4A	1104	4JA211A
1104	18110A	1104	183016(RECT)	1104	28J60A	1104	4JA2FX355
1104	18111	1104	183016R			1104	4JA2X355

RS 276-	Industry Standard No.	RS 276-	Industry Standard No.	RS 276-	Industry Standard No.	RS 276-	Industry Standard No.
1104	4JA4DRT00	1104	8-905-198-007	1104	10D-6	1104	13-17596-2(RECT)
1104	4JA4DX520	1104	8-905-198-008	1104	10D-V	1104	13-17825-1
1104	4JA6MRT00	1104	8-905-198-010	1104	10D0.5	1104	13-18458-1
1104	4T501	1104	8-905-198-010	1104	10D05	1104	13-22452-0
1104	4T502	1104	8-905-305-400	1104	10D1	1104	13-22463-1
1104	4T503	1104	8-905-405-002	1104	10D3	1104	13-29867-2(SUPP)
1104	4T504	1104	8-905-405-026	1104	10D3G	1104	13-31013-6
1104	4T505	1104	8-905-405-069	1104	10D4	1104	13-31014-1
1104	4T506	1104	8-905-405-134	1104	10D4C	1104	13-31014-3
1104	05-000104	1104	8-905-405-146	1104	10D4D	1104	13-31014-6
1104	05-03016-01	1104	8-905-405-206	1104	10D4E	1104	13-33376-1
1104	05-04001-02	1104	8-905-413-092	1104	10D5	1104	013-339
1104	05-141025	1104	8A11667	1104	10D5B	1104	13-34057-1
1104	05-174004	1104	8C015	1104	10D5C	1104	13-39073-1
1104	05-190061	1104	8C015RE	1104	10D5E	1104	13-39860-1
1104	5-30086.1	1104	8D4	1104	10D5F	1104	13-41122-1
1104	5-30088.1	1104	8D6	1104	10D6	1104	13-41122-2
1104	5-30088.2	1104	8Q-7-01	1104	10D6B	1104	13-41122-4
1104	5-30094.1	1104	8Q-7-02	1104	10DB	1104	13-41123-4
1104	5-30095.1	1104	8Q-7-03	1104	10DBF	1104	13-43766-1
1104	5-30098.1	1104	09-306-083	1104	10DC	1104	13-55029-1
1104	5-30099.1	1104	09306033	1104	10DC-1	1104	13-87539-4
1104	5-30099.3	1104	09306042	1104	10DC-1R	1104	13-88302
1104	5-30099.4	1104	09-306042	1104	10DC-2	1104	0013-911
1104	5-30106.1	1104	09-306042(RECT)	1104	10DC-2B	1104	13D4
1104	5-30109.1	1104	09-306046	1104	10DC-2C	1104	13J2
1104	5-30111.1	1104	09-306050	1104	10DC-2F	1104	13J2F
1104	5-30113.1	1104	09-306054	1104	10DC-2J	1104	13P1
1104	5-30120.1	1104	09-306059	1104	10DC-4	1104	14-0072-1
1104	5-30122.1	1104	09-306063	1104	10DC-4R	1104	14-0072-1(PHILCO)
1104	05-540001	1104	09-306083	1104	10DC0.5	1104	14-0072-2
1104	05-540112	1104	09-306088	1104	10DC05(RED)	1104	14-0072-2(PHILCO)
1104	05-750010	1104	09-306100	1104	10DC05R	1104	14-0072-3
1104	05-931601	1104	09-306103	1104	10DC0B	1104	14-0072-3(PHILCO)
1104	05-931609	1104	09-306104	1104	10DC0H	1104	14-503-01
1104	05-931971	1104	09-306110	1104	000010DC1	1104	14-503-03
1104	05-935201	1104	09-306115	1104	10DC1(BLACK)	1104	14-503-0B
1104	5-830082.1003	1104	09-306119	1104	10DC1(RED)	1104	14-504-04
1104	5A-D	1104	09-306125	1104	10DC1BLACK	1104	14-514-75
1104	05AO7	1104	09-306138	1104	10DC1R	1104	14J2
1104	5A1	1104	09-306157	1104	10DC2	1104	14J2F
1104	5A2	1104	09-306160	1104	10DC2F	1104	14P2
1104	5A3	1104	09-306162	1104	10DC4	1104	015-002
1104	5A4	1104	09-306169	1104	10DC4R	1104	015-006
1104	5A4D	1104	09-306172	1104	10DC5	1104	15-033-0
1104	5A4D-C	1104	09-306176	1104	10DC05R	1104	15-085006
1104	5A5	1104	09-306177	1104	10DC0B	1104	15-085007
1104	5A5D	1104	09-306192	1104	10DC0H	1104	15-085015
1104	5A6	1104	09-306205	1104	10DG	1104	15-085016
1104	5A6D	1104	09-306213	1104	10D0.5	1104	15-085040
1104	5A6D-C	1104	09-306214	1104	10DRV	1104	15-085043
1104	5B-1	1104	09-306224	1104	10DY	1104	15-100001
1104	5B-15H	1104	09-306245	1104	10DX2	1104	15-100002
1104	5B-2	1104	09-306249	1104	10DZ	1104	15-100004
1104	5B-2-H5W	1104	09-306250	1104	10E-1	1104	15-103022
1104	5B3	1104	09-306254	1104	10E-2	1104	15-108003
1104	5B4	1104	09-306255	1104	10E-4D	1104	15-108004
1104	5D1	1104	09-306258	1104	10E-7L	1104	15-108005
1104	5D2	1104	09-306259	1104	10E1	1104	15-108006
1104	5E1	1104	09-306263	1104	10E1LF	1104	15-108010
1104	5E2	1104	09-306264	1104	10E2	1104	15-108011
1104	5E3	1104	09-306285	1104	10E6	1104	15-108015
1104	5E4	1104	09-306300	1104	10H	1104	15-108016
1104	5E5	1104	09-306302	1104	10J2	1104	15-108020
1104	5E6	1104	09-306303	1104	10J2F	1104	15-108021
1104	5G-D	1104	09-306312	1104	10K	1104	15-108022
1104	5GA	1104	09-306315	1104	10K-1	1104	15-108024
1104	5GB	1104	09-306323	1104	10M	1104	15-108031
1104	5GD	1104	09-306333	1104	10N1	1104	15-108032
1104	5GPH	1104	09-3063353	1104	10R1B	1104	15-108034
1104	5GJ/FR1N	1104	09-306341	1104	10R2B	1104	15-108036
1104	5GJFR1N	1104	09-306350	1104	10R3B	1104	15-108037
1104	5GL	1104	09-306365	1104	10R4B	1104	15-108049
1104	5H	1104	09-306366	1104	10R5B	1104	15-108050
1104	5H4D1	1104	09-306376	1104	10R6B	1104	15-123105
1104	5H750M	1104	09-306384	1104	10T200	1104	15-123243
1104	5J-P1	1104	09-306389	1104	11-0429	1104	0015-911
1104	5MA2	1104	09-306394	1104	11-0769	1104	15B2
1104	5MA4	1104	09-306417	1104	11-0771	1104	15BD11
1104	5MA5	1104	09-306421	1104	11-085003	1104	15J2
1104	5MA6	1104	09-306422	1104	11-085013	1104	15J2F
1104	5MAK	1104	09-306431	1104	11-085024	1104	15N1
1104	5MP1	1104	09-306432	1104	11-102001	1104	15S6
1104	5M310	1104	09-306433	1104	11-120007	1104	16-2
1104	5M320	1104	09-307043	1104	11-1592	1104	16C-4
1104	5M330	1104	09-307084	1104	11/1	1104	16C4
1104	5M340	1104	9D13	1104	11/1&12/1	1104	16G5
1104	5M35	1104	9LR2-2	1104	11/1+12/1	1104	16J2G(DIODE)
1104	5M350	1104	90069-1	1104	11/15	1104	16J2P
1104	5N1	1104	9RE1	1104	11/1592	1104	16X10
1104	05V-50	1104	10-012	1104	11J2F	1104	17-10
1104	6-59010	1104	10-085006	1104	12-085031	1104	17-410
1104	608	1104	10-085009	1104	12-085040	1104	018-00006
1104	6GA1750	1104	10-085010	1104	12-100001	1104	018-00007
1104	6GA175D	1104	10-085026	1104	12-100003	1104	018-0000B
1104	6GD1	1104	10-085030	1104	12-100008	1104	018-00009
1104	6M4	1104	10-12	1104	12-102001	1104	018-0009
1104	6M404-1	1104	10-42	1104	012-1022-002	1104	18-085030
1104	6M404-2	1104	10-7	1104	0012-911	1104	18-22-17
1104	6M404-3	1104	10-D1	1104	12B-2	1104	18P2
1104	6M404-4	1104	10A590B	1104	12B-2B1P-M	1104	019-002935
1104	6M404-5	1104	10AG2	1104	1202	1104	019-003420
1104	6M404-6	1104	10AG4	1104	12C0ZP-114	1104	019-003870-013
1104	6M404-7	1104	10AG6	1104	12J2	1104	019-003870-02Q
1104	6R522PC7BAD1	1104	10AL2	1104	12J2F	1104	19-040-002
1104	6RW62HY	1104	10AL6	1104	13-0015	1104	19-040-003
1104	7-0002	1104	10A8	1104	13-085039	1104	19-040-004
1104	7-0004	1104	10AT2	1104	13-087027	1104	19-060-002
1104	7-0008	1104	10AT4	1104	13-102001	1104	19-085010
1104	7D	1104	10AT6	1104	13-102001	1104	19-085022
1104	7L6-0495-14	1104	10B-2-B1W	1104	13-10321-3	1104	19A115024-P4
1104	7MA60	1104	10B-2-N1W	1104	13-14094-12	1104	19A115100-P1
1104	08-0040	1104	10B-Y	1104	13-14094-16	1104	19A115145-P3
1104	008-024-00	1104	10B1	1104	13-14094-17	1104	19A115145-P4
1104	08-08122	1104	10C	1104	13-14094-24	1104	19A115569-P1
1104	08-0821	1104	0005	1104	13-14094-38	1104	19A115569-P2
1104	8-22	1104	10C1	1104	13-14094-39	1104	19AR11
1104	8-25	1104	1002	1104	13-14094-42	1104	19A112
1104	8-38	1104	1003	1104	13-14094-54	1104	19AR17
1104	8-619-030-012	1104	10C4D	1104	13-14261-1	1104	19AR2
1104	8-639-001-095	1104	1005	1104	13-14627-1	1104	19AR34
1104	8-710-222-21	1104	1006	1104	13-14627-4	1104	19AR4
1104	8-719-200-02	1104	10D	1104	13-14858-1	1104	19A85
1104	8-719-205-10	1104	10D-02	1104	13-16104-8	1104	192200011-P5
1104	8-719-900-63	1104	10D-05	1104	13-16104-9	1104	19C300076-P1
1104	8-719-901-02	1104	10D-06	1104	13-16247-3	1104	19C300076-P2
1104	8-719-901-13	1104	10D-1	1104	13-17174-1	1104	19C300076-P3
1104	8-719-908-03	1104	10D-2B(-4)	1104	13-17174-2	1104	19C300076-P4
1104	8-905-013-752	1104	10D-2B(4)	1104	13-17174-3	1104	19C300076-P5
1104	8-905-013-759	1104	10D-2B-4	1104	13-17557-1	1104	19C300076-P6
1104	8-905-013-760			1104	13-17596-1	1104	19P2
1104	8-905-198-001					1104	20-1680-143
1104	8-905-198-004					1104	20-22-08
1104	8-905-198-005					1104	20A0054

RS 276-	Industry Standard No.	RS 276-	Industry Standard No.	RS 276-	Industry Standard No.	RS 276-	Industry Standard No.
1104	20A8	1104	28-22-06	1104	42-23350	1104	46BR7
1104	20B8	1104	28-22-07	1104	42-23350A	1104	46BR9
1104	20C	1104	28-22-10	1104	42-23975	1104	46BX2
1104	20H	1104	28-22-12	1104	42-27202	1104	46BX3
1104	20K	1104	28-22-14	1104	42-27278	1104	48-01
1104	20M	1104	28-22-15	1104	42-27463	1104	48-03005A07
1104	20N1	1104	28-22-17	1104	42-28058	1104	48-05-001
1104	21-810	1104	28-22-21	1104	42-28202	1104	48-10001-A01
1104	21-810-2	1104	28-25-01	1104	42-7	1104	48-10001-A03
1104	21/3	1104	28-254566-1	1104	42A11	1104	48-10001-A030-1
1104	21A001-00	1104	28-29-01	1104	42A23	1104	48-1005
1104	21A001-000	1104	28-6-01	1104	42B16	1104	48-10062A01
1104	21A006-000	1104	28-65-01	1104	42B2	1104	48-10062A01A
1104	21A007-000	1104	28-7-01	1104	42J2	1104	48-10062A04
1104	21A008-000	1104	28J2	1104	42X244	1104	48-10062A05
1104	21A008-001	1104	30-8054-7	1104	42X244B	1104	48-134769
1104	21A008-003	1104	30A8	1104	42X245	1104	48-134790
1104	21A020-001	1104	30B5	1104	42X245B	1104	48-134958
1104	21A020-006	1104	30B8	1104	42X25	1104	48-134959
1104	21A040-44	1104	30C	1104	42X32	1104	48-134990
1104	21A079-000	1104	30H	1104	43-540012	1104	48-137029
1104	21A101-001(RECT)	1104	30K	1104	44-530	1104	48-137074
1104	21A102-001	1104	30M	1104	44T-300-95	1104	48-137143
1104	21A102-002	1104	31-195	1104	45A2FX355	1104	48-137198
1104	21A103-007	1104	32-0026	1104	046-0909	1104	48-137208
1104	21A103-018	1104	32-0037	1104	46-16261	1104	48-137291
1104	21A103-021	1104	32-0038	1104	46-16261-3	1104	48-155041
1104	21A103-044	1104	32-0042	1104	46-34-3	1104	48-155063
1104	21A103-050	1104	32-0045	1104	46-61249-3	1104	48-155099
1104	21A103-058	1104	32-0046	1104	46-6661-2	1104	48-155126
1104	21A103-070	1104	32-0047	1104	46-67120A13	1104	48-155136
1104	21A110-001	1104	32-0048	1104	46-86-3	1104	48-155178
1104	21A110-002	1104	32-0050	1104	46-8601-3	1104	48-155193
1104	21A110-007	1104	32-0059	1104	46-8611-4	1104	48-155235
1104	21A110-010	1104	32-0060	1104	46-861148-3	1104	48-155236
1104	21A110-012	1104	32-0061	1104	46-86116-3	1104	48-191807
1104	21A119-030	1104	33-0002	1104	46-86139-3	1104	48-191A01
1104	21A6	1104	33-0006	1104	46-86148-3	1104	48-191A01-9
1104	21A7	1104	33-0023	1104	46-86150-3	1104	48-191A02
1104	21B-14	1104	33-0024	1104	46-86151-3	1104	48-191A03
1104	21B-17	1104	33-0026	1104	46-86177-3	1104	48-191A04
1104	21M248	1104	33-0029	1104	46-8619	1104	48-191A06
1104	21M283	1104	33-0030	1104	46-86199-3	1104	48-191A07
1104	21M302	1104	33-0031	1104	46-86201-3	1104	48-191A07A
1104	21M312	1104	33059024	1104	46-86201-3(POWER)	1104	48-191A08
1104	21M315	1104	33059121	1104	46-86212-3	1104	48-191A09
1104	21M517	1104	33059122	1104	46-86261-3	1104	48-191A11
1104	21M386	1104	34-8003	1104	46-86264-3	1104	48-355016
1104	21M386(PWR)	1104	34-8026-1	1104	46-86267-3	1104	48-355023
1104	21M416	1104	34-8026-2	1104	46-86270-3	1104	48-355025
1104	21M417	1104	34-8026-3	1104	46-86271-3	1104	48-355045
1104	21M419	1104	34-8026-4	1104	46-86290-3	1104	48-355046
1104	21M433	1104	34-8034-1	1104	46-86503-3	1104	48-355049
1104	21M434	1104	34-8034-2	1104	46-86304-3	1104	48-40235901
1104	21M435	1104	34-8034-3	1104	46-86307-3	1104	48-40235902
1104	21M436	1104	34-8034-4	1104	46-86321-3	1104	48-40739901
1104	21M437	1104	34-8036-1	1104	46-86322-3	1104	48-41508A01
1104	21M469	1104	34-8036-2	1104	46-86326-3	1104	48-41508A02
1104	21M487	1104	34-8036-3	1104	46-86337-3	1104	48-43265901
1104	21M519	1104	34-8036-4	1104	46-86339-3	1104	48-44887G0
1104	21M545	1104	34-8040-2	1104	46-86351-3	1104	48-57120A01
1104	21M590	1104	34-8042-1	1104	46-86355-2	1104	48-6002249B
1104	22-1-075	1104	34-8042-2	1104	46-86355-3	1104	48-63086A16
1104	22-1-138	1104	34-8042-3	1104	46-86358-3	1104	48-64169
1104	022-2823-011	1104	34-8047-2	1104	46-86364-3	1104	48-646954
1104	022-3905-001	1104	34-8048-1	1104	46-86365-3	1104	48-647829
1104	022-3905-001	1104	34-8048-2	1104	46-86383-3	1104	48-6514SA74
1104	22A001-17	1104	34-8048-3	1104	46-86402-3	1104	48-660370A05
1104	022D	1104	34-8048-4	1104	46-86416-3	1104	48-66037A03
1104	23-0003	1104	34-8048-5	1104	46-86430-3	1104	48-66037A04
1104	23-0004	1104	34-8050-13	1104	46-86433-3	1104	48-66037A05
1104	23-0010	1104	34-8050-14	1104	46-86435-3	1104	48-66037A08
1104	23-0017	1104	34-8050-2	1104	46-86438-3	1104	48-66037A10
1104	23B8C101	1104	34-8050-5	1104	46-86463-3	1104	48-66037A12
1104	24-198	1104	34-8050-6	1104	46-8647-3	1104	48-66629A01
1104	24-DP1	1104	34-8050-7	1104	46-86488-3	1104	48-66629A03
1104	24MW1066	1104	34-8050-8	1104	46-86497-3	1104	48-66629A05
1104	24MW1071	1104	34-8050-9	1104	46-86503-3	1104	48-66629A06
1104	24MW1108	1104	34-8051	1104	46-8661-3	1104	48-66653A001
1104	24MW1113	1104	34-8054-1	1104	46-86807-3	1104	48-66653A002
1104	24MW1123	1104	34-8054-10	1104	46-8687-3	1104	48-66654A02
1104	24MW1144	1104	34-8054-11	1104	46-8689-3	1104	48-67120A01
1104	24MW1146	1104	34-8054-12	1104	46-8697-3	1104	48-67120A06
1104	24MW1162	1104	34-8054-13	1104	46AR1	1104	48-67120A0607
1104	24MW175	1104	34-8054-15	1104	46AR10	1104	48-67120A07
1104	24MW196	1104	34-8054-16	1104	46AR11	1104	48-67926A01
1104	24MW197	1104	34-8054-17	1104	46AR12	1104	48-68688A79
1104	24MW208	1104	34-8054-18	1104	46AR2	1104	48-68688A79B
1104	24MW227	1104	34-8054-2	1104	46AR21	1104	48-733746
1104	24MW241	1104	34-8054-23	1104	46AR28	1104	48-741752
1104	24MW267	1104	34-8054-27	1104	46AR29	1104	48-746831
1104	24MW268	1104	34-8054-3	1104	46AR3	1104	48-752497
1104	24MW269	1104	34-8054-4	1104	46AR35	1104	48-754153
1104	24MW619	1104	34-8054-5	1104	46AR4	1104	48-82095C01
1104	24MW669	1104	34-8054-6	1104	46AR5	1104	48-82095C02
1104	24MW671	1104	34-8054-7	1104	46AR7	1104	48-82095C03
1104	24MW721	1104	34-8055-2	1104	46AR8	1104	48-82095C64
1104	24MW768	1104	34-8055-3	1104	46AR9	1104	48-8240006
1104	24MW772	1104	34-8057-11	1104	46AX1	1104	48-8240008
1104	24MW851	1104	34-8057-18	1104	46AX10	1104	48-82466B01
1104	24MW862	1104	34-8057-28	1104	46AX115	1104	48-82466B02
1104	24MW867	1104	34-8057-45	1104	46AX16	1104	48-82466B03
1104	24MW871	1104	35-003-001	1104	46AX19	1104	48-82466B04
1104	24MW974	1104	35-1004	1104	46AX21	1104	48-82466B06
1104	24MW975	1104	35-1005	1104	46AX23	1104	48-82466B07
1104	24MW995	1104	39-02	1104	46AX3	1104	48-86148
1104	025-100016	1104	40-0502	1104	46AX30	1104	48-90229A01
1104	025-100024	1104	40A8	1104	46AX34	1104	48-90233A04
1104	025-100028	1104	40B5	1104	46AX4	1104	48-90234A01
1104	025-100029	1104	40B8	1104	46AX5	1104	48-90234A02
1104	025-100035	1104	40C	1104	46AX52	1104	48-90343A64
1104	25-5	1104	40D6665A03	1104	46AX54	1104	48-90343A91
1104	26D00505	1104	40H	1104	46AX7	1104	48-90343A92
1104	027-000296	1104	40J2	1104	46AX8	1104	48-90343A93
1104	027-000306	1104	40JZ	1104	46AX9	1104	48-90420A79
1104	027-000312	1104	40K	1104	46BD1	1104	48-90420A80
1104	28-1-01	1104	40KR	1104	46BD12	1104	48-90420A97
1104	28-1-02	1104	40M	1104	46BD14	1104	48-97048A04
1104	28-13-01	1104	40N1	1104	46BD19	1104	48-97048A07
1104	28-14-01	1104	40NJ	1104	46BD2	1104	48-97048A10
1104	28-15-01	1104	40Y3P	1104	46BD25	1104	48-97048A18
1104	28-15-02	1104	41-032	1104	46BD5	1104	48-97127A01
1104	28-18-01	1104	41-J2	1104	46BD8	1104	48-97168A02
1104	28-19-01	1104	41J2	1104	46BD9	1104	48-97168A10
1104	28-20-01	1104	42-14027	1104	46BR10	1104	48-97168A14
1104	28-20-02	1104	42-17443	1104	46BR11	1104	48-97168A14
1104	28-21-01	1104	42-17443A	1104	46BR15	1104	48-97172A01
1104	28-22-01	1104	42-19865	1104	46BR17	1104	48-97222A02
1104	28-22-02	1104	42-21400	1104	46BR5	1104	48-97270A01
1104	28-22-03	1104	42-21408			1104	48A41508A01
1104	28-22-04	1104	42-21866			1104	48A41508A02
1104	28-22-05	1104	42-22835				

RS 276-	Industry Standard No.	RS 276-	Industry Standard No.	RS 276-	Industry Standard No.	RS 276-	Industry Standard No.	RS 276-	Industry Standard No.
1104	48B43265Q01	1104	53-0099-3	1104	62A05	1104	86-0013	1104	93B1-11
1104	48B66629A03	1104	53-0099-4	1104	62J2	1104	86-0016-01		
1104	48B66629A05	1104	53-0106-1	1104	63-10064	1104	86-0511		
1104	48C40235-602	1104	53-0109-1	1104	63-10709	1104	86-0516		
1104	48C40235C01	1104	53-0111-1	1104	63-11215	1104	86-1-3		
1104	48C40235G01	1104	53-0113-1	1104	63-11291	1104	86-111-3		
1104	48C66037A03	1104	53-0120-1	1104	63-11881	1104	86-139-3		
1104	48C66037A04	1104	53-0122-1	1104	63-11957	1104	86-146-3		
1104	48C66037A05	1104	53-1086	1104	63-12077	1104	86-147-3		
1104	48C66037A10	1104	53A001-12	1104	63-13903	1104	86-21-1		
1104	48C66037A12	1104	53A001-3	1104	63-13919	1104	86-22-3		
1104	48C66629A02	1104	53A001-35	1104	63-14195	1104	86-23-1		
1104	48C66653A02	1104	53A001-9	1104	63-15483	1104	86-23-3		
1104	48C67120A05	1104	53A002-1	1104	63-18135	1104	86-26-1		
1104	48C67926A01	1104	53A010-1	1104	63-26597	1104	86-28-3		
1104	48D66037A03	1104	53A011-1	1104	63-27483	1104	86-3-1		
1104	48D66037A04	1104	53A022-1	1104	63-27622	1104	86-3-3		
1104	48D66037A05	1104	53A022-4	1104	63-29383	1104	86-30-1		
1104	48D66037A08	1104	53B001-5	1104	63-7435	1104	86-30-3		
1104	48D66654A02	1104	53B001-6	1104	63-8685	1104	86-32-1		
1104	48D67120A01	1104	53B003-1	1104	63-8819	1104	86-34-3		
1104	48D67120A02	1104	53B003-2	1104	63-8824	1104	86-35-3		
1104	48D67120A05	1104	53B006-1	1104	63-9787	1104	86-36-1		
1104	48D67120A06	1104	53B010-3	1104	63J2	1104	86-38-1		
1104	48D67120A07	1104	53B010-4	1104	64-8054-6	1104	86-4-1		
1104	48D67120H02	1104	53B010-9	1104	64-J2	1104	86-40-3		
1104	48K64169	1104	53B011-1	1104	64J2	1104	86-42-3		
1104	48K646954	1104	53B011-2	1104	65(DIODE)	1104	86-46-3		
1104	48K647829	1104	53B019-2	1104	065-015	1104	86-49-3		
1104	48K746831	1104	53B020-2	1104	65-085013	1104	86-50-3		
1104	48K752497	1104	53C0003-1	1104	65-744238	1104	86-5000-3		
1104	48M355016	1104	53C0005-3	1104	65J2	1104	86-5001-3		
1104	48M355023	1104	53C007-1	1104	65P117	1104	86-5002-3		
1104	48M355025	1104	53C009-1	1104	65P124	1104	86-5003-3		
1104	48M355045	1104	53C012-1	1104	65P124-1	1104	86-5006-3		
1104	48M355046	1104	53C014-1	1104	65P124-2	1104	86-5009-3		
1104	48M355048	1104	53C015-1	1104	65P153	1104	86-5012-3		
1104	48M355049	1104	53C016-1	1104	65P155	1104	86-5024-3		
1104	48R10001-A01	1104	53C017-1	1104	65P206	1104	86-5029-3		
1104	48R10001-A03	1104	53C022-1	1104	65P284	1104	86-51-2		
1104	48R10001-A030-1	1104	53D001-7	1104	65P297	1104	86-51-3		
1104	48R100620A02	1104	53D003-1	1104	66-2246	1104	86-52-1		
1104	48R100620A04	1104	53D003-2	1104	66-6030-00	1104	86-54-3		
1104	48R100620A05	1104	53E010-1	1104	66-6031-00	1104	86-56-3		
1104	48R10062A01	1104	53F001-1	1104	66P-001	1104	86-57-3		
1104	48R10062A02	1104	53F002-1	1104	66P-001-1	1104	86-58-3		
1104	48R134671	1104	53H001-1	1104	66F001	1104	86-59-1		
1104	48R660370A05	1104	53J002-2	1104	66F001-1	1104	86-59-3		
1104	48R10062A01	1104	53J003-1	1104	66J2	1104	86-591-3		
1104	48R10062A02	1104	53K001-11	1104	66X0023-001	1104	86-60-3		
1104	48R10062A05	1104	53K001-6	1104	66X0023-002	1104	86-62-3		
1104	48R10062A05A	1104	53K001-6(5,7)	1104	66X0023-003	1104	86-63-3		
1104	48R134736	1104	53K001-7	1104	66X0023-004	1104	86-65-3		
1104	48R134790	1104	53K001-7(6,8)	1104	66X0023-005	1104	86-65-9		
1104	48R134921	1104	53K001-8	1104	66X0023-006	1104	86-67-0		
1104	48R134939	1104	53K001-9	1104	66X0023-007	1104	86-67-3		
1104	48R134958	1104	53N001-2	1104	66X0023-008	1104	86-67-8		
1104	48R134959	1104	53N001-3	1104	66X0023-1	1104	86-67-9		
1104	48R134990	1104	53N002-2	1104	66X0028-001	1104	86-7-1		
1104	48R137029	1104	53N004-7	1104	66X0033-000	1104	86-75-3		
1104	48R137040	1104	53N004-8	1104	66X0033-001	1104	86-75-3(SEARS)		
1104	48R137074	1104	53N004-9	1104	66X0037-001	1104	86-78-3		
1104	48R137208	1104	53T001-3	1104	66X041-001	1104	86-80-1		
1104	48R137347	1104	53T001-1	1104	66X053-001	1104	86-80-3		
1104	48R137348	1104	56-33	1104	66X19	1104	86-84-1		
1104	48R155037	1104	56-4839	1104	66X23	1104	86-85-3		
1104	48R155040	1104	56-52	1104	66X24	1104	86-89-1		
1104	48R155041	1104	56-78	1104	66X26	1104	86-89-3		
1104	48R155063	1104	56-8097	1104	67-1000-00	1104	86-9-3		
1104	48R155079	1104	56-8198	1104	67-1003-00	1104	87-67-3		
1104	48R155099	1104	56-8199	1104	69-1823	1104	88-125		
1104	48R155100	1104	57-0006	1104	69-2246	1104	88-3		
1104	48R155106	1104	57-1	1104	070-019	1104	88-831		
1104	48R155126	1104	57-12	1104	70P40	1104	88-832		
1104	48R191A02	1104	57-13	1104	70840	1104	88-833		
1104	48R191A04	1104	57-15	1104	70P40	1104	089-262		
1104	48R191A04(A)	1104	57-17	1104	72-11	1104	91A01		
1104	48R191A05	1104	57-2	1104	72-15	1104	91A018		
1104	48R191A06	1104	57-20	1104	72Z	1104	91A02		
1104	48R191A07	1104	57-21	1104	75	1104	91A03		
1104	48R191A08	1104	57-22	1104	75D1	1104	91A05		
1104	48R40235G02	1104	57-23	1104	75D2	1104	91A06		
1104	48R41508A01	1104	57-24	1104	75E1	1104	91A08		
1104	48X90233A04	1104	57-25	1104	75E2	1104	91A11		
1104	48X90234A01	1104	57-26	1104	75E3	1104	92A11-1		
1104	48X90234A02	1104	57-27	1104	75E4	1104	92A11-1		
1104	48X90234A03	1104	57-28	1104	75E5	1104	92B12-2		
1104	48X97048A04	1104	57-29	1104	75E6	1104	92L102-2		
1104	48X97048A07	1104	57-31	1104	75P05	1104	93-302		
1104	48X97048A10	1104	57-33	1104	75R1B	1104	93A1-20		
1104	48X97127A01	1104	57-38	1104	75R2B	1104	93A1-21		
1104	48X97168A06	1104	57-43	1104	75R3B	1104	93A10-1		
1104	48X97168A10	1104	57-46	1104	75R4B	1104	93A11		
1104	48X97168A16	1104	57-49	1104	75R5B	1104	93A12-1		
1104	48X97172A01	1104	57-58	1104	75R6B	1104	93A12-3		
1104	48X97222A02	1104	57-59	1104	75W-005	1104	93A120		
1104	48X97271A0.	1104	57-6	1104	76Q145OJ	1104	93A15-1		
1104	49-1042	1104	57-60	1104	77-270993-1	1104	93A1D-1		
1104	49-3112	1104	57-65	1104	77-271374-1	1104	93A2		
1104	050-0011-0Q	1104	57A1-63	1104	078-0016	1104	93A27-2		
1104	50A	1104	57D1-63	1104	078-1696	1104	93A30-1		
1104	50A8	1104	57X14	1104	078-2400	1104	93A30-3		
1104	50B5	1104	60-3	1104	78-254566-1	1104	93A3D-2		
1104	50C	1104	60C	1104	78-254566-4	1104	93A4-2		
1104	50D2	1104	60D	1104	78-271030-1	1104	93A42-1		
1104	50D4	1104	60H	1104	78-271143-1	1104	93A42-2		
1104	50E05	1104	60J2	1104	78-272160-1	1104	93A42-7		
1104	50E1	1104	60M	1104	78-273008	1104	93A42-8		
1104	50E2	1104	61-1320	1104	78-273008-1	1104	93A45-1		
1104	50E3	1104	61-1765	1104	78-273085	1104	93A45-2		
1104	50E4	1104	61-756	1104	078-5001	1104	93A47		
1104	50E5	1104	61-7728	1104	81-27123100-3	1104	93A48-2		
1104	50E6	1104	61-820	1104	81-27123300-3	1104	93A51-3		
1104	50M	1104	61-926	1104	81-27123307-3	1104	93A52-1		
1104	051-0006	1104	61J2	1104	81-46123004-5	1104	93A53-3		
1104	051-0016-0Q	1104	62-113998	1104	81-46123004-7	1104	93A56-1		
1104	51-03007-06	1104	62-126856	1104	81-46123005	1104	93A58-1		
1104	52-1	1104	62-13261	1104	81-46123011-2	1104	93A6-1		
1104	52A011	1104	62-13477	1104	81-46123012-2	1104	93A6-2		
1104	52ABIA	1104	62-15483	1104	81-46123014-6	1104	93A60-14		
1104	52BBLA	1104	62-16711	1104	81-46123018-7	1104	93A60-2		
1104	52DS-18	1104	62-18135	1104	81-46123022-9	1104	93A60-7		
1104	52SD-1	1104	62-18431	1104	81-46123034-4	1104	93A60-8		
1104	52SD1	1104	62-18434	1104	81-46123035-1	1104	93A60-80		
1104	53-0051-2	1104	62-18436	1104	81-46123043-5	1104	93A60-9		
1104	53-0082-1003	1104	62-18438	1104	82-4	1104	93A67-1		
1104	53-0086-1	1104	62-19115	1104	83-829	1104	93A78-1		
1104	53-0088-1	1104	62-19749	1104	83-880	1104	93A97-1		
1104	53-0088-2	1104	62-19814	1104	83A30-1	1104	93A97-2		
1104	53-0094-1	1104	62-21369	1104	83C30-1	1104	93A97-3		
1104	53-0095-1	1104	62A01	1104	85-5	1104	93A97-37		
1104	53-0098-1	1104	62A02	1104	86-0006	1104	93B1-1		
1104	53-0099-1	1104	62A04	1104	86-0010	1104	93B1-10		
				1104	86-0012				

RS 276-	Industry Standard No.
1104	93B1-12
1104	93B1-13
1104	93B1-14
1104	93B1-15
1104	93B1-16
1104	93B1-17
1104	93B1-18
1104	93B1-2
1104	93B1-20
1104	93B1-21
1104	93B1-3
1104	93B1-4
1104	93B1-5
1104	93B1-6
1104	93B1-7
1104	93B1-8
1104	93B1-9
1104	93B12-1
1104	93B12-2
1104	93B12-3
1104	93B122
1104	93B123
1104	93B2-1
1104	93B20-1
1104	93B20-2
1104	93B20-3
1104	93B24-2
1104	93B24-3
1104	93B27-2
1104	93B30-1
1104	93B30-3
1104	93B41-12
1104	93B41-14
1104	93B41-20
1104	93B41-4
1104	93B41-6
1104	93B41-8
1104	93B42-1
1104	93B42-10
1104	93B42-11
1104	93B42-12
1104	93B42-2
1104	93B42-3
1104	93B42-4
1104	93B42-5
1104	93B42-6
1104	93B42-7
1104	93B42-8
1104	93B42-9
1104	93B45-1
1104	93B45-2
1104	93B47-1
1104	93B51-3
1104	93B97-1
1104	93C1-20
1104	93C1-21
1104	93C103-4
1104	93C12-1
1104	93C12-2
1104	93C12-3
1104	93C16-2
1104	93C19-1
1104	93C24-1
1104	93C24-2
1104	93C24-3
1104	93C24-4
1104	93C25-4
1104	93C26-1
1104	93C26-2
1104	93C26-3
1104	93C26-4
1104	93C26-5
1104	93C26-8
1104	93C267
1104	93C27-4
1104	93C28-4
1104	93C30-1
1104	93C30-3
1104	93C42-2
1104	93C42-7
1104	93C5-3
1104	93C51-3
1104	93C60-7
1104	93C60-9
1104	93CRH-DX0155//
1104	93CRH-DX0156//
1104	93L101-2
1104	93L102-2
1104	93L103-4
1104	93L107-2
1104	93L3-1
1104	93L3-2
1104	93L5-1
1104	93L5-2
1104	93L5-4
1104	93L5-6
1104	93L5-7
1104	93M8-1
1104	094-010
1104	94-1066-1
1104	96-5022-01
1104	96-5022-1
1104	96-5023-01
1104	96-5046-01
1104	96-5082-01
1104	96-5088-01
1104	96-5096-01
1104	96-5103-01
1104	96-5109-01
1104	96-5109-02
1104	96-5113-01
1104	96-5119-01
1104	96-5121-01
1104	96-5166-01
1104	96-5196-01
1104	96-5333-01
1104	98-301
1104	98-302
1104	98A12518
1104	0100
1104	100-001-01/2228-3
1104	100-10121-05
1104	100-10132-00
1104	100-13
1104	100-132-00
1104	100-13B
1104	100-15
1104	100-162
1104	100-217
1104	100-438
1104	100-520
1104	100-527
1104	100A
1104	100C-4R

RS 276-	Industry Standard No.
1104	100R1B
1104	100R2B
1104	100R3B
1104	100R4B
1104	100R5B
1104	100R6B
1104	101B6
1104	102B6
1104	102D
1104	103-191
1104	103-203
1104	103-216
1104	103-228
1104	103-245
1104	103-254
1104	103-254-01
1104	103-261
1104	103-261-0
1104	103-261-01
1104	103-315-03A
1104	103-59
1104	103-76
1104	103-82
1104	103-90
1104	103B6
1104	10303125
1104	10303125A
1104	104B6
1104	105(JULIETTEERECT)
1104	105B6
1104	106
1104	106-011
1104	106-1
1104	106-111
1104	106B6
1104	107B6
1104	108
1104	108A4
1104	108B4
1104	108B6
1104	108E-E2
1104	109B6
1104	0110
1104	0110-0011
1104	0110-0141
1104	0110-0141/4460
1104	0110-0209
1104	110-629
1104	110-635
1104	110-672
1104	110-684
1104	0111
1104	111-1
1104	111-4-2020-0800
1104	111-4-2020-08000
1104	111-4-2020-14500
1104	111-4-2020-14600
1104	0112
1104	112-601-0-102
1104	112-826
1104	113-039
1104	113-321
1104	113-392
1104	113-998
1104	113A7739
1104	114-013
1104	114-4-2020-14500
1104	114-42020-14500
1104	115-039
1104	115-559
1104	115-599
1104	115-867
1104	116-052
1104	117-134-11
1104	119-6511
1104	120-001-300
1104	120-001300
1104	120-00195
1104	120-002012
1104	120-002013
1104	120-002014
1104	120-003148
1104	120-004503
1104	120-004878
1104	121-14(ZENITH)
1104	122-80
1104	123-021
1104	124-0178
1104	124J490
1104	126-4
1104	126-7
1104	126-7(ARVIN)
1104	130-30192
1104	130-30313
1104	130-338-00
1104	130-398-99
1104	130A17268
1104	130YE
1104	0131-0026
1104	0131-0035
1104	0131-0044
1104	0131-0053
1104	0131-026
1104	0131-053
1104	131A
1104	135-3
1104	135P4
1104	137-718
1104	140-013
1104	140-8
1104	140-9
1104	142-013
1104	142-014
1104	145A9254
1104	145A9786
1104	150
1104	150-018-9-001
1104	150-022-9-001
1104	151-011-9-011
1104	151-018-9-001
1104	151-023-9-001
1104	151-025-9-001
1104	151-029-9-001
1104	151-030-9-005
1104	151-040-7-003
1104	151-040-9-005
1104	151-045-9-001
1104	151-046-9-001
1104	151-1
1104	152-047
1104	154A3992
1104	162-005A
1104	162J2

RS 276-	Industry Standard No.
1104	163J2
1104	164J2
1104	165J2
1104	166J2
1104	173A3981
1104	173A3981-1
1104	173A4393
1104	174-1
1104	174-2
1104	174-3
1104	180-1
1104	185-014
1104	185-6
1104	185-6(RCA)
1104	185-6(RECT)
1104	186-014
1104	196-654
1104	0200
1104	200-6582-22
1104	200X7120-224
1104	201X3120-255
1104	202-5-2300-01710
1104	203
1104	212-18
1104	212-192
1104	212-21
1104	212-22
1104	212-23
1104	212-25
1104	212-254
1104	212-27
1104	212-33
1104	212-35
1104	212-36
1104	212-37
1104	212-38
1104	212-39
1104	212-41
1104	212-42
1104	212-47
1104	212-49
1104	212-50
1104	212-51
1104	212-57
1104	212-58
1104	212-59
1104	212-60
1104	212-61
1104	212-62
1104	212-64
1104	212-65
1104	212-7
1104	212-70
1104	212-75
1104	212-76
1104	212-76-02
1104	212-77
1104	212-79
1104	212-92
1104	212-94
1104	212-94B
1104	215-51
1104	215-52
1104	215-76(GE)
1104	215-76GE
1104	217-76-02
1104	220-003001
1104	229-0102
1104	229-0117
1104	229-0119
1104	229-0120
1104	229-0135
1104	229-0147
1104	229-0162
1104	229-1054-5
1104	229-1054-82
1104	229-1054-85
1104	229-1054-9
1104	229-5100-232
1104	229-5100-233
1104	229-5100-234
1104	232-0001
1104	232-0006
1104	232-1009
1104	232-1011
1104	0234
1104	0244
1104	247-255
1104	247-621
1104	260-10-033
1104	264D00505
1104	264D01112
1104	264F00601
1104	264F01011
1104	264F01012
1104	264F02001
1104	264F02402
1104	264P03001
1104	264P03607
1104	264P03701
1104	264P04206
1104	264P04301
1104	264P04402
1104	264P04702
1104	264P04703
1104	264P04801
1104	264P04901
1104	264P05001
1104	264P05002
1104	264P06601
1104	264P06606
1104	264P08002
1104	264P08801
1104	264P09801
1104	264P10102
1104	264P10103
1104	264P10105
1104	264Z00201
1104	265P03301
1104	276-1104
1104	288(SEARS)
1104	291-04
1104	291-20
1104	292-10
1104	295L001H01
1104	295L001M01
1104	295L001M02
1104	295L002H01
1104	295L002M03
1104	295L003H01
1104	295V002H01
1104	295V005H02

RS 276-	Industry Standard No.
1104	295V006H01
1104	295V006H02
1104	295V006H03
1104	295V006H05
1104	295V006H06
1104	295V006H07
1104	295V006H09
1104	295V007H02
1104	295V008H01
1104	295V012H01
1104	295V012H02
1104	295V012H03
1104	295V012H06
1104	295V014H01
1104	295V014H7
1104	295V015H02
1104	295V016H01
1104	295V017H01
1104	295V020H01
1104	295V020H01
1104	295V023H01
1104	295V027C01
1104	295V027C01-1
1104	295V028C01
1104	295V028C02
1104	295V028C03
1104	295V029C01
1104	295V029C02
1104	295V035C01
1104	296V006H03
1104	296V006H07
1104	296V020B02
1104	297V027C01
1104	0300
1104	301
1104	0304
1104	304B
1104	307A
1104	307B
1104	307C
1104	307D
1104	307F
1104	307H
1104	307K
1104	307M
1104	309-327-910
1104	309-327-927
1104	309-327-932
1104	310
1104	310-4
1104	0311
1104	0312
1104	0314
1104	315
1104	0320
1104	320A
1104	320B
1104	320C
1104	320D
1104	320F
1104	320H
1104	320K
1104	320M
1104	0321
1104	0322
1104	322-0147
1104	324
1104	324-0102
1104	324-0117
1104	324-0119
1104	324-0135
1104	324-0147
1104	324-0162
1104	325-0028-89
1104	325-0031-338
1104	325-0047-517
1104	325-0054-312
1104	325-0076-5
1104	325-0081-110
1104	325-0135-B
1104	325-0141-23
1104	325-0670-16
1104	325-1441-10
1104	325-1441-11
1104	325-1446-29
1104	354-9001-001
1104	354-9101-006
1104	354-9102-001
1104	354-9110-001
1104	359A
1104	359B
1104	359C
1104	359D
1104	359F
1104	359H
1104	359K
1104	359M
1104	366B
1104	369-2
1104	369-3
1104	380H61
1104	380K62
1104	380M63
1104	384A
1104	384B
1104	384C
1104	384D
1104	384F
1104	384H
1104	385A
1104	385B
1104	385C
1104	385D
1104	385P
1104	385H
1104	385K
1104	385KW
1104	385M
1104	386-1AY
1104	386-1CY
1104	386-1FY
1104	386AK
1104	386AW
1104	386AX
1104	386AY
1104	386BW
1104	386BY
1104	386CW
1104	386CX
1104	386CY
1104	386DW

RS 276-	Industry Standard No.	RS 276-	Industry Standard No.	RS 276-	Industry Standard No.	RS 276-	Industry Standard No.
1104	386DY	1104	630-052	1104	977-18B	1104	2093A12-1
1104	386FW	1104	642-219	1104	977-19	1104	2093A38-13
1104	386FX	1104	642-304	1104	977-19B	1104	2093A38-25
1104	386FY	1104	650-110	1104	977-20B	1104	2093A38-28
1104	386K	1104	656-141	1104	977-218	1104	2093A38-4
1104	386KX	1104	660-230	1104	977-22B	1104	2093A3D-2
1104	386KY	1104	684-652423-1	1104	977-23B	1104	2093A4-2
1104	386MW	1104	690O092B88	1104	977-25B	1104	2093A41-103
1104	386MY	1104	690Y031H33	1104	977-27B	1104	2093A41-104
1104	0400	1104	690Y039B52	1104	977-28B	1104	2093A41-108
1104	4000-11958	1104	690Y041H08	1104	977-2B	1104	2093A41-110
1104	0401	1104	690Y053B57	1104	977-8B	1104	2093A41-115
1104	0402	1104	690Y069B39	1104	991-00-1449	1104	2093A41-116
1104	410	1104	690Y080B47	1104	991-00-1449-1	1104	2093A41-126
1104	0411	1104	690Y080H52	1104	991-00-2440	1104	2093A41-131
1104	421-4027	1104	690Y080H91	1104	991-00-2440-1	1104	2093A41-139
1104	421-7443	1104	690Y082B40	1104	991-00-2440-2	1104	2093A41-151
1104	421-7443A	1104	690Y086H91	1104	991-00-7394	1104	2093A41-152
1104	421-9865	1104	690Y086B92	1104	991-00-7776	1104	2093A41-173
1104	422-1362	1104	690Y089B91	1104	992-10	1104	2093A41-180
1104	422-1866	1104	690Y092B88	1104	1000-129	1104	2093A41-182
1104	422-2540	1104	690Y094B55	1104	1001-11	1104	2093A41-185
1104	429-0989-68	1104	690Y098B53	1104	1002-0219	1104	2093A41-189
1104	430-31	1104	690Y105B31	1104	1002-6219	1104	2093A41-197
1104	435-40012	1104	690Y109B44	1104	1003-13	1104	2093A41-45
1104	0450	1104	690Y114H36	1104	1005-20	1104	2093A41-49
1104	454-A2534-1	1104	690Y116B41	1104	1006-24	1104	2093A41-51
1104	454-A2534-10	1104	690Y118B58	1104	1007-0951	1104	2093A41-54
1104	454A25	1104	690Y119H14	1104	1009-09	1104	2093A41-55
1104	0460	1104	0700	1104	1010(JULIETTE)	1104	2093A41-57
1104	460-1013	1104	700-137	1104	1010-9486	1104	2093A41-61
1104	464-280-19	1104	700A858-285	1104	1010-9494	1104	2093A41-62
1104	464-285-15	1104	700A858-286	1104	1011(VO-6C)	1104	2093A41-66
1104	464-285-19	1104	0701	1104	1012	1104	2093A41-75
1104	471-010	1104	0702	1104	1014	1104	2093A41-78
1104	474-004	1104	710	1104	1016-17	1104	2093A41-8
1104	474-025	1104	0727-50	1104	1016-78	1104	2093A41-81
1104	486-40235-002	1104	750-141	1104	1016-79	1104	2093A41-84
1104	511-898	1104	750A858-285	1104	1018-6963	1104	2093A41-86
1104	513-891	1104	750M63-149	1104	1020	1104	2093A41-88
1104	515-299	1104	751-2001-212	1104	1022-5548	1104	2093A41-89
1104	517-0021	1104	751-6300-001	1104	1024	1104	2093A42-7
1104	517-0025	1104	751-9001-124	1104	1033-17	1104	2093A5-2
1104	517-0031	1104	754-2000-002	1104	1035-8	1104	2093A6-1
1104	517-0033	1104	754-2000-005	1104	1038-1697	1104	2093A6-2
1104	518-499	1104	754-5000-021	1104	1040	1104	2093B38-14
1104	521-094	1104	754-5750-284	1104	1040-10	1104	2093B4-6
1104	522-726	1104	754-900-124	1104	1041-109	1104	2093B41-10
1104	522-998	1104	754-9000-953	1104	1041-63	1104	2093B41-12
1104	523-0001-001	1104	754-9001-124	1104	1041-64	1104	2093B41-14
1104	523-0001-002	1104	756	1104	1042-17	1104	2093B41-18
1104	523-0001-003	1104	792-238	1104	1043-1534	1104	2093B41-20
1104	523-0001-004	1104	800-006-00	1104	1043-7582	1104	2093B41-21
1104	523-0001-005	1104	800-013-00	1104	1044	1104	2093B41-22
1104	523-0001-006	1104	800-01300-504	1104	1045-0518	1104	2093B41-25
1104	523-0013-002	1104	800-014-00	1104	1045-0534	1104	2093B41-28
1104	523-0501-002	1104	800-014-01	1104	1048-9946	1104	2093B41-32
1104	523-0501-003	1104	800-01400	1104	1048-9995	1104	2093B41-34
1104	523-1000-001	1104	800-017-00	1104	1049-3435	1104	2093B41-35
1104	523-1000-882	1104	800-024-00	1104	1050	1104	2093B41-37
1104	523-1000882	1104	800-032-00	1104	1050-64	1104	2093B41-6
1104	523-1500-002	1104	800-038-00	1104	1059-7961	1104	2093B41-62
1104	525-24	1104	800-041-00	1104	1060	1104	2093B41-8
1104	525-498	1104	808-206	1104	1061-8379	1104	2093B41-9
1104	526-376	1104	822-1	1104	1061-8387	1104	2102
1104	527-798	1104	822-2	1104	1061-8916	1104	2102-014
1104	528-325	1104	822-3	1104	1061-8924	1104	2102-017
1104	530-082-1003	1104	822-4	1104	1063-79	1104	2102-074
1104	530-086-1	1104	822-5	1104	1065-9928	1104	2110R-42
1104	530-088-1	1104	851-0372-130	1104	1074-119	1104	2116-17
1104	530-088-2	1104	854-0372-020	1104	1077-0261	1104	2148-17
1104	530-094-1	1104	880-207-000	1104	1077-2366	1104	2152
1104	530-095-1	1104	903-00594	1104	1077-3836	1104	2182-17
1104	530-098-1	1104	903-104B	1104	1080-08	1104	2198-17
1104	530-099-1	1104	903-105B	1104	1095J2F	1104	2199-17
1104	530-099-3	1104	903-117B	1104	1096J2F	1104	2200-17
1104	530-099-4	1104	903-13B	1104	1103-88	1104	2206
1104	530-106-1	1104	903-14B	1104	1104-96	1104	2206-17
1104	530-109-1	1104	903-156B	1104	1106-29	1104	2211B
1104	530-111-1	1104	903-15B	1104	1106-36	1104	2231-17
1104	530-116-1	1104	903-164B	1104	1108-73	1104	2252
1104	530-120-1	1104	903-169B	1104	1110-86	1104	2283-17
1104	530-122-1	1104	903-197	1104	1115-15	1104	2285-17
1104	532-341	1104	903-28B	1104	1116-42	1104	2302
1104	532-341A	1104	903-303	1104	1117-76	1104	2315-046
1104	536J2F	1104	903-330	1104	1118-20	1104	2319-17
1104	537J2F	1104	903-49B	1104	1120-18	1104	2400-23
1104	538J2F	1104	903-52B	1104	1121-17	1104	2400-27
1104	539J2F	1104	903-69B	1104	1123(JULIETTE)	1104	2402
1104	540-008	1104	903-79B	1104	1138	1104	2402-461
1104	540-014	1104	903X00212	1104	1139-17	1104	2402-462
1104	540J2F	1104	916-33003-2	1104	1263A	1104	2402-463
1104	0557-010	1104	919-00-2440-1	1104	1289	1104	2405-459
1104	575-028	1104	919-00-2440-2	1104	01339	1104	2405-462
1104	575-042	1104	919-00-5045	1104	1405-11	1104	2405-463
1104	575-048	1104	919-00-7109	1104	1410-102(ROGERS)	1104	2408-331
1104	575-050	1104	919-00-7766	1104	1410-167	1104	2410-17
1104	580-029	1104	919-00-7776	1104	1412-170	1104	2411
1104	600	1104	919-001172	1104	1488(SEARS)	1104	2460-13
1104	600X0099-066	1104	919-001172-1	1104	1562	1104	2510-32
1104	601	1104	919-004326	1104	1611-17	1104	2522
1104	601X0048-066	1104	919-004799	1104	1674-17	1104	2523
1104	601X0152-066	1104	919-007109	1104	1713-17	1104	2606-299
1104	601X0224-066	1104	919-007109RA	1104	1827-17	1104	2703-390
1104	601X0227-066	1104	919-007394	1104	1848-17	1104	2704-389
1104	602X0019-000	1104	919-007776	1104	1849	1104	2762
1104	604B	1104	919-007776RA	1104	1849R	1104	2763
1104	606-113	1104	919-009459RA	1104	1850	1104	2766
1104	607-113	1104	919-01-0459	1104	1850R	1104	2777
1104	608	1104	919-01-0623	1104	1851-17	1104	2779
1104	608-030	1104	919-01-0829	1104	1854-17	1104	2784
1104	608-101	1104	919-01-0829-1	1104	1855-17	1104	2786
1104	608-113	1104	919-01-1211	1104	1859R	1104	2794
1104	609-030	1104	919-01-1212	1104	1885-17(RECT.)	1104	3001A
1104	609-113	1104	919-01-1339	1104	1941-17	1104	003007
1104	610	1104	919-010873	1104	1949-17	1104	003015
1104	610-001-103	1104	919-011212	1104	1950-17	1104	3023
1104	610C	1104	919-011215	1104	1970-16	1104	3069
1104	612	1104	919-011307	1104	1970-17	1104	3069(ARVIN)
1104	614	1104	919-013044	1104	1982-17	1104	3074A20
1104	616	1104	919-013060	1104	2000-304	1104	3074A21
1104	617-162	1104	919-013072	1104	2000-320	1104	3377
1104	617-163	1104	919-013081	1104	2010-02	1104	3529
1104	617-46	1104	936-10	1104	2017-111	1104	3700-153
1104	617-53	1104	936-20	1104	2017-114	1104	4013(RECTIFIER)
1104	617-62	1104	964-17443	1104	2021-05	1104	4014
1104	617-66	1104	964-174443	1104	2022-08	1104	4041-000-10150
1104	618	1104	964-19865	1104	2031-17	1104	04049B
1104	620	1104	964-21866	1104	2045-17	1104	4051-300-10150
1104	622	1104	972D7	1104	2052	1104	4101-685
1104	624	1104	977-10B	1104	2060-024	1104	4354-001C
1104	624-0009	1104	977-11B	1104	2061A45-72	1104	4403
1104	624-0010	1104	977-13B	1104	2061B45-35	1104	004567
1104	629A02	1104		1104		1104	4686-116-3
1104		1104		1104		1104	4686-147-3

RS 276-	Industry Standard No.	RS 276-	Industry Standard No.	RS 276-	Industry Standard No.	RS 276-	Industry Standard No.
1104	4686-148-3	1104	013339	1104	72113A	1104	130389-00
1104	4686-149-3	1104	13782	1104	72119	1104	130607
1104	4686-150-3	1104	14027	1104	72123	1104	131050
1104	4686-151-3	1104	14126-1	1104	72123A	1104	131245
1104	4686-177-3	1104	14588-9	1104	72130-1	1104	131476
1104	4686-87-3	1104	16681	1104	72145	1104	131501
1104	4800-224	1104	17002	1104	72158	1104	131502(RCA)
1104	4802-2000-012	1104	17443	1104	72171-1	1104	131950
1104	4802-2000-019	1104	19042	1104	72176	1104	132149
1104	4806-0000-004	1104	19865	1104	74200-8	1104	132814
1104	4822-130-30192	1104	20810-21	1104	74200-9	1104	133390(PWR. RECT)
1104	4822-130-30256	1104	20810-22	1104	75230-9	1104	133390(PWR.-RECT)
1104	4822-130-30259	1104	021154	1104	75700-14-02	1104	133543
1104	4822-130-50221	1104	24198(RECT.)	1104	75702-15-21	1104	133950
1104	5001-117	1104	025026	1104	75702-15-24	1104	135281
1104	5001-129	1104	025056	1104	77190-7	1104	135284
1104	5001-135	1104	25115-115	1104	77271-3	1104	135386
1104	5001-163	1104	25202-002	1104	78894	1104	136635
1104	5051-300-10150	1104	25810-51	1104	080040	1104	137652
1104	5205NI	1104	25840-51	1104	81513-8	1104	138172
1104	5205NL	1104	25840-52	1104	81514-2	1104	138173
1104	5300-82-1003	1104	26235-1	1104	83008	1104	139605
1104	5300-86-1	1104	27113-100	1104	88641	1104	140693
1104	5300-88-1	1104	27123-050	1104	88999-205	1104	140694
1104	5300-88-2	1104	27123-070	1104	90203-6	1104	141849
1104	5300-94-1	1104	27123-100	1104	91001	1104	141872-6
1104	5300-95-1	1104	27123-120	1104	93005	1104	143595
1104	5300-98-1	1104	27840-41	1104	93006	1104	144052
1104	5300-99-1	1104	28810-61	1104	93007	1104	144860
1104	5300-99-3	1104	031034	1104	93011	1104	145839
1104	5300-99-4	1104	34174	1104	93012	1104	147478-8-2
1104	5301-06-1	1104	34394	1104	93022	1104	161029
1104	5301-09-1	1104	34894	1104	93022A	1104	161030
1104	5301-11-1	1104	35287	1104	93025	1104	161031
1104	5301-13-1	1104	35287A	1104	93027	1104	161037
1104	5301-20-1	1104	35289	1104	93028	1104	165597
1104	5301-22-1	1104	35306	1104	95015	1104	165739
1104	5380-21	1104	35500	1104	95016	1104	166040
1104	5416	1104	35860	1104	99023-3	1104	166593
1104	5601-MS	1104	36201	1104	99203-005	1104	166726
1104	5601-NI	1104	36503	1104	99203-006	1104	166881
1104	5601-NL	1104	36503(AIRCASTLE)	1104	0099203-007	1104	166920
1104	5632-81B01-02	1104	36535	1104	99203-5	1104	166922
1104	5632-W03B	1104	36537	1104	99203-6	1104	167034
1104	5632-W06A	1104	36549	1104	99203-6(RCA)	1104	167413
1104	5632-W06B	1104	36554	1104	99203-9(RCA)	1104	167543
1104	5641-MV11	1104	36555	1104	100412	1104	167544
1104	5923	1104	36564	1104	100520	1104	168339
1104	6171-28	1104	37126-1	1104	100617	1104	168692
1104	6523-34	1104	37680	1104	100624	1104	169113
1104	7211-9	1104	37987	1104	101405	1104	169116
1104	7212-3	1104	38174	1104	103318	1104	169363
1104	7212-3A	1104	40024	1104	104081	1104	169565
1104	7213(DIODE)	1104	40024VM	1104	104213	1104	169570
1104	7213-0	1104	40013	1104	104273	1104	169590
1104	7213-7	1104	40265	1104	104325	1104	169765
1104	72148	1104	40383	1104	104615	1104	170133
1104	7314	1104	41001-4	1104	106379	1104	170297
1104	7342-1	1104	42221	1104	107540	1104	170750
1104	7568	1104	44002(RCA)	1104	110043	1104	170856
1104	7962	1104	44465-3	1104	110388	1104	170955
1104	8000-00003-044	1104	44465-4	1104	110496	1104	170967
1104	8000-00004-044	1104	44465-5	1104	110629	1104	170967(RECTIFIER)
1104	8000-00004-064	1104	44465-6	1104	110636	1104	170970-1
1104	8000-00004-068	1104	44936	1104	110875	1104	171019
1104	8000-00005-022	1104	44937	1104	111086	1104	171149-024
1104	8000-00005-152	1104	44938	1104	111516	1104	171162-252
1104	8000-00006-201	1104	45810-51	1104	111642	1104	171305
1104	8000-00006-231	1104	46140-2	1104	111776	1104	171416
1104	8000-00011-041	1104	47126	1104	111820	1104	171561
1104	8000-00011-044	1104	47126-003	1104	112017	1104	171657
1104	8000-00011-104	1104	47126-1	1104	112018	1104	172551
1104	8000-00030-010	1104	47126-12	1104	112329	1104	172721
1104	8000-00035-001	1104	47126-1A	1104	112531	1104	185704
1104	8000-00035-002	1104	47126-2	1104	112826	1104	194917
1104	8000-0004-044	1104	47126-3	1104	113039	1104	195648-6
1104	8000-0004-P-061	1104	47126-3A	1104	113392	1104	196184-3
1104	8000-0004-P067	1104	47126-4	1104	113398	1104	196259-4
1104	8000-0004-P068	1104	47126-4A	1104	114013	1104	200062-5-36
1104	8000-0004-P067	1104	47126-6	1104	115039	1104	200062-6-32
1104	8000-0004-P068	1104	47126-7	1104	115559	1104	200064-6-115
1104	8000-00049-010	1104	47126-9	1104	115559A	1104	200064-6-119
1104	8000-0005-022	1104	47127-3	1104	115599	1104	200064-6-120
1104	8003-115	1104	48106	1104	115867	1104	202463
1104	8020-203	1104	50745-01	1104	116052	1104	214105
1104	08050	1104	50745-02	1104	116054	1104	215669
1104	8102-206	1104	50745-06	1104	117145	1104	216014
1104	8300-8	1104	50745-07	1104	117145A	1104	216020
1104	8500-206	1104	50745-08	1104	118244	1104	216449
1104	8710-55	1104	53099-1	1104	118825	1104	216817
1104	8800-201	1104	53099-3	1104	118873	1104	218612
1104	8800-206	1104	53300-41	1104	120231	1104	219245
1104	8840-56	1104	55206	1104	120503	1104	219935
1104	8864-1	1104	055210	1104	120544	1104	221128
1104	8910-55	1104	58810-80	1104	121468	1104	222611
1104	8999-201	1104	58810-85	1104	121680	1104	222315
1104	8999-205	1104	58840-111	1104	122129	1104	223216
1104	9100-1	1104	58840-116	1104	122788	1104	223323
1104	9300-5	1104	59557-49	1104	123004	1104	223358
1104	9300-6	1104	59844-1	1104	123296	1104	223462
1104	9300-7	1104	59844-3	1104	123702	1104	223467
1104	9301-1	1104	61011-2	1104	123804	1104	223489
1104	9302-2	1104	66682-42	1104	123804-1	1104	223720
1104	9302-2A	1104	68177	1104	124098	1104	223724
1104	9302-3	1104	68177A	1104	124812	1104	223753
1104	9330-006-11112	1104	68504-78	1104	125105	1104	224159
1104	9330-229-20112	1104	68645-03	1104	125127	1104	224597
1104	9348-3	1104	69111	1104	125458	1104	224774
1104	9650-001	1104	69213-78	1104	125471	1104	225200
1104	10180	1104	70064-7	1104	125529	1104	225297
1104	10386	1104	70066-3	1104	125549	1104	225592
1104	10909	1104	70066-4	1104	125835	1104	226058
1104	11282-6	1104	70270-23	1104	125856	1104	226182
1104	11303-9	1104	70432-16	1104	125964	1104	226237
1104	11332-1	1104	71006-28	1104	125993	1104	226546
1104	11339-2	1104	71006-30	1104	126131	1104	226788
1104	11399-8	1104	71588-1	1104	126148	1104	227015
1104	11401-3	1104	71588-2	1104	126176	1104	227348
1104	11503-9	1104	71588-5	1104	126320	1104	227565
1104	11555-9	1104	71588-6	1104	126855	1104	227720
1104	11559-9	1104	71779	1104	126861	1104	227744
1104	11586-7	1104	000072020	1104	127102	1104	229522
1104	11605-2	1104	72025	1104	127176	1104	230218
1104	11746	1104	72026	1104	127396	1104	231150
1104	12602	1104	72041	1104	127695	1104	231339
1104	12720	1104	000072050	1104	128256	1104	231665
1104	12736	1104	72058	1104	129029	1104	231669
1104	12746	1104	72058A	1104	129029(DIO)	1104	231923
1104	12768	1104	72097	1104	129213	1104	232203
1104	12786	1104	72101	1104	129334	1104	233011
1104	12788	1104	72103	1104	129494	1104	233561
1104	12803	1104	72109	1104	129759	1104	233597
1104	12837	1104	72113	1104	130052	1104	234552
1104	12844			1104	130338-01	1104	234565

RS 276-	Industry Standard No.	RS 276-	Industry Standard No.	RS 276-	Industry Standard No.	RS 276-	Industry Standard No.	RS 276-	Industry Standard No.
1104	234611	1104	530127-6	1104	924801-5	1104	2002402-29	1114	A-118038
1104	234761	1104	530135-1003	1104	924805-5	1104	2003069	1114	A04093-A
1104	235313	1104	530135-3	1104	924805-8	1104	2003069-5	1114	A04093-X
1104	235545	1104	530151-1	1104	945820-4	1104	2003073-67	1114	A04093A
1104	235546	1104	530151-1001	1104	971457	1104	2003073-68	1114	A04093A2
1104	237227	1104	530162-1	1104	971458	1104	2004358-142	1114	A04093X
1104	237453	1104	530162-1001	1104	972216	1104	2006422-133	1114	A04212-A
1104	237929	1104	530171-1	1104	972571-4	1104	2006436-38	1114	A04230
1104	259221	1104	530171-1001	1104	972571-6	1104	2006436-89	1114	A04230-A
1104	240456	1104	530171-1002	1104	972571-7	1104	2006441-91	1114	A04241-A
1104	240594	1104	530171-1003	1104	972571-20	1104	2006463-89	1114	A04294
1104	240603	1104	530171-3	1104	973925-20	1104	2006512-40	1114	A04294-1
1104	242029	1104	530180-1001	1104	973936-1	1104	2006582	1114	A059-118
1104	242226	1104	533034	1104	973936-10	1104	2006582-22	1114	A068-104
1104	243364	1104	533058	1104	973936-11	1104	2006582-23	1114	A08
1104	248917	1104	535151-1001	1104	973936-12	1104	2006582-24	1114	A1000
1104	249217	1104	0537640	1104	973936-13	1104	2006623-49	1114	A10P
1104	252817	1104	0537640(DIO)	1104	973936-14	1104	2008292-89	1114	A114C
1104	258842	1104	551026	1104	973936-15	1104	2008293-9	1114	A13M2
1104	259315	1104	0552006	1104	973936-16	1104	2008302-41	1114	A14PD1
1104	259368	1104	0552006H	1104	973936-17	1104	2010097-49	1114	A14PD2
1104	259878	1104	0552007H	1104	973936-18	1104	2013019-117	1114	A14PD3
1104	260429	1104	0552010	1104	973936-19	1104	2057062-0702	1114	A1H1
1104	261465	1104	0552010H	1104	973936-2	1104	2060041	1114	A1H5
1104	261596	1104	575028	1104	973936-21	1104	2206582-22	1114	A1H9
1104	261898	1104	575042	1104	973936-3	1104	2327031	1114	A1K1
1104	261975	1104	0575050	1104	973936-4	1104	2327041	1114	A1K5
1104	262112	1104	575051	1104	973936-5	1104	2330201	1114	A1K9
1104	262872	1104	0576054	1104	973936-6	1104	2330251	1114	A1M1
1104	265164	1104	0576054(BIAS)	1104	973936-7	1104	2330251H	1114	A1M5
1104	265235	1104	580029	1104	973936-8	1104	2330252	1114	A1M9
1104	265236	1104	603114	1104	973936-9	1104	2330254	1114	A2458
1104	265634	1104	606113	1104	973962-1	1104	2330256	1114	A2H1
1104	266585	1104	606131	1104	980143	1104	2330562	1114	A2H4
1104	276097	1104	608030	1104	980164	1104	2330553	1114	A2H5
1104	300233	1104	608101	1104	980540	1104	2330561	1114	A2H9
1104	300315	1104	608113	1104	980964	1104	2330562	1114	A2K1
1104	300524	1104	609030	1104	981445	1104	2330611	1114	A2K4
1104	300550	1104	609113	1104	981739	1104	2330612	1114	A2K5
1104	301586	1104	610020	1104	981952	1104	2330721	1114	A2K9
1104	330005	1104	610112	1104	981953	1104	2330773	1114	A2M1
1104	330018	1104	611111	1104	981954	1104	2331141	1114	A2M4
1104	330019	1104	612112	1104	981955	1104	2331142	1114	A2M5
1104	348048-2	1104	612130	1104	981956	1104	232141	1114	A3H1
1104	348053-3	1104	612132	1104	982214	1104	2337071	1114	A3H3
1104	348054-1	1104	613020	1104	982253	1104	2347021	1114	A3H5
1104	348054-10	1104	613330	1104	982377	1104	2486836	1114	A3H9
1104	348054-11	1104	614010	1104	982526	1104	2487305	1114	A3K1
1104	348054-14	1104	619130	1104	982823	1104	2498513	1114	A3K3
1104	348054-2	1104	651038	1104	983101	1104	3430063	1114	A3K5
1104	348054-5	1104	652092	1104	983413	1104	1N2090	1114	A3K9
1104	348054-6	1104	652615	1104	984182	1114	.5B10	1114	A3M1
1104	348054-9	1104	699739	1104	984183	1114	.5B12	1114	A3M3
1104	348055-2	1104	700043-00	1104	984184	1114	.5B7	1114	A3M5
1104	348055-3	1104	700063-00	1104	984189	1114	.5B8	1114	A3M9
1104	464010	1104	700647	1104	984254	1114	.5J10	1114	A4212A
1104	464070	1104	700663	1104	984522	1114	.5J12	1114	A4H1
1104	489752-005	1104	700664	1104	984594	1114	.5J7	1114	A4H5
1104	489752-013	1104	740183	1104	984713	1114	.5J8	1114	A4H9
1104	489752-014	1104	740289	1104	984882	1114	.7B10	1114	A4K1
1104	489752-015	1104	740570	1104	985104	1114	.7B12	1114	A4K5
1104	489752-017	1104	740630	1104	985105	1114	.7B7	1114	A4K9
1104	489752-022	1104	741101	1104	985218	1114	.7B8	1114	A4M1
1104	489752-025	1104	741740	1104	985472	1114	.7J10	1114	A4M5
1104	489752-027	1104	741865	1104	986414	1114	.7J12	1114	A4M9
1104	489752-028	1104	742008	1104	986637	1114	.7J7	1114	A5H1
1104	489752-035	1104	742009	1104	986935	1114	.7J8	1114	A5H2
1104	489752-037	1104	746003	1104	988049				
1104	489752-043	1104	746004	1104	988051				
1104	489752-050	1104	752309	1104	992171				
1104	489752-051	1104	760202-0003	1104	1043176-1				
1104	489752-066	1104	761113	1104	1043176-2				
1104	489752-073	1104	771907	1104	1043176-3				
1104	489752-075	1104	772713	1104	1043176-4				
1104	489752-092	1104	801707	1104	1043176-5				
1104	489752-094	1104	801711	1104	1045013				
1104	489752-097	1104	801714	1104	1045154-1				
1104	489752-108	1104	801715	1104	1045154-2				
1104	489752-123	1104	801716	1104	1045154-3				
1104	500859	1104	801723	1104	1045494-1				
1104	501010	1104	815138	1104	1223772				
1104	501152	1104	815142	1104	1223928				
1104	0510006	1104	817042	1104	1223929				
1104	0517550-3	1104	817043	1104	1464778-9				
1104	0517750	1104	817044	1104	1471858-1				
1104	0517750H	1104	817053	1104	1471872-001				
1104	530045-3	1104	817064	1104	1471872-004				
1104	530051-2	1104	817066	1104	1471872-006				
1104	530057-1	1104	817067	1104	1471872-008				
1104	530063-11	1104	817068	1104	1471872-6				
1104	530063-12	1104	817068P	1104	1472171-32				
1104	530063-6	1104	817079	1104	1472460-1				
1104	530063-7	1104	817088	1104	1472460-16				
1104	530071-1	1104	817104	1104	1472460-4				
1104	530071-2	1104	817109	1104	1472460-5				
1104	530071-3	1104	817111	1104	1471478-10				
1104	530072-1	1104	817112	1104	1474777-8				
1104	530072-10	1104	817114	1104	1474778				
1104	530072-1017	1104	817117	1104	1474778-013				
1104	530072-1019	1104	817120	1104	1474778-10				
1104	530072-11	1104	817121	1104	1474778-11				
1104	530072-14	1104	817122	1104	1474778-2				
1104	530072-15	1104	817128	1104	1474778-3				
1104	530072-2	1104	817130	1104	1474778-9				
1104	530072-4	1104	817133	1104	1474788-10				
1104	530072-5	1104	817134	1104	1476161-13				
1104	530072-9	1104	817135	1104	1476171-1				
1104	530082-1	1104	817138	1104	1476171-11				
1104	530082-1003	1104	817140	1104	1476171-19				
1104	530082-1004	1104	817141	1104	1476171-27				
1104	530084-4	1104	817143	1104	1476171-28				
1104	530086-1	1104	817148	1104	1476171-29				
1104	530088-1003	1104	817156	1104	1476171-31				
1104	530088-1004	1104	817157	1104	1476171-8				
1104	530088-4	1104	817161	1104	1800006				
1104	530093-1001	1104	817164	1104	1800018				
1104	530093-3	1104	817166	1104	1811119				
1104	530094-1	1104	817167	1104	1956486				
1104	530095-1	1104	817180	1104	1961843				
1104	530098-1	1104	817195	1104	1962594				
1104	530098-1001	1104	817196	1104	1966808				
1104	530099-1	1104	817962	1104	1967813				
1104	530099-3	1104	852158-7-1	1104	1969497				
1104	530099-4	1104	852158-7.1	1104	2000625-36				
1104	530099-6	1104	921608	1104	2000626-32				
1104	530109-1	1104	922092	1104	2000646-115				
1104	530111-1	1104	922094	1104	2000646-119				
1104	530111-1001	1104	922183	1104	2000646-120				
1104	530116-1003	1104	922311	1104	2000648-120				
1104	530122-1	1104	922360	1104	2001786-141				
1104	530124-1	1104	922567	1104	2001786-207				
1104	530124-3	1104	922799	1104	2002332-57				
1104	530126-1	1104	922860	1104	2002332-58				
1104	530127-1	1104	922969	1104	2002336-115				
1104	530127-4	1104	924605-3						

RS 276-	Industry Standard No.	RS 276-	Industry Standard No.	RS 276-	Industry Standard No.	RS 276-	Industry Standard No.	RS 276-	Industry Standard No.
1114	A5H5	1114	BBGS	1114	DB210	1114	DD058	1114	E8KEB125C500Q
1114	A5H9	1114	BR-108	1114	DD058	1114		1114	ET57X39
1114	A5K1	1114	BXY10	1114	DD206B	1114		1114	ET57X40
1114	A5K2	1114	BT100	1114	DD268	1114		1114	EU16X20
1114	A5K5	1114	BT1001	1114	DER1	1114		1114	EV16X20
1114	A5K9	1114	BT1002	1114	DG-1N	1114		1114	F10
1114	A5M1	1114	BY103	1114	DG-1NR	1114		1114	F11034
1114	A5M2	1114	BY104	1114	DG1K	1114		1114	F114E
1114	A5M5	1114	BY105	1114	DG1N	1114		1114	F14C
1114	A5M9	1114	BY108	1114	DG1NR	1114		1114	F14D
1114	A6N1	1114	BY109	1114	DI-650	1114		1114	F14E
1114	A6N5	1114	BY1101	1114	DI650	1114		1114	F14F
1114	A6N9	1114	BY1102	1114	DICR1	1114		1114	F14H
1114	A725EH2AB1	1114	BY118	1114	DID	1114		1114	F14J
1114	A7568300	1114	BY119	1114	DIJ72163	1114		1114	F8
1114	A75779500	1114	BY12 00	1114	DIJ72167	1114		1114	FA-4
1114	A7579500	1114	BY1200	1114	DIJ72170	1114		1114	FA8
1114	A758011	1114	BY1201	1114	DIJ72171	1114		1114	FDH600
1114	A7580111	1114	BY1202	1114	DIJ72174	1114		1114	FG-2N
1114	A7580910	1114	BY127	1114	DIJ72290	1114		1114	FG-2N/10D-4
1114	A7582000	1114	BY127/500	1114	DIJ72291	1114		1114	FG-2N2
1114	A759500	1114	BY127/600	1114	DIK	1114		1114	FG2N
1114	A7978850	1114	BY127/700	1114	DP100	1114		1114	FG2PC
1114	A7978855	1114	BY128	1114	DR1000	1114		1114	FR-10
1114	A7N1	1114	BY129	1114	DR210	1114		1114	FR2-005
1114	A7N5	1114	BY133	1114	DR700	1114		1114	FR2-02(RECT)
1114	A7N9	1114	BY158	1114	DR800	1114		1114	FR2-02C
1114	A800	1114	BYX10	1114	DR900	1114		1114	FR2-04
1114	AA1000	1114	BYX12/400	1114	DR8107	1114		1114	FR2-10
1114	AA800	1114	BYX13/600	1114	DR8108	1114		1114	FT-1M
1114	AD-29B	1114	BYX22-1200	1114	DS-113A	1114		1114	FT-1P
1114	AD-29S	1114	BYX22/800	1114	DS-113B	1114		1114	FT10
1114	AD29S	1114	BYX55009	1114	DS-15A(SANYO)	1114		1114	FT1M
1114	AD4002	1114	BYX57-500	1114	DS-15B(SANYO)	1114		1114	FT1P
1114	AD4003	1114	BYX57-600	1114	DS-16A(SANYO)	1114		1114	FU1M
1114	AD4004	1114	BYX60-700	1114	DS-1K	1114		1114	FU1K
1114	AD4005	1114	BYY37	1114	DS-2M	1114		1114	FU10
1114	AD4006	1114	BYY91	1114	DS113B	1114		1114	FU1K
1114	AD4007	1114	C-20	1114	DS135B	1114		1114	FU1NA
1114	AD8	1114	C0410	1114	DS15A	1114		1114	G10
1114	AP1(DIODE)	1114	C1.0EO2	1114	DS16NA	1114		1114	G100K
1114	AH1005	1114	CA102PA	1114	DS16NB(SONY)	1114		1114	G100M
1114	AH1010	1114	CA102RA	1114	DS1M	1114		1114	G8
1114	AH1015	1114	CA102VA	1114	DS2M	1114		1114	GE-505
1114	AH14	1114	CA152VA	1114	DT230B	1114		1114	GE-509
1114	AH805	1114	CC102PA	1114	DU1000	1114		1114	GE-509A
1114	AH810	1114	CC102RA	1114	DU800	1114		1114	GE-510
1114	AH814	1114	CC102RA	1114	E.857X9	1114		1114	GE-511
1114	AH815	1114	CC102VA	1114	E10	1114		1114	GE-531
1114	AM-G-5A	1114	CC152VA	1114	E108(ELCOM)	1114		1114	GE505
1114	AM-G-5C	1114	CE508	1114	E1145R	1114		1114	GEBR-1000
1114	AN-05B	1114	CE510	1114	E1145RED	1114		1114	GER4001
1114	A04093-X	1114	CERT2	1114	E1145RJH	1114		1114	GER4002
1114	A04230-A	1114	CERT720	1114	E1M3	1114		1114	GER4003
1114	A04241-A	1114	CERT720A	1114	E1M5	1114		1114	GER4004
1114	AR-23(DIO)	1114	CERT720B	1114	E758(ELCOM)	1114		1114	GER4005
1114	AR-24	1114	CERT720C	1114	E760(ELCOM)	1114		1114	GER4006
1114	AR23	1114	CERT72A	1114	E8	1114		1114	GER4007
1114	AR24	1114	CERT72C	1114	E9	1114		1114	GH3P
1114	B1H1	1114	CERT72D	1114	EA080	1114		1114	GI237
1114	B1H5	1114	CERT72F	1114	EA100	1114		1114	GM-1C
1114	B1H9	1114	CERT3	1114	EC100	1114		1114	GMIJ2
1114	B1K1	1114	CERT730	1114	ECG125	1114		1114	GRU2A
1114	B1K5	1114	CERT730A	1114	ECG506	1114		1114	GSR1
1114	B1K9	1114	CERT730B	1114	ED23(ELCOM)	1114		1114	H-881
1114	B1M1	1114	CERT730C	1114	ED2847	1114		1114	H1000
1114	B1M5	1114	CERT73A	1114	ED2848	1114		1114	H621
1114	B1M9	1114	CERT73B	1114	ED2849	1114		1114	H800
1114	B24-06B	1114	CERT73C	1114	ED2910	1114		1114	H815
1114	B24-06C	1114	CERT73D	1114	ED2911	1114		1114	HA-1
1114	B2620	1114	CERT73F	1114	ED2912	1114		1114	HA1000
1114	B2H1	1114	CP102PA	1114	ED2913	1114		1114	HA800
1114	B2H5	1114	CP102RA	1114	ED2922	1114		1114	HC72
1114	B2H9	1114	CGJ-1	1114	ED2923	1114		1114	HC720
1114	B2K1	1114	COD1537	1114	ED2924	1114		1114	HC73
1114	B2K5	1114	COD1538	1114	ED3007	1114		1114	HC730
1114	B2K9	1114	COD1617	1114	ED3007A	1114		1114	HD-2000108
1114	B2M1	1114	COD1618	1114	ED3007B	1114		1114	HD-20008
1114	B2M5	1114	CODI537	1114	ED3008	1114		1114	HEP159
1114	B2M9	1114	CODI538	1114	ED3008A	1114		1114	HEP160
1114	B3H1	1114	CODI617	1114	ED3008B	1114		1114	HEP170
1114	B3H5	1114	CP102	1114	ED3009	1114		1114	HEPR0055
1114	B3H9	1114	CP102PA	1114	ED3009A	1114		1114	HEPR0056
1114	B3K1	1114	CP102RA	1114	ED3009B	1114		1114	HEPR0600
1114	B3K5	1114	CP102VA	1114	ED3010	1114		1114	HEPR0602
1114	B3K9	1114	CP103	1114	ED3010A	1114		1114	HEPR0604
1114	B3M1	1114	CP152VA	1114	ED3010B	1114		1114	HEPR0606
1114	B3M5	1114	CTN100	1114	ED8-0004	1114		1114	HEPR3010
1114	B3M9	1114	CTN1000	1114	EP100	1114		1114	HP-1
1114	B3N1	1114	CTN300	1114	EM507	1114		1114	HP-17
1114	B3N5	1114	CTN400	1114	EM508	1114		1114	HP-1A
1114	B3N9	1114	CTN50	1114	EM510	1114		1114	HP-1B
1114	B4H1	1114	CTN500	1114	EP1000	1114		1114	HP-1C
1114	B4H5	1114	CTN600	1114	EP16X10	1114		1114	HP-1Z
1114	B4H9	1114	CTN800	1114	EP16X11	1114		1114	HP-20071
1114	B4K1	1114	CTP100	1114	EP16X24	1114		1114	HP-SD1/BB-4
1114	B4K5	1114	CTP1000	1114	EP57X5	1114		1114	HP1
1114	B4K9	1114	CTP200	1114	EP6X11	1114		1114	HP1A
1114	B4M1	1114	CTP300	1114	EP800	1114		1114	HP1B
1114	B4M5	1114	CTP400	1114	EPB22-15	1114		1114	HP1C
1114	B4M9	1114	CTP50	1114	ER1001	1114		1114	HP1Z
1114	B5H1	1114	CTP500	1114	ER107D	1114		1114	HPSD-1
1114	B5H5	1114	CTP600	1114	ER108D	1114		1114	HPSD-14
1114	B5H9	1114	CTP800	1114	ER186	1114		1114	HPSD-1A
1114	B5K1	1114	CXL506	1114	ER187	1114		1114	HPSD-1B
1114	B5K5	1114	CXL507	1114	ER308	1114		1114	HPSD-1C
1114	B5K9	1114	CY100	1114	ER310	1114		1114	HPSD-1C(SEARS)
1114	B5M1	1114	D1000	1114	ER801	1114		1114	HPSD-1Z
1114	B5M5	1114	D105C	1114	ER81	1114		1114	HPSD1
1114	B5M9	1114	D108	1114	ERB24-06	1114		1114	HPSD1Z
1114	B6N1	1114	D1201A	1114	ERB24-04A	1114		1114	HS310S
1114	B6N5	1114	D1201B	1114	ERB24-04B	1114		1114	HS31100
1114	B6N9	1114	D1201M	1114	ERB24-04C	1114		1114	HSC1
1114	B7579500	1114	D1201N	1114	ERB24-04D	1114		1114	HSPD-1A(SONY)
1114	B7978850	1114	D1201P	1114	ERB24-06B	1114		1114	HSPD-1A(SONY-5B)
1114	BTN1	1114	D172-P	1114	ERB24-06C	1114		1114	HSPD1
1114	BTN5	1114	D1D	1114	ERB24-06D	1114		1114	IP20-0026
1114	BTN9	1114	D1F	1114	ERC01-06	1114		1114	IR10D3L
1114	BA133	1114	D1K	1114	ERC04-10	1114		1114	IR1P
1114	BA145	1114	D431-P	1114	ERC26-131	1114		1114	IR3D
1114	BAY15	1114	D4L	1114	ERC26-13L	1114		1114	IR1F
1114	BAY16	1114	D5H	1114	ERC27-13	1114		1114	J-10
1114	BAY23	1114	D800	1114	ERCC4-06	1114		1114	J-8
1114	BAY90	1114	D85C	1114	ERD1000	1114		1114	J24H183
1114	BB-10	1114	D88	1114	ERD700	1114		1114	J24I215
1114	BB-108	1114	D8G	1114	ERD800	1114		1114	JC-00017
1114	BB-2	1114	D8HZ	1114	ERD900	1114		1114	JCN7
1114	BB-4	1114	D8L	1114	ES-16X16	1114		1114	JCV7
1114	BB-6	1114	D8M	1114	ES16X16	1114		1114	K1K5
1114	BB10	1114	DA058	1114	ES16X20	1114		1114	K1M5
1114	BB2	1114	DA206B	1114	ES16X28	1114		1114	K2H5
1114	BB4	1114		1114	ES16X31	1114		1114	K2K5
1114	BB6	1114		1114	ES57X8	1114		1114	K2M5
1114	BB6S	1114		1114	ES57X9	1114		1114	

RS 276-	Industry Standard No.	RS 276-	Industry Standard No.	RS 276-	Industry Standard No.	RS 276-	Industry Standard No.	RS 276-	Industry Standard No.
1114	K3H5	1114	R5B	1114	SB-2	1114	SLA19C		
1114	K3K5	1114	R5C	1114	SB-2(CENTERING)	1114	SLA2616		
1114	K5M5	1114	R6	1114	SB-2B	1114	SLA2617		
1114	K4H5	1114	R8	1114	SB-2C	1114	SLA3196		
1114	K4K5	1114	R8/2H	1114	SB-2CGL	1114	SLA560		
1114	K4M5	1114	RA-1C	1114	SB-2CH	1114	SLA561		
1114	K5H5	1114	RA-1Z	1114	SB-2T	1114	SLA01		
1114	K5K5	1114	RA-1ZC	1114	SB2C	1114	SM-150		
1114	K5M5	1114	RA-2	1114	SB2CH	1114	SM-150-005		
1114	KBF-02	1114	RA-2C	1114	SB2G	1114	SM-150-02		
1114	KCD-80P11/1+12/1	1114	RA-Z'	1114	SC10	1114	SM-150-02(BOOST)		
1114	KCD-80P11/1912/1	1114	RA1	1114	SC10A	1114	SM-150-10		
1114	KDC.80P11/1&12/1	1114	RA1Z	1114	SC8	1114	SM-150-6(FOCUS)		
1114	KDC.80P11/1+12/1	1114	RA2C	1114	SC8A	1114	SM-150-D		
1114	KSKE12C5200	1114	RC080	1114	SCA10	1114	SM-150A		
1114	KSKE12C500	1114	RE-51	1114	SCA8	1114	SM100		
1114	L82H	1114	RE-90	1114	SCBR05F	1114	SM101		
1114	LCO.09M1113	1114	RE51	1114	SCBR1F	1114	SM103		
1114	LM1932	1114	RE55	1114	SCBR35F	1114	SM150		
1114	M102	1114	RE90	1114	SCBR6F	1114	SM150-01		
1114	M41032A	1114	REN125	1114	SCE10	1114	SM150-02		
1114	M702	1114	REN506	1114	SCB8	1114	SM150-11		
1114	M702B	1114	REN90	1114	SCT1	1114	SM150-6		
1114	M72	1114	RF-1A	1114	SCT2	1114	SM150A		
1114	MT20	1114	RP10K35	1114	SCT3	1114	SM150B		
1114	MT20A	1114	RPJ70149	1114	SCT4	1114	SM150C		
1114	MT20B	1114	RPJ70971	1114	SCT5	1114	SM150D		
1114	MT20C	1114	RPJ70976	1114	SD-1	1114	SM150S		
1114	MT2A	1114	RH-1	1114	SD-1(BOOST)	1114	SM150SB		
1114	MT2B	1114	RH-1B	1114	SD-1-30DA	1114	SM170		
1114	MT2C	1114	RH-1C	1114	SD-1A4F	1114	SM180		
1114	MT3	1114	RH-1V	1114	SD-1B(HF)	1114	SM200		
1114	MT30	1114	RH-DX0004TAZZ	1114	SD-1BHF	1114	SM270		
1114	MT30A	1114	RH-DX0017CEZZ	1114	SD-1C	1114	SM280		
1114	MT30B	1114	RH-DX0028CEZZ	1114	SD-1VHF	1114	SM300		
1114	MT30C	1114	RH-DX0029CEZZ	1114	SD-1Z	1114	SM517		
1114	MT3A	1114	RH-DX0041CEZZ	1114	SD-22	1114	SM518		
1114	MT3B	1114	RH-DX0042CEZZ	1114	SD-6AUF	1114	SM520		
1114	MT3C	1114	RH-DX0045CEZZ	1114	SD-ERB24-04D	1114	SM70		
1114	M82	1114	RH-DX0051CEZZ	1114	SD-ERC26-13	1114	SM71		
1114	M8HZ	1114	RH-DX0056TAZZ	1114	SD1-11	1114	SM73		
1114	M91A06	1114	RH-DX0062CEZZ	1114	0000008D1A	1114	SM73(DIODE)		
1114	MB-1D	1114	RH-DX0065CEZZ	1114	SD1BHF	1114	SM770		
1114	MB-1P	1114	RH-DX0065CZZ	1114	SD1C	1114	SM780		
1114	MC070	1114	RH-DX0073CEZZ	1114	SD2	1114	SMB0		
1114	MC070A	1114	RH-DX0081CEZZ	1114	SD22	1114	SMB00		
1114	MC080	1114	RH-DX0090CEZZ	1114	SD2C	1114	SMB1		
1114	MC080A	1114	RH-DX0096CEZZ	1114	SD7(RECT)	1114	SMB3		
1114	MC090	1114	RH1	1114	SD8	1114	SO10G		
1114	MC090A	1114	RH1B	1114	SD800	1114	SOA		
1114	MC100	1114	RL005	1114	SD910	1114	SR-1K-2A		
1114	MC1527	1114	RL010	1114	SD910A	1114	SR-1K-2B		
1114	MC1528	1114	RL020	1114	SD910S	1114	SR-1K-2C		
1114	MC1529	1114	RL040	1114	SD98	1114	SR-1K-2D		
1114	MH72	1114	RL060	1114	SD98A	1114	SR-2		
1114	MH720	1114	RM-2AV	1114	SD98S	1114	SR-35		
1114	MH730	1114	RM-2C	1114	SE-05-D	1114	SR-9001		
1114	MM10	1114	RM1C	1114	SE0	1114	SR-9007		
1114	MM7	1114	RM2C	1114	SE05C	1114	SR1156		
1114	MM8	1114	RO2A	1114	SEI1730	1114	SR150		
1114	MM9	1114	RS1296(SEARS)	1114	SELEN-92	1114	SR1C-20		
1114	MP549	1114	RSLND0003CEZZ	1114	SF-1	1114	SR1EM-20		
1114	MR-1C	1114	RT-210	1114	SF1	1114	SR1FM-12		
1114	MR-1M	1114	RT1686	1114	SF1CN1	1114	SR1FM-20		
1114	MR-852	1114	RT210	1114	SF3CN1	1114	SR1FM-20		
1114	MR1031	1114	RT8665	1114	SF4	1114	SR1FM-4		
1114	MR1M	1114	RU-IN	1114	SF4CN1	1114	SR1FM20		
1114	MR2266	1114	RU1C	1114	SF5	1114	SR1HM-8		
1114	MR2272	1114	RU2	1114	SF6	1114	SR1K-20		
1114	MR2273	1114	RU2V	1114	SFR25B	1114	SR25		
1114	MR801	1114	S-05-10	1114	SFR26B	1114	SR2A-12		
1114	MR811	1114	S-05/01	1114	SH1B	1114	SR2A-8		
1114	MR812	1114	S-1.5-10	1114	SH1C	1114	SR3AM		
1114	MR814	1114	S-15H	1114	SH4DB	1114	SR3AM-4		
1114	MR816	1114	S-34	1114	SI-05A	1114	SR3AM-4		
1114	MR990	1114	S-4C	1114	SI-10A	1114	SRIFM-12		
1114	MT84	1114	S-500	1114	SI-1A	1114	SRIFM-4		
1114	NA104	1114	S010G	1114	SI-2A	1114	SRIHM-8		
1114	NA105	1114	S100	1114	SI-3A	1114	ST18		
1114	NA74	1114	S10110	1114	SI-4A	1114	SW-05-D		
1114	NA75	1114	S10120	1114	SI-5A	1114	SW-1-10		
1114	NA76	1114	S10AR1	1114	SI-6A	1114	SW05C		
1114	NA84	1114	S10AR2	1114	SI-7A	1114	SW05D		
1114	NA85	1114	S1201F	1114	SI-8A	1114	SW1C		
1114	NA86	1114	S12110	1114	SI-RECT-102A	1114	SW1D		
1114	N81585541	1114	S12120	1114	SI-RECT-114	1114	SWC		
1114	NSS1021	1114	S12130	1114	SI-RECT-124	1114	SWD		
1114	NTC-17	1114	S12140	1114	SI-RECT-136	1114	T-1000X		
1114	NTC-18	1114	S154	1114	SI-RECT-158	1114	T-1000X		
1114	O766	1114	S1A100	1114	SI-RECT-162	1114	T-2062		
1114	OV02	1114	S1B-0306	1114	SI-RECT-170	1114	T-B1145		
1114	OY5067	1114	S1B01-04	1114	SI-RECT-180	1114	T-B1146		
1114	P1000	1114	S1C	1114	SI-RECT-182	1114	T-B1153		
1114	P10115A	1114	S1D	1114	SI-RECT-204	1114	T1000X		
1114	P10156A	1114	S1D23-13	1114	SI-RECT-206	1114	T800		
1114	P5100	1114	S1D23-15	1114	SI-RECT-224	1114	T800X		
1114	P580	1114	S1D26	1114	SI-RECT-36	1114	TA1000		
1114	P6RP10	1114	S1D30-13	1114	SI-RECT-44	1114	TA7051/01		
1114	P6RP8	1114	S1D30-15	1114	SI1000E	1114	TA7805		
1114	P800	1114	S1P	1114	SI10A	1114	TA7806		
1114	PA380	1114	S1R80	1114	SI8A	1114	TA800		
1114	PD114	1114	S20	1114	SIB-0306	1114	TC-0-2P11/1		
1114	PD115	1114	S20B	1114	SIB0-1	1114	TC-0-.2		
1114	PD116	1114	S210	1114	SIB01	1114	TCO-0.2P11/1		
1114	PD913	1114	S24	1114	SIB01-04	1114	TCO.1P		
1114	PD914	1114	S257	1114	SIB03-10	1114	TD-13		
1114	PD915	1114	S258	1114	SID01K	1114	TD-15		
1114	PD916	1114	S260	1114	SID02K	1114	TD15		
1114	PE408	1114	S28	1114	SID23-13	1114	TD15M		
1114	PE410	1114	S2E100	1114	SID23-15	1114	TE1064		
1114	PE508	1114	S2Q100	1114	SID30-13	1114	TFR105		
1114	PE510	1114	S2V	1114	SID30-132	1114	TFR110		
1114	PH108	1114	S2VC	1114	SID30-15	1114	TFR140		
1114	PH109	1114	S2VC10	1114	SIL	1114	TH1000		
1114	PH9D5	1114	S2VC10R	1114	SISD-1	1114	TH800		
1114	PL-151-045-9-005	1114	S304	1114	SISD-1HF	1114	TH808		
1114	PS1140	1114	S5295G	1114	SK-218	1114	TH810		
1114	PS2416	1114	S5295J	1114	SK3032	1114	TIR07		
1114	PS2417	1114	S61	1114	SK3033	1114	TIR08		
1114	PT580	1114	S62	1114	SK3033A	1114	TIR09		
1114	PT9C22/1	1114	S63	1114	SK3043	1114	TIR10		
1114	PTC203	1114	S750	1114	SK3080	1114	TK1000		
1114	PTC204	1114	S750C	1114	SK3081	1114	TK800		
1114	PTC205	1114	S7AR1	1114	SK3081/125	1114	TKP100		
1114	PTC216	1114	S8AR1	1114	SK3130	1114	TKP80		
1114	QRT-210	1114	S8AR2	1114	SK3175	1114	TM125		
1114	R0606	1114	S9AR1	1114	SL60B	1114	TM506		
1114	R080	1114	SA-2	1114	SL610	1114	TM86		
1114	R170	1114	SA-2C	1114	SL70B	1114	TRBR3AM		
1114	R1Z	1114	SA2H	1114	SL710	1114	TS3		
1114	R210	1114	SA2Z	1114	SLA01	1114	TS8		
1114	R250	1114	SB-1	1114	SLA18AB	1114	TSTD-15		
1114	R250F	1114	SB-1000	1114	SLA18C	1114	TV24167		
1114	R5	1114	SB-1B	1114	SLA19AB				

RS 276-	Industry Standard No.	RS 276-	Industry Standard No.	RS 276-	Industry Standard No.	RS 276-	Industry Standard No.
1114	TV24169	1114	1N2224A	1114	1N525	1114	1S2312
1114	TV24217	1114	1N2225	1114	1N526	1114	1S2314
1114	TV24219	1114	1N2225A	1114	1N5392	1114	1S2315
1114	TV24221	1114	1N2327	1114	1N5392GP	1114	1S2319
1114	TV24617	1114	1N2357	1114	1N5398	1114	1S2324
1114	TV24648	1114	1N2372	1114	1N5418	1114	1S2353
1114	TV24942	1114	1N2374	1114	1N5419	1114	1S2354
1114	TV34	1114	1N2397	1114	1N543	1114	1S2364
1114	TV8	1114	1N2398	1114	1N5433	1114	1S2365
1114	TVM535	1114	1N2406	1114	1N5434	1114	1S2399
1114	TVM537	1114	1N2407	1114	1N543A	1114	1S2400
1114	TV8-0V-02	1114	1N2415	1114	1N548	1114	1S2403
1114	TV8-2B-2C	1114	1N2416	1114	1N549	1114	1S2405
1114	TV8-FR10	1114	1N2424	1114	1N5552	1114	1S2406
1114	TV8-FT-1N	1114	1N2425	1114	1N5553	1114	1S2593
1114	TV8-FT10	1114	1N2501	1114	1N5554	1114	1S2608
1114	TV8-SB-2C	1114	1N2502	1114	1N560	1114	1S2609
1114	TV8-SB-20	1114	1N2503	1114	1N561	1114	1S2610
1114	TV8-SD-1B	1114	1N2504	1114	1N5616	1114	1S2711
1114	TV8-SD1B	1114	1N2505	1114	1N5617	1114	1S2756
1114	TV8IP80	1114	1N2506	1114	1N5618	1114	1S312(HITACHI)
1114	TV8S24-06Q	1114	1N2507	1114	1N5619	1114	1S315(HITACHI)
1114	TV8RB10	1114	1N2508	1114	1N562	1114	1S39
1114	TV8BB2	1114	1N2617	1114	1N5620	1114	1S397
1114	TV8C041O	1114	1N2618	1114	1N5621	1114	1S398
1114	TV8D8-2K	1114	1N2619	1114	1N563	1114	1S401
1114	TV8MR-1M	1114	1N2653	1114	1N597	1114	1S402
1114	TV8MR1C	1114	1N2772	1114	1N598	1114	1S406
1114	TV8RF-1A	1114	1N2773	1114	1N693	1114	1S47
1114	TV8S3G4	1114	1N2774	1114	1N853	1114	1S557
1114	TV8SA-2B	1114	1N2775	1114	1N854	1114	1S557H
1114	TV8SF-1	1114	1N2776	1114	1N855	1114	1S663
1114	TV8SID30-15	1114	1N2777	1114	1N856	1114	1S686
1114	TW100	1114	1N2778	1114	1N864	1114	1S687
1114	TW80	1114	1N2779	1114	1N865	1114	1S814
1114	U07L	1114	1N2852	1114	1N866	1114	1S819
1114	U119	1114	1N2865	1114	1N867	1114	1S848
1114	U120	1114	1N2866	1114	1N875	1114	1S849R
1114	U361	1114	1N2867	1114	1N876	1114	1S850
1114	UP-1A	1114	1N2868	1114	1N877	1114	1S850N
1114	UP-1C	1114	1N2878	1114	1N878	1114	1S871C
1114	UP01	1114	1N2879	1114	1N886	1114	1S872C
1114	UP8D-18	1114	1N2880	1114	1N887	1114	1S880C
1114	UP8D-1C	1114	1N2881	1114	1N888	1114	1S882C
1114	UT345	1114	1N2882	1114	1N889	1114	1S950
1114	V-11N	1114	1N2883	1114	1N893	1114	1S96
1114	VO-6A	1114	1N2884	1114	1N923	1114	1S96R
1114	VO1G	1114	1N2885	1114	1N931	1114	1S97
1114	VO3	1114	1N3107	1114	1NJ27	1114	1S97R
1114	VO3-C	1114	1N3185	1114	1NJ70976	1114	1S98
1114	VO3E	1114	1N3186	1114	1N261	1114	1S98R
1114	VO3E(HITACHI)	1114	1N3196	1114	1R10D3L	1114	1S99
1114	VO50	1114	1N3521A	1114	1R31	1114	1S99A
1114	VO6-G	1114	1N3522	1114	1R3B	1114	1S99R
1114	VO6C	1114	1N3221	1114	1R5GZ61	1114	1SN1835
1114	VO6C(BOOST)	1114	1N3222A	1114	1R5GZ61 PA-1	1114	1S038
1114	VO6C(HITACHI)	1114	1N3231	1114	1R5ZH61	1114	1S058
1114	VO9	1114	1N3232	1114	1S96	1114	1SR12-1000
1114	VO9A	1114	1N3233	1114	1S038	1114	1SR13-1000
1114	VO9B	1114	1N3242	1114	1S058	1114	1SR31-1000
1114	VO9C	1114	1N3243	1114	1S1015	1114	1SR77-1000
1114	V11N	1114	1N3244	1114	1S1024	1114	1SR79-1000
1114	V30N	1114	1N3251	1114	1S1030	1114	1T507
1114	VF-8D1A	1114	1N3252	1114	1S1046	1114	1T508
1114	VHD181834//-1	1114	1N3256	1114	1S1065	1114	1T509
1114	VHD182230//1B	1114	1N328	1114	1S1066	1114	1T510
1114	VO-3C	1114	1N3280	1114	1S107	1114	1T861
1114	VO-9C	1114	1N3281	1114	1S107R	1114	2-64701508
1114	VO3-G	1114	1N3282	1114	1S108	1114	2B-2
1114	VO3C	1114	1N328A	1114	1S108R	1114	2F5CN1
1114	VO9C	1114	1N329	1114	1S109	1114	2HR3J
1114	V8-8D-1B	1114	1N329A	1114	1S109R	1114	2HR3M
1114	V8-8D1	1114	1N3487	1114	1S117	1114	2K02
1114	V8-8D1B	1114	1N3563	1114	1S1223	1114	2K8M-1
1114	V89-0003-911	1114	1N364	1114	1S1225	1114	2SB-2C
1114	V89-0004-911	1114	1N364A	1114	1S1225A	1114	28C-2C
1114	WEP158A	1114	1N365	1114	1S1226	1114	2SJ6A
1114	WEP160	1114	1N365A	1114	1S1233	1114	2T507
1114	WEP170	1114	1N3723	1114	1S1234	1114	2T508
1114	WEP172	1114	1N3724	1114	1S1234H	1114	2T509
1114	WEP186	1114	1N3725	1114	1S1234A	1114	2T510
1114	WEP81000-1	1114	1N3729	1114	1S1237	1114	2W6A
1114	XV604	1114	1N3731	1114	1S1238	1114	2W7A
1114	Z1A103-018	1114	1N3751	1114	1S1255A	1114	2W9A
1114	Z878	1114	1N3752	1114	1S1345	1114	03-936011
1114	Z878A	1114	1N3760	1114	1S1347	1114	3A-200
1114	Z878B	1114	1N3761	1114	1S1348	1114	3A1510
1114	001-0072-00	1114	1N3869	1114	1S1349	1114	3A1510
1114	1.5E12	1114	1N3915	1114	1S1363	1114	3A2510
1114	1.5U12	1114	1N3929	1114	1S1363A	1114	3A258
1114	1C10	1114	1N3957	1114	1S1369	1114	3G1510
1114	1C8	1114	1N4006	1114	1S1369A	1114	3G158
1114	1D10	1114	1N4007	1114	1S138(RECT)	1114	3G2510
1114	1D8	1114	1N4011	1114	1S1471	1114	3G258
1114	1E10	1114	1N4146	1114	1S151	1114	3T507
1114	1E7	1114	1N4155	1114	1S1517	1114	3T508
1114	1E8	1114	1N4248	1114	1S1517A	1114	3T509
1114	1ET10	1114	1N4249	1114	1S1608	1114	3T510
1114	1ET7	1114	1N4250	1114	1S1669	1114	0004-003500
1114	1ET8	1114	1N4251	1114	1S1683	1114	4-2020-03-700
1114	1F8	1114	1N4252	1114	1S1689	1114	4-2020-03700
1114	1G2C1	1114	1N4361	1114	1S1695	1114	4-2020-03800
1114	1G2Z1	1114	1N4514	1114	1S1732	1114	4-2020-03900
1114	1G8	1114	1N4585	1114	1S1732R	1114	4-2020-06500
1114	1HC-100F	1114	1N4817	1114	1S1829	1114	4-2020-06700
1114	1HC-100R	1114	1N4818	1114	1S1830	1114	4-2020-07800
1114	1HY100	1114	1N4826	1114	1S1834	1114	4-2020-07801
1114	1HY80	1114	1N487	1114	1S1835	1114	4-2020-07802
1114	1N1095	1114	1N487A	1114	1S1845D	1114	4-2020-08000
1114	1N1096	1114	1N487B	1114	1S1846D	1114	4-2020-14400
1114	1N1108	1114	1N488	1114	1S1855	1114	4-2020-14600
1114	1N1224A	1114	1N488A	1114	1S1882	1114	4-2020-15100
1114	1N1224B	1114	1N488B	1114	1S1892	1114	4-2020-8000
1114	1N1225A	1114	1N4934	1114	1S1920	1114	4-2020-8700
1114	1N1226A	1114	1N4935	1114	1S1921A	1114	4-2021-04470
1114	1N1258	1114	1N4936	1114	1S1921B	1114	4-2021-04570
1114	1N1259	1114	1N4937	1114	1S1921C	1114	4-2021-04770
1114	1N1260	1114	1N4945	1114	1S1921D	1114	4-2021-06470
1114	1N1261	1114	1N4946	1114	1S1921D	1114	4-2021-06970
1114	1N1407	1114	1N4947	1114	1S1921E	1114	4-2021-07970
1114	1N1408	1114	1N4948	1114	1S1921F	1114	4-202104570
1114	1N1409	1114	1N5052	1114	1S210	1114	4-202104870
1114	1N1443	1114	1N5053	1114	1S210/28J6A	1114	4-686105-3
1114	1N1443A	1114	1N5054	1114	1S211	1114	4-686106-3
1114	1N1730	1114	1N5054A	1114	1S211/28J8A	1114	4-686139-3
1114	1N1730A	1114	1N509	1114	1S2231	1114	4-686179-3
1114	1N1914	1114	1N510	1114	1S2306	1114	4-686184-3
1114	1N1915	1114	1N517	1114	1S2307	1114	4-686186-3
1114	1N1916	1114	1N518	1114	1S2308	1114	4-686189-3
1114	1N2222	1114	1N5188	1114	1S2309	1114	4-686199-3
1114	1N2222A	1114	1N5189	1114	1S2311	1114	4-686201-3
1114	1N2223	1114	1N5190	1114		1114	4-686212-3
1114	1N2223A	1114	1N5214	1114		1114	4-68689-3
1114	1N2224	1114	1N5218	1114		1114	4-68697-3

RS 276-	Industry Standard No.	RS 276-	Industry Standard No.	RS 276-	Industry Standard No.	RS 276-	Industry Standard No.
1114	4G8	1114	13-55031-2	1114	46AX55	1114	67P2
1114	4JA16MR700M	1114	13-55031-3	1114	46AX56	1114	67P2C
1114	4T507	1114	13-55078-1	1114	46AX59	1114	67P3
1114	4T508	1114	132(RECTIFIER)	1114	46AX70	1114	67P3C
1114	4T509	1114	14-514-03	1114	46AX82	1114	070-004
1114	4T510	1114	14-514-03/53	1114	46AX84	1114	070-005
1114	05-110046	1114	14-514-07	1114	46AX85	1114	070-006
1114	05-112404	1114	14-514-07/57	1114	46BD101	1114	070-007
1114	05-112406	1114	14-514-14	1114	46BD27	1114	070-008
1114	5-3/830136.?	1114	14-514-17/67	1114	46BD30	1114	070-009
1114	5-30082.3	1114	14-514-18	1114	46BD32	1114	070-010
1114	5-30082.4	1114	14-514-18/68	1114	46BD33	1114	070-013
1114	5-30088.3	1114	14-514-20/70	1114	46BD34	1114	070-014
1114	05-800015	1114	14-514-23	1114	46BD38	1114	070-015
1114	05-860002	1114	14-514-23/73	1114	46BD39	1114	070-016
1114	05-936010	1114	14-514-66	1114	46BD52	1114	070-017
1114	5-830088.1002	1114	14-514-74	1114	46BR18	1114	070-028
1114	5-830088.1003	1114	14/514-12/62	1114	46BR21	1114	070-030
1114	5-830088.1004	1114	14/514-14/64	1114	46BR27	1114	070-032
1114	5-830106.1001	1114	14/514-19/69	1114	46BR62	1114	070-033
1114	5-830111.1001	1114	15-085033	1114	46BR63	1114	70-270050
1114	5A10	1114	15-085041	1114	46BR64	1114	72-9
1114	5A10C	1114	15-08015	1114	46BR68	1114	73A60-11
1114	5A10D	1114	15-108026	1114	48-10577AO4	1114	75D10
1114	5A10D-C	1114	15-108027	1114	48-115107	1114	75D8
1114	5A8	1114	15-108035	1114	48-134921	1114	75E10
1114	5A8D	1114	15-108040	1114	48-134939	1114	75E7
1114	5A8DC	1114	15-108041	1114	48-134978	1114	75E8
1114	5B-15H(SHARP)	1114	15-108043	1114	48-137112	1114	75R10B
1114	5B-2-H5W(SHARP)	1114	15-108044	1114	48-137205	1114	75R11B
1114	5D10	1114	15-108046	1114	48-137212	1114	75R12B
1114	5D8	1114	15-108047	1114	48-137290	1114	75R13B
1114	5E8	1114	15-108048	1114	48-137301	1114	75R14B
1114	5G8	1114	15-123102	1114	48-137302	1114	75R7B
1114	5GF	1114	15-123103	1114	48-137316	1114	75R8B
1114	5GJ	1114	15-123242	1114	48-137347	1114	75R9B
1114	5K10	1114	16-501190016	1114	48-137348	1114	80-001300
1114	5MA10	1114	18J2	1114	48-137533	1114	80-001400
1114	5MA8	1114	18J2P	1114	48-137546	1114	80A5
1114	7-11(STANDEL)	1114	19-100001	1114	48-137551	1114	80A8
1114	7-15(STANDEL)	1114	19A115024-P6	1114	48-155083	1114	80H
1114	7D210	1114	21A004-000	1114	48-155107	1114	83-2
1114	7D210A	1114	21A008-002	1114	48-155108	1114	86-104-3
1114	8-0001300	1114	21A103-012	1114	48-155125	1114	86-105-3
1114	8-0001400	1114	21A103-013	1114	48-155198	1114	86-116-3
1114	8-719-320-11	1114	21A103-104	1114	48-155223	1114	86-117-3
1114	8-719-320-31	1114	21A110-004	1114	48-191A05	1114	86-118-3
1114	8-719-901-19	1114	21A110-005	1114	48-191A05A	1114	86-128-3
1114	8-719-903-09	1114	21A110-006	1114	48-355013	1114	86-67-2
1114	8-719-906-15	1114	21A110-008	1114	48-355048	1114	86-72-3
1114	8-905-405-105	1114	21A110-009	1114	48-41266G01	1114	87-104-3
1114	8-905-405-160	1114	21A110-013	1114	48-44125AO7	1114	088-2
1114	8-905-405-170	1114	21A110-014	1114	48-67120AO4	1114	91A04
1114	8D10	1114	21A110-071	1114	48-67120AO5	1114	93.20.709
1114	8D8	1114	21A110-072	1114	48-67120A10	1114	93.20.714
1114	8G7	1114	21A112-006	1114	48-8240C10	1114	93A40-1
1114	8GA	1114	21A119-068	1114	48-8240C12	1114	93A52-7
1114	09-306030	1114	22-004004	1114	48-82466H	1114	93A60-1
1114	09-306031	1114	23-0018	1114	48-82466H12	1114	93A60-10
1114	09-306042(DIO)	1114	24MW246	1114	48-82466H13	1114	93A60-11
1114	09-306102	1114	24MW602	1114	48-82466H14	1114	93A60-3
1114	09-306141	1114	24MW605	1114	48-82466H15	1114	93A69-1
1114	09-306144	1114	24MW779	1114	48-82466H16	1114	93A71-1
1114	09-306212(RECT)	1114	24MW829	1114	48-82466H17	1114	93A79-6
1114	09-306225	1114	24MW864	1114	48-82466H18	1114	93B45-3
1114	09-306237	1114	24MW924	1114	48-82466H19	1114	93B52-1
1114	09-306260	1114	025-10029	1114	48-82466H21	1114	93B52-2
1114	09-306274	1114	25-7	1114	48-82466H25	1114	93B60-1
1114	09-306311	1114	25D10	1114	48-82466H27	1114	93B60-10
1114	09-306392	1114	25K10	1114	48-82466H28	1114	93B60-11
1114	09-306418	1114	26-470150B	1114	48-86308-3	1114	93B60-3
1114	09-306423	1114	28-22-13	1114	48-90158A01	1114	93B64-2
1114	09-306424	1114	28-22-22	1114	48-90234A12	1114	93B65-1
1114	09-306425	1114	030-007-0	1114	48-90343A53	1114	93B65-2
1114	9-511511500	1114	30D2	1114	48-90343A54	1114	93B67-1
1114	10-102005	1114	31-194	1114	48-90343A62	1114	93B71-1
1114	10AG10	1114	33-0036	1114	48-90420A98	1114	93C118-1
1114	10AG8	1114	34-8050-10	1114	48-97048A20	1114	93C118-2
1114	10AT10	1114	34-8054-24	1114	48-97768A06	1114	93C18-1
1114	10AT8	1114	34-8054-25	1114	48B41226G01	1114	93C18-2
1114	10B8	1114	34-8054-8	1114	48B41266G01	1114	93C52-1
1114	10C10	1114	34-8057-22	1114	48C40235G02	1114	93C60-10
1114	10C8	1114	34-8057-24	1114	48M355013	1114	93C60-3
1114	10D-2	1114	34-8057-39	1114	48M355021	1114	93C69-1
1114	10D-2(CROWN)	1114	35-1000	1114	48B137533	1114	93C09-1
1114	10D-2B	1114	35-1003	1114	48B137546	1114	094-011
1114	10D-4	1114	35-1029	1114	48B137551	1114	094-012
1114	10D-5	1114	42-051	1114	48B155056	1114	95-11511500
1114	10D-7K	1114	42-19645	1114	48B155083	1114	96-5178-01
1114	10D10	1114	46-86106-3	1114	48B155085	1114	96-5345-01
1114	10D2(CROWN)	1114	46-861149-3	1114	48B155107	1114	098O1219
1114	10D2(DAMPER)	1114	46-861179-3	1114	48B155108	1114	100C
1114	10D2L	1114	46-86141-3	1114	48B191A05A	1114	100C(ADMIRAL)
1114	10D4L	1114	46-86146-3	1114	50D10	1114	100R10
1114	10D5A	1114	46-86147-3	1114	50D8	1114	100K10
1114	10D5D	1114	46-86148	1114	50E7	1114	100R10B
1114	10D7	1114	46-86173-3	1114	50E8	1114	100R11B
1114	10D7P	1114	46-8620-3	1114	52A011-1	1114	100R12B
1114	10D8	1114	46-86280-3	1114	53-0082-3	1114	100R13B
1114	10DC8	1114	46-86281-3	1114	53-0088-1002	1114	100R14B
1114	10DC8R	1114	46-86282-3	1114	53-0088-1003	1114	100R7B
1114	10DI	1114	46-86308-3	1114	53-0088-1004	1114	100R8B
1114	10G4	1114	46-86320-3	1114	53-0088-3	1114	100R9B
1114	10G4A	1114	46-86324-3	1114	53-0106-1001	1114	103-1
1114	10I10	1114	46-86328-3	1114	53-0111-1001	1114	103-160
1114	10R10B	1114	46-86331-3	1114	53-0136T	1114	103-193
1114	10R11B	1114	46-86382-3	1114	53A001-34	1114	103-196
1114	10R12B	1114	46-86391-3	1114	53A001-4	1114	103-244
1114	10R13B	1114	46-86393-3	1114	53A014-1	1114	103-247
1114	10R14B	1114	46-86393-3	1114	53A015-1	1114	103-265
1114	10R7B	1114	46-86395-3	1114	53A017-1	1114	103-287
1114	10R8B	1114	46-86440-3	1114	53A022-5	1114	110-636
1114	10R9B	1114	46-86489-3	1114	53B010-8	1114	110B6
1114	11/110	1114	46-86494	1114	53B001-14	1114	111-4-2020-06200
1114	12B-2-BIP-M	1114	46-86494-3	1114	53B001-5	1114	111-4-2020-14400
1114	1204	1114	46-86507-3	1114	55-001	1114	111-4-2020-15100
1114	13-0050	1114	46-86511-3	1114	56-15	1114	120-003149
1114	13-085024	1114	46-86519-3	1114	60DE10	1114	120-004061
1114	13-085027	1114	46-86525-1	1114	62-118825	1114	121G1241
1114	13-14094-6	1114	46-86525-3	1114	63-11762	1114	121G1241
1114	13-17174-5	1114	46-86553-3	1114	63-12366	1114	121GL241
1114	13-29663-1	1114	46-86562-3	1114	66-B504	1114	123-018
1114	13-33172-1	1114	46AR13	1114	66X0023-009	1114	124-0028
1114	13-33172-2	1114	46AR15	1114	66X0028-008	1114	130-50256
1114	13-34591-1	1114	46AR16	1114	66X0036-001	1114	137-684
1114	13-37868-1	1114	46AR18	1114	66X0036-002	1114	137-828
1114	13-39072-1	1114	46AR50	1114	66X0038-001	1114	142-002(RECT.)
1114	13-39072-2	1114	46AR52	1114	66X0055-001	1114	148-3
1114	13-39867-2	1114	46AR59	1114	66X36-1	1114	151-045-9-003
1114	13-41123-2	1114	46AX12	1114	67I2	1114	167J2
1114	13-43956-1	1114	46AX13	1114	67J2A	1114	200X8130-171
1114	13-55030-1	1114	46AX14	1114	67P1	1114	201X2120-009
1114	13-55031-1	1114		1114	67P1C	1114	201X3130-109
						1114	212-80

RS 276-	Industry Standard No.	RS 276-	Industry Standard No.	RS 276-	Industry Standard No.	RS 276-	Industry Standard No.
1114	221-64(RECT)	1114	4686-106-3	1114	147187-2-14	1122	A-1901-0096-1
1114	229-1301-25	1114	4686-139-3	1114	147477-2-10	1122	A-1901-0150-1
1114	229-5100-235	1114	4686-179-3	1114	147477-8-2	1122	A-1901-0156-1
1114	264-701508	1114	4686-184-3	1114	147477-8-3	1122	A-1901-0196-1
1114	264P00602	1114	4686-186-3	1114	147477-8-7	1122	A-1901-044-1
1114	264P01701	1114	4686-189-3	1114	147477-8-8	1122	A-1901-0461-1
1114	264P02301	1114	4686-199-3	1114	147617-11	1122	A-1901-1067-1
1114	264P03601	1114	4686-201-3	1114	147617-12	1122	A-2008-9140
1114	264P03603	1114	4686-212-3	1114	151476	1122	A-36617
1114	264P03604	1114	4686-89-3	1114	161032	1122	A01
1114	264P03605	1114	4686-97-3	1114	161033	1122	A01(MOTOROLA)
1114	264P03606	1114	04970	1114	185022	1122	A054-228
1114	264P04303	1114	5001-089	1114	222867	1122	A066-119(1S1555)
1114	264P04701	1114	5300-82-3	1114	223357	1122	A066-119(8I)
1114	264P04705	1114	5300-82-4	1114	226922	1122	A066-12
1114	264P06603	1114	5300-88-1002	1114	228560	1122	A11159761
1114	264P09001	1114	5300-88-1003	1114	233025	1122	A42200390-01
1114	264P09101	1114	5300-88-1004	1114	233062	1122	A488-A0001
1114	264P09501	1114	5300-88-3	1114	234553	1122	A488-A0060
1114	264P10201	1114	5301-06-1001	1114	235382	1122	A5010005
1114	264P13002	1114	5301-11-1001	1114	239429	1122	A65-P11305-0001A
1114	264P14701	1114	5301-36-T	1114	241295	1122	A65-P11311-0001
1114	276-1114	1114	5861	1114	241302	1122	A65-P11324-0001
1114	290V034001	1114	6171-17	1114	242141	1122	A66X0043-001
1114	295V028004	1114	6171-18	1114	270642	1122	A691M4
1114	0507	1114	6629	1114	275831	1122	A691M5
1114	309-327-803	1114	6629-A05	1114	280217	1122	A691M5-2
1114	320P	1114	7214-6	1114	348054-15	1122	A691N4
1114	3208	1114	7215-1	1114	348054-7	1122	A72-49-600
1114	320Z	1114	7215-2	1114	348057-8	1122	A7246602
1114	325-4610-100	1114	7510	1114	348057-9	1122	A7246711
1114	0327	1114	7701	1114	348058-2	1122	A7246727
1114	359P	1114	7702-1A	1114	476171-18	1122	A7246901
1114	359S	1114	7704-1	1114	489752-016	1122	A7275400
1114	359V	1114	7706-1	1114	489752-026	1122	A73-16-179
1114	359Z	1114	7708-1	1114	489752-096	1122	A9218
1114	384K	1114	7710-1	1114	500003	1122	A9228-3
1114	384M	1114	7711-1	1114	530073-8	1122	A050(DIODE)
1114	384P	1114	7712-1	1114	530082-1002	1122	AD100(DIODE)
1114	384S	1114	7713-1	1114	530082-2	1122	AD150(DIODE)
1114	384V	1114	8000-00006-147	1114	530088-1	1122	AD200(DIODE)
1114	384Z	1114	8000-004-P061	1114	530088-2	1122	AD50
1114	385P	1114	8000-004-P064	1114	530097-3	1122	AB10
1114	385S	1114	8000-004-P067	1114	530106-1001	1122	AB100
1114	385Z	1114	8000-004-P068	1114	530111-1002	1122	AB150
1114	388B	1114	9203-8	1114	530113-1001	1122	AB200
1114	400(QUASAR)	1114	9920-3-6	1114	530113-2	1122	AB50
1114	420-2003-173	1114	10100	1114	530122-2	1122	A250
1114	420-2104-570	1114	10110	1114	530123-3	1122	A350
1114	422-1408	1114	12685	1114	530123-4	1122	A30
1114	429-0093-71	1114	013339(RECTIFIER)	1114	530123-5	1122	AM300
1114	500B10	1114	0015107	1114	530125-7	1122	AM300A
1114	500B10	1114	21008-002	1114	530135-1	1122	AM301
1114	523-0001-007	1114	23115-042	1114	530136-T	1122	AM301A
1114	523-0001-008	1114	23115-072	1114	530144	1122	AM302
1114	523-0001-009	1114	025072	1114	530148-1003	1122	AM302A
1114	523-0001-010	1114	34424	1114	530148-1004	1122	AM303
1114	525-212	1114	35333	1114	530148-3	1122	AM303A
1114	525-26	1114	40808	1114	530184-4	1122	AM304
1114	530-082-3	1114	40809	1114	530184-1002	1122	AM304A
1114	530-082-4	1114	41020-618	1114	530972-14	1122	AM305
1114	530-088-1002	1114	41023-224	1114	552005	1122	AM305A
1114	530-088-1003	1114	41023-225	1114	552007	1122	AM306
1114	530-088-1004	1114	42021	1114	575047	1122	AM306A
1114	530-088-3	1114	44001	1114	0575047H	1122	AM307
1114	530-106-1001	1114	44003	1114	0575049	1122	AM307A
1114	530-111-1001	1114	44004	1114	0575049H	1122	AM308
1114	530-136-T	1114	44005	1114	0575054	1122	AM308A
1114	540-010	1114	44006	1114	0575066	1122	AM620
1114	547J2F	1114	44007	1114	759500	1122	AM620A
1114	601X0049-066	1114	53088-4	1114	765713	1122	AM626
1114	601X0375-006	1114	53300-31	1114	817193	1122	AM626A
1114	626(RECT.)	1114	53301-01	1114	972575-5	1122	AM632
1114	690V080H49	1114	55207	1114	973936-20	1122	AM632A
1114	690V080H50	1114	055228	1114	984260	1122	AV0000105-0
1114	690V080H51	1114	055228H	1114	991064	1122	B-1599
1114	690V081H91	1114	59844-2	1114	991129	1122	B-1702
1114	750D858-211	1114	62140	1114	991421	1122	B269-3345
1114	800-029-30	1114	70205-8A	1114	991422	1122	B614 007 0
1114	800-282	1114	70270-05	1114	991499	1122	B614-007-0
1114	903-35B	1114	70270-38	1122	1045154-4	1122	BA114
1114	903-36	1114	72137	1122	1390655	1122	BA147
1114	903-36A	1114	72161-1	1122	1415721-1	1122	BA167
1114	903-36B	1114	72172-1	1122	1415721-35	1122	BA168
1114	903-36C	1114	75700-14-05	1122	1445470-502	1122	BA170
1114	903-36D	1114	75700-24-02	1122	1446149-1	1122	BA174
1114	903-36F	1114	75700-12-15-34	1122	1446149-2	1122	BA182
1114	903-68B	1114	77190-8	1122	1471872-14	1122	BA187
1114	977-3B	1114	77271-4	1122	1471872-3	1122	BA202
1114	1006-17	1114	78524-39-01	1122	1472460-13	1122	BA216
1114	1010-8116	1114	99203-7	1122	1474778-14	1122	BA219
1114	1019-1385	1114	99203-8	1122	1474778-21	1122	BA243
1114	1038-1788	1114	113998	1122	1474778-5	1122	BA243A
1114	1050(GE)	1114	115524	1122	1474778-6	1122	BA244
1114	1051	1114	115529	1122	1476049-2	1122	BA316
1114	1051(GE)	1114	120471	1122	1476161-12	1122	BA317
1114	1059-2848	1114	125488	1122	1476171-13	1122	BAC SHIMI
1114	1061-6290	1114	125844	1122	1476171-15(DIODE)	1122	BACSHIMI
1114	1061-8561	1114	125848	1122	1476171-17	1122	BACSIMI
1114	1063-8971	1114	126527	1122	1476171-18	1122	BAW10TP20
1114	1070	1114	126826(DIODE)	1122	1476171-22	1122	BAW11TP21
1114	1070-0623	1114	126856	1122	1476171-25	1122	BAW12TP22
1114	1076-1674	1114	127379	1122	1476171-26	1122	BAW13TP23
1114	1084	1114	129348	1122	1476171-32	1122	BAW16
1114	1090	1114	130110	1122	1476171-34	1122	BAW17
1114	1872-3	1114	131148	1122	1476172-12	1122	BAW18
1114	1971-17	1114	131318	1122	1476183-2	1122	BAW24
1114	2084-17	1114	132418	1122	1476183-5	1122	BAW24A
1114	2093A41-28	1114	132418	1122	1476183-6	1122	BAW24B
1114	2093A41-42	1114	132501	1122	1476183-8	1122	BAW25
1114	2093A41-43	1114	132509	1122	2001786-142	1122	BAW25A
1114	2093A41-6	1114	132537	1122	2003069-1	1122	BAW25B
1114	2093A41-65	1114	132548	1122	2006582-21	1122	BAW45
1114	2093A41-67	1114	132549	1122	2311598	1122	BAW53
1114	2093A41-69	1114	133615	1122	2330101	1122	BAW59
1114	2093A41-72	1114	133616	1122	2330191	1122	BAW59A
1114	2093A41-73	1114	135341	1122	2330211	1122	BAW59B
1114	2093A41-77	1114	135380	1122	2330253	1122	BAW62
1114	2093A41-96	1114	135734	1122	2330356	1122	BAW63A
1114	2093A52-1	1114	136605	1122	2330551	1122	BAW63B
1114	2093A69-1	1114	136606	1122	2330564	1122	BAW99
1114	2093A71-1	1114	137399	1122	2331121	1122	BAX-13
1114	2093A79-1	1114	137606	1122	2331581	1122	BAX-16
1114	2093A79-6	1114	138736	1122	2331991	1122	BAX15
1114	2330-191	1114	140503	1122	2352152	1122	BAX28
1114	2330-201	1114	140971	1122	A-113367(DIODE)	1122	BAX30
1114	2330-252	1114	140972	1122	A-11790169	1122	BAX74
1114	2411-17	1114	141489	1122	A-120125(DIODE)	1122	BAX87
1114	2454-17	1114	142569	1122	A-1634 1125	1122	BAX87A
1114	2495-489	1114	143594	1122	A-1634-1125	1122	BAX88TP11
1114	2498-513	1114	144050	1122	A-1901-0025-1	1122	BAX89A
1114	2802	1114	144051	1122	A-1901-0033-1	1122	BAX91C/TP102
1114	4001-151	1114	147015	1122	A-1901-0050-1	1122	BAX95TP600
1114	4686-105-3	1114		1122	A-1901-0053-1	1122	BAY17

RS 276-	Industry Standard No.	RS 276-	Industry Standard No.	RS 276-	Industry Standard No.	RS 276-	Industry Standard No.
1122	BAY18	1122	DIJ70545	1122	FD1708	1122	JM-40
1122	BAY19	1122	DIJ71273	1122	FD1843	1122	JM40
1122	BAY20	1122	DIJ71711	1122	FD200	1122	JM401
1122	BAY31	1122	DIJ71960	1122	FD222	1122	KO1208A
1122	BAY36	1122	DIJ72164	1122	FD600	1122	K1192804-5
1122	BAY41	1122	DND800	1122	FD6451	1122	KB-162
1122	BAY41A	1122	DR1575	1122	FD6489	1122	KB-162C5
1122	BAY41B	1122	D8-113	1122	FD777	1122	KB-162C5A
1122	BAY52	1122	D8-117	1122	FDH-A	1122	KB-165
1122	BAY68	1122	D8-31	1122	FDH444	1122	KB-262
1122	BAY72	1122	D8-31(DELCO)	1122	FDH6229	1122	KB-265
1122	BAY73	1122	D8-410	1122	FDH666	1122	KB-269
1122	BP8-8-50	1122	D8-410(AMPEX)	1122	FDH694	1122	KB102
1122	BRN-8PEC-24-12	1122	D8-442	1122	FDH900	1122	KB162N
1122	B8AO1	1122	D8-442(SEARS)	1122	FDH999	1122	KB165
1122	B8AO2	1122	D8-97	1122	FDM1006	1122	KB169
1122	B8A11	1122	D81-002-0	1122	FDM1007	1122	KB265
1122	BYX58-100	1122	D81005-1X8628	1122	FDN600	1122	KB265A
1122	BYX58-200	1122	D8117	1122	FDN666	1122	KB269
1122	BYX58-50	1122	D8410	1122	FQ-12377	1122	KD300A
1122	BZ1021V4	1122	D8410(COURIER)	1122	FR1D	1122	KGE46109
1122	C-4401	1122	D8410(EMERSON)	1122	FT1	1122	KGE46465
1122	C-505	1122	D8410(PANON)	1122	FT1(SHARP)	1122	KLH4763
1122	C23018	1122	D8410(OLYMPIC)	1122	G00-012-A	1122	KR-162
1122	C255110-011	1122	D8410R	1122	G00-012A	1122	KX-1
1122	C256125-011	1122	D8441	1122	G00-014-A	1122	LD34R
1122	C6141990	1122	D8442	1122	G00-803-A	1122	LM-1159
1122	CA90	1122	D8442FM	1122	G01-083-A	1122	LM1159
1122	CD-00-9	1122	D8443	1122	G01-209-B	1122	M-8489A
1122	CD-0021	1122	D8448	1122	G01-209B	1122	M-8513A
1122	CD-20	1122	D897(DELCO)	1122	G01-217-A	1122	M1-301
1122	CD-37	1122	D8A150	1122	G01-803A	1122	M150-1
1122	CD-37A	1122	D8I-104-2	1122	G01209	1122	M26
1122	CD-5038	1122	D8S953	1122	G01209B	1122	M8489
1122	CD-84857	1122	DT230A	1122	G01803	1122	M8489A
1122	CDOOOONC	1122	DT230F	1122	G01803A	1122	M8513
1122	CDO014	1122	DT230Q	1122	G101OA	1122	M8513(LAFAYETTE)
1122	CDO014(MORSE)	1122	DX-0150	1122	G657061	1122	M8513-0
1122	CD1224	1122	DX-0255	1122	G657123	1122	M8513A
1122	CD37	1122	DX-0270	1122	GD-510	1122	M8513A-R
1122	CD37A	1122	DX-0273	1122	GD101	1122	M8513AO
1122	CD5002	1122	DX-0299	1122	GD102	1122	M8513B
1122	CD5003	1122	DX-0543	1122	GD3638	1122	M8640E(C-M)
1122	CD6016	1122	DX7429	1122	GE-300	1122	M8640E(DIO)
1122	CD6016-013-689	1122	E09-306112	1122	GE-514	1122	MA-25A
1122	CD6161P1NO13-75	1122	E1121R	1122	GE414	1122	MA-26
1122	CD8457	1122	E1138R	1122	G86063	1122	MA150
1122	CD8547	1122	E13-017-01	1122	G01209	1122	MA150TA
1122	CD86003	1122	EA1405	1122	G01211	1122	MA161
1122	CD860037	1122	EA1661	1122	G01803	1122	MA162
1122	CD84857	1122	EA16X101	1122	GP2-345	1122	MA26
1122	CDG-00	1122	EA16X110	1122	GP2-345/MA161	1122	MA26A
1122	CDG-20	1122	EA16X122	1122	GV5760	1122	MA26WA
1122	CDG-20/494	1122	EA16X134	1122	GVL20226	1122	MB513AR
1122	CDG-21	1122	EA16X135	1122	GVL20265	1122	MC-301
1122	CDG-22	1122	EA16X146	1122	GVL2O327-1	1122	MC2
1122	CDG-24	1122	EA16X152	1122	H623	1122	MC2326
1122	CD900	1122	EA16X171	1122	H8513	1122	MC301
1122	CDG025	1122	EA16X20	1122	H889	1122	MC5321
1122	CD922	1122	EA16X39	1122	HD-2000106	1122	MDP173
1122	CD924	1122	EA16X49	1122	HD2000106	1122	ME-1
1122	CD924/3490	1122	EA16X60	1122	HD2000121O	1122	MER-65-L11324
1122	CD927	1122	EA16X61	1122	HD2000206	1122	MI-301
1122	CDJ-00	1122	EA16X68	1122	HD2001105	1122	MI301
1122	CE37	1122	EA16X69	1122	HD2001105Q	1122	ML2812
1122	CG23018	1122	EA16X73	1122	HD601	1122	MP4300158
1122	CGB-500	1122	EA16X75	1122	HD125	1122	MPC3500
1122	CR353	1122	EA16X84	1122	HE-10030	1122	MPN-3401
1122	CX-0055	1122	EA2607	1122	HE-M8489	1122	MPN3401
1122	CXL177	1122	EA3447	1122	HEPR9137	1122	MP89444
1122	CXL178MP	1122	EC9177	1122	HF-20034	1122	MP89606
1122	CXL519	1122	EC9519	1122	HF-20034	1122	MP89606(H,I)
1122	D-00184R	1122	ED21(ELCOM)	1122	HF-2004B	1122	MP89061
1122	D-12	1122	ED31(ELCOM)	1122	HF-20060	1122	MP89606H
1122	D-215-1	1122	ED32(ELCOM)	1122	HF-20061	1122	MP89065
1122	D1-20	1122	E8514721	1122	HF-20065	1122	MP89644
1122	D1J70545	1122	E8515790	1122	HF-20084	1122	MP89646
1122	D2-77-1	1122	E8516420	1122	HF-20095	1122	MP89646I
1122	D200(IR)	1122	E8536062	1122	HF-20124	1122	MP89646G
1122	D200MP	1122	E8560913	1122	HF-D8410	1122	MP89648
1122	D200MP(IR)	1122	EDH6023	1122	HF-MV2	1122	MP89646I
1122	D3356	1122	ED8-0001	1122	HF20052	1122	MP89646J
1122	D34003220-001	1122	ED8-0014	1122	HF20060	1122	MP89646M
1122	D353	1122	ED8-1	1122	HN-00002	1122	M8S1000
1122	D355	1122	ED8-25	1122	HN-0047	1122	MV-12
1122	D3A	1122	EH16X20	1122	HP5082-2800	1122	MV-13(DIO)
1122	D5Y	1122	EM8 72272	1122	HF-23G	1122	MV-2
1122	D6462	1122	EM872258	1122	HV-23	1122	MV-3
1122	D6726	1122	EM872272	1122	HV-23G(BL)	1122	MV3(DIODE)
1122	D72	1122	EP16X15	1122	HV-25	1122	MV4
1122	D77	1122	EP16X20	1122	HV-25(DIODE)	1122	MZ-00
1122	D77(DIODE)	1122	EP16X22	1122	HV-25(RCA)	1122	MZ207
1122	D7E	1122	EP16X23	1122	HV-27	1122	MZ2360
1122	D7Z	1122	EP16X30	1122	HV-80	1122	MZ2361
1122	D919039-2	1122	EP16X4	1122	HV000001050	1122	ON120623
1122	DA101	1122	EP2798	1122	HV0000206	1122	ON143285
1122	DA101A	1122	EP8X10	1122	HV0000502	1122	ON206068
1122	DA101B	1122	E815057	1122	HV23	1122	N5406
1122	DA102	1122	E81627	1122	HV23(DIODE)	1122	N5406(RCA)
1122	DA102A	1122	E816X23	1122	HV25(HITACHI)	1122	N8-446AQ
1122	DA102B	1122	E816X24	1122	HV460R	1122	NT0-15
1122	DA205	1122	E816X27	1122	HV80	1122	09-306113
1122	DAAYOO3002	1122	E816X30	1122	I95436	1122	09-306195
1122	DAAY004001	1122	E816X32	1122	I964(6	1122	0A-95
1122	DA8Z158800	1122	E816X4	1122	IC743049	1122	0A200
1122	DC8457	1122	E816X40	1122	IC743051	1122	0A2002
1122	DDAY-048001	1122	E857X12	1122	INJ61433	1122	0A2005
1122	DDAY004001	1122	ETD-8G9150	1122	INJ61677	1122	0A202
1122	DDAY047001	1122	ETD182788	1122	INJ61725	1122	0A205
1122	DDAY047005	1122	ETDCDG21	1122	IP20-0021	1122	0F156
1122	DDAY048001	1122	EU16X11	1122	IP20-0023	1122	0F162
1122	DDAY048007	1122	EVD-3	1122	IP20-0061	1122	P-6006
1122	DDAY048008	1122	EX142-X	1122	IP20-0145	1122	P04-41-0025-001
1122	DDAY048012	1122	EX16X10	1122	IP20-0184	1122	P04-42-0011
1122	DDAY048013	1122	EX16X27	1122	IP20-0216	1122	P04410025-003
1122	DDAY048014	1122	EYV-320D1R2J	1122	IP20-0282	1122	P04410042-001
1122	DDAY069001	1122	EY2P-384	1122	IR1026X	1122	P04410042-002
1122	DDAY090001	1122	0F162	1122	IR914	1122	P1172
1122	DDFY004002	1122	FA111	1122	ITT200	1122	P1172-1
1122	DE104	1122	FA2310E	1122	ITT413	1122	P4326
1122	DE110	1122	FA2310U	1122	ITT7215	1122	P7394
1122	DE111	1122	FA2320B	1122	ITT73N	1122	PD137
1122	DE112	1122	FA2320U	1122	ITT921	1122	PL-150-040-9-002
1122	DE113	1122	FA2330E	1122	IX8055-379005N	1122	PL-151-050-9-001
1122	DE114	1122	FA2330U	1122	J241182	1122	PL-151-032-9-004
1122	DE115	1122	FA2330U	1122	J241213	1122	PL-151-035-9-001
1122	DFFT007001	1122	FCDOOO3PC	1122	J241234	1122	PL-151-040-9-001
1122	DHD800	1122	FCDOO14NC8	1122	J241235	1122	PL-151-040-9-002
1122	DHD8001	1122	FD-1029-GP	1122	J241242	1122	PL-151-045-9-002
1122	DHD805	1122	FD-1029-MC	1122	J24755	1122	P830802
1122	DHD806	1122	FD01880	1122	J4-1610	1122	P863A205
1122	D1-20	1122	FD06193	1122	JL-40A	1122	P8700
1122	D1-3	1122	FD100	1122	JL40A		
1122	DIJ70486	1122	FD111				

RS 276-	Industry Standard No.	RS 276-	Industry Standard No.	RS 276-	Industry Standard No.	RS 276-	Industry Standard No.
1122	P9720	1122	SG3432	1122	WG1010A	1122	1N3147
1122	PT4-2268-011	1122	SG3516	1122	WG1010B	1122	1N3197
1122	PT4-2268-01B	1122	SG3583	1122	WG1012	1122	1N3206
1122	PT4-2287-01	1122	SG5028	1122	WG1014A	1122	1N3207
1122	PT4-2311-011	1122	SG5392	1122	WG1021	1122	1N3223
1122	P740063	1122	SG5400	1122	WG4599	1122	1N3257
1122	PTC214	1122	SG9150	1122	WG713	1122	1N3258
1122	PTC214M	1122	SI-RECT-152	1122	WG714	1122	1N330
1122	PTC215	1122	SI-RECT-35	1122	WG851	1122	1N331
1122	Q-1N914	1122	SIB-C1-02	1122	WH1012	1122	1N3471
1122	Q-21115C	1122	SID50B94	1122	WS100	1122	1N350
1122	Q-23115C	1122	SID50B951	1122	WS100A	1122	1N351
1122	Q-24115C	1122	SK3100	1122	WS100B	1122	1N352
1122	QD-818953XA	1122	SK3100/519	1122	WS100C	1122	1N353
1122	QD-8MA150XN	1122	SL103	1122	WS200	1122	1N3550
1122	QD-8S1555XT	1122	SM-C-706156	1122	WS200A	1122	1N3575
1122	QD-8S852XXA	1122	SMB-541191	1122	WS200B	1122	1N3576
1122	QRT-218	1122	SN4448	1122	WS200C	1122	1N3577
1122	R-7026	1122	SP101	1122	WS300	1122	1N3593
1122	R-7027	1122	SR-34	1122	WS300A	1122	1N3594
1122	R-7092	1122	SR0004	1122	WS300B	1122	1N3595
1122	R14010658	1122	ST/123/CR	1122	WS300C	1122	1N3596
1122	R3410-P1	1122	ST/146/CR	1122	WSD002C	1122	1N3598
1122	R8022	1122	ST22546-1	1122	X-19031-A	1122	1N3599
1122	R8023	1122	ST32012-0037	1122	X1022220-1	1122	1N3600
1122	R6314	1122	STB01-02	1122	X72A42416	1122	1N3601
1122	RA1A	1122	STB576	1122	X925940-5018	1122	1N3602
1122	RD1343	1122	SV-04	1122	X925940-501B	1122	1N3603
1122	RE-94	1122	SV-3	1122	Y56001-21	1122	1N3604
1122	RB52	1122	SV-3A	1122	YAAD010	1122	1N3605
1122	RE94	1122	SV-3B	1122	YAAD018	1122	1N3606
1122	RBJ71253	1122	SV-5C	1122	YD121	1122	1N3607
1122	REN177	1122	SV-8	1122	Z-175-011	1122	1N3608
1122	REN94	1122	SV14B	1122	ZA29312	1122	1N3609
1122	RER-023	1122	SV14C	1122	ZC18358S	1122	1N3625
1122	RF34661	1122	SV9	1122	ZS142	1122	1N3653
1122	RH-DX0033TAZZ	1122	SVDMA26-2	1122	ZS40	1122	1N3654
1122	RH-DX0046CEZZ	1122	SVDVD1121	1122	ZS4P	1122	1N3666
1122	RH-DX0048CEZZ	1122	SX780	1122	1-0002-0001	1122	1N3668
1122	RH-DX0054CEZZ	1122	T-10010	1122	1-0002-001	1122	1N3722
1122	RH-VX0004TAZZ	1122	T-E1098	1122	1-001/2207	1122	1N379
1122	RHDX0033TAZZ	1122	T-E1118	1122	001-0095-00	1122	1N380
1122	R14010658	1122	T-E1119	1122	001-0112-00	1122	1N381
1122	RKZ12003	1122	T-E1121	1122	001-0125-00	1122	1N382
1122	RKZ120101	1122	T12-1	1122	1-014/2207	1122	1N383
1122	RS-253C	1122	T12-2	1122	001-0151-00	1122	1N384
1122	RS-253B	1122	T12A	1122	001-0151-01	1122	1N385
1122	RS-267S	1122	T12B	1122	001-026010	1122	1N386
1122	RS-272US	1122	T12C	1122	001-026030	1122	1N3864
1122	RS-280S	1122	T151	1122	001-026060	1122	1N387
1122	RS1428	1122	T152	1122	1-13989	1122	1N3872
1122	RT-218	1122	T153	1122	01-1501	1122	1N3873
1122	RT1669	1122	T154	1122	1-16549	1122	1N388
1122	RT2061(G.E.)	1122	T155	1122	001-226010	1122	1N389
1122	RT218	1122	T16	1122	01-2601	1122	1N390
1122	RT5554	1122	T21600	1122	01-2601-0	1122	1N391
1122	RT5909	1122	TA198785-2	1122	01-2603-0	1122	1N392
1122	RT7946	1122	TAB532787	1122	1-DI-007	1122	1N393
1122	RV1017	1122	TD-15-BL	1122	1A16549	1122	1N394
1122	RV1226	1122	TD81515	1122	1A16551	1122	1N3953
1122	RV2071	1122	TD81518	1122	1A35	1122	1N3954
1122	RVDKB265J2	1122	TD96001 6-1M	1122	1A6942S-1	1122	1N3956
1122	RVDVD1150L	1122	TE-1029	1122	1A99812-1001	1122	1N4009
1122	RVDVD1210L	1122	TF54	1122	1B2992	1122	1N4043
1122	RVDVD1210M	1122	TF44	1122	1D098-001V-022	1122	1N4048
1122	RVDVD1212L	1122	TF44J	1122	1N35	1122	1N4087
1122	RVDVD1213	1122	TI-51	1122	1N37A	1122	1N4092
1122	RVDVD1250M	1122	TI-UG-1888	1122	1N37B	1122	1N4147
1122	S-2064-G	1122	TI51	1122	1N38	1122	1N4148
1122	S-3016R	1122	TM177	1122	1N138A	1122	1N4149
1122	S04	1122	TM519	1122	1N138B	1122	1N4148
1122	S04-1	1122	TR2083-42	1122	1N448	1122	1N4150
1122	S04A-1	1122	TR2083-44	1122	1N630	1122	1N4151
1122	S04B-1	1122	TR2337011	1122	1N638	1122	1N4152
1122	S074-005-0001	1122	TSC156	1122	1N1839	1122	1N4153
1122	S074-007-0001	1122	TV24554	1122	1N1840	1122	1N4154
1122	S074-007-001	1122	TVS-181211	1122	1N1841	1122	1N4156
1122	S1-RECT-35	1122	TVS-0A81	1122	1N1842	1122	1N4157
1122	S11	1122	TVS-0A95	1122	1N1843	1122	1N4242
1122	S1428	1122	TVS182076	1122	1N1844	1122	1N4243
1122	S160	1122	TVS18954	1122	1N1845	1122	1N4305
1122	S180	1122	TVSB01-2	1122	1N1846	1122	1N4306
1122	S1D50B851-A	1122	TVSBAX-13	1122	1N1847	1122	1N4307
1122	S1R20	1122	TVSBAX13	1122	1N194	1122	1N4308
1122	S182	1122	TVS9P2-354	1122	1N194A	1122	1N4309
1122	S20870	1122	TVSMA26	1122	1N195	1122	1N431
1122	S3004-1716	1122	TVSRA1A	1122	1N196	1122	1N4311
1122	S3016	1122	TVSS1R20	1122	1N200	1122	1N4315
1122	S3016R	1122	TZ1153	1122	1N201	1122	1N4318
1122	S30720	1122	UO-5E	1122	1N202	1122	1N432
1122	084.000654	1122	UG1888	1122	1N203	1122	1N4322
1122	S502	1122	UO-5B	1122	1N204	1122	1N432A
1122	S502A	1122	US1555	1122	1N205	1122	1N432B
1122	S502B	1122	V1112	1122	1N206	1122	1N433
1122	S506	1122	V50260-10	1122	1N207	1122	1N433A
1122	S506A	1122	V50260-36	1122	1N2075K	1122	1N433B
1122	S506B	1122	VAR	1122	1N208	1122	1N434
1122	S509	1122	VAR-1R2	1122	1N209	1122	1N434A
1122	S509A	1122	VD-1121	1122	1N210	1122	1N434B
1122	S509B	1122	VD-1123	1122	1N211	1122	1N4363
1122	S7-8	1122	VD-1124	1122	1N212	1122	1N4375
1122	SA-93792	1122	VD-S-7-013626-001	1122	1N213	1122	1N4376
1122	SC-1016	1122	VD1121	1122	1N214	1122	1N4382
1122	SC12(DIODE)	1122	VD1123	1122	1N215	1122	1N4389
1122	SC1431(GE)	1122	VD1124	1122	1N216	1122	1N4392
1122	SD-110	1122	VD1127	1122	1N217	1122	1N4395
1122	SD-1AUF	1122	VD1150M	1122	1N218	1122	1N4395A
1122	SD-181555	1122	VD1150M	1122	1N2473	1122	1N4444
1122	SD-34	1122	VD1210L	1122	1N251	1122	1N4446
1122	SD-43	1122	VD1212	1122	1N251A	1122	1N4447
1122	SD-5	1122	VHD181553//-1	1122	1N252	1122	1N4448
1122	SD-5(PHILCO)	1122	VHD181555-R-1	1122	1N252A	1122	1N4449
1122	SD-630	1122	VHD181555//1A	1122	1N300	1122	1N4450
1122	SD-7(PHILCO)	1122	VHD182076-//-1	1122	1N300A	1122	1N4451
1122	SD100	1122	VHD182076-1	1122	1N300B	1122	1N4453
1122	SD110	1122	V8V181209-1	1122	1N301	1122	1N4454
1122	SD12(PHILCO)	1122	VS9-0008-911	1122	1N301A	1122	1N4455
1122	SD165	1122	VS9-0014-911	1122	1N301B	1122	1N4531
1122	SD43	1122	W1R	1122	1N303	1122	1N4532
1122	SD500	1122	W1R(DIODE)	1122	1N303A	1122	1N4533
1122	SD600	1122	WA-26	1122	1N303B	1122	1N4534
1122	SD630	1122	WAOA-90	1122	1N3062	1122	1N4536
1122	SD701-02	1122	WD4	1122	1N3063	1122	1N4547
1122	SD974	1122	WEP1060	1122	1N3064	1122	1N4548
1122	SE5-0247-C	1122	WEP925	1122	1N3065	1122	1N456
1122	SE5-0456	1122	WG-1010-A	1122	1N3066	1122	1N456A
1122	SE5-0966	1122	WG-1012	1122	1N3067	1122	1N457
1122	SPD43	1122	WG-101DA	1122	1N3068	1122	1N457A
1122	SPD83	1122	WG-599	1122	1N3069	1122	1N457M
1122	SG-9150	1122	WG-713	1122	1N3069M	1122	1N458
1122	SG3182	1122	WGOA-90	1122	1N3070	1122	1N458A
1122	SG3183	1122	000WG1010	1122	1N3071	1122	1N458M
1122	SG3198			1122	1N3123	1122	1N459
				1122	1N3124	1122	1N459A

RS 276-	Industry Standard No.	RS 276-	Industry Standard No.	RS 276-	Industry Standard No.	RS 276-	Industry Standard No.
1122	1N459M	1122	1N799	1122	1S1216	1122	1S560
1122	1N460	1122	1N800	1122	1S1217	1122	1S560H
1122	1N4606	1122	1N801	1122	1S1218	1122	1S642
1122	1N4607	1122	1N802	1122	1S1218GR	1122	1S693
1122	1N4608	1122	1N803	1122	1S1219	1122	1S731
1122	1N460A	1122	1N804	1122	1S1219H	1122	1S773
1122	1N460B	1122	1N806	1122	1S1220	1122	1S773A
1122	1N461	1122	1N807	1122	1S130	1122	1S775
1122	1N4610	1122	1N808	1122	1S1301	1122	1S776
1122	1N461A	1122	1N809	1122	1S1302	1122	1S801
1122	1N462	1122	1N810	1122	1S1303	1122	1S802
1122	1N462A	1122	1N811	1122	1S1305	1122	1S803
1122	1N463	1122	1N811M	1122	1S131	1122	1S804
1122	1N463A	1122	1N812	1122	1S132	1122	1S805
1122	1N464	1122	1N812M	1122	1S1420H	1122	1S806
1122	1N464A	1122	1N813	1122	1S144	1122	1S807
1122	1N4726	1122	1N813M	1122	1S1473	1122	1S808
1122	1N4727	1122	1N814	1122	1S1514	1122	1S84
1122	1N482	1122	1N814M	1122	1S1515	1122	1S89
1122	1N4827	1122	1N815	1122	1S1516	1122	1S90R
1122	1N4828	1122	1N815M	1122	1S1532	1122	1S920
1122	1N4829	1122	1N817	1122	1S1544	1122	1S921
1122	1N482A	1122	1N818	1122	1S1544A	1122	1S922(DIODE)
1122	1N482B	1122	1N818M	1122	1S1545	1122	1S923
1122	1N482C	1122	1N837	1122	1S155	1122	1S941
1122	1N483	1122	1N837A	1122	1S155-1	1122	1S942
1122	1N4830	1122	1N838	1122	1S1553	1122	1S951
1122	1N483A	1122	1N839	1122	1S1554	1122	1S952
1122	1N483AM	1122	1N840	1122	000001S1555	1122	1S953
1122	1N483B	1122	1N840M	1122	1S1555-1	1122	1S954
1122	1N483BM	1122	1N841	1122	1S1555-8	1122	1S955
1122	1N483C	1122	1N842	1122	1S1555-Z	1122	1S977
1122	1N484	1122	1N843	1122	1S1555I	1122	1S978
1122	1N484A	1122	1N844	1122	1S1555FA1	1122	1S981
1122	1N484C	1122	1N845	1122	1S1555V	1122	1S982
1122	1N485	1122	1N890	1122	1S1555Z	1122	1S983
1122	1N485A	1122	1N891	1122	1S1580	1122	1S984
1122	1N485B	1122	1N892	1122	1S1585	1122	1S985
1122	1N485C	1122	1N897	1122	1S1586	1122	1S986
1122	1N486	1122	1N898	1122	1S1587	1122	1S987
1122	1N4861	1122	1N899	1122	1S1588	1122	1S988
1122	1N4862	1122	1N900	1122	1S1588V	1122	000001S990
1122	1N4863	1122	1N901	1122	1S1589	1122	1S990-AM
1122	1N4864	1122	1N902	1122	1S1621	1122	1S990A
1122	1N486A	1122	1N903	1122	1S1621-0	1122	1S990S(DIODE)
1122	1N486B	1122	1N903/A	1122	1S1621-R	1122	1S994A
1122	1N4938	1122	1N903A	1122	1S1621-Y	1122	1SC2567
1122	1N4949	1122	1N903AM	1122	1S1650	1122	1S830
1122	1N4950	1122	1N903M	1122	1S1651	1122	1S831
1122	1N5062(SEARS)	1122	1N904	1122	1S1710	1122	1S849
1122	1N5179	1122	1N904/A	1122	1S1711	1122	1S850
1122	1N5194	1122	1N904A	1122	1S1712	1122	1S851
1122	1N5195	1122	1N904AM	1122	1S1712A	1122	1S853
1122	1N5196	1122	1N904M	1122	1S1713	1122	1S855
1122	1N5208	1122	1N905	1122	1S1714	1122	1S868
1122	1N5209	1122	1N905/A	1122	1S180	1122	1S881
1122	1N5210	1122	1N905A	1122	1S180B	1122	1S934
1122	1N5219	1122	1N905AM	1122	1S181	1122	1W35
1122	1N5220	1122	1N905M	1122	1S181PA	1122	1Z243
1122	1N5282	1122	1N906	1122	1S182	1122	1T40
1122	1N5315	1122	1N906/A	1122	1S1825	1122	1X9179
1122	1N5316	1122	1N906A	1122	1S1973	1122	1X9805
1122	1N5317	1122	1N906AM	1122	1S1992	1122	1X9809
1122	1N5318	1122	1N906M	1122	1S1993	1122	02-1006-2/2221-3
1122	1N5319	1122	1N907	1122	1S1994	1122	2-1K60
1122	1N5320	1122	1N907/A	1122	1S1995	1122	2D165
1122	1N5413	1122	1N907A	1122	1S2070	1122	2P2150M
1122	1N5414	1122	1N907AM	1122	1S2071	1122	2S100J2
1122	1N5426	1122	1N907M	1122	1S2074H	1122	2SC1124-OR
1122	1N5605	1122	1N908	1122	1S2075	1122	03-0063-04
1122	1N5606	1122	1N908/A	1122	1S2075IC	1122	03-034-042
1122	1N5607	1122	1N908A	1122	1S2075K	1122	03-931641
1122	1N5610	1122	1N908AM	1122	1S2076	1122	03-931642
1122	1N568	1122	1N908M	1122	1S2076-27	1122	03-931645
1122	1N5711	1122	1N914	1122	1S2076-TP1	1122	3L4-2001-3
1122	1N5712	1122	1N914/A/B	1122	1S2076-TPI	1122	3L4-2001-4
1122	1N5713	1122	1N914A	1122	1S2076A	1122	3L4-2003-3
1122	1N5719	1122	1N914B	1122	1S2076A-07	1122	3L4-3001-1
1122	1N5767	1122	1N914M	1122	1S2091	1122	3L4-3001-7
1122	1N619	1122	1N915	1122	1S2091(DIODE)	1122	3L4-3002-01
1122	1N622	1122	1N916	1122	1S2091-BK	1122	3L4-3002-10
1122	1N625	1122	1N916A	1122	1S2091-BL	1122	3L4-3002-25
1122	1N625A	1122	1N916B	1122	1S2091-W	1122	3L4-3002-32
1122	1N625M	1122	1N917	1122	1S2091BK	1122	3L4-3002-7
1122	1N626	1122	1N919	1122	1S2091BL	1122	04-000653
1122	1N626A	1122	1N920	1122	1S2091W	1122	04-000655-1
1122	1N626M	1122	1N921	1122	1S2092	1122	004-00900
1122	1N627	1122	1N922	1122	1S2097	1122	4-1724
1122	1N627A	1122	1N924	1122	1S2098	1122	4-1726
1122	1N628	1122	1N925	1122	1S2099	1122	4-2020-05800
1122	1N628A	1122	1N926	1122	1S2134	1122	4-2020-06100
1122	1N629	1122	1N927	1122	1S2144A	1122	4-2020-06200
1122	1N629A	1122	1N928	1122	1S2144A	1122	4-2020-06400
1122	1N653	1122	1N929	1122	1S2144Z	1122	4-2020-08001
1122	1N643	1122	1N930	1122	1S2145	1122	4-2020-08200
1122	1N643A	1122	1N931	1122	1S218	1122	4-2020-10100
1122	1N658	1122	1N932	1122	1S2186	1122	4-2020-15600
1122	1N658A	1122	1N933	1122	1S2186GR	1122	4-2020-16200
1122	1N658M	1122	1N934	1122	1S2212	1122	4-2021-07470
1122	1N659	1122	1N948	1122	1S2276	1122	4-2021-07670
1122	1N659/A	1122	1N993	1122	1S2460	1122	4-2021 04970
1122	1N659A	1122	1N995	1122	1S2461	1122	4-2040-08000
1122	1N660	1122	1N997	1122	1S2462	1122	4-3034
1122	1N660A	1122	1N999	1122	1S2471	1122	05-02160-01
1122	1N661	1122	1NJ61433	1122	1S2472	1122	05-110442
1122	1N661A	1122	1NJ61677	1122	1S2473	1122	05-180053
1122	1N662	1122	1NJ61725	1122	1S2473-T72	1122	05-180753
1122	1N662A	1122	1N70980	1122	1S2473H	1122	05-180953
1122	1N663	1122	1NJ71224	1122	1S2473HC	1122	05-18155
1122	1N663A	1122	1R0	1122	1S2473K	1122	05-181555
1122	1N663M	1122	1R0E	1122	1S2473T	1122	05-182076
1122	1N690	1122	1R10D3K	1122	1S2473VE	1122	05-320301
1122	1N696	1122	1R2	1122	1S2597	1122	05-330150
1122	1N697	1122	1R3A	1122	1S2692	1122	05-330161
1122	1N70980	1122	1R4	1122	1S2692AB	1122	05-931642
1122	1N760	1122	1R0E	1122	1S2788	1122	05-931645
1122	1N778	1122	1S-1555-Z	1122	1S2788B	1122	05-936470
1122	1N779	1122	1S-1555V	1122	1S306	1122	05A01
1122	1N788	1122	1S-180	1122	1S306M	1122	05A05
1122	1N789	1122	1S-2144Z	1122	1S307	1122	05A06
1122	1N789M	1122	1S-2472	1122	1S3076	1122	006-0000004
1122	1N790	1122	1S1001	1122	1S322	1122	006-6400902
1122	1N790M	1122	1S1052	1122	1S324	1122	6X97174A01
1122	1N791	1122	1S1053	1122	1S325	1122	6X97174XA08
1122	1N791M	1122	1S1147	1122	1S38	1122	7-0013
1122	1N792	1122	1S1155	1122	1S444	1122	007-25005-01
1122	1N792M	1122	1S1941	1122	1S460	1122	007-25013-01
1122	1N793	1122	1S1207	1122	1S461	1122	007-25016-01
1122	1N793M	1122	1S1210	1122	1S465	1122	007-6016-00
1122	1N794	1122	1S1212	1122	1S500	1122	007-6060-00
1122	1N795	1122	1S1212A	1122	1S501	1122	08-08117
1122	1N796	1122	1S1213	1122	1S501M	1122	08-08119
1122	1N797	1122	1S1214	1122	1S548	1122	8-719-815-55
1122	1N798	1122	1S1215	1122	1S555	1122	8-719-923-76

RS 276-	Industry Standard No.	RS 276-	Industry Standard No.	RS 276-	Industry Standard No.	RS 276-	Industry Standard No.
1122	8-905-405-098	1122	14-514-70	1122	48-10577A01	1122	65-P11308-0001
1122	8-905-406-020	1122	0014-911	1122	48-134781	1122	65-P11311-0001
1122	8-905-421-300	1122	15-085005	1122	48-134816	1122	65-P11324-0001
1122	08A159-007	1122	15-085008	1122	48-137585	1122	65-P11325-0001
1122	08A165-001	1122	15-108008	1122	48-137514	1122	66A10319
1122	09-302090(DIODE)	1122	15-108025	1122	48-137573	1122	66P016-1
1122	09-3060111	1122	15-108042	1122	48-155047	1122	66P016-2
1122	09-306060	1122	15-123101	1122	48-155060	1122	66P018
1122	09-306062	1122	15P2	1122	48-155077	1122	66X0043-001
1122	09-306110	1122	16028	1122	48-155159	1122	66X0044-001
1122	09-306111	1122	16028A	1122	48-34816	1122	66X0044-100
1122	09-306113	1122	16P2	1122	48-355035	1122	66X0046-001
1122	09-306129	1122	16X39	1122	48-355036	1122	66X0048-001
1122	09-306134	1122	17P2	1122	48-40738P01	1122	66X0054-001
1122	09-306135	1122	019-002964	1122	48-42899J01	1122	66X0056-001
1122	09-306145	1122	019-003676-334	1122	48-46669H01	1122	66X0062-001
1122	09-306148	1122	19-080-001	1122	48-67120A02	1122	66X29
1122	09-306151	1122	19-080-008	1122	48-67120A03	1122	68A7252
1122	09-306154	1122	19-130-001(DIODE)	1122	48-67120A11	1122	69-1822
1122	09-306159	1122	19A115250	1122	48-67120A13	1122	69A49728-P1
1122	09-306161	1122	19A115371-1	1122	48-67120AB	1122	070-022
1122	09-3061633	1122	19A115661P1	1122	48-6712A11	1122	070-047
1122	09-306170	1122	19A116081	1122	48-6720A02	1122	70-943-083-002
1122	09-306171	1122	19B200249-P1	1122	48-82392B01	1122	70-943-083-003
1122	09-306178	1122	19B200249-P2	1122	48-82392B02	1122	71-13/51/60
1122	09-306195	1122	19B200249-P3	1122	48-82392B03	1122	72-18
1122	09-306198	1122	19P1	1122	48-82392B04	1122	77A01
1122	09-306199	1122	19Q820	1122	48-82392B05	1122	77A02
1122	09-306202	1122	20-161002	1122	48-82392B06	1122	77A11
1122	09-306206	1122	20A11	1122	48-82392B07	1122	77A13
1122	09-306211	1122	20A13	1122	48-82392B08	1122	77E1017
1122	09-306211(RECT)	1122	21-606-0001	1122	48-82392B09	1122	79E104-2
1122	09-306219	1122	21-606-0001-00H	1122	48-82392B10	1122	81-46123023-7
1122	09-306220	1122	21-608-4148-006	1122	48-82392B11	1122	81-46123038-5
1122	09-306221	1122	21-609-3595-009	1122	48-82392B12	1122	84A20
1122	09-306223	1122	21A008-008	1122	48-82392B13	1122	86-0515
1122	09-306231	1122	21A009	1122	48-82392B14	1122	86-0515
1122	09-306233	1122	21A009-008	1122	48-82392B15	1122	86-100013
1122	09-306236	1122	21A009-009	1122	48-8240007	1122	86-100014
1122	09-306244	1122	21A103-065	1122	48-8240009	1122	86-100014
1122	09-306248	1122	21A108-001	1122	48-8240011	1122	86-147-1
1122	09-306257	1122	21A108-003	1122	48-8240013	1122	86-27-1
1122	09-306266	1122	21A108-004	1122	48-8240014	1122	86-41-1(SEARS)
1122	09-306276	1122	21A111-001	1122	48-8240015	1122	86-5010-3
1122	09-306283	1122	21A111-002	1122	48-8240016	1122	86-5011-3
1122	09-306288	1122	21M330	1122	48-8240017	1122	86-5015-3
1122	09-306291	1122	21M415	1122	48-8240018	1122	86-5027-3
1122	09-306309	1122	21M562	1122	48-82420001	1122	86-5037-3
1122	09-306313	1122	21M562(DIODE)	1122	48-82420002	1122	86-574-2
1122	09-306326	1122	022	1122	48-82420003	1122	86-62-1
1122	09-306368	1122	022-0163-00	1122	48-82420004	1122	86-65-4
1122	09-306373	1122	22-1-044	1122	48-82420005	1122	86-70-1
1122	09-306390	1122	22-1-70	1122	48-86289-3	1122	86-74-1
1122	09-306426	1122	022-2823-004	1122	48-90420A03	1122	86-74-9
1122	09-307039	1122	022-2823-005	1122	48-90420A04	1122	86-76-1
1122	09-307045	1122	23-001R03A10	1122	48-97048A01	1122	86-77-1
1122	09-307055	1122	23-001R03AA10	1122	48-97048A17	1122	86-78-1
1122	09-307075	1122	23-P2274-125	1122	48-V34816	1122	87-190XX-001
1122	09-307080	1122	23-P2283-125	1122	48C134816	1122	88-77-1
1122	09-307081	1122	23J2	1122	48C0524A02	1122	089-241
1122	09-307083	1122	24-282201	1122	48C67120A02	1122	92A64-1
1122	09-307085	1122	24B-001	1122	48D67	1122	93-13
1122	09-307089	1122	24B-006	1122	48D67120A13	1122	93A27-3
1122	9D1	1122	24J2	1122	48D67120A13(C)	1122	93A27-8
1122	9D11	1122	24MW1109	1122	48D82420001	1122	93A31-1
1122	9D14	1122	24MW667	1122	48D82420002	1122	93A3912
1122	9DI	1122	24MW744	1122	48D82420003	1122	93A48-1
1122	9DI1	1122	24MW785	1122	48D82420005	1122	93A52-2
1122	9DI3	1122	24MW825	1122	48O10546A02	1122	93A60-5
1122	10-010	1122	24MW858	1122	48M355014	1122	93A60-6
1122	10-085005	1122	24MW861	1122	48M355035	1122	93A64-1
1122	10-10112	1122	24MW894	1122	48810577A01	1122	93A64-2
1122	11-0430	1122	24MW956	1122	48810577A02	1122	93A64-3
1122	11-0781	1122	25J2	1122	48810577A11	1122	93A64-5
1122	11-085012	1122	26J2	1122	48810577A13	1122	93A64-7
1122	11-9-156438	1122	27J2	1122	488134816	1122	93A69-2
1122	012-0121-001	1122	31-011	1122	488137133	1122	93B27-4
1122	12-087004	1122	31-093	1122	488155047	1122	93B27-5
1122	012-1020-005	1122	32-0022	1122	488155060	1122	93B27-8
1122	012-1021-001	1122	32-0057	1122	488155077	1122	93B48-1
1122	012-1024-001	1122	32-0063	1122	48844887G01	1122	93B48-2
1122	12-200009	1122	34-0037-102	1122	48867120A13	1122	93B48-3
1122	12A1027-P5	1122	034-032-0	1122	4981	1122	93B48-4
1122	12P2	1122	34-8057-13	1122	051-0003	1122	93B60-8
1122	13-0003	1122	34-8057-40	1122	051-0020	1122	93B64-1
1122	13-0003(PACE)	1122	34-8057-5	1122	51-08001-11	1122	93C027-2
1122	13-085002	1122	34-8057-6	1122	51-13-14	1122	93C027-3
1122	13-085015	1122	34P4	1122	51-47-28	1122	93C027-7
1122	13-085022	1122	35-010-04	1122	52A4	1122	93C060-5
1122	13-085023	1122	35-1014	1122	53A001-32	1122	93C060-6
1122	13-085026	1122	35P4	1122	53A001-36	1122	93C064-1
1122	13-087005	1122	36D-32	1122	53A018-1	1122	93C064-2
1122	13-10321-34	1122	38-8057-6	1122	53A019-1	1122	93C064-3
1122	13-10321-54	1122	39A69-2	1122	53A020-1	1122	93C7-2
1122	13-10321-55	1122	40-09297	1122	53A030-1	1122	93C7-3
1122	13-14094-33	1122	42-22558	1122	53B001-7	1122	93C60-5
1122	13-14097-7	1122	42-24387	1122	53B001-9	1122	93EVHD1N4148//
1122	13-17204-1	1122	42-28201	1122	53B016-6	1122	094-007
1122	13-17569-1	1122	43A113534	1122	53B015-1	1122	96-51-518-01
1122	13-17569-1	1122	43A114346-1	1122	53B014-1	1122	96-5254-01
1122	13-17569-2	1122	43A114346-P1	1122	53D001-4	1122	100-00310-09
1122	13-17595-2	1122	43A114832-3	1122	53D001-8	1122	100-007-10/2228/3
1122	13-17596-10	1122	43A167229-01	1122	53D002-3	1122	100-011-20
1122	13-17596-2	1122	43A175989P1	1122	53J003-2	1122	100-011-50/2228-3
1122	13-17596-3	1122	447-300-96	1122	53L001-15	1122	100-01110-01
1122	13-17596-4	1122	046-0134	1122	53N001-5	1122	100-01120-09
1122	13-17596-5	1122	46-61267-3	1122	53N004-12	1122	100-11
1122	13-17596-7	1122	46-61307-3	1122	53T001-7	1122	100-120
1122	13-17596-8	1122	46-80309-3	1122	53U001-2	1122	100-125
1122	13-17596-9	1122	46-836380-3	1122	56-24	1122	100-130
1122	13-22017-0	1122	46-861187-3	1122	56-27	1122	100-137
1122	13-22606-0	1122	46-86168-3	1122	56-28	1122	100-161
1122	13-22609-0	1122	46-86184	1122	56-4832	1122	100-184
1122	13-23917-1	1122	46-86184-3	1122	56-5	1122	100-216
1122	13-27596-5	1122	46-86186-3	1122	56-56	1122	100-435
1122	13-29687-2	1122	46-86187-3	1122	56-73	1122	0000101
1122	13-29867-1	1122	46-86250-3	1122	56-59-2867	1122	102-339
1122	13-29867-2	1122	46-86289-3	1122	59B001-1	1122	102-412
1122	13-29867-2(SW)	1122	46-86309-3	1122	62-125528	1122	103-131
1122	13-34056-1	1122	46-86335-3	1122	62-126521	1122	103-141
1122	13-43250-1	1122	46-86343-3	1122	62-130045	1122	103-142
1122	13-547604	1122	46-86422-3	1122	62-130046	1122	103-142(DET)
1122	13-55333-1	1122	46-86428-3	1122	62-20223	1122	103-145
1122	13-67590-1	1122	46-86431-3	1122	62-20319	1122	103-145-01
1122	13-67590-1/3464	1122	46-86436-3	1122	62-20437	1122	103-145-1
1122	13P2	1122	46-86481-3	1122	62-20597	1122	103-159
1122	14-507-01	1122	46-86483-3	1122	62-21552	1122	103-178
1122	14-514-02	1122	46-8676-3	1122	62-26597	1122	103-178-01
1122	14-514-12	1122	047(DIODE)	1122	065-012	1122	103-222
1122	14-514-15	1122	472102536-P1	1122	65-K11305-0001	1122	103-240
1122	14-514-17	1122	48-03005A01	1122	65-L11305-0001	1122	103-261-02
1122	14-514-19	1122	48-03005A05	1122	65-L11324-0001	1122	103-295-02
1122	14-514-62	1122	48-03005A06	1122	65-P11305-0001	1122	103-42
1122	14-514-64	1122	48-05-011	1122		1122	103-51

RS 276-	Industry Standard No.	RS 276-	Industry Standard No.	RS 276-	Industry Standard No.	RS 276-	Industry Standard No.	RS 276-	Industry Standard No.
1122	106-007	1122	332-4009	1122	919-013059	1122	5084-600-4521-0	1122	127993
1122	106-009	1122	344-6005-6	1122	919-013067	1122	5084-600-45ZI-0		
1122	111-4-2020-06100	1122	349-0002-001	1122	919-013082	1122	5217		
1122	111-4-2020-15600	1122	349-0002-002	1122	921-342B	1122	5631-MA150		
1122	112-500-0-50	1122	351-3031	1122	9260206P1	1122	6019		
1122	112-500-0-501	1122	353-2575-00	1122	931	1122	6129-P1		
1122	116,666	1122	353-2655-000	1122	939-12	1122	7112		
1122	118-02900	1122	353-3024-000	1122	941-026-0001	1122	7344		
1122	118-02902	1122	353-3083-000	1122	943-086	1122	8000-00003-046		
1122	118-030	1122	353-3273-00	1122	943-087	1122	8000-00004-061		
1122	120-003147	1122	353-3289-000	1122	943-087-1	1122	8000-00004-062		
1122	120-004877	1122	353-3338-000	1122	943-105-001	1122	8000-00004-066		
1122	120A02	1122	353-3339-000	1122	976-0036-921	1122	8000-00004-067		
1122	120A11	1122	353-3627-010	1122	977-1	1122	8000-00004-184		
1122	120A13	1122	353-3687-010	1122	991-00-1172	1122	8000-00006-008		
1122	123-016	1122	353-3687-020	1122	991-00-1172-1	1122	8000-00006-281		
1122	123-017(DIODE)	1122	360-32	1122	1000-130	1122	8000-00010-109		
1122	123-022	1122	380-1000	1122	1003-11	1122	8000-00011-042		
1122	123-024	1122	380-1001	1122	1009-07	1122	8000-00011-045		
1122	123-025	1122	394-1571-1	1122	1009-08	1122	8000-00028-045		
1122	123B-003	1122	394-1592-1	1122	1010-143	1122	8000-00058-008		
1122	130-30189	1122	394-1602-1	1122	1010A	1122	8000-0004-066		
1122	130-30265	1122	400-1417-101	1122	1011-0302	1122	8000-0004-P061		
1122	130-30266	1122	400-1596	1122	1018-3259	1122	8000-0004-P062		
1122	130-30274	1122	402-004-02	1122	1018-9884	1122	8000-0004-P064		
1122	130-30702	1122	413(CONCERT HALL)	1122	1033-0911	1122	8000-0004-P066		
1122	141-078-0001	1122	413(CONCERT-HALL)	1122	1033-0983	1122	8000-0004-P061		
1122	142-006(DIODE)	1122	413(TRUETONE)	1122	1033-0991	1122	8000-0004-P062		
1122	142-009	1122	429-10036-0A	1122	1040-07	1122	8000-0004-P066		
1122	142-010	1122	429-10036-0B	1122	1040-9332	1122	8000-00041-019		
1122	144-1	1122	429-10054-0A	1122	1041-66	1122	8000-00042-007		
1122	144-3	1122	429-20004-0B	1122	1042-15	1122	8010-52		
1122	146D-1	1122	460-1009	1122	1042-16	1122	8020-202		
1122	149-142-01	1122	479-0547-001	1122	1042-23	1122	8710-52		
1122	150-030-9-002	1122	479-0663-005	1122	1043-0049	1122	8840-53		
1122	150-040-9-002	1122	479-1012-001	1122	1043-10	1122	8910-52		
1122	150-066-9-001	1122	479-1013-001	1122	1044-8983	1122	9128-1503-001		
1122	150-1	1122	479-1013-002	1122	1045-7802	1122	9330-228-60112		
1122	151-001-9-001	1122	479-1055-001	1122	1048-6421	1122	9440		
1122	151-002-9-001	1122	479-1163-001	1122	1048-9987	1122	9644		
1122	151-011-9-001	1122	479-1229-001	1122	1062OE	1122	9646		
1122	151-014-9-1	1122	479-1248-001	1122	1074-118	1122	9803		
1122	151-015-9-001	1122	0500	1122	1076-1484	1122	10895		
1122	151-021-001	1122	514-042791	1122	1077-2341	1122	11305-0001		
1122	151-021-9-001	1122	523-0006-002	1122	1079-01	1122	11352-78		
1122	151-022-9-001	1122	523-0007-001	1122	1079-89	1122	011950		
1122	151-024-9-001	1122	523-0013-201	1122	1092-16	1122	011956		
1122	151-029-9-001	1122	523-1000-881	1122	1100-9487	1122	12011		
1122	151-029-9-002	1122	523-1000-883	1122	1120-17	1122	12101		
1122	151-030-9-001	1122	523-1500-803	1122	1212	1122	12255-235		
1122	151-030-9-006	1122	523-1500-883	1122	1212(R088)	1122	13150L		
1122	151-030-9-009	1122	540-02B	1122	1410-102	1122	014002-1		
1122	151-032-004	1122	540-028-00	1122	1410-169	1122	014007-1		
1122	151-032-9-001	1122	540-033-00	1122	1410-169(ROGERS)	1122	014007-2		
1122	151-032-9-002	1122	596-2	1122	1543-00	1122	014024		
1122	151-032-9-004	1122	596-5	1122	1627 1843	1122	15122		
1122	151-034-9-001	1122	600C	1122	1627-1843	1122	18600-52		
1122	151-035-0-003	1122	600X0100-066	1122	1699-17	1122	22164-000		
1122	151-035-9-001	1122	601-1(DIODE)	1122	1703-8662	1122	25260-61-067		
1122	151-035-9-003	1122	601C	1122	1750-103	1122	25810-52		
1122	151-035-9-004	1122	604C	1122	1756-17	1122	27840-43		
1122	151-040-9-001	1122	606-6003-101	1122	1809	1122	28810-63		
1122	151-040-9-002	1122	606-6003-102	1122	1818	1122	37510-52		
1122	151-042-9-001	1122	606-6021-101	1122	1846-17	1122	38510-330		
1122	151-045-9-001	1122	610-017-706	1122	1872-6	1122	38510-54		
1122	151-049-9-001	1122	612C	1122	1937-17	1122	39042		
1122	151-049-9-002	1122	614-118	1122	1937-17(DIODE)	1122	41173		
1122	151-051-9-001	1122	614C	1122	1979-17	1122	041200-30110		
1122	151-059-9-001	1122	616C	1122	1981-17	1122	41616		
1122	151-060-9-001	1122	618C	1122	2000-125	1122	43062		
1122	151-064-9-001	1122	620C	1122	2000-302	1122	44002		
1122	151-064-9-002	1122	622C	1122	2000-303	1122	50212-19		
1122	151-066-9-001	1122	642-126	1122	2000-317	1122	50212-28		
1122	151-067-9-001	1122	642-275	1122	2000-332	1122	50212-30		
1122	151-069-9-001	1122	642-281	1122	2000-345	1122	50213-5		
1122	151-072-9-001	1122	690V088H20	1122	2004-67	1122	53016R		
1122	151-267-9-001	1122	690V102H40	1122	2005-2981	1122	057293		
1122	151-32-9-004	1122	690V118H57	1122	2010-01	1122	58810-82		
1122	152-061-00	1122	690V119H15	1122	2010-03	1122	58840-113		
1122	152-0242-00	1122	700A858-322	1122	2019-45	1122	59840		
1122	152-0333-00	1122	754-2000-001	1122	2041-05	1122	67055		
1122	152-141-1	1122	754-2000-011	1122	2061A45-38	1122	70055		
1122	165-432-2-48-1	1122	754-2009-150	1122	2061A45-93	1122	70372-2		
1122	165A4378P2	1122	754-2509-150	1122	2065-08	1122	70581-1		
1122	168-107-001	1122	754-2720-021	1122	2065-17	1122	71119-1		
1122	170-1	1122	754-5750-283	1122	2066-17	1122	71467-1		
1122	171-1	1122	754-9000-473	1122	2076	1122	71667-1		
1122	173A4394-1	1122	754-9052-473	1122	2093A38-11	1122	000072130		
1122	185-011	1122	755-845049	1122	2093A38-37	1122	72146		
1122	185-012	1122	863-254B	1122	2093A41-102	1122	72147		
1122	186-011	1122	863-567B	1122	2093A41-105	1122	72152(TELEDYNE)		
1122	186-011(DIODE)	1122	863-776B	1122	2093A41-129	1122	72197		
1122	186-012	1122	903-00393	1122	2093A41-155	1122	085003		
1122	186-015	1122	903-100B	1122	2093A41-156	1122	87756		
1122	200X8010-165	1122	903-112B	1122	2093A41-158	1122	88060-52		
1122	201X2010-144	1122	903-116B	1122	2093A41-164	1122	88060-55		
1122	201X2010-159	1122	903-118B	1122	2093A41-165	1122	88510-52		
1122	202-5-9531-01010	1122	903-121B	1122	2093A41-171	1122	88510-55		
1122	204-1	1122	903-177B	1122	2093A41-172	1122	95005		
1122	209-32	1122	903-178	1122	2093A41-76	1122	95012		
1122	212-142	1122	903-20	1122	2093B38-4	1122	95015(EICO)		
1122	212-72	1122	903-25B	1122	2101-03	1122	95332-1		
1122	229-1301-19	1122	903-311	1122	2102-029	1122	99203-3		
1122	229-1301-20	1122	903-332	1122	2110N-41	1122	99203-5(DIODE)		
1122	230-014	1122	903-41B	1122	2180-41	1122	100828-003		
1122	232-0006-02	1122	903-42B	1122	2209-17	1122	101154		
1122	241ABNG414B	1122	903-48B	1122	2304	1122	102009		
1122	260-10-011	1122	903-58B	1122	2328	1122	103514		
1122	260-10-032	1122	903-72B	1122	2328-17	1122	104762		
1122	260-10-047	1122	903-82B	1122	2400-17	1122	110667		
1122	260-10-049	1122	903-84B	1122	2403	1122	112529(AMPHENOL)		
1122	260-10-057	1122	903-95B	1122	2405	1122	112529(AMPHENOL)		
1122	260-61-011	1122	914-006-0-00	1122	2409-17	1122	115060-102		
1122	260-61-047	1122	914-001-1-00	1122	2477-173	1122	116021		
1122	260-61-067	1122	900-0-4326	1122	3008	1122	116021(SEARS)		
1122	264D00101	1122	919-00-4799	1122	003507	1122	118335		
1122	264D00209	1122	919-00-7394	1122	4001(MAGNAVOX)	1122	118527-04		
1122	264P04501	1122	919-00-9929	1122	4011(PENNCREST)	1122	119264-001		
1122	264P04502	1122	919-01-0873	1122	4014-200-30110	1122	119507		
1122	264P04507	1122	919-01-1172-1	1122	004763	1122	119596		
1122	265D00702	1122	919-01-1215	1122	4801-00154	1122	119597		
1122	265200101	1122	919-01-1307	1122	4805-1241-200	1122	119697		
1122	269M01201	1122	919-01-3072	1122	4820-0201	1122	119956		
1122	290-1003	1122	919-010873-050	1122	4822-130-40182	1122	120068		
1122	296-42-9-3			1122	5001-083	1122	123805(DIODE)		
1122	296I002B01			1122	5001-107	1122	123813		
1122	296V020B01			1122	5001-120	1122	125397		
1122	296V024B02			1122	5001-128	1122	125528		
1122	300-0003-002			1122	5001-144	1122	125588		
1122	300-0003-003			1122	5001-145	1122	125884		
1122	311-0126-001			1122	5001-146	1122	126321		
1122	311-0139-001			1122	5001-156	1122	127474		
1122	311D883-P01			1122	5001-162				
				1122	5001-164				

RS 276-	Industry Standard No.
1122	126A74
1122	129095
1122	129375
1122	129475
1122	129682-001
1122	129708-101
1122	130045
1122	130046
1122	131501(DIODE)
1122	131502
1122	132553
1122	132645
1122	133390
1122	135571
1122	136162
1122	136163
1122	136688
1122	137028
1122	137028-001
1122	137057-001
1122	138175
1122	138196
1122	139029
1122	139328
1122	139706
1122	143837
1122	144581
1122	144691
1122	146571
1122	146576
1122	150834
1122	152473K
1122	161021
1122	168706
1122	169114
1122	169117
1122	169558
1122	0170301
1122	170370
1122	170857
1122	171149-016
1122	171149-017
1122	171162-196
1122	171162-197
1122	171162-271
1122	171217-4
1122	171840
1122	171841
1122	171843
1122	172165
1122	172201
1122	172253
1122	172547
1122	172722
1122	181073
1122	181619
1122	181681
1122	183033
1122	184798
1122	185335
1122	185411
1122	188056
1122	193717
1122	197464
1122	198773-1
1122	198775-1
1122	198776-1
1122	198779-1
1122	198785
1122	198785-1
1122	198785-2
1122	198799-4
1122	198809-1
1122	198810-1
1122	198813-1
1122	199596
1122	202315(THOMAS)
1122	202862-518
1122	206617
1122	211083
1122	240564
1122	245192
1122	248817
1122	249508-3
1122	267611
1122	269922-001
1122	280017
1122	281101-97
1122	304281-P1
1122	309481
1122	326309-10A
1122	338307
1122	379005N
1122	462580-1
1122	476690-2
1122	497442
1122	497616
1122	497616-1
1122	497616-2
1122	501343
1122	504720
1122	504720-1
1122	0517132
1122	0517133
1122	0518926
1122	0926232
1122	529657
1122	529658
1122	530063-13
1122	530072-1002
1122	530072-1006
1122	530072-1008
1122	530072-1009
1122	530072-1010
1122	530072-1011
1122	530072-1015
1122	530072-1018
1122	530072-6
1122	530072-8
1122	530082-3
1122	530092-1002
1122	530116-1001
1122	530116-3
1122	530135-2
1122	530144-1
1122	530144-1001
1122	530144-1002
1122	530144-1003
1122	530144-1004
1122	530144-3
1122	530150-1
1122	530154-1
1122	530170-1
1122	530179-1
1122	530179-1001
1122	530179-1002
1122	530179-2
1122	530179-3
1122	530181-1
1122	530181-1-1
1122	530181-1001
1122	530181-1003
1122	535013
1122	534001H
1122	562654
1122	0576054(SW)
1122	604407
1122	615004-8
1122	615154-1
1122	618150-3
1122	618639-1
1122	619011
1122	619087-1
1122	619526-1
1122	650845
1122	650854
1122	651030
1122	654032
1122	701662-00
1122	710838-1
1122	720608-13
1122	720609-1
1122	726654
1122	740828
1122	741100
1122	741741
1122	741864
1122	760007
1122	760037
1122	800743-001
1122	801712
1122	801724
1122	801728
1122	802008
1122	817172
1122	817190
1122	860001-153
1122	860011
1122	900545-2
1122	900546-2
1122	900546-20
1122	908703-1
1122	908705-1
1122	908705-2
1122	908721-1
1122	908742-1
1122	921150-021
1122	922433
1122	922873
1122	922943-1
1122	923147
1122	925075-501B
1122	925252-1
1122	925252-102
1122	925252-2
1122	925252-4
1122	925253-1
1122	925253-3
1122	925297
1122	925297-1
1122	925521-1B
1122	925521-1C
1122	925915-101
1122	925939-101
1122	925940-1
1122	925940-501B
1122	930293
1122	932022-0001
1122	932033-1
1122	932050-1
1122	942677-9
1122	943502
1122	969099-1
1122	983689
1122	984162
1122	984880
1122	985106
1122	986577
1122	986934
1122	988995
1122	988996
1122	992150
1122	992157-1
1122	1223771
1122	1223930
1122	1417872-11
1122	1421207-1
1122	1470872-6
1122	1471072-4
1122	1471872-1
1122	1471872-11
1122	1471872-12
1122	1471872-13
1122	1471872-16
1122	1471872-18
1122	1471872-2
1122	1471872-4
1122	1471872-5
1122	1471872-8
1122	1472460-2
1122	1472460-6
1122	1472460-7
1122	1472460-8
1122	1472872-4
1122	1472778-16
1122	1472872-1
1122	1476171-24
1122	1476171-33
1122	1476690-1
1122	1476690-2
1122	1477022-1
1122	1588035-42
1122	1638776-103J81659
1122	1641141-101
1122	1641141-102
1122	1680008-01
1122	1780169-1
1122	1780174-1
1122	1800009
1122	1819393-1
1122	1846794-1
1122	1905490-1
1122	1908519
1122	1950003
1122	1950060
1122	2002402
1122	2003069-2
1122	2003069-6
1122	2006441-132
1122	2006623-48
1122	2006627-54
1122	2008292-88
1122	2010967-83
1122	2014400
1122	2132524-1
1122	2182124-1
1122	2185434-1
1122	2185494-2
1122	2330351
1122	2330352
1122	2331351
1122	2337011
1122	2485080
1122	2546059
1122	2610032
1122	2621499-1
1122	2621786-1
1122	2635012
1122	2760438-1
1122	2865141
1122	3110883-P01
1122	3160000-00-08A
1122	3180006
1122	3195638
1122	3263029-1Q
1122	3311133
1122	3464611
1122	3596061
1122	5596559
1122(2)	ALC-1A
1122(2)	B-46-110
1122(2)	C-10-20A
1122(2)	C10-20A
1122(2)	C10-22C
1122(2)	C1A
1122(2)	CDB60011
1122(2)	D30
1122(2)	D1J71895-1
1122(2)	B1176R
1122(2)	ECG178MP
1122(2)	EP16X1
1122(2)	G01-803-A
1122(2)	IS-446D
1122(2)	M8569
1122(2)	PTC406
1122(2)	RF32426-7
1122(2)	S1B02-CR1
1122(2)	SD-7
1122(2)	SD7
1122(2)	SIB02-CR1
1122(2)	SR-13
1122(2)	SR-20
1122(2)	TM178MP
1122(2)	TR-C72(DIODE)
1122(2)	TVM-530
1122(2)	TVM-K-112C
1122(2)	TVM554A
1122(2)	Y80-V81-2-3
1122(2)	ZEN432
1122(2)	1N4093
1122(2)	1N4951
1122(2)	1N4952
1122(2)	181579
1122(2)	181701
1122(2)	18184
1122(2)	18200
1122(2)	182210
1122(2)	182211
1122(2)	18824
1122(2)	18825
1122(2)	18S26
1122(2)	18S27
1122(2)	18828
1122(2)	18829
1122(2)	4-2021-05000
1122(2)	09-306168
1122(2)	12/1N10
1122(2)	13-15465-1
1122(2)	14-514-20
1122(2)	14-514-65
1122(2)	15-100003
1122(2)	15-108001
1122(2)	21N586
1122(2)	30-8057-13
1122(2)	34-8047-1
1122(2)	34-8057-34
1122(2)	34-8057-8
1122(2)	46-861-78-3
1122(2)	46-86254-3
1122(2)	46-86380-3
1122(2)	48-134916
1122(2)	48-134917
1122(2)	48-137167
1122(2)	48-355014
1122(2)	48-6712A02
1122(2)	48-741255
1122(2)	48-741656
1122(2)	48-741724
1122(2)	488134917
1122(2)	488137167
1122(2)	53A022-3
1122(2)	530008-1
1122(2)	667017
1122(2)	73A64-1
1122(2)	86-141-1
1122(2)	86-574-2(DIODE)
1122(2)	92-64-1
1122(2)	93A64-2(APC)
1122(2)	103-101
1122(2)	103-192
1122(2)	120-004629
1122(2)	146D1
1122(2)	146D1B112
1122(2)	151-013-9-001
1122(2)	151-030-9-002
1122(2)	205-142-G1
1122(2)	601X0225-066
1122(2)	1042-14
1122(2)	2231-17
1122(2)	2289-13
1122(2)	3505(AIRLINE)
1122(2)	5203N1
1122(2)	06102
1122(2)	72152
1122(2)	72163-1
1122(2)	139065
1122(2)	139605(APC)
1122(2)	170733-1
1122(2)	503146-1
1122(2)	530093-1
1122(2)	530146-1
1122(2)	530146-2
1122(2)	740629
1122(2)	983744
1122(2)	1471872-1Q
1122(2)	1471876-6
1123	A02
1123	A054-103
1123	A054-105
1123	A054-187
1123	A066-119
1123	A066-119(GE)
1123	A066-120
1123	A066-121
1123	A068-101
1123	A069-109
1123	A069-111
1123	A069-115
1123	A07
1123	A090
1123	A15-1007
1123	A20571
1123	A2419
1123	A2420
1123	A2473
1123	A2476
1123	A25-1007
1123	A30
1123	A30(DIODE)
1123	A36508
1123	A42X00340-01
1123	A48-63078A52
1123	A514-022057
1123	A514-042791
1123	A556-142
1123	A615-1012
1123	A692X13-4
1123	0A7
1123	A7001800
1123	A73
1123	0A81
1123	0A9
1123	0A90
1123	A909-1015
1123	A909-1017
1123	A90M
1123	0A90Z
1123	0A90ZA
1123	0A91
1123	A9M
1123	AA111
1123	AA112
1123	AA112P
1123	AA113
1123	AA114
1123	AA116
1123	AA117
1123	AA118
1123	AA119
1123	AA120
1123	AA121
1123	AA123
1123	AA130
1123	AA131
1123	AA132
1123	AA134
1123	AA135
1123	AA136
1123	AA137
1123	AA138
1123	AA139
1123	AA140
1123	AA142
1123	AA143
1123	AA143B
1123	AA144
1123	AA218
1123	AA779
1123	AAY139
1123	AAY15
1123	AAY18
1123	AAY22
1123	AAY27
1123	AAY30
1123	AAY33
1123	AAY46
1123	AA210
1123	AA218
1123	ACR810-107
1123	ACR83-1007
1123	AFS-160-1017
1123	B-28
1123	B-30
1123	B-30P
1123	B-3P
1123	B28
1123	B30
1123	B601-1011
1123	B692X13
1123	BD1
1123	BD2
1123	BD3
1123	BD4
1123	BD5
1123	BD6
1123	BD7
1123	BE-090(DIODE)
1123	C21480
1123	C5005
1123	C60
1123	C60(DIODE)
1123	C8106
1123	CB163
1123	CB393
1123	CD-0000
1123	CD-0000N
1123	CD-0014N
1123	CD000
1123	CD0000
1123	CD0000/7825B
1123	CD101
1123	CR0495/7839
1123	C012-E
1123	C012E
1123	C064H
1123	C065H
1123	C065H
1123	C074H
1123	C086H
1123	CQD1029
1123	CQD462
1123	CQD591
1123	CQD685
1123	CK-706

RS 276-	Industry Standard No.	RS 276-	Industry Standard No.	RS 276-	Industry Standard No.	RS 276-	Industry Standard No.	RS 276-	Industry Standard No.
1123	CK705	1123	ED4(ELCOM)	1123	GD5E	1123	K52	1123	
1123	CK706	1123	ED46	1123	GD663	1123	K6	1123	
1123	CK706A	1123	ED6(ELCOM)	1123	GD6E	1123	K60	1123	
1123	CK706P	1123	ED60	1123	GD72E/3	1123	K882	1123	
1123	CK715	1123	ED9(ELCOM)	1123	GD72E/4	1123	KD27	1123	
1123	CR0000	1123	EDG-0003	1123	GD72E/5	1123	KG841959	1123	
1123	CR101/6515	1123	EDG-0006	1123	GD73E/4	1123	K0825201-1	1123	
1123	CR102/6515	1123	EDG-1	1123	GD73E/5	1123	K0825201-2	1123	
1123	CT-2002	1123	EDG-3	1123	GD74E/3	1123	KR-Q0005	1123	
1123	CT-2005	1123	EDG-6	1123	GD74E/4	1123	M,KIM,I-/U	1123	
1123	CT2002	1123	EE16X3	1123	GD74E/5	1123	M-8513R	1123	
1123	CT2007	1123	ES10189	1123	GD8E	1123	M34A	1123	
1123	CT461	1123	ES10224	1123	GE-X66	1123	M51	1123	
1123	CTP-2001-1010	1123	ES10225	1123	GE3638	1123	M60	1123	
1123	CTP-2006-1004	1123	ES15054	1123	GED05B850	1123	M8489-A	1123	
1123	CTP-461	1123	ES16X103	1123	G00-003-A	1123	M95	1123	
1123	CTP461	1123	ES16X12	1123	GP-354	1123	MA-23B	1123	
1123	CTP573	1123	ES16X14	1123	GP2354	1123	MA-900	1123	
1123	CV425	1123	ES16X2	1123	GP2364	1123	MA21	1123	
1123	CV442	1123	ES16X3	1123	GPM1NA	1123	MA23(B)	1123	
1123	CX-0031	1123	ES16X5	1123	GPM1NB	1123	MA25A	1123	
1123	CX-0033	1123	ES16X6	1123	GPM1NC	1123	MA3	1123	
1123	CX-0036	1123	ES16X7	1123	GPM2NA	1123	MA47	1123	
1123	CX-0042	1123	ES356X103	1123	G873E/3	1123	MA51A	1123	
1123	CX-0045	1123	ET16X1	1123	H087	1123	MA55	1123	
1123	CX-0047	1123	ET16X19	1123	H091	1123	MA790	1123	
1123	CX0036	1123	ET16X20	1123	H316	1123	MA8	1123	
1123	CX0041	1123	ET16X21	1123	H614	1123	MA90	1123	
1123	CXL109	1123	ET41X37	1123	H624	1123	MC2526	1123	
1123	CXL110	1123	ETD-1N60	1123	H8287	1123	MC308	1123	
1123	D-00169C	1123	ETD-8D46	1123	H8287-4	1123	MD-34	1123	
1123	D-00204R	1123	EU16X1	1123	HD-1000101	1123	MD-60A	1123	
1123	D-00269C	1123	EU16X2	1123	HD10-001-01	1123	MD34	1123	
1123	D-00284R	1123	EW166	1123	HD10000101	1123	MD34A	1123	
1123	D-00669C	1123	EW167	1123	HD1000101	1123	MD35	1123	
1123	D-2	1123	EXY420DIR5JB	1123	HD1000101-0	1123	MD46	1123	
1123	D-GM-2	1123	EYV420DIR5JB	1123	HD1001010	1123	MD54	1123	
1123	D093	1123	EYY420DIR5JB	1123	HD1000105-0	1123	MD56	1123	
1123	D1-2	1123	F136	1123	HD1000105-0	1123	MD60	1123	
1123	D1J70542	1123	0P173	1123	HD1000301	1123	MD604A	1123	
1123	D1J70543	1123	F20-1010	1123	HD1000302	1123	MD60A	1123	
1123	D1R35	1123	F20-1012	1123	HD1000303	1123	MN34A	1123	
1123	D1R39	1123	F20-1013	1123	HD1000303-0	1123	MN51	1123	
1123	D286	1123	F20-1014	1123	HD1468	1123	MN60(DIODE)	1123	
1123	D2R31	1123	F20303	1123	HD2149	1123	MT1889	1123	
1123	D2U	1123	F215-1010	1123	HD2155	1123	N2A	1123	
1123	D4R	1123	F215-1012	1123	HD4000109	1123	N2A(DIODE)	1123	
1123	D8410	1123	F215-1013	1123	HE-0A90	1123	N48	1123	
1123	DA90	1123	F215-1014	1123	HE-10001	1123	NC29	1123	
1123	DAAY001002	1123	FA-1	1123	HE-10002	1123	NGP3002	1123	
1123	DANZ00600	1123	FA-1(DIODE)	1123	HE-10003	1123	NTC-13	1123	
1123	DANZ006000	1123	FB1043	1123	HE-10024	1123	NTC-14	1123	
1123	DANZ00600000	1123	FD1980	1123	HE-10025	1123	NU34	1123	
1123	DC-13	1123	PPR40-1006	1123	HE-10027	1123	OA-90	1123	
1123	DDAY001001	1123	F819	1123	HE-10040	1123	OA-90(G)	1123	
1123	DDAY001002	1123	FV-23	1123	HE-1024	1123	OA-900	1123	
1123	DDAY001004	1123	FV23	1123	HE-1N34	1123	OA-91	1123	
1123	DDAY001010	1123	G-1010	1123	HE-1N34A	1123	OA134Q	1123	
1123	DDAY001022	1123	G00-003-A	1123	HE-1N60	1123	OA150	1123	
1123	DDAY101001	1123	G00-004-A	1123	HE-1N60P	1123	OA159	1123	
1123	DDMV-1	1123	G00-008-A	1123	HE-18188	1123	OA160	1123	
1123	DDMV-2	1123	G00-009-A	1123	HE-18426	1123	OA161	1123	
1123	DG1N60	1123	G00-013-B	1123	HE-18446	1123	OA172	1123	
1123	DG1834	1123	G00003A	1123	HE-CD0000	1123	OA174	1123	
1123	DGM-2	1123	G00009A	1123	HE-0A90	1123	OA47	1123	
1123	DGM-3	1123	G100	1123	HEP-R9134	1123	OA50	1123	
1123	DGM3	1123	G1010	1123	HEP-R9135	1123	OA541	1123	
1123	DI-1	1123	G1288	1123	HEP134	1123	OA6	1123	
1123	DI-2	1123	G156	1123	HEP135	1123	OA7	1123	
1123	DI-8(DIODE)	1123	G157	1123	HEPR9134	1123	OA70	1123	
1123	DIJ61224	1123	G158	1123	HEPR9134A	1123	OA71	1123	
1123	DIJ70542	1123	G159	1123	HEPR9135	1123	OA71C	1123	
1123	DIJ70645	1123	G198	1123	HP-10024	1123	OA72	1123	
1123	DIJ70646	1123	G199	1123	HF-20008	1123	OA73	1123	
1123	DIJ71776	1123	G1HA	1123	H01090	1123	OA73C	1123	
1123	DIJ72166	1123	G200	1123	HG5002	1123	OA74	1123	
1123	DIJ72294	1123	G297	1123	HG5004	1123	OA74A	1123	
1123	DIJ72349	1123	G409	1123	HG5006	1123	OA79	1123	
1123	DK19	1123	G498	1123	HG5007	1123	OA81	1123	
1123	DK20	1123	G580	1123	HG5008	1123	OA81C	1123	
1123	DK21	1123	G5C	1123	HG5009	1123	OA85	1123	
1123	DR291	1123	G5P	1123	HG5078	1123	OA85C	1123	
1123	DR351	1123	G5K	1123	HG5079	1123	OA9	1123	
1123	DR352	1123	G766	1123	HG5085	1123	OA90	1123	
1123	DR365	1123	G769	1123	HG5088	1123	OA90-FM	1123	
1123	DR385	1123	G770	1123	HG5808	1123	OA909	1123	
1123	DR426	1123	G788	1123	HN-00003	1123	OA90Q	1123	
1123	DR434	1123	G789	1123	H091	1123	OA90QA	1123	
1123	DR449	1123	G790	1123	HV-15	1123	OA90LF	1123	
1123	DR464	1123	G7D	1123	HV-23BL	1123	OA90M	1123	
1123	DS-18(DELCO)	1123	G7E	1123	HV-801	1123	OA90MLF	1123	
1123	DS-32	1123	G7F	1123	HVO000202	1123	OA90E	1123	
1123	DS-33	1123	G7G	1123	HV15	1123	OA90ZA	1123	
1123	DS-39	1123	G814	1123	IC743050	1123	OA91	1123	
1123	DS2-27	1123	G815	1123	IN60-1	1123	OA92	1123	
1123	DS27	1123	G816	1123	INJ33349	1123	OA95	1123	
1123	DS31	1123	G820	1123	INJ60034	1123	OA99	1123	
1123	DS31(DELCO)	1123	G821	1123	INJ60284	1123	OA9D	1123	
1123	DS32	1123	G822	1123	INJ61224	1123	OA9Z	1123	
1123	DS33	1123	G823	1123	INJ61675	1123	OP173	1123	
1123	DS33(DELCO)	1123	G824	1123	INJ70973	1123	ON67A	1123	
1123	DS410(G.E.)	1123	G844	1123	INJ70980	1123	O870	1123	
1123	DS410R(G.E.)	1123	G845	1123	IP20-0015	1123	OSS-16685	1123	
1123	DS97	1123	G846	1123	IP20-0016	1123	OSS16308	1123	
1123	DS816685	1123	G847	1123	IP20-0020	1123	OSS16685	1123	
1123	DX-0161	1123	G869	1123	IP20-0060	1123	P10155	1123	
1123	DX-0162	1123	G8-1	1123	IP20-0179	1123	PBB3322	1123	
1123	DX-0241	1123	GC5012	1123	IP20-0283	1123	PL-150-001-9-005	1123	
1123	DX-0725	1123	GD-25	1123	IR5JA	1123	PL-150-006-9-001	1123	
1123	DX6873	1123	GD-26	1123	IT22	1123	P093	1123	
1123	E0018	1123	GD-29	1123	IT23	1123	PTC206	1123	
1123	E1031RXT	1123	GD-30	1123	IT23G	1123	PTC206M	1123	
1123	E20030	1123	GD10	1123	ITT102	1123	PTC207	1123	
1123	E21430	1123	GD1001	1123	ITT301	1123	PTC207M	1123	
1123	E295ZZ01	1123	GD11E	1123	ITT718	1123	Q-22115C	1123	
1123	E30060	1123	GD12E	1123	J241245	1123	Q49	1123	
1123	EA1123	1123	GD13E	1123	J242	1123	Q50	1123	
1123	EA16X1	1123	GD1E	1123	J243	1123	Q51	1123	
1123	EA16X11	1123	GD1P	1123	J2441	1123	QD-01N60PXT	1123	
1123	EA16X140	1123	GD3638-00	1123	J24628	1123	QD-01N60XXT	1123	
1123	EA16X22	1123	GD400	1123	J24643	1123	QD-01832XXT	1123	
1123	EA16X27	1123	GD401	1123	J24820	1123	QRT-200	1123	
1123	EA16X48	1123	GD402	1123	J24838	1123	QVD1KF114	1123	
1123	EA16X5	1123	GD403	1123	J24911	1123	QVM800B	1123	
1123	EA16X9	1123	GD404	1123	J24913	1123	R-18188	1123	
1123	EA16X97	1123	GD405	1123	J320041	1123	R-7029	1123	
1123	EA2502	1123	GD406	1123	J685	1123	R-7051	1123	
1123	EA3127	1123	GD409	1123	JT-E1014	1123	R1106	1123	
1123	EA3718	1123	GD4E	1123	JT-E1031	1123	R1107	1123	
1123	EC0109	1123	GD5004	1123	K115J511-2	1123	R1109	1123	
1123	ED-46	1123	GD510	1123	K3	1123	R112524	1123	
1123	ED-60	1123	GD556	1123	K4-550	1123	R1667	1123	
1123	ED12(ELCOM)							1123	R1889
1123	ED219464							1123	R2164

RS 276-	Industry Standard No.	RS 276-	Industry Standard No.	RS 276-	Industry Standard No.	RS 276-	Industry Standard No.
1123	R2334	1123	SFT108	1123	ZTR-1N60	1123	1N3146
1123	R5096	1123	SI811	1123	001-0000-00	1123	1N31A
1123	R5522	1123	SI82	1123	001-0010-00	1123	1N3204
1123	R60-1007	1123	SI820	1123	001-0022-00	1123	1N3287
1123	R7028	1123	SK-1W8Q	1123	001-0081	1123	1N3287N
1123	R7029	1123	SK30B7	1123	001-01501-Q	1123	1N3287W
1123	R7743	1123	SK30B8	1123	001-015010	1123	1N34
1123	R7892	1123	SK3090	1123	001-015011	1123	1N3465
1123	R7893	1123	SK3091	1123	1-016	1123	1N3466
1123	R8060	1123	SQ46	1123	1-017	1123	1N3467
1123	R8061	1123	SS0007	1123	1-037/2207	1123	1N3468
1123	R8219	1123	SS0008	1123	1-12689	1123	1N3469
1123	R8257	1123	SU-31	1123	1-425-636	1123	1N3470
1123	R8475	1123	SV-31	1123	1-DI-009	1123	1N3482
1123	R8887	1123	SV-31(DIODE)	1123	1A11306	1123	1N3483
1123	R8970	1123	SV30	1123	1A12689	1123	1N3484
1123	R9590	1123	SVD0A70	1123	1A14584	1123	1N34A
1123	R847	1123	SVD0A79	1123	1C0029	1123	1N34A-Z
1123	RE86	1123	SVD0A90	1123	1C0039	1123	1N34AM
1123	RE8109	1123	SVD20A70	1123	1DG2	1123	1N34AS
1123	RP1811	1123	SVD20A79	1123	1G01	1123	1N34Q
1123	RP33550-1	1123	SVDMA26-1	1123	1G02	1123	1N34QA
1123	RP35123	1123	SVD0A70	1123	1G25	1123	1N34Z
1123	RP60034	1123	SVD0A90	1123	1G86	1123	1N35
1123	RPJ60172	1123	SVDSC20	1123	1GD1Q	1123	1N355
1123	RPJ60614	1123	SX-990	1123	1GD2	1123	1N3564
1123	RL232G	1123	T-700-709	1123	1GD4	1123	1N3592
1123	RL246	1123	T-750-713	1123	1GD5X	1123	1N36
1123	RL252	1123	T-E1014	1123	1GD6	1123	1N367
1123	RL51	1123	T-E1031	1123	1K110	1123	1N367B
1123	RL52	1123	T-E1116	1123	1K188FM	1123	1N3753
1123	RL32G	1123	T-G1173B	1123	1K188FM-1	1123	1N3773
1123	RL34	1123	T11	1123	1K261	1123	1N38
1123	RL34G	1123	T12	1123	1K34A	1123	1N38A
1123	RL41	1123	T12G	1123	1K60	1123	1N38B
1123	RL41G	1123	T13	1123	1K60A	1123	1N3991
1123	RL42	1123	T13G	1123	1K90	1123	1N40
1123	RL52	1123	T14	1123	1N-22	1123	1N4088
1123	RS1811	1123	T14G	1123	1N10	1123	1N41
1123	RS2801	1123	T17	1123	1N100	1123	1N417
1123	RSWY/16L7Z	1123	T18	1123	1N100A	1123	1N418
1123	RT-1689	1123	T1G	1123	1N103	1123	1N419
1123	RT-200	1123	T20	1123	1N104	1123	1N419B
1123	RT-2016	1123	T20G	1123	1N105	1123	1N42
1123	RT1008	1123	T21	1123	1N107	1123	1N43
1123	RT1106	1123	T21237	1123	1N108	1123	1N435
1123	RT1108	1123	T21238	1123	1N109	1123	1N44
1123	RT1184	1123	T21271	1123	1N1093	1123	1N447
1123	RT1667	1123	T21313	1123	1N111	1123	1N449
1123	RT2334	1123	T21G	1123	1N112	1123	1N45
1123	RT2452	1123	T22	1123	1N113	1123	1N4502
1123	RT2694	1123	T22G	1123	1N114	1123	1N452
1123	RT3072	1123	T23	1123	1N115	1123	1N4523
1123	RT3099	1123	T23G	1123	1N116	1123	1N454
1123	RT3233	1123	T24G	1123	1N116A	1123	1N455
1123	RT3336	1123	T26G	1123	1N117	1123	1N45A
1123	RT3469	1123	T27G	1123	1N117A	1123	1N46
1123	RT4293	1123	T2G(DIODE)	1123	1N118	1123	1N46A
1123	RT4644	1123	T3G	1123	1N118A	1123	1N47
1123	RT5213	1123	T7	1123	1N119	1123	1N476
1123	RT5214	1123	T700-709	1123	1N119A	1123	1N477
1123	RT5379	1123	T8G	1123	1N120	1123	1N478
1123	RT5908	1123	T9	1123	1N120A	1123	1N479
1123	RT5939	1123	T9G	1123	1N125	1123	1N48
1123	RT61012	1123	T0311200600	1123	1N126	1123	1N480
1123	RT6119	1123	T03112006000	1123	1N126A	1123	1N48A
1123	RT6179	1123	TE-1014	1123	1N128	1123	1N49
1123	RT6180	1123	TE-1031	1123	1N128A	1123	1N497
1123	RT6181	1123	TE1014	1123	1N132	1123	1N498
1123	RT6182	1123	TE1031	1123	1N133	1123	1N50
1123	RT6183	1123	TE1098	1123	1N134	1123	1N500
1123	RT6184	1123	TE1105	1123	1N136	1123	1N51
1123	RT6189	1123	TG-28	1123	1N137	1123	1N52
1123	RT6619	1123	TG-48	1123	1N139	1123	1N527
1123	RT7330	1123	TM109	1123	1N1391	1123	1N52A
1123	RT7636	1123	TP34	1123	1N140	1123	1N542A
1123	RT7689	1123	TP34A	1123	1N142	1123	1N542MP
1123	RT7851	1123	TRO575002	1123	1N143	1123	1N54A
1123	RT8671	1123	TRO575005	1123	1N144	1123	1N54G
1123	RV1479	1123	TR12001-4	1123	1N145	1123	1N54GA
1123	RVD1K110	1123	TR2083-41	1123	1N148	1123	1N56
1123	RVD1N34A	1123	TR228736002003	1123	1N1561	1123	1N564
1123	RVD2-1K110	1123	TR320008	1123	1N1562	1123	1N569
1123	S-21271	1123	TR320039	1123	1N191	1123	1N56A
1123	S-320F	1123	TR320041	1123	1N192	1123	1N57
1123	OS-90	1123	TR320048	1123	1N193	1123	1N571
1123	S1811	1123	TR48	1123	1N198A	1123	1N57A
1123	S1820	1123	TV241013	1123	1N198B	1123	1N58
1123	S2085	1123	TV24273	1123	1N22	1123	1N58A
1123	S21271	1123	TV8-0A70	1123	1N265	1123	0000N60
1123	0822.3901-001	1123	TV8-0A81	1123	1N266	1123	1N60(7V)(PA-1)
1123	0822.3902-001	1123	TV8-0A90	1123	1N267	1123	1N60-1
1123	830300	1123	TV8-0A91	1123	1N268	1123	1N60-PA1
1123	S3577G	1123	TV8-0A95	1123	1N270	1123	1N60-M3
1123	S3603G	1123	TV8-0A70	1123	1N273	1123	1N60-P
1123	S3776B	1123	TV8-0A90	1123	1N276	1123	1N60-S
1123	S3838GA	1123	TV8-0A91	1123	1N277	1123	1N60-T
1123	S3885G	1123	TV80A70	1123	1N278	1123	1N60-Z
1123	S93.24.401	1123	TV80A90	1123	1N2801	1123	1N60/3490
1123	S93.24.601	1123	TV80A91	1123	1N281	1123	1N60/4454C
1123	S93.24.604	1123	TV80A90	1123	1N282	1123	1N60/7825B
1123	SC-6	1123	UF-8D1	1123	1N283	1123	1N60A
1123	SC20	1123	V-10916-3	1123	1N285	1123	1N60AM
1123	SC54	1123	V-210C	1123	1N287	1123	1N60B
1123	SD-12	1123	V10916-1	1123	1N288	1123	1N60C
1123	SD-13(PHILCO)	1123	V10916-3	1123	1N289	1123	1N60D
1123	SD-14	1123	V115	1123	1N290	1123	1N60P
1123	SD-150	1123	V117	1123	1N292	1123	1N60PA1
1123	SD-16	1123	V135	1123	1N294	1123	1N60FD1
1123	SD-1N60	1123	V50260-16	1123	1N294A	1123	1N60FM
1123	SD-1N60B	1123	V50A260-36	1123	1N295	1123	1N60FMX
1123	SD-46	1123	V74	1123	1N295A	1123	1N60G
1123	SD-46-2	1123	VD11	1123	1N295B	1123	1N60GA
1123	SD-56	1123	VD12	1123	1N295X	1123	1N60GB
1123	SD-60	1123	VD13	1123	1N296	1123	1N60M
1123	SD12	1123	VHD1N34A///-1	1123	1N297	1123	1N60M3
1123	SD12B	1123	VHD1N34A///1	1123	1N297A	1123	1N60P
1123	SD12E	1123	VHD1N60-1	1123	1N298	1123	1N60R
1123	SD12M	1123	VHD1N60////-1	1123	1N298A	1123	1N60S
1123	SD14	1123	VHD1S34///-1	1123	1N304	1123	1N60SD60
1123	SD15	1123	V8-0A70	1123	1N305	1123	1N60T
1123	SD16	1123	WD1	1123	1N306	1123	1N60TV
1123	SD21A	1123	WRP134	1123	1N307	1123	1N60TV-TOOL
1123	SD34	1123	WX6	1123	1N309	1123	1N60TVGL
1123	SD46	1123	X16	1123	1N310	1123	1N60Z
1123	SD46-2	1123	X18	1123	1N3110	1123	1N616
1123	SD461	1123	YAAD009	1123	1N312	1123	1N617
1123	SD56	1123	YBAD032	1123	1N3121	1123	1N618
1123	SD60	1123	YBAD1N60P	1123	1N3122	1123	1N62
1123	SDH-2	1123	Y8G-V139-2-2	1123	1N3125	1123	1N63
1123	SFD107	1123	Y8G-V139-22	1123	1N314	1123	1N631
1123	SFD111	1123	ZC1N34A	1123		1123	1N632
1123	SFD112	1123	ZE-1.5	1123		1123	1N636
1123	SFT104	1123	ZEN430				

RS 276-	Industry Standard No.	RS 276-	Industry Standard No.	RS 276-	Industry Standard No.	RS 276-	Industry Standard No.
1123	1N636A	1123	18354	1123	05-490095	1123	14-514-01/51
1123	1N63A	1123	18355	1123	05-610046	1123	14-514-05
1123	1N64	1123	18356	1123	05-931771	1123	14-514-05/55
1123	1N64A	1123	18357	1123	05-932510	1123	14-514-06
1123	1N64B	1123	18358S	1123	05A03	1123	14-514-08
1123	1N64G	1123	18426	1123	50	1123	14-514-08/58
1123	1N64GA	1123	18266	1123	7-0005	1123	14-514-09
9123	1N64P	1123	18266FM	1123	7-0006	1123	14-514-10
1123	1N65	1123	18426FM	1123	07-5134-14	1123	14-514-10/60
1123	1N65A	1123	18426G	1123	07-5134-14A	1123	14-514-11
1123	1N66	1123	18426GPM	1123	07-5134-14B	1123	14-514-21
1123	1N66A	1123	18428	1123	07-5134-14C	1123	14-514-22
1123	1N67	1123	18441	1123	7-59-001/3477	1123	14-514-22/72
1123	1N67A	1123	18442	1123	7-59-0013477	1123	14-514-55
1123	1N67D	1123	18446	1123	08-08111	1123	14-514-61
1123	1N68	1123	000018446D	1123	08-08112	1123	14-514-72
1123	1N68A	1123	000018446D	1123	8-619-030-011	1123	14-515-06/66
1123	1N69	1123	18447	1123	8-697-020-571	1123	15-085002
1123	1N695	1123	18447P	1123	8-719-026-11	1123	15-085003
1123	1N695A	1123	18448	1123	8-719-422-21	1123	15-085009
1123	1N698	1123	18449	1123	8-905-305-007	1123	15-085027
1123	1N69A	1123	18451	1123	8-905-305-020	1123	15-085032
1123	1N70	1123	18452	1123	8-905-305-055	1123	15-085037
1123	1N70A	1123	18453	1123	8-905-305-318	1123	15-085061
1123	1N71	1123	18454	1123	8-905-305-327	1123	15-20A-90M
1123	1N72	1123	18455	1123	8-905-305-336	1123	019-001918
1123	1N72G	1123	18456	1123	8-905-305-338	1123	019-001980
1123	1N73	1123	18467R	1123	8-905-305-339	1123	019-002718
1123	1N74	1123	18542	1123	8-905-305-342	1123	019-00301980
1123	1N75A	1123	18545A	1123	8-905-305-348	1123	019-005043
1123	1N76	1123	1858	1123	8-905-305-405	1123	19-080-009
1123	1N76A	1123	18589	1123	8-905-305-555	1123	19-085005
1123	1N76C	1123	1860P	1123	8-905-305-561	1123	19-085018
1123	1N76G	1123	1872	1123	8-905-305-580	1123	019-301980
1123	1N770	1123	1873	1123	8-905-305-635	1123	19A115086-P1
1123	1N771	1123	1873A	1123	8-905-313-010	1123	19AR3
1123	1N771A	1123	1874	1123	8-905-313-011	1123	020-00030
1123	1N771B	1123	18744	1123	8-905-313-100	1123	20-16-3
1123	1N772A	1123	18745	1123	8-905-313-101	1123	20-1680-175
1123	1N773	1123	18746	1123	8-905-313-120	1123	20A-70
1123	1N773A	1123	18747	1123	8-905-405-077	1123	20A-90
1123	1N774	1123	1875	1123	8-905-405-838	1123	20A-90H
1123	1N774A	1123	1876	1123	8A01	1123	20A-90M
1123	1N775	1123	1877	1123	09-306002	1123	20A70
1123	1N776	1123	1877H	1123	09-306009	1123	20A79
1123	1N777	1123	1878	1123	09-306010	1123	20A90
1123	1N781	1123	1878B	1123	09-306012	1123	20A90LF
1123	1N805	1123	1878H	1123	09-306020	1123	20A90M
1123	1N81	1123	1879	1123	09-306024	1123	20A90MLF
1123	1N81A	1123	1879H	1123	09-306036	1123	20A9M
1123	1N835	1123	1880	1123	09-306037	1123	21A009-000
1123	1N84	1123	1882	1123	09-306039	1123	21A009-002
1123	1N86	1123	18851	1123	09-306040	1123	21A020-005
1123	1N86AG	1123	1887	1123	09-306047	1123	21A040-001
1123	1N87	1123	1888	1123	09-306049	1123	21A103-010
1123	1N87A	1123	18926G	1123	09-306051	1123	21A103-011
1123	1N87G	1123	18990B	1123	09-306058	1123	21A103-016
1123	1N87GA	1123	18990AM	1123	09-306061	1123	21A103-017
1123	1N87S	1123	1P188	1123	09-306064	1123	21A103-019
1123	1N87T	1123	12213	1123	09-306093	1123	21A103-022
1123	1N88	1123	1P22	1123	09-306107	1123	21A103-046
1123	1N89	1123	1P22A	1123	09-306108	1123	21A103-052
1123	1N90	1123	1P22AJ	1123	09-306149	1123	21A103-055
1123	1N909	1123	1P22B	1123	09-306200	1123	21A109-001
1123	1N90G	1123	1P22G	1123	09-306222	1123	21A109-002
1123	1N90GA	1123	1P23	1123	09-306270	1123	21A109-022
1123	1N910	1123	1P231	1123	09-306290	1123	21A119-005
1123	1N911	1123	1236	1123	09-306331	1123	21K60
1123	1N949	1123	1P23A	1123	09-306334	1123	21M288
1123	1N95	1123	1P23B	1123	09-306336	1123	21M288(DIODE)
1123	1N96	1123	1P23G	1123	09-306339	1123	21M323
1123	1N96A	1123	1P23J	1123	09-306349	1123	21M432
1123	1N97	1123	1P23M	1123	09-306370	1123	21M594
1123	1N97A	1123	1P240	1123	9D12	1123	22-004003
1123	1N98	1123	1P240A	1123	9D16	1123	22-1-005
1123	1N98A	1123	1P26	1123	9D12	1123	22-1-129
1123	1N99	1123	1P26-2	1123	9Df1100310	1123	22-1-5
1123	1N994	1123	1P261	1123	10-085001	1123	022-2823-003
1123	1N995M	1123	1P262	1123	10-085004	1123	022-2823-006
1123	1N996	1123	1V9002	1123	10-085005	1123	022-2823-007
1123	1N99A	1123	2-0A90	1123	10-085018	1123	022-2823-008
1123	1NA4	1123	02-1001-1/2221-3	1123	10-085025	1123	022-3901-001
1123	1NA4G	1123	02-1001-1221-3	1123	100-202	1123	022-3902-001
1123	1NJ33233	1123	2-0A90	1123	11-085001	1123	022.3901-001
1123	1NJ60284	1123	2AA119	1123	11-085004	1123	022.3902-001
1123	1NJ61224	1123	2MW665	1123	11-085005	1123	24MW1029
1123	1NJ61675	1123	22D-358	1123	11-085007	1123	24MW1030
1123	1NJ70973	1123	2XAA111	1123	11-085008	1123	24MW1043
1123	1NJ71185	1123	2XAA112	1123	11-085014	1123	24MW1051
1123	1P541	1123	2XAA113	1123	11-085015	1123	24MW1067
1123	1P542	1123	03-0021-0	1123	11-085022	1123	24MW1092
1123	13-18B	1123	003-004200	1123	12-085005	1123	24MW199
1123	13-188AM	1123	003-005400	1123	12-085006	1123	24MW243
1123	18-446D	1123	003-006700	1123	12-085009	1123	24MW603
1123	181006	1123	003-007500	1123	12-085029	1123	24MW771
1123	181007	1123	003-009000	1123	12-085034	1123	24MW860
1123	18.1007B	1123	003-009600	1123	12-085035	1123	24MW987
1123	181008	1123	03-160	1123	12-085038	1123	24MW967
1123	181009	1123	03-931051	1123	12-087003	1123	025-100027
1123	181010	1123	03-931771	1123	13-004	1123	25-3
1123	1811	1123	3A90	1123	13-004	1123	27P1
1123	1812	1123	3B-4	1123	13-085012	1123	28P1
1123	181239	1123	3L4-2091	1123	13-12001-0	1123	29-505
1123	181244	1123	3L4-2001-1	1123	13-12002-0	1123	30P1
1123	18127	1123	3L4-2001-1A	1123	13-12003-0	1123	31-0039
1123	18128	1123	3L4-2003-1	1123	13-14094-1	1123	32-0000
1123	18129	1123	3L4-2003-4	1123	13-14094-11	1123	32-0001
1123	1813	1123	004-009200	1123	13-14094-14	1123	32-0002
1123	1814	1123	4-2020	1123	13-14094-15	1123	32-0003
1123	1815	1123	4-2020-03500	1123	13-14094-2	1123	32-0008
1123	1816	1123	4-2020-03571	1123	13-14094-3	1123	32-0013
1123	181690	1123	4-2020-03600	1123	13-14094-5	1123	32-0023
1123	1817D1	1123	4-202003500	1123	13-14094-9	1123	32-0029
1123	18185	1123	4-202003571	1123	13-14890-1	1123	32-0036
1123	18186	1123	4-2021-05870	1123	13-16235-8	1123	32-0039
1123	18186(PM)	1123	4-202A16	1123	13-30281	1123	32-16557
1123	18187	1123	4-282	1123	13-35621-1	1123	034-001-0
1123	18187S	1123	4-852	1123	13-55046-1	1123	34-028-0
1123	18188	1123	4-853	1123	13-55166-1	1123	34-2001-1
1123	18188AM	1123	4-854	1123	13-55166-1/2459-2	1123	34-3022-7
1123	18188PM	1123	4-855	1123	13-55166-1/3464	1123	34-8002-1
1123	18189	1123	4-857	1123	13-67590-1/2439-2	1123	34-8002-2
1123	1819	1123	4-882	1123	13-89562-1	1123	34-8002-3
1123	1820	1123	48D46-2	1123	13Z-1005	1123	34-8002-4
1123	182479	1123	05-00000-00	1123	14-10	1123	34-8002-5
1123	18318	1123	05-00060-00	1123	14-502-01	1123	34-8002-6
1123	18319	1123	05-00060-01	1123	14-504-01	1123	34-8002-7
1123	1832	1123	05-00160-01	1123	14-510-01	1123	34-8022
1123	18323	1123	05-170034	1123	14-511-01	1123	34-8022-1
1123	1833	1123	05-170060	1123	14-512-01	1123	34-8022-2
1123	1834	1123	05-180034	1123	14-513-01	1123	34-8022-3
1123	1834A	1123	05-180188	1123	14-514-01	1123	34-8022-4
1123	18348	1123	05-490090	1123		1123	34-8022-5
1123	1835	1123	05-490091				

RS 276-	Industry Standard No.	RS 276-	Industry Standard No.	RS 276-	Industry Standard No.	RS 276-	Industry Standard No.
1123	34-8022-6	1123	48K640675	1123	63-12754	1123	100-007-13/2228-3
1123	34-8022-7	1123	48K644681	1123	63-12755	1123	100-007-13/228-3
1123	34-8022-77	1123	48K647311	1123	63-12757	1123	100-00910-07
1123	34-8057-23	1123	48K647713	1123	63-13080	1123	100-00914-10
1123	34-8057-25	1123	48K647769	1123	63-13842	1123	100-0124
1123	34-8057-26	1123	48K863030	1123	63-22724	1123	100-0125
1123	34-8057-29	1123	48K867716	1123	63-25973	1123	100-12
1123	34-8057-30	1123	48M355008	1123	63-25982	1123	100-136
1123	34-8057-52	1123	48M355009	1123	63-28250	1123	100-160
1123	36E004-1	1123	48P60022A97	1123	63-28888	1123	100-180
1123	41E3-1	1123	48P60077A06	1123	63-8381	1123	100-181
1123	42-22537	1123	48P63006A56	1123	63-9523	1123	100-215
1123	42-22539	1123	48P63077A11	1123	64-1	1123	100-436
1123	42-22539(DIODE)	1123	48P63077A32	1123	065-013	1123	102-02
1123	42-22755	1123	48P63077A52	1123	65-085002	1123	102-207
1123	42-23969	1123	48R134587	1123	65-085003	1123	103-114
1123	42-27378	1123	48810346A02	1123	65-085010	1123	103-19
1123	42-27380	1123	48B134587	1123	65-085012	1123	103-202-GE
1123	42-27381	1123	48B137101	1123	66X0020-000	1123	103-22
1123	42-27543	1123	48B137299	1123	66X0020-001	1123	103-23
1123	42-28199	1123	48B137495	1123	66X0039-001	1123	103-23-01
1123	42A14	1123	48B155039	1123	66X0047-001	1123	103-3
1123	44E-300-97	1123	48B155061	1123	66X0047-901	1123	103-31
1123	46-8619-3	1123	48B155078	1123	66X0049-002	1123	103-34
1123	46-86200-3	1123	48B623334A01	1123	66X0049-100	1123	103-44
1123	46-86214-3	1123	48K644681	1123	66X0051-001	1123	103-73
1123	46-8623-3	1123	48X90233A01	1123	66X260	1123	103-74
1123	46-86253-3	1123	48X90233A02	1123	69-1820	1123	103-79
1123	46-86266-3	1123	48X90233A06	1123	69-2922	1123	103-87
1123	46-8630	1123	48X97048A02	1123	76-12965-26	1123	103-Z9001
1123	46-8644-3	1123	48X97048A05	1123	77-271051-1	1123	105-02
1123	46-8646-3	1123	48X97048A06	1123	77-271032-1	1123	105-03
1123	46-86484-3	1123	48X97048A19	1123	78-271199-1	1123	106-008
1123	03005A03	1123	48X97168A01	1123	78-271228-1	1123	108(FARFISA)
1123	48-06-001	1123	48X97168A03	1123	78-273002-1	1123	108(FARTISA)
1123	48-134387	1123	48X97168A04	1123	79F015	1123	108-74
1123	48-134537	1123	48X97168A07	1123	80-60-1	1123	109-036500
1123	48-134587	1123	48X97177A15	1123	81-27123150-8	1123	109-192
1123	48-134588	1123	48X97222A01	1123	81-46123001-3	1123	110-763
1123	48-137299	1123	48X97239A01	1123	81-46123006-2	1123	111-4-2020-08600
1123	48-137495	1123	48X97271A02	1123	81-46123015-3	1123	0112-0019
1123	48-137497	1123	51-04001-01	1123	81-46123015-3	1123	0112-0026
1123	48-155039	1123	51IN60P	1123	83B38-1	1123	0112-0028
1123	48-155061	1123	51IS348	1123	86-0002	1123	0112-0028-6438
1123	48-155114	1123	51LN60P	1123	86-0008	1123	0112-0028/4460
1123	48-355009	1123	51L8348	1123	86-001	1123	0112-0037
1123	48-41768G01	1123	5I81834	1123	86-0013	1123	0112-0045
1123	48-60022A97	1123	52-050-021-0	1123	86-10-1	1123	0112-0075
1123	48-60077A06	1123	52-051-017-0	1123	86-125-1	1123	0112-0082
1123	48-61074B01	1123	53-1519	1123	86-14-1	1123	120-004498
1123	48-61767B01	1123	53A001-1	1123	86-146-1	1123	120-004499
1123	48-62334A01	1123	53A006-1	1123	86-20-1	1123	120-00470-0
1123	48-62334A02	1123	53A008-1	1123	86-22-1	1123	120-004730
1123	48-63006A56	1123	53A0081	1123	86-45-1	1123	120-005299
1123	48-63029A20	1123	53A022-2	1123	86-49-1	1123	121-31
1123	48-63075A78	1123	53B001-1	1123	86-5007-3	1123	123-013
1123	48-63077A11	1123	53B001-3	1123	86-60-1	1123	123-015
1123	48-63077A32	1123	53B004-1	1123	86-64-1	1123	130-30281
1123	48-63084A06	1123	53B005-2	1123	86-88-3	1123	130-30301
1123	48-63590A01	1123	53B007-1	1123	87-10-0	1123	130-40229
1123	48-644587	1123	53C001	1123	87-10-1	1123	137-824
1123	48-644681	1123	53C000-1	1123	089-256	1123	141-003
1123	48-647311	1123	53C0006-1	1123	089-248	1123	142-011
1123	48-647313	1123	53C0006-2	1123	089-293	1123	150-001-005
1123	48-647713	1123	53C0006-52	1123	89WJ75/46N	1123	150-001-9-005
1123	48-65837A02	1123	53C009-2	1123	090A64-1	1123	150-001-9-007
1123	48-65937A02	1123	53C020-1	1123	91-3	1123	150-002-9-001
1123	48-67020A11	1123	53CD01-1	1123	91-46	1123	150-004-9-001
1123	48-67120A09	1123	53D002-1	1123	092-1	1123	150-005-9-001
1123	48-711052	1123	53D002-2	1123	92-1001	1123	150-006-9-001
1123	48-739300	1123	53E001-1	1123	93.24.401	1123	150-012-9-001
1123	48-741280	1123	53E003-1	1123	93.24.601	1123	150-013-9-001
1123	48-82139901	1123	53K0006-1	1123	93.24.604	1123	150-014-9-001
1123	48-82139902	1123	53H001-2	1123	93A105--1	1123	150-015-9-001
1123	48-82178A01	1123	53K001-3	1123	93A110-1	1123	185-015
1123	48-82178A02	1123	53K001-5	1123	93A25-1	1123	186-013
1123	48-82178A03	1123	53L001-10	1123	93A25-3	1123	200X8000-026
1123	48-82178A04	1123	53L001-10	1123	93A27-1	1123	201X2000-118
1123	48-82178A05	1123	53N001-1	1123	93A33-1	1123	209-31
1123	48-82178A06	1123	53N001-7	1123	93A38-1	1123	229-0014
1123	48-82178A07	1123	53N003-1	1123	93A41-2	1123	229-0035
1123	48-82178A08	1123	53N003-2	1123	93A41-3	1123	229-0037
1123	48-82178A09	1123	53N004-11	1123	93A77-1	1123	229-0049
1123	48-82178A10	1123	53N004-14	1123	93A8-1	1123	229-0057
1123	48-82178A11	1123	53N004-5	1123	93B25-1	1123	229-0093
1123	48-82178A12	1123	53N004-6	1123	93B25-2	1123	229-0105
1123	48-82178A13	1123	53T001-1	1123	93B25-3	1123	229-0107
1123	48-82292A01	1123	53T001-4	1123	93B27-1	1123	229-0108
1123	48-82292A02	1123	53T001-5	1123	93B38-1	1123	229-0141
1123	48-82292A03	1123	53T001-6	1123	93B38-5	1123	229-0182-65
1123	48-82292A04	1123	53U001-1	1123	93B41-1	1123	229-5100-231
1123	48-82292A05	1123	54V001-2	1123	93B41-2	1123	230-0006
1123	48-82292B03	1123	55-1	1123	93B41-29	1123	230-006
1123	48-86168-3	1123	56-1	1123	93B46-1	1123	260-10-025
1123	48-8619-3	1123	56-10	1123	93B77-1	1123	260-10-04B
1123	48-86200-3	1123	56-11	1123	93B8-1	1123	260D00507
1123	48-863030	1123	56-2	1123	9302-6	1123	264D00612
1123	48-8643-3	1123	56-20	1123	93C21-1	1123	264D00701
1123	48-867716	1123	56-26	1123	93C21-2	1123	264D00901
1123	48-90210A01	1123	56-3	1123	93C21-3	1123	264D01001
1123	48-90222A0B	1123	56-4	1123	93C21-4	1123	264P00401
1123	48-90233A01	1123	56-4886	1123	93C21-5	1123	264P0080
1123	48-90233A06	1123	56-7	1123	93C21-6	1123	264P00801
1123	48-90233A08	1123	56-8	1123	93C21-7	1123	264P01301
1123	48-97048A06	1123	56-8093	1123	93C21-8	1123	264P01305
1123	48-97048A19	1123	56-89	1123	93C21B	1123	264P01306
1123	48-97168A01	1123	57A1-1	1123	93C25-3	1123	264P01350
1123	48-97168A04	1123	57A1-2	1123	93C27-1	1123	264P03802
1123	48-97168A07	1123	57A1-54	1123	93C27-5	1123	264200701
1123	48-97168A09	1123	57A1-62	1123	93C27-6	1123	276-1122
1123	48-97168A13	1123	57A1-64	1123	93C055-1	1123	294-42-9
1123	48-97177A15	1123	57D1-1	1123	93C7-1	1123	295V005H03
1123	48-97222A01	1123	57D1-2	1123	93C77-1	1123	296V002H01
1123	48-97239A01	1123	57D1-54	1123	9308-1	1123	296V002H02
1123	48-97270A02	1123	57D1-62	1123	094-014	1123	296V002H03
1123	48B41768G01	1123	57D1-64	1123	94-42-9	1123	296V002H05
1123	48B62334A01	1123	61-259	1123	96-0008	1123	296V002H06
1123	48B6629A01	1123	61-5595	1123	96-5007-01	1123	296V002H07
1123	48B6629A02	1123	62-10234	1123	96-5059-01	1123	296V002H08
1123	48B67020A11	1123	62-10655	1123	96-5087-01	1123	296V002H01
1123	48C134587	1123	62-12524	1123	96-5363-01	1123	296V002K01
1123	48C42428A01	1123	62-12034	1123	96XZ0778-44N	1123	296V006H02
1123	48C61074B01	1123	62-13518	1123	96XZ077844N	1123	296V007H02
1123	48C61767B01	1123	62-16769	1123	96XZ0868/15X	1123	296V012H01
1123	48C65832A02	1123	62-16841	1123	96XZ778/21N	1123	296V015H01
1123	48C65837A01	1123	62-19846	1123	96XZ778/44N	1123	296V015H01
1123	48C65837A02	1123	63-10739	1123	100-003-40/2228-3	1123	296V02H01
1123	48C63590A01	1123	63-11074	1123	100-003-40/228-3	1123	309-324-616
1123	48D67120A11	1123	63-11879	1123	100-00340-00	1123	309-327-916
1123	48D67120A11(DIO)	1123	63-12158	1123	100-0051	1123	324-0014
1123	48G10346A01	1123	63-12607	1123		1123	324-0035
1123	48K134587	1123	63-12645	1123		1123	324-0037
1123	48K544539	1123		1123		1123	324-0049
1123		1123		1123		1123	324-0057
1123		1123		1123		1123	324-0093(PHILCO)
1123		1123		1123		1123	324-0105

RS 276-	Industry Standard No.	RS 276-	Industry Standard No.	RS 276-	Industry Standard No.	RS 276-	Industry Standard No.
1123	324-0107-01	1123	903-45B	1123	3002	1123	76675A
1123	324-0108	1123	903-51B	1123	003016	1123	76675B
1123	324-0141	1123	903-54B	1123	3322-6	1123	79985
1123	324-0160	1123	903-65B	1123	3505	1123	085002
1123	325-0025-327	1123	903-83B	1123	03571	1123	085004
1123	325-0028-86	1123	903-92B	1123	4001-230	1123	085006
1123	325-0028-877	1123	903-9B	1123	4002(SEARS)	1123	085016
1123	325-0031-335	1123	904-97B	1123	4003(SEARS)	1123	86001
1123	325-0036-562	1123	908D1	1123	4004(SEARS)	1123	95000
1123	325-1376-60	1123	914-000-4-00	1123	4005(SEARS)	1123	95001
1123	339-529-001	1123	914-001-7-00	1123	4006(SEARS)	1123	95002
1123	344-6006-6	1123	916-32000-7	1123	4007(SEARS)	1123	95004
1123	400A	1123	916-32003-2	1123	4008(SEARS)	1123	95007
1123	400D	1123	916-32006-2	1123	4009(SEARS)	1123	95008
1123	403-1	1123	919-01-0867	1123	4010(PENNCREST)	1123	95014
1123	429-0092-56	1123	919-010867	1123	4010(SEARS)	1123	95017
1123	429-0910-54	1123	1000-131	1123	4011(SEARS)	1123	95018
1123	464-100-19	1123	1000-17	1123	4012(SEARS)	1123	95170-2(DIO)
1123	464-103-19	1123	1001-10	1123	4041-000-1018	1123	100017
1123	464-106-19	1123	1002-08	1123	4041-200-10140100	1123	100844
1123	464-110-19	1123	1002-09	1123	4041-200-10180	1123	102989
1123	464-111-19	1123	1002-17	1123	4041-200-40100	1123	104152
1123	464-113-19	1123	1002-17(DIODE)	1123	4351-.0012	1123	105517
1123	464-119-19	1123	1004(DIODE)	1123	4351-.0013	1123	110610
1123	503-721271	1123	1005	1123	4351-.0031	1123	111207
1123	503-721472	1123	1006(DIODE)	1123	4354	1123	111605
1123	510A90	1123	1006-9292	1123	04770	1123	112530
1123	510ED46	1123	1007	1123	4801-00628	1123	112524
1123	510IN60	1123	1007-1124	1123	4801-00629	1123	112526
1123	510I834	1123	1007-17(DIODE)	1123	4822-130-30281	1123	112529
1123	510LN60	1123	1010-145	1123	4822-130-30301	1123	115101
1123	510L834	1123	1010-8173	1123	4822-130-30311	1123	116048
1123	5118345	1123	1011	1123	4822-130-30312	1123	116273
1123	521-145	1123	1012-17	1123	4822-130-40229	1123	117659
1123	523-1000-067	1123	1013	1123	4828-4	1123	117760
1123	523-1000-294	1123	1016-77	1123	5001-080	1123	119199
1123	523-1000-295	1123	1019-6699	1123	5001-134	1123	119919
1123	523-1000-326	1123	1030-17	1123	5001-141	1123	120617
1123	523-1002-326	1123	1033-1916	1123	5001-161	1123	122166
1123	523-1500-067	1123	1040-08	1123	5101834	1123	122520
1123	523-1500-881	1123	1041-65	1123	5120A90	1123	126177
1123	523-2003-0-1	1123	1042-12	1123	5631-1N34A	1123	127017
1123	523-2003-001	1123	1042-13	1123	5631-1N60	1123	127532
1123	524-457	1123	1042-3	1123	5631-1N60P	1123	127784
1123	525-877	1123	1048-9870	1123	5631-20A90	1123	129028
1123	0575-005	1123	1048-9888	1123	8000-00003-045	1123	129157
1123	600X0096-066	1123	1063-3553	1123	8000-00004-045	1123	129158
1123	600X0097-066	1123	1063-8591	1123	8000-00004-060	1123	129348(DIO)
1123	601X0150-066	1123	1074-124	1123	8000-00004-063	1123	129474
1123	601X0151-066	1123	1074-24	1123	8000-00005-015	1123	132912
1123	617-15	1123	1077-2325	1123	8000-00005-016	1123	132915
1123	617-156	1123	1077-2760	1123	8000-00005-017	1123	134180
1123	617-17	1123	1077-9296	1123	8000-00005-018	1123	134587
1123	624-0011	1123	1118-17	1123	8000-00005-023	1123	135872
1123	630-002	1123	1119-17	1123	8000-00006-007	1123	137876
1123	630-079	1123	0112-0073	1123	8000-00011-046	1123	139634
1123	642-028	1123	1206-17	1123	8000-00011-060	1123	144585
1123	642-102	1123	1207-17	1123	8000-00038-009	1123	146575
1123	642-119	1123	1410-171	1123	8000-0004-042	1123	161001
1123	642-132	1123	1489-17	1123	8000-0004-063	1123	161006
1123	642-199	1123	1512	1123	8000-0004-P060	1123	161016
1123	642-221	1123	1550	1123	8000-0004-P063	1123	161038
1123	650	1123	1778-17	1123	8000-0004-P063	1123	162002-039
1123	656-142	1123	1956-17	1123	8000-0004-P063	1123	162002-39
1123	675-158	1123	1977-17	1123	8000-00041-015	1123	165572
1123	690V034H32	1123	1980-17	1123	8000-00041-016	1123	166273
1123	690V034H74	1123	2000-501	1123	8000-0005-015	1123	167562
1123	690V037H91	1123	2000-318	1123	8000-0005-016	1123	168910
1123	690V040H63	1123	2022-07	1123	8000-0005-017	1123	169115
1123	690V047H61	1123	2077-07	1123	8000-0005-018	1123	169362
1123	690V052H50	1123	2093A25-3	1123	8000-0005-023	1123	169501
1123	690V052H68	1123	2093A33-1	1123	8000-004-P063	1123	170575
1123	690V059H63	1123	2093A38-1	1123	8010-53	1123	171162-042
1123	690V066H48	1123	2093A38-10	1123	8710-53	1123	171162-269
1123	690V067H09	1123	2093A38-14	1123	8840-54	1123	175043-066
1123	690V068H32	1123	2093A38-15	1123	9861B-43	1123	175043-068
1123	690V073H60	1123	2093A38-21	1123	10181	1123	190716
1123	690V083H89	1123	2093A38-22	1123	11252	1123	195617
1123	690V092H85	1123	2093A38-27	1123	0012060	1123	200648-26
1123	690V098H52	1123	2093A38-30	1123	12808	1123	202315
1123	690V103H54	1123	2093A38-31	1123	12850	1123	216001
1123	690V110H88	1123	2093A38-32	1123	15027(DIODE)	1123	225410
1123	690V68H32	1123	2093A38-33	1123	18410-42	1123	226334
1123	690V73H60	1123	2093A38-34	1123	18600-53	1123	226344
1123	690V037H91	1123	2093A38-40	1123	25201-001	1123	245517
1123	754-1003-030	1123	2093A38-5	1123	25810-53	1123	320007
1123	754-2000-009	1123	2093A41	1123	25810-55	1123	489751-001
1123	754-4000-088	1123	2093A41-14	1123	25840-53	1123	489752-001
1123	754-4000-188	1123	2093A41-141	1123	25840-55	1123	489752-003
1123	754-4000-410	1123	2093A41-148	1123	26810-51	1123	489752-031
1123	754-9000-090	1123	2093A41-154	1123	27840-42	1123	489752-042
1123	754-9900-460	1123	2093A41-167	1123	28287	1123	489752-049
1123	792-292	1123	2093A41-169	1123	031033	1123	489752-076
1123	800-002-00	1123	2093A41-181	1123	031040	1123	489752-08
1123	800-003-00	1123	2093A41-187	1123	36508	1123	489850-004
1123	800-005-00	1123	2093A41-196	1123	58510-52	1123	500000
1123	800-020-00	1123	2093A41-2	1123	42020	1123	500009G
1123	800-022-00	1123	2093A41-38	1123	43959	1123	0517022
1123	800-039-00	1123	2093A41-50	1123	45810-53	1123	0517826
1123	800-517-00	1123	2093A41-59	1123	46287-4	1123	0517828
1123	805-1060	1123	2093A41-68	1123	48287	1123	0517829
1123	805-10600	1123	2093A41-92	1123	48287-4	1123	0525002
1123	808-312	1123	2093A77-1	1123	55005-01	1123	0525002H
1123	903-00390	1123	2093A8-1	1123	53092-1	1123	0526224
1123	903-105B	1123	2093B11-21	1123	55166	1123	530015-2
1123	903-10B	1123	2093B41-11	1123	55810-51	1123	530063-1
1123	903-108B	1123	2093B41-29	1123	55810-52	1123	530063-10
1123	903-10B	1123	2102-010	1123	057001	1123	530063-2
1123	903-113B	1123	2102-025	1123	057001H	1123	530063-4
1123	903-114B	1123	2102-028	1123	057005	1123	530063-5
1123	903-115B	1123	2106-124	1123	057005H	1123	530065-1002
1123	903-12B	1123	2122-17	1123	057500	1123	530065-1002A
1123	903-168B	1123	2151-17	1123	58810-83	1123	530065-1003
1123	903-16B	1123	2196-17	1123	059395	1123	530065-3
1123	903-18B	1123	2232-17	1123	59395(RCA)	1123	530072-7
1123	903-212	1123	2279-13	1123	59840-1	1123	530085-2
1123	903-23	1123	2280-13	1123	60215-1	1123	530092-1
1123	903-23A	1123	2281-13	1123	67590	1123	530092-1001
1123	903-23B	1123	2282-13	1123	68504-76	1123	530092-2
1123	903-23C	1123	2282-17	1123	68504-77	1123	530105-1
1123	903-23D	1123	2283-13	1123	71119-2	1123	530105-1001
1123	903-23B	1123	2284-13	1123	71778	1123	530115-1001
1123	903-27	1123	2290-13	1123	72013	1123	530123-1
1123	903-27B	1123	2402-459	1123	72060	1123	530127-1
1123	903-29B	1123	2405-458	1123	72080	1123	0537820
1123	903-30B	1123	2408-330	1123	72089	1123	568101
1123	903-34	1123	2510-31	1123	000072090	1123	0570519
1123	903-34A	1123	2603-186	1123	72104	1123	0573024
1123	903-34B	1123	2606-296	1123	72128A	1123	0575001
1123	903-34C	1123	2703-389	1123	72129	1123	0575001H
1123	903-34D	1123	2704	1123	72129A	1123	0575002
1123	903-37B	1123	2704-388	1123	000072160	1123	0575002H
1123	903-43B	1123	2789	1123	075005	1123	0575004
		1123	2796	1123	76675		

RS 276-	Industry Standard No.	RS 276-	Industry Standard No.	RS 276-	Industry Standard No.	RS 276-	Industry Standard No.	RS 276-	Industry Standard No.
1123	0575005	1123(2)	D010	1123(2)	264D00801	1142	UR205		
1123	0575005H	1123(2)	DIJ70543	1123(2)	264D0701	1142	UR210		
1123	0575007	1123(2)	DIJ70644	1123(2)	324-0107	1142	UT211		
1123	575009	1123(2)	DIJ71778	1123(2)	354-9101-002	1142	UT112		
1123	0575067	1123(2)	D839	1123(2)	429-0910-53	1142	UT2005		
1123	575091	1123(2)	E21135	1123(2)	903-167B	1142	UT2010		
1123	0575099	1123(2)	E2484	1123(2)	903-8B	1142	UT236		
1123	576063	1123(2)	EA2137	1123(2)	1947-17	1142	UT249		
1123	0577001	1123(2)	EA2606	1123(2)	2093A38-35	1142	UT251		
1123	601030	1123(2)	EG8110	1123(2)	2093A41-29	1142	UT261		
1123	610030	1123(2)	EF16X21	1123(2)	2282-17(RAT.DET)	1142	UT3005		
1123	614020	1123(2)	E816X21	1123(2)	5101N60	1142	UT3010		
1123	615010	1123(2)	E816X70	1123(2)	5361-1N60P	1142	WR-006		
1123	700180-00	1123(2)	E816X19	1123(2)	8910-53	1142	WR-030		
1123	740402	1123(2)	EW168	1123(2)	011119	1142	WR-040		
1123	740952	1123(2)	G000004A	1123(2)	28810-64	1142	ZR10		
1123	740954	1123(2)	HD1000105	1123(2)	37510-53	1142	ZR11		
1123	741866	1123(2)	HE-10044	1123(2)	58840-114	1142	ZR601		
1123	742004	1123(2)	HP-20088	1123(2)	72128	1142	Z8270		
1123	742730	1123(2)	J241	1123(2)	085005	1142	Z8271		
1123	755722	1123(2)	J24567	1123(2)	085026	1142	1.5B05		
1123	760101-0005	1123(2)	JP575005	1123(2)	80060-53	1142	1.5B1		
1123	760101-0006	1123(2)	JP575995	1123(2)	88510-53	1142	1.5J05		
1123	765722	1123(2)	N-EA16X27	1123(2)	117730	1142	1.5J1		
1123	771909	1123(2)	OA9OFM	1123(2)	161015	1142	1N1556		
1123	771910	1123(2)	REN110	1123(2)	167572	1142	1N1563		
1123	771911	1123(2)	RT2451	1123(2)	171162-270	1142	1N1617		
1123	772712	1123(2)	RT4880	1123(2)	216003	1142	1N1644		
1123	785278-01	1123(2)	RT5212	1123(2)	489752-036	1142	1N1645		
1123	801722	1123(2)	RT5470	1123(2)	489752-125	1142	1N1692		
1123	817032	1123(2)	RT5738	1123(2)	5115348	1142	1N1701		
1123	817077	1123(2)	RT5912	1123(2)	530063-14	1142	1N1702		
1123	817082	1123(2)	RT6728	1123(2)	530072-1001	1142	1N1707		
1123	817125	1123(2)	RT7538	1123(2)	0575019	1142	1N1708		
1123	817158	1123(2)	S1384	1123(2)	0575019H	1142	1N1907		
1123	817159	1123(2)	SD020	1123(2)	772740	1142	1N1908		
1123	817194	1123(2)	SD46(4)	1123(2)	817160	1142	1N2013		
1123	817199	1123(2)	SD46R	1123(2)	817177	1142	1N2014		
1123	871125	1123(2)	S046	1123(2)	984881	1142	1N2390		
1123	922021	1123(2)	STL128	1123(2)	1471822-11	1142	1N2391		
1123	922604	1123(2)	T-E1105	1123(2)	2495083	1142	1N2399		
1123	942677-1	1123(2)	T-E1177	1141	A150(RECTIFIER)	1142	1N2400		
1123	942677-2	1123(2)	T4590	1141	HEPRO080	1142	1N2408		
1123	942677-3	1123(2)	TM110	1141	HEPRO090	1142	1N2409		
1123	942677-4	1123(2)	TR228T36002004	1141	V330	1142	1N2609		
1123	942677-6	1123(2)	TR320007	1141	1N4719R	1142	1N2610		
1123	942677-7	1123(2)	TY24122	1141	1N4997R	1142	1N2858		
1123	964298	1123(2)	XD2A	1141	192760	1142	1N2858A		
1123	972258-1	1123(2)	001-0020-00	1141	18R15-50	1142	1N2859		
1123	972258-2	1123(2)	1N34M	1141	18R16-50	1142	1N2859A		
1123	972258-3	1123(2)	1N541	1141	18R17-50	1142	1N3072		
1123	972258-4	1123(2)	1N542	1141	276-1141	1142	1N3073		
1123	972258-5	1123(2)	1N60-5	1142	A15A	1142	1N3227		
1123	972258-6	1123(2)	1N60MP	1142	A15F	1142	1N3237		
1123	972258-8	1123(2)	1818PM	1142	A150(DIODE)	1142	1N3238		
1123	972258-9	1123(2)	1816PM	1142	A050(RECTIFIER)	1142	1N3246		
1123	972259-8	1123(2)	1818A	1142	AE3A	1142	1N3247		
1123	980514	1123(2)	1818AR	1142	AE3B	1142	1N3486		
1123	981150	1123(2)	1818PM-1	1142	A83A	1142	1N3544		
1123	981153	1123(2)	1818PM1	1142	A83B	1142	1N3754		
1123	981207	1123(2)	1818PM1A	1142	B56-33	1142	1N4139		
1123	981522	1123(2)	1818PM2	1142	BA136A	1142	1N4140		
1123	981676	1123(2)	1818PMA	1142	BA178	1142	1N4364		
1123	982065	1123(2)	1818PMI	1142	BTM-50	1142	1N4719		
1123	982270	1123(2)	1818Q	1142	EC05800	1142	1N4720		
1123	982271	1123(2)	1818MPX	1142	EC05801	1142	1N4816		
1123	982275	1123(2)	1818P	1142	HEP-R0090	1142	1N4997		
1123	982290	1123(2)	1818PM	1142	HEP-R0091	1142	1N4998		
1123	982822	1123(2)	1818S	1142	HEP161	1142	1N5004		
1123	983099	1123(2)	1818TV	1142	OA10	1142	1N5055		
1123	983239	1123(2)	1850	1142	PF505	1142	1N5171		
1123	984163	1123(2)	1860	1142	QDSSR3AMBE	1142	1N5197		
1123	984200	1123(2)	1223S	1142	RS-05	1142	1N5391		
1123	984226	1123(2)	2A119	1142	RT7848	1142	1N5400		
1123	984666	1123(2)	20A90	1142	S1010	1142	1643		
1123	984667	1123(2)	4-2020-05600	1142	S105A	1142	1644		
1123	985621	1123(2)	4-2020-08600	1142	S1210	1142	18020		
1123	988994	1123(2)	4I29200602	1142	S1GR2	1142	18030		
1123	988997	1123(2)	07-5160-15	1142	S1M1	1142	182408		
1123	988998	1123(2)	09-306019	1142	S1M2	1142	182415		
1123	992143	1123(2)	09-306057	1142	S2A06	1142	182416		
1123	1121008/7611	1123(2)	09-306229	1142	S2M1	1142	1840		
1123	1223770	1123(2)	09-306335	1142	S2M2	1142	18430		
1123	1223921	1123(2)	09-30636	1142	S3A1	1142	18431		
1123	1408649-1	1123(2)	12-085041	1142	S3BA05	1142	2P05		
1123	1471872-15	1123(2)	13-085029	1142	S91A	1142	3A15		
1123	1471872-17	1123(2)	13-14094-8	1142	S91B	1142	3A30		
1123	1473619-3	1123(2)	14-514-13	1142	S91H	1142	3B05		
1123	1476179-001	1123(2)	14-514-13/63	1142	S92A	1142	S805E		
1123	1476179-1	1123(2)	15-20A70	1142	S92H	1142	SB1B		
1123	1800002	1123(2)	15-20A90	1142	SA1M1	1142	7-6006-00		
1123	2000433-150	1123(2)	020-00011	1142	SB1Z	1142	13-18481-2		
1123	2000648-26	1123(2)	020-00012	1142	SD1X	1142	15C05		
1123	2000757-18	1123(2)	20A90Z	1142	0000008D1Y	1142	15C1		
1123	2001786-134	1123(2)	21A103-006	1142	SD91	1142	15805		
1123	2002151-020	1123(2)	21A103-048	1142	S105	1142	1581		
1123	2002151-20	1123(2)	21A109-003	1142	S105A	1142	20A1		
1123	2002336-20	1123(2)	21M289	1142	SI1	1142	20C1		
1123	2003069-4	1123(2)	21M525	1142	SJ053P	1142	30R1		
1123	2004357-106	1123(2)	1M568	1142	SJ103P	1142	30805		
1123	2006422-132	1123(2)	24MW122	1142	SLA21A	1142	3081		
1123	2006431-50	1123(2)	24MW244	1142	SLA21B	1142	48-13709B		
1123	2006441-122	1123(2)	24MW665	1142	SLA21C	1142	0135-1		
1123	2006582-20	1123(2)	24MW820	1142	SLA22A	1142	250R1B		
1123	2006299-2	1123(2)	32-0007	1142	SLA22B	1142	300R1B		
1123	2092055-0001	1123(2)	32-18539	1142	SLA22C	1142	39GB		
1123	2092055-0007	1123(2)	34-8022-6(PHILCO)	1142	SLA5197	1142	399A		
1123	2092055-001	1123(2)	34-8057-3	1142	SLA5198	1142	5210DC-1		
1123	2092055-0010	1123(2)	42-19681	1142	SOD30AL	1142	40266		
1123	2092055-0711	1123(2)	42-21362	1142	SOD30BL	1142	61088		
1123	2092055-0713	1123(2)	42-23972	1142	SOD50BL	1142	720453		
1123	2092055-1	1123(2)	42-27544	1142	SOD50AL	1142	720454		
1123	2092055-7	1123(2)	48-134954	1142	SOD50BL	1142	3460758-2		
1123	2092405-7	1123(2)	48-355008	1142	SOD50CL	1143	A-132591		
1123	2095083	1123(2)	48-60154A01	1142	SOD50DL	1143	A15B		
1123	2485080(DIODE)	1123(2)	48-97048A05	1142	SP-1.5A	1143	A83C		
1123	2495083-2	1123(2)	48-97168A03	1142	SR132-1	1143	A83C		
1123	2495084	1123(2)	53A001-2	1142	SR1422	1143	B56-15		
1123	2495380	1123(2)	53B010-7	1142	SR1598	1143	BYX30/150		
1123	2495383	1123(2)	53B003-3	1142	SR1643A	1143	D311BA		
1123	2495529(DIODE)	1123(2)	53N004-10	1142	SR1DM-2	1143	E302(RLCOM)		
1123	2496436	1123(2)	56-8095	1142	SR1PM2	1143	EC05802		
1123	2498530	1123(2)	065-014	1142	SR3AM-3	1143	GM-3Y		
1123	3596062	1123(2)	76-14196-1	1142	SR3AM1	1143	GM-38		
1123(2)	A054-226	1123(2)	86-15-1	1142	SR3AM2	1143	GU-38Z		
1123(2)	A069-118	1123(2)	86-48-1	1142	SR475	1143	HEP-R0092		
1123(2)	A909-1016	1123(2)	91N1	1142	SW-05	1143	HEP162		
1123(2)	OA95	1123(2)	93A25-2	1142	TB5	1143	HEPRO070		
1123(2)	APB-160-1020	1123(2)	93A83-1	1142	TC50	1143	HEPRO071		
1123(2)	CI270645	1123(2)	93B27-3	1142	TI-56	1143	HEPRO080		
1123(2)	CT2005	1123(2)	93025-2	1142	TMS800	1143	HEPRO081		
1123(2)	CT2008	1123(2)	103-102	1142	TV24651	1143	HEPRO082		
1123(2)	CX0042	1123(2)	103-271	1142	TVS-181893	1143	HEPRO091		
1123(2)	D-05	1123(2)	120-001301	1142	TVS-181922	1143	HEPRO092		
						1143	HR-200		

RS 276-	Industry Standard No.	RS 276-	Industry Standard No.	RS 276-	Industry Standard No.	RS 276-	Industry Standard No.
1143	MR501	1143	1N3757	1144	PT525	1144	1N3240
1143	MR502	1143	1N3866	1144	PT540	1144	1N3249
1143	NGF3003	1143	1N3938	1144	PTC201	1144	1N3254
1143	PGR-24	1143	1N3952	1144	R-1	1144	1N3546
1143	PT520	1143	1N3981	1144	R-4A	1144	1N3547
1143	R-1A	1143	1N4141	1144	R-81805	1144	1N3612
1143	R-1B	1143	1N4245	1144	RPJ70147	1144	1N3640
1143	R-5B	1143	1N4365	1144	RV1476	1144	1N3657
1143	R-5C	1143	1N4383	1144	S1030	1144	1N3749
1143	RO2Z	1143	1N4517	1144	S1040	1144	1N3756
1143	S-12	1143	1N4720R	1144	S1230	1144	1N3758
1143	S-12A	1143	1N4721	1144	S1240	1144	1N3867
1143	S1020	1143	1N4721R	1144	S13A	1144	1N3959
1143	S1220	1143	1N4998R	1144	S14A	1144	1N3982
1143	S229	1143	1N4999	1144	S2A30	1144	1N4089
1143	S236	1143	1N4999R	1144	S2A40	1144	1N4142
1143	S2A20	1143	1N5005	1144	S2240	1144	1N4246
1143	S2CN1	1143	1N5056	1144	S3A5	1144	1N4367
1143	S2GR2	1143	1N5059	1144	S3A4	1144	1N4384
1143	S2220	1143	1N5199	1144	S3CN1	1144	1N4722
1143	S3A2	1143	1N5393	1144	S3G40	1144	1N4722R
1143	S3G10	1143	1N5402	1144	S3G40Z	1144	1N4819
1143	S3G20	1143	1N5624	1144	S3G82	1144	1N4820
1143	S3M1	1143	1R5B261	1144	S4M1	1144	1N5000
1143	S4CN1	1143	1R5D261	1144	S500B	1144	1N5000R
1143	S4GR2	1143	18021	1144	S500C	1144	1N5006
1143	SA-2Z	1143	18032	1144	S5CN1	1144	1N5057
1143	SA2M1	1143	181071	1144	S5M1	1144	1N5058
1143	SD1Z	1143	181072	1144	SA3M1	1144	1N5060
1143	SD6(DUAL)	1143	181236	1144	SD101	1144	1N5173
1143	SD91A	1143	181474	1144	SD93	1144	1N5174
1143	SD91B	1143	181488	1144	SD93A	1144	1N5200
1143	SD92	1143	181914	1144	SD94	1144	1N5394
1143	SD92A	1143	181915	1144	SD94A	1144	1N5395
1143	SD92B	1143	181991	1144	SE-0.5A	1144	1N5403
1143	SG205	1143	182409	1144	SE05A	1144	1N5404
1143	SH18	1143	182410	1144	SG005	1144	1N5625
1143	SI1A	1143	182417	1144	SG305	1144	1Z645
1143	SIG1/100	1143	182455	1144	SJ403P	1144	1Z646
1143	SIG1/200	1143	182456	1144	SLA1103	1144	1Z647
1143	SJ203P	1143	182596	1144	SLA24A	1144	18023
1143	SL-5	1143	182761	1144	SLA24B	1144	181039
1143	SLA23B	1143	182762	1144	SLA24C	1144	181073
1143	SLA23C	1143	182778	1144	SLA25A	1144	181475
1143	SLA5199	1143	182827A	1144	SLA25B	1144	181489
1143	SOD100AL	1143	18432	1144	SLA25C	1144	181916
1143	SOD100BL	1143	18921(RECT)	1144	SLA5200	1144	182350
1143	SOD100CL	1143	18956	1144	SR1378-2	1144	182379
1143	SOD100DB	1143	1SR15-100	1144	SR1FM6	1144	182411
1143	SOD200AL	1143	1SR15-200	1144	SR1FM8	1144	182412
1143	SOD200BL	1143	1SR16-100	1144	SR3AM-6	1144	182418
1143	SOD200CL	1143	1SR16-200	1144	SR3AM6	1144	182419
1143	SOD200DL	1143	1SR17-100	1144	SR3AM8	1144	182457
1143	SR-100	1143	1SR17-200	1144	SSI1120	1144	182763
1143	SR-114	1143	3B261	1144	SSIC0820	1144	182764
1143	SR1762	1143	3D261	1144	SSIC1220	1144	182779
1143	SR210	1143	3E2	1144	SX-1.5-04	1144	182828A
1143	SR220	1143	382E	1144	TC0-09M22/1	1144	1842
1143	SR3AM4	1143	1082	1144	TM5804	1144	1843
1143	SR507	1143	13-18481-1	1144	T84	1144	1844
1143	SSIC0810	1143	13-18481-3	1144	U05E	1144	18853
1143	SSIC1110	1143	13-34368-1	1144	U06E	1144	18854
1143	SSIC1210	1143	15B1	1144	U0225	1144	18922
1143	SV03	1143	1502	1144	U0114	1144	18922(RECT)
1143	SW-1-01	1143	1502D	1144	UT115	1144	18957
1143	SW-1-02	1143	1582	1144	UT2040	1144	1SR15-400
1143	SX-1.5-01	1143	20A2	1144	UT211	1144	1SR16-400
1143	SX-1.5-02	1143	30D1	1144	UT212	1144	1SR17-400
1143	TB100	1143	3082	1144	UT213	1144	1W84
1143	TB200	1143	35-1008	1144	UT235	1144	3-1477
1143	TC100	1143	86-5028-3	1144	UT244	1144	3A300
1143	TC200	1143	86-5032-3	1144	UT254	1144	3AP4
1143	TM5802	1143	96-5106-01	1144	UT262	1144	30261
1143	U05B	1143	96-5149-01	1144	UT5040	1144	383E
1143	U05C	1143	96-5184-01	1144	VO1E	1144	384E
1143	U06C	1143	137-737	1144	V334	1144	38M4
1143	UR215	1143	150R2B	1144	WRR1955	1144	10B3
1143	UR220	1143	250R2B	1144	WRR1956	1144	10B4
1143	UT113	1143	276-1142	1144	X-23305-3	1144	15B4
1143	UT12	1143	276-1143	1144	X-23305-4	1144	1504
1143	UT13	1143	300R2B	1144	ZR13	1144	1584
1143	UT2020	1143	399B	1144	ZR14	1144	20A3
1143	UT234	1143	399C	1144	ZR604	1144	20A4
1143	UT242	1143	399D	1144	ZR606	1144	2003
1143	UT252	1143	475-018	1144	1.5E4	1144	2004
1143	UT262	1143	501-363-2	1144	1.5J3	1144	30R3
1143	UT3020	1143	540-015	1144	1.5J4	1144	30B5
1143	VO1C	1143	919-005045	1144	1N1221	1144	30B4
1143	V331	1143	1412-182	1144	1N1221A	1144	48-137327
1143	V332	1143	1901-0045	1144	1N1221B	1144	48-137340
1143	ZR12	1143	5522-8	1144	1N1222	1144	150R3B
1143	ZR602	1143	7921	1144	1N1558	1144	150R4B
1143	ZS2T2	1143	031450	1144	1N1559	1144	152-0047
1143	ZS701	1143	34405	1144	1N1565	1144	152-0047-00
1143	ZS702	1143	36147	1144	1N1566	1144	200R3B
1143	1.5E2	1143	36591	1144	1N1619	1144	200RB4
1143	1.5J2	1143	38052	1144	1N1620	1144	211-58
1143	1K2	1143	39804	1144	1N1648	1144	250R3B
1143	1N1055	1143	40267	1144	1N1649	1144	250R4B
1143	1N1220	1143	46914	1144	1N1650	1144	276-1144
1143	1N1220A	1143	61807	1144	1N1651	1144	300R3B
1143	1N1220B	1143	71449	1144	1N1694	1144	1901-0028
1143	1N1557	1143	71449-1	1144	1N1695	1144	1901-0036
1143	1N1564	1143	121180	1144	1N1704	1144	1901-0388
1143	1N1618	1143	206180	1144	1N1705	1144	1901-0389
1143	1N1646	1143	206185	1144	1N1710	1144	3583
1143	1N1647	1143	206190	1144	1N1711	1144	35604
1143	1N1693	1143	475018	1144	1N1763A	1144	38074
1143	1N1703	1143	506911	1144	1N1910	1144	720456
1143	1N1709	1143	630063	1144	1N1911	1144	720458
1143	1N1909	1143	654420	1144	1N2017	1144	3430063-1
1143	1N2015	1143	720455	1144	1N2018	1144	3430063-11
1143	1N2016	1143	801730	1144	1N2019	1151	DP13B00-2
1143	1N2392	1144	A15C	1144	1N2020	1151	HEPR0801
1143	1N2401	1144	A15D	1144	1N2393	1151	HEPR0841
1143	1N2410	1144	AC300	1144	1N2394	1151	HEPR0851
1143	1N2482	1144	AC400	1144	1N2402	1151	MDA100
1143	1N2485	1144	AB3D	1144	1N2403	1151	MDA942-1
1143	1N2611	1144	A83D	1144	1N2411	1151	MDA942A-1
1143	1N2860	1144	D8118B	1144	1N2412	1151	VM08
1143	1N2860A	1144	D813B	1144	1N2483	1151	VM25
1143	1N3074	1144	D82P	1144	1N2487	1151	276-1151
1143	1N3075	1144	E304(ELCOM)	1144	1N2612	1152	A04210-A
1143	1N3082	1144	EC05803	1144	1N2613	1152	A04231-A
1143	1N3228	1144	EC05804	1144	1N2861	1152	A04283-A
1143	1N3239	1144	GM-3	1144	1N2861A	1152	A04284-A
1143	1N3248	1144	HEP-R0094	1144	1N2862	1152	A04716
1143	1N3253	1144	HEPRO072	1144	1N2862A	1152	AM-G-11
1143	1N3545	1144	HEPRO084	1144	1N3076	1152	B0043
1143	1N3611	1144	HEPRO094	1144	1N3077	1152	B50C1000
1143	1N3639	1144	J4-1602	1144	1N3078	1152	BD-3A184
1143	1N3656	1144	MR504	1144	1N3079	1152	BD3A-1B4
1143	1N3748	1144	PT-72130-1	1144	1N3083	1152	BD0A
1143	1N3755					1152	BR-1

RS 276-	Industry Standard No.
1152	BS-B1
1152	BY122
1152	BY164
1152	CXL166
1152	D236(RECT.)
1152	DP13B01-2
1152	EB1(ELCOM)
1152	EB3(ELCOM)
1152	EB9(ELCOM)
1152	ECG166
1152	FW50
1152	HEP175
1152	HEPR0802
1152	HEPR0842
1152	HEPR0852
1152	JB00036
1152	K1.3022-1A
1152	K2CDP221B
1152	KCO.8C22/1
1152	KC08C2219
1152	KC2AP22/1B
1152	KC2AP221B
1152	KC2DD22/1
1152	KC2DP22/1B
1152	KC2DP221C
1152	M604
1152	MB-01
1152	MB-4-01
1152	MDA101
1152	MDA920-1
1152	MDA942-2
1152	MDA942A-2
1152	PA9D522/1
1152	PD1011
1152	PH25C221
1152	PH9-221
1152	PH9D5221
1152	PH9D522M
1152	PH9DS221
1152	PT6D22-1
1152	PT6D22/1
1152	R154B
1152	R204B
1152	REN255
1152	REN166
1152	RT7402
1152	RVD12DP22/1C
1152	RVD2DP221B
1152	RVD8MB4
1152	RVDD1245
1152	RVDD124B
1152	RVDR154B
1152	S1B0201B
1152	S1QB10
1152	S1RB
1152	S1RB-10
1152	S1RB10
1152	S1RBA10
1152	S2PB
1152	S2PB10
1152	S2TB10
1152	S4VB10
1152	S6-3
1152	SB-01
1152	SB309A
1152	SB309C
1152	SD-19
1152	SELEN-44
1152	SEN2A1
1152	SIB020-1B
1152	SIB508794-1
1152	SIRB-10
1152	SK3105
1152	SP2-01
1152	SR1B
1152	SVD12B2B1P-M
1152	SVD81RB10
1152	T-E1042
1152	T10195
1152	TB-1
1152	T00.09M22/1
1152	TI-365A
1152	TM166
1152	TV24285
1152	TVM-PT6D22/1
1152	TVM529
1152	TVSW04
1152	TVSW04M
1152	VM18
1152	W-005
1152	WRO11
1152	WRO30
1152	WRO40
1152	ZA150
1152	1B05
1152	1B05J05
1152	1B08T05
1152	1B1
1152	1N1054
1152	181121
1152	181670
1152	18237I
1152	182371A
1152	18R70-100
1152	2B2DM
1152	23B-8851
1152	33B-20B01
1152	4-2021-04170
1152	7-0003
1152	7-59-005/3477
1152	10B2-B1W
1152	10DB1
1152	11-102-001
1152	11-102003
1152	11-108002
1152	12C2P-114
1152	13-67539-1
1152	13-67539-1/3464
1152	16C-4P
1152	16C4B1P
1152	24MW192
1152	46-86421-3
1152	48-97305A03
1152	53B018-1
1152	53E011-1
1152	86-68-3
1152	86-68-3A
1152	130-30261
1152	276-1152
1152	325-0042-311
1152	429-0958-41
1152	642-216
1152	690V081H40
1152	977-42B
1152	977-4B
1152	977-6B
1152	992-531-01
1152	1808T10
1152	04170
1152	4822-130-30414
1152	72174-1
1152	0112945
1152	0112945(ELGIN)
1152	125787
1152	126413
1152	489752-038
1152	530088-1002
1152	530120-1
1152	530152-1
1152	0551029
1152	0551029H
1152	771908
1152	772714
1152	984690
1152	2330011
1152	2330011H
1152	2330361
1172	B004C
1172	B30C250KP
1172	B30C50KP
1172	BS-1
1172	BS-B2
1172	BY123
1172	C674(JAPAN)
1172	CXL167
1172	DP13B02-2
1172	E03090-002
1172	EB10(ELCOM)
1172	EB4(ELCOM)
1172	EC0167
1172	FWB3001
1172	FWB3002
1172	FWLC100
1172	FWLC200
1172	FWLC50
1172	HEP176
1172	HEPR0803
1172	HEPR0843
1172	HEPR0853
1172	J4-1605
1172	KBU02
1172	MB-02
1172	MB-4-02
1172	MDA102
1172	MDA920-4
1172	MDA942-3
1172	MDA942A-3
1172	MP-02
1172	QRT-230
1172	RT-230
1172	S1PB15
1172	S1Q20Z
1172	S1QB20
1172	S1QB20Z
1172	S1RB20
1172	S1RB20Z
1172	S1RBA20
1172	S2PB20
1172	S2PB20Z
1172	S4VB20
1172	S93.20.709
1172	S93.20.714
1172	SB-02
1172	SD-W04
1172	SEN2A2
1172	SP2-02
1172	SR50253-2
1172	TB-2
1172	TM167
1172	VM28
1172	W04
1172	ZEN433
1172	ZEN434
1172	1-531-024
1172	1308T20
1172	181122
1172	181671
1172	182372A
1172	18R70-200
1172	18RBA20Z
1172	2B4DM
1172	2N6214
1172	10DB2
1172	10DB2A
1172	18DB2A
1172	18DB2A-C
1172	25F-B1P
1172	35B611
1172	35EL611
1172	65-11148
1172	63-12287
1172	93A104-1
1172	325-0081-109
1172	690VO80H53
1172	2093A41-112
1172	4822-130-30261
1172	126849
1172	199919
1172	1471393-4
1173	B004B
1173	BS-B4
1173	BY179
1173	CXL168
1173	DP13B04-2
1173	EB11(ELCOM)
1173	EB2(ELCOM)
1173	EB6(ELCOM)
1173	ECG168
1173	FB-200
1173	FB200
1173	FW500
1173	FWB3003
1173	FWB3004
1173	FWLC300
1173	FWLC400
1173	GMO378
1173	HEP177
1173	HEPR0804
1173	HEPR0845
1173	HEPR0855
1173	MB-04
1173	MB-4-04
1173	MDA920-6
1173	MDA942-4
1173	MDA942-5
1173	MDA942A-4
1173	MDA942A-5
1173	MP-04
1173	PTC401
1173	S1QB40
1173	S1QB40Z
1173	S1RB40
1173	S1RB40Z
1173	S1RBA40
1173	S1RBA40Z
1173	S2PB40
1173	S2RB40
1173	S2TB40
1173	S4VB40
1173	SB-04
1173	SEN2A4
1173	SK3106
1173	SK3106/5304
1173	SP2-04
1173	TB-4
1173	TM168
1173	VM48
1173	1B4
1173	181123
1173	181124
1173	181672
1173	182373A
1173	182375A
1173	18R70-400
1173	2B6DM
1173	2B8DM
1173	7-8(STANDEL)
1173	10DB4
1173	10DB4A
1173	10DB6A
1173	10DB6A-C
1173	13-33190-1
1173	18DB4A
1173	18DB4A-C
1173	4822-130-50228
1701	EC0075491B
1702	EC0075492B
1711	EC0987
1713	EC0992
1723	CA555CE
1723	EC0955M
1723	GEIC-269
1723	IC-520(ELCOM)
1723	J1000-NE555
1723	J4-1555
1723	M51841P
1723	MC1455P
1723	MC1455P1
1723	NE555V
1723	SK3564
1723	SK3564/955M
1723	SN72555JP
1723	SN72555P
1723	TN955M
1723	555
1728	EC0978
1731	EC0823
1740	AMU6A7723393
1740	EC0923D
1740	FU6A7723393
1740	FU9A7723393
1740	G390507S2
1740	GEIC-260
1740	HA17723
1740	IC-20
1740	IC-20(PHILCO)
1740	IC-53(ELCOM)
1740	IC-53I(ELCOM)
1740	I053(ELCOM)
1740	ITT723(D.I.P.)
1740	ITT723-5(D.I.P.)
1740	LM723C
1740	LM723CD
1740	MC1723CL
1740	MIC723-5(D.I.P.)
1740	N5723A
1740	P8-801
1740	RB6-P8-801
1740	SG723CD
1740	SG723CN
1740	SK3165
1740	SK3165/923D
1740	SL21385
1740	SL22910
1740	SL22935
1740	SL23325
1740	TDB0723A
1740	Z10003
1740	007-1669901
1740	13-5020-6
1740	19-10415-00
1740	44A393611
1740	44A393611-001
1740	46-136284-P2
1740	46-5002-20
1740	156-0071-00
1740	221-Z9020
1740	224HA2A0723
1740	398-15222-1
1740	723DC
1740	723PC
1740	1677-1149
1740	142349
1740	157800
1740	157800-2760
1740	930419
1740	1802677-1
1740	2899414
1740	3008340
1771	7812
1801	A00
1801	C3000P
1801	DDBY090001
1801	DMT400
1801	DMT400N
1801	ECG7400
1801	FD-1073-BF
1801	FJH131
1801	FJH101
1801	GE-7400
1801	HC10004110
1801	HD2503
1801	HD2503P
1801	HD7400
1801	HD7400P
1801	HEP-C3000P
1801	HEPC3000L
1801	HEPC3000P
1801	HEPC7400P
1801	HL18998
1801	HL56420
1801	IC-801(ELCOM)
1801	I220-0205
1801	ITT7400N
1801	ITT7400N
1801	J1000-7400
1801	J4-1000
1801	K820967-L1
1801	LB3000
1801	M53200
1801	M53200P
1801	MB400
1801	MB601
1801	MC7400
1801	MC7400L
1801	MC7400N
1801	MC7400P
1801	MIC7400J
1801	MIC7400N
1801	N-7400A
1801	N7400A
1801	N7400F
1801	N7400N
1801	N7401A
1801	N7401F
1801	PA7001/521
1801	REN7400
1801	R87400
1801	SK7400
1801	SL16793
1801	SN7400
1801	SN7400A
1801	SN7400N
1801	SN7400N-10
1801	T7400B1
1801	TD1401
1801	TD1401P
1801	TD3400A
1801	TD3400AP
1801	TD3400N
1801	TL7400N
1801	US7400A
1801	WEP7400
1801	ZN7400E
1801	006-0000146
1801	007-1695001
1801	09-308022
1801	9NOO
1801	9NOODC
1801	9NOOPC
1801	19A116180P1
1801	51-10611A11
1801	51B10611A11
1801	68A9025
1801	78A200010P4
1801	225A6946-P000
1801	236-0005
1801	301-576-4
1801	398-13223-1
1801	435-21026-0A
1801	443-1
1801	1065-4861
1801	1741-0051
1801	1820-0054
1801	7400
1801	7400-6A
1801	7400-9A
1801	7400/9NOO
1801	7400A
1801	7400PC
1801	8000-00038-004
1801	10302-04
1801	11216-1
1801	43200
1801	55001
1801	138311
1801	339300
1801	339300-2
1801	373401-1
1801	558875
1801	611563
1801	760011
1801	800024-001
1801	930347-3
1801	2610786
1801	3520041-001
1802	A03
1802	C3004P
1802	DMT404N
1802	ECG7404
1802	FD-1073-BJ
1802	FJH211
1802	FLH211
1802	GE-7404
1802	HD2522
1802	HD2522P
1802	HD7404
1802	HD7404P
1802	HEP-C3004P
1802	HEPC3004L
1802	HEPC3004P
1802	HEPC7404P
1802	HLI9000
1802	HL55862
1802	hL56421
1802	IC-84(ELCOM)
1802	ITT7404N
1802	J1000-7404
1802	J4-1004
1802	K820967-L2
1802	LB3006
1802	M53204
1802	M53204P
1802	MB418
1802	MC7404L
1802	MC7404P
1802	MIC7404J
1802	MIC7404N
1802	N-7404A
1802	N7404A
1802	N7404F
1802	PA7001/527
1802	R87404
1802	SK7404
1802	SL16796
1802	SN7404N
1802	SN7404N-1Q
1802	T7404B1
1802	TD3404A
1802	TD3404AP
1802	TL7404N
1802	US7404A
1802	US7404A
1802	ZN7404E

RS 276-	Industry Standard No.	RS 276-	Industry Standard No.	RS 276-	Industry Standard No.	RS 276-	Industry Standard No.
1802	007-1695301	1804	TD3441AP	1807	MC7410F	1809	EC07420
1802	9N04	1804	US7441A	1807	MC7410L	1809	EP84X11
1802	9N04DC	1804	ZN7441AE	1807	MC7410P	1809	FD-1073-BR
1802	9N04PC	1804	007-1697801	1807	MIC7410J	1809	PJH111
1802	19A116180P20	1804	376-0099	1807	MIC7410N	1809	PLH21
1802	51-10611A12	1804	1741-1190	1807	N7410A	1809	GE-7420
1802	51810611A12	1804	7441	1807	N7410P	1809	HD2504
1802	68A9028	1804	7441-6A	1807	N7410N	1809	HD2504P
1802	156-0148-00	1804	7441-9A	1807	PA7001/520	1809	HD7420
1802	225A6946-P004	1804	7441DC	1807	RLH111	1809	HD7420P
1802	236-0007	1804	7441PC	1807	R87410(IC)	1809	HEP-C3020P
1802	398-13224-1	1804	9315DC	1807	SK7410	1809	HEP3020L
1802	398-13632-1	1804	9315PC	1807	SL16801	1809	HEPC3020P
1802	435-21028-0A	1804	DM7447N	1807	SMC7410N	1809	HEPC7420P
1802	443-18	1805	EC07447	1807	SN7410J	1809	HL19003
1802	1741-0143	1805	FLL121T	1807	SN7410N	1809	HL56422
1802	1806	1805	FLL121V	1807	SN7410N-10	1809	IC-87(ELCOM)
1802	1820-0174	1805	GE-7447	1807	T7410B1	1809	IC7420
1802	1820-0894	1805	HD2532	1807	T7410D1	1809	ITT7420N
1802	7404	1805	HD2532P	1807	T7410D2	1809	LB3002
1802	7404-6A	1805	HEPC7447AP	1807	TD1402	1809	M53220
1802	7404-9A	1805	IC-101(ELCOM)	1807	TD1402P	1809	M53220P
1802	7404A	1805	ITT7447AN	1807	TD3410A	1809	MB402
1802	7404PC	1805	J1000-7447	1807	TD3410P	1809	MB603
1802	8000-00028-042	1805	J4-1047	1807	TL7410N	1809	MC7420P
1802	8000-0028-042	1805	M53247	1807	UPB202C	1809	MC7420L
1802	11202-1	1805	M53247P	1807	UPB202D	1809	MC7420P
1802	015040/7	1805	MC7447L	1807	UPB7410C	1809	MIC7420J
1802	43201	1805	MC7447P	1807	UST410A	1809	MIC7420N
1802	55003	1805	N7447B	1807	UST410J	1809	N7420A
1802	138314	1805	N7447P	1807	ZN7410E	1809	N7420P
1802	373404-1	1805	RS7447	1807	ZN7410F	1809	N7420N
1802	508590	1805	SK7447	1807	006-0000147	1809	PA7001/519
1802	611565	1805	SN7447AN	1807	007-1695901	1809	RS7420
1802	800387-001	1805	SN7447N-10	1807	9N10	1809	SK7420
1802	801806	1805	TD3447A	1807	9N10DC	1809	SL16800
1802	930347-13	1805	TD3447AP	1807	9N10PC	1809	SMC7420N
1802	3520048-001	1805	TL7447AN	1807	19A116180P4	1809	SN7420J
1803	B0075660	1805	TVCM-503	1807	49A0005-000	1809	SN7420N
1803	C3075P	1805	UST447A	1807	68A9030	1809	SN7420N-10
1803	DM7473N	1805	443-36	1807	225A6946-P010	1809	T7420B1
1803	EA33X8385	1805	7447	1807	435-21030-0A	1809	T7420D1
1803	EC07473	1805	7447BDC	1807	443-12	1809	T7420D2
1803	PJJ121	1805	7447BPC	1807	1741-0234	1809	TD1403
1803	PLJ121	1805	7447DC	1807	1820-0068	1809	TD1403P
1803	GE-7473	1805	7447PC	1807	7410	1809	TD3420A
1803	HD2515	1805	9357B	1807	7410-6A	1809	TD3420AP
1803	HD2515P	1805	9357BDC	1807	7410-9A	1809	TD3420P
1803	HD7473AP	1805	9357BPC	1807	7410DC	1809	TL7420N
1803	HD7473P	1806	B01	1807	7410P	1809	UPB203D
1803	HEP-C3073P	1806	C3075P	1807	7410PC	1809	UPBT420C
1803	HEPC3073L	1806	DM7475N	1807	10302-03	1809	UST420A
1803	HEPC3073P	1806	EC07475	1807	11200-1	1809	UST420J
1803	HEPC7473P	1806	PJJ181	1807	55005	1809	ZN7420E
1803	HL19002	1806	PJJ151	1807	55006	1809	ZN7420P
1803	IC-95(ELCOM)	1806	GE-7475	1807	373405-1	1809	007-1695101
1803	ITT7473N	1806	HD2517	1807	558877	1809	9N20
1803	M53273	1806	HD2517P	1807	611566	1809	9N20DC
1803	M53275P	1806	HD7475	1807	800025-001	1809	9N20PC
1803	M53842P	1806	HD7475P	1808	3520042-001	1809	19A116180P5
1803	N7641	1806	HEP-C3075P	1808	A17(I.C.)	1809	49A0006-000
1803	MC7473L	1806	HEPC3075L	1808	C3800P	1809	68A9033
1803	MC7473P	1806	HEPC3075P	1808	DDEY029001	1809	225A6946-P020
1803	MIC7473J	1806	HEPC7475P	1808	DM7490	1809	435-21033-0A
1803	MIC7473N	1806	HL19012	1808	DM7490N	1809	443-2
1803	N-7473A	1806	IC-96(ELCOM)	1808	EC07490	1809	1741-0325
1803	N7473	1806	ITT7475N	1808	PJJ141	1809	1820-0069
1803	N7473A	1806	J4-1075	1808	PLJ161	1809	7420(IC)
1803	N7473P	1806	M53275	1808	GE-7490	1809	7420-6A
1803	PA7001/531	1806	M53275P	1808	HD2519	1809	7420-9A
1803	REN7473	1806	MC7475L	1808	HD2519P	1809	7420DC
1803	RS7473	1806	MC7475P	1808	HD7490A	1809	7420PC
1803	SK7473	1806	MIC7475J	1808	HD7490AP	1809	10302-02
1803	SL16806	1806	MIC7475N	1808	HEP-C3800P	1809	11205-1
1803	SL17242	1806	N7475B	1808	HEPC3800L	1809	55007
1803	SN7473	1806	RS7475	1808	HEPC3800P	1809	138318
1803	SN7473N	1806	SK7475	1808	HEPC7490AP	1809	373406-1
1803	SN7473N-10	1806	SM63	1808	HL19015	1809	558878
1803	T7473B1	1806	SMT75(I.C.)	1808	IC-98(ELCOM)	1809	611567
1803	TD1409	1806	SN7475N	1808	ITT7490N	1809	800020-001
1803	TD1409P	1806	SN7475N-10	1808	J1000-7490	1809	930347-1
1803	TD3473A	1806	T7475B1	1808	J4-1090	1809	930347-10
1803	TD3473AP	1806	TD3475A	1808	LB3150	1811	C3002P
1803	TL7473N	1806	TD3475AP	1808	M53290	1811	DM7402N
1803	UST473A	1806	TL7475N	1808	M53290P	1811	EC07402
1803	UST473J	1806	UST475A	1808	MC7490L	1811	FD-1073-BG
1803	WBP7473	1806	UST475J	1808	MC7490P	1811	FJH221
1803	XAA107	1806	ZN7475E	1808	MIC7490J	1811	FLH191
1803	ZN7473B	1806	40-065-19-027	1808	MIC7490N	1811	GE-7402
1803	9N73	1806	49A0000	1808	N7490A	1811	HD2511
1803	9N73DC	1806	51-10611A16	1808	N7490P	1811	HD2511P
1803	9N73PC	1806	51810611A16	1808	R87490	1811	HD7402
1803	19A116180P15	1806	68A9041	1808	SK7490	1811	HD7402P
1803	43A223025	1806	443-13	1808	SN7490AN	1811	HEP-C3002P
1803	49A0002-000	1806	1741-0747	1808	SN7490N	1811	HEPC3002P
1803	236-0009	1806	1820-0301	1808	SN7490N-10	1811	HEPC7402P
1803	435-23006-0A	1806	7475	1808	T7490B1	1811	HL19004
1803	443-5	1806	7475-6A	1808	TD3490A	1811	IC-82(ELCOM)
1803	477-0412-004	1806	7475-9A	1808	TD3490AP	1811	ITT7402N
1803	1030-25	1806	7475PC	1808	TD3490P	1811	J1000-7402
1803	1820-0075	1806	7475PC	1808	TL7490N	1811	J4-1002
1803	7473	1806	9375DC	1808	TVCM-505	1811	LB3008
1803	7473-6A	1806	9375PC	1808	UST490A	1811	M53202
1803	7473-9A	1806	373713-1	1808	UST490C	1811	M53202P
1803	7473PC	1806	611065	1808	WBP7490	1811	MB417
1803	43205	1806	800382-001	1808	XAA109	1811	MC-7402
1803	138403	1807	A05(I.C.)	1808	1-000-099-00	1811	MC7402L
1803	558881	1807	C3010P	1808	19-130-005	1811	MC7402P
1803	930347-7	1807	DM7410J	1808	19A116180-24	1811	MIC7402J
1803	3520043-001	1807	DM7410N	1808	443-7	1811	MIC7402N
1804	C3041P	1807	EC07410	1808	443-7-16088	1811	N7402A
1804	DM7441N	1807	FD-1073-BN	1808	733W00039	1811	N7402P
1804	EC07441	1807	PJH121	1808	905-102	1811	PA7001/525
1804	PJL101	1807	GE-7410	1808	1808	1811	RS7402
1804	GE-7441	1807	HD2507	1808	1820-0055	1811	SK7402
1804	HD2518	1807	HD2507P	1808	7490	1811	SL16795
1804	HD2518P	1807	HD7410	1808	7490-6A	1811	SN7402
1804	HEP-C3041P	1807	HD7410P	1808	7490-9A	1811	SN7402N
1804	HEPC3041L	1807	HEP-C3010P	1808	7490DC	1811	SN7402N-10
1804	HEPC3041P	1807	HEPC3010L	1808	7490PC	1811	T7402B1
1804	HEPC7441AP	1807	HEPC3010P	1808	9390DC	1811	TD3402A
1804	IC-89(ELCOM)	1807	HEPC7410P	1808	9390PC	1811	TD3402AP
1804	M53241	1807	HL19001	1808	16088	1811	TL7402N
1804	M53241P	1807	HL56999	1808	102005	1811	UST402A
1804	MC7441AL	1807	IC-7410	1808	373427-1	1811	UST402J
1804	MIC7441AJ	1807	IC-86(ELCOM)	1808	558803	1811	WBP7402
1804	MIC7441AN	1807	ITT7410N	1808	611572	1811	XAA104
1804	N7441B	1807	J1000-7410	1808	760013	1811	ZN7402E
1804	N7441F	1807	JA-1010	1808	801808	1811	007-1696201
1804	RS7441	1807	LB3001	1809	A06(I.C.)	1811	9N02
1804	SK7441	1807	M53210	1809	C3020P	1811	9N02DC
1804	SN7441N-1Q	1807	M53210P	1809	DM7420J	1811	9N02PC
1804	T7441AB1	1807	MB401	1809	DM7420N	1811	19A116180P3
1804	TD3441A	1807	MB602			1811	43A223009

RS 276-	Industry Standard No.
1811	68A9027
1811	435-21027-0A
1811	443-46
1811	1741-0119
1811	1820-0328
1811	7402
1811	7402-6A
1811	7402-9A
1811	7402PC
1811	11207-1
1811	55002
1811	138313
1811	558876
1811	611564
1811	800080-001
1811	930547-11
1811	2610783
1811	B02
1813	DM7476N
1813	ECG7476
1813	FJJ191
1813	FLJ131
1813	GE-7476
1813	HD2516
1813	HD2516P
1813	HEPC7476P
1813	HL19010
1813	IC-99(ELCOM)
1813	ITT7476N
1813	J1000-7476
1813	J4-1076
1813	M53276
1813	M53276P
1813	MC7476L
1813	MC7476P
1813	MIC7476J
1813	MIC7476N
1813	N7476B
1813	N7476P
1813	RS7476
1813	SK7476
1813	SL16808
1813	SN7476B
1813	SN7476N
1813	SN7476N-10
1813	T7476B1
1813	TL7476N
1813	US7476A
1813	WEP7476
1813	XAA108
1813	ZN7476E
1813	9N76
1813	9N76DC
1813	9N76PC
1813	43A223028
1813	6849042
1813	443-16
1813	1348A14H01
1813	7476
1813	7476-6A
1813	7476-9A
1813	7476PC
1813	55012
1813	72185
1813	373414-1
1813	611870
1813	760015
1815	DM7413N
1815	ECG7413
1815	FLH351
1815	GE-7413
1815	HD2545
1815	HD2545P
1815	HEPC7413P
1815	IC-103(ELCOM)
1815	ITT7413N
1815	MIC7413N
1815	MIC77413J
1815	N7413A
1815	N7413P
1815	RSD7413(IC)
1815	SK7413
1815	SN7413N
1815	SN7413N-10
1815	TL7413N
1815	UPB7413C
1815	9N13
1815	9N13DC
1815	9N13PC
1815	443-44
1815	443-44-2854
1815	601-0100865
1815	3531-021-000
1815	7413(IC)
1815	7413PC
1815	WEP7413
1816	DM7448N
1816	ECG7448
1816	FLH551
1816	GE-7448
1816	HEPC7448P
1816	ITT7448N
1816	M53248
1816	M53248A
1816	M53248P
1816	MC7448L
1816	MC7448P
1816	N7448B
1816	RS7448
1816	SK7448
1816	SN7448N
1816	SN7448N-10
1816	TL7448N
1816	US7448A
1816	7448
1816	7448DC
1816	7448PC
1816	9358
1816	9358DC
1816	9358PC
1817	DM74123N
1817	ECG74123
1817	EP84X19
1817	FLK121
1817	GE-74123
1817	HD2561
1817	HD2561P
1817	HEPC74123P
1817	IC-74123
1817	ITT74123N
1817	N74122A
1817	N74123B
1817	N74123F
1817	REN74123
1817	RH-IX0041PAZZ
1817	RS74123
1817	SK74123
1817	SN74123N
1817	SN74123N-10
1817	TL74123N
1817	UPB74123C
1817	9N123
1817	9N123DC
1817	9N123PC
1817	1462
1817	74123
1817	74123PC
1818	A15
1818	DM7474N
1818	ECG7474
1818	F7474PC
1818	FJJ131
1818	FJJ141
1818	GE-7474
1818	GEIC-193
1818	HD2510
1818	HD2510P
1818	HD7474
1818	HD7474P
1818	HEPC7474P
1818	HL18999
1818	HL56425
1818	IC-97(ELCOM)
1818	IP20-0206
1818	IP20-0316
1818	M53274
1818	M53274P
1818	M5374
1818	M5374P
1818	MB420
1818	MIC7474J
1818	MIC7474N
1818	N7474A
1818	N7474P
1818	PA7001/529
1818	REN7474
1818	RS7474
1818	SL16807
1818	SN-7474
1818	SN7474
1818	SN7474M
1818	SN7474N-10
1818	T7474B1
1818	TD3474A
1818	TD3474AP
1818	TD3474P
1818	TT7474N
1818	TVCM-502
1818	US7474A
1818	US7474J
1818	WBP7474
1818	YBAM53274P
1818	ZN7474E
1818	007-1699801
1818	9N74
1818	9N74DC
1818	9N74PC
1818	19A116180P16
1818	43A225026P1
1818	49A0012-000
1818	075-045037
1818	90-39
1818	90-67
1818	435-23007-0A
1818	443-6
1818	1820-0077
1818	7474
1818	7474-6A
1818	7474-9A
1818	7474/9N74
1818	7474PC
1818	8000-00038-007
1818	K821282-L3
1818	373409-1
1818	558882
1818	611571
1818	800400-001
1818	281916
1818	2868556-1
1818	3520046-001
1819	C3801P
1819	ECG7492
1819	PJJ152
1819	FLJ171
1819	GE-7492
1819	HD2521
1819	HD2521P
1819	HD7492A
1819	HD7492AP
1819	HBP-C3801P
1819	HEPC3801L
1819	HEPC3801P
1819	HEPC7492AP
1819	IC-100(ELCOM)
1819	ITT7492N
1819	J-1000-7492
1819	J4-1092
1819	K820969-L3
1819	M53292
1819	M53292P
1819	MC7492L
1819	MC7492P
1819	MIC7492J
1819	MIC7492N
1819	N7492A
1819	N7492P
1819	RS7492
1819	SK7492
1819	SL16809
1819	SN7492AN
1819	SN7492AN-10
1819	TD3492A
1819	TD3492AP
1819	TD3492P
1819	TL7492N
1819	US7492A
1819	US7492J
1819	19A116180-27
1819	436-10010-0A
1819	7492
1819	7492-6A
1819	7492-9A
1819	7492DC
1819	7492PC
1819	9392DC
1819	9392PC
1819	373712-1
1819	611573
1819	1000100
1819	1000100-000
1820	DM74193N
1820	ECG74193
1820	FLJ251
1820	GE-74193
1820	HD2542
1820	HD2542P
1820	HEPC74193P
1820	HL56430
1820	ITT74193N
1820	M53593
1820	M53593P
1820	MC74193P
1820	N74193B
1820	N74193P
1820	RS74193
1820	SK74193
1820	SN74193N
1820	SN74193N-10
1820	T74193B1
1820	TL74193N
1820	007-1698401
1820	43C216447
1820	43C216447P1
1820	443-162
1820	9366DC
1820	9366PC
1820	11204-1
1820	74193
1820	74193DC
1820	74193PC
1820	138320
1820	611730
1820	800386-001
1821	DM7406N
1821	ECG7406
1821	ELN481
1821	GE-7406
1821	HD7406
1821	HD7406P
1821	HEPC7406P
1821	IC-104(ELCOM)
1821	ITT7406N
1821	MC7406L
1821	MC7406P
1821	N7406A
1821	N7406P
1821	RH-IX0038PAZZ
1821	RS7406(IC)
1821	SK7406
1821	SN7406N
1821	SN7406N-10
1821	T7406B1
1821	TL7406N
1821	007-1696901
1821	9N06
1821	9N06DC
1821	9N06PC
1821	68A9032
1821	1607A80
1821	7406
1821	7406PC
1821	55936
1821	373429-1
1821	800651-001
1822	DM7408N
1822	ECG7408
1822	FD-1073-BM
1822	FLH381
1822	GE-7408
1822	HD2550
1822	HD2550P
1822	HEPC7408P
1822	HL55763
1822	IC-102(ELCOM)
1822	ITT7408N
1822	K821282-L1
1822	L-612099
1822	MC7408L
1822	MC7408P
1822	N-7408A
1822	N7408A
1822	RS7408(IC)
1822	SK7408
1822	SL14971
1822	SL16798
1822	SL17869
1822	SN7408N
1822	SN7408N-10
1822	US7408A
1822	US7408J
1822	WEP7408
1822	007-1699301
1822	9N08
1822	9N08DC
1822	9N08PC
1822	435-21029-0A
1822	443-45
1822	1741-0200
1822	1820-0870
1822	2473-2109
1822	7408
1822	7408-6A
1822	7408-9A
1822	7408A
1822	7408DC
1822	43202
1822	94152
1822	138315
1822	310254
1822	374109-1
1823	DM7427J
1823	DM7427N
1823	DM7427W
1823	ECG7427
1823	FLH66
1823	GE-7427
1823	HD7427
1823	HD7427P
1823	HEPC7427P
1823	N7427A
1823	N7427F
1823	N7427N
1823	RS7427
1823	SK7427
1823	SN7427J
1823	SN7427N
1823	SN7427N-10
1823	US7427A
1823	ZN7427E
1823	ZN7427J
1823	9N27
1823	9N27DC
1823	9N27PC
1823	443-65
1823	7427(IC)
1823	7427DC
1823	7427PC
1824	DM7432J
1824	DM7432N
1824	ECG7432
1824	FLH631
1824	GE-7432
1824	HD7432
1824	HD7432P
1824	HEPC7432P
1824	ITT7432N
1824	K821282-L3
1824	L-612107
1824	N7432A
1824	N7432P
1824	N7432N
1824	RS7432
1824	SK7432
1824	SN7432J
1824	SN7432N
1824	SN7432N-10
1824	US7432A
1824	US7432J
1824	ZN7432E
1824	ZN7432J
1824	9N32
1824	9N32DC
1824	9N32PC
1824	7432(IC)
1824	7432DC
1824	7432PC
1824	138381
1824	1383SH
1825	A12(I.C.)
1825	DM7451J
1825	DM7451N
1825	ECG7451
1825	FD-1073-BW
1825	FJH161
1825	FLH161
1825	GE-7451
1825	HD2505
1825	HD2505P
1825	HD7451
1825	HD7451P
1825	HEPC7451P
1825	IC-91(ELCOM)
1825	ITT7451N
1825	MC7451J
1825	MC7451L
1825	MC7451P
1825	MIC7451J
1825	MIC7451N
1825	N7451A
1825	N7451A
1825	N7451P
1825	N7451N
1825	PA7001/523
1825	RS7451
1825	SK7451
1825	SNC7451N
1825	SN7451
1825	SN7451J
1825	SN7451N-10
1825	T7451B1
1825	T7451N1
1825	T7451N2
1825	TD1419
1825	TD1419P
1825	TD3451A
1825	TD3451AP
1825	TD3451P
1825	TL7451N
1825	UPB207D
1825	UPB7451C
1825	US7451A
1825	US7451J
1825	ZN7451E
1825	ZN7451F
1825	9N51
1825	9N51DC
1825	9N51PC
1825	435-21034-0A
1825	1741-0564
1825	1820-0063
1825	7451
1825	7451-6A
1825	7451-9A
1825	7451DC
1825	7451PC
1825	10302-05
1825	373715-1
1825	930547-12
1825	3520045-001
1826	DM7485N
1826	ECG7485
1826	FLH431
1826	GE-7485
1826	HD7485
1826	HD7485P
1826	HEPC7485P
1826	HL56426
1826	MB448
1826	N7485A
1826	N7485B
1826	N7485P
1826	N7485N
1826	RS7485
1826	SK7485
1826	SN7485J
1826	SN7485N
1826	SN7485N-10
1826	TL7485N
1826	UPB208SD
1826	UPB7485C
1826	007-1696001
1826	7485
1826	7485DC
1826	7485PC
1826	9385DC
1826	9385PC
1827	DM7486J
1827	DM7486N
1827	ECG7486
1827	FD-1073-CA
1827	FLH341
1827	GE-7486

RS 276-	Industry Standard No.
1827	HD2526
1827	HD2526P
1827	HD7486
1827	HD7486P
1827	HEPC7486P
1827	HL19014
1827	ITT7486N
1827	KS20967-L3
1827	M53286
1827	M53286P
1827	MB449
1827	MC7486F
1827	MC7486L
1827	MC7486P
1827	MIC7486J
1827	MIC7486N
1827	N7486A
1827	N7486P
1827	N7486N
1827	RS7486
1827	SK7486
1827	SN7486J
1827	SN7486N
1827	SN7486N-1Q
1827	T7486B1
1827	T7486D1
1827	T7486D2
1827	TL7486N
1827	UPB2086D
1827	UPB7486
1827	US7486A
1827	US7486J
1827	ZN7486E
1827	ZN7486J
1827	9N86
1827	9N86DC
1827	9N86PC
1827	19A116180-18
1827	19A116180P18
1827	40-065-19-029
1827	435-21035-0A
1827	1741-0804
1827	7486
1827	7486DC
1827	7486PC
1827	339486
1827	373410-1
1827	611066
1827	611844
1828	DM74145N
1828	EC974145
1828	FLL111T
1828	GB-74145
1828	HD2555
1828	HD2555P
1828	HEPC74145P
1828	ITT74145N
1828	MB443
1828	MC74145P
1828	MIC74145J
1828	MIC74145N
1828	N74145
1828	N74145B
1828	PA7001/593
1828	RS74145
1828	SK74145
1828	SN74145N
1828	SN74145N-10
1828	TL74145N
1828	US74145A
1828	WEP74145
1828	007-1696801
1828	19-130-004
1828	276-1828
1828	443-87
1828	1542
1828	74145
1828	74145DC
1828	74145PC
1828	93145DC
1828	93145PC
1829	DM74150N
1829	EC974150
1829	FLY111
1829	GE-74150
1829	HD2548
1829	HD2548P
1829	HD74150
1829	HD74150P
1829	HEPC74150P
1829	HL55861
1829	ITT74150N
1829	MC74150P
1829	MIC74150J
1829	MIC74150N
1829	N74150B
1829	N74150F
1829	N74150N
1829	RS74150
1829	SK74150
1829	SN74150N
1829	SN74150N-1Q
1829	TL74150N
1829	UPB2150D
1829	UPB74150C
1829	1741-1042
1829	74150
1829	74150DC
1829	74150PC
1829	93150DC
1829	93150PC
1831	DM74192N
1831	EC974192
1831	FLW241
1831	GE-74192
1831	HD2541
1831	HD2541P
1831	HEPC74192P
1831	HL56429
1831	ITT74192N
1831	M53392
1831	M53392P
1831	MC74192P
1831	N74192B
1831	N74192P
1831	RS74192
1831	SK74192
1831	SN74192
1831	SN74192N
1831	SN74192N-1Q
1831	TD34192A
1831	TD34192AP
1831	TL74192N
1831	007-1698301
1831	443-66
1831	9360DC
1831	9360PC
1831	74192
1831	74192DC
1831	74192PC
1831	611731
1833	DM74196N
1833	EC974196
1833	FLJ381
1833	GE-74196
1833	HD2572
1833	HD2572P
1833	HEPC74196P
1833	N74196A
1833	RS74196
1833	SK74196
1833	SN74196J
1833	SN74196N
1833	SN74196N-1Q
1833	TL74196N
1833	443-628
1833	74196
1833	74196DC
1833	74196PC
1833	93196DC
1833	93196PC
1834	DM74154N
1834	EC974154
1834	FLY141
1834	GE-74154
1834	HD2580
1834	HD2580P
1834	HEPC74154P
1834	ITT74154N
1834	MIC74154J
1834	MIC74154N
1834	N74154P
1834	N74154N
1834	RS74154
1834	SK74154
1834	SN74154N
1834	SN74154N-10
1834	TL74154N
1834	US74154A
1834	74154
1834	74154DC
1834	74154PC
1834	93154DC
1834	93154PC
1835	EC974LS367
1835	SK74LS367
1835	SN74LS367N
1900	DM74LS00N
1900	EC974LS00
1900	N74LS00
1900	SN74LS00N
1900	TISM74LS00N
1900	74LS00N
1902	EC974LS02
1902	SK74LS02
1902	SN74LS02N
1904	EC974LS04
1904	SK74LS04
1904	SN74LS04N
1908	EC974LS08
1908	SN74LS08N
1910	EC974LS10
1910	SK74LS10N
1910	SN74LS10N
1912	EC974LS20
1912	SK74LS20
1912	SN74LS20N
1913	EC974LS27
1913	SK74LS27
1913	SN74LS27N
1914	EC974LS30
1914	SK74LS30
1914	SN74LS30N
1915	EC974LS32
1915	SK74LS32
1915	SN74LS32N
1917	EC974LS51
1917	SK74LS51
1917	SN74LS51N
1918	EC974LS73
1918	SK74LS73
1918	SN74LS73N
1919	EC974LS74A
1919	SK74LS74
1919	SN74LS74
1920	MC74LS75
1920	SK74LS75
1920	SN74LS75N
1922	EC974LS85
1922	SN74LS85N
1925	EC974LS93
1925	SK74LS93
1925	SN74LS93
1925	SN74LS93N
1925	74LS93
1926	EC974LS123
1926	SK74LS123
1926	SN74LS123N
1929	EC974LS151
1929	SK74LS151
1929	SN74LS151N
1930	EC974LS157
1930	SK74LS157
1930	SN74LS157N
1931	EC974LS161A
1932	EC974LS174
1932	SK74LS174
1932	SN74LS174N
1932	EC974LS175
1933	SK74LS175
1934	SN74LS175N
1934	EC974LS193
1936	SK74LS193
1936	SN74LS193N
2002	A-567
2002	A121-1410
2002	A121-15
2002	A121-16
2002	A121-17
2002	A121-21
2002	A121-50
2002	A121-762
2002	A122-1962
2002	A127-7
2002	A129-30
2002	A13-86416-1
2002	A13-86420-1
2002	A13-87433-1
2002	A1396
2002	A1465-29
2002	A1465-4
2002	A1465-49
2002	A1465-2
2002	A1641
2002	A16A2
2002	A1858
2002	A198794-1
2002	A2039-2
2002	A2092418
2002	A2092418-0711
2002	A27682
2002	A3607
2002	A3609
2002	A37201
2002	A37202
2002	A37203
2002	A42X210
2002	A46-8614-3
2002	A4700
2002	A48-124216
2002	A48-124217
2002	A48-124218
2002	A48-124220
2002	A48-124221
2002	A48-125233
2002	A48-125234
2002	A48-125235
2002	A48-125236
2002	A48-128239
2002	A48-134520
2002	A48-134700
2002	A48-134931
2002	A48-869254
2002	A48-869283
2002	A48-869476
2002	A48-869476A
2002	A4JD3B1
2002	A4JX2A822
2002	A4L
2002	A514-023553
2002	A514-032815
2002	A5705
2002	A65-2-70
2002	A65-2-705
2002	A65-2A9G
2002	A65-4-70
2002	A65-4-705
2002	A65-4A9G
2002	A86-10-2
2002	A86-44-2
2002	A95115
2002	A95211
2002	A99807
2002	A998K5
2002	A998K7
2002	AA2
2002	AC-127
2002	AC-157
2002	AC-172
2002	AC127
2002	AC127-01
2002	AC127-132
2002	AC130
2002	AC141
2002	AC141B
2002	AC141K
2002	AC157
2002	AC172
2002	AC175
2002	AC175A
2002	AC175B
2002	AC175P
2002	AC176
2002	AC179
2002	AC181
2002	AC181K
2002	AC183
2002	AC185
2002	AC186
2002	AC187
2002	AC187/01
2002	AC187K
2002	AC187R
2002	AC187U
2002	AC187U/U
2002	AC194
2002	AC194K
2002	AF192
2002	A07
2002	AS3428
2002	ASY-62
2002	ASY-72
2002	ASY29
2002	ASY53
2002	ASY61
2002	ASY62
2002	ASY72
2002	ASY73
2002	ASY74
2002	ASY75
2002	ASY86
2002	ASY87
2002	ASY88
2002	ASY89
2002	AT521
2002	AT53
2002	AT551
2002	AT71
2002	AT72
2002	AT73R
2002	AT75R
2002	AT76R
2002	AT77
2002	B65-2-A-21
2002	B65-2A21
2002	B65-4-A-21
2002	B65-4A21
2002	B92-1-A-21
2002	BD-00072
2002	BTX071
2002	C128
2002	C129
2002	C13(TRANSISTOR)
2002	C14
2002	C173
2002	C175
2002	C176
2002	C177
2002	C178
2002	C179
2002	C180(TRANSISTOR)
2002	C181
2002	C277C
2002	C34(TRANSISTOR)
2002	C35(TRANSISTOR)
2002	C36(TRANSISTOR)
2002	C50(TRANSISTOR)
2002	C50A
2002	C60(TRANSISTOR)
2002	C71
2002	C72
2002	C73(JAPAN)
2002	C75(JAPAN)
2002	C76(JAPAN)
2002	C77
2002	C77C
2002	C78
2002	C89
2002	C90
2002	C91
2002	CK261
2002	CK262
2002	CS1759
2002	CXL101
2002	CXL103
2002	D-F1
2002	D083
2002	D085
2002	D100A
2002	D101
2002	D104
2002	D105
2002	D11
2002	D127
2002	D127A
2002	D128
2002	D128A
2002	D162
2002	D167
2002	D170
2002	D170A
2002	D170AA
2002	D170AB
2002	D170AC
2002	D170B
2002	D170BC
2002	D170C
2002	D170PB
2002	D178
2002	D178A
2002	D178Q
2002	D178T
2002	D186
2002	D186A
2002	D186B
2002	D187
2002	D187A
2002	D187R
2002	D187Y
2002	D19
2002	D192
2002	D193
2002	D194
2002	D195
2002	D195A
2002	D20
2002	D21
2002	D215
2002	D22
2002	D23
2002	D30-0
2002	D30-N
2002	D31
2002	D31D
2002	D32
2002	D33
2002	D33C
2002	D34
2002	D35
2002	D352D
2002	D352E
2002	D352F
2002	D36
2002	D367
2002	D367A
2002	D367B
2002	D367C
2002	D367D
2002	D367E
2002	D367F
2002	D37
2002	D37A
2002	D37B
2002	D37C
2002	D38
2002	D43
2002	D43A
2002	D44
2002	D61
2002	D62
2002	D63
2002	D64
2002	D65-1
2002	D72-2C
2002	D72-3C
2002	D72-4C
2002	D720D
2002	D720E
2002	D72A
2002	D72B
2002	D72C
2002	D72RE
2002	D75
2002	D75A
2002	D75AH
2002	D75B
2002	D75C
2002	D75H
2002	D77(TRANSISTOR)
2002	D77A
2002	D77AH
2002	D77B
2002	D77C
2002	D77D
2002	D77H
2002	D77F
2002	D96
2002	DF1
2002	D083
2002	D085
2002	DS11

RS 276-	Industry Standard No.	RS 276-	Industry Standard No.	RS 276-	Industry Standard No.	RS 276-	Industry Standard No.
2002	D812	2002	HT400721C	2002	R62	2002	SYL1329
2002	D82	2002	HT400721D	2002	R63	2002	SYL1380
2002	D84	2002	HT400721E	2002	R7362	2002	SYL1396
2002	D85	2002	HT400723	2002	R79	2002	SYL1408
2002	D86	2002	HT400723A	2002	R80	2002	SYL1454
2002	D87	2002	HT400723B	2002	RCA40231	2002	SYL1468
2002	D88	2002	HT400770B	2002	RCA40396N	2002	SYL1524
2002	D89	2002	HT403523A	2002	RE-5	2002	SYL1536
2002	E-2466	2002	IE-850	2002	RE5	2002	SYL1537
2002	E24105	2002	IP20-0008	2002	RE6	2002	SYL1538
2002	E2427	2002	IP20-0076	2002	REN101	2002	SYL1539
2002	E2428	2002	IP20-016Q	2002	REN103	2002	SYL1547
2002	E2429	2002	IRTR08	2002	REN103A	2002	SYL1591
2002	E2447	2002	IRTR09	2002	RS-104	2002	SYL1617
2002	E2466	2002	IRTR10	2002	RS-1524	2002	SYL1750
2002	E4002	2002	ISD-162	2002	RS-1536	2002	SYL1941
2002	EA15X8443	2002	ISD162	2002	RS-1537	2002	SYL1987
2002	ECG101	2002	J241111(28D170)	2002	RS-1538	2002	SYL2130
2002	ECG101L	2002	J241185	2002	RS-1545	2002	SYL2131
2002	ECG103	2002	J24868	2002	RS-1547	2002	SYL2132
2002	ECG103A	2002	K4-501	2002	RS-1553	2002	SYL2134
2002	EGH-1	2002	KD2124	2002	RS-2001	2002	SYL2135
2002	ES15X48	2002	M2R168A	2002	RS-2359	2002	SYL2136
2002	ES15X71	2002	M4700	2002	RS-2360	2002	SYL2245
2002	ES15X74	2002	M8120	2002	RS-2364	2002	SYL2246
2002	ES38(ELCOM)	2002	M9092	2002	RS-2365	2002	SYL2650
2002	ES5	2002	M9093	2002	RS-2366	2002	SYL4315
2002	ES5(ELCOM)	2002	MHT2002	2002	RS-2373	2002	SYL4339
2002	ES6(ELCOM)	2002	MHT2003	2002	RS-2374	2002	SYL792
2002	ET10	2002	MHT2004	2002	RS-2375	2002	T-81
2002	ET11	2002	MHT2008	2002	RS1513	2002	T-Q5031
2002	ET8	2002	MHT2009	2002	RS1524	2002	T-Q5039
2002	ET9	2002	MHT2010	2002	RS1530	2002	T-Q5050
2002	G101079	2002	MIS-14150-18A	2002	RS1531	2002	T59276
2002	G16	2002	MIS14150-18	2002	RS1532	2002	T59277
2002	G16506	2002	N57B2-8	2002	RS1533	2002	T81
2002	G17	2002	NA20	2002	RS1534	2002	TA-1620A
2002	G18	2002	NA30	2002	RS1536	2002	TA-1620B
2002	GA53270	2002	NC33	2002	RS1537	2002	TA-1759
2002	GC1034	2002	NK65-2A19	2002	RS1538	2002	TA-1767
2002	GC1035	2002	NK65-4A19	2002	RS1547	2002	TA-1771
2002	GC1036	2002	NKT701	2002	RS1549	2002	TA-1772
2002	GC1137	2002	NKT703	2002	RS1553	2002	TA1620A
2002	GC1185	2002	NKT713	2002	RS2356	2002	TA1620B
2002	GC1423	2002	NKT717	2002	RS2359	2002	TA1759
2002	GC148	2002	NKT732	2002	RS2360	2002	TA1767
2002	GC285	2002	NKT734	2002	RS2364	2002	TA1771
2002	GC286	2002	NKT736	2002	RS2365	2002	TA1772
2002	GC452	2002	NKT751	2002	RS2366	2002	TF70
2002	GC453	2002	NKT752	2002	RS2375	2002	TF71
2002	GC454	2002	NKT753	2002	RS3306	2002	TF72
2002	GC465	2002	NKT773	2002	RS8931	2002	TIX896
2002	GC467	2002	NKT781	2002	RS8407	2002	TK33C
2002	GC608	2002	NR-10	2002	RS8420	2002	TM101
2002	GC609	2002	NR05	2002	RS8441	2002	TM103
2002	GE-5	2002	NR10	2002	RS8443	2002	TM103A
2002	GE-5(59)	2002	NR20	2002	RS8445	2002	TNJ61671
2002	GE-6	2002	NR30	2002	RT-119	2002	TNJ61671(28D72)
2002	GE-7	2002	NR5	2002	RT-122	2002	TNJ61734
2002	GE-8	2002	NR700	2002	RT3096	2002	TNJ72284
2002	GE-X8	2002	OO139	2002	RT7400	2002	TP4274
2002	GEX8	2002	OO140	2002	RT7944	2002	TQ5031
2002	G15	2002	OO141	2002	S-1453	2002	TQ5032
2002	G16	2002	PBE3020-2	2002	S-95202	2002	TQ5039
2002	G16506	2002	PTC108	2002	S028	2002	TQ5044
2002	G17	2002	PTC134	2002	S2042634	2002	TQ5050
2002	G18	2002	Q-2	2002	S65-2-A-3P	2002	TQ5062
2002	GT-1200	2002	Q-5	2002	S65-4-A-3P	2002	TR-08
2002	GT-35	2002	Q-9	2002	SA-7	2002	TR-08C
2002	GT-903	2002	QOV60527	2002	SA354B	2002	TR-09
2002	GT-905	2002	QOV60537	2002	SA7	2002	TR-09C
2002	GT-947	2002	Q2N4105	2002	SC-56	2002	TR-10
2002	GT1200	2002	Q5039	2002	SC56	2002	TR-10C
2002	GT1201	2002	Q5050	2002	SE-7001	2002	TR-159(OLSON)
2002	GT1202	2002	QOV60527	2002	SF.T184	2002	TR-160(OLSON)
2002	GT1608	2002	QOV60527	2002	SF.T377	2002	TRO1065
2002	GT1609	2002	QQV61772	2002	SPT-184	2002	TRO8
2002	GT1658	2002	QRT-119	2002	SPT-298	2002	TRO8C
2002	GT167	2002	QRT-122	2002	SPT184	2002	TRO9
2002	GT229	2002	R-125	2002	SPT259	2002	TRO9C
2002	GT2765	2002	R-135	2002	SPT260	2002	TR10
2002	GT2766	2002	R-136	2002	SPT261	2002	TR10C
2002	GT2767	2002	R-137	2002	SPT298	2002	TR167
2002	GT2768	2002	R-1533	2002	SPT377	2002	TR182
2002	GT2884	2002	R-202	2002	SK-7	2002	TR183
2002	GT2886	2002	R-203	2002	SK3010	2002	TR184
2002	GT2888	2002	R-28D187	2002	SK3011	2002	TR193
2002	GT2906	2002	R-33	2002	SK7	2002	TR194
2002	GT3150	2002	R-34	2002	SN60	2002	TR211
2002	GT336	2002	R-62	2002	SN80	2002	TR212
2002	GT35	2002	R-63	2002	SNW-Q-2	2002	TR213
2002	GT364	2002	R117	2002	SNW-Q-3	2002	TR214
2002	GT365	2002	R12	2002	SNW-Q-5	2002	TR216
2002	GT366	2002	R125	2002	SP2158	2002	TR310160
2002	GT792	2002	R135	2002	SP2188	2002	TR310236
2002	GT903	2002	R136	2002	SQ-7	2002	TR335
2002	GT904	2002	R137	2002	SQ7	2002	TR336
2002	GT905	2002	R14	2002	ST-172	2002	TR337
2002	GT905R	2002	R1530	2002	ST172	2002	TR338
2002	GT947	2002	R1531	2002	SY101	2002	TV-27
2002	GT948	2002	R1532	2002	SYL-101	2002	TV24143
2002	GT948R	2002	R1533	2002	SYL-102	2002	TV24983
2002	GT949	2002	R1534	2002	SYL-103	2002	TV27
2002	GT949R	2002	R1537	2002	SYL-104	2002	TVS28C647A
2002	HA5001	2002	R1538	2002	SYL-1297	2002	W2
2002	HA5002	2002	R1545	2002	SYL-1310	2002	W4
2002	HA5003	2002	R1547	2002	SYL-1311	2002	WC19862
2002	HA5005	2002	R1549	2002	SYL-1329	2002	WEP641
2002	HA5009	2002	R1553	2002	SYL-1396	2002	WEP641A
2002	HA5010	2002	R177	2002	SYL-152	2002	WEP641B
2002	HA5011	2002	R202	2002	SYL-1524	2002	WTV-L6
2002	HA5012	2002	R203	2002	SYL-1987	2002	WTVL6
2002	HA5014	2002	R2356	2002	SYL-2130	2002	WTVSA7
2002	HA5016	2002	R2359	2002	SYL-2131	2002	WTVSK7
2002	HA5020	2002	R2360	2002	SYL-2132	2002	WTVSQ7
2002	HA5021	2002	R2364	2002	SYL-2134	2002	XAT01
2002	HA5022	2002	R2365	2002	SYL-2135	2002	XAT02
2002	HA5023	2002	R2374	2002	SYL-2136	2002	XAT03
2002	HA5024	2002	R2375	2002	SYL-2245	2002	XB4
2002	HA5025	2002	R3293	2002	SYL-2246	2002	XG28
2002	HA5026	2002	R33	2002	SYL-4315	2002	XG29
2002	HC-00730	2002	R34	2002	SYL-4339	2002	XG33
2002	HC00730	2002	R3573-1	2002	SYL102	2002	XNC101
2002	HD-00072	2002	R41	2002	SYL103	2002	XS26
2002	HD-187	2002	R5050	2002	SYL104	2002	ZEN315
2002	HEP641	2002	R5054	2002	SYL1279	2002	001-011101-Q
2002	HEP00011	2002	R5056	2002	SYL1297	2002	001-011010
2002	HTO40519C	2002	R5179	2002	SYL1310	2002	1-801-309
2002	HTO40519C(28D72)	2002	R5180	2002	SYL1311	2002	1-TR-112
2002	HT340519C	2002	R592	2002	SYL1312	2002	002-011700
2002	HT400721	2002	R61	2002	SYL1313	2002	002-11700
2002	HT400721A			2002	SYL1326	2002	2-56
2002	HT400721B			2002	SYL1327	2002	2G339
						2002	2G339A

RS 276-	Industry Standard No.	RS 276-	Industry Standard No.	RS 276-	Industry Standard No.	RS 276-	Industry Standard No.
2002	2N78	2002	2N306#	2002	2SC129Y	2002	2SD178K
2002	2N100	2002	2N306A	2002	2SC13	2002	2SD178L
2002	2N1000	2002	2N312	2002	2SC14	2002	2SD178M
2002	2N1010	2002	2N313	2002	2SC173	2002	2SD178Q
2002	2N1010#	2002	2N314	2002	2SC175	2002	2SD178T
2002	2N1012	2002	2N35	2002	2SC175B	2002	2SD178X
2002	2N102	2002	2N356	2002	2SC176	2002	2SD178Y
2002	2N103	2002	2N356A	2002	2SC177	2002	2SD186
2002	2N1058	2002	2N357	2002	2SC178	2002	2SD186A
2002	2N1059	2002	2N357A	2002	2SC179	2002	2SD186B
2002	2N1059-1	2002	2N358	2002	2SC179A	2002	2SD187
2002	2N1086	2002	2N358A	2002	2SC179B	2002	2SD187-OR
2002	2N1086A	2002	2N364	2002	2SC179C	2002	2SD187A
2002	2N1087	2002	2N364#	2002	2SC179D	2002	2SD187B
2002	2N1090	2002	2N364#	2002	2SC179F	2002	2SD187C
2002	2N1091	2002	2N365	2002	2SC179G	2002	2SD187D
2002	2N1095	2002	2N365#	2002	2SC179GN	2002	2SD187E
2002	2N1096	2002	2N365#	2002	2SC179H	2002	2SD187F
2002	2N1101	2002	2N366	2002	2SC179J	2002	2SD187G
2002	2N1102	2002	2N366#	2002	2SC179K	2002	2SD187GN
2002	2N1102/5	2002	2N366#	2002	2SC179L	2002	2SD187H
2002	2N1112	2002	2N377	2002	2SC179M	2002	2SD187J
2002	2N1114	2002	2N377A	2002	2SC179OR	2002	2SD187K
2002	2N1169	2002	2N385	2002	2SC179R	2002	2SD187L
2002	2N1170	2002	2N385A	2002	2SC179X	2002	2SD187M
2002	2N1173	2002	2N388	2002	2SC179Y	2002	2SD187OR
2002	2N1173W	2002	2N388A	2002	2SC180A	2002	2SD187R
2002	2N1198	2002	2N4105	2002	2SC180B	2002	2SD187X
2002	2N1217	2002	2N438	2002	2SC180C	2002	2SD187Y
2002	2N124	2002	2N439	2002	2SC180D	2002	2SD19
2002	2N125	2002	2N440	2002	2SC180F	2002	2SD191
2002	2N1251	2002	2N444	2002	2SC180G	2002	2SD192
2002	2N126	2002	2N444#	2002	2SC180GN	2002	2SD193
2002	2N127	2002	2N444#	2002	2SC180H	2002	2SD194
2002	2N1288	2002	2N444A	2002	2SC180J	2002	2SD195
2002	2N1289	2002	2N445	2002	2SC180K	2002	2SD195A
2002	2N1299	2002	2N445#	2002	2SC180L	2002	2SD20
2002	2N1302	2002	2N445#	2002	2SC180M	2002	2SD21
2002	2N1304	2002	2N445A	2002	2SC180OR	2002	2SD215
2002	2N1306	2002	2N446	2002	2SC180R	2002	2SD22
2002	2N1308	2002	2N446A	2002	2SC180X	2002	2SD23
2002	2N1310	2002	2N447	2002	2SC180Y	2002	2SD25
2002	2N1311	2002	2N447#	2002	2SC181	2002	2SD30
2002	2N1312	2002	2N447A	2002	2SC181A	2002	2SD30-O
2002	2N1366	2002	2N447B	2002	2SC181B	2002	2SD30-OR
2002	2N1367	2002	2N448	2002	2SC181C	2002	2SD30-N
2002	2N1391	2002	2N449	2002	2SC181D	2002	2SD30-O
2002	2N1431	2002	2N507	2002	2SC181E	2002	2SD30-OR
2002	2N145	2002	2N515	2002	2SC181F	2002	2SD30A
2002	2N146	2002	2N516	2002	2SC181G	2002	2SD30B
2002	2N147	2002	2N517	2002	2SC181GN	2002	2SD30C
2002	2N1473	2002	2N529/N	2002	2SC181H	2002	2SD30D
2002	2N148A	2002	2N530/N	2002	2SC181J	2002	2SD30E
2002	2N148B	2002	2N531/N	2002	2SC181L	2002	2SD30F
2002	2N149	2002	2N532/N	2002	2SC181M	2002	2SD30G
2002	2N149A	2002	2N533/N	2002	2SC181OR	2002	2SD30GN
2002	2N150	2002	2N556	2002	2SC181R	2002	2SD30H
2002	2N150A	2002	2N557	2002	2SC181X	2002	2SD30J
2002	2N1510	2002	2N558	2002	2SC181Y	2002	2SD30K
2002	2N1585	2002	2N567	2002	2SC187(I)	2002	2SD30L
2002	2N1605	2002	2N576	2002	2SC187-OR	2002	2SD30M
2002	2N1605-A	2002	2N576A	2002	2SC187A	2002	2SD30N
2002	2N1605A	2002	2N585	2002	2SC187B	2002	2SD30OR
2002	2N1622	2002	2N587	2002	2SC187C	2002	2SD30P
2002	2N1624	2002	2N594	2002	2SC187D	2002	2SD30R
2002	2N164	2002	2N595	2002	2SC187E	2002	2SD30X
2002	2N164A	2002	2N596	2002	2SC187F	2002	2SD30Y
2002	2N165	2002	2N625	2002	2SC187G	2002	2SD31
2002	2N166	2002	2N634	2002	2SC187H	2002	2SD31D
2002	2N167	2002	2N634A	2002	2SC187I	2002	2SD32
2002	2N1672	2002	2N635	2002	2SC187J	2002	2SD33
2002	2N1672A	2002	2N635A	2002	2SC187K	2002	2SD34
2002	2N167A	2002	2N636	2002	2SC187L	2002	2SD35
2002	2N168	2002	2N636A	2002	2SC187OR	2002	2SD352
2002	2N1685	2002	2N646	2002	2SC187R	2002	2SD352B
2002	2N168A	2002	2N647	2002	2SC187X	2002	2SD352E
2002	2N169	2002	2N647/22	2002	2SC187Y	2002	2SD352F
2002	2N169A	2002	2N648	2002	2SC277C	2002	2SD36
2002	2N1694	2002	2N649	2002	2SC34	2002	2SD367
2002	2N169A	2002	2N649/22	2002	2SC35	2002	2SD367A
2002	2N170	2002	2N649/22	2002	2SC36	2002	2SD367B
2002	2N171	2002	2N649/5	2002	2SC50	2002	2SD367C
2002	2N172	2002	2N679	2002	2SC50A	2002	2SD367D
2002	2N1730	2002	2N78	2002	2SC60	2002	2SD367E
2002	2N1752	2002	2N78A	2002	2SC71	2002	2SD367F
2002	2N1779	2002	2N797	2002	2SC72	2002	2SD367P
2002	2N1780	2002	2N821	2002	2SC73	2002	2SD37
2002	2N1781	2002	2N822	2002	2SC75	2002	2SD37A
2002	2N1783	2002	2N823	2002	2SC76	2002	2SD37B
2002	2N1808	2002	2N824	2002	2SC77	2002	2SD37C
2002	2N182	2002	2N94	2002	2SC77B	2002	2SD58
2002	2N182#	2002	2N94A	2002	2SC77C	2002	2SD58Y
2002	2N183	2002	2N955	2002	2SC79	2002	2SD43
2002	2N185#	2002	2N955A	2002	2SC89	2002	2SD43A
2002	2N184	2002	2N97	2002	2SC90	2002	2SD44
2002	2N1858	2002	2N97A	2002	2SC91	2002	2SD61
2002	2N1891	2002	2N98	2002	2SD-P1	2002	2SD62
2002	2N1892	2002	2N98A	2002	2SD100	2002	2SD63
2002	2N193	2002	2N99	2002	2SD100A	2002	2SD64
2002	2N194	2002	2SC11	2002	2SD101	2002	2SD65
2002	2N194A	2002	2SC128	2002	2SD104	2002	2SD65-1
2002	2N1993	2002	2SC128A	2002	2SD105	2002	2SD65A
2002	2N1994	2002	2SC128B	2002	2SD127	2002	2SD65B
2002	2N1995	2002	2SC128C	2002	2SD127A	2002	2SD65C
2002	2N1996	2002	2SC128D	2002	2SD128	2002	2SD65D
2002	2N2085	2002	2SC128E	2002	2SD128A	2002	2SD65E
2002	2N211	2002	2SC128F	2002	2SD161	2002	2SD65F
2002	2N212	2002	2SC128G	2002	2SD162	2002	2SD65G
2002	2N213	2002	2SC128GN	2002	2SD167	2002	2SD65GN
2002	2N213A	2002	2SC128H	2002	2SD170	2002	2SD65H
2002	2N214	2002	2SC128J	2002	2SD170A	2002	2SD65J
2002	2N214A	2002	2SC128K	2002	2SD170AA	2002	2SD65K
2002	2N216	2002	2SC128L	2002	2SD170AB	2002	2SD65L
2002	2N228	2002	2SC128M	2002	2SD170AC	2002	2SD65M
2002	2N229	2002	2SC128OR	2002	2SD170B	2002	2SD65OR
2002	2N233	2002	2SC128R	2002	2SD170BC	2002	2SD65R
2002	2N233#	2002	2SC128X	2002	2SD170P	2002	2SD65X
2002	2N233A	2002	2SC128Y	2002	2SD170PB	2002	2SD65Y
2002	2N233A#	2002	2SC129	2002	2SD178	2002	2SD66
2002	2N2345	2002	2SC129A	2002	2SD178B	2002	2SD72
2002	2N2354	2002	2SC129B	2002	2SD178D	2002	2SD72-2C
2002	2N2426	2002	2SC129C	2002	2SD178F	2002	2SD72-3C
2002	2N2430	2002	2SC129D	2002	2SD178G	2002	2SD72-4C
2002	2N2482	2002	2SC129E	2002	2SD178	2002	2SD72-6
2002	2N253	2002	2SC129F	2002	2SD178B	2002	2SD72-OR
2002	2N254	2002	2SC129GN	2002	2SD178D	2002	2SD720D
2002	2N2699	2002	2SC129H	2002	2SD178F	2002	2SD720E
2002	2N28	2002	2SC129J	2002	2SD178	2002	2SD72A
2002	2N29	2002	2SC129K	2002	2SD178F	2002	2SD72B
2002	2N292	2002	2SC129L	2002	2SD178G	2002	2SD72BR
2002	2N292A	2002	2SC129M	2002	2SD178GN	2002	2SD72C
2002	2N293	2002	2SC129OR	2002	2SD178H	2002	2SD72D
2002	2N294A	2002	2SC129R	2002	2SD178J	2002	2SD72E
2002	2N306	2002	2SC129X	2002		2002	2SD72F

RS 276-	Industry Standard No.	RS 276-	Industry Standard No.	RS 276-	Industry Standard No.	RS 276-	Industry Standard No.
2002	28D72G	2002	8-905-605-384	2002	63-7248	2002	99807
2002	28D72GA	2002	8-905-605-390	2002	63-7549	2002	998A7
2002	28D72H	2002	8-905-613-015	2002	63-7565	2002	998K5
2002	28D72J	2002	8-905-613-062	2002	63-8473	2002	998K7
2002	28D72K	2002	8Q-3-04	2002	63-8705	2002	998Q7
2002	28D72L	2002	09-033006	2002	63-9340	2002	114N4U
2002	28D72L	2002	09-303006	2002	065-2	2002	116-687
2002	28D72M	2002	09-303012	2002	065-2-12	2002	120-004888
2002	28D72P	2002	09-303013	2002	65-2-70	2002	121-100
2002	28D72R	2002	09-303023	2002	65-2-70-12	2002	121-1410
2002	28D72RE	2002	09-303030	2002	65-2-70-12-7	2002	121-15
2002	28D72X	2002	09-309075	2002	65-20	2002	121-16
2002	28D72Y	2002	9-5112	2002	65-21	2002	121-17
2002	28D735	2002	9-5113	2002	65-22	2002	121-21
2002	28D75	2002	9-5114	2002	65-23	2002	121-22
2002	28D75A	2002	9-5202	2002	65-24	2002	121-237
2002	28D75AH	2002	9-5222-2	2002	65-25	2002	121-238
2002	28D75B	2002	9-5224-2	2002	65-26	2002	121-24
2002	28D75C	2002	12AA2	2002	65-27	2002	121-247
2002	28D75H	2002	13-14279-1	2002	65-28	2002	121-248
2002	28D77	2002	13-18654-1	2002	65-29	2002	121-25
2002	28D77-A	2002	13-27050-1	2002	65-2A	2002	121-26
2002	28D77A	2002	13-86416-1	2002	65-2AO	2002	121-302
2002	28D77AH	2002	13-86420-1	2002	65-2AOR	2002	121-50
2002	28D77B	2002	13-87433-1	2002	65-2A1	2002	121-51
2002	28D77C	2002	14-602-21	2002	65-2A19	2002	121-557
2002	28D77D	2002	14-602-52	2002	65-2A2	2002	121-558
2002	28D77E	2002	16A1	2002	65-2A21	2002	121-59
2002	28D77F	2002	16A2	2002	65-2A3	2002	121-6
2002	28D77G	2002	019-003317	2002	65-2A3P	2002	121-60
2002	28D77GN	2002	019-003318	2002	65-2A4	2002	121-641
2002	28D77H	2002	019-003319	2002	65-2A4-7	2002	121-650
2002	28D77J	2002	19A115103-P1	2002	65-2A4-7B	2002	121-673
2002	28D77L	2002	19A115129-2	2002	65-2A5	2002	121-7
2002	28D77M	2002	19A115129-P1	2002	65-2A5L	2002	121-70
2002	28D77OR	2002	19A115201-P1	2002	65-2A6	2002	121-71
2002	28D77P	2002	19A115201-P2	2002	65-2A6-1	2002	121-762
2002	28D77R	2002	19A115546-P1	2002	65-2A7	2002	121-762CL
2002	28D77X	2002	19A115546-P2	2002	65-2A7-1	2002	121-8
2002	28D77Y	2002	19A115673-P1	2002	65-2A8	2002	121-Z9031
2002	28D96	2002	19A115673-P2	2002	65-2A82	2002	122-1962
2002	28DF1	2002	19AR6-1	2002	65-2A9	2002	124-#16
2002	28DF1A	2002	19AR6-2	2002	65-2A9Q	2002	124N1
2002	2T51	2002	19AR6-3	2002	65-4	2002	126N1
2002	2T513	2002	19B200065-P1	2002	65-4-70	2002	126N2
2002	2T52	2002	19B200065-P2	2002	65-4-70-12	2002	127-7
2002	2T520	2002	21A015-005	2002	65-4-70-12-7	2002	129-30
2002	2T521	2002	022-2876-002	2002	65-40	2002	130-40089
2002	2T522	2002	24MW1116(NPN)	2002	65-41	2002	130-40096
2002	2T523	2002	24MW130	2002	65-42	2002	130-40314
2002	2T524	2002	27T408	2002	65-43	2002	130-40347
2002	2T53	2002	27T410	2002	65-44	2002	151-0040
2002	2T54	2002	030-034-0	2002	65-45	2002	151-0040-00
2002	2T55	2002	34-6001-84	2002	65-46	2002	151-0238
2002	2T551	2002	42-19862	2002	65-47	2002	165-2A82
2002	2T552	2002	42-21404	2002	65-48	2002	165-4A82
2002	2T56	2002	42X210	2002	65-49	2002	174-002-9-001
2002	2T57	2002	42X310	2002	65-4A	2002	186-4-127
2002	2T58	2002	45N1	2002	65-4AO	2002	200A
2002	2T61	2002	45N2	2002	65-4AOR	2002	202A
2002	2T62	2002	45N2A	2002	65-4A1	2002	247-256
2002	2T63	2002	46-86115-3	2002	65-4A19	2002	251M1
2002	2T64	2002	46-8614-3	2002	65-4A2	2002	297L001H01
2002	2T64R	2002	46-86211-3	2002	65-4A21	2002	297L001H02
2002	2T65	2002	46-8671-3	2002	65-4A3	2002	297L001M01
2002	2T650	2002	48-124216	2002	65-4A3P	2002	297V002H03
2002	2T65R	2002	48-124217	2002	65-4A4	2002	297V002H04
2002	2T66	2002	48-124218	2002	65-4A4-7	2002	297V002H05
2002	2T66R	2002	48-124220	2002	65-4A4-7B	2002	297V002M04
2002	2T67	2002	48-124221	2002	65-4A5	2002	297V002M05
2002	2T681	2002	48-125233	2002	65-4A5L	2002	324-0122
2002	2T682	2002	48-125234	2002	65-4A6	2002	324-0134
2002	2T69	2002	48-125235	2002	65-4A6-2	2002	324-3102-1
2002	2T71	2002	48-125236	2002	65-4A7	2002	417-121
2002	2T72	2002	48-128239	2002	65-4A7-1	2002	421-9862
2002	2T73	2002	48-134520	2002	65-4A8	2002	465-2A5
2002	2T73R	2002	48-13466	2002	65-4A82	2002	465-4A5
2002	2T74	2002	48-134700	2002	65-4A9	2002	576-0002-013
2002	2T75	2002	48-134931	2002	65-4A9Q	2002	601-065
2002	2T75R	2002	48-869092	2002	80-050300	2002	614X8
2002	2T76	2002	48-869093	2002	80-050600	2002	642-277
2002	2T76R	2002	48-869254	2002	80-052800	2002	650-109
2002	2T77	2002	48-869283	2002	86-10-2	2002	660-228
2002	2T77R	2002	48-869476	2002	86-11-2	2002	665-2A5L
2002	2T78	2002	48-869476A	2002	86-12-2	2002	665-4A5L
2002	2T78R	2002	48C125233	2002	86-13-2	2002	690V067H35
2002	2T82	2002	48C125235	2002	86-14-2	2002	690V081B96
2002	2T83	2002	48C125236	2002	86-24-2	2002	690V102H39
2002	2T84	2002	48K45N2	2002	86-25-2	2002	698V102H39
2002	2T85	2002	48P63078A86(NPN)	2002	86-26-2	2002	800-310(CP)(NPN)
2002	2T85A	2002	48R134665	2002	86-301-2	2002	800-310(CP)(PNP)
2002	2T86	2002	48R869092	2002	86-31-2	2002	800-50300
2002	2T89	2002	48R869093	2002	86-35-2	2002	800-506-00
2002	3N23	2002	48R869254	2002	86-4-2	2002	800-50600
2002	3N23A	2002	48R869283	2002	86-44-2	2002	800-528
2002	3N23B	2002	48R869476	2002	86-5-2	2002	800-528-00
2002	3N23C	2002	48R869476A	2002	86-5003-2	2002	800-52800
2002	3N29	2002	55-1027	2002	86-5004-2	2002	800-537-01
2002	3N30	2002	55-1032	2002	86-5005-2	2002	808-309
2002	3N31	2002	56-8100	2002	86-5007-2	2002	921-46B
2002	3N36	2002	56-8100A	2002	86-5008-2	2002	921-54B
2002	3N37	2002	56-8100B	2002	86-5011-2	2002	921-5A
2002	3T201	2002	56-8100C	2002	86-5012-2	2002	921-5B
2002	3T202	2002	56-8100D	2002	86-5013-2	2002	921-6A
2002	3T203	2002	57A1-3	2002	86-5015-2	2002	921-6B
2002	04-00072-01	2002	57A1-4	2002	86-5016-2	2002	921-71
2002	4-65-2A7-1	2002	57A1-5	2002	86-5017-2	2002	921-71A
2002	4-65-4A7-1	2002	57A1-6	2002	86-5026-2	2002	921-71B
2002	4JD5B1	2002	57A1-78	2002	86-5029-2	2002	921-7B
2002	4JX16A569	2002	57A5	2002	86-5034-2	2002	965-2A6-1
2002	4JX1EB50	2002	57A6-20	2002	86-5047-2	2002	965-4A6-2
2002	4JX2816	2002	57A6-21	2002	86-5048-2	2002	1033-6
2002	4JX2825	2002	57A6-5	2002	86-5060-2	2002	1034-43
2002	4JX2A601	2002	57A6-6	2002	86-5061-2	2002	1040-80
2002	4JX2A616	2002	57B2-5	2002	86-5062-2	2002	1102-63
2002	4JX2A801	2002	57B2-8	2002	86-5080-2	2002	1104-95
2002	4JX2A816	2002	57C5	2002	86-5086-2	2002	1119-58
2002	4JX2A822	2002	57C6-20	2002	86-5087-2	2002	1344-3767
2002	6-89X	2002	57C6-21	2002	86-6-2	2002	1349-17
2002	6A12992	2002	57C6-5	2002	86-76-2	2002	1371-17
2002	6A12993	2002	57C6-6	2002	86-81-2	2002	1396
2002	07-07167	2002	57D1-3	2002	86X037-001	2002	1465-2
2002	8-0050300	2002	57D1-4	2002	089-233	2002	1465-2-12
2002	8-0050600	2002	57D1-5	2002	95-112	2002	1465-2-12-8
2002	8-0052800	2002	57D1-6	2002	95-113	2002	1465-4
2002	8-723-650	2002	57D1-78	2002	95-114	2002	1465-4-12
2002	8-905-605-105	2002	57M1-16	2002	95-202	2002	1465-4-12-8
2002	8-905-605-108	2002	57M1-18	2002	95-222-2	2002	1515(PNP)
2002	8-905-605-109	2002	57M2-1	2002	95-224-2	2002	1858
2002	8-905-605-110	2002	57M2-2	2002	96-5205-01	2002	1865-2
2002	8-905-605-111	2002	57M2-6	2002	97N2	2002	1865-2-12
2002	8-905-605-112	2002	57M2-9	2002	97N2U	2002	1865-2L
2002	8-905-605-113	2002	62-13259	2002	99K7	2002	1865-2L8
2002	8-905-605-365	2002	63-10583	2002	99L6	2002	1865-4

RS 276-	Industry Standard No.	RS 276-	Industry Standard No.	RS 276-	Industry Standard No.	RS 276-	Industry Standard No.
2002	1865-4-12	2002	610126-2	2004	A376	2004	ACZ
2002	1865-4-127	2002	650860	2004	A377	2004	ACZ21
2002	1865-4L	2002	723000-18	2004	A378	2004	AF-105A
2002	1865-4LB	2002	723001-19	2004	A379	2004	AF-106
2002	1906-17	2002	815026	2004	A401	2004	AF-109
2002	1958-17	2002	815026A	2004	A403	2004	AF-121
2002	2013(E.F.JOHNSON)	2002	815026B	2004	A404	2004	AF-137
2002	2039-2	2002	815026C	2004	A405	2004	AF-166
2002	2057A100-30(NPN)	2002	815026D	2004	A408	2004	AF-182
2002	2057A100-35(NPN)	2002	815075	2004	A409	2004	AF101
2002	2057A2-167	2002	815218-4	2004	A41	2004	AF102
2002	2057B100-4	2002	815232	2004	A410	2004	AF105A
2002	2057B2-46	2002	985735	2004	A411	2004	AF106
2002	3607	2002	985735A	2004	A413	2004	AF106A
2002	3609	2002	985735A-1	2004	A417(JAPAN)	2004	AF107
2002	4700	2002	1475573-1	2004	A419(JAPAN)	2004	AF109
2002	4822-130-40096	2002	1815139	2004	A42	2004	AF109R
2002	4822-130-40314	2002	1817017	2004	A420(JAPAN)	2004	AF110
2002	5001-512	2002	2002153-83	2004	A421	2004	AF111
2002	5614-77C	2002	2010952-14	2004	A422	2004	AF112
2002	8000-00004-086	2002	2091260	2004	A425	2004	AF113
2002	8000-00011-086	2002	2091260-1	2004	A426	2004	AF114
2002	8000-00030-009	2002	2091260-1(NPN)	2004	A430(JAPAN)	2004	AF114N
2002	8000-00032-027	2002	2091260-2	2004	A431	2004	AF115
2002	8000-0004-086	2002	2091260-2(NPN)	2004	A431A	2004	AF115N
2002	8000-0004-P086	2002	2091260-3	2004	A432	2004	AF116
2002	11668-7	2002	2091260-3(NPN)	2004	A432A	2004	AF116N
2002	27125-310	2002	2092418-0711	2004	A433	2004	AF117
2002	27125-490	2002	2320331	2004	A434	2004	AF117C
2002	030812-1	2002-	28C180	2004	A435	2004	AF117N
2002	37279	2004	A-1384	2004	A435A	2004	AF118
2002	37552	2004	A01(TRANSISTOR)	2004	A435B	2004	AF119
2002	38175	2004	A059-100	2004	A440	2004	AF121
2002	38199	2004	A059-101	2004	A440A	2004	AF121B
2002	38200	2004	A059-102	2004	A448	2004	AF122
2002	40037	2004	A059-103	2004	A450	2004	AF124
2002	40037-1	2004	A105	2004	A450H	2004	AF125
2002	40037-2	2004	A107(JAPAN)	2004	A451(JAPAN)	2004	AF126
2002	40037-3	2004	A108(JAPAN)	2004	A451H	2004	AF127
2002	40037VM	2004	A117	2004	A452	2004	AF127/01
2002	40396	2004	A118	2004	A452H	2004	AF128
2002	40396N	2004	A122	2004	A460	2004	AF137A
2002	43992-2	2004	A1220	2004	A461	2004	AF138/20
2002	45495-2	2004	A126	2004	A462	2004	AF139
2002	46490-2	2004	A127	2004	A463	2004	AF142
2002	46590-2	2004	A13(TRANSISTOR)	2004	A464	2004	AF143
2002	46591-2	2004	A130(JAPAN)	2004	A506	2004	AF146
2002	46592-2	2004	A135	2004	A507	2004	AF147
2002	46593-2	2004	A1377	2004	A508	2004	AF148
2002	46631-2	2004	A1378	2004	A51	2004	AF149
2002	46774-1	2004	A14(TRANSISTOR)	2004	A525	2004	AF150
2002	46775-2	2004	A1462-19	2004	A525A	2004	AF164
2002	47645-2	2004	A1465-1	2004	A525B	2004	AF165
2002	48385-2	2004	A1465-19	2004	A56	2004	AF166
2002	49058-2	2004	A1465A	2004	A61	2004	AF167
2002	49138-2	2004	A1465A9	2004	A615-1008	2004	AF168
2002	61012-4-1	2004	A1465B	2004	A615-1009	2004	AF169
2002	72191	2004	A1465B9	2004	A65-1-1A9G	2004	AF170
2002	80817VM	2004	A1488B	2004	A65-1-70	2004	AF171
2002	81502-6	2004	A1488B9	2004	A65-1-705	2004	AF172
2002	81502-6A	2004	A14A8-1	2004	A65-1A9G	2004	AF178
2002	81502-6B	2004	A14A8-19	2004	A65A-70	2004	AF179
2002	81502-6C	2004	A14A8-19G	2004	A65A-705	2004	AF180
2002	81502-6D	2004	A151(JAPAN)	2004	A65A19G	2004	AF181
2002	81507-5	2004	A152(JAPAN)	2004	A70	2004	AF182
2002	81507-6	2004	A163	2004	A70F	2004	AF185
2002	91021	2004	OA180	2004	A70L	2004	AF186
2002	94023	2004	A215	2004	A70MA	2004	AF186A
2002	94029	2004	A218	2004	A74	2004	AF186W
2002	95112	2004	A219	2004	A79	2004	AF193
2002	95113	2004	A220(TRANSISTOR)	2004	OA8-1	2004	AF200
2002	95114	2004	A221	2004	OA8-1-12	2004	AF201
2002	95115	2004	A222	2004	OA8-1-12-7	2004	AF201C
2002	95117	2004	A223	2004	A8-1-70	2004	AF202
2002	95202	2004	A224	2004	A8-1-70-1	2004	AF202L
2002	95211	2004	A225	2004	A8-1-70-12	2004	AF202S
2002	95222-2	2004	A226	2004	A8-1-70-12-7	2004	AF239
2002	95224-2	2004	A227	2004	A8-1-A-4-7B	2004	AF239B
2002	95224-4	2004	A228	2004	A8-10	2004	AF240
2002	101678	2004	A229	2004	A8-11	2004	AF251
2002	103443	2004	A230	2004	A8-12	2004	AF252
2002	104080	2004	A233	2004	A8-13	2004	AF253
2002	110495	2004	A233A	2004	A8-14	2004	AF256
2002	110958	2004	A233B	2004	A8-15	2004	AF267
2002	111958	2004	A233C	2004	A8-16	2004	AF279
2002	116204	2004	A234	2004	A8-17	2004	AF306
2002	116687	2004	A234A	2004	A8-18	2004	AF712
2002	122061	2004	A235	2004	A8-19	2004	AFY14
2002	122111	2004	A235B	2004	A8-1A	2004	AFY15
2002	122112	2004	A239	2004	A8-1AO	2004	AFY16
2002	130400-96	2004	A240	2004	A8-1AOR	2004	AFY18C
2002	130403-47	2004	A240A	2004	A8-1A1	2004	AFY18D
2002	168953	2004	A240B	2004	A8-1A19	2004	AFY18E
2002	170132-1(NPN)	2004	A240B2	2004	A8-1A2	2004	AFY19
2002	170783-3	2004	A240BL	2004	A8-1A21	2004	AFY34
2002	171005(TOSHIBA)	2004	A241	2004	A8-1A3	2004	AFY37
2002	198794-1	2004	A242	2004	A8-1A3P	2004	AFY39
2002	219016	2004	A243	2004	A8-1A4	2004	AFY40
2002	221601	2004	A244(JAPAN)	2004	A8-1A4-7	2004	AFY40K
2002	221924	2004	A245(JAPAN)	2004	A8-1A4-7B	2004	AFY40R
2002	222367	2004	A276(JAPAN)	2004	A8-1A5	2004	AFY41
2002	223568	2004	A285	2004	A8-1A5L	2004	AFY42
2002	223570	2004	A286	2004	A8-1A6	2004	AFZ11
2002	223482	2004	A287	2004	A8-1A6-4	2004	AFZ12
2002	223684	2004	A296	2004	A8-1A7	2004	AL210
2002	224820	2004	A297	2004	A8-1A7-1	2004	APB-11H-1001
2002	225300	2004	A298	2004	A8-1A8	2004	APB-11H-1004
2002	226441	2004	A306(JAPAN)	2004	A8-1A82	2004	AR103
2002	226791	2004	A307(JAPAN)	2004	A8-1A9	2004	AR104
2002	230209	2004	A321(JAPAN)	2004	A8-1A9G	2004	AR105
2002	230256	2004	A322	2004	A84	2004	AS34280
2002	231140-45	2004	A322(JAPAN)	2004	A86	2004	ASY-24
2002	232949	2004	A323(JAPAN)	2004	A88-70	2004	ASY63
2002	236265	2004	A324(JAPAN)	2004	A88-705	2004	ASY63N
2002	243837	2004	A327(TRANSISTOR)	2004	A88B-70	2004	ASY67
2002	250400	2004	A340	2004	A88B-705	2004	AS220
2002	256127	2004	A341	2004	A88B19G	2004	AS220N
2002	257385	2004	A341-0A	2004	A89	2004	AS221
2002	258993	2004	A341-0B	2004	A909-1008	2004	AT-1
2002	260468	2004	A341-0B	2004	AA3	2004	AT-14
2002	269367	2004	A342	2004	AA8-1-70	2004	AT-2
2002	270781	2004	A342A	2004	AA8-1-705	2004	AT-3
2002	279317	2004	A345(JAPAN)	2004	AA8-1A9G	2004	AT-4
2002	300486	2004	A346(JAPAN)	2004	AC107	2004	AT-6
2002	300536	2004	A347	2004	AC107M	2004	AT-8
2002	300542	2004	A348	2004	AC107N	2004	AT-9
2002	300774	2004	A349	2004	AC129	2004	AT/RP1
2002	0573037	2004	A360	2004	AC150-0RN	2004	AT/RP2
2002	0573037M	2004	A361	2004	AC150-YEL	2004	AT/813
2002	0573139	2004	A362	2004	AC160-0RN	2004	AT13
2002	581024	2004	A363	2004	AC160-RED	2004	AT14
2002	601065	2004	A372	2004	AC160-VIO	2004	AT15
2002	607030	2004	A373	2004	AC160-YEL	2004	AT16
2002	610124-1	2004	A375	2004	ACY24	2004	AT17

RS 276-	Industry Standard No.	RS 276-	Industry Standard No.	RS 276-	Industry Standard No.	RS 276-	Industry Standard No.
2004	AT4	2004	FS2299	2004	HJ60A	2004	MD836
2004	AT5	2004	G0002	2004	HJ60C	2004	MD837
2004	B240	2004	013	2004	HJ70	2004	MD838
2004	B240A	2004	GC1003	2004	HJ75	2004	MD839
2004	B51	2004	GC1004	2004	HR-20	2004	MD840
2004	B601-1006	2004	GC1005	2004	HR-20A	2004	MM1139
2004	B601-1007	2004	GC1006	2004	HR-21	2004	MM1199
2004	B601-1008	2004	GC1007	2004	HR-21A	2004	MM2503
2004	B65-1-A-21	2004	GC1092	2004	HR-22	2004	MM2550
2004	B65-1A21	2004	GC1093	2004	HR-22A	2004	MM2552
2004	B65A-1-21	2004	GC1093X3	2004	HR-22B	2004	MM2554
2004	B65B-1-21	2004	GC1142	2004	HR-24	2004	MM2894
2004	B74	2004	GC1146	2004	HR-24A	2004	MM500
2004	B88B-1-21	2004	GC1148	2004	HR-25	2004	MM5000
2004	BA8-1A-21	2004	GC1149	2004	HR-25A	2004	MM5001
2004	BCM1002-1	2004	GC1155	2004	HR-26	2004	MM5002
2004	BCY501	2004	GC1182	2004	HR-26A	2004	MPS1097
2004	BCY511	2004	GC1573	2004	HR-27	2004	MP99600-5
2004	BF5263	2004	GC282	2004	HR-27A	2004	MT102351A
2004	C10215-2	2004	GC283	2004	HR-40	2004	N-020
2004	C10258	2004	GC284	2004	HR-43	2004	N57B2-11
2004	C10260	2004	GC387	2004	HR-45	2004	N57B2-13
2004	C10261	2004	GC388	2004	HR2	2004	N57B2-14
2004	C10262	2004	GC389	2004	HR20	2004	N57B2-22
2004	C125	2004	GC460	2004	HR20A	2004	NA-1114-1001
2004	CB156	2004	GC461	2004	HR21	2004	NA-1114-1002
2004	CB244	2004	GC462	2004	HR21A	2004	NA1022-1001
2004	CB254	2004	GC630	2004	HR22	2004	NA5018-1001
2004	CGB-50	2004	GC630A	2004	HR22A	2004	NK1302
2004	CGB-51	2004	GC631	2004	HR22B	2004	NK1404
2004	CXL160	2004	GE-208	2004	HR24	2004	NK65-1A19
2004	DO63	2004	GE-245	2004	HR24A	2004	NK65A119
2004	DO73	2004	GE-50	2004	HR25	2004	NK65B119
2004	DO79	2004	GE-51	2004	HR25A	2004	NK48-1A19
2004	DO86	2004	GE-9	2004	HR26	2004	NK121
2004	D149	2004	GE-9A	2004	HR26A	2004	NK2122
2004	D172	2004	GE-M100	2004	HR27	2004	NK2124
2004	D173	2004	GER-A-D	2004	HR27A	2004	NK2125
2004	D174	2004	GET-672	2004	HR40	2004	NK2251
2004	DO63	2004	GET-672A	2004	HR40836	2004	NK2674P
2004	DO79	2004	GET-673	2004	HR40837	2004	NK2677P
2004	DO86	2004	GET-692	2004	HR41	2004	OC-169
2004	D8-34	2004	GET-873A	2004	HR42	2004	OC-170
2004	D8-35	2004	GET-883	2004	HR43	2004	OC-171
2004	D8-36	2004	GET5116	2004	HR43835	2004	OC-615
2004	D8-37	2004	GET5117	2004	HR44	2004	OC169
2004	D8-41	2004	GET671	2004	HR448636	2004	OC169R
2004	D8-42	2004	GET672	2004	HR45	2004	OC170
2004	D8-51	2004	GET672A	2004	HR45838	2004	OC170M
2004	D8-52	2004	GET673	2004	HR45910	2004	OC170R
2004	D8-56	2004	GET691	2004	HR45913	2004	OC170V
2004	D8-62	2004	GET692	2004	HR46	2004	OC171
2004	D8-63	2004	GET693	2004	HR50	2004	OC171N
2004	D8-64	2004	GPT44	2004	HR51	2004	OC171R
2004	D8-65	2004	GPT45	2004	HR52	2004	OC171V
2004	D834	2004	GGE-51	2004	HR7	2004	OC320
2004	D835	2004	GM0290	2004	HT102341	2004	OC400
2004	D837	2004	GM0375	2004	HT102341A	2004	OC410
2004	D838(DELCO)	2004	GM0376	2004	HT102341B	2004	OC53
2004	D841	2004	GM0377	2004	HT102341C	2004	OC54
2004	D842	2004	GM290	2004	HT102341UB	2004	OC55
2004	D856	2004	GM290A	2004	HT102351A	2004	OC615N
2004	D862	2004	GM2351	2004	HT103501	2004	ON174
2004	D863	2004	GM378A	2004	HT2005410	2004	PA10880
2004	D864	2004	GM378RED	2004	I12032	2004	PA9154
2004	D865	2004	GM656A	2004	IRTR06	2004	PA9155
2004	DU1	2004	GM875	2004	IRTR07	2004	PADT20
2004	DU12	2004	GM876	2004	IRTR12	2004	PADT21
2004	DU2	2004	GM877	2004	IRTR17	2004	PADT22
2004	E-2462	2004	GM878	2004	IRTR18	2004	PADT23
2004	E070	2004	GM878A	2004	IT205A	2004	PADT24
2004	E2438	2004	GM878B	2004	J24620	2004	PADT25
2004	E2439	2004	GMO290	2004	J24621	2004	PADT26
2004	E2440	2004	GMO375	2004	J24623	2004	PADT27
2004	E2450	2004	GMO376	2004	J310251	2004	PADT28
2004	E2462	2004	GMO377	2004	J5062	2004	PADT30
2004	E2474	2004	GMO378	2004	JP5062	2004	PADT31
2004	E2475	2004	GT1	2004	JR10	2004	PADT35
2004	E2477	2004	GT2	2004	JR100	2004	PADT40
2004	E2478	2004	GT20	2004	JR200	2004	PADT51
2004	E2479	2004	GT24H	2004	JR30	2004	PIL/4956
2004	EA0002	2004	GT3	2004	JR30X	2004	PQ27
2004	EA0007	2004	GT34	2004	K75508-1	2004	PQ30
2004	EA0053	2004	GT40	2004	K0825642-10	2004	PT2A
2004	EA1337	2004	GT760	2004	L2091241-2	2004	PT28
2004	EA1338	2004	HA-00354	2004	L2091241-3	2004	PT855
2004	EA1339	2004	HA-234	2004	L5108	2004	PT856
2004	EA1340	2004	HA-234B	2004	L5121	2004	PTC07
2004	EA1342	2004	HA-235A	2004	L5122	2004	PTC160
2004	EA15X11	2004	HA-235C	2004	L5181	2004	Q40359
2004	EA15X13	2004	HA-268	2004	M351	2004	Q5044
2004	EA15X27	2004	HA1040	2004	M4439	2004	QA-1
2004	EA15X29	2004	HA12	2004	M4484	2004	R-28A222
2004	EA15X30	2004	HA2190	2004	M4485	2004	R1539
2004	EA15X40	2004	HA2356	2004	M4486	2004	R1550
2004	EA15X41	2004	HA235A	2004	M4504	2004	R1554
2004	EA15X43	2004	HA240	2004	M4506	2004	R2683
2004	EA15X5	2004	HA266	2004	M4507	2004	R2664
2004	EA15X66	2004	HA267	2004	M4524	2004	R2685
2004	EA18Y11	2004	HA269	2004	M4526	2004	R2686
2004	ECG160	2004	HA30	2004	M4697	2004	R2687
2004	ED51	2004	HA3210	2004	M4860	2004	R2688
2004	ER15X11	2004	HA3480	2004	M75516-2	2004	R2694
2004	ER15X12	2004	HA350	2004	M75516-2P	2004	R2695
2004	ER15X13	2004	HA353	2004	M75516-2R	2004	R2696
2004	ER15X14	2004	HA354	2004	M75516-P	2004	R2697
2004	ER15X15	2004	HA3670	2004	M75162-P	2004	R3277
2004	ER15X16	2004	HA4400	2004	M75162-R	2004	R3278
2004	ER15X19	2004	HA525	2004	M76	2004	R3279
2004	ER15X20	2004	HA70	2004	M77	2004	R3287
2004	ER15X21	2004	HEP3	2004	M78	2004	R3288
2004	ER15X24	2004	HEP637	2004	M78A	2004	R3309
2004	ER15X25	2004	HEPG0002	2004	M78B	2004	R339
2004	ER15X26	2004	HEPG0003	2004	M78BLK	2004	R339
2004	ER15X4	2004	HP-35	2004	M78C	2004	R341
2004	ER15X5	2004	HP12H	2004	M78D	2004	R424
2004	ER15X6	2004	HP12M	2004	M78GRN	2004	R424-1
2004	ES31(ELCOM)	2004	HP12N	2004	M78RED	2004	R425
2004	ES311(ELCOM)	2004	HP20H	2004	M78YEL	2004	R497
2004	ES314(ELCOM)	2004	HP20M	2004	M8124	2004	R5102
2004	ES315(ELCOM)	2004	HP3H	2004	MA1	2004	R5103
2004	ES319(ELCOM)	2004	HP3M	2004	MA23A	2004	R515
2004	ES323(ELCOM)	2004	HP50H	2004	MC101	2004	R515A
2004	ES325(ELCOM)	2004	HP50M	2004	MC103	2004	R516(T.I.)
2004	ES341(ELCOM)	2004	HP6H	2004	MD420	2004	R516A
2004	ES88(ELCOM)	2004	HP6M	2004	MD831	2004	R539
2004	ET1	2004	HJ15D	2004	MD832	2004	R558
2004	ET15X29	2004	HJ32	2004	MD833	2004	R563
2004	ET2	2004	HJ34	2004	MD833A	2004	R564
2004	F73216	2004	HJ34A	2004	MD833C	2004	R565
2004	FB401	2004	HJ37	2004	MD833D	2004	R579
2004	FB402	2004	HJ55	2004	MD834	2004	R593
2004	FB403	2004	HJ56	2004		2004	R593A
2004	FB440	2004	HJ57	2004		2004	R60-1002

RS 276-	Industry Standard No.	RS 276-	Industry Standard No.	RS 276-	Industry Standard No.	RS 276-	Industry Standard No.
2004	R60-1003	2004	SB101	2004	T2028	2004	TIXM08
2004	R7885	2004	SB102	2004	T2029	2004	TIXM10
2004	R7886	2004	SB103	2004	T2030	2004	TIXM101
2004	R7891	2004	SC-71	2004	T2191	2004	TIXM103
2004	R7962	2004	SC-72	2004	T2364	2004	TIXM104
2004	R8240	2004	SC-74	2004	T2379	2004	TIXM105
2004	R8241	2004	SC-78	2004	T2384	2004	TIXM106
2004	R8242	2004	SC-79	2004	T253	2004	TIXM107
2004	R8559	2004	SC-80	2004	T253(SEARS)	2004	TIXM108
2004	R8685	2004	SC1007	2004	T278	2004	TIXM1
2004	R8686	2004	SC71	2004	T2788	2004	TIXM13
2004	R8692	2004	SC72	2004	T279	2004	TIXM14
2004	R8693	2004	SC74	2004	T280	2004	TIXM15
2004	R8694	2004	SC78	2004	T281	2004	TIXM16
2004	R8703	2004	SC79	2004	T282	2004	TIXM17
2004	R8704	2004	SC80	2004	T2878	2004	TIXM18
2004	R8705	2004	SS2400	2004	T2896	2004	TIXM9
2004	R8881	2004	SF.T163	2004	T2945	2004	TIXM201
2004	R8882	2004	SF.T316	2004	T2946	2004	TIXM202
2004	R9531	2004	SF.T317	2004	T348	2004	TIXM203
2004	R9532	2004	SF.T319	2004	T367	2004	TIXM204
2004	R9601	2004	SF.T320	2004	T368	2004	TIXM205
2004	R9602	2004	SF.T354	2004	T373	2004	TIXM206
2004	RE27	2004	SF.T357	2004	T374	2004	TIXM207
2004	REN160	2004	SF.T358	2004	T449	2004	TK1228-001
2004	REN69	2004	SFT-163	2004	T6058	2004	TK410
2004	R8-101	2004	SFT-358	2004	T811	2004	TK42C
2004	R8-103	2004	SFT120	2004	TA-1628	2004	TM160
2004	R8-2002	2004	SFT162	2004	TA-1658	2004	TNJ-60067
2004	R8-3892	2004	SFT163	2004	TA-1659	2004	TNJ-60068
2004	R8-3898	2004	SFT171	2004	TA-1660	2004	TNJ-60069
2004	R8-3900	2004	SFT172	2004	TA-1662	2004	TNJ-60071
2004	R8-3901	2004	SFT173	2004	TA-1731	2004	TNJ-60073
2004	R8-3902	2004	SFT174	2004	TA-1757	2004	TNJ-60077
2004	R8-3903	2004	SFT268	2004	TA-1796	2004	TNJ-60279
2004	R8-3911	2004	SFT315	2004	TA-1797	2004	TNJ-60280
2004	R8-3926	2004	SFT316	2004	TA-1798	2004	TNJ-60281
2004	R8-5208	2004	SFT358	2004	TA-1828	2004	TNJ-60362
2004	R81539	2004	SK3006	2004	TA-1829	2004	TNJ-60363
2004	R81550	2004	SK3007	2004	TA-1846	2004	TNJ-60364
2004	R81554	2004	SK3730	2004	TA-1847	2004	TNJ-60365
2004	R82679	2004	SK3770	2004	TA-1860	2004	TNJ-60608
2004	R82680	2004	SM1297	2004	TA-1861	2004	TNJ60063
2004	R82683	2004	SM1600	2004	TA-1990	2004	TNJ60064
2004	R82684	2004	SM217	2004	TA2322	2004	TNJ60065
2004	R82685	2004	SM2491	2004	TQ28A201(C)	2004	TNJ60067
2004	R82686	2004	SM2492	2004	TI-338	2004	TNJ60068
2004	R82687	2004	SM249I	2004	TI-387	2004	TNJ60069
2004	R82688	2004	SM3014	2004	TI-388	2004	TNJ60071
2004	R82694	2004	SMB454760	2004	TI-389	2004	TNJ60073
2004	R82695	2004	SO-1	2004	TI-400	2004	TNJ60077
2004	R83277	2004	SO-2	2004	TI-401	2004	TNJ60279
2004	R83278	2004	SO-3	2004	TI-403	2004	TNJ60280
2004	R83279	2004	SO-65A	2004	TI3010	2004	TNJ60281
2004	R83288	2004	S01	2004	TI3011	2004	TNJ60450
2004	R83309	2004	S02	2004	TI338	2004	TNJ60456
2004	R83322	2004	S03	2004	TI363	2004	TNJ70641
2004	R83323	2004	SP85328	2004	TI364	2004	TO-003
2004	R83324	2004	S8155	2004	TI365	2004	TO-004
2004	R83668	2004	ST-125	2004	TI387	2004	TS-5034
2004	R83862	2004	ST0T279	2004	TI388	2004	TQ5021
2004	R83863	2004	STX0036	2004	TI389	2004	TQ5022
2004	R83864	2004	STX0085	2004	TI390	2004	TQ5034
2004	R83868	2004	STX0087	2004	TI391	2004	TQ5035
2004	R83898	2004	STX0089	2004	TI393	2004	TQ5038
2004	R83900	2004	STX0090	2004	TI395	2004	TR-07
2004	R83901	2004	SWT1728	2004	TI396	2004	TR-11C
2004	R83902	2004	SWT3588	2004	TI397	2004	TR-12
2004	R83903	2004	STL2189	2004	TI398	2004	TR-12C
2004	R83905	2004	T-163	2004	TI399	2004	TR-13(RP)
2004	R83906	2004	T-2028	2004	TI400	2004	TR-13C
2004	R83907	2004	T-2029	2004	TI401	2004	TR-161(OLSON)
2004	R83911	2004	T-2030	2004	TI402	2004	TR-166(OLSON)
2004	R83912	2004	T-278	2004	TI403	2004	TR-17
2004	R83929	2004	T-279	2004	TIM-01	2004	TR-17A
2004	R83986	2004	T-99	2004	TIM-10	2004	TR-17C
2004	R83995	2004	T-Q5021	2004	TIM-11	2004	TR-18
2004	R85101	2004	T-Q5022	2004	TIX-M01	2004	TR-18C
2004	R85106	2004	T-Q5034	2004	TIX-M02	2004	TR-1R26
2004	R85107	2004	T-Q5035	2004	TIX-M03	2004	TR-2R26
2004	R85108	2004	T-Q5038	2004	TIX-M04	2004	TR-3R26
2004	R85109	2004	T1011	2004	TIX-M05	2004	TR-4R26
2004	R85201	2004	T1012	2004	TIX-M06	2004	TR-8001
2004	R85204	2004	T1028	2004	TIX-M07	2004	TR-8002
2004	R85205	2004	T1032	2004	TIX-M08	2004	TR-8003
2004	R85206	2004	T1033	2004	TIX-M101	2004	TRO-2012
2004	R85207	2004	T1034	2004	TIX-M11	2004	TR06
2004	R85208	2004	T1058	2004	TIX-M17	2004	TR06C
2004	R85301	2004	T1166	2004	TIX-M201	2004	TR07
2004	R85305	2004	T1224	2004	TIX-M202	2004	TR105
2004	R85306	2004	T1225	2004	TIX-M203	2004	TR11C
2004	R85311	2004	T1232	2004	TIX-M204	2004	TR12
2004	R85312	2004	T1233	2004	TIX-M205	2004	TR12C
2004	R85313	2004	T1250	2004	TIX-M206	2004	TR13
2004	R85314	2004	T1298	2004	TIX-M207	2004	TR139
2004	R85317	2004	T1299	2004	TIX3016	2004	TR13C
2004	R85818	2004	T1305	2004	TIX3016A	2004	TR14
2004	R8593	2004	T1306	2004	TIX3032	2004	TR17
2004	R8684	2004	T1314	2004	TIX316	2004	TR18
2004	R8685	2004	T1387	2004	TIX91	2004	TR18C
2004	R8686	2004	T1388	2004	TIX92	2004	TR1R26
2004	R8687	2004	T1389	2004	TIXM-201	2004	TR218
2004	RT3466	2004	T1390	2004	TIXM-203	2004	TR2R26
2004	RT4525	2004	T1391	2004	TIXM-205	2004	TR310019
2004	RT6988	2004	T1400	2004	TIXM-206	2004	TR310025
2004	S-1640	2004	T1401	2004	TIXM01	2004	TR310065
2004	S-371	2004	T1402	2004	TIXM02	2004	TR310068
2004	S-873TB	2004	T1403	2004	TIXM03	2004	TR310069
2004	S-874TB	2004	T1454	2004	TIXM04	2004	TR310123
2004	S01	2004	T1459	2004	TIXM05	2004	TR310124
2004	S02	2004	T1461	2004	TIXM06	2004	TR310139
2004	S03	2004	T1548	2004	TIXM07	2004	TR310147
2004	S1640	2004	T1618			2004	TR310150
2004	0822.3511-770	2004	T163			2004	TR310155
2004	0822.3511-780	2004	T1657			2004	TR310156
2004	0822.3511-790	2004	T1690			2004	TR310157
2004	S65-1-A-3P	2004	T1691			2004	TR310158
2004	S65A-1-3P	2004	T1692			2004	TR310193
2004	8684	2004	T1737			2004	TR310224
2004	S70T	2004	T1738			2004	TR310232
2004	S87TB	2004	T1788			2004	TR331
2004	S88B-1-3P	2004	T1814			2004	TR3R26
2004	S88TB	2004	T1831			2004	TR4R26
2004	S95101	2004	T1XM05			2004	TR51
2004	S95102	2004	T2015			2004	TR52
2004	S95103	2004	T2016			2004	TR62
2004	S95104	2004	T2017			2004	TR77
2004	S95106	2004	T2019			2004	TR8001
2004	S9516	2004	T2020			2004	TR8002
2004	S99101	2004	T2021			2004	TR8003
2004	S99102	2004	T2022			2004	TR87
2004	S99103	2004	T2024			2004	TR88
2004	S99104	2004	T2025			2004	TRA10R
2004	SA8-1-A-3P	2004	T2026			2004	TRA11R

RS 276-	Industry Standard No.	RS 276-	Industry Standard No.	RS 276-	Industry Standard No.	RS 276-	Industry Standard No.
2004	TRA12R	2004	1-522211021	2004	2N1782	2004	2N3280
2004	TRA22A	2004	1-522211921	2004	2N1784	2004	2N3281
2004	TRA22B	2004	1-522214400	2004	2N1785	2004	2N3282
2004	TRA23	2004	1-522214411	2004	2N1786	2004	2N3283
2004	TRA23A	2004	1-522214435	2004	2N1787	2004	2N3284
2004	TRA23B	2004	1-522214821	2004	2N1788	2004	2N3285
2004	TRA24	2004	1-522214831	2004	2N1789	2004	2N3286
2004	TRA24A	2004	1-522216600	2004	2N1790	2004	2N331
2004	TRA24C	2004	1-522217400	2004	2N1853/18	2004	2N3320
2004	TRM13	2004	1A8-1A82	2004	2N1864	2004	2N3321
2004	TRM14	2004	1220SA	2004	2N1865	2004	2N3322
2004	TRM81	2004	002-007100	2004	2N1866	2004	2N3323
2004	TR02012	2004	002-007200	2004	2N1867	2004	2N3324
2004	T8-615	2004	002-007400	2004	2N1868	2004	2N3325
2004	T8-620	2004	002-00900	2004	2N1872	2004	2N3328
2004	T8-621	2004	002-009000	2004	2N1873	2004	2N3371
2004	T8-627A	2004	2P	2004	2N1874	2004	2N3399
2004	T8-627B	2004	2G101	2004	2N1875	2004	2N3400
2004	T8-630	2004	2G102	2004	2N1960	2004	2N3412
2004	T8-672A	2004	2G103	2004	2N1960/46	2004	2N3443
2004	T8-672B	2004	2G104	2004	2N1961	2004	2N3449
2004	T8-673A	2004	2G106	2004	2N19616	2004	2N3588
2004	T8-673B	2004	2G108	2004	2N1999	2004	2N36
2004	T8630	2004	2G109	2004	2N2009	2004	2N36#
2004	T8669C	2004	2G110	2004	2N2022	2004	2N37
2004	T8672B	2004	2G201	2004	2N2048	2004	2N3770
2004	T8673A	2004	2G202	2004	2N2048A	2004	2N3783
2004	T8673B	2004	2G301	2004	2N2059	2004	2N3784
2004	TV24137	2004	2G345	2004	2N2083	2004	2N3785
2004	TV24158	2004	2G382	2004	2N2084	2004	2N38
2004	TV24166	2004	2G401	2004	2N2091	2004	2N38#
2004	TV24172	2004	2G402	2004	2N2093	2004	2N384
2004	TV24229	2004	2G403	2004	2N2097	2004	2N384/33
2004	TV24230	2004	2G404	2004	2N2098	2004	2N3883
2004	TV24239	2004	2G413	2004	2N2100	2004	2N3995
2004	TV24351	2004	2G414	2004	2N2168	2004	2N509
2004	TV2455	2004	2G415	2004	2N2170	2004	2N529/P
2004	TV2479	2004	2G416	2004	2N218	2004	2N530/P
2004	TV8-2SA103	2004	2G417	2004	2N2188	2004	2N532/P
2004	V120	2004	2J72	2004	2N2189	2004	2N533/P
2004	V205	2004	2J73	2004	2N219	2004	2N537
2004	V58	2004	2K48	2004	2N2190	2004	2N559
2004	V75	2004	2N1305	2004	2N2191	2004	2N601
2004	VPL2744K	2004	2N1017	2004	2N2220	2004	2N63#
2004	VFW2745D	2004	2N1018	2004	2N2220#	2004	2N65
2004	V8-2SA103	2004	2N1023	2004	2N2200	2004	2N6365A
2004	V8-2SA385L	2004	2N106	2004	2N2225	2004	2N64#
2004	V8-2SA71	2004	2N106#	2004	2N2238	2004	2N65#
2004	V8-2SA71B	2004	2N1066	2004	2N2258	2004	2N700
2004	V8-2SA71B8	2004	2N107	2004	2N2259	2004	2N700/18
2004	V8-2SC385L	2004	2N1093	2004	2N2273	2004	2N700A
2004	V82A71B8	2004	2N1094	2004	2N231	2004	2N700A/18
2004	V82SA103	2004	2N1115	2004	2N232	2004	2N705
2004	V82SA378	2004	2N1115A	2004	2N2360	2004	2N705A
2004	V82SA379	2004	2N1122	2004	2N2361	2004	2N710
2004	V82SA71B	2004	2N1122A	2004	2N2362	2004	2N710A
2004	V82SA71B8	2004	2N1141	2004	2N238	2004	2N711
2004	V82SC385L	2004	2N1141A	2004	2N2382	2004	2N711A
2004	W12	2004	2N1142	2004	2N2398	2004	2N711B
2004	W22	2004	2N1142A	2004	2N2399	2004	2N725
2004	W83	2004	2N1143	2004	2N240	2004	2N741
2004	WEP637	2004	2N1143A	2004	2N2400	2004	2N741A
2004	WTV12MC	2004	2N1144	2004	2N2401	2004	2N768
2004	WTV20MC	2004	2N1145	2004	2N2402	2004	2N769
2004	WTV3MC	2004	2N1158	2004	2N2415	2004	2N779
2004	WTV6MC	2004	2N1158A	2004	2N2416	2004	2N779A
2004	WTVAT6A	2004	2N1177	2004	2N2455	2004	2N779B
2004	WTVB5	2004	2N1178	2004	2N2495	2004	2N781
2004	WTVBA6	2004	2N1195	2004	2N2496	2004	2N782
2004	WTVBB6	2004	2N123/5	2004	2N2512	2004	2N794
2004	X32C4293	2004	2N12345	2004	2N252#	2004	2N795
2004	X42	2004	2N1266	2004	2N2587	2004	2N796
2004	XA101	2004	2N1266#	2004	2N2621	2004	2N827
2004	XA102	2004	2N1282	2004	2N2622	2004	2N828
2004	XA103	2004	2N1285	2004	2N2623	2004	2N828A
2004	XA104	2004	2N1300	2004	2N2624	2004	2N829
2004	XA111	2004	2N1301	2004	2N2625	2004	2N837
2004	XA112	2004	2N1309A	2004	2N2626	2004	2N838
2004	XA123	2004	2N132#	2004	2N2627	2004	2N84
2004	XA124	2004	2N1385	2004	2N2628	2004	2N846
2004	XA126	2004	2N1394	2004	2N2629	2004	2N846A
2004	XA131	2004	2N1403	2004	2N2630	2004	2N846B
2004	XA141	2004	2N1404A	2004	2N2635	2004	2N934
2004	XA142	2004	2N1405	2004	2N2654	2004	2N960
2004	XA143	2004	2N1406	2004	2N267	2004	2N960/46
2004	XA161	2004	2N1407	2004	2N2671	2004	2N961
2004	XB10	2004	2N1408	2004	2N2672	2004	2N961/46
2004	XB8	2004	2N1409	2004	2N2672A	2004	2N962
2004	XB9	2004	2N1410	2004	2N2672BLK	2004	2N962/46
2004	XG1	2004	2N1411	2004	2N2672GRN	2004	2N963
2004	XG10	2004	2N1436	2004	2N269	2004	2N964
2004	XG11	2004	2N1494	2004	2N2717	2004	2N964/46
2004	XG12	2004	2N1495	2004	2N274	2004	2N964A
2004	XG2	2004	2N1500/18	2004	2N274BLU	2004	2N965
2004	XG24	2004	2N1515	2004	2N274WHT	2004	2N966
2004	XG3	2004	2N1516	2004	2N276	2004	2N967
2004	XG5	2004	2N1517	2004	2N2786	2004	2N968
2004	XJ71	2004	2N1517A	2004	2N2786A	2004	2N969
2004	XJ72	2004	2N1524	2004	2N2795	2004	2N970
2004	XJ73	2004	2N1524-1	2004	2N2796	2004	2N971
2004	2C2SA101	2004	2N1524-2	2004	2N2797	2004	2N972
2004	2C2SA101BA	2004	2N1525	2004	2N2798	2004	2N973
2004	2C2SA102	2004	2N1526/33	2004	2N2799	2004	2N974
2004	2C2SA102CA	2004	2N1527	2004	2N286	2004	2N975
2004	2C2SA103	2004	2N1561	2004	2N2860	2004	2N976
2004	2C2SA103CA	2004	2N1562	2004	2N2873	2004	2N977
2004	2C2SA377	2004	2N1625	2004	2N289	2004	2N979
2004	2C2SA70	2004	2N1633	2004	2N2928	2004	2N980
2004	2C2SA700A	2004	2N1634	2004	2N2929	2004	2N982
2004	2C2SA700B	2004	2N1637/33	2004	2N2942	2004	2N983
2004	2C2SA71	2004	2N1638	2004	2N2943	2004	2N984
2004	2C2SA71A	2004	2N1639	2004	2N2955	2004	2N985
2004	2EN300	2004	2N1639/33	2004	2N2956	2004	2N987
2004	2EN301	2004	2N1646	2004	2N2957	2004	2N990
2004	ZJ72	2004	2N1665	2004	2N299	2004	2N991
2004	ZJ73	2004	2N1670	2004	2N2996	2004	2N992
2004	1-21-135	2004	2N1678	2004	2N2997	2004	2N993
2004	1-21-137	2004	2N1713	2004	2N2998	2004	2N994
2004	1-21-139	2004	2N1726	2004	2N2999	2004	2S144
2004	1-21-150	2004	2N1727	2004	2N300	2004	2S148
2004	1-21-157	2004	2N1728	2004	2N3074	2004	2S175
2004	1-21-190	2004	2N1742	2004	2N308#	2004	2S201
2004	1-21-228	2004	2N1745	2004	2N309#	2004	2S32D
2004	1-21-229	2004	2N1746	2004	2N3127	2004	2S471-1
2004	1-21-230	2004	2N1747	2004	2N3148	2004	2S57
2004	1-21-231	2004	2N1748	2004	2N3153	2004	2SA-4551
2004	1-21-233	2004	2N1748A	2004	2N316	2004	2SA-4561
2004	1-21-256	2004	2N1749	2004	2N3216	2004	2SA076F
2004	1-21-258	2004	2N1750	2004	2N3267	2004	2SA098R
2004	1-21-259	2004	2N1752	2004	2N3279	2004	2SA100D
2004	1-21-260	2004	2N1753			2004	2SA100B
2004	1-522210131	2004	2N1754			2004	2SA100F
2004	1-522210300					2004	2SA100G
2004	1-522210921					2004	2SA100H

RS 276-	Industry Standard No.	RS 276-	Industry Standard No.	RS 276-	Industry Standard No.	RS 276-	Industry Standard No.
2004	2SA100J	2004	2SA112K	2004	2SA122L	2004	2SA134H
2004	2SA100K	2004	2SA112L	2004	2SA122M	2004	2SA134J
2004	2SA100M	2004	2SA112M	2004	2SA220R	2004	2SA134K
2004	2SA100OR	2004	2SA112OR	2004	2SA122R	2004	2SA134L
2004	2SA100R	2004	2SA112R	2004	2SA122X	2004	2SA134OR
2004	2SA100X	2004	2SA112X	2004	2SA122Y	2004	2SA134R
2004	2SA100Y	2004	2SA112Y	2004	2SA123A	2004	2SA134X
2004	2SA1018	2004	2SA113A	2004	2SA123B	2004	2SA134Y
2004	2SA103(CA)	2004	2SA113B	2004	2SA123C	2004	2SA135
2004	2SA103CB	2004	2SA113C	2004	2SA123D	2004	2SA135A
2004	2SA103CB	2004	2SA113D	2004	2SA123E	2004	2SA135B
2004	2SA103D	2004	2SA113E	2004	2SA123F	2004	2SA135C
2004	2SA103E	2004	2SA113F	2004	2SA123G	2004	2SA135D
2004	2SA103F	2004	2SA113G	2004	2SA123GN	2004	2SA135E
2004	2SA103G	2004	2SA113GN	2004	2SA123H	2004	2SA135F
2004	2SA103GA	2004	2SA113H	2004	2SA123J	2004	2SA135G
2004	2SA103L	2004	2SA113J	2004	2SA123K	2004	2SA135GN
2004	2SA103M	2004	2SA113L	2004	2SA123L	2004	2SA135H
2004	2SA103OR	2004	2SA113M	2004	2SA123M	2004	2SA135J
2004	2SA103R	2004	2SA113R	2004	2SA123OR	2004	2SA135K
2004	2SA103X	2004	2SA113X	2004	2SA123R	2004	2SA135L
2004	2SA103Y	2004	2SA113Y	2004	2SA123X	2004	2SA135M
2004	2SA104A	2004	2SA114A	2004	2SA123Y	2004	2SA135OR
2004	2SA104B	2004	2SA114B	2004	2SA124A	2004	2SA135R
2004	2SA104C	2004	2SA114C	2004	2SA124B	2004	2SA135X
2004	2SA104E	2004	2SA114D	2004	2SA124C	2004	2SA135Y
2004	2SA104F	2004	2SA114E	2004	2SA124D	2004	2SA136A
2004	2SA104G	2004	2SA114F	2004	2SA124E	2004	2SA136B
2004	2SA104H	2004	2SA114G	2004	2SA124F	2004	2SA136C
2004	2SA104K	2004	2SA114H	2004	2SA124G	2004	2SA136D
2004	2SA104L	2004	2SA114K	2004	2SA124GN	2004	2SA136E
2004	2SA104M	2004	2SA114L	2004	2SA124H	2004	2SA136F
2004	2SA104OR	2004	2SA114M	2004	2SA124J	2004	2SA136G
2004	2SA104R	2004	2SA114OR	2004	2SA124K	2004	2SA136GN
2004	2SA104X	2004	2SA114R	2004	2SA124L	2004	2SA136H
2004	2SA104Y	2004	2SA114X	2004	2SA124M	2004	2SA136J
2004	2SA105	2004	2SA114Y	2004	2SA124OR	2004	2SA136K
2004	2SA106	2004	2SA115A	2004	2SA124R	2004	2SA136L
2004	2SA106A	2004	2SA115B	2004	2SA124X	2004	2SA136M
2004	2SA106B	2004	2SA115C	2004	2SA124Y	2004	2SA136OR
2004	2SA106C	2004	2SA115D	2004	2SA125A	2004	2SA136R
2004	2SA106D	2004	2SA115E	2004	2SA125B	2004	2SA136X
2004	2SA106E	2004	2SA115F	2004	2SA125C	2004	2SA137A
2004	2SA106F	2004	2SA115G	2004	2SA125D	2004	2SA137B
2004	2SA106G	2004	2SA115GN	2004	2SA125E	2004	2SA137C
2004	2SA106H	2004	2SA115H	2004	2SA125F	2004	2SA137D
2004	2SA106K	2004	2SA115J	2004	2SA125G	2004	2SA137E
2004	2SA106L	2004	2SA115K	2004	2SA125GN	2004	2SA137F
2004	2SA106M	2004	2SA115L	2004	2SA125H	2004	2SA137G
2004	2SA106OR	2004	2SA115M	2004	2SA125J	2004	2SA137GN
2004	2SA106R	2004	2SA115OR	2004	2SA125K	2004	2SA137H
2004	2SA106X	2004	2SA115R	2004	2SA125L	2004	2SA137J
2004	2SA106Y	2004	2SA115X	2004	2SA125M	2004	2SA137K
2004	2SA107	2004	2SA115Y	2004	2SA125OR	2004	2SA137L
2004	2SA107A	2004	2SA116A	2004	2SA125R	2004	2SA137M
2004	2SA107B	2004	2SA116B	2004	2SA125X	2004	2SA137OR
2004	2SA107C	2004	2SA116C	2004	2SA125Y	2004	2SA137R
2004	2SA107D	2004	2SA116D	2004	2SA126	2004	2SA137X
2004	2SA107E	2004	2SA116E	2004	2SA127	2004	2SA137Y
2004	2SA107F	2004	2SA116F	2004	2SA13	2004	2SA139A
2004	2SA107G	2004	2SA116G	2004	2SA130	2004	2SA139B
2004	2SA107H	2004	2SA116GN	2004	2SA130A	2004	2SA139C
2004	2SA107K	2004	2SA116H	2004	2SA130B	2004	2SA139D
2004	2SA107L	2004	2SA116J	2004	2SA130C	2004	2SA139E
2004	2SA107M	2004	2SA116K	2004	2SA130D	2004	2SA139F
2004	2SA107OR	2004	2SA116L	2004	2SA130E	2004	2SA139G
2004	2SA107R	2004	2SA116M	2004	2SA130F	2004	2SA139GN
2004	2SA107X	2004	2SA116OR	2004	2SA130G	2004	2SA139J
2004	2SA107Y	2004	2SA116R	2004	2SA130GN	2004	2SA139K
2004	2SA108A	2004	2SA116X	2004	2SA130H	2004	2SA139L
2004	2SA108B	2004	2SA116Y	2004	2SA130J	2004	2SA139M
2004	2SA108C	2004	2SA117	2004	2SA130K	2004	2SA139OR
2004	2SA108D	2004	2SA117A	2004	2SA130L	2004	2SA139R
2004	2SA108E	2004	2SA117B	2004	2SA130M	2004	2SA139X
2004	2SA108F	2004	2SA117C	2004	2SA130OR	2004	2SA139Y
2004	2SA108G	2004	2SA117D	2004	2SA130R	2004	2SA13A
2004	2SA108H	2004	2SA117E	2004	2SA130X	2004	2SA13B
2004	2SA108K	2004	2SA117F	2004	2SA130Y	2004	2SA13C
2004	2SA108L	2004	2SA117G	2004	2SA131A	2004	2SA13D
2004	2SA108M	2004	2SA117GN	2004	2SA131B	2004	2SA13G
2004	2SA108OR	2004	2SA117H	2004	2SA131C	2004	2SA13L
2004	2SA108R	2004	2SA117J	2004	2SA131D	2004	2SA13M
2004	2SA108X	2004	2SA117K	2004	2SA131E	2004	2SA13OR
2004	2SA108Y	2004	2SA117L	2004	2SA131F	2004	2SA13R
2004	2SA109A	2004	2SA117M	2004	2SA131G	2004	2SA13X
2004	2SA109B	2004	2SA117OR	2004	2SA131GN	2004	2SA13Y
2004	2SA109C	2004	2SA117R	2004	2SA131H	2004	2SA14
2004	2SA109D	2004	2SA117X	2004	2SA131J	2004	2SA141A
2004	2SA109E	2004	2SA117Y	2004	2SA131K	2004	2SA141D
2004	2SA109F	2004	2SA118	2004	2SA131L	2004	2SA141E
2004	2SA109G	2004	2SA118A	2004	2SA131M	2004	2SA141F
2004	2SA109K	2004	2SA118B	2004	2SA131OR	2004	2SA141G
2004	2SA109L	2004	2SA118C	2004	2SA131R	2004	2SA141H
2004	2SA109M	2004	2SA118D	2004	2SA131X	2004	2SA141K
2004	2SA109OR	2004	2SA118E	2004	2SA131Y	2004	2SA141L
2004	2SA109R	2004	2SA118F	2004	2SA132A	2004	2SA141M
2004	2SA109X	2004	2SA118G	2004	2SA132B	2004	2SA141OR
2004	2SA109Y	2004	2SA118GN	2004	2SA132C	2004	2SA141R
2004	2SA110A	2004	2SA118H	2004	2SA132D	2004	2SA141X
2004	2SA110B	2004	2SA118J	2004	2SA132E	2004	2SA141Y
2004	2SA110C	2004	2SA118K	2004	2SA132F	2004	2SA143A
2004	2SA110D	2004	2SA118L	2004	2SA132G	2004	2SA143B
2004	2SA110E	2004	2SA118M	2004	2SA132H	2004	2SA143C
2004	2SA110F	2004	2SA118OR	2004	2SA132J	2004	2SA143D
2004	2SA110G	2004	2SA118R	2004	2SA132K	2004	2SA143E
2004	2SA110K	2004	2SA118X	2004	2SA132L	2004	2SA143F
2004	2SA110L	2004	2SA118Y	2004	2SA132M	2004	2SA143G
2004	2SA110M	2004	2SA121A	2004	2SA132OR	2004	2SA143GN
2004	2SA110OR	2004	2SA121B	2004	2SA132R	2004	2SA143H
2004	2SA110R	2004	2SA121C	2004	2SA132X	2004	2SA143J
2004	2SA110X	2004	2SA121D	2004	2SA132Y	2004	2SA143K
2004	2SA110Y	2004	2SA121E	2004	2SA133A	2004	2SA143L
2004	2SA111A	2004	2SA121F	2004	2SA133B	2004	2SA143M
2004	2SA111B	2004	2SA121G	2004	2SA133C	2004	2SA143OR
2004	2SA111C	2004	2SA121GN	2004	2SA133D	2004	2SA143R
2004	2SA111D	2004	2SA121H	2004	2SA133E	2004	2SA143X
2004	2SA111E	2004	2SA121J	2004	2SA133F	2004	2SA144A
2004	2SA111F	2004	2SA121K	2004	2SA133G	2004	2SA144D
2004	2SA111G	2004	2*A121L	2004	2SA133GN	2004	2SA144E
2004	2SA111K	2004	2SA121M	2004	2SA133H	2004	2SA144F
2004	2SA111L	2004	2SA121OR	2004	2SA133J	2004	2SA144G
2004	2SA111M	2004	2SA121R	2004	2SA133K	2004	2SA144GN
2004	2SA111OR	2004	2SA121X	2004	2SA133L	2004	2SA144H
2004	2SA111R	2004	2SA121Y	2004	2SA133M	2004	2SA144J
2004	2SA111X	2004	2SA122A	2004	2SA133OR	2004	2SA144L
2004	2SA111Y	2004	2SA122B	2004	2SA133R	2004	2SA144M
2004	2SA112A	2004	2SA122C	2004	2SA133X	2004	2SA144OR
2004	2SA112B	2004	2SA122D	2004	2SA133Y	2004	2SA144R
2004	2SA112C	2004	2SA122E	2004	2SA134A	2004	2SA144X
2004	2SA112D	2004	2SA122F	2004	2SA134B		
2004	2SA112E	2004	2SA122G	2004	2SA134C		
2004	2SA112F	2004	2SA122GN	2004	2SA134D		
2004	2SA112G	2004	2SA122H	2004	2SA134E		
2004	2SA112GN	2004	2SA122K	2004	2SA134F		
2004	2SA112H			2004	2SA134G		

RS 276-	Industry Standard No.
2004	2SA144Y
2004	2SA145B
2004	2SA145D
2004	2SA145E
2004	2SA145G
2004	2SA145GN
2004	2SA145K
2004	2SA145M
2004	2SA145OR
2004	2SA145R
2004	2SA145X
2004	2SA145Y
2004	2SA146A
2004	2SA146B
2004	2SA146C
2004	2SA146D
2004	2SA146E
2004	2SA146F
2004	2SA146G
2004	2SA146GN
2004	2SA146H
2004	2SA146J
2004	2SA146K
2004	2SA146L
2004	2SA146M
2004	2SA146OR
2004	2SA146X
2004	2SA146Y
2004	2SA147A
2004	2SA147B
2004	2SA147C
2004	2SA147D
2004	2SA147E
2004	2SA147F
2004	2SA147G
2004	2SA147H
2004	2SA147J
2004	2SA147K
2004	2SA147L
2004	2SA147M
2004	2SA147OR
2004	2SA147R
2004	2SA147X
2004	2SA147Y
2004	2SA148A
2004	2SA148B
2004	2SA148C
2004	2SA148D
2004	2SA148E
2004	2SA148F
2004	2SA148G
2004	2SA148GN
2004	2SA148H
2004	2SA148J
2004	2SA148K
2004	2SA148L
2004	2SA148M
2004	2SA148OR
2004	2SA148R
2004	2SA148X
2004	2SA148Y
2004	2SA149A
2004	2SA149B
2004	2SA149C
2004	2SA149D
2004	2SA149E
2004	2SA149F
2004	2SA149G
2004	2SA149GN
2004	2SA149H
2004	2SA149J
2004	2SA149K
2004	2SA149L
2004	2SA149M
2004	2SA149OR
2004	2SA149R
2004	2SA149X
2004	2SA149Y
2004	2SA14A
2004	2SA14B
2004	2SA14C
2004	2SA14D
2004	2SA14E
2004	2SA14G
2004	2SA14L
2004	2SA14M
2004	2SA14OR
2004	2SA14R
2004	2SA14X
2004	2SA14Y
2004	2SA15-6
2004	2SA151
2004	2SA151A
2004	2SA151B
2004	2SA151C
2004	2SA151D
2004	2SA151E
2004	2SA151F
2004	2SA151G
2004	2SA151GN
2004	2SA151H
2004	2SA151J
2004	2SA151K
2004	2SA151L
2004	2SA151M
2004	2SA151OR
2004	2SA151R
2004	2SA151X
2004	2SA151Y
2004	2SA152
2004	2SA153A
2004	2SA153B
2004	2SA153C
2004	2SA153D
2004	2SA153E
2004	2SA153F
2004	2SA153G
2004	2SA153GN
2004	2SA153H
2004	2SA153J
2004	2SA153K
2004	2SA153L
2004	2SA153M
2004	2SA153OR
2004	2SA153R
2004	2SA153X
2004	2SA153Y
2004	2SA154A
2004	2SA154B
2004	2SA154C
2004	2SA154D
2004	2SA154E
2004	2SA154F
2004	2SA154G
2004	2SA154GN
2004	2SA154H
2004	2SA154J
2004	2SA154K
2004	2SA154L
2004	2SA154M
2004	2SA154OR
2004	2SA154R
2004	2SA154X
2004	2SA154Y
2004	2SA155A
2004	2SA155B
2004	2SA155C
2004	2SA155D
2004	2SA155E
2004	2SA155F
2004	2SA155G
2004	2SA155GN
2004	2SA155H
2004	2SA155J
2004	2SA155K
2004	2SA155L
2004	2SA155M
2004	2SA155OR
2004	2SA155R
2004	2SA155Y
2004	2SA156A
2004	2SA156B
2004	2SA156C
2004	2SA156D
2004	2SA156E
2004	2SA156F
2004	2SA156G
2004	2SA156GN
2004	2SA156H
2004	2SA156J
2004	2SA156K
2004	2SA156L
2004	2SA156M
2004	2SA156OR
2004	2SA156X
2004	2SA156Y
2004	2SA157A
2004	2SA157B
2004	2SA157C
2004	2SA157D
2004	2SA157E
2004	2SA157F
2004	2SA157G
2004	2SA157GN
2004	2SA157H
2004	2SA157J
2004	2SA157K
2004	2SA157L
2004	2SA157M
2004	2SA157OR
2004	2SA157R
2004	2SA157X
2004	2SA157Y
2004	2SA159A
2004	2SA159B
2004	2SA159C
2004	2SA159D
2004	2SA159E
2004	2SA159F
2004	2SA159G
2004	2SA159GN
2004	2SA159H
2004	2SA159J
2004	2SA159K
2004	2SA159L
2004	2SA159M
2004	2SA159OR
2004	2SA159R
2004	2SA159X
2004	2SA159Y
2004	2SA15A
2004	2SA15B
2004	2SA15C
2004	2SA15D
2004	2SA15E
2004	2SA15F
2004	2SA15G
2004	2SA15L
2004	2SA15M
2004	2SA15OR
2004	2SA15RD
2004	2SA15X
2004	2SA160A
2004	2SA160B
2004	2SA160C
2004	2SA160D
2004	2SA160E
2004	2SA160F
2004	2SA160GN
2004	2SA160H
2004	2SA160J
2004	2SA160K
2004	2SA160L
2004	2SA160M
2004	2SA160OR
2004	2SA160R
2004	2SA160X
2004	2SA160Y
2004	2SA161A
2004	2SA161B
2004	2SA161C
2004	2SA161D
2004	2SA161E
2004	2SA161F
2004	2SA161G
2004	2SA161GN
2004	2SA161H
2004	2SA161J
2004	2SA161K
2004	2SA161L
2004	2SA161M
2004	2SA161OR
2004	2SA161R
2004	2SA161X
2004	2SA161Y
2004	2SA162A
2004	2SA162B
2004	2SA162C
2004	2SA162D
2004	2SA162E
2004	2SA162F
2004	2SA162G
2004	2SA162GN
2004	2SA162H
2004	2SA162J
2004	2SA162K
2004	2SA162L
2004	2SA162N
2004	2SA162OR
2004	2SA162R
2004	2SA162X
2004	2SA162Y
2004	2SA163A
2004	2SA163B
2004	2SA163C
2004	2SA163D
2004	2SA163E
2004	2SA163F
2004	2SA163G
2004	2SA163GN
2004	2SA163H
2004	2SA163J
2004	2SA163K
2004	2SA163L
2004	2SA163M
2004	2SA163OR
2004	2SA163R
2004	2SA163X
2004	2SA163Y
2004	2SA164A
2004	2SA164B
2004	2SA164C
2004	2SA164D
2004	2SA164E
2004	2SA164F
2004	2SA164G
2004	2SA164GN
2004	2SA164H
2004	2SA164J
2004	2SA164K
2004	2SA164L
2004	2SA164M
2004	2SA164OR
2004	2SA164R
2004	2SA164X
2004	2SA164Y
2004	2SA165A
2004	2SA165B
2004	2SA165C
2004	2SA165D
2004	2SA165E
2004	2SA165F
2004	2SA165G
2004	2SA165GN
2004	2SA165H
2004	2SA165J
2004	2SA165K
2004	2SA165M
2004	2SA165OR
2004	2SA165R
2004	2SA165X
2004	2SA165Y
2004	2SA166A
2004	2SA166B
2004	2SA166C
2004	2SA166D
2004	2SA166E
2004	2SA166F
2004	2SA166G
2004	2SA166GN
2004	2SA166H
2004	2SA166J
2004	2SA166K
2004	2SA166L
2004	2SA166M
2004	2SA166OR
2004	2SA166R
2004	2SA166X
2004	2SA166Y
2004	2SA167A
2004	2SA167B
2004	2SA167C
2004	2SA167D
2004	2SA167F
2004	2SA167P
2004	2SA167GN
2004	2SA167H
2004	2SA167J
2004	2SA167K
2004	2SA167L
2004	2SA167M
2004	2SA167OR
2004	2SA167R
2004	2SA167X
2004	2SA167Y
2004	2SA168A
2004	2SA168B
2004	2SA168C
2004	2SA168D
2004	2SA168E
2004	2SA168F
2004	2SA168G
2004	2SA168GN
2004	2SA168H
2004	2SA168J
2004	2SA168K
2004	2SA168L
2004	2SA168M
2004	2SA168OR
2004	2SA168R
2004	2SA168X
2004	2SA168Y
2004	2SA169A
2004	2SA169B
2004	2SA169C
2004	2SA169D
2004	2SA169E
2004	2SA169F
2004	2SA169G
2004	2SA169GN
2004	2SA169H
2004	2SA169J
2004	2SA169K
2004	2SA169L
2004	2SA169M
2004	2SA169OR
2004	2SA169R
2004	2SA169X
2004	2SA169Y
2004	2SA16A
2004	2SA16B
2004	2SA16C
2004	2SA16D
2004	2SA16E
2004	2SA16F
2004	2SA16L
2004	2SA16M
2004	2SA16OR
2004	2SA16R
2004	2SA16X
2004	2SA16Y
2004	2SA170A
2004	2SA170B
2004	2SA170C
2004	2SA170D
2004	2SA170E
2004	2SA170F
2004	2SA170G
2004	2SA170GN
2004	2SA170H
2004	2SA170J
2004	2SA170K
2004	2SA170L
2004	2SA170M
2004	2SA170OR
2004	2SA170R
2004	2SA170X
2004	2SA170Y
2004	2SA171A
2004	2SA171B
2004	2SA171C
2004	2SA171D
2004	2SA171E
2004	2SA171F
2004	2SA171G
2004	2SA171GN
2004	2SA171H
2004	2SA171J
2004	2SA171K
2004	2SA171L
2004	2SA171M
2004	2SA171OR
2004	2SA171R
2004	2SA171X
2004	2SA171Y
2004	2SA172B
2004	2SA172C
2004	2SA172D
2004	2SA172E
2004	2SA172F
2004	2SA172G
2004	2SA172GN
2004	2SA172H
2004	2SA172J
2004	2SA172K
2004	2SA172L
2004	2SA172M
2004	2SA172OR
2004	2SA172R
2004	2SA172X
2004	2SA172Y
2004	2SA173A
2004	2SA173C
2004	2SA173D
2004	2SA173E
2004	2SA173F
2004	2SA173G
2004	2SA173GN
2004	2SA173H
2004	2SA173J
2004	2SA173K
2004	2SA173L
2004	2SA173M
2004	2SA173OR
2004	2SA173R
2004	2SA173X
2004	2SA173Y
2004	2SA174A
2004	2SA174B
2004	2SA174C
2004	2SA174D
2004	2SA174E
2004	2SA174F
2004	2SA174G
2004	2SA174GN
2004	2SA174H
2004	2SA174J
2004	2SA174K
2004	2SA174L
2004	2SA174M
2004	2SA174OR
2004	2SA174R
2004	2SA174X
2004	2SA174Y
2004	2SA175A
2004	2SA175B
2004	2SA175C
2004	2SA175D
2004	2SA175E
2004	2SA175F
2004	2SA175G
2004	2SA175GN
2004	2SA175H
2004	2SA175J
2004	2SA175K
2004	2SA175L
2004	2SA175M
2004	2SA175OR
2004	2SA175R
2004	2SA175X
2004	2SA175Y
2004	2SA17A
2004	2SA17B
2004	2SA17C
2004	2SA17D
2004	2SA17E
2004	2SA17F
2004	2SA17G
2004	2SA17L
2004	2SA17OR
2004	2SA17R
2004	2SA17X
2004	2SA17Y
2004	2SA180
2004	2SA180A
2004	2SA180B
2004	2SA180C
2004	2SA180D
2004	2SA180E
2004	2SA180F
2004	2SA180G
2004	2SA180GN
2004	2SA180H
2004	2SA180J
2004	2SA180K
2004	2SA180L
2004	2SA180M
2004	2SA180OR
2004	2SA180R
2004	2SA180X
2004	2SA180Y
2004	2SA181A

RS 276-	Industry Standard No.	RS 276-	Industry Standard No.	RS 276-	Industry Standard No.	RS 276-	Industry Standard No.	RS 276-	Industry Standard No.
2004	2SA181B	2004	2SA198M	2004	2SA211H	2004	2SA219D	2004	2SA234D
2004	2SA181C	2004	2SA198OR	2004	2SA211J	2004	2SA219E		
2004	2SA181D	2004	2SA198R	2004	2SA211K	2004	2SA219F		
2004	2SA181E	2004	2SA198X	2004	2SA211L	2004	2SA219G		
2004	2SA181F	2004	2SA198Y	2004	2SA211M	2004	2SA219GN		
2004	2SA181G	2004	2SA199	2004	2SA211OR	2004	2SA219H		
2004	2SA181GN	2004	2SA200	2004	2SA211R	2004	2SA219J		
2004	2SA181H	2004	2SA204A	2004	2SA211X	2004	2SA219K		
2004	2SA181J	2004	2SA204B	2004	2SA211Y	2004	2SA219L		
2004	2SA181K	2004	2SA204C	2004	2SA212A	2004	2SA219M		
2004	2SA181L	2004	2SA204D	2004	2SA212C	2004	2SA219OR		
2004	2SA181M	2004	2SA204E	2004	2SA212D	2004	2SA219R		
2004	2SA181OR	2004	2SA204F	2004	2SA212E	2004	2SA219X		
2004	2SA181R	2004	2SA204G	2004	2SA212F	2004	2SA219Y		
2004	2SA181X	2004	2SA204GN	2004	2SA212G	2004	2SA22		
2004	2SA181Y	2004	2SA204H	2004	2SA212GN	2004	2SA220		
2004	2SA182A	2004	2SA204J	2004	2SA212H	2004	2SA221		
2004	2SA182B	2004	2SA204K	2004	2SA212J	2004	2SA222		
2004	2SA182C	2004	2SA204L	2004	2SA212K	2004	2SA223		
2004	2SA182D	2004	2SA204M	2004	2SA212L	2004	2SA223A		
2004	2SA182E	2004	2SA204OR	2004	2SA212M	2004	2SA223B		
2004	2SA182F	2004	2SA204R	2004	2SA212OR	2004	2SA223C		
2004	2SA182G	2004	2SA204X	2004	2SA212R	2004	2SA223D		
2004	2SA182GN	2004	2SA204Y	2004	2SA212X	2004	2SA223E		
2004	2SA182H	2004	2SA205A	2004	2SA212Y	2004	2SA223G		
2004	2SA182J	2004	2SA205B	2004	2SA213A	2004	2SA223GN		
2004	2SA182K	2004	2SA205C	2004	2SA213B	2004	2SA223H		
2004	2SA182L	2004	2SA205D	2004	2SA213C	2004	2SA223J		
2004	2SA182M	2004	2SA205E	2004	2SA213D	2004	2SA223K		
2004	2SA182OR	2004	2SA205F	2004	2SA213E	2004	2SA223L		
2004	2SA182R	2004	2SA205G	2004	2SA213F	2004	2SA223M		
2004	2SA182X	2004	2SA205GN	2004	2SA213G	2004	2SA223OR		
2004	2SA182Y	2004	2SA205H	2004	2SA213GN	2004	2SA223R		
2004	2SA183A	2004	2SA205J	2004	2SA213H	2004	2SA223X		
2004	2SA183B	2004	2SA205K	2004	2SA213J	2004	2SA223Y		
2004	2SA183C	2004	2SA205L	2004	2SA213K	2004	2SA224		
2004	2SA183D	2004	2SA205M	2004	2SA213L	2004	2SA225		
2004	2SA183E	2004	2SA205OR	2004	2SA213M	2004	2SA225A		
2004	2SA183F	2004	2SA205R	2004	2SA213OR	2004	2SA225B		
2004	2SA183G	2004	2SA205X	2004	2SA213R	2004	2SA225C		
2004	2SA183GN	2004	2SA205Y	2004	2SA213X	2004	2SA225D		
2004	2SA183H	2004	2SA206A	2004	2SA213Y	2004	2SA225E		
2004	2SA183J	2004	2SA206B	2004	2SA214A	2004	2SA225F		
2004	2SA183K	2004	2SA206C	2004	2SA214B	2004	2SA225G		
2004	2SA183L	2004	2SA206D	2004	2SA214C	2004	2SA225GN		
2004	2SA183M	2004	2SA206E	2004	2SA214D	2004	2SA225H		
2004	2SA183OR	2004	2SA206F	2004	2SA214E	2004	2SA225J		
2004	2SA183R	2004	2SA206G	2004	2SA214F	2004	2SA225K		
2004	2SA183X	2004	2SA206GN	2004	2SA214G	2004	2SA225L		
2004	2SA183Y	2004	2SA206H	2004	2SA214GN	2004	2SA225M		
2004	2SA184	2004	2SA206J	2004	2SA214H	2004	2SA225OR		
2004	2SA188A	2004	2SA206K	2004	2SA214J	2004	2SA225R		
2004	2SA188B	2004	2SA206L	2004	2SA214K	2004	2SA225X		
2004	2SA188C	2004	2SA206M	2004	2SA214L	2004	2SA225Y		
2004	2SA188D	2004	2SA206OR	2004	2SA214M	2004	2SA226		
2004	2SA188E	2004	2SA206R	2004	2SA214OR	2004	2SA227		
2004	2SA188F	2004	2SA206X	2004	2SA214R	2004	2SA227A		
2004	2SA188G	2004	2SA206Y	2004	2SA214X	2004	2SA227B		
2004	2SA188GN	2004	2SA207A	2004	2SA214Y	2004	2SA227C		
2004	2SA188H	2004	2SA207B	2004	2SA215	2004	2SA227D		
2004	2SA188J	2004	2SA207C	2004	2SA215A	2004	2SA227E		
2004	2SA188K	2004	2SA207D	2004	2SA215B	2004	2SA227F		
2004	2SA188L	2004	2SA207E	2004	2SA215C	2004	2SA227G		
2004	2SA188M	2004	2SA207F	2004	2SA215D	2004	2SA227GN		
2004	2SA188OR	2004	2SA207G	2004	2SA215E	2004	2SA227H		
2004	2SA188R	2004	2SA207GN	2004	2SA215F	2004	2SA227J		
2004	2SA188X	2004	2SA207H	2004	2SA215G	2004	2SA227K		
2004	2SA188Y	2004	2SA207J	2004	2SA215GN	2004	2SA227L		
2004	2SA189A	2004	2SA207K	2004	2SA215H	2004	2SA227M		
2004	2SA189B	2004	2SA207L	2004	2SA215J	2004	2SA227OR		
2004	2SA189C	2004	2SA207M	2004	2SA215K	2004	2SA227R		
2004	2SA189D	2004	2SA207OR	2004	2SA215L	2004	2SA227X		
2004	2SA189E	2004	2SA207R	2004	2SA215M	2004	2SA227Y		
2004	2SA189F	2004	2SA207X	2004	2SA215OR	2004	2SA228		
2004	2SA189G	2004	2SA207Y	2004	2SA215R	2004	2SA229		
2004	2SA189GN	2004	2SA208A	2004	2SA215X	2004	2SA229A		
2004	2SA189H	2004	2SA208B	2004	2SA215Y	2004	2SA229B		
2004	2SA189J	2004	2SA208C	2004	2SA216A	2004	2SA229C		
2004	2SA189K	2004	2SA208D	2004	2SA216B	2004	2SA229D		
2004	2SA189L	2004	2SA208E	2004	2SA216C	2004	2SA229E		
2004	2SA189M	2004	2SA208F	2004	2SA216D	2004	2SA229F		
2004	2SA189OR	2004	2SA208G	2004	2SA216E	2004	2SA229G		
2004	2SA189R	2004	2SA208GN	2004	2SA216F	2004	2SA229GN		
2004	2SA189X	2004	2SA208H	2004	2SA216G	2004	2SA229H		
2004	2SA189Y	2004	2SA208J	2004	2SA216GN	2004	2SA229J		
2004	2SA18A	2004	2SA208K	2004	2SA216H	2004	2SA229K		
2004	2SA18B	2004	2SA208L	2004	2SA216J	2004	2SA229L		
2004	2SA18C	2004	2SA208M	2004	2SA216K	2004	2SA229OR		
2004	2SA18D	2004	2SA208OR	2004	2SA216L	2004	2SA229R		
2004	2SA18E	2004	2SA208R	2004	2SA216M	2004	2SA229X		
2004	2SA18F	2004	2SA208X	2004	2SA216OR	2004	2SA229Y		
2004	2SA18G	2004	2SA208Y	2004	2SA216R	2004	2SA23		
2004	2SA18L	2004	2SA209A	2004	2SA216X	2004	2SA230		
2004	2SA18M	2004	2SA209B	2004	2SA217A	2004	2SA230A		
2004	2SA18OR	2004	2SA209C	2004	2SA217B	2004	2SA230B		
2004	2SA18R	2004	2SA209D	2004	2SA217C	2004	2SA230C		
2004	2SA18X	2004	2SA209E	2004	2SA217D	2004	2SA230D		
2004	2SA18Y	2004	2SA209F	2004	2SA217E	2004	2SA230E		
2004	2SA190	2004	2SA209G	2004	2SA217F	2004	2SA230F		
2004	2SA191	2004	2SA209GN	2004	2SA217G	2004	2SA230G		
2004	2SA192	2004	2SA209H	2004	2SA217GN	2004	2SA230GN		
2004	2SA193	2004	2SA209J	2004	2SA217H	2004	2SA230H		
2004	2SA194	2004	2SA209K	2004	2SA217J	2004	2SA230J		
2004	2SA195	2004	2SA209L	2004	2SA217K	2004	2SA230K		
2004	2SA196	2004	2SA209M	2004	2SA217L	2004	2SA230L		
2004	2SA197A	2004	2SA209OR	2004	2SA217M	2004	2SA230M		
2004	2SA197B	2004	2SA209R	2004	2SA217OR	2004	2SA230OR		
2004	2SA197C	2004	2SA209X	2004	2SA217R	2004	2SA230R		
2004	2SA197D	2004	2SA209Y	2004	2SA217X	2004	2SA230X		
2004	2SA197E	2004	2SA210A	2004	2SA217Y	2004	2SA230Y		
2004	2SA197F	2004	2SA210B	2004	2SA218	2004	2SA233		
2004	2SA197G	2004	2SA210C	2004	2SA218A	2004	2SA233A		
2004	2SA197GN	2004	2SA210D	2004	2SA218B	2004	2SA233B		
2004	2SA197H	2004	2SA210E	2004	2SA218C	2004	2SA233C		
2004	2SA197J	2004	2SA210F	2004	2SA218D	2004	2SA233D		
2004	2SA197K	2004	2SA210G	2004	2SA218E	2004	2SA233E		
2004	2SA197L	2004	2SA210GN	2004	2SA218F	2004	2SA233F		
2004	2SA197M	2004	2SA210H	2004	2SA218G	2004	2SA233G		
2004	2SA197OR	2004	2SA210J	2004	2SA218GN	2004	2SA233GN		
2004	2SA197R	2004	2SA210K	2004	2SA218H	2004	2SA233H		
2004	2SA197X	2004	2SA210L	2004	2SA218J	2004	2SA233J		
2004	2SA197Y	2004	2SA210M	2004	2SA218K	2004	2SA233K		
2004	2SA198A	2004	2SA210OR	2004	2SA218L	2004	2SA233L		
2004	2SA198B	2004	2SA210R	2004	2SA218M	2004	2SA233M		
2004	2SA198C	2004	2SA210X	2004	2SA218OR	2004	2SA233OR		
2004	2SA198D	2004	2SA210Y	2004	2SA218R	2004	2SA233R		
2004	2SA198E	2004	2SA211A	2004	2SA218X	2004	2SA233X		
2004	2SA198F	2004	2SA211B	2004	2SA218Y	2004	2SA233Y		
2004	2SA198G	2004	2SA211C	2004	2SA219	2004	2SA234		
2004	2SA198GN	2004	2SA211D	2004	2SA219A	2004	2SA234A		
2004	2SA198H	2004	2SA211E	2004	2SA219B	2004	2SA234B		
2004	2SA198J	2004	2SA211F	2004	2SA219C	2004	2SA234C		
2004	2SA198K	2004	2SA211G						
2004	2SA198L	2004	2SA211GN						

RS 276-	Industry Standard No.	RS 276-	Industry Standard No.	RS 276-	Industry Standard No.	RS 276-	Industry Standard No.
2004	2SA234E	2004	2SA246L	2004	2SA257K	2004	2SA267K
2004	2SA234F	2004	2SA246M	2004	2SA257L	2004	2SA267L
2004	2SA234G	2004	2SA246OR	2004	2SA257OR	2004	2SA267M
2004	2SA234GN	2004	2SA246R	2004	2SA257R	2004	2SA267OR
2004	2SA234H	2004	2SA246X	2004	2SA257X	2004	2SA267R
2004	2SA234J	2004	2SA246Y	2004	2SA257Y	2004	2SA267X
2004	2SA234K	2004	2SA247A	2004	2SA258A	2004	2SA267Y
2004	2SA234L	2004	2SA247B	2004	2SA258B	2004	2SA268A
2004	2SA234M	2004	2SA247C	2004	2SA258C	2004	2SA268B
2004	2SA234OR	2004	2SA247D	2004	2SA258D	2004	2SA268C
2004	2SA234R	2004	2SA247E	2004	2SA258E	2004	2SA268D
2004	2SA234X	2004	2SA247F	2004	2SA258F	2004	2SA268E
2004	2SA234Y	2004	2SA247G	2004	2SA258G	2004	2SA268F
2004	2SA235	2004	2SA247GN	2004	2SA258GN	2004	2SA268G
2004	2SA235A	2004	2SA247H	2004	2SA258H	2004	2SA268GN
2004	2SA235B	2004	2SA247J	2004	2SA258J	2004	2SA268H
2004	2SA235C	2004	2SA247K	2004	2SA258K	2004	2SA268J
2004	2SA235D	2004	2SA247L	2004	2SA258L	2004	2SA268K
2004	2SA235E	2004	2SA247M	2004	2SA258N	2004	2SA268L
2004	2SA235F	2004	2SA247OR	2004	2SA258OR	2004	2SA268M
2004	2SA235G	2004	2SA247R	2004	2SA258R	2004	2SA268OR
2004	2SA235GN	2004	2SA247X	2004	2SA258X	2004	2SA268R
2004	2SA235H	2004	2SA247Y	2004	2SA258Y	2004	2SA268X
2004	2SA235K	2004	2SA248A	2004	2SA260A	2004	2SA268Y
2004	2SA235M	2004	2SA248B	2004	2SA260B	2004	2SA26A
2004	2SA235OR	2004	2SA248C	2004	2SA260C	2004	2SA26B
2004	2SA235R	2004	2SA248D	2004	2SA260D	2004	2SA26C
2004	2SA235X	2004	2SA248E	2004	2SA260E	2004	2SA26D
2004	2SA235Y	2004	2SA248F	2004	2SA260F	2004	2SA26E
2004	2SA236A	2004	2SA248G	2004	2SA260G	2004	2SA26F
2004	2SA236B	2004	2SA248GN	2004	2SA260GN	2004	2SA26G
2004	2SA236C	2004	2SA248H	2004	2SA260H	2004	2SA26L
2004	2SA236D	2004	2SA248J	2004	2SA260J	2004	2SA26M
2004	2SA236E	2004	2SA248K	2004	2SA260K	2004	2SA260R
2004	2SA236F	2004	2SA248L	2004	2SA260L	2004	2SA26R
2004	2SA236G	2004	2SA248M	2004	2SA260M	2004	2SA26X
2004	2SA236GN	2004	2SA248OR	2004	2SA260OR	2004	2SA26Y
2004	2SA236H	2004	2SA248R	2004	2SA260R	2004	2SA27
2004	2SA236J	2004	2SA248X	2004	2SA260X	2004	2SA270A
2004	2SA236K	2004	2SA248T	2004	2SA260Y	2004	2SA270B
2004	2SA236L	2004	2SA249	2004	2SA261A	2004	2SA270C
2004	2SA236M	2004	2SA25	2004	2SA261B	2004	2SA270D
2004	2SA236OR	2004	2SA251A	2004	2SA261C	2004	2SA270E
2004	2SA236R	2004	2SA251B	2004	2SA261D	2004	2SA270F
2004	2SA236X	2004	2SA251C	2004	2SA261E	2004	2SA270G
2004	2SA236Y	2004	2SA251D	2004	2SA261F	2004	2SA270GN
2004	2SA238A	2004	2SA251E	2004	2SA261G	2004	2SA270H
2004	2SA238B	2004	2SA251F	2004	2SA261GN	2004	2SA270J
2004	2SA238C	2004	2SA251G	2004	2SA261H	2004	2SA270K
2004	2SA238D	2004	2SA251H	2004	2SA261K	2004	2SA270M
2004	2SA238E	2004	2SA251J	2004	2SA261L	2004	2SA270OR
2004	2SA238F	2004	2SA251K	2004	2SA261M	2004	2SA270R
2004	2SA238G	2004	2SA251L	2004	2SA261OR	2004	2SA270X
2004	2SA238GN	2004	2SA251M	2004	2SA261R	2004	2SA270Y
2004	2SA238H	2004	2SA251OR	2004	2SA261X	2004	2SA276
2004	2SA238J	2004	2SA251R	2004	2SA261Y	2004	2SA277A
2004	2SA238K	2004	2SA251X	2004	2SA262A	2004	2SA277B
2004	2SA238L	2004	2SA251Y	2004	2SA262B	2004	2SA277C
2004	2SA239	2004	2SA252A	2004	2SA262C	2004	2SA277D
2004	2SA239A	2004	2SA252B	2004	2SA262D	2004	2SA277E
2004	2SA239B	2004	2SA252C	2004	2SA262E	2004	2SA277F
2004	2SA239C	2004	2SA252D	2004	2SA262F	2004	2SA277G
2004	2SA239D	2004	2SA252E	2004	2SA262G	2004	2SA277GN
2004	2SA239E	2004	2SA252F	2004	2SA262GN	2004	2SA277H
2004	2SA239F	2004	2SA252G	2004	2SA262H	2004	2SA277J
2004	2SA239G	2004	2SA252H	2004	2SA262J	2004	2SA277K
2004	2SA239GN	2004	2SA252J	2004	2SA262K	2004	2SA277L
2004	2SA239GREEN	2004	2SA252K	2004	2SA262L	2004	2SA277M
2004	2SA239H	2004	2SA252L	2004	2SA262M	2004	2SA277OR
2004	2SA239J	2004	2SA252M	2004	2SA262OR	2004	2SA277X
2004	2SA239K	2004	2SA252OR	2004	2SA262R	2004	2SA277Y
2004	2SA239L	2004	2SA252R	2004	2SA262X	2004	2SA278A
2004	2SA239M	2004	2SA252X	2004	2SA262Y	2004	2SA278B
2004	2SA239OR	2004	2SA252Y	2004	2SA263A	2004	2SA278C
2004	2SA239R	2004	2SA253A	2004	2SA263B	2004	2SA278D
2004	2SA239RED	2004	2SA253B	2004	2SA263C	2004	2SA278E
2004	2SA239X	2004	2SA253C	2004	2SA263D	2004	2SA278F
2004	2SA239Y	2004	2SA253D	2004	2SA263E	2004	2SA278G
2004	2SA24	2004	2SA253E	2004	2SA263F	2004	2SA278GN
2004	2SA240	2004	2SA253F	2004	2SA263G	2004	2SA278J
2004	2SA240A	2004	2SA253G	2004	2SA263GN	2004	2SA278K
2004	2SA240B	2004	2SA253GN	2004	2SA263H	2004	2SA278L
2004	2SA240B2	2004	2SA253H	2004	2SA263J	2004	2SA278M
2004	2SA240BL	2004	2SA253J	2004	2SA263K	2004	2SA278OR
2004	2SA240C	2004	2SA253K	2004	2SA263L	2004	2SA278R
2004	2SA240D	2004	2SA253L	2004	2SA263M	2004	2SA278X
2004	2SA240E	2004	2SA253M	2004	2SA263OR	2004	2SA278Y
2004	2SA240F	2004	2SA253OR	2004	2SA263R	2004	2SA279A
2004	2SA240G	2004	2SA253R	2004	2SA263X	2004	2SA279B
2004	2SA240GN	2004	2SA253X	2004	2SA264(1)	2004	2SA279C
2004	2SA240GREEN	2004	2SA253Y	2004	2SA264A	2004	2SA279D
2004	2SA240H	2004	2SA254A	2004	2SA264B	2004	2SA279E
2004	2SA240J	2004	2SA254B	2004	2SA264C	2004	2SA279F
2004	2SA240K	2004	2SA254C	2004	2SA264D	2004	2SA279G
2004	2SA240L	2004	2SA254D	2004	2SA264E	2004	2SA279GN
2004	2SA240M	2004	2SA254F	2004	2SA264F	2004	2SA279H
2004	2SA240OR	2004	2SA254G	2004	2SA264G	2004	2SA279J
2004	2SA240R	2004	2SA254H	2004	2SA264GN	2004	2SA279K
2004	2SA240RED	2004	2SA254J	2004	2SA264H	2004	2SA279L
2004	2SA240X	2004	2SA254K	2004	2SA264K	2004	2SA279OR
2004	2SA240Y	2004	2SA254L	2004	2SA264L	2004	2SA279R
2004	2SA241	2004	2SA254M	2004	2SA264M	2004	2SA279X
2004	2SA241A	2004	2SA254OR	2004	2SA264O	2004	2SA279Y
2004	2SA241B	2004	2SA254R	2004	2SA264R	2004	2SA280A
2004	2SA241C	2004	2SA254X	2004	2SA264X	2004	2SA280B
2004	2SA241D	2004	2SA254Y	2004	2SA264Y	2004	2SA280C
2004	2SA241E	2004	2SA255A	2004	2SA265A	2004	2SA280E
2004	2SA241F	2004	2SA255B	2004	2SA265B	2004	2SA280F
2004	2SA241G	2004	2SA255C	2004	2SA265C	2004	2SA280G
2004	2SA241GN	2004	2SA255D	2004	2SA265D	2004	2SA280GN
2004	2SA241H	2004	2SA255E	2004	2SA265E	2004	2SA280H
2004	2SA241J	2004	2SA255F	2004	2SA265F	2004	2SA280J
2004	2SA241K	2004	2SA255G	2004	2SA265G	2004	2SA280K
2004	2SA241L	2004	2SA255O	2004	2SA265O	2004	2SA280L
2004	2SA241M	2004	2SA255GN	2004	2SA265GN	2004	2SA280M
2004	2SA241OR	2004	2SA255H	2004	2SA265H	2004	2SA280OR
2004	2SA241R	2004	2SA255J	2004	2SA265J	2004	2SA280R
2004	2SA241X	2004	2SA255K	2004	2SA265K	2004	2SA280X
2004	2SA241Y	2004	2SA255L	2004	2SA265L	2004	2SA280Y
2004	2SA242	2004	2SA255M	2004	2SA265M	2004	2SA281A
2004	2SA243	2004	2SA255OR	2004	2SA265OR	2004	2SA281B
2004	2SA244	2004	2SA255R	2004	2SA265R	2004	2SA281C
2004	2SA245	2004	2SA255X	2004	2SA265X	2004	2SA281D
2004	2SA246A	2004	2SA255Y	2004	2SA265Y	2004	2SA281E
2004	2SA246B	2004	2SA257A	2004	2SA267A	2004	2SA281F
2004	2SA246C	2004	2SA257B	2004	2SA267B	2004	2SA281G
2004	2SA246D	2004	2SA257C	2004	2SA267C	2004	2SA281GN
2004	2SA246E	2004	2SA257D	2004	2SA267D	2004	2SA281H
2004	2SA246F	2004	2SA257E	2004	2SA267E	2004	2SA281J
2004	2SA246G	2004	2SA257F	2004	2SA267F	2004	2SA281K
2004	2SA246GN	2004	2SA257G	2004	2SA267G	2004	2SA281L
2004	2SA246H	2004	2SA257GN	2004	2SA267GN	2004	2SA281M
2004	2SA246J	2004	2SA257H	2004	2SA267H	2004	2SA281OR
2004	2SA246K	2004	2SA257J	2004	2SA267J		

RS 276-	Industry Standard No.	RS 276-	Industry Standard No.	RS 276-	Industry Standard No.	RS 276-	Industry Standard No.
2004	2SA281R	2004	2SA292Y	2004	2SA309OR	2004	2SA329E
2004	2SA281X	2004	2SA293A	2004	2SA309R	2004	2SA329F
2004	2SA281Y	2004	2SA293B	2004	2SA309X	2004	2SA329G
2004	2SA282A	2004	2SA293C	2004	2SA309Y	2004	2SA329GN
2004	2SA282B	2004	2SA293D	2004	2SA30A	2004	2SA329H
2004	2SA282C	2004	2SA293E	2004	2SA30B	2004	2SA329J
2004	2SA282D	2004	2SA293G	2004	2SA30C	2004	2SA329K
2004	2SA282E	2004	2SA293GN	2004	2SA30D	2004	2SA329L
2004	2SA282P	2004	2SA293H	2004	2SA30F	2004	2SA329M
2004	2SA282G	2004	2SA293J	2004	2SA30G	2004	2SA329OR
2004	2SA282GN	2004	2SA293K	2004	2SA30L	2004	2SA329R
2004	2SA282H	2004	2SA293L	2004	2SA30M	2004	2SA329X
2004	2SA282J	2004	2SA293M	2004	2SA30OR	2004	2SA329Y
2004	2SA282K	2004	2SA293OR	2004	2SA30X	2004	2SA32A
2004	2SA282L	2004	2SA293R	2004	2SA30Y	2004	2SA32B
2004	2SA282M	2004	2SA293X	2004	2SA310A	2004	2SA32C
2004	2SA282OR	2004	2SA293Y	2004	2SA310B	2004	2SA32E
2004	2SA282R	2004	2SA294A	2004	2SA310C	2004	2SA32F
2004	2SA282X	2004	2SA294B	2004	2SA310D	2004	2SA32G
2004	2SA282Y	2004	2SA294C	2004	2SA310E	2004	2SA32L
2004	2SA283A	2004	2SA294D	2004	2SA310F	2004	2SA32M
2004	2SA283B	2004	2SA294E	2004	2SA310G	2004	2SA32OR
2004	2SA283C	2004	2SA294F	2004	2SA310GN	2004	2SA32X
2004	2SA283D	2004	2SA294GN	2004	2SA310H	2004	2SA32Y
2004	2SA283E	2004	2SA294H	2004	2SA310J	2004	2SA330A
2004	2SA283F	2004	2SA294J	2004	2SA310K	2004	2SA330B
2004	2SA283G	2004	2SA294K	2004	2SA310L	2004	2SA330C
2004	2SA283GN	2004	2SA294L	2004	2SA310M	2004	2SA330D
2004	2SA283H	2004	2SA294M	2004	2SA310OR	2004	2SA330E
2004	2SA283J	2004	2SA294OR	2004	2SA310R	2004	2SA330F
2004	2SA283K	2004	2SA294R	2004	2SA310X	2004	2SA330G
2004	2SA283L	2004	2SA294X	2004	2SA310Y	2004	2SA330GN
2004	2SA283M	2004	2SA294Y	2004	2SA311A	2004	2SA330H
2004	2SA283OR	2004	2SA295A	2004	2SA311B	2004	2SA330J
2004	2SA283R	2004	2SA295B	2004	2SA311C	2004	2SA330K
2004	2SA283X	2004	2SA295C	2004	2SA311D	2004	2SA330L
2004	2SA283Y	2004	2SA295D	2004	2SA311E	2004	2SA330M
2004	2SA284A	2004	2SA295E	2004	2SA311F	2004	2SA330OR
2004	2SA284B	2004	2SA295F	2004	2SA311G	2004	2SA330X
2004	2SA284C	2004	2SA295G	2004	2SA311L	2004	2SA330Y
2004	2SA284D	2004	2SA295GN	2004	2SA311M	2004	2SA331A
2004	2SA284E	2004	2SA295H	2004	2SA311OR	2004	2SA331B
2004	2SA284F	2004	2SA295J	2004	2SA311X	2004	2SA331C
2004	2SA284G	2004	2SA295K	2004	2SA311Y	2004	2SA331D
2004	2SA284GN	2004	2SA295L	2004	2SA321	2004	2SA331E
2004	2SA284J	2004	2SA295OR	2004	2SA321-1	2004	2SA331F
2004	2SA284K	2004	2SA295R	2004	2SA321A	2004	2SA331G
2004	2SA284L	2004	2SA295Y	2004	2SA321B	2004	2SA331GN
2004	2SA284OR	2004	2SA296	2004	2SA321C	2004	2SA331H
2004	2SA284R	2004	2SA296A	2004	2SA321D	2004	2SA331J
2004	2SA284X	2004	2SA296B	2004	2SA321E	2004	2SA331K
2004	2SA284Y	2004	2SA296C	2004	2SA321F	2004	2SA331L
2004	2SA285	2004	2SA296D	2004	2SA321G	2004	2SA331M
2004	2SA286	2004	2SA296E	2004	2SA321GN	2004	2SA331OR
2004	2SA287	2004	2SA296F	2004	2SA321H	2004	2SA331R
2004	2SA289A	2004	2SA296G	2004	2SA321J	2004	2SA331X
2004	2SA289B	2004	2SA296GN	2004	2SA321K	2004	2SA331Y
2004	2SA289C	2004	2SA296H	2004	2SA321L	2004	2SA335A
2004	2SA289D	2004	2SA296J	2004	2SA321M	2004	2SA335B
2004	2SA289E	2004	2SA296K	2004	2SA321OR	2004	2SA335C
2004	2SA289F	2004	2SA296L	2004	2SA321R	2004	2SA335D
2004	2SA289G	2004	2SA296M	2004	2SA321X	2004	2SA335E
2004	2SA289GN	2004	2SA296OR	2004	2SA321Y	2004	2SA335F
2004	2SA289H	2004	2SA296R	2004	2SA322	2004	2SA335G
2004	2SA289J	2004	2SA296X	2004	2SA323	2004	2SA335GN
2004	2SA289K	2004	2SA296Y	2004	2SA323A	2004	2SA335H
2004	2SA289L	2004	2SA297	2004	2SA323B	2004	2SA335J
2004	2SA289M	2004	2SA297A	2004	2SA323C	2004	2SA335K
2004	2SA289OR	2004	2SA297B	2004	2SA323D	2004	2SA335L
2004	2SA289R	2004	2SA297C	2004	2SA323E	2004	2SA335M
2004	2SA289X	2004	2SA297D	2004	2SA323F	2004	2SA335OR
2004	2SA289Y	2004	2SA297E	2004	2SA323G	2004	2SA335R
2004	2SA28A	2004	2SA297F	2004	2SA323GN	2004	2SA335X
2004	2SA28B	2004	2SA297G	2004	2SA323J	2004	2SA335Y
2004	2SA28C	2004	2SA297H	2004	2SA323K	2004	2SA337A
2004	2SA28D	2004	2SA297J	2004	2SA323L	2004	2SA337B
2004	2SA28E	2004	2SA297K	2004	2SA323M	2004	2SA337C
2004	2SA28F	2004	2SA297L	2004	2SA323X	2004	2SA337D
2004	2SA28G	2004	2SA297M	2004	2SA323Y	2004	2SA337F
2004	2SA28L	2004	2SA297OR	2004	2SA324	2004	2SA337G
2004	2SA28M	2004	2SA297R	2004	2SA324A	2004	2SA337GN
2004	2SA28OR	2004	2SA297X	2004	2SA324B	2004	2SA337H
2004	2SA28X	2004	2SA297Y	2004	2SA324C	2004	2SA337J
2004	2SA28Y	2004	2SA298	2004	2SA324D	2004	2SA337K
2004	2SA290A	2004	2SA299	2004	2SA324E	2004	2SA337L
2004	2SA290B	2004	2SA300	2004	2SA324G	2004	2SA337M
2004	2SA290C	2004	2SA306	2004	2SA324GN	2004	2SA337OR
2004	2SA290D	2004	2SA307	2004	2SA324H	2004	2SA337R
2004	2SA290E	2004	2SA307A	2004	2SA324K	2004	2SA337X
2004	2SA290F	2004	2SA307B	2004	2SA324L	2004	2SA337Y
2004	2SA290G	2004	2SA307C	2004	2SA324M	2004	2SA339A
2004	2SA290GN	2004	2SA307D	2004	2SA324OR	2004	2SA339B
2004	2SA290H	2004	2SA307E	2004	2SA324R	2004	2SA339C
2004	2SA290J	2004	2SA307F	2004	2SA324X	2004	2SA339D
2004	2SA290K	2004	2SA307G	2004	2SA324Y	2004	2SA339E
2004	2SA290L	2004	2SA307GN	2004	2SA325A	2004	2SA339F
2004	2SA290M	2004	2SA307J	2004	2SA325B	2004	2SA339G
2004	2SA290OR	2004	2SA307K	2004	2SA325D	2004	2SA339GN
2004	2SA290R	2004	2SA307L	2004	2SA325E	2004	2SA339H
2004	2SA290X	2004	2SA307M	2004	2SA325F	2004	2SA339J
2004	2SA290Y	2004	2SA307OR	2004	2SA325GN	2004	2SA339K
2004	2SA291A	2004	2SA307R	2004	2SA325H	2004	2SA339L
2004	2SA291B	2004	2SA307X	2004	2SA325J	2004	2SA339M
2004	2SA291C	2004	2SA307Y	2004	2SA325K	2004	2SA339OR
2004	2SA291D	2004	2SA308A	2004	2SA325L	2004	2SA339X
2004	2SA291E	2004	2SA308B	2004	2SA325M	2004	2SA339Y
2004	2SA291F	2004	2SA308C	2004	2SA325OR	2004	2SA340
2004	2SA291G	2004	2SA308D	2004	2SA325R	2004	2SA341
2004	2SA291GN	2004	2SA308E	2004	2SA325X	2004	2SA341-OA
2004	2SA291H	2004	2SA308F	2004	2SA326A	2004	2SA341-OB
2004	2SA291J	2004	2SA308G	2004	2SA326B	2004	2SA341OA
2004	2SA291K	2004	2SA308GN	2004	2SA326C	2004	2SA341OB
2004	2SA291L	2004	2SA308H	2004	2SA326D	2004	2SA342
2004	2SA291M	2004	2SA308J	2004	2SA326E	2004	2SA342A
2004	2SA291R	2004	2SA308K	2004	2SA326F	2004	2SA342B
2004	2SA291X	2004	2SA308L	2004	2SA326G	2004	2SA342C
2004	2SA291Y	2004	2SA308M	2004	2SA326GN	2004	2SA342D
2004	2SA292A	2004	2SA308OR	2004	2SA326H	2004	2SA342E
2004	2SA292B	2004	2SA308X	2004	2SA326J	2004	2SA342F
2004	2SA292C	2004	2SA308Y	2004	2SA326K	2004	2SA342G
2004	2SA292D	2004	2SA309A	2004	2SA326L	2004	2SA342GN
2004	2SA292E	2004	2SA309B	2004	2SA326M	2004	2SA342H
2004	2SA292F	2004	2SA309C	2004	2SA326OR	2004	2SA342J
2004	2SA292G	2004	2SA309D	2004	2SA326Q	2004	2SA342K
2004	2SA292GN	2004	2SA309E	2004	2SA326X	2004	2SA342L
2004	2SA292H	2004	2SA309F	2004	2SA326Y	2004	2SA342M
2004	2SA292J	2004	2SA309G	2004	2SA327	2004	2SA342OR
2004	2SA292K	2004	2SA309GN	2004	2SA328	2004	2SA342X
2004	2SA292L	2004	2SA309H	2004	2SA329C	2004	2SA342Y
2004	2SA292M	2004	2SA309J	2004	2SA329D	2004	2SA343A
2004	2SA292OR	2004	2SA309K			2004	2SA343B
2004	2SA292R	2004	2SA309L				
2004	2SA292X	2004	2SA309M				

RS 276-2004	Industry Standard No.	RS 276-2004	Industry Standard No.	RS 276-2004	Industry Standard No.	RS 276-2004	Industry Standard No.
2004	28A343C	2004	28A355F	2004	28A367H	2004	28A381M
2004	28A343D	2004	28A355G	2004	28A367J	2004	28A381OR
2004	28A343E	2004	28A355H	2004	28A367K	2004	28A381R
2004	28A343P	2004	28A355J	2004	28A367L	2004	28A381X
2004	28A343G	2004	28A355K	2004	28A367M	2004	28A381Y
2004	28A343H	2004	28A355L	2004	28A367OR	2004	28A382A
2004	28A343J	2004	28A355M	2004	28A367R	2004	28A382C
2004	28A343K	2004	28A355OR	2004	28A367X	2004	28A382D
2004	28A343L	2004	28A355R	2004	28A367Y	2004	28A382E
2004	28A343M	2004	28A355X	2004	28A368A	2004	28A382F
2004	28A343OR	2004	28A355Y	2004	28A368B	2004	28A382G
2004	28A343R	2004	28A356A	2004	28A368C	2004	28A382GN
2004	28A343X	2004	28A356B	2004	28A368D	2004	28A382H
2004	28A343Y	2004	28A356C	2004	28A368E	2004	28A382J
2004	28A344A	2004	28A356D	2004	28A368F	2004	28A382K
2004	28A344B	2004	28A356E	2004	28A368G	2004	28A382L
2004	28A344C	2004	28A356F	2004	28A368GN	2004	28A382M
2004	28A344D	2004	28A356G	2004	28A368H	2004	28A382OR
2004	28A344E	2004	28A356GN	2004	28A368J	2004	28A382R
2004	28A344P	2004	28A356H	2004	28A368K	2004	28A382X
2004	28A344GN	2004	28A356J	2004	28A368L	2004	28A382Y
2004	28A344H	2004	28A356K	2004	28A368M	2004	28A383A
2004	28A344J	2004	28A356L	2004	28A368OR	2004	28A383B
2004	28A344K	2004	28A356M	2004	28A368X	2004	28A383C
2004	28A344L	2004	28A356OR	2004	28A368Y	2004	28A383D
2004	28A344M	2004	28A356R	2004	28A369A	2004	28A383E
2004	28A344OR	2004	28A358-3	2004	28A369B	2004	28A383F
2004	28A344R	2004	28A358A	2004	28A369C	2004	28A383G
2004	28A344X	2004	28A358B	2004	28A369D	2004	28A383GN
2004	28A344Y	2004	28A358C	2004	28A369E	2004	28A383H
2004	28A345	2004	28A358D	2004	28A369F	2004	28A383J
2004	28A346	2004	28A358E	2004	28A369G	2004	28A383K
2004	28A347	2004	28A358F	2004	28A369GN	2004	28A383L
2004	28A348	2004	28A358G	2004	28A369H	2004	28A383M
2004	28A348A	2004	28A358GN	2004	28A369J	2004	28A383OR
2004	28A348B	2004	28A358H	2004	28A369K	2004	28A383R
2004	28A348C	2004	28A358J	2004	28A369L	2004	28A383X
2004	28A348D	2004	28A358K	2004	28A369M	2004	28A383Y
2004	28A348E	2004	28A358L	2004	28A369OR	2004	28A384A
2004	28A348F	2004	28A358M	2004	28A369R	2004	28A384B
2004	28A348G	2004	28A358OR	2004	28A369X	2004	28A384C
2004	28A348GN	2004	28A358R	2004	28A369Y	2004	28A384D
2004	28A348H	2004	28A358X	2004	28A36A	2004	28A384F
2004	28A348J	2004	28A358Y	2004	28A36B	2004	28A384G
2004	28A348K	2004	28A35A	2004	28A36C	2004	28A384GN
2004	28A348L	2004	28A35B	2004	28A36D	2004	28A384H
2004	28A348M	2004	28A35C	2004	28A36E	2004	28A384J
2004	28A348OR	2004	28A35D	2004	28A36F	2004	28A384K
2004	28A348R	2004	28A35E	2004	28A36G	2004	28A384L
2004	28A348X	2004	28A35F	2004	28A36L	2004	28A384M
2004	28A348Y	2004	28A35G	2004	28A36M	2004	28A384OR
2004	28A349	2004	28A35L	2004	28A36OR	2004	28A384R
2004	28A350AV	2004	28A35M	2004	28A36R	2004	28A384X
2004	28A350B	2004	28A35OR	2004	28A36X	2004	28A384Y
2004	28A350BK	2004	28A35X	2004	28A36Y	2004	28A385B
2004	28A350D	2004	28A35Y	2004	28A372	2004	28A385C
2004	28A350E	2004	28A360	2004	28A373	2004	28A385E
2004	28A350P	2004	28A360A	2004	28A375	2004	28A385F
2004	28A350G	2004	28A360B	2004	28A376A	2004	28A385G
2004	28A350GN	2004	28A360C	2004	28A376B	2004	28A385GN
2004	28A350C	2004	28A360E	2004	28A376C	2004	28A385H
2004	28A350L	2004	28A360F	2004	28A376D	2004	28A385K
2004	28A350M	2004	28A360G	2004	28A376E	2004	28A385L
2004	28A350OR	2004	28A360GN	2004	28A376F	2004	28A385M
2004	28A350X	2004	28A360H	2004	28A376GN	2004	28A385OR
2004	28A351A-2	2004	28A360J	2004	28A376H	2004	28A385X
2004	28A351C	2004	28A360K	2004	28A376J	2004	28A385Y
2004	28A351D	2004	28A360L	2004	28A376K	2004	28A38A
2004	28A351E	2004	28A360M	2004	28A376L	2004	28A38B
2004	28A351F	2004	28A360OR	2004	28A376OR	2004	28A38C
2004	28A351G	2004	28A360X	2004	28A376R	2004	28A38D
2004	28A351GN	2004	28A360Y	2004	28A376X	2004	28A38F
2004	28A351K	2004	28A361	2004	28A376Y	2004	28A38G
2004	28A351L	2004	28A361A	2004	28A377	2004	28A38L
2004	28A351M	2004	28A361B	2004	28A377A	2004	28A38M
2004	28A351OR	2004	28A361C	2004	28A377B	2004	28A38OR
2004	28A351R	2004	28A361D	2004	28A377C	2004	28A38R
2004	28A351X	2004	28A361E	2004	28A377D	2004	28A38X
2004	28A351Y	2004	28A361GN	2004	28A377E	2004	28A38Y
2004	28A352C	2004	28A361H	2004	28A377F	2004	28A391A
2004	28A352D	2004	28A361J	2004	28A377G	2004	28A391B
2004	28A352E	2004	28A361K	2004	28A377GN	2004	28A391C
2004	28A352F	2004	28A361L	2004	28A377H	2004	28A391D
2004	28A352G	2004	28A361M	2004	28A377J	2004	28A391E
2004	28A352GN	2004	28A361OR	2004	28A377K	2004	28A391F
2004	28A352H	2004	28A361R	2004	28A377L	2004	28A391G
2004	28A352J	2004	28A361X	2004	28A377M	2004	28A391GN
2004	28A352K	2004	28A361Y	2004	28A377OR	2004	28A391H
2004	28A352L	2004	28A362	2004	28A377R	2004	28A391J
2004	28A352M	2004	28A363	2004	28A377X	2004	28A391K
2004	28A352OR	2004	28A364A	2004	28A377Y	2004	28A391L
2004	28A352R	2004	28A364B	2004	28A378	2004	28A391M
2004	28A352X	2004	28A364C	2004	28A379	2004	28A391OR
2004	28A352Y	2004	28A364D	2004	28A37A	2004	28A391R
2004	28A353AL	2004	28A364F	2004	28A37B	2004	28A391X
2004	28A353B	2004	28A364G	2004	28A37C	2004	28A391Y
2004	28A353CL	2004	28A364GN	2004	28A37D	2004	28A392A
2004	28A353D	2004	28A364H	2004	28A37E	2004	28A392B
2004	28A353E	2004	28A364J	2004	28A37F	2004	28A392C
2004	28A353F	2004	28A364K	2004	28A37G	2004	28A392D
2004	28A353G	2004	28A364L	2004	28A37L	2004	28A392E
2004	28A353GN	2004	28A364M	2004	28A37M	2004	28A392F
2004	28A353H	2004	28A364OR	2004	28A37OR	2004	28A392G
2004	28A353J	2004	28A364R	2004	28A37R	2004	28A392GN
2004	28A353K	2004	28A364X	2004	28A37X	2004	28A392H
2004	28A353L	2004	28A364Y	2004	28A37Y	2004	28A392J
2004	28A353M	2004	28A365A	2004	28A380A	2004	28A392K
2004	28A353OR	2004	28A365B	2004	28A380B	2004	28A392L
2004	28A353R	2004	28A365C	2004	28A380C	2004	28A392M
2004	28A353X	2004	28A365D	2004	28A380E	2004	28A392OR
2004	28A353Y	2004	28A365F	2004	28A380F	2004	28A392R
2004	28A354-B	2004	28A365G	2004	28A380G	2004	28A392X
2004	28A354BK	2004	28A365GN	2004	28A380GN	2004	28A392Y
2004	28A354C	2004	28A365H	2004	28A380H	2004	28A394A
2004	28A354D	2004	28A365J	2004	28A380J	2004	28A394B
2004	28A354E	2004	28A365K	2004	28A380K	2004	28A394C
2004	28A354F	2004	28A365L	2004	28A380L	2004	28A394D
2004	28A354G	2004	28A365M	2004	28A380M	2004	28A394E
2004	28A354GN	2004	28A365OR	2004	28A380OR	2004	28A394F
2004	28A354H	2004	28A365R	2004	28A380R	2004	28A394G
2004	28A354J	2004	28A365X	2004	28A380X	2004	28A394GN
2004	28A354K	2004	28A365Y	2004	28A380Y	2004	28A394H
2004	28A354L	2004	28A367A	2004	28A381A	2004	28A394J
2004	28A354M	2004	28A367B	2004	28A381B	2004	28A394K
2004	28A354OR	2004	28A367C	2004	28A381D	2004	28A394L
2004	28A354R	2004	28A367D	2004	28A381E	2004	28A394M
2004	28A354X	2004	28A367E	2004	28A381F	2004	28A394OR
2004	28A354Y	2004	28A367F	2004	28A381G	2004	28A394R
2004	28A355B	2004	28A367GN	2004	28A381GN	2004	28A394X
2004	28A355C			2004	28A381H		
2004	28A355D			2004	28A381K		
2004	28A355E			2004	28A381L		

RS 276-	Industry Standard No.	RS 276-	Industry Standard No.	RS 276-	Industry Standard No.	RS 276-	Industry Standard No.
2004	2SA394Y	2004	2SA408	2004	2SA428J	2004	2SA440GN
2004	2SA395A	2004	2SA409	2004	2SA428K	2004	2SA440H
2004	2SA395B	2004	2SA40A	2004	2SA428L	2004	2SA440J
2004	2SA395C	2004	2SA40B	2004	2SA428M	2004	2SA440K
2004	2SA395D	2004	2SA40C	2004	2SA428OR	2004	2SA440L
2004	2SA395E	2004	2SA40D	2004	2SA428R	2004	2SA440M
2004	2SA395F	2004	2SA40E	2004	2SA428X	2004	2SA440OR
2004	2SA395G	2004	2SA40F	2004	2SA428Y	2004	2SA440R
2004	2SA395GN	2004	2SA40G	2004	2SA430	2004	2SA440X
2004	2SA395H	2004	2SA40L	2004	2SA431	2004	2SA440Y
2004	2SA395J	2004	2SA40M	2004	2SA431A	2004	2SA448
2004	2SA395K	2004	2SA40OR	2004	2SA432	2004	2SA44A
2004	2SA395L	2004	2SA40R	2004	2SA432A	2004	2SA44B
2004	2SA395M	2004	2SA40X	2004	2SA432B	2004	2SA44C
2004	2SA395OR	2004	2SA40Y	2004	2SA432C	2004	2SA44D
2004	2SA395R	2004	2SA41	2004	2SA432D	2004	2SA44E
2004	2SA395X	2004	2SA410	2004	2SA432E	2004	2SA44F
2004	2SA395Y	2004	2SA411	2004	2SA432F	2004	2SA44G
2004	2SA398A	2004	2SA412A	2004	2SA432G	2004	2SA44L
2004	2SA398B	2004	2SA412B	2004	2SA432GN	2004	2SA44M
2004	2SA398C	2004	2SA412C	2004	2SA432H	2004	2SA44OR
2004	2SA398D	2004	2SA412D	2004	2SA432K	2004	2SA44R
2004	2SA398E	2004	2SA412E	2004	2SA432L	2004	2SA44Y
2004	2SA398F	2004	2SA412F	2004	2SA432M	2004	2SA450
2004	2SA398G	2004	2SA412G	2004	2SA432OR	2004	2SA450H
2004	2SA398GN	2004	2SA412GN	2004	2SA432R	2004	2SA451
2004	2SA398H	2004	2SA412H	2004	2SA432X	2004	2SA451H
2004	2SA398J	2004	2SA412J	2004	2SA432Y	2004	2SA452
2004	2SA398K	2004	2SA412K	2004	2SA433	2004	2SA452H
2004	2SA398L	2004	2SA412L	2004	2SA433A	2004	2SA453A
2004	2SA398M	2004	2SA412M	2004	2SA433B	2004	2SA453B
2004	2SA398OR	2004	2SA412OR	2004	2SA433C	2004	2SA453C
2004	2SA398X	2004	2SA412R	2004	2SA433D	2004	2SA453D
2004	2SA398Y	2004	2SA412X	2004	2SA433E	2004	2SA453E
2004	2SA399A	2004	2SA412Y	2004	2SA433F	2004	2SA453F
2004	2SA399B	2004	2SA413	2004	2SA433G	2004	2SA453G
2004	2SA399C	2004	2SA414A	2004	2SA433GN	2004	2SA453GN
2004	2SA399D	2004	2SA414B	2004	2SA433H	2004	2SA453H
2004	2SA399E	2004	2SA414C	2004	2SA433K	2004	2SA453J
2004	2SA399F	2004	2SA414D	2004	2SA433L	2004	2SA453K
2004	2SA399G	2004	2SA414E	2004	2SA433M	2004	2SA453L
2004	2SA399GN	2004	2SA414F	2004	2SA433OR	2004	2SA453M
2004	2SA399H	2004	2SA414G	2004	2SA433R	2004	2SA453OR
2004	2SA399J	2004	2SA414GN	2004	2SA433X	2004	2SA453R
2004	2SA399K	2004	2SA414H	2004	2SA433Y	2004	2SA453X
2004	2SA399L	2004	2SA414J	2004	2SA434	2004	2SA453Y
2004	2SA399M	2004	2SA414K	2004	2SA434A	2004	2SA454A
2004	2SA399OR	2004	2SA414L	2004	2SA434B	2004	2SA454B
2004	2SA399X	2004	2SA414M	2004	2SA434C	2004	2SA454C
2004	2SA399Y	2004	2SA414OR	2004	2SA434D	2004	2SA454D
2004	2SA39A	2004	2SA414R	2004	2SA434E	2004	2SA454E
2004	2SA39B	2004	2SA414X	2004	2SA434F	2004	2SA454F
2004	2SA39C	2004	2SA415A	2004	2SA434G	2004	2SA454G
2004	2SA39D	2004	2SA415B	2004	2SA434GN	2004	2SA454GN
2004	2SA39E	2004	2SA415C	2004	2SA434H	2004	2SA454H
2004	2SA39F	2004	2SA415D	2004	2SA434J	2004	2SA454J
2004	2SA39G	2004	2SA415E	2004	2SA434K	2004	2SA454K
2004	2SA39L	2004	2SA415F	2004	2SA434L	2004	2SA454L
2004	2SA39M	2004	2SA415G	2004	2SA434M	2004	2SA454OR
2004	2SA39OR	2004	2SA415H	2004	2SA434R	2004	2SA454X
2004	2SA39R	2004	2SA415J	2004	2SA434X	2004	2SA454Y
2004	2SA39X	2004	2SA415K	2004	2SA434Y	2004	2SA455A
2004	2SA39Y	2004	2SA415L	2004	2SA435	2004	2SA455B
2004	2SA401	2004	2SA415M	2004	2SA435A	2004	2SA455C
2004	2SA403	2004	2SA415OR	2004	2SA435B	2004	2SA455D
2004	2SA403A	2004	2SA415R	2004	2SA435C	2004	2SA455E
2004	2SA403B	2004	2SA415X	2004	2SA435D	2004	2SA455F
2004	2SA403C	2004	2SA415Y	2004	2SA435E	2004	2SA455G
2004	2SA403D	2004	2SA417	2004	2SA435F	2004	2SA455GN
2004	2SA403E	2004	2SA419	2004	2SA435G	2004	2SA455J
2004	2SA403F	2004	2SA41A	2004	2SA435GN	2004	2SA455K
2004	2SA403G	2004	2SA41B	2004	2SA435H	2004	2SA455L
2004	2SA403GN	2004	2SA41C	2004	2SA435J	2004	2SA455M
2004	2SA403H	2004	2SA41D	2004	2SA435L	2004	2SA455OR
2004	2SA403J	2004	2SA41E	2004	2SA435M	2004	2SA455R
2004	2SA403K	2004	2SA41F	2004	2SA435OR	2004	2SA455X
2004	2SA403L	2004	2SA41G	2004	2SA435R	2004	2SA455Y
2004	2SA403M	2004	2SA41L	2004	2SA435X	2004	2SA456A
2004	2SA403OR	2004	2SA41M	2004	2SA435Y	2004	2SA456B
2004	2SA403R	2004	2SA41OR	2004	2SA436A	2004	2SA456C
2004	2SA403X	2004	2SA41R	2004	2SA436B	2004	2SA456D
2004	2SA403Y	2004	2SA41X	2004	2SA436C	2004	2SA456E
2004	2SA404	2004	2SA41Y	2004	2SA436D	2004	2SA456F
2004	2SA404A	2004	2SA42	2004	2SA436E	2004	2SA456G
2004	2SA404B	2004	2SA420	2004	2SA436F	2004	2SA456GN
2004	2SA404C	2004	2SA420A	2004	2SA436G	2004	2SA456H
2004	2SA404D	2004	2SA420B	2004	2SA436GN	2004	2SA456J
2004	2SA404E	2004	2SA420C	2004	2SA436H	2004	2SA456K
2004	2SA404F	2004	2SA420D	2004	2SA436J	2004	2SA456L
2004	2SA404G	2004	2SA420E	2004	2SA436K	2004	2SA456M
2004	2SA404GN	2004	2SA420F	2004	2SA436L	2004	2SA456OR
2004	2SA404H	2004	2SA420G	2004	2SA437A	2004	2SA456R
2004	2SA404K	2004	2SA420H	2004	2SA437B	2004	2SA456X
2004	2SA404L	2004	2SA420J	2004	2SA437C	2004	2SA457A
2004	2SA404M	2004	2SA420K	2004	2SA437D	2004	2SA457B
2004	2SA404OR	2004	2SA420L	2004	2SA437E	2004	2SA457C
2004	2SA404R	2004	2SA420M	2004	2SA437F	2004	2SA457D
2004	2SA404X	2004	2SA420OR	2004	2SA437G	2004	2SA457E
2004	2SA404Y	2004	2SA420R	2004	2SA437GN	2004	2SA457F
2004	2SA405	2004	2SA420X	2004	2SA437H	2004	2SA457G
2004	2SA405-0	2004	2SA420Y	2004	2SA437J	2004	2SA457GN
2004	2SA406A	2004	2SA421	2004	2SA437K	2004	2SA457H
2004	2SA406B	2004	2SA422	2004	2SA437L	2004	2SA457J
2004	2SA406C	2004	2SA423	2004	2SA437M	2004	2SA457K
2004	2SA406D	2004	2SA424	2004	2SA437OR	2004	2SA457L
2004	2SA406E	2004	2SA425	2004	2SA437R	2004	2SA457M
2004	2SA406F	2004	2SA426	2004	2SA437X	2004	2SA457OR
2004	2SA406G	2004	2SA426GN	2004	2SA437Y	2004	2SA457R
2004	2SA406GN	2004	2SA427A	2004	2SA438A	2004	2SA457X
2004	2SA406H	2004	2SA427B	2004	2SA438B	2004	2SA457Y
2004	2SA406J	2004	2SA427C	2004	2SA438C	2004	2SA460
2004	2SA406K	2004	2SA427D	2004	2SA438D	2004	2SA461
2004	2SA406L	2004	2SA427E	2004	2SA438E	2004	2SA462
2004	2SA406M	2004	2SA427F	2004	2SA438F	2004	2SA463
2004	2SA406OR	2004	2SA427G	2004	2SA438G	2004	2SA464
2004	2SA406R	2004	2SA427GN	2004	2SA438GN	2004	2SA466A
2004	2SA406X	2004	2SA427H	2004	2SA438H	2004	2SA466B
2004	2SA406Y	2004	2SA427J	2004	2SA438J	2004	2SA466C
2004	2SA407A	2004	2SA427K	2004	2SA438K	2004	2SA466D
2004	2SA407B	2004	2SA427L	2004	2SA438L	2004	2SA466E
2004	2SA407C	2004	2SA427M	2004	2SA438M	2004	2SA466F
2004	2SA407D	2004	2SA427OR	2004	2SA438OR	2004	2SA466G
2004	2SA407E	2004	2SA427R	2004	2SA438R	2004	2SA466GN
2004	2SA407F	2004	2SA427X	2004	2SA438X	2004	2SA466J
2004	2SA407GN	2004	2SA427Y	2004	2SA438Y	2004	2SA466K
2004	2SA407H	2004	2SA428A	2004	2SA440	2004	2SA466L
2004	2SA407J	2004	2SA428B	2004	2SA440A	2004	2SA466M
2004	2SA407K	2004	2SA428C	2004	2SA440AL	2004	2SA466OR
2004	2SA407L	2004	2SA428D	2004	2SA440C	2004	2SA466R
2004	2SA407M	2004	2SA428E	2004	2SA440D		
2004	2SA407OR	2004	2SA428F	2004	2SA440F		
2004	2SA407R	2004	2SA428GN	2004	2SA440G		
2004	2SA407X	2004	2SA428H				
2004	2SA407Y						

RS 276-	Industry Standard No.	RS 276-	Industry Standard No.	RS 276-	Industry Standard No.	RS 276-	Industry Standard No.
2004	2SA466X	2004	2SA490R	2004	2SA55M	2004	2SA70L
2004	2SA466Y	2004	2SA49R	2004	2SA550R	2004	2SA70MA
2004	2SA468A	2004	2SA49X	2004	2SA55R	2004	2SA700A
2004	2SA468B	2004	2SA49Y	2004	2SA55X	2004	2SA700R
2004	2SA468C	2004	2SA506	2004	2SA55Y	2004	2SA70R
2004	2SA468D	2004	2SA507	2004	2SA56	2004	2SA70X
2004	2SA468E	2004	2SA507A	2004	2SA581	2004	2SA70Y
2004	2SA468F	2004	2SA507B	2004	2SA58A	2004	2SA71C
2004	2SA468G	2004	2SA507C	2004	2SA58B	2004	2SA71E
2004	2SA468GN	2004	2SA507D	2004	2SA58C	2004	2SA71F
2004	2SA468H	2004	2SA507E	2004	2SA58D	2004	2SA71G
2004	2SA468J	2004	2SA507F	2004	2SA58E	2004	2SA71H
2004	2SA468K	2004	2SA507G	2004	2SA58F	2004	2SA71K
2004	2SA468L	2004	2SA507GN	2004	2SA58G	2004	2SA71L
2004	2SA468M	2004	2SA507H	2004	2SA58H	2004	2SA71M
2004	2SA468OR	2004	2SA507J	2004	2SA58J	2004	2SA71OR
2004	2SA468R	2004	2SA507K	2004	2SA58K	2004	2SA71R
2004	2SA468X	2004	2SA507L	2004	2SA58L	2004	2SA71Y
2004	2SA468Y	2004	2SA507OR	2004	2SA58M	2004	2SA71YA
2004	2SA469A	2004	2SA507R	2004	2SA580R	2004	2SA73
2004	2SA469B	2004	2SA507X	2004	2SA58R	2004	2SA73A
2004	2SA469C	2004	2SA507Y	2004	2SA58X	2004	2SA73B
2004	2SA469D	2004	2SA50B	2004	2SA58Y	2004	2SA73C
2004	2SA469E	2004	2SA50A	2004	2SA59A	2004	2SA73D
2004	2SA469F	2004	2SA50B	2004	2SA59B	2004	2SA73E
2004	2SA469G	2004	2SA50C	2004	2SA59C	2004	2SA73F
2004	2SA469GN	2004	2SA50D	2004	2SA59D	2004	2SA73G
2004	2SA469H	2004	2SA50E	2004	2SA59E	2004	2SA73H
2004	2SA469J	2004	2SA50F	2004	2SA59F	2004	2SA73K
2004	2SA469K	2004	2SA50G	2004	2SA59G	2004	2SA73L
2004	2SA469L	2004	2SA50L	2004	2SA59L	2004	2SA73M
2004	2SA469M	2004	2SA50M	2004	2SA59M	2004	2SA730R
2004	2SA469OR	2004	2SA500R	2004	2SA590R	2004	2SA73R
2004	2SA469R	2004	2SA50R	2004	2SA59R	2004	2SA73X
2004	2SA469X	2004	2SA50X	2004	2SA59X	2004	2SA73Y
2004	2SA469Y	2004	2SA50Y	2004	2SA59Y	2004	2SA74
2004	2SA470A	2004	2SA51	2004	2SA60A	2004	2SA74A
2004	2SA470B	2004	2SA518-Q	2004	2SA60B	2004	2SA74B
2004	2SA470C	2004	2SA518A	2004	2SA60C	2004	2SA74C
2004	2SA470D	2004	2SA518B	2004	2SA60D	2004	2SA74D
2004	2SA470E	2004	2SA518C	2004	2SA60E	2004	2SA74F
2004	2SA470F	2004	2SA518D	2004	2SA60F	2004	2SA74G
2004	2SA470G	2004	2SA518E	2004	2SA60G	2004	2SA74H
2004	2SA470GN	2004	2SA518F	2004	2SA60H	2004	2SA74K
2004	2SA470H	2004	2SA518G	2004	2SA60K	2004	2SA74L
2004	2SA470J	2004	2SA518GN	2004	2SA60L	2004	2SA74M
2004	2SA470K	2004	2SA518H	2004	2SA60M	2004	2SA740R
2004	2SA470L	2004	2SA518J	2004	2SA600R	2004	2SA74R
2004	2SA470M	2004	2SA518K	2004	2SA60R	2004	2SA74X
2004	2SA470OR	2004	2SA518L	2004	2SA60X	2004	2SA74Y
2004	2SA470R	2004	2SA518M	2004	2SA60Y	2004	2SA76A
2004	2SA470X	2004	2SA518OR	2004	2SA61	2004	2SA76B
2004	2SA470Y	2004	2SA518R	2004	2SA61A	2004	2SA76C
2004	2SA471A	2004	2SA518X	2004	2SA61C	2004	2SA76D
2004	2SA471B	2004	2SA518Y	2004	2SA61D	2004	2SA76E
2004	2SA471C	2004	2SA51A	2004	2SA61E	2004	2SA76F
2004	2SA471D	2004	2SA51B	2004	2SA61G	2004	2SA76G
2004	2SA471F	2004	2SA51C	2004	2SA61K	2004	2SA76H
2004	2SA471G	2004	2SA51D	2004	2SA61L	2004	2SA76K
2004	2SA471GN	2004	2SA51E	2004	2SA61OR	2004	2SA760R
2004	2SA471H	2004	2SA51F	2004	2SA61R	2004	2SA76R
2004	2SA471J	2004	2SA51G	2004	2SA61X	2004	2SA76X
2004	2SA471K	2004	2SA51L	2004	2SA61Y	2004	2SA76Y
2004	2SA471L	2004	2SA51M	2004	2SA65A	2004	2SA77E
2004	2SA471M	2004	2SA51OR	2004	2SA65B	2004	2SA77F
2004	2SA471OR	2004	2SA51R	2004	2SA65C	2004	2SA77G
2004	2SA471R	2004	2SA51X	2004	2SA65D	2004	2SA77H
2004	2SA471X	2004	2SA51Y	2004	2SA65E	2004	2SA77K
2004	2SA471Y	2004	2SA525	2004	2SA65F	2004	2SA77L
2004	2SA472-1	2004	2SA525A	2004	2SA65G	2004	2SA77M
2004	2SA472-2	2004	2SA525B	2004	2SA65K	2004	2SA770R
2004	2SA472-3	2004	2SA525C	2004	2SA65L	2004	2SA77R
2004	2SA472-4	2004	2SA525D	2004	2SA65M	2004	2SA77X
2004	2SA472-5	2004	2SA525E	2004	2SA650R	2004	2SA77Y
2004	2SA472-6	2004	2SA525F	2004	2SA65R	2004	2SA79
2004	2SA472BB	2004	2SA525GN	2004	2SA65X	2004	2SA81
2004	2SA472F	2004	2SA525H	2004	2SA65Y	2004	2SA83B
2004	2SA472G	2004	2SA525J	2004	2SA663-R	2004	2SA83A
2004	2SA472GN	2004	2SA525K	2004	2SA66A	2004	2SA83B
2004	2SA472H	2004	2SA525L	2004	2SA66B	2004	2SA83C
2004	2SA472J	2004	2SA525M	2004	2SA66C	2004	2SA83D
2004	2SA472K	2004	2SA525OR	2004	2SA66D	2004	2SA83E
2004	2SA472L	2004	2SA525R	2004	2SA66E	2004	2SA83F
2004	2SA472M	2004	2SA525X	2004	2SA66F	2004	2SA83G
2004	2SA472OR	2004	2SA525Y	2004	2SA66G	2004	2SA83H
2004	2SA472R	2004	2SA52A	2004	2SA66K	2004	2SA83K
2004	2SA472X	2004	2SA52B	2004	2SA66L	2004	2SA83L
2004	2SA472Y	2004	2SA52C	2004	2SA66M	2004	2SA83M
2004	2SA475	2004	2SA52D	2004	2SA660R	2004	2SA830R
2004	2SA478A	2004	2SA52E	2004	2SA66R	2004	2SA83R
2004	2SA478B	2004	2SA52G	2004	2SA66X	2004	2SA83X
2004	2SA478C	2004	2SA52L	2004	2SA66Y	2004	2SA83Y
2004	2SA478D	2004	2SA52M	2004	2SA67A	2004	2SA84
2004	2SA478E	2004	2SA520R	2004	2SA67B	2004	2SA84A
2004	2SA478F	2004	2SA52R	2004	2SA67C	2004	2SA84B
2004	2SA478GN	2004	2SA52X	2004	2SA67D	2004	2SA84C
2004	2SA478H	2004	2SA52Y	2004	2SA67F	2004	2SA84D
2004	2SA478J	2004	2SA536	2004	2SA67G	2004	2SA84E
2004	2SA478K	2004	2SA53A	2004	2SA67H	2004	2SA84F
2004	2SA478L	2004	2SA53B	2004	2SA67K	2004	2SA84G
2004	2SA478M	2004	2SA53C	2004	2SA67L	2004	2SA84H
2004	2SA478OR	2004	2SA53D	2004	2SA670R	2004	2SA84K
2004	2SA478R	2004	2SA53E	2004	2SA67R	2004	2SA84L
2004	2SA478X	2004	2SA53F	2004	2SA67X	2004	2SA84M
2004	2SA478Y	2004	2SA53G	2004	2SA67Y	2004	2SA840R
2004	2SA479A	2004	2SA53L	2004	2SA69A	2004	2SA84R
2004	2SA479B	2004	2SA53M	2004	2SA69B	2004	2SA84X
2004	2SA479C	2004	2SA530R	2004	2SA69C	2004	2SA84Y
2004	2SA479D	2004	2SA53R	2004	2SA69D	2004	2SA86
2004	2SA479E	2004	2SA53X	2004	2SA69E	2004	2SA87A
2004	2SA479F	2004	2SA53Y	2004	2SA69F	2004	2SA87B
2004	2SA479G	2004	2SA54	2004	2SA69G	2004	2SA87C
2004	2SA479GN	2004	2SA54A	2004	2SA69H	2004	2SA87D
2004	2SA479H	2004	2SA54B	2004	2SA69K	2004	2SA87E
2004	2SA479J	2004	2SA54C	2004	2SA69L	2004	2SA87F
2004	2SA479K	2004	2SA54D	2004	2SA69M	2004	2SA87G
2004	2SA479L	2004	2SA54E	2004	2SA690R	2004	2SA87H
2004	2SA479M	2004	2SA54F	2004	2SA69R	2004	2SA87K
2004	2SA479OR	2004	2SA54G	2004	2SA69X	2004	2SA87L
2004	2SA479R	2004	2SA54L	2004	2SA69Y	2004	2SA87M
2004	2SA479X	2004	2SA54M	2004	2SA70	2004	2SA870R
2004	2SA479Y	2004	2SA540R	2004	2SA70-0S	2004	2SA87R
2004	2SA48	2004	2SA54R	2004	2SA70-0B	2004	2SA87X
2004	2SA49A	2004	2SA54X	2004	2SA70-0B	2004	2SA87Y
2004	2SA49B	2004	2SA54Y	2004	2SA70A	2004	2SA88
2004	2SA49C	2004	2SA55A	2004	2SA70B	2004	2SA89
2004	2SA49D	2004	2SA55B	2004	2SA70C	2004	2SA92A
2004	2SA49E	2004	2SA55C	2004	2SA70D	2004	2SA92B
2004	2SA49F	2004	2SA55D	2004	2SA70E	2004	2SA92C
2004	2SA49G	2004	2SA55E	2004	2SA70F	2004	2SA92D
2004	2SA49L	2004	2SA55F	2004	2SA70G	2004	2SA92E
2004	2SA49M	2004	2SA55G	2004	2SA70H	2004	2SA92F
2004		2004	2SA55L	2004	2SA70K	2004	

RS 276-	Industry Standard No.	RS 276-	Industry Standard No.	RS 276-	Industry Standard No.	RS 276-	Industry Standard No.
2004	2SA92G	2004	2T201	2004	9-9120	2004	31-0161
2004	2SA92H	2004	2T203	2004	9-9121	2004	31-0163
2004	2SA92K	2004	2T204	2004	9A8-1A64	2004	31-0165
2004	2SA92L	2004	2T204A	2004	10A	2004	31-0166
2004	2SA92M	2004	2T205	2004	10B	2004	31-0168
2004	2SA92OR	2004	2T205A	2004	12-1-135	2004	31-0170
2004	2SA92R	2004	2Y485	2004	12-1-137	2004	31-0171
2004	2SA92X	2004	2Y559	2004	12-1-138	2004	31-0180
2004	2SA92Y	2004	2Y560	2004	12-1-139	2004	31-0181
2004	2SA93A	2004	2Y561	2004	12-1-150	2004	31-0241
2004	2SA93B	2004	2Y562	2004	12-1-157	2004	31-0241-1
2004	2SA93C	2004	2Y563	2004	12-1-190	2004	31-21004900
2004	2SA93E	2004	03-57-003	2004	12-1-228	2004	31-21007744
2004	2SA93F	2004	03-57-102	2004	12-1-229	2004	31-21024033
2004	2SA93G	2004	03-57-200	2004	12-1-230	2004	31-21024044
2004	2SA93H	2004	03-57-202	2004	12-1-231	2004	31-21047111
2004	2SA93K	2004	3MC	2004	12-1-233	2004	31-2104733
2004	2SA93L	2004	3N25/501	2004	12-1-256	2004	31-21050611
2004	2SA93M	2004	3N34	2004	12-1-258	2004	31-21050622
2004	2SA93OR	2004	3B35A	2004	12-1-259	2004	32-12066-10
2004	2SA93R	2004	3BA324	2004	12-1-260	2004	34-119
2004	2SA93X	2004	4-2073	2004	12A9244-1	2004	34-220
2004	2SA93Y	2004	4-432	2004	12A9244-P2	2004	34-221
2004	2SAJ150M	2004	4-44-0012-PT2	2004	12MC	2004	34-298
2004	2SAU/5H	2004	4-65-1A7-1	2004	12MZ	2004	34-6
2004	2SAUO5H	2004	4-65A17-1	2004	13-14085-28	2004	34-6000-10
2004	2SBB120	2004	4-65B17-1	2004	13-14085-30	2004	34-6000-11
2004	2SBB120A	2004	4-88B17-1	2004	13-18946-1	2004	34-6000-12
2004	2SBB120B	2004	4A8-1A5	2004	13-18946-2	2004	34-6000-13
2004	2SBB120C	2004	4A8-1A7-1	2004	13-18947-1	2004	34-6000-14
2004	2SBB120D	2004	4JX1A813	2004	13-18948-1	2004	34-6000-16
2004	2SBB120E	2004	4JX1C707	2004	13-18948-2	2004	34-6000-17
2004	2SBB120F	2004	6-1260039	2004	13-18950-2	2004	34-6000-18
2004	2SBB120G	2004	6-1260039A	2004	14-569-09	2004	34-6000-19
2004	2SBB120GN	2004	6-60A	2004	14-580-01	2004	34-6000-20
2004	2SBB120H	2004	6-60C	2004	14-581-01	2004	34-6000-26
2004	2SBB120J	2004	6-60B	2004	14-585-01	2004	34-6000-3
2004	2SBB120K	2004	6-60F	2004	14-587-01	2004	34-6000-58
2004	2SBB120L	2004	6-60T	2004	14-588-01	2004	34-6000-59
2004	2SBB120M	2004	6-60X	2004	14-591-01	2004	34-6000-60
2004	2SBB120OR	2004	6-61A	2004	14-600-01	2004	34-6000-61
2004	2SBB120R	2004	6-61E	2004	14-600-02	2004	34-6000-62
2004	2SBB120X	2004	6-61F	2004	14-600-10	2004	34-6000-63
2004	2SBB120Y	2004	6-61T	2004	14-600-11	2004	34-6000-65
2004	2SB158A	2004	6-61X	2004	14-600-13	2004	34-6000-66
2004	2SB158B	2004	6-62C	2004	14-600-16	2004	34-6000-67
2004	2SB158C	2004	6-62E	2004	14-600-19	2004	34-6000-68
2004	2SB158D	2004	6-69	2004	14-600-20	2004	34-6000-76
2004	2SB158E	2004	6-69X	2004	14-600-22	2004	34-6000-77
2004	2SB158F	2004	6-84F	2004	14A8-1	2004	34-6000-78
2004	2SB158G	2004	6-85F	2004	14A8-1-12	2004	34-6000-79
2004	2SB158GN	2004	6A12889	2004	15-2210921	2004	34-6000-80
2004	2SB158H	2004	6A8-1A5L	2004	15-22210131	2004	34-6000-81
2004	2SB158J	2004	6MC	2004	15-22210300	2004	34-6000-82
2004	2SB158K	2004	07-3080-06	2004	15-22210921	2004	34-6000-9
2004	2SB158L	2004	07-4233-19	2004	15-22211021	2004	34-6005-1
2004	2SB158M	2004	07-4235-13	2004	15-22211921	2004	34-6015-32
2004	2SB158OR	2004	07-4235-73	2004	15-22214400	2004	34-6015-33
2004	2SB158R	2004	8-0050400	2004	15-22214411	2004	34-6015-34
2004	2SB158X	2004	8-0050500	2004	15-22214435	2004	34-6015-35
2004	2SB158Y	2004	8-0104900	2004	15-22214821	2004	34-6015-36
2004	2SB159A	2004	8-0105200	2004	15-22214831	2004	34-6015-39
2004	2SB159B	2004	8-0105300	2004	15-22216600	2004	34-6015-40
2004	2SB159C	2004	8-905-605-320	2004	15-22217400	2004	34-6016-28
2004	2SB159D	2004	8-905-606-001	2004	18A8-1	2004	34-6016-29
2004	2SB159E	2004	8-905-606-003	2004	18A8-1-12	2004	36(SEARS)
2004	2SB159F	2004	8-905-606-007	2004	18A8-1-127	2004	40D1547
2004	2SB159G	2004	8-905-606-008	2004	18A8-1L	2004	42-19682
2004	2SB159GN	2004	8-905-606-010	2004	18A8-1L8	2004	42-19792
2004	2SB159H	2004	8-905-606-016	2004	019-003315	2004	42-22778
2004	2SB159J	2004	8-905-606-051	2004	019-003778	2004	42-22779
2004	2SB159K	2004	8-905-606-075	2004	19-020-031	2004	42-22780
2004	2SB159L	2004	8-905-606-077	2004	19-020-032	2004	42-22781
2004	2SB159M	2004	8-905-606-090	2004	19A115140-P1	2004	42-22784
2004	2SB159OR	2004	8-905-606-105	2004	19A115140-P2	2004	42-23965P
2004	2SB159R	2004	8-905-606-106	2004	19A115192-P1	2004	43A111449
2004	2SB159X	2004	8-905-606-120	2004	19A115192-P2	2004	44P1
2004	2SB159Y	2004	8-905-606-142	2004	19A115553-P1	2004	46-8612-3
2004	2SB160A	2004	8-905-606-152	2004	19A115553-P2	2004	46-86123-3
2004	2SB160B	2004	8-905-606-153	2004	19A115554-P1	2004	46-8613-3(SYNC)
2004	2SB160C	2004	8-905-606-154	2004	19A115554-P2	2004	46-862-3
2004	2SB160D	2004	8-905-606-155	2004	19A115567-P1	2004	46-86300-3
2004	2SB160E	2004	8-905-606-158	2004	19A115567-P2	2004	46-865-3
2004	2SB160F	2004	8-905-606-165	2004	19A115628-P1	2004	46-866-3
2004	2SB160G	2004	8-905-606-168	2004	19A115628-P2	2004	46-868-3
2004	2SB160GN	2004	8-905-606-180	2004	19A115635-1	2004	48-10079A01
2004	2SB160H	2004	8-905-606-211	2004	19A115635-P1	2004	48-10079A02
2004	2SB160J	2004	8-905-606-225	2004	19A115636-P1	2004	48-124296
2004	2SB160K	2004	8-905-606-241	2004	19A115665-P1	2004	48-124363
2004	2SB160L	2004	8-905-606-255	2004	19A115665-P2	2004	48-124368
2004	2SB160M	2004	8-905-606-256	2004	19A126265-1	2004	48-128219
2004	2SB160OR	2004	8-905-606-349	2004	19A126265-2	2004	48-134411
2004	2SB160R	2004	8-905-606-350	2004	19AR13-1	2004	48-134412
2004	2SB160X	2004	8-905-606-351	2004	19AR13-2	2004	48-134413
2004	2SB160Y	2004	8-905-606-352	2004	19AR13-3	2004	48-134439
2004	2SB179	2004	8-905-606-360	2004	19AR13-4	2004	48-134484
2004	2SB182	2004	8-905-606-375	2004	19AR18	2004	48-134485
2004	2SB240	2004	8-905-606-390	2004	19AR24	2004	48-134486
2004	2SB240A	2004	8-905-606-391	2004	19B2000130-P1	2004	48-134504
2004	2SB277	2004	8-905-606-392	2004	19B2000130-P2	2004	48-134506
2004	2SB302A	2004	8-905-606-405	2004	19B200130-P1	2004	48-134507
2004	2SB302B	2004	8-905-606-419	2004	19B200130-P2	2004	48-134524
2004	2SB302C	2004	8-905-606-420	2004	19O300216-P1	2004	48-134526
2004	2SB302D	2004	8-905-606-423	2004	19O300216-P2	2004	48-134579
2004	2SB302E	2004	8-905-706-790	2004	020-1110-006	2004	48-134676
2004	2SB302F	2004	8D	2004	20MC	2004	48-134677
2004	2SB302G	2004	8E	2004	21A045-000	2004	48-134678
2004	2SB302J	2004	8F	2004	21A048-000	2004	48-134679
2004	2SB302K	2004	8L	2004	21A049-000	2004	48-134693
2004	2SB302L	2004	8P	2004	21A050-000	2004	48-134694
2004	2SB302M	2004	09-300021	2004	21A050-001	2004	48-63029A16
2004	2SB302OR	2004	09-300024	2004	022-3511-770	2004	48-63029A60
2004	2SB302R	2004	09-300028	2004	022-3511-780	2004	48-63075A72
2004	2SB302X	2004	09-300084	2004	022-3511-790	2004	48-63075A73
2004	2SB302Y	2004	09-30012	2004	022.3511790	2004	48-63075A75
2004	2SB74	2004	09-304011	2004	022.3511-770	2004	48-63078A63
2004	2SC101B	2004	9-5108	2004	022.3511-780	2004	48-63081A82
2004	2SC101C	2004	9-511	2004	022.3511-790	2004	48-644676
2004	2SC101D	2004	9-5110	2004	23-PT284-122	2004	48-644677
2004	2SC101E	2004	9-5111	2004	23-PT284-123	2004	48-645867
2004	2SC101F	2004	9-5116	2004	24MW157	2004	48-64978A27
2004	2SC101G	2004	9-5117	2004	24MW351	2004	48-64978A28
2004	2SC101GN	2004	9-5118	2004	24MW44	2004	48-64978A29
2004	2SC101H	2004	9-5119	2004	24MW55	2004	48-65132A79
2004	2SC101J	2004	9-5120	2004	24M959	2004	48-669040F
2004	2SC101K	2004	9-5121	2004	31-0001	2004	48-971A203
2004	2SC101L	2004	9-5122	2004	31-0002	2004	48-97271A05
2004	2SC101M	2004	9-5123	2004	31-0003	2004	48-97271A06
2004	2SC101OH	2004	9-5124	2004	31-0004	2004	48K134494
2004	2SC101R	2004	9-905-606-001	2004	31-0015	2004	48K134495
2004	2SC101XL	2004	9-9105	2004	31-0016	2004	48K134496
2004	2SC101Y	2004	9-9106	2004	31-0141	2004	48K134601
2004	2SC125	2004	9-9107	2004	31-0150	2004	48K134796
2004	2T20	2004	9-9108				

RS 276-	Industry Standard No.	RS 276-	Industry Standard No.	RS 276-	Industry Standard No.	RS 276-	Industry Standard No.
2004	48K134798	2004	57B2-30	2004	63-18424	2004	86-150-2
2004	48B64978A27	2004	57B2-31	2004	63-25726	2004	86-151-2
2004	48B64978A28	2004	57B2-35	2004	63-26850	2004	86-162-2
2004	48B64978A29	2004	57B2-37	2004	63-27366	2004	86-163-2
2004	48R134545	2004	57B2-40	2004	63-27500	2004	86-164-2
2004	48B134405	2004	57B2-41	2004	63-28348	2004	86-179-2
2004	48B134406	2004	57B2-42	2004	63-28358	2004	86-18-2
2004	48B134407	2004	57B2-48	2004	63-29661	2004	86-180-2
2004	48B134956	2004	57B2-50	2004	63-29819	2004	86-181-2
2004	48X97046A16	2004	57B2-51	2004	63-29820	2004	86-253-2
2004	051-0062	2004	57B2-65	2004	63-29821	2004	86-254-2
2004	051-0063	2004	57B2-66	2004	63-29862	2004	86-278-2
2004	56P1	2004	57B2-67	2004	63-3954	2004	86-279-2
2004	56P2	2004	57B2-68	2004	63-7538	2004	86-296-2
2004	56P3	2004	57B2-75	2004	63-7541	2004	86-312-2
2004	56P4	2004	57B2-77	2004	63-7548	2004	86-320-2
2004	56P4P	2004	57B2-80	2004	63-7579	2004	86-321-2
2004	57A1-10	2004	57B2-89	2004	63-7580	2004	86-322-2
2004	57A1-100	2004	57B2-9	2004	63-7581	2004	86-347-2
2004	57A1-101	2004	57B2-90	2004	63-7582	2004	86-348-2
2004	57A1-102	2004	57B2-93	2004	63-7660	2004	86-36-2
2004	57A1-103	2004	57D5-2	2004	63-8119	2004	86-363-2
2004	57A1-107	2004	57D1-10	2004	63-8376	2004	86-366-2
2004	57A1-108	2004	57D1-100	2004	63-8377	2004	86-367-2
2004	57A1-109	2004	57D1-101	2004	63-8378	2004	86-368-2
2004	57A1-110	2004	57D1-102	2004	63-8379	2004	86-37-2
2004	57A1-112	2004	57D1-103	2004	63-8699	2004	86-373-2
2004	57A1-113	2004	57D1-107	2004	63-8700	2004	86-374-2
2004	57A1-114	2004	57D1-108	2004	63-8954	2004	86-376-2
2004	57A1-115	2004	57D1-109	2004	63-9072	2004	86-38-2
2004	57A1-12	2004	57D1-110	2004	63-9517	2004	86-449-9
2004	57A1-120	2004	57D1-112	2004	63-9664	2004	86-86-2
2004	57A1-13	2004	57D1-113	2004	63-9665	2004	86-87-2
2004	57A1-15	2004	57D1-114	2004	63-9876	2004	86-88-2
2004	57A1-16	2004	57D1-115	2004	63-9877	2004	86-89-2
2004	57A1-22	2004	57D1-12	2004	63-9941	2004	86-90-2
2004	57A1-24	2004	57D1-120	2004	065-1-12	2004	86-91-2
2004	57A1-25	2004	57D1-13	2004	065-1-1-12-7	2004	86-92-2
2004	57A1-30	2004	57D1-15	2004	65-1-70-12	2004	86-99-2
2004	57A1-31	2004	57D1-16	2004	65-1-70-12-7	2004	088B
2004	57A1-32	2004	57D1-22	2004	65-10	2004	088B-12
2004	57A1-33	2004	57D1-25	2004	65-11	2004	088B-12-7
2004	57A1-35	2004	57D1-30	2004	65-12	2004	88B-70
2004	57A1-36	2004	57D1-31	2004	65-13	2004	88B-70-12
2004	57A1-37	2004	57D1-32	2004	65-14	2004	88B-70-12-7
2004	57A1-41	2004	57D1-33	2004	65-15	2004	88B0
2004	57A1-45	2004	57D1-35	2004	65-16	2004	88B1
2004	57A1-46	2004	57D1-36	2004	65-17	2004	88B10
2004	57A1-47	2004	57D1-37	2004	65-18	2004	88B10R
2004	57A1-48	2004	57D1-39	2004	65-19	2004	88B11
2004	57A1-49	2004	57D1-41	2004	65-1A	2004	88B119
2004	57A1-50	2004	57D1-45	2004	65-1A0	2004	88B12
2004	57A1-57	2004	57D1-46	2004	65-1A0R	2004	88B121
2004	57A1-61	2004	57D1-47	2004	65-1A1	2004	88B13P
2004	57A1-67	2004	57D1-48	2004	65-1A19	2004	88B14
2004	57A1-69	2004	57D1-49	2004	65-1A2	2004	88B14-7B
2004	57A1-72	2004	57D1-50	2004	65-1A21	2004	88B15
2004	57A1-73	2004	57D1-57	2004	65-1A3	2004	88B16
2004	57A1-74	2004	57D1-61	2004	65-1A3P	2004	88B16E
2004	57A1-80	2004	57D1-67	2004	65-1A4	2004	88B17
2004	57A1-81	2004	57D1-69	2004	65-1A4-7	2004	88B17-1
2004	57A1-84	2004	57D1-72	2004	65-1A4-7B	2004	88B18
2004	57A1-85	2004	57D1-73	2004	65-1A5	2004	88B182
2004	57A1-86	2004	57D1-74	2004	65-1A5L	2004	88B19
2004	57A1-87	2004	57D1-80	2004	65-1A6	2004	88B19Q
2004	57A1-89	2004	57D1-81	2004	65-1A6-5	2004	88B2
2004	57A1-9	2004	57D1-84	2004	65-1A7	2004	88B3
2004	57A1-94	2004	57D1-85	2004	65-1A7-1	2004	88B4
2004	57A1-96	2004	57D1-86	2004	65-1A8	2004	88B5
2004	57A1-98	2004	57D1-87	2004	65-1A82	2004	88B6
2004	57A1-99	2004	57D1-88	2004	65-1A9	2004	88B7
2004	57A1130	2004	57D1-89	2004	65-1A9G	2004	88B8
2004	57A1131	2004	57D1-9	2004	65-1A0	2004	88B9
2004	57A1132	2004	57D1-94	2004	65-1A0R	2004	90-37
2004	57A1186	2004	57D1-96	2004	065A-12	2004	90-37(TRANSISTOR)
2004	57A132-9	2004	57D1-98	2004	065A-12-7	2004	90-54
2004	57A180	2004	57D1-99	2004	65A-70	2004	90-59
2004	57A184	2004	57D1130	2004	65A-70-12	2004	91-4
2004	57A187	2004	57D1131	2004	65A-70-12-7	2004	95-108
2004	57A188	2004	57D1132	2004	65A0	2004	95-110
2004	57A2-104	2004	57D1186	2004	65A1	2004	95-116
2004	57A2-105	2004	57D132-9	2004	65A10	2004	95-117
2004	57A2-149	2004	57D9-1	2004	65A10R	2004	95-118
2004	57A2-157	2004	57L1-1	2004	65A119	2004	95-119
2004	57A2-158	2004	57L1-10	2004	65A12	2004	95-120
2004	57A2-159	2004	57L1-11	2004	65A121	2004	95-121
2004	57A2-22	2004	57L1-12	2004	65A13	2004	95-122
2004	57A2-26	2004	57L1-2	2004	65A13P	2004	95-123
2004	57A2-30	2004	57L1-3	2004	65A14	2004	95-124
2004	57A2-31	2004	57L1-4	2004	65A14-7	2004	96-5062-01
2004	57A2-35	2004	57L1-9	2004	65A14-7B	2004	96-5095-01
2004	57A2-37	2004	57M1-1	2004	65A15	2004	96-5099-01
2004	57A2-40	2004	57M1-10	2004	65A15L	2004	96-5138-01
2004	57A2-41	2004	57M1-11	2004	65A16	2004	96-5139-01
2004	57A2-42	2004	57M1-12	2004	65A16-3	2004	96-5140-01
2004	57A2-48	2004	57M1-17	2004	65A17	2004	96-5141-01
2004	57A2-50	2004	57M1-2	2004	65A17-1	2004	96X2801/37N
2004	57A2-51	2004	57M1-3	2004	65A18	2004	99-105
2004	57A2-65	2004	57M1-4	2004	65A182	2004	99-106
2004	57A2-66	2004	57M1-9	2004	65A19	2004	99-107
2004	57A2-67	2004	58B2-14	2004	65A19Q	2004	99-108
2004	57A2-68	2004	61B002-1	2004	65A2	2004	99-120
2004	57A2-75	2004	61P1D	2004	65A3	2004	99-121
2004	57A2-77	2004	62-17390	2004	65A4	2004	998006
2004	57A2-80	2004	62-17391	2004	65A5	2004	998007
2004	57A2-89	2004	62-18418	2004	65A6	2004	101A
2004	57A2-90	2004	62-18419	2004	65A7	2004	101B
2004	57A2-93	2004	62-18422	2004	65A8	2004	101M
2004	57A9-1	2004	62-26851	2004	65A9	2004	107A
2004	57B1130	2004	62-8781	2004	65C01	2004	107B
2004	57B1131	2004	63-10035	2004	65C02	2004	107M
2004	57B1186	2004	63-10036	2004	65C03	2004	108-1
2004	57B180	2004	63-10145	2004	070-001	2004	108-2
2004	57B184	2004	63-10146	2004	70.00.730	2004	108-3
2004	57B186	2004	63-10148	2004	74Q1262	2004	108-4
2004	57B187	2004	63-10149	2004	77-271029-1	2004	111-6910
2004	57B188	2004	63-10150	2004	77-271029-2	2004	112-000267
2004	57B2-1	2004	63-10195	2004	77-271038-1	2004	115-227
2004	57B2-104	2004	63-10196	2004	77-271166-2	2004	115-228
2004	57B2-105	2004	63-10375	2004	77-271166-3	2004	115-229
2004	57B2-11	2004	63-10376	2004	77-271166-2	2004	116-202
2004	57B2-13	2004	63-11055	2004	77-273001-3	2004	116-207
2004	57B2-14	2004	63-11496	2004	80-050400	2004	116-209
2004	57B2-149	2004	63-11582	2004	80-050500	2004	116-683
2004	57B2-157	2004	63-11584	2004	86-100-2	2004	116-684
2004	57B2-158	2004	63-12610	2004	86-101-2	2004	120-000190
2004	57B2-159	2004	63-13025	2004	86-102-2	2004	120-001190
2004	57B2-17	2004	63-13839	2004	86-105-2	2004	120-002213
2004	57B2-18	2004	63-13899	2004	86-112-2	2004	120-002214
2004	57B2-19	2004	63-17390	2004	86-117-2	2004	120-002216
2004	57B2-2	2004	63-18418	2004	86-135-2		
2004	57B2-20	2004	63-18419	2004	86-136-2		
2004	57B2-22	2004	63-18423	2004	86-149-2		
2004	57B2-26						

RS 276-	Industry Standard No.
2004	120-004722
2004	120-02213
2004	121-101
2004	121-102
2004	121-103
2004	121-104
2004	121-105
2004	121-119
2004	121-132
2004	121-134
2004	121-135
2004	121-136
2004	121-137
2004	121-138
2004	121-139
2004	121-157
2004	121-179
2004	121-180
2004	121-181
2004	121-185
2004	121-242
2004	121-243
2004	121-244
2004	121-268
2004	121-269
2004	121-284
2004	121-304
2004	121-312
2004	121-313
2004	121-329
2004	121-330
2004	121-331
2004	121-332
2004	121-333
2004	121-334
2004	121-335
2004	121-336
2004	121-349
2004	121-350
2004	121-351
2004	121-352
2004	121-353
2004	121-356
2004	121-357
2004	121-358
2004	121-359
2004	121-384
2004	121-385
2004	121-411
2004	121-412
2004	121-413
2004	121-414
2004	121-415
2004	121-415B
2004	121-426
2004	121-427
2004	121-428
2004	121-429
2004	121-432
2004	121-48
2004	121-49
2004	121-538
2004	121-538B
2004	121-539
2004	121-540
2004	121-540B
2004	121-541
2004	121-541B
2004	121-542
2004	121-542B
2004	121-552
2004	121-553
2004	121-601
2004	121-63
2004	121-697
2004	121-698
2004	121-705
2004	121-714
2004	121-73
2004	121-74
2004	121-75
2004	121-76
2004	121-78
2004	121-79
2004	0131-000418
2004	0131-000419
2004	0131-000498
2004	0131-000802
2004	0131-000859
2004	0131-000862
2004	0131-000863
2004	0131-001182
2004	0131-001314
2004	0131-001332
2004	0131-001433
2004	0131-001434
2004	0131-001435
2004	0131-001436
2004	0131-001697
2004	0131-003029
2004	132-019
2004	132-020
2004	132-027
2004	145T1B
2004	154T1
2004	154T1A
2004	154T1B
2004	155T1
2004	156T1
2004	157T1
2004	157T1A
2004	159T1
2004	160T1
2004	161T1
2004	162T1
2004	165-1A82
2004	165A-182
2004	165B-182
2004	188B-1-82
2004	201A
2004	201B
2004	207A
2004	207A1
2004	207B
2004	207M
2004	229-0026
2004	229-0038
2004	229-0077
2004	229-0079
2004	229-0082
2004	229-0083
2004	229-0086
2004	229-0087
2004	229-0089
2004	229-0090
2004	229-0095
2004	229-0098
2004	229-0099
2004	229-0106
2004	229-0110
2004	229-0111
2004	229-0112
2004	229-0121
2004	229-0123
2004	229-0129
2004	229-0131
2004	229-0132
2004	229-0136
2004	229-0145
2004	241-15A
2004	296-46-9
2004	297C011H01
2004	297V008M01
2004	297V012H04
2004	297V012H07
2004	297V012H11
2004	297V012H12
2004	297V012H13
2004	297V020H01
2004	297V024H01
2004	297V024H03
2004	297V038H02
2004	297V038H03
2004	297V038H04
2004	297V063H01
2004	297V064H01
2004	297V064H01
2004	324-0016
2004	324-0026
2004	324-0077
2004	324-0082
2004	324-0083
2004	324-0086
2004	324-0087
2004	324-0095
2004	324-0110
2004	324-0111
2004	324-0112
2004	324-012
2004	324-0123
2004	324-0131
2004	324-0132
2004	324-0155
2004	324-0187
2004	324-132
2004	325-0028-85
2004	353-9301-002
2004	367(SEARS)
2004	417-11
2004	417-12
2004	417-13
2004	417-14
2004	417-143
2004	417-16
2004	417-2
2004	417-22
2004	417-23
2004	417-25
2004	417-26
2004	417-27
2004	417-31
2004	417-33
2004	417-35
2004	417-36
2004	417-37
2004	417-38
2004	417-39
2004	417-50
2004	417-53
2004	417-54
2004	417-56
2004	417-57
2004	417-58
2004	417-60
2004	417-66
2004	417-68
2004	417-70
2004	417-71
2004	417-72
2004	417-76
2004	417-79
2004	421-6
2004	421-6B
2004	421-7
2004	421-7B
2004	421-8
2004	421-8B
2004	421-9792
2004	422-2778
2004	422-2779
2004	422-2780
2004	429-0092-1
2004	429-0910-50
2004	444-0012-PT2
2004	444-012-P1
2004	465-032-19
2004	465-042-19
2004	465-045-19
2004	465-049-19
2004	465-061-19
2004	465-066-19
2004	465-146-19
2004	465-1A5
2004	465-223-19
2004	465A-15
2004	465B-15
2004	488B15
2004	501T1
2004	503T1
2004	504T1
2004	505T1
2004	506T1
2004	507T1
2004	508T1
2004	573-518
2004	576-0003-010
2004	576-0003-025
2004	576-003-009
2004	601-040
2004	602-075
2004	603-020
2004	603-030
2004	603-040
2004	604-030
2004	604-080
2004	609-020
2004	610-050
2004	610-050-1
2004	610-050-2
2004	610-050-3
2004	610-051
2004	610-051-1
2004	610-051-2
2004	610-051-4
2004	610-053
2004	610-053-1
2004	610-053-2
2004	610-055
2004	610-055-1
2004	610-055-2
2004	610-055-3
2004	617-54
2004	617-55
2004	617-57
2004	617-58
2004	642-116
2004	642-147
2004	642-173
2004	642-202
2004	642-207
2004	665-1A5L
2004	665A-1-5L
2004	688B-1-5L
2004	690L297H01
2004	690V040H56
2004	690V052H63
2004	690V056H27
2004	690V056H29
2004	690V056H30
2004	690V056H31
2004	690V056H32
2004	690V057H25
2004	690V066H89
2004	690V081H0B
2004	690V105H-21
2004	690V105H21
2004	690V105H26
2004	690V119H95
2004	690V119H96
2004	690V119H97
2004	750M63-104
2004	750M63-116
2004	750M63-117
2004	800-504-00
2004	866-20-2
2004	921-10B
2004	921-11B
2004	921-12B
2004	921-13B
2004	921-15B
2004	921-16B
2004	921-17B
2004	921-1A
2004	921-1B
2004	921-25B
2004	921-26B
2004	921-2A
2004	921-2B
2004	921-3A
2004	921-4A
2004	921-9B
2004	955-1
2004	955-2
2004	955-3
2004	965-1A6-5
2004	965A16-3
2004	972X1
2004	972X2
2004	972X3
2004	972X4
2004	972X5
2004	991-011221
2004	1006-78
2004	1010-78
2004	1010-87
2004	1040-59
2004	1111-17
2004	1111-18
2004	1113-13
2004	1119-54
2004	1119-55
2004	1119-56
2004	1301-1
2004	1501-2
2004	1449
2004	1465-1
2004	1465-1-12
2004	1465-1-12-8
2004	1465A
2004	1465A-12
2004	1465A-12-8
2004	1488B
2004	1488B-12
2004	1488B-12-8
2004	1526
2004	1865-1
2004	1865-1-12
2004	1865-1-127
2004	1865-1L
2004	1865A
2004	1865A-12
2004	1865A-127
2004	1865AL8
2004	1888B
2004	1888B-12
2004	1888B-127
2004	1888BL
2004	1888BL8
2004	2015
2004	2020
2004	2021
2004	2057A2-149
2004	2057A2-159
2004	2057A2-166
2004	2057A2-205
2004	2057A2-231
2004	2057A2-232
2004	2057A2-252
2004	2057A2-49
2004	2057B186
2004	2057B2-149
2004	2057B2-157
2004	2057B2-158
2004	2057B2-159
2004	2057B2-89
2004	2057B2-90
2004	2057B2-93
2004	2057B2A2-118
2004	2106-120
2004	2215-17
2004	2487B
2004	2489A
2004	2495-078
2004	2495-079
2004	2495-082
2004	2495-376
2004	2495-377
2004	2495-378
2004	2495-488-1
2004	2495-488-2
2004	2606-295
2004	2700
2004	2797
2004	3008(E.F.JOHNSON)
2004	3009(E.F.JOHNSON)
2004	3024
2004	3024(E.F.JOHNSON)
2004	3025
2004	3458
2004	3504(RCA)
2004	3534
2004	3534(RCA)
2004	3603
2004	3617
2004	3652
2004	3750
2004	3907/2N404A
2004	3961(G.E.)
2004	4001-222
2004	4001-223
2004	4006
2004	4368
2004	4822-130-40252
2004	4822-130-40255
2004	4822-130-40441
2004	5085
2004	5722-3000
2004	06008
2004	6008(E.F.JOHNSON)
2004	6313
2004	6990
2004	9403-3
2004	9403-6
2004	9510-1
2004	9510-2
2004	9510-7
2004	11522-7
2004	11522-8
2004	11522-9
2004	11620-2
2004	11620-7
2004	11620-8
2004	11620-9
2004	11668-3
2004	11668-4
2004	12113-5
2004	12113-7
2004	12113-8
2004	12113-9
2004	12115-0
2004	12115-7
2004	12119-0
2004	12122-A8
2004	12122-9
2004	12123-0
2004	12123-1
2004	12123-3
2004	12125-6
2004	12125-8
2004	12125-9
2004	12126-0
2004	18540
2004	18541
2004	30218(RCA)
2004	30238
2004	30239
2004	30240
2004	30273
2004	34118
2004	34342
2004	34389
2004	34425
2004	34553
2004	34675
2004	34942
2004	35070
2004	35169
2004	35170
2004	35815
2004	37278
2004	38680
2004	38681
2004	39053
2004	40004
2004	40005
2004	40265(RCA)
2004	40268
2004	40487
2004	40489
2004	49939-2
2004	61012-5-1
2004	65804-62
2004	65804-62
2004	69107-43
2004	72797-81
2004	72925-08
2004	77271-8
2004	080026
2004	080027
2004	080028
2004	080061
2004	80071
2004	80114
2004	080244
2004	080245
2004	080258
2004	080266
2004	080267
2004	080269
2004	94007
2004	094013
2004	94028
2004	94033
2004	94035
2004	94036
2004	95101
2004	95102
2004	95103
2004	95107
2004	95108
2004	95110
2004	95111
2004	95121
2004	95124
2004	99103

RS 276-	Industry Standard No.	RS 276-	Industry Standard No.	RS 276-	Industry Standard No.	RS 276-	Industry Standard No.
2004	99104	2004	815067C	2005	TI-402	2005	2SB157OR
2004	99105	2004	815193	2005	TR123	2005	2SB157R
2004	99106	2004	815197	2005	TR20	2005	2SB157X
2004	99107	2004	815234	2005	TR269	2005	2SB157Y
2004	99108	2004	851759-3	2005	TR382	2005	2SB158
2004	101078	2004	853640	2005	TR383	2005	2SB159
2004	101087	2004	853864-0	2005	TR803	2005	2SB160
2004	111117	2004	902521	2005	T8619	2005	2SB267
2004	111118	2004	980140	2005	T8620	2005	2SB378
2004	111313	2004	980142	2005	TV825A103	2005	2SB389
2004	111954	2004	980146	2005	UPI1301	2005	2SB436
2004	111955	2004	980435	2005	UPI1352	2005	2SB73B
2004	111956	2004	980441	2005	UPI404	2005	2SB75H
2004	112001	2004	980505	2005	1T264	2005	2SB93
2004	112002	2004	980506	2005	2N1108#	2005	2SC101
2004	112011	2004	980507	2005	2N1109#	2005	2SC101A
2004	112032	2004	980509	2005	2N1110#	2005	2SC101X
2004	112296	2004	980514A	2005	2N131#	2005	2T321
2004	115227	2004	980545A	2005	2N133#	2005	2T322
2004	115228	2004	980636A	2005	2N135#	2005	6-62D
2004	115229	2004	981143	2005	2N136#	2005	121-10
2004	115504	2004	981144	2005	2N137#	2005	121-163
2004	116202	2004	981145	2005	2N207B#	2005	121-164
2004	116207	2004	981146	2005	2N207B#	2005	121-319
2004	116208	2004	981205	2005	2N2169	2005	121-320
2004	116209	2004	981204	2005	2N2224#	2005	121-9
2004	116683	2004	981959	2005	2N224#	2005	121-29013
2004	116684	2004	982322	2005	2N226#	2005	276-2003
2004	117618	2004	982374	2005	2N280	2005	276-2003/RS2003
2004	117658	2004	982497	2005	2N291#	2005	383
2004	117725	2004	985445A	2005	2N291#	2005	916-31001-1B
2004	117726	2004	985446A	2005	2N409#	2005	916-31026-9B
2004	117866	2004	1222136	2005	2N409#	2005	95109
2004	119013	2004	1222314	2005	2N410#	2005	99201
2004	124625	2004	1222371	2005	2N410#	2005	236288
2004	125790	2004	2000646-104	2005	2N411#	2006	A059-108
2004	125972	2004	2000648-21	2005	2N411#	2006	A059-115
2004	126184	2004	2000648-22	2005	2N412#	2006	A061-116
2004	126186	2004	2001653-58	2005	2N412#	2006	A065-111
2004	129347	2004	2001653-59	2005	2N502BRN	2006	A1124C
2004	129389	2004	2002151-18	2005	2N502BRN	2006	A12153
2004	171016	2004	2002151-18A	2005	2N502GRN	2006	A1217B
2004	171039	2004	2002153-58	2005	2N502GRN	2006	A14-586-01
2004	171039(SEARS)	2004	2002153-59	2005	2N502RED	2006	A1414A
2004	171039(TOSHIBA)	2004	2002153-60	2005	2N502RED	2006	A1414A9
2004	171162-169	2004	2002153-76	2005	2SA100A	2006	A144A-1
2004	175006-181	2004	2003073-11	2005	2SA101D	2006	A144A-19
2004	175006-182	2004	2003073-12	2005	2SA101P	2006	A1465C
2004	175006-183	2004	2003073-13	2005	2SA101Q	2006	A146509
2004	175006-184	2004	2076403	2005	2SA101H	2006	A1477C
2004	175006-185	2004	2076403-0703	2005	2SA101K	2006	A1477O9
2004	190427	2004	2091217-0014	2005	2SA101L	2006	A1484A
2004	200064-6-104	2004	2091241-0005	2005	2SA101M	2006	A1484A9
2004	200064-6-106	2004	2091247-005	2005	2SA101R	2006	A14A-70
2004	215002	2004	2092417-005	2005	2SA101XHX	2006	A14A-705
2004	215008	2004	2092417-0704	2005	2SA101YA	2006	A14A10Q
2004	215031	2004	2092417-0707	2005	2SA114	2006	A15927
2004	215038	2004	2092417-0708	2005	2SA115	2006	A2090056-1
2004	223369	2004	2092417-0709	2005	2SA116	2006	A2090056-27
2004	223474	2004	2092417-0710	2005	2SA12	2006	A2090056-5
2004	224586	2004	2092417-0717	2005	2SA12A	2006	A2091859-0025
2004	224587	2004	2092417-1	2005	2SA12B	2006	A2091859-0720
2004	225311	2004	2092417-2	2005	2SA12C	2006	A2091859-10
2004	225594	2004	2092417-3	2005	2SA12D	2006	A2091859-11
2004	225600	2004	2092417-4	2005	2SA12H	2006	A21801208
2004	229133	2004	2092417-5	2005	2SA12V	2006	A2SB240A
2004	232676	2004	2092417-6	2005	2SA138	2006	A28B242A
2004	232680	2004	2092417-7	2005	2SA15	2006	A28B248A
2004	234015	2004	2092417-8	2005	2SA15BK	2006	A30302
2004	234631	2004	2092417-9	2005	2SA15BLU	2006	A34-6001-1
2004	235200	2004	2092418-0708	2005	2SA15H	2006	A34-6002-17
2004	261586	2004	2092418-071	2005	2SA15R	2006	A34715
2004	265771	2004	2092418-0710	2005	2SA15V	2006	A35084
2004	297065003	2004	2092418-0712	2005	2SA15Y	2006	A35201
2004	310030	2004	2092418-1	2005	2SA16	2006	A35260
2004	310132	2004	2092418-10	2005	2SA17	2006	A36896
2004	310162	2004	2092418-11	2005	2SA173	2006	A416
2004	310204	2004	2092418-2	2005	2SA173B	2006	A417-62
2004	346607-4	2004	2092418-5	2005	2SA17H	2006	A4247
2004	454760	2004	2092418-6	2005	2SA18	2006	A4347
2004	0510079	2004	2092418-7	2005	2SA183	2006	A48-134727
2004	0510079H	2004	2092418-8	2005	2SA18H	2006	A48-134731
2004	537428	2004	2092693-2	2005	2SA243A	2006	A48-134907
2004	0573335	2004	2092693-3	2005	2SA244A	2006	A48-137102
2004	573336	2004	2092693-4	2005	2SA280D	2006	A48-137214
2004	573366	2004	2092693-9	2005	2SA304	2006	A48-137215
2004	573371	2004	2320161(RCA)	2005	2SA313-BLUE	2006	A48-137216
2004	573405	2004	2320514	2005	2SA313-GREEN	2006	A48-137217
2004	573406	2004	2495078	2005	2SA313-RED	2006	A48-137218
2004	601032	2004	2495079	2005	2SA313-YELLOW	2006	A48-137219
2004	601040	2004	2495082	2005	2SA314-RED	2006	A48-137220
2004	601052	2004	2495376	2005	2SA314-YELLOW	2006	A48-63076A81
2004	602075	2004	2495377	2005	2SA315-RED	2006	A4A-1-70
2004	603020	2004	2498837	2005	2SA315-YELLOW	2006	A4A-1-705
2004	603030	2004	3460550-1	2005	2SA316-RED	2006	A4A-1A9Q
2004	603040	2004	3460550-3	2005	2SA316-YELLOW	2006	A5255
2004	603112	2004	3460550-4	2005	2SA32	2006	A578124-10
2004	603312	2004&	28A407G	2005	2SA333	2006	A57L5-1
2004	604030	2004'	28A361G	2005	2SA334	2006	A57M3-7
2004	604040	2005	AA4	2005	2SA336	2006	A57M3-8
2004	604080	2005	AA5	2005	2SA373A	2006	A62-18427
2004	605112	2005	ACY34	2005	2SA374	2006	A63-18427
2004	609020	2005	ACY35	2005	2SA386	2006	A650-190
2004	610050	2005	AP188	2005	2SA387	2006	A650-70
2004	610050-1	2005	A8Y26-RT	2005	2SA406	2006	A650-705
2004	610050-2	2005	A8Y32	2005	2SA407	2006	A650190
2004	610050-3	2005	CK83	2005	2SA46	2006	A660097
2004	610051	2005	E83120	2005	2SA47	2006	A690V081H97
2004	610051-1	2005	E83121	2005	2SA50	2006	A7279039
2004	610051-2	2005	E83122	2005	2SA65	2006	A7279049
2004	610051-4	2005	E83123	2005	2SA66	2006	A7285774
2004	610053	2005	E83124	2005	2SA68	2006	A7285778
2004	610053-1	2005	GT14H	2005	2SA723	2006	A7289047
2004	610053-2	2005	GT20H	2005	2SA95	2006	A7290594
2004	610055	2005	GT2696	2005	2SA96	2006	A7291252
2004	610055-1	2005	GT33	2005	2SA97	2006	A7297043
2004	610055-2	2005	HEP-G0003	2005	2SA98	2006	A7297092
2004	610055-3	2005	HJ22D	2005	2SA99	2006	A7297093
2004	610056-1	2005	HJ25D	2005	2SB12	2006	A76-11770
2004	610056-2	2005	HJ71	2005	2SB14	2006	A770-70
2004	610056-3	2005	HJ72	2005	2SB157	2006	A770-705
2004	610056-4	2005	RE-27	2005	2SB157A	2006	A77C19G
2004	610061-1	2005	RB69	2005	2SB157B	2006	A84A-70
2004	660064	2005	RS2003	2005	2SB157C	2006	A84A-705
2004	660084	2005	RS5209	2005	2SB157D	2006	A84A190
2004	660085	2005	SF.T351	2005	2SB157E	2006	A8P-404-ORN
2004	741050	2005	SK3006/160	2005	2SB157F	2006	A8P404-ORN
2004	772716	2005	T0003	2005	2SB157G	2006	A8P404F
2004	772717	2005	T0004	2005	2SB157GN	2006	A94004
2004	772719	2005	T0005	2005	2SB157H	2006	A964-17887
2004	785897-01	2005	T1327	2005	2SB157J	2006	A97A83
2004	815067	2005	T1328	2005	2SB157K	2006	A992-00-1192
2004	815067A	2005	T1342	2005	2SB157L	2006	AA2SB240A
2004	815067B	2005	T1930	2005	2SB157M	2006	AC148

RS 276-	Industry Standard No.
2006	ACY16
2006	AD-140
2006	AD-148
2006	AD-149
2006	AD-150
2006	AD-152
2006	AD-156
2006	AD-157
2006	AD-159
2006	AD-162
2006	AD103
2006	AD104
2006	AD105
2006	AD130
2006	AD130-III
2006	AD130-IV
2006	AD130-V
2006	AD131
2006	AD131-III
2006	AD131-IV
2006	AD131-V
2006	AD132
2006	AD138
2006	AD138/50
2006	AD138S0
2006	AD139
2006	AD140
2006	AD143R
2006	AD145
2006	AD148
2006	AD149
2006	AD149-01
2006	AD149-02
2006	AD149-IV
2006	AD149-V
2006	AD149B
2006	AD149C
2006	AD150
2006	AD150-IV
2006	AD150-V
2006	AD152
2006	AD153
2006	AD155
2006	AD156
2006	AD159
2006	AD162
2006	AD164
2006	AD169
2006	AD262
2006	AD263
2006	ADY22
2006	ADY23
2006	ADY24
2006	ADY27
2006	ADY28
2006	AP280
2006	AL101
2006	AR-10
2006	AR-11
2006	AR-12
2006	AR-13
2006	AR-14
2006	AR-4
2006	AR-5
2006	AR-6
2006	AR-7
2006	AR-8
2006	AR-9
2006	AR10
2006	AR11
2006	AR12
2006	AR13
2006	AR14
2006	AR4
2006	AR5
2006	AR6
2006	AR7
2006	AR8
2006	AR8P404R
2006	AR9
2006	AS215
2006	ASZ15
2006	ASZ16
2006	ASZ17
2006	ASZ18
2006	ATC-TR-14
2006	ATC-TR-5
2006	ATC-TR-6
2006	AU102
2006	AUY-21
2006	AUY10
2006	AUY19
2006	AUY20
2006	AUY21
2006	AUY21A
2006	AUY22
2006	AUY22A
2006	AUY31
2006	AUY33
2006	AV105
2006	B-1511
2006	B-1914
2006	B10064
2006	B10069
2006	B10162
2006	B10163
2006	B10474
2006	B10475
2006	B107
2006	B107A
2006	B1085
2006	B10912
2006	B10913
2006	B1151
2006	B1151A
2006	B1151B
2006	B1152
2006	B1152A
2006	B1152B
2006	B1181
2006	B119
2006	B119A
2006	B1215
2006	B122
2006	B123
2006	B123A
2006	B124
2006	B126
2006	B126A
2006	B126F
2006	B126V
2006	B127
2006	B1274
2006	B1274A
2006	B1274B
2006	B127A
2006	B128
2006	B128A
2006	B128V
2006	B129
2006	B130
2006	B130A
2006	B131(JAPAN)
2006	B131A
2006	B132(JAPAN)
2006	B132A
2006	B134
2006	B134A
2006	B134C
2006	B136B
2006	B136BA
2006	B136BB
2006	B136BC
2006	B136BD
2006	B136BE
2006	B136BF
2006	B137
2006	B138
2006	B140
2006	B141
2006	B142
2006	B142B
2006	B142C
2006	B143
2006	B143P
2006	B144
2006	B144P
2006	B145
2006	B146
2006	B147
2006	B149
2006	B14A-1-21
2006	B151(TRANSISTOR)
2006	B152(TRANSISTOR)
2006	B177
2006	B178
2006	B179
2006	B19
2006	B1904
2006	B20
2006	B20-001
2006	B21
2006	B215
2006	B216
2006	B216A
2006	B217
2006	B217A
2006	B217Q
2006	B217U
2006	B224
2006	B228
2006	B229
2006	B230
2006	B233
2006	B239
2006	B239A
2006	B246
2006	B247
2006	B248
2006	B248A
2006	B249
2006	B249A
2006	B25
2006	B250
2006	B250A
2006	B254
2006	B255
2006	B256
2006	B25B
2006	B26(JAPAN)
2006	B26A
2006	B27
2006	B28(JAPAN)
2006	B282
2006	B283
2006	B284
2006	B285
2006	B29
2006	B295
2006	B28B241
2006	B28B244
2006	B2W
2006	B30(JAPAN)
2006	B31(TRANSISTOR)
2006	B337
2006	B337A
2006	B337B
2006	B337BK
2006	B337H
2006	B337HA
2006	B337HB
2006	B338
2006	B338H
2006	B338HA
2006	B338HB
2006	B355
2006	B356
2006	B367
2006	B367(A)
2006	B367A
2006	B367B
2006	B367C
2006	B367H
2006	B368
2006	B368A
2006	B368B
2006	B368H
2006	B391
2006	B407
2006	B407-0
2006	B407-0
2006	B407TV
2006	B41
2006	B413
2006	B414
2006	B42
2006	B424
2006	B426
2006	B426BL
2006	B426R
2006	B426Y
2006	B445
2006	B446
2006	B448
2006	B449
2006	B449F
2006	B449P
2006	B458
2006	B458A
2006	B462
2006	B463
2006	B463BL
2006	B463E
2006	B463R
2006	B463Y
2006	B466
2006	B467
2006	B471
2006	B471-2
2006	B471A
2006	B471B
2006	B472
2006	B472A
2006	B472B
2006	B473
2006	B473D
2006	B473F
2006	B473H
2006	B481
2006	B481D
2006	B481E
2006	B4A-1-A-21
2006	B62
2006	B63
2006	B64
2006	B65C-1-21
2006	B69
2006	B77B-1-21
2006	B80
2006	B81
2006	B82
2006	B83
2006	B84
2006	B84A-1-21
2006	BC1073
2006	BC1073A
2006	BC1274
2006	BC1274A
2006	BC1274B
2006	BCM1002-6
2006	BDY62
2006	BF5
2006	C337B
2006	C463(Y)
2006	C50BA042
2006	CDT1309
2006	CDT1310
2006	CDT1311
2006	CDT1319
2006	CDT1320
2006	CDT1321
2006	CDT1349
2006	CDT1349A
2006	CDT1350
2006	CDT1350A
2006	CM2550
2006	CQT1075
2006	CQT1076
2006	CQT1077
2006	CQT1110
2006	CQT1110A
2006	CQT1111
2006	CQT1111A
2006	CQT1112
2006	CQT1129
2006	CQT940A
2006	CQT940B
2006	CQT940BA
2006	CRT1544
2006	CRT1545
2006	CRT1552
2006	CRT1553
2006	CRT1602
2006	CRT3602A
2006	CST1739
2006	CST1740
2006	CST1741
2006	CST1742
2006	CST1743
2006	CST1744
2006	CST1745
2006	CST1746
2006	CT1122
2006	CT1124
2006	CT1124A
2006	CT1124AB
2006	CTP1104
2006	CTP1106
2006	CTP1108
2006	CTP1109
2006	CTP1111
2006	CTP1117
2006	CTP1119
2006	CTP1124
2006	CTP1133
2006	CTP1135
2006	CTP1136
2006	CTP1137
2006	CTP1265
2006	CTP1266
2006	CTP1306
2006	CTP1307
2006	CTP1500
2006	CTP1503
2006	CTP1508
2006	CTP1511
2006	CTP1513
2006	CTP1514
2006	CTP1550
2006	CTP1551
2006	CTP1728
2006	CTP1729
2006	CTP1730
2006	CTP1731
2006	CTP1732
2006	CTP1733
2006	CTP1735
2006	CTP1736
2006	CTP1739
2006	CTP1740
2006	CTP3500
2006	CTP3503
2006	CTP3504
2006	CTP3550B
2006	CXL104
2006	CXL121
2006	CXL121MP
2006	CXL131
2006	CXL131MP
2006	D080
2006	D081
2006	DS-515
2006	DS-520
2006	DS503
2006	DS515
2006	DS520
2006	DT104Q
2006	DT401
2006	DT41
2006	DT6110
2006	DTA1011
2006	DTG-110
2006	DTG1011
2006	DTG104Q
2006	DTG110
2006	DTG110(SEARS)
2006	DTG110(WARDS)
2006	DU6
2006	E24107
2006	EA1082
2006	EA1341
2006	EA15X10
2006	EA15X12
2006	EA15X139
2006	EA15X15
2006	EA15X154
2006	EA15X173
2006	EA15X226
2006	EA15X33
2006	EA15X35
2006	EA15X38
2006	EA15X53
2006	EA15X88
2006	EA1700
2006	ECG104
2006	ECG131
2006	EQ0-6
2006	EQ0-8
2006	EH15X10
2006	ES10110
2006	ES13
2006	ES13(ELCOM)
2006	ES15X17
2006	ES15X43
2006	ES15X45
2006	ES15X51
2006	ES15X78
2006	ES18
2006	ES18(ELCOM)
2006	ES21(ELCOM)
2006	ES29(ELCOM)
2006	ES50(ELCOM)
2006	ES503
2006	ES503(ELCOM)
2006	ES7
2006	ES7(ELCOM)
2006	ES9
2006	ES9(ELCOM)
2006	ET15X17
2006	ET15X4
2006	ET15X43
2006	ET15X5
2006	ET6
2006	ETFB-367B
2006	F-67-E
2006	F20
2006	F67E
2006	FD-1029-ET
2006	FD-1029ET
2006	G04-701-A
2006	G04-704-A
2006	G19
2006	G6013
2006	G6016
2006	GC4045
2006	GC4057
2006	GC4062
2006	GC4087
2006	GC4094
2006	GC4097
2006	GC4111
2006	GC4125
2006	GC4156
2006	GC4251
2006	GC4267-2
2006	GC641
2006	GC691
2006	GC692
2006	GB-16
2006	GE-239
2006	GE-3
2006	GE-30
2006	GE-44
2006	GET-572
2006	GET572
2006	GM428
2006	G04-701-A
2006	G04-704-A
2006	GP1432
2006	GP1493
2006	GP1494
2006	GP1882
2006	GP420
2006	GPT-16
2006	GPT16
2006	GTX-2001
2006	GTX2001
2006	H103A
2006	HB367
2006	HEP-6005
2006	HEP200
2006	HEP230
2006	HEP232
2006	HEP623
2006	HEP624
2006	HEP626
2006	HEP628
2006	HEP642
2006	HEP643
2006	HEPG6000
2006	HEPG6000P/Q
2006	HEPG6003
2006	HEPG6003P/Q
2006	HEPG6005
2006	HEPG6005P/Q
2006	HEPG6013
2006	HEPG6013P/Q
2006	HEPG6016
2006	HF19
2006	HF20
2006	HJ35
2006	HR-101A
2006	HR101
2006	HR101A

RS 276-	Industry Standard No.	RS 276-	Industry Standard No.	RS 276-	Industry Standard No.	RS 276-	Industry Standard No.
2006	HR102	2006	NKT-416	2006	PT-235	2006	8C4274
2006	HR102C	2006	NKT-451	2006	PT-235A	2006	SC70
2006	HR103	2006	NKT-452	2006	PT-236	2006	SDT-3048
2006	HR103A	2006	NKT-453	2006	PT-236A	2006	SE-40022
2006	HR105	2006	NKT-454	2006	PT-236B	2006	SE40022
2006	HR105A	2006	NKT-501	2006	PT-242	2006	SF.T191
2006	HR105B	2006	NKT-503	2006	PT-25	2006	SFT190
2006	HT2040710A	2006	NKT-504	2006	PT-255	2006	SFT191
2006	HT2040071D	2006	NKT401	2006	PT-256	2006	SFT192
2006	HT20436A	2006	NKT402	2006	PT-285	2006	SFT212
2006	HT2046710	2006	NKT403	2006	PT-285A	2006	SFT213
2006	HT204736	2006	NKT404	2006	PT-301	2006	SFT214
2006	I//L	2006	NKT405	2006	PT-301A	2006	SFT238
2006	I472446-I	2006	NKT406	2006	PT-307	2006	SFT239
2006	IP20-0032	2006	NKT415	2006	PT-307A	2006	SFT240
2006	IR-RE50	2006	NKT416	2006	PT-3A	2006	SFT250
2006	IRTR-50	2006	NKT450	2006	PT-40	2006	SK3009
2006	IRTR01	2006	NKT451	2006	PT-554	2006	SK3014
2006	IRTR16	2006	NKT452	2006	PT-6	2006	SK3052
2006	JP40	2006	NKT452-81	2006	PT06	2006	SP-1108
2006	JR40	2006	NKT453	2006	PT12	2006	SP-148-3
2006	K04774	2006	NKT454	2006	PT155	2006	SP-1482-5
2006	K4-520	2006	NKT501	2006	PT176	2006	SP-1483
2006	K4-521	2006	NKT503	2006	PT234	2006	SP-1484
2006	K04774	2006	NKT504	2006	PT235	2006	SP-1556-2
2006	KT1017	2006	OC-16	2006	PT235A	2006	SP-1603
2006	L-417-29BLK	2006	OC-22	2006	PT236	2006	SP-1603-1
2006	L-417-29GRN	2006	OC-23	2006	PT236A	2006	SP-1603-2
2006	L-417-29WHT	2006	OC-24	2006	PT236B	2006	SP-404T
2006	L-417-60	2006	OC-25	2006	PT236C	2006	SP-441
2006	L852	2006	OC-26	2006	PT242	2006	SP-485
2006	M4331	2006	OC-28	2006	PT25	2006	SP-486
2006	M4463	2006	OC-29	2006	PT255	2006	SP-486W
2006	M4570	2006	OC-30A	2006	PT256	2006	SP-634
2006	M4582	2006	OC-35	2006	PT285	2006	SP-649
2006	M4582BRN	2006	OC-36	2006	PT285A	2006	SP-649-1
2006	M4583	2006	OC16	2006	PT3	2006	SP-834
2006	M4583RED	2006	OC19	2006	PT30	2006	SP-880
2006	M4584	2006	OC20	2006	PT301	2006	SP-880-1
2006	M4584GRN	2006	OC22	2006	PT301A	2006	SP-880-3
2006	M4606	2006	OC23	2006	PT307	2006	SP-891
2006	M4608	2006	OC24	2006	PT307A	2006	SP-891B
2006	M4619	2006	OC25	2006	PT32	2006	SP-491W
2006	M4619RED	2006	OC26	2006	PT336B	2006	SP1013A
2006	M4620	2006	OC27	2006	PT366B	2006	SP1013B
2006	M4620GRN	2006	OC28	2006	PT3A	2006	SP1108
2006	M4649	2006	OC29	2006	PT40	2006	SP1118
2006	M4722	2006	OC30	2006	PT50	2006	SP1137
2006	M4722BLU	2006	OC30A	2006	PT501	2006	SP1271
2006	M4722GRN	2006	OC30B	2006	PT554	2006	SP1323
2006	M4722PUR	2006	OC35	2006	PT555	2006	SP1405
2006	M4722RED	2006	OC36	2006	PT6	2006	SP148-3
2006	M4722YEL	2006	P-31898	2006	PTC105	2006	SP1481
2006	M4727	2006	P-T-30	2006	PTC114	2006	SP1481-1
2006	M4730	2006	P1A	2006	PTC120	2006	SP1481-2
2006	M4766	2006	P1E-1	2006	PTO-6	2006	SP1481-3
2006	M4767	2006	P1G	2006	Q-01084R	2006	SP1481-4
2006	M4887A	2006	P1K	2006	QG-0074	2006	SP1481-5
2006	M4888	2006	P1KBLK	2006	Q00074	2006	SP1482
2006	M4888A	2006	P1KBLU	2006	QP1	2006	SP1482-2
2006	M4888B	2006	P1KBRN	2006	QP1A	2006	SP1482-3
2006	M4974	2006	P1KGRN	2006	QP2	2006	SP1482-4
2006	M4974/P1R	2006	P1KORN	2006	QP6	2006	SP1482-5
2006	M501	2006	P1KRED	2006	QP7	2006	SP1482-6
2006	M7031	2006	P1KYEL	2006	R0092	2006	SP1482-7
2006	M84	2006	P1R	2006	R102-41	2006	SP1483
2006	M84B	2006	P1T	2006	R2446	2006	SP1483-1
2006	M9141	2006	P2C	2006	R2446-1	2006	SP1483-2
2006	M9142	2006	P2D	2006	R265A	2006	SP1483-3
2006	M9202	2006	P2DBLU	2006	R3512-1	2006	SP1484
2006	M9237	2006	P2DBRN	2006	R3515	2006	SP1550-3
2006	M9241	2006	P2DGRN	2006	R3515(RCA)	2006	SP1556
2006	M9255	2006	P2DORN	2006	R516	2006	SP1556-1
2006	M9263	2006	P2DRED	2006	R7167	2006	SP1556-2
2006	M9342	2006	P2DYEL	2006	R7253	2006	SP1556-3
2006	M9436	2006	P2R	2006	R7620	2006	SP1556-4
2006	M9550	2006	P31898	2006	R8313	2006	SP1556-2
2006	MA4670	2006	P3E	2006	R8659	2006	SP1595BLK
2006	MF-55-62	2006	P3EBLK	2006	RE11	2006	SP1595BLU
2006	MN194	2006	P3EBLU	2006	RE11MP	2006	SP1595GRN
2006	MN22	2006	P3EGRN	2006	RE20	2006	SP1595RED
2006	MN23	2006	P3ERED	2006	RE20MP	2006	SP1596BLK
2006	MN24	2006	P3R	2006	RE7	2006	SP1596BLU
2006	MN25	2006	P3R-1	2006	REN104	2006	SP1596GRN
2006	MN26	2006	P3R-2	2006	REN121	2006	SP1596RED
2006	MN29	2006	P3R-3	2006	REN131	2006	SP1600
2006	MN29BLK	2006	P3R-4	2006	RS-105	2006	SP1603
2006	MN29GRN	2006	P3T	2006	RS-1055	2006	SP1603-1
2006	MN29PUR	2006	P3T-1	2006	RS-2006	2006	SP1603-2
2006	MN29WHT	2006	P3T-2	2006	RS-3858-1	2006	SP1603-3
2006	MN32	2006	P4D	2006	RS-5613	2006	SP1619
2006	MN46	2006	P4L	2006	RS-5835	2006	SP1651
2006	MN48	2006	P4M	2006	RS-5855	2006	SP1657
2006	MN49	2006	P4N	2006	RS3358-1	2006	SP176
2006	MN73	2006	P6480001	2006	RS3359-1	2006	SP1801
2006	MN73BLK	2006	P75534	2006	RS3858	2006	SP1844
2006	MN73WHT	2006	P75534-1	2006	RS3858-1	2006	SP1927
2006	MN76	2006	P75534-2	2006	RS3959	2006	SP1938
2006	MP1014	2006	P75534-3	2006	RS3959-1	2006	SP1950
2006	MP1509-1	2006	P75534-4	2006	RS5612	2006	SP2045
2006	MP1509-2	2006	P75534-5	2006	RS5613	2006	SP2046
2006	MP1509-3	2006	P8870	2006	RS5614	2006	SP2048
2006	MP2060	2006	P8890	2006	RS5616	2006	SP2072
2006	MP2060-1	2006	P8890A	2006	RS5835	2006	SP2076
2006	MP2061	2006	P8890L	2006	RT-124	2006	SP2094
2006	MP2062	2006	PA-10889-1	2006	RT-127	2006	SP2155
2006	MP2137A	2006	PA-10889-2	2006	RT4762	2006	SP2234
2006	MP2138A	2006	PA-10890	2006	RT4762MHF25	2006	SP2247
2006	MP2139A	2006	PA-10890-1	2006	S-1556-2	2006	SP230
2006	MP2142A	2006	PA10890	2006	S-39T	2006	SP2341
2006	MP2143A	2006	PA10890-1	2006	S-40T	2006	SP2358
2006	MP2144A	2006	PADT50	2006	S-40TB	2006	SP2361
2006	MP3611	2006	PAR-12	2006	S-41T	2006	SP2361BLU
2006	MP3612	2006	PAR12	2006	S-42T	2006	SP2361BRN
2006	MP3613	2006	PB110	2006	S-43T	2006	SP2361GRN
2006	MP3614	2006	PC3004	2006	S-46T	2006	SP2361ORN
2006	MP3615	2006	PC3010	2006	S-48T	2006	SP2361RED
2006	MP3617	2006	PIK	2006	S-49T	2006	SP2361YEL
2006	MP825	2006	POWER-12	2006	S-58TB	2006	SP2395
2006	MZ9	2006	POWER-25	2006	S-95253	2006	SP2431
2006	N57B4-2	2006	POWER-299	2006	S-95253-1	2006	SP2493
2006	N57B4-4	2006	POWER-99	2006	S14A-1-3P	2006	SP2541
2006	NAP-T2-8	2006	POWER12	2006	S1556-2	2006	SP2551
2006	NK14A119	2006	POWER25	2006	0S22.3640-050	2006	SP26
2006	NK4A-1A19	2006	POWER299	2006	S4A-1-A-3P	2006	SP334
2006	NK65C119	2006	POWER99	2006	S650-1-3P	2006	SP404
2006	NK77C119	2006	P231	2006	867809	2006	SP404T
2006	NK84AA19	2006	PN5-3	2006	S770-1-3P	2006	SP441
2006	NKT-401	2006	PS-1	2006	S84A-1-3P	2006	SP441D
2006	NKT-402	2006	PS1	2006	S95253	2006	SP441G
2006	NKT-403	2006	PT-12	2006	S95253-1	2006	SP441S
2006	NKT-404	2006	PT-155	2006	SC-70	2006	SP47
2006	NKT-405	2006	PT-176			2006	SP485
2006	NKT-415	2006	PT-234			2006	SP485B

RS 276-	Industry Standard No.	RS 276-	Industry Standard No.	RS 276-	Industry Standard No.	RS 276-	Industry Standard No.
2006	SP485BLK	2006	TR-16	2006	001-01252	2006	2N1531A
2006	SP485BLU	2006	TR-16C	2006	001-01253	2006	2N1532
2006	SP485BRN	2006	TR-172(OLSON)	2006	001-01254	2006	2N1532A
2006	SP485W	2006	TR-178(OLSON)	2006	001-01255	2006	2N1533
2006	SP485WHT	2006	TR-184(OLSON)	2006	001-01256	2006	2N1534
2006	SP486	2006	TR-3	2006	001-01257	2006	2N1534A
2006	SP486W	2006	TR-43B	2006	001-01258	2006	2N1535
2006	SP486WHT	2006	TR-5	2006	001-01259	2006	2N1535A
2006	SP634	2006	TR-50	2006	1-21-270	2006	2N1535B
2006	SP649	2006	TR-8006	2006	1-21-271	2006	2N1536
2006	SP649-1	2006	TR01	2006	1R9-1	2006	2N1536A
2006	SP744	2006	TR01C	2006	002-007000	2006	2N1537
2006	SP819R	2006	TR01MP	2006	002-008100	2006	2N1537A
2006	SP834	2006	TR02	2006	002-008800	2006	2N1538
2006	SP875	2006	TR02C	2006	002-009700	2006	2N1538A
2006	SP880	2006	TR16C	2006	002-009701	2006	2N1539
2006	SP880-1	2006	TR333	2006	002-010100	2006	2N1539A
2006	SP880-3	2006	TR35144	2006	002-012700	2006	2N1540
2006	SP891	2006	TR35144A	2006	002-9700	2006	2N1540A
2006	SP891-B	2006	TR35524	2006	2.4341.0018	2006	2N1541
2006	SP891B	2006	TR50	2006	2AD140	2006	2N1541A
2006	SP891BLU	2006	TR56	2006	2D001	2006	2N1542
2006	SP891G	2006	TR8006	2006	2D004	2006	2N1542A
2006	SP891GRN	2006	TRA-7R	2006	2D004-9	2006	2N1543
2006	SP891R	2006	TRA-7RM	2006	2D015	2006	2N1543A
2006	SP891W	2006	TRA-8R	2006	2G210	2006	2N1544
2006	SP891WHT	2006	TRA7R	2006	2G220	2006	2N1544A
2006	SP891	2006	TRA8R	2006	2G222	2006	2N1545
2006	S81606A	2006	TS-1657	2006	2G223	2006	2N1545A
2006	ST-235	2006	TS-173	2006	2G224	2006	2N1546
2006	ST235	2006	TS-176	2006	2G225	2006	2N1546A
2006	SYL109	2006	TS-610	2006	2G240	2006	2N1547
2006	T-101	2006	TS-612	2006	2N1007	2006	2N1547A
2006	T-127	2006	TS-613	2006	2N1011	2006	2N1548
2006	T-235	2006	TS-614	2006	2N1014	2006	2N1548A
2006	T-Q5028	2006	TS1657	2006	2N1020	2006	2N1549
2006	T-Q5036	2006	TS173	2006	2N1021	2006	2N1549A
2006	T1040	2006	TS176	2006	2N1021A	2006	2N155
2006	T1041	2006	TS610	2006	2N1022	2006	2N1550
2006	T1167	2006	TS612	2006	2N1022A	2006	2N1550A
2006	T1168	2006	TS613	2006	2N1029	2006	2N1551
2006	T13029	2006	TS614	2006	2N1029A	2006	2N1551A
2006	T1366	2006	TT-1083	2006	2N1029B	2006	2N1552
2006	T1366A	2006	TT1083	2006	2N1029C	2006	2N1552A
2006	T1367	2006	TV24337	2006	2N1030	2006	2N1553
2006	T1367A	2006	TV24678	2006	2N1030A	2006	2N1553A
2006	T1368	2006	TV28B126P	2006	2N1030B	2006	2N1554
2006	T1368A	2006	TV8-28B126	2006	2N1030C	2006	2N1554A
2006	T1369	2006	TV8-28B126P	2006	2N1031	2006	2N1555
2006	T1369A	2006	TV8-28B449P	2006	2N1031A	2006	2N1555A
2006	T1370	2006	TV828B126	2006	2N1031B	2006	2N1556
2006	T1370A	2006	TV828B449	2006	2N1031C	2006	2N1556A
2006	T139	2006	TV828B449(P)	2006	2N1032	2006	2N1557
2006	T142	2006	UTRA-7RM	2006	2N1032A	2006	2N1557A
2006	T1601	2006	V145	2006	2N1032B	2006	2N1558
2006	T1602	2006	V15/10DP	2006	2N1032C	2006	2N1558A
2006	TA-1614	2006	V15/20DP	2006	2N1033	2006	2N1559
2006	TA-1682	2006	V15/30DP	2006	2N1039	2006	2N1559A
2006	TA-1682A	2006	V152A	2006	2N1040	2006	2N156
2006	TA-1705	2006	V162A	2006	2N1041	2006	2N1560
2006	TA-1765	2006	V30/10DP	2006	2N1046	2006	2N1560A
2006	TA-1766	2006	V30/20DP	2006	2N1046B	2006	2N157
2006	TA-1773	2006	V30/30DP	2006	2N1073	2006	2N157A
2006	TA-1794	2006	V60/10DP	2006	2N1073A	2006	2N158
2006	TA-1881	2006	V60/20DP	2006	2N1120	2006	2N158A
2006	TA-1890	2006	V60/30DP	2006	2N1136	2006	2N1609
2006	TA-1891	2006	V60/30P	2006	2N1136A	2006	2N1610
2006	TA-2	2006	VP0-274513	2006	2N1136C	2006	2N1611
2006	TA1614	2006	VP02745B	2006	2N1137	2006	2N1612
2006	TA1682	2006	VFP-2746C	2006	2N1137A	2006	2N1652
2006	TA1682A	2006	VFP-6537C	2006	2N1138	2006	2N1653
2006	TA1705	2006	VPP2746C	2006	2N1138A	2006	2N1666
2006	TA1765	2006	VP26357C	2006	2N1138B	2006	2N1667
2006	TA1766	2006	VP66537C	2006	2N1146	2006	2N1668
2006	TA1773	2006	VFU-2746B	2006	2N1146A	2006	2N1669
2006	TA1794	2006	VFU2746B	2006	2N1146B	2006	2N1755
2006	TA1881	2006	VP065326B	2006	2N1146C	2006	2N1756
2006	TA1890	2006	VM-30203	2006	2N1147	2006	2N1757
2006	TA1891	2006	VM30203	2006	2N1147A	2006	2N1758
2006	TA2301	2006	VS-28B126	2006	2N1147B	2006	2N1759
2006	TA2672	2006	VS28B126	2006	2N1147C	2006	2N176
2006	TP-80/30	2006	VS28B126P	2006	2N115	2006	2N176-1
2006	TP78	2006	VS28B126V	2006	2N1159	2006	2N176-1BLU
2006	TP78/30	2006	W17	2006	2N1160	2006	2N176-1WHT
2006	TP78/30Z	2006	W5	2006	2N1162	2006	2N176-1YEL
2006	TP78/60	2006	W9	2006	2N1162A	2006	2N176-3PUR
2006	TP80/30	2006	WEP230	2006	2N1163	2006	2N176-4PUR
2006	TP80/30Z	2006	WEP230MP	2006	2N1163A	2006	2N176-5WHT
2006	TI-1A6	2006	WEP232	2006	2N1164	2006	2N176-6WHT
2006	TI-266A	2006	WEP367	2006	2N1164A	2006	2N1760
2006	TI-269	2006	WEP481	2006	2N1165	2006	2N1761
2006	TI-3029	2006	WEP624	2006	2N1165A	2006	2N1762
2006	TI-366	2006	WEP628	2006	2N1166	2006	2N176A
2006	TI-366A	2006	WEP628MP	2006	2N1166A	2006	2N176BLK
2006	TI-367	2006	WEP642	2006	2N1168	2006	2N176BLU
2006	TI-367A	2006	WEP642MP	2006	2N1172	2006	2N176G
2006	TI-368	2006	WEP643	2006	2N1182	2006	2N176GRN
2006	TI-368A	2006	WTV12PWR	2006	2N1218	2006	2N176M
2006	TI-369	2006	WTV199PWR	2006	2N1227	2006	2N176PUR
2006	TI-369A	2006	WTV25PWR	2006	2N1227-3	2006	2N176RED
2006	TI-370	2006	WTV40PWR	2006	2N1227-4	2006	2N176W
2006	TI-370A	2006	WTV6PWR	2006	2N1227-4A	2006	2N176WHT
2006	TI-7A	2006	WTV99PWR	2006	2N1227-4R	2006	2N176YEL
2006	TI266A	2006	X1005	2006	2N1227A	2006	2N178
2006	TI269	2006	XB-5	2006	2N1245	2006	2N179
2006	TI3012	2006	XB-7	2006	2N1246	2006	2N1905
2006	TI3027	2006	XB14	2006	2N1291	2006	2N1971
2006	TI3028	2006	XB5	2006	2N1292	2006	2N2061
2006	TI3029	2006	XB7	2006	2N1293	2006	2N2061A
2006	TI366	2006	XC141	2006	2N1294	2006	2N2062
2006	TI366A	2006	XC142	2006	2N1295	2006	2N2062A
2006	TI367	2006	XC155	2006	2N1296	2006	2N2063
2006	TI367A	2006	XC156	2006	2N1297	2006	2N2063A
2006	TI368	2006	XN12A	2006	2N1314	2006	2N2064
2006	TI368A	2006	XN12B	2006	2N1314R	2006	2N2064A
2006	TI369	2006	XN12C	2006	2N1359	2006	2N2065
2006	TI369A	2006	XN12E	2006	2N1360	2006	2N2065A
2006	TI370	2006	XN12F	2006	2N1362	2006	2N2066
2006	TI370A	2006	Y410	2006	2N1363	2006	2N2066A
2006	TM121	2006	Z20	2006	2N1364	2006	2N2067
2006	TM131	2006	ZEN325	2006	2N1365	2006	2N2067B
2006	TNJ60454	2006	ZEN326	2006	2N141	2006	2N2067G
2006	TNJ70483	2006	ZEN330	2006	2N1419	2006	2N2067W
2006	TNJ70541	2006	ZEN331	2006	2N143	2006	2N2069
2006	TNJ72318	2006	001-01204-0	2006	2N1430	2006	2N2070
2006	TO-012	2006	001-012040	2006	2N1437	2006	2N2071
2006	TO-015	2006	001-01205-0	2006	2N1438	2006	2N2072
2006	TQ-5064	2006	001-01205-1	2006	2N1501	2006	2N2137
2006	TQ5036	2006	001-01250	2006	2N1502	2006	2N2137A
2006	TQ5064	2006	001-012051	2006	2N1529	2006	2N2138
2006	TR-01	2006	001-01250	2006	2N1529A	2006	2N2138A
2006	TR-01C	2006	001-01251	2006	2N1530	2006	2N2139
2006	TR-01MP			2006	2N1530A	2006	2N2140
2006	TR-02			2006	2N1531	2006	2N2140A

RS 276-	Industry Standard No.	RS 276-	Industry Standard No.	RS 276-	Industry Standard No.	RS 276-	Industry Standard No.
2006	2N2141	2006	2N511B	2006	2SB248A	2006	2SB414B
2006	2N2141A	2006	2N512	2006	2SB249	2006	2SB414C
2006	2N2142	2006	2N512A	2006	2SB249A	2006	2SB414D
2006	2N2142A	2006	2N512B	2006	2SB25	2006	2SB414E
2006	2N2143	2006	2N513	2006	2SB250	2006	2SB414P
2006	2N2143A	2006	2N513A	2006	2SB250A	2006	2SB414G
2006	2N2144	2006	2N513B	2006	2SB254	2006	2SB414GN
2006	2N2144A	2006	2N514	2006	2SB255	2006	2SB414H
2006	2N2145	2006	2N514A	2006	2SB256	2006	2SB414J
2006	2N2145A	2006	2N514B	2006	2SB25B	2006	2SB414K
2006	2N2146	2006	2N538	2006	2SB26	2006	2SB414L
2006	2N2146A	2006	2N538A	2006	2SB26A	2006	2SB414M
2006	2N2147	2006	2N539	2006	2SB27	2006	2SB414OR
2006	2N2148	2006	2N539A	2006	2SB28	2006	2SB414R
2006	2N2212	2006	2N540	2006	2SB282	2006	2SB414X
2006	2N2282	2006	2N540A	2006	2SB283	2006	2SB414Y
2006	2N2287	2006	2N553	2006	2SB284	2006	2SB42
2006	2N2288	2006	2N554	2006	2SB285	2006	2SB424
2006	2N2289	2006	2N555	2006	2SB29	2006	2SB426
2006	2N2291	2006	2N561	2006	2SB295	2006	2SB426BL
2006	2N2292	2006	2N5618	2006	2SB30	2006	2SB426R
2006	2N2293	2006	2N57	2006	2SB31	2006	2SB426Y
2006	2N2294	2006	2N5887	2006	2SB337	2006	2SB445
2006	2N2295	2006	2N5888	2006	2SB337-OR	2006	2SB445A
2006	2N2296	2006	2N5889	2006	2SB337A	2006	2SB445B
2006	2N230	2006	2N589	2006	2SB337B	2006	2SB445C
2006	2N234	2006	2N5893	2006	2SB337BK	2006	2SB445D
2006	2N234A	2006	2N5897	2006	2SB337C	2006	2SB445E
2006	2N235	2006	2N5901	2006	2SB337D	2006	2SB445F
2006	2N2357	2006	2N618	2006	2SB337E	2006	2SB445G
2006	2N2358	2006	2N627	2006	2SB337F	2006	2SB445GN
2006	2N235A	2006	2N628	2006	2SB337G	2006	2SB445H
2006	2N235B	2006	2N629	2006	2SB337GN	2006	2SB445J
2006	2N236	2006	2N637	2006	2SB337H	2006	2SB445K
2006	2N236A	2006	2N637A	2006	2SB337HA	2006	2SB445L
2006	2N236B	2006	2N637B	2006	2SB337HB	2006	2SB445M
2006	2N242	2006	2N638	2006	2SB337J	2006	2SB445OR
2006	2N2423	2006	2N638A	2006	2SB337LB	2006	2SB445R
2006	2N2446	2006	2N638B	2006	2SB337M	2006	2SB445X
2006	2N250	2006	2N639	2006	2SB337R	2006	2SB445Y
2006	2N250A	2006	2N639A	2006	2SB337X	2006	2SB446
2006	2N251	2006	2N639B	2006	2SB337Y	2006	2SB446A
2006	2N251A	2006	2N66	2006	2SB338	2006	2SB446B
2006	2N255	2006	2N663	2006	2SB338H	2006	2SB446C
2006	2N255A	2006	2N665	2006	2SB338HA	2006	2SB446D
2006	2N256	2006	2N669	2006	2SB338HB	2006	2SB446E
2006	2N256A	2006	2N67	2006	2SB555	2006	2SB446F
2006	2N257	2006	2N677C	2006	2SB567	2006	2SB446G
2006	2N257A	2006	2N678B	2006	2SB567(A)	2006	2SB446GN
2006	2N257B	2006	2N678C	2006	2SB567(B)P	2006	2SB446H
2006	2N257G	2006	2N68	2006	2SB567-4	2006	2SB446J
2006	2N257W	2006	2826	2006	2SB567-5	2006	2SB446K
2006	2N2612	2006	2826A	2006	2SB567-OR	2006	2SB446L
2006	2N268	2006	2841	2006	2SB567A	2006	2SB446M
2006	2N268A	2006	2841A	2006	2SB567AL	2006	2SB446OR
2006	2N275W	2006	2842	2006	2SB567B	2006	2SB446R
2006	2N275W	2006	2SA416	2006	2SB567B-2	2006	2SB446X
2006	2N2832	2006	2SA416A	2006	2SB567BL	2006	2SB446Y
2006	2N2835	2006	2SA416B	2006	2SB567BP	2006	2SB449
2006	2N2836	2006	2SA416C	2006	2SB567C	2006	2SB449F
2006	2N285	2006	2SA416D	2006	2SB567D	2006	2SB449P
2006	2N285A	2006	2SA416E	2006	2SB567E	2006	2SB458
2006	2N285B	2006	2SA416F	2006	2SB567F	2006	2SB458A
2006	2N2869	2006	2SA416G	2006	2SB567G	2006	2SB458B
2006	2N2869/2N301	2006	2SA416GN	2006	2SB567H	2006	2SB458BC
2006	2N2870	2006	2SA416H	2006	2SB567J	2006	2SB458BL
2006	2N290	2006	2SA416J	2006	2SB567K	2006	2SB458C
2006	2N296	2006	2SA416K	2006	2SB567L	2006	2SB462
2006	2N297	2006	2SA416L	2006	2SB567M	2006	2SB463
2006	2N297A	2006	2SA416M	2006	2SB567OR	2006	2SB463-0
2006	2N301	2006	2SA416OR	2006	2SB567P	2006	2SB463A
2006	2N301A	2006	2SA416R	2006	2SB567R	2006	2SB463B
2006	2N301B	2006	2SA416X	2006	2SB567X	2006	2SB463BLU
2006	2N301W	2006	2SA416Y	2006	2SB567Y	2006	2SB463BLU-Q
2006	2N307	2006	2SB107	2006	2SB568	2006	2SB463C
2006	2N307A	2006	2SB107A	2006	2SB568-OR	2006	2SB463D
2006	2N307B	2006	2SB119	2006	2SB568A	2006	2SB463F
2006	2N3125	2006	2SB119A	2006	2SB568B	2006	2SB463G
2006	2N3126	2006	2SB122	2006	2SB568C	2006	2SB463G-BL
2006	2N3132	2006	2SB123	2006	2SB568D	2006	2SB463G-R
2006	2N3146	2006	2SB123A	2006	2SB568E	2006	2SB463G-Y
2006	2N3147	2006	2SB124	2006	2SB568F	2006	2SB463GN
2006	2N3212	2006	2SB126	2006	2SB568G	2006	2SB463J
2006	2N3213	2006	2SB126A	2006	2SB568GN	2006	2SB463K
2006	2N3214	2006	2SB126P	2006	2SB568H	2006	2SB463L
2006	2N3215	2006	2SB126V	2006	2SB568J	2006	2SB463M
2006	2N325	2006	2SB127	2006	2SB568K	2006	2SB463R
2006	2N350	2006	2SB127A	2006	2SB568L	2006	2SB463RED
2006	2N350A	2006	2SB128	2006	2SB568M	2006	2SB463RED-Q
2006	2N351	2006	2SB128A	2006	2SB568X	2006	2SB463X
2006	2N351A	2006	2SB128V	2006	2SB568Y	2006	2SB463XL
2006	2N352	2006	2SB129	2006	2SB591	2006	2SB463Y
2006	2N353	2006	2SB130	2006	2SB407	2006	2SB463YEL
2006	2N3611	2006	2SB130A	2006	2SB407-0	2006	2SB463YEL-Q
2006	2N3612	2006	2SB131	2006	2SB407-OR	2006	2SB466
2006	2N3613	2006	2SB131A	2006	2SB407A	2006	2SB467
2006	2N3614	2006	2SB132	2006	2SB407B	2006	2SB471
2006	2N3615	2006	2SB132A	2006	2SB407BK	2006	2SB471-2
2006	2N3616	2006	2SB137	2006	2SB407C	2006	2SB471-0
2006	2N3617	2006	2SB138	2006	2SB407D	2006	2SB471A
2006	2N3618	2006	2SB140	2006	2SB407E	2006	2SB471B
2006	2N375	2006	2SB141	2006	2SB407F	2006	2SB471D
2006	2N376	2006	2SB142	2006	2SB407G	2006	2SB471E
2006	2N376A	2006	2SB142A	2006	2SB407GN	2006	2SB471F
2006	2N378	2006	2SB142B	2006	2SB407H	2006	2SB471G
2006	2N379	2006	2SB142C	2006	2SB407J	2006	2SB471GN
2006	2N380	2006	2SB143	2006	2SB407K	2006	2SB471H
2006	2N386	2006	2SB143P	2006	2SB407M	2006	2SB471J
2006	2N387	2006	2SB144	2006	2SB407R	2006	2SB471K
2006	2N392	2006	2SB144P	2006	2SB407TV	2006	2SB471L
2006	2N399	2006	2SB145	2006	2SB407TV-2	2006	2SB471M
2006	2N400	2006	2SB146	2006	2SB407X	2006	2SB471R
2006	2N401	2006	2SB147	2006	2SB407Y	2006	2SB471X
2006	2N4078	2006	2SB149	2006	2SB41	2006	2SB471Y
2006	2N418	2006	2SB151	2006	2SB413	2006	2SB472
2006	2N419	2006	2SB152	2006	2SB413A	2006	2SB472A
2006	2N420	2006	2SB19	2006	2SB413B	2006	2SB472B
2006	2N420A	2006	2SB20	2006	2SB413C	2006	2SB473
2006	2N4241	2006	2SB21	2006	2SB413D	2006	2SB473D
2006	2N4244	2006	2SB215	2006	2SB413E	2006	2SB473F
2006	2N4247	2006	2SB216	2006	2SB413F	2006	2SB473H
2006	2N456	2006	2SB216A	2006	2SB413G	2006	2SB481-OR
2006	2N456A	2006	2SB217	2006	2SB413GN	2006	2SB481A
2006	2N456B	2006	2SB217A	2006	2SB413H	2006	2SB481B
2006	2N457	2006	2SB217U	2006	2SB413J	2006	2SB481C
2006	2N457A	2006	2SB217U	2006	2SB413K	2006	2SB481D
2006	2N457B	2006	2SB22B	2006	2SB413L	2006	2SB481E
2006	2N458	2006	2SB229	2006	2SB413M	2006	2SB481F
2006	2N458A	2006	2SB230	2006	2SB413OR	2006	2SB481G
2006	2N458B	2006	2SB231	2006	2SB413R	2006	2SB481GN
2006	2N459	2006	2SB235	2006	2SB413X		
2006	2N459A	2006	2SB239	2006	2SB413Y		
2006	2N511	2006	2SB239A	2006	2SB414		
2006	2N511A	2006	2SB246				
		2006	2SB247				
		2006	2SB248				

RS 276-	Industry Standard No.	RS 276-	Industry Standard No.	RS 276-	Industry Standard No.	RS 276-	Industry Standard No.
2006	28B481H	2006	8A10521	2006	21A097-000	2006	48-137026
2006	28B481J	2006	8A10625	2006	22-002001	2006	48-137031
2006	28B481K	2006	8A11083	2006	22-002008	2006	48-137078
2006	28B481L	2006	8A11721	2006	22-002009	2006	48-137102
2006	28B481M	2006	8A12359	2006	022-3640-050	2006	48-137118
2006	28B481OR	2006	8A12991	2006	022.3640-050	2006	48-137119
2006	28B481R	2006	8A13164	2006	23-5309	2006	48-137120
2006	28B481X	2006	8H303	2006	23-5042	2006	48-137122
2006	28B62	2006	8L201	2006	23B-210-025	2006	48-137123
2006	28B63	2006	8L201B	2006	24MW994	2006	48-137124
2006	28B64	2006	8L201C	2006	026-1000-20	2006	48-137213
2006	28B69	2006	8L201R	2006	026-100003	2006	48-137214
2006	28B80	2006	8L301V	2006	026-100028	2006	48-137215
2006	28B81	2006	8L301V	2006	27T406	2006	48-137216
2006	28B82	2006	8L404	2006	31-0192	2006	48-137217
2006	28B83	2006	8P-404	2006	31-0196	2006	48-137218
2006	28B84	2006	8P-404R	2006	31-0240	2006	48-137219
2006	28B84A	2006	8P-505	2006	33-1004-00	2006	48-137220
2006	28B84B	2006	8P202	2006	34-6001-1	2006	48-137267
2006	28B84C	2006	8P40	2006	34-6001-79	2006	48-137268
2006	28B84D	2006	8P404	2006	34-6002-10	2006	48-137269
2006	28B84E	2006	8P404B	2006	34-6002-11	2006	48-137270
2006	28B84F	2006	8P404F	2006	34-6002-13	2006	48-137271
2006	28B84G	2006	8P404M	2006	34-6002-14	2006	48-137308
2006	28B84GN	2006	8P404M-1	2006	34-6002-17	2006	48-137329
2006	28B84H	2006	8P404N	2006	34-6002-18	2006	48-137978
2006	28B84J	2006	8P404ORN	2006	34-6002-18A	2006	48-57B2
2006	28B84K	2006	8P404R	2006	34-6002-19	2006	48-40172901
2006	28B84L	2006	8P404V	2006	34-6002-2	2006	48-57B2
2006	28B84M	2006	8P415C	2006	34-6002-20	2006	48-57B42
2006	28B84OR	2006	8P416C	2006	34-6002-22	2006	48-869087B
2006	28B84R	2006	8P505	2006	34-6002-22A	2006	48-869099B
2006	28B84X	2006	8P508	2006	34-6002-3	2006	48-869141
2006	28B84Y	2006	8P060	2006	34-6002-34	2006	48-869142
2006	28BF5	2006	8P860	2006	34-6002-4	2006	48-869182
2006	28C357	2006	09-300037A	2006	34-6002-5	2006	48-869202
2006	28C357B	2006	09-301010	2006	34-6002-6	2006	48-869237
2006	28C467	2006	09-301024	2006	34-6002-7	2006	48-869241
2006	28F.T212	2006	09-301034	2006	34-6002-8	2006	48-869255
2006	28FT212	2006	09-301052	2006	34-6002-9	2006	48-869342
2006	2T3011	2006	09-301075	2006	38P1	2006	48-869436
2006	2T3021	2006	9-51141400	2006	38P1C	2006	48-869550
2006	2T3022	2006	9-5250	2006	39P1	2006	48-97046A15
2006	2T3030	2006	9-5251	2006	39P1C	2006	48-97046A18
2006	2T3031	2006	11-0399	2006	39PC1	2006	48-97046A31
2006	2T3032	2006	11-0400	2006	42-16599	2006	48-97258A06
2006	2T3033	2006	12-1-270	2006	42-23968	2006	48K125250
2006	2T3041	2006	12-1-271	2006	42-23968P	2006	48K134583
2006	2T3042	2006	12M2	2006	43-025834	2006	48K134584
2006	2T3043	2006	13-14735	2006	44A-1A5	2006	48X39P1
2006	03-57-501	2006	13-14735-1	2006	46-86125-3	2006	48K57B2
2006	4-142	2006	13-14735A	2006	46-86135-3	2006	48K57B42
2006	4-14A17-1	2006	13-15806-1	2006	46-86136-3	2006	48K869342
2006	4-435	2006	13-18034	2006	46-8615-3	2006	48N8P1035
2006	4-4A-1A7-1	2006	13-18034-1	2006	46-8617-3	2006	48Q134722
2006	4-65017-1	2006	13-18034A	2006	46-86213-3	2006	48R134582
2006	4-686213-3	2006	13-22741	2006	46-8638-3	2006	48R134606
2006	4-77017-1	2006	13-26377-1	2006	47P1	2006	48R134722
2006	004-8000	2006	14-573-10	2006	48-10103A06	2006	48R869141
2006	4-88A17-1	2006	14-574-10	2006	48-10103A11	2006	48R869142
2006	4A-1	2006	14-578-10	2006	48-124204	2006	48R869202
2006	04A-1-1-12-7	2006	14-579-10	2006	48-124246	2006	48R869205
2006	4A-1-70	2006	14-586-01	2006	48-124247	2006	48R869241
2006	4A-1-70-12	2006	14-589-01	2006	48-124285	2006	48R869255
2006	4A-1-70-12-7	2006	14-590-01	2006	48-124302	2006	48R869436
2006	4A-1-A-7B	2006	14-601-01	2006	48-124332	2006	48R869550
2006	4A-10	2006	14-601-03	2006	48-124356	2006	48S132270
2006	4A-11	2006	14-601-04	2006	48-125204	2006	48S134695
2006	4A-12	2006	14-601-05	2006	48-125208	2006	48S134746
2006	4A-13	2006	14-601-06	2006	48-125267	2006	48S134747
2006	4A-14	2006	14-601-07	2006	48-125288	2006	48S134751
2006	4A-15	2006	14-601-08	2006	48-125332	2006	48S134758
2006	4A-16	2006	14-601-09	2006	48-129934	2006	48S134759
2006	4A-17	2006	14-601-11	2006	48-129935	2006	48S134760
2006	4A-18	2006	14-604-03(PNP)	2006	48-129936	2006	48S134761
2006	4A-19	2006	14-604-07	2006	48-129937	2006	48S134766
2006	4A-1A	2006	14-604-08	2006	48-134302	2006	48S134767
2006	4A-1AO	2006	14A	2006	48-134447	2006	48S134888
2006	4A-1AOR	2006	014A-12	2006	48-134448	2006	48S134947
2006	4A-1A1	2006	014A-12-7	2006	48-134449	2006	48S134974
2006	4A-1A19	2006	14AO	2006	48-134465	2006	48S137031
2006	4A-1A2	2006	14A1	2006	48-134487	2006	48S372270
2006	4A-1A21	2006	14A1-A82	2006	48-134488	2006	48S40172901
2006	4A-1A3	2006	14A10	2006	48-134493	2006	48X97046A15
2006	4A-1A3P	2006	14A1OR	2006	48-134519	2006	48X97046A31
2006	4A-1A4	2006	14A11	2006	48-134560	2006	49P1C
2006	4A-1A4-7	2006	14A12	2006	48-134570	2006	53P153
2006	4A-1A5	2006	14A13	2006	48-134574	2006	57A11-119
2006	4A-1A5L	2006	14A13P	2006	48-134575	2006	57A100-11
2006	4A-1A6	2006	14A14	2006	48-134582	2006	57A124-10
2006	4A-1A6-4	2006	14A14-7	2006	48-134583	2006	57A4-1
2006	4A-1A7	2006	14A14-7B	2006	48-134592	2006	57A4-2
2006	4A-1A7-1	2006	14A15	2006	48-134606	2006	57A4-4
2006	4A-1A8	2006	14A15L	2006	48-134611	2006	57A6-12
2006	4A-1A82	2006	14A16	2006	48-134612	2006	57A6-2
2006	4A-1A9	2006	14A16-5	2006	48-134613	2006	57A6-23
2006	4A-1A9G	2006	14A17	2006	48-134634	2006	57A6-3
2006	4A-1AO	2006	14A17-1	2006	48-134638	2006	57A6-8
2006	4A-1AOR	2006	14A18	2006	48-134639	2006	57A9-2
2006	04A1	2006	14A19	2006	48-134644	2006	57B100-11
2006	04A1-12	2006	14A19G	2006	48-134645	2006	57B124-10
2006	4JXBD404	2006	14A2	2006	48-134646	2006	57B4-1
2006	4JXBD404	2006	14A3	2006	48-134647	2006	57B4-2
2006	4JXBD409	2006	14A4	2006	48-134649	2006	57B4-4
2006	6-0000158	2006	14A5	2006	48-134651	2006	57B6-12
2006	6-158	2006	14A6	2006	48-134670	2006	5706-12
2006	6-88(AUTOMATIC)	2006	14A7	2006	48-134672	2006	5706-2
2006	6A10229	2006	14A8	2006	48-134696	2006	5706-23
2006	06P1C	2006	14A9	2006	48-134722	2006	5706-3
2006	7-1(STANDEL)	2006	14AO	2006	48-134723	2006	5706-8
2006	7-2	2006	17A4422-1	2006	48-134727	2006	5709-2
2006	8-0050700	2006	18AA-1-82	2006	48-134730	2006	57D1-119
2006	8-619-030-015	2006	19A115101	2006	48-134731	2006	57D4-1
2006	8-905-605-607	2006	19A115101-P1	2006	48-134738	2006	57D4-2
2006	8-905-605-624	2006	19A115184-P1	2006	48-134744	2006	57D6-12
2006	8-905-605-635	2006	19A15267P1	2006	48-134746	2006	57D9-2
2006	8-905-605-636	2006	19A115268	2006	48-134750	2006	57DQ-23
2006	8-905-605-637	2006	19A15341P1	2006	48-134751	2006	57DQ-32
2006	8-905-605-650	2006	19A115361-P1	2006	48-134757	2006	57L5-1
2006	8-905-606-720	2006	19A115376	2006	48-134758	2006	57M3-10P
2006	8-905-613-210	2006	19A115385-P1	2006	48-134759	2006	57M3-12
2006	8-905-613-215	2006	19A115561	2006	48-134763	2006	57M3-7
2006	8-905-613-240	2006	19A115561-1	2006	48-134764	2006	57M3-8
2006	8-905-613-241	2006	19AR31	2006	48-134766	2006	57M3-9P
2006	8-905-613-242	2006	19C300113-P1	2006	48-134767	2006	61-1906
2006	8-905-613-245	2006	20A0017	2006	48-134888	2006	61-782
2006	8-905-613-250	2006	20A0041	2006	48-134907	2006	610009-1
2006	8-905-613-265	2006	20A0042	2006	48-134930	2006	62-18427
2006	8-905-613-266	2006	20A0074	2006	48-134938	2006	62-18428
2006	8-905-613-277	2006	21-28	2006	48-134947	2006	63-10378
2006	8-905-613-282	2006	21A015-003	2006	48-134974	2006	63-18427
2006	8-905-613-283	2006	21A015-022	2006	48-134977	2006	63-29451
2006	8-905-613-284	2006	21A064-000	2006	48-137025	2006	63-29459
2006	8-905-613-555					2006	63-8590

RS 276-	Industry Standard No.	RS 276-	Industry Standard No.	RS 276-	Industry Standard No.	RS 276-	Industry Standard No.
2006	63-8706	2006	86-142-2	2006	173A4419-4	2006	1124C
2006	65C	2006	86-146-2	2006	173A4419-5	2006	1415-168
2006	065C-12	2006	86-147-2	2006	173A4419-6	2006	1415-172
2006	065C-12-7	2006	86-173-2	2006	173A4419-7	2006	1415-178
2006	65C-70	2006	86-173-9	2006	173A4419-8	2006	1414A
2006	65C-70-12	2006	86-19-2	2006	173A4419-9	2006	1414A-12
2006	65C-70-12-7	2006	86-230-2	2006	173A4420	2006	1414A-12-8
2006	65C0	2006	86-231-2	2006	173A4420-1	2006	1465C
2006	65C10	2006	86-232-2	2006	173A4420-5	2006	1465C-12
2006	65C10R	2006	86-235-2	2006	173A4421-1	2006	1465C-12-8
2006	65C11	2006	86-248-2	2006	173A4422-1	2006	1484A
2006	65C119	2006	86-313-2	2006	173A436	2006	1484A-12
2006	65C12	2006	86-317-2	2006	173A469	2006	1484A-12-8
2006	65C121	2006	86-319-2	2006	184A-1	2006	1559-17
2006	65C13	2006	86-353-2	2006	184A-1-12	2006	1559-17A
2006	65C13P	2006	86-354-2	2006	184A-1-12-7	2006	1561-17
2006	65C14	2006	86-370-2	2006	184A-1L	2006	1814A
2006	65C14-7	2006	86-370-2(GRN)	2006	184A-1L8	2006	1814A-12
2006	65C14-7B	2006	86-370-2(ORN)	2006	188-826	2006	1814A-127
2006	65C15	2006	86-370-2(VIO)	2006	199-POWER	2006	1814AL
2006	65C15L	2006	86-570-2YEL	2006	207A20	2006	1814AL-8
2006	65C16	2006	86-480-9	2006	207A20A	2006	1840-17
2006	65C16-4	2006	86-5039-2	2006	220-002001	2006	1859-14
2006	65C17	2006	86-5043-2	2006	229-0116	2006	1859-16
2006	65C17-1	2006	86-5057-2	2006	231-0006B	2006	1865C
2006	65C18	2006	86-5058-2	2006	231-0011	2006	1865C-12
2006	65C18Z	2006	86-5083-2	2006	231-0015	2006	1865C-127
2006	65C19	2006	86-5088-2	2006	231-006B	2006	1865CL
2006	65C19G	2006	86-5089-2	2006	247-624	2006	1865CL8
2006	65C2	2006	86-5090-2	2006	260P01209	2006	1884A
2006	65C3	2006	86-5113-2	2006	260P07502	2006	1884A-12
2006	65C4	2006	86-5125-2	2006	295V041H04	2006	1884A-127
2006	65C5	2006	86-5943-2	2006	296-61-9	2006	1884AL
2006	65C6	2006	86-62-2	2006	297V040H15	2006	1884AL8
2006	65C7	2006	86-63-2	2006	297V041H02	2006	1945-17
2006	65C8	2006	86-8-2	2006	297V041H03	2006	1955-17
2006	65C9	2006	86X0009-001	2006	297V041H04	2006	2002
2006	68P1	2006	86X0015-001	2006	297V041H05	2006	2002(E.F.JOHNSON)
2006	68P1B	2006	86X0033-001	2006	297V041H06	2006	2005
2006	76-11770	2006	86X2	2006	297V041H07	2006	2007-01
2006	77-270877-2	2006	0087	2006	297V041H15	2006	2057A2-211
2006	77-270878-2	2006	094-013	2006	297V062001	2006	2057A2-302
2006	77-271491-1	2006	94A-1A6-4	2006	297V062C05	2006	2057B100-11
2006	77C	2006	95-250	2006	310-192	2006	2057B124-10
2006	077C-12	2006	95-251	2006	324-0126	2006	2057B2-133
2006	077C-12-7	2006	96-5026-01	2006	414A-15	2006	2093A3D-20
2006	77C-70	2006	96-5045-01	2006	417-141	2006	2243
2006	77C-70-12	2006	96-5064-01	2006	417-160	2006	2347-17
2006	77C-70-12-7	2006	96-5081-01	2006	417-20	2006	2402-456
2006	77C0	2006	96-5086-02	2006	417-216	2006	2446-1(RCA)
2006	77C1	2006	96-5100-01	2006	417-30	2006	2577
2006	77C10	2006	96-5100-03	2006	417-32	2006	2780
2006	77C10R	2006	96-5125-01	2006	417-44	2006	2780-4
2006	77C11	2006	96-5143-01	2006	417-45	2006	2780-5
2006	77C119	2006	96-5143-02	2006	417-62	2006	2901-010
2006	77C12	2006	96-5430-02	2006	417-90	2006	2904-008
2006	77C121	2006	96-5148-01	2006	417-99	2006	2904-014
2006	77C13	2006	96-5155-01	2006	421-24	2006	3107-204-90070
2006	77C13P	2006	96-5192-01	2006	421-25	2006	3107-204-90140
2006	77C14	2006	96-5378-01	2006	421-6599	2006	3107-204-90190
2006	77C14-7	2006	96XZ801/06N	2006	430-25834	2006	3512
2006	77C14-7B	2006	96XZ801/10N	2006	465-137-19	2006	3514
2006	77C15	2006	96XZ801/34X	2006	465-206-19	2006	003515
2006	77C15L	2006	97A83	2006	465C-15	2006	3515(RCA)
2006	77C16	2006	98P1P	2006	473A3	2006	003516
2006	77C16-3	2006	99-PWR	2006	480=6(SEARS)	2006	3618-1
2006	77C17	2006	998001	2006	484A15	2006	4082-501-0001
2006	77C17-1	2006	998014	2006	572-0040-051	2006	0004203
2006	77C18	2006	998014A	2006	576-0002-002	2006	4247
2006	77C18Z	2006	998015	2006	576-0002-005	2006	4331
2006	77C19	2006	106P1	2006	576-0040-051	2006	4347
2006	77C19G	2006	106P1AG	2006	576-0040-254	2006	4465
2006	77C2	2006	106P1T	2006	602-032	2006	4570
2006	77C3	2006	111P5C	2006	610-039	2006	4582BRN
2006	77C4	2006	111P7C	2006	610-039-1	2006	4583RED
2006	77C5	2006	112-202147	2006	610-067	2006	4584GRN
2006	77C6	2006	112-524	2006	610-067-1	2006	4608
2006	77C7	2006	114A-1-82	2006	610-067-2	2006	4619RED
2006	77C8	2006	115-063	2006	610-067-3	2006	4620GRN
2006	77C9	2006	115-268	2006	610-068	2006	4649
2006	78-272212-1	2006	115-269	2006	610-068-1	2006	4686-213-3
2006	78-5009	2006	120-003150	2006	614A-1-5L	2006	4722
2006	80-050700	2006	120-004887	2006	614A1-5L	2006	4722BLU
2006	81-27126130-7	2006	121-1124	2006	642-152	2006	4722GRN
2006	81-27126130-7A	2006	121-1134	2006	642-176	2006	4722ORN
2006	81-27126130-7B	2006	121-171	2006	642-206	2006	4722PUR
2006	81-46125028-4	2006	121-270	2006	642-217	2006	4722RED
2006	83-1056	2006	121-271	2006	642-264	2006	4722YEL
2006	84	2006	121-308	2006	642-272	2006	4727
2006	84A	2006	121-363	2006	642-316	2006	4730
2006	084A-12	2006	121-371	2006	650=1-5L	2006	4801-1100-011
2006	084A-12-7	2006	121-382	2006	684A-1-5L	2006	4822-130-40213
2006	84A-70	2006	121-389	2006	690L270H02	2006	4822-130-40233
2006	84A-70-12	2006	121-398	2006	690L287H02	2006	4888A
2006	84A-70-12-7	2006	121-793	2006	690V001H09	2006	4888B
2006	84A0	2006	127-29006	2006	690V081H97	2006	5253
2006	84A1	2006	122-1625	2006	690V098H51	2006	6377-1(SYLVANIA)
2006	84A10	2006	124-1	2006	753-2100-002	2006	7219-3
2006	84A1OR	2006	125-402	2006	753-4001-474	2006	8000-00003-040
2006	84A12	2006	125-403	2006	800-329	2006	8000-00006-190
2006	84A121	2006	129-10	2006	800-507-00	2006	8500-204
2006	84A13	2006	129-13	2006	800-507OO	2006	8883-2
2006	84A13P	2006	129-5	2006	800-518-00	2006	8999-115
2006	84A14	2006	129-6	2006	880-250-001	2006	9005-0
2006	84A14-7	2006	129-7	2006	880-250-010	2006	9403-2
2006	84A14-7B	2006	129-9	2006	884-250-001	2006	9404-0
2006	84A15	2006	130-104	2006	884-250-010	2006	9925-0
2006	84A15L	2006	0131-000192	2006	914A-1-6-5	2006	11252-4
2006	84A16	2006	0131-000336	2006	924-17945	2006	11506-3
2006	84A16B	2006	0131-000337	2006	964-16599	2006	11526-8
2006	84A17	2006	131-000562	2006	964-17887	2006	11526-9
2006	84A17-1	2006	0131-001425	2006	964-17945	2006	12127-0
2006	84A18	2006	143	2006	965C-16-4	2006	12127-1
2006	84A182	2006	144A-1	2006	984A-1-6B	2006	12163
2006	84A19	2006	144A-1-12	2006	991-01-0099	2006	12178
2006	84A19G	2006	144A-1-12-3	2006	991-01-1216	2006	014382
2006	84A3	2006	144A-1-12-8	2006	992-00-1192	2006	15024
2006	84A4	2006	146-T1	2006	992-00-8870	2006	15027
2006	84A5	2006	146T1	2006	992-001192	2006	15354-3
2006	84A6	2006	147-T1	2006	992-008870	2006	15927
2006	84A7	2006	147T1	2006	992-008890	2006	16599
2006	84A8	2006	154A3680	2006	992-01-1216	2006	16959
2006	84A9	2006	165C-182	2006	992-01-1218	2006	17887
2006	84AA1	2006	171-001-9-001	2006	992-011218	2006	17945
2006	84AA19	2006	171-015-9-001	2006	992-08890	2006	0022481
2006	84B	2006	171-016-9-001	2006	1008(JULIETTE)	2006	23311-006
2006	85-370-2 BLU	2006	173A3936	2006	1008(POWER)	2006	25658-120
2006	85-370-2(BLU)	2006	173A3963	2006	1008-17	2006	25658-121
2006	85-5058-2	2006	173A419-2	2006	1024-17	2006	25661-020
2006	0086	2006	173A4419	2006	1105-15	2006	25661-022
2006	86-120-2	2006	173A4419-1	2006	1124	2006	27126-060
2006	86-127-2	2006	173A4419-10	2006	1124A	2006	27126-090
2006	86-141-2	2006	173A4419-2	2006	1124B	2006	30203
		2006	173A4419-3			2006	30211

RS 276-	Industry Standard No.
2006	30215(RCA)
2006	30216(RCA)
2006	30246
2006	30246A
2006	30302
2006	35989-2069
2006	34022
2006	34298
2006	34315
2006	34425
2006	34525
2006	34715
2006	35044
2006	35084
2006	35144
2006	55201
2006	55231
2006	55260
2006	55349
2006	35728
2006	35885A
2006	35885B
2006	35951
2006	36203
2006	36303
2006	36304
2006	36304-4
2006	36312
2006	36395
2006	36477
2006	36687
2006	36800-2
2006	36800-3
2006	36800-4
2006	36800-5
2006	36800-6
2006	36800-7
2006	36896
2006	36910
2006	36971
2006	39893
2006	40022
2006	40050
2006	40051
2006	40051-2
2006	40254
2006	40462
2006	40612
2006	40626
2006	43046
2006	43074
2006	49751-163
2006	50447-4
2006	50477-4
2006	51650
2006	057040
2006	59990-1
2006	60770
2006	61010-6
2006	61010-6-1
2006	62177
2006	66009-5
2006	66010-3
2006	70434
2006	000071090
2006	71448
2006	71448-1
2006	71448-2
2006	71448-3
2006	71448-4
2006	71448-5
2006	71448-6
2006	71448-7
2006	71488-4
2006	71488-5
2006	71488-6
2006	72193
2006	72856-63
2006	75700-03-01
2006	080048
2006	080050
2006	80416C
2006	81513-7
2006	084001C
2006	88832
2006	90050
2006	94004
2006	94024
2006	94025
2006	94026
2006	94032
2006	94034
2006	94040
2006	95250
2006	95250-1
2006	95251
2006	95253
2006	95250
2006	95250-1
2006	110515
2006	115268
2006	115269
2006	115281
2006	115282
2006	115283
2006	115284
2006	116093
2006	119721
2006	121243
2006	122792
2006	123792
2006	123808
2006	125703
2006	125761
2006	145134-526
2006	147351-5-1
2006	162002-033
2006	162002-062
2006	162002-062A
2006	162002-095
2006	167285
2006	170307-1
2006	170376
2006	170376-1
2006	170407-1
2006	170479-1
2006	171004(SEARS)
2006	171162-082
2006	171162-083
2006	171162-086
2006	171162-089
2006	171217-1
2006	175027-022
2006	175043-023
2006	175043-065
2006	175043-81
2006	190425
2006	190425A
2006	194474-8
2006	196058-4
2006	196148-0
2006	196183-5
2006	196501-7
2006	196607-9
2006	209185-962
2006	214396
2006	215089
2006	216986
2006	217892
2006	21801DB
2006	219301
2006	219361
2006	219440
2006	219940
2006	221602
2006	221605
2006	221940
2006	221941
2006	222915
2006	223365
2006	223430
2006	223576
2006	224503
2006	224873
2006	225595
2006	225596
2006	225925
2006	225927
2006	226634
2006	226999
2006	227566
2006	227804
2006	228229
2006	228230
2006	228558
2006	228559
2006	230208
2006	230523
2006	231140-11
2006	231140-33
2006	231672
2006	231797
2006	232194
2006	232674
2006	232675
2006	233509
2006	234077
2006	234178
2006	234566
2006	235312
2006	236935
2006	237452
2006	242183
2006	242838
2006	256068
2006	256071
2006	257341
2006	257403
2006	257536
2006	258990
2006	261970
2006	262114
2006	262370
2006	263856
2006	270744
2006	270745
2006	270746
2006	270780
2006	270785
2006	275612
2006	309412
2006	322968-140
2006	489751-163
2006	0573030
2006	0573030-14
2006	0573166
2006	0573205
2006	0573212
2006	602032
2006	603031
2006	605122
2006	610039
2006	610039-1
2006	610049-1
2006	610067
2006	610067-1
2006	610067-2
2006	610067-D
2006	610068
2006	610106
2006	610106-1
2006	610111-5
2006	610111-7
2006	610152-1
2006	617871-1
2006	618139-1
2006	650970
2006	651202
2006	652085
2006	652086
2006	655319
2006	660077
2006	660094
2006	660095
2006	660097
2006	660103
2006	702885
2006	702885-00
2006	740471
2006	801518
2006	801519
2006	801522
2006	801523
2006	801538
2006	815137B
2006	815137
2006	815246-2
2006	880092
2006	980132
2006	980134
2006	980135
2006	980155
2006	980437
2006	982307
2006	983036
2006	983795
2006	983874
2006	983975
2006	984261
2006	984431
2006	984521
2006	985036
2006	985103
2006	985431
2006	985443
2006	985447
2006	985449
2006	985453
2006	985455
2006	985686
2006	988080
2006	988336
2006	988413
2006	988468
2006	989387
2006	995001
2006	995014
2006	995015
2006	1221615
2006	1221625
2006	1407205-1
2006	1407206-1
2006	1471036-14
2006	1471036-20
2006	1471102-41
2006	1472446
2006	1472446-1
2006	1473512-1
2006	1473515-1
2006	1476171-11(TRANS)
2006	1476171-13(TRANS)
2006	1960584
2006	1961479
2006	1961480
2006	1961835
2006	1965017
2006	1966079
2006	2000646-113
2006	2006511-59
2006	2006607-59
2006	2090056-1
2006	2090056-27
2006	2090056-5
2006	2091858-0712
2006	2091858-11
2006	2091859-0008
2006	2091859-0011
2006	2091859-0025
2006	2091859-0712
2006	2091859-0713
2006	2091859-0714
2006	2091859-0715
2006	2091859-0716
2006	2091859-0717
2006	2091859-0718
2006	2091859-0720
2006	2091859-0723
2006	2091859-10
2006	2091859-11
2006	2091859-16
2006	2091859-2
2006	2091859-25
2006	2091859-4
2006	2091859-8
2006	2091859-8
2006	2091859-9
2006	2091959-16
2006	2320092
2006	2320541
2006	2904014
2006	3130006
2006	3130109
2006	3460553-2
2006	3460555-4
2006	3462221-1
2006	3462306-1
2006(2)	A13-14604-1A
2006(2)	A13-14604-1B
2006(2)	A13-14604-1C
2006(2)	A13-14604-1D
2006(2)	A13-14604-1E
2006(2)	A13-14777-1
2006(2)	A13-14777-1A
2006(2)	A13-14777-1B
2006(2)	A13-14777-1C
2006(2)	A13-14777-1D
2006(2)	A13-14778-1A
2006(2)	A13-14778-1B
2006(2)	A13-14778-1C
2006(2)	A13-14778-1D
2006(2)	A13-22741-2
2006(2)	A146B-3
2006(2)	A146B-39
2006(2)	A1477B
2006(2)	A1477B9
2006(2)	A168P1
2006(2)	A48-10075A01
2006(2)	A48-10075A02
2006(2)	A48-10075A03
2006(2)	A48-10075A04
2006(2)	A48-10075A05
2006(2)	A48-10075A06
2006(2)	A48-10075A07
2006(2)	A48-10075A08
2006(2)	A48-10103A01
2006(2)	A48-10103A02
2006(2)	A48-10103A03
2006(2)	A48-10103A04
2006(2)	A48-10103A05
2006(2)	A48-10103A06
2006(2)	A48-10103A07
2006(2)	A48-10103A08
2006(2)	A48-10103A09
2006(2)	A48-10103A10
2006(2)	A48-10103A11
2006(2)	A48-64978A10
2006(2)	A48-64978A11
2006(2)	A48-64978A24
2006(2)	A642-271
2006(2)	A660031
2006(2)	A6B-3-70
2006(2)	A6B-3-705
2006(2)	A6B-3A9G
2006(2)	A77B-70
2006(2)	A77B-705
2006(2)	A77B19G
2006(2)	A815203-5
2006(2)	A86X0030-100
2006(2)	B6B-3-A-21
2006(2)	B6B-3A21
2006(2)	B77C-1-21
2006(2)	CXL104MP
2006(2)	EC0104MP
2006(2)	EC0121MP
2006(2)	EC0131MP
2006(2)	GE-13MP
2006(2)	GE-31MP
2006(2)	J24566
2006(2)	N-8A15X139
2006(2)	NK6B-3A19
2006(2)	NK77B119
2006(2)	REN121MP
2006(2)	REN131MP
2006(2)	RB5855
2006(2)	86B-3-A-3P
2006(2)	S77B-1-3P
2006(2)	SK3013
2006(2)	SK3015
2006(2)	TM104MP
2006(2)	TM121MP
2006(2)	TM131MP
2006(2)	W17MP
2006(2)	W9MP
2006(2)	24D149
2006(2)	28B473(H)
2006(2)	28B481
2006(2)	4-1848
2006(2)	4-6B-3A7-1
2006(2)	4-77B17-1
2006(2)	6B-3
2006(2)	06B-3-12
2006(2)	06B-3-12-7
2006(2)	6B-3-70
2006(2)	6B-3-70-12
2006(2)	6B-30
2006(2)	6B-31
2006(2)	6B-32
2006(2)	6B-33
2006(2)	6B-34
2006(2)	6B-35
2006(2)	6B-36
2006(2)	6B-37
2006(2)	6B-38
2006(2)	6B-39
2006(2)	6B-3A
2006(2)	6B-3A0
2006(2)	6B-3A19
2006(2)	6B-3A2
2006(2)	6B-3A21
2006(2)	6B-3A3
2006(2)	6B-3A3F
2006(2)	6B-3A4
2006(2)	6B-3A4-7
2006(2)	6B-3A4-7B
2006(2)	6B-3A5
2006(2)	6B-3A5L
2006(2)	6B-3A6
2006(2)	6B-3A7
2006(2)	6B-3A7-1
2006(2)	6B-3A8
2006(2)	6B-3A82
2006(2)	6B-3A9
2006(2)	6B-3A9G
2006(2)	6B-3A0R
2006(2)	7-2(STANDEL)
2006(2)	9-5250-1
2006(2)	9-5257
2006(2)	13-14604-1
2006(2)	13-14604-1A
2006(2)	13-14604-1B
2006(2)	13-14604-1C
2006(2)	13-14604-1D
2006(2)	13-14604-1E
2006(2)	13-14777-1
2006(2)	13-14777-1A
2006(2)	13-14777-1B
2006(2)	13-14777-1C
2006(2)	13-14777-1D
2006(2)	13-14778-1
2006(2)	13-14778-1A
2006(2)	13-14778-1B
2006(2)	13-14778-1C
2006(2)	13-14778-1D
2006(2)	13-14778-1E
2006(2)	13-22739-1
2006(2)	13-22741-1
2006(2)	13-22741-2
2006(2)	16B-3A82
2006(2)	24NW618
2006(2)	026-100004
2006(2)	026-100020
2006(2)	31-0248
2006(2)	32-16591
2006(2)	32-16599
2006(2)	33-1002-00
2006(2)	42-21443
2006(2)	42-22834
2006(2)	46B-3A5
2006(2)	48-10075A01
2006(2)	48-10075A02
2006(2)	48-10075A03
2006(2)	48-10075A04
2006(2)	48-10075A05
2006(2)	48-10075A06
2006(2)	48-10075A07
2006(2)	48-10075A08
2006(2)	48-10103A01
2006(2)	48-10103A02
2006(2)	48-10103A03
2006(2)	48-10103A04
2006(2)	48-10103A05
2006(2)	48-10103A07
2006(2)	48-10103A08
2006(2)	48-10103A09
2006(2)	48-10103A10
2006(2)	48-134747
2006(2)	48-64978A10
2006(2)	48-64978A11
2006(2)	48-64978A24
2006(2)	48X97238A06
2006(2)	57A3-10
2006(2)	57A3-11
2006(2)	57A3-12
2006(2)	57A3-7
2006(2)	57A3-8
2006(2)	57A3-9
2006(2)	57B3-10
2006(2)	57B3-11
2006(2)	57B3-12
2006(2)	57B3-7
2006(2)	57B3-8
2006(2)	57B3-9
2006(2)	66B-3A5L
2006(2)	077B
2006(2)	077B-12

RS 276-	Industry Standard No.	RS 276-	Industry Standard No.	RS 276-	Industry Standard No.	RS 276-	Industry Standard No.
2006(2)	077B-12-7	2007	A065-106	2007	A514-027662	2007	AC151RIV
2006(2)	77B-70	2007	A065-112	2007	A538	2007	AC151RV
2006(2)	77B-70-12	2007	A069-105	2007	A55	2007	AC151RVI
2006(2)	77B-70-12-7	2007	A069-107	2007	A595	2007	AC151V
2006(2)	77B0	2007	A114(TRANSISTOR)	2007	A615-1010	2007	AC151VI
2006(2)	77B1	2007	A115(JAPAN)	2007	A615-1011	2007	AC151VII
2006(2)	77B10	2007	A116(JAPAN)	2007	A64	2007	AC152
2006(2)	77B10R	2007	A12	2007	A65	2007	AC152-IV
2006(2)	77B11	2007	A1243	2007	A66	2007	AC152-V
2006(2)	77B119	2007	A128(JAPAN)	2007	A66-1-70	2007	AC152-VI
2006(2)	77B12	2007	A129(TRANSISTOR)	2007	A66-1-705	2007	AC152IV
2006(2)	77B121	2007	A12A	2007	A66-1A9G	2007	AC152V
2006(2)	77B13	2007	A12B	2007	A66-2-70	2007	AC152VI
2006(2)	77B13P	2007	A12C	2007	A66-2-705	2007	AC153
2006(2)	77B14	2007	A12D	2007	A66-2A9G	2007	AC153K
2006(2)	77B14-7	2007	A12H	2007	A66-3-3A9G	2007	AC154
2006(2)	77B14-7B	2007	A12V	2007	A66-3-70	2007	AC155
2006(2)	77B15	2007	A138(JAPAN)	2007	A66-3-705	2007	AC156
2006(2)	77B15L	2007	A14-1004	2007	A66-3A9G	2007	AC160
2006(2)	77B16	2007	A14-1005	2007	A74-3-3A9G	2007	AC160A
2006(2)	77B16-2	2007	A14-1006	2007	A74-3-70	2007	AC160B
2006(2)	77B17	2007	A14-1007	2007	A74-3-705	2007	AC160GRN
2006(2)	77B17-1	2007	A14-1008	2007	A74-3A9G	2007	AC160RED
2006(2)	77B18	2007	A14-1009	2007	A88C-70	2007	AC160YEL
2006(2)	77B182	2007	A14-1010	2007	A88C-705	2007	AC161
2006(2)	77B19	2007	A146	2007	A88019G	2007	AC162
2006(2)	77B19Q	2007	A1466-1	2007	A8P-2-70	2007	AC163
2006(2)	77B2	2007	A1466-19	2007	A8P-2-705	2007	AC165
2006(2)	77B3	2007	A1466-19-	2007	A8P-2A9G	2007	AC166
2006(2)	77B4	2007	A1466-2	2007	A909-1011	2007	AC167
2006(2)	77B5	2007	A1466-29	2007	A909-1012	2007	AC168
2006(2)	77B6	2007	A1466-3	2007	A909-1013	2007	AC170
2006(2)	77B7	2007	A1466-39	2007	A9L-4-70	2007	AC171
2006(2)	77B8	2007	A147	2007	A9L-4-705	2007	AC173
2006(2)	77B9	2007	A1474-3	2007	A9L-4A9G	2007	AC176K
2006(2)	86-0033-007	2007	A1474-39	2007	AA1	2007	AC178
2006(2)	86X0030-001	2007	A148	2007	AC-107	2007	AC180
2006(2)	86X0030-100	2007	A1488C	2007	AC-113	2007	AC180K
2006(2)	95-250-1	2007	A1488C9	2007	AC-113A	2007	AC182
2006(2)	95-257	2007	A148P-2	2007	AC-114	2007	AC184
2006(2)	96B-3A65	2007	A148P-29	2007	AC-116	2007	AC188
2006(2)	96X26054/45X	2007	A148P2	2007	AC-117	2007	AC188/01
2006(2)	998013	2007	A148P2-29	2007	AC-117A	2007	AC188O1
2006(2)	998013A	2007	A149	2007	AC-117P	2007	AC188K
2006(2)	115-281	2007	A149L-4	2007	AC-121IV	2007	AC191
2006(2)	115-282	2007	A149L-49	2007	AC-122	2007	AC192
2006(2)	115-283	2007	A15(TRANSISTOR)	2007	AC-123	2007	AC193
2006(2)	115-284	2007	A15-1004	2007	AC-125	2007	AC193K
2006(2)	146B-3	2007	A15-1005	2007	AC-126	2007	ACR810-104
2006(2)	146B-3-12	2007	A15BK	2007	AC-128	2007	ACR810-105
2006(2)	146B-3-12-8	2007	A15BL	2007	AC-132	2007	ACR810-106
2006(2)	168P1	2007	A15BLU	2007	AC-150	2007	ACR83-1004
2006(2)	177B-1-82	2007	A15H	2007	AC-151	2007	ACR83-1005
2006(2)	177C-1-82	2007	A15K	2007	AC-152	2007	ACR83-1006
2006(2)	186B-3	2007	A15R	2007	AC-154	2007	ACY-17
2006(2)	186B-3-12	2007	A15U(TRANSISTOR)	2007	AC-155	2007	ACY-18
2006(2)	186B-3-127	2007	A15V	2007	AC-156	2007	ACY-19
2006(2)	186B-3L	2007	A15VR	2007	AC-161	2007	ACY-20
2006(2)	186B-3L8	2007	A15Y	2007	AC-162	2007	ACY-21
2006(2)	297V041H01	2007	A16	2007	AC-165	2007	ACY-22
2006(2)	553-9201-001	2007	A160(JAPAN)	2007	AC-166	2007	ACY-23
2006(2)	0418	2007	A167	2007	AC-167	2007	ACY-32
2006(2)	422-1443	2007	A168(JAPAN)	2007	AC-168	2007	ACY17
2006(2)	465-166-19	2007	A168A	2007	AC-169	2007	ACY17-1
2006(2)	477B15	2007	A169	2007	AC-N7B	2007	ACY18
2006(2)	477015	2007	A17	2007	AC105	2007	ACY18-1
2006(2)	642-271	2007	A170(JAPAN)	2007	AC106	2007	ACY19
2006(2)	675-206	2007	A171(JAPAN)	2007	AC108	2007	ACY19-1
2006(2)	677B-1-5L	2007	A172	2007	AC109	2007	ACY20
2006(2)	677C-1-5L	2007	A172A	2007	AC110	2007	ACY20-1
2006(2)	755-2000-463	2007	A173	2007	AC113	2007	ACY21
2006(2)	800-196	2007	A173B	2007	AC114	2007	ACY21-1
2006(2)	800-253	2007	A174	2007	AC115	2007	ACY22
2006(2)	977B1-6-2	2007	A17H	2007	AC116	2007	ACY22-1
2006(2)	977C1-6-3	2007	A181	2007	AC117	2007	ACY23
2006(2)	992-008-890	2007	A182	2007	AC117A	2007	ACY23-V
2006(2)	1477B	2007	A183	2007	AC117B	2007	ACY23-VI
2006(2)	1477B-12	2007	A187TV	2007	AC117P	2007	ACY23V
2006(2)	1477B-12-8	2007	A18H	2007	AC118	2007	ACY23VI
2006(2)	1477C	2007	A197	2007	AC119	2007	ACY27
2006(2)	1477C-12	2007	A198	2007	AC120	2007	ACY28
2006(2)	1477C-12-8	2007	A203(PNP)	2007	AC121	2007	ACY29
2006(2)	1877B	2007	A203AA	2007	AC121-IV	2007	ACY30
2006(2)	1877B-12	2007	A203B	2007	AC121-VI	2007	ACY31
2006(2)	1877B-127	2007	A203P	2007	AC121-VII	2007	ACY32
2006(2)	1877BL	2007	A204	2007	AC121IV	2007	ACY32-V
2006(2)	1877BL8	2007	A205	2007	AC121V	2007	ACY32-VI
2006(2)	1877C	2007	A206	2007	AC121VI	2007	ACY32V
2006(2)	1877C-12	2007	A207	2007	AC121VII	2007	ACY32VI
2006(2)	1877C-127	2007	A208(JAPAN)	2007	AC122	2007	ACY33
2006(2)	1877CL	2007	A209	2007	AC122-30	2007	ACY33-VII
2006(2)	1877CL8	2007	A210(JAPAN)	2007	AC122-GRN	2007	ACY33-VIII
2006(2)	2780(AIRLINE)	2007	A211(JAPAN)	2007	AC122-RED	2007	ACY33VI
2006(2)	2780-3	2007	A212	2007	AC122-VIO	2007	ACY33VII
2006(2)	4800-223	2007	A217	2007	AC122-YEL	2007	ACY36
2006(2)	8102-210	2007	A231	2007	AC122GRN	2007	ACY38
2006(2)	11528-1	2007	A232	2007	AC122RED	2007	ACY40
2006(2)	11528-2	2007	A2414	2007	AC122YEL	2007	ACY41
2006(2)	11528-3	2007	A248(JAPAN)	2007	AC123	2007	ACY41-1
2006(2)	11528-4	2007	A25-1004	2007	AC124	2007	ACY44
2006(2)	36359-4	2007	A25-1005	2007	AC125	2007	ACY44-1
2006(2)	38094	2007	A25-1006	2007	AC126	2007	ADY-27
2006(2)	40623	2007	A26	2007	AC128	2007	AE-50
2006(2)	67085-0	2007	A277(JAPAN)	2007	AC128-01	2007	AF-101
2006(2)	67085-0-1	2007	A278(JAPAN)	2007	AC128/01	2007	AF138/290
2006(2)	081042	2007	A279(JAPAN)	2007	AC128O1	2007	AF187
2006(2)	95257	2007	A282	2007	AC128K	2007	AF200U
2006(2)	115063	2007	A283	2007	AC131	2007	AF201U
2006(2)	170666-1	2007	A284	2007	AC131-30	2007	AFZ23
2006(2)	170668-1	2007	A28A666PQR	2007	AC132	2007	ALZ10
2006(2)	170850-1	2007	A302	2007	AC132-01	2007	APB-11A-1008
2006(2)	171162-090	2007	A303	2007	AC133A	2007	APB-11H-1008
2006(2)	175043-081	2007	A304	2007	AC134	2007	APB-11H-101Q
2006(2)	215071	2007	A305	2007	AC135	2007	AR-102
2006(2)	560004	2007	A31	2007	AC136	2007	AR-103
2006(2)	0573031	2007	A311(JAPAN)	2007	AC137	2007	AR-104
2006(2)	0573040	2007	A312	2007	AC138	2007	AR-105
2006(2)	610067-3	2007	A32	2007	AC139	2007	AR102
2006(2)	610068-1	2007	A33	2007	AC142	2007	AS33867
2006(2)	660031	2007	A330	2007	AC142K	2007	AS3386B
2006(2)	670850	2007	A332	2007	AC150	2007	ASY-26
2006(2)	670850-1	2007	A350A	2007	AC150GRN	2007	ASY-27
2006(2)	740247	2007	A36	2007	AC150YBL	2007	ASY12-1
2006(2)	740443	2007	A374	2007	AC151	2007	ASY12-2
2006(2)	815137	2007	A396(TRANSISTOR)	2007	AC151-RIV	2007	ASY13-1
2006(2)	815203-3	2007	A397	2007	AC151-RV	2007	ASY13-2
2006(2)	815203-5	2007	A40	2007	AC151-RVI	2007	ASY14
2007	A-514-027662	2007	A406	2007	AC151-V	2007	ASY14-1
2007	A059-106	2007	A407	2007	AC151-VI	2007	ASY14-2
2007	A059-116	2007	A414	2007	AC151-VII	2007	ASY14-3
2007	A061-114	2007	A415(JAPAN)	2007	AC151IV	2007	ASY24
2007	A061-115	2007	A42X00286-01	2007	AC151R	2007	ASY26
2007	A065-105	2007	A44	2007		2007	ASY27
		2007	A50(TRANSISTOR)				

RS 276-	Industry Standard No.	RS 276-	Industry Standard No.	RS 276-	Industry Standard No.	RS 276-	Industry Standard No.
2007	A3Y31	2007	B171	2007	B266P	2007	B389
2007	A3Y48	2007	B171(JAPAN)	2007	B266Q	2007	B39
2007	A3Y48-IV	2007	B171A	2007	B267	2007	B392
2007	A3Y48-V	2007	B171B	2007	B268	2007	B393
2007	A3Y48-VI	2007	B172	2007	B269	2007	B394
2007	A3Y49	2007	B172A	2007	B270	2007	B395
2007	A3Y50	2007	B172AF	2007	B270A	2007	B396
2007	A3Y51	2007	B172B	2007	B270B	2007	B40
2007	A3Y52	2007	B172C	2007	B270C	2007	B400
2007	A3Y54	2007	B172D	2007	B270D	2007	B4004B
2007	A3Y55	2007	B172E	2007	B270E	2007	B400A
2007	A3Y56	2007	B172F	2007	B271	2007	B400B
2007	A3Y56N	2007	B172H	2007	B272	2007	B400K
2007	A3Y57N	2007	B172P	2007	B273	2007	B401
2007	A3Y58N	2007	B172R	2007	B290	2007	B402
2007	A3Y70	2007	B173	2007	B291	2007	B403
2007	A3Y70-IV	2007	B173A	2007	B292	2007	B405
2007	A3Y70-VI	2007	B173B	2007	B292A	2007	B405-2C
2007	A3Y70IV	2007	B173C	2007	B293	2007	B405-3C
2007	A3Y70V	2007	B173L	2007	B294(TRANSISTOR)	2007	B405-4C
2007	A3Y70VI	2007	B174	2007	B299	2007	B405A
2007	A3Y71	2007	B175	2007	B302	2007	B405B
2007	A3Y76	2007	B175A	2007	B303(TRANSISTOR)	2007	B405C
2007	A3Y77	2007	B175B	2007	B303-0	2007	B405D
2007	A3Y80	2007	B175C	2007	B3030	2007	B405E
2007	A3Y90	2007	B175E	2007	B303A	2007	B405G
2007	A3Y91	2007	B176	2007	B303B	2007	B405H
2007	AT-15	2007	B176-0	2007	B303C	2007	B405K
2007	AT-5	2007	B176-P	2007	B303H	2007	B405R
2007	AT-50	2007	B176-PR	2007	B303K	2007	B405RE
2007	AT-6A	2007	B176B	2007	B304	2007	B406
2007	AT100H	2007	B176M	2007	B304A	2007	B415
2007	AT100M	2007	B176P	2007	B314	2007	B415A
2007	AT100N	2007	B176PRC	2007	B315	2007	B415B
2007	AT10H	2007	B176R	2007	B316	2007	B416
2007	AT10M	2007	B177(JAPAN)	2007	B317	2007	B417
2007	AT10N	2007	B178(JAPAN)	2007	B32	2007	B421
2007	AT20H	2007	B178-0	2007	B32-0	2007	B422
2007	AT20M	2007	B178-8	2007	B32-1	2007	B423
2007	AT20N	2007	B178A	2007	B32-2	2007	B427
2007	AT30H	2007	B178C	2007	B32-4	2007	B428
2007	AT30M	2007	B178D	2007	B321	2007	B43
2007	AT30N	2007	B178M	2007	B322	2007	B431
2007	AT50	2007	B178N	2007	B323	2007	B439
2007	AT6	2007	B178T	2007	B324	2007	B439A
2007	AT6A	2007	B178U	2007	B324A	2007	B43A
2007	AT74	2007	B178V	2007	B324B	2007	B440
2007	AT74S	2007	B178X	2007	B324D	2007	B443
2007	AT874	2007	B178Y	2007	B324E	2007	B443A
2007	ATAP1	2007	B180	2007	B324E-1	2007	B443B
2007	ATAP2	2007	B180A	2007	B324F	2007	B450
2007	ATGP	2007	B181	2007	B324G	2007	B450A
2007	AU100N	2007	B181A	2007	B324H	2007	B451
2007	B-105B	2007	B183	2007	B324I	2007	B452
2007	B-22-3	2007	B184	2007	B324J	2007	B452A
2007	B-22-4	2007	B185	2007	B324K	2007	B453
2007	B-23	2007	B185(0)	2007	B324L	2007	B454
2007	B-23-1	2007	B185(0)	2007	B324N	2007	B455
2007	B-23-2	2007	B185AA	2007	B324P	2007	B457
2007	B-24-1	2007	B185F	2007	B324S	2007	B457-C
2007	B-26	2007	B185P	2007	B324V	2007	B457A
2007	B-26-1	2007	B186	2007	B326	2007	B459
2007	B-315-1	2007	B186(0)	2007	B327	2007	B459-0
2007	B-324	2007	B186(SANYO)	2007	B328	2007	B459-0
2007	B-P1A	2007	B186-1	2007	B329	2007	B459A
2007	B100	2007	B186-K	2007	B329K	2007	B459B
2007	B101	2007	B186A	2007	B32N	2007	B459C
2007	B102	2007	B186AG	2007	B33	2007	B459D
2007	B1022	2007	B186B	2007	B33-4	2007	B46
2007	B1022-1	2007	B186BY	2007	B335	2007	B460
2007	B103	2007	B186G	2007	B336	2007	B460A
2007	B104	2007	B186H	2007	B33C	2007	B460B
2007	B105	2007	B186L	2007	B33D	2007	B47
2007	B105B	2007	B186Y	2007	B33E	2007	B470
2007	B105B-1	2007	B187	2007	B33F	2007	B475
2007	B106	2007	B187(1)	2007	B34	2007	B475A
2007	B108	2007	B187(SANYO)	2007	B345	2007	B475B
2007	B108A	2007	B187AA	2007	B346	2007	B475D
2007	B108B	2007	B187B	2007	B346K	2007	B475E
2007	B109	2007	B187C	2007	B346Q	2007	B475F
2007	B10(TRANSISTOR)	2007	B187D	2007	B347	2007	B475G
2007	B111	2007	B187G	2007	B348	2007	B475P
2007	B111K	2007	B187K	2007	B348Q	2007	B475Q
2007	B112	2007	B187R	2007	B348R	2007	B476
2007	B113	2007	B187RED	2007	B349	2007	B48
2007	B114	2007	B187S	2007	B34N	2007	B482
2007	B115	2007	B187Y	2007	B350	2007	B486
2007	B1154	2007	B187YEL	2007	B364	2007	B49
2007	B1154-1	2007	B188	2007	B365	2007	B494
2007	B116	2007	B189	2007	B365B	2007	B495A
2007	B117	2007	B199	2007	B366	2007	B495C
2007	B117K	2007	B200	2007	B37	2007	B495D
2007	B120	2007	B200A	2007	B370	2007	B495T
2007	B134(JAPAN)	2007	B201	2007	B370A	2007	B496
2007	B134-D	2007	B202	2007	B370AA	2007	B497
2007	B134-E	2007	B203AA	2007	B370AB	2007	B498
2007	B135	2007	B218	2007	B370AC	2007	B5
2007	B135B	2007	B219	2007	B370AHA	2007	B50
2007	B135C	2007	B22	2007	B370AHB	2007	B51(JAPAN)
2007	B135E	2007	B22-3	2007	B370B	2007	B516C
2007	B136	2007	B22-4	2007	B370C	2007	B516CD
2007	B136-2	2007	B220	2007	B370D	2007	B516D
2007	B136-3	2007	B220A	2007	B370P	2007	B516P
2007	B136A	2007	B221	2007	B370PB	2007	B52
2007	B136B	2007	B221A	2007	B370V	2007	B53
2007	B136C	2007	B222	2007	B371	2007	B534
2007	B136U	2007	B223	2007	B371D	2007	B534A
2007	B153	2007	B224(JAPAN)	2007	B372	2007	B535
2007	B154	2007	B225	2007	B373	2007	B54
2007	B155	2007	B226	2007	B376	2007	B54B
2007	B155A	2007	B227	2007	B376G	2007	B54E
2007	B155B	2007	B22A	2007	B377	2007	B54P
2007	B156	2007	B22B	2007	B377B	2007	B54Y
2007	B156A	2007	B22I	2007	B378	2007	B55(TRANSISTOR)
2007	B156AA	2007	B22R	2007	B378A	2007	B56
2007	B156AB	2007	B22Y	2007	B379	2007	B560
2007	B156AC	2007	B23	2007	B379-2	2007	B56A
2007	B156B	2007	B23-1	2007	B379A	2007	B56B
2007	B156C	2007	B23-2	2007	B379B	2007	B56C
2007	B156D	2007	B238	2007	B37A	2007	B57
2007	B156F	2007	B238-12A	2007	B37B	2007	B58
2007	B157	2007	B238-12B	2007	B37C	2007	B59
2007	B158	2007	B238-12C	2007	B37E	2007	B5A
2007	B159	2007	B24	2007	B37F	2007	B60
2007	B160	2007	B24-1	2007	B38	2007	B601-1009
2007	B161	2007	B241	2007	B380	2007	B601-1010
2007	B162	2007	B257	2007	B380A	2007	B60A
2007	B163	2007	B26	2007	B381	2007	B61
2007	B164	2007	B261	2007	B382	2007	B65
2007	B165	2007	B262	2007	B383	2007	B66
2007	B166	2007	B263	2007	B383-1	2007	B66-1-A-21
2007	B167	2007	B264	2007	B383-2	2007	B66-1A21
2007	B168	2007	B265	2007	B386		
2007	B170	2007	B266	2007	B387		

RS 276-	Industry Standard No.	RS 276-	Industry Standard No.	RS 276-	Industry Standard No.	RS 276-	Industry Standard No.
2007	B66-2-A-21	2007	CK721	2007	E241B	2007	G04-711-H
2007	B66-2A21	2007	CK722	2007	E2445	2007	G11
2007	B66-3-A-21	2007	CK725	2007	E2448	2007	G12
2007	B66-3A21	2007	CK727	2007	E2453	2007	G14
2007	B66H	2007	CK751	2007	E2465	2007	GA52829
2007	B67	2007	CK754	2007	E2467	2007	GA53149
2007	B67A	2007	CK759	2007	E2476	2007	GA53242
2007	B71	2007	CK759A	2007	E2480	2007	GC1097
2007	B72	2007	CK760	2007	E2481	2007	GC1134
2007	B73	2007	CK760A	2007	E2482	2007	GC1136
2007	B73A	2007	CK761	2007	E4	2007	GC1143
2007	B73B	2007	CK768	2007	EA0009	2007	GC1145
2007	B73C	2007	CK776	2007	EA1081	2007	GC1150
2007	B73GR	2007	CK776A	2007	EA1346	2007	GC1159
2007	B74-3-A-21	2007	CK790	2007	EA15X164	2007	GC1183
2007	B74-3A21	2007	CK791	2007	EA15X19	2007	GC1184
2007	B75	2007	CK793	2007	EA15X2	2007	GC1186
2007	B75A	2007	CK794	2007	EA15X203	2007	GC1187
2007	B75AH	2007	CK870	2007	EA15X207	2007	GC1257
2007	B75B	2007	CK871	2007	EA15X212	2007	GC1302
2007	B75C	2007	CK872	2007	EA15X23	2007	GC1422
2007	B75P	2007	CK875	2007	EA15X25	2007	GC181
2007	B75H	2007	CK878	2007	EA15X257	2007	GC182
2007	B75LB	2007	CK879	2007	EA15X28	2007	GC250
2007	B76	2007	CK882	2007	EA15X3	2007	GC31
2007	B77	2007	CK888	2007	EA15X326	2007	GC32
2007	B77(B)	2007	CK891	2007	EA15X36	2007	GC33
2007	B77A	2007	CK892	2007	EA15X4	2007	GC34
2007	B77AA	2007	CM8640E	2007	EA15X67	2007	GC343
2007	B77AB	2007	CP800	2007	EA15X684Q	2007	GC35
2007	B77AC	2007	CP801	2007	EA15X7	2007	GC360
2007	B77AD	2007	CP802	2007	EA15X8442	2007	GC4022
2007	B77AH	2007	CP803	2007	EA15X8444	2007	GC408
2007	B77AP	2007	CQ1	2007	EA2134	2007	GC144
2007	B77B	2007	CS1758	2007	EA2135	2007	GC464
2007	B77B-11	2007	CT1009	2007	EA2136	2007	GC466
2007	B77C	2007	CT1017	2007	EA2176	2007	GC5000
2007	B77D	2007	CTP-2001-1001	2007	EB0001	2007	GC5010
2007	B77H	2007	CTP-2001-1002	2007	EB0003	2007	GC520
2007	B77V	2007	CTP-2001-1003	2007	ECG100	2007	GC521
2007	B77VRED	2007	CTP-2001-1004	2007	ECG102	2007	GC532
2007	B78	2007	CTP-2001-1009	2007	ECG102A	2007	GC551
2007	B79	2007	CTP-2006-1001	2007	ECG158	2007	GC552
2007	B85	2007	CTP-2006-1002	2007	ED52	2007	GC578
2007	B87	2007	CTP-2006-1003	2007	ED53	2007	GC579
2007	B88C-1-21	2007	CTP1032	2007	ED54B	2007	GC580
2007	B89	2007	CTP1033	2007	ED55	2007	GC581
2007	B89A	2007	CTP1034	2007	ED56	2007	GC588
2007	B89AH	2007	CTP1035	2007	ED57	2007	GC60
2007	B89H	2007	CTP1036	2007	EE100	2007	GC61
2007	B8P-2-A-21	2007	CTP2076-1001	2007	EK136	2007	GC639
2007	B8P-2A21	2007	CTP2076-1002	2007	EK159	2007	GC640
2007	B90	2007	CTP2076-1003	2007	EO-44A	2007	GC680
2007	B91	2007	CTP2076-1004	2007	EO105	2007	GC681
2007	B92	2007	CTP2076-1005	2007	EO44A	2007	GC682
2007	B94	2007	CTP2076-1006	2007	EO65	2007	GC733B
2007	B95	2007	CTP2076-1007	2007	EO66	2007	GC856
2007	B97	2007	CTP2076-1008	2007	EO67	2007	GC864
2007	B98	2007	CTP2076-1009	2007	EO68	2007	GE-1
2007	B9L-4-A-21	2007	CTP2076-1010	2007	EQ0-15	2007	GE-2
2007	B9L-4A21	2007	CTP2076-1011	2007	EQ0-9	2007	GE-52
2007	BA6	2007	CTP2076-1012	2007	ER15X17	2007	GE-57
2007	BA6A	2007	CXL100	2007	ER15X18	2007	GE-X9
2007	BCM1002-18	2007	CXL102	2007	ER15X22	2007	GER-A
2007	BCM1002-3	2007	CXL158	2007	ER15X23	2007	GET-103
2007	BCM1002-4	2007	D-P1A	2007	ER15X7	2007	GET-113
2007	BCM1002-5	2007	D008	2007	ER15X9	2007	GET-113A
2007	BCY91B	2007	D018	2007	ES15X10Q	2007	GET-114
2007	BB6	2007	D019	2007	ES15X31	2007	GETO-50P
2007	BB6A	2007	D021	2007	ES15X32	2007	GET103
2007	BP1	2007	D030	2007	ES15X4	2007	GET113
2007	BP1A	2007	D031	2007	ES15X49	2007	GET113A
2007	BP2	2007	D038	2007	ES15X50	2007	GET114
2007	BTX070	2007	D043	2007	ES15X53	2007	GET880
2007	BX-324	2007	D059	2007	ES15X55	2007	GET881
2007	C10227	2007	D078	2007	ES15X63	2007	GET882
2007	C10230-3	2007	D101B	2007	ES15X72	2007	GET887
2007	C11021	2007	D105B	2007	ES15X75	2007	GET888
2007	C1437	2007	D117	2007	ES15X8	2007	GET889
2007	C1438	2007	D135	2007	ES15X99	2007	GET890
2007	C175B	2007	D156	2007	ES17(ELCOM)	2007	GET891
2007	C73	2007	D171	2007	ES19	2007	GET892
2007	C75	2007	D180	2007	ES2(ELCOM)	2007	GET895
2007	C76	2007	D352	2007	ES23	2007	GET896
2007	CA2D2	2007	DC-10	2007	ES25	2007	GET897
2007	CB-103	2007	DC-12	2007	ES26	2007	GET898
2007	CB161	2007	DC-9	2007	ES26(ELCOM)	2007	GETO-50P
2007	CB248	2007	DB-3181	2007	ES3(ELCOM)	2007	GF20
2007	CB249	2007	D019	2007	ES3126	2007	GF21
2007	CB0360/7839	2007	D038	2007	ES37(ELCOM)	2007	GF32
2007	CB0360/7839	2007	D043	2007	ES4(ELCOM)	2007	GFT3008/40
2007	CB0361/7839	2007	D078	2007	ET15X1	2007	G11
2007	CB0362/7839	2007	D80-81252	2007	ET15X1	2007	G12
2007	CB0363/7839	2007	D8-14	2007	ET15X25	2007	G14
2007	CE213811	2007	D8-19	2007	ET15X31	2007	G04-711-E
2007	CGB-52	2007	D8-22	2007	ET15X32	2007	G04-711-F
2007	CGB-53	2007	D8-26	2007	ET3	2007	G04-711-G
2007	CJ5204	2007	D8-28(DELCO)	2007	ET4	2007	G04-711-H
2007	CK13	2007	D8-8	2007	ET5	2007	GP139
2007	CK13A	2007	D813	2007	ETTB-28B176	2007	GP139A
2007	CK14	2007	D816	2007	ETTB-28B176A	2007	GP139B
2007	CK14A	2007	D821	2007	ETTB-28B176B	2007	GT-109
2007	CK16	2007	D822	2007	ETTB-28B176R	2007	GT-269
2007	CK16A	2007	D823	2007	ETTB-75LB	2007	GT-348
2007	CK17	2007	D826	2007	EX15X25	2007	GT-759R
2007	CK17A	2007	D829	2007	F20-1006	2007	GT-760R
2007	CK22	2007	D83	2007	F20-1007	2007	GT-761R
2007	CK22A	2007	D853	2007	F20-1008	2007	GT-762R
2007	CK22B	2007	DU3	2007	F20-1009	2007	GT100
2007	CK22C	2007	DU4	2007	F215-1006	2007	GT109
2007	CK25	2007	DU5	2007	F215-1007	2007	GT109R
2007	CK25A	2007	E-044A	2007	F215-1008	2007	GT11
2007	CK26	2007	E-070	2007	F215-1009	2007	GT2
2007	CK26A	2007	E-158	2007	F2480	2007	GT122
2007	CK27	2007	E-2465	2007	FB420	2007	GT1223
2007	CK27A	2007	E0105	2007	FB421	2007	GT123
2007	CK64	2007	E044A	2007	FBN-CP2293	2007	GT13
2007	CK64A	2007	E066	2007	FD-1029-BG	2007	GT132
2007	CK64B	2007	E067	2007	FD-1029-EE	2007	GT14
2007	CK64C	2007	E068	2007	FD1029EE	2007	GT153
2007	CK65	2007	E132	2007	FPR40-1004	2007	GT1604
2007	CK65A	2007	E158	2007	FPR40-1005	2007	GT1605
2007	CK65B	2007	E181	2007	FPR50-1005	2007	GT1606
2007	CK65C	2007	E181A	2007	FPR50-1006	2007	GT1607
2007	CK66	2007	E181B	2007	FV2747C	2007	GT1665
2007	CK661	2007	E181C	2007	G0005	2007	GT18
2007	CK662	2007	E181D	2007	G0006	2007	GT20R
2007	CK66A	2007	E214B	2007	G0007	2007	GT210H
2007	CK66B	2007	E241	2007	G004	2007	GT222
2007	CK66C	2007	E24104	2007	G04-711-E	2007	GT269
2007	CK67	2007	E24106	2007	G04-711-F	2007	GT2693
2007	CK67A	2007	E2412	2007	G04-711-Q	2007	GT2694
2007	CK67B	2007	E241A			2007	GT2695
2007	CK67C					2007	GT2883

RS 276-	Industry Standard No.	RS 276-	Industry Standard No.	RS 276-	Industry Standard No.	RS 276-	Industry Standard No.
2007	GT2885	2007	HJ22	2007	K14-0066-4	2007	MA883
2007	GT2887	2007	HJ226	2007	K4-500	2007	MA884
2007	GT31	2007	HJ228	2007	KD2101	2007	MA885
2007	GT32	2007	HJ25	2007	K081000	2007	MA886
2007	GT34HV	2007	HJ230	2007	KO825642-20	2007	MA887
2007	GT348	2007	HJ315	2007	KO825642-40	2007	MA888
2007	GT41	2007	HJ41	2007	KO825643-10	2007	MA889
2007	GT42	2007	HJ43	2007	KO825643-15	2007	MA890
2007	GT44	2007	HJ50	2007	KO825651-20	2007	MA891
2007	GT45	2007	HJ51	2007	KO825657-53	2007	MA892
2007	GT74	2007	HJ54	2007	KR-Q0001	2007	MA893
2007	GT75	2007	HJ60	2007	KR-Q0002	2007	MA894
2007	GT751	2007	HJ606	2007	KR-Q0004	2007	MA895
2007	GT758	2007	HJ62	2007	KR-Q1010	2007	MA896
2007	GT759	2007	HJ73	2007	KR-Q1011	2007	MA897
2007	GT759R	2007	HJ74	2007	KR-Q1012	2007	MA898
2007	GT760R	2007	HJX2	2007	KV-1	2007	MA899
2007	GT761	2007	HM-00049	2007	KV-2	2007	MA900
2007	GT761R	2007	HM-08014	2007	KV-4A	2007	MA901
2007	GT762	2007	HR-1	2007	KV-4D	2007	MA902
2007	GT762R	2007	HR-2	2007	KV1	2007	MA903
2007	GT763	2007	HR-3	2007	KV2	2007	MA904
2007	GT764	2007	HR-39	2007	KV4	2007	MA909
2007	GT766	2007	HR-4	2007	L5021	2007	MA910
2007	GT766A	2007	HR-4A	2007	L5022	2007	MD501
2007	GT81	2007	HR-5	2007	L5022A	2007	MD501B
2007	GT81H	2007	HR-6	2007	L5025	2007	MM1151
2007	GT81H8	2007	HR-61	2007	L5025A	2007	MM1152
2007	GT81R	2007	HR-7	2007	M-75517-1	2007	MM1153
2007	GT82	2007	HR-7A	2007	M-8641A	2007	MM1154
2007	GT83	2007	HR-8	2007	M-P3D	2007	MM1742
2007	GT832	2007	HR-8A	2007	M108	2007	MN-53
2007	GT87	2007	HR-9	2007	M4313	2007	MN52
2007	GT88	2007	HR-9A	2007	M4315	2007	MN53
2007	GTB-2	2007	HR2A	2007	M4327	2007	MN53BLU
2007	GT81	2007	HR3	2007	M4389	2007	MN53GRN
2007	GT82	2007	HR30	2007	M4398	2007	MN53RED
2007	GTJ33231	2007	HR4	2007	M4450	2007	MN60
2007	GTJ33232	2007	HR4A	2007	M4462	2007	MP1014-1
2007	GTV	2007	HR5	2007	M4466	2007	MP1014-4
2007	H10	2007	HR53	2007	M4466ORN	2007	MP1014-5
2007	HA-00102	2007	HR6	2007	M4468	2007	MP1014-6
2007	HA-12	2007	HR61	2007	M4468BRN	2007	N-EA2136
2007	HA-201	2007	HR7A	2007	M4469	2007	N57B2-15
2007	HA-350	2007	HR8	2007	M4469RED	2007	N57B2-25
2007	HA-52	2007	HR8A	2007	M4470	2007	N57B2-3
2007	HA-53	2007	HR9	2007	M4470ORN	2007	N57B2-6
2007	HA00052	2007	HR9A	2007	M4471	2007	N57B2-7
2007	HA00053	2007	H8-15	2007	M4471YEL	2007	NA-1114-1004
2007	HA1350	2007	H8-22D	2007	M4472	2007	NA-1114-1005
2007	HA1360	2007	HS102	2007	M4472GRN	2007	NA-1114-1006
2007	HA15	2007	HS15	2007	M4473	2007	NA-1114-1007
2007	HA202	2007	HS170	2007	M4474	2007	NA-1114-1008
2007	HA49	2007	HS17D	2007	M4474YEL	2007	NA-1114-1009
2007	HA54	2007	HS23D	2007	M4475	2007	NA-1114-1010
2007	HA56	2007	HS290	2007	M4475ORN	2007	NA-1114-1011
2007	HAM-1	2007	HS5	2007	M4476	2007	NA1022-1007
2007	HB-00054	2007	HT040519C(28B405)	2007	M4476BLU	2007	NA5018-1013
2007	HB-00056	2007	HT1001510	2007	M4477	2007	NA5018-1014
2007	HB-00156	2007	HT2000771C	2007	M4477PUR	2007	NA5018-1015
2007	HB-00171	2007	HT200540	2007	M4482	2007	NA5018-1016
2007	HB-00172	2007	HT200540A	2007	M4483	2007	NAP-T-Z-10
2007	HB-00173	2007	HT200541	2007	M4510	2007	NAP-TZ-10
2007	HB-00175	2007	HT200541A	2007	M4553	2007	NO30
2007	HB-00176	2007	HT200541B	2007	M4553BLU	2007	NO32
2007	HB-00178	2007	HT200541B-0	2007	M4553BRN	2007	NB269
2007	HB-00186	2007	HT200561	2007	M4553GRN	2007	NJ181B
2007	HB-00187	2007	HT200561A	2007	M4553ORN	2007	NK66-1A19
2007	HB-00303	2007	HT200561B	2007	M4553RED	2007	NK66-2A19
2007	HB-00324	2007	HT200561C	2007	M4553YEL	2007	NK66-3A19
2007	HB-00370	2007	HT200561C-Q	2007	M4562	2007	NK74-3A19
2007	HB-00405	2007	HT200751B	2007	M4563	2007	NK88C119
2007	HB-156	2007	HT200770B	2007	M4564	2007	NK8P-2A19
2007	HB-172	2007	HT200771	2007	M4565	2007	NK91-4A19
2007	HB-173	2007	HT200771A	2007	M4567	2007	NKT102
2007	HB-175	2007	HT200771B	2007	M4573	2007	NKT104
2007	HB-475	2007	HT200771C	2007	M4595	2007	NKT105
2007	HB-54	2007	HT201721A	2007	M4596	2007	NKT106
2007	HB-56	2007	HT201721D	2007	M4597	2007	NKT107
2007	HB-75B	2007	HT201725A	2007	M4597GRN	2007	NKT108
2007	HB-77C	2007	HT201782A	2007	M4597RED	2007	NKT109
2007	HB-85	2007	HT201861A	2007	M4607	2007	NKT211
2007	HB156	2007	HT201871L	2007	M4627	2007	NKT212
2007	HB156C	2007	HT203243A	2007	M5285	2007	NKT125
2007	HB171	2007	HT203701	2007	M75517-1	2007	NKT126
2007	HB172	2007	HT203701A	2007	M75517-2	2007	NKT128
2007	HB175	2007	HT203701B	2007	M75561-7	2007	NKT129
2007	HB176	2007	HT204051	2007	M8014(TRANSISTOR)	2007	NKT133
2007	HB178	2007	HT204051B	2007	M8062A	2007	NKT141
2007	HB186	2007	HT204051C	2007	M8062B	2007	NKT142
2007	HB187	2007	HT204051D	2007	M8062C	2007	NKT143
2007	HB263	2007	HT204051E	2007	M8604	2007	NKT144
2007	HB270	2007	HT204053	2007	M8604A	2007	NKT153/25
2007	HB32	2007	HT204053A	2007	M8640	2007	NKT154/25
2007	HB324	2007	HT204053B	2007	M8640A	2007	NKT162
2007	HB33	2007	HT204051A	2007	M8640B(XSTR)	2007	NKT163
2007	HB365	2007	HV0000102	2007	M9002	2007	NKT163/25
2007	HB415	2007	HV0000405	2007	M9148	2007	NKT163325
2007	HB499	2007	HV12	2007	M9198	2007	NKT164
2007	HB475	2007	HV15(TRANSISTOR)	2007	M9249	2007	NKT164/25
2007	HB54	2007	HV16	2007	M9250	2007	NKT16425
2007	HB55	2007	HV17	2007	MA100	2007	NKT202
2007	HB75	2007	HV17B	2007	MA112	2007	NKT203
2007	HB75	2007	HV19	2007	MA113	2007	NKT204
2007	HB75C	2007	IF-65	2007	MA114	2007	NKT205
2007	HB77	2007	IF65	2007	MA115	2007	NKT206
2007	HB77B	2007	IRTR04	2007	MA116	2007	NKT207
2007	HB77C	2007	IRTR05	2007	MA117	2007	NKT208
2007	HC-00176	2007	IRTR11	2007	MA1318	2007	NKT211
2007	HEP-0004	2007	IRTR14	2007	MA1700	2007	NKT212
2007	HEP2	2007	IRTR84	2007	MA1702	2007	NKT213
2007	HEP250	2007	IRTR85	2007	MA1703	2007	NKT214
2007	HEP251	2007	ISBP1	2007	MA1704	2007	NKT215
2007	HEP252	2007	J241111(28B370)	2007	MA1706	2007	NKT216
2007	HEP253	2007	J241164	2007	MA1707	2007	NKT217
2007	HEP254	2007	J241178	2007	MA1708	2007	NKT218
2007	HEP280	2007	J241190	2007	MA206	2007	NKT219
2007	HEP281	2007	J24626	2007	MA23	2007	NKT221
2007	HEP629	2007	J24639	2007	MA23B	2007	NKT222
2007	HEP630	2007	J24833	2007	MA240	2007	NKT22281
2007	HEP631	2007	J24834	2007	MA25	2007	NKT22282
2007	HEP632	2007	J24869	2007	MA286	2007	NKT223
2007	HEP633	2007	J24870	2007	MA287	2007	NKT223A
2007	HEP634	2007	J24934	2007	MA288	2007	NKT224
2007	HEP00004	2007	J310199	2007	MA393	2007	NKT224A
2007	HEP00005	2007	J310224	2007	MA393A	2007	NKT225
2007	HEP00005P/Q	2007	J310252	2007	MA393B	2007	NKT225A
2007	HEP00006	2007	J5063	2007	MA393E	2007	NKT226
2007	HEP00006P/Q	2007	J5064	2007	MA393GRN	2007	NKT226A
2007	HEP00007	2007	JP5063	2007	MA393G	2007	NKT227
2007	HEP00007P/Q	2007	JP5064	2007	MA393R	2007	NKT227A
2007	HJ15	2007	JR15	2007	MA815	2007	NKT228
2007	HJ17	2007	JR5	2007	MA881	2007	NKT228A
2007	HJ17D	2007	K14-0066-4	2007	MA882	2007	NKT229

RS 276-	Industry Standard No.	RS 276-	Industry Standard No.	RS 276-	Industry Standard No.	RS 276-	Industry Standard No.
2007	NKT231	2007	0056	2007	R-120	2007	R64
2007	NKT231A	2007	0057	2007	R-152	2007	R65
2007	NKT232	2007	0058	2007	R-16	2007	R6553
2007	NKT232A	2007	0059	2007	R-163	2007	R66
2007	NKT24	2007	0060	2007	R-164	2007	R67
2007	NKT243	2007	0C601	2007	R-186	2007	R6922
2007	NKT244	2007	0C602	2007	R-227	2007	R7048
2007	NKT245	2007	0C602SP	2007	R-23-1003	2007	R7124
2007	NKT246	2007	0C602SQ	2007	R-23-1004	2007	R7127
2007	NKT247	2007	0C603	2007	R-24-1001	2007	R7164
2007	NKT247A	2007	0C604	2007	R-24-1002	2007	R7166
2007	NKT25	2007	0C604SP	2007	R-242	2007	R7363
2007	NKT261	2007	0C65	2007	R-244	2007	R7489
2007	NKT262	2007	0C66	2007	R-245	2007	R7490
2007	NKT263	2007	0C70	2007	R-291	2007	R7491
2007	NKT264	2007	0C70N	2007	R-28B186	2007	R7612
2007	NKT271	2007	0C71	2007	R-28B187	2007	R7888
2007	NKT272	2007	0C711	2007	R-28B303	2007	R7889
2007	NKT273	2007	0C71A	2007	R-28B405	2007	R8121
2007	NKT274	2007	0C71N	2007	R-424	2007	R83
2007	NKT275	2007	0C72	2007	R-425	2007	R8310
2007	NKT275A	2007	0C73	2007	R-488	2007	R8311
2007	NKT275B	2007	0C74	2007	R-506	2007	R868
2007	NKT275J	2007	0C74N	2007	R-56	2007	R8687
2007	NKT278	2007	0C75	2007	R-593	2007	R8688
2007	NKT281	2007	0C75N	2007	R-608A	2007	R8695
2007	NKT303	2007	0C76	2007	R-64	2007	R8697
2007	NKT308	2007	0C77	2007	R-66	2007	R87
2007	NKT32	2007	0C77M	2007	R-67	2007	R8706
2007	NKT33	2007	0C78	2007	R-83	2007	R8707
2007	NKT4	2007	0C79	2007	R06-1007	2007	R8883
2007	NKT42	2007	0C80	2007	R06-1008	2007	R8884
2007	NKT43	2007	0C81	2007	R06-1009	2007	R8885
2007	NKT5	2007	0C810	2007	R06-1010	2007	R8886
2007	NKT52	2007	0C81D	2007	R100-1	2007	R8971
2007	NKT53	2007	0C81DD	2007	R100-8	2007	R9533
2007	NKT54	2007	0C81DN	2007	R100-9	2007	R9534
2007	NKT62	2007	0C81N	2007	R101-2	2007	R9603
2007	NKT63	2007	0C83	2007	R101-3	2007	R9604
2007	NKT64	2007	0C83N	2007	R101-4	2007	R98
2007	NKT72	2007	0C84	2007	R119	2007	RCA34101
2007	NKT73	2007	0C84N	2007	R120	2007	RCA34106
2007	NKT74	2007	0F-129	2007	R1273	2007	RCA3517
2007	N8121	2007	0F129	2007	R1274	2007	RCA35953
2007	N832	2007	P1L	2007	R152	2007	RCA35954
2007	00-130	2007	P1L4956	2007	R1540	2007	RCA3858
2007	00-140	2007	P3B	2007	R1541	2007	RCA40395
2007	00-304/1	2007	P3D	2007	R1542	2007	RCA40395P
2007	00-304/2	2007	P6460006	2007	R1543	2007	RB1
2007	00-304/3	2007	P6460037	2007	R1544	2007	RB25
2007	00-305/1	2007	PA10889-1	2007	R1546	2007	REN100
2007	00-305/2	2007	PA10889-2	2007	R1548	2007	REN102
2007	00-306/1	2007	PA9156	2007	R1555	2007	REN102A
2007	00-306/2	2007	PA9157	2007	R16	2007	RBN158
2007	00-306/3	2007	PA9158	2007	R163	2007	RS-102
2007	00-307	2007	PBE3014-1	2007	R164	2007	RS-1192
2007	00-308	2007	PBE3014-2	2007	R186	2007	RS-1539
2007	00-318	2007	PBE3020-1	2007	R227	2007	RS-1540
2007	00-330	2007	PBE3162	2007	R23-1003	2007	RS-1541
2007	00-34	2007	PBE3162-1	2007	R23-1004	2007	RS-1542
2007	00-340	2007	PBE3162-2	2007	R2350	2007	RS-1543
2007	00-341	2007	PBX103	2007	R2351	2007	RS-1544
2007	00-342	2007	PBX113	2007	R2352	2007	RS-1546
2007	00-343	2007	PC1066T	2007	R2353	2007	RS-1548
2007	00-350	2007	PC1067T	2007	R2355	2007	RS-1550
2007	00-351	2007	PC1068T	2007	R2366	2007	RS-1555
2007	00-360	2007	PC3002	2007	R2367	2007	RS-2004
2007	00-362	2007	PC3003	2007	R2373	2007	RS-2005
2007	00-363	2007	PC3005	2007	R24-1001	2007	RS-2007
2007	00-364	2007	PC3006	2007	R24-1002	2007	RS-2350
2007	00-38	2007	PC3007	2007	R24-1003	2007	RS-2351
2007	00-410	2007	PC3009	2007	R24-1004	2007	RS-2352
2007	00-44	2007	PQ28	2007	R242	2007	RS-2353
2007	00-45	2007	PQ29	2007	R244	2007	RS-2354
2007	00-46	2007	PT-530A	2007	R245	2007	RS-2355
2007	00-47	2007	PT0-139	2007	R2482-1	2007	RS-2367
2007	00-602	2007	PT530	2007	R255	2007	RS-2675
2007	00-604	2007	PT530A	2007	R258	2007	RS-2677
2007	00-65	2007	PT0102	2007	R2675	2007	RS-2683
2007	00-66	2007	PT0109	2007	R2677	2007	RS-2684
2007	00-70	2007	PTC135	2007	R2689	2007	RS-2685
2007	00-71	2007	PT0156	2007	R2749	2007	RS-2686
2007	00-71A	2007	PT0159	2007	R2749M	2007	RS-2687
2007	00-71N	2007	PXB-103	2007	R289	2007	RS-2688
2007	00-72	2007	PXB-113	2007	R290	2007	RS-2689
2007	00-73	2007	PXB103	2007	R291	2007	RS-2690
2007	00-74	2007	PXB113	2007	R28B492	2007	RS-2691
2007	00-75	2007	PXC-101	2007	R324	2007	RS-2692
2007	00-75N	2007	PXC-101AB	2007	R3275	2007	RS-2694
2007	00-77	2007	PXC101	2007	R3276	2007	RS-2695
2007	00-79	2007	PXC101A	2007	R3280(RCA)	2007	RS-2696
2007	00-81DD	2007	PXC101AB	2007	R3282	2007	RS-2697
2007	0C110	2007	Q-1	2007	R3284	2007	RS-3275
2007	0C120	2007	Q-16	2007	R3286	2007	RS-3276
2007	0C122	2007	Q-1A	2007	R3299	2007	RS-3277
2007	0C123	2007	Q-4	2007	R3301	2007	RS-3278
2007	0C302	2007	Q-6	2007	R35	2007	RS-3279
2007	0C303	2007	Q-7	2007	R3578-1	2007	RS-3282
2007	0C304	2007	Q-8	2007	R3598-2	2007	RS-3283
2007	0C304-1	2007	Q0V60526	2007	R364	2007	RS-3284
2007	0C304-2	2007	Q0V60528	2007	R428	2007	RS-3285
2007	0C304-3	2007	Q1	2007	R4348	2007	RS-3286
2007	0C304N	2007	Q1-7C	2007	R4349	2007	RS-3288
2007	0C305	2007	Q11/6515	2007	R46	2007	RS-3289
2007	0C305-1	2007	Q12/6515	2007	R488	2007	RS-3299
2007	0C305-2	2007	Q16	2007	R5051	2007	RS-3301
2007	0C306	2007	Q2-7C	2007	R5052	2007	RS-3308
2007	0C306-1	2007	Q2N2428	2007	R5053	2007	RS-3309
2007	0C306-2	2007	Q2N2613	2007	R5055	2007	RS-3310
2007	0C306-3	2007	Q2N406	2007	R506	2007	RS-3316
2007	0C307	2007	Q2N4106	2007	R5097	2007	RS-3316-1
2007	0C307-1	2007	Q35218	2007	R5098	2007	RS-3316-2
2007	0C307-2	2007	Q40263	2007	R5099	2007	RS-3318
2007	0C307-3	2007	Q6	2007	R5100	2007	RS-3867
2007	0C308	2007	Q7	2007	R5101	2007	RS-3868
2007	0C309	2007	Q7/6515	2007	R5181	2007	RS-3904
2007	0C309-1	2007	Q8	2007	R5182	2007	RS-3907
2007	0C309-2	2007	Q8/6515	2007	R52	2007	RS-3913
2007	0C309-3	2007	Q9-0076	2007	R530	2007	RS-3914
2007	0C318	2007	Q00076	2007	R537	2007	RS-3915
2007	0C32	2007	QN2613	2007	R5523	2007	RS-3925
2007	0C33	2007	Q0V60526	2007	R5524	2007	RS-3929
2007	0C330	2007	Q0V60528	2007	R5525	2007	RS-5008
2007	0C340	2007	Q0V60538	2007	R558(T.I.)	2007	RS-5008
2007	0C350	2007	Q0V60539	2007	R56	2007	RS-5104
2007	0C360	2007	QQ061209	2007	R563(T.I.)	2007	RS-5105
2007	0C364	2007	QQ061210	2007	R5708	2007	RS-5106
2007	0C38	2007	QQV60526	2007	R579(T.I.)	2007	RS-5206
2007	0C41	2007	Q0V60528	2007	R60-1004	2007	RS-5311
2007	0C41A	2007	QQV60538	2007	R60-1005	2007	RS-5401
2007	0C45	2007	QQV60539	2007	R60-1006	2007	RS-5406
2007	0C46	2007	QR2378	2007	R608	2007	RS-5502
2007	0C47	2007	R-119	2007	R608A	2007	RS-5504
						2007	RS-5505

RS 276-	Industry Standard No.	RS 276-	Industry Standard No.	RS 276-	Industry Standard No.	RS 276-	Industry Standard No.
2007	RS-5506	2007	R85507	2007	RT7401	2007	8FT131
2007	RS-5511	2007	R85511	2007	RT7558	2007	8FT131P
2007	RS-5530	2007	R85530	2007	RT7846	2007	8FT143
2007	RS-5531	2007	R85531	2007	RT8442	2007	8FT144
2007	RS-5532	2007	R85532	2007	RT8602	2007	8FT145
2007	RS-5533	2007	R85533	2007	RT8842	2007	8FT146
2007	RS-5534	2007	R85534	2007	RV1180	2007	8FT151
2007	RS-5535	2007	R85535	2007	RV1475	2007	8FT152
2007	RS-5536	2007	R85536	2007	S-1348	2007	8FT221
2007	RS-5540	2007	R85540	2007	S-1349	2007	8FT221A
2007	RS-5541	2007	R85541	2007	S-1639	2007	8FT222
2007	RS-5542	2007	R85542	2007	S-95201	2007	8FT222A
2007	RS-5544	2007	R85543	2007	S-95204	2007	8FT223
2007	RS-5551	2007	R85544	2007	S025	2007	8FT226
2007	RS-5552	2007	R85545	2007	S065	2007	8FT227
2007	RS-5553	2007	R85551	2007	S065A	2007	8FT228
2007	RS-5554	2007	R85552	2007	S088	2007	8FT229
2007	RS-5555	2007	R85553	2007	S1348	2007	8FT232
2007	RS-5556	2007	R85554	2007	S1349	2007	8FT237
2007	RS-5557	2007	R85555	2007	S1639	2007	8FT241
2007	RS-5558	2007	R85556	2007	S1672	2007	8FT242
2007	RS-5602	2007	R85557	2007	S1802C	2007	8FT243
2007	RS-5704	2007	R85558	2007	S2041635	2007	8FT251
2007	RS-57042	2007	R85563	2007	0822.3504-040	2007	8FT252
2007	RS-57062	2007	R85564	2007	0822.3504-060	2007	8FT253
2007	RS-5708	2007	R85565	2007	0822.3505-910	2007	8FT288
2007	RS-5708-2	2007	R85566	2007	8413796	2007	8FT306
2007	RS-5709	2007	R85567	2007	8A248	2007	8FT321
2007	RS-5711	2007	R85568	2007	866-1-A-3P	2007	8FT322
2007	RS-5717	2007	R85602	2007	866-2-A-3P	2007	8FT323
2007	RS-5717-3	2007	R85603	2007	866-3-A-3P	2007	8FT327
2007	RS-5717-6	2007	R85605	2007	8685	2007	8FT337
2007	RS-5720	2007	R85607	2007	8686	2007	8FT337B
2007	RS-5731	2007	R85608	2007	8687	2007	8FT337V
2007	RS-5733	2007	R85610	2007	874-3-A-3P	2007	8FT351
2007	RS-5734	2007	R85704	2007	888C-1-3P	2007	8FT352
2007	RS-5736	2007	R85704-2	2007	88F-2-A-3P	2007	8FT353
2007	RS-5737	2007	R857042	2007	895201	2007	8FT367
2007	RS-5742	2007	R857062	2007	895203	2007	8FT523
2007	RS-5743	2007	R85708	2007	895204	2007	8FT526
2007	RS-5743-1	2007	R85708-2	2007	895206	2007	8K3003
2007	RS-5743-2	2007	R85709	2007	895207	2007	8K3004
2007	RS-5743-3	2007	R85711	2007	895214	2007	8K3005
2007	RS-5744	2007	R85717	2007	895218	2007	8M-217
2007	RS-5744-3	2007	R85717-1	2007	89524	2007	8MO843
2007	RS-5749	2007	R85717-3	2007	899201	2007	8MB341
2007	RS-5753-2	2007	R85717-6	2007	899203	2007	8M843
2007	RS-5852	2007	R85720	2007	899218	2007	8MB447610
2007	RS-5854	2007	R85731	2007	89923	2007	8MB447610A
2007	RS-686	2007	R85732	2007	89L-4-A-3P	2007	8MB454549
2007	RS-687	2007	R85733	2007	8A128	2007	8MB621960
2007	RS1192	2007	R85734	2007	8A128-1	2007	8NW-Q-1
2007	RS1540	2007	R85735	2007	8A15V	2007	8NW-Q-4
2007	RS1541	2007	R85736	2007	8A197	2007	8NW-Q-6
2007	RS1543	2007	R85737	2007	8A197-1	2007	8O-25
2007	RS1544	2007	R85738	2007	8A197-2	2007	8O-88
2007	RS1545	2007	R85740	2007	8A197-3	2007	8O25
2007	RS1546	2007	R85740-1	2007	8A204	2007	8O65
2007	RS1548	2007	R85742	2007	8A205BLU	2007	8O88
2007	RS1555	2007	R85743	2007	8A205BRN	2007	8R1
2007	RS2350	2007	R85743-1	2007	8A205GRN	2007	8S0001
2007	RS2351	2007	R85743-2	2007	8A205ORN	2007	8S0001A
2007	RS2352	2007	R85743-3	2007	8A205RED	2007	8S0002
2007	RS2353	2007	R85743.3	2007	8A205VIO	2007	8S0002A
2007	RS2354	2007	R857433	2007	8A205WHT	2007	8S0003
2007	RS2355	2007	R85744	2007	8A205YEL	2007	8S0003A
2007	RS2367	2007	R85744-3	2007	8A240	2007	8S0004
2007	RS2373	2007	R85745	2007	8A29	2007	8S0004A
2007	RS2374	2007	R85746	2007	8A318-2	2007	8S0005
2007	RS2675	2007	R85747	2007	8A318-3	2007	8S0005A
2007	RS2677	2007	R85748	2007	8A33	2007	8S0001
2007	RS2689	2007	R85749	2007	8A33BRN	2007	8T-122
2007	RS2690	2007	R85750	2007	8A33RED	2007	8T-123
2007	RS2691	2007	R85751	2007	8A529	2007	8T-28B
2007	RS2692	2007	R85752	2007	8A565	2007	8T-28C
2007	RS2696	2007	R85765	2007	8A646	2007	8T-301
2007	RS2697	2007	R85766	2007	8A681	2007	8T-302
2007	RS2867	2007	R85767	2007	8A853	2007	8T-304
2007	RS3211	2007	R85768	2007	8B168	2007	8T-332
2007	RS322	2007	R85825	2007	8B169	2007	8T-370
2007	RS3275	2007	R85852	2007	8C-12	2007	8T-37C
2007	RS3276	2007	R85854	2007	8C-63	2007	8T-37D
2007	RS3280	2007	R86824	2007	8C-65	2007	8T122
2007	RS3281	2007	R86840	2007	8C-68	2007	8T123
2007	RS3282	2007	R86843	2007	8C-69	2007	8T28A
2007	RS3283	2007	R86843A	2007	8C-73	2007	8T28B
2007	RS3284	2007	R86846	2007	8C12	2007	8T28C
2007	RS3285	2007	R86846A	2007	8C43	2007	8T301
2007	RS3286	2007	R87568	2007	8C44	2007	8T302
2007	RS3287	2007	R88406	2007	8C45	2007	8T303
2007	RS3289	2007	R88421	2007	8C46	2007	8T304
2007	RS3293	2007	R88424	2007	8C65	2007	8T332
2007	RS3299	2007	R88444	2007	8C66	2007	8T370
2007	RS3301	2007	R88446	2007	8C68	2007	8T37C
2007	RS3308	2007	R81350	2007	8C69	2007	8T37D
2007	RS3310	2007	RT-185	2007	8C73	2007	8T37E
2007	RS3316	2007	RT-4625	2007	8B-5-0819	2007	8T382
2007	RS3316-1	2007	RT-61015	2007	8F.T124	2007	8TX0096
2007	RS3316-2	2007	RT-61016	2007	8F.T130	2007	8TX0099
2007	RS3318	2007	RT121	2007	8F.T131P	2007	8TX0104
2007	RS3717	2007	RT185	2007	8F.T171	2007	8TX0105
2007	RS3726	2007	RT2230	2007	8F.T172	2007	8TX0110
2007	RS3857	2007	RT2329	2007	8F.T173	2007	8TX0114
2007	RS3866	2007	RT2330	2007	8F.T174	2007	8TX0121
2007	RS3867	2007	RT2331	2007	8F.T221	2007	8TX0123
2007	RS3880	2007	RT2709	2007	8F.T222	2007	8TX0224
2007	RS3892	2007	RT3097	2007	8F.T223	2007	8TX0260
2007	RS3897	2007	RT3098	2007	8F.T227	2007	8TX0263
2007	RS3904	2007	RT3229	2007	8F.T237	2007	8TX0264
2007	RS3913	2007	RT3230	2007	8F.T251	2007	8TX0265
2007	RS3914	2007	RT3231	2007	8F.T252	2007	8TX0268
2007	RS3915	2007	RT3363	2007	8F.T253	2007	8TX0269
2007	RS3925	2007	RT3364	2007	8F.T306	2007	000000BV31
2007	RS3926	2007	RT3365	2007	8F.T318	2007	8YL-105
2007	RS406	2007	RT3449	2007	8F.T321	2007	8YL-106
2007	RS5008	2007	RT3467	2007	8F.T322	2007	8YL-107
2007	RS5102	2007	RT3468	2007	8F.T337	2007	8YL-108
2007	RS5103	2007	RT3566	2007	8F.T352	2007	8YL-1583
2007	RS5104	2007	RT3568	2007	8FT-306	2007	8YL-160
2007	RS5105	2007	RT4624	2007	8FT-307	2007	8YL-1608
2007	RS5202	2007	RT4625	2007	8FT-319	2007	8YL-1668
2007	RS5203	2007	RT4568	2007	8FT-322	2007	8YL-2248
2007	RS5243-2	2007	RT5521	2007	8FT-323	2007	8YL-2249
2007	RS5302	2007	RT5522	2007	8FT-327	2007	8YL-2250
2007	RS5303	2007	RT5637	2007	8FT-337	2007	8YL105
2007	RS5401	2007	RT61015	2007	8FT-352	2007	8YL106
2007	RS5402	2007	RT61016	2007	8FT-353	2007	8YL107
2007	RS5403	2007	RT6205	2007	8FT121	2007	8YL107A
2007	RS5406	2007	RT6604	2007	8FT122	2007	8YL108
2007	RS5502	2007	RT6734	2007	8FT123	2007	8YL108A
2007	RS5503	2007	RT6735	2007	8FT124	2007	8YL1430
2007	RS5504	2007	RT6736	2007	8FT125	2007	8YL1535
2007	RS5505	2007	RT6990	2007	8FT125P	2007	8YL1583
2007	RS5506			2007	8FT130	2007	8YL1583A

RS 276-	Industry Standard No.	RS 276-	Industry Standard No.	RS 276-	Industry Standard No.	RS 276-	Industry Standard No.
2007	SYL1588	2007	T1546	2007	TIA04	2007	TR15C
2007	SYL160	2007	T1559	2007	TIA05	2007	TR2083-75
2007	SYL1608	2007	T1573	2007	TIA05A	2007	TR215
2007	SYL1655	2007	T1574	2007	TIX895	2007	TR217
2007	SYL1655A	2007	T1577	2007	TIX90	2007	TR2N2614C
2007	SYL1665	2007	T1583	2007	TIXA-03	2007	TR310011
2007	SYL1665A	2007	T1593	2007	TIXA-04	2007	TR310012
2007	SYL1668	2007	T1594	2007	TIXA-05	2007	TR310015
2007	SYL1668A	2007	T1595	2007	TIXA01	2007	TR310017
2007	SYL1690	2007	T1596	2007	TIXA02	2007	TR310018
2007	SYL1697	2007	T1597	2007	TIXA03	2007	TR310026
2007	SYL1717	2007	T1598	2007	TIXA04	2007	TR310075
2007	SYL2120	2007	T1599	2007	TIXA05	2007	TR310107
2007	SYL2247	2007	T160	2007	TK-23C	2007	TR310125
2007	SYL2248	2007	T1740	2007	TK-40C	2007	TR310136
2007	SYL2248A	2007	T1877	2007	TK-41C	2007	TR310149
2007	SYL2249	2007	T1902	2007	TK-42C	2007	TR310153
2007	SYL2249A	2007	T1903	2007	TK-45C	2007	TR310159
2007	SYL2250	2007	T1904	2007	TK1228-1002	2007	TR310161
2007	SYL2300	2007	T1961	2007	TK1228-1003	2007	TR310164
2007	SYL2300A	2007	T2038	2007	TK1228-1004	2007	TR310225
2007	SYL3613	2007	T2039	2007	TK1228-1005	2007	TR310227
2007	SYL5613A	2007	T2040	2007	TK1228-1006	2007	TR310235
2007	T-00014	2007	T2091	2007	TK1228-1007	2007	TR310251
2007	T-109	2007	T2122	2007	TK23C	2007	TR310252
2007	T-116	2007	T2159	2007	TK40	2007	TR310255
2007	T-126	2007	T2172	2007	TK40A	2007	TR320
2007	T-129	2007	T2173	2007	TK40C	2007	TR320A
2007	T-130	2007	T2256	2007	TK400A	2007	TR320AN
2007	T-131	2007	T2257	2007	TK41	2007	TR321
2007	T-152148	2007	T2258	2007	TK41A	2007	TR321(HPGH1)
2007	T-1877	2007	T2259	2007	TK42	2007	TR321A
2007	T-2038	2007	T2260	2007	TK42A	2007	TR323
2007	T-2039	2007	T2261	2007	TK45C	2007	TR323A
2007	T-2040	2007	T2439	2007	TK49C	2007	TR323AN
2007	T-2091	2007	T2440	2007	TM100	2007	TR332
2007	T-2122	2007	T2441	2007	TM102	2007	TR381
2007	T-23	2007	T2515	2007	TM102A	2007	TR38117
2007	T-2439	2007	T2517	2007	TM158	2007	TR383(HGPH-2)
2007	T-2440	2007	T282(SEARS)	2007	TN591	2007	TR43
2007	T-2441	2007	T3005	2007	TNJ-60070	2007	TR44
2007	T-3321	2007	T3321	2007	TNJ-60074	2007	TR45
2007	T-3322	2007	T3322	2007	TNJ-60079	2007	TR482
2007	T-3321	2007	T3321	2007	TNJ-60282	2007	TR482A
2007	T-3323	2007	T3323	2007	TNJ-60283	2007	TR508
2007	T-39	2007	T346	2007	TNJ-60610	2007	TR508A
2007	T-46	2007	T39	2007	TNJ-60611	2007	TR508AN
2007	T-47	2007	T45	2007	TNJ-60612	2007	TR53
2007	T-48	2007	T46	2007	TNJ-60728	2007	TR54
2007	T-52148Z	2007	T47	2007	TNJ60070	2007	TR55
2007	T-52149	2007	T48	2007	TNJ60074	2007	TR5R26
2007	T-52149Z	2007	T50	2007	TNJ60079	2007	TR64
2007	T-52150	2007	T50039A	2007	TNJ60282	2007	TR65
2007	T-52150Z	2007	T50631	2007	TNJ60283	2007	TR650
2007	T-52151	2007	T50931B	2007	TNJ60365	2007	TR650A
2007	T-52151Z	2007	T50933B	2007	TNJ60608	2007	TR653
2007	T-78	2007	T50944	2007	TNJ60610	2007	TR653A
2007	T-82	2007	T51573A	2007	TNJ60611	2007	TR6R26
2007	T-95	2007	T52147	2007	TNJ60612	2007	TR71
2007	T-Q5020	2007	T52147Z	2007	TNJ60728	2007	TR72
2007	T-Q5023	2007	T52148Z	2007	TNJ61221	2007	TR721
2007	T-Q5025	2007	T52149	2007	TNJ61222	2007	TR721A
2007	T-Q5026	2007	T52149Z	2007	TNJ61282	2007	TR722
2007	T-Q5027	2007	T52150	2007	TNJ61673(2SB186)	2007	TR722A
2007	TO-101	2007	T52150Z	2007	TNJ61674	2007	TR758A
2007	TO-102	2007	T52151	2007	TNJ70634	2007	TR759
2007	TO00014	2007	T52151Z	2007	TNJ70635	2007	TR763
2007	TO012	2007	T52159	2007	TNJ70688	2007	TR763A
2007	TO014	2007	T59	2007	TNJ72278	2007	TR764
2007	TO015	2007	T59247	2007	TNJ72283	2007	TR792
2007	TO031	2007	T59249	2007	TNJ72285	2007	TR8007
2007	TO033	2007	T60	2007	TNJ72287	2007	TR801
2007	TO038	2007	T61	2007	TNJ72289	2007	TR802
2007	TO039	2007	T72	2007	TO-005	2007	TR81
2007	TO040	2007	T74	2007	TO-014	2007	TR84
2007	TO041	2007	T77	2007	TO-041	2007	TR85
2007	TO051	2007	T78	2007	TO-101	2007	TRA-32
2007	TO101	2007	T814	2007	TO-102	2007	TRA-33
2007	TO102	2007	T815	2007	TO-103	2007	TRA32
2007	TO33Z3	2007	T82	2007	TO-104	2007	TRA33
2007	T100	2007	T83	2007	TO101	2007	TRC44
2007	T1000	2007	T84	2007	TO102	2007	TRC44A
2007	T1001	2007	T87	2007	TO103	2007	TRC45
2007	T10010	2007	T95	2007	TO104	2007	TRC45A
2007	T1002	2007	T99	2007	TQ-5051	2007	TRC70
2007	T1002A	2007	TA-1575	2007	TQ-5061	2007	TRC71
2007	T1003	2007	TA-1575B	2007	TQ5020	2007	TRC72
2007	T1004	2007	TA-1655B	2007	TQ5023	2007	TRM15
2007	T1005	2007	TA-1697	2007	TQ5025	2007	TRM21
2007	T1006	2007	TA-1704	2007	TQ5026	2007	TS-1
2007	T1007	2007	TA-1706	2007	TQ5051	2007	TS-1007
2007	T1008	2007	TA-1730	2007	TQ5061	2007	TS-1266
2007	T10085	2007	TA-1763	2007	TQ8A0-222	2007	TS-13
2007	T1009	2007	TA-1763A	2007	TR-04	2007	TS-14
2007	T101	2007	TA-1778	2007	TR-05	2007	TS-15
2007	T1010	2007	TA-1782	2007	TR-06	2007	TS-162
2007	T1013	2007	TA-1783	2007	TR-14	2007	TS-163
2007	T1023	2007	TA-4	2007	TR-14C	2007	TS-164
2007	T102A	2007	TA1575	2007	TR-15	2007	TS-165
2007	T1036	2007	TA1575B	2007	TR-157(OLSON)	2007	TS-166
2007	T1037	2007	TA1655B	2007	TR-158(OLSON)	2007	TS-1727
2007	T1042	2007	TA1697	2007	TR-15C	2007	TS-1728
2007	T1043	2007	TA1704	2007	TR-169(OLSON)	2007	TS-2
2007	T1046	2007	TA1706	2007	TR-170(OLSON)	2007	TS-3
2007	T1047	2007	TA1730	2007	TR-2N2641C	2007	TS-601
2007	T1076	2007	TA1763	2007	TR-320	2007	TS-602
2007	T108	2007	TA1763A	2007	TR-320A	2007	TS-603
2007	T109	2007	TA1778	2007	TR-321	2007	TS-604
2007	T116	2007	TA1782	2007	TR-321A	2007	TS-616
2007	T11618	2007	TA1783	2007	TR-323A	2007	TS-617
2007	T1202	2007	TC3123041557	2007	TR-482A	2007	TS-618
2007	T1203	2007	TF-30	2007	TR-508A	2007	TS-627
2007	T1251	2007	TF-65	2007	TR-5R26	2007	TS-629
2007	T126	2007	TF-66	2007	TR-6R26	2007	TS-739
2007	T127	2007	TF30	2007	TR-84	2007	TS-739B
2007	T1289	2007	TF49	2007	TR-85	2007	TS-740
2007	T129	2007	TF65	2007	TR-C044	2007	TS-765
2007	T1291	2007	TF65/30	2007	TR-C044A	2007	TS1007
2007	T130	2007	TF65/M	2007	TR-C45	2007	TS1266
2007	T1300	2007	TF65/8/30	2007	TR-C45A	2007	TS13
2007	T13000	2007	TF65M	2007	TR-C70	2007	TS14
2007	T131	2007	TF66	2007	TR-C71	2007	TS15
2007	T1310	2007	TF66/30	2007	TR-C72	2007	TS162
2007	T1312	2007	TF66/60	2007	TR04	2007	TS163
2007	T1322	2007	TF75	2007	TR04C	2007	TS164
2007	T1326	2007	TF77	2007	TR05	2007	TS165
2007	T1334	2007	TF80/302	2007	TR05C	2007	TS166
2007	T1346	2007	TG48	2007	TR07C	2007	TS1727
2007	T1352	2007	THU60U	2007	TR104	2007	TS1728
2007	T1363	2007	TI-363	2007	TR109	2007	TS1792
2007	T1364	2007	TI-364	2007	TR11	2007	TS601
2007	T1474	2007	TIA-01	2007	TR14C	2007	TS602
2007	T1510	2007	TIA01	2007	TR15	2007	TS603
2007	T152148	2007	TIA03				

RS 276-	Industry Standard No.	RS 276-	Industry Standard No.	RS 276-	Industry Standard No.	RS 276-	Industry Standard No.	RS 276-	Industry Standard No.
2007	TS604	2007	WTV20MG	2007	1-52221011	2007	1-52221011	2007	2N114
2007	TS615	2007	WTV20VH6	2007	1-522210011	2007	1-522210011	2007	2N1171
2007	TS616	2007	WTV20VHG	2007	1-522211200	2007	1-522211200	2007	2N1174
2007	TS617	2007	WTV20VMG	2007	1-522211328	2007	1-522211328	2007	2N1175
2007	TS618	2007	WTV30VH6	2007	1-522216500	2007	1-522216500	2007	2N1175A
2007	TS621	2007	WTV30VHG	2007	1-6207190405	2007	1-6207190405	2007	2N1176
2007	TS627A	2007	WTV30VLG	2007	1-801-005	2007	1-801-005	2007	2N1176A
2007	TS627B	2007	WTV30VMG	2007	1-801-005-23	2007	1-801-005-23	2007	2N1176B
2007	TS629	2007	WTVA76	2007	1-801-006	2007	1-801-006	2007	2N1185
2007	TS669A	2007	WTVB9A	2007	1-801-006-12	2007	1-801-006-12	2007	2N1186
2007	TS669B	2007	WTVB6	2007	1-801-006-14	2007	1-801-006-14	2007	2N1187
2007	TS669D	2007	WTVB6A	2007	1-801-308	2007	1-801-308	2007	2N1188
2007	TS669E	2007	WTVBA6A	2007	1-801-308-24	2007	1-801-308-24	2007	2N1189
2007	TS669F	2007	WTVB86A	2007	1-801-310	2007	1-801-310	2007	2N1190
2007	TS672A	2007	WTVBMC	2007	1-TR-111	2007	1-TR-111	2007	2N1191
2007	TS739	2007	X-78	2007	1/4L£3.6D	2007	1/4L£3.6D	2007	2N1192
2007	TS739B	2007	X1C1644	2007	1/4L£3.6D5	2007	1/4L£3.6D5	2007	2N1193
2007	TS740	2007	X45C-H06	2007	1A0055	2007	1A0055	2007	2N1194
2007	TS765	2007	X78	2007	1A0056	2007	1A0056	2007	2I23
2007	TU000	2007	XA122	2007	1T495	2007	1T495	2007	2N1264
2007	TV-61	2007	XA151	2007	002-005100	2007	002-005100	2007	2N1265
2007	TV-61A	2007	XA152	2007	002-006600	2007	002-006600	2007	2N1265/5
2007	TV24115	2007	XB1	2007	002-006800	2007	002-006800	2007	2N1265A
2007	TV24152	2007	XB102	2007	002-006900	2007	002-006900	2007	2N1273
2007	TV24154	2007	XB103	2007	002-007300	2007	002-007300	2007	2N1273BLU
2007	TV24156	2007	XB104	2007	002-00840	2007	002-00840	2007	2N1273GRN
2007	TV24189	2007	XB112	2007	002-008400	2007	002-008400	2007	2N1273ORN
2007	TV24194	2007	XB113	2007	002-011000	2007	002-011000	2007	2N1273RED
2007	TV2428	2007	XB114	2007	002-011900	2007	002-011900	2007	2N1273YEL
2007	TV2429	2007	XB13	2007	002-011800	2007	002-011800	2007	2N1274
2007	TV2434	2007	XB1A	2007	002-11900	2007	002-11900	2007	2N1274BLU
2007	TV24370	2007	XB2	2007	2B	2007	2B	2007	2N1274BRN
2007	TV24599	2007	XB2A	2007	2C	2007	2C	2007	2N1274ORN
2007	TV24945	2007	XB3	2007	2D	2007	2D	2007	2N1274OHN
2007	TV24984	2007	XB3A	2007	2D013	2007	2D013	2007	2N1274PUR
2007	TV4152	2007	XB3B	2007	2D013-109	2007	2D013-109	2007	2N1274RED
2007	TV47	2007	XB3BN	2007	2D013-13	2007	2D013-13	2007	2N1280
2007	TV61	2007	XB3C	2007	2D013-160	2007	2D013-160	2007	2N1281
2007	TV8-28172P	2007	XB3C-1	2007	2D013-54	2007	2D013-54	2007	2N1284
2007	TV8-28A171	2007	XB4-1	2007	2D016	2007	2D016	2007	2N1287
2007	TV8-28A385	2007	XC101	2007	2D016-45	2007	2D016-45	2007	2N1287A
2007	TV8-28A385A	2007	XC121	2007	2D016-54	2007	2D016-54	2007	2N30
2007	TV8-28A385L	2007	XC131	2007	2D021	2007	2D021	2007	2N1303
2007	TV8-28A71B	2007	XC171	2007	2D021-11	2007	2D021-11	2007	2N1305
2007	TV8-28B171	2007	XP-14B217	2007	2D021-56	2007	2D021-56	2007	2N1307
2007	TV8-28B171A	2007	XG32	2007	2D021-8	2007	2D021-8	2007	2N1309
2007	TV8-28B172A	2007	XG8	2007	2D023	2007	2D023	2007	2N130A
2007	TV8-28B172A	2007	XJ13	2007	2D036	2007	2D036	2007	2N131
2007	TV8-28B172P	2007	XJ13-1	2007	2D039	2007	2D039	2007	2N1316
2007	TV8-28B176	2007	XS101	2007	2D021-56	2007	2D021-56	2007	2N1317
2007	TV8-28B234	2007	XS104	2007	2D023	2007	2D023	2007	2N1318
2007	TV8-28B324	2007	XS121	2007	2G1024	2007	2G1024	2007	2N131A
2007	TV828A171	2007	XT100	2007	2G1025	2007	2G1025	2007	2N132
2007	TV828A71B	2007	XT200	2007	2G1026	2007	2G1026	2007	2N132A
2007	TV828B171	2007	Y363	2007	2G1027	2007	2G1027	2007	2N133
2007	TV828B171A	2007	Y633	2007	2G138	2007	2G138	2007	2N133A
2007	TV828B171B	2007	YV1	2007	2G139	2007	2G139	2007	2N1343
2007	TV828B172	2007	YV1A	2007	2G140	2007	2G140	2007	2N1344
2007	TV828B172A	2007	YV2	2007	2G270	2007	2G270	2007	2N1345
2007	TV828B324	2007	ZA105604	2007	2G271	2007	2G271	2007	2N1346
2007	UPI1303	2007	ZC28B172	2007	2G302	2007	2G302	2007	2N1347
2007	UPI1305	2007	ZC28B172A	2007	2G303	2007	2G303	2007	2N1348
2007	UPI1307	2007	ZEN303	2007	2G304	2007	2G304	2007	2N35
2007	UPI1309	2007	ZEN304	2007	2G308	2007	2G308	2007	2N1350
2007	UPI1345	2007	ZEN305	2007	2G309	2007	2G309	2007	2N1351
2007	UPI1347	2007	ZEN306	2007	2G319	2007	2G319	2007	2N1352
2007	UPI1353	2007	ZEN307	2007	2G320	2007	2G320	2007	2N354
2007	V10/15A	2007	ZEN308	2007	2G321	2007	2G321	2007	2N1354
2007	V10/15A18	2007	ZEN309	2007	2G322	2007	2G322	2007	2N1355
2007	V10/28	2007	ZEN310	2007	2G323	2007	2G323	2007	2N1356
2007	V10/28J	2007	ZJ13	2007	2G324	2007	2G324	2007	2N1357
2007	V10/30A	2007	ZS38	2007	2G371	2007	2G371	2007	2N136
2007	V10/30A18	2007	ZS56	2007	2G371A	2007	2G371A	2007	2N1361
2007	V10/50A	2007	ZT2102	2007	2G374	2007	2G374	2007	2N1361A
2007	V10/50A18	2007	ZTR-B54	2007	2G374A	2007	2G374A	2007	2N137
2007	V13/11	2007	ZTR-B56	2007	2G376	2007	2G376	2007	2N1370
2007	V51	2007	001-012011	2007	2G377	2007	2G377	2007	2N1371
2007	V6/2RC	2007	001-01202-1	2007	2G381	2007	2G381	2007	2N1372
2007	V6/2RJ	2007	001-012021	2007	2G381A	2007	2G381A	2007	2N1373
2007	V6/4RC	2007	001-01203-1	2007	2G383	2007	2G383	2007	2N1374
2007	V6/4RJ	2007	001-012031	2007	2G384	2007	2G384	2007	2N1375
2007	V6/8RJ	2007	001-012060	2007	2G385	2007	2G385	2007	2N1376
2007	V6/RC	2007	001-012060	2007	2G386	2007	2G386	2007	2N1377
2007	VB11	2007	001-02010	2007	2G387	2007	2G387	2007	2N1378
2007	VFQ-2745P	2007	01-1-201-0	2007	2G394	2007	2G394	2007	2N1379
2007	VFQ2745F	2007	1-21-100	2007	2G395	2007	2G395	2007	2N38
2007	VF8-2745	2007	1-21-102	2007	2G396	2007	2G396	2007	2N1380
2007	VF8-2745J	2007	1-21-103	2007	2G397	2007	2G397	2007	2N1381
2007	VF82745	2007	1-21-104	2007	2G508	2007	2G508	2007	2N1382
2007	VF82745J	2007	1-21-105	2007	2G509	2007	2G509	2007	2N1383
2007	VFT-2745H	2007	1-21-106	2007	2G524	2007	2G524	2007	2N38A
2007	VFT2745H	2007	1-21-107	2007	2G525	2007	2G525	2007	2N1383
2007	VFY-2745E	2007	1-21-120	2007	2G526	2007	2G526	2007	2N392
2007	VFY2745E	2007	1-21-128	2007	2G527	2007	2G527	2007	2N393
2007	VL/8RJ	2007	1-21-148	2007	2G577	2007	2G577	2007	2N395
2007	VM-30244	2007	1-21-161	2007	2G603	2007	2G603	2007	2N1404
2007	VM30244	2007	1-21-162	2007	2G604	2007	2G604	2007	2N1413
2007	VS-28B171	2007	1-21-164	2007	2G605	2007	2G605	2007	2N1414
2007	VS-28B172	2007	1-21-179	2007	2M1127	2007	2M1127	2007	2N1415
2007	VS-28B172FN	2007	1-21-180	2007	2M1303	2007	2M1303	2007	2N1416
2007	VS-28B176	2007	1-21-184	2007	2N1008	2007	2N1008	2007	2N1432
2007	VS-28B178	2007	1-21-186	2007	2N1008A	2007	2N1008A	2007	2N1446
2007	VS-28B178A	2007	1-21-191	2007	2N1008B	2007	2N1008B	2007	2N1447
2007	VS-28B324	2007	1-21-192	2007	2N1009	2007	2N1009	2007	2N1448
2007	V828A385	2007	1-21-225	2007	2N104	2007	2N104	2007	2N1449
2007	V828B171	2007	1-21-226	2007	2N1044	2007	2N1044	2007	2N1450
2007	V828B172	2007	1-21-227	2007	2N1045	2007	2N1045	2007	2N1451
2007	V828B172F	2007	1-21-232	2007	2N105	2007	2N105	2007	2N1452
2007	V828B172FN	2007	1-21-234	2007	2N1056	2007	2N1056	2007	2N1469
2007	V828B176	2007	1-21-235	2007	2N1057	2007	2N1057	2007	2N1470
2007	V828B178	2007	1-21-236	2007	2N108	2007	2N108	2007	2N1471
2007	V828B178A	2007	1-21-240	2007	2N109	2007	2N109	2007	2N149B
2007	VSF2745	2007	1-21-241	2007	2N109/5	2007	2N109/5	2007	2N151
2007	W1	2007	1-21-246	2007	2N1097	2007	2N1097	2007	2N1570
2007	W25	2007	1-21-254	2007	2N1098	2007	2N1098	2007	2N1581
2007	W3	2007	1-21-266	2007	2N109BLU	2007	2N109BLU	2007	2N1583
2007	WC19862A	2007	1-21-267	2007	2N109GRN	2007	2N109GRN	2007	2N1584
2007	WC19863	2007	1-21-272	2007	2N109M1	2007	2N109M1	2007	2N159
2007	WC19864	2007	1-21-273	2007	2N109M2	2007	2N109M2	2007	2N1614
2007	WEP1860	2007	1-21-274	2007	2N109WHT	2007	2N109WHT	2007	2N1664
2007	WEP187	2007	1-21-275	2007	2N109YEL	2007	2N109YEL	2007	2N1673
2007	WEP2	2007	1-21-289	2007	2N11	2007	2N11	2007	2N1681
2007	WEP250	2007	1-21-73	2007	2N111A	2007	2N111A	2007	2N1683
2007	WEP253	2007	1-21-74	2007	2N123	2007	2N123	2007	2N1684
2007	WEP254	2007	1-21-75	2007	2N124	2007	2N124	2007	2N1705
2007	WEP405	2007	1-21-76	2007	2N125	2007	2N125	2007	2N1706
2007	WEP630	2007	1-21-78	2007	2N126	2007	2N126	2007	2N1707
2007	WEP631	2007	1-21-83	2007	2N127	2007	2N127	2007	2N1729
2007	WEP632	2007	1-21-91	2007	2N128	2007	2N128	2007	2N1731
2007	WRT1114	2007	1-21-92	2007	2N129	2007	2N129	2007	2N1743
2007	WTV-BMC	2007	1-21-93	2007	2N112A	2007	2N112A	2007	2N1744
2007	WTV15MG	2007	1-21-95	2007	2N113	2007	2N113	2007	2N175
2007	WTV15VMG	2007	1-21-96	2007	2N1130	2007	2N1130		

RS 276-	Industry Standard No.	RS 276-	Industry Standard No.	RS 276-	Industry Standard No.	RS 276-	Industry Standard No.
2007	2N180	2007	2N319	2007	2N523	2007	2N825
2007	2N1800	2007	2N32	2007	2N523A	2007	2N826
2007	2N181	2007	2N320	2007	2N524	2007	2N83
2007	2N185	2007	2N321	2007	2N524A	2007	2N96
2007	2N1853	2007	2N322	2007	2N525	2007	2NJ5A
2007	2N1854	2007	2N323	2007	2N525A	2007	2NJ5D
2007	2N185BLU	2007	2N324	2007	2N526	2007	2NJ6
2007	2N186	2007	2N327	2007	2N526A	2007	2NJ8A
2007	2N186A	2007	2N327A	2007	2N527	2007	2NJ9A
2007	2N187	2007	2N32A	2007	2N527A	2007	2NJ9D
2007	2N187A	2007	2N331M	2007	2N529	2007	2N8121
2007	2N188	2007	2N34	2007	2N53	2007	2N831
2007	2N188A	2007	2N3427	2007	2N530	2007	2N832
2007	2N189	2007	2N3428	2007	2N531	2007	2O072
2007	2N189#	2007	2N34A	2007	2N532	2007	2N8187
2007	2N189#	2007	2N359	2007	2N533	2007	28111
2007	2N190	2007	2N360	2007	2N534	2007	2812
2007	2N191	2007	2N361	2007	2N535	2007	2813
2007	2N192	2007	2N362	2007	2N535A	2007	2814
2007	2N1924	2007	2N362B	2007	2N535B	2007	2815
2007	2N1925	2007	2N363	2007	2N536	2007	28155
2007	2N1926	2007	2N367	2007	2N54	2007	28159
2007	2N1940	2007	2N368	2007	2N55	2007	2815A
2007	2N195	2007	2N369	2007	2N56	2007	28160
2007	2N1954	2007	2N381	2007	2N563	2007	28163
2007	2N1955	2007	2N382	2007	2N564	2007	28167
2007	2N1956	2007	2N383	2007	2N565	2007	28174
2007	2N1957	2007	2N38A	2007	2N566	2007	281760
2007	2N196	2007	2N39	2007	2N568	2007	28178
2007	2N1961/46	2007	2N394	2007	2N569	2007	28179
2007	2N1969	2007	2N394A	2007	2N570	2007	28189
2007	2N197	2007	2N395	2007	2N571	2007	2822
2007	2N198	2007	2N396	2007	2N572	2007	2824
2007	2N199	2007	2N396A	2007	2N573	2007	2825
2007	2N1997	2007	2N397	2007	2N573BRN	2007	28273
2007	2N1998	2007	2N40	2007	2N573ORN	2007	2830
2007	2N200	2007	2N402	2007	2N573RED	2007	2831
2007	2N2000	2007	2N403	2007	2N578	2007	2832
2007	2N2001	2007	2N404	2007	2N579	2007	2833
2007	2N204	2007	2N404A	2007	2N58	2007	2834
2007	2N205	2007	2N405	2007	2N580	2007	2837
2007	2N206	2007	2N406	2007	2N581	2007	2838
2007	2N207	2007	2N406BLU	2007	2N582	2007	2839
2007	2N207A	2007	2N406BRN	2007	2N586	2007	2840
2007	2N207B	2007	2N406GRN	2007	2N59	2007	2843
2007	2N207BLU	2007	2N406GRN-YEL	2007	2N591	2007	2844
2007	2N21	2007	2N406ORN	2007	2N591/5	2007	2845
2007	2N215	2007	2N406RED	2007	2N591A	2007	2846
2007	2N217	2007	2N407	2007	2N592	2007	2847
2007	2N2171	2007	2N407BLK	2007	2N593	2007	2849
2007	2N2172	2007	2N407GRN	2007	2N597	2007	2851
2007	2N2173	2007	2N407J	2007	2N598	2007	2852
2007	2N217A	2007	2N407RED	2007	2N599	2007	2853
2007	2N217RED	2007	2N407WHT	2007	2N59A	2007	2854
2007	2N217WHT	2007	2N407YEL	2007	2N59B	2007	2856
2007	2N217YEL	2007	2N408	2007	2N59C	2007	2860
2007	2N22	2007	2N408J	2007	2N59D	2007	2891
2007	2N2209	2007	2N408WHT	2007	2N60	2007	2892
2007	2N223	2007	2N41	2007	2N600	2007	2892A
2007	2N224	2007	2N106	2007	2N609	2007	2893
2007	2N225	2007	2N107	2007	2N60A	2007	2893A
2007	2N226	2007	2N413	2007	2N60B	2007	28A007H
2007	2N-027	2007	2N413A	2007	2N60C	2007	28A081C
2007	2N2271	2007	2N414	2007	2N60R	2007	28A101-OR
2007	2N228A	2007	2N414A	2007	2N61	2007	28A105A
2007	2N23	2007	2N414B	2007	2N610	2007	28A105B
2007	2N237	2007	2N414C	2007	2N611	2007	28A105C
2007	2N2374	2007	2N415	2007	2N612	2007	28A105D
2007	2N2375	2007	2N415A	2007	2N613	2007	28A105E
2007	2N2376	2007	2N416	2007	2N614	2007	28A105G
2007	2N238-ORN	2007	2N417	2007	2N615	2007	28A105H
2007	2N238D	2007	2N42	2007	2N616	2007	28A105K
2007	2N238B	2007	2N422	2007	2N617	2007	28A105L
2007	2N238P	2007	2N422A	2007	2N61A	2007	28A105M
2007	2N24	2007	2N425	2007	2N61B	2007	28A105OR
2007	2N241	2007	2N426	2007	2N61C	2007	28A105R
2007	2N241A	2007	2N427	2007	2N62	2007	28A105X
2007	2N242B	2007	2N427A	2007	2N63	2007	28A105Y
2007	2N2429	2007	2N428	2007	2N631	2007	28128
2007	2N2431	2007	2N428A	2007	2N632	2007	28A128A
2007	2N2431B	2007	2N43	2007	2N633	2007	28A128B
2007	2N2447	2007	2N435	2007	2N633B	2007	28A128C
2007	2N2448	2007	2N438	2007	2N64	2007	28A128D
2007	2N2449	2007	2N438A	2007	2N65	2007	28A128E
2007	2N2450	2007	2N439A	2007	2N650	2007	28A128F
2007	2N2468	2007	2N43A	2007	2N650A	2007	28A128G
2007	2N2469	2007	2N44	2007	2N651	2007	28A128GN
2007	2N249	2007	2N440A	2007	2N651A	2007	28A128H
2007	2N25	2007	2N444A	2007	2N652	2007	28A128J
2007	2N2564	2007	2N45	2007	2N652A	2007	28A128K
2007	2N2565	2007	2N450	2007	2N653	2007	28A128L
2007	2N26	2007	2N450A	2007	2N654	2007	28A128M
2007	2N2613	2007	2N46	2007	2N655	2007	28A128OR
2007	2N2614	2007	2N460	2007	2N655ORN	2007	28A128R
2007	2N262	2007	2N461	2007	2N655RED	2007	28A128Y
2007	2N265	2007	2N462	2007	2N670	2007	28A129
2007	2N266	2007	2N464	2007	2N670	2007	28A129A
2007	2N27	2007	2N465	2007	2N671	2007	28A129B
2007	2N270	2007	2N466	2007	2N673	2007	28A129C
2007	2N270-5E	2007	2N467	2007	2N673	2007	28A129D
2007	2N2706	2007	2N468	2007	2N674	2007	28A129E
2007	2N2707	2007	2N469	2007	2N675	2007	28A129F
2007	2N270A	2007	2N47	2007	2N71	2007	28A129GN
2007	2N271	2007	2N48	2007	2N73	2007	28A129H
2007	2N271A	2007	2N481	2007	2N74	2007	28A129J
2007	2N272	2007	2N482	2007	2N76	2007	28A129K
2007	2N273	2007	2N483	2007	2N79	2007	28A129L
2007	2N279	2007	2N483-6M	2007	2N799	2007	28A129M
2007	2N281	2007	2N483B	2007	2N80	2007	28A129OR
2007	2N282	2007	2N484	2007	2N800	2007	28A129R
2007	2N283	2007	2N485	2007	2N801	2007	28A129X
2007	2N284	2007	2N486	2007	2N802	2007	28A129Y
2007	2N284A	2007	2N486B	2007	2N803	2007	28A146
2007	2N291	2007	2N487	2007	2N804	2007	28A147
2007	2N2930	2007	2N48A	2007	2N805	2007	28A148
2007	2N2953	2007	2N49	2007	2N806	2007	28A149
2007	2N2966	2007	2N50	2007	2N807	2007	28A150A
2007	2N2Q374	2007	2N503	2007	2N808	2007	28A150B
2007	2N30	2007	2N505	2007	2N809	2007	28A150C
2007	2N302	2007	2N508	2007	2N81	2007	28A150D
2007	2N303	2007	2N508A	2007	2N810	2007	28A150E
2007	2N3075	2007	2N51	2007	2N811	2007	28A150F
2007	2N308	2007	2N518	2007	2N812	2007	28A150G
2007	2N309	2007	2N519	2007	2N813	2007	28A150GN
2007	2N31	2007	2N519A	2007	2N814	2007	28A150H
2007	2N310	2007	2N52	2007	2N815	2007	28A150J
2007	2N311	2007	2N520	2007	2N816	2007	28A150K
2007	2N315	2007	2N5201	2007	2N817	2007	28A150L
2007	2N315A	2007	2N520A	2007	2N818	2007	28A150M
2007	2N315B	2007	2N521	2007	2N819	2007	28A150OR
2007	2N316A	2007	2N521A	2007	2N82	2007	28A150R
2007	2N317	2007	2N522	2007	2N820	2007	28A150X
2007	2N317A	2007	2N522A			2007	28A150Y
						2007	28A15BL

RS 276-	Industry Standard No.	RS 276-	Industry Standard No.	RS 276-	Industry Standard No.	RS 276-	Industry Standard No.
2007	2SA15K	2007	2SA222OR	2007	2SA271K	2007	2SA288M
2007	2SA15U	2007	2SA222R	2007	2SA271L	2007	2SA288OR
2007	2SA15VR	2007	2SA222X	2007	2SA271M	2007	2SA288X
2007	2SA16O	2007	2SA222Y	2007	2SA271OR	2007	2SA288Y
2007	2SA167	2007	2SA224A	2007	2SA271R	2007	2SA298A
2007	2SA168	2007	2SA224B	2007	2SA271X	2007	2SA298B
2007	2SA168A	2007	2SA224C	2007	2SA271Y	2007	2SA298C
2007	2SA169	2007	2SA224D	2007	2SA272A	2007	2SA298D
2007	2SA170	2007	2SA224E	2007	2SA272B	2007	2SA298E
2007	2SA171	2007	2SA224F	2007	2SA272C	2007	2SA298F
2007	2SA172	2007	2SA224G	2007	2SA272G	2007	2SA298G
2007	2SA172A	2007	2SA224GN	2007	2SA272P	2007	2SA298GN
2007	2SA174	2007	2SA224H	2007	2SA272G	2007	2SA298H
2007	2SA176A	2007	2SA224J	2007	2SA272GN	2007	2SA298J
2007	2SA176B	2007	2SA224K	2007	2SA272H	2007	2SA298K
2007	2SA176C	2007	2SA224L	2007	2SA272J	2007	2SA298L
2007	2SA176D	2007	2SA224M	2007	2SA272K	2007	2SA298M
2007	2SA176E	2007	2SA224OR	2007	2SA272M	2007	2SA298OR
2007	2SA176F	2007	2SA224R	2007	2SA272OR	2007	2SA298R
2007	2SA176G	2007	2SA224X	2007	2SA272R	2007	2SA298X
2007	2SA176GN	2007	2SA224Y	2007	2SA272X	2007	2SA298Y
2007	2SA176H	2007	2SA231	2007	2SA272Y	2007	2SA29A
2007	2SA176J	2007	2SA232	2007	2SA273A	2007	2SA29B
2007	2SA176K	2007	2SA237A	2007	2SA273B	2007	2SA29C
2007	2SA176L	2007	2SA237B	2007	2SA273C	2007	2SA29D
2007	2SA176M	2007	2SA237C	2007	2SA273D	2007	2SA29E
2007	2SA176OR	2007	2SA237D	2007	2SA273E	2007	2SA29F
2007	2SA176R	2007	2SA237E	2007	2SA273F	2007	2SA29G
2007	2SA176X	2007	2SA237F	2007	2SA273G	2007	2SA29L
2007	2SA176Y	2007	2SA237G	2007	2SA273GN	2007	2SA29R
2007	2SA181	2007	2SA237GN	2007	2SA273H	2007	2SA29X
2007	2SA182	2007	2SA237H	2007	2SA273J	2007	2SA29Y
2007	2SA186	2007	2SA237J	2007	2SA273K	2007	2SA301A
2007	2SA187TV	2007	2SA237K	2007	2SA273L	2007	2SA301B
2007	2SA197	2007	2SA237L	2007	2SA273M	2007	2SA301C
2007	2SA198	2007	2SA237M	2007	2SA273OR	2007	2SA301D
2007	2SA19A	2007	2SA237OR	2007	2SA273R	2007	2SA301E
2007	2SA19B	2007	2SA237R	2007	2SA273X	2007	2SA301F
2007	2SA19C	2007	2SA237X	2007	2SA273Y	2007	2SA301G
2007	2SA19D	2007	2SA237Y	2007	2SA274A	2007	2SA301GN
2007	2SA19E	2007	2SA248	2007	2SA274C	2007	2SA301H
2007	2SA19F	2007	2SA250A	2007	2SA274D	2007	2SA301J
2007	2SA19G	2007	2SA250B	2007	2SA274E	2007	2SA301K
2007	2SA19L	2007	2SA250C	2007	2SA274F	2007	2SA301L
2007	2SA19M	2007	2SA250D	2007	2SA274G	2007	2SA301M
2007	2SA19OR	2007	2SA250E	2007	2SA274GN	2007	2SA301OR
2007	2SA19R	2007	2SA250F	2007	2SA274H	2007	2SA301R
2007	2SA19X	2007	2SA250G	2007	2SA274J	2007	2SA301X
2007	2SA19Y	2007	2SA250GN	2007	2SA274K	2007	2SA301Y
2007	2SA202	2007	2SA250H	2007	2SA274L	2007	2SA302
2007	2SA204	2007	2SA250J	2007	2SA274M	2007	2SA303
2007	2SA205	2007	2SA250L	2007	2SA274OR	2007	2SA305
2007	2SA206	2007	2SA250M	2007	2SA274R	2007	2SA311
2007	2SA207	2007	2SA250OR	2007	2SA274X	2007	2SA311A
2007	2SA208	2007	2SA250X	2007	2SA274Y	2007	2SA311B
2007	2SA209	2007	2SA250Y	2007	2SA275A	2007	2SA311C
2007	2SA20A	2007	2SA256A	2007	2SA275B	2007	2SA311D
2007	2SA20B	2007	2SA256B	2007	2SA275C	2007	2SA311E
2007	2SA20C	2007	2SA256C	2007	2SA275D	2007	2SA311F
2007	2SA20D	2007	2SA256D	2007	2SA275E	2007	2SA311G
2007	2SA20E	2007	2SA256E	2007	2SA275F	2007	2SA311GN
2007	2SA20F	2007	2SA256F	2007	2SA275G	2007	2SA311H
2007	2SA20G	2007	2SA256G	2007	2SA275GN	2007	2SA311J
2007	2SA20L	2007	2SA256GN	2007	2SA275H	2007	2SA311K
2007	2SA20M	2007	2SA256H	2007	2SA275J	2007	2SA311L
2007	2SA20OR	2007	2SA256J	2007	2SA275L	2007	2SA311M
2007	2SA20R	2007	2SA256K	2007	2SA275M	2007	2SA311OR
2007	2SA20X	2007	2SA256L	2007	2SA275OR	2007	2SA311R
2007	2SA20Y	2007	2SA256M	2007	2SA275R	2007	2SA311X
2007	2SA210	2007	2SA256OR	2007	2SA275X	2007	2SA311Y
2007	2SA212	2007	2SA256R	2007	2SA275Y	2007	2SA312
2007	2SA217	2007	2SA256X	2007	2SA277	2007	2SA312A
2007	2SA21A	2007	2SA256Y	2007	2SA278	2007	2SA312B
2007	2SA21B	2007	2SA259B	2007	2SA279	2007	2SA312C
2007	2SA21C	2007	2SA259C	2007	2SA282	2007	2SA312D
2007	2SA21D	2007	2SA259D	2007	2SA283	2007	2SA312F
2007	2SA21E	2007	2SA259E	2007	2SA284	2007	2SA312G
2007	2SA21F	2007	2SA259F	2007	2SA285B	2007	2SA312GN
2007	2SA21G	2007	2SA259G	2007	2SA285C	2007	2SA312H
2007	2SA21L	2007	2SA259GN	2007	2SA285E	2007	2SA312J
2007	2SA21M	2007	2SA259H	2007	2SA285F	2007	2SA312K
2007	2SA21OR	2007	2SA259J	2007	2SA285G	2007	2SA312L
2007	2SA21R	2007	2SA259L	2007	2SA285GN	2007	2SA312M
2007	2SA21X	2007	2SA259M	2007	2SA285H	2007	2SA312OR
2007	2SA21Y	2007	2SA259OR	2007	2SA285J	2007	2SA312R
2007	2SA220A	2007	2SA259R	2007	2SA285L	2007	2SA312X
2007	2SA220B	2007	2SA259X	2007	2SA285M	2007	2SA312Y
2007	2SA220C	2007	2SA259Y	2007	2SA285OR	2007	2SA313A
2007	2SA220D	2007	2SA26	2007	2SA285R	2007	2SA313B
2007	2SA220E	2007	2SA266A	2007	2SA285X	2007	2SA313C
2007	2SA220F	2007	2SA266B	2007	2SA285Y	2007	2SA313D
2007	2SA220G	2007	2SA266C	2007	2SA286A	2007	2SA313F
2007	2SA220GN	2007	2SA266D	2007	2SA286B	2007	2SA313G
2007	2SA220H	2007	2SA266E	2007	2SA286C	2007	2SA313GN
2007	2SA220J	2007	2SA266F	2007	2SA286D	2007	2SA313H
2007	2SA220K	2007	2SA266G	2007	2SA286E	2007	2SA313J
2007	2SA220L	2007	2SA266GN	2007	2SA286F	2007	2SA313K
2007	2SA220M	2007	2SA266GREEN	2007	2SA286G	2007	2SA313L
2007	2SA220OR	2007	2SA266H	2007	2SA286GN	2007	2SA313M
2007	2SA220R	2007	2SA266J	2007	2SA286H	2007	2SA313OR
2007	2SA220X	2007	2SA266K	2007	2SA286K	2007	2SA313R
2007	2SA220Y	2007	2SA266L	2007	2SA286M	2007	2SA313X
2007	2SA221-OR	2007	2SA266M	2007	2SA286OR	2007	2SA313Y
2007	2SA221A	2007	2SA266OR	2007	2SA286R	2007	2SA314A
2007	2SA221B	2007	2SA266R	2007	2SA286X	2007	2SA314B
2007	2SA221C	2007	2SA266X	2007	2SA286Y	2007	2SA314C
2007	2SA221D	2007	2SA266Y	2007	2SA287A	2007	2SA314D
2007	2SA221E	2007	2SA269A	2007	2SA287B	2007	2SA314F
2007	2SA221F	2007	2SA269B	2007	2SA287C	2007	2SA314G
2007	2SA221G	2007	2SA269C	2007	2SA287D	2007	2SA314GN
2007	2SA221GN	2007	2SA269D	2007	2SA287E	2007	2SA314H
2007	2SA221H	2007	2SA269E	2007	2SA287F	2007	2SA314J
2007	2SA221J	2007	2SA269F	2007	2SA287G	2007	2SA314K
2007	2SA221K	2007	2SA269GN	2007	2SA287GN	2007	2SA314L
2007	2SA221L	2007	2SA269H	2007	2SA287H	2007	2SA314M
2007	2SA221M	2007	2SA269J	2007	2SA287J	2007	2SA314OR
2007	2SA221OR	2007	2SA269K	2007	2SA287K	2007	2SA314R
2007	2SA221R	2007	2SA269L	2007	2SA287L	2007	2SA314X
2007	2SA221X	2007	2SA269M	2007	2SA287M	2007	2SA314Y
2007	2SA221Y	2007	2SA269OR	2007	2SA287OR	2007	2SA315A
2007	2SA222A	2007	2SA269X	2007	2SA287X	2007	2SA315B
2007	2SA222B	2007	2SA269Y	2007	2SA287Y	2007	2SA315C
2007	2SA222C	2007	2SA271(2)	2007	2SA288B	2007	2SA315D
2007	2SA222D	2007	2SA271(3)	2007	2SA288C	2007	2SA315E
2007	2SA222E	2007	2SA271A	2007	2SA288D	2007	2SA315F
2007	2SA222F	2007	2SA271B	2007	2SA288E	2007	2SA315G
2007	2SA222G	2007	2SA271C	2007	2SA288F	2007	2SA315GN
2007	2SA222GN	2007	2SA271D	2007	2SA288G	2007	2SA315H
2007	2SA222H	2007	2SA271E	2007	2SA288GN	2007	2SA315J
2007	2SA222J	2007	2SA271F	2007	2SA288J	2007	2SA315J
2007	2SA222K	2007	2SA271G	2007	2SA288K	2007	
2007	2SA222L	2007	2SA271GN	2007	2SA288L	2007	
2007	2SA222M	2007	2SA271H				
		2007	2SA271J				

RS 276-	Industry Standard No.	RS 276-	Industry Standard No.	RS 276-	Industry Standard No.	RS 276-	Industry Standard No.
2007	28A315K	2007	28A400A	2007	28B-P1A	2007	28B110G
2007	28A315L	2007	28A400B	2007	28B100	2007	28B110GN
2007	28A315M	2007	28A400C	2007	28B100A	2007	28B110H
2007	28A315OR	2007	28A400D	2007	28B100B	2007	28B110J
2007	28A315R	2007	28A400E	2007	28B100C	2007	28B110K
2007	28A315X	2007	28A400F	2007	28B100D	2007	28B110L
2007	28A315Y	2007	28A400G	2007	28B100E	2007	28B110M
2007	28A316A	2007	28A400GN	2007	28B100F	2007	28B110OR
2007	28A316B	2007	28A400H	2007	28B100G	2007	28B110X
2007	28A316C	2007	28A400J	2007	28B100GN	2007	28B110Y
2007	28A316D	2007	28A400K	2007	28B100H	2007	28B111
2007	28A316E	2007	28A400L	2007	28B100J	2007	28B111A
2007	28A316P	2007	28A400M	2007	28B100K	2007	28B111B
2007	28A316Q	2007	28A400OR	2007	28B100L	2007	28B111C
2007	28A316GN	2007	28A400R	2007	28B100M	2007	28B111D
2007	28A316H	2007	28A400X	2007	28B100OR	2007	28B111E
2007	28A316J	2007	28A400Y	2007	28B100R	2007	28B111F
2007	28A316K	2007	28A414	2007	28B100X	2007	28B111G
2007	28A316L	2007	28A415	2007	28B100Y	2007	28B111GN
2007	28A316M	2007	28A44	2007	28B101	2007	28B111H
2007	28A316OR	2007	28A446A	2007	28B101A	2007	28B111J
2007	28A316R	2007	28A446B	2007	28B101B	2007	28B111K
2007	28A316X	2007	28A446C	2007	28B101C	2007	28B111L
2007	28A316Y	2007	28A446D	2007	28B101D	2007	28B111M
2007	28A322A	2007	28A446E	2007	28B101E	2007	28B111OR
2007	28A322B	2007	28A446P	2007	28B101F	2007	28B111R
2007	28A322C	2007	28A446G	2007	28B101G	2007	28B111X
2007	28A322D	2007	28A446GN	2007	28B101GN	2007	28B111Y
2007	28A322E	2007	28A446H	2007	28B101H	2007	28B112
2007	28A322P	2007	28A446J	2007	28B101J	2007	28B112A
2007	28A322G	2007	28A446K	2007	28B101K	2007	28B112B
2007	28A322GN	2007	28A446L	2007	28B101L	2007	28B112C
2007	28A322H	2007	28A446M	2007	28B101M	2007	28B112D
2007	28A322K	2007	28A446OR	2007	28B101OR	2007	28B112E
2007	28A322L	2007	28A446R	2007	28B101R	2007	28B112F
2007	28A322M	2007	28A446X	2007	28B101X	2007	28B112G
2007	28A322OR	2007	28A446Y	2007	28B101Y	2007	28B112H
2007	28A322R	2007	28A447A	2007	28B102	2007	28B112J
2007	28A322X	2007	28A447B	2007	28B102A	2007	28B112K
2007	28A322Y	2007	28A447C	2007	28B102B	2007	28B112L
2007	28A33	2007	28A447D	2007	28B102C	2007	28B112M
2007	28A330	2007	28A447E	2007	28B102D	2007	28B112OR
2007	28A332	2007	28A447P	2007	28B102E	2007	28B112R
2007	28A332A	2007	28A447GN	2007	28B102F	2007	28B112X
2007	28A332B	2007	28A447H	2007	28B102G	2007	28B112Y
2007	28A332C	2007	28A447J	2007	28B102GN	2007	28B113
2007	28A332D	2007	28A447K	2007	28B102H	2007	28B113A
2007	28A332E	2007	28A447L	2007	28B102J	2007	28B113B
2007	28A332P	2007	28A447M	2007	28B102K	2007	28B113C
2007	28A332Q	2007	28A447OR	2007	28B102L	2007	28B113D
2007	28A332GN	2007	28A447R	2007	28B102M	2007	28B113E
2007	28A332H	2007	28A447X	2007	28B102OR	2007	28B113F
2007	28A332J	2007	28A447Y	2007	28B102R	2007	28B113G
2007	28A332K	2007	28A538	2007	28B102X	2007	28B113GN
2007	28A332L	2007	28A55	2007	28B102Y	2007	28B113H
2007	28A332M	2007	28A57A	2007	28B103	2007	28B113J
2007	28A332OR	2007	28A57B	2007	28B103A	2007	28B113K
2007	28A332R	2007	28A57C	2007	28B103B	2007	28B113L
2007	28A332X	2007	28A57D	2007	28B103C	2007	28B113M
2007	28A332Y	2007	28A57E	2007	28B103D	2007	28B113OR
2007	28A338A	2007	28A57P	2007	28B103E	2007	28B113R
2007	28A338B	2007	28A57G	2007	28B103F	2007	28B113X
2007	28A338C	2007	28A57L	2007	28B103G	2007	28B113Y
2007	28A338D	2007	28A57M	2007	28B103GN	2007	28B114
2007	28A338E	2007	28A57OR	2007	28B103H	2007	28B114A
2007	28A338P	2007	28A57R	2007	28B103J	2007	28B114B
2007	28A338Q	2007	28A57X	2007	28B103K	2007	28B114C
2007	28A338GN	2007	28A57Y	2007	28B103L	2007	28B114D
2007	28A338H	2007	28A64	2007	28B103M	2007	28B114E
2007	28A338J	2007	28A64A	2007	28B103OR	2007	28B114F
2007	28A338K	2007	28A64B	2007	28B103R	2007	28B114G
2007	28A338L	2007	28A64C	2007	28B103X	2007	28B114GN
2007	28A338M	2007	28A64D	2007	28B103Y	2007	28B114H
2007	28A338OR	2007	28A64E	2007	28B104	2007	28B114J
2007	28A338R	2007	28A64P	2007	28B104A	2007	28B114K
2007	28A338X	2007	28A64G	2007	28B104B	2007	28B114L
2007	28A338Y	2007	28A64GN	2007	28B104C	2007	28B114M
2007	28A341A	2007	28A64H	2007	28B104D	2007	28B114OR
2007	28A341B	2007	28A64J	2007	28B104E	2007	28B114R
2007	28A341C	2007	28A64K	2007	28B104F	2007	28B114X
2007	28A341D	2007	28A64L	2007	28B104G	2007	28B114Y
2007	28A341E	2007	28A64M	2007	28B104GN	2007	28B115
2007	28A341P	2007	28A64OR	2007	28B104H	2007	28B115A
2007	28A341Q	2007	28A64R	2007	28B104J	2007	28B115B
2007	28A341GN	2007	28A64X	2007	28B104K	2007	28B115C
2007	28A341H	2007	28A64Y	2007	28B104L	2007	28B115D
2007	28A341J	2007	28A80A	2007	28B104M	2007	28B115E
2007	28A341K	2007	28A80B	2007	28B104OR	2007	28B115F
2007	28A341L	2007	28A80D	2007	28B104R	2007	28B115G
2007	28A341M	2007	28A80E	2007	28B104Y	2007	28B115GN
2007	28A341OR	2007	28A80F	2007	28B105	2007	28B115H
2007	28A341R	2007	28A80G	2007	28B105A	2007	28B115J
2007	28A341X	2007	28A80H	2007	28B105B	2007	28B115K
2007	28A341Y	2007	28A80K	2007	28B105C	2007	28B115L
2007	28A357A	2007	28A80L	2007	28B105D	2007	28B115M
2007	28A357B	2007	28A80M	2007	28B105E	2007	28B115OR
2007	28A357C	2007	28A80OR	2007	28B105F	2007	28B115R
2007	28A357D	2007	28A80R	2007	28B105G	2007	28B115X
2007	28A357E	2007	28A80X	2007	28B105GN	2007	28B115Y
2007	28A357P	2007	28A80Y	2007	28B105H	2007	28B116
2007	28A357G	2007	28A82A	2007	28B105K	2007	28B116A
2007	28A357GN	2007	28A82B	2007	28B105L	2007	28B116B
2007	28A357H	2007	28A82C	2007	28B105M	2007	28B116C
2007	28A357J	2007	28A82D	2007	28B105OR	2007	28B116D
2007	28A357K	2007	28A82E	2007	28B105X	2007	28B116E
2007	28A357L	2007	28A82F	2007	28B105Y	2007	28B116F
2007	28A357M	2007	28A82G	2007	28B106	2007	28B116G
2007	28A357OR	2007	28A82H	2007	28B108	2007	28B116GN
2007	28A357R	2007	28A82J	2007	28B108A	2007	28B116H
2007	28A357X	2007	28A82M	2007	28B108B	2007	28B116J
2007	28A357Y	2007	28A82OR	2007	28B108C	2007	28B116K
2007	28A359A	2007	28A82R	2007	28B108D	2007	28B116L
2007	28A359B	2007	28A82X	2007	28B108E	2007	28B116M
2007	28A359C	2007	28A82Z	2007	28B108F	2007	28B116OR
2007	28A359D	2007	28A85L	2007	28B108G	2007	28B116X
2007	28A359E	2007	28A945	2007	28B108GN	2007	28B116Y
2007	28A359P	2007	28A94A	2007	28B108H	2007	28B117
2007	28A359Q	2007	28A94C	2007	28B108J	2007	28B117A
2007	28A359GN	2007	28A94D	2007	28B108K	2007	28B117B
2007	28A359H	2007	28A94E	2007	28B108L	2007	28B117C
2007	28A359J	2007	28A94F	2007	28B108M	2007	28B117D
2007	28A359K	2007	28A94G	2007	28B108OR	2007	28B117E
2007	28A359L	2007	28A94H	2007	28B108X	2007	28B117F
2007	28A359M	2007	28A94K	2007	28B108Y	2007	28B117G
2007	28A359OR	2007	28A94L	2007	28B109	2007	28B117GN
2007	28A359R	2007	28A94M	2007	28B110	2007	28B117H
2007	28A359X	2007	28A94OR	2007	28B110A	2007	28B117J
2007	28A359Y	2007	28A94R	2007	28B110B	2007	28B117K
2007	28A36	2007	28A94X	2007	28B110C	2007	28B117L
2007	28A370	2007	28A94Y	2007	28B110D	2007	28B117M
2007	28A371	2007	28B-3783	2007	28B110E	2007	28B117OR
2007	28A396	2007	28B-3812	2007	28B110P	2007	28B117R
2007	28A397	2007	28B-3813				
2007	28A40						

RS 276-	Industry Standard No.	RS 276-	Industry Standard No.	RS 276-	Industry Standard No.	RS 276-	Industry Standard No.
2007	2SB117X	2007	2SB155R	2007	2SB167A	2007	2SB174
2007	2SB117Y	2007	2SB155X	2007	2SB167B	2007	2SB174A
2007	2SB118	2007	2SB155Y	2007	2SB167BK	2007	2SB174B
2007	2SB13	2007	2SB156	2007	2SB167C	2007	2SB174C
2007	2SB134	2007	2SB156/7825B	2007	2SB167D	2007	2SB174D
2007	2SB134-D	2007	2SB156A	2007	2SB167E	2007	2SB174E
2007	2SB134-E	2007	2SB156AA	2007	2SB167F	2007	2SB174F
2007	2SB134A	2007	2SB156AB	2007	2SB167G	2007	2SB174G
2007	2SB134B	2007	2SB156AC	2007	2SB167GN	2007	2SB174GN
2007	2SB134C	2007	2SB156B	2007	2SB167H	2007	2SB174H
2007	2SB134D	2007	2SB156BK	2007	2SB167J	2007	2SB174K
2007	2SB134E	2007	2SB156C	2007	2SB167K	2007	2SB174L
2007	2SB134F	2007	2SB156D	2007	2SB167L	2007	2SB174M
2007	2SB134G	2007	2SB156E	2007	2SB167M	2007	2SB174OR
2007	2SB134GN	2007	2SB156F	2007	2SB167OR	2007	2SB174R
2007	2SB134H	2007	2SB156R	2007	2SB167R	2007	2SB174X
2007	2SB134J	2007	2SB156GN	2007	2SB167X	2007	2SB174Y
2007	2SB134K	2007	2SB156H	2007	2SB167Y	2007	2SB175
2007	2SB134L	2007	2SB156J	2007	2SB168	2007	2SB175(A)
2007	2SB134M	2007	2SB156K	2007	2SB168A	2007	2SB175(B)
2007	2SB134OR	2007	2SB156L	2007	2SB168B	2007	2SB175(C)
2007	2SB134R	2007	2SB156M	2007	2SB168C	2007	2SB175A
2007	2SB134X	2007	2SB156OR	2007	2SB168D	2007	2SB175B
2007	2SB134Y	2007	2SB156P	2007	2SB168E	2007	2SB175B-1
2007	2SB135	2007	2SB156R	2007	2SB168F	2007	2SB175BL
2007	2SB135(C)	2007	2SB156X	2007	2SB168G	2007	2SB175C
2007	2SB135A	2007	2SB156Y	2007	2SB168GN	2007	2SB175CL
2007	2SB135B	2007	2SB161	2007	2SB168H	2007	2SB175D
2007	2SB135C	2007	2SB161A	2007	2SB168J	2007	2SB175E
2007	2SB135D	2007	2SB161B	2007	2SB168K	2007	2SB175F
2007	2SB135E	2007	2SB161C	2007	2SB168L	2007	2SB175G
2007	2SB135F	2007	2SB161D	2007	2SB168M	2007	2SB175GN
2007	2SB135G	2007	2SB161E	2007	2SB168OR	2007	2SB175H
2007	2SB135GN	2007	2SB161F	2007	2SB168R	2007	2SB175L
2007	2SB135H	2007	2SB161G	2007	2SB168X	2007	2SB175M
2007	2SB135J	2007	2SB161GN	2007	2SB168Y	2007	2SB175OR
2007	2SB135K	2007	2SB161H	2007	2SB169	2007	2SB175R
2007	2SB135L	2007	2SB161J	2007	2SB169A	2007	2SB175X
2007	2SB135M	2007	2SB161K	2007	2SB169B	2007	2SB175Y
2007	2SB135OR	2007	2SB161L	2007	2SB169C	2007	2SB176
2007	2SB135R	2007	2SB161M	2007	2SB169D	2007	2SB176-O
2007	2SB135X	2007	2SB161OR	2007	2SB169E	2007	2SB176-P
2007	2SB135Y	2007	2SB161R	2007	2SB169F	2007	2SB176-PR
2007	2SB36	2007	2SB161X	2007	2SB169G	2007	2SB176A
2007	2SB36(C)	2007	2SB161Y	2007	2SB169GN	2007	2SB176B
2007	2SB36-2	2007	2SB162	2007	2SB169H	2007	2SB176C
2007	2SB36-3	2007	2SB162A	2007	2SB169J	2007	2SB176D
2007	2SB36A	2007	2SB162B	2007	2SB169K	2007	2SB176E
2007	2SB36B	2007	2SB162C	2007	2SB169L	2007	2SB176F
2007	2SB36C	2007	2SB162D	2007	2SB169M	2007	2SB176G
2007	2SB36D	2007	2SB162F	2007	2SB169OR	2007	2SB176GN
2007	2SB36E	2007	2SB162G	2007	2SB169R	2007	2SB176H
2007	2SB36F	2007	2SB162GN	2007	2SB169X	2007	2SB176J
2007	2SB36G	2007	2SB162H	2007	2SB169Y	2007	2SB176K
2007	2SB36GN	2007	2SB162J	2007	2SB170	2007	2SB176L
2007	2SB36H	2007	2SB162K	2007	2SB170A	2007	2SB176M
2007	2SB36J	2007	2SB162L	2007	2SB170B	2007	2SB176O
2007	2SB36K	2007	2SB162M	2007	2SB170C	2007	2SB176OR
2007	2SB36L	2007	2SB162OR	2007	2SB170D	2007	2SB176P
2007	2SB36M	2007	2SB162R	2007	2SB170E	2007	2SB176PL
2007	2SB36OR	2007	2SB162X	2007	2SB170F	2007	2SB176PR
2007	2SB36R	2007	2SB162Y	2007	2SB170G	2007	2SB176PRC
2007	2SB36U	2007	2SB163	2007	2SB170GN	2007	2SB176R
2007	2SB36X	2007	2SB163A	2007	2SB170H	2007	2SB176R(1)
2007	2SB5	2007	2SB163B	2007	2SB170J	2007	2SB176RG
2007	2SB50	2007	2SB163C	2007	2SB170K	2007	2SB176X
2007	2SB50A	2007	2SB163D	2007	2SB170L	2007	2SB176Y
2007	2SB50B	2007	2SB163E	2007	2SB170M	2007	2SB177
2007	2SB50C	2007	2SB163F	2007	2SB170OR	2007	2SB177A
2007	2SB50D	2007	2SB163G	2007	2SB170R	2007	2SB177B
2007	2SB50E	2007	2SB163GN	2007	2SB170X	2007	2SB177C
2007	2SB50F	2007	2SB163H	2007	2SB170Y	2007	2SB177D
2007	2SB50G	2007	2SB163J	2007	2SB171	2007	2SB177E
2007	2SB50GN	2007	2SB163K	2007	2SB171A	2007	2SB177F
2007	2SB50H	2007	2SB163L	2007	2SB171B	2007	2SB177G
2007	2SB50J	2007	2SB163M	2007	2SB171C	2007	2SB177GN
2007	2SB50K	2007	2SB163OR	2007	2SB171D	2007	2SB177H
2007	2SB50L	2007	2SB163R	2007	2SB171E	2007	2SB177K
2007	2SB50M	2007	2SB163X	2007	2SB171F	2007	2SB177L
2007	2SB50OR	2007	2SB163Y	2007	2SB171G	2007	2SB177M
2007	2SB50X	2007	2SB164	2007	2SB171GN	2007	2SB177OR
2007	2SB50Y	2007	2SB164A	2007	2SB171H	2007	2SB177R
2007	2SB153	2007	2SB164B	2007	2SB171J	2007	2SB177X
2007	2SB153A	2007	2SB164C	2007	2SB171K	2007	2SB177Y
2007	2SB153B	2007	2SB164D	2007	2SB171L	2007	2SB178
2007	2SB153C	2007	2SB164E	2007	2SB171M	2007	2SB178-O
2007	2SB153D	2007	2SB164F	2007	2SB171OR	2007	2SB178-OR
2007	2SB153E	2007	2SB164G	2007	2SB171X	2007	2SB178-S
2007	2SB153F	2007	2SB164GN	2007	2SB171Y	2007	2SB178O
2007	2SB153G	2007	2SB164H	2007	2SB172	2007	2SB178OA
2007	2SB153GN	2007	2SB164J	2007	2SB172A	2007	2SB178OB
2007	2SB153H	2007	2SB164L	2007	2SB172A-1	2007	2SB178OC
2007	2SB153J	2007	2SB164M	2007	2SB172A-P	2007	2SB178OD
2007	2SB153K	2007	2SB164OR	2007	2SB172AF	2007	2SB178OE
2007	2SB153L	2007	2SB164R	2007	2SB172B	2007	2SB178OF
2007	2SB153M	2007	2SB164X	2007	2SB172C	2007	2SB178OG
2007	2SB153OR	2007	2SB164Y	2007	2SB172D	2007	2SB178OGN
2007	2SB153R	2007	2SB165	2007	2SB172E	2007	2SB178OH
2007	2SB153X	2007	2SB165A	2007	2SB172F	2007	2SB178OJ
2007	2SB153Y	2007	2SB165B	2007	2SB172FN	2007	2SB178OK
2007	2SB154	2007	2SB165C	2007	2SB172G	2007	2SB178OL
2007	2SB154A	2007	2SB165D	2007	2SB172GN	2007	2SB178OM
2007	2SB154B	2007	2SB165E	2007	2SB172H	2007	2SB178OOR
2007	2SB154C	2007	2SB165F	2007	2SB172J	2007	2SB178OR
2007	2SB154D	2007	2SB165G	2007	2SB172K	2007	2SB178OX
2007	2SB154E	2007	2SB165GN	2007	2SB172L	2007	2SB178OY
2007	2SB154F	2007	2SB165H	2007	2SB172M	2007	2SB178A
2007	2SB154G	2007	2SB165J	2007	2SB172OR	2007	2SB178B
2007	2SB154GN	2007	2SB165K	2007	2SB172P	2007	2SB178BU
2007	2SB154H	2007	2SB165M	2007	2SB172R	2007	2SB178C
2007	2SB154J	2007	2SB165OR	2007	2SB172X	2007	2SB178D
2007	2SB154K	2007	2SB165R	2007	2SB172T	2007	2SB178E
2007	2SB154L	2007	2SB165X	2007	2SB173	2007	2SB178F
2007	2SB154M	2007	2SB165Y	2007	2SB173(C)	2007	2SB178G
2007	2SB154OR	2007	2SB166	2007	2SB173A	2007	2SB178GN
2007	2SB154R	2007	2SB166A	2007	2SB173B	2007	2SB178H
2007	2SB154X	2007	2SB166B	2007	2SB173BL	2007	2SB178J
2007	2SB154Y	2007	2SB166C	2007	2SB173C	2007	2SB178K
2007	2SB155	2007	2SB166D	2007	2SB173CL	2007	2SB178L
2007	2SB155A	2007	2SB166E	2007	2SB173D	2007	2SB178M
2007	2SB155B	2007	2SB166F	2007	2SB173E	2007	2SB178N
2007	2SB155C	2007	2SB166G	2007	2SB173F	2007	2SB178OR
2007	2SB155D	2007	2SB166GN	2007	2SB173G	2007	2SB178R
2007	2SB155E	2007	2SB166H	2007	2SB173GN	2007	2SB178S
2007	2SB155F	2007	2SB166J	2007	2SB173H	2007	2SB178T
2007	2SB155G	2007	2SB166K	2007	2SB173J	2007	2SB178TC
2007	2SB155GN	2007	2SB166L	2007	2SB173K	2007	2SB178TS
2007	2SB155H	2007	2SB166M	2007	2SB173L	2007	2SB178V
2007	2SB155J	2007	2SB166OR	2007	2SB173M	2007	2SB178X
2007	2SB155K	2007	2SB166R	2007	2SB173OR	2007	2SB178Y
2007	2SB155L	2007	2SB166X	2007	2SB173R	2007	2SB180
2007	2SB155M	2007	2SB166Y	2007	2SB173X	2007	2SB180A
2007	2SB155OR	2007	2SB167	2007	2SB173Y	2007	2SB181
						2007	2SB181A

RS 276-	Industry Standard No.	RS 276-	Industry Standard No.	RS 276-	Industry Standard No.	RS 276-	Industry Standard No.
2007	28B183	2007	28B188R	2007	28B220	2007	28B22E
2007	28B183A	2007	28B188X	2007	28B220A	2007	28B22F
2007	28B183B	2007	28B188Y	2007	28B220B	2007	28B22G
2007	28B183C	2007	28B189	2007	28B220C	2007	28B22GN
2007	28B183D	2007	28B189A	2007	28B220D	2007	28B22H
2007	28B183E	2007	28B189B	2007	28B220E	2007	28B22I
2007	28B183F	2007	28B189C	2007	28B220F	2007	28B22J
2007	28B183G	2007	28B189D	2007	28B220GN	2007	28B22K
2007	28B183GN	2007	28B189E	2007	28B220H	2007	28B22L
2007	28B183H	2007	28B189F	2007	28B220J	2007	28B22M
2007	28B183J	2007	28B189G	2007	28B220K	2007	28B220R
2007	28B183K	2007	28B189GN	2007	28B220L	2007	28B22P
2007	28B183L	2007	28B189H	2007	28B220M	2007	28B22R
2007	28B183M	2007	28B189J	2007	28B2200R	2007	28B22X
2007	28B183OR	2007	28B189K	2007	28B220R	2007	28B22Y
2007	28B183R	2007	28B189L	2007	28B220X	2007	28B23
2007	28B183X	2007	28B189M	2007	28B220Y	2007	28B23B
2007	28B183Y	2007	28B189OR	2007	28B221	2007	28B23B-12A
2007	28B184	2007	28B189R	2007	28B221A	2007	28B23B-12B
2007	28B184A	2007	28B189X	2007	28B221B	2007	28B23B-12C
2007	28B184B	2007	28B189Y	2007	28B221C	2007	28B23A
2007	28B184C	2007	28B190	2007	28B221D	2007	28B23B
2007	28B184D	2007	28B192	2007	28B221E	2007	28B23C
2007	28B184E	2007	28B193	2007	28B221F	2007	28B23D
2007	28B184F	2007	28B194	2007	28B221G	2007	28B23E
2007	28B184G	2007	28B195	2007	28B221GN	2007	28B23F
2007	28B184GN	2007	28B196	2007	28B221H	2007	28B23G
2007	28B184H	2007	28B197	2007	28B221J	2007	28B23GN
2007	28B184J	2007	28B198	2007	28B221K	2007	28B23H
2007	28B184K	2007	28B199	2007	28B221L	2007	28B23J
2007	28B184L	2007	28B199A	2007	28B221M	2007	28B23K
2007	28B184M	2007	28B199B	2007	28B221OR	2007	28B23L
2007	28B184OR	2007	28B199C	2007	28B221R	2007	28B23M
2007	28B184R	2007	28B199D	2007	28B221X	2007	28B230R
2007	28B184X	2007	28B199E	2007	28B221Y	2007	28B23R
2007	28B184Y	2007	28B199F	2007	28B222	2007	28B23X
2007	28B185	2007	28B199G	2007	28B222A	2007	28B23Y
2007	28B185(O)	2007	28B199GN	2007	28B222B	2007	28B24
2007	28B185(O)	2007	28B199H	2007	28B222C	2007	28B241
2007	28B185(P)	2007	28B199J	2007	28B222D	2007	28B241A
2007	00028B185-0	2007	28B199L	2007	28B222F	2007	28B241B
2007	0028B18500	2007	28B199M	2007	28B222G	2007	28B241C
2007	28B185A	2007	28B199OR	2007	28B222GN	2007	28B241D
2007	0028B185AA	2007	28B199R	2007	28B222H	2007	28B241E
2007	28B185B	2007	28B199X	2007	28B222J	2007	28B241F
2007	28B185C	2007	28B199Y	2007	28B222K	2007	28B241G
2007	28B185D	2007	28B200	2007	28B222L	2007	28B241GN
2007	28B185E	2007	28B200A	2007	28B222M	2007	28B241H
2007	28B185F	2007	28B200B	2007	28B222R	2007	28B241J
2007	28B185G	2007	28B200C	2007	28B222X	2007	28B241K
2007	28B185GN	2007	28B200D	2007	28B222Y	2007	28B241L
2007	28B185H	2007	28B200E	2007	28B223	2007	28B241M
2007	28B185I	2007	28B200F	2007	28B223A	2007	28B2410R
2007	28B185J	2007	28B200G	2007	28B223B	2007	28B241R
2007	28B185L	2007	28B200GN	2007	28B223C	2007	28B241V
2007	28B185M	2007	28B200H	2007	28B223D	2007	28B241X
2007	28B185OR	2007	28B200J	2007	28B223E	2007	28B241Y
2007	28B185P	2007	28B200K	2007	28B223F	2007	28B242
2007	28B185R	2007	28B200L	2007	28B223Y	2007	28B242A
2007	28B185X	2007	28B200M	2007	28B224	2007	28B242B
2007	28B185Y	2007	28B2000R	2007	28B224A	2007	28B242C
2007	00028B186	2007	28B200R	2007	28B224B	2007	28B242D
2007	28B186(O)	2007	28B200X	2007	28B224C	2007	28B242E
2007	28B186(O)	2007	28B200Y	2007	28B224D	2007	28B242F
2007	00028B186-0	2007	28B201	2007	28B224E	2007	28B242G
2007	28B186-1	2007	28B202	2007	28B224F	2007	28B242GN
2007	28B186-7	2007	28B202A	2007	28B224G	2007	28B242H
2007	28B186-K	2007	28B202B	2007	28B224GN	2007	28B242J
2007	28B186-0	2007	28B202C	2007	28B224H	2007	28B242K
2007	28B186-OR	2007	28B202D	2007	28B224K	2007	28B242L
2007	00028B1860	2007	28B202E	2007	28B224L	2007	28B242M
2007	28B186A	2007	28B202F	2007	28B224M	2007	28B2420R
2007	28B186AG	2007	28B202G	2007	28B2240R	2007	28B242R
2007	28B186B	2007	28B202GN	2007	28B224R	2007	28B242X
2007	28B186BY	2007	28B202H	2007	28B224X	2007	28B242Y
2007	28B186C	2007	28B202J	2007	28B224Y	2007	28B24A
2007	28B186D	2007	28B202L	2007	28B225	2007	28B24B
2007	28B186E	2007	28B202M	2007	28B225A	2007	28B24C
2007	28B186F	2007	28B2020R	2007	28B225B	2007	28B24D
2007	28B186G	2007	28B202R	2007	28B225C	2007	28B24E
2007	28B186GN	2007	28B202X	2007	28B225D	2007	28B24F
2007	28B186H	2007	28B202Y	2007	28B225E	2007	28B24G
2007	28B186J	2007	28B203A	2007	28B225F	2007	28B24GN
2007	28B186K	2007	28B203AA	2007	28B225G	2007	28B24H
2007	28B186L	2007	28B203B	2007	28B225GN	2007	28B24J
2007	28B186M	2007	28B2030D	2007	28B225H	2007	28B24K
2007	28B186O	2007	28B203E	2007	28B225K	2007	28B24L
2007	28B186OR	2007	28B203F	2007	28B225M	2007	28B24M
2007	28B186R	2007	28B203G	2007	28B2250R	2007	28B240R
2007	28B186X	2007	28B203GN	2007	28B225R	2007	28B24R
2007	28B186Y	2007	28B203H	2007	28B225X	2007	28B24X
2007	00028B187	2007	28B203J	2007	28B225Y	2007	28B24Y
2007	28B187(1)	2007	28B203K	2007	28B2250R	2007	28B257
2007	28B187(K)	2007	28B203L	2007	28B225R	2007	28B257A
2007	00028B187(RED)	2007	28B2030R	2007	28B225X	2007	28B257B
2007	28B187-1	2007	28B2030R	2007	28B225Y	2007	28B257C
2007	28B187-OR	2007	28B203R	2007	28B226	2007	28B257D
2007	28B187A	2007	28B203X	2007	28B226A	2007	28B257E
2007	28B187AA	2007	28B203Y	2007	28B226B	2007	28B257F
2007	28B187B	2007	28B218	2007	28B226C	2007	28B257GN
2007	28B187BK	2007	28B218A	2007	28B226E	2007	28B257H
2007	28B187C	2007	28B218B	2007	28B226F	2007	28B257K
2007	28B187D	2007	28B218C	2007	28B226G	2007	28B257L
2007	28B187E	2007	28B218D	2007	28B226GN	2007	28B257M
2007	28B187F	2007	28B218E	2007	28B226J	2007	28B2570R
2007	28B187G	2007	28B218F	2007	28B226K	2007	28B257R
2007	28B187GN	2007	28B218G	2007	28B226L	2007	28B257X
2007	28B187H	2007	28B218GN	2007	28B226M	2007	28B257Y
2007	28B187K	2007	28B218H	2007	28B2260R	2007	28B258A
2007	28B187L	2007	28B218J	2007	28B226R	2007	28B258B
2007	28B187OR	2007	28B218K	2007	28B226X	2007	28B258C
2007	28B187R	2007	28B218L	2007	28B226Y	2007	28B258D
2007	28B187RED	2007	28B218M	2007	28B227	2007	28B258E
2007	28B187S	2007	28B218OR	2007	28B227A	2007	28B258F
2007	28B187TV	2007	28B218R	2007	28B227B	2007	28B258G
2007	28B187X	2007	28B218X	2007	28B227C	2007	28B258GN
2007	28B187Y	2007	28B218Y	2007	28B227D	2007	28B258H
2007	28B187YEL	2007	28B219	2007	28B227F	2007	28B258J
2007	28B188	2007	28B219A	2007	28B227G	2007	28B258K
2007	28B188A	2007	28B219B	2007	28B227GN	2007	28B258L
2007	28B188B	2007	28B219E	2007	28B227H	2007	28B258M
2007	28B188C	2007	28B219F	2007	28B227J	2007	28B2580R
2007	28B188D	2007	28B219G	2007	28B227K	2007	28B258R
2007	28B188E	2007	28B219H	2007	28B227L	2007	28B258X
2007	28B188F	2007	28B219J	2007	28B227M	2007	28B258Y
2007	28B188G	2007	28B219K	2007	28B2270R	2007	28B261
2007	28B188GN	2007	28B219L	2007	28B227R	2007	28B261A
2007	28B188H	2007	28B219R	2007	28B227X	2007	28B261B
2007	28B188J	2007	28B219X	2007	28B227Y	2007	28B261C
2007	28B188K	2007	28B219Y	2007	28B22A	2007	28B261D
2007	28B188L	2007	28B22	2007	28B22B	2007	28B261E
2007	28B188M	2007	28B22-0	2007	28B22C	2007	28B261F
2007	28B188OR	2007	28B22/09-30100	2007	28B22D	2007	28B261G

RS 276-	Industry Standard No.	RS 276-	Industry Standard No.	RS 276-	Industry Standard No.	RS 276-	Industry Standard No.
2007	2SB261GN	2007	2SB268X	2007	2SB292M	2007	2SB32-0
2007	2SB261H	2007	2SB268Y	2007	2SB292OR	2007	2SB32-1
2007	2SB261J	2007	2SB269	2007	2SB292R	2007	2SB32-2
2007	2SB261K	2007	2SB269A	2007	2SB292X	2007	2SB32-4
2007	2SB261L	2007	2SB269C	2007	2SB292Y	2007	2SB321
2007	2SB261M	2007	2SB269D	2007	2SB293	2007	2SB322
2007	2SB261OR	2007	2SB269E	2007	2SB293A	2007	2SB323
2007	2SB261R	2007	2SB269F	2007	2SB293B	2007	2SB324
2007	2SB261X	2007	2SB269G	2007	2SB293C	2007	2SB324(E)
2007	2SB261Y	2007	2SB269GN	2007	2SB293D	2007	2SB324(F)
2007	2SB262	2007	2SB269J	2007	2SB293E	2007	2SB324(I)
2007	2SB262A	2007	2SB269K	2007	2SB293F	2007	2SB324(L)
2007	2SB262B	2007	2SB269L	2007	2SB293G	2007	2SB324(N)
2007	2SB262C	2007	2SB269M	2007	2SB293GN	2007	2SB324,K
2007	2SB262D	2007	2SB269OR	2007	2SB293H	2007	2SB324-OR
2007	2SB262E	2007	2SB269R	2007	2SB293J	2007	2SB324/4454C
2007	2SB262F	2007	2SB269X	2007	2SB293K	2007	2SB324O
2007	2SB262G	2007	2SB269Y	2007	2SB293L	2007	2SB324OA
2007	2SB262GN	2007	2SB270	2007	2SB293M	2007	2SB324OB
2007	2SB262H	2007	2SB270A	2007	2SB293OR	2007	2SB324OC
2007	2SB262J	2007	2SB270B	2007	2SB293R	2007	2SB324OD
2007	2SB262K	2007	2SB270C	2007	2SB293X	2007	2SB324OE
2007	2SB262L	2007	2SB270D	2007	2SB293Y	2007	2SB324OF
2007	2SB262M	2007	2SB270E	2007	2SB294	2007	2SB324OG
2007	2SB262OR	2007	2SB271	2007	2SB294A	2007	2SB324OGN
2007	2SB262R	2007	2SB272	2007	2SB294B	2007	2SB324OH
2007	2SB262X	2007	2SB272A	2007	2SB294C	2007	2SB324OJ
2007	2SB262Y	2007	2SB272B	2007	2SB294D	2007	2SB324OK
2007	2SB263	2007	2SB272C	2007	2SB294E	2007	2SB324OL
2007	2SB263A	2007	2SB272D	2007	2SB294F	2007	2SB324OM
2007	2SB263B	2007	2SB272E	2007	2SB294G	2007	2SB324OOR
2007	2SB263C	2007	2SB272F	2007	2SB294GN	2007	2SB324OR
2007	2SB263D	2007	2SB272G	2007	2SB294H	2007	2SB324OX
2007	2SB263E	2007	2SB272GN	2007	2SB294J	2007	2SB324OY
2007	2SB263F	2007	2SB272J	2007	2SB294K	2007	2SB324AH
2007	2SB263G	2007	2SB272K	2007	2SB294L	2007	2SB324A
2007	2SB263GN	2007	2SB272L	2007	2SB294M	2007	2SB324B
2007	2SB263H	2007	2SB272M	2007	2SB294OR	2007	2SB324C
2007	2SB263J	2007	2SB272OR	2007	2SB294R	2007	2SB324D
2007	2SB263K	2007	2SB272R	2007	2SB294X	2007	2SB324E
2007	2SB263L	2007	2SB272X	2007	2SB294Y	2007	2SB324E-1
2007	2SB263M	2007	2SB272Y	2007	2SB299	2007	2SB324E-L
2007	2SB263OR	2007	2SB273	2007	2SB299A	2007	2SB324F
2007	2SB263R	2007	2SB273A	2007	2SB299B	2007	2SB324G
2007	2SB263X	2007	2SB273B	2007	2SB299C	2007	2SB324GN
2007	2SB263Y	2007	2SB273C	2007	2SB299D	2007	2SB324H
2007	2SB264	2007	2SB273D	2007	2SB299E	2007	2SB324I
2007	2SB264A	2007	2SB273E	2007	2SB299F	2007	2SB324J
2007	2SB264B	2007	2SB273F	2007	2SB299G	2007	2SB324K
2007	2SB264C	2007	2SB273G	2007	2SB299GN	2007	2SB324L
2007	2SB264D	2007	2SB273GN	2007	2SB299H	2007	2SB324M
2007	2SB264E	2007	2SB273H	2007	2SB299J	2007	2SB324N
2007	2SB264F	2007	2SB273J	2007	2SB299K	2007	2SB324OR
2007	2SB264G	2007	2SB273K	2007	2SB299L	2007	2SB324P
2007	2SB264GN	2007	2SB273L	2007	2SB299M	2007	2SB324R
2007	2SB264H	2007	2SB273M	2007	2SB299OR	2007	2SB324S
2007	2SB264J	2007	2SB273OR	2007	2SB299X	2007	2SB324V
2007	2SB264K	2007	2SB273R	2007	2SB299Y	2007	2SB324X
2007	2SB264L	2007	2SB273X	2007	2SB302	2007	2SB324Y
2007	2SB264M	2007	2SB273Y	2007	000028SB303	2007	2SB326A
2007	2SB264OR	2007	2SB274A	2007	2SB303(0)	2007	2SB326B
2007	2SB264R	2007	2SB274B	2007	0002SB303-0	2007	2SB326C
2007	2SB264X	2007	2SB274C	2007	0002SB3030	2007	2SB326D
2007	2SB264Y	2007	2SB274D	2007	2SB303A	2007	2SB326E
2007	2SB265	2007	2SB274E	2007	2SB303B	2007	2SB326F
2007	2SB265A	2007	2SB274F	2007	2SB303BK	2007	2SB326G
2007	2SB265B	2007	2SB274G	2007	2SB303C	2007	2SB326GN
2007	2SB265C	2007	2SB274H	2007	2SB303D	2007	2SB326H
2007	2SB265D	2007	2SB274J	2007	2SB303E	2007	2SB326J
2007	2SB265E	2007	2SB274K	2007	2SB303F	2007	2SB326K
2007	2SB265F	2007	2SB274L	2007	2SB303G	2007	2SB326L
2007	2SB265GN	2007	2SB274M	2007	2SB303GN	2007	2SB326M
2007	2SB265H	2007	2SB274OR	2007	2SB303H	2007	2SB326OR
2007	2SB265J	2007	2SB274R	2007	2SB303J	2007	2SB326R
2007	2SB265K	2007	2SB274V	2007	2SB303K	2007	2SB326X
2007	2SB265L	2007	2SB274X	2007	2SB303L	2007	2SB326Y
2007	2SB265M	2007	2SB274Y	2007	2SB303M	2007	2SB327
2007	2SB265OR	2007	2SB290	2007	2SB303OR	2007	2SB327A
2007	2SB265X	2007	2SB290A	2007	2SB303R	2007	2SB327B
2007	2SB265Y	2007	2SB290B	2007	2SB303X	2007	2SB327C
2007	2SB266	2007	2SB290C	2007	2SB303Y	2007	2SB327D
2007	2SB266A	2007	2SB290D	2007	2SB304	2007	2SB327E
2007	2SB266B	2007	2SB290E	2007	2SB304A	2007	2SB327F
2007	2SB266C	2007	2SB290F	2007	2SB314	2007	2SB327G
2007	2SB266D	2007	2SB290G	2007	2SB314A	2007	2SB327GN
2007	2SB266E	2007	2SB290GN	2007	2SB314C	2007	2SB327H
2007	2SB266F	2007	2SB290H	2007	2SB314D	2007	2SB327J
2007	2SB266G	2007	2SB290J	2007	2SB314E	2007	2SB327K
2007	2SB266GN	2007	2SB290K	2007	2SB314F	2007	2SB327L
2007	2SB266H	2007	2SB290L	2007	2SB314GN	2007	2SB327OR
2007	2SB266J	2007	2SB290M	2007	2SB314H	2007	2SB327R
2007	2SB266K	2007	2SB290OR	2007	2SB314J	2007	2SB327X
2007	2SB266L	2007	2SB290R	2007	2SB314K	2007	2SB327Y
2007	2SB266M	2007	2SB290X	2007	2SB314L	2007	2SB328
2007	2SB266OR	2007	2SB290Y	2007	2SB314M	2007	2SB328A
2007	2SB266P	2007	2SB291	2007	2SB314OR	2007	2SB328B
2007	2SB266Q	2007	2SB291-GREEN	2007	2SB314R	2007	2SB328C
2007	2SB266R	2007	2SB291-RED	2007	2SB314X	2007	2SB328D
2007	2SB266X	2007	2SB291-YELLOW	2007	2SB314Y	2007	2SB328E
2007	2SB266Y	2007	2SB291A	2007	2SB315	2007	2SB328F
2007	2SB267A	2007	2SB291B	2007	2SB315A	2007	2SB328G
2007	2SB267B	2007	2SB291C	2007	2SB315B	2007	2SB328GN
2007	2SB267C	2007	2SB291D	2007	2SB315C	2007	2SB328H
2007	2SB267D	2007	2SB291E	2007	2SB315D	2007	2SB328J
2007	2SB267E	2007	2SB291F	2007	2SB315E	2007	2SB328K
2007	2SB267F	2007	2SB291G	2007	2SB315F	2007	2SB328L
2007	2SB267G	2007	2SB291GN	2007	2SB315G	2007	2SB328M
2007	2SB267GN	2007	2SB291H	2007	2SB315GN	2007	2SB328OR
2007	2SB267H	2007	2SB291J	2007	2SB315H	2007	2SB328R
2007	2SB267J	2007	2SB291K	2007	2SB315J	2007	2SB328X
2007	2SB267K	2007	2SB291L	2007	2SB315L	2007	2SB328Y
2007	2SB267L	2007	2SB291M	2007	2SB315M	2007	2SB329
2007	2SB267M	2007	2SB291OR	2007	2SB315R	2007	2SB329A
2007	2SB267OR	2007	2SB291R	2007	2SB315X	2007	2SB329B
2007	2SB267R	2007	2SB291X	2007	2SB315Y	2007	2SB329C
2007	2SB267X	2007	2SB291Y	2007	2SB316	2007	2SB329D
2007	2SB267Y	2007	2SB292	2007	2SB316A	2007	2SB329E
2007	2SB268	2007	2SB292-BLUE	2007	2SB316B	2007	2SB329F
2007	2SB268A	2007	2SB292-GREEN	2007	2SB316C	2007	2SB329G
2007	2SB268B	2007	2SB292-ORANGE	2007	2SB316D	2007	2SB329GN
2007	2SB268C	2007	2SB292-RED	2007	2SB316E	2007	2SB329H
2007	2SB268D	2007	2SB292-YELLOW	2007	2SB316G	2007	2SB329J
2007	2SB268E	2007	2SB292A	2007	2SB316GN	2007	2SB329K
2007	2SB268F	2007	2SB292B	2007	2SB316H	2007	2SB329L
2007	2SB268G	2007	2SB292C	2007	2SB316OR	2007	2SB329M
2007	2SB268GN	2007	2SB292D	2007	2SB316R	2007	2SB329OR
2007	2SB268H	2007	2SB292E	2007	2SB316Y	2007	2SB329R
2007	2SB268J	2007	2SB292F	2007	2SB317	2007	2SB329X
2007	2SB268K	2007	2SB292G	2007	2SB32	2007	2SB32A
2007	2SB268L	2007	2SB292GN			2007	2SB32B
2007	2SB268M	2007	2SB292H			2007	2SB32C
2007	2SB268OR	2007	2SB292J				
2007	2SB268R	2007	2SB292K				
		2007	2SB292L				

RS 276-	Industry Standard No.	RS 276-	Industry Standard No.	RS 276-	Industry Standard No.	RS 276-	Industry Standard No.
2007	2SB32D	2007	2SB35	2007	2SB3760R	2007	2SB381L
2007	2SB32E	2007	2SB350	2007	2SB376R	2007	2SB381M
2007	2SB32F	2007	2SB350A	2007	2SB376X	2007	2SB381OR
2007	2SB32G	2007	2SB350B	2007	2SB376Y	2007	2SB381R
2007	2SB32GN	2007	2SB350C	2007	2SB377	2007	2SB381X
2007	2SB32H	2007	2SB350D	2007	2SB377B	2007	2SB381Y
2007	2SB32J	2007	2SB350E	2007	2SB37783A	2007	2SB382
2007	2SB32K	2007	2SB350F	2007	2SB37783B	2007	2SB382A
2007	2SB32M	2007	2SB350G	2007	2SB37783C	2007	2SB382B
2007	2SB32N	2007	2SB350GN	2007	2SB37783D	2007	2SB382BK
2007	2SB32OR	2007	2SB350H	2007	2SB37783E	2007	2SB382BN
2007	2SB32R	2007	2SB350J	2007	2SB37783F	2007	2SB382C
2007	2SB32X	2007	2SB350K	2007	2SB37783G	2007	2SB382D
2007	2SB32Y	2007	2SB350L	2007	2SB37783GN	2007	2SB382E
2007	2SB33	2007	2SB350M	2007	2SB37783H	2007	2SB382F
2007	2SB33(5)	2007	2SB350OR	2007	2SB37783J	2007	2SB382G
2007	2SB33-4	2007	2SB350R	2007	2SB37783K	2007	2SB382GN
2007	2SB33-5	2007	2SB350X	2007	2SB37783L	2007	2SB382H
2007	2SB335	2007	2SB350Y	2007	2SB37783M	2007	2SB382J
2007	2SB336	2007	2SB35A	2007	2SB37830R	2007	2SB382K
2007	2SB33A	2007	2SB35B	2007	2SB37783R	2007	2SB382L
2007	2SB33B	2007	2SB35C	2007	2SB37783X	2007	2SB382OR
2007	2SB33BK	2007	2SB35D	2007	2SB37783Y	2007	2SB382R
2007	2SB33C	2007	2SB35E	2007	2SB378A	2007	2SB382X
2007	2SB33D	2007	2SB35F	2007	2SB378B	2007	2SB382Y
2007	2SB33E	2007	2SB35G	2007	2SB378C	2007	2SB383
2007	2SB33F	2007	2SB35GN	2007	2SB378D	2007	2SB383-1
2007	2SB33G	2007	2SB35H	2007	2SB378E	2007	2SB383-2
2007	2SB33GN	2007	2SB35J	2007	2SB378F	2007	2SB383A
2007	2SB33H	2007	2SB35K	2007	2SB378GN	2007	2SB383B
2007	2SB33J	2007	2SB35L	2007	2SB378J	2007	2SB383C
2007	2SB33K	2007	2SB35M	2007	2SB378K	2007	2SB383D
2007	2SB33L	2007	2SB35OR	2007	2SB378L	2007	2SB383E
2007	2SB33M	2007	2SB35R	2007	2SB378M	2007	2SB383F
2007	2SB33OR	2007	2SB35X	2007	2SB378OR	2007	2SB383G
2007	2SB33R	2007	2SB35Y	2007	2SB378R	2007	2SB383GN
2007	2SB33X	2007	2SB364-OR	2007	2SB378X	2007	2SB383H
2007	2SB33Y	2007	2SB364A	2007	2SB378Y	2007	2SB383J
2007	2SB4	2007	2SB364B	2007	2SB379	2007	2SB383K
2007	2SB345	2007	2SB364C	2007	2SB379-2	2007	2SB383L
2007	2SB345A	2007	2SB364D	2007	2SB379A	2007	2SB383M
2007	2SB345B	2007	2SB364E	2007	2SB379B	2007	2SB38330R
2007	2SB345C	2007	2SB364F	2007	2SB379C	2007	2SB383R
2007	2SB345D	2007	2SB364G	2007	2SB379D	2007	2SB383X
2007	2SB345E	2007	2SB364GN	2007	2SB379E	2007	2SB383Y
2007	2SB345F	2007	2SB364H	2007	2SB379F	2007	2SB384A
2007	2SB345G	2007	2SB364J	2007	2SB379GN	2007	2SB384B
2007	2SB345GN	2007	2SB364K	2007	2SB379H	2007	2SB384C
2007	2SB345H	2007	2SB364L	2007	2SB379J	2007	2SB384D
2007	2SB345J	2007	2SB364M	2007	2SB379K	2007	2SB384E
2007	2SB345K	2007	2SB364R	2007	2SB379L	2007	2SB384F
2007	2SB345L	2007	2SB364X	2007	2SB379M	2007	2SB384G
2007	2SB345M	2007	2SB364Y	2007	2SB379OR	2007	2SB384GN
2007	2SB345OR	2007	2SB365	2007	2SB379R	2007	2SB384H
2007	2SB345R	2007	2SB365A	2007	2SB379Y	2007	2SB384J
2007	2SB345X	2007	2SB365B	2007	2SB37A	2007	2SB384K
2007	2SB345Y	2007	2SB365C	2007	2SB37B	2007	2SB384L
2007	2SB346	2007	2SB365D	2007	2SB37C	2007	2SB384M
2007	2SB346(Q)	2007	2SB365F	2007	2SB37D	2007	2SB3840R
2007	2SB346A	2007	2SB365G	2007	2SB37E	2007	2SB384R
2007	2SB346B	2007	2SB365GN	2007	2SB37G	2007	2SB384X
2007	2SB346C	2007	2SB365H	2007	2SB37GN	2007	2SB384Y
2007	2SB346D	2007	2SB365J	2007	2SB37H	2007	2SB385A
2007	2SB346E	2007	2SB365K	2007	2SB37J	2007	2SB385B
2007	2SB346F	2007	2SB365L	2007	2SB37K	2007	2SB385C
2007	2SB346G	2007	2SB365M	2007	2SB37L	2007	2SB385D
2007	2SB346GN	2007	2SB365OR	2007	2SB37M	2007	2SB385E
2007	2SB346H	2007	2SB365R	2007	2SB37OR	2007	2SB385F
2007	2SB346J	2007	2SB365X	2007	2SB37R	2007	2SB385G
2007	2SB346K	2007	2SB365Y	2007	2SB37X	2007	2SB385GN
2007	2SB346L	2007	2SB366	2007	2SB37Y	2007	2SB385H
2007	2SB346M	2007	2SB37	2007	2SB38	2007	2SB385J
2007	2SB346OR	2007	2SB370-Q	2007	2SB380	2007	2SB385K
2007	2SB346Q	2007	2SB370A	2007	2SB380A	2007	2SB385L
2007	2SB346R	2007	2SB370AA	2007	2SB380B	2007	2SB385M
2007	2SB346X	2007	2SB370AB	2007	2SB380C	2007	2SB3850R
2007	2SB346Y	2007	2SB370AC	2007	2SB380D	2007	2SB385R
2007	2SB347	2007	2SB370AHA	2007	2SB380E	2007	2SB385X
2007	2SB347A	2007	2SB370AHB	2007	2SB380F	2007	2SB385Y
2007	2SB347B	2007	2SB370B	2007	2SB380GN	2007	2SB386
2007	2SB347C	2007	2SB370C	2007	2SB380H	2007	2SB387
2007	2SB347D	2007	2SB370D	2007	2SB380K	2007	2SB387A
2007	2SB347E	2007	2SB370E	2007	2SB380L	2007	2SB387B
2007	2SB347F	2007	2SB370F	2007	2SB380OR	2007	2SB387C
2007	2SB347G	2007	2SB370G	2007	2SB380R	2007	2SB387D
2007	2SB347GN	2007	2SB370GN	2007	2SB380X	2007	2SB387E
2007	2SB347H	2007	2SB370H	2007	2SB381	2007	2SB387F
2007	2SB347J	2007	2SB370J	2007	2SB3812A	2007	2SB387G
2007	2SB347K	2007	2SB370K	2007	2SB3812B	2007	2SB387GN
2007	2SB347L	2007	2SB370L	2007	2SB3812C	2007	2SB387H
2007	2SB347M	2007	2SB370M	2007	2SB3812D	2007	2SB387J
2007	2SB347OR	2007	2SB370OR	2007	2SB3812F	2007	2SB387K
2007	2SB347R	2007	2SB370P	2007	2SB3812G	2007	2SB387L
2007	2SB347X	2007	2SB370PB	2007	2SB3812GN	2007	2SB387M
2007	2SB347Y	2007	2SB370R	2007	2SB3812H	2007	2SB387OR
2007	2SB348	2007	2SB370V	2007	2SB3812J	2007	2SB387R
2007	2SB348(Q)	2007	2SB370X	2007	2SB3812K	2007	2SB387X
2007	2SB348A	2007	2SB370Y	2007	2SB3812L	2007	2SB387Y
2007	2SB348B	2007	2SB371	2007	2SB3812M	2007	2SB389A
2007	2SB348C	2007	2SB371A	2007	2SB3812OR	2007	2SB389B
2007	2SB348D	2007	2SB371B	2007	2SB3812R	2007	2SB389BK
2007	2SB348E	2007	2SB371C	2007	2SB3812X	2007	2SB389C
2007	2SB348F	2007	2SB371D	2007	2SB3812Y	2007	2SB389D
2007	2SB348G	2007	2SB371E	2007	2SB3813A	2007	2SB389E
2007	2SB348GN	2007	2SB371F	2007	2SB3813B	2007	2SB389F
2007	2SB348H	2007	2SB371G	2007	2SB3813C	2007	2SB389G
2007	2SB348J	2007	2SB371GN	2007	2SB3813D	2007	2SB389GN
2007	2SB348K	2007	2SB371H	2007	2SB3813E	2007	2SB389H
2007	2SB348L	2007	2SB371J	2007	2SB3813F	2007	2SB389J
2007	2SB348M	2007	2SB371K	2007	2SB3813G	2007	2SB389K
2007	2SB348OR	2007	2SB371L	2007	2SB3813GN	2007	2SB389L
2007	2SB348Q	2007	2SB371M	2007	2SB3813H	2007	2SB389M
2007	2SB348R	2007	2SB371OR	2007	2SB3813J	2007	2SB3890R
2007	2SB348X	2007	2SB371R	2007	2SB3813K	2007	2SB389R
2007	2SB348Y	2007	2SB371X	2007	2SB3813L	2007	2SB389X
2007	2SB349	2007	2SB371Y	2007	2SB3813M	2007	2SB389Y
2007	2SB34A	2007	2SB372	2007	2SB38130R	2007	2SB38A
2007	2SB34B	2007	2SB373	2007	2SB3813R	2007	2SB38B
2007	2SB34C	2007	2SB376	2007	2SB3813X	2007	2SB38C
2007	2SB34D	2007	2SB376A	2007	2SB3813Y	2007	2SB38D
2007	2SB34E	2007	2SB376B	2007	2SB381A	2007	2SB38E
2007	2SB34F	2007	2SB376C	2007	2SB381B	2007	2SB38F
2007	2SB34G	2007	2SB376D	2007	2SB381C	2007	2SB38G
2007	2SB34GN	2007	2SB376E	2007	2SB381D	2007	2SB38GN
2007	2SB34H	2007	2SB376F	2007	2SB381E	2007	2SB38H
2007	2SB34J	2007	2SB376G	2007	2SB381F	2007	2SB38J
2007	2SB34K	2007	2SB376GN	2007	2SB381G	2007	2SB38K
2007	2SB34L	2007	2SB376J	2007	2SB381GN	2007	2SB38L
2007	2SB34M	2007	2SB376K	2007	2SB381H	2007	2SB38M
2007	2SB34N	2007	2SB376L	2007	2SB381K	2007	2SB380R
2007	2SB34OR	2007	2SB376M				2SB38R
2007	2SB34R					2007	2SB38X
2007	2SB34X					2007	2SB38Y
2007	2SB34Y					2007	2SB39
						2007	2SB392

RS 276-	Industry Standard No.	RS 276-	Industry Standard No.	RS 276-	Industry Standard No.	RS 276-	Industry Standard No.
2007	2SB392A	2007	2SB402F	2007	2SB417	2007	2SB455
2007	2SB392B	2007	2SB402G	2007	2SB417A	2007	2SB457
2007	2SB392C	2007	2SB402GN	2007	2SB417B	2007	2SB457-C
2007	2SB392E	2007	2SB402H	2007	2SB417C	2007	2SB457A
2007	2SB392F	2007	2SB402J	2007	2SB417D	2007	2SB457AC
2007	2SB392G	2007	2SB402K	2007	2SB417E	2007	2SB457B
2007	2SB392GN	2007	2SB402L	2007	2SB417F	2007	2SB457C
2007	2SB392H	2007	2SB402M	2007	2SB417G	2007	2SB457D
2007	2SB392J	2007	2SB402OR	2007	2SB417GN	2007	2SB457E
2007	2SB392K	2007	2SB402R	2007	2SB417H	2007	2SB457F
2007	2SB392L	2007	2SB402X	2007	2SB417J	2007	2SB457G
2007	2SB392M	2007	2SB402Y	2007	2SB417K	2007	2SB457GN
2007	2SB392OR	2007	2SB403	2007	2SB417L	2007	2SB457H
2007	2SB392R	2007	2SB403A	2007	2SB417M	2007	2SB457J
2007	2SB392X	2007	2SB403B	2007	2SB417OR	2007	2SB457K
2007	2SB392Y	2007	2SB403C	2007	2SB417R	2007	2SB457L
2007	2SB393	2007	2SB403D	2007	2SB417X	2007	2SB457M
2007	2SB393A	2007	2SB403E	2007	2SB417Y	2007	2SB457OR
2007	2SB393B	2007	2SB403F	2007	2SB421	2007	2SB457R
2007	2SB393C	2007	2SB403G	2007	2SB422	2007	2SB457X
2007	2SB393D	2007	2SB403GN	2007	2SB422A	2007	2SB457Y
2007	2SB393E	2007	2SB403H	2007	2SB422B	2007	2SB459
2007	2SB393F	2007	2SB403J	2007	2SB422C	2007	2SB459-0
2007	2SB393G	2007	2SB403K	2007	2SB422D	2007	2SB459A
2007	2SB393GN	2007	2SB403L	2007	2SB422E	2007	2SB459B
2007	2SB393H	2007	2SB403M	2007	2SB422F	2007	2SB459C
2007	2SB393J	2007	2SB403OR	2007	2SB422G	2007	2SB459C-2
2007	2SB393K	2007	2SB403R	2007	2SB422GN	2007	2SB459D
2007	2SB393L	2007	2SB403X	2007	2SB422H	2007	2SB459E
2007	2SB393M	2007	2SB403Y	2007	2SB422J	2007	2SB459F
2007	2SB393OR	2007	0000 2SB405	2007	2SB422K	2007	2SB459G
2007	2SB393R	2007	2SB405(K)	2007	2SB422L	2007	2SB459GN
2007	2SB393X	2007	2SB405-0	2007	2SB422M	2007	2SB459H
2007	2SB393Y	2007	2SB405-1	2007	2SB422OR	2007	2SB459J
2007	2SB394	2007	2SB405-2C	2007	2SB422R	2007	2SB459K
2007	2SB394A	2007	2SB405-3C	2007	2SB422X	2007	2SB459L
2007	2SB394B	2007	2SB405-4C	2007	2SB422Y	2007	2SB459M
2007	2SB394C	2007	2SB405-0	2007	2SB423	2007	2SB459OR
2007	2SB394D	2007	2SB405-OR	2007	2SB423A	2007	2SB459R
2007	2SB394E	2007	2SB405-R	2007	2SB423B	2007	2SB459X
2007	2SB394F	2007	2SB405A	2007	2SB423C	2007	2SB459Y
2007	2SB394G	2007	2SB405AG	2007	2SB423D	2007	2SB46
2007	2SB394GN	2007	2SB405B	2007	2SB423E	2007	2SB460A
2007	2SB394H	2007	2SB405BR	2007	2SB423F	2007	2SB460B
2007	2SB394K	2007	2SB405C	2007	2SB423G	2007	2SB461A
2007	2SB394L	2007	2SB405D	2007	2SB423GN	2007	2SB461B
2007	2SB394M	2007	2SB405DK	2007	2SB423H	2007	2SB461BL
2007	2SB394OR	2007	2SB405E	2007	2SB423J	2007	2SB461C
2007	2SB394R	2007	2SB405EK	2007	2SB423K	2007	2SB461D
2007	2SB394X	2007	2SB405F	2007	2SB423L	2007	2SB461E
2007	2SB394Y	2007	2SB405G	2007	2SB423M	2007	2SB461F
2007	2SB395	2007	2SB405GN	2007	2SB423OR	2007	2SB461G
2007	2SB395A	2007	2SB405H	2007	2SB423R	2007	2SB461H
2007	2SB395B	2007	2SB405J	2007	2SB423X	2007	2SB461J
2007	2SB395C	2007	2SB405K	2007	2SB423Y	2007	2SB461K
2007	2SB395D	2007	2SB405L	2007	2SB427	2007	2SB461L
2007	2SB395E	2007	2SB405M	2007	2SB428	2007	2SB461M
2007	2SB395F	2007	2SB405OR	2007	2SB43	2007	2SB461OR
2007	2SB395G	2007	2SB405P	2007	2SB431	2007	2SB461X
2007	2SB395GN	2007	2SB405R	2007	2SB439	2007	2SB461Y
2007	2SB395H	2007	2SB405RE	2007	2SB439A	2007	2SB47
2007	2SB395J	2007	2SB405X	2007	2SB439B	2007	2SB470
2007	2SB395K	2007	2SB405Y	2007	2SB439C	2007	2SB470A
2007	2SB395M	2007	2SB408	2007	2SB439D	2007	2SB470B
2007	2SB395OR	2007	2SB408A	2007	2SB439F	2007	2SB470C
2007	2SB395R	2007	2SB408B	2007	2SB439G	2007	2SB470D
2007	2SB395X	2007	2SB408C	2007	2SB439GN	2007	2SB470E
2007	2SB395Y	2007	2SB408D	2007	2SB439H	2007	2SB470F
2007	2SB396	2007	2SB408E	2007	2SB439J	2007	2SB470G
2007	2SB396A	2007	2SB408F	2007	2SB439L	2007	2SB470GN
2007	2SB396B	2007	2SB408G	2007	2SB439M	2007	2SB470H
2007	2SB396C	2007	2SB408GN	2007	2SB439OR	2007	2SB470J
2007	2SB396D	2007	2SB408H	2007	2SB439R	2007	2SB470K
2007	2SB396E	2007	2SB408J	2007	2SB439X	2007	2SB470L
2007	2SB396G	2007	2SB408K	2007	2SB439Y	2007	2SB470M
2007	2SB396GN	2007	2SB408L	2007	2SB43A	2007	2SB470OR
2007	2SB396H	2007	2SB408M	2007	2SB44	2007	2SB470R
2007	2SB396J	2007	2SB408OR	2007	2SB440	2007	2SB470X
2007	2SB396K	2007	2SB408R	2007	2SB440A	2007	2SB470Y
2007	2SB396L	2007	2SB408X	2007	2SB440B	2007	2SB475
2007	2SB396M	2007	2SB408Y	2007	2SB440C	2007	2SB475A
2007	2SB396OR	2007	2SB40A	2007	2SB440E	2007	2SB475B
2007	2SB396R	2007	2SB40B	2007	2SB440G	2007	2SB475C
2007	2SB396X	2007	2SB40C	2007	2SB440GN	2007	2SB475D
2007	2SB396Y	2007	2SB40D	2007	2SB440H	2007	2SB475E
2007	2SB398	2007	2SB40E	2007	2SB440K	2007	2SB475F
2007	2SB39A	2007	2SB40F	2007	2SB440L	2007	2SB475G
2007	2SB39B	2007	2SB40G	2007	2SB440M	2007	2SB475GN
2007	2SB39C	2007	2SB40GN	2007	2SB440OR	2007	2SB475H
2007	2SB39D	2007	2SB40H	2007	2SB440R	2007	2SB475J
2007	2SB39E	2007	2SB40J	2007	2SB440X	2007	2SB475K
2007	2SB39F	2007	2SB40K	2007	2SB440Y	2007	2SB475L
2007	2SB39G	2007	2SB40L	2007	2SB443A	2007	2SB475M
2007	2SB39GN	2007	2SB40M	2007	2SB443B	2007	2SB475OR
2007	2SB39H	2007	2SB40OR	2007	2SB443C	2007	2SB475P
2007	2SB39J	2007	2SB40R	2007	2SB443D	2007	2SB475PL
2007	2SB39K	2007	2SB40X	2007	2SB443E	2007	2SB475Q
2007	2SB39L	2007	2SB40Y	2007	2SB443F	2007	2SB475R
2007	2SB39M	2007	2SB415	2007	2SB443G	2007	2SB475X
2007	2SB390R	2007	2SB415-OR	2007	2SB443GN	2007	2SB475Y
2007	2SB39R	2007	2SB415A	2007	2SB443H	2007	2SB476
2007	2SB39X	2007	2SB415B	2007	2SB443J	2007	2SB477(C)
2007	2SB39Y	2007	2SB415C	2007	2SB443K	2007	2SB48
2007	2SB40	2007	2SB415D	2007	2SB443L	2007	2SB482
2007	2SB400	2007	2SB415E	2007	2SB443M	2007	2SB482A
2007	2SB400A	2007	2SB415F	2007	2SB443OR	2007	2SB482B
2007	2SB400B	2007	2SB415G	2007	2SB443R	2007	2SB482C
2007	2SB400BL	2007	2SB415GN	2007	2SB443X	2007	2SB482E
2007	2SB400K	2007	2SB415H	2007	2SB443Y	2007	2SB482F
2007	2SB401	2007	2SB415J	2007	2SB444OD	2007	2SB482G
2007	2SB401A	2007	2SB415K	2007	2SB444C	2007	2SB482GN
2007	2SB401B	2007	2SB415L	2007	2SB444D	2007	2SB482H
2007	2SB401C	2007	2SB415M	2007	2SB444E	2007	2SB482J
2007	2SB401D	2007	2SB415R	2007	2SB444F	2007	2SB482K
2007	2SB401E	2007	2SB415X	2007	2SB444G	2007	2SB482L
2007	2SB401F	2007	2SB415Y	2007	2SB444GN	2007	2SB482M
2007	2SB401G	2007	2SB416	2007	2SB444H	2007	2SB482V
2007	2SB401GN	2007	2SB416A	2007	2SB444K	2007	2SB482X
2007	2SB401H	2007	2SB416B	2007	2SB444L	2007	2SB482Y
2007	2SB401K	2007	2SB416D	2007	2SB444M	2007	2SB486
2007	2SB401L	2007	2SB416E	2007	2SB444OR	2007	2SB486A
2007	2SB401M	2007	2SB416F	2007	2SB444R	2007	2SB486C
2007	2SB401OR	2007	2SB416G	2007	2SB444X	2007	2SB486D
2007	2SB401R	2007	2SB416H	2007	2SB444Y	2007	2SB486E
2007	2SB401X	2007	2SB416J	2007	2SB450	2007	2SB486F
2007	2SB401Y	2007	2SB416K	2007	2SB450A	2007	2SB486G
2007	2SB402	2007	2SB416L	2007	2SB451	2007	2SB486GN
2007	2SB402A	2007	2SB416M	2007	2SB452	2007	2SB486H
2007	2SB402B	2007	2SB416OR	2007	2SB452A	2007	2SB486J
2007	2SB402C	2007	2SB416X	2007	2SB453	2007	2SB486K
2007	2SB402D	2007	2SB416Y	2007	2SB454	2007	2SB486L
2007	2SB402E						

RS 276-	Industry Standard No.	RS 276-	Industry Standard No.	RS 276-	Industry Standard No.	RS 276-	Industry Standard No.
2007	2SB486M	2007	2SB520R	2007	2SB68C	2007	2SB77L
2007	2SB486OR	2007	2SB52R	2007	2SB68D	2007	2SB77M
2007	2SB486R	2007	2SB52X	2007	2SB68E	2007	2SB77P
2007	2SB486X	2007	2SB52Y	2007	2SB68F	2007	2SB77PD
2007	2SB486Y	2007	2SB53	2007	2SB68G	2007	2SB77R
2007	2SB49	2007	2SB534	2007	2SB68GN	2007	2SB77RED
2007	2SB492A	2007	2SB534(A)	2007	2SB68H	2007	2SB77V
2007	2SB492C	2007	2SB534A	2007	2SB68J	2007	2SB77VRED
2007	2SB492D	2007	2SB535	2007	2SB68K	2007	2SB77X
2007	2SB492E	2007	2SB53A	2007	2SB68L	2007	2SB77Y
2007	2SB492F	2007	2SB53B	2007	2SB68M	2007	2SB78
2007	2SB492G	2007	2SB53C	2007	2SB68OR	2007	2SB78A
2007	2SB492GN	2007	2SB53D	2007	2SB68R	2007	2SB78B
2007	2SB492H	2007	2SB53E	2007	2SB68X	2007	2SB78C
2007	2SB492J	2007	2SB53F	2007	2SB68Y	2007	2SB78D
2007	2SB492K	2007	2SB53G	2007	2SB70	2007	2SB78E
2007	2SB492L	2007	2SB53GN	2007	2SB71	2007	2SB78F
2007	2SB492M	2007	2SB53H	2007	2SB71A	2007	2SB78G
2007	2SB492OR	2007	2SB53J	2007	2SB71B	2007	2SB78GN
2007	2SB492R	2007	2SB53K	2007	2SB71C	2007	2SB78H
2007	2SB492X	2007	2SB53L	2007	2SB71D	2007	2SB78J
2007	2SB492Y	2007	2SB53M	2007	2SB71E	2007	2SB78K
2007	2SB494	2007	2SB530R	2007	2SB71F	2007	2SB78L
2007	2SB495	2007	2SB53R	2007	2SB71G	2007	2SB78M
2007	2SB495A	2007	2SB53X	2007	2SB71GN	2007	2SB78OR
2007	2SB495B	2007	2SB53Y	2007	2SB71H	2007	2SB78R
2007	2SB495C	2007	2SB54	2007	2SB71J	2007	2SB78X
2007	2SB495D	2007	2SB54B	2007	2SB71K	2007	2SB78Y
2007	2SB495E	2007	2SB54E	2007	2SB71L	2007	2SB79
2007	2SB495F	2007	2SB54F	2007	2SB71M	2007	2SB79A
2007	2SB495G	2007	2SB54Y	2007	2SB71OR	2007	2SB79B
2007	2SB495GN	2007	2SB55	2007	2SB71R	2007	2SB79C
2007	2SB495H	2007	2SB56	2007	2SB71X	2007	2SB79D
2007	2SB495J	2007	2SB560	2007	2SB71Y	2007	2SB79E
2007	2SB495K	2007	2SB565	2007	2SB72	2007	2SB79F
2007	2SB495L	2007	2SB565A	2007	2SB72A	2007	2SB79G
2007	2SB495M	2007	2SB565B	2007	2SB72B	2007	2SB79GN
2007	2SB495OR	2007	2SB565C	2007	2SB72C	2007	2SB79H
2007	2SB495R	2007	2SB565D	2007	2SB72D	2007	2SB79K
2007	2SB495T	2007	2SB565E	2007	2SB72E	2007	2SB79L
2007	2SB495X	2007	2SB565F	2007	2SB72F	2007	2SB79M
2007	2SB495Y	2007	2SB565G	2007	2SB72G	2007	2SB790R
2007	2SB496	2007	2SB565GN	2007	2SB72GN	2007	2SB79X
2007	2SB496A	2007	2SB565H	2007	2SB72H	2007	2SB79Y
2007	2SB496B	2007	2SB565J	2007	2SB72J	2007	2SB85
2007	2SB496C	2007	2SB565K	2007	2SB72K	2007	2SB85A
2007	2SB496D	2007	2SB565L	2007	2SB72L	2007	2SB85B
2007	2SB496E	2007	2SB565M	2007	2SB72M	2007	2SB85C
2007	2SB496F	2007	2SB565OR	2007	2SB720R	2007	2SB85D
2007	2SB496G	2007	2SB565R	2007	2SB72R	2007	2SB85E
2007	2SB496GN	2007	2SB565X	2007	2SB72X	2007	2SB85F
2007	2SB496H	2007	2SB565Y	2007	2SB72Y	2007	2SB85G
2007	2SB496J	2007	2SB56A	2007	2SB73	2007	2SB85GN
2007	2SB496K	2007	2SB56B	2007	2SB73A-1	2007	2SB85H
2007	2SB496L	2007	2SB56C	2007	2SB73C	2007	2SB85J
2007	2SB496M	2007	2SB57	2007	2SB73D	2007	2SB85K
2007	2SB496OR	2007	2SB58	2007	2SB73E	2007	2SB85L
2007	2SB496R	2007	2SB59	2007	2SB73F	2007	2SB85M
2007	2SB496X	2007	2SB60	2007	2SB73G	2007	2SB850R
2007	2SB496Y	2007	2SB60A	2007	2SB73GN	2007	2SB85R
2007	2SB497	2007	2SB61	2007	2SB73GR	2007	2SB85X
2007	2SB497A	2007	2SB61A	2007	2SB73H	2007	2SB85Y
2007	2SB497B	2007	2SB61B	2007	2SB73J	2007	2SB87
2007	2SB497C	2007	2SB61C	2007	2SB73K	2007	2SB87A
2007	2SB497D	2007	2SB61D	2007	2SB73L	2007	2SB87B
2007	2SB497E	2007	2SB61E	2007	2SB73M	2007	2SB87C
2007	2SB497F	2007	2SB61F	2007	2SB730R	2007	2SB87D
2007	2SB497G	2007	2SB61G	2007	2SB73R	2007	2SB87E
2007	2SB497GN	2007	2SB61GN	2007	2SB73S	2007	2SB87F
2007	2SB497H	2007	2SB61H	2007	2SB73X	2007	2SB87G
2007	2SB497J	2007	2SB61J	2007	2SB73Y	2007	2SB87GN
2007	2SB497K	2007	2SB61K	2007	2SB74A	2007	2SB87H
2007	2SB497L	2007	2SB61L	2007	2SB74B	2007	2SB87J
2007	2SB497M	2007	2SB61M	2007	2SB74C	2007	2SB87K
2007	2SB497OR	2007	2SB61OR	2007	2SB74D	2007	2SB87L
2007	2SB497R	2007	2SB61R	2007	2SB74E	2007	2SB87M
2007	2SB497X	2007	2SB61X	2007	2SB74F	2007	2SB87OR
2007	2SB497Y	2007	2SB61Y	2007	2SB74G	2007	2SB87R
2007	2SB498	2007	2SB65	2007	2SB74GN	2007	2SB87X
2007	2SB498A	2007	2SB65A	2007	2SB74H	2007	2SB87Y
2007	2SB498B	2007	2SB65B	2007	2SB74J	2007	2SB89
2007	2SB498C	2007	2SB65C	2007	2SB74K	2007	2SB89AH
2007	2SB498D	2007	2SB65D	2007	2SB74L	2007	2SB89C
2007	2SB498E	2007	2SB65F	2007	2SB74M	2007	2SB89D
2007	2SB498F	2007	2SB65G	2007	2SB740R	2007	2SB89E
2007	2SB498G	2007	2SB65GN	2007	2SB74R	2007	2SB89F
2007	2SB498GN	2007	2SB65H	2007	2SB74X	2007	2SB89G
2007	2SB498H	2007	2SB65J	2007	2SB75	2007	2SB89GN
2007	2SB498J	2007	2SB65K	2007	2SB75A	2007	2SB89H
2007	2SB498K	2007	2SB65L	2007	2SB75AH	2007	2SB89J
2007	2SB498L	2007	2SB65M	2007	2SB75B	2007	2SB89K
2007	2SB498M	2007	2SB650R	2007	2SB75C	2007	2SB89L
2007	2SB498OR	2007	2SB65R	2007	2SB75F	2007	2SB89M
2007	2SB498R	2007	2SB65X	2007	2SB75LB	2007	2SB890R
2007	2SB498X	2007	2SB65Y	2007	2SB76	2007	2SB89R
2007	2SB498Y	2007	2SB66	2007	2SB76A	2007	2SB89X
2007	2SB50	2007	2SB66A	2007	2SB76B	2007	2SB89Y
2007	2SB51	2007	2SB66B	2007	2SB76C	2007	2SB90
2007	2SB516	2007	2SB66C	2007	2SB76D	2007	2SB91
2007	2SB516C	2007	2SB66D	2007	2SB76E	2007	2SB92
2007	2SB516CD	2007	2SB66E	2007	2SB76F	2007	2SB92A
2007	2SB516D	2007	2SB66F	2007	2SB76G	2007	2SB92B
2007	2SB516P	2007	2SB66G	2007	2SB76GN	2007	2SB92C
2007	2SB51A	2007	2SB66GN	2007	2SB76H	2007	2SB92D
2007	2SB51B	2007	2SB66J	2007	2SB76J	2007	2SB92E
2007	2SB51C	2007	2SB66K	2007	2SB76K	2007	2SB92F
2007	2SB51D	2007	2SB66L	2007	2SB76L	2007	2SB92G
2007	2SB51E	2007	2SB66M	2007	2SB76M	2007	2SB92GN
2007	2SB51F	2007	2SB660R	2007	2SB760R	2007	2SB92H
2007	2SB51G	2007	2SB66R	2007	2SB76R	2007	2SB92J
2007	2SB51GN	2007	2SB66X	2007	2SB76X	2007	2SB92K
2007	2SB51H	2007	2SB66Y	2007	2SB76Y	2007	2SB92L
2007	2SB51J	2007	2SB67	2007	2SB77	2007	2SB92M
2007	2SB51K	2007	2SB67A	2007	2SB77-OR	2007	2SB92X
2007	2SB51L	2007	2SB67B	2007	2SB77A/P	2007	2SB92Y
2007	2SB51M	2007	2SB67C	2007	2SB77AA	2007	2SB94
2007	2SB510R	2007	2SB67D	2007	2SB77AB	2007	2SB94A
2007	2SB51R	2007	2SB67E	2007	2SB77AC	2007	2SB94B
2007	2SB51X	2007	2SB67F	2007	2SB77AD	2007	2SB94C
2007	2SB51Y	2007	2SB67G	2007	2SB77AH	2007	2SB94D
2007	2SB52	2007	2SB67GN	2007	2SB77AP	2007	2SB94E
2007	2SB52A	2007	2SB67H	2007	2SB77B	2007	2SB94F
2007	2SB52B	2007	2SB67J	2007	2SB77B-11	2007	2SB94G
2007	2SB52C	2007	2SB67K	2007	2SB77C	2007	2SB94GN
2007	2SB52D	2007	2SB67L	2007	2SB77D	2007	2SB94H
2007	2SB52E	2007	2SB67M	2007	2SB77F	2007	2SB94J
2007	2SB52F	2007	2SB670R	2007	2SB77G	2007	2SB94K
2007	2SB52G	2007	2SB67R	2007	2SB77GN	2007	2SB94M
2007	2SB52GN	2007	2SB67X	2007	2SB77H	2007	2SB940R
2007	2SB52H	2007	2SB67Y	2007	2SB77K	2007	2SB94R
2007	2SB52J	2007	2SB68				
2007	2SB52K	2007	2SB68A				2SB94X
2007	2SB52L	2007	2SB68B				
2007	2SB52M						

RS 276-	Industry Standard No.
2007	28B94Y
2007	28B95
2007	28B95A
2007	28B95B
2007	28B95C
2007	28B95D
2007	28B95E
2007	28B95F
2007	28B95G
2007	28B95GN
2007	28B95H
2007	28B95J
2007	28B95L
2007	28B95M
2007	28B95OR
2007	28B95R
2007	28B95X
2007	28B95Y
2007	28B97
2007	28B97A
2007	28B97B
2007	28B97C
2007	28B97D
2007	28B97P
2007	28B97G
2007	28B97GN
2007	28B97GP
2007	28B97K
2007	28B97L
2007	28B97M
2007	28B97OR
2007	28B97R
2007	28B97Y
2007	28B98
2007	28B98A
2007	28B98B
2007	28B98C
2007	28B98D
2007	28B98E
2007	28B98F
2007	28B98G
2007	28B98GN
2007	28B98H
2007	28B98J
2007	28B98K
2007	28B98L
2007	28B98M
2007	28B98OR
2007	28B98R
2007	28B98X
2007	28B98Y
2007	28B99
2007	28B99A
2007	28B99B
2007	28B99C
2007	28B99D
2007	28B99H
2007	28B99P
2007	28B99G
2007	28B99GN
2007	28B99H
2007	28B99J
2007	28B99K
2007	28B99L
2007	28B99M
2007	28B99OR
2007	28B99R
2007	28B99X
2007	28B99Y
2007	28BP1
2007	28BP1A
2007	28BP2
2007	28BP2A
2007	28BM77
2007	28D-P1A
2007	28H203
2007	2T11
2007	2T12
2007	2T13
2007	2T14
2007	2T15
2007	2T16
2007	2T17
2007	2T2001
2007	2T21-
2007	2T22
2007	2T23
2007	2T230
2007	2T231
2007	2T24
2007	2T25
2007	2T26
2007	2T29
2007	2T3
2007	2T311
2007	2T312
2007	2T313
2007	2T314
2007	2T315
2007	2T323
2007	2T324
2007	2T383
2007	2TN33
2007	2TN32
2007	2TN45A
2007	2TN48
2007	2TN49
2007	2TN52
2007	2TN53
2007	2TN56
2007	2TN95
2007	2TN95A
2007	2V362
2007	2V363
2007	2V464
2007	2V465
2007	2V466
2007	2V467
2007	2V482
2007	2V483
2007	2V484
2007	2V486
2007	2V631
2007	2V632
2007	2V633
2007	03-0020-0
2007	03-0022
2007	03-035-0
2007	03-156B
2007	03-57-301
2007	03-57-302
2007	03-57-304
2007	003-H03
2007	3B15
2007	3B15-1
2007	3B-27
2007	3E-28
2007	3E-29
2007	003H03
2007	38B347
2007	04-00156-03
2007	4-201104770
2007	04-57-303
2007	4-66-1A7-1
2007	4-66-2A7-1
2007	4-66-3A7-1
2007	4-686163-3
2007	4-686195-3
2007	4-686196-3
2007	4-68681-2
2007	4-68681-3
2007	4-74-3A7-1
2007	4-88C17-1
2007	4-8P-2A7-567
2007	4-9L-4A7-1
2007	4JD1A17
2007	4JD1A73
2007	4JX1A520
2007	4JX1A520B
2007	4JX1A520C
2007	4JX1A520D
2007	4JX1A520E
2007	4JX1C1224
2007	4JX1C0850
2007	4JX1C0850A
2007	4JX1C0925
2007	4JX1B821
2007	4JX2A60
2007	6-0000155A
2007	6-000105
2007	6-13
2007	6-155
2007	6-53
2007	6-53/63
2007	6-5363
2007	6-53A
2007	6-53P
2007	6-60
2007	6-60-P
2007	6-60B
2007	6-60D
2007	6-60P
2007	6-61
2007	6-61-P
2007	6-61B
2007	6-61D
2007	6-61P
2007	6-62
2007	6-62-P
2007	6-62A
2007	6-62B
2007	6-63
2007	6-63A
2007	6-63T
2007	6-65
2007	6-65T
2007	6-66
2007	6-66T
2007	6-67
2007	6-67T
2007	6-87
2007	6-88
2007	6A10622
2007	6A10624
2007	6A11301
2007	6A11665
2007	6A11666B
2007	6A12515
2007	6A12516
2007	6A12517
2007	6A12678
2007	6A12684
2007	6A12685
2007	6A12989
2007	6A12990
2007	6D0000105
2007	6D122
2007	6D122R
2007	6D122T
2007	6D122TC
2007	6D122TH
2007	6D122U
2007	6D122V
2007	6D122W
2007	6D122Y
2007	6L122
2007	6X97047A02
2007	07-07119
2007	07-1075-01
2007	07-1075-02
2007	07-1156-03
2007	07-2012-04
2007	07-3012-04
2007	07-3015-05
2007	7-59-010/3477
2007	7-59-0103477
2007	7-59-029/3477
2007	7-59-0293477
2007	7-59-060/3477
2007	7-59-0603477
2007	07-6015-16
2007	8-0060
2007	8-0062
2007	8-0205400
2007	8-0205600
2007	8-022651U
2007	8-0236400
2007	8-0236430
2007	8-0243900
2007	8-619-030-007
2007	8-619-030-00B
2007	8-619-030-009
2007	8-619-030-014
2007	8-619-030-016
2007	8-619-030-017
2007	8-697-020-567
2007	8-697-020-568
2007	8-697-020-569
2007	8-721-323-00
2007	8-729-447-53
2007	8-905-605-016
2007	8-905-605-030
2007	8-905-605-032
2007	8-905-605-050
2007	8-905-605-051
2007	8-905-605-075
2007	8-905-605-090
2007	8-905-605-091
2007	8-905-605-120
2007	8-905-605-123
2007	8-905-605-125
2007	8-905-605-126
2007	8-905-605-127
2007	8-905-605-128
2007	8-905-605-129
2007	8-905-605-230
2007	8-905-605-232
2007	8-905-605-234
2007	8-905-605-250
2007	8-905-605-255
2007	8-905-605-260
2007	8-905-605-264
2007	8-905-605-266
2007	8-905-605-268
2007	8-905-605-269
2007	8-905-605-292
2007	8-905-605-305
2007	8-905-605-750
2007	8-905-606-800
2007	8-905-606-815
2007	8-905-606-817
2007	8-905-606-885
2007	8-905-613-010
2007	8-905-613-070
2007	8-905-613-071
2007	8-905-613-131
2007	8-905-613-132
2007	8-905-613-133
2007	8-905-613-160
2007	8-905-613-640
2007	8-905-613-710
2007	8-905-613-955
2007	8-905-615-156
2007	8A13718
2007	08P-12-12
2007	08P-2
2007	08P-2-12-7
2007	8P-2-70
2007	8P-2-70-12
2007	8P-2-70-12-7
2007	8P-20
2007	8P-21
2007	8P-22
2007	8P-23
2007	8P-24
2007	8P-25
2007	8P-26
2007	8P-27
2007	8P-28
2007	8P-29
2007	8P-2A
2007	8P-2AO
2007	8P-2AOR
2007	8P-2A1
2007	8P-2A19
2007	8P-2A2
2007	8P-2A21
2007	8P-2A3
2007	8P-2A3P
2007	8P-2A4
2007	8P-2A4-7
2007	8P-2A4-7B
2007	8P-2A5
2007	8P-2A5L
2007	8P-2A6
2007	8P-2A6-2
2007	8P-2A7
2007	8P-2A7-1
2007	8P-2A82
2007	8P-2A9
2007	8P-2A90
2007	8P-2AO
2007	8P-2AOR
2007	09-300005
2007	09-300017
2007	09-301001
2007	09-301002
2007	09-301002-6
2007	09-301003
2007	09-301004
2007	09-301005
2007	09-301006
2007	09-301007
2007	09-301008
2007	09-301008-18
2007	09-301009
2007	09-301012
2007	09-301014
2007	09-301016
2007	09-301019
2007	09-301020
2007	09-301022
2007	09-301023
2007	09-301025
2007	09-301025-6
2007	09-301026
2007	09-301027
2007	09-301032
2007	09-301036
2007	09-301048
2007	09-301048B
2007	09-301054
2007	09-301066
2007	09-301066
2007	09-301072
2007	09-301074
2007	09-30126
2007	09-30313
2007	09-30131
2007	9-511410100
2007	9-511410200
2007	9-511410900
2007	9-511413500
2007	9-5120A
2007	9-5201
2007	9-5203
2007	9-5204
2007	9-5209
2007	9-5212
2007	9-5213
2007	9-5214
2007	9-5217
2007	9-5218
2007	9-5222-1
2007	9-5224-1
2007	9-9104
2007	9-9201
2007	9-9202
2007	9-9203
2007	9.8011
2007	9L-4
2007	09L-4-12
2007	09L-4-12-7
2007	9L-4-70
2007	9L-4-70-12
2007	9L-4-70-12-7
2007	9L-40
2007	9L-41
2007	9L-42
2007	9L-43
2007	9L-44
2007	9L-45
2007	9L-46
2007	9L-47
2007	9L-48
2007	9L-49
2007	9L-4A
2007	9L-4AO
2007	9L-4AOR
2007	9L-4A1
2007	9L-4A2
2007	9L-4A21
2007	9L-4A3
2007	9L-4A3P
2007	9L-4A4-7
2007	9L-4A4-7B
2007	9L-4A5
2007	9L-4A5L
2007	9L-4A6
2007	9L-4A6-1
2007	9L-4A7
2007	9L-4A7-1
2007	9L-4A8
2007	9L-4A82
2007	9L-4A9
2007	9L-4A90
2007	98011
2007	10-28B54
2007	10-28B56
2007	011-H01
2007	011H01
2007	12-1-100
2007	12-1-102
2007	12-1-103
2007	12-1-104
2007	12-1-105
2007	12-1-106
2007	12-1-107
2007	12-1-120
2007	12-1-128
2007	12-1-148
2007	12-1-161
2007	12-1-162
2007	12-1-164
2007	12-1-179
2007	12-1-180
2007	12-1-184
2007	12-1-186
2007	12-1-191
2007	12-1-226
2007	12-1-227
2007	12-1-232
2007	12-1-234
2007	12-1-235
2007	12-1-236
2007	12-1-240
2007	12-1-241
2007	12-1-246
2007	12-1-254
2007	12-1-266
2007	12-1-267
2007	12-1-272
2007	12-1-273
2007	12-1-274
2007	12-1-275
2007	12-1-289
2007	12-1-73
2007	12-1-74
2007	12-1-75
2007	12-1-76
2007	12-1-78
2007	12-1-83
2007	12-1-91
2007	12-1-92
2007	12-1-93
2007	12-1-95
2007	12-1-96
2007	012-H02
2007	12A6240
2007	12A7239P1
2007	12A9275
2007	12A9275-1
2007	012H01
2007	13-14085-10
2007	13-14085-11
2007	13-14085-12
2007	13-14085-18
2007	13-14085-23
2007	13-14085-25
2007	13-14085-35
2007	13-14085-60
2007	13-14085-71
2007	13-14085-9
2007	13-14888-3
2007	13-15805-1
2007	13-15836-1
2007	13-18032-1
2007	13-18033-1
2007	13-18304-1
2007	13-18671-1
2007	13-18671-1A
2007	13-18671-1B
2007	13-18671-1C
2007	13-18944-1
2007	13-18944-2
2007	13-23785-1
2007	13-35792-1
2007	13-50484-1
2007	13-50486-1
2007	13-50631-1
2007	13-50944-1
2007	13-67599-3
2007	13-67599-3/2439-2
2007	13-67599-3/3464
2007	13-94096-2
2007	14-557-10
2007	14-564-08
2007	14-576-10
2007	14-577-10
2007	14-584-01

RS 276-	Industry Standard No.	RS 276-	Industry Standard No.	RS 276-	Industry Standard No.	RS 276-	Industry Standard No.
2007	14-602-04	2007	022-3504-060	2007	34-6000-4	2007	46-8681-2
2007	14-602-05	2007	022-3505-910	2007	34-6000-5	2007	46-8681-3
2007	14-602-05A	2007	022-3640-253	2007	34-6000-6	2007	46-869-3
2007	14-602-06	2007	022.3504-040	2007	34-6000-7	2007	48-10073A01
2007	14-602-07	2007	022-3504-060	2007	34-6000-8	2007	48-10073A02
2007	14-602-08	2007	022.3505-910	2007	34-6000-83	2007	48-10074A01
2007	14-602-09	2007	22N1319	2007	34-6000-84	2007	48-10074A02
2007	14-602-10	2007	23-5014	2007	34-6000-85	2007	48-10074A03
2007	14-602-15	2007	23-5017	2007	34-6001-10	2007	48-124158
2007	14-602-51	2007	23-6001-16	2007	34-6001-13	2007	48-124159
2007	14MW69	2007	23-6001-17	2007	34-6001-14	2007	48-124175
2007	15-2221011	2007	23-6001-20	2007	34-6001-16	2007	48-124219
2007	15-22210111	2007	23-6001-21	2007	34-6001-17	2007	48-124258
2007	15-22211200	2007	23-6001-23	2007	34-6001-18	2007	48-124259
2007	15-22211328	2007	24MW1040	2007	34-6001-19	2007	48-124275
2007	15-22216500	2007	24MW107	2007	34-6001-20	2007	48-124279
2007	16-207190405	2007	24MW1083	2007	34-6001-21	2007	48-124286
2007	16A787	2007	24MW1084	2007	34-6001-22	2007	48-124297
2007	18P-2A82	2007	24MW11	2007	34-6001-23	2007	48-124300
2007	019-003324	2007	24MW111	2007	34-6001-26	2007	48-124303
2007	019-003342	2007	24MW1115	2007	34-6001-29	2007	48-124304
2007	019-003343	2007	24MW1116(PNP)	2007	34-6001-30	2007	48-124306
2007	019-003415	2007	24MW115	2007	34-6001-31	2007	48-124307
2007	019-003416	2007	24MW116	2007	34-6001-33	2007	48-124308
2007	19-020-003	2007	24MW132	2007	34-6001-41	2007	48-124309
2007	19-020-007	2007	24MW15	2007	34-6001-42	2007	48-124314
2007	19-020-015	2007	24MW16	2007	34-6001-43	2007	48-124315
2007	19-020-033	2007	24MW178	2007	34-6001-44	2007	48-124318
2007	19-020-034	2007	24MW179	2007	34-6001-47	2007	48-124319
2007	19-020-035	2007	24MW185	2007	34-6001-66	2007	48-124322
2007	19-020-036	2007	24MW187	2007	34-6001-7	2007	48-124327
2007	19-3415	2007	24MW256	2007	34-6001-72	2007	48-124328
2007	19-3416	2007	24MW263	2007	34-6001-76	2007	48-124343
2007	19A115077-P1	2007	24MW27	2007	34-6001-8	2007	48-124544
2007	19A115077-P2	2007	24MW28	2007	34-6001-9	2007	48-124545
2007	19A115208	2007	24MW29	2007	34-6008	2007	48-124553
2007	19A115208-P1	2007	24MW34	2007	34-6009	2007	48-124554
2007	19A115208-P2	2007	24MW370	2007	34-6015-44A	2007	48-124555
2007	19A115281-P1	2007	24MW384	2007	34-6016-11	2007	48-124557
2007	19A115301-P1	2007	24MW43	2007	34-6016-50	2007	48-124558
2007	19A115301-P2	2007	24MW441	2007	35P1	2007	48-124559
2007	19A115548-P1	2007	24MW598	2007	35P2	2007	48-124570
2007	19A115556-P1	2007	24MW599	2007	35P2C	2007	48-124571
2007	19A115674-P1	2007	24MW60	2007	35T1	2007	48-124573
2007	19A115674-P2	2007	24MW600	2007	36J003-1	2007	48-124578
2007	19AR14-1	2007	24MW601	2007	36P1	2007	48-124579
2007	19AR14-2	2007	24MW608	2007	36P1C	2007	48-124580
2007	19AR16-1	2007	24MW613	2007	36P1F	2007	48-124589
2007	19AR16-2	2007	24MW614	2007	36P2F	2007	48-124598
2007	19AR19-1	2007	24MW615	2007	36P3	2007	48-12443
2007	19AR19-2	2007	24MW69	2007	36P3A	2007	48-124444
2007	19AR25	2007	24MW70	2007	36P3C	2007	48-124445
2007	19AR26	2007	24MW741	2007	36P4	2007	48-124446
2007	19AR27	2007	24MW77	2007	36P4C	2007	48-125229
2007	19AR32	2007	24MW777	2007	36P5	2007	48-125230
2007	19AR7-1	2007	24MW778	2007	36P5C	2007	48-125231
2007	19AR7-2	2007	24MW780	2007	36P6C	2007	48-125232
2007	19B2000129-P1	2007	24MW781	2007	36P7	2007	48-125237
2007	19B2000132-P1	2007	24MW782	2007	36P7C	2007	48-125238
2007	19B2000132-P2	2007	24MW783	2007	36P7T	2007	48-125239
2007	19B2000132-P3	2007	24MW789	2007	36P8	2007	48-125240
2007	19B2000132-P4	2007	24MW799	2007	36P8C	2007	48-125242
2007	19B200054-P1	2007	24MW819	2007	36T1	2007	48-125271
2007	19B200061-P1	2007	24MW824	2007	38T1	2007	48-125276
2007	19B200061-P2	2007	24MW83	2007	39A9	2007	48-125282
2007	19B200061-P3	2007	24MW84	2007	39T1	2007	48-125285
2007	19B200061-P4	2007	24MW853	2007	40-601	2007	48-125294
2007	19B200063-P1	2007	24MW856	2007	40P1	2007	48-125296
2007	19B200129-P1	2007	24MW857	2007	40P2	2007	48-128094
2007	19B200132-P1	2007	24MW892	2007	42-17143	2007	48-128303
2007	19B200132-P2	2007	24MW893	2007	42-19671	2007	48-13407
2007	19B200132-P3	2007	24MW973	2007	42-19862A	2007	48-134408
2007	19B200132-P4	2007	025-100031	2007	42-19863	2007	48-134415
2007	19B200210-P1	2007	025-100031	2007	42-19863A	2007	48-134416
2007	19B200210-P2	2007	25B378	2007	42-19864	2007	48-134417
2007	19B200210-P3	2007	25C1	2007	42-19864A	2007	48-134419
2007	19O300073-P1	2007	026-100005	2007	42-20222	2007	48-134420
2007	19O300073-P2	2007	026-100018	2007	42-21405	2007	48-134421
2007	19O300073-P3	2007	026-100012	2007	42-21406	2007	48-134422
2007	19O300073-P4	2007	26MW613	2007	42-22534	2007	48-134423
2007	19O300073-P5	2007	26C1	2007	42-22535	2007	48-134424
2007	19O300073-P6	2007	277401	2007	42-22535Q	2007	48-134425
2007	19O300074-P2	2007	277402	2007	42-23622	2007	48-134426
2007	19O300128-P1	2007	277403	2007	42-23967	2007	48-134427
2007	19O300128-P2	2007	277404	2007	42-23967P	2007	48-134429
2007	19O300128-P3	2007	277405	2007	42-27376	2007	48-134432
2007	19O300128-P4	2007	29V0058H03	2007	42-28057	2007	48-134433
2007	19O300128-P5	2007	29V008M01	2007	42X230	2007	48-134443
2007	19O300128-P6	2007	29V011H01	2007	42X233	2007	48-134444
2007	19O300128-P7	2007	29V012H01	2007	42X308	2007	48-134445
2007	19O300128-P8	2007	29V12H01	2007	42X309	2007	48-134446
2007	19O300138-P4	2007	30V-H6	2007	42X311	2007	48-134450
2007	19O300138-P8	2007	30V-HG	2007	43P1	2007	48-134458
2007	19U-4A82	2007	31-0006	2007	43P2	2007	48-134462
2007	020-1110-025	2007	31-0008	2007	43P3	2007	48-134466
2007	20-1680-189	2007	31-0017	2007	43P4	2007	48-134468
2007	20A0007	2007	31-0018	2007	43P4C	2007	48-134469
2007	20A0009	2007	31-0025	2007	43P6	2007	48-134470
2007	20A0015	2007	31-0026	2007	43P6A	2007	48-134471
2007	20C71	2007	31-0033	2007	43P6C	2007	48-134472
2007	20C72	2007	31-0035	2007	43P7	2007	48-134473
2007	20V-HG	2007	31-0053	2007	43P7A	2007	48-134474
2007	21-34	2007	31-0070	2007	43P7C	2007	48-134475
2007	21-36	2007	31-0075	2007	45X1A502C	2007	48-134476
2007	21-37	2007	31-0105	2007	45X1A520C	2007	48-134477
2007	21A005-000	2007	31-0107	2007	45X2	2007	48-134482
2007	21A015-001	2007	31-0148	2007	46-163-3	2007	48-134483
2007	21A015-006	2007	31-0153	2007	46-840-3	2007	48-134494
2007	21A038-000	2007	31-0172	2007	46-8610-3	2007	48-134495
2007	21A039-000	2007	31-0182	2007	46-8611	2007	48-134496
2007	21A040-000	2007	31-0183	2007	46-8611-3	2007	48-134499
2007	21A040-005	2007	31-0188	2007	46-8616-3	2007	48-134500
2007	21A040-014	2007	31-0189	2007	46-8165-3	2007	48-134501
2007	21A040-021	2007	31-0205	2007	46-8166-3	2007	48-134509
2007	21A040-022	2007	31-0229	2007	46-8195-3	2007	48-134510
2007	21A040-036	2007	31-0247	2007	46-8196-3	2007	48-134512
2007	21A040-057	2007	31-025	2007	46-8256-3	2007	48-134535
2007	21A040-058	2007	31-0255	2007	46-8631-3	2007	48-134538
2007	21A040-060	2007	31-22005400	2007	46-8673-3	2007	48-134539
2007	21A040-061	2007	33-1000-00	2007	46-8430-3	2007	48-134540
2007	21A040-079	2007	33-1001-00	2007	46-8660-3	2007	48-134541
2007	21A040-081	2007	33-1009-01	2007	46-8664-3	2007	48-134542
2007	21A040-36	2007	33-1019-00	2007	46-8665-3	2007	48-134543
2007	21A051-000	2007	33-1020-00	2007	46-8666-3	2007	48-134544
2007	21A053-000	2007	33-1021-00	2007	46-8668-3	2007	48-134553
2007	21A054-000	2007	34-6000-15	2007	46-8679-1	2007	48-134554
2007	21A055-000	2007	34-6000-27	2007	46-8679-2	2007	48-134555
2007	21A063-000	2007	34-6000-28	2007	46-8679-3	2007	48-134556
2007	21A074-000	2007	34-6000-29	2007	46-8680-1	2007	48-134557
2007	21MO07	2007	34-6000-30	2007	46-8680-2	2007	48-134558
2007	21MW132	2007	34-6000-31	2007	46-8680-3		
2007	22-002006	2007	34-6000-32	2007	46-8681-1		
2007	22-002007	2007	34-6000-33				
2007	022-3504-040	2007	34-6000-34				

RS 276-	Industry Standard No.	RS 276-	Industry Standard No.	RS 276-	Industry Standard No.	RS 276-	Industry Standard No.
2007	48-134559	2007	48P63077A03	2007	57A2-60	2007	59P2C
2007	48-134562	2007	48P63078A62	2007	57A2-61	2007	61-1130
2007	48-134563	2007	48P63078A86(PNP)	2007	57A2-72	2007	61-1131
2007	48-134564	2007	48R134407	2007	57A2-78	2007	61-1215
2007	48-134565	2007	48R134573	2007	57A2-83	2007	61-1907
2007	48-134567	2007	48R134621	2007	57A2-88	2007	61-1934
2007	48-134572	2007	48R134632	2007	57A2.4	2007	61-1935
2007	48-134573	2007	48R869148	2007	57A3-4	2007	61-607
2007	48-134603	2007	48R869249	2007	57A3-5	2007	61-608
2007	48-134604	2007	48R869250	2007	57A3-6	2007	61-654
2007	48-134610	2007	48R869253	2007	57A5-3	2007	61-655
2007	48-134621	2007	48R869282	2007	57A5-5	2007	61-656
2007	48-134625	2007	48R869475A	2007	57A6(PNP)	2007	61-928
2007	48-134626	2007	48S134404	2007	57A6-1	2007	61-929
2007	48-134631	2007	48S134408	2007	57A6-22	2007	61B0015-1
2007	48-134632	2007	48S134860	2007	57A6-25	2007	61B004-1
2007	48-134633	2007	48S134861	2007	57A6-6A	2007	61B005-1
2007	48-134636	2007	48S134862	2007	57A6-6B	2007	61B006-1
2007	48-134637	2007	48X97046A34	2007	57A6-6C	2007	61B009-1
2007	48-134641	2007	48X97046A48	2007	57A6B	2007	61B015-1
2007	48-134655	2007	48X97046A53	2007	57B100-7	2007	61B016-1
2007	48-134656	2007	48X97046A54	2007	57B1127	2007	61B017-1
2007	48-134657	2007	48X97046A55	2007	57B1143	2007	61B018-1
2007	48-134956	2007	48X97162A06	2007	57B168	2007	61B019-1
2007	48-137193	2007	48X97162A07	2007	57B169	2007	61B020-1
2007	48-137199	2007	48X97162A36	2007	57B170	2007	61B021-1
2007	48-171-A06	2007	48X97162A37	2007	57B2-10	2007	61B022-2
2007	48-17162A06	2007	48X97238A05	2007	57B2-12	2007	61B022-3
2007	48-17162A10	2007	49L-4A5	2007	57B2-15	2007	61B023-1
2007	48-17162A17	2007	50B173-C	2007	57B2-16	2007	61B026-1
2007	48-17162A22	2007	50B173-8	2007	57B2-21	2007	61B027-1
2007	48-17271A03	2007	50B175A	2007	57B2-23	2007	61B45-14
2007	48-21598B01	2007	50B175B	2007	57B2-25	2007	61C002-1
2007	48-35P1	2007	50B175C	2007	57B2-29	2007	61C005-1
2007	48-36P1	2007	50B324	2007	57B2-3	2007	61L004-1
2007	48-36P3	2007	50B364	2007	57B2-32	2007	62-128543
2007	48-39P3	2007	50B415	2007	57B2-34	2007	62-13258
2007	48-43351A01	2007	50B423	2007	57B2-36	2007	62-13494
2007	48-43354A83	2007	50B54	2007	57B2-39	2007	62-16918
2007	48-43P3	2007	50BU75-C	2007	57B2-4	2007	62-18415
2007	48-43P4	2007	51D170	2007	57B2-43	2007	62-18416
2007	48-56P1	2007	51D176	2007	57B2-44	2007	62-18417
2007	48-63029A17	2007	51D188	2007	57B2-45	2007	62-18420
2007	48-63029A18	2007	51D189	2007	57B2-52	2007	62-18421
2007	48-63029A19	2007	52004	2007	57B2-6	2007	62-18423
2007	48-63029A91	2007	52D189	2007	57B2-60	2007	62-18424
2007	48-63029A92	2007	53P157	2007	57B2-7	2007	62-18430
2007	48-63029A93	2007	55-1016	2007	57B2-72	2007	63-10037
2007	48-63029A94	2007	55-1029	2007	57B2-78	2007	63-10098
2007	48-63044A05	2007	55-1031	2007	57B2-83	2007	63-10147
2007	48-63075A76	2007	56-8091	2007	57B2-88	2007	63-10151
2007	48-63077A03	2007	56-8091A	2007	57B3-4	2007	63-10152
2007	48-63078A59	2007	56-8091B	2007	57B3-5	2007	63-10153
2007	48-63078A60	2007	56-8091C	2007	57B3-6	2007	63-10154
2007	48-63078A61	2007	56-8091D	2007	57B6	2007	63-10156
2007	48-63078A62	2007	56-8092	2007	5705-3	2007	63-10158
2007	48-63078A64	2007	56-8092A	2007	5705-5	2007	63-10159
2007	48-63078A66	2007	56-8092B	2007	5706-1	2007	63-10200
2007	48-63078A68	2007	56-8099	2007	5706-22	2007	63-10384
2007	48-63078A69	2007	56-8101	2007	5706-25	2007	63-10408
2007	48-63082A15	2007	57-4015-27	2007	5706-6A	2007	63-11073
2007	48-63082A16	2007	57-4015-28	2007	5706-6B	2007	63-11144
2007	48-63084A03	2007	57A1-104	2007	5706-6C	2007	63-11474
2007	48-63084A04	2007	57A1-105	2007	5706B	2007	63-11497
2007	48-63084A05	2007	57A1-106	2007	57D1-104	2007	63-11585
2007	48-63086A19	2007	57A1-11	2007	57D1-105	2007	63-11586
2007	48-644678	2007	57A1-116	2007	57D1-106	2007	63-11661
2007	48-644679	2007	57A1-117	2007	57D1-11	2007	63-12316
2007	48-869001	2007	57A1-118	2007	57D1-111	2007	63-12317
2007	48-869148	2007	57A1-121	2007	57D1-116	2007	63-12669
2007	48-869198	2007	57A1-14	2007	57D1-117	2007	63-12670
2007	48-869249	2007	57A1-17	2007	57D1-118	2007	63-12698
2007	48-869250	2007	57A1-18	2007	57D1-121	2007	63-12876
2007	48-869253	2007	57A1-19	2007	57D1-14	2007	63-12880
2007	48-869282	2007	57A1-20	2007	57D1-17	2007	63-12881
2007	48-869475	2007	57A1-21	2007	57D1-18	2007	63-12945
2007	48-869475A	2007	57A1-23	2007	57D1-19	2007	63-12947
2007	48-97046A02	2007	57A1-26	2007	57D1-20	2007	63-13323
2007	48-97046A03	2007	57A1-27	2007	57D1-21	2007	63-13840
2007	48-97046A10	2007	57A1-28	2007	57D1-23	2007	63-16918
2007	48-97046A32	2007	57A1-34	2007	57D1-26	2007	63-18416
2007	48-97046A33	2007	57A1-38	2007	57D1-27	2007	63-18420
2007	48-97046A34	2007	57A1-42	2007	57D1-28	2007	63-18421
2007	48-97046A37	2007	57A1-43	2007	57D1-34	2007	63-18430
2007	48-97046A53	2007	57A1-44	2007	57D1-38	2007	63-25041
2007	48-97046A54	2007	57A1-53	2007	57D1-40	2007	63-25179
2007	48-97046A55	2007	57A1-56	2007	57D1-42	2007	63-25180
2007	48-97046A56	2007	57A1-58	2007	57D1-43	2007	63-25181
2007	48-97046A57	2007	57A1-59	2007	57D1-44	2007	63-25182
2007	48-97127A05	2007	57A1-60	2007	57D1-53	2007	63-25261
2007	48-97127A09	2007	57A1-66	2007	57D1-56	2007	63-25281
2007	48-97127A20	2007	57A1-7	2007	57D1-58	2007	63-25282
2007	48-97127A22	2007	57A1-70	2007	57D1-59	2007	63-25720
2007	48-97127A30	2007	57A1-71	2007	57D1-66	2007	63-25727
2007	48-97127A31	2007	57A1-77	2007	57D1-68	2007	63-25728
2007	48-97127A32	2007	57A1-79	2007	57D1-7	2007	63-25729
2007	48-97162A06	2007	57A1-8	2007	57D1-70	2007	63-25942
2007	48-97162A07	2007	57A1-82	2007	57D1-71	2007	63-25944
2007	48-97162A08	2007	57A1-83	2007	57D1-77	2007	63-25946
2007	48-97162A11	2007	57A1-90	2007	57D1-79	2007	63-26849
2007	48-97162A16	2007	57A1-91	2007	57D1-8	2007	63-26851
2007	48-97162A18	2007	57A1-92	2007	57D1-82	2007	63-27278
2007	48-97162A19	2007	57A1-93	2007	57D1-83	2007	63-27279
2007	48-97162A20	2007	57A1-95	2007	57D1-90	2007	63-27280
2007	48-97162A24	2007	57A1-97	2007	57D1-91	2007	63-27281
2007	48-97162A25	2007	57A100-7	2007	57D1-92	2007	63-27367
2007	48-97162A34	2007	57A1127	2007	57D1-93	2007	63-28390
2007	48-97221A01	2007	57A1143	2007	57D1-95	2007	63-28399
2007	48-97221A02	2007	57A126	2007	57D1-97	2007	63-29662
2007	48-97221A03	2007	57A143	2007	57D1127	2007	63-29663
2007	48-97221A04	2007	57A168-1	2007	57D1143	2007	63-29664
2007	48-97221A05	2007	57A169	2007	57D126	2007	63-29665
2007	48-97238A05	2007	57A170	2007	57D143	2007	63-29666
2007	48-97238A07	2007	57A189	2007	57D156	2007	63-29863
2007	48-97271A01	2007	57A2-15	2007	57D169	2007	63-7246
2007	48-97271A02	2007	57A2-23	2007	57D170	2007	63-7247
2007	48-97271A03	2007	57A2-24	2007	57D180	2007	63-7396
2007	48-97271A04	2007	57A2-25	2007	57D184	2007	63-7397
2007	48-97271A2	2007	57A2-29	2007	57D189	2007	63-7398
2007	48-97271A3	2007	57A2-32	2007	57D3-6	2007	63-7399
2007	48-97271A4	2007	57A2-33	2007	57D68	2007	63-7420
2007	48-97271A5	2007	57A2-34	2007	57L1-5	2007	63-7547
2007	48-97271A6	2007	57A2-36	2007	57L1-8	2007	63-7564
2007	48A124315	2007	57A2-39	2007	57M1-35	2007	63-7596
2007	48A124327	2007	57A2-4	2007	57M1-5	2007	63-7871
2007	48K134450	2007	57A2-43	2007	57M1-6	2007	63-7872
2007	48K134458	2007	57A2-44	2007	57M1-8	2007	63-7873
2007	48K134482	2007	57A2-45	2007	57M2-3	2007	63-8120
2007	48K134483	2007	57A2-46	2007	57M2-4	2007	63-8380
2007	48K56P1	2007	57A2-52	2007	57M2-8	2007	63-8703
2007	48K869001			2007	57P1	2007	63-8704
2007	48P-2A5					2007	63-9519
2007	48P1					2007	63-9520
						2007	63-9521

RS 276-	Industry Standard No.	RS 276-	Industry Standard No.	RS 276-	Industry Standard No.	RS 276-	Industry Standard No.
2007	63-9659	2007	74-33	2007	86-98-2	2007	0000104
2007	63P3	2007	74-34	2007	86A318	2007	110-01563-00
2007	64T1	2007	74-35	2007	86X00011-001	2007	112-001
2007	65T1	2007	74-36	2007	86X0014-001	2007	112-003
2007	66-1	2007	74-37	2007	86X0017-001	2007	112-004
2007	066-1-12	2007	74-38	2007	86X0018-001	2007	112-034923
2007	066-1-12-7	2007	74-39	2007	86X0037-002	2007	112-200525
2007	66-1-70	2007	74-3A	2007	86X3	2007	114P3U
2007	66-1-70-12	2007	74-3AO	2007	87-0018	2007	116-091
2007	66-1-70-12-7	2007	74-3AOR	2007	87-0019	2007	116-201
2007	66-10	2007	74-3A1	2007	87-0020	2007	116-203
2007	66-11	2007	74-3A19	2007	87-0021	2007	116-206
2007	66-12	2007	74-3A2	2007	88C	2007	116-685
2007	66-13	2007	74-3A21	2007	088C-12	2007	116-686
2007	66-14	2007	74-3A3	2007	088C-12-7	2007	116-757
2007	66-15	2007	74-3A3P	2007	88C-70	2007	116-997
2007	66-16	2007	74-3A4	2007	88C-70-12	2007	117-2
2007	66-17	2007	74-3A4-7	2007	88C-70-12-7	2007	120-00-19
2007	66-18	2007	74-3A4-7B	2007	8800	2007	120-001192
2007	66-19	2007	74-3A5	2007	8801	2007	120-001195
2007	66-1A	2007	74-3A5L	2007	88010	2007	120-002520
2007	66-1AO	2007	74-3A6	2007	88010R	2007	120-002521
2007	66-1AOR	2007	74-3A6-3	2007	88011	2007	120-002748
2007	66-1A1	2007	74-3A7	2007	88C119	2007	120-004493
2007	66-1A19	2007	74-3A7-1	2007	88012	2007	120-004494
2007	66-1A2	2007	74-3A8	2007	88C121	2007	120-004495
2007	66-1A21	2007	74-3A82	2007	88013	2007	120-004727
2007	66-1A3	2007	74-3A9	2007	88C13P	2007	120-004728
2007	66-1A3P	2007	74-3A9G	2007	88014	2007	120-004729
2007	66-1A4	2007	75-461	2007	88C14-7	2007	120-01193
2007	66-1A4-7	2007	77	2007	88C14-7B	2007	120-190
2007	66-1A4-7B	2007	77-271025-1	2007	88015	2007	121-1030
2007	66-1A5	2007	77-271026-1	2007	88C15L	2007	121-1033
2007	66-1A5L	2007	77-271027-1	2007	88016	2007	121-1034
2007	66-1A6	2007	77-271036-1	2007	88C16D	2007	121-1035
2007	66-1A6-3	2007	77-271037-1	2007	88017	2007	121-1036
2007	66-1A7	2007	77-271039-1	2007	88C17-1	2007	121-106
2007	66-1A7-1	2007	77-273004-1	2007	88018	2007	121-107
2007	66-1A8	2007	78BLK	2007	88C182	2007	121-11
2007	66-1A82	2007	78GRN	2007	88019	2007	121-12
2007	66-1A9	2007	78RED	2007	88C19G	2007	121-120
2007	66-1A9G	2007	78YEL	2007	88C2	2007	121-128
2007	66-1AA19	2007	79P1	2007	88C3	2007	121-1330
2007	66-2	2007	80-205400	2007	88C4	2007	121-1350
2007	066-2-12	2007	80-205600	2007	88C5	2007	121-1360
2007	066-2-12-7	2007	80-2226314	2007	88C6	2007	121-1390
2007	66-2-70	2007	80-236400	2007	88C7	2007	121-14
2007	66-2-70-12	2007	80-236430	2007	88C8	2007	121-1400
2007	66-2-70-12-7	2007	80-243900	2007	88C9	2007	121-145
2007	66-20	2007	80P1	2007	88C-70-12	2007	121-146
2007	66-21	2007	81-46125002-9	2007	089-222	2007	121-147
2007	66-22	2007	81-46125003-7	2007	089-231	2007	121-148
2007	66-23	2007	81-46125004-5	2007	90-56	2007	121-151
2007	66-24	2007	81-46125005-2	2007	90-58	2007	121-152
2007	66-25	2007	81-46125009-4	2007	90-60	2007	121-160
2007	66-26	2007	81-46125010-2	2007	93A9	2007	121-167
2007	66-27	2007	81-46125011-0	2007	93A9-1	2007	121-184
2007	66-28	2007	81-46125018-5	2007	93A9-2	2007	121-19
2007	66-29	2007	81-46125029-2	2007	93A9-3	2007	121-190
2007	66-2A	2007	86-103-2	2007	93A9-4	2007	121-191
2007	66-2AO	2007	86-114-2	2007	94T1	2007	121-192
2007	66-2AOR	2007	86-115-2	2007	95-11410100	2007	121-193
2007	66-2A1	2007	86-126-2	2007	95-11410200	2007	121-200
2007	66-2A19	2007	86-128-2	2007	95-11410900	2007	121-205
2007	66-2A2	2007	86-129-2	2007	95-11413500	2007	121-206
2007	66-2A3	2007	86-130-2	2007	95-11414000	2007	121-207
2007	66-2A3P	2007	86-131-2	2007	95-114140000	2007	121-208
2007	66-2A4-7	2007	86-132-2	2007	95-120A	2007	121-209
2007	66-2A4-7B	2007	86-133-2	2007	95-201	2007	121-210
2007	66-2A5	2007	86-152-2	2007	95-203	2007	121-211
2007	66-2A5L	2007	86-156-2	2007	95-204	2007	121-212
2007	66-2A6	2007	86-156-2A	2007	95-208	2007	121-213
2007	66-2A6-4	2007	86-159-2	2007	95-209	2007	121-219
2007	66-2A7	2007	86-16-2	2007	95-212	2007	121-220
2007	66-2A7-1	2007	86-169-2	2007	95-213	2007	121-221
2007	66-2A8	2007	86-172-2	2007	95-214	2007	121-222
2007	66-2A82	2007	86-176-2	2007	95-217	2007	121-225
2007	66-2A9	2007	86-21-2	2007	95-218	2007	121-226
2007	66-2A9G	2007	86-22-2	2007	95-222-1	2007	121-227
2007	66-3	2007	86-23-2	2007	95-224-1	2007	121-234
2007	066-3-12	2007	86-249-2	2007	96-5032-01	2007	121-235
2007	066-3-12-7	2007	86-27-2	2007	96-5033-01	2007	121-236
2007	66-3-70	2007	86-28-2	2007	96-5033-02	2007	121-239
2007	66-3-70-12	2007	86-283-2	2007	96-5033-03	2007	121-240X
2007	66-3-70-12-7	2007	86-29-2	2007	96-5033-04	2007	121-245
2007	66-30	2007	86-295-2	2007	96-5076-01	2007	121-246
2007	66-31	2007	86-297-2	2007	96-5085-01	2007	121-254
2007	66-32	2007	86-30-2	2007	96-5085-02	2007	121-266
2007	66-33	2007	86-300-2	2007	96-5098-01	2007	121-267
2007	66-34	2007	86-303-2	2007	96-5101-01	2007	121-27
2007	66-35	2007	86-304-2	2007	96-5102-01	2007	121-272
2007	66-36	2007	86-305-2	2007	96XZ6053-09N	2007	121-274
2007	66-37	2007	86-32-2	2007	96XZ6053-27N	2007	121-275
2007	66-38	2007	86-33-2	2007	96XZ6053-51N	2007	121-287
2007	66-3A	2007	86-39-2	2007	96XZ6053/24N	2007	121-291
2007	66-3AO	2007	86-392-2	2007	96XZ6053/27N	2007	121-300
2007	66-3AOR	2007	86-419-2	2007	96XZ6053/51N	2007	121-301
2007	66-3A1	2007	86-421-2	2007	96XZ801/50N	2007	121-305
2007	66-3A19	2007	86-45-2	2007	97P1	2007	121-306
2007	66-3A2	2007	86-46-2	2007	97P1U	2007	121-307
2007	66-3A21	2007	86-47-2	2007	98-1	2007	121-309
2007	66-3A3	2007	86-476-2	2007	98-2	2007	121-310
2007	66-3A3P	2007	86-48-2	2007	98-3	2007	121-311
2007	66-3A4	2007	86-49-2	2007	98-4	2007	121-311C
2007	66-3A4-7	2007	86-497-2	2007	98-5	2007	121-311D
2007	66-3A4-7B	2007	86-50-2	2007	98-6	2007	121-311B
2007	66-3A5	2007	86-5000-2	2007	98-7	2007	121-311F
2007	66-3A5L	2007	86-5001-2	2007	98P-2A6-2	2007	121-314
2007	66-3A6	2007	86-5006-2	2007	99-104	2007	121-327
2007	66-3A6C	2007	86-5027-2	2007	99-201	2007	121-328
2007	66-3A7	2007	86-5042-2	2007	99-202	2007	121-33
2007	66-3A7-1	2007	86-5052-2	2007	99-203	2007	121-34
2007	66-3A8	2007	86-5063-2	2007	99AT6	2007	121-347
2007	66-3A82	2007	86-5067-2	2007	99B5	2007	121-348
2007	66-3A9	2007	86-509-2	2007	99BA6	2007	121-354
2007	66-3A9G	2007	86-5091-2	2007	99BB6	2007	121-362
2007	66-6023	2007	86-5342-2	2007	99J-4A6-1	2007	121-368
2007	66-6023-00	2007	86-54-2	2007	99P3	2007	121-372
2007	66-6024-00	2007	86-59-2	2007	99P3AA	2007	121-373
2007	66-6025-00	2007	86-61-2	2007	998002	2007	121-373B
2007	66-6026-00	2007	86-72-2	2007	998003	2007	121-373C
2007	66-6027-00	2007	86-73-2	2007	998004	2007	121-373D
2007	66-6028-00	2007	86-74-2	2007	998004A	2007	121-373E
2007	66-6033	2007	86-75-2	2007	998005	2007	121-373F
2007	68P-2A5L	2007	86-77-2	2007	998010	2007	121-373G
2007	69M-4A5L	2007	86-78-2	2007	998010A	2007	121-373H
2007	070-020	2007	86-79-2	2007	998011	2007	121-373I
2007	74-3	2007	86-80-2	2007	998011A	2007	121-373J
2007	74-3-70	2007	86-82-2	2007	100B63	2007	121-373K
2007	74-3-70-12	2007	86-83-2	2007	101-1	2007	121-373L
2007	74-3-70-12-7	2007	86-84-2	2007	101-15	2007	121-373M
2007	74-30	2007	86-93-2	2007	0102-0371		
2007	74-31	2007	86-95-2	2007	0000103		
2007	74-32			2007	0103-95		

RS 276-	Industry Standard No.	RS 276-	Industry Standard No.	RS 276-	Industry Standard No.	RS 276-	Industry Standard No.
2007	121-373N	2007	175-007-9-001	2007	297V040H08	2007	417-21
2007	121-373O	2007	175-008-9-001	2007	297V040H09	2007	417-28
2007	121-373P	2007	188G-1-82	2007	297V040H10	2007	417-40
2007	121-373Q	2007	188F-2	2007	297V040H11	2007	417-41
2007	121-374	2007	188F-2-12	2007	297V040H12	2007	417-47
2007	121-375	2007	188F-2-127	2007	297V040H13	2007	417-48
2007	121-388	2007	188F-2L	2007	297V040H16	2007	417-5
2007	121-395	2007	188F-2L8	2007	297V042C01	2007	417-51
2007	121-396	2007	189I-4	2007	297V042C02	2007	417-52
2007	121-397	2007	189I-4-12	2007	297V042C03	2007	417-6
2007	121-399	2007	189I-4-127	2007	297V042C04	2007	417-73
2007	121-400	2007	189I-4L	2007	297V042H01	2007	417-74
2007	121-401	2007	189I-4L8	2007	297V042H02	2007	417-75
2007	121-403	2007	201-15	2007	297V043H01	2007	417-78
2007	121-408	2007	202	2007	297V043H02	2007	420T1
2007	121-409	2007	205A2-210	2007	297V044H01	2007	421-10
2007	121-410	2007	207A3	2007	297V050C02	2007	421-11
2007	121-416	2007	207AT	2007	297V050H03	2007	421-11B
2007	121-420	2007	229-0027	2007	297V051C03	2007	421-12
2007	121-421	2007	229-0028	2007	297V051C04	2007	421-12B
2007	121-425	2007	229-0029	2007	297V051H01	2007	421-13
2007	121-425A	2007	229-0030	2007	297V051H02	2007	421-13B
2007	121-425B	2007	229-0055	2007	297V051H03	2007	421-13C
2007	121-425C	2007	229-0056	2007	297V051H04	2007	421-14
2007	121-425D	2007	229-0062	2007	297V052C01	2007	421-14B
2007	121-425E	2007	229-0080	2007	297V052H01	2007	421-15
2007	121-425F	2007	229-0085	2007	297V052H02	2007	421-15B
2007	121-425G	2007	229-0088	2007	297V052H04	2007	421-19
2007	121-425H	2007	229-0091	2007	297V053C01	2007	421-7143
2007	121-425I	2007	229-0092	2007	297V053H01	2007	421-8109
2007	121-425J	2007	229-0097	2007	297V053H02	2007	421-9
2007	121-425K	2007	229-0100	2007	297V054C01	2007	421-9671
2007	121-425L	2007	229-0124	2007	297V054C02	2007	421-9682
2007	121-425M	2007	229-0125	2007	297V055C01	2007	421-9862A
2007	121-425N	2007	229-0130	2007	297V057H02	2007	421-9863
2007	121-425O	2007	229-0133	2007	297V065H01	2007	421-9863A
2007	121-425P	2007	229-0137	2007	297V065H02	2007	421-9864
2007	121-425Q	2007	229-0138	2007	297V065H03	2007	421-9864A
2007	121-43	2007	229-0159	2007	297V076B01	2007	421T1
2007	121-437	2007	229-0140	2007	297V076C01	2007	422-0222
2007	121-46	2007	229-0142	2007	297V081C01	2007	422-2535
2007	121-47	2007	229-0143	2007	302	2007	429-0092-2
2007	121-490	2007	229-0146	2007	309-327-931	2007	429-0092-5
2007	121-52	2007	231-000-001	2007	310-188	2007	429-0910-51
2007	121-53	2007	231-0000-01	2007	310-189	2007	429-0910-52
2007	121-543	2007	231-0009	2007	322T1	2007	430-85
2007	121-544	2007	235	2007	323T1	2007	465-005-19
2007	121-61	2007	247-623	2007	324-0029	2007	465-036-19
2007	121-632	2007	260D00401	2007	324-0041	2007	465-067-19
2007	121-633	2007	260D00402	2007	324-0055	2007	465-072-19
2007	121-634	2007	260D00403	2007	324-0056	2007	465-073-19
2007	121-635	2007	260D02501	2007	324-0062	2007	465-075-19
2007	121-636	2007	260D02601	2007	324-0074	2007	465-080-19
2007	121-636B	2007	260D02701	2007	324-0080	2007	465-082-19
2007	121-636C	2007	260D04701	2007	324-0085	2007	465-108-19
2007	121-636D	2007	260D08514	2007	324-0088	2007	465-115-19
2007	121-64	2007	260D08912	2007	324-0089	2007	465-132-19
2007	121-640	2007	260D09413	2007	324-0090	2007	465-163-19
2007	121-663	2007	260D13704	2007	324-0091	2007	465-165-19
2007	121-674	2007	260P03001	2007	324-0092	2007	465-191-15
2007	121-68	2007	260P07601	2007	324-0093	2007	466-1A5
2007	121-69	2007	260P2100	2007	324-0097	2007	466-2A5
2007	121-72	2007	260P21001	2007	324-0100	2007	466-3A5
2007	121-734	2007	260P21002	2007	324-0124	2007	473B5
2007	121-80	2007	260Z00703	2007	324-0125	2007	473B6-2
2007	121-81	2007	260Z01201	2007	324-0133	2007	473B6-2A
2007	121-82	2007	264P06301	2007	324-0139	2007	473B6-4
2007	121-83	2007	296-18-9	2007	324-0140	2007	473B6-7
2007	121-830	2007	296-19-9	2007	324-0142	2007	473B6-7
2007	121-84	2007	296-60-9	2007	324-0143	2007	474-3A5
2007	121-85	2007	296-62-9	2007	324-0144	2007	488C15
2007	121-86	2007	296-64-9-1	2007	324-0146	2007	509B8025P
2007	121-87	2007	297L001M02	2007	324-144	2007	520T1
2007	121-88	2007	297L002H01	2007	324-6011(PNP)	2007	521T1
2007	121-89	2007	297L005H01	2007	324T1	2007	573-529
2007	121-90	2007	297V003	2007	324T2	2007	576-0001-003
2007	121-91	2007	297V003H03	2007	325-0025-329	2007	576-0001-009
2007	121-92	2007	297V003H09	2007	325-0025-330	2007	576-0001-014
2007	121-93	2007	297V003M01	2007	325-0025-331	2007	576-0002-004
2007	121-94	2007	297V003M07	2007	325-0028-79	2007	576-0002-012
2007	121-95	2007	297V004H01	2007	325-0028-80	2007	576-001
2007	121-96	2007	297V004H010	2007	325-0028-81	2007	576-002-004
2007	121-29004	2007	297V004H03	2007	325-0028-83	2007	576-0040-253
2007	122-229	2007	297V004H04	2007	325-0030-315	2007	576-005
2007	125-425C	2007	297V004H06	2007	325-0030-317	2007	600-104-308
2007	125AJ34	2007	297V004H08	2007	325-0030-318	2007	602-040
2007	126P1	2007	297V004H09	2007	325-0030-319	2007	605-030
2007	129-11	2007	297V004H10	2007	325-0031-306	2007	606-020
2007	129-17	2007	297V004H11	2007	325-0036-536	2007	610-035
2007	129-18	2007	297V004H14	2007	325-0047-516	2007	610-035-1
2007	129-31	2007	297V004H15	2007	325-0054-310	2007	610-036
2007	129-32	2007	297V004H16	2007	325-0054-311	2007	610-036-1
2007	129-8	2007	297V004M01	2007	325-0081-102	2007	610-036-2
2007	129-8-1	2007	297V005H01	2007	325-0670	2007	610-036-3
2007	129-8-1A	2007	297V008H01	2007	325-0670-1	2007	610-036-4
2007	129-8-2	2007	297V011	2007	325-0670-7	2007	610-036-5
2007	130-40095	2007	297V011H02	2007	325-0670A	2007	610-036-8
2007	130-40236	2007	297V012	2007	325-1370-18	2007	610-040
2007	130-40352	2007	297V012H02	2007	325-1376-53	2007	610-040-1
2007	130-40456	2007	297V012H03	2007	325-1376-56	2007	610-040-2
2007	0131-000100	2007	297V012H05	2007	325-1376-57	2007	610-043
2007	0131-000101	2007	297V012H06	2007	325-1376-58	2007	610-043-1
2007	0131-000102	2007	297V012H08	2007	325-1378-20	2007	610-043-2
2007	0131-000563	2007	297V012H09	2007	325-1378-21	2007	610-043-3
2007	0131-001050	2007	297V012H14	2007	325-1378-22	2007	610-043-4
2007	0131-001056	2007	297V01TH01	2007	325-1442-8	2007	610-043-6
2007	0131-001419	2007	297V017H02	2007	325T1	2007	610-043-7
2007	0131-001426	2007	297V018H01	2007	326T1	2007	610-046-7
2007	0131-002656	2007	297V019H01	2007	350	2007	610-079
2007	132-001	2007	297V020H02	2007	352	2007	610-079-1
2007	132-010	2007	297V020M01	2007	353	2007	610-080
2007	132-090	2007	297V021H01	2007	353-9012-001	2007	610-080-1
2007	132-90	2007	297V021H02	2007	353-9312-001	2007	614X1
2007	148P-2	2007	297V021H03	2007	354-3052	2007	614X10
2007	148P-2-12	2007	297V022H01	2007	365T1	2007	614X2
2007	148P-2-12-8	2007	297V025H02	2007	394-3074-2	2007	614X3
2007	149L-4	2007	297V025H04	2007	394-3074-5	2007	614X4
2007	149L-4-12	2007	297V025H05	2007	394-3097-1	2007	614X5
2007	149L-4-12-8	2007	297V025H15	2007	394-3097-2	2007	614X6
2007	152-221011	2007	297V026H03	2007	410-012-0150	2007	614X7
2007	154A3675-105	2007	297V027H01	2007	410-013-0240	2007	614X9
2007	154A3676-205	2007	297V032H01	2007	412	2007	617-50
2007	154A3679	2007	297V033H01	2007	417-103	2007	617-52
2007	154A8681	2007	297V037H01	2007	417-122	2007	617-69
2007	166-1A82	2007	297V037B02	2007	417-146	2007	617-70
2007	166-2A82	2007	297V037H01	2007	417-147	2007	642-117
2007	166-3A82	2007	297V037H02	2007	417-148	2007	642-150
2007	173A3970	2007	297V038H01	2007	417-149	2007	642-151
2007	173A4348	2007	297V038H05	2007	417-150	2007	650-105
2007	173A4349	2007	297V038H07	2007	417-151	2007	650-106
2007	173A4389-1	2007	297V038H09	2007	417-152	2007	650-107
2007	173A4390	2007	297V040H01	2007	417-17	2007	650-108
2007	174-3A82			2007	417-18		
2007	175-006-9-001						

RS 276-	Industry Standard No.	RS 276-	Industry Standard No.	RS 276-	Industry Standard No.	RS 276-	Industry Standard No.
2007	656-136	2007	921-147B	2007	1148-17	2007	2057B2-142
2007	656-137	2007	921-148B	2007	1166-7821	2007	2057B2-206
2007	656-138	2007	921-14B	2007	1178-3453	2007	2057B2-23
2007	656-139	2007	921-150B	2007	1180-0182	2007	2057B2-28
2007	660-224	2007	921-153B	2007	1192	2007	2057B2-29
2007	660-227	2007	921-216B	2007	1192(XSTR)	2007	2057B2-32
2007	660B	2007	921-217B	2007	1241A	2007	2057B2-34
2007	662A21	2007	921-222B	2007	124B	2007	2057B2-4
2007	662A4	2007	921-223B	2007	1277-17	2007	2057B2-43
2007	666-1-A-5L	2007	921-224B	2007	1316-17	2007	2057B2-44
2007	666-2-A-5L	2007	921-227B	2007	1317-17	2007	2057B2-45
2007	666-3A-5L	2007	921-238B	2007	1320	2007	2057B2-49
2007	674-3A5L	2007	921-242B	2007	1321-7724	2007	2057B2-52
2007	675-153	2007	921-243B	2007	1321-7732	2007	2057B2-57
2007	675-154	2007	921-244B	2007	1329	2007	2057B2-61
2007	675-155	2007	921-24B	2007	1330	2007	2057B2-72
2007	675-156	2007	921-256X	2007	1340	2007	2057B2-78
2007	688C-1-5L	2007	921-27	2007	1344-7321	2007	2057B2-83
2007	690V034H30	2007	921-273B	2007	1347-17	2007	2057B2-86
2007	690V034H31	2007	921-274B	2007	1350	2007	2057B2-94
2007	690V034H39	2007	921-27A	2007	1360	2007	2057B2-99
2007	690V040H61	2007	921-282B	2007	1362-17	2007	2057B206
2007	690V040H62	2007	921-318B	2007	1362-17A	2007	2057B45-14
2007	690V043H62	2007	921-319B	2007	1364-17	2007	2061A45-47
2007	690V043H63	2007	921-327B	2007	1390	2007	2061B45-14
2007	690V047H56	2007	921-35	2007	1400	2007	2093A38-23
2007	690V047H57	2007	921-35A	2007	1410	2007	2093A41-40
2007	690V047H58	2007	921-35B	2007	1413-160	2007	2093A41-41
2007	690V047H59	2007	921-36B	2007	1413-175	2007	2093A41-87
2007	690V047H60	2007	921-37B	2007	1436-17	2007	2093A9-3
2007	690V052H23	2007	921-38	2007	1459	2007	2093A9-4
2007	690V052H24	2007	921-38A	2007	1466-1	2007	2106-119
2007	690V054H20	2007	921-38B	2007	1466-1-12	2007	2106-121
2007	690V054H21	2007	921-39	2007	1466-1-12-8	2007	2106-122
2007	690V056H33	2007	921-39A	2007	1466-2	2007	2106-123
2007	690V056H34	2007	921-39B	2007	1466-2-12	2007	2112-17
2007	690V056H90	2007	921-40	2007	1466-2-12-8	2007	2114-17
2007	690V057H27	2007	921-40A	2007	1466-3	2007	2402-453
2007	690V057H28	2007	921-40B	2007	1466-3-12	2007	2402-454
2007	690V059H20	2007	921-41B	2007	1466-3-12-8	2007	2402-455
2007	690V059H21	2007	921-44B	2007	1474-3	2007	2402-457
2007	690V059H52	2007	921-45	2007	1474-3-12	2007	2405-453
2007	690V059H55	2007	921-45A	2007	1474-3-12-8	2007	2405-454
2007	690V061H98	2007	921-45B	2007	1476-17	2007	2405-455
2007	690V061H99	2007	921-51B	2007	1476-17-6	2007	2405-456
2007	690V062H47	2007	921-52B	2007	1488C	2007	2405-457
2007	690V063H16	2007	921-53B	2007	1488C-12	2007	2408-326
2007	690V063H17	2007	921-67	2007	1488C-12-8	2007	2408-328
2007	690V063H50	2007	921-67A	2007	1510-2718	2007	2408-329
2007	690V063H51	2007	921-67B	2007	1777-17	2007	2490
2007	690V066H46	2007	921-68	2007	1850-0040	2007	2490A
2007	690V066H47	2007	921-68A	2007	1850-0040-1	2007	2495-014
2007	690V068H30	2007	921-68B	2007	1850-0060	2007	2495-080
2007	690V068H31	2007	921-69	2007	1850-0062	2007	2495-388
2007	690V077H37	2007	921-69A	2007	1850-0062-1	2007	2495-567-2
2007	690V077H73	2007	921-69B	2007	1850-0101	2007	2495-567-3
2007	690V080H39	2007	924-16598	2007	1850-0184	2007	2495-586-2
2007	690V080H44	2007	930X10	2007	1850-0184-1	2007	2497-473
2007	690V082H47	2007	930X7	2007	1866-1	2007	2497-496
2007	690V084H61	2007	930X8	2007	1866-1-12	2007	2502(RCA)
2007	690V084H63	2007	930X9	2007	1866-1-127	2007	2603-180
2007	690V085	2007	951-1	2007	1866-1L	2007	2603-181
2007	690V085H44	2007	964-17142	2007	1866-1L8	2007	2603-182
2007	690V086H39	2007	964-19862A	2007	1866-2	2007	2606-286
2007	690V088H52	2007	964-19863	2007	1866-2-12	2007	2606-287
2007	690V089H90	2007	964-19864	2007	1866-2-127	2007	2606-291
2007	690V094H18	2007	965T1	2007	1866-2L	2007	2612
2007	690V094H19	2007	966-1A6-3	2007	1866-2L8	2007	2703-384
2007	690V094H20	2007	966-2A6-4	2007	1866-3	2007	2703-385
2007	690V097H59	2007	966-3A6C	2007	1866-3-12	2007	2703-386
2007	690V099H59	2007	972X10	2007	1866-3-127	2007	2704-384
2007	690V104H53	2007	972X11	2007	1866-3L	2007	2704-385
2007	690V104H54	2007	972X12	2007	1866-3L8	2007	2704-386
2007	690V105H24	2007	972X9	2007	1874-3	2007	2704-387
2007	690V118H62	2007	974-3A6-3	2007	1874-3-12	2007	2781
2007	690V119H54	2007	987T1	2007	1874-3-127	2007	2904-038H05
2007	720-35019	2007	988T1	2007	1874-3L	2007	3004-856
2007	720-35019A	2007	989T1	2007	1874-3L8	2007	3009(SEARS)
2007	750-137	2007	990T1	2007	1888C	2007	3010(SEARS)
2007	750-138	2007	991-00-1221	2007	1888C-12	2007	3014(SEARS)
2007	750-139	2007	991-00-1222	2007	1888C-127	2007	3112
2007	750-140	2007	991-01-1217	2007	1888CL	2007	3425
2007	750-35019	2007	991-01-1221	2007	1888CL8	2007	3434
2007	750M63-105	2007	991-01-1222	2007	1917-17	2007	3435
2007	750M63-115	2007	991-01-1223	2007	1919-17	2007	3500
2007	750M63-147	2007	991-01-1224	2007	1919-17A	2007	3504
2007	750M63-148	2007	991-011217	2007	1946-17	2007	3507(RCA)
2007	792-286	2007	991-011222	2007	1954-17	2007	3517
2007	792-287	2007	991-011223	2007	1960-17	2007	3544
2007	792-288	2007	991T1	2007	1973-17	2007	3550
2007	792-289	2007	992T1	2007	1974-17	2007	3578-1
2007	792-290	2007	1001	2007	2004	2007	3600
2007	800-205	2007	1001-7663	2007	2004(TRANSISTOR)	2007	3686(RCA)
2007	800-502-00	2007	1002(SQUELCH)	2007	2012	2007	3746
2007	800-505-00	2007	1002-05	2007	2012(E.F.JOHNSON)	2007	3852
2007	800-537-03	2007	1003	2007	2042-17	2007	3907
2007	802-05400	2007	1005(E.F.JOHNSON)	2007	2047A2-288	2007	3970
2007	802-05600	2007	1004-01	2007	2057A100-21	2007	3970(G.E.)
2007	802-22631U	2007	1006(JULIETTE)	2007	2057A100-23	2007	3970CL
2007	802-36400	2007	1005-17	2007	2057A100-24	2007	4001-224
2007	802-36430	2007	1006(JULIETTE)	2007	2057A100-30(PNP)	2007	4001-225
2007	802-43900	2007	1006-93	2007	2057A100-34	2007	4001-226
2007	808-304	2007	1007(28B405)	2007	2057A100-35(PNP)	2007	4004(PENNCREST)
2007	808-305	2007	1007-17	2007	2057A100-41	2007	4009(PENNCREST)
2007	808-306	2007	1008(E.F.JOHNSON)	2007	2057A100-44	2007	4014-000-10160
2007	808-307	2007	1009	2007	2057A100-48	2007	4041-000-10160
2007	808-308	2007	1009(E.F.JOHNSON)	2007	2057A100-55	2007	4041-000-20120
2007	808-310	2007	1009(SEARS)	2007	2057A100-8	2007	4041-000-30180
2007	808-311	2007	1014(E.F.JOHNSON)	2007	2057A100-9	2007	0004201
2007	815-1810	2007	1019-74	2007	2057A2-147	2007	0004202
2007	815-181D	2007	1021-17	2007	2057A2-148	2007	4315
2007	830	2007	1023-17	2007	2057A2-165	2007	4316
2007	901-000-6-51	2007	1032	2007	2057A2-206	2007	4348
2007	910X1	2007	1033-5	2007	2057A2-210	2007	4349
2007	910X10	2007	1033-7	2007	2057A2-241	2007	4398
2007	910X2	2007	1034	2007	2057A2-28	2007	4450
2007	910X3	2007	1035	2007	2057A2-288	2007	4451
2007	910X4	2007	1036	2007	2057A2-317	2007	4462
2007	910X5	2007	1052-17	2007	2057A2-329	2007	4460ORN
2007	910X6	2007	1057-17	2007	2057A2-61	2007	4468BRN
2007	910X7	2007	1060-17	2007	2057A2-988	2007	4469RED
2007	910X8	2007	1060-17(DRIVER)	2007	2057B100-1	2007	4471ORN
2007	910X9	2007	1060-17(PRE-AMP)	2007	2057B100-13	2007	4471YEL
2007	916-31001-1	2007	1102-17	2007	2057B100-6	2007	4472GRN
2007	916-31001-7B	2007	1102-17A	2007	2057B100-7	2007	4473
2007	916-31003-5B	2007	1104-94	2007	2057B100-8	2007	4474YEL
2007	916-31007-5	2007	1119-2875	2007	2057B100-9	2007	4475GRN
2007	916-31007-5B	2007	1119-57	2007	2057B169	2007	4476BLU
2007	916-31012-6	2007	1119-59	2007	2057B169	2007	4477PUR
2007	916-31012-6B	2007	1128-17	2007	2057B2-120	2007	4477V10
2007	921-100B	2007	1140-17	2007	2057B2-124	2007	4484
2007	921-105B	2007	1145	2007	2057B2-129	2007	4485
2007	921-118B	2007	1145-17	2007	2057B2-135	2007	4486
2007	921-140B	2007	1146	2007	2057B2-137		

RS 276-	Industry Standard No.	RS 276-	Industry Standard No.	RS 276-	Industry Standard No.	RS 276-	Industry Standard No.
2007	4510	2007	12196	2007	41570	2007	81504-1A
2007	4553BLU	2007	15009	2007	42304	2007	81504-1B
2007	4553BRN	2007	16598	2007	42305	2007	81504-1C
2007	4553GRN	2007	16958	2007	42322	2007	81504-3
2007	4553ORN	2007	17047-1	2007	42324	2007	81504-3A
2007	4553RED	2007	17142	2007	44616-1	2007	81504-3B
2007	4553V10	2007	17143	2007	44967-2	2007	81504-3C
2007	4553YEL	2007	18109	2007	46776-2	2007	81505-8
2007	4562	2007	18529	2007	47994-2	2007	81505-8A
2007	4563	2007	18530	2007	47737-2	2007	81505-8B
2007	4564	2007	18601	2007	49139-2	2007	81505-8C
2007	4565	2007	18611	2007	49341	2007	81505-8X
2007	4595	2007	18731	2007	53201-01	2007	81506-5
2007	4607	2007	020156	2007	53201-11	2007	81506-5A
2007	4627	2007	23114-061	2007	59557-48	2007	81506-5B
2007	4686-163-3	2007	23785(SYLVANIA)	2007	59987-1	2007	81506-5C
2007	4686-195-3	2007	23785-1(SYLVANIA)	2007	61003-4	2007	81506-6
2007	4686-196-3	2007	24785	2007	61008-8	2007	81506-6A
2007	4686-81-2	2007	25114-101	2007	61008-8-1	2007	81506-6B
2007	4686-81-3	2007	25114-102	2007	61008-8-2	2007	81506B
2007	4800-200	2007	25114-103	2007	61009-9-3	2007	81506-9
2007	4800-220	2007	25114-104	2007	62032	2007	81506-9A
2007	4800-221	2007	25651-020	2007	64071-1	2007	81506-9B
2007	4800-222	2007	25651-021	2007	65804-63	2007	81506-9C
2007	4822-130-40095	2007	25651-033	2007	66008-2	2007	81507-0
2007	4822-130-40235	2007	25655-055	2007	67193-82	2007	81507-0A
2007	4822-130-40236	2007	25655-056	2007	67193-85	2007	81507-0B
2007	4822-130-40456	2007	25657-050	2007	67599	2007	81507-0C
2007	4907-976	2007	27125-120	2007	68504-63	2007	81507-0D
2007	5052	2007	27125-150	2007	68895-13	2007	81507-4
2007	5464	2007	27125-170	2007	69107-45	2007	81510-3
2007	5612-370	2007	27125-330	2007	71193-2	2007	81510-4
2007	5612-370C	2007	27125-340	2007	72117	2007	81510-5
2007	5612-75(C)	2007	27125-350	2007	72784-21	2007	81511-4
2007	5612-75C	2007	27125-360	2007	72784-22	2007	81511-5
2007	5612-77C	2007	27125-480	2007	72784-23	2007	81511-6
2007	5766-25	2007	27125-540	2007	72799-41	2007	81511-7
2007	6100	2007	27125-550	2007	72813-10	2007	81511-8
2007	6100-35	2007	27910-12153	2007	72847-51	2007	81512-0
2007	6440	2007	30201	2007	72941-33	2007	81512-0A
2007	6445	2007	30202	2007	73100-9	2007	81512-0B
2007	6452	2007	30204	2007	75960CH	2007	81512-0C
2007	7215-0	2007	30206	2007	77052-3	2007	81512-0D
2007	7239	2007	30207	2007	77052-4	2007	81512-0E
2007	7993	2007	30208	2007	77053-2	2007	81513-6
2007	8000-00001-068	2007	30208-1	2007	77272-0	2007	81513-9
2007	8000-00003-038	2007	30208-2	2007	77272-1	2007	81515-8
2007	8000-00003-039	2007	30216	2007	77272-5	2007	81516-0
2007	8000-00004-088	2007	30218	2007	77272-7	2007	81516-0A
2007	8000-00004-P088	2007	30231	2007	77272-9	2007	81516-0B
2007	8000-0004-P088	2007	30244	2007	77273-2	2007	81516-0C
2007	8000-00043-020	2007	30263	2007	77273-5	2007	81516-0D
2007	8070-4	2007	30293	2007	77273-6	2007	81516-0E
2007	8071-4	2007	30302(RCA)	2007	77273-7	2007	81516-0F
2007	8072-4	2007	34119	2007	78527-75-01	2007	81516-0G
2007	8073-4	2007	34219	2007	78527-76-01	2007	81516-0H
2007	8500-201	2007	34220	2007	78527-78-01	2007	81516-0I
2007	8500-202	2007	34221	2007	78527-79-01	2007	81516-0J
2007	8500-203	2007	34262	2007	080003	2007	86452
2007	8999-202	2007	34493P	2007	080040	2007	86632
2007	8999-203	2007	34871	2007	080043	2007	86842
2007	9012HE	2007	34923	2007	080052	2007	087003
2007	9012HP	2007	35045	2007	80073	2007	94000
2007	9330-011-70112	2007	35086	2007	80818VM	2007	94001
2007	9390	2007	35452-2	2007	081001	2007	94002
2007	9391	2007	35454	2007	081018	2007	94003
2007	9400-8	2007	35454-1	2007	081019	2007	94005
2007	9400-9	2007	35454-2	2007	081026	2007	94006
2007	9401-7	2007	35454-3	2007	081029	2007	94008
2007	9403-7	2007	35590	2007	081038	2007	94009
2007	9403-9	2007	35628	2007	081046	2007	94014
2007	9564	2007	35677	2007	081047	2007	94015
2007	9920-4	2007	35678	2007	081048	2007	94016
2007	9920-5	2007	35792	2007	081049	2007	94017
2007	9921-7	2007	35816	2007	081050	2007	94018
2007	9921-8	2007	35817	2007	081056	2007	94019
2007	10032	2007	35819	2007	081059	2007	94020
2007	10036	2007	35820	2007	81404-4A	2007	94021
2007	10037	2007	35820-1	2007	81500-3	2007	94022
2007	10038	2007	35820-2	2007	81501-5	2007	94030
2007	10039	2007	35820-3	2007	81502-0	2007	94037(EICO)
2007	11609-1	2007	35824	2007	81502-0A	2007	94038(EICO)
2007	11620-3	2007	35950	2007	81502-0B	2007	94039
2007	11620-6	2007	35952	2007	81502-1	2007	95120A
2007	11668-5	2007	35953	2007	81502-1A	2007	95172-2
2007	11668-6	2007	35954	2007	81502-1B	2007	95173-1
2007	11675-7	2007	35955	2007	81502-2	2007	95201
2007	11699-7	2007	36534	2007	81502-2A	2007	95203
2007	12110-0	2007	36557	2007	81502-2B	2007	95204
2007	12110-3	2007	36558	2007	81502-3B	2007	95208
2007	12110-4	2007	36816	2007	81502-4	2007	95209
2007	12110-5	2007	37549	2007	81502-4A	2007	95212
2007	12110-6	2007	37550	2007	81502-4B	2007	95213
2007	12110-7	2007	37551	2007	81502-5	2007	95214
2007	12112-0	2007	37677	2007	81502-5A	2007	95217
2007	12112-8	2007	37833	2007	81502-5B	2007	95218
2007	12114-8	2007	38057	2007	81502-7	2007	95219
2007	12116-1	2007	38091	2007	81502-7A	2007	95222-1
2007	12116-2	2007	38176	2007	81502-7B	2007	95224-1
2007	12116-4	2007	38177	2007	81502-7C	2007	95224-3
2007	12117-9	2007	38209	2007	81502-8	2007	95255-000
2007	12118-0	2007	38269	2007	81502-8A	2007	99202
2007	12118-4	2007	38685	2007	81502-8B	2007	99203
2007	12118-6	2007	40034	2007	81502-8C	2007	99204
2007	12119-1	2007	40034-1	2007	81502-9A	2007	99205
2007	12119-2	2007	40034-2	2007	81502-9B	2007	99217
2007	12122-5	2007	40034-3	2007	81502-9C	2007	99218
2007	12122-6	2007	40034VM	2007	81503-0	2007	100693
2007	12122-7	2007	40035	2007	81503-0A	2007	101973
2007	12123-2	2007	40035-1	2007	81503-0B	2007	101974
2007	12123-4	2007	40035-2	2007	81503-1	2007	103562
2007	12123-5	2007	40035-3	2007	81503-1A	2007	104444
2007	12123-6	2007	40036	2007	81503-1B	2007	107274
2007	12124-0	2007	40036-1	2007	81503-3	2007	110263
2007	12124-1	2007	40036-2	2007	81503-3A	2007	110494
2007	12124-6	2007	40036-3	2007	81503-4	2007	110957
2007	12125-4	2007	40038	2007	81503-4A	2007	110959
2007	12126-6	2007	40038-1	2007	81503-4B	2007	111001
2007	12126-7	2007	40038-2	2007	81503-4C	2007	111011
2007	12127-2	2007	40038-3	2007	81503-6	2007	111012
2007	12127-3	2007	40038VM	2007	81503-6A	2007	111013
2007	12127-4	2007	40253	2007	81503-6B	2007	111959
2007	12127-5	2007	40269	2007	81503-6C	2007	112071
2007	12128-9	2007	40329	2007	81503-7	2007	112297
2007	12173	2007	40359	2007	81503-7A	2007	116084
2007	12174	2007	40395	2007	81503-7B	2007	116091
2007	12175	2007	40396/P	2007	81503-7C	2007	116201
2007	12176	2007	40396P	2007	81503-8	2007	116203
2007	12183	2007	40403	2007	81503-8B	2007	116205
2007	12191	2007	40490	2007	81503-8C	2007	116206
2007	12192	2007	40763	2007	81504-1	2007	116286
2007	12193	2007	40828			2007	116662B
2007	12195	2007	40852(VM)			2007	116685
		2007	40853(VM)				

RS 276-2007	Industry Standard No.	RS 276-2007	Industry Standard No.	RS 276-2007	Industry Standard No.	RS 276-2007	Industry Standard No.
2007	116686	2007	226338	2007	610036-1	2007	815037C
2007	116757	2007	226924	2007	610036-2	2007	815038
2007	116988	2007	227752	2007	610036-3	2007	815038A
2007	116996	2007	228287	2007	610036-4	2007	815038B
2007	116997	2007	230253	2007	610036-5	2007	815038C
2007	116998	2007	230259	2007	610036-6	2007	815041
2007	117208	2007	230524	2007	610036-7	2007	815041A
2007	117209	2007	230525	2007	610036-8	2007	815041B
2007	117210	2007	231140-21	2007	610040	2007	815041C
2007	117616	2007	231588	2007	610040-1	2007	815043
2007	117617	2007	231706	2007	610040-2	2007	815043A
2007	117727	2007	233507	2007	610043	2007	815043B
2007	117728	2007	233945	2007	610043-1	2007	815043C
2007	117867	2007	234076	2007	610043-2	2007	815055
2007	119727	2007	234630	2007	610043-3	2007	815056
2007	120075	2007	235194	2007	610043-4	2007	815057
2007	120143	2007	236709	2007	610043-6	2007	815058
2007	120144	2007	238417	2007	610043-7	2007	815058A
2007	120545	2007	238418	2007	610052-1	2007	815058B
2007	120546	2007	240003	2007	610059-2	2007	815058C
2007	120909-24.4	2007	240006	2007	610074-2	2007	815058X
2007	121151	2007	242221	2007	610079	2007	815065
2007	121152	2007	243939	2007	610079-1	2007	815065A
2007	121153	2007	244007	2007	610080	2007	815065B
2007	121154	2007	255728	2007	610080-1	2007	815065C
2007	122243	2007	256126	2007	610088	2007	815066
2007	122244	2007	257340	2007	610088-1	2007	815066A
2007	122901	2007	257470	2007	610088-2	2007	815066B
2007	123379	2007	257473	2007	610099-3	2007	815066C
2007	123791	2007	262113	2007	610126-1	2007	815069
2007	123805	2007	266686	2007	611020	2007	815069A
2007	123806	2007	266702	2007	650196	2007	815069B
2007	123809	2007	269374	2007	650859-1	2007	815069C
2007	123877	2007	297240-1	2007	650859-2	2007	815070
2007	124626	2007	300008	2007	650859-3	2007	815070A
2007	126093	2007	300538	2007	651012	2007	815070B
2007	126093-1	2007	300540	2007	651236	2007	815070C
2007	126093-2	2007	300541	2007	660030	2007	815070D
2007	126093-3	2007	309421	2007	660059	2007	815074
2007	126093-4	2007	310017	2007	660060	2007	815082
2007	126187	2007	310035	2007	660072	2007	815083
2007	126276	2007	310159	2007	660082	2007	815101
2007	126697	2007	310160	2007	731009	2007	815103
2007	126945	2007	310201	2007	740417	2007	815104
2007	127112	2007	310223	2007	770501	2007	815105
2007	127114	2007	310225	2007	770524	2007	815107
2007	127297	2007	322968-167	2007	770525	2007	815108
2007	127303	2007	322968-17	2007	770730	2007	815109
2007	127397	2007	346016-1	2007	772718	2007	815114
2007	127589	2007	346016-11	2007	772720	2007	815115
2007	127962	2007	373003	2007	772721	2007	815116
2007	128940	2007	373117	2007	772722	2007	815117
2007	129286	2007	373119	2007	772723	2007	815118
2007	129348(XSTR)	2007	454549	2007	772724	2007	815120
2007	129507	2007	489751-045	2007	772725	2007	815120A
2007	129508	2007	489751-108	2007	772726	2007	815120B
2007	129802	2007	489751-109	2007	772727	2007	815120C
2007	130200-00	2007	489751-113	2007	772729	2007	815120D
2007	130200-02	2007	489751-114	2007	772732	2007	815120E
2007	130400-95	2007	510007	2007	772733	2007	815122
2007	130402-36	2007	537200	2007	772736	2007	815136
2007	130403-52	2007	0537640(XSTR)	2007	772737	2007	815139
2007	137093	2007	537790	2007	800747	2007	815158
2007	161705	2007	551015	2007	801500	2007	815160
2007	165976	2007	551051	2007	801501	2007	815160-C
2007	166400	2007	0563012H	2007	801507	2007	815160-I
2007	166882	2007	0573001-14	2007	801509	2007	815160-J
2007	166883	2007	0573001H	2007	801510	2007	815160-K
2007	166997	2007	0573003	2007	801511	2007	815160-L
2007	167679	2007	0573003H	2007	801520	2007	815160-O
2007	167680	2007	0573004	2007	802032-2	2007	815160-P
2007	167998	2007	0573005	2007	802032-4	2007	815160-Q
2007	167999	2007	0573005-14	2007	802033-3	2007	815160A
2007	168906	2007	0573005H	2007	802054-0	2007	815160B
2007	168907	2007	0573011	2007	802056-0	2007	815160C
2007	168954	2007	0573012	2007	802189-7	2007	815160D
2007	168983	2007	0573012H	2007	802189-8	2007	815160E
2007	168984	2007	0573018	2007	802263-0	2007	815160F
2007	169175	2007	0573018H	2007	802263-1	2007	815160H
2007	169359	2007	0573022	2007	802389-2	2007	815177
2007	169360	2007	0573022H	2007	802415-2	2007	815178
2007	169361	2007	0573023	2007	802439-0	2007	815179
2007	169773	2007	0573023A	2007	802560	2007	815181
2007	170132-1(PNP)	2007	0573023H	2007	814044A	2007	815181-B
2007	171016(SEARS)	2007	0573024-14	2007	815003	2007	815181A
2007	171017	2007	0573025	2007	815015	2007	815181B
2007	171018	2007	573029	2007	815020	2007	815181C
2007	171026(SEARS)	2007	0573034	2007	815020A	2007	815181D
2007	171049	2007	0573036	2007	815020B	2007	815195
2007	171162-072	2007	0573036H	2007	815021	2007	81596
2007	171162-073	2007	0573055	2007	815021A	2007	815216-3
2007	171162-074	2007	0573056	2007	815021B	2007	815228A
2007	171162-075	2007	573103	2007	815022	2007	815228A01
2007	171162-076	2007	573110	2007	815022A	2007	815228A1
2007	171162-080	2007	573114	2007	815022B	2007	815228B
2007	171162-081	2007	0573114H	2007	815023	2007	815228B1
2007	171162-120	2007	0573117	2007	815023A	2007	815308A
2007	171162-121	2007	0573117-14	2007	815023B	2007	825065
2007	171522	2007	573118	2007	815024	2007	910050-2
2007	171916	2007	573119	2007	815024A	2007	910062-1
2007	171917	2007	573125	2007	815024B	2007	910070-6
2007	172816	2007	0573131	2007	815025	2007	910094-4
2007	175006-186	2007	0573142	2007	815025A	2007	922896
2007	177105	2007	0573142H	2007	815025B	2007	971477
2007	195601-6	2007	0573152	2007	815027	2007	980156
2007	196064-3	2007	0573153	2007	815027B	2007	980144
2007	196183-7	2007	0573153H	2007	815027C	2007	980148
2007	200028-7-28	2007	573184	2007	815028	2007	980149
2007	200062-5-31	2007	0573187	2007	815028A	2007	980153
2007	200062-5-32	2007	0573200	2007	815028B	2007	980316
2007	200062-5-33	2007	0573204	2007	815028C	2007	980375
2007	200062-5-34	2007	573528	2007	815029	2007	980376
2007	200064-6-108	2007	573556	2007	815029A	2007	980426
2007	200064-6-111	2007	0573432	2007	815029B	2007	980432
2007	215053	2007	0573422H	2007	815029C	2007	980434
2007	218502	2007	0573429	2007	815030	2007	980438
2007	218505	2007	573432	2007	815030A	2007	980439
2007	222509	2007	0573529	2007	815030B	2007	980508
2007	223124	2007	0573742	2007	815031	2007	980510
2007	223366	2007	574003	2007	815031A	2007	980511
2007	223571	2007	576001	2007	815031B	2007	980836
2007	223572	2007	576005	2007	815033	2007	980837
2007	223473	2007	581005	2007	815034	2007	980960
2007	223475	2007	581042	2007	815034A	2007	980961
2007	223483	2007	602040	2007	815034B	2007	981147
2007	223484	2007	602051	2007	815034C	2007	981148
2007	223485	2007	606030	2007	815036	2007	981149
2007	223486	2007	606020	2007	815036A	2007	981206
2007	223810	2007	606112	2007	815036B	2007	981672
2007	224584	2007	610035	2007	815036C	2007	981673
2007	224696	2007	610035-1	2007	815037	2007	981674
2007	224857	2007	610035-2	2007	815037A	2007	981675
2007	225593	2007	610036	2007	815037B	2007	982151
2007	226181						

RS 276-	Industry Standard No.	RS 276-	Industry Standard No.	RS 276-	Industry Standard No.	RS 276-	Industry Standard No.
2007	982152	2007	2091578-1	2009	28C1293B(3RD IP)	2009	28C710F
2007	982244	2007	2092693-1	2009	28C1293B(3RD-IP)	2009	28C710G
2007	982283	2007	2092693-8	2009	28C1293B(LAST IP)	2009	28C710GN
2007	982284	2007	2243255-1	2009	28C1293B(LAST-IP)	2009	28C710H
2007	982285	2007	2320011	2009	28C1293C	2009	28C710K
2007	982375	2007	2320154	2009	28C1293C(3RD IP)	2009	28C710L
2007	982531	2007	2320261	2009	28C1293C(3RD-IP)	2009	28C710M
2007	982532	2007	2320302	2009	28C1293D	2009	28C710OR
2007	982820	2007	2320302H	2009	28C1390I(W)	2009	28C710R
2007	983237	2007	2320422	2009	28C1390J(X)	2009	28C710XL
2007	983238	2007	2320422-1	2009	28C1687	2009	28C710Y
2007	983405	2007	2320423	2009	28C1688	2009	28C715
2007	983406	2007	2320492	2009	28C372	2009	28C715-0R
2007	983407	2007	2320512	2009	28C372(3RD-IP)	2009	28C715A
2007	983408	2007	2320513	2009	28C372(H)	2009	28C715B
2007	983409	2007	2320514-1	2009	28C372(O)	2009	28C715C
2007	983411	2007	2320515	2009	28C372(Y)	2009	28C715D
2007	984160	2007	2495014	2009	28C372-0	2009	28C715E
2007	984161	2007	2495080	2009	28C372-1	2009	28C715EJ
2007	984221	2007	2495388	2009	28C372-2	2009	28C715EV
2007	984228	2007	2495388-1	2009	28C372-0R	2009	28C715F
2007	984746	2007	2495388-2	2009	28C372-0RG	2009	28C715G
2007	985216	2007	2495567-2	2009	28C372-0RG-G	2009	28C715GN
2007	985217	2007	2495567-3	2009	28C372-R	2009	28C715H
2007	985468	2007	2495568	2009	28C372-WORR	2009	28C715J
2007	985468A	2007	2495568-2	2009	28C372-Y	2009	28C715K
2007	985469	2007	2497473	2009	28C372-YEL-Q	2009	28C715L
2007	985469A	2007	2497473-1	2009	28C372-Z	2009	28C715M
2007	985470	2007	2497496	2009	28C372/4454C	2009	28C715N
2007	985470A	2007	2497888	2009	28C372A	2009	28C715X
2007	985471	2007	2498837-4	2009	28C372AR	2009	28C715XL
2007	985609	2007	2855296-01	2009	28C372B	2009	28C715Y
2007	985610	2007	2970038805	2009	28C372BL	2009	28CF14
2007	986302	2007	3004856	2009	28C372C	2009	28C227D
2007	986305	2007	3130011	2009	28C372D	2009	28D636
2007	986543	2007	3130025	2009	28C372E	2009	28D636-0
2007	986766	2007	3130060	2009	28C372F	2009	28D636-0
2007	986779	2007	3404114-1	2009	28C372G	2009	004-00(LAST IP)
2007	988005	2007	3404114-2	2009	28C372G-0	2009	09-302004
2007	992289	2007	3460679-1	2009	28C372G-Y	2009	09-302016
2007	995002	2007	3464482-1	2009	28C372GN	2009	09-302200
2007	995003	2009	A1U(3RDIF)	2009	28C372GR	2009	13-15842-1
2007	1211016/7603/7608	2009	A1U(LAST IP)	2009	28C372H	2009	13-23160-2
2007	1211017/7603/7608	2009	A1Z	2009	28C372J	2009	13-23824-1(3RDIP)
2007	1211019/7603/7608	2009	BC758	2009	28C372K	2009	13-23824-3
2007	1221648	2009	BP249	2009	28C372L	2009	21A040-046
2007	1221649	2009	C1128(3RD IP)	2009	28C372M	2009	21A040-047
2007	1420427-1	2009	C1293A(LAST IP)	2009	28C372OR	2009	34-6000-72
2007	1420427-2	2009	C1293B(3RD IP)	2009	28C372R	2009	34-6015-37
2007	1420427-3	2009	C1293B(3RD-IP)	2009	28C372X	2009	34-6015-38
2007	1443200-3	2009	C1293B(LAST IP)	2009	28C372Y	2009	34-6015-61
2007	1471100-1	2009	C1293B(LAST-IP)	2009	28C372X(3RD IP)	2009	34-6015-62
2007	1471100-8	2009	C388(TRANSISTOR)	2009	28C372YEL	2009	46-86131-3
2007	1471100-9	2009	C388A	2009	28C372YEL-G	2009	46-86531-3
2007	1471101-15	2009	C388ATV	2009	28C372Z	2009	48-134932
2007	1471101-2	2009	C460A	2009	28C383(3RD IP)	2009	48-134932(3RD LP)
2007	1471101-3	2009	C460B	2009	28C383(PINAL IP)	2009	48-4937
2007	1471101-4	2009	C460C	2009	28C383T(LAST IP)	2009	57A138-4(3RD LP)
2007	1472482-1	2009	C460D	2009	28C388	2009	57A138-4(LAST LP)
2007	1473578-1	2009	C460G	2009	28C388-0R	2009	57A142-4
2007	1473598-2	2009	C460GB	2009	28C388A	2009	57A142-4(3RDLP)
2007	1956016	2009	C460H	2009	28C388A(3RD IP)	2009	57A180-4
2007	1960643	2009	C460K	2009	28C388A(3RD-IP)	2009	57A180-4(3RDIP)
2007	1961837	2009	C460L	2009	28C388ATV	2009	57A180-4(3RDLP)
2007	2002887-28	2009	C577	2009	28C388B	2009	57B138-4(LAST LP)
2007	2000625-31	2009	C563A(3RDIP)	2009	28C388C	2009	57B142-4
2007	2000625-32	2009	C715	2009	28C388D	2009	86-513-2
2007	2000625-33	2009	C717(FINAL IP)	2009	28C388E	2009	86X0034-001
2007	2000625-34	2009	CF14	2009	28C388F	2009	998032(3RD LP)
2007	2000646-108	2009	CG8-61	2009	28C388G	2009	121-522
2007	2000646-111	2009	C89012HF(LAST IP)	2009	28C388GN	2009	121-524
2007	2000604-9	2009	C89021HF(LAST IP)	2009	28C388H	2009	121-524A
2007	2001653	2009	C89021HG(LAST IP)	2009	28C388J	2009	121-526
2007	2001653-22	2009	DDBY224003	2009	28C388K	2009	121-713
2007	2001653-23	2009	DDBY273001	2009	28C388L	2009	121-951
2007	2001653-24	2009	D8-94	2009	28C388M	2009	145N1(LAST LP)
2007	2001809-47	2009	E13-004-00	2009	28C388OR	2009	180N1
2007	2001809-48	2009	ECG233	2009	28C388R	2009	186N1(LAST LP)
2007	2001809-48B	2009	EP15X20	2009	28C388S	2009	260P11102
2007	2001812-65	2009	EP15X54	2009	28C388X	2009	576-1
2007	2002151-19	2009	ES15X106	2009	28C536A(3RD IP)	2009	676-1
2007	2002152-14	2009	ES15X127	2009	28C537(P)	2009	1010(GE)
2007	2002153-71	2009	ES15X22	2009	28C537(Q)	2009	1075
2007	2002153-78	2009	LM1133	2009	28C537-EH	2009	1373-17
2007	2002210-110	2009	M4937(3RD IP)	2009	28C537-EV	2009	2502-17
2007	2002211-24	2009	M89002	2009	28C537ALC	2009	3576
2007	2002211-25	2009	NPC1075	2009	28C537B	2009	3676
2007	2003073-14	2009	PE5025	2009	28C537BK	2009	3676(RCA)
2007	2003073-15	2009	Q-00384R	2009	28C537C	2009	8000-00003-034
2007	2003073-8	2009	Q-00584R	2009	28C537C7	2009	12454
2007	2004358-123	2009	Q50787	2009	28C537D	2009	61086-1
2007	2004358-168	2009	Q50787Z	2009	28C537D1	2009	082020
2007	2006226-14	2009	R3576-1	2009	28C537D2	2009	124754
2007	2006334-31	2009	R3676-1	2009	28C537E	2009	137338
2007	2006441-113	2009	REN233	2009	28C537EF	2009	137338(3RD IP)
2007	2006513-133	2009	S0002	2009	28C537EH	2009	137338(3RD-IP)
2007	2006681-120	2009	S0024	2009	28C537EJ	2009	147676-1
2007	2008293-109	2009	SK3132	2009	28C537EK	2009	0573486
2007	2008293-111	2009	SK3551	2009	28C537EL	2009	05734BT
2007	2047102	2009	T-Q5078	2009	28C537F	2009	610145-1
2007	2076945-0701	2009	TD102	2009	28C537F-C7	2009	610180-1(LAST IP)
2007	2090924-0008	2009	TD201	2009	28C537F1	2009	741856
2007	2090924-008	2009	TD202	2009	28C537F2	2009	1473576-1
2007	2090924-48	2009	TD2219	2009	28C537FC	2009	1473676-1(3RD LP)
2007	2090924-48A	2009	T628C1293	2009	28C537FC7	2009	2320041
2007	2090924B	2009	T628C1293(A)	2009	28C537FJ	2009	2320041(LAST IP)
2007	2091211-0014	2009	T628C1293-	2009	28C537FK	2010	A-1853-0024-1
2007	2091241-0013	2009	T628C1293-A-A	2009	28C537FV	2010	A-1854-0023-1
2007	2091241-0014	2009	T628C1293-B-A	2009	28C5370	2010	A-1854-0084-1
2007	2091241-0015	2009	T628C1293-C-A	2009	28C537G1	2010	A-1854-0284-1
2007	2091241-0018	2009	T628C1293-D-A	2009	28C537G2	2010	A066-143
2007	2091241-1	2009	T628C1293A	2009	28C537GFL	2010	A110
2007	2091241-11	2009	TM233	2009	28C537GI	2010	A110(JAPAN)
2007	2091241-12	2009	TNJ72276	2009	28C537H	2010	A1238
2007	2091241-13	2009	TV-20	2009	28C537HT	2010	A139
2007	2091241-13A	2009	TV-54	2009	28C537W	2010	A139
2007	2091241-14	2009	TV-55	2009	28C537WF	2010	A1460
2007	2091241-15	2009	TV-77	2009	28C705TV(3RD IP)	2010	A2370773
2007	2091241-15A	2009	TV37	2009	28C710	2010	A28C538R
2007	2091241-2	2009	TV81	2009	28C710(B)	2010	A3J
2007	2091241-3	2009	01-030458	2009	28C710(C)	2010	A3K
2007	2091241-4	2009	01-030710	2009	28C710(D)	2010	A4B
2007	2091241-6	2009	01-031293	2009	28C710-1	2010	A4F
2007	2091241-7	2009	2N6004	2009	28C710-2	2010	A523707
2007	2091241-8	2009	28C1128(3RD IP)	2009	28C710-4	2010	A523708
2007	2091241-9	2009	28C1128(FINAL IP)	2009	28C710-0R	2010	A523709
2007	2091260-1(PNP)	2009	28C1293	2009	28C710AL	2010	A523710
2007	2091260-2(PNP)	2009	28C1293(3RD IP)	2009	28C710B2	2010	A523711
2007	2091260-3(PNP)	2009	28C1293(3RD-IP)	2009	28C710BC	2010	A642L
2007	2091578-0702	2009	28C1293(A)	2009	28C710D	2010	A642L(NPN)
		2009	28C1293A	2009	28C710DE	2010	A6428
		2009	28C1293A(LAST IP)	2009	28C710E	2010	A6428(NPN)
		2009	28C1293A(LAST-IP)			2010	A644L
		2009	28C1293B				

RS 276-	Industry Standard No.	RS 276-	Industry Standard No.	RS 276-	Industry Standard No.	RS 276-	Industry Standard No.
2010	A644S	2010	C1312H	2010	C693GH	2010	CQB-62
2010	A645L	2010	C1312Y	2010	C693GS	2010	CK420
2010	A645S	2010	C1312YF	2010	C693GU	2010	CK421
2010	A667-GRN	2010	C1312YG	2010	C693GZ	2010	CS4003
2010	A667-ORG	2010	C1312YH	2010	C693H	2010	CS4061
2010	A667-RED	2010	C1313	2010	C694	2010	CS9104
2010	A667-YEL	2010	C1313F	2010	C694E	2010	CXL199
2010	A668-GRN	2010	C1313G	2010	C694F	2010	D057
2010	A668-ORG	2010	C1313H	2010	C694G	2010	D232
2010	A668-YEL	2010	C1313Y	2010	C694Z	2010	D24A33391
2010	A669-GRN	2010	C1313YF	2010	C7076	2010	D24A33391A
2010	A669-YEL	2010	C1313YG	2010	C711	2010	D24A33392
2010	A67-07-244	2010	C1313YH	2010	C711(E)	2010	D24A33393
2010	A67-33-340	2010	C1327	2010	C711A	2010	D24A3900
2010	A6754194H	2010	C1327FS	2010	C711AE	2010	D24A3900A
2010	A747B	2010	C1327U	2010	C711D	2010	D2604
2010	A748C	2010	C1328	2010	C711E	2010	D2605
2010	A749	2010	C1328T	2010	C711F	2010	D26B-1
2010	A749C	2010	C1328U	2010	C711FG	2010	D26B-5
2010	A76228	2010	C1335	2010	C711G	2010	D26B-7
2010	A8R	2010	C1335A	2010	C732	2010	D30B
2010	A9-175	2010	C1335B	2010	C732BL	2010	D32P1
2010	AR213(V)	2010	C1335C	2010	C732GR	2010	D32P2
2010	AR213VIOLET	2010	C1335D	2010	C732S	2010	D32P3
2010	AR218(RO)	2010	C1335E	2010	C732Y	2010	D32P4
2010	B-1910	2010	C1335F	2010	C732Y	2010	DBC2037300
2010	BC109B	2010	C1344	2010	C733	2010	DBC2094504
2010	BC127	2010	C1344C	2010	C733-O	2010	DBC2373000
2010	BC128	2010	C1344D	2010	C733-O	2010	DDBY224001
2010	BC155C	2010	C1344E	2010	C733BL	2010	DDBY224004
2010	BC156C	2010	C1344F	2010	C733GR	2010	DDBY224006
2010	BC170	2010	C1345	2010	C733R	2010	DDBY262001
2010	BC170A	2010	C1345C	2010	C733V	2010	DDBY299001
2010	BC170B	2010	C1345F	2010	C733Y	2010	DDBY4233001
2010	BC170C	2010	C1416BL	2010	C828	2010	EA15X152
2010	BC173	2010	C1537	2010	C828-O	2010	EA15X161
2010	BC173A	2010	C1537-3	2010	C828-OP	2010	EA15X245
2010	BC173B	2010	C1537-O	2010	C828A	2010	EA15X251
2010	BC173C	2010	C1537B	2010	C828AP	2010	EA15X258
2010	BC182K	2010	C1537FB	2010	C828AQ	2010	EA15X259
2010	BC182KA	2010	C1538	2010	C828AR	2010	EA15X264
2010	BC182KB	2010	C1538A	2010	C828B	2010	EA15X288
2010	BC183K	2010	C1538B	2010	C828BP	2010	EA15X325
2010	BC183KA	2010	C1538BA	2010	C828FR	2010	EA15X352
2010	BC183KB	2010	C1633	2010	C828H	2010	EA15X353
2010	BC183KC	2010	C1648	2010	C828K	2010	EA15X354
2010	BC184K	2010	C1681BL	2010	C828LR	2010	EA15X386
2010	BC184KB	2010	C1684	2010	C828LS	2010	EA15X404
2010	BC184KC	2010	C1685Q	2010	C828N	2010	EA15X7245
2010	BC209/7825B	2010	C335	2010	C828P	2010	EA15X7583
2010	B20C	2010	C568	2010	C828Q	2010	EA15X7639
2010	BC238C	2010	C368BL	2010	C828QRS	2010	EA2429
2010	BC239A	2010	C368GR	2010	C828R	2010	EA2738
2010	BC239B	2010	C368Y	2010	C828R/494	2010	EA2740
2010	BC239C	2010	C2771-R-1	2010	C828T	2010	EA2771
2010	BC382B	2010	C373	2010	C828W	2010	EA3211
2010	BC382C	2010	C373BL	2010	C828Y	2010	EA3763
2010	BC383B	2010	C373G	2010	C833BL	2010	EA4025
2010	BC383C	2010	C373GR	2010	C838(H)	2010	EC9199
2010	BC384B	2010	C373W	2010	C838(J)	2010	ED1702N
2010	BC384C	2010	C374	2010	C838(K)	2010	EDC-CR-11-1
2010	BC385A	2010	C374-BL	2010	C838(M)	2010	EN956
2010	BC385B	2010	C374-V	2010	C858	2010	EP15X3
2010	BC386A	2010	C374JA	2010	C858B	2010	EQ8-0061
2010	BC386B	2010	C400-GR	2010	C858P	2010	EQ8-10
2010	BC408B	2010	C536	2010	C858PG	2010	EQ8-11
2010	BC408C	2010	C536A	2010	C858Q	2010	EQ8-131
2010	BC413B	2010	C536AG	2010	C859	2010	EQ8-78
2010	BC413C	2010	C536B	2010	C859C	2010	ES10231
2010	BC414B	2010	C536C	2010	C859F	2010	ES15049
2010	BC414C	2010	C536D	2010	C859PG	2010	ES15052
2010	BC520	2010	C536DK	2010	C859G	2010	ET379262
2010	BC520B	2010	C536E	2010	C859GK	2010	ET379462
2010	BC520C	2010	C536ED	2010	C900	2010	ET380834
2010	BC521	2010	C536EH	2010	C900(L)	2010	ET398711
2010	BC521C	2010	C536EJ	2010	C900A	2010	ET398777
2010	BC521D	2010	C536EN	2010	C900B	2010	ET517263
2010	BC522	2010	C536ER	2010	C900C	2010	ET517994
2010	BC522C	2010	C536ET	2010	C900D	2010	ETTC-945
2010	BC522D	2010	C536EZ	2010	C900E	2010	EW8-78
2010	BC522E	2010	C536F	2010	C900F	2010	F079
2010	BC523	2010	C536F1	2010	C900L	2010	F08100
2010	BC523B	2010	C536F2	2010	C900M	2010	F08101
2010	BC523C	2010	C536FC	2010	C900U	2010	FSE4002
2010	BC546	2010	C536FP	2010	C907	2010	FT107B
2010	BCW87	2010	C536FR	2010	C907A	2010	G05-012-Q
2010	BCW98A	2010	C536FS6	2010	C907AC	2010	G05-035-D,E
2010	BCW98B	2010	C536FZ	2010	C907AD	2010	G05-035-E
2010	BCW98C	2010	C536G	2010	C907AH	2010	G05-035D
2010	BCW98D	2010	C536GF	2010	C907C	2010	G05-706-D
2010	BCY55	2010	C536GK	2010	C907D	2010	G05-706-E
2010	BF340	2010	C536GT	2010	C907H	2010	G05035D
2010	BF341	2010	C536GV	2010	C907HA	2010	G212
2010	BF342	2010	C536GY	2010	C923	2010	GE-10A
2010	BF343	2010	C536H	2010	C923A	2010	GE-212
2010	BF597	2010	C536W	2010	C923B	2010	GE-62
2010	BFR25	2010	C644	2010	C923C	2010	GE-85
2010	BF855A	2010	C644C	2010	C923D	2010	GE12483
2010	BPT55	2010	C644F	2010	C923F	2010	GE13562
2010	BFV37	2010	C644F/494	2010	C930CK	2010	GI-2924
2010	BFV38	2010	C644FR	2010	C930DH	2010	GI-2925
2010	BFV62	2010	C644FS	2010	C930DZ	2010	GI-3391
2010	BFV89	2010	C644H	2010	C945	2010	GI-3391A
2010	BFV89A	2010	C644HR	2010	C945(R)	2010	GI-3392
2010	BFY39-1	2010	C644P	2010	C945-O	2010	GI-3393
2010	BFY39-2	2010	C644PJ	2010	C945A	2010	GI-3394
2010	BFY39-3	2010	C644Q	2010	C945AP	2010	GI-3395
2010	BTX-068	2010	C644R	2010	C945AQ	2010	GI-3396
2010	C1000	2010	C644RST	2010	C945B	2010	GI-3397
2010	C1000-BL	2010	C644S/494	2010	C945C	2010	GI-3398
2010	C1000-GR	2010	C644ST	2010	C945D	2010	GI-3707
2010	C1000Y	2010	C644T	2010	C945E	2010	GI-3708
2010	C1002	2010	C648	2010	C945F	2010	GI-3709
2010	C1006	2010	C648H	2010	C945G	2010	GI-3710
2010	C1006A	2010	C650	2010	C945H	2010	GI-3711
2010	C1006B	2010	C650B	2010	C945K	2010	GI-3721
2010	C1006C	2010	C644H	2010	C945L	2010	GI-3900
2010	C1010	2010	C693	2010	C945M	2010	GI-3900A
2010	C1010A	2010	C693(JAPAN)	2010	C945O	2010	G05-035-D,E
2010	C1010B	2010	C693A	2010	C945P	2010	G05-035-D,E
2010	C1010C	2010	C693B	2010	C945Q	2010	G05-035-E
2010	C1204	2010	C693C	2010	C945QL	2010	HC-00373
2010	C1204B	2010	C693D	2010	C945QP	2010	HC-00536
2010	C1204C	2010	C693E	2010	C945R	2010	HC-00711
2010	C1204D	2010	C693E(JAPAN)	2010	C945S	2010	HC-00732
2010	C1215	2010	C693EB	2010	C945T	2010	HC-00900
2010	C1222	2010	C693ET	2010	C945TQ	2010	HC-00925
2010	C1222A	2010	C693F	2010	C945TR	2010	HC-00929
2010	C1222B	2010	C693FC	2010	C945X	2010	HC-00930
2010	C1222C	2010	C693FL	2010	CC82004B	2010	HC-01000
2010	C1222D	2010	C693FU	2010	CC82004D303	2010	HC-01355
2010	C1312	2010	C693G	2010	CD441	2010	HC-537
2010	C1312F	2010	C693G(JAPAN)	2010	CD562	2010	HE-00930
2010	C1312Q					2010	HEP726

RS 276-	Industry Standard No.	RS 276-	Industry Standard No.	RS 276-	Industry Standard No.	RS 276-	Industry Standard No.
2010	HEP730	2010	PL4062	2010	TE3899	2010	2N3397-RED
2010	HEP737	2010	PN929	2010	TE3860	2010	2N3397-WHT
2010	HEP80023	2010	PN930	2010	TE3900	2010	2N3397-YEL
2010	HEP80024	2010	P86010-1	2010	TE3900A	2010	2N3398-BLU
2010	HEP80030	2010	Q-00469	2010	TE3901	2010	2N3398-ORG
2010	HP17	2010	Q-00469A	2010	TE4256	2010	2N3398-RED
2010	HP309301E	2010	Q-00469C	2010	TE5310	2010	2N3398-WHT
2010	HR-14	2010	Q-00569C	2010	TE5311	2010	2N3708-BLU
2010	HR-15	2010	Q-08115C	2010	TG280536	2010	2N3708-BRN
2010	HR-47	2010	Q-09115C	2010	TG280536(C)	2010	2N3708-GRN
2010	HR-75	2010	Q-13115C	2010	TG280536(E)	2010	2N3708-ORG
2010	HR47	2010	Q5053	2010	TG280536-D-A	2010	2N3708-RED
2010	HST5001	2010	Q5053D	2010	TG280536-D-B	2010	2N3708-VIO
2010	HT306442A	2010	Q5053E	2010	TG280536-E	2010	2N3708-YEL
2010	HT306442B	2010	Q5053F	2010	TG280536-E-A	2010	2N4086
2010	HT30733100	2010	Q5053G	2010	TG280536-E-B	2010	2N4087
2010	HT308281D	2010	Q5078Z	2010	TG280536-F	2010	2N4087A
2010	HT308281H	2010	Q5121	2010	TG280536-F-A	2010	2N4138
2010	HT308282B	2010	Q51210	2010	TG280536C	2010	2N5249
2010	HT309002A0	2010	Q51210	2010	TG280536E	2010	2N5249A
2010	HT309301C	2010	Q5121Q	2010	TI-415	2010	2N5824
2010	HT309301E	2010	Q5121R	2010	TI-416	2010	2N5825
2010	HT309301F	2010	Q5120R	2010	TI-418	2010	2N5826
2010	HT309451L0	2010	Q5180	2010	TI-419	2010	2N5827
2010	HT310001F	2010	Q5183	2010	TI-421	2010	2N5827A
2010	HT310002A	2010	Q5183P	2010	TI-92	2010	2N5828
2010	HT313271T	2010	QRT105	2010	TI54D	2010	2N5828A
2010	HT313272B	2010	QT-C0828XAN	2010	TIS-94	2010	2N5852
2010	HT36441B	2010	QT-C0828XDN	2010	TIS-97	2010	2N6112
2010	HX-50107	2010	QT-C0900XBA	2010	TM199	2010	2N841/46
2010	IP20-0034	2010	QT-C0900XBD	2010	TN28C945-Q	2010	28C-929
2010	IRTR-51	2010	QT-C0900XCA	2010	TN28C945-R	2010	28C-NJ-100
2010	IRTR51	2010	QT-C0945ACA	2010	TN28C945R	2010	28C-NJ100
2010	IT122	2010	QT-C0945AGA	2010	TNJ1034	2010	2SC1000
2010	IT918A	2010	QT-C1359XAN	2010	TNJ70691	2010	2SC1000(GR)
2010	ITC918A	2010	R8192	2010	TNJ71034	2010	2SC1000-BL
2010	J24812	2010	R84001	2010	TNJ71271	2010	2SC1000-GR
2010	J24875	2010	R84002	2010	TNJ71277	2010	2SC1000-Y
2010	J24932	2010	R84010	2010	TNJ71965	2010	2SC1000A
2010	JA1350	2010	R864	2010	TP4067-410	2010	2SC1000B
2010	JA1350B	2010	REN199	2010	TP4067-411	2010	2SC1000BL
2010	JA1350W	2010	REN64	2010	TR 19A	2010	2SC1000C
2010	KGB46146	2010	REN67	2010	TR-1993	2010	2SC1000D
2010	KSC945Y	2010	RLB-17	2010	TR-8034	2010	2SC1000E
2010	L4	2010	RQ-4448	2010	TR-8040	2010	2SC1000F
2010	L5	2010	RS-279U8	2010	TR-BC149C	2010	2SC1000G
2010	L6	2010	RT-104	2010	TRO1014	2010	2SC1000G-BL
2010	L7	2010	RT-105	2010	TRO1015	2010	2SC1000G-GR
2010	LDA410	2010	RT-107	2010	TRO1040	2010	2SC1000GN
2010	LDS207	2010	RT-108	2010	TR105(SPRAGUE)	2010	2SC1000GR
2010	LID929	2010	RT-112	2010	TR19A	2010	2SC1000H
2010	LID930	2010	RT2309	2010	TR2327443	2010	2SC1000J
2010	LM1117	2010	RT5208	2010	TR2327444	2010	2SC1000K
2010	LM1117C	2010	RT5435	2010	TR281570LH	2010	2SC1000L
2010	LM1540C	2010	RT7557	2010	TR33	2010	2SC1000M
2010	LM1540C	2010	RT7559	2010	TR4010	2010	2SC1000OR
2010	LS-0031-AR-218	2010	RT8047	2010	TRO1037	2010	2SC1000R
2010	LS-0095-AR-213	2010	RT8337	2010	TRPLC711	2010	2SC1000X
2010	LS-0095-AR-213	2010	RT8666	2010	TSC-499	2010	2SC1000Y
2010	M9197	2010	RT8863	2010	TSC767	2010	2SC1006
2010	M9269	2010	RY2070	2010	VS28C1335D/-1	2010	2SC1006A
2010	M9293	2010	RY2248	2010	VS28C1335D/-1E	2010	2SC1006B
2010	M9329	2010	RY2354	2010	VS28C16810/1E	2010	2SC1006C
2010	M9338	2010	RVTC81473	2010	VS28C1890A/1E	2010	2SC1010
2010	M9384	2010	S001465	2010	VS28C371-R-1	2010	2SC1010A
2010	M9409	2010	S0023	2010	VS28C373-/1E	2010	2SC1010B
2010	M9416	2010	S006793	2010	VS28C373//-1E	2010	2SC1010C
2010	M9447	2010	S006927	2010	VS28C373//-1E	2010	2SC103B
2010	M9474	2010	S007764	2010	VS28C730-1	2010	2SC103C
2010	M9486	2010	S2058	2010	VS28C374-B-1	2010	2SC103D
2010	M9547	2010	S24592	2010	VS28C383-W/1E	2010	2SC103E
2010	M9594	2010	S29956	2010	VS28C732-V1F	2010	2SC103F
2010	MMCM930	2010	SC147A	2010	VS28C733E-1	2010	2SC103G
2010	MM270	2010	SC147B	2010	VS28C945LK-1	2010	2SC103GN
2010	MPS-5172	2010	SC148	2010	VS28C945LP-1	2010	2SC103H
2010	MPS-A05	2010	SC148A	2010	VS9-0003-913	2010	2SC103K
2010	MPS-A09	2010	SC148B	2010	WEP1945	2010	2SC103L
2010	MPS3710	2010	SC148C	2010	WEP373	2010	2SC103M
2010	MPS6564	2010	SC149	2010	WEP403	2010	2SC103X
2010	MPS6572	2010	SC149B	2010	WEP536	2010	2SC103Y
2010	MPS89433J	2010	SCA3244	2010	WEP537	2010	2SC104B
2010	MPS89630J	2010	SE5-0127	2010	WEP634	2010	2SC104C
2010	MPS89630K	2010	SE5-0565	2010	WEP644	2010	2SC104D
2010	MPS89633C	2010	SE5-0569	2010	X19001-D	2010	2SC104F
2010	MPS89633D	2010	SE5-0938	2010	ZEN116	2010	2SC104G
2010	MPS8634	2010	SE5-0938-55	2010	ZT930	2010	2SC104GN
2010	MPS89634B	2010	SE5-0938-56	2010	ZTX107	2010	2SC104H
2010	MPS89634C	2010	SE5-0938-57	2010	ZTX108	2010	2SC104K
2010	MPSA09	2010	SKA-4802	2010	ZTX109	2010	2SC104L
2010	MPSA10	2010	S019806	2010	01-030373	2010	2SC104M
2010	MPSA10-BLU	2010	S025094	2010	01-030711	2010	2SC104R
2010	MPSA10-GRN	2010	SPB-1476	2010	01-030732	2010	2SC104OR
2010	MPSA10-RED	2010	SPB-1539(WT)	2010	01-030784	2010	2SC104X
2010	MPSA10-WHT	2010	SPS2216	2010	01-030900	2010	2SC104Y
2010	MPSA10-YEL	2010	SPS2217	2010	01-030945	2010	2SC104Z
2010	MPSA16	2010	SPS2271	2010	01-031327	2010	2SC1204B
2010	MPSA17	2010	SPS4272	2010	01-031685	2010	2SC1204C
2010	MPSA18	2010	SPS4814	2010	01-031815	2010	2SC1204D
2010	MPSA20-BLU	2010	SPS952	2010	1A0013	2010	2SC120A
2010	MPSA20-GRN	2010	ST-28C383W	2010	002-104-000	2010	2SC120B
2010	MPSA20-RED	2010	ST.082.114.016	2010	002-9501	2010	2SC120C
2010	MPSA20-WHT	2010	STZT003	2010	002-9502	2010	2SC120D
2010	MPSA20-YEL	2010	S133026	2010	002-9502-12	2010	2SC120E
2010	MPSD06	2010	S77100	2010	28537FC	2010	2SC120F
2010	MPSK20	2010	T-Q5093	2010	2N1268	2010	2SC120G
2010	MPSK21	2010	T01-047	2010	2N1269	2010	2SC120GN
2010	MPSK22	2010	T1007(ZENITH)	2010	2N1270	2010	2SC120H
2010	NB013	2010	T1341A3K	2010	2N1271	2010	2SC120J
2010	NPO069	2010	T82711	2010	2N1272	2010	2SC120K
2010	NPO069-98	2010	T82712	2010	2N1390	2010	2SC120L
2010	NPSA20	2010	T82921	2010	2N2432A	2010	2SC120M
2010	NR-421AS	2010	T82922	2010	2N2639	2010	2SC120R
2010	NTC-7	2010	T82923	2010	2N2640	2010	2SC120Y
2010	P1901-48	2010	T82924	2010	2N2641	2010	2SC121A
2010	P69941	2010	T82925	2010	2N2642	2010	2SC121B
2010	PA8260	2010	T82926	2010	2N2643	2010	2SC121C
2010	PA8543	2010	TE3390	2010	2N2644	2010	2SC121D
2010	PA9004	2010	TE3391	2010	2N2926-BRN	2010	2SC121E
2010	PA9005	2010	TE3391A	2010	2N2926-GRN	2010	2SC121F
2010	PBC107	2010	TE3392	2010	2N2926-ORG	2010	2SC121G
2010	PBC107A	2010	TE3393	2010	2N2926-RED	2010	2SC121GN
2010	PBC107B	2010	TE3394	2010	2N2926-YEL	2010	2SC121H
2010	PBC108	2010	TE3395	2010	2N3390-U29	2010	2SC121J
2010	PBC108A	2010	TE3396	2010	2N3391-U29	2010	2SC121L
2010	PBC108B	2010	TE3397	2010	2N3392-U29	2010	2SC121M
2010	PBC108C	2010	TE3398	2010	2N3393-U29	2010	2SC121OR
2010	PBC109	2010	TE3843	2010	2N3394-U29	2010	2SC121R
2010	PBC109B	2010	TE3844	2010	2N3395-WHT	2010	2SC121X
2010	PBC109C	2010	TE3845	2010	2N3395-YEL	2010	2SC1212
2010	PET4003	2010	TE3854	2010	2N3396-ORG	2010	2SC1222
2010	PIT-37	2010	TE3854A	2010	2N3396-WHT	2010	2SC1222A
2010	PL-176-025-9-001	2010	TE3855	2010	2N3396-YEL	2010	2SC1222B
2010	PL-176-031-9-002	2010	TE3855A	2010	2N3397-ORG		
2010	PL4061						

RS 276-	Industry Standard No.	RS 276-	Industry Standard No.	RS 276-	Industry Standard No.	RS 276-	Industry Standard No.
2010	2SC1222C	2010	2SC1380	2010	2SC196F	2010	2SC323N
2010	2SC1222D	2010	2SC1380-BL	2010	2SC196G	2010	2SC323OR
2010	2SC1222E	2010	2SC1380-GR	2010	2SC196GN	2010	2SC323R
2010	2SC1222U	2010	2SC1380A	2010	2SC196H	2010	2SC323X
2010	2SC122A	2010	2SC1380A-BL	2010	2SC196J	2010	2SC323Y
2010	2SC122B	2010	2SC1380A-GR	2010	2SC196K	2010	2SC355
2010	2SC122C	2010	2SC1399	2010	2SC196L	2010	2SC350A
2010	2SC122D	2010	2SC1399B	2010	2SC196M	2010	2SC350B
2010	2SC122E	2010	2SC1416	2010	2SC196OR	2010	2SC350C
2010	2SC122F	2010	2SC1453	2010	2SC196R	2010	2SC350D
2010	2SC122G	2010	2SC1538	2010	2SC196X	2010	2SC350E
2010	2SC122GN	2010	2SC1538A	2010	2SC196Y	2010	2SC350F
2010	2SC122H	2010	2SC1538S	2010	2SC197A	2010	2SC350G
2010	2SC122J	2010	2SC1538S(A)	2010	2SC197B	2010	2SC350GN
2010	2SC122K	2010	2SC1538SA	2010	2SC197C	2010	2SC350J
2010	2SC122L	2010	2SC1623	2010	2SC197D	2010	2SC350K
2010	2SC122M	2010	2SC1632	2010	2SC197E	2010	2SC350L
2010	2SC122OR	2010	2SC1633	2010	2SC197F	2010	2SC350M
2010	2SC122R	2010	2SC1647	2010	2SC197G	2010	2SC350OR
2010	2SC122X	2010	2SC1647Q	2010	2SC197GN	2010	2SC350R
2010	2SC122Y	2010	2SC1647RY	2010	2SC197H	2010	2SC350X
2010	2SC123A	2010	2SC1648	2010	2SC197J	2010	2SC350Y
2010	2SC123B	2010	2SC1648E	2010	2SC197K	2010	2SC368
2010	2SC123C	2010	2SC1648S	2010	2SC197L	2010	2SC368-BL
2010	2SC123D	2010	2SC1648SH	2010	2SC197OR	2010	2SC368-GR
2010	2SC123E	2010	2SC1681	2010	2SC197R	2010	2SC368A
2010	2SC123F	2010	2SC1681-GR	2010	2SC197X	2010	2SC368B
2010	2SC123G	2010	2SC1681BL	2010	2SC197Y	2010	2SC368BL
2010	2SC123GN	2010	2SC1681GR	2010	2SC281	2010	2SC368C
2010	2SC123H	2010	2SC1681V	2010	2SC281(B)	2010	2SC368D
2010	2SC123J	2010	2SC1682	2010	2SC281-OR	2010	2SC368E
2010	2SC123K	2010	2SC1684	2010	2SC281A	2010	2SC368F
2010	2SC123L	2010	2SC1684BL	2010	2SC281B	2010	2SC368G
2010	2SC123M	2010	2SC1684P	2010	2SC281BL	2010	2SC368GN
2010	2SC123OR	2010	2SC1684Q	2010	2SC281C	2010	2SC368GR
2010	2SC123R	2010	2SC1684R	2010	2SC281C-EP	2010	2SC368H
2010	2SC123X	2010	2SC1684S	2010	2SC281D	2010	2SC368J
2010	2SC123Y	2010	2SC1684T	2010	2SC281E	2010	2SC368K
2010	2SC124A	2010	2SC1685	2010	2SC281EP	2010	2SC368L
2010	2SC124B	2010	2SC1685-O	2010	2SC281F	2010	2SC368M
2010	2SC124C	2010	2SC1685P	2010	2SC281G	2010	2SC368OR
2010	2SC124D	2010	2SC1685Q	2010	2SC281GN	2010	2SC368R
2010	2SC124E	2010	2SC1685R	2010	2SC281H	2010	2SC368V
2010	2SC124F	2010	2SC1685S	2010	2SC281HA	2010	2SC368X
2010	2SC124G	2010	2SC1685T	2010	2SC281HB	2010	2SC368Y
2010	2SC124GN	2010	2SC1787	2010	2SC281HC	2010	2SC369-BL
2010	2SC124H	2010	2SC191A	2010	2SC281J	2010	2SC369-BLU-G
2010	2SC124J	2010	2SC191B	2010	2SC281K	2010	2SC369-GR
2010	2SC124K	2010	2SC191C	2010	2SC281L	2010	2SC369-GRN-Q
2010	2SC124L	2010	2SC191D	2010	2SC281M	2010	2SC369A
2010	2SC124M	2010	2SC191E	2010	2SC281OR	2010	2SC369B
2010	2SC124OR	2010	2SC191F	2010	2SC281R	2010	2SC369C
2010	2SC124R	2010	2SC191G	2010	2SC281X	2010	2SC369D
2010	2SC124X	2010	2SC191GN	2010	2SC283A	2010	2SC369E
2010	2SC124Y	2010	2SC191H	2010	2SC283B	2010	2SC369F
2010	2SC1285	2010	2SC191J	2010	2SC283C	2010	2SC369G-V
2010	2SC1310	2010	2SC191K	2010	2SC283D	2010	2SC369G/BL
2010	2SC1311	2010	2SC191L	2010	2SC283E	2010	2SC369G/GR
2010	2SC1312	2010	2SC191M	2010	2SC283F	2010	2SC369GN
2010	2SC1312A	2010	2SC191OR	2010	2SC283G	2010	2SC369H
2010	2SC1312C	2010	2SC191R	2010	2SC283GN	2010	2SC369J
2010	2SC1312D	2010	2SC191X	2010	2SC283H	2010	2SC369K
2010	2SC1312E	2010	2SC191Y	2010	2SC283J	2010	2SC369L
2010	2SC1312F	2010	2SC192A	2010	2SC283K	2010	2SC369M
2010	2SC1312G	2010	2SC192B	2010	2SC283L	2010	2SC369OR
2010	2SC1312GN	2010	2SC192C	2010	2SC283M	2010	2SC369R
2010	2SC1312H	2010	2SC192D	2010	2SC283OR	2010	2SC369X
2010	2SC1312J	2010	2SC192E	2010	2SC283R	2010	2SC369Y
2010	2SC1312K	2010	2SC192F	2010	2SC283X	2010	2SC3721
2010	2SC1312L	2010	2SC192G	2010	2SC283Y	2010	2SC3721GR
2010	2SC1312M	2010	2SC192GN	2010	2SC284A	2010	2SC3721(GR)
2010	2SC1312OR	2010	2SC192H	2010	2SC284B	2010	2SC373-14
2010	2SC1312R	2010	2SC192J	2010	2SC284C	2010	2SC373-O
2010	2SC1312X	2010	2SC192K	2010	2SC284D	2010	2SC373-OR
2010	2SC1312Y	2010	2SC192L	2010	2SC284E	2010	2SC373A
2010	2SC1312YP	2010	2SC192M	2010	2SC284F	2010	2SC373AL
2010	2SC1312YG	2010	2SC192OR	2010	2SC284G	2010	2SC373B
2010	2SC1312YH	2010	2SC192R	2010	2SC284GN	2010	2SC373BL
2010	2SC1313	2010	2SC192X	2010	2SC284J	2010	2SC373C
2010	2SC1313B	2010	2SC192Y	2010	2SC284K	2010	2SC373D
2010	2SC1313F	2010	2SC193A	2010	2SC284L	2010	2SC373F
2010	2SC1313G	2010	2SC193B	2010	2SC284M	2010	2SC373G
2010	2SC1313H	2010	2SC193C	2010	2SC284OR	2010	2SC373GN
2010	2SC1313Y	2010	2SC193D	2010	2SC284R	2010	2SC373GR
2010	2SC1313YE	2010	2SC193E	2010	2SC284X	2010	2SC373H
2010	2SC1313YG	2010	2SC193F	2010	2SC284Y	2010	2SC373J
2010	2SC1313YH	2010	2SC193G	2010	2SC28A	2010	2SC373K
2010	2SC1327	2010	2SC193GN	2010	2SC28B	2010	2SC373L
2010	2SC1327PS	2010	2SC193H	2010	2SC28C	2010	2SC373M
2010	2SC1327S	2010	2SC193J	2010	2SC28D	2010	2SC373OR
2010	2SC1327T	2010	2SC193K	2010	2SC28E	2010	2SC373R
2010	2SC1327TU	2010	2SC193L	2010	2SC28F	2010	2SC373W
2010	2SC1327TV	2010	2SC193M	2010	2SC28G	2010	2SC373X
2010	2SC1327U	2010	2SC193OR	2010	2SC28GN	2010	2SC373Y
2010	2SC1328	2010	2SC193R	2010	2SC28H	2010	2SC374
2010	2SC1328(U)	2010	2SC193X	2010	2SC28J	2010	2SC374(BL)
2010	2SC1328(U)(T)	2010	2SC194A	2010	2SC28K	2010	2SC374(V)
2010	2SC1328T	2010	2SC194B	2010	2SC28L	2010	2SC374-BL
2010	2SC1328U	2010	2SC194C	2010	2SC28M	2010	2SC374-OR
2010	2SC1335	2010	2SC194D	2010	2SC28OR	2010	2SC374-V
2010	2SC1335(E)	2010	2SC194E	2010	2SC28P	2010	2SC374A
2010	2SC1335-OR	2010	2SC194F	2010	2SC28X	2010	2SC374B
2010	2SC1335A	2010	2SC194G	2010	2SC318AB	2010	2SC374BLK
2010	2SC1335B	2010	2SC194GN	2010	2SC318B	2010	2SC374C
2010	2SC1335C	2010	2SC194H	2010	2SC318C	2010	2SC374D
2010	2SC1335D	2010	2SC194J	2010	2SC318D	2010	2SC374E
2010	2SC1335E	2010	2SC194K	2010	2SC318E	2010	2SC374F
2010	2SC1335F	2010	2SC194L	2010	2SC318F	2010	2SC374G
2010	2SC1335G	2010	2SC194M	2010	2SC318G	2010	2SC374GN
2010	2SC1335GN	2010	2SC194OR	2010	2SC318GN	2010	2SC374H
2010	2SC1335H	2010	2SC194X	2010	2SC318H	2010	2SC374J
2010	2SC1335J	2010	2SC194Y	2010	2SC318J	2010	2SC374JA
2010	2SC1335K	2010	2SC195A	2010	2SC318K	2010	2SC374K
2010	2SC1335L	2010	2SC195B	2010	2SC318L	2010	2SC374L
2010	2SC1335M	2010	2SC195C	2010	2SC318M	2010	2SC374M
2010	2SC1335OR	2010	2SC195D	2010	2SC318OR	2010	2SC374OR
2010	2SC1335R	2010	2SC195E	2010	2SC318R	2010	2SC374R
2010	2SC1335X	2010	2SC195F	2010	2SC318X	2010	2SC374V
2010	2SC1335Y	2010	2SC195G	2010	2SC318Y	2010	2SC374X
2010	2SC1344	2010	2SC195GN	2010	2SC323A	2010	2SC374Y
2010	2SC1344(E)	2010	2SC195H	2010	2SC323B	2010	2SC376A
2010	2SC1344C	2010	2SC195J	2010	2SC323C	2010	2SC376B
2010	2SC1344D	2010	2SC195K	2010	2SC323D	2010	2SC376C
2010	2SC1344E	2010	2SC195L	2010	2SC323E	2010	2SC376D
2010	2SC1344F	2010	2SC195M	2010	2SC323F	2010	2SC376E
2010	2SC1345	2010	2SC195OR	2010	2SC323G	2010	2SC376F
2010	2SC1345(E)	2010	2SC195R	2010	2SC323GN	2010	2SC376G
2010	2SC1345C	2010	2SC195X	2010	2SC323H	2010	2SC376GN
2010	2SC1345D	2010	2SC195Y	2010	2SC323J	2010	2SC376H
2010	2SC1345E	2010	2SC196A	2010	2SC323K	2010	2SC376J
2010	2SC1345F	2010	2SC196B	2010	2SC323L	2010	2SC376K
2010	2SC1359C	2010	2SC196C				
2010	2SC1359Q	2010	2SC196D				
2010	2SC1362	2010	2SC196E			2010	2SC376L

RS 276-	Industry Standard No.	RS 276-	Industry Standard No.	RS 276-	Industry Standard No.	RS 276-	Industry Standard No.	RS 276-	Industry Standard No.
2010	28C376M	2010	28C475R	2010	28C644(R)	2010	2SC711A(E)	2010	2SC828G
2010	28C3760R	2010	28C475X	2010	28C644(R,S)	2010	2SC711AE		
2010	28C376R	2010	28C475Y	2010	28C644(S)	2010	2SC711AF		
2010	28C376X	2010	28C476A	2010	28C644-OR	2010	2SC711AG		
2010	28C376Y	2010	28C476B	2010	28C644A	2010	2SC711AN		
2010	28C378-0	2010	28C476C	2010	28C644B	2010	2SC711B		
2010	28C378-ORG	2010	28C476D	2010	28C644C	2010	2SC711C		
2010	28C378-R	2010	28C476E	2010	28C644D	2010	2SC711D		
2010	28C378-RED	2010	28C476F	2010	28C644E	2010	2SC711E		
2010	28C378-Y	2010	28C476GN	2010	28C644F	2010	2SC711F		
2010	28C378-YEL	2010	28C476H	2010	28C644F(H)(S)	2010	2SC711PG		
2010	28C378A	2010	28C476J	2010	28C644FH	2010	2SC711G		
2010	28C378B	2010	28C476K	2010	28C644FHS	2010	2SC711GN		
2010	28C378C	2010	28C476L	2010	28C644FR	2010	2SC711H		
2010	28C378D	2010	28C476M	2010	28C644FS	2010	2SC711J		
2010	28C378E	2010	28C4760R	2010	28C644G	2010	2SC711M		
2010	28C378F	2010	28C476R	2010	28C644GN	2010	2SC711OR		
2010	28C378G	2010	28C476X	2010	28C644H	2010	2SC711R		
2010	28C378GN	2010	28C476Y	2010	28C644H(S)	2010	2SC711X		
2010	28C378H	2010	28C528	2010	28C644HR	2010	2SC711Y		
2010	28C378J	2010	28C536	2010	28C644HS	2010	2SC732		
2010	28C378K	2010	28C536(C)	2010	28C644J	2010	2SC732(BL)		
2010	28C378L	2010	28C536(D)	2010	28C644K	2010	2SC732-B		
2010	28C378M	2010	28C536(E)	2010	28C644L	2010	2SC732-BL		
2010	28C3780R	2010	28C536-D	2010	28C644M	2010	2SC732-BLU		
2010	28C378R	2010	28C536-E	2010	28C6440R	2010	2SC732-G		
2010	28C378X	2010	28C536-F	2010	28C644P	2010	2SC732-GR		
2010	28C378Y	2010	28C536-G	2010	28C644Q	2010	2SC732-GRN		
2010	28C379A	2010	28C536-OR	2010	28C644R	2010	2SC732-OR		
2010	28C379B	2010	28C536A	2010	28C644RST	2010	2SC732-V		
2010	28C379C	2010	28C536A(3RD-IP)	2010	28C644S	2010	2SC732-V10		
2010	28C379D	2010	28C536AG	2010	0002C8C644S,R,Q	2010	2SC732-VIO		
2010	28C379E	2010	28C536B	2010	28C644T	2010	2SC732A		
2010	28C379F	2010	28C536C	2010	28C644X	2010	2SC732B		
2010	28C379G	2010	28C536D	2010	28C644Y	2010	2SC732BL		
2010	28C379GN	2010	28C536DK	2010	28C648	2010	2SC732BL-1		
2010	28C379H	2010	28C536E	2010	28C648H	2010	2SC732BLU		
2010	28C379J	2010	28C536ED	2010	28C648Y-RST	2010	2SC732C		
2010	28C379K	2010	28C536EH	2010	28C650	2010	2SC732D		
2010	28C379L	2010	28C536EJ	2010	28C650-OR	2010	2SC732E		
2010	28C379M	2010	28C536EN	2010	28C650-Y	2010	2SC732F		
2010	28C3790R	2010	28C536EP	2010	28C650A	2010	2SC732G		
2010	28C379X	2010	28C536ER	2010	28C650B	2010	2SC732GN		
2010	28C379Y	2010	28C536ET	2010	28C650C	2010	2SC732GR		
2010	28C383	2010	28C536EZ	2010	28C650D	2010	2SC732GR/44540		
2010	28C383(3RD-IP)	2010	28C536P	2010	28C650E	2010	2SC732GRB		
2010	28C383(FINAL-IP)	2010	28C536P1	2010	28C650F	2010	2SC732GRN		
2010	28C383(T)	2010	28C536P2	2010	28C650G	2010	2SC732H		
2010	28C383-OR	2010	28C536PC	2010	28C650GN	2010	2SC732J		
2010	28C383A	2010	28C536PP	2010	28C650H	2010	2SC732L		
2010	28C383B	2010	28C536PS	2010	28C650J	2010	2SC732M		
2010	28C383C	2010	28C536PS6	2010	28C650K	2010	2SC732OR		
2010	28C383D	2010	28C536PZ	2010	28C650L	2010	2SC732R		
2010	28C383E	2010	28C536Q	2010	28C650M	2010	2SC732S		
2010	28C383F	2010	28C536Q-1	2010	28C6500R	2010	2SC732V		
2010	28C383GN	2010	28C536QG1	2010	28C650R	2010	2SC732V10		
2010	28C383H	2010	28C536QG2	2010	28C650X	2010	2SC732VIO		
2010	28C383J	2010	28C536QGP	2010	28C650Y	2010	2SC732X		
2010	28C383K	2010	28C536QGK	2010	28C655A	2010	2SC732Y		
2010	28C383L	2010	28C536QGL	2010	28C655B	2010	2SC733		
2010	28C383M	2010	28C536QGM	2010	28C655C	2010	2SC733(GR)		
2010	28C3830R	2010	28C536QGN	2010	28C655E	2010	2SC733-0		
2010	28C383R	2010	28C536QGP	2010	28C655F	2010	2SC733-B		
2010	28C383T	2010	28C536QGT	2010	28C655G	2010	2SC733-BL		
2010	28C383T(LAST-IP)	2010	28C536QGV	2010	28C655GN	2010	2SC733-BLU		
2010	28C383W	2010	28C536QGY	2010	28C655H	2010	2SC733-G		
2010	28C383X	2010	28C536QGZ	2010	28C655J	2010	2SC733-GR		
2010	28C383Y	2010	28C536H	2010	28C655K	2010	2SC733-GRN		
2010	28C400-GR	2010	28C536J	2010	28C655M	2010	2SC733-0		
2010	28C400-0	2010	28C536K	2010	28C6550R	2010	2SC733-OR		
2010	28C400A	2010	28C536L	2010	28C655R	2010	2SC733-ORG		
2010	28C400B	2010	28C536M	2010	28C655S	2010	2SC733-Y		
2010	28C400C	2010	28C536NP	2010	28C655T	2010	2SC733-YEL		
2010	28C400D	2010	28C5360R	2010	28C655X	2010	2SC733A		
2010	28C400E	2010	28C536Q	2010	28C689A	2010	2SC733B		
2010	28C400F	2010	28C536R	2010	28C689B	2010	2SC733BL		
2010	28C400G	2010	28C536W	2010	28C689C	2010	2SC733BLK		
2010	28C400GN	2010	28C536X	2010	28C689D	2010	2SC733BLU		
2010	28C400H	2010	28C536XL	2010	28C689E	2010	2SC733C		
2010	28C400J	2010	28C536Y	2010	28C689G	2010	2SC733D		
2010	28C400K	2010	28C539(L)(K)	2010	28C689GN	2010	2SC733E		
2010	28C400L	2010	28C539(R)	2010	28C689J	2010	2SC733ER		
2010	28C400M	2010	28C539T	2010	28C689K	2010	2SC733F		
2010	28C4000R	2010	28C540A	2010	28C689L	2010	2SC733GN		
2010	28C400R	2010	28C540B	2010	28C689M	2010	2SC733GR		
2010	28C400X	2010	28C540C	2010	28C6890R	2010	2SC733GRN		
2010	28C400Y	2010	28C540D	2010	28C689R	2010	2SC733H		
2010	28C402B	2010	28C540E	2010	28C689X	2010	2SC733J		
2010	28C402C	2010	28C540F	2010	28C689Y	2010	2SC733K		
2010	28C402D	2010	28C540G	2010	28C693	2010	2SC733L		
2010	28C402E	2010	28C540GN	2010	28C693-OR	2010	2SC733M		
2010	28C402F	2010	28C540H	2010	28C693A	2010	2SC7330		
2010	28C402G	2010	28C540J	2010	28C693B	2010	2SC7330R		
2010	28C402GN	2010	28C540K	2010	28C693C	2010	2SC733Q		
2010	28C402H	2010	28C540L	2010	28C693D	2010	2SC733R		
2010	28C402J	2010	28C540M	2010	28C693E	2010	2SC733S		
2010	28C402K	2010	28C5400R	2010	28C693EB	2010	2SC733S-BL		
2010	28C402L	2010	28C540R	2010	28C693ET	2010	2SC733V		
2010	28C402M	2010	28C540X	2010	28C693F	2010	2SC733X		
2010	28C4020R	2010	28C540Y	2010	28C693FC	2010	2SC733Y		
2010	28C402R	2010	28C587C	2010	28C693FL	2010	2SC733YEL		
2010	28C402X	2010	28C587C	2010	28C693FP	2010	2SC828(H)		
2010	28C402Y	2010	28C587D	2010	28C693FU	2010	2SC828(N)		
2010	28C404A	2010	28C587E	2010	28C693G	2010	2SC828(0)		
2010	28C404B	2010	28C587F	2010	28C693GL	2010	2SC828(P)		
2010	28C404C	2010	28C587G	2010	28C693GN	2010	2SC828(P)(Q)		
2010	28C404D	2010	28C587GN	2010	28C693GS	2010	2SC828(Q)		
2010	28C404E	2010	28C587H	2010	28C693GU	2010	2SC828(R)		
2010	28C404F	2010	28C587J	2010	28C693GZ	2010	2SC828(R)(S)		
2010	28C404GN	2010	28C587K	2010	28C693H	2010	2SC828(R,Q,P)		
2010	28C404H	2010	28C587L	2010	28C693J	2010	2SC828(R,S,T)		
2010	28C404J	2010	28C587M	2010	28C693K	2010	2SC828(S)		
2010	28C404K	2010	28C5870R	2010	28C693M	2010	2SC828(T)		
2010	28C404L	2010	28C587R	2010	28C693P	2010	2SC828-0		
2010	28C404M	2010	28C587X	2010	28C6930R	2010	2SC828-OP		
2010	28C4040R	2010	28C587Y	2010	28C693R	2010	2SC828-OR		
2010	28C404R	2010	28C640A	2010	28C693U	2010	2SC828A		
2010	28C404X	2010	28C640C	2010	28C693X	2010	2SC828A(P)		
2010	28C404Y	2010	28C640Y	2010	28C693Y	2010	2SC828A(Q)		
2010	28C475A	2010	28C640B	2010	28C694	2010	2SC828A(R)		
2010	28C475B	2010	28C640F	2010	28C694E	2010	2SC828A0		
2010	28C475C	2010	28C640Q	2010	28C694F	2010	2SC828AF		
2010	28C475D	2010	28C640GN	2010	28C694G	2010	2SC828AQ		
2010	28C475E	2010	28C640J	2010	28C694Z	2010	2SC828AR		
2010	28C475F	2010	28C640K	2010	2SC711	2010	2SC828AB		
2010	28C475G	2010	28C640L	2010	2SC711(D)	2010	2SC828B		
2010	28C475GN	2010	28C640M	2010	2SC711(E)	2010	2SC828C		
2010	28C475H	2010	28C640R	2010	2SC711(F)	2010	2SC828D		
2010	28C475J	2010	28C640X	2010	2SC711-A,P,Q	2010	2SC828F		
2010	28C475L	2010	28C644	2010	2SC711-OR	2010	2SC828FR		
2010	28C475M	2010	28C644(F)	2010	2SC711A	2010	2SC828G		
2010	28C4750R	2010	28C644(H)						

RS 276-	Industry Standard No.
2010	2SC828GN
2010	2SC828HR
2010	2SC828K
2010	2SC828L
2010	2SC828LR
2010	2SC828LS
2010	2SC828M
2010	2SC828N
2010	2SC828OR
2010	2SC828P
2010	2SC828PQ
2010	2SC828Q-6
2010	2SC828QRS
2010	2SC828R
2010	2SC828R-1
2010	2SC828RA
2010	2SC828RH
2010	2SC828RS
2010	2SC828RST
2010	2SC828S
2010	2SC828T
2010	2SC828W
2010	2SC828X
2010	2SC828Y
2010	2SC828YL
2010	0000 2SC858
2010	2SC859
2010	2SC859A
2010	2SC859B
2010	2SC859D
2010	2SC859E
2010	2SC859F
2010	2SC859PQ
2010	2SC859G
2010	2SC859GK
2010	2SC859GL
2010	2SC859GM
2010	2SC859GN
2010	2SC859H
2010	2SC859J
2010	2SC859K
2010	2SC859L
2010	2SC859M
2010	2SC859OR
2010	2SC859X
2010	2SC859Y
2010	2SC900
2010	2SC900(E)(L)
2010	2SC900(P)
2010	2SC900(L)
2010	2SC900(U)
2010	2SC900-OR
2010	2SC900A
2010	2SC900B
2010	2SC900C
2010	2SC900D
2010	2SC900E
2010	2SC900F
2010	2SC900G
2010	2SC900J
2010	2SC900K
2010	2SC900M
2010	2SC900OM
2010	2SC900OR
2010	2SC900R
2010	2SC900S
2010	2SC900SA
2010	2SC900SB
2010	2SC900SC
2010	2SC900SD
2010	2SC900U
2010	2SC900VE
2010	2SC900X
2010	2SC900Y
2010	2SC907
2010	2SC907A
2010	2SC907AC
2010	2SC907AD
2010	2SC907AH
2010	2SC907C
2010	2SC907D
2010	2SC907HA
2010	2SC912M
2010	2SC923
2010	2SC923(E)(F)
2010	2SC923A
2010	2SC923B
2010	2SC923C
2010	2SC923D
2010	2SC923F
2010	2SC923G
2010	2SC923GN
2010	2SC923H
2010	2SC923K
2010	2SC923L
2010	2SC923M
2010	2SC923OR
2010	2SC923R
2010	2SC923X
2010	2SC923Y
2010	2SC929
2010	2SC930
2010	2SC930D
2010	2SC930E
2010	0000 2SC945
2010	2SC945(K)
2010	2SC945(L)
2010	2SC945(P)
2010	2SC945(Q)
2010	2SC945(R)
2010	2SC945(TK)
2010	2SC945(TK,P)
2010	2SC945(TP)
2010	2SC945(TQ)
2010	2SC945(TQ,Q)
2010	2SC945-0
2010	2SC945-O
2010	2SC945-OR
2010	2SC945-R
2010	2SC945A
2010	2SC945A/D
2010	2SC945AK
2010	2SC945AP
2010	2SC945AQ
2010	2SC945AR
2010	2SC945B
2010	2SC945C
2010	2SC945CK
2010	2SC945D
2010	2SC945E
2010	2SC945F
2010	2SC945G
2010	2SC945GN
2010	2SC945H
2010	2SC945J
2010	2SC945K
2010	2SC945L
2010	2SC945LP
2010	2SC945LPQ
2010	2SC945LQ
2010	2SC945M
2010	2SC945O
2010	2SC945OR
2010	2SC945P
2010	2SC945PJ
2010	2SC945PO
2010	2SC945PQ
2010	2SC945Q
2010	2SC945QL
2010	2SC945QP
2010	2SC945QR
2010	2SC945R
2010	2SC945RA
2010	2SC945S
2010	2SC945T
2010	2SC945TK
2010	2SC945TP
2010	2SC945TQ
2010	2SC945TR
2010	2SC945X
2010	2SC945Y
2010	2SC985A
2010	2SCNJ100
2010	28D591
2010	28D591R
2010	28D599
2010	28E4002
2010	3L4-6007-15
2010	3L4-6007-16
2010	3N74
2010	3N75
2010	3N76
2010	3N77
2010	3N78
2010	4-1544
2010	4-1791
2010	4-46
2010	4-47
2010	07-07166
2010	08-B1250108
2010	8Q-3-01
2010	8Q-3-10
2010	09-302038
2010	09-302053
2010	09-302074(SHARP)
2010	09-302080
2010	09-302086
2010	09-302093
2010	09-302097
2010	09-302107
2010	09-302125
2010	09-302127
2010	09-302139
2010	09-302194
2010	09-305001
2010	09-305048
2010	09-305052
2010	09-305123
2010	09-305126
2010	09-305059
2010	09-305070
2010	9TR3
2010	10-003
2010	11-27070
2010	11-B551-2
2010	11-B551-3
2010	11-B552-2
2010	11-B552-3
2010	11-B554-2
2010	11-B554-3
2010	11-B555-2
2010	11-B555-3
2010	11-0551-2
2010	11-0551-3
2010	11-0553-2
2010	11-0553-3
2010	13-14085-49
2010	13-14085-95
2010	13-14085-96
2010	13-14085-97
2010	13-18365-1
2010	13-23327-4
2010	13-23338-3
2010	13-23338-4
2010	13-23339-2
2010	13-23339-3
2010	13-23510-4
2010	13-29033-3
2010	13-29033-4
2010	13-29033-5
2010	13-32362-1
2010	13-34381-1
2010	13-35550-1
2010	14-03512-1
2010	13-4085-121
2010	13-4085-122
2010	13-4085-41
2010	14-602-25
2010	14-602-46
2010	14-602-46A
2010	14-602-63
2010	14-801-23
2010	014-862
2010	15-02757-00
2010	15-09587
2010	16A667-QRN
2010	16A667-ORG
2010	16A667-RED
2010	16A667-YEL
2010	16A668-QRN
2010	16A668-ORG
2010	16A668-YEL
2010	16A669-QRN
2010	16A669-YEL
2010	16L24
2010	16L25
2010	16L4
2010	16L45
2010	16L65
2010	019-0-003675-205
2010	019-004094
2010	020-1110-010
2010	020-1110-012
2010	020-1110-013
2010	21A040-064
2010	21A040-065
2010	21A040-066
2010	21A040-067
2010	21A040-082
2010	21A040-082
2010	21A040-083
2010	21A112-019
2010	21A112-045
2010	21A112-046
2010	21A112-058
2010	21A112-062
2010	21A112-063
2010	21A404-066
2010	21MO87
2010	21MO91
2010	21MO95
2010	21MO96
2010	21M124
2010	21M137
2010	21M160
2010	21M161
2010	21M170
2010	21M174
2010	21M408
2010	21M446
2010	21M550
2010	21M603
2010	24-001326
2010	24-001354
2010	24M125
2010	24MW1022
2010	24MW1025
2010	24MW1060
2010	24MW826
2010	24MW964
2010	24MW965
2010	24MW990
2010	24T-009
2010	025-100017
2010	27A10489-101-11
2010	31-0012
2010	31-0013
2010	31-0052
2010	31-0099
2010	31-0100
2010	033A
2010	34-1009
2010	34-1019
2010	34-6007-14
2010	37-193MP
2010	37-21401
2010	42-27277
2010	42-27533
2010	42-27534
2010	42-27535
2010	42-30092
2010	42-9029-31B
2010	42-9029-31L
2010	42-9029-31P
2010	42-9029-31Q
2010	42-9029-60D
2010	43A162445P1
2010	43A212090P1
2010	43B168613-1
2010	44T-300-106
2010	44T-300-111
2010	44T-300-112
2010	46-86434-3
2010	46-86513-3
2010	46-86535-3
2010	46-86536-3
2010	46-86543-3
2010	48-01-049
2010	48-134810
2010	48-134813
2010	48-134823
2010	48-134997
2010	48-137015
2010	48-137300
2010	48-137325
2010	48-137390
2010	48-155073
2010	48-155134
2010	48-155154
2010	48-355004
2010	48-40170-Q01
2010	48-401700-Q01
2010	48-40246-Q01
2010	48-40247Q02
2010	48-40606J02
2010	48-40607J01
2010	48-41816J02
2010	48-44885Q01
2010	48-45N2
2010	48-869197
2010	48-869269
2010	48-869293
2010	48-869338
2010	48-869384
2010	48-869409
2010	48-869416
2010	48-869447
2010	48-869474
2010	48-869486
2010	48-869547
2010	48-869594
2010	48-90343A06
2010	48K869269
2010	48K869293
2010	48K869409
2010	48K869447
2010	48K869474
2010	48K869486
2010	48K869197
2010	48K869338
2010	48K869384
2010	48K869416
2010	48K869547
2010	48K869594
2010	488155073
2010	488155089
2010	48840170001
2010	48840241001
2010	48840247001
2010	48840606002
2010	48840606J02
2010	48840607001
2010	48840607J01
2010	56-4826
2010	56-4827
2010	56-8196
2010	56-8197
2010	57A126-1
2010	57A142-2
2010	57A143-12
2010	57A144-12
2010	57A6(NFN)
2010	57B203-14
2010	57B5-6
2010	57B5-7
2010	57B5-8
2010	57C6
2010	61-309686
2010	065-006
2010	65A11573
2010	66-P11112-0001
2010	66P026-1
2010	67A9060
2010	68A7366-1
2010	69-1816
2010	70.01.704
2010	70N1M
2010	74-01-772
2010	76N1(VID)
2010	77-271798-2
2010	77-27198-3
2010	78B67-2
2010	81-46125053-2
2010	81-46125063-1
2010	86-005135-2
2010	86-400-2
2010	86-444-2
2010	86-526-2
2010	86A327
2010	87-0005
2010	87-0009
2010	87-0013
2010	87-0212-1
2010	87-0218-U
2010	87-0227
2010	87-0230
2010	87-0230-1
2010	87-0231
2010	87-0235
2010	87-0235B
2010	87-0238
2010	88-1250108
2010	90-140
2010	90-33
2010	90-614
2010	90-70
2010	9176
2010	9226
2010	9376
2010	96-5346-01
2010	998033
2010	102-0373-00
2010	102-0732-28
2010	102-0945-16
2010	102-0945-17
2010	102-0945-58
2010	102-0945-59
2010	0103-0088/4460
2010	0103-0492
2010	0103-93
2010	106-003
2010	112-203391
2010	121-767
2010	121-876
2010	121-928
2010	122-7
2010	122-7(RCA)
2010	123-005
2010	123-007
2010	127-115
2010	130-40216
2010	130-40883
2010	130-40901
2010	0131-004792
2010	0131-005347
2010	0131-005348
2010	0131-005349
2010	0131-005350
2010	134B1038-13
2010	134B1038-22
2010	134B1038-4
2010	134B1038-8
2010	0140-5
2010	142-002
2010	142-002(REGENCY)
2010	142-003
2010	147-7016-01
2010	151-0341-00
2010	151-0341-00-A
2010	151-0456-00
2010	154A5944-413
2010	154A5945-519
2010	173A4473-5
2010	176-006-9-002
2010	176-025-9-001
2010	176-031-9-002
2010	176-037-9-003
2010	176-037-9-004
2010	176-042-9-003
2010	176-042-9-004
2010	176-060-9-003
2010	176-062-9-001
2010	185-005
2010	185-010
2010	186-003
2010	186-004
2010	186-005
2010	195
2010	207A17
2010	215-37567
2010	229-1301-23
2010	250-0373
2010	250-0700
2010	250-0711
2010	250-1312
2010	260-10-023
2010	260-10-042
2010	260D13701
2010	260D15901
2010	260P17501
2010	260P17502
2010	260P17701
2010	260P17704
2010	260P19503
2010	296-55-9
2010	296-98-9
2010	297L007C
2010	297L007C003
2010	297L015C01
2010	311D916P01
2010	325-0500-12

RS 276-	Industry Standard No.	RS 276-	Industry Standard No.	RS 276-	Industry Standard No.	RS 276-	Industry Standard No.
2010	325-0500-13	2010	2057A2-275	2010	100292	2012	A127712
2010	352-0549-000	2010	2057A2-290	2010	133218	2012	A130
2010	352-0638	2010	2057A2-333	2010	136282	2012	A130-ORN
2010	353-9306-006	2010	2057A2-334	2010	136430	2012	A130-V10
2010	353-9306-007	2010	2057A2-352	2010	138001	2012	A132(TRANSISTOR)
2010	353-9318-001	2010	2057A2-370	2010	138019-001	2012	A1409
2010	353-9318-002	2010	2057A2-373	2010	138789-4	2012	A1A
2010	386-7178-P001	2010	2057A2-385	2010	138789-5	2012	A1M
2010	400-1569-101	2010	2057A2-391	2010	147115-7	2012	A1S
2010	417-108-13163	2010	2057A2-404	2010	171175-8	2012	A2057B104-8
2010	417-126-12903	2010	2057A2-405	2010	147122P7	2012	A2090
2010	417-226-13163	2010	2057A2-428	2010	150714-1	2012	A23114130
2010	417-244-12903	2010	2057A2-436	2010	162002-085	2012	A247
2010	417-283-13271	2010	2057A2-454	2010	168716	2012	A247(AMC)
2010	434	2010	2057A2-475	2010	171162-004	2012	A24114130
2010	462-1038-01	2010	2057A2-502	2010	171162-095	2012	A25762-010
2010	462-1066-01	2010	2057A2-518	2010	171162-100	2012	A25762-012
2010	462-2004	2010	2057A2-542	2010	171162-180	2012	A2620
2010	536P(JVC)	2010	2057A2-543	2010	171162-204	2012	A2A
2010	536OT	2010	2057B-59	2010	171162-247	2012	A2K
2010	576-0003-022	2010	2063-17	2010	171162-285	2012	A2S
2010	599P3430	2010	2065-54	2010	171162-288	2012	A310
2010	600-207-801	2010	2101	2010	171553	2012	A3170717
2010	600-224-605	2010	2121-17	2010	171676	2012	A35
2010	602X0008-002	2010	2132-17	2010	171678	2012	A35(JAPAN)
2010	602X0018-002	2010	2195-17	2010	181012	2012	A3M
2010	686-257-0	2010	2207-17	2010	200200	2012	A3MA
2010	693GV	2010	2212-17	2010	200200-700	2012	A417-115
2010	700-155	2010	2271	2010	202617	2012	A46-867-3
2010	700-156	2010	2338-17	2010	204117	2012	A4648
2010	700A-858-318	2010	2510-103	2010	213217	2012	A48-134819
2010	700A-858-328	2010	2904-045	2010	256417	2012	A48-134843
2010	700A858-319	2010	2904-054	2010	257017	2012	A48-134853
2010	733-0	2010	3107-204-90150	2010	268003	2012	A48-134898
2010	750A858-328	2010	3391	2010	281001-53	2012	A48-134919
2010	753-1644-100	2010	3391(SEARS)	2010	281001-83	2012	A48-134927
2010	753-2000-100	2010	3391A	2010	323954	2012	A48-137002
2010	753-4000-010	2010	3391A(SEARS)	2010	506902	2012	A48-137035
2010	753-6000-002	2010	03460	2010	510584	2012	A4819
2010	753-8400-230	2010	003461	2010	530130-1	2012	A4838
2010	755-8500-380	2010	3519-1(RCA)	2010	531841-002	2012	A4843
2010	753-9000-839	2010	4039-00	2010	532775	2012	A4853
2010	772-101-00	2010	4039-01	2010	610079-2	2012	A4H
2010	853-0300-644	2010	4801-0000-003	2010	610094-3	2012	A573501
2010	858	2010	4801-0000-010	2010	610128-1	2012	A57012-1
2010	859K	2010	4824-0014	2010	610128-2	2012	A57012-2
2010	880-250108	2010	4824-0014-02	2010	610128-5	2012	A57D1-122
2010	881-250108	2010	5001-038	2010	610128-6	2012	A57M2-16
2010	902-002-3-06	2010	5001-043	2010	610151-1	2012	A57M2-17
2010	902-003-3-17	2010	5001-074	2010	610151-3	2012	A575058
2010	902-003-6-006	2010	5001-505	2010	610151-5	2012	A610075-1
2010	903Y002150	2010	5001-511	2010	610224-1	2012	A6181-1
2010	904-95B	2010	5001-541	2010	610226-1	2012	A63-18426
2010	921-1022	2010	5001-544	2010	618165-1	2012	A7253
2010	921-111B	2010	05206-00	2010	618181-1	2012	A86-213-2
2010	921-114B	2010	5613-1327T	2010	699291	2012	A86-214-2
2010	921-115B	2010	5613-1335(E)	2010	740306	2012	A86-215-2
2010	921-116B	2010	5613-1359B	2010	740439	2012	A86-316-2
2010	921-133	2010	5613-1684T	2010	740440	2012	A8V
2010	921-133B	2010	5613-7111E	2010	740442	2012	A8VA
2010	921-154B	2010	5613-871	2010	740886	2012	AM9
2010	921-196	2010	5613-871(P)	2010	740887	2012	AT350
2010	921-196B	2010	5613-871P	2010	741114	2012	AT351
2010	921-202B	2010	7991	2010	741737	2012	B5D
2010	921-205B	2010	8000-00004-301	2010	741857	2012	BC100
2010	921-206	2010	8000-00005-003	2010	742512	2012	BC117
2010	921-206B	2010	8000-00005-004	2010	742513	2012	BC145
2010	921-207B	2010	8000-00005-005	2010	742548	2012	BC394
2010	921-208B	2010	8000-0009-089	2010	801527	2012	BD115
2010	921-209B	2010	8000-00041-041	2010	805696	2012	BP108
2010	921-239B	2010	8000-00041-042	2010	815173	2012	BP109
2010	921-240B	2010	8000-0005-004	2010	908844-1	2012	BP110
2010	921-281B	2010	8000-0009-089	2010	916031(CARTAPE)	2012	BP111
2010	921-407	2010	8020-204	2010	916033	2012	BP114
2010	921-408	2010	8102-208	2010	916052	2012	BP117
2010	921-92B	2010	8200-202	2010	916055	2012	BP118
2010	929(O)	2010	8200-203	2010	916100	2012	BP119
2010	929C	2010	8394	2010	964547-2	2012	BP140
2010	929CA	2010	8840-126	2010	971460	2012	BP140A
2010	929CU	2010	8868-7	2010	992066	2012	BP140R
2010	929D(JVC)	2010	8910-144	2010	1223915	2012	BP140S
2010	929DX	2010	9033-2	2010	1417308-1	2012	BP155R
2010	930(OV)	2010	9033-3	2010	1417308-2	2012	BP155S
2010	930(OY)	2010	9033-4	2010	1443024-1	2012	BP156
2010	930B	2010	9033-5	2010	1471115-1	2012	BP157
2010	930C	2010	9033GREEN	2010	1471115-2	2012	BP157B
2010	930C(WARDS)	2010	9033ORANGE	2010	1471115-3	2012	BP174
2010	930D	2010	9033Q	2010	1471115-4	2012	BP177
2010	930D(WARDS)	2010	9033RED	2010	1471120-1	2012	BP178
2010	930DU	2010	9033WHITE	2010	1471122	2012	BP179
2010	930DX	2010	9330-688-30112	2010	1471122-6	2012	BP179A
2010	930E	2010	09500	2010	1471122-7	2012	BP179B
2010	930E(JVC)	2010	09501	2010	1472475-1	2012	BP179C
2010	930E(WARDS)	2010	9502	2010	1473506-1	2012	BP186
2010	930EX	2010	16237	2010	1473614-2	2012	BP257
2010	943-742-002	2010	18410-144	2010	1612738-1	2012	BP258
2010	991-013057	2010	19500-253	2010	2125310	2012	BP259
2010	1003-6754	2010	20103	2010	2498665-1	2012	BP292A
2010	1004-0780	2010	20810-91	2010	2498665-2	2012	BP292B
2010	1030-21	2010	20810-92	2010	2498665-3	2012	BP292C
2010	1035-80	2010	22810-174	2010	2499950	2012	BP294
2010	1040-01	2010	25840-162	2010	2505207	2012	BP305
2010	1042-03	2010	26810-153	2011	BFY167	2012	BP336
2010	1048-9920	2010	27125-460	2011	Q4	2012	BP337
2010	1049-0092	2010	27125-470	2011	SE5-0250	2012	BP358
2010	1050-21	2010	030828	2011	SK3716	2012	BP355
2010	1074-03	2010	37510-161	2011	SK3716/161	2012	BFW37
2010	1074-115	2010	38478	2011	TM-3200	2012	BFW45
2010	1080-03	2010	38510-164	2011	TR-95(IR)	2012	BFX34
2010	1080-21	2010	38510-350	2011	TVB-28C948	2012	BFX98
2010	1096-12	2010	41175	2011	2N117#	2012	BFY43
2010	1472 8349	2010	45337-A	2011	2N118#	2012	BFY45
2010	1472-8349	2010	45810-162	2011	2N118A#	2012	BFY57
2010	1479 7963	2010	45810-166	2011	2N119#	2012	BF65
2010	1479-7963	2010	51429	2011	2N120#	2012	BF780
2010	1515(NPN)	2010	53200-74	2011	2N902	2012	BN7253
2010	1751-17	2010	58810-164	2011	2N903	2012	BSW32
2010	1835-17	2010	58810-167	2011	2N904	2012	BSW69
2010	1854-0060	2010	000073070	2011	2N905	2012	BSW70
2010	1854-0387	2010	000073080	2011	28C1054	2012	BSX21
2010	1854-SRK1-1	2010	000073350	2011	28C1331	2012	C1012
2010	2020-05	2010	000073351	2011	28C618	2012	C1012A
2010	2035-5100-53660	2010	000073360	2011	28C618A	2012	C1048
2010	2035-5100-69372	2010	000073361	2011	28C660	2012	C1048B
2010	2041-02	2010	000073373	2011	28C661	2012	C1048C
2010	2048-17	2010	000073374	2011	28C706	2012	C1048D
2010	2057A2-212	2010	75613-1	2011	28C811	2012	C1048E
2010	2057A2-249	2010	75613-2	2011	28D33C	2012	C1048F
2010	2057A2-257	2010	81410-145	2012	A1112	2012	C1056
2010	2057A2-260	2010	88060-144	2012	A116081	2012	C1103
2010	2057A2-262	2010	88510-174	2012	A121-361	2012	C1103(A)
2010	2057A2-272	2010	90934-35	2012	A12546	2012	C1103A
2010	2057A2-273	2010	99240-269	2012	A126705	2012	C1103B
2010	2057A2-274					2012	C1103C

RS 276-	Industry Standard No.	RS 276-	Industry Standard No.	RS 276-	Industry Standard No.	RS 276-	Industry Standard No.
2012	C1103L	2012	MT1893	2012	2N3700	2012	2SC58E
2012	C1116	2012	MT698	2012	2N3701	2012	2SC58F
2012	C1116-0	2012	MT699	2012	2N3712	2012	2SC58G
2012	C273	2012	MT870	2012	2N3742	2012	2SC58GN
2012	C407	2012	MT871	2012	2N3743	2012	2SC58H
2012	C470	2012	MT910	2012	2N3923	2012	2SC58J
2012	C472Y	2012	MT911	2012	2N4068	2012	2SC58K
2012	C500	2012	MT912	2012	2N4269	2012	2SC58L
2012	C500R	2012	N1X	2012	2N4270	2012	2SC58M
2012	C500Y	2012	N2XA	2012	2N4410	2012	2SC58OR
2012	C505	2012	N820-42	2012	2N4924	2012	2SC58R
2012	C505-0	2012	N848004	2012	2N4925	2012	2SC58X
2012	C505-R	2012	N86212	2012	2N4926	2012	2SC58Y
2012	C506	2012	PET1075A	2012	2N4927	2012	2SC627
2012	C506-0	2012	PRT101	2012	2N5058	2012	2SC64
2012	C506-R	2012	PTC117	2012	2N5059	2012	2SC65
2012	C507	2012	Q5217	2012	2N5174	2012	2SC65-0
2012	C507-0	2012	R2474-2	2012	2N5175	2012	2SC65-OR
2012	C507-R	2012	R823	2012	2N5176	2012	2SC65A
2012	C507-Y	2012	RRN154	2012	2N5184	2012	2SC65B
2012	C526	2012	RS-2008	2012	2N5185	2012	2SC65C
2012	C558	2012	RT-110	2012	2N699AB	2012	2SC65D
2012	C589	2012	S1366	2012	2N699AB	2012	2SC65E
2012	C58A	2012	S1407	2012	2N738	2012	2SC65F
2012	C627	2012	S1769	2012	2N739	2012	2SC65G
2012	C64(JAPAN)	2012	S17862	2012	2N739A	2012	2SC65GN
2012	C65	2012	S2986	2012	2N740	2012	2SC65H
2012	C65B	2012	S3002	2012	2N740A	2012	2SC65K
2012	C65N	2012	S3033	2012	2N743	2012	2SC65L
2012	C65Y	2012	S3034	2012	2N743A	2012	2SC65M
2012	C65YA	2012	S3035	2012	2N745A	2012	2SC65N
2012	C65YB	2012	S39560	2012	2N746	2012	2SC65OR
2012	C65YTV	2012	S40205	2012	2N746A	2012	2SC65R
2012	C66	2012	SC843(T.I.)	2012	2SC1012	2012	2SC65X
2012	C686	2012	SE7001	2012	2SC1012A	2012	2SC65Y
2012	C70	2012	SE7002	2012	2SC1012B	2012	2SC65Y(B)
2012	C727	2012	SE7010	2012	2SC1012C	2012	2SC65YA
2012	C728	2012	SE7016	2012	2SC1012E	2012	2SC65YB
2012	C743A	2012	SE7017	2012	2SC1012F	2012	2SC65YTV
2012	C746A	2012	SE7050	2012	2SC1012G	2012	2SC65YTV1
2012	C788	2012	SE7055	2012	2SC1012GN	2012	2SC66
2012	C805	2012	SE7056	2012	2SC1012H	2012	2SC686
2012	C818	2012	SFT186	2012	2SC1012J	2012	2SC686A
2012	C856	2012	SFT187	2012	2SC1012K	2012	2SC686B
2012	C856-02	2012	SK3040	2012	2SC1012L	2012	2SC686C
2012	C856C	2012	SK3044	2012	2SC1012M	2012	2SC686D
2012	C857	2012	SK8261	2012	2SC1012OR	2012	2SC686E
2012	C857H	2012	SM6727	2012	2SC1012R	2012	2SC686F
2012	C868	2012	SP8400	2012	2SC1012X	2012	2SC686G
2012	C869	2012	SP8401	2012	2SC1012Y	2012	2SC686GN
2012	C88	2012	S31912	2012	2SC104B	2012	2SC686H
2012	C88A	2012	S8524	2012	2SC104BB	2012	2SC686J
2012	C926	2012	S86111	2012	2SC104C	2012	2SC686K
2012	C926A	2012	STX0028	2012	2SC104D	2012	2SC686L
2012	C995	2012	SX60M	2012	2SC104BE	2012	2SC686M
2012	CDC744	2012	T-481	2012	2SC104BF	2012	2SC686OR
2012	CDQ10013	2012	T-Q5082	2012	2SC104BN	2012	2SC686R
2012	CDQ10015	2012	T481(SEARS)	2012	2SC1056	2012	2SC686X
2012	CDQ10034	2012	TA7292	2012	2SC1062	2012	2SC686Y
2012	CDQ10037	2012	TA7293	2012	2SC1090	2012	2SC70
2012	CDQ1004	2012	TB80143	2012	2SC1090D	2012	2SC727
2012	CDQ10045	2012	TG28C65	2012	2SC1103	2012	2SC728
2012	CDQ10046	2012	TG28C65Y	2012	2SC1103(A)	2012	2SC743A
2012	CDQ10047	2012	TI-722	2012	2SC1103A	2012	2SC746A
2012	CDQ10049	2012	TI8100	2012	2SC1103B	2012	2SC788
2012	CS1347	2012	TI8101	2012	2SC1103C	2012	2SC788B
2012	CT6776	2012	TI8102	2012	2SC1103L	2012	2SC788C
2012	CX1154	2012	TI8103	2012	2SC1116	2012	2SC788D
2012	DT1003	2012	TM154	2012	2SC1116-0	2012	2SC788E
2012	DT1602	2012	TNJ60072	2012	2SC154	2012	2SC788F
2012	DT1603	2012	TNJ72282	2012	2SC154(C)	2012	2SC788G
2012	DT1612	2012	TQ5063	2012	2SC154-OR	2012	2SC788GN
2012	DT1613	2012	TR301	2012	2SC546	2012	2SC788J
2012	EC01S4	2012	TR8100	2012	2SC154A	2012	2SC788K
2012	EP15X107	2012	TR8101	2012	2SC154B	2012	2SC788L
2012	EP15X18	2012	TR8120	2012	2SC154C	2012	2SC788M
2012	ERS8140	2012	TR8140	2012	2SC154D	2012	2SC788OR
2012	ERS8160	2012	TR8160	2012	2SC154E	2012	2SC788R
2012	ERS8180	2012	TR8180	2012	2SC154F	2012	2SC788X
2012	ERS8200	2012	TR8200	2012	2SC154G	2012	2SC788Y
2012	ERS8225	2012	TR8225	2012	2SC154GN	2012	2SC805
2012	ERS8250	2012	TR8250	2012	2SC154H	2012	2SC805A
2012	ERS8275	2012	TR8275	2012	2SC154J	2012	2SC805B
2012	ERS8301	2012	TR8301	2012	2SC154K	2012	2SC805C
2012	ERS8325	2012	TR83011	2012	2SC154L	2012	2SC805D
2012	ES15X107	2012	TR83012	2012	2SC154M	2012	2SC805E
2012	ES15X59	2012	TS2776	2012	2SC1540R	2012	2SC805F
2012	ES15X89	2012	TV-19	2012	2SC154R	2012	2SC805G
2012	ES32(ELCOM)	2012	TV-49	2012	2SC154X	2012	2SC805GN
2012	ET15X34	2012	TV-70	2012	2SC154Y	2012	2SC805H
2012	ETP2008	2012	TV19	2012	2SC273	2012	2SC805J
2012	ETP3923	2012	TV24164	2012	2SC470	2012	2SC805L
2012	EX743-X	2012	TV24435	2012	2SC470-3	2012	2SC805OR
2012	FM1893	2012	TV24499	2012	2SC470-4	2012	2SC805R
2012	FT34C	2012	TV70	2012	2SC470-5	2012	2SC805X
2012	FT34D	2012	TV8-28C526	2012	2SC470-6	2012	2SC805Y
2012	FT5641	2012	TV8-28C58	2012	2SC470A	2012	2SC818
2012	GE-235	2012	TV8-28C58A	2012	2SC470B	2012	2SC856
2012	GE-40	2012	V8-28C58	2012	2SC470C	2012	2SC856-02
2012	H932	2012	V8-28C58A	2012	2SC470D	2012	2SC856-OR
2012	HEP706	2012	V8-28C58B	2012	2SC470E	2012	2SC856A
2012	HEP712	2012	V8-28C58C	2012	2SC470F	2012	2SC856B
2012	HEP713	2012	VS28C1921//1E	2012	2SC470G	2012	2SC856C
2012	HEP83033	2012	W28	2012	2SC470GN	2012	2SC856D
2012	HEP83034	2012	WEP712	2012	2SC470H	2012	2SC856E
2012	HEP83035	2012	ZEN205	2012	2SC470J	2012	2SC856F
2012	HEP85024	2012	001-021100	2012	2SC470K	2012	2SC856G
2012	HEP85025	2012	01-571941	2012	2SC470L	2012	2SC856GN
2012	HT8000101.	2012	01-572831	2012	2SC470M	2012	2SC856J
2012	IRHR78	2012	2N1052	2012	2SC470OR	2012	2SC856K
2012	M4648	2012	2N1053	2012	2SC470R	2012	2SC856L
2012	M4819	2012	2N1054	2012	2SC470X	2012	2SC856M
2012	M4838	2012	2N1207	2012	2SC470Y	2012	2SC856OR
2012	M4839	2012	2N1495	2012	2SC470YY	2012	2SC856R
2012	M4843	2012	2N1572	2012	2SC500	2012	2SC856X
2012	M4853	2012	2N1573	2012	2SC500R	2012	2SC856Y
2012	M4927	2012	2N1574	2012	2SC500Y	2012	2SC857
2012	M7002	2012	2N1613A	2012	2SC505	2012	2SC857H
2012	M819	2012	2N1613B	2012	2SC505-0	2012	2SC857X
2012	ME1110	2012	2N2102A	2012	2SC505-R	2012	2SC868
2012	ME1120	2012	2N2509	2012	2SC506	2012	2SC869
2012	MHT4402	2012	2N2510	2012	2SC506-0	2012	2SC88
2012	MI814150/37	2012	2N2511	2012	2SC506-R	2012	2SC88A
2012	MJ420	2012	2N2618	2012	2SC507	2012	2SC926
2012	MJ421	2012	2N2726	2012	2SC507-0	2012	2SC926(A)
2012	MM2260	2012	2N3114	2012	2SC507-R	2012	2SC926-OR
2012	MM3000	2012	2N3388	2012	2SC507-Y	2012	2SC926A
2012	MM3002	2012	2N3389	2012	2SC58	2012	2SC926B
2012	MM3005	2012	2N340	2012	2SC589	2012	2SC926C
2012	MM3009	2012	2N340A	2012	2SC58A	2012	2SC926D
2012	MM3100	2012	2N341	2012	2SC58AC	2012	2SC926E
2012	MM3101	2012	2N341A	2012	2SC58B	2012	2SC926F
2012	MM7087	2012	2N342A	2012	2SC58D	2012	2SC926G
2012	MM7088	2012	2N3526			2012	2SC926GN

RS 276-	Industry Standard No.	RS 276-	Industry Standard No.	RS 276-	Industry Standard No.	RS 276-	Industry Standard No.
2012	280926H	2012	260D07201	2015	A069-101	2015	AT345
2012	280926J	2012	260D7201	2015	A069-102	2015	AT346
2012	280926K	2012	260P1030	2015	A069-103	2015	AZG
2012	280926L	2012	260P10301	2015	A069-114	2015	AZY
2012	280926M	2012	260P10801	2015	A069-116	2015	B-75568-2
2012	280926OR	2012	260P22101	2015	A069-119	2015	B-T1000-139
2012	280926R	2012	321-264	2015	A121-585	2015	B-T1000-139
2012	280926X	2012	352-0403-010	2015	A121-585B	2015	B1H
2012	280926Y	2012	417-115	2015	A121-687	2015	B1H(XSTR)
2012	28095	2012	576-0004-001	2015	A1G	2015	BC121
2012	28095	2012	613-4	2015	A1G-1	2015	BC122
2012	28C1090B	2012	656-4	2015	A1M-1	2015	BC123
2012	4-686145-3	2012	690V080H38	2015	A1P	2015	BC155
2012	4-686232-3	2012	743	2015	A1P-1	2015	BC156
2012	6A12988	2012	992-01-3684	2015	A1P-1A	2015	BC188
2012	09-302099	2012	1018	2015	A1P/4922	2015	BC189
2012	010-694(AMPEX)	2012	1067	2015	A1P/4923	2015	BC442
2012	12-1	2012	1067(GE)	2015	A1R	2015	BC510C
2012	13-15809-1	2012	1076-1559	2015	A1R-1	2015	BCM1002-2
2012	13-23825-1	2012	1169	2015	A1R-2	2015	BE173
2012	13-28432-2	2012	1804-17	2015	A1R/4924	2015	BP-200(PENNCREST)
2012	13-55062-1	2012	1885-17	2015	A1R/4925	2015	BF121
2012	14-602-19	2012	2017-110	2015	A1R/4926	2015	BF123
2012	14-602-36	2012	2057A2-261	2015	A2057B2-115	2015	BF125
2012	14-602-37	2012	2057B104-8	2015	A2352	2015	BF127
2012	19-020-046	2012	2057B2-140	2015	A244	2015	BF155
2012	34-6001-65	2012	2090(CROWN)	2015	A244(AMC)	2015	BF161
2012	34-6015-59	2012	2114-0	2015	A2464	2015	BF166
2012	46-86173-3	2012	2474	2015	A2465	2015	BF167
2012	46-86182-3	2012	2474(RCA)	2015	A2479	2015	BF168
2012	46-86183-3	2012	2620	2015	A248	2015	BF169
2012	46-86210-3	2012	3476	2015	A2480	2015	BF173
2012	46-86232-3	2012	3532	2015	A2746	2015	BF175
2012	46-86318-3	2012	3545	2015	A2M	2015	BF180
2012	46-867-3	2012	3545(RCA)	2015	A2M-1	2015	BF181
2012	48-134648	2012	3552(RCA)	2015	A2N	2015	BF182
2012	48-134819	2012	3552-1	2015	A2N-1	2015	BF183
2012	48-134843	2012	3553(RCA)	2015	A2N-2	2015	BF184
2012	48-134853	2012	3582(RCA)	2015	A2P	2015	BF185
2012	48-134898	2012	3687	2015	A2T	2015	BF194
2012	48-134919	2012	3878	2015	A3772.01	2015	BF194B
2012	48-134927	2012	4001	2015	A3T918	2015	BF195
2012	48-137002	2012	4648	2015	A417-154	2015	BF195C
2012	48-137035	2012	4686-232-3	2015	A417-174	2015	BF195D
2012	48-137364	2012	4819	2015	A417-190	2015	BF196
2012	48-137415	2012	4858	2015	A417-205	2015	BF197
2012	48-155006	2012	4843	2015	A429-0981-12	2015	BF198
2012	48M355012	2012	4853	2015	A430	2015	BF199
2012	488134919	2012	6181-1	2015	A451	2015	BP200
2012	488137364(H)	2012	7253(LOWREY)	2015	A46-86101-3	2015	BP200(ZENITH)
2012	488137415(I)	2012	8000-00043-064	2015	A46-86109-3	2015	BP206
2012	488155006	2012	12546	2015	A46-86110-3	2015	BP207
2012	57A1-122	2012	014558	2015	A46-86133-3	2015	BP208
2012	57A104-1	2012	23114-052	2015	A46-86301-3	2015	BP209
2012	57A104-2	2012	25672-016	2015	A46-86302-3	2015	BP212
2012	57A104-3	2012	37725	2015	A46-86303-3	2015	BP213
2012	57A104-4	2012	38120	2015	A465-181-19	2015	BP214
2012	57A104-5	2012	38121	2015	A466A	2015	BP215
2012	57A104-6	2012	38996	2015	A481	2015	BP216
2012	57A104-7	2012	40354	2015	A482	2015	BP217
2012	57A104-8	2012	40355	2015	A483	2015	BP218
2012	57A12-1	2012	40459	2015	A484	2015	BP219
2012	57A12-2	2012	60684	2015	A484(ADMIRAL)	2015	BP220
2012	57A136-1	2012	94051(EICO)	2015	A484(ZENITH)	2015	BP222
2012	57A136-10	2012	116081	2015	A486	2015	BP226
2012	57A136-11	2012	123944	2015	A489	2015	BP227
2012	57A136-2	2012	126705	2015	A492	2015	BP229
2012	57A136-3	2012	126709	2015	A495	2015	BP230
2012	57A136-4	2012	126710	2015	A496	2015	BP232
2012	57A136-5	2012	127712	2015	A4E	2015	BP240
2012	57A136-6	2012	136066	2015	A4Y-1	2015	BP241
2012	57A136-7	2012	171162-124	2015	A6V	2015	BP251
2012	57A136-8	2012	171162-125	2015	A772B1	2015	BP252
2012	57A160-8	2012	171162-126	2015	A772EH	2015	BP253
2012	57A194-11	2012	231140-28	2015	A772FE	2015	BP254B
2012	57A194-11(PULSE)	2012	236282	2015	A1E	2015	BP255C
2012	57A194-1L	2012	0320051	2015	AR111	2015	BP255D
2012	57A207-8	2012	540205	2015	AR200(GREEN)	2015	BP260
2012	57A261-10	2012	0573501	2015	AR209	2015	BP261
2012	57A283-11	2012	0573519	2015	AR210	2015	BP270
2012	57B136-1	2012	610075-1	2015	AR211	2015	BP271
2012	57B136-10	2012	610135-1	2015	AR213	2015	BP273
2012	57B136-11	2012	610144-3	2015	AR213(VIOLET)	2015	BP273C
2012	57B136-2	2012	610144-4	2015	AR213V	2015	BP273D
2012	57B136-3	2012	984227	2015	AR218	2015	BP274
2012	57B136-4	2012	1449098-1	2015	AR218(ORANGE)	2015	BP274B
2012	57B136-5	2012	1472474-2	2015	AR218(RED)	2015	BP274C
2012	57B136-6	2012	1473541-1	2015	AR218RO	2015	BP279
2012	57B136-7	2012	1473552-1	2015	AR219	2015	BP287
2012	57B136-8	2012	1473656-4	2015	AR219(YY)	2015	BP288
2012	57B136-9	2012	1473679-2	2015	AR219YY	2015	BP290
2012	57B207-8	2012	1960085-2	2015	AR220	2015	BP302
2012	57B283-11	2012	2320051	2015	AR220(GY)	2015	BP304
2012	57C12-1	2012	2320051H	2015	AR220(YELLOW)	2015	BP310
2012	57C12-2	2012	2320191	2015	AR220GREEN	2015	BP311
2012	57D1-122	2012	2320892	2015	AR220GY	2015	BP329
2012	57M2-16	2012	2321101	2015	AR220YELLOW	2015	BP332
2012	57M2-17	2012	3170717	2015	AR222	2015	BP333
2012	62-18426	2012	3170757	2015	AR222(BLUE)	2015	BP333C
2012	63-28426	2012	3450842-10	2015	AR222(BY)	2015	BP333D
2012	73C180497-4	2012	3450842-20	2015	AR222(YELLOW)	2015	BP334
2012	73C180499-5	2012	3450842-30	2015	AR222BLUE	2015	BP335
2012	73C180499-6	2012	BF254(SIEG)	2015	AR222BY	2015	BF344
2012	73C182080-33	2013	QRT-107	2015	AR222YELLOW	2015	BF345
2012	73C182088-31	2013	28C1319	2015	AR224	2015	BF362
2012	75N1	2013	28C1989	2015	AR224(WHITE)	2015	BF363
2012	86-213-2	2013	28C2210	2015	AR224(YELLOW)	2015	BF516
2012	86-214-2	2013	28C380A(R)	2015	AR224WHITE	2015	BF811
2012	86-215-2	2014	BC130A	2015	AR224YELLOW	2015	BF813E
2012	86-316-2	2014	J4-1626	2015	AT311	2015	BF813F
2012	86-612-2	2014	QRT-187	2015	AT312	2015	BF813G
2012	86-629-2	2014	RT-187	2015	AT313	2015	BF814E
2012	998087-1	2014	2N2885	2015	AT314	2015	BF814F
2012	998101-1	2014	28C1276	2015	AT315	2015	BF814G
2012	104-8	2014	28C1428	2015	AT316	2015	BF815B
2012	111T2	2014	28C1999	2015	AT318	2015	BF815F
2012	121-241	2014	28C2308	2015	AT319	2015	BF815G
2012	121-361	2014	28C2309	2015	AT321	2015	BF816E
2012	121-445	2014	28C2310	2015	AT322	2015	BF816F
2012	121-473	2014	28C812	2015	AT323	2015	BF816G
2012	121-600(ZENITH)	2014	28C963	2015	AT324	2015	BF818
2012	121-743	2014	28C964	2015	AT325	2015	BF818CA
2012	121-743-01(T05)	2014	28C965	2015	AT326	2015	BF818R
2012	121-744	2014	16B670-GRN	2015	AT327	2015	BF819
2012	121-776	2014	121-1004	2015	AT328	2015	BF819R
2012	121-777-01	2015	A-1854-0092-1	2015	AT330	2015	BF820
2012	121-792	2015	A054-148	2015	AT338	2015	BF820R
2012	121-868-01	2015	A054-170	2015	AT340	2015	BF827E
2012	135N1	2015	A060-100	2015	AT341	2015	BF827P
2012	135N1M	2015	A061-106	2015	AT342	2015	BF827Q
2012	144-4	2015	A061-108	2015	AT343	2015	BFV27
2012	144N4	2015	A061-109	2015	AT344	2015	BFV28
2012	260-10-052	2015	A061-112			2015	BFV59
						2015	BFV60

RS 276-	Industry Standard No.	RS 276-	Industry Standard No.	RS 276-	Industry Standard No.	RS 276-	Industry Standard No.
2015	BPV61	2015	C663	2015	CS9017F	2015	G04041B
2015	BPW80	2015	C674B	2015	CS89017O	2015	G05-003-A
2015	BPW41	2015	C674C	2015	CS89017H	2015	G05-003-B
2015	BPW63	2015	C674CV	2015	CS929	2015	G05-050-C
2015	BPX31	2015	C674D	2015	CS93	2015	GE-11
2015	BPX32	2015	C674E	2015	CT1012	2015	GE-39
2015	BPX47	2015	C674F	2015	CT1013	2015	GE-60
2015	BPX60	2015	C674G	2015	CXL1071	2015	GE129
2015	BPX62	2015	C682	2015	CXL161	2015	GI-2715
2015	BPX73	2015	C682A	2015	DO06	2015	GI-2716
2015	BPX77	2015	C682B	2015	D10B1051	2015	GI-2921
2015	BPX89	2015	C683	2015	D10B1055	2015	GI-2922
2015	BFY47	2015	C683A	2015	D10Q1051	2015	GI-2923
2015	BFY48	2015	C683B	2015	D10Q1052	2015	GI-3605
2015	BFY49	2015	C683V	2015	D16O	2015	GI-3606
2015	BFY69	2015	C695	2015	D16O6	2015	GI-3607
2015	BFY69A	2015	C705	2015	D16K1	2015	GI3793
2015	BFY69B	2015	C705B	2015	D16K2	2015	GM-770
2015	BFY79	2015	C705C	2015	D16K3	2015	GM308
2015	BFY87	2015	C705E	2015	D16K4	2015	GM770
2015	BO-71	2015	C705F	2015	D1Y	2015	GMB6001
2015	BSV53P	2015	C705TV	2015	D26B1	2015	GMB6002
2015	BSV54P	2015	C707	2015	D26B2	2015	GMO3601
2015	BT929	2015	C707H	2015	D26C1	2015	GO-05-004A
2015	C1023(JAPAN)	2015	C717	2015	D26C2	2015	HC-00380
2015	C1023-O	2015	C717B	2015	D26C3	2015	HC-00784
2015	C1023-Y	2015	C717BK	2015	D26E-4	2015	HC-00829
2015	C1023Q	2015	C717BLK	2015	D26E2	2015	HC-00920
2015	C1026	2015	C717C	2015	D26Q-1	2015	HC-01047
2015	C1026Q	2015	C717E	2015	D26OI	2015	HC-01359
2015	C1026Y	2015	C738	2015	D292(CHAN.MASTER)	2015	HC-668
2015	C1032	2015	C738C	2015	D4D22	2015	HC380
2015	C1032Q	2015	C738D	2015	DDBY277002	2015	HC381
2015	C1032Y	2015	C739	2015	DO87	2015	HC394
2015	C1035	2015	C739C	2015	DO88	2015	HC454
2015	C1035C	2015	C761	2015	DS-71	2015	HC460
2015	C1035D	2015	C761Y	2015	DS-72	2015	HC461
2015	C1035E	2015	C761Z	2015	DS-73	2015	HC535
2015	C1036	2015	C762	2015	DS-74	2015	HC535A
2015	C1044	2015	C763(C)	2015	DS-78	2015	HC535B
2015	C1070	2015	C785	2015	DS-781	2015	HC537
2015	C1123	2015	C786R	2015	DS-85	2015	HC545
2015	C1126	2015	C787	2015	DS71	2015	HC784
2015	C1128	2015	C800	2015	DS72	2015	HE-00829
2015	C1128(S)	2015	C835	2015	DS73	2015	HEP709
2015	C1129	2015	C837	2015	DS74	2015	HEP719
2015	C1129(R)	2015	C837F	2015	DS75	2015	HEP721
2015	C1159	2015	C837H	2015	DS78	2015	HEP731
2015	C1182B	2015	C837K	2015	DS81	2015	HEP732
2015	C1182C	2015	C837L	2015	DS85	2015	HEP733
2015	C1182D	2015	C837WF	2015	E1A	2015	HEP734
2015	C1360	2015	C860	2015	EA15X134	2015	HR-58
2015	C1390A	2015	C860C	2015	EA15X135	2015	HR-59
2015	C1394	2015	C860D	2015	EA15X351	2015	HR76
2015	C1417	2015	C860E	2015	EA15X4064	2015	HR79
2015	C1417C	2015	C863	2015	EA15X7125	2015	HR80
2015	C1417D	2015	C864	2015	EA15X7140	2015	HB40021
2015	C1417D(1)	2015	C918	2015	EA15X7215	2015	HB40022
2015	C1417D(U)	2015	C920E	2015	EA15X7231	2015	HB40023
2015	C1417DU	2015	C921	2015	EA15X7264	2015	HB40024
2015	C1417F	2015	C921C1	2015	EA2131	2015	HB40025
2015	C1417G	2015	C921K	2015	EA2132	2015	HT303711A
2015	C1417H	2015	C921L	2015	EA2494	2015	HT303711B
2015	C1417U	2015	C921M	2015	EA2496	2015	HT303711C
2015	C1417V	2015	C922	2015	EA2812	2015	HT303T310
2015	C1417VW	2015	C922A	2015	EA3713	2015	HT303801
2015	C1417W	2015	C922B	2015	EC0107	2015	HT303801-B
2015	C171	2015	C922C	2015	EC0161	2015	HT303801A
2015	C174	2015	C922L	2015	EP200	2015	HT303801A0
2015	C174A	2015	C922M	2015	EL231	2015	HT303801B
2015	C186	2015	C927CJ	2015	EL434	2015	HT303801B-0
2015	C187	2015	C928	2015	ES5172	2015	HT303801B0
2015	C1908E	2015	C928B	2015	EP15X106	2015	HT303801C
2015	C250	2015	C928C	2015	EP15X123	2015	HT303801C0
2015	C251	2015	C928D	2015	EP15X37	2015	HT304601B0
2015	C251A	2015	C928E	2015	EP15X38	2015	HT306451A
2015	C252	2015	C930DE	2015	EQS-18	2015	HT307720B
2015	C253	2015	C930ET	2015	ES10186	2015	HT307721C
2015	C266	2015	C947	2015	ES10188	2015	HT307721D
2015	C313(JAPAN)	2015	C948	2015	ES15046	2015	HT308291A
2015	C313C	2015	C957	2015	ES15047	2015	HT308291A-0
2015	C313H	2015	C997	2015	ES15102	2015	HT308291A-0
2015	C329	2015	CA05028A	2015	ES15X102	2015	HT308291B
2015	C329B	2015	CCS6225F	2015	ES15X104	2015	HT308291B-0
2015	C329C	2015	CCS6226G	2015	ES15X105	2015	HT308291C
2015	C348	2015	CCS6227F	2015	ES15X119	2015	HT309291E
2015	C351	2015	CS89017	2015	ES15X120	2015	HT309301D
2015	C351(PA)	2015	CS89017O925	2015	ES15X121	2015	HT8001610
2015	C361	2015	CS89018H924	2015	ES15X122	2015	HT8001710
2015	C362	2015	CDC12030B	2015	ES15X123	2015	HV0000302
2015	C365T6	2015	CDC5000	2015	ES15X56	2015	HX50003
2015	C375-O	2015	CS-6225E	2015	ES15X57	2015	IP20-0037
2015	C381	2015	CS-9014	2015	ES15X65	2015	IP20-0038
2015	C381-O	2015	CS-9016D	2015	ES15X66	2015	IRTR-71
2015	C381-O	2015	C81014G	2015	ES15X67	2015	IRTR70
2015	C381-R	2015	C81014H	2015	ES15X79	2015	IRTR71
2015	C381BN	2015	C81120C	2015	ES15X80	2015	IT918
2015	C381R	2015	C81120D	2015	ES15X81	2015	IT929
2015	C385(TRANSISTOR)	2015	C81120E	2015	ES15X82	2015	IT930
2015	C385A	2015	C81120F	2015	ES15X87	2015	J107
2015	C389	2015	C81120H	2015	ES15X88	2015	J241177
2015	C389-O	2015	C81284B	2015	ES15X96	2015	J241188
2015	C389R	2015	C81284F	2015	ES15X97	2015	J241189
2015	C390(TRANSISTOR)	2015	C81284G	2015	ES54(ELCOM)	2015	J24561
2015	C394	2015	C81284H	2015	ES73(ELCOM)	2015	J24562
2015	C398(PA-1)	2015	C81330	2015	ES86(ELCOM)	2015	J24563
2015	C398(TRANSISTOR)	2015	C81460E	2015	ET15X2	2015	J24701
2015	C399	2015	C81460H	2015	ET15X3	2015	J24903
2015	C464	2015	C81461J	2015	EW163	2015	K1214688-1
2015	C464C	2015	C81461X	2015	EW164	2015	K2001
2015	C465	2015	C81462F	2015	EW165	2015	K2119
2015	C466	2015	C81585H	2015	EW212	2015	K2120
2015	C466H	2015	C81589E	2015	F24T-011-013	2015	K2121
2015	C545	2015	C81589F	2015	F24T-011-015	2015	K2122
2015	C545A	2015	C81589B	2015	F24T-016-024	2015	K2123
2015	C545B	2015	C81594E	2015	F3535	2015	K2124
2015	C545C	2015	C81596E	2015	F3574	2015	K2125
2015	C545D	2015	C82715	2015	F501	2015	K2126
2015	C545E	2015	C82716	2015	F501(ZENITH)	2015	K2127
2015	C56	2015	C83662	2015	F501-16	2015	K2501
2015	C562	2015	C83663	2015	F50116	2015	K2502
2015	C562Y	2015	C83707	2015	F502	2015	K2503
2015	C563	2015	C83708	2015	F502(ZENITH)	2015	K2509
2015	C563A	2015	C83709	2015	F523	2015	K2601C
2015	C5b7	2015	C83710	2015	FI 1023	2015	K2602C
2015	C629	2015	C83711	2015	FI-1023	2015	K2603C
2015	C645	2015	C84001	2015	FS1308	2015	K2604
2015	C645A	2015	C84021	2015	FS1682	2015	K2604C
2015	C645B	2015	C84060	2015	FS32669	2015	K2615
2015	C645C	2015	C84193	2015	FSP1	2015	K2616
2015	C645G	2015	C84194	2015	FT118	2015	K2857C
2015	C645N	2015	C89016	2015	FT45	2015	K2857P
2015	C656			2015	FX709	2015	K3683C
2015	C657			2015	G04-041B		

RS 276-	Industry Standard No.	RS 276-	Industry Standard No.	RS 276-	Industry Standard No.	RS 276-	Industry Standard No.
2015	K3683P	2015	RE5001	2015	SPS-2265	2015	TV35
2015	K3880C	2015	RE5002	2015	SPS-2266	2015	TV50
2015	K4002	2015	RE9	2015	SPS-856	2015	TV54
2015	KLH4792	2015	REN107	2015	SPS-860	2015	TV57
2015	LM1110B	2015	RF8161	2015	SPS2167	2015	TV58
2015	LM1120B	2015	RF200	2015	SPS2425	2015	TV60
2015	LM1120C	2015	RS-2011	2015	SPS3787	2015	TV8-28C288A
2015	LTI1016	2015	RS-7201	2015	SPS4143	2015	TV8-28C183P
2015	M-128J509-1	2015	RS-7212	2015	SPS4168	2015	TV8-28C183Q
2015	M-128J510-1	2015	RS7143	2015	SPS429	2015	TV8-28C313
2015	M-128J511-3	2015	RS7222	2015	SPS43-1	2015	TV8-28C429A
2015	M140-1	2015	RS7523	2015	SPS4343	2015	TV8-28C466
2015	M1400-1	2015	RS9510	2015	SPS5569	2015	TV8-28C469A
2015	M401	2015	RS9511	2015	SPS6682	2015	TV8-28C562
2015	M4746	2015	RS9512	2015	SPS816	2015	TV8-28C563
2015	M4756	2015	RT-113	2015	SPS856	2015	TV8-28C563A
2015	M4757	2015	RT112	2015	SPS860	2015	TV8-28C644
2015	M4789	2015	RT5464	2015	SPS906	2015	TV8-28C645
2015	M4825	2015	RT5465	2015	SPS919	2015	TV8-28C645A
2015	M4840A	2015	RT6204	2015	SS4042	2015	TV8-28C645B
2015	M4904	2015	RT6601	2015	ST11	2015	TV8-28C645C
2015	M546	2015	RT6602	2015	ST5641	2015	TV8-28C683
2015	M612	2015	RT6732	2015	ST6510	2015	TV8-28C683V
2015	M613	2015	RT6787	2015	T-203	2015	TV8-28C684
2015	M614	2015	RT7703	2015	T-3568	2015	TV8-28C762
2015	M75545-1	2015	RT7704	2015	T-Q5049	2015	TV8280562
2015	M9032	2015	RT8668	2015	T-Q5071	2015	TV8280563
2015	M91	2015	RT8669	2015	T-Q5079	2015	TV8280829B
2015	M9266	2015	S-1019	2015	T-Q5086	2015	V118
2015	M9450	2015	S-1019(UHP)	2015	T-Q5106	2015	V129
2015	M9481	2015	S-1313	2015	T1202(QE)	2015	V143
2015	MF1161	2015	S-2617	2015	T1886	2015	V8-28C446
2015	MF1162	2015	S1041	2015	T1X-M14	2015	V8-28C683
2015	MF1163	2015	S1041-16GN	2015	T1X-M15	2015	V8280394-0-1
2015	MF1164	2015	S1122	2015	T1X-M16	2015	V8280684
2015	MM8006	2015	S1126	2015	T3568	2015	V8280717///-1
2015	MM8007	2015	S1286	2015	T3601(RCA)	2015	V8280784R1F
2015	MMT8015	2015	S130-138	2015	T381	2015	W23
2015	MPS2369	2015	S130-251	2015	T381(SEARS)	2015	W29
2015	MPS2823	2015	S1308	2015	T576-1	2015	WRP56
2015	MPS2894	2015	S15650	2015	TA2554	2015	WRP784
2015	MPS4145	2015	S15657	2015	TA2555	2015	XC371
2015	MPS6528	2015	S15658	2015	TEG0129	2015	X836
2015	MPS6529	2015	S15659	2015	TG28C927	2015	X837
2015	MPS6531	2015	S1897	2015	TG28C927(C)	2015	X838
2015	MPS6532	2015	S2002	2015	TG28C927A	2015	X839
2015	MPS654	2015	S2131	2015	TG28C927C	2015	X840
2015	MPS6540	2015	S2132	2015	TH-H28C313	2015	ZEN105
2015	MPS6569	2015	S2133	2015	TI-3016	2015	01-030682
2015	MPS6570	2015	S2134	2015	TI-407	2015	01-031560
2015	MPS834	2015	S21648	2015	TI-408	2015	01-571794
2015	MPS918	2015	S25261	2015	TI-409	2015	1-801-304-15
2015	MPS9623C	2015	S25941	2015	TI-492	2015	1-TR-048
2015	MPSH08	2015	S2617(UHP)	2015	TI-493	2015	1 SS35B
2015	MPSH09	2015	S27604	2015	TI-494	2015	180683
2015	MPSH24	2015	S35232	2015	TIS-18	2015	002-011400
2015	MPSH34	2015	S4002	2015	TIS-1B	2015	002-011500
2015	MRD150	2015	S40204	2015	TIS108	2015	2AG
2015	MRD450	2015	S40545	2015	TIS18	2015	2AH
2015	MRF502	2015	S5020	2015	TIS412	2015	2N1586
2015	MRS6548	2015	S5021	2015	TIS86	2015	2N1589
2015	MT3001	2015	SAB1044	2015	TIS87	2015	2N1592
2015	MT3002	2015	SAB3469	2015	TIS97	2015	2N162
2015	MT4104	2015	SAO-1843	2015	TIS98	2015	2N162A
2015	N-EA15X130	2015	SDB820	2015	TIS99	2015	2N222
2015	N-EA15X134	2015	SDB240	2015	TIX-M14	2015	2N2865
2015	N-EA15X135	2015	SDB420	2015	TIX-M15	2015	2N2901
2015	NC89018D	2015	SE-1010	2015	TIX-M16	2015	2N3082
2015	NJ100A	2015	SE-5023	2015	TM107	2015	2N3083
2015	NJ202B	2015	SE-5024	2015	TM161	2015	2N3137
2015	NKT16229	2015	SE-5025	2015	TN3200	2015	2N3287
2015	NKT35219	2015	SEO02(1)	2015	TNJ-60605	2015	2N3288
2015	NL100B	2015	SE1002	2015	TNJ1173	2015	2N3289
2015	NPC167	2015	SE1002-1	2015	TNJ60069(28C74)	2015	2N3290
2015	NR421	2015	SE1002-2	2015	TNJ60447	2015	2N3291
2015	NR421DG	2015	SE1010	2015	TNJ60448	2015	2N3292
2015	NR461AA	2015	SE2001	2015	TNJ60449	2015	2N3293
2015	NR461AF	2015	SE2002	2015	TNJ60604	2015	2N3294
2015	NS1510	2015	SE2020	2015	TNJ60605	2015	2N3337
2015	NS3039	2015	SE2397	2015	TNJ60606	2015	2N3338
2015	NS3040	2015	SE3001	2015	TNJ60607	2015	2N3339
2015	NS3041	2015	SE3005	2015	TNJ61217	2015	2N33681
2015	NS3455	2015	SE3019	2015	TNJ61218	2015	2N337
2015	NS406	2015	SE3646	2015	TNJ61671(28C688)	2015	2N338
2015	P/N14-603-02	2015	SE4020	2015	TNJ61722(28C722)	2015	2N3493
2015	PB5031	2015	SE4020/6-04	2015	TNJ61730	2015	2N3633/46
2015	PL-176-026-9-001	2015	SE4021	2015	TNJ61731	2015	2N3633/TNT
2015	PL-176-037-9-001	2015	SE4022	2015	TNJ71629	2015	2N3662
2015	PL-176-042-9-003	2015	SE5-0249	2015	TNJ71957	2015	2N3663
2015	PL-176-049-9-002	2015	SE5001	2015	TNJ71963	2015	2N3721
2015	PL1021	2015	SE5002	2015	TNJ71964	2015	2N3825
2015	PL1022	2015	SE5003	2015	TNJ72150	2015	2N3826
2015	PL1023	2015	SE5004	2015	TNJ72151	2015	2N3845
2015	PL1024	2015	SE5021	2015	TNJ72275	2015	2N3845A
2015	PL1025	2015	SE5022	2015	TNJ72277	2015	2N3846
2015	PL1026	2015	SE5023	2015	TNJ72279	2015	2N3855
2015	PL1066	2015	SE5024	2015	TNJ72368	2015	2N3855A
2015	PL1067	2015	SE5025	2015	TNJ72701	2015	2N3856
2015	PL1111	2015	SE5032	2015	TR-171	2015	2N3856A
2015	PL1112	2015	SE5050	2015	TR-8010	2015	2N3860
2015	PL1113	2015	SE5051	2015	TR-8043	2015	2N3932
2015	PM194	2015	SE5052	2015	TR-95(XSTR)	2015	2N3933
2015	PM195	2015	SE6002	2015	TRO1042	2015	2N4081
2015	PT4816	2015	SF167	2015	TR2083-74	2015	2N4254
2015	PT4830	2015	SF173	2015	TR228735046011	2015	2N4255
2015	PTC126	2015	SF194	2015	TRO1042	2015	2N4295
2015	PTC132	2015	SF194B	2015	TV-15A	2015	2N4597
2015	PTG132	2015	SF195	2015	TV-16	2015	2N4433
2015	Q-00584	2015	SF195C	2015	TV-24399	2015	2N4434
2015	Q-01115C	2015	SF195D	2015	TV-36	2015	2N471
2015	Q1/6515	2015	SF196	2015	TV-37	2015	2N476A
2015	Q2	2015	SF197	2015	TV-38	2015	2N5031
2015	Q2/6515	2015	SF294	2015	TV-39	2015	2N5032
2015	Q3	2015	SF294B	2015	TV15	2015	2N5126
2015	Q301	2015	SF295	2015	TV15A	2015	2N5130
2015	Q35259	2015	SF295C	2015	TV15B	2015	2N5132
2015	Q4/6515	2015	SF295D	2015	TV15C	2015	2N5133
2015	Q5	2015	SF310	2015	TV20	2015	2N5181
2015	Q6/6515	2015	SF314	2015	TV24160	2015	2N5182
2015	QSE5020	2015	SF334	2015	TV24161	2015	2N5650
2015	QT-C1047XAA	2015	SF334B	2015	TV24209	2015	2N5651
2015	QT-C1047XBN	2015	SF335	2015	TV24210	2015	2N5652
2015	R-28C535	2015	SF335C	2015	TV24380	2015	2N708/TNT
2015	R-28C772	2015	SF335D	2015	TV24382	2015	2N709/TNT
2015	R-28C858	2015	SG887231	2015	TV24383	2015	2N780
2015	R118	2015	SK8937	2015	TV24385	2015	2N917
2015	RS529	2015	SKA-4075	2015	TV24399	2015	2N917/51
2015	RCA40245	2015	SKA-5541	2015	TV24436	2015	2N917A
2015	RCA40246	2015	SKA1416	2015	TV24437	2015	000028606
2015	RE2001	2015	SKA4768	2015	TV24438	2015	2382
2015	RE2002	2015	SM-4304-8	2015	TV24571	2015	23B460
2015	RE28	2015	SM-A-595830-12	2015	TV24806	2015	23B683
2015	RE3001	2015	SP4436	2015	TV33	2015	28C-313
2015	RE3002	2015	SPS-1473RT				

RS 276-	Industry Standard No.	RS 276-	Industry Standard No.	RS 276-	Industry Standard No.	RS 276-	Industry Standard No.
2015	2SC1026	2015	2SC2009	2015	2SC383	2015	2SC674CZ
2015	0002SC1026A	2015	2SC2012	2015	2SC3854	2015	2SC674D
2015	0002SC1026C	2015	2SC250	2015	2SC385A	2015	2SC674EB
2015	2SC1026G	2015	2SC251	2015	2SC389	2015	2SC674F
2015	2SC1026Y	2015	2SC251A	2015	2SC389-0	2015	2SC674G
2015	2SC1032	2015	2SC252	2015	2SC389R	2015	2SC674V
2015	2SC1032(Y)	2015	2SC253	2015	2SC390	2015	2SC682
2015	0002SC1032A	2015	2SC266	2015	2SC398	2015	2SC682(B)
2015	0002SC1032B	2015	2SC2884	2015	2SC399	2015	2SC682-OR
2015	2SC1032BL	2015	2SC313	2015	2SC463A	2015	2SC682A
2015	0002SC1032C	2015	2SC313-OR	2015	2SC463B	2015	2SC682B
2015	2SC1032D	2015	2SC313A	2015	2SC463C	2015	2SC682C
2015	2SC1032E	2015	2SC313B	2015	2SC463D	2015	2SC682D
2015	2SC1032GN	2015	2SC313C	2015	2SC463E	2015	2SC682E
2015	2SC1032H	2015	2SC313D	2015	2SC463F	2015	2SC682F
2015	2SC1032J	2015	2SC313E	2015	2SC463G	2015	2SC682G
2015	2SC1032L	2015	2SC313F	2015	2SC463GN	2015	2SC682GN
2015	2SC1032M	2015	2SC313G	2015	2SC463J	2015	2SC682H
2015	2SC1032OR	2015	2SC313GN	2015	2SC463K	2015	2SC682J
2015	2SC1032R	2015	2SC313H	2015	2SC463M	2015	2SC682L
2015	2SC1032X	2015	2SC313J	2015	2SC463OR	2015	2SC682M
2015	2SC1032Y	2015	2SC313K	2015	2SC463R	2015	2SC682OR
2015	2SC1035	2015	2SC313L	2015	2SC463X	2015	2SC682R
2015	2SC1035C	2015	2SC313M	2015	2SC463Y	2015	2SC682X
2015	2SC1035D	2015	2SC313OR	2015	2SC464	2015	2SC682Y
2015	2SC1035E	2015	2SC313R	2015	2SC464C	2015	2SC684-OR
2015	2SC1036	2015	2SC313X	2015	2SC465	2015	2SC684C
2015	2SC1044	2015	2SC313Y	2015	2SC466	2015	2SC684D
2015	2SC1048DC	2015	2SC329	2015	2SC466B	2015	2SC684E
2015	2SC1070	2015	2SC329B	2015	2SC527	2015	2SC684G
2015	2SC1117	2015	2SC329C	2015	2SC545	2015	2SC684GN
2015	2SC1123	2015	2SC344	2015	2SC545A	2015	2SC684H
2015	2SC1126	2015	2SC344(Y)	2015	2SC545B	2015	2SC684J
2015	2SC1128	2015	2SC344Y	2015	2SC545C	2015	2SC684L
2015	2SC1128(3RD-IP)	2015	2SC348	2015	2SC545D	2015	2SC684M
2015	2SC1128(FINAL-IP)	2015	2SC349	2015	2SC545E	2015	2SC684R
2015	2SC1128(M)	2015	2SC349R	2015	2SC546	2015	2SC684X
2015	2SC1128(S)	2015	2SC351	2015	2SC546K	2015	2SC695
2015	2SC1128-0	2015	2SC351(PA)	2015	2SC562	2015	2SC705
2015	2SC1128A	2015	2SC361	2015	2SC562	2015	2SC705B
2015	2SC1128B	2015	2SC362	2015	2SC562-0	2015	2SC705C
2015	2SC1128BL	2015	2SC375-0	2015	2SC562-OR	2015	2SC705D
2015	2SC1128C	2015	2SC380	2015	2SC562A	2015	2SC705E
2015	2SC1128D	2015	2SC380(R)	2015	2SC562C	2015	2SC705F
2015	2SC1128G	2015	2SC380-0	2015	2SC562D	2015	2SC705TV
2015	2SC1128H	2015	2SC380-0/4454C	2015	2SC562E	2015	2SC707
2015	2SC1128M	2015	2SC380-BRN	2015	2SC562F	2015	2SC707H
2015	2SC1128R	2015	2SC380-0	2015	2SC562G	2015	2SC717
2015	2SC1128S(3RDIP)	2015	2SC380-0/4454C	2015	2SC562GN	2015	2SC717(3RD-IP)
2015	2SC1128Y	2015	2SC380-OR	2015	2SC562H	2015	2SC717(LAST-IP)
2015	2SC1129	2015	2SC380-ORG	2015	2SC562J	2015	2SC717B
2015	2SC1129(M)	2015	2SC380-R	2015	2SC562K	2015	2SC717BK
2015	2SC1129(R)	2015	2SC380-RED	2015	2SC562L	2015	2SC717BLK
2015	2SC1129-0	2015	2SC380-Y	2015	2SC562M	2015	2SC717C
2015	2SC1129A	2015	2SC380-YEL	2015	2SC562OR	2015	2SC717E
2015	2SC1129B	2015	2SC380/4454C	2015	2SC562R	2015	2SC717F
2015	2SC1129BL	2015	2SC380A	2015	2SC562X	2015	2SC717GN
2015	2SC1129C	2015	2SC380A(D)	2015	2SC562Y	2015	2SC717H
2015	2SC1129G	2015	2SC380A(0)	2015	2SC563	2015	2SC717K
2015	2SC1129M	2015	2SC380A(Y)	2015	2SC563(3RDIP)	2015	2SC717L
2015	2SC1129R	2015	2SC380A+0)	2015	2SC563-F	2015	2SC717M
2015	2SC1129Y	2015	2SC380A-0	2015	2SC563-G	2015	2SC717X
2015	2SC1159	2015	2SC380A-0(TV)	2015	2SC563-OR	2015	2SC739
2015	2SC1180	2015	2SC380A-0	2015	2SC563A	2015	2SC739C
2015	2SC1182B	2015	2SC380A-0(TV)	2015	2SC563A(3RDIP)	2015	2SC761
2015	2SC1182C	2015	2SC380A-R	2015	2SC563B	2015	2SC761(Y)
2015	2SC1182D	2015	2SC380A-R(TV)	2015	2SC563C	2015	2SC761A
2015	2SC1215	2015	2SC380A-Y	2015	2SC563D	2015	2SC761B
2015	2SC1215C	2015	2SC380A0	2015	2SC563E	2015	2SC761C
2015	2SC1215D	2015	2SC380AR	2015	2SC563F	2015	2SC761D
2015	2SC1215E	2015	2SC380ATV	2015	2SC563G	2015	2SC761E
2015	2SC1215F	2015	2SC380AY	2015	2SC563GN	2015	2SC761F
2015	2SC1215G	2015	2SC380B	2015	2SC563H	2015	2SC761G
2015	2SC1215GN	2015	2SC380B-Y	2015	2SC563J	2015	2SC761GN
2015	2SC1215H	2015	2SC380BY	2015	2SC563K	2015	2SC761H
2015	2SC1215J	2015	2SC380C	2015	2SC563L	2015	2SC761J
2015	2SC1215K	2015	2SC380C-Y	2015	2SC563M	2015	2SC761K
2015	2SC1215L	2015	2SC380CY	2015	2SC563OR	2015	2SC761L
2015	2SC1215M	2015	2SC380D	2015	2SC563R	2015	2SC761M
2015	2SC1215OR	2015	2SC380D-Y	2015	2SC563X	2015	2SC761OR
2015	2SC1215R	2015	2SC380DY	2015	2SC563Y	2015	2SC761R
2015	2SC1215X	2015	2SC380E	2015	2SC567	2015	2SC761X
2015	2SC1215Y	2015	2SC380E-Y	2015	2SC591	2015	2SC761Y
2015	2SC1254	2015	2SC380EY	2015	2SC602	2015	2SC761Z
2015	2SC1320K	2015	2SC380F	2015	2SC604	2015	2SC762
2015	2SC1321	2015	2SC380F-Y	2015	2SC629	2015	2SC762B
2015	2SC1360	2015	2SC380FY	2015	2SC645	2015	2SC762C
2015	2SC1394	2015	2SC380G	2015	2SC645-OR	2015	2SC762D
2015	2SC1395	2015	2SC380GN	2015	2SC645A	2015	2SC762E
2015	2SC1417(V,G)	2015	2SC380H	2015	2SC645B	2015	2SC762F
2015	2SC1417VW	2015	2SC380J	2015	2SC645B-1	2015	2SC762G
2015	2SC1585F	2015	2SC380K	2015	2SC645C	2015	2SC762GN
2015	2SC1585F,H	2015	2SC380L	2015	2SC645E	2015	2SC762H
2015	2SC1585H	2015	2SC380M	2015	2SC645F	2015	2SC762J
2015	2SC1622	2015	2SC380O	2015	2SC645G	2015	2SC762K
2015	2SC166A	2015	2SC380OR	2015	2SC645GN	2015	2SC762L
2015	2SC166B	2015	2SC380R	2015	2SC645GR	2015	2SC762M
2015	2SC166C	2015	2SC380R/4454C	2015	2SC645H	2015	2SC762R
2015	2SC166D	2015	2SC380RED	2015	2SC645J	2015	2SC762X
2015	2SC166E	2015	2SC380X	2015	2SC645K	2015	2SC762Y
2015	2SC166F	2015	2SC380Y	2015	2SC645L	2015	0002SC772
2015	2SC166G	2015	2SC380YEL	2015	2SC645M	2015	0002SC772C
2015	2SC166GN	2015	2SC381	2015	2SC645N	2015	2SC772CA
2015	2SC166H	2015	2SC381(BN)	2015	2SC645OR	2015	2SC786
2015	2SC166J	2015	2SC381-0	2015	2SC645R	2015	2SC786A
2015	2SC166K	2015	2SC381-BN	2015	2SC645V	2015	2SC786B
2015	2SC166L	2015	2SC381-BRN	2015	2SC645X	2015	2SC786C
2015	2SC166OR	2015	2SC381-0	2015	2SC645Y	2015	2SC786D
2015	2SC166OR	2015	2SC381-OR	2015	2SC653	2015	2SC786E
2015	2SC166R	2015	2SC381-ORG	2015	2SC656	2015	2SC786F
2015	2SC166X	2015	2SC381-R	2015	2SC657	2015	2SC786G
2015	2SC166Y	2015	2SC381-RED	2015	2SC663	2015	2SC786H
2015	2SC167A	2015	2SC381A	2015	2SC667	2015	2SC786J
2015	2SC167B	2015	2SC381B	2015	0002SC668D	2015	2SC786K
2015	2SC167C	2015	2SC381BN	2015	2SC673	2015	2SC786L
2015	2SC167D	2015	2SC381BN-1	2015	2SC673(B)	2015	2SC786M
2015	2SC167E	2015	2SC381BRN	2015	2SC673B	2015	2SC786OR
2015	2SC167F	2015	2SC381C	2015	2SC673C	2015	2SC786R
2015	2SC167G	2015	2SC381D	2015	2SC673C2	2015	2SC786X
2015	2SC167GN	2015	2SC381E	2015	2SC673D	2015	2SC787
2015	2SC167H	2015	2SC381F	2015	2SC674	2015	2SC787A
2015	2SC167K	2015	2SC381G	2015	2SC674(D)	2015	2SC787B
2015	2SC167L	2015	2SC381GN	2015	2SC674(P)	2015	2SC787C
2015	2SC167M	2015	2SC381H	2015	2SC674(G)	2015	2SC787D
2015	2SC167OR	2015	2SC381J	2015	2SC674-B	2015	2SC787E
2015	2SC167X	2015	2SC381K	2015	2SC674-F	2015	2SC787F
2015	2SC167Y	2015	2SC381L	2015	2SC674B	2015	2SC787G
2015	2SC171	2015	2SC381M	2015	2SC674C	2015	2SC787GN
2015	2SC174	2015	2SC381O	2015	2SC674CK		
2015	2SC174A	2015	2SC381OR	2015	2SC674CL		
2015	2SC186	2015	2SC381RED	2015	2SC674CV		
2015	2SC1919C	2015	2SC381RL				
		2015	2SC381X				
		2015	2SC381Y				

RS 276-	Industry Standard No.	RS 276-	Industry Standard No.	RS 276-	Industry Standard No.	RS 276-	Industry Standard No.
2015	28C787H	2015	28D33J	2015	13-10321-47	2015	24T-013-003
2015	28C787K	2015	28D33K	2015	13-10321-50	2015	24T-016
2015	28C787L	2015	28D33L	2015	13-10321-65	2015	24T-016-010
2015	28C787M	2015	28D330R	2015	13-10321-66	2015	24T-016-024
2015	28C787OR	2015	28D33X	2015	13-10321-67	2015	24T-016-0B
2015	28C787R	2015	28D33Y	2015	13-10321-71	2015	24T002
2015	28C787X	2015	28D829	2015	13-10321-72	2015	24T011-012
2015	28C787Y	2015	2T15X3	2015	13-10321-75	2015	24T013003
2015	28C800	2015	2T919	2015	13-10321-78	2015	24T013005
2015	000028C829	2015	03A03	2015	13-10321-79	2015	24T016
2015	28C82BN	2015	3L4-6007-10	2015	13-103176	2015	24T016001
2015	28C82R	2015	3L4-6007-11	2015	13-15021-15	2015	24T016005
2015	28C83	2015	3L4-6007-12	2015	13-14085-16	2015	24T021
2015	28C835	2015	3L4-6007-13	2015	13-14085-17	2015	025-009600
2015	28C837	2015	3L4-6007-14	2015	13-14085-26	2015	025-100012
2015	28C837F	2015	3L4-6007-17	2015	13-14085-3	2015	025-100026
2015	28C837H	2015	3L4-6007-19	2015	13-14085-4	2015	025-100036
2015	28C837K	2015	3L4-6007-20	2015	13-15808-2	2015	025-100037
2015	28C837L	2015	3L4-6007-21	2015	13-15835-1	2015	025-100038
2015	28C837WF	2015	3L4-6007-22	2015	13-23013-2	2015	25T-002
2015	28C852	2015	3L4-6007-23	2015	13-23822-1	2015	28-819-172
2015	28C852A	2015	3L4-6007-35	2015	13-23824-1	2015	31-0048
2015	28C860	2015	3N120	2015	13-23824-2	2015	31-0049
2015	28C860C	2015	3N121	2015	13-2384-1	2015	31-0050
2015	28C860D	2015	3N127	2015	13-2384-2	2015	31-0054
2015	28C860E	2015	3N71	2015	13-26009-1	2015	31-0098
2015	28C863	2015	3N72	2015	13-26576-1	2015	31-0206
2015	28C864	2015	3N73	2015	13-26576-2	2015	31-10
2015	28C879	2015	3N87	2015	13-26577-1	2015	34-3015-46
2015	28C918	2015	3N88	2015	13-26577-2	2015	34-3015-47
2015	28C921	2015	04-00461-02	2015	13-26577-3	2015	34-3015-49
2015	28C921(L)	2015	04-00535-02	2015	13-29947-1	2015	34-6007-10
2015	28C921(VMF)	2015	04-01585-08	2015	13-29974-1	2015	34-6007-11
2015	28C921A	2015	04-15850-06	2015	13-370103-5	2015	34-6007-12
2015	28C921B	2015	4-850	2015	13-67583-6/3464	2015	34-6007-13
2015	28C921C	2015	6A11223	2015	13-67585-4/3464	2015	34-6015-18
2015	28C921C1	2015	7-10	2015	14-0110-1	2015	34-6015-19
2015	28C921D	2015	7-10(SARKES)	2015	14-0110-2	2015	34-6015-20
2015	28C921E	2015	7-11(SARKES)	2015	14-602-31	2015	34-6015-22
2015	28C921F	2015	7-16	2015	14-602-34	2015	34-6015-31
2015	28C921G	2015	7-20	2015	14-602-45	2015	34-6015-50
2015	28C921GN	2015	7-21(SARKES)	2015	14-602-77	2015	34-6015-8
2015	28C921H	2015	7-22	2015	14-602-77B	2015	36-1
2015	28C921J	2015	7-22(SARKES)	2015	14-603-02	2015	36-6015-46
2015	28C921K	2015	7-23	2015	14-603-05	2015	41M-19
2015	28C921L	2015	7-23(SARKES)	2015	14-603-05-2	2015	41N
2015	28C921M	2015	7-24(SARKES)	2015	14-603-06	2015	42-21401
2015	28C921OR	2015	7-25(SARKES)	2015	14-603-08	2015	42-21402
2015	28C921R	2015	7-26	2015	14-603-09	2015	42-22532
2015	28C921W	2015	7-28	2015	14-603-13	2015	42-23960
2015	28C921X	2015	7-36	2015	14-652-12	2015	42-23960P
2015	28C921Y	2015	7-39	2015	14-653-21	2015	42-23961
2015	28C927	2015	7-43	2015	14-661-21	2015	42-23961P
2015	28C927(D)	2015	7-45	2015	14-850-12	2015	42-23962
2015	28C927(E)	2015	7-9	2015	14-851-12	2015	42-23962P
2015	28C927B	2015	7-9(SARKES)	2015	15-03051-00	2015	42-23963
2015	28C927C(E)	2015	7A30	2015	15-088003	2015	42-23963P
2015	28C927C(K)	2015	7A31	2015	15-166N	2015	42-27372
2015	28C927CJ	2015	7A32	2015	16-21426	2015	42-27573
2015	28C927CK	2015	8-722-923-00	2015	16GN	2015	045-1
2015	28C927CT	2015	8-724-034-00	2015	16L64	2015	045-2
2015	28C927CU	2015	8-905-605-644	2015	018B	2015	46-06311-3
2015	28C927CW	2015	8-905-706-044	2015	019-005157	2015	46-119-3
2015	28C927D	2015	8-905-706-055	2015	19-020-071	2015	46-6829
2015	28C927E	2015	8-905-706-060	2015	19A115342-2	2015	46-86101-3
2015	28C927E,Z	2015	8-905-706-070	2015	19Q017	2015	46-86107-3
2015	28C927F	2015	8-905-706-071	2015	020-00024	2015	46-86108-3
2015	28C927G	2015	8-905-706-075	2015	020-00025	2015	46-86109-3
2015	28C927GN	2015	8-905-706-080	2015	020-00028	2015	46-86110-3
2015	28C927H	2015	8-905-706-101	2015	020-1112-003	2015	46-86112-2
2015	28C927J	2015	8-905-706-110	2015	020-1112-004	2015	46-86112-3
2015	28C927K	2015	8-905-706-112	2015	21-2	2015	46-86113-3
2015	28C927L	2015	8-905-706-730	2015	21-4	2015	46-86114-3
2015	28C927M	2015	9-003	2015	21-6	2015	46-86117-3
2015	28C927OR	2015	09-301039	2015	21-7	2015	46-86118-3
2015	28C927R	2015	09-302014	2015	21A040-023	2015	46-86119-3
2015	28C927X	2015	09-302032	2015	21A040-024	2015	46-86120-3
2015	28C927XL	2015	09-302044	2015	21A040-025	2015	46-86126-3
2015	28C927Y	2015	09-302063	2015	21A040-054	2015	46-86127-3
2015	28C927Z	2015	09-302072	2015	21A040-055	2015	46-86133-3
2015	28C928	2015	09-302073	2015	21A040-063	2015	46-86172-3
2015	28C928B	2015	09-302079	2015	21A112-007	2015	46-86207-3
2015	28C928C	2015	09-302092	2015	21A112-010	2015	46-86209-3
2015	28C928D	2015	09-302103	2015	21M095	2015	46-86224-3
2015	28C928E	2015	09-302114	2015	21M094	2015	46-86239-3
2015	28C947	2015	09-302128	2015	21M099	2015	46-86251-3
2015	28C947A	2015	09-302129	2015	21M140	2015	46-86265-3
2015	28C947B	2015	09-302138	2015	21M151	2015	46-86285-3
2015	28C947C	2015	09-302142	2015	21M152	2015	46-8629
2015	28C947D	2015	09-302143	2015	21M153	2015	46-86295-3
2015	28C947E	2015	09-302144	2015	21M154	2015	46-86299-3
2015	28C947F	2015	09-302145	2015	21M178	2015	46-86301-3
2015	28C947G	2015	09-302151	2015	21M179	2015	46-86302-3
2015	28C947GN	2015	09-302152	2015	21M182	2015	46-86311-3
2015	28C947J	2015	09-302174	2015	21M188	2015	46-86352-3
2015	28C947K	2015	09-302199	2015	22-1	2015	46-86353-3
2015	28C947L	2015	09-302201	2015	23-PT274-120	2015	46-86354-3
2015	28C947M	2015	09-302205	2015	23-PT274-123	2015	46-86435-3
2015	28C947OR	2015	09-302216	2015	23-PT275-121	2015	46-8647
2015	28C967R	2015	09-302240	2015	23-PT283-122	2015	46-8650
2015	28C947X	2015	09-304019	2015	23-PT283-124	2015	46-8652
2015	28C947Y	2015	09-305006	2015	24-016	2015	46-8653
2015	28C94B	2015	09-305007	2015	24-016-005	2015	46-86558-3
2015	28C948A	2015	09-305011	2015	24MW 656	2015	46-8654
2015	28C948B	2015	09-305033	2015	24MW1057	2015	46-8672-3
2015	28C948D	2015	09-305036	2015	24MW1058	2015	46-8677-3
2015	28C948E	2015	09-305041	2015	24MW1081	2015	46-96434-3
2015	28C948F	2015	09-305132	2015	24MW1106	2015	47-(X88R)
2015	28C948G	2015	09-308072	2015	24MW287	2015	48-01-017
2015	28C948GN	2015	09-309069	2015	24MW361	2015	48-01-027
2015	28C948H	2015	09-309072	2015	24MW535	2015	48-01-031
2015	28C948J	2015	09-309672	2015	24MW593	2015	48-134190A1Q
2015	28C948K	2015	9TR1	2015	24MW594	2015	48-134756
2015	28C948L	2015	9TR11001-01	2015	24MW595	2015	48-134789
2015	28C948M	2015	10-1	2015	24MW596	2015	48-134837
2015	28C948OR	2015	10-2	2015	24MW597	2015	48-134845
2015	28C948R	2015	10-28C380	2015	24MW653	2015	48-134904
2015	28C948X	2015	10B551-1	2015	24MW656	2015	48-134904A1Q
2015	28C948Y	2015	10B551-2	2015	24MW673	2015	48-134904F
2015	28C957	2015	10B553-2	2015	24MW700	2015	48-134922
2015	28C987	2015	10B553-3	2015	24MW737	2015	48-134923
2015	28C987A	2015	100573-2	2015	24MW738	2015	48-134924
2015	28C988	2015	100573-3	2015	24MW793	2015	48-134925
2015	28C988A	2015	100574-2	2015	24MW805	2015	48-134926
2015	28C988B	2015	100574-3	2015	24MW812	2015	48-134945
2015	28C997	2015	11C557-2	2015	24MW813	2015	48-134961
2015	28D33A	2015	11C557-3	2015	24MW814	2015	48-134962
2015	28D33B	2015	13-10320-14	2015	24MW815	2015	48-134963
2015	28D33D	2015	13-10321-29	2015	24MW863	2015	48-134964
2015	28D33E	2015	13-10321-31	2015	24MW953	2015	48-134965
2015	28D33F	2015	13-10321-32	2015	24MW957	2015	48-134966
2015	28D33G	2015	13-10321-35	2015	24MW958	2015	48-134981
2015	28D330N	2015	13-10321-36	2015	24T-011-001	2015	48-137040
2015	28D33H	2015	13-10321-42	2015	24T-011-003	2015	48-137059
				2015	24T-011-015	2015	48-137071

RS 276-	Industry Standard No.
2015	48-137075
2015	48-137076
2015	48-137077
2015	48-137197
2015	48-137400
2015	48-137491
2015	48-3003A03
2015	48-40247G01
2015	48-41815J02
2015	48-41816J01
2015	48-43351A02
2015	48-63076A81
2015	48-63076A82
2015	48-63078A52
2015	48-63078A54
2015	48-65112A68
2015	48-65146A61
2015	48-65146A62
2015	48-65146A63
2015	48-65173A78
2015	48-65174A24
2015	48-869266
2015	48-869450
2015	48-869481
2015	48-90232A01
2015	48-90232A17
2015	48-90232A18
2015	48-90343A66
2015	48-97046A05
2015	48-97046A06
2015	48-97046A07
2015	48-97046A17
2015	48-97046A20
2015	48-97046A21
2015	48-97046A25
2015	48-97046A45
2015	48-97127A06
2015	48-97162A26
2015	48-97162A28
2015	48-97162A30
2015	48-97162A31
2015	48-97162A32
2015	48-97177A02
2015	48-97177A03
2015	48P63078A69
2015	48P65123A67
2015	48P65123A95
2015	48P65144A72
2015	48P65146A61
2015	48P65146A62
2015	48P65174A24
2015	48P65193A56
2015	48P65194A92
2015	488134756
2015	488134932
2015	488134981
2015	488157158
2015	48865123A67
2015	488P134826
2015	488P134837
2015	488P134857
2015	488P134904
2015	488P134937
2015	48X134902
2015	48X90232A17
2015	48X90232A18
2015	48X90232A19
2015	48X90232A20
2015	500380-0
2015	500380-0R
2015	500380-0R
2015	500394-0
2015	500394-R
2015	500784-R
2015	051-0049
2015	51A180-4
2015	56-8086
2015	56-8086A
2015	56-8086B
2015	56-8086C
2015	56-8087
2015	56-8087B
2015	56-8087C
2015	56-8088
2015	56-8088A
2015	56-8088C
2015	57A102-4
2015	57A103-4
2015	57A119-2
2015	57A126-12
2015	57A139-1
2015	57A139-2
2015	57A139-3
2015	57A141-4
2015	57A142-7
2015	57A164-4
2015	57A177-12
2015	57A179-4
2015	57A21
2015	57A21-11
2015	57A21-12
2015	57A21-13
2015	57A21-14
2015	57A21-16
2015	57A21-17
2015	57A21-18
2015	57A21-21
2015	57A21-26
2015	57A21-45
2015	57A24
2015	57A249-4
2015	57B102-4
2015	57B103-4
2015	57B104-8
2015	57B119-2
2015	57B141-4
2015	57B21-17
2015	57B21-6
2015	57B21-7
2015	57B21-9
2015	57B249-4
2015	57B30-12
2015	057B474H
2015	57C104-8
2015	57C142-4
2015	57C164-4
2015	57C21
2015	57C21-5
2015	57C24
2015	62-114267
2015	62-129604
2015	62-19581
2015	65(TRANSISTOR)
2015	66F021-1
2015	66F022-1
2015	66F042-1
2015	70N3
2015	73N1
2015	76-13866-19(VHP)
2015	81-46125007-8
2015	81-46125012-8
2015	81-46125013-6
2015	81-46125030-0
2015	81-46125032-6
2015	81-461250324
2015	86-138-3
2015	86-186-2
2015	86-204-2
2015	86-205-2
2015	86-262-2
2015	86-263-2
2015	86-289-2
2015	86-290-2
2015	86-381-2
2015	86-442-2
2015	86-525-2
2015	86-596-2
2015	86-597-2
2015	86X0019-001
2015	86X0052-001
2015	86X0060-001
2015	86X0061-001
2015	86X0062-001
2015	86X7-6
2015	87-593-2
2015	90-45
2015	90-46
2015	90-604
2015	90T2
2015	91N1B
2015	93A39-17
2015	96-5131-01
2015	96-5163-01
2015	96-5174-01
2015	96-5175-01
2015	96-5198-01
2015	96-5199-01
2015	96-5295-01
2015	96-5236-01
2015	96-5259-01
2015	96-5260-01
2015	96-5334-01
2015	96X26050/25N
2015	998031
2015	998032
2015	998044
2015	998067-v
2015	100-4107
2015	100N1
2015	100N1A8
2015	100N1P
2015	100N1P(VID)
2015	100N3P
2015	101-1(ADMIRAL)
2015	102-1047-03
2015	102-4
2015	0103-0531/4460
2015	0103-0568
2015	0103-0568/4460
2015	105-00107-09
2015	105-02005-07
2015	105-02006-05
2015	105-02008-01
2015	121-283
2015	121-377
2015	121-378
2015	121-379
2015	121-380
2015	121-383
2015	121-453
2015	121-460
2015	121-461
2015	121-462
2015	121-470
2015	121-471
2015	121-480
2015	121-500
2015	121-501
2015	121-502
2015	121-505(3LEADS)
2015	121-509
2015	121-510
2015	121-521
2015	121-523
2015	121-580
2015	121-583
2015	121-585
2015	121-585B
2015	121-612
2015	121-645
2015	121-651
2015	121-653
2015	121-687
2015	121-692
2015	121-704
2015	121-723
2015	121-735B
2015	121-739
2015	121-760
2015	121-761
2015	121-775
2015	121-779
2015	121-802
2015	121-807
2015	121-823
2015	121-824
2015	121-847
2015	121-849
2015	121-883
2015	121-885
2015	121-895
2015	121-924
2015	121-946
2015	121-974
2015	121-983
2015	121-984
2015	121-995
2015	121J688-1
2015	121J688-2
2015	129-27
2015	129N1
2015	130-112
2015	130-152
2015	130-185
2015	130-240
2015	130-V10
2015	130-VLO
2015	1300RN
2015	132-008
2015	132-009
2015	132-076
2015	132-082
2015	132-087
2015	132-185
2015	139N2
2015	141 402
2015	141-402
2015	142N1P
2015	145N1
2015	145N1P
2015	150-1N
2015	150N1
2015	150N3
2015	151-0259-00
2015	151-0427-00
2015	151-0471-00
2015	170(RCA)
2015	176-025-9-002
2015	176-056-9-001
2015	176-056-9-003
2015	176-060-9-001
2015	176-075-9-001
2015	181N1
2015	181N1D
2015	181N2
2015	181N2D
2015	186N1
2015	201-254343-13
2015	201-254343-22
2015	201-254343-26
2015	201-254343-28
2015	201-254343-30
2015	201-254343-33
2015	201-254343-34
2015	207A25
2015	207A27
2015	217(RCA)
2015	229-0-180-33
2015	229-0-180-34
2015	229-0180-119
2015	229-0180-32
2015	229-0180-33
2015	229-0190-30
2015	229-0190-31
2015	229-0191-29
2015	229-0191-30
2015	229-0204-6
2015	229-0204-6(VHP)
2015	229-0210-19
2015	229-0240-25
2015	229-0260-18
2015	229-1301-22
2015	229-180-32
2015	229-485-2
2015	229-5100-31V
2015	229-5100-32
2015	229-5100-32V
2015	260D05704
2015	260D08013
2015	260P03201
2015	260P03201A
2015	260P05402
2015	260P05402A
2015	260P05901
2015	260P05901A
2015	260P09201
2015	260P0601
2015	260P1601
2015	260P1610
2015	260P16101
2015	260P1760
2015	260P17702
2015	260P24901
2015	297L011C01
2015	297V07401C01
2015	325-1378-18
2015	325-1378-19
2015	325-1771-16
2015	344-6015-8
2015	344-6015-9
2015	352-0630
2015	352-0630-010
2015	352-0653-010
2015	352-0653-020
2015	352-0658-010
2015	352-0658-020
2015	352-0658-030
2015	352-0658-040
2015	352-0658-050
2015	355D6
2015	355DB
2015	386-7243-P001
2015	404-2
2015	417-125-12903
2015	417-154
2015	417-19
2015	417-190
2015	417-205
2015	417-243
2015	417-258
2015	417-262
2015	429-0981-12
2015	429-0985-12
2015	465-181-19
2015	522(ZENITH)
2015	527-1
2015	535A
2015	536-1
2015	536-1(RCA)
2015	536-2
2015	576-0003-023
2015	576-0003-026
2015	576-0003-029
2015	576-1(RCA)
2015	600X0093-086
2015	600X0094-086
2015	600X0141-000
2015	600X0143-000
2015	600X0175-000
2015	600X0195-000
2015	615-1
2015	617-65
2015	634-1
2015	642-229
2015	642-230
2015	642-254
2015	642-260
2015	642-268
2015	642-269
2015	642-270
2015	642-274
2015	642A84076-101
2015	642A84076-101
2015	657-31
2015	680-1
2015	680-1(RCA)
2015	680-1(TRANSISTOR)
2015	690V010H40
2015	690V080H36
2015	690V086H94
2015	690V086H95
2015	690V088H46
2015	690V089H46
2015	690V089H86
2015	690V103H28
2015	690V109H46
2015	690V116H20
2015	702-0002
2015	715FB
2015	742
2015	753-2000-460
2015	753-4000-668
2015	753-4000-929
2015	753-5751-359
2015	753-5851-359
2015	753-9000-922
2015	753-9020-784
2015	753-9050-785
2015	772A
2015	772B
2015	772B1
2015	772BJ
2015	772BL
2015	772BM
2015	772BN
2015	772BY
2015	772C
2015	772C0C
2015	772D
2015	772D1
2015	772DC
2015	772DG
2015	772E
2015	772EH
2015	772F
2015	772FE
2015	772G
2015	800-557-00
2015	822-1(SYLVANIA)
2015	824-1(SYLVANIA)
2015	903
2015	904
2015	904A
2015	905
2015	909(RCA)
2015	909-27125-160
2015	910
2015	916-31024-3B
2015	916-31024-5
2015	916-31025-4
2015	916-31025-4B
2015	916-31025-5B
2015	921-1014
2015	921-102
2015	921-102A
2015	921-102B
2015	921-141B
2015	921-142B
2015	921-143B
2015	921-145B
2015	921-1528
2015	921-1528
2015	921-176B
2015	921-177B
2015	921-204B
2015	921-210B
2015	921-211B
2015	921-213B
2015	921-226B
2015	921-232B
2015	921-233B
2015	921-235B
2015	921-257B
2015	921-258B
2015	921-55B
2015	921-56B
2015	921-57B
2015	921-58B
2015	921-59B
2015	921-62
2015	921-62A
2015	921-62B
2015	921-63
2015	921-63A
2015	921-63B
2015	921-64
2015	921-64A
2015	921-64B
2015	921-64C
2015	921-84
2015	921-84A
2015	921-84B
2015	921-85
2015	921-85A
2015	921-85B
2015	921-86
2015	921-86A
2015	921-86B
2015	930DZ
2015	947-1
2015	974-1(SYLVANIA)
2015	991-01-1316
2015	1000-135
2015	1001-02(COURIER)
2015	1002A(JULIETTE)
2015	1003-01
2015	1004(G.E.)
2015	1006(G.E.)
2015	1009(G.E.)
2015	1039-0482
2015	1080-01
2015	1080-20
2015	1300RN
2015	1373-17A
2015	1420-1-1
2015	1420-2-2
2015	1455-7-4
2015	1585H
2015	1687-17
2015	1801
2015	1843-17
2015	1845-17
2015	1854-0231

RS 276-	Industry Standard No.	RS 276-	Industry Standard No.	RS 276-	Industry Standard No.	RS 276-	Industry Standard No.
2015	1854-0417	2015	4822-130-40318	2015	171338-3	2015	2006681-95
2015	1931-17	2015	4851	2015	171983	2015	2008292-56
2015	1952-17	2015	5001-543	2015	172761	2015	2008299-1
2015	1998-17	2015	5613-1342	2015	183015	2015	2092117-0018
2015	1999-17	2015	5613-1342C	2015	183016	2015	2092417-0017
2015	2004-01	2015	5613-460	2015	183018	2015	2092417-0018
2015	2020-02	2015	5613-460A	2015	183019	2015	2092417-0019
2015	2057A100-53	2015	5613-460B	2015	190714	2015	2092417-0711
2015	2057A2-109	2015	5613-460C	2015	212717	2015	2092417-0712
2015	2057A2-110	2015	5613-461	2015	231140-36	2015	2092417-0713
2015	2057A2-116	2015	5613-461C	2015	231140-37	2015	2092417-0714
2015	2057A2-117	2015	5613-535	2015	231140-43	2015	2092417-0715
2015	2057A2-120	2015	5613-535A	2015	236039	2015	2092417-0716
2015	2057A2-128	2015	5613-535B	2015	256617	2015	2092418-0022
2015	2057A2-157	2015	5613-535C	2015	489751-039	2015	2092418-0023
2015	2057A2-158	2015	5862	2015	489751-047	2015	2092418-0024
2015	2057A2-180	2015	72368	2015	489751-048	2015	2092418-0716
2015	2057A2-181	2015	8000-00006-005	2015	489751-049	2015	2092418-0717
2015	2057A2-185	2015	8000-00032-026	2015	489751-052	2015	2092418-0718
2015	2057A2-187	2015	8000-00041-040	2015	489751-058	2015	2092418-0719
2015	2057A2-192	2015	8000-00043-019	2015	489751-120	2015	2092418-0720
2015	2057A2-193	2015	8000-0005-002	2015	489751-121	2015	2092418-0721
2015	2057A2-195	2015	8000-0005-003	2015	489751-127	2015	2092693-0724
2015	2057A2-196	2015	8010-174	2015	489751-128	2015	2092693-0725
2015	2057A2-197	2015	8010-175	2015	489751-173	2015	2093308-0704
2015	2057A2-202	2015	8503	2015	489752-095	2015	2093308-0705
2015	2057A2-204	2015	8840-162	2015	514045	2015	2093308-0706
2015	2057A2-207	2015	9330-229-60112	2015	5150418	2015	2093308-0725
2015	2057A2-216	2015	9330-229-70112	2015	540204	2015	2316177
2015	2057A2-217	2015	18410-142	2015	0573468(HITACHI)	2015	2316183
2015	2057A2-218	2015	18600-151	2015	0573474	2015	2320031
2015	2057A2-219	2015	0023829	2015	0573474H	2015	2320041H
2015	2057A2-220	2015	24002	2015	0573475	2015	2320042
2015	2057A2-221	2015	25671-020	2015	0573485	2015	2320043
2015	2057A2-237	2015	25671-021	2015	0573487H	2015	2320141
2015	2057A2-251	2015	25671-023	2015	0573507	2015	2320141H
2015	2057A2-258	2015	28810-172	2015	0573508	2015	2320471-1
2015	2057A2-259	2015	29076-005	2015	0573509H	2015	2320471H
2015	2057A2-304	2015	29076-006	2015	0573510	2015	2320981
2015	2057A2-305	2015	33563	2015	0573510H	2015	2495520
2015	2057A2-322	2015	35259	2015	0573511H	2015	2495521
2015	2057A2-323	2015	37334	2015	0573607	2015	2495522-1
2015	2057A2-325	2015	37986-3563	2015	610041-2	2015	2495523-1
2015	2057A2-326	2015	37986-4040	2015	610042	2015	2497094-1
2015	2057A2-331	2015	37986-4046	2015	610073-1	2015	2497094-2
2015	2057A2-356	2015	38787	2015	610100-1	2015	2498456-2
2015	2057A2-395	2015	040001	2015	610100-2	2015	2498482-2
2015	2057A2-466	2015	40231	2015	610139-2	2015	2498508-2
2015	2057A2-501	2015	40233	2015	610150-1	2015	2498508-3
2015	2057A2-503	2015	40234	2015	610150-3	2015	2498902-1
2015	2057A2-539	2015	40235	2015	610181-2	2015	2498902-2
2015	2057A2-540	2015	40236	2015	610186-1	2015	2498903-1
2015	2057B2-101	2015	40237	2015	701678-00	2015	2498903-3
2015	2057B2-102	2015	40238	2015	740949	2015	3459332-1
2015	2057B2-114	2015	40239	2015	741862	2015	3463609-2
2015	2057B2-115	2015	40240	2015	742537	2015	3468068-1
2015	2057B2-116	2015	40242	2015	742547	2015	3468068-2
2015	2057B2-117	2015	40243	2015	749002	2015	3468068-3
2015	2057B2-125	2015	40244	2015	749014	2015	3468068-4
2015	2057B2-138	2015	40246	2015	760213-0002	2015	3559307-001
2015	2057B2-139	2015	40259	2015	760213-0005	2015	3559307-002
2015	2057B2-143	2015	40260	2015	760249	2015	3596440
2015	2057B2-192	2015	40296	2015	760253	2015	BFX59
2015	2110N-132	2015	40897	2015	772738	2016	C182
2015	2110N-133	2015	55810-161	2015	772739	2016	C182Q
2015	2127-17	2015	55810-163	2015	815206	2016	C183
2015	2197-17	2015	55810-164	2015	916049	2016	C183E
2015	2213-17	2015	58810-161	2015	916068	2016	C183J
2015	2284-17	2015	58840-192	2015	960201	2016	C183K
2015	2291-17	2015	61558	2015	960202	2016	C183L
2015	2427	2015	61755	2015	965632	2016	C183M
2015	2443	2015	75568-3	2015	970311	2016	C183P
2015	2448	2015	88510-172	2015	970939	2016	C183Q
2015	2475	2015	119412	2015	970962	2016	C183R
2015	2495	2015	119823	2015	971035	2016	C183W
2015	2546	2015	119824	2015	971459	2016	C184
2015	2634	2015	119825	2015	971526	2016	C184H
2015	3003	2015	122902	2015	971904	2016	C184J
2015	3003(SEARS)	2015	124024	2015	972305	2016	C184L
2015	3004(SEARS)	2015	124623	2015	972306	2016	C185
2015	3027	2015	124624	2015	972507	2016	C185(TRANSISTOR)
2015	3029	2015	125144	2015	972417	2016	C185A
2015	3107-204-90100	2015	125329	2015	972418	2016	C185J
2015	3301(SEARS)	2015	125390(RCA)	2015	972419	2016	C185M
2015	3476(RCA)	2015	125994(RCA)	2015	972420	2016	C185Q
2015	3507(SEARS)	2015	126025	2015	983233	2016	C185R
2015	3508(SEARS)	2015	129392	2015	983234	2016	C287(TRANSISTOR)
2015	3514(SEARS)	2015	129509	2015	983235	2016	C287A
2015	3516(RCA)	2015	129573	2015	983742	2016	C288
2015	3518	2015	129604	2015	984191	2016	C288A
2015	3518(RCA)	2015	130793	2015	984192	2016	C382
2015	3519-2	2015	131543	2015	984876	2016	C382-BK(1)
2015	3521-1	2015	131544	2015	984877	2016	C382-BK(2)
2015	3524-2	2015	131545	2015	984878	2016	C382-GR
2015	3527-1	2015	133171	2015	985442	2016	C382-GY
2015	3530	2015	134144-1	2015	985444	2016	C382BK
2015	3530-1	2015	137383	2015	985619	2016	C382BL
2015	3530-2	2015	137588	2015	986576	2016	C382BN
2015	3535(RCA)	2015	141402	2015	986693	2016	C382BR
2015	3536-2	2015	147115-9	2015	986694	2016	C382G
2015	3560-1(RCA)	2015	147353-0-1	2015	992108	2016	C382R
2015	3568	2015	165392	2015	1223781	2016	C382V
2015	3569	2015	165931	2015	1223919	2016	C383
2015	3571-1	2015	165932	2015	1471115-10	2016	C383G
2015	3572	2015	166906	2015	1472633	2016	C383P
2015	3579	2015	168567	2015	1472676-1	2016	C383W
2015	3579(RCA)	2015	169194	2015	1473521-1	2016	C383Y
2015	3586-2	2015	169196	2015	1473524-1	2016	C463
2015	3610	2015	169505	2015	1473524-3	2016	C463R
2015	3676-1	2015	170755-1	2015	1473529-1	2016	C469
2015	3680	2015	170756-1	2015	1473535-1	2016	C469A
2015	3680-1	2015	170906	2015	1473535-1(RCA)	2016	C469F
2015	3881	2015	170906-1	2015	1473536-2	2016	C469K
2015	4041-000-40270	2015	171009	2015	1473543-1	2016	C469Q
2015	4041-000-40300	2015	171029	2015	1473544-1	2016	C605
2015	4041-000-60170	2015	171029(TOSHIBA)	2015	1473558-1	2016	C605(NEC)
2015	4041-000-60200	2015	171030	2015	1473571-1	2016	C605(Q)
2015	4167(AIRLINE)	2015	171031	2015	1473576	2016	C605K
2015	4167(PENNCREST)	2015	171031(TOSHIBA)	2015	1473577-1	2016	C605L
2015	4168(PENNCREST)	2015	171044	2015	1473579-1	2016	C605M
2015	4168(WARDS)	2015	171144-1	2015	1473586	2016	C605Q
2015	4169(PENNCREST)	2015	171162-118	2015	1473610-1	2016	C605TW
2015	4169(WARDS)	2015	171206-10	2015	1473617-2	2016	C606
2015	4756	2015	171206-11	2015	1473676-1	2016	C606(NEC)
2015	4799	2015	171206-13	2015	1473680-1	2016	C927
2015	4792	2015	171206-7	2015	2003342-244	2016	C927A
2015	4822-130-40214	2015	171206-9	2015	2003779-22	2016	C927B
2015	4822-130-40215	2015	171207-3	2015	2003779-23	2016	C927C
2015	4822-130-40216	2015	171318	2015	2003779-24	2016	C927C1
2015	4822-130-40304	2015	171319	2015	2003779-25	2016	C927CW
2015	4822-130-40311	2015	171320	2015	2006513-39	2016	C927D
2015	4822-130-40312	2015	171338-1	2015	2006582-101	2016	C927E
2015	4822-130-40313	2015	171338-2	2015	2006681-93	2016	CF11
2015	4822-130-40317			2015	2006681-94		

RS 276-	Industry Standard No.	RS 276-	Industry Standard No.	RS 276-	Industry Standard No.	RS 276-	Industry Standard No.
2016	CGE-60	2016	2SC185M	2016	2SC392J	2016	2SC928Y
2016	DDBT209003	2016	2SC185Q	2016	2SC392K	2016	2SCF11
2016	EA15X336	2016	2SC185R	2016	2SC392L	2016	13-21860-3
2016	EA15X365	2016	2SC185V	2016	2SC392M	2016	022-2876-007
2016	EA15X367	2016	2SC2057	2016	2SC392OR	2016	34-6015-15
2016	EA15X7173	2016	2SC2057-C	2016	2SC392R	2016	34-6015-16
2016	EA15X7174	2016	2SC2057C	2016	2SC392X	2016	34-6015-17
2016	EA15X7178	2016	2SC2057D	2016	2SC463	2016	34-6015-29
2016	EA15X7179	2016	2SC2057E	2016	2SC463H	2016	46-86112-3
2016	ECG313	2016	2SC2057F	2016	2SC469	2016	50-40102-05
2016	ECG319	2016	2SC278-0R	2016	2SC469-0R	2016	57A139-4
2016	EQ3-0196	2016	2SC287	2016	2SC469A	2016	57B139-4
2016	EQ8-19	2016	2SC287-0R	2016	2SC469B	2016	86-185-2
2016	EQ8-20	2016	2SC287A	2016	2SC469C	2016	86-607-2
2016	FCS-9016F	2016	2SC287B	2016	2SC469D	2016	90-602
2016	FCS-9016G	2016	2SC287C	2016	2SC469E	2016	100N3
2016	FC890016F	2016	2SC287D	2016	2SC469F	2016	121-503
2016	FC890016G	2016	2SC287E	2016	2SC469G	2016	121-504
2016	GE-211	2016	2SC287F	2016	2SC469GN	2016	121-505
2016	GE-278	2016	2SC287G	2016	2SC469H	2016	121-506
2016	GE-283	2016	2SC287GN	2016	2SC469J	2016	121-507
2016	GI-2926	2016	2SC287H	2016	2SC469K	2016	121-508
2016	H9-40054	2016	2SC287J	2016	2SC469L	2016	184-999
2016	I20-0029	2016	2SC287K	2016	2SC469M	2016	184R
2016	NS45006	2016	2SC287L	2016	2SC469OR	2016	0516-3101-250
2016	PL-176-047-9-002	2016	2SC287M	2016	2SC469Q	2016	0516-3101-406
2016	QT-C0710XAE	2016	2SC287OR	2016	2SC469R	2016	1016-85
2016	SE5055	2016	2SC287R	2016	2SC469X	2016	1210-17
2016	SPS-952-2	2016	2SC287X	2016	2SC469Y	2016	1634-17
2016	TG2SC2057-C	2016	2SC287Y	2016	2SC605	2016	2134-17
2016	TG2SC2057-D	2016	2SC288	2016	2SC605(L)	2016	2448-17
2016	TG2SC2057C	2016	2SC288A	2016	2SC605-Q	2016	4801-0000-015
2016	TG2SC927-C-A	2016	2SC348A	2016	2SC605-QR	2016	8000-00004-003
2016	TG2SC927-D-A	2016	2SC348B	2016	2SC605A	2016	124757
2016	TG2SC927-E-A	2016	2SC348C	2016	2SC605B	2016	227517
2016	TM313	2016	2SC348D	2016	2SC605C	2016	244817
2016	TM319	2016	2SC348E	2016	2SC605D	2016	610181-1
2016	TV-33	2016	2SC348F	2016	2SC605E	2016	1223911
2016	TV-34	2016	2SC348G	2016	2SC605F	2016	23204T1
2016	TV-35	2016	2SC348GN	2016	2SC605G	2017	A201
2016	TV-50	2016	2SC348H	2016	2SC605GN	2017	A203(NPN)
2016	WEP710	2016	2SC348J	2016	2SC605H	2017	A208
2016	WEP921	2016	2SC348K	2016	2SC605J	2017	A208(NPN)
2016	WEP924	2016	2SC348L	2016	2SC605K	2017	A210
2016	01-032057	2016	2SC348M	2016	2SC605L	2017	A211
2016	2SC1035A	2016	2SC348OR	2016	2SC605M	2017	A253
2016	2SC1035B	2016	2SC348X	2016	2SC605OR	2017	A271
2016	2SC1035F	2016	2SC348Y	2016	2SC605Q	2017	A275
2016	2SC1035GN	2016	2SC382	2016	2SC605R	2017	A4K
2016	2SC1035H	2016	2SC382(BL)	2016	2SC605TW	2017	A514-047830
2016	2SC1035J	2016	2SC382(BN)	2016	2SC605X	2017	A5A-1
2016	2SC1035L	2016	2SC382(R)	2016	2SC605Y	2017	A5A-2
2016	2SC1035M	2016	2SC382-BK	2016	2SC606(VHF)	2017	A5A-3
2016	2SC1035OR	2016	2SC382-BK(1)	2016	2SC606A	2017	A5A-4
2016	2SC1035R	2016	2SC382-BK(2)	2016	2SC606B	2017	A5A-5
2016	2SC1035X	2016	2SC382-G	2016	2SC606C	2017	A5G
2016	2SC1035Y	2016	2SC382-GR	2016	2SC606D	2017	A5T
2016	2SC1036A	2016	2SC382-GY	2016	2SC606E	2017	A6C
2016	2SC1036B	2016	2SC382-0R	2016	2SC606F	2017	A6C(GRN)
2016	2SC1036C	2016	2SC382-R	2016	2SC606G	2017	A6C-1
2016	2SC1036D	2016	2SC382-V	2016	2SC606GN	2017	A6C-1(RED)
2016	2SC1036E	2016	2SC382A	2016	2SC606H	2017	A6C-1-RED
2016	2SC1036F	2016	2SC382B	2016	2SC606J	2017	A6C-2
2016	2SC1036G	2016	2SC382BK	2016	2SC606K	2017	A6C-2(BLK)
2016	2SC1036GN	2016	2SC382BK1	2016	2SC606L	2017	A6C-2-BLACK
2016	2SC1036H	2016	2SC382BK2	2016	2SC606M	2017	A6C-3
2016	2SC1036J	2016	2SC382BL	2016	2SC606N	2017	A6C-3(WHT)
2016	2SC1036K	2016	2SC382BN	2016	2SC606OR	2017	A6C-3-WHITE
2016	2SC1036L	2016	2SC382BR	2016	2SC606R	2017	A6C-4
2016	2SC1036M	2016	2SC382C	2016	2SC606X	2017	A6C-GREEN
2016	2SC1036OR	2016	2SC382D	2016	2SC606Y	2017	A6W
2016	2SC1036R	2016	2SC382E	2016	2SC663A	2017	A7C(MOTOROLA)
2016	2SC1036X	2016	2SC382F	2016	2SC663B	2017	A7Z
2016	2SC1036Y	2016	2SC382G	2016	2SC663C	2017	A3F
2016	2SC1779	2016	2SC382GN	2016	2SC663D	2017	AR22(TRANSISTOR)
2016	2SC1790	2016	2SC382GR	2016	2SC663E	2017	AR24(PHILCO)
2016	2SC182	2016	2SC382GY	2016	2SC663F	2017	AR24(TRANSISTOR)
2016	2SC182Q	2016	2SC382H	2016	2SC663G	2017	B-1790
2016	2SC183	2016	2SC382J	2016	2SC663GN	2017	B143000
2016	2SC183(P)	2016	2SC382K	2016	2SC663H	2017	B143001
2016	2SC183(Q)(R)	2016	2SC382L	2016	2SC663J	2017	B143003
2016	2SC183(R)	2016	2SC382M	2016	2SC663K	2017	B143009
2016	2SC183-1	2016	2SC382OR	2016	2SC663L	2017	B143010
2016	2SC183-0R	2016	2SC382R	2016	2SC663M	2017	B143015
2016	2SC183A	2016	2SC382V	2016	2SC663OR	2017	B143016
2016	2SC183AP	2016	2SC382W	2016	2SC663R	2017	B143024
2016	2SC183B	2016	2SC382W,R	2016	2SC663X	2017	B143025
2016	2SC183BK	2016	2SC382X	2016	2SC663Y	2017	B1C
2016	2SC183C	2016	2SC382Y	2016	2SC694A	2017	B1C-1
2016	2SC183D	2016	2SC383G	2016	2SC694B	2017	B1C-2
2016	2SC183E	2016	2SC390A	2016	2SC694C	2017	B1D-1
2016	2SC183F	2016	2SC390B	2016	2SC694D	2017	B1F
2016	2SC183G	2016	2SC390C	2016	2SC694GN	2017	B1U148
2016	2SC183GN	2016	2SC390D	2016	2SC694H	2017	B2J
2016	2SC183H	2016	2SC390E	2016	2SC694J	2017	B2V
2016	2SC183J	2016	2SC390F	2016	2SC694K	2017	B3531
2016	2SC183K	2016	2SC390G	2016	2SC694L	2017	B3533
2016	2SC183L	2016	2SC390GN	2016	2SC694M	2017	B3537
2016	2SC183M	2016	2SC694M	2016	2SC694OR	2017	B3538
2016	2SC183OR	2016	2SC390H	2016	2SC694R	2017	B3540
2016	2SC183P	2016	2SC390J	2016	2SC694X	2017	B3541
2016	2SC183Q	2016	2SC390K	2016	2SC694Y	2017	B3542
2016	2SC183R	2016	2SC390L	2016	2SC863A	2017	B3570
2016	2SC183S	2016	2SC390M	2016	2SC863B	2017	B3576
2016	2SC183W	2016	2SC390OR	2016	2SC863C	2017	B3606
2016	2SC183X	2016	2SC390R	2016	2SC863D	2017	B3607
2016	2SC183Y	2016	2SC390X	2016	2SC863E	2017	B3608
2016	2SC184	2016	2SC390Y	2016	2SC863F	2017	B3609
2016	2SC184(R)	2016	2SC391	2016	2SC863G	2017	B3610
2016	2SC184-0R	2016	2SC391A	2016	2SC863GN	2017	B3611
2016	2SC184A	2016	2SC391B	2016	2SC863H	2017	B3612
2016	2SC184AP	2016	2SC391C	2016	2SC863J	2017	B3613
2016	2SC184B	2016	2SC391D	2016	2SC863K	2017	B3614
2016	2SC184BK	2016	2SC391E	2016	2SC863L	2017	B3747
2016	2SC184C	2016	2SC391F	2016	2SC863M	2017	B3748
2016	2SC184D	2016	2SC391G	2016	2SC863OR	2017	B3750
2016	2SC184E	2016	2SC391GN	2016	2SC863R	2017	B5001
2016	2SC184F	2016	2SC391H	2016	2SC863X	2017	BD107
2016	2SC184G	2016	2SC391J	2016	2SC863Y	2017	BD131
2016	2SC184GN	2016	2SC391K	2016	2SC927A	2017	BD135
2016	2SC184H	2016	2SC391L	2016	2SC927C	2017	BD137
2016	2SC184J	2016	2SC391M	2016	2SC928	2017	BD139
2016	2SC184K	2016	2SC391OR	2016	2SC928F	2017	BD149
2016	2SC184L	2016	2SC391R	2016	2SC928G	2017	BD153
2016	2SC184M	2016	2SC391X	2016	2SC928H	2017	BD154
2016	2SC184OR	2016	2SC391Y	2016	2SC928I	2017	BD155
2016	2SC184P	2016	2SC392A	2016	2SC928K	2017	BD165
2016	2SC184Q	2016	2SC392B	2016	2SC928L	2017	BD167
2016	2SC184R	2016	2SC392C	2016	2SC928M	2017	BD175
2016	2SC184X	2016	2SC392D	2016	2SC928OR	2017	BD177
2016	2SC184Y	2016	2SC392E	2016	2SC928R	2017	BD185
2016	2SC185	2016	2SC392G	2016	2SC928R	2017	BD187
2016	2SC185A	2016	2SC392GN	2016	2SC928R	2017	BD189
2016	2SC185J	2016	2SC392H	2016	2SC928X	2017	BD195

RS 276-	Industry Standard No.
2017	BD197
2017	BD205
2017	BD207
2017	BD561
2017	BF850
2017	BF851
2017	BFX55
2017	BFX96A
2017	BLY20
2017	BLY3
2017	BLY37
2017	BLY38
2017	BLY53
2017	BLY62
2017	BLY78
2017	BLY91
2017	BR100B
2017	B8V60
2017	C1162
2017	C1162A
2017	C1162B
2017	C1162C
2017	C1162CP
2017	C1162D
2017	C1162MP
2017	C1162WB
2017	C1162WTB
2017	C1162WTC
2017	C1162WTD
2017	C1212
2017	C1212A
2017	C1212AA
2017	C1212AB
2017	C1212AC
2017	C1212AD
2017	C1212AWT
2017	C1212AWTA
2017	C1212AWTB
2017	C1212AWTC
2017	C1212AWTD
2017	C1212B
2017	C1212C
2017	C1212D
2017	C1212WT
2017	C1212WTA
2017	C1212WTB
2017	C1212WTC
2017	C1212WTD
2017	C1368
2017	C1368C
2017	C1386C
2017	C495
2017	C495-0
2017	C495-R
2017	C496T
2017	CS9013(HG)
2017	CXL184
2017	CXL186
2017	D18A12
2017	D28D8
2017	D4201
2017	D4202
2017	D4203
2017	D4204
2017	D4205
2017	D4206
2017	D4207
2017	D4208
2017	D4408B
2017	EA1-380
2017	EA15X160
2017	EA15X244
2017	EA15X248
2017	EA15X269
2017	EA2488
2017	EA3674
2017	EA4055
2017	EA4085
2017	EAI-380
2017	EC0184
2017	EC0186
2017	EP100
2017	EP15X25
2017	EP15X32
2017	EP15X34
2017	EQ8-66
2017	EQ8-67
2017	EQ8-89
2017	E815227
2017	E858
2017	E858(ELCOM)
2017	ET495371
2017	G03-007C
2017	G03007C
2017	G06-711-B
2017	G06-714C
2017	G06-717-B
2017	G06-717-C
2017	G06-717-D
2017	G06-717B
2017	G06714C
2017	GE-28
2017	GE-57
2017	GE-83
2017	GE-X18
2017	GEMR-6
2017	G03-007C
2017	G06-714C
2017	G06-717-B
2017	HC-00496
2017	HC-01060
2017	HC-01096
2017	HC-01098
2017	HC-01226
2017	HC-495
2017	HD-00471
2017	HEP245
2017	HEP701
2017	HEP85000
2017	HEP85003
2017	HP11
2017	HP15
2017	HR-69
2017	HT104961C
2017	HT304961B
2017	HT304961C
2017	HT304961C-0
2017	HT304961C-0
2017	HT307902B
2017	HT313681BO
2017	IRTR-55
2017	IRTR55
2017	J241250
2017	KGB41061
2017	M0TMJE521
2017	M3567-2
2017	M844
2017	M852
2017	M9134
2017	M9556
2017	M9582
2017	M9618
2017	MJE-220
2017	MJE180
2017	MJE200
2017	MJE200E
2017	MJE220
2017	MJE221
2017	MJE222
2017	MJE223
2017	MJE224
2017	MJE225
2017	MJE3520
2017	MJE3521
2017	MJE482
2017	MJE483
2017	MJE488
2017	MJE520
2017	MJE521
2017	MJE720
2017	MJE721
2017	MJE9400
2017	MM3004
2017	MM4429
2017	MM4430
2017	M0TMJE521
2017	MSA8508
2017	MT1070
2017	PA890Q
2017	PL-172-024-9-001
2017	PL-176-042-9-005
2017	PLE52
2017	PLE552
2017	PPR1006
2017	PPR1008
2017	PT2620
2017	PT2640
2017	PT2660
2017	PT2690
2017	PT4690
2017	PT600
2017	PT601
2017	PT6618
2017	PT6669
2017	PT6696
2017	PTC110
2017	PTC143
2017	PTC193
2017	Q-14
2017	QT-D0325XAC
2017	RE40
2017	RE42
2017	REN184
2017	REN186
2017	RS-2017
2017	RS-2020
2017	RT-152
2017	RT152
2017	S-85
2017	S-86
2017	S10153
2017	S32903
2017	S715
2017	S801
2017	S8660
2017	S8660121-808
2017	SC4133
2017	SD1023
2017	SDT3326
2017	SDT3421
2017	SDT3425
2017	SDT3426
2017	SDT4301
2017	SDT4302
2017	SDT4304
2017	SDT4305
2017	SDT4307
2017	SDT4308
2017	SDT4310
2017	SDT4311
2017	SDT4455
2017	SDT4483
2017	SDT4551
2017	SDT4553
2017	SDT4583
2017	SDT4611
2017	SDT4612
2017	SDT4614
2017	SDT4615
2017	SDT5001
2017	SDT5006
2017	SDT5501
2017	SDT5506
2017	SDT5511
2017	SDT5901
2017	SDT5902
2017	SDT5906
2017	SDT6101
2017	SDT6102
2017	SDT6104
2017	SDT6105
2017	SDT6106
2017	SDT7401
2017	SDT7411
2017	SDT7414
2017	SDT9001
2017	SDT9002
2017	SDT9003
2017	SDT9005
2017	SDT9006
2017	SDT9007
2017	SDT9008
2017	SJE-5038
2017	SJE-5402
2017	SJE-649
2017	SJE100
2017	SJE106
2017	SJE113
2017	SJE133
2017	SJE1519
2017	SJE1520
2017	SJE203
2017	SJE211
2017	SJE222
2017	SJE228
2017	SJE229
2017	SJE237
2017	SJE242
2017	SJE244
2017	SJE246
2017	SJE248
2017	SJE253
2017	SJE254
2017	SJE255
2017	SJE261
2017	SJE262
2017	SJE271
2017	SJE272
2017	SJE274
2017	SJE278
2017	SJE280
2017	SJE284
2017	SJE289
2017	SJE305
2017	SJE320
2017	SJE401
2017	SJE402
2017	SJE404
2017	SJE405
2017	SJE407
2017	SJE527
2017	SJE5402
2017	SJE583
2017	SJE634
2017	SJE649
2017	SJE669
2017	SJE721
2017	SJE724
2017	SJE737
2017	SJE769
2017	SJE781
2017	SJE784
2017	SJE785
2017	SK3190
2017	SP-8660
2017	SP8918
2017	STC1800
2017	STC1862
2017	STD9007
2017	STT4451
2017	STX0013
2017	STX0026
2017	T-342
2017	T-344
2017	T1P33
2017	T1P31A
2017	T344
2017	T611-1(RCA)
2017	T612-1(RCA)
2017	TIP14
2017	TIP31B
2017	TIS82
2017	TM184
2017	TM186
2017	TNJ70450
2017	TNJ70540
2017	TNJ72775
2017	TR-1037-3
2017	TR-55
2017	TR-76
2017	TR01045
2017	TR01056-5
2017	TR01057-3
2017	TR2327203
2017	TR76
2017	TRAPLC1013
2017	TV-73
2017	TV-75
2017	TV-82
2017	V176
2017	VX3375
2017	WEP1096
2017	WEP245
2017	WEP701
2017	WEP908
2017	WEP85003
2017	X713
2017	XB401
2017	YAAN28C1096
2017	YAAN28C1096K
2017	YAAN28C1096L
2017	YAAN28C1096M
2017	YAAN28C1096N
2017	YAAN128C1096
2017	YAAN128C1096Q
2017	YAAN128C1096R
2017	YAANZ
2017	YAAN28I096
2017	YAAN228C1096
2017	YAAN228C1096-BLUE
2017	YAAN228C1096-RED
2017	YAAN228C1096L
2017	YAAN228C1096M
2017	YANZ28C1096
2017	YANZ28C1096
2017	YANZ28C1096K
2017	YANZ28C1096L
2017	YANZ28C1096M
2017	ZEN202
2017	ZEN210
2017	ZT3375
2017	ZT600
2017	002-012200
2017	2N1067
2017	2N1068
2017	2N1710
2017	2N3296
2017	2N3375
2017	2N3418
2017	2N3420
2017	2N3525A
2017	2N3619
2017	2N3620
2017	2N3623
2017	2N3624
2017	2N3628
2017	2N3852
2017	2N3853
2017	2N3919
2017	2N3925
2017	2N3926
2017	2N3961
2017	2N4012
2017	2N4040
2017	2N4041
2017	2N4225
2017	2N4226
2017	2N4350
2017	2N4428
2017	2N4429
2017	2N4430
2017	2N4440
2017	2N4921
2017	2N4922
2017	2N4923
2017	2N4976
2017	2N5079
2017	2N5080
2017	2N5190
2017	2N5191
2017	2N5192
2017	2N5422
2017	2N5423
2017	2N5481
2017	2N5482
2017	2N5489
2017	2N5608
2017	2N5610
2017	2N5612
2017	2N5644
2017	2N5645
2017	2N5688
2017	2N5697
2017	2N5698
2017	2N5699
2017	2N5703
2017	2N5764
2017	2N5766
2017	2N5767
2017	2N6037
2017	2N6038
2017	2N6288
2017	2N6289
2017	2N6290
2017	2N6291
2017	2N6293
2017	2SC0148
2017	2SC1061
2017	2SC11070
2017	2SC1107YG
2017	2SC1162
2017	2SC1212
2017	2SC1212A
2017	2SC1212AA
2017	2SC1212AB
2017	2SC1212ABWT
2017	2SC1212AC
2017	2SC1212ACWT
2017	2SC1212AD
2017	2SC1212AWT
2017	2SC1212AWTA
2017	2SC1212AWTB
2017	2SC1212AWTC
2017	2SC1212AWTD
2017	2SC1212B
2017	2SC1212C
2017	2SC1212D
2017	2SC1212WT
2017	2SC1212WTA
2017	2SC1212WTB
2017	2SC1212WTC
2017	2SC1212WTD
2017	2SC1368
2017	2SC1368(B)
2017	2SC1368B
2017	2SC1368C
2017	2SC1449
2017	2SC1449(CB)
2017	2SC1449CB
2017	2SC1449M
2017	2SC1568
2017	2SC1568(R)
2017	2SC1568R
2017	2SC2222
2017	2SC2223A
2017	2SC2223B
2017	2SC2223C
2017	2SC2223D
2017	2SC2223E
2017	2SC2223F
2017	2SC2223G
2017	2SC2223GN
2017	2SC2223H
2017	2SC2223J
2017	2SC2223L
2017	2SC2223M
2017	2SC2230R
2017	2SC2223K
2017	2SC2223X
2017	2SC2223Y
2017	2SC224A
2017	2SC224B
2017	2SC224C
2017	2SC224D
2017	2SC224E
2017	2SC224F
2017	2SC224G
2017	2SC224H
2017	2SC224J
2017	2SC224K
2017	2SC224L
2017	2SC224M
2017	2SC2240R
2017	2SC224R
2017	2SC224X
2017	2SC224Y
2017	2SC2225A
2017	2SC2225B
2017	2SC2225C
2017	2SC2225D
2017	2SC2225E
2017	2SC2225G
2017	2SC2225GN
2017	2SC2225H
2017	2SC2225J
2017	2SC2225L
2017	2SC2225M
2017	2SC2250R
2017	2SC2225R
2017	2SC2225X
2017	2SC2225Y
2017	2SC2229A
2017	2SC2229B
2017	2SC2229C
2017	2SC2229D
2017	2SC2229F
2017	2SC2229P
2017	2SC2229G
2017	2SC2229GN
2017	2SC2229H
2017	2SC2229J

RS 276-	Industry Standard No.	RS 276-	Industry Standard No.	RS 276-	Industry Standard No.	RS 276-	Industry Standard No.
2017	28C229K	2017	13-34046-1	2017	260P21102	2017	610162-2
2017	28C229L	2017	13-34046-2	2017	260P21308	2017	610162B2
2017	28C229M	2017	13-34046-3	2017	260P28701	2017	610190-3
2017	28C229OR	2017	13-34046-4	2017	314-6005-1	2017	651956
2017	28C229R	2017	13-34046-5	2017	361-1	2017	657180
2017	28C229X	2017	13-34372-2	2017	364-6004	2017	741115
2017	28C229Y	2017	13-36444-3	2017	417-144	2017	760275
2017	28C2456	2017	13-39046-3	2017	449	2017	916046
2017	28C2497Q	2017	13-39074-1	2017	576-0002-011	2017	916114
2017	28C2497R	2017	13-39098-1	2017	576-002-001	2017	986932
2017	28C2568K	2017	13-40344-1	2017	612-1	2017	1471132-2
2017	28C2568L	2017	13-43790-1	2017	625-1	2017	1471140-1
2017	28C2568M	2017	14-601-17	2017	642-261	2017	1473567-2
2017	28C291A	2017	14-608-01	2017	690V110HB9	2017	1473632-2
2017	28C291B	2017	14-608-02	2017	690V116H26	2017	2006436-35
2017	28C291C	2017	14-608-03	2017	700-110	2017	2006436-3C
2017	28C291D	2017	14-608-04	2017	753-0101-226	2017	2006623-88
2017	28C291E	2017	14-906-13	2017	753-4001-931	2017	2320843
2017	28C291F	2017	14-907-13	2017	753-9000-096	2017	2320845
2017	28C291G	2017	14-910-13	2017	0772	2017	2327152
2017	28C291GN	2017	14-911-13	2017	800-533-01	2017	2327153
2017	28C291H	2017	15-39098-1	2017	853-0301-096	2017	2327203
2017	28C291J	2017	018-00005	2017	880-250-011	2018	A5E
2017	28C291K	2017	19-020-101	2017	884-250-011	2018	A6D
2017	28C291L	2017	19A115300-P1	2017	903Y002152	2018	A6D-1
2017	28C291M	2017	19A115300-P2	2017	921-01131	2018	A6D-2
2017	28C291OR	2017	19A115300-P3	2017	921-163B	2018	A6D-3
2017	28C291R	2017	21A112-049	2017	936 NPN	2018	A6Q
2017	28C291X	2017	21M180	2017	936(NPN)	2018	A8J
2017	28C291Y	2017	21M181	2017	01057-1	2018	AG6
2017	28C292A	2017	21M183	2017	1062-0615	2018	B1M
2017	28C292B	2017	21M184	2017	1074-116	2018	B2M
2017	28C292C	2017	21M286	2017	1080-06	2018	BC337
2017	28C292D	2017	21M367	2017	1097(GE)	2018	BC338
2017	28C292E	2017	21M369	2017	1116(GE)	2018	BC430
2017	28C292F	2017	21M466	2017	1125	2018	C1018B
2017	28C292G	2017	21M556	2017	1125(GE)	2018	C1243-24
2017	28C292GN	2017	24MW1143	2017	1132-2(RCA)	2018	C1429
2017	28C292H	2017	24MW978	2017	1269	2018	D3288
2017	28C292J	2017	30-090	2017	1282	2018	D40D1
2017	28C292K	2017	31-0066	2017	1345(GE)	2018	D40D10
2017	28C292L	2017	33-090	2017	1358GE	2018	D40D11
2017	28C292M	2017	34-6002-41	2017	2011	2018	D40D14
2017	28C292OR	2017	42-28212	2017	2017-108	2018	D40D2
2017	28C292X	2017	046-1	2017	2057A100-45(NPN)	2018	D40D3
2017	28C292Y	2017	046-1(SYLVANIA)	2017	2057A100-47(NPN)	2018	D40D4
2017	28C297	2017	46-86234-3	2017	2057A100-49	2018	D40D5
2017	28C541	2017	46-86360-3	2017	2057A100-58(NPN)	2018	D40D7
2017	28C542	2017	48-134145	2017	2057A100-66(NPN)	2018	D40D8
2017	28C547	2017	48-137037	2017	2081-17	2018	EC0188
2017	28C548	2017	48-137095	2017	2085-17	2018	EC0210
2017	28C550	2017	48-137128	2017	2132	2018	ES82(ELCOM)
2017	28C572	2017	48-137145	2017	2160-17	2018	ET392927
2017	28C592	2017	48-137146	2017	2168-17	2018	GE-217
2017	28C598	2017	48-137147	2017	2202-17	2018	GE-226
2017	28C637	2017	48-137148	2017	2245-17	2018	GE-252
2017	28C691	2017	48-137200	2017	2312-17	2018	H7618
2017	28C692	2017	48-137211	2017	2334-17	2018	HEP83023
2017	28C893	2017	48-137277	2017	2510-105	2018	HEP83024
2017	28C909	2017	48-137323	2017	2853-1	2018	HEP83025
2017	28C911	2017	48-137473	2017	2853-3	2018	HEP83026
2017	28C932(E)	2017	48-137549	2017	2854-1	2018	HR-73
2017	28C932A	2017	48-40382J01	2017	2854-3	2018	HR85
2017	28C932B	2017	48-41784J03	2017	2855-1	2018	IRTR72
2017	28C932BK	2017	48-41884J03	2017	2855-3	2018	M9640
2017	28C932C	2017	48-86961B	2017	2856-1	2018	M3-5
2017	28C932D	2017	48-90254A36	2017	2856-3	2018	MPS-U01
2017	28C932F	2017	48-97162A35	2017	2904-057	2018	MPS-U07
2017	28C932G	2017	48-97162A69	2017	003526	2018	MPSU01
2017	28C932GN	2017	48M359039	2017	3526(SEARS)	2018	MPSU01A
2017	28C932H	2017	488686961B	2017	3611	2018	MPSU02
2017	28C932J	2017	488137093	2017	3612	2018	MPSU05
2017	28C932K	2017	488137145	2017	3632-2(RCA)	2018	MPSU06
2017	28C932L	2017	488137473	2017	4491-6	2018	MU9610
2017	28C932M	2017	48840382J01	2017	4491-9	2018	MU9610F
2017	28C932OR	2017	53-1362	2017	4802-0000-002	2018	MU9610T
2017	28C932R	2017	53-1967	2017	4822-130-40537	2018	MU9611
2017	28C932X	2017	56-86412-3	2017	5001-064	2018	MU9611G
2017	28C932Y	2017	57A214-2	2017	5001-075	2018	MU9611T
2017	28D317A(F)	2017	69-1818	2017	5847(RCA)	2018	P218-1
2017	28D317A(F)(P)	2017	072	2017	7316-1	2018	REN188
2017	28D317AF	2017	074-1(PHILCO)	2017	7413	2018	REN210
2017	28D317AP	2017	77-271798-1	2017	7513	2018	RS-2018
2017	28D318-O	2017	77-271798-3	2017	8000-00004-185	2018	SPS1107
2017	28D318Q	2017	080	2017	8000-0004-P185	2018	T422
2017	28D325	2017	081	2017	8000-00041-043	2018	TM188
2017	28D325-OR	2017	86-344-2	2017	8471	2018	TM210
2017	28D325A	2017	86-506-2	2017	8710-170	2018	TV73
2017	28D325B	2017	86-51100-7	2017	8840-169	2018	TV75
2017	28D325C	2017	86X00059-001	2017	12539	2018	WEP83023
2017	28D325D	2017	86X0074-001	2017	20810-94	2018	ZZX360
2017	28D325E	2017	90-38	2017	30272	2018	2N2853-1
2017	28D325F	2017	90-75	2017	38448	2018	2N2854-1
2017	28D325G	2017	998105-1	2017	39458	2018	2N2855-1
2017	28D325GN	2017	102-1061-01	2017	39750(ORRTRONICS)	2018	2SC1429
2017	28D325H	2017	0103-0419	2017	40279	2018	2SC1429-1
2017	28D325J	2017	0103-0419A	2017	40281	2018	2SC1429-2
2017	28D325K	2017	0103-512M	2017	40305	2018	2SC1663
2017	28D325L	2017	114N2P	2017	40306	2018	2SC1663H
2017	28D325M	2017	121-708	2017	40347V1	2018	2SC2194
2017	28D325OR	2017	121-710	2017	40389	2018	2SD3288
2017	28D325R	2017	121-805	2017	40605	2018	13-33180-1
2017	28D325X	2017	121-808	2017	40666	2018	13-33185-1
2017	28D325Y	2017	121-853X	2017	41178	2018	13-33742-1PLASTIC
2017	28D343A	2017	121-29002	2017	41344	2018	13-34003-1
2017	28D343B	2017	0122(AIRLINE)	2017	58810-169	2018	13-34372-1
2017	28D343C	2017	123-011	2017	58840-200	2018	13-36508-1
2017	28D343D	2017	123-011A	2017	60132	2018	13-37526-1
2017	28D343H	2017	123B-002	2017	60407	2018	34-6016-31
2017	28D343J	2017	0124	2017	60413	2018	34-6016-53
2017	28D343K	2017	0124(HOFFMAN)	2017	61242	2018	46-86585-3
2017	28D390Q	2017	0124(WARDS)	2017	62221	2018	48-137091
2017	28D612	2017	128-853	2017	62950	2018	48-137092
2017	28D612E	2017	131-2	2017	000073280	2018	48-137149
2017	28D612K	2017	132-046	2017	000073320	2018	48-137169
2017	314-6005-2	2017	132-072	2017	000073380	2018	48-137202
2017	314-6005-3	2017	132-080	2017	88060-146	2018	48-137319
2017	716-0531-19(NPN)	2017	132-081	2017	88510-176	2018	48-137505
2017	8-905-706-801	2017	0140-7	2017	95262-1	2018	48-869640
2017	8-905-713-110	2017	149N1	2017	95263-1	2018	48-97177A10
2017	09-302113	2017	149N2	2017	127978	2018	488869640
2017	09-309071	2017	149N4	2017	132495	2018	488137169
2017	11-0772	2017	162N2	2017	132499	2018	488137572
2017	13-0049	2017	172-006-9-001	2017	135573	2018	48X9717A10
2017	13-14085-88	2017	173A4490-2	2017	134280	2018	57A214-12
2017	13-23507-0	2017	176-024-9-004	2017	135739	2018	66R058-2
2017	13-28336-2	2017	176-042-9-005	2017	137369	2018	66P069-1
2017	13-32634-1	2017	185-001	2017	144856	2018	81-23860400-3
2017	13-32636-1	2017	186-008	2017	166918	2018	81-23860400A
2017	13-32642-1	2017	186-009	2017	167542	2018	81-23860400B
2017	13-33742-1	2017	190N1C	2017	171053-2(NPN)	2018	81-27125530-9
2017	13-33925-1	2017	190N3	2017	171175-1(NPN)	2018	81-27125530-9A
		2017	190N3C	2017	570031	2018	81-27125530-9B
		2017	229-1301-34				

RS 276-	Industry Standard No.	RS 276-	Industry Standard No.	RS 276-	Industry Standard No.	RS 276-	Industry Standard No.	RS 276-	Industry Standard No.
2018	81-27126100-0	2020	BD587	2020	EP15X43	2020	SDT6905		
2018	86-284-2	2020	BD589	2020	ES103	2020	SE3040		
2018	86-422-3	2020	BD70	2020	ES113	2020	SE3041		
2018	86-660-2	2020	BDX71	2020	ES16X98	2020	SE9060		
2018	86-673-2	2020	BDX72	2020	ES36(ELCOM)	2020	SE9061		
2018	86X0073-001	2020	BDX73	2020	ES44(ELCOM)	2020	SE9062		
2018	86X0073-002	2020	BDY16	2020	ES45(ELCOM)	2020	SE9063		
2018	87B02	2020	BDY71	2020	ET8-017	2020	SJ1172		
2018	96-5263-01	2020	BDY72	2020	ETEC-28C490	2020	SJ2095		
2018	99S091-1	2020	BDY78	2020	EW185B	2020	SJ3408		
2018	112-203053	2020	BDY79	2020	FA-1(TRANSISTOR)	2020	SJ3447		
2018	121-1020	2020	BLY15A	2020	FBN-35469	2020	SJ3648		
2018	131N2(MAGNAVOX)	2020	BLY47A	2020	FBN-35903	2020	SJ3680		
2018	133-1	2020	BLY48A	2020	FBN-36486	2020	SJ811		
2018	157N3	2020	BLY49A	2020	FBN-36488	2020	SJE5439		
2018	258-1	2020	BN7168	2020	FRB-564	2020	SJE5441		
2018	260D08214	2020	BRC-5496	2020	G8-23	2020	SK3026		
2018	260Z1106	2020	BUY38	2020	G8-241	2020	SK3538		
2018	276-2018	2020	C1024	2020	G8-246	2020	SK3558		
2018	276-2018/RS2018	2020	C1024-D2	2020	HEP241	2020	SM-A-618687-1		
2018	297Y087B02	2020	C1024C	2020	HEP703	2020	STC-4401		
2018	471-1(PLASTIC)	2020	C1024D	2020	HEP703X2	2020	STC4401		
2018	628-3	2020	C1024E	2020	HEP85012	2020	STX0010		
2018	632-1(SYLVANIA)	2020	C1024P	2020	HEP85019	2020	STX0016		
2018	742-1	2020	C1025	2020	HR-107(PHILCO)	2020	STX3326		
2018	964-27986	2020	C1025CTV	2020	HR-68	2020	T-Q5080		
2018	1043-1328	2020	C1160	2020	HR107H	2020	T1233		
2018	2904-058	2020	C1160K	2020	HT304911	2020	T1233A		
2018	3107-204-90180	2020	C1160L	2020	HT304911A	2020	T25-93		
2018	3628-3	2020	C1161	2020	HT304911B	2020	T271		
2018	3742	2020	C1448A	2020	HT308301B0	2020	TA-7155		
2018	4442-3	2020	C1828	2020	HT401301B0	2020	TA2402		
2018	7316	2020	C254	2020	HT403131E0	2020	TA2402A		
2018	7553	2020	C487	2020	HT403151B	2020	TA7155		
2018	8471(SYLVANIA)	2020	C488	2020	HT403152A	2020	TA7362		
2018	36213	2020	C489	2020	HT403152B	2020	TA7782		
2018	40311	2020	C490	2020	I51-0141-00	2020	TA7783		
2018	40311A	2020	C491	2020	I6114	2020	TA7784		
2018	40311B	2020	C491BL	2020	INTRON-127	2020	TA8231		
2018	40314V1	2020	C491R	2020	IP20-0212	2020	TA8232		
2018	40314V2	2020	C491Y	2020	IR-TR57	2020	TA8233		
2018	40315V1	2020	C680	2020	IRTR-57	2020	TG28C1025		
2018	40315V2	2020	C680A	2020	IRTR57	2020	TG28C1025D		
2018	40317V1	2020	C680R	2020	IRTR57X2	2020	TRT251		
2018	40317V2	2020	C791	2020	IRTR92	2020	TIP-24		
2018	40319V1	2020	C830	2020	J241231	2020	TIP-29		
2018	40319V2	2020	C830A	2020	J241252	2020	TIP-31		
2018	40320V1	2020	C830B	2020	J24642	2020	TIP-31B		
2018	40320V2	2020	C830C	2020	KLH4745	2020	TIP29XA		
2018	40321V1	2020	C840	2020	M-75543-1	2020	TIP4		
2018	40321V2	2020	C840A	2020	M4936	2020	TIP41		
2018	40323V1	2020	C840AC	2020	M9225	2020	TIP41A		
2018	40323V2	2020	C840H	2020	M9274	2020	TIP41B		
2018	40326V2	2020	C840PQ	2020	M9301	2020	TIP503		
2018	40327V2	2020	C895	2020	M9309	2020	TM175		
2018	40360V1	2020	CS93-1007	2020	M9316	2020	TM196		
2018	40360V2	2020	CXL175	2020	M9593	2020	TM241		
2018	40361V2	2020	CXL196	2020	M9610	2020	TN60451		
2018	40361V3	2020	DO250	2020	M9676	2020	TN60453		
2018	40385V1	2020	D102	2020	M9676/NPN	2020	TNJ72286		
2018	40385V2	2020	D102-0	2020	MHT5901	2020	TR-03(PENNCREST)		
2018	40437V1	2020	D102-0	2020	MHT5911	2020	TR-180(OLSON)		
2018	40437V2	2020	D102-R	2020	MJ2249	2020	TR-180(OLSON)		
2018	40594	2020	D102-Y	2020	MJ2250	2020	TR-57		
2018	44763	2020	D103	2020	MJ3101	2020	TR-81		
2018	60457	2020	D103-0	2020	MJ4101	2020	TR-81MP		
2018	95258-1	2020	D103-0	2020	MJ4102	2020	TRI591		
2018	95285-1	2020	D103-R	2020	MJ5202	2020	TRI593		
2018	112525	2020	D103-Y	2020	MJ5203	2020	TR2327852		
2018	130474	2020	D129	2020	MJ5204	2020	TR26-1		
2018	132445	2020	D129-BL	2020	MJE2021	2020	TV-109		
2018	132446	2020	D129-R	2020	MJE205K	2020	TV-112		
2018	137066	2020	D129-Y	2020	MJE2380	2020	TV109		
2018	139266	2020	D130	2020	MJE2381	2020	TV112		
2018	171162-164	2020	D130-R	2020	MJE2382	2020	TV117		
2018	171162-235	2020	D130-Y	2020	MJE2383	2020	TV24211		
2018	610157-3	2020	D130BL	2020	MJE2480	2020	TV24487		
2018	610228-1	2020	D142	2020	MJE2481	2020	TVS-28C840		
2018	702415-00	2020	D142M	2020	MJE2482	2020	UPE221		
2018	1417316-1	2020	D143	2020	MJE2483	2020	VS28D476-C/-1		
2018	1445829-503	2020	D144	2020	MJE2520	2020	WEP241		
2018	1445829-504	2020	D145	2020	MJE2521	2020	WEP703		
2018	1471133-1	2020	D150	2020	MJE2522	2020	X1-549		
2018	1473628-1	2020	D155	2020	MJE2523	2020	X7338934		
2018	1473628-3	2020	D155H	2020	MJE29	2020	X735-40C		
2020	A-120304	2020	D155K	2020	MJE29A	2020	XI-549		
2020	A-1854-0449-1	2020	D155L	2020	MJE29B	2020	XT-549A		
2020	A054-224	2020	D226	2020	MJE3054	2020	YAAN28D141		
2020	A068-114	2020	D226-0	2020	MJE31	2020	ZSD235Y		
2020	A14743	2020	D226A	2020	MJE31A	2020	01-040313		
2020	A1Y	2020	D226AP	2020	MJE31B	2020	1A0013(YAMAHA)		
2020	A2415	2020	D226B	2020	MJE371	2020	1A0027		
2020	A27(RCA)	2020	D226BP	2020	MJE41	2020	1A0027(YAMAHA)		
2020	A2SD2260P	2020	D226P	2020	MJE41A	2020	1A0048		
2020	A2SD226PQ	2020	D226Q	2020	MJE41B	2020	1A0048(YAMAHA)		
2020	A5A	2020	D236(TRANSISTOR)	2020	MJE520K	2020	002-1(SYLVANIA)		
2020	A6843401	2020	D258	2020	MJE521K	2020	002-2(SYLVANIA)		
2020	A8E	2020	D258P	2020	MJE591	2020	002-3(SYLVANIA)		
2020	A8P	2020	D254	2020	P3172	2020	002-4(SYLVANIA)		
2020	A8K	2020	D255	2020	P5148	2020	2N3054		
2020	A8Y	2020	D257	2020	P6022A	2020	2N3483		
2020	AD160	2020	D29	2020	P6128	2020	2N3766		
2020	ALB6494612	2020	D290	2020	P6804	2020	2N3767		
2020	AR-50	2020	D290L	2020	PP-AR18	2020	2N3779		
2020	AR17	2020	D291	2020	PN26	2020	2N3878		
2020	AR17(PHILCO)	2020	D292	2020	PTC112	2020	2N3879		
2020	AR17A	2020	D297	2020	Q-00769	2020	2N4231		
2020	AR17B	2020	D313C	2020	Q-00769A	2020	2N4232		
2020	AR18(PHILCO)	2020	D313D	2020	Q-00769B	2020	2N4233		
2020	AR24(RED)	2020	D313F	2020	Q-00769C	2020	2N4910		
2020	AR24RED	2020	D315	2020	Q-01269	2020	2N4911		
2020	AR28	2020	D34014094	2020	Q-01269B	2020	2N4912		
2020	AR28(RED)	2020	D49	2020	Q-01269C	2020	2N5202		
2020	AR28RED	2020	D56	2020	Q5101D	2020	2N5490		
2020	AR30	2020	D57	2020	Q51342	2020	2N5491		
2020	AR35	2020	D58	2020	R2096	2020	2N5493		
2020	AR38	2020	D70	2020	R623-1	2020	2N5494		
2020	AR38(RED)	2020	D71	2020	R640-1	2020	2N5495		
2020	AR38RED	2020	D92	2020	RCA1C10	2020	2N5496		
2020	B133823	2020	D92D	2020	RCA1C14	2020	2N5497		
2020	B1U	2020	DDBY407004	2020	RCA40250	2020	2N5598		
2020	B1Z	2020	DT3301	2020	RE34	2020	2N5600		
2020	B23-82	2020	DT3302	2020	RBN175	2020	2N5601		
2020	B2B	2020	E570022-01	2020	RBN196	2020	2N5602		
2020	B5C	2020	EA15X121	2020	S21520	2020	2N5604		
2020	BD-131	2020	EC0175	2020	S2321	2020	2N6233		
2020	BD273	2020	EC0196	2020	S2486	2020	2N6260		
2020	BD275	2020	EC0241	2020	S20153	2020	2N6261		
2020	BD278	2020	EC0390	2020	S306A	2020	2N6263		
2020	BD441	2020	EP-276	2020	S354	2020	2N6292		
2020	BD537	2020	EP15X11	2020	S37162	2020	2N6500		
2020	BD575	2020	EP15X14	2020	SD141	2020	2SC1024		
2020	BD577	2020	EP15X30	2020	SDT6901	2020	2SC1024(D)		
2020	BD585					2020	2SC1024(F)		

RS 276-	Industry Standard No.	RS 276-	Industry Standard No.	RS 276-	Industry Standard No.	RS 276-	Industry Standard No.
2020	2SC1024-D2	2020	2SD103-0	2020	2SD314E	2020	48-60022A14
2020	2SC1024-E	2020	2SD103-R	2020	2SD315	2020	48-869225
2020	2SC1024A	2020	2SD103-Y	2020	2SD315C	2020	48-869274
2020	2SC1024B	2020	2SD129	2020	2SD315D	2020	48-869301
2020	2SC1024C	2020	2SD129-BL	2020	2SD315E	2020	48-869309
2020	2SC1024D	2020	2SD129-R	2020	2SD470A	2020	48-869316
2020	2SC1024E	2020	2SD129-Y	2020	2SD476	2020	48-869393
2020	2SC1024F	2020	2SD130	2020	2SD476A	2020	48-869610
2020	2SC1024G	2020	2SD130(BL)	2020	2SD476B	2020	48-869676
2020	2SC1024L	2020	2SD130(Y)	2020	2SD476C	2020	48-90234A11
2020	2SC1024Y	2020	2SD130-BLU	2020	2SD476D	2020	48-90343A56
2020	2SC1025	2020	2SD130-R	2020	2SD476TL	2020	48X869309
2020	2SC1025CTV	2020	2SD130-RED	2020	2SD49	2020	48X355042
2020	2SC1025D	2020	2SD130-Y	2020	2SD526	2020	48R869225
2020	2SC1025E	2020	2SD130-YEL	2020	2SD526-0	2020	48B869301
2020	2SC1025J	2020	2SD130A	2020	2SD546	2020	48B869676
2020	2SC1025MT	2020	2SD130B	2020	2SD56	2020	48B134936
2020	2SC106A	2020	2SD130BL	2020	2SD57	2020	48B137323
2020	2SC106B	2020	2SD130C	2020	2SD570	2020	48B155005
2020	2SC106C	2020	2SD130D	2020	2SD58	2020	48B155044
2020	2SC106G	2020	2SD130E	2020	2SD586	2020	051-0151
2020	2SC106GN	2020	2SD130P	2020	2SD586R	2020	57A12-3
2020	2SC106H	2020	2SD130G	2020	2SD656	2020	57A12-5
2020	2SC106J	2020	2SD130GN	2020	2SD70	2020	57A123-10
2020	2SC106K	2020	2SD130H	2020	2SD71	2020	57A155-10
2020	2SC106L	2020	2SD130J	2020	2SD92	2020	57A2-47
2020	2SC106M	2020	2SD130K	2020	2SD92D	2020	57A2-58
2020	2SC106OR	2020	2SD130OR	2020	3L4-6OR4	2020	57A2-84
2020	2SC106R	2020	2SD130R	2020	3L4-6011-1	2020	57A244-14
2020	2SC106X	2020	2SD130X	2020	3L4-6012-06	2020	57B123-10
2020	2SC106Y	2020	2SD130Y	2020	3L4-6012-1	2020	57B155-10
2020	2SC1085	2020	2SD142	2020	3L4-6012-5	2020	57B158-10
2020	2SC1160	2020	2SD142A	2020	3L4-6012-55	2020	57B2-47
2020	2SC1160K	2020	2SD142B	2020	3L4-6012-56	2020	57B2-58
2020	2SC1160L	2020	2SD142C	2020	3L4-6012-58	2020	57B2-84
2020	2SC1161	2020	2SD142D	2020	3L4-6012-6	2020	57B244-14
2020	2SC1664A	2020	2SD142E	2020	3L4-6012-8	2020	57C12-3
2020	2SC1826	2020	2SD142F	2020	3SD313	2020	57C12-5
2020	2SC1826F	2020	2SD142G	2020	3SD313C	2020	57M3-5
2020	2SC1826Q	2020	2SD142GN	2020	4-003721-00	2020	61-1053-1
2020	2SC1826R	2020	2SD142H	2020	4-265	2020	62-16919
2020	2SC254	2020	2SD142J	2020	4-686130-3	2020	63-10062
2020	2SC487	2020	2SD142K	2020	4-686226-3	2020	63-11290
2020	2SC488	2020	2SD142L	2020	4-686234-3	2020	63-11878
2020	2SC489	2020	2SD142M	2020	6-000140	2020	63-11938
2020	2SC489Y	2020	2SD142OR	2020	6-000555-2	2020	63-12954
2020	2SC490	2020	2SD142R	2020	6-140	2020	63-8512
2020	2SC491	2020	2SD142X	2020	6-24	2020	63-8945
2020	2SC491BL	2020	2SD142Y	2020	6-490001	2020	065
2020	2SC491R	2020	2SD143	2020	6-5552	2020	76-0105
2020	2SC491Y	2020	2SD144	2020	6-6490001	2020	86-271-2
2020	2SC508	2020	2SD145	2020	007-0030	2020	86-272-2
2020	2SC680	2020	2SD150	2020	7-0050-00	2020	86-412-2
2020	2SC680(A)	2020	2SD152(L)	2020	007-0074	2020	86-5085-2
2020	2SC680A	2020	2SD152(M)	2020	7-29(STANDEL)	2020	86-5106-2
2020	2SC680B	2020	2SD152A	2020	7L6-0105	2020	86-568-2
2020	2SC680C	2020	2SD152B	2020	7L6-0444-1(NPN)	2020	86A338
2020	2SC680G	2020	2SD152C	2020	7L6-0531-1(NPN)	2020	86X0042-002
2020	2SC680GN	2020	2SD152D	2020	8-1	2020	86X0080-002
2020	2SC680H	2020	2SD152E	2020	8-1(BENDIX)	2020	91AJ150
2020	2SC680J	2020	2SD152F	2020	09-302218	2020	96-5190-01
2020	2SC680K	2020	2SD152G	2020	09-303005	2020	96-5191-01
2020	2SC680L	2020	2SD152GN	2020	09-303021	2020	96-5232-01
2020	2SC680M	2020	2SD152H	2020	10-13159-002	2020	96-5232-03
2020	2SC680R	2020	2SD152J	2020	10-13159-002	2020	96-5245-01
2020	2SC680X	2020	2SD152K	2020	10-3159-002	2020	96-5267-01
2020	2SC791	2020	2SD152L	2020	13-0014	2020	96-5285-01
2020	2SC791-OR	2020	2SD152M	2020	13-1847-1	2020	98-24320-2
2020	2SC791A	2020	2SD152OR	2020	13-34002-1	2020	099-1&58PHILCO)
2020	2SC791B	2020	2SD152R	2020	13-34002-2	2020	99-8-075
2020	2SC791C	2020	2SD152X	2020	13-34002-3	2020	998047
2020	2SC791D	2020	2SD152Y	2020	13-34002-4	2020	998075
2020	2SC791E	2020	2SD155	2020	13-34002-5	2020	998085-1
2020	2SC791F	2020	2SD155H	2020	13-34616-1	2020	998100
2020	2SC791FA1	2020	2SD155K	2020	13-34838-1	2020	100-5765-001
2020	2SC791GN	2020	2SD155L	2020	13-36314-1	2020	111NSC
2020	2SC791H	2020	2SD226	2020	13-39004-1	2020	119-0068
2020	2SC791J	2020	2SD226-0	2020	13-40340-1	2020	121-588
2020	2SC791K	2020	2SD226A	2020	13-55064-1(SYL)	2020	121-770CL
2020	2SC791M	2020	2SD226AP	2020	14-602-67	2020	121-772CL
2020	2SC791OR	2020	2SD226B	2020	14-608-02A	2020	121-853
2020	2SC791R	2020	2SD226BP	2020	14-609-03A	2020	121-880
2020	2SC791X	2020	2SD226P	2020	14-609-09	2020	121-967
2020	2SC791Y	2020	2SD226Q	2020	1414-180	2020	121-987-02
2020	2SC804H	2020	2SD236	2020	15-0009-05	2020	125-410
2020	2SC830	2020	2SD236(02Y)	2020	18-177-1	2020	125A8251
2020	2SC830A	2020	2SD236A	2020	19-00-3485	2020	125B410
2020	2SC830B	2020	2SD236B	2020	019-003485	2020	129-23
2020	2SC830C	2020	2SD236C	2020	19-020-045	2020	0131-04560
2020	2SC830D	2020	2SD236D	2020	19-020-045	2020	132-065
2020	2SC830E	2020	2SD236E	2020	19A115527	2020	132-5
2020	2SC830F	2020	2SD236F	2020	19A115527-1	2020	149N2B
2020	2SC830G	2020	2SD236G	2020	19A115527-P1	2020	149N4B
2020	2SC830GN	2020	2SD236GN	2020	19A115527-1	2020	151-0141
2020	2SC830H	2020	2SD236H	2020	22-001008	2020	151-0141-00
2020	2SC830J	2020	2SD236J	2020	22-001009	2020	151-0148
2020	2SC830K	2020	2SD236K	2020	23	2020	151-0149
2020	2SC830L	2020	2SD236L	2020	23-5031	2020	151-0149-1A
2020	2SC830M	2020	2SD236M	2020	31-0010	2020	151-0217
2020	2SC830OR	2020	2SD236R	2020	33-00234-B	2020	151-0217-1
2020	2SC830R	2020	2SD236X	2020	33-050	2020	151-0I49
2020	2SC830X	2020	2SD236Y	2020	34-1026	2020	151-148-00-BC
2020	2SC830Y	2020	2SD238	2020	34-6002-1	2020	153N1
2020	2SC840	2020	2SD238F	2020	34-6002-28	2020	155N1
2020	2SC840(P)	2020	2SD254	2020	34-6002-29	2020	161-5
2020	2SC840A	2020	2SD255	2020	34-6002-43	2020	185-736
2020	2SC840AC	2020	2SD257	2020	34-6005-2	2020	200-018
2020	2SC840B	2020	2SD28	2020	34-6005-3	2020	207A16
2020	2SC840C	2020	2SD29	2020	41-0909	2020	207A16A
2020	2SC840D	2020	2SD290	2020	41B5H144-P001	2020	211A6381-1
2020	2SC840E	2020	2SD290L	2020	42-20960	2020	211A6381-I
2020	2SC840F	2020	2SD291	2020	44A353980-002	2020	220
2020	2SC840G	2020	2SD291(R)	2020	44A354637-001	2020	00234-B
2020	2SC840GN	2020	2SD291-0	2020	44A350245-001	2020	239A7920
2020	2SC840H	2020	2SD291B	2020	44B238208-001	2020	0243-001
2020	2SC840HP	2020	2SD291BL	2020	44B238208-002	2020	260P08601
2020	2SC840J	2020	2SD291C	2020	44B238208002	2020	260P14202
2020	2SC840K	2020	2SD291D	2020	46-86130-3	2020	260P16009
2020	2SC840L	2020	2SD291E	2020	46-86189-3	2020	260P20101
2020	2SC840M	2020	2SD291G	2020	46-86226-3	2020	260P24803
2020	2SC840OR	2020	2SD291GA	2020	46-86347-3	2020	260P34202
2020	2SC840P	2020	2SD291GN	2020	46-86349-3	2020	314-6006
2020	2SC840PQ	2020	2SD291H	2020	46-86349-9	2020	322-1
2020	2SC840Q	2020	2SD291J	2020	46-86350-3	2020	331-1
2020	2SC840X	2020	2SD291K	2020	46-86459-3	2020	352-0581-011
2020	2SC840Y	2020	2SD291L	2020	46-86476-3	2020	352-0581-020
2020	2SC895	2020	2SD291M	2020	46-86496-3	2020	352-0581-021
2020	2SD102	2020	2SD291OR	2020	46-86568-3	2020	352-0581-030
2020	2SD102-0	2020	2SD291R	2020	46-88676-3	2020	352-0581-031
2020	2SD102-R	2020	2SD291X	2020	48-134936	2020	352-0606-011
2020	2SD102-Y	2020	2SD291Y	2020	48-137030	2020	413(E.F.JOHNSON)
2020	2SD103	2020	2SD292	2020	48-137309	2020	417-104
		2020	2SD297	2020	48-137311	2020	417-175
				2020	48-137369	2020	417-199
				2020	48-137526		
				2020	48-355042		

RS 276-	Industry Standard No.	RS 276-	Industry Standard No.	RS 276-	Industry Standard No.	RS 276-	Industry Standard No.
2020	417-203	2020	40373	2022	REN744	2023	BCY78VIII
2020	417-298	2020	40424	2022	SK3171	2023	BCY79A
2020	417-4-00226	2020	40624	2022	SK3171/744	2023	BCY79B
2020	422-0960	2020	40627	2022	SL23971	2023	BCY79VII
2020	424-9001	2020	40632	2022	SN76635N	2023	BCY79VIII
2020	576-0002-001	2020	40664	2022	TM744	2023	BPR61
2020	576-0002-026	2020	40816(RCA)	2022	TVCM-19	2023	BPR80
2020	576-0002-029	2020	40910	2022	U6A7720354	2023	BPR80T05
2020	632-3	2020	41500	2022	U6A7720395	2023	BPS34
2020	638H	2020	42065	2022	U9A7720354	2023	BPS44
2020	638HJ	2020	42342	2022	U9A7720395	2023	BPS45
2020	686-0210	2020	50200-8	2022	UA720	2023	BPS96
2020	686-0210-0	2020	50280-3	2022	UA720DC	2023	BPS97
2020	707	2020	51300	2022	UA720PC	2023	BPT70
2020	750-045	2020	55810-166	2022	ULN2137A	2023	BFW20
2020	753-2000-006	2020	60201	2022	ULN2137N	2023	BFW31
2020	817-275	2020	60219	2022	28A854	2023	BSV43B
2020	991-01-3170	2020	60237	2022	28A854Q	2023	BSV44B
2020	991-01-5001	2020	60416	2022	28B642	2023	BSV45B
2020	991-01-5063	2020	60632	2022	15-36995-1	2023	BSV47B
2020	992-003172	2020	60678	2022	88-9302	2023	BSV48B
2020	992-01-3705	2020	60977	2022	88-9302R	2023	BSV49B
2020	1009-05A	2020	61035	2022	88-9302RS	2023	CDC9000-1C
2020	1040-02	2020	61252	2022	88-9302S	2023	CDC90001B
2020	1061-8338	2020	61534	2022	221-107	2023	CGE-67
2020	1064-4417	2020	61926	2022	720DC	2023	CS-2142
2020	1065-9944	2020	61997	2022	720PC	2023	CS125SHF
2020	1109(GE)	2020	62142	2022	7208DC	2023	CS2082
2020	1123-3335	2020	62144	2022	7208PC	2023	CS2142
2020	1239 5752	2020	62511	2022	889302	2023	CXL193
2020	1239-5752	2020	62660	2023	A3T2907A	2023	D29B08
2020	1384(GB)	2020	62763	2023	A545	2023	D29B08J1
2020	1414-180	2020	70019-1	2023	A545GRN	2023	D29B09
2020	1507-7183	2020	70019-5	2023	A545K	2023	D29B09J1
2020	1534-8931	2020	75700-22-01	2023	A545KLM	2023	D29B10
2020	1548-17	2020	79922	2023	A545L	2023	D29B10J1
2020	1714-0402	2020	80131	2023	A545LM	2023	D29B1J1
2020	1714-0602	2020	88801-3-1	2023	A547	2023	D29B2J1
2020	1714-0605	2020	94049	2023	A547A	2023	D29B4J1
2020	1714-0802	2020	94094	2023	A575K	2023	D29B5J1
2020	1714-0805	2020	115527-1	2023	A575L	2023	D29B6J1
2020	1714-1002	2020	131095-2	2023	A5T3505	2023	D29B7J1
2020	1714-1005	2020	131161	2023	A5T4026	2023	D29B8J1
2020	1854-0265	2020	131239	2023	A5T4028	2023	D29B9
2020	1936-17	2020	133823	2023	A643	2023	D29B9J1
2020	2008	2020	134771	2023	A643A	2023	EA15X242
2020	2026-4	2020	135735	2023	A643B	2023	EA15X273
2020	2026-5	2020	139270	2023	A643C	2023	EC0193
2020	2057B155-10	2020	140626	2023	A643D	2023	EC0193-1
2020	2057B158-10	2020	140979	2023	A643F	2023	EP15X13
2020	2110N-134	2020	141020	2023	A643L	2023	EP15X21
2020	2163-17	2020	150060	2023	A643R	2023	EP15X51
2020	2180-155	2020	165758	2023	A643S	2023	EP35
2020	2295	2020	167958	2023	A643V	2023	F2041
2020	02375-A	2020	171033-1	2023	A643W	2023	FT3644
2020	2842-056	2020	171174-3	2023	A707	2023	FX3962
2020	2957A100-25	2020	198038-1	2023	A707V	2023	GB-67
2020	3054	2020	198038-3	2023	A751	2023	GB-89
2020	3152-159	2020	198038-4	2023	A751Q	2023	HA-00643
2020	3632-1	2020	198039-1	2023	A751QR	2023	HA7520
2020	3632-3	2020	198039-3	2023	A751R	2023	HA7521
2020	3638	2020	198039-501	2023	A751S	2023	HA7522
2020	3640(RCA)	2020	198039-503	2023	A752	2023	HA7523
2020	3721	2020	198049-1	2023	A752P	2023	HA7524
2020	3843(SEARS)	2020	230084	2023	A752Q	2023	HA7526
2020	3980-002	2020	232268	2023	A752R	2023	HA7527
2020	4080-838-0001	2020	238208-002	2023	A752S	2023	HA7528
2020	4080-838-1	2020	239517	2023	A8T4026	2023	HA7725
2020	4080-838-2	2020	262116	2023	A8T4028	2023	HA7730
2020	4080-838-3	2020	309689	2023	AR304GREEN	2023	HA7732
2020	4080-866-0006	2020	400127	2023	AR304RED	2023	HA7734
2020	4080-866-1	2020	489751-174	2023	AR308	2023	HA7735
2020	4080-866-2	2020	505256	2023	BC161-06	2023	HA7736
2020	4080-879-0001	2020	505257	2023	BC212A	2023	HA7737
2020	4442-366	2020	505434	2023	BC212B	2023	HEP76
2020	4686-130-3	2020	505469	2023	BC213B	2023	HR-72
2020	4686-226-3	2020	532003	2023	BC214B	2023	HR10S611C
2020	4686-234-3	2020	0573525	2023	BC224	2023	HR72
2020	4745	2020	610111-8	2023	BC231B	2023	HS5811
2020	4823-0018	2020	610155-1	2023	BC257A	2023	HS5813
2020	5259	2020	610195-1	2023	BC257B	2023	HS5815
2020	5320-003	2020	619009-1	2023	BC258A	2023	HS5817
2020	5380-73(POWER)	2020	619361-1	2023	BC258B	2023	HS5819
2020	6099-2	2020	700191	2023	BC259A	2023	HS5821
2020	6381-1	2020	700195	2023	BC259B	2023	HS5823
2020	7312	2020	705784-1	2023	BC297	2023	HS6011
2020	7322	2020	723060-29	2023	BC298	2023	HS6013
2020	7564-1	2020	723423-16	2023	BC304-4	2023	HS6015
2020	7510(TRANSISTOR)	2020	723423-20	2023	BC304-5	2023	HS6017
2020	7920-1	2020	723423-7	2023	BC304-6	2023	HX-50104
2020	8000-00005-008	2020	723423-9	2023	BC307B	2023	IP20-0211
2020	8000-00005-012	2020	910807-11	2023	BC309C	2023	IRTR-52
2020	8620	2020	932081-1	2023	BC327-16	2023	J241015(2SA695C)
2020	9404-9	2020	982523	2023	BC361-06	2023	MD86518
2020	9409-4	2020	1417322-1	2023	BC405	2023	ME513
2020	12044-0021	2020	1417364-1	2023	BC406	2023	MH1502
2020	012085	2020	1471132-002	2023	BC461-4	2023	MM4008
2020	12538	2020	1471132-3	2023	BC461-5	2023	MM7379B
2020	13159-2	2020	1471132-5	2023	BC461-6	2023	MP8-A56
2020	16029	2020	1473612-1	2023	BC512B	2023	MP84354
2020	16039	2020	1473621-1	2023	BC513B	2023	MP84355
2020	16114	2020	1473623-1	2023	BC514	2023	MP86535M
2020	16115	2020	1473632-3	2023	BC514B	2023	NN7500
2020	16163	2020	1473640-1	2023	BCW45	2023	NN7501
2020	16165	2020	1851516	2023	BCW52	2023	NN7502
2020	16277	2020	1851518	2023	BCW61A	2023	NN7503
2020	16341	2020	1971296	2023	BCW61B	2023	NN7504
2020	20294	2020	2320084	2023	BCW61BA	2023	NN7505
2020	020425-3	2020	3130057	2023	BCW61BB	2023	NN7511
2020	23754	2020	3130093	2023	BCW63B	2023	NR-671ET
2020	30236	2020	3152159	2023	BCW64	2023	P4B
2020	30256	2020	3463100-1	2023	BCW64B	2023	PTC177
2020	35002	2020	3463604-1	2023	BCW70	2023	RB197
2020	35405	2020	3463604-2	2023	BCW70R	2023	REN193
2020	36274	2020	3468071-1	2023	BCW75-10	2023	RT-121
2020	36320	2020	3468841-1	2023	BCW75-16	2023	RV2260
2020	36370-05490	2020	3596091	2023	BCW76-10	2023	SC256B
2020	37077	2020	3596092	2023	BCW76-16	2023	SHA7520
2020	37484	2022	BC320	2023	BCW86	2023	SHA7521
2020	37599	2022	CA3123E	2023	BCW92	2023	SHA7522
2020	37763	2022	DM-20	2023	BCW92K	2023	SHA7523
2020	37900	2022	DM-32	2023	BCW93	2023	SHA7524
2020	37913	2022	DM20	2023	BCW93A	2023	SHA7526
2020	38443	2022	DM32	2023	BCW93B	2023	SHA7527
2020	39285	2022	EC0744	2023	BCW93K	2023	SHA7528
2020	40250	2022	GEIC-215	2023	BCW93KA	2023	SHA7597
2020	40282	2022	GEIC-24	2023	BCW93KB	2023	SHA7598
2020	40310	2022	HA1139A	2023	BCX71BG	2023	SHA7599
2020	40310V1	2022	IC-607(ELCOM)	2023	BCX71BH	2023	SK3138
2020	40312	2022	LM1820	2023	BCY78A	2023	ST61000
2020	40312V1	2022	LM1820A	2023	BCY78B	2023	TC8101
2020	40316	2022	LM1820N	2023	BCY78VII	2023	TC8103
2020	40324	2022	PTC734			2023	TE3702
2020	40364	2022	RE362-IC			2023	TE3703
2020	40372						

RS 276-	Industry Standard No.
2023	TE5086
2023	TE5087
2023	TE5088
2023	TE5089
2023	TE5249
2023	TE5309
2023	TE5367
2023	TI-891
2023	TM193
2023	TNJ70481
2023	TNJ72152
2023	TP363dA
2023	TQ53
2023	TQ54
2023	TQ55
2023	TQ57
2023	TQ58
2023	TQ59
2023	TQ59A
2023	TQ60
2023	TQ60A
2023	TV82C81256HG
2023	V654
2023	V655
2023	V741
2023	001-02117-2
2023	2N1132B/51
2023	2N1132B46
2023	2N1242
2023	2N1243
2023	2N3081/46
2023	2N3244
2023	2N4971A
2023	2N4972A
2023	2N4982A
2023	2N5055
2023	2N5086A
2023	2N5087A
2023	2N5138A
2023	2N5139A
2023	2N5142A
2023	2N5143A
2023	2N5226A
2023	2N5227A
2023	2N5354A
2023	2N5355A
2023	2N5356A
2023	2N5365A
2023	2N5366A
2023	2N5367A
2023	2N5372A
2023	2N5373A
2023	2N5374A
2023	2N5375A
2023	2N5378A
2023	2N5379A
2023	2N5382A
2023	2N5383A
2023	2N5447A
2023	2N5448A
2023	2N5821H8
2023	2N5823
2023	2N5855
2023	2N6015
2023	2S3020A
2023	2S3021A
2023	2S3030A
2023	2S3040A
2023	2S307A
2023	2S310A
2023	2S321A
2023	2S3220A
2023	2S3221A
2023	2S3222AB
2023	2S3230A
2023	2S323A
2023	2S3240A
2023	2S324A
2023	2S326A
2023	2S327A
2023	2S494-YE
2023	2S4930
2023	2S4940
2023	2S4950R
2023	2SA510-OR
2023	2SA510-RD
2023	2SA510A
2023	2SA510B
2023	2SA510C
2023	2SA510D
2023	2SA510E
2023	2SA510F
2023	2SA510G
2023	2SA510GN
2023	2SA510H
2023	2SA510J
2023	2SA510K
2023	2SA510L
2023	2SA510M
2023	2SA510OR
2023	2SA510R
2023	2SA510X
2023	2SA510Y
2023	2SA511-G
2023	2SA511-OR
2023	2SA511-RD
2023	2SA511A
2023	2SA511B
2023	2SA511C
2023	2SA511D
2023	2SA511E
2023	2SA511F
2023	2SA511G
2023	2SA511GN
2023	2SA511H
2023	2SA511J
2023	2SA511K
2023	2SA511L
2023	2SA511M
2023	2SA511OR
2023	2SA511R
2023	2SA511X
2023	2SA511Y
2023	2SA512-OR
2023	2SA512-OR1
2023	2SA512-ORG
2023	2SA512-RD
2023	2SA512-RED
2023	2SA512A
2023	2SA512B
2023	2SA512C
2023	2SA512D
2023	2SA512E
2023	2SA512F
2023	2SA512G
2023	2SA512GN
2023	2SA512H
2023	2SA512J
2023	2SA512K
2023	2SA512L
2023	2SA512M
2023	2SA512OR
2023	2SA512R
2023	2SA512X
2023	2SA512Y
2023	2SA513-OR
2023	2SA513-ORG
2023	2SA513-RD
2023	2SA513-RED
2023	2SA513A
2023	2SA513B
2023	2SA513C
2023	2SA513D
2023	2SA513E
2023	2SA513F
2023	2SA513G
2023	2SA513GN
2023	2SA513H
2023	2SA513J
2023	2SA513K
2023	2SA513L
2023	2SA513M
2023	2SA513OR
2023	2SA513R
2023	2SA513X
2023	2SA513Y
2023	2SA530A
2023	2SA530B
2023	2SA530C
2023	2SA530D
2023	2SA530E
2023	2SA530F
2023	2SA530G
2023	2SA530GN
2023	2SA530GR
2023	2SA530H1
2023	2SA530J
2023	2SA530K
2023	2SA530L
2023	2SA530M
2023	2SA530OR
2023	2SA530R
2023	2SA530X
2023	2SA530Y
2023	2SA544A
2023	2SA544B
2023	2SA544C
2023	2SA544D
2023	2SA544E
2023	2SA544F
2023	2SA544G
2023	2SA544GN
2023	2SA544H
2023	2SA544J
2023	2SA544K
2023	2SA544L
2023	2SA544M
2023	2SA544OR
2023	2SA544R
2023	2SA544X
2023	2SA544Y
2023	2SA545
2023	2SA545(K)
2023	2SA545(L)
2023	2SA545A
2023	2SA545B
2023	2SA545C
2023	2SA545D
2023	2SA545E
2023	2SA545F
2023	2SA545G
2023	2SA545GN
2023	2SA545GRN
2023	2SA545H
2023	2SA545J
2023	2SA545K
2023	2SA545KLM
2023	2SA545L
2023	2SA545M
2023	2SA545OR
2023	2SA545R
2023	2SA545X
2023	2SA545Y
2023	2SA547
2023	2SA547A
2023	2SA620
2023	2SA567O
2023	2SA567OR
2023	2SA568O
2023	2SA568OR
2023	2SA569O
2023	2SA569OR
2023	2SA569A
2023	2SA569B
2023	2SA569C
2023	2SA569D
2023	2SA569E
2023	2SA569F
2023	2SA569G
2023	2SA569GN
2023	2SA569H
2023	2SA569K
2023	2SA569L
2023	2SA569M
2023	2SA569OR
2023	2SA569R
2023	2SA569X
2023	2SA569Y
2023	2SA575K
2023	2SA575L
2023	2SA599
2023	2SA599(Y)
2023	2SA599Y
2023	2SA607E
2023	2SA607F
2023	2SA607G
2023	2SA607GN
2023	2SA607H
2023	2SA607J
2023	2SA607K
2023	2SA607L
2023	2SA607M
2023	2SA607OR
2023	2SA607R
2023	2SA607X
2023	2SA607Y
2023	2SA609GN
2023	2SA609J
2023	2SA609K
2023	2SA609L
2023	2SA609M
2023	2SA609OR
2023	2SA609R
2023	2SA609Y
2023	2SA643
2023	2SA643(R)
2023	2SA643(V,R)
2023	2SA643(W)
2023	2SA643A
2023	2SA643B
2023	2SA643C
2023	2SA643D
2023	2SA643E
2023	2SA643F
2023	2SA643R
2023	2SA643V
2023	2SA643W
2023	2SA707
2023	2SA707V
2023	2SA728A
2023	2SA751
2023	2SA751(P)
2023	2SA751P
2023	2SA751Q
2023	2SA751QR
2023	2SA751R
2023	2SA751S
2023	2SA752
2023	2SA752P
2023	2SA752Q
2023	2SA752R
2023	2SB623
2023	2SB677
2023	09-300072
2023	09-302161
2023	09-30930
2023	13-28393-3
2023	13-32631
2023	13-32631-2
2023	13-32631-4
2023	13-33178-1
2023	13-39115-1
2023	13-39115-3
2023	14-602-54
2023	14-602-71
2023	14-602-75
2023	21A112-004
2023	21A118-032
2023	21M027
2023	21M459
2023	24NW1062(PNP)
2023	31-0055
2023	31-0102
2023	34-6016-59
2023	34-6016-59(PNP)
2023	36-6016-59
2023	48-437321
2023	48-355055
2023	488134989
2023	48X97046A36
2023	56-4857
2023	57A168-9
2023	86-5104-2
2023	86-602-2
2023	87-0028
2023	87-0029
2023	0101-0439
2023	134P1(REMOTE)
2023	168-9
2023	178-1(PHILCO)
2023	229-1301-28
2023	229-1301-38
2023	393-1
2023	643RIX
2023	643RLX
2023	853-0300-643
2023	991-011319
2023	1072
2023	1072Q
2023	1072K
2023	1072K(GE)
2023	1080Q(GE)
2023	1184Q(GE)
2023	2057A100-40(PNP)
2023	2057A2-480
2023	3524
2023	5611-695
2023	5611-695C
2023	40319L
2023	40319B
2023	40406L
2023	40406B
2023	71687
2023	71687-1
2023	71687-101
2023	95226-003
2023	183035
2023	255821H8
2023	5140688
2023	988992
2023	2320632
2024	A-102
2024	A059-104
2024	A059-105
2024	A061-107
2024	A061-110
2024	A061-111
2024	A069-121
2024	A1-3
2024	A100(TRANSISTOR)
2024	A100A
2024	A100B
2024	A100C
2024	A101
2024	A101A(TRANSISTOR)
2024	A101AA
2024	A101AY
2024	A101B
2024	A101BA
2024	A101BB
2024	A101BC
2024	A101BX
2024	A101C
2024	A101CA
2024	A101CV
2024	A101CX
2024	A101E
2024	A101QA
2024	A101V
2024	A101X
2024	A101Y
2024	A101Z
2024	A102
2024	A102(JAPAN)
2024	A102A
2024	A102AA
2024	A102AB
2024	A102B
2024	A102BA
2024	A102BN
2024	A102CA
2024	A102TV
2024	A103
2024	A103A
2024	A103B
2024	A103C
2024	A103CA
2024	A103CAK
2024	A103CQ
2024	A103DA
2024	A104
2024	A104(JAPAN)
2024	A104A
2024	A104B
2024	A104D
2024	A104P
2024	A104Y
2024	A109
2024	A111(JAPAN)
2024	A112
2024	A113
2024	A121
2024	A122(JAPAN)
2024	A125
2024	A124
2024	A125
2024	A131
2024	A132(JAPAN)
2024	A133(JAPAN)
2024	A134
2024	A137(JAPAN)
2024	A1383
2024	A1384
2024	A139(JAPAN)
2024	A14-1001
2024	A14-1002
2024	A14-1003
2024	A141(JAPAN)
2024	A141B
2024	A141C
2024	A142(JAPAN)
2024	A142A
2024	A142B
2024	A142C
2024	A143(JAPAN)
2024	A144
2024	A144C
2024	A145
2024	A145A
2024	A145C
2024	A15-1001
2024	A15-1002
2024	A15-1003
2024	A150
2024	A153(JAPAN)
2024	A154(PNP)
2024	A155(PNP)
2024	A156(JAPAN)
2024	A157(JAPAN)
2024	A164(PNP)
2024	A165(PNP)
2024	A166
2024	A175
2024	A176
2024	A178A
2024	A178B
2024	A179A
2024	A179B
2024	A188
2024	A189
2024	A19
2024	A20
2024	A201-Q
2024	A201A
2024	A201B
2024	A201E
2024	A201TVO
2024	A201TVO
2024	A202
2024	A202(JAPAN)
2024	A202A
2024	A202B
2024	A202C
2024	A202D
2024	A203
2024	A203A
2024	A203AA(PNP)
2024	A203B(PNP)
2024	A205P(PNP)
2024	A21
2024	A213
2024	A214
2024	A216
2024	A234B
2024	A234C
2024	A235A
2024	A235C
2024	A236
2024	A237
2024	A238
2024	A246(JAPAN)
2024	A246V
2024	A247(JAPAN)
2024	A25-1001
2024	A25-1002
2024	A25-1003

RS 276-	Industry Standard No.	RS 276-	Industry Standard No.	RS 276-	Industry Standard No.	RS 276-	Industry Standard No.	RS 276-	Industry Standard No.
2024	A250	2024	A466YEL	2024	BC259	2024	HA201		
2024	A251	2024	A468	2024	BC260A	2024	HA235		
2024	A252	2024	A469	2024	BC260B	2024	HA330		
2024	A253(JAPAN)	2024	A470	2024	BC260C	2024	HA342		
2024	A254	2024	A471	2024	BC262A	2024	HA471		
2024	A255	2024	A471-1	2024	BC262B	2024	HA52		
2024	A256	2024	A471-2	2024	BC263A	2024	HA53		
2024	A257	2024	A471-3	2024	BC263B	2024	HA9048		
2024	A258	2024	A472(JAPAN)	2024	BC263C	2024	HA9049		
2024	A259	2024	A472A	2024	BC308	2024	HA9054		
2024	A260	2024	A472B	2024	BC308V1	2024	HA9055		
2024	A261	2024	A472C	2024	BC309	2024	HA9079		
2024	A262	2024	A472D	2024	BC309(ALGG)	2024	HB-186		
2024	A263	2024	A472E	2024	BC309(SIEG)	2024	HB-187		
2024	A264	2024	A474	2024	BC309A	2024	HB-32		
2024	A265	2024	A476	2024	BC309B	2024	HB-33		
2024	A266	2024	A477	2024	BC418B	2024	HB422		
2024	A267	2024	A478	2024	BC419	2024	HEP-80006		
2024	A268	2024	A479	2024	BC419A	2024	HEP1		
2024	A269	2024	A49	2024	BC419B	2024	HEP57		
2024	A270(JAPAN)	2024	A517	2024	BC772	2024	HEP635		
2024	A271(JAPAN)	2024	A518	2024	BFV82	2024	HEP636		
2024	A272(JAPAN)	2024	A52	2024	BFV82A	2024	HEP638		
2024	A273(JAPAN)	2024	A53	2024	BFV82B	2024	HEP639		
2024	A274(PNP)	2024	A57	2024	BFV82C	2024	HEP640		
2024	A275(JAPAN)	2024	A58	2024	B8Y40	2024	HEPG0001		
2024	A28	2024	A59	2024	B8Y41	2024	HEPG0008		
2024	A280	2024	A60	2024	C10291	2024	HEPG0009		
2024	A281	2024	A65B-70	2024	CB103	2024	HEPG0010		
2024	A288	2024	A65B-705	2024	CB157	2024	HR-30		
2024	A288A	2024	A65B19G	2024	CB158	2024	HR15		
2024	A289	2024	A67	2024	CDC-9000	2024	H322D		
2024	A29	2024	A69	2024	CK28	2024	H329D		
2024	A290	2024	A71	2024	CK28A	2024	HT100		
2024	A291(TRANSISTOR)	2024	A71A(TRANSISTOR)	2024	CK4	2024	H2101		
2024	A292	2024	A71AB	2024	CK4A	2024	HT101011X		
2024	A293	2024	A71AC	2024	CK762	2024	HT101021A		
2024	A294	2024	A71B(TRANSISTOR)	2024	CK766	2024	HT103501A		
2024	A295	2024	A71BS	2024	CK766A	2024	HT103531C		
2024	A30(TRANSISTOR)	2024	A71D(TRANSISTOR)	2024	CXL102A	2024	HT103541B		
2024	A30(JAPAN)	2024	A71Y	2024	CXL126	2024	HV0000405-0		
2024	A308	2024	A72	2024	D020	2024	I9A115180-2		
2024	A309	2024	A72BLU	2024	D026	2024	JR05		
2024	A310(JAPAN)	2024	A72BRN	2024	D134	2024	K417-68		
2024	A313	2024	A72ORN	2024	D65	2024	LU2N544		
2024	A314	2024	A72WHT	2024	D66	2024	M4363		
2024	A315	2024	A75	2024	DAT1A	2024	M4363BLU		
2024	A316	2024	A759	2024	DAT2	2024	M4363GRN		
2024	A325	2024	A75B	2024	DS-24	2024	M4363ORN		
2024	A326	2024	A76	2024	DS-25	2024	M4363WHT		
2024	A329	2024	A77	2024	DS14	2024	M4364		
2024	A329A	2024	A77A	2024	D824	2024	M4365		
2024	A329B	2024	A77B	2024	D825	2024	M4366		
2024	A331	2024	A77C	2024	D836	2024	M4367		
2024	A335	2024	A77D	2024	D851	2024	M4368		
2024	A337	2024	A78	2024	D852	2024	M4388		
2024	A338	2024	A78B	2024	E-065	2024	M4454		
2024	A339	2024	A78C	2024	E-066	2024	M4456		
2024	A343	2024	A78D	2024	E-067	2024	M4457		
2024	A344(JAPAN)	2024	A80	2024	E-068	2024	M4501		
2024	A350	2024	A82	2024	E102	2024	M4509		
2024	A350C	2024	A83	2024	E105	2024	M4545		
2024	A350H	2024	A85	2024	E2451	2024	M4545BLU		
2024	A350R	2024	A87	2024	EA15X133	2024	M4545WHT		
2024	A350T	2024	A90(TRANSISTOR)	2024	EA15X140	2024	M4586		
2024	A350TY	2024	A909-1009	2024	EA15X141	2024	M4589		
2024	A350Y	2024	A909-1010	2024	EA2133	2024	M4603		
2024	A351	2024	A92	2024	EA2491	2024	M4604		
2024	A351A	2024	A93	2024	EA2497	2024	M4605		
2024	A351B	2024	A94	2024	EA2498	2024	M4605RED		
2024	A352	2024	AC164	2024	EO0126	2024	M4621		
2024	A352A	2024	AC169	2024	E070	2024	M4652		
2024	A352B	2024	ACR810-101	2024	EP20	2024	M8116		
2024	A353	2024	ACR810-102	2024	E814	2024	M8640E		
2024	A353A	2024	ACR810-103	2024	ES15X61	2024	MA0414		
2024	A353C	2024	ACR83-1001	2024	ES15X73	2024	MA1705		
2024	A354	2024	ACR83-1002	2024	E83	2024	MA820		
2024	A354A	2024	ACR83-1003	2024	E83110	2024	MA821		
2024	A354B	2024	AF105	2024	E83111	2024	MA822		
2024	A355	2024	AF108	2024	E83112	2024	MA823		
2024	A355A	2024	AF120	2024	E83113	2024	MD835		
2024	A356	2024	AF129	2024	E83114	2024	MMT71		
2024	A357	2024	AF130	2024	E83115	2024	MPS404		
2024	A358	2024	AF131	2024	E83116	2024	MPS6076		
2024	A359	2024	AF132	2024	E841	2024	MTO404		
2024	A364	2024	AF133	2024	ESA213	2024	N-EA15X133		
2024	A365	2024	AF134	2024	ESA233	2024	N020		
2024	A366	2024	AF135	2024	ET12	2024	N57B2-17		
2024	A367	2024	AF136	2024	G0008	2024	N57B2-18		
2024	A368	2024	AF137	2024	G0010	2024	N57B2-19		
2024	A369	2024	AF138	2024	GA-52829	2024	N57B2-25		
2024	A37	2024	AF144	2024	GA-53149	2024	NA5015-1012		
2024	A376(JAPAN)	2024	AFY10	2024	GA-53242	2024	NA5018-1002		
2024	A38	2024	AFY11	2024	GC1081	2024	NA5018-1003		
2024	A380	2024	AFY18	2024	GET871	2024	NA5018-1004		
2024	A381	2024	A0134	2024	GET872	2024	NA5018-1005		
2024	A382	2024	A01	2024	GET873	2024	NA5018-1006		
2024	A383	2024	A04091A	2024	GET873A	2024	NA5018-1007		
2024	A384	2024	APB-11B-1007	2024	GET874	2024	NA5018-1008		
2024	A385	2024	ASY30	2024	GET875	2024	NA5018-1009		
2024	A385A	2024	ASY57	2024	GET885	2024	NA5018-1010		
2024	A385D	2024	ASY58	2024	GI-3638	2024	NA5018-1011		
2024	A39	2024	ASY59	2024	GI-3638A	2024	NA5018-1022		
2024	A391	2024	ASZ10	2024	G13	2024	NA5018-1219		
2024	A392	2024	ASZ11	2024	GM580	2024	NA5018-1220		
2024	A393	2024	ASZ30	2024	GT-34HV	2024	NK65B119		
2024	A393A	2024	ATRF1	2024	GT-66	2024	NKT103		
2024	A394	2024	ATRF2	2024	GT43	2024	NKT127		
2024	A395	2024	ATS13	2024	GT46	2024	NKT131		
2024	A398	2024	B384	2024	GT47	2024	NKT132		
2024	A399	2024	B385	2024	GT5116	2024	NKT151		
2024	A400(TRANSISTOR)	2024	B444	2024	GT5117	2024	NKT152		
2024	A412	2024	B444A	2024	GT5148	2024	NKT15325		
2024	A428	2024	B444B	2024	GT5149	2024	NKT15425		
2024	A43	2024	BC158VI	2024	GT5151	2024	NKT242		
2024	A436	2024	BC159	2024	GT5153	2024	NKT249		
2024	A437	2024	BC159A	2024	GT66	2024	NKT252		
2024	A438	2024	BC159B	2024	GTJ33141	2024	NKT253		
2024	A446	2024	BC201	2024	GTJ33229	2024	NKT254		
2024	A447	2024	BC205	2024	GTJ33250	2024	NKT255		
2024	A45	2024	BC205A	2024	HA-101	2024	NKT265		
2024	A45-1	2024	BC205V	2024	HA-15	2024	NKT270		
2024	A45-2	2024	BC205VI	2024	HA-269	2024	NKT618		
2024	A45-3	2024	BC250A	2024	HA-350	2024	NKT675		
2024	A453	2024	BC250B	2024	HA-350A	2024	NKT676		
2024	A454(JAPAN)	2024	BC250C	2024	HA-353	2024	NKT677		
2024	A455(TRANSISTOR)	2024	BC252	2024	HA-353C	2024	OC-390		
2024	A456	2024	BC252A	2024	HA-354	2024	OC-612		
2024	A457	2024	BC252B	2024	HA-354B	2024	OC-613		
2024	A466(JAPAN)	2024	BC253	2024	HA49	2024	OC-614		
2024	A466-2	2024	BC253A	2024	HA102	2024	OC-Y75		
2024	A466-3	2024	BC253B	2024	HA103	2024	OC130		
2024	A466BLK	2024	BC253C	2024	HA104	2024	OC201		
2024	A466BLU	2024	BC258VI					2024	OC202

RS 276-	Industry Standard No.	RS 276-	Industry Standard No.	RS 276-	Industry Standard No.	RS 276-	Industry Standard No.
2024	00331	2024	SPT317	2024	2N111M1	2024	2N643
2024	00341	2024	SPT318	2024	2N111M2	2024	2N644
2024	00342	2024	SPT319	2024	2N112	2024	2N645
2024	00343	2024	SPT320	2024	2N112M1	2024	2N680
2024	00351	2024	SPT354	2024	2N1180	2024	2N694
2024	00361	2024	SPT357	2024	2N1204	2024	2N695
2024	00362	2024	SPT357P	2024	2N1204A	2024	2N72
2024	00363	2024	SK3008	2024	2N1213	2024	2N75
2024	00390	2024	SM862	2024	2N1214	2024	2N85
2024	0040	2024	S065A	2024	2N1216	2024	2N86
2024	0041N	2024	SP8871	2024	2N1224	2024	2N87
2024	0042	2024	ST-103	2024	2N1225	2024	2N88
2024	0042N	2024	STX0033	2024	2N1226	2024	2N89
2024	0043	2024	STX0034	2024	2N123A	2024	2N90
2024	0043N	2024	T-1363	2024	2N28	2024	2N933
2024	0044	2024	T-1364	2024	2N129	2024	2N986
2024	0044N	2024	T-1460	2024	2N1313	2024	2NJ50
2024	0045N	2024	T-348	2024	2N1384	2024	2NJ51
2024	0046N	2024	T-6028	2024	2N139	2024	2NJ52
2024	0047N	2024	T-6029	2024	2N1396	2024	2NJ53
2024	0050	2024	T-6030	2024	2N1397	2024	2NJ59D
2024	00612	2024	T-6031	2024	2N1398	2024	2S109
2024	00613	2024	T-6032	2024	2N1399	2024	2S110
2024	00614	2024	T1460	2024	2N140	2024	2S112
2024	00615	2024	T1524	2024	2N1400	2024	2S141
2024	PB-998005	2024	T1524BRN	2024	2N1401	2024	2S142
2024	PN72	2024	T1524BRN/RED	2024	2N1401A	2024	2S143
2024	Q2N1526	2024	T1654	2024	2N1402	2024	2S145
2024	Q9/6515	2024	T1654BLU	2024	2N140M1	2024	2S146
2024	R-539	2024	T2322	2024	2N140M2	2024	2N76
2024	R104-5	2024	T2323	2024	2N1425	2024	2S3040
2024	R104-6	2024	T2324	2024	2N1426	2024	2S3370
2024	R104-7	2024	T280(SEARS)	2024	2N1427	2024	2S35
2024	R104-8	2024	T281(SEARS)	2024	2N1499	2024	2S36
2024	R336	2024	T449(SEARS)	2024	2N1499A	2024	2S58
2024	R337	2024	T50818	2024	2N1499B	2024	2S96
2024	R581	2024	T52054	2024	2N1500	2024	2S97
2024	R60-1001	2024	T602B	2024	2N1526	2024	2S98
2024	R684	2024	T6029	2024	2N1631	2024	2SA100
2024	R714	2024	T6050	2024	2N1632	2024	2SA100B
2024	R715	2024	T6031	2024	2N1635	2024	2SA100C
2024	RCA34098	2024	T6032	2024	2N1636	2024	2SA101
2024	RCA34099	2024	TA-1650A	2024	2N1637	2024	2SA101OV
2024	RCA34100	2024	TA-1755	2024	2N1638/33	2024	2SA101A
2024	RCA44098	2024	TA-1756	2024	2N1662	2024	2SA101AA
2024	RB15	2024	TA-1830+	2024	2N1699	2024	2SA101AY
2024	RB4	2024	TA-5	2024	2N2089	2024	2SA101B
2024	REN126	2024	TG28A201	2024	2N2090	2024	2SA101BA
2024	RS-1554	2024	TG28A201-0	2024	2N2092	2024	2SA101BB
2024	RS-2003	2024	TG28A201-N	2024	2N2180	2024	2SA101BC
2024	RS-3322	2024	TG28A201C	2024	2N2207	2024	2SA101BX
2024	RS-3523	2024	TL-890	2024	2N2208	2024	2SA101C
2024	RS-3324	2024	TK1228-1001	2024	2N231-YEL-RED	2024	2SA101CA
2024	RS-3862	2024	TM126	2024	2N231BLU	2024	2SA101CV
2024	RS-3863	2024	TNJ60362	2024	2N231RED	2024	2SA101E
2024	RS-3866	2024	TNJ60363	2024	2N231YEL	2024	2SA101OR
2024	RS-5107	2024	TNJ60364	2024	2N2363	2024	2SA101QA
2024	RS-5108	2024	TNJ71248	2024	2N2391	2024	2SA101V
2024	RS-5109	2024	TR-11	2024	2N2392	2024	2SA101X
2024	RS-5201	2024	TR-168(OLSON)	2024	2N2411	2024	2SA101Y
2024	RS-5205	2024	TR02012	2024	2N2412	2024	2SA101Z
2024	RS-5207	2024	TR02063-1	2024	2N2451	2024	2SA102
2024	RS-5209	2024	TR63	2024	2N2456	2024	2SA102(BA)
2024	RS-5301	2024	TR760	2024	2N247	2024	2SA102-0R
2024	RS-5305	2024	TR761	2024	2N247/33	2024	2SA1028A
2024	RS-5306	2024	TR762	2024	2N248	2024	2SA102A
2024	RS-5312	2024	TRA-10R	2024	2N2487	2024	2SA102AA
2024	RS-5313	2024	TRA-11R	2024	2N2488	2024	2SA102AB
2024	RS-5752	2024	TRA-12R	2024	2N2489	2024	2SA102B
2024	RS-5753	2024	TRA-2	2024	2N2494	2024	2SA102BA
2024	RS-5754	2024	TRA-22	2024	2S252	2024	2SA102BA-2
2024	RS-5755	2024	TRA-22A	2024	2N2588	2024	2SA102BN
2024	RS-5756	2024	TRA-22B	2024	2N275	2024	2SA102C
2024	RS-5757	2024	TRA-23	2024	2N2783	2024	2SA102CA
2024	RS-5758	2024	TRA-23A	2024	2N2862	2024	2SA102CA-1
2024	RS-5759	2024	TRA-23B	2024	2N2894	2024	2SA102D
2024	RS-5760	2024	TRA-24	2024	2N2894A	2024	2SA102E
2024	RS-5761	2024	TRA-24A	2024	2N301G	2024	2SA102F
2024	RS-5762	2024	TRA-24B	2024	2N318	2024	2SA102G
2024	RS-5802	2024	TRM16	2024	2N33	2024	2SA102H
2024	RS-5818	2024	TRM17	2024	2N344	2024	2SA102K
2024	RS-684	2024	TV24157	2024	2N345	2024	2SA102L
2024	RS-685	2024	V10/18	2024	2N346	2024	2SA102N
2024	RS5753	2024	V10/18J	2024	2N370	2024	2SA102OR
2024	RS5753-2	2024	V15/20R	2024	2N370/33	2024	2SA102TV
2024	RS5754	2024	V6/2R	2024	2N370A	2024	2SA102TV-2
2024	RS5755	2024	V6/4R	2024	2N371	2024	2SA102X
2024	RS5756	2024	V6/8R	2024	2N371/33	2024	2SA102Y
2024	RS5757	2024	VFL-2744K	2024	2N372	2024	2SA103
2024	RS5758	2024	VP85K	2024	2N372/33	2024	2SA103A
2024	RS5759	2024	VPW-27450	2024	2N373	2024	2SA103B
2024	RS5760	2024	VL18RJ	2024	2N374	2024	2SA103C
2024	RS5761	2024	VS-28A358	2024	2N393	2024	2SA103CA
2024	RS5762	2024	VS-28A378	2024	2N409	2024	2SA103CAK
2024	RS5802	2024	VS-28A379	2024	2N410	2024	2SA103OG
2024	RS6821	2024	VS-28A385	2024	2N411	2024	2SA103DA
2024	RS6822	2024	WRF635	2024	2N412	2024	2SA103K
2024	RT-61014	2024	XT300	2024	2N499	2024	2SA104
2024	RT3361	2024	XT400	2024	2N499A	2024	2SA104D
2024	RT3362	2024	ZEN311	2024	2N500	2024	2SA104P
2024	RT5063	2024	ZEN312	2024	2N500BLU	2024	2SA108
2024	RT5466	2024	ZEN313	2024	2N500RED	2024	2SA109
2024	RT5467	2024	ZEN314	2024	2N500WHT	2024	2SA110
2024	RT5520	2024	1-21-138	2024	2N501	2024	2SA111
2024	RT61014	2024	1-21-189	2024	2N501/18	2024	2SA112
2024	S-55TB	2024	1-21-242	2024	2N501A	2024	2SA113
2024	S-70T	2024	1-21-243	2024	2N502	2024	2SA121
2024	S-80T	2024	1-21-244	2024	2N502A	2024	2SA122
2024	S-87TB	2024	1-21-257	2024	2N502B	2024	2SA123
2024	S-88TB	2024	002-006300	2024	2N504	2024	2SA124
2024	S-95101	2024	002-011600	2024	2N544	2024	2SA125
2024	S-95102	2024	2A	2024	2N544/33	2024	2SA131
2024	S-95103	2024	2AC128	2024	2N583	2024	2SA132
2024	S1332	2024	2E	2024	2N584	2024	2SA133
2024	0822.3516-380	2024	2G	2024	2N588	2024	2SA134
2024	865B-1-3P	2024	20344	2024	2N588A	2024	2SA136
2024	S70.00.730	2024	20601	2024	2N591-6M	2024	2SA137
2024	SA102	2024	20602	2024	2N602	2024	2SA139
2024	SAW-28B56	2024	2MC	2024	2N602A	2024	2SA141
2024	SB-100	2024	2N1003	2024	2N603	2024	2SA141B
2024	SB100	2024	2N1004	2024	2N603A	2024	2SA141C
2024	SB200	2024	2N1042	2024	2N604	2024	2SA142
2024	SB5122	2024	2N1043	2024	2N604A	2024	2SA142A
2024	SC159	2024	2N1107	2024	2N605	2024	2SA142B
2024	SC159A	2024	2N1108	2024	2N606	2024	2SA142C
2024	SC259	2024	2N1108RED	2024	2N607	2024	2SA142D
2024	SC259A	2024	2N1109	2024	2N608	2024	2SA142E
2024	SPT-315	2024	2N1110	2024	2N623	2024	2SA142F
2024	SPT-316	2024	2N1111	2024	2N624	2024	2SA142G
2024	SPT-317	2024	2N1111A	2024	2N640	2024	2SA142GN
2024	SPT-320	2024	2N1111B	2024	2N641	2024	2SA142H
2024	SPT-357	2024	2N1111M1	2024	2N641REDM/P	2024	2SA142J
2024	SPT307	2024	2N1111RED	2024	2N642	2024	2SA142K
2024	SPT308	2024	2N111B				

RS 276-	Industry Standard No.	RS 276-	Industry Standard No.	RS 276-	Industry Standard No.	RS 276-	Industry Standard No.
2024	2SA142L	2024	2SA272	2024	2SA436	2024	2SA71Y
2024	2SA142M	2024	2SA273	2024	2SA437	2024	2SA72
2024	2SA142OR	2024	2SA274	2024	2SA438	2024	2SA72BLU
2024	2SA142R	2024	2SA275	2024	2SA446	2024	2SA72BLU-BLU
2024	2SA142X	2024	2SA28	2024	2SA447	2024	2SA72BRN
2024	2SA142Y	2024	2SA280	2024	2SA45	2024	2SA72ORN
2024	2SA143	2024	2SA281	2024	2SA45-1	2024	2SA72WHT
2024	2SA144	2024	2SA288	2024	2SA45-2	2024	2SA75
2024	2SA144C	2024	2SA288A	2024	2SA45-3	2024	2SA75A
2024	2SA145	2024	2SA289	2024	2SA453	2024	2SA75B
2024	2SA145A	2024	2SA29	2024	2SA454	2024	2SA75C
2024	2SA145C	2024	2SA290	2024	2SA455	2024	2SA75E
2024	2SA150	2024	2SA291	2024	2SA456	2024	2SA75F
2024	2SA152A	2024	2SA292	2024	2SA457	2024	2SA75G
2024	2SA152B	2024	2SA293	2024	2SA466	2024	2SA75H
2024	2SA152C	2024	2SA294	2024	2SA466-2	2024	2SA75K
2024	2SA152D	2024	2SA295	2024	2SA466-3	2024	2SA75L
2024	2SA152E	2024	2SA30	2024	2SA466BLK	2024	2SA75M
2024	2SA152F	2024	2SA301	2024	2SA466BLU	2024	2SA75OR
2024	2SA152G	2024	2SA305-RED	2024	2SA466YEL	2024	2SA75R
2024	2SA152GN	2024	2SA305-YELLOW	2024	2SA468	2024	2SA75X
2024	2SA152H	2024	2SA305A	2024	2SA469	2024	2SA75Y
2024	2SA152J	2024	2SA305B	2024	2SA470	2024	2SA76
2024	2SA152K	2024	2SA305C	2024	2SA471	2024	2SA77
2024	2SA152L	2024	2SA305D	2024	2SA471-1	2024	2SA774
2024	2SA152M	2024	2SA305E	2024	2SA471-2	2024	2SA777A
2024	2SA152OR	2024	2SA305F	2024	2SA471-3	2024	2SA777B
2024	2SA152X	2024	2SA305G	2024	2SA472	2024	2SA777C
2024	2SA152Y	2024	2SA305GN	2024	2SA472A	2024	2SA777D
2024	2SA153	2024	2SA305H	2024	2SA472B	2024	2SA78
2024	2SA154	2024	2SA305J	2024	2SA472C	2024	2SA789
2024	2SA155	2024	2SA305K	2024	2SA472D	2024	2SA78B
2024	2SA156	2024	2SA305L	2024	2SA472E	2024	2SA78C
2024	2SA157	2024	2SA305M	2024	2SA474	2024	2SA78D
2024	2SA159	2024	2SA305OR	2024	2SA474A	2024	2SA80
2024	2SA161	2024	2SA305R	2024	2SA474B	2024	2SA82
2024	2SA162	2024	2SA305X	2024	2SA474C	2024	2SA83
2024	2SA163	2024	2SA305Y	2024	2SA474D	2024	2SA85
2024	2SA164	2024	2SA308	2024	2SA474E	2024	2SA866
2024	2SA165	2024	2SA309	2024	2SA474F	2024	2SA87
2024	2SA166	2024	2SA310	2024	2SA474G	2024	2SA871
2024	2SA175	2024	2SA313	2024	2SA474GN	2024	2SA880
2024	2SA176	2024	2SA314	2024	2SA474H	2024	2SA888
2024	2SA188	2024	2SA315	2024	2SA474J	2024	2SA889
2024	2SA189	2024	2SA316	2024	2SA474K	2024	2SA890
2024	2SA19	2024	2SA325	2024	2SA474L	2024	2SA891
2024	2SA20	2024	2SA326	2024	2SA474M	2024	2SA90
2024	2SA201	2024	2SA329	2024	2SA474OR	2024	2SA92
2024	2SA201-0	2024	2SA329A	2024	2SA474R	2024	2SA93
2024	2SA201-N	2024	2SA329B	2024	2SA474AX	2024	2SA94
2024	2SA201-0	2024	2SA331	2024	2SA474Y	2024	2SB185-0
2024	2SA201-OR	2024	2SA335	2024	2SA476	2024	2SB384
2024	2SA201A	2024	2SA337	2024	2SA476A	2024	2SB385
2024	2SA201B	2024	2SA338	2024	2SA476B	2024	2SB400BK
2024	2SA201CL	2024	2SA339	2024	2SA476C	2024	2SB400C
2024	2SA201D	2024	2SA343	2024	2SA476D	2024	2SB400D
2024	2SA201E	2024	2SA344	2024	2SA476E	2024	2SB400E
2024	2SA201F	2024	2SA35	2024	2SA476F	2024	2SB400F
2024	2SA201G	2024	2SA350	2024	2SA476G	2024	2SB400G
2024	2SA201GN	2024	2SA350A	2024	2SA476GN	2024	2SB400GN
2024	2SA201H	2024	2SA350C	2024	2SA476H	2024	2SB400H
2024	2SA201J	2024	2SA350H	2024	2SA476J	2024	2SB400J
2024	2SA201K	2024	2SA350OR	2024	2SA476K	2024	2SB400L
2024	2SA201L	2024	2SA350T	2024	2SA476L	2024	2SB400M
2024	2SA201M	2024	2SA350TY	2024	2SA476M	2024	2SB400OR
2024	2SA201N	2024	2SA350Y	2024	2SA476OR	2024	2SB400R
2024	2SA201O	2024	2SA351	2024	2SA476R	2024	2SB400X
2024	2SA201OR	2024	2SA351A	2024	2SA476X	2024	2SB400Y
2024	2SA201R	2024	2SA351B	2024	2SA476Y	2024	2SB43B
2024	2SA201TV	2024	2SA351GR	2024	2SA477	2024	2SB43C
2024	2SA201TVO	2024	2SA352	2024	2SA477A	2024	2SB43E
2024	2SA201TVO	2024	2SA352A	2024	2SA477B	2024	2SB43F
2024	2SA201X	2024	2SA352B	2024	2SA477C	2024	2SB43G
2024	2SA201Y	2024	2SA353	2024	2SA477D	2024	2SB43GN
2024	2SA202-OR	2024	2SA353-AC	2024	2SA477E	2024	2SB43H
2024	2SA202A	2024	2SA353A	2024	2SA477F	2024	2SB43J
2024	2SA202AP	2024	2SA353C	2024	2SA477G	2024	2SB43K
2024	2SA202B	2024	2SA354	2024	2SA477GN	2024	2SB43L
2024	2SA202C	2024	2SA354A	2024	2SA477H	2024	2SB43M
2024	2SA202D	2024	2SA354B	2024	2SA477J	2024	2SB430R
2024	2SA202D-4	2024	2SA355	2024	2SA477K	2024	2SB43R
2024	2SA202E	2024	2SA355A	2024	2SA477L	2024	2SB43X
2024	2SA202F	2024	2SA356	2024	2SA477M	2024	2SB43Y
2024	2SA202G	2024	2SA357	2024	2SA477OR	2024	2SB444
2024	2SA202GN	2024	2SA358	2024	2SA477R	2024	2SB444A
2024	2SA202H	2024	2SA359	2024	2SA477X	2024	2SB444B
2024	2SA202J	2024	2SA364	2024	2SA477Y	2024	2SB44A
2024	2SA202K	2024	2SA365	2024	2SA478	2024	2SB44C
2024	2SA202L	2024	2SA366	2024	2SA479	2024	2SB44D
2024	2SA202M	2024	2SA366A	2024	2SA49	2024	2SB44E
2024	2SA202OR	2024	2SA366B	2024	2SA517	2024	2SB44F
2024	2SA202R	2024	2SA366C	2024	2SA518	2024	2SB44G
2024	2SA202X	2024	2SA366D	2024	2SA52	2024	2SB44GN
2024	2SA202Y	2024	2SA366E	2024	2SA522	2024	2SB44H
2024	2SA203	2024	2SA366F	2024	2SA524	2024	2SB44J
2024	2SA203A	2024	2SA366G	2024	2SA53	2024	2SB44K
2024	2SA203AA	2024	2SA366GN	2024	00002SA550	2024	2SB44L
2024	2SA203B	2024	2SA366H	2024	2SA556	2024	2SB44M
2024	2SA203C	2024	2SA366J	2024	2SA565K	2024	2SB440R
2024	2SA203D	2024	2SA366K	2024	2SA567	2024	2SB44X
2024	2SA203P	2024	2SA366L	2024	2SA567A	2024	2SB44Y
2024	2SA21	2024	2SA366M	2024	2SA567B	2024	2SB46A
2024	2SA213	2024	2SA366OR	2024	2SA567C	2024	2SB46B
2024	2SA214	2024	2SA366R	2024	2SA567D	2024	2SB46C
2024	2SA216	2024	2SA366X	2024	2SA567E	2024	2SB46D
2024	2SA236	2024	2SA366Y	2024	2SA567F	2024	2SB46E
2024	2SA237	2024	2SA367	2024	2SA567G	2024	2SB46F
2024	2SA238	2024	2SA368	2024	2SA567GN	2024	2SB46G
2024	2SA246	2024	2SA369	2024	2SA567GR	2024	2SB46GN
2024	2SA246V	2024	2SA37	2024	2SA567H	2024	2SB46H
2024	2SA247	2024	2SA376	2024	2SA567J	2024	2SB46J
2024	2SA250	2024	2SA38	2024	2SA567K	2024	2SB46K
2024	2SA251	2024	2SA380	2024	2SA567L	2024	2SB46L
2024	2SA252	2024	2SA381	2024	2SA567M	2024	2SB46M
2024	2SA253	2024	2SA382	2024	2SA567OR	2024	2SB46OR
2024	2SA254	2024	2SA383	2024	2SA567R	2024	2SB46R
2024	2SA255	2024	2SA384	2024	2SA567X	2024	2SB46X
2024	2SA256	2024	2SA385	2024	2SA567Y	2024	2SB46Y
2024	2SA257	2024	2SA385A	2024	2SA56a	2024	2SB47C
2024	2SA258	2024	2SA385D	2024	2SA57	2024	2SB47A
2024	2SA259	2024	2SA39	2024	2SA58	2024	2SB47B
2024	2SA260	2024	2SA391	2024	2SA59	2024	2SB47C
2024	2SA261	2024	2SA392	2024	2SA60	2024	2SB47E
2024	2SA262	2024	2SA393	2024	2SA67	2024	2SB47F
2024	2SA263	2024	2SA393A	2024	2SA677	2024	2SB47G
2024	2SA264	2024	2SA394	2024	2SA69	2024	2SB47GN
2024	2SA265	2024	2SA395	2024	2SA704	2024	2SB47H
2024	2SA266	2024	2SA398	2024	2SA71	2024	2SB47J
2024	2SA267	2024	2SA399	2024	2SA71A	2024	2SB47K
2024	2SA268	2024	2SA400	2024	2SA71AB		
2024	2SA269	2024	2SA412	2024	2SA71AC		
2024	2SA270	2024	2SA427	2024	2SA71B		
2024	2SA271	2024	2SA42B	2024	2SA71BS		
		2024	2SA43	2024	2SA71D		

RS 276-	Industry Standard No.	RS 276-	Industry Standard No.	RS 276-	Industry Standard No.	RS 276-	Industry Standard No.
2024	28B47L	2024	23B59J	2024	19-020-001	2024	48-134859
2024	28B47M	2024	23B59K	2024	19-020-002	2024	48-134860
2024	28B47OR	2024	23B59L	2024	19-020-005	2024	48-134861
2024	28B47R	2024	23B59M	2024	19A115087-P1	2024	48-134862
2024	28B47X	2024	23B59OR	2024	19A115098	2024	48-134880
2024	28B47Y	2024	23B59R	2024	19A115098-P1	2024	48-63029A90
2024	28B48A	2024	23B59X	2024	19A115099-P1	2024	48-63075A74
2024	28B48B	2024	23B59Y	2024	020-00023	2024	48-63078A65
2024	28B48C	2024	28B60B	2024	21-32	2024	48-63082A24
2024	28B48D	2024	28B60C	2024	21-33	2024	48-8613-3
2024	28B48E	2024	28B60D	2024	21A040-031	2024	48-97046A08
2024	28B48F	2024	28B60E	2024	21M006	2024	48-97046A09
2024	28B48G	2024	28B60F	2024	022-3516-380	2024	48-97162A03
2024	28B48GN	2024	28B60G	2024	022-5311-770	2024	48-97238A01
2024	28B48H	2024	28B60GN	2024	022-5311-780	2024	48-97238A02
2024	28B48J	2024	28B60H	2024	022-5311-798	2024	48-97238A03
2024	28B48K	2024	28B60J	2024	022.3516-380	2024	48C125237
2024	28B48L	2024	28B60K	2024	22N1215	2024	48K35P1
2024	28B48M	2024	28B60L	2024	24NW152	2024	48K36P1
2024	28B48OR	2024	28B60M	2024	24MW205	2024	48K36P3
2024	28B48R	2024	28B60OR	2024	24MW271	2024	48K43P3
2024	28B48X	2024	28B60R	2024	24MW303	2024	48K63P4
2024	28B48Y	2024	28B60X	2024	24MW352	2024	48R10073A02
2024	28B49A	2024	28B60Y	2024	24MW353	2024	48R10074A02
2024	28B49B	2024	28B75O-4	2024	24MW368	2024	488134458
2024	28B49C	2024	28B75D	2024	24MW61	2024	48X97238A01
2024	28B49D	2024	28B75E	2024	24MW74	2024	48X97238A02
2024	28B49E	2024	28B75G	2024	24MW816	2024	48X97238A03
2024	28B49F	2024	28B75GN	2024	24MW991	2024	50A102
2024	28B49G	2024	28B75J	2024	27T412	2024	50A103K
2024	28B49GN	2024	28B75L	2024	28A477	2024	50A52
2024	28B49H	2024	28B75M	2024	31-0005	2024	50P2
2024	28B49J	2024	28B75OR	2024	31-0041	2024	50P3
2024	28B49K	2024	28B75R	2024	31-0042	2024	051-0079
2024	28B49L	2024	28B75X	2024	31-0065	2024	51P2
2024	28B49M	2024	28B75Y	2024	31-0108	2024	51P4
2024	28B49OR	2024	28B9OA	2024	31-0123	2024	55P2
2024	28B49R	2024	28B9OB	2024	31-0124	2024	55P3
2024	28B49X	2024	28B9OC	2024	31-0132	2024	57A130
2024	28B49Y	2024	28B9OD	2024	31-0134	2024	57A131
2024	28B503-R	2024	28B9OE	2024	31-0139	2024	57A132
2024	28B503A-R	2024	28B9OF	2024	31-0178	2024	57A16
2024	28B50A	2024	28B9OG	2024	31-0184	2024	57A168
2024	28B50B	2024	28B90GN	2024	31-0190	2024	57A186
2024	28B50C	2024	28B9OH	2024	31-0191	2024	57A2-1
2024	28B50D	2024	28B9OJ	2024	31-0217	2024	57A2-19
2024	28B50E	2024	28B9OK	2024	31-0228	2024	57A5-1
2024	28B50F	2024	28B9OL	2024	54-8001-43	2024	57A5-10
2024	28B50G	2024	28B9OM	2024	37?1	2024	57A5-2
2024	28B50GN	2024	28B900R	2024	42-0094-399	2024	57A5-4
2024	28B50H	2024	28B9OR	2024	42-21403	2024	57A5-9
2024	28B50J	2024	28B9OX	2024	42-23965	2024	57A6-16
2024	28B50K	2024	28B9OY	2024	44-13	2024	57016
2024	28B50L	2024	28B91A	2024	46-86102-3	2024	5705-1
2024	28B50M	2024	28B91B	2024	46-8613-3	2024	5705-10
2024	28B50OR	2024	28B91C	2024	46-8625-1	2024	5705-4
2024	28B50R	2024	28B91D	2024	46-8625-2	2024	5705-9
2024	28B50X	2024	28B91E	2024	46-8625-3	2024	5706-16
2024	28B50Y	2024	28B91F	2024	46-8625-4	2024	57D130
2024	28B516(C)	2024	28B91G	2024	46-8625-5	2024	57D131
2024	28B516(D)	2024	28B91GN	2024	46-8625-6	2024	57D132
2024	28B516CD(P)	2024	28B91H	2024	46-86256-1	2024	57D168
2024	28B54A	2024	28B91J	2024	46-86256-2	2024	57D186
2024	28B54BA	2024	28B91K	2024	46-8636-3	2024	57D187
2024	28B54C	2024	28B91L	2024	48-123522	2024	57D188
2024	28B54D	2024	28B91M	2024	48-123536	2024	57D5-1
2024	28B54G	2024	28B910R	2024	48-124255	2024	57D5-2
2024	28B54GN	2024	28B91R	2024	48-124256	2024	57D5-4
2024	28B54H	2024	28B91X	2024	48-124305	2024	61-260039
2024	28B54J	2024	28B91Y	2024	48-124310	2024	61-260039A
2024	28B54K	2024	0028C203AA	2024	48-124311	2024	61B003-1
2024	28B54L	2024	2T14A	2024	48-124312	2024	61P1
2024	28B54L1	2024	03-57-001	2024	48-124316	2024	61P10
2024	28B54M	2024	03-57-002	2024	48-124346	2024	65B
2024	28B54OR	2024	03-57-101	2024	48-124347	2024	065B-12
2024	28B54R	2024	03-57-201	2024	48-124348	2024	065B-12-7
2024	28B54X	2024	3N129	2024	48-124349	2024	65B-70
2024	28B556-R	2024	3N134	2024	48-124350	2024	65B-70-12
2024	28B55A	2024	3N21	2024	48-124351	2024	65B-70-12-7
2024	28B55B	2024	4-279	2024	48-124352	2024	65B1
2024	28B55C	2024	4-280	2024	48-124360	2024	65B10
2024	28B55D	2024	4-686256-1	2024	48-124364	2024	65B10R
2024	28B55E	2024	4-686256-2	2024	48-124365	2024	65B11
2024	28B55F	2024	4-686256-3	2024	48-124366	2024	65B119
2024	28B55G	2024	4J56	2024	48-124367	2024	65B121
2024	28B55GN	2024	6-50	2024	48-124377	2024	65B13
2024	28B55H	2024	6-62F	2024	48-124388	2024	65B13P
2024	28B55J	2024	6-62P	2024	48-125228	2024	65B14
2024	28B55K	2024	6-62X	2024	48-125278	2024	65B14-7B
2024	28B55L	2024	6-70	2024	48-128093	2024	65B15
2024	28B55M	2024	6-71	2024	48-128095	2024	65B15L
2024	28B55OR	2024	6-72	2024	48-128096	2024	65B16
2024	28B55R	2024	6-89	2024	48-134101	2024	65B16-2
2024	28B55X	2024	6412680	2024	48-134372	2024	65B17
2024	28B55Y	2024	07-3350-57	2024	48-134405	2024	65B17-1
2024	28B56CK	2024	7-3401A	2024	48-134406	2024	65B18
2024	28B56D	2024	7-7340102	2024	48-134414	2024	65B182
2024	28B56E	2024	8E(AUTOMATIC)	2024	48-134434	2024	65B19
2024	28B56F	2024	8P(AUTOMATIC)	2024	48-134454	2024	65B19Q
2024	28B56G	2024	09-300002	2024	48-134456	2024	65B3
2024	28B56GN	2024	09-300006	2024	48-134457	2024	65B4
2024	28B56H	2024	09-300007	2024	48-134479	2024	65B5
2024	28B56J	2024	09-300011	2024	48-134480	2024	65B6
2024	28B56K	2024	09-300012	2024	48-134481	2024	65B7
2024	28B56L	2024	09-300015	2024	48-134508	2024	65B9
2024	28B56OR	2024	09-300016	2024	48-134514	2024	65B9G
2024	28B56R	2024	09-300027	2024	48-134521	2024	65B0
2024	28B56X	2024	09-300029	2024	48-134522	2024	73A03
2024	28B57A	2024	09-300078	2024	48-134536	2024	76
2024	28B57B	2024	09-300079	2024	48-134545	2024	80-104900
2024	28B57C	2024	09-30011	2024	48-134547	2024	80-105200
2024	28B57D	2024	9-9101	2024	48-134561	2024	80-105300
2024	28B57E	2024	9-9102	2024	48-134576	2024	81-46125001-1
2024	28B57F	2024	9-9103	2024	48-134577	2024	86-108-2
2024	28B57G	2024	10-28A49	2024	48-134578	2024	86-109-2
2024	28B57GN	2024	12-1-189	2024	48-134600	2024	86-111-2
2024	28B57H	2024	12-1-242	2024	48-134601	2024	86-116-2
2024	28B57J	2024	12-1-243	2024	48-134602	2024	86-280-2
2024	28B57K	2024	12-1-244	2024	48-134605	2024	86-281-2
2024	28B57L	2024	12-1-257	2024	48-134635	2024	86-282-2
2024	28B57OR	2024	13-14085-31	2024	48-134680	2024	86-311-2
2024	28B57R	2024	13-14085-32	2024	48-134681	2024	86-449-2
2024	28B57X	2024	13-14085-33	2024	48-134682	2024	86X0011-001
2024	28B57Y	2024	13-14085-93	2024	48-134683	2024	86X0013-001
2024	28B59A	2024	13-14986-1	2024	48-134684	2024	87-0015
2024	28B59B	2024	13-14887-1	2024	48-134697	2024	87-0016
2024	28B59D	2024	13-14889-1	2024	48-134711	2024	87-0017
2024	28B59E	2024	13-18951-1	2024	48-134796	2024	089-220
2024	28B59F	2024	13-18951-2	2024	48-134797	2024	95-111
2024	28B59G	2024	13-67599-9	2024	48-134798	2024	96-5094-01(TRANS)
2024	28B59GN	2024	14-566-08			2024	96XZ6051-28N
		2024	14-568-04				
		2024	14-582-01				
		2024	019-003777				

RS 276-	Industry Standard No.	RS 276-	Industry Standard No.	RS 276-	Industry Standard No.	RS 276-	Industry Standard No.
2024	96XZ6051-35N	2024	324-0145	2024	2057B2-42	2024	40488
2024	96XZ6051-36N	2024	325-0028-82	2024	2057B2-48	2024	42302
2024	96XZ6053-10N	2024	325-0036-564	2024	2057B2-50	2024	42311
2024	96XZ6053-24N	2024	325-0036-565	2024	2057B2-51	2024	48937-2
2024	99-101	2024	325-1375-10	2024	2057B2-60	2024	48939-2
2024	99-102	2024	325-1375-11	2024	2057B2-65	2024	61010-2-1
2024	99-103	2024	325-1375-12	2024	2057B2-66	2024	61102-0
2024	0101-0034	2024	325-1376-54	2024	2057B2-67	2024	69107-42
2024	0101-0222	2024	325-1376-55	2024	2057B2-68	2024	69107-44
2024	0102	2024	353-9001-001	2024	2057B2-70	2024	72797-80
2024	104-17	2024	353-9001-002	2024	2057B2-71	2024	72879-39
2024	104-19	2024	353-9001-003	2024	2057B2-77	2024	72879-40
2024	104-21	2024	353-9002-002	2024	2057B2-79	2024	080001
2024	111-6935	2024	421-16	2024	2057B2-80	2024	080072
2024	112-002	2024	421-17	2024	2057B2-81	2024	080206
2024	115-275	2024	421-20	2024	2057B2-88	2024	080224
2024	116-072	2024	421-20B	2024	2058	2024	080225
2024	116-756	2024	421-21	2024	2478	2024	080228
2024	120-00190	2024	421-21B	2024	2478A	2024	080236
2024	120-00213	2024	421-22	2024	2478B	2024	080253
2024	120-002513	2024	421-22B	2024	2482	2024	080274
2024	120-002515	2024	421-26	2024	2488	2024	080275
2024	120-002656	2024	422-1403	2024	2488A	2024	080276
2024	120-004048	2024	429-0093-69	2024	2489	2024	080277
2024	120-004492	2024	429-0094-39	2024	2495-012	2024	81506-4
2024	121-1005	2024	455-1	2024	2495-013	2024	81506-4A
2024	121-1019	2024	576-0003-008	2024	2495-200	2024	81506-4B
2024	121-150	2024	576-0003-009	2024	2603-183	2024	81506-4C
2024	121-153	2024	576-0003-013	2024	2606-292	2024	81506-7
2024	121-154	2024	576-0003-014	2024	2791	2024	81506-7A
2024	121-161	2024	576-0003-015	2024	2799	2024	81506-7B
2024	121-162	2024	576-0003-024	2024	2904-016	2024	81506-7C
2024	121-186	2024	576-2000-990	2024	2904-029	2024	81506-8
2024	121-187	2024	576-2000-993	2024	3008(CB)	2024	81506-8A
2024	121-189	2024	601-054	2024	3009	2024	81506-8B
2024	121-228	2024	610-052	2024	3009(MIXER)	2024	81506-8C
2024	121-229	2024	610-052-1	2024	3010	2024	94038
2024	121-230	2024	610-056	2024	3010(E.F.JOHNSON)	2024	95116
2024	121-231	2024	610-056-1	2024	3010(IF)	2024	95118
2024	121-232	2024	610-056-2	2024	3019(EFJOHNSON)	2024	95119
2024	121-233	2024	610-056-3	2024	3551	2024	95120
2024	121-240	2024	610-056-4	2024	3551A	2024	95122
2024	121-256	2024	610-074	2024	3551A(BLU)	2024	95123
2024	121-257	2024	610-074-1	2024	3551A-BLU	2024	99101
2024	121-258	2024	612-60039	2024	3551A-GRN	2024	99120
2024	121-259	2024	612-60039A	2024	3564	2024	99121
2024	121-260	2024	617-56	2024	3586	2024	100678
2024	121-261	2024	665B-1-5L	2024	3637	2024	101089
2024	121-262	2024	690V010H42	2024	3686	2024	104009
2024	121-263	2024	690V034E29	2024	3851	2024	104059
2024	121-289	2024	690V040H57	2024	3907(2N404A)	2024	112041
2024	121-290	2024	690V040H58	2024	3961	2024	115275
2024	121-292	2024	690V040H59	2024	4041-000-80100	2024	116072
2024	121-293	2024	690V040H60	2024	4363	2024	116756
2024	121-294	2024	690V047H54	2024	4363BLU	2024	117724
2024	121-295	2024	690V047H55	2024	4363GRN	2024	117824
2024	121-296	2024	690V056H89	2024	4363ORN	2024	119526
2024	121-297	2024	690V057H59	2024	4363WHT	2024	122725
2024	121-298	2024	690V057H62	2024	4364	2024	123244
2024	121-299	2024	690V063H14	2024	4365	2024	123511
2024	121-360	2024	690V063H15	2024	4366	2024	124097
2024	121-381	2024	690V066H44	2024	4367	2024	125330
2024	121-44	2024	690V066H45	2024	4454	2024	126185
2024	121-45	2024	690V066H49	2024	4456	2024	128543
2024	121-491	2024	690V068H29	2024	4457	2024	128938
2024	121-492	2024	690V073H59	2024	4501	2024	138946
2024	121-493	2024	690V073H85	2024	4509	2024	162002-040
2024	121-494	2024	690V077H34	2024	4545	2024	162002-041
2024	121-54	2024	690V077H35	2024	4545BLU	2024	162002-042
2024	121-554	2024	690V077H36	2024	4545WHT	2024	162002-40
2024	121-555	2024	690V080H57	2024	4586	2024	162002-41
2024	121-62	2024	690V080H40	2024	4589	2024	165667
2024	121-65	2024	690V084H60	2024	4595	2024	166908
2024	121-66	2024	690V085H42	2024	4603	2024	166909
2024	121-67	2024	690V102H96	2024	4604	2024	171005
2024	122-1648	2024	690V105H-24	2024	4605	2024	198010-1
2024	145-T1B	2024	690V119H94	2024	4605RED	2024	221856
2024	151-045	2024	800-50500	2024	4621	2024	223487
2024	151-1002	2024	801-04900	2024	4632	2024	225594A
2024	154A3676	2024	801-05200	2024	4677	2024	232681
2024	154A3677	2024	801-05300	2024	4686-256-1	2024	249588
2024	154A3679-5110	2024	916-31019-3	2024	4686-256-2	2024	310157
2024	229-5100-227	2024	916-31019-3B	2024	4686-256-3	2024	310158
2024	260D0404	2024	921-3B	2024	6154	2024	310221
2024	260D04501	2024	921-4B	2024	6155	2024	310224
2024	260P1300	2024	921-65	2024	6162	2024	0573002
2024	260P13001	2024	921-65A	2024	9403-8	2024	573303
2024	297V011H01	2024	921-65B	2024	9510-3	2024	573329
2024	297V012H01	2024	921-66	2024	11527-5	2024	573330
2024	297V012H10	2024	921-66A	2024	11607-2	2024	573398
2024	297V012H15	2024	921-66B	2024	11675-6	2024	573402
2024	297V026H01	2024	930X6	2024	12118-9	2024	0573427
2024	297V035H01	2024	951-1(SYLVANIA)	2024	12124-2	2024	0573428
2024	297V036H01	2024	964-16598	2024	12124-3	2024	0573471
2024	297V036H02	2024	965B16-2	2024	12124-4	2024	0573518
2024	297V038H06	2024	972X6	2024	12125-7	2024	601054
2024	297V038H10	2024	972X7	2024	12180	2024	604112
2024	297V038H11	2024	972X8	2024	18493	2024	610052
2024	297V038H12	2024	1003(JULIETTE)	2024	23785	2024	610056
2024	297V042H03	2024	1007(JULIETTE)	2024	23785-1	2024	610074
2024	297V042H04	2024	1010-89	2024	24198	2024	740946
2024	297V045H01	2024	1013-16	2024	25566-01	2024	740947
2024	297V045H02	2024	1033-1	2024	25642-020	2024	740948
2024	297V054H01	2024	1033-2	2024	25642-030	2024	772768
2024	297V054H02	2024	1033-3	2024	25642-031	2024	81506A
2024	297V055H01	2024	1033-4	2024	25642-040	2024	815064A
2024	297V063C01	2024	1122-96	2024	25642-041	2024	815064B
2024	297V065C01	2024	1241-719	2024	25642-110	2024	815064C
2024	297V065C02	2024	1465B	2024	25642-115	2024	81506B
2024	297V065C03	2024	1465B-12	2024	25642-120	2024	815068A
2024	297V070C01	2024	1465B-12-8	2024	30213	2024	815068B
2024	297V077C01	2024	1865D	2024	30214	2024	815068C
2024	310-068	2024	1865B-12	2024	30215	2024	980372
2024	310-123	2024	1865B-127	2024	30217	2024	980373
2024	310-124	2024	1865BL	2024	30221	2024	980374
2024	310-139	2024	1865BLB	2024	30222	2024	980626
2024	310-190	2024	1951-17	2024	30223	2024	980833
2024	310-191	2024	2015-00	2024	30230	2024	980834
2024	310-68	2024	2020-00	2024	30247	2024	980855
2024	324-0027	2024	2021-00	2024	30274	2024	980958
2024	324-0028	2024	2057A100-10	2024	34048	2024	980959
2024	324-0038	2024	2057A2-263	2024	34098	2024	982150
2024	324-0079	2024	2057A2-37	2024	34099	2024	982267
2024	324-0098	2024	2057A2-60	2024	34100	2024	982289
2024	324-0099	2024	2057A2-65	2024	35168	2024	983256
2024	324-0106	2024	2057A2-66	2024	35818	2024	983271
2024	324-0121	2024	2057A2-80	2024	36559	2024	983272
2024	324-0129	2024	2057B2-104	2024	36560	2024	984685
2024	324-0130	2024	2057B2-105	2024	36563	2024	985445
2024	324-0136	2024	2057B2-118	2024	38093	2024	985446
2024	324-0137	2024	2057B2-141	2024	38533	2024	985611
2024	324-0138	2024	2057B2-35	2024	40261	2024	1061854-2
		2024	2057B2-37	2024	40262	2024	1471104-5
		2024	2057B2-41	2024	40263		

RS 276-	Industry Standard No.	RS 276-	Industry Standard No.	RS 276-	Industry Standard No.	RS 276-	Industry Standard No.
2024	1471104-6	2025	HP51	2025	TR-1036-3	2025	13-39047-3
2024	1471104-7	2025	HR-70	2025	TR-182(OLSON)	2025	13-40345-1
2024	1471104-8	2025	HR104961C	2025	TR-56	2025	13-41629-4
2024	1473598-1	2025	HT104961B	2025	TR-77(XSTR)	2025	13-43791-1
2024	1473686-1	2025	I-B198	2025	TR02057-3	2025	018-00004
2024	2000646-106	2025	IB198	2025	TR2327393	2025	19-020-102
2024	2000646-109	2025	IR-TR56	2025	TV-74	2025	21A112-048
2024	2000648-23	2025	IRTR-56	2025	WEP246	2025	21M025
2024	2001653-20	2025	IRTR56	2025	WEP474	2025	21M026
2024	2001653-21	2025	M0TMJE371	2025	WEP474MP	2025	21M028
2024	2002336-19	2025	MJB100	2025	WEP700	2025	21M345
2024	2002403-19	2025	MJE-371	2025	WEP85007	2025	21M395
2024	2008292-87	2025	MJE230	2025	ZEN203	2025	21M443
2024	2076393	2025	MJE231	2025	ZEN211	2025	21M465
2024	2091241-005	2025	MJE232	2025	002-012100	2025	21M028
2024	2091241-0719)	2025	MJE233	2025	2N4918	2025	33-096
2024	2091241-5	2025	MJE234	2025	2N4919	2025	34-6002-42
2024	2091241-5A	2025	MJE235	2025	2N4920	2025	34-6016-13
2024	2495012	2025	MJE3370	2025	2N5193	2025	34-6016-65
2024	2495013	2025	MJE33371	2025	2N5194	2025	42-28213
2024	2495200	2025	MJE370	2025	2N5195	2025	46-86235-3
2024	2495378	2025	MJE492	2025	2N5782	2025	46-86411-3
2024	2495379	2025	MJE493	2025	2N5783	2025	047-1
2024	2495488-1	2025	MJE711	2025	2SA496	2025	047-1(SYLVANIA)
2024	2495488-2	2025	MJE9450	2025	2SA496(0)	2025	48-134987
2024	5404520-81	2025	M0TMJE371	2025	2SA496-0	2025	48-137153
2025	A496(JAPAN)	2025	P1V-	2025	2SA496-0	2025	48-137154
2025	A496-0	2025	P1V-1	2025	2SA496-ORG	2025	48-137155
2025	A496-R	2025	P1V-2	2025	2SA496-R	2025	48-137156
2025	A496-Y	2025	P1V-3	2025	2SA496-RED	2025	48-137157
2025	A496Y	2025	P2T	2025	2SA496-Y	2025	48-137256
2025	A505	2025	P2T-1	2025	2SA496-YEL	2025	48-137258
2025	A505-0	2025	P2T-2	2025	2SA4964	2025	48-137303
2025	A505-R	2025	P2T-3	2025	2SA4960	2025	48-137304
2025	A505-Y	2025	P2T-4	2025	2SA496ORG	2025	48-137550
2025	A715	2025	P3M	2025	2SA496R	2025	48-40383J01
2025	A715A	2025	P3N	2025	2SA496RED	2025	48-41785J03
2025	A715B	2025	P3N-1	2025	2SA496Y	2025	48-869582
2025	A715C	2025	P3N-2	2025	2SA496YEL	2025	48M355055
2025	A715D	2025	P3N-3	2025	2SA505	2025	48M869582
2025	A715WTA	2025	P3N-4	2025	2SA505-0	2025	488137308
2025	A715WTB	2025	P3P	2025	2SA505-ORG	2025	488137331
2025	A715WTC	2025	P3P-1	2025	2SA505-R	2025	488137472
2025	A715WTD	2025	P3P-2	2025	2SA505-RED	2025	488155116
2025	A738C	2025	P3P-3	2025	2SA505-Y	2025	48840383J01
2025	A738D	2025	P3P-4	2025	2SA505-YEL	2025	63-11990
2025	A743	2025	P3P-5	2025	2SA5050	2025	65-80001
2025	A743A	2025	P58	2025	2SA505R	2025	69-1819
2025	A743AA	2025	P48	2025	2SA505Y	2025	079
2025	A743AB	2025	P48	2025	2SA636(4)K	2025	86-345-2
2025	A743AC	2025	P4U	2025	2SA636(4)L	2025	86-396-2
2025	A743AD	2025	P4V	2025	2SA636(L)	2025	86-507-2
2025	A743B	2025	P4V-1	2025	2SA636(M)	2025	86004
2025	A743C	2025	P4W-1	2025	2SA636I	2025	86X0075-001
2025	A743D	2025	P4W-2	2025	2SA6361,K	2025	0101-0448
2025	A8C	2025	P5L	2025	2SA636UK	2025	0101-0448A
2025	AR-29	2025	P5R	2025	2SA700(B)	2025	0101-0466
2025	AR23(TRANSISTOR)	2025	PLE37	2025	2SA715	2025	114P1P
2025	AR25	2025	PTC111	2025	2SA715A	2025	0121(AIRLINE)
2025	AR25G	2025	PTC142	2025	2SA715B	2025	121-707
2025	B-1695	2025	QP-13	2025	2SA715C	2025	121-709
2025	B474	2025	RE313-IC	2025	2SA715D	2025	121-992(V.REG)
2025	B474-2	2025	RE41	2025	2SA715WB	2025	0123
2025	B474-3	2025	RE43	2025	2SA715WT	2025	0123(WARDS)
2025	B474-4	2025	RE83	2025	2SA715WT(C,B)	2025	132-024
2025	B474-6D	2025	RE83MP	2025	2SA715WTA	2025	132-079
2025	B474MP	2025	REN185	2025	2SA715WTB	2025	149P1
2025	B474S	2025	REN187	2025	2SA715WTC	2025	149P3
2025	B474V10	2025	REN226	2025	2SA715WTD	2025	162P1
2025	B474Y4	2025	R8-2025	2025	2SA738	2025	260P26301
2025	B474Y	2025	RV2356	2025	2SA738B	2025	297V086C04
2025	BD136	2025	SDT3325	2025	2SA738C	2025	417-145
2025	BD138	2025	SDT3505	2025	2SA743	2025	417-225
2025	BD140	2025	SDT3506	2025	2SA743A	2025	624-1
2025	BD151	2025	SDT3550	2025	2SA743AA	2025	642-266
2025	BD152	2025	SDT3552	2025	2SA743AB	2025	644-1
2025	BD156	2025	SDT3553	2025	2SA743AC	2025	690V116H25
2025	BD166	2025	SDT3775	2025	2SA743AD	2025	733-0100-699
2025	BD168	2025	SDT3776	2025	2SA743B	2025	753-0100-699
2025	BD176	2025	SDT3778	2025	2SA743C	2025	753-0160-699
2025	BD178	2025	SDT9004	2025	2SA743D	2025	0770
2025	BD186	2025	SE9570	2025	2SA886	2025	0773
2025	BD188	2025	SE9571	2025	2SA886V	2025	853-0300-634
2025	BD190	2025	SE9572	2025	2SA886VR	2025	9037002151
2025	BD196	2025	SE9573	2025	2SB474	2025	936PNP
2025	BD206	2025	SJ285	2025	2SB474-2	2025	1100
2025	BD208	2025	SJE108	2025	2SB474-3	2025	1100(GE)
2025	BD562	2025	SJE111	2025	2SB474-4	2025	1285
2025	C636	2025	SJE112	2025	2SB474-6D	2025	1285(GE)
2025	CXL185	2025	SJE114	2025	2SB474-OR	2025	02057-1
2025	CXL187	2025	SJE1518	2025	2SB474A	2025	2057A100-45(PNP)
2025	CXL226	2025	SJE202	2025	2SB474B	2025	2057A100-47(PNP)
2025	CXL226MP	2025	SJE210	2025	2SB474C	2025	2057A100-58(PNP)
2025	D27D1	2025	SJE221	2025	2SB474D	2025	2057A100-66(PNP)
2025	D27D2	2025	SJE227	2025	2SB474E	2025	2244-1
2025	D27D3	2025	SJE231	2025	2SB474F	2025	3525
2025	D27D4	2025	SJE241	2025	2SB474G	2025	3525(SEARS)
2025	D43C1	2025	SJE243	2025	2SB474N	2025	3682(RCA)
2025	D43C2	2025	SJE245	2025	2SB474H	2025	7317-1
2025	D43C3	2025	SJE256	2025	2SB474J	2025	7351
2025	D43C4	2025	SJE257	2025	2SB474K	2025	7359-1
2025	D43C5	2025	SJE265	2025	2SB474L	2025	8200-204
2025	D43C6	2025	SJE267	2025	2SB474M	2025	9582
2025	D43C8	2025	SJE273	2025	2SB474MP	2025	30271
2025	EA15X243	2025	SJE275	2025	2SB474OR	2025	040048
2025	EA15X270	2025	SJE276	2025	2SB474R	2025	41179
2025	EA15X8605	2025	SJE277	2025	2SB474S	2025	41342
2025	EC0185	2025	SJE279	2025	2SB474V10	2025	61239
2025	EC0187	2025	SJE283	2025	2SB474V4	2025	000071120
2025	EC0226	2025	SJE288	2025	2SB474X	2025	95262-2
2025	EP101A	2025	SJE403	2025	2SB474Y	2025	95263-2
2025	EP15X24	2025	SJE408	2025	2SB474YE1	2025	132571
2025	ES15051	2025	SJE584	2025	2SB474YEL	2025	134279
2025	E360(ELCOM)	2025	SJE633	2025	2SB632	2025	138193
2025	E861(ELCOM)	2025	SJE723	2025	2SB632K	2025	144033
2025	E868(ELCOM)	2025	SJE736	2025	2SC549	2025	146569
2025	G03-406-C	2025	SJE743	2025	4-283	2025	166919
2025	G03703C	2025	SJE768	2025	7L6-0531-19(PNP)	2025	171033-2(PNP)
2025	GB-29	2025	SJE797	2025	09-300067	2025	171175-1(PNP)
2025	GE-49	2025	SJE799	2025	09-300068	2025	570030
2025	GE-58	2025	SK3082	2025	09-302170	2025	572001
2025	GE-84	2025	SK3191	2025	11-0770	2025	610227-1
2025	GEA-121	2025	SPS4237	2025	11-0773	2025	657179
2025	HA-00496	2025	STX0020	2025	13-14085-86	2025	657181
2025	HA-00634	2025	T-345	2025	13-23508-0	2025	760276
2025	HA-00636	2025	T-396	2025	13-28336-1	2025	910952
2025	HA-505	2025	T345	2025	13-32635-1	2025	986933
2025	HB-00564	2025	T396	2025	13-34040-1A	2025	1417359-1
2025	HEP246	2025	TA7520	2025	13-34047-1	2025	1471141-1
2025	HEP700	2025	TA7741	2025	13-34047-2	2025	1473682-1
2025	HEPS5006	2025	TM185	2025	13-34047-3	2025	2320855
2025	HEPS5007	2025	TM187	2025	13-34047-4	2025(2)	D116
2025	HP12	2025	TM226	2025	13-34373-2	2025(2)	EC0226MP
2025	HP16	2025	TNJ72774	2025	13-36445-3	2025(2)	IRTR94

RS 276-	Industry Standard No.	RS 276-	Industry Standard No.	RS 276-	Industry Standard No.	RS 276-	Industry Standard No.	RS 276-	Industry Standard No.
2025(2)	IRTR94MP	2026	488137160	2027	A546H	2027	BD-132		
2025(2)	R-2SB474	2026	488137168	2027	A548	2027	BD132		
2025(2)	REN226MP	2026	48X97177A11	2027	A551	2027	BD170		
2025(2)	SX3086	2026	66FO69-2	2027	A551C	2027	BD180		
2025(2)	TM226MP	2026	86-423-3	2027	A551D	2027	BD200		
2025(2)	4-48	2026	86-659-2	2027	A551E	2027	BD274		
2025(2)	4-48(SEARS)	2026	96-5262-01	2027	A552	2027	BD276		
2025(2)	09-301030	2026	96-5320-01	2027	A560	2027	BD442		
2025(2)	65-080001	2026	998092-1	2027	A571	2027	BD462		
2025(2)	74Q22881	2026	121-1007	2027	A594	2027	BD464		
2025(2)	1042-10	2026	121-1021	2027	A594-0	2027	BD538		
2025(2)	5001-049	2026	121-980	2027	A594-R	2027	BD576		
2025(2)	22881	2026	121-980-01	2027	A594-Y	2027	BD578		
2026	A486(JAPAN)	2026	0126(WARD8)	2027	A595C	2027	BD580		
2026	A486EBJAPAN)	2026	134-1	2027	A597	2027	BD586		
2026	A645	2026	157P4	2027	A604	2027	BD588		
2026	A717	2026	202-1	2027	A606	2027	BD590		
2026	AR26	2026	202N1	2027	A6068	2027	BDY70		
2026	AR29	2026	202P1	2027	A6532921	2027	BFT60		
2026	BC327	2026	202P2	2027	A6532941	2027	BFT61		
2026	BC328	2026	258-2	2027	A677	2027	BFT62		
2026	BC460	2026	276-2026	2027	A708	2027	BFT79		
2026	BC461	2026	276-2026/RS2026	2027	A708A	2027	BFT80		
2026	BCW79-25	2026	417-221	2027	A708B	2027	BFT81		
2026	BCW80-25	2026	629-3	2027	A708C	2027	BFW44		
2026	BCX10	2026	921-1021	2027	A736	2027	BFW91		
2026	BFW87	2026	3629-3	2027	A742	2027	BFX38		
2026	BFW88	2026	4099A	2027	A800-511-00	2027	BFX39		
2026	CXL189	2026	13299	2027	A800-516-00	2027	BFX40		
2026	CXL211	2026	40362V1	2027	A8015613	2027	BFX41		
2026	D41D1	2026	40362V2	2027	A815185	2027	BFX74A		
2026	D41D10	2026	40595	2027	A815185E	2027	BFX87		
2026	D41D11	2026	44764	2027	A836	2027	BFX88		
2026	D41D14	2026	60458	2027	A836D	2027	BFY18		
2026	D41D2	2026	64074	2027	A836E	2027	BFY64		
2026	D41D4	2026	95258-2	2027	A83P2B	2027	BRC6109		
2026	D41D5	2026	132447	2027	A844	2027	BS817		
2026	D41D7	2026	132488	2027	A844D	2027	BS818		
2026	D41D8	2026	137065	2027	A844E	2027	BSV15		
2026	EC9189	2026	138121	2027	A880-250-107	2027	BSV16		
2026	EC9211	2026	139267	2027	A94063	2027	BSV82		
2026	EP-976	2026	145172	2027	AEX-85715	2027	BSV83		
2026	ES81	2026	171162-195	2027	AMP2970-2	2027	B8W23		
2026	ES83(ELCOM)	2026	610157-4	2027	AR23(PHILCO)	2027	B8W40		
2026	GE-218	2026	610202-1	2027	AR27	2027	B8X40		
2026	GE-253	2026	610202-2	2027	AR304	2027	B8X41		
2026	HEP-83027	2026	610361-5	2027	AR304(RED)	2027	BTX-071		
2026	HEP-83028	2026	1417317-1	2027	AR308VIOLET	2027	BTX-097		
2026	HEP-83029	2026	1445829-501	2027	AR37	2027	C102		
2026	HEP-83030	2026	1445829-502	2027	AT2848	2027	C106(PNP)		
2026	HEP-83031	2026	1471134-1	2027	AT391	2027	C112(PNP)		
2026	HEP-83032	2026	1473629-1	2027	AT392	2027	C118		
2026	HEP83027	2026	1473629-3	2027	AT393	2027	C119		
2026	HEP83028	2026	A-1853-0041-1	2027	AT394	2027	C201		
2026	HEP83029	2027	A054-109	2027	AT395	2027	C202		
2026	HEP83030	2027	A1016	2027	AT396	2027	C301A(PNP)		
2026	HEP83031	2027	A116084	2027	AT397	2027	C302(PNP)		
2026	HEP83032	2027	A116284	2027	AT398	2027	C36577		
2026	HR-74	2027	A119983	2027	AT460	2027	C402		
2026	HR86	2027	A126724	2027	AT461	2027	C41001		
2026	IRTR73	2027	A1473549-1	2027	AT462	2027	C502		
2026	M9641	2027	A1473616-1	2027	AT463	2027	C5620		
2026	MPS-U51	2027	A170	2027	AT464	2027	C5620		
2026	MPS-U55	2027	A1853-0233-1	2027	AT465	2027	C9080		
2026	MPS-U56	2027	A2057B110-9	2027	AT466	2027	C9081		
2026	MPSU51	2027	A2057B112-9	2027	AT467	2027	CDC-9000-1B		
2026	MPSU51A	2027	A2057B114-9	2027	AT468	2027	CDC9000-1B		
2026	MPSU52	2027	A2057B115-9	2027	AT480	2027	CDC9002		
2026	MPSU55	2027	A2057B116-9	2027	AT481	2027	CK4002D		
2026	MPSU56	2027	A2057B121-9	2027	AT482	2027	CJ-5209		
2026	MU9660	2027	A2057B122-9	2027	AT483	2027	C8-6228G		
2026	MU9660B	2027	A2057B145-12	2027	AT484	2027	C8-9012HH		
2026	MU9660T	2027	A2057B163-12	2027	AT485	2027	CK1237		
2026	MU9661	2027	A2482	2027	AT9156	2027	C81256H		
2026	MU9661T	2027	A297074C11	2027	B23-79	2027	C81312G		
2026	P218-2	2027	A297V073C01	2027	B502	2027	C81369		
2026	P2U	2027	A297V073C02	2027	B503	2027	C81465H		
2026	P2U-1	2027	A297V082B03	2027	B510	2027	C83V280		
2026	P2U-2	2027	A29V082B03	2027	B5108	2027	C89012/3490		
2026	P2V	2027	A2M9	2027	B5493957-4	2027	C89012I		
2026	P3K	2027	A28A550P	2027	B5493957-5	2027	C89012F		
2026	P3Y	2027	A28A564P	2027	B5493957-6	2027	C89012FC		
2026	PTC141	2027	A30278	2027	B550	2027	C89012PB		
2026	RE76	2027	A3523	2027	BC139	2027	C89012HB		
2026	RE81	2027	A353-9008-001	2027	BC143	2027	C89012HF		
2026	REN189	2027	A3533	2027	BC160	2027	C89012HG/3490		
2026	REN211	2027	A3533-1	2027	BC160-10	2027	C89012I		
2026	RS-2026	2027	A3616-1	2027	BC160-16	2027	C89011HF		
2026	RT-157	2027	A36577	2027	BC160-6	2027	C89102B		
2026	839509	2027	A4037764-2	2027	BC161	2027	CXL129		
2026	SJE687	2027	A40410	2027	BC161-10	2027	CXL183		
2026	SP84092	2027	A417-138	2027	BC161-16	2027	CXL197		
2026	SP84099	2027	A417-170	2027	BC161-6	2027	D2H		
2026	SP8837	2027	A417-234	2027	BC287	2027	D45H1		
2026	T423	2027	A417-43	2027	BC303	2027	D45H10		
2026	TIP30B	2027	A417756	2027	BC311	2027	D45H2		
2026	TM189	2027	A4478	2027	BC360-10	2027	D45H4		
2026	TM211	2027	A497(JAPAN)	2027	BC360-16	2027	D45H5		
2026	TRO2053-5	2027	A498(JAPAN)	2027	BC360-6	2027	D45H7		
2026	TRO2053-7	2027	A98Y	2027	BC361-10	2027	D45H8		
2026	WEP83027	2027	A501	2027	BC361-6	2027	DDBYO08001		
2026	WEP83031	2027	A503	2027	BC396	2027	D8-86		
2026	X43C248	2027	A503-0	2027	BC404A	2027	E2498		
2026	02P1BC	2027	A503-R	2027	BC404V1	2027	EA0086		
2026	28A503GR	2027	A503-Y	2027	BC477	2027	EA0087		
2026	28A645	2027	A503GR	2027	BC477A	2027	EA15X194		
2026	28A706	2027	A504	2027	BC477V1	2027	EC9129		
2026	28A717	2027	A504-R	2027	BC527	2027	EC9183		
2026	28A835	2027	A504-Y	2027	BC528	2027	EC9197		
2026	28A835H	2027	A504GR	2027	BC534	2027	EC9242		
2026	28A962	2027	A516	2027	BC727	2027	EL264		
2026	3L4-6011-2	2027	A516A	2027	BC728	2027	EN2905		
2026	3L4-6011-3	2027	A527	2027	BCY17	2027	EN3502		
2026	3L4-6011-52	2027	A528	2027	BCY18	2027	EP15X4		
2026	3L4-6011-53	2027	A532	2027	BCY19	2027	EP15X44		
2026	004-1	2027	A532A	2027	BCY21	2027	E8104		
2026	13-34004-1	2027	A532B	2027	BCY22	2027	E8114		
2026	13-34373-1	2027	A532C	2027	BCY23	2027	E851(ELCOM)		
2026	13-36509-1	2027	A532D	2027	BCY24	2027	E867(ELCOM)		
2026	13-37527-1	2027	A532E	2027	BCY25	2027	ET15X33		
2026	20A0059	2027	A532F	2027	BCY26	2027	ET15X38		
2026	20A0060	2027	A537	2027	BCY27	2027	ET15X39		
2026	34-6016-54	2027	A537A	2027	BCY28	2027	ET8-069		
2026	045	2027	A537AA	2027	BCY29	2027	ET8-071		
2026	46-86584-3	2027	A537AB	2027	BCY30	2027	F3549		
2026	48-137160	2027	A537AC	2027	BCY31	2027	FC81795D		
2026	48-137168	2027	A537AH	2027	BCY32	2027	FC89012		
2026	48-137240	2027	A537B	2027	BCY33	2027	FC89012HG		
2026	48-137314	2027	A537C	2027	BCY34	2027	FD-1029-MB		
2026	48-137320	2027	A537H	2027	BCY167	2027	FD-1029-ML		
2026	48-137610	2027	A546	2027	BCY96	2027	FD-1029-RB		
2026	48-869641	2027	A546A	2027	BCY96B	2027	G03-404-B		
2026	48-97177A11	2027	A546E	2027	BCY97	2027	G03-404-C		
2026	48R869641			2027	BCY97B				

RS 276-	Industry Standard No.	RS 276-	Industry Standard No.	RS 276-	Industry Standard No.	RS 276-	Industry Standard No.
2027	GE-21A	2027	Q5100A	2027	TA8210	2027	2N3345
2027	GB-244	2027	QA-11	2027	TA8211	2027	2N3467
2027	GE-250	2027	QA-17	2027	TA8212	2027	2N346B
2027	GE-34	2027	QA-9	2027	TI808E	2027	2N3494
2027	GE-56	2027	QP-31	2027	TIP32B	2027	2N3502
2027	GE56	2027	QT-A0733X0N	2027	TIP34	2027	2N3503
2027	903-404-B	2027	R2270-7693	2027	TIP34A	2027	2N3645
2027	HA7530	2027	R2270-76963	2027	TIP34B	2027	2N3671
2027	HA7531	2027	R264-1	2027	TIP3A	2027	2N3677
2027	HA7630	2027	RCA1AQ2	2027	TIP42	2027	2N3719
2027	HA7631	2027	RCA1A05	2027	TIP42A	2027	2N3720
2027	HA7632	2027	RCA1A08	2027	TIP42B	2027	2N3762
2027	HEP-85002	2027	RCA1A16	2027	TIS61A	2027	2N3763
2027	HEP-85005	2027	RCA1A19	2027	TIS61B	2027	2N3764
2027	HEP-85008	2027	RCA1C06	2027	TIS61C	2027	2N3765
2027	HEP-85009	2027	RCA1C11	2027	TIS61D	2027	2N3774
2027	HEP-85010	2027	RE39	2027	TIS61E	2027	2N3775
2027	HEP242	2027	REN129	2027	TIS61M	2027	2N3776
2027	HEP710	2027	REN183	2027	TIS93M	2027	2N3778
2027	HEPS3012	2027	REN197	2027	TM129	2027	2N3780
2027	HEP85002	2027	RS-2027	2027	TM183	2027	2N3781
2027	HEP85005	2027	R88100	2027	TM197	2027	2N3782
2027	HEP85008	2027	R88102	2027	TM242	2027	2N3867
2027	HEP85009	2027	R88104	2027	TQ63	2027	2N3868
2027	HEP85010	2027	R88106	2027	TQ63A	2027	2N4026
2027	HEP85013	2027	R88108	2027	TQ64	2027	2N4027
2027	HEP85023	2027	R88110	2027	TQ64A	2027	2N4028
2027	HP10	2027	R88112	2027	TR-04C	2027	2N4029
2027	HP12(PHILCO)	2027	RT5230	2027	TR-04C(PENNCREST)	2027	2N4030
2027	HS40032	2027	RV1472	2027	TR-08(PENNCREST)	2027	2N4031
2027	HT104971A	2027	RV2069	2027	TR-1036-1	2027	2N4032
2027	HT104971A-0	2027	S-437	2027	TR-1036-2	2027	2N4033
2027	HT104971A-0	2027	S-437F	2027	TR-165(OLSON)	2027	2N4036
2027	HT104971AO	2027	S1430	2027	TR-28	2027	2N4037
2027	HT104971AO	2027	S1431	2027	TR-8020	2027	2N4234
2027	HT70011100	2027	S1520	2027	TR-88	2027	2N4235
2027	I6342	2027	S1698	2027	TRO-202B-5	2027	2N4236
2027	IP20-0217	2027	S1863	2027	TRO-2057-3	2027	2N4314
2027	IRTR88	2027	S1983	2027	TRO-2057-4	2027	2N4404
2027	J241015	2027	S2117	2027	TRO-2058-1	2027	2N4405
2027	J24908	2027	S2274	2027	TRO-2058-5	2027	2N4406
2027	JSF6009	2027	S2368	2027	TRO2054-7	2027	2N4407
2027	K071818-001	2027	S2370	2027	TR1000	2027	2N4412
2027	K071962-001	2027	S2398C	2027	TR1002	2027	2N4412A
2027	K1181	2027	S24594	2027	TR1012	2027	2N4414
2027	K1181	2027	S24597	2027	TR1036-2	2027	2N4414A
2027	KD2120	2027	S24612	2027	TR1036-3	2027	2N4890
2027	KB1007-0004-00	2027	S24612A	2027	TR1037-1	2027	2N4928
2027	KLH4781	2027	S24615	2027	TR1038-5	2027	2N5022
2027	KLH5353	2027	S2771	2027	TR1058-6	2027	2N5023
2027	KLH5808	2027	S2991	2027	TR22	2027	2N5040
2027	LDA450	2027	S2993	2027	TR22A	2027	2N5041
2027	LN76963	2027	S2994	2027	TR23A	2027	2N5042
2027	M447B	2027	S2995	2027	TR8037	2027	2N5110
2027	M652P1C	2027	S3012	2027	TRO2054-1	2027	2N5111
2027	M652PIC	2027	S3386	2027	TV-29	2027	2N5160
2027	M7310	2027	S33886	2027	TV116	2027	2N5242
2027	M75561-17	2027	S33886A	2027	TV29	2027	2N5243
2027	M75561-8	2027	S504-0	2027	TVS28A546	2027	2N5252
2027	M828	2027	SC1294H	2027	TVS28A546B	2027	2N5821
2027	M9145	2027	SC365	2027	TVS28A546B	2027	2N5865
2027	M9257	2027	SC843	2027	TVS28A546H	2027	2N5954
2027	M9308	2027	SCI444204037	2027	V180	2027	2N5955
2027	M9400	2027	SCI444291004	2027	V410A	2027	2N5956
2027	M9426	2027	SDT3321	2027	VS-C81256HG	2027	2N6036
2027	M9432	2027	SDT3322	2027	VS2S8A844-D/-1	2027	2N6106
2027	M9435	2027	SDT3501	2027	W15	2027	2N6107
2027	M9520	2027	SDT3502	2027	WEP242	2027	2N6108
2027	M9677	2027	SDT3503	2027	WEP85005	2027	2N6109
2027	M9677/PNP	2027	SE8540	2027	XA-495C	2027	2N6110
2027	M9701	2027	SE8541	2027	XA1199	2027	2N6111
2027	MH9460A	2027	SE8542	2027	XA492D	2027	2N6132
2027	MJB-2020	2027	SFD-23	2027	XA495D	2027	2N6133
2027	MJE-32B	2027	SJ2031	2027	Y56601-46	2027	2N6134
2027	MJE-42	2027	SJE517	2027	Y56601-76	2027	28021
2027	MJE-521	2027	SJE5442	2027	ZDP-D22-69 54	2027	28022
2027	MJE104	2027	SK16510006-2	2027	ZDP-D22-69-54	2027	28023
2027	MJE105	2027	SK16510006-4	2027	ZN 35024712	2027	28501
2027	MJE105K	2027	SK3025	2027	ZN-35024712	2027	28010
2027	MJE1290	2027	SK3083	2027	ZT286	2027	28302
2027	MJE1291	2027	SK3084	2027	ZTX500	2027	2S302A
2027	MJE170	2027	SK3085	2027	01-020566	2027	28303
2027	MJE2010	2027	SKA1079	2027	01-030733	2027	28304
2027	MJE2020	2027	SKA4621	2027	002-010300-6	2027	2SA57AC
2027	MJE210	2027	SL3101	2027	002-010700	2027	2SA497
2027	MJE2370	2027	SL3111	2027	002-012600	2027	2SA497-0
2027	MJE2371	2027	SM2718	2027	002-9800-12	2027	2SA497-ORG
2027	MJE2490	2027	SM3987	2027	2A12	2027	2SA497-R
2027	MJE2491	2027	SM6728	2027	2N1034	2027	2SA497-RED
2027	MJE2901	2027	SMC449077	2027	2N1035	2027	2SA497-Y
2027	MJE2955	2027	S080121	2027	2N1036	2027	2SA497-YEL
2027	MJE30	2027	S080121	2027	2N1197	2027	2SA497R
2027	MJE30A	2027	SPS-0121	2027	2N1197	2027	2SA497RED
2027	MJE30B	2027	SPS-29	2027	2N1228	2027	2SA497Y
2027	MJE32	2027	SPS-4076	2027	2N1229	2027	2SA498
2027	MJE32A	2027	SPS-4078	2027	2N1230	2027	2SA498Y
2027	MJE370K	2027	SPS0121	2027	2N1231	2027	2SA501
2027	MJE371K	2027	SPS2131	2027	2N1232	2027	2SA503
2027	MJE3740	2027	SPS2226	2027	2N1233	2027	2SA503-0
2027	MJE3741	2027	SPS4010	2027	2N1234	2027	2SA503-R
2027	MJE42A	2027	SPS4301	2027	2N1254	2027	2SA503-Y
2027	MJE42B	2027	SPS4310	2027	2N1255	2027	2SA504
2027	MJE710	2027	SPS4312	2027	2N1256	2027	2SA504-0
2027	MJE712	2027	SPS4462	2027	2N1257	2027	2SA504-R
2027	MJ05194	2027	SPS4477	2027	2N1258	2027	2SA504-Y
2027	MM4005	2027	SPS4492	2027	2N1259	2027	2SA504GR
2027	MM4006	2027	SPS4497	2027	2N1275	2027	2SA516
2027	MM4019	2027	SPS6125	2027	2N1429	2027	2SA516A
2027	MPS9460A	2027	ST-021660	2027	2N1606	2027	2SA527
2027	MPS9460H	2027	ST72039	2027	2N1623	2027	2SA528
2027	MR3934	2027	ST72040	2027	2N1922	2027	2SA532
2027	M3S75006G	2027	STC5610	2027	2N2904	2027	2SA532A
2027	M3S77505	2027	STC5611	2027	2N2904A	2027	2SA532B
2027	P1M	2027	STC5612	2027	2N2905	2027	2SA532C
2027	P3A	2027	STH7251	2027	2N2905A	2027	2SA532D
2027	P3N-5	2027	STX0011	2027	2N3120	2027	2SA532E
2027	P3U	2027	SX61	2027	2N3133	2027	2SA532F
2027	P3UA	2027	T-340	2027	2N3134	2027	2SA537
2027	P3V	2027	T-482	2027	2N3202	2027	2SA537A
2027	P4E-1	2027	T1275	2027	2N3203	2027	2SA537AA
2027	P4E-2	2027	T1276	2027	2N3208	2027	2SA537AB
2027	P4E-3	2027	T1808	2027	2N327B	2027	2SA537AH
2027	P4E-4	2027	T1808A	2027	2N328	2027	2SA537B
2027	P4J	2027	T1808B	2027	2N328A	2027	2SA537C
2027	P4T	2027	T1808C	2027	2N328B	2027	2SA537H
2027	P4V-2	2027	T1808D	2027	2N329	2027	2SA546
2027	P5F	2027	T1808E	2027	2N329A	2027	2SA546A
2027	P58	2027	T1936A	2027	2N329B	2027	2SA546B
2027	P6009	2027	T246	2027	2N330	2027	2SA546E
2027	PIV	2027	T334-2	2027	2N330A	2027	2SA546H
2027	PIV-1	2027	T459	2027	2N3343	2027	2SA548
2027	PIV-2	2027	T475	2027	2N3344	2027	2SA550AB
2027	PIV-3	2027	TA7742	2027		2027	2SA550AR
2027	PMC-Q8-0280	2027	TA7743	2027		2027	2SA550AY

RS 276-	Industry Standard No.	RS 276-	Industry Standard No.	RS 276-	Industry Standard No.	RS 276-	Industry Standard No.
2027	28A550B	2027	21A040-050	2027	57B148-6	2027	317-0139-001
2027	28A550BC	2027	21A112-002	2027	57B148-7	2027	344-6012-1
2027	28A550BL	2027	24MW727	2027	57B148-8	2027	344-6012-3
2027	28A550C	2027	026-100019	2027	57B148-9	2027	344-6014-1B
2027	28A550D	2027	29A4	2027	57B163-12	2027	353-9008-001
2027	28A550Y	2027	32E64	2027	57B163-12A	2027	353-9301-004
2027	28A551	2027	33-00742	2027	57B166-9A	2027	353-9304-004
2027	28A551C	2027	34-6001-86	2027	57B168-9	2027	353-9317-001
2027	28A551D	2027	34-6016-12	2027	57B168-99	2027	380-0171-000
2027	28A551E	2027	34-6016-23	2027	57B171-9	2027	386-7184P1
2027	28A552	2027	34-6016-23A	2027	57B245-14	2027	386-7254-P202
2027	28A560	2027	34-6016-32A	2027	57C110-9	2027	393-1(SYLVANIA)
2027	28A57	2027	039	2027	57C122-9	2027	404B(NCR)
2027	28A594	2027	41-0500	2027	57C148-12	2027	417-111
2027	28A594-0	2027	41-0500A	2027	57C148-12A	2027	417-138
2027	28A594-R	2027	42-19643	2027	57C23	2027	417-170
2027	28A594-Y	2027	42-20739	2027	57C23-1	2027	417-181
2027	28A595C	2027	42-21232	2027	57C23-2	2027	417-234
2027	28A597	2027	42-9029-40U	2027	57C23-3	2027	417-255
2027	28A604	2027	42-9029-60W	2027	57C6-15	2027	417-260
2027	28A605	2027	42-9029-70G	2027	57C6-26	2027	417-289
2027	28A606	2027	43-0203845	2027	57C6-26A	2027	417-43
2027	28A606B	2027	43-023222	2027	57C6-31	2027	417-822-13262
2027	28A612	2027	44A358624-003	2027	57C6-31A	2027	422-0739
2027	28A70B	2027	44A391505	2027	57C6-33	2027	422-1232
2027	28A70BA	2027	44A417756-1	2027	57C6-33A	2027	422-1405
2027	28A70BB	2027	46-86403-3	2027	57M1-13	2027	422-2008
2027	28A70BC	2027	46-86406-3	2027	57M1-13A	2027	430-203845
2027	28A734	2027	46-86425-3	2027	57M1-21	2027	430-23222
2027	28A736	2027	46-86517-3	2027	57M1-21A	2027	433M852
2027	28A742	2027	46-86565-3	2027	57M1-22	2027	461-1014
2027	28A773	2027	46-86574-3	2027	57M1-25	2027	461-1048
2027	28A844	2027	48-03-10111102	2027	57M1-25A	2027	610-083
2027	28A844C	2027	48-03-10744802	2027	57M2-10	2027	610-083-1
2027	28A844D	2027	48-134478	2027	57M2-10A	2027	610-083-2
2027	28A844E	2027	48-134478A	2027	57M2-15	2027	610-083-3
2027	28A950	2027	48-134951	2027	57M2-15A	2027	631-1
2027	28A950-0	2027	48-134951A	2027	61-3096-90	2027	631-3(SYLVANIA)
2027	28A950-Q	2027	48-137080	2027	61-309688	2027	690V086HB9
2027	28A950Y	2027	48-137185	2027	000062	2027	690V110H55
2027	28A952	2027	48-137259	2027	66-P11139	2027	700-136
2027	28B502	2027	48-137310	2027	66-P11141	2027	762-105-00
2027	28B503	2027	48-137331	2027	69BP112	2027	762-120
2027	28B510	2027	48-137370	2027	74P1M	2027	774CL
2027	28B510S	2027	48-137472	2027	77-271818-1	2027	800-101-114-1
2027	28B550	2027	48-137527	2027	79P114-2A	2027	800-284
2027	28B596	2027	48-137562	2027	79P114-4	2027	800-511-00
2027	28B596-0	2027	48-37312	2027	79P114-4A	2027	800-516-00
2027	28B596-Q	2027	48-65177A77	2027	80-051600	2027	800-51600
2027	28B598	2027	48-65177A77A	2027	80-052700	2027	800-525-03
2027	28B598F	2027	48-869145	2027	80P2	2027	800-52700
2027	28C844	2027	48-869257	2027	80P2B	2027	824-10300
2027	3-041	2027	48-869308	2027	80P3	2027	880-250-107
2027	3-215	2027	48-869400	2027	80P3B	2027	881-250-107
2027	3-233	2027	48-869426	2027	81P3	2027	921-110B
2027	3-7	2027	48-869432	2027	82-410300	2027	921-182B
2027	314-6010-4	2027	48-869435	2027	83	2027	921-270B
2027	314-6011-02	2027	48-869520	2027	83P2B	2027	921-315B
2027	314-6013-4	2027	48-869677	2027	86-329-2	2027	921-47B
2027	314-6013-6	2027	48-869681	2027	86-334-2	2027	964-20739
2027	314-6013-8	2027	48-869701	2027	86-431-8	2027	972-659R-0
2027	4-30203845	2027	48-90232A06	2027	86-431-9	2027	977-64197
2027	4-3023222	2027	48-90232A06A	2027	86-5064-2	2027	978-1923
2027	4-686170-3	2027	48-90232A09	2027	86-5064-2A	2027	991-01-1315
2027	4-686229-3	2027	48-90232A09A	2027	86-5082-2	2027	991-01-2686
2027	4-686230-3	2027	48-90232A12	2027	86-5082-2A	2027	991-01-5000
2027	4-686235-3	2027	48-90232A12A	2027	86-616-2	2027	991-01-5062
2027	4-686238-3	2027	48-90232A15	2027	86X0042-001	2027	991-2?
2027	006-0004956	2027	48-90232A15A	2027	86X0080-001	2027	1002A-1
2027	006-0005191	2027	48-97046A36	2027	95-226-003	2027	1004P
2027	6-9029-20J	2027	48-97046A39	2027	95-226-1	2027	1011M57P01
2027	7-0012-00	2027	48R869145	2027	95-226-3	2027	1011M62P01
2027	7-14A	2027	48R869257	2027	96-5165-01	2027	10800
2027	7-3(STANDEL)	2027	48R869432	2027	96-5176-01	2027	1184G
2027	7-4(STANDEL)	2027	48R869520	2027	96-5209-01	2027	1385(GE)
2027	716-0444-1(PNP)	2027	48R969681	2027	96-5220-01(PNP)	2027	1414-185
2027	716-0531-1(PNP)	2027	50A103	2027	96-5230-01	2027	1414-187
2027	8-0051600	2027	53P166	2027	96-5316-01	2027	1428
2027	8-052700	2027	53P170	2027	99P117	2027	1853-0041
2027	8-2410300	2027	55-643	2027	99P2	2027	1853-0045
2027	8-905-706-545	2027	57A10B-6-8	2027	99P2B	2027	1853-0215
2027	8-905-713-810	2027	57A110-9	2027	99P3C	2027	1889-17
2027	09-300043	2027	57A112-9	2027	99PLM	2027	2057A2-150
2027	09-305075	2027	57A112-9A	2027	998073	2027	2057B1B-9
2027	09-305134	2027	57A114-9	2027	998099	2027	2057B110-9
2027	9-5226-003	2027	57A114-9A	2027	0101-439	2027	2057B112-9
2027	9-5226-1	2027	57A115-9	2027	106KB0	2027	2057B114-9
2027	9-5226-3	2027	57A116-9	2027	106KBA	2027	2057B115-9
2027	12-100027	2027	57A116-9A	2027	112-2	2027	2057B116-9
2027	13-14085-92	2027	57A122-9	2027	112-2A	2027	2057B121-9
2027	13-28222-1	2027	57A130-9	2027	121-765	2027	2057B122-9
2027	13-283361-1	2027	57A132-10	2027	121-774CL	2027	2057B145-12
2027	13-28386-1	2027	57A148-10	2027	121-845	2027	2057B163-12
2027	13-28393-1	2027	57A148-11	2027	121-879	2027	2482(RCA)
2027	13-28393-1A	2027	57A148-12A	2027	121-973	2027	2502
2027	13-28393-2	2027	57A148-2	2027	121-988-02	2027	3222
2027	13-28393-2A	2027	57A148-4	2027	0131-001427	2027	3523(RCA)
2027	13-28394-4	2027	57A148-5	2027	0131-001428	2027	3533(RCA)
2027	13-28394-4A	2027	57A148-6	2027	131-005-353-1	2027	3553-1
2027	13-28394-5	2027	57A148-7	2027	131-005-808	2027	3574-1
2027	13-28394-5A	2027	57A148-9	2027	131-005353-1	2027	3590
2027	13-28394-6	2027	57A23	2027	131-005808	2027	3592(RCA)
2027	13-32631-1	2027	57A23-1	2027	131-04561	2027	3598-2
2027	13-32631-3	2027	57A23-2	2027	131-04561	2027	3616-1(RCA)
2027	13-34003-1(METAL)	2027	57A23-3	2027	132-007	2027	3746-00
2027	13-34004-1(METAL)	2027	57A245-14	2027	132-032	2027	3746-01
2027	13-34617-1	2027	57A6-15	2027	132-039	2027	4219
2027	13-34859-1	2027	57A6-26	2027	134P11AA	2027	4367-001
2027	13-36443-1	2027	57A6-26A	2027	134P2	2027	4478
2027	13-4800869145	2027	57A6-31	2027	149P1B	2027	4686-170-3
2027	14-0086-1	2027	57A6-31A	2027	149P4B	2027	4686-229-3
2027	14-0104-3	2027	57A6-33	2027	151-0208	2027	4686-230-3
2027	14-602-28	2027	57A6-33A	2027	151-0208-00-AA	2027	4686-235-3
2027	14-602-28A	2027	57B110-9	2027	154A5946-667	2027	4686-238-3
2027	14-602-60	2027	57B112-9	2027	158P1M	2027	4781
2027	14-602-66	2027	57B112-9A	2027	171-9	2027	5680
2027	14-602-73	2027	57B114-9	2027	174-25566-01	2027	6285
2027	14-602-79	2027	57B114-9A	2027	174-25566-62	2027	7301-1
2027	14-602-79A	2027	57B115-9	2027	177-001	2027	750B
2027	14-602-85	2027	57B115-9A	2027	177-001-9-001	2027	7921-1
2027	14-607-29	2027	57B116-9	2027	183P1	2027	8000-004-P089
2027	14-607-29A	2027	57B116-9A	2027	195P2C	2027	8303
2027	14-957-32	2027	57B148-1	2027	226-3	2027	8400-1
2027	16-19(SYMPHONIC)	2027	57B148-10	2027	239AT921-1	2027	8400-1A
2027	17-458	2027	57B148-11	2027	260P13704	2027	8400-1B
2027	19-020-100	2027	57B148-12	2027	297V073001	2027	8624-003
2027	19A115180-2	2027	57B148-12A	2027	297V073002	2027	9405-2
2027	19A115976P1	2027	57B148-3	2027	297V074011	2027	09800-12
2027	020-1110-009	2027	57B148-4	2027	297V080001	2027	10036-001
2027	020-1110-014	2027	57B148-5	2027	297V082B02	2027	10300-12
2027	020-1111-017			2027	297V082B03	2027	12048-0011
2027	20A0075			2027	297V083C03	2027	012099-1
2027	021-0224-00			2027	303-2	2027	16167
2027	21A040-049					2027	16169
						2027	16175

RS 276-	Industry Standard No.	RS 276-	Industry Standard No.	RS 276-	Industry Standard No.	RS 276-	Industry Standard No.
2027	16239	2027	325077	2028	IP20-0010	2028	NF522
2027	16279	2027	335613	2028	IRTRPE100	2028	NF523
2027	16306	2027	386726-1	2028	JNU61673	2028	NF531
2027	16342	2027	405965-35A	2028	K10	2028	NF533
2027	018069	2027	5140678	2028	K11(1-GATE)	2028	NF5485
2027	020426-3	2027	5140728	2028	K11-0(1-GATE)	2028	NF5486
2027	20739	2027	564671	2028	K11-R(1-GATE)	2028	NKT800112
2027	23114-050	2027	0573542	2028	K11-Y(1-GATE)	2028	NKT800113
2027	23114-051	2027	0573559	2028	K12(1-GATE)	2028	NKT80111
2027	25566-62	2027	0573560	2028	K12-3(1-GATE)	2028	NKT80211
2027	30278	2027	602909-3A	2028	K12-GR(1-GATE)	2028	NKT80212
2027	31006	2027	610099	2028	K12-R(1-GATE)	2028	NKT80213
2027	31032-0	2027	610099-1	2028	K120T(1-GATE)	2028	NKT80214
2027	035571	2027	610099-2	2028	K13(1-GATE)	2028	NKT80215
2027	36577	2027	610099-5	2028	K15-0(1-GATE)	2028	NKT80216
2027	37269	2027	610110	2028	K15-GR-(1-GATE)	2028	NPC211N
2027	37664	2027	610129-1	2028	K15-R(1-GATE)	2028	NPC212N
2027	37740	2027	610158-1	2028	K15-Y(1-GATE)	2028	NPC214N
2027	37764	2027	610195-2	2028	K17	2028	NPC215N
2027	37793	2027	650175	2028	K17(1-GATE)	2028	NPC216N
2027	37918	2027	681266	2028	K170(1-GATE)	2028	NPC312N
2027	37966	2027	681266-1	2028	K170R(1-GATE)	2028	P1087
2027	38588	2027	723043-1	2028	K174(1-GATE)	2028	QA-18
2027	38458	2027	779821	2028	K17B(1-GATE)	2028	QA-20
2027	38496	2027	810002-733	2028	K17BL(1-GATE)	2028	RE46
2027	38654	2027	815185	2028	K17GR(1-GATE)	2028	REN133
2027	38734	2027	815185E	2028	K17R(1-GATE)	2028	RT-176
2027	38737	2027	970107	2028	K19	2028	R2176
2027	38870	2027	970762-6	2028	K19(1-GATE)	2028	S1211N
2027	39114	2027	986030	2028	K19(GR)	2028	S1212N
2027	39250	2027	986931	2028	K19BL	2028	S1213N
2027	39440	2027	993570-4	2028	K19BL(1-GATE)	2028	S1214N
2027	39618	2027	126195-191	2028	K19GC	2028	S1215N
2027	39619	2027	1417303-1	2028	K19GC(1-GATE)	2028	S1216N
2027	39853	2027	1417303-2	2028	K19GE(1-GATE)	2028	S1221N
2027	39865	2027	1471112-12	2028	K19GR	2028	S1222N
2027	40319	2027	1471112-3	2028	K19GR(1-GATE)	2028	S1223N
2027	40362	2027	1471112-8-9	2028	K19Y	2028	S1224N
2027	40406	2027	1473516-1	2028	K19Y(1-GATE)	2028	S1225N
2027	40407	2027	1473592-1	2028	K22(1-GATE)	2028	S1226N
2027	40410	2027	1473616-1	2028	K22-Y(1-GATE)	2028	S1231N
2027	40537	2027	1473624-1	2028	K23(1-GATE)	2028	S1232N
2027	40537L	2027	1473666-1	2028	K23A	2028	S1233N
2027	40537B	2027	1827322	2028	K24(1 GATE)	2028	S1234N
2027	40538	2027	1835667	2028	K24C(1 GATE)	2028	S1235N
2027	40538L	2027	1965016	2028	K24D(1 GATE)	2028	S1236N
2027	40538B	2027	2004746-116	2028	K24DR(1 GATE)	2028	S83819
2027	40595VX	2027	2004746-117	2028	K24E(1 GATE)	2028	SB5-0996
2027	40634	2027	2006436-37	2028	K24F(1 GATE)	2028	SPE145
2027	41052	2027	2096700	2028	K24G(1 GATE)	2028	SJ1925
2027	41501	2027	2096700-TM18	2028	K25(1-GATE)	2028	SK3112
2027	41503	2027	2320242	2028	K25C(1-GATE)	2028	SPP024
2027	43089	2027	2320243	2028	K25D(1-GATE)	2028	S3-3704
2027	43107	2027	2321351	2028	K25E(1-GATE)	2028	S83586
2027	43116	2027	2327282	2028	K25ET(1-GATE)	2028	S83672
2027	59989-1	2027	2327283	2028	K25F(1-GATE)	2028	S83735
2027	60154	2027	2412949-0001	2028	K25G(1-GATE)	2028	SU2076
2027	60701	2027	3596100	2028	K30-0(1-GATE)	2028	SU2077
2027	60947	2027	3596101	2028	K30-0(1-GATE)	2028	SU2080
2027	61009-9	2027	3596340	2028	K30A(1-GATE)	2028	SU2081
2027	61009-9-1	2027	3596341	2028	K30AGR(1-GATE)	2028	SX-58Y
2027	61009-9-2	2027	A04201	2028	K30B(1-GATE)	2028	T01-044
2027	61011-0	2028	A194	2028	K30C(1-GATE)	2028	T1814
2027	61011-0-1	2028	A195	2028	K30D(1-GATE)	2028	TAA320
2027	61013-4-1	2028	A196	2028	K30R(1-GATE)	2028	TI814
2027	61244	2028	A2652-919	2028	K30Y	2028	TI878
2027	61371/4561	2028	A514-040296	2028	K30Y(1-GATE)	2028	TI879
2027	61666	2028	B-6001	2028	K31(1-GATE)	2028	TM135
2027	61774	2028	B0264C	2028	K31C	2028	TNJ61673
2027	61937	2028	BP244	2028	K31C(1-GATE)	2028	TNJ61673(2SK24)
2027	62204	2028	BP245	2028	K33(1-GATE)	2028	V159
2027	62277	2028	BP245A	2028	K33(E)	2028	V16502-1
2027	62279	2028	BP245B	2028	K33E	2028	V16503-4
2027	62512	2028	BP245C	2028	K33F	2028	V16533B
2027	62584	2028	BP246	2028	K33F(1-GATE)	2028	WEP801
2027	62708	2028	BP247	2028	K33GR	2028	WEP802
2027	62759	2028	BP256A	2028	K33GR(1-GATE)	2028	WP1
2027	70158-9-00	2028	BP256B	2028	K34	2028	001-02701-1
2027	70260-19	2028	BP256C	2028	K34(1-GATE)	2028	001-02702-0
2027	70260-29	2028	BF348	2028	K34(E)	2028	1W11706
2027	071818	2028	BPS21	2028	K34B(1-GATE)	2028	1W11706
2027	75700-13-01	2028	BPS21A	2028	K34C	2028	2D031
2027	84001A	2028	BFW15	2028	K34D(1-GATE)	2028	2N296BA
2027	84001B	2028	C-36582	2028	K34D	2028	2N3066
2027	86257	2028	C674	2028	K34D1(1-GATE)	2028	2N3067
2027	94063	2028	C764	2028	K34E	2028	2N3068
2027	94064	2028	CXL132	2028	K35(1 GATE)	2028	2N3069
2027	94067	2028	CXL133	2028	K35-0(1 GATE)	2028	2N3070
2027	94068	2028	D1101	2028	K35-1(1 GATE)	2028	2N3071
2027	95239-1	2028	D1102	2028	K35-2(1 GATE)	2028	2N3084
2027	111945	2028	D1177	2028	K35A(1 GATE)	2028	2N3085
2027	116084(RCA)	2028	D1178	2028	K35BL(1 GATE)	2028	2N3086
2027	116284	2028	D1180	2028	K35A(1 GATE)	2028	2N3087
2027	119298-001	2028	D1181	2028	K350N(1 GATE)	2028	2N3088
2027	119983	2028	D1301	2028	K35R(1 GATE)	2028	2N3088A
2027	124616	2028	D1302	2028	K35Y(1 GATE)	2028	2N3089
2027	126724	2028	D1303	2028	K37(1-GATE)	2028	2N3089A
2027	130404-29	2028	D1421	2028	K37E(1-GATE)	2028	2N3365
2027	131242-12	2028	E103	2028	K37L(1-GATE)	2028	2N3366
2027	132448	2028	EA15X169	2028	K42	2028	2N3367
2027	132574	2028	EA15X446	2028	K47	2028	2N3368
2027	136423	2028	EA3278	2028	K47M	2028	2N3369
2027	137155	2028	EG0133	2028	K47M(1-GATE)	2028	2N3370
2027	140625	2028	EP1(ELCOM)	2028	K49	2028	2N3436
2027	147112-7	2028	EL131	2028	K49F	2028	2N3437
2027	147357-0-1	2028	EQF-3	2028	K49F(1-GATE)	2028	2N3438
2027	147357-4-1	2028	ET491051	2028	K49H	2028	2N3452
2027	147359-0-1	2028	FE100	2028	K49H(1-GATE)	2028	2N3453
2027	147624-1	2028	FE100A	2028	K49HK(1-GATE)	2028	2N3454
2027	157004	2028	FE102	2028	K49I	2028	2N3455
2027	171033-1(PNP)	2028	FE102A	2028	K49I(1-GATE)	2028	2N3457
2027	171162-193	2028	FE104A	2028	K49M(1-GATE)	2028	2N3458
2027	172336	2028	FE400	2028	M100	2028	2N3459
2027	188226	2028	FE402	2028	M101	2028	2N3460
2027	198065-1	2028	FE402A	2028	MK-10	2028	2N3465
2027	198065-3	2028	FE404A	2028	MK-10-2	2028	2N3466
2027	198074-1	2028	FF400	2028	MPF-102	2028	2N3684
2027	198078-1	2028	G08-005L	2028	MPF101	2028	2N3684A
2027	202917-137	2028	G08005L	2028	MPF103	2028	2N3685
2027	203718	2028	GE-FET-1	2028	MPF104	2028	2N3685A
2027	218537	2028	G08-005L	2028	MPF105	2028	2N3686
2027	233969	2028	HA2001	2028	MPF108	2028	2N3686A
2027	236433	2028	HA2010	2028	MPF298	2028	2N3687
2027	240402	2028	HEF801	2028	MPF112	2028	2N3821
2027	241052	2028	HEPF0015	2028	MPS566B	2028	2N3822
2027	242422	2028	HP200191A/	2028	NE4304	2028	2N3823
2027	242460	2028	HP200191B0	2028	NF4302	2028	2N3966
2027	242958	2028	HP200301B	2028	NF4303	2028	2N3967
2027	262638	2028	HP200301B0	2028	NF520	2028	2N3967A
2027	268717	2028	HP200301B0			2028	2N3968
2027	297074011	2028	HP200301C-0			2028	2N3968A
2027	302865	2028	HP200301C-0			2028	2N3969
2027	309690	2028	HK-00530				
2027	319304						

RS 276-	Industry Standard No.	RS 276-	Industry Standard No.	RS 276-	Industry Standard No.	RS 276-	Industry Standard No.
2028	2N5969A	2028	4-324	2029	ZEN129	2030	AT442
2028	2N4117	2028	006-0004443	2029	2D022	2030	AT443
2028	2N4117A	2028	007-0214-00	2029	2D022-211	2030	AT444
2028	2N4118	2028	007-0214-01	2029	2D030	2030	AT445
2028	2N4119	2028	07-07159	2029	2N1671	2030	AT446
2028	2N4119A	2028	8-723-302-00	2029	2N2160	2030	AT470
2028	2N4220	2028	8-905-706-901	2029	2N2646	2030	AT471
2028	2N4222	2028	8P111	2029	2N2647	2030	AT472
2028	2N4222A	2028	09-305021	2029	28H11	2030	AT473
2028	2N4223	2028	09-305023	2029	28H12	2030	AT474
2028	2N4224	2028	09-305031	2029	28H18	2030	AT475
2028	2N4302	2028	09-305133	2029	28H18K	2030	AT476
2028	2N4303	2028	9ACW	2029	28H18L	2030	AT477
2028	2N4338	2028	13-28654-4	2029	28H18M	2030	AT478
2028	2N4339	2028	13-28654-5	2029	28H18N	2030	AT479
2028	2N4340	2028	13-44290	2029	28H19	2030	AT7
2028	2N4341	2028	13-44291	2029	28H19K	2030	ATC-TR-13
2028	2N4416	2028	14-2002-01	2029	28H19L	2030	ATC-TR-4
2028	2N4416A	2028	14-700-02	2029	28H19M	2030	ATC-TR-7
2028	2N4417	2028	16027	2029	28H19N	2030	B-6340
2028	2N4867	2028	020-1110-016	2029	28H20	2030	B274(SYLVANIA)
2028	2N4867A	2028	020-1110-018	2029	28H21	2030	B3746
2028	2N4868	2028	020-1110-021	2029	28H22	2030	B87J0007
2028	2N4868A	2028	020-1110-022	2029	4JD5B29	2030	BC-119
2028	2N4869	2028	020-1112-005	2029	4JX5B670	2030	BC-138
2028	2N4869A	2028	020-1112-007	2029	14-593-01	2030	BC-140
2028	2N5045	2028	020-1112-008	2029	14-593-03	2030	BC-140A
2028	2N5046	2028	020-1112-009	2029	48-134792	2030	BC-140B
2028	2N5047	2028	022-2876-005	2029	48-137058	2030	BC-140C
2028	2N5267	2028	24T-026-001	2029	48-137165	2030	BC-140D
2028	2N5268	2028	34-6018-2	2029	48-137282	2030	BC-141
2028	2N5269	2028	48-134944	2029	48-869206	2030	BC-142
2028	2N5270	2028	48-137023	2029	48-869256	2030	BC103
2028	2N5277	2028	48-97177HO1	2029	48-869264	2030	BC103C
2028	2N5358	2028	48X90232A14	2029	48R869206	2030	BC119
2028	2N5359	2028	48X97177HO1	2029	48R869256	2030	BC120
2028	2N5360	2028	57A31-4	2029	48R869264	2030	BC138
2028	2N5361	2028	57L106-9	2029	86-668-2	2030	BC140
2028	2N5362	2028	81-46125065-6	2029	96-5269-01	2030	BC140-10
2028	2N5363	2028	86-477-2	2029	107-0021-00	2030	BC140-16
2028	2N5364	2028	86-500-2	2029	242-997	2030	BC140C
2028	2N5391	2028	86-5095-2	2029	417-183	2030	BC140D
2028	2N5392	2028	86-5096-2	2029	417-187	2030	BC141
2028	2N5393	2028	86-5122-2	2029	417-81	2030	BC141-10
2028	2N5394	2028	108-0068-12	2029	576-0005-001	2030	BC141-16
2028	2N5395	2028	121-860	2029	753-9010-021	2030	BC141-6
2028	2N5396	2028	123-003	2029	945	2030	BC142
2028	2N5452	2028	149-12	2029	2160	2030	BC144
2028	2N5453	2028	150-12	2029	5001	2030	BC173,B
2028	2N5454	2028	173-1	2029	70399-1	2030	BC174
2028	2N5457	2028	182-014-9-002	2029	70399-2	2030	BC174A
2028	2N5458	2028	182-014-9-003	2029	132650	2030	BC174B
2028	2N5459	2028	182-046-9-001	2029	654001	2030	BC190A
2028	2N5543	2028	276-2028	2029	656064	2030	BC190B
2028	2N5544	2028	276-2035	2029	740855	2030	BC211
2028	2N5556	2028	364-10048	2029	801525	2030	BC216
2028	2N5557	2028	417-140	2029	801551	2030	BC216A
2028	2N5558	2028	417-167	2030	A-1141-5932	2030	BC216B
2028	2N5561	2028	417-169	2030	A-120018	2030	BC226
2028	2N5562	2028	417-194	2030	A-1854-0022-1	2030	BC232A
2028	2N5563	2028	417-231	2030	A-1854-0087-1	2030	BC232B
2028	2N5648	2028	417-241	2030	A-1854-0090-1	2030	BC254
2028	2N5649	2028	417-246	2030	A0-54-175	2030	BC255
2028	2N5668	2028	417-252	2030	A054-156	2030	BC286
2028	2N5716	2028	417-253	2030	A054-160	2030	BC288
2028	2N5717	2028	576-0006-003	2030	A054-186	2030	BC301
2028	2N5718	2028	690V116H22	2030	A054-206	2030	BC302
2028	2SJ11	2028	700-04	2030	A059-110	2030	BC302-4
2028	2SJ12	2028	753-4000-024	2030	A107	2030	BC302-5
2028	2SK106	2028	921-1019	2030	A115	2030	BC302-6
2028	2SK17(O)	2028	991-01-1706	2030	A1314	2030	BC310
2028	2SK17-0R	2028	991-01-3055	2030	A1341	2030	BC313
2028	2SK24	2028	991-011576	2030	A188103	2030	BC337
2028	2SK24C	2028	991-011706	2030	A2471	2030	BC337-16
2028	2SK24D	2028	1002-01	2030	A249(AMC)	2030	BC338-16
2028	2SK24DR	2028	1042-02	2030	A25A305020101	2030	BC340-06
2028	2SK24E	2028	1612SK24E	2030	A3011112	2030	BC340-10
2028	2SK24F	2028	2000-104	2030	A311	2030	BC340-16
2028	2SK24G	2028	2010-17	2030	A32-2805-50-1	2030	BC340-6
2028	2SK24SET	2028	2020-04	2030	A322805-50-1	2030	BC341-06
2028	2SK35	2028	2057B149-12	2030	A3T2484	2030	BC341-10
2028	2SK35-0	2028	5001-047	2030	A417034	2030	BC341-6
2028	2SK35-1	2028	5459	2030	A466	2030	BC377
2028	2SK35-2	2028	6003	2030	A47392R-0	2030	BC378
2028	2SK35A	2028	8000-00004-080	2030	A490	2030	BC429
2028	2SK35BL	2028	8000-00010-017	2030	A543	2030	BC508
2028	2SK35C	2028	8000-0004-P080	2030	A5B	2030	BC509
2028	2SK35GN	2028	8000-0004-P081	2030	A5T2192	2030	BC510
2028	2SK35R	2028	8000-0004-P080	2030	A5T3391	2030	BC535
2028	2SK35Y	2028	40461	2030	A5T3391A	2030	BC537
2028	2SK35Q	2028	40468	2030	A5T5210	2030	BC538
2028	2SK40	2028	42396	2030	A5T6222	2030	BC582
2028	2SK40(C)	2028	44766	2030	A66-P11138-0001	2030	BC582A
2028	2SK40-3	2028	71686-4	2030	A67-70-960	2030	BC582B
2028	2SK40A	2028	71686-5	2030	A7T3391	2030	BC583A
2028	2SK40B	2028	71748-1	2030	A7T3391A	2030	BC583B
2028	2SK40C	2028	95133	2030	A8D	2030	BC584
2028	2SK40D	2028	95228	2030	A8T3391	2030	BC737
2028	2SK43	2028	95231	2030	A8T3391A	2030	BC738
2028	2SK43-0R	2028	226517	2030	A8T3704	2030	BCW44
2028	2SK43A	2028	452077	2030	A8T3705	2030	BCW46
2028	2SK43B	2028	618580	2030	A8T3706	2030	BCW47
2028	2SK43C	2028	3412004-1912	2030	AM	2030	BCW48
2028	2SK43D	2029	C-6BX212	2030	AMP-2971-4	2030	BCW49
2028	2SK43E	2029	CXL6400	2030	AQ2	2030	BCW51
2028	2SK43F	2029	CXL6401	2030	AQ3	2030	BCW65EC
2028	2SK43GN	2029	CXL6409	2030	AQ5	2030	BCW66EF
2028	2SK43H	2029	D5E-37	2030	AR-203(R)	2030	BCW66EG
2028	2SK43J	2029	D5E-43	2030	AR203	2030	BCW66EH
2028	2SK43K	2029	D5E-44	2030	AR203(R)	2030	BCW66EW
2028	2SK43L	2029	D5E-45	2030	AR203(RED)	2030	BCW73-16
2028	2SK43M	2029	D5E37	2030	AR203R	2030	BCW74-16
2028	2SK43OR	2029	D5E43	2030	AR203RED	2030	BCW77-16
2028	2SK43R	2029	D5E44	2030	AR207	2030	BCW78-16
2028	2SK43X	2029	D5E45	2030	AR218ORANGE	2030	BCW82
2028	2SK43Y	2029	ECG392	2030	AR218RED	2030	BCW82A
2028	2SK44	2029	ECG6400	2030	AR306BLUE	2030	BCW82B
2028	2SK44(D)	2029	ECG6401	2030	AR306ORANGE	2030	BCW83A
2028	2SK44C	2029	ECG6409	2030	AT-12	2030	BCW83B
2028	2SK44D	2029	ES47(ELCOM)	2030	AT-12(PHILCO)	2030	BCW84
2028	2SK491	2029	GE-X10	2030	AT-7	2030	BCW90
2028	2SK68	2029	GET4870	2030	AT12	2030	BCW90A
2028	2SK68-L	2029	GET4871	2030	AT339	2030	BCW90B
2028	2SK68A	2029	HEP-89002	2030	AT380	2030	BCW90C
2028	2SK68AL	2029	HEP310	2030	AT381	2030	BCW90K
2028	2SK68AM	2029	HEPS9002	2030	AT382	2030	BCW90KA
2028	2SK68L	2029	IR2160	2030	AT383	2030	BCW90KB
2028	2SK68M	2029	M9256	2030	AT384	2030	BCW90KC
2028	2SK68Q	2029	M9264	2030	AT385	2030	BCW91
2028	2SK68Y	2029	REN6401	2030	AT386	2030	BCW91A
2028	2SR24	2029	RS-2029	2030	AT387	2030	BCW91B
2028	33K34	2029	SU110	2030	AT388	2030	BCW91K
2028	33K34C	2029	SU44	2030	AT440	2030	BCW91KA
		2029	WEP510	2030	AT441	2030	BCW91KB

RS 276-	Industry Standard No.	RS 276-	Industry Standard No.	RS 276-	Industry Standard No.	RS 276-	Industry Standard No.
2030	BCW94K	2030	BSX61	2030	C215	2030	C853
2030	BCW95	2030	BSX62	2030	C216	2030	C853A
2030	BCW95K	2030	BSX62B	2030	C217	2030	C853B
2030	BCX58IX	2030	BSX62C	2030	C22	2030	C853C
2030	BCX58X	2030	BSX62D	2030	C220	2030	C853KLM
2030	BCX59IX	2030	BSX63	2030	C221	2030	C853L
2030	BCX59VII	2030	BSX63B	2030	C222	2030	C875
2030	BCX59VIII	2030	BSX63C	2030	C223(TRANSISTOR)	2030	C875-1
2030	BCX59X	2030	BSX70	2030	C224	2030	C875-1C
2030	BCX73-16	2030	BSX71	2030	C225	2030	C875-1D
2030	BCX73-25	2030	BSX72	2030	C226	2030	C875-1E
2030	BCX73-40	2030	BSX95	2030	C227	2030	C875-1F
2030	BCX74-16	2030	BSX96	2030	C228	2030	C875-2
2030	BCX74-25	2030	BSY25	2030	C229	2030	C875-2C
2030	BCX74-40	2030	BSY44	2030	C230	2030	C875-2D
2030	BCY443	2030	BSY45	2030	C231	2030	C875-2E
2030	BCY46	2030	BSY46	2030	C232	2030	C875-2F
2030	BCY47	2030	BSY51	2030	C233	2030	C875-3
2030	BCY48	2030	BSY52	2030	C234	2030	C875-3C
2030	BCY49	2030	BSY53	2030	C235	2030	C875-3D
2030	BCY58VII	2030	BSY54	2030	C235-0	2030	C875-3E
2030	BCY58VIII	2030	BSY55	2030	C235-0	2030	C875-3F
2030	BCY59VIII	2030	BSY56	2030	C236	2030	C875BR
2030	BCY59VIII	2030	BSY62B	2030	C238	2030	C875C
2030	BCY65	2030	BSY71	2030	C24	2030	C875D
2030	BCY66	2030	BSY77	2030	C247	2030	C875E
2030	BF71	2030	BSY78	2030	C248	2030	C875F
2030	BF387	2030	BSY79	2030	C2485076-3	2030	C876
2030	BFR16	2030	BSY81	2030	C2485077-2	2030	C876C
2030	BFR36	2030	BSY82	2030	C249	2030	C876D
2030	BFR40	2030	BSY83	2030	C261	2030	C876E
2030	BFR40T05	2030	BSY84	2030	C268	2030	C876F
2030	BFR41	2030	BSY85	2030	C268A	2030	C876TV
2030	BFR41T05	2030	BSY86	2030	C27	2030	C876TVD
2030	BFR51	2030	BSY87	2030	C291(TRANSISTOR)	2030	C876TVE
2030	BFR52	2030	BSY88	2030	C292	2030	C876TVEF
2030	BF829	2030	BSY90	2030	C30	2030	C881
2030	BF829P	2030	BSY91	2030	C306	2030	C881A
2030	BF830	2030	BSY92	2030	C307	2030	C881B
2030	BF830P	2030	C0013000-1C	2030	C308	2030	C881C
2030	BF836	2030	C1008	2030	C309	2030	C881D
2030	BF839	2030	C1072	2030	C31	2030	C881K
2030	BF859	2030	C1072A	2030	C310	2030	C881L
2030	BF860	2030	C108	2030	C32	2030	C934
2030	BFT30	2030	C109	2030	C32A	2030	C959
2030	BFT31	2030	C109A	2030	C352A(JAPAN)	2030	C959A
2030	BF254	2030	C112(JAPAN)	2030	C353	2030	C959B
2030	BFV14	2030	C113	2030	C353A	2030	C959C
2030	BFV52	2030	C114	2030	C36579	2030	C959D
2030	BFW29	2030	C115	2030	C376	2030	C959M
2030	BFW33	2030	C1166	2030	C403B(SONY)	2030	C959S
2030	BFX17	2030	C1166-0	2030	C403C(SONY)	2030	C959SA
2030	BFX50	2030	C11660	2030	C420	2030	C959SB
2030	BFX51	2030	C1166D	2030	C426	2030	C959SC
2030	BFX52	2030	C11660R	2030	C428	2030	C959SD
2030	BFX53	2030	C11660	2030	C443	2030	C97
2030	BFX59F	2030	C1166R	2030	C459	2030	C971
2030	BFX61	2030	C1166Y	2030	C459B	2030	C972
2030	BFX68	2030	C117	2030	C46	2030	C972C
2030	BFX68A	2030	C118(JAPAN)	2030	C479	2030	C972D
2030	BFX69	2030	C119(JAPAN)	2030	C49	2030	C972B
2030	BFX69A	2030	C12(TRANSISTOR)	2030	C497	2030	C97A
2030	BFX74	2030	C120	2030	C497-0	2030	C993
2030	BFX84	2030	C1209	2030	C497-R	2030	C99CD
2030	BFX85	2030	C1209C	2030	C497-Y	2030	C99P-10A580
2030	BFX86	2030	C121	2030	C498	2030	CC86168G
2030	BFX92A	2030	C1211	2030	C498-0	2030	CC86229H
2030	BFX94	2030	C1211C	2030	C498-R	2030	CD6153
2030	BFX95	2030	C1211D	2030	C498-Y	2030	CD6153-2
2030	BFX96	2030	C1211E	2030	C503(TRANSISTOR)	2030	CDC-8000-1D
2030	BFX97	2030	C1218	2030	C503	2030	CDC120700
2030	BFX97A	2030	C122(TRANSISTOR)	2030	C503-0	2030	CDC5008
2030	BFY10	2030	C1220E	2030	C503-Y	2030	CDC5028A
2030	BFY11	2030	C123(TRANSISTOR)	2030	C503GR	2030	CDC587
2030	BFY12	2030	C124	2030	C504	2030	CDC745
2030	BFY13	2030	C130	2030	C504-0	2030	CDC8000-1
2030	BFY14	2030	C1318	2030	C504-Y	2030	CDC8000-1C
2030	BFY15	2030	C1318C	2030	C504GR	2030	CDC8000-1D
2030	BFY16	2030	C1318Q	2030	C509(0)	2030	CDC8000-CM
2030	BFY17	2030	C1318R	2030	C509(0)	2030	CD8002
2030	BFY19	2030	C1318B	2030	C509G	2030	CD8002-1
2030	BFY27	2030	C1330	2030	C51	2030	CDC9002-18
2030	BFY34	2030	C1330A	2030	C512	2030	CDC9002-1C
2030	BFY40	2030	C1330B	2030	C512-0	2030	CDQ10011
2030	BFY44	2030	C1330C	2030	C512-R	2030	CDQ10012
2030	BFY46	2030	C1330D	2030	C513	2030	CDQ10014
2030	BFY50	2030	C1330L	2030	C513-0	2030	CDQ10033
2030	BFY51	2030	C1330R	2030	C513R	2030	CDQ10044
2030	BFY52	2030	C1346	2030	C516	2030	CDQ10048
2030	BFY53	2030	C1346R	2030	C560	2030	CDQ10051
2030	BFY55	2030	C1346B	2030	C564	2030	CDQ10052
2030	BFY56	2030	C1347	2030	C564A	2030	CDQ10053
2030	BFY56A	2030	C1347Q	2030	C564P	2030	CDQ10057
2030	BFY66	2030	C1347R	2030	C564Q	2030	CDQ10058
2030	BFY67	2030	C1347B	2030	C564U	2030	CD215000
2030	BFY67A	2030	C1383	2030	C564B	2030	CB4003D
2030	BFY67C	2030	C1383P	2030	C564T	2030	CF-1295H
2030	BFY68	2030	C1383Q	2030	C59	2030	CGB-63
2030	BFY68A	2030	C1383R	2030	C594	2030	CI3704
2030	BFY70	2030	C1383R/494	2030	C61	2030	CI3705
2030	BFY72	2030	C1383S	2030	C628	2030	CI3706
2030	BFY99	2030	C1383X	2030	C644HS	2030	CJ-5210
2030	BLY27	2030	C1384	2030	C69	2030	CJ5206A
2030	BLY28	2030	C1384Q	2030	C708	2030	CJ5210
2030	BLY61	2030	C1384R	2030	C708A	2030	CJ5213
2030	BLY93	2030	C1384S	2030	C708AA	2030	CJ5214
2030	BS823	2030	C140(TRANSISTOR)	2030	C708AB	2030	CJ5215
2030	BS840	2030	C1407	2030	C708AC	2030	CK419
2030	BS841	2030	C1407P	2030	C708AH	2030	CK422
2030	BSV51	2030	C1407R	2030	C708AHA	2030	CK474
2030	BSV69	2030	C1407S	2030	C708AHB	2030	CK475
2030	BSV85	2030	C1429-1	2030	C708AHC	2030	CMO770
2030	BSW10	2030	C1429-2	2030	C708B	2030	CN2484
2030	BSW26	2030	C147	2030	C708C	2030	C0013000-1C
2030	BSW27	2030	C1509	2030	C731R	2030	CP2357
2030	BSW28	2030	C1509P	2030	C734	2030	CP409
2030	BSW29	2030	C1509Q	2030	C734-0	2030	CS-1352
2030	BSW35	2030	C1509R	2030	C734-0	2030	CS-1372
2030	BSW49	2030	C1509S	2030	C734-R	2030	CS-2143
2030	BSW66	2030	C150T	2030	C734-Y	2030	CS-9013HG
2030	BSW66	2030	C152(TRANSISTOR)	2030	C734GR	2030	CS1129E
2030	BSX12A	2030	C188	2030	C7340	2030	CS1225H
2030	BSX22	2030	C188A	2030	C734Y	2030	CS1225HF
2030	BSX23	2030	C188AB	2030	C744	2030	CS1229K
2030	BSX33	2030	C189	2030	C797	2030	CS1229N
2030	BSX45	2030	C19	2030	C798	2030	CS1248
2030	BSX45-10	2030	C190	2030	C803	2030	CS1248I
2030	BSX45-16	2030	C20	2030	C814	2030	CS1248T
2030	BSX45-6	2030	C20B	2030	C815	2030	CS1250F
2030	BSX46	2030	C210	2030	C816K	2030	CS1255H
2030	BSX46-10	2030	C211	2030	C826	2030	CS1255M
2030	BSX46-16	2030	C213	2030	C827	2030	CS1256HG
2030	BSX46-6	2030	C214	2030	C839	2030	CS1295H
2030	BSX60					2030	CS1305

RS 276-	Industry Standard No.	RS 276-	Industry Standard No.	RS 276-	Industry Standard No.	RS 276-	Industry Standard No.	RS 276-	Industry Standard No.
2030	C81352	2030	E-01381	2030	HC-01209	2030	HC-01209	2030	MA8001
2030	C81453F	2030	E-167-228	2030	HC-01317	2030	HC-01317	2030	MB1075
2030	C81453G	2030	E-185B121712	2030	HC-01318	2030	HC-01318	2030	MB8001
2030	C81462I	2030	E-2491B	2030	HD-00261	2030	HD-00261	2030	MB8002
2030	C81464H	2030	E2441	2030	HEP243	2030	HEP243	2030	MB8003
2030	C81591LE	2030	E2449	2030	HEP711	2030	HEP711	2030	MH1501
2030	C81609F	2030	E318-1	2030	HEP736	2030	HEP736	2030	MH9410A
2030	C81613	2030	EA-15X8517	2030	HEP75	2030	HEP75	2030	MHT2414
2030	C81664	2030	EA0081	2030	HEP80003	2030	HEP80003	2030	MHT2418
2030	C81711	2030	EA0090	2030	HEP80014	2030	HEP80014	2030	MHT4401
2030	C81893	2030	EA1549	2030	HEP85026	2030	HEP85026	2030	MHT4411
2030	C81909B	2030	EA15X102	2030	HP11(PHILCO)	2030	HP11(PHILCO)	2030	MHT4412
2030	C81990	2030	EA15X144	2030	HP9	2030	HP9	2030	MHT4413
2030	C82023	2030	EA15X249	2030	HR-67	2030	HR-67	2030	MHT4451
2030	C82484	2030	EA15X250	2030	HR12A	2030	HR12A	2030	MHT4483
2030	C83704	2030	EA15X2522	2030	HR12B	2030	HR12B	2030	MHT4511
2030	C83705	2030	EA15X274	2030	HR12C	2030	HR12C	2030	MHT4512
2030	C83706	2030	EA15X349	2030	HR12D	2030	HR12D	2030	MHT4513
2030	C85449	2030	EA15X378	2030	HR12E	2030	HR12E	2030	MHT7401
2030	C85450	2030	EA15X397	2030	HR12F	2030	HR12F	2030	MHT7411
2030	C85451	2030	EA15X413	2030	HR28	2030	HR28	2030	MHT7412
2030	C86168G	2030	EA15X4531	2030	HR29	2030	HR29	2030	MHT7414
2030	C87229F	2030	EA15X57	2030	HR38	2030	HR38	2030	MHT7417
2030	C88050	2030	EA15X8517	2030	HR67	2030	HR67	2030	MH9001
2030	C89013/3490	2030	EA15X8521	2030	HR81	2030	HR81	2030	MHT9002
2030	C89013E	2030	EA1684	2030	HR82	2030	HR82	2030	MHT9004
2030	C89013E-F	2030	EA1698	2030	HR83	2030	HR83	2030	MHT9005
2030	C89013F	2030	EA1873	2030	HS40026	2030	HS40026	2030	MM1809
2030	C89013G	2030	EA3990	2030	HS5810	2030	HS5810	2030	MM1809A
2030	C89013H	2030	EC0128	2030	HS5812	2030	HS5812	2030	MM1810A
2030	C89013HG/3490	2030	EC0192	2030	HS5814	2030	HS5814	2030	MM1943
2030	C89022LE	2030	EC0192-1	2030	HS5816	2030	HS5816	2030	MM2193A
2030	C89103B	2030	EC0293	2030	HS5818	2030	HS5818	2030	MM2261
2030	C89103C	2030	EC0297	2030	HS5820	2030	HS5820	2030	MM2266
2030	C89417	2030	EC0297MP	2030	HS5822	2030	HS5822	2030	MM2270
2030	C8956	2030	ED1702M	2030	HS6010	2030	HS6010	2030	MM306
2030	CXL128	2030	EDC TR11-4	2030	HS6012	2030	HS6012	2030	MM486
2030	CXL192	2030	EDC-TR11-4	2030	HS6014	2030	HS6014	2030	MM487
2030	DOOD	2030	EDC-TR11-5	2030	HS6016	2030	HS6016	2030	MM488
2030	D11C10B1	2030	ELZ14	2030	HT104861	2030	HT104861	2030	MS511
2030	D11C11B1	2030	EM873278	2030	HT104861A	2030	HT104861A	2030	MS512
2030	D11C1B1	2030	EM873279	2030	HT104861B	2030	HT104861B	2030	MS513
2030	D11C3B1	2030	EN1613	2030	HT304861	2030	HT304861	2030	MP4906063
2030	D11C5B1	2030	EN1711	2030	HT304861A	2030	HT304861A	2030	MPS-A06
2030	D11C7B1	2030	EN3904	2030	HT304971C	2030	HT304971C	2030	MPS3642
2030	D182	2030	EP15X33	2030	HT3049714	2030	HT3049714	2030	MP89410A
2030	D183	2030	EP15X5	2030	HT3049714 AO	2030	HT3049714 AO	2030	MP89410AJ
2030	D204	2030	EP16X7	2030	HT3049714 AO	2030	HT3049714 AO	2030	MP89410AK
2030	D204L	2030	E28131	2030	HT3049714B	2030	HT3049714B	2030	MP89410H
2030	D219	2030	ER8100	2030	HT306441 AO	2030	HT306441 AO	2030	MP89416
2030	D221	2030	ER8120	2030	HT306441 AO	2030	HT306441 AO	2030	MP89416A
2030	D227	2030	ES15226	2030	HT306441B-0	2030	HT306441B-0	2030	MP89416AT
2030	D228	2030	ES15X93	2030	HT306441B0	2030	HT306441B0	2030	MP89416BT
2030	D233	2030	ES22(ELCOM)	2030	HT306441C-0	2030	HT306441C-0	2030	MP89417A
2030	D261	2030	ES51X65	2030	HT307541	2030	HT307541	2030	MP89417AT
2030	D261A	2030	ES56(ELCOM)	2030	HT307341A	2030	HT307341A	2030	MP89417T
2030	D261B	2030	ES57(ELCOM)	2030	HT307341C-0	2030	HT307341C-0	2030	MP894188
2030	D261C	2030	ES62(ELCOM)	2030	HT307342B	2030	HT307342B	2030	MP89418T
2030	D261D	2030	ES8964P	2030	HT307342C	2030	HT307342C	2030	MP89466A
2030	D261E	2030	ET15136	2030	HT082810	2030	HT082810	2030	MP89416A
2030	D261F	2030	ET15X8	2030	HT309680B	2030	HT309680B	2030	MP89616J
2030	D261L	2030	ET8-070	2030	HT309714A-0	2030	HT309714A-0	2030	MP89633
2030	D261P	2030	EU15X27	2030	HT309841B0	2030	HT309841B0	2030	MP89696
2030	D261R	2030	EU15X34	2030	HT309841B0	2030	HT309841B0	2030	MPSA06
2030	D261V	2030	EX-141216	2030	HT309842A-0	2030	HT309842A-0	2030	MX89410H
2030	D261W	2030	EX744-X	2030	HT313181C	2030	HT313181C	2030	MX3933
2030	D28D07	2030	F318-1	2030	HT31383X	2030	HT31383X	2030	MS1010
2030	D327	2030	F3560	2030	HT31383ZC	2030	HT31383ZC	2030	MS2991
2030	D336	2030	F3561	2030	HT313841R	2030	HT313841R	2030	MS510
2030	D336R	2030	F3565	2030	HT3140771Q	2030	HT3140771Q	2030	MST506H
2030	D336Y	2030	F3589	2030	HT31909100	2030	HT31909100	2030	MST506J
2030	D33D21J1	2030	F361	2030	HT404001E	2030	HT404001E	2030	MS07506
2030	D33D22J1	2030	F4709	2030	HT800012F	2030	HT800012F	2030	MSK5405
2030	D33D22J2	2030	P625-1	2030	HT8001210	2030	HT8001210	2030	MSP999058-1
2030	D33D24J1	2030	P69916	2030	HT8001310	2030	HT8001310	2030	MST-10
2030	D33D25J1	2030	FBN-CP34634	2030	HT8001191H	2030	HT8001191H	2030	MT1613
2030	D33D26J1	2030	FBN-L109	2030	HX-50103	2030	HX-50103	2030	MT1T11
2030	D33D27J1	2030	FBN-L113	2030	I473608-2	2030	I473608-2	2030	ON020540
2030	D33D28J1	2030	FBN-L115	2030	I473679-1	2030	I473679-1	2030	NB211
2030	D33D29	2030	FBN-L148	2030	I6191	2030	I6191	2030	NB211E1
2030	D33D29J1	2030	FBTX070	2030	I9631	2030	I9631	2030	NCR046
2030	D33D30	2030	FC81229	2030	INTRON-108	2030	INTRON-108	2030	NCR047
2030	D33D30J1	2030	FC89013HG	2030	IP20-0230	2030	IP20-0230	2030	NJ107
2030	D35SC	2030	FC89013HH	2030	IR-TR53	2030	IR-TR53	2030	NN7000
2030	D355E	2030	FC89014C	2030	IRTR-53	2030	IRTR-53	2030	NN7001
2030	D400(TRANSISTOR)	2030	FD-1029-FT	2030	IRTR21	2030	IRTR21	2030	NN7002
2030	D400D	2030	FD-1029-GE	2030	IRTR33	2030	IRTR33	2030	NN7003
2030	D400E	2030	FD-1029-GM	2030	IRTR53	2030	IRTR53	2030	NN7004
2030	D400P	2030	FD-1029-JN	2030	IRTR87	2030	IRTR87	2030	NN7005
2030	D400P1	2030	FD-1029-MM	2030	J2410F5(28C1209C)	2030	J2410F5(28C1209C)	2030	NPC115
2030	D400P1D	2030	FD-1029-NS	2030	J241255	2030	J241255	2030	NPO187
2030	D400P1E	2030	FD-1029-PA	2030	J241256	2030	J241256	2030	NPO189
2030	D400P1F	2030	F81331	2030	K071961-001	2030	K071961-001	2030	NR-141E8
2030	D4028	2030	F82042	2030	K14-0066-13	2030	K14-0066-13	2030	NR-141ET
2030	D4029	2030	F827233	2030	KA1225	2030	KA1225	2030	NS1355
2030	D4C30	2030	FT001	2030	KD2118	2030	KD2118	2030	NS1900
2030	D4C31	2030	FT0019H	2030	KGE41054	2030	KGE41054	2030	NS1960
2030	D4D20	2030	FT002	2030	KGE41414	2030	KGE41414	2030	NS2100
2030	D4D21	2030	FT003	2030	KLH5807	2030	KLH5807	2030	NS2101
2030	D7A30	2030	FT004	2030	KS2018O-L1	2030	KS2018O-L1	2030	NS9400
2030	D7A31	2030	FT004A	2030	KT600	2030	KT600	2030	NS9420
2030	D7A32	2030	FT027	2030	KT600F	2030	KT600F	2030	NS950
2030	D911138-1	2030	FT3642	2030	KT600T	2030	KT600T	2030	NS9500
2030	D911138-2	2030	FX3724	2030	L532 008 012	2030	L532 008 012	2030	NS9540
2030	D911138-3	2030	FX4960	2030	L532-008-012	2030	L532-008-012	2030	NS9728
2030	D911138-4	2030	G05-055-C	2030	L532000162	2030	L532000162	2030	NS9729
2030	D911138-5	2030	G05-055-E	2030	LDA404	2030	LDA404	2030	NS9730
2030	D911138-6	2030	G05-413A	2030	LDA405	2030	LDA405	2030	NS9731
2030	D911138-7	2030	G05-413C	2030	LDA406	2030	LDA406	2030	NTC-11
2030	DBRZ272001	2030	G05-413D	2030	LDA408	2030	LDA408	2030	OA-10
2030	DDBY410001	2030	G05-415-B	2030	LD8200	2030	LD8200	2030	00V60529
2030	DDBY41002	2030	G05413A	2030	LD8201	2030	LD8201	2030	ORP-2
2030	DN20-00453	2030	G05413B	2030	M-75536-1	2030	M-75536-1	2030	P-11748-1
2030	D8-512	2030	G05415	2030	M-75536-2	2030	M-75536-2	2030	P-11903-1
2030	D8512	2030	G23-46	2030	M1X	2030	M1X	2030	P04-45-0026-P5
2030	DT1110	2030	G5A7A66-2	2030	M300-1300A	2030	M300-1300A	2030	P0445O026P5
2030	DT1111	2030	GC1615-1	2030	M4689	2030	M4689	2030	P4069
2030	DT1112	2030	GE-18	2030	M4918	2030	M4918	2030	P480A0018
2030	DT1120	2030	GB-243	2030	M4919	2030	M4919	2030	P4Z
2030	DT1121	2030	GB-271	2030	M530	2030	M530	2030	P6533024
2030	DT1122	2030	GB-47	2030	M9138	2030	M9138	2030	P6330240
2030	DT1311	2030	GB-63	2030	M9170	2030	M9170	2030	P6450026
2030	DT1321	2030	GB-663	2030	M9184	2030	M9184	2030	P6786
2030	DT1510	2030	GB-88	2030	M9209	2030	M9209	2030	P9962-1
2030	DT1511	2030	GET929	2030	M9221	2030	M9221	2030	P9962-2
2030	DT1512	2030	GET930	2030	M9228	2030	M9228	2030	P9962-4
2030	DT1520	2030	G05-015-D	2030	M9380	2030	M9380	2030	P9962-5
2030	DT1521	2030	G05-034-D	2030	M9491	2030	M9491	2030	PA9483
2030	DT1522	2030	G05-036D	2030	M9519	2030	M9519	2030	PBC184
2030	DT1621	2030	G05-055-D	2030	M9521	2030	M9521	2030	PC4900-1
2030	DTIN206	2030	G05-413A	2030	M9562	2030	M9562	2030	PF2001
2030	DTH206	2030	G05415	2030	M9591	2030	M9591	2030	PET1001
2030	DW-6982	2030	HC-00644	2030	M9631	2030	M9631	2030	PET9021
2030	DW6195			2030	M9703	2030	M9703	2030	PET9022

RS 276-	Industry Standard No.	RS 276-	Industry Standard No.	RS 276-	Industry Standard No.	RS 276-	Industry Standard No.
2030	PG330024Q	2030	S1683	2030	ST4203	2030	TV-70B
2030	PIT-74	2030	S1689	2030	ST4204	2030	TV21
2030	PL1083	2030	S1762	2030	ST4341	2030	TV23
2030	PL1084	2030	S1773	2030	ST5061	2030	TV25
2030	PMC-Q8-0320	2030	S1777	2030	ST6573	2030	TV26
2030	PMC-Q8-0400	2030	S17900	2030	ST6574	2030	TV28
2030	PN2221	2030	S18000	2030	STC1336	2030	TV41
2030	PN2222	2030	S18200	2030	SYL4280	2030	TV42
2030	PRT-104-4	2030	S1864	2030	T-04689	2030	TV45
2030	PT1544	2030	S1874	2030	T-291	2030	TV51
2030	PT1545	2030	S19386	2030	T-339	2030	TV52
2030	PT2540	2030	S2104	2030	T-Q5033	2030	TV53
2030	PT3502	2030	S2118	2030	T-Q5055	2030	TV59
2030	PT612	2030	S21549	2030	T-Q5032	2030	TV6080
2030	PT850	2030	S2209	2030	T-Q5063	2030	TV8-28C582
2030	PT850A	2030	S2369	2030	T-Q5081	2030	TV8-28C582A
2030	PT888	2030	S2371	2030	T-Q5099	2030	TV828A543
2030	PT896	2030	S2400A	2030	T01-022	2030	TV828C696
2030	PTC101	2030	S2400B	2030	T13015	2030	UPI1613
2030	PTC178	2030	S2401	2030	T1340A3H	2030	UPI2217
2030	Q-0-172	2030	S2401A	2030	T164213	2030	UPI2218
2030	Q-00869	2030	S2401B	2030	T1706	2030	UPI4046
2030	Q-00869A	2030	S2401C	2030	T1706A	2030	V120RH
2030	Q-00869B	2030	S2402	2030	T1706B	2030	V139
2030	Q-00969	2030	S2402A	2030	T1706C	2030	V142
2030	Q-00969A	2030	S2402B	2030	T1811	2030	V154
2030	Q-00969B	2030	S2402C	2030	T1811E	2030	V166
2030	Q-00969C	2030	S2427	2030	T1811G	2030	V172
2030	Q-01169	2030	S2459B	2030	T23-94	2030	V177
2030	Q-01169A	2030	S2461A	2030	T247	2030	V8-C81255H
2030	Q-01169B	2030	S24616	2030	T336-2	2030	V8-C81255HP
2030	Q-01169C	2030	S2487	2030	T4205L1	2030	V828C1166-Y-1
2030	Q-0172	2030	S2526	2030	T452	2030	V828C1627A
2030	Q5099E	2030	S2648	2030	T9418	2030	V828C1627Y
2030	Q5099F	2030	S27233	2030	T9631	2030	V828C1741-1
2030	Q5124	2030	S2794	2030	TA198035-1	2030	V828D467-0/-1
2030	Q5160	2030	S2992	2030	TA1Z	2030	V8C81255H
2030	Q7-C1318XDN	2030	S409F	2030	TA6200	2030	W14
2030	QA-10	2030	S6804	2030	TCS100	2030	WEP53
2030	QA-8	2030	SC1229E	2030	TCS102	2030	WEP553
2030	QA8	2030	SC144191005	2030	TB1990	2030	WEP912
2030	QQV60529	2030	SC1444103053	2030	TE3416	2030	X16E3890
2030	QRF-2	2030	SDD1220	2030	TE3417	2030	X19001-B
2030	Q8C509	2030	SDD420	2030	TE3859A	2030	X19001-C
2030	QT-C1318XDN	2030	SE-4020	2030	TE4425	2030	X300-1300A
2030	R-3552-1	2030	SE-8001	2030	TP101-A	2030	X325099
2030	R-3553-1	2030	SE-8010	2030	TP101-B	2030	X32C5111
2030	R-3555	2030	SE1012	2030	TP101-D	2030	XA-1095
2030	R-3580-1	2030	SE5-0452	2030	TG28C1175C	2030	XA1018
2030	R123	2030	SE5-0958	2030	TG28C65(Y)	2030	XB12
2030	R123-2	2030	SE6001	2030	TI-424	2030	X830
2030	R123-3	2030	SE6006	2030	TI-425	2030	Y56601-47
2030	R123-4	2030	SE6020	2030	TI-475	2030	ZT1479
2030	R123-5	2030	SE6020A	2030	TI-480	2030	ZT1481
2030	R15003	2030	SE6021	2030	TI-482	2030	ZT1613
2030	R15003P1	2030	SE6021A	2030	TI-483	2030	ZT1700
2030	R2270-60106	2030	SE6022	2030	TI-484	2030	ZT190
2030	R2270-77873D	2030	SE6023	2030	TI-485	2030	ZT191
2030	R227077873D	2030	SE7005	2030	TI-496	2030	ZT192
2030	R3508	2030	SE7015	2030	T13015	2030	ZT193
2030	R3555-3	2030	SE8001	2030	T164213	2030	ZT210
2030	R3593	2030	SE8002	2030	TI8110	2030	ZT211
2030	R3608	2030	SE8010	2030	TIS107	2030	ZT2270
2030	R3608-1	2030	SE8012	2030	TIS133	2030	ZT3053
2030	R3608-2	2030	SE8041	2030	TIS134	2030	ZT3512
2030	R3679	2030	SE8042	2030	TI860A	2030	ZT3866
2030	R5048	2030	SE8510	2030	TI860B	2030	ZT66A
2030	R7613	2030	SE8520	2030	TI860C	2030	ZT66B
2030	R8915	2030	SE8521	2030	TI860D	2030	ZT66C
2030	RC2270	2030	SEC1078	2030	TI860E	2030	ZT90
2030	RCA1A01	2030	SEC1079	2030	TI860M	2030	ZT93
2030	RCA1A06	2030	SEC1477	2030	TI892M	2030	ZT94
2030	RCA1A07	2030	SEC1479	2030	TM128	2030	ZTX302
2030	RCA1A15	2030	SF.T440	2030	TM1613	2030	ZTX303
2030	RCA1A17	2030	SF.T443	2030	TM1711	2030	ZTX331
2030	RCA1A18	2030	SF.T443A	2030	TM192	2030	001-021219-0
2030	RE17	2030	SF.T445	2030	TM293	2030	001-021190
2030	RE196	2030	SF.T714	2030	TM297	2030	001-021221
2030	RE70	2030	SFT443	2030	TN421	2030	001-021290
2030	REN128	2030	SFT443A	2030	TN79	2030	01-030734
2030	REN192	2030	SFT445	2030	TN81	2030	01-031166
2030	REN293	2030	S05013	2030	TNJ70482	2030	01-031318
2030	REN297	2030	S9C7202	2030	TNJ71035	2030	1-035/2207
2030	REN70	2030	SJ2032	2030	TNJ71234	2030	01-2110
2030	R8-2014	2030	SK3024	2030	TNJ71252	2030	01-2111
2030	R8132	2030	SK3047	2030	TNJ72288	2030	01-2114
2030	R8T672	2030	SK3137	2030	TQ-5055	2030	1-2114-Q
2030	R8T678	2030	SK5512	2030	TQ5055	2030	1A0066
2030	R88101	2030	SKA4410	2030	TQ5081	2030	1A34
2030	R88103	2030	SKA4616	2030	TQPD3053	2030	1A34(R)
2030	R88105	2030	SL301C	2030	TR-01(PENNCREST)	2030	1A348
2030	R88107	2030	SL301CE	2030	TR-01R	2030	1A348(R)
2030	R88109	2030	SM2716	2030	TR-01E(PENNCREST)	2030	1A348R
2030	R88111	2030	SM3978	2030	TR-07(PENNCREST)	2030	1A34R
2030	R88113	2030	SM6251	2030	TR-1000-3	2030	1A38
2030	RT-114	2030	SM7991	2030	TR-13	2030	1A38(R)
2030	RT-141	2030	SM-706009D	2030	TR-164(OLSON)	2030	1A38R
2030	RT-188	2030	SMC-583259	2030	TR-25	2030	1N3112
2030	RT141	2030	SMC-620774-1	2030	TR-28C482	2030	1WB995A
2030	RT154	2030	SN166	2030	TR-31B	2030	002-010600
2030	RT188	2030	SN167	2030	TR-33	2030	002-012500
2030	RT482	2030	SPB402	2030	TR-4R31	2030	02-33379-6
2030	RT483	2030	SPD-80123	2030	TR-53	2030	2C291(TRANSISTOR)
2030	RT484	2030	SPS-0122	2030	TR-5R31	2030	2E1A20A22AAB
2030	RT5151	2030	SPS-4077	2030	TR-7R31	2030	2N1092
2030	RT5152	2030	SS80122	2030	TR-8021	2030	2N1154
2030	RT5203	2030	SS82130	2030	TR-8023	2030	2N1154/951
2030	RT5204	2030	SS83912	2030	TR-8024	2030	2N1155
2030	RT5401	2030	SS83914	2030	TR-8036	2030	2N1155/952
2030	RT5402	2030	SS84038	2030	TR-87	2030	2N1156
2030	RT5403	2030	SS84300	2030	TRO1054-1	2030	2N1156/953
2030	RT5404	2030	SS84309	2030	TRO1062-1	2030	2N1206
2030	RT5905	2030	SS84311	2030	TRO1062-7	2030	2N1252
2030	RT5906	2030	SS84361	2030	TR1001	2030	2N1253
2030	RT5907	2030	SS84391	2030	TR1003	2030	2N1253A
2030	RT699M	2030	SS84461	2030	TR1005	2030	2N1335
2030	RT7945	2030	SS84490	2030	TR23	2030	2N1336
2030	RV1473	2030	SS84495	2030	TR25	2030	2N1337
2030	RV2068	2030	SS849	2030	TR28C482	2030	2N1339
2030	RYN12104	2030	SS85809	2030	TR36643	2030	2N1340
2030	S-522	2030	SS86124	2030	TR8036	2030	2N1341
2030	S001683	2030	ST 254 Q	2030	TRO1054-1	2030	2N1409A
2030	S007220	2030	ST-201	2030	TRO1054-7	2030	2N1410A
2030	S1368	2030	ST-213	2030	TS2218	2030	2N1444
2030	S1421	2030	ST-254-Q	2030	TS2219	2030	2N1472
2030	S1514	2030	ST-LM2682	2030	TSC-722	2030	2N1479
2030	S1516	2030	ST1504	2030	TSC722	2030	2N1480
2030	S1517	2030	ST1505	2030	TV-26	2030	2N1481
2030	S1523	2030	ST213	2030	TV-28	2030	2N1482
2030	S1525	2030	ST402	2030	TV-41	2030	2N1613
2030	S15660	2030	ST4150	2030	TV-43	2030	2N1613L
2030	S1642	2030	ST4201	2030	TV-45	2030	2N1613B
2030	S1644	2030	ST4202	2030	TV-59	2030	2N1700
2030	S1671			2030	TV-67	2030	2N1711

RS 276-	Industry Standard No.	RS 276-	Industry Standard No.	RS 276-	Industry Standard No.	RS 276-	Industry Standard No.
2030	2N1711A	2030	2N3981	2030	2SC1072J	2030	2SC114R
2030	2N1711L	2030	2N3982	2030	2SC1072K	2030	2SC114X
2030	2N1711S	2030	2N4046	2030	2SC1072L	2030	2SC114Y
2030	2N1714	2030	2N4237	2030	2SC1072M	2030	2SC115
2030	2N1716	2030	2N4383	2030	2SC1072R	2030	2SC115-1
2030	2N1764	2030	2N4384	2030	2SC1072R	2030	2SC115-2
2030	2N1889	2030	2N4385	2030	2SC1072X	2030	2SC115-3
2030	2N1890	2030	2N4386	2030	2SC1072Y	2030	2SC115-43
2030	2N1943	2030	2N4425	2030	2SC108	2030	2SC115A
2030	2N1958A	2030	2N4427	2030	2SC108-R	2030	2SC115B
2030	2N1958A/51	2030	2N4943	2030	2SC108A	2030	2SC115C
2030	2N1959A	2030	2N4944	2030	2SC108A-O	2030	2SC115D
2030	2N1959A/51	2030	2N4946	2030	2SC108A-R	2030	2SC115E
2030	2N1962/46	2030	2N4960	2030	2SC108B	2030	2SC115F
2030	2N1964/46	2030	2N4961	2030	2SC108C	2030	2SC115G
2030	2N1965/46	2030	2N4962	2030	2SC108D	2030	2SC115GN
2030	2N1990	2030	2N4963	2030	2SC108E	2030	2SC115H
2030	2N2017	2030	2N4964	2030	2SC108F	2030	2SC115J
2030	2N2086	2030	2N497	2030	2SC108G	2030	2SC115K
2030	2N2102	2030	2N497A	2030	2SC108GN	2030	2SC115M
2030	2N2102L	2030	2N498	2030	2SC108H	2030	2SC115OR
2030	2N2102B	2030	2N5106	2030	2SC108J	2030	2SC115R
2030	2N2106	2030	2N5120	2030	2SC108K	2030	2SC115X
2030	2N2107	2030	2N5129	2030	2SC108L	2030	2SC116
2030	2N2108	2030	2N5135	2030	2SC108M	2030	2SC116-O
2030	2N2192	2030	2N5136	2030	2SC108OR	2030	2SC116-QR
2030	2N2192A	2030	2N5188	2030	2SC108R	2030	2SC116-O
2030	2N2192B	2030	2N5211	2030	2SC108X	2030	2SC116-R
2030	2N2193	2030	2N5233	2030	2SC108Y	2030	2SC116-Y
2030	2N2195A	2030	2N5234	2030	2SC109	2030	2SC116D
2030	2N2224	2030	2N5235	2030	2SC109A	2030	2SC116QR
2030	2N2239	2030	2N5300	2030	2SC109A-O	2030	2SC116O
2030	2N2270	2030	2N5309	2030	2SC109A-R	2030	2SC116R
2030	2N2270L	2030	2N5310	2030	2SC109A-Y	2030	2SC116Y
2030	2N2270B	2030	2N5311	2030	2SC109B	2030	2SC117
2030	2N2279/51	2030	2N5320	2030	2SC109C	2030	2SC117A
2030	2N2297	2030	2N5320HB	2030	2SC109D	2030	2SC117B
2030	2N2350	2030	2N5421	2030	2SC109E	2030	2SC117C
2030	2N2350A	2030	2N5449	2030	2SC109F	2030	2SC117D
2030	2N2351	2030	2N545	2030	2SC109G	2030	2SC117P
2030	2N2352	2030	2N5450	2030	2SC109G1	2030	2SC117Q
2030	2N2352A	2030	2N5451	2030	2SC109GN	2030	2SC117GN
2030	2N2368/51	2030	2N5784	2030	2SC109J	2030	2SC117H
2030	2N2369/51	2030	2N5812	2030	2SC109K	2030	2SC117J
2030	2N2403	2030	2N5820	2030	2SC109L	2030	2SC117K
2030	2N2404	2030	2N5820HB	2030	2SC109OR	2030	2SC117L
2030	2N2410	2030	2N5822	2030	2SC109R	2030	2SC117M
2030	2N2410/51	2030	2N5856	2030	2SC109X	2030	2SC117R
2030	2N243	2030	2N5859	2030	2SC109Y	2030	2SC117X
2030	2N244	2030	2N5881	2030	2SC110A	2030	2SC117Y
2030	2N2443	2030	2N5882	2030	2SC110B	2030	2SC118
2030	2N245	2030	2N5964	2030	2SC110C	2030	2SC118A
2030	2N246	2030	2N5965	2030	2SC110D	2030	2SC118B
2030	2N2478	2030	2N6002	2030	2SC110E	2030	2SC118C
2030	2N2484A	2030	2N6006	2030	2SC110F	2030	2SC118D
2030	2N2537	2030	2N6010	2030	2SC110G	2030	2SC118E
2030	2N2538	2030	2N6014	2030	2SC110GN	2030	2SC118F
2030	2N2594	2030	2N655A	2030	2SC110H	2030	2SC118G
2030	2N2788	2030	2N656	2030	2SC110J	2030	2SC118GN
2030	2N2789	2030	2N656A	2030	2SC110K	2030	2SC118H
2030	2N2846	2030	2N657	2030	2SC110L	2030	2SC118J
2030	2N2848	2030	2N696A	2030	2SC110M	2030	2SC118L
2030	2N2849	2030	2N697	2030	2SC110OR	2030	2SC118M
2030	2N2850	2030	2N697L	2030	2SC110R	2030	2SC118OR
2030	2N2851	2030	2N697B	2030	2SC110X	2030	2SC118R
2030	2N2852	2030	2N698	2030	2SC110Y	2030	2SC118X
2030	2N2853	2030	2N699A	2030	2SC111A	2030	2SC118Y
2030	2N2854	2030	2N699B	2030	2SC111B	2030	2SC119
2030	2N2855	2030	2N706M-JAN	2030	2SC111C	2030	2SC119A
2030	2N2856	2030	2N709/46	2030	2SC111D	2030	2SC119B
2030	2N2863	2030	2N709/51	2030	2SC111E	2030	2SC119C
2030	2N2864	2030	2N709A/51	2030	2SC111F	2030	2SC119D
2030	2N2886	2030	2N720	2030	2SC111G	2030	2SC119E
2030	2N2890	2030	2N720A	2030	2SC111GN	2030	2SC119F
2030	2N2891	2030	2N743/46	2030	2SC111H	2030	2SC119G
2030	2N2895	2030	2N743/51	2030	2SC111J	2030	2SC119GN
2030	2N2897	2030	2N744/46	2030	2SC111K	2030	2SC119H
2030	2N2900	2030	2N744/51	2030	2SC111L	2030	2SC119J
2030	2N2926G	2030	2N753/51	2030	2SC111M	2030	2SC119K
2030	2N2926R	2030	2N784/51	2030	2SC111OR	2030	2SC119L
2030	2N2J324	2030	2N834/51	2030	2SC111R	2030	2SC119M
2030	2N2J374	2030	2N835/51	2030	2SC111X	2030	2SC119OR
2030	2N3019	2030	2N870	2030	2SC111Y	2030	2SC119R
2030	2N3020	2030	2N871	2030	2SC112	2030	2SC119Y
2030	2N3036	2030	2N907	2030	2SC112A	2030	2SC12
2030	2N3038	2030	2N908	2030	2SC112B	2030	2SC120
2030	2N3052	2030	2N910	2030	2SC112C	2030	2SC1206
2030	2N3053	2030	2N911	2030	2SC112D	2030	2SC1209
2030	2N3053/40053	2030	2N912	2030	2SC112E	2030	2SC1209(C)
2030	2N3056	2030	2N918/46	2030	2SC112F	2030	2SC1209C
2030	2N3057	2030	2N981	2030	2SC112G	2030	2SC1209D
2030	2N3077	2030	2NL48	2030	2SC112GN	2030	2SC1209E
2030	2N3078	2030	2S014	2030	2SC112H	2030	2SC121
2030	2N3107	2030	2S017	2030	2SC112J	2030	2SC1211
2030	2N3108	2030	2S018	2030	2SC112K	2030	2SC1211C
2030	2N3109	2030	2S019	2030	2SC112L	2030	2SC1211D
2030	2N3110	2030	2S020	2030	2SC112M	2030	2SC1211E
2030	2N3117	2030	2S103	2030	2SC112OR	2030	2SC1218
2030	2N3118	2030	2S104	2030	2SC112R	2030	2SC122
2030	2N3252	2030	2S644(S)	2030	2SC112X	2030	2SC1220
2030	2N3262	2030	2S741	2030	2SC112Y	2030	2SC1220(E)
2030	2N3295	2030	2S742	2030	2SC113	2030	2SC1220-003
2030	2N3299	2030	2S742A	2030	2SC113A	2030	2SC1220A
2030	2N3374	2030	2S744	2030	2SC113B	2030	2SC1220A(QPR)
2030	2N3339	2030	2S745	2030	2SC113C	2030	2SC1220AP
2030	2N3339A	2030	2SC-4012	2030	2SC113D	2030	2SC1220AQ
2030	2N3403	2030	2SC-NJ-107	2030	2SC113E	2030	2SC1220AR
2030	2N3404	2030	2SC-NJ107	2030	2SC113F	2030	2SC1220E
2030	2N3405	2030	2SC1008	2030	2SC113G	2030	2SC1220P
2030	2N342	2030	2SC1010D	2030	2SC113GN	2030	2SC1220Q
2030	2N3426	2030	2SC1010E	2030	2SC113H	2030	2SC1220R
2030	2N342B	2030	2SC1010F	2030	2SC113J	2030	2SC123
2030	2N343	2030	2SC1010G	2030	2SC113K	2030	2SC124
2030	2N3435	2030	2SC1010GN	2030	2SC113L	2030	2SC1247
2030	2N3439	2030	2SC1010H	2030	2SC113OR	2030	2SC12A
2030	2N343A	2030	2SC1010J	2030	2SC113R	2030	2SC12B
2030	2N343B	2030	2SC1010K	2030	2SC113X	2030	2SC12C
2030	2N3469	2030	2SC1010L	2030	2SC113Y	2030	2SC12E
2030	2N3499	2030	2SC1010M	2030	2SC114	2030	2SC12F
2030	2N3501	2030	2SC1010R	2030	2SC114A	2030	2SC12G
2030	2N3506	2030	2SC1010R	2030	2SC114B	2030	2SC12GN
2030	2N3507	2030	2SC1010X	2030	2SC114C	2030	2SC12H
2030	2N3512	2030	2SC1010Y	2030	2SC114D	2030	2SC12J
2030	2N3567	2030	2SC1063	2030	2SC114E	2030	2SC12K
2030	2N3568	2030	2SC1072	2030	2SC114F	2030	2SC12L
2030	2N3569	2030	2SC1072A	2030	2SC114G	2030	2SC12M
2030	2N3633/52	2030	2SC1072B	2030	2SC114GN	2030	2SC12R
2030	2N3665	2030	2SC1072C	2030	2SC114H	2030	2SC12X
2030	2N3666	2030	2SC1072D	2030	2SC114J	2030	2SC12Y
2030	2N3724	2030	2SC1072E	2030	2SC114K		
2030	2N3725	2030	2SC1072F	2030	2SC114L		
2030	2N3877	2030	2SC1072G	2030	2SC114M		
2030	2N3881	2030	2SC1072GN	2030	2SC114OR		
2030	2N3948	2030	2SC1072H				

RS 276-	Industry Standard No.
2030	2SC130
2030	2SC130A
2030	2SC130B
2030	2SC130C
2030	2SC130D
2030	2SC130E
2030	2SC130F
2030	2SC130GN
2030	2SC130H
2030	2SC130J
2030	2SC130K
2030	2SC130L
2030	2SC130M
2030	2SC130OR
2030	2SC130R
2030	2SC130X
2030	2SC130Y
2030	2SC1318
2030	2SC1318(P)
2030	2SC1318(P,R)
2030	2SC1318(Q)
2030	2SC1318A
2030	2SC1318B
2030	2SC1318E
2030	2SC1318E
2030	2SC1318P
2030	2SC1318G
2030	2SC1318GN
2030	2SC1318H
2030	2SC1318J
2030	2SC1318K
2030	2SC1318L
2030	2SC1318M
2030	2SC1318P
2030	2SC1318PR
2030	2SC1318Q
2030	2SC1318QP
2030	2SC1318QR
2030	2SC1318R
2030	2SC1318S
2030	2SC1318S,R
2030	2SC1318Y
2030	2SC1318X
2030	2SC1330
2030	2SC1330A
2030	2SC1330B
2030	2SC1330C
2030	2SC1330D
2030	2SC1330L
2030	2SC1330R
2030	2SC346
2030	2SC346(R)
2030	2SC346Q
2030	2SC346R
2030	2SC346S
2030	2SC347
2030	2SC347(Q)
2030	2SC347A
2030	2SC347F
2030	2SC347P
2030	2SC347L
2030	2SC347Q
2030	2SC347R
2030	2SC347RQ
2030	2SC347S
2030	2SC347X
2030	2SC347Y
2030	2SC383
2030	2SC383(P,Q,R)
2030	2SC383(S)
2030	2SC383,RS
2030	2SC383B
2030	2SC383P
2030	2SC383Q
2030	2SC383R
2030	2SC383RS
2030	2SC383S
2030	2SC383X
2030	2SC384
2030	2SC384(Q)
2030	2SC384-OR
2030	2SC384A
2030	2SC384B
2030	2SC384C
2030	2SC384D
2030	2SC384E
2030	2SC384F
2030	2SC384G
2030	2SC384GN
2030	2SC384H
2030	2SC384J
2030	2SC384K
2030	2SC384L
2030	2SC384M
2030	2SC384OR
2030	2SC384Q
2030	2SC384Q,R
2030	2SC384R
2030	2SC384S
2030	2SC384X
2030	2SC384Y
2030	2SC385H
2030	2SC385H
2030	2SC138B
2030	2SC138C
2030	2SC138D
2030	2SC138E
2030	2SC138F
2030	2SC138G
2030	2SC138GN
2030	2SC138H
2030	2SC138J
2030	2SC138L
2030	2SC138M
2030	2SC138OR
2030	2SC138R
2030	2SC138X
2030	2SC139A
2030	2SC139B
2030	2SC139C
2030	2SC139D
2030	2SC139E
2030	2SC139F
2030	2SC139G
2030	2SC139GN
2030	2SC139H
2030	2SC139J
2030	2SC139K
2030	2SC139M
2030	2SC139OR
2030	2SC139J
2030	2SC139X
2030	2SC139Y
2030	2SC140
2030	2SC1406
2030	2SC1406(P)
2030	2SC1406P
2030	2SC1406Q
2030	2SC1407
2030	2SC1407(Q)
2030	2SC1407O
2030	2SC1407B
2030	2SC1407P
2030	2SC1407Q
2030	2SC1407R
2030	2SC1407S
2030	2SC1407X
2030	2SC140A
2030	2SC140C
2030	2SC140D
2030	2SC140E
2030	2SC140F
2030	2SC140G
2030	2SC140GN
2030	2SC140H
2030	2SC140J
2030	2SC140K
2030	2SC140L
2030	2SC140M
2030	2SC140OR
2030	2SC140R
2030	2SC140X
2030	2SC140Y
2030	2SC147
2030	2SC1474
2030	2SC1474-3
2030	2SC1474-4
2030	2SC1474J
2030	2SC1474S
2030	2SC147A
2030	2SC147B
2030	2SC147C
2030	2SC147D
2030	2SC147P
2030	2SC147Q
2030	2SC147GN
2030	2SC147H
2030	2SC147J
2030	2SC147K
2030	2SC147L
2030	2SC147M
2030	2SC147R
2030	2SC147X
2030	2SC147Y
2030	2SC15-Q
2030	2SC150
2030	2SC150-OR
2030	2SC1509
2030	2SC1509P
2030	2SC1509Q
2030	2SC1509R
2030	2SC1509S
2030	2SC150A
2030	2SC150B
2030	2SC150C
2030	2SC150D
2030	2SC150B
2030	2SC150F
2030	2SC150G
2030	2SC150GN
2030	2SC150H
2030	2SC150J
2030	2SC150K
2030	2SC150L
2030	2SC150M
2030	2SC1500R
2030	2SC150R
2030	2SC150T
2030	2SC150X
2030	2SC150Y
2030	2SC151B
2030	2SC151A
2030	2SC151B
2030	2SC151C
2030	2SC151D
2030	2SC151E
2030	2SC151G
2030	2SC151GN
2030	2SC151H
2030	2SC151J
2030	2SC151K
2030	2SC151L
2030	2SC151M
2030	2SC1510R
2030	2SC151R
2030	2SC151X
2030	2SC151Y
2030	2SC152
2030	2SC152A
2030	2SC152B
2030	2SC152C
2030	2SC152D
2030	2SC152E
2030	2SC152F
2030	2SC152G
2030	2SC152GN
2030	2SC152H
2030	2SC152J
2030	2SC152K
2030	2SC152L
2030	2SC152M
2030	2SC152OR
2030	2SC152R
2030	2SC152X
2030	2SC152Y
2030	2SC1570
2030	2SC1570LH
2030	2SC1571
2030	2SC1571G
2030	2SC1571L
2030	2SC157A
2030	2SC157B
2030	2SC157C
2030	2SC157D
2030	2SC157E
2030	2SC157P
2030	2SC157G
2030	2SC157GN
2030	2SC157J
2030	2SC157K
2030	2SC157L
2030	2SC157R
2030	2SC157OR
2030	2SC157R
2030	2SC157Y
2030	2SC158A
2030	2SC158B
2030	2SC158C
2030	2SC158D
2030	2SC158E
2030	2SC158F
2030	2SC158G
2030	2SC158GN
2030	2SC158H
2030	2SC158K
2030	2SC158L
2030	2SC158M
2030	2SC158OR
2030	2SC158R
2030	2SC158X
2030	2SC158Y
2030	2SC15A
2030	2SC15B
2030	2SC15C
2030	2SC15D
2030	2SC15E
2030	2SC15P
2030	2SC15G
2030	2SC15GN
2030	2SC15H
2030	2SC15J
2030	2SC15K
2030	2SC15L
2030	2SC15M
2030	2SC15OR
2030	2SC15R
2030	2SC15X
2030	2SC15Y
2030	2SC1627
2030	2SC1627-Q
2030	2SC1627Y
2030	2SC163A
2030	2SC163B
2030	2SC163C
2030	2SC163D
2030	2SC163E
2030	2SC163F
2030	2SC163G
2030	2SC163GN
2030	2SC163H
2030	2SC163J
2030	2SC163K
2030	2SC163L
2030	2SC163M
2030	2SC163OR
2030	2SC163R
2030	2SC163X
2030	2SC163Y
2030	2SC164
2030	2SC164A
2030	2SC164B
2030	2SC164C
2030	2SC164D
2030	2SC164E
2030	2SC164F
2030	2SC164G
2030	2SC164GN
2030	2SC164H
2030	2SC164J
2030	2SC164K
2030	2SC164L
2030	2SC164OR
2030	2SC164R
2030	2SC164X
2030	2SC165
2030	2SC1682-BL
2030	2SC1682-GR
2030	2SC1682V
2030	2SC16B
2030	2SC16C
2030	2SC16D
2030	2SC16E
2030	2SC16F
2030	2SC16G
2030	2SC16GN
2030	2SC16H
2030	2SC16J
2030	2SC16K
2030	2SC16M
2030	2SC16OR
2030	2SC16R
2030	2SC16X
2030	2SC16Y
2030	2SC1741
2030	2SC1788
2030	2SC1788R
2030	2SC182V
2030	2SC1851
2030	2SC188
2030	2SC188A
2030	2SC188AB
2030	2SC188B
2030	2SC188C
2030	2SC188D
2030	2SC188E
2030	2SC188F
2030	2SC188G
2030	2SC188GN
2030	2SC188J
2030	2SC188K
2030	2SC188L
2030	2SC188M
2030	2SC188OR
2030	2SC188R
2030	2SC188X
2030	2SC188Y
2030	2SC189
2030	2SC189A
2030	2SC189B
2030	2SC189C
2030	2SC189D
2030	2SC189E
2030	2SC189F
2030	2SC189G
2030	2SC189GN
2030	2SC189H
2030	2SC189J
2030	2SC189K
2030	2SC189M
2030	2SC189OR
2030	2SC189R
2030	2SC189X
2030	2SC189Y
2030	2SC18A
2030	2SC18B
2030	2SC18C
2030	2SC18D
2030	2SC18K
2030	2SC18F
2030	2SC18G
2030	2SC18GN
2030	2SC18H
2030	2SC18J
2030	2SC18K
2030	2SC18M
2030	2SC18OR
2030	2SC18R
2030	2SC18X
2030	2SC19
2030	2SC19O
2030	2SC190A
2030	2SC190B
2030	2SC190C
2030	2SC190D
2030	2SC190E
2030	2SC190P
2030	2SC190GN
2030	2SC190H
2030	2SC190K
2030	2SC190L
2030	2SC190M
2030	2SC1900R
2030	2SC190R
2030	2SC190X
2030	2SC190Y
2030	2SC1973
2030	2SC19A
2030	2SC19B
2030	2SC19C
2030	2SC19D
2030	2SC19E
2030	2SC19F
2030	2SC19G
2030	2SC19GN
2030	2SC19H
2030	2SC19J
2030	2SC19K
2030	2SC19L
2030	2SC19M
2030	2SC19OR
2030	2SC19R
2030	2SC19X
2030	2SC19Y
2030	2SC20
2030	2SC200A
2030	2SC200B
2030	2SC200C
2030	2SC200D
2030	2SC200P
2030	2SC200G
2030	2SC200GN
2030	2SC200H
2030	2SC200J
2030	2SC200K
2030	2SC200L
2030	2SC200M
2030	2SC2000R
2030	2SC200R
2030	2SC200X
2030	2SC200Y
2030	2SC201A
2030	2SC201B
2030	2SC201C
2030	2SC201D
2030	2SC201E
2030	2SC201F
2030	2SC201G
2030	2SC201GN
2030	2SC201H
2030	2SC201J
2030	2SC201K
2030	2SC201L
2030	2SC201M
2030	2SC201N
2030	2SC201X
2030	2SC201Y
2030	2SC2028
2030	2SC2028-2
2030	2SC2028/2
2030	2SC2028B
2030	2SC2028B/20
2030	2SC2028B2Q
2030	2SC202A
2030	2SC202B
2030	2SC202C
2030	2SC202D
2030	2SC202E
2030	2SC202F
2030	2SC202G
2030	2SC202GN
2030	2SC202H
2030	2SC202J
2030	2SC202K
2030	2SC202L
2030	2SC202M
2030	2SC202OR
2030	2SC202R
2030	2SC202X
2030	2SC202Y
2030	002SC203A
2030	2SC203B
2030	2SC203C
2030	2SC203D
2030	2SC203E
2030	2SC203F
2030	2SC203G
2030	2SC203GN
2030	2SC203H
2030	2SC203J
2030	2SC203K
2030	2SC203L
2030	2SC203M
2030	2SC203OR
2030	2SC203R
2030	2SC203X
2030	2SC203Y
2030	2SC204A
2030	2SC204B
2030	2SC204C
2030	2SC204D
2030	2SC204E
2030	2SC204F
2030	2SC204G
2030	2SC204GN
2030	2SC204H
2030	2SC204J
2030	2SC204K
2030	2SC204L
2030	2SC204M

RS 276-	Industry Standard No.
2030	2SC2040R
2030	2SC204R
2030	2SC204X
2030	2SC204Y
2030	2SC205A
2030	2SC205B
2030	2SC205C
2030	2SC205D
2030	2SC205F
2030	2SC205G
2030	2SC205GN
2030	2SC205H
2030	2SC205J
2030	2SC205K
2030	2SC205L
2030	2SC205M
2030	2SC205OR
2030	2SC205R
2030	2SC205X
2030	2SC205Y
2030	2SC206O
2030	2SC206OQ
2030	2SC208
2030	2SC2081
2030	2SC2086
2030	2SC208A
2030	2SC208B
2030	2SC208C
2030	2SC208D
2030	2SC208E
2030	2SC208F
2030	2SC208G
2030	2SC208GN
2030	2SC208H
2030	2SC208J
2030	2SC208K
2030	2SC208L
2030	2SC208M
2030	2SC208OR
2030	2SC208R
2030	2SC208X
2030	2SC208Y
2030	2SC20A
2030	2SC20B
2030	2SC20C
2030	2SC20D
2030	2SC20E
2030	2SC20F
2030	2SC20G
2030	2SC20GN
2030	2SC20H
2030	2SC20J
2030	2SC20K
2030	2SC20L
2030	2SC20M
2030	2SC200R
2030	2SC20R
2030	2SC20X
2030	2SC20Y
2030	2SC210
2030	2SC210A
2030	2SC210B
2030	2SC210C
2030	2SC210D
2030	2SC210E
2030	2SC210F
2030	2SC210GN
2030	2SC210H
2030	2SC210J
2030	2SC210K
2030	2SC210L
2030	2SC210M
2030	2SC2100R
2030	2SC210R
2030	2SC210X
2030	2SC210Y
2030	2SC211
2030	2SC211O
2030	2SC212O
2030	2SC2120-O
2030	2SC2120Y
2030	2SC213
2030	2SC213A
2030	2SC213B
2030	2SC213C
2030	2SC213D
2030	2SC213E
2030	2SC213F
2030	2SC213G
2030	2SC213GN
2030	2SC213H
2030	2SC213J
2030	2SC213K
2030	2SC213L
2030	2SC213M
2030	2SC2130R
2030	2SC213R
2030	2SC213X
2030	2SC213Y
2030	2SC214
2030	2SC214A
2030	2SC214B
2030	2SC214C
2030	2SC214D
2030	2SC214E
2030	2SC214F
2030	2SC214G
2030	2SC214GN
2030	2SC214H
2030	2SC214J
2030	2SC214K
2030	2SC214L
2030	2SC214M
2030	2SC2140R
2030	2SC214X
2030	2SC214Y
2030	2SC215
2030	2SC215A
2030	2SC215B
2030	2SC215C
2030	2SC215E
2030	2SC215F
2030	2SC215G
2030	2SC215GN
2030	2SC215H
2030	2SC215J
2030	2SC215K
2030	2SC215L
2030	2SC215M
2030	2SC215OR
2030	2SC215R
2030	2SC215X
2030	2SC215Y
2030	2SC216
2030	2SC216A
2030	2SC216B
2030	2SC216C
2030	2SC216D
2030	2SC216E
2030	2SC216F
2030	2SC216G
2030	2SC216GN
2030	2SC216H
2030	2SC216J
2030	2SC216K
2030	2SC216L
2030	2SC216M
2030	2SC216OR
2030	2SC216R
2030	2SC216X
2030	2SC216Y
2030	2SC217
2030	2SC217A
2030	2SC217B
2030	2SC217C
2030	2SC217D
2030	2SC217E
2030	2SC217F
2030	2SC217G
2030	2SC217GN
2030	2SC217H
2030	2SC217J
2030	2SC217K
2030	2SC217L
2030	2SC217M
2030	2SC217OR
2030	2SC217R
2030	2SC217X
2030	2SC217Y
2030	2SC218B
2030	2SC218C
2030	2SC218D
2030	2SC218E
2030	2SC218F
2030	2SC218G
2030	2SC218GN
2030	2SC218H
2030	2SC218J
2030	2SC218L
2030	2SC218M
2030	2SC2180R
2030	2SC218R
2030	2SC218X
2030	2SC218Y
2030	2SC22
2030	2SC220
2030	2SC220A
2030	2SC220B
2030	2SC220C
2030	2SC220D
2030	2SC220E
2030	2SC220F
2030	2SC220G
2030	2SC220GN
2030	2SC220H
2030	2SC220K
2030	2SC220L
2030	2SC220M
2030	2SC2200R
2030	2SC220R
2030	2SC220X
2030	2SC220Y
2030	2SC221
2030	2SC221A
2030	2SC221B
2030	2SC221C
2030	2SC221D
2030	2SC221E
2030	2SC221F
2030	2SC221G
2030	2SC221GN
2030	2SC221H
2030	2SC221J
2030	2SC221K
2030	2SC221L
2030	2SC221M
2030	2SC2210R
2030	2SC221R
2030	2SC221X
2030	2SC221Y
2030	2SC222
2030	2SC222A
2030	2SC222B
2030	2SC222C
2030	2SC222D
2030	2SC222E
2030	2SC222F
2030	2SC222G
2030	2SC222GN
2030	2SC222H
2030	2SC222J
2030	2SC222M
2030	2SC2220R
2030	2SC222R
2030	2SC222X
2030	2SC222Y
2030	2SC223
2030	2SC2236
2030	2SC22236-O
2030	2SC22236-O
2030	2SC2236Y
2030	2SC224
2030	2SC225
2030	2SC226
2030	2SC226A
2030	2SC226B
2030	2SC226C
2030	2SC226D
2030	2SC226E
2030	2SC226F
2030	2SC226G
2030	2SC226GN
2030	2SC226H
2030	2SC226J
2030	2SC226K
2030	2SC226M
2030	2SC226OR
2030	2SC226R
2030	2SC226X
2030	2SC227
2030	2SC227A
2030	2SC227B
2030	2SC227C
2030	2SC227D
2030	2SC227E
2030	2SC227F
2030	2SC227G
2030	2SC227GN
2030	2SC227H
2030	2SC227J
2030	2SC227K
2030	2SC227L
2030	2SC227M
2030	2SC227OR
2030	2SC227R
2030	2SC227X
2030	2SC227Y
2030	2SC228
2030	2SC228A
2030	2SC228B
2030	2SC228C
2030	2SC228D
2030	2SC228E
2030	2SC228F
2030	2SC228G
2030	2SC228GN
2030	2SC228H
2030	2SC228J
2030	2SC228K
2030	2SC228L
2030	2SC228M
2030	2SC228OR
2030	2SC228R
2030	2SC228X
2030	2SC228Y
2030	2SC229
2030	2SC22A
2030	2SC22B
2030	2SC22D
2030	2SC22E
2030	2SC22P
2030	2SC22G
2030	2SC22GN
2030	2SC22H
2030	2SC22J
2030	2SC22K
2030	2SC22L
2030	2SC22M
2030	2SC22OR
2030	2SC22R
2030	2SC22X
2030	2SC22Y
2030	2SC23
2030	2SC230A
2030	2SC230B
2030	2SC230C
2030	2SC230D
2030	2SC230F
2030	2SC230G
2030	2SC230GN
2030	2SC230H
2030	2SC230K
2030	2SC230L
2030	2SC230M
2030	2SC2300R
2030	2SC230X
2030	2SC230Y
2030	2SC231
2030	2SC231A
2030	2SC231B
2030	2SC231C
2030	2SC231D
2030	2SC231E
2030	2SC231F
2030	2SC231G
2030	2SC231GN
2030	2SC231H
2030	2SC231J
2030	2SC231K
2030	2SC231L
2030	2SC231M
2030	2SC231OR
2030	2SC231R
2030	2SC231X
2030	2SC231Y
2030	2SC232
2030	2SC232A
2030	2SC232B
2030	2SC232C
2030	2SC232D
2030	2SC232E
2030	2SC232F
2030	2SC232G
2030	2SC232GN
2030	2SC232H
2030	2SC232J
2030	2SC232K
2030	2SC232L
2030	2SC232M
2030	2SC2320R
2030	2SC232R
2030	2SC232X
2030	2SC232Y
2030	2SC233
2030	2SC233A
2030	2SC233B
2030	2SC233C
2030	2SC233D
2030	2SC233F
2030	2SC233G
2030	2SC233GN
2030	2SC233H
2030	2SC233J
2030	2SC233M
2030	2SC2330R
2030	2SC233X
2030	2SC233Y
2030	2SC235-O
2030	2SC235A
2030	2SC235B
2030	2SC235C
2030	2SC235E
2030	2SC235F
2030	2SC235GN
2030	2SC235H
2030	2SC235J
2030	2SC235K
2030	2SC235L
2030	2SC235M
2030	2SC2350R
2030	2SC235R
2030	2SC235X
2030	2SC235Y
2030	2SC236
2030	2SC236A
2030	2SC236B
2030	2SC236C
2030	2SC236D
2030	2SC236E
2030	2SC236F
2030	2SC236G
2030	2SC236GN
2030	2SC236H
2030	2SC236K
2030	2SC236L
2030	2SC236M
2030	2SC236OR
2030	2SC236R
2030	2SC236X
2030	2SC236Y
2030	2SC238
2030	2SC238A
2030	2SC238B
2030	2SC238C
2030	2SC238D
2030	2SC238E
2030	2SC238F
2030	2SC238G
2030	2SC238GN
2030	2SC238H
2030	2SC238J
2030	2SC238K
2030	2SC238L
2030	2SC238M
2030	2SC238OR
2030	2SC238R
2030	2SC238X
2030	2SC238Y
2030	2SC23A
2030	2SC23B
2030	2SC23D
2030	2SC23E
2030	2SC23F
2030	2SC23G
2030	2SC23GN
2030	2SC23H
2030	2SC23J
2030	2SC23K
2030	2SC23L
2030	2SC23M
2030	2SC23OR
2030	2SC23R
2030	2SC23X
2030	2SC23Y
2030	2SC24
2030	2SC247
2030	2SC247A
2030	2SC247B
2030	2SC247C
2030	2SC247D
2030	2SC247E
2030	2SC247F
2030	2SC247G
2030	2SC247GN
2030	2SC247H
2030	2SC247J
2030	2SC247L
2030	2SC247M
2030	2SC247OR
2030	2SC247R
2030	2SC247X
2030	2SC247Y
2030	2SC248
2030	2SC248A
2030	2SC248B
2030	2SC248C
2030	2SC248D
2030	2SC248E
2030	2SC248F
2030	2SC248G
2030	2SC248J
2030	2SC248K
2030	2SC248L
2030	2SC248M
2030	2SC248OR
2030	2SC248R
2030	2SC248X
2030	2SC248Y
2030	2SC249
2030	2SC24A
2030	2SC24B
2030	2SC24C
2030	2SC24D
2030	2SC24E
2030	2SC24F
2030	2SC24G
2030	2SC24GN
2030	2SC24H
2030	2SC24J
2030	2SC24K
2030	2SC24L
2030	2SC24M
2030	2SC240R
2030	2SC24R
2030	2SC24X
2030	2SC24Y
2030	2SC260
2030	2SC261
2030	2SC262
2030	2SC267B
2030	2SC267C
2030	2SC267D
2030	2SC267E
2030	2SC267F
2030	2SC267G
2030	2SC267GN
2030	2SC267H
2030	2SC267J
2030	2SC267K
2030	2SC267L
2030	2SC267M
2030	2SC267OR
2030	2SC267R
2030	2SC267X
2030	2SC267Y
2030	2SC268
2030	2SC268-OR
2030	2SC268A
2030	2SC268B
2030	2SC268C
2030	2SC268D
2030	2SC268E
2030	2SC268F

RS 276-	Industry Standard No.
2030	28C268G
2030	28C268GN
2030	28C268H
2030	28C268J
2030	28C268K
2030	28C268L
2030	28C268M
2030	28C2680R
2030	28C268R
2030	28C268X
2030	28C268Y
2030	28C26A
2030	28C26B
2030	28C26C
2030	28C26D
2030	28C26E
2030	28C26F
2030	28C26G
2030	28C26GN
2030	28C26H
2030	28C26J
2030	28C26K
2030	28C26L
2030	28C26M
2030	28C26OR
2030	28C26R
2030	28C26X
2030	28C26Y
2030	28C27
2030	28C27A
2030	28C27B
2030	28C27D
2030	28C27E
2030	28C27F
2030	28C27G
2030	28C27GN
2030	28C27H
2030	28C27J
2030	28C27K
2030	28C27L
2030	28C27M
2030	28C27OR
2030	28C27R
2030	28C27X
2030	28C27Y
2030	28C285B
2030	28C285C
2030	28C285D
2030	28C285E
2030	28C285F
2030	28C285G
2030	28C285GN
2030	28C285H
2030	28C285J
2030	28C285K
2030	28C285L
2030	28C285M
2030	28C2850R
2030	28C285R
2030	28C285X
2030	28C285Y
2030	28C288A1
2030	28C288AB
2030	28C288B
2030	28C288C
2030	28C288D
2030	28C288E
2030	28C288F
2030	28C288G
2030	28C288GN
2030	28C288H
2030	28C288J
2030	28C288K
2030	28C288L
2030	28C288M
2030	28C2880R
2030	28C288R
2030	28C288X
2030	28C288Y
2030	28C291
2030	28C292
2030	28C293A
2030	28C293B
2030	28C293C
2030	28C293D
2030	28C293E
2030	28C293F
2030	28C293G
2030	28C293GN
2030	28C293H
2030	28C293J
2030	28C293K
2030	28C293L
2030	28C293M
2030	28C2930R
2030	28C293R
2030	28C293X
2030	28C293Y
2030	28C298
2030	28C298-4
2030	28C300A
2030	28C300B
2030	28C300C
2030	28C300D
2030	28C300E
2030	28C300F
2030	28C300G
2030	28C300GN
2030	28C300H
2030	28C300J
2030	28C300K
2030	28C300L
2030	28C300M
2030	28C3000R
2030	28C300R
2030	28C300X
2030	28C300Y
2030	28C301A
2030	28C301B
2030	28C301C
2030	28C301D
2030	28C301E
2030	28C301F
2030	28C301G
2030	28C301GN
2030	28C301H
2030	28C301J
2030	28C301K
2030	28C301L
2030	28C301M
2030	28C3010R
2030	28C301R
2030	28C301X
2030	28C301Y
2030	28C302A
2030	28C302B
2030	28C302C
2030	28C302D
2030	28C302E
2030	28C302F
2030	28C302G
2030	28C302GN
2030	28C302H
2030	28C302J
2030	28C302K
2030	28C302L
2030	28C302M
2030	28C3020R
2030	28C302R
2030	28C302X
2030	28C302Y
2030	28C303
2030	28C304
2030	28C305
2030	28C305A
2030	28C305B
2030	28C305C
2030	28C305D
2030	28C305E
2030	28C305F
2030	28C305G
2030	28C305GN
2030	28C305H
2030	28C305J
2030	28C305K
2030	28C305L
2030	28C305M
2030	28C3050R
2030	28C305R
2030	28C305X
2030	28C305Y
2030	28C306
2030	28C306A
2030	28C306B
2030	28C306C
2030	28C306D
2030	28C306E
2030	28C306F
2030	28C306G
2030	28C306GN
2030	28C306H
2030	28C306J
2030	28C306K
2030	28C306L
2030	28C306M
2030	28C3060R
2030	28C306R
2030	28C306X
2030	28C306Y
2030	28C307
2030	28C307(CB-FINAL)
2030	28C307-OR
2030	28C307A
2030	28C307B
2030	28C307C
2030	28C307D
2030	28C307E
2030	28C307F
2030	28C307G
2030	28C307GN
2030	28C307H
2030	28C307J
2030	28C307K
2030	28C307L
2030	28C307M
2030	28C3070R
2030	28C307R
2030	28C307T
2030	28C307X
2030	28C307Y
2030	28C30B
2030	28C308A
2030	28C308B
2030	28C308C
2030	28C308D
2030	28C308E
2030	28C308F
2030	28C308G
2030	28C308GN
2030	28C308H
2030	28C308J
2030	28C308K
2030	28C308L
2030	28C308M
2030	28C3080R
2030	28C308R
2030	28C308X
2030	28C308Y
2030	28C309
2030	28C309A
2030	28C309B
2030	28C309C
2030	28C309D
2030	28C309E
2030	28C309F
2030	28C309G
2030	28C309GN
2030	28C309H
2030	28C309J
2030	28C309K
2030	28C309L
2030	28C309M
2030	28C3090R
2030	28C309R
2030	28C309X
2030	28C309Y
2030	28C31
2030	28C310
2030	28C310A
2030	28C310B
2030	28C310C
2030	28C310D
2030	28C310E
2030	28C310F
2030	28C310G
2030	28C310GN
2030	28C310H
2030	28C310J
2030	28C310K
2030	28C310L
2030	28C310M
2030	28C3100R
2030	28C310R
2030	28C310X
2030	28C310Y
2030	28C317A
2030	28C317B
2030	28C317D
2030	28C317E
2030	28C317F
2030	28C317G
2030	28C317GN
2030	28C317H
2030	28C317J
2030	28C317K
2030	28C317M
2030	28C3170R
2030	28C317R
2030	28C317X
2030	28C317Y
2030	28C31A
2030	28C31B
2030	28C31C
2030	28C31D
2030	28C31E
2030	28C31F
2030	28C31G
2030	28C31GN
2030	28C31H
2030	28C31J
2030	28C31K
2030	28C31L
2030	28C31M
2030	28C310R
2030	28C31R
2030	28C31X
2030	28C31Y
2030	28C32
2030	28C32A
2030	28C32B
2030	28C32C
2030	28C32E
2030	28C32F
2030	28C32G
2030	28C32GN
2030	28C32H
2030	28C32J
2030	28C32K
2030	28C32L
2030	28C32M
2030	28C320R
2030	28C32R
2030	28C32X
2030	28C32Y
2030	28C33A
2030	28C33B
2030	28C33C
2030	28C33E
2030	28C33F
2030	28C33G
2030	28C330N
2030	28C33H
2030	28C33J
2030	28C33K
2030	28C33L
2030	28C33M
2030	28C330R
2030	28C33X
2030	28C33Y
2030	28C352
2030	28C352-OR
2030	28C352A
2030	28C352AC
2030	28C352B
2030	28C352C
2030	28C352D
2030	28C352E
2030	28C352F
2030	28C352G
2030	28C352GN
2030	28C352H
2030	28C352J
2030	28C352L
2030	28C352M
2030	28C3520R
2030	28C352R
2030	28C352X
2030	28C352Y
2030	28C353
2030	28C353A
2030	28C353AC
2030	28C353C
2030	28C353D
2030	28C353E
2030	28C353F
2030	28C353G
2030	28C353H
2030	28C353J
2030	28C353K
2030	28C353L
2030	28C353M
2030	28C3530R
2030	28C353R
2030	28C353X
2030	28C353Y
2030	28C376
2030	28C396A
2030	28C396B
2030	28C396C
2030	28C396D
2030	28C396E
2030	28C396F
2030	28C396G
2030	28C396GN
2030	28C396H
2030	28C396J
2030	28C396K
2030	28C396L
2030	28C396M
2030	28C3960R
2030	28C396R
2030	28C396X
2030	28C396Y
2030	28C4012A
2030	28C4012B
2030	28C4012C
2030	28C4012E
2030	28C4012P
2030	28C4012G
2030	28C4012GN
2030	28C4012H
2030	28C4012J
2030	28C4012L
2030	28C4012M
2030	28C4012X
2030	28C4012Y
2030	28C425A
2030	28C425G
2030	28C425GN
2030	28C425H
2030	28C425K
2030	28C425L
2030	28C425M
2030	28C4250R
2030	28C425R
2030	28C425X
2030	28C425Y
2030	28C441A
2030	28C441C
2030	28C441D
2030	28C441E
2030	28C441F
2030	28C441G
2030	28C441GN
2030	28C441H
2030	28C441J
2030	28C441K
2030	28C441L
2030	28C4410R
2030	28C441R
2030	28C441X
2030	28C441Y
2030	28C442A
2030	28C442B
2030	28C442C
2030	28C442D
2030	28C442E
2030	28C442F
2030	28C442G
2030	28C442GN
2030	28C442H
2030	28C442J
2030	28C442K
2030	28C442L
2030	28C442M
2030	28C4420R
2030	28C442R
2030	28C442X
2030	28C442Y
2030	28C443
2030	28C443A
2030	28C443B
2030	28C443C
2030	28C443D
2030	28C443E
2030	28C443F
2030	28C443G
2030	28C443GN
2030	28C443H
2030	28C443J
2030	28C443K
2030	28C443L
2030	28C4430R
2030	28C443R
2030	28C443X
2030	28C443Y
2030	28C444
2030	28C456B
2030	28C456C
2030	28C456E
2030	28C456G
2030	28C456GN
2030	28C456H
2030	28C456J
2030	28C456K
2030	28C456L
2030	28C456M
2030	28C4560R
2030	28C456R
2030	28C456X
2030	28C456Y
2030	28C459
2030	28C459B
2030	28C46
2030	28C46A
2030	28C46B
2030	28C46C
2030	28C46E
2030	28C46F
2030	28C46G
2030	28C46GN
2030	28C46H
2030	28C46J
2030	28C46K
2030	28C46L
2030	28C46M
2030	28C460R
2030	28C46X
2030	28C46Y
2030	28C471
2030	28C479A
2030	28C479B
2030	28C479C
2030	28C479D
2030	28C479E
2030	28C479F
2030	28C479G
2030	28C479GN
2030	28C479H
2030	28C479J
2030	28C479K
2030	28C479L
2030	28C479M
2030	28C4790R
2030	28C479R
2030	28C479X
2030	28C479Y
2030	28C47A
2030	28C47B
2030	28C47C
2030	28C47D
2030	28C47E
2030	28C47G
2030	28C47GN
2030	28C47H
2030	28C47J
2030	28C47K
2030	28C47L
2030	28C47M
2030	28C470R
2030	28C47R
2030	28C47X
2030	28C47Y

RS 276-	Industry Standard No.
2030	2SC486-R
2030	2SC486-RED
2030	2SC486-Y
2030	2SC486-YEL
2030	2SC486A
2030	2SC486B
2030	2SC486C
2030	2SC486D
2030	2SC486E
2030	2SC486F
2030	2SC486G
2030	2SC486GN
2030	2SC486H
2030	2SC486J
2030	2SC486K
2030	2SC486L
2030	2SC486M
2030	2SC486OR
2030	2SC486R
2030	2SC486X
2030	2SC48A
2030	2SC48B
2030	2SC48D
2030	2SC48E
2030	2SC48F
2030	2SC48G
2030	2SC48GN
2030	2SC48H
2030	2SC48J
2030	2SC48K
2030	2SC48L
2030	2SC48M
2030	2SC48OR
2030	2SC48R
2030	2SC48X
2030	2SC48Y
2030	2SC49
2030	2SC497-O
2030	2SC497-0
2030	2SC497-OR
2030	2SC497-ORG
2030	2SC497-R
2030	2SC497-RED
2030	2SC497-Y
2030	2SC497A
2030	2SC497B
2030	2SC497C
2030	2SC497D
2030	2SC497E
2030	2SC497F
2030	2SC497G
2030	2SC497GN
2030	2SC497H
2030	2SC497J
2030	2SC497K
2030	2SC497L
2030	2SC497M
2030	2SC497OR
2030	2SC497R
2030	2SC497RED
2030	2SC497X
2030	2SC497Y
2030	2SC498
2030	2SC498-O
2030	2SC498-0
2030	2SC498-OR
2030	2SC498-ORG
2030	2SC498-R
2030	2SC498-RED
2030	2SC498-Y
2030	2SC498-YEL
2030	2SC498A
2030	2SC498B
2030	2SC498C
2030	2SC498D
2030	2SC498E
2030	2SC498F
2030	2SC498G
2030	2SC498GN
2030	2SC498H
2030	2SC498J
2030	2SC498K
2030	2SC498L
2030	2SC498M
2030	2SC498OR
2030	2SC498R
2030	2SC498RED
2030	2SC498X
2030	2SC498Y
2030	2SC498YEL
2030	2SC49A
2030	2SC49B
2030	2SC49C
2030	2SC49D
2030	2SC49E
2030	2SC49F
2030	2SC49G
2030	2SC49GN
2030	2SC49H
2030	2SC49J
2030	2SC49K
2030	2SC49L
2030	2SC49M
2030	2SC49OR
2030	2SC49X
2030	2SC49Y
2030	2SC501
2030	2SC501-O
2030	2SC501-ORG
2030	2SC501-R
2030	2SC501-RED
2030	2SC501-Y
2030	2SC501-YEL
2030	2SC501A
2030	2SC501B
2030	2SC501C
2030	2SC501D
2030	2SC501E
2030	2SC501F
2030	2SC501G
2030	2SC501GN
2030	2SC501H
2030	2SC501J
2030	2SC501K
2030	2SC501L
2030	2SC501M
2030	2SC501OR
2030	2SC501R
2030	2SC501X
2030	2SC501Y
2030	2SC502A
2030	2SC502B
2030	2SC502C
2030	2SC502D
2030	2SC502E
2030	2SC502F
2030	2SC502G
2030	2SC502GN
2030	2SC502H
2030	2SC502J
2030	2SC502K
2030	2SC502L
2030	2SC502M
2030	2SC502OR
2030	2SC502R
2030	2SC502X
2030	2SC502Y
2030	2SC503
2030	2SC503-GR
2030	2SC503-O
2030	2SC503-Y
2030	2SC503A
2030	2SC503B
2030	2SC503C
2030	2SC503D
2030	2SC503E
2030	2SC503F
2030	2SC503G
2030	2SC503GN
2030	2SC503GR
2030	2SC503H
2030	2SC503J
2030	2SC503K
2030	2SC503L
2030	2SC503M
2030	2SC503OR
2030	2SC503R
2030	2SC503X
2030	2SC504
2030	2SC504-GR
2030	2SC504-O
2030	2SC504-Y
2030	2SC504A
2030	2SC504B
2030	2SC504C
2030	2SC504D
2030	2SC504E
2030	2SC504F
2030	2SC504G
2030	2SC504GN
2030	2SC504GR
2030	2SC504H
2030	2SC504J
2030	2SC504K
2030	2SC504L
2030	2SC504M
2030	2SC504OR
2030	2SC504R
2030	2SC504X
2030	2SC50B
2030	2SC50C
2030	2SC50E
2030	2SC50F
2030	2SC50G
2030	2SC50GN
2030	2SC50H
2030	2SC50J
2030	2SC50K
2030	2SC50L
2030	2SC50M
2030	2SC50OR
2030	2SC50R
2030	2SC50X
2030	2SC50Y
2030	2SC51
2030	2SC512
2030	2SC512-O
2030	2SC512-0
2030	2SC512-ORG
2030	2SC512-R
2030	2SC512-RED
2030	2SC512A
2030	2SC512B
2030	2SC512C
2030	2SC512D
2030	2SC512E
2030	2SC512F
2030	2SC512G
2030	2SC512GN
2030	2SC512H
2030	2SC512J
2030	2SC512K
2030	2SC512L
2030	2SC512M
2030	2SC512O
2030	2SC512OR
2030	2SC512X
2030	2SC512Y
2030	2SC513
2030	2SC513-O
2030	2SC513-0
2030	2SC513-ORG
2030	2SC513-R
2030	2SC513-RED
2030	2SC513A
2030	2SC513B
2030	2SC513C
2030	2SC513D
2030	2SC513E
2030	2SC513F
2030	2SC513G
2030	2SC513GN
2030	2SC513H
2030	2SC513J
2030	2SC513K
2030	2SC513L
2030	2SC513M
2030	2SC513O
2030	2SC513OR
2030	2SC513R
2030	2SC513X
2030	2SC513Y
2030	2SC516
2030	2SC516A
2030	2SC516B
2030	2SC516C
2030	2SC516D
2030	2SC516E
2030	2SC516F
2030	2SC516G
2030	2SC516GN
2030	2SC516H
2030	2SC516J
2030	2SC516K
2030	2SC516L
2030	2SC516M
2030	2SC516OR
2030	2SC516R
2030	2SC516X
2030	2SC516Y
2030	2SC51A
2030	2SC51B
2030	2SC51C
2030	2SC51D
2030	2SC51E
2030	2SC51F
2030	2SC51G
2030	2SC51H
2030	2SC51J
2030	2SC51K
2030	2SC51L
2030	2SC51M
2030	2SC510R
2030	2SC51R
2030	2SC51X
2030	2SC51Y
2030	2SC52A
2030	2SC52B
2030	2SC52C
2030	2SC52D
2030	2SC52E
2030	2SC52F
2030	2SC52G
2030	2SC52GN
2030	2SC52H
2030	2SC52J
2030	2SC52K
2030	2SC52L
2030	2SC52M
2030	2SC52R
2030	2SC52X
2030	2SC52Y
2030	2SC5370
2030	2SC5370A
2030	2SC5370B
2030	2SC5370C
2030	2SC5370D
2030	2SC5370E
2030	2SC5370F
2030	2SC5370G
2030	2SC5370GN
2030	2SC5370H
2030	2SC5370J
2030	2SC5370K
2030	2SC5370L
2030	2SC5370M
2030	2SC5370OR
2030	2SC5370X
2030	2SC5370Y
2030	2SC53B
2030	2SC53C
2030	2SC53D
2030	2SC53F
2030	2SC53G
2030	2SC53GN
2030	2SC53H
2030	2SC53J
2030	2SC53L
2030	2SC530R
2030	2SC53R
2030	2SC53X
2030	2SC53Y
2030	2SC54A
2030	2SC54B
2030	2SC54C
2030	2SC54D
2030	2SC54E
2030	2SC54F
2030	2SC54G
2030	2SC54GN
2030	2SC54H
2030	2SC54J
2030	2SC54K
2030	2SC54L
2030	2SC54M
2030	2SC540R
2030	2SC54X
2030	2SC54Y
2030	2SC559
2030	2SC560
2030	2SC564
2030	2SC564(Q)
2030	2SC564(Q)(R)
2030	2SC564A
2030	2SC564B
2030	2SC564C
2030	2SC564D
2030	2SC564E
2030	2SC564F
2030	2SC564G
2030	2SC564GN
2030	2SC564H
2030	2SC564J
2030	2SC564K
2030	2SC564L
2030	2SC564M
2030	2SC564P
2030	2SC564PL
2030	2SC564Q
2030	2SC564QC
2030	2SC564R
2030	2SC564S
2030	2SC564T
2030	2SC564X
2030	2SC564Y
2030	2SC566A
2030	2SC566B
2030	2SC566C
2030	2SC566D
2030	2SC566E
2030	2SC566F
2030	2SC566G
2030	2SC566GN
2030	2SC566H
2030	2SC566J
2030	2SC566K
2030	2SC566L
2030	2SC566M
2030	2SC566OR
2030	2SC566R
2030	2SC566X
2030	2SC56A
2030	2SC56B
2030	2SC56C
2030	2SC56D
2030	2SC56E
2030	2SC56F
2030	2SC56G
2030	2SC56H
2030	2SC56J
2030	2SC56K
2030	2SC56L
2030	2SC56OR
2030	2SC56R
2030	2SC56X
2030	2SC56Y
2030	2SC588B
2030	2SC588C
2030	2SC588D
2030	2SC588E
2030	2SC588F
2030	2SC588G
2030	2SC588GN
2030	2SC588H
2030	2SC588J
2030	2SC588K
2030	2SC588L
2030	2SC588M
2030	2SC588OR
2030	2SC588X
2030	2SC588Y
2030	2SC59
2030	2SC590A
2030	2SC590B
2030	2SC590C
2030	2SC590D
2030	2SC590E
2030	2SC590F
2030	2SC590G
2030	2SC590GN
2030	2SC590H
2030	2SC590J
2030	2SC590K
2030	2SC590L
2030	2SC590M
2030	2SC590OR
2030	2SC590X
2030	2SC594
2030	2SC594-O
2030	2SC594-R
2030	2SC594-Y
2030	2SC594A
2030	2SC594B
2030	2SC594C
2030	2SC594D
2030	2SC594E
2030	2SC594F
2030	2SC594G
2030	2SC594H
2030	2SC594J
2030	2SC594K
2030	2SC594L
2030	2SC594OR
2030	2SC594R
2030	2SC594X
2030	2SC594Y
2030	2SC595A
2030	2SC595B
2030	2SC595C
2030	2SC595D
2030	2SC595E
2030	2SC595F
2030	2SC595G
2030	2SC595GN
2030	2SC595H
2030	2SC595J
2030	2SC595K
2030	2SC595L
2030	2SC595M
2030	2SC595OR
2030	2SC595R
2030	2SC595X
2030	2SC595Y
2030	2SC596A
2030	2SC596B
2030	2SC596C
2030	2SC596D
2030	2SC596E
2030	2SC596F
2030	2SC596G
2030	2SC596GN
2030	2SC596H
2030	2SC596J
2030	2SC596K
2030	2SC596L
2030	2SC596M
2030	2SC596OR
2030	2SC596R
2030	2SC596X
2030	2SC596Y
2030	2SC59A
2030	2SC59B
2030	2SC59C
2030	2SC59D
2030	2SC59E
2030	2SC59F
2030	2SC59G
2030	2SC59GN
2030	2SC59H
2030	2SC59J
2030	2SC59K
2030	2SC59L
2030	2SC59M
2030	2SC59OR
2030	2SC59R
2030	2SC59X
2030	2SC59Y
2030	2SC61
2030	2SC610A
2030	2SC610B
2030	2SC610C
2030	2SC610D
2030	2SC610E
2030	2SC610F
2030	2SC610G
2030	2SC610GN
2030	2SC610H
2030	2SC610J
2030	2SC610K
2030	2SC610L
2030	2SC610M
2030	2SC6100H

| 2030 | 2SC588A |

RS 276-	Industry Standard No.	RS 276-	Industry Standard No.	RS 276-	Industry Standard No.	RS 276-	Industry Standard No.
2030	28C610R	2030	28C68C	2030	28C734-GRN	2030	28C81J
2030	28C610X	2030	28C68D	2030	28C734-O	2030	28C81K
2030	28C610Y	2030	28C68E	2030	28C734-OR	2030	28C81L
2030	28C614A	2030	28C68F	2030	28C734-ORG	2030	28C81M
2030	28C614B	2030	28C68G	2030	28C734-OY	2030	28C810R
2030	28C614GN	2030	28C68GN	2030	28C734-R	2030	28C81R
2030	28C614H	2030	28C68H	2030	28C734-RED	2030	28C826
2030	28C614J	2030	28C68K	2030	28C734-Y	2030	28C826A
2030	28C614K	2030	28C68L	2030	28C734-YEL	2030	28C826B
2030	28C614L	2030	28C68L	2030	28C734A	2030	28C826C
2030	28C614M	2030	28C68M	2030	28C734B	2030	28C826D
2030	28C6140R	2030	28C680R	2030	28C734C	2030	28C826E
2030	28C614R	2030	28C68R	2030	28C734D	2030	28C826F
2030	28C614X	2030	28C68X	2030	28C734E	2030	28C826G
2030	28C614Y	2030	28C68Y	2030	28C734F	2030	28C826H
2030	28C61A	2030	28C69	2030	28C734G	2030	28C826J
2030	28C61B	2030	28C699	2030	28C734GN	2030	28C826K
2030	28C61C	2030	28C69A	2030	28C734GR	2030	28C826L
2030	28C61D	2030	28C69B	2030	28C734GRN	2030	28C826M
2030	28C61E	2030	28C69C	2030	28C734H	2030	28C8260R
2030	28C61F	2030	28C69D	2030	28C734J	2030	28C826R
2030	28C61G	2030	28C69E	2030	28C734K	2030	28C826X
2030	28C61GN	2030	28C69F	2030	28C734K/GR	2030	28C826Y
2030	28C61H	2030	28C69G	2030	28C734L	2030	28C827
2030	28C61J	2030	28C69GN	2030	28C734M	2030	28C827A
2030	28C61K	2030	28C69H	2030	28C7340	2030	28C827B
2030	28C61L	2030	28C69J	2030	28C7340R	2030	28C827C
2030	28C61M	2030	28C69K	2030	28C734R	2030	28C827D
2030	28C610R	2030	28C69L	2030	28C734RED	2030	28C827E
2030	28C61R	2030	28C69M	2030	28C734X	2030	28C827F
2030	28C61Y	2030	28C690R	2030	28C734Y	2030	28C827G
2030	28C621A	2030	28C69R	2030	28C734YEL	2030	28C827GN
2030	28C621B	2030	28C69X	2030	28C741H	2030	28C827H
2030	28C621C	2030	28C69Y	2030	28C741J	2030	28C827J
2030	28C621D	2030	28C700	2030	28C741K	2030	28C827K
2030	28C621E	2030	28C701A	2030	28C741L	2030	28C827M
2030	28C621F	2030	28C701B	2030	28C741M	2030	28C8270R
2030	28C621G	2030	28C701C	2030	28C7410R	2030	28C827X
2030	28C621GN	2030	28C701D	2030	28C741R	2030	28C827Y
2030	28C621H	2030	28C701E	2030	28C741X	2030	28C839
2030	28C621J	2030	28C701F	2030	28C741Y	2030	28C839(P)
2030	28C621K	2030	28C701G	2030	28C744	2030	28C839(H)
2030	28C621L	2030	28C701GN	2030	28C744A	2030	28C839(J)
2030	28C621M	2030	28C701H	2030	28C796A	2030	28C839(JI)
2030	28C6210R	2030	28C701J	2030	28C796B	2030	28C839(L)
2030	28C621R	2030	28C701K	2030	28C796C	2030	28C839(M)
2030	28C621X	2030	28C701L	2030	28C796D	2030	28C839-E
2030	28C621Y	2030	28C701M	2030	28C796E	2030	28C839-P
2030	28C622A	2030	28C7010R	2030	28C796F	2030	28C839A
2030	28C622B	2030	28C701R	2030	28C796G	2030	28C839B
2030	28C622C	2030	28C701X	2030	28C796GN	2030	28C839C
2030	28C622D	2030	28C701Y	2030	28C796H	2030	28C839D
2030	28C622E	2030	28C702A	2030	28C796J	2030	28C839F
2030	28C622F	2030	28C702B	2030	28C796K	2030	28C839G
2030	28C622G	2030	28C702C	2030	28C796L	2030	28C839GN
2030	28C622GN	2030	28C702D	2030	28C796M	2030	28C839H
2030	28C622H	2030	28C702E	2030	28C7960R	2030	28C839J
2030	28C622J	2030	28C702F	2030	28C796X	2030	28C839JH
2030	28C622K	2030	28C702G	2030	28C796Y	2030	28C839JI
2030	28C622L	2030	28C702GN	2030	28C797	2030	28C839K
2030	28C622M	2030	28C702H	2030	28C798	2030	28C839L
2030	28C6220R	2030	28C702J	2030	28C798A	2030	28C839M
2030	28C622R	2030	28C702K	2030	28C798B	2030	28C839N
2030	28C622Y	2030	28C702L	2030	28C798C	2030	28C8390R
2030	28C62B	2030	28C702M	2030	28C798D	2030	28C839S
2030	28C62A	2030	28C7020R	2030	28C798E	2030	28C839X
2030	28C62B	2030	28C702R	2030	28C798G	2030	28C839Y
2030	28C62C	2030	28C702X	2030	28C798GN	2030	28C847A
2030	28C62D	2030	28C702Y	2030	28C798H	2030	28C847B
2030	28C62E	2030	28C708	2030	28C798J	2030	28C847C
2030	28C62F	2030	28C708(A)	2030	28C798K	2030	28C847D
2030	28C62G	2030	28C708(B)	2030	28C798L	2030	28C847E
2030	28C62GN	2030	28C708(C)	2030	28C798M	2030	28C847F
2030	28C62H	2030	28C708-OR	2030	28C7980R	2030	28C847G
2030	28C62J	2030	28C708A	2030	28C798R	2030	28C847GN
2030	28C62L	2030	28C708AA	2030	28C798X	2030	28C847H
2030	28C62M	2030	28C708AB	2030	28C798Y	2030	28C847J
2030	28C62OR	2030	28C708AC	2030	28C802	2030	28C847K
2030	28C62R	2030	28C708AH	2030	28C803	2030	28C847L
2030	28C62X	2030	28C708AHA	2030	28C81	2030	28C847M
2030	28C62Y	2030	28C708AHB	2030	28C814	2030	28C847R
2030	28C631B	2030	28C708AHC	2030	28C8146	2030	28C847X
2030	28C631C	2030	28C708B	2030	28C815	2030	28C848A
2030	28C631D	2030	28C708C	2030	28C815(M)	2030	28C848B
2030	28C631E	2030	28C708D	2030	28C815-1	2030	28C848D
2030	28C631F	2030	28C708E	2030	28C815A	2030	28C848E
2030	28C631G	2030	28C708F	2030	28C815B	2030	28C848F
2030	28C6310N	2030	28C708G	2030	28C815BK	2030	28C848G
2030	28C631H	2030	28C708GN	2030	28C815C	2030	28C848GN
2030	28C631J	2030	28C708H	2030	28C815D	2030	28C848H
2030	28C631K	2030	28C708HA	2030	28C815E	2030	28C848J
2030	28C631L	2030	28C708HB	2030	28C815F	2030	28C848K
2030	28C631M	2030	28C708L	2030	28C815G	2030	28C848L
2030	28C6310R	2030	28C708M	2030	28C815GN	2030	28C848M
2030	28C631R	2030	28C7080R	2030	28C815H	2030	28C8480R
2030	28C631X	2030	28C708R	2030	28C815J	2030	28C848X
2030	28C631Y	2030	28C708X	2030	28C815K	2030	28C848Y
2030	28C654A	2030	28C708Y	2030	28C815K,L	2030	28C849A
2030	28C654B	2030	28C716A	2030	28C815L	2030	28C849B
2030	28C654C	2030	28C716GN	2030	28C815LJ	2030	28C849C
2030	28C654D	2030	28C716H	2030	28C815M	2030	28C849D
2030	28C654E	2030	28C716J	2030	28C8150R	2030	28C849E
2030	28C654F	2030	28C716K	2030	28C815R	2030	28C849F
2030	28C654G	2030	28C716L	2030	28C815SA	2030	28C849G
2030	28C654GN	2030	28C716M	2030	28C815SC	2030	28C849GN
2030	28C654H	2030	28C7160R	2030	28C815X	2030	28C849H
2030	28C654J	2030	28C716R	2030	28C815Y	2030	28C849J
2030	28C654K	2030	28C716X	2030	28C816	2030	28C849K
2030	28C654L	2030	28C716Y	2030	28C816A	2030	28C849L
2030	28C654M	2030	28C727A	2030	28C816B	2030	28C849M
2030	28C6540R	2030	28C727B	2030	28C816C	2030	28C8490R
2030	28C654R	2030	28C727C	2030	28C816D	2030	28C849R
2030	28C654X	2030	28C727D	2030	28C816E	2030	28C849Y
2030	28C654Y	2030	28C727E	2030	28C816F	2030	28C850A
2030	28C67A	2030	28C727F	2030	28C816G	2030	28C850B
2030	28C67B	2030	28C727G	2030	28C816GN	2030	28C850C
2030	28C67C	2030	28C727GN	2030	28C816H	2030	28C850D
2030	28C67D	2030	28C727H	2030	28C816K	2030	28C850E
2030	28C67E	2030	28C727J	2030	28C816L	2030	28C850F
2030	28C67F	2030	28C727K	2030	28C816M	2030	28C850G
2030	28C67G	2030	28C727L	2030	28C8160R	2030	28C850GN
2030	28C67GN	2030	28C727M	2030	28C816R	2030	28C850H
2030	28C67H	2030	28C727R	2030	28C816X	2030	28C850J
2030	28C67J	2030	28C727X	2030	28C816Y	2030	28C850K
2030	28C67K	2030	28C727Y	2030	28C81A	2030	28C850L
2030	28C67L	2030	28C731	2030	28C81B	2030	28C850M
2030	28C67M	2030	28C731R	2030	28C81C	2030	28C8500R
2030	28C670R	2030	28C734	2030	28C81D		
2030	28C67R	2030	28C734(O)	2030	28C81E		
2030	28C67X	2030	28C734(R)	2030	28C81F		
2030	28C67Y	2030	28C734(Y)	2030	28C81G		
2030	28C68A	2030	28C734-O	2030	28C81GN		
2030	28C68B	2030	28C734-G	2030	28C81H		

RS 276-	Industry Standard No.	RS 276-	Industry Standard No.	RS 276-	Industry Standard No.	RS 276-	Industry Standard No.
2030	2SC850R	2030	2SC925(M)	2030	2SC991Y	2030	2SD182C
2030	2SC850X	2030	2SC925A	2030	2SC992A	2030	2SD182D
2030	2SC850Y	2030	2SC925B	2030	2SC992B	2030	2SD182E
2030	2SC853	2030	2SC925C	2030	2SC992C	2030	2SD182F
2030	2SC853-OR	2030	2SC925D	2030	2SC992D	2030	2SD182G
2030	2SC853A	2030	2SC925E	2030	2SC992E	2030	2SD182GN
2030	2SC853B	2030	2SC925F	2030	2SC992F	2030	2SD182H
2030	2SC853C	2030	2SC925G	2030	2SC992G	2030	2SD182J
2030	2SC853D	2030	2SC925GN	2030	2SC992GN	2030	2SD182K
2030	2SC853E	2030	2SC925H	2030	2SC992H	2030	2SD182L
2030	2SC853F	2030	2SC925J	2030	2SC992J	2030	2SD182OR
2030	2SC853G	2030	2SC925K	2030	2SC992K	2030	2SD182R
2030	2SC853GN	2030	2SC925L	2030	2SC992L	2030	2SD182X
2030	2SC853H	2030	2SC925M	2030	2SC992M	2030	2SD182Y
2030	2SC853J	2030	2SC925OR	2030	2SC992OR	2030	2SD183
2030	2SC853K	2030	2SC925R	2030	2SC992R	2030	2SD183B
2030	2SC853KLM	2030	2SC925X	2030	2SC992X	2030	2SD183E
2030	2SC853L	2030	2SC925Y	2030	2SC992Y	2030	2SD183GN
2030	2SC853M	2030	2SC938-O	2030	2SC993	2030	2SD183K
2030	2SC853OR	2030	2SC938D	2030	2SC993D	2030	2SD183L
2030	2SC853R	2030	2SC938E	2030	2SC96A	2030	2SD183OR
2030	2SC853X	2030	2SC938F	2030	2SC96B	2030	2SD183R
2030	2SC853Y	2030	2SC938G	2030	2SC96C	2030	2SD204
2030	2SC870FL	2030	2SC938GN	2030	2SC96D	2030	2SD204(L)
2030	2SC870R	2030	2SC938H	2030	2SC96E	2030	2SD204A
2030	2SC875	2030	2SC938J	2030	2SC96F	2030	2SD204B
2030	2SC875(D)	2030	2SC938K	2030	2SC96G	2030	2SD204BL
2030	2SC875(E)	2030	2SC938L	2030	2SC96GN	2030	2SD204C
2030	2SC875(F)	2030	2SC938M	2030	2SC96H	2030	2SD204D
2030	2SC875-1	2030	2SC938OR	2030	2SC96J	2030	2SD204E
2030	2SC875-1C	2030	2SC938R	2030	2SC96K	2030	2SD204F
2030	2SC875-1D	2030	2SC938X	2030	2SC96L	2030	2SD204G
2030	2SC875-1E	2030	2SC938Y	2030	2SC96M	2030	2SD204GA
2030	2SC875-1F	2030	2SC959	2030	2SC96OR	2030	2SD204GN
2030	2SC875-2	2030	2SC959A	2030	2SC96R	2030	2SD204H
2030	2SC875-2C	2030	2SC959B	2030	2SC96X	2030	2SD204J
2030	2SC875-2D	2030	2SC959C	2030	2SC96Y	2030	2SD204K
2030	2SC875-2E	2030	2SC959D	2030	2SCM98P	2030	2SD204L
2030	2SC875-2F	2030	2SC959M	2030	2SCNJ107	2030	2SD204M
2030	2SC875-3	2030	2SC959S	2030	2SC8183A	2030	2SD204R
2030	2SC875-3C	2030	2SC959SA	2030	2SC8183B	2030	2SD204Y
2030	2SC875-3D	2030	2SC959SC	2030	2SC8183D	2030	2SD205(L)
2030	2SC875-3E	2030	2SC959SD	2030	2SC8183F	2030	2SD205(M)
2030	2SC875-3F	2030	2SC966A	2030	2SC8183G	2030	2SD205A
2030	2SC875B	2030	2SC966B	2030	2SC8183GN	2030	2SD205B
2030	2SC875BR	2030	2SC966C	2030	2SC8183H	2030	2SD205C
2030	2SC875C	2030	2SC966D	2030	2SC8183J	2030	2SD205D
2030	2SC875D	2030	2SC966E	2030	2SC8183K	2030	2SD205E
2030	2SC875DL	2030	2SC966F	2030	2SC8183L	2030	2SD205F
2030	2SC875E	2030	2SC966G	2030	2SC8183M	2030	2SD205G
2030	2SC875EL	2030	2SC966GN	2030	2SC8183OR	2030	2SD205GN
2030	2SC875F	2030	2SC966H	2030	2SC8183R	2030	2SD205H
2030	2SC875G	2030	2SC966J	2030	2SC8183Y	2030	2SD205J
2030	2SC875GN	2030	2SC966K	2030	2SC8184A	2030	2SD205L
2030	2SC875J	2030	2SC966L	2030	2SC8184B	2030	2SD205M
2030	2SC875K	2030	2SC966M	2030	2SC8184C	2030	2SD205OR
2030	2SC875L	2030	2SC966OR	2030	2SC8184D	2030	2SD205R
2030	2SC875M	2030	2SC966R	2030	2SC8184F	2030	2SD205X
2030	2SC875OR	2030	2SC966X	2030	2SC8184G	2030	2SD205Y
2030	2SC875X	2030	2SC966Y	2030	2SC8184GN	2030	2SD219
2030	2SC875Y	2030	2SC967A	2030	2SC8184H	2030	2SD219A
2030	2SC876	2030	2SC967B	2030	2SC8184K	2030	2SD219B
2030	2SC876(F)	2030	2SC967C	2030	2SC8184L	2030	2SD219C
2030	2SC876A	2030	2SC967D	2030	2SC8184M	2030	2SD219D
2030	2SC876B	2030	2SC967E	2030	2SC8184OR	2030	2SD219E
2030	2SC876C	2030	2SC967F	2030	2SC8184R	2030	2SD219F
2030	2SC876D	2030	2SC967G	2030	2SC8184X	2030	2SD219G
2030	2SC876E	2030	2SC967GN	2030	2SC8184Y	2030	2SD219GN
2030	2SC876F	2030	2SC967H	2030	2SD102A	2030	2SD219H
2030	2SC876G	2030	2SC967J	2030	2SD102B	2030	2SD219J
2030	2SC876GN	2030	2SC967K	2030	2SD102C	2030	2SD219K
2030	2SC876H	2030	2SC967L	2030	2SD102D	2030	2SD219L
2030	2SC876J	2030	2SC967M	2030	2SD102E	2030	2SD219M
2030	2SC876K	2030	2SC967OR	2030	2SD102F	2030	2SD219OR
2030	2SC876L	2030	2SC967R	2030	2SD102G	2030	2SD219R
2030	2SC876M	2030	2SC967X	2030	2SD102GN	2030	2SD219X
2030	2SC876OR	2030	2SC967Y	2030	2SD102H	2030	2SD219Y
2030	2SC876R	2030	0002SC968P	2030	2SD102J	2030	2SD220
2030	2SC876TV(D)	2030	2SC968Y	2030	2SD102K	2030	2SD220A
2030	2SC876TV(E)	2030	2SC97	2030	2SD102L	2030	2SD220B
2030	2SC876TVD	2030	2SC971	2030	2SD102M	2030	2SD220C
2030	2SC876TVE	2030	2SC972	2030	2SD1020R	2030	2SD220D
2030	2SC876TVEF	2030	2SC972A	2030	2SD102R	2030	2SD220E
2030	2SC876TVF	2030	2SC972B	2030	2SD102X	2030	2SD220F
2030	2SC876X	2030	2SC972C	2030	2SD102Y	2030	2SD220G
2030	2SC876Y	2030	2SC972D	2030	2SD121C	2030	2SD220GN
2030	2SC87A	2030	2SC972E	2030	2SD121D	2030	2SD220H
2030	2SC87B	2030	2SC972F	2030	2SD121E	2030	2SD220J
2030	2SC87C	2030	2SC972G	2030	2SD121F	2030	2SD220K
2030	2SC87D	2030	2SC972GN	2030	2SD121G	2030	2SD220L
2030	2SC87E	2030	2SC972H	2030	2SD121GN	2030	2SD220M
2030	2SC87F	2030	2SC972J	2030	2SD121J	2030	2SD220OR
2030	2SC87G	2030	2SC972K	2030	2SD121K	2030	2SD220R
2030	2SC87GN	2030	2SC972L	2030	2SD121L	2030	2SD220X
2030	2SC87H	2030	2SC972M	2030	2SD121M	2030	2SD220Y
2030	2SC87J	2030	2SC972OR	2030	2SD121OR	2030	2SD221
2030	2SC87K	2030	2SC972X	2030	2SD121R	2030	2SD222
2030	2SC87L	2030	2SC972Y	2030	2SD121X	2030	2SD222A
2030	2SC87M	2030	2SC976TV	2030	2SD121Y	2030	2SD222B
2030	2SC87OR	2030	2SC97A	2030	2SD131B	2030	2SD222C
2030	2SC87R	2030	2SC97B	2030	2SD1347	2030	2SD222D
2030	2SC87X	2030	2SC97C	2030	2SD134A	2030	2SD222E
2030	2SC87Y	2030	2SC97D	2030	2SD134C	2030	2SD222F
2030	2SC881	2030	2SC97E	2030	2SD134D	2030	2SD222G
2030	2SC881A	2030	2SC97F	2030	2SD134F	2030	2SD222GN
2030	2SC881B	2030	2SC97G	2030	2SD134G	2030	2SD222H
2030	2SC881C	2030	2SC97GN	2030	2SD134GN	2030	2SD222J
2030	2SC881D	2030	2SC97H	2030	2SD134H	2030	2SD222K
2030	2SC881E	2030	2SC97J	2030	2SD134J	2030	2SD222L
2030	2SC881F	2030	2SC97K	2030	2SD134K	2030	2SD222M
2030	2SC881G	2030	2SC97L	2030	2SD134L	2030	2SD222OR
2030	2SC881GN	2030	2SC97M	2030	2SD134M	2030	2SD222R
2030	2SC881H	2030	2SC97OR	2030	2SD134OR	2030	2SD222X
2030	2SC881K	2030	2SC97R	2030	2SD134R	2030	2SD222Y
2030	2SC881L	2030	2SC97X	2030	2SD134X	2030	2SD227
2030	2SC881M	2030	2SC97Y	2030	2SD134Y	2030	2SD227(PANASONIC)
2030	2SC881OR	2030	2SC991A	2030	2SD136D	2030	2SD227(R)
2030	2SC881R	2030	2SC991B	2030	2SD136E	2030	2SD227-175
2030	2SC881X	2030	2SC991C	2030	2SD136F	2030	2SD227-OR
2030	2SC881Y	2030	2SC991D	2030	2SD136G	2030	2SD227A
2030	2SC907B	2030	2SC991E	2030	2SD136GN	2030	2SD227B
2030	2SC907E	2030	2SC991F	2030	2SD136J	2030	2SD227C
2030	2SC907F	2030	2SC991G	2030	2SD136L	2030	2SD227E
2030	2SC907G	2030	2SC991GN	2030	2SD136M	2030	2SD227F
2030	2SC907GN	2030	2SC991H	2030	2SD136OR	2030	2SD227G
2030	2SC907J	2030	2SC991J	2030	2SD136R	2030	2SD227GN
2030	2SC907K	2030	2SC991K	2030	2SD136X	2030	2SD227H
2030	2SC907L	2030	2SC991L	2030	2SD136Y	2030	2SD227J
2030	2SC907M	2030	2SC991M	2030	2SD182	2030	2SD227K
2030	2SC907OR	2030	2SC991OR	2030	2SD182A	2030	2SD227L
2030	2SC907R	2030	2SC991R	2030	2SD182B	2030	2SD227LF
2030	2SC907X	2030	2SC991X			2030	2SD227M
2030	2SC907Y					2030	2SD227OR
						2030	2SD227R

RS 276-	Industry Standard No.
2030	28D227S
2030	28D227V
2030	28D227W
2030	28D227X
2030	28D227Y
2030	28D228
2030	000028D261
2030	28D261(L)
2030	28D261(O)
2030	28D261(Q)
2030	28D261(R)
2030	28D261(U)
2030	28D261(V)
2030	28D261-0
2030	28D261-0
2030	28D261A
2030	28D261B
2030	28D261C
2030	28D261D
2030	28D261E
2030	28D261F
2030	28D261G
2030	28D261GN
2030	28D261H
2030	28D261J
2030	28D261K
2030	28D261L
2030	28D261M
2030	28D261O
2030	28D261OR
2030	28D261P
2030	28D261Q
2030	28D261R
2030	28D261S
2030	28D261U
2030	28D261V
2030	28D261W
2030	28D261X
2030	28D261Y
2030	28D327
2030	28D327A
2030	28D327B
2030	28D327C
2030	28D327D
2030	28D327E
2030	28D327F
2030	28D327OR
2030	28D327R
2030	28D327Y
2030	28D336
2030	28D336R
2030	28D336Y
2030	28D355
2030	28D555C
2030	28D555D
2030	28D555E
2030	28D567H
2030	28D567J
2030	28D567K
2030	28D567L
2030	28D567M
2030	28D567OR
2030	28D567R
2030	28D567X
2030	28D567Y
2030	28D400
2030	28D400D
2030	28D400E
2030	28D400P
2030	28D400P1
2030	28D400P1D
2030	28D400P1E
2030	28D400P1F
2030	28D400P2
2030	28D438
2030	28D438E
2030	28D467
2030	28D467B
2030	28D467C
2030	28D468
2030	28D468A
2030	28D468AC
2030	28D468B
2030	28D468C
2030	28D468D
2030	28D468E
2030	28D468F
2030	28D468G
2030	28D468GN
2030	28D468L
2030	28D468LN
2030	28D468Y
2030	28D545
2030	28D545F
2030	28D545G
2030	28D773K
2030	28DU780R
2030	28DU84GN
2030	2E2102
2030	03A02
2030	3L4-6007-4
2030	3L4-6007-41
2030	3LA-6007-4
2030	3TE150
2030	3TE160
2030	3TE250
2030	3TE260
2030	4-049B
2030	4-1001
2030	4-1792
2030	4-274
2030	4-288
2030	4-3023223
2030	4-3023844
2030	4-397
2030	4-398
2030	4-686182-3
2030	4C28
2030	4C29
2030	4C30
2030	4C31
2030	4JX11C2848
2030	5E4850/56-0002
2030	6-0000139
2030	6-139
2030	6-856
2030	6A10228
2030	6X4850/56-0002
2030	610148-1
2030	7-0002-00
2030	7-0011-00
2030	007-0051
2030	7-0051-00
2030	007-74659-01
2030	007-74659-04
2030	007-74659-06
2030	7A1011
2030	7A1011(GE)
2030	7A1011(SHERWOOD)
2030	7A30(GE)
2030	7A31(GE)
2030	7A32(GE)
2030	7A32(SHERWOOD)
2030	7A35
2030	7A35(GE)
2030	7A995
2030	7A995(GE)
2030	7A995(SHERWOOD)
2030	7013
2030	7D13
2030	7B13
2030	7Y13
2030	7D13
2030	8-0053300
2030	8-726-357-10
2030	8-760-335-10
2030	8-760-343-10
2030	8-905-705-410
2030	8Q-3-13
2030	09-302075
2030	09-302090
2030	09-302116
2030	09-302155
2030	09-302171
2030	09-302222
2030	09-302226
2030	09-303019
2030	09-30319
2030	09-305076
2030	09-305092
2030	09-309031
2030	09-309063
2030	9-5220
2030	9-5226-004
2030	9-5226-4
2030	9TR61001-07
2030	11B556
2030	11B560
2030	11C1536
2030	11C211B20
2030	11C702
2030	13-0026666-001
2030	13-105698-1
2030	13-14085-15A
2030	13-14085-91
2030	13-14085-94
2030	13-14605-1
2030	13-415833-1
2030	13-18927-1A
2030	13-23840-1
2030	13-23892-5
2030	13-26666
2030	13-26666-1
2030	13-27432-1
2030	13-27443-1
2030	13-28392-1
2030	13-28392-2
2030	13-28392-3
2030	13-28394-1
2030	13-28394-2
2030	13-28394-3
2030	13-28471-1
2030	13-28471-1(METAL)
2030	13-32630-1
2030	13-32630-2
2030	13-32630-3
2030	13-32630-4
2030	13-34371-1
2030	13-39114-3
2030	13-55018-4
2030	13-55061-1(AGC)
2030	13-55061-2(WARDS)
2030	13-55064-1
2030	13-55064-1(WARDS)
2030	13-55066-2(STYL)
2030	13-55066-2(WARDS)
2030	14-0104-4
2030	14-602-18
2030	14-602-24
2030	14-602-29
2030	14-602-30
2030	14-602-43
2030	14-602-49
2030	14-602-55
2030	14-602-59
2030	14-602-65
2030	14-602-70
2030	14-602-72
2030	14-602-72
2030	14-603-01
2030	14-603-02A
2030	14-603-07
2030	14-654-21
2030	15-021215
2030	15-09650
2030	15-10062-0
2030	16-201(SYMPHONIC)
2030	17-443
2030	18-3539-1
2030	019-003637
2030	19-020-038
2030	19-020-050
2030	19-020-075
2030	19-15840
2030	19-3349
2030	19-3692
2030	19-3934-643
2030	19A115238-2
2030	19A115238-P1
2030	19A115300
2030	19A115300-1
2030	19A115300-2
2030	19A115304-2
2030	19A115889-P1
2030	19A115889-P2
2030	19A115889-P3
2030	19C300115-1
2030	19C300115-P1
2030	19C300LL5-1
2030	020-1110-008
2030	020-1111-018
2030	020-1111-038
2030	20-1680-174
2030	20A0055
2030	20A0076
2030	021
2030	21-0101
2030	021-0137-00
2030	21A015-018
2030	21A015-019
2030	21A015-026
2030	21A040-051
2030	21A040-051(XSTR)
2030	21A105-004
2030	21A105-006
2030	21M185
2030	21M192
2030	21M193
2030	21M228
2030	21M387
2030	21M448
2030	21M455
2030	21N541
2030	21M606
2030	23-5039
2030	23B114053
2030	23B114054
2030	24-000452
2030	24MW1062(NPN)
2030	24MW1152
2030	24MW1161
2030	24MW660
2030	24MW663
2030	24MW674
2030	24MW677
2030	24MW714
2030	026-100013
2030	26/P19503
2030	31-0083
2030	31-0101
2030	33-048
2030	03-3
2030	34-6001-50
2030	34-6001-51
2030	34-6001-51MR-12
2030	34-6001-78
2030	34-6001-80
2030	34-6001-82
2030	34-6001-83
2030	34-6001-85
2030	34-6007-4
2030	34-6007-8
2030	34-6015-23
2030	34-6015-24
2030	34-6015-25
2030	34-6015-30
2030	34-6015-51
2030	34-6016-22
2030	34-6016-27
2030	34-6016-30
2030	34-6016-33
2030	34-6016-44
2030	34-6016-51
2030	34-6016-59(NPN)
2030	35-020-21
2030	58
2030	40-0068-2
2030	41B581014
2030	42-19642
2030	42-21233
2030	42-22154
2030	42-9029-40T
2030	42-9029-70J
2030	42-9029-70K
2030	43-023223
2030	43-023844
2030	43-D-09
2030	43A126932
2030	43A162455-1
2030	43A223060-1
2030	43A219819-1
2030	444A399497-001
2030	444A399497-002
2030	444A390264-001
2030	444A395992-001
2030	444A395992-002
2030	444A417034
2030	44A417063-001
2030	46-86233-3
2030	46-86315-3
2030	46-86571-3
2030	46-86381
2030	46-86381-3
2030	46-8638P-3
2030	46-86398-3
2030	46-86405-3
2030	46-86426-3
2030	46-86427-3
2030	46-86485-3
2030	46-86550-3
2030	46-86589-3
2030	46-86591-3
2030	48-01-003
2030	48-01-007
2030	48-03-04046103
2030	48-03-05013702
2030	48-03-10744702
2030	48-134689
2030	48-134838
2030	48-134953
2030	48-134982
2030	48-137005
2030	48-137041
2030	48-137088
2030	48-137307
2030	48-137315
2030	48-137988
2030	48-155046
2030	48-3003A02
2030	48-355037
2030	48-40212
2030	48-859248
2030	48-859428
2030	48-869138
2030	48-869170
2030	48-869184
2030	48-869209
2030	48-869221
2030	48-869228
2030	48-869263
2030	48-869380
2030	48-869464
2030	48-869491
2030	48-869519
2030	48-869521
2030	48-869562
2030	48-869591
2030	48-869599
2030	48-869631
2030	48-869703-3
2030	48-90232A08
2030	48-97046A29
2030	48-97046A40
2030	48-97127A019
2030	48-97127A04
2030	48K869228
2030	48M355037
2030	48P63082A27
2030	48R134666
2030	488859428
2030	488869138
2030	488869170
2030	488869209
2030	488869464
2030	488869491
2030	48U869519
2030	488869521
2030	488869562
2030	488134768
2030	488134853
2030	488134935
2030	488134942
2030	488134953
2030	488134988
2030	488137007
2030	488137014
2030	488137022
2030	488137041
2030	488137047
2030	488137106
2030	488137174
2030	488137206
2030	488155046
2030	488155121
2030	488155123
2030	48840171Q01
2030	48844886001
2030	48X90232A05
2030	48X90232A07
2030	48X97046A04
2030	48X97046A52
2030	48X97177A12
2030	49-62139
2030	50-40306-07
2030	51-47-20
2030	53P169
2030	55-1084
2030	55-642
2030	56-4830
2030	57-01491-C
2030	57-01494C
2030	57A104-8-6
2030	57A109-9
2030	57A113-9
2030	57A12-4
2030	57A128-9
2030	57A129-9
2030	57A131-10
2030	57A14-1
2030	57A14-2
2030	57A14-3
2030	57A156
2030	57A167-9
2030	57A2-38
2030	57A2141-14
2030	57A7-8
2030	57B109-9
2030	57B113-9
2030	57B12-4
2030	57B128-9
2030	57B162-12
2030	57B165-11
2030	57B167-9
2030	57B170-9
2030	57B2-38
2030	57B2-57
2030	57C109-9
2030	57C12-4
2030	57C14-1
2030	57C14-2
2030	57C14-3
2030	57C7-8
2030	57L105-12
2030	57L2-1
2030	57M1-33
2030	57M1-34
2030	57M2-11
2030	57M2-14
2030	57M2-18
2030	57M3-7
2030	57M3-3
2030	61-1764
2030	61-309-458
2030	61-309687
2030	61-747
2030	61-813
2030	61C004-1
2030	63-11936
2030	63-11989
2030	63-12989
2030	63-12990
2030	63-13214
2030	63-13215
2030	065-008
2030	066
2030	66A10298
2030	66F025-1
2030	66800000A
2030	68A7368
2030	68A7380-2
2030	69-1814
2030	72N1B
2030	73B140385-001
2030	76-1
2030	76-14090-1
2030	76N1B
2030	76N2B
2030	76N3B
2030	7681050-000
2030	77-271490
2030	77-271490-1
2030	77-272999-1
2030	77N2B
2030	79P114-1
2030	79P114-3
2030	80-053300
2030	81T2
2030	86-100010
2030	86-10010
2030	86-161-2
2030	86-170-2
2030	86-207-2
2030	86-208-2
2030	86-210-2

RS 276-	Industry Standard No.	RS 276-	Industry Standard No.	RS 276-	Industry Standard No.	RS 276-	Industry Standard No.	RS 276-	Industry Standard No.
2030	86-211-2	2030	151-211	2030	422-2158	2030	1087	2030	
2030	86-234-2	2030	153-9	2030	430-10047-0C	2030	1087-2380		
2030	86-266-2	2030	154A5946-622	2030	430-23223	2030	1099-0950		
2030	86-267-2	2030	158-045-0027	2030	430-23844	2030	1106-99		
2030	86-273-2	2030	167-9	2030	432-1	2030	1112-78		
2030	86-291-2	2030	170-9	2030	448A662	2030	1116-6535		
2030	86-306-2	2030	173A-4490-5	2030	454A104	2030	1157		
2030	86-330-2	2030	173A4391	2030	462-1007	2030	1187		
2030	86-336-2	2030	173A4489-2	2030	462-1016	2030	1187(GE)		
2030	86-393-2	2030	173A4490-5	2030	462-1019	2030	1204(GE)		
2030	86-428-2	2030	174-25566-21	2030	462-I007	2030	1206(GE)		
2030	86-428-9	2030	174-25566-50	2030	469-646-3	2030	1207(GE)		
2030	86-440-2	2030	174-25566-63	2030	472-0309-001	2030	1300-1		
2030	86-441-2	2030	174-25566-76	2030	472-0445-001	2030	1300A		
2030	86-452-2	2030	176-004	2030	481-201-A	2030	1414-157		
2030	86-463-2	2030	176-056-9-005	2030	481-201-B	2030	1414-173		
2030	86-463-3 (SEARS)	2030	176-074-9-004	2030	491A948	2030	1414-184		
2030	86-5070-2	2030	182B2003JDP1	2030	510E8030M	2030	1414-186		
2030	86-5073-2	2030	200-011	2030	510E8031M	2030	1414-189		
2030	86-5074-2	2030	200-12	2030	511E8035P	2030	1473-4255		
2030	86-5075-2	2030	218-26	2030	511E8036P	2030	1476 111B		
2030	86-5093-2	2030	226-4	2030	520-301	2030	1476-111B		
2030	86-510-2	2030	229-1301-37	2030	522E8075M	2030	1483		
2030	86-543-2	2030	229-1513-46	2030	522E8076M	2030	1573-00		
2030	86-567-2	2030	229-6011	2030	523E8077M	2030	1573-01		
2030	86-601-2	2030	231-0013	2030	523E8078M	2030	1673-0475		
2030	86-609-9	2030	233(SEARS)	2030	536GU(WARDS)	2030	1676-1991		
2030	86-609-2	2030	241B	2030	542-1033	2030	1705-5351		
2030	86-650-2	2030	247-625	2030	555-3	2030	1789-17		
2030	86X0071-001	2030	247-626	2030	555-3(RCA)	2030	1854-0022-1		
2030	86X483-2	2030	248-38104-1	2030	555-4	2030	1854-0090		
2030	87-0006	2030	260-10-041	2030	559-1516-001	2030	1854-0090-1		
2030	87-0236Q	2030	260D07901	2030	565-072	2030	1854-0274		
2030	87-0008	2030	260D08601	2030	565-1	2030	1854-0332		
2030	87-0236R	2030	260D08701	2030	573-532	2030	1854-0498		
2030	90-181	2030	260D09612	2030	576-0004-013	2030	1879-17A		
2030	90-31	2030	260P10003	2030	576-0004-02	2030	1933-17		
2030	90-66	2030	260P10003A	2030	576-0005-004	2030	1972-17		
2030	94N1V	2030	260P1005	2030	577R8H9H01	2030	2000-209		
2030	95-220	2030	260P21003	2030	600-188-1-21	2030	2000-276		
2030	95-226-004	2030	260P12401	2030	601-0100792	2030	2003JDP1		
2030	95-226-4	2030	260P12481	2030	608-2	2030	2057A100-17		
2030	96-5107-01	2030	260P14104	2030	608-3	2030	2057A100-40(NPN)		
2030	96-5107-02	2030	260P16304	2030	614-1	2030	2057A100-62(NPN)		
2030	96-5161-01	2030	260P017002	2030	622-1	2030	2057A100-67		
2030	96-5170-01	2030	260P31303	2030	622-1(RCA)	2030	2057A2-151		
2030	96-5180-01	2030	260P31402	2030	622-2	2030	2057A2-156		
2030	96-5180-02	2030	260P200401	2030	625-1(RCA)	2030	2057A2-199		
2030	96-5203-01	2030	261RAX	2030	638-1	2030	2057A2-230		
2030	96-5204-01	2030	270-950-037-02	2030	639GL	2030	2057A2-234		
2030	96-5208-01	2030	296-58-9	2030	642-306	2030	2057A2-284		
2030	96-5231-01	2030	296-81-9	2030	660-125	2030	2057A2-295		
2030	96-5244-01	2030	297V049H06	2030	660-128	2030	2057A2-468		
2030	96-5252-01	2030	297V061005	2030	660-145	2030	2057A2-484		
2030	96-5256-01	2030	297V062006	2030	666-1(RCA)	2030	2057A2-486		
2030	96-5364-01	2030	297V072005	2030	686-0012	2030	2057A2-524		
2030	96XZ6053/38N	2030	297V074010	2030	686-0112	2030	2057A2-530		
2030	99L6(SHARP)	2030	297V074012	2030	686-0130	2030	2057B100-17		
2030	998036(TELEDYNE)	2030	297V082B01	2030	686-0165	2030	2057B107-8		
2030	998074	2030	297V083C04	2030	686-229-0	2030	2057B109-3		
2030	100-4846-001	2030	300D043	2030	690V075H62	2030	2057B13-9		
2030	100-5338	2030	318-2	2030	690V086H90	2030	2057B126-12		
2030	100-5338-001	2030	324-019	2030	690V103H30	2030	2057B129-9		
2030	102-1384-17	2030	324-6011	2030	690V110H34	2030	2057B144-12		
2030	0103-0014(R,S)	2030	324-6011(NPN)	2030	690V111H66	2030	2057B153-9		
2030	0103-0014R	2030	324-6013	2030	690V116H24	2030	2057B156-9		
2030	0103-0014S	2030	325-1513-46	2030	691C844	2030	2093A3B-24		
2030	0103-0014T	2030	342-1	2030	700-135	2030	2208-17		
2030	0103-0014U	2030	344-6001-1	2030	703-1	2030	2246-17		
2030	0103-0051	2030	344-6001-2	2030	703-2	2030	2270-5		
2030	0103-0503	2030	344-6011-1	2030	703B	2030	2498-163		
2030	0103-05038	2030	344-6011-2	2030	729-3	2030	2904-032		
2030	0103-0607	2030	344-6013-1B	2030	0731	2030	3022		
2030	0103-0616	2030	344-6015-4	2030	744	2030	3034		
2030	0104-0015	2030	344-6017-4	2030	753-2000-107	2030	3034(RCA)		
2030	104H01	2030	345-2	2030	753-8510-470	2030	3202-51-01		
2030	107N1	2030	349-1	2030	763-1	2030	3202-5H01		
2030	112-361	2030	349-2	2030	772-120-00	2030	3223		
2030	112-525	2030	352-0092-020	2030	772-121-00	2030	003501		
2030	114-1(PHILCO)	2030	352-0364-000	2030	773CL	2030	3507		
2030	119-0077	2030	352-0364-010	2030	800-001-106-1	2030	3508		
2030	121-14(COLUMBIA)	2030	352-043-010	2030	800-101-101-2	2030	3521		
2030	121-431	2030	352-0479-010	2030	800-512-00	2030	3523		
2030	121-639CL	2030	352-0766-010	2030	800-513-00	2030	3532-1		
2030	121-664	2030	352-0783-020	2030	800-515-00	2030	3539		
2030	121-676	2030	352-0816-010	2030	800-521-02	2030	3541		
2030	121-703	2030	352-9014-00	2030	800-521-03	2030	3544(RCA)		
2030	121-722	2030	353-9301-001	2030	800-522-03	2030	3553-3		
2030	121-737CL	2030	560-1	2030	800-53300	2030	3554		
2030	121-766	2030	560-1(RCA)	2030	800-669B	2030	3555		
2030	121-773CL	2030	362A10	2030	800-669B(NPN)	2030	3555-1(RCA)		
2030	121-843A	2030	385-1	2030	819-1	2030	3555-3		
2030	121-844	2030	386-7181P2	2030	824-1	2030	3556		
2030	121-878	2030	386-7316-P2	2030	914F298-1	2030	3560-1		
2030	122N2	2030	386-7316-PU	2030	921-188B	2030	3560-2		
2030	1238-437	2030	392-1	2030	921-230B	2030	3561-1		
2030	1238425	2030	392-1(SYLVANIA)	2030	921-236B	2030	3561-1(GE)		
2030	125-121	2030	394-3005-2	2030	921-337B	2030	3565(RCA)		
2030	125A137	2030	394-3127-1	2030	921-410	2030	3566(RCA)		
2030	125A137A	2030	394-3127-2	2030	922-50B	2030	3591		
2030	125B159	2030	394-3127-3	2030	943-728-001	2030	3601		
2030	128N2	2030	394-3141-1	2030	951	2030	360B(RCA)		
2030	130	2030	400-1362-101	2030	952	2030	3608-1		
2030	130-191-00	2030	400-1362-102	2030	953	2030	3608-2		
2030	0131-000561	2030	400-1362-201	2030	964-22158	2030	3608-2(RCA)		
2030	0131-001429	2030	411-237	2030	964-24387	2030	3615		
2030	0131-001430	2030	417-100	2030	964-25046	2030	3615(RCA)		
2030	131-005-807	2030	417-109	2030	991-00-2888	2030	3615-1		
2030	131-00561	2030	417-114	2030	991-01-1305	2030	3622(RCA)		
2030	131-000807	2030	417-115=13173	2030	991-01-1314	2030	3622-1		
2030	131-045-60	2030	417-128	2030	991-011305	2030	3622-2		
2030	132-021B	2030	417-133	2030	991-011313	2030	3622-2(RCA)		
2030	132-022	2030	417-136	2030	991-011314	2030	3625-1		
2030	132-038	2030	417-137	2030	991-3N	2030	3666(RCA)		
2030	132-066	2030	417-155	2030	999-4601	2030	4006(E.F.JOHNSON)		
2030	132N1	2030	417-178	2030	1001-01	2030	4008(E.F.JOHNSON)		
2030	134B1040-7	2030	417-180	2030	1002A-2	2030	4012		
2030	136-12	2030	417-193	2030	1008	2030	4080-187-0507		
2030	147-7031-01	2030	417-224	2030	1009-03-17	2030	4218		
2030	147N1	2030	417-233	2030	1010-7951	2030	4306		
2030	148-1	2030	417-237	2030	1010-7993	2030	4473-M-12		
2030	148N1	2030	417-237-13163	2030	1010-8041	2030	4473-M3		
2030	148N3	2030	417-247	2030	1027G	2030	4473-N		
2030	151-1096-00	2030	417-250	2030	1027(GE)	2030	4483		
2030	151-0136	2030	417-257	2030	1038(G.E.)	2030	4490		
2030	151-0136-00	2030	417-49	2030	1039-0458	2030	4686-182-3		
2030	151-0136-02	2030	417-59	2030	1040-11	2030	4689		
2030	151-0211	2030	417-821-13163	2030	1040-7	2030	4822-130-40356		
2030	151-0211-00	2030	417-87	2030	1044-9544	2030	4824-33		
2030	151-0211-01	2030	417-88	2030	1045-5082	2030	5258		
2030	151-096-1C	2030	417-89	2030	1049-0167	2030	5380-73		
2030	151-150	2030	422-1233	2030	1080	2030	5553		
2030	151-150-1B	2030	422-1404	2030	1080-7584	2030	5613-1209		
				2030	1081-3087	2030	5613-1209C		

RS 276-	Industry Standard No.	RS 276-	Industry Standard No.	RS 276-	Industry Standard No.	RS 276-	Industry Standard No.
2030	5613-1213D	2030	39311	2030	62612	2030	196779-9-1
2030	5613-1788R	2030	39329	2030	63900-229	2030	196780-1
2030	6151	2030	39443	2030	66007-4	2030	198014-1
2030	6151(RCA)	2030	39462	2030	67001	2030	198020-1
2030	6284	2030	39485	2030	67003	2030	198020-2
2030	6651-486	2030	39486	2030	70087-31	2030	198020-3
2030	7325-1	2030	39561	2030	71447-1	2030	198023-2
2030	7334	2030	39587	2030	71447-3	2030	198035-1
2030	7344-1	2030	39617	2030	000073110	2030	198035-3
2030	7345-2	2030	39705	2030	000073300	2030	198045-4
2030	7349	2030	39713	2030	000073301	2030	198047-1
2030	7381-2	2030	39741	2030	000073302	2030	198047-3
2030	7507	2030	39835	2030	000073303	2030	198047-5
2030	7909	2030	39842	2030	000073305	2030	198047-6
2030	8000-00004-P082	2030	39863	2030	75145-3	2030	198048-1
2030	0000-00012-040	2030	39864	2030	75561-20	2030	198048-2
2030	8000-0004-P083	2030	39868	2030	75561-32	2030	198072-1
2030	8000-0004-P084	2030	39876	2030	80902-1	2030	198077-1
2030	8000-0049-059	2030	39919	2030	80904-1	2030	200252
2030	8000-0049-061	2030	39920	2030	87532	2030	202913-057
2030	8000-0005-010	2030	39940	2030	88060-145	2030	204201-001
2030	8000-0005-011	2030	40053	2030	88803	2030	210575
2030	8304	2030	40080	2030	88803-2-1	2030	210511
2030	8517	2030	40309	2030	88803-3-1	2030	221158
2030	8554-9	2030	40309V1	2030	88834	2030	230214
2030	8868-6	2030	40309V2	2030	91271	2030	230233
2030	8880-3	2030	40314	2030	91272	2030	231375
2030	8883-4	2030	40315	2030	91273	2030	232841
2030	90530	2030	40317	2030	91411	2030	233735
2030	9279	2030	40317L	2030	94041	2030	233944
2030	9367-1	2030	403178	2030	94042	2030	234024
2030	9404-2	2030	40320	2030	94050	2030	236287
2030	9405-0	2030	40320L	2030	94051	2030	239612
2030	9405-1	2030	403203	2030	94052	2030	248017
2030	9617K	2030	40323	2030	94062	2030	262417
2030	9696H	2030	40326	2030	94066	2030	262417-1
2030	10416-010	2030	40326L	2030	94070	2030	267704
2030	11236-1	2030	40326B	2030	94835-145-00	2030	276160
2030	11252-5	2030	40327	2030	95220	2030	276413
2030	012015	2030	40347	2030	95226-004	2030	276415
2030	12047-0023	2030	40350	2030	95226-1	2030	301606
2030	13213E4252	2030	40360	2030	95233	2030	310010
2030	14692	2030	40361	2030	95241-3	2030	315930
2030	14995	2030	40366	2030	96481	2030	315932
2030	14996-1	2030	40367	2030	101435	2030	321145
2030	14996-2	2030	40385	2030	102209	2030	321166
2030	15486	2030	40407L	2030	104719	2030	325101
2030	16065	2030	40407B	2030	105180	2030	331383
2030	16082	2030	40409	2030	110669	2030	340866-2
2030	16191	2030	40450	2030	110699	2030	346015-23
2030	16254	2030	40451	2030	111278	2030	346015-30
2030	018077	2030	40452	2030	112360	2030	346016-16
2030	20810-93	2030	40453	2030	112361	2030	346016-17
2030	21201	2030	40454	2030	113942	2030	346016-27
2030	21221	2030	40455	2030	115300-1	2030	348846-1
2030	21290	2030	40456	2030	115810P2	2030	00351980
2030	21676A	2030	40457	2030	119728	2030	400108
2030	23648	2030	40458	2030	119822	2030	400909
2030	25566-21	2030	40461-2	2030	123139	2030	409965-8A
2030	25566-50	2030	40501	2030	123243	2030	425411-01
2030	25566-63	2030	40539	2030	123703	2030	0440002-003
2030	25566-76	2030	40539L	2030	126703	2030	450826-1
2030	25840-163	2030	40539B	2030	126720	2030	481335
2030	26666-1	2030	40578	2030	126721	2030	489751-037
2030	27125-110	2030	40611	2030	126725	2030	489751-038
2030	27125-380	2030	40616	2030	127376	2030	489751-129
2030	28977	2030	40635	2030	127845	2030	489751-144
2030	030011	2030	40637A	2030	129051	2030	500879
2030	030011-1	2030	41053	2030	130172	2030	505287
2030	030011-2	2030	41502	2030	130174	2030	511806
2030	30254	2030	43021-198	2030	132175	2030	515039B
2030	30291(NPN)	2030	43082	2030	132327	2030	531972
2030	34044	2030	43088	2030	132328	2030	0573480H
2030	34588A	2030	43115	2030	132500	2030	0573517
2030	35210	2030	43117	2030	133176	2030	0573527
2030	35212	2030	43122	2030	133177	2030	0573557
2030	35303	2030	43165	2030	133275	2030	602122
2030	35383	2030	43354	2030	133876	2030	602909-7A
2030	35888	2030	45810-165	2030	134155	2030	605113
2030	36145	2030	49092	2030	134989	2030	608122
2030	36387	2030	50137-2	2030	136424	2030	610107-2
2030	36466	2030	50200-12	2030	136696	2030	610148-1
2030	36579	2030	50202-12	2030	137241	2030	610148-3
2030	36682	2030	50202-24	2030	137648	2030	610213-1
2030	36748	2030	51194	2030	138035-001	2030	618136-1
2030	37280	2030	51194-01	2030	138763	2030	618197
2030	37287	2030	51194-02	2030	139268	2030	619006-1
2030	37393	2030	51194-03	2030	139696	2030	652231
2030	37445	2030	53203-72	2030	140501	2030	652321
2030	37464	2030	57001-01	2030	141003	2030	653406
2030	37649	2030	59988-1	2030	141008	2030	656746
2030	37694	2030	60031	2030	141019	2030	660070
2030	37741	2030	60091	2030	141335	2030	660074
2030	37767	2030	60106	2030	141355	2030	660100
2030	37800	2030	60172	2030	141767	2030	696575-198
2030	37806	2030	60194	2030	141783	2030	699410-140
2030	37840	2030	60228	2030	143042	2030	702407-00
2030	37847	2030	60294	2030	143793	2030	729020-41
2030	37899	2030	60335	2030	143796	2030	742729
2030	37975	2030	60417	2030	143797	2030	800019-001
2030	37982	2030	60423	2030	143798	2030	800073-6
2030	38045	2030	60428	2030	145258	2030	800073-7
2030	38058	2030	60597	2030	147355	2030	800946-001
2030	38182	2030	60659	2030	150045	2030	810002-736
2030	38270	2030	60677	2030	150046A	2030	860001-8
2030	38271	2030	60680	2030	150095	2030	870006
2030	38334	2030	60682	2030	150730	2030	900201-104
2030	38354	2030	60697	2030	150787	2030	900201-105
2030	38361	2030	60700	2030	150796	2030	900201-167
2030	38393	2030	60703	2030	161919-29	2030	900201-81
2030	38424	2030	60720	2030	162002-081	2030	910088
2030	38432	2030	60994	2030	162002-082	2030	910634
2030	38468	2030	61209	2030	165735	2030	910634
2030	38475	2030	61219	2030	165736	2030	922114
2030	38476	2030	61275	2030	165737	2030	922125
2030	38495	2030	61359	2030	166917	2030	928291-101
2030	38497	2030	61370/4560	2030	167569	2030	928291-102
2030	38510-165	2030	61538	2030	168660	2030	932017-0001
2030	38551	2030	61562	2030	171038	2030	932055-1
2030	38588	2030	61667	2030	171044(SEARS)	2030	960494-1
2030	38659	2030	61733	2030	171162-026	2030	960494-2
2030	38716	2030	61828	2030	171162-163	2030	970108
2030	38725	2030	61841	2030	171162-188	2030	982300
2030	38735	2030	61917	2030	171557	2030	984196
2030	38736	2030	62019	2030	172763	2030	984229
2030	38837	2030	62185	2030	175007-277	2030	987030
2030	38869	2030	62192	2030	181515-4	2030	988993
2030	38916	2030	62203	2030	181515-6	2030	995624-2
2030	39097	2030	62243	2030	181515-7	2030	995928 1
2030	39231	2030	62398	2030	181515-9	2030	995928-1
2030	39238	2030	62404	2030	183034	2030	1288055
2030	39248	2030	62446	2030	186342A	2030	1417318-2
2030	39252	2030	62452	2030	194243	2030	1417325-1
2030	39255	2030	62540	2030	196779-9		

RS 276-	Industry Standard No.	RS 276-	Industry Standard No.	RS 276-	Industry Standard No.	RS 276-	Industry Standard No.
2030	1417338-5	2032	A068-109A	2032	A467G-0(JAPAN)	2032	A5T404A
2030	1417342-1	2032	A1030	2032	A467G-R	2032	A5T4125
2030	1417344-1	2032	A112-000172	2032	A467G-R(JAPAN)	2032	A5T4126
2030	1417345-2	2032	A112-000185	2032	A467G-Y	2032	A5T4248
2030	1417349-2	2032	A112-000187	2032	A467G-Y(JAPAN)	2032	A5T4249
2030	1417381-2	2032	A11200482	2032	A4745	2032	A5T5086
2030	1471120-14	2032	A116078	2032	A480(JAPAN)	2032	A5T5221
2030	1471123-2	2032	A118284	2032	A4802-00004	2032	A5T5226
2030	1471123-3	2032	A119730	2032	A4815	2032	A5T5227
2030	1471123-4	2032	A120P1	2032	A481AO030	2032	A603
2030	1471123-5	2032	A121-1	2032	A482(JAPAN)	2032	A608-C
2030	1473508-1	2032	A121-1RED	2032	A4822-130-40348	2032	A608-D
2030	1473545-1	2032	A121-444	2032	A4844	2032	A608-E
2030	1473547-1	2032	A121-446	2032	A489751-028	2032	A608-F
2030	1473553-1	2032	A121-495	2032	A489751-031	2032	A608A
2030	1473555-3	2032	A121-496	2032	A490(JAPAN)	2032	A608B
2030	1473555-I	2032	A121-497	2032	A490Y	2032	A608C
2030	1473560	2032	A121-497WHT	2032	A493	2032	A608D
2030	1473560-1	2032	A121-602	2032	A493-0	2032	A608E
2030	1473560-2	2032	A121-603	2032	A4930R	2032	A608F
2030	1473561-1	2032	A121-679	2032	A493Y	2032	A608G
2030	1473565-001	2032	A121-699	2032	A494-GR	2032	A609
2030	1473565-1	2032	A121-746	2032	A494-0	2032	A609A
2030	1473566-1	2032	A121-774	2032	A494-Y	2032	A609B
2030	1473572-3	2032	A1214	2032	A495(JAPAN)	2032	A609C
2030	1473580-1	2032	A121467	2032	A495-G	2032	A609D
2030	1473593-1	2032	A121659	2032	A495-0	2032	A609E
2030	1473608-002	2032	A1220RN	2032	A495-R	2032	A609F
2030	1473608-1	2032	A122YEL	2032	A495-Y	2032	A609G
2030	1473608-2	2032	A124047	2032	A4950	2032	A610
2030	1473608-3	2032	A124755	2032	A495A	2032	A610074-1
2030	1473613-5	2032	A12594	2032	A495D	2032	A610083
2030	1473615-1	2032	A126524	2032	A495G-GR	2032	A610083-1
2030	1473622	2032	A126700	2032	A4950-0	2032	A610083-2
2030	1473622-002	2032	A126707	2032	A4950-R	2032	A610083-3
2030	1473622-2	2032	A126715	2032	A4950-Y	2032	A610110-1
2030	1473625-001	2032	A126718	2032	A495W	2032	A610120-1
2030	1473625-1	2032	A126719	2032	A495Y	2032	A610B
2030	1473679-1	2032	A12888	2032	A499	2032	A611
2030	1502039	2032	A129-34	2032	A499-0	2032	A611-4E
2030	1815153	2032	A129697	2032	A499-R	2032	A617K
2030	1815154-9	2032	A129699	2032	A499-Y	2032	A618K
2030	1815154/7	2032	A130-149	2032	A500(JAPAN)	2032	A628
2030	1815156	2032	A130-40315	2032	A500-0	2032	A628A
2030	1815157	2032	A130-40429	2032	A500-R	2032	A628D
2030	1815157-9	2032	A130139	2032	A500-Y	2032	A628B
2030	1815159	2032	A137(PNP)	2032	A502	2032	A628F
2030	1817006	2032	A1471114-1	2032	A509	2032	A629
2030	1820829	2032	A1473563-1	2032	A509-0	2032	A637
2030	1943313A1	2032	A1473570-1	2032	A509-Y	2032	A640
2030	1960083-1	2032	A1473574-1	2032	A509GR	2032	A640(JAPAN)
2030	1967799	2032	A1473590-1	2032	A509R	2032	A640A
2030	1967799-1	2032	A1473591-1	2032	A509Y	2032	A640B
2030	1967801	2032	A1473597-1	2032	A510	2032	A640C
2030	1968959	2032	A1558-17	2032	A510-0	2032	A640D
2030	1971489	2032	A160	2032	A510-R	2032	A640E
2030	2000752-80	2032	A161	2032	A511	2032	A640M
2030	2006334-155	2032	A162	2032	A511-0	2032	A642
2030	2006436-40	2032	A171	2032	A511-R	2032	A642(JAPAN)
2030	2006623-148	2032	A177	2032	A512	2032	A642A
2030	2097013-0702	2032	A177(A)	2032	A512-0	2032	A642B
2030	2320052	2032	A177A	2032	A512-R	2032	A642C
2030	2320233	2032	A177AB	2032	A513	2032	A642D
2030	2320664	2032	A178AB	2032	A513-0	2032	A642E
2030	2320946	2032	A178BA	2032	A513-R	2032	A642F
2030	2321881	2032	A179AC	2032	A514-044910	2032	A642W
2030	2327022	2032	A179BB	2032	A522(JAPAN)	2032	A65-09-220
2030	2327292	2032	A1844-17	2032	A5226-1	2032	A650238A
2030	2327293	2032	A1867-17	2032	A522A	2032	A650372D
2030	2327332	2032	A1901-5338	2032	A530	2032	A650923F
2030	2327403	2032	A190429	2032	A530H	2032	A650925H
2030	2469936-1	2032	A200-052	2032	A5320111	2032	A659
2030	2485076-2	2032	A20030703-0702	2032	A539	2032	A659C
2030	2485076-3	2032	A2057013-0004	2032	A539L	2032	A659D
2030	2485077-2	2032	A2057013-0701	2032	A539B	2032	A659B
2030	2485077-3	2032	A2057013-0702	2032	A54-96-003	2032	A659P
2030	2487424	2032	A2057013-0703	2032	A542	2032	A6661QRS
2030	2487424(NPN)	2032	A2057A2-198	2032	A544	2032	A669
2030	2495529(ARVIN)	2032	A2057B106-12	2032	A550	2032	A672
2030	2498163	2032	A2057B108-6	2032	A550A	2032	A672A
2030	2545989-2	2032	A20K	2032	A550AQ	2032	A672B
2030	2605022	2032	A20KA	2032	A550Q	2032	A672C
2030	2608169-1	2032	A22008	2032	A550R	2032	A673A
2030	2666307-1	2032	A23114050	2032	A550B	2032	A673AA
2030	2777301	2032	A23114051	2032	A561	2032	A673AB
2030	2865101	2032	A23114550	2032	A561-0	2032	A673AC
2030	2928054-1	2032	A2428	2032	A561-R	2032	A673AD
2030	3130053	2032	A2448	2032	A561-Y	2032	A673AS
2030	3130092	2032	A2498512	2032	A561GR	2032	A673ABC
2030	3403866-3	2032	A2798	2032	A562	2032	A673B
2030	3404520-301	2032	A297L012C01	2032	A562-0	2032	A673C
2030	3412907-1	2032	A297V073C03	2032	A562-R	2032	A673C2
2030	3457633-1	2032	A297V073C04	2032	A562-Y	2032	A673D
2030	3457633-2	2032	A28A564FR	2032	A5620	2032	A673WT
2030	3458267-1	2032	A30270	2032	A5620R	2032	A675
2030	3458573-1	2032	A30290	2032	A562R	2032	A675A
2030	3463099-1	2032	A3513	2032	A562Y	2032	A675B
2030	3596116	2032	A3540	2032	A565	2032	A675C
2030(2)	EC0295MP	2032	A3549	2032	A565A	2032	A676
2030(2)	GE-47NP	2032	A3559	2032	A565B	2032	A678(SONY)
2030(2)	TM293MP	2032	A3562	2032	A565C	2032	A690D
2031	C710(C)	2032	A3563	2032	A565D	2032	A701
2031	C710B2	2032	A3574	2032	A565K	2032	A701584-00
2031	C710DB	2032	A3581	2032	A567(JAPAN)	2032	A701589-00
2031	SK3356	2032	A312894	2032	A568	2032	A701P
2031	2801727	2032	A312906	2032	A569	2032	A701PJ
2031	2801898	2032	A312906A	2032	A569J	2032	A701PO
2031	2801899	2032	A312907	2032	A570(TRANSISTOR)	2032	A71687-1
2032	A-1005-725	2032	A402	2032	A576-0001-002	2032	A718
2032	A-113110	2032	A4086	2032	A576-0001-013	2032	A719
2032	A-120417	2032	A4087	2032	A592Y	2032	A719P
2032	A-120526	2032	A4126	2032	A59625-1	2032	A719Q
2032	A-1853-0016-1	2032	A41440	2032	A59625-10	2032	A719R
2032	A-1853-0020-1	2032	A417-116	2032	A59625-11	2032	A719RB
2032	A-1853-0027-1	2032	A417-132	2032	A59625-12	2032	A719B
2032	A-1853-0036-1	2032	A417-153	2032	A59625-2	2032	A723
2032	A-1853-0039-1	2032	A417-176	2032	A59625-3	2032	A723A
2032	A-1853-0049-1	2032	A417-182	2032	A59625-4	2032	A723B
2032	A-1853-0058-1	2032	A417-184	2032	A59625-5	2032	A723C
2032	A-1853-0062-1	2032	A417-196	2032	A59625-6	2032	A723D
2032	A-1853-0065-1	2032	A417-200	2032	A59625-7	2032	A723E
2032	A-1853-0092-1	2032	A417-201	2032	A59625-8	2032	A723F
2032	A-1853-0099-1	2032	A417-235	2032	A59625-9	2032	A730
2032	A-1853-0285-1	2032	A3025845	2032	A5T2604	2032	A731
2032	A-1853-0321-1	2032	A4310	2032	A5T2605	2032	A753-4004-248
2032	A-195C	2032	A4442	2032	A5T2907	2032	A7570013-01
2032	A054-223	2032	A467(JAPAN)	2032	A5T3638	2032	A7576004-01
2032	A066-113	2032	A467-0	2032	A5T3638A	2032	A759A
2032	A066-113A	2032	A467-0(JAPAN)	2032	A5T3644	2032	A759B
2032	A066-113AB	2032	A467-Y	2032	A5T3645	2032	A78931
2032	A066-118	2032	A467-Y(JAPAN)	2032	A5T3905	2032	A800-523-01
2032	A066-118A	2032	A467G	2032	A5T3906	2032	A800-523-02
2032	A066-118B	2032	A467G(JAPAN)	2032	A5T404	2032	A800-527-00
2032	A068-109	2032	A467G-0				

RS 276-	Industry Standard No.	RS 276-	Industry Standard No.	RS 276-	Industry Standard No.	RS 276-	Industry Standard No.
2032	A815199	2032	B2E	2032	BC214KC	2032	BC513A
2032	A815199-6	2032	B2Q	2032	BC214KC-1	2032	BC514A
2032	A815211	2032	B2M-1	2032	BC214L	2032	BC556
2032	A815213	2032	B2M-2	2032	BC214L-1	2032	BC556A
2032	A815229	2032	B2M-3	2032	BC214LA	2032	BC556VI
2032	A815247	2032	B28	2032	BC214LA-1	2032	BC557
2032	A829	2032	B2Y	2032	BC214LB	2032	BC557A
2032	A829A	2032	B54731-3Q	2032	BC214LB-1	2032	BC557VI
2032	A829B	2032	B561	2032	BC214LC	2032	BC558
2032	A829C	2032	B561B	2032	BC214LC-1	2032	BC558VI
2032	A829D	2032	B561C	2032	BC221	2032	BC559
2032	A829E	2032	BC-261	2032	BC221-1	2032	BCW35
2032	A829F	2032	BO116	2032	BC225-1	2032	BCW37
2032	A833	2032	BO116A	2032	BC231A	2032	BCW37A
2032	A8405	2032	B126	2032	BC238-16	2032	BCW56
2032	A8540	2032	BC126-1	2032	BC250	2032	BCW56A
2032	A8867	2032	BC137	2032	BC250-1	2032	BCW57
2032	A823702	2032	BC137-1	2032	BC250A-1	2032	BCW57A
2032	A823703	2032	BC153-1	2032	BC250B-1	2032	BCW58
2032	A8T404	2032	BC154-1	2032	BC250C-1	2032	BCW58A
2032	A8T404A	2032	BC157	2032	BC251	2032	BCW59
2032	A8T4058	2032	BC157-1	2032	BC251A	2032	BCW59A
2032	A8T4059	2032	BC157A	2032	BC251A-1	2032	BCW61
2032	A8T4060	2032	BC157B	2032	BC251B	2032	BCW61C
2032	A8T4061	2032	BC158	2032	BC251B-1	2032	BCW61D
2032	A8T4062	2032	BC158-1	2032	BC251C	2032	BCW62
2032	A921-70B	2032	BC158A	2032	BC251C-1	2032	BCW62A
2032	A94057	2032	BC158A-1	2032	BC252A	2032	BCW63
2032	A945-0	2032	BC158B	2032	BC252A-1	2032	BCW63A
2032	A95227	2032	BC158B-1	2032	BC252B-1	2032	BCW64A
2032	A95232	2032	BC159-1	2032	BC252C	2032	BCW69
2032	A970246	2032	BC159A-1	2032	BC253-1	2032	BCW69R
2032	A970248	2032	BC159B-1	2032	BC253A-1	2032	BCW79-10
2032	A970251	2032	BC177-1	2032	BC253B-1	2032	BCW79-16
2032	A970254	2032	BC177A-1	2032	BC253C-1	2032	BCW80-10
2032	A984193	2032	BC177B-1	2032	BC256	2032	BCW80-16
2032	A991-01-0098	2032	BC177V	2032	BC256-1	2032	BCW96
2032	A991-01-1225	2032	BC177V-1	2032	BC256A	2032	BCW96A
2032	A991-01-1319	2032	BC177V1	2032	BC256A-1	2032	BCW96B
2032	A991-01-3058	2032	BC177V1-1	2032	BC256B	2032	BCW96K
2032	A09082	2032	BC178-1	2032	BC256B-1	2032	BCW96KA
2032	A09083	2032	BC178A-1	2032	BC257	2032	BCW96KB
2032	A09084	2032	BC178B-1	2032	BC257-1	2032	BCW97
2032	A09085	2032	BC178D	2032	BC257VI	2032	BCW97A
2032	ACDC09000-1	2032	BC178D-1	2032	BC258	2032	BCW97B
2032	AD29E-1	2032	BC178V	2032	BC258-1	2032	BCW97K
2032	AD29E-2	2032	BC178V-1	2032	BC259-1	2032	BCW97KA
2032	AD29B10	2032	BC178V1	2032	BC25BB	2032	BCW97KB
2032	AD29B4	2032	BC178V1-1	2032	BC260	2032	BCX17
2032	AD29B5	2032	BC179-1	2032	BC260-1	2032	BCX17R
2032	AD29B6	2032	BC179A-1	2032	BC260A-1	2032	BCY10
2032	AD29B7	2032	BC179B-1	2032	BC260B-1	2032	BCY10A
2032	AD29B9	2032	BC17LB-1	2032	BC260C-1	2032	BCY11
2032	AD29B9	2032	BC181	2032	BC261	2032	BCY11A
2032	AD30A1	2032	BC181A	2032	BC261A	2032	BCY12
2032	AD30A2	2032	BC181A-1	2032	BC261A-1	2032	BCY12A
2032	AD30A3	2032	BC187	2032	BC261B	2032	BCY35
2032	AD30A4	2032	BC187-1	2032	BC261B-1	2032	BCY35A
2032	AD30A5	2032	BC192	2032	BC261C	2032	BCY37
2032	AF21490	2032	BC192-1	2032	BC261C-1	2032	BCY37A
2032	AF3570	2032	BC196	2032	BC262	2032	BCY38
2032	AF3590	2032	BC196-1	2032	BC262-1	2032	BCY38A
2032	AF699	2032	BC196A	2032	BC262A-1	2032	BCY39
2032	AF031170P	2032	BC196A-1	2032	BC262B-1	2032	BCY39A
2032	AF824226	2032	BC196B	2032	BC262C	2032	BCY40
2032	AFT0019M	2032	BC196B-1	2032	BC262C-1	2032	BCY40A
2032	AFT052	2032	BC196V1	2032	BC263	2032	BCY54
2032	AFT1341	2032	BC196V1-1	2032	BC263A-1	2032	BCY54A
2032	AFT1746	2032	BC200	2032	BC263B-1	2032	BCY70
2032	AN4101	2032	BC200-1	2032	BC263C-1	2032	BCY70A
2032	AR304(GREEN)	2032	BC201-1	2032	BC266	2032	BCY71
2032	AR308(VIOLET)	2032	BC202	2032	BC266-1	2032	BCY71A
2032	AT-11	2032	BC202-1	2032	BC266A	2032	BCY72A
2032	AT331	2032	BC203	2032	BC266B	2032	BCY77IX
2032	AT331A	2032	BC203-1	2032	BC281	2032	BCY77VII
2032	AT332	2032	BC204-1	2032	BC281A	2032	BCY77VIII
2032	AT332A	2032	BC204A-1	2032	BC281A-1	2032	BCY90
2032	AT333	2032	BC204B-1	2032	BC281B	2032	BCY90A
2032	AT333A	2032	BC204V	2032	BC281B-1	2032	BCY90B
2032	AT410	2032	BC204V-1	2032	BC281C	2032	BCY90B-1
2032	AT410-1	2032	BC204V1	2032	BC281C-1	2032	BCY91
2032	AT412	2032	BC204V1-1	2032	BC283	2032	BCY91A
2032	AT412-1	2032	BC205-1	2032	BC283-1	2032	BCY91B-1
2032	AT413	2032	BC205A-1	2032	BC291	2032	BCY92
2032	AT413-1	2032	BC205B-1	2032	BC291-1	2032	BCY92A
2032	AT414	2032	BC205V-1	2032	BC291A	2032	BCY92B-1
2032	AT414-1	2032	BC205V1	2032	BC291A-1	2032	BCY93
2032	AT415	2032	BC205V1-1	2032	BC291D	2032	BCY93A
2032	AT415-1	2032	BC206-1	2032	BC291D-1	2032	BCY93B
2032	AT416	2032	BC206B-1	2032	BC292	2032	BCY93B-1
2032	AT416-1	2032	BC212	2032	BC292A	2032	BCY94
2032	AT417	2032	BC212-1	2032	BC292A-1	2032	BCY94A
2032	AT417-1	2032	BC212K	2032	BC292D	2032	BCY94B
2032	AT418	2032	BC212K-1	2032	BC292D-1	2032	BCY94B-1
2032	AT418-1	2032	BC212KA	2032	BC307	2032	BCY95
2032	AT419	2032	BC212KA-1	2032	BC307A	2032	BCY95A
2032	AT419-1	2032	BC212KB	2032	BC307V1	2032	BCY95B
2032	AT430	2032	BC212KB-1	2032	BC307VI	2032	BCY95B-1
2032	AT430-1	2032	BC212L	2032	BC308A	2032	BCY98
2032	AT431	2032	BC212L-1	2032	BC308B	2032	BCY98A
2032	AT431-1	2032	BC212LA	2032	BC308V1	2032	BCY98B
2032	AT432	2032	BC212LA-1	2032	BC315	2032	BCY98B-1
2032	AT432-1	2032	BC212LB	2032	BC325	2032	BCZ10
2032	AT433	2032	BC212LB-1	2032	BC325A	2032	BCZ10A
2032	AT433-1	2032	BC212V1	2032	BC326	2032	BCZ10B
2032	AT434	2032	BC212VI	2032	BC326A	2032	BCZ10C
2032	AT434-1	2032	BC213	2032	BC328-16	2032	BCZ11
2032	AT435	2032	BC213A	2032	BC381	2032	BCZ11A
2032	AT435-1	2032	BC213K	2032	BC400	2032	BCZ12
2032	AT436	2032	BC213K-1	2032	BC404	2032	BCZ12A
2032	AT436-1	2032	BC213KA	2032	BC404VI	2032	BCZ12B
2032	AT437	2032	BC213KA-1	2032	BC405	2032	BCZ12C
2032	AT437-1	2032	BC213KB	2032	BC405A	2032	BCZ13
2032	AT438	2032	BC213KB-1	2032	BC405B	2032	BCZ13A
2032	AT438-1	2032	BC213KC	2032	BC406B	2032	BCZ13B
2032	AT451	2032	BC213KC-1	2032	BC415A	2032	BCZ13C
2032	AT451-1	2032	BC213L	2032	BC415B	2032	BCZ13E
2032	AT452	2032	BC213L-1	2032	BC416A	2032	BCZ13F
2032	AT452-1	2032	BC213LA	2032	BC416B	2032	BCZ13G
2032	AT453	2032	BC213LA-1	2032	BC417	2032	BCZ13H
2032	AT453-1	2032	BC213LB	2032	BC418	2032	BCZ14
2032	AT454	2032	BC213LB-1	2032	BC418A	2032	BCZ14A
2032	AT454-1	2032	BC213LC	2032	BC432	2032	BCZ14B
2032	AT455	2032	BC213LC-1	2032	BC478	2032	BCZ14C
2032	AT455-1	2032	BC214	2032	BC478A	2032	BCZ14D
2032	B-1426	2032	BC214-1	2032	BC478B	2032	BCZ14E
2032	B-6288	2032	BC214A	2032	BC479	2032	BCZ14F
2032	B-75561-31	2032	BC214K	2032	BC479B	2032	BF-832
2032	B1J	2032	BC214K-1	2032	BC512	2032	BF340A
2032	B1N-1	2032	BC214KA	2032	BC512A	2032	BF340B
2032	B1N-2	2032	BC214KA-1	2032	BC513	2032	BF340C
2032	B266A-1	2032	BC214KB			2032	BF340D
2032	B266B-1	2032	BC214KB-1				
2032	B2A						

RS 276-	Industry Standard No.
2032	BF341A
2032	BF341B
2032	BF341C
2032	BF341D
2032	BF342A
2032	BF342B
2032	BF342C
2032	BF342D
2032	BF814A
2032	BF814B
2032	BF814C
2032	BF814D
2032	BF816
2032	BF816A
2032	BF816B
2032	BF816C
2032	BF816D
2032	BF826
2032	BF826A
2032	BF826B
2032	BF826C
2032	BF826D
2032	BF826E
2032	BF826F
2032	BF826G
2032	BF831
2032	BF832
2032	BF832P
2032	BF833
2032	BF833P
2032	BF834P
2032	BF837
2032	BF837A
2032	BF840
2032	BF840A
2032	BF841
2032	BF869
2032	BFV20
2032	BFV21
2032	BFV22
2032	BFV25
2032	BFV26
2032	BFV29
2032	BFV30
2032	BFV33
2032	BFV34
2032	BFV86
2032	BFV86A
2032	BFV86B
2032	BFV86C
2032	BFW89
2032	BFW90
2032	BFX29
2032	BFX30
2032	BFX55
2032	BFX65
2032	BFY94
2032	BI-82
2032	BJ10
2032	BJ11
2032	BJ11A
2032	BJ11B
2032	BJ12
2032	BJ12B
2032	BJ12C
2032	BJ13
2032	BJ13A
2032	BJ13B
2032	BJ14
2032	BJ14A
2032	BJ15
2032	BJ160
2032	BJ161
2032	BJ161A
2032	BJ161B
2032	BJ161C
2032	BJ1A
2032	BJ1B
2032	BJ1C
2032	BJ2
2032	BJ2A
2032	BJ2B
2032	BJ2C
2032	BJ2D
2032	BJ3
2032	BJ3A
2032	BJ3B
2032	BJ4
2032	BJ4A
2032	BJ4B
2032	BJ4C
2032	BJ4D
2032	BJ5
2032	BJ56
2032	BJ5A
2032	BJ5B
2032	BJ6
2032	BJ6A
2032	BJ6B
2032	BJ6C
2032	BJ6D
2032	BJ7
2032	BJ7A
2032	BJ7B
2032	BJ7C
2032	BJ7D
2032	BJ8
2032	BJ8A
2032	BJ8B
2032	BJ8C
2032	BJ8D
2032	BJ9
2032	BJ9A
2032	BJ9B
2032	BJ9C
2032	BJ9D
2032	BMT1991
2032	BMT2303
2032	BMT2411
2032	BMT2412
2032	B08875/2
2032	BR-82
2032	BR-832
2032	BSV21
2032	BSV21A
2032	BSV43A
2032	BSV44A
2032	BSV45A
2032	BSV47A
2032	BSV48A
2032	BSV49A
2032	BSV55A
2032	BSV55AP
2032	BSV55P
2032	BSV96
2032	BSV97
2032	BSV98
2032	BSW-21A
2032	BSW-22
2032	BSW-22A
2032	BSW-24
2032	BSW-44
2032	BSW-44A
2032	BSW-45
2032	BSW-45A
2032	BSW-72
2032	BSW-73
2032	BSW-74
2032	BSW-75
2032	BSW21
2032	BSW21A
2032	BSW22
2032	BSW22A
2032	BSW24
2032	BSW44
2032	BSW44A
2032	BSW45
2032	BSW45A
2032	BSW72
2032	BSW73
2032	BSW74
2032	BSW75
2032	BSX36
2032	BSY-40
2032	BSY-41
2032	BT82
2032	BT832
2032	C00-686-0241
2032	C00686-0258-0
2032	C0068602720
2032	C51909B
2032	C610
2032	C673C2
2032	C673D
2032	C686-248-0
2032	C9082
2032	C9083
2032	C9084
2032	C9085
2032	CC82005B
2032	CC86228F
2032	CC89015
2032	CD10000-1E
2032	CD437
2032	CD445
2032	CDC-9000-1D
2032	CDC10000-1E
2032	CD496
2032	CDC746(ZENITH)
2032	CDC9000-1
2032	CDC9000-1D
2032	CE4005C
2032	CK942
2032	CS-2005B
2032	CS-2005C
2032	CS-9012HF
2032	CS-9015D
2032	CS-901ZHF
2032	C81121G
2032	C81170F
2032	C81221F
2032	C81228
2032	C81228B
2032	C81251E
2032	C81251F
2032	C81294E
2032	C81294H
2032	C81298
2032	C81303
2032	C81308
2032	C81354
2032	C81627
2032	C82941
2032	C83702
2032	C83703
2032	C83906
2032	C85447
2032	C85448
2032	C86228F
2032	C89012
2032	C89012E
2032	C89012E-F
2032	C89012FG
2032	C89012HG
2032	C89012HH
2032	C89012HH/3490
2032	C89015C
2032	C89015C2
2032	C89015D
2032	C8901ZHF
2032	C89020E
2032	C89020F
2032	C89128-B2
2032	C89128C1
2032	C89129
2032	C89129B
2032	C89129B1
2032	C89129B2
2032	D-50492-01
2032	D29A4
2032	D29A5
2032	D29A6
2032	D29A9
2032	D29E1
2032	D29E2
2032	D29E4
2032	D29E5
2032	D29E6
2032	D29E7
2032	D29E8
2032	D29F5
2032	D29F6
2032	D29F7
2032	D30A1
2032	D30A2
2032	D30A3
2032	D30A4
2032	D30A5
2032	DDBY003001
2032	DDBY004001
2032	DDCY007002
2032	DED4191
2032	D8-82
2032	D8-83
2032	D882
2032	D883
2032	E13-001-02
2032	E13-001-03
2032	E13-001-04
2032	E13-006-02
2032	E211
2032	E213
2032	EA15X185
2032	EA15X233
2032	EA15X395
2032	EA3714
2032	EC0159
2032	EC0290
2032	ED1602C
2032	ED1602D
2032	E13-006-02
2032	EN1132
2032	EN2894
2032	EN2894A
2032	EN2907
2032	EN3250
2032	EN3504
2032	EN3905
2032	EN3906
2032	EN3962
2032	EN722
2032	EP15X17
2032	EP15X26
2032	EP15X48
2032	EP15X48(PNP)
2032	EP15X53
2032	EP15X60
2032	EP25
2032	EPX15X17
2032	ES15X101
2032	ES15X128
2032	ES15X9
2032	ES15X90
2032	ES34(ELCOM)
2032	ES65(ELCOM)
2032	EWQ202
2032	EX746-X
2032	EY2P-546
2032	EY2P-623
2032	EY2P623
2032	F209
2032	F21490
2032	F3559
2032	F3570
2032	F3590
2032	F3597
2032	F549-1
2032	F699
2032	FCS-9012-HH
2032	FCS-9012F
2032	FCS-9012G
2032	F81170P
2032	FC88550
2032	FC88550C
2032	FC89012H
2032	FC89012HE
2032	FC89015B
2032	FD-1029-4&
2032	FD-1029-JP
2032	FI-1007
2032	FI-1008
2032	FS24954
2032	FSI2IIL
2032	FTO019M
2032	FTO52
2032	FT1341
2032	FT1746
2032	FT19B
2032	FT19M
2032	FT3638
2032	FT5040
2032	FT5041
2032	G03-407-Y
2032	G03007
2032	G13638
2032	GE-21
2032	GE-22
2032	GE-269
2032	GE-334
2032	GE-82
2032	GET3638
2032	GET3638A
2032	GI3638
2032	GI3638A
2032	GI3644
2032	GI3702
2032	GI3703
2032	GM760
2032	GME040-1
2032	GO3007
2032	GO3014
2032	G89015H
2032	G89015I
2032	G89015J
2032	GT1644
2032	HA-00495
2032	HA-00610
2032	HAO0562
2032	HAT501
2032	HAT502
2032	HAT506
2032	HAT507
2032	HAT510
2032	HAT533
2032	HAT534
2032	HAT536
2032	HAT537
2032	HAT538
2032	HAT543
2032	HAT597
2032	HAT598
2032	HAT599
2032	HAT633
2032	HAT804
2032	HAT806
2032	HAT808
2032	HAT810
2032	HAT815
2032	HEP51
2032	HEP52
2032	HEP708
2032	HEP715
2032	HEP716
2032	HEP717
2032	HEP739
2032	HEP80026
2032	HR-71
2032	HR84(PNP)
2032	HS-40027
2032	HS-40031
2032	HS-40040
2032	HS-40050
2032	HS-40057
2032	HS-90028
2032	HT104941B-0
2032	HT104941B-0
2032	HT104941C-0
2032	HT104941C0
2032	HT104942A
2032	HT104951A-Q
2032	HT104951B
2032	HT104951B-0
2032	HT104951C
2032	HT105611
2032	HT105611A
2032	HT105611B
2032	HT105611B0
2032	HT105611C
2032	HT105611C0
2032	HT105612B
2032	HT105621B0
2032	HT105621B0
2032	HT106731B
2032	HT106731B0
2032	HT6000011P
2032	HT6000011H
2032	HT6000210
2032	HX-50105
2032	HX-50112
2032	I81C-30
2032	19680
2032	IC743043
2032	IP20-0009
2032	IP20-00192
2032	IRTR20
2032	IRTR31
2032	IRTR52
2032	J241225
2032	J241226
2032	J241253
2032	J241259
2032	J24640
2032	J9680
2032	J9697
2032	JA1050
2032	JA1050G
2032	JA1050GL
2032	K4-505
2032	K9682
2032	KLH4746
2032	KSA495Y
2032	LJ-152
2032	LJ152
2032	LJ152(0)
2032	LJ152-0
2032	LJ152B
2032	LM-1149
2032	LM-1150
2032	LM-1151
2032	LM-1153
2032	LM-2589
2032	LM1153
2032	LM1404
2032	LM1795
2032	L8-0079-01
2032	L8-0079-02
2032	M094-585-46
2032	M4442
2032	M446
2032	M4525
2032	M4590
2032	M4745
2032	M4815
2032	M4815D
2032	M4943
2032	M4989
2032	M644
2032	M65A
2032	M65B
2032	M65C
2032	M65D
2032	M65E
2032	M65F
2032	M6931
2032	M694
2032	M7127/P28
2032	M829A
2032	M829B
2032	M829C
2032	M829D
2032	M829E
2032	M829F
2032	M833
2032	M9334
2032	M9514
2032	M9526
2032	M9527
2032	M9531
2032	M9571
2032	M9649
2032	MAO404
2032	MEO404
2032	MEO404-1
2032	MEO404-2
2032	ME501
2032	ME502
2032	MF3304
2032	MM3726
2032	MM3905
2032	MM3906
2032	MM4048
2032	MM4052
2032	MM999
2032	MPL1000
2032	MPS-3638A
2032	MPS-3702
2032	MPS1572
2032	MPS3638
2032	MPS3638A
2032	MPS3640
2032	MPS3702
2032	MPS3703
2032	MPS404A
2032	MPS5086
2032	MPS6134
2032	MPS6516
2032	MPS6517
2032	MPS6518
2032	MPS6519
2032	MPS6522
2032	MPS6523
2032	MPS6524
2032	MPS6533
2032	MPS6534

RS 276-	Industry Standard No.	RS 276-	Industry Standard No.	RS 276-	Industry Standard No.	RS 276-	Industry Standard No.
2032	MP86534M	2032	0C445K	2032	RS-2021	2032	SL119
2032	MP86535	2032	0C449	2032	RS-2022	2032	SM-A-726658
2032	MP86562	2032	0C449K	2032	RS-2023	2032	SM-B-523974
2032	MP86563	2032	0C450	2032	RS-2024	2032	SM-B-574495
2032	MP88598	2032	0C460	2032	RS-7665A	2032	SM1507
2032	MP88599	2032	0C460K	2032	RST665	2032	SM4547
2032	MP89467A-T	2032	0C463	2032	RT-101	2032	SM4574A
2032	MP89666	2032	0C463K	2032	RT-103	2032	SM4719
2032	MP89680	2032	0C465	2032	RT3065	2032	SN-400-319-P1
2032	MP89680O	2032	0C465K	2032	RT3065A	2032	SNT204
2032	MP89680H	2032	0C466	2032	RT3071	2032	SNT204A
2032	MP89680H/I	2032	0C466K	2032	RT3071A	2032	SP70
2032	MP89680I	2032	0C467	2032	RT8670	2032	SP90
2032	MP89680I/J	2032	0C467K	2032	RT8895	2032	SPS-1539
2032	MP89680J	2032	0C468	2032	RV2351	2032	SPS-1539WT
2032	MP89680T	2032	0C468K	2032	RV2355	2032	SPS-952
2032	MP89681	2032	0C469	2032	RVTS281382	2032	SP8169
2032	MP89681I	2032	0C469K	2032	RVTS22411	2032	SP81097A
2032	MP89681J	2032	0C470	2032	S-1367A	2032	SP812
2032	MP89681K	2032	0C470K	2032	S0026	2032	SP81523
2032	MP89681T	2032	0C700	2032	S017446	2032	SP81523A
2032	MP89682	2032	0C700A	2032	S019843	2032	SP81593WT
2032	MP89682(I)	2032	0C700B	2032	S022012	2032	SP822
2032	MP89682-1	2032	0C702	2032	S023735	2032	SP82269
2032	MP89682I	2032	0C702A	2032	S1047	2032	SP82272
2032	MP89682J	2032	0C702Q	2032	0S133.0008T	2032	SP82279
2032	MP89682K	2032	0C704	2032	0S133.1003T	2032	SP83329
2032	MP89750D	2032	0C740	2032	S1350	2032	SP83724
2032	MP89750F	2032	0C7400	2032	S1350A	2032	SP83724A
2032	MP89750O	2032	0C740Q	2032	S1367	2032	SP83786
2032	MP89750OP	2032	0C740M	2032	S1477	2032	SP83786A
2032	MP8A55	2032	0C7400	2032	S18100	2032	SP83924
2032	MP8A56	2032	0C742	2032	S1889	2032	SP83924A
2032	MP8A70	2032	0C7420	2032	S2091	2032	SP83927
2032	MP8D55	2032	0C74200	2032	S2128	2032	SP83927A
2032	MP89681J	2032	0C742Q	2032	S2129	2032	SP83931
2032	M97505	2032	0C742M	2032	S23130	2032	SP83931A
2032	M89667	2032	P00347100	2032	S24226	2032	SP83987
2032	M89681	2032	P00347101	2032	S2525	2032	SP83988
2032	M8M63	2032	P04-45-0015-P1	2032	S2645	2032	SP83988A
2032	M8M631	2032	P04-45-0016-P1	2032	S2645A	2032	SP83990
2032	MT0404-1	2032	P04-450016-002	2032	S2988	2032	SP83990A
2032	MT0404-2	2032	P04450016-004	2032	S2988A	2032	SP84000
2032	MT0411	2032	P1000A	2032	S3004	2032	SP84007
2032	MT0412	2032	P1901-70	2032	S31551	2032	SP84007A
2032	MT0413	2032	P1B	2032	S3639	2032	SP84013
2032	MT1131	2032	P1C	2032	S3640A	2032	SP84013A
2032	MT1131A	2032	P1CQ	2032	S3655	2032	SP84014
2032	MT1132	2032	P1D	2032	S3655A	2032	SP84014A
2032	MT1132A	2032	P1H	2032	S39094	2032	SP84018
2032	MT1132B	2032	P1J	2032	S4249	2032	SP84018A
2032	MT1254	2032	P1N	2032	S20	2032	SP84019
2032	MT1255	2032	P1N-1	2032	S608	2032	SP84019A
2032	MT1256	2032	P1N-2	2032	SA310	2032	SP84401K
2032	MT1257	2032	P1N-3	2032	SA310A	2032	SPS4025
2032	MT1258	2032	P1P	2032	SA311	2032	SPS4025A
2032	MT1259	2032	P1P-1	2032	SA311A	2032	SP84026
2032	MT1420	2032	P1W	2032	SA312	2032	SP84026A
2032	MT1991	2032	P2A	2032	SA312A	2032	SP84027
2032	MT2303	2032	P2E	2032	SA313	2032	SPS4027A
2032	MT2411	2032	P2G	2032	SA314	2032	SP84028
2032	MT2412	2032	P20E	2032	SA314A	2032	SP84028A
2032	MT726	2032	P2H	2032	SA315	2032	SP84031
2032	MT869	2032	P2L	2032	SA315A	2032	SP84031A
2032	NB121	2032	P2M-1	2032	SA316	2032	SPS4054
2032	NJ101B	2032	P2M-2	2032	SA316A	2032	SP84054A
2032	NKT20329	2032	P2M-3	2032	SA410	2032	SPS4056
2032	NKT20329A	2032	P2P	2032	SA410A	2032	SP84056A
2032	NKT20339	2032	P2B	2032	SA411	2032	SP84064
2032	NN650	2032	P2W	2032	SA411A	2032	SP84064A
2032	NP8404	2032	P2Y	2032	SA412	2032	SP84072
2032	NP8404A	2032	P3C	2032	SA412A	2032	SP84072A
2032	NR-601AT	2032	P3CA	2032	SA413	2032	SP84073
2032	NR601BT	2032	P3Z	2032	SA413A	2032	SP84073A
2032	NR621AT	2032	P480A0022	2032	SA414	2032	SP84076
2032	NR621BU	2032	P480A0023	2032	SA414A	2032	SP84076A
2032	NR631AY	2032	P480A0027	2032	SA415	2032	SP84078
2032	NS1000	2032	P4C	2032	SA415A	2032	SP84078A
2032	NS1000A	2032	P4G	2032	SA416	2032	SP84082
2032	NS1001	2032	P4K	2032	SA416A	2032	SP84082A
2032	NS1001A	2032	P4P	2032	SA50	2032	SP84086
2032	NS1672	2032	P4R	2032	SA50A	2032	SP84086A
2032	NS1672A	2032	P4Y	2032	SA51	2032	SP84087
2032	NS1673	2032	P5B	2032	SA52	2032	SP84087A
2032	NS1674	2032	P5C	2032	SA52A	2032	SP84090
2032	NS1674A	2032	P5D	2032	SA52AC	2032	SP84090A
2032	NS1675	2032	PA1000	2032	SA52B	2032	SP842
2032	NS1675A	2032	PA1001	2032	SA52BC	2032	SP842A
2032	NS1861	2032	PA1001A	2032	SA53	2032	SPS4302
2032	NS1861A	2032	PI-10,131	2032	SA53A	2032	SPS4314
2032	NS1862	2032	PIC	2032	SA54	2032	SP84314A
2032	NS1863	2032	PIT-50	2032	SA55A	2032	SPS4330
2032	NS1863A	2032	PIT-79	2032	SA56	2032	SP84338
2032	NS1864	2032	PIT-81	2032	SA56A	2032	SPS4348
2032	NS1864A	2032	PL-177-006-9-001	2032	SA70	2032	SP84348A
2032	NS404	2032	PL-177-006-9-002	2032	SA70A	2032	SPS4354
2032	NS6001	2032	PL1031	2032	SCD-7326	2032	SP84354A
2032	NS6062	2032	PL1033	2032	SB5-0370	2032	SPS4355
2032	NS6062A	2032	PL1034	2032	SB5-0798	2032	SP84365
2032	NS6063	2032	PL1101	2032	SB5-0831	2032	SPS4365A
2032	NS6063A	2032	PL1102	2032	SB5-0949	2032	SP84397
2032	NS6064	2032	PL1103	2032	SB5-1057	2032	SP84452
2032	NS6065	2032	PL1104	2032	SB5-1223	2032	SP84452A
2032	NS6065A	2032	PN	2032	SP8014	2032	SPS4458
2032	NS6211	2032	PTC103	2032	SHA7530	2032	SPS4458A
2032	NS6211A	2032	PTC131	2032	SHA7531	2032	SP84460
2032	NS6241	2032	Q-00984R	2032	SHA7532	2032	SP84460A
2032	NS661	2032	Q-36	2032	SHA7533	2032	SP84473
2032	NS662	2032	Q-36A	2032	SHA7534	2032	SPS4473A
2032	NS663	2032	Q0-419	2032	SHA7536	2032	SP84480
2032	NS664	2032	Q0415	2032	SHA7537	2032	SPS4480A
2032	NS665	2032	Q5087Z	2032	SHA7538	2032	SP84489
2032	NS666	2032	Q5116C	2032	SK1639	2032	SP84489A
2032	NS667	2032	Q5135	2032	SK1639D	2032	SP847
2032	NS668	2032	Q5205	2032	SK1640	2032	SP84813
2032	NS732	2032	QA-21	2032	SK1856	2032	SP84815
2032	NS732A	2032	QA-21A	2032	SK1856A	2032	SP85007
2032	NTC-10	2032	Q0-419	2032	SK2604	2032	SPS5007-1
2032	NTC-6	2032	QQC61689	2032	SK2604A	2032	SP85007-1A
2032	00415	2032	QQC61689A	2032	SK3114	2032	SP85007-2
2032	0C200	2032	QRT-101	2032	SK3118	2032	SP85007-2A
2032	0C202	2032	Q8316	2032	SK5797	2032	SP85007A
2032	0C204	2032	QT-A0719AXN	2032	SK5798	2032	SP85008
2032	0C205	2032	QT-A0719XHN	2032	SK6345	2032	SP514
2032	0C206	2032	R8967	2032	SK6346	2032	SP8514A
2032	0C207	2032	R8967A	2032	SK6347	2032	SPS545B
2032	0C430	2032	R8969	2032	SK6347A	2032	SP86109
2032	0C430K	2032	R8969A	2032	SK7664	2032	SP86109A
2032	0C440	2032	RE26	2032	SKA-4061	2032	SP668
2032	0C440K	2032	RE63	2032	SKA1279	2032	SP86953
2032	0C443	2032	REN159	2032	SKA4129	2032	SS1606
2032	0C443K	2032	REN290	2032	SKA6250	2032	SS1906
2032	0C445	2032	REN63	2032	SKWHG7006	2032	SS1906A
		2032	RS-110				

RS 276-	Industry Standard No.	RS 276-	Industry Standard No.	RS 276-	Industry Standard No.	RS 276-	Industry Standard No.
2032	SS2503	2032	TI893-VIO	2032	V761	2032	002-9800-A
2032	SS2503A	2032	TI893-YEL	2032	V763	2032	2.01.03.02
2032	SS6724	2032	TI8991	2032	VS28A495-0/1B	2032	2CT38
2032	SSA43	2032	TI8993	2032	VS28A495-0/-1	2032	2CT39
2032	SSA43A	2032	TIX804	2032	VS28A495-Y/1B	2032	2D017-165
2032	SSA43A-1	2032	TIX805	2032	VS28A562-0/1E	2032	2D017-166
2032	SSA46	2032	TIX890	2032	VS28A562-Y/1E	2032	2D017-167
2032	SSA46A	2032	TIX891	2032	VS28A673-B/1K	2032	2D017-169
2032	SSA48	2032	TM159	2032	VS28A673-C/1A	2032	2D027
2032	SSA48A	2032	TM1614	2032	VS28B561-C/1	2032	2N1025
2032	ST-MP89682J	2032	TM1712	2032	VS28B561-C/-1	2032	2N1026
2032	ST-MP89750D	2032	TM2614	2032	VSOS1256HG	2032	2N1026A
2032	ST.082.115.015	2032	TM2712	2032	W21	2032	2N2712
2032	ST129-1	2032	TM290	2032	WEP717	2032	2N1028
2032	ST6110	2032	TN061690	2032	WEP911	2032	2N1118
2032	ST62180	2032	TN061703	2032	X29A829	2032	2N1118A
2032	ST8014	2032	TN061690	2032	X32A1389	2032	2N1119
2032	ST8033	2032	TN061705	2032	X34B1226	2032	2N1131
2032	ST8033A	2032	TN71037	2032	XA-1072	2032	2N1131A
2032	ST8034	2032	TNJ71773	2032	XA-1140	2032	2N1132
2032	ST8034A	2032	TNJ71774	2032	XA-495	2032	2N1132/46
2032	ST8035	2032	TN72154	2032	XA494	2032	2N1132A
2032	ST8035A	2032	TP3638	2032	XA495	2032	2N1132B
2032	ST8036	2032	TP4125	2032	XA495(C)	2032	2N1135
2032	ST8036A	2032	TP4126	2032	XA495AC	2032	2N1135A
2032	ST8065	2032	TP4257	2032	XA495C	2032	2N1196
2032	ST8065A	2032	TP4258	2032	XN-400-319-P2	2032	2N1219
2032	ST8190	2032	TP5142	2032	Y56601-50	2032	2N1220
2032	ST8500	2032	TP86516	2032	Y56601-63	2032	2N1221
2032	ST8500A	2032	TP86517	2032	Y56601-74	2032	2N1222
2032	ST8509	2032	TP86518	2032	Y56601-79	2032	2N1223
2032	ST8509A	2032	TP86519	2032	Y56601-82	2032	2N1428
2032	STMT3Q	2032	TP86522	2032	Y56601-84	2032	2N1439
2032	SX3702	2032	TP86523	2032	ZAG-9673	2032	2N1440
2032	SX3702A	2032	TQ-63	2032	ZEN101	2032	2N1441
2032	SX61M	2032	TQ-64	2032	ZEN106	2032	2N1442
2032	SX61MA	2032	TQ61	2032	ZEN107	2032	2N1443
2032	SYL4275	2032	TQ61A	2032	ZEN122	2032	2N1474
2032	T 112	2032	TQ62	2032	ZT131	2032	2N1474A
2032	T-246	2032	TQ62A	2032	ZT131A	2032	2N1475
2032	T-251	2032	TR-1000-2	2032	ZT152	2032	2N1607
2032	T-Q5077	2032	TR-1030-1	2032	ZT152A	2032	2N1608
2032	T-Q5087	2032	TR-1030-2	2032	ZT153	2032	2N1643
2032	T01-023	2032	TR-1032-2	2032	ZT153A	2032	2N1676
2032	T1-503	2032	TR-167(OLSON)	2032	ZT154	2032	2N1677
2032	T1-503A	2032	TR-19	2032	ZT154A	2032	2N1917
2032	T1-743	2032	TR-19A	2032	ZT180	2032	2N1918
2032	T1-743A	2032	TR-20	2032	ZT180A	2032	2N1919
2032	T1-744	2032	TR-20A	2032	ZT181	2032	2N1920
2032	T1-744A	2032	TR-30	2032	ZT181A	2032	2N1921
2032	T1-752	2032	TR-30A	2032	ZT182	2032	2N1991
2032	T1-752A	2032	TR-31	2032	ZT182A	2032	2N2002
2032	T1-906	2032	TR-4R38	2032	ZT183	2032	2N2003
2032	T112	2032	TR-6R35	2032	ZT183A	2032	2N2004
2032	T1803	2032	TR-6R35A	2032	ZT184	2032	2N2005
2032	T1803A	2032	TR-8007	2032	ZT184A	2032	2N2006
2032	T1804	2032	TR-8007(FISHER)	2032	ZT187	2032	2N2007
2032	T1804A	2032	TR-8026	2032	ZT187A	2032	2N2104
2032	T1837A	2032	TR-8037	2032	ZT189	2032	2N2105
2032	T1838	2032	TR0055	2032	ZT280	2032	2N2121
2032	T1838A	2032	TRO1053-1	2032	ZT280A	2032	2N2162
2032	T1853	2032	TRO2051-1	2032	ZT281	2032	2N2163
2032	T1853A	2032	TRO2051-3	2032	ZT281A	2032	2N2164
2032	T1854	2032	TRO2051-5	2032	ZT282	2032	2N2165
2032	T185A	2032	TRO2051-6	2032	ZT282A	2032	2N2166
2032	T1861A	2032	TRO2062-1	2032	ZT283	2032	2N2167
2032	T1891	2032	TRO2062-6	2032	ZT283A	2032	2N2178
2032	T1891A	2032	TRO2063-8	2032	ZT284	2032	2N2181
2032	T1893	2032	TR1000A	2032	ZT284A	2032	2N2182
2032	T1893A	2032	TR1030	2032	ZT287	2032	2N2183
2032	T276	2032	TR1030-1	2032	ZT287A	2032	2N2184
2032	T340	2032	TR1030-2	2032	ZTX501	2032	2N2185
2032	T3570	2032	TR1030A	2032	ZTX502	2032	2N2186
2032	T3570A	2032	TR1032	2032	ZTX503	2032	2N2187
2032	T407185152	2032	TR1032-1	2032	ZTX504	2032	2N2274
2032	T460(SEARS)	2032	TR1032A	2032	ZTX510	2032	2N2275
2032	T475(SEARS)	2032	TR1034	2032	ZTX530	2032	2N2276
2032	T482(SEARS)	2032	TR1034A	2032	ZTX530A	2032	2N2277
2032	T597-1	2032	TR19	2032	ZTX530B	2032	2N2278
2032	T900	2032	TR2327743	2032	ZTX530C	2032	2N2279
2032	T9681	2032	TR28	2032	ZTX530D	2032	2N2280
2032	TA198036-2	2032	TR30	2032	ZTX531	2032	2N2281
2032	TCH98	2032	TR31	2032	ZTX531A	2032	2N2299
2032	TCH99	2032	TR8007A	2032	ZTX531B	2032	2N2303
2032	TCH99B	2032	TR8020	2032	001	2032	2N2303/46
2032	TD2905	2032	TR8026A	2032	1,000,111-00	2032	2N2332
2032	TD400	2032	TR8055	2032	1-0006-0023	2032	2N2333
2032	TD401	2032	TS-21756640	2032	01-010495	2032	2N2334
2032	TD402	2032	TS2904	2032	01-010562	2032	2N2335
2032	TD500	2032	TS2905	2032	01-010673	2032	2N2336
2032	TD501	2032	TS2906	2032	01-010719	2032	2N2337
2032	TD502	2032	TS2907	2032	001-02	2032	2N2370
2032	TD550	2032	T897-1	2032	001-02201-0	2032	2N2371
2032	TE3905	2032	TT28A495-0	2032	001-022010	2032	2N2372
2032	TE4125	2032	TT28A495-0-A	2032	001-022020	2032	2N2373
2032	TE4126	2032	TT28A495-0-A	2032	001-03	2032	2N2377
2032	TE5365	2032	TT28A495-Y	2032	01-030509	2032	2N2378
2032	TE5366	2032	TT28A495-Y-A	2032	01-030536	2032	2N2393
2032	TE5378	2032	TV-44	2032	1-034/2207	2032	2N2394
2032	TE5379	2032	TV-47	2032	001-04	2032	2N2424
2032	TE5447	2032	TV-47A	2032	1-043/2207	2032	2N2425
2032	TE5448	2032	TV-57	2032	001-533-00	2032	2N2474
2032	TG28A608	2032	TV-87	2032	01-572588	2032	2N2595
2032	TG28A608-D	2032	TV-93	2032	01-572811	2032	2N2596
2032	TG28A608-B	2032	TV24214	2032	1B3096-1	2032	2N2597
2032	TG28A608C	2032	TV24214A	2032	1V68611A47	2032	2N2598
2032	TI-428	2032	TV24363	2032	1V68611A47A	2032	2N2599
2032	TI-429	2032	TV24363A	2032	1W11700	2032	2N2600
2032	TI-503	2032	TV24495	2032	1W11700A	2032	2N2601
2032	TI-743	2032	TV24495A	2032	1W11702	2032	2N2602
2032	TI-744	2032	TV44	2032	1W11702A	2032	2N2603
2032	TI-752	2032	TV72	2032	1W11711	2032	2N2604
2032	TI-905	2032	TV8-28A564	2032	1W8337	2032	2N2605
2032	TI-906	2032	TV8-28A564A	2032	1W9148	2032	2N2605A
2032	TI503	2032	TV8-28A564P	2032	1W9640	2032	2N2695
2032	TI743	2032	TV8-28C564	2032	1W9640A	2032	2N2696
2032	TI744	2032	TV828A564	2032	1W9728	2032	2N2709
2032	TI752	2032	TV828A564-0	2032	1W9728A	2032	2N2800
2032	TI8-03	2032	TV828A564A	2032	1W9782	2032	2N2800/46
2032	TI803	2032	TV828A564C	2032	1W9782A	2032	2N2801
2032	TI804	2032	TV828A564P	2032	1W9810	2032	2N2837
2032	TI8104	2032	TV828A564PY	2032	1W9810A	2032	2N2838
2032	TI8112	2032	TV828A564Q	2032	1W9810B	2032	2N2906
2032	TI837	2032	TV828A607	2032	1W98108A	2032	2N2906A
2032	TI838	2032	TV828A609	2032	002-009800	2032	2N2907
2032	TI850	2032	TV828C564	2032	002-010300	2032	2N2907A
2032	TI853	2032	TV828C564-3	2032	002-010300A	2032	2N2927
2032	TI854	2032	TV828C564-0	2032	002-010500	2032	2N2927/46
2032	TI861	2032	TV828C5640	2032	002-010500A	2032	2N2944
2032	TI891	2032	TV828C564R	2032	002-010900	2032	2N2945
2032	TI893	2032	V52	2032	002-010900A	2032	2N2945A
2032	TI893-BLU	2032	V162	2032	002-012800	2032	2N2946
2032	TI893-GRN	2032	V410	2032	002-012800A	2032	2N2946A
2032	TI893-GRY	2032	V435A				

RS 276-	Industry Standard No.
2032	2N3012
2032	2N3039
2032	2N3040
2032	2N3058
2032	2N3059
2032	2N3060
2032	2N3061
2032	2N3062
2032	2N3072
2032	2N3073
2032	2N3081
2032	2N3121
2032	2N3135
2032	2N3136
2032	2N3217
2032	2N3218
2032	2N3219
2032	2N3248
2032	2N3249
2032	2N3250
2032	2N3250A
2032	2N3251
2032	2N3251A
2032	2N3305
2032	2N3306
2032	2N3307
2032	2N3308
2032	2N3317
2032	2N3318
2032	2N3319
2032	2N3341
2032	2N3346
2032	2N3401
2032	2N3451
2032	2N3464
2032	2N3485
2032	2N3485A
2032	2N3486
2032	2N3486A
2032	2N3496
2032	2N3504
2032	2N3505
2032	2N3527
2032	2N3545
2032	2N3546
2032	2N3547
2032	2N3548
2032	2N3549
2032	2N3550
2032	2N3579
2032	2N3580
2032	2N3581
2032	2N3582
2032	2N3638
2032	2N3638A
2032	2N3639
2032	2N3640
2032	2N3644
2032	2N3672
2032	2N3673
2032	2N3702
2032	2N3703
2032	2N3798
2032	2N3799
2032	2N3829
2032	2N3840
2032	2N3841
2032	2N3842
2032	2N3857
2032	2N3905
2032	2N3906
2032	2N3910
2032	2N3911
2032	2N3912
2032	2N3913
2032	2N3914
2032	2N3930
2032	2N3931
2032	2N3962
2032	2N3963
2032	2N3964
2032	2N3965
2032	2N3977
2032	2N3978
2032	2N3979
2032	2N4006
2032	2N4007
2032	2N4008
2032	2N4121
2032	2N4125
2032	2N4126
2032	2N4149
2032	2N4207
2032	2N4208
2032	2N4209
2032	2N4221
2032	2N4248
2032	2N4249
2032	2N4250
2032	2N4250A
2032	2N4257
2032	2N4257A
2032	2N4258
2032	2N4258A
2032	2N4284
2032	2N4285
2032	2N4288
2032	2N4289
2032	2N4290
2032	2N4291
2032	2N4354
2032	2N4355
2032	2N4356
2032	2N4359
2032	2N4402
2032	2N4403
2032	2N4411
2032	2N4413
2032	2N4413A
2032	2N4415
2032	2N4415A
2032	2N4452
2032	2N4916
2032	2N4917
2032	2N4965
2032	2N4971
2032	2N4972
2032	2N4980
2032	2N4981
2032	2N4982
2032	2N5086
2032	2N5087
2032	2N5138
2032	2N5139
2032	2N5142
2032	2N5143
2032	2N5221
2032	2N5226
2032	2N5227
2032	2N5229
2032	2N5230
2032	2N5231
2032	2N5354
2032	2N5355
2032	2N5356
2032	2N5365
2032	2N5366
2032	2N5367
2032	2N5372
2032	2N5373
2032	2N5374
2032	2N5375
2032	2N5378
2032	2N5379
2032	2N5382
2032	2N5383
2032	2N5447
2032	2N5448
2032	2N5910
2032	2N5999
2032	2N6003
2032	2N6067
2032	2N6076
2032	2N721
2032	2N721A
2032	2N722
2032	2N722A
2032	2N858
2032	2N859
2032	2N860
2032	2N861
2032	2N862
2032	2N863
2032	2N864
2032	2N923
2032	2N924
2032	2N925
2032	2N926
2032	2N927
2032	2N928
2032	2N935
2032	2N936
2032	2N937
2032	2N938
2032	2N939
2032	2N940
2032	2N941
2032	2N942
2032	2N943
2032	2N944
2032	2N945
2032	2N946
2032	02P1B
2032	2S3020
2032	2S3021
2032	2S3030
2032	2S305
2032	2S306
2032	2S307
2032	2S321
2032	2S3210
2032	2S322
2032	2S3220
2032	2S3221
2032	2S3222A
2032	2S323
2032	2S3230
2032	2S324
2032	2S3240
2032	2S326
2032	2S327
2032	2S673C
2032	2SA-NJ-101
2032	2SA-NJ101
2032	2SA1029
2032	2SA1029C
2032	2SA119
2032	2SA402
2032	2SA467
2032	2SA467-O
2032	2SA467-Y
2032	2SA467G
2032	2SA467G-O
2032	2SA467G-R
2032	2SA467G-Y
2032	2SA480
2032	2SA480-OR
2032	2SA480A
2032	2SA480B
2032	2SA480C
2032	2SA480D
2032	2SA480E
2032	2SA480F
2032	2SA480G
2032	2SA480GN
2032	2SA480H
2032	2SA480J
2032	2SA480K
2032	2SA480L
2032	2SA480M
2032	2SA480OR
2032	2SA480R
2032	2SA480X
2032	2SA480Y
2032	2SA482
2032	2SA482A
2032	2SA482B
2032	2SA482C
2032	2SA482D
2032	2SA482E
2032	2SA482F
2032	2SA482GN
2032	2SA482H
2032	2SA482J
2032	2SA482K
2032	2SA482L
2032	2SA482M
2032	2SA482OR
2032	2SA482R
2032	2SA482X
2032	2SA482Y
2032	2SA493
2032	2SA493-O
2032	2SA493R
2032	2SA493Y
2032	2SA494
2032	2SA494-GR
2032	2SA494-O
2032	2SA494-Y
2032	2SA495
2032	2SA495(O)
2032	2SA495(R)
2032	2SA495(Y)
2032	2SA495-O
2032	2SA495-1
2032	2SA495-G
2032	2SA495-GN
2032	2SA495-O
2032	2SA495-OF
2032	2SA495-OR
2032	2SA495-ORG
2032	2SA495-ORG-Q
2032	2SA495-R
2032	2SA495-RD
2032	2SA495-RED
2032	2SA495-RED-Q
2032	2SA495-Y
2032	2SA495-YEL
2032	2SA495-YEL-G
2032	2SA495-YL
2032	2SA495O
2032	2SA495A
2032	2SA495B
2032	2SA495C
2032	2SA495D
2032	2SA495E
2032	2SA495F
2032	2SA495G
2032	2SA495G-GR
2032	2SA495G-O
2032	2SA495G-R
2032	2SA495G-Y
2032	2SA495GN
2032	2SA495GR
2032	2SA495H
2032	2SA495J
2032	2SA495K
2032	2SA495L
2032	2SA495M
2032	2SA495O
2032	2SA495OF
2032	2SA495OR
2032	2SA495R
2032	2SA495RD
2032	2SA495RED
2032	2SA495RED-Q
2032	2SA495W
2032	2SA495W1
2032	2SA495WI
2032	2SA495X
2032	2SA495Y
2032	2SA495YEL
2032	2SA495YEL-G
2032	2SA495YL
2032	2SA499
2032	2SA499-O
2032	2SA499-ORG
2032	2SA499-R
2032	2SA499-RED
2032	2SA499-Y
2032	2SA499-YEL
2032	2SA499A
2032	2SA499B
2032	2SA499C
2032	2SA499D
2032	2SA499B
2032	2SA499F
2032	2SA499G
2032	2SA499GN
2032	2SA499H
2032	2SA499J
2032	2SA499K
2032	2SA499L
2032	2SA499M
2032	2SA499O
2032	2SA499R
2032	2SA499X
2032	2SA499Y
2032	2SA500
2032	2SA500-O
2032	2SA500-ORG
2032	2SA500-R
2032	2SA500-RED
2032	2SA500-Y
2032	2SA500-YEL
2032	2SA500A
2032	2SA500B
2032	2SA500C
2032	2SA500D
2032	2SA500E
2032	2SA500F
2032	2SA500G
2032	2SA500GN
2032	2SA500H
2032	2SA500J
2032	2SA500K
2032	2SA500L
2032	2SA500M
2032	2SA500OR
2032	2SA500R
2032	2SA500X
2032	2SA500Y
2032	2SA502
2032	2SA502-O
2032	2SA502-OR
2032	2SA502-R
2032	2SA502-Y
2032	2SA502A
2032	2SA502B
2032	2SA502C
2032	2SA502D
2032	2SA502E
2032	2SA502F
2032	2SA502G
2032	2SA502GN
2032	2SA502H
2032	2SA502J
2032	2SA502K
2032	2SA502L
2032	2SA502M
2032	2SA5020R
2032	2SA502R
2032	2SA502X
2032	2SA502Y
2032	2SA509
2032	2SA509(A)
2032	2SA509(O)
2032	2SA509-O
2032	2SA509-O
2032	2SA509-OR
2032	2SA509-R
2032	2SA509-RD
2032	2SA509-Y
2032	2SA509-YE
2032	2SA509A
2032	2SA509B
2032	2SA509BL
2032	2SA509C
2032	2SA509D
2032	2SA509E
2032	2SA509F
2032	2SA509G
2032	2SA509G-O
2032	2SA509G-Y
2032	2SA509GN
2032	2SA509GR
2032	2SA509GR-1
2032	2SA509H
2032	2SA509J
2032	2SA509K
2032	2SA509L
2032	2SA509M
2032	2SA509OR
2032	2SA509Q
2032	2SA509R
2032	2SA509RD
2032	2SA509Y
2032	2SA509X
2032	2SA509Y
2032	2SA509YE
2032	2SA510
2032	2SA510-O
2032	2SA510-R
2032	2SA511
2032	2SA511-O
2032	2SA511-R
2032	2SA512
2032	2SA512-O
2032	2SA512-R
2032	2SA513
2032	2SA513-O
2032	2SA513-R
2032	2SA522A
2032	2SA522AL
2032	2SA522B
2032	2SA522C
2032	2SA522D
2032	2SA522E
2032	2SA522F
2032	2SA522G
2032	2SA522GN
2032	2SA522H
2032	2SA522J
2032	2SA522K
2032	2SA522L
2032	2SA522M
2032	2SA522OR
2032	2SA522R
2032	2SA522X
2032	2SA522Y
2032	2SA530
2032	2SA530H
2032	2SA539
2032	2SA539(K)
2032	2SA539(L)
2032	2SA539(M)
2032	2SA539A
2032	2SA539B
2032	2SA539C
2032	2SA539D
2032	2SA539E
2032	2SA539F
2032	2SA539G
2032	2SA539GN
2032	2SA539H
2032	2SA539J
2032	2SA539K
2032	2SA539L
2032	2SA539M
2032	2SA539OR
2032	2SA539R
2032	2SA539B
2032	2SA539X
2032	2SA539Y
2032	2SA542
2032	2SA542A
2032	2SA542C
2032	2SA542D
2032	2SA542E
2032	2SA542F
2032	2SA542G
2032	2SA542GN
2032	2SA542H
2032	2SA542J
2032	2SA542L
2032	2SA542M
2032	2SA542OR
2032	2SA542R
2032	2SA542X
2032	2SA542Y
2032	2SA544
2032	2SA550A
2032	2SA550A(Q)
2032	2SA550A(R)
2032	2SA550A(R,Q,S)
2032	2SA550A(S)
2032	2SA550AQ
2032	2SA550P
2032	2SA550Q
2032	2SA550R
2032	2SA550B
2032	2SA560A
2032	2SA561
2032	2SA561(O)
2032	2SA561-O
2032	2SA561-GR
2032	2SA561-GRN
2032	2SA561-O
2032	2SA561-OR
2032	2SA561-ORG
2032	2SA561-R
2032	2SA561-RD
2032	2SA561-RED
2032	2SA561-Y
2032	2SA561-YEL
2032	2SA561-YL
2032	2SA561GR
2032	2SA561GRN
2032	2SA561R
2032	2SA561RD
2032	2SA561RED
2032	2SA561Y
2032	2SA561YEL

RS 276-	Industry Standard No.
2032	28A561YL
2032	28A562
2032	28A562(O)
2032	28A562(Y)
2032	28A562-0
2032	28A562-GR
2032	28A562-GRN
2032	28A562-0
2032	28A562-OR
2032	28A562-ORG
2032	28A562-R
2032	28A562-RD
2032	28A562-RED
2032	28A562-Y
2032	28A562-YE
2032	28A562-YEL
2032	28A562D
2032	28A562B
2032	28A562G
2032	28A562GR
2032	28A562GRN
2032	28A562R
2032	28A562RD
2032	28A562RED
2032	28A562V
2032	28A562VO
2032	28A562Y
2032	28A562YE
2032	28A562YEL
2032	28A565
2032	28A565A
2032	28A565B
2032	28A565C
2032	28A565D
2032	28A568
2032	28A568A
2032	28A568B
2032	28A568C
2032	28A568D
2032	28A568E
2032	28A568F
2032	28A568G
2032	28A568GN
2032	28A568H
2032	28A568J
2032	28A568K
2032	28A568L
2032	28A568M
2032	28A568OR
2032	28A568R
2032	28A568X
2032	28A568Y
2032	28A569
2032	28A569J
2032	28A570
2032	28A570A
2032	28A570B
2032	28A570C
2032	28A570D
2032	28A570E
2032	28A570F
2032	28A570G
2032	28A570GN
2032	28A570H
2032	28A570J
2032	28A570K
2032	28A570L
2032	28A570M
2032	28A570OR
2032	28A570R
2032	28A570X
2032	28A570Y
2032	28A592Y
2032	28A603
2032	28A603A
2032	28A603B
2032	28A603C
2032	28A603D
2032	28A603E
2032	28A603F
2032	28A603G
2032	28A603GN
2032	28A603H
2032	28A603J
2032	28A603K
2032	28A603L
2032	28A603M
2032	28A603OR
2032	28A603R
2032	28A603X
2032	28A603Y
2032	28A608
2032	28A608(C)
2032	28A608(D)
2032	28A608(F)
2032	28A608-0
2032	28A608-C
2032	28A608-D
2032	28A608-E
2032	28A608-F
2032	28A608-0
2032	28A608-OR
2032	28A608A
2032	28A608B
2032	28A608BL
2032	28A608C
2032	28A608D
2032	28A608D(F)
2032	28A608E
2032	28A608F
2032	28A608G
2032	28A608GN
2032	28A608H
2032	28A608J
2032	28A608K
2032	28A608L
2032	28A608M
2032	28A608OR
2032	28A608P
2032	28A608R
2032	28A608X
2032	28A608Y
2032	000028A609
2032	28A609A
2032	28A609B
2032	28A609C
2032	28A609D
2032	28A609E
2032	28A609F
2032	28A609G
2032	28A609H
2032	28A610
2032	28A610B
2032	28A611
2032	28A611-4B
2032	28A617K
2032	28A618K
2032	28A628
2032	28A628(EP)
2032	28A628(P)
2032	28A628-0
2032	28A628-OR
2032	28A628A
2032	28A628AA
2032	28A628AD
2032	28A628AE
2032	28A628B
2032	28A628C
2032	28A628D
2032	28A628E
2032	28A628E,F
2032	28A628EF
2032	28A628F
2032	28A628G
2032	28A628GN
2032	28A628H
2032	28A628J
2032	28A628K
2032	28A628L
2032	28A628M
2032	28A628OR
2032	28A628R
2032	28A628X
2032	28A628Y
2032	28A629
2032	28A629A
2032	28A629B
2032	28A629C
2032	28A629D
2032	28A629E
2032	28A629F
2032	28A629G
2032	28A629GN
2032	28A629H
2032	28A629J
2032	28A629K
2032	28A629L
2032	28A629M
2032	28A629OR
2032	28A629R
2032	28A629X
2032	28A629Y
2032	28A638E,F
2032	28A641
2032	28A641A
2032	28A641B
2032	28A641BL
2032	28A641C
2032	28A641D
2032	28A641G
2032	28A641GR
2032	28A641K
2032	28A641L
2032	28A641M
2032	28A641O
2032	28A641OR
2032	28A641R
2032	28A641Y
2032	28A642
2032	28A642A
2032	28A642B
2032	28A642C
2032	28A642D
2032	28A642F
2032	28A642G
2032	28A642GN
2032	28A642H
2032	28A642J
2032	28A642K
2032	28A642L
2032	28A642M
2032	28A642OR
2032	28A642R
2032	28A642S
2032	28A642V
2032	28A642W
2032	28A642X
2032	28A642Y
2032	28A659
2032	28A659(D)
2032	28A659(E)
2032	28A659A
2032	28A659B
2032	28A659C
2032	28A659D
2032	28A659E
2032	28A659F
2032	28A659G
2032	28A659L
2032	28A659P
2032	28A659R
2032	28A659Y
2032	28A661QRS
2032	28A672
2032	28A672A
2032	28A672B
2032	28A672C
2032	28A673(B)
2032	28A673(D)
2032	28A673-OR
2032	28A673A
2032	28A673A(C)
2032	28A673AA
2032	28A673AB
2032	28A673AC
2032	28A673AD
2032	28A673AE
2032	28A673AK
2032	28A673AKA
2032	28A673AKB
2032	28A673AKC
2032	28A673AS
2032	28A673AS(C)
2032	28A673ASC
2032	28A673B
2032	28A673C
2032	28A673C2
2032	28A673D
2032	28A673E
2032	28A673F
2032	28A673G
2032	28A673GN
2032	28A673H
2032	28A673J
2032	28A673K
2032	28A673L
2032	28A673M
2032	28A673OR
2032	28A673R
2032	28A673WT
2032	28A673X
2032	28A673Y
2032	28A675
2032	28A675A
2032	28A675B
2032	28A675C
2032	28A677-0
2032	28A677-0
2032	28A677-OR
2032	28A677A
2032	28A677B
2032	28A677C
2032	28A677D
2032	28A677E
2032	28A677F
2032	28A677G
2032	28A677GN
2032	28A677H
2032	28A677HL
2032	28A677J
2032	28A677K
2032	28A677L
2032	28A677M
2032	28A677OR
2032	28A677R
2032	28A677X
2032	28A677Y
2032	28A678
2032	28A678(C)
2032	28A678(SONY)
2032	28A678-0
2032	28A678-OR
2032	28A678A
2032	28A678B
2032	28A678C
2032	28A678D
2032	28A678E
2032	28A678F
2032	28A678G
2032	28A678H
2032	28A678J
2032	28A678K
2032	28A678L
2032	28A678M
2032	28A678OR
2032	28A678R
2032	28A678X
2032	28A678Y
2032	28A685
2032	28A690D
2032	28A693
2032	28A693C
2032	28A701
2032	28A701F
2032	28A701FJ
2032	28A701PO
2032	28A704A
2032	28A704B
2032	28A704C
2032	28A704D
2032	28A704E
2032	28A704F
2032	28A704G
2032	28A704GN
2032	28A704H
2032	28A704J
2032	28A704K
2032	28A704L
2032	28A704M
2032	28A704OR
2032	28A704R
2032	28A704X
2032	28A704Y
2032	28A705
2032	28A705A
2032	28A705B
2032	28A705C
2032	28A705D
2032	28A705E
2032	28A705F
2032	28A705G
2032	28A705GN
2032	28A705H
2032	28A705J
2032	28A705K
2032	28A705L
2032	28A705M
2032	28A705OR
2032	28A705R
2032	28A705X
2032	28A705Y
2032	28A718
2032	28A719
2032	28A719(Q)
2032	28A719,R
2032	28A719K
2032	28A719P
2032	28A719PQR
2032	28A719Q
2032	28A719QR
2032	28A719R
2032	28A719RS
2032	28A719S
2032	28A7200
2032	28A723A
2032	28A723B
2032	28A723C
2032	28A723D
2032	28A723E
2032	28A723F
2032	28A723K
2032	28A730
2032	28A731
2032	28A735
2032	28A786
2032	28A701FJ
2032	28A811
2032	28A812
2032	28A826
2032	28A826P
2032	28A826Q
2032	28A826R
2032	28A826RY
2032	28A828A
2032	28A838B
2032	28A838C
2032	28A841
2032	28A841-GR
2032	28A918
2032	28A945-0
2032	28A945Y
2032	28ANJ101
2032	28B561
2032	28B561B
2032	28B561C
2032	28B641
2032	28B641S
2032	28B643
2032	28B643Q
2032	28B643R
2032	28B643S
2032	28B741
2032	28C610
2032	28C677
2032	28C906(P)
2032	28M610B
2032	03A06
2032	3L4-6007-34
2032	3L4-6010-8
2032	3L4-6017-01
2032	3N114
2032	3N115
2032	3N116
2032	3N117
2032	3N118
2032	3N119
2032	3N130
2032	3N131
2032	3N132
2032	3N133
2032	3N135
2032	3N136
2032	3N90
2032	3N91
2032	3N92
2032	3N93
2032	3N94
2032	3N95
2032	04-440032-002
2032	04-440032-008
2032	04-67000-01
2032	4JX29A529
2032	4JX29A826
2032	4JX29A829
2032	006-0000135
2032	006-02
2032	6-31
2032	6-31A
2032	6-38
2032	6-38A
2032	7-0014
2032	07-07113
2032	007-74004-01
2032	007-74008-01
2032	8-905-706-247
2032	8-905-706-251
2032	8-905-706-253
2032	8-905-706-254
2032	8-905-706-255
2032	8-905-706-256
2032	8-905-706-280
2032	8-905-706-286
2032	8-905-706-287
2032	8-905-706-288
2032	8-905-706-289
2032	8-905-706-290
2032	8-905-713-058
2032	80200
2032	80202
2032	80203
2032	80204
2032	80205
2032	80206
2032	80207
2032	80430
2032	80430K
2032	80440K
2032	80440
2032	80443
2032	80443K
2032	80445
2032	80445K
2032	80447
2032	80449K
2032	80450
2032	80460
2032	80460K
2032	80463
2032	80463K
2032	80465
2032	80465K
2032	80466
2032	80466K
2032	80467
2032	80467K
2032	80468
2032	80468K
2032	80469
2032	80469K
2032	80470
2032	80470K
2032	8C700
2032	8C700A
2032	8C700B
2032	8C702
2032	8C702A
2032	8C702B
2032	8C704
2032	8C7400
2032	8C7400
2032	8C7400G
2032	8C7400M
2032	8C742
2032	8C7420
2032	8C7420
2032	8C742M
2032	8Q-3-11
2032	8Q-3-14
2032	09-300037
2032	09-300059
2032	09-300061
2032	09-300062
2032	09-300063
2032	09-300064
2032	09-300070
2032	09-300074
2032	09-300077
2032	09-300080
2032	09-300081
2032	09-303301
2032	09-30063
2032	09-304012
2032	09-304047

RS 276-	Industry Standard No.	RS 276-	Industry Standard No.	RS 276-	Industry Standard No.	RS 276-	Industry Standard No.
2032	09-304049	2032	19-10	2032	46-86238-3A	2032	488134913
2032	09-304050	2032	19-2	2032	46-86263-3	2032	488134915
2032	09-304051	2032	19-20	2032	46-86293-3	2032	488134943
2032	09-305024	2032	19-3	2032	46-86377-3	2032	488137032
2032	09-305073	2032	19-30	2032	46-86399-3	2032	488137127
2032	09-305149	2032	19A115178-P1	2032	46-86412-3	2032	488137173
2032	09-309038	2032	19A115178-P2	2032	46-86424-3	2032	488137314
2032	09-309042	2032	19A115458-P1	2032	46-86514-3	2032	488137321
2032	10P1	2032	19A115458-P2	2032	46-86546-3	2032	488155035
2032	10P1A	2032	19A115562P2	2032	46-86578-3	2032	50-40205-09
2032	11-691504	2032	19A115653-P1	2032	48-03-002	2032	051-0107
2032	12-0AM	2032	19A115653-P2	2032	48-134525	2032	51-47-21
2032	13-0006	2032	19A115654-P1	2032	48-134525A	2032	52-010-109-0
2032	13-0006A	2032	19A115654-P2	2032	48-134702	2032	53-1516
2032	13-0043	2032	19A115688-P1	2032	48-134702A	2032	55-1083
2032	13-0043A	2032	19A115688-P2	2032	48-134745	2032	55-1083A
2032	13-0044	2032	19A115706-1	2032	48-134745A	2032	55-1085
2032	13-0044A	2032	19A115706-2	2032	48-134815	2032	55-1085A
2032	13-006	2032	19A115706-P1	2032	48-134815A	2032	55-152579
2032	13-0061	2032	19A115706-P2	2032	48-134829	2032	056
2032	13-0061A	2032	19A115768-1	2032	48-134829A	2032	56-8098
2032	13-13532-1	2032	19A115768-2	2032	48-134833	2032	56-8098A
2032	13-14085-13	2032	19A115768-3	2032	48-134833A	2032	56-8098B
2032	13-14085-87	2032	19A115768-P1	2032	48-134865	2032	56-8098C
2032	13-16570-1	2032	19A115768-P2	2032	48-134865A	2032	57A1-76
2032	13-16570-1A	2032	19A115779P1	2032	48-134866	2032	57A1-76A
2032	13-16570-2	2032	19A115852P1	2032	48-134866A	2032	57A106-12
2032	13-16570-2A	2032	19A116223P1	2032	48-134867	2032	57A106-6A
2032	13-19776-1	2032	19A116408-1	2032	48-134867A	2032	57A122-9A
2032	13-22582-1	2032	020-1110-004C	2032	48-134868	2032	57A130-9A
2032	13-22582-1A	2032	20-JLM	2032	48-134868A	2032	57A133-12
2032	13-23325-5	2032	21A015-008	2032	48-134869	2032	57A137-12
2032	13-23826-1	2032	21A015-008A	2032	48-134869A	2032	57A137-12A
2032	13-23826-1A	2032	21A015-009	2032	48-134870	2032	57A145-12
2032	13-23826-2	2032	21A015-009A	2032	48-134870A	2032	57A145-12A
2032	13-23826-2A	2032	21A015-011	2032	48-134871	2032	57A147-12
2032	13-23826-3	2032	21A015-011A	2032	48-134871A	2032	57A147-12A
2032	13-23826-3A	2032	21A015-012	2032	48-134909	2032	57A148-12
2032	13-26386-1	2032	21A015-012A	2032	48-134909A	2032	57A15-5
2032	13-26386-1A	2032	21A015-025	2032	48-134910	2032	57A15-50
2032	13-26386-2	2032	21A040-059	2032	48-134910A	2032	57A157
2032	13-26386-2A	2032	21A112-001	2032	48-134910P	2032	57A157-9
2032	13-26386-3	2032	21A112-003	2032	48-134911	2032	57A157-90
2032	13-28391-1	2032	21A112-047	2032	48-134913	2032	57A157-9A
2032	13-28391-1A	2032	21A112-065	2032	48-134913A	2032	57A159-12
2032	13-28391-2	2032	21A112-075	2032	48-134914	2032	57A159-12A
2032	13-28391-2A	2032	21A112-093	2032	48-134914A	2032	57A174-8
2032	13-29776-1	2032	21A112-100	2032	48-134915	2032	57A175-12
2032	13-29776-2	2032	21A112-102	2032	48-134915A	2032	57A178-12
2032	13-29776-3	2032	21M020	2032	48-134940	2032	57A185-12
2032	13-31013-1(SYLV.)	2032	21M022	2032	48-134940A	2032	57A189-8
2032	13-31013-1/2	2032	21M355	2032	48-134943	2032	57A19
2032	13-31013-2	2032	21M5B1	2032	48-134943A	2032	57A19-1
2032	13-32364-1	2032	22-001010	2032	48-134967	2032	57A19-10
2032	13-34367-1	2032	022-1110-005C	2032	48-134967A	2032	57A19-1A
2032	13-34367-3	2032	23-1	2032	48-134973	2032	57A19-2
2032	13-36386-1	2032	23-10	2032	48-134973A	2032	57A19-20
2032	13-39970-1	2032	23-2	2032	48-134975	2032	57A19-30
2032	13-43634-1	2032	23-20	2032	48-134975A	2032	57A19-3
2032	13-55069-1	2032	23-3	2032	48-134989	2032	57A197-12
2032	13-55069-1A	2032	23-30	2032	48-134989A	2032	57A2-70
2032	14-602-11	2032	23-5045	2032	48-137020	2032	57A2-70A
2032	14-602-11A	2032	23-LLB	2032	48-137020A	2032	57A2-71
2032	14-602-20	2032	23-PT274-122	2032	48-137021	2032	57A2-71A
2032	14-602-20A	2032	24-AWH	2032	48-137032	2032	57A201-14
2032	14-602-32	2032	24MW1031	2032	48-137032A	2032	57A215-12
2032	14-602-32A	2032	24MW1049	2032	48-137033	2032	57A216-12
2032	14-602-42	2032	24MW1061	2032	48-137045A	2032	57A220-14
2032	14-602-42A	2032	24MW661	2032	48-137046	2032	57A235-12
2032	14-602-44	2032	24MW976	2032	48-137061	2032	57A240-14
2032	14-602-44A	2032	24T-011-011	2032	48-137067	2032	57A258-8
2032	14-602-47	2032	25-000453	2032	48-137067A	2032	57A281-14
2032	14-602-47A	2032	25-000462	2032	48-137068	2032	57B106-12
2032	14-602-54A	2032	25-MEF	2032	48-137068A	2032	57B106-6
2032	14-602-56	2032	25A561Y	2032	48-137069	2032	57B108-12
2032	14-602-56A	2032	27A10533	2032	48-137069A	2032	57B108-6A
2032	14-602-58	2032	29-HCL	2032	48-137090	2032	57B122-9
2032	14-602-580	2032	32-207739	2032	48-137090A	2032	57B12-9A
2032	14-602-58A	2032	33-016	2032	48-137127	2032	57B130-9
2032	14-602-600	2032	33-086	2032	48-137127A	2032	57B130-9A
2032	14-602-68	2032	34-1013	2032	48-137-12	2032	57B133-12
2032	14-602-88	2032	34-1022	2032	48-137173	2032	57B137-12
2032	14-602-90	2032	34-143-12	2032	48-137173A	2032	57B137-12A
2032	014-611	2032	34-3015-28	2032	48-137176	2032	57B145-12
2032	014-652	2032	34-6001-15	2032	48-137176A	2032	57B145-12A
2032	014-652C	2032	34-6015-26	2032	48-137195	2032	57B147-12
2032	014-772	2032	34-6015-42	2032	48-137318	2032	57B147-12A
2032	14-803 12	2032	34-6016-15	2032	48-137324	2032	57B157-9
2032	14-803-12	2032	34-6016-15A	2032	48-137366	2032	57B157-9A
2032	14-804-12	2032	34-6016-32	2032	48-137379	2032	57B159-12
2032	14-804-12	2032	34-6016-47	2032	48-137380	2032	57B159-12A
2032	14-855-32	2032	34-6016-60	2032	48-137381	2032	57B175-12
2032	14-856-23	2032	34P1AA	2032	48-137382	2032	57B178-12
2032	14-857-12	2032	35(RCA)	2032	48-137383	2032	57B185-12
2032	14-857-79	2032	35-ALD	2032	48-137391	2032	57B189-8
2032	14-863-23	2032	0036-001	2032	48-137502	2032	57B197-12
2032	14-864-23	2032	42-22008	2032	48-137504	2032	57B2-70
2032	14-867-32	2032	42-22008A	2032	48-155035	2032	57B2-70A
2032	15-01742	2032	42-22810	2032	48-155055	2032	57B2-71
2032	15-01913-00	2032	42-22810A	2032	48-155156	2032	57B2-71A
2032	15-02762-00	2032	42-23541	2032	48-40118B01	2032	57B201-14
2032	15-02762-1	2032	42-23541A	2032	48-40118B01A	2032	57B216-12
2032	15-02762-2	2032	42-27536	2032	48-42098B01	2032	57B220-14
2032	15-02979	2032	42-28208	2032	48-42098B01A	2032	57B235-12
2032	15-03099	2032	42-28211	2032	48-43258B01	2032	57B240-14
2032	15-03409-0	2032	42-9029-40X	2032	48-64978A40	2032	57B258-8
2032	15-03409-02	2032	42-9029-60Q	2032	48-64978A40A	2032	57B281-14
2032	15-03409-1	2032	42-9029-70D	2032	48-64978A41	2032	57C1-5-5
2032	15-088002	2032	42-9029-70E	2032	48-64978A41A	2032	57C15-50
2032	15-088002A	2032	43A145291-1	2032	48-869334	2032	57C157-9
2032	15-09090-01	2032	43A145291-2	2032	48-869413	2032	57C157-90
2032	15-3	2032	43A16720TP1	2032	48-869526	2032	57C19-1
2032	15-30	2032	43A16720TP2	2032	48-869571	2032	57C19-1A
2032	15-4	2032	43A168064-1	2032	48-869649	2032	57D1-76
2032	15-40	2032	43A168064P1	2032	48-90165A01	2032	57D1-76A
2032	15-5	2032	43A176002	2032	48-90165A01A	2032	57D19
2032	15-50	2032	43B168450-1	2032	48-90234A14	2032	57D19-1
2032	15-875-075-001	2032	43B168495-1	2032	48-90343A68	2032	57D19-10
2032	1523	2032	43B168566-P1	2032	48-97046A26	2032	57D19-2
2032	17-459	2032	44A333464	2032	48-97046A27	2032	57D19-20
2032	17-459A	2032	44A333464-1	2032	48-97177A14	2032	57D19-3
2032	018-00001	2032	44A390248-001	2032	48-97177A14A	2032	57D19-30
2032	018-00002	2032	44A390256-001	2032	48B869334	2032	62-19452
2032	019-003675-231	2032	44A390261	2032	48B869413	2032	62-20154
2032	019-003675-232	2032	44A397905	2032	48B869526	2032	62-20154A
2032	019-003675-234	2032	44A417031-001	2032	48B869571	2032	62-20244
2032	019-003675-257	2032	44A417041-001	2032	48B869649	2032	62-20244A
2032	019-003931	2032	44A238203-1	2032	48B134814	2032	62A11871
2032	019-004558	2032	44A238246	2032	48B134815	2032	63-12154
2032	019-005010	2032	46-86170-3	2032	48B134830	2032	63-12154A
2032	019-005179	2032	46-86229-3	2032	48B134831	2032	63-12156
2032	19-020-114	2032	46-86229-3A	2032	48B134832	2032	63-12156A
2032	19-1	2032	46-86230-3			2032	63-12157
2032		2032	46-86238-3			2032	63-12157A
						2032	63-13322

RS 276-	Industry Standard No.	RS 276-	Industry Standard No.	RS 276-	Industry Standard No.	RS 276-	Industry Standard No.
2032	63-13322A	2032	101P10	2032	223P1	2032	921-70A
2032	65A	2032	102P1	2032	260-10-016	2032	921-70B
2032	65C1	2032	102P10	2032	260-10-027	2032	943-721-001
2032	65D	2032	103P(AIRLINE)	2032	260-10-039	2032	958-023
2032	65D1	2032	103P935	2032	260P08201	2032	991-01-0098
2032	65E	2032	103P935A	2032	260P11403	2032	991-01-0462
2032	65E1	2032	103PA	2032	260P15201	2032	991-01-1225
2032	65P	2032	103PNAIRLINEE	2032	260P15202	2032	991-01-1319
2032	65P1	2032	104-17(RCA)	2032	260P15203	2032	991-01-2328
2032	66-&11I20	2032	104-170	2032	260P16502	2032	991-01-3058
2032	66-P11120	2032	104-17NRCAE	2032	260P16503	2032	991-01-3599
2032	66F023-1	2032	105	2032	260P16504	2032	991-010098
2032	66F024-1	2032	106-12	2032	260P16603	2032	991-011225
2032	66F041-1	2032	106-120	2032	260P36001	2032	991-012686
2032	68 A 8318-P1	2032	106RED	2032	294	2032	991-015614
2032	68-110-02	2032	108-6	2032	297L012C01	2032	1005M19
2032	68A7370-1	2032	108-60	2032	297L013B02	2032	1010-7738
2032	68A7370-P3	2032	108GRN	2032	297V073C003	2032	1012(G.B.)
2032	68A7382-P1	2032	110P1	2032	297V073C04	2032	1012(GE)
2032	68A7754P1	2032	110P1AA	2032	297V083C01	2032	1013(E.F.JOHNSON)
2032	68A8318-P1	2032	110P1M	2032	297V086C01	2032	1013(GE)
2032	73B-140-005-1	2032	112-000172	2032	309(CATALINA)	2032	1016(GE)
2032	73B-140005-4	2032	112-000185	2032	311D589-P2	2032	1030
2032	73C180831-1	2032	112-000187	2032	317-0083-001	2032	1039-0060
2032	73C180831-2	2032	112-10	2032	344-6017-1	2032	1042-06
2032	73C182082-31	2032	112-7	2032	352-0219-000	2032	1043-1278
2032	74P1	2032	112-8	2032	352-0551-010	2032	1044-0295
2032	76-1(SYLVANIA)	2032	119-0055	2032	352-0551-021	2032	1061-0807
2032	76-1(SYLVANIA)	2032	120-006604	2032	352-0610-030	2032	1062-6018
2032	81-46125071-4	2032	120P1	2032	352-0610-040	2032	1063-5423
2032	082.115.015	2032	120P1M	2032	352-0636-010	2032	1063-5431
2032	83P1	2032	121-1016	2032	352-0636-020	2032	1063-5449
2032	83P1A	2032	121-170	2032	352-0754-020	2032	1063-6926
2032	83P1B	2032	121-1RED	2032	352-0778-010	2032	1079-85
2032	83P1BC	2032	121-417	2032	352-0848-020	2032	1081-4000
2032	83P1M	2032	121-444	2032	352-0959-010	2032	1081-4010
2032	83P1MC	2032	121-446	2032	352-0959-020	2032	1084-9784
2032	83P2	2032	121-495	2032	352-0959-030	2032	1089 6199
2032	83P2A	2032	121-496	2032	353-9304-001	2032	1089-6199
2032	83P2AA	2032	121-497	2032	364-1(SYLVANIA)	2032	1112-8
2032	83P2AA1	2032	121-497WHT	2032	386-1	2032	1147(GE)
2032	83P2M	2032	121-602	2032	386-1(SYLVANIA)	2032	1186
2032	83P2M1	2032	121-603	2032	394-3145	2032	1186(GE)
2032	83P2N	2032	121-608	2032	00415	2032	1214
2032	83P3	2032	121-654	2032	417-116	2032	1236-3750
2032	83P3A	2032	121-661	2032	417-116-13165	2032	1254(GE)
2032	83P3AA	2032	121-667	2032	417-132	2032	1294(GE)
2032	83P3AA1	2032	121-679	2032	417-153	2032	1314
2032	83P3B	2032	121-679GREEN	2032	417-168	2032	1314(GE)
2032	83P3B1	2032	121-679YELLOW	2032	417-176	2032	1414-158(ROGERS)
2032	83P3M	2032	121-680	2032	417-182	2032	1414-176
2032	83P3M1	2032	121-681	2032	417-184	2032	1479-8029
2032	83P4	2032	121-682	2032	417-196	2032	1553-17
2032	83PS	2032	121-683	2032	417-200	2032	1582
2032	86-0036-001	2032	121-699	2032	417-201	2032	1679 7391
2032	86-100009	2032	121-699-02	2032	417-234-13165	2032	1679-7391
2032	86-10009	2032	121-725	2032	417-235	2032	1844-17
2032	86-178-2	2032	121-746	2032	417-235-13262	2032	1850-17
2032	86-178-20	2032	121-765-01	2032	417-242-8181	2032	1853-0001-1
2032	86-183-2	2032	121-774	2032	417-260-50127	2032	1853-0081
2032	86-183-20	2032	121-777	2032	430-20013-0B	2032	1853-0089
2032	86-216-2	2032	121-801	2032	430-20018-0A	2032	1867-17
2032	86-217-2	2032	121-838	2032	430-20021	2032	1940-17
2032	86-217-20	2032	121-861	2032	430-20023-0A	2032	1979-808-10
2032	86-218-2	2032	121-865	2032	430-20026-0	2032	2000-201
2032	86-218-20	2032	121-875	2032	436-404-002	2032	2000-218
2032	86-219-2	2032	121-952	2032	450-1167-1	2032	2004-06
2032	86-233-2	2032	121-978	2032	461-1006	2032	2017-107
2032	86-246-2	2032	121-986	2032	461-1055-01	2032	2020-07
2032	86-246-20	2032	121-29003	2032	509R	2032	2032-35
2032	86-251-2	2032	122GRN	2032	509Y	2032	2043-17
2032	86-251-20	2032	122YEL	2032	549-2	2032	2057A100-51
2032	86-276-2	2032	123-006	2032	550-027-00	2032	2057A2-182
2032	86-276-20	2032	0124(KNIGHT)	2032	559-1	2032	2057A2-183
2032	86-286-2	2032	0124A	2032	570-1	2032	2057A2-198
2032	86-286-20	2032	125B133	2032	574	2032	2057A2-277
2032	86-294-2	2032	125P1	2032	576-0001-002	2032	2057A2-298
2032	86-294-20	2032	125P116	2032	576-0001-013	2032	2057A2-307
2032	86-298-2	2032	125P1M	2032	576-0002-008	2032	2057A2-343
2032	86-298-20	2032	125PI	2032	576-0003-017	2032	2057A2-353
2032	86-340-2	2032	125PL	2032	576-0003-019	2032	2057A2-359
2032	86-340-20	2032	129-20	2032	576-0003-12(PNP)	2032	2057A2-397
2032	86-406-2	2032	129-34	2032	5808304H01	2032	2057A2-400
2032	86-407-2	2032	129-34(PILOT)	2032	597-1(RCA)	2032	2057A2-403
2032	86-423-2	2032	130-149	2032	600X0095-086	2032	2057A2-406
2032	86-459-2	2032	130-40315	2032	601X0417-086	2032	2057A2-430
2032	86-475-2	2032	130-40429	2032	602-56	2032	2057A2-457
2032	86-482-2	2032	0131-000335	2032	602-60	2032	2057A2-489
2032	86-501-2	2032	0131-001328	2032	620-1	2032	2057B106-12
2032	86-527-2	2032	0131-001329	2032	627-1	2032	2057B108-6
2032	86-528-2	2032	0131-001420	2032	635	2032	2057B147-12
2032	86-533-2	2032	0131-001439	2032	669	2032	2057B159-12
2032	86-547-2	2032	0131-004746	2032	686-0325-0	2032	2158-1558
2032	86-552-2	2032	0131-005351	2032	686-2700	2032	2181-17
2032	86-555-2	2032	0131-4328	2032	690V086H86	2032	2220-17
2032	86-600-2	2032	132-056	2032	690V116H23	2032	2269
2032	86-610-9	2032	132-074	2032	690V118H60	2032	2272
2032	86-622-2	2032	0133	2032	690V118H61	2032	2300.036.096
2032	86-669-2	2032	134P1A	2032	700-133	2032	2381-17
2032	86A335	2032	134P1M	2032	753-2000-101	2032	2448(RCA)
2032	86P1AA	2032	134P4	2032	753-4004-248	2032	2798
2032	86X0016-001	2032	134P4AA	2032	755-422494	2032	2904-038
2032	86X0016-001A	2032	134P4M	2032	774(ZENITH)	2032	3012(PNP)
2032	86X0036-001	2032	137(ADMIRAL)	2032	0776-0195	2032	3017
2032	86X0036-001A	2032	147-7009-01	2032	776-1(SYLVANIA)	2032	3017(E.F.JOHNSON)
2032	86X0041-001	2032	151-0124-00	2032	800-001-051-1	2032	3019
2032	86X0041-001A	2032	151-0188-00	2032	800-10-108-1	2032	3513
2032	86X0044-001	2032	151-0221-00	2032	800-523-01	2032	3513(SEARS)
2032	86X0044-001A	2032	151-0221-02	2032	800-523-02	2032	3513(WARDS)
2032	86X0046-001	2032	151-0325-00	2032	800-525-04	2032	3522(SEARS)
2032	86X0047-001	2032	151-0458-00	2032	800-527-00	2032	3524(SEARS)
2032	86X0066-001	2032	151-0459-00	2032	800-547-00	2032	3540(RCA)
2032	86X0072-001	2032	157	2032	826-1	2032	3540-1
2032	86X46	2032	157YEL	2032	829	2032	3540-1(GE)
2032	86X47	2032	158P2	2032	829B	2032	3549(RCA)
2032	87-423-2	2032	158P2M	2032	829C	2032	3549-1(RCA)
2032	93P1AA	2032	161-012J	2032	829E	2032	3549-2(RCA)
2032	93P-5215-01	2032	173A4483-1	2032	829EB	2032	3559(RCA)
2032	96-5282-01	2032	173A4483-2	2032	829E	2032	3559-1
2032	96-5283-01	2032	177-006-9-001	2032	829F	2032	3562(RCA)
2032	96-5365-01	2032	177-006-9-002	2032	833	2032	3563(RCA)
2032	98P1	2032	177-012-9-001	2032	921-1016	2032	3570
2032	98P10	2032	177-025-9-001	2032	921-160B	2032	3570(RCA)
2032	99P1	2032	200-052	2032	921-197B	2032	3570-1
2032	99P10	2032	0201	2032	921-254B	2032	3570P
2032	99P1M	2032	207V073C04	2032	921-292B	2032	3574
2032	99P5	2032	209-1	2032	921-296B	2032	3574(RCA)
2032	998039	2032	209P1	2032	921-29B	2032	3574-1(RCA)
2032	998039A	2032	210ATTF3638	2032	921-308	2032	3581(RCA)
2032	998062-1	2032	211ATPE3391	2032	921-308B	2032	3597
2032	998084-1	2032	212-699	2032	921-332B	2032	3597-1
2032	100-0673-04			2032	921-343B	2032	3597-1(RCA)
2032	100-4790			2032	921-405	2032	3597-2
2032	101P1			2032	921-70	2032	3616-1

RS 276-	Industry Standard No.	RS 276-	Industry Standard No.	RS 276-	Industry Standard No.	RS 276-	Industry Standard No.
2032	3620(RCA)	2032	121659	2032	815247	2034	CS1369D
2032	3620-1	2032	123940	2032	838105	2034	CS2005
2032	3627	2032	123971	2032	891008	2034	CS2005B
2032	3627(RCA)	2032	123991	2032	900552-17	2034	CS9102
2032	3627-1	2032	124047	2032	908864-2	2034	CS9123C1
2032	3631	2032	124755	2032	916051	2034	CS9128
2032	3634.2011	2032	125142	2032	916062	2034	CS9129(B)
2032	4086	2032	126524	2032	928408-101	2034	CS9129-B1
2032	4087	2032	126700	2032	932040	2034	CXL106
2032	4126(E.F.JOHNSON)	2032	126707	2032	932107-1	2034	D14
2032	4151-01	2032	126715	2032	960106-3	2034	DS-68
2032	4310 (AIRLINE)	2032	126718	2032	970246	2034	D068
2032	4310(AIRLINE)	2032	126719	2032	970248	2034	D886
2032	4442	2032	129699	2032	970251	2034	DS96
2032	004746	2032	130139	2032	970254	2034	EA0088
2032	4802-00004	2032	130536	2032	970663	2034	EA15X69
2032	4815	2032	131241	2032	970762	2034	EA15X70
2032	4822-130-40315	2032	131242	2032	971059	2034	EA15X71
2032	4844	2032	131647	2032	971905	2034	ECG106
2032	4856-0106	2032	132176	2032	984193	2034	ED1802-0
2032	5001-048	2032	132285	2032	988990	2034	ED1802N
2032	5001-066	2032	132498	2032	1223914	2034	ED1802N,M
2032	5001-509	2032	132830	2032	1417330-3	2034	EP15X90
2032	5059-0236	2032	133182	2032	1417330-4	2034	EQR-0038
2032	5226-1	2032	133253	2032	1417339-1	2034	EQR-1
2032	5611-628	2032	135286	2032	1417347-1	2034	EX699-X
2032	5611-628F	2032	138776	2032	1417363-1	2034	F103P
2032	5611-628P	2032	139455	2032	1471112-7	2034	F121-60216
2032	5611-673	2032	140290	2032	1471112-8	2034	F121-603
2032	5611-673D	2032	140371	2032	1471114-1	2034	FC8-9015C
2032	5701	2032	140372	2032	1472501-1	2034	FC89012HH
2032	6201	2032	140623	2032	1473501-1	2034	FC89015D
2032	6762	2032	141018	2032	1473523-1	2034	FC89015D
2032	6854K90-062	2032	141227	2032	1473540-1	2034	FI-1019
2032	7303-1	2032	141343	2032	1473549-1	2034	FK2894
2032	7303-2	2032	141344	2032	1473549-2	2034	FM2894
2032	7313(TRANSISTOR)	2032	141345	2032	1473559-1	2034	FS-2299
2032	7363-1	2032	141421	2032	1473562-1	2034	FS26382
2032	7503	2032	141711	2032	1473563-1	2034	FT1702
2032	8000-00003-037	2032	141738P63-1	2032	1473570-1	2034	FV2894
2032	8000-00004-089	2032	142838	2032	1473570-2	2034	GET2904
2032	8000-00004-P089	2032	142839	2032	1473574-1	2034	GET2905
2032	8000-0006-004	2032	143802	2032	1473581-1	2034	GET3905
2032	8000-0004-004	2032	143803	2032	1473591-1	2034	GET3906
2032	8000-0004-P089	2032	143806	2032	1473597-1	2034	GMB0404
2032	8000-0004-P089	2032	143807	2032	1473597-2	2034	GMB0404-1
2032	8000-00049-056	2032	143963	2032	1473599-1	2034	GMB0404-2
2032	8301	2032	144031	2032	1473620-1	2034	GMB404-1
2032	8405	2032	144051	2032	1473627-1	2034	HA-0054
2032	8540	2032	145410	2032	1503097-0	2034	HA9500
2032	8601	2032	147549-1	2032	1616226-1	2034	HA9501
2032	8710-169	2032	147549-2	2032	1617032	2034	HA9502
2032	8867	2032	150742	2032	1700001	2034	HBP80012
2032	90150	2032	150753	2032	1700034	2034	HBP80013
2032	9330-767-60112	2032	150758	2032	1780142	2034	HBP80019
2032	9330-908-10112	2032	150762	2032	1780522-1	2034	HF-47
2032	9652H	2032	150771	2032	1780522-2	2034	HR-84
2032	09800	2032	167690	2032	1780522-2-001	2034	HR84
2032	10300	2032	170128	2032	1861223-1	2034	H8-40035
2032	010562	2032	171555	2032	1945294	2034	H8-40053
2032	12594	2032	181015	2032	1950052	2034	HT10494IC-0
2032	12888	2032	181030	2032	1950056-1	2034	HT10494ICO
2032	13162	2032	181034	2032	1969281	2034	HT10494S1A-0
2032	17045	2032	185032	2032	2003073-0702	2034	HT105621B-0
2032	19680	2032	187217	2032	2056606-0701	2034	I50865
2032	20011	2032	188180	2032	2057013-0004	2034	IF20-0046
2032	2200B	2032	190429	2032	2057013-0007	2034	IRTR-54
2032	22595-000	2032	198024	2032	2057013-0008	2034	IRTR19
2032	22605-005	2032	198036=1	2032	2057013-0012	2034	IRTR28
2032	23826	2032	198050	2032	2057013-0701	2034	IRTR30
2032	23826(SYLVANIA)	2032	200067	2032	2057013-0702	2034	J24832
2032	26810-152	2032	200220	2032	2057013-0703	2034	K071687
2032	29076-023	2032	200433	2032	2092609-0025	2034	LDA452
2032	30270	2032	202909-577	2032	2092693-0734	2034	LDS202
2032	30290	2032	202909-587	2032	2132523-1	2034	LDS203
2032	31005	2032	202911-737	2032	2320161	2034	LJ152(O)
2032	33509-1	2032	203364	2032	2320162	2034	LJ1528
2032	033589	2032	205032	2032	2320631	2034	M4910
2032	37486	2032	205048	2032	2320671	2034	M652/PIC
2032	38095	2032	205049	2032	2320681	2034	MA0401
2032	41177	2032	205367	2032	2327262	2034	MA0402
2032	41440	2032	210076	2032	2327387	2034	MA0404-1
2032	43127	2032	232631	2032	2487340	2034	MA0404-2
2032	45122	2032	241517	2032	2487341	2034	MA0413
2032	45337-C	2032	256319	2032	2487424(PNP)	2034	MD3133P
2032	50203-12	2032	267838	2032	2498512	2034	MD3134P
2032	50203-8	2032	309684	2032	250134-105	2034	ME0401
2032	59625-1	2032	320280	2032	2521108-1	2034	ME0402
2032	59625-10	2032	322165	2032	2621570	2034	ME0463
2032	59625-11	2032	324144	2032	2903993-1	2034	ME503
2032	59625-12	2032	333060-1029	2032	2903993-1	2034	M8511
2032	59625-2	2032	337342	2032	3468183-1	2034	M8512
2032	59625-3	2032	401003-001 O	2032	3468242-1	2034	MMCM2907
2032	59625-4	2032	401003-0010	2032	3596063	2034	MMT3905
2032	59625-5	2032	436119-002	2032	3596118	2034	MPS-3640
2032	59625-6	2032	489751-028	2032	3650238A	2034	MPS-A55
2032	59625-7	2032	489751-031	2033	A640L	2034	MPS3534
2032	59625-8	2032	489751-042	2033	A640R	2034	MPS3644
2032	59625-9	2032	489751-097	2033	BP255(SIEG)	2034	MPS3645
2032	60719-1	2032	489751-124	2033	QRT-104	2034	MPS6534
2032	63282	2032	489751-130	2033	2N2711#	2034	MPS6533M
2032	000071150	2032	489751-146	2033	2SC1637	2034	MPS6580
2032	000071151	2032	543995	2034	A-1853-0009-1	2034	MPS9682T
2032	71818-1	2032	592503	2034	A-1853-0010-1	2034	MPSBB1
2032	75561-1	2032	610074-1	2034	A-1853-0034-1	2034	MT0465
2032	75561-2	2032	610083	2034	A5E4402	2034	NP36516
2032	75561-31	2032	610083-3	2034	A5E4403	2034	NP86517
2032	75617-1	2032	610093-1	2034	A311	2034	N83905
2032	75617-2	2032	610099-6	2034	BC126A	2034	N83906
2032	75561-27	2032	610110-1	2034	BC177VI	2034	P67
2032	78331	2032	610110-2	2034	BC178VI	2034	PL4031
2032	83272	2032	610120-1	2034	BC204VI	2034	PL4032
2032	84001	2032	610125-1	2034	BC205L	2034	PL4033
2032	87758	2032	610147-2	2034	BC415	2034	PL4034
2032	87759	2032	610158-2	2034	BC416	2034	PN2904
2032	90330-001	2032	610209-1	2034	BCW29R	2034	PN2905
2032	90432	2032	615180-1	2034	BCW29R	2034	PN2906
2032	94037	2032	615180-2	2034	BCW30	2034	PN2906A
2032	95227	2032	615180-3	2034	BCW30R	2034	PN2907
2032	95227-1	2032	615180-4	2034	BCX78VII	2034	PN2907A
2032	95232	2032	650060	2034	BCX79VII	2034	PN70
2032	95240-1	2032	698941-1	2034	BF315	2034	PN71
2032	96458-1	2032	701584-00	2034	BF316	2034	Q5102Q
2032	101497	2032	701589-00	2034	BF339	2034	Q5116CA
2032	102001	2032	721272	2034	BFX12	2034	QT-A0719XCN
2032	102260	2032	741729	2034	BFX13	2034	RB18
2032	102263	2032	742546	2034	BFX48	2034	RB53
2032	113182	2032	760269	2034	BSX29	2034	REN106
2032	115517-001	2032	801540	2034	C902	2034	RT-115
2032	116078(RCA)	2032	815199	2034	C401	2034	RT-126
2032	116090	2032	815199-6	2034	CDC-9002-IC	2034	RT115
2032	118284	2032	815211	2034	CB-1294F	2034	S2
2032	119228-001	2032	815213	2034	CB-9013HH	2034	S3
2032	119730	2032	815229	2034	CS1124G		
2032	121467	2032	815236				

RS 276-	Industry Standard No.	RS 276-	Industry Standard No.	RS 276-	Industry Standard No.	RS 276-	Industry Standard No.
2034	84	2034	2N626	2034	136P1	2035	BFW54
2034	S500	2034	2N723	2034	137-12	2035	BFW55
2034	S501	2034	2N726	2034	145-12	2035	BFW56
2034	SA495	2034	2N727	2034	148-12	2035	BFW61
2034	SA495A	2034	2N864A	2034	151-0417-00	2035	CDC731
2034	SA496	2034	2N865	2034	154A5947-7732	2035	DDCY001002
2034	SA496A	2034	2N865A	2034	159-12	2035	DDCYO02002
2034	SA496B	2034	2N869	2034	177-025-9-002	2035	DDCYO06001
2034	SA537	2034	2N869A	2034	210BWTF4121	2035	DS-88
2034	SA538	2034	2N978	2034	246P1	2035	DS88
2034	SA539	2034	2N995	2034	297L012C-01	2035	E300
2034	SA540	2034	2N995A	2034	333-1	2035	EA15X165
2034	SAC40	2034	2N996	2034	352-0950-010	2035	EA15X192
2034	SAC40A	2034	28564A	2034	352-0950-020	2035	EA15X193
2034	SAC40B	2034	28A465	2034	364-1	2035	EA15X394
2034	SAC42	2034	28A558-G	2034	417-102	2035	EA15X400
2034	SAC42A	2034	28A548G	2034	417-284	2035	EA15X401
2034	SAC42B	2034	28A548GN	2034	508ES020P	2035	ECG132
2034	SAC44	2034	28A548OR	2034	508ES021P	2035	ECG312
2034	SI341P	2034	28A548R	2034	515ES045M	2035	EP2(ELCOM)
2034	SI342P	2034	28A548Y	2034	516ES046M	2035	EP3
2034	SI343P	2034	28A565(D,C)	2034	516ES047M	2035	EP15X92
2034	SI351P	2034	28A565AB	2034	516ES048M	2035	EQP-0009
2034	SI352P	2034	28A565BA	2034	520ES070M	2035	EQF-4
2034	SI353P	2034	28A565E	2034	521ES071M	2035	ES15X92
2034	SL200	2034	28A565F	2034	549-1	2035	P1462
2034	SL201	2034	28A565G	2034	574-1	2035	P1463
2034	SM62186	2034	28A565GN	2034	576-0003-012	2035	FE-100
2034	SP12271	2034	28A565H	2034	576-001-013	2035	FE5245
2034	SPS2526	2034	28A565J	2034	597-1	2035	FT34G
2034	SPS4401	2034	28A565K	2034	616-1	2035	FT34Y
2034	SP2-2517	2034	28A565M	2034	620-56	2035	G08-007-B
2034	T-2357	2034	28A565OR	2034	666-1	2035	GB-FET-2
2034	T-HU6OU	2034	28A565R	2034	774	2035	GRF-3
2034	T2357	2034	28A565X	2034	921-103B	2035	HEP802
2034	T40	2034	28A565Y	2034	921-112B	2035	HEPF0010
2034	TGZSA608(C)	2034	28A572Y	2034	921-333B	2035	HEPF0021
2034	TIS128	2034	28A594N	2034	921-333P	2035	HEP2000411B
2034	TIS138	2034	28A6111	2034	991-013058	2035	HP200191A
2034	TM-22	2034	28A666I	2034	1012GE	2035	HP200191A-0
2034	TM106	2034	28A672(B)	2034	1043-7374	2035	HP200191A0
2034	TR-1032-1	2034	28A701FO	2034	1050B	2035	HP200191A0
2034	TRO1053-1	2034	28A720Q,R,8	2034	1147	2035	HP200191B-0
2034	TRO2051-1	2034	28A742H	2034	1254	2035	HP200191B0
2034	TRO2062-1	2034	28A813	2034	1294	2035	HP200411B
2034	TV-71	2034	28C182(Q)	2034	1853-0069	2035	HP200411CO
2034	TV24848	2034	28C182(V)	2034	2000-202	2035	HK-00049
2034	W7	2034	28C182A	2034	2022-01	2035	IC743046
2034	WEP51	2034	28C182B	2034	2057A-430	2035	IP20-0011
2034	WEP52	2034	28C182C	2034	2057A2-200	2035	IP20-0012
2034	WEP715	2034	28C182D	2034	2057A2-203	2035	IP20-0035
2034	WEP716	2034	28C182E	2034	2057A2-229	2035	IP20-0078
2034	002-0105-00	2034	28C182F	2034	2057A2-561	2035	IT10B
2034	2H1254	2034	28C182G	2034	3012	2035	ITE4416
2034	2H1255	2034	28C182GN	2034	3012(EPJOHNSON)	2035	J308
2034	2H1256	2034	28C182H	2034	003552	2035	JF1033B
2034	2H1257	2034	28C182J	2034	3540	2035	JF-1033
2034	2H1258	2034	28C182K	2034	3549	2035	JF1033
2034	2H1259	2034	28C182L	2034	3549-1	2035	JF1033B
2034	2N1024	2034	28C182M	2034	3549-2	2035	JF1033G
2034	2N1131/51	2034	28C182R	2034	3559	2035	JF1033S
2034	2N1131A/51	2034	28C182X	2034	3562	2035	K11(1GATE)
2034	2N1132/51	2034	28C182Y	2034	3563	2035	K11-0(1 GATE)
2034	2N1132A/51	2034	28C719	2034	3581	2035	K11-R(1 GATE)
2034	2N1132A46	2034	28C719Q	2034	3616	2035	K11-Y(1 GATE)
2034	2N1238	2034	28C9012-HG	2034	3620	2035	K12(1 GATE)
2034	2N1239	2034	28C9012-HH	2034	3651	2035	K12-GR(1 GATE)
2034	2N1240	2034	28H678	2034	3666	2035	K12-0(1 GATE)
2034	2N1241	2034	03A04	2034	4484-1	2035	K12-R(1 GATE)
2034	2N1640	2034	3H112	2034	4484-2	2035	K12OT(1 GATE)
2034	2N1641	2034	3H113	2034	4485-1	2035	K13(1 GATE)
2034	2N1642	2034	8G-3-02	2034	4590	2035	K15-GR(1 GATE)
2034	2N2175	2034	11RT1	2034	4801-0000-001	2035	K15-0(1 GATE)
2034	2N2176	2034	13-26386-4	2034	4801-0000-060	2035	K15-R(1 GATE)
2034	2N2177	2034	13-34369-1	2034	4822-130-40348	2035	K15-Y(1 GATE)
2034	2N2395	2034	13-34940-1	2034	4822-130-40369	2035	K17-0(1 GATE)
2034	2N2591	2034	13-40083-1	2034	4822-130-40477	2035	K170(1 GATE)
2034	2N2592	2034	13-40083-2	2034	4822-130-40508	2035	K170R(1 GATE)
2034	2N2593	2034	14-807-12	2034	4822-130-40614	2035	K17A(1 GATE)
2034	2N2800/51	2034	14-861-12	2034	5001-540	2035	K17B(1 GATE)
2034	2N2801/51	2034	15-008-1	2034	5611-628(P)	2035	K17BL(1 GATE)
2034	2N2802	2034	020-1110-004	2034	5611-6730	2035	K170R(1 GATE)
2034	2N2803	2034	21A062-000	2034	6855K90	2035	K17R(1 GATE)
2034	2N2804	2034	24-016-001	2034	7303	2035	K17Y(1 GATE)
2034	2N2805	2034	25-001328	2034	7340	2035	K19(1 GATE)
2034	2N2806	2034	34-1011	2034	9012HG	2035	K19BL(1 GATE)
2034	2N2807	2034	4002FW8V18P	2034	10508	2035	K19GC(1 GATE)
2034	2N2927/51	2034	46-86342-3	2034	30291(PNP)	2035	K19GB(1 GATE)
2034	2N2968	2034	47C23-3	2034	48751-028	2035	K19GR(1 GATE)
2034	2N2969	2034	48-137066	2034	95229	2035	K19Y(1 GATE)
2034	2N2970	2034	48-3003A04	2034	97680	2035	K22(1 GATE)
2034	2N2971	2034	48-97127A015	2034	105468	2035	K22-Y(1 GATE)
2034	2N3049	2034	48-97127A15	2034	115270-101	2035	K23(1 GATE)
2034	2N3050	2034	48M355007	2034	124634	2035	K25(1 GATE)
2034	2N3051	2034	48R869426	2034	125707	2035	K25C(1 GATE)
2034	2N3209	2034	488134821	2034	129697	2035	K25D(1 GATE)
2034	2N3224	2034	488134909	2034	130215	2035	K25E(1 GATE)
2034	2N3245	2034	488134910	2034	131262	2035	K25ET(1 GATE)
2034	2N3304	2034	488137045	2034	135347	2035	K25P(1 GATE)
2034	2N3342	2034	488155095	2034	135766	2035	K25Q(1 GATE)
2034	2N3576	2034	48X90232A06	2034	150865	2035	K30-0(1 GATE)
2034	2N3838	2034	48X90232A09	2034	160196	2035	K30A(1 GATE)
2034	2N3915	2034	48X90232A12	2034	405192	2035	K30AD(1 GATE)
2034	2N4034	2034	48X90232A15	2034	531298-001	2035	K30AGR(1 GATE)
2034	2N4035	2034	48X97177A14	2034	547684	2035	K30B(1 GATE)
2034	2N4058	2034	50-40106-09	2034	610024-1	2035	K30C(1 GATE)
2034	2N4059	2034	50-40204-10	2034	610102-1	2035	K30D(1 GATE)
2034	2N4060	2034	57A1-52	2034	610136-1	2035	K30GR(1 GATE)
2034	2N4061	2034	57A108-1	2034	610246-1	2035	K30R(1 GATE)
2034	2N4062	2034	57A108-2	2034	989015	2035	K30Y(1 GATE)
2034	2N4080	2034	57A108-3	2034	1471112-9	2035	K31-(1 GATE)
2034	2N4122	2034	57A108-4	2034	1473559-001	2035	K31C(1 GATE)
2034	2N4142	2034	57A108-5	2034	1473597	2035	K33(1 GATE)
2034	2N4228	2034	57A108-6	2034	1473620-001	2035	K33B(1 GATE)
2034	2N4313	2034	57A108-7	2034	1473651-1	2035	K33P(1 GATE)
2034	2N4389	2034	57A108-8	2034	2621811	2035	K33GR(1 GATE)
2034	2N4423	2034	57D1-52	2035	A066-110	2035	K34(1 GATE)
2034	2N4451	2034	62-130139	2035	A068-106	2035	K34B(1GATE)
2034	2N4453	2034	62-132497	2035	A068-107	2035	K34C(1 GATE)
2034	2N4872	2034	62-21683	2035	A11744	2035	K34D(1 GATE)
2034	2N4937	2034	62-22524	2035	A11745	2035	K37(1 GATE)
2034	2N4938	2034	62-22529	2035	A192	2035	K37H(1 GATE)
2034	2N4940	2034	62-3597-1	2035	A5T3821	2035	K37L(1 GATE)
2034	2N4941	2034	62-3597-2	2035	B9TUI	2035	K49P(1 GATE)
2034	2N495	2034	69-1815	2035	BP244A	2035	K49H(1 GATE)
2034	2N495/18	2034	69-1817	2035	BP244B	2035	K49HK(1 GATE)
2034	2N496	2034	86X0066-003	2035	BP244C	2035	K49I(1 GATE)
2034	2N5140	2034	110P2	2035	BP256	2035	K49M(1 GATE)
2034	2N5141	2034	121-615	2035	BF868	2035	K55(1 GATE)
2034	2N5208	2034	125-1	2035	BF868P	2035	K55(1-GATE)
2034	2N5228	2034	0131-001438	2035	BFW10	2035	K55D(1 GATE)
2034	2N5352	2034	134-1P	2035	BFW11		
2034	2N5763	2034	134P1(VID)	2035	BFW12		
2034	2N5857	2034	134P6				

RS 276-	Industry Standard No.
2035	K55D(1-GATE)
2035	K55E(1 GATE)
2035	K55E(1-GATE)
2035	KA4559
2035	KE4416
2035	KE5105
2035	LS5484
2035	LS5485
2035	MJF10335
2035	MJF1033G
2035	MK-10-E
2035	MK10
2035	MK10-2
2035	MK102
2035	MK5485
2035	MPF-106
2035	MPF102
2035	MPF106
2035	MPF107
2035	MPF111
2035	NH7916
2035	N8316
2035	PL-182-009-9-001
2035	PL1091
2035	PL1092
2035	PL1093
2035	PL1094
2035	PTC151
2035	PTC152
2035	PTC161
2035	Q-00169
2035	Q-00169A
2035	Q-00169B
2035	Q-00169C
2035	Q-00184R
2035	QKT-0033XBE
2035	QKT0033XBE
2035	QRF-3
2035	QRF3
2035	QRG-3
2035	QT-K0023AAS
2035	QT-K0033XBE
2035	R45
2035	REN132
2035	RS-2028
2035	R87916
2035	RT-175
2035	RT-8667
2035	RT175
2035	RT8331
2035	RT8667
2035	RVTMK10-2
2035	RVTMK10-E
2035	S1241N
2035	S1242N
2035	S383819
2035	SK19
2035	SK3116
2035	S83534-4
2035	S83704
2035	SX3819
2035	T1-741
2035	T1208
2035	T1834
2035	T1858
2035	T1859
2035	T1888
2035	TB-500-E
2035	TB500
2035	TI-741
2035	TI741
2035	T1S-88
2035	T1834
2035	T1842
2035	T1858
2035	T1859
2035	T1888
2035	TM132
2035	TM312
2035	TNJ61672(28K25)
2035	TR-8027
2035	TR-FE100
2035	TR-FET-1
2035	TR-U1650E
2035	TR-U1650E-1
2035	TR-U1835E
2035	TR06011
2035	TR06014
2035	TR2083-70
2035	TR228735120325
2035	TR28K55
2035	TR5528
2035	TR06011
2035	TU834
2035	TV-83
2035	TV80
2035	V183
2035	VS28K49F-1
2035	W1P
2035	WEP920
2035	WF2
2035	WG-10AS
2035	ZEN123
2035	001-02703-Q
2035	001-027030
2035	01-070030
2035	2N3819
2035	2N4093
2035	2N5078
2035	2N5078A
2035	2N5103A
2035	2N5104A
2035	2N5105A
2035	2N5163
2035	2N5163A
2035	2N5245
2035	2N5245A
2035	2N5246
2035	2N5246A
2035	2N5247
2035	2N5247A
2035	2N5248
2035	2N5248A
2035	2N5258
2035	2N5278A
2035	2N5360A
2035	2N5397
2035	2N5398
2035	2N5484
2035	2N5485
2035	2N5485-1
2035	2N5486
2035	2N5555
2035	2N5592
2035	2N5593
2035	2N5594
2035	2N5669
2035	2N5670
2035	2N5949
2035	2N5950
2035	2N5951
2035	2N5952
2035	2N5953
2035	2NJ253B
2035	2N241P
2035	28K1033B
2035	28K104
2035	28K104H
2035	28K11
2035	28K11-0
2035	28K11-R
2035	28K11-Y
2035	28K12
2035	28K12-GR
2035	28K12-0
2035	28K12-R
2035	28K12-Y
2035	28K13
2035	28K15-GR
2035	28K15-0
2035	28K15-R
2035	28K15-Y
2035	28K17
2035	28K17-Q
2035	28K170
2035	28K170R
2035	28K17A
2035	28K17B
2035	28K17BL
2035	28K170R
2035	28K17R
2035	28K17Y
2035	28K19
2035	28K19(BL)
2035	28K19(GR)
2035	28K19-14
2035	28K19-BL
2035	28K19-GR
2035	28K19-Y
2035	28K19A
2035	28K19B
2035	28K19BB
2035	28K19BL
2035	28K19FET
2035	28K19GB
2035	28K19GC
2035	28K19GE
2035	28K19GR
2035	28K19H
2035	28K19K
2035	28K19V
2035	28K19Y
2035	28K22Y
2035	28K23
2035	28K23A
2035	28K23A540
2035	28K25
2035	28K25C
2035	28K25D
2035	28K25E
2035	28K25ET
2035	28K25F
2035	28K25G
2035	28K30
2035	28K30(0)
2035	28K30-0
2035	28K30-0
2035	28K30-R
2035	28K30-Y
2035	28K304
2035	28K30A
2035	28K30A(D)
2035	28K30A-Y
2035	28K30AD
2035	28K30AGR
2035	28K30A0
2035	28K30AY
2035	28K30D
2035	28K30GR
2035	28K30R
2035	28K30Y
2035	28K31
2035	28K31(C)
2035	28K31C
2035	28K32B
2035	28K33
2035	28K33(E)
2035	28K33D
2035	28K33E
2035	28K33F
2035	28K33GR
2035	28K33H
2035	28K34
2035	28K34A
2035	28K34B
2035	28K34C
2035	28K34D
2035	28K34E
2035	28K37
2035	28K37(K)
2035	28K37H
2035	28K37K
2035	28K37L
2035	28K41
2035	28K41E
2035	28K41E2
2035	28K41F
2035	28K42
2035	28K42-CM1
2035	28K42-CMI
2035	28K42CM1
2035	28K47
2035	28K47M
2035	28K49
2035	28K49B2
2035	28K49F
2035	28K49H
2035	28K49H1
2035	28K49H2
2035	28K49HK
2035	28K49I
2035	28K49M
2035	28K54
2035	28K55
2035	28K55D
2035	28K55DE
2035	28K55E
2035	28K55R
2035	28K61
2035	28K61GR
2035	28K61Y
2035	28K83
2035	28K84
2035	28K68AM
2035	03A09
2035	3Q2
2035	3830B
2035	38K23
2035	38K30
2035	38K30A
2035	38K30B
2035	38K30C
2035	04-58190-01
2035	07-07158
2035	09-304017
2035	09-305014
2035	09-305032
2035	09-305135
2035	09-309074
2035	10-004
2035	13-22690-1
2035	13-22692-1
2035	13-22692-2
2035	13-28654-1
2035	13-28654-2
2035	13-28654-3
2035	13-33173-1
2035	13-34375-1
2035	13-34375-2
2035	13-34378-1
2035	13-34378-2
2035	13-34378-3
2035	13-34312-1
2035	14-700-01
2035	14-700-03
2035	14-700-04
2035	14-700-05
2035	14-700-06
2035	14-710-21
2035	14-713-31
2035	14-713-32
2035	14-714-13
2035	16-17
2035	19-020-115
2035	020-1112-002
2035	020-1112-006
2035	21A040-015
2035	21A113-002
2035	21M196
2035	21M224
2035	21M412
2035	21M534
2035	022-2876-006
2035	022-2876-009
2035	24KW652
2035	24MW723
2035	24MW736
2035	24MW989
2035	33K59
2035	44T-300-105
2035	44T-300-113
2035	46-86316-3
2035	46-8638
2035	46-8648
2035	46-8649
2035	48-137343
2035	48-3003A09
2035	48-4A567J01
2035	48-90232A14
2035	48-97046A47
2035	48-97046A48
2035	48-97177A01
2035	48-97177A06
2035	48B134944
2035	48B137070
2035	48B137343
2035	57A149-12
2035	57A150-12
2035	57A31-1
2035	57A31-2
2035	57A31-3
2035	57B149-12
2035	57B150-12
2035	57B150-12
2035	63-13926
2035	065-001
2035	065-002
2035	66F-084-1
2035	87-0001
2035	90-179
2035	90-50
2035	90-55
2035	90-606
2035	90-607
2035	90-608
2035	90-613
2035	90-62
2035	0105-0012
2035	106M
2035	108-0049-08
2035	111-731
2035	121-731
2035	121-756
2035	121-858
2035	132-049
2035	173-1(SYLVANIA)
2035	182-009-9-001
2035	182-015-9-001
2035	182-021-9-001
2035	182-029-9-001
2035	182-039-9-001
2035	182-044-9-001
2035	182-044-9-002
2035	182-056-9-001
2035	200-053
2035	200-064
2035	220-008001
2035	229-5192-20
2035	260-10-006
2035	260P22001
2035	260P22002
2035	260P22003
2035	417-25
2035	430-25762
2035	576-0004-004
2035	588U
2035	654-1
2035	654-1(SYLVANIA)
2035	734EU
2035	753-4000-025
2035	753-6000-019
2035	753-5000-019
2035	800-535-00
2035	800-535-01
2035	914-000-2-00
2035	921-203B
2035	921-231B
2035	1009-01
2035	1009-127
2035	1021
2035	1041-70
2035	1095-01
2035	1859-17
2035	1934-17
2035	2000-101
2035	2000-105
2035	2000-107
2035	2020-01
2035	2032-36
2035	2056-75
2035	2057A2-445
2035	2058-02
2035	2074-17
2035	2335-17
2035	2336-17
2035	2450-17
2035	3511(SEARS)
2035	3511(WARDS)
2035	3512(RCA)
2035	3512(SEARS)
2035	3512(WARDS)
2035	3819
2035	3819(RCA)
2035	4802-00010
2035	4811-0000-015
2035	4811-0000-025
2035	5096
2035	6013
2035	8000-00004-081
2035	8000-00004-P080
2035	8000-00004-P081
2035	8000-00005-001
2035	8000-00009-178
2035	8000-00011-054
2035	8000-00011-055
2035	8000-00004-P081
2035	8000-00004-062
2035	8000-0005-001
2035	8010-173
2035	8710-161
2035	8840-121
2035	8840-161
2035	8910-141
2035	18410-141
2035	23606
2035	28810-171
2035	36582
2035	38510-161
2035	43296
2035	45810-161
2035	48009
2035	48009(1)
2035	58810-160
2035	58840-191
2035	71686-6
2035	88060-141
2035	88510-171
2035	127214
2035	171207-4
2035	207417
2035	489751-208
2035	516009S
2035	610164-1
2035	760268
2035	916082
2035	970253
2035	989715
2035	2006623-47
2035	2327132
2035	2327142
2035	2327142(JFET)
2035	A054-142
2035	A498(F.E.T.)
2036	AR501
2036	AR502
2036	BF828
2036	BF828R
2036	CXL220
2036	CXL221
2036	CXL222
2036	DDCY105001
2036	DDCY104001
2036	DDCY104003
2036	DS-102
2036	DS-105
2036	DS-106
2036	EA15X238
2036	EA15X402
2036	EA15X405
2036	ECG220
2036	ECG222
2036	EP4(ELCOM)
2036	EP5(ELCOM)
2036	EP15X36
2036	EP15X40
2036	EP15X64
2036	FT0601
2036	GE-FET-3
2036	GE-FET-4
2036	HEP-F2005
2036	HEPF2004
2036	HEPF2005
2036	HEPF2007
2036	HEPF2007A
2036	HF200301E
2036	IP20-0157
2036	IP20-0218
2036	IP20-0305
2036	K32A(2 GATE)
2036	K32B(2 GATE)
2036	K32C(2 GATE)
2036	K32D(2 GATE)
2036	K35G(1 GATE)
2036	K39Q(2 GATE)
2036	K39R(2 GATE)
2036	K40(2 GATE)
2036	K40A(2 GATE)
2036	K40AK(2 GATE)
2036	K40C(2 GATE)
2036	K40D(2 GATE)
2036	K40M(2 GATE)
2036	K45(2 GATE)
2036	K45B(2 GATE)
2036	KD2130

RS 276-	Industry Standard No.	RS 276-	Industry Standard No.	RS 276-	Industry Standard No.	RS 276-	Industry Standard No.
2036	M75561-10RK	2036	38K39E	2036	60337	2038	A32-2809
2036	M75561-23	2036	38K39F	2036	60793	2038	A321
2036	M75561-23RN	2036	38K39Q	2036	95132	2038	A3A
2036	MEM564C	2036	38K39R	2036	114267	2038	A3C
2036	MEM650	2036	38K5E	2036	127980	2038	A3D
2036	MEM680Y	2036	38K40	2036	129424	2038	A3H
2036	MPE-3008	2036	38K40I	2036	129980	2038	A3N71
2036	MPE121	2036	38K40M	2036	134450	2038	A3N72
2036	MPE130	2036	38K41	2036	135324	2038	A3N73
2036	MPE130-712	2036	38K41(L)	2036	135963	2038	A3P
2036	MPE131	2036	38K411	2036	141332	2038	A3R
2036	MPE3004	2036	38K41C	2036	171206-6	2038	A38B
2036	MPE3005	2036	38K41L	2036	258017	2038	A417
2036	MPE3006	2036	38K41M	2036	610166-1	2038	A418
2036	MPE3007	2036	38K44	2036	610203-1	2038	A419
2036	MPE3008	2036	38K45	2036	610203-2	2038	A420
2036	MPP121	2036	38K45-B-09	2036	610203-3	2038	A427
2036	MO83635	2036	38K45B	2036	610203-4	2038	A427(JAPAN)
2036	MPF-121	2036	38K45B09	2036	610203-5	2038	A467
2036	MPF121	2036	38K49	2036	610203-6	2038	A473
2036	P2C181	2036	38K49E2	2036	610358-1	2038	A4789
2036	R3651-1	2036	38K49Q	2036	610358-2	2038	A48-134789
2036	R8199	2036	38K59	2036	610358-3	2038	A48-134837
2036	REN220	2036	38K59BL	2036	741860	2038	A48-134845
2036	REN221	2036	38K59GR	2036	760005	2038	A48-134902
2036	REN222	2036	7-117-02	2036	985175	2038	A48-134904
2036	RT-180	2036	7-40	2036	986930	2038	A48-134922
2036	RT180	2036	09-305040	2036	1408694-1	2038	A48-134923
2036	SR5315	2036	13-0118	2036	1417372-1	2038	A48-134924
2036	SPB8970	2036	13-0165	2036	1473588-2	2038	A48-134925
2036	SPC-1616	2036	13-10321-37	2036	1473588-8	2038	A48-134926
2036	SPC-1617	2036	13-10321-53	2036	1473618-1	2038	A48-134945
2036	SPC8999	2036	13-28583-1	2036	1473635	2038	A48-134961
2036	SPD2285	2036	13-37900-1	2036	1473635-1	2038	A48-134962
2036	SPE253	2036	022-2876-008	2036	1473635-2	2038	A48-134963
2036	SPE303	2036	24MW1122	2036	2068491-704	2038	A48-134964
2036	SPE305424	2036	033-014-0	2036	2327111	2038	A48-134965
2036	SPE425	2036	041A	2036	2327142(MOSFET)	2038	A48-134966
2036	SPE427	2036	46-86396-3	2036	2327232	2038	A48-134981
2036	SK3050/221	2036	46-8646	2036	2327431	2038	A48-137071
2036	SK3065	2036	047	2036	3404520-601	2038	A48-137075
2036	SK3065/222	2036	048	2036	3596401	2038	A48-137076
2036	SK3531	2036	48-13707	2036	3596402	2038	A48-137077
2036	SPP215	2036	48-137070	2037	EC0326	2038	A48-137197
2036	SPP274	2036	48-137488	2037	REN85	2038	A48-40247G01
2036	SPP512	2036	48-137567	2038	A-1854-0485	2038	A48-43351A02
2036	SPP609	2036	48-5005A10	2038	A-1854-JBD1	2038	A48-63076A82
2036	T1731	2036	48-64978A39	2038	A-473	2038	A48-97046A05
2036	T464	2036	57A267-4	2038	A054-157	2038	A48-97046A06
2036	T513	2036	86-464-2	2038	A054-158	2038	A48-97046A07
2036	TA2644	2036	86-540-2	2038	A054-159	2038	A48-97127A06
2036	TA2840	2036	86-605-2	2038	A054-163	2038	A48-97127A12
2036	TA7149	2036	86-606-2	2038	A054-164	2038	A48-97127A18
2036	TA7150	2036	86-625-2	2038	A054-470	2038	A480
2036	TA7151	2036	86-632-2	2038	A066-111	2038	A485
2036	TA7189	2036	90-178	2038	A066-115	2038	A4851
2036	TA7262(RCA)	2036	998041	2038	A068-111	2038	A497
2036	TA7274	2036	998045	2038	A068-112	2038	A48-5
2036	TA7374	2036	998045A	2038	A1109	2038	A49
2036	TA7399	2036	998046	2038	A1170	2038	A4Z
2036	TA7669	2036	0117-02	2038	A121-480	2038	A4Z-1A
2036	TA7684	2036	121-1024	2038	A124623	2038	A57A144-12
2036	TA8242	2036	121-1030	2038	A124624	2038	A57A145-12
2036	TM220	2036	121-782	2038	A125329	2038	A5Z
2036	TM221	2036	121-783	2038	A129509	2038	A62-19581
2036	TM222	2036	121-784	2038	A129510	2038	A642-254
2036	TR08004	2036	121-785	2038	A129511	2038	A642-260
2036	TR2327431	2036	121-786	2038	A129512	2038	A642-268
2036	TR39453	2036	121-787	2038	A129513	2038	A667RED
2036	TR40603	2036	121-826	2038	A129571	2038	A67-33-540
2036	VOC256-001	2036	123-002	2038	A129572	2038	A67-37-940
2036	VOC256-00I	2036	132-045	2038	A129573	2038	A670720K
2036	W1E	2036	132-047	2038	A129574	2038	A6708850
2036	W1U	2036	132-048	2038	A1300RN	2038	A673351K
2036	W1U-1	2036	182-038-9-001	2038	A130V100	2038	A673354K
2036	WBP903	2036	182-158-9-001	2038	A14-602-63	2038	A673355H
2036	WBP904	2036	201-283818-1	2038	A14-603-05	2038	A673355K
2036	WBP905	2036	201-283818-2	2038	A14-603-06	2038	A6B
2036	WIE	2036	201-283818-3	2038	A1418	2038	A6B
2036	YBAN38K39Q	2036	203-1	2038	A1462	2038	A6F
2036	ZEN124	2036	203-4	2038	A1518	2038	A6T
2036	01-080045	2036	229-0192-18	2038	A1519	2038	A6U
2036	28K39	2036	355D7	2038	A1520	2038	A6V-5
2036	28K39B	2036	417-206	2038	A1521	2038	A715FB
2036	28K39F	2036	417-207	2038	A154	2038	A772T38
2036	28K45B	2036	417-225	2038	A154(NPN)	2038	A772T39
2036	03A10	2036	417-240	2038	A155	2038	A772D1
2036	3L4-6503-1	2036	417-863	2038	A155(NPN)	2038	A7A30
2036	3L4-6503-2	2036	576-0003-224	2038	A164	2038	A7A31
2036	3N128	2036	576-0006-221	2038	A164(NPN)	2038	A7A32
2036	3N140	2036	576-0006-222	2038	A165	2038	A7N
2036	3N142	2036	576-0006-227	2038	A165(NPN)	2038	A7P
2036	3N143	2036	635-1	2038	A176-025-9-002	2038	A7U
2036	3N152	2036	635-1(RCA)	2038	A19-020-072	2038	A7V
2036	3N200	2036	921-157B	2038	A1E	2038	A7W
2036	3N201	2036	1042-01	2038	A1G-1A	2038	A8N
2036	3N201A	2036	1251-1-1	2038	A1K	2038	A909-27125-160
2036	3N202	2036	2000-102	2038	A1P-/4923	2038	A90Z2
2036	3N205	2036	2000-103	2038	A1P-5	2038	A916-31025-58
2036	3N213	2036	2022-06	2038	A1P/4923-1	2038	A916-31025-5B
2036	38K-30B	2036	2065-55	2038	A1R-1/4925	2038	A921-59B
2036	38K-35	2036	2079-40	2038	A1R-1A	2038	A921-62B
2036	38K-39	2036	2359-17	2038	A1R-2/4926	2038	A921-63B
2036	38K14	2036	3588(RCA)	2038	A1R-24926	2038	A921-64B
2036	38K20	2036	3588-2	2038	A1R-2A	2038	A991-01-1316
2036	38K20H	2036	3618	2038	A1R-5	2038	A9A
2036	38K20HW	2036	3635	2038	A1R/4925A	2038	A9D
2036	38K20HY	2036	3635(RCA)	2038	A1R/4926A	2038	AR200
2036	38K21	2036	3635-1	2038	A2006681-95	2038	AR200W
2036	38K21H	2036	3635-2	2038	A2092693-0724	2038	AR201
2036	38K22	2036	5001-046	2038	A2092693-0725	2038	AR201(YELLOW)
2036	38K22-Y	2036	6227	2038	A24	2038	AR201Y
2036	38K22GR	2036	6351	2038	A245	2038	AR202
2036	38K22Y	2036	7372-1	2038	A245(AMC)	2038	AR202(GREEN)
2036	38K29	2036	8000-00011-053	2038	A249B	2038	AR202G
2036	38K32	2036	8000-00042-013	2038	A24MW594	2038	AR212
2036	38K32(B)	2036	38378	2038	A24MW595	2038	AR221
2036	38K32A	2036	38563	2038	A24MW596	2038	AR25(ORANGE)
2036	38K32B	2036	40467A	2038	A24MW597	2038	AR25(WHITE)
2036	38K32B-6	2036	40468A	2038	A24T-016-016	2038	AR313
2036	38K32C	2036	40559A	2038	A24T-016-01L	2038	AT959A
2036	38K32D	2036	40600	2038	A2C	2038	AX91770
2036	38K32E	2036	40601	2038	A2D	2038	B2
2036	38K32E-4	2036	40602	2038	A2F	2038	B3
2036	38K33	2036	40603	2038	A2G	2038	B4
2036	38K35	2036	40604	2038	A2H	2038	B9426
2036	38K35-BL	2036	40673	2038	A2J	2038	BC111
2036	38K35-GR	2036	40819	2038	A2N-2A	2038	BC112
2036	38K35-Y	2036	40820	2038	A2P-5	2038	BC155A
2036	38K35BL	2036	40821	2038	A2T	2038	BC156A
2036	38K35G	2036	40822	2038	A2T919	2038	BC194
2036	38K37	2036	40823	2038	A2V	2038	BC194B
2036	38K39	2036	40841	2038	A2W	2038	BC195
				2038	A31-0206		

RS 276-	Industry Standard No.	RS 276-	Industry Standard No.	RS 276-	Industry Standard No.	RS 276-	Industry Standard No.
2038	BC195CD	2038	C1213AKC	2038	C458GLB	2038	C772DX
2038	BC295	2038	C1213AKD	2038	C458K	2038	C772DY
2038	BC456	2038	C1213B	2038	C458KA	2038	C772E
2038	BC6500	2038	C1213BC	2038	C458KB	2038	C772P
2038	BCW31	2038	C1213C	2038	C458KC	2038	C772K
2038	BCW31R	2038	C1213CD	2038	C458KD	2038	C772KB
2038	BCW32	2038	C1213D	2038	C458L	2038	C772KC
2038	BCW32R	2038	C1214	2038	C458LB	2038	C772KD
2038	BCW71	2038	C1214A	2038	C458LG	2038	C772KD1
2038	BCW71R	2038	C1214B	2038	C458LGA	2038	C772KD2
2038	BCW72	2038	C1214C	2038	C458LGB	2038	C772R(JAPAN)
2038	BCW72R	2038	C1214D	2038	C458LGBM	2038	C772RB-D
2038	BCY87	2038	C27	2038	C458LGC	2038	C772RD
2038	BCY88	2038	C1317	2038	C458LGD	2038	C784
2038	BCY89	2038	C1317B	2038	C458LGQ	2038	C784-0
2038	BD71	2038	C1317P	2038	C458LGS	2038	C784-0
2038	BF-115	2038	C1317Q	2038	C458M	2038	C7840
2038	BF152	2038	C1317R	2038	C458P	2038	C784A
2038	BF153	2038	C1317S	2038	C458RGS	2038	C784BN
2038	BF154	2038	C1317T	2038	C458TOK	2038	C784R
2038	BF158	2038	C1342	2038	C458VC	2038	C784Y
2038	BF159	2038	C1342A	2038	C459D	2038	C785(0)
2038	BF160	2038	C1342B	2038	C460	2038	C785BN
2038	BF162	2038	C1342C	2038	C461	2038	C785D
2038	BF163	2038	C1359	2038	C461A	2038	C785R
2038	BF164	2038	C1359A	2038	C461B	2038	C785R
2038	BF165	2038	C1359B	2038	C461C	2038	C786
2038	BF173A	2038	C1359C	2038	C461E	2038	C79
2038	BF176	2038	C148(TRANSISTOR)	2038	C461L	2038	C80
2038	BF187	2038	C155	2038	C468(LGR)	2038	C828
2038	BF188	2038	C156	2038	C477	2038	C903D
2038	BF194A	2038	C1674	2038	C509	2038	C912
2038	BF223	2038	C1674K	2038	C509(TRANSISTOR)	2038	C917
2038	BF224	2038	C1674L	2038	C509Y	2038	C917K
2038	BF225	2038	C1674M	2038	C535	2038	C920
2038	BF233-2	2038	C1675	2038	C5359	2038	C920Q
2038	BF233-3	2038	C1675K	2038	C535A	2038	C920R
2038	BF233-4	2038	C1675L	2038	C535B	2038	C924
2038	BF233-5	2038	C1675M	2038	C535C	2038	C924E
2038	BF234	2038	C1686	2038	C535G	2038	C924F
2038	BF235	2038	C17	2038	C544	2038	C924M
2038	BF236	2038	C170	2038	C544C	2038	C929
2038	BF237	2038	C172	2038	C544D	2038	C929-0
2038	BF238	2038	C172A	2038	C544E	2038	C929B
2038	BF240B	2038	C17A	2038	C561	2038	C929C
2038	BF241C	2038	C185V	2038	C568	2038	C929C1
2038	BF241D	2038	C199	2038	C590	2038	C929D
2038	BF254	2038	C206	2038	C590Y	2038	C929D1
2038	BF255	2038	C2475078-3	2038	C611	2038	C929DE
2038	BF262	2038	C2485078-1	2038	C611(TRANSISTOR)	2038	C929DF
2038	BF263	2038	C2485079-1	2038	C612	2038	C929DU
2038	BF264	2038	C263	2038	C613	2038	C929DV
2038	BF357	2038	C269	2038	C63(JAPAN)	2038	C929E
2038	BF364	2038	C271	2038	C641	2038	C929ED
2038	BF365	2038	C272	2038	C641B	2038	C929P
2038	BF494	2038	C282(TRANSISTOR)	2038	C649	2038	C929FK
2038	BF495	2038	C286(TRANSISTOR)	2038	C658	2038	C930
2038	BF594	2038	C289	2038	C658A	2038	C930B
2038	BF595	2038	C296	2038	C659	2038	C930BK
2038	BFR28	2038	C3123	2038	C662	2038	C930BV
2038	BF817	2038	C313	2038	C668	2038	C930C
2038	BF817R	2038	C316	2038	C668-0	2038	C930CL
2038	BF858	2038	C318	2038	C668A	2038	C930CS
2038	BF862	2038	C318A	2038	C668B	2038	C930D
2038	BFW83	2038	C370	2038	C668B1	2038	C930DS
2038	BFW83A	2038	C370P	2038	C668BC2	2038	C930DT
2038	BFW85D	2038	C370Q	2038	C668C	2038	C930DT-2
2038	BFW85E	2038	C370H	2038	C668C1	2038	C930DX
2038	BFW85F	2038	C370J	2038	C668CD	2038	C930E
2038	BFW85G	2038	C370K	2038	C668D	2038	C930EP
2038	BFW64	2038	C380	2038	C668D1	2038	C930EW
2038	BFX18	2038	C380(TRANSISTOR)	2038	C668DE	2038	C930EX
2038	BFX19	2038	C380-0	2038	C668DQ	2038	C930F
2038	BFX20	2038	C380-0	2038	C668DV	2038	C939D
2038	BFX21	2038	C380-0/4454C	2038	C668DX	2038	C9426
2038	BFX43	2038	C380A	2038	C668DZ	2038	C9634
2038	BFX44	2038	C380A(0)	2038	C668E	2038	C98
2038	BFX45	2038	C380A-0(TV)	2038	C668E1	2038	C99
2038	BFY-37	2038	C380A-R	2038	C668E2	2038	CC82006D
2038	BFY-47	2038	C380A-R(TV)	2038	C668EP	2038	CC82008P015
2038	BFY-48	2038	C380A0	2038	C668EV	2038	CC89016D
2038	BFY78	2038	C380ATV	2038	C668EX	2038	CC89016E
2038	BFY90	2038	C380AY	2038	C668F	2038	CC89018F
2038	BFY90B	2038	C380D	2038	C684	2038	CDC12112C
2038	B8826	2038	C380R	2038	C684(JAPAN)	2038	CDC12112D
2038	BSY35	2038	C380R/4454C	2038	C684A	2038	CDC12112E
2038	BSY35B	2038	C380Y	2038	C684A(JAPAN)	2038	CDC12112F
2038	BSY35C	2038	C384	2038	C684B	2038	CDC5000-1B
2038	BSY35D	2038	C384-0	2038	C684BK	2038	CDC5038A
2038	BSY52	2038	C384Y	2038	C684F	2038	CDC5071A
2038	BSY52R	2038	C386	2038	C688	2038	CDC5075B
2038	BSX12	2038	C386A-0(TV)	2038	C722	2038	CE4001C
2038	BSX26	2038	C387	2038	C735	2038	CE4008B
2038	BSX27	2038	C387A	2038	C735(PA-3)	2038	CE4008C
2038	BSX28	2038	C387A(PA-3)	2038	C735(0)	2038	CE4010D
2038	BSX35	2038	C387Q	2038	C735-0	2038	CE4010E
2038	BSX87A	2038	C391(TRANSISTOR)	2038	C735B	2038	CF1
2038	BSX88	2038	C392	2038	C735P	2038	CIL511
2038	BSX88A	2038	C394(TRANSISTOR)	2038	C735PA3	2038	CIL512
2038	BSX92	2038	C394-0	2038	C735GR	2038	CIL513
2038	BSX93	2038	C394-0	2038	C735H	2038	CIL521
2038	BSY-62	2038	C394GR	2038	C735J	2038	CIL522
2038	BSY-72	2038	C394R	2038	C735K	2038	CIL523
2038	BSY-73	2038	C394W	2038	C735Y	2038	CIL531
2038	BSY-74	2038	C394Y	2038	C74	2038	CIL532
2038	BSY-80	2038	C397(TRANSISTOR)	2038	C740	2038	CIL533
2038	BSY-95	2038	C39A	2038	C748	2038	CK476
2038	BSY22	2038	C40	2038	C763	2038	CK477
2038	BSY23	2038	C405	2038	C763B	2038	CS-2004C
2038	BSY70	2038	C406	2038	C763C	2038	CS-2007O
2038	BSY72	2038	C424	2038	C763CD	2038	CS-2007H
2038	BT930	2038	C429	2038	C763D	2038	CS-2008P
2038	C1023	2038	C429J	2038	C772	2038	CS-3001B
2038	C1047	2038	C429X	2038	C772B	2038	CS-461B
2038	C1047B	2038	C430	2038	C772BG	2038	CS-6227G
2038	C1047C	2038	C430H	2038	C772BH	2038	CS-9016
2038	C1047D	2038	C430W	2038	C772BV	2038	CS-9016P
2038	C1047E	2038	C454	2038	C772BX	2038	CS-9018
2038	C111B	2038	C455	2038	C772BY	2038	CS-9018E
2038	C1158	2038	C458	2038	C7721	2038	CS-9018F
2038	C1187	2038	C458(C)	2038	C772C1	2038	CS-9018G
2038	C1205	2038	C458A	2038	C772C2	2038	CS-9018H
2038	C1205A	2038	C458AD	2038	C772CK	2038	CS1014
2038	C1205B	2038	C458B	2038	C772CL	2038	CS1014D
2038	C1205C	2038	C458B-D	2038	C772CS	2038	CS1014E
2038	C1213	2038	C458BC	2038	C772CU	2038	CS1014F
2038	C1213A	2038	C458BK	2038	C772CV	2038	CS1018
2038	C1213AB	2038	C458BL	2038	C772CX	2038	CS1168E
2038	C1213AC	2038	C458BLG	2038	C772D	2038	CS1225D
2038	C1213AD	2038	C458C	2038	C772DJ	2038	CS1225E
2038	C1213AK	2038	C458CLG	2038	C772DU	2038	CS1225F
2038	C1213AKA	2038	C458CM	2038	C772DV	2038	CS1226
2038	C1213AKB	2038	C458D	2038		2038	CS1226E
2038		2038	C458Q	2038		2038	

RS 276-	Industry Standard No.	RS 276-	Industry Standard No.	RS 276-	Industry Standard No.	RS 276-	Industry Standard No.
2038	CS1226F	2038	EA15X393	2038	FC89016D	2038	HT3037201A
2038	CS1226G	2038	EA15X396	2038	FC89018D	2038	HT3037201B
2038	CS1226H	2038	EA15X415	2038	FC89018E	2038	HT3037201A
2038	CS1227	2038	EA15X48	2038	FC89018F	2038	HT303941
2038	CS1227D	2038	EA15X49	2038	FC89018G	2038	HT303941A
2038	CS1227E	2038	EA15X50	2038	FC89018H	2038	HT303941B
2038	CS1227F	2038	EA15X51	2038	FC89066	2038	HT30454OA0
2038	CS1227G	2038	EA15X54	2038	FC89166P	2038	HT30460100
2038	CS1238	2038	EA15X55	2038	FJ1033	2038	HT304611B
2038	CS1238G	2038	EA15X7141	2038	FK2369A	2038	HT305351B0
2038	CS1238H	2038	EA15X7177	2038	FK2484	2038	HT305551C0
2038	CS1238I	2038	EA15X7228	2038	FK3014	2038	HT306451
2038	CS1243E	2038	EA15X7233	2038	FK3299	2038	HT306451B
2038	CS1243H	2038	EA15X7234	2038	FK3300	2038	HT306681C
2038	CS1244H	2038	EA15X7235	2038	FK914	2038	HT306962A-0
2038	CS1244J	2038	EA15X7236	2038	FK918	2038	HT308291B0
2038	CS1244X	2038	EA15X7243	2038	FM1613	2038	HT308291D0
2038	CS1252B	2038	EA15X7244	2038	FM1711	2038	HT309291C
2038	CS1252C	2038	EA15X7263	2038	FM2368	2038	HT312131C0
2038	CS1293	2038	EA15X7587	2038	FM2369	2038	HT313171R
2038	CS1330A	2038	EA15X7722	2038	FM2846	2038	HT313172A
2038	CS1330B	2038	EA15X8589	2038	FM3014	2038	HT313592B
2038	CS1330C	2038	EA15X8601	2038	FM708	2038	HT316751M0
2038	CS1340D	2038	EA15X8608	2038	FM709	2038	HX50001
2038	CS1340E	2038	EA15X8609	2038	FM720A	2038	I/O,
2038	CS1340F	2038	EA15X8610	2038	FM870	2038	I9623
2038	CS1340G	2038	EA15X94	2038	FM871	2038	IC743040
2038	CS1340H	2038	EA1733	2038	FM910	2038	IC743041
2038	CS1350	2038	EA1793	2038	FM911	2038	IC743044
2038	CS1351	2038	EA2493	2038	FM914	2038	IP20-0040
2038	CS1359	2038	EA2495	2038	FPR40-1003	2038	IP20-006
2038	CS1360	2038	EA2600	2038	FPR50-1003	2038	IP20-0110
2038	CS1361E	2038	EA2601	2038	FPR50-1004	2038	IP20-0191
2038	CS1361F	2038	EA2602	2038	F83266	2038	IP20-0251
2038	CS1386H	2038	EA2603	2038	F8326690	2038	IRTR-70
2038	CS1508G	2038	EA2604	2038	F855529	2038	IRTR80
2038	CS1509B	2038	EA2605	2038	F85683	2038	IRTR83
2038	CS1509F	2038	EA3406	2038	F881001	2038	IRTR95
2038	CS1518E	2038	ECG108	2038	F8E3001	2038	J108
2038	CS1555	2038	ECG229	2038	F8E5002	2038	J187
2038	CS1661	2038	ECG289	2038	FSP-1	2038	J24596
2038	CS1834	2038	ED-1502C	2038	FSP-164	2038	J24635
2038	CS184J	2038	ED1502	2038	FSP-165	2038	J24636
2038	CS2006F	2038	ED1502A	2038	FSP-166	2038	J24637
2038	CS2006G	2038	ED1502B	2038	FSP-166-1	2038	J24813
2038	CS2008	2038	ED1502C	2038	FSP-215	2038	J24814
2038	CS2008G	2038	ED1502D	2038	FSP-242-1	2038	J24844
2038	CS2008H	2038	ED1502E	2038	FSP-270-1	2038	J24852
2038	CS2008H552	2038	ED1502F	2038	FSP-289-1	2038	J24863
2038	CS429J	2038	ED592K	2038	FSP-42	2038	J24904
2038	CS430H	2038	ED592M	2038	FSP-42-1	2038	J24905
2038	CS461F	2038	EL403	2038	FT1315	2038	J24915
2038	CS469F	2038	EL75	2038	FT1324B	2038	J24921
2038	CS6225F	2038	EN3011	2038	FT1324C	2038	J24923
2038	CS6226F	2038	EN718A	2038	FT709	2038	J24933
2038	CS6226T E	2038	EN744	2038	FV2369A	2038	J8P7001
2038	CS6227F	2038	EN914	2038	FV3014	2038	J8P7005
2038	CS9001	2038	EN916	2038	FV3299	2038	J8P7006
2038	CS9016/3490	2038	EN918	2038	FV3300	2038	K14-0066-6
2038	CS9016E	2038	EP15X41	2038	FV914	2038	K2109
2038	CS9016F	2038	EP15X42	2038	FV918	2038	K2110
2038	CS9016G	2038	EP15X55	2038	FX3013	2038	K2111
2038	CS9016G/3490	2038	EP15X6	2038	FX3014	2038	K2112
2038	CS9016H	2038	EQS-0018	2038	FX914	2038	K2113
2038	CS9018	2038	EQS-0100	2038	FX918	2038	K2114
2038	CS9018/3490	2038	EQS-0198	2038	G05-004A	2038	K2115
2038	CS9018D	2038	EQS-159	2038	G05-037-D	2038	K2116
2038	CS9018E	2038	EQS-21	2038	G05-055-D	2038	K2117
2038	CS9018F	2038	ES10187	2038	G05-063-R	2038	K2118
2038	CS9018F/3490	2038	ES15X10	2038	G05-065-A	2038	K3880P
2038	CS9018G	2038	ES15X18	2038	G05-066A	2038	K4-510
2038	CS9021G-I	2038	ES15X19	2038	G05004A	2038	KD2119
2038	CS9124-G2	2038	ES15X2	2038	G9423	2038	KD5000
2038	CS9124B1	2038	ES15X3	2038	G9625	2038	K0825671-20
2038	CS9125-B1	2038	ES15X30	2038	GE-214	2038	K0825671-21
2038	CS918	2038	ES15X6	2038	GE-268	2038	K0825671-23
2038	CT1300	2038	ES15X60	2038	GE-327	2038	K8C11870
2038	CT1500	2038	ES15X69	2038	GE-61	2038	K8C11877R
2038	CXL108	2038	ES1X122	2038	GE-86	2038	K8C16740
2038	D009	2038	ES3266	2038	GET3563	2038	K8C16774R
2038	D058	2038	ET15X18	2038	GMO580	2038	L842
2038	D069	2038	ET15X19	2038	GME.022	2038	LM1110A
2038	D072	2038	ET15X21	2038	GME3001	2038	LM1123H
2038	D087	2038	ET15X23	2038	GME3002	2038	LM1138
2038	D088	2038	ET15X30	2038	GME9001	2038	LM1138E
2038	D133(CHAN.MASTER)	2038	ET15X7	2038	GME9002	2038	LM1138E/F
2038	D141(CHAN.MASTER)	2038	ET15X9	2038	GMB9021	2038	LM1138F
2038	D1666	2038	ET234854	2038	H1V	2038	LM1138G
2038	D24A3394	2038	ET32321B	2038	H442	2038	LM1138G/F
2038	D26E-2	2038	ET352146	2038	H9623	2038	LM1138G/H
2038	D26B-3	2038	ETTC-930D	2038	H9625	2038	LM1138H
2038	D26B-6	2038	EU15X2	2038	HC-00372	2038	LM1138H/I
2038	D471	2038	EU15X3	2038	HC-00394	2038	LM1138I
2038	D471L	2038	EU15X6	2038	HC-00668	2038	LT1016
2038	D562	2038	EW162	2038	HC-00772	2038	LT1016D
2038	D71L	2038	F15810	2038	HC-00839	2038	LT1016E
2038	D9634	2038	F15835	2038	HC-01830	2038	LT1016H
2038	DBCZ039404	2038	F15841	2038	HC009930	2038	LT1016I
2038	DBCZ073503	2038	F20-1001	2038	HC206	2038	LTE1016
2038	DBCZ083905	2038	F20-1002	2038	HC645	2038	LTE1016(G.E.)
2038	DDBY216002	2038	F20-1003	2038	HC668	2038	LTH1016
2038	DDBY219001	2038	F20-1004	2038	HC772	2038	LTH1016(G.E.)
2038	DDBY259001	2038	F20-1005	2038	HC829	2038	M012
2038	DDBY259002	2038	F215-1001	2038	HEP56	2038	M024
2038	DDBY261002	2038	F215-1002	2038	HEP718	2038	M4709
2038	DDBY287001	2038	F215-1003	2038	HEP720	2038	M4733
2038	DS-81	2038	F215-1004	2038	HEP722	2038	M4820
2038	DX1018	2038	F215-1005	2038	HEP723	2038	M4826
2038	E2434	2038	F2427	2038	HEP727	2038	M4857
2038	E2435	2038	F2450	2038	HEP80008	2038	M4845
2038	E629	2038	F2633	2038	HEP80010	2038	M4855
2038	E842	2038	F2634	2038	HEP80016	2038	M4857
2038	E843	2038	F2656	2038	HEP80017	2038	M75547-1
2038	E844	2038	F3530	2038	HEP80020	2038	M75547-2
2038	E9625	2038	F4706	2038	HEP80021	2038	M9010
2038	EA0013	2038	F9623(G.E.)	2038	HEP80033	2038	M9389
2038	EA0091	2038	F9625	2038	HR58	2038	M9482
2038	EA0093	2038	F96N	2038	HR59	2038	M9575
2038	EA0094	2038	FC5006	2038	HR60	2038	MA4101
2038	EA0095	2038	FC81168E	2038	HR77	2038	MA4103
2038	EA1343	2038	FC81168B641	2038	HR78	2038	MA4104
2038	EA1562	2038	FC81225E	2038	HR87	2038	ME1108
2038	EA1563	2038	FC81227E	2038	HS-1225	2038	MB3001
2038	EA15X113	2038	FC81227FB814	2038	HS-1226	2038	MB3002
2038	EA15X130	2038	FC81227F	2038	HS-1227	2038	MB3011
2038	EA15X131	2038	FC81227FP743	2038	HS-40014	2038	MB5001
2038	EA15X132	2038	FC81227G	2038	HS-40019	2038	MB8201
2038	EA15X180	2038	FC81227G810	2038	HS-40045	2038	MB9003
2038	EA15X256	2038	FC89011E	2038	HS-40047	2038	MB9021
2038	EA15X267	2038	FC89011H	2038	HS40049	2038	MB9022
2038	EA15X350	2038	FC89014	2038	HT3037010	2038	MHM1001
2038	EA15X360	2038	FC89014B	2038	HT303711A-0	2038	MHM1101
2038	EA15X374	2038	FC89014D	2038	HT3037201	2038	MI1546
2038	EA15X376					2038	MM1367/28C684

RS 276-	Industry Standard No.	RS 276-	Industry Standard No.	RS 276-	Industry Standard No.	RS 276-	Industry Standard No.
2038	MM1382	2038	R-28C668	2038	RT6158	2038	SE3001R
2038	MM1387	2038	R-3530-1	2038	RT6159	2038	SE3001Y
2038	MM1803	2038	R06-1001	2038	RT6160	2038	SE3002
2038	MM1941	2038	R06-1002	2038	RT6201	2038	SE3003
2038	MM1945	2038	R06-1003	2038	RT6202	2038	SE50
2038	MM709	2038	R06-1004	2038	RT6203	2038	SE5006
2038	MMCM918	2038	R06-1005	2038	RT6600	2038	SE5009
2038	MMT918	2038	R06-1006	2038	RT6733	2038	SE5010
2038	MPS-H17	2038	R2473	2038	RT6737	2038	SE5015
2038	MPS3536	2038	R2476	2038	RT5991	2038	SE5020
2038	MPS3563	2038	R2477	2038	RT7320	2038	SE5029
2038	MPS3693	2038	R62194	2038	RT7321	2038	SE5030
2038	MPS3694	2038	R9600	2038	RT7323	2038	SE5031
2038	MPS6507	2038	RB1001	2038	RT7324	2038	SE5035
2038	MPS6511	2038	RB1002	2038	RT8193	2038	SE5036
2038	MPS6511-8	2038	REN108	2038	RT8333	2038	SE504
2038	MPS6512	2038	REN229	2038	RT8527	2038	SE5040
2038	MPS6513	2038	REN289	2038	RT930H	2038	SE5056
2038	MPS6515	2038	RR8070	2038	RV1068	2038	SE521
2038	MPS6539	2038	RR8116	2038	RV1467	2038	SP115
2038	MPS6541	2038	RR8118	2038	RV1468	2038	SP115A
2038	MPS6542	2038	RR8119	2038	RV1469	2038	SP115B
2038	MPS6543	2038	RR8989	2038	RV1470	2038	SP115C
2038	MPS6546	2038	RR8999	2038	RV8280645	2038	SP115D
2038	MPS6547	2038	RS-109	2038	RV8C81384	2038	SP115B
2038	MPS6548	2038	RS-2015	2038	S-1037	2038	8GB-9742
2038	MPS6568A	2038	RS-1037	2038	S-1041	2038	SK-31024-3
2038	MPS8706	2038	RS-7102	2038	S-1058	2038	SK1320
2038	MPS8706A	2038	RS-7104	2038	S-1059	2038	SK3018
2038	MPS7513	2038	RS-7106	2038	S-1060	2038	SK3019
2038	MPS805	2038	RS-7107	2038	S-1062	2038	SK3039
2038	MPS89418	2038	RS-7108	2038	S-1078	2038	SK3117
2038	MPS894231	2038	RS-7109	2038	S-1079	2038	SK7181
2038	MPS89423F	2038	RS-7110	2038	S-1155	2038	SKA-4074
2038	MPS894230	2038	RS-7113	2038	S-1227	2038	SKA-4076
2038	MPS89423H	2038	RS-7114	2038	S-1276	2038	SKA-4590
2038	MPS89423I	2038	RS-7115	2038	S-1286	2038	SKA-5248
2038	MPS89426	2038	RS-7202	2038	S-1296	2038	SKA-5886
2038	MPS89426B	2038	RS-7512	2038	S-1316	2038	SKA4074
2038	MPS89426BC	2038	RS-805UB	2038	S-1317	2038	SKA4075
2038	MPS89426C	2038	RS1726	2038	S-1318	2038	SKA4076
2038	MPS89427	2038	RS6523	2038	S-1360	2038	SKA4525
2038	MPS89427B	2038	RS7101	2038	S-1361	2038	SKA5248
2038	MPS89600G/H	2038	RS7102	2038	S-1362	2038	SKA5886
2038	MPS89600H	2038	RS7104	2038	S-1408	2038	SKA9015
2038	MPS89601	2038	RS7106	2038	S-1409	2038	SKA9096
2038	MPS89604	2038	RS7107	2038	S-5328E	2038	SL-100
2038	MPS89625	2038	RS7109	2038	S-5670-E	2038	SL100
2038	MPS89625C	2038	RS7110	2038	S-95125	2038	SL403
2038	MPS89625D	2038	RS7112	2038	S-95125A	2038	SM4304-8
2038	MPS89625E	2038	RS7113	2038	S-95126	2038	SM5796
2038	MPS89625F	2038	RS7114	2038	S-95126A	2038	SP4168
2038	MPS89625G	2038	RS7115	2038	S0016	2038	SPP530
2038	MPS89625H	2038	RS7116	2038	S0020	2038	SPS-1351
2038	MPS89696G	2038	RS7117	2038	S0021	2038	SPS-1352
2038	MPSH10	2038	RS7118	2038	S1009	2038	SPS-1353
2038	MPSH11	2038	RS7119	2038	S1019	2038	SPS-1473
2038	MPSH19	2038	RS7120	2038	S1037	2038	SPS-2111
2038	MPSH30	2038	RS7122	2038	S1044	2038	SPS-2520
2038	MPSH31	2038	RS7123	2038	S1058	2038	SPS-4145
2038	MPSH37	2038	RS7124	2038	S1059	2038	SPS-4423
2038	MRP501	2038	RS7125	2038	S1062	2038	SPS-917
2038	MS701T	2038	RS7126	2038	S1076	2038	SPS1351
2038	MS7501B	2038	RS7128	2038	S1078	2038	SPS1352
2038	MS7501T	2038	RS7135	2038	S1079	2038	SPS1353
2038	MS7502B	2038	RS7138	2038	S1142	2038	SPS1846
2038	MS7502T	2038	RS7139	2038	S1153	2038	SPS820
2038	MSR7502	2038	RS7140	2038	S1227	2038	SPS2110
2038	MS87501	2038	RS7141	2038	S1276	2038	SPS2111
2038	MS87502	2038	RS7142	2038	S1296	2038	SPS2135
2038	MST7501	2038	RS7144	2038	S1313	2038	SPS220
2038	MT100	2038	RS7145	2038	S1316	2038	SPS2224
2038	MT101	2038	RS7161	2038	S1317	2038	SPS2265
2038	MT102	2038	RS7162	2038	S1318	2038	SPS2265-2
2038	MT106	2038	RS7163	2038	S1360	2038	SPS2266
2038	MT107	2038	RS7164	2038	S1362	2038	SPS2320
2038	MT3011	2038	RS7166	2038	S1408	2038	SPS2664
2038	MT743	2038	RS7167	2038	S1409	2038	SPS3003
2038	MT744	2038	RS7168	2038	S1636	2038	SPS3370
2038	MT753	2038	RS7169	2038	S1674	2038	SPS838
2038	MT9003	2038	RS7170	2038	S1674A	2038	SPS3929
2038	N-EA15X131	2038	RS7173	2038	S1682	2038	SPS3937
2038	N-EA15X132	2038	RS7174	2038	S2159	2038	SPS3948
2038	NPC173	2038	RS7175	2038	0S22.3640-080	2038	SPS3952
2038	NPC188	2038	RS7176	2038	S2224	2038	SPS3968
2038	NS1356	2038	RS7177	2038	S2438	2038	SPS3971
2038	NS381	2038	RS7201	2038	S25805	2038	SPS84
2038	NS3300	2038	RS7202	2038	S2617	2038	SPS840
2038	NS382	2038	RS7209	2038	S2716	2038	SPS84002
2038	NS6112	2038	RS7210	2038	S2719	2038	SPS84005
2038	NS6113	2038	RS7211	2038	S27893	2038	SPS84008
2038	NS7261	2038	RS7212	2038	S5019	2038	SPS84016
2038	NS7267	2038	RS7214	2038	S32020	2038	SPS84030
2038	NS9710	2038	RS7215	2038	S32417	2038	SPS84043
2038	NTC-4	2038	RS7216	2038	S32669	2038	SPS84050
2038	NTC-5	2038	RS7217	2038	S326690	2038	SPS84051
2038	P346	2038	RS7218	2038	S33990	2038	SPS84068
2038	PA-10556	2038	RS7219	2038	S35233	2038	SPS84079
2038	PB0182	2038	RS7220	2038	S36951	2038	SPS84080
2038	PC-20066	2038	RS7221	2038	S5327E	2038	SPS84091
2038	PEP1001	2038	RS7225	2038	S5328E	2038	SPS84145
2038	PET-101-1	2038	RS7227	2038	S5670E	2038	SPS84167
2038	PET1075	2038	RS7228	2038	S95125	2038	SPS84428
2038	PET3001	2038	RS7229	2038	S95125A	2038	SPS84351
2038	PET8201	2038	RS7230	2038	S95126	2038	SPS84399
2038	PET8250	2038	RS7231	2038	S95126A	2038	SPS84423
2038	PET8251	2038	RS7233	2038	SATC0E8339	2038	SPS84436
2038	PET8300	2038	RS7237	2038	SC1227P	2038	SPS84610
2038	PL-176-031-9-001	2038	RS7333	2038	SC1227G	2038	SPS6155
2038	PL-176-042-9-001	2038	RS7334	2038	SCA3021	2038	SPS915
2038	PL-176-047-9-001	2038	RS7511	2038	SE-1001	2038	SPS917
2038	PL1051	2038	RS7512	2038	SE-1019	2038	SPS920
2038	PL1053	2038	RS7520	2038	SE-1044	2038	SR130-1
2038	PL1055	2038	RS7522	2038	SE-1419	2038	SR20226
2038	PL1061	2038	RS7524	2038	SE-3001	2038	ST10
2038	PL1062	2038	RS7532	2038	SE-3002	2038	ST1026
2038	PL1063	2038	RS7533	2038	SE-3005	2038	ST1050
2038	PL1064	2038	RT2915	2038	SE-3019	2038	ST1051
2038	PL1065	2038	RT3069	2038	SE-5001	2038	ST12
2038	PL1081	2038	RT3070	2038	SE-5002	2038	ST13
2038	PL1082	2038	RT3095	2038	SE-5003	2038	ST1336
2038	PM195A	2038	RT3225	2038	SE-5006-14	2038	ST14
2038	PMT1767	2038	RT3226	2038	SE-5020	2038	ST15
2038	Q-00284R	2038	RT3227	2038	SE-5021	2038	ST1694
2038	Q-00284R-3	2038	RT3232	2038	SE-5050	2038	ST29
2038	Q-00384R-3	2038	RT5061	2038	SE0001	2038	ST30
2038	Q-00484R-1	2038	RT5200	2038	SE1001	2038	ST3030
2038	Q-00584R-3	2038	RT5201	2038	SE1001-1	2038	ST3031
2038	Q-00784R	2038	RT5205	2038	SE1001-2	2038	ST31
2038	QBE3001	2038	RT5900	2038	SE1019	2038	ST32
2038	QT-00829XBN	2038	RT5901	2038	SE1044	2038	ST33
2038	QT-00839XDA	2038	RT5902	2038	SE1419	2038	ST34
2038	QT-00460CBB	2038	RT5903			2038	ST35
2038	R-280545	2038	RT5904			2038	ST40
		2038	RT6157				

RS 276-	Industry Standard No.	RS 276-	Industry Standard No.	RS 276-	Industry Standard No.	RS 276-	Industry Standard No.
2038	ST41	2038	TNJ60066	2038	TV828C828Q	2038	2N1505
2038	ST415	2038	TNJ61729	2038	UPI706	2038	2N1506
2038	ST42	2038	TNJ70478	2038	UPI706A	2038	2N1506A
2038	ST43	2038	TNJ70478-1	2038	UPI706B	2038	2N1507
2038	ST44	2038	TNJ70479	2038	V120PH	2038	2N1941
2038	ST45	2038	TNJ70480	2038	V220	2038	2N1989
2038	ST47025	2038	TNJ70484	2038	V221	2038	2N2032
2038	ST60	2038	TNJ71173	2038	V222	2038	2N2193A
2038	ST61	2038	TNJ71498	2038	V405	2038	2N2193B
2038	ST6120	2038	TN7-839	2038	V415	2038	2N2195
2038	ST62	2038	TN7-840	2038	V417	2038	2N2195B
2038	ST70	2038	TN7-841	2038	V435	2038	2N2197
2038	ST71	2038	TN7839	2038	V8-28C206	2038	2N2475
2038	ST72	2038	TN7840	2038	V8-28C208	2038	2N2475/46
2038	ST80	2038	TN7841	2038	V8-28C288A	2038	2N2475/51
2038	ST82	2038	TN7843	2038	V8-28C324	2038	2N2489A
2038	ST9	2038	TP4275	2038	V8-28C371-R-1	2038	2N2615
2038	ST903	2038	TQ1	2038	V8-28C466	2038	2N2616
2038	ST904	2038	TQ2	2038	V8-28C563	2038	2N2651
2038	ST904A	2038	TQ3	2038	V8-28C645	2038	2N2655
2038	ST905	2038	TQ5	2038	V8-28C645A	2038	2N2708
2038	ST910	2038	TQ5049	2038	V8-28C645B	2038	2N2710
2038	STE400	2038	TQ6	2038	V8-28C645C	2038	2N2711
2038	SX-3825	2038	TQ7	2038	V8-28C683Y	2038	2N2712
2038	SX3825	2038	TQ8	2038	V8-28C684	2038	2N2712BLUE
2038	SX3826	2038	TQ9	2038	V8-28C762	2038	2N2715
2038	SX3827	2038	TR-016	2038	V828C1213-C/1A	2038	2N2716
2038	SX408	2038	TR-01B	2038	V828C1213AC/1A	2038	2N2729
2038	SYL-2300	2038	TR-163(OLSON)	2038	V828C16742/-1	2038	2N2784
2038	SYL-4131	2038	TR-1831	2038	V828C16741/-1	2038	2N2784/52
2038	SYL4131	2038	TR-1835	2038	V828C1675M-1	2038	2N2883
2038	T-483	2038	TR-24	2038	V828C1855//-1	2038	2N2884
2038	T-484	2038	TR-2831	2038	V828C394-0-1	2038	2N2921
2038	T-486	2038	TR-2833	2038	V828C394-Y-1	2038	2N2922
2038	T-H28C313	2038	TR-2835	2038	V828C454-B/1E	2038	2N3010
2038	T-H28C387	2038	TR-28C371	2038	V828C454-C/1A	2038	2N3035
2038	T-Q5055	2038	TR-28C372	2038	V828C454-C/1E	2038	2N3227
2038	T1003-521	2038	TR-28C384	2038	V828C454-C/3A	2038	2N33682
2038	T1003521	2038	TR-3831	2038	V828C458-C/1E	2038	2N337A
2038	T1408	2038	TR-3R33	2038	V828C460B-1	2038	2N3407
2038	T18-18	2038	TR-3R55	2038	V828C735-Y-1	2038	2N3508
2038	T1894	2038	TR-3R03	2038	V828C784-R1P	2038	2N3509
2038	T1898	2038	TR-4R35	2038	V828D471-8/-1	2038	2N3511
2038	T1828	2038	TR-70	2038	W8	2038	2N3544
2038	T1XM15	2038	TR-80	2038	WEP371	2038	2N3562
2038	T1XM17	2038	TR-8004-4	2038	WEP380	2038	2N3563
2038	T2634	2038	TR-8004-5	2038	WEP394	2038	2N3563-1
2038	T508	2038	TR-8028	2038	WEP460	2038	2N3570
2038	T3550	2038	TR-8029	2038	WEP535	2038	2N3571
2038	T3535	2038	TR-8030	2038	WEP709	2038	2N3572
2038	T3536	2038	TR-8031	2038	WEP719	2038	2N3600
2038	T3539	2038	TR-8032	2038	WEP720	2038	2N3633
2038	T3568(RCA)	2038	TR-8038	2038	WEP772	2038	2N3648
2038	T586(SEARS)	2038	TR-83	2038	WEP828	2038	2N3683
2038	T9011CD	2038	TR01026	2038	WEP910	2038	2N3831
2038	T9011EF	2038	TR010602-1	2038	WEP956	2038	2N3852
2038	T9011G	2038	TR0573486	2038	X32P5660	2038	2N3854
2038	T9011EP	2038	TR0573507	2038	X81	2038	2N3854A
2038	T9011GH	2038	TR112	2038	X814	2038	2N3953
2038	T9011H	2038	TR1512-80	2038	X815	2038	2N3983
2038	T9011HEF	2038	TR21C	2038	X82	2038	2N3984
2038	T9011I	2038	TR228735045311	2038	X83	2038	2N3985
2038	T9011J	2038	TR228735048617	2038	X84	2038	2N4072
2038	T9016P	2038	TR228735048618	2038	X86	2038	2N4073
2038	T9016H	2038	TR22C	2038	X215X3	2038	2N4134
2038	T9423	2038	TR24	2038	ZA100962E	2038	2N4251
2038	TA-7	2038	TR28C1342	2038	ZDT-30	2038	2N4252
2038	TA2401	2038	TR28C371	2038	ZDT-31	2038	2N4253
2038	TA2503	2038	TR28C372	2038	ZDT10	2038	2N4292
2038	TA7303	2038	TR28C384	2038	ZDT11	2038	2N4293
2038	TA7319	2038	TR28C535	2038	ZDT20	2038	2N4418
2038	TC-0918	2038	TR310230	2038	ZDT21	2038	2N4419
2038	TC0914	2038	TR310244	2038	ZDT30	2038	2N4873
2038	TC0918	2038	TR310249	2038	ZDT31	2038	2N4934
2038	TC2369A	2038	TR310250	2038	ZEN104	2038	2N4935
2038	TC2483	2038	TR38	2038	ZEN108	2038	2N4936
2038	TC2484	2038	TR5320326	2038	ZEN109	2038	2N4996
2038	TE2484	2038	TR70	2038	ZEN117	2038	2N4997
2038	TE2715	2038	TR80	2038	ZEN118	2038	2N5024
2038	TE2716	2038	TR8004	2038	ZEN121	2038	2N5053
2038	TE3707	2038	TR8042	2038	Z140	2038	2N5054
2038	TE3708	2038	TR83	2038	Z8C535B	2038	2N5058B
2038	TE3709	2038	T89013	2038	ZT2368	2038	2N5059B
2038	TE3710	2038	T8C614	2038	ZT2369	2038	2N5131
2038	TE3711	2038	TT-204	2038	ZT2369A	2038	2N5180
2038	TE706	2038	TT-204A	2038	ZT2475	2038	2N5200
2038	TI-410	2038	TT-204AB	2038	ZT2708	2038	2N5220
2038	TI-417	2038	TT-204B	2038	ZT2857	2038	2N5292
2038	TI-420	2038	TT-204C	2038	ZT2938	2038	2N5304
2038	TI-430	2038	TT204	2038	ZT3269A	2038	2N5399
2038	TI-431	2038	TT204A	2038	ZT3600	2038	2N5487-1
2038	TI-474	2038	TT204AB	2038	ZT709	2038	2N5487-3
2038	TI-490	2038	TT204B	2038	ZT917	2038	2N5770
2038	TI-495	2038	TT204C	2038	ZT918	2038	2N5772
2038	TI25A	2038	TV-15	2038	ZTX490	2038	2N5830A
2038	TI25B	2038	TV-15B	2038	ZTX321	2038	2N6045
2038	TI3016	2038	TV-22	2038	ZTX3IU	2038	2N619
2038	TI407	2038	TV-32	2038	01-030380	2038	2N620
2038	TI408	2038	TV-48	2038	01-030394	2038	2N621
2038	TI409	2038	TV-7	2038	01-030735	2038	2N6218
2038	TI410	2038	TV1000	2038	01-030763	2038	2N6220
2038	TI431	2038	TV115	2038	01-030930	2038	2N6221
2038	TI8105	2038	TV16	2038	01-031047	2038	2N6222
2038	TI8125	2038	TV17A	2038	01-031317	2038	2N6232
2038	TI824	2038	TV18	2038	01-031359	2038	2N6232-4
2038	TI862	2038	TV2403	2038	01-031674	2038	2N6367
2038	TI863	2038	TV2404	2038	01-031675	2038	2N6370
2038	TI864	2038	TV24102	2038	01-031686	2038	2N6372
2038	TI884	2038	TV24148	2038	1-041/2207	2038	2N6373
2038	TI885	2038	TV24203	2038	01-117005	2038	2N702
2038	TI898A	2038	TV24204	2038	01-117006	2038	2N703
2038	TIX876	2038	TV24313	2038	01-30828	2038	2N706
2038	TIX880	2038	TV24387	2038	01-572814	2038	2N706A
2038	TIX809	2038	TV24573	2038	1-801-003-12	2038	2N706B
2038	TIX810	2038	TV24574	2038	1-801-003-13	2038	2N706C
2038	TIX828	2038	TV24589	2038	1-801-003-14	2038	2N707
2038	TIX829	2038	TV24684	2038	1-801-003-15	2038	2N708/51
2038	TIX830	2038	TV32	2038	1-801-305-13	2038	2N709
2038	TIX831	2038	TV55	2038	1-801-306	2038	2N709/52
2038	TM108	2038	TV8-1818	2038	1-801-306-13	2038	2N709A
2038	TM229	2038	TV8-28C185A	2038	1-801-306-15	2038	2N717
2038	TM289	2038	TV8-28C208	2038	01-9016-42221-3	2038	2N717A
2038	TMT-2427	2038	TV8-28C287A	2038	01-9018-62221-3	2038	2N718
2038	TMT2427	2038	TV8-28C446	2038	1B535A	2038	2N718A
2038	TMT696	2038	TV8-28C605	2038	1B535A/7825B	2038	2N743
2038	TMT697	2038	TV8-28C606	2038	002-009600	2038	2N743A
2038	TMT839	2038	TVS28C288A	2038	002-009601	2038	2N744
2038	TMT840	2038	TV8280466	2038	002-9601	2038	2N752
2038	TMT841	2038	TV828C538	2038	002-9601-12	2038	2N756
2038	TMT842	2038	TV828C645	2038	2M6219	2038	2N757
2038	TMT843	2038	TV828C645A	2038	2M6374	2038	2N758
2038	TNJ-60066	2038	TV828C645C	2038	2N1005		
		2038	TV828C683	2038	2N1060		
		2038	TV828C762	2038	2N1085		

RS 276-	Industry Standard No.	RS 276-	Industry Standard No.	RS 276-	Industry Standard No.	RS 276-	Industry Standard No.
2038	2N759	2038	28C1126A	2038	28C1342M	2038	28C1700R
2038	2N760	2038	28C1126B	2038	28C1420R	2038	28C170R
2038	2N761	2038	28C1126E	2038	28C1342R	2038	28C170X
2038	2N762	2038	28C1126F	2038	28C1342X	2038	28C170Y
2038	2N834	2038	28C1126GN	2038	28C1342Y	2038	28C171A
2038	2N841	2038	28C1126J	2038	28C1359	2038	28C171B
2038	2N843	2038	28C1126K	2038	28C1359(A)	2038	28C171C
2038	2N849	2038	28C1126L	2038	28C1359(B)	2038	28C171D
2038	2N850	2038	28C1126M	2038	28C1359(C)	2038	28C171F
2038	2N851	2038	28C1126OR	2038	28C1359(C,B)	2038	28C171G
2038	2N852	2038	28C1126R	2038	28C1359A	2038	28C171GN
2038	2N913	2038	28C1126X	2038	28C1359B	2038	28C171H
2038	2N914	2038	28C1126Y	2038	28C1359C	2038	28C171J
2038	2N914/51	2038	28C1198	2038	28C1417	2038	28C171K
2038	2N914A	2038	28C1182	2038	28C1417(W)	2038	28C171L
2038	2N915	2038	28C1187	2038	28C1417O	2038	28C171M
2038	2N915A	2038	28C1189L	2038	28C1417D	2038	28C171OR
2038	2N916	2038	28C1205	2038	28C1417D(U)	2038	28C171R
2038	2N916A	2038	28C1205A	2038	28C1417DU	2038	28C171X
2038	2N918	2038	28C1205B	2038	28C1417F	2038	28C171Y
2038	2N918/51	2038	28C1205C	2038	28C1417G	2038	28C172
2038	2N988	2038	28C1213	2038	28C1417H	2038	28C172A
2038	2N989	2038	28C1213(B)	2038	28C1417V	2038	28C172B
2038	28006	2038	28C1213-OR	2038	28C1417VP	2038	28C172C
2038	28512	2038	28C1213A	2038	28C1417W	2038	28C172D
2038	28C-4033	2038	28C1213A(C)	2038	28C148	2038	28C172E
2038	28C-P11	2038	28C1213AA	2038	28C148A	2038	28C172F
2038	28C-P11A	2038	28C1213AB	2038	28C148B	2038	28C172GN
2038	28C-P11B	2038	28C1213AC	2038	28C148D	2038	28C172H
2038	28C-P11C	2038	28C1213AD	2038	28C148E	2038	28C172J
2038	28C-P11D	2038	28C1213AK	2038	28C148G	2038	28C172K
2038	28C-P11E	2038	28C1213AKA	2038	28C148GN	2038	28C172L
2038	28C-P11F	2038	28C1213AKB	2038	28C148H	2038	28C172M
2038	28C-P11G	2038	28C1213AKC	2038	28C148J	2038	28C172OR
2038	28C-P11GN	2038	28C1213AKD	2038	28C148K	2038	28C172R
2038	28C-P11H	2038	28C1213B	2038	28C148L	2038	28C172T
2038	28C-P11J	2038	28C1213BC	2038	28C148M	2038	28C174B
2038	28C-P11K	2038	28C1213CD	2038	28C148OR	2038	28C174C
2038	28C-P11L	2038	28C1213D	2038	28C148R	2038	28C174D
2038	28C-P11M	2038	28C1213E	2038	28C148X	2038	28C174E
2038	28C-P11OR	2038	28C1213F	2038	28C148Y	2038	28C174F
2038	28C-P11R	2038	28C1213G	2038	28C155	2038	28C174G
2038	28C-P11X	2038	28C1213GN	2038	28C155A	2038	28C174GN
2038	28C-P11Y	2038	28C1213H	2038	28C155B	2038	28C174H
2038	28C-P14	2038	28C1213J	2038	28C155C	2038	28C174J
2038	28C-P14A	2038	28C1213K	2038	28C155D	2038	28C174K
2038	28C-P14B	2038	28C1213L	2038	28C155E	2038	28C174L
2038	28C-P14C	2038	28C1213M	2038	28C155F	2038	28C174M
2038	28C-P14D	2038	28C1213OR	2038	28C155GN	2038	28C174R
2038	28C-P14E	2038	28C1213X	2038	28C155H	2038	28C174X
2038	28C-P14F	2038	28C1213Y	2038	28C155K	2038	28C174Y
2038	28C-P14G	2038	28C1214	2038	28C155L	2038	28C177B
2038	28C-P14H	2038	28C1214(B)	2038	28C155M	2038	28C177S9
2038	28C-P14J	2038	28C1214A	2038	28C155OR	2038	28C17A
2038	28C-P14K	2038	28C1214B	2038	28C155R	2038	28C17B
2038	28C-P14L	2038	28C1214C	2038	28C155X	2038	28C17C
2038	28C-P14M	2038	28C1214D	2038	28C155Y	2038	28C17D
2038	28C-P14OR	2038	28C1216	2038	28C156	2038	28C17E
2038	28C-P14R	2038	28C127	2038	28C156A	2038	28C17F
2038	28C-P14X	2038	28C127A	2038	28C156B	2038	28C17G
2038	28C-P14Y	2038	28C127B	2038	28C156C	2038	28C17GN
2038	28C1009	2038	28C127C	2038	28C156D	2038	28C17H
2038	28C1026(G)	2038	28C127D	2038	28C156E	2038	28C17J
2038	28C1026-O	2038	28C127E	2038	28C156F	2038	28C17K
2038	28C1026-R	2038	28C127F	2038	28C156G	2038	28C17L
2038	0002SC1026B	2038	28C127G	2038	28C156GN	2038	28C17M
2038	28C1026BL	2038	28C127GN	2038	28C156H	2038	28C17OR
2038	28C1026D	2038	28C127H	2038	28C156J	2038	28C17R
2038	28C1026E	2038	28C127J	2038	28C156K	2038	28C17X
2038	28C1026F	2038	28C127K	2038	28C156L	2038	28C17Y
2038	28C1026GN	2038	28C127L	2038	28C156M	2038	28C1852
2038	28C1026GR	2038	28C127OR	2038	28C156OR	2038	28C1854C
2038	28C1026H	2038	28C127R	2038	28C156R	2038	28C1854S
2038	28C1026J	2038	28C127X	2038	28C156X	2038	28C1855
2038	28C1026K	2038	28C127Y	2038	28C156Y	2038	28C1856
2038	28C1026L	2038	28C1317(P)	2038	28C159A	2038	28C185B
2038	28C1026M	2038	28C1317(R)	2038	28C159B	2038	28C185C
2038	28C1026OR	2038	28C1317(S)	2038	28C159C	2038	28C185E
2038	28C1026R	2038	28C1317-OR	2038	28C159D	2038	28C185P
2038	28C1026X	2038	28C1317A	2038	28C159E	2038	28C185G
2038	28C1032P	2038	28C1317B	2038	28C159F	2038	28C1850N
2038	28C1047	2038	28C1317BC	2038	28C159GN	2038	28C185H
2038	28C1047A	2038	28C1317C	2038	28C159H	2038	28C185L
2038	28C1047B	2038	28C1317D	2038	28C159J	2038	28C1850R
2038	28C1047BC	2038	28C1317E	2038	28C159K	2038	28C185X
2038	28C1047BCD	2038	28C1317G	2038	28C159L	2038	28C185Y
2038	28C1047C	2038	28C1317GR	2038	28C159M	2038	28C186A
2038	28C1047D	2038	28C1317L	2038	28C159OR	2038	28C186B
2038	28C1047E	2038	28C1317OR	2038	28C159X	2038	28C186C
2038	28C1047F	2038	28C1317P	2038	28C159Y	2038	28C186D
2038	28C1047G	2038	28C1317Q	2038	28C160A	2038	28C186E
2038	28C1047GN	2038	28C1317R	2038	28C160B	2038	28C186F
2038	28C1047GR	2038	28C1317S	2038	28C160C	2038	28C186GN
2038	28C1047H	2038	28C1317T	2038	28C160D	2038	28C186H
2038	28C1047J	2038	28C1317Y	2038	28C160E	2038	28C186J
2038	28C1047K	2038	28C1320(K)	2038	28C160F	2038	28C186K
2038	28C1047L	2038	28C1320A	2038	28C160G	2038	28C186L
2038	28C1047M	2038	28C1320B	2038	28C160GN	2038	28C186M
2038	28C1047R	2038	28C1320C	2038	28C160H	2038	28C186R
2038	28C1047X	2038	28C1320D	2038	28C160J	2038	28C186X
2038	28C1047Y	2038	28C1320E	2038	28C160K	2038	28C186Y
2038	28C1117A	2038	28C1320F	2038	28C160L	2038	28C1906
2038	28C1117B	2038	28C1320G	2038	28C160M	2038	28C1908
2038	28C1117C	2038	28C1320GN	2038	28C160OR	2038	28C1908E
2038	28C1117D	2038	28C1320H	2038	28C160X	2038	28C1908H
2038	28C1117E	2038	28C1320J	2038	28C160Y	2038	28C1923
2038	28C1117F	2038	28C1320L	2038	28C1621	2038	28C1923A
2038	28C1117G	2038	28C1320M	2038	28C1636	2038	28C1923BN
2038	28C1117GN	2038	28C1320OR	2038	28C1674	2038	28C199
2038	28C1117H	2038	28C1320R	2038	28C1674K	2038	28C1990
2038	28C1117J	2038	28C1320X	2038	28C1674L	2038	28C1990B
2038	28C1117K	2038	28C1320Y	2038	28C1674M	2038	28C2001
2038	28C1117L	2038	28C1336	2038	28C1686	2038	28C2001L
2038	28C1117M	2038	28C1336JK	2038	28C1686B	2038	28C206
2038	28C1117OR	2038	28C1342	2038	2817	2038	28C206-OR
2038	28C1117R	2038	28C1342(A)	2038	28C170	2038	28C206A
2038	28C1117X	2038	28C1342(B)	2038	28C170A	2038	28C206B
2038	28C1117Y	2038	28C1342(C)	2038	28C170B	2038	28C206C
2038	28C1123A	2038	28C1342-OR	2038	28C170C	2038	28C206D
2038	28C1123B	2038	28C1342A	2038	28C170D	2038	28C206E
2038	28C1123C	2038	28C1342B	2038	28C170E	2038	28C206G
2038	28C1123D	2038	28C1342C	2038	28C170F	2038	28C206GN
2038	28C1123E	2038	28C1342D	2038	28C170G	2038	28C206H
2038	28C1123F	2038	28C1342E	2038	28C170GN	2038	28C206J
2038	28C1123GN	2038	28C1342F	2038	28C170H	2038	28C206K
2038	28C1123H	2038	28C1342G	2038	28C170J	2038	28C206L
2038	28C1123J	2038	28C1342GN	2038	28C170K	2038	28C206M
2038	28C1123K	2038	28C1342H	2038	28C170L	2038	28C2060R
2038	28C1123L	2038	28C1342J	2038	28C170M		
2038	28C1123M	2038	28C1342K				
2038	28C1123OR	2038	28C1342L				
2038	28C1123R						
2038	28C1123X						
2038	28C1123Y						

RS 276-	Industry Standard No.
2038	2SC206R
2038	2SC206RED
2038	2SC206WHITE
2038	2SC206K
2038	2SC206Y
2038	2SC250A
2038	2SC250B
2038	2SC250C
2038	2SC250D
2038	2SC250E
2038	2SC250F
2038	2SC250G
2038	2SC250GN
2038	2SC250H
2038	2SC250J
2038	2SC250K
2038	2SC250L
2038	2SC250M
2038	2SC250OR
2038	2SC250R
2038	2SC250X
2038	2SC250Y
2038	2SC251B
2038	2SC251C
2038	2SC251D
2038	2SC251E
2038	2SC251F
2038	2SC251G
2038	2SC251GN
2038	2SC251H
2038	2SC251J
2038	2SC251K
2038	2SC251L
2038	2SC251M
2038	2SC251OR
2038	2SC251R
2038	2SC251X
2038	2SC251Y
2038	2SC252A
2038	2SC252B
2038	2SC252C
2038	2SC252D
2038	2SC252E
2038	2SC252F
2038	2SC252G
2038	2SC252GN
2038	2SC252H
2038	2SC252J
2038	2SC252K
2038	2SC252L
2038	2SC252M
2038	2SC252OR
2038	2SC252R
2038	2SC252X
2038	2SC252Y
2038	2SC253A
2038	2SC253B
2038	2SC253C
2038	2SC253D
2038	2SC253E
2038	2SC253F
2038	2SC253G
2038	2SC253GN
2038	2SC253H
2038	2SC253J
2038	2SC253K
2038	2SC253L
2038	2SC253M
2038	2SC253OR
2038	2SC253R
2038	2SC253X
2038	2SC253Y
2038	2SC263
2038	2SC263A
2038	2SC263B
2038	2SC263C
2038	2SC263D
2038	2SC263E
2038	2SC263F
2038	2SC263G
2038	2SC263GN
2038	2SC263H
2038	2SC263J
2038	2SC263K
2038	2SC263L
2038	2SC263M
2038	2SC263OR
2038	2SC263R
2038	2SC263X
2038	2SC263Y
2038	2SC266A
2038	2SC266B
2038	2SC266C
2038	2SC266D
2038	2SC266E
2038	2SC266F
2038	2SC266G
2038	2SC266GN
2038	2SC266H
2038	2SC266J
2038	2SC266K
2038	2SC266L
2038	2SC266M
2038	2SC266OR
2038	2SC266R
2038	2SC266X
2038	2SC266Y
2038	2SC269
2038	2SC269A
2038	2SC269B
2038	2SC269C
2038	2SC269D
2038	2SC269E
2038	2SC269F
2038	2SC269G
2038	2SC269GN
2038	2SC269H
2038	2SC269J
2038	2SC269K
2038	2SC269M
2038	2SC269OR
2038	2SC269R
2038	2SC269X
2038	2SC269Y
2038	2SC271
2038	2SC271A
2038	2SC271B
2038	2SC271C
2038	2SC271D
2038	2SC271E
2038	2SC271F
2038	2SC271G
2038	2SC271GN
2038	2SC271H
2038	2SC271J
2038	2SC271K
2038	2SC271L
2038	2SC271M
2038	2SC271OR
2038	2SC271R
2038	2SC271X
2038	2SC271Y
2038	2SC272
2038	2SC272A
2038	2SC272B
2038	2SC272C
2038	2SC272D
2038	2SC272E
2038	2SC272F
2038	2SC272G
2038	2SC272GN
2038	2SC272H
2038	2SC272J
2038	2SC272K
2038	2SC272L
2038	2SC272M
2038	2SC272OR
2038	2SC272R
2038	2SC272X
2038	2SC272Y
2038	2SC282A
2038	2SC282B
2038	2SC282C
2038	2SC282D
2038	2SC282E
2038	2SC282F
2038	2SC282G
2038	2SC282GN
2038	2SC282J
2038	2SC282K
2038	2SC282L
2038	2SC282M
2038	2SC282OR
2038	2SC282R
2038	2SC282X
2038	2SC282Y
2038	2SC286
2038	2SC286A
2038	2SC286B
2038	2SC286C
2038	2SC286D
2038	2SC286E
2038	2SC286F
2038	2SC286G
2038	2SC286GN
2038	2SC286H
2038	2SC286J
2038	2SC286K
2038	2SC286L
2038	2SC286M
2038	2SC286OR
2038	2SC286R
2038	2SC286X
2038	2SC286Y
2038	2SC289
2038	2SC289A
2038	2SC289B
2038	2SC289C
2038	2SC289D
2038	2SC289E
2038	2SC289F
2038	2SC289G
2038	2SC289GN
2038	2SC289H
2038	2SC289J
2038	2SC289K
2038	2SC289L
2038	2SC289M
2038	2SC289OR
2038	2SC289X
2038	2SC289Y
2038	2SC296
2038	2SC29A
2038	2SC29B
2038	2SC29C
2038	2SC29D
2038	2SC29E
2038	2SC29F
2038	2SC29G
2038	2SC29GN
2038	2SC29H
2038	2SC29J
2038	2SC29K
2038	2SC29L
2038	2SC29M
2038	2SC29OR
2038	2SC29R
2038	2SC29X
2038	2SC29Y
2038	2SC316
2038	2SC316A
2038	2SC316B
2038	2SC316C
2038	2SC316D
2038	2SC316E
2038	2SC316F
2038	2SC316G
2038	2SC316GN
2038	2SC316H
2038	2SC316J
2038	2SC316K
2038	2SC316L
2038	2SC316M
2038	2SC316OR
2038	2SC316R
2038	2SC316X
2038	2SC316Y
2038	2SC324B
2038	2SC324C
2038	2SC324D
2038	2SC324E
2038	2SC324F
2038	2SC324G
2038	2SC324GN
2038	2SC324J
2038	2SC324K
2038	2SC324L
2038	2SC324M
2038	2SC324OR
2038	2SC324R
2038	2SC324X
2038	2SC324Y
2038	2SC351A
2038	2SC351B
2038	2SC351C
2038	2SC351D
2038	2SC351E
2038	2SC351F
2038	2SC351FA1
2038	2SC351G
2038	2SC351GN
2038	2SC351H
2038	2SC351J
2038	2SC351K
2038	2SC351L
2038	2SC351M
2038	2SC351OR
2038	2SC351R
2038	2SC351X
2038	2SC351Y
2038	2SC361A
2038	2SC361B
2038	2SC361C
2038	2SC361D
2038	2SC361E
2038	2SC361F
2038	2SC361G
2038	2SC361GN
2038	2SC361H
2038	2SC361J
2038	2SC361K
2038	2SC361L
2038	2SC361M
2038	2SC361OR
2038	2SC361R
2038	2SC361X
2038	2SC361Y
2038	2SC362A
2038	2SC362B
2038	2SC362C
2038	2SC362D
2038	2SC362E
2038	2SC362F
2038	2SC362G
2038	2SC362GN
2038	2SC362H
2038	2SC362J
2038	2SC362K
2038	2SC362L
2038	2SC362OR
2038	2SC362R
2038	2SC362X
2038	2SC362Y
2038	2SC370
2038	2SC370-O
2038	2SC370-Q
2038	2SC370-0
2038	2SC370-T
2038	2SC370A
2038	2SC370B
2038	2SC370C
2038	2SC370D
2038	2SC370E
2038	2SC370P
2038	2SC370G
2038	2SC370GN
2038	2SC370H
2038	2SC370J
2038	2SC370K
2038	2SC370L
2038	2SC370M
2038	2SC370OR
2038	2SC370R
2038	2SC370X
2038	2SC370Y
2038	2SC720
2038	2SC720A
2038	2SC720B
2038	2SC720C
2038	2SC720D
2038	2SC720F
2038	2SC720GN
2038	2SC720H
2038	2SC720J
2038	2SC720K
2038	2SC720L
2038	2SC720OR
2038	2SC720R
2038	2SC720X
2038	2SC720Y
2038	2SC375A
2038	2SC375B
2038	2SC375C
2038	2SC375D
2038	2SC375E
2038	2SC375F
2038	2SC375G
2038	2SC375GN
2038	2SC375H
2038	2SC375J
2038	2SC375K
2038	2SC375L
2038	2SC375R
2038	2SC375OR
2038	2SC375R
2038	2SC375X
2038	2SC374A
2038	2SC374B
2038	2SC374C
2038	2SC374D
2038	2SC374E
2038	2SC374F
2038	2SC374G
2038	2SC374GN
2038	2SC374H
2038	2SC374J
2038	2SC374K
2038	2SC374L
2038	2SC374M
2038	2SC374OR
2038	2SC374R
2038	2SC374X
2038	2SC374Y
2038	2SC3800
2038	2SC3800A
2038	2SC3800B
2038	2SC3800C
2038	2SC3800D
2038	2SC3800E
2038	2SC3800F
2038	2SC3800G
2038	2SC3800GN
2038	2SC3800H
2038	2SC3800K
2038	2SC3800L
2038	2SC3800M
2038	2SC38000R
2038	2SC3800OR
2038	2SC3800X
2038	2SC3800Y
2038	2SC384
2038	2SC384(O)
2038	2SC384(Y)
2038	2SC384-O
2038	2SC384-O
2038	2SC384A
2038	2SC384B
2038	2SC384C
2038	2SC384D
2038	2SC384E
2038	2SC384F
2038	2SC384G
2038	2SC384GN
2038	2SC384H
2038	2SC384J
2038	2SC384K
2038	2SC384L
2038	2SC384M
2038	2SC384OR
2038	2SC384R
2038	2SC384X
2038	2SC384Y
2038	2SC385C
2038	2SC385D
2038	2SC385E
2038	2SC385F
2038	2SC385G
2038	2SC385GN
2038	2SC385H
2038	2SC385J
2038	2SC385K
2038	2SC385L
2038	2SC385M
2038	2SC385OR
2038	2SC385R
2038	2SC385X
2038	2SC385Y
2038	2SC386
2038	2SC386-O
2038	2SC386A
2038	2SC386A-O(TV)
2038	2SC386AO
2038	2SC386B
2038	2SC386C
2038	2SC386D
2038	2SC386E
2038	2SC386G
2038	2SC386GN
2038	2SC386H
2038	2SC386J
2038	2SC386K
2038	2SC386L
2038	2SC386M
2038	2SC386OR
2038	2SC386R
2038	2SC386X
2038	2SC386Y
2038	2SC389-O
2038	2SC389-OR
2038	2SC389A
2038	2SC389B
2038	2SC389C
2038	2SC389D
2038	2SC389E
2038	2SC389F
2038	2SC389G
2038	2SC389GN
2038	2SC389H
2038	2SC389J
2038	2SC389K
2038	2SC389L
2038	2SC389M
2038	2SC389X
2038	2SC389Y
2038	2SC39
2038	2SC392
2038	2SC394
2038	2SC394(O)
2038	2SC394(0)
2038	2SC394-O
2038	2SC394-GR
2038	2SC394-GRN
2038	2SC394-O
2038	2SC394-OR
2038	2SC394-ORG
2038	2SC394-R
2038	2SC394-RBD
2038	2SC394-Y
2038	2SC394-YEL
2038	2SC3940
2038	2SC3940A
2038	2SC3940B
2038	2SC3940C
2038	2SC3940D
2038	2SC3940F
2038	2SC3940G
2038	2SC3940GN
2038	2SC3940H
2038	2SC3940J
2038	2SC3940K
2038	2SC3940L
2038	2SC3940M
2038	2SC3940OR
2038	2SC3940R
2038	2SC3940X
2038	2SC3940Y
2038	2SC394A
2038	2SC394AP
2038	2SC394B
2038	2SC394C
2038	2SC394D
2038	2SC394E
2038	2SC394F
2038	2SC394G
2038	2SC394GN
2038	2SC394GR
2038	2SC394GRN
2038	2SC394H
2038	2SC394J
2038	2SC394K
2038	2SC394L
2038	2SC394M
2038	2SC394O
2038	2SC394OR
2038	2SC394R
2038	2SC394RED
2038	2SC394W
2038	2SC394X

RS 276-	Industry Standard No.	RS 276-	Industry Standard No.	RS 276-	Industry Standard No.	RS 276-	Industry Standard No.
2038	2SC394Y	2038	2SC403R	2038	2SC458R	2038	00002SC535
2038	2SC394YEL	2038	2SC403X	2038	2SC458RGS	2038	2SC535(B)
2038	2SC397	2038	2SC403Y	2038	2SC458TOK	2038	2SC535-OR
2038	2SC397A	2038	2SC405	2038	2SC458Y	2038	2SC5359
2038	2SC397B	2038	2SC406	2038	2SC458VC	2038	2SC535A
2038	2SC397C	2038	2SC40A	2038	2SC458X	2038	2SC535ABC
2038	2SC397D	2038	2SC40B	2038	2SC458Y	2038	2SC535AL
2038	2SC397E	2038	2SC40C	2038	2SC460A	2038	2SC535B
2038	2SC397F	2038	2SC40D	2038	00002SC460C	2038	2SC535C
2038	2SC397G	2038	2SC40E	2038	00002SC461	2038	2SC535D
2038	2SC397GN	2038	2SC40F	2038	2SC461(8F)	2038	2SC535E
2038	2SC397H	2038	2SC40G	2038	2SC461-8F	2038	2SC535F
2038	2SC397J	2038	2SC40GN	2038	2SC461-A	2038	2SC535G
2038	2SC397K	2038	2SC40H	2038	2SC461-B	2038	2SC535GN
2038	2SC397L	2038	2SC40J	2038	2SC461 A	2038	2SC535H
2038	2SC397M	2038	2SC40K	2038	2SC461AL	2038	2SC535J
2038	2SC397R	2038	2SC40L	2038	2SC461B	2038	2SC535L
2038	2SC397OR	2038	2SC40OR	2038	2SC461BF	2038	2SC535M
2038	2SC397X	2038	2SC40R	2038	2SC461BK	2038	2SC535OR
2038	2SC397Y	2038	2SC40X	2038	2SC461BL	2038	2SC535R
2038	2SC398(FA-1)	2038	2SC40Y	2038	2SC461C	2038	2SC535X
2038	2SC398A	2038	2SC429	2038	2SC461E	2038	2SC535Y
2038	2SC398B	2038	2SC429A	2038	2SC461EF	2038	2SC535T
2038	2SC398C	2038	2SC429B	2038	2SC461L	2038	2SC535OOR
2038	2SC398D	2038	2SC429C	2038	2SC464A	2038	2SC543A
2038	2SC398E	2038	2SC429D	2038	2SC464B	2038	2SC543B
2038	2SC398F	2038	2SC429E	2038	2SC464D	2038	2SC543C
2038	2SC398FA1	2038	2SC429F	2038	2SC464E	2038	2SC543D
2038	2SC398G	2038	2SC429G	2038	2SC464F	2038	2SC543E
2038	2SC398GN	2038	2SC429GN	2038	2SC464G	2038	2SC543G
2038	2SC398H	2038	2SC429H	2038	2SC464GN	2038	2SC543GN
2038	2SC398J	2038	2SC429J	2038	2SC464H	2038	2SC543H
2038	2SC398K	2038	2SC429K	2038	2SC464K	2038	2SC543J
2038	2SC398L	2038	2SC429L	2038	2SC464L	2038	2SC543K
2038	2SC398M	2038	2SC429M	2038	2SC464M	2038	2SC543L
2038	2SC398OR	2038	2SC4290R	2038	2SC464OR	2038	2SC543M
2038	2SC398R	2038	2SC429R	2038	2SC464R	2038	2SC543OR
2038	2SC398X	2038	2SC429X	2038	2SC464X	2038	2SC543R
2038	2SC398Y	2038	2SC429Y	2038	2SC464Y	2038	2SC543X
2038	2SC399A	2038	2SC430	2038	2SC465A	2038	2SC543Y
2038	2SC399B	2038	2SC430A	2038	2SC465B	2038	2SC544
2038	2SC399C	2038	2SC430B	2038	2SC465C	2038	2SC544A
2038	2SC399D	2038	2SC430C	2038	2SC465D	2038	2SC544AG
2038	2SC399E	2038	2SC430D	2038	2SC465E	2038	2SC544B
2038	2SC399F	2038	2SC430E	2038	2SC465F	2038	2SC544C
2038	2SC399FA1	2038	2SC430F	2038	2SC465G	2038	2SC544D
2038	2SC399G	2038	2SC430G	2038	2SC465GN	2038	2SC544D(VHF)
2038	2SC399GN	2038	2SC430GN	2038	2SC465H	2038	2SC544E
2038	2SC399H	2038	2SC430H	2038	2SC465J	2038	2SC544F
2038	2SC399J	2038	2SC430J	2038	2SC465K	2038	2SC544G
2038	2SC399K	2038	2SC430K	2038	2SC465L	2038	2SC544GN
2038	2SC399L	2038	2SC430L	2038	2SC465M	2038	2SC544H
2038	2SC399M	2038	2SC430M	2038	2SC465OR	2038	2SC544J
2038	2SC399OR	2038	2SC430OR	2038	2SC465R	2038	2SC544K
2038	2SC399R	2038	2SC430R	2038	2SC465X	2038	2SC544L
2038	2SC399X	2038	2SC430W	2038	2SC465Y	2038	2SC544M
2038	2SC399Y	2038	2SC430X	2038	2SC466A	2038	2SC544OR
2038	2SC39A	2038	2SC430Y	2038	2SC466C	2038	2SC544R
2038	2SC39B	2038	2SC455	2038	2SC466D	2038	2SC544X
2038	2SC39C	2038	2SC455-OR	2038	2SC466E	2038	2SC544Y
2038	2SC39D	2038	2SC455A	2038	2SC466F	2038	2SC556
2038	2SC39E	2038	2SC455B	2038	2SC466G	2038	2SC55A
2038	2SC39F	2038	2SC455C	2038	2SC466GN	2038	2SC55B
2038	2SC39G	2038	2SC455D	2038	2SC466J	2038	2SC55C
2038	2SC39GN	2038	2SC455E	2038	2SC466K	2038	2SC55D
2038	2SC39H	2038	2SC455F	2038	2SC466L	2038	2SC55E
2038	2SC39J	2038	2SC455G	2038	2SC466M	2038	2SC55F
2038	2SC39K	2038	2SC455GN	2038	2SC466OR	2038	2SC55G
2038	2SC39L	2038	2SC455H	2038	2SC466R	2038	2SC55GN
2038	2SC39M	2038	2SC455J	2038	2SC466X	2038	2SC55H
2038	2SC390R	2038	2SC455K	2038	2SC472	2038	2SC55J
2038	2SC39R	2038	2SC455L	2038	2SC472A	2038	2SC55K
2038	2SC39X	2038	2SC455M	2038	2SC472B	2038	2SC55L
2038	2SC39Y	2038	2SC455OR	2038	2SC472C	2038	2SC55M
2038	2SC40	2038	2SC455R	2038	2SC472D	2038	2SC550OR
2038	2SC401A	2038	2SC455X	2038	2SC472E	2038	2SC55R
2038	2SC401B	2038	2SC455Y	2038	2SC472F	2038	2SC55X
2038	2SC401C	2038	2SC458	2038	2SC472G	2038	2SC55Y
2038	2SC401D	2038	2SC458(C)	2038	2SC472GN	2038	2SC561
2038	2SC401E	2038	2SC458(C,D)	2038	2SC472H	2038	2SC561A
2038	2SC401F	2038	2SC458(D)	2038	2SC472J	2038	2SC561B
2038	2SC401G	2038	2SC458(LG)	2038	2SC472K	2038	2SC561C
2038	2SC401GN	2038	2SC458-4	2038	2SC472L	2038	2SC561D
2038	2SC401H	2038	2SC458-5	2038	2SC472M	2038	2SC561E
2038	2SC401J	2038	2SC458-0	2038	2SC472OR	2038	2SC561F
2038	2SC401K	2038	2SC458-OR	2038	2SC472R	2038	2SC561G
2038	2SC401L	2038	2SC458A	2038	2SC472X	2038	2SC561GN
2038	2SC401M	2038	2SC458AD	2038	2SC477	2038	2SC561H
2038	2SC401OR	2038	2SC458AK	2038	2SC477A	2038	2SC561J
2038	2SC401R	2038	2SC458B-D	2038	2SC477B	2038	2SC561K
2038	2SC401X	2038	2SC458BC	2038	2SC477C	2038	2SC561L
2038	2SC401Y	2038	2SC458BD	2038	2SC477D	2038	2SC561M
2038	2SC403	2038	2SC458BK	2038	2SC477E	2038	2SC561OR
2038	2SC403(C)	2038	2SC458BL	2038	2SC477F	2038	2SC561R
2038	2SC403(SONY)	2038	2SC458BLG	2038	2SC477G	2038	2SC561X
2038	2SC403,A,B	2038	2SC458BM	2038	2SC477GN	2038	2SC561Y
2038	2SC403-OR	2038	2SC458C/L6	2038	2SC477H	2038	2SC567A
2038	2SC4033A	2038	2SC458CL	2038	2SC477J	2038	2SC567B
2038	2SC4033B	2038	2SC458CLG	2038	2SC477K	2038	2SC567C
2038	2SC4033C	2038	2SC458CM	2038	2SC477L	2038	2SC567D
2038	2SC4033D	2038	2SC458D	2038	2SC477M	2038	2SC567E
2038	2SC4033E	2038	2SC458E	2038	2SC477OR	2038	2SC567F
2038	2SC4033F	2038	2SC458F	2038	2SC477R	2038	2SC567G
2038	2SC4033G	2038	2SC458G	2038	2SC477X	2038	2SC567GN
2038	2SC4033GN	2038	2SC458GLB	2038	2SC509	2038	2SC567H
2038	2SC4033H	2038	2SC458GN	2038	2SC509(0)	2038	2SC567J
2038	2SC4033J	2038	2SC458H	2038	2SC509(O)	2038	2SC567K
2038	2SC4033K	2038	2SC458J	2038	2SC509(Y)	2038	2SC567L
2038	2SC4033L	2038	2SC458K	2038	2SC509-0	2038	2SC567M
2038	2SC4033M	2038	2SC458KA	2038	2SC509-Y	2038	2SC567OR
2038	2SC4033OR	2038	2SC458KB	2038	2SC509G	2038	2SC567R
2038	2SC4033R	2038	2SC458KC	2038	2SC509G-0	2038	2SC567X
2038	2SC4033X	2038	2SC458KD	2038	2SC509G-Y	2038	2SC567Y
2038	2SC4033Y	2038	2SC458L	2038	2SC529B	2038	2SC568
2038	2SC403A	2038	2SC458L(C)	2038	2SC529B	2038	2SC568A
2038	2SC403AL	2038	2SC458L6	2038	2SC529C	2038	2SC568B
2038	2SC403B	2038	2SC458LB	2038	2SC529D	2038	2SC568C
2038	2SC403B(SONY)	2038	2SC458LC	2038	2SC529E	2038	2SC568D
2038	2SC403C	2038	2SC458LD	2038	2SC529F	2038	2SC568E
2038	2SC403C(SONY)	2038	2SC458LG	2038	2SC529G	2038	2SC568F
2038	2SC403C,A	2038	2SC458LG(B)	2038	2SC529GN	2038	2SC568G
2038	2SC403C,B,A	2038	2SC458LG(C)	2038	2SC529H	2038	2SC568GN
2038	2SC403CG	2038	2SC458LG(D)	2038	2SC529J	2038	2SC568H
2038	2SC403D	2038	2SC458LGA	2038	2SC529K	2038	2SC568J
2038	2SC403E	2038	2SC458LGB	2038	2SC529L	2038	2SC568K
2038	2SC403F	2038	2SC458LGBM	2038	2SC529M	2038	2SC568L
2038	2SC403G	2038	2SC458LGC-6	2038	2SC529OR	2038	2SC568M
2038	2SC403H	2038	2SC458LGD	2038	2SC529R	2038	2SC568OR
2038	2SC403J	2038	2SC458LGO	2038	2SC529X	2038	2SC568R
2038	2SC403K	2038	2SC458LGR	2038	2SC529Y	2038	2SC568X
2038	2SC403L	2038	2SC458LGS			2038	2SC568Y
2038	2SC403M	2038	2SC458M			2038	2SC590
2038	2SC4030R	2038	2SC458OR			2038	2SC590Y
		2038	2SC458P				

RS 276-	Industry Standard No.	RS 276-	Industry Standard No.	RS 276-	Industry Standard No.	RS 276-	Industry Standard No.
2038	2SC606	2038	2SC656H	2038	2SC683R	2038	2SC748
2038	2SC611	2038	2SC656J	2038	2SC683S	2038	2SC748A
2038	2SC611A	2038	2SC656K	2038	2SC683V	2038	2SC748C
2038	2SC611B	2038	2SC656L	2038	2SC683X	2038	2SC748D
2038	2SC611C	2038	2SC656M	2038	2SC683Y	2038	2SC748E
2038	2SC611D	2038	2SC6560R	2038	2SC684	2038	2SC748F
2038	2SC611E	2038	2SC656R	2038	2SC684A	2038	2SC748G
2038	2SC611F	2038	2SC656X	2038	2SC684B	2038	2SC748GN
2038	2SC611G	2038	2SC656Y	2038	2SC684BK	2038	2SC748H
2038	2SC611GN	2038	2SC657A	2038	2SC684F	2038	2SC748J
2038	2SC611H	2038	2SC657B	2038	2SC688	2038	2SC748K
2038	2SC611J	2038	2SC657C	2038	2SC688A	2038	2SC748L
2038	2SC611K	2038	2SC657D	2038	2SC688B	2038	2SC748M
2038	2SC611L	2038	2SC657E	2038	2SC688C	2038	2SC7480R
2038	2SC611M	2038	2SC657F	2038	2SC688D	2038	2SC748R
2038	2SC611OR	2038	2SC657G	2038	2SC688E	2038	2SC748X
2038	2SC611R	2038	2SC657GN	2038	2SC688F	2038	2SC748Y
2038	2SC611X	2038	2SC657H	2038	2SC688G	2038	2SC74A
2038	2SC611Y	2038	2SC657J	2038	2SC688GN	2038	2SC74B
2038	2SC612	2038	2SC657K	2038	2SC688H	2038	2SC74C
2038	2SC612A	2038	2SC657L	2038	2SC688J	2038	2SC74D
2038	2SC612B	2038	2SC657M	2038	2SC688K	2038	2SC74E
2038	2SC612C	2038	2SC6570R	2038	2SC688M	2038	2SC74F
2038	2SC612E	2038	2SC657R	2038	2SC6880R	2038	2SC74G
2038	2SC612F	2038	2SC657X	2038	2SC688R	2038	2SC74GN
2038	2SC612G	2038	2SC657Y	2038	2SC688X	2038	2SC74H
2038	2SC612GN	2038	2SC658	2038	2SC705A	2038	2SC74J
2038	2SC612H	2038	2SC658A	2038	2SC705G	2038	2SC74K
2038	2SC612J	2038	2SC658B	2038	2SC705GN	2038	2SC74L
2038	2SC612K	2038	2SC658C	2038	2SC705J	2038	2SC74M
2038	2SC612L	2038	2SC658D	2038	2SC705K	2038	2SC74R
2038	2SC612M	2038	2SC658E	2038	2SC705L	2038	2SC74X
2038	2SC6120R	2038	2SC658F	2038	2SC705M	2038	2SC74Y
2038	2SC612R	2038	2SC658G	2038	2SC7050R	2038	2SC7580R
2038	2SC612X	2038	2SC658GN	2038	2SC705R	2038	2SC763
2038	2SC612Y	2038	2SC658H	2038	2SC705TVV	2038	2SC763(C)
2038	2SC613	2038	2SC658J	2038	2SC705TW	2038	2SC763-OR
2038	2SC613A	2038	2SC658K	2038	2SC705X	2038	2SC763A
2038	2SC613B	2038	2SC658L	2038	2SC707A	2038	2SC763B
2038	2SC613C	2038	2SC658M	2038	2SC707B	2038	2SC763C
2038	2SC613D	2038	2SC6580R	2038	2SC707C	2038	2SC763CD
2038	2SC613E	2038	2SC658R	2038	2SC707D	2038	2SC763D
2038	2SC613F	2038	2SC658X	2038	2SC707F	2038	2SC763E
2038	2SC613G	2038	2SC658Y	2038	2SC707G	2038	2SC763F
2038	2SC613GN	2038	2SC659	2038	2SC707GN	2038	2SC763G
2038	2SC613H	2038	2SC659A	2038	2SC707K	2038	2SC763GN
2038	2SC613J	2038	2SC659B	2038	2SC707L	2038	2SC763H
2038	2SC613K	2038	2SC659C	2038	2SC707M	2038	2SC763J
2038	2SC613L	2038	2SC659D	2038	2SC7070R	2038	2SC763K
2038	2SC613M	2038	2SC659E	2038	2SC707R	2038	2SC763L
2038	2SC6130R	2038	2SC659F	2038	2SC707X	2038	2SC7630R
2038	2SC613R	2038	2SC659G	2038	2SC707Y	2038	2SC763X
2038	2SC613X	2038	2SC659GN	2038	2SC722	2038	2SC763Y
2038	2SC613Y	2038	2SC659H	2038	2SC735(O)	2038	2SC771
2038	2SC629-31	2038	2SC659J	2038	2SC735(PA-3)	2038	2SC771A
2038	2SC629-41	2038	2SC659K	2038	2SC735(O)	2038	2SC771B
2038	2SC629A	2038	2SC659L	2038	2SC735(Y)	2038	2SC771BX
2038	2SC629B	2038	2SC659M	2038	2SC735-O	2038	2SC771C
2038	2SC629C	2038	2SC6590R	2038	2SC735-GRN	2038	2SC771D
2038	2SC629D	2038	2SC659R	2038	2SC735-O	2038	2SC771E
2038	2SC629E	2038	2SC659X	2038	2SC735-OR	2038	2SC771F
2038	2SC629F	2038	2SC659Y	2038	2SC735-ORG	2038	2SC771G
2038	2SC629G	2038	2SC662	2038	2SC735-ORN	2038	2SC771GN
2038	2SC629GN	2038	2SC662A	2038	2SC735-OY	2038	2SC771H
2038	2SC629H	2038	2SC662B	2038	2SC735-R	2038	2SC771J
2038	2SC629J	2038	2SC662C	2038	2SC735-RED	2038	2SC771K
2038	2SC629K	2038	2SC662D	2038	2SC735-Y	2038	2SC771L
2038	2SC629L	2038	2SC662E	2038	2SC735-YEL	2038	2SC771M
2038	2SC629M	2038	2SC662F	2038	2SC735/4454C	2038	2SC7710R
2038	2SC6290R	2038	2SC662G	2038	002SC7350Y	2038	2SC771R
2038	2SC629R	2038	2SC662GN	2038	2SC735A	2038	2SC771X
2038	2SC629X	2038	2SC662H	2038	2SC735B	2038	2SC771Y
2038	2SC629Y	2038	2SC662J	2038	2SC735C	2038	2SC772-OR
2038	2SC63	2038	2SC662K	2038	2SC735D	2038	2SC772A
2038	2SC63A	2038	2SC662L	2038	2SC735E	2038	2SC772B
2038	2SC63B	2038	2SC662M	2038	2SC735F	2038	2SC772BG
2038	2SC63D	2038	2SC6620R	2038	2SC735FA3	2038	2SC772BH
2038	2SC63E	2038	2SC662R	2038	2SC735G	2038	2SC772BV
2038	2SC63F	2038	2SC662X	2038	2SC735GN	2038	2SC772BX
2038	2SC63G	2038	2SC662Y	2038	2SC735GR	2038	2SC772BY
2038	2SC6630N	2038	2SC668	2038	2SC735GRN	2038	2SC772C1
2038	2SC63H	2038	2SC668(C)	2038	2SC735H	2038	2SC772C2
2038	2SC63J	2038	2SC668(D)	2038	2SC735J	2038	2SC772CK
2038	2SC63K	2038	2SC668-0	2038	2SC735K	2038	2SC772CL
2038	2SC63L	2038	2SC668-OR	2038	2SC735L	2038	2SC772CU
2038	2SC63M	2038	2SC668A	2038	2SC735M	2038	2SC772CV
2038	2SC6630R	2038	2SC668B	2038	2SC7350	2038	2SC772CX
2038	2SC63R	2038	2SC668B1	2038	2SC7350R	2038	2SC772D
2038	2SC63X	2038	2SC668BC2	2038	2SC7350RN	2038	2SC772DJ
2038	2SC63Y	2038	2SC668C1	2038	2SC735R	2038	2SC772DV
2038	2SC641	2038	2SC668C2	2038	2SC735RED	2038	2SC772DX
2038	2SC641B	2038	2SC668CD	2038	2SC735X	2038	2SC772DY
2038	2SC648A	2038	2SC668D0	2038	2SC735Y	2038	2SC772E
2038	2SC648B	2038	2SC668D1	2038	2SC735Y/4454C	2038	2SC772F
2038	2SC648C	2038	2SC668DE	2038	2SC735YEL	2038	2SC772G
2038	2SC648D	2038	2SC668DO	2038	2SC738	2038	2SC772GN
2038	2SC648E	2038	2SC668DV	2038	2SC738A	2038	2SC772H
2038	2SC648F	2038	2SC668DX	2038	2SC738B	2038	2SC772J
2038	2SC648G	2038	2SC668DZ	2038	2SC738C	2038	2SC772K
2038	2SC648J	2038	2SC668E	2038	2SC738D	2038	2SC772KB
2038	2SC648K	2038	2SC668E1	2038	2SC738E	2038	2SC772KC
2038	2SC648L	2038	2SC668E2	2038	2SC738F	2038	2SC772KD
2038	2SC648M	2038	2SC668EP	2038	2SC738G	2038	2SC772KD1
2038	2SC6480R	2038	2SC668EV	2038	2SC738H	2038	2SC772KD2
2038	2SC648R	2038	2SC668EX	2038	2SC738J	2038	2SC772L
2038	2SC648X	2038	2SC668F	2038	2SC738K	2038	2SC772M
2038	2SC648Y	2038	2SC668G	2038	2SC738L	2038	2SC7720R
2038	2SC649	2038	2SC668GN	2038	2SC738M	2038	2SC772R
2038	2SC649A	2038	2SC668H	2038	2SC7380R	2038	2SC772RB-D
2038	2SC649B	2038	2SC668K	2038	2SC738R	2038	2SC772RD
2038	2SC649C	2038	2SC668L	2038	2SC738X	2038	2SC772RS-D
2038	2SC649D	2038	2SC668M	2038	2SC738Y	2038	2SC772X
2038	2SC649E	2038	2SC6680R	2038	2SC739A	2038	2SC772Y
2038	2SC649F	2038	2SC668R	2038	2SC739B	2038	2SC784
2038	2SC649G	2038	2SC668X	2038	2SC739D	2038	2SC784(BN)
2038	2SC649GN	2038	2SC668Y	2038	2SC739E	2038	2SC784-0
2038	2SC649H	2038	2SC683	2038	2SC739F	2038	2SC784-6
2038	2SC649J	2038	2SC683(B)	2038	2SC739G	2038	2SC784-B
2038	2SC649K	2038	2SC683-OR	2038	2SC739GN	2038	2SC784-BN
2038	2SC649L	2038	2SC683A	2038	2SC739H	2038	2SC784-BRN
2038	2SC649M	2038	2SC683B	2038	2SC739K	2038	2SC784-O
2038	2SC6490R	2038	2SC683C	2038	2SC739L	2038	2SC784-OR
2038	2SC649R	2038	2SC683D	2038	2SC739M	2038	2SC784-ORG
2038	2SC649X	2038	2SC683E	2038	2SC7390R	2038	2SC784-R
2038	2SC649Y	2038	2SC683F	2038	2SC739R	2038	2SC784-RED
2038	2SC656A	2038	2SC6830	2038	2SC739Y	2038	2SC784-Y
2038	2SC656B	2038	2SC6830N	2038	2SC74	2038	2SC784A
2038	2SC656C	2038	2SC683H	2038	2SC74-O	2038	2SC784B
2038	2SC656D	2038	2SC683J	2038	2SC74-R	2038	2SC784BN
2038	2SC656E	2038	2SC683K			2038	2SC784BN-1
2038	2SC656F	2038	2SC683L			2038	2SC784BRN
2038	2SC656G	2038	2SC6830R				
2038	2SC656GN						

RS 276-	Industry Standard No.	RS 276-	Industry Standard No.	RS 276-	Industry Standard No.	RS 276-	Industry Standard No.
2038	2SC784C	2038	2SC917D	2038	2SC930K	2038	2SC8461X
2038	2SC784D	2038	2SC917E	2038	2SC930L	2038	2SC8461Y
2038	2SC784E	2038	2SC917F	2038	2SC930N	2038	2SC8469
2038	2SC784F	2038	2SC917G	2038	2SC930NP	2038	2SC8469A
2038	2SC784G	2038	2SC917GN	2038	2SC930OR	2038	2SC8469B
2038	2SC784GN	2038	2SC917H	2038	2SC930R	2038	2SC8469C
2038	2SC784H	2038	2SC917J	2038	2SC930X	2038	2SC8469D
2038	2SC784J	2038	2SC917K	2038	2SC9307	2038	2SC8469E
2038	2SC784K	2038	2SC917L	2038	2SC947H	2038	2SC8469F
2038	2SC784L	2038	2SC917M	2038	2SC957A	2038	2SC8469G
2038	2SC784M	2038	2SC917OR.	2038	2SC957AL	2038	2SC8469GN
2038	2SC784O	2038	2SC917R	2038	2SC957B	2038	2SC8469H
2038	2SC784OR	2038	2SC917X	2038	2SC957C	2038	2SC8469J
2038	2SC784Q	2038	2SC917Y	2038	2SC957D	2038	2SC8469K
2038	2SC784R	2038	2SC918A	2038	2SC957E	2038	2SC8469L
2038	2SC784R/4454C	2038	2SC918AL	2038	2SC957F	2038	2SC8469M
2038	2SC784RA	2038	2SC918B	2038	2SC957G	2038	2SC8469R
2038	2SC784RED	2038	2SC918C	2038	2SC957GN	2038	2SC8469X
2038	2SC784X	2038	2SC918D	2038	2SC957H	2038	2SC8469Y
2038	2SC784Y	2038	2SC918E	2038	2SC957J	2038	2SD471
2038	2SC785	2038	2SC918F	2038	2SC957K	2038	2SD471K
2038	2SC785(E)(D)	2038	2SC918G	2038	2SC957L	2038	2SD471L
2038	2SC785(O)	2038	2SC918GN	2038	2SC957M	2038	2SD471M
2038	2SC785-O	2038	2SC918H	2038	2SC957OR	2038	2SD562
2038	2SC785-B	2038	2SC918J	2038	2SC957R	2038	2SD771L
2038	2SC785-BN	2038	2SC918K	2038	2SC957X	2038	2SB629
2038	2SC785-BRN	2038	2SC918L	2038	2SC957XL	2038	2SQ371
2038	2SC785-O	2038	2SC918LF	2038	2SC957Y	2038	003-01
2038	2SC785-ORG	2038	2SC918M	2038	2SC98	2038	03-460C
2038	2SC785-R	2038	2SC918OR	2038	2SC985	2038	03-461B
2038	2SC785-RED	2038	2SC918R	2038	2SC99	2038	03-535A
2038	2SC785-Y	2038	2SC918X	2038	2SC997A	2038	3C38(TRANSISTOR)
2038	2SC785-YEL	2038	2SC918XL	2038	2SC997B	2038	3L4-6007-51
2038	2SC785A	2038	2SC918Y	2038	2SC997C	2038	04-00355-06
2038	2SC785B	2038	2SC920	2038	2SC997D	2038	04-0485
2038	2SC785BL	2038	2SC920-OQ	2038	2SC997E	2038	4-3022861
2038	2SC785BN	2038	2SC920-OR	2038	2SC997F	2038	4-3025763
2038	2SC785BR	2038	2SC920A	2038	2SC997G	2038	4-3025764
2038	2SC785BRN	2038	2SC920B	2038	2SC997GN	2038	4-3025765
2038	2SC785C	2038	2SC920C	2038	2SC997H	2038	4-3025767
2038	2SC785D	2038	2SC920CL	2038	2SC997K	2038	4-399
2038	2SC785E	2038	2SC920D	2038	2SC997L	2038	4-400
2038	2SC785F	2038	2SC920E	2038	2SC997M	2038	4-433
2038	2SC785G	2038	2SC920P	2038	2SC997N	2038	4-434
2038	2SC785GN	2038	2SC920Q	2038	2SC997OR	2038	4-443
2038	2SC785GR	2038	2SC920GN	2038	2SC997R	2038	04-46000-02
2038	2SC785H	2038	2SC920H	2038	2SC997X	2038	4-684120-3
2038	2SC785K	2038	2SC920L	2038	2SC997Y	2038	4-685285-3
2038	2SC785L	2038	2SC920M	2038	2SCP1	2038	4-686107-3
2038	2SC785M	2038	2SC920OR	2038	2SCP1A	2038	4-686108-3
2038	2SC785O	2038	2SC920Q	2038	2SCP1B	2038	4-686112-3
2038	2SC785R	2038	2SC920R	2038	2SCP1C	2038	4-686114-3
2038	2SC785RA	2038	2SC920X	2038	2SCP1D	2038	4-686118-3
2038	2SC785RED	2038	2SC920Y	2038	2SCP1E	2038	4-686119-3
2038	2SC785V	2038	2SC921CL	2038	2SCP1F	2038	4-686120-3
2038	2SC785X	2038	2SC924	2038	2SCP1G	2038	4-686126-3
2038	2SC785Y	2038	2SC924A	2038	2SCP1GN	2038	4-686127-3
2038	2SC785YEL	2038	2SC924B	2038	2SCP1H	2038	4-686131-3
2038	2SC79	2038	2SC924C	2038	2SCP1J	2038	4-686140-3
2038	2SC79A	2038	2SC924D	2038	2SCP1K	2038	4-686169-3
2038	2SC79B	2038	2SC924E	2038	2SCP1L	2038	4-686171-3
2038	2SC79C	2038	2SC924F	2038	2SCP1M	2038	4-686172-3
2038	2SC79D	2038	2SC924G	2038	2SCP1OR	2038	4-686207-3
2038	2SC79E	2038	2SC924GN	2038	2SCP1R	2038	4-686208-3
2038	2SC79F	2038	2SC924H	2038	2SCP1X	2038	4-686209-3
2038	2SC79G	2038	2SC924J	2038	2SCP1Y	2038	4-686224-3
2038	2SC79GN	2038	2SC924K	2038	2SCP2A	2038	4-686228-3
2038	2SC79H	2038	2SC924L	2038	2SCP2B	2038	4-686244-3
2038	2SC79J	2038	2SC924M	2038	2SCP2C	2038	4-686251-3
2038	2SC79K	2038	2SC924OR	2038	2SCP2D	2038	4-68695-3
2038	2SC79L	2038	2SC924R	2038	2SCP2E	2038	4-JBD1
2038	2SC79M	2038	2SC924X	2038	2SCP2G	2038	6-2708
2038	2SC79OR	2038	2SC924Y	2038	2SCP2GN	2038	6A12677
2038	2SC79R	2038	2SC929(O)	2038	2SCP2H	2038	6A12679
2038	2SC79X	2038	2SC929(B)	2038	2SCP2J	2038	07-07129
2038	2SC79Y	2038	2SC929-O	2038	2SCP2L	2038	07-07163
2038	2SC80	2038	2SC929-O	2038	2SCP2M	2038	7-4
2038	2SC80A	2038	2SC929A	2038	2SCP2R	2038	7-44
2038	2SC80B	2038	2SC929B	2038	2SCP2X	2038	7-59-0193477
2038	2SC80C	2038	2SC929C	2038	2SCP2Y	2038	7-59-0203477
2038	2SC80D	2038	2SC929C1	2038	2SCM39J	2038	7-59-0213477
2038	2SC80E	2038	2SC929D	2038	2SCM39X	2038	7-59-0223477
2038	2SC80F	2038	2SC929D1	2038	2SC8164J	2038	7-59-0233477
2038	2SC80G	2038	2SC929DE	2038	2SC8429	2038	7-6
2038	2SC80GN	2038	2SC929DP	2038	2SC8429A	2038	7-7
2038	2SC80H	2038	2SC929DU	2038	2SC8429B	2038	7-8
2038	2SC80J	2038	2SC929DV	2038	2SC8429C	2038	8-0024-1
2038	2SC80K	2038	2SC929E	2038	2SC8429D	2038	8-0024-2
2038	2SC80L	2038	2SC929ED	2038	2SC8429E	2038	8-0053600
2038	2SC80M	2038	2SC929EZ	2038	2SC8429F	2038	8-0338050
2038	2SC80OR	2038	2SC929F	2038	2SC8429G	2038	8-0338040
2038	2SC80R	2038	2SC929FK	2038	2SC8429GN	2038	8-0339430
2038	2SC80X	2038	2SC929G	2038	2SC8429H	2038	8-0339440
2038	2SC80Y	2038	2SC929GN	2038	2SC8429J	2038	8-0389840
2038	2SC837(K)	2038	2SC929H	2038	2SC8429K	2038	8-0389930
2038	2SC837(KL)	2038	2SC929J	2038	2SC8429L	2038	08-302152
2038	2SC837(L)	2038	2SC929K	2038	2SC8429M	2038	8-729-803-04
2038	2SC837A	2038	2SC929L	2038	2SC8429OR	2038	09-3002006
2038	2SC837B	2038	2SC929M	2038	2SC8429R	2038	09-302002
2038	2SC837C	2038	2SC929NP	2038	2SC8429X	2038	09-302003
2038	2SC837D	2038	2SC929OR	2038	2SC8429Y	2038	09-302005
2038	2SC837E	2038	2SC929R	2038	2SC8430	2038	09-302006
2038	2SC837G	2038	2SC929X	2038	2SC8430A	2038	09-302009
2038	2SC837GN	2038	2SC930(D)	2038	2SC8430A	2038	09-302010
2038	2SC837J	2038	2SC930(E)	2038	2SC8430B	2038	09-302017
2038	2SC837KL	2038	2SC930-OR	2038	2SC8430C	2038	09-302020
2038	2SC837M	2038	2SC930A	2038	2SC8430D	2038	09-302036
2038	2SC837OR	2038	2SC930B	2038	2SC8430E	2038	09-302037
2038	2SC837R	2038	2SC930BB	2038	2SC8430F	2038	09-302040
2038	2SC837X	2038	2SC930BK	2038	2SC8430GN	2038	09-302060
2038	2SC837Y	2038	2SC930BV	2038	2SC8430H	2038	09-302095
2038	2SC8380	2038	2SC930C	2038	2SC8430J	2038	09-302115
2038	2SC860A	2038	2SC930CK	2038	2SC8430K	2038	09-302141
2038	2SC860B	2038	2SC930CL	2038	2SC8430L	2038	09-302149
2038	2SC860F	2038	2SC930CS	2038	2SC8430M	2038	09-302162
2038	2SC860G	2038	2SC930DB	2038	2SC8430R	2038	09-302173
2038	2SC860GN	2038	2SC930DC	2038	2SC8430X	2038	09-302190
2038	2SC860H	2038	2SC930DE	2038	2SC8430Y	2038	09-302206
2038	2SC860J	2038	2SC930DH	2038	2SC8461	2038	09-302207
2038	2SC860K	2038	2SC930DK	2038	2SC8461A	2038	09-302224
2038	2SC860L	2038	2SC930DS	2038	2SC8461B	2038	09-302225
2038	2SC860M	2038	2SC930DT	2038	2SC8461C	2038	09-302241
2038	2SC860OR	2038	2SC930DT-2	2038	2SC8461D	2038	09-302242
2038	2SC860R	2038	2SC930DX	2038	2SC8461E	2038	09-304042
2038	2SC860X	2038	2SC930DZ	2038	2SC8461F	2038	09-304042
2038	2SC860Y	2038	2SC930DP	2038	2SC8461G	2038	09-304043
2038	2SC899E	2038	2SC930ET	2038	2SC8461GN	2038	09-305050
2038	2SC912	2038	2SC930EV	2038	2SC8461H	2038	09-305051
2038	2SC917	2038	2SC930EX	2038	2SC8461J	2038	09-305069
2038	2SC917(K)	2038	2SC930F	2038	2SC8461K	2038	09-305070
2038	2SC917A	2038	2SC930G	2038	2SC8461L	2038	09-305071
2038	2SC917B	2038	2SC930GN	2038	2SC8461M	2038	09-305072
2038	2SC917C	2038	2SC930H	2038	2SC8461OR	2038	09-305074
2038		2038	2SC930J	2038	2SC8461R	2038	09-305093

RS 276-	Industry Standard No.
2038	09-305094
2038	09-305096
2038	09-309007
2038	09-309013
2038	09-309024
2038	09-309027
2038	09-309028
2038	09-309032
2038	09-309061
2038	09-309065
2038	09-309073
2038	09-32124
2038	9-5125
2038	9-5126
2038	9-5127
2038	9-5128
2038	9-5129
2038	9-5130
2038	9-5131
2038	9-5223
2038	9TR9
2038	10-002
2038	10-008
2038	10-080009
2038	10-28C080
2038	10-28C094
2038	10B1051
2038	10B1055
2038	10B551
2038	10B553
2038	10B555
2038	10B555-2
2038	10B555-3
2038	10B556
2038	10B556-2
2038	10B556-3
2038	100573
2038	100574
2038	10Q1051
2038	10Q1052
2038	10H1051
2038	10H1053
2038	10B551
2038	10B553
2038	11B1052
2038	11B1055
2038	11B551
2038	11B552
2038	11B554
2038	11B555
2038	11Q1051
2038	11Q1053
2038	11Q1057
2038	110551
2038	110553
2038	110557
2038	12-23163-3
2038	13-0009
2038	13-0010
2038	13-0020
2038	13-0040
2038	13-0062
2038	13-0063
2038	13-0065
2038	13-0178
2038	13-0321-14
2038	13-0321-15
2038	13-0321-16
2038	13-0321-17
2038	13-0321-21
2038	13-1032-5
2038	13-10321-1
2038	13-10321-10
2038	13-10321-11
2038	13-10321-12
2038	13-10321-14
2038	13-10321-15
2038	13-10321-16
2038	13-10321-17
2038	13-10321-2
2038	13-10321-20
2038	13-10321-21
2038	13-10321-26
2038	13-10321-30
2038	13-10321-41
2038	13-10321-43
2038	13-10321-46
2038	13-10321-5
2038	13-10321-51
2038	13-10321-6
2038	13-10321-62
2038	13-10321-7
2038	13-10321-77
2038	13-10321-8
2038	13-10321-9
2038	13-14085-1
2038	13-14085-2
2038	13-14085-24
2038	13-14085-27
2038	13-14085-74
2038	13-14085-75
2038	13-14085-76
2038	13-14085-77
2038	13-15810-1
2038	13-15841-1
2038	13-16744-1
2038	13-18949-1
2038	13-18950-1
2038	13-23001-2
2038	13-23002-2
2038	13-23160-5
2038	13-23163-2
2038	13-23822
2038	13-28584
2038	13-28584-1
2038	13-29033-6
2038	13-31013-4
2038	13-32366-1
2038	13-32366-2
2038	13-34045-1
2038	13-34045-2
2038	13-35550
2038	13-55020-1
2038	13-55065-1
2038	13-55065-1
2038	13-67583-5
2038	13-67585-5/2439-2
2038	13-67585-6/2439-2
2038	14-32430
2038	14-602-41
2038	14-603-12
2038	14-609-49A
2038	015
2038	15-088004
2038	15-08800U
2038	16-736
2038	16J1
2038	16J2
2038	16K1
2038	16K2
2038	16K3
2038	16L2
2038	16L22
2038	16L23
2038	16L3
2038	16L5
2038	019-003929
2038	19-020-037
2038	19-020-044
2038	19-020-048
2038	19-020-052
2038	19-020-070
2038	19-020-44
2038	19-19420
2038	19A115249-1
2038	19A115342-1
2038	19A115440-1
2038	19A115440-2
2038	19A115441-1
2038	19A115666-1
2038	19A115925-1
2038	19A123160-1
2038	19A123160-2
2038	020-00026
2038	020-00027
2038	20-00229-001
2038	20-00444-001
2038	20-1
2038	21A015-004
2038	21A015-014
2038	21A015-016
2038	21A040-003
2038	21A040-004
2038	21A040-007
2038	21A040-010
2038	21A040-016
2038	21A040-017
2038	21A040-019
2038	21A040-045
2038	21A040-053
2038	21A040-54
2038	21A050-004
2038	21A105-001
2038	21A112-084
2038	21A112-086
2038	21A112-087
2038	21A112-101
2038	21A118-031
2038	21M476
2038	21M481
2038	21M577
2038	22-001002
2038	22-001003
2038	22-001004
2038	22-001005
2038	022-2876-003
2038	022-3640-080
2038	022.3640-080
2038	23-PT275-122
2038	24-3564
2038	24A1
2038	24MW1038
2038	24MW1082
2038	24MW654
2038	24MW657
2038	24MW675
2038	24MW724
2038	24MW725
2038	24MW739
2038	24MW827
2038	24MW852
2038	24MW865
2038	24T-002
2038	24T-011-008
2038	24T-011-013
2038	24T-013-005
2038	24T-016-001
2038	24T-016-005
2038	24T-016-013
2038	24T-016-015
2038	24T-016-016
2038	24T011-008
2038	025-100003
2038	025-100004
2038	025-100009
2038	025-100013
2038	025-100014
2038	025-100015
2038	25A
2038	25A1262-005
2038	25A1281-001
2038	25AM624
2038	25B-1
2038	25B1
2038	25B2
2038	25C206
2038	25R
2038	31-0051
2038	31-0097
2038	31-0103
2038	31-0242
2038	31-0243
2038	33H50
2038	34-6001-3
2038	34-6001-6
2038	34-6015-27
2038	34-6015-47
2038	34-6015-48
2038	34-6015-49
2038	34-6015-52
2038	34-6016-17
2038	34B31
2038	34B3L
2038	34H31
2038	41N1
2038	41N2
2038	41N2A
2038	41N2AA
2038	41N2B
2038	41N2M
2038	41N3
2038	42-19683
2038	42-22785
2038	42-27529
2038	42-27530
2038	42-27537
2038	42-28203
2038	42-28204
2038	42-28206
2038	43-022861
2038	43-025763
2038	43-025764
2038	43-025765
2038	43-025767
2038	44-44886901
2038	44T-300-104
2038	44T-300-110
2038	045-1(SYLVANIA)
2038	045-2(SYLVANIA)
2038	45N7
2038	46-84120-3
2038	46-85285-3
2038	46-86132-3
2038	46-86140-3
2038	46-86208-3
2038	46-86240-3
2038	46-86244-3
2038	46-86262-3
2038	46-86269-3
2038	46-86314-3
2038	46-86314-3A
2038	46-86357-3
2038	46-86376-3
2038	46-86397-3
2038	46-864-3
2038	46-8677-2
2038	47-2(BRADFORD)
2038	48-01-004
2038	48-01-010
2038	48-124804
2038	48-124805
2038	48-124808
2038	48-13470
2038	48-134706
2038	48-134709
2038	48-134713
2038	48-134717
2038	48-134719
2038	48-134724
2038	48-134725
2038	48-134772
2038	48-134773
2038	48-134774
2038	48-134777
2038	48-134780
2038	48-134783
2038	48-134785
2038	48-134786
2038	48-134787
2038	48-134800
2038	48-134805
2038	48-134806
2038	48-134814
2038	48-134818
2038	48-134820
2038	48-134821
2038	48-134825
2038	48-134826
2038	48-134827
2038	48-134828
2038	48-134855
2038	48-134857
2038	48-134879
2038	48-134891
2038	48-134892
2038	48-134893
2038	48-134902
2038	48-134908
2038	48-134937
2038	48-134946
2038	48-134948
2038	48-134949
2038	48-134950
2038	48-134960
2038	48-134979
2038	48-134983
2038	48-134985
2038	48-137004
2038	48-137006
2038	48-137033
2038	48-137055
2038	48-137104
2038	48-137105
2038	48-137126
2038	48-137136
2038	48-137140
2038	48-137144
2038	48-137158
2038	48-137166
2038	48-137190
2038	48-137191
2038	48-137194
2038	48-137196
2038	48-137339
2038	48-137351
2038	48-137352
2038	48-137371
2038	48-137372
2038	48-137375
2038	48-137376
2038	48-137388
2038	48-137483
2038	48-155087
2038	48-155119
2038	48-155053
2038	48-43351A03
2038	48-43351A04
2038	48-43351A05
2038	48-43992J01
2038	48-44886001
2038	48-63026A46
2038	48-63077A29
2038	48-65112A65
2038	48-65112A67
2038	48-65113A88
2038	48-65118A64
2038	48-65123A67
2038	48-65123A95
2038	48-65144A72
2038	48-869589
2038	48-90232A03
2038	48-90232A04
2038	48-90232A10
2038	48-90232A13
2038	48-90232A19
2038	48-97046A04
2038	48-97046A18
2038	48-97046A51
2038	48-97127A02
2038	48-97127A03
2038	48-97162A01
2038	48-97162A02
2038	48-97177A04
2038	48-97177A07
2038	48-97177A08
2038	48-971A04
2038	48-97762A02
2038	48-K8669575
2038	48N355004
2038	48P63082A45
2038	48P63082A71
2038	48P65146A63
2038	48P65173A78
2038	48R869589
2038	48S134902
2038	48S134946
2038	48S134960
2038	48S134970
2038	48S134979
2038	48S137006
2038	48S43991J01
2038	48S43992J01
2038	48X97046A51
2038	48X97162A01
2038	48X97162A02
2038	48X97162A04
2038	48X97162A09
2038	48X97162A10
2038	49-1
2038	50-40101-04
2038	50-40101-05
2038	50C1047
2038	50C784
2038	500829
2038	500829B
2038	500829C
2038	51
2038	54BLK
2038	54BRN
2038	540RN
2038	54RED
2038	56-234
2038	56-4829
2038	57A10-1
2038	57A10-2
2038	57A101-4
2038	57A107-1
2038	57A107-2
2038	57A107-3
2038	57A107-4
2038	57A107-5
2038	57A107-6
2038	57A107-8
2038	57A10A-8-6
2038	57A34-12
2038	57A138-4
2038	57A138-4-6
2038	57A139-4-6
2038	57A141-1
2038	57A141-2
2038	57A141-3
2038	57A142-1
2038	57A143-1
2038	57A143-10
2038	57A143-11
2038	57A143-2
2038	57A143-4
2038	57A143-5
2038	57A143-6
2038	57A143-8
2038	57A143-9
2038	57A146-12
2038	57A151-6
2038	57A152-12
2038	57A160-1
2038	57A160-2
2038	57A160-3
2038	57A160-4
2038	57A20-1
2038	57A21-1
2038	57A21-10
2038	57A21-15
2038	57A21-2
2038	57A21-3
2038	57A21-4
2038	57A21-5
2038	57A21-6
2038	57A21-7
2038	57A21-9
2038	57A219-14
2038	57A241-14
2038	57A27-2
2038	57A280-14
2038	57A5-6
2038	57A5-7
2038	57A5-8
2038	57A7-1
2038	57A7-2
2038	57A7-3
2038	57A7-4
2038	57A7-5
2038	57A7-6
2038	57A7-7
2038	57B101-4
2038	57B134-12
2038	57B141-1
2038	57B141-2
2038	57B141-3
2038	57B142-1
2038	57B142-2
2038	57B142-3
2038	57B143-1
2038	57B143-10
2038	57B143-11
2038	57B143-2
2038	57B143-3
2038	57B143-5
2038	57B143-6
2038	57B143-7
2038	57B143-8
2038	57B143-9
2038	57B151-6
2038	57B152-12
2038	57B160-1
2038	57B160-2
2038	57B160-3
2038	57B160-4
2038	57B160-5
2038	57B160-6

RS 276-	Industry Standard No.	RS 276-	Industry Standard No.	RS 276-	Industry Standard No.	RS 276-	Industry Standard No.	RS 276-	Industry Standard No.
2038	57B160-7	2038	998017	2038	121-841	2038	260P08001		
2038	57B160-8	2038	998018	2038	121-846	2038	260P08401		
2038	57B166-12	2038	998018A	2038	121-848	2038	260P10403		
2038	57B21	2038	998019	2038	121-851	2038	260P10501		
2038	57B21-1	2038	998019A	2038	121-855	2038	260P10502		
2038	57B21-12	2038	998019B	2038	121-857	2038	260P1060		
2038	57B21-13	2038	998037	2038	121-869	2038	260P10602		
2038	57B21-14	2038	998055	2038	121-872	2038	260P11101		
2038	57B21-15	2038	998056	2038	121-884	2038	260P11101A		
2038	57B21-16	2038	998090-1	2038	121-895A	2038	260P16301		
2038	57B21-18	2038	100W1	2038	121-899	2038	260P16302		
2038	57B21-2	2038	0101-0060A	2038	121-900	2038	260P17201		
2038	57B21-3	2038	0101-0531	2038	121-907	2038	260P17601		
2038	57B21-4	2038	101-2(ADMIRAL)	2038	121-909	2038	260P17602		
2038	57B21-5	2038	101-3(ADMIRAL)	2038	121-910	2038	260P17603		
2038	57B819-14	2038	101-4(ADMIRAL)	2038	121-925	2038	260P28107		
2038	57B241-14	2038	102-0394-25	2038	121-929	2038	260P36501		
2038	57B280-14	2038	102-0460-02	2038	121-930	2038	260P70403		
2038	57C10-1	2038	102-0461-02	2038	121-943	2038	260P70501		
2038	57C10-2	2038	102-0535-02	2038	121-945	2038	260P70502		
2038	57C20-1	2038	102-0735-25	2038	121-953	2038	260Z00109		
2038	57C27-2	2038	102-0828-17	2038	121-954	2038	260Z00209		
2038	5705-6	2038	102-1342-02	2038	122-A484	2038	260Z00309		
2038	5705-7	2038	102-1675-11	2038	123-012	2038	290V02H69		
2038	5705-8	2038	102-1675-12	2038	125-655	2038	296-86		
2038	57C7-1	2038	0103-0060	2038	128	2038	297V070H49		
2038	57C7-2	2038	0103-0060B	2038	128N4	2038	297V072C01		
2038	57C7-3	2038	0103-0191	2038	130-138	2038	297V072C03		
2038	57C7-4	2038	0103-0389	2038	130-40304	2038	297V072C04		
2038	57C7-5	2038	0103-0521	2038	130-40362	2038	297V074C09		
2038	57C7-6	2038	0103-0521B	2038	130-40421	2038	297V078C01		
2038	57C7-7	2038	0103-0531	2038	130-40459	2038	297V078C02		
2038	57D107-8	2038	0103-389	2038	131(ARVIN)	2038	324-0149		
2038	57D24-1	2038	103-4	2038	131(SEARS)	2038	324-0150		
2038	57D24-2	2038	0103-9531/4460	2038	132-015	2038	324-1		
2038	57D24-3	2038	105-001-08	2038	139-4	2038	325-0028-84		
2038	58-1	2038	105-00106-00	2038	139N1	2038	330-1304-8		
2038	61B007-1	2038	105-00108-07	2038	139N1D	2038	344-6000-3		
2038	61B007-2	2038	105-005-04/2228-3	2038	0142	2038	344-6000-3A		
2038	66X0007-104	2038	105-005-12	2038	142-005	2038	344-6015-10		
2038	69N1	2038	105-008-04/2228-3	2038	142B1	2038	344-6015-11		
2038	72N1	2038	105-009-21/2228-3	2038	142N6	2038	344-6015-7		
2038	72N2	2038	105-02004-09	2038	151-0138	2038	344-6015-7A		
2038	74	2038	105-06004-00	2038	151M11	2038	344-6017-6		
2038	76-042-9-006	2038	105-24191-04	2038	151N1	2038	366-1(SYLVANIA)		
2038	76-13570-39	2038	105-941-97/2228-3	2038	151N11	2038	366-2(SYLVANIA)		
2038	76-13570-59	2038	106-001	2038	151N116	2038	386-7118P1		
2038	76-13866-17	2038	106-002	2038	161T2	2038	386-7188P1		
2038	76-13866-18	2038	106-351	2038	162T2	2038	396-717BP1		
2038	76-13866-19	2038	108-002	2038	173A04490-1	2038	417-124		
2038	76-13866-20	2038	112-520	2038	173A04490-2	2038	417-125		
2038	76-13866-59	2038	112-521	2038	176-003	2038	417-83		
2038	76-13866-62	2038	112-522	2038	176-003-9-001	2038	417-84		
2038	78C01	2038	113-398	2038	176-004-9-001	2038	417-85		
2038	78C02	2038	113-958	2038	176-005	2038	421-9685		
2038	80-053600	2038	114-118	2038	176-005-9-001	2038	422-1401		
2038	80-338030	2038	114-267	2038	176-006	2038	422-1402		
2038	80-338040	2038	116-073	2038	176-006-9-001	2038	422-2532		
2038	80-339430	2038	116-079	2038	176-007	2038	429-0986-12		
2038	80-339440	2038	116-080	2038	176-007-9-001	2038	430-22861		
2038	80-383940	2038	116-082	2038	176-016-9-001	2038	430-25763		
2038	80-383930	2038	116-083	2038	176-026-9-001	2038	430-25764		
2038	81-46125006-0	2038	116-198	2038	176-029-9-001	2038	430-25765		
2038	86-100002	2038	116-199	2038	176-031-9-001	2038	430-25767		
2038	86-100004	2038	116-200	2038	176-037-9-001	2038	488-2(SEARS)		
2038	86-100006	2038	118-1	2038	176-039-9-001	2038	490-2(SEARS)		
2038	86-100007	2038	118-2	2038	176-042-9-001	2038	491-2(SEARS)		
2038	86-10003	2038	118-3	2038	176-042-9-006	2038	499-1		
2038	86-10006	2038	118-4	2038	176-047-9-001	2038	501ES001M		
2038	86-10000	2038	120-004496	2038	176-072-9-005	2038	515-521		
2038	86-138-2	2038	120-004497	2038	176-074-9-001	2038	537FS		
2038	86-243-2	2038	120-004723	2038	180N1P	2038	546		
2038	86-244-2	2038	120-004724	2038	185-002	2038	551		
2038	86-245-2	2038	120-004725	2038	185-006	2038	573-472		
2038	86-386-2	2038	120-004881	2038	186-001	2038	573-474		
2038	86-416-2	2038	120-005291	2038	186-006	2038	573-474A		
2038	86-417-2	2038	120-005292	2038	189	2038	573-475		
2038	86-422-2	2038	120-005293	2038	200-007	2038	573-491		
2038	86-467-2	2038	120-005294	2038	200-010	2038	573-494		
2038	86-488-2	2038	120-005295	2038	200-015	2038	573-495		
2038	86-490-2	2038	120-005296	2038	200-055	2038	573-507		
2038	86-491-2	2038	120-005297	2038	200-056	2038	573-509		
2038	86-511-9	2038	120-005298	2038	201-25-4343-12	2038	576-0001-006		
2038	86-593-2	2038	121-1010	2038	201-254323-12	2038	576-0003-001		
2038	86-593-9	2038	121-113	2038	201-254323-13	2038	576-0003-002		
2038	86-594-2	2038	121-303	2038	201-254343-12	2038	576-0003-003		
2038	86-619-2	2038	121-316	2038	201-254343-49	2038	576-0003-004		
2038	86-620-2	2038	121-317	2038	207A9	2038	576-0003-005		
2038	86-621-2	2038	121-318	2038	217-1	2038	576-0003-006		
2038	86X0007-004	2038	121-318L	2038	220-001011	2038	576-0003-007		
2038	86X0007-204	2038	121-321	2038	220-001012	2038	576-0003-018		
2038	86X0038-001	2038	121-345	2038	223	2038	576-0003-020		
2038	86X0043-001	2038	121-434	2038	229-0151-3	2038	576-0003-021		
2038	86X6029-001	2038	121-434H	2038	229-0180-124	2038	576-0003-027		
2038	86X7-6013	2038	121-472	2038	229-0180-149	2038	576-0003-028		
2038	87-0002	2038	121-481	2038	229-0180-34	2038	576-0006-011		
2038	87-0002-1	2038	121-482	2038	229-0185-2	2038	576-0036-918		
2038	87-0003	2038	121-483	2038	229-0185-3	2038	576-0036-919		
2038	87-0023-7	2038	121-498	2038	229-0190-29	2038	600X0092-086		
2038	87-0027	2038	121-520	2038	229-0192-19	2038	601-113		
2038	87-0023-1	2038	121-546	2038	229-0204-23	2038	602-113		
2038	87-0235A	2038	121-546B	2038	229-0204-4	2038	602-61		
2038	87-0235C	2038	121-547	2038	229-0210-14	2038	603-113		
2038	089-214	2038	121-551	2038	229-0214-40	2038	604-113		
2038	089-215	2038	121-560	2038	229-0220-19	2038	605-113		
2038	089-216	2038	121-612-16	2038	229-0220-9	2038	610-041		
2038	90-452	2038	121-613	2038	229-0248-45	2038	610-041-1		
2038	90-453	2038	121-613-16	2038	229-0250-10	2038	610-041-2		
2038	90-457	2038	121-614	2038	229-5100-15U	2038	610-041-3		
2038	90-49	2038	121-614-9	2038	229-5100-15V	2038	610-042		
2038	90-601	2038	121-616	2038	229-5100-224	2038	610-042-1		
2038	90-612	2038	121-630	2038	229-5100-225	2038	610-045		
2038	91A	2038	121-637	2038	229-5100-226	2038	610-045-1		
2038	91B	2038	121-638	2038	229-5100-228	2038	610-045-2		
2038	91B0RN	2038	121-638B	2038	229-5100-33V	2038	610-069		
2038	91P	2038	121-642	2038	247-016-013	2038	610-069-1		
2038	92N1	2038	121-643	2038	249-1L	2038	610-072		
2038	92N1B	2038	121-644	2038	249N1	2038	610-072-1		
2038	95-125	2038	121-655	2038	250-0380	2038	610-072-2		
2038	95-126	2038	121-656	2038	250-1213	2038	610-073		
2038	95-127	2038	121-658	2038	260-10-021	2038	610-073-1		
2038	95-128	2038	121-659	2038	260-10-026	2038	612-16(ZENITH)		
2038	95-129	2038	121-684	2038	260-10-040	2038	612-16A		
2038	95-130	2038	121-733	2038	260-10-051	2038	613(ZENITH)		
2038	95-131	2038	121-742	2038	260D05701	2038	613-72		
2038	95-223	2038	121-753	2038	260D05707	2038	614(ZENITH)		
2038	96-056-234	2038	121-754	2038	260D15902	2038	614-12		
2038	96-138-2	2038	121-819	2038	260P05801	2038	653-202		
2038	96N(AIRLINE)	2038	121-825	2038	260P06901	2038	660-127		
2038	96N927	2038	121-827	2038	260P06902	2038	668CS		
2038	96N932	2038	121-834	2038	260P06903	2038	690V010H41		
2038	96NPT	2038	121-835	2038	260P07004	2038	690V028H2B		
2038	998016	2038	121-840	2038	260P07901	2038	690V028H48		
2038	998016-1					2038	690V028H69		

RS 276-	Industry Standard No.	RS 276-	Industry Standard No.	RS 276-	Industry Standard No.	RS 276-	Industry Standard No.
2038	690V028B89	2038	1006	2038	2495-523-1	2038	7131
2038	690V02H69	2038	1007-3054	2038	2498-507-2	2038	7132
2038	690V049HB1	2038	1007-3062	2038	2498-507-3	2038	7133
2038	690V060B58	2038	1011-11(R.P.)	2038	2498-508-2	2038	7134
2038	690V060C59	2038	1015-15	2038	2498-508-3	2038	7173
2038	690V070E49	2038	1016-83	2038	2498-903-2	2038	7174
2038	690V070E98	2038	1016-84	2038	2498-903-3	2038	7175
2038	690V075H68	2038	1026(GE)	2038	2510-101	2038	7177
2038	690V081H07	2038	1027(G.E.)	2038	2510-102	2038	7178
2038	690V084H94	2038	1039-0441	2038	2606-294	2038	7214
2038	690V084H95	2038	1039-0961	2038	2633(RCA)	2038	7215
2038	690V084H96	2038	1041-71	2038	2634(RCA)	2038	7216
2038	690V086H52	2038	1042-05	2038	2634-1	2038	7217
2038	690V086H87	2038	1049-0060	2038	2636	2038	7218
2038	690V086H96	2038	1049-0100	2038	2667	2038	7219
2038	690V088H44	2038	1049-1744	2038	2900-007	2038	7220
2038	690V088H45	2038	1061-8320	2038	2904-033	2038	7221
2038	690V088H48	2038	1071-4913	2038	2904-053	2038	7232
2038	690V103H23	2038	1076-1377	2038	3001	2038	7233
2038	690V103H24	2038	1080-07	2038	3012(NPN)	2038	7234
2038	690V103H25	2038	1106-97	2038	3018	2038	7235
2038	690V103H26	2038	1123-55	2038	3020	2038	7236
2038	690V103H27	2038	1123-56	2038	3021	2038	7237
2038	690V110H30	2038	1123-57	2038	3028	2038	7238
2038	690V110H31	2038	1123-58	2038	3107-204-90080	2038	7261
2038	690V110H32	2038	1123-59	2038	3227-E	2038	7262
2038	690V110H33	2038	1229H	2038	3370	2038	7264
2038	690V114H29	2038	1284	2038	003449	2038	7425
2038	690V114H31	2038	1373-17AL	2038	3508(WARDS)	2038	7426
2038	690V116H19	2038	1374-17	2038	3509	2038	7427
2038	690V118H59	2038	1374-17A	2038	3510(SEARS)	2038	7428
2038	0703	2038	1501	2038	3510(WARDS)	2038	7593-2
2038	750D858-213	2038	1502B	2038	3511	2038	7642
2038	753-0101-047	2038	1502D	2038	3516(WARDS)	2038	7810
2038	753-1303-801	2038	1634-17-14A	2038	3524(RCA)	2038	7811
2038	753-1372-100	2038	1710	2038	3524-1	2038	7813
2038	753-2000-007	2038	1761-17	2038	3524-1(RCA)	2038	7814
2038	753-2000-535	2038	1792-17	2038	3524-2(RCA)	2038	7815
2038	753-2000-710	2038	1852-17	2038	3527(RCA)	2038	8000-00003-035
2038	753-3000-535	2038	1880-17	2038	3530(RCA)	2038	8000-00003-036
2038	753-5751359	2038	1881-17	2038	3537(RCA)	2038	8000-00004-085
2038	753-9001-674	2038	1890-17	2038	3539(RCA)	2038	8000-00004-242
2038	753-9001-675	2038	1923-17	2038	3539-307-001	2038	8000-00004-298
2038	773RED	2038	1923-17-1	2038	3539-307-002	2038	8000-00004-299
2038	7740RN	2038	1925-17	2038	3568(RCA)	2038	8000-00004-300
2038	775BRN	2038	1931-17A	2038	3568(WARDS)	2038	8000-00004-P079
2038	779BLU	2038	1983-17	2038	3572-3	2038	8000-00004-P085
2038	785RED	2038	2000-204	2038	3576(RCA)	2038	8000-00009-177
2038	7840RN	2038	2000-205	2038	3598(RCA)	2038	8000-00011-049
2038	786	2038	2000-213	2038	3603(RCA)	2038	8000-00028-037
2038	787BLU	2038	2004-02	2038	3604(RCA)	2038	8000-00028-038
2038	800-536-00	2038	2022-03	2038	3610(RCA)	2038	8000-00028-206
2038	800-53600	2038	2028	2038	3618(RCA)	2038	8000-00055-003
2038	803-38030	2038	2028-00	2038	3646-2(RCA)	2038	8000-00041-046
2038	803-38040	2038	2032-33	2038	3652-2	2038	8000-00049-053
2038	803-39430	2038	2032-34	2038	3657-1	2038	8000-00049-054
2038	803-39440	2038	2041-01	2038	3657-2	2038	8606
2038	803-83840	2038	2057A-120	2038	3693(ARVINE)	2038	8607
2038	803-83930	2038	2057A-429	2038	4010(E.F.JOHNSON)	2038	8609
2038	85-0301-317	2038	2057A2-119	2038	4167(SEARS)	2038	8611
2038	902-003-0-012	2038	2057A2-127	2038	4168(SEARS)	2038	8710-162
2038	916-31024-5B	2038	2057A2-163	2038	4169(SEARS)	2038	8840-123
2038	921-1013	2038	2057A2-179	2038	4473-11	2038	8910-142
2038	921-106B	2038	2057A2-201	2038	4473-2	2038	9011E
2038	921-119B	2038	2057A2-224	2038	4473-3	2038	9011F
2038	921-129B	2038	2057A2-309	2038	4473-6	2038	9011F(TUNER)
2038	921-158B	2038	2057A2-310	2038	4473-7	2038	9011G
2038	921-170B	2038	2057A2-311	2038	4473-8	2038	9011H
2038	921-171B	2038	2057A2-313	2038	4490-1	2038	9016
2038	921-172B	2038	2057A2-314	2038	4587	2038	9016D
2038	921-173B	2038	2057A2-342	2038	4684-120-3	2038	9016E
2038	921-174B	2038	2057A2-386	2038	4685-285-3	2038	9016F
2038	921-181B	2038	2057A2-392	2038	4686-107-3	2038	9016G
2038	921-20	2038	2057A2-393	2038	4686-108-3	2038	09018
2038	921-20A	2038	2057A2-394	2038	4686-112-3	2038	9018D
2038	921-20B	2038	2057A2-402	2038	4686-114-3	2038	9018E
2038	921-21	2038	2057A2-427	2038	4686-118-3	2038	9018F
2038	921-212B	2038	2057A2-432	2038	4686-120-3	2038	9018G
2038	921-21A	2038	2057A2-448	2038	4686-126-3	2038	9300
2038	921-21B	2038	2057A2-465	2038	4686-127-3	2038	9300A
2038	921-21BK	2038	2057A2-477	2038	4686-131-3	2038	9300B
2038	921-22	2038	2057A2-478	2038	4686-140-3	2038	9300Z
2038	921-22A	2038	2057A2-483	2038	4686-169-3	2038	9314
2038	921-22B	2038	2057A2-504	2038	4686-171-3	2038	9426B
2038	921-23	2038	2057A2-505	2038	4686-172-3	2038	9426C
2038	921-23A	2038	2057A2-507	2038	4686-207-3	2038	9513
2038	921-23B	2038	2057A2-508	2038	4686-208-3	2038	9600C
2038	921-264B	2038	2057A2-509	2038	4686-209-3	2038	9600F
2038	921-265	2038	2057A2-526	2038	4686-224-3	2038	9600G
2038	921-265B	2038	2057A2-527	2038	4686-228-3	2038	9600H
2038	921-266	2038	2057A2-541	2038	4686-244-3	2038	9601
2038	921-266B	2038	2057A2-87	2038	4686-251-3	2038	9601-12
2038	921-267	2038	2057A42-477	2038	4686-95-3	2038	9604F
2038	921-267B	2038	2057B-113	2038	4706	2038	9618
2038	921-275B	2038	2057B100-12	2038	4709	2038	9623F
2038	921-30	2038	2057B2-103	2038	4802-00002	2038	9623G
2038	921-301B	2038	2057B2-108	2038	4802-00014	2038	9623H
2038	921-30A	2038	2057B2-109	2038	4820	2038	9625F
2038	921-30B	2038	2057B2-110	2038	4821	2038	9625E
2038	921-31	2038	2057B2-111	2038	4825	2038	9630C
2038	921-312B	2038	2057B2-112	2038	4826	2038	11252-0
2038	921-313B	2038	2057B2-119	2038	4837	2038	11252-1
2038	921-31A	2038	2057B2-120	2038	4845	2038	11252-2
2038	921-31B	2038	2057B2-127	2038	4855	2038	11339-8
2038	921-32	2038	2057B2-128	2038	4857	2038	11395-8
2038	921-325B	2038	2057B2-14	2038	5001-021	2038	11426-7
2038	921-32A	2038	2057B2-160	2038	5001-032	2038	11607-3
2038	921-32B	2038	2057B2-161	2038	5001-070	2038	11607-9
2038	921-33	2038	2057B2-162	2038	5001-506	2038	11608-0
2038	921-334B	2038	2057B2-64	2038	5001-510	2038	11608-2
2038	921-335B	2038	2057B2-85	2038	5001-545	2038	11608-3
2038	921-336B	2038	2057B2-87	2038	5093	2038	11619-8
2038	921-338B	2038	2065-04	2038	5313-461B	2038	11619-9
2038	921-33B	2038	2093A2-289	2038	5613-46B	2038	11620-0
2038	921-34	2038	2180-151	2038	5613-8288	2038	11620-1
2038	921-349	2038	2180-152	2038	5710	2038	16190
2038	921-349B	2038	2214-17	2038	6158	2038	16194
2038	921-34A	2038	2224-17	2038	6158-3	2038	19420
2038	921-34B	2038	2225-17	2038	6185-3	2038	23114-056
2038	921-350B	2038	2226-17	2038	6507(AIRLINE)	2038	23114-057
2038	921-379	2038	2362-1(SYLVANIA)	2038	7112(TRANSISTOR)	2038	23114-060
2038	921-428	2038	2427(RCA)	2038	7115	2038	23114-078
2038	921-72B	2038	2445	2038	7116	2038	23114-104
2038	921-97B	2038	2450(RCA)	2038	7117(GE)	2038	23125-037
2038	921-98B	2038	2473(RCA)	2038	7118	2038	25114-121
2038	930X4	2038	2476(RCA)	2038	7122	2038	26810-151
2038	930X5	2038	2477(RCA)	2038	7123	2038	26810-155
2038	947-1(SYLVANIA)	2038	2495-166-1	2038	7124	2038	30292
2038	1002-02	2038	2495-166-4	2038	7125	2038	030930
2038	1002-02A	2038	2495-166-8	2038	7126	2038	35004
2038	1002-04-1	2038	2495-166-9	2038	7127	2038	35449
2038	1002-68	2038	2495-520	2038	7128	2038	36212
2038	1004(JULIETTE)	2038	2495-521			2038	36212V1
2038	1004-17	2038	2495-522-1			2038	36578

RS 276-	Industry Standard No.	RS 276-	Industry Standard No.	RS 276-	Industry Standard No.	RS 276-	Industry Standard No.	RS 276-	Industry Standard No.
2038	36581	2038	122517	2038	231140-31	2038	970244	2038	970244
2038	36847	2038	122518	2038	231140-34	2038	970245		
2038	36918	2038	122904	2038	231140-44	2038	970249		
2038	36919	2038	123160	2038	232017	2038	970309		
2038	37383	2038	123429	2038	232840	2038	970309-1		
2038	37384	2038	123430	2038	233117	2038	970309-12		
2038	37694A	2038	123431	2038	236251	2038	970309-2		
2038	37694B	2038	124263	2038	236706	2038	970309-3		
2038	38207	2038	124412	2038	256907	2038	970309-4		
2038	38208	2038	125137	2038	237020	2038	970309-5		
2038	38246	2038	125138	2038	237021	2038	970310		
2038	38246A	2038	125263	2038	237024	2038	970310-1		
2038	38510-162	2038	125264	2038	237026	2038	970310-12		
2038	38511	2038	125392	2038	237785	2038	970310-2		
2038	38511A	2038	125475-14	2038	237840	2038	970310-3		
2038	38785	2038	125944	2038	241249	2038	970310-4		
2038	38786	2038	125994	2038	241778	2038	970310-5		
2038	38920	2038	125994-14	2038	241960	2038	970332		
2038	38921	2038	125995	2038	242590	2038	970332-12		
2038	39331	2038	126023	2038	242960	2038	970911		
2038	39730	2038	126024	2038	243318	2038	980138		
2038	39731	2038	126670	2038	243645	2038	980139		
2038	39789	2038	126698	2038	245078-3	2038	982268		
2038	40294	2038	127693	2038	256817	2038	982269		
2038	40295	2038	127792	2038	257540	2038	982321		
2038	40351	2038	127793	2038	260565	2038	982815		
2038	40352	2038	127794	2038	265074	2038	982816		
2038	40413	2038	129050	2038	265241	2038	982817		
2038	40414	2038	129144	2038	267797	2038	982818		
2038	40469	2038	129392-14	2038	304900	2038	982819		
2038	40470	2038	129393-14	2038	346015-15	2038	983095		
2038	40472	2038	129394	2038	346015-16	2038	983096		
2038	40475	2038	129394-14	2038	346015-17	2038	984156		
2038	40478	2038	129574	2038	346015-18	2038	984158		
2038	40479	2038	129574	2038	346015-19	2038	984159		
2038	40480	2038	129897	2038	346015-20	2038	984194		
2038	40481	2038	129979	2038	346015-21	2038	984195		
2038	40482	2038	130278	2038	346015-22	2038	984577		
2038	40894	2038	130403-04	2038	346015-25	2038	984743		
2038	40895	2038	130403-62	2038	346015-37	2038	984744		
2038	40896	2038	130404-21	2038	489751-027	2038	984851		
2038	41689	2038	130404-59	2038	489751-131	2038	984852		
2038	41694	2038	131221	2038	489751-137	2038	984853		
2038	043001	2038	131648	2038	489751-143	2038	984875		
2038	45022-860	2038	131844	2038	489751-145	2038	985096		
2038	45810-163	2038	131848	2038	489751-147	2038	985097		
2038	50957-03	2038	134142	2038	489751-148	2038	985215		
2038	55170-1	2038	134144	2038	489751-162	2038	985442A		
2038	57000-5452	2038	134263	2038	489751-165	2038	985443A		
2038	60048	2038	134417	2038	489751-167	2038	985444A		
2038	60314	2038	134419	2038	489751-168	2038	986634		
2038	61009-1	2038	134442	2038	489751-169	2038	986635		
2038	61009-1-1	2038	134857	2038	489751-171	2038	988000		
2038	61009-1-2	2038	136165	2038	489751-206	2038	988001		
2038	61009-2	2038	136168	2038	573101	2038	988002		
2038	61009-2-1	2038	136239	2038	573472	2038	988985		
2038	61009-6	2038	136240	2038	573479	2038	988986		
2038	61009-6-1	2038	137127	2038	0573491	2038	988987		
2038	61010-0	2038	145595	2038	0573492	2038	988988		
2038	61010-0-1	2038	147245-0-1	2038	573494	2038	988989		
2038	61010-7-1	2038	147356-9-1	2038	0573495	2038	992052		
2038	61013-9-1	2038	147357-2-1	2038	0573506	2038	994634		
2038	61015-0-1	2038	147357-9-1	2038	0573506H	2038	1222463		
2038	61133	2038	148751-147	2038	0573507H	2038	1223920		
2038	61661	2038	150117	2038	0573509	2038	1408615-1		
2038	61663	2038	156931	2038	0573511	2038	1408640-1		
2038	62449	2038	162002-090	2038	0573570	2038	1471115-13		
2038	67802	2038	165995	2038	601113	2038	1471115-14		
2038	70167-8-00	2038	166272	2038	602113	2038	1472450-1		
2038	70231	2038	168657	2038	603113	2038	1472634-1		
2038	70260-11	2038	168658	2038	604113	2038	1473524-2		
2038	70260-12	2038	168659	2038	610041	2038	1473530-1		
2038	70260-13	2038	169195	2038	610041-1	2038	1473530-2		
2038	72949-10	2038	170398	2038	610041-3	2038	1473533-1		
2038	72951-95	2038	170794	2038	610042-1	2038	1473537-1		
2038	72951-96	2038	171003(SEARS)	2038	610045	2038	1473568-1		
2038	72979-80	2038	171009(SEARS)	2038	610045-1	2038	1473603-1		
2038	000073140	2038	171028	2038	610045-2	2038	1473604-3		
2038	75616-6	2038	171029(SEARS)	2038	610046-7	2038	1473606-1		
2038	75810-17	2038	171030(SEARS)	2038	610069	2038	1473617-1		
2038	79855	2038	171031(SEARS)	2038	610069-1	2038	1473652-1		
2038	79856	2038	171032	2038	610072	2038	1473657-1		
2038	080006	2038	171033	2038	610072-2	2038	1473657-2		
2038	080021	2038	171034	2038	610072-2	2038	1700020		
2038	080022	2038	171045	2038	610073	2038	1700032		
2038	080023	2038	171048	2038	610091	2038	1700033		
2038	080041	2038	171052	2038	610091-1	2038	1700039		
2038	080042	2038	171054	2038	610091-2	2038	1810037		
2038	080059	2038	171090-1	2038	610092	2038	1810038		
2038	080060	2038	171139-1	2038	610092-1	2038	1810039		
2038	080028	2038	171140-1	2038	610092-3	2038	1815056		
2038	82716	2038	171141-1	2038	610096	2038	1815037		
2038	88045-142	2038	171162-027	2038	610096-1	2038	1815039		
2038	94044	2038	171162-128	2038	610100	2038	1815045		
2038	95125	2038	171162-129	2038	610100-3	2038	1815047		
2038	95126	2038	171162-130	2038	610107-1	2038	1815067		
2038	95127	2038	171162-131	2038	610128-4	2038	1815068		
2038	95128	2038	171162-186	2038	610129-D	2038	1817004		
2038	95129	2038	171162-187	2038	610139-1	2038	1817005-3		
2038	95130	2038	171162-278	2038	610142-1	2038	1817006-3		
2038	95131	2038	171162-279	2038	610142-6	2038	1817008		
2038	95170-1	2038	171162-280	2038	610150	2038	1817045		
2038	95170-2	2038	171206-1	2038	610180-1	2038	1819045		
2038	95170-2(XSTR)	2038	171206-2	2038	610232-2	2038	2000646-105		
2038	95171-1	2038	171206-4	2038	610249-1	2038	2000757-80		
2038	95171-3	2038	171206-5	2038	613112	2038	2000804-7		
2038	95171-4	2038	171206-8	2038	740950	2038	2000804-8		
2038	95221	2038	171207-1	2038	740951	2038	2002332-53		
2038	95242-1	2038	171207-2	2038	742549	2038	2002332-54		
2038	101434	2038	171217-3	2038	760142	2038	2002332-55		
2038	111943	2038	171915	2038	815164	2038	2002332-56		
2038	112355	2038	175006-187	2038	815165	2038	2002620-18		
2038	113938	2038	175043-062	2038	815170	2038	2002620-19		
2038	114143-1	2038	175043-063	2038	815172	2038	2003542-109		
2038	114525	2038	175043-064	2038	815172A	2038	2004746-114		
2038	115440	2038	175043-100	2038	815173A	2038	2004746-115		
2038	115910	2038	175043-107	2038	815173C	2038	2006431-44		
2038	115925	2038	181003-7	2038	815173P	2038	2006513-19		
2038	116073	2038	181003-8	2038	815209	2038	2091859-0711		
2038	116079	2038	181003-9	2038	824960-Q	2038	2092418-0715		
2038	116080	2038	181503-6	2038	910799	2038	2092418-0724		
2038	116082	2038	181503-7	2038	916029	2038	2093308-070		
2038	116083	2038	181503-7	2038	916060	2038	2093308-0711		
2038	116119	2038	181503-9	2038	916069	2038	2093308-0700		
2038	116198	2038	181504-1	2038	964634	2038	2093308-0704A		
2038	116199	2038	181504-7	2038	964713	2038	2093308-0705A		
2038	116200	2038	181506-7	2038	965074	2038	2093308-0706A		
2038	117823	2038	200064-6-103	2038	965633	2038	2093308-1		
2038	118822	2038	200064-6-105	2038	965634	2038	2093308-2		
2038	119414	2038	209417-0714	2038	970046	2038	2093308-3		
2038	119554	2038	227000	2038	970046-1	2038	2320062		
2038	119555	2038	229392	2038	970046-2	2038	2320063		
2038	119556	2038	231140-01	2038	970046A	2038	2320073		
2038	119557	2038	231140-07			2038	2320591		
		2038	231140-23			2038	2320596		
						2038	2320598		

RS 276-	Industry Standard No.
2038	2320643
2038	2320644
2038	2320646
2038	2320646-1
2038	2320647
2038	2320647-1
2038	2321511
2038	2495166-1
2038	2498507-1
2038	2498507-2
2038	2498507-3
2038	2498903-2
2038	2596071
2038	3181972
2038	3596067
2038	3596068
2038	3596069
2038	3596070
2038	3596071
2038	3596072
2038	3596260
2038	3596261
2038	3597103
2038	3597104
2038	3673351K
2038(2)	EC0289MP
2038(2)	GE-268MP
2041	2SD1470R
2041	A-11166527
2041	A-120327
2041	A-140605
2041	A-18
2041	A-1854-0291-1
2041	A-1854-0294-1
2041	A-1854-0458
2041	A-6-67703
2041	A-6-67703-A-7
2041	A-PS-1509730-0-0
2041	A-PS-1510196-0-0
2041	A054-154
2041	A08-1050115
2041	A08-105018
2041	A1-44
2041	A112363
2041	A13-0032
2041	A13-17918-1
2041	A13-23594-1
2041	A13-33188-2
2041	A14-601-10
2041	A14-601-12
2041	A14-601-13
2041	A18
2041	A18-4
2041	A2418
2041	A2B
2041	A2B-2
2041	A2EBLK
2041	A2EBRN
2041	A2EBRN-1
2041	A28
2041	A28-3
2041	A3902441
2041	A391593
2041	A3L4-6001
2041	A3L4-6001-01
2041	A3TE120
2041	A3TE230
2041	A3TE240
2041	A3TX003
2041	A3TX004
2041	A3U
2041	A3U-4
2041	A417014
2041	A417033
2041	A43023843
2041	A417014
2041	A4J
2041	A4J(RED)
2041	A4JBLK
2041	A4JBRN
2041	A4JRED-1
2041	A4S
2041	A4S-1
2041	A4Z
2041	A515
2041	A522
2041	A522-3
2041	A523
2041	A572
2041	A572-1
2041	A580-040215
2041	A580-040315
2041	A580-040515
2041	A580-080215
2041	A580-080315
2041	A580-080515
2041	A5V
2041	A6L
2041	A6LBLK
2041	A6LBLK-1
2041	A6LBRN
2041	A6LBRN-1
2041	A6LRED
2041	A6LRED-1
2041	A6N
2041	A6N-6
2041	A7-12
2041	A7-13
2041	A7M
2041	A7M-5(TRANSISTOR)
2041	A80052402
2041	A80414120
2041	A80414130
2041	A8P
2041	A8U
2041	A8W
2041	A9N
2041	AEX-82308
2041	AEX79846
2041	AEX9846
2041	AM3235
2041	AMF-121
2041	AMP104
2041	AMP105
2041	AMP115
2041	AMP116
2041	AMP117
2041	AMP117A
2041	AMP118
2041	AMP118A
2041	AMP119
2041	AMP119A
2041	AMP120
2041	AMP120A
2041	AMP201
2041	AMP201B
2041	AMP201C
2041	AMP210
2041	AMP210A
2041	AMP210B
2041	AMP2919-2
2041	AR15
2041	AR15-L8-0026
2041	AT-10
2041	AT-1856
2041	AT1856
2041	AT3260
2041	ATC-TR-15
2041	B-12822-2
2041	B-12822-4
2041	B0301-049
2041	B133550
2041	B133577
2041	B133684
2041	B133685
2041	B170000
2041	B170000-ORG
2041	B170000-ORN
2041	B170000-RED
2041	B170000-YEL
2041	B170000BLK
2041	B170000BRN
2041	B170001
2041	B170001-BLK
2041	B170001-BRN
2041	B170001-ORG
2041	B170001-RED
2041	B170001-YEL
2041	B170001BLK
2041	B170001BRN
2041	B170002
2041	B170002-ORG
2041	B170002-RED
2041	B170002-YEL
2041	B170003
2041	B170003-BLK
2041	B170003-ORG
2041	B170003-RED
2041	B170003-YEL
2041	B170004
2041	B170004-BLK
2041	B170004-BRN
2041	B170004-ORG
2041	B170004-YEL
2041	B170005
2041	B170005-BLK
2041	B170005-BRN
2041	B170005-ORG
2041	B170005-YEL
2041	B170006
2041	B170006-BLK
2041	B170006-ORG
2041	B170006-RED
2041	B170006-YEL
2041	B170007
2041	B170007-BLK
2041	B170007-BRN
2041	B170007-ORG
2041	B170007-RED
2041	B170007-YEL
2041	B170008-BLK
2041	B170008-BRN
2041	B170008-RED
2041	B170008-YEL
2041	B170009
2041	B170010
2041	B170011
2041	B170012
2041	B170013
2041	B170014
2041	B170015
2041	B170016
2041	B170017
2041	B170018
2041	B170019
2041	B170020
2041	B170022
2041	B170023
2041	B170024
2041	B170025
2041	B170026
2041	B17307
2041	B177000
2041	B5020
2041	B66X0040-006
2041	BD111
2041	BD111A
2041	BD112
2041	BD113
2041	BD116
2041	BD118
2041	BD121
2041	BD123
2041	BD130
2041	BD141
2041	BD142
2041	BD145
2041	BD181
2041	BD182
2041	BD183ELK
2041	BD184
2041	BD245
2041	BD245A
2041	BD249
2041	BD249A
2041	BD249B
2041	BD249C
2041	BDX10
2041	BDX11
2041	BDX13
2041	BDX24
2041	BDX40
2041	BDX41
2041	BDY-10
2041	BDY10
2041	BDY11
2041	BDY17
2041	BDY23
2041	BDY38
2041	BDY39
2041	BDY53
2041	BDY57
2041	BDY58
2041	BDY63
2041	BDY74
2041	BDY76
2041	BDY90(AUDIO)
2041	BDY91(AUDIO)
2041	BDY92(AUDIO)
2041	BLY10
2041	BLY11
2041	BLY12
2041	BLY15
2041	BLY47
2041	BLY48
2041	BN7133
2041	BN7214
2041	BRC-116
2041	BUY10
2041	BUY11
2041	BUY43
2041	BUY46
2041	C1030
2041	C1030A
2041	C1030B
2041	C1030C
2041	C1030D
2041	C1030P
2041	C1051
2041	C1051C
2041	C1051D
2041	C1051E
2041	C1051F
2041	C1051LC
2041	C1051LD
2041	C1051LE
2041	C1051LF
2041	C1079
2041	C1079R
2041	C1079Y
2041	C1080
2041	C1080R
2041	C1080Y
2041	C1111
2041	C1115
2041	C1343
2041	C1343A
2041	C1343B
2041	C1343C
2041	C1343H
2041	C1343HA
2041	C1343HB
2041	C1402
2041	C21
2041	C244
2041	C36566
2041	C493
2041	C493-BL
2041	C493-R
2041	C493-Y
2041	C494
2041	C494-BL
2041	C494-R
2041	C494-Y
2041	C494BL
2041	C49Y
2041	C520(TRANSISTOR)
2041	C520A
2041	C521
2041	C521A
2041	C646
2041	C647
2041	C647Q
2041	C647R
2041	C664
2041	C664B
2041	C664C
2041	C665
2041	C665H
2041	C665HA
2041	C665HB
2041	C736
2041	C765
2041	C768
2041	C793
2041	C793R
2041	C793Y
2041	C794R
2041	C79BL
2041	C851
2041	C897
2041	C897A
2041	C897B
2041	C897C
2041	C898
2041	C898A
2041	C898B
2041	C898C
2041	CD461
2041	CD461-014-614
2041	C11-225-Q
2041	CP400
2041	CP401
2041	CP404
2041	CP405
2041	CP406
2041	CP407
2041	CP408
2041	CXL130
2041	CXL181
2041	CXL223
2041	D113
2041	D113-0
2041	D113-0
2041	D113-R
2041	D113-Y
2041	D114
2041	D114-0
2041	D114-0
2041	D114-R
2041	D114-Y
2041	D118
2041	D118BL
2041	D118R
2041	D118Y
2041	D119
2041	D119BL
2041	D119R
2041	D119Y
2041	D12
2041	D124(TRANSISTOR)
2041	D124A
2041	D124AH
2041	D124AHA
2041	D124B
2041	D125
2041	D125A
2041	D125AH
2041	D125AHA
2041	D125AHB
2041	D126
2041	D126A
2041	D126AH
2041	D126AHA
2041	D126AHB
2041	D126H
2041	D126HA
2041	D126HB
2041	D132
2041	D146
2041	D146UK
2041	D147
2041	D15
2041	D151
2041	D16
2041	D163
2041	D164
2041	D175
2041	D180A
2041	D180B
2041	D180C
2041	D180D
2041	D180M
2041	D188
2041	D188A
2041	D188B
2041	D188C
2041	D201
2041	D201(0)
2041	D201(0)
2041	D201M
2041	D201Y
2041	D211
2041	D212
2041	D241H
2041	D26A
2041	D26B
2041	D26C
2041	D3005VN
2041	D319
2041	D322
2041	D322A
2041	D322B
2041	D322C
2041	D323
2041	D323A
2041	D323B
2041	D323C
2041	D41
2041	D425
2041	D4250
2041	D4250
2041	D53
2041	D55
2041	D55A
2041	D68B
2041	D68C
2041	D68D
2041	D68E
2041	D69
2041	D73
2041	D73A
2041	D73B
2041	D73C
2041	D73D
2041	D73E
2041	D74
2041	D74A
2041	D74B
2041	D74C
2041	D74D
2041	D74E
2041	D80
2041	D81
2041	D82
2041	D82A
2041	DD-79D107-1
2041	DF-2
2041	DS-509
2041	DS-514
2041	DS-519
2041	DS509
2041	DS514
2041	DS519
2041	DT4011
2041	DT4110
2041	DT4111
2041	DT4120
2041	DT4126
2041	EA15X100
2041	EA15X123
2041	EA1740
2041	EC961
2041	ECG130
2041	ECG181
2041	ECG223
2041	ECG280
2041	EP15X29
2041	E816(ELCOM)
2041	E831(ELCOM)
2041	E843(ELCOM)
2041	E869(ELCOM)
2041	ETS-003
2041	ETS-005
2041	EX524-X
2041	FBN-36220
2041	FBN-36485
2041	FBN-36603
2041	FBN-36972
2041	FBN-36973
2041	FBN-38022
2041	FD-1029-UU
2041	FD4500AL
2041	F82003-1
2041	G181-725-001
2041	G23-45
2041	G23-67
2041	G23-76
2041	GE-14
2041	GE-19
2041	GE-255
2041	GE-262
2041	GE-75
2041	GRA83-R2982
2041	HEP247
2041	HEP704
2041	HEP705
2041	HEP87000
2041	HEP87002
2041	HEP87004

RS 276-	Industry Standard No.	RS 276-	Industry Standard No.	RS 276-	Industry Standard No.	RS 276-	Industry Standard No.
2041	HP22	2041	R227077499	2041	STC4255	2041	2N3714
2041	HP24	2041	R227078533	2041	STT2300	2041	2N3715
2041	HST-9201	2041	R2982	2041	STT3500	2041	2N3771
2041	HST-9205	2041	R4369	2041	STX0014	2041	2N3772
2041	HST-9206	2041	RC-1700	2041	STX0027	2041	2N3863
2041	HST-9210	2041	RC1700	2041	STX0032	2041	2N3864
2041	HT30494	2041	RCA1806	2041	T-23-71	2041	2N3917
2041	HT304941X	2041	RC8242	2041	T-Q5105	2041	2N3918
2041	HT30494X	2041	RE19	2041	T1P3055	2041	2N4111
2041	HT401191	2041	RE37	2041	T841	2041	2N4112
2041	HT401191A	2041	REN130	2041	T842	2041	2N4113
2041	HT401191B	2041	REN181	2041	T843	2041	2N4114
2041	HT401193AO	2041	REN223	2041	T844	2041	2N4130
2041	HT9000410-0	2041	REN280	2041	TA2577A	2041	2N4347
2041	HT9000410-0	2041	RT-131	2041	TA7068	2041	2N4348
2041	IP20-0028	2041	RT-149	2041	TA7069	2041	2N4395
2041	TR-TR59	2041	RT-154	2041	TA7199	2041	2N4396
2041	IRTR-59	2041	R-305	2041	TA7200	2041	2N4913
2041	IRTR-61	2041	S-305-PD	2041	TA7201	2041	2N4914
2041	IRTR26	2041	S-305A	2041	TA7202	2041	2N4915
2041	IRTR36	2041	S-356	2041	TK30551	2041	2N5034
2041	IRTR59	2041	S124AHB	2041	TK30552	2041	2N5055
2041	IRTR61	2041	S1685	2041	TK30555	2041	2N5036
2041	K071964-001	2041	S1691	2041	TK30556	2041	2N5037
2041	K4-525	2041	S1692	2041	TK30557	2041	2N5067
2041	KB-1007	2041	S1865	2041	TK30560	2041	2N5068
2041	KS-19938	2041	S1905	2041	TK9201	2041	2N5069
2041	KSD1051	2041	S1905A	2041	TM130	2041	2N5301
2041	KSD1052	2041	S1907	2041	TM181	2041	2N5302
2041	KSD1055	2041	S1977634	2041	TM223	2041	2N5303
2041	KSD1056	2041	S2003-1	2041	TM280	2041	2N5614
2041	KSD1057	2041	S2241	2041	TNJ72148	2041	2N5616
2041	KSD2203	2041	S2392	2041	TQ-PD-3055	2041	2N5622
2041	KSD3055	2041	S2403B	2041	TR-1000-7	2041	2N5629
2041	KSD3771	2041	S2403C	2041	TR-1039-4	2041	2N5632
2041	KSD3772	2041	S2471	2041	TR-176(OLSON)	2041	2N5750
2041	KSD9701	2041	S2741	2041	TR-26	2041	2N5885
2041	KSD9701A	2041	S305	2041	TR-36	2041	2N5886
2041	KSD9704	2041	S305A	2041	TR-36MP	2041	2N5970
2041	KSD9707	2041	S305D	2041	TR-59	2041	2N6253
2041	LM51116Q	2041	S353	2041	TR-8018	2041	2N6254
2041	LNT5116	2041	S35487	2041	TRO1060-7	2041	2N6257
2041	LNT5497	2041	S356	2041	TR1007	2041	2N6258
2041	LN78533	2041	S3771	2041	TR1009A	2041	2N6326
2041	M18-12795B	2041	89S8133	2041	TR1025	2041	2N6327
2041	M4715	2041	89S8165	2041	TR1039-4	2041	2N6328
2041	M4882	2041	SAB-1	2041	TR1039-6	2041	2N6559
2041	M7543-1	2041	SB5	2041	TR1077	2041	2N6371
2041	M75549-2	2041	SB6	2041	TR1490	2041	28033
2041	M9244	2041	SB7	2041	TR1491	2041	28034
2041	M9259	2041	SC0321	2041	TR1492	2041	2SC1030
2041	M9278	2041	SC0328	2041	TR1493	2041	2SC1030-0R
2041	M9302	2041	SC0321	2041	TR2327574	2041	2SC1030A
2041	M9321	2041	SCD-T32Q	2041	TR26	2041	2SC1030B
2041	M9480	2041	SCD321	2041	TR26C	2041	2SC1030B2C
2041	M9515	2041	SCE321	2041	TR271 TR26	2041	2SC1030C
2041	M9628	2041	SDI1621	2041	TR36	2041	2SC1030D
2041	M9639	2041	SDI1622	2041	TR59	2041	2SC1030E
2041	M9666	2041	SDI1623	2041	TR8018	2041	2SC1030P
2041	M9715	2041	SDI1631	2041	TS-1193-736	2041	2SC1030G
2041	MHT7601	2041	SDI1632	2041	TVS-28C646	2041	2SC1030H
2041	MHT7602	2041	SDI1633	2041	TVS28C1629A	2041	2SC1030K
2041	MHT7603	2041	SDT9201	2041	TVS28C647	2041	2SC1030X
2041	MHT7607	2041	SDT9205	2041	VS-28C41	2041	2SC1030L
2041	MHT7608	2041	SDT9206	2041	W16	2041	2SC1030M
2041	MHT7609	2041	SDT9210	2041	WEP247	2041	2SC1030R
2041	MJ2800	2041	SDT9261	2041	WEP247MP	2041	2SC1030X
2041	MJ2801	2041	SDT9303	2041	WEP704	2041	2SC1030X
2041	MJ2802	2041	SDT9306	2041	WEP87000	2041	2SC1030Y
2041	MJ2840	2041	SDT9309	2041	X1-548	2041	2SC1051
2041	MJ2841	2041	SB-3033	2041	X194-3005829A	2041	2SC1051C
2041	MJ3771	2041	SE3032	2041	XA-1078	2041	2SC1051D
2041	MJ3772	2041	SE3033	2041	XA-1161	2041	2SC1051B
2041	MJ480	2041	SE3035	2041	XC723	2041	2SC1051F
2041	MJ481	2041	SE3036	2041	XI-548	2041	2SC1051LC
2041	MJ5257	2041	SE9002	2041	XT-548A	2041	2SC1051LD
2041	MJ6257	2041	SE9080	2041	ZT1487	2041	2SC1051LE
2041	MJ802	2041	SE8632	2041	ZT1488	2041	2SC1051LF
2041	MJ2940	2041	SE8881	2041	ZT1489	2041	2SC1079
2041	MU-26-1C	2041	SJ1106	2041	ZT1490	2041	2SC1079R
2041	N-121122	2041	SJ1470	2041	ZT1702	2041	2SC1079Y
2041	N-52329	2041	SJ2000	2041	1-002	2041	2SC1080
2041	N0282CT	2041	SJ2008	2041	1-003	2041	2SC1080R
2041	N121122	2041	SJ2047	2041	001-021270	2041	2SC1080Y
2041	P-10954-1	2041	SJ2064	2041	001-021280	2041	2SC1111
2041	P-10954-2	2041	SJ3464	2041	01-040201	2041	2SC1115
2041	P-11810-1	2041	SJ3519	2041	1A1123100-1	2041	2SC1322
2041	P-11901-1	2041	SJ3604	2041	1S1-0356-00-A	2041	2SC1343
2041	P-11901-3	2041	SJ3678	2041	2-G-3055	2041	2SC1343A
2041	P0445-0034-1	2041	SJ619	2041	2CD1988	2041	2SC1343B
2041	P0445-0034-2	2041	SJ619-1	2041	2D010	2041	2SC1343BL
2041	P04450034-1	2041	SJ820	2041	203055	2041	2SC1343C
2041	P04450034-2	2041	SJ8701	2041	2N1069	2041	2SC1343D
2041	P04450037	2041	SJ9110	2041	2N1070	2041	2SC1343E
2041	P04450040-002	2041	SK3027	2041	2N1072	2041	2SC1343F
2041	P10619-1	2041	SK3036	2041	2N1422	2041	2SC1343G
2041	P2271	2041	SK3510	2041	2N1423	2041	2SC1343G-R
2041	P3139	2041	SK3511	2041	2N1487	2041	2SC1343GN
2041	P50200-11	2041	SK3535	2041	2N1488	2041	2SC1343H
2041	P5034	2041	SK3535/181	2041	2N1489	2041	2SC1343HA
2041	P5149	2041	SK3561	2041	2N1490	2041	2SC1343HB
2041	P6500A	2041	SK3563	2041	2N1702	2041	2SC1343J
2041	PP-AR15	2041	SK4231	2041	2N1703	2041	2SC1343K
2041	PMC-QP0010	2041	SPC40411	2041	2N2305	2041	2SC1343L
2041	PMC-QP0012	2041	SPD-80059	2041	2N2383	2041	2SC1343M
2041	PMC-QP0040	2041	SPD-80060	2041	2N2384	2041	2SC1343O
2041	PN350	2041	SPD-80061	2041	2N2948	2041	2SC1343OR
2041	PP3000	2041	SPD-80062	2041	2N3055	2041	2SC1343R
2041	PP3001	2041	SPT3713	2041	2N3055-1	2041	2SC1343X
2041	PP3003	2041	ST101	2041	2N3055-10	2041	2SC1343Y
2041	PP3004	2041	STC-1035	2041	2N3055-2	2041	2SC1402
2041	PP3006	2041	STC-1035A	2041	2N3055-3	2041	2SC1618
2041	PP3007	2041	STC-1036	2041	2N3055-4	2041	2SC1618B
2041	PT1941	2041	STC-1036A	2041	2N3055-5	2041	2SC1629
2041	PT7930	2041	STC-1085	2041	2N3055-6	2041	2SC1629A
2041	PT7931	2041	STC1035	2041	2N3055-7	2041	2SC1629AO
2041	PTC116	2041	STC1035A	2041	2N3055-8	2041	2SC1629M
2041	PTC119	2041	STC1036	2041	2N3055-9	2041	2SC1777
2041	PTC140	2041	STC1036A	2041	2N3055B	2041	2SC198
2041	PTC173	2041	STC1080	2041	2N3226	2041	2SC198B
2041	Q5085B	2041	STC1081	2041	2N3232	2041	2SC988
2041	Q5110Z	2041	STC1082	2041	2N3233	2041	2SC21
2041	QP-11	2041	STC1083	2041	2N3234	2041	2SC21A
2041	QP-12	2041	STC1084	2041	2N3236	2041	2SC21B
2041	QP-8	2041	STC1094	2041	2N3237	2041	2SC21C
2041	QP-8-P	2041	STC2220	2041	2N3238	2041	2SC21D
2041	QP001200A	2041	STC2221	2041	2N3239	2041	2SC21E
2041	QP8	2041	STC2224	2041	2N3297	2041	2SC21F
2041	R135-1	2041	STC2225	2041	2N3445	2041	2SC21G
2041	R2270-75116	2041	STC2228	2041	2N3446	2041	2SC21GN
2041	R2270-75497	2041	STC2229	2041	2N3447	2041	2SC21H
2041	R2270-78399	2041	STC4252	2041	2N3448	2041	2SC21J
2041	R22707-8399	2041	STC4253	2041	2N3667	2041	2SC21K
2041	R227075497	2041	STC4254	2041	2N3713	2041	2SC21L

RS 276-	Industry Standard No.	RS 276-	Industry Standard No.	RS 276-	Industry Standard No.	RS 276-	Industry Standard No.	RS 276-	Industry Standard No.
2041	2SC221M	2041	2SC765F	2041	2SD119-YEL	2041	2SD163G	2041	2SD176A
2041	2SC221OR	2041	2SC765G	2041	2SD119A	2041	2SD163GN	2041	2SD176B
2041	2SC221R	2041	2SC765GN	2041	2SD119B	2041	2SD163H	2041	2SD176C
2041	2SC221X	2041	2SC765H	2041	2SD119BL	2041	2SD165J	2041	2SD176D
2041	2SC221Y	2041	2SC765J	2041	2SD119C	2041	2SD163K	2041	2SD176E
2041	2SC240A	2041	2SC765K	2041	2SD119D	2041	2SD163L	2041	2SD176F
2041	2SC240B	2041	2SC765L	2041	2SD119R	2041	2SD163M	2041	2SD176G
2041	2SC240C	2041	2SC765M	2041	2SD119Y	2041	2SD163OR	2041	2SD176GN
2041	2SC240D	2041	2SC765OR	2041	2SD12	2041	2SD163R	2041	2SD176H
2041	2SC240R	2041	2SC765R	2041	2SD124	2041	2SD163X	2041	2SD176J
2041	2SC240P	2041	2SC765X	2041	2SD124A	2041	2SD163Y	2041	2SD176K
2041	2SC240G	2041	2SC765Y	2041	2SD124AH	2041	2SD164	2041	2SD176L
2041	2SC240GN	2041	2SC766	2041	2SD124AHA	2041	2SD164A	2041	2SD176OR
2041	2SC240H	2041	2SC767	2041	2SD124AHB	2041	2SD164B	2041	2SD176R
2041	2SC240J	2041	2SC768	2041	2SD124B	2041	2SD164C	2041	2SD176X
2041	2SC240K	2041	2SC768A	2041	2SD124C	2041	2SD164D	2041	2SD176Y
2041	2SC240L	2041	2SC768B	2041	2SD124E	2041	2SD164E	2041	2SD17A
2041	2SC240M	2041	2SC768C	2041	2SD124F	2041	2SD164F	2041	2SD17C
2041	2SC240OR	2041	2SC768D	2041	2SD124GN	2041	2SD164G	2041	2SD17D
2041	2SC240R	2041	2SC768E	2041	2SD124H	2041	2SD164GN	2041	2SD17E
2041	2SC240X	2041	2SC768F	2041	2SD124J	2041	2SD164H	2041	2SD17F
2041	2SC240Y	2041	2SC768G	2041	2SD124K	2041	2SD164J	2041	2SD17GN
2041	2SC241A	2041	2SC768GN	2041	2SD124L	2041	2SD164K	2041	2SD17H
2041	2SC241B	2041	2SC768H	2041	2SD124M	2041	2SD164L	2041	2SD17J
2041	2SC241C	2041	2SC768J	2041	2SD124OR	2041	2SD164M	2041	2SD17K
2041	2SC241D	2041	2SC768K	2041	2SD124R	2041	2SD164OR	2041	2SD17L
2041	2SC241E	2041	2SC768L	2041	2SD124X	2041	2SD164R	2041	2SD17M
2041	2SC241F	2041	2SC768M	2041	2SD124T	2041	2SD164X	2041	2SD17R
2041	2SC241G	2041	2SC768OR	2041	2SD125	2041	2SD164Y	2041	2SD17X
2041	2SC241GN	2041	2SC768R	2041	2SD125A	2041	2SD16A	2041	2SD17Y
2041	2SC241H	2041	2SC768X	2041	2SD125AH	2041	2SD16B	2041	2SD180
2041	2SC241K	2041	2SC768Y	2041	2SD125AHB	2041	2SD16C	2041	2SD180A
2041	2SC241L	2041	2SC793	2041	2SD125B	2041	2SD16D	2041	2SD180B
2041	2SC241M	2041	2SC793-BLU	2041	2SD125C	2041	2SD16E		
2041	2SC241OR	2041	2SC793-R	2041	2SD125E	2041	2SD16F		
2041	2SC241R	2041	2SC793-RED	2041	2SD125P	2041	2SD16G		
2041	2SC241X	2041	2SC793-YEL	2041	2SD125G	2041	2SD16GN		
2041	2SC241Y	2041	2SC793A	2041	2SD125GN	2041	2SD16H		
2041	2SC242A	2041	2SC793B	2041	2SD125H	2041	2SD16J		
2041	2SC242B	2041	2SC793BL	2041	2SD125J	2041	2SD16K		
2041	2SC242C	2041	2SC793C	2041	2SD125K	2041	2SD16L		
2041	2SC242D	2041	2SC793E	2041	2SD125L	2041	2SD16M		
2041	2SC242E	2041	2SC793F	2041	2SD125M	2041	2SD16OR		
2041	2SC242F	2041	2SC793G	2041	2SD125OR	2041	2SD16R		
2041	2SC242G	2041	2SC793GN	2041	2SD125R	2041	2SD16X		
2041	2SC242GN	2041	2SC793H	2041	2SD125X	2041	2SD16Y		
2041	2SC242H	2041	2SC793J	2041	2SD125Y	2041	2SD17		
2041	2SC242J	2041	2SC793K	2041	2SD126	2041	2SD172		
2041	2SC242K	2041	2SC793L	2041	2SD126A	2041	2SD172A		
2041	2SC242L	2041	2SC793M	2041	2SD126AH	2041	2SD172B		
2041	2SC242M	2041	2SC793OR	2041	2SD126AHA	2041	2SD172C		
2041	2SC242OR	2041	2SC793R	2041	2SD126AHB	2041	2SD172D		
2041	2SC242R	2041	2SC793X	2041	2SD126H	2041	2SD172E		
2041	2SC242X	2041	2SC793Y	2041	2SD126HA	2041	2SD172F		
2041	2SC242Y	2041	2SC794	2041	2SD126HB	2041	2SD172G		
2041	2SC244	2041	2SC794A	2041	2SD132	2041	2SD172GN		
2041	2SC244A	2041	2SC794B	2041	2SD146	2041	2SD172H		
2041	2SC244B	2041	2SC794C	2041	2SD146A	2041	2SD172J		
2041	2SC244C	2041	2SC794D	2041	2SD146B	2041	2SD172K		
2041	2SC244D	2041	2SC794E	2041	2SD146C	2041	2SD172L		
2041	2SC244E	2041	2SC794F	2041	2SD146D	2041	2SD172M		
2041	2SC244F	2041	2SC794G	2041	2SD146E	2041	2SD172OR		
2041	2SC244G	2041	2SC794GN	2041	2SD146F	2041	2SD172R		
2041	2SC244GN	2041	2SC794H	2041	2SD146G	2041	2SD172X		
2041	2SC244J	2041	2SC794J	2041	2SD146GN	2041	2SD172Y		
2041	2SC244K	2041	2SC794K	2041	2SD146H	2041	2SD173		
2041	2SC244L	2041	2SC794L	2041	2SD146J	2041	2SD173A		
2041	2SC244M	2041	2SC794OR	2041	2SD146K	2041	2SD173B		
2041	2SC244OR	2041	2SC794R	2041	2SD146L	2041	2SD173C		
2041	2SC244R	2041	2SC794RA	2041	2SD146OR	2041	2SD173D		
2041	2SC244X	2041	2SC794X	2041	2SD146R	2041	2SD173E		
2041	2SC244Y	2041	2SC794Y	2041	2SD146VK	2041	2SD173F		
2041	2SC493	2041	2SC851	2041	2SD146VK	2041	2SD173G		
2041	2SC493-BL	2041	2SC851A	2041	2SD146X	2041	2SD173GN		
2041	2SC493-R	2041	2SC851B	2041	2SD146Y	2041	2SD173H		
2041	2SC493-Y	2041	2SC851C	2041	2SD147	2041	2SD173J		
2041	2SC493A	2041	2SC851D	2041	2SD147A	2041	2SD173K		
2041	2SC493B	2041	2SC851E	2041	2SD147C	2041	2SD173M		
2041	2SC493C	2041	2SC851F	2041	2SD147D	2041	2SD173OR		
2041	2SC493D	2041	2SC851G	2041	2SD147E	2041	2SD173X		
2041	2SC493E	2041	2SC851GN	2041	2SD147F	2041	2SD173Y		
2041	2SC493F	2041	2SC851H	2041	2SD147GN	2041	2SD174		
2041	2SC493G	2041	2SC851K	2041	2SD147H	2041	2SD175		
2041	2SC493GN	2041	2SC851L	2041	2SD147J	2041	2SD175A		
2041	2SC493H	2041	2SC851M	2041	2SD147K	2041	2SD175B		
2041	2SC493J	2041	2SC851OR	2041	2SD147M	2041	2SD175D		
2041	2SC493K	2041	2SC851R	2041	2SD147R	2041	2SD175E		
2041	2SC493L	2041	2SC851X	2041	2SD147X	2041	2SD175F		
2041	2SC493M	2041	2SC851Y	2041	2SD147Y	2041	2SD175G		
2041	2SC493OR	2041	2SC897	2041	2SD15	2041	2SD175GN		
2041	2SC493R	2041	2SC897A	2041	2SD151	2041	2SD175H		
2041	2SC493X	2041	2SC897B	2041	2SD151A	2041	2SD175J		
2041	2SC493Y	2041	2SC897C	2041	2SD151B	2041	2SD175K		
2041	2SC494	2041	2SC898	2041	2SD151C	2041	2SD175L		
2041	2SC494-BL	2041	2SC898A	2041	2SD151E	2041	2SD175M		
2041	2SC494-R	2041	2SC898B	2041	2SD151F	2041	2SD175OR		
2041	2SC494-Y	2041	2SC898C	2041	2SD151G	2041	2SD175R		
2041	2SC520	2041	2SCM93D	2041	2SD151GN	2041	2SD175X		
2041	2SC520A	2041	2SD111-ORG	2041	2SD151H	2041	2SD175Y		
2041	2SC521	2041	2SD111-RED	2041	2SD151J	2041	2SD176		
2041	2SC521A	2041	2SD111-YEL	2041	2SD151K				
2041	2SC646	2041	2SD113	2041	2SD151L				
2041	2SC647	2041	2SD113-O	2041	2SD151M				
2041	2SC647Q	2041	2SD113-ORG	2041	2SD151OR				
2041	2SC647R	2041	2SD113-R	2041	2SD151R				
2041	2SC665	2041	2SD113-RED	2041	2SD151T				
2041	2SC665H	2041	2SD113-Y	2041	2SD15A				
2041	2SC665HA	2041	2SD113-YEL	2041	2SD15B				
2041	2SC665HB	2041	2SD114	2041	2SD15C				
2041	2SC7	2041	2SD114-O	2041	2SD15D				
2041	2SC736	2041	2SD114-ORG	2041	2SD15E				
2041	2SC736A	2041	2SD114-R	2041	2SD15F				
2041	2SC736B	2041	2SD114-RED	2041	2SD15G				
2041	2SC736C	2041	2SD114-Y	2041	2SD15GN				
2041	2SC736D	2041	2SD114-YEL	2041	2SD15H				
2041	2SC736E	2041	2SD118	2041	2SD15J				
2041	2SC736F	2041	2SD118-B	2041	2SD15K				
2041	2SC736G	2041	2SD118-BLU	2041	2SD15L				
2041	2SC736GN	2041	2SD118-R	2041	2SD15M				
2041	2SC736H	2041	2SD118-RED	2041	2SD15OR				
2041	2SC736J	2041	2SD118-Y	2041	2SD15R				
2041	2SC736K	2041	2SD118-YEL	2041	2SD15X				
2041	2SC736L	2041	2SD118A	2041	2SD15Y				
2041	2SC736M	2041	2SD118BL	2041	2SD16				
2041	2SC736OR	2041	2SD118C	2041	2SD163				
2041	2SC736R	2041	2SD118D	2041	2SD163A				
2041	2SC736X	2041	2SD118R	2041	2SD163B				
2041	2SC736Y	2041	2SD118Y	2041	2SD163C				
2041	2SC765	2041	2SD119	2041	2SD163D				
2041	2SC765A	2041	2SD119-BL	2041	2SD163E				
2041	2SC765B	2041	2SD119-BLU						
2041	2SC765C	2041	2SD119-R						
2041	2SC765D	2041	2SD119-RED						
2041	2SC765E	2041	2SD119-Y						

RS 276-	Industry Standard No.	RS 276-	Industry Standard No.	RS 276-	Industry Standard No.	RS 276-	Industry Standard No.
2041	2SD180C	2041	2SD335	2041	2SD80E	2041	14-40363A
2041	2SD180D	2041	2SD338	2041	2SD80F	2041	14-40369A
2041	2SD180P	2041	2SD339	2041	2SD80G	2041	14-40421A
2041	2SD180GN	2041	2SD41	2041	2SD80GN	2041	14-40464A
2041	2SD180H	2041	2SD41A	2041	2SD80H	2041	14-40465A
2041	2SD180J	2041	2SD41B	2041	2SD80J	2041	14-40466A
2041	2SD180K	2041	2SD41C	2041	2SD80K	2041	14-40471
2041	2SD180L	2041	2SD41D	2041	2SD80L	2041	14-40934-1
2041	2SD180M	2041	2SD41E	2041	2SD80OR	2041	14-601-10
2041	2SD180OR	2041	2SD41F	2041	2SD80R	2041	14-601-12
2041	2SD180R	2041	2SD41G	2041	2SD80X	2041	14-601-13
2041	2SD180X	2041	2SD41GN	2041	2SD80Y	2041	14-601-15
2041	2SD180Y	2041	2SD41H	2041	2SD81	2041	14-601-15A
2041	2SD188	2041	2SD41J	2041	2SD81A	2041	14-601-16
2041	2SD188A	2041	2SD41K	2041	2SD81B	2041	14-601-16A
2041	2SD188B	2041	2SD41L	2041	2SD81C	2041	14-601-18
2041	2SD188C	2041	2SD41M	2041	2SD81E	2041	14-601-20
2041	2SD188D	2041	2SD41OR	2041	2SD81F	2041	14-601-23
2041	2SD188E	2041	2SD41R	2041	2SD81G	2041	14-601-24
2041	2SD188F	2041	2SD41X	2041	2SD81GN	2041	14-601-26
2041	2SD188G	2041	2SD41Y	2041	2SD81H	2041	15-03068
2041	2SD188GN	2041	2SD425	2041	2SD81K	2041	17-50(FISHER)
2041	2SD188H	2041	2SD425-R	2041	2SD81L	2041	18/P2C
2041	2SD188J	2041	2SD425O	2041	2SD81M	2041	019-003935
2041	2SD188L	2041	2SD426-R	2041	2SD81OR	2041	19A115818
2041	2SD188N	2041	2SD428	2041	2SD81R	2041	19A116753P1
2041	2SD188OR	2041	2SD50	2041	2SD81Y	2041	19A116761P1
2041	2SD188R	2041	2SD50A	2041	2SD82	2041	19A126813
2041	2SD188X	2041	2SD50B	2041	2SD82-OR	2041	19A126813A
2041	2SD188Y	2041	2SD50C	2041	2SD82A	2041	19A126826-P2
2041	2SD189B	2041	2SD50D	2041	2SD82B	2041	20-1111-002
2041	2SD189C	2041	2SD50E	2041	2SD82C	2041	20-1111-003
2041	2SD189D	2041	2SD50F	2041	2SD82D	2041	020-1111-007
2041	2SD189E	2041	2SD50G	2041	2SD82E	2041	020-1111-008
2041	2SD189F	2041	2SD50GN	2041	2SD82F	2041	020-1111-019
2041	2SD189G	2041	2SD50H	2041	2SD82G	2041	21A112-124
2041	2SD189GN	2041	2SD50J	2041	2SD82GN	2041	23-5035
2041	2SD189H	2041	2SD50K	2041	2SD82H	2041	23-5038
2041	2SD189J	2041	2SD50L	2041	2SD82J	2041	23-5041
2041	2SD189K	2041	2SD50M	2041	2SD82K	2041	025
2041	2SD189L	2041	2SD50OR	2041	2SD82L	2041	30-005072
2041	2SD189M	2041	2SD50R	2041	2SD82M	2041	33-052
2041	2SD189OR	2041	2SD50X	2041	2SD82OR	2041	33-108
2041	2SD189R	2041	2SD50Y	2041	2SD82R	2041	34-1000
2041	2SD189X	2041	2SD51	2041	2SD82X	2041	34-1000A
2041	2SD189Y	2041	2SD51A	2041	2SD82Y	2041	34-1028
2041	2SD201	2041	2SD51B	2041	2SDF2A	2041	34-1028A
2041	2SD201(O)	2041	2SD51C	2041	2SDF2B	2041	34-6002-27
2041	2SD201-O	2041	2SD51D	2041	2SDF2C	2041	34-6002-32
2041	2SD201A	2041	2SD51E	2041	2SDF2D	2041	34-6002-32A
2041	2SD201B	2041	2SD51F	2041	2SDF2E	2041	34-6002-35
2041	2SD201C	2041	2SD51G	2041	2SDF2F	2041	34-6002-37
2041	2SD201P	2041	2SD51GN	2041	2SDF2G	2041	41-0318
2041	2SD201G	2041	2SD51H	2041	2SDF2GN	2041	41-0318A
2041	2SD201GN	2041	2SD51J	2041	2SDF2H	2041	42-20961
2041	2SD201H	2041	2SD51K	2041	2SDF2J	2041	42-20961A
2041	2SD201J	2041	2SD51L	2041	2SDF2K	2041	43-023190
2041	2SD201K	2041	2SD51M	2041	2SDF2L	2041	43-023843
2041	2SD201L	2041	2SD51OR	2041	2SDF2M	2041	43A165137P1
2041	2SD201M	2041	2SD51R	2041	2SDF2OR	2041	43A165137P3
2041	2SD201M(O)	2041	2SD51X	2041	2SDF2R	2041	43A165137P4
2041	2SD201M(O,Y)	2041	2SD51Y	2041	2SDF2X	2041	43A167885-P1
2041	2SD201M(Y)	2041	2SD53	2041	2SDF2Y	2041	43A167885-P2
2041	2SD201MO	2041	2SD53A	2041	2SDU047X	2041	43A167885-P3
2041	2SD201MY	2041	2SD53B	2041	28D33	2041	43A212067
2041	2SD201O	2041	2SD53C	2041	28D34	2041	44A-417014-001
2041	2SD201Q	2041	2SD53E	2041	28D35	2041	44A355565-001
2041	2SD201X	2041	2SD53F	2041	28D36	2041	44A355565001
2041	2SD201Y	2041	2SD53G	2041	3-30173	2041	44A390244-001
2041	2SD211	2041	2SD53GN	2041	314-6001-01	2041	44A391593-001
2041	2SD211-OR	2041	2SD53H	2041	3TE120	2041	44A395909-1
2041	2SD211A	2041	2SD53J	2041	3TE140	2041	44A417033-001
2041	2SD211B	2041	2SD53K	2041	3TE230	2041	44A417716
2041	2SD211C	2041	2SD53L	2041	3TE240	2041	44A417716-001
2041	2SD211D	2041	2SD53M	2041	3TX003	2041	44A4417714
2041	2SD211E	2041	2SD53OR	2041	3TX004	2041	44A4417714-001
2041	2SD211F	2041	2SD53R	2041	4-0294	2041	46-86388-3
2041	2SD211G	2041	2SD53X	2041	4-0563	2041	48-03-04093403
2041	2SD211GN	2041	2SD53Y	2041	4-245	2041	48-03-041840-2
2041	2SD211H	2041	2SD55	2041	4-3023190	2041	48-12091A
2041	2SD211J	2041	2SD55A	2041	4-3025843	2041	48-134701
2041	2SD211K	2041	2SD582	2041	4-458	2041	48-134701A
2041	2SD211L	2041	2SD582A	2041	4-490	2041	48-134715
2041	2SD211M	2041	2SD68	2041	6-137	2041	48-134715A
2041	2SD211OR	2041	2SD68A	2041	6-138(PWR)	2041	48-134882
2041	2SD211R	2041	2SD68B	2041	6-64990004	2041	48-134884
2041	2SD211X	2041	2SD68C	2041	007-0040-00	2041	48-134969
2041	2SD211Y	2041	2SD68D	2041	7-0115	2041	48-137008
2041	2SD212	2041	2SD68E	2041	7-0115-000	2041	48-137008A
2041	2SD212A	2041	2SD68F	2041	7-0197	2041	48-137027
2041	2SD212B	2041	2SD68G	2041	7-0197-00	2041	48-137027A
2041	2SD212C	2041	2SD68GN	2041	7-12	2041	48-137036
2041	2SD212D	2041	2SD68H	2041	7-12(STANDEL)	2041	48-137036A
2041	2SD212E	2041	2SD68J	2041	7-13	2041	48-137053
2041	2SD212F	2041	2SD68L	2041	7-13(STANDEL)	2041	48-137053A
2041	2SD212G	2041	2SD68M	2041	7-466201	2041	48-137079
2041	2SD212GN	2041	2SD68OR	2041	007-7450301	2041	48-137079A
2041	2SD212H	2041	2SD68R	2041	7-7466201	2041	48-137116
2041	2SD212J	2041	2SD68X	2041	8-0052402	2041	48-137175
2041	2SD212K	2041	2SD68Y	2041	8-0414120	2041	48-137175A
2041	2SD212L	2041	2SD69	2041	8-0414130	2041	48-137180
2041	2SD212M	2041	2SD70A	2041	8-1075	2041	48-137180A
2041	2SD212OR	2041	2SD70B	2041	8-905-706-555	2041	48-137251
2041	2SD212R	2041	2SD70C	2041	8-905-706-556	2041	48-137333
2041	2SD212X	2041	2SD70D	2041	8-905-706-557	2041	48-137344
2041	2SD212Y	2041	2SD70E	2041	8-905-713-101	2041	48-137368
2041	2SD241H	2041	2SD70F	2041	8-905-713-556	2041	48-15
2041	2SD26	2041	2SD70G	2041	09-302122	2041	48-155140
2041	2SD26A	2041	2SD70GN	2041	09-305138	2041	48-217241
2041	2SD26B	2041	2SD70H	2041	9TR91001-09	2041	48-232796
2041	2SD26C	2041	2SD70J	2041	10-13-002-003	2041	48-869244
2041	2SD26D	2041	2SD70K	2041	10-13-002-004	2041	48-869259
2041	2SD26E	2041	2SD70L	2041	10-13-002-3	2041	48-869278
2041	2SD26F	2041	2SD70M	2041	10-13-002-4	2041	48-869302
2041	2SD26G	2041	2SD70OR	2041	10-13002-003	2041	48-869321
2041	2SD26GN	2041	2SD70R	2041	10-13002-004	2041	48-869480
2041	2SD26H	2041	2SD70X	2041	10-13030-004	2041	48-869515
2041	2SD26J	2041	2SD70Y	2041	10-13030-005	2041	48-869628
2041	2SD26K	2041	2SD73	2041	10-374101	2041	48-869639
2041	2SD26L	2041	2SD73A	2041	10-13002-004	2041	48-869660
2041	2SD26M	2041	2SD73B	2041	11-11911-1	2041	48K869515
2041	2SD26OR	2041	2SD73C	2041	12-1000047	2041	48B217241
2041	2SD26R	2041	2SD74	2041	13-0032	2041	48B232796
2041	2SD26X	2041	2SD74A	2041	13-17918-1	2041	48B869244
2041	2SD26Y	2041	2SD74B	2041	13-23594-1	2041	48B869302
2041	2SD319	2041	2SD74C	2041	13-28396-1	2041	48B869321
2041	2SD322	2041	2SD74D	2041	13-28396-I	2041	48B137344
2041	2SD322A	2041	2SD74E	2041	13-33188-2	2041	48B155053
2041	2SD322B	2041	2SD80	2041	13-34374-1	2041	48B155067
2041	2SD322C	2041	2SD80A	2041	13-34684-1	2041	500401
2041	2SD323	2041	2SD80B	2041	13-41738-1	2041	52-025-004-Q
2041	2SD323A	2041	2SD80C	2041	14-40325A	2041	57-4018-60
2041	2SD323B	2041	2SD80D	2041		2041	57A196-10
2041	2SD323C	2041		2041		2041	57A256-10

RS 276-	Industry Standard No.	RS 276-	Industry Standard No.	RS 276-	Industry Standard No.	RS 276-	Industry Standard No.
2041	57B175-9	2041	260D09301	2041	1561-0815	2041	13002-4
2041	57B175-9A	2041	260724008	2041	1561-1004	2041	13030-4
2041	57B196-10	2041	297T061001	2041	1561-1005	2041	13298
2041	57B256-10	2041	297T061001A	2041	1561-1010	2041	16001
2041	57M3-1	2041	297T061002	2041	1561-1015	2041	16083
2041	57M3-1A	2041	297T061002A	2041	1561A603	2041	16176
2041	57M3-2	2041	0501-3055-00	2041	1561A608	2041	16201
2041	57M3-4	2041	332-2911	2041	1561A615	2041	16230
2041	57M3-4A	2041	332-2912	2041	1582-0408	2041	16234
2041	57M3-6	2041	332-3562	2041	1582-0410	2041	16235
2041	57M3-6A	2041	352-0583-011	2041	1582-0415	2041	16240
2041	60-211040	2041	352-0677-010	2041	1582-0508	2041	16261
2041	61-309449	2041	352-0677-011	2041	1582-0510	2041	16266
2041	63-11991	2041	352-0677-020	2041	1582-0608	2041	16267
2041	63-11991A	2041	352-0677-021	2041	1582-0610	2041	16287
2041	63-8707	2041	352-0677-030	2041	1582-0615	2041	16292
2041	63-8707A	2041	352-0677-031	2041	1582-0803	2041	16299
2041	68A8319001	2041	352-0677-040	2041	1582-0804	2041	16319
2041	070	2041	352-0677-041	2041	1582-0805	2041	16320
2041	80-052402	2041	352-0677-051	2041	1582-0810	2041	16338
2041	80-414120	2041	352-0677-40	2041	1582-0815	2041	21280
2041	80-414130	2041	352-0749-010	2041	1582-1003	2041	23114-070
2041	085	2041	378-44	2041	1582-1004	2041	023762
2041	86-5084-2	2041	378-44A	2041	1582-1005	2041	27126-10Q
2041	86-5084-2A	2041	386-40	2041	1582-1008	2041	28396
2041	86-5101-2	2041	386-7183P1	2041	1582-1010	2041	28474
2041	86-5101-2A	2041	386-7183P1A	2041	1582-1015	2041	30276
2041	86-5112-2	2041	386-7270-P2	2041	1723-0405	2041	030539-1
2041	86-5112-2(THOMAS)	2041	394-3127-4	2041	1723-0605	2041	33188-2
2041	86-665-2	2041	394-3135A	2041	1723-0610	2041	34208
2041	86-665-9	2041	403-009/07	2041	1723-0805	2041	35001
2041	86A332	2041	417-139	2041	1723-0810	2041	36545
2041	93-SB-124	2041	417-139-13286	2041	1723-1005	2041	36846
2041	938C165	2041	417-139A	2041	1723-1010	2041	36855
2041	938C165133	2041	417-162	2041	1723-1015	2041	36892
2041	938C165133A	2041	417-212	2041	1723-1210	2041	36946
2041	938E165	2041	417-212A	2041	1723-1405	2041	36953
2041	96-5117-01	2041	417-214	2041	1723-1410	2041	037085
2041	96-5117-91	2041	417-214-13286	2041	1723-1605	2041	37267
2041	96-5162-03	2041	417-215	2041	1723-1610	2041	37475
2041	96-5162-04	2041	417-215-13286	2041	1723-1805	2041	37476
2041	96-5164-02	2041	417-215A	2041	1723-1810	2041	37563
2041	96-5164-03	2041	417-254	2041	1763-0415	2041	37663
2041	96-5201-01	2041	417-273	2041	1763-0420	2041	37888
2041	96-5207-01	2041	417-273-13286	2041	1763-0425	2041	37967
2041	96-5310-01	2041	417-282	2041	1763-0615	2041	37974
2041	96-5315-01	2041	417-286	2041	1763-0620	2041	38049
2041	998103-1	2041	422-0961	2041	1763-0625	2041	38137
2041	998103-2	2041	430-23190	2041	1763-0815	2041	38138
2041	998103-3	2041	430-23843	2041	1763-0825	2041	38166
2041	100T2	2041	445-0023	2041	1763-1015	2041	38267
2041	100T2A	2041	445-0023-P1	2041	1763-1020	2041	38268
2041	100T2	2041	445-0023-P3	2041	1763-1025	2041	38272
2041	100X6	2041	445-0023-P4	2041	1763-1215	2041	38335
2041	100X6A	2041	445-0034-1	2041	1763-1220	2041	38397
2041	104T2	2041	472-0946-001	2041	1763-1225	2041	38473
2041	104T2A	2041	472-0946-002	2041	1763-1415	2041	38474
2041	1113N2C	2041	514-054214	2041	1763-1420	2041	38491
2041	111N4	2041	542-1034	2041	1763-1425	2041	38494
2041	111N4A	2041	571-844	2041	1806-1	2041	38513-50360C
2041	111N4B	2041	571-844A	2041	1806-17A	2041	38626
2041	111N4C	2041	576-0002-003	2041	1854-0245	2041	38731(KALOP)
2041	111N6	2041	576-0040-251	2041	1854-0490-1	2041	38804
2041	112-203055	2041	686-0243	2041	1854-0563	2041	38897
2041	112-363	2041	686-0243-0	2041	2000-221	2041	38965
2041	11603475	2041	686-143	2041	2003	2041	39127
2041	119-0075	2041	686-143A	2041	2015-1	2041	39140
2041	121-665	2041	690V094H17	2041	2015-1A	2041	39148
2041	121-726	2041	700-113	2041	2015-2	2041	39196
2041	121-726A	2041	711-001	2041	2015-2A	2041	39213
2041	123B-001	2041	742C02030-020	2041	2015-3	2041	39251
2041	128-9050	2041	766-100999	2041	2015-3A	2041	39343
2041	130-146	2041	800-510-00	2041	2015-4	2041	39369
2041	131-001-007	2041	800-510-01	2041	2015-5	2041	39414
2041	131-001007	2041	800-524-02	2041	2015-6	2041	39455
2041	0131-001597	2041	800-524-02A	2041	2015-7	2041	39465
2041	0131-002068	2041	800-524-03	2041	2057A100	2041	39466
2041	0131-004367	2041	800-524-03A	2041	2071	2041	39492
2041	131-043-67	2041	800-524-04A	2041	2842-875	2041	39581
2041	131-04367	2041	800-52402	2041	3055-1	2041	39616
2041	0132	2041	800-525-04A	2041	3055-3	2041	39635
2041	132-070	2041	804-14120	2041	3146-977	2041	39751
2041	132-085	2041	804-14130	2041	3152-170	2041	39803A
2041	132-541	2041	860-022-01	2041	3373	2041	39819
2041	140N1	2041	866-6	2041	3499	2041	39921
2041	140N1C	2041	866-6(BENDIX)	2041	3637-1	2041	39922
2041	140N2	2041	921-1011	2041	3665	2041	39954
2041	151-0140	2041	991-01-1317	2041	3665-2	2041	40151
2041	151-0140-00	2041	991-01-3063	2041	3683	2041	40251
2041	151-0275	2041	991-013063	2041	3683-1	2041	40325
2041	151-0275-00	2041	992-00-2271	2041	3714H1	2041	40363
2041	151-0336-00	2041	992-00-3139	2041	3714H1A	2041	40369
2041	151-0336-00-A	2041	992-00-3139A	2041	3771	2041	40411
2041	151-0337-00	2041	992-00-4091	2041	3772-1	2041	40421
2041	151-0337-00-A	2041	992-00-4092	2041	3772-2	2041	40444
2041	152N-2	2041	992-002271	2041	4216	2041	40464
2041	152N2	2041	992-00271	2041	4491-4	2041	40465
2041	152N2C	2041	992-00271A	2041	4491-7	2041	40466
2041	156-032	2041	992-003139	2041	4701	2041	40471
2041	156-043	2041	992-004091	2041	4715	2041	40513
2041	156-043A	2041	992-004092	2041	4715A	2041	40514
2041	156-053	2041	992-01-1317	2041	4802-00005	2041	40542
2041	156-063	2041	992-011317	2041	4802-00005A	2041	40543
2041	156-083	2041	992-017169	2041	4822-130-40132	2041	40633
2041	156-084	2041	992-02271	2041	4882	2041	40934-1
2041	156-104	2041	1001-09	2041	5001-508	2041	43060
2041	156-83	2041	1045-7844	2041	5036-1	2041	43095
2041	161-2NC	2041	1045-7851	2041	5036-2	2041	43114
2041	161-LNC	2041	1064-6032	2041	5504	2041	43168
2041	161-LNC	2041	1069-7032	2041	5505	2041	52215-00
2041	161N1C	2041	1071-3642	2041	5565-001	2041	52329
2041	161N4	2041	1074-117	2041	005575	2041	52560
2041	161N4C	2041	1080-5364	2041	5909-001	2041	60024
2041	172-003-9-001	2041	1081-7104	2041	6380-1	2041	60041
2041	172-003-9-001A	2041	1092	2041	7214(LOWREY)	2041	60046
2041	172-024-9-005	2041	1092(GE)	2041	7214A	2041	60047
2041	173A4491-2	2041	1116-6527	2041	7320-1	2041	60076
2041	173A4491-2A	2041	1119-4628	2041	7356-2	2041	60085
2041	173A4491-4	2041	1359(GE)	2041	7514	2041	60115
2041	173A4491-7	2041	1414-179	2041	7885-1	2041	60127
2041	174-20989-22	2041	1414-188	2041	7885-2	2041	60130
2041	176-040-9-001	2041	1431-7184	2041	7885-3	2041	60142
2041	180T2	2041	1471-4802	2041	8000-00004-241	2041	60192
2041	180T2A	2041	1561-0408	2041	8000-00006-006	2041	60205
2041	180T2B	2041	1561-0410	2041	8000-0005-006	2041	60213(TRANSISTOR)
2041	180T2C	2041	1561-0604	2041	8005(FERNCREST)	2041	60234
2041	181T2A	2041	1561-0608	2041	8319-001	2041	60243
2041	181T2B	2041	1561-0610	2041	10003	2041	60339
2041	181T2C	2041	1561-0615	2041	11236-3	2041	60350
2041	211-40140-18	2041	1561-0803	2041	12536	2041	60380
2041	211A6380-1	2041	1561-0804	2041	12536A	2041	60465
2041	211AESU3055	2041	1561-0805	2041	13002-3	2041	60710
2041	229-1301-64	2041	1561-0808			2041	60810
2041	231-0004	2041	1561-0810			2041	60837

RS 276-	Industry Standard No.	RS 276-	Industry Standard No.	RS 276-	Industry Standard No.	RS 276-	Industry Standard No.
2041	60885	2041	2057199-0700	2043	HEP248	2043	2N6246
2041	60944	2041	2057199-0701	2043	HEP87001	2043	2N6247
2041	60973	2041	2057199-070BA	2043	HEP87003	2043	2N6329
2041	60991	2041	2057199-701	2043	HP23	2043	2N6330
2041	61007	2041	2057323-0500	2043	HP25	2043	2N6331
2041	61012	2041	2057323-0501	2043	K8-20033L2	2043	2SA626
2041	61019	2041	2321652	2043	M9344	2043	2SA626L
2041	61173	2041	2327052	2043	M9359	2043	2SA627
2041	61234	2041	2327053	2043	MJ2267	2043	2SA648
2041	61367	2041	2327172	2043	MJ2268	2043	2SA648A
2041	61369/4367	2041	3020061	2043	MJ2901	2043	2SA648B
2041	61451	2041	3130058	2043	MJ2940	2043	2SA648C
2041	61456	2041	3130090	2043	MJ2941	2043	2SA658
2041	61868	2041	3130091	2043	MJ2955	2043	2SA663
2041	62004	2041	3130104	2043	MJ450	2043	2SA679
2041	62005-1	2041	3146977	2043	MJ4502	2043	2SA679-R
2041	62143	2041	3146977A	2043	MJ490	2043	2SA679R
2041	62287	2041	3152170	2043	MJ491	2043	2SA679Y
2041	62382	2041	3152170A	2043	NKT4055	2043	2SA680
2041	62792	2041	3463101-1	2043	P1E	2043	2SA680-R
2041	70008-0	2041	3464648-1	2043	P1E-1BLK	2043	2SA680R
2041	70008-3	2041	3464648-2	2043	P1E-1BLU	2043	2SA680T
2041	70260-18	2041(2)	CXL130MP	2043	P1E-1GRN	2043	2SA744
2041	74662	2041(2)	ECG130MP	2043	P1E-1RED	2043	2SA746
2041	76251	2041(2)	ECG280MP	2043	P1E-1V10	2043	2SA753
2041	78399	2041(2)	EB95(ELCOM)	2043	P1E-1VIO	2043	2SA753A
2041	80807	2041(2)	GE-15MP	2043	P1E-2BLK	2043	2SA753B
2041	91274	2041(2)	GE-262MP	2043	P1E-2BLU	2043	2SA753C
2041	94065	2041(2)	HEP704X2	2043	P1E-2GRN	2043	2SA756
2041	94065A	2041(2)	IRTR61X2	2043	P1E-2RED	2043	2SA756A
2041	94094(EICO)	2041(2)	S93SB140	2043	P1E-2V10	2043	2SA756B
2041	98484-001	2041(2)	SK3029	2043	P1E-3BLK	2043	2SA756C
2041	101568	2041(2)	TM130MP	2043	P1E-3BLU	2043	2SA757
2041	113875	2041(2)	W16MP	2043	P1E-3GRN	2043	2SA757A
2041	113876	2041(2)	2SC494A	2043	P1E-3RED	2043	2SA757B
2041	124511	2041(2)	2SC494B	2043	P1E-3V10	2043	2SA757C
2041	130014	2041(2)	2SC494BL	2043	P1E-3VIO	2043	2SA758
2041	131257	2041(2)	2SC494C	2043	P2J	2043	2SA758A
2041	132776	2041(2)	2SC494D	2043	P3W	2043	2SA758B
2041	133550	2041(2)	2SC494E	2043	PTC172	2043	2SA758C
2041	133684	2041(2)	2SC494F	2043	RE-61	2043	2SA877
2041	133685	2041(2)	2SC494G	2043	RE-82	2043	2SB506
2041	133923	2041(2)	2SC494GN	2043	RB61	2043	2SB506A
2041	133925	2041(2)	2SC494H	2043	RET4	2043	2SB506B
2041	134282	2041(2)	2SC494J	2043	REB2	2043	2SB506C
2041	138194	2041(2)	2SC494K	2043	REN180	2043	2SB506D
2041	138195	2041(2)	2SC494L	2043	REN219	2043	2SB518
2041	140612	2041(2)	2SC494M	2043	REN61	2043	2SB519
2041	146466-1	2041(2)	2SC494OR	2043	RT-148	2043	2SB541
2041	150070	2041(2)	2SC494X	2043	S35486	2043	2SB555
2041	162002-071	2041(2)	2SC494Z	2043	SAB-2	2043	2SB555-0
2041	162002-101	2041(2)	8-0053702	2043	SC0421	2043	2SD341
2041	162002-71	2041(2)	80-053702	2043	SC0428	2043	2SD341H
2041	170890-1(NPN)	2041(2)	132-542	2043	SC0421	2043	13-40346-1
2041	170891-1	2041(2)	161N2	2043	SC0421	2043	13-40347-1
2041	188165	2041(2)	800-537-02	2043	SCBA21	2043	020-1111-009
2041	198034-1	2041(2)	800-53702	2043	SDT3760	2043	020-1111-080
2041	198034-2	2041(2)	2057A100-16	2043	SDT3764	2043	34-1001
2041	198034-3	2041(2)	2057A100-26	2043	SDT3765	2043	34-1029
2041	198034-4	2041(2)	2057B100-16	2043	SDT3766	2043	34-6002-23
2041	198034-5	2041(2)	171174-1	2043	SDT3826	2043	34-6002-36
2041	198039-0507	2043	A626	2043	SDT3827	2043	34-6002-38
2041	198039-506	2043	A627	2043	SDT3875	2043	48-137049
2041	198039-507	2043	A648	2043	SDT3876	2043	48-137313
2041	198039-6	2043	A648A	2043	SDT3877	2043	51N3M
2041	198039-7	2043	A648B	2043	SDT9202	2043	111P5
2041	198064-1	2043	A648C	2043	SDT9207	2043	0137
2041	198079-1	2043	A658	2043	SDT9701	2043	152P1C
2041	198079-2	2043	A663	2043	SDT9704	2043	2549
2041	202909-827	2043	A679	2043	SDT9707	2043	4419-4
2041	204211-001	2043	A679R	2043	SJ1272	2043	5118
2041	211040-1	2043	A679Y	2043	SJ2001	2043	12537
2041	231378	2043	A680	2043	SJ2023	2043	99240-292
2041	232359	2043	A680R	2043	SJ2024	2043	134281
2041	232359A	2043	A680T	2043	SJ3507	2043	170890-1(PNP)
2041	236854	2043	A744	2043	SJ3520	2043	171053-1
2041	239713	2043	A746	2043	SJ3636	2043	181038
2041	241657	2043	A753	2043	SJ3637	2043	202925-047
2041	267791	2043	A753A	2043	SJ3679	2043	332762
2041	267878	2043	A753B	2043	SJ821	2043	559557
2041	282217	2043	A753C	2043	SJ8264	2043	659141
2041	291509	2043	A756	2043	SJ8764	2043	760021
2041	309449	2043	A756A	2043	SK3173	2043	1471139-1
2041	318835	2043	A756B	2043	SK3183	2047	BD183
2041	395253-1	2043	A756C	2043	ST29045	2047	BD245B
2041	445023-P1	2043	A757	2043	ST29046	2047	BD245C
2041	489751-119	2043	A757A	2043	ST29047	2047	BDX50
2041	502349	2043	A757B	2043	T1P2955	2047	BDX51
2041	505198	2043	A757C	2043	TIP2955	2047	BDX60
2041	505568	2043	A758	2043	TIP544	2047	BDX61
2041	570029	2043	A758A	2043	TM180	2047	BDY18
2041	0573562	2043	A758B	2043	TM219	2047	BDY19
2041	581070	2043	A758C	2043	TM281	2047	BDY20
2041	600115-413-001	2043	A837	2043	TR-1038-4	2047	BDY55
2041	610111-2	2043	B506	2043	TR-29	2047	BDY56
2041	610111-4	2043	B506A	2043	TRO2060-7	2047	BDY73
2041	610140-1	2043	B506C	2043	TRO259-6	2047	BDY77
2041	610161-2	2043	B506D	2043	TR1038-4	2047	C1667
2041	610161-4	2043	B541	2043	TR2327841	2047	D287
2041	618955-2	2043	B555	2043	TR29	2047	D379
2041	649002	2043	B555-0	2043	WEP87001	2047	D379P
2041	659174	2043	B555-0	2043	2N3171	2047	D379Q
2041	700080	2043	BD246	2043	2N3172	2047	D379R
2041	700080A	2043	BD246A	2043	2N3173	2047	D3798
2041	700083	2043	BD246B	2043	2N3183	2047	D424
2041	700083A	2043	BD246C	2043	2N3184	2047	D424-0
2041	793356-1	2043	BD250	2043	2N3185	2047	D424-R
2041	801537	2043	BD250A	2043	2N3186	2047	ECG284
2041	815246-1	2043	BD250B	2043	2N3195	2047	GE-265
2041	882028	2043	BD250C	2043	2N3196	2047	MJ6302
2041	891032	2043	BDX18	2043	2N3197	2047	RC41301
2041	983055	2043	BDX18N	2043	2N3198	2047	SJ2519
2041	984259	2043	BDX20	2043	2N3789	2047	TM284
2041	984259A	2043	BDY69	2043	2N3790	2047	2N3442
2041	1289050	2043	CXL180	2043	2N3791	2047	2N3716
2041	1417320	2043	CXL219	2043	2N3792	2047	2N3773
2041	1417320-1	2043	D341	2043	2N5621	2047	2N5630
2041	1417356-2	2043	D341H	2043	2N5623	2047	2N5634
2041	1471132-6	2043	EA15X124	2043	2N5625	2047	2N5878
2041	1471135-001	2043	ECG180	2043	2N5738	2047	2N6259
2041	1471135-1	2043	ECG219	2043	2N5741	2047	2N6262
2041	1471135-2	2043	ECG281	2043	2N5742	2047	2N6354
2041	1473665-2	2043	ES64(ELCOM)	2043	2N5745	2047	2N6496
2041	1473683-1	2043	ES74(ELCOM)	2043	2N5879	2047	2SC1586
2041	1701790-1	2043	ES90	2043	2N5883	2047	2SC1667
2041	1702601-1	2043	ES90(ELCOM)	2043	2N5884	2047	2SD287
2041	1833404	2043	PD-1029-LW	2043	2N6025	2047	2SD379
2041	1851517	2043	GE-263	2043	2N6226	2047	2SD379P
2041	1967784	2043	GE-74	2043	2N6227	2047	2SD379Q
2041	1968977	2043	HEP-248	2043	2N6228	2047	2SD379R
2041	1971487	2043	HEP-705	2043	2N6229	2047	2SD3798
2041	1971503	2043	HEP-37001	2043	2N6231	2047	2SD424
2041	2003073-91	2043	HEP-37001	2043	2N6231	2047	2SD424-0
2041	2003073-91A	2043	HEP-37003	2043	2N6231	2047	2SD424-R

RS 276-	Industry Standard No.	RS 276-	Industry Standard No.	RS 276-	Industry Standard No.	RS 276-	Industry Standard No.
2047	13-36440-1	2048	C154	2048	EP-100	2048	TA2911
2047	13-40342-1	2048	C154B	2048	EP-422	2048	TA7137
2047	46-86551-3	2048	C154C	2048	EP-797	2048	TA7156
2047	96-5370-01	2048	C154H	2048	EP-801	2048	TA7262
2047	40656	2048	C325	2048	EP-944	2048	TA7363
2047	43104	2048	C325A	2048	EP15X22	2048	TA7554
2048	A-1854-0420-1	2048	C325E	2048	EP15X68	2048	TA7555
2048	A-1854-0464	2048	C36583	2048	EP3053	2048	T1486
2048	A066-114	2048	C789	2048	EQ8-140	2048	T1487
2048	A272	2048	C789-0	2048	EQ8140	2048	TIP-14
2048	A273	2048	C789-0	2048	EQ15X86	2048	TIP-31A
2048	A276	2048	C789-R	2048	E88O(ELCOM)	2048	TIP24
2048	A277	2048	C789-Y	2048	ETTD-235	2048	TIP29
2048	A28D2260P	2048	C932	2048	G05705	2048	TIP29A
2048	A417032	2048	C932E	2048	GE-3229	2048	TIP31
2048	A54-3	2048	CGE-66	2048	GE-66	2048	TIP31A
2048	A5A-1B	2048	CII73Y	2048	G05705	2048	TIP31A
2048	A5A-IB	2048	CXL152	2048	HC495	2048	TM152
2048	A68-23-560	2048	D141	2048	HP57	2048	TR28D330E
2048	A9V	2048	D141HO1	2048	HR69	2048	TV-115
2048	AR-17	2048	D141H9Z	2048	HT308301BO	2048	TV-117
2048	AR-22(XSTR)	2048	D154	2048	HT311621B	2048	V409
2048	AR17(GREY)	2048	D184	2048	HT402352B	2048	V828C1173-Y-3
2048	AR17GREY	2048	D234	2048	IP20-0007	2048	V828C1983//-1
2048	ATC-TR-19	2048	D234-0	2048	IP20-0036	2048	VX3733
2048	B-1823	2048	D234-0	2048	IP20-0083	2048	W18
2048	B131	2048	D234-R	2048	IP20-0323	2048	X44C358
2048	B143004	2048	D234-Y	2048	IRTR76	2048	XA-1160
2048	B143011	2048	D235	2048	J241241	2048	XB404
2048	B143012	2048	D235-0	2048	L8-0066	2048	XB408
2048	B143018	2048	D235-0	2048	M75543-1	2048	XB476
2048	B143019	2048	D235-R	2048	M9576	2048	ZC-140
2048	B143026	2048	D235-Y	2048	M9661	2048	ZT1483
2048	B143027	2048	D235D	2048	MHT5906	2048	ZT1484
2048	B1D	2048	D235G	2048	MJE201	2048	ZT1485
2048	B3547	2048	D235GR	2048	MJE202	2048	ZT1486
2048	B3548	2048	D235R	2048	MM1619	2048	ZT1701
2048	B3550	2048	D235Y	2048	MPS111	2048	ZT2876
2048	B3551	2048	D27C1	2048	MPS112	2048	01-040243
2048	B3577	2048	D27C2	2048	MSA7505	2048	01-040389
2048	B3578	2048	D27C3	2048	MSA8505	2048	01-572784
2048	B3580	2048	D27C4	2048	P/FEV/117	2048	01-572791
2048	B3584	2048	D288	2048	P4J148	2048	01-572861
2048	B3585	2048	D288A	2048	PL-172-010-9-001	2048	01-572291
2048	B3586	2048	D288B	2048	PL-172-014-9-001	2048	1A0046
2048	B3588	2048	D288C	2048	PLE-48	2048	1A0058
2048	B3589	2048	D288L	2048	PN66	2048	1A0059
2048	B5000	2048	D289	2048	PP3250	2048	002-012400
2048	B5002	2048	D289A	2048	PP3310	2048	002D235RY
2048	B5021	2048	D289B	2048	PP3312	2048	2N1047
2048	B5022	2048	D289C	2048	PT2635	2048	2N1049
2048	B5031	2048	D28A1	2048	PT5693	2048	2N1483
2048	B5032	2048	D28A10	2048	PT665	2048	2N1484
2048	B5E	2048	D28A12	2048	Q-11115C	2048	2N1485
2048	BD106	2048	D28A13	2048	Q-12115C	2048	2N1486
2048	BD106A	2048	D28A2	2048	QT-D0313XAC	2048	2N1701
2048	BD106B	2048	D28A3	2048	R3611-1	2048	2N1709
2048	BD107A	2048	D28A4	2048	R3681-1	2048	2N1718
2048	BD107B	2048	D28A5	2048	R612-1	2048	2N1720
2048	BD109	2048	D28A6	2048	R621-1	2048	2N1768
2048	BD109-6	2048	D28A7	2048	R632-1	2048	2N1769
2048	BD124	2048	D28A9	2048	R632-2	2048	2N1886
2048	BD162	2048	D28D1	2048	RCA29	2048	2N2033
2048	BD163	2048	D28D10	2048	RCA29/SDH	2048	2N2034
2048	BD220	2048	D28D2	2048	RCA29A	2048	2N2035
2048	BD221	2048	D28D3	2048	RCA29A/SDH	2048	2N2036
2048	BD222	2048	D28D4	2048	RE21	2048	2N2339
2048	BD231A	2048	D28D5	2048	REN152	2048	2N2632
2048	BD239	2048	D28D7	2048	RT-133	2048	2N2657
2048	BD239A	2048	D317	2048	RT-150	2048	2N2828
2048	BD241	2048	D317A	2048	RT150	2048	2N2829
2048	BD243	2048	D317P	2048	RT8335	2048	2N2877
2048	BD243B	2048	D317P	2048	S-310E	2048	2N2878
2048	BD271	2048	D318	2048	S12020-04	2048	2N2947
2048	BD433	2048	D318A	2048	S1D153	2048	2N3138
2048	BD435	2048	D325	2048	S2042	2048	2N3140
2048	BD437	2048	D325C	2048	S33530	2048	2N3142
2048	BD439	2048	D325D	2048	S37166	2048	2N3144
2048	BD533	2048	D325E	2048	S39262	2048	2N3229
2048	BD535	2048	D325F	2048	SC0519	2048	2N3621
2048	BDX74	2048	D325D	2048	SC4303	2048	2N3622
2048	BDX75	2048	D343	2048	SC4303-1	2048	2N3625
2048	BDY12	2048	D359	2048	SC4303-2	2048	2N3626
2048	BDY13	2048	D359C	2048	SC4308	2048	2N3629
2048	BDY34	2048	D359C2	2048	SCD-D330	2048	2N3630
2048	BLY21	2048	D359D	2048	SD1345	2048	2N3632
2048	BLY35	2048	D359D1	2048	SD345	2048	2N3675
2048	BLY36	2048	D359D2	2048	SDA345	2048	2N3744
2048	BLY63	2048	D359E	2048	SDB345	2048	2N3745
2048	BLY79	2048	D360	2048	SDC345	2048	2N3747
2048	BLY88	2048	D360C	2048	SDD345	2048	2N3748
2048	BLY89	2048	D360E	2048	SDK345	2048	2N3818
2048	BLY92	2048	D365H	2048	SDL345	2048	2N3927
2048	BR101B	2048	D366-0	2048	SDM345	2048	2N4127
2048	BRC5296	2048	D366P	2048	SDN345	2048	2N4128
2048	BUY24	2048	D366Q	2048	SDT3422	2048	2N4307
2048	C1060	2048	D382	2048	SDT5102	2048	2N4308
2048	C1060A	2048	D382LM	2048	SDT5907	2048	2N4311
2048	C1060B	2048	D389	2048	SDT6001	2048	2N4312
2048	C1060BM	2048	D389-0	2048	SDT6011	2048	2N4877
2048	C1060C	2048	D389-OP	2048	SDT6013	2048	2N5293
2048	C1060D	2048	D389A	2048	SDT6031	2048	2N5294
2048	C1061	2048	D389AFP	2048	SDT6103	2048	2N5295
2048	C1061A	2048	D389B	2048	SDT7402	2048	2N5296
2048	C1061B	2048	D389BL	2048	SDT7412	2048	2N5297
2048	C1061C	2048	D389LB	2048	SDT7415	2048	2N5298
2048	C1061D	2048	D44C2	2048	SDT7511	2048	2N5334
2048	C1061T	2048	D44C3	2048	SDT7512	2048	2N5424
2048	C1061T-B	2048	D44C4	2048	SDT7514	2048	2N5483
2048	C1061TB	2048	D44C5	2048	SDT7515	2048	2N5492
2048	C1173	2048	D44C6	2048	SDT9009	2048	2N5606
2048	C1173-0	2048	D44C7	2048	SE5-0963	2048	2N5637
2048	C1173-GR	2048	D44C8	2048	SJE-513	2048	2N5642
2048	C1173-0	2048	D44C8B	2048	SJE-515	2048	2N5689
2048	C1173-R	2048	D44C9	2048	SJ842	2048	2N5690
2048	C1173-Y	2048	D44C8B	2048	SJE513	2048	2N5700
2048	C1173C	2048	D513	2048	SJE515	2048	2N5701
2048	C1173R	2048	D90	2048	SJE678	2048	2N5704
2048	C1173X	2048	D91	2048	SJE694	2048	2N5705
2048	C1173X0	2048	D91P	2048	SJE785	2048	2N5712
2048	C1173T	2048	DDBY278001	2048	SK3041	2048	2N5713
2048	C1398	2048	DDBY278002	2048	SP8416	2048	2N5765
2048	C1398Q	2048	DDBY407001	2048	SP8660	2048	2N5768
2048	C1418	2048	DS-513	2048	ST81436	2048	2N6121
2048	C1418A	2048	DS513	2048	STC1300	2048	2N6122
2048	C1418B	2048	DE496	2048	STC1850	2048	2N6312
2048	C1418C	2048	EA15X327	2048	STC1860	2048	2SA542B
2048	C1418D	2048	EA15X333	2048	T01-030	2048	2SB389-J
2048	C1419	2048	EA15X7121	2048	T1486	2048	2SC1019C
2048	C1419A	2048	EA15X8119	2048	T1487	2048	2SC1060
2048	C1419B	2048	EA15X8602	2048	T1P29X	2048	2SC1060(C,D)
2048	C1419C	2048	EA15X99	2048	T1P29X	2048	2SC1060A
2048	C1419D	2048	EA3716	2048	T611-1	2048	2SC1060B
2048	C1450B	2048	ECG152	2048	T612-1	2048	2SC1060BL

RS 276-	Industry Standard No.	RS 276-	Industry Standard No.	RS 276-	Industry Standard No.	RS 276-	Industry Standard No.
2048	28C1060BM	2048	28C553	2048	28D235-RED	2048	28D365A
2048	28C1060BY	2048	28C554	2048	28D235-Y	2048	28D365B
2048	28C1060C	2048	28C575	2048	28D2350	2048	28D365H
2048	28C1060CR	2048	28C585	2048	28D235A	2048	28D365P
2048	28C1060E	2048	28C599	2048	28D235B	2048	28D365Q
2048	28C1060F	2048	28C600	2048	28D235C	2048	28D366
2048	28C1060G	2048	28C636	2048	28D235D	2048	28D366-0
2048	28C1060GN	2048	28C638	2048	28D235E	2048	28D366P
2048	28C1060H	2048	28C638C	2048	28D235F	2048	28D366Q
2048	28C1060J	2048	28C690	2048	28D235G	2048	28D389
2048	28C1060K	2048	28C703	2048	28D235GN	2048	28D389(0)
2048	28C1060L	2048	28C704	2048	28D235GR	2048	28D389(LP)
2048	28C1060M	2048	28C789	2048	28D235H	2048	28D389(O)
2048	28C1060N	2048	28C789-0	2048	28D235J	2048	28D389(O,LP,P)
2048	28C1060X	2048	28C789-0	2048	28D235L	2048	28D389(P)
2048	28C1060Y	2048	28C789-R	2048	28D235LBY	2048	28D389-0
2048	28C1061(B)	2048	28C789-Y	2048	28D235M	2048	28D389-0P
2048	28C1061(C)	2048	28C789A	2048	28D2350R	2048	28D389AP
2048	0002BC1061A	2048	28C789B	2048	28D235R	2048	28D389APO
2048	28C1061B	2048	28C789C	2048	28D235RED	2048	28D389APO
2048	28C1061BM	2048	28C789D	2048	0028D235RY	2048	28D389APP
2048	28C1061BT	2048	28C789E	2048	28D235X	2048	28D389AP
2048	28C1061C	2048	28C789F	2048	28D235Y	2048	28D389AQ
2048	28C1061D	2048	28C789G	2048	28D288	2048	28D389B
2048	28C1061KA	2048	28C789GN	2048	28D288A	2048	28D389BLB
2048	28C1061KB	2048	28C789H	2048	28D288B	2048	28D389BLB-0
2048	28C1061KC	2048	28C789J	2048	28D288C	2048	28D389BLB-P
2048	28C1061T	2048	28C789K	2048	28D288K	2048	28D389BP
2048	28C1061T-B	2048	28C789L	2048	28D288L	2048	28D389L
2048	28C1061TB	2048	28C789M	2048	28D289	2048	28D389LB
2048	28C107	2048	28C7890R	2048	28D289A	2048	28D389LBP
2048	28C1398	2048	28C789R	2048	28D289B	2048	28D389LP
2048	28C1398P	2048	28C789X	2048	28D289C	2048	28D389P
2048	28C1398Q	2048	28C789Y	2048	28D28A	2048	28D389Q
2048	28C1409(B)	2048	28C790	2048	28D28B	2048	28D390
2048	28C1418	2048	28C790(0)	2048	28D28C	2048	28D390(0)
2048	28C1418A	2048	28C790-0	2048	28D28D	2048	28D390P
2048	28C1418B	2048	28C790Y	2048	28D28E	2048	28D477
2048	28C1418C	2048	28C892	2048	28D28F	2048	28D48
2048	28C1418D	2048	28C916	2048	28D28G	2048	28D49A
2048	28C1419	2048	28C92	2048	28D28GN	2048	28D49B
2048	28C1419A	2048	28C93	2048	28D28H	2048	28D49C
2048	28C1419B	2048	28C932	2048	28D28J	2048	28D49D
2048	28C1419C	2048	28C932E	2048	28D28K	2048	28D49E
2048	28C1419D	2048	28C94	2048	28D28L	2048	28D49F
2048	28C1450	2048	28C990	2048	28D28M	2048	28D49G
2048	28C14505	2048	28CD78	2048	28D280R	2048	28D49GN
2048	28C14508	2048	28D141	2048	28D28R	2048	28D49J
2048	28C1600R	2048	28D141A	2048	28D28X	2048	28D49K
2048	28C1983	2048	28D141B	2048	28D28Y	2048	28D49L
2048	28C2317	2048	28D141C	2048	28D292-0	2048	28D49M
2048	28C234	2048	28D141E	2048	28D292A	2048	28D490R
2048	28C234A	2048	28D141F	2048	28D292B	2048	28D49R
2048	28C234B	2048	28D141G	2048	28D292EL	2048	28D49X
2048	28C234C	2048	28D141GN	2048	28D292C	2048	28D49Y
2048	28C234D	2048	28D141H	2048	28D292D	2048	28D525
2048	28C234E	2048	28D141HO1	2048	28D292E	2048	28D525-0
2048	28C234F	2048	28D141H9Z	2048	28D292F	2048	28D525-0
2048	28C2340Q	2048	28D141J	2048	28D292G	2048	28D525Y
2048	28C2340GN	2048	28D141K	2048	28D292GA	2048	28D56A
2048	28C234H	2048	28D141L	2048	28D292GN	2048	28D56B
2048	28C234J	2048	28D141M	2048	28D292H	2048	28D56C
2048	28C234K	2048	28D141OR	2048	28D292J	2048	28D56D
2048	28C234L	2048	28D141R	2048	28D292K	2048	28D56E
2048	28C234M	2048	28D141X	2048	28D292L	2048	28D56F
2048	28C2340R	2048	28D141Y	2048	28D292M	2048	28D56G
2048	28C234R	2048	28D154	2048	28D2920R	2048	28D56GN
2048	28C234X	2048	28D184	2048	28D292R	2048	28D56H
2048	28C234Y	2048	28D184A	2048	28D292X	2048	28D56J
2048	28C234Z	2048	28D184B	2048	28D292Y	2048	28D56K
2048	28C2350	2048	28D184C	2048	28D29A	2048	28D56L
2048	28C325	2048	28D184D	2048	28D29B	2048	28D56M
2048	28C325A	2048	28D184E	2048	28D29E	2048	28D560R
2048	28C325C	2048	28D184F	2048	28D29F	2048	28D56R
2048	28C325B	2048	28D184H	2048	28D29GN	2048	28D56X
2048	28C4116	2048	28D184J	2048	28D29J	2048	28D56T
2048	28C4116A	2048	28D184K	2048	28D29K	2048	28D804HP
2048	28C4116B	2048	28D184L	2048	28D29L	2048	28D90
2048	28C4116C	2048	28D184M	2048	28D29M	2048	28D90A
2048	28C4116D	2048	28D1840R	2048	28D2900R	2048	28D90B
2048	28C4116E	2048	28D184Y	2048	28D29R	2048	28D90C
2048	28C4116F	2048	28D185	2048	28D29X	2048	28D90D
2048	28C4116GN	2048	28D226-0	2048	28D29Y	2048	28D90F
2048	28C4116H	2048	28D226A(0)	2048	28D313	2048	28D90G
2048	28C4116J	2048	28D226C	2048	28D313(D,E)	2048	28D900GN
2048	28C4116K	2048	28D226E	2048	28D313(DE)	2048	28D90H
2048	28C4116L	2048	28D226F	2048	28D313A	2048	28D90J
2048	28C4116M	2048	28D226G	2048	28D313B	2048	28D90K
2048	28C41160R	2048	28D226GN	2048	28D313C	2048	28D90L
2048	28C4116R	2048	28D226H	2048	28D313D	2048	28D90M
2048	28C4116X	2048	28D226J	2048	28D313DE	2048	28D900R
2048	28C4116Y	2048	28D226K	2048	28D313E	2048	28D90X
2048	28C489-R	2048	28D226L	2048	28D313F	2048	28D91
2048	28C489-Y	2048	28D226M	2048	28D313G	2048	28D91A
2048	28C489A	2048	28D2260R	2048	28D313GN	2048	28D91B
2048	28C489B	2048	28D226R	2048	28D313H	2048	28D91C
2048	28C489C	2048	28D226X	2048	28D313K	2048	28D91D
2048	28C489D	2048	28D226Y	2048	28D313L	2048	28D91E
2048	28C489E	2048	28D234	2048	28D313M	2048	28D91F
2048	28C489F	2048	28D234-0	2048	28D313N	2048	28D91GN
2048	28C489G	2048	28D234-0	2048	28D313R	2048	28D91H
2048	28C489GN	2048	28D234-ORG	2048	28D313Y	2048	28D91J
2048	28C489H	2048	28D234-R	2048	28D314	2048	28D91L
2048	28C489J	2048	28D234-RED	2048	28D314A	2048	28D91M
2048	28C489K	2048	28D234-Y	2048	28D314B	2048	28D910R
2048	28C489L	2048	28D2340	2048	28D314C	2048	28D91R
2048	28C489M	2048	28D234A	2048	28D314D	2048	28D91X
2048	28C4890R	2048	28D234B	2048	28D314F	2048	28F248T
2048	28C489R	2048	28D234C	2048	28D314GN	2048	28012
2048	28C489X	2048	28D234D	2048	28D314H	2048	3L4-6005-1
2048	28C491-BL	2048	28D234E	2048	28D314L	2048	3L4-6005-5
2048	28C491-BLU	2048	28D234F	2048	28D314M	2048	3L4-6005-55
2048	28C491-R	2048	28D234GA	2048	28D314N	2048	3L4-6012-02
2048	28C491-RED	2048	28D234GN	2048	28D314R	2048	3L4-6012-2
2048	28C491-Y	2048	28D234GR	2048	28D314Y	2048	3L4-6012-3
2048	28C491-YEL	2048	28D234H	2048	28D317	2048	4-464
2048	28C491A	2048	28D234J	2048	28D317A	2048	6-0005193
2048	28C491B	2048	28D234K	2048	28D317P	2048	6-5193
2048	28C491C	2048	28D234L	2048	28D317P	2048	007-0112
2048	28C491D	2048	28D234M	2048	28D318	2048	7-0112-00
2048	28C491E	2048	28D234N	2048	28D318(0)	2048	007-0112-03
2048	28C491F	2048	28D2340R	2048	28D318A	2048	7-0112-04
2048	28C491G	2048	28D234R	2048	28D318BA	2048	7-0112-05
2048	28C491GN	2048	28D234X	2048	28D318BB	2048	007-112-04
2048	28C491H	2048	28D234Y	2048	28D318P	2048	8-1074
2048	28C491J	2048	28D235	2048	28D330	2048	8-729-316-12
2048	28C491K	2048	28D235(0)	2048	28D330D	2048	8F345
2048	28C491L	2048	28D235(Y)	2048	28D330E		
2048	28C491M	2048	28D235-0	2048	28D343		
2048	28C4910R	2048	28D235-0	2048	28D360		
2048	28C491X	2048	28D235-OR	2048	28D360C		
2048	28C543	2048	28D235-ORG	2048	28D360D		
2048	28C551	2048	28D235-R	2048	28D360E		
2048	28C552			2048	28D365		
				2048	28D365-0		

RS 276-	Industry Standard No.	RS 276-	Industry Standard No.	RS 276-	Industry Standard No.	RS 276-	Industry Standard No.
2048	8P73ELU	2048	96-5225-01	2048	2856-2	2048	610162-4
2048	8P73GRN	2048	96-5252-02	2048	3107-204-90182	2048	610162-8
2048	8P73YEL	2048	96-5348-01	2048	3567-2	2048	610199-3
2048	8P70BLU	2048	96-5357-01	2048	3612(RCA)	2048	702886
2048	09-302083	2048	099-1(PHILCO)	2048	3621(RCA)	2048	717101
2048	09-302132	2048	099-1(SYL)	2048	3631(RCA)	2048	740856
2048	09-302164	2048	998100-1	2048	3631-1(RCA)	2048	996817
2048	09-302236	2048	103-0235-85	2048	3681	2048	1223783
2048	09-303018	2048	111N6C	2048	3843	2048	1417358-1
2048	09-303022	2048	121-1006	2048	4360D	2048	1471132-4
2048	09-303031	2048	121-719	2048	4490-7	2048	1473567-4
2048	09-303032	2048	121-770	2048	4491-5	2048	1473611
2048	09-303033	2048	121-772	2048	4491-8	2048	1473611-1
2048	09-304140	2048	121-804	2048	5001-053	2048	1473612
2048	09-305140	2048	121-806	2048	5847	2048	1473612-11
2048	97E8	2048	121-809	2048	6380-3	2048	1473681-1
2048	10-26-123-313	2048	121-854	2048	6381-2	2048	1473681-I
2048	012-103002	2048	121-874	2048	6382-2	2048	1700036
2048	13-14085-121	2048	121-887	2048	7213	2048	1950160
2048	13-28469-2	2048	121-927	2048	7252	2048	2320083
2048	13-2P64	2048	121-966	2048	7358-1	2048	2320432
2048	13-2P64-1	2048	121-966-01	2048	7414	2048	2320482
2048	13-32636-1	2048	121-976	2048	7423	2048	2320482H
2048	13-32640-1	2048	121-992-01	2048	8000-00011-050	2048	2320483
2048	13-39004-2	2048	125-B415	2048	8000-00028-041	2048	2320485
2048	13-39099-1	2048	129-33	2048	8020-206	2048	2320486
2048	13-39884-1	2048	0131-005352	2048	8800-205	2048	2320602
2048	13-39884-2	2048	131N2G	2048	12020-02	2048	2320651
2048	13-39884-3	2048	132-3	2048	16113	2048	2320652
2048	13-41628-2	2048	132-4	2048	16164	2048	2321302
2048	13-41628-3	2048	133-3	2048	16166	2048	2327206
2048	13-43635-1	2048	142-004	2048	16181	2048	2875493
2048	13-4P63	2048	142-008	2048	16182	2048	3438095
2048	13-4P63-1	2048	149N2002D	2048	16207	2048	3438867
2048	13-4P63-I	2048	149N2004	2048	16241	2048	3596446
2048	14-0104-1	2048	149N2D	2048	16305	2048	3596447
2048	14-609-00	2048	149P1D	2048	16334	2048	3596448
2048	14-609-03	2048	153N1C	2048	16335	2048	3596449
2048	14-609-04	2048	153N2C	2048	16336	2049	A-1853-0233-1
2048	14-609-06	2048	153N4C	2048	20295	2049	A-1853-0234-1
2048	14-609-08	2048	153N5C	2048	30294	2049	A-1853-0254-1
2048	14-902-23	2048	153N6	2048	38733	2049	A473(JAPAN)
2048	14-903-23	2048	172-010-9-001	2048	39302	2049	A473-0R
2048	14-905-23	2048	172-014-9-001	2048	39750	2049	A473-0
2048	14-908-23	2048	172-014-9-003	2048	39767	2049	A473-R
2048	14-909-23	2048	172-014-9-007	2048	37789(POWER)	2049	A473-Y
2048	15-123065	2048	172-024-9-004	2048	39824	2049	A473Y
2048	19-020-056	2048	172-031-9-003	2048	39948	2049	A489(JAPAN)
2048	19-020-066	2048	172-044-9-001	2048	39981	2049	A489-0
2048	19A115200-P1	2048	173A-4490-7	2048	40292	2049	A489-R
2048	19A116118-1	2048	173A-4491-5	2048	40307	2049	A489-Y
2048	19A116118-2	2048	173A4490-7	2048	40368	2049	A490(POWER)
2048	19A116118-I	2048	173A4491-5	2048	40613	2049	A670
2048	19A116118P1	2048	173A4491-8	2048	40618	2049	A670A
2048	19A116118P2	2048	176-042-9-007	2048	40621	2049	A670B
2048	020-1111-005	2048	176-005-9-004	2048	40622	2049	A670C
2048	21A040-052	2048	185-007	2048	40629	2049	A748Q
2048	21A112-095	2048	195N1	2048	40630	2049	A754
2048	21A112095	2048	195N1C	2048	40631	2049	A754A
2048	21A118-029	2048	195N1D	2048	40665	2049	A754B
2048	24MW662	2048	195N3	2048	41504	2049	A754C
2048	24MW778	2048	200-076	2048	42942	2049	A754D
2048	24MW977	2048	207A30	2048	43163	2049	A755
2048	34-1002	2048	207A33	2048	44208	2049	A755A
2048	34-6002-50	2048	209-30	2048	44699	2049	A755B
2048	34-6002-52	2048	211A6380-3	2048	50200-1B	2049	A755C
2048	34-6002-56	2048	211A6381-2	2048	50200-24	2049	A755D
2048	34-6006-1	2048	211A6382-2	2048	50200-9	2049	AR-23(XSTR)
2048	34-6016-45	2048	216-001-001	2048	50308-0100	2049	AR-25
2048	42-42459	2048	229-1301-35	2048	60133	2049	AR25(G)
2048	42-27539	2048	250-0359	2048	60175	2049	AR25(GREEN)
2048	44A395986-001	2048	260-10-024	2048	60216	2049	AR27(GREEN)
2048	44A417032-001	2048	260-10-054	2048	60408	2049	AR37(GREEN)
2048	46-86317-3	2048	260P12701	2048	60679	2049	AR37GREEN
2048	46-86335-3	2048	260P28401	2048	60719	2049	AR44
2048	46-86348-3	2048	276-2017	2048	60835	2049	B132
2048	46-86374-3	2048	276-2017/RS2017	2048	60838	2049	B434
2048	46-86400-3	2048	296(REGENCY)	2048	60886	2049	B434-0
2048	46-86516-3	2048	353-9203-001	2048	60966	2049	B434-0
2048	48-15309X3	2048	353-9502-001	2048	60987	2049	B434-R
2048	48-137396	2048	413	2048	61102	2049	B434-Y
2048	48-137437	2048	417-175-12993	2048	61193	2049	B435
2048	48-137506	2048	417-852	2048	61285	2049	B435-0
2048	48-355040	2048	474A410BEP2	2048	61286	2049	B435-0
2048	48-355059	2048	474A410BW-2	2048	61418	2049	B435-R
2048	48-44885901	2048	499(CHRYSLER)	2048	61772	2049	B435-Y
2048	48-869576	2048	514-047830	2048	61866	2049	B507
2048	48-869661	2048	565-073	2048	61875	2049	B508
2048	48-97046A30	2048	617-117	2048	61958	2049	B509
2048	48-97046A58	2048	623(RCA)	2048	61981	2049	B511
2048	48K355040	2048	640-1(SYLVANIA)	2048	62156	2049	B511C
2048	48K355059	2048	690L-021H25	2048	62571	2049	B511D
2048	488137309	2048	750A858-448	2048	62676	2049	B512
2048	488137311	2048	753-2001-173	2048	62681	2049	B512A
2048	488137369	2048	753-4001-932	2048	64076	2049	B513
2048	488155013	2048	755-9010-235	2048	79992	2049	B513A
2048	488155062	2048	770-045	2048	95261-1	2049	B514
2048	48840662904	2048	800-533-00	2048	99252-4	2049	B515
2048	48843240901	2048	800-546-00	2048	116118	2049	B537LM
2048	48843241901	2048	834-250-011	2048	116118-2	2049	BD223
2048	48844885901	2048	921-1009	2048	127798	2049	BD224
2048	57A195-10	2048	995-01-6131	2048	128056	2049	BD225
2048	57A250-14	2048	1000-138	2048	131075	2049	BD240
2048	57A278-14	2048	1000-142	2048	131848(RCA)	2049	BD240A
2048	57A279-14	2048	1010-7936	2048	131849	2049	BD242
2048	57A286-10	2048	1016-81	2048	132575	2049	BD242A
2048	57B131-10	2048	1041-74	2048	132697	2049	BD242B
2048	57B205-14	2048	1043-1286	2048	136648	2049	BD244
2048	57B250-14	2048	1043-7358	2048	137527	2049	BD244A
2048	57B278-14	2048	1045-7828	2048	137692	2049	BD244B
2048	57B279-14	2048	1065-5225	2048	138579	2049	BD262A
2048	57B286-10	2048	1069	2048	142691	2049	BD272
2048	61-309689	2048	1069(GE)	2048	146139	2049	BD434
2048	62B046-1	2048	1080-130	2048	153107	2049	BD436
2048	62B046-2	2048	1081-3368	2048	167692	2049	BD438
2048	62B046-3	2048	1098-14	2048	171162-265	2049	BD440
2048	62B046-4	2048	1111(GE)	2048	171162-291	2049	BD534
2048	66P-010	2048	1714-0405	2048	172463	2049	BD536
2048	73C18028-11	2048	1968-17	2048	172643	2049	BDX27
2048	73C18028-12	2048	1969-17	2048	242102	2049	BDX27-10
2048	730180829-11	2048	2000-216	2048	279517	2049	BDX27-6
2048	730180829-12	2048	2001	2048	309441	2049	BDX28
2048	77-272913-1	2048	2001(E.P.JOHNSON)	2048	309459	2049	BDX28-10
2048	77-273715-1	2048	2017-109	2048	489751-033	2049	BDX28-6
2048	77-273738-1	2048	2036-09	2048	489751-044	2049	BDY82
2048	86-5102-2	2048	2057A2-301	2048	610111-6	2049	BDY83
2048	86-5107-2	2048	2057B2-151	2048	610149-2	2049	CUB-69
2048	86-5108-2	2048	2082-6	2048	610153-1	2049	CXL153
2048	86-529-2	2048	2210-17	2048	610153-2	2049	D4C7
2048	86-544-2	2048	2382-17	2048	610153-3	2049	D4C1
2048	86-604-2	2048	2853-2	2048	610153-4	2049	D4C2
2048	86-663-2	2048	2854-2	2048	610153-5	2049	D4C3
2048	90-600	2048	2855-2	2048	610153-6	2049	D4C4

RS 276-	Industry Standard No.	RS 276-	Industry Standard No.	RS 276-	Industry Standard No.	RS 276-	Industry Standard No.
2049	D4505	2049	01-572774	2049	3L4-6013-2	2050	A564AS
2049	D45C6	2049	002-012300	2049	3L4-6013-3	2050	A564AT
2049	D45C7	2049	2N2875	2049	3L4-6013-55	2050	A564F
2049	D45C8	2049	2N3021	2049	3L4-6013-56	2050	A564FQ
2049	D45C9	2049	2N3022	2049	3L4-6013-58	2050	A564FR
2049	EA15X3118	2049	2N3023	2049	8P445	2050	A564J
2049	EA15X328	2049	2N3024	2049	09-300090	2050	A564P
2049	EA15X334	2049	2N3025	2049	09-301077	2050	A564POR
2049	EA15X8118	2049	2N3026	2049	13-39100-1	2050	A564Q
2049	EA15X8130	2049	2N3199	2049	13-39819-1	2050	A564QHD
2049	EA3715	2049	2N3200	2049	13-39819-2	2050	A564QR
2049	ECG153	2049	2N3205	2049	14-0104-2	2050	A564R
2049	EP-421	2049	2N3206	2049	19A116375	2050	A564S
2049	EP-802	2049	2N3740A	2049	19A116375P1	2050	A564T
2049	EP-943	2049	2N4387	2049	020-1111-004	2050	A565Q
2049	EP15X15	2049	2N4388	2049	21A112-094	2050	A578
2049	EP15X23	2049	2N5112	2049	25A473Y	2050	A578A
2049	ES81(ELCOM)	2049	2N5161	2049	34-1003	2050	A578B
2049	EW183	2049	2N5597	2049	34-6002-57	2050	A578C
2049	GE-26	2049	2N6021	2049	34-6016-46	2050	A579
2049	GE-69	2049	2N6022	2049	42-27538	2050	A579A
2049	HA-00699	2049	2N6023	2049	46-86515-3	2050	A579B
2049	HA505	2049	2N6024	2049	46-86581-3	2050	A579C
2049	HF58	2049	2N6026	2049	48-137312	2050	A5T405B
2049	HR70	2049	2N6124	2049	48-137501	2050	A5T4059
2049	IRTR77	2049	2N6125	2049	48-137507	2050	A5T4060
2049	L8-0067	2049	2N6126	2049	48-137540	2050	A5T4061
2049	M9348	2049	28A473	2049	48-137566	2050	A5T4062
2049	MJE101	2049	28A473(O)	2049	48-43241901	2050	A641
2049	MJE102	2049	28A473(O)	2049	48-44884901	2050	A641(JAPAN)
2049	MJE103	2049	28A473-GR	2049	48-90420A06	2050	A641(PBP)
2049	MM4020	2049	28A473-O	2049	48B137310	2050	A641(PNP)
2049	NPS6518	2049	28A473-R	2049	48B137312	2050	A641A
2049	P1V-4	2049	28A473-Y	2049	48B137370	2050	A641B
2049	P2B	2049	28A473GR	2049	488155014	2050	A641C
2049	P2K	2049	28A473R	2049	488155066	2050	A641L
2049	P4W	2049	28A473Y	2049	48844884901	2050	A641M
2049	P5H	2049	28A486	2049	57A205-14	2050	A641B
2049	P5U	2049	28A486-R	2049	57A206-14	2050	A666
2049	P8B	2049	28A486-RED	2049	57A251-14	2050	A666A
2049	P8H	2049	28A486-Y	2049	57A277-14	2050	A666H
2049	RCA30	2049	28A486-YEL	2049	57B132-10	2050	A666HR
2049	RCA30A	2049	28A489	2049	57B206-14	2050	A666HRS
2049	RE22	2049	28A489-O	2049	57B251-14	2050	A666R
2049	REN153	2049	28A489-R	2049	57B277-14	2050	A666S
2049	RT-151	2049	28A489-Y	2049	61-309690	2050	A721
2049	RT-155	2049	28A490	2049	730180830-11	2050	A721B
2049	RT8336	2049	28A490(POWER)	2049	730180830-12	2050	A721T
2049	RT8336	2049	28A490-O	2049	77-272914-1	2050	A721U
2049	S2041	2049	28A490-O	2049	77-273716-1	2050	A722
2049	S33529	2049	28A490-Y	2049	77-273739-1	2050	A722B
2049	S37165	2049	28A490A	2049	86-530-2	2050	A722R
2049	S39261	2049	28A490B	2049	96-5303-01	2050	A722U
2049	SCD T334	2049	28A490C	2049	96-5349-01	2050	A725
2049	SCD-T334	2049	28A490D	2049	96-5356-01	2050	A725F
2049	SD1445	2049	28A490E	2049	998099-1	2050	A725G
2049	SD445	2049	28A490F	2049	100-1(PHILCO)	2050	A725H
2049	SDA445	2049	28A490G	2049	121-803	2050	A726
2049	SDB445	2049	28A490GN	2049	121-873	2050	A726F
2049	SDI445	2049	28A490H	2049	121-886	2050	A726G
2049	SDJ445	2049	28A490J	2049	121-926	2050	A726H
2049	SDK445	2049	28A490K	2049	121-969-02	2050	A825
2049	SDL445	2049	28A490L	2049	121-977	2050	BC153
2049	SDM445	2049	28A490LBG1	2049	121-997	2050	BC154
2049	SDN445	2049	28A490M	2049	0131-002049	2050	BC158C
2049	SDT-445	2049	28A490R	2049	0131-005353	2050	BC159C
2049	SDT3509	2049	28A490X	2049	149F2001D	2050	BC177
2049	SDT3510	2049	28A490Y	2049	149R3003	2050	BC177A
2049	SDT3513	2049	28A490YA	2049	177-023-9-001	2050	BC177B
2049	SDT3514	2049	28A490YLBG11	2049	195P2	2050	BC178
2049	SDT3701	2049	28A670	2049	195P4	2050	BC178A
2049	SDT3702	2049	28A670A	2049	229-1-301-36	2050	BC178B
2049	SDT3703	2049	28A670B	2049	276-2025	2050	BC178C
2049	SDT3704	2049	28A670C	2049	276-2025/R82025	2050	BC179
2049	SDT3706	2049	28A748Q	2049	624(RCA)	2050	BC179A
2049	SDT3707	2049	28A748R	2049	690L-021H26	2050	BC179B
2049	SDT3708	2049	28A754	2049	995-01-6130	2050	BC179C
2049	SDT3709	2049	28A754A	2049	1043-1294	2050	BC204
2049	SDT3710	2049	28A754B	2049	1045-7836	2050	BC204A
2049	SDT3711	2049	28A754C	2049	1071	2050	BC205B
2049	SDT3712	2049	28A754D	2049	1071(GE)	2050	BC205C
2049	SDT3713	2049	28A755	2049	1113(GE)	2050	BC206
2049	SDT3715	2049	28A755A	2049	2036-58	2050	BC206A
2049	SDT3716	2049	28A755B	2049	2057B2-150	2050	BC206B
2049	SDT3717	2049	28A755C	2049	2081-6	2050	BC206C
2049	SDT3720	2049	28A755D	2049	2211-17	2050	BC225
2049	SDT3721	2049	28A766	2049	3682	2050	BC258C
2049	SDT3722	2049	28A766S	2049	7420	2050	BC307C
2049	SDT3725	2049	28A768	2049	31004-1	2050	BC308C
2049	SDT3726	2049	28B434	2049	44209	2050	BC320A
2049	SDT3727	2049	28B434(O)	2049	50201-4	2050	BC320C
2049	SDT3729	2049	28B434-O	2049	61865	2050	BC321
2049	SDT3730	2049	28B434-R	2049	64075	2050	BC321A
2049	SDT3733	2049	28B434-Y	2049	64197	2050	BC321B
2049	SDT445	2049	28B435	2049	75803-1	2050	BC321C
2049	SDT5112	2049	28B435-O	2049	75803-2	2050	BC322
2049	SE5-0964	2049	28B435-O	2049	75803-3	2050	BC322B
2049	SJ1152	2049	28B435-R	2049	115792	2050	BC322C
2049	SJ1171	2049	28B435-Y	2049	120057	2050	BC526B
2049	SJ1284	2049	28B435R	2049	139069	2050	BC526B
2049	SJE514	2049	0028B435RY	2049	144076	2050	BC526C
2049	SJE677	2049	28B435Y	2049	146081	2050	BC565B
2049	SJE695	2049	28B507	2049	198065-1	2050	BCW59B
2049	SPS1437	2049	28B507E	2049	489751-032	2050	BCW59B
2049	ST27020	2049	28B508	2049	610112-1	2050	BCW61BC
2049	STC5202	2049	28B509	2049	610149-1	2050	BCW61BD
2049	STC5203	2049	28B511	2049	610149-3	2050	BCW86
2049	STC5205	2049	28B511C	2049	610162-7	2050	BCX71BJ
2049	STC5206	2049	28B511D	2049	610195-4	2050	BCX71BK
2049	STC5303	2049	28B511E	2049	802037-001	2050	BCY78
2049	STC5802	2049	28B512	2049	1471134	2050	BCY79
2049	STC5803	2049	28B512A	2049	2320884	2050	BF243
2049	STC5805	2049	28B512P	2049	2321281	2050	BF65
2049	STC5806	2049	28B513	2049	3596451	2050	BFW22
2049	STX0029	2049	28B513A	2049	3596452	2050	BFX37
2049	STX0030	2049	28B513P	2049	3596453	2050	BTX-084
2049	TA7555	2049	28B513Q	2049	3596454	2050	C532000585
2049	TA7557	2049	28B513R	2050	A-1853-0050-1	2050	C673
2049	TIP-30	2049	28B514	2050	A-1853-0066-1	2050	C7
2049	TIP30	2049	28B515	2050	A-1853-0077-1	2050	C8
2049	TIP30A	2049	28B537	2050	A-1853-0086-1	2050	CD500
2049	TIP32	2049	28B537LM	2050	A-1853-0098-1	2050	D29P1
2049	TIP32A	2049	28B699	2050	A-1853-0300-1	2050	D29P2
2049	TM153	2049	28B699Q	2050	A2311	2050	D29P3
2049	TR-8019	2049	28C831	2050	A564	2050	D29P4
2049	TR2327723	2049	3L4-6011-11	2050	A564-0	2050	DDBY003003
2049	TR8019	2049	3L4-6011-12	2050	A564A	2050	DW-7655
2049	TV-116	2049	3L4-6011-14	2050	A5640R	2050	DW-7655-LV00223
2049	TV828A483	2049	3L4-6011-9	2050	A564A	2050	DW7655
2049	W19	2049	3L4-6013-02	2050	A564ABQ	2050	ECG234
2049	X45C359	2049	3L4-6013-15	2050	A564A0	2050	E8112
2049	01-010473			2050	A564AP	2050	ET350335
				2050	A564AQ	2050	ET35035

RS 276-	Industry Standard No.
2050	EYZP-808
2050	EYZP808
2050	FI-1021
2050	F31990
2050	FX3964
2050	GE-65
2050	HA-00564
2050	HA-00733
2050	HEP80006
2050	HEP80031
2050	HF47
2050	HT105641C
2050	HT105641D
2050	HT105641H
2050	HT105642B
2050	HT107211T
2050	HT305642B
2050	LDA454
2050	LDA455
2050	LDB257
2050	M7
2050	M9412
2050	M9467
2050	MI813674/47
2050	MPS3639
2050	MPS9680H/E
2050	MPSA70-YEL
2050	MPSD56
2050	MPSKT0
2050	MPSKT1
2050	MPSKT2
2050	NR-621AU
2050	P/N297L01OCO1
2050	P12407-1
2050	PM1120
2050	Q5102
2050	Q5102P
2050	Q5102Q
2050	Q5102R
2050	QRT106
2050	RB195
2050	REN234
2050	REN62
2050	RV2353
2050	SC25094
2050	S1990
2050	SC158B
2050	SC159B
2050	SC258B
2050	SC259B
2050	SE5-0909
2050	SS829A4
2050	SS829A5
2050	T309
2050	TM234
2050	TR02020-2
2050	TR106(SPRAGUE)
2050	TR28A763
2050	V328A854-Q/1E
2050	V39-0004-923
2050	V39-0008-923
2050	WEP907
2050	XA-1164
2050	Y56601-44
2050	Y56601-4B
2050	07722
2050	01-010564
2050	01-010844
2050	001-021170
2050	001-021171
2050	001-021172
2050	001-021173
2050	28A494(Y)
2050	28A494-GR-1
2050	28A494-OR
2050	28A494A
2050	28A494B
2050	28A494C
2050	28A494D
2050	28A494E
2050	28A494F
2050	28A494G
2050	28A494GN
2050	28A494GR
2050	28A494H
2050	28A494J
2050	28A494K
2050	28A494L
2050	28A494M
2050	28A4940
2050	28A494OR
2050	28A494R
2050	28A494X
2050	28A494Y
2050	28A543
2050	28A564
2050	28A564(O)
2050	28A564(O)
2050	28A564(P)
2050	28A564(Q)
2050	28A564(R)
2050	28A564(S,T)
2050	28A564(T)
2050	28A564-O
2050	28A564-0
2050	28A564-OGD
2050	28A564-OR
2050	28A564-P
2050	28A564-Q
2050	28A564-R
2050	28A564A
2050	28A564A(P)
2050	28A564A(R)
2050	28A564A(S)
2050	28A564ABQ
2050	28A564ABQ-1
2050	28A564AG
2050	28A564AK
2050	28A564AL
2050	28A564AO
2050	28A564AP
2050	28A564AQ
2050	28A564AR
2050	28A564AT
2050	28A564B
2050	28A564C
2050	28A564D
2050	28A564E
2050	28A564F
2050	28A564FQ
2050	28A564FQ-1
2050	28A564FR
2050	28A564FR-1
2050	28A564G
2050	28A564GN
2050	28A564H
2050	28A564J
2050	28A564K
2050	28A564L
2050	28A564M
2050	28A564OR
2050	28A564P
2050	28A564P.A
2050	28A564POR
2050	28A564Q
2050	28A564QQD
2050	28A564QHD
2050	28A564QP
2050	28A564QR
2050	28A564R
2050	28A564S
2050	28A564T
2050	28A564X
2050	28A564XL
2050	28A564Y
2050	28A572
2050	28A578
2050	28A578A
2050	28A578B
2050	28A578C
2050	28A579
2050	28A579A
2050	28A579B
2050	28A579C
2050	28A640
2050	28A640(M)
2050	28A640A
2050	28A640B
2050	28A640C
2050	28A640D
2050	28A640E
2050	28A640L
2050	28A640M
2050	28A640S
2050	28A666
2050	28A666QRS
2050	28A666A
2050	28A666B
2050	28A666BL
2050	28A666C
2050	28A666D
2050	28A666E
2050	28A666H
2050	28A666HR
2050	28A666IQRS
2050	28A666Q
2050	28A666QRS
2050	28A666R
2050	28A666S
2050	28A666Y
2050	28A702
2050	28A721
2050	28A721Q
2050	28A721R
2050	28A721B
2050	28A721T
2050	28A721U
2050	07722
2050	28L.22S
2050	28A722T
2050	28A722U
2050	28A725
2050	28A725P
2050	28A725Q
2050	28A725H
2050	28A726
2050	28A726F
2050	28A726Q
2050	28A726E
2050	28A726Y
2050	28A728
2050	28A741H
2050	28A763
2050	28A763-W
2050	28A763-WL-3
2050	28A763-WL-4
2050	28A763-WL-5
2050	28A763-WL-6
2050	28A763-WN
2050	28A763-WN-3
2050	28A763-WN-4
2050	28A763-WN-5
2050	28A763-WN-6
2050	28A763-Y
2050	28A763-YL-3
2050	28A763-YL-4
2050	28A763-YL-5
2050	28A763-YL-6
2050	28A763-YN
2050	28A763-YN-3
2050	28A763-YN-4
2050	28A763-YN-6
2050	28A787
2050	28A825
2050	28A825Q
2050	28A825R
2050	28A836
2050	28A836D
2050	28A836E
2050	28A836F
2050	28A842-BL
2050	28A842-GR
2050	28A642Q
2050	28B642R
2050	28B774
2050	28B774H
2050	28B774S
2050	28C641H
2050	3N123
2050	006-00055182
2050	6-138
2050	6-49
2050	8Q-000003-11
2050	09-300026
2050	09-300036
2050	09-30055B
2050	11-691501
2050	11-691502
2050	12-21-050
2050	12-PG 01
2050	12-PG-01
2050	13-343617-30
2050	14-602-64
2050	014-754
2050	14-803-32
2050	17-RLB
2050	19QC19
2050	020-1110-005
2050	020-1110-027
2050	21M158
2050	21N604
2050	25-000456-1
2050	34-1007
2050	34-1008
2050	34-6016-64
2050	37-19201
2050	40-09437
2050	40-09952
2050	40-11253
2050	42-9029-40P
2050	42-9029-40W
2050	42-9029-60B
2050	42-9029-70Q
2050	46-86429-3
2050	48-134830A
2050	48-134831
2050	48-134831A
2050	48-134832
2050	48-134832A
2050	48-155045
2050	48-355006
2050	48-86429-3
2050	48-869412
2050	48-869467
2050	48-90234A38
2050	48-90234A99
2050	48-90343A59
2050	48B869412
2050	48B869467
2050	48B155045
2050	51-47-54
2050	66A10310
2050	67A8926
2050	68A7355P1
2050	70-943-773-001
2050	730180831-3
2050	86-608-2
2050	96-5258-01
2050	121-716
2050	132-029
2050	132-031
2050	134P1
2050	151-0219-00
2050	151-0342-00
2050	151-0410-00
2050	151-0453-00
2050	177-007-9-001
2050	177-018-9-001
2050	200X4085-415
2050	207114
2050	297L010C
2050	297L010001
2050	352-0773-010
2050	352-0773-020
2050	352-0773-030
2050	417-153-1341
2050	417-153-13431
2050	417-196-13262
2050	461-2001
2050	514-044910
2050	601-010081 0
2050	762-110
2050	0831
2050	991-012328
2050	1002
2050	1016
2050	1038-9922
2050	1063-8435
2050	1152
2050	1341
2050	1574-01
2050	1935-17
2050	2057A2-529
2050	4053
2050	4825-0031-01
2050	11000
2050	31015
2050	57009
2050	75561-18
2050	75561-21
2050	75596-1
2050	75596-2
2050	75596-3
2050	75596-4
2050	90029-246
2050	130253
2050	131710
2050	136281
2050	138049-001
2050	158049-004
2050	144030
2050	144858
2050	144863
2050	146902
2050	309683
2050	309685
2050	610083-1
2050	610083-2
2050	610083-4
2050	610134-1
2050	610134-2
2050	610134-4
2050	610134-5
2050	610134-6
2050	610223-1
2050	611233
2050	741863
2050	1471112-10
2050	3457936-1
2051	28C1348
2051	A-106
2051	A-11095924
2051	A-11237336
2051	A-1141 6062
2051	A-1141-6062
2051	A-120278
2051	A-125332
2051	A-128
2051	A-1379
2051	A-1380
2051	A-156
2051	A-1567
2051	A-158B
2051	A-158C
2051	A-168
2051	A-1854-0003-1
2051	A-1854-0019-1
2051	A-1854-0025-1
2051	A-1854-0027-1
2051	A-1854-0071-1
2051	A-1854-0094-1
2051	A-1854-0099-1
2051	A-1854-0201-1
2051	A-1854-0215-1
2051	A-1854-0241-1
2051	A-1854-0246-1
2051	A-1854-0251-1
2051	A-1854-0354-1
2051	A-1854-0408-1
2051	A-1854-0434-1
2051	A-1854-0471-1
2051	A-1854-0492-1
2051	A-1854-0541-1
2051	A-1854-0554-1
2051	A-415
2051	A-494
2051	A-567A
2051	A.184/5
2051	A0-54-195
2051	A054-108
2051	A054-114
2051	A054-115
2051	A054-155
2051	A054-173
2051	A054-175
2051	A054-195
2051	A054-221
2051	A054-222
2051	A054-225
2051	A054-233
2051	A054-234
2051	A059-109
2051	A06-1-12
2051	A065-102
2051	A065-103
2051	A065-104
2051	A065-108
2051	A065-109
2051	A065-110
2051	A065-113
2051	A066-109
2051	A066-112
2051	A066-113(28C7321)
2051	A068-108
2051	A068-113
2051	A069-102/103
2051	A069-104
2051	A069-104/106
2051	A069-106
2051	A069-120
2051	A069-122
2051	A106
2051	A106(JAPAN)
2051	A108
2051	A1086
2051	A1087
2051	A108A
2051	A108B
2051	A10005-010-A
2051	A10005-011-A
2051	A10005-015-D
2051	A2011
2051	A1414257
2051	A116
2051	A12-1-70
2051	A12-1-705
2051	A12-1A9G
2051	A128
2051	A128A
2051	A137
2051	A137(NPN)
2051	A1379
2051	A1380
2051	A13N1
2051	A141
2051	A1412-1
2051	A142
2051	A143
2051	A1472-19
2051	A151
2051	A152
2051	A153
2051	A156
2051	A1567
2051	A1567-1
2051	A157
2051	A157A
2051	A157B
2051	A157C
2051	A158
2051	A158A
2051	A158B
2051	A158C
2051	A159
2051	A159A
2051	A159B
2051	A159C
2051	A168
2051	A1B
2051	A1F
2051	A1H
2051	A1U
2051	A1L
2051	A1P
2051	A12-1
2051	A1U(LABTIP)
2051	A1V
2051	A1VE
2051	A1W
2051	A2019ZC
2051	A20372
2051	A2410
2051	A24100
2051	A2411
2051	A2412
2051	A2413
2051	A2434
2051	A246
2051	A246(AMC)
2051	A2466
2051	A2468
2051	A2469
2051	A2470
2051	A248(AMC)
2051	A249
2051	A2499
2051	A25A509-015-101
2051	A2B
2051	A2BRN
2051	A2PGRN
2051	A2J

RS 276-	Industry Standard No.	RS 276-	Industry Standard No.	RS 276-	Industry Standard No.	RS 276-	Industry Standard No.
2051	A2SC538PQR	2051	AR-108	2051	BC134	2051	BC547B
2051	A301	2051	AR-200	2051	BC135	2051	BC548
2051	A306	2051	AR-201	2051	BC136	2051	BC548VI
2051	A307	2051	AR-202	2051	BC146	2051	BC583
2051	A323	2051	AR107	2051	BC147	2051	BC71
2051	A324	2051	AR108	2051	BC147B	2051	BCW34
2051	A344	2051	AR200(W)	2051	BC147A	2051	BCW36
2051	A345	2051	AR200WHITE	2051	BC147B	2051	BCW48A
2051	A346	2051	AR201(Y)	2051	BC148	2051	BCW60A
2051	A3B	2051	AR201YELLOW	2051	BC148A	2051	BCW60AA
2051	A3F	2051	AR202GREEN	2051	BC148B	2051	BCW60AB
2051	A3G	2051	AR204	2051	BC148C	2051	BCW60AC
2051	A3N	2051	AR205	2051	BC149	2051	BCW60B
2051	A3S	2051	AR206	2051	BC149A	2051	BCW60C
2051	A3T	2051	AR208	2051	BC149B	2051	BCW65BA
2051	A3T2221	2051	AR303	2051	BC149G	2051	BCW65EB
2051	A3T2221A	2051	AR306	2051	BC150	2051	BCW83
2051	A3T2222	2051	AR306(BLUE)	2051	BC151	2051	BCW94
2051	A3T2222A	2051	AR306(ORANGE)	2051	BC152	2051	BCW94A
2051	A3T3011	2051	AT329	2051	BC155B	2051	BCW94B
2051	A3T929	2051	AT335	2051	BC156B	2051	BCW94C
2051	A3T930	2051	AT336	2051	BC167	2051	BCW94KA
2051	A3W	2051	AT337	2051	BC167A	2051	BCW94KB
2051	A3Z	2051	AT347	2051	BC167B	2051	BCW94KC
2051	A415	2051	AT348	2051	BC168	2051	BCW95A
2051	A42X00434-01	2051	AT349	2051	BC168A	2051	BCW95B
2051	A43021415	2051	AT370	2051	BC168B	2051	BCW95KA
2051	A454	2051	AT400	2051	BC168C	2051	BCW95KB
2051	A455	2051	AT401	2051	BC169	2051	BCX19
2051	A472	2051	AT402	2051	BC169A	2051	BCX19R
2051	A481A0028	2051	AT403	2051	BC169B	2051	BCX20
2051	A481A0031	2051	AT404	2051	BC169C	2051	BCX20R
2051	A494	2051	AT405	2051	BC169CL	2051	BCX58VII
2051	A494(JAPAN)	2051	AT406	2051	BC171	2051	BCX58VIII
2051	A4A	2051	AT407	2051	BC171A	2051	BCX70AG
2051	A4M	2051	AT420	2051	BC171B	2051	BCX70AH
2051	A4N	2051	AT421	2051	BC172	2051	BCX70AJ
2051	A4P	2051	AT422	2051	BC172A	2051	BCY-50
2051	A4R	2051	AT423	2051	BC172B	2051	BCY-58
2051	A4U	2051	AT424	2051	BC172C	2051	BCY13
2051	A4V	2051	AT425	2051	BC175	2051	BCY15
2051	A4Y-2	2051	AT426	2051	BC180	2051	BCY16
2051	A514-033338	2051	AT427	2051	BC180B	2051	BCY36
2051	A54-96-001	2051	AT490	2051	BC182	2051	BCY42
2051	A54-96-002	2051	AT491	2051	BC182A	2051	BCY43
2051	A567	2051	AT492	2051	BC182L	2051	BCY50
2051	A567A	2051	AT493	2051	BC183	2051	BCY51
2051	A593	2051	AT494	2051	BC183A	2051	BCY511
2051	A5C	2051	AT495	2051	BC183B	2051	BCY56
2051	A5H	2051	AWH-24	2051	BC183L	2051	BCY57
2051	A5K	2051	B 722246-2	2051	BC184	2051	BCY58B
2051	A5L	2051	B-1338	2051	BC184B	2051	BCY58C
2051	A5M	2051	B-1421	2051	BC184L	2051	BCY58D
2051	A5N	2051	B-1433	2051	BC185	2051	BCY59
2051	A5P	2051	B-1666	2051	BC186	2051	BCY59A
2051	A5R	2051	B-169	2051	BC197	2051	BCY59B
2051	A5S	2051	B-1842	2051	BC197A	2051	BCY59C
2051	A5T2222	2051	B-1872	2051	BC197B	2051	BCY59D
2051	A5T3392	2051	B-66	2051	BC198	2051	BCY69
2051	A5T3565	2051	B-722246-2	2051	BC198A	2051	BCY84A
2051	A5T3903	2051	B-75583-1	2051	BC199	2051	BE-66
2051	A5T3904	2051	B-75583-2	2051	BC207	2051	BF-180
2051	A5T4123	2051	B-75583-202	2051	BC207A	2051	BF-214
2051	A5T4124	2051	B-75583-I02	2051	BC207B	2051	BF-215
2051	A5T4409	2051	B-75589-13	2051	BC207BL	2051	BF-226
2051	A5T5172	2051	B-75589-3	2051	BC208	2051	BF-255
2051	A5T5209	2051	B-75608-3	2051	BC208A	2051	BF115
2051	A5T5219	2051	B12-1-A-21	2051	BC208AL	2051	BF183A
2051	A5T5220	2051	B12-1A21	2051	BC208B	2051	BF189
2051	A5T5223	2051	B133578	2051	BC208C	2051	BF224J
2051	A5T5225	2051	B169	2051	BC208CL	2051	BF225J
2051	A5U	2051	B169(JAPAN)	2051	BC209	2051	BF248
2051	A5W	2051	B1K	2051	BC209A	2051	BF250
2051	A641(NPN)	2051	B1N	2051	BC209B	2051	BF291
2051	A649L	2051	B1P7201	2051	BC209BL	2051	BF291A
2051	A649B	2051	B1R	2051	BC209C	2051	BF291B
2051	A670722D	2051	B1W	2051	BC209CL	2051	BF293
2051	A670729B	2051	B2D	2051	BC210	2051	BF293A
2051	A6H	2051	B2Z	2051	BC220	2051	BF293D
2051	A6HD	2051	B539R	2051	BC222	2051	BF321A
2051	A6J	2051	B6P	2051	BC223A	2051	BF321B
2051	A6K	2051	B8780010	2051	BC223B	2051	BF321C
2051	A6R	2051	BA67	2051	BC233A	2051	BF321D
2051	A6S	2051	BA71	2051	BC237	2051	BF321E
2051	A747	2051	BAC8H2M1	2051	BC237A	2051	BF321F
2051	A747A	2051	BAC8H2M2	2051	BC237B	2051	BF440
2051	A748	2051	BAC8H2M3	2051	BC238	2051	BF441
2051	A748B	2051	BACT2F	2051	BC238A	2051	BF596
2051	A749B	2051	BC-107	2051	BC238B	2051	BF71
2051	A7A	2051	BC-1072	2051	BC239	2051	BFR11
2051	A7B(TRANSISTOR)	2051	BC-107A	2051	BC267	2051	BFR26
2051	A7R	2051	BC-108	2051	BC268	2051	BFS19CB
2051	A7S	2051	BC-1082	2051	BC269	2051	BFS31P
2051	A7T	2051	BC-1086	2051	BC270	2051	BFS36A
2051	A7T3392	2051	BC-108B	2051	BC280	2051	BFS36B
2051	A7T5172	2051	BC-1096	2051	BC280A	2051	BFS36C
2051	A7Y	2051	BC-109B	2051	BC280B	2051	BFS38
2051	A88	2051	BC-114	2051	BC280C	2051	BFS58A
2051	A88(JAPAN)	2051	BC-121	2051	BC282	2051	BFS42
2051	A8B	2051	BC-122	2051	BC284	2051	BFS42A
2051	A8G	2051	BC-123	2051	BC284A	2051	BFS42B
2051	A8L	2051	BC-148A	2051	BC284B	2051	BFS42C
2051	A8T3392	2051	BC-148B	2051	BC285	2051	BFS43
2051	A8T3707	2051	BC-148C	2051	BC289	2051	BFS43A
2051	A8T3708	2051	BC-167-B	2051	BC289A	2051	BFS43B
2051	A8T3709	2051	BC-169-C	2051	BC289B	2051	BFS43C
2051	A8T3710	2051	BC-1690	2051	BC290	2051	BFV10
2051	A8T3711	2051	BC-169B	2051	BC290B	2051	BFV11
2051	A8T5172	2051	BC-169C	2051	BC290C	2051	BFV12
2051	A937	2051	BC-71	2051	BC317	2051	BFV40
2051	A937-1	2051	BC107	2051	BC317A	2051	BFV41
2051	A937-3	2051	BC107A	2051	BC317B	2051	BFV42
2051	A9B	2051	BC107B	2051	BC318	2051	BFV43
2051	A9E	2051	BC108	2051	BC318A	2051	BFV44
2051	A9F	2051	BC108A	2051	BC318B	2051	BFV45
2051	A9G	2051	BC108B	2051	BC318C	2051	BFV46
2051	A9H	2051	BC108C	2051	BC319	2051	BFV47
2051	A9J	2051	BC109	2051	BC319B	2051	BFV49
2051	A9S	2051	BC1096	2051	BC319C	2051	BFV50
2051	A9T	2051	BC109BP	2051	BC408	2051	BFV51
2051	A9U	2051	BC109C	2051	BC409	2051	BFV53
2051	A9W	2051	BC110	2051	BC431	2051	BFV54
2051	A9Y	2051	BC113	2051	BC507A	2051	BFV55
2051	AC-175A	2051	BC114	2051	BC507B	2051	BFV83B
2051	AC-175B	2051	BC114TR	2051	BC508A	2051	BFV83C
2051	AC-175P	2051	BC115	2051	BC508B	2051	BFV85
2051	APC3527	2051	BC118	2051	BC508C	2051	BFV85A
2051	ALD-3141	2051	BC125	2051	BC509B	2051	BFV85B
2051	ALD-35	2051	BC125B	2051	BC509C	2051	BFV85C
2051	AL8-8922	2051	BC129	2051	BC510B	2051	BFV87
2051	AN	2051	BC130	2051	BC546A	2051	BFV88
2051	AQ4	2051	BC131	2051	BC546B	2051	BFV88A
2051	AQ6	2051	BC132			2051	BFV88B
2051	AR-107					2051	BFV88C

RS 276-	Industry Standard No.
2051	BPW32
2051	BPW46
2051	BPW59
2051	BPW60
2051	BPW68
2051	BPX92
2051	BPX93
2051	BPX95A
2051	BPY
2051	BPY-22
2051	BPY-23
2051	BPY-23A
2051	BPY-24
2051	BPY-29
2051	BPY-30
2051	BPY-39
2051	BPY22
2051	BPY23
2051	BPY23A
2051	BPY24
2051	BPY25
2051	BPY26
2051	BPY28
2051	BPY29
2051	BPY30
2051	BPY33
2051	BPY37
2051	BPY371
2051	BPY37I
2051	BPY39
2051	BPY39/1
2051	BPY39/2
2051	BPY39/3
2051	BPY391
2051	BPY39I
2051	BPY63
2051	BPY73
2051	BPY74
2051	BPY75
2051	BPY76
2051	BPY77
2051	BQ-66
2051	BQ-94
2051	BQ71
2051	BH71
2051	BI71
2051	BIP7201
2051	BN-66
2051	BN7517
2051	BN751B
2051	BP67
2051	BQ-94
2051	BQ67
2051	BR-66
2051	BR67
2051	BS-66
2051	BS-94
2051	BS475
2051	BS67
2051	BS90116
2051	BS810
2051	BS821
2051	BSY35A
2051	BSY40
2051	BSY41
2051	BSY53
2051	BSY54
2051	BSY59
2051	BSY65PA
2051	BSY65PB
2051	BSY84
2051	BSY86
2051	BSY87
2051	BSY88
2051	BSY89
2051	BSY90
2051	BSY91
2051	BSW11
2051	BSW12
2051	BSW19
2051	BSW33
2051	BSW34
2051	BSW39
2051	BSW41
2051	BSW42
2051	BSW42A
2051	BSW43
2051	BSW43A
2051	BSW51
2051	BSW52
2051	BSW53
2051	BSW58
2051	BSW82
2051	BSW83
2051	BSW84
2051	BSW85
2051	BSW88
2051	BSW88A
2051	BSW88B
2051	BSW89
2051	BSW89A
2051	BSW89B
2051	BSW92
2051	BSX19
2051	BSX20
2051	BSX24
2051	BSX25
2051	BSX30
2051	BSX38
2051	BSX38A
2051	BSX38B
2051	BSX39
2051	BSX44
2051	BSX48
2051	BSX49
2051	BSX51
2051	BSX51A
2051	BSX52
2051	BSX52A
2051	BSX53
2051	BSX54
2051	BSX66
2051	BSX67
2051	BSX68
2051	BSX69
2051	BSX75
2051	BSX76
2051	BSX77
2051	BSX78
2051	BSX79
2051	BSX79A
2051	BSX79B
2051	BSX80
2051	BSX81
2051	BSX81A
2051	BSX81B
2051	BSX87
2051	BSX89
2051	BSX90
2051	BSX91
2051	BSX94A
2051	BSX97
2051	BSY10
2051	BSY11
2051	BSY165
2051	BSY168
2051	BSY17
2051	BSY18
2051	BSY19
2051	BSY20
2051	BSY21
2051	BSY24
2051	BSY26
2051	BSY27
2051	BSY28
2051	BSY29
2051	BSY34
2051	BSY38
2051	BSY39
2051	BSY48
2051	BSY49
2051	BSY58
2051	BSY59
2051	BSY61
2051	BSY62
2051	BSY62A
2051	BSY63
2051	BSY73
2051	BSY74
2051	BSY75
2051	BSY76
2051	BSY80
2051	BSY89
2051	BSY93
2051	BSY95
2051	BSY95A
2051	BT-94
2051	BT67
2051	BT71
2051	BTX-070
2051	BTX-094
2051	BTX-095
2051	BTX-096
2051	BTX-2367B
2051	BTXO68
2051	BTX2367B
2051	BU67
2051	BU71
2051	BUC 97704-2
2051	BUC97704-2
2051	BV67
2051	BV71
2051	BW67
2051	BW71
2051	BX67
2051	BX71
2051	BX67
2051	BY71
2051	BZ67
2051	BZ71
2051	C00 68602300
2051	C00-68602300
2051	C100
2051	C100-OY
2051	C1000-Y
2051	C1003
2051	C1007
2051	C10279-1
2051	C10279-3
2051	C103
2051	C103(JAPAN)
2051	C1033
2051	C1033A
2051	C103A(TRANSISTOR)
2051	C104
2051	C104A
2051	C105
2051	C1071
2051	C110
2051	C111
2051	C111E
2051	C1128(3RD-IF)
2051	C1128D
2051	C1175
2051	C1175C
2051	C1175D
2051	C1175E
2051	C1175F
2051	C122
2051	C1244
2051	C12711
2051	C1293A(LAST-IF)
2051	C1293B
2051	C131
2051	C132
2051	C133
2051	C134
2051	C134B
2051	C135
2051	C136
2051	C1361
2051	C1362
2051	C1363
2051	C1364
2051	C1364A
2051	C137(TRANSISTOR)
2051	C1372Y
2051	C138(TRANSISTOR)
2051	C138A
2051	C139(TRANSISTOR)
2051	C13901
2051	C1390I
2051	C1390J
2051	C1390K
2051	C1390V
2051	C1390W
2051	C1390WH
2051	C1390WI
2051	C1390WX
2051	C1390WY
2051	C1390X
2051	C1390XJ
2051	C1390XK
2051	C1390YM
2051	C1393
2051	C1416
2051	C15(TRANSISTOR)
2051	C15-1
2051	C15-2
2051	C15-3
2051	C151
2051	C1542
2051	C157
2051	C158
2051	C159
2051	C16
2051	C160
2051	C1639
2051	C166
2051	C167
2051	C1687
2051	C1688
2051	C16A
2051	C1759
2051	C18
2051	C191
2051	C192
2051	C193
2051	C194
2051	C1945295DY1
2051	C195
2051	C196
2051	C197
2051	C200(TRANSISTOR)
2051	C201(JAPAN)
2051	C202(JAPAN)
2051	C203
2051	C204
2051	C205
2051	C218
2051	C218A
2051	C230
2051	C2300.037-096
2051	C237
2051	C239
2051	C2538-11
2051	C26
2051	C267
2051	C267A
2051	C28
2051	C281
2051	C281A
2051	C281B
2051	C281C
2051	C281C-EP
2051	C281D
2051	C281EP
2051	C281H
2051	C281HA
2051	C281HB
2051	C281HC
2051	C282
2051	C282H
2051	C282HA
2051	C282HB
2051	C282HC
2051	C283
2051	C283(TRANSISTOR)
2051	C284
2051	C284H
2051	C284HA
2051	C284HB
2051	C285
2051	C285A
2051	C29
2051	C300
2051	C301
2051	C302(JAPAN)
2051	C315
2051	C317
2051	C317-0
2051	C317C
2051	C318(JAPAN)
2051	C318A(JAPAN)
2051	C319
2051	C320
2051	C321
2051	C321H
2051	C321HA
2051	C321HB
2051	C321HC
2051	C323
2051	C324
2051	C324A
2051	C324H
2051	C324HA
2051	C328A
2051	C33
2051	C350(TRANSISTOR)
2051	C350H
2051	C352
2051	C352(JAPAN)
2051	C352A
2051	C356
2051	C360
2051	C360D
2051	C363
2051	C36580
2051	C366
2051	C367
2051	C369
2051	C369BL
2051	C3690
2051	C3690-BL
2051	C3690-GR
2051	C3690-V
2051	C3690BL
2051	C3690GR
2051	C3690R
2051	C369V
2051	C37
2051	C37(TRANSISTOR)
2051	C371
2051	C371(0)
2051	C371-0
2051	C371-R
2051	C3710
2051	C371B
2051	C371G
2051	C371R
2051	C372(0)
2051	C372-0
2051	C372-1
2051	C372-2
2051	C372-0
2051	C372-R
2051	C372-Y
2051	C372-Z
2051	C372GR
2051	C372H
2051	C372Y
2051	C377
2051	C378
2051	C379
2051	C38
2051	C38(TRANSISTOR)
2051	C388
2051	C39-207
2051	C395
2051	C395A
2051	C395B
2051	C396
2051	C400
2051	C400-0
2051	C400-R
2051	C400-Y
2051	C401(JAPAN)
2051	C403(C)
2051	C403C
2051	C404
2051	C423
2051	C423B
2051	C423C
2051	C423D
2051	C423E
2051	C423F
2051	C424(JAPAN)
2051	C424D
2051	C425
2051	C425B
2051	C425C
2051	C425D
2051	C425E
2051	C425F
2051	C440
2051	C441
2051	C442
2051	C444
2051	C45
2051	C450
2051	C454(A)
2051	C454A
2051	C454B
2051	C454C
2051	C454D
2051	C454L
2051	C454LA
2051	C458LG(B)
2051	C460(A)
2051	C460(B)
2051	C468
2051	C468A
2051	C47
2051	C475
2051	C475K
2051	C476
2051	C48
2051	C48C
2051	C52(TRANSISTOR)
2051	C523383
2051	C529
2051	C529A
2051	C53
2051	C537(F)
2051	C537(G)
2051	C537-01
2051	C537A
2051	C537B
2051	C537C
2051	C537D
2051	C537D2
2051	C537B
2051	C537EF
2051	C537EH
2051	C537EI
2051	C537EK
2051	C537F
2051	C537P1
2051	C537P2
2051	C537PC
2051	C537PV
2051	C537G
2051	C537G1
2051	C537GF
2051	C537GI
2051	C537HI
2051	C537HT
2051	C537W
2051	C538
2051	C538A
2051	C538AQ
2051	C538P
2051	C538Q
2051	C538R
2051	C538S
2051	C538T
2051	C539
2051	C539K
2051	C539L
2051	C539R
2051	C539S
2051	C54
2051	C540
2051	C55
2051	C566
2051	C587
2051	C587A
2051	C588
2051	C593
2051	C595
2051	C596
2051	C602E
2051	C619
2051	C619B
2051	C619C
2051	C619D
2051	C62(TRANSISTOR)
2051	C620
2051	C620C
2051	C620CD
2051	C620D
2051	C620DE
2051	C620E
2051	C621
2051	C622
2051	C63
2051	C631
2051	C631A
2051	C632
2051	C632A
2051	C633
2051	C633-7
2051	C633A
2051	C633G

RS 276-	Industry Standard No.	RS 276-	Industry Standard No.	RS 276-	Industry Standard No.	RS 276-	Industry Standard No.
2051	C633H	2051	C870	2051	CDC8021	2051	CS1166D
2051	C634	2051	C870BL	2051	CDC8054	2051	CS1166D-Q
2051	C634A	2051	C870E	2051	CDC82201	2051	CS1166E
2051	C64	2051	C870F	2051	CDC86X7-5	2051	CS1166F
2051	C640	2051	C871	2051	CDQ10001	2051	CS1166G
2051	C640B	2051	C871BL	2051	CDQ10002	2051	CS1166H
2051	C654	2051	C871D	2051	CDQ10003	2051	CS1166H/P
2051	C655	2051	C871E	2051	CDQ10004	2051	CS1168F
2051	C66-P11111-0001	2051	C871F	2051	CDQ10005	2051	CS1168G
2051	C66-P11150-00001	2051	C871G	2051	CDQ10006	2051	CS1168H
2051	C67	2051	C894	2051	CDQ10007	2051	C8226N
2051	C68	2051	C896	2051	CDQ10008	2051	C8229
2051	C686240Q	2051	C899	2051	CDQ10009	2051	C8229A
2051	C689	2051	C899K	2051	CDQ10010	2051	C8229B
2051	C689H	2051	C913	2051	CDQ10016	2051	C8229C
2051	C694D	2051	C923E	2051	CDQ10017	2051	C8229D
2051	C701(TRANSISTOR)	2051	C925	2051	CDQ10018	2051	C8229E
2051	C702	2051	C933	2051	CDQ10019	2051	C8229F
2051	C709	2051	C933BB	2051	CDQ10020	2051	C8229G
2051	C709B	2051	C933C	2051	CDQ10021	2051	C8229H
2051	C709C	2051	C933D	2051	CDQ10022	2051	C8235C
2051	C709CD	2051	C933E	2051	CDQ10023	2051	C8235E
2051	C709D	2051	C933F	2051	CDQ10024	2051	C8235G
2051	C710	2051	C933FP	2051	CDQ10025	2051	C8236D
2051	C710(B)	2051	C933PPC	2051	CDQ10026	2051	C8236H
2051	C710(D)	2051	C933PPD	2051	CDQ10027	2051	C8238P
2051	C710-1	2051	C933PPE	2051	CDQ10028	2051	C8238P
2051	C710-2	2051	C933PPP	2051	CDQ10032	2051	C8245P
2051	C710-4	2051	C933PPG	2051	CDQ10035	2051	C8245G
2051	C710B	2051	C933Q	2051	CDQ10036	2051	C8245H
2051	C710BC	2051	C934(TRANSISTOR)	2051	CDQ1018	2051	C8245I
2051	C710C	2051	C934C	2051	CDQ1021	2051	C8245T
2051	C710D	2051	C934D	2051	CDQ1024	2051	C8250E
2051	C710E	2051	C934E	2051	CE4001B	2051	C8257
2051	C712	2051	C934F	2051	CE4001E	2051	C8258
2051	C712A	2051	C934Q	2051	CE4003E	2051	C8259
2051	C712C	2051	C934P	2051	CE4004C	2051	C8283A
2051	C712D	2051	C938	2051	CE4013E	2051	C8286
2051	C712E	2051	C938A	2051	CF-2	2051	C8288
2051	C712W	2051	C938B	2051	CF2	2051	C8289
2051	C713	2051	C938C	2051	CF5	2051	C8295E
2051	C714	2051	C941	2051	CG1	2051	C8295G
2051	C715(JAPAN)	2051	C941-0	2051	CI2711	2051	C81330D
2051	C715A	2051	C941-0	2051	CI2712	2051	C813401
2051	C715B	2051	C941-R	2051	CI2713	2051	C81340I
2051	C715C	2051	C941-Y	2051	CI2714	2051	C81544
2051	C715D	2051	C941R	2051	CI2923	2051	C81545
2051	C715E	2051	C943	2051	CI2924	2051	C81548
2051	C715EJ	2051	C943A	2051	CI2925	2051	C81549
2051	C715EV	2051	C943B	2051	CI2926	2051	C81553
2051	C715P	2051	C943C	2051	CI3390	2051	C81361G
2051	C715XL	2051	C944	2051	CI3391	2051	C81362
2051	C716	2051	C944A	2051	CI3391A	2051	C81363
2051	C716B	2051	C944B	2051	CI3392	2051	C81368
2051	C716C	2051	C944C	2051	CI3393	2051	C81568A
2051	C716D	2051	C944D	2051	CI3394	2051	C81568B
2051	C716E	2051	C944K	2051	CI3395	2051	C81568D
2051	C716F	2051	C960	2051	CI3396	2051	C81370
2051	C716G	2051	C9604	2051	CI3397	2051	C81371
2051	C717(FINAL-IF)	2051	C966	2051	CI3398	2051	C81372
2051	C720	2051	C967	2051	CI3402	2051	C81383
2051	C725	2051	C968	2051	CI3403	2051	C81420
2051	C725-0	2051	C968P	2051	CI3404	2051	C81453E
2051	C7335-BL	2051	C984	2051	CI3405	2051	C81463A
2051	C735-0	2051	C984A	2051	CI3414	2051	C81585
2051	C735-R	2051	C984B	2051	CI3415	2051	C81585E/F
2051	C7350RN	2051	C984C	2051	CI3416	2051	C81585G
2051	C741	2051	C991	2051	CI3417	2051	C81625
2051	C742	2051	C992	2051	CI3900	2051	C81665
2051	C752	2051	CA-9011H	2051	CI3900A	2051	C81683E
2051	C752G	2051	CAM-12	2051	CI3901	2051	C8184E
2051	C760	2051	C8246	2051	CI4256	2051	C82001
2051	C773	2051	CC1168F	2051	CI4424	2051	C82001H
2051	C773C	2051	CC59018F	2051	CI4425	2051	C82004
2051	C773D	2051	CC8-2006D	2051	CIL-531	2051	C82004C
2051	C773E	2051	CC812359	2051	CIL-532	2051	C82004D
2051	C796	2051	CC82001H	2051	CJ-5206	2051	C82006
2051	C815(M)	2051	C82004	2051	CJ-5207	2051	C82007G
2051	C815A	2051	CC84004	2051	CJ-5208	2051	C82007H
2051	C815B	2051	CC86168	2051	CJ-5211	2051	C82218
2051	C815C	2051	CC86168F	2051	CJ-5212	2051	C82219
2051	C815F	2051	CC89018E	2051	CJ5201	2051	C82221
2051	C815K	2051	CD0014NA	2051	CJ5202	2051	C82222
2051	C815L	2051	CD0014NG	2051	CJ5203	2051	C82569
2051	C815M	2051	CD0015N	2051	CJ5206	2051	C82481
2051	C8158	2051	CD0021	2051	CJ5207	2051	C82711
2051	C8158A	2051	CD1200Q	2051	CJ5211	2051	C82712
2051	C8158C	2051	CD38	2051	CJ5212	2051	C82713
2051	C825	2051	CD446	2051	CM7163	2051	C82714
2051	C828A8	2051	CD6019	2051	CMO354-423	2051	C82922
2051	C829	2051	CD6150	2051	CN0770	2051	C82923
2051	C829A	2051	CD6157	2051	CS-1120C1	2051	C82924
2051	C829B	2051	CD6375	2051	CS-1120C2	2051	C82925
2051	C829BC	2051	CD8000	2051	CS-1120D1	2051	C83001B
2051	C829C	2051	CD8000-1	2051	CS-1120H	2051	C83390
2051	C829D	2051	CD9525	2051	CS-1120I	2051	C83391
2051	C829R	2051	CDC-13000-1	2051	CS-123EP	2051	C83391A
2051	C829X	2051	CDC-13000-1D	2051	CS-1238P	2051	C83392
2051	C829Y	2051	CDC-8000-1	2051	CS-1238P	2051	C83393
2051	C856M	2051	CDC-8001	2051	CS-1244X	2051	C83394
2051	C858	2051	CDC12000	2051	CS-1258	2051	C83395
2051	C858A	2051	CDC12000-1C	2051	CS-1259	2051	C83396
2051	C858B	2051	CDC12018C	2051	CS-12941	2051	C83397
2051	C858C	2051	CDC1201BC	2051	CS-1305	2051	C83398
2051	C858D	2051	CDC12077F	2051	CS-1330	2051	C83402
2051	C858E	2051	CDC1210B	2051	CS-1359	2051	C83403
2051	C858F	2051	CDC13000	2051	CS-1561E	2051	C83404
2051	C858H	2051	CDC13000-1	2051	CS-1561F	2051	C83405
2051	C858J	2051	CDC13000-1B	2051	CS-1561G	2051	C83414
2051	C858K	2051	CDC13000-1B	2051	CS-1386E	2051	C83415
2051	C858L	2051	CDC13000-1C	2051	CS-1386H	2051	C83416
2051	C858M	2051	CDC13000-1D	2051	CS-460B	2051	C83417
2051	C858R	2051	CDC13000C	2051	CS-6168F	2051	C8360
2051	C859(H)	2051	CDC13016A	2051	CS-6168G	2051	C83605
2051	C859(J)	2051	CDC13019B	2051	CS-6168H	2051	C83606
2051	C859(M)	2051	CDC3500-1	2051	CS-6225F	2051	C83607
2051	C859A	2051	CDC15018	2051	CS-6225G	2051	C83843
2051	C859B	2051	CDC2010	2051	CS-6227E	2051	C83844
2051	C859C	2051	CDC2010C	2051	CS-6227F	2051	C83845
2051	C859D	2051	CDC2010D	2051	CS-9011	2051	C83854
2051	C859E	2051	CDC25100-6	2051	CS-9011I	2051	C83854A
2051	C859F	2051	CDC25100C	2051	CS-9011F	2051	C83855
2051	C859H	2051	CDC25100-Q	2051	CS-9011G	2051	C83855A
2051	C859J	2051	CDC2510G	2051	CS-9011L	2051	C83859
2051	C859L	2051	CDC40023A130	2051	CS-9013	2051	C83859A
2051	C859N	2051	CDC430	2051	CS-9013HE	2051	C83860
2051	C859S	2051	CDC4306813	2051	CS-9014B	2051	C83900
2051	C844	2051	CDC60132	2051	CS-9014D	2051	C83900A
2051	C847	2051	CDC745(ZENITH)	2051	CS-9018D	2051	C83901
2051	C848	2051	CDC746	2051	CS-9104	2051	C83903
2051	C849	2051	CDC8000	2051	CS-9125B	2051	C83904
2051	C850	2051	CDC8000-1B	2051	C81068	2051	C84424
2051	C87	2051	CDC8001	2051	CS1120I	2051	C84425
		2051	CDC8011B	2051	CS1166		

RS 276-	Industry Standard No.	RS 276-	Industry Standard No.	RS 276-	Industry Standard No.	RS 276-	Industry Standard No.
2051	C85088	2051	DS-46	2051	EA15X7635	2051	ET15X15
2051	C85369	2051	DS-47	2051	EA15X7658	2051	ET15X16
2051	C86168F	2051	DS-64	2051	EA15X7643	2051	ET15X20
2051	C86225E	2051	DS-66L	2051	EA15X77	2051	ET15X24
2051	C86229F	2051	DS-66W	2051	EA15X8122	2051	ET15X27
2051	C86229G	2051	DS-67	2051	EA15X83	2051	ET15X37
2051	C8696	2051	DS-67W	2051	EA15X84	2051	ET15X41
2051	C8697	2051	DS-75	2051	EA15X85	2051	ET15X42
2051	C8706	2051	DS-76	2051	EA15X8502	2051	ET15X45
2051	C8718	2051	DS-77	2051	EA15X8511	2051	ET15X54
2051	C8718A	2051	DS1B	2051	EA15X8518	2051	ET234843
2051	C8720A	2051	DS44	2051	EA15X8529	2051	ET238894
2051	C8722Q	2051	DS45	2051	EA15X86	2051	ET368021
2051	C89011	2051	DS46	2051	EA15X89	2051	ET412626
2051	C89011(E)(F)	2051	DS47	2051	EA15X9	2051	ET8-068
2051	C89011(EF)	2051	DS66	2051	EA15X90	2051	ETT-CDC-12000
2051	C89011(GH)	2051	DS67	2051	EA15X91	2051	ETTC-458LG
2051	C89011/3490	2051	DS67W	2051	EA15X96	2051	ETTC-CD12000
2051	C89011I	2051	DS76	2051	EA15X98	2051	ETTC-CD13000
2051	C89011D	2051	DS77	2051	EA1628	2051	ETTC-CD8000
2051	C89011E	2051	DT161	2051	EA1629	2051	ETX18
2051	C89011F	2051	DT1610	2051	EA1630	2051	EW165V
2051	C89011G	2051	DW-6505	2051	EA1638	2051	EW181
2051	C89011G/3490	2051	DW-7375	2051	EA1695	2051	EW182
2051	C89011H	2051	DW6034/M	2051	EA1696	2051	EX499-X
2051	C89011I	2051	E13-000-03	2051	EA1697	2051	EX500-X
2051	C89011N	2051	E13-000-04	2051	EA1703	2051	EX695-X
2051	C89012HP(LAST-IP)	2051	E13-002-03	2051	EA1716	2051	EX748-X
2051	C89013	2051	E13-003-00	2051	EA1718	2051	EX888-X
2051	C89013A	2051	E13-003-01	2051	EA1735	2051	EYZP-632
2051	C89013B	2051	E13-005-02	2051	EA1778	2051	EYZP-791
2051	C89013C	2051	E210	2051	EA1872	2051	EYZP632
2051	C89013D	2051	E212	2051	EA2271	2051	EYZP791
2051	C89013HE	2051	E24103	2051	EA2489	2051	F-302-1
2051	C89013HF	2051	E2430	2051	EA2490	2051	F-302-1532
2051	C89013HG	2051	E2431	2051	EA2739	2051	F-302-2532
2051	C89013HH	2051	E2436	2051	EA2770	2051	F121-433804
2051	C89014	2051	E2444	2051	EA2770(N)	2051	F121-546
2051	C89014(C)	2051	E2452	2051	EA3149	2051	F15840
2051	C89014/3490	2051	E2454	2051	EA4112	2051	F15840-1
2051	C89014A	2051	E2455	2051	EC0123	2051	F222
2051	C89014B	2051	E2459	2051	EC0123A	2051	F2443
2051	C89014B-C	2051	E2461	2051	EC0123AP	2051	F2448
2051	C89014C	2051	E2497	2051	ED-1402	2051	F2584
2051	C89014C/3490	2051	E2499	2051	ED1402A	2051	F302-1
2051	C89014D	2051	EA0092	2051	ED1402A/09-305066	2051	F302-2
2051	C89014G	2051	EA1080	2051	ED1402B	2051	F302-2532
2051	C89015	2051	EA1128	2051	ED1402C	2051	F306-001
2051	C89015B	2051	EA1129	2051	ED1402D	2051	F306-022
2051	C89016(G)	2051	EA1135	2051	ED1402E	2051	F3519
2051	C89016D	2051	EA1145	2051	ED1502R	2051	F3532
2051	C89016EF	2051	EA1344	2051	ED150Z	2051	F3569
2051	C89016EF(TRUETONE)	2051	EA1345	2051	ED1702L	2051	F3571
2051	C89016F(WESTGHSE)	2051	EA1406	2051	ED1702L/09-305068	2051	F366
2051	C89016FG	2051	EA1407	2051	ED1704L	2051	F572-1
2051	C89018EF	2051	EA1408	2051	EDC-Q10-1	2051	F587
2051	C89018FG	2051	EA1451	2051	EDO 219	2051	F7316
2051	C89021HP(LAST-IP)	2051	EA1452	2051	EDO-219	2051	F75116
2051	C89021HG(LAST-IP)	2051	EA1499	2051	ED8-100	2051	F9600
2051	C89101B	2051	EA1564	2051	EL232	2051	F9623
2051	C89125B	2051	EA1578	2051	EL238	2051	F9623F
2051	C89126	2051	EA1581	2051	EL642	2051	PA-1(SEARS)
2051	C89025M	2051	EA15X1	2051	EMS-73500	2051	F36853
2051	C89600-4	2051	EA15X101	2051	EN10	2051	FBC237
2051	C89600-5	2051	EA15X103	2051	EN2219	2051	FC8-9013F
2051	CTP-2001-1007	2051	EA15X111	2051	EN2222	2051	FC8-9013Q
2051	CTP-2001-1008	2051	EA15X112	2051	EN2484	2051	FC8-9016F
2051	CXL123A	2051	EA15X136	2051	EN3	2051	FC8116B9F813
2051	CXL159	2051	EA15X137	2051	EN3009	2051	FC8116BG
2051	D031(CHAN.MASTER)	2051	EA15X142	2051	EN3013	2051	FC8116BG704
2051	D048	2051	EA15X143	2051	EN3014	2051	FC81229F
2051	D053	2051	EA15X153	2051	EN3903	2051	FC81229G
2051	D1101P1	2051	EA15X157	2051	EN60	2051	FC88050C
2051	D1102P1B20	2051	EA15X162	2051	EN697	2051	FC89011F
2051	D1105P1	2051	EA15X163	2051	EN706	2051	FC89011G
2051	D133	2051	EA15X167	2051	EN708	2051	FC89013
2051	D16E7	2051	EA15X168	2051	EN930	2051	FC89013F
2051	D16E9	2051	EA15X18	2051	EP15X1	2051	FC89013G
2051	D16EC18	2051	EA15X189	2051	EP15X2	2051	FC89013H
2051	D1A	2051	EA15X190	2051	EP15X39	2051	FC89014(B)
2051	D1R38	2051	EA15X20	2051	EP15X47	2051	FC89016
2051	D227A	2051	EA15X213	2051	EP15X48(NPN)	2051	FC89016E
2051	D227B	2051	EA15X22	2051	EP15X49	2051	FD-1029-JA
2051	D227C	2051	EA15X237	2051	EP15X7	2051	FD-1029-LL
2051	D227D	2051	EA15X239	2051	EP15X8	2051	FD-1029-NG
2051	D227E	2051	EA15X24	2051	EP15X86	2051	FD-1029-PP
2051	D227F	2051	EA15X240	2051	EP15X88	2051	FD-1029-PT
2051	D227L	2051	EA15X241	2051	EP15X9	2051	FK3484
2051	D227R	2051	EA15X246	2051	EPX2	2051	FK3494
2051	D227S	2051	EA15X272	2051	EQ8-0165	2051	FMPS-A20
2051	D227W	2051	EA15X31	2051	EQ8-0192	2051	FMPSA20
2051	D294	2051	EA15X330	2051	EQ8-100	2051	FN-51-1A
2051	D2R38	2051	EA15X331	2051	EQ8-13	2051	FO810A
2051	D327A	2051	EA15X355	2051	EQ8-22	2051	FPR40-1001
2051	D327B	2051	EA15X356	2051	EQ8-5	2051	FPR50-1001
2051	D327C	2051	EA15X361	2051	EQ8-61	2051	FPS50-1002
2051	D327D	2051	EA15X364	2051	EQ8-62	2051	FR83693
2051	D327E	2051	EA15X37	2051	EQ8-64	2051	FS-1133
2051	D327F	2051	EA15X370	2051	EQ8-9	2051	FS8116B9641
2051	D328	2051	EA15X371	2051	ES10222	2051	FS8116B9813
2051	D32K1	2051	EA15X373	2051	ES10223	2051	FS1221
2051	D32K2	2051	EA15X379	2051	ES10232	2051	FS1974
2051	D33D21	2051	EA15X408	2051	ES15048	2051	FS2043
2051	D33D22	2051	EA15X437	2051	ES15050	2051	FS27604
2051	D33D24	2051	EA15X44	2051	ES15X1	2051	FS36999
2051	D33D25	2051	EA15X441	2051	ES15X11	2051	FT005
2051	D33D26	2051	EA15X45	2051	ES15X12	2051	FT006
2051	D33D27	2051	EA15X52	2051	ES15X14	2051	FT008
2051	D33D28	2051	EA15X56	2051	ES15X16	2051	FT008A
2051	D342	2051	EA15X58	2051	ES15X20	2051	FT023
2051	D372BL	2051	EA15X59	2051	ES15X23	2051	FT024
2051	D4D24	2051	EA15X60	2051	ES15X24	2051	FT025
2051	D4D25	2051	EA15X63	2051	ES15X37	2051	FT026
2051	D4D26	2051	EA15X68	2051	ES15X42	2051	FT053
2051	D912	2051	EA15X7112	2051	ES15X58	2051	FT3567
2051	D917254-2	2051	EA15X7113	2051	ES15X62	2051	FT3568
2051	D921881-1	2051	EA15X7115	2051	ES15X64	2051	FT3569
2051	D926640-1	2051	EA15X7117	2051	ES15X68	2051	FT3643
2051	D928121	2051	EA15X7118	2051	ES15X7	2051	FT40
2051	DBCZ073304	2051	EA15X7119	2051	ES15X70	2051	FX2368
2051	DBCZ073504	2051	EA15X7120	2051	ES15X76	2051	FX3300
2051	DBCZ083906	2051	EA15X7175	2051	ES15X83	2051	FX4046
2051	DBCZ136406	2051	EA15X7176	2051	ES15X84	2051	FZ101
2051	DDBY222002	2051	EA15X72	2051	ES15X85	2051	G005-036C
2051	DDBY233001	2051	EA15X7232	2051	ES20(ELCOM)	2051	G005-036E
2051	DDBY27001	2051	EA15X7262	2051	ES46	2051	G05-010-A
2051	DDBY283001	2051	EA15X73	2051	ES46(ELCOM)	2051	G05-011-A
2051	DDBT301001	2051	EA15X7514	2051	ES95(ELCOM)	2051	G05-015-D
2051	DN	2051	EA15X7517	2051	ESB5(ELCOM)	2051	G05-015C
2051	DO31	2051	EA15X7586	2051	ET15X10	2051	G05-034-D
2051	DRC-87540	2051	EA15X7588	2051	ET15X11	2051	G05-035-D
2051	D8-410(MOTOROLA)	2051	EA15X7589	2051	ET15X12	2051	G05-035E
2051	D8-44	2051	EA15X76	2051	ET15X13	2051	G05-036-B
2051	D8-45			2051	ET15X14		

RS 276-	Industry Standard No.	RS 276-	Industry Standard No.	RS 276-	Industry Standard No.	RS 276-	Industry Standard No.
2051	G05-036-C	2051	HC-00461	2051	HT303801A0	2051	J310250
2051	G05-036-C,D,E	2051	HC-00509	2051	HT303801B-0	2051	J961B(G.E.)
2051	G05-036-D	2051	HC-00535	2051	HT303801B0	2051	JA-H
2051	G05-036-E	2051	HC-00537	2051	HT303801O0	2051	JA-L
2051	G05-036B	2051	HC-00693	2051	HT304508K	2051	JA1200
2051	G05-036C	2051	HC-00735	2051	HT304531	2051	JE9011G
2051	G05-036D	2051	HC-00828	2051	HT304531A	2051	JE9011H
2051	G05-037-A	2051	HC-00838	2051	HT304531B	2051	JLM-20
2051	G05-037-B	2051	HC-00871	2051	HT304531C	2051	J8271
2051	G05-037B	2051	HC-00921	2051	HT304540A0	2051	JSP7001B
2051	G05-064-A	2051	HC-00924	2051	HT304540B0	2051	JT-1601-40
2051	G05-413B	2051	HC-00945	2051	HT304540B0	2051	K14-0066-12
2051	G05015C	2051	HC-01390	2051	HT304580	2051	K4-506
2051	G05035E	2051	HC-01417	2051	HT304580A	2051	K88339
2051	G05036	2051	HC-01820	2051	HT304580B	2051	KB8416
2051	G05036B	2051	HC-373	2051	HT304580O0	2051	KD2102
2051	G05036C	2051	HC-535	2051	HT304580K	2051	KGE41055
2051	G05036D	2051	HC-56	2051	HT304580Y0	2051	KLH1422
2051	G05036E	2051	HC-561	2051	HT304580Z	2051	KLH704
2051	G05037B	2051	HC-772	2051	HT304581	2051	KM917P
2051	G05059	2051	HC00838	2051	HT304581A	2051	KM917Q
2051	G395967	2051	HC01820	2051	HT304581B	2051	KP66682
2051	G395967-2	2051	HC01830	2051	HT304581B-0	2051	KR-401013
2051	G9600	2051	HC371	2051	HT304581B-0	2051	KR8417
2051	G9600(G.E.)	2051	HC372	2051	HT304581C	2051	KSC815-0
2051	G9623	2051	HC373	2051	HT304601B0	2051	KSC815-0
2051	G9696	2051	HC458	2051	HT304601C0	2051	KT218
2051	GC1144	2051	HC559	2051	HT304861B	2051	LDA400
2051	GC783	2051	HC561	2051	HT305351C0	2051	LDA400MP
2051	GC784	2051	HCL-29	2051	HT305361E	2051	LDA401
2051	GE-10	2051	HCL-6066	2051	HT305361G	2051	LDA401MP
2051	GE-17	2051	HD-00227	2051	HT305371E	2051	LDA402
2051	GE-20	2051	HEP50	2051	HT306441	2051	LD8206
2051	GE-210	2051	HEP53	2051	HT306441A	2051	LD8210
2051	GE-3265	2051	HEP54	2051	HT306441B	2051	LLB-23
2051	GE-81	2051	HEP55	2051	HT306441B-0	2051	LM-1129
2051	GE-X16A1938	2051	HEP724	2051	HT306441B0	2051	LM-1130
2051	GE3265	2051	HEP725	2051	HT306441C	2051	LM-1132
2051	GET2221	2051	HEP728	2051	HT306441C-0	2051	LM-1133
2051	GET2221A	2051	HEP729	2051	HT306451H	2051	LM-1147
2051	GET2222	2051	HEP735	2051	HT306962A-0	2051	LM-1148
2051	GET2222A	2051	HEP738	2051	HT307321A	2051	LM-1155
2051	GET2369	2051	HEP80002	2051	HT307321B-0	2051	LM1090E
2051	GET3015	2051	HEP80004	2051	HT307321B-0	2051	LM1090F
2051	GET3014	2051	HEP80011	2051	HT307322A	2051	LM1090G
2051	GET3646	2051	HEP80015	2051	HT307331O	2051	LM1117D
2051	GET3903	2051	HEP80022	2051	HT307331B	2051	LM1403
2051	GET3904	2051	HEP80025	2051	HT307331C	2051	LM1415-6
2051	GET706	2051	HF-40	2051	HT307331C0	2051	LM1415-7
2051	GET708	2051	HP2	2051	HT307341B	2051	LM1540
2051	GET914	2051	HP3	2051	HT307341C-0	2051	LM1566F
2051	GI-2711	2051	HP4	2051	HT308281B	2051	LM1614D
2051	GI-2712	2051	HP5	2051	HT308281C	2051	LM1614M
2051	GI-2714	2051	HP50	2051	HT308281G	2051	LM1818
2051	GI-3403	2051	HP6	2051	HT308281O	2051	LM2152
2051	GI-3405	2051	HP7	2051	HT308282A	2051	LRQ849
2051	GI-3415	2051	HP8	2051	HT308282A-0	2051	L8-0085-Q1
2051	GI-3416	2051	HKT-158	2051	HT308282A-0	2051	L8TG05
2051	GI-3417	2051	HKT-161	2051	HT308291B-0	2051	LT1016(B)
2051	GI-3641	2051	HR-11	2051	HT308291B0	2051	LT10161,H
2051	GI-3642	2051	HR-11A	2051	HT309842A-0	2051	LT1016Y,H
2051	GI-3643	2051	HR-11B	2051	HT38281D	2051	M-1002-17-NC
2051	GI-3704	2051	HR-13	2051	HT400	2051	M-1002-17NC
2051	GI-3705	2051	HR-13A	2051	HT401	2051	M-1002-2
2051	GI-3706	2051	HR-14A	2051	HT800011P	2051	M-4721
2051	GI10	2051	HR-15A	2051	HT800011G	2051	M-75557-1
2051	GI2711	2051	HR-16	2051	HT80011H	2051	M-75557-2
2051	GI2712	2051	HR-16A	2051	HT80011K	2051	M-75557-3
2051	GI2713	2051	HR-17	2051	HT8001810	2051	M-75557-4
2051	GI2714	2051	HR-17A	2051	HV23(TRANSISTOR)	2051	M-75557-5
2051	GI2715	2051	HR-18	2051	HV25	2051	M-75557-6
2051	GI2716	2051	HR-18A	2051	HX-50063	2051	M-8641
2051	GI2921	2051	HR-19	2051	HX-50072	2051	M140-3
2051	GI2922	2051	HR-19A	2051	HX-50092	2051	M24
2051	GI2923	2051	HR-32	2051	HX-50097	2051	M24A
2051	GI2924	2051	HR-36	2051	HX-50110	2051	M24B
2051	GI3641	2051	HR-37	2051	HX-50113	2051	M25
2051	GI3643	2051	HR-38	2051	HX-50161	2051	M25A
2051	GI3704	2051	HR-48	2051	HX50002	2051	M25A2
2051	GI3705	2051	HR-60	2051	HY3045801C	2051	M25B
2051	GI3706	2051	HR11A	2051	19A115728-2	2051	M25B2
2051	GI3707	2051	HR11B	2051	IC743042	2051	M31001
2051	GI3708	2051	HR13A	2051	IE460B	2051	M3519
2051	GI3709	2051	HR14	2051	IE850	2051	M4464
2051	GI3710	2051	HR14A	2051	IP20-0001	2051	M4465
2051	GI3711	2051	HR15A	2051	IP20-0002	2051	M447
2051	GMB1001	2051	HR16	2051	IP20-0003	2051	M4594
2051	GMB1002	2051	HR16A	2051	IP20-0006	2051	M4624
2051	GME2001	2051	HR17	2051	IP20-0039	2051	M4630
2051	GMB2002	2051	HR17A	2051	IP20-0041	2051	M4705
2051	GME4001	2051	HR18	2051	IP20-0122	2051	M4706
2051	GMB4002	2051	HR18A	2051	IP20-0165	2051	M4714
2051	GMB4003	2051	HR19A	2051	IP20-0214	2051	M4732
2051	GMB6003	2051	HR32	2051	IRTR-62	2051	M4734
2051	GMO-380	2051	HR36	2051	IRTR22	2051	M4737
2051	G04-041B	2051	HR37	2051	IRTR24	2051	M4739
2051	G05-003-A	2051	HR48	2051	IRTR62	2051	M4765
2051	G05-003-B	2051	HR62	2051	IRTR63	2051	M4768
2051	G05-010-A	2051	HR63	2051	IRTR86	2051	M4821
2051	G05-011-A	2051	HR64	2051	IT120	2051	M4834
2051	G05-015-C	2051	HR65	2051	IT121	2051	M484
2051	G05-036-C,D,E	2051	HR66	2051	IT2218	2051	M4840
2051	G05036B	2051	HR84(NPN)	2051	IT2219	2051	M4841
2051	G05036C	2051	HS-1168	2051	IT2221	2051	M4842
2051	G05036E	2051	HS-1229	2051	IT2222	2051	M4842A
2051	G05037B	2051	HS-40016	2051	J139A	2051	M4842C
2051	G05-050-C	2051	HS-40017	2051	J241054	2051	M4844
2051	G05036	2051	HS-40020	2051	J241099	2051	M4852
2051	G05059	2051	HS-40030	2051	J241230	2051	M4854
2051	G89014	2051	HS-40037	2051	J241251	2051	M4898
2051	G890141	2051	HS-40039	2051	J24458	2051	M4906
2051	G89014J	2051	HS-40044	2051	J24564	2051	M4926
2051	G89014K	2051	HS-40046	2051	J24565	2051	M4933
2051	G89014F	2051	HS-40055	2051	J24624	2051	M4935
2051	G89023H	2051	HS40046	2051	J24625	2051	M4937
2051	G89023I	2051	HT3036201	2051	J24641	2051	M4937(3RD-IP)
2051	G89023J	2051	HT303620B	2051	J24658	2051	M4941
2051	G89023K	2051	HT3036210	2051	J24752	2051	M4952
2051	GV6063	2051	HT303711A-0	2051	J24753	2051	M4953
2051	GVL 20077	2051	HT303711A0	2051	J24817	2051	M4970
2051	GVL20077	2051	HT303711AO	2051	J24842	2051	M54
2051	H-1567	2051	HT303711B-0	2051	J24843	2051	M54A
2051	H102	2051	HT303711B-0	2051	J24845	2051	M54B
2051	H104	2051	HT 303711B0	2051	J24846	2051	M54BLK
2051	H1567	2051	HT303711BO	2051	J24855	2051	M54BLU
2051	H931	2051	HT303721-0	2051	J24874	2051	M54BRN
2051	H933	2051	HT3037210	2051	J24878	2051	M54C
2051	H934	2051	HT303721O-0	2051	J24906	2051	M54G
2051	H9423	2051	HT303721D	2051	J24907	2051	M54E
2051	H9618	2051	HT303730	2051	J24909	2051	M54GRN
2051	H9696	2051	HT303730A	2051	J24916	2051	M54ORN
2051	HC-00458	2051	HT30373100	2051	J310249	2051	M54RED
2051	HC-00460					2051	M54WHT
						2051	M54YEL

RS 276-	Industry Standard No.	RS 276-	Industry Standard No.	RS 276-	Industry Standard No.	RS 276-	Industry Standard No.	RS 276-	Industry Standard No.
2051	M671	2051	MPS2716	2051	MP89696F	2051	P5152
2051	M7003	2051	MPS2923	2051	MP89696H	2051	P5153
2051	M7006	2051	MPS2924	2051	MP89696I	2051	P633567
2051	M7014	2051	MPS2925	2051	MP89700D	2051	P64447
2051	M7015	2051	MPS2926	2051	MP89700E	2051	P8393
2051	M7033	2051	MPS2926-BRN	2051	MP89700F	2051	P8394
2051	M7108	2051	MPS2926-ORG	2051	MP8A05	2051	P9623
2051	M7108/A5N	2051	MPS2926-RED	2051	MP8A2O	2051	PA7001/0001
2051	M7109	2051	MPS2926-YEL	2051	MP8D05	2051	PA9006
2051	M7109/A5P	2051	MPS2926BRN	2051	MP8EL239	2051	PBC183
2051	M7171	2051	MPS2926GRN	2051	MP8H02	2051	PE3001
2051	M75565-1	2051	MPS2926ORN	2051	MP8H17	2051	PE5015
2051	M773	2051	MPS2926RED	2051	MP8H17	2051	PE5010
2051	M773RED	2051	MPS2926YEL	2051	MP8H20	2051	PE5013
2051	M774	2051	MPS3392	2051	MP8H32	2051	PEP2
2051	M740RN	2051	MPS3393	2051	MPX9623	2051	PEP5
2051	M775	2051	MPS3394	2051	MPX9623H	2051	PEP6
2051	M775BRN	2051	MPS3395	2051	MPX9623H/I	2051	PEP7
2051	M776	2051	MPS3396	2051	MPX9623I	2051	PEP8
2051	M776GRN	2051	MPS3397	2051	MPX9623OI	2051	PEP9
2051	M779BLU	2051	MPS3398	2051	MQ2	2051	PET1002
2051	M780WHT	2051	MPS3643	2051	MR3932	2051	PET2001
2051	M783	2051	MPS3646	2051	MR9604	2051	PET2002
2051	M783RED	2051	MPS3704	2051	MS22B	2051	PET3704
2051	M784	2051	MPS3705	2051	MS3694	2051	PET3705
2051	M7840RN	2051	MPS3706	2051	MS7502R	2051	PET3706
2051	M785	2051	MPS3707	2051	MS7503R	2051	PET4001
2051	M785YEL	2051	MPS3708	2051	M8R7503	2051	PET4002
2051	M787BLU	2051	MPS3709	2051	MT104	2051	PET6001
2051	M791	2051	MPS3711	2051	MT4101	2051	PET6002
2051	M8105	2051	MPS3721	2051	MT4102	2051	PET8000
2051	M818	2051	MPS3826	2051	MT4102A	2051	PET8001
2051	M818WHT	2051	MPS3827	2051	MT4103	2051	PET8002
2051	M822	2051	MPS393	2051	MT6001	2051	PET8003
2051	M8221	2051	MPS3992	2051	MT6002	2051	PET8004
2051	M822A	2051	MPS5172	2051	MT6003	2051	PET9002
2051	M822A-BLU	2051	MPS6351	2051	MT696	2051	PL-176-029-9-001
2051	M822B	2051	MPS6413	2051	MT697	2051	PL-176-042-9-002
2051	M823	2051	MPS6514	2051	MT706	2051	PL-176-042-9-004
2051	M823B	2051	MPS6520	2051	MT706A	2051	PL-176-042-9-006
2051	M823WHT	2051	MPS6521	2051	MT706B	2051	PL-182-014-9-002
2051	M827	2051	MPS6530	2051	MT707	2051	PL1052
2051	M827BRN	2051	MPS6544	2051	MT708	2051	PL1054
2051	M828GRN	2051	MPS6545	2051	MT9001	2051	PL4021
2051	M847	2051	MPS6552	2051	MT9002	2051	PL4051
2051	M847BLK	2051	MPS6553	2051	N-EA15X136	2051	PL4052
2051	M9095	2051	MPS6554	2051	N-EA15X137	2051	PL4053
2051	M9159	2051	MPS6555	2051	N-EA15X138	2051	PL4054
2051	M91A	2051	MPS6556	2051	0N047204-2	2051	PL4055
2051	M91B	2051	MPS6561	2051	N201AY	2051	PM1121
2051	M91B0RN	2051	MPS6611	2051	N271	2051	PN107
2051	M91C	2051	MPS6565	2051	N5563	2051	PN108
2051	M91CM624	2051	MPS6566	2051	0N47204-1	2051	PN109
2051	M91D	2051	MPS6567	2051	N4T	2051	PN2369
2051	M91E	2051	MPS6568	2051	NB011(NPN)	2051	PN2369A
2051	M91F	2051	MPS6571	2051	NO207AL	2051	PN2484
2051	M91FM624	2051	MPS6573	2051	NJ100B	2051	PRT-101
2051	M9226	2051	MPS6574	2051	NJ102C	2051	PRT-104
2051	M924	2051	MPS6575	2051	NK12-1A19	2051	PRT-104-1
2051	M9248	2051	MPS6576	2051	NKT10359	2051	PRT-104-2
2051	M9282	2051	MPS6590	2051	NKT10419	2051	PRT-104-3
2051	M9475	2051	MS8001	2051	NKT10439	2051	PS209800
2051	M9525	2051	MS8097	2051	NKT10519	2051	PT1558
2051	M9532	2051	MS8098	2051	NKT12329	2051	PT1559
2051	M9563	2051	MS8099	2051	NKT12429	2051	PT1610
2051	M9568	2051	MP89185	2051	NKT13329	2051	PT1835
2051	M9570	2051	MP89417A-T	2051	NKT13429	2051	PT1836
2051	MA6001	2051	MP89423	2051	NL-102	2051	PT1837
2051	MA6002	2051	MP89426A	2051	NN9017	2051	PT2760
2051	MA6003	2051	MP89426A.B	2051	NPC737	2051	PT2896
2051	MA9426	2051	MP89427B.C	2051	N86512	2051	PT3141
2051	MAQ7786	2051	MP89427C	2051	N86513	2051	PT3141A
2051	MC9427	2051	MP89433	2051	N86514	2051	PT3141B
2051	ME-2	2051	MP89433K	2051	N86520	2051	PT3151A
2051	ME-3	2051	MP89434J	2051	NR-071AU	2051	PT3151B
2051	MEI001	2051	MP89434K	2051	NR-261A8	2051	PT3151C
2051	ME1002	2051	MP89600	2051	NR-431A8	2051	PT3500
2051	ME2001	2051	MP89600(G)	2051	NR-431A8	2051	PT3501
2051	ME2002	2051	MP89600F	2051	NR-461A8	2051	PT4-7158
2051	ME213	2051	MP89600G	2051	NR041E	2051	PT4-7158-012
2051	ME213A	2051	MP89604I	2051	NR041E	2051	PT4-7158-013
2051	ME216	2051	MP89604D	2051	NR041A	2051	PT4-7158-01A
2051	ME217	2051	MP89604B	2051	NR071AU	2051	PT4-7158-021
2051	ME4001	2051	MP89604F	2051	NR091ET	2051	PT4-7158-022
2051	ME4002	2051	MP89604FG	2051	NR201AY	2051	PT4-7158-023
2051	ME4003	2051	MP89604I	2051	NR261A8	2051	PT4-7158-02A
2051	ME4003C	2051	MP89604R	2051	NR271AY	2051	PT4800
2051	ME4101	2051	MP89600T	2051	NR461	2051	PT627
2051	ME4102	2051	MP89611-5	2051	NR461EH	2051	PT703
2051	ME4103	2051	MP89616	2051	NS1500	2051	PT720
2051	ME4104	2051	MP89618	2051	NS1972	2051	PT851
2051	ME6001	2051	MP89618(J)	2051	NS1973	2051	PT886
2051	ME6002	2051	MP89618H	2051	NS1974	2051	PT887
2051	ME6003	2051	MP89618I	2051	NS1975	2051	PT897
2051	ME900	2051	MP89618J	2051	NS3903	2051	PT898
2051	ME9001	2051	MP89623	2051	NS3923	2051	PTC115
2051	ME9005A	2051	MP89623C(P)	2051	NS475	2051	PTC121
2051	ME900A	2051	MP89623E	2051	NS476	2051	PTC136
2051	ME901	2051	MP89623E.G	2051	NS477	2051	PTC139
2051	ME901A	2051	MP89623F	2051	NS478	2051	Q-00269
2051	MEF-25	2051	MP89623G	2051	NS479	2051	Q-00269A
2051	MG9623	2051	MP89623G/H	2051	NS480	2051	Q-00269B
2051	MI9623	2051	MP89623H	2051	N86114	2051	Q-00269C
2051	MI9630	2051	MP89623H/I	2051	N86115	2051	Q-00369
2051	MI9623	2051	MP89623I	2051	N86207	2051	Q-00369A
2051	MI9630	2051	MP89623I/J	2051	N86260	2051	Q-00369B
2051	MJ89411T	2051	MP89623J	2051	NS7262	2051	Q-00369C
2051	MM1755	2051	MP8962G	2051	NS731	2051	Q-00484R
2051	MM1756	2051	MP89626Q	2051	NS731A	2051	Q-00569
2051	MM1757	2051	MP89626H	2051	NS733	2051	Q-00569A
2051	MM1758	2051	MP89626I	2051	NS733A	2051	Q-00569B
2051	MM3903	2051	MP89630	2051	NS734	2051	Q-00669
2051	MM3904	2051	MP89630H	2051	NS734A	2051	Q-00669A
2051	MMT2222	2051	MP89630H.I	2051	NS949	2051	Q-00669B
2051	MMT3014	2051	MP89630I	2051	09-309060	2051	Q-00669C
2051	MMT3903	2051	MP89630T	2051	0N047204-2	2051	Q-00684R
2051	MMT3904	2051	MP89631	2051	0N271	2051	Q-0115C
2051	MMT72	2051	MP89631(I)	2051	0N274	2051	Q-02115C
2051	MN54	2051	MP89631I	2051	0N47204-1	2051	Q-03115C
2051	MP1014-2	2051	MP89631J	2051	0S536Q	2051	Q-04115C
2051	MPM5006	2051	MP89631K	2051	P-8393	2051	Q-05115C
2051	MP8 9623G	2051	MP89631S	2051	P/810000020	2051	Q-06115C
2051	MP8-2716	2051	MP89631F	2051	P04-444-0028	2051	Q-07115C
2051	MP8-3563	2051	MP89632	2051	P04-45-0014-P2	2051	Q-10115C
2051	MP8-3705	2051	MP89632(I)	2051	P04-45-0014-P5	2051	Q-14115C
2051	MP8-6571	2051	MP89632(K)	2051	P0444O028-001	2051	Q-15115C
2051	MP8-706	2051	MP89632I	2051	P0444OO28-009	2051	Q-16115C
2051	MP8-9630I	2051	MP89632H	2051	P0444O028-014	2051	Q-2N5225
2051	MP8-A10	2051	MP89632I	2051	P04440028-8	2051	Q-35
2051	MP8-H32	2051	MP89632J	2051	P04440032-001	2051	Q-RP-2
2051	MPS2711	2051	MP89632K	2051	P15153	2051	Q-3E1001
2051	MPS2712	2051	MP89632L	2051	P1901-50	2051	Q0V60O529
2051	MPS2713	2051	MP89632O	2051	P1901-50	2051	Q0V60530
2051	MPS2714	2051	MP89633O	2051	P480A0028	2051	Q0V60530
2051	MPS2715	2051	MP89634D	2051	P480A0029	2051	Q0V60538

RS 276-2051	Industry Standard No.	RS 276-2051	Industry Standard No.	RS 276-2051	Industry Standard No.	RS 276-2051	Industry Standard No.
2051	Q3/6515	2051	RS-7607	2051	RT7325	2051	S1891A
2051	Q35242	2051	RS-7609	2051	RT7326	2051	S1891B
2051	Q5053C	2051	RS-7610	2051	RT7327	2051	S1955
2051	Q5073D	2051	RS-7611	2051	RT7511	2051	S1993
2051	Q5073E	2051	RS-7612	2051	RT7514	2051	S2034
2051	Q5073F	2051	RS-7613	2051	RT7515	2051	S2043
2051	Q5123E	2051	RS-7614	2051	RT7517	2051	S2044
2051	Q5123F	2051	RS-7622	2051	RT7518	2051	S2121
2051	Q5182	2051	RS-7623	2051	RT7528	2051	S2122
2051	QA-12	2051	RS1049	2051	RT7638	2051	S2123
2051	QA-13	2051	RS1059	2051	RT7645	2051	S2124
2051	QA-14	2051	RS128	2051	RT7943	2051	S2171
2051	QA-15	2051	RS136	2051	RT8195	2051	S2172
2051	QA-16	2051	RS1504B	2051	RT8197	2051	S2225
2051	QA-19	2051	RS2914	2051	RT8198	2051	S2543
2051	QO0254	2051	RS5851	2051	RT8201	2051	S23579
2051	QOV60529	2051	RS5853	2051	RT8330	2051	S2397
2051	QOV60530	2051	RS5856	2051	RT8332	2051	S24591
2051	Q8-0254	2051	RS5857	2051	RT929H	2051	S24596
2051	Q8O254	2051	RS7103	2051	RV11471	2051	S2581
2051	Q8054	2051	RS7105	2051	RV1474	2051	S2582
2051	Q8O380	2051	RS7108	2051	RV2249	2051	S2590
2051	Q8E1001	2051	RS7111	2051	RVTC81381	2051	S2593
2051	QT-C0372XAT	2051	RS7121	2051	RVTC81383	2051	S2635
2051	QT-C0710XBE	2051	RS7127	2051	RVTS22410	2051	S26636
2051	QT-C0710XEE	2051	RS7129	2051	RYN121105	2051	S26822
2051	QT-C0735XBT	2051	RS7132	2051	RYN121105-3	2051	S2935
2051	QT-C0829XAN	2051	RS7133	2051	RYN121105-4	2051	S2944
2051	QT-C131BXDN	2051	RS7136	2051	S-1061	2051	S29445
2051	QT-C168TXAN	2051	RS7160	2051	S-1065	2051	S2984
2051	QT-CBC546AA	2051	RS7223	2051	S-1066	2051	S2985
2051	R-28C537	2051	RS7224	2051	S-1068	2051	S2989
2051	R3273-P1	2051	RS7226	2051	S-1128	2051	S2996
2051	R3273-P2	2051	RS7232	2051	S-1143	2051	S2997
2051	R3283	2051	RS7234	2051	S-1221	2051	S2998
2051	R3293(GE)	2051	RS7235	2051	S-1221A	2051	S2999
2051	R34-6016-58	2051	RS7236	2051	S-1245	2051	S31866
2051	R340	2051	RS7238	2051	S-1331W	2051	S32550
2051	R4057	2051	RS7241	2051	S-1363	2051	S33755
2051	R582	2051	RS7242	2051	S-1364	2051	S34540
2051	R7163	2051	RS7405	2051	S-1403	2051	S36999
2051	R7165	2051	RS7406	2051	S-1512	2051	S37182
2051	R7249	2051	RS7407	2051	S-1533	2051	S37214
2051	R7343	2051	RS7408	2051	S-1559	2051	S37423
2051	R7359	2051	RS7409	2051	S001466	2051	S38763
2051	R7360	2051	RS7410	2051	S0015	2051	S38787
2051	R7361	2051	RS7411	2051	S0022	2051	S38854
2051	R7582	2051	RS7412	2051	S0025	2051	0S5360
2051	R7887	2051	RS7413	2051	S022010	2051	S6801
2051	R7953	2051	RS7415	2051	S022011	2051	S70.01.704
2051	R8066	2051	RS7421	2051	S024428	2051	S95202
2051	R8067	2051	RS7504	2051	S024987	2051	S9631
2051	R8068	2051	RS7510	2051	S025232	2051	SAW-28C372GR
2051	R8069	2051	RS7513	2051	S025289	2051	SAW-28C372Y
2051	R8070	2051	RS7513-15	2051	S031A	2051	SAW-28C945R
2051	R8115	2051	RS7514	2051	S037	2051	SC-4044
2051	R8116	2051	RS7515	2051	S0704	2051	SC-4244
2051	R8117	2051	RS7516	2051	S1016	2051	SC-65
2051	R8118	2051	RS7517	2051	S1061	2051	SC-832
2051	R8119	2051	RS7517-19	2051	S1065	2051	SC1001
2051	R8120	2051	RS7518	2051	S1066	2051	SC1010
2051	R8223	2051	RS7519	2051	S1068	2051	SC108
2051	R8224	2051	RS7521	2051	S1069	2051	SC108A
2051	R8225	2051	RS7525	2051	S1074	2051	SC108B
2051	R8243	2051	RS7526	2051	S1074(R)	2051	SC109A
2051	R8244	2051	RS7527	2051	S1074R	2051	SC1168Q
2051	R8259	2051	RS7528	2051	S1080(TRANSISTOR)	2051	SC1168H
2051	R8260	2051	RS7529	2051	S1128	2051	SC1229Q
2051	R8261	2051	RS7530	2051	S1143	2051	SC350
2051	R8305	2051	RS7542	2051	S12-1-A-3P	2051	SC4010
2051	R8312	2051	RS7543	2051	S1221	2051	SC4044
2051	R8528	2051	RS7544	2051	S1221A	2051	SC65
2051	R8530	2051	RS7555	2051	S1226	2051	SC785
2051	R8543	2051	RS7606	2051	S1241	2051	SC832
2051	R8551	2051	RS7607	2051	S1242	2051	SC842
2051	R8552	2051	RS7609	2051	S1243	2051	SCD-T322
2051	R8553	2051	RS7610	2051	S1245	2051	SCDT323
2051	R8554	2051	RS7611	2051	S1272	2051	SD-109
2051	R8555	2051	RS7612	2051	S1307	2051	SD109
2051	R8556	2051	RS7613	2051	S1309	2051	SDD3000
2051	R8557	2051	RS7614	2051	S133-1	2051	SDD421
2051	R8620	2051	RS7620	2051	S1331	2051	SDD821
2051	R8646	2051	RS7621	2051	S1331N	2051	SE-0566
2051	R8647	2051	RS7622	2051	S1331W	2051	SE-1002
2051	R8648	2051	RS7623	2051	S1363	2051	SE-1331
2051	R8658	2051	RS7624	2051	S1364	2051	SE-2001
2051	R8889	2051	RS7625	2051	S1369	2051	SE-3646
2051	R8914	2051	RS7626	2051	S1373	2051	SE-4001
2051	R8916	2051	RS7627	2051	S1374	2051	SE-4002
2051	R8963	2051	RS7628	2051	S1403	2051	SE-4010
2051	R8964	2051	RS7634	2051	S1405	2051	SE-5006
2051	R8965	2051	RS7635	2051	S1419	2051	SE-6001
2051	R8966	2051	RS7636	2051	S1420	2051	SE-6002
2051	R8968	2051	RS7637	2051	S1429-3	2051	SE1331
2051	R9004	2051	RS7638	2051	S1432	2051	SE2401
2051	R9005	2051	RS7639	2051	S1443	2051	SE2402
2051	R9006	2051	RS7640	2051	S1453	2051	SE4001
2051	R9025	2051	RS7641	2051	S1475	2051	SE4002
2051	R9071	2051	RS7642	2051	S1476	2051	SE4010
2051	R9384	2051	RS7643	2051	S1487	2051	SE4172
2051	R9385	2051	RS7814	2051	S1502	2051	SE5-0128
2051	R9483	2051	RS8442	2051	S1510	2051	SE5-0253
2051	R810	2051	RS8503	2051	S1512	2051	SE5-0274
2051	R812	2051	RS86057332	2051	S1526	2051	SE5-0567
2051	R813	2051	RT-100	2051	S1527	2051	SE5-0608
2051	R866	2051	RT-102	2051	S1529	2051	SE5-0848
2051	R867	2051	RT-929-H	2051	S1530	2051	SE5-0854
2051	REN123	2051	RT-929H	2051	S1533	2051	SE5-0855
2051	REN123A	2051	RT-930H	2051	S1559	2051	SE5-0887
2051	RH120	2051	RT100	2051	S15649	2051	SE5-0888
2051	RR7504	2051	RT114	2051	S1568	2051	SE5-0938-54
2051	RR8068	2051	RT2016	2051	S1570	2051	SE5030A
2051	RR8914	2051	RT2332	2051	S1619	2051	SE5030B
2051	RS-107	2051	RT2914	2051	S1620	2051	SE5151
2051	RS-108	2051	RT304	2051	S1629	2051	SE6010
2051	RS-2009	2051	RT3063	2051	S1697	2051	SE8040
2051	RS-2013	2051	RT3064	2051	S169N	2051	SEK-0367
2051	RS-2016	2051	RT3228	2051	S1761	2051	SF1001
2051	RS-5851	2051	RT3565	2051	S1761A	2051	SF1713
2051	RS-5853	2051	RT3567	2051	S1761B	2051	SF1714
2051	RS-5856	2051	RT476	2051	S1761C	2051	SF1726
2051	RS-5857	2051	RT4760	2051	S1764	2051	SF1730
2051	RS-7103	2051	RT4761	2051	S1765	2051	SF7713
2051	RS-7105	2051	RT5202	2051	S1766	2051	SF7714
2051	RS-7124	2051	RT5206	2051	S1768	2051	SG0-7202
2051	RS-7127	2051	RT5207	2051	S1770	2051	SH1064
2051	RS-7129	2051	RT5551	2051	S1772	2051	SJ3629
2051	RS-7409	2051	RT6600MHF25	2051	S1784	2051	SJ570
2051	RS-7411	2051	RT6921	2051	S1785	2051	SK1640A
2051	RS-7412	2051	RT6921MHF25	2051	S1788	2051	SK1641
2051	RS-7413	2051	RT6921	2051	S1855	2051	SK3020
2051	RS-7504	2051	RT697M	2051	S1871	2051	SK3038
2051	RS-7511	2051	RT6989	2051	S1878	2051	SK3046
2051	RS-7606	2051	RT7322	2051	S1891	2051	SK3122

RS 276-	Industry Standard No.
2051	8K3124
2051	8K3434A
2051	8K3444/123A
2051	8K5801
2051	8K5915
2051	8K6215
2051	8K8251
2051	8KA-6256
2051	8KA-6437
2051	8KA-8105
2051	8KA0030
2051	8KA1080
2051	8KA1117
2051	8KA1395
2051	8KA4141
2051	8KB8339
2051	8L300
2051	8L7990
2051	8M-4508-B
2051	8M-5564
2051	8M-5643
2051	8M-716
2051	8M-7815
2051	8M-7836
2051	8M-A-726655
2051	8M-A-726664
2051	8M-B-610342
2051	8M-B-686767
2051	8M-C-583256
2051	8M07275
2051	8M07286
2051	8M2700
2051	8M2701
2051	8M3104
2051	8M3117A
2051	8M3505
2051	8M3986
2051	8M4508-B
2051	8M5379
2051	8M5564
2051	8M5643
2051	8M576-1
2051	8M576-2
2051	8M5981
2051	8M6762
2051	8M6773
2051	8M716
2051	8M7545
2051	8M7815
2051	8M7836
2051	8M8112
2051	8M8113
2051	8M8978
2051	8M9008
2051	8M9135
2051	8M9253
2051	8PC040
2051	8PC042
2051	8PC50
2051	8PC51
2051	8PC052
2051	8PS-1475
2051	8PS-1475(YT)
2051	8PS-1475YT
2051	8PS-4075
2051	8PS-41
2051	8PS-4396
2051	8PS-934
2051	8PS1045
2051	8PS1082
2051	8PS1475
2051	8PS1802
2051	8PS1817
2051	8PS1977
2051	8PS2104
2051	8PS2129
2051	8PS2142
2051	8PS2164
2051	8PS2194
2051	8PS2225
2051	8PS2270
2051	8PS2415
2051	8PS3015
2051	8PS3735
2051	8PS3751
2051	8PS3900
2051	8PS3907
2051	8PS3908
2051	8PS3909
2051	8PS3915
2051	8PS3923
2051	8PS3925
2051	8PS3926
2051	8PS3930
2051	8PS3936
2051	8PS3938
2051	8PS3940
2051	8PS3951
2051	8PS3957C
2051	8PS3967
2051	8PS3972
2051	8PS3973
2051	8PS3999
2051	8PS4003
2051	8PS4004
2051	8PS4006
2051	8PS4009
2051	8PS4017
2051	8PS4020
2051	8PS4029
2051	8PS4032
2051	8PS4034
2051	8PS4037
2051	8PS4039
2051	8PS4040
2051	8PS4041
2051	8PS4042
2051	8PS4044
2051	8PS4045
2051	8PS4049
2051	8PS4052
2051	8PS4053
2051	8PS4055
2051	8PS4059
2051	8PS4060
2051	8PS4061
2051	8PS4062
2051	8PS4063
2051	8PS4066
2051	8PS4067
2051	8PS4069
2051	8PS4074
2051	8PS4075
2051	8PS4077
2051	8PS4081
2051	8PS4083
2051	8PS4084
2051	8PS4085
2051	8PS4088
2051	8PS4089
2051	8PS4095
2051	8PS41
2051	8PS4169
2051	8PS4199
2051	8PS4236
2051	8PS4303
2051	8PS4313
2051	8PS4344
2051	8PS4345
2051	8PS4356
2051	8PS4359
2051	8PS4360
2051	8PS4363
2051	8PS4367
2051	8PS4368
2051	8PS4382
2051	8PS4390
2051	8PS4392
2051	8PS4443
2051	8PS4446
2051	8PS4450
2051	8PS4451
2051	8PS4453
2051	8PS4455
2051	8PS4456
2051	8PS4457
2051	8PS4459
2051	8PS4472
2051	8PS4476
2051	8PS4478
2051	8PS4491
2051	8PS4493
2051	8PS4494
2051	8PS4498
2051	8PS4477
2051	8PS4920
2051	8PS4942
2051	8PS5000
2051	8PS5006
2051	8PS5006-1
2051	8PS5006-2
2051	8PS5457
2051	8PS6111
2051	8PS6112
2051	8PS6113
2051	8PS627
2051	8PS6571
2051	8PS8699
2051	8PS7652
2051	8PS817
2051	8PS817N
2051	8PS868
2051	8PS907
2051	8PS918
2051	8QD-2170
2051	8R20234
2051	8R75844
2051	8S1-145128
2051	8S2308
2051	8S2504
2051	8S3694
2051	8S9328
2051	8T-1242
2051	8T-1243
2051	8T-1244
2051	8T-1290
2051	8T-LM2152
2051	8T-MPS9433
2051	8T-MPS9700D
2051	8T-MPS9700E
2051	8T-MPS9700F
2051	8T.082.112.005
2051	8T.082.114.015
2051	8T/217/Q
2051	8T01
2051	8T02
2051	8T03
2051	8T04
2051	8T05
2051	8T06
2051	8T1242
2051	8T1243
2051	8T1244
2051	8T1290
2051	8T1402D
2051	8T1402E
2051	8T150
2051	8T1506
2051	8T151
2051	8T152
2051	8T153
2051	8T154
2051	8T155
2051	8T156
2051	8T157
2051	8T160
2051	8T1607
2051	8T161
2051	8T162
2051	8T163
2051	8T1702M
2051	8T1702N
2051	8T175
2051	8T176
2051	8T177
2051	8T178
2051	8T180
2051	8T181
2051	8T182
2051	8T250
2051	8T251
2051	8T25A
2051	8T25B
2051	8T25C
2051	8T403
2051	8T50
2051	8T501
2051	8T502
2051	8T503
2051	8T504
2051	8T5060
2051	8T51
2051	8T53
2051	8T54
2051	8T55
2051	8T56
2051	8T57
2051	8T58
2051	8T59
2051	8T63
2051	8T64
2051	8T6511
2051	8T6512
2051	8X3709
2051	8X3711
2051	8X55
2051	8YL-1182
2051	8YL1182
2051	8YL152
2051	8YL3460
2051	T-1416
2051	T-255
2051	T-256
2051	T-399
2051	T-H28C536
2051	T-H28C693
2051	T-H28C715
2051	T-S6433
2051	T-Q5053C
2051	T-Q5073
2051	T01-013
2051	T01-014
2051	T01-101
2051	T01-104
2051	T01-105
2051	T1-1A6
2051	T1004671
2051	T1008-834
2051	T1008834
2051	T1340A31
2051	T1340A3I
2051	T1340A3J
2051	T1340A3K
2051	T1413
2051	T1414
2051	T1415
2051	T1416
2051	T1417
2051	T143
2051	T1495
2051	T157
2051	T158
2051	T1642B
2051	T170
2051	T171
2051	T1746
2051	T1746A
2051	T1746B
2051	T1746C
2051	T1748
2051	T1748A
2051	T1748B
2051	T1748C
2051	T1802
2051	T1802A
2051	T1802B
2051	T1804
2051	T1805
2051	T1810
2051	T1810B
2051	T185
2051	T1909
2051	T1855
2051	T1895
2051	T235A013-2
2051	T237
2051	T2446
2051	T255
2051	T256
2051	T2277
2051	T291
2051	T327
2051	T327-2
2051	T328
2051	T339
2051	T342
2051	T3565
2051	T35A-5
2051	T3601
2051	T386
2051	T399
2051	T416-16(SEARS)
2051	T417
2051	T457-16
2051	T457-16(SEARS)
2051	T458-16
2051	T459(SEARS)
2051	T460
2051	T461-16
2051	T461-16(SEARS)
2051	T462
2051	T462(SEARS)
2051	T472
2051	T472(SEARS)
2051	T483(SEARS)
2051	T484(SEARS)
2051	T485(SEARS)
2051	T486(SEARS)
2051	T59235A
2051	T615A002
2051	T615A006-1
2051	T650(TRANSISTOR)
2051	T6565
2051	T76
2051	T9011A1C
2051	T9011A1G
2051	T9011AZ
2051	T9011G(CD)
2051	T9011G(EF)
2051	T9011H(EF)
2051	T9011I(EF)
2051	T9011J(GH)
2051	TA-6
2051	TA198030-4
2051	TA2A
2051	TA6
2051	TA7
2051	TAC-047
2051	TAC047
2051	TBRC147B
2051	TC3123036722
2051	TC3123036900
2051	TC3123037111
2051	TC3123037222
2051	TC3123037412
2051	TC3123307222
2051	TD100
2051	TD101
2051	TD200
2051	TD250
2051	TE1420
2051	T82369
2051	T83414
2051	T83415
2051	T83605
2051	T83605A
2051	T83606
2051	T83606A
2051	T83607
2051	T83704
2051	T83705
2051	T83903
2051	T83904
2051	T83906
2051	T84123
2051	T84124
2051	T84424
2051	T84951
2051	T84952
2051	T84953
2051	T84954
2051	T85309A
2051	T85311A
2051	T85368
2051	T85369
2051	T85370
2051	T85371
2051	T85376
2051	T85377
2051	T85449
2051	T85450
2051	T85697
2051	TBH0147
2051	TP-78
2051	TG28C1175
2051	TG28C1175(C)
2051	TG28C1175-
2051	TG28C1175-D
2051	TG28C1175-E
2051	TGB0331
2051	TH28C536
2051	TH28C693
2051	TH28C715
2051	TI-412
2051	TI-413
2051	TI-414
2051	TI-422
2051	TI-425
2051	TI-432
2051	TI-433
2051	TI-481
2051	TI-714
2051	TI-714A
2051	TI-751
2051	TI-806G
2051	TI-907
2051	TI-908
2051	TI1A6
2051	TI24A
2051	TI24B
2051	TI415
2051	TI416
2051	TI417
2051	TI418
2051	TI419
2051	TI420
2051	TI422
2051	TI424
2051	TI430
2051	TI432
2051	TI480
2051	TI481
2051	TI482
2051	TI483
2051	TI484
2051	TI485
2051	TI492
2051	TI493
2051	TI494
2051	TI495
2051	TI496
2051	TI54A
2051	TI54B
2051	TI54C
2051	TI54E
2051	TI642B
2051	TI714
2051	TI751
2051	TI8003B
2051	TI803B
2051	TI810B
2051	TI904
2051	TI907
2051	TI908
2051	TIA06
2051	TIA102
2051	TI8-125
2051	TI8-62
2051	TI8113
2051	TI8114
2051	TI822
2051	TI823
2051	TI844
2051	TI845
2051	TI846
2051	TI847
2051	TI848
2051	TI849
2051	TI851
2051	TI852
2051	TI855
2051	TI856
2051	TI857
2051	TI860
2051	TI871
2051	TI872
2051	TI893
2051	TI890
2051	TI890-2
2051	TI892
2051	TI892-BLU
2051	TI892-GRN
2051	TI892-GRY
2051	TI892-VIO
2051	TI892-YEL
2051	TI894
2051	TI894(AFAMP)
2051	TI894(XSTR)
2051	TI895
2051	TI896
2051	TIS992
2051	TIX712

RS 276-	Industry Standard No.	RS 276-	Industry Standard No.	RS 276-	Industry Standard No.	RS 276-	Industry Standard No.
2051	TIXS12	2051	TR310243	2051	VS-28C324H	2051	ZT22B
2051	TIXS13	2051	TR310245	2051	VS-28C458	2051	ZT220
2051	TK1228-1008	2051	TR4010-2	2051	VS-28C53B	2051	ZT22C
2051	TK1228-1009	2051	TR601	2051	VS28B324	2051	ZT23
2051	TK1228-1010	2051	TR8004-4	2051	VS28C1166-0-1	2051	ZT23-1
2051	TK1228-1011	2051	TR8010	2051	VS28C1166Y-1	2051	ZT23-12
2051	TK1228-1012	2051	TR8014	2051	VS28C1335D/1	2051	ZT23A
2051	TM123	2051	TR8021	2051	VS28C206	2051	ZT23B
2051	TM123A	2051	TR8025	2051	VS28C208	2051	ZT23C
2051	TM123AP	2051	TR8028	2051	VS28C288A	2051	ZT24
2051	TM2613	2051	TR8029	2051	VS28C324H	2051	ZT24-1
2051	TM2711	2051	TR8030	2051	VS28C372-Y/1E	2051	ZT24-12
2051	TMT-1543	2051	TR8031	2051	VS28C458	2051	ZT24-55
2051	TMT1543	2051	TR8034	2051	VS28C538	2051	ZT2476
2051	TN237	2051	TR8035	2051	VS28C645A	2051	ZT2477
2051	TN53	2051	TR8038	2051	VS28C732-VIP	2051	ZT24A
2051	TN55	2051	TR8039	2051	VS28D227V-1	2051	ZT24B
2051	TN56	2051	TR8040	2051	VS9-0005-913	2051	ZT24C
2051	TN59	2051	TR8043	2051	VS9-0006-913	2051	ZT23.55
2051	TN60	2051	TR8330	2051	W10	2051	ZT40
2051	TN61	2051	TR86	2051	W20	2051	ZT402
2051	TN62	2051	TR9100	2051	W24	2051	ZT403
2051	TN63	2051	TRA-34	2051	WEP1717	2051	ZT404
2051	TN64	2051	TRA-36	2051	WEP454	2051	ZT406
2051	TN80	2051	TRA-4	2051	WEP458	2051	ZT41
2051	TNC61689	2051	TRA-4A	2051	WEP50	2051	ZT42
2051	TNC61702	2051	TRA-4B	2051	WEP54	2051	ZT43
2051	TNJ-60076	2051	TRA-9R	2051	WEP55	2051	ZT44
2051	TNJ-60604	2051	TRA34	2051	WEP723	2051	ZT50
2051	TNJ-60606	2051	TRA36	2051	WEP724	2051	ZT60
2051	TNJ-60607	2051	TRA4	2051	WEP728	2051	ZT60-1
2051	TNJ1036	2051	TRA4A	2051	WEP729	2051	ZT60-12
2051	TNJ60076	2051	TRA4B	2051	WEP735	2051	ZT60-55
2051	TNJ61219	2051	TRA9R	2051	WEP735A	2051	ZT60A
2051	TNJ61220	2051	TRAPLC711	2051	WEP736	2051	ZT60B
2051	TNJ70479-1	2051	TRAPLC871	2051	WEP773	2051	ZT60C
2051	TNJ70537	2051	TRAPLC871A	2051	WEP829	2051	ZT61
2051	TNJ70539	2051	TRBC147B	2051	WEP838	2051	ZT61-1
2051	TNJ70637	2051	TRO1026	2051	WEP906	2051	ZT61-12
2051	TNJ70638	2051	TRO10602-1	2051	WRR1952	2051	ZT61-55
2051	TNJ70639	2051	TS2221	2051	WRR1953	2051	ZT61A
2051	TNJ71036	2051	TS2222	2051	WRR1954	2051	ZT61B
2051	TNJ72280	2051	TSC499	2051	X1G01829	2051	ZT61C
2051	TNJ72281	2051	TSC695	2051	X16A1938	2051	ZT62-1
2051	TNJ72783	2051	TST705899A	2051	X16A545-7	2051	ZT62-12
2051	TNJ72784	2051	TT-1097	2051	X16E3860	2051	ZT62-15
2051	TNT-843	2051	TT1097	2051	X16E3960	2051	ZT62C
2051	TNT842	2051	TV-17	2051	X16N1485	2051	ZT63
2051	TO-033	2051	TV-18	2051	X19001-A	2051	ZT63-1
2051	TO-038	2051	TV-21	2051	X3204211	2051	ZT63-12
2051	TO-039	2051	TV-23	2051	X3204296	2051	ZT63-55
2051	TO-040	2051	TV-40	2051	X3205198	2051	ZT63A
2051	TO1-101	2051	TV-42	2051	X3206105	2051	ZT63B
2051	TO1-104	2051	TV-46	2051	X3205422	2051	ZT63C
2051	TO1-105	2051	TV-51	2051	X32M5026	2051	ZT64
2051	TP4123	2051	TV-52	2051	X34E2111	2051	ZT64-1
2051	TP4124	2051	TV-53	2051	X6584-C	2051	ZT64-12
2051	TP86512	2051	TV-56	2051	X735-41	2051	ZT64-5
2051	TP86513	2051	TV-58	2051	XA-1071	2051	ZT64-55
2051	TP86514	2051	TV-6	2051	XA-1139	2051	ZT64A
2051	TP86515	2051	TV-60	2051	XC372	2051	ZT64B
2051	TP86520	2051	TV-62	2051	XC373	2051	ZT64C
2051	TP86521	2051	TV-65	2051	XC374	2051	ZT66
2051	TQ-5052	2051	TV-66	2051	XEA040017	2051	ZT68
2051	TQ-5053	2051	TV-68	2051	XG30	2051	ZT696
2051	TQ-5054	2051	TV-84	2051	XN-400-318-P1	2051	ZT697
2051	TQ-5060	2051	TV-92	2051	X821	2051	ZT706
2051	TQ4	2051	TV17	2051	X822	2051	ZT706A
2051	TQ5052	2051	TV16B	2051	Y49001-21	2051	ZT708
2051	TQ5053	2051	TV241077	2051	Y56001-86	2051	ZT80
2051	TQ5054	2051	TV241078	2051	Y56601-08	2051	ZT81
2051	TQ5060	2051	TV24215	2051	Y56601-45	2051	ZT82(TRANSISTOR)
2051	TR-01B(PENNCREST)	2051	TV24216	2051	Y56601-49	2051	ZT83
2051	TR-01C(PENNCREST)	2051	TV24281	2051	Y56601-51	2051	ZT84
2051	TR-06(PENNCREST)	2051	TV24372	2051	Y56601-73	2051	ZT86
2051	TR-1033-1	2051	TV24453	2051	Y56601-75	2051	ZT87
2051	TR-1033-2	2051	TV24454	2051	Y56601-80	2051	ZT88
2051	TR-1033-3	2051	TV24458	2051	Y56601-86-AD	2051	ZT89
2051	TR-1347	2051	TV24576	2051	Y56601-93	2051	ZTX300
2051	TR-162(OLSON)	2051	TV24655	2051	YBAN28C941	2051	ZTX301
2051	TR-1R33	2051	TV28C208	2051	Z-28058-1	2051	ZTX304
2051	TR-21	2051	TV36	2051	Z4MW333	2051	ZTX310
2051	TR-21-6	2051	TV38	2051	ZDT	2051	ZTX311
2051	TR-21C	2051	TV39	2051	ZBN100	2051	ZTX312
2051	TR-22	2051	TV40	2051	ZEN102	2051	ZTX330
2051	TR-22C	2051	TV43	2051	ZEN103	2051	001-00
2051	TR-24(PHILCO)	2051	TV46	2051	ZEN110	2051	1-0006-0021
2051	TR-28C367	2051	TV48	2051	ZEN111	2051	1-0006-0022
2051	TR-28C373	2051	TV56	2051	ZEN112	2051	1-001-02415
2051	TR-28C735	2051	TV57A	2051	ZEN113	2051	001-02011
2051	TR-3R38	2051	TV58A	2051	ZEN114	2051	02020
2051	TR-4R33	2051	TV59A	2051	ZEN115	2051	001-02101-0
2051	TR-51	2051	TV60A	2051	ZEN119	2051	001-02101-1
2051	TR-5R33	2051	TV65	2051	ZEN120	2051	001-021010
2051	TR-5R35	2051	TV71	2051	ZEN126	2051	001-02102-0
2051	TR-5R38	2051	TV92	2051	ZEN127	2051	001-021020
2051	TR-62	2051	TVS-2C8645A	2051	ZT-110	2051	001-02103-0
2051	TR-6R33	2051	TVS-28C206	2051	ZT-62	2051	001-021030
2051	TR-7R35	2051	TVS-28C208A	2051	ZT-82	2051	001-02104-0
2051	TR-8004	2051	TVS-28C538	2051	ZT111	2051	001-021040
2051	TR-8014	2051	TVS-28C538A	2051	ZT112	2051	001-021050-0
2051	TR-8025	2051	TVS-28C828	2051	ZT113	2051	001-021050
2051	TR-8035	2051	TVS-28C828A	2051	ZT114	2051	001-021060
2051	TR-8039	2051	TVS-28C828Q	2051	ZT116	2051	001-021070
2051	TR-8042	2051	TVS28C538A	2051	ZT117	2051	001-021070
2051	TR-86	2051	TVS28C58A	2051	ZT118	2051	001-0210B-0
2051	TR-8R35	2051	TVS28C644	2051	ZT119	2051	001-0210B0
2051	TR-9100-18	2051	TVS28C645B	2051	ZT1420	2051	001-021090-0
2051	TR-BC147B	2051	TVS28C684	2051	ZT1708	2051	001-021090
2051	TR-BRC149C	2051	TVS28C828	2051	ZT1711	2051	001-021110-0
2051	TR-RR38	2051	TVS28C828(Q)	2051	ZT20	2051	001-021111-0
2051	TR-TR38	2051	TVS28C828A	2051	ZT20-1	2051	001-021111-1
2051	TRO1037	2051	TVS28C828BP	2051	ZT20-12	2051	001-021110
2051	TRO1073	2051	TVS28C828BR	2051	ZT20-55	2051	001-021111
2051	TRO1074	2051	TVS28C829(B)	2051	ZT202	2051	001-021113-2
2051	TRO573491	2051	11585E	2051	ZT203	2051	001-021113-3
2051	TR1011	2051	UPI2222	2051	ZT204	2051	001-021113-4
2051	TR1031	2051	UPI2222B	2051	ZT20A	2051	001-021113-5
2051	TR1033	2051	UPI2222F	2051	ZT20B	2051	001-021132
2051	TR1993-2	2051	UPI4046-46	2051	ZT20C	2051	001-021133
2051	TR2083-71	2051	UPI4047-46	2051	ZT21	2051	001-021134
2051	TR2083-72	2051	UPI718A	2051	ZT21-1	2051	001-021121-0
2051	TR2083-73	2051	UPI956	2051	ZT21-12	2051	001-021210
2051	TR21	2051	V119	2051	ZT21-55	2051	001-021232
2051	TR2320063	2051	V169	2051	ZT21A	2051	01-030829
2051	TR2327293	2051	V297	2051	ZT21B	2051	01-031175
2051	TR2327333	2051	VM-30209	2051	ZT21C	2051	01-031364
2051	TR2327363	2051	VM-30241	2051	ZT22	2051	01-031687
2051	TR28C3677	2051	VM-30242	2051	ZT22-1	2051	01-031688
2051	TR28C373	2051	VM30209	2051	ZT22-12	2051	01-032076
2051	TR28C735	2051	VM30241	2051	ZT22-55	2051	1-042/2207
2051	TR302	2051	VM30242	2051	ZT2205	2051	1-044/2207
2051	TR310231	2051	VS-28A288A	2051	ZT2206	2051	01-201-0
		2051	VS-28C-458	2051	ZT22A		

RS 276-	Industry Standard No.	RS 276-	Industry Standard No.	RS 276-	Industry Standard No.	RS 276-	Industry Standard No.	RS 276-	Industry Standard No.
2051	1-21-276	2051	2N1338	2051	2N2318	2051	2N3241	2051	2N3241
2051	1-21-277	2051	2N1386	2051	2N2319	2051	2N3241A	2051	2N3241A
2051	1-21-278	2051	2N1387	2051	2N2320	2051	2N3242	2051	2N3242
2051	1-21-279	2051	2N1388	2051	2N2330	2051	2N3242A	2051	2N3242A
2051	01-2101	2051	2N1389	2051	2N2353	2051	2N3246	2051	2N3246
2051	01-2101-0	2051	2N1417	2051	2N2349	2051	2N3247	2051	2N3247
2051	001-21011	2051	2N1418	2051	2N2353	2051	2N3261	2051	2N3261
2051	01-2102	2051	2N1420	2051	2N2353A	2051	2N3268	2051	2N3268
2051	01-2104	2051	2N1420A	2051	2N2368	2051	2N3298	2051	2N3298
2051	01-2105	2051	2N148	2051	2N2368A	2051	2N3300	2051	2N3300
2051	01-2106	2051	2N148C	2051	2N2369	2051	2N3301	2051	2N3301
2051	01-2107	2051	2N148C/D	2051	2N2369A	2051	2N3302	2051	2N3302
2051	01-2108	2051	2N148D	2051	2N2380	2051	2N3310	2051	2N3310
2051	01-2109	2051	2N1528	2051	2N2380A	2051	2N332	2051	2N332
2051	01-30829	2051	2N1564	2051	2N2387	2051	2N3326	2051	2N3326
2051	01-349418	2051	2N1565	2051	2N2388	2051	2N332A	2051	2N332A
2051	1-522223720	2051	2N1566	2051	2N2389	2051	2N333	2051	2N333
2051	01-571811	2051	2N1566A	2051	2N2390	2051	2N333A	2051	2N333A
2051	01-571821	2051	2N1587	2051	2N2396	2051	2N334	2051	2N334
2051	01-6147191229	2051	2N1588	2051	2N2397	2051	2N3340	2051	2N3340
2051	1-6171191356B	2051	2N1590	2051	2N2413	2051	2N334A	2051	2N334A
2051	01-680815	2051	2N1591	2051	2N2417	2051	2N335	2051	2N335
2051	1-801-003	2051	2N1593	2051	2N2427	2051	2N335A	2051	2N335A
2051	1-801-004	2051	2N1594	2051	2N2432	2051	2N335B	2051	2N335B
2051	1-801-004-17	2051	2N160	2051	2N2466	2051	2N336	2051	2N336
2051	1-801-314	2051	2N160A	2051	2N2472	2051	2N336A	2051	2N336A
2051	1-801-314-15	2051	2N161	2051	2N2473	2051	2N338A	2051	2N338A
2051	1-801-314-16	2051	2N1613/46	2051	2N2476	2051	2N3390	2051	2N3390
2051	01-9011-5/2221-3	2051	2N161A	2051	2N2477	2051	2N3391	2051	2N3391
2051	01-9011-52221-3	2051	2N163	2051	2N2479	2051	2N3391A	2051	2N3391A
2051	01-9013-7/2221-3	2051	2N163A	2051	2N2481	2051	2N3392	2051	2N3392
2051	01-9013-72221-3	2051	2N1644	2051	2N2483	2051	2N3393	2051	2N3393
2051	01-9014-2/2221-3	2051	2N1644A	2051	2N2484	2051	2N3394	2051	2N3394
2051	01-9014-22221-3	2051	2N1663	2051	2N2501	2051	2N3395	2051	2N3395
2051	01-9016-4/2221-3	2051	2N1674	2051	2N2514	2051	2N3396	2051	2N3396
2051	01-9018-6/2221-3	2051	2N1682	2051	2N2515	2051	2N3397	2051	2N3397
2051	1A0020	2051	2N1704	2051	2N2520	2051	2N3398	2051	2N3398
2051	1A0021	2051	2N1708	2051	2N2521	2051	2N3402	2051	2N3402
2051	1A0022	2051	2N1708A	2051	2N2522	2051	2N3409	2051	2N3409
2051	1A0024	2051	2N1711/46	2051	2N2523	2051	2N3410	2051	2N3410
2051	1A0025	2051	2N1763	2051	2N2524	2051	2N3411	2051	2N3411
2051	1A0029	2051	2N1837	2051	2N2530	2051	2N3414	2051	2N3414
2051	1A0032	2051	2N1838	2051	2N2531	2051	2N3415	2051	2N3415
2051	1A0033	2051	2N1839	2051	2N2532	2051	2N3416	2051	2N3416
2051	1A0034	2051	2N1840	2051	2N2533	2051	2N3417	2051	2N3417
2051	1A0035	2051	2N1944	2051	2N2534	2051	2N3423	2051	2N3423
2051	1A0037	2051	2N1945	2051	2N2539	2051	2N3424	2051	2N3424
2051	1A0043	2051	2N1946	2051	2N2540	2051	2N3462	2051	2N3462
2051	1A0044	2051	2N1947	2051	2N2569	2051	2N3463	2051	2N3463
2051	1A0045	2051	2N1948	2051	2N2570	2051	2N3498	2051	2N3498
2051	1A0051	2051	2N1949	2051	2N2571	2051	2N3510	2051	2N3510
2051	1A0063	2051	2N1950	2051	2N2572	2051	2N3542	2051	2N3542
2051	1A0067	2051	2N1951	2051	2N2586	2051	2N3564	2051	2N3564
2051	1A0070	2051	2N1952	2051	2N2610	2051	2N3565	2051	2N3565
2051	1A0076	2051	2N1953	2051	2N2617	2051	2N3566	2051	2N3566
2051	1A0077	2051	2N1958	2051	2N2618/46	2051	2N3605	2051	2N3605
2051	1A0078	2051	2N1959	2051	2N263	2051	2N3605A	2051	2N3605A
2051	1A0079	2051	2N1962	2051	2N264	2051	2N3606	2051	2N3606
2051	1A0080	2051	2N1963	2051	2N2645	2051	2N3606A	2051	2N3606A
2051	1A0081	2051	2N1964	2051	2N2656	2051	2N3607	2051	2N3607
2051	1A0083	2051	2N1965	2051	2N2673	2051	2N3641	2051	2N3641
2051	1A0084	2051	2N1972	2051	2N2674	2051	2N3642	2051	2N3642
2051	1A4757-1	2051	2N1973	2051	2N2675	2051	2N3643	2051	2N3643
2051	1C5576	2051	2N1974	2051	2N2676	2051	2N3646	2051	2N3646
2051	1J1	2051	2N1975	2051	2N2677	2051	2N3647	2051	2N3647
2051	1U585F	2051	2N1983	2051	2N2692	2051	2N3678	2051	2N3678
2051	1U585F/7825B	2051	2N1984	2051	2N2693	2051	2N3688	2051	2N3688
2051	1W8358	2051	2N1985	2051	2N2713	2051	2N3689	2051	2N3689
2051	1W9723	2051	2N1986	2051	2N2714	2051	2N3690	2051	2N3690
2051	1W9787	2051	2N1987	2051	2N2719	2051	2N3691	2051	2N3691
2051	002-006500	2051	2N1988	2051	2N2720	2051	2N3692	2051	2N3692
2051	002-008300	2051	2N1992	2051	2N2721	2051	2N3693	2051	2N3693
2051	002-009500	2051	2N2038	2051	2N2722	2051	2N3694	2051	2N3694
2051	002-009501	2051	2N2039	2051	2N2784/TNT	2051	2N3704	2051	2N3704
2051	002-009502	2051	2N2040	2051	2N2787	2051	2N3705	2051	2N3705
2051	002-009900	2051	2N2041	2051	2N2790	2051	2N3706	2051	2N3706
2051	002-010400	2051	2N2049	2051	2N2791	2051	2N3707	2051	2N3707
2051	002-010800	2051	2N2087	2051	2N2792	2051	2N3708	2051	2N3708
2051	002-012000	2051	2N2094	2051	2N2831	2051	2N3709	2051	2N3709
2051	002-03	2051	2N2094A	2051	2N2845	2051	2N3710	2051	2N3710
2051	02-1078-01	2051	2N2095	2051	2N2847	2051	2N3711	2051	2N3711
2051	002-12000	2051	2N2095A	2051	2N2868	2051	2N3723	2051	2N3723
2051	2-8454-031	2051	2N2096	2051	2N2910	2051	2N3736	2051	2N3736
2051	2C8900	2051	2N2097A	2051	2N2913	2051	2N3793	2051	2N3793
2051	2D002	2051	2N2161	2051	2N2914	2051	2N3794	2051	2N3794
2051	2D002-168	2051	2N2194	2051	2N2917	2051	2N3828	2051	2N3828
2051	2D002-169	2051	2N2194A	2051	2N2918	2051	2N3843	2051	2N3843
2051	2D002-170	2051	2N2194B	2051	2N2923	2051	2N3843A	2051	2N3843A
2051	2D002-171	2051	2N2198	2051	2N2924	2051	2N3844	2051	2N3844
2051	2D002-175	2051	2N2205	2051	2N2925	2051	2N3844A	2051	2N3844A
2051	2D002-41	2051	2N2206	2051	2N2926	2051	2N3858	2051	2N3858
2051	2D017	2051	2N2217	2051	2N2926-6	2051	2N3858A	2051	2N3858A
2051	2D026	2051	2N2217/51	2051	2N2926ORN	2051	2N3859	2051	2N3859
2051	2D026-274	2051	2N2218	2051	2N2926G	2051	2N3859A	2051	2N3859A
2051	2D033	2051	2N2218/51	2051	2N2926GRN	2051	2N3862	2051	2N3862
2051	2D038	2051	2N2218A	2051	2N2926ORN	2051	2N3869	2051	2N3869
2051	2D033	2051	2N2219	2051	2N2926Y	2051	2N3877A	2051	2N3877A
2051	2JMW961	2051	2N2219/51	2051	2N2931	2051	2N3880	2051	2N3880
2051	2N1006	2051	2N2219A	2051	2N2932	2051	2N3885	2051	2N3885
2051	2N1051	2051	2N2220	2051	2N2933	2051	2N3900	2051	2N3900
2051	2N1081	2051	2N2221	2051	2N2934	2051	2N3900A	2051	2N3900A
2051	2N1082	2051	2N2221A	2051	2N2935	2051	2N3901	2051	2N3901
2051	2N1103	2051	2N2222	2051	2N2936	2051	2N3903	2051	2N3903
2051	2N1104	2051	2N2222A	2051	2N2937	2051	2N3904	2051	2N3904
2051	2N1116	2051	2N2222A	2051	2N2938	2051	2N3924	2051	2N3924
2051	2N1117	2051	2N2234	2051	2N2951	2051	2N3946	2051	2N3946
2051	2N1139	2051	2N2235	2051	2N2952	2051	2N3947	2051	2N3947
2051	2N1140	2051	2N2236	2051	2N2954	2051	2N3973	2051	2N3973
2051	2N1149	2051	2N2237	2051	2N2958	2051	2N3974	2051	2N3974
2051	2N1150	2051	2N2240	2051	2N2959	2051	2N3975	2051	2N3975
2051	2N1150/904	2051	2N2241	2051	2N2960	2051	2N3976	2051	2N3976
2051	2N1151	2051	2N2242	2051	2N2961	2051	2N4013	2051	2N4013
2051	2N1151/904A	2051	2N2244	2051	2N2974	2051	2N4014	2051	2N4014
2051	2N1152	2051	2N2245	2051	2N3009	2051	2N4063	2051	2N4063
2051	2N1153	2051	2N2246	2051	2N3011	2051	2N4064	2051	2N4064
2051	2N1153/910	2051	2N2247	2051	2N3013	2051	2N4074	2051	2N4074
2051	2N117	2051	2N2248	2051	2N3015	2051	2N4100	2051	2N4100
2051	2N118	2051	2N2249	2051	2N3043	2051	2N4104	2051	2N4104
2051	2N118A	2051	2N2250	2051	2N3044	2051	2N4123	2051	2N4123
2051	2N119	2051	2N2251	2051	2N3045	2051	2N4124	2051	2N4124
2051	2N199	2051	2N2252	2051	2N3046	2051	2N4137	2051	2N4137
2051	2N199A	2051	2N2253	2051	2N3047	2051	2N4140	2051	2N4140
2051	2N120	2051	2N2254	2051	2N3048	2051	2N4141	2051	2N4141
2051	2N1200	2051	2N2255	2051	2N3115	2051	2N4227	2051	2N4227
2051	2N1201	2051	2N2256	2051	2N3116	2051	2N4256	2051	2N4256
2051	2N1205	2051	2N2257	2051	2N3122	2051	2N4259	2051	2N4259
2051	2N1247	2051	2N2272	2051	2N3128	2051	2N4264	2051	2N4264
2051	2N1248	2051	2N2309	2051	2N3129	2051	2N4265	2051	2N4265
2051	2N1249	2051	2N2310	2051	2N3130	2051	2N4274	2051	2N4274
2051	2N1267	2051	2N2312	2051	2N3210	2051	2N4275	2051	2N4275
2051	2N1276	2051	2N2314	2051	2N3211	2051	2N4286	2051	2N4286
2051	2N1277	2051	2N2315					2051	2N4287
2051	2N1278	2051	2N2317					2051	2N4294
2051	2N1279							2051	2N431

RS 276-	Industry Standard No.	RS 276-	Industry Standard No.	RS 276-	Industry Standard No.	RS 276-	Industry Standard No.	RS 276-	Industry Standard No.
2051	2N432	2051	2N736	2051	2SC1033GN	2051	2SC137		
2051	2N433	2051	2N736A	2051	2SC1033H	2051	2SC1372Y		
2051	2N4400	2051	2N736B	2051	2SC1033J	2051	2SC137A		
2051	2N4401	2051	2N742	2051	2SC1033K	2051	2SC137B		
2051	2N4420	2051	2N742A	2051	2SC1033L	2051	2SC137C		
2051	2N4421	2051	2N745	2051	2SC1033OR	2051	2SC137D		
2051	2N4422	2051	2N746	2051	2SC1033R	2051	2SC137E		
2051	2N4424	2051	2N747	2051	2SC1033X	2051	2SC137F		
2051	2N4432	2051	2N748	2051	2SC1033Y	2051	2SC137G		
2051	2N4432A	2051	2N749	2051	2SC104	2051	2SC137GN		
2051	2N4436	2051	2N750	2051	2SC104A	2051	2SC137H		
2051	2N4437	2051	2N751	2051	2SC105	2051	2SC137J		
2051	2N4450	2051	2N753	2051	2SC1071	2051	2SC137K		
2051	2N470	2051	2N753/46	2051	2SC110	2051	2SC137L		
2051	2N471A	2051	2N754	2051	2SC111	2051	2SC137M		
2051	2N472	2051	2N756A	2051	2SC1175	2051	2SC137OR		
2051	2N472A	2051	2N757A	2051	2SC1175(D,E,F)	2051	2SC137R		
2051	2N473	2051	2N758A	2051	2SC1175C	2051	2SC137X		
2051	2N474	2051	2N758B	2051	2SC1175CTV	2051	2SC137Y		
2051	2N474A	2051	2N759A	2051	2SC1175D	2051	2SC138		
2051	2N475	2051	2N759B	2051	2SC1175E	2051	2SC138A		
2051	2N475A	2051	2N760A	2051	2SC1175E,F,D	2051	2SC138B		
2051	2N476	2051	2N760B	2051	2SC1175F	2051	2SC139		
2051	2N477	2051	2N770	2051	2SC1244	2051	2SC1390		
2051	2N478	2051	2N771	2051	2SC131	2051	2SC1390(L,Y)		
2051	2N479	2051	2N772	2051	2SC1317	2051	2SC1390(V)		
2051	2N479A	2051	2N773	2051	2SC131A	2051	2SC1390(W)		
2051	2N480	2051	2N774	2051	2SC131B	2051	2SC1390(X)		
2051	2N480A	2051	2N775	2051	2SC131C	2051	2SC1390(Y)		
2051	2N480B	2051	2N776	2051	2SC131D	2051	2SC13901		
2051	2N4950	2051	2N777	2051	2SC131E	2051	2SC1390I		
2051	2N4951	2051	2N778	2051	2SC131F	2051	2SC1390IW		
2051	2N4952	2051	2N783	2051	2SC131G	2051	2SC1390J		
2051	2N4953	2051	2N784	2051	2SC131GN	2051	2SC1390JX		
2051	2N4954	2051	2N784A	2051	2SC131H	2051	2SC1390K		
2051	2N4966	2051	2N784A/46	2051	2SC131J	2051	2SC1390L		
2051	2N4967	2051	2N784A/51	2051	2SC131L	2051	2SC1390V		
2051	2N4968	2051	2N789	2051	2SC131M	2051	2SC1390W		
2051	2N4969	2051	2N790	2051	2SC131OR	2051	2SC1390WH		
2051	2N4970	2051	2N791	2051	2SC131R	2051	2SC1390WI		
2051	2N4994	2051	2N792	2051	2SC131T	2051	2SC1390WX		
2051	2N4995	2051	2N793	2051	2SC131Y	2051	2SC1390X		
2051	2N5027	2051	2N834/46	2051	2SC132	2051	2SC1390XJ		
2051	2N5028	2051	2N834A	2051	2SC1324	2051	2SC1390XK		
2051	2N5030	2051	2N835	2051	2SC1324(C)	2051	2SC1390Y		
2051	2N5066	2051	2N835/46	2051	2SC1324C	2051	2SC1390YM		
2051	2N5081	2051	2N839	2051	2SC132A	2051	2SC1393		
2051	2N5082	2051	2N840	2051	2SC132B	2051	2SC1393K		
2051	2N5088	2051	2N842	2051	2SC132C	2051	2SC1393M		
2051	2N5089	2051	2N844	2051	2SC132D	2051	2SC141		
2051	2N5103	2051	2N866	2051	2SC132E	2051	2SC1416A		
2051	2N5107	2051	2N867	2051	2SC132F	2051	2SC1416BL		
2051	2N5127	2051	2N909	2051	2SC132G	2051	2SC142		
2051	2N5128	2051	2N914/46	2051	2SC132GN	2051	2SC143		
2051	2N5134	2051	2N915	2051	2SC132H	2051	2SC144		
2051	2N5137	2051	2N919	2051	2SC132J	2051	2SC144A		
2051	2N5144	2051	2N920	2051	2SC132K	2051	2SC15		
2051	2N5172	2051	2N921	2051	2SC132L	2051	2SC15-1		
2051	2N5183	2051	2N922	2051	2SC132M	2051	2SC15-2		
2051	2N5186	2051	2N929	2051	2SC132OR	2051	2SC15-3		
2051	2N5187	2051	2N929/46	2051	2SC132R	2051	2SC151		
2051	2N5209	2051	2N929A	2051	2SC132X	2051	2SC1537		
2051	2N5210	2051	2N930	2051	2SC132Y	2051	2SC1537(S)		
2051	2N5219	2051	2N930/46	2051	2SC133	2051	2SC1537-0		
2051	2N5223	2051	2N930/TNT	2051	2SC133A	2051	2SC1537-0		
2051	2N5224	2051	2N930A	2051	2SC133C	2051	2SC1537B		
2051	2N5225	2051	2N930A/46	2051	2SC133D	2051	2SC1537S		
2051	2N5232	2051	2N930B	2051	2SC133E	2051	2SC1542		
2051	2N5232A	2051	2N947	2051	2SC133F	2051	2SC157		
2051	2N5368	2051	2N951	2051	2SC133G	2051	2SC158		
2051	2N5369	2051	2N956	2051	2SC133GN	2051	2SC159		
2051	2N5370	2051	2N957	2051	2SC133H	2051	2SC16		
2051	2N5371	2051	2S001	2051	2SC133J	2051	2SC160		
2051	2N5380	2051	2S002	2051	2SC133K	2051	2SC162		
2051	2N5381	2051	2S003	2051	2SC133L	2051	2SC1634		
2051	2N541	2051	2S004	2051	2SC133M	2051	2SC1639		
2051	2N5417	2051	2S005	2051	2SC133OR	2051	2SC1641		
2051	2N5418	2051	2S095A	2051	2SC133X	2051	2SC1641Q		
2051	2N5419	2051	2S101	2051	2SC133Y	2051	2SC1641R		
2051	2N542	2051	2S102	2051	2SC134	2051	2SC166		
2051	2N5420	2051	2S11182D	2051	2SC134A	2051	2SC167		
2051	2N542A	2051	2S131	2051	2SC134B	2051	2SC1675		
2051	2N543	2051	2S363	2051	2SC134C	2051	2SC1675K		
2051	2N543A	2051	2S501	2051	2SC134D	2051	2SC1675L		
2051	2N5456	2051	2S502	2051	2SC134E	2051	2SC1675M		
2051	2N546	2051	2S503	2051	2SC134F	2051	2SC16A		
2051	2N547	2051	2S645	2051	2SC134G	2051	2SC1739		
2051	2N548	2051	2S701	2051	2SC134GN	2051	2SC1740		
2051	2N549	2051	2S702	2051	2SC134J	2051	2SC1740L		
2051	2N550	2051	2S703	2051	2SC134K	2051	2SC1740P		
2051	2N551	2051	2S711	2051	2SC134L	2051	2SC1740Q		
2051	2N552	2051	2S712	2051	2SC134M	2051	2SC1740QH		
2051	2N5581	2051	2S731	2051	2SC134OR	2051	2SC1740QJ		
2051	2N5582	2051	2S732	2051	2SC134R	2051	2SC1740OR		
2051	2N560	2051	2S733	2051	2SC134X	2051	2SC1740HH		
2051	2N5735	2051	2S741A	2051	2SC134Y	2051	2SC1740B		
2051	2N5736	2051	2S744A	2051	2SC135	2051	2SC1766		
2051	2N5769	2051	2S95A	2051	2SC135A	2051	2SC1766C		
2051	2N5810	2051	2SB645E	2051	2SC135B	2051	2SC1775		
2051	2N5814	2051	2SC-P6	2051	2SC135C	2051	2SC1775E		
2051	2N5816	2051	2SC100	2051	2SC135D	2051	2SC1775F		
2051	2N5818	2051	2SC100-OY	2051	2SC135E	2051	2SC18		
2051	2N5830	2051	2SC1007	2051	2SC135F	2051	2SC1815		
2051	2N5845	2051	2SC1023	2051	2SC135G	2051	2SC1815-0		
2051	2N5845A	2051	2SC1023(O)	2051	2SC135GN	2051	2SC1815GR		
2051	2N5961	2051	2SC1023(0)	2051	2SC135H	2051	2SC1815Y		
2051	2N5998	2051	2SC1023-0	2051	2SC135J	2051	2SC1815YW		
2051	2N6000	2051	2SC1023-Y	2051	2SC135L	2051	2SC1853		
2051	2N6008	2051	2SC1023A	2051	2SC135M	2051	2SC191		
2051	2N622	2051	2SC1023B	2051	2SC135OR	2051	2SC192		
2051	2N696	2051	2SC1023C	2051	2SC135R	2051	2SC193		
2051	2N697A	2051	2SC1023D	2051	2SC135X	2051	2SC194		
2051	2N701	2051	2SC1023E	2051	2SC135Y	2051	2SC195		
2051	2N706/46	2051	2SC1023F	2051	2SC136	2051	2SC1959Y		
2051	2N706/51	2051	2SC1023G	2051	2SC1361	2051	2SC196		
2051	2N706A/46	2051	2SC1023GN	2051	2SC1363	2051	2SC197		
2051	2N706A/51	2051	2SC1023H	2051	2SC1364	2051	2SC199A		
2051	2N706B/46	2051	2SC1023J	2051	2SC1364-6	2051	2SC199B		
2051	2N706B/51	2051	2SC1023K	2051	2SC1364-OR	2051	2SC199C		
2051	2N707A	2051	2SC1023M	2051	2SC1364A	2051	2SC199D		
2051	2N708	2051	2SC1023OR	2051	2SC1364B	2051	2SC199E		
2051	2N708/46	2051	2SC1023R	2051	2SC1364C	2051	2SC199F		
2051	2N708A	2051	2SC1023X	2051	2SC1364D	2051	2SC199G		
2051	2N709A46	2051	2SC1023Y	2051	2SC1364D	2051	2SC199GN		
2051	2N715	2051	2SC103	2051	2SC1364E	2051	2SC199H		
2051	2N716	2051	2SC1033	2051	2SC1364H	2051	2SC199J		
2051	2N728	2051	2SC1033A	2051	2SC1364K	2051	2SC199K		
2051	2N729	2051	2SC1033B	2051	2SC1364L	2051	2SC199L		
2051	2N730	2051	2SC1033D	2051	2SC1364M	2051	2SC199M		
2051	2N731	2051	2SC1033E	2051	2SC1364OR	2051	2SC199OR		
2051	2N734	2051	2SC1033F	2051	2SC1364R	2051	2SC199R		
2051	2N734A	2051	2SC1033G	2051	2SC1364X	2051	2SC199X		
2051	2N735			2051	2SC1364Y	2051	2SC199Y		
2051	2N735A			2051	2SC136D	2051	2SC200		

RS 276-	Industry Standard No.	RS 276-	Industry Standard No.	RS 276-	Industry Standard No.	RS 276-	Industry Standard No.
2051	28C2000	2051	28C30B	2051	28C363H	2051	28C405J
2051	28C2000L	2051	28C30C	2051	28C363J	2051	28C405K
2051	28C201	2051	28C30D	2051	28C363K	2051	28C405L
2051	28C202	2051	28C30E	2051	28C363L	2051	28C405M
2051	28C2021	2051	28C30F	2051	28C363M	2051	28C405OR
2051	28C2021Q	2051	28C30G	2051	28C363OR	2051	28C405R
2051	28C2021R	2051	28C30GN	2051	28C363R	2051	28C405X
2051	28C2021S	2051	28C30H	2051	28C363X	2051	28C405Y
	0028C203	2051	28C30J	2051	28C363Y	2051	28C406A
2051	28C204	2051	28C30L	2051	28C366	2051	28C406B
2051	28C205	2051	28C30M	2051	28C367	2051	28C406C
2051	28C2076	2051	28C300R	2051	28C369	2051	28C406D
2051	28C2076B	2051	28C30R	2051	28C369BL	2051	28C406E
2051	28C2076C	2051	28C30X	2051	28C369G	2051	28C406F
2051	28C2076CB	2051	28C30Y	2051	28C369G-BL	2051	28C406G
2051	28C2076CD	2051	28C317	2051	28C369G-OR	2051	28C406GN
2051	28C2076D	2051	28C317C	2051	28C369GBL	2051	28C406H
2051	28C211A	2051	28C318	2051	28C369GGR	2051	28C406J
2051	28C211B	2051	28C318A	2051	28C369GR	2051	28C406K
2051	28C211C	2051	28C319	2051	28C369V	2051	28C406L
2051	28C211D	2051	28C319A	2051	28C37	2051	28C406M
2051	28C211F	2051	28C319B	2051	28C371(O)	2051	28C406OR
2051	28C211G	2051	28C319C	205	28C371(O)	2051	28C406R
2051	28C211GN	2051	28C319D	2051	28C371-O	2051	28C406X
2051	28C211H	2051	28C319F	2051	28C371-O	2051	28C406Y
2051	28C211J	2051	28C319G	2051	28C371-OR	2051	28C423
2051	28C211K	2051	28C319GN	2051	28C371-ORG-G	2051	28C423-O
2051	28C211L	2051	28C319H	2051	28C371-R	2051	28C423A
2051	28C211M	2051	28C319J	2051	28C371-R-1	2051	28C423B
2051	28C211OR	2051	28C319K	2051	28C371-RED-G	2051	28C423C
2051	28C211R	2051	28C319L	2051	28C371-T	2051	28C423D
2051	28C211X	2051	28C319M	2051	28C371O	2051	28C423E
2051	28C211Y	2051	28C319OR	2051	28C371A	2051	28C423F
2051	28C212A	2051	28C319R	2051	28C371B	2051	28C423G
2051	28C212B	2051	28C319X	2051	28C371C	2051	28C423H
2051	28C212C	2051	28C319Y	2051	28C371D	2051	28C423J
2051	28C212D	2051	28C320	2051	28C371E	2051	28C423K
2051	28C212E	2051	28C320A	2051	28C371F	2051	28C423L
2051	28C212F	2051	28C320B	2051	28C371G	2051	28C423M
2051	28C212G	2051	28C320C	2051	28C371G-Q	2051	28C423OR
2051	28C212GN	2051	28C320D	2051	28C371G-R	2051	28C423R
2051	28C212H	2051	28C320E	2051	28C371GN	2051	28C423X
2051	28C212J	2051	28C320F	2051	28C371H	2051	28C423Y
2051	28C212K	2051	28C320G	2051	28C371J	2051	28C424
2051	28C212L	2051	28C320GN	2051	28C371K	2051	28C424D
2051	28C212M	2051	28C320H	2051	28C371L	2051	28C425
2051	28C212OR	2051	28C320J	2051	28C371M	2051	28C425B
2051	28C212R	2051	28C320K	2051	28C371O	2051	28C425C
2051	28C212X	2051	28C320L	2051	28C371OR	2051	28C425D
2051	28C212Y	2051	28C320M	2051	28C371R	2051	28C425E
2051	28C218	2051	28C320OR	2051	28C371R-1	2051	28C425P
2051	28C218A	2051	28C320R	2051	28C371RED-G	2051	28C426
2051	28C230	2051	28C320X	2051	28C371T	2051	28C427
2051	28C237	2051	28C320Y	2051	28C371Y	2051	28C428
2051	28C237A	2051	28C321	2051	28C37J24	2051	28C433
2051	28C237B	2051	28C321A	2051	28C37J20	2051	28C440
2051	28C237C	2051	28C321B	2051	28C373	2051	28C441
2051	28C237D	2051	28C321C	2051	28C377	2051	28C442
2051	28C237E	2051	28C321D	2051	28C377-BN	2051	28C45
2051	28C237F	2051	28C321E	2051	28C377-BRN	2051	28C454
2051	28C237G	2051	28C321F	2051	28C377-O	2051	28C454(A)
2051	28C237GN	2051	28C321G	2051	28C377-OR	2051	28C454(B)
2051	28C237H	2051	28C321GN	2051	28C377-ORG	2051	28C454-3
2051	28C237J	2051	28C321H	2051	28C377-R	2051	28C454-5
2051	28C237K	2051	28C321HA	2051	28C377-RED	2051	28C454-OR
2051	28C237L	2051	28C321HB	2051	28C377A	2051	28C454A
2051	28C237M	2051	28C321HC	2051	28C377B	2051	28C454B
2051	28C237OR	2051	28C321J	2051	28C377BN	2051	28C454B-6
2051	28C237R	2051	28C321K	2051	28C377BRN	2051	28C454BL
2051	28C237X	2051	28C321L	2051	28C377C	2051	28C454C
2051	28C237Y	2051	28C321M	2051	28C377D	2051	28C454D
2051	28C239	2051	28C321OR	2051	28C377E	2051	28C454E
2051	28C239A	2051	28C321R	2051	28C377F	2051	28C454F
2051	28C239B	2051	28C321X	2051	28C377G	2051	28C454G
2051	28C239C	2051	28C321Y	2051	28C377GN	2051	28C454GN
2051	28C239D	2051	28C323	2051	28C377H	2051	28C454H
2051	28C239E	2051	28C324	2051	28C377J	2051	28C454J
2051	28C239F	2051	28C324A	2051	28C377K	2051	28C454K
2051	28C239G	2051	28C324H	2051	28C377L	2051	28C454L
2051	28C239GN	2051	28C324HA	2051	28C377M	2051	28C454LA
2051	28C239H	2051	28C328A	2051	28C377O	2051	28C454M
2051	28C239J	2051	28C32D	2051	28C377OR	2051	28C454OR
2051	28C239K	2051	28C33	2051	28C377R	2051	28C454R
2051	28C239L	2051	28C350	2051	28C377RED	2051	28C454X
2051	28C239M	2051	28C350H	2051	28C377X	2051	28C454Y
2051	28C239OR	2051	28C356	2051	28C377Y	2051	28C458B
2051	28C239R	2051	28C356A	2051	28C378	2051	28C458C
2051	28C239X	2051	28C356B	2051	28C379	2051	28C458LGC
2051	28C239Y	2051	28C356C	2051	28C38	2051	28C459A
2051	28C249A	2051	28C356D	2051	28C395	2051	28C459C
2051	28C249B	2051	28C356E	2051	28C395A	2051	28C459D
2051	28C249C	2051	28C356F	2051	28C395A-O	2051	28C459B
2051	28C249D	2051	28C356G	2051	28C395A-ORG	2051	28C459F
2051	28C249E	2051	28C356GN	2051	28C395A-R	2051	28C459G
2051	28C249F	2051	28C356H	2051	28C395A-RED	2051	28C459GN
2051	28C249G	2051	28C356J	2051	28C395A-Y	2051	28C459H
2051	28C249GN	2051	28C356K	2051	28C395A-YEL	2051	28C459J
2051	28C249H	2051	28C356L	2051	28C395B	2051	28C459K
2051	28C249J	2051	28C356M	2051	28C395C	2051	28C459L
2051	28C249K	2051	28C356OR	2051	28C395D	2051	28C459M
2051	28C249L	2051	28C356X	2051	28C395E	2051	28C459OR
2051	28C249M	2051	28C356Y	2051	28C395F	2051	28C459R
2051	28C249OR	2051	28C360	2051	28C395G	2051	28C459X
2051	28C249R	2051	28C360-OR	2051	28C395GN	2051	28C459Y
2051	28C249X	2051	28C360A	2051	28C395H	2051	000028C460
2051	28C249Y	2051	28C360B	2051	28C395J	2051	28C460(A)
2051	28C26	2051	28C360C	2051	28C395K	2051	28C460(B)
2051	28C264	2051	28C360D	2051	28C395L	2051	28C460-5
2051	28C265	2051	28C360E	2051	28C395M	2051	28C460-B
2051	28C267	2051	28C360F	2051	28C395OR	2051	28C460-C
2051	28C267A	2051	28C360G	2051	28C395R	2051	28C460-OR
2051	28C28	2051	28C360GN	2051	28C395X	2051	28C460B
2051	28C280AO	2051	28C360H	2051	28C395Y	2051	28C460D
2051	28C2818	2051	28C360J	2051	28C396	2051	28C460E
2051	28C282	2051	28C360K	2051	28C3Y	2051	28C460F
2051	28C282H	2051	28C360L	2051	28C400	2051	28C460G
2051	28C282HA	2051	28C360M	2051	28C400-O	2051	28C460GB
2051	28C282HB	2051	28C360OR	2051	28C400-R	2051	28C460GN
2051	28C282HC	2051	28C360R	2051	28C400-Y	2051	28C460H
2051	28C283	2051	28C360X	2051	28C401	2051	28C460J
2051	28C284	2051	28C360Y	2051	28C402	2051	28C460K
2051	28C284H	2051	28C363	2051	28C402A	2051	28C460L
2051	28C284HA	2051	28C363-OR	2051	28C404	2051	28C460M
2051	28C284HB	2051	28C363A	2051	28C405A	2051	28C460OR
2051	28C284HC	2051	28C363B	2051	28C405B	2051	28C460R
2051	28C285	2051	28C363C	2051	28C405C	2051	28C460X
2051	28C285A	2051	28C363D	2051	28C405D	2051	28C460Y
2051	28C29	2051	28C363E	2051	28C405E	2051	28C462
2051	28C30	2051	28C363F	2051	28C405F	2051	28C468
2051	28C30-OR	2051	28C363G	2051	28C405GN	2051	28C468A
2051	28C300	2051	28C363GN	2051	28C405H	2051	28C47
2051	28C301					2051	28C474
2051	28C302					2051	28C475
2051	28C30A						

RS 276-	Industry Standard No.	RS 276-	Industry Standard No.	RS 276-	Industry Standard No.	RS 276-	Industry Standard No.
2051	2SC475K	2051	2SC634	2051	2SC716C	2051	2SC829GN
2051	2SC476	2051	2SC634(2)	2051	2SC716D	2051	2SC829H
2051	2SC48	2051	2SC634-0	2051	2SC716E	2051	2SC829K
2051	2SC48C	2051	2SC634-OR	2051	2SC716F	2051	2SC829L
2051	2SC52	2051	2SC634A	2051	2SC716G	2051	2SC829M
2051	2SC529	2051	2SC634AK	2051	2SC723	2051	2SC829OR
2051	2SC529A	2051	2SC634AJ	2051	2SC723BL	2051	2SC829R
2051	2SC53	2051	2SC634AL	2051	2SC725	2051	2SC829X
2051	2SC531	2051	2SC634AXL	2051	2SC725-Q	2051	2SC829Y
2051	0002SC531P	2051	2SC634B	2051	2SC728A	2051	2SC833
2051	2SC537	2051	2SC634C	2051	2SC728B	2051	2SC833BL
2051	2SC538	2051	2SC634D	2051	2SC728C	2051	2SC834L
2051	2SC538(P)	2051	2SC634E	2051	2SC728D	2051	2SC836M
2051	2SC538(R)	2051	2SC634F	2051	2SC728E	2051	2SC838
2051	2SC538-Q	2051	2SC634G	2051	2SC728F	2051	2SC838(A)
2051	2SC538A	2051	2SC634GN	2051	2SC728G	2051	2SC838(B)
2051	2SC538A(Q)	2051	2SC634J	2051	2SC728GN	2051	2SC838(F)
2051	2SC538A-P	2051	2SC634K	2051	2SC728H	2051	2SC838(H)
2051	2SC538A-R	2051	2SC634L	2051	2SC728J	2051	2SC838(J)
2051	2SC538AQ	2051	2SC634M	2051	2SC728K	2051	2SC838(K)
2051	2SC538AR	2051	2SC634OR	2051	2SC728L	2051	2SC838(M)
2051	2SC538AB	2051	2SC634R	2051	2SC728OR	2051	2SC838(O)
2051	2SC538K	2051	2SC634X	2051	2SC728X	2051	2SC838-2
2051	2SC538P	2051	2SC634Y	2051	2SC728Y	2051	2SC838-O
2051	2SC538Q	2051	2SC635	2051	2SC735-BL	2051	2SC838A
2051	2SC538R	2051	2SC635A	2051	2SC735	2051	2SC838B
2051	2SC538S	2051	2SC640	2051	00-2SC735-OY	2051	2SC838BL
2051	2SC538T	2051	2SC640B	2051	2SC7354	2051	2SC838C
2051	2SC539	2051	2SC641A	2051	000-2SC7350Y	2051	2SC838D
2051	2SC539K	2051	2SC641C	2051	2SC737	2051	2SC838E
2051	2SC539L	2051	2SC641D	2051	2SC737Y	2051	2SC838F
2051	2SC539R	2051	2SC641E	2051	2SC752	2051	2SC838H
2051	2SC539S	2051	2SC641F	2051	2SC752-ORG-G	2051	2SC838HF
2051	2SC54	2051	2SC641G	2051	2SC752-RED-G	2051	2SC838J
2051	2SC540	2051	2SC641GN	2051	2SC752-YEL-G	2051	2SC838K
2051	2SC55	2051	2SC641H	2051	2SC752A	2051	2SC838L
2051	2SC566	2051	2SC641J	2051	2SC752B	2051	2SC838M
2051	2SC587	2051	2SC641K	2051	2SC752C	2051	2SC838S
2051	2SC587A	2051	2SC641L	2051	2SC752D	2051	2SC847
2051	2SC588	2051	2SC641M	2051	2SC752E	2051	2SC848
2051	2SC593	2051	2SC641OR	2051	2SC752F	2051	2SC849
2051	2SC595	2051	2SC641X	2051	2SC752G	2051	2SC850
2051	2SC596	2051	2SC641Y	2051	2SC752-O	2051	2SC858A
2051	2SC619	2051	2SC654	2051	2SC752-0-R	2051	2SC858B
2051	2SC619(B)	2051	2SC655	2051	2SC752-0-Y	2051	2SC858C
2051	2SC619(C)	2051	2SC67	2051	2SC752GA	2051	2SC858D
2051	2SC619A	2051	2SC68	2051	2SC752H	2051	2SC858E
2051	2SC619B	2051	2SC689	2051	2SC752J	2051	2SC858F
2051	2SC619C	2051	2SC689H	2051	2SC752K	2051	2SC858FG
2051	2SC619D	2051	2SC702	2051	2SC752L	2051	2SC858G
2051	2SC619E	2051	2SC705TV(3RD-IP)	2051	2SC752M	2051	2SC858GA
2051	2SC619F	2051	2SC709	2051	2SC752OR	2051	2SC858GN
2051	2SC619G	2051	2SC709(B)(C)	2051	2SC752R	2051	2SC858H
2051	2SC619GN	2051	2SC709(C)	2051	2SC752X	2051	2SC858K
2051	2SC619H	2051	2SC709A	2051	2SC752Y	2051	2SC858L
2051	2SC619J	2051	2SC709B	2051	2SC773	2051	2SC858M
2051	2SC619K	2051	2SC709C	2051	2SC773(E)	2051	2SC858OR
2051	2SC619L	2051	2SC709CD	2051	2SC773A	2051	2SC858R
2051	2SC619M	2051	2SC709D	2051	2SC773B	2051	2SC858X
2051	2SC619R	2051	2SC709E	2051	2SC773C	2051	2SC859R
2051	2SC619X	2051	2SC709F	2051	2SC773D	2051	2SC87
2051	2SC619Y	2051	2SC709G	2051	2SC773E	2051	2SC870
2051	2SC62	2051	2SC709GN	2051	2SC773F	2051	000-2SC870A
2051	2SC620	2051	2SC709H	2051	2SC773G	2051	000-2SC870B
2051	2SC620(C)	2051	2SC709J	2051	2SC773GN	2051	2SC870BL
2051	2SC620(D)	2051	2SC709L	2051	2SC773H	2051	2SC870C
2051	2SC620-OR	2051	2SC709M	2051	2SC773J	2051	2SC870D
2051	2SC620A	2051	2SC709OR	2051	2SC773K	2051	2SC870F
2051	2SC620B	2051	2SC709X	2051	2SC773L	2051	2SC870G
2051	2SC620C	2051	2SC709Y	2051	2SC773M	2051	2SC870GN
2051	2SC620CD	2051	2SC710BB	2051	2SC773OR	2051	2SC870H
2051	2SC620D	2051	0002SC710B	2051	2SC773R	2051	2SC870J
2051	2SC620DE	2051	0002SC710C	2051	2SC773X	2051	2SC870K
2051	2SC620E	2051	2SC712	2051	2SC773Y	2051	2SC870L
2051	2SC620F	2051	2SC712(D)	2051	2SC796	2051	2SC870M
2051	2SC620G	2051	2SC712-CD	2051	2SC797A	2051	2SC870OR
2051	2SC620GN	2051	2SC712A	2051	2SC797B	2051	2SC870X
2051	2SC620H	2051	2SC712B	2051	2SC797C	2051	2SC870Y
2051	2SC620J	2051	2SC712C	2051	2SC797D	2051	2SC871
2051	2SC620K	2051	2SC712CD	2051	2SC797E	2051	2SC871-G
2051	2SC620L	2051	2SC712D	2051	2SC797F	2051	2SC871A
2051	2SC620M	2051	2SC712DC	2051	2SC797G	2051	2SC871AM
2051	2SC620OR	2051	2SC712E	2051	2SC797GN	2051	2SC871B
2051	2SC620X	2051	2SC712F	2051	2SC797H	2051	2SC871BL
2051	2SC620Y	2051	2SC712G	2051	2SC797J	2051	2SC871C
2051	2SC621	2051	2SC712GN	2051	2SC797K	2051	2SC871D
2051	2SC622	2051	2SC712H	2051	2SC797L	2051	2SC871E
2051	2SC626	2051	2SC712J	2051	2SC797M	2051	2SC871F
2051	2SC628E	2051	2SC712K	2051	2SC797OR	2051	2SC871G
2051	2SC628F	2051	2SC712L	2051	2SC797R	2051	2SC871GN
2051	2SC631	2051	2SC712M	2051	2SC797X	2051	2SC871H
2051	2SC631A	2051	2SC712OR	2051	2SC814A	2051	2SC871J
2051	2SC632	2051	2SC712R	2051	2SC814B	2051	2SC871K
2051	2SC632(1)	2051	2SC712W	2051	2SC814C	2051	2SC871L
2051	2SC632-OR	2051	2SC712X	2051	2SC814D	2051	2SC871M
2051	2SC632A	2051	2SC712Y	2051	2SC814E	2051	2SC871OR
2051	2SC632B	2051	2SC713	2051	2SC814F	2051	2SC871R
2051	2SC632C	2051	2SC713A	2051	2SC814GN	2051	2SC871X
2051	2SC632D	2051	2SC713B	2051	2SC814H	2051	2SC871Y
2051	2SC632E	2051	2SC713C	2051	2SC814J	2051	2SC894
2051	2SC632F	2051	2SC713D	2051	2SC814K	2051	2SC894A
2051	2SC632G	2051	2SC713E	2051	2SC814M	2051	2SC894B
2051	2SC632GN	2051	2SC713F	2051	2SC814OR	2051	2SC894C
2051	2SC632H	2051	2SC713G	2051	2SC814R	2051	2SC894D
2051	2SC632J	2051	2SC713GN	2051	2SC814X	2051	2SC894E
2051	2SC632K	2051	2SC713H	2051	2SC814Y	2051	2SC894F
2051	2SC632L	2051	2SC713J	2051	2SC815S	2051	2SC894G
2051	2SC632M	2051	2SC713K	2051	2SC828	2051	2SC894GN
2051	2SC632OR	2051	2SC713L	2051	2SC828H	2051	2SC894H
2051	2SC632R	2051	2SC713M	2051	2SC828Q	2051	2SC894J
2051	2SC632X	2051	2SC713OR	2051	2SC829(Y)	2051	2SC894K
2051	2SC632Y	2051	2SC713R	2051	2SC829-OR	2051	2SC894L
2051	2SC633	2051	2SC713X	2051	2SC829/4454C	2051	2SC894M
2051	2SC633-7	2051	2SC713Y	2051	2SC829	2051	2SC894OR
2051	2SC633-OR	2051	2SC714	2051	2SC829A	2051	2SC894X
2051	2SC633A	2051	2SC714A	2051	2SC829AK	2051	2SC894Y
2051	2SC633B	2051	2SC714B	2051	2SC829B	2051	2SC896
2051	2SC633C	2051	2SC714C	2051	2SC829B/4454C	2051	2SC899
2051	2SC633D	2051	2SC714D	2051	2SC829BG	2051	2SC899A
2051	2SC633E	2051	2SC714E	2051	2SC829BJ	2051	2SC899B
2051	2SC633F	2051	2SC714F	2051	2SC829BK	2051	2SC899C
2051	2SC633G	2051	2SC714G	2051	2SC829BY	2051	2SC899D
2051	2SC633GN	2051	2SC714GN	2051	2SC829C	2051	2SC899F
2051	2SC633H	2051	2SC714H	2051	2SC829CL	2051	2SC899G
2051	2SC633J	2051	2SC714J	2051	2SC829D	2051	2SC899GN
2051	2SC633K	2051	2SC714K	2051	2SC829E	2051	2SC899H
2051	2SC633L	2051	2SC714L	2051	2SC829F	2051	2SC899J
2051	2SC633M	2051	2SC714M	2051	2SC829G	2051	2SC899K
2051	2SC633OR	2051	2SC714OR				2SC899L
2051	2SC633R	2051	2SC714R				
2051	2SC633X	2051	2SC714X				
2051	2SC633Y	2051	2SC714Y				
		2051	2SC716				
		2051	2SC716B				

RS 276-	Industry Standard No.	RS 276-	Industry Standard No.	RS 276-	Industry Standard No.	RS 276-	Industry Standard No.
2051	280899M	2051	280984M	2051	04-01585-06	2051	7-6(SARKES)
2051	280899O	2051	280984OR	2051	04-01585-07	2051	7-7(SARKES)
2051	280899R	2051	280984R	2051	04-02090-02	2051	007-74655-02
2051	280899X	2051	280984X	2051	4-12-1A7-1	2051	007-74655-06
2051	280899Y	2051	280984Y	2051	4-1545	2051	007-74661-01
2051	28C9011N	2051	28098A	2051	4-1790	2051	007-74661 01
2051	28C9011F	2051	28098B	2051	4-3023212	2051	7-8(SARKES)
2051	28C9011H	2051	28098C	2051	4-3023221	2051	7A30(SHERWOOD)
2051	280906	2051	28098D	2051	4-3025766	2051	7A31(SHERWOOD)
2051	280906P	2051	28098E	2051	4-47(SEARS)	2051	8-0024-3
2051	280913	2051	28098F	2051	4-5145	2051	8-00243
2051	280922	2051	28098G	2051	4-686132-3	2051	8-0050100
2051	280922A	2051	28098GN	2051	4-686143-3	2051	8-0051500
2051	280922B	2051	28098H	2051	4-686144-3	2051	8-005202
2051	280922C	2051	28098J	2051	4-686173-3	2051	8-0052102
2051	280922K	2051	28098K	2051	4-686183-3	2051	8-0052502
2051	280922L	2051	28098L	2051	4-686231-3	2051	8-0052600
2051	280922M	2051	28098M	2051	4-686257-3	2051	8-0053001
2051	280925	2051	28098OR	2051	4-68682-3	2051	8-0053400
2051	280933	2051	28098R	2051	4-B51	2051	8-0318250
2051	280933(D)	2051	28098X	2051	4C43	2051	8-0337390
2051	280933(P)	2051	28098Y	2051	4D20	2051	8-0383940
2051	280933(G)	2051	280991	2051	4D21	2051	8-0389910
2051	280933A	2051	280992	2051	4D22	2051	8-0389930
2051	280933B	2051	280993E	2051	4D24	2051	8-0421980
2051	280933BB	2051	280994	2051	4D25	2051	8-2409501
2051	280933C	2051	28099A	2051	4D26	2051	8-4(BENDIX)
2051	280933D	2051	28099B	2051	4JX16A567	2051	8-697-020-570
2051	280933D(F)	2051	28099C	2051	4JX16A667	2051	8-724-733-30
2051	280933E	2051	28099D	2051	4JX16A667/G	2051	8-729-663-47
2051	280933E(F)	2051	28099E	2051	4JX16A667/O	2051	8-81250109
2051	280933F	2051	28099F	2051	4JX16A667/R	2051	8-902-0706-071
2051	280933FP	2051	28099G	2051	4JX16A667/Y	2051	8-905-014-017
2051	280933FPC	2051	28099GN	2051	4JX16A667O	2051	8-905-705-112
2051	280933FPD	2051	28099H	2051	4JX16A667O	2051	8-905-705-403
2051	280933FPB	2051	28099J	2051	4JX16A667O	2051	8-905-705-405
2051	280933FPF	2051	28099K	2051	4JX16A667R	2051	8-905-706-104
2051	280933FPG	2051	28099L	2051	4JX16A667Y	2051	8-905-706-201
2051	280933G	2051	28099M	2051	4JX16A668	2051	8-905-706-202
2051	280933GN	2051	28099OR	2051	4JX16A668/G	2051	8-905-706-203
2051	280933H	2051	28099R	2051	4JX16A668/O	2051	8-905-706-206
2051	280933J	2051	28099X	2051	4JX16A668/Y	2051	8-905-706-208
2051	280933K	2051	28099Y	2051	4JX16A668O	2051	8-905-706-211
2051	280933L	2051	28CF-2	2051	4JX16A668G	2051	8-905-706-215
2051	280933M	2051	28CP2	2051	4JX16A668Y	2051	8-905-706-235
2051	280933OR	2051	28CP5	2051	4JX16A669	2051	8-905-706-236
2051	280933R	2051	28CP5A	2051	4JX16A669G	2051	8-905-706-238
2051	280933X	2051	28CP5B	2051	4JX16A669Y	2051	8-905-706-239
2051	280933Y	2051	28CP5C	2051	4JX16A670	2051	8-905-706-240
2051	280934	2051	28CP5D	2051	4JX16A670G	2051	8-905-706-242
2051	280934-0	2051	28CP5E	2051	4JX16B670/B	2051	8-905-706-245
2051	280934A	2051	28CP5F	2051	4JX16B670/R	2051	8-905-706-246
2051	280934B	2051	28CP5G	2051	4JX16B670/Y	2051	8-905-706-250
2051	280934C	2051	28CP5GN	2051	4JX16B670B	2051	8-905-706-257
2051	280934D	2051	28CP5H	2051	4JX16B670G	2051	8-905-706-260
2051	280934E	2051	28CP5J	2051	4JX16B670R	2051	8-905-706-263
2051	280934F	2051	28CP5K	2051	4JX16B670Y	2051	8-905-706-336
2051	280934G	2051	28CP5L	2051	4JX16E3860	2051	8-905-706-606
2051	280934GN	2051	28CP5M	2051	4JX16E3890	2051	8-905-707-254
2051	280934H	2051	28CP5OR	2051	4JX16E3960	2051	8-905-707-265
2051	280934J	2051	28CP5R	2051	4JX18596	2051	8-905-707-315
2051	280934K	2051	28CP5X	2051	4JX7A972	2051	8A12789
2051	280934L	2051	28CP5Y	2051	005-02	2051	08A8300-2
2051	280934M	2051	28CS183E	2051	5-70004503	2051	8Q-3-12
2051	280934OR	2051	28CS184E	2051	5-70005452	2051	09-002012
2051	280934P	2051	28D134	2051	5-70005503	2051	9-006
2051	280934Q	2051	28D228A	2051	5-7000901504	2051	09-302007
2051	280934R	2051	28D228B	2051	5-8	2051	09-302012
2051	280934X	2051	28D228C	2051	006-0000134	2051	09-302019
2051	280934Y	2051	28D228D	2051	6-04	2051	09-302033
2051	280938	2051	28D228E	2051	6-0451	2051	09-302034
2051	280938A	2051	28D228F	2051	6-0452	2051	09-302039
2051	280938B	2051	28D228G	2051	6-04GRN	2051	09-302041
2051	280938C	2051	28D228GN	2051	6-040RN	2051	09-302045
2051	280941	2051	28D228H	2051	6-040RNN	2051	09-302045-12
2051	280941(O),(R)	2051	28D228J	2051	6-0431	2051	09-302054
2051	280941-0	2051	28D228K	2051	6-0482	2051	09-302058
2051	280941-0Y	2051	28D228L	2051	6-05	2051	09-302062
2051	280941-R	2051	28D228M	2051	6-05P	2051	09-302074
2051	280941-Y	2051	28D228OR	2051	6-05TEL	2051	09-302078
2051	280941K	2051	28D228R	2051	6-11	2051	09-302101
2051	280941R	2051	28D228X	2051	6-19	2051	09-302106
2051	280941Y	2051	28D228Y	2051	6-30	2051	09-302118
2051	280943	2051	28D592	2051	6-4	2051	09-302124
2051	280943A	2051	28D592R	2051	006-6450032	2051	09-302131
2051	280943B	2051	28D636P	2051	6-6450036	2051	09-302140
2051	280943C	2051	28D636Q	2051	6-90	2051	09-302148
2051	280943D	2051	28D636R	2051	6-9029-15D	2051	09-302153
2051	280943E	2051	28D639	2051	6-9029-15E	2051	09-302165
2051	280943F	2051	28D5360N	2051	6-93	2051	09-302172
2051	280943G	2051	2T172	2051	6A10227	2051	09-302175
2051	280943GN	2051	2T202	2051	6A10422	2051	09-302189
2051	280943H	2051	2T270B	2051	6A10423	2051	09-302191
2051	280943J	2051	2T2785	2051	6A10520	2051	09-302204
2051	280943K	2051	2T2857	2051	6A10851	2051	09-302215
2051	280943L	2051	2T40	2051	6A10855	2051	09-302227
2051	280943M	2051	2T402	2051	6A11180	2051	09-302244
2051	280943OR	2051	2T403	2051	6A12681	2051	09-303025
2051	280943R	2051	2T404	2051	6A12682	2051	09-304044
2051	280943X	2051	2T41	2051	6A12683	2051	09-304045
2051	280943Y	2051	2T42	2051	6A12725	2051	09-304048
2051	280944	2051	2T43	2051	6A12788	2051	09-304058
2051	280944K	2051	2T44	2051	6A12789	2051	09-305034
2051	280966	2051	2T918	2051	6A16399	2051	09-305062
2051	280967	2051	3-0033	2051	68A690/56-0001	2051	09-305063
2051	280968	2051	03-1585/Q	2051	6X97047A01	2051	09-305064
2051	280968A	2051	03A05	2051	7-0015	2051	09-305065
2051	280968B	2051	03A11	2051	07-07124	2051	09-305066
2051	280968C	2051	03A12	2051	07-07125	2051	09-305067
2051	280968D	2051	3E-1	2051	07-07139	2051	09-305068
2051	280968E	2051	3E-2	2051	07-07156	2051	09-305077
2051	280968F	2051	3E-3	2051	7-1(SARKES)	2051	09-305139
2051	280968G	2051	3I4-5007-3	2051	07-1458-85	2051	09-305148
2051	280968GN	2051	3I4-6007-02	2051	7-15(SARKES)	2051	09-305152
2051	280968H	2051	3I4-6007-03	2051	7-16(SARKES)	2051	09-309006
2051	280968J	2051	3I4-6007-08	2051	7-17	2051	09-309012
2051	280968K	2051	3I4-6007-09	2051	7-17(SARKES)	2051	09-309023
2051	280968L	2051	3I4-6007-1	2051	7-18(SARKES)	2051	09-309049
2051	280968M	2051	3I4-6007-2	2051	7-19(SARKES)	2051	09-309050
2051	280968N	2051	3I4-6007-37	2051	7-2(SARKES)	2051	09-309060
2051	280968OR	2051	3I4-6007-38	2051	7-3(SARKES)	2051	09-309064
2051	280968R	2051	3I4-6010-03	2051	7-4(SARKES)	2051	09-309076
2051	280968X	2051	3I4-6010-3	2051	7-5	2051	9-5216
2051	280984	2051	3I4-6010-6	2051	7-5(SARKES)	2051	9-5221
2051	280984A	2051	3I4-6015-01	2051	7-59-019/3477	2051	9-5225
2051	280984B	2051	3I46007-1	2051	7-59-020/3477	2051	9-5226-2
2051	280984C	2051	3I46007-2	2051	7-59-021/3477	2051	9-5227
2051	280984D	2051	3N35	2051	7-59-022/3477	2051	9-5296
2051	280984E	2051	38004	2051	7-59-023/3477	2051	9-9109-1
2051	280984F	2051	004-00	2051	7-59-024/3477	2051	9-9109-2
2051	280984G	2051	004-00(LAST-IF)	2051	7-59-024 3477	2051	9.3037
2051	280984GN	2051	04-00460-03	2051	7-59-068	2051	9QR2
2051	280984H					2051	09N1
2051	280984L						

RS 276-	Industry Standard No.	RS 276-	Industry Standard No.	RS 276-	Industry Standard No.	RS 276-	Industry Standard No.
2051	98037	2051	13-34381	2051	019-003349	2051	22-001007
2051	9TR10	2051	13-34381-2	2051	019-003675-196	2051	022-006500
2051	9TR2	2051	13-35226-1	2051	019-003675-203	2051	022-009600
2051	9TRZ1001-02	2051	13-35807-2	2051	019-003675-207	2051	23-5020
2051	9TR31001-03	2051	13-39114-1	2051	019-003675-246	2051	23-5021
2051	9TR7	2051	13-39114-2	2051	019-003932	2051	23-5022
2051	9TRZ1001-02	2051	13-47773-1	2051	019-003934	2051	23-5023
2051	10-080010	2051	13-55061-1	2051	019-004111	2051	23-5024
2051	11-0422	2051	13-55061-1(SOUND)	2051	019-004428-002	2051	23-5025
2051	11-0774	2051	13-55061-2	2051	019-005006	2051	23-5026
2051	11-0778	2051	13-55061-2(SYL)	2051	019-005021	2051	23-5027
2051	11-085010	2051	13-55066-1	2051	19-020-043	2051	23-5029
2051	012-1-12	2051	13-55066-2	2051	19-020-043A	2051	23-5033
2051	012-1-12-7	2051	13-55067-1	2051	19-020-058	2051	23-5052
2051	12-1-276	2051	13-55068-1	2051	19-020-067	2051	23-PE274-121
2051	12-1-277	2051	13-67583-6	2051	19-020-073	2051	23B114044
2051	12-1-278	2051	13-67585-4	2051	19-020-074	2051	23B001-1
2051	12-1-279	2051	13-67585-4/2439-2	2051	19-020071	2051	24-0003714-1
2051	12-1-70	2051	13-67585-5	2051	19-2-02616	2051	24-000451
2051	12-1-70-12	2051	13-67585-5/3464	2051	19A115061-P1	2051	24-000457
2051	12-1-70-12-7	2051	13-67585-7	2051	19A115061-P2	2051	24-000653-1
2051	12-10	2051	13-67585-7/2439-2	2051	19A115102-P1	2051	24-001327-1
2051	12-101001	2051	13-67586-3	2051	19A115108-P1	2051	24-002
2051	12-11	2051	13-67586-5/3464	2051	19A115108-P2	2051	24-602-25
2051	12-12	2051	13-68617-1	2051	19A115123-2	2051	24A
2051	12-13	2051	14 806 12	2051	19A115123-P1	2051	24B
2051	12-14	2051	14-0104-7	2051	19A115123-P2	2051	24B1
2051	12-15	2051	14-1	2051	19A115142-P1	2051	24MW1023
2051	12-16	2051	14-2	2051	19A115142-P2	2051	24MW1024
2051	12-17	2051	14-3	2051	19A115157-1	2051	24MW1059
2051	12-18	2051	14-572-10	2051	19A115167-2	2051	24MW1068
2051	12-19	2051	14-575-10	2051	19A115245-P1	2051	24MW1069
2051	12-1A	2051	14-583-01	2051	19A115245-P2	2051	24MW1089
2051	12-1AO	2051	14-601-28	2051	19A115253-P1	2051	24MW1096
2051	12-1AOR	2051	14-601-29	2051	19A115253-P2	2051	24MW1120
2051	12-1A1	2051	14-602-01	2051	19A115315-P1	2051	24MW1141
2051	12-1A19	2051	14-602-02	2051	19A115315-P2	2051	24MW1147
2051	12-1A2	2051	14-602-03	2051	19A115328-1	2051	24MW119
2051	12-1A21	2051	14-602-12	2051	19A115330	2051	24MW333
2051	12-1A3	2051	14-602-13	2051	19A115342-P1	2051	24MW372
2051	12-1A3P	2051	14-602-14	2051	19A115342-P2	2051	24MW454
2051	12-1A4	2051	14-602-16	2051	19A115359-P1	2051	24MW458
2051	12-1A4-7	2051	14-602-17	2051	19A115359-P2	2051	24MW460
2051	12-1A4-7B	2051	14-602-23	2051	19A115362-P1	2051	24MW461
2051	12-1A5	2051	14-602-26	2051	19A115410-P1	2051	24MW609
2051	12-1A5L	2051	14-602-35	2051	19A115410-P2	2051	24MW655
2051	12-1A6	2051	14-602-48	2051	19A115552-P1	2051	24MW658
2051	12-1A6A	2051	14-602-55A	2051	19A115552-P2	2051	24MW659
2051	12-1A7	2051	14-602-61	2051	19A115552P1	2051	24MW676
2051	12-1A7-1	2051	14-602-62	2051	19A115552P2	2051	24MW740
2051	12-1AB	2051	14-602-69	2051	19A115591P1	2051	24MW760
2051	12-1AB2	2051	14-602-78	2051	19A115591P2	2051	24MW773
2051	12-1A9	2051	14-602-80	2051	19A115720-1	2051	24MW774
2051	12-1A9Q	2051	14-602-81	2051	19A115720-2	2051	24MW775
2051	12-1AO	2051	14-602-87	2051	19A115728-1	2051	24MW776
2051	12-1AOR	2051	14-602-89	2051	19A115728-2	2051	24MW796
2051	12-4	2051	14-603-03	2051	19A115786	2051	24MW790
2051	12CLN	2051	14-603-04	2051	19A115786A	2051	24MW796
2051	012E	2051	14-603-10	2051	19A115910P1	2051	24MW797
2051	12X047	2051	14-603-11	2051	19A115910P2	2051	24MW801
2051	13-0021	2051	14-651-12	2051	19A115944P1	2051	24MW807
2051	13-0022	2051	14-655-13	2051	19A115944P2	2051	24MW808
2051	13-0024	2051	14-656-21	2051	19A116631P1	2051	24MW809
2051	13-0041	2051	14-659-12	2051	19A116755P1	2051	24MW817
2051	13-0048	2051	14-660-12	2051	19A116774-P1	2051	24MW818
2051	13-0058	2051	014-680	2051	19A116865	2051	24MW823
2051	13-0062-1	2051	014-686	2051	19A129207P1	2051	24MW854
2051	13-0321-10	2051	014-698	2051	19AR20	2051	24MW855
2051	13-0321-11	2051	014-784	2051	19AR36	2051	24MW874
2051	13-0321-12	2051	14-800-32	2051	19C300114-P1	2051	24MW899
2051	13-0321-5	2051	14-801-12	2051	19C300114P1	2051	24MW954
2051	13-0321-6	2051	14-802-12	2051	19C300114P2	2051	24MW961
2051	13-0321-7	2051	14-805-12	2051	19C300114P3	2051	24MW988
2051	13-0321-8	2051	14-806-12	2051	19C300141-4P2	2051	24MW992
2051	13-0321-81	2051	14-806-23	2051	020-1110-011	2051	24R
2051	13-0321-9	2051	14-807-23	2051	020-1110-017	2051	24T-015-013
2051	13-10321-59	2051	14-809-25	2051	020-1112-001	2051	25-0060-4
2051	13-10321-70	2051	14-809-32	2051	20A0053	2051	025-100018
2051	13-10321-76	2051	14-851-32	2051	20A0073	2051	025-100030
2051	13-14065-77	2051	14-853-23	2051	20A10849	2051	025-100040
2051	13-14085-122	2051	14-854-12	2051	021-0121-00	2051	025-10030
2051	13-14085-15	2051	14-858-12	2051	21-1	2051	25A1
2051	13-14085-34	2051	14-862-23	2051	21-1L	2051	25A1273-001
2051	13-14085-41	2051	14-862-32	2051	21A015-013	2051	25A2
2051	13-14085-50	2051	14-864-12	2051	21A015-020	2051	25B
2051	13-14085-54	2051	14-865-12	2051	21A015-027	2051	025B-YEL
2051	13-14085-6	2051	14-866-32	2051	21A040-020	2051	25B21
2051	13-14085-7	2051	15-01093	2051	21A040-032	2051	25DB58LGBM
2051	13-14085-72	2051	15-03014-00	2051	21A040-033	2051	25B2
2051	13-14085-83	2051	15-03100	2051	21A040-033A	2051	026-100017
2051	13-14085-84	2051	15-05302	2051	21A040-034	2051	026-100026
2051	13-14085-85	2051	15-05369	2051	21A040-037	2051	277409
2051	13-14085-89	2051	15-05650	2051	21A040-056	2051	277411
2051	13-14606-1	2051	15-082019	2051	21A040-077	2051	31-0007
2051	13-15804-1	2051	15-09358	2051	21A040-078	2051	31-0009
2051	13-15808-1	2051	15-09980	2051	21A040-092	2051	31-002-Q
2051	13-15840-1	2051	15-1	2051	21A040-37	2051	31-0068
2051	13-15840-2	2051	15-2	2051	21A112-013	2051	31-0069
2051	13-15865-1	2051	15-22223720	2051	21A112-015	2051	31-0080
2051	13-16769-1	2051	15-875-075-003	2051	21A112-017	2051	31-0081
2051	13-17-6(SEAR)	2051	16-147191229	2051	21A112-018	2051	31-0082
2051	13-17-6(SEARS)	2051	16-471119136B	2051	21A112-020	2051	31-0084
2051	13-18087-1	2051	16A1(FLEETWOOD)	2051	21A112-050	2051	31-0085
2051	13-18087-2	2051	16A1938	2051	21A112-085	2051	31-0104
2051	13-18158-1	2051	16A2(FLEETWOOD)	2051	21A112-088	2051	31-0106
2051	13-1836-1	2051	16A545-7	2051	21A112-089	2051	31-0115
2051	13-18363	2051	016B12	2051	21A112-090	2051	31-0116
2051	13-18363-1	2051	16B2	2051	21A112-091	2051	31-0177
2051	13-18363-1A	2051	016B810	2051	21A112-092	2051	31-0187
2051	13-18364-1	2051	016B812	2051	21A112-104	2051	31-0230
2051	13-18927-1	2051	16E1330	2051	21M084	2051	31-0239
2051	13-22581	2051	16E1330(GE)	2051	21M085	2051	31-0246
2051	13-22581-1	2051	16O2	2051	21M086	2051	31-025-Q
2051	13-23160-4	2051	16J3	2051	21M122	2051	31-058
2051	13-23309-5	2051	16J42	2051	21M123	2051	31-1
2051	13-23323-4	2051	16L43	2051	21M125	2051	31-16
2051	13-23323-6	2051	16L44	2051	21M139	2051	031A
2051	13-23824	2051	16L62	2051	21M146	2051	32-13843-2
2051	13-23916-1	2051	16L63	2051	21M149	2051	32-20738
2051	13-25226-1	2051	16U1	2051	21M150	2051	33-00706A
2051	13-27404-1	2051	16U1(HEATH KIT)	2051	21M186	2051	33-070
2051	13-27404-2	2051	16U1(HEATH-KIT)	2051	21M200	2051	33-0706
2051	13-27432-2	2051	16X1	2051	21M205	2051	33-084
2051	13-27435-1	2051	16X2	2051	21M366	2051	033-31(SYLVANIA)
2051	13-29033-1	2051	17-451	2051	21M488	2051	34-1010
2051	13-29033-2	2051	17-457	2051	21M520	2051	34-34-6015-43
2051	13-29392-2	2051	017EB24	2051	21M563	2051	34-6000-64
2051	13-29432-1	2051	001E	2051	21M578	2051	34-6000-69
2051	13-31013-1	2051	018-00003	2051	21M579	2051	34-6000-70
2051	13-33350-1	2051	18-148A	2051	21M60	2051	34-6000-71
2051	13-33595-1	2051	019-00009	2051	21M605	2051	34-60001-63
2051	13-33595-2	2051	019-0001Q	2051	22-001001	2051	34-6001-54
2051	13-33595-3			2051	22-001006		

RS 276-	Industry Standard No.
2051	34-6001-48
2051	34-6001-49
2051	34-6001-5
2051	34-6001-52
2051	34-6001-53
2051	34-6001-54
2051	34-6001-55
2051	34-6001-56
2051	34-6001-57
2051	34-6001-58
2051	34-6001-60
2051	34-6001-61
2051	34-6001-62
2051	34-6001-63
2051	34-6001-69
2051	34-6001-70
2051	34-6001-71
2051	34-6001-73
2051	34-6001-74
2051	34-6001-77
2051	34-6002-54
2051	34-6002-55
2051	34-6007-1
2051	34-6007-2
2051	34-6007-3
2051	34-6007-5
2051	34-6007-6
2051	34-6007-7
2051	34-6007-9
2051	34-6015-1
2051	34-6015-10
2051	34-6015-11
2051	34-6015-12
2051	34-6015-13
2051	34-6015-14
2051	34-6015-2
2051	34-6015-21
2051	34-6015-3
2051	34-6015-4
2051	34-6015-41
2051	34-6015-42A
2051	34-6015-43
2051	34-6015-43A
2051	34-6015-44
2051	34-6015-46
2051	34-6015-5
2051	34-6015-54
2051	34-6015-6
2051	34-6015-60
2051	34-6015-63
2051	34-6015-7
2051	34-6015-80
2051	34-6015-9
2051	34-6016-14
2051	34-6016-16
2051	34-6016-18
2051	34-6016-19
2051	34-6016-2
2051	34-6016-24
2051	34-6016-25
2051	34-6016-26
2051	34-6016-3
2051	34-6016-4
2051	34-6016-49
2051	34-6016-49A
2051	34-6016-6
2051	34-6016-63
2051	34-6016-7
2051	34-6016-8
2051	34-6075-46
2051	35-39306001
2051	35-39306002
2051	35-39306003
2051	037
2051	041
2051	042
2051	41-0499
2051	42-17444
2051	42-18111
2051	42-19644
2051	42-19670
2051	42-19840
2051	42-20738
2051	42-21234
2051	42-21407
2051	42-22158
2051	42-22553
2051	42-22786
2051	42-22787
2051	42-22809
2051	42-22811
2051	42-22812
2051	42-22847
2051	42-23348
2051	42-23349
2051	42-23542
2051	42-23964
2051	42-23964P
2051	42-23966
2051	42-23966P
2051	42-27374
2051	42-27375
2051	42-28056
2051	42-28205
2051	42-28207
2051	42-28210
2051	42-9029-31M
2051	42-9029-31R
2051	42-9029-31V
2051	42-9029-40C
2051	42-9029-40L
2051	42-9029-40X
2051	42-9029-60A
2051	42-9029-60C
2051	42-9029-70C
2051	42-9029-70P
2051	42-9029-70P
2051	43-023212
2051	43-023221
2051	43-025766
2051	43A128340-1
2051	43A128340-2
2051	43A128340-3
2051	43A128340-4
2051	43A128342-1
2051	43A128342-2
2051	43A128342-3
2051	43A128342-4
2051	43A128342-5
2051	43A128342-6
2051	43A128342-7
2051	43A167851
2051	43A168016P1
2051	43A180002-P1
2051	43A180002-P2
2051	43B140883-1
2051	43B168610
2051	43C168567
2051	43N5
2051	43N6
2051	45X16A567
2051	044-9667-02
2051	44A333463-001
2051	44A390247
2051	44A390249
2051	44A390251
2051	44A390251-001
2051	44A395994-001
2051	44B311097
2051	44L
2051	442-300-103
2051	45AN4AA
2051	45N2M
2051	45N3
2051	45N4
2051	45N4M
2051	46-8257-3
2051	46-86121-3
2051	46-86122-3
2051	46-86143-3
2051	46-86144-3
2051	46-86145-3
2051	46-86152-3
2051	46-86169-3
2051	46-86171-3
2051	46-86192-3
2051	46-86228-3
2051	46-86231-3
2051	46-8624-3
2051	46-86247-2
2051	46-86247-3
2051	46-86252-3
2051	46-86257-3
2051	46-86268-3
2051	46-86274-3
2051	46-86310-3
2051	46-86375-3
2051	46-86378-3
2051	46-86404-3
2051	46-86407-3
2051	46-86408-3
2051	46-86409-3
2051	46-86419-3
2051	46-8651
2051	46-86512-3
2051	46-86540-3
2051	46-8656
2051	46-8682-2
2051	46-8695-3
2051	46-01-005
2051	48-123175
2051	48-123802
2051	48-123803
2051	48-134173
2051	48-134464
2051	48-134465
2051	48-134654
2051	48-134664
2051	48-134665
2051	48-134666
2051	48-134667
2051	48-134668
2051	48-134669
2051	48-134673
2051	48-134674
2051	48-134675
2051	48-134690
2051	48-134691
2051	48-134703
2051	48-134705
2051	48-134714
2051	48-134718
2051	48-134720
2051	48-134721
2051	48-134726
2051	48-134732
2051	48-134733
2051	48-134733A
2051	48-134734
2051	48-134734A
2051	48-134737
2051	48-134739
2051	48-134765
2051	48-134775
2051	48-134776
2051	48-134782
2051	48-134785
2051	48-134791
2051	48-134801
2051	48-134802
2051	48-134803
2051	48-134804
2051	48-134808
2051	48-134809
2051	48-134811
2051	48-134817
2051	48-134822
2051	48-134824
2051	48-134839
2051	48-134840
2051	48-134841
2051	48-134844
2051	48-134847
2051	48-134848
2051	48-134852
2051	48-134854
2051	48-134889
2051	48-134894
2051	48-134895
2051	48-134897
2051	48-134903
2051	48-134905
2051	48-134906
2051	48-134918
2051	48-134928
2051	48-134929
2051	48-134932(3RD-IP)
2051	48-134933
2051	48-134933E
2051	48-134935
2051	48-134942
2051	48-134952
2051	48-134970
2051	48-134980
2051	48-134988
2051	48-134992
2051	48-134994
2051	48-134996
2051	48-156665
2051	48-137003
2051	48-137007
2051	48-137010
2051	48-137013
2051	48-137014
2051	48-137019
2051	48-137022
2051	48-137043
2051	48-137044
2051	48-137047
2051	48-137056
2051	48-137057
2051	48-137072
2051	48-137073
2051	48-137083
2051	48-137089
2051	48-137096
2051	48-137101
2051	48-137106
2051	48-137107
2051	48-137108
2051	48-137109
2051	48-137110
2051	48-137111
2051	48-137115
2051	48-137137
2051	48-137138
2051	48-137139
2051	48-137171
2051	48-137171D
2051	48-137172
2051	48-137174
2051	48-137192
2051	48-137206
2051	48-137257
2051	48-137260
2051	48-137265
2051	48-137336
2051	48-137350
2051	48-137353
2051	48-137354
2051	48-137373
2051	48-137374
2051	48-137377
2051	48-137378
2051	48-137384
2051	48-137398
2051	48-137399
2051	48-137499
2051	48-137500
2051	48-137509
2051	48-137530
2051	48-137543
2051	48-137612
2051	48-137955
2051	48-137998
2051	48-155088
2051	48-3003A05
2051	48-3003A11
2051	48-3003A12
2051	48-355002
2051	48-355052
2051	48-40170Q01
2051	48-40171Q01
2051	48-40246Q01
2051	48-40246Q02
2051	48-40606J01
2051	48-43354A81
2051	48-43354A82
2051	48-44885Q02
2051	48-60022A13
2051	48-63005A66
2051	48-63005A72
2051	48-63026A47
2051	48-63026A48
2051	48-63076A52
2051	48-63076A83
2051	48-63077A10
2051	48-63077A30
2051	48-63077A31
2051	48-63078A70
2051	48-63078A71
2051	48-63079A97
2051	48-63082A25
2051	48-63082A26
2051	48-63082A27
2051	48-63082A45
2051	48-63082A71
2051	48-65123A94
2051	48-65147A72
2051	48-69394A01
2051	48-83750Q01
2051	48-86376-3
2051	48-869226-0
2051	48-869248
2051	48-869312
2051	48-869325
2051	48-869329
2051	48-869444
2051	48-869525
2051	48-869563
2051	48-869568
2051	48-869570
2051	48-90172A01
2051	48-90232A05
2051	48-90232A11
2051	48-90232A13
2051	48-97046A22
2051	48-97046A23
2051	48-97046A28
2051	48-97046A42
2051	48-97046A43
2051	48-97046A46
2051	48-97046A50
2051	48-97046A52
2051	48-971-A95
2051	48-97127A012
2051	48-97127A013
2051	48-97127A01B
2051	48-97127A12
2051	48-97127A13
2051	48-97127A18
2051	48-97127A19
2051	48-97127A24
2051	48-97127A29
2051	48-97127A33
2051	48-97162A04
2051	48-97162A05
2051	48-97162A09
2051	48-97162A12
2051	48-97162A15
2051	48-97162A21
2051	48-97162A23
2051	48-97162A33
2051	48-97177A09
2051	48-97177A12
2051	48-97177A13
2051	48-971A05
2051	48-97238A04
2051	48-PO2597A
2051	48B40762A P1
2051	48D67120A11(XSTR)
2051	48N355Q02
2051	48N355052
2051	48N355053
2051	48K63005A72
2051	48K63076A81
2051	48K63076A82
2051	48K63077A31
2051	48K63078A52
2051	48K63078A70
2051	48K63078A71
2051	48K63079A97
2051	48K63082A24
2051	48K63082A25
2051	48K63082A26
2051	48K63086A18
2051	48K65112A65
2051	48K65113A88
2051	48K65118A64
2051	48R869248
2051	48R869312
2051	48R869325
2051	48R869329
2051	48R869444
2051	48R869525
2051	48R869563
2051	48R869568
2051	48R869570
2051	48R869767
2051	48S134665
2051	48S134718
2051	48S134719
2051	48S134720
2051	48S134721
2051	48S134732
2051	48S134733
2051	48S134733A
2051	48S134734
2051	48S134734A
2051	48S134737
2051	48S134765
2051	48S134773
2051	48S134774
2051	48S134776
2051	48S134783
2051	48S134784
2051	48S134785
2051	48S134789
2051	48S134797
2051	48S134804
2051	48S134805
2051	48S134807
2051	48S134809
2051	48S134810
2051	48S134811
2051	48S134820
2051	48S134823
2051	48S134825
2051	48S134826
2051	48S134827
2051	48S134837
2051	48S134838
2051	48S134840
2051	48S134841
2051	48S134844
2051	48S134845
2051	48S134846
2051	48S134854
2051	48S134857
2051	48S134887
2051	48S134889
2051	48S134894
2051	48S134899
2051	48S134899
2051	48S134903
2051	48S134904
2051	48S134905
2051	48S134908
2051	48S134918
2051	48S134922
2051	48S134923
2051	48S134924
2051	48S134925
2051	48S134926
2051	48S134927
2051	48S134933
2051	48S134937
2051	48S134941
2051	48S134945(A2C
2051	48S134948
2051	48S134949
2051	48S134950
2051	48S134952
2051	48S134961
2051	48S134962
2051	48S134963
2051	48S134964
2051	48S134990
2051	48S134997
2051	48S137003
2051	48S137033
2051	48S137044
2051	48S137055
2051	48S137056
2051	48S137057
2051	48S137107
2051	48S137108
2051	48S137109
2051	48S137110
2051	48S137115
2051	48S137171
2051	48S137171(D)
2051	48S137172

RS 276-	Industry Standard No.	RS 276-	Industry Standard No.	RS 276-	Industry Standard No.	RS 276-	Industry Standard No.
2051	488137190	2051	57A140-12	2051	57B2-126	2051	63-11143
2051	488137191	2051	57A142-4(3RDIP)	2051	57B2-153	2051	63-11289
2051	488137192	2051	57A15-1	2051	57B2-192	2051	63-11468
2051	488137260	2051	57A15-2	2051	57B2-27	2051	63-11469
2051	488137300	2051	57A15-3	2051	57B2-28	2051	63-11470
2051	488137315	2051	57A15-4	2051	57B2-59	2051	63-11471
2051	488137350	2051	57A152-1	2051	57B2-62	2051	63-11472
2051	488137351	2051	57A152-10	2051	57B2-63	2051	63-11660
2051	488137476	2051	57A152-11	2051	57B2-64	2051	63-11757
2051	488137498	2051	57A152-2	2051	57B2-73	2051	63-11758
2051	488137512	2051	57A152-3	2051	57B2-85	2051	63-11759
2051	488137530	2051	57A152-4	2051	57B2-87	2051	63-11825
2051	488137543	2051	57A152-5	2051	57B2-97	2051	63-11831
2051	488137855	2051	57A152-6	2051	57B200-12	2051	63-11833
2051	488155087	2051	57A152-7	2051	57B202-13	2051	63-11916
2051	488155088	2051	57A152-8	2051	57B253-14	2051	63-11932
2051	488155094	2051	57A152-9	2051	57B27-2	2051	63-11934
2051	488155117	2051	57A153-1	2051	57B282-12	2051	63-11935
2051	48840246901	2051	57A153-2	2051	57B6-11	2051	63-11937
2051	48840247902	2051	57A153-3	2051	57B6-19	2051	63-12003
2051	48840606901	2051	57A153-4	2051	57B6-4	2051	63-12004
2051	48844885901	2051	57A153-5	2051	57B6-9	2051	63-12062
2051	48844885902	2051	57A153-6	2051	57D105-12	2051	63-12272
2051	488P134804	2051	57A153-7	2051	57D11-1	2051	63-12605
2051	488P134855	2051	57A153-8	2051	57D121-9	2051	63-12608
2051	488P134894	2051	57A153-9	2051	57D15-1	2051	63-12609
2051	488P134897	2051	57A156-9	2051	57D15-3	2051	63-12641
2051	488P134903	2051	57A16-1	2051	57D15-4	2051	63-12642
2051	488P134905	2051	57A166-12	2051	57D156-9	2051	63-12696
2051	488P134906	2051	57A181-12	2051	57D16-1	2051	63-12697
2051	488P134933	2051	57A182-10	2051	57D24-1	2051	63-12706
2051	48X134970	2051	57A182-12	2051	57D24-2	2051	63-12707
2051	48X90232A01	2051	57A184-12	2051	57D24-3	2051	63-12750
2051	48X90232A02	2051	57A191-12	2051	57D24-4	2051	63-12751
2051	48X90232A03	2051	57A193-12	2051	57D27-1	2051	63-12752
2051	48X90232A04	2051	57A199-4	2051	57D6-11	2051	63-12753
2051	48X90232A08	2051	57A2-101	2051	57D6-17	2051	63-12874
2051	48X90232A10	2051	57A2-102	2051	57D6-19	2051	63-12875
2051	48X90232A11	2051	57A2-103	2051	57D6-29	2051	63-12877
2051	48X90232A13	2051	57A2-113	2051	57D6-30	2051	63-12878
2051	48X97046A17	2051	57A2-116	2051	57D6-32	2051	63-12879
2051	48X97046A18	2051	57A2-126	2051	57D6-7	2051	63-12933
2051	48X97046A19	2051	57A2-153	2051	57D6-9	2051	63-12940
2051	48X97046A20	2051	57A2-192	2051	57D7-10	2051	63-12941
2051	48X97046A21	2051	57A2-27	2051	57D7-15	2051	63-12942
2051	48X97046A22	2051	57A2-28	2051	57D7-18	2051	63-12943
2051	48X97046A23	2051	57A2-59	2051	57D7-20	2051	63-12944
2051	48X97046A25	2051	57A2-62	2051	57D7-9	2051	63-12946
2051	48X97046A50	2051	57A2-63	2051	57D1-123	2051	63-12948
2051	48X97046A60	2051	57A2-64	2051	57D1-124	2051	63-12949
2051	48X97046A61	2051	57A2-73	2051	57D1-51	2051	63-12950
2051	48X97046A62	2051	57A2-85	2051	57D1-75	2051	63-12951
2051	48X97048A18	2051	57A2-87	2051	57D136-12	2051	63-12952
2051	48X97162A05	2051	57A2-97	2051	57D14-1	2051	63-12953
2051	48X97162A21	2051	57A200-12	2051	57D14-2	2051	63-13419
2051	48X97177A03	2051	57A201-13	2051	57D14-3	2051	63-13438
2051	48X97177A13	2051	57A202-13	2051	57D6-19	2051	63-13440
2051	48X97238A04	2051	57A203-14	2051	57D6-4	2051	63-13441
2051	49X90232A05	2051	57A204-14	2051	57T12-2	2051	63-13864
2051	50-40102-04	2051	57A21-8	2051	57T3-1	2051	63-13927
2051	50-40105-08	2051	57A24-1	2051	57T3-4	2051	63-14032
2051	50-40201-08	2051	57A24-2	2051	57M1-14	2051	63-14051
2051	50-40201-09	2051	57A24-3	2051	57M1-15	2051	63-14052
2051	50-40201-10	2051	57A24-4	2051	57M1-19	2051	63-14057
2051	50C371	2051	57A252-1	2051	57M1-20	2051	63-18643
2051	50C372	2051	57A253-14	2051	57M1-23	2051	63-19280
2051	50C373	2051	57A265-4	2051	57M1-24	2051	63-19282
2051	50C374	2051	57A268-9	2051	57M1-26	2051	63-29461
2051	50C538	2051	57A27-1	2051	57M1-27	2051	63-7421
2051	50C644	2051	57A282-12	2051	57M1-28	2051	63-7567
2051	50C828	2051	57A6-11	2051	57M1-29	2051	63-7670
2051	50C838	2051	57A6-17	2051	57M1-30	2051	63-8555
2051	50CJ139	2051	57A6-27	2051	57M1-31	2051	63-8701
2051	051-0046	2051	57A6-29	2051	57M1-32	2051	63-8702
2051	051-0047	2051	57A6-30	2051	58-(TRUETONE)	2051	63-9337
2051	051-0155	2051	57A6-32	2051	61-1400	2051	63-9338
2051	51-47-23	2051	57A6-4	2051	61-1401	2051	63-9339
2051	51-47-24	2051	57A6-7	2051	61-1402	2051	63-9341
2051	52-020-108-0	2051	57A6-9	2051	61-1403	2051	63-9516
2051	53-1110	2051	57A7-10	2051	61-1404	2051	63-9518
2051	53P151	2051	57A7-15	2051	61-1763	2051	63-9829
2051	53P158	2051	57A7-17	2051	61-746	2051	63-9830
2051	53P159	2051	57A7-18	2051	61-751	2051	63-9831
2051	53P161	2051	57A7-20	2051	61-754	2051	63-9832
2051	53P162	2051	57A7-9	2051	61-755	2051	63-9833
2051	53P163	2051	57B-102-4	2051	61-814	2051	63-9847
2051	53P165	2051	57B105-12	2051	61-815	2051	065-004
2051	054	2051	57B107-8	2051	610001-1	2051	65-1
2051	54-1	2051	57B117-9	2051	61J001-1	2051	66-127119
2051	54A	2051	57B118-12	2051	61J002-1	2051	66-F29-1
2051	54B	2051	57B119-12	2051	61J003-1	2051	66A00008A
2051	54BLU	2051	57B120-12	2051	62-16905	2051	66A00010A
2051	54C	2051	57B121-9	2051	62-17550	2051	66P027-1
2051	54D	2051	57B125-9	2051	62-18425	2051	66P028-1
2051	54F	2051	57B129-9	2051	62-18641	2051	66P028-1
2051	54GRN	2051	57B135-12	2051	62-18642	2051	66P029-1
2051	54WHT	2051	57B139-12	2051	62-18643	2051	66P057-1
2051	54YEL	2051	57B138-4	2051	62-18828	2051	66P057-2
2051	055	2051	57B138-4(LAST-IP)	2051	62-19280	2051	66M
2051	55-1026	2051	57B140-12	2051	62-19516	2051	68A7380-1
2051	55-1034	2051	57B143-12	2051	62-19548	2051	68A7715P1
2051	55-1082	2051	57B144-12	2051	62-19837	2051	68A8321
2051	56-35	2051	57B146-12	2051	62-19838	2051	069
2051	56-8089	2051	57B152-1	2051	62-20155	2051	69-1810
2051	56-8089A	2051	57B152-10	2051	62-20240	2051	69-1811
2051	56-8089C	2051	57B152-11	2051	62-20241	2051	69-1812
2051	56-8090	2051	57B152-2	2051	62-20242	2051	69-1813
2051	56-8090A	2051	57B152-3	2051	62-20243	2051	70-943-722-001
2051	56-8090C	2051	57B152-4	2051	62-20360	2051	70-943-754-002
2051	56A22-1	2051	57B152-5	2051	62-21496	2051	70-943-762-001
2051	56B22-1	2051	57B152-6	2051	62-22098	2051	70-943-772-002
2051	057	2051	57B152-7	2051	62-22039	2051	70N1
2051	57-0004503	2051	57B152-8	2051	62-22250	2051	70N2
2051	57-0005452	2051	57B152-9	2051	62-22251	2051	70N4
2051	57-0005503	2051	57B153-1	2051	62-7567	2051	71-12626B
2051	57-000901504	2051	57B153-2	2051	62-8555	2051	71N1B
2051	57-00901504	2051	57B153-3	2051	62A11868	2051	72N2B
2051	57-01491-B	2051	57B153-4	2051	63-10188	2051	73B-140-003-5
2051	57A1-123	2051	57B153-5	2051	63-10377	2051	73C182081-31
2051	57A1-124	2051	57B153-6	2051	63-10708	2051	73N1B
2051	57A1-51	2051	57B153-7	2051	63-10732	2051	74N1
2051	57A1-75	2051	57B153-8	2051	63-10733	2051	75N5AA
2051	57A105-12	2051	57B153-9	2051	63-10734	2051	76N1
2051	57A11-1	2051	57B156-9	2051	63-10735	2051	76N1(REMOTE)
2051	57A117-9	2051	57B182-12	2051	63-10736	2051	76N1M
2051	57A118-12	2051	57B184-12	2051	63-10737	2051	76N2
2051	57A119-12	2051	57B191-12	2051	63-10860	2051	76N2369-000
2051	57A120-12	2051	57B194-11	2051	63-11025	2051	76N2369-001
2051	57A121-9	2051	57B2-101			2051	77-271453-1
2051	57A125-9	2051	57B2-102			2051	77-271819-1
2051	57A135-12	2051	57B2-103			2051	77-271967-1
2051	57A136-12	2051	57B2-113			2051	77-273001-2
2051	57A138-4(3RD-IP)	2051	57B2-116			2051	77N1
2051	57A138-4(LAST-IP)					2051	77N2
						2051	77N3

RS 276-	Industry Standard No.	RS 276-	Industry Standard No.	RS 276-	Industry Standard No.	RS 276-	Industry Standard No.
2051	77N4	2051	86-5046-2	2051	96-5220-01(NPN)	2051	121-430CL
2051	77N5	2051	86-5049-2	2051	96-5221-01	2051	121-433
2051	77N6	2051	86-5050-2	2051	96-5228-01	2051	121-433CL
2051	78N1	2051	86-5051-2	2051	96-5229-01	2051	121-435
2051	78N2B	2051	86-5055-2	2051	96-5257-01	2051	121-442
2051	80-050100	2051	86-5056-2	2051	96-5255-01	2051	121-447
2051	80-051500	2051	86-5065-2	2051	96-5257-01	2051	121-448
2051	80-052102	2051	86-5081-2	2051	96-5281-01	2051	121-450
2051	80-052202	2051	86-5097-2	2051	96-5290-01	2051	121-499
2051	80-052302	2051	86-5103-2	2051	96-5314-01	2051	121-499-01
2051	80-052600	2051	86-5110-2	2051	96X26052-52N	2051	121-505501
2051	80-053001	2051	86-5111-2	2051	96X26052/52N	2051	121-5065
2051	80-053400	2051	86-5114-2	2051	96X26053-11N	2051	121-521(MOTOROLA)
2051	80-308-2	2051	86-5117-2	2051	96X26053/11N	2051	121-581
2051	80-318250	2051	86-514-2	2051	96X26053/35N	2051	121-587
2051	80-337390	2051	86-515-2(SEARS)	2051	96X26053/36N	2051	121-600
2051	80-383940	2051	86-520-2	2051	96X2801/14N	2051	121-610
2051	80-389910	2051	86-534-2	2051	98T2	2051	121-629
2051	80-389930	2051	86-536-2	2051	99-109-1	2051	121-639
2051	80-421980	2051	86-537-2	2051	99-109-2	2051	121-646
2051	81-27125140-7	2051	86-538-2	2051	998012	2051	121-647
2051	81-27125140-7A	2051	86-539-2	2051	998012A	2051	121-648
2051	81-27125140-7B	2051	86-548-2	2051	998012E	2051	121-649
2051	81-27125160-5	2051	86-551-2	2051	998020	2051	121-652
2051	81-27125160-5A	2051	86-554-2	2051	998025	2051	121-657
2051	81-27125160-5B	2051	86-559-2	2051	998025A	2051	121-660
2051	81-27125270-2	2051	86-560-2	2051	998032(3RD-IP)	2051	121-662
2051	81-27125270-2A	2051	86-561-2	2051	998033A	2051	121-668
2051	81-27125270-2B	2051	86-564-2	2051	998035	2051	121-671
2051	81-27125300-7	2051	86-565-2	2051	998036	2051	121-672
2051	81-46125016-9	2051	86-573-2	2051	998038	2051	121-675
2051	81-46125019-3	2051	86-58-2	2051	998085	2051	121-677
2051	81-46125026-8	2051	86-595-2	2051	100N1P(SOUND)	2051	121-678
2051	81-46125027-6	2051	86-598-2	2051	0101-0491	2051	121-678GREEN
2051	82-409501	2051	86-599-2	2051	0101-0540	2051	121-678YELLOW
2051	85-5056-2	2051	86-646-2	2051	102-0454-02	2051	121-695
2051	85004	2051	86-655-2	2051	102-1166-25	2051	121-701
2051	86 A 86A327	2051	86-661-2	2051	102-1317-18	2051	121-702
2051	86-0007-004	2051	86-675-2	2051	102-1335-04	2051	121-706
2051	86-0012-001	2051	86-702-2	2051	0103-0014	2051	121-711
2051	86-0022-001	2051	86A334	2051	0103-0014/4460	2051	121-730
2051	86-0029-001	2051	86A336	2051	0103-0060(B)	2051	121-737
2051	86-0031-001	2051	86A350	2051	0103-0060A	2051	121-745
2051	086-005132-02	2051	86A86A327	2051	0103-0088	2051	121-748
2051	86-100003	2051	86X0006-001	2051	0103-0088H	2051	121-751
2051	86-100005	2051	86X0007-001	2051	0103-0088R	2051	121-764
2051	86-100008	2051	86X0007-104	2051	0103-0088B	2051	121-767CL
2051	86-110-2	2051	86X0008-001	2051	0103-0227-18	2051	121-768
2051	86-119-2	2051	86X0012-001	2051	0103-0473	2051	121-768CL
2051	86-123-2	2051	86X0022-001	2051	0103-0482	2051	121-773
2051	86-139-2	2051	86X0025-001	2051	0103-0491	2051	121-787-01
2051	86-1392	2051	86X0025-001(TIK)	2051	0103-0491/4460	2051	121-812
2051	86-143-2	2051	86X0029-001	2051	0103-0504	2051	121-836
2051	86-144-2	2051	86X0031-001	2051	0103-0521(B)	2051	121-837
2051	86-155-2	2051	86X0031-002	2051	0103-0540	2051	121-850
2051	86-157-2	2051	86X0031-003	2051	0103-94	2051	121-856
2051	86-158-2	2051	86X0032-001	2051	105(ADMIRAL)	2051	121-862
2051	86-166-2	2051	86X0035-001	2051	105-001-04	2051	121-863
2051	86-171-2	2051	86X0040-00	2051	105-001-05	2051	121-877
2051	86-175-2	2051	86X0040-001	2051	105-001-07	2051	121-881
2051	86-182-2	2051	86X0045-001	2051	105-003-06	2051	121-888
2051	86-188-2	2051	86X0048-001	2051	105-003-09	2051	121-889
2051	86-189-2	2051	86X0050-001	2051	105-006-08	2051	121-913
2051	86-190-2	2051	86X0051-001	2051	105-060-09	2051	121-914
2051	86-191-2	2051	86X0054-001	2051	105-06007-05	2051	121-915
2051	86-192-2	2051	86X0058-001	2051	105-08243-05	2051	121-916
2051	86-193-2	2051	86X0058-002	2051	105-085-33	2051	121-931
2051	86-194-2	2051	86X0058-003	2051	105-085-54	2051	121-944
2051	86-195-2	2051	86X006-001	2051	105-12	2051	121-950
2051	86-196-2	2051	86X0063-001	2051	105-28196-07	2051	121-972
2051	86-197-2	2051	86X007-004	2051	105-904-85	2051	121-972-01
2051	86-198-2	2051	86X007-034	2051	105-904-86	2051	121-975
2051	86-199-2	2051	86X0079-001	2051	105-904-87	2051	121-982
2051	86-201-2	2051	86X0090-001	2051	105-931-91/2228-3	2051	121-29000
2051	86-202-2	2051	86X54-1	2051	105NADMIRALE	2051	121-29000A
2051	86-237-2	2051	86X6	2051	107-3088	2051	12103019
2051	86-238-2	2051	86X6-1	2051	107-8	2051	12103020
2051	86-247-2	2051	86X6-4-518	2051	107BRN	2051	122-1
2051	86-250-2	2051	86X7-2	2051	107R2	2051	122-2
2051	86-255-2	2051	86X7-3	2051	109	2051	122-6
2051	86-256-2	2051	86X7-4	2051	109-1	2051	123-004
2051	86-264-2	2051	86X8-1	2051	109-1(RCA)	2051	123-010
2051	86-265-2	2051	86X8-2	2051	112-000088	2051	124N16
2051	86-277-2	2051	86X8-3	2051	112-0011 A	2051	0125
2051	86-291-9	2051	86X8-4	2051	112-011A	2051	125B132
2051	86-293-2	2051	87-0014	2051	112-1A82	2051	125Q211
2051	86-308-2	2051	88-1250109	2051	112-523	2051	0126
2051	86-309-2	2051	089-223	2051	113-118	2051	126-12
2051	86-310-2	2051	089-226	2051	115-1	2051	127
2051	86-323-2	2051	90-180	2051	115-13	2051	128WHT
2051	86-324-2	2051	90-2213-00-18	2051	115-225	2051	129-14
2051	86-327-2	2051	90-30	2051	115-4	2051	129-15
2051	86-328-2	2051	90-32	2051	115-875	2051	129-16
2051	86-339-2	2051	90-455	2051	116-074	2051	129-21
2051	86-339-9	2051	90-458	2051	116-078	2051	129-33(PILOT)
2051	86-342-2	2051	90-459	2051	116-085	2051	129BRN
2051	86-359-2	2051	90-47	2051	116-092	2051	129WHT
2051	86-362-2	2051	90-48	2051	116-588	2051	130-40214
2051	86-365-2	2051	90-57	2051	116-875	2051	130-40215
2051	86-379-2	2051	90-603	2051	119-0054	2051	130-40294
2051	86-389-2	2051	90-605	2051	119-0056	2051	130-40311
2051	86-390-2	2051	90-61	2051	120-004480	2051	130-40312
2051	86-391-2	2051	90-65	2051	120-004482	2051	130-40313
2051	86-399-1	2051	90-69	2051	120-004483	2051	130-40317
2051	86-399-2	2051	90-71	2051	120-004880	2051	130-40318
2051	86-399-9	2051	91C	2051	120-004882	2051	130-40357
2051	86-403-2	2051	91D	2051	120-004883	2051	130-40896
2051	86-42-1	2051	91E	2051	120-1	2051	130-40922
2051	86-420-2	2051	92-30942	2051	120-2	2051	0131
2051	86-445-2	2051	93A39-15	2051	120-3	2051	0131-000473
2051	86-457-2	2051	93D39-11	2051	120-7	2051	0131-000704
2051	86-458-2	2051	94N1	2051	120-8	2051	0131-001417
2051	86-460-2	2051	94N1B	2051	120-8A	2051	0131-001418
2051	86-461-2	2051	94N1R	2051	120BLU	2051	0131-001421
2051	86-462-2	2051	94N2	2051	121	2051	0131-001422
2051	86-465-2	2051	95-216	2051	121(SEARS)	2051	0131-001423
2051	86-472-2	2051	95-221	2051	121-1040	2051	0131-001424
2051	86-481-1	2051	95-225	2051	121-195B	2051	0131-001464
2051	86-481-2	2051	95-226-2	2051	121-276	2051	0131-001864
2051	86-483-2	2051	95-227	2051	121-277	2051	0131-004323
2051	86-483-3	2051	95-296	2051	121-278	2051	132-002
2051	86-484-2	2051	96-5080-02	2051	121-286	2051	132-004
2051	86-485-2	2051	96-5115-01	2051	121-288	2051	132-005
2051	86-486-2	2051	96-5115-02	2051	121-364	2051	132-011
2051	86-493-2	2051	96-5115-03	2051	121-365	2051	132-014
2051	86-494-2	2051	96-5115-04	2051	121-366	2051	132-018
2051	86-495-2	2051	96-5115-05	2051	121-367	2051	132-021
2051	86-496-2	2051	96-5152-01	2051	121-369	2051	132-023
2051	86-5018-2	2051	96-5152-03	2051	121-404	2051	132-026
2051	86-502-2	2051	96-5153-03	2051	121-422	2051	132-030
2051	86-5040-2	2051	96-5153-05	2051	121-423	2051	132-041
2051	86-5041-2	2051	96-5177-01	2051	121-430	2051	132-042
2051	86-5044-2	2051	96-5187-01	2051	121-430	2051	132-050
2051	86-5045-2	2051	96-5213-01	2051	121-430B	2051	132-051

RS 276-	Industry Standard No.	RS 276-	Industry Standard No.	RS 276-	Industry Standard No.	RS 276-	Industry Standard No.
2051	132-054	2051	229-0154	2051	314-6010-3	2051	417-135
2051	132-055	2051	229-0180-123	2051	317-8504-001	2051	417-155-13163
2051	132-057	2051	229-0190-90	2051	319C	2051	417-171
2051	132-062	2051	229-1200-36	2051	322(CATALINA)	2051	417-171-13163
2051	132-063	2051	229-1301-24	2051	324-0151	2051	417-172
2051	132-069	2051	231-0004-01	2051	324-0152	2051	417-172-13271
2051	132-075	2051	231-0004-03	2051	324-0154	2051	417-185
2051	132-077	2051	232N1	2051	324-6005-5	2051	417-192
2051	132-501	2051	232N2	2051	325-0031-303	2051	417-197
2051	132-502	2051	247-257	2051	325-0031-304	2051	417-213
2051	132-503	2051	247-629	2051	325-0031-305	2051	417-217
2051	132-504	2051	247ABc21249-001	2051	325-0031-310	2051	417-218
2051	132-539	2051	250-0712	2051	325-0042-351	2051	417-219
2051	132-540	2051	260-10-020	2051	325-0076-306	2051	417-226-1
2051	0134	2051	260-10-20	2051	325-0076-307	2051	417-228
2051	134B1038-21	2051	260D05709	2051	325-0076-308	2051	417-229
2051	135044322-542	2051	260D07412	2051	325-0081-100	2051	417-233-13163
2051	136GRN	2051	260D08201	2051	325-0081-101	2051	417-244
2051	136RED	2051	260D08801	2051	325-0574-30	2051	417-283
2051	138-4	2051	260D09001	2051	325-0574-31	2051	417-67
2051	140-0007	2051	260D09314	2051	325-1370-19	2051	417-69
2051	0140-6	2051	260D106A1	2051	325-1370-20	2051	417-7
2051	141-4	2051	260D13702	2051	325-1446-26	2051	417-77
2051	141-430	2051	260P02903	2051	325-1446-27	2051	417-801
2051	142-001	2051	260P029033	2051	325-1446-28	2051	417-801-12903
2051	142-3	2051	260P02903A	2051	325-1513-29	2051	417-91
2051	142-4	2051	260P02903B	2051	325-1513-30	2051	417-92
2051	142N3	2051	260P04001	2051	325-1771-15	2051	417-93
2051	142N3P	2051	260P04002	2051	344-6000-2	2051	417-93-12903
2051	142N3T	2051	260P04003	2051	344-6000-4	2051	417-94
2051	142N4	2051	260P04004	2051	344-6000-5	2051	421-7444
2051	142N5	2051	260P04502	2051	344-6000-5A	2051	421-8111
2051	143-12	2051	260P04503	2051	344-6000-3	2051	421-9644
2051	144-12	2051	260P04505	2051	344-6005-1	2051	421-9670
2051	145N1(LAST-IF)	2051	260P06904	2051	344-6005-2	2051	421-9840
2051	145NIP	2051	260P07001	2051	344-6005-5	2051	422-0738
2051	146-12	2051	260P07002	2051	344-6017-2	2051	422-1234
2051	146N3	2051	260P07301	2051	344-6017-3	2051	422-1407
2051	146N5	2051	260P0770	2051	344-6017-5	2051	422-2533
2051	148N2	2051	260P07701	2051	352-0195-000	2051	422-2534
2051	148N212	2051	260P07702	2051	352-0197-000	2051	429-0958-42
2051	150N2	2051	260P07703	2051	352-0197-001	2051	430
2051	151-0103-00	2051	260P07704	2051	352-0206-001	2051	430(ZENITH)
2051	151-0127-00	2051	260P07705	2051	352-0316-00	2051	430-10034
2051	151-0190-00	2051	260P07707	2051	352-0318-00	2051	430-10034-06
2051	151-0223-00	2051	260P08801	2051	352-0318-001	2051	430-10053-0
2051	151-0224-00	2051	260P08801A	2051	352-0319-000	2051	430-10053-0A
2051	151-0302-00	2051	260P09902	2051	352-0322-010	2051	430-1044-0A
2051	151-0424-00	2051	260P11302	2051	352-0349-000	2051	430-23212
2051	151-6	2051	260P11303	2051	352-0365-000	2051	430-23221
2051	151N2	2051	260P11304	2051	352-0400-00	2051	430-25766
2051	151N4	2051	260P11305	2051	352-0400-010	2051	430-86
2051	151N5	2051	260P11502	2051	352-0400-030	2051	430-87
2051	152-12	2051	260P11503	2051	352-0433-00	2051	430CL
2051	154A5941	2051	260P11504	2051	352-0477-00	2051	433
2051	154A5944-410	2051	260P11505	2051	352-0506-000	2051	433(ZENITH)
2051	154A5946	2051	260P12001	2051	352-0519-00	2051	433-1
2051	154A5946-624	2051	260P12002	2051	352-0546-00	2051	433-1(SYLVANIA)
2051	156	2051	260P13701	2051	352-0569-00	2051	433CL
2051	156WHT	2051	260P13702	2051	352-0569-010	2051	436-403-001
2051	158(SEARS)	2051	260P14101	2051	352-0569-020	2051	444(SEARS)
2051	160-8	2051	260P14102	2051	352-0579-00	2051	447-00
2051	161-001I	2051	260P14103	2051	352-0579-010	2051	447(ZENITH)
2051	161-011J	2051	260P14105	2051	352-0579-020	2051	450-1167-2
2051	161-016H	2051	260P141103	2051	352-0596-010	2051	450-1261
2051	161-016K	2051	260P17101	2051	352-0596-020	2051	462-0119
2051	165A4383	2051	260P17102	2051	352-0596-030	2051	462-1000
2051	165N1	2051	260P17103	2051	352-0629-010	2051	462-1009-01
2051	167N1	2051	260P17104	2051	352-0661-010	2051	462-1061
2051	167N2	2051	260P17105	2051	352-0661-020	2051	462-1063
2051	168N1	2051	260P17106	2051	352-0667-010	2051	462-2002
2051	173A4057	2051	260P17503	2051	352-0675-000	2051	465-106-19
2051	173A4399	2051	260P19101	2051	352-0675-020	2051	472-0491-001
2051	173A4416	2051	260P19103	2051	352-0675-030	2051	472-1198-001
2051	173A4470-11	2051	260P19501	2051	352-0675-040	2051	483-3141
2051	173A4470-13	2051	260P4002	2051	352-0675-050	2051	486-1551
2051	173A4470-32	2051	260Z00402	2051	352-0680-010	2051	499
2051	176-008-9-001	2051	266P00101(XSTR)	2051	352-0680-020	2051	499(ZENITH)
2051	176-014-9-001	2051	270-950-030	2051	352-0713-030	2051	502
2051	176-017-9-001	2051	276-2009	2051	352-0809	2051	509
2051	176-024-9-001	2051	276-2009/RS2009	2051	352-7500-010	2051	509(SEARS)
2051	176-042-9-002	2051	276-2014	2051	352-7500-450	2051	511-515
2051	176-043-9-002	2051	276-2014/RS2014	2051	352-8000-010	2051	511-519
2051	176-047-9-002	2051	276-2030	2051	352-8000-020	2051	512RED
2051	176-047-9-003	2051	276-2030/RS2030	2051	352-8000-030	2051	514-033338
2051	176-048-9-002	2051	276-2033	2051	352-8000-040	2051	515
2051	176-049-9-002	2051	276-2033/RS2033	2051	352-9056-00	2051	515-074
2051	176-054-9-001	2051	281	2051	352-9079-00	2051	515ORN
2051	176-060-9-002	2051	284HC	2051	352-9103-000	2051	516
2051	176-060-9-004	2051	296-50-9	2051	353-9306-001	2051	517-518
2051	176-073-9-001	2051	296-51-9	2051	353-9306-002	2051	519-1
2051	185-003	2051	296-56-9	2051	353-9306-003	2051	519-1(RCA)
2051	185-004	2051	296-59-9	2051	353-9306-004	2051	519ES067M
2051	185-009	2051	296-77-9	2051	353-9306-005	2051	519ES068M
2051	186-002	2051	297L006H01	2051	353-9310-001	2051	522
2051	186-007	2051	297L006H02	2051	353-9315-001	2051	524WHT
2051	186N1(LAST-IF)	2051	297L007C02	2051	353-9315-001	2051	526-2(RCA)
2051	200-016	2051	297L007H01	2051	353-9319-001	2051	536D
2051	200-057	2051	297L007H02	2051	353-9319-002	2051	536D9
2051	200-058	2051	297L007H03/C03	2051	354-3127-1	2051	536F
2051	200-846	2051	297L013B01	2051	355D9	2051	536FS
2051	200-862	2051	297V049H01	2051	365-1	2051	536FU
2051	200-863	2051	297V049H03	2051	366-1	2051	536G(WARDS)
2051	200X3174-006	2051	297V049H04	2051	366-2	2051	537D
2051	200X3174-014	2051	297V049H05	2051	375-1005	2051	537E
2051	200X3174-021	2051	297V059H01	2051	386-1102-P1	2051	537FV
2051	207A10	2051	297V059H02	2051	386-1102-P2	2051	537FY
2051	207A29	2051	297V059H03	2051	386-1102-P3	2051	000646-1
2051	207A31	2051	297V061C03	2051	386-7178P1	2051	550-026-00
2051	207A35	2051	297V061C04	2051	386-7185P1	2051	560-2
2051	209-846	2051	297V061C06	2051	394-3003-1	2051	565-074
2051	209-862	2051	297V061H01	2051	394-3003-3	2051	567-0003-011
2051	209-863	2051	297V061H02	2051	394-3003-7	2051	570-004503
2051	211AVPF3415	2051	297V061H03	2051	394-3003-9	2051	570-005503
2051	211AVTEA275	2051	297V072C06	2051	400-1371-101	2051	572-1
2051	212-505	2051	297V074C02	2051	400-2023-101	2051	572-683
2051	212-507	2051	297V074C03	2051	400-2023-201	2051	573-469
2051	212-695	2051	297V074C04	2051	404-2(SYLVANIA)	2051	573-479
2051	218-22	2051	297V074C06	2051	412-1A5	2051	573-480
2051	218-23	2051	297V074C07	2051	414	2051	573-481
2051	218-24	2051	297V074C08	2051	417-105	2051	576-0001-004
2051	218-25	2051	297V083C02	2051	417-106	2051	576-0001-005
2051	220-001001	2051	297V085C01	2051	417-107	2051	576-0001-008
2051	220-001002	2051	297V085C02	2051	417-108	2051	576-0001-012
2051	221(SEARS)	2051	297V085C03	2051	417-109-13163	2051	576-0001-018
2051	226-1(SYLVANIA)	2051	297V085C04	2051	417-110	2051	576-0002-006
2051	229-0050-13	2051	297V086C01	2051	417-110-13163	2051	576-0003-011
2051	229-0050-14	2051	297V086C02	2051	417-114-13163	2051	576-0003-12(NPN)
2051	229-0050-15	2051	297V086C03	2051	417-118	2051	576-0004-010
2051	229-0144	2051	306-1	2051	417-126	2051	576-0036-212
2051	229-0149	2051	309-327-926	2051	417-127	2051	576-0036-847
2051	229-0150	2051	310-187	2051	417-129	2051	576-0036-916
2051	229-0151	2051	314-6007-1	2051	417-134	2051	576-0036-917
2051	229-0152	2051	314-6007-2	2051	417-134-13271	2051	576-0036-920
		2051	314-6007-3				

RS 276-	Industry Standard No.
2051	576-0036-921
2051	586-2
2051	590-593031
2051	593D742-1
2051	595-1
2051	595-1(SYLVANIA)
2051	595-2
2051	595-2(SYLVANIA)
2051	600-188-1-13
2051	600-188-1-20
2051	600-188-1-23
2051	600-229-201
2051	600-301-801
2051	600X0091-086
2051	601-0100793
2051	601-1
2051	601-1(RCA)
2051	601-1(TRANSISTOR)
2051	601-2
2051	601X0149-086
2051	602X0018-000
2051	604
2051	604(SEARS)
2051	605
2051	606-9601-101
2051	606-9602-101
2051	607-030
2051	609-112
2051	610-045-3
2051	610-045-4
2051	610-070
2051	610-070-1
2051	610-070-2
2051	610-070-3
2051	610-076
2051	610-076-1
2051	610-076-2
2051	610-077
2051	610-077-1
2051	610-077-2
2051	610-077-3
2051	610-077-4
2051	610-077-5
2051	610-078
2051	610-078-1
2051	612-16
2051	612-1A5L
2051	613
2051	614-2
2051	614-3
2051	617-10
2051	617-161
2051	617-29
2051	617-63
2051	617-64
2051	617-67
2051	617-68
2051	617-71
2051	617-87
2051	625-1
2051	626
2051	626-1
2051	630-076
2051	638
2051	639
2051	639(ZENITH)
2051	642-174
2051	642-242
2051	642-246
2051	642-319
2051	000653
2051	660-126
2051	660-131
2051	660-134
2051	660-220
2051	660-221
2051	660-222
2051	660-225
2051	690Y0103H27
2051	690Y080H41
2051	690Y086H51
2051	690Y086H88
2051	690Y088H50
2051	690Y088H51
2051	690Y089H89
2051	690Y090H36
2051	690Y092H52
2051	690Y092H54
2051	690Y092H81
2051	690Y092H84
2051	690Y092H96
2051	690Y092H97
2051	690Y094H21
2051	690Y097H62
2051	690Y098H48
2051	690Y098H49
2051	690Y098H50
2051	690Y099H79
2051	690Y102H71
2051	690Y103H29
2051	690Y103H31
2051	690Y103H32
2051	690Y103H33
2051	690Y105H-25
2051	690Y105H19
2051	690Y105H22
2051	690Y105H25
2051	690Y105H27
2051	690Y105H29
2051	690Y114H30
2051	690Y114H33
2051	690Y116H21
2051	690Y120H89
2051	693BP
2051	693PB
2051	693Q
2051	693QT
2051	694D
2051	694E
2051	699
2051	700-134
2051	700-154
2051	700-325
2051	700A858-318
2051	700A858-328
2051	705 056 (4)
2051	703-056(4)
2051	0704
2051	715BN
2051	737
2051	737(ZENITH)
2051	739H01
2051	748
2051	748(ZENITH)
2051	750A858-319
2051	7500858-123
2051	7500858-124
2051	7500858-125
2051	7500858-212
2051	750M63-119
2051	750M63-120
2051	750M63-146
2051	753-1828-001
2051	753-2000-003
2051	753-2000-004
2051	753-2000-008
2051	753-2000-009
2051	753-2000-011
2051	753-2000-711
2051	753-2000-735
2051	753-2000-870
2051	753-2000-871
2051	753-2100-001
2051	753-2100-008
2051	753-4000-011
2051	753-4000-101
2051	753-4000-537
2051	767
2051	767(ZENITH)
2051	767CL
2051	772-110
2051	773(ZENITH)
2051	0776-0160
2051	776-151
2051	776-183
2051	776-2(PHILCO)
2051	776GRN
2051	780WHT
2051	785YEL
2051	791
2051	800-001-034
2051	800-101-101-1
2051	800-101-102-1
2051	800-250-102
2051	800-501-00
2051	800-501-01
2051	800-501-02
2051	800-501-03
2051	800-501-11
2051	800-501-22
2051	800-50100
2051	800-508-00
2051	800-509-00
2051	800-514-00
2051	800-51500
2051	800-521-01
2051	800-52102
2051	800-522-01
2051	800-522-02
2051	800-522-04
2051	800-52202
2051	800-52302
2051	800-526-00
2051	800-52600
2051	800-529-00
2051	800-530-00
2051	800-530-01
2051	800-53001
2051	800-534-00
2051	800-534-00(XSTR)
2051	800-534-01
2051	800-53400
2051	800-538-00
2051	800-544-00
2051	800-544-10
2051	800-544-20
2051	800-544-30
2051	800-548-00
2051	800-550-00
2051	801B
2051	803-18250
2051	803-37390
2051	803-83940
2051	803-89910
2051	803-89930
2051	804
2051	804-21980
2051	818WHT
2051	822
2051	822A
2051	822ABLU
2051	822B
2051	823B
2051	823WHT
2051	824-09501
2051	827BRN
2051	828GRN
2051	8288
2051	834-6066
2051	847BLK
2051	853-0300-632
2051	853-0300-900
2051	853-0300-923
2051	853-0373-110
2051	85808
2051	880-250-102
2051	880-250-108
2051	880-250-109
2051	881-250-102
2051	881-250-108
2051	881-250-109
2051	902-000-2-04
2051	902-000-8-04
2051	903-3
2051	903-30
2051	903Y002149
2051	904-95
2051	904-95A
2051	904-96B
2051	909-27125-140
2051	912-1A6A
2051	916-31024-3
2051	916-31025-5
2051	916-31026-8B
2051	921-01122
2051	921-01123
2051	921-01124
2051	921-01127
2051	921-1017
2051	921-1109B
2051	921-117B
2051	921-120B
2051	921-123B
2051	921-124B
2051	921-125B
2051	921-127B
2051	921-128B
2051	921-155B
2051	921-159B
2051	921-161B
2051	921-189B
2051	921-191B
2051	921-195B
2051	921-200B
2051	921-20BK
2051	921-214B
2051	921-215B
2051	921-225B
2051	921-228B
2051	921-229B
2051	921-22BG
2051	921-234B
2051	921-237B
2051	921-238K
2051	921-252B
2051	921-255B
2051	921-26
2051	921-268B
2051	921-269B
2051	921-26A
2051	921-272B
2051	921-275R
2051	921-276B
2051	921-27B
2051	921-28
2051	921-28A
2051	921-28B
2051	921-28BLU
2051	921-291B
2051	921-301
2051	921-303B
2051	921-304B
2051	921-305B
2051	921-306B
2051	921-307
2051	921-307B
2051	921-309
2051	921-314B
2051	921-326B
2051	921-339B
2051	921-345B
2051	921-351
2051	921-351B
2051	921-352B
2051	921-353
2051	921-353B
2051	921-354B
2051	921-355B
2051	921-360B
2051	921-369
2051	921-43B
2051	921-43R
2051	921-449
2051	921-450
2051	921-46
2051	921-462
2051	921-463
2051	921-464
2051	921-46A
2051	921-46BK
2051	921-47
2051	921-470
2051	921-47A
2051	921-47BL
2051	921-49B
2051	921-50B
2051	921-7
2051	921-73B
2051	921-77B
2051	921-8
2051	921-93B
2051	921-99B
2051	9260193-1
2051	9260193-2
2051	9260193-P1M4165
2051	930X1
2051	930X2
2051	930X3
2051	935-1
2051	964-17444
2051	964-20738
2051	964-20733B
2051	964-24584
2051	991-00-1219
2051	991-00-2232
2051	991-00-2248
2051	991-00-2298
2051	991-00-2356
2051	991-00-2356/K
2051	991-00-2873
2051	991-00-3144
2051	991-00-3304
2051	991-00-8393
2051	991-00-8393A
2051	991-00-8393M
2051	991-00-8394
2051	991-00-8394A
2051	991-00-8394AH
2051	991-00-8395
2051	991-002232
2051	991-002298
2051	991-002356
2051	991-002873
2051	991-003304
2051	991-008393
2051	991-01-1219
2051	991-01-1220
2051	991-01-1306
2051	991-01-1312
2051	991-01-1318
2051	991-01-1705
2051	991-01-3044
2051	991-01-3056
2051	991-01-3057
2051	991-01-3068
2051	991-01-3544
2051	991-01-3685
2051	991-01-3740
2051	991-010462
2051	991-011219
2051	991-011220
2051	991-011306
2051	991-011312
2051	991-011318
2051	991-013044
2051	991-013056
2051	991-013068
2051	991-013544
2051	991-015587
2051	991-015615
2051	991-016274
2051	992-00-2298
2051	992-00-3144
2051	992-01-3738
2051	998-0061114
2051	998-0200816
2051	1000-136
2051	1000-137
2051	1001(JULIETTE)
2051	1001-02
2051	1001-03
2051	1001-04
2051	1001-05
2051	1001-06
2051	1002-03
2051	1002-04
2051	1004
2051	1004(28C537)
2051	1004-03
2051	1005(28C537)
2051	1005-03
2051	1005-3
2051	1006(JULIETTE)
2051	1006-48
2051	1007-3153
2051	1008-02
2051	1009-02
2051	1009-02-16
2051	1009-17
2051	1009-2
2051	1010-7928
2051	1010-8066
2051	1010-8082
2051	1010-8090
2051	1019-3852
2051	1020-17
2051	1023Q
2051	1023Q(GE)
2051	1024Q
2051	1024Q(GE)
2051	1025Q
2051	1025Q(GE)
2051	1026Q
2051	1026Q(GE)
2051	1028
2051	1028(G.E.)
2051	1028Q
2051	1028Q(GE)
2051	1029
2051	1029(G.E.)
2051	1029Q
2051	1029Q(GE)
2051	1034-17
2051	1038-17
2051	1038-1-10
2051	1038-10
2051	1038-15
2051	1038-15CL
2051	1038-18
2051	1038-18CL
2051	1038-2
2051	1038-23
2051	1038-23CL
2051	1038-24
2051	1038-6
2051	1038-6CL
2051	1038-8
2051	1039-01
2051	1040-03
2051	1040-155
2051	1040-2
2051	1041-72
2051	1041-73
2051	1041-75
2051	1042-04
2051	1042-07
2051	1042-7
2051	1043-07
2051	1043-1229
2051	1043-1260
2051	1045-2951
2051	1048-9904
2051	1048-9912
2051	1063-5381
2051	1063-8963
2051	1077-07
2051	1080-6396
2051	1081-3301
2051	1081-3319
2051	1081-3475
2051	1081-9464
2051	1087
2051	1100-9461
2051	1100-9479
2051	1113-03
2051	1117(MC)
2051	1119-8132
2051	1123-60
2051	1202
2051	1203(GE)
2051	1203-169
2051	1205(GE)
2051	1208(GE)
2051	1210-17(MIXER)
2051	1210-17(OSC)
2051	1210-17B
2051	1227-17
2051	1228-17
2051	1236-3776
2051	1272
2051	1315
2051	1316
2051	1368C/D
2051	1374-17AC
2051	1402B
2051	1402C
2051	1402E
2051	1412-1
2051	1412-1-12
2051	1412-1-12-8
2051	1414-174
2051	1414-183
2051	1415
2051	001422
2051	1424
2051	1431 8349
2051	1431-8349
2051	1463
2051	1465
2051	1471-4778
2051	1479-7989

RS 276-	Industry Standard No.
2051	1482
2051	1493-17
2051	1515
2051	1524
2051	1540
2051	1567
2051	1567-0
2051	1567-2
2051	1702M
2051	1705-4834
2051	1711
2051	1711-17
2051	1711MC
2051	1712-17
2051	1723-17
2051	1751G036
2051	1799-17
2051	1800-17
2051	1812-1
2051	1812-1-12
2051	1812-1-127
2051	1812-1L
2051	1812-1L8
2051	1841-17
2051	1854-0003
2051	1854-0005
2051	1854-0033
2051	1854-0353
2051	1854-0432
2051	1866-17
2051	1879-17
2051	1882-17
2051	1883-17
2051	1884-17
2051	1893-17
2051	1915-17
2051	1929-17
2051	1932-17
2051	1961-17
2051	1966-17
2051	1984-17
2051	1999
2051	2000-203
2051	2000-206
2051	2000-210
2051	2000-214
2051	2001-17
2051	2003-17
2051	2004-03
2051	2004-04
2051	2004-05
2051	2004-14
2051	2006
2051	2008-17
2051	2017-115
2051	2018-01
2051	2020-06
2051	2022-05
2051	2022-244
2051	2026
2051	2026-00
2051	2027
2051	2027(E.F.JOHNSON)
2051	2027-00
2051	2027R.F.
2051	2032-37
2051	2032-40
2051	2044-17
2051	2057A10-64
2051	2057A2-103
2051	2057A2-113
2051	2057A2-117(OSC)
2051	2057A2-121
2051	2057A2-122
2051	2057A2-131
2051	2057A2-143
2051	2057A2-145
2051	2057A2-146
2051	2057A2-152
2051	2057A2-153
2051	2057A2-154
2051	2057A2-155
2051	2057A2-184
2051	2057A2-208
2051	2057A2-209
2051	2057A2-215
2051	2057A2-222
2051	2057A2-225
2051	2057A2-226
2051	2057A2-227
2051	2057A2-228
2051	2057A2-264
2051	2057A2-27
2051	2057A2-276
2051	2057A2-278
2051	2057A2-279
2051	2057A2-280
2051	2057A2-281
2051	2057A2-285
2051	2057A2-289
2051	2057A2-294
2051	2057A2-296
2051	2057A2-297
2051	2057A2-300
2051	2057A2-303
2051	2057A2-306
2051	2057A2-316
2051	2057A2-319
2051	2057A2-324
2051	2057A2-332
2051	2057A2-341
2051	2057A2-374
2051	2057A2-387
2051	2057A2-390
2051	2057A2-396
2051	2057A2-398
2051	2057A2-399
2051	2057A2-401
2051	2057A2-412
2051	2057A2-433
2051	2057A2-434
2051	2057A2-449
2051	2057A2-452
2051	2057A2-463
2051	2057A2-464
2051	2057A2-479
2051	2057A2-487
2051	2057A2-510
2051	2057A2-511
2051	2057A2-558
2051	2057A2-559
2051	2057A2-62
2051	2057A2-64
2051	2057B-85
2051	2057B101-4
2051	2057B102-4
2051	2057B103-4
2051	2057B117-9
2051	2057B118-12
2051	2057B119-2
2051	2057B120-12
2051	2057B125-9
2051	2057B141-4
2051	2057B142-4
2051	2057B143-12
2051	2057B146-12
2051	2057B151-6
2051	2057B152-12
2051	2057B2-113
2051	2057B2-121
2051	2057B2-122
2051	2057B2-123
2051	2057B2-130
2051	2057B2-152
2051	2057B2-153
2051	2057B2-154
2051	2057B2-155
2051	2057B2-27
2051	2057B2-38
2051	2057B2-59
2051	2057B2-62
2051	2057B2-63
2051	2057B2-69
2051	2057B2-73
2051	2057B2-97
2051	2057B2-276
2051	2063-17-12
2051	2064
2051	2064(CROWN)
2051	2065-03
2051	2065-07
2051	2132(GE)
2051	2158-1541
2051	2180-153
2051	2180-154
2051	2204-17
2051	2263
2051	2270
2051	2275-17
2051	2290-17
2051	2320-17
2051	2321-17
2051	2337-17
2051	2361-17
2051	2362-1
2051	2443(RCA)
2051	2446
2051	2446(RCA)
2051	2447
2051	2447(RCA)
2051	2449-17
2051	2450
2051	2472-5632
2051	2473
2051	2475(RCA)
2051	2476
2051	2477
2051	2495(RCA)
2051	2495-166-2
2051	2495-522-4
2051	2495-529
2051	2496-125-2
2051	2510-104
2051	2546(RCA)
2051	2584
2051	2603-184
2051	2787
2051	2854
2051	2904-003
2051	2904-034
2051	2904-035
2051	2925
2051	3005(SEARS)
2051	3005-861
2051	3006(SEARS)
2051	3007(SEARS)
2051	03008-1
2051	3011
2051	3026
2051	3107-204-9000
2051	3107-204-90010
2051	3107-204-90020
2051	3111
2051	3113
2051	3505(RCA)
2051	3506(RCA)
2051	3507(WARDS)
2051	3508(RCA)
2051	3509(SEARS)
2051	3509(WARDS)
2051	3510
2051	3510(RCA)
2051	3513(RCA)
2051	3514(WARDS)
2051	3519
2051	3519(RCA)
2051	3519-1
2051	3521(SEARS)
2051	3523(SEARS)
2051	3525(RCA)
2051	3526(RCA)
2051	3527
2051	3532(RCA)
2051	3535
2051	003536
2051	3536(RCA)
2051	3536-1
2051	3537
2051	3538
2051	3538(RCA)
2051	3541(RCA)
2051	3543
2051	3543(RCA)
2051	3544-1
2051	3546
2051	3546(RCA)
2051	3546-1(RCA)
2051	3546-2(RCA)
2051	3548
2051	3548(RCA)
2051	3551(RCA)
2051	3551A(RCA)
2051	3554(RCA)
2051	3555(RCA)
2051	3556-1
2051	3558
2051	3558(RCA)
2051	3560
2051	3560(RCA)
2051	3560-2(RCA)
2051	3561
2051	3561(RCA)
2051	3561-1(RCA)
2051	3565-1
2051	3569(RCA)
2051	3569-1
2051	3571(RCA)
2051	3571R
2051	3572(RCA)
2051	3577
2051	3577(RCA)
2051	3577-1
2051	3586(RCA)
2051	3588
2051	3589
2051	3601(RCA)
2051	3601-1
2051	3603-1
2051	3604-3
2051	3614-1
2051	3614-3
2051	3625
2051	3625(RCA)
2051	3626
2051	3631-1
2051	3634.0011
2051	3706
2051	3867
2051	3999
2051	4002(PACE)
2051	4002(PENNCREST)
2051	4003E
2051	4010
2051	4021
2051	4022
2051	4046(SEARS)
2051	4057
2051	4066
2051	4085
2051	4150-01
2051	4167
2051	4168
2051	4169
2051	4309(AIRLINE)
2051	4322-542
2051	4464
2051	4465
2051	4470
2051	4470-31
2051	4470-32
2051	4470-33
2051	4470M-32
2051	4473-12
2051	4473-4
2051	4473-5
2051	4473-5X
2051	4473-9
2051	4473-M-3
2051	4594
2051	4624
2051	4630
2051	4686-132-3
2051	4686-143-3
2051	4686-144-3
2051	4686-173-3
2051	4686-183-3
2051	4686-251-3
2051	4686-257-3
2051	4686-82-3
2051	4705
2051	4714
2051	4732
2051	4733
2051	4734
2051	4737
2051	4765
2051	4768
2051	4801-0000-035
2051	4801-1100-0011
2051	4802-00003
2051	4802-0009
2051	4802-00009
2051	4802-00012
2051	4802-00015
2051	4822-130-40184
2051	4822-130-40333
2051	4822-130-40343
2051	4822-130-40354
2051	4822-130-40361
2051	4822-130-40454
2051	4839
2051	4840
2051	4841
2051	4842
2051	4852
2051	4854
2051	4856-0101
2051	4856-0107
2051	4856-0109
2051	4856-0110
2051	4927
2051	5001-002
2051	5001-014
2051	5001-020
2051	5001-037
2051	5001-069
2051	5001-072
2051	5001-539
2051	5001-542
2051	5065
2051	5226-2
2051	5380-71
2051	5380-72
2051	5613-1335
2051	5613-1335D
2051	5613-458
2051	5613-4581C
2051	5613-458B
2051	5613-458C
2051	5613-458D
2051	5613-458LQC
2051	5613-461(B)
2051	5613-535(B)
2051	5613-558C
2051	5613-710B
2051	5613-711(E)
2051	5613-870
2051	5613-870(F)
2051	5613-870F
2051	5721
2051	6136
2051	06246-00
2051	6343-1
2051	6367
2051	6367-1
2051	6514
2051	6517
2051	6854K90-074
2051	6954K90-074
2051	7001
2051	7001(E.F.JOHNSON)
2051	7002
2051	7005G(LOWREY)
2051	7006
2051	7014
2051	7015
2051	7113
2051	7117
2051	7122-5
2051	7129
2051	7171
2051	7171(GE)
2051	7172
2051	7176
2051	7233B
2051	7302
2051	7306
2051	7306-1
2051	7306-4
2051	7306-5
2051	7318
2051	7318-1
2051	7318-2
2051	7321-1
2051	7340-2
2051	7398-6117P1
2051	7398-6118P1
2051	7398-6119P
2051	7398-6119P1
2051	7429
2051	7430
2051	7431
2051	7432
2051	7433
2051	7501
2051	7502
2051	7505
2051	7506
2051	7509
2051	7515
2051	7516
2051	7517
2051	7518
2051	7519
2051	7585
2051	7586
2051	7586(GE)
2051	7587
2051	7587(GE)
2051	7588
2051	7588(GE)
2051	7589
2051	7590
2051	7590(GE)
2051	7591(GE)
2051	7637
2051	7639
2051	7641
2051	7675
2051	7676
2051	7816
2051	7817
2051	7818
2051	7992
2051	8000-00003-033
2051	8000-00004-079
2051	8000-00004-082
2051	8000-00004-243
2051	8000-00004-B5
2051	8000-00004-P086
2051	8000-00005-002
2051	8000-00005-007
2051	8000-00005-005
2051	8000-00006-280
2051	8000-00009-174
2051	8000-00011-004
2051	8000-00011-047
2051	8000-00011-048
2051	8000-00012-059
2051	8000-00029-006
2051	8000-00029-007
2051	8000-00030-007
2051	8000-00032-025
2051	8000-0004-P079
2051	8000-0004-P082
2051	8000-0004-P085
2051	8000-0004-P086
2051	8000-00049-055
2051	8000-00049-057
2051	8000-00005-009
2051	8003-114
2051	8010-176
2051	8020-205
2051	8074-4
2051	8075-4
2051	8102-207
2051	8102-209
2051	8210-1203
2051	8281
2051	8281-1
2051	8302
2051	8504
2051	8509
2051	8600
2051	8602
2051	8614 007 0
2051	8710-163
2051	8710-164
2051	8710-165
2051	8710-166
2051	8710-167
2051	8710-168
2051	8800-202
2051	8800-203
2051	8800-204
2051	8840-122
2051	8840-124
2051	8840-163
2051	8840-164
2051	8840-165

RS 276-	Industry Standard No.	RS 276-	Industry Standard No.	RS 276-	Industry Standard No.	RS 276-	Industry Standard No.	RS 276-	Industry Standard No.
2051	8840-166	2051	30289	2051	68617	2051	116076		
2051	8840-167	2051	030512-1	2051	70023-0-00	2051	116077		
2051	8840-168	2051	030512-2	2051	70023-1-00	2051	116078		
2051	8868-8	2051	030515	2051	70260-14	2051	116085		
2051	8886-2	2051	030515-4	2051	70260-15	2051	116092		
2051	8910-143	2051	030527	2051	70260-16	2051	116148		
2051	8910-145	2051	030536	2051	70260-20	2051	116588		
2051	09011	2051	030536-1	2051	70398-1	2051	116875		
2051	9011(G)	2051	030537	2051	70511	2051	118200		
2051	9013G	2051	030537-1	2051	71226-1	2051	118713		
2051	9013H	2051	030537-2	2051	71226-10	2051	119232-001		
2051	9013HF	2051	030538	2051	71226-15	2051	119258-001		
2051	9013HG	2051	030542	2051	71226-2	2051	119635		
2051	9013HH	2051	030542-1	2051	71226-3	2051	119636		
2051	9014	2051	030543	2051	71226-4	2051	119724		
2051	9014(D)	2051	030543-1	2051	71226-5	2051	119725		
2051	9014B	2051	030543-2	2051	71226-6	2051	119726		
2051	9014C	2051	030548	2051	71266-4	2051	119982		
2051	9014D	2051	31003	2051	71412-4	2051	120073		
2051	9016(P)	2051	31009	2051	71412-5	2051	120074		
2051	9016(G)	2051	35242	2051	71447-2	2051	120085		
2051	9033	2051	36580	2051	71819-1	2051	120481		
2051	9033(SYLVANIA)	2051	36917	2051	71963-1	2051	120482		
2051	9033-1	2051	36920	2051	72114	2051	120483		
2051	9033BROWN	2051	36921	2051	72115	2051	121655		
2051	9033G(SYLVANIA)	2051	37510-162	2051	72116	2051	121658		
2051	9410A	2051	37510-163	2051	72151	2051	121660		
2051	09502-8	2051	37585	2051	72204	2051	121661(RCA)		
2051	9600	2051	37884	2051	72206	2051	121662		
2051	9600-5	2051	38178	2051	72207	2051	121663		
2051	9623	2051	38283	2051	72874-52	2051	121664		
2051	9920-6-1	2051	38510-163	2051	72963-14	2051	122074		
2051	9920-6-2	2051	38510-166	2051	000073090	2051	122519		
2051	9920-7-2	2051	38510-167	2051	000073100	2051	122519(IFAMP)		
2051	10226/2	2051	38788	2051	000073120	2051	122519(SW)		
2051	10416-009	2051	39034	2051	000073130	2051	122664		
2051	11252-3	2051	39096	2051	000073230	2051	122665		
2051	11522-5	2051	40084	2051	000073251	2051	123274		
2051	11587-5	2051	40217	2051	000075290	2051	123807		
2051	11607-4	2051	40218	2051	000073310	2051	123941		
2051	11607-8	2051	40219	2051	000073332	2051	124557		
2051	11608-5	2051	40220	2051	000073370	2051	124753		
2051	11609-2	2051	40221	2051	000073390	2051	124756		
2051	11658-8	2051	40222	2051	000073391	2051	124759		
2051	11687-5	2051	40232	2051	74651-02	2051	125135		
2051	12112-C	2051	40280	2051	75561-16	2051	125139		
2051	12112-D	2051	40283	2051	75561-28	2051	125140		
2051	12112-E	2051	40290	2051	75561-3	2051	125141		
2051	12112-F	2051	40397	2051	75561-333	2051	125143		
2051	12112C	2051	40398	2051	75614-1	2051	125389		
2051	12112D	2051	40399	2051	75700-04	2051	125389-14		
2051	12112E	2051	40400	2051	75700-04-01	2051	125390		
2051	12112P	2051	40404	2051	75700-05	2051	125394		
2051	12127-6	2051	40405	2051	75700-05-01	2051	125474		
2051	12127-7	2051	40432	2051	75700-05-02	2051	125475		
2051	12127-8	2051	40456(RCA)	2051	75700-05-03	2051	125519		
2051	12127-9	2051	40473	2051	75700-08	2051	126150		
2051	12593	2051	40474	2051	75700-08-02	2051	126156		
2051	14303	2051	40476	2051	75700-09-01	2051	126331		
2051	15809-1	2051	40477	2051	75700-09-21	2051	126534		
2051	15810-1	2051	40500	2051	76236	2051	126525		
2051	15820-1	2051	40517	2051	80540	2051	126526		
2051	15835-1	2051	40518	2051	80544	2051	126699		
2051	15840-1	2051	40519	2051	80545	2051	126701		
2051	15841-1	2051	40577	2051	80813VM	2051	126702		
2051	17144	2051	40637	2051	80814VM	2051	126704		
2051	17412-5	2051	41051	2051	80815VM	2051	126706		
2051	17444	2051	41176	2051	80816VM	2051	126708		
2051	18410-143	2051	42464	2051	81170-6	2051	126711		
2051	18410-145	2051	43021-017	2051	81513-3	2051	126712		
2051	18410-146	2051	43044	2051	082006	2051	126713		
2051	18509	2051	43045	2051	082019	2051	126714		
2051	18555	2051	43054	2051	85549	2051	126716		
2051	18600-152	2051	43055	2051	86287	2051	126717		
2051	18600-153	2051	43139	2051	86812	2051	126863		
2051	19645	2051	45184	2051	86822	2051	127263		
2051	20738	2051	45810-164	2051	87757	2051	127354		
2051	22158	2051	48004-07	2051	88060-143	2051	127355		
2051	22635-002	2051	48004-08	2051	88510-173	2051	127393		
2051	22635-003	2051	50202-1	2051	88510-175	2051	127529		
2051	22810-173	2051	50202-13	2051	88686	2051	127899		
2051	23114-046	2051	50202-14	2051	88687	2051	129029(XSTR)		
2051	23114-053	2051	50202-2	2051	88688	2051	129049		
2051	23114-054	2051	50202-23	2051	88862	2051	129145		
2051	23114-082	2051	51213	2051	90209-172	2051	129146		
2051	23114-095	2051	51213-01	2051	90209-182	2051	129147		
2051	23115-057	2051	51213-02	2051	90326-001	2051	129393		
2051	23115-058	2051	51213-03	2051	90429	2051	129425		
2051	23316	2051	51213-2	2051	91605	2051	129510		
2051	0023645	2051	51428-01	2051	94027	2051	129511		
2051	0023828	2051	51429-02	2051	94047	2051	129512		
2051	25011(HONEYWELL)	2051	51429-03	2051	94048	2051	129513		
2051	25114-116	2051	51429-3	2051	95171-2	2051	129698		
2051	25114-130	2051	51441	2051	95216	2051	129899		
2051	25114-161	2051	51441-01	2051	95216RED	2051	129949		
2051	25810-161	2051	51441-02	2051	95216YEL	2051	130013		
2051	25810-162	2051	51441-03	2051	95223	2051	130403-13		
2051	25810-163	2051	51442	2051	95225	2051	130403-17		
2051	25840-161	2051	51442-01	2051	95226-2	2051	130403-18		
2051	026237	2051	51442-02	2051	96457-1	2051	130537		
2051	26810-154	2051	51442-03	2051	99109-1	2051	131240		
2051	27125-080	2051	51545	2051	99109-2	2051	131243		
2051	27125-090	2051	51547	2051	99206-1	2051	131243-12		
2051	27125-140	2051	53200-22	2051	99206-2	2051	131311		
2051	27125-160	2051	53200-23	2051	99207-2	2051	132329		
2051	27125-270	2051	53200-51	2051	100092	2051	132642		
2051	27125-300	2051	53400-01	2051	100093	2051	132643		
2051	27125-370	2051	053492	2051	100802	2051	132823		
2051	27125-500	2051	55606	2051	100119	2051	133178		
2051	27125-530	2051	55810-162	2051	101185	2051	133249		
2051	27127-550	2051	57000-5503	2051	102002	2051	133690		
2051	27840-161	2051	58215-01	2051	104389	2051	133743		
2051	27840-162	2051	58810-162	2051	105432	2051	134143		
2051	27910-12150	2051	58810-163	2051	110697	2051	137339		
2051	28810-173	2051	58810-165	2051	111303	2051	137614		
2051	28810-174	2051	58810-166	2051	112356	2051	137875		
2051	030010	2051	58810-168	2051	112357	2051	138191		
2051	030010-1	2051	58840-193	2051	112358	2051	138378		
2051	30210	2051	58840-194	2051	112359	2051	138789-1		
2051	30219	2051	58840-195	2051	112520	2051	138789-2		
2051	30224	2051	58840-196	2051	112521	2051	138789-3		
2051	30226	2051	58840-197	2051	112522	2051	139362		
2051	30227	2051	58840-198	2051	112523	2051	139569		
2051	30228	2051	58840-199	2051	113348	2051	140622		
2051	30229	2051	60395	2051	113438	2051	140858-12		
2051	30235	2051	61009-4	2051	113524	2051	141330		
2051	30241	2051	61009-4-1	2051	115167	2051	141331		
2051	30242	2051	61010-7-2	2051	115225	2051	141558		
2051	30243	2051	61011-3-2	2051	115720	2051	142683		
2051	30248	2051	61013-2-1	2051	115728	2051	142684		
2051	30253	2051	61049	2051	115875	2051	142686		
2051	30259	2051	061366	2051	116074	2051	142711		
2051	30268	2051	67586					2051	143316
2051	30269								

RS 276-	Industry Standard No.	RS 276-	Industry Standard No.	RS 276-	Industry Standard No.	RS 276-	Industry Standard No.
2051	143792	2051	205782-297	2051	600096-413	2051	810000-373
2051	143794	2051	210074	2051	600098-413-001	2051	815133
2051	143795	2051	215072	2051	601122	2051	815134
2051	143804	2051	215074	2051	602113(SHARP)	2051	815171
2051	143805	2051	215081	2051	602909-2A	2051	815171D
2051	145173	2051	216445-2	2051	603122	2051	815174
2051	145398	2051	221600	2051	604122	2051	815174L
2051	146141	2051	221857	2051	609112	2051	815182
2051	146142	2051	221897	2051	610045-3	2051	815183
2051	146144-2	2051	221918	2051	610045-4	2051	815184
2051	146153-1	2051	222131	2051	610045-5	2051	815184E
2051	147115	2051	224506	2051	610070	2051	815186
2051	147115-5	2051	228417	2051	610070-1	2051	815186C
2051	147115-6	2051	229017	2051	610070-2	2051	815186L
2051	147357-1-1	2051	231017	2051	610070-3	2051	815190
2051	147357-7-1	2051	231140-15	2051	610070-4	2051	815191
2051	147360-1	2051	231374	2051	610076	2051	815198
2051	147363-1	2051	232678	2051	610076-1	2051	815201
2051	147513	2051	234612	2051	610076-2	2051	815202
2051	147555-1	2051	234758	2051	610077	2051	815210
2051	147664	2051	234763	2051	610077-1	2051	815212
2051	147665	2051	235192	2051	610077-2	2051	815227
2051	150741	2051	235205	2051	610077-4	2051	815233
2051	150763	2051	235206	2051	610077-5	2051	815237
2051	150768	2051	236285	2051	610077-6	2051	0820220
2051	157008	2051	236286	2051	610078	2051	845050
2051	161918-28	2051	237025	2051	610078-1	2051	848082
2051	165668	2051	237223	2051	610078-2	2051	851881
2051	165827	2051	238368	2051	610094	2051	883802
2051	165828	2051	239970	2051	610094-1	2051	894876
2051	167263	2051	240401	2051	610094-2	2051	900552-20
2051	167540	2051	242758	2051	610132	2051	900552-30
2051	167541	2051	242759	2051	610132-1	2051	900552-6
2051	167688	2051	256217	2051	610142-2	2051	900552-8
2051	167956	2051	256517	2051	610142-3	2051	911743-1
2051	167957	2051	256917	2051	610142-4	2051	916009
2051	168405	2051	262066	2051	610142-5	2051	916028
2051	168651	2051	265240	2051	610142-7	2051	916030
2051	169197	2051	266685	2051	610142-8	2051	916031
2051	169574	2051	267898	2051	610143-3	2051	916031(PENNYS)
2051	169679	2051	267899	2051	610146-3	2051	916050
2051	169680	2051	268044L	2051	610146-5	2051	916059
2051	169771	2051	270819	2051	610147-1	2051	916091
2051	170294	2051	275131	2051	610148-2	2051	928103-1
2051	170308	2051	276331	2051	610148-2A	2051	930236
2051	170967-1	2051	279417	2051	610150-2	2051	941295-2
2051	170968-1	2051	279917	2051	610151-2	2051	941295-3
2051	171003(TOSHIBA)	2051	282317	2051	610151-4	2051	943720-001
2051	171009(TOSHIBA)	2051	290458LGD	2051	610165-1	2051	954330-2
2051	171026	2051	299371-1	2051	610165-2	2051	959492-2
2051	171026(TOSHIBA)	2051	300113	2051	610167-1	2051	961544-1
2051	171027	2051	301591	2051	610167-2	2051	965000
2051	171030(TOSHIBA)	2051	302342	2051	610168-1	2051	970247
2051	171040	2051	3045581B	2051	610168-2	2051	970250
2051	171040(SEARS)	2051	308449	2051	610180-1(LAST-IP)	2051	970252
2051	171040(TOSHIBA)	2051	309442	2051	610232-1	2051	970659
2051	171044(TOSHIBA)	2051	313909-1	2051	611428	2051	970660
2051	171046	2051	0320031	2051	615093-2	2051	970661
2051	171162-005	2051	320529	2051	615179-1	2051	970662
2051	171162-006	2051	321517	2051	615179-2	2051	970916
2051	171162-008	2051	321573	2051	618072	2051	970916-6
2051	171162-009	2051	325079	2051	618126-1	2051	970940
2051	171162-113	2051	328785	2051	618217-2	2051	972155
2051	171162-119	2051	330803	2051	618810-2	2051	972156
2051	171162-132	2051	333241	2051	619006	2051	972214
2051	171162-143	2051	334724-1	2051	619006-7	2051	972215
2051	171162-161	2051	335288-4	2051	651891	2051	980147
2051	171162-162	2051	335774	2051	651955	2051	980440
2051	171162-190	2051	346015-24	2051	651955-1	2051	982231
2051	171162-191	2051	346016-14	2051	651955-2	2051	982510
2051	171162-202	2051	346016-18	2051	651955-3	2051	982511
2051	171162-286	2051	346016-19	2051	651995-1	2051	982512
2051	171162-U90	2051	346016-25	2051	651995-2	2051	983097
2051	171554	2051	346016-26	2051	651995-3	2051	983743
2051	171558	2051	00352080	2051	652072	2051	984197
2051	171559	2051	379101K	2051	652091	2051	984198
2051	171677	2051	379102	2051	652230	2051	984222
2051	175007-275	2051	388060	2051	654000	2051	984224
2051	175007-276	2051	405457	2051	656204	2051	984286
2051	175027-021	2051	433836	2051	656524	2051	984590
2051	175043-058	2051	00444028-010	2051	656719	2051	984591
2051	175043-059	2051	00444028-014	2051	658577	2051	984593
2051	175043-060	2051	489751-025	2051	658578	2051	984686
2051	181023	2051	489751-026	2051	658657	2051	984687
2051	181214	2051	489751-029	2051	671077-6	2051	984745
2051	181504-2	2051	489751-030	2051	700047-47	2051	984854
2051	183017	2051	489751-040	2051	700047-49	2051	984879
2051	183030	2051	489751-041	2051	700181	2051	985087
2051	183031	2051	489751-107	2051	700230-00	2051	985098
2051	185236	2051	489751-122	2051	700231-00	2051	985099
2051	187218	2051	489751-123	2051	702884	2051	985100
2051	190426	2051	489751-125	2051	720236	2051	985101
2051	190428	2051	489751-164	2051	720240	2051	985102
2051	190715	2051	489751-166	2051	740437	2051	985543
2051	196023-1	2051	489751-172	2051	740441	2051	986542
2051	196023-2	2051	508762	2051	740461	2051	986636
2051	198003-1	2051	514023	2051	740462	2051	987010
2051	198003-2	2051	5150438	2051	740463	2051	988003
2051	198007-3	2051	5150458	2051	740466	2051	988991
2051	198013-P1	2051	533802	2051	740470	2051	992129
2051	198023-1	2051	552308	2051	740857	2051	995016
2051	198023-3	2051	555606	2051	741726	2051	995017
2051	198023-4	2051	567312	2051	741731	2051	995870-1
2051	198023-5	2051	570000-5452	2051	741855	2051	995870-3
2051	198030	2051	570000-5503	2051	741861	2051	996746
2051	198030-2	2051	570004-503	2051	742728	2051	1020612
2051	198030-3	2051	570005-452	2051	757008-02	2051	1127859
2051	198030-4	2051	570005-503	2051	760236	2051	1221962
2051	198030-6	2051	570009-01-504	2051	760251	2051	1222123
2051	198030-7	2051	572683	2051	760259	2051	1222133
2051	198031-1	2051	0573066	2051	789278-101	2051	1222424
2051	198031-2	2051	0573202	2051	800132-001	2051	1223782
2051	198042-2	2051	0573418	2051	801512	2051	1223912
2051	198042-3	2051	0573430	2051	801513	2051	1223913
2051	198051-1	2051	0573460	2051	801514	2051	1223916
2051	198051-2	2051	573467	2051	801515	2051	1261915-383
2051	198051-3	2051	0573468	2051	801516	2051	1320135
2051	198051-4	2051	0573469	2051	801517	2051	1320135A
2051	198067-1	2051	0573469H	2051	801524	2051	1320135BC
2051	198081-2	2051	0573479H	2051	801529	2051	1320135C
2051	198581-2	2051	0573480	2051	801530	2051	1417302-1
2051	198581-3	2051	0573481	2051	801532	2051	1417306-1
2051	200064-6-107	2051	0573481H	2051	801534	2051	1417306-2
2051	200076	2051	0573490	2051	801536	2051	1417306-4
2051	200251-5377	2051	0573491H	2051	801543	2051	1417312-1
2051	202609-0713	2051	0573523	2051	801729	2051	1417312-2
2051	202862-947	2051	573932	2051	803182-5	2051	1417318-1
2051	202907-047P1	2051	0573556	2051	803369-6	2051	1417324-1
2051	202914-417	2051	0573981	2051	803372-0	2051	1417340-2
2051	202915-627	2051	581034A	2051	803733-0	2051	1471113-2
2051	202922-237	2051	581054	2051	803733-3	2051	1471113-3
2051	204210-002	2051	581055	2051	803735-3	2051	1471115-1
2051	204969	2051	600080-413-001			2051	1471115-12
2051	205782-103	2051	600080-413-002				

RS 276-	Industry Standard No.	RS 276-	Industry Standard No.	RS 276-	Industry Standard No.	RS 276-	Industry Standard No.
2051	1471120-15	2051	2320111	2052	28B523	2052	488155042
2051	1471120-7	2051	2320123	2052	28C1095	2052	488155074
2051	1471120-8	2051	2320413	2052	28C1095(6)	2052	065-007
2051	1471120-8-9	2051	2320441	2052	28C1095L	2052	90-111
2051	1472495-1	2051	2320471(LAST-IP)	2052	28C1095M	2052	90-176
2051	1473500-1	2051	2320591-1	2052	28C1096	2052	90-451
2051	1473505-1	2051	2320595	2052	28C1096(M)	2052	106-004
2051	1473519-1	2051	2320696	2052	28C1096-32M	2052	142-006
2051	1473527-1	2051	2320696-1	2052	28C1096-4ZL	2052	142-006(REGENCY)
2051	1473532-1	2051	2321541	2052	28C1096-OR	2052	172-011-9-001
2051	1473556-001	2051	2326953	2052	28C109632M	2052	172-038-9-003
2051	1473536-1	2051	2327023	2052	28C10964ZL	2052	260P22801
2051	1473538-1	2051	2327122	2052	28C1096A	2052	1039-0433
2051	1473539-1	2051	2327363	2052	28C1096B	2052	2041-03
2051	1473546-1	2051	2360924-5601	2052	28C1096C	2052	2408-17
2051	1473546-2	2051	2469749	2052	28C1096D	2052	5614-359C
2051	1473546-3	2051	2469755	2052	28C1096F	2052	8910-146
2051	1473548-1	2051	2479692	2052	28C1096G	2052	18410-147
2051	1473550-1	2051	2479836	2052	28C1096GN	2052	38510-168
2051	1473551-1	2051	2485078-1	2052	28C1096H	2052	000073381
2051	1473554-1	2051	2485078-2	2052	28C1096J	2052	2320846
2051	1473555-1	2051	2485078-2(SEARS)	2052	28C1096K	2053	A67-76-200
2051	1473555-2	2051	2485078-3	2052	28C1096L	2053	0117S(RF-PWR)
2051	1473556-1	2051	2485079-1	2052	28C1096LM	2053	C1237E
2051	1473557-1	2051	2485079-2	2052	28C1096M	2053	C1307
2051	1473560-002	2051	2485079-3	2052	28C1096N	2053	C1377
2051	1473561-Y	2051	2495166-2	2052	28C1096OR	2053	C1816
2051	1473569-1	2051	2495166-4	2052	28C1096Q	2053	C1964
2051	1473572-1	2051	2495166-8	2052	28C1096R	2053	C799
2051	1473582-1	2051	2495166-9	2052	28C1096W	2053	DDBY228001
2051	1473586-2	2051	2495521-1	2052	28C1096X	2053	DDDY231002
2051	1473589-1	2051	2495522-4	2052	28C1096Y	2053	DDBY307003
2051	1473595-1	2051	2496125-2	2052	28C1098	2053	E00236
2051	1473601-001	2051	2498457-2	2052	28C1098(4)K	2053	EQ8-0159
2051	1473601-1	2051	2498904-3	2052	28C1098(4)L	2053	EQ8-141
2051	1473601-2	2051	2498904-4	2052	28C1098(L)	2053	ESQ-141
2051	1473614-1	2051	2498904-6	2052	28C1098(M)	2053	GB-216
2051	1473614-3	2051	2505209	2052	28C1098A	2053	GB-332
2051	1473622-1	2051	2520063	2052	28C1098B	2053	HST5906
2051	1473626-1	2051	2520733	2052	28C1098C	2053	IP20-0154
2051	1473631-1	2051	2621567-1	2052	28C1098D	2053	PL-172-013-9001
2051	1473676-1(3RD-IP)	2051	2621764	2052	28C1098L	2053	PL-172-024-9-003
2051	1476188-1	2051	2622284	2052	28C1098M	2053	PTC186
2051	1501883	2051	2640830-1	2052	28C1107Q	2053	QT-C1307XZA
2051	1522237-20	2051	2640843-1	2052	28C1162(C)	2053	RE201
2051	1563295-101	2051	2712080	2052	28C1162(RF-PWR)	2053	REN236
2051	1596408	2051	3005861	2052	28C1162C	2053	STT2405
2051	1611708-2	2051	3068305-2	2052	28C1162CB	2053	STT2406
2051	1617510-1	2051	3201104-10	2052	28C1162C	2053	STT4483
2051	1690019-01	2051	3403787	2052	28C1162C(RF-PWR)	2053	STT9001
2051	1700008	2051	3457107-1	2052	28C1162CP	2053	STT9002
2051	1700019	2051	3457632-5	2052	28C1162D	2053	STT9004
2051	1780145-1	2051	3468182-1	2052	28C1162MP	2053	STT9005
2051	1780145-2	2051	3468182-2	2052	28C1162WB	2053	TM236
2051	1780145-2-001	2051	3468242-2	2052	28C1162WBP	2053	UP1121
2051	1780724-1	2051	3596117	2052	28C1162WT	2053	UP1611
2051	1780738-1	2051	3596338	2052	28C1162WTA	2053	UP1621
2051	1815041	2051	3596339	2052	28C1162WTB	2053	V82SC1237-1
2051	1815042	2051	3596570	2052	28C1162WTC	2053	WEP1307
2051	1815043	2051	3597114	2052	28C1162WTD	2053	01-031173
2051	1815054	2051	3598070	2052	28C1226	2053	2N2876
2051	1815154	2051	3670724H	2052	28C1226(A)	2053	2N3253
2051	1817005	2052	C1095	2052	28C1226(AP)	2053	2N5785
2051	1817007	2052	C1095(6)	2052	28C1226(P)	2053	28C-1307
2051	1817108	2052	C1095L	2052	28C1226(R)	2053	28C1173X(RF-PWR)
2051	1851515	2052	C1095M	2052	28C1226-0	2053	28C1307
2051	1950039	2052	C1096	2052	28C1226A	2053	28C1307-1
2051	1960023	2052	C1096(M)	2052	28C1226A(P)	2053	28C1377
2051	1960177-2	2052	C109632M	2052	28C1226A(Q)	2053	28C1816
2051	1968958	2052	C10964ZL	2052	28C1226A(QPR)	2053	28C1969
2051	2000646-103	2052	C1096A	2052	28C1226A(R)	2053	28C1969B
2051	2000646-107	2052	C1096B	2052	28C1226AC	2053	28C1969BH
2051	2002153-77	2052	C1096C	2052	28C1226ACP	2053	28C1969H
2051	2002621-2	2052	C1096D	2052	28C1226AF	2053	28C2043
2051	2003073-0701	2052	C1096K	2052	28C1226AP	2053	28C2393D
2051	2003073-10	2052	C1096L	2052	28C1226AQ	2053	28C2394D
2051	2003073-9	2052	C1096M	2052	28C1226AR	2053	28C490-BLU
2051	2003168-135	2052	C1096N	2052	28C1226ARL	2053	28C490-R
2051	2003168-136	2052	C1096W	2052	28C1226B	2053	28C490-RED
2051	2003229-25	2052	C1098	2052	28C1226BL	2053	28C490-Y
2051	2003239-65	2052	C1098A	2052	28C1226C	2053	28C490-YEL
2051	2006227-51	2052	C1098B	2052	28C1226CP	2053	2809
2051	2006534-115	2052	C1098C	2052	28C1226D	2053	28C256
2051	2006431-45	2052	C1098D	2052	28C1226E	2053	28D562A
2051	2006431-46	2052	C1098L	2052	28C1226F	2053	28D562B
2051	2006431-49	2052	C1098M	2052	28C1226G	2053	28D562C
2051	2006514-60	2052	C1107Q	2052	28C1226H	2053	28D562J
2051	2006582-25	2052	C1162WBP	2052	28C1226L	2053	28D562K
2051	2006607-60	2052	C1226	2052	28C1226O	2053	28D562L
2051	2006613-77	2052	C1226-0	2052	28C1226OR	2053	28D5620R
2051	2006623-145	2052	C1226A	2052	28C1226P	2053	28D562R
2051	2006681-96	2052	C1226AC	2052	28C1226Q	2053	28D562X
2051	2010088-49	2052	C1226AO	2052	28C1226QR	2053	28D562Y
2051	2010499-52	2052	C1226AP	2052	28C1226R	2053	09-302192
2051	2041614	2052	C1226AQ	2052	28C1226RL	2053	09-302193
2051	2092055-0714	2052	C1226AR	2052	28C1226RLP	2053	09-305137
2051	2092417-0719	2052	C1226C	2052	28C1226RLQ	2053	21A118-049
2051	2092417-0720	2052	C1226CF	2052	28C1226RLR	2053	022-2876-012
2051	2092417-0721	2052	C1226F	2052	28C1226SC	2053	31-091-3
2051	2092417-0724	2052	C1226P	2052	28C1226Y	2053	48-43240G01
2051	2092417-0725	2052	C1226Q	2052	28C1368D	2053	065-210
2051	2092417-17	2052	C1226QR	2052	28C1848	2053	90-610
2051	2092417-18	2052	C1226R	2052	28C1848Q	2053	123-008
2051	2092417-19	2052	C1368D	2052	28C1848R	2053	172-013-9-001
2051	2092605-0705	2052	C931	2052	28C1848V	2053	172-024-9-003
2051	2092608-22	2052	C931C	2052	28C931	2053	172-029-9-001
2051	2092609	2052	C931D	2052	28C931C	2053	176-024-9-001
2051	2092609-0001	2052	C931E	2052	28C931D	2053	176-044-9-001
2051	2092609-0002	2052	D44C1	2052	28C931D	2053	176-073-9-002
2051	2092609-001	2052	DBBY003001	2052	28C931E	2053	176-073-9-012
2051	2092609-0022	2052	DBBY003002	2052	28D1173	2053	260-10-043
2051	2092609-0023	2052	DBBY227001	2052	28D1368D	2053	260-10-053
2051	2092609-0024	2052	DBBY227004	2052	28D359C	2053	260-10-055
2051	2092609-0026	2052	EC0186A	2052	28D359C2	2053	1042-09
2051	2092609-0027	2052	ET453611	2052	28D359D	2053	2000-212
2051	2092609-0705	2052	G05-416-C	2052	28D359D1	2053	2437-17
2051	2092609-0706	2052	GB-247	2052	28D359D2	2053	5001-071
2051	2092609-0707	2052	KSC1096-0	2052	28D359E	2053	8000-00028-039
2051	2092609-0713	2052	KSC1096-Y	2052	04-11620-01	2053	8000-00032-028
2051	2092609-0715	2052	MJB-200E	2052	4ZL	2053	8000-00043-065
2051	2092609-0718	2052	NC8V14	2052	07-07165	2301	HD1-74000
2051	2092609-0720	2052	NSB-181	2052	09-3-2123	2301	HD9-74000
2051	2092609-0721	2052	PL-172-024-9-004	2052	09-302080	2301	MM74U0ON
2051	2092609-1	2052	PT-029	2052	09-302121	2302	EC0074002
2051	2092609-2	2052	REN186A	2052	09-302123	2302	HD1-74002
2051	2092609-3	2052	TN186A	2052	09-302126	2302	HD9-74002
2051	2092609-5	2052	WEP900	2052	10-0009	2302	MM74C02N
2051	2093308-0701	2052	YAANL28C1096K	2052	13-41628-5	2302	EC0074004
2051	2093308-0702	2052	YAANL28C1096KLMN	2052	48-155042	2305	EC0074008
2051	2093308-0703	2052	YAANL28C1096L	2052	48-155074	2305	HD1-74008
2051	2093308-0708	2052	YAANL28C1096M	2052	48-155097	2305	HD9-74008
2051	2093308-8708	2052	YAANL28C1096N	2052	48-355039	2305	MM74C08N
2051	2320022	2052	01-051096	2052	48-41784J04		
				2052	48-90343A76		

RS 276-	Industry Standard No.	RS 276-	Industry Standard No.	RS 276-	Industry Standard No.	RS 276-	Industry Standard No.
2310	ECG74C74	2420	F4020				
2310	HD1-74C74	2420	GE-4020				
2310	HD9-74C74	2420	HBF4020				
2310	MM74C74N	2420	HD4020				
2312	ECG74C76	2420	HEPC4020P				
2312	HD1-74C76	2420	MC14020				
2312	HD9-74C76	2420	MM4020(IC)				
2312	MM74C76N	2420	RS4020				
2315	HD1-74C90	2420	SCL4020				
2315	HD9-74C90	2420	SK4020				
2315	MM74C90N	2420	SW4020				
2321	ECG74C192	2420	TP4020				
2321	HD1-74C192	2420	4020(CMOS)				
2321	HD9-74C192	2421	CD4021BE				
2321	MM74C192N	2421	CM4021				
2322	ECG74C193	2421	ECG4021				
2322	HD1-74C193	2421	ECG4021B				
2322	HD9-74C193	2421	F4021				
2322	MM74C193N	2421	GE-4021				
2401	CD4001	2421	HBF4021				
2401	CD4001BE	2421	HD4021				
2401	CM4001	2421	HEPC4021P				
2401	ECG4001	2421	MC14021				
2401	ECG4001B	2421	MM4021				
2401	F4001	2421	N4021				
2401	GE-4001	2421	SCL4021				
2401	HBF4001	2421	SK4021				
2401	HD4001	2421	SW4021				
2401	HEPC4001P	2421	TP4021				
2401	MB84011V	2421	4021(IC)				
2401	MC14001	2423	CD4023BE				
2401	MEM4001	2423	CM4023				
2401	MM4001	2423	ECG4023				
2401	N4001	2423	ECG4023B				
2401	RS4001	2423	F4023				
2401	SCL4001	2423	GE-4023				
2401	SW4001	2423	HBF4023				
2401	TC4001P	2423	HD4023				
2401	TP4001	2423	MC14023				
2401	51-10655A17	2423	MM4023				
2401	544-3001-103	2423	N4023				
2401	905-125	2423	SCL4023				
2401	4001(IC)	2423	SW4023				
2411	CD4011	2423	TP4023				
2411	CD4011BE	2423	4023				
2411	CM4011	2427	CD4027BE				
2411	DDEY089001	2427	CM4027				
2411	ECG4011	2427	ECG4027				
2411	ECG4011B	2427	ECG4027B				
2411	F4011	2427	F4027				
2411	F4011PC	2427	GE-4027				
2411	GE-4011	2427	HBF4027				
2411	HBF4011	2427	HD4027				
2411	HD4011	2427	MC14027				
2411	MB84011	2427	MM4027				
2411	MB84011-0	2427	N4027				
2411	MB84011U	2427	RS4027				
2411	MC-14011CP	2427	SCL4027				
2411	MC14011	2427	SW4027				
2411	MC14011B	2427	TP4027				
2411	MC14011CP	2427	4027				
2411	MEM4011	2428	ECG4028				
2411	MM4011	2428	ECG4028B				
2411	N4011	2446	ECG980				
2411	RS4011	2446	SK4046/980				
2411	SCL4011	2447	ECG4511B				
2411	SW4011	2449	CD4049UBE				
2411	TC4011BP	2449	CM4049				
2411	TC4011P	2449	ECG4049				
2411	TP4011	2449	ECG4049B				
2411	UPD4011C	2449	F4049				
2411	307-113-9-001	2449	GE-4049				
2411	307-152-9-012	2449	HD4049				
2411	905-126	2449	MC14049				
2411	1147-09	2449	MC14049B				
2411	4011(IC)	2449	MC14049CP				
2411	4011-PC	2449	MEM4049				
2411	4011PC	2449	MM4049				
2411	84011	2449	N4049				
2411	84011U	2449	RS4049				
2412	CD4012BE	2449	SCL4049				
2412	CM4012	2449	SK4049				
2412	ECG4012	2449	SW4049				
2412	ECG4012B	2449	TP4049				
2412	F4012	2449	4049				
2412	GE-4012	2449	4049(IC)				
2412	HBF4012	2450	CD4050BE				
2412	HD4012	2450	CM4050				
2412	MC14012	2450	ECG4050				
2412	MM4012	2450	ECG4050B				
2412	N4012	2450	F4050				
2412	SCL4012	2450	GE-4050				
2412	SW4012	2450	HD4050				
2412	TP4012	2450	HEPC4050P				
2412	4012(IC)	2450	MC14050				
2413	CD4013	2450	MEM4050				
2413	CD4013AE	2450	MM4050				
2413	CD4013BE	2450	N4050				
2413	CM4013	2450	RS4050				
2413	ECG4013	2450	SCL4050				
2413	ECG4013B	2450	SK4050				
2413	F4013	2450	SW4050				
2413	GE-4013	2450	TP4050				
2413	HBF4013	2450	4050				
2413	HD4013	2451	CD4051BE				
2413	MC14013	2451	CM4051				
2413	MEM4013	2451	ECG4051				
2413	MM4013	2451	ECG4051B				
2413	N4013	2451	F4051				
2413	RS4013	2451	GE-4051				
2413	SCL4013	2451	HEPC4051P				
2413	SW4013	2451	MC14051				
2413	TP4013	2451	MEM4051				
2413	51-10655A19	2451	MM4051				
2413	4013(IC)	2451	SCL4051				
2417	CD4017BE	2451	SK4051				
2417	CM4017	2451	TP4051				
2417	ECG4017	2451	4051				
2417	ECG4017B	2466	ECG4066B				
2417	F4017	2490	CD4518BE				
2417	GE-4017	2490	ECG4518				
2417	HBF4017	2490	ECG4518B				
2417	HD4017	2490	F4518				
2417	MC14017	2490	GE-4518				
2417	MM4017	2490	HD4518				
2417	RS4017	2490	MC14518				
2417	SCL4017	2490	MM4518				
2417	SW4017	2490	RS4518				
2417	TP4017	2490	SCL4518				
2417	4017	2490	SK4518				
2420	CD4020BE	2490	TP4518				
2420	CM4020	2490	4518				
2420	ECG4020						
2420	ECG4020B						

MOTOR.	Industry Standard No.	MOTOR.	Industry Standard No.	MOTOR.	Industry Standard No.	MOTOR.	Industry Standard No.
(2) 2N6297	MJ4211	BC415	BC253	BU205	IR801	C230D3	C230D3
(2) 2N6298	MJ920	BC415	BC336	BU205	MJ105	C230E	C230E
(2) 2N6299	MJ921	BC415	BC415	BU205	MJ205	C230E3	C230E3
(2) 2N6300	MJ1200	BC416	BC416	BU205	TIP550	C230F	C230F
(2) 2N6301	MJ1201	BC445	BC445	BU205	TIP551	C230F3	C230F3
.4M10525	1N1326	BC446	BC446	BU205	TIP69	C230M	C230M
.4M10525	1N1326A	BC447	BC447	BU205	TIP70	C230M3	C230M3
.4M127.5Z5	1N1327A	BC448	BC448	BU205	TIP71	C231A	C231A
.4M140Z10	MZ292-140	BC449	BC449	BU205	TIP72	C231A3	C231A3
.4M40Z10	1N5891	BC450	BC450	BU205	2SC1004A	C231B	C231B
.4M170Z10	MZ92-170	BC485	BC485	BU205	2SC1892	C231B3	C231B3
.4M170Z10	1N5894	BC485A	BF860	BU205	2SC999	C231C	C231C
.4M190Z10	MZ92-190	BC486	BC486	BU207	BU207	C231C3	C231C3
.4M190Z10	1N5896	BC486	BF597	BU207	D56W3	C231D	C231D
.4M33Z	CD3909732	BC487	BC487	BU207	D56W4	C231D3	C231D3
.4M8.2Z	CD3907562	BC488	BC488	BU207	DTS712	C231E	C231E
.5M110Z10	04Z110	BC489	BC489	BU207	DTS802	C231E3	C231E3
.5M11AZ10	MZ92-11	BC490	BC490	BU207	DTS812	C231F	C231F
.5M11AZ10	1N5857	BC490	BF397	BU207	TIP552	C231F3	C231F3
.5M12AZ	GA4Z12.0	BC546	BC546	BU207	2SC1005	C231M	C231M
.5M14Z10	MZ92-14	BC547	BC547	BU207	2SC1046	C231M3	C231M3
.5M14Z10	1N5860	BC548	BC548	BU207	2SC1086	C232A	C232A
.5M16Z10,5	1S2160,A	BC549	BC549	BU207	2SC1099	C232B	C232B
.5M16Z10,5	1S7160,A	BC550	BC550	BU207	2SC1100	C232C	C232C
.5M17Z10	MZ92-17	BC556	BC556	BU207	2SC1132	C232D	C232D
.5M17Z10	1N5863	BC557	BC557	BU207	2SC1170	C232E	C232E
.5M19Z10	MZ92-19	BC558	BC558	BU207	2SC1309	C232F	C232F
.5M19Z10	1N5865	BC559	BC559	BU207	2SC1348	C232M	C232M
.5M2.4AZ	GA4Z2.4	BC560	BC560	BU207	2SC1413	C233A	C233A
.5M2.5AZ10	MZ92-2.5	BCX25	BCX25	BU208	BU108	C233B	C233B
.5M2.5AZ10	1N5858	BCX26	BCX26	BU208	BU208	C233C	C233C
.5M2.8AZ10	MZ92-2.8	BCX27	BCX27	BU208	D56W2	C233D	C233D
.5M2.8AZ10	1N5840	BCX27	BSS38	BU208	DTS714	C233E	C233E
.5M25Z10	MZ92-25	BCX28	BCX28	BU208	DTS804	C233F	C233F
.5M25Z10	1N5869	BCX29	BCX29	BU208	DTS814	C233M	C233M
.5M28Z10	MZ92-28	BCX29	BF899	BU208	MJ3480	C35A	C35A
.5M28Z10	1N5871	BCX30	BCX30	BU208	TIP309	C35B	C35B
.5M3.3AZ5	BZT88-C3V3	BCX45	BCX45	BU208	TIP310	C35C	C35C
.5M3025	BZT88-C30	BCX46	BCX46	BU208	TIP553	C35D	C35D
.5M6.0AZ10	MZ92-6.0	BCX47	BCX47	BU208	2SC1170A	C35E	C35E
.5M6.0AZ10	1N5849	BCX47	BF861	BU208	2SC1172	C35F	C35F
.5M60Z10	MZ92-60	BCX48	BCX48	BU208	2SC1358	C35G	C35G
.5M60Z10	1N5880	BCX48	BFS98	BU208	2SC1894	C35H	C35H
.5M7.5Z10	G4Z7.5	BCX49	BCX49	BU208	2SD246	C35M	C35M
.5M8.7AZ10	MZ92-8.7	BCX50	BCX50	BZX79-C10	BZX79-C10	C35N	C35N
.5M8.7AZ10	1N5854	BCX58	BCX58	BZX79-C4V7	BZX79-C4V7	C358	C358
.5M87Z10	MZ92-87	BCX59	BCX59	BZX79-C75	BZX79-C75	C35U	C35U
.5M87Z10	1N5885	BF198	BF198	BZX79-C9V1	BZX79-C9V1	CA3054	CA3026
AM26LS31DC	AM26LS31DC	BF199	BF199	C106A1	C107A	CA3054	CA3054
AM26LS31PC	AM26LS31PC	BF254	BF254	C106A1	S106A	CA3054	LM3026
BB105A	BB105A	BF254	BF494	C106A1	S107A	CA3054	LM3054
BB105B	BB105 B	BF255	BF255	C106B	C107B	CA3054	JA3026HM
BB105Q	MA320B	BF255	BF495	C106B1	S106B	CA3054P	JA3054DC
BB105Q	BB105Q	BF362	BF362	C106B1	S107B	CA3059	CA3058
BB105Q	RF400	BF363	BF363	C106C	C107C	CA3059	CA3059
BB105QW	1S220B	BF366	BF314	C106C1	C106C1	CA3059	CA3079
BB205B	BB205 B	BF366	BF366	C106C1	S107C	CA3059	JA742DC
BB205Q	BB205Q	BF367	BF367	C106D1	C107D	CA3139	CA3139
BC174	BC174	BF368	BF368	C106D1	S106D	C8122A	C8122A
BC182	BC182	BF369	BF369	C106D1	S107D	C8122B	C8122B
BC183	BC183	BF371	BF371	C106E	C107E	C8122C	C8122C
BC184	BC184	BF373	BF373	C106E1	S106E	C8122D	C8122D
BC213	BC213	BF374	BF374	C106E1	S107E	C8122E	C8122E
BC214	BC214	BF374	BF384	C106F	C107F	C8122M	C8122M
BC237	BC167	BF375	BF375	C106F1	C106F1	C8122N	C8122N
BC237	BC171	BF375	BF385	C106F1	S106F	C8122S	C8122S
BC237	BC237	BF391	BF297	C106F1	S107F	C8220-5	C8220-5
BC237	BC332	BF391	BF391	C106M	C107M	C8220-7	C8220-7
BC237	BC382	BF392	BF298	C106M1	C106M1	C8220-9	C8220-9
BC237	BC582	BF392	BF392	C106M1	S106M	C8221-5	C8221-5
BC237A	BC385	BF393	BF299	C106M1	S107M	C8221-7	C8221-7
BC238	BC168	BF393	BF393	C106Q	C107Q	C8221-9	C8221-9
BC238	BC170	BF394	BF394	C106Y	C107Y	C86395	C86395
BC238	BC172	BF395	BF395	C106Y1	C106Y1	C86396	C86396
BC238	BC238	BF479	BF479	C122A1	C122A	C86397	C86397
BC238	BC333	BF506	BF324	C122A1	T1C116A	C86398	C86398
BC238	BC383	BF506	BF414	C122B1	C122B1	C86399	C86399
BC238	BC386	BF506	BF450	C122B1	S2600B	C86401	C86401
BC238	BC583	BF506	BF451	C122B1	T1C116B	C86402	C86402
BC238B	BSW92	BF506	BF506	C122C1	C122C1	C86403	C86403
BC238B	BSW42	BF506	BF706	C122C1	T1C116C	C86404	C86404
BC239	BC169	BF506	BF906	C122D1	C122D1	C86405	C86405
BC239	BC173	BF509	BF500	C122D1	S2600D	C86505	C86505
BC239	BC239	BF509	BF509	C122D1	T1C116D	C86506	C86506
BC239	BC584	BF509	BF709	C122E1	C122E1	C86507	C86507
BC239	BCW87	BF679	BF679	C122E1	T1C116E	C86508	C86508
BC239	BCW88	BF680	BF680	C122F1	C122F1	C86509	C86509
BC307	BC307	BFR90	AT1425	C122F1	T1C116F	C872-3	C872-3
BC307	BC512	BFR90	AT50	C122M1	C122M1	C872-4	C872-4
BC308	BC181	BFR90	AT51	C122M1	S2600M	C872-5	C872-5
BC308	BC250	BFR90	AT52	C122M1	T1C116M	C872-6	C872-6
BC308	BC308	BFR90	BFR90	C122N1	C122N1	C872-7	C872-7
BC308	BC334	BFR90	NE73435	C1228I	C1228I	C872-8	C872-8
BC308	BC513	BFR90	2SC1090	C228A	C228A	CT15-10	CT15-10
BC308A	BSW44	BFR91	BFR91	C228A3	C228A3	CT15-3	CT15-3
BC309	BC251	BFR91	TP393	C228B	C228B	CT15-4	CT15-4
BC309	BC252	BFR91	TP491	C228B3	C228B3	CT15-5	CT15-5
BC309	BC309	BFR96	BFR96	C228C	C228C	CT15-6	CT15-6
BC309	BC514	BFR96	TP312	C228C3	C228C3	CT15-7	CT15-7
BC317	BC317	BFR96	TP394	C228D	C228D	CT15-9	CT15-9
BC318	BC318	BFW92A	A486	C228D3	C228D3	CT15A10	CT15A10
BC319	BC319	BFW92A	BFW92	C228E	C228E	CT15A3	CT15A3
BC320	BC320	BFW92A	BFW93	C228E3	C228E3	CT15A4	CT15A4
BC321	BC321	BFW92A	TP390	C228F	C228F	CT15A5	CT15A5
BC322	BC322	BFX89	A490	C228F3	C228F3	CT15A6	CT15A6
BC327	BC327	BFX89	BFX89	C228M	C228M	CT15A7	CT15A7
BC328	BC381	BFY90	BFY90	C228M3	C228M3	CT15A8	CT15A8
BC328	BF896	BFY90	SD1300	C229A	C229A	CT15A9	CT15A9
BC330	BC330	BFY90	2SC1807	C229A3	C229A3	CT220-3	CT220-3
BC337	BC337	BU204	BU204	C229B	C229B	CT220-5	CT220-5
BC338	BC338	BU204	DTS701	C229B3	C229B3	CT220-7	CT220-7
BC338	BF859	BU204	IR701	C229C	C229C	CT220-9	CT220-9
BC347	BC331	BU204	2SC1004	C229C3	C229C3	CT221-3	CT221-3
BC347	BC347	BU204	2SC1034	C229D	C229D	CT221-5	CT221-5
BC348	BC348	BU204	2SC1078	C229D3	C229D3	CT221-7	CT221-7
BC348	BC358	BU204	2SC1101	C229E	C229E	CT221-9	CT221-9
BC349	BC349	BU204	2SC1151	C229E3	C229E3	CT223-10	CT223-10
BC350	BC257	BU204	2SC1153	C229F	C229F	CT223-3	CT223-3
BC350	BC350	BU204	2SC1167	C229F3	C229F3	CT223-4	CT223-4
BC351	BC224	BU204	2SC1171	C229M	C229M	CT223-5	CT223-5
BC351	BC258	BU204	2SC1184	C229M3	C229M3	CT223-6	CT223-6
BC351	BC351	BU204	2SC1295	C230A	C230A	CT223-7	CT223-7
BC351	BC354	BU204	2SC1367	C230A3	C230A3	CT223-8	CT223-8
BC351	BC355	BU204	2SC1891	C230B	C230B	CT223-9	CT223-9
BC351	BC357	BU204	2SC642	C230B3	C230B3	CT223A10	CT223A10
BC352	BC352	BU204	2SC643			CT223A3	CT223A3
BC413	BC335	BU204	2SC936			CT223A4	CT223A4
BC413	BC384	BU204	2SC937			CT223A5	CT223A5
BC414	BC384	BU204	2SD199			CT223A6	CT223A6
		BU204	2SD200			CT223A7	CT223A7
		BU205	BU105				

MOTOR.	Industry Standard No.
CT223A8	CT223A8
CT225A9	CT225A9
CT6342	CT6342
CT6342	CT6342
CT6342A	CT6342A
CT6343	CT6343
CT6343A	CT6343A
CT6344	CT6344
CT6344A	CT6344A
CT6345	CT6345
CT6345A	CT6345A
CT6346	CT6346
CT6346A	CT6346A
CT6347	CT6347
CT6347A	CT6347A
CT6348	CT6348
CT6348A	CT6348A
CT6349	CT6349
CT6349A	CT6349A
D40C1	D40C1
D40C2	D40C2
D40C4	D40C4
D40C5	D40C5
D40D1	D40D1
D40D10	D40D10
D40D11	D40D11
D40D13	D40D13
D40D13	2SD1155
D40D14	D40D14
D40D2	D40D2
D40D2	D40D3
D40D4	D40D4
D40D7	D40D5
D40D7	D40D7
D40D8	D40D8
D40E1	D40E1
D40E5	D40E5
D40E7	D40E7
D40E7	2SC1848
D40K1	D40K1
D40K2	D40K2
D40K3	D40K3
D40K3	2SD1243
D40K4	D40K4
D40N1	D40N1
D40N2	D40N2
D40N3	D40N3
D40N4	D40N4
D40N4	2SC2068
D40P1	D40P1
D40P3	D40P3
D40P5	D40P5
D41D1	D41D1
D41D10	D41D10
D41D10	2SA645
D41D11	D41D11
D41D13	D41D13
D41D14	D41D14
D41D2	D41D2
D41D4	D41D4
D41D5	D41D5
D41D7	D41D7
D41D8	2SA635
D41E1	D41E1
D41E1	2SA623
D41E1	2SA633
D41E1	2SA703
D41E5	D41E5
D41E5	2SA624
D41E5	2SA634
D41E5	2SA699
D41E7	D41E7
D41E7	2SA887
D41K1	D41K1
D41K2	D41K2
D41K3	D41K3
D41K4	D41K4
D44H10	D44H10
D44H10	D44H4
D44H10	D44H7
D44H10	2SD717
D44H11	D44H11
D44H11	D44H2
D44H11	D44H5
D44H11	D44H8
D45H10	D45H1
D45H10	D45H10
D45H10	D45H4
D45H10	D45H7
D45H11	D45H11
D45H11	D45H12
D45H11	D45H5
D45H11	D45H8
D45H11	D45H9
DAC-08AQ	ADDAC-08AD
DAC-08AQ	DAC-08AQ
DAC-08AQ	JA0801ADM
DAC-08AQ	JA0801APM
DAC-08CP	DAC-08CP
DAC-08CP	JA0801CPC
DAC-08CQ	ADDAC-08CD
DAC-08CQ	DAC-08CQ
DAC-08CQ	JA0801CDC
DAC-08EP	DAC-08EP
DAC-08EP	JA0801EPC
DAC-08EQ	ADDAC-08ED
DAC-08EQ	DAC-08EQ
DAC-08EQ	JA0801EDC
DAC-08HP	DAC-08HP
DAC-08HP	JA0801HPC
DAC-08HQ	ADDAC-08HD
DAC-08HQ	DAC-08HQ
DAC-08HQ	JA0801HDC
DAC-08Q	ADDAC-08D
DAC-08Q	DAC-08BM
DAC-08Q	DAC-08Q
DAC-08Q	JA0801DM
DAC-08Q	JA0801FM
DS80026CP1	DS80026CN
DS80026G	DS80026H
DS80026L	DS80026J
DS8641N/J	DS8641N/J
FT2955	FT2955
FT3055	FT3055
FT317	FT317
FT317	RCA1003
FT317	2N6473
FT317A	FT317A
FT317A	RCA1C12
FT317A	2N6474
FT317A	2N6477

MOTOR.	Industry Standard No.
FT317B	FT317B
FT317B	RCA3441
FT317B	RCA6263
FT317B	2N6478
FT359	FT359
FT401	FT401
FT402	FT402
FT417	FT417
FT417	RCA1004
FT417	2N6475
FT417A	FT417A
FT417A	RCA1013
FT417A	2N6476
FT417B	FT417B
H11A1	H11A1
H11A2	H11A2
H11A3	H11A3
H11A4	H11A4
H11A5	H11A5
H11A5100	H11A5100
H11A520	H11A520
H11A55Q	H11A55Q
H11B1	H11B1
H11B2	H11B2
H11B3	H11B3
H11C1	H11C1
H11C2	H11C2
H11C3	H11C3
H11C4	H11C4
H11C5	H11C5
H11C6	H11C6
H74C1	H74C1
H74C2	H74C2
HA1199	HA1199
ICTE-10	ICT-10
ICTE-12	ICT-12
ICTE-15	ICT-15
ICTE-18	ICT-18
ICTE-22	ICT-22
ICTE-36	ICT-36
ICTE-45	ICT-45
ICTE-5	ICT-5
ICTE-8	ICT-8
IN4004	SEN140
IN4936	4FC30
IN5221,A,B	MZP5221,A,B
IN5270,A,B	MZP5270,A,B
J308	J308
J309	J309
J310	J310
J310	K300
JAN2N3330	2N3330,J
L	AMLM101AD
L	AMLM101AP
L1444	L1444
L14H1	BPW24
L14H2	L14H2
L14H4	L14H3
LF155AH	LF155AH
LF155AH	LF155AL
LF155AH	JAF155AHM
LF155AJ	LF155AJ
LF155H	LF155H
LF155H	LF155L
LF155H	LH740AH
LF155H	JA740HM
LF155H	JAF155HM
LF155J	LF155J
LF155J	LF155JG
LF156AH	LF156AH
LF156AH	LF156AL
LF156AJ	JAF156AHM
LF156H	LF156H
LF156H	LF156L
LF156J	LF156JG
LF156J	JAF156HM
LF157AH	LF157AH
LF157AH	LF157AL
LF157AJ	JAF157AJG
LF157H	LF157H
LF157H	LF157L
LF157H	JAF157HM
LF157J	LF157JG
LF255H	LF255H
LF255H	LF255L
LF255J	LF255D
LF255J	LF255P
LF256H	LF256H
LF256H	LF256L
LF256J	LF256JG
LF256J	LF256P
LF257H	LF257H
LF257H	LF257L
LF257J	LF257P
LF355AH	LF355AH
LF355AH	LF355AL
LF355AH	JAF355AHC
LF355AJ	LF355AJG
LF355AN	LF355AP
LF355BH	LF355BH
LF355BN	LF355BN
LF355H	LF355H
LF355H	LF355L
LF355H	LH740ACH
LF355H	JA740HC
LF355H	JAF355HC
LF355J	LF355JG
LF355J	LF355JG
LF355N	LF355N
LF355N	LF355P
LF356AH	LF356AH
LF356AH	LF356AL
LF356AH	JAF356AHC
LF356AJ	LF356AJG
LF356AJ	LF356AJG
LF356BH	LF356BH
LF356BN	LF356BN
LF356H	LF356H
LF356H	LF356L
LF356H	JAF356HC
LF356J	LF356JG
LF356N	LF356N
LF356N	LF356P
LF357AH	LF357AH
LF357AH	JAF357AHC
LF357AJ	LF357BH
LF357BJ	LF357BJ

MOTOR.	Industry Standard No.
LF357BN	LF357BN
LF357H	LF357H
LF357H	LF357L
LF357H	JAF357HC
LF357J	LF357JG
LF357N	LF357N
LF357N	LF357P
LM408N	SG1760M
LM101AH	AD5098
LM101AH	AD518S
LM101AH	AMLM101
LM101AH	AMLM101A
LM101AH	AMLM101D
LM101AH	AMLM101F
LM101AH	CA101AT
LM101AH	CA101T
LM101AH	ICL101ALNDP
LM101AH	ICL101ALNPB
LM101AH	ICL101ALNTY
LM101AH	LM101AD
LM101AH	LM101AF
LM101AH	LM101AL
LM101AH	LM101P
LM101AH	LM101H
LM101AH	ML101AF
LM101AH	ML101AM
LM101AH	ML101AT
LM101AH	ML101P
LM101AH	ML101M
LM101AH	ML101T
LM101AH	SG101AD
LM101AH	SG101AT
LM101AH	SG101J
LM101AH	SG101T
LM101AH	SN52101AL
LM101AH	SSS101AJ
LM101AH	SSS101AL
LM101AH	JA101AH
LM101AH	JA101H
LM101AJ	LM101AJ
LM101AJ	LM101AJ-14
LM101AJ	LM101AJG
LM101AJ	LM101D
LM101AJ	LM101AJ-14
LM101AJ	JA101AD
LM101AJ	JA101AP
LM101AJ	LM101D
LM101H	SN52104L
LM104H	LM104F
LM104H	LM104H
LM104H	LM104J
LM104H	LM104M
LM104H	SG104T
LM104H	JA104HM
LM105H	AMLM105
LM105H	AMLM105A
LM105H	AMLM105H
LM105H	LM100F
LM105H	LM100H
LM105H	LM105H
LM105H	LM105JG
LM105H	LM105L
LM105H	SG105T
LM105H	SN52105L
LM105H	JA105HM
LM107H	AMLM107
LM107H	AMLM107D
LM107H	AMLM107P
LM107H	CA107T
LM107H	LM107F
LM107H	LM107H
LM107H	LM107L
LM107H	ML107F
LM107H	ML107M
LM107H	ML107P
LM107H	SG107J
LM107H	SN52107L
LM107H	SSS107J
LM107H	SSS107P
LM107H	JA107H
LM108AF	LM108AF
LM108AF	JA108AF
LM108AF	CA108AT
LM108AH	LM108AH
LM108AH	ML108AT
LM108AH	SG1118AT
LM108AH	SG7777T
LM108AH	SN52108AL
LM108AH	JA108AH
LM108AH	JA725AHM
LM108AH	JA725HM
LM108AJ	JA777ML
LM108AJ	LM108AD
LM108AJ	ML108AM
LM108AJ	SG1118AJ
LM108AJ	SG7777J
LM108AJ	JA108AQ
LM108AJ-8	CA108A8
LM108AJ-8	JA777MJ
LM108AJ-8	JA777MJG
LM108F	LM108AF
LM108F	JA108F
LM108H	CA108T
LM108H	ML108T
LM108H	SG108T
LM108H	SG1118T
LM108H	SN52108L
LM108H	JA108L
LM108J	LM108D
LM108J	ML108M
LM108J	SG108J
LM108J	SG1118J
LM108J-8	CA108S
LM108J-8	JA108D
LM109H	LM109H
LM109H	SG109T
LM109H	SN52109L
LM109K	LM109K
LM109K	LM109LA
LM109K	SG109K
LM109K	JA109KM
LM111H	AMLM111H
LM111H	LM111H
LM111H	ML111H
LM111H	SG111T

MOTOR.	Industry Standard No.
LM111J	AMLM111D
LM111J	ICB8000C
LM111J	ICB8001C
LM111J	ICL8001CTZ
LM111J	ICL8001MTZ
LM111J	LM111D
LM111J	ML111M
LM111J	ML111B
LM111J	SG111D
LM117H	LM117H
LM117K	LM117K
LM117K	JA760HM
LM117K	JA789KC
LM117K	JA789KM
LM120H-12	LM120H-12
LM120H-15	LM120H-15
LM120H-18	LM120H-18
LM120H-24	LM120H-24
LM120H-5.0	LM120H-5.0
LM120H-6.0	LM120H-6.0
LM20K-05	SG120K-05
LM20K-12	LM120K-12
LM120K-15	SG120K-15
LM120K-15	LM120K-15
LM120K-18	LM120K-18
LM120K-24	LM120K-24
LM120K-5.0	LM120K-5.0
LM120K-6.0	LM120K-8.0
LM120T-05	LM120T-05
LM120T-12	SG120T-12
LM120T-15	SG120T-15
LM124J	LM124AD
LM124J	LM124AJ
LM124J	LM124D
LM124J	LM124F
LM124J	LM124J
LM124J	TL044MJ
LM139AJ	CA139AJ
LM139AJ	LM139AD
LM139J	CA139J
LM139J	LM139D
LM139J	LM139J
LM140K-12	LM140K-12
LM140K-12	SG140K-12
LM140K-15	LM140K-15
LM140K-15	SG140K-15
LM140K-18	LM140K-18
LM140K-18	SG140K-18
LM140K-24	LM140K-24
LM140K-24	SG140K-24
LM140K-5.0	LM140K-5.0
LM140K-5.0	SG140K-05
LM140K-6.0	LM140K-6.0
LM140K-6.0	SG140K-06
LM140K-8.0	LM140K-8.0
LM140K-8.0	SG140K-08
LM148J	LM148J
LM148J	LM148J
LM158H	LM158AH
LM158H	LM158H
LM158H	LM158L
LM158H	TL022ML
LM158J	LM158JG
LM158J	TL022MJG
LM741CJ	LM741CD
LM193AH	LM193AH
LM193H	LM193H
LM193H	LM193JG
LM193H	LM193U
LM201AH	AMLM201
LM201AH	AMLM201A
LM201AH	AMLM201AP
LM201AH	AMLM201P
LM201AH	CA201AT
LM201AH	CA201T
LM201AH	LM201AF
LM201AH	LM201AL
LM201AH	LM201F
LM201AH	LM201H
LM201AH	ML201AF
LM201AH	ML201AM
LM201AH	ML201AT
LM201AH	ML201F
LM201AH	ML201M
LM201AH	ML201P
LM201AH	SG201AD
LM201AH	SG201AT
LM201AH	SG201J
LM201AH	SG201T
LM201AH	SSS201AJ
LM201AH	SSS201AL
LM201AH	JA201H
LM201AH	JA201H
LM201AJ	LM201AD
LM201AJ	LM201AJ-14
LM201AJ	LM201AJG
LM201AJ	LM201D
LM201AJ	LM201J-14
LM201AJ	JA201AD
LM201AJ	JA201AP
LM201AJ	JA201D
LM201AN	LM201AP
LM201AN	AMLM201AD
LM201AN	LM201AN
LM201AN	LM201AP
LM201AN	SG201AN
LM201AN	SG201M
LM201AN	SG201N
LM201AN	SSS201AP
LM204H	LM204F
LM204H	LM204H
LM205H	SG204T
LM205H	AMLM205
LM205H	AMLM205P
LM205H	AMLM205H
LM205H	LM205F
LM205H	LM200H
LM205H	LM205P
LM205H	LM205H
LM205H	SG205N
LM205H	SG205T
LM207H	AMLM207D
LM207H	AMLM207D

MOTOR.	Industry Standard No.	MOTOR.	Industry Standard No.	MOTOR.	Industry Standard No.	MOTOR.	Industry Standard No.
LM207H	AMLM207F	LM301AN	ML301T	LM311N	SNT2311T	MAC15A4	T4700T
LM207H	CA207T	LM301AN	ML748CP	LM311N	JA311P	MAC15A4	T4706B
LM207H	LM207F	LM301AN	ML748CS	LM317H	LM317H	MAC15A4	T6000B
LM207H	LM207H	LM301AN	SG301AM	LM317H	JA78MGHC	MAC15A4	T6001B
LM207H	LM207P	LM301AN	SG301AN	LM317K	LM317K	MAC15A4	T6006B
LM207H	ML207M	LM301AN	SNT2301AP	LM317T	LM317F	MAC15A5	MAC15A5
LM207H	ML207T	LM301AN	SSS301AP	LM317T	LM317T	MAC15A5	T6000C
LM207H	SG207J	LM301AN	JA301AT	LM317T	JA78GU1C	MAC15A5	T6001C
LM207H	SG207M	LM304H	LM304F	LM317T	JA78MGT2C	MAC15A5	T6006C
LM207H	SG207N	LM304H	LM304H	LM317T	JA78MGU1C	MAC15A6	MAC15A6
LM207H	SG207T	LM304H	LM304J	LM320H-12	LM320H-12	MAC15A6	T4700D
LM207H	SSS207J	LM304H	LM304L	LM320H-15	LM320H-15	MAC15A6	T4706D
LM207H	SSS207P	LM304N	LM304N	LM320H-18	LM320H-18	MAC15A6	T6000D
LM207H	JA207H	LM304T	SG504T	LM320H-24	LM320H-24	MAC15A6	T6001D
LM208AF	LM208AF	LM304T	SNT2304L	LM320H-5.0	LM320H-5.0	MAC15A6	T6006D
LM208AF	JA208AF	LM304T	JA304HC	LM320H-6.0	LM320H-6.0	MAC15A7	MAC15A7
LM208AH	CA208AT	LM305H	AMLM305	LM320K-12	LM320K-12	MAC15A7	T4700E
LM208AH	LM208AH	LM305H	AMLM305A	LM320K-12	SG320K-12	MAC15A7	T6000E
LM208AH	ML208AT	LM305H	AMLM305F	LM320K-15	LM320K-15	MAC15A7	T6001E
LM208AH	SG208AT	LM305H	AMLM305H	LM320K-15	SG320K-15	MAC15A7	T6006E
LM208AH	SG2118AT	LM305H	LM305F	LM320K-18	LM320K-18	MAC15A8	MAC15A8
LM208AH	JA208AH	LM305H	LM305H	LM320K-18	SG320K-18	MAC15A8	T6000M
LM208AJ	LM208AD	LM305H	LM305AH	LM320K-24	LM320K-24	MAC15A8	T6001M
LM208AJ	ML208AM	LM305H	LM305AJG	LM320K-5.0	LM320K-5.0	MAC15A8	T6006M
LM208AJ	SG208AJ	LM305H	LM305AL	LM320K-5.0	SG320K-05	MAC15A9	MAC15A9
LM208AJ	SG2118AJ	LM305H	LM305AP	LM320K-6.0	LM320K-6.0	MAC20-4	MAC20-4
LM208AJ	JA208AD	LM305H	LM305F	LM320K-8.0	LM320K-8.0	MAC20-4	TIC253B
LM208AJ-8	LM208AJ	LM305H	LM305H	LM320T-12	LM320T-12	MAC20-5	MAC20-5
LM208AJ-8	SG208AM	LM305H	LM305JG	LM320T-12	SG320T-12	MAC20-6	MAC20-6
LM208AJ-8	SG2118AM	LM305H	LM305L	LM320T-15	LM320T-15	MAC20-6	TIC253D
LM208F	LM208P	LM305H	LM305P	LM320T-18	LM320T-18	MAC20-7	MAC20-7
LM208F	JA208F	LM305H	LM376JG	LM320T-18	SG320T-18	MAC20-7	TIC253E
LM208H	CA208T	LM305H	LM376L	LM320T-24	LM320T-24	MAC20-8	MAC20-8
LM208H	LM208H	LM305H	LM376N	LM320T-5.0	LM320T-5.0	MAC20-8	TIC253M
LM208H	LM208T	LM305H	LM376P	LM320T-5.0	SG320T-05	MAC20-9	MAC20-9
LM208H	SG208T	LM305H	SG305AT	LM320T-6.0	LM320T-6.0	MAC20A10	MAC20A10
LM208H	SG2118T	LM305H	SG305N	LM320T-8.0	LM320T-8.0	MAC20A4	MAC20A4
LM208H	JA208H	LM305H	SG305T	LM324J	LM324AJ	MAC20A5	MAC20A5
LM208J	ML208M	LM305H	SNT72305AL	LM324J	LM324J	MAC20A6	MAC20A6
LM208J	SG208J	LM305H	SNT72305L	LM324J	SG324J	MAC20A7	MAC20A7
LM208J	SG2118J	LM305H	SNT72376L	LM324N	TLO440J	MAC20A8	MAC20A8
LM208J	JA208D	LM305H	JA305HC	LM324N	L144AP	MAC20A9	MAC20A9
LM208J-8	CA208B	LM305H	JA376TC	LM324N	LM324AN	MAC220-2	MAC220-2
LM208J-8	LM208D	LM307H	CA307T	LM324N	LM324N	MAC220-3	MAC220-3
LM208J-8	SG208M	LM307H	LM307F	LM324N	SG324N	MAC220-5	MAC220-5
LM208J-8	SG2118M	LM307H	LM307H	LM324N	SNT2L044JA	MAC220-7	MAC220-7
LM209H	LM209H	LM307H	LM307L	LM324N	SNT2L044N	MAC220-9	MAC220-9
LM209T	SG209T	LM307H	ML307P	LM324N	TLO44CN	MAC221-2	MAC221-2
LM209K	SG209K	LM307H	ML307P	LM339AJ	CA339AG	MAC221-5	MAC221-5
LM209K	SG209K	LM307H	SG1760D	LM339AJ	LM339AD	MAC221-5	MAC221-5
LM209K	JA209KM	LM307H	SG1760F	LM339AN	CA339AG	MAC221-7	MAC221-7
LM211H	AMLM211H	LM307H	SG307T	LM339AN	LM339AN	MAC221-9	MAC221-9
LM211H	LM211H	LM307H	SNT2307L	LM339J	CA339G	MAC222-1	MAC222-1
LM211H	ML211T	LM307N	JA307H	LM339J	JA775DC	MAC222-10	MAC222-10
LM211H	SG211T	LM307N	LM307N	LM339N	LM339N	MAC222-2	MAC222-2
LM211J	AMLM211D	LM307N	LM307P	LM339N	LM339N	MAC222-3	MAC222-3
LM211J	LM211D	LM307N	ML307S	LM339N	JA775PC	MAC222-4	MAC222-4
LM211J	ML211M	LM307N	SG307J	LM340K-12	LM340K-12	MAC222-4	SC143B
LM211J	SG211D	LM307N	SG307M	LM340K-15	LM340K-15	MAC222-5	MAC222-5
LM211N	SG211S	LM307T	JA307T	LM340K-18	LM340K-18	MAC222-6	MAC222-6
LM217H	LM217H	LM308AH	CA308AT	LM340K-24	LM340K-24	MAC222-6	SC143D
LM217K	LM217K	LM308AH	LM308AH	LM340K-5.0	LM340K-5.0	MAC222-7	MAC222-7
LM224J	LM224AD	LM308AH	LM308AH-1	LM340K-6.0	LM340K-6.0	MAC222-7	SC143E
LM224J	LM224AP	LM308AH	LM308AH-2	LM340K-8.0	LM340K-8.0	MAC222-8	MAC222-8
LM224J	LM224D	LM308AH	SG308AT	LM348J	LM348D	MAC222-8	SC143M
LM224J	LM224J	LM308AH	SG308AT	LM348J	LM348J	MAC222-9	MAC222-9
LM224J	SG224J	LM308AH	SG9777CT	LM348N	LM348N	MAC222A1	MAC222A1
LM224L	LM224F	LM308AH	SNT2308AL	LM358AH	LM358AH	MAC222A10	MAC222A10
LM224N	SG224N	LM308AH	SG308AH	LM358H	LM358H	MAC222A2	MAC222A2
LM239AJ	CA239AG	LM308AH	JA725BHC	LM358H	LM358L	MAC222A3	MAC222A3
LM239AJ	LM239AD	LM308AH	JA725HC	LM358H	TLO22CL	MAC222A4	MAC222A4
LM239AJ	LM239AJ	LM308AH	JA777CL	LM358J	LM358JG	MAC222A5	MAC222A5
LM239AN	CA239AE	LM308AH	JA777HC	LM358J	TLO22CJG	MAC222A6	MAC222A6
LM239J	CA239G	LM308AJ	LM308AF	LM358N	LM358AN	MAC222A7	MAC222A7
LM239J	LM239D	LM308AJ	ML308AM	LM358N	LM358N	MAC222A8	MAC222A8
LM239J	LM239J	LM308AJ	SG308AJ	LM358N	LM358P	MAC222A9	MAC222A9
LM239N	CA239B	LM308AJ	SG777CJ	LM358N	SNT2LO22P	MAC223-1	MAC223-1
LM248J	LM248D	LM308AJ	LM308AJ	LM358N	TLO22CP	MAC223-10	MAC223-10
LM248J	LM248J	LM308AJ-8	LM308AJ	LM393AH	LM393AH	MAC223-2	MAC223-2
LM258H	LM258AH	LM308AJ-8	JA777CJ	LM393AN	LM393AN	MAC223-3	MAC223-3
LM258H	LM258H	LM308AJ-8	JA777CJG	LM393H	LM393H	MAC223-4	MAC223-4
LM2901N	LM2901N	LM308AJ-8	JA777DC	LM393JG	LM393JG	MAC223-5	MAC223-5
LM2902J	LM2902J	LM308AN	SG308AM	LM393N	LM393N	MAC223-6	MAC223-6
LM2902N	LM2902N	LM308AN	SG777CM	LM393N	LM393P	MAC223-7	MAC223-7
LM2903N	LM2903	LM308AN	SG777CN	LM393N	LM393U	MAC223-8	MAC223-8
LM2903N	LM2903JG	LM308AN	JA777CN	LM520B-8.0	LM320B-8.0	MAC223-9	MAC223-9
LM2903S	LM2903P	LM308AN	JA777CN	LM723CH	LM723CH	MAC223A1	MAC223A1
LM2903S	LM2903P	LM308AN	JA777CP	LM723CD	LM723CD	MAC223A10	MAC223A10
LM2904N	LM2904N	LM308AN	JA777TC	LM723CJ	LM723CJ	MAC223A2	MAC223A2
LM293AH	LM293AH	LM308H	CA308B	LM723CN	LM723CN	MAC223A3	MAC223A3
LM293H	LM293H	LM308H	LM308H	LM723H	LM723H	MAC223A4	MAC223A4
LM293H	LM293P	LM308H	ML308P	LM723J	LM723D	MAC223A5	MAC223A5
LM293H	LM293U	LM308H	SG1660T	LM723J	LM723J	MAC223A6	MAC223A6
LM301AH	AD301AL	LM308H	SG1760T	LM741CP	LM741CP	MAC223A7	MAC223A7
LM301AH	AD509J	LM308H	SG308T	LM741CH	LM741CH	MAC223A8	MAC223A8
LM301AH	AD509K	LM308H	SNT2308L	LM741CJ	LM741CJ	MAC223A9	MAC223A9
LM301AH	AD518J	LM308H	JA308H	LM741CJ-14	LM741CJ-14	MAC228-1	MAC228-1
LM301AH	AD518K	LM308J	LM308D	LM741CN	LM741CN	MAC228-2	MAC228-2
LM301AH	AM166039F	LM308J	ML308M	LM741CN-14	LM741CN-14	MAC228-3	MAC228-3
LM301AH	AM166039T	LM308J	SG1660J	LM741F	LM741F	MAC228-4	MAC228-4
LM301AH	AMLM301	LM308J	SG1760J	LM741H	LM741H	MAC228-5	MAC228-5
LM301AH	AMLM301A	LM308J	SG308J	LM741J	LM741J	MAC228-6	MAC228-6
LM301AH	CA301AT	LM308J	JA308D	LM741J-14	LM741J-14	MAC228-7	MAC228-7
LM301AH	ICL301ALNPA	LM308N	CA308AS	LM747CF	LM747CF	MAC228-8	MAC228-8
LM301AH	ICL301ALNTY	LM308N	LM308N	LM747CH	LM747CH	MAC228-9	MAC228-9
LM301AF	LM301AF	LM308N	SG1660M	LM747CD	LM747CD	MAC228A1	MAC228A1
LM301AH	LM301AH	LM308N	SG308M	LM747CJ	LM747CJ	MAC228A10	MAC228A10
LM301AH	LM301AL	LM309H	LM309H	LM747CN	LM747CN	MAC228A2	MAC228A2
LM301AH	ML301AT	LM309H	SG309T	LM747F	LM747F	MAC228A3	MAC228A3
LM301AH	SG1660D	LM309H	SNT2309L	LM747H	LM747H	MAC228A4	T1C216A
LM301AD	SG301AD	LM309K	LM309K	LM747J	LM747D	MAC228A4	MAC228A4
LM301AH	SG301AT	LM309K	LM309KC	LM747J	LM747J	MAC228A5	T1C216B
LM301AH	SNT72301AL	LM309K	LM309LA	MAC15-10	MAC15-10	MAC228A5	MAC228A5
LM301AH	SSS301AJ	LM309K	JA309KC	MAC15-4	MAC15-4	MAC228A6	T1C216C
LM301AH	SSS301AL	LM310H	LM302H	MAC15-4	SC153B	MAC228A6	MAC228A6
LM301AH	JA301AH	LM311H	AMLM311H	MAC15-5	T1C246B	MAC228A7	MAC228A7
LM301AJ	AMLM301AD	LM311H	LM311H	MAC15-5	MAC15-5	MAC228A8	MAC228A8
LM301AJ	AMLM301D	LM311H	ML311T	MAC15-6	SC151D	MAC228A9	MAC228A9
LM301AJ	LM301AD	LM311H	SG311T	MAC15-6	T1C246D	MAC25-10	MAC25-10
LM301AJ	LM301AJ	LM311H	SNT2311L	MAC15-7	MAC15-7	MAC25-4	MAC25-4
LM301AJ	LM301AJG	LM311H	JA734HC	MAC15-7	SC151E	MAC25-5	TIC253B
LM301AD	JA301AD	LM311H	JA734HM	MAC15-8	MAC15-8	MAC25-5	MAC25-5
LM301AN	ICL8008CPA	LM311J	LM311D	MAC15-8	SC151M	MAC25-6	MAC25-6
LM301AN	ICL8008CTY	LM311J	LM311N-14	MAC15-9	MAC15-9	MAC25-7	MAC25-7
LM301AN	ICL8017CTW	LM311J	ML311N	MAC15A10	MAC15A10	MAC25-7	TIC253E
LM301AN	ICL8017MTW	LM311J	ML311P	MAC15A2	T4700F	MAC25-8	MAC25-8
LM301AN	LM301AN	LM311J	SG311D	MAC15A2	T6000F	MAC25-8	TIC253E
LM301AN	LM301AP	LM311J	JA734DC	MAC15A2	T6001F	MAC25-9	MAC25-9
LM301AN	ML301AD	LM311J	JA734DM	MAC15A3	T6000A	MAC25A10	MAC25A10
LM301AN	ML301AS	LM311J-8	AMLM311D	MAC15A3	T6001A	MAC25A4	MAC25A4
LM301AN	ML301P	LM311N	LM311N	MAC15A3	T6006A	MAC25A5	MAC25A5
LM301AN	ML301S	LM311N	ML311S	MAC15A4	MAC15A4		
		LM311N	SG311M				

MOTOR.	Industry Standard No.	MOTOR.	Industry Standard No.	MOTOR.	Industry Standard No.	MOTOR.	Industry Standard No.
MAC25A6	MAC25A6	MAC95A4	MAC95A4	MC10506	SP10106	MC1327P	TAA630
MAC25A7	MAC25A7	MAC95A5	MAC95A5	MC10507	P10107	MC1327P	TBA520
MAC25A8	MAC25A8	MAC95A6	L400E5	MC10507	SP10107	MC1327P	TBA920
MAC25A9	MAC25A9	MAC95A6	MAC95A6	MC10509	P10109	MC1327P	jA78kDC
MAC3010-15	MAC3010-15	MAC95A7	MAC95A7	MC10509	SP10109	MC1330P	ITT1330
MAC3010-25	MAC3010-25	MAC95A8	MAC95A8	MC10513	P10113	MC1344P	CA3120E
MAC3010-4	MAC3010-4	MAC96-1	MAC96-1	MC10513	SP10113	MC1344P	LM1845N
MAC3010-40	MAC3010-40	MAC96-2	MAC96-2	MC10514	P10114	MC1344P	TBA940
MAC3010-8	MAC3010-8	MAC96-3	MAC96-3	MC10514	SP10114	MC1344P	TBA950
MAC3020-15	MAC3020-15	MAC96-4	MAC96-4	MC10515	P10115	MC1344P	ULN2125A
MAC3020-25	MAC3020-25	MAC96-5	MAC96-5	MC10515	SP10115	MC1350P	LM703LN
MAC3020-4	MAC3020-4	MAC96-6	MAC96-6	MC10516	P10116	MC1350P	SN76600P
MAC3020-40	MAC3020-40	MAC96-7	MAC96-7	MC10516	SP10116	MC1350P	A757DC
MAC3020-8	MAC3020-8	MAC96-8	MAC96-8	MC10517	P10117	MC1350P	jA757DM
MAC3030-15	MAC3030-15	MAC96A1	MAC96A1	MC10517	SP10117	MC1351P	CA3041
MAC3030-25	MAC3030-25	MAC96A2	MAC96A2	MC10518	P10118	MC1351P	LM1351N
MAC3030-4	MAC3030-4	MAC96A3	MAC96A3	MC10518	SP10118	MC1351P	SN76651N
MAC3030-40	MAC3030-40	MAC96A4	L200E5	MC10519	P10119	MC1352P	CA1352E
MAC3030-8	MAC3030-8	MAC96A4	MAC96A4	MC10519	SP10119	MC1352P	CA3035
MAC3040-15	MAC3040-15	MAC96A5	MAC96A5	MC10521	P10121	MC1352P	CA3035V1
MAC3040-25	MAC3040-25	MAC96A6	L400E5	MC10521	SP10121	MC1352P	CA3068
MAC3040-4	MAC3040-4	MAC96A6	MAC96A6	MC10524	P10124	MC1352P	ITT1352
MAC3040-40	MAC3040-40	MAC96A7	MAC96A7	MC10524	SP10124	MC1352P	SN76644N
MAC3040-8	MAC3040-8	MAC96A8	MAC96A8	MC10525	P10125	MC1352P	SN76653N
MAC50-10	MAC50-10	MBD101	MBD101	MC10525	SP10125	MC1352P	TBA440
MAC50-4	MAC50-4	MBD102	MBD102	MC10530	P10130	MC1355P	SN76678P
MAC50-5	MAC50-5	MBD201	MBD201	MC10530	SP10130	MC1355P	jA754HC
MAC50-6	MAC50-6	MBD201	1N5712	MC10531	P10131	MC1355P	jA754TC
MAC50-7	MAC50-7	MBD201	1N5713	MC10531	SP10131	MC1356P	LM1841N
MAC50-8	MAC50-8	MBD201	1N5767	MC10533	P10133	MC1356P	SN76669N
MAC50-9	MAC50-9	MBD301	MBD301	MC10535	P10135	MC1356P	ULN2136A
MAC50A10	MAC50A10	MBD301	1N5765	MC10535	SP10135	MC1356P	ULN2209A
MAC50A4	MAC50A4	MBD301	1N5766	MC10536	P10136	MC1356P	jA2136PC
MAC50A5	MAC50A5	MBD501	MBD501	MC10536	SP10136	MC1356P	jA753TC
MAC50A6	MAC50A6	MBD502	MBD502	MC10537	P10137	MC1357P	CA2111AE
MAC50A7	MAC50A7	MBD701	MBD701	MC10537	SP10137	MC1357P	CA3013
MAC50A8	MAC50A8	MBD701/702	1N2663	MC10538	SP10138	MC1357P	CA3014
MAC50A9	MAC50A9	MBD701/702	1N5711	MC10541	P10141	MC1357P	CA3042
MAC525-2	Q2025P	MBD702	MBD702	MC10541	SP10141	MC1357P	CA3043
MAC525-4	SC160B	MBR1530CT	12CTQ030	MC10553	P10153	MC1357P	LM2111N
MAC525-4	Q4025P	MBR1535CT	12CTQ035	MC10558	P10158	MC1357P	LM2113N
MAC525-6	SC160D	MBR1540CT	12CTQ040	MC10559	P10159	MC1357P	SN76642N
MAC525-6	Q6025P	MBR1545CT	12CTQ045	MC10560	P10160	MC1357P	ULN2111A
MAC525-7	SC160B	MBR2530CT	20CTQ030	MC10560	SP10160	MC1357P	ULN2113A
MAC525-7	Q5025P	MBR2535CT	20CTQ035	MC10561	P10161	MC1357P	ULN2113N
MAC525-8	SC160B	MBR2545CT	20CTQ040	MC10561	SP10161	MC1357PQ	CA2111AQ
MAC525-8	Q6025P	MBR3020CT	USD320C	MC10562	P10162	MC1357PQ	ULN2111N
MAC525-8	SC160M	MBR3020CT	VSK3020C	MC10562	SP10162	MC1358P	CA3065
MAC91-1	MAC91-1	MBR3035CT	USD335C	MC10564	P10164	MC1358P	ITT3065
MAC91-2	MAC91-2	MBR3035CT	VSK3030T	MC10564	SP10164	MC1358P	LM3065N
MAC91-3	MAC91-3	MBR3045CT	USD345C	MC10565	P10165	MC1358P	N5065A
MAC91-4	MAC91-4	MBR3045CT	VSK3040T	MC10565	SP10165	MC1358P	SN76666N
MAC91-5	MAC91-5	MBR3520	USD420	MC10568	P10168	MC1358P	TBA1208
MAC91-6	MAC91-6	MBR3520	VSK1520	MC10570	P10170	MC1358P	ULN2165A
MAC91-7	MAC91-7	MBR3520	VSK3020S	MC10570	SP10170	MC1358P	A5065PC
MAC91-8	MAC91-8	MBR3520	20PQ020	MC10571	P10171	MC1358PQ	ULN2165N
MAC91A1	MAC91A1	MBR3535	VSK1535	MC10571	SP10171	MC1364P	CA3044
MAC91A2	MAC91A2	MBR3535	VSK1530	MC10572	P10172	MC1364P	CA3044V1
MAC91A3	MAC91A3	MBR3535	VSK3030S	MC10572	SP10172	MC1364P	CA3064
MAC91A4	MAC91A4	MBR3535	20PQ030	MC10574	P10174	MC1364P	CA3064E
MAC91A5	MAC91A5	MBR3535	20PQ035	MC10574	SP10174	MC1364P	ITT3064
MAC91A6	MAC91A6	MBR3535	21PQ030	MC10575	P10175	MC1364P	LM3064N
MAC91A7	MAC91A7	MBR3535	21PQ035	MC10575	SP10175	MC1364P	SN76564N
MAC91A8	MAC91A8	MBR3535	30PQ030	MC10576	P10176	MC1364P	SN76565N
MAC92-1	MAC92-1	MBR3545	SD31	MC10576	SP10176	MC1364P	SN76665N
MAC92-1	SC92P	MBR3545	SD32	MC10578	SP10178	MC1364P	ULN2264A
MAC92-2	MAC92-2	MBR3545	USD445	MC10579	P10179	MC1364P	A3064PC
MAC92-2	SC92A	MBR3545	VSK1540	MC10579	SP10179	MC1375P	LM3075N
MAC92-3	MAC92-3	MBR3545	VSK3040S	MC10580	P10180	MC1375P	SN76675N
MAC92-4	MAC92-4	MBR3545	20PQ040	MC10580	SP10180	MC1385P	ITT641
MAC92-4	SC92B	MBR3545	20PQ045	MC10581	P10181	MC1385P	LM1805
MAC92-5	MAC92-5	MBR3545	21PQ040	MC10581	SP10181	MC1391P	CA1391E
MAC92-6	MAC92-6	MBR3545	21PQ045	MC10586	SP10186	MC1391P	ITT3710
MAC92-6	SC92D	MBR3545	30PQ045	MC10610	P10210	MC1391P	LM1391N
MAC92-7	MAC92-7	MBR6020	VSK4020	MC10611	P10211	MC1391P	SN76591P
MAC92-8	MAC92-8	MBR6035	VSK4030	MC10611	SP10211	MC1391P	TBA920
MAC92A1	MAC92A1	MBR6045	VSK4040	MC10612	P10212	MC1391P	TBA920B
MAC92A2	MAC92A2	MBR7520	USD520	MC10612	SP10212	MC1391P	A1391PC
MAC92A3	MAC92A3	MBR7535	USD535	MC10616	SP10216	MC1394P	CA1394E
MAC92A4	MAC92A4	MBR7545	SD-71	MC10631	P10231	MC1394P	ITT3714
MAC92A5	MAC92A5	MBR7545	SD-72	MC10631	SP10231	MC1394P	LM1394N
MAC92A6	MAC92A6	MBR7545	SD-75	MC1303P	FA239A	MC1394P	SN76594P
MAC92A7	MAC92A7	MBR7545	USD545	MC1303P	SN76130N	MC1394P	A1394PC
MAC92A8	MAC92A8	MBS4991	D130500	MC1305P	SN76131N	MC1398P	CA1398E
MAC93-1	MAC93-1	MBS4991	MBS4991	MC1305P	SN76149N	MC1398P	SN76298N
MAC93-2	MAC93-2	MBS4991	MUS4987	MC1306P	LM386N	MC1398P	ULN2298A
MAC93-3	MAC93-3	MBS4991	1N5779	MC1306P	SN76000P	MC1398P	CA3066
MAC93-4	MAC93-4	MBS4991	1N5780	MC1310P	CA3110B	MC1398P	CA3070
MAC93-5	MAC93-5	MBS4991	1N5781	MC1310P	CA3090AQ	MC1399P	CA3071
MAC93-6	MAC93-6	MBS4991	1N5782	MC1310P	CAT58E	MC1399P	ITT3066
MAC93-7	MAC93-7	MBS4991	1N5783	MC1310P	LM1310N	MC1399P	ITT3707
MAC93-8	MAC93-8	MBS4991	1N5784	MC1310P	LM1800AN	MC1399P	LM3066N
MAC93A1	MAC93A1	MBS4991	1N5785	MC1310P	LM1800N	MC1399P	LM3070N
MAC93A2	MAC93A2	MBS4991	1N5786	MC1310P	MC1310A	MC1399P	LM3071N
MAC93A3	MAC93A3	MBS4991	1N5787	MC1310P	SN76104N	MC1399P	LM5126
MAC93A4	MAC93A4	MBS4991	1N5788	MC1310P	SN76105N	MC1399P	N5070B
MAC93A5	MAC93A5	MBS4991	1N5789	MC1310P	SN76111N	MC1399P	N5071A
MAC93A6	MAC93A6	MBS4991	1N5790	MC1310P	SN76113N	MC1399P	SN76242N
MAC93A7	MAC93A7	MBS4991	1N5791	MC1310P	SN76115N	MC1399P	SN76243N
MAC93A8	MAC93A8	MBS4991	1N5792	MC1310P	SN76116N	MC1399P	ULN2124A
MAC94-1	MAC94-1	MBS4991	1N5793	MC1310P	SN76117N	MC1399P	ULN2127A
MAC94-2	MAC94-2	MBS4991	2N4991	MC1310P	ULN2120A	MC1399P	ULN2262A
MAC94-4	MAC94-3	MBS4991	2N4993	MC1310P	ULN2121A	MC1399P	A780DC
MAC94-4	L200E9	MBS4992	MBS4992	MC1310P	ULN2122A	MC1399P	A780PC
MAC94-4	MAC94-4	MBS4992	MUS4988	MC1310P	ULN2128A	MC1399P	A781DC
MAC94-5	Q200E4	MBS4992	2N4992	MC1310P	ULN2210A	MC1399P	A781PC
MAC94-5	MAC94-5	MC10110	P10110	MC1310P	ULN2244A	MC1399P	A787PC
MAC94-6	L400E9	MC10110	SP10110	MC1310P	jA732DC	MC14000UB	CD40000UB
MAC94-6	MAC94-6	MC10111	P10111	MC1310P	jA732PC	MC14000UB	SCL4000B
MAC94-8	Q400E4	MC10111	SP10111	MC1310P	jA758DC	MC14001B	CD4001B
MAC94-7	MAC94-7	MC10123	P10123	MC1310P	jA758PC	MC14001B	F4001B
MAC94-8	MAC94-8	MC10128	SP10128	MC1310P	jA767DC	MC14001B	MS4002
MAC94A1	MAC94A1	MC10129	SP10129	MC1312P	jA767PC	MC14001B	MM74C02
MAC94A2	MAC94A2	MC10132	P10132	MC1314P	jA1312PC	MC14001B	SCL4001B
MAC94A3	MAC94A3	MC10134	P10134	MC1314P	jA1314PC	MC14001UB	CD4001UB
MAC94A3	L200E7	MC10134	SP10134	MC1315P	jA1315PC	MC14001UB	SCL4001UB
MAC94A4	MAC94A4	MC10166	P10166	MC1323P	CA3067	MC14002B	CD4002B
MAC94A4	Q200E3	MC10173	P10173	MC1323P	CA3072	MC14002B	F4002B
MAC94A5	MAC94A5	MC10173	SP10173	MC1323P	CA3125E	MC14002B	SCL4002B
MAC94A5	L400E7	MC10177	P10177	MC1323P	CA3137E	MC14002UB	CD4002UB
MAC94A6	MAC94A6	MC10500	P10500	MC1323P	LM1828N	MC14006B	CD4006B
MAC94A6	Q400E3	MC10500	SP10100	MC1323P	LM1848N	MC14006B	F4006B
MAC94A7	MAC94A7	MC10501	P10101	MC1323P	LM5067N	MC14006B	SCL4006B
MAC94A8	MAC94A8	MC10501	SP10101	MC1323P	LM746N	MC14007B	F4007UB
MAC95-1	MAC95-1	MC10502	P10102	MC1323P	N5072A	MC14007UB	CD4007UB
MAC95-2	MAC95-2	MC10502	SP10102	MC1323P	SN76246N	MC14007UB	SCL4007UB
MAC95-3	MAC95-3	MC10503	P10103	MC1323P	ULN2114A	MC14008B	CD4008B
MAC95-4	MAC95-4	MC10503	SP10103	MC1323P	ULN2114K	MC14008B	F4008B
MAC95-5	MAC95-5	MC10504	P10104	MC1323P	ULN2114N	MC14008B	MM54C83
MAC95-6	MAC95-6	MC10504	SP10104	MC1323P	ULN2228A	MC14008B	MM74C83
MAC95-7	MAC95-7	MC10505	P10105	MC1323P	ULN2267A	MC14008B	SCL4008B
MAC95-8	MAC95-8	MC10505	SP10105	MC1323P	jA746DC	MC1400AU10	REF-01EJ
MAC95A1	MAC95A1	MC10506	P10106	MC1323P	jA746HC	MC1400AU5	REF-02EJ
MAC95A2	MAC95A2			MC1324P	ULN2224A		
MAC95A3	MAC95A3						
MAC95A4	L200E5						

MOTOR.	Industry Standard No.
MC1400U10	REF-01HJ
MC1400U10	REF-01HP
MC1400U5	REF-02HJ
MC1400U5	REF-02HP
MC14011B	CD4011B
MC14011B	F4011B
MC14011B	MM54C00
MC14011B	MM74C00
MC14011B	SCL4011B
MC14011UB	CD4011UB
MC14011UB	SCL4011UB
MC14012B	CD4012UB
MC14012B	F4012B
MC14012B	MM54C20
MC14012B	MM74C20
MC14012B	SCL4012B
MC14013B	CD4013B
MC14013B	F4013B
MC14013B	MM54C74
MC14013B	MM74C74
MC14013B	SCL4013B
MC14014B	CD4014B
MC14014B	F4014B
MC14014B	SCL4014B
MC14015B	CD4015B
MC14015B	F4015B
MC14015B	MM54C164
MC14015B	MM74C164
MC14015B	SCL4015B
MC14016B	CD4016B
MC14016B	F4016B
MC14016B	SCL4016B
MC14017B	CD4017B
MC14017B	F4017B
MC14017B	SCL4017B
MC14018B	CD4018B
MC14018B	F4018B
MC14018B	SCL4018B
MC14019B	SCL4019B
MC14020B	CD4020B
MC14020B	F4020B
MC14020B	SCL4020B
MC14021B	CD4021B
MC14021B	F4021B
MC14021B	MM54C165
MC14021B	MM74C165
MC14021B	SCL4021B
MC14022B	CD4022B
MC14022B	F4022B
MC14022B	SCL4022B
MC14023B	CD4023B
MC14023B	F4023B
MC14023B	MM54C10
MC14023B	MM74C10
MC14023B	SCL4023B
MC14023UB	CD4023UB
MC14024B	CD4024B
MC14024B	F4024B
MC14024B	SCL4024B
MC14025B	F4025B
MC14025B	SCL4025B
MC14025UB	CD4025UB
MC14027B	CD4027B
MC14027B	F4027B
MC14027B	MM54C107
MC14027B	MM54C73
MC14027B	MM54C76
MC14027B	MM74C107
MC14027B	MM74C73
MC14027B	MM74C76
MC14027B	SCL4027B
MC14028B	CD4028B
MC14028B	F4028B
MC14028B	MM54C42
MC14028B	MM74C42
MC14028B	SCL4028B
MC14029B	CD4029B
MC14029B	F4029B
MC14029B	SCL4029B
MC14032B	CD4032B
MC14034B	CD4034B
MC14034B	SCL4034AB
MC14035B	CD4035B
MC14035B	F4035B
MC14035B	MM54C195
MC14035B	MM54C95
MC14035B	MM74C195
MC14035B	MM74C95
MC14035B	SCL4035B
MC14038B	CD4038B
MC1403AP1	AD580M
MC1403AU	SG2503
MC1403P1	AD580K
MC1403U	AD1403AN
MC1403U	AD1403N
MC1403U	AD580J
MC1403U	SG3503
MC14040B	CD4040B
MC14040B	F4040B
MC14040B	SCL4040B
MC14042B	CD4042B
MC14042B	F4042B
MC14042B	SCL4042B
MC14043B	CD4043B
MC14043B	F4043B
MC14043B	SCL4043B
MC14044B	CD4044B
MC14044B	F4044B
MC14044B	SCL4044B
MC14046B	CD4046B
MC14046B	F4046B
MC14046B	SCL4046B
MC14049UB	CD4009UB
MC14049UB	CD4049UB
MC14049UB	F4049B
MC14049UB	MM54C901
MC14049UB	MM74C901
MC14049UB	SCL4009UB
MC14049UB	SCL4049UB
MC1404U10	REF-01CJ
MC1404U10	REF-01CP
MC1404U10	REF-01DJ
MC1404U10	REF-01PP
MC1404U5	REF-02CJ
MC1404U5	REF-02CP
MC1404U5	REF-02DJ
MC1404U5	REF-02DP
MC14050B	CD4010B
MC14050B	CD4050B
MC14050B	F4050B
MC14050B	MM54C902
MC14050B	MM74C902
MC14050B	SCL4010B
MC14050B	SCL4050B

MOTOR.	Industry Standard No.
MC14051B	CD4051B
MC14051B	F4051B
MC14051B	SCL4051B
MC14052B	CD4052B
MC14052B	F4052B
MC14052B	SCL4052B
MC14053B	CD4053B
MC14053B	F4053B
MC14053B	SCL4053B
MC1405L	LD111CJ
MC14060B	CD4060B
MC14060B	SCL4060B
MC14066B	CD4066B
MC14066B	F4066B
MC14066B	SCL4066B
MC14068B	CD4068B
MC14068B	F4068B
MC14068B	MM54C30
MC14068B	MM74C30
MC14068B	SCL4068B
MC14069UB	CD4069UB
MC14069UB	F4069UB
MC14069UB	MM54C04
MC14069UB	MM74C04
MC14069UB	SCL4069UB
MC14070B	CD4030B
MC14070B	CD4070B
MC14070B	F4030B
MC14070B	F4070B
MC14070B	MM54C86
MC14070B	MM74C86
MC14070B	SCL4030B
MC14070B	SCL4070B
MC14071B	CD4071B
MC14071B	F4071B
MC14071B	MM54C32
MC14071B	MM74C32
MC14071B	SCL4071B
MC14072B	CD4072B
MC14072B	SCL4072B
MC14073B	CD4073B
MC14073B	F4073B
MC14073B	SCL4073B
MC14075B	CD4075B
MC14075B	F4075B
MC14075B	SCL4075B
MC14076B	CD4076B
MC14076B	F4076B
MC14076B	MM54C173
MC14076B	MM74C173
MC14076B	SCL4076B
MC14077B	CD4077B
MC14077B	F4077B
MC14077B	SCL4077B
MC14078B	CD4078B
MC14078B	F4078B
MC14078B	SCL4078B
MC1408-L7	AD1408-7D
MC14081B	CD4081B
MC14081B	F4081B
MC14081B	MM54C08
MC14081B	MM74C08
MC14081B	SCL4081B
MC14082B	CD4082B
MC14082B	SCL4082B
MC14085B	SCL4085B
MC14086B	SCL4086B
MC1408L6	LM1408J6
MC1408L6	SSS1408A-6Z
MC1408L6	]A0802DC-3
MC1408L7	LM1408J7
MC1408L7	SSS1408A-7Z
MC1408L7	]A0802DC-2
MC1408L8	AD1408-8D
MC1408L8	AD559JD
MC1408L8	AD559K
MC1408L8	AD559KD
MC1408L8	LM1408J8
MC1408L8	MC1408F
MC1408L8	SSS1408A-8Z
MC1408L8	]A0802DC-1
MC1408P6	LM1408N6
MC1408P6	]A0802PC-3
MC1408P7	LM1408N7
MC1408P7	]A0802PC-2
MC1408P8	LM1408N8
MC1408P8	MC1408B
MC1408P8	]A0802PC-1
MC14093B	CD4093B
MC14093B	F4093B
MC14093B	SCL4093B
MC14094B	CD4094B
MC14094B	SCL4094B
MC14099B	CD4099B
MC14099B	SCL4099B
MC1411	ULN2001A
MC1411L	SN75466J
MC1411P	ITT652
MC1411P	L201
MC1411P	SN75466N
MC1412	ULN2002A
MC1412L	SN75467J
MC1412P	ITT654
MC1412P	L202
MC1412P	SN75467N
MC1413	ULN2003A
MC1413L	SN75468J
MC1413P	ITT656
MC1413P	L203
MC1413P	SN75468N
MC1414L	LM1414J
MC1414L	RC1414DC
MC1414L	SN72514J
MC1414P	LM1414N
MC1414P	RC1414DP
MC1414P	SN72514N
MC1416	ULN2004A
MC14160B	CD40160B
MC14160B	F40160B
MC14160B	MM54C160
MC14160B	MM74C160
MC14160B	SCL4160B
MC14161B	CD40161B
MC14161B	F40161B
MC14161B	MM54C161
MC14161B	MM74C161
MC14161B	SCL4161B
MC14162B	CD40162B
MC14162B	F40162B
MC14162B	MM54C162
MC14162B	MM74C162
MC14162B	SCL4162B
MC14163B	CD40163B
MC14163B	F40163B

MOTOR.	Industry Standard No.
MC14163B	MM54C163
MC14163B	MM74C163
MC14163B	SCL4163B
MC14174B	CD40174B
MC14174B	F40174B
MC14174B	MM54C174
MC14174B	MM74C174
MC14174B	SCL4174B
MC14175B	CD40175B
MC14175B	F40175B
MC14175B	MM54C175
MC14175B	MM74C175
MC14192B	SCL4192B
MC14193B	SCL4193B
MC14194B	CD40194B
MC14194B	F40194B
MC1420G	NE515A
MC1420G	NE515K
MC1420G	NE516A
MC1420G	NE516K
MC1420G	]A727HC
MC1420G	]A730HC
MC142100	CD22100
MC1433G	SG2401N
MC1433G	SG3401N
MC1433G	SG3401T
MC1433L	CA3047
MC1433L	CA3047A
MC1435G	]A749DHC
MC1435G	]A749HC
MC1435L	]A749DC
MC1436CG	SG1436CT
MC1436G	LH0004CH
MC1436G	LM343D
MC1436G	LM343H
MC1436G	ML1436T
MC1436G	SG1436T
MC1436U	SG1436M
MC1437L	LH2301AD
MC1437L	LH2301AF
MC1437L	ML7503M
MC1437L	RC1437D
MC1437P	ML1437P
MC1437P	RC1437DP
MC1438R	]A791KC
MC1438R	]A791P5
MC1439G	NE531G
MC1439G	NE531T
MC1439G	ULN2139D
MC1439G	ULN2139G
MC1439P	NE531V
MC1439P1	ULN2139M
MC1439P1	ULS2139M
MC1439P2	ULN2139H
MC14404	CD4404
MC14406	CD4406
MC14407	CD4407
MC14413	CD4413
MC14414	CD4414
MC14416	CD4416
MC14418	CD4418
MC14502B	CD4502B
MC14502B	SCL4502B
MC14503B	CD4503B
MC14503B	MM80C97
MC14504B	CD40109B
MC14506B	CD4085B
MC14506B	CD4086B
MC14506B	F4085B
MC14506B	F4086B
MC14508B	CD4508B
MC14508B	SCL4508B
MC145104	MM55104
MC145104	MM55114
MC145106	MM55106
MC145106	MM55116
MC145106	MM55126
MC145107	MM55107
MC14510B	CD40192B
MC14510B	CD4510B
MC14510B	F40192B
MC14510B	F4510B
MC14510B	MM54C192
MC14510B	MM74C192
MC14510B	SCL4510B
MC14511B	CD4511B
MC14511B	F4511B
MC14511B	SCL4511B
MC14512B	CD4512B
MC14512B	F4512B
MC14512B	MM54C151
MC14512B	MM74C151
MC14512B	SCL4512B
MC14514B	CD4514B
MC14514B	F4514B
MC14514B	MM54C154
MC14514B	SCL4514B
MC14515B	CD4515B
MC14515B	F4515B
MC14515B	MM74C154
MC14515B	SCL4515B
MC14516B	CD40193B
MC14516B	CD4516B
MC14516B	F40193B
MC14516B	F4516B
MC14516B	MM54C193
MC14516B	MM74C193
MC14516B	SCL4516B
MC14517B	CD4517B
MC14517B	SCL4517B
MC14518B	CD4518B
MC14518B	F4518B
MC14518B	MM54C90
MC14518B	MM74C90
MC14518B	SCL4518B
MC14519B	CD4019B
MC14519B	F4019B
MC14519B	MM54C157
MC14519B	MM74C157
MC14520B	CD4520B
MC14520B	F4520B
MC14520B	MM54C93
MC14520B	MM74C93
MC14520B	SCL4520B
MC14522B	F4522B
MC14522B	SCL4522B
MC14526B	F4526B
MC14526B	SCL4526B
MC14527B	CD4527B
MC14527B	SCL4527B
MC14528B	F4528B
MC14528B	SCL4528B
MC14529B	CD4067B
MC14529B	F4067B

MOTOR.	Industry Standard No.
MC14531B	CD40101B
MC14531B	F4531B
MC14531B	SCL4531B
MC14532B	CD4532B
MC14532B	F4532B
MC14532B	SCL4532B
MC14538B	MM54C221
MC14538B	MM74C221
MC14539B	F4539B
MC14543B	CD4055B
MC14543B	CD4056B
MC14543B	SCL4543B
MC14549B	MM74C905
MC1454G	CA3020A
MC14555B	CD4555B
MC14555B	F4555B
MC14555B	SCL4555B
MC14556B	CD4556B
MC14556B	F4556B
MC14556B	SCL4556B
MC14557B	F4031B
MC14558B	MM54C48
MC14558B	MM74C48
MC14559B	MM54C905
MC1455G	LM322H
MC1455G	LM555CH
MC1455G	NE555L
MC1455G	NE555T
MC1455G	SG555CT
MC1455G	SN72555L
MC1455G	]A555HC
MC1455P1	LM2905N
MC1455P1	LM322N
MC1455P1	LM3905N
MC1455P1	LM555CN
MC1455P1	NE555P
MC1455P1	NE555V
MC1455P1	SG555CM
MC1455P1	SN72555P
MC1455P1	]A2240PC
MC1455P1	]A555TC
MC1455U	NE555JG
MC1455U	]A2240DC
MC14562B	CD4062A
MC1456CG	RC1556T
MC1456CG	SG1456CT
MC1456G	LM212H
MC1456G	LM312H
MC1456G	N5556T
MC1456G	NE537G
MC1456G	NE537T
MC1456G	SG1456T
MC1456G	SN72770L
MC1456G	SN72771L
MC1456G	ULN2156D
MC1456G	ULN2156G
MC1456G	ULN2156H
MC1456G	ULN2156M
MC1456L	LM312D
MC1456L	LM312F
MC1456P1	N5556V
MC14580B	CD40108B
MC14581B	CD40181B
MC14581B	F4581B
MC14581B	SCL4581B
MC14582B	CD40182B
MC14582B	F4582B
MC14582B	SCL4582B
MC14583B	F4583B
MC14584B	CD40106B
MC14584B	MM54C14
MC14584B	MM74C14
MC14584B	SCL4584B
MC14585B	CD4063B
MC14585B	MM54C85
MC14585B	MM74C85
MC14585B	SCL4585B
MC1458CG	]A1458CHC
MC1458CP1	CA1458S
MC1458CP1	]A1458CP
MC1458CP1	]A1458CTC
MC1458CU	]A1458CRC
MC1458G	CA1458T
MC1458G	LM1458H
MC1458G	MC1458L
MC1458G	ML1458T
MC1458G	N5558T
MC1458G	RC1458T
MC1458G	SG1458T
MC1458G	SN72558L
MC1458G	SSS1458J
MC1458G	ULN2157K
MC1458G	]A1458E
MC1458L	N5558F
MC1458P1	LM1458N
MC1458P1	MC1458P
MC1458P1	ML1458S
MC1458P1	N5558V
MC1458P1	RC1458DN
MC1458P1	SG1458M
MC1458P1	SN72558P
MC1458P1	]A1458P
MC1458P1	]A1458TC
MC1458P2	LM1458N-14
MC1458P2	ML1458P
MC1458P2	ULN2157A
MC1458P2	ULN2157H
MC1458U	LM1458J
MC1458U	MC1458JG
MC1458U	]A1458RC
MC1468G	LM325H
MC1468G	LM326H
MC1468G	LM328H
MC1468G	RC4195T
MC1468G	SG1468T
MC1468G	SG2501AT
MC1468G	SG2501T
MC1468G	SG2502T
MC1468G	SG3501AT
MC1468G	SG3501T
MC1468G	SG3502G
MC1468G	SG4501T
MC1468L	LM325AN
MC1468L	LM325N
MC1468L	LM326N
MC1468L	LM328AN
MC1468L	LM328N
MC1468L	SG1468J
MC1468L	SG1468N
MC1468L	SG2501AD
MC1468L	SG2501D
MC1468L	SG2501N
MC1468L	SG2502D
MC1468L	SG2502N

MOTOR.	Industry Standard No.
MC1468L	SG3501AD
MC1468L	SG3501D
MC1468L	SG3501N
MC1468L	SG3502D
MC1468L	SG3502N
MC1468L	SG4501D
MC1468L	SG4501N
MC1468R	RC4195TK
MC1471P1	UDN5711M
MC1471SCP1	RC4131DP
MC1472P1	DS3612N
MC1472P1	DS3632N
MC1472P1	SN75475P
MC1472P1	UDN5712M
MC1472U	DS3612H
MC1472U	DS3632H
MC1472U	DS3632J
MC1472U	SN75475JG
MC1473P1	UDN5713M
MC1474P1	UDN5714M
MC1488L	DS1488J
MC1488L	LM1488J
MC1488L	N8215A
MC1488L	N8215P
MC1488L	RC1488DC
MC1488L	SN75150J
MC1488L	SN75188J
MC1488P	DS1488N
MC1488P	LM1488N
MC1488P	SN75150N
MC1488P	SN75188N
MC1489AL	DM7822J
MC1489AL	DM8822J
MC1489AL	DS1489AJ
MC1489AL	LM1489AJ
MC1489AL	ML1489AM
MC1489AL	RC1489ADC
MC1489AL	SN75189AJ
MC1489AP	DM8822N
MC1489AP	DS1489AN
MC1489AP	LM1489AN
MC1489AP	SN75189AN
MC1489L	DS1489J
MC1489L	LM1489J
MC1489L	ML1489M
MC1489L	N8216A
MC1489L	RC1489DC
MC1489L	SN75154J
MC1489L	SN75189J
MC1489P	DS1489N
MC1489P	LM1489N
MC1489P	SN75154N
MC1489P	SN75189N
MC1494L	SG2402N
MC1494L	SG2402T
MC1494L	SG3402N
MC1494L	SG3402T
MC1495L	N5595A
MC1495L	N5595P
MC1495L	SG1495D
MC1495L	SG1495N
MC1496G	LM1496H
MC1496G	N5596A
MC1496G	N5596K
MC1496G	SG1496T
MC1496G	SN76514L
MC1496G	JA796HC
MC1496L	LM1496J
MC1496L	N5596A
MC1496L	SG1496D
MC1496L	SG1496N
MC1496L	JA796DC
MC1496P	LM1496N
MC1496P	SN76514N
MC1500AU10	REF-01AJ
MC1500AU10	REF-01J
MC1500AU5	REF-02AJ
MC1500AU5	REF-02J
MC1503AU	AD580T
MC1503U	AD580S
MC1503U	SG1503
MC1504L	DAC-01
MC1508L8	AD1508-8D
MC1508L8	AD559S
MC1508L8	AD559SD
MC1508L8	SH8090PM
MC1508L8	SS81508A-8Z
MC1508L8	JA0802DM-1
MC1510P	SN5510FA
MC1510G	CA3040
MC1510G	SN5510L
MC1514L	HM1514J
MC1514L	RM1514DC
MC1514L	SN52514J
MC1520P	NE515G
MC1520P	NE516G
MC1520P	SE515G
MC1520P	SE516G
MC1520G	SE515K
MC1520G	SE516K
MC1520G	JA727EM
MC1520G	JA730HM
MC1533G	SG1401N
MC1533G	SG1401T
MC1533L	CA3033
MC1533L	CA3033A
MC1533L	JA749DM
MC1536	OP-01C
MC1536	OP-01G
MC1536	OP-01H
MC1536G	LH0004H
MC1536G	LM143D
MC1536G	LM143P
MC1536G	LM243H
MC1536G	ML1536T
MC1536G	OP-01J
MC1536G	OP-01L
MC1536G	SG1536T
MC1536G	OP-01P
MC1537L	LH2101AD
MC1537L	LH2101AP
MC1537L	LH2201AD
MC1537L	LH2201AP
MC1537M	RM1537D
MC1537L	M16503M
MC1537L	RM1537D
MC1538R	LH0002CH
MC1538R	LH0002H
MC1538R	JA791KM
MC1539G	SE531G
MC1539G	SE531T
MC1539G	UL82139D

MOTOR.	Industry Standard No.
MC1539G	UL82139G
MC1539G	UL82139H
MC1544L	SE528E
MC1544L	SE528R
MC1544L	SN55244J
MC1545G	IH5101IIE
MC1545G	IH5101MIE
MC1550G	CA3000
MC1550G	CA3001
MC1550G	CA3002
MC1550G	CA3004
MC1550G	CA3005
MC1550G	CA3006
MC1550G	CA3007
MC1550G	CA3028A
MC1550G	CA3028AF
MC1550G	CA3028A3
MC1550G	CA3028B
MC1550G	CA3028BF
MC1550G	CA3028BS
MC1550G	CA3053
MC1550G	CA3053P
MC1550G	CA3053S
MC1550G	LM3011H
MC1554G	CA3020
MC1554G	NE540L
MC1555G	D555GJ
MC1555G	LM122P
MC1555G	LM122H
MC1555G	LM222H
MC1555G	LM555H
MC1555G	SE555L
MC1555G	SE555T
MC1555G	SG555T
MC1555G	SN52555L
MC1555G	JA2240DM
MC1555G	JA555HM
MC1555U	SE555JG
MC1556G	AM725A31T
MC1556G	CMP-01CJ
MC1556G	LM112H
MC1556G	ML108AP
MC1556G	ML208AP
MC1556G	ML741AF
MC1556G	ML741AM
MC1556G	ML741AM
MC1556G	S5556T
MC1556G	SE537G
MC1556G	SE537T
MC1556G	SG1556T
MC1556G	SN52770L
MC1556G	SN52771L
MC1556G	UL82156D
MC1556G	UL82156G
MC1556G	UL82156H
MC1556G	UL82156M
MC1556L	LM112D
MC1556L	LM112P
MC1556L	LM212D
MC1556L	LM212P
MC1556P	CMP-01CP
MC1558G	CA1558T
MC1558G	LM1558H
MC1558G	MC1558L
MC1558G	ML1558T
MC1558G	RC1558T
MC1558G	S5558T
MC1558G	SG1558T
MC1558G	SN52558L
MC1558G	SS81558J
MC1558G	UL82157K
MC1558G	JA1458HC
MC1558G	JA1558E
MC1558G	JA1558HM
MC1558L	ML1558M
MC1558L	S5558R
MC1558L	UL82157A
MC1558L	UL82157H
MC1558U	CA1558S
MC1558U	LM1558J
MC1558U	MC1558JG
MC1568G	LM125H
MC1568G	LM126H
MC1568G	LM128H
MC1568G	LM225H
MC1568G	LM226H
MC1568G	LM228H
MC1568G	RM4195T
MC1568G	SG1501AT
MC1568G	SG1501T
MC1568L	SG1501AD
MC1568L	SG1501D
MC1568L	SG1502D
MC1568L	SG1502N
MC1568R	RM4195TK
MC1590G	CA3011
MC1590G	CA3012
MC1590G	CA3021
MC1590G	CA3022
MC1590G	CA3023
MC1590G	CA3076
MC1590G	LM171H
MC1590G	LM271H
MC1590G	LM711H
MC1594G	ICL8013A
MC1594G	ICL8013B
MC1594G	ICL8013C
MC1594L	CA5091D
MC1594L	SG1402N
MC1594L	SG1402T
MC1595G	AD532J
MC1595L	AD530
MC1595L	AD531
MC1595L	SG1595D
MC1596G	LM1596H
MC1596G	S5596K
MC1596G	SG1596T
MC1596L	JA796HM
MC1596L	LM1596J
MC1596L	S5596P
MC1596L	SG1596D
MC1596L	JA796DM
MC1709AF	LM709AF
MC1709AF	SN52709APA
MC1709AF	JA709AFM
MC1709AG	LM709AH
MC1709AG	ML709AT
MC1709AG	SN52709AL
MC1709AG	JA709AHM
MC1709AG	JA709AML
MC1709AL	LM709AJ
MC1709AL	SN52709AJ
MC1709AL	JA709ADM

MOTOR.	Industry Standard No.
MC1709AL	JA709AMJ
MC1709AU	JA709AMJG
MC1709CF	N5709G
MC1709CG	ICL8007CTA
MC1709CG	ICL8007MTA
MC1709CG	LM709CH
MC1709CG	MIC709-5
MC1709CG	ML709CT
MC1709CG	N5709T
MC1709CG	RC709T
MC1709CG	SG72709L
MC1709CG	JA709CL
MC1709CG	JA709HC
MC1709CL	LM709CJ
MC1709CL	RC709D
MC1709CL	SN72709J
MC1709CL	JA709CJ
MC1709CL	JA709DC
MC1709CP1	LM709CN-8
MC1709CP1	N5709V
MC1709CP1	RC709DN
MC1709CP1	SN72709P
MC1709CP1	JA709CP
MC1709CP1	JA709TC
MC1709CP2	LM709CN
MC1709CP2	MIC709CP
MC1709CP2	N5709A
MC1709CP2	RC709DP
MC1709CP2	SN72709N
MC1709CT	JA709CN
MC1709CT	JA709CP
MC1709CU	JA709CJG
MC1709F	CA3008
MC1709F	CA3008A
MC1709F	CA3016
MC1709F	CA3016A
MC1709F	ML709F
MC1709F	RM709Q
MC1709F	S5709G
MC1709F	SN52709FA
MC1709G	CA3010
MC1709G	CA3010A
MC1709G	CA3015
MC1709G	CA3015A
MC1709G	LM709H
MC1709G	MIC709-1
MC1709G	ML709T
MC1709G	RM709T
MC1709G	S5709T
MC1709G	SN52709L
MC1709G	JA709HM
MC1709G	JA709ML
MC1709L	CA3037
MC1709L	CA3037A
MC1709L	CA3038
MC1709L	CA3038A
MC1709L	LM709J
MC1709L	ML709M
MC1709L	RM709D
MC1709L	SN52709J
MC1709L	JA709DM
MC1709L	JA709MJ
MC1709P2	CA3029A
MC1709P2	CA3029A
MC1709P2	CA3030
MC1709P2	CA3030A
MC1709U	JA709MJG
MC1710CG	LM306H
MC1710CG	LM306H
MC1710CG	LM710CH
MC1710CG	MIC710-5C
MC1710CG	S5710T
MC1710CG	RC710T
MC1710CG	SG710CT
MC1710CG	SN72506L
MC1710CG	SN72510L
MC1710CG	SN72710L
MC1710CG	SN72810L
MC1710CL	JA710DC
MC1710CL	RC710DC
MC1710CL	SG710CD
MC1710CL	SN72306J
MC1710CL	SN72510J
MC1710CL	SN72710J
MC1710CL	SN72720J
MC1710CL	SN72810J
MC1710CL	JA710CD
MC1710CP	LM710CN
MC1710CP	N5710A
MC1710CP	RC710DP
MC1710CP	SG710CN
MC1710CP	SN72306N
MC1710CP	SN72510N
MC1710CP	SN72710N
MC1710CP	SN72720N
MC1710CP	SN72810N
MC1710CP	JA710PC
MC1710F	SN52810FA
MC1710F	SN52810PA
MC1710G	LM306H
MC1710G	MIC710-1C
MC1710G	RM710T
MC1710G	S5710T
MC1710G	SN52106L
MC1710G	SN52510L
MC1710G	SN52710L
MC1710G	SN52810L
MC1710G	JA710HM
MC1710L	RM710D
MC1710L	SN52106J
MC1710L	SN52510J
MC1710L	SN52710J
MC1710L	SN52810J
MC1710L	JA710DM
MC1711CG	MIC711-5C
MC1711CG	N5711K
MC1711CG	RC711T
MC1711CG	SG711CT
MC1711CG	SN72711L
MC1711CG	SN72811L
MC1711CG	JA711HC
MC1711CL	RC711DC
MC1711CL	SG711CD
MC1711CL	SN72811J
MC1711CL	SN72811J
MC1711CL	JA711DC
MC1711CP	LM711CN

MOTOR.	Industry Standard No.
MC1711CP	N5711A
MC1711CP	RC711DP
MC1711CP	SG711CN
MC1711CP	SN72711N
MC1711CP	SN72811N
MC1711CP	JA711PC
MC1711F	SN52711FA
MC1711F	SN52811FA
MC1711H	LM711H
MC1711G	JA711H-1G
MC1711G	RM711T
MC1711K	S5711K
MC1711G	SG711T
MC1711L	SN52711L
MC1711L	S5281L
MC1711L	JA711HM
MC1711L	RM711DC
MC1711L	SG711D
MC1711J	SN52711J
MC1711J	SN52811J
MC1711L	SN52711J
MC1712CF	MIC712-5B
MC1712CG	CA3032
MC1712CG	MIC712-5C
MC1712CG	RC702T
MC1712CG	SN72702L
MC1712CG	JA702HC
MC1712CL	MIC712-5D
MC1712CL	SN72702J
MC1712CL	JA702EC
MC1712F	MIC712-1B
MC1712F	RM702Q
MC1712F	SN52702APA
MC1712F	SN52702FA
MC1712G	CA3051
MC1712G	MIC712-1C
MC1712G	SN52702AL
MC1712G	SN52702L
MC1712G	JA702HM
MC1712G	JA702ML
MC1712L	MIC712-1D
MC1712L	SN52702AJ
MC1712L	SN52702M
MC1712L	JA702DM
MC1712L	JA702MJ
MC1721G	RM702T
MC1723CG	AMU5R7723393
MC1723CG	MIC723-5
MC1723CG	MIC723-5
MC1723CO	MIC723-5
MC1723CG	N5723T
MC1723CG	NE550L
MC1723CG	RC723T
MC1723CG	SG300T
MC1723CG	SG723CT
MC1723CG	SN72723L
MC1723CG	JA723CL
MC1723CL	AMU6A7723393
MC1723CL	MLT723CF
MC1723CL	MLT723CM
MC1723CL	MLT723CP
MC1723CL	RC723D
MC1723CL	SG723CD
MC1723CL	SN72723J
MC1723CL	JA723CL
MC1723CP	CA723CK
MC1723CP	N5723A
MC1723CP	NE550A
MC1723CP	SG300N
MC1723CP	SG723CN
MC1723CP	JA723CN
MC1723F	SN52723FA
MC1723G	AMU5R7723312
MC1723G	CA3085
MC1723G	CA3085A
MC1723G	CA3085AB
MC1723G	CA3085B
MC1723G	CA3085BS
MC1723G	CA3085S
MC1723G	MIC723-1
MC1723G	MLT723T
MC1723G	RM723T
MC1723G	S5723T
MC1723G	SE550L
MC1723G	SG100T
MC1723G	SG200T
MC1723G	SG723T
MC1723G	SN52723L
MC1723G	JA723HM
MC1723G	JA723ML
MC1723L	AMU6A7723312
MC1723L	CA3085AF
MC1723L	CA3085BF
MC1723L	CA3085F
MC1723L	MLT723F
MC1723L	MLT723M
MC1723L	RM723D
MC1723L	SN52723J
MC1723L	JA723DM
MC1723L	JA723MJ
MC1733CG	AMU5B7733393
MC1733CG	LMT733CH
MC1733CG	N5733K
MC1733CG	NE501K
MC1733CG	RC733T
MC1733CG	SG733CT
MC1733CG	SN72733L
MC1733CG	JA733CL
MC1733CG	JA733HC
MC1733CL	AMU5B7733393
MC1733CL	AMU6A7733393
MC1733CL	LMT733CD
MC1733CL	NE501A
MC1733CL	RC733D
MC1733CL	SG733CD
MC1733CL	JA733CD
MC1733CL	JA733CJ
MC1733CP	LMT733CN
MC1733CP	SG733CN
MC1733F	JA733CN
MC1733F	AMU5B7733312
MC1733H	LM733H
MC1733G	RM733T
MC1733G	S5733K
MC1733G	SE501K
MC1733G	SG733T
MC1733G	SN52733L
MC1733G	JA733HM
MC1733G	JA733ML

MOTOR.	Industry Standard No.
MC1733L	AMU3F7733312
MC1733L	AMU6A7733312
MC1733L	LM733D
MC1733L	LM733J
MC1733L	RM733D
MC1733L	SG733D
MC1733L	SG733N
MC1733L	SN52733J
MC1733L	μA733DM
MC1733L	μA733MJ
MC1741CF	RC741Q
MC1741CF	SG741CF
MC1741CF	SN72741FA
MC1741CF	SSS741CL
MC1741CF	ULN2151G
MC1741CF	μA741FC
MC1741CG	AD741CJ
MC1741CG	AMU5B7741393
MC1741CG	CA3056
MC1741CG	CA741CT
MC1741CG	ICB8741C
MC1741CG	LM741EH
MC1741CG	MIC741-5C
MC1741CG	ML741CT
MC1741CG	N5741T
MC1741CG	RC741T
MC1741CG	SG318T
MC1741CG	SG741CT
MC1741CG	SN74741L
MC1741CG	SSS741CJ
MC1741CG	ULN2151D
MC1741CG	ULN2741D
MC1741CG	μA741CL
MC1741CL	AMU3I7741393
MC1741CL	AMU6A7741393
MC1741CL	LM741ED
MC1741CL	LM741EJ-14
MC1741CL	MIC741-5D
MC1741CL	RC741D
MC1741CL	SG741CD
MC1741CL	SN72741J
MC1741CL	μA741CJ
MC1741CL	μA747CJ
MC1741CP1	CA741CS
MC1741CP1	ICL741CLNPA
MC1741CP1	ICL741CLNTY
MC1741CP1	LM741EN
MC1741CP1	ML741CS
MC1741CP1	N5741V
MC1741CP1	RC741DN
MC1741CP1	SG318M
MC1741CP1	SG741CM
MC1741CP1	SN72741P
MC1741CP1	ULN2151M
MC1741CP1	ULS2151M
MC1741CP1	μA741CP
MC1741CP2	ML741CP
MC1741CP2	N5741A
MC1741CP2	RC741DP
MC1741CP2	SG741CN
MC1741CP2	SN72741N
MC1741CP2	SSS741CP
MC1741CP2	ULN2151H
MC1741CP2	μA741CN
MC1741CP2	μA741PC
MC1741CU	LM741EJ
MC1741CU	μA741CJG
MC1741CU	μA741RC
MC1741F	AMU3I7741312
MC1741F	LH101F
MC1741F	LH201F
MC1741F	LM741AF
MC1741F	ML741F
MC1741F	RM741Q
MC1741F	SG741F
MC1741F	SN52741FA
MC1741F	SSS741BL
MC1741F	SSS741L
MC1741F	ULS2151G
MC1741F	μA741AFM
MC1741F	μA741FM
MC1741G	AD741J
MC1741G	AD741K
MC1741G	AD741L
MC1741G	AMU5B7741312
MC1741G	CA3056A
MC1741G	CA741T
MC1741G	LH101H
MC1741G	LH201H
MC1741G	LM741AH
MC1741G	MIC741-1C
MC1741G	ML741T
MC1741G	RM741T
MC1741G	S5741T
MC1741G	SG1217
MC1741G	SG741T
MC1741G	SN52741L
MC1741G	SSS741BJ
MC1741G	SSS741J
MC1741G	ULS2151D
MC1741G	μA741AHM
MC1741G	μA741EHC
MC1741G	μA741HM
MC1741G	μA741ML
MC1741G	μA799HC
MC1741G	μA799HM
MC1741L	AMU6A7741312
MC1741L	ICL741LNDP
MC1741L	ICL741LNFB
MC1741L	ICL741LNTY
MC1741L	LM741AD
MC1741L	LM741AJ-14
MC1741L	MIC741-1D
MC1741L	ML741M
MC1741L	RM741D
MC1741L	SG741D
MC1741L	SN52741J
MC1741L	ULS2151H
MC1741L	μA741ADM
MC1741L	μA741DM
MC1741L	μA741EDC
MC1741L	μA741MJ
MC1741P	RM741DP
MC1741P2	SSS741BP
MC1741P2	SSS741P
MC1741S	μA772
MC1741SCG	LM318H
MC1741SCG	ML318T
MC1741SCG	SG741SCT
MC1741SCG	μA715HC
MC1741SCL	LM318D
MC1741SCL	LM318F
MC1741SCL	SG318J
MC1741SCL	μA715DC
MC1741SCP1	LM318N
MC1741SCP1	ML318M
MC1741SCP1	SG741SCM
MC1741SG	AD741S
MC1741SG	LM118H
MC1741SG	LM218H
MC1741SG	ML118F
MC1741SG	ML118M
MC1741SG	ML118T
MC1741SG	ML218F
MC1741SG	ML218M
MC1741SG	ML218T
MC1741SG	RC4131T
MC1741SG	SG118T
MC1741SG	SG1217T
MC1741SG	SG218T
MC1741SG	SG741ST
MC1741SG	SSS741GJ
MC1741SG	SSS741GP
MC1741SG	μA715HM
MC1741SL	LM118D
MC1741SL	LM118F
MC1741SL	LM218D
MC1741SL	LM218F
MC1741SL	SG118J
MC1741SL	SG1217J
MC1741SL	SG218J
MC1741SL	SG218M
MC1741SL	μA715DM
MC1741U	CA741S
MC1741U	μA741MJG
MC1741U	μA741RM
MC1747CCBM	μA747EDC
MC1747CF	SN72747FA
MC1747CF	SSS747CM
MC1747CG	AMU5B7747393
MC1747CG	CA747CT
MC1747CG	ML747CT
MC1747CG	N5748A
MC1747CG	RC747T
MC1747CG	SG747CT
MC1747CG	SN72747L
MC1747CG	SSS747CK
MC1747CG	μA747CL
MC1747CG	μA747HC
MC1747CICM	μA747EHC
MC1747CL	AMU6W7747393
MC1747CL	CA747CE
MC1747CL	CA747CF
MC1747CL	ML747CP
MC1747CL	N5747A
MC1747CL	N5747F
MC1747CL	RC747D
MC1747CL	SG747CJ
MC1747CL	SN72747J
MC1747CL	SSS747CP
MC1747CL	ULN2747A
MC1747CL	μA747DC
MC1747CP2	SG747CN
MC1747CP2	SN72747N
MC1747CP2	μA747CN
MC1747CP2	μA747PC
MC1747F	ML747F
MC1747F	SN52747FA
MC1747F	SSS747B2
MC1747F	SSS747GM
MC1747F	SSS747L
MC1747G	AMU5B7747312
MC1747G	CA747T
MC1747G	ML747T
MC1747G	RM747T
MC1747G	SG747T
MC1747G	SN52747L
MC1747G	SSS747GK
MC1747G	μA747AHM
MC1747G	μA747HM
MC1747G	μA747ML
MC1747L	AMU6W7747312
MC1747L	CA747E
MC1747L	CA747F
MC1747L	ML747M
MC1747L	RM747D
MC1747L	SG747J
MC1747L	SN52747J
MC1747L	SSS747BP
MC1747L	SSS747GP
MC1747L	SSS747P
MC1747L	μA747ADM
MC1747L	μA747DM
MC1747L	μA747MJ
MC1748CG	AMU5B7748393
MC1748CG	CA748CT
MC1748CG	LM748CH
MC1748CG	ML748CT
MC1748CG	N5748T
MC1748CG	RC748T
MC1748CG	SG748CT
MC1748CG	SN72748L
MC1748CG	μA748CL
MC1748CG	μA748HC
MC1748CL	μA748CJ
MC1748CL	μA748DC
MC1748CP1	AMU6A7748393
MC1748CP1	CA748CS
MC1748CP1	LM748CN
MC1748CP1	SG748CD
MC1748CP1	SG748CM
MC1748CP1	SG748CN
MC1748CP1	SN72748P
MC1748CP1	μA748CP
MC1748CP1	μA748TC
MC1748CP2	μA748CN
MC1748CU	LM748CJ
MC1748CU	μA748CJG
MC1748F	μA748AFM
MC1748F	μA748FM
MC1748G	AMU3F7748312
MC1748G	AMU5B7748312
MC1748G	AMU6A7748312
MC1748G	CA748T
MC1748G	LM748H
MC1748G	ML748F
MC1748G	ML748M
MC1748G	ML748T
MC1748G	RM748T
MC1748G	SG748D
MC1748G	SG748T
MC1748G	SN52748L
MC1748G	μA748AHM
MC1748G	μA748HM
MC1748G	μA748ML
MC1748L	μA748DM
MC1748L	μA748MJ
MC1748U	CA748S
MC1748U	LM748J
MC1748U	μA748MJG
MC174L	RF103DC
MC174L	SF103-01
MC1776	OP-08
MC1776	OP-08A
MC1776	OP-08B
MC1776	OP-08C
MC1776	OP-08E
MC1776CG	AD505J
MC1776CG	AD505K
MC1776CG	CA3078S
MC1776CG	CA3078T
MC1776CG	ICH8500ATV
MC1776CG	ICH8500TV
MC1776CG	LH0001ACD
MC1776CG	LH0001ACF
MC1776CG	LH0001ACH
MC1776CG	LM4250CH
MC1776CG	ML4250CS
MC1776CG	ML4250CT
MC1776CG	ML4251CS
MC1776CG	ML4251CT
MC1776CG	NE533G
MC1776CG	NE533T
MC1776CG	NE533V
MC1776CG	SG4250CT
MC1776CG	μA776DC
MC1776CG	μA776HC
MC1776CP1	LM4250CN
MC1776CP1	SG4250CM
MC1776CP1	μA776TC
MC1776G	AD505S
MC1776G	CA3078AS
MC1776G	CA3078AT
MC1776G	CA6078AS
MC1776G	CA6078AT
MC1776G	CA6741S
MC1776G	CA6741T
MC1776G	ICL8021C
MC1776G	ICL8021M
MC1776G	ICL8022C
MC1776G	ICL8022M
MC1776G	ICL8043CDE
MC1776G	ICL8043CPE
MC1776G	ICL8043MDE
MC1776G	ICL8048CDE
MC1776G	ICL8048DPE
MC1776G	LH0001AD
MC1776G	LH0001AF
MC1776G	LH0001AH
MC1776G	LH0042CH
MC1776G	LM4250H
MC1776G	ML4250T
MC1776G	ML4251T
MC1776G	SE533G
MC1776G	SE533T
MC1776G	SG1250T
MC1776G	SG2250T
MC1776G	SG3250T
MC1776G	SG4250T
MC1776G	μA776DM
MC1776G	μA776HM
MC1800F	SN151800U
MC1800F	SW1800-2F
MC1800F	U3I180059
MC1800L	ITT1800-5D
MC1800L	SN151800J
MC1800L	SW1800-2P
MC1800L	U6A180059
MC1800P	DM1800N
MC1800P	ITT1800-5N
MC1800P	SN151800N
MC1800P	SW1800-2M
MC1800P	U9A180059
MC1801F	SN151801U
MC1801F	SW1801-2F
MC1801F	U3I180159
MC1801L	SN151801J
MC1801L	SW1801-2P
MC1801L	U6A180159
MC1801P	DM1801N
MC1801P	SN151801N
MC1801P	SW1801-2M
MC1801P	U9A180159
MC1802F	SN151802U
MC1802F	SW1802-2F
MC1802F	U3I180259
MC1802L	SN151802J
MC1802L	SW1802-2P
MC1802L	U6A180259
MC1802P	SN151802N
MC1802P	SW1802-2M
MC1802P	U9A180259
MC1803F	SN151803U
MC1803F	U3I180359
MC1803L	U6A180359
MC1803P	SN151803N
MC1803P	U9A180359
MC1804F	SN151804U
MC1804F	U3I180459
MC1804L	SN151804J
MC1804L	U6A180459
MC1804P	SN151804N
MC1804P	U9A180459
MC1805F	SN151805U
MC1805F	SW1805-2F
MC1805F	U3I180559
MC1805L	SN151805J
MC1805L	SW1805-2P
MC1805L	U6A180559
MC1805P	SN151805N
MC1805P	SW1805-2M
MC1805P	U9A180559
MC1806F	SN151806U
MC1806F	SW1806-2F
MC1806F	U3I109659
MC1806L	ITT1806-5D
MC1806L	SN151806J
MC1806L	SW1806-2P
MC1806L	U6A180659
MC1806P	ITT1806-5N
MC1806P	SN151806N
MC1806P	SW1806-2M
MC1806P	U9A180659
MC1807F	SN151807U
MC1807F	SW1807-2F
MC1807F	U3I180759
MC1807L	ITT1807-5D
MC1807L	SN151807J
MC1807L	SW1807-2P
MC1807L	SW1808-2P
MC1807L	U6A180759
MC1807P	ITT1807-5N
MC1807P	SN151807N
MC1807P	SW1807-2M
MC1807P	U9A180759
MC1808F	SN151808U
MC1808F	SW1808-2F
MC1808F	U3I180859
MC1808L	ITT1808-5D
MC1808L	SN151808J
MC1808L	U6A180859
MC1808P	ITT1808-5N
MC1808P	SN151808N
MC1808P	SW1808-2M
MC1808P	U9A180859
MC1809F	SN151809U
MC1809F	U3I180959
MC1809L	SN151809J
MC1809L	U6A180959
MC1809P	SN151809N
MC1809P	U9A180959
MC1810F	SN151810U
MC1810F	SW1810-2F
MC1810F	U3I181059
MC1810L	SN151810J
MC1810L	SW1810-2P
MC1810L	U6A181059
MC1810P	SN151810N
MC1810P	SW1810-2M
MC1810P	U9A181059
MC1811F	SN151811U
MC1811F	U3I181159
MC1811L	SN151811J
MC1811L	U6A181159
MC1811P	SN151811N
MC1811P	U9A181159
MC1812F	SN151812U
MC1812F	SW1812-2F
MC1812F	U3I181259
MC1812L	SN151812J
MC1812L	SW1812-2P
MC1812L	U6A181259
MC1812P	SN151812N
MC1812P	SW1812-2M
MC1812P	U9A181259
MC1814F	SW1814-2F
MC1814F	U3I181459
MC1814L	SW1814-2P
MC1814L	U6A181459
MC1814P	SW1814-2M
MC1814P	U9A181459
MC1816F	U3I181659
MC1816L	U6A181659
MC1816P	U9A181659
MC1820F	SN151820U
MC1820L	SN151820J
MC1820P	SN151820N
MC1900F	SN151900U
MC1900F	SW1900-1F
MC1900F	U3I180051
MC1900L	ITT1800-1D
MC1900L	SN151900J
MC1900L	SW1900-1P
MC1900L	U6A180051
MC1901F	SN151901U
MC1901F	SW1901-1F
MC1901F	U3I180151
MC1901L	SN151901J
MC1901L	SW1901-1P
MC1901L	U6A180151
MC1902F	SN151902U
MC1902F	SW1902-1F
MC1902F	U3I180251
MC1902L	SN151902J
MC1902L	SW1902-1P
MC1902L	U6A180251
MC1903F	U3I180351
MC1903L	SN151803J
MC1903L	U6A180351
MC1904F	SN151904U
MC1904F	U3I180451
MC1904L	SN151904J
MC1904L	U6A180451
MC1905F	SN151905U
MC1905F	SW1905-1F
MC1905F	U3I180551
MC1905L	SN151905J
MC1905L	SW1905-1P
MC1905L	U6A180551
MC1906F	SN151906U
MC1906F	SW1906-1F
MC1906F	U3I180651
MC1906L	ITT1806-1D
MC1906L	SN151906J
MC1906L	SW1906-1P
MC1906L	U6A180651
MC1907F	SN151907U
MC1907F	SW1907-1F
MC1907F	U3I180751
MC1907L	ITT1807-1D
MC1907L	SN151907J
MC1907L	SW1907-1P
MC1907L	U6A180751
MC1908F	SN151908U
MC1908F	SW1908-1F
MC1908F	U3I180851
MC1908L	ITT1808-1D
MC1908L	SN151908J
MC1908L	SW1908-1P
MC1908L	U6A180851
MC1909F	SN151909U
MC1909F	U3I180951
MC1909L	SN151909J
MC1909L	U6A180951
MC1910F	SN151910U
MC1910F	SW1910-1F
MC1910F	U3I181051
MC1910L	SN151910J
MC1910L	SW1910-1P
MC1910L	U6A181051
MC1911F	SN151911U
MC1911F	U3I181151
MC1911L	SN151911J
MC1911L	U6A181151
MC1912F	SN151912U
MC1912F	SW1912-1F
MC1912F	U3I181251
MC1912L	SN151912J
MC1912L	SW1912-1P
MC1912L	U6A181251
MC1914F	SW1914-1F
MC1914F	U3I181451
MC1914L	SW1914-1P
MC1914L	U6A181451
MC1916F	U3I181651
MC1916L	U6A181651

MOTOR.	Industry Standard No.	MOTOR.	Industry Standard No.	MOTOR.	Industry Standard No.	MOTOR.	Industry Standard No.
MC1920F	SN151920U	MC2123L	RF120DC	MC34002BG	TL082ACL	MC3520L	SG1524J
MC1920L	SN151920J	MC2123L	SF120-01	MC34002BF	LF353N	MC3520L	SG2524J
MC2000P	SG212-02	MC2124P	RP130K	MC34002BP	TL082ACP	MC3520L	TL497MJ
MC2000L	SG212-01	MC2124P	SF130-02	MC34002BP	TL082ACP	MC3556L	LM556J
MC2001P	SG222-02	MC2124L	RP130DC	MC34002BU	TL072ACJG	MC3556L	LM556A
MC2001L	SG222-01	MC2124L	SF130-01	MC34002BU	TL082ACJG	MC3556L	SE556A
MC2001L	SG222-01	MC2150P	RG211K	MC34002G	LF353N	MC3556L	SG556J
MC2002P	SG232-02	MC2150P	SG211-02	MC34002G	TL072CL	MC3556L	SG556N
MC2002L	SG232DC	MC2150L	RG211DC	MC34002G	TL082CL	MC3556L	JA556DM
MC2002L	SG232-01	MC2150L	SG211-01	MC34002P	LF353N	MC3558G	JA798HM
MC2003F	SG242-02	MC2151P	RG221K	MC34002P	TL072CP	MC3558U	JA798RM
MC2003L	SG242DC	MC2151P	SG221-02	MC34002P	TL082CP	MC4004F	SM82-02
MC2003L	SG242-01	MC2151L	RG221DC	MC34002U	TL072CJG	MC4004L	RL82DC
MC2004F	SG251-02	MC2151L	SG221-01	MC34002U	TL082CJG	MC4004L	SM82-01
MC2004L	RG252DC	MC2152P	RG231K	MC34004A	TL074BCJ	MC4005F	SM83-02
MC2004L	SG252-01	MC2152P	SG231-02	MC34004AL	TL084BCJ	MC4005L	RL83DC
MC2005F	SG262-02	MC2152L	RG231DC	MC34004AP	LF347AN	MC4005L	SM83-01
MC2005L	SG262-01	MC2152L	SG231DC	MC34004AP	TL074BCN	MC400F	SG42-02
MC2006F	SG272-02	MC2153P	RG241K	MC34004AP	TL084BCN	MC400L	RG42DC
MC2006L	RG262DC	MC2153P	SG241-02	MC34004BL	TL074ACJ	MC401F	SG52-02
MC2006L	SG272-01	MC2153L	RG241DC	MC34004BL	TL084ACJ	MC401L	RG52DC
MC2007F	SG322-02	MC2153L	SG241-01	MC34004BP	LF347AN	MC401L	SG52-01
MC2007L	RG322DC	MC2154F	RG251K	MC34004BP	TL074ACN	MC4026F	SM12-02
MC2007L	SG322-01	MC2154L	RG251DC	MC34004BP	TL084ACN	MC4026L	RL12DC
MC2011P	SG202-02	MC2154L	SG251-01	MC34004L	TL074CJ	MC4026L	SM12-01
MC2011L	SG202DC	MC2155P	RG261K	MC34004L	TL084CJ	MC4027F	SM13-02
MC2011L	SG202-01	MC2155P	SG261-02	MC34004P	LF347N	MC4027L	RL13DC
MC2012P	SG302-02	MC2155L	RG261DC	MC34004P	TL074CN	MC4027L	SM13-01
MC2012L	SG302DC	MC2155L	SG261-01	MC34004P	TL084CN	MC4028F	SM22-01
MC2012L	SG302-01	MC2156P	RG271K	MC3401P	CA3401E	MC4028F	RL22DC
MC2013F	SG312-02	MC2156P	SG271-02	MC3401P	LM3401N	MC4028L	RL22DC
MC2013L	SG312DC	MC2156L	RG271DC	MC3401P	LM3900N	MC4029F	SM23-02
MC2013L	SG312-01	MC2156L	SG271-01	MC3401P	JA3401P	MC4029L	RL23DC
MC2016F	SG382-02	MC2157P	RG321K	MC3403L	RC4136D	MC4029L	SM23-01
MC2016L	RG382DC	MC2157P	SG321-02	MC3403L	RC4136J	MC402F	SG62-02
MC2016L	SG382-01	MC2157L	RG321DC	MC3403L	JA3403D	MC402L	RG62DC
MC2023F	SF122-02	MC2157L	SG321-01	MC3403P	RC4136DP	MC402L	SG62-01
MC2023L	RF122DC	MC2161P	RG201K	MC3403P	RC4136N	MC4030F	SM32-02
MC2023L	SF122-01	MC2161P	SG201-02	MC3403P	JA3403P	MC4030L	RL32DC
MC2024F	SF132-02	MC2161L	RG201DC	MC3410OL	AD7520D	MC4030L	SM32-01
MC2024L	RF132DC	MC2161L	SG201-01	MC3410OL	AD7520P	MC4031F	SM33-02
MC2024L	SF132-01	MC2162P	RG301K	MC3410OL	AD7520ON	MC4031L	RL33DC
MC2050F	SG213-02	MC2162P	SG301-02	MC3410L	DAC-1C10BC	MC4031L	SM33-01
MC2050L	RG213DC	MC2162L	RG301DC	MC3416L	RC4444R	MC4032F	SM43-02
MC2050L	SG213-01	MC2162L	SG301-01	MC3420OL	SG3524J	MC4032L	RL43DC
MC2051F	SG223-02	MC2163F	RG311K	MC3420L	TL497CJ	MC4032L	SM43-01
MC2051L	RG223DC	MC2163F	SG311-02	MC3420P	TL497CN	MC4035F	SM63-02
MC2051L	SG223-01	MC2163L	RG311DC	MC3426L	LM1850N	MC4035L	RL63DC
MC2052F	SG233-02	MC2165P	RG351K	MC3430L	DS3651J	MC4035L	SM63-01
MC2052L	RG233DC	MC2165P	SG351-02	MC3430P	DS3651N	MC4037F	SM73-02
MC2052L	SG233-01	MC2165L	SG351-01	MC3432L	DS3653J	MC4037L	RL73DC
MC2053F	SG243-02	MC2166P	RG381K	MC3432P	DS3653N	MC4037L	SM73-01
MC2053L	RG243DC	MC2166P	SG381-02	MC3437L	DS7837J	MC403F	SG92-02
MC2053L	SG243-01	MC2166L	RG381DC	MC3437L	DS7837J	MC403L	RG92DC
MC2054F	SG253-02	MC2166L	SG381-01	MC3437L	DS7837W	MC403L	SG92-01
MC2054L	RG253DC	MC2173F	RF121K	MC3437P	DS8837J	MC404F	SG102-02
MC2054L	SG253-01	MC2173F	SF121-02	MC3437P	DM8837N	MC404L	RG102DC
MC2055F	RG253K	MC2173L	RF121DC	MC3437P	DS8837N	MC404L	SG102-01
MC2055F	SG263-02	MC2173L	SF121-01	MC3437P	N8837A	MC405F	SG112-02
MC2055L	RG263DC	MC2174P	RF131K	MC3438L	DM7838J	MC405L	RG112DC
MC2055L	SG263-01	MC2174P	SF131-02	MC3438L	DS7838J	MC405L	SG112-01
MC2056L	SG273DC	MC2174L	RF131DC	MC3438L	DS7838W	MC406F	SG122-02
MC2056L	SG273-02	MC2174L	SF131-01	MC3438L	DS8838J	MC406L	RG122DC
MC2057F	SG323-02	MC26810L	AM26810DC	MC3438P	DM8838N	MC406L	SG122-01
MC2057L	RG323DC	MC26810P	AM26810PC	MC3438P	DS8838N	MC407F	SG132-02
MC2057L	SG323-01	MC26811L	AM26811DC	MC3438P	N8T38A	MC407L	RG132DC
MC2061F	SG203-02	MC26811P	AM26811PC	MC3443P	SN75138J	MC407L	SG132-01
MC2061L	RG203DC	MC3232AF	D2532	MC3443P	SN75138N	MC408F	SG142-02
MC2061L	SG203-01	MC3242AF	D3242	MC3450L	DS3650J	MC408F	SG142-01
MC2062F	SG303-02	MC3301L	LM1900D	MC3450L	LM163J	MC408L	RG142DC
MC2062L	RG303DC	MC3301L	LM2900J	MC3450L	LM363AJ	MC408L	SG142-01
MC2062L	SG303-01	MC3301P	CA3048	MC3450L	LM363J	MC409F	SG152-02
MC2063F	SG313-02	MC3301P	CA3052	MC3450P	DS3650N	MC409L	RG152DC
MC2063L	SG313DC	MC3301P	LM2900N	MC3450P	LM363AN	MC409L	SG152-01
MC2063L	SG313-01	MC3301P	LM3301N	MC3450P	LM363N	MC410F	RG172K
MC2065F	SG353-02	MC3301P	RV3301DB	MC3452L	DS3652J	MC410P	SG172-02
MC2065L	SG353-01	MC3301P	JA3301P	MC3452P	DS3652N	MC410L	RG172DC
MC2066F	SG383-02	MC3302L	LM3302J	MC3456L	LM556CD	MC410L	SG172-01
MC2066L	RG383DC	MC3302P	CA3302E	MC3456L	LM556CJ	MC411F	SG182-02
MC2066L	SG383-01	MC3302P	LM3302N	MC3456L	N5561	MC411L	RG182DC
MC2073F	SF123-02	MC3302P	RC3302DB	MC3456L	SG556CJ	MC411L	SG182-01
MC2073L	RF123DC	MC3303P	JA3302P	MC3456L	JA556DC	MC412F	SG192-02
MC2073L	SF123-01	MC3303P	JA3302P	MC3456L	LM556CN	MC412L	RG192DC
MC2076F	SG273-02	MC3346P	CA5045	MC3456P	N5556A	MC412L	SG192-01
MC2100P	RG210K	MC3346P	CA3045F	MC3456P	SG556CN	MC413F	SF12-02
MC2100P	SG210-02	MC3346P	CA3046	MC3456P	JA556PC	MC413L	SF12-01
MC2100L	RG210DC	MC3346P	CA3086F	MC3458G	JA798HC	MC414F	SP22-02
MC2100L	SG210-01	MC3346P	CA3136A	MC3458P1	JA798HC	MC414L	SP22-01
MC2101F	RG220K	MC3346P	CA3146	MC3458U	JA798RC	MC415F	SP52-02
MC2101P	SG220-02	MC3346P	LM3045	MC3460L	DS3674J	MC415L	RP52DC
MC2101L	RG220DC	MC3346P	LM3046N	MC3460P	DS3674N	MC415L	SP52-01
MC2101L	SG220-01	MC3346P	LM3146	MC3481/5L	SN75126J	MC416F	SP62-02
MC2102F	RG230K	MC3346P	LM3146A	MC3481/5P	SN75126N	MC416L	RP62DC
MC2102P	SG230-02	MC3346P	MLM046P	MC3486L	AM26LS32DC	MC416L	SP62-01
MC2102L	SG230-01	MC3346P	JA3045	MC3486L	AM26LS33DC	MC419F	SG162-02
MC2103F	SG240-02	MC3346P	JA3046DC	MC3486L	DS3486J	MC419L	RG162DC
MC2103F	SG240-02	MC3370P	SN72440J	MC3486P	AM26LS32PC	MC419L	SG162-01
MC2103L	SG240DC	MC3370P	SN72440N	MC3486P	AM26LS33PC	MC420F	SG72-02
MC2103L	SG240-01	MC3386F	CA3086	MC3486P	DS3486N	MC420L	RG72DC
MC2104F	RG250K	MC3386F	LM3086N	MC3487L	AM26LS30DC	MC420L	SG72-01
MC2104F	SG250-02	MC34001AG	LF351AH	MC3487L	DS3487J	MC421F	RF32K
MC2104L	RG250DC	MC34001AG	TL074BCL	MC3487P	AM26LS30PC	MC421F	SF32-02
MC2104L	SG250-01	MC34001AG	TL081BCL	MC3487P	DS3487N	MC421L	RP32DC
MC2105F	RG260K	MC34001AP	LF351AN	MC3490P	DM7887J	MC421L	SF32-01
MC2105P	SG260-02	MC34001AP	TL071BCP	MC3490P	DM7887N	MC422F	SF82-02
MC2105L	SG260DC	MC34001AP	TL081BCP	MC3490P	DM8887J	MC422L	SF82-01
MC2105L	SG260-01	MC34001AU	TL071BCJG	MC3490P	DS7887J	MC423F	SP102-02
MC2106F	RG270K	MC34001AU	TL081BCJG	MC3490P	DM8887N	MC423L	RP102DC
MC2106F	SG270-02	MC34001BG	LF351BH	MC3490P	DS8887J	MC423L	SP102-01
MC2106L	RG270DC	MC34001BG	TL071ACL	MC3490P	DS8887N	MC424F	SP112-02
MC2106L	SG270-01	MC34001BG	TL081ACL	MC3490P	UDN-6144A	MC424L	RP112DC
MC2107F	RG320K	MC34001BP	LF351BN	MC3490P	UDN-6164A	MC424L	SP112-01
MC2107F	SG320-02	MC34001BP	TL071ACP	MC3490P	UDN-6184A	MC426F	SG82-02
MC2107L	RG320DC	MC34001BP	TL081ACP	MC3491P	UHP-495	MC426L	RG82DC
MC2107L	SG320-01	MC34001BU	TL071ACJG	MC3491P	DM7889J	MC426L	SG82-01
MC2111F	RG200K	MC34001BU	TL081ACJG	MC3491P	DM7889N	MC427F	SG282-02
MC2111F	SG200-02	MC34001G	LF351H	MC3491P	DM8889J	MC427F	RG282DC
MC2111L	RG200DC	MC34001G	TL071CL	MC3491P	DS7889J	MC427L	SG282-01
MC2111L	SG200-01	MC34001G	TL081CL	MC3491P	DM8889N	MC428F	SG292-02
MC2112F	RG300K	MC34001P	LF351N	MC3491P	DS8889J	MC428L	RG292DC
MC2112F	SG300-02	MC34001P	TL081CP	MC3491P	DM8889N	MC428L	SG292-01
MC2112L	RG300DC	MC34001U	TL071CJG	MC3491P	UDN-7183A	MC429F	SG372-02
MC2112L	SG300-01	MC34001U	TL081CJG	MC3491P	UDN-7184A	MC429L	RG372DC
MC2113F	RG310K	MC34002AG	LF353AH	MC3491P	UDN-7186A	MC429L	SG372-01
MC2113F	SG310-02	MC34002AG	TL072BCL	MC3494P	DM7897J	MC4304F	RL80K
MC2113L	RG310DC	MC34002AG	TL082BCL	MC3494P	DM7897J	MC4304F	SM80-02
MC2113L	SG310-01	MC34002AP	LF353AN	MC3494P	DM8897J	MC4304L	RL80DC
MC2116F	RG380K	MC34002AP	TL072BCP	MC3494P	DM8897N	MC4304L	SM80-01
MC2116F	SG380-02	MC34002AP	TL082BCP	MC3494P	DS8897N	MC4305F	RL81K
MC2116L	RG380DC	MC34002AU	TL072BCJG	MC3494P	UHD-490	MC4305F	SM81-02
MC2116L	SG380-01	MC34002AU	TL082BCJG	MC3494P	UHD-491	MC4305L	RL81DC
MC2123F	RF120K	MC34002BG	LF353BH	MC3494P	UHP-490	MC4305L	SM81-01
MC2123F	SF120-02	MC34002BG	TL072ACL	MC3494P	UHP-491	MC4326F	RL10K
				MC3503L	RM4136D	MC4326F	SM10-02
				MC3503L	RM4136J		

MOTOR.	Industry Standard No.	MOTOR.	Industry Standard No.	MOTOR.	Industry Standard No.	MOTOR.	Industry Standard No.
MC4326L	RL10DC	MC500F	SG40-02	MC528F	SG290-02	MC528F	TD200BP
MC4326L	SM10-01	MC500L	RG40DC	MC528L	RP100DC	MC660P	TD660P
MC4327F	RL11K	MC500L	SG40-01	MC528L	RG290DC	MC660TL	PZH135
MC4327F	SM11-02	MC501F	RG50K	MC528L	SF100-01	MC660TL	H104D2
MC4327L	RL11DC	MC501F	SG50-02	MC529F	RG370K	MC661L	H124D6
MC4327L	SM11-01	MC501L	RG50DC	MC529F	SG370-02	MC661L	SG370-01
MC4328F	RL20K	MC501L	SG50-01	MC529L	RG370DC	MC667F	TD2001P
MC4328F	SM20-02	MC502F	RG60K	MC529L	SG370-01	MC661P	TL661P
MC4328L	RI20DC	MC502F	SG60-02	MC550F	RG41K	MC661TL	H124D2
MC4328L	SM20-01	MC502L	RG60DC	MC550L	RG41DC	MC662L	PZH141
MC4329F	RL21K	MC502L	SG60-01	MC550L	SG41-01	MC662L	HNIL301AL
MC4329F	SM21-02	MC503F	RP90K	MC55107L	D855107J	MC662L	HNIL301CL
MC4329L	RL21DC	MC503F	SG90-02	MC55107L	LM55107AJ	MC662L	TL662L
MC4329L	SM21-01	MC503L	RG90DC	MC55107L	RM55107AD	MC662P	HNIL301AJ
MC4330F	RL30K	MC503L	SG90-01	MC55107L	SN55107AJ	MC662P	HNIL301CJ
MC4330F	SM30-02	MC504F	RG100K	MC55107L	SN55107BJ	MC662P	TD2011P
MC4330F	RL30DC	MC504F	SG100-02	MC55108L	D855108J	MC662P	TL662P
MC4330L	SM30-01	MC504L	RG100DC	MC55108L	D855108W	MC662TL	PZH145
MC4331F	RL31K	MC504L	SG100-01	MC55108L	LM55108AJ	MC663L	PZJ121
MC4331L	RL31DC	MC505F	RG110K	MC55108L	SN55108AJ	MC663L	H110D1
MC4331L	SM31-01	MC505F	SG110-02	MC551F	RG51K	MC663L	H110D6
MC4332F	RL41K	MC505L	RG110DC	MC551F	SG51-02	MC663P	TD2005P
MC4332F	SM41-02	MC505L	SG110-01	MC551L	SG51-01	MC663TL	PZJ125
MC4332L	RL41DC	MC506F	RG120K	MC5525L	LM5525J	MC663TL	H110D2
MC4332L	SM41-01	MC506F	SG120-02	MC5528L	LM552BJ	MC664L	HNIL311AL
MC4335F	RL61K	MC506L	RG120DC	MC552F	R061K	MC664L	HNIL311CL
MC4335F	SM61-02	MC506L	SG120-01	MC552F	SG61-02	MC664L	TL664L
MC4335L	RL61DC	MC507F	RG130K	MC552L	RG61DC	MC664P	HNIL311AJ
MC4335L	SM61-01	MC507F	SG130-02	MC55325L	D855325J	MC664P	HNIL311CJ
MC4337F	RL71K	MC507L	RG130DC	MC55325L	LM55325N	MC664P	TD2004P
MC4337F	SM71-02	MC507L	SG130-01	MC55325L	RM55325DD	MC664P	TL664P
MC4337L	RL71DC	MC508F	RG140K	MC5534L	LM5534J	MC665L	HNIL361AL
MC4337L	SM71-01	MC508F	SG140-02	MC5535L	LM5535J	MC665L	HNIL361CL
MC450F	S943-02	MC508L	RG140DC	MC5538L	LM5538AJ	MC665L	TL665L
MC450L	RG43DC	MC508L	SG140-01	MC5539L	LM5529J	MC665P	HNIL361AJ
MC450L	SG43-01	MC5090F	SM90-02	MC553F	RG91K	MC665P	HNIL361CJ
MC451F	S053-02	MC5090L	SM90-01	MC553L	SG91-01	MC665P	TD2016P
MC451L	RG53DC	MC5092F	SM92-02	MC554F	RG101K	MC665P	TL665P
MC451L	S053-01	MC5092L	SM92-01	MC554F	SG101-02	MC666L	HNIL362AL
MC452F	SG63-02	MC509F	RG150K	MC554L	SG101-01	MC666L	HNIL362CL
MC452L	RG63DC	MC509F	SG150-02	MC555F	RG111K	MC666L	TL666L
MC452L	SG63-01	MC509L	RG150DC	MC555F	SG111-02	MC666P	HNIL362AJ
MC453F	SG93-02	MC509L	SG150-01	MC555L	SG111-01	MC666P	HNIL362CJ
MC453L	RG93DC	MC510F	RG170K	MC555L	SG111DC	MC666P	TL666P
MC453L	SG93-01	MC510F	SG170-02	MC556L	RG121DC	MC667	H117D1
MC454F	SG103-02	MC510L	RG170DC	MC556L	SG121-01	MC667	H117D6
MC454L	RG103DC	MC510L	SG170-01	MC557F	RG131K	MC667	P2K101
MC454L	SG103-01	MC5111F	RL111K	MC557F	SG131-02	MC667L	HNIL342AL
MC4558CG	RC4558L	MC5111F	SM111-02	MC557L	SG131DC	MC667L	HNIL342CL
MC4558CG	RC4558T	MC5111L	RL111DC	MC557L	SG31-01	MC667P	HNIL342AJ
MC4558CG	jA4558BC	MC5111L	SM111-01	MC558F	RG141K	MC667P	HNIL342CJ
MC4558CP1	RC4558DN	MC5113F	SM113-02	MC558L	SG141-02	MC667P	TD2015P
MC4558CP1	RC4558P	MC5113L	RL113DC	MC558L	SG141DC	MC667TL	P2K105
MC4558CP1	JA4558TC	MC5113L	SM113-01	MC559F	SG141-01	MC668L	H122D1
MC4558CU	RC4558JG	MC511F	RG180K	MC559F	RG151K	MC668L	H122D6
MC4558G	RM4558L	MC511L	RG180DC	MC559F	SG151-02	MC668L	HNIL303AL
MC4558G	RM4558T	MC511L	SM180-01	MC559L	SG151-01	MC668L	HNIL303CL
MC4558G	jA4558BM	MC5121F	RL121K	MC560F	RG171K	MC668L	TL668L
MC4558U	RM4558D	MC5121F	SM121-02	MC560L	SG171-02	MC668P	HNIL303AJ
MC4558U	RM4558JG	MC5121L	RL121DC	MC560L	SG171DC	MC668P	HNIL303CJ
MC455F	SG113-02	MC5121L	SM121-01	MC561F	RG181K	MC668P	HNIL324CJ
MC455L	RG113DC	MC5123F	SM123-02	MC561F	SG181-02	MC668P	TD2003P
MC455L	SG113-01	MC5123L	RL123DC	MC561L	SG181DC	MC668P	TL668P
MC456F	SG123-02	MC5123L	SM123-01	MC561L	SG181-01	MC668TL	H122D2
MC456L	RG123DC	MC512F	RG190K	MC562F	RG191K	MC669L	HNIL331AL
MC456L	SG123-01	MC512F	SG190-02	MC562F	SG191-02	MC669L	HNIL331CL
MC457F	SG133-02	MC512L	RG190DC	MC562L	RG191DC	MC669P	HNIL331AJ
MC457L	RG133DC	MC512L	SG190-01	MC562L	SG191-01	MC669P	HNIL331CJ
MC457L	SG133-01	MC5131F	SM131-02	MC563F	SP11-02	MC669P	TD2006P
MC458F	SG143-02	MC5131L	SM131-01	MC564L	SP21-02	MC670L	HNIL326AL
MC458L	RG143DC	MC5133F	SM133-02	MC564L	SP21-01	MC670L	HNIL326CL
MC458L	SG143-01	MC5133L	SM133-01	MC565F	RF51K	MC670L	TL670L
MC459F	SG153-02	MC513F	SF10-02	MC565F	SP51-02	MC670P	HNIL326AJ
MC459L	RG153DC	MC513L	SF10-01	MC565L	RF51DC	MC670P	HNIL326CJ
MC459L	SG153-01	MC5141F	SM141-02	MC566F	SP61-02	MC670P	TD2002P
MC460F	SG173-02	MC5141L	SM141-01	MC566L	RG61DC	MC670P	TL670P
MC460L	RG173DC	MC5143F	SM143-02	MC566L	SP61-01	MC671L	PZH191
MC460L	SG173-01	MC5143L	SM143-01	MC569F	RG161K	MC671L	H103D1
MC461F	SG183-02	MC514F	SP20-02	MC569F	SG161-02	MC671L	H103D6
MC461L	RG183DC	MC514L	SP20-01	MC569L	RG161DC	MC671L	HNIL325AL
MC461L	SG183-01	MC5151F	SM151-02	MC569L	SG161-01	MC671L	HNIL325CL
MC462F	SG193-02	MC5151L	SM151-01	MC570F	RG71K	MC671L	TL671L
MC462L	RG193DC	MC5153F	SM153-02	MC570F	SG71-02	MC671P	HNIL325AJ
MC462L	SG193-01	MC5153L	SM153-01	MC570L	RF71DC	MC671P	HNIL325CJ
MC463F	SF13-02	MC515F	RP50K	MC570L	SG71-01	MC671P	TD2009P
MC463L	SF13-01	MC515F	SP50-02	MC571F	RF31K	MC671P	TL671P
MC464F	SP23-02	MC515L	RD50DC	MC571F	SP31DC	MC671TL	PZH195
MC464L	SP23-01	MC515L	SP50-01	MC571L	SF31-01	MC671TL	H103D2
MC465F	SF53-02	MC5163F	SM163-02	MC572F	SP81-02	MC672L	PZH111
MC465L	RD53DC	MC5163L	SM163-01	MC572F	SG280-02	MC672L	H102D6
MC465L	SF53-01	MC516F	RP60K	MC572L	SP81-01	MC672L	HNIL321AL
MC466F	SF63-02	MC516F	SF60-02	MC573F	RP101K	MC672L	HNIL321CL
MC466L	RF63DC	MC516L	RP60DC	MC573F	RP101K	MC672L	TL672L
MC466L	SF63-01	MC516L	SF60-01	MC573L	RP101DC	MC672P	HNIL321AJ
MC469F	SG163-02	MC5173F	SM173-02	MC573L	SP101-01	MC672P	HNIL321CJ
MC469L	RG163DC	MC5173L	SM173-01	MC574F	RP111K	MC672P	TD2010P
MC469L	SG163-01	MC5181F	SM181-02	MC574F	SP111-02	MC672P	TL672P
MC470F	SG73-02	MC5181L	SM181-01	MC574L	RP111DC	MC672TL	PZH115
MC470L	RG73DC	MC5183F	SM183-02	MC574L	SP111-01	MC672TL	H102D2
MC470L	SG73-01	MC5183L	SM183-01	MC576F	SG81-02	MC673L	H105D1
MC471F	SF33-02	MC5191F	SM191-02	MC576L	SG81-01	MC673L	H105D6
MC471L	RF53DC	MC5193F	MJ3772	MC577F	RG281K	MC673L	HNIL341AL
MC471L	SF33-01	MC5193F	SM193-02	MC577F	SG281-02	MC673L	HNIL341CL
MC472F	SF63-02	MC5193L	SM193-01	MC577L	RG281DC	MC673P	HNIL341AJ
MC472L	SF63-01	MC5195L	SM191-01	MC577L	SG281-01	MC673P	HNIL341CJ
MC473F	SF103-02	MC519F	RG160K	MC578F	RG291K	MC673TL	H105D2
MC47741CL	LM349D	MC519F	SG160-02	MC578F	SG291-02	MC675L	HNIL347AL
MC47741CL	LM349J	MC519L	RG160DC	MC578L	RG291DC	MC675L	HNIL347CL
MC47741CL	LM349N	MC519L	SG160-01	MC578L	SG291-01	MC675L	HNIL347AJ
MC47741CL	jA4136DC	MC520F	RF70K	MC579F	RG371K	MC675P	HNIL347CJ
MC47741CP	jA4136PC	MC520L	SG70-02	MC579F	SG371-02	MC675P	H158D1
MC47741L	LM148F	MC520L	RG70DC	MC579L	RG371DC	MC676L	H158D6
MC47741L	LM149D	MC520L	SG70-01	MC579L	SG371-01	MC676L	HNIL380AL
MC47741L	LM149F	MC521F	RF30K	MC653L	SP11-01	MC676L	HNIL380CL
MC47741L	LM249D	MC521F	SP30-02	MC660L	PZH131	MC676L	HNIL381AL
MC47741L	LM249J	MC521F	SP30DC	MC660L	H104D1	MC676L	HNIL381CL
MC47741L	jA4136DM	MC521L	SP30-01	MC660L	H104D6	MC676L	HNIL382AL
MC474F	SF113-02	MC522F	SF80-02	MC660L	HNIL322AL	MC676L	HNIL382CL
MC474L	RF113DC	MC522L	SF80-01	MC660L	HNIL322CL	MC676P	HNIL380AJ
MC474L	SF113-01	MC524F	RF110K	MC660L	TL660L	MC676P	HNIL380CJ
MC475F	SF83-02	MC524F	SF110-02	MC660P	HNIL322AJ	MC676P	HNIL381AJ
MC476L	RG83DC	MC524L	RF110DC	MC660P	HNIL322CJ	MC676P	HNIL381CJ
MC476L	SG83-01	MC524L	SF110-01			MC676P	HNIL382AJ
MC477F	SG283-02	MC526F	RG80K			MC676P	HNIL382CJ
MC477L	RG283DC	MC526F	SG80-02			MC676P	TD2018P
MC477L	SG283-01	MC526L	RG80DC			MC676TL	H158D2
MC478F	SG293-02	MC526L	SG80-01			MC677L	PZH201
MC478L	RG293DC	MC527F	RG280K			MC677L	H119D1
MC478L	SG293-01	MC527L	RG280DC			MC677L	H119D6
MC479F	SG373-02	MC527L	SG280-01			MC677L	HNIL335CL
MC479L	RF373DC	MC528F	RP100K			MC677L	HNIL335AJ
MC479L	SG373-01	MC528F	SP100-02			MC677L	HNIL335AL
MC500F	HG40K					MC677P	HNIL335CJ

MOTOR.	Industry Standard No.
MC6677TL	F2H205
MC6677TL	H119D2
MC678L	H115D1
MC678L	H115D6
MC678L	HNIL334AL
MC678L	HNIL334AJ
MC678P	HNIL334AJ
MC678P	HNIL334CJ
MC678TL	H115D2
MC680L	H118D1
MC680L	H118D6
MC680P	TD2013P
MC680TL	H118D2
MC681L	H112D1
MC681L	H112D6
MC681L	HNIL332AL
MC681L	HNIL336AL
MC681L	HNIL336CL
MC681L	TL681L
MC681P	HNIL332AJ
MC681P	HNIL332CJ
MC681P	HNIL336AJ
MC681P	HNIL336CJ
MC681P	TL681P
MC681TL	H112D2
MC682L	FZJ131
MC682L	HNIL370AL
MC682L	HNIL370CL
MC682P	HNIL370AJ
MC682P	HNIL370CJ
MC682TL	FZJ135
MC684L	FZJ141
MC684L	H157D1
MC684L	H157D6
MC684L	HNIL371AL
MC684L	HNIL371CL
MC684P	HNIL371AJ
MC684P	HNIL371CJ
MC684TL	FZJ145
MC684TL	H157D2
MC685L	FZJ151
MC685L	H156D1
MC685L	H156D6
MC685L	HNIL372AL
MC685P	HNIL372CL
MC685P	HNIL372AJ
MC685P	HNIL372CJ
MC685TL	FZJ155
MC685TL	H156D2
MC686L	HNIL375AL
MC686P	HNIL375CL
MC686P	HNIL375CJ
MC688L	H111D1
MC688L	H111D6
MC688L	HNIL312AL
MC688L	HNIL312CL
MC688L	HNIL324AL
MC688L	HNIL324CL
MC688L	HNIL312AJ
MC688P	HNIL312CJ
MC688TL	H111D2
MC689L	H113D1
MC689L	H113D6
MC689L	H113D2
MC691L	H114D1
MC691L	H114D6
MC691L	HNIL363AL
MC691L	HNIL363CL
MC691P	HNIL363AJ
MC691P	HNIL363CJ
MC691P	TD2017P
MC691TL	H114D2
MC693L	HNIL302AL
MC693L	HNIL302CL
MC693P	HNIL302AJ
MC693P	HNIL302CJ
MC694L	HNIL368AL
MC694L	HNIL368CL
MC694P	HNIL368AJ
MC694P	HNIL368CJ
MC697L	HNIL333AL
MC697L	HNIL333CL
MC697L	HNIL333AJ
MC697P	HNIL333CJ
MC697P	TD2012P
MC697P	SN76660N
MC75107L	D855107W
MC75107J	D875107J
MC75107L	D875207J
MC75107L	LM75107AJ
MC75107L	LM75207L
MC75107L	RC75107AD
MC75107L	SN75107AJ
MC75107L	SN75107BJ
MC75107L	SN75207J
MC75107P	D875107N
MC75107AN	LM75207N
MC75107P	LM75207N
MC75107P	RC75107ADP
MC75107P	SN75107AN
MC75107P	SN75107BN
MC75107L	SN75207N
MC75108L	D875108J
MC75108L	D875208J
MC75108L	LM75108AJ
MC75108L	LM75208J
MC75108L	RC75108AD
MC75108L	SN75108AJ
MC75108L	SN55108BJ
MC75108L	SN75108BJ
MC75108L	SN75208J
MC75108P	D875108N
MC75108P	D875208N
MC75108P	LM75108AN
MC75108P	LM75208N
MC75108P	RC75108ADP
MC75108P	SN75108AN
MC75108P	SN75108BN
MC75108P	SN75208N
MC75125/27	D875125/27
MC75125L	SN75125J
MC75125P	SN75125N
MC75127L	SN75127J
MC75127P	SN75127N
MC75128/29	D875128/29
MC75128L	SN75128J
MC75128P	SN75128N
MC75129L	SN75129J
MC75129P	SN75129N
MC75140P1	SN75140P
MC75524L	LM75524J
MC7524P	LM7524N
MC7525L	LM75J
MC75325L	D875325J
MC75325L	LM75324J
MC75325L	LM75325N
MC75325L	RC75325DD
MC75325P	D875325R
MC75325P	LM75324N
MC75325P	LM75325J
MC75491P	DM7491N
MC75491P	DN8861N
MC75491P	D875491J
MC75491P	SN75461N
MC75491P	SN75491N
MC75492P	DN7492N
MC75492P	DN8863N
MC75492P	D875492J
MC75492P	D875492N
MC75492P	SN75492N
MC758110L	D855110J
MC758110L	S875110J
MC758110L	LM55110J
MC758110L	SN55110J
MC758110L	D875110N
MC758110L	SN55110J
MC758110N	LM75110N
MC758110P	RC75110DP
MC7802ACP	JA78L26AWC
MC7805CK	LM340KC-5.0
MC7805CK	LM7805KC
MC7805CK	SG340K-05
MC7805CK	SG7805K
MC7805CK	LM7805KC
MC7805CK	JA7805KC
MC7805CT	LM340T-5.0
MC7805CT	JA7805CKC
MC7805CT	JA7805UC
MC7805K	JA7805KM
MC7806CK	LM340KC-6.0
MC7806CK	LM7806KC
MC7806CK	SG340K-06
MC7806CK	SG7806K
MC7806CK	JA7806KC
MC7806CT	LM340T-6.0
MC7806CT	JA7806CKC
MC7806CT	JA7806UC
MC7806K	JA7806KM
MC7808CK	LM340KC-8.0
MC7808CK	LM7808KC
MC7808CK	SG340K-08
MC7808CK	SG7808K
MC7808CK	JA7808K
MC7808CK	JA7808KC
MC7808CT	LM340T-8.0
MC7808CT	JA7808CKC
MC7808CT	JA7808UC
MC7808K	JA7808HM
MC7812CK	LM340KC-12
MC7812CK	LM7812KC
MC7812CK	SG340K-12
MC7812CK	SG7812CK
MC7812CT	JA7812KC
MC7812CT	JA7812CKC
MC7812CT	JA7812UC
MC7812K	JA7812KM
MC7815CK	LM340KC-15
MC7815CK	JA7815KC
MC7815CK	SG340K-15
MC7815CK	SG7815CK
MC7815CK	JA7815KC
MC7815CT	LM340T-15
MC7815CT	JA7815CKC
MC7815CT	JA7815UC
MC7815CT	JA7815KM
MC7818CK	LM340KC-18
MC7818CK	JA7818KC
MC7818CK	SG340K-18
MC7818CK	SG7818K
MC7818CT	JA7818KC
MC7818CT	LM340T-18
MC7818CT	JA7818CKC
MC7818CT	JA7818UC
MC7818K	JA7818KM
MC7824CK	LM340KC-24
MC7824CK	JA7824KC
MC7824CK	SG340K-24
MC7824CK	SG7824CK
MC7824CK	JA7824KC
MC7824CT	JA7824CKC
MC7824CT	JA7824UC
MC7824K	JA7824KM
MC78L02ACG	JA78L02ACJG
MC78L05ACG	LM140LAH-5.0
MC78L05ACG	LM340LAH-5.0
MC78L05ACG	LM7805ACH
MC78L05ACG	JA78L05ACJG
MC78L05ACG	JA78L05AHC
MC78L05ACP	LM240LAZ-5.0
MC78L05ACP	LM340LAZ-5.0
MC78L05ACP	JA78L05ACLP
MC78L05ACP	JA78L05AWC
MC78L05CG	LM78L05CH
MC78L05CG	LM78L05CJG
MC78L05CG	JA78L05HC
MC78L05CP	LM78L05CZ
MC78L05CP	JA78L05CLP
MC78L05CP	JA78L05WC
MC78L06ACG	LM140LAH-6.0
MC78L06ACG	LM340LAH-6.0
MC78L06ACG	JA78L06ACJG
MC78L06ACP	LM240LAZ-6.0
MC78L06ACP	LM340LAZ-6.0
MC78L06ACP	JA78L06ACLP
MC78L06CG	LM240LAH-6.0
MC78L06CG	JA78L06CJG
MC78L08ACG	LM140LAH-8.0
MC78L08ACG	LM240LAH-8.0
MC78L08ACG	LM340LAH-8.0
MC78L08ACG	LM340LAH-8.0
MC78L08ACG	JA78L08ACH
MC78L08ACG	JA78L08ACJG
MC78L08ACP	LM240LAZ-8.0
MC78L08ACP	LM340LAZ-8.0
MC78L08ACP	LM78L08ACZ
MC78L08ACP	JA78L08ACLP
MC78L08CG	LM78L08CH
MC78L08CG	JA78L08CJG
MC78L08CP	LM78L08CZ
MC78L08CP	JA78L08CLP
MC78L12ACG	LM140LAH-12
MC78L12ACG	LM340LAH-12
MC78L12ACG	LM78L12ACH
MC78L12ACG	JA78L12ACJG
MC78L12ACG	JA78L12AHC
MC78L12ACP	LM240LAZ-12
MC78L12ACP	LM340LAZ-12
MC78L12ACP	JA78L12ACLP
MC78L12ACP	JA78L12AWC
MC78L12CG	LM78L12CH
MC78L12CG	LM78L12CJG
MC78L12CG	JA78L12HC
MC78L12CP	LM78L12CZ
MC78L12CP	JA78L12CLP
MC78L15ACG	LM140LAH-15
MC78L15ACG	LM340LAH-15
MC78L15ACG	LM340LAH-15
MC78L15ACG	JA78L15ACH
MC78L15ACG	JA78L15ACJG
MC78L15ACG	JA78L15AHC
MC78L15ACP	LM240LAZ-15
MC78L15ACP	LM340LAZ-15
MC78L15ACP	JA78L15ACZ
MC78L15ACP	JA78L15ACLP
MC78L15ACP	JA78L15AWC
MC78L15CG	LM78L15CH
MC78L15CG	JA78L15HC
MC78L15CG	LM78L15CZ
MC78L15CP	JA78L15CLP
MC78L15CP	JA78L15WC
MC78L18ACG	LM140LAH-18
MC78L18ACG	LM240LAH-18
MC78L18ACG	LM340LAH-18
MC78L18ACP	LM240LAZ-18
MC78L18ACP	LM340LAZ-18
MC78L18ACP	JA78L18ACZ
MC78L18CG	LM78L18CH
MC78L18CP	LM78L18CZ
MC78L24ACG	LM140LAH-24
MC78L24ACG	LM240LAH-24
MC78L24ACG	LM340LAH-24
MC78L24ACP	LM240LAZ-24
MC78L24ACP	LM340LAZ-24
MC78L24ACP	JA78L24ACZ
MC78L24CG	LM78L24CH
MC78L24CP	LM78L24CZ
MC78M05CG	JA78M05HC
MC78M05CG	JA78M05HM
MC78M05CT	LM341P-5.0
MC78M05CT	LM342P-5.0
MC78M05CT	JA78M05KC
MC78M05CT	JA78M05UC
MC78M06CG	JA78M06HC
MC78M06CG	JA78M06HM
MC78M06CT	LM341P-6.0
MC78M06CT	LM342P-6.0
MC78M06CT	JA78M06CKC
MC78M06CT	JA78M06UC
MC78M08CG	JA78M08HC
MC78M08CG	JA78M08HM
MC78M08CT	LM342P-8.0
MC78M08CT	LM342P-8.0
MC78M08CT	JA78M08CKC
MC78M08CT	JA78M08UC
MC78M12CG	JA78M12HC
MC78M12CG	LM341P-12
MC78M12CT	LM342P-12
MC78M12CT	JA78M12CKC
MC78M12CT	JA78M12UC
MC78M15CG	JA78M15HC
MC78M15CG	JA78M15HM
MC78M15CT	LM341P-15
MC78M15CT	LM342P-15
MC78M15CT	JA78M15CKC
MC78M15UG	JA78M15UG
MC78M18CG	JA78M18HC
MC78M18CG	JA78M18HM
MC78M18CT	LM341P-18
MC78M18CT	LM342P-18
MC78M18UG	JA78M18UG
MC78M20CG	JA78M20HC
MC78M20CG	JA78M20HM
MC78M20CT	JA78M20CKC
MC78M20CG	JA78M20HM
MC78M24CG	JA78M24HC
MC78M24CG	JA78M24HM
MC78M24CT	LM341P-24
MC78M24CT	LM342P-24
MC78M24CT	JA78M24CKC
MC78M24CT	JA78M24UC
MC7902CT	JA7902UC
MC7902K	JA7902KM
MC7902CK	JA7902KM
MC7905.2CK	LM120H-5.2
MC7905.2CK	LM220H-5.2
MC7905.2CK	LM220K-5.2
MC7905.2CK	SG320K-5.2
MC7905.2CK	SG120H-5.2
MC7905.2CK	SG320K-5.2
MC7905.2CT	LM320MP-5.2
MC7905.2CT	SG320T-5.2
MC7905CK	LM145K
MC7905CK	LM220H-5.0
MC7905CK	LM220K-5.0
MC7905CK	LM245K
MC7905CK	LM145K
MC7905CK	JA7905KC
MC7905CK	JA7905KM
MC7905CK	JA79M05AHM
MC7905CK	JA79M05HM
MC7905CT	LM320MP-5.0
MC7905CT	SN72905
MC7905CT	JA7905UC
MC7905CT	JA79M05AUC
MC7905CT	JA79M05CKC
MC7905CT	JA79M05UC
MC7906CK	LM220H-6.0
MC7906CK	LM220K-6.0
MC7906CK	JA7906KC
MC7906CK	JA7906KM
MC7906CK	JA79M06AHM
MC7906CK	JA79M06HM
MC7906CT	LM320MP-6.0
MC7906CT	SN72906
MC7906CT	JA7906UC
MC7906CT	JA79M06AUC
MC7906CT	JA79M06CKC
MC7906CT	JA79M06UC
MC7908CK	LM220H-8.0
MC7908CK	LM220K-8.0
MC7908CK	JA7908KC
MC7908CK	JA7908KM
MC7908CK	JA79M08AHM
MC7908CK	JA79M08M
MC7908CT	LM320MP-8.0
MC7908CT	SN72908
MC7908CT	JA7908UC
MC7908CT	JA79M08AUC
MC7908CT	JA79M08CKC
MC7908CT	JA79M08UC
MC7912CK	LM220H-12
MC7912CK	LM220K-12
MC7912CK	JA7912KC
MC7912CK	JA7912KM
MC7912CK	JA79M12AHM
MC7912CK	JA79M12HM
MC7912CT	LM320MP-12
MC7912CT	SN72912
MC7912CT	JA7912UC
MC7912CT	JA79M12AUC
MC7912CT	JA79M12CKC
MC7912CT	JA79M12UC
MC7915CK	LM220H-15
MC7915CK	LM220K-15
MC7915CK	JA7915KC
MC7915CK	JA7915KM
MC7915CK	JA79M15AHM
MC7915CK	JA79M15HM
MC7915CT	LM320MP-15
MC7915CT	SN72915
MC7915CT	JA7915UC
MC7915CT	JA79M15AUC
MC7915CT	JA79M15CKC
MC7915CT	JA79M15UC
MC7918CK	LM220H-18
MC7918CK	LM220K-18
MC7918CK	JA7918KC
MC7918CK	JA7918KM
MC7918CK	JA79M18AHM
MC7918CK	JA79M18HM
MC7918CT	LM320MP-18
MC7918CT	JA7918CKC
MC7918CT	JA79M18UC
MC7918CT	JA79M18AUC
MC7918CT	JA79M18UC
MC7924CK	LM220H-24
MC7924CK	LM220K-24
MC7924CK	JA7924KC
MC7924CK	JA7924KM
MC7924CK	JA79M24AHM
MC7924CK	JA79M24HM
MC7924CT	LM320MP-24
MC7924CT	JA7924CKC
MC7924CT	JA7924UC
MC7924CT	JA79M24AUC
MC7924CT	JA79M24UC
MC79L05ACG	JA79L05AHC
MC79L05ACG	JA79L05AWC
MC79L05CG	JA79L05HC
MC79L05CP	JA79L05WC
MC79L12ACG	JA79L12AHC
MC79L12ACP	JA79L12AWC
MC79L12CG	JA79L12HC
MC79L12CP	JA79L12WC
MC79L15ACG	JA79L15AHC
MC79L15ACP	JA79L15AWC
MC79L15CG	JA79L15HC
MC79L15CP	JA79L15WC
MC830L	SN75905-5D
MC830L	SN15830J
MC830L	SW950-2P
MC830L	U6A993059
MC830P	DM930N
MC830P	ITT903-5N
MC830P	SN15830N
MC830P	SW950-2M
MC830P	U9A993059
MC832L	ITT932-5D
MC832L	SN15832J
MC832L	SW932-2P
MC832L	U6A993259
MC832P	DM932N
MC832P	ITT932-5N
MC832P	SN15832N
MC832P	SW932-2M
MC832P	U9A993259
MC833L	ITT933-5D
MC833L	SN15833J
MC833L	U6A993359
MC833P	DM933N
MC833P	ITT933-5N
MC833P	SN15833N
MC833P	SW933-2M
MC833P	U9A993359
MC833P	DM933N
MC834L	SN15834J
MC834L	SN15844J
MC834P	SN15834N
MC835L	ITT938-5D
MC835L	SN15858J
MC835L	U6A913559
MC835P	ITT938-5N
MC835P	SN15858N
MC835P	U9A913559
MC836L	ITT936-5D
MC836L	SN15836J
MC836L	SW936-2P
MC836L	U6A993659
MC836P	DM936N
MC836P	ITT936-5N
MC836P	SN15836N
MC836P	SW936-2M
MC836P	U9A993659
MC837L	ITT937-5D
MC837L	SN15837J
MC837L	SW937-2P
MC837L	SW937-2P

MOTOR.	Industry Standard No.	MOTOR.	Industry Standard No.	MOTOR.	Industry Standard No.	MOTOR.	Industry Standard No.
MC837L	U6A993759	MC863L	SN158063J	MC935L	SN15935J	MC955L	U6A909351
MC837P	DM937N	MC863L	SW963-2P	MC935L	U6A913551	MC955F	MC855F
MC837P	ITT937-5N	MC863L	U6A996359	MC936F	MC836F	MC955F	SN15809U
MC837P	SN15837N	MC863P	DM963N	MC936F	SN15836U	MC955F	SN150097U
MC837P	SW937-2M	MC863P	ITT963-5N	MC936F	SN15936U	MC955F	SW709-2P
MC837P	U9A993759	MC863P	SN15863N	MC936F	SW936-1P	MC955F	U31909751
MC840L	ITT935-5D	MC863P	SW963-2M	MC936F	SW936-2P	MC955F	U31909759
MC840L	SN15835J	MC863P	U9A996359	MC936F	U31993651	MC955L	SN15909J
MC840P	DM935N	MC8713L	DS55121J	MC936L	U31993659	MC955L	SW709-1P
MC840P	ITT935-5N	MC8713L	DS55121N	MC936L	ITT936-1D	MC955L	U6A909751
MC840P	SN15835M	MC8713L	DS75121J	MC936L	SN15936J	MC956F	MC856F
MC840P	SW935-2M	MC8713L	LM55121J	MC936L	SW936-1P	MC956F	SN15809U
MC840P	U9A993559	MC8713L	LM75121J	MC936L	U6A993651	MC956F	SN150094U
MC844L	ITT944-5D	MC8713L	N8713P	MC937P	MC837P	MC956F	SW708-1P
MC844L	SW944-2P	MC8713L	RC8713DD	MC937P	SN15837U	MC956F	U31909451
MC844L	U6A994459	MC8713L	S8713E	MC937P	SN15937U	MC956F	U31909459
MC844P	DM944N	MC8713L	SN75121J	MC937P	SW937-1P	MC956L	SN15909J
MC844P	ITT944-5N	MC8713L	JA8713DC	MC937P	U31993751	MC956L	SW708-1P
MC844P	SN15844N	MC8713P	DS75121N	MC937P	U31993759	MC956L	U6A909451
MC844P	SW944-2M	MC8713P	LM75121N	MC937L	ITT937-1D	MC957F	MC857F
MC844P	U9A994459	MC8713P	N8713B	MC937L	SN15937J	MC957F	SN15857U
MC845L	ITT945-5D	MC8713P	RC8713MP	MC937L	SW937-1P	MC957F	SN15957U
MC845L	SW945-2P	MC8713P	SN75121N	MC937L	U6A993751	MC957F	SW957-1P
MC845L	U6A994559	MC8713P	JA8713PC	MC938F	MC838F	MC957F	U31915751
MC845P	DM945N	MC8714L	DS55122J	MC939F	MC839F	MC957L	U31915759
MC845P	ITT945-5N	MC8714L	DS55122N	MC940F	MC840F	MC957L	SN15957J
MC845P	SN15845N	MC8714L	DS75122J	MC940F	SN15935U	MC957L	SW957-1P
MC845P	SW945-2M	MC8714L	LM55122J	MC940F	SN15935U	MC957L	U6A915751
MC845P	U9A994559	MC8714L	LM75122J	MC940P	SW935-2P	MC958F	MC858F
MC846L	ITT946-5D	MC8714L	N8714E	MC940P	U31993551	MC958F	SN15858U
MC846L	SN15846J	MC8714L	RC8714DD	MC940P	U31993559	MC958F	SN15958U
MC846L	SW946-2P	MC8714L	S8714B	MC940P	U6A993559	MC958F	SW958-1P
MC846L	U6A994659	MC8714L	SN75122J	MC940L	ITT935-1D	MC958F	U31915851
MC846N	DM946N	MC8714L	JA8714DC	MC940L	SN15935J	MC958P	U31915859
MC846P	ITT946-5N	MC8714P	DS75122N	MC940L	U6A993551	MC958L	SN15958J
MC846P	SN15846N	MC8714P	LM75122N	MC941F	MC841F	MC958L	SW958-1P
MC846P	SW946-2M	MC8714P	N8714B	MC944F	MC844F	MC958L	U6A915851
MC846P	U9A994659	MC8714P	RC8714MP	MC944F	SN15944U	MC961F	MC861F
MC848L	ITT948-5D	MC8714P	SN75122N	MC944F	SW709-2P	MC961F	SN15861U
MC848L	SN15845J	MC8714P	JA8714PC	MC944L	SW944-1P	MC961P	SN15861U
MC848L	SW948-2P	MC8723L	DS216	MC944L	SW944-2P	MC961P	SW961-1P
MC848L	U6A994859	MC8723L	LM55123J	MC944L	U31994451	MC961P	U31915951
MC848P	DM948N	MC8723L	LM75123J	MC944L	U31994459	MC961P	U31915959
MC848P	ITT948-5N	MC8723L	N8723E	MC944L	ITT944-1D	MC961L	ITT961-1D
MC848P	SN15848N	MC8723L	RC8723DD	MC944L	SN15944J	MC961L	SN15961J
MC848P	SW948-2M	MC8723L	SN75123J	MC944L	SW944-1P	MC961L	SW961-1P
MC848P	U9A994859	MC8723L	JA8723DC	MC944L	U6A994451	MC961L	U6A916151
MC849L	ITT949-5D	MC8723P	DS75123N	MC945F	SN15945U	MC962P	MC862P
MC849L	SN15849J	MC8723P	LM75123N	MC945F	SN15945U	MC962P	SN15862U
MC849L	SW949-2P	MC8723P	N8723B	MC945P	SW945-1P	MC962P	SN15962U
MC849L	U6A994959	MC8723P	RC8723MP	MC945P	SW945-2P	MC962P	SW962-1P
MC849P	DM949N	MC8723P	SN75123N	MC945P	U31994551	MC962P	U31996251
MC849P	ITT949-5N	MC8723P	JA8723PC	MC945P	U31994559	MC962P	U31996259
MC849P	SN15849N	MC8724L	DS75124J	MC945L	ITT945-1D	MC962L	ITT962-1D
MC849P	SW949-2M	MC8724L	LM55124J	MC945L	SN15945J	MC962L	SN15962J
MC849P	U9A994959	MC8724L	LM75124J	MC945L	SW945-1P	MC962L	SW962-1P
MC850L	SN15850J	MC8724L	N8724E	MC945L	U6A994551	MC962L	U6A996251
MC850L	SW950-2M	MC8724L	RC8724DD	MC946F	MC846F	MC963P	MC863P
MC850L	U6A995059	MC8724L	SN75124J	MC946F	SN15846U	MC963P	SN16963U
MC850P	SN15850N	MC8724L	JA8724DC	MC946L	SN15946U	MC963P	SW963-1P
MC850P	SW950-2M	MC8724P	DS75124N	MC946L	SW946-1P	MC963P	SW963-2P
MC850P	U9A995059	MC8724P	LM75124N	MC946L	SW946-2P	MC963P	U31963551
MC851L	ITT951-5D	MC8724P	N8724B	MC946L	U31994651	MC963P	U31996359
MC851L	SN15851J	MC8724P	RC8724MP	MC946L	U31994659	MC963L	SN15963J
MC851L	SW951-2P	MC8724P	SN75124N	MC946L	ITT946-1D	MC963L	ITT963-1D
MC851L	U6A995159	MC8724P	JA8724PC	MC946L	SN15946J	MC963L	SW963-1P
MC851P	ITT951-5N	MC8726AL	DS216	MC946L	SW946-1P	MC963L	U6A996351
MC851P	SN15851N	MC8726AL	DS8834J	MC946L	U6A994651	MCA230	MCA230
MC851P	SW951-2M	MC8726AL	DS8835J	MC947P	MC847P	MCA231	MCA231
MC851P	U9A995159	MC8726AE	RC8726AE	MC948F	MC848F	MCA255	MCA255
MC852L	SN158099J	MC8726AJ/N	DS8826AJ/N	MC948F	SN15848U	MCM10147	F10405
MC852L	SW706-2P	MC8726AL/P	DS8726AMJ/N	MC948F	SN15948U	MCM10544	SP10144
MC852L	U6A909959	MC8726AP	DS8834N	MC948L	SW948-1P	MCM10544	F10410
MC852P	DM909N	MC8726AP	DS8835N	MC948L	SW948-2P	MCM10545	F10145A
MC852P	SN158099N	MC8726AP	N8726AB	MC948L	U31994851	MCM10545	SP10145
MC852P	SW706-2M	MC8726AP	S8726B	MC948L	U31994859	MCM10546	F10415
MC852P	U9A909959	MC8728J/P	DS8728J/N	MC948L	ITT948-1D	MCM14505	MM54089
MC853L	SN158093J	MC828L	DS226	MC948L	SN15948J	MCM14505	MM74089
MC853L	SW705-2N	MC828L	DS8833J	MC948L	SW948-1P	MCM14537	MM54O200
MC853P	DM9093N	MC828L	DS8833N	MC948L	U6A994851	MCM14537	MM74C200
MC853P	SN158093N	MC828LP	DS8728MJ/N	MC949F	MC849F	MCM2114	AM9114
MC853P	SW705-2M	MC828P	DS8833N	MC949F	SN15949U	MCM2114	HM472114
MC855P	U9A909959	MC828P	N8728B	MC949F	SW949-1P	MCM2114	I2114
MC855L	SN158097J	MC8795P	N8795B	MC949P	SW949-2P	MCM2114	IM2114
MC855L	SW709-2P	MC8796L	N8796P	MC949P	U31994951	MCM2114	IM7114
MC855P	DM9097N	MC8796P	N8796B	MC949P	U31994959	MCM2114	MBB114
MC855P	SN158097N	MC8797L	N8797PB	MC949L	ITT949-1D	MCM2114	MM2114
MC855P	SW709-2M	MC8797P	N8797B	MC949L	SN15949J	MCM2114	S2114
MC855P	U9A909959	MC898L	N8798P	MC949L	SW949-1P	MCM2114	SY2114
MC856L	SN158094J	MC898P	N8798B	MC949L	U6A994951	MCM2114	TMM314
MC856L	SW708-2P	MC830P	RC830P	MC950F	MC850F	MCM2114	TMS2114
MC856L	U6A909459	MC930F	SN15830U	MC950F	SN15850U	MCM2115	S4015
MC856P	DM9094N	MC930F	SN15930U	MC950F	SN15950U	MCM2125	S4025
MC856P	SN158094N	MC930P	SW930-1P	MC950P	SW950-1P	MCM2147	AM9147
MC856P	SW708-2M	MC930P	SW930-2P	MC950P	SW950-2P	MCM2147	HM4847
MC856P	U9A909459	MC930P	U31993051	MC950P	U31995051	MCM2147	IM2147
MC857L	SN15857J	MC930P	U31993059	MC950P	U31995059	MCM2147	MB2147
MC857L	SW957-2P	MC930L	ITT903-1D	MC950L	SN15950J	MCM2147	MK2147
MC857L	U6A915759	MC930L	SN15930J	MC950L	SW950-1P	MCM2147	MM2147
MC857P	DM957N	MC930L	SW930-1P	MC950L	U6A995051	MCM2147	S2147
MC857P	SN15857N	MC930L	U6A993051	MC951P	MC851P	MCM2147	SY2147
MC857P	SW957-2M	MC932F	MC832F	MC951P	SN15851U	MCM2147	TMS2147
MC857P	U9A915759	MC932F	SN15832U	MC951P	SN15951U	MCM2147	JPD2147
MC858L	SN15858J	MC932F	SN15932U	MC951P	SW951-1P	MCM2148	MM2148
MC858L	SW958-2P	MC932P	SW932-1P	MC951P	SW951-2P	MCM2L14	AM9L114
MC858L	U6A915859	MC932P	SW932-2P	MC951P	U31995151	MCM2L14	IM2114L
MC858P	DM958N	MC932P	U31993251	MC951P	U31995159	MCM2L14	S2114L
MC858P	SN15858N	MC932P	U31993259	MC951L	ITT951-1D	MCM2L14	JPD2114L
MC858P	SW958-2M	MC932L	ITT932-1D	MC951L	SN15951J	MCM2532	HN462532
MC858P	U9A915859	MC932L	SN15932J	MC951L	SW951-1P	MCM2532	MM2532
MC861L	ITT961-5D	MC932L	SW932-1P	MC951L	U6A995151	MCM2532	TMS2532
MC861L	SN15861J	MC932L	U6A993251	MC952P	MC852P	MCM2708	AM2708
MC861L	SW961-2P	MC933F	SN15833U	MC952P	SN15809U	MCM2708	EA2708
MC861L	U6A916159	MC933F	SN15933U	MC952P	SN150099U	MCM2708	F2708
MC861P	DM961N	MC933P	SW933-1P	MC952P	SW706-1P	MCM2708	I2708
MC861P	ITT961-5N	MC933P	SW933-2P	MC952P	SW706-2P	MCM2708	MBB518H
MC861P	SN15861N	MC933P	U31993551	MC952P	U31909951	MCM2708	MM2708
MC861P	SW961-2M	MC933L	ITT933-1D	MC952P	U31909959	MCM2708	TMM322
MC861P	U9A996159	MC933L	SN15933J	MC952L	SN15909J	MCM2708	TMS2708
MC862L	ITT962-5D	MC933L	SW933-1P	MC952L	SW706-1P	MCM2716	AM2716
MC862L	SN15862J	MC933L	U6A993351	MC952L	U6A909951	MCM2716	EA2716
MC862L	SW962-2P	MC934F	MC835F	MC953F	MC853F	MCM2716	HN462716
MC862L	U6A916259	MC934F	SN15834U	MC953F	SN15809U	MCM2716	I2716
MC862P	DM962N	MC934F	SN15934U	MC953F	SN150093U	MCM2716	MBM2716
MC862P	ITT962-5N	MC934L	SN15934J	MC953P	SW705-2P	MCM2716	MK2716
MC862P	SN15862N	MC935F	MC835F	MC953P	U31909351	MCM2716	MM2716
MC862P	SW962-2M	MC935F	SN15835U	MC953P	U31909359	MCM2716	SY2716
MC862P	U9A996259	MC935F	SN15935U	MC953L	RO91DC	MCM2716	TMM323
MC863F	SN158063U	MC935F	U31913551	MC953L	SN150093J	MCM2716	TMS2516
MC863L	ITT963-5D	MC935F	U31913559	MC953L	SW705-1P		
		MC935L	ITT938-1D				

MOTOR.	Industry Standard No.
MCM2716	TMS2716
MCM2716	JPD2716
MCM27A08	P2708-1
MCM2816	HN48016
MCM2816	I2816
MCM4016	MK4802
MCM4016	TC5516
MCM4027	MW4104
MCM4027	TMM415
MCM4027A	TMS4027
MCM4027A	FM4027
MCM4027A	IM7027
MCM4027A	MB8227
MCM4027A	MK4027
MCM4027A	JPD414A
MCM4116	AM9016
MCM4116	P16K
MCM4116	HM4716
MCM4116	IM4116
MCM4116	MB8116
MCM4116	MB8216
MCM4116	MK4116
MCM4116	MM5290
MCM4116	TMM416
MCM4116	JPD416
MCM4332	MK4332
MCM4516	MK4516
MCM4517	HM4816
MCM4517	MM5295
MCM4517	JPD2118
MCM5101	HM435101
MCM5101	IM6551
MCM5101	85101
MCM5101	SY5101
MCM5101	TC5501
MCM6508	IM6508
MCM6508	MM6508
MCM6508	86508
MCM6508	TC5508B
MCM6508	JPD443
MCM65116	HM6116P
MCM65147	MK4104
MCM65148	HM6148P
MCM6518	IM6518
MCM6518	MM6518
MCM6518	86518
MCM6641	IM7141
MCM6641	MB4044
MCM6641	MM5257
MCM6641	TMS4044
MCM6664	MK4164
MCM6665	HM4864
MCM6665	MB8264
MCM6665	MM4164
MCM6665	TMM4164
MCM6665	JPD4164
MCM66L41	AM4044
MCM66L41	IM7141L
MCM66L41	JPD4104
MCM6810	86810
MCM68308	AM9208
MCM68316E	EA2316E/8316E
MCM68316E	RO-3-9316B
MCM68316E	S68316B
MCM68365	HM46364
MCM68365	IM6364
MCM68365	84264
MCM68365	SY2364
MCM68365	TMS4764
MCM68700	F68708
MCM68A10	HM468A10
MCM68A30A	86830
MCM68A316E	EA2308/8308
MCM68A316E	F5516B
MCM68A316E	HM462316EP
MCM68A316E	IM6516
MCM68A316E	MK34000
MCM68A316E	RO-3-9316A
MCM68A316E	RO-3-9316C
MCM68A316E	SY2316B
MCM68A332	AM9232
MCM68A332	EA8332
MCM68A332	HM46332
MCM68A332	MM52132
MCM68A332	RO-3-9332C
MCM68A332	S68332
MCM68A332	SY2332
MCM68A332	TMS4732
MCM68A332	JPD2332
MCM68A364	MM52164
MCM68A364	JPD2364
MCM68B10	F68B10
MCM68B308	F68B308
MCM68B364	MK36000-4
MCM7641	AM27831
MCM7641	HMT641
MCM7641	IM5625
MCM7641	SN74S474
MCM7641	AM27S33
MCM7643	HM7643
MCM7643	IM5626
MCM7643	IM56826
MCM7643	SN74S476
MCM7681	HM7681
MCM7681	IM56818
MCM7681	SN74S478
MCM7685	HM7685
MCM93415	DM93415
MCM93415	IM55808
MCM93415	SN748314
MCM93425	DM93425
MCM93425	IM55818
MCM93425	SN748214
MCR100-1	MCR100-1
MCR100-2	MCR100-2
MCR100-3	MCR100-3
MCR100-4	MCR100-4
MCR100-4	EC103C
MCR100-5	MCR100-5
MCR100-5	T1C39C
MCR100-5	2N5722
MCR100-5	2N5727
MCR100-6	2N882
MCR100-6	2N890
MCR100-6	EC103D
MCR100-6	MCR100-6
MCR100-6	2N5723
MCR100-6	2N5728
MCR100-6	2N883
MCR100-6	2N891
MCR100-7	MCR100-7
MCR100-7	T1C39E
MCR100-8	MCR100-8
MCR101	MCR101
MCR101	2N4144
MCR102	2N4145
MCR102	EC103J
MCR102	MCR102
MCR102	MCR600-1
MCR102	2N2679
MCR102	2N2683
MCR102	2N2687
MCR102	2N3001
MCR102	2N3005
MCR102	2N3027
MCR102	2N3030
MCR102	2N3254
MCR102	2N3255
MCR102	2N3257
MCR102	2N3258
MCR102	2N4146
MCR102	2N4332
MCR102	2N876
MCR102	2N877
MCR102	2N884
MCR102	2N885
MCR102	2N948
MCR103	MCR103
MCR103	MCR600-2
MCR103	2N2680
MCR103	2N2684
MCR103	2N2688
MCR103	2N3002
MCR103	2N3006
MCR103	2N3028
MCR103	2N3031
MCR103	2N3256
MCR103	2N3259
MCR103	2N4096
MCR103	2N4108
MCR103	2N4147
MCR103	2N5719
MCR103	2N5724
MCR103	2N878
MCR103	2N886
MCR103	2N949
MCR104	EC103A
MCR104	MCR104
MCR104	MCR600-3
MCR104	2N2681
MCR104	2N2685
MCR104	2N2689
MCR104	2N3003
MCR104	2N3007
MCR104	2N3029
MCR104	2N3032
MCR104	2N4097
MCR104	2N4109
MCR104	2N4148
MCR104	2N4334
MCR104	2N5720
MCR104	2N5725
MCR104	2N879
MCR104	2N887
MCR104	2N950
MCR106-1	MCR106-1
MCR106-1	MCR107-1
MCR106-1	SO303L82
MCR106-1	SO303L83
MCR106-2	MCR106-2
MCR106-2	MCR107-2
MCR106-2	SO501L82
MCR106-2	SO503L83
MCR106-3	MCR106-3
MCR106-3	MCR107-3
MCR106-3	S1003L82
MCR106-3	S1003L83
MCR106-4	MCR106-4
MCR106-4	MCR107-4
MCR106-4	S2003L82
MCR106-4	S2003L83
MCR106-5	MCR106-5
MCR106-5	MCR107-5
MCR106-5	S2060C
MCR106-5	S2061C
MCR106-5	S2062C
MCR106-5	2N6336
MCR106-6	MCR106-6
MCR106-6	MCR107-6
MCR106-7	MCR106-7
MCR106-7	MCR107-7
MCR106-7	S2060E
MCR106-7	S2061E
MCR106-7	S2062E
MCR106-8	MCR106-8
MCR106-8	MCR107-8
MCR106-9	MCR106-9
MCR115	MCR115
MCR115	MCR600-3.5
MCR115	2N4335
MCR115	2N880
MCR115	2N888
MCR120	EC103B
MCR120	MCR120
MCR120	MCR600-4
MCR120	2N2682
MCR120	2N2686
MCR120	2N2690
MCR120	2N3004
MCR120	2N3008
MCR120	2N4098
MCR120	2N4110
MCR120	2N4149
MCR120	2N4336
MCR120	2N5721
MCR120	2N5726
MCR120	2N881
MCR120	2N889
MCR120	2N951
MCR1718-5	MCR1718-5
MCR1718-6	MCR1718-6
MCR1718-7	MCR1718-7
MCR1718-8	MCR1718-8
MCR1906-1	C6U
MCR1906-1	C7U
MCR1906-1	MCR1906-1
MCR1906-1	SO301M83
MCR1906-1	2N1869A
MCR1906-1	2N1870A
MCR1906-1	2N1875A
MCR1906-1	2N1876A
MCR1906-1	2N1881
MCR1906-1	2N2344
MCR1906-2	C6P
MCR1906-2	MCR1906-2
MCR1906-2	SO501M83
MCR1906-2	2N1871A
MCR1906-2	2N1877A
MCR1906-2	2N1882
MCR1906-2	2N2345
MCR1906-3	C6A
MCR1906-3	C7A
MCR1906-3	MCR1906-3
MCR1906-3	S1001M83
MCR1906-3	2N1878A
MCR1906-3	2N1883
MCR1906-3	2N2346
MCR1906-4	C6B
MCR1906-4	C6G
MCR1906-4	C7B
MCR1906-4	C7G
MCR1906-4	MCR1906-4
MCR1906-4	S2001M83
MCR1906-4	2N1873A
MCR1906-4	2N1874A
MCR1906-4	2N1879A
MCR1906-4	2N1880A
MCR1906-4	2N1884
MCR1906-4	2N1885
MCR1906-4	2N2347
MCR1906-4	2N2348
MCR1906-5	MCR1906-5
MCR1906-6	MCR1906-6
MCR1906-6	S4001M83
MCR1906-7	MCR1906-7
MCR1906-8	MCR1906-8
MCR201	MCR201
MCR202	MCR202
MCR203	MCR203
MCR204	MCR204
MCR205	MCR205
MCR206	MCR206
MCR22-6	C205D
MCR220-5	C126C
MCR220-5	MCR220-5
MCR220-5	S6000C
MCR220-5	T1C126C
MCR220-7	C126E
MCR220-7	MCR220-7
MCR220-7	S6000E
MCR220-7	T1C126E
MCR220-9	MCR220-9
MCR220-9	S60003
MCR221-5	C127C
MCR221-5	MCR221-5
MCR221-5	S6100C
MCR221-7	C127E
MCR221-7	MCR221-7
MCR221-7	S6100E
MCR221-9	MCR221-9
MCR221-9	S61003
MCR3000-1	MCR3000-1C
MCR3000-1	SO308P1
MCR3000-1	SO308P3
MCR3000-10	MCR3000-10C
MCR3000-3	MCR3000-3C
MCR3000-3	S1008P1
MCR3000-3	S1008P3
MCR3000-5	MCR3000-5C
MCR3000-5	S1008L
MCR3000-6	MCR3000-6L
MCR3000-6	S4008L
MCR3000-7	MCR3000-7C
MCR3000-7	S5008L
MCR3000-9	MCR3000-9C
MCR3818-1	C33U
MCR3818-1	MCR2818-1
MCR3818-1	MCR808-1
MCR3818-10	MCR3818-10
MCR3818-2	MCR2818-2
MCR3818-3	C33A
MCR3818-3	MCR2818-3
MCR3818-3	MCR3818-3
MCR3818-4	MCR3818-4
MCR3818-5	C33C
MCR3818-5	MCR2818-5
MCR3818-5	MCR3818-5
MCR3818-5	MCR808-5
MCR3818-6	MCR3818-6
MCR3818-7	MCR2818-7
MCR3818-7	MCR3818-7
MCR3818-7	MCR808-7
MCR3818-8	MCR3818-8
MCR3818-8	MCR808-8
MCR3818-9	MCR808-9
MCR3835-1	MCR3835-1
MCR3835-1	SO335G
MCR3835-10	MCR3835-10
MCR3835-2	MCR2835-2
MCR3835-2	SO535G
MCR3835-5	MCR2835-5
MCR3835-5	MCR3835-5
MCR3835-7	MCR2835-7
MCR3835-7	850250
MCR3835-7	950350
MCR3835-9	MCR3835-9
MCR3918-1	C37U
MCR3918-1	MCR1308-1
MCR3918-1	MCR2918-1
MCR3918-10	MCR3918-10
MCR3918-2	MCR3918-2
MCR3918-3	C37A
MCR3918-3	MCR1308-3
MCR3918-3	MCR3918-3
MCR3918-4	MCR3918-4
MCR3918-5	C37C
MCR3918-5	MCR1308-5
MCR3918-5	MCR2918-5
MCR3918-5	MCR3918-5
MCR3918-6	MCR3918-6
MCR3918-7	C37E
MCR3918-7	MCR1308-7
MCR3918-7	MCR2918-7
MCR3918-7	MCR3918-7
MCR3918-8	MCR3918-8
MCR3918-9	MCR1308-9
MCR3918-9	MCR3918-9
MCR3935-1	C58U
MCR3935-1	MCR2935-1
MCR3935-1	MCR3935-1
MCR3935-1	SO335H
MCR3935-10	C137N
MCR3935-10	MCR3935-10
MCR3935-10	2N5205
MCR3935-2	C135P
MCR3935-2	C36P
MCR3935-2	MCR3935-2
MCR3935-2	S0535H
MCR3935-5	C135C
MCR3935-5	C58C
MCR3935-5	C38H
MCR3935-5	MCR2935-5
MCR3935-5	MCR3935-5
MCR3935-7	C135E
MCR3935-7	C137E
MCR3935-7	C38E
MCR3935-7	MCR2935-7
MCR3935-7	MCR3935-7
MCR3935-7	S6025H
MCR3935-9	C137E
MCR3935-9	MCR3935-9
MCR63-1	MCR63-1
MCR63-10	MCR63-10
MCR63-2	MCR63-2
MCR63-3	MCR63-3
MCR63-4	MCR63-4
MCR63-5	MCR63-5
MCR63-6	MCR63-6
MCR63-7	MCR63-7
MCR63-8	MCR63-8
MCR63-9	MCR63-9
MCR64-1	MCR64-1
MCR64-10	MCR64-10
MCR64-2	MCR64-2
MCR64-3	MCR64-3
MCR64-4	MCR64-4
MCR64-5	MCR64-5
MCR64-6	MCR64-6
MCR64-7	MCR64-7
MCR64-8	MCR64-8
MCR64-9	MCR64-9
MCR649-1	SO325C
MCR649-2	SO525C
MCR649-3	S1025C
MCR649-4	S2025C
MCR649-6	S4025C
MCR649-7	S5025C
MCR649AP1	MCR649AP1
MCR649AP10	MCR649AP10
MCR649AP2	MCR649AP2
MCR649AP3	MCR649AP3
MCR649AP4	MCR649AP4
MCR649AP5	MCR649AP5
MCR649AP6	MCR649AP6
MCR649AP7	MCR649AP7
MCR649AP8	MCR649AP8
MCR649AP9	MCR649AP9
MCR649P1	MCR649P1
MCR649P10	MCR649P10
MCR649P2	MCR649P2
MCR649P3	MCR649P3
MCR649P4	MCR649P4
MCR649P5	MCR649P5
MCR649P6	MCR649P6
MCR649P7	MCR649P7
MCR649P8	MCR649P8
MCR649P9	MCR649P9
MCR65-1	MCR65-1
MCR65-10	MCR65-10
MCR65-3	MCR65-3
MCR65-4	MCR65-4
MCR65-5	MCR65-5
MCR65-6	MCR65-6
MCR65-8	MCR65-8
MCR65-9	MCR65-9
MCR67-1	MCR67-1
MCR67-2	MCR67-2
MCR67-3	MCR67-3
MCR67-6	MCR67-6
MCR68-1	MCR68-1
MCR68-3	MCR68-3
MCR68-6	MCR68-6
MCR69-1	MCR69-1
MCR69-2	MCR69-2
MCR69-3	MCR69-3
MCR69-6	MCR69-6
MCR70-1	MCR70-1
MCR70-2	MCR70-2
MCR70-3	MCR70-3
MCR70-6	MCR70-6
MCR71-1	MCR71-1
MCR71-3	MCR71-3
MCR71-6	MCR71-6
MCR72-1	SO506P821
MCR72-1	SO306P831
MCR72-1	SO308P831
MCR72-1	SO308P84
MCR72-2	SO505P821
MCR72-2	SO505P831
MCR72-2	SO506P821
MCR72-2	SO506P831
MCR72-3	S1006P821
MCR72-3	S1006P831
MCR72-3	S1008P821
MCR72-3	S1008P831
MCR72-4	S2006P821
MCR72-4	S2006P831
MCR72-4	S2008P821
MCR72-4	S2008P831
MCR72-5	MCR72-5
MCR72-6	MCR72-6
MCR72-6	S4006P821
MCR72-6	S4006P831
MCR72-6	S4008P821
MCR72-6	S4008P831
MCR72-7	MCR72-7
MCR72-8	MCR72-8
MCR729-10	MCR729-10
MCR729-5	MCR729-5
MCR729-6	MCR729-6
MCR729-6	S4006G
MCR729-7	MCR729-7
MCR729-7	S5006B
MCR729-8	MCR729-8
MCR729-9	MCR729-9

MOTOR.	Industry Standard No.
MC82	MC82
MC82400	MC82400
MCT2	MCT2
MCT26	MCT26
MCT271	MCT271
MCT272	MCT272
MCT273	MCT273
MCT274	MCT274
MCT275	MCT275
MCT276	MCT277
MCT2E	MCT2E
MD1120	MD1120
MD1120F	MD1120F
MD1120P	SP2946F
MD1120P	SP30QP
MD1121	MD1121
MD1121F	MD1121F
MD1122	MD1122
MD1122F	MD1122F
MD1123	MD1123
MD1129	MD1129
MD1129F	MD1129F
MD1130	MD1130
MD1130F	MD1130F
MD1132	MD1132
MD2060F	MD2060F
MD2218	D2T2218
MD2218	MD2218
MD2218A	D2T2218A
MD2218A	MD2218A
MD2218AF	MD2218AF
MD2218F	MD2218F
MD2219	D2T2219
MD2219	MD2219
MD2219	2N2910
MD2219A	D2T2219A
MD2219A	MD2219A
MD2219AF	MD2219AF
MD2219AF	SP2222AF
MD2219AF	SP2223AF
MD2219AF	2N3515
MD2219AF	2N3517
MD2219F	2N3518
MD2219F	MD2219F
MD2369	MD1127
MD2369	MD1128
MD2369	MD1134
MD2369	MD2369
MD2369A	MD2369A
MD2369AF	MD2369AF
MD2369B	MD2369B
MD2369BF	MD2369BF
MD2369F	MD1128F
MD2369F	MD2369F
MD2369F	2N3052
MD2904	D2T2904
MD2904	MD2904
MD2904	MD2906
MD2904	MD3133
MD2904	MD990
MD2904A	D2T2904A
MD2904A	MD2904A
MD2904AF	MD2904AF
MD2904F	MD3133F
MD2904F	D2T2905
MD2905	MD2905
MD2905	SP2905
MD2905A	D2T2905A
MD2905A	MD2905A
MD2905A	MD3134
MD2905AF	MD2905AF
MD2905AF	MD3134F
MD2905AF	SP2907AF
MD2905F	MD2905F
MD3250	FT1718B
MD3250	FT1718D
MD3250	MD3250
MD3250A	MD1124
MD3250A	MD1125
MD3250A	MD1126
MD3250A	MD3250A
MD3250A	2N5843
MD3250AF	MD1123F
MD3250AF	MD1124F
MD3250AF	MD1125F
MD3250AF	MD1126F
MD3250AF	MD3250AF
MD3250AF	2N3049
MD3250AF	2N3050
MD3250AF	2N3051
MD3250F	MD3250F
MD3250P	MD984F
MD3251	FT1718
MD3251	FT1718A
MD3251	FT1718C
MD3251	FT1718E
MD3251	MD3251
MD3251A	MD3251A
MD3251A	2N5844
MD3251AF	MD3251AF
MD3251F	MD3251F
MD3409	MD3409
MD3409	2N3409
MD3410	MD3410
MD3410	2N3410
MD3410	2N3411
MD3467	MD3467
MD3467F	MD3467F
MD3467F	SP3467F
MD3725	MD1133
MD3725	MD3725
MD3725F	MD1133F
MD3725F	MD3725F
MD3725F	SP3725F
MD3762	MD3762
MD3762F	MD3762F
MD4049	MD4049
MD4957	MD4957
MD5000	MD5000
MD5000A	MD5000A
MD5000B	MD5000B
MD5000B	MD548
MD6001	MD3001
MD6001	MD6001
MD6001F	MD6001F
MD6002	MD3002
MD6002F	MD6002F
MD6003	MD6003
MD6003	MD7011
MD6003F	MD6003F

MOTOR.	Industry Standard No.
MD6100	MD6100
MD6100F	MD6100F
MD7000	MD7000
MD7001	MD7001
MD7001F	MD7001F
MD7001F	SP5136F
MD7002	MD7002
MD7002A	MD7002A
MD7002B	MD7002B
MD7003	MD7003
MD7003	MD7006
MD7003A	MD7003A
MD7003A	MD7006A
MD7003AF	MD7003AF
MD7003B	MD7003B
MD7003B	MD7006B
MD7003F	MD7003F
MD7004	MD7004
MD7005	MD7005
MD7005F	MD7005F
MD7007	MD7007
MD7007A	MD7007A
MD7007B	MD7007B
MD7007BF	MD7007BF
MD7007F	MD7007F
MD7021	MD7021
MD7021F	MD7021F
MD708	MD708
MD708A	MD708A
MD708AF	MD708AF
MD708B	MD708B
MD708BF	MD708BF
MD708F	MD708F
MD8001	MD7008
MD8001	MD8001
MD8002	MD8002
MD8003	MD7091
MD8003	MD8003
MD8003	MD981F
MD8003	MD983
MD918	D2T918
MD918	MD1131
MD918	MD918
MD918	2N3423
MD918	2N3424
MD918A	MD918A
MD918AF	MD1132F
MD918AF	MD918AF
MD918AF	SP918AF
MD918B	MD918B
MD918BF	MD918BF
MD918BF	SP918BF
MD918F	MD1131F
MD918F	MD918F
MD930	SP528QF
MD930F	SP529QF
MD982	MD982
MD982F	MD982F
MD984	MD984
MD985	MD985
MD985F	MD985F
MD986	MD986
MD986F	MD986F
MDA 3508	KBH08
MDA 3508	KBH2508
MDA 3508	KBH508
MDA 3508	KBHG08
MDA 3508	KBHG2508
MDA 3508	KBHG508
MDA 3510	KBH10
MDA 3510	KBH2510
MDA 3510	KBH510
MDA 3510	KBHG10
MDA 3510	KBHG2510
MDA 3510	KBHG510
MDA100A	FWL50
MDA100A	KBP005
MDA100A	NSS3058A
MDA100A	W005
MDA100A	W005M
MDA101A	FWB3001A
MDA101A	FWL100
MDA101A	NSS3059A
MDA101A	10DB1F
MDA102A	FWB3002A
MDA102A	FWL200
MDA102A	KBP02
MDA102A	NSS3060A
MDA102A	W02
MDA102A	W02M
MDA102A	WL02M
MDA102A	10DB2P
MDA104A	FWB3004A
MDA104A	FWL300
MDA104A	FWL400
MDA104A	KBP04
MDA104A	NSS3061A
MDA104A	NSS3062A
MDA104A	W04
MDA104A	W04M
MDA104A	WL04M
MDA104A	10DB3P
MDA104A	10DB4P
MDA106A	FWB3006A
MDA106A	FWL500
MDA106A	FWL600
MDA106A	KBP06
MDA106A	NSS3063A
MDA106A	NSS3064A
MDA106A	W06
MDA106A	W06M
MDA106A	WL06M
MDA108A	FWB3008A
MDA108A	FWL700
MDA108A	FWL800
MDA108A	KBP08
MDA108A	NSS3065A
MDA108A	W08
MDA108A	W08M
MDA108A	WL08M
MDA110A	FWB3010A
MDA110A	FWL1000
MDA110A	KBP10
MDA110A	NSS3066A
MDA110A	W10
MDA110A	W10M
MDA1200	KBPC12-005
MDA1202	KBPC12-02
MDA1204	KBPC12-04
MDA1206	KBPC12-06
MDA200	FWLA50
MDA200	2FB050
MDA200	2KBP005

MOTOR.	Industry Standard No.
MDA200	2W005
MDA201	FWLA100
MDA201	SEN3A1
MDA201	2FB100
MDA202	FWLA200
MDA202	SEN3A2
MDA202	2FB200
MDA202	2KBP002
MDA202	2W02
MDA204	FWLA300
MDA204	FWLA400
MDA204	SEN3A4
MDA204	2FB400
MDA204	2KBP004
MDA204	2W04
MDA206	FWLA500
MDA206	FWLA600
MDA206	SEN3A6
MDA206	2FB600
MDA206	2KBP006
MDA206	2W06
MDA208	FWLA700
MDA208	FWLA800
MDA208	SEN3A8
MDA208	2FB800
MDA208	2KBP008
MDA208	2W08
MDA210	FWLA1000
MDA210	2FB1000
MDA210	2KBP010
MDA210	2W10
MDA2500	A4B6-50E
MDA2500	A4B6-50F
MDA2500	KBPC10005
MDA2500	KBPC1005
MDA2500	KBPC25-005
MDA2500	KBPC25005
MDA2500	KBPC6005
MDA2500	KBPC8005
MDA2500	NB12-50
MDA2500	NB25-50
MDA2500	SCAJ05
MDA2500	SCBA05
MDA2500	SDA129A
MDA2500	SDA130A
MDA2500	SDA985-1
MDA2500	SLBA05
MDA2500	5SCBR05
MDA2500	10HB050
MDA2500,F	SDA980-1,F
MDA2501	A4B6-100E
MDA2501	A4B6-100F
MDA2501	EGF211B1
MDA2501	EGL211B1
MDA2501	EGT211B1
MDA2501	MPO12JBB
MDA2501	MPO13MBB
MDA2501	MPO13RBB
MDA2501	NB812-100
MDA2501	NB820-100
MDA2501	S-5745
MDA2501	SBR1OA1
MDA2501	SBR6A1
MDA2501	SCAJ1
MDA2501	SCBA1
MDA2501	SDA129B
MDA2501	SDA130B
MDA2501	SDA985-2
MDA2501	SLBA1
MDA2501	5SCBR1
MDA2501	10HB100
MDA2501,F	SDA980-2,F
MDA2502	A4B6-200E
MDA2502	A4B6-200F
MDA2502	EGF212B1
MDA2502	EGL212B1
MDA2502	EGT212B1
MDA2502	KBPC02
MDA2502	KBPC1002
MDA2502	KBPC102
MDA2502	KBPC25-02
MDA2502	KBPC2502
MDA2502	KBPC602
MDA2502	KBPC802
MDA2502	MPO12JBD
MDA2502	MPO13MBD
MDA2502	MPO13RBD
MDA2502	NB312-200
MDA2502	NBS25-200
MDA2502	S-5745-1
MDA2502	SBR1OA2
MDA2502	SBR6A2
MDA2502	SCAJ2
MDA2502	SCBA2
MDA2502	SDA129C
MDA2502	SDA130C
MDA2502	SDA985-3
MDA2502	SLBA2
MDA2502	5SCBR2
MDA2502	10DB200
MDA2502,F	SDA980-3,F
MDA2504	A4B6-400E
MDA2504	A4B6-400F
MDA2504	EGF213B1
MDA2504	EGF214B1
MDA2504	EGL213B1
MDA2504	EGL214B1
MDA2504	EGT213B1
MDA2504	EGT214B1
MDA2504	KBPC04
MDA2504	KBPC1004
MDA2504	KBPC104
MDA2504	KBPC25-04
MDA2504	KBPC2504
MDA2504	KBPC604
MDA2504	KBPC804
MDA2504	MPO12JBH
MDA2504	MPO13MBF
MDA2504	MPO13MBH
MDA2504	MPO13RBF
MDA2504	MPO13RBH
MDA2504	NB312-300
MDA2504	NBS12-400
MDA2504	NBS25-300
MDA2504	NBS25-400
MDA2504	S-5745-2
MDA2504	S-5745-3
MDA2504	SBR10A3
MDA2504	SBR10A4
MDA2504	SBR6A3
MDA2504	SBR6A4
MDA2504	SCAJ3
MDA2504	SCAJ4
MDA2504	SCBA3

MOTOR.	Industry Standard No.
MDA2504	SCBA4
MDA2504	SDA129D
MDA2504	SDA130D
MDA2504	SDA985-4
MDA2504	SDA985-5
MDA2504	SLBA3
MDA2504	SLBA4
MDA2504	5SCBR4
MDA2504	10HB400
MDA2504,F	SDA980-4,F
MDA2504,F	SDA980-5,F
MDA2506	KBPC25-06
MDA2506	NB812-600
MDA2506	NBS25-600
MDA2506	SBR10A5
MDA2506	SBR10A6
MDA2506	SBR6A5
MDA2506	SBR6A6
MDA2506	SDA129E
MDA2506	SDA130E
MDA2506	5SCBR6
MDA2506	10HB600
MDA2506,F	SDA980-6,F
MDA2508	SDA129F
MDA2508	SDA130F
MDA2510	SDA129G
MDA2510	SDA130G
MDA3500	KBPC35-005
MDA3500	NBS30-50
MDA3500	SDA990-1
MDA3500	35MB5A
MDA3501	NBS30-100
MDA3501	SDA990-2
MDA3501	35MB1A
MDA3502	KBPC35-02
MDA3502	NBS30-200
MDA3502	SDA990-3
MDA3502	VT200-8
MDA3502	VT200-T
MDA3502	35MB2A
MDA3504	KBPC35-04
MDA3504	NBS30-300
MDA3504	NBS30-400
MDA3504	SDA990-4
MDA3504	SDA990-5
MDA3504	VT400-8
MDA3504	VT400-T
MDA3504	35MB4A
MDA3506	A4B6-600E
MDA3506	A4B6-600F
MDA3506	EGF215B1
MDA3506	EGF216B1
MDA3506	EGL215B1
MDA3506	EGL216B1
MDA3506	EGT215B1
MDA3506	EGT216B1
MDA3506	KBPC06
MDA3506	KBPC1006
MDA3506	KBPC106
MDA3506	KBPC2506
MDA3506	KBPC35-06
MDA3506	KBPC606
MDA3506	KBPC806
MDA3506	MPO12JBM
MDA3506	MPO13MBK
MDA3506	MPO13MBM
MDA3506	MPO13RBK
MDA3506	MPO13RBM
MDA3506	NBS30-600
MDA3506	S-5745-4
MDA3506	S-5745-5
MDA3506	SCAJ6
MDA3506	SCBA6
MDA3506	SDA990-6
MDA3506	SLBA6
MDA3506	VT600-8
MDA3506	VT600-T
MDA3508	35MB6A
MDA3508	KBPC08
MDA3508	KBPC108
MDA3508	KBPC2508
MDA3508	KBPC35-08
MDA3508	NBS30-800
MDA3508	SDA990-8
MDA3508	30DB8T
MDA3508	35MB8A
MDA3510	KBPC010
MDA3510	KBPC110
MDA3510	KBPC2510
MDA3510	KBPC35-10
MDA3510	NBS30-1000
MDA3510	SDA990-10
MDA3510	30DB10T
MDA3510	35MB10A
MDA800	KBPC10-005
MDA801	MPO12EBB
MDA802	KBPC10-02
MDA802	MPO12HBB
MDA804	KBPC10-04
MDA804	MPO12BBH
MDA806	KBPC10-06
MDA806	MPO12EBM
MDA920A2	18DB10A
MDA920A2	A4B.5-50A
MDA920A2	A4B1-50B
MDA920A2	A4B2-50C
MDA920A2	A4B2-50D
MDA920A2	BR505
MDA920A2	KBP8005
MDA920A2	SCBR05
MDA920A3	A4B.5-100A
MDA920A3	A4B1-100B
MDA920A3	A4B2-100C
MDA920A3	A4B2-100D
MDA920A3	BR51
MDA920A3	KBP802
MDA920A3	MPO10ABB
MDA920A3	SB1
MDA920A3	SCBR1
MDA920A3	V318
MDA920A4	A4B.5-200A
MDA920A4	A4B1-200B
MDA920A4	A4B2-200C
MDA920A4	A4B2-200D
MDA920A4	BR22
MDA920A4	BR42
MDA920A4	BR52
MDA920A4	MPO10ABQ
MDA920A4	S-6211
MDA920A4	SB2
MDA920A4	SCBR2
MDA920A4	SLA200
MDA920A4	VX28
MDA920A4	VS244

MOTOR.	Industry Standard No.
MDA920A4	VS248
MDA920A4	18DB2A
MDA920A5	SLA300
MDA920A6	A4B.5-400A
MDA920A6	A4B1-400B
MDA920A6	A4B2-400C
MDA920A6	A4B2-400D
MDA920A6	AR44
MDA920A6	BR24
MDA920A6	BR54
MDA920A6	KBPS04
MDA920A6	MPO10ABH
MDA920A6	S-6211-1
MDA920A6	SB4
MDA920A6	SCBR4
MDA920A6	SLA400
MDA920A6	VE48
MDA920A6	VS444
MDA920A6	VS448
MDA920A6	18DB4A
MDA920A7	A4B.5-600A
MDA920A7	A4B1-600B
MDA920A7	A4B2-600C
MDA920A7	A4B2-600D
MDA920A7	BR26
MDA920A7	BR46
MDA920A7	BR56
MDA920A7	KBPS06
MDA920A7	MP010ABM
MDA920A7	S-6211-2
MDA920A7	SB6
MDA920A7	SCBR6
MDA920A7	SLA500
MDA920A7	SLA600
MDA920A7	VE68
MDA920A7	VS644
MDA920A7	VS648
MDA920A7	10DB6P
MDA920A7	18DB6A
MDA920A8	A4B.5-800A
MDA920A8	A4B1-800B
MDA920A8	A4B2-800C
MDA920A8	A4B2-800D
MDA920A8	BR48
MDA920A8	BR58
MDA920A8	KBPS08
MDA920A8	MP010AB8
MDA920A8	S-6211-3
MDA920A8	SB8
MDA920A8	SLA700
MDA920A8	SLA800
MDA920A8	18DB8A
MDA920A9	BR410
MDA920A9	BR510
MDA920A9	KBPS10
MDA920A9	MP010ABZ
MDA920A9	S-6211-4
MDA920A9	SB10
MDA920A9	SCBR10
MDA920A9	SLA1000
MDA920A9	SLA900
MDA960A2	MP011BBB
MDA960A3	MP011BBD
MDA960A3	S-5959
MDA970A1	A4B3-50D
MDA970A1	A4B3-50F
MDA970A1	EJL1B1
MDA970A1	EJT1B1
MDA970A1	FWLC50
MDA970A1	FWLD50
MDA970A1	KBLO05
MDA970A1	KBPC005
MDA970A1	KBPC2005
MDA970A1	KB8005
MDA970A1	38CBR05
MDA970A2	A4B3-100D
MDA970A2	A4B3-100F
MDA970A2	FWLC100
MDA970A2	FWLD100
MDA970A2	MPO12FBB
MDA970A2	2KBB10R
MDA970A2	38CBR1
MDA970A3	A4B3-200D
MDA970A3	A4B3-200F
MDA970A3	EJL2B1
MDA970A3	EJT2B1
MDA970A3	FWLC200
MDA970A3	FWLD200
MDA970A3	KBLO2
MDA970A3	KBPC202
MDA970A3	KB802
MDA970A3	MPO1FBD
MDA970A3	2KBB20R
MDA970A3	38CBR2
MDA970A3	30DB2T
MDA970A4	FWLC300
MDA970A4	FWLD300
MDA970A4	2KBB40R
MDA970A5	FWLC400
MDA970A5	FWLD400
MDA970A5	KBLO4
MDS20	MD820
MDS21	MD820
MDS26	D4201
MDS26	D4202
MDS26	D4203
MDS26	MD826
MDS26	2SC1013
MDS26	2SC1096
MDS27	D4204
MDS27	D4205
MDS27	D4206
MDS27	D4207
MDS27	D4208
MDS27	D4209
MDS27	MD827
MDS27	NSD102
MDS27	NSD103
MDS27	2SC1014
MDS60	MD860
MDS60	2SA698
MDS76	D4301
MDS76	D4302
MDS76	D4303
MDS76	MD876
MDS77	D4304
MDS77	D4305
MDS77	D4306
MDS77	D4307
MDS77	D43CB
MDS77	D4309
MDS77	MD877
MDS77	NSD202
MDS77	NSD203

MOTOR.	Industry Standard No.
MFE120	MFE120
MFE121	MFE121
MFE122	MFE122
MFE130	MFE130
MFE131	MEM620
MFE131	MEM621
MFE131	MEM622
MFE131	MEM630
MFE131	MEM631
MFE131	MEM632
MFE131	MEM633
MFE131	MEM643
MFE131	MEM644
MFE131	MEM645
MFE131	MFE131
MFE132	MFE132
MFE140	MFE140
MFE2000	MFE2000Q
MFE2000	U183
MFE2000	UC707
MFE2001	MFE2001
MFE2001	0184
MFE2004	MFE2004
MFE2004	TIX842
MFE2004	U200
MFE2005	MFE2005
MFE2005	U201
MFE2006	MFE2006
MFE2006	U202
MFE2010	MFE2010
MFE2011	MFE2011
MFE2011	NF583
MFE2011	NF585
MFE2011	U243
MFE2012	CM650
MFE2012	CM651
MFE2012	MFE2012
MFE2012	NF4445
MFE2012	NF580
MFE2012	NF581
MFE2012	NF584
MFE2012	U240
MFE2012	U241
MFE2012	U242
MFE2012	U244
MFE2012	U290
MFE2012	U291
MFE2012	2N4445
MFE2012	2N4446
MFE2012	2N4447
MFE2012	2N4448
MFE2012	2N4977
MFE2012	2N5158
MFE2012	2N5159
MFE2012	2N5432
MFE2012	2N5433
MFE2012	2N5434
MFE2013	MFE2013
MFE2093	C614
MFE2093	C620
MFE2093	C621
MFE2093	C622
MFE2093	C623
MFE2093	C624
MFE2093	C625
MFE2093	C650
MFE2093	C651
MFE2093	C652
MFE2093	C653
MFE2093	C680
MFE2093	C680A
MFE2093	C681
MFE2093	C681A
MFE2093	D1103
MFE2093	D1179
MFE2093	D1203
MFE2093	DN3068A
MFE2093	DN3370A
MFE2093	DNX3
MFE2093	DNX3A
MFE2093	DNX6
MFE2093	DNX6A
MFE2093	FB104
MFE2093	FB104A
MFE2093	FB202
MFE2093	FB204
MFE2093	TN4117
MFE2093	TN4117A
MFE2093	U1179
MFE2093	U1182
MFE2093	U1279
MFE2093	U1285
MFE2093	2N5647
MFE2093/4	U1178
MFE2093/5	MFE2093/5
MFE2094	C682
MFE2094	C682A
MFE2094	C683
MFE2094	C683A
MFE2094	D1102
MFE2094	D1178
MFE2094	DN3067A
MFE2094	DNX2
MFE2094	DNX2A
MFE2094	FB102
MFE2094	FB102A
MFE2094	TN4118
MFE2094	TN4118A
MFE2094	U1278
MFE2094	2N5648
MFE2094	2N5649
MFE2095	C611
MFE2095	C612
MFE2095	C613
MFE2095	C615
MFE2095	C692
MFE2095	C684
MFE2095	C684A
MFE2095	C685
MFE2095	C685A
MFE2095	D1177
MFE2095	DN3066A
MFE2095	FB100
MFE2095	FB100A
MFE2095	TN4119,A
MFE2095	U1277
MFE3001	K1004
MFE3001	MEM668
MFE3001	MEM670
MFE3001	MFE3001
MFE3002	MEM563

MOTOR.	Industry Standard No.
MFE3002	MFE3002
MFE3003	DE1004
MFE3003	PT703
MFE3003	HRN1020
MFE3003	HRN1030
MFE3003	HRN8318D
MFE3003	HRN8346D
MFE3003	HRN8350
MFE3003	HRN8353
MFE3003	IT1700
MFE3003	IT1701
MFE3003	IT1702
MFE3003	M103
MFE3003	M511
MFE3003	M5111A
MFE3003	MEM561C
MFE3003	MFE3003
MFE3003	PT320
MFE3003	2N4268
MFE3003	2N5548
MFE3004	K1202
MFE3004	MEM655
MFE3004	MFE3004
MFE3004	MPP157
MFE3004	PT200
MFE3004	PT201
MFE3004	3N98
MFE3004	3N99
MFE3005	K1001
MFE3005	K1003
MFE3005	K1201
MFE3005	MFE3005
MFE3005	MPP158
MFE3006	PT57
MFE3021	FI0049
MFE3021	PT701
MFE4001	UC420
MFE4007	MFE4007
MFE4007	2N3697
MFE4008	MFE4008
MFE4009	MFE4009
MFE4009	UC410
MFE4009	2N3695
MFE4009	2N3696
MFE4010	MFE4010
MFE4011	MFE4011
MFE4012	MFE4012
MFE4012	UC400
MFE4012	2N3331
MFE4856	2N4978
MFE4857	2N4979
MFE5000	MEM857
MFE521	MFE521
MHQ2221	MHQ2221
MHQ2221	SP2221QD
MHQ2222	MHQ2222
MHQ2222	SP2222QD
MHQ2369	MHQ2369
MHQ2369	SP2369QD
MHQ2483	MHQ2483
MHQ2484	MHQ2484
MHQ2484	SP2484QD
MHQ2906	MHQ2906
MHQ2906	SP2906QD
MHQ2907	MHQ2907
MHQ2907	SP2907QD
MHQ3251	SP3251AQD
MHQ3467	MHQ3467
MHQ3467	SP3467QD
MHQ3546	MHQ3546
MHQ3798	SP3762QD
MHQ3799	MHQ3799
MHQ4001A	MHQ4001A
MHQ4002A	MHQ4002A
MHQ4013	MHQ4013
MHQ4013	SP3724QD
MHQ4014	MHQ4014
MHQ4014	SP3725QD
MHQ6001	MHQ6001
MHQ6002	MHQ6002
MHQ6100	MHQ6100
MHQ918	MHQ918
MHW1121	BGY50
MHW1122	BGY51
MHW1171	BGY52
MHW1171	BGY54
MHW1171	CA100
MHW1171	CA2100
MHW1171R	MHW594
MHW1172	CA2100R
MHW1172	BGY53
MHW1172	BGY55
MHW1172	CA200
MHW1172	CA2200
MHW1172	CA2800
MHW1172	MHW595
MHW1172R	CA2200R
MHW1182	CA2418
MHW1182	CA2818
MHW1182	CA2850
MHW1182	CA2875
MHW1182	CA401B
MHW1182	CA416
MHW1182	CA418
MHW1221	CA2300
MHW1221	CA2876
MHW1221	BGY57
MHW1222	CA2301
MHW1222	CA2400
MHW1342	BGY59
MHW1342	CA2600
MHW1342	CA2810
MHW1342	CA2870
MHW1342	CA601B/U
MHW1342	CA636
MHW1343	CA2601BU
MHW1344	CA2603
MHW1392	CA2830
MHW401	BGY22
MHW401	BGY22A
MHW401	MX1.5
MHW590	CA2820
MHW590	CA801
MHW590	CA804
MHW590	CA860
MHW590	CA870
MHW592	CA2830
MHW593	CA2812

MOTOR.	Industry Standard No.
MHW612	M57706
MHW612	M57715
MHW612	VP20A
MHW612	VP20B
MHW612A	BGY36
MHW612A	MV20
MHW613	M57710
MHW613	M57712
MHW613A	MV30
MHW709	BGY23
MHW709	BGY23A
MHW709	MX7.5
MHW710	R47M10
MHW710	MX12
MHW710	MX15
MHW710	R47M13
MHW710	R47M17
MHW710-1	M57704L
MHW710-1	M57704M
MHW710-2	M57704H
MJ1000	IR1000
MJ1000	IR1060
MJ1000	PMD12K-40
MJ1000	PMD12K-60
MJ1000	RCA1000
MJ10000	DTS4039
MJ10000	DTS4040
MJ10000	DTS4041
MJ10000	DTS4045
MJ10000	DTS4059
MJ10000	DTS4061
MJ10000	DTS4067
MJ10000	GE5060
MJ10000	GE5061
MJ10000	IR4039
MJ10000	IR4040
MJ10000	IR4041
MJ10000	IR4045
MJ10000	IR4050
MJ10000	IR4055
MJ10000	IR4059
MJ10000	IR4061
MJ10000	IR5000
MJ10000	IR5001
MJ10000	IR5060
MJ10000	MJ10000
MJ10000	SDN4040
MJ10000	SDN4045
MJ10000	SDN6000
MJ10000	SDN6001
MJ10000	SDN6060
MJ10000	SDN6061
MJ10000	SDN6062
MJ10001	2SC435
MJ10001	2SC450
MJ10001	DTS4060
MJ10001	DTS4065
MJ10001	GE5062
MJ10001	IR4060
MJ10001	IR4065
MJ10001	IR5002
MJ10001	IR5062
MJ10001	MJ10001
MJ10001	SDN6002
MJ10002	IR5261
MJ10002	MJ10002
MJ10002	RCA8766
MJ10002	RCA8766A
MJ10002	SDN6251
MJ10002	SDN6252
MJ10002	TIP660
MJ10002	TIP661
MJ10002	TIP662
MJ10003	IR5252
MJ10003	MJ10003
MJ10003	RCA8766B
MJ10003	RCA8766C
MJ10003	RCA8766D
MJ10003	RCA8766E
MJ10003	SDN6253
MJ10004	DTS4074
MJ10004	DTS4075
MJ10004	GE6060
MJ10004	GE6061
MJ10004	IR6000
MJ10004	IR6001
MJ10004	IR6060
MJ10004	IR6061
MJ10004	MJ10004
MJ10004	SVT6000
MJ10004	SVT6001
MJ10004	SVT6060
MJ10004	SVT6061
MJ10004	TIP663
MJ10004	TIP664
MJ10004	2SD710
MJ10005	GE6062
MJ10005	IR6002
MJ10005	IR6062
MJ10005	MJ10005
MJ10005	SVT6002
MJ10005	SVT6062
MJ10005	TIP665
MJ10006	IR6251
MJ10006	MJ10006
MJ10006	SVT6251
MJ10006	SVT6252
MJ10006	TIP666
MJ10006	TIP667
MJ10006	2SC1227
MJ10006	2SC1229
MJ10006	2SC1477
MJ10007	IR6252
MJ10007	MJ10007
MJ10007	SVT6253
MJ10007	TIP668
MJ10008	MJ10008
MJ10009	MJ10009
MJ10009	2SC1832
MJ1001	IR1001
MJ1001	IR1061
MJ1001	PMD12K-80
MJ1001	RCA1001
MJ1001	2SC1629
MJ10011	MJ10011
MJ10012	DTS4026
MJ10012	MJ10012
MJ10012	SDM6000
MJ10012	SDM6001
MJ10012	SDM6002
MJ10012	SDN6003
MJ10012	2SD626

MOTOR.	Industry Standard No.	MOTOR.	Industry Standard No.	MOTOR.	Industry Standard No.	MOTOR.	Industry Standard No.	MOTOR.	Industry Standard No.
MJ10012	28D693	MJ13015	28C1142	MJ3001	SDN1020	MJ4247	MJ4247	MJE1000	MJE1000
MJ10012	28D705	MJ13015	28D519	MJ3001	TIP641	MJ4247	TIP509	MJE1001	MJE1001
MJ10013	MJ10013	MJ13015	28D539	MJ3026	MJ3026	MJ4247	TIP511	MJE105	MJE105
MJ10013	28C1579	MJ13015	28D555	MJ3027	MJ3027	MJ4247	28D322	MJE1090	MJE1090
MJ10013	28D572	MJ13330	MJ13330	MJ3028	MJ3028	MJ4247	28D388	MJE1091	MJE1091
MJ10013	28D706	MJ13330	28D375	MJ3029	MJ3029	MJ4248	MJ4248	MJE1092	MJE1092
MJ10013	28D707	MJ13330	28D434	MJ3029	TIP575A	MJ4248	TIP510	MJE1093	MJE1093
MJ10013	28D708	MJ13331	MJ13331	MJ3029	28C558	MJ4248	TIP512	MJE1100	MJE1100
MJ10014	MJ10014	MJ13331	28D376	MJ3029	28D861	MJ4248	28D523	MJE1101	MJE1101
MJ10014	28C1578	MJ13332	MJ13332	MJ3030	BU126	MJ431	DT8431	MJE1102	MJE1102
MJ10014	28C1580	MJ13332	2N6653	MJ3030	MJ3030	MJ431	PT431	MJE1103	MJE1103
MJ10014	28C2123	MJ13332	2N6654	MJ3030	MJ3760	MJ431	IR431	MJE12007	MJE12007
MJ10014	28C2151	MJ13332	2N6676	MJ3030	MJ3761	MJ431	KDT431	MJE1290	MJE1290
MJ10014	28D573	MJ13332	2N6677	MJ3030	28C862	MJ431	MJ431	MJE1291	MJE1291
MJ10014	28D606	MJ13332	28D435	MJ3040	MJ3040	MJ431	RCA431	MJE13002	MJE13002
MJ10015	MJ10015	MJ13333	MJ13333	MJ3041	DT84010	MJ431	SDT431	MJE13003	MJX86
MJ10015	2N6322	MJ13333	2N6244	MJ3041	DT84025	MJ431	28C2122	MJE13003	BUX87
MJ10015	2N6323	MJ13333	2N6655	MJ3041	MJ3041	MJ431	28C806	MJE13003	MJE13003
MJ10015	2N6324	MJ13333	2N6678	MJ3041	28C1768	MJ4360	MJ4360	MJE13004	MJE13004
MJ10015	2N6325	MJ13333	28D436	MJ3041	28C1829	MJ4361	MJ4361	MJE13004	TIP75A
MJ10015	28C2128	MJ13334	MJ13334	MJ3041	28D604	MJ4380	MJ4380	MJE13004	TIP75B
MJ10015	28C2147	MJ13334	2N6243	MJ3042	28D709	MJ4380	28C2126	MJE13004	UMT1203
MJ10015	28C2159	MJ13334	2N6245	MJ3042	MJ3042	MJ4380	28D422	MJE13004	28C2331
MJ10015	28C2249	MJ13334	2N6581	MJ3042	28D605	MJ4380	28D423	MJE13004	28D386
MJ10015	28C2403	MJ13334	2N6584	MJ3042	28D650	MJ4380	28D518	MJE13004	28D387
MJ10015	28D363	MJ13334	28D377	MJ3042	28D663	MJ4381	MJ4381	MJE13004	28D724
MJ10015	28D372	MJ13335	BUX81	MJ3237	MJ3237	MJ4381	28C1316	MJE13005	BUX84
MJ10015	28D373	MJ13335	MJ13335	MJ3237	2N6467	MJ4381	28C1467	MJE13005	BUX85
MJ10015	28D457	MJ13335	28D295	MJ3237	2N6468	MJ4381	28C1504	MJE13005	MJE13005
MJ10015	28D643	MJ13335	28D296	MJ3238	MJ3238	MJ4381	28C2559	MJE13005	TIP75
MJ10015	28D694	MJ14000	MJ14000	MJ3238	TIP514	MJ4381	28D622	MJE13005	TIP75C
MJ10015	28D695	MJ14001	MJ14001	MJ3238	28A969	MJ4400	MJ4400	MJE13005	UMT1204
MJ10015	28D696	MJ14002	MJ14002	MJ3247	MJ3247	MJ4401	MJ4401	MJE13005	28C2333
MJ10015	28D702	MJ14003	MJ14003	MJ3247	2N6466	MJ4401	28D272	MJE13006	MJE13006
MJ10016	MJ10016	MJ15001	MJ15001	MJ3247	28C1113	MJ4502	IB4502	MJE13006	TIP150
MJ10016	28C2204	MJ15001	28C1079	MJ3247	28D243	MJ4502	MJ4502	MJE13006	28C2373
MJ10016	28C2220	MJ15001	28C1080	MJ3247	28D258	MJ4502	2N5744	MJE13007	MJE13007
MJ10016	28C2250	MJ15001	28C1440	MJ3247	28D283	MJ4645	MJ4645	MJE13007	TIP151
MJ10016	28C2366	MJ15001	28C1782	MJ3247	28D284	MJ4646	MJ4646	MJE13007	TIP152
MJ10016	28C2442	MJ15001	28C1784	MJ3247	28D285	MJ4647	MJ4647	MJE13007	28C2335
MJ10016	28C2443	MJ15001	28C2189	MJ3248	MJ3248	MJ4647	MJ4648	MJE13008	MJE13008
MJ10016	28D364	MJ15001	28C2323	MJ3248	28D244	MJ6502	MJ6502	MJE13008	TIP55A
MJ10016	28D374	MJ15001	28D939	MJ3248	28D259	MJ6502	28A739	MJE13008	TIP56A
MJ10016	28D642	MJ15001	28D181	MJ3248	28D297	MJ6503	MJ6503	MJE13009	MJE13009
MJ10016	28D644	MJ15001	28D424	MJ4030	MJ4030	MJ6700	MJ6700	MJE13009	TIP57A
MJ10016	28D645	MJ15001	28D733	MJ4030	SDM21301	MJ802	IR802	MJE13009	TIP58A
MJ10016	28D646	MJ15001	28D752	MJ4030	SDM21302	MJ802	MJ802	MJE15028	MJE15028
MJ10016	28D703	MJ15002	MJ15002	MJ4030	SDM21311	MJ802	28D113	MJE15028	2N6669
MJ10016	28D805	MJ15002	28A908	MJ4030	SDM21312	MJ804	MJ804	MJE15028	28D718
MJ10020	MJ10020	MJ15002	28B655	MJ4030	SB9406	MJ8100	MJ8100	MJE15028	28D643
MJ10021	MJ10021	MJ15002	28B656	MJ4031	MJ4031	MJ8100	SDT3321	MJE15029	MJE15029
MJ10022	MJ10022	MJ15002	28B681	MJ4031	SDM21303	MJ8100	SDT3322	MJE15029	28A965
MJ10023	MJ10023	MJ15002	28B697	MJ4031	SDM21313	MJ8100	SDT3325	MJE15029	28D727
MJ10024	MJ10024	MJ15002	28B722	MJ4031	SB9407	MJ8100	SDT3326	MJE15030	MJE15030
MJ10025	MJ10025	MJ15003	MJ15003	MJ4032	MJ4032	MJ8500	MJ8500	MJE15030	28C2167
MJ11011	MJ11011	MJ15003	RCA1B06	MJ4032	SDM21304	MJ8501	MJ8501	MJE15030	28C2168
MJ11012	MJ11012	MJ15003	2N5972	MJ4032	SDM21314	MJ8502	MJ8502	MJE15030	28C2334
MJ11013	MJ11013	MJ15004	MJ15004	MJ4032	SB9408	MJ8503	MJ8503	MJE15030	28D381
MJ11014	MJ11014	MJ15011	MJ15011	MJ4033	MJ4033	MJ8504	MJ8504	MJE15030	28D382
MJ11015	MJ11015	MJ15011	TIP525	MJ4033	SDM20301	MJ8505	MJ8505	MJE15030	28D823
MJ11015	28B694	MJ15011	TIP526	MJ4033	SDM20302	MJ900	IR900	MJE15031	MJE15031
MJ11016	MJ11016	MJ15011	28C1116	MJ4033	SDM20311	MJ900	MJ900	MJE15031	28A1011
MJ11016	28C2433	MJ15011	28C1343	MJ4033	SDM20312	MJ900	PMD13K-40	MJE15031	28A1111
MJ11016	28D574	MJ15011	28C407	MJ4033	SDM20321	MJ900	PMD13K-60	MJE15031	28A1112
MJ11018	28C2199	MJ15011	28C408	MJ4033	SDM20322	MJ9000	MJ9000	MJE15031	28A843
MJ11028	MJ11028	MJ15011	28C681	MJ4033	SB9306	MJ901	IR901		
MJ11029	MJ11029	MJ15011	28C770	MJ4034	MJ4034	MJ901	MJ901		
MJ11030	MJ11030	MJ15011	28C771	MJ4034	SDM20303	MJ901	PMD13K-80		
MJ11031	MJ11031	MJ15011	28D166	MJ4034	SDM20313				
MJ11032	MJ11032	MJ15011	28D18	MJ4034	SDM20323				
MJ11033	MJ11033	MJ15011	28D286	MJ4034	SB9307				
MJ12002	MJ12002	MJ15011	28D287	MJ4035	MJ4035				
MJ12002	MJ701	MJ15011	28D433	MJ4035	SDM20304				
MJ12002	MJ702	MJ15011	28D84	MJ4035	SDM20314				
MJ12002	MJ704	MJ15012	MJ15012	MJ4035	SDM20324				
MJ12002	MJ721	MJ15012	TIP513	MJ4035	SB9308				
MJ12002	MJ723	MJ15012	TIP523	MJ410	DT8410				
MJ12002	MJ764	MJ15012	TIP527	MJ410	PT410				
MJ12002	IR665	MJ15012	TIP528	MJ410	IR410				
MJ12003	MJ12003	MJ15012	28B555	MJ410	IR660				
MJ12003	28C1154	MJ15012	28B556	MJ410	KDT410				
MJ12003	28C1174	MJ15012	28B600	MJ410	MJ410				
MJ12003	28C1875	MJ15015	MJ15015	MJ410	RCA410				
MJ12003	28C1893	MJ15015	28C2322	MJ410	SDT410				
MJ12003	28C1922	MJ15015	28C2431	MJ410	TIP575				
MJ12003	28C1942	MJ15015	28D340	MJ410	28C243				
MJ12003	28D517	MJ15015	28D341	MJ410	28C246				
MJ12003	28D765	MJ15016	MJ15016	MJ410	28C42				
MJ12004	MJ12004	MJ15016	2N6248	MJ410	28C42A				
MJ12004	MJ8020	MJ15016	28A679	MJ410	28C586				
MJ12004	MJ8400	MJ15016	28A907	MJ410	28C687				
MJ12004	28D299	MJ15022	MJ15022	MJ410	28C887				
MJ12004	28D300	MJ15022	RCA1B04	MJ410	28C888				
MJ12004	28D350	MJ15022	28C433	MJ410	28C889				
MJ12004	28D577	MJ15022	28C434	MJ411	DT8411				
MJ12004	28D627	MJ15023	MJ15023	MJ411	PT411				
MJ12004	28D649	MJ15023	28A909	MJ411	IR411				
MJ12004	28D950	MJ15023	28B552	MJ411	KDT411				
MJ12004	28D951	MJ15023	28B554	MJ411	MJ411				
MJ12004	28D952	MJ15023	28B723	MJ411	RCA411				
MJ12005	MJ12005	MJ15024	MJ15024	MJ411	SDT411				
MJ12005	28C1325	MJ15024	RCA1B05	MJ411	28C1050				
MJ12005	28C1895	MJ15024	RCA1B09	MJ411	28C1433				
MJ12005	28C1896	MJ15025	MJ15025	MJ411	28C1454				
MJ12005	28C2027	MJ16010	MJ16010	MJ411	28C1617				
MJ12005	28D348	MJ16012	MJ16012	MJ411	28C2121				
MJ12005	28D368	MJ1800	MJ1800	MJ411	28C270				
MJ12005	28D380	MJ2252	MJ2252	MJ411	28C808				
MJ12005	28D416	MJ2300	MJ2300	MJ411	28D383				
MJ12005	28D418	MJ2305	MJ2305	MJ411	28D461				
MJ12005	28D589	MJ2500	IR2500	MJ413	DT8413				
MJ12005	28D725	MJ2500	IR645	MJ413	PT413				
MJ12005	28D903	MJ2500	MJ2500	MJ413	IR413				
MJ12005	28D953	MJ2500	TIP645	MJ413	KDT413				
MJ12010	MJ12010	MJ2501	TIP2501	MJ413	MJ413				
MJ12010	28C2357	MJ2501	IR646	MJ413	RCA413				
MJ12010	28C2358	MJ2501	MJ2501	MJ413	SDT413				
MJ12010	28D811	MJ2501	TIP646	MJ413	28C807				
MJ13014	MJ13014	MJ2955	MJ2268	MJ423	DT8423				
MJ13014	MJ7160	MJ2955	MJ2955	MJ423	PT423				
MJ13014	TIP564	MJ2955A	MJ2955A	MJ423	IR423				
MJ13014	UMT1008	MJ3000	IR3000	MJ423	IR663				
MJ13014	2N6579	MJ3000	IR640	MJ423	SDT423				
MJ13014	2N6674	MJ3000	MJ3000	MJ423	MJ423				
MJ13014	28C1143	MJ3000	MJ3520	MJ423	RCA423				
MJ13015	MJ13015	MJ3000	TIP640	MJ423	SDT423				
MJ13015	MJ7161	MJ3001	DT81020	MJ423	28D533				
MJ13015	PM26K380	MJ3001	IR1020	MJ423	28D741				
MJ13015	TIP565	MJ3001	IR3001	MJ4237	MJ4237				
MJ13015	UMT1009	MJ3001	MJ3001	MJ4238	MJ4238				
MJ13015	WT5100	MJ3001	MJ3001	MJ4238	TIP519				
MJ13015	2N6242	MJ3001	MJ3521	MJ4238	TIP520				
MJ13015	2N6580	MJ3001	MJ3801						
MJ13015	2N6675	MJ3001	MJ3802						

MOTOR.	Industry Standard No.
MJE15031	2SA940
MJE15031	2SA949
MJE15031	2SA957
MJE15031	2SA958
MJE15031	2SA968
MJE15031	2SB546
MJE15031	2SB547
MJE15031	2SB567
MJE15031	2SB568
MJE15031	2SB628
MJE15031	2SB630
MJE15031	2SB719
MJE15031	2SB720
MJE1660	MJE1660
MJE1661	MJE1661
MJE170	D45C1
MJE170	D45C2
MJE170	D45C3
MJE170	MJE170
MJE170	2N6414
MJE170	2SA715
MJE170	2SA738
MJE170	2SB743
MJE170	2SB772
MJE171	D45C4
MJE171	D45C5
MJE171	D45C6
MJE171	D45C7
MJE171	D45C8
MJE171	D45C9
MJE171	MJE171
MJE171	2N6406
MJE171	2N6415
MJE171	2SA963
MJE172	D45C10
MJE172	D45C11
MJE172	D45C12
MJE172	MJE172
MJE172	2N6407
MJE172	2SB744
MJE180	D44C1
MJE180	D44C2
MJE180	D44C3
MJE180	MJE180
MJE180	2N6412
MJE180	2SC1162
MJE180	2SC1449
MJE180	2SC1846
MJE180	2SC2080
MJE180	2SD793
MJE180	2SD882
MJE181	D44C4
MJE181	D44C5
MJE181	D44C6
MJE181	D44C7
MJE181	D44C8
MJE181	D44C9
MJE181	MJE181
MJE181	2N6413
MJE181	2SC1847
MJE181	2SC2209
MJE182	D44C10
MJE182	D44C11
MJE182	D44C12
MJE182	MJE182
MJE182	2SC1381
MJE182	2SC1382
MJE182	2SD794
MJE200	MJE200
MJE200	MJE2050
MJE205	MJE205
MJE205	2SC931
MJE210	MJE210
MJE210	MJE2150
MJE210	2N6411
MJE210	2SA900
MJE220	MJE220
MJE221	MJE221
MJE222	MJE222
MJE223	MJE223
MJE224	MJE224
MJE225	MJE225
MJE230	MJE230
MJE231	MJE231
MJE232	MJE232
MJE233	MJE233
MJE234	MJE234
MJE235	MJE235
MJE2360T	MJE2360T
MJE2360T	MJE2360T
MJE2360T	2SC1755
MJE2360T	2SC1756
MJE2360T	2SC1757
MJE2361T	MJE2361
MJE2361T	MJE2361T
MJE2361T	2SC1819
MJE2361T	2SC1905
MJE2361T	2SC2085
MJE2361T	2SC2242
MJE2361T	2SC2482
MJE240	MJE240
MJE241	MJE241
MJE242	MJE242
MJE243	MJE243
MJE243	2N6416
MJE243	2N6417
MJE244	MJE244
MJE250	MJE250
MJE251	MJE251
MJE252	MJE252
MJE253	MJE253
MJE253	2N6418
MJE253	2N6419
MJE253	2SA681
MJE253	2SA682
MJE253	2SA794
MJE253	2SA795
MJE270	MJE270
MJE270	2SC2298
MJE271	MJE271
MJE2801T	MJE2801
MJE2801T	MJE2801K
MJE2801T	MJE2801T
MJE2901T	MJE2901
MJE2901T	MJE2901K
MJE2901T	MJE2901T
MJE2955T	MJE2955
MJE2955T	2SB578
MJE2955T	MJE2955K
MJE2955T	MJE2955T
MJE2955T	RCA1008
MJE2955T	TIP2955
MJE3055	MJE2055
MJE3055	MJE3055
MJE3055	2SD491
MJE3055	2SD499
MJE3055	2SD500
MJE3055T	MJE3055K
MJE3055T	MJE3055T
MJE3055T	NSP3055
MJE3055T	RCA1007
MJE3055T	RCA1C09
MJE3055T	TIP3055
MJE3055T	2SC2397
MJE31	MJE31
MJE31A	MJE31A
MJE31B	MJE31B
MJE31C	MJE31C
MJE32	MJE32
MJE32A	MJE32A
MJE32B	MJE32B
MJE32C	MJE32C
MJE3300	MJE3300
MJE3300	2SC1516
MJE3301	MJE3301
MJE3302	MJE3302
MJE3310	MJE3310
MJE3310	2SB569
MJE3311	MJE3311
MJE3311	2SB570
MJE3312	MJE3312
MJE3312	2SB571
MJE340	MJE340
MJE340	2SC1749
MJE341	MJE341
MJE341	2SC1903
MJE341	2SC1904
MJE341	2SC2590
MJE341	2SD414
MJE341	2SD415
MJE3439	MJE3439
MJE3439	MJE345
MJE3439	2SC1088
MJE3439	2SC1089
MJE3439	2SC1501
MJE3439	2SC1514
MJE3439	2SC2278
MJE3439	2SC2371
MJE344	MJE344
MJE344	2SB668
MJE344	2SB669
MJE3440	MJE3440
MJE3440	2SC2071
MJE3440	2SD757
MJE3440	2SD758
MJE350	MJE350
MJE350	2SA1110
MJE350	2SA898
MJE350	2SA899
MJE350	2SA939
MJE350	2SB648
MJE350	2SB649
MJE350	2SB717
MJE350	2SB718
MJE370	MJE370
MJE370	2SB370
MJE370	MJE3370
MJE371	MJE371
MJE4341	MJE4340
MJE4341	MJE4341
MJE4342	MJE4342
MJE4350	MJE4350
MJE4351	MJE4351
MJE4352	MJE4352
MJE4352	2SB691
MJE4352	2SB695
MJE4352	2SB713
MJE5170	MJE5170
MJE5171	MJE5171
MJE5172	MJE5172
MJE5180	MJE5180
MJE5181	MJE5181
MJE5182	MJE5182
MJE51T	MJE51
MJE51T	MJE51T
MJE520	MJE520
MJE520	2SD612
MJE521	MJE521
MJE52T	MJE52T
MJE53T	MJE53
MJE53T	MJE53T
MJE5730	MJE5730
MJE5731	MJE5731
MJE5732	MJE5732
MJE5740	MJE5740
MJE5740	TIP160
MJE5741	BU180
MJE5741	MJE5741
MJE5741	TIP161
MJE5742	BU180A
MJE5742	MJE5742
MJE5742	TIP162
MJE5780	MJE5780
MJE5781	MJE5781
MJE5782	MJE5782
MJE5850	MJE5850
MJE5851	MJE5851
MJE5852	MJE5852
MJE6040	MJE6040
MJE6040	2SB582
MJE6041	MJE6041
MJE6041	2SB583
MJE6042	MJE6042
MJE6042	2SB584
MJE6043	MJE6043
MJE6043	2SD496
MJE6044	MJE6044
MJE6044	2SB497
MJE6045	MJE6045
MJE6045	2SD498
MJE700	MJE700
MJE700T	MJE700T
MJE701	MJE701
MJE701T	MJE701T
MJE702	MJE702
MJE702T	MJE702T
MJE703	MJE703
MJE703T	MJE703T
MJE703T	2SB751
MJE710	MJE710
MJE711	MJE711
MJE712	MJE712
MJE720	MJE720
MJE721	MJE721
MJE722	MJE722
MJE800	MJE800
MJE800T	MJE800T
MJE800T	2SD687
MJE800T	2SD839
MJE801	MJE801
MJE801T	MJE801T
MJE802	MJE802
MJE802	2SD797
MJE802T	MJE802T
MJE802T	2SD840
MJE803	MJE803
MJE803T	MJE803T
MJE8500	MJE8500
MJE8501	MJE8501
MJE8502	MJE8502
MJE8503	MJE8503
MLED60	EE100
MLED60	EE60
MLED60	IRL60
MLED60	LD261
MLED60	ME60
MLED60	ME61
MLED60	TIL26
MLED900	ME702
MLED900	OP160
MLED930	CL100
MLED930	CL110
MLED930	CL110A
MLED930	CL110B
MLED930	CQY10
MLED930	CQY11,B,C
MLED930	CQY12,B
MLED930	CQY31
MLED930	CQY32
MLED930	FPE100
MLED930	FPE510
MLED930	IRL40
MLED930	LBD56,P
MLED930	OP130
MLED930	OP131
MLED930	SE1450 SERIES
MLED930	SSL34,54
MLED930	SSL4,P
MLED930	TIL31
MLED930	TIL33
MLED930	TIL34
MLM308AG	S03118AT
MLM308AL	S03118AJ
MLM308AP1	S03118AM
MLM308G	S03118T
MLM308L	S03118J
MLM308P1	S03118M
MLM565CP	SR565A
MLM565CP	SR565K
MM1505	2N3010
MM1748	2N709
MM2258	2N2258
MM2259	2N2259
MM2260	2N2260
MM3006	2N1654
MM3007	2N1655
MM3220	2N3725
MM3726	2S3726
MM4049	A440
MM4257	2N4257
MM4258	2N4258
MMBR901	NB0Z133
MMBR920	BPR53
MMBR920	BPR92
MMBR930	BPR93
MMHO026C1	MHO026CP
MMHO026CG	BPR92
MMHO026CG	DS00026CP
MMHO026CG	DS00056CH
MMHO026CG	DS00056GU
MMHO026CG	MHO026CG
MMHO026CG	2SC2398
MMHO026CG	MHO026CH
MMHO026CG	MHO026G
MMHO026CG	MHO026H
MMHO026CL	SHO013HC
MMHO026CL	DS00026CJ
MMHO026CL	DS00560QL
MMHO026CL	MHO026P
MMHO026CP1	SN75362P
MMHO026CP1	MHO026CN
MMHO026CP1	SN75369P
MMHO026G	DS00026G
MMHO026G	DS00056G
MMHO026G	DS00056H
MMHO026LG	SHO013HM
MOC1005	CQY80
MOC1005	FCD820,C,D
MOC1005	FCD830,C,D
MOC1006	OPI110
MOC1006	FCD830,B,C,D
MOC1006	FCD831,C,D
MOC1006	FCD836C,D
MOC1006	OPI2150
MOC1006	OPI2250
MOC1006	OPI2251
MOC3000	OPI4402
MOC3000	SCS11C6
MOC3001	OPI4401
MOC3001	SCS11C4
MOC3002	OPI4202
MOC3002	SCS11C3
MOC3003	OPI4201
MOC3003	SCS11C1
MP5918	SE3002
MPE256	PN4416
MPP102	E100
MPP102	K84223
MPP102	K84224
MPP102	NF500
MPP102	NF501
MPP102	NF506
MPP102	2N5248
MPP108	MPP108
MPP108/112	2N3819
MPP109	MPP109
MPP110	MPP110
MPP111	MPP111
MPP112	MPP112
MPP161	MPP161
MPP161	2N3820
MPP161	2N4088
MPP161	2N4089
MPP161	2N4090
MPP256	A6108
MPP256	MPP256
MPP4391	C413E
MPF4391	CMX740
MPF4391	CP650
MPF4391	CP651
MPF4391	E105
MPF4391	E106
MPF4391	E108
MPF4391	E109
MPF4391	E110
MPF4391	E111
MPF4391	ITE4391
MPF4391	K84391
MPF4391	K84856
MPF4392	CM697
MPF4392	CP652
MPF4392	CP653
MPF4392	E112
MPF4392	ITE4392
MPF4392	K84392
MPF4392	K84857
MPF4393	E113
MPF4393	ITE4393
MPF4393	K84393
MPF4393	K84858
MPF970	E174
MPF970	E175
MPF970	E270
MPF970	E271
MPF970	P1069E
MPF970	P1086E
MPF970	P1087E
MPF971	P1117E
MPF971	P1118E
MPF971	P1119E
MPN3401	MPN3401
MPN3401	18222
MPN3402	MPN3402
MPN3403	MPN3403
MPN3404	MPN3404
MPN3500	MPN3500
MPN3503	MPN3503
MPN3504	MPN3504
MPQ1000	MPQ1000
MPQ2001	MPQ2001
MPQ2221	MPQ2221
MPQ2222	2N2222
MPQ2369	MPQ2369
MPQ2483	MPQ2483
MPQ2484	MPQ2484
MPQ2906	MPQ2906
MPQ2907	MPQ2907
MPQ2907	Q2T2905
MPQ2907	Q2T5244
MPQ3303	MPQ3303
MPQ3467	FPQ3467
MPQ3467	FPQ3468
MPQ3467	MPQ3467
MPQ3546	MPQ3546
MPQ3724	FPQ3724
MPQ3725	MPQ3725
MPQ3725	MPQ3725
MPQ3725A	MPQ3725A
MPQ3762	MPQ3762
MPQ3798	MPQ3798
MPQ3799	MPQ3799
MPQ3904	MPQ3904
MPQ3906	MPQ3906
MPQ6001	MPQ6001
MPQ6002	MPQ6002
MPQ6100	MPQ6100
MPQ6100A	MPQ6100A
MPQ6426	MPQ6426
MPQ6427	MPQ6427
MPQ6501	MPQ6501
MPQ6502	MPQ6502
MPQ6600	MPQ6600
MPQ6600A	MPQ6600A
MPQ6700	MPQ6700
MPQ7041	MPQ7041
MPQ7042	MPQ7042
MPQ7043	MPQ7043
MPQ7051	MPQ7051
MPQ7052	MPQ7052
MPQ7053	MPQ7053
MPQ7091	MPQ7091
MPQ7092	MPQ7092
MPQ7093	MPQ7093
MPQ918	MPQ918
MPS2222	2N2222
MPS2222	2N2222
MPS2222	TIB109
MPS2222	TIB110
MPS2222A	PN2222A
MPS2222A	TIB111
MPS2369	2N2369A
MPS2369	EN9009
MPS2369	EN3014
MPS2369	TIB50
MPS2369	TIB51
MPS2369	2N5769
MPS2369	2N5771
MPS2369	2N5910
MPS2711	PN5135
MPS2714	PN5133
MPS2907	PN2907
MPS2907	TIB112
MPS2907A	PN2907A
MPS3392	A5T3392
MPS3392	A7T3392
MPS3392	A8T3392
MPS3563	PN5130
MPS3638	A5T3638
MPS3638	A5T4260
MPS3638	PN5142
MPS3638	PN5143
MPS3638A	A5T3638A
MPS3640	A5T4261
MPS3640	PN5910
MPS3640	TIB55
MPS3640	TIB54
MPS3646	EN3012
MPS3646	EN3013
MPS3646	PN5131
MPS3646	2N5772
MPS3702	A8T3702
MPS3703	A8T3703
MPS3703	EN722
MPS3704	A8T3704
MPS3705	A8T3705
MPS3706	A8T3706

MOTOR.	Industry Standard No.
MPS3706	PN5136
MPS3707	A5T3707
MPS3708	A5T3708
MPS3708	A8T3708
MPS3709	A5T3709
MPS3709	A8T3709
MPS3710	A5T3710
MPS3710	A8T3710
MPS3711	A8T3711
MPS3826	EN915
MPS404	A5T404
MPS404	A8T404
MPS404A	A5T404A
MPS404A	A8T404A
MPS4257	PN4257
MPS4257	PN4257A
MPS4258	PN4258
MPS4258A	PN4258A
MPS5172	A5T5172
MPS5172	A7T5172
MPS5172	A8T5172
MPS5172	EN5172
MPS5551	TIS101
MPS6512	EN916
MPS6513	PN3691
MPS6513	PN3692
MPS6514	A5T3565
MPS6514	EN930
MPS6514	PN5565
MPS6514	PN5128
MPS6514	PN5129
MPS6515	2N6008
MPS6516	A5T3644
MPS6516	A5T4059
MPS6516	A5T4060
MPS6516	A8T4059
MPS6516	A8T4060
MPS6516	EN3011
MPS6516	EN3250
MPS6516	PN3250
MPS6516	PN3644
MPS6516	PN5138
MPS6516	PN5139
MPS6517	A5T4061
MPS6517	A8T4061
MPS6518	A5T4062
MPS6518	A8T4062
MPS6521	TI894
MPS6522	A5T4058
MPS6522	A8T4058
MPS6530	A5T2243
MPS6530	EN1613
MPS6530	EN1711
MPS6530	EN697
MPS6530	PN1613
MPS6530	PN1711
MPS6530	PN3567
MPS6530	PN3641
MPS6530	PN3642
MPS6531	A5T2192
MPS6531	A5T2193
MPS6531	EN956
MPS6531	PN3643
MPS6531	TI860
MPS6534	TI861
MPS6539	PN5132
MPS6560	PN5137
MPS6566	PN5855
MPS6595	A5T3571
MPS6595	A5T3572
MPS8706	
MPS8092	TI890
MPS8092	TI892
MPS8093	TI891
MPS8093	TI893
MPS8097	SE4001
MPS8097	SE4002
MPS8097	SE4010
MPS8097	TI897
MPS8098	BC254
MPS8098	TI895
MPS8098	TI898
MPS8098	2N6014
MPS8099	BSW32
MPS8099	TI896
MPS8099	TI899
MPS8834	EN708
MPS8598	BC256
MPS8598	2N6017
MPS918	C89018
MPS918	EN918
MPS918	SE3001
MPS918	TIS105
MPS918	TI862
MPS918	TI863
MPS918	TI864
MPSA05	BC255
MPSA05	BC425
MPSA05	BC431
MPSA05	BC635
MPSA05	BC637
MPSA05	BPR40
MPSA05	BPR41
MPSA05	EN2484
MPSA05	PB6020
MPSA05	PB6022
MPSA05	PN2484
MPSA05	PN5569
MPSA05	TC8100
MPSA05	TC8102
MPSA06	BC424
MPSA06	BC639
MPSA06	BC640
MPSA06	BPR39
MPSA06	B8834
MPSA06	PB6021
MPSA06	PB6023
MPSA06	PN3568
MPSA06	PN5856
MPSA13	BC517
MPSA18	SE1001
MPSA18	SE1002
MPSA42	PN5964
MPSA43	TI8100
MPSA55	A5T3645
MPSA55	A5T4026
MPSA55	A5T4028
MPSA55	A8T4026
MPSA55	A8T4028
MPSA55	BC427
MPSA55	BC432
MPSA55	BC527
MPSA55	BC535
MPSA55	BC636
MPSA55	BC638
MPSA55	BPR80
MPSA55	BPR81
MPSA55	PN3645
MPSA55	TC8101
MPSA55	TC8103
MPSA56	A5T4027
MPSA56	A5T4029
MPSA56	A8T4027
MPSA56	A8T4029
MPSA56	BC426
MPSA56	BC534
MPSA56	BC640
MPSA56	PN5857
MPSA56	PN5858
MPSA63	BC516
MPSA93	BP398
MPSH02	SE5035
MPSH04	BF310
MPSH04	EN718A
MPSH07	PE5010
MPSH07	PE5013
MPSH07	PE5015
MPSH08	PTR168
MPSH08	PE5029
MPSH08	TIS125
MPSH10	PE3100
MPSH10	PN5126
MPSH10	TIS129
MPSH11	PTR129A
MPSH24	PN3690
MPSH24	BF311
MPSH24	BF523
MPSH24	PTR129
MPSH24	PN3688
MPSH24	PN3689
MPSH24	TIS104
MPSH30	TI886
MPSH30	C89016
MPSH30	C89019
MPSH30	SE5020
MPSH30	SE5021
MPSH30	SE5022
MPSH30	SE5023
MPSH30	SE5024
MPSH30	SE5050
MPSH30	SE5051
MPSH30	SE5052
MPSH32	C89010
MPSH32	EN914
MPSH32	PTR118
MPSH32	PTR158
MPSH32	SE5055
MPSH32	TIS108
MPSH32	TI856
MPSH32	TI857
MPSH32	TI884
MPSH33	BF357
MPSH34	PE5031
MPSH34	TIS126
MPSH34	TI887
MPSH37	PE5030B
MPSH54	BP243
MPSH54	BP340
MPSH54	BF341
MPSH54	BF540
MPSH54	BF541
MPSH54	TIS137
MPSH54	TIS138
MPSH83	TIS128
MPSU01	MPSU01
MPSU01	2BC1429
MPSU01	2BC1761
MPSU01A	MPSU01A
MPSU02	MPSU02
MPSU03	MPSU03
MPSU04	2BC1124
MPSU04	2BC1628
MPSU04	2BC1630
MPSU05	MPSU05
MPSU05	2N6179
MPSU06	MPSU06
MPSU06	2N6178
MPSU07	MPSU07
MPSU07	2BC1728
MPSU07	2BC1760
MPSU07	2BD558
MPSU10	MPSU10
MPSU10	MPSU11
MPSU10	2N6175
MPSU10	2N6176
MPSU10	2BC1125
MPSU31	MPSU31
MPSU31	MPSU47
MPSU45	MPSU12
MPSU45	MPSU45
MPSU51	2BA861
MPSU51A	MPSU51A
MPSU52	MPSU52
MPSU55	MPSU55
MPSU55	2N6181
MPSU55	2BA706
MPSU55	2BA897
MPSU55	2BA962
MPSU56	MPSU56
MPSU56	2N6180
MPSU57	MPSU57
MPSU60	MPSU60
MPSU60	2BA818
MPSU60	2BA835
MPSU95	MPSU95
MPTE-10	MPT-10
MPTE-12	MPT-12
MPTE-15	MPT-15
MPTE-18	MPT-18
MPTE-22	MPT-22
MPTE-36	MPT-36
MPTE-45	MPT-45
MPT-5	MPT-5
MPTE-8	MPT-8
MPU131	MCR1330
MPU131	MCR1331
MPU132	MCR1350
MPU132	MPU132
MPU133	MPU133
MPU6027	D1371
MPU6027	MPU6027
MPU6028	D1372
MPU6028	MPU6028
MQ1120	MQ1120
MQ2218	MQ2218
MQ2218	MQ2221
MQ2218	SP2221QF
MQ2219A	MQ2219A
MQ2219A	MQ2222A
MQ2219A	SP2222AQF
MQ2219A	SP2222QF
MQ2369	MQ2369
MQ2484	MQ2484
MQ2484	SP2483QF
MQ2484	SP2484QF
MQ2904	MQ2904
MQ2904	MQ2906
MQ2904	SP2906QF
MQ2905A	MQ2905A
MQ2905A	MQ2907A
MQ2905A	SP2906AQF
MQ2905A	SP2907AQF
MQ3251	MQ3251
MQ3467	FQ3467
MQ3467	FQ3468
MQ3467	MQ3467
MQ3467	SP3762QF
MQ3467	SP3763QF
MQ3725	FQ3724
MQ3725	FQ3725
MQ3725	MQ3725
MQ3725	MQ3762
MQ3725	SP3725QF
MQ3798	MQ3798
MQ3799	MQ3799
MQ3799A	MQ3799A
MQ6001	MQ6001
MQ6002	MQ6002
MQ6100	MQ6100
MQ7001	MQ7001
MQ7003	MQ7003
MQ7004	MQ7004
MQ7005	MQ7005
MQ7007	MQ7007
MQ7021	MQ7021
MQ918	MQ918
MQ930	MQ930
MQ982	MQ982
MR1-1000	G11-1000
MR1-1000	1B3929
MR1-1200	A1425MH2AB1
MR1-1200	A1425MH2AB1
MR1-1200	BF4-120L
MR1-1200	E12
MR1-1200	E12
MR1-1200	G12
MR1-1200	G11-1200
MR1-1200	M1200
MR1-1200	MR1212
MR1-1200	MR1200
MR1-1200	MV-12A,B,C
MR1-1200	NC1200
MR1-1200	S10110
MR1-1200	S10120
MR1-1200	S112
MR1-1200	SLA-20
MR1-1200	TA120
MR1-1200	TW120
MR1-1200	1N2503
MR1-1200	1N2507
MR1-1200	1B2618
MR1-1200	1N3107
MR1-1200	1N3487
MR1-1400	A114PD2
MR1-1400	A14PD
MR1-1400	E127
MR1-1400	E14
MR1-1400	F14
MR1-1400	G11-1400
MR1-1600	1B2557
MR1-1600	BY184
MR1-1600	BYX10
MR1-1600	G11-1600
MR1-1600	MPR15
MR1-1600	MR1500
MR1-1600	MV-16A,B,C
MR1-1600	S115
MR1-1600	1N2358
MR1-1600	1N2359
MR1-1600	1N2504
MR1-1600	1N2508
MR1-1600	1N2619
MR1-1600	1N2866
MR1-1600	1N3870
MR1-1600	1N3990
MR1-1600	1N4374
MR1-1600	3CF815
MR1120	SL50
MR1120	TR1120
MR1120	1N2348
MR1121	SL100
MR1121	ST210P
MR1121	TR1121
MR1121	1N3569
MR1121	3P10
MR1121	10B
MR1122	10B3P
MR1122	BYX38-300,R
MR1122	BYX42-300,R
MR1122	SL200
MR1122	ST220P
MR1122	TR1122
MR1122	1N1124,A
MR1122	1N2350
MR1122	1N3570
MR1122	3P20
MR1122	20B
MR1122	20B3P
MR1123	SL3
MR1123	SL300
MR1123	ST230P
MR1123	TR1123
MR1123	30B
MR1123	30B3P
MR1124	BYX42-600,R
MR1124	BYX48/300
MR1124	SL400
MR1124	ST240P
MR1124	TR1124
MR1124	1N1125,A
MR1124	1N1126,A
MR1124	1N3571
MR1124	1N3572
MR1124	3P30
MR1124	3P40
MR1124	40B
MR1124	40B3P
MR1125	SL5
MR1125	SL500
MR1125	ST250P
MR1125	TR1125
MR1126	BYX38-600,R
MR1126	BYX42-900,R
MR1126	BYX48/600
MR1126	SL600
MR1126	ST260P
MR1126	TR1126
MR1126	1N1127,A
MR1126	1N1128,A
MR1126	1N3573
MR1126	1N3574
MR1126	3P50
MR1126	3P60
MR1128	BYX42-1200,R
MR1128	SL8
MR1128	SL800
MR1128	SL800X
MR1128	ST280P
MR1128	TM74
MR1128	TM75
MR1128	TM76
MR1128	TM78
MR1128	TM79
MR1128	TM84
MR1128	TM85
MR1128	TM86
MR1128	TM88
MR1128	TM89
MR1128	TR1128
MR1128	1N1235,A,B
MR1128	1N1236,A,B
MR1128	1N2240,A
MR1128	1N2260A
MR1128	1N3649
MR1128	1N3650
MR1128	1N3670,A
MR1128	1N3671,A
MR1128	1N3987
MR1128	1N3988
MR1128	1N4012
MR1128	1N4013
MR1128	1N562
MR1128	3P80
MR1128	6A700
MR1128	6A800
MR1128	6F70A,B
MR1128	6F80A,B
MR1128	12A700
MR1128	12A800
MR1128	12P80B
MR1130	BYX38-1200,R
MR1130	BYX38-900,R
MR1130	BYX48/900
MR1130	F12100B
MR1130	SL10
MR1130	SL1000
MR1130	SL1000X
MR1130	ST2100P
MR1130	TM104
MR1130	TM105
MR1130	TM105
MR1130	TR1130
MR1130	1N1444,A,B
MR1130	1N2224,A
MR1130	1N2242,A
MR1130	1N2262A
MR1130	1N3672,A
MR1130	1N3673,A
MR1130	1N3924
MR1130	1N3989
MR1130	1N3990
MR1130	1N4014
MR1130	1N4015
MR1130	1N563
MR1130	3P100
MR1130	6A1000
MR1130	6F100A,B
MR1130	6F90A,B
MR1130	12A1000
MR1130	12A900
MR1130	12P100B
MR1215PL	1N3288
MR1215PL	1N3289
MR1215PL	1N3290
MR1219SL	1N3291
MR1219SL	1N3292
MR1219SL	1N3293
MR1366	D2406M
MR1366	ED8307
MR1366	N86005
MR1366	R3020606
MR1366	R302506
MR1366	6A6F
MR1366	6FL50
MR1366	6FL60
MR1366	6FT50
MR1366	6FV50
MR1366	6FV60
MR1376	A129R
MR1376	A129M
MR1376	D2412M
MR1376	ED8310
MR1376	N812006
MR1376	R3020612
MR1376	R302512
MR1376	RT60
MR1376	ST2FR60P
MR1376	12A6F
MR1376	12FL60,502
MR1376	12FL60,502
MR1376	12FT50
MR1376	12FT60
MR1376	12FV50
MR1376	12FV60
MR1386	A139R
MR1386	A139M
MR1386	BYX38-500,R
MR1386	BYX36-600,R
MR1386	2S520M
MR1386	R4020620
MR1386	20A6P
MR1396	D2540M
MR1396	R4020530
MR1396	R4020630
MR1396	30A6P
MR20008	A327P
MR20008	1N3615
MR20008	16P5

MOTOR.	Industry Standard No.	MOTOR.	Industry Standard No.	MOTOR.	Industry Standard No.	MOTOR.	Industry Standard No.
MR2001S	A327A	MR502	G3D	MR506	3A500	MR817	T880
MR2001S	1N3616	MR502	GI502	MR506	3A6	MR817	1N4947
MR2001S	16P10	MR502	GP25D	MR506	3A600	MR817	1N5621GP
MR2001S	20P10	MR502	GP30D	MR506	3AP6	MR818	D2601N
MR2002S	A327B	MR502	HB200	MR506	3BP6	MR818	GI818
MR2002S	1N3617	MR502	MB229	MR506	3E6	MR818	RG1-M
MR2002S	1N3618	MR502	P300D	MR506	3S16	MR818	RG1M
MR2002S	16P15	MR502	S-3A2	MR506	3SM6	MR818	RGP10H
MR2002S	16P20	MR502	S5A2	MR506	30S5	MR818	RGP15M
MR2002S	A327C	MR502	SEN220	MR506	30S6	MR818	RGP20M
MR2004S	1N3619	MR502	SEN320	MR5060	1N5060GP	MR818	RL100
MR2004S	1N3620	MR502	SI-2A	MR5061	1N5061GP	MR818	RMC100
MR2004S	16P30	MR502	SLA-23	MR5062	1N5062GP	MR818	S0F
MR2004S	16P40	MR502	SLA5199	MR508	A15N	MR818	S1A10F
MR2004S	20P30	MR502	SRS220	MR508	AB800	MR818	SRSFR1100
MR2004S	20P40	MR502	SRB320	MR508	AC800	MR818	1N4948
MR2006S	1N3621	MR502	UT2020	MR508	AC880	MR818	1N5623GP
MR2006S	1N3622	MR502	UT262	MR508	BP5-80L	MR820	GI820
MR2006S	16P50	MR502	UT4020	MR508	BP6-80L	MR821	GI821
MR2006S	16P60	MR502	V332	MR508	G3K	MR821	6ALR1
MR2008S	1N3623	MR502	1N5199	MR508	GI508	MR822	GI822
MR2008S	16P80	MR502	1N5624,GP	MR508	GP25K	MR822	6ALR2
MR2010S	1N3624	MR502	2AP2	MR508	GP30K	MR824	GI824
MR2010S	16P100	MR502	3A2	MR508	HB800	MR824	6ALR3
MR2400P	BYW29-50	MR502	3A200	MR508	MB234	MR824	6ALR4
MR2401P	BYW29-100	MR502	3AP2	MR508	P300K	MR826	GI826
MR2402P	BYW29-150	MR502	3BP2	MR508	S-3A8	MR826	6ALR6
MR250-1	1N1730,A	MR502	3E2	MR508	S5A8	MR850	A115P
MR250-2	G2	MR502	3S12	MR508	SEN280	MR850	GI850
MR250-2	HG-2	MR502	3SM2	MR508	SEN380	MR850	NS2000
MR250-3	HG-3	MR502	30S2	MR508	SI-10A	MR850	N53000
MR250-4	HG-4	MR5020	SR5020	MR508	SI-8A	MR850	RG3-A
MR250-5	HG-5	MR5030	SR5030	MR508	SLA-28	MR850	RG3A
MR2500	AR25A	MR504	A15C	MR508	SRS280	MR850	RGP25A
MR2500	GI2500	MR504	A15D	MR508	SRS380	MR850	RGP30A
MR2501	AR25B	MR504	AB300	MR508	UT268	MR850	SEN205FR
MR2501	GI2501	MR504	AB400	MR508	V338	MR850	SEN305FR
MR2502	AR25D	MR504	AC300	MR508	1N5627GP	MR850	SRBFR205
MR2502	GI2502	MR504	AC400	MR508	2AP8	MR850	SRSFR305
MR2504	AR25F	MR504	BP5-40L	MR508	3A8	MR850	UTR2305
MR2504	AR25G	MR504	BP6-40L	MR508	3A800	MR850	UTR3305
MR2504	GI2504	MR504	DSR1203	MR508	3A88	MR850	UT4305
MR2506	AR25H	MR504	ER2003	MR508	3BP8	MR850	UTX3105
MR2506	AR25J	MR504	ER2004	MR508	3B8	MR850	UTX4105
MR2506	GI2506	MR504	G3F	MR508	3SM8	MR850	V330X
MR2508	AR25K	MR504	G3G	MR508	30S8	MR850	1N5185,GP
MR2508	GI2508	MR504	GI504	MR510	AB1000	MR850	1N5415
MR2510	AR25M	MR504	GP25G	MR510	AC1000	MR850	3L05
MR2510	GI2510	MR504	GP30G	MR510	BP5-100L	MR850	3L05
MR327	A44E	MR504	HB300	MR510	BP6-100L	MR851	A115A
MR328	A44M	MR504	HB400	MR510	G3M	MR851	GI851
MR328	PZ-140P	MR504	MB230	MR510	GI510	MR851	HRF100
MR328	V500	MR504	MB231	MR510	GP25M	MR851	N52001
MR328	V600	MR504	P300F	MR510	GP30M	MR851	RG3B
MR328	25PW50	MR504	P300G	MR510	HB1000	MR851	RGP25B
MR328	25PW60	MR504	S-3A3	MR510	MB235	MR851	RGP30B
MR330	SV800	MR504	S-3A4	MR510	P300M	MR851	S5A1P
MR330	V800	MR504	S5A3	MR510	S-3A10	MR851	SEN210FR
MR331	SV1000	MR504	S5A4	MR510	S5A10	MR851	SEN310FR
MR331	V1000	MR504	SEN240	MR510	SEN2100	MR851	SRSFR210
MR500	03A	MR504	SEN300	MR510	SEN3100	MR851	SRBFR210
MR500	GI500	MR504	SEN340	MR510	SLA-29	MR851	UTR2310
MR500	GP25A	MR504	SI-3A	MR510	SRS2100	MR851	UTR3310
MR500	GP30A	MR504	SI-4A	MR510	SRS3100	MR851	UTR4310
MR500	P300A	MR504	SLA-24	MR510	V3310	MR851	UTX3110
MR500	S5A025	MR504	SLA-25	MR510	2AP10	MR851	UTX4110
MR500	V330	MR504	SLA5200	MR510	3A1000	MR851	V331X
MR500	1N5197	MR504	SRS240	MR510	3AP10	MR851	1N5186,GP
MR5005	SR5005	MR504	UT2040	MR510	3BP10	MR851	1N5416
MR501	A15A	MR504	UT264	MR510	3E10	MR851	2AFR1
MR501	A15F	MR504	UT4040	MR510	3SM0	MR851	3AFR1
MR501	AB100	MR504	V334	MR510	30S10	MR851	3BP1
MR501	AB50	MR504	1N5200	MR750	GI750	MR851	3SP1
MR501	AC100	MR504	1N5625,GP	MR750	R3400106	MR852	A115B
MR501	A050	MR504	2AP3	MR751	GI751	MR852	GI852
MR501	BP5-05L	MR504	2AP4	MR751	R3400106	MR852	HRF200
MR501	BP5-10L	MR504	3A300	MR751	S6A1	MR852	NS2002
MR501	BP6-05L	MR504	3A400	MR751	6AL1	MR852	N53002
MR501	BP6-10L	MR504	3AP3	MR752	GI752	MR852	RG3D
MR501	DSR1201	MR504	3AP4	MR752	R3400206	MR852	RGP25D
MR501	ER2000	MR504	3BP3	MR752	S-5A2	MR852	RGP30D
MR501	ER2001	MR504	3E4	MR752	S6A2	MR852	RIV020
MR501	G3B	MR504	3S14	MR752	6AL2	MR852	S3A2P
MR501	GI501	MR504	3SM4	MR754	GI754	MR852	SEN220FR
MR501	GP25B	MR504	30S3	MR754	R3400306	MR852	SEN320FR
MR501	GP30B	MR504	30S4	MR754	R3400406	MR852	SRBFR320
MR501	HB100	MR5040	SR5040	MR754	R3400506	MR852	UTR2320
MR501	HB50	MR5059	1N5059GP	MR754	S-5A3	MR852	UTR3320
MR501	MB228	MR506	A15E	MR754	S-5A4	MR852	UTR4320
MR501	P300B	MR506	A15M	MR754	S6A3	MR852	3S115
MR501	S-3A1	MR506	AB500	MR754	S6A4	MR852	UTX3120
MR501	S5A1	MR506	AB600	MR754	6AL3	MR852	UTX4115
MR501	SEN205	MR506	AC500	MR754	6AL4	MR852	UTX4120
MR501	SEN210	MR506	AC600	MR756	GI756	MR852	V332X
MR501	SEN310	MR506	BP5-60L	MR756	R3400606	MR852	1N5187,GP
MR501	SI-1A	MR506	BP6-60L	MR756	R3400706	MR852	1N5417
MR501	SLA-21	MR506	DSR1205	MR756	S-5A5	MR852	2AFR2
MR501	SLA-22	MR506	ER2005	MR756	S-5A6	MR852	3AFR2
MR501	SLA519I	MR506	ER2006	MR756	S6A5	MR852	3BFR2
MR501	SLA519B	MR506	G3H	MR756	S6A6	MR852	3SP2
MR501	SRB205	MR506	G3J	MR756	6AL6	MR854	A115C
MR501	SRB210	MR506	GI506	MR758	GI758	MR854	A115D
MR501	SRB305	MR506	GP25J	MR758	R3400806	MR854	GI854
MR501	SRB310	MR506	GP30J	MR758	S6A8	MR854	HRF400
MR501	UT2005	MR506	HB500	MR760	R3400906	MR854	NS2003
MR501	UT2010	MR506	HB600	MR760	R3401006	MR854	NS2004
MR501	UT261	MR506	MB232	MR760	S6A10	MR854	N53003
MR501	UT4005	MR506	MB233	MR810	D2601F	MR854	N53004
MR501	UT4010	MR506	P300H	MR810	GI810	MR854	RG3F
MR501	V331	MR506	P300J	MR811	D2601A	MR854	RG3G
MR501	1N5198	MR506	S-3A5	MR811	GI811	MR854	RGP25F
MR501	2AP1	MR506	S-3A6	MR812	D2601B	MR854	RGP25G
MR501	03A05	MR506	S5A5	MR812	GI812	MR854	RGP30F
MR501	3A1	MR506	S5A6	MR814	D2601D	MR854	RGP30G
MR501	3A100	MR506	SEN260	MR814	GI814	MR854	RIV040
MR501	3A15	MR506	SEN350	MR816	D2201N	MR854	S3A3P
MR501	3A30	MR506	SEN360	MR816	D2601M	MR854	S3A4P
MR501	3A50	MR506	SI-5A	MR816	GI816	MR854	SEN230FR
MR501	3AP1	MR506	SI-6A	MR817	A114N	MR854	SEN240FR
MR501	3BP1	MR506	SLA-26	MR817	GI817	MR854	SEN330FR
MR501	3E05	MR506	SLA-27	MR817	MB220	MR854	SEN340FR
MR501	3E1	MR506	SLA5201	MR817	RG1-K	MR854	SRBFR230
MR501	3S105	MR506	SRS260	MR817	RG1K	MR854	SRBFR330
MR501	3S11	MR506	SRS360	MR817	RGP10K	MR854	SEN340FR
MR501	30S1	MR506	UT2060	MR817	RGP15K	MR854	UTR2340
MR5010	SR5010	MR506	UT265	MR817	RGP20K	MR854	UTR3340
MR502	A15B	MR506	UT267	MR817	RL080	MR854	UTR4340
MR502	AB200	MR506	UT4060	MR817	RMC080	MR854	V334X
MR502	AC200	MR506	V336	MR817	S1ABF	MR854	1N5188,GP
MR502	BP5-20L	MR506	1N5201	MR817	S8F	MR854	1N5418
MR502	BP6-20L	MR506	1N5626,GP	MR817	SRSFR180	MR854	2AFR3
MR502	ER2002	MR506	2AP6	MR817	TKP100		
				MR817	TKP80		

MOTOR.	Industry Standard No.
MR854	2AFR4
MR854	3AFR3
MR854	3AFR4
MR854	3BPR3
MR854	3BPR4
MR854	38P4
MR856	A115E
MR856	A115M
MR856	G1856
MR856	HRP600
MR856	NS2005
MR856	NS2006
MR856	NS3005
MR856	NS3006
MR856	RG3H
MR856	RG3J
MR856	RGP25H
MR856	RGP25J
MR856	RGP30H
MR856	RGP30J
MR856	RIVO60
MR856	S3A5F
MR856	S3A6F
MR856	SEN250FR
MR856	SEN260FR
MR856	SEN350FR
MR856	SEN360FR
MR856	SRSFR250
MR856	SRSFR260
MR856	SRSFR350
MR856	SRSFR360
MR856	UTR2350
MR856	UTR2360
MR856	UTR3350
MR856	UTR3360
MR856	UTR4350
MR856	UTR4360
MR856	V336X
MR856	1N5189,GP
MR856	1N5190,GP
MR856	1N5419
MR856	1N5420
MR856	2AFR6
MR856	3AFR6
MR856	3BPR6
MR861	ST4FR10P
MR862	ST4FR20P
MR864	ST4FR30P
MR864	ST4FR40P
MR866	ST4FR60P
MR910	G1910
MR911	G1911
MR912	G1912
MR914	G1914
MR916	G1916
MR917	G1917
MR917	RG3K
MR917	RGP25K
MR917	RGP30K
MR917	S3A8P
MR917	S5A8P
MR918	G1918
MR918	RG3M
MR918	RGP25M
MR918	RGP30M
MR918	S3A10P
MR918	S5A10P
MRD160	BPW16
MRD160	BPW17
MRD160	BPX81
MRD160	FPT100,B
MRD160	FPT131
MRD160	FPT132
MRD160	FPT220
MRD160	M-161
MRD160	M-162
MRD160	STPT10
MRD160	STPT15
MRD300	BPW14
MRD300	BPX25
MRD300	BPX37
MRD300	BPX43
MRD300	BPX58
MRD300	CLR2150
MRD300	CLR2160
MRD300	FPT120,C
MRD300	FPT450A
MRD300	FPT500,A
MRD300	FPT520
MRD300	FPT520A
MRD300	FPT530A
MRD300	FPT550A
MRD300	FPT560
MRD300	QG686
MRD300	G8600,3,6,9,10
MRD300	G8680
MRD300	G8683
MRD300	G8686
MRD300	L1401
MRD300	OP803
MRD300	OP804
MRD300	OP805
MRD300	OP830
MRD300	SD5443-2
MRD300	SD5443-3
MRD300	SD5443-4
MRD300	STPT300
MRD300	TIL66
MRD300	TIL67
MRD300	TIL81
MRD3050	BP101
MRD3050	BP102
MRD3050	BPY62
MRD3050	G8612
MRD3050	G8670
MRD3050	OP801
MRD3050	STPT51
MRD3054	STPT83
MRD3055	BPX38
MRD3055	EPY62-1
MRD3055	FPT510,A
MRD3055	OP800
MRD3056	EPY62-2
MRD3056	STPT53
MRD3056	STPT80
MRD3056	STPT84
MRD310	BPX29
MRD310	CLR2110
MRD310	CLR2140
MRD310	EPY62-3
MRD310	L1402
MRD310	L1403
MRD310	OP802
MRD310	SD5443-1
MRD310	TIL63
MRD310	TIL64
MRD310	TIL65
MRD36/	SD5400-2
MRD36/	SD5400-3
MRD360	BPW30
MRD360	BPX59
MRD360	CLR2060
MRD360	CLR2180
MRD360	FPT400
MRD360	FPT570
MRD360	L14P1
MRD360	STPT260
MRD360	STPT310
MRD370	BPX25A
MRD370	BPX29A
MRD370	CLR2050
MRD370	CLR2170
MRD370	L14F2
MRD370	SD5400-1
MRD450	BPX70,C,D,E
MRD450	BPX72,C,D,E
MRD450	LPT
MRD450	LPT100A
MRD450	LPT100B
MRD450	M-163
MRD450	M-164
MRD450	M-165
MRD450	OP500
MRD450	STPT45
MRD450	TIL78
MRD500	SD5420-2
MRD510	SD3420-2
MRP1008MB	DME7
MRP1015MB	DM10P
MRP1015MB	DME10
MRP1015MB	TPR10
MRP1035MB	DM30P
MRP1035MB	DME25
MRP1090MB	DM50P
MRP1090MB	DME50
MRP1090MB	TPR50
MRP1150MB	DME150
MRP1150MB	TPR150
MRP1250M	DME250
MRP1250M	MSC1250M
MRP1325M	DME375
MRP1325M	MSC1325M
MRP2001	CTC2001
MRP2001	MSC2001
MRP2001	MSC82001
MRP2001	MSC82201
MRP2001	TRW2001
MRP2003	CTC2002
MRP2003	CTC2003
MRP2003	MSC2003
MRP2003	MSC2302
MRP2003	MSC82003
MRP2003	MSC82203
MRP2003	TRW2003
MRP2003M	MRA1720-2
MRP2003M	MRAL1720-2
MRP2003M	MRAL2023-1.5
MRP2003M	MRAL2023-3
MRP2005	CTC2005
MRP2005	MSC2005
MRP2005	MSC2304
MRP2005	MSC82005
MRP2005	TRW2005
MRP2005M	MRA1720-5
MRP2005M	MRAL1720-5
MRP2005M	MSC82005M
MRP2005M	MSC82304M
MRP2010	CTC2010
MRP2010	MSC2010
MRP2010	MSC2307
MRP2010	MSC82010
MRP2010	TRW2010
MRP2010M	MRA1720-9
MRP2010M	MRAL1720-9
MRP2010M	MRAL2023-6
MRP2010M	MSC82012M
MRP2010M	MSC82310M
MRP2016M	MRA1720-20
MRP2016M	MRAL1720-20
MRP2016M	MRAL2023-12
MRP2016M	MSC82020M
MRP2016M	MSC82313M
MRP207	SD1080
MRP208	BLW64
MRP209	CD1803
MRP209	SD1218
MRP212	B8-12
MRP212	BLY87A
MRP212	PT4544
MRP212	PT8828
MRP212	SD1133
MRP212	SD1133-1
MRP212	SD1143
MRP212	2N5995
MRP215	BM15-12
MRP215	J04020
MRP216	BM30-12
MRP216	J04030
MRP216	J04036
MRP216	J04040
MRP216	J04045
MRP216	RF2081
MRP216	SD1415
MRP216	SD1428
MRP221	PT8875P
MRP221	RP1003
MRP222	SD1229-1
MRP222	2SC1946
MRP223	RP1004
MRP223	RF2135
MRP224	PT8874F
MRP224	RF2144
MRP224	SD1018-4
MRP224	SD1134-1
MRP225	2SC628
MRP225	BLW75
MRP226	CD1802
MRP226	SD1262
MRP226	2SC1605A
MRP226	2SC1729
MRP227	2SC1589
MRP227	2SC1970
MRP230	PT3501
MRP230	SD1068
MRP231	PT8717
MRP231	2N5846
MRP231	2N6366
MRP232	PT3503
MRP232	PT8861
MRF232	PT8861A
MRF232	2N5992
MRF233	PT5695
MRF233	PT8769
MRF233	SD1014
MRF234	PT4555
MRF234	PT8863
MRF234	PT8863A
MRF234	2N5993
MRF234	2SC1329
MRF237	PT8866
MRF237	PT8877
MRF238	RF2123
MRF239	SD1272
MRF243	BLY90
MRF243	BM45-12
MRF243	SD1427
MRF245	BM70-12
MRF245	BM80-12
MRF245	J04080
MRF245	MRP203
MRF245	RF2127
MRF245	SD1124
MRF247	J04070
MRF247	J04075
MRF247	SD1416
MRF247	2SC2097
MRF260	2SC1590
MRF262	2SC1591
MRF313	C1-28
MRF313	SD1020-7
MRF313A	BLX91
MRF313A	SD1020-6
MRF313A	SD1461
MRF313A	SD1462
MRF314	A25-28
MRF314	B12-28
MRF314	B25-28
MRF314	BAM20
MRF314	PT9731
MRF314	PT9734
MRF315	B40-28
MRF315	BAM40
MRF315	CD3400
MRF315	CD3550
MRF315	J01006
MRF315	PT9733
MRF315	SD1015
MRF316	BAM80
MRF316	BM80-28
MRF316	CD3401
MRF316	SD1219
MRF316	SD1487
MRF317	BAM120
MRF317	BM100-28
MRF317	CD1752
MRF317	CD3403
MRF317	PT9782
MRF317	PT9782A
MRF317	SD1019
MRF317	SD1438
MRF317	SD1480
MRF321	BLX93
MRF321	BLX96
MRF321	BLX97
MRF321	C12-28
MRF321	CD1979
MRF321	CD2088
MRF321	CD2810
MRF321	CD5918
MRF321	MRF5175
MRF321	PHO412H
MRF321	PHO506H
MRF321	PHO512H
MRF321	PT9735
MRF321	SD1148
MRF321	SD1245
MRF321	SD5918
MRF321	2SC1804
MRF321	BLX98
MRF323	C25-28
MRF323	CD2089
MRF323	CD5919A
MRF323	MRF5176
MRF323	PT9702
MRF323	SD1149
MRF323	2N5919A
MRF323	2SC1805
MRF325	CM25-28
MRF325	J02009
MRF325	MRP305
MRF325	PHO425H
MRF325	PHO525H
MRF325	SD1464
MRF326	2N6361
MRF326	BLX95
MRF326	C40-28
MRF326	CM45-28
MRF326	J02014
MRF326	J02401
MRF326	SD1299
MRF327	SD1466
MRF327	SD1467
MRF327	CM100-28
MRF327	C2N70-28R
MRF327	CM80-28
MRF327	CM80-28R
MRF327	J02015A
MRF327	J02016
MRF327	SD1468
MRF328	2SC1606
MRF328	SD1469
MRF338A	SD1409
MRF340	SD1013
MRF342	2SC1945
MRF342	2SC2207
MRF401	PT9788
MRF401	PT9788A
MRF401	2N5707
MRF402	PT5701
MRF402	SD1020
MRF403	SD1347-7
MRF403	A3-12
MRF406	SD1285
MRF406	2N6455
MRF406	2N6456
MRF412	PT9847
MRF412	2SC2100
MRF421	PT9785
MRF421	S100-12
MRF421	SD1295
MRF421	SD1449
MRF421	2SC2290
MRF422	S100-28
MRF422	S175-28
MRF422	SD1407
MRF422	SD1450
MRF426A	BLX13
MRF427	S15-50
MRF427	SD1404
MRF428	PT9790
MRF428	S100-50
MRF428	S175-50
MRF428	SD1403
MRF428A	BLX15
MRF428A	CTC015
MRF428A	PT9790A
MRF433	PT9795
MRF433	S10-12
MRF433	2N6367
MRF433	2SC2395
MRF449	PT9796
MRF449A	PT9796A
MRF449A	SD1424
MRF450	CD2545
MRF450	PT9797
MRF450	RP2125
MRF450	S50-12
MRF450A	BLW60
MRF450A	PT4556
MRF450A	PT9797A
MRF453	MRF451
MRF453	MRF452
MRF453	SD1074
MRF453	SD1289
MRF453	SD1451
MRF453A	SD1288
MRF454	CD7012
MRF454	MRF420
MRF454	RP2143
MRF454	S80-12
MRF454	SD1076
MRF454	SD1078
MRF454	2N5460
MRF455	2N6459
MRF455	PT9776
MRF455	PT9784
MRF455A	PT9776A
MRF455A	PT9784A
MRF458	SD1405
MRF458	SD1452
MRF460	MRF418
MRF460	RP2092
MRF460	2N6368
MRF460	2N6457
MRF460	2N6458
MRF463	2N4130
MRF463	2N5942
MRF464	PT9780
MRF464	S50-28
MRF464	2N6025
MRF464A	BLX14
MRF464A	CTC14
MRF464A	PT9780A
MRF465	PT9785
MRF466	SD1224-4
MRF466	2N5708
MRF466	2N5941
MRF475	RP2147
MRF475	SD1214-4
MRF475	2S6688
MRF475	2SC1969
MRF476	RP2146
MRF476	SD1212-4
MRF485	2SC1307
MRF492	PT8854
MRF492A	PT8854A
MRF511	BFR63
MRF511	BFR64
MRF511	BFR65
MRF511	BFR94
MRF511	LT2001
MRF511	MM8003
MRF511	MRF504
MRF511	NE22120
MRF511	NE22154
MRF511	NET4020
MRF511	NET4054
MRF511	PT4574
MRF511	SD1005
MRF511	SD1315
MRF511	2N6135
MRF511	2SC1043
MRF511	2SC1251
MRF515	2SC2040
MRF515	2N5697
MRF515	2SC890
MRF517	A210
MRF517	BFR95
MRF517	BFW16A
MRF517	BFW17A
MRF517	LT1001
MRF517	MRF519
MRF517	NET4113
MRF517	NET4114
MRF517	PT4578
MRF517	SD1232
MRF517	2SC1252
MRF517	2SC1365
MRF517	2SC1366
MRF5174	BLX92
MRF5174	PHO401H
MRF5174	PT9700
MRF5174	2N5773
MRF5174	2N5917
MRF5174	2N6202
MRF5175	C3-28
MRF5175	CD2035
MRF5175	CD2087
MRF5175	PHO403H
MRF5175	PHO406H
MRF5175	PHO501H
MRF5175	PHO503H
MRF5175	PT9701
MRF5175	SD1147
MRF5175	2N6203
MRF5177	MM5177
MRF5177	2N5775
MRF5177	2N5916
MRF5177	2N6104
MRF5177	2N6205
MRF5177	BLX94A
MRF5177A	CD6105
MRF5177A	CD6105A

MOTOR.	Industry Standard No.
MRF5177A	J02000
MRF5177A	J02005
MRF5177A	PT9704
MRF5177A	P9704A
MRF5177A	SD1465
MRF5177A	2N6105
MRF5177A	28C1806
MRP526	PT8889
MRP531	2N1491
MRP531	2N3119
MRP531	2N4192
MRP531	2N4193
MRP531	2N5262
MRP559	SD1402
MRP604	SD1080-4
MRP606	2N5422
MRP607	PT8740A
MRP607	SD1115-4
MRP607	2N5913
MRP607	28C320
MRP607	28C822
MRP616	C1-12
MRP616	C1/2-12
MRP626	CHE
MRP626	SD1080-7
MRP627	SD1080-6
MRP628	SD1080-2
MRP628	28C1604
MRP629	BLX65
MRP629	PT8740
MRP629	SD1131
MRP629	SD1444
MRP641	CM10-12A
MRP641	MRP618
MRP641	SD1087
MRP641	SD1429
MRP641	28C1968
MRP644	CM20-12A
MRP644	J03020
MRP644	J03025
MRP644	MRP619
MRP644	MRP620
MRP646	SD1088
MRP646	CM30-12A
MRP646	CM45-12A
MRP646	J03030
MRP646	J03037
MRP646	J03040
MRP646	MRP621
MRP646	SD1089
MRP646	SD1434
MRP646	28C2132
MRP648	CM50-12A
MRP648	CM60-12A
MRP648	J03055
MRP648	SD1460
MRP8004	MM8004
MRP8004	SD1077
MRP8004	SD1425
MRP8004	28C1239
MRP816	D1/2-12
MRP817	D1-12B
MRP817	D2-12B
MRP838	N80801
MRP838A	D1-12E
MRP838A	DHE
MRP840	DMB10-12BA
MRP840	DMB5-12BA
MRP840	N80803
MRP840	N80810
MRP840	SD1095
MRP840	SD1410
MRP840	8MOB10
MRP840	8MOB5
MRP841	SD1412
MRP842	DMB20-12BA
MRP842	SD1096
MRP842	SD1421
MRP842	8MOB15
MRP844	DM30-12BA
MRP844	SD1098
MRP844	8MOB30
MRP846	DMB45-12BA
MRP846	SD1099
MRP846	8MOB45
MRP870	8MOB2
MRP870A	D2-12E
MRP901	AT25
MRP901	AT25A
MRP901	AT25B
MRP901	NE62135
MRP901	NE32730
MRP901	NE41735
MRP902	AT2625
MRP902	28C1119
MRP902	28C1336
MRP904	A400
MRP904	AT0017
MRP904	AT0017A
MRP904	AT004
MRP904	AT0045
MRP904	LT3072
MRP904	NE02112
MRP904	NE32712
MRP904	NE41712
MRP904	2N6595
MRP904	2N6596
MRP904	28C1988
MRP904	28C988A
MRP905	MM1500
MRP905	MM1500A
MRP905	MM1501
MRP905	MM1501A
MRP905	MM8008
MRP905	MM8010
MRP905	MM8011
MRP905	NE57510
MRP905	NE64310
MRP905	PH8193
MRP905	SD1308
MRP905	28C1600
MRP914	A401
MRP914	2N6597
MRP914	2N6598
MRP971	BPF24
MRP962	A561
MRP962	AT2715
MRP962	NE41603
MRP962	NE41607
MRP962	28C1949
MRP965	A406
MRP965	NE41610
MRP965	2N6599
MRP965	2N6600
MRB5100	FPA103,4,5
MRB5100	H15B1,2
MTM1224	MTM1224
MTM1225	MTM1225
MTM474	MTM474
MTM475	MTM475
MTM564	MTM564
MTM565	MTM565
MTP1224	MTP1224
MTP1225	MTP1225
MTP474	MTP474
MTP475	MTP475
MTP564	MTP564
MTP565	MTP565
MU10	MU10
MU20	MU20
MU2646	MU2646
MU4891	MU4891
MU4891	2N4891
MU4892	MU4892
MU4892	2N4892
MU4893	MU4893
MU4893	2N4893
MU4894	MU4894
MU4894	2N4894
MV104	MV104
MV104Q	BB104Q
MV104Q	MV104Q
MV109	MV109
MV1401	MV1401
MV1403	MV1403
MV1403H	MV1403H
MV1404	MV1404
MV1404H	MV1404H
MV1405	MV1405
MV1405H	MV1405H
MV1620	MV1620
MV1622	MV1622
MV1624	MV1624
MV1626	MV1626
MV1628	MV1628
MV1630	MV1630
MV1632	MV1632
MV1632	1N3945
MV1634	MV1634
MV1636	BA102
MV1636	MV1636
MV1638	MV1638
MV1640	MV1640
MV1642	MV1642
MV1642	1N3551
MV1644	BA111
MV1644	MV1644
MV1646	MV1646
MV1648	MV1648
MV1650	MV1650
MV1652	MV1652
MV1654	MV1654
MV1666	MV1666
MV1866	MV1866
MV1866	PG210
MV1866	PG310
MV1868	MV1868
MV1868	PG215
MV1868	PG315
MV1870	MV1870
MV1871	MV1871
MV1872	MV1872
MV1872	PG222
MV1872	PG322
MV1874	MV1874
MV1876	MV1876
MV1876	PG233
MV1876	PG333
MV1877	MV1877
MV1877	PG239
MV1877	PG339
MV1878	MV1878
MV1878	PG247
MV1878	PG347
MV209	MV209
MV2101	MV2101
MV2102	MV2102
MV2103	MV2103
MV2104	MV2104
MV2105	MV2105
MV2106	MV2106
MV2107	MV2107
MV2108	MV2108
MV2109	MV2109
MV2110	MV2110
MV2111	MV2111
MV2112	MV2112
MV2113	MV2113
MV2114	MV2114
MV2115	MV2115
MV2201	MV2201
MV2203	MV2203
MV2205	MV2205
MV2209	MV2209
MV2301	MV2301
MV2302	MV2302
MV2303	MV2303
MV2304	MV2304
MV2305	MV2305
MV2306	MV2306
MV2307	MV2307
MV2308	MV2308
MV309	MV309
MV310	MV310
MV3102	MV3102
MV3103	MV3103
MV3140	MV3140
MV3141	MV3141
MV3142	MV3142
MV830	V15
MV831	MV831
MV832	MV832
MV833	MV833
MV833	V27
MV834	MV834
MV834	V35
MV835	MV835
MV835	V39
MV836	MV836
MV836	V47
MV837	MV837
MV838	MV838
MV838	V68
MV839	MV839
MV839	V82
MV840	MV840
MV840	V100
MV857	V56
MVAM108	MVAM108
MVAM109	MVAM109
MVAM115	BB113
MVAM115	MVAM115
MVAM125	MVAM125
MZ2360	SG1910
MZ2360	SG1912
MZ2360	SG1
MZ2361	SG1920
MZ2361	SS1
MZ2361	SS1-2
MZ2361	ST8567
MZ605	PRD105
MZ605	SV7401
MZ605	1N3504
MZ610	PRD110
MZ610	1N3505
MZ610	1N4894
MZ610	1N4894A
MZ610	1N4895
MZ610	1N4895A
MZ620	PRD120
MZ620	1N3502
MZ620	1N4892
MZ620	1N4892A
MZ620	1N4893
MZ620	1N4893A
MZ640	PRD140
MZ640	PRD160
MZ640	1N3501
MZ640	1N4890
MZ640	1N4890A
MZ640	1N4891
MZ640	1N4891A
MZC2.7A10	C6012
MZC47A10	C6032
NB565N	LM565CH
NB565N	LM565CN
NB565N	LM565H
NB565N	NB565A
NB565N	NB565K
NB592A	NB592A
NB592K	NB592K
NOTE 1	SN54LS00 SERIES
NOTE 1	SN74LS00 SERIES
OPI4501	OPI4501
OPI4502	OPI4502
R711X	R711X
R712X	R712X
R714X	R714X
S2800A1	S2800A1
S2800B	S2610B
S2800B	S2710B
S2800B1	2N3228
S2800B1	2N3528
S2800B1	S2800B1
S2800C1	S2800C1
S2800D	S2610D
S2800D	S2620D
S2800D	S2710D
S2800D	2N3525
S2800D	2N3529
S2800D1	S2800D1
S2800E1	S2800E1
S2800F1	S2800F1
S2800M	S2610M
S2800M	S2620M
S2800M	S2710M
S2800M	2N4101
S2800N1	S2800N1
S2800N1	S2800N1
S2800S1	S2800S1
S6200A	S6200A
S6200B	S6200B
S6200D	S6200D
S6200M	S6200M
S6210A	S6210A
S6210B	S6210B
S6210D	S6210D
S6210M	S6210M
S6220A	S6220A
S6220B	S6220B
S6220D	S6220D
S6220M	S6220M
S6401B	TIC250B
S6401D	TIC250D
S6401E	TIC250E
S6401M	TIC250M
SC136A	SC136A
SC136A	T2503P
SC136A	T2506A
SC136A	T2313A
SC136A	T2313F
SC136A	2N5754
SC136B	L2001P91
SC136B	L2001M9
SC136B	L2003M9
SC136B	Q2001M4
SC136B	Q2003P4
SC136B	Q2004P41
SC136B	Q2006P31
SC136B	Q2006P41
SC136B	SC136B
SC136B	T2306B
SC136B	T2313B
SC136B	2N5755
SC136C	SC136C
SC136C	2N5756
SC136D	L4001P91
SC136D	L4001M9
SC136D	L4003M9
SC136D	Q4001M4
SC136D	Q4003P4
SC136D	Q4006P41
SC136D	SC136D
SC136D	T2306D
SC136D	T2313D
SC136D	2N5757
SC136E	Q5004P41
SC136E	SC136E
SC136M	Q6004P41
SC136M	Q6006P51
SC136M	SC136M
SC136M	T2313M
SC141A	SC141A
SC141A1	SC141A1
SC141B	TIC226B
SC141B1	SC141B
SC141B2	SC141B2
SC141B3	SC141B3
SC141B4	SC141B4
SC141B5	SC141B5
SC141B6	SC141B6
SC141C1	SC141C
SC141C2	SC141C2
SC141C3	SC141C3
SC141C4	SC141C4
SC141C5	SC141C5
SC141C6	SC141C6
SC141D	TIC226D
SC141D1	SC141D
SC141D2	SC141D2
SC141D3	SC141D3
SC141D4	SC141D4
SC141D5	SC141D5
SC141D6	SC141D6
SC141E1	SC141E
SC141E2	SC141E2
SC141E3	SC141E3
SC141E4	SC141E4
SC141E5	SC141E5
SC141E6	SC141E6
SC141M1	SC141M
SC141M2	SC141M2
SC141M3	SC141M3
SC141M4	SC141M4
SC141M5	SC141M5
SC141M6	SC141M6
SC146A1	SC146A
SC146A2	SC146A2
SC146A3	SC146A3
SC146A4	SC146A4
SC146A5	SC146A5
SC146A6	SC146A6
SC146B	Q2010P41
SC146B1	SC146B
SC146B2	SC146B2
SC146B3	SC146B3
SC146B4	SC146B4
SC146B5	SC146B5
SC146B6	SC146B6
SC146C1	SC146C
SC146C2	SC146C2
SC146C3	SC146C3
SC146C4	SC146C4
SC146C5	SC146C5
SC146C6	SC146C6
SC146D	Q4010P41
SC146D1	SC146D
SC146D2	SC146D2
SC146D3	SC146D3
SC146D4	SC146D4
SC146D6	SC146D6
SC146E1	SC146E
SC146E2	SC146E2
SC146E3	SC146E3
SC146E4	SC146E4
SC146E5	SC146E5
SC146E6	SC146E6
SC146M	Q6010P51
SC146M1	SC146M
SC146M2	SC146M2
SC146M3	SC146M3
SC146M4	SC146M4
SC146M5	SC146M5
SC146N1	SC146N
SC146N2	SC146N2
SC146N3	SC146N3
SC146N4	SC146N4
SC146N5	SC146N5
SC146N6	SC146N6
SC146S1	SC146S
SC146S2	SC146S2
SC146S3	SC146S3
SC146S4	SC146S4
SC146S5	SC146S5
SC146S6	SC146S6
SC245A	SC245A
SC245B	SC245B
SC245C	SC245C
SC245D	SC245D
SC245E	SC245E
SC245M	SC245M
SC245N	SC245N
SC245S	SC245S
SC246A	SC246A
SC246B	SC246B
SC246C	SC246C
SC246D	SC246D
SC246E	SC246E
SC246M	SC246M
SC246N	SC246N
SC246S	SC246S
SC250A	SC250A
SC250B	Q2015H
SC250B	SC250B
SC250C	SC250C
SC250D	Q4015H
SC250D	SC250D
SC250E	Q6015H
SC250E	SC250E
SC250M	Q6015H
SC250M	SC250M
SC250N	SC250N
SC250S	SC250S
SC251A	SC251A
SC251B	Q2015Q
SC251B	SC251B
SC251C	SC251C
SC251D	Q4015Q
SC251D	SC251D
SC251E	Q5015Q
SC251E	SC251E
SC251M	Q6015Q
SC251M	SC251M
SC251N	SC251N
SC251S	SC251S
SC260A	SC260A
SC260B	Q2025H
SC260B	SC260B
SC260C	SC260C
SC260D	Q4025H
SC260D	SC260D
SC260E	Q5025H
SC260E	SC260E
SC260M	Q6025H
SC260M	SC260M
SC261A	SC261A
SC261B	Q2025Q
SC261B	SC261B

MOTOR.	Industry Standard No.
SC261C	SC261C
SC261D	Q4025G
SC261D	SC261D
SC261E	Q5025G
SC261E	SC261E
SC261M	Q6025G
SC261M	SC261M
SD241	SD241
SD41	SD-41
SD51	SD-51
SD51	VSK51
SD5171	SD-5171
SD5171	VSK51B
SDT13304	SDT13304
SDT13304	2SC1468
SDT13304	2SC1576
SDT13304	2SC1868
SDT13304	2SC2138
SDT13304	2SC2140
SDT13304	2SC2245
SDT13305	SDT13305
SDT13305	2SC1228
SDT13305	2SC1469
SDT13305	2SC1577
SDT13305	2SC2139
SDT13305	2SC2148
SDT13305	2SC2292
SDT13305	2SC2293
SDT13305	2SC2356
SDT13305	2SC2448
SDT13305	2SC2449
SDT13305	2SC2450
SDT13305	2SC2451
SDT13305	2SC2452
SDT13305	2SC2453
SDT13305	2SC2541 (T0218)
SDT13305	2SD393
SDT13305	2SD394
SDT13305	2SD395
SDT13305	2SD437
SDT13305	2SD458
SDT13305	2SD538
SE592G	SE592K
SE592L	SE592A
SE9300	SE9300
SE9301	SE9301
SE9302	SE9302
SE9303	SE9303
SE9304	SE9304
SE9305	SE9305
SE9400	SE9400
SE9401	SE9401
SE9402	SE9402
SE9403	SE9403
SE9404	SE9404
SE9405	SE9405
SN54LS00J	DM54LS00J
SN54LS00J	S54LS00F
SN54LS00W	DM54LS00W
SN54LS00W	S54LS00W
SN54LS01J	DM54LS01J
SN54LS01J	S54LS01F
SN54LS01W	DM54LS01W
SN54LS01W	S54LS01W
SN54LS02J	DM54LS02J
SN54LS02J	S54LS02F
SN54LS02W	DM54LS02W
SN54LS02W	S54LS02W
SN54LS03J	DM54LS03J
SN54LS03J	S54LS03F
SN54LS03W	DM54LS03W
SN54LS03W	S54LS03W
SN54LS04J	DM54LS04J
SN54LS04J	S54LS04F
SN54LS04W	DM54LS04W
SN54LS04W	S54LS04W
SN54LS05J	DM54LS05J
SN54LS05J	S54LS05F
SN54LS05W	DM54LS05W
SN54LS05W	S54LS05W
SN54LS08J	DM54LS08J
SN54LS08J	S54LS08F
SN54LS08W	DM54LS08W
SN54LS08W	S54LS08W
SN54LS09J	DM54LS09J
SN54LS09J	S54LS09F
SN54LS09W	DM54LS09W
SN54LS09W	S54LS09W
SN54LS10J	DM54LS10J
SN54LS10J	S54LS10F
SN54LS10W	DM54LS10W
SN54LS10W	S54LS10W
SN54LS11J	DM54LS11J
SN54LS11J	S54LS11F
SN54LS11W	DM54LS11W
SN54LS11W	S54LS11W
SN54LS12J	DM54LS12J
SN54LS12J	S54LS12F
SN54LS12W	DM54LS12W
SN54LS12W	S54LS12W
SN54LS132J	DM54LS132J
SN54LS132J	S54LS132F
SN54LS132W	DM54LS132W
SN54LS132W	S54LS132W
SN54LS136J	DM54LS136J
SN54LS136J	S54LS136F
SN54LS136W	DM54LS136W
SN54LS136W	S54LS136W
SN54LS138J	DM54LS138J
SN54LS138J	S54LS138F
SN54LS138W	DM54LS138W
SN54LS138W	S54LS138W
SN54LS139J	DM54LS139J
SN54LS139J	S54LS139F
SN54LS139W	DM54LS139W
SN54LS139W	S54LS139W
SN54LS13J	DM54LS13J
SN54LS13J	S54LS13F
SN54LS13W	DM54LS13W
SN54LS13W	S54LS13W
SN54LS145J	S54LS145F
SN54LS145W	S54LS145W
SN54LS14J	DM54LS14J
SN54LS14J	S54LS14F
SN54LS14W	DM54LS14W
SN54LS14W	S54LS14W
SN54LS151J	DM54LS151J
SN54LS151J	S54LS151F
SN54LS151W	DM54LS151W
SN54LS151W	S54LS151W
SN54LS153J	DM54LS153J
SN54LS153J	S54LS153F
SN54LS153W	DM54LS153W
SN54LS153W	S54LS153W
SN54LS155J	DM54LS155J
SN54LS155W	DM54LS155W
SN54LS156J	DM54LS156J
SN54LS156W	DM54LS156W
SN54LS157J	AM54LS157J
SN54LS157J	DM54LS157J
SN54LS157J	S54LS157F
SN54LS157W	AM54LS157W
SN54LS157W	DM54LS157W
SN54LS157W	S54LS157W
SN54LS158J	AM54LS158J
SN54LS158J	DM54LS158J
SN54LS158J	S54LS158F
SN54LS158W	AM54LS158W
SN54LS158W	DM54LS158W
SN54LS158W	S54LS158W
SN54LS15J	DM54LS15J
SN54LS15J	S54LS15F
SN54LS15W	DM54LS15W
SN54LS15W	S54LS15W
SN54LS164J	AM54LS164J
SN54LS164J	DM54LS164J
SN54LS164J	S54LS164F
SN54LS164W	AM54LS164W
SN54LS164W	DM54LS164W
SN54LS164W	S54LS164W
SN54LS168J	DM54LS168J
SN54LS168W	DM54LS168W
SN54LS169J	DM54LS169J
SN54LS169W	DM54LS169W
SN54LS170J	DM54LS170J
SN54LS170J	S54LS170F
SN54LS170W	DM54LS170W
SN54LS170W	S54LS170W
SN54LS173J	DM54LS173J
SN54LS173W	DM54LS173W
SN54LS174J	AM54LS174J
SN54LS174J	DM54LS174J
SN54LS174J	S54LS174F
SN54LS174W	AM54LS174W
SN54LS174W	DM54LS174W
SN54LS174W	S54LS174W
SN54LS175J	AM54LS175J
SN54LS175J	DM54LS175J
SN54LS175J	S54LS175F
SN54LS175W	AM54LS175W
SN54LS175W	DM54LS175W
SN54LS175W	S54LS175W
SN54LS181J	AM54LS181J
SN54LS181J	S54LS181F
SN54LS181W	AM54LS181W
SN54LS181W	S54LS181W
SN54LS189J	DM54LS189J
SN54LS189W	DM54LS189W
SN54LS190J	AM54LS190J
SN54LS190J	DM54LS190J
SN54LS190J	S54LS190F
SN54LS190W	AM54LS190W
SN54LS190W	DM54LS190W
SN54LS190W	S54LS190W
SN54LS191J	AM54LS191J
SN54LS191J	DM54LS191J
SN54LS191J	S54LS191F
SN54LS191W	AM54LS191W
SN54LS191W	DM54LS191W
SN54LS191W	S54LS191W
SN54LS192J	AM54LS192J
SN54LS192J	DM54LS192J
SN54LS192J	S54LS192F
SN54LS192W	AM54LS192W
SN54LS192W	DM54LS192W
SN54LS192W	S54LS192W
SN54LS193J	AM54LS193J
SN54LS193J	DM54LS193J
SN54LS193J	S54LS193F
SN54LS193W	AM54LS193W
SN54LS193W	DM54LS193W
SN54LS193W	S54LS193W
SN54LS196J	DM54LS196J
SN54LS196J	S54LS196F
SN54LS196W	DM54LS196W
SN54LS196W	S54LS196W
SN54LS197J	DM54LS197J
SN54LS197J	S54LS197F
SN54LS197W	DM54LS197W
SN54LS197W	S54LS197W
SN54LS20J	DM54LS20J
SN54LS20J	S54LS20F
SN54LS20W	DM54LS20W
SN54LS20W	S54LS20W
SN54LS21J	DM54LS21J
SN54LS21J	S54LS21F
SN54LS21W	DM54LS21W
SN54LS21W	S54LS21W
SN54LS221J	S54LS221F
SN54LS221W	S54LS221W
SN54LS22J	DM54LS22J
SN54LS22J	S54LS22F
SN54LS22W	DM54LS22W
SN54LS22W	S54LS22W
SN54LS248J	DM54LS248J
SN54LS248W	DM54LS248W
SN54LS249J	DM54LS249J
SN54LS249W	DM54LS249W
SN54LS251J	AM54LS251J
SN54LS251J	S54LS251F
SN54LS251W	AM54LS251W
SN54LS251W	S54LS251W
SN54LS253J	AM54LS253J
SN54LS253J	DM54LS253J
SN54LS253J	S54LS253F
SN54LS253W	AM54LS253W
SN54LS253W	DM54LS253W
SN54LS253W	S54LS253W
SN54LS260J	S54LS260F
SN54LS260W	S54LS260W
SN54LS266J	DM54LS266J
SN54LS266J	S54LS266F
SN54LS266W	DM54LS266W
SN54LS266W	S54LS266W
SN54LS26J	DM54LS26J
SN54LS26J	S54LS26F
SN54LS26W	DM54LS26W
SN54LS26W	S54LS26W
SN54LS273J	AM54LS273J
SN54LS273W	AM54LS273W
SN54LS279J	DM54LS279J
SN54LS279W	DM54LS279W
SN54LS27J	DM54LS27J
SN54LS27J	S54LS27F
SN54LS27W	DM54LS27W
SN54LS27W	S54LS27W
SN54LS283J	DM54LS283J
SN54LS283J	S54LS283F
SN54LS283W	DM54LS283W
SN54LS283W	S54LS283W
SN54LS28J	S54LS28F
SN54LS28W	S54LS28W
SN54LS293J	S54LS293F
SN54LS293W	S54LS293W
SN54LS299J	AM54LS299J
SN54LS299W	AM54LS299W
SN54LS30J	DM54LS30J
SN54LS30J	S54LS30F
SN54LS30W	DM54LS30W
SN54LS30W	S54LS30W
SN54LS322J	AM25LS22DM
SN54LS322W	AM25LS22FM
SN54LS323J	AM25LS23DM
SN54LS323W	AM25LS23FM
SN54LS32J	DM54LS32J
SN54LS32J	S54LS32F
SN54LS32W	DM54LS32W
SN54LS32W	S54LS32W
SN54LS33J	S54LS33F
SN54LS33W	S54LS33W
SN54LS353J	DM54LS353J
SN54LS353W	DM54LS353W
SN54LS374J	AM54LS374J
SN54LS374W	AM54LS374W
SN54LS377J	AM54LS377J
SN54LS377W	AM54LS377W
SN54LS378J	AM25LS07DM
SN54LS378W	AM25LS07FM
SN54LS379J	AM25LS08DM
SN54LS379W	AM25LS08FM
SN54LS37J	DM54LS37J
SN54LS37J	S54LS37F
SN54LS37W	DM54LS37W
SN54LS37W	S54LS37W
SN54LS384J	AM25LS14DM
SN54LS384W	AM25LS14FM
SN54LS385J	AM25LS15DM
SN54LS385W	AM25LS15FM
SN54LS386J	DM54LS386J
SN54LS386J	S54LS386F
SN54LS386W	DM54LS386W
SN54LS386W	S54LS386W
SN54LS388J	AM25LS2518DM
SN54LS388W	AM25LS2518FM
SN54LS38J	DM54LS38J
SN54LS38J	S54LS38F
SN54LS38W	DM54LS38W
SN54LS38W	S54LS38W
SN54LS399J	AM25LS09DM
SN54LS399W	AM25LS09FM
SN54LS40J	DM54LS40J
SN54LS40J	S54LS40F
SN54LS40W	DM54LS40W
SN54LS40W	S54LS40W
SN54LS42J	DM54LS42J
SN54LS42J	S54LS42F
SN54LS42W	DM54LS42W
SN54LS42W	S54LS42W
SN54LS47J	DM54LS47J
SN54LS47W	DM54LS47W
SN54LS48J	DM54LS48J
SN54LS48W	DM54LS48W
SN54LS49J	DM54LS49J
SN54LS49W	DM54LS49W
SN54LS51J	DM54LS51J
SN54LS51J	S54LS51F
SN54LS51W	DM54LS51W
SN54LS51W	S54LS51W
SN54LS54J	DM54LS54J
SN54LS54J	S54LS54F
SN54LS54W	DM54LS54W
SN54LS54W	S54LS54W
SN54LS55J	DM54LS55J
SN54LS55J	S54LS55F
SN54LS55W	DM54LS55W
SN54LS55W	S54LS55W
SN54LS670J	DM54LS670J
SN54LS670J	S54LS670F
SN54LS670W	DM54LS670W
SN54LS670W	S54LS670W
SN54LS73AJ	DM54LS73J
SN54LS73AJ	S54LS73F
SN54LS73AW	DM54LS73W
SN54LS73AW	S54LS73W
SN54LS74AJ	DM54LS74J
SN54LS74AJ	S54LS74F
SN54LS74AW	DM54LS74W
SN54LS74AW	S54LS74W
SN54LS75J	DM54LS75J
SN54LS75J	S54LS75F
SN54LS75W	DM54LS75W
SN54LS75W	S54LS75W
SN54LS76AJ	DM54LS76J
SN54LS76AJ	S54LS76F
SN54LS76AW	DM54LS76W
SN54LS76AW	S54LS76W
SN54LS77J	DM54LS77J
SN54LS77W	DM54LS77W
SN54LS78AJ	DM54LS78J
SN54LS78AJ	S54LS78F
SN54LS78AW	DM54LS78W
SN54LS78AW	S54LS78W
SN54LS795J	DM71LS95J
SN54LS795W	DM71LS95W
SN54LS796J	DM71LS96J
SN54LS796W	DM71LS96W
SN54LS797J	DM71LS97J
SN54LS797W	DM71LS97W
SN54LS798J	DM71LS98J
SN54LS798W	DM71LS98W
SN54LS83AJ	DM54LS83AJ
SN54LS83AW	DM54LS83AW
SN54LS85J	DM54LS85J
SN54LS85J	S54LS85F
SN54LS85W	DM54LS85W
SN54LS85W	S54LS85W
SN54LS86J	DM54LS86J
SN54LS86J	S54LS86F
SN54LS86W	DM54LS86W
SN54LS86W	S54LS86W
SN54LS90J	S54LS90F
SN54LS90W	S54LS90W
SN54LS92J	S54LS92F
SN54LS92W	S54LS92W
SN54LS93J	S54LS93F
SN54LS93W	S54LS93W
SN54LS95BJ	S54LS95BF
SN54LS95BW	S54LS95BW
SN74LS00J	DM74LS00J
SN74LS00J	N74LS00F
SN74LS00N	DM74LS00N
SN74LS00N	N74LS00N
SN74LS01J	DM74LS01J
SN74LS01J	N74LS01F
SN74LS01N	DM74LS01N
SN74LS01N	N74LS01N
SN74LS02J	DM74LS02J
SN74LS02J	N74LS02F
SN74LS02N	DM74LS02N
SN74LS02N	N74LS02N
SN74LS03J	DM74LS03J
SN74LS03J	N74LS03F
SN74LS03N	DM74LS03N
SN74LS03N	N74LS03N
SN74LS04J	DM74LS04J
SN74LS04J	N74LS04F
SN74LS04N	DM74LS04N
SN74LS04N	N74LS04N
SN74LS05J	DM74LS05J
SN74LS05J	N74LS05F
SN74LS05N	DM74LS05N
SN74LS05N	N74LS05N
SN74LS08J	DM74LS08J
SN74LS08J	N74LS08F
SN74LS08N	DM74LS08N
SN74LS08N	N74LS08N
SN74LS09J	DM74LS09J
SN74LS09J	N74LS09F
SN74LS09N	DM74LS09N
SN74LS09N	N74LS09N
SN74LS10J	DM74LS10J
SN74LS10J	N74LS10F
SN74LS10N	DM74LS10N
SN74LS10N	N74LS10N
SN74LS11J	DM74LS11J
SN74LS11J	N74LS11F
SN74LS11N	DM74LS11N
SN74LS11N	N74LS11N
SN74LS12J	DM74LS12J
SN74LS12J	N74LS12F
SN74LS12N	DM74LS12N
SN74LS12N	N74LS12N
SN74LS132J	DM74LS132J
SN74LS132J	N74LS132F
SN74LS132N	DM74LS132N
SN74LS132N	N74LS132N
SN74LS136J	DM74LS136J
SN74LS136J	N74LS136F
SN74LS136N	DM74LS136N
SN74LS136N	N74LS136N
SN74LS138J	AM74LS138J
SN74LS138J	DM74LS138J
SN74LS138J	N74LS138F
SN74LS138N	AM74LS138N
SN74LS138N	DM74LS138N
SN74LS138N	N74LS138N
SN74LS139J	AM74LS139J
SN74LS139J	DM74LS139J
SN74LS139J	N74LS139F
SN74LS139N	AM74LS139N
SN74LS139N	DM74LS139N
SN74LS139N	N74LS139N
SN74LS13J	DM74LS13J
SN74LS13J	N74LS13F
SN74LS13N	DM74LS13N
SN74LS13N	N74LS13N
SN74LS145J	N74LS145F
SN74LS145N	N74LS145N
SN74LS14J	DM74LS14J
SN74LS14J	N74LS14F
SN74LS14N	DM74LS14N
SN74LS14N	N74LS14N
SN74LS151J	AM74LS151J
SN74LS151J	DM74LS151J
SN74LS151J	N74LS151F
SN74LS151N	AM74LS151N
SN74LS151N	DM74LS151N
SN74LS151N	N74LS151N
SN74LS153J	AM74LS153J
SN74LS153J	DM74LS153J
SN74LS153J	N74LS153F
SN74LS153N	AM74LS153N
SN74LS153N	DM74LS153N
SN74LS153N	N74LS153N
SN74LS155J	DM74LS155J
SN74LS155N	DM74LS155N
SN74LS156J	DM74LS156J
SN74LS156N	DM74LS156N
SN74LS157J	AM74LS157J
SN74LS157J	DM74LS157J
SN74LS157J	N74LS157F
SN74LS157N	AM74LS157N
SN74LS157N	DM74LS157N
SN74LS157N	N74LS157N
SN74LS158J	AM74LS158J
SN74LS158J	DM74LS158J
SN74LS158J	N74LS158F
SN74LS158N	AM74LS158N
SN74LS158N	DM74LS158N
SN74LS158N	N74LS158N
SN74LS15J	DM74LS15J
SN74LS15J	N74LS15F
SN74LS15N	DM74LS15N
SN74LS15N	N74LS15N
SN74LS164J	AM74LS164J
SN74LS164J	DM74LS164J
SN74LS164J	N74LS164F
SN74LS164N	AM74LS164N
SN74LS164N	DM74LS164N
SN74LS164N	N74LS164N
SN74LS168J	DM74LS168J
SN74LS168N	DM74LS168N
SN74LS169J	DM74LS169J
SN74LS169N	DM74LS169N
SN74LS170J	DM74LS170J
SN74LS170J	N74LS170F
SN74LS170N	DM74LS170N
SN74LS170N	N74LS170N
SN74LS173J	DM74LS173J
SN74LS173N	AM74LS173N
SN74LS174J	AM74LS174J
SN74LS174J	DM74LS174J
SN74LS174J	N74LS174F
SN74LS174N	AM74LS174N
SN74LS174N	DM74LS174N
SN74LS174N	N74LS174N
SN74LS175J	AM74LS175J
SN74LS175J	DM74LS175J
SN74LS175J	N74LS175F
SN74LS175N	AM74LS175N
SN74LS175N	DM74LS175N
SN74LS175N	N74LS175N
SN74LS181J	AM74LS181J
SN74LS181J	N74LS181F
SN74LS181N	AM74LS181N
SN74LS181N	N74LS181N

MOTOR.	Industry Standard No.
SN74LS189J	DM74LS189J
SN74LS189N	DM74LS189N
SN74LS190J	AM74LS190J
SN74LS190J	DM74LS190J
SN74LS190J	N74LS190F
SN74LS190N	AM74LS190N
SN74LS190N	DM74LS190N
SN74LS190N	N74LS190N
SN74LS191J	AM74LS191J
SN74LS191J	DM74LS191J
SN74LS191J	N74LS191F
SN74LS191N	AM74LS191N
SN74LS191N	DM74LS191N
SN74LS191N	N74LS191N
SN74LS192J	AM74LS192J
SN74LS192J	DM74LS192J
SN74LS192J	N74LS192F
SN74LS192N	AM74LS192N
SN74LS192N	DM74LS192N
SN74LS192N	N74LS192N
SN74LS193J	AM74LS193J
SN74LS193J	DM74LS193J
SN74LS193J	N74LS193F
SN74LS193N	AM74LS193N
SN74LS193N	DM74LS193N
SN74LS193N	N74LS193N
SN74LS196J	DM74LS196J
SN74LS196J	N74LS196F
SN74LS196N	DM74LS196N
SN74LS196N	N74LS196N
SN74LS197J	DM74LS197J
SN74LS197J	N74LS197F
SN74LS197N	DM74LS197N
SN74LS197N	N74LS197N
SN74LS20J	DM74LS20J
SN74LS20J	N74LS20F
SN74LS20N	DM74LS20N
SN74LS20N	N74LS20N
SN74LS21J	DM74LS21J
SN74LS21J	N74LS21F
SN74LS21N	DM74LS21N
SN74LS21N	N74LS21N
SN74LS221J	N74LS221F
SN74LS221N	N74LS221N
SN74LS22J	DM74LS22J
SN74LS22J	N74LS22F
SN74LS22N	DM74LS22N
SN74LS22N	N74LS22N
SN74LS248J	DM74LS248J
SN74LS248N	DM74LS248N
SN74LS249J	DM74LS249J
SN74LS249N	DM74LS249N
SN74LS251J	AM74LS251J
SN74LS251J	N74LS251F
SN74LS251N	AM74LS251N
SN74LS251N	N74LS251N
SN74LS253J	AM74LS253J
SN74LS253J	DM74LS253J
SN74LS253J	N74LS253F
SN74LS253N	AM74LS253N
SN74LS253N	DM74LS253N
SN74LS253N	N74LS253N
SN74LS260J	N74LS260F
SN74LS260N	N74LS260N
SN74LS266J	DM74LS266J
SN74LS266J	N74LS266F
SN74LS266N	DM74LS266N
SN74LS266N	N74LS266N
SN74LS26J	DM74LS26J
SN74LS26J	N74LS26F
SN74LS26N	DM74LS26N
SN74LS26N	N74LS26N
SN74LS273J	AM74LS273J
SN74LS273N	AM74LS273N
SN74LS279J	DM74LS279J
SN74LS279N	DM74LS279N
SN74LS27J	DM74LS27J
SN74LS27J	N74LS27F
SN74LS27N	DM74LS27N
SN74LS27N	N74LS27N
SN74LS283J	DM74LS283J
SN74LS283J	N74LS283F
SN74LS283N	DM74LS283N
SN74LS283N	N74LS283N
SN74LS289J	DM74LS289J
SN74LS289N	DM74LS289N
SN74LS28J	N74LS28F
SN74LS28N	N74LS28N
SN74LS293J	N74LS293F
SN74LS293N	N74LS293N
SN74LS299J	AM74LS299J
SN74LS299N	AM74LS299N
SN74LS30J	DM74LS30J
SN74LS30J	N74LS30F
SN74LS30N	DM74LS30N
SN74LS30N	N74LS30N
SN74LS322J	AM25LS22DC
SN74LS322N	AM25LS22PC
SN74LS323J	AM25LS23DC
SN74LS323N	AM25LS23PC
SN74LS32J	DM74LS32J
SN74LS32J	N74LS32F
SN74LS32N	DM74LS32N
SN74LS32N	N74LS32N
SN74LS33J	N74LS33F
SN74LS33N	N74LS33N
SN74LS353J	DM74LS353J
SN74LS353N	DM74LS353N
SN74LS374J	AM74LS374J
SN74LS374N	AM74LS374N
SN74LS377J	AM74LS377J
SN74LS377N	AM74LS377N
SN74LS378J	AM25LS07DC
SN74LS378N	AM25LS07PC
SN74LS379J	AM25LS08DC
SN74LS379N	AM25LS08PC
SN74LS37J	DM74LS37J
SN74LS37J	N74LS37F
SN74LS37N	DM74LS37N
SN74LS37N	N74LS37N
SN74LS384J	AM25LS14DC
SN74LS384N	AM25LS14PC
SN74LS385J	AM25LS15DC
SN74LS385N	AM25LS15PC
SN74LS386J	DM74LS386J
SN74LS386J	N74LS386F
SN74LS386N	DM74LS386N
SN74LS386N	N74LS386N
SN74LS388J	AM25LS2518DC
SN74LS388N	AM25LS2518PC
SN74LS38J	DM74LS38J
SN74LS38J	N74LS38F
SN74LS38N	DM74LS38N
SN74LS38N	N74LS38N
SN74LS399J	AM25LS09DC
SN74LS399N	AM25LS09PC
SN74LS40J	DM74LS40J
SN74LS40J	N74LS40F
SN74LS40N	DM74LS40N
SN74LS40N	N74LS40N
SN74LS42J	DM74LS42J
SN74LS42J	N74LS42F
SN74LS42N	DM74LS42N
SN74LS42N	N74LS42N
SN74LS47J	DM74LS47J
SN74LS47N	DM74LS47N
SN74LS48J	DM74LS48J
SN74LS48N	DM74LS48N
SN74LS49J	DM74LS49J
SN74LS49N	DM74LS49N
SN74LS51J	DM74LS51J
SN74LS51J	N74LS51F
SN74LS51N	DM74LS51N
SN74LS51N	N74LS51N
SN74LS54J	DM74LS54J
SN74LS54J	N74LS54F
SN74LS54N	DM74LS54N
SN74LS54N	N74LS54N
SN74LS55J	DM74LS55J
SN74LS55J	N74LS55F
SN74LS55N	DM74LS55N
SN74LS55N	N74LS55N
SN74LS670J	DM74LS670J
SN74LS670J	N74LS670F
SN74LS670N	DM74LS670N
SN74LS670N	N74LS670N
SN74LS73AJ	DM74LS73J
SN74LS73AJ	N74LS73F
SN74LS73AN	DM74LS73N
SN74LS73AN	N74LS73N
SN74LS74AJ	DM74LS74J
SN74LS74AJ	N74LS74F
SN74LS74AN	DM74LS74N
SN74LS74AN	N74LS74N
SN74LS75J	DM74LS75J
SN74LS75J	N74LS75F
SN74LS75N	DM74LS75N
SN74LS75N	N74LS75N
SN74LS76AJ	DM74LS76J
SN74LS76AJ	N74LS76F
SN74LS76AN	DM74LS76N
SN74LS76AN	N74LS76N
SN74LS77J	DM74LS77J
SN74LS77N	DM74LS77N
SN74LS78AJ	DM74LS78J
SN74LS78AJ	N74LS78F
SN74LS78AN	DM74LS78N
SN74LS78AN	N74LS78N
SN74LS795J	DM81LS95J
SN74LS795N	DM81LS95N
SN74LS796J	DM81LS96J
SN74LS796N	DM81LS96N
SN74LS797J	DM81LS97J
SN74LS797N	DM81LS97N
SN74LS798J	DM81LS98J
SN74LS798N	DM81LS98N
SN74LS83AJ	DM74LS83AJ
SN74LS83AN	DM74LS83AN
SN74LS85J	DM74LS85J
SN74LS85J	N74LS85F
SN74LS85N	DM74LS85N
SN74LS85N	N74LS85N
SN74LS86J	DM74LS86J
SN74LS86J	N74LS86F
SN74LS86N	DM74LS86N
SN74LS86N	N74LS86N
SN74LS90J	N74LS90F
SN74LS90N	N74LS90N
SN74LS92J	N74LS92F
SN74LS92N	N74LS92N
SN74LS93J	N74LS93F
SN74LS93N	N74LS93N
SN74LS95BJ	N74LS95BF
SN74LS95BN	N74LS95BN
SPECIAL	BYW30-100
SPECIAL	BYW30-150
SPECIAL	BYW30-50
SPECIAL	BYW31-100
SPECIAL	BYW31-150
SPECIAL	BYW31-50
SPECIAL	C118M
SPECIAL	FWLC1000
SPECIAL	FWLC500
SPECIAL	FWLC600
SPECIAL	FWLC800
SPECIAL	FWLD1000
SPECIAL	FWLD500
SPECIAL	FWLD600
SPECIAL	FWLD800
SPECIAL	NSR7140
SPECIAL	NSR7141
SPECIAL	NSR7142
SPECIAL	NSR7143
SPECIAL	NSR8140
SPECIAL	NSR8141
SPECIAL	NSR8142
SPECIAL	NSR8143
SPECIAL	S0303M
SPECIAL	S0304F1
SPECIAL	S0306F1
SPECIAL	S0310F1
SPECIAL	S0310F3
SPECIAL	S0315G
SPECIAL	S0315H
SPECIAL	S0325G
SPECIAL	S0503M
SPECIAL	S0504F1
SPECIAL	S0506F1
SPECIAL	S0510F1
SPECIAL	S0510F3
SPECIAL	S0515G
SPECIAL	S0515H
SPECIAL	S0525G
SPECIAL	S1003M
SPECIAL	S1004F1
SPECIAL	S1006F1
SPECIAL	S1010F1
SPECIAL	S1010F3
SPECIAL	S1015G
SPECIAL	S1015H
SPECIAL	S1025G
SPECIAL	S1A12F
SPECIAL	S2003M
SPECIAL	S2004F1
SPECIAL	S2006F1
SPECIAL	S2010F1
SPECIAL	S2010F3
SPECIAL	S2015G
SPECIAL	S2015H
SPECIAL	S2025G
SPECIAL	S3A12F
SPECIAL	S4003M
SPECIAL	S4004F1
SPECIAL	S4006F1
SPECIAL	S4010F1
SPECIAL	S4010F3
SPECIAL	S4015G
SPECIAL	S4015H
SPECIAL	S4025G
SPECIAL	S5A12F
SPECIAL	S6001M
SPECIAL	S6003M
SPECIAL	S6004F1
SPECIAL	S6006F1
SPECIAL	S6010F1
SPECIAL	S6010F3
SPECIAL	S6015G
SPECIAL	S6015H
SPECIAL	S6025G
SPECIAL	S6230A
SPECIAL	S6230B
SPECIAL	S6230D
SPECIAL	S6230M
SPECIAL	S6240A
SPECIAL	S6240B
SPECIAL	S6240D
SPECIAL	S6240M
SPECIAL	S6250A
SPECIAL	S6250B
SPECIAL	S6250D
SPECIAL	S6250M
SPECIAL	S6430A
SPECIAL	S6430B
SPECIAL	S6430D
SPECIAL	S6430M
SPECIAL	S6493M
SPECIAL	SBR10A10
SPECIAL	SBR10A8
SPECIAL	SBR6A10
SPECIAL	SBR6A8
SPECIAL	SC149E
SPECIAL	SDA980-10
SPECIAL	SDA980-8
SPECIAL	SEN2A1FR
SPECIAL	SEN2A2FR
SPECIAL	SEN2A4FR
SPECIAL	SEN2A6FR
SPECIAL	SEN2A8FR
SPECIAL	SEN3A1FR
SPECIAL	SEN3A2FR
SPECIAL	SEN3A4FR
SPECIAL	SEN3A6FR
SPECIAL	SEN3A8FR
SPECIAL	SES5001
SPECIAL	SES5002
SPECIAL	SES5003
SPECIAL	SES5301
SPECIAL	SES5302
SPECIAL	SES5303
SPECIAL	SES5401
SPECIAL	SES5401C
SPECIAL	SES5402
SPECIAL	SES5402C
SPECIAL	SES5403
SPECIAL	SES5403C
SPECIAL	SES5601C
SPECIAL	SES5602C
SPECIAL	SES5603C
SPECIAL	SES5701
SPECIAL	SES5702
SPECIAL	SES5703
SPECIAL	SES5801
SPECIAL	SES5802
SPECIAL	SES5803
SPECIAL	SF10
SPECIAL	SF11
SPECIAL	SF12
SPECIAL	SF315
SPECIAL	SF3151
SPECIAL	SF3152
SPECIAL	SF330
SPECIAL	SF3301
SPECIAL	SF3302
SPECIAL	SF415
SPECIAL	SF4151
SPECIAL	SF4152
SPECIAL	SF46
SPECIAL	SF461
SPECIAL	SF462
SPECIAL	SF515
SPECIAL	SF5151
SPECIAL	SF5152
SPECIAL	SF530
SPECIAL	SF5301
SPECIAL	SF5302
SPECIAL	SR710
SPECIAL	SR710F
SPECIAL	SR711
SPECIAL	SR711F
SPECIAL	SR712
SPECIAL	SR712F
SPECIAL	SR713
SPECIAL	SR713F
SPECIAL	SR714
SPECIAL	SR714F
SPECIAL	SR716
SPECIAL	SR716F
SPECIAL	T12A6F
SPECIAL	T1C35
SPECIAL	T1C36
SPECIAL	T1C67
SPECIAL	T1C68
SPECIAL	T20A6F
SPECIAL	T30A6F
SPECIAL	T3889
SPECIAL	T3890
SPECIAL	T3891
SPECIAL	T3892
SPECIAL	T3893
SPECIAL	T3899
SPECIAL	T3900
SPECIAL	T3901
SPECIAL	T3902
SPECIAL	T3903
SPECIAL	T3909
SPECIAL	T3910
SPECIAL	T3911
SPECIAL	T3912
SPECIAL	T3913
SPECIAL	T4130A
SPECIAL	T4130B
SPECIAL	T4130C
SPECIAL	T4130D
SPECIAL	T4130E
SPECIAL	T4130F
SPECIAL	T4130M
SPECIAL	T4131A
SPECIAL	T4131B
SPECIAL	T4131C
SPECIAL	T4131D
SPECIAL	T4131E
SPECIAL	T4131F
SPECIAL	T4131M
SPECIAL	T4140A
SPECIAL	T4140B
SPECIAL	T4140C
SPECIAL	T4140D
SPECIAL	T4140E
SPECIAL	T4140F
SPECIAL	T4140M
SPECIAL	T4141A
SPECIAL	T4141B
SPECIAL	T4141C
SPECIAL	T4141D
SPECIAL	T4141E
SPECIAL	T4141F
SPECIAL	T4141M
SPECIAL	T4150A
SPECIAL	T4150B
SPECIAL	T4150C
SPECIAL	T4150D
SPECIAL	T4150E
SPECIAL	T4150F
SPECIAL	T4150M
SPECIAL	T4151A
SPECIAL	T4151B
SPECIAL	T4151C
SPECIAL	T4151D
SPECIAL	T4151E
SPECIAL	T4151F
SPECIAL	T4151M
SPECIAL	T6430B
SPECIAL	T6430D
SPECIAL	T6430M
SPECIAL	T6431A
SPECIAL	T6431B
SPECIAL	T6431C
SPECIAL	T6431D
SPECIAL	T6431E
SPECIAL	T6431F
SPECIAL	T6431M
SPECIAL	T6440A
SPECIAL	T6440B
SPECIAL	T6440C
SPECIAL	T6440D
SPECIAL	T6440E
SPECIAL	T6440F
SPECIAL	T6440M
SPECIAL	T6441A
SPECIAL	T6441B
SPECIAL	T6441C
SPECIAL	T6441D
SPECIAL	T6441E
SPECIAL	T6441F
SPECIAL	T6441M
SPECIAL	T6450A
SPECIAL	T6450B
SPECIAL	T6450C
SPECIAL	T6450D
SPECIAL	T6450E
SPECIAL	T6450F
SPECIAL	T6450M
SPECIAL	T6451A
SPECIAL	T6451B
SPECIAL	T6451C
SPECIAL	T6451D
SPECIAL	T6451E
SPECIAL	T6451F
SPECIAL	T6451M
SPECIAL	TIR101A
SPECIAL	TIR101B
SPECIAL	TIR101C
SPECIAL	TIR101D
SPECIAL	TIR102A
SPECIAL	TIR102B
SPECIAL	TIR102C
SPECIAL	TIR102D
SPECIAL	TIR201A
SPECIAL	TIR201B
SPECIAL	TIR201C
SPECIAL	TIR201D
SPECIAL	TIR202A
SPECIAL	TIR202B
SPECIAL	TIR202C
SPECIAL	TIR202D
SPECIAL	UES1001
SPECIAL	UES1002
SPECIAL	UES1003
SPECIAL	UES1101
SPECIAL	UES1102
SPECIAL	UES1103
SPECIAL	UES1301
SPECIAL	UES1302
SPECIAL	UES1303
SPECIAL	UES2401
SPECIAL	UES2402
SPECIAL	UES2403
SPECIAL	UES2601
SPECIAL	UES2602
SPECIAL	UES2603
SPECIAL	UES701
SPECIAL	UES702
SPECIAL	UES703
SPECIAL	UES801
SPECIAL	UES802
SPECIAL	UES803
SPECIAL	1N2226,A
SPECIAL	1N2244,A
SPECIAL	1N3192
SPECIAL	1N3938
SPECIAL	1N3939
SPECIAL	1N3940
SPECIAL	1N4506
SPECIAL	1N4507
SPECIAL	1N4508
SPECIAL	1N5812
SPECIAL	1N5813
SPECIAL	1N5814
SPECIAL	1N5815
SPECIAL	1N5816
SPECIAL	2FB050R
SPECIAL	2FB100R
SPECIAL	2FB200R
SPECIAL	2FB400R
SPECIAL	2FB600R
SPECIAL	6A10F

MOTOR.	Industry Standard No.
SPECIAL	6A8F
SPECIAL	6HB050R
SPECIAL	6HB100R
SPECIAL	6HB200R
SPECIAL	6HB400R
SPECIAL	6HB600R
SPECIAL	12A10F
SPECIAL	12A8F
SPECIAL	20A10F
SPECIAL	20A8F
SPECIAL	30A10F
SPECIAL	30A8F
SPECIAL	30CTQ030
SPECIAL	30CTQ035
SPECIAL	30CTQ040
SPECIAL	30CTQ045
SPECIAL	30FQ30A
SPECIAL	30FQ35A
SPECIAL	30FQ40A
SPECIAL	30FQ45A
SPECIAL	30QHC030
SPECIAL	30QHC045
T2300PA	T2300PA
T2300PA	T2310A
T2300PA5	T2300A
T2300PB	L2001F31
T2300PB	L2001M3
T2300PB	L2003M3
T2300PB	T2300PB
T2300PB	T2310B
T2300PB5	T2300B
T2300PC	T2300PC
T2300PC5	T2300C
T2300PD	L4001F31
T2300PD	L4001M3
T2300PD	L4003M3
T2300PD	T2300PD
T2300PD	T2302PD
T2300PD	T2310D
T2300PD5	T2300D
T2300PE	T2300PE
T2300PE5	T2300E
T2300PF	T2300PF
T2300PF	T2301PF
T2300PF	T2310P
T2300PF5	T2300P
T2300PM	T2300PM
T2300PM5	T2300M
T2301PA	T1C205A
T2301PA	T2301PA
T2301PA	T2311A
T2301PA5	T2301A
T2301PB	L2001F51
T2301PB	L2003M5
T2301PB	T1C205B
T2301PB	T2301PB
T2301PB	T2311B
T2301PB5	T2301B
T2301PC	T2301PC
T2301PC5	T2301C
T2301PD	L4001F51
T2301PD	L4001M5
T2301PD	L4003M5
T2301PD	T1C205D
T2301PD	T2301PD
T2301PD5	T2301D
T2301PE	T2301PE
T2301PE5	T2311P
T2301PF	T2301PF
T2301PF5	T2301P
T2301PM	T2301PM
T2301PM5	T2301M
T2302A	T2312A
T2302B	T2312B
T2302D	T2312D
T2302F	T2312F
T2302PA	T2302PA
T2302PA5	T2302A
T2302PB	L2001F71
T2302PB	L2001M7
T2302PB	L2003M7
T2302PB	Q2001M3
T2302PB	Q2003F3
T2302PB	Q2004F31
T2302PB	T2302PB
T2302PB5	T2302B
T2302PC	T2302PC
T2302PC5	T2302C
T2302PD	L4001F71
T2302PD	L4001M7
T2302PD	L4003M7
T2302PD	Q4001M3
T2302PD	Q4003F3
T2302PD	Q4004F31
T2302PD5	T2302D
T2302PE	T2302PE
T2302PE5	T2302E
T2302PF	T2302PF
T2302PF5	T2302P
T2302PM	T2302PM
T2302PM5	T2302M
T2320A	T2320A
T2320B	T2320B
T2320C	T2320C
T2320D	T2320D
T2320E	T2320E
T2320F	T2320F
T2322A	T2322A
T2322B	T2322B
T2322C	T2322C
T2322D	T2322D
T2322E	T2322E
T2322F	T2322F
T2323A	T2323A
T2323B	T2323B
T2323C	T2323C
T2323D	T2323D
T2323E	T2323E
T2323F	T2323F
T2327A	T2327A
T2327B	T2327B
T2327C	T2327C
T2327D	T2327D
T2327E	T2327E
T2327F	T2327F
T2500A	T2500A
T2500B	T2500B
T2500C	T2500C
T2500D	T2500D
T2500E	T2500E
T2500M	T2500M
T2500N	T2500N
T2500S	T2500S
T2800B	Q2008F41
T2800B	T2800B
T2800B	T2806B
T2800B	T2856B
T2800C	T2800C
T2800C	T2806C
T2800C	T2856C
T2800C	Q4008F41
T2800D	T2800D
T2800D	T2806D
T2800D	T2856D
T2800E	T2800E
T2800M	T2800M
T2800N	T2806M
T2800N	T2806N
T2801B	T2801B
T2801B	T2851B
T2801C	T2801C
T2801D	T2801D
T2801D	T2851D
T2801E	T2801E
T2801E	T2851E
T2801M	T2801M
T2802B	T2802B
T2802C	T2802C
T2802D	T2802D
T2802E	T2802E
T2802M	Q6008F51
T2802M	T2802M
T2850E	T2850E
T4100B	T4100B
T4100C	T4100C
T4100D	T4100D
T4100E	T4100E
T4100M	T4100M
T4106M	T4106M
T4100M	TIC240E
T4101B	SC241B
T4101B	T4101B
T4101C	T4101C
T4101D	SC241D
T4101D	T4101D
T4101E	SC241E
T4101E	T4101E
T4101M	MAC40795
T4101M	SC241M
T4101M	T4101M
T4101N	T4107M
T4101N	TIC220E
T4101N	TIC230E
T4110B	T4110B
T4110C	T4110C
T4110D	T4110D
T4110E	T4110E
T4110M	T4110M
T4110M	TIC242E
T4111B	T4111B
T4111C	T4111C
T4111D	T4111D
T4111E	T4111E
T4111M	MAC40796
T4111M	T4111M
T4111M	TIC222E
T4111N	TIC232E
T4120A	T4120A
T4120B	T4120B
T4120C	T4120C
T4120D	T4120D
T4120D	T4126D
T4120D	TIC241D
T4120E	T4120E
T4120M	T4120M
T4120M	T4126M
T4120N	TIC241E
T4120N	T4120N
T4120S	T4120S
T4121A	MAC40799
T4121B	SC240B
T4121B	T4121B
T4121B	T4127B
T4121B	TIC221B
T4121B	TIC231B
T4121C	T4121C
T4121C	MAC40800
T4121D	SC240D
T4121D	T4121D
T4121D	T4127D
T4121D	TIC221D
T4121D	TIC231D
T4121E	SC240E
T4121E	T4121E
T4121E	MAC40801
T4121M	SC240M
T4121M	T4121M
T4121M	T4127M
T4121M	TIC221E
T4121M	TIC231E
T4121N	T4121N
T4121S	T4121S
T6400A	T6400A
T6400B	T6400B
T6400C	T6400C
T6400D	T6400D
T6400E	SC266E
T6400E	T6400E
T6400M	T6400M
T6400N	T6400N
T6400S	T6400S
T6401A	T6401A
T6401B	T6401B
T6401B	TIC260B
T6401C	T6401C
T6401D	T6401D
T6401D	T6407D
T6401D	TIC260D
T6401E	T6401E
T6401E	T6407E
T6401E	TIC260E
T6401M	T6401M
T6401M	T6407M
T6401M	TIC260M
T6401N	T6401N
T6401S	T6401S
T6410A	T6410A
T6410B	T6410B
T6410C	T6410C
T6410D	T6410D
T6410D	SC265E
T6410E	T6410E
T6410M	T6410M
T6410N	T6410N
T6410S	T6410S
T6411A	T6411A
T6411B	T6411B
T6411B	TIC252B
T6411C	T6411C
T6411D	T6411D
T6411D	TIC252D
T6411D	TIC262D
T6411E	T6411E
T6411E	TIC252E
T6411E	TIC262E
T6411M	T6411M
T6411M	TIC252M
T6411M	TIC262M
T6411N	T6411N
T6411S	T6411S
T6420A	T6420A
T6420B	MAC40688
T6420B	Q2040D
T6420B	T6420B
T6420B	T6426B
T6420C	T6420C
T6420D	MAC40689
T6420D	T6420D
T6420D	T6426D
T6420E	T6420E
T6420M	MAC40690
T6420M	T6420M
T6420M	T6426M
T6420N	T6420N
T6420S	T6420S
T6421A	T6421A
T6421B	T6421B
T6421C	T6421C
T6421D	T6421D
T6421D	T6427D
T6421E	T6421E
T6421M	T6421M
T6421M	T6427M
T6421S	T6421S
TBA1190Z	TBA1190Z
TDA1190Z	CA3134E
TDA1190Z	CA3134EM
TDA1190Z	CA3134QM
TDA1190Z	ITT3701
TDA1190Z	LM1808N
TDA1190Z	TDA1190Z
TDA2002	TDA2002
TIL111	TIL111
TIL112	TIL112
TIL113	TIL113
TIL114	TIL114
TIL115	TIL115
TIL116	TIL116
TIL117	TIL117
TIL118	TIL118
TIL119	TIL119
TIP100	NSP695A
TIP100	NSP697A
TIP100	TIP100
TIP101	TIP101
TIP101	2N6530
TIP102	TIP102
TIP102	2N6531
TIP102	2N6532
TIP102	2N6535
TIP102	2N6536
TIP105	NSP696A
TIP105	NSP698A
TIP105	TIP105
TIP106	NSP700A
TIP106	TIP106
TIP107	TIP107
TIP110	TIP110
TIP110	2SC1881
TIP110	2SD678
TIP110	SC246
TIP111	TIP111
TIP111	2SC1983
TIP111	2SD679
TIP112	TIP112
TIP112	2SC1880
TIP112	2SC1984
TIP112	2SD689
TIP115	TIP115
TIP116	2SB750
TIP117	TIP117
TIP117	2SB679
TIP120	MJE2100
TIP120	MJE2101
TIP120	NSP2100
TIP120	NSP2101
TIP120	NSP695
TIP120	NSP697
TIP120	RCA120
TIP120	TIP120
TIP120	2SD404
TIP120	2SD635
TIP120	2SD837
TIP121	MJE2102
TIP121	MJE2103
TIP121	NSP2102
TIP121	NSP2103
TIP121	NSP699
TIP121	RCA121
TIP121	TIP121
TIP121	2SD459
TIP121	2SD634
TIP121	NSP701
TIP122	RCA122
TIP122	TIP122
TIP122	2SC1883
TIP122	2SD460
TIP122	2SD633
TIP122	2SD686
TIP125	D45Z1
TIP125	D45Z2
TIP125	MJE2090
TIP125	MJE2091
TIP125	NSP2090
TIP125	NSP2091
TIP125	NSP696
TIP125	NSP698
TIP125	RCA125
TIP125	TIP125
TIP125	2SB677
TIP126	D45E3
TIP126	MJE2092
TIP126	MJE2093
TIP126	NSP2092
TIP126	NSP2093
TIP126	NSP700
TIP126	RCA126
TIP126	TIP126
TIP127	NSP702
TIP127	TIP127
TIP127	2SB676
TIP140	TIP140
TIP141	TIP141
TIP142	TIP142
TIP145	TIP145
TIP146	TIP146
TIP147	TIP147
TIP29	MJE29
TIP29	MJE4921
TIP29	NSP4921
TIP29	RCA29
TIP29	TIP29
TIP29A	MJE29A
TIP29A	MJE4922
TIP29A	NSP4922
TIP29A	NSP575
TIP29A	NSP577
TIP29A	NSP585
TIP29A	NSP597
TIP29A	RCA29A
TIP29A	TIP29A
TIP29B	MJE29B
TIP29B	MJE4923
TIP29B	NSP579
TIP29B	NSP589
TIP29B	RCA29B
TIP29B	TIP29B
TIP29C	MJE29C
TIP29C	NSP581
TIP29C	RCA29C
TIP29C	TIP29C
TIP30	MJE30
TIP30	MJE4918
TIP30	NSP4918
TIP30	RCA30
TIP30	TIP30
TIP30	2SA700
TIP30A	MJE30A
TIP30A	MJE4919
TIP30A	NSP576
TIP30A	NSP578
TIP30A	NSP586
TIP30A	NSP588
TIP30A	RCA30A
TIP30A	TIP30A
TIP30B	MJE30B
TIP30B	MJE4920
TIP30B	NSP580
TIP30B	NSP590
TIP30B	RCA30B
TIP30B	TIP30B
TIP30B	2SA816
TIP30C	MJE30C
TIP30C	NSP582
TIP30C	RCA30C
TIP30C	TIP30C
TIP30C	2SA775
TIP30C	2SA814
TIP30C	2SA815
TIP31	MJE2480
TIP31	MJE520K
TIP31	MJB520K
TIP31	MJB521K
TIP31	NSP2520
TIP31	NSP520
TIP31	NSP521
TIP31	RCA31
TIP31	TIP31
TIP31	2SC1173
TIP31	2SC1418
TIP31	2SC1419
TIP31	2SC2236
TIP31	2SC2500
TIP31	2SD325
TIP31A	MJE2481
TIP31A	NSP2481
TIP31A	NSP3054
TIP31A	NSP595
TIP31A	NSP597
TIP31A	RCA31A
TIP31A	TIP31A
TIP31A	2SC1060
TIP31A	2SC1061
TIP31A	2SC790
TIP31A	2SD224
TIP31A	2SB235
TIP31A	2SD313
TIP31A	2SD314
TIP31A	2SD317
TIP31A	2SD318
TIP31A	2SD330
TIP31A	2SD331
TIP31A	2SD365
TIP31A	2SD366
TIP31A	2SD389
TIP31A	2SD390
TIP31A	2SD762
TIP31B	NSP599
TIP31B	RCA31B
TIP31B	TIP31B
TIP31B	2SC1237
TIP31B	2SD288
TIP31B	2SD289
TIP31B	2SD342
TIP31B	2SD343
TIP31B	2SD344
TIP31B	2SD345
TIP31C	RCA31C
TIP31C	TIP31C
TIP31C	2SD723
TIP31C	2SD726
TIP32	MJE2370
TIP32	MJE2490
TIP32	MJE370K
TIP32	MJE371K
TIP32	NSP2370

MOTOR.	Industry Standard No.	MOTOR.	Industry Standard No.	MOTOR.	Industry Standard No.	MOTOR.	Industry Standard No.
TIP32	NSP2490	TIP42C	2SA1010	1M120ZSB5	1N5106	1M8.2ZZ10	LPZ78.2
TIP32	NSP370	TIP42C	2SB595	1M130ZS10	1N1798	1M80ZS5	N5093
TIP32	NSP371	TIP42C	2SB633	1M130ZS10	1N3706	1M862G10	MZ1000-35
TIP32	RCA32	TIP42C	2SB689	1M130ZS10	1N3706A	1N183	S25A05
TIP32	TIP32	TIP42C	2SB690	1M130ZS10	1N4352	1N183	1N2154
TIP32	2SA1020	TIP42C	2SB753	1M130ZS10	1N4354A	1N183	1N2446
TIP32	2SA966	TIP47	D44R1	1M130ZS10	1N4431	1N183	1N2458
TIP32	2SB511	TIP47	D44R2	1M130ZS10	1N4859	1N183A	TR53
TIP32A	MJE2371	TIP47	D44R5	1M130ZS10	1N4859A	1N183A	1N1301
TIP32A	MJE2491	TIP47	FP47	1M130ZS	1EZ130D5	1N183A	1N1434
TIP32A	NSP596	TIP47	MJE341K	1M130ZS	1N1798A	1N183A	40A50
TIP32A	NSP598	TIP47	MJE344K	1M130ZS	1N3706B	1N183A	40HP5
TIP32A	RCA32A	TIP47	MJE3738	1M130ZS	1N4354B	1N184	S25A1
TIP32A	TIP32A	TIP47	MJE47	1M130ZS	1N4492	1N184	TR103
TIP32A	2SB514	TIP47	MJB5655	1M130ZS	1N4859B	1N184	1N2155
TIP32A	2SB515	TIP47	TIP47	1M130ZS	1N5098	1N184	1N2447
TIP32A	2SB668	TIP47	2SC1409	1M130ZSB5	1N5107	1N184	1N2459
TIP32A	2SB724	TIP47	2SC1410	1M135ZSB5	1N5108	1N184A	S40A1
TIP32B	NSP600	TIP47	2SC1447	1M140ZS	1EZ140D5	1N184A	ST410P
TIP32B	RCA32B	TIP47	2SC1448	1M140ZS	1N5099	1N184A	1N1302
TIP32B	TIP32B	TIP47	2SC1669	1M140ZSB5	1N5109	1N184A	1N1435
TIP32B	2SB669	TIP47	2SC1683	1M142Z10,5	Z4Z14B,A	1N184A	40A100
TIP32C	RCA32C	TIP47	2SC2073	1M142Z10	1N5024	1N184A	40HP10
TIP32C	TIP32C	TIP47	2SC2229	1M142S	1N5024A	1N186	1N2021
TIP32C	2SA1008	TIP47	2SC2235	1M142S	1N5070	1N186	1N2156
TIP32C	2SA839	TIP47	2SC2258	1M150ZS10	1N1799	1N186	1N2448
TIP35	TIP35	TIP47	2SC2344	1M150ZS10	1N1890	1N186	1N2449
TIP35A	TIP35A	TIP47	2SD401	1M150ZS10	1N3461	1N186	1N2460
TIP35B	TIP35B	TIP47	2SD402	1M150ZS10	1N3707	1N186	1N2461
TIP35C	TIP35C	TIP47	2SD478	1M150ZS10	1N3707A	1N186	1N2786
TIP36	TIP36	TIP47	2SD608	1M150ZS10	1N4355	1N186	1N2788
TIP36A	TIP36A	TIP47	2SD610	1M150ZS10	1N4355A	1N186A	S40A2
TIP36B	TIP36B	TIP47	2SD759	1M150ZS10	1N4432	1N186A	ST420P
TIP36C	TIP36C	TIP47	2SD760	1M150ZS10	1N4860	1N186A	TR153
TIP41	MJE2020	TIP47	2SD761	1M150ZS10	1N4860A	1N186A	1N1304
TIP41	MJE33	TIP48	D44R3	1M150ZS	1EZ150D5	1N186A	1N1436
TIP41	MJE41	TIP48	D44R4	1M150ZS	1N4433	1N186A	40A200
TIP41	MJE5977	TIP48	D44R6	1M150ZS	1N1799A	1N186A	40HP15
TIP41	NSP41	TIP48	FP48	1M150ZS	1N3707B	1N186A	40HP20
TIP41	NSP5977	TIP48	MJE2160	1M150ZS	1N4098	1N187	S25A3
TIP41	RCA41	TIP48	MJE340K	1M150ZS	1N4355B	1N187	TR303
TIP41	TIP33	TIP48	MJE3739	1M150ZS	1N4493	1N187A	S40A3
TIP41	TIP41	TIP48	MJE48	1M150ZS	1N4860B	1N187A	ST430P
TIP41	2N1483	TIP48	MJB5656	1M150ZSB5	1N5110	1N188	S25A4
TIP41	2N1485	TIP48	TIP48	1M160ZS10	1N1800	1N188	1N2022
TIP41A	MJE2021	TIP48	2SC1505	1M160ZS10	1N3708	1N188	1N2023
TIP41A	MJE205K	TIP48	2SC1506	1M160ZS10	1N3708A	1N188	1N2024
TIP41A	MJE33A	TIP48	2SC1507	1M160ZS10	1N4356	1N188	1N2025
TIP41A	MJE41A	TIP48	2SC1722	1M160ZS10	1N4356A	1N188	1N2157
TIP41A	MJE5978	TIP48	2SC1723	1M160ZS10	1N4433	1N188	1N2282
TIP41A	NSP2021	TIP48	2SC1929	1M160ZS	1EZ160D5	1N188	1N2283
TIP41A	NSP205	TIP49	FP49	1M160ZS	1N1800A	1N188	1N2450
TIP41A	NSP41A	TIP49	MJE49	1M160ZS	1N3708B	1N188	1N2451
TIP41A	NSP5978	TIP49	MJB5657	1M160ZS	1N4356B	1N188	1N2452
TIP41A	RCA41A	TIP49	TIP49	1M160ZS	1N4494	1N188	1N2453
TIP41A	SDT5101	TIP50	FP50	1M160ZS	1N5100	1N188	1N2462
TIP41A	SDT5102	TIP50	TIP50	1M160ZSB5	1N5111	1N188	1N2463
TIP41A	SDT5103	TIP61	TIP61	1M165ZSB5	1N5112	1N188	1N2464
TIP41A	TIP33A	TIP61A	TIP61A	1M16ZS10,5	SZ16.0,A	1N188	1N2465
TIP41A	TIP41A	TIP61B	TIP61B	1M170ZS	1EZ170D5	1N188	1N2787
TIP41A	2N1484	TIP61C	TIP61C	1M170ZS	1N5101	1N188	1N2789
TIP41A	2N1486	TIP62	TIP62	1M170ZSB5	1N5113	1N188A	ST440P
TIP41A	2SD346	TIP62A	TIP62A	1M17ZS10	1N5027	1N188A	TR203
TIP41A	2SD347	TIP62B	TIP62B	1M17ZS	1N5027A	1N188A	TR253
TIP41B	MJE33B	TIP62C	TIP62C	1M180ZS10	1N180	1N188A	TR353
TIP41B	MJE41B	TL494CJ	TL494CJ	1M180ZS10	1N3462	1N188A	TR403
TIP41B	MJE5979	TL494CN	TL494CN	1M180ZS10	1N3709	1N188A	1N1306
TIP41B	NSP41B	TL495CJ	TL495CJ	1M180ZS10	1N3709A	1N188A	1N1437
TIP41B	NSP5979	TL495CN	TL495CN	1M180ZS10	1N4357	1N188A	40A400
TIP41B	RCA41B	U308	E308	1M180ZS10	1N4357A	1N188A	40HP30
TIP41B	TIP33B	U308	U308	1M180ZS10	1N4434	1N188A	40HP40
TIP41B	TIP41B	U308	2N5078	1M180ZS	1EZ180D5	1N189	TR503
TIP41B	2SC1826	U309	E309	1M180ZS	1N1801A	1N189A	S40A5
TIP41B	2SC1985	U309	U309	1M180ZS	1N3709B	1N189A	ST450P
TIP41B	2SB526	U309	2N5397	1M180ZS	1N4357B	1N190	S25A6
TIP41B	2SB553	U309	2N5398	1M180ZS	1N4495	1N190	S40A6
TIP41C	MJE33C	U310	E310	1M180ZS	1N5102	1N190	TR603
TIP41C	MJE41C	U310	U310	1M180ZSB5	1N5114	1N190	1N2159
TIP41C	NSP41C	JA723DC	JA723DC	1M190ZS	1EZ190D5	1N190	1N2160
TIP41C	RCA41C	JA723HC	JA723HC	1M190ZS	1N5103	1N190	1N2284
TIP41C	TIP33C	JA723PC	JA723PC	1M190ZSB5	1N5115	1N190	1N2285
TIP41C	TIP41C	JA741DC	JA741DC	1M195ZSB5	1N5116	1N190	1N2454
TIP41C	2SC1827	JA741HC	JA741HC	1M19ZS	1N5029	1N190	1N2455
TIP41C	2SC1986	JA741TC	JA741TC	1M19ZS	1N5029A	1N190	1N2466
TIP41C	2SB531	.5M2525	N5951	1M200Z10,5	LPZ200,A	1N190	1N2467
TIP41C	2SD544	1/4M10.5Z5	1N1314A	1M200ZS10	U28220	1N190A	ST460P
TIP41C	2SD613	1/4M23.5Z5	1N1318A	1M200ZS10	VR200	1N190A	1N1438
TIP41C	2SD716	1/4M28.5Z5	1N1319A	1M200ZS10	1N1802	1N190A	40A500
TIP42	MJE2010	1/4M34.5Z5	1N1320A	1M200ZS10	1N3710	1N190A	40HP50
TIP42	MJE34	1/4M41Z10	1N321	1M200ZS10	1N3710A	1N190A	40HP60
TIP42	MJE42	1/4M4125	1N321A	1M200ZS10	1N4358	1N194	TR200
TIP42	MJE5974	1/4M48.5Z5	1N1322A	1M200ZS10	1N4358A	1N196	TR352
TIP42	NSP2010	1/4M58Z10	1N323	1M200ZS10	1N4435	1N196	TR400
TIP42	NSP42	1/4M58Z5	1N1323A	1M200ZS10	2VR200	1N196	TR402
TIP42	NSP5974	1/4M71Z10	1N324	1M200ZS	U28120	1N198	TR600
TIP42	RCA42	1/4M7125	1N1324A	1M200ZS	1EZ200D5	1N198	TR601
TIP42	TIP34	1/4M87.5Z5	1N1325A	1M200ZS	1N1802A	1N198	TR602
TIP42	TIP42	1M100Z	1T100	1M200ZS	1N3463	1N199C	1N1227,A,B
TIP42B	2SA1069	1M100ZS	1ZS100	1M200ZS	1N3710B	1N199C	1N1537
TIP42A	MJE105K	1M110ZB10	1N1796	1M200ZS	1N4358B	1N199C	1N1581
TIP42A	MJE2011	1M110ZB10	1N3704	1M200ZS	1N4496	1N199C	1N2026
TIP42A	MJE34A	1M110ZB10	1N3704A	1M200ZS	1N5104	1N199C	1N2216
TIP42A	MJE42A	1M110ZB10	1N4352	1M252ZS10	1N5033	1N199C	1N2228,A
TIP42A	MJE5975	1M110ZB10	1N4429	1M25ZS5	1N5033A	1N199C	1N2246A
TIP42A	NSP105	1M110ZS	1N4857	1M3.3ZS	1ZS3.3	1N199C	1N2266
TIP42A	NSP2011	1M110ZB5	1N4857A	1M30Z,10,5	1Z30T20,10,5	1N199C	1N2491
TIP42A	NSP42A	1M110ZS	1EZ110D5	1M30ZS	SZ30	1N199C	1N607,A
TIP42A	NSP5975	1M110ZS	1N1796A	1M33ZS	CD3211062	1N199C	1N1115
TIP42A	RCA42A	1M110ZS	1N3704B	1M33ZS	CD3214752	1N200C	1N1228,A,B
TIP42A	SDT5111	1M110ZB5	1N4352B	1M33ZZ10	LPZ733	1N200C	1N1538
TIP42A	SDT5112	1M110ZS	1N4490	1M4.7AZS	BZY96-C4V7	1N200C	1N1551
TIP42A	SDT5113	1M110ZS	1N4857B	1M4005	1N2614	1N200C	1N1582
TIP42A	TIP34A	1M110ZS	1N5096	1M40ZS	1N5081	1N200C	1N2248A
TIP42A	TIP42A	1M110ZS	1N5105	1M45ZS10	1N5040	1N200C	1N2492
TIP42A	2SA1012	1M120ZB10	1N1797	1M45ZS10	1N5040A	1N200C	1N2512
TIP42A	2SB521	1M120ZB10	1N1889	1M45ZS	1N5083	1N200C	1N253
TIP42A	2SB522	1M120ZB10	1N3460	1M47ZS10,5	MC6130,A	1N200C	1N338
TIP42A	2SC2562	1M120ZB10	1N3705	1M5.6AZ	1T5.6	1N200C	1N339
TIP42B	MJE34B	1M120ZB10	1N3705A	1M50ZS10	1N5042	1N200C	1N340
TIP42B	MJE42B	1M120ZB10	1N4353	1M50ZS	1N5042A	1N200C	1N347
TIP42B	MJE5976	1M120ZB10	1N4353A	1M52ZS10	1N5044	1N200C	1N348
TIP42B	NSP42B	1M120ZB10	1N4430	1M52ZS10	1N5044A	1N200C	1N349
TIP42B	NSP5976	1M120ZB10	1N4858	1M6.2ZS10	VR6.2	1N200C	1N350
TIP42B	RCA42B	1M120ZB10	1N4858A	1M6.2ZS10	2VR6.2	1N200C	1N608,A
TIP42B	TIP34B	1M120ZS	1E120D5	1M6.2ZS10	U28806	1N202C	1N116
TIP42B	TIP42B	1M120ZS	1N1797A	1M6.8ZS	U28706	1N202C	1N1229,A,B
TIP42C	MJE34C	1M120ZS	1N3705B	1M60ZS	1N5088	1N202C	1N1230,A,B
TIP42C	MJE42C	1M120ZS	1N4353B	1M7.5ZS	BZX61-C7V5	1N202C	1N1539
TIP42C	NSP42C	1M120ZS	1N4491	1M70ZS	1N5091	1N202C	1N1552
TIP42C	RCA42C	1M120ZS	1N4858B	1M75ZS	BZY96-C75	1N202C	1N1583
TIP42C	TIP34C	1M120ZB5	1N5097	1M75ZS	BZX61-C75	1N202C	1N2027
TIP42C	TIP42C			1M8.2ZS	CD3212048		
				1M8.2ZS	CD3214738		

MOTOR.	Industry Standard No.	MOTOR.	Industry Standard No.	MOTOR.	Industry Standard No.	MOTOR.	Industry Standard No.
1N1202C	1N2230,A	1N2974A	1N2498	1N2984B	1N4025B	1N2995B	1N1829A
1N1202C	1N2250A	1N2974A	1N4018	1N2984B	1N4208B	1N2995B	1N4034B
1N1202C	1N2493	1N2974A	1N4018A	1N2984B	1N4269B	1N2995B	1N4219B
1N1202C	1N2513	1N2974A	1N4198	1N2985A	1N1359	1N2995B	1N4278B
1N1202C	1N254	1N2974A	1N4198A	1N2985A	1N1821	1N2996A	1N2937
1N1202C	1N336	1N2974A	1N4262	1N2985A	1N1896	1N2996A	1N4220
1N1202C	1N337	1N2974A	1N4262A	1N2985A	1N4026	1N2996A	1N4220A
1N1202C	1N345	1N2974B	1N1351A	1N2985A	1N4026A	1N2996B	1N4220B
1N1202C	1N346	1N2974B	1N2498A	1N2985A	1N4209	1N2997A	1N1368
1N1202C	1N351	1N2974B	1N4018B	1N2985A	1N4270	1N2997A	1N1830
1N1202C	1N609,A	1N2974B	1N4198B	1N2985A	1N4270A	1N2997A	1N4035
1N1202C	1N610,A	1N2974B	1N4262B	1N2985B	1N1359A	1N2997A	1N4035A
1N1204C	1N353	1N2974RA	1N1593	1N2985B	1N1420	1N2997A	1N4221
1N1204C	1N1117	1N2974RA	1N1604	1N2985B	1N1821A	1N2997A	1N4221A
1N1204C	1N1118	1N2974RB	1N1593A	1N2985B	1N4026B	1N2997A	1N4279
1N1204C	1N1231,A,B	1N2974RB	1N1604A	1N2985B	1N4209B	1N2997B	1N4279A
1N1204C	1N1232,A,B	1N2974RB	1N2045	1N2985B	1N4270B	1N2997B	1N1368A
1N1204C	1N1541	1N2975A	1N1352	1N2985RA	1N1608	1N2997B	1N1830A
1N1204C	1N1542	1N2975A	1N2499	1N2985RB	1N1597A	1N2997B	1N4035B
1N1204C	1N1553	1N2975A	1N4019	1N2985RB	1N1608A	1N2997B	1N4221B
1N1204C	1N1554	1N2975A	1N4019A	1N2986A	1N1360	1N2997B	1N4279B
1N1204C	1N1584	1N2975A	1N199	1N2986A	1N1822	1N2998A	1N4222
1N1204C	1N1585	1N2975A	1N4199A	1N2986A	1N4027	1N2998A	1N4222B
1N1204C	1N2028	1N2975A	1N4263	1N2986A	1N4027A	1N2999A	1N1369
1N1204C	1N2029	1N2975B	1N1352A	1N2986A	1N4210	1N2999A	1N1831
1N1204C	1N2232,A	1N2975B	1N2499A	1N2986A	1N4210A	1N2999A	1N1901
1N1204C	1N2234,A	1N2975B	1N4019B	1N2986A	1N4271	1N2999A	1N4036
1N1204C	1N2252A	1N2975B	1N4199B	1N2986B	1N1360A	1N2999A	1N4036A
1N1204C	1N2254A	1N2975B	1N4263B	1N2986B	1N1822A	1N2999A	1N4223
1N1204C	1N2494	1N2976A	1N1353	1N2986B	1N4027B	1N2999A	1N4223A
1N1204C	1N2495	1N2976A	1N1893	1N2986B	1N4210B	1N2999A	1N4280
1N1204C	1N2514	1N2976A	1N2500	1N2986RA	1N2049	1N2999A	1N4280A
1N1204C	1N2515	1N2976A	1N4020	1N2987A	1N4211	1N2999B	1N1369A
1N1204C	1N255	1N2976A	1N4020A	1N2987A	1N4211A	1N2999B	1N1831A
1N1204C	1N332	1N2976A	1N4200	1N2987B	1N4211B	1N2999B	1N4036B
1N1204C	1N333	1N2976A	1N4200A	1N2988A	1N1361	1N2999B	1N4223B
1N1204C	1N334	1N2976A	1N4264	1N2988A	1N1823	1N2999B	1N4280B
1N1204C	1N335	1N2976A	1N4264A	1N2988A	1N1897	1N3000A	1N1370
1N1204C	1N341	1N2976B	1N1353A	1N2988A	1N4028	1N3000A	1N1832
1N1204C	1N342	1N2976B	1N1417	1N2988A	1N4028A	1N3000A	1N1785
1N1204C	1N343	1N2976B	1N2500A	1N2988A	1N4212	1N3000A	1N4037
1N1204C	1N344	1N2976B	1N4020B	1N2988A	1N4212A	1N3000A	1N4037A
1N1204C	1N352	1N2976B	1N4200B	1N2988B	1N1361A	1N3000A	1N4224
1N1204C	1N611,A	1N2976B	1N4264B	1N2988B	1N1421	1N3000A	1N4224A
1N1204C	1N612,A	1N2976RA	1N1594	1N2988B	1N1823A	1N3000A	1N4281
1N1206C	1N1119	1N2976RA	1N1605	1N2988B	1N4028B	1N3000A	1N4281A
1N1206C	1N1120	1N2976RB	1N1594A	1N2988B	1N4212B	1N3000B	1N1370A
1N1206C	1N1233,A,B	1N2976RB	1N1605A	1N2988B	1N1598	1N3000B	1N1832A
1N1206C	1N1234,A,B	1N2977A	1N1354	1N2988RA	1N1609	1N3000B	1N4037B
1N1206C	1N1543	1N2977A	1N1816	1N2988RA	1N1598A	1N3000B	1N4224B
1N1206C	1N1544	1N2977A	1N4021	1N2988RB	1N1609A	1N3000B	1N4281B
1N1206C	1N1555	1N2977A	1N4021A	1N2989A	1N1362	1N3000B	1N4889
1N1206C	1N1586	1N2977A	1N4201	1N2989A	1N1824	1N3001A	1N1371
1N1206C	1N1587	1N2977A	1N4201A	1N2989A	1N4029	1N3001A	1N1833
1N1206C	1N2030	1N2977A	1N4265	1N2989A	1N4029A	1N3001A	1N1902
1N1206C	1N2031	1N2977A	1N4265A	1N2989A	1N4213	1N3001A	1N4038
1N1206C	1N2218	1N2977A	1N1816A	1N2989A	1N4213A	1N3001A	1N4038A
1N1206C	1N2220	1N2977B	1N3154A	1N2989A	1N4275	1N3001A	1N4225
1N1206C	1N2236,A	1N2977B	1N4021B	1N2989A	1N4275A	1N3001A	1N4225A
1N1206C	1N2238,A	1N2977B	1N4201B	1N2989B	1N1362A	1N3001A	1N4282
1N1206C	1N2256A	1N2977B	1N4265B	1N2989B	1N1824A	1N3001A	1N4282A
1N1206C	1N2258A	1N2977RA	1N2046	1N2989B	1N4029B	1N3001A	1N1971
1N1206C	1N2268	1N2978A	1N4202	1N2989B	1N4213B	1N3001B	1N1422
1N1206C	1N2270	1N2978A	1N4202A	1N2989B	1N4275B	1N3001B	1N1833A
1N1206C	1N2496	1N2978B	1N4202B	1N2990A	1N1363	1N3001B	1N4038B
1N1206C	1N2497	1N2979A	1N1817	1N2990A	1N1825	1N3001B	1N4225B
1N1206C	1N2516	1N2979A	1N1894	1N2990A	1N1898	1N3001B	1N4282B
1N1206C	1N2517	1N2979A	1N4022	1N2990A	1N4030	1N3002A	1N1372
1N1206C	1N256	1N2979A	1N4022A	1N2990A	1N4030A	1N3002A	1N1834
1N1206C	1N354	1N2979A	1N4203	1N2990A	1N4214	1N3002A	1N4039
1N1206C	1N355	1N2979A	1N4203A	1N2990A	1N4214A	1N3002A	1N4039A
1N1206C	1N613,A	1N2979A	1N4266	1N2990A	1N4274	1N3002A	1N4226
1N1206C	1N614,A	1N2979A	1N4266A	1N2990A	1N4274A	1N3002A	1N4226A
1N1736A	1N615,A	1N2979B	1N1418	1N2990B	1N1363A	1N3002A	1N4283
1N1736A	1N2766	1N2979B	1N1817A	1N2990B	1N1825A	1N3002A	1N4283A
1N1736A	1N2766A	1N2979B	1N4022B	1N2990B	1N4030B	1N3002B	1N1372A
1N248B	TR50	1N2979B	1N4203B	1N2990B	1N4214B	1N3002B	1N1834A
1N249B	TR100	1N2979B	1N4266B	1N2990B	1N4274B	1N3002B	1N4039B
1N250B	TR150	1N2979RA	1N1595	1N2991A	1N1364	1N3002B	1N4226B
1N250B	TR152	1N2979RA	1N1606	1N2991A	1N1826	1N3002B	1N4283B
1N2624B	1N4094	1N2979RB	1N1595A	1N2991A	1N4031	1N3003A	1N1373
1N2970A	1N1805	1N2979RB	1N1606A	1N2991A	1N4031A	1N3003A	1N1835
1N2970A	1N4194	1N2980A	1N1356	1N2991A	1N4215	1N3003A	1N1903
1N2970A	1N4194A	1N2980A	1N1818	1N2991A	1N4215A	1N3003A	1N4040
1N2970A	1N4258	1N2980A	1N4023	1N2991A	1N4275	1N3003A	1N4040A
1N2970A	1N4258A	1N2980A	1N4023A	1N2991A	1N1364A	1N3003A	1N4227
1N2970A,B	U03470,A,B	1N2980A	1N4204	1N2991B	1N1826A	1N3003A	1N4227A
1N2970B	1N1805A	1N2980A	1N4204A	1N2991B	1N4031B	1N3003A	1N4284
1N2970B	1N4194B	1N2980A	1N4267	1N2991B	1N4215B	1N3003A	1N4284A
1N2970B	1N4258B	1N2980A	1N4267A	1N2991B	1N4275B	1N3003B	1N1373A
1N2970RA	1N1591	1N2980A	1N1356A	1N2992A	1N1365	1N3003B	1N1835A
1N2970RA	1N1602	1N2980A	1N1818A	1N2992A	1N1827	1N3003B	1N4040B
1N2970RA	1N2043	1N2980A	1N4023B	1N2992A	1N1899	1N3003B	1N4227B
1N2970RB	1N1591A	1N2980A	1N4204B	1N2992A	1N4032	1N3003B	1N4284B
1N2970RB	1N1602A	1N2980A	1N4267B	1N2992A	1N4032A	1N3004A	1N1374
1N2971A	1N1806	1N2981A	1N2047	1N2992A	1N4216	1N3004A	1N1836
1N2971A	1N4195	1N2981A	1N4205	1N2992A	1N4216A	1N3004A	1N4041
1N2971A	1N4195A	1N2981A	1N4205A	1N2992A	1N4276	1N3004A	1N4041A
1N2971A	1N4259	1N2981B	1N4205B	1N2992A	1N1365A	1N3004A	1N4228
1N2971A	1N4259A	1N2982A	1N1357	1N2992B	1N1827A	1N3004A	1N4228A
1N2971B	1N1806A	1N2982A	1N1819	1N2992B	1N4032B	1N3004A	1N4285
1N2971B	1N4195B	1N2982A	1N1895	1N2992B	1N4216B	1N3004B	1N4285A
1N2971B	1N4259B	1N2982A	1N4024	1N2992B	1N4276B	1N3004B	1N1374A
1N2972A	1N1807	1N2982A	1N4024A	1N2993A	1N1366	1N3004B	1N1836A
1N2972A	1N1891	1N2982A	1N4206	1N2993A	1N1828	1N3004B	1N4041B
1N2972A	1N4016	1N2982A	1N4206A	1N2993A	1N4033	1N3004B	1N4228B
1N2972A	1N4016A	1N2982A	1N4268	1N2993A	1N4033A	1N3004B	1N4285B
1N2972A	1N4196	1N2982A	1N4268A	1N2993A	1N4217	1N3005A	1N1375
1N2972A	1N4196A	1N2982B	1N1357A	1N2993A	1N4217A	1N3005A	1N1904
1N2972A	1N4260	1N2982B	1N1419	1N2993A	1N4277	1N3005A	1N2008
1N2972A	1N4260A	1N2982B	1N1819A	1N2993A	1N4277A	1N3005A	1N4042
1N2972B	1N1416	1N2982B	1N4024B	1N2993B	1N1366A	1N3005A	1N4042A
1N2972B	1N1807A	1N2982B	1N4206B	1N2993B	1N1828A	1N3005A	1N4229
1N2972B	1N4016B	1N2982B	1N4268B	1N2993B	1N4033B	1N3005A	1N4229A
1N2972B	1N4196B	1N2982RA	1N1596	1N2993B	1N4217B	1N3005A	1N4286
1N2972B	1N4260B	1N2982RA	1N1607	1N2993B	1N4277B	1N3005B	1N4286A
1N2972RA	1N1592	1N2982RB	1N1596A	1N2994A	1N4218	1N3005B	1N1375A
1N2972RA	1N1603	1N2982RB	1N1607A	1N2994B	1N4218A	1N3005B	1N1423
1N2972RB	1N1592A	1N2983A	1N4207	1N2994B	1N4218B	1N3005B	1N4042B
1N2972RB	1N1603A	1N2983A	1N4207A	1N2995A	1N1367	1N3005B	1N4229B
1N2973A	1N1808	1N2983B	1N4207B	1N2995A	1N1829	1N3005B	1N4286B
1N2973A	1N4017	1N2983RA	1N204B	1N2995A	1N1900	1N3006A	1N4230
1N2973A	1N4017A	1N2984A	1N1358	1N2995A	1N4034	1N3006A	1N4230A
1N2973A	1N4197	1N2984A	1N1820	1N2995A	1N4034A	1N3006A	1N4230B
1N2973A	1N4197A	1N2984A	1N4025	1N2995A	1N4219	1N3006A	1N1809
1N2973A	1N4261	1N2984A	1N4025A	1N2995A	1N4219A	1N3007A	1N2009
1N2973A	1N4261A	1N2984A	1N4208	1N2995A	1N4278	1N3007A	1N4231
1N2973B	1N1808A	1N2984A	1N4208A	1N2995A	1N4278A	1N3007A	1N4231A
1N2973B	1N4017B	1N2984A	1N4269	1N2995B	1N1367A	1N3007A	1N4287
1N2973B	1N4197B	1N2984B	1N4269A			1N3007A	1N4287A
1N2973B	1N4261B	1N2984B	1N1358A			1N3007B	1N1809A
1N2973RA	1N2044	1N2984B	1N1820A			1N3007B	1N4231B
1N2974A	1N1351	1N2984B	1N3949			1N3007B	1N4287B
1N2974A	1N743						
1N2974A	1N1892						

MOTOR.	Industry Standard No.	MOTOR.	Industry Standard No.	MOTOR.	Industry Standard No.	MOTOR.	Industry Standard No.
1N3008A	1N1810	1N3766	1N2457	1N4001	A50	1N4002	10B1
1N3008A	1N1905	1N3766	1N2468	1N4001	AA50	1N4002	20A1
1N3008A	1N2010	1N3766	1N2469	1N4001	AR16	1N4003	A14B
1N3008A	1N3102,A	1N3768	S25A10	1N4001	B50	1N4003	AA200
1N3008A	1N4232	1N3768	S40A10	1N4001	BA50	1N4003	AR18
1N3008A	1N4232A	1N3768	1N2287	1N4001	BF4-05L	1N4003	B200
1N3008A	1N4288	1N3796B	1N3950	1N4001	BY111	1N4003	BA200
1N3008A	1N4288A	1N3879	D2406F	1N4001	BY121	1N4003	BF4-20L
1N3008B	1N1810A	1N3879	NS6000	1N4001	BY141	1N4003	BY102
1N3008B	1N4232B	1N3879	TPR105	1N4001	CER67,A,B,C	1N4003	BY113
1N3008B	1N4288B	1N3879	TPR305	1N4001	D1201F	1N4003	BY123
1N3009A	1N1811	1N3879	TPR605	1N4001	D50	1N4003	BYX36350
1N3009A	1N2011	1N3879	6FL5	1N4001	DT230F	1N4003	BYX36300
1N3009A	1N4233	1N3879	6PT5	1N4001	ED3100	1N4003	BYY31
1N3009A	1N4233A	1N3879	6FV5	1N4001	ER1	1N4003	BYY32
1N3009A	1N4289	1N3880	D2406A	1N4001	ER181	1N4003	CER69,A,B,Q
1N3009A	1N4289A	1N3880	NS6001	1N4001	G100A	1N4003	D1201B
1N3009B	1N1811A	1N3880	TPR110	1N4001	G1A	1N4003	DI-42
1N3009B	1N4233B	1N3880	TPR310	1N4001	GER4001	1N4003	DI-52
1N3009B	1N4289B	1N3880	TPR610	1N4001	GP10A	1N4003	DI-72
1N3010A	1N4234	1N3880	6FL10	1N4001	HC67	1N4003	DT230Q
1N3010A	1N4234A	1N3880	6PT10	1N4001	HGR-5	1N4003	E2
1N3010B	1N4234B	1N3880	6FV10	1N4001	J-05	1N4003	ED3102
1N3011A	1N1812	1N3880	10BR	1N4001	M100A	1N4003	EM502
1N3011A	1N1906	1N3880	10HR3P	1N4001	M67,A,B,Q	1N4003	ER183
1N3011A	1N2012	1N3881	D2406B	1N4001	PA305	1N4003	ER21
1N3011A	1N3103,A	1N3881	NS6002	1N4001	PS405	1N4003	G100D
1N3011A	1N4235	1N3881	TPR120	1N4001	PT505	1N4003	G1D
1N3011A	1N4235A	1N3881	TPR320	1N4001	RG1122	1N4003	GER4003
1N3011A	1N4290	1N3881	TPR620	1N4001	SD-05	1N4003	GP10D
1N3011A	1N4290A	1N3881	6FL20	1N4001	SEN105	1N4003	HC69
1N3011B	1N1424	1N3881	6PT20	1N4001	SI-50E	1N4003	HGR-2Q
1N3011B	1N1812A	1N3881	6FV20	1N4001	SLA-11	1N4003	J-2
1N3011B	1N2012A,AH	1N3881	20BR	1N4001	SR3512	1N4003	M100D
1N3011B	1N4290B	1N3881	20HR3P	1N4001	SR6523	1N4003	M2
1N3012A	1N1813	1N3882	BY18	1N4001	SR8105	1N4003	M69,A,B,Q
1N3012A	1N4236	1N3882	D2406C	1N4001	TA5	1N4003	MB237
1N3012A	1N4236A	1N3882	NS6003	1N4001	TA50	1N4003	MB245
1N3012A	1N4291	1N3882	30BR	1N4001	TK5	1N4003	PA315
1N3012A	1N4291A	1N3882	30HR3P	1N4001	TS-05	1N4003	PA320
1N3012B	1N1813A	1N3883	D2406D	1N4001	TW5	1N4003	PS415
1N3012B	1N4236B	1N3883	NS6004	1N4001	UT111	1N4003	PS420
1N3012B	1N4291B	1N3883	TPR140	1N4001	1N1217,A,B	1N4003	PT515
1N3013A	1N4237	1N3883	TPR340	1N4001	1N1251	1N4003	PT520
1N3013A	1N4237A	1N3883	TPR640	1N4001	1N1644	1N4003	R200
1N3013B	1N4237B	1N3883	6FL30	1N4001	1N1701	1N4003	S102Q
1N3014A	1N1814	1N3883	6FL40	1N4001	1N1707	1N4003	S2M
1N3014A	1N3104,A	1N3883	6FT30	1N4001	1N1907	1N4003	SD-2
1N3014A	1N4238	1N3883	6FT40	1N4001	1N2013	1N4003	SEN120
1N3014A	1N4238A	1N3883	6FV30	1N4001	1N2072	1N4003	SGR200A
1N3014A	1N4292	1N3883	6FV40	1N4001	1N2080	1N4003	SI-200E
1N3014A	1N4292A	1N3883	40BR	1N4001	1N2103	1N4003	SL92
1N3014B	1N1814A	1N3883	40HR3P	1N4001	1N2609	1N4003	SLA-13
1N3014B	1N4238B	1N3889	A28P	1N4001	1N3072	1N4003	SR613
1N3014B	1N4292B	1N3889	D2412F	1N4001	1N316,A	1N4003	SR614
1N3015A	1N1815	1N3889	RT05	1N4001	1N323,A	1N4003	SR6585
1N3015A	1N3105,A	1N3889	TPR1205	1N4001	1N359,A	1N4003	SRS12Q
1N3015A	1N4239	1N3889	12FL5,502	1N4001	1N536	1N4003	TA20
1N3015A	1N4239A	1N3889	12PT5	1N4002	A100	1N4003	TA200
1N3015A	1N4293	1N3889	12FV5	1N4002	A14A	1N4003	TK20
1N3015A	1N4293A	1N3890	A18A	1N4002	AA100	1N4003	TK21
1N3015A,B	UZ3015,A,B	1N3890	A28A	1N4002	AR17	1N4003	TS-2
1N3015B	1N1815A	1N3890	D2412A	1N4002	B100	1N4003	TW20
1N3015B	1N4239B	1N3890	RT10	1N4002	BA100	1N4003	UT113
1N3015B	1N4293B	1N3890	ST2FR10P	1N4002	BF4-10L	1N4003	U234
1N3016A,B	UZ3016,A,B	1N3890	TPR1210	1N4002	CER68,A,B,Q	1N4003	UT244
1N3046A	1N3098,A	1N3890	12FL10,502	1N4002	D100	1N4003	UT252
1N3048A	1N3099,A	1N3890	12FT10	1N4002	D1201A	1N4003	1N1101
1N3050A	1N3100,A	1N3890	12FV10	1N4002	DT230A	1N4003	1N1219,A,B
1N3051A	1N3101,A	1N3891	A28B	1N4002	E1	1N4003	1N1220,A,B
1N3051A,B	UZ3051,A,B	1N3891	D2412B	1N4002	ED3101	1N4003	1N1253
1N3154	CD4112	1N3891	RT20	1N4002	EM501	1N4003	1N1488
1N3155	1N199	1N3891	ST2FR20P	1N4002	ER11	1N4003	1N1557
1N3155A	1N3148	1N3891	TPR1220	1N4002	ER182	1N4003	1N1645
1N3156	1N1530	1N3891	12FL20,502	1N4002	G1	1N4003	1N1646
1N3156	1N2790	1N3891	12FT20	1N4002	G100B	1N4003	1N1693
1N3156	1N3200	1N3891	12FV20	1N4002	G1B	1N4003	1N1703
1N3156	1N3201	1N3892	A28C	1N4002	GER4002	1N4003	1N1709
1N3156	1N430	1N3892	D2412C	1N4002	GP10B	1N4003	1N1909
1N3156	CD4115	1N3892	RT30	1N4002	HC68	1N4003	1N2015
1N3157	1N1530A	1N3892	ST2FR30P	1N4002	HGR-10	1N4003	1N2016
1N3157	1N3202	1N3893	A28D	1N4002	J-1	1N4003	1N2069,A
1N3157	1N430A	1N3893	D2412D	1N4002	M100B	1N4003	1N2074
1N3157A	1N430B	1N3893	RT40	1N4002	M68,A,B,C	1N4003	1N2075
1N3182	1N3182	1N3893	ST2FR40P	1N4002	MB236	1N4003	1N2082
1N3208	A40F	1N3893	TPR1240	1N4002	MB244	1N4003	1N2105
1N3209	A40A	1N3893	12FL30,502	1N4002	PA310	1N4003	1N2482
1N3209	ST210E	1N3893	12FL40,502	1N4002	PS410	1N4003	1N2611
1N3210	A40B	1N3893	12FT30	1N4002	PT510	1N4003	1N3074
1N3210	ST220E	1N3893	12FV30	1N4002	RG1123	1N4003	1N3075
1N3210	TR151	1N3893	12FV40	1N4002	S1010	1N4003	1N318,A
1N3211	A40C	1N3899	D2520F	1N4002	SD-1	1N4003	1N3189
1N3211	ST230E	1N3900	D2520A	1N4002	SEN110	1N4003	1N3193
1N3211	TR251	1N3901	BYX30-200,H	1N4002	SGR100	1N4003	1N325,A
1N3211	TR252	1N3901	D2520B	1N4002	SI-100E	1N4003	1N3253
1N3211	TR300	1N3902	BYX30-300,H	1N4002	SL91	1N4003	1N361,A
1N3211	TR301	1N3902	D2520C	1N4002	SLA-12	1N4003	1N3611
1N3211	TR302	1N3903	BYX30-400,R	1N4002	SR3502	1N4003	1N3866
1N3212	A40D	1N3903	D2520D	1N4002	SR6560	1N4003	1N3981
1N3212	ST240E	1N3909	D2540F	1N4002	SR8110	1N4003	1N4245
1N3212	TR351	1N3909	NS3000Q	1N4002	TA10	1N4003	1N4365
1N3212	TR401	1N3910	D2540A	1N4002	TK10	1N4003	1N441,B
1N3213	A40E	1N3910	NS30001	1N4002	TK11	1N4003	1N531
1N3213	ST250E	1N3911	D2540B	1N4002	TS-1	1N4003	1N538
1N3214	A40M	1N3911	NS30002	1N4002	TW10	1N4003	1N5614GP
1N3214	ST260E	1N3912	D2540C	1N4002	UT112	1N4003	1N601,A
1N3491	A44F	1N3912	NS30003	1N4002	UT236	1N4003	1N602,A
1N3491	25PW5	1N3913	D2540D	1N4002	UT249	1N4003	5A2
1N3491	A44A	1N3913	NS30004	1N4002	UT251	1N4003	10B2
1N3492	BYX21100	1N3993A	1N1588	1N4002	1N1110	1N4004	20A2
1N3492	BYX21L100	1N3993A	1N1588A	1N4002	1N1218,A,B	1N4004	A14C
1N3492	25PW10	1N3993A	1N1599	1N4002	1N1252	1N4004	A14D
1N3493	A44B	1N3993A	1N1599A	1N4002	1N1487	1N4004	A300
1N3493	BYX21200	1N3995A	1N1482	1N4002	1N1556	1N4004	AA300
1N3493	BYX21L200	1N3995A	1N1589	1N4002	1N1692	1N4004	AA400
1N3493	PZ-140B	1N3995A	1N1589A	1N4002	1N1702	1N4004	AR19
1N3493	25PW20	1N3995A	1N1600	1N4002	1N1708	1N4004	AR20
1N3493R	BYX20200R	1N3995A	1N1600A	1N4002	1N1908	1N4004	B300
1N3493R	BYX21200R	1N3995A	1N2041	1N4002	1N2014	1N4004	B400
1N3493R	BYY20	1N3997A	1N1590	1N4002	1N2073	1N4004	BA300
1N3493R	BYY20/200	1N3997A	1N1590A	1N4002	1N2081	1N4004	BA400
1N3493R	BYY21/200	1N3997A	1N1601	1N4002	1N2104	1N4004	BF4-40L
1N3494	A44C	1N3997A	1N1601A	1N4002	1N2610	1N4004	BY112
1N3494	25PW30	1N3997RA	1N3984	1N4002	1N3073	1N4004	BY116
1N3495	A44D	1N3997RA	1N1803	1N4002	1N317,A	1N4004	BY124
1N3495	BYX216400	1N3997RA	1N1803A	1N4002	1N324,A	1N4004	BY125
1N3495	PZ-140D	1N3998A	1N1483	1N4002	1N360,A	1N4004	BYX36600Q
1N3495	25PW40	1N3998A	1N1591	1N4002	1N4364	1N4004	BYY33
1N3495R	BYX21L400R	1N3998A	1N3986	1N4002	1N440,B	1N4004	BYY34
1N3766	S25A8	1N3998RA	1N1804	1N4002	1N530	1N4004	CER70,A,B,Q
1N3766	S40A8	1N3998RA	1N1804A	1N4002	1N537	1N4004	D1201D
1N3766	1N2286	1N4001	A14F	1N4002	1N600,A	1N4004	D300
1N3766	1N2456			1N4002	5A1	1N4004	DI-44
						1N4004	DI-54
						1N4004	DI-74

MOTOR.	Industry Standard No.	MOTOR.	Industry Standard No.	MOTOR.	Industry Standard No.	MOTOR.	Industry Standard No.
1N4004	DT230H	1N4004	1N604,A	1N4005	1N4247	1N4007	R1000
1N4004	E3	1N4004	2/A4	1N4005	1N4368	1N4007	S0M
1N4004	E4	1N4004	4D4	1N4005	1N4369	1N4007	S10100
1N4004	ED3104	1N4004	5A	1N4005	1N444,B	1N4007	S1090
1N4004	EM503	1N4004	5A3	1N4005	1N445,B	1N4007	SEN1100
1N4004	EM504	1N4004	5A4	1N4005	1N554	1N4007	SGR1000A
1N4004	ER184	1N4004	5D4	1N4005	1N555	1N4007	SI-1000E
1N4004	ER31	1N4004	10B3	1N4005	1N547	1N4007	SL610
1N4004	ER41	1N4004	10B4	1N4005	1N5618GP	1N4007	SL710
1N4004	F3	1N4004	20A3	1N4005	1N596	1N4007	SLA-19
1N4004	G100F	1N4004	20A4	1N4005	1N605,A	1N4007	SR6593
1N4004	G100G	1N4004	30C	1N4005	1N606,A	1N4007	SRB1100
1N4004	G1F	1N4004	40C	1N4006	4D6	1N4007	T1000
1N4004	G1G	1N4005	A14E	1N4006	5A5	1N4007	TA100
1N4004	GER4004	1N4005	A14M	1N4006	5A6	1N4007	TA1000
1N4004	GP10G	1N4005	AA500	1N4006	5D6	1N4007	TW100
1N4004	HC70	1N4005	AA600	1N4006	10B5	1N4007	UT347
1N4004	HGR-30	1N4005	AR21	1N4006	10B6	1N4007	UT363
1N4004	HGR-40	1N4005	AR22	1N4006	20A5	1N4007	UT364
1N4004	J-4	1N4005	B500	1N4006	20A6	1N4007	1N1260
1N4004	M100F	1N4005	B600	1N4006	A14N	1N4007	1N1261
1N4004	M100G	1N4005	BA500	1N4006	A800	1N4007	1N1443,A,B
1N4004	M4	1N4005	BA600	1N4006	AA800	1N4007	1N1916
1N4004	M70,A,B,C	1N4005	BF4-60L	1N4006	AR23	1N4007	1N2502
1N4004	MB238	1N4005	CER500,A,B,C	1N4006	B800	1N4007	1N2506
1N4004	MB239	1N4005	CER71,A,B,C	1N4006	BA800	1N4007	1N2617
1N4004	MB246	1N4005	D1201M	1N4006	BF4-80L	1N4007	1N2865
1N4004	MB247	1N4005	D500	1N4006	BY126	1N4007	1N321,A
1N4004	PA325	1N4005	DI-46	1N4006	CER72,A,B,C,D	1N4007	1N322,A
1N4004	PA330	1N4005	DI-56	1N4006	D1201N	1N4007	1N328,A
1N4004	PA340	1N4005	DI-76	1N4006	D800	1N4007	1N329,A
1N4004	PS425	1N4005	E6	1N4006	DI-48	1N4007	1N3486
1N4004	PS430	1N4005	ED3106	1N4006	DI-58	1N4007	1N3563
1N4004	PS435	1N4005	EM505	1N4006	DI-78	1N4007	1N364,A
1N4004	PS440	1N4005	EM506	1N4006	E8	1N4007	1N365,A
1N4004	PT525	1N4005	ER185	1N4006	ED3108	1N4007	1N3869
1N4004	PT530	1N4005	ER51	1N4006	EM508	1N4007	1N4249
1N4004	PT540	1N4005	ER61	1N4006	ER186	1N4007	1N561
1N4004	R400	1N4005	F6	1N4006	ER81	1N4007	1N5622GP
1N4004	S1030	1N4005	G100H	1N4006	F8	1N4007	1N598
1N4004	S1040	1N4005	G100J	1N4006	G100K	1N4007	3CP810
1N4004	S4M	1N4005	G1H	1N4006	G1K	1N4007	5A10
1N4004	SD-4	1N4005	G1J	1N4006	G8	1N4007	10B10
1N4004	SEN130	1N4005	G6	1N4006	GER4006	1N4007	20A10
1N4004	SGR400A	1N4005	GER4005	1N4006	GP10K	1N4102	1N1313
1N4004	SI-300E	1N4005	G210J	1N4006	H800	1N4102	1N1315A
1N4004	SI-400E	1N4005	HC71	1N4006	HC72	1N4113	1N1317
1N4004	SL93	1N4005	HGR-60	1N4006	J-8	1N4113	1N1317A
1N4004	SLA-14	1N4005	J-6	1N4006	M100K	1N4370	MZ92-2.4
1N4004	SLA-15	1N4005	M100H	1N4006	M72,A,B,C	1N4370	1N5837
1N4004	SR462	1N4005	M100J	1N4006	M8	1N4370A	MLV4370A
1N4004	SR6569	1N4005	M500,A,B,C	1N4006	MB242	1N4370A	1N4360
1N4004	SRB140	1N4005	M6	1N4006	MB250	1N4371	MZ92-2.7
1N4004	TA300	1N4005	M71,A,B,C	1N4006	PT580	1N4371	1N5839
1N4004	TA40	1N4005	MB240	1N4006	R800	1N4372	MZ92-3.0
1N4004	TA400	1N4005	MB241	1N4006	S1070	1N4372	1N5841
1N4004	TK30	1N4005	MB248	1N4006	S1080	1N4372A	MLV4372A
1N4004	TK40	1N4005	MB249	1N4006	S8M	1N4565A	PS3546
1N4004	TK41	1N4005	PA350	1N4006	SD-8	1N4568A	PS3549
1N4004	TS-4	1N4005	PA360	1N4006	SEN180	1N4570A	PS3555
1N4004	TW30	1N4005	PS450	1N4006	SGR800A	1N4573A	PS3539
1N4004	TW40	1N4005	PS460	1N4006	SI-800E	1N4576A	1N4611
1N4004	UT114	1N4005	PT550	1N4006	SL608	1N4577A	1N4611A
1N4004	UT115	1N4005	PT560	1N4006	SL708	1N4578A	1N4611B
1N4004	UT211	1N4005	R600	1N4006	SLA-18	1N4579A	1N4611C
1N4004	UT212	1N4005	S1050	1N4006	SR6404	1N4581A	1N4613
1N4004	UT213	1N4005	S1060	1N4006	SR6592	1N4581A	1N4612A
1N4004	UT235	1N4005	S6M	1N4006	SRB180	1N4582A	1N4613A
1N4004	UT244	1N4005	SD-6	1N4006	T800	1N4583A	1N4612B
1N4004	UT254	1N4005	SEN150	1N4006	TA80	1N4583A	1N4613B
1N4004	1N1102	1N4005	SEN160	1N4006	TA800	1N4584A	1N4612C
1N4004	1N1103	1N4005	SGR600A	1N4006	TS-8	1N4584A	1N4613C
1N4004	1N1169,A	1N4005	SI-500E	1N4006	TW80	1N4722	HC300
1N4004	1N1221,A,B	1N4005	SI-600E	1N4006	UT119	1N4723	HC500
1N4004	1N1222,A,B	1N4005	SLA-16	1N4006	UT258	1N4728	CD3110001
1N4004	1N1254	1N4005	SLA-17	1N4006	UT361	1N4728	MZ1000-1
1N4004	1N1255,A	1N4005	SR3946	1N4006	UT362	1N4728	1N5008
1N4004	1N1489	1N4005	SRB160	1N4006	1N1105	1N4728A	1N4649
1N4004	1N1490	1N4005	TA500	1N4006	1N1225,A,B	1N4728A	1N5008A
1N4004	1N1558	1N4005	TA60	1N4006	1N1226,A,B	1N4729	MZ1000-2
1N4004	1N1559	1N4005	TA600	1N4006	1N1258	1N4729	1N5009
1N4004	1N1647	1N4005	TK50	1N4006	1N1259	1N4729A	1N4650
1N4004	1N1648	1N4005	TK60	1N4006	1N1914	1N4729A	1N5009A
1N4004	1N1649	1N4005	TK61	1N4006	1N1915	1N4730	MZ1000-3
1N4004	1N1650	1N4005	TS-6	1N4006	1N2501	1N4730	1N507
1N4004	1N1694	1N4005	TW50	1N4006	1N2505	1N4730	1N518
1N4004	1N1695	1N4005	TW60	1N4006	1N2616	1N4730A	1N5010
1N4004	1N1704	1N4005	UT117	1N4006	1N3106	1N4730A	1N507A
1N4004	1N1705	1N4005	UT118	1N4006	1N196	1N4730A	1N518A
1N4004	1N1710	1N4005	UT214	1N4006	1N3256	1N4730A	1N4651
1N4004	1N1711	1N4005	UT215	1N4006	1N327,A	1N4730A	1N5010A
1N4004	1N1763	1N4005	UT225	1N4006	1N3614	1N4731	MZ1000-4
1N4004	1N1910	1N4005	UT237	1N4006	1N363,A	1N4731	1N5011
1N4004	1N1911	1N4005	UT245	1N4006	1N560	1N4731A	1N4652
1N4004	1N2017	1N4005	UT247	1N4006	1N56200P	1N4731A	1N5011A
1N4004	1N2018	1N4005	UT257	1N4006	1N597	1N4732	MZ1000-5
1N4004	1N2019	1N4005	UT358	1N4006	5A8	1N4732	1N508
1N4004	1N2020	1N4005	1N1095	1N4006	10B8	1N4732	1N519
1N4004	1N2070,A	1N4005	1N1096	1N4006	20A8	1N4732	1N2032
1N4004	1N2076	1N4005	1N1104	1N4007	A1000	1N4732	1N5012
1N4004	1N2077	1N4005	1N1223,A,B	1N4007	A14P	1N4732A	1N1484
1N4004	1N2078	1N4005	1N1224,A,B	1N4007	AA1000	1N4732A	1N508A
1N4004	1N2083	1N4005	1N1256	1N4007	AR24	1N4732A	1N519A
1N4004	1N2084	1N4005	1N1257	1N4007	B1000	1N4732A	1N4653
1N4004	1N2106	1N4005	1N1486	1N4007	BA1000	1N4732A	1N5012A
1N4004	1N2107	1N4005	1N1491	1N4007	BF4-100L	1N4733	MZ1000-6
1N4004	1N2116	1N4005	1N1492	1N4007	BY128	1N4733	1N5013
1N4004	1N2483	1N4005	1N1560	1N4007	CER75,A,B,C,D	1N4733A	1N4654
1N4004	1N2612	1N4005	1N1651	1N4007	D1000	1N4733A	1N5013A
1N4004	1N2613	1N4005	1N1652	1N4007	D1201P	1N4734	MZ1000-7
1N4004	1N3076	1N4005	1N1653	1N4007	DI-410	1N4734	1N509
1N4004	1N3077	1N4005	1N1696	1N4007	DI-510	1N4734	1N520
1N4004	1N3078	1N4005	1N1697	1N4007	DI-710	1N4734	1N2033
1N4004	1N3079	1N4005	1N1706	1N4007	E10	1N4734	1N5014
1N4004	1N3519,A	1N4005	1N1712	1N4007	ED3110	1N4734A	1N509A
1N4004	1N3190	1N4005	1N1764	1N4007	EM510	1N4734A	1N520A
1N4004	1N3194	1N4005	1N1912	1N4007	ER187	1N4734A	1N765A
1N4004	1N3254	1N4005	1N1913	1N4007	F10	1N4734A	1N4655
1N4004	1N326,A	1N4005	1N2071,A	1N4007	G10	1N4734A	1N5014A
1N4004	1N3612	1N4005	1N2079	1N4007	G100M	1N4735	MZ1000-8
1N4004	1N362,A	1N4005	1N2085	1N4007	G1M	1N4735	1N766
1N4004	1N3867	1N4005	1N2086	1N4007	GER4007	1N4735	1N3443
1N4004	1N3982	1N4005	1N2108	1N4007	GP10M	1N4735	1N5015
1N4004	1N4246	1N4005	1N2484	1N4007	H1000	1N4735	1N485
1N4004	1N4366	1N4005	1N2615	1N4007	HC75	1N4735A	1N766A
1N4004	1N4367	1N4005	1N3080	1N4007	J-10	1N4735A	1N460
1N4004	1N442,B	1N4005	1N3081	1N4007	M0	1N4735A	1N4499
1N4004	1N443,B	1N4005	1N3191	1N4007	M100M	1N4735A	1N4656
1N4004	1N532	1N4005	1N3195	1N4007	M73,A,B,C	1N4736	CD3112016
1N4004	1N533	1N4005	1N320,A	1N4007	MB243	1N4736	MZ1000-9
1N4004	1N539	1N4005	1N3255	1N4007	MB251	1N4736	1N1510
1N4004	1N540	1N4005	1N3613	1N4007	MPR10		
1N4004	1N5616GP	1N4005	1N3868				
1N4004	1N603,A	1N4005	1N3983				

MOTOR.	Industry Standard No.	MOTOR.	Industry Standard No.	MOTOR.	Industry Standard No.	MOTOR.	Industry Standard No.
1N4736	1N1521	1N4741	1N4833A	1N4746	1N4168A	1N4751	1N1782
1N4736	1N1767	1N4741	1N5021	1N4746	1N4333	1N4751	1N2387
1N4736	1N2054	1N4741A	1N1772A	1N4746	1N4333A	1N4751	1N3452
1N4736	1N3444	1N4741A	1N3680B	1N4746	1N4410	1N4751	1N3690
1N4736	1N3675	1N4741A	1N4163B	1N4746	1N4838	1N4751	1N3690A
1N4736	1N3675A	1N4741A	1N4328B	1N4746	1N4838A	1N4751	1N4173
1N4736	1N4158	1N4741A	1N4466	1N4746	1N5028	1N4751	1N4173A
1N4736	1N4158A	1N4741A	1N4633	1N4746A	MZ2623-14	1N4751	1N4338
1N4736	1N4323	1N4741A	1N4662	1N4746A	MZ2623-14A	1N4751	1N4338A
1N4736	1N4323A	1N4741A	1N4833B	1N4746A	MZ2623-14B	1N4751	1N4415
1N4736	1N4400	1N4741A	1N5021A	1N4746A	1N1428	1N4751	1N4843
1N4736	1N5016	1N4741A	1N5068	1N4746A	1N1515A	1N4751	1N4843A
1N4736,A	U24736,A	1N4742	MZ1000-15	1N4746A	1N1526A	1N4751	1N5035
1N4736A	1N1510A	1N4742	1N1513	1N4746A	1N1777A	1N4751A	1N1782A
1N4736A	1N1521A	1N4742	1N1524	1N4746A	1N5285B	1N4751A	1N3690B
1N4736A	1N1767A	1N4742	1N1773	1N4746A	1N4168B	1N4751A	1N4173B
1N4736A	1N4158B	1N4742	1N1877	1N4746A	1N4333B	1N4751A	1N4338B
1N4736A	1N4323B	1N4742	1N3435	1N4746A	1N4471	1N4751A	1N4476
1N4736A	1N4461	1N4742	1N3447	1N4746A	1N4638	1N4751A	1N4643
1N4736A	1N4628	1N4742	1N3681	1N4746A	1N4667	1N4751A	1N4672
1N4736A	1N4657	1N4742	1N3681A	1N4746A	1N4838B	1N4751A	1N4843B
1N4736A	1N5016A	1N4742	1N4164	1N4746A	1N5028A	1N4751A	1N5035A
1N4736A	1N5063	1N4742	1N4164A	1N4746A	1N5073	1N4751A	1N5077
1N4737	MZ1000-10	1N4742	1N4329	1N4747	MZ1000-20	1N4752	CD3112032
1N4737	1N1768	1N4742	1N4329A	1N4747	1N1778	1N4752	MZ1000-25
1N4737	1N3676	1N4742	1N4406	1N4747	1N2039	1N4752	1N1783
1N4737	1N3676A	1N4742	1N4834	1N4747	1N3686	1N4752	1N1882
1N4737	1N4159A	1N4742	1N4834A	1N4747	1N3686A	1N4752	1N3440
1N4737	1N4324	1N4742	1N5022	1N4747	1N4169	1N4752	1N3453
1N4737	1N4324A	1N4742A	1N1426	1N4747	1N4169A	1N4752	1N3691
1N4737	1N4401	1N4742A	1N1513A	1N4747	1N4334	1N4752	1N3691A
1N4737	1N5017	1N4742A	1N1524A	1N4747	1N4334A	1N4752	1N4174
1N4737A	1N1768A	1N4742A	1N1773A	1N4747	1N4411	1N4752	1N4174A
1N4737A	1N3112	1N4742A	1N3681B	1N4747	1N4839	1N4752	1N4339
1N4737A	1N3676B	1N4742A	1N4164B	1N4747	1N4839A	1N4752	1N4339A
1N4737A	1N4159B	1N4742A	1N4329B	1N4747	1N4881	1N4752	1N4416
1N4737A	1N4324B	1N4742A	1N4467	1N4747	1N5030	1N4752	1N4503
1N4737A	1N4462	1N4742A	1N4634	1N4747A	1N1778A	1N4752	1N4844
1N4737A	1N4629	1N4742A	1N4663	1N4747A	1N3686B	1N4752	1N4844A
1N4737A	1N4658	1N4742A	1N4834B	1N4747A	1N4169B	1N4752A	1N5036
1N4737A	1N5017A	1N4742A	1N4883	1N4747A	1N4334B	1N4752A	1N3783
1N4737A	1N5064	1N4742A	1N5022A	1N4747A	1N4472	1N4752A	1N3691B
1N4738	MZ1000-11	1N4743	MZ1000-16	1N4747A	1N4639	1N4752A	1N4174B
1N4738	1N1511	1N4743	1N1774	1N4747A	1N4669	1N4752A	1N4339B
1N4738	1N1522	1N4743	1N2037	1N4747A	1N4859B	1N4752A	1N4477
1N4738	1N1769	1N4743	1N3682	1N4747A	1N4884	1N4752A	1N4644
1N4738	1N3433	1N4743	1N3682A	1N4747A	1N5030A	1N4752A	1N4673
1N4738	1N3445	1N4743	1N4165	1N4748	MZ1000-21	1N4752A	1N4844B
1N4738	1N3677	1N4743	1N4165A	1N4748	1N1516	1N4752A	1N5036A
1N4738	1N3677A	1N4743	1N4330	1N4748	1N1527	1N4752A	1N5078
1N4738	1N4160	1N4743	1N4330A	1N4748	1N1779	1N4753	CD3100025
1N4738	1N4160A	1N4743	1N4407	1N4748	1N1880	1N4753	MZ1000-26
1N4738	1N4325	1N4743	1N4835	1N4748	1N3438	1N4753	1N1784
1N4738	1N4325A	1N4743	1N4835A	1N4748	1N3450	1N4753	1N3692
1N4738	1N5018	1N4743	1N5023	1N4748	1N3687	1N4753	1N3692A
1N4738	1N1425	1N4743A	M2623-9	1N4748	1N3687A	1N4753	1N4175
1N4738A	1N1511A	1N4743A	MZ2623-9A	1N4748	1N4170	1N4753	1N4175A
1N4738A	1N1522A	1N4743A	MZ2623-9B	1N4748	1N4170A	1N4753	1N4340
1N4738A	1N1769A	1N4743A	1N1774A	1N4748	1N4335	1N4753	1N4340A
1N4738A	1N3677B	1N4743A	1N3682B	1N4748	1N4335A	1N4753	1N4417
1N4738A	1N4160B	1N4743A	1N4165B	1N4748	1N4412	1N4753	1N4845
1N4738A	1N4325B	1N4743A	1N4330B	1N4748	1N4840	1N4753	1N4845A
1N4738A	1N4463	1N4743A	1N4468	1N4748	1N4840A	1N4753	1N4882
1N4738A	1N4630	1N4743A	1N4664	1N4748	1N5031	1N4753	1N5037
1N4738A	1N4659	1N4743A	1N4835B	1N4748A	1N1429	1N4753A	1N1784A
1N4738A	1N5018A	1N4743A	1N5023A	1N4748A	1N1516A	1N4753A	1N3692B
1N4738A	1N5065	1N4743A	1N5069	1N4748A	1N1527A	1N4753A	1N4175B
1N4739	MZ1000-12	1N4743A	MZ1000-17	1N4748A	1N1779A	1N4753A	1N4340B
1N4739	1N1770	1N4744	1N1514	1N4748A	1N3687B	1N4753A	1N4478
1N4739	1N2055	1N4744	1N1525	1N4748A	1N4170B	1N4753A	1N4645
1N4739	1N3678	1N4744	1N1775	1N4748A	1N4335B	1N4753A	1N4674
1N4739	1N3678A	1N4744	1N1878	1N4748A	1N4473	1N4753A	1N4845B
1N4739	1N4161	1N4744	1N3436	1N4748A	1N4640	1N4753A	1N5037A
1N4739	1N4161A	1N4744	1N3448	1N4748A	1N4840B	1N4753A	1N5079
1N4739	1N4326	1N4744	1N3683	1N4748A	1N5031A	1N4754	MZ1000-27
1N4739	1N4326A	1N4744	1N3683A	1N4748A	1N5074	1N4754	1N1785
1N4739	1N4403	1N4744	1N4166	1N4749	MZ1000-22	1N4754	1N1883
1N4739	1N4831	1N4744	1N4166A	1N4749	1N1780	1N4754	1N3441
1N4739	1N4831A	1N4744	1N4331	1N4749	1N2040	1N4754	1N3454
1N4739	1N5019	1N4744	1N4331A	1N4749	1N3688	1N4754	1N3693
1N4739A	1N1770A	1N4744	1N4408	1N4749	1N3688A	1N4754	1N3693A
1N4739A	1N3678B	1N4744	1N4836	1N4749	1N4171	1N4754	1N4176
1N4739A	1N4161B	1N4744	1N4836A	1N4749	1N4171A	1N4754	1N4176A
1N4739A	1N4326B	1N4744	1N5025	1N4749	1N4336	1N4754	1N4341
1N4739A	1N4464	1N4744A	1N1427	1N4749	1N4336A	1N4754	1N4341A
1N4739A	1N4631	1N4744A	1N1514A	1N4749	1N4413	1N4754	1N4418
1N4739A	1N4660	1N4744A	1N1525A	1N4749	1N4841	1N4754	1N4846
1N4739A	1N4831B	1N4744A	1N1775A	1N4749	1N5032	1N4754	1N5038
1N4739A	1N5019A	1N4744A	1N3683B	1N4749A	MZ2623-18	1N4754A	1N1785A
1N4739A	1N5066	1N4744A	1N4166B	1N4749A	MZ2623-18A	1N4754A	1N3693B
1N4740	MZ1000-13	1N4744A	1N4331B	1N4749A	MZ2623-18B	1N4754A	1N4176B
1N4740	MZ1000-14	1N4744A	1N4469	1N4749A	1N1780A	1N4754A	1N4341B
1N4740	1N1512	1N4744A	1N4636	1N4749A	1N3688B	1N4754A	1N4479
1N4740	1N1523	1N4744A	1N4665	1N4749A	1N4171B	1N4754A	1N4646
1N4740	1N1744	1N4744A	1N4836B	1N4749A	1N4336B	1N4754A	1N4675
1N4740	1N1771	1N4744A	1N5025A	1N4749A	1N4474	1N4754A	1N4846B
1N4740	1N1876	1N4744A	1N5071	1N4749A	1N4670	1N4754A	1N5038A
1N4740	1N2036	1N4745	MZ1000-18	1N4749A	1N4841B	1N4754A	1N5080
1N4740	1N3434	1N4745	1N1776	1N4749A	1N5032A	1N4755	MZ1000-28
1N4740	1N3446	1N4745	1N2038	1N4749A	1N5075	1N4755	1N1786
1N4740	1N3679	1N4745	1N3684	1N4750	MZ1000-23	1N4755	1N3694
1N4740	1N3679A	1N4745	1N3684A	1N4750	1N1517	1N4755	1N3694A
1N4740	1N4162	1N4745	1N4167	1N4750	1N1528	1N4755	1N4177
1N4740	1N4162A	1N4745	1N4167A	1N4750	1N1781	1N4755	1N4177A
1N4740	1N4327	1N4745	1N4332	1N4750	1N1881	1N4755	1N4342
1N4740	1N4327A	1N4745	1N4332A	1N4750	1N3439	1N4755	1N4342A
1N4740	1N4404	1N4745	1N4409	1N4750	1N3451	1N4755	1N4419
1N4740	1N4832	1N4745	1N4837	1N4750	1N3689	1N4755	1N4847
1N4740	1N4832A	1N4745	1N4837A	1N4750	1N3689A	1N4755	1N4847A
1N4740	1N5020	1N4745	1N5026	1N4750	1N4172	1N4755	1N5039
1N4740A	1N1512A	1N4745A	MZ2623-12	1N4750	1N4337	1N4755A	MZ2623-25
1N4740A	1N1523A	1N4745A	MZ2623-12A	1N4750	1N4337A	1N4755A	MZ2623-25A
1N4740A	1N1771A	1N4745A	MZ2623-12B	1N4750	1N4414	1N4755A	MZ2623-25B
1N4740A	1N3679B	1N4745A	1N1776A	1N4750	1N4842	1N4755A	1N1786A
1N4740A	1N4162B	1N4745A	1N3684B	1N4750	1N4842A	1N4755A	1N3694B
1N4740A	1N4327B	1N4745A	1N4167B	1N4750	1N5034	1N4755A	1N4177B
1N4740A	1N4465	1N4745A	1N4332B	1N4750A	1N1430	1N4755A	1N4342B
1N4740A	1N4632	1N4745A	1N4470	1N4750A	1N1517A	1N4755A	1N4480
1N4740A	1N4661	1N4745A	1N4637	1N4750A	1N1528A	1N4755A	1N4647
1N4740A	1N4832B	1N4745A	1N4666	1N4750A	1N1781A	1N4755A	1N4847B
1N4740A	1N5020A	1N4745A	1N4837B	1N4750A	1N3689B	1N4755A	1N5039A
1N4740A	1N5067	1N4745A	1N5026A	1N4750A	1N4172B	1N4755A	1N5082
1N4741	1N1772	1N4745A	1N5072	1N4750A	1N4337B	1N4756	MZ1000-29
1N4741	1N3680	1N4746	MZ1000-19	1N4750A	1N4475	1N4756	1N1787
1N4741	1N3680A	1N4746	1N1515	1N4750A	1N4642	1N4756	1N1884
1N4741	1N4163	1N4746	1N1526	1N4750A	1N4671	1N4756	1N3442
1N4741	1N4163A	1N4746	1N1777	1N4750A	1N4842B	1N4756	1N3455
1N4741	1N4328	1N4746	1N1879	1N4750A	1N5034A	1N4756	1N3695
1N4741	1N4328A	1N4746	1N3437	1N4750A	1N5072	1N4756	1N3695A
1N4741	1N4405	1N4746	1N3449	1N4751	MZ1000-24	1N4756	1N4178
1N4741	1N4833	1N4746	1N3685			1N4756	1N4178A
		1N4746	1N3685A			1N4756	1N4343
		1N4746	1N4168				

MOTOR.	Industry Standard No.
1N4756	1N4343A
1N4756	1N4420
1N4756	1N4848
1N4756	1N4848A
1N4756	1N5041
1N4756A	1N1787A
1N4756A	1N3695B
1N4756A	1N4178B
1N4756A	1N4343B
1N4756A	1N4481
1N4756A	1N4648
1N4756A	1N4677
1N4756A	1N4848B
1N4756A	1N5041A
1N4756A	1N5084
1N4757	MZ1000-30
1N4757	1N1788
1N4757	1N3696
1N4757	1N3696A
1N4757	1N4179
1N4757	1N4179A
1N4757	1N4344
1N4757	1N4344A
1N4757	1N4421
1N4757	1N4849
1N4757	1N4849A
1N4757	1N5043
1N4757A	1N1788A
1N4757A	1N3696B
1N4757A	1N4179B
1N4757A	1N4344B
1N4757A	1N4482
1N4757A	1N4849B
1N4757A	1N5043A
1N4757A	1N5086
1N4758	MZ1000-31
1N4758	1N1789
1N4758	1N1885
1N4758	1N3456
1N4758	1N3697
1N4758	1N3697A
1N4758	1N4180
1N4758	1N4180A
1N4758	1N4345
1N4758	1N4345A
1N4758	1N4422
1N4758	1N4850
1N4758	1N4850A
1N4758	1N5045
1N4758A	1N1789A
1N4758A	1N3697B
1N4758A	1N4180B
1N4758A	1N4345B
1N4758A	1N4483
1N4758A	1N4850B
1N4758A	1N5045A
1N4758A	1N5087
1N4759	MZ1000-32
1N4759	1N1790
1N4759	1N3698
1N4759	1N3698A
1N4759	1N4181
1N4759	1N4181A
1N4759	1N4346
1N4759	1N4346A
1N4759	1N4423
1N4759	1N4851
1N4759	1N4851A
1N4759	1N5046
1N4759A	1N1790A
1N4759A	1N3698B
1N4759A	1N4181B
1N4759A	1N4346B
1N4759A	1N4484
1N4759A	1N4851B
1N4759A	1N5046A
1N4759A	1N5089
1N4760	MZ1000-33
1N4760	1N1791
1N4760	1N1886
1N4760	1N3457
1N4760	1N3699
1N4760	1N3699A
1N4760	1N4182
1N4760	1N4182A
1N4760	1N4347
1N4760	1N4347A
1N4760	1N4424
1N4760	1N4852
1N4760	1N4852A
1N4760	1N5047
1N4760A	1N1431
1N4760A	1N1791A
1N4760A	1N3699B
1N4760A	1N4182B
1N4760A	1N4347B
1N4760A	1N4485
1N4760A	1N4852B
1N4760A	1N5047A
1N4760A	1N5090
1N4761	MZ1000-34
1N4761	1N1792
1N4761	1N3700
1N4761	1N3700A
1N4761	1N4183
1N4761	1N4183A
1N4761	1N4348
1N4761	1N4348A
1N4761	1N4425
1N4761	1N4853
1N4761	1N4853A
1N4761	1N5048
1N4761A	1N1792A
1N4761A	1N3700B
1N4761A	1N4183B
1N4761A	1N4348B
1N4761A	1N4486
1N4761A	1N4853B
1N4761A	1N5048A
1N4761A	1N5092
1N4762	1N1793
1N4762	1N1887
1N4762	1N3458
1N4762	1N3701
1N4762	1N3701A
1N4762	1N4184
1N4762	1N4184A
1N4762	1N4349
1N4762	1N4349A
1N4762	1N4426
1N4762	1N4854
1N4762	1N4854A
1N4762	1N5049
1N4762A	1N1793A
1N4762A	1N3701B
1N4762A	1N4184B
1N4762A	1N4349B
1N4762A	1N4487
1N4762A	1N4854B
1N4762A	1N5049A
1N4762A	1N5094
1N4763	MZ1000-36
1N4763	1N1794
1N4763	1N3702
1N4763	1N3702A
1N4763	1N4185
1N4763	1N4185A
1N4763	1N4350
1N4763	1N4350A
1N4763	1N4427
1N4763	1N4855
1N4763	1N4855A
1N4763	1N5050
1N4763A	1N1794A
1N4763A	1N3702B
1N4763A	1N4096
1N4763A	1N4185B
1N4763A	1N4350B
1N4763A	1N4488
1N4763A	1N4855B
1N4763A	1N5050A
1N4763A	1N5095
1N4764	MZ1000-37
1N4764	1N1795
1N4764	1N1888
1N4764	1N3459
1N4764	1N3703
1N4764	1N3703A
1N4764	1N4186
1N4764	1N4186A
1N4764	1N4351
1N4764	1N4351A
1N4764	1N4428
1N4764	1N4856
1N4764	1N4856A
1N4764	1N5051
1N4764,A	U24764,A
1N4764A	1N1432
1N4764A	1N1795A
1N4764A	1N3703B
1N4764A	1N4097
1N4764A	1N4186B
1N4764A	1N4351B
1N4764A	1N4489
1N4764A	1N4856B
1N4764A	1N5051A
1N4933	A114F
1N4933	D2201F
1N4933	HPR-5
1N4933	N31000
1N4933	N3500
1N4933	R142001Q
1N4933	RG1A
1N4933	RGP10A
1N4933	RGP15A
1N4933	RGP20A
1N4933	RL005
1N4933	RMC005
1N4933	SEN105FR
1N4933	SRSFR105
1N4933	TKP5
1N4933	TS3
1N4933	TS5
1N4933	TSV
1N4933	UTR01
1N4933	UTR02
1N4933	1N4933GP
1N4933	4FB5
1N4933	4PC5
1N4933	18FA5
1N4933	18FB5
1N4933	18FC5
1N4934	A114A
1N4934	D2201A
1N4934	P1
1N4934	FR1
1N4934	GR1
1N4934	HPR-1Q
1N4934	MB214
1N4934	MB221
1N4934	MRP100
1N4934	MRP200
1N4934	N31001
1N4934	N3501
1N4934	R142011Q
1N4934	RG1B
1N4934	RGP10B
1N4934	RGP15B
1N4934	RGP20B
1N4934	RL010
1N4934	RMC010
1N4934	S1A1F
1N4934	SEN110FR
1N4934	SRSFR110
1N4934	TKP10
1N4934	TS10
1N4934	UTR10
1N4934	UTR11
1N4934	UTR12
1N4934	1N4934GP
1N4934	1N5055
1N4934	4FB10
1N4934	4FC
1N4934	4FC10
1N4934	18FA10
1N4934	18FB10
1N4934	18FC10
1N4935	A114B
1N4935	D2201B
1N4935	BR2
1N4935	F2
1N4935	FR2
1N4935	GR2
1N4935	HPR-150
1N4935	HPR-200
1N4935	MB215
1N4935	MB222
1N4935	N31002
1N4935	N3502
1N4935	R142021Q
1N4935	RG1-D
1N4935	RG1D
1N4935	RGP10D
1N4935	RGP15D
1N4935	RGP20D
1N4935	RL020
1N4935	RMC020
1N4935	S1A2F
1N4935	32F
1N4935	SRSFR120
1N4935	TKP20
1N4935	TS20
1N4935	UTR20
1N4935	UTR21
1N4935	UTR22
1N4935	1N4935GP
1N4935	1N4942
1N4935	1N5056
1N4935	1N5615GP
1N4935	4FB20
1N4935	4FC20
1N4935	18FA20
1N4935	18FB20
1N4935	18FC20
1N4936	A114C
1N4936	A114D
1N4936	D2201D
1N4936	ER4
1N4936	P4
1N4936	FR3
1N4936	FR4
1N4936	GR4
1N4936	MB217
1N4936	MB224
1N4936	MRP400
1N4936	N31004
1N4936	N3504
1N4936	R142041Q
1N4936	RG1F
1N4936	RG1G
1N4936	RGP10F
1N4936	RGP10G
1N4936	RGP15F
1N4936	RGP15G
1N4936	RGP20F
1N4936	RGP20G
1N4936	RL040
1N4936	RMC040
1N4936	S1A3F
1N4936	S1A4F
1N4936	S4F
1N4936	SEN120FR
1N4936	SEN140FR
1N4936	SRSFR140
1N4936	TKP40
1N4936	TS40
1N4936	UTR40
1N4936	UTR41
1N4936	UTR42
1N4936	1N4936GP
1N4936	1N4943
1N4936	1N4944
1N4936	1N5057
1N4936	1N5206
1N4936	1N5617GP
1N4936	4FB30
1N4936	4FB40
1N4936	4FC40
1N4936	18FA30
1N4936	18FA40
1N4936	18FB30
1N4936	18FB40
1N4936	18FC30
1N4936	18FC40
1N4937	A114E
1N4937	A114M
1N4937	D2201M
1N4937	ER6
1N4937	P5
1N4937	FR6
1N4937	GR6
1N4937	MB218
1N4937	MB219
1N4937	MB225
1N4937	MB226
1N4937	MRP600
1N4937	N31005
1N4937	N31006
1N4937	N3505
1N4937	N3506
1N4937	R142061Q
1N4937	RG1H
1N4937	RG1J
1N4937	RGP10J
1N4937	RGP15H
1N4937	RGP15J
1N4937	RGP20H
1N4937	RGP20J
1N4937	RL060
1N4937	RMC060
1N4937	S1A5F
1N4937	S1AGF
1N4937	S6F
1N4937	SEN150FR
1N4937	SEN160FR
1N4937	SRSFR150
1N4937	SRSFR160
1N4937	TKP50
1N4937	TKP60
1N4937	TS50
1N4937	TS60
1N4937	UTR50
1N4937	UTR52
1N4937	UTR60
1N4937	UTR61
1N4937	UTR62
1N4937	1N4937GP
1N4937	1N4945
1N4937	1N4946
1N4937	1N5058
1N4937	1N5619GP
1N5138	V7B
1N5139	BA149
1N5139	PC139
1N5139	PC141
1N5139	PG207
1N5139	PG307
1N5139	V907B
1N5139	1N4801
1N5139,A	1N5139,A
1N5140	PC112
1N5140	PC115
1N5140	PC135
1N5140	V10E
1N5140	V910E
1N5140,A	1N4803
1N5141	V12E
1N5141	1N4804
1N5141,A	1N5141,A
1N5141A	1N3554
1N5142	PC124
1N5142	PC125
1N5142	PC126
1N5142	V15E
1N5142	1N4085
1N5142,A	1N5142,A
1N5143	1N4806
1N5143,A	1N5143,A
1N5144	PC113
1N5144	PC116
1N5144	PC136
1N5144	1N3555
1N5144	1N3557
1N5144	1N4807
1N5144,A	1N5144,A
1N5145	V27E
1N5145	1N4808
1N5145,A	1N5145,A
1N5146	PC128
1N5146	PC129
1N5146	PC130
1N5146	V33E
1N5146	1N4809
1N5146,A	1N5146,A
1N5147	V34E
1N5147	1N4810
1N5147,A	1N5147,A
1N5148	PC114
1N5148	PC117
1N5148	PC122
1N5148	PC137
1N5148	V47E
1N5148	1N3556
1N5148	1N4811
1N5148	1N4812
1N5148,A	1N5148,A
1N5221	1N5985
1N5221A	MZ500-1
1N5221A	1N3477
1N5221A	1N371
1N5221B	1N3198
1N5221B	1N3477A
1N5221B	1N370
1N5223	1N3898
1N5223	1N5986
1N5223A	MZ500-2
1N5223A	1N465
1N5223B	1N465A
1N5225	1N5987
1N5225A	MZ500-3
1N5225A	1N372
1N5226	1N5988
1N5226A	MZ500-4
1N5226A	1N466
1N5226B	1N5506
1N5226B	1N466A
1N5227	1N5989
1N5227A	MZ500-5
1N5227A	1N373
1N5227B	1N5990
1N5228	1N5990
1N5228A	MZ500-6
1N5228A	1N1927
1N5228A	1N1954
1N5228A	1N981
1N5228B	1N5508
1N5228B	1N467
1N5228B	1N467A
1N5229	1N5991
1N5229A	MZ500-7
1N5229B	1N574
1N5229B	1N5509
1N5230	1N5992
1N5230A	MZ500-8
1N5230A	1N1928
1N5230A	1N1985
1N5230A	1N1982
1N5230A	1N375
1N5230A	1N468
1N5230A	1N674
1N5230B	1N5510
1N5230B	1N468A
1N5230B	1N5728
1N5231	1N5993
1N5231A	MZ500-9
1N5231A	1N4095
1N5231B	1N5511
1N5231B	1N5729
1N5232	1N5994
1N5232A	MZ500-10
1N5232A	1N1929
1N5232A	1N1956
1N5232A	1N1983
1N5232B	1N3512
1N5232B	1N469
1N5232B	1N5730
1N5233A	1N376
1N5234	1N5995
1N5234A	MZ500-11
1N5234A	1N3411
1N5234B	1N5513
1N5234B	1N5731
1N5234B	1N675
1N5235	1N5996
1N5235,A,B	U23235,A,B
1N5235A	MZ500-12
1N5235A	1N1930
1N5235A	1N1957
1N5235A	1N984
1N5235A	1N3412
1N5235A	1N5514
1N5235B	1N470
1N5235B	1N470A
1N5235B	1V752B
1N5236	1N5997
1N5236A	MZ500-13
1N5236A	1N3413
1N5236A	1N377
1N5236B	1N5515
1N5236B	1N5733B
1N5237	1N5998
1N5237A	MZ500-14
1N5237A	1N1931
1N5237A	1N1958
1N5237A	1N985
1N5237A	1N5181
1N5237A	1N3414
1N5237A	1N664
1N5237B	1N3516
1N5237B	1N5734B
1N5238A	1N78
1N5239	1N5999
1N5239A	MZ500-15

MOTOR.	Industry Standard No.
1N5239B	1N5517
1N5239B	1N5735B
1N5240	1N6000
1N5240A	MZ500-16
1N5240A	1N1932
1N5240A	1N1959
1N5240A	1N1986
1N5240A	1N3415
1N5240A	1N579
1N5240B	1N3518
1N5240B	1N5736B
1N5241	1N6001
1N5241A	MZ500-17
1N5241A	1N5519
1N5242	1N6002
1N5242A	MZ500-18
1N5242A	1N1933
1N5242A	1N1960
1N5242A	1N1987
1N5242A	1N3416
1N5242A	1N665
1N5242B	1N3520
1N5242B	1N5738B
1N5243	1N6003
1N5243A	MZ500-19
1N5243A	1N580
1N5243A	1N3521
1N5243B	1N5739B
1N5245	1N6004
1N5245A	MZ500-20
1N5245A	1N1934
1N5245A	1N1961
1N5245A	1N1988
1N5245A	1N3417
1N5245B	1N3522
1N5245B	1N5740B
1N5245B	1N666
1N5246	1N6005
1N5246A	MZ500-21
1N5246A	1N581
1N5246B	1N3523
1N5246B	1N5741B
1N5248	1N6006
1N5248A	MZ500-22
1N5248A	1N1935
1N5248A	1N1962
1N5248A	1N1989
1N5248A	1N3418
1N5248A	1N667
1N5248B	1N3524
1N5248B	1N5742B
1N5249A	1N582
1N5250	1N6007
1N5250A	MZ500-23
1N5250B	1N3525
1N5250B	1N5743B
1N5251	1N6008
1N5251A	MZ500-24
1N5251A	1N1936
1N5251A	1N1963
1N5251A	1N1990
1N5251A	1N3419
1N5251A	1N668
1N5251B	1N3526
1N5251B	1N5744B
1N5252	1N6009
1N5252A	MZ500-25
1N5252A	1N583
1N5252B	1N3527
1N5252B	1N5745B
1N5254	1N6010
1N5254A	MZ500-26
1N5254A	1N1937
1N5254A	1N1964
1N5254A	1N1991
1N5254A	1N3420
1N5254A	1N669
1N5254B	1N3528
1N5254B	1N5746B
1N5255A	1N584
1N5256	1N6011
1N5256A	MZ500-27
1N5256A	1N3421
1N5256B	1N3529
1N5256B	1N5747B
1N5257	1N6012
1N5257A	MZ500-28
1N5257A	1N1938
1N5257A	1N1965
1N5257A	1N1992
1N5257A	1N3422
1N5257B	1N3530
1N5257B	1N5748B
1N5258	1N6013
1N5258A	MZ500-29
1N5258A	1N585
1N5258B	1N3531
1N5258B	1N5749B
1N5259	1N6014
1N5259A	MZ500-30
1N5259A	1N1939
1N5259A	1N1966
1N5259A	1N1993
1N5259A	1N3423
1N5259B	1N3532
1N5259B	1N5750B
1N5260	1N6015
1N5260A	MZ500-31
1N5260A	1N586
1N5260B	1N3533
1N5261	1N6016
1N5261A	MZ500-32
1N5261A	1N1940
1N5261A	1N1967
1N5261A	1N1994
1N5261A	1N3424
1N5261A	1N587
1N5261B	1N3534
1N5261B	1N5752
1N5262	CD3168
1N5262	1N6017
1N5262A	MZ500-33
1N5262A	1N5753
1N5263	1N6018
1N5263A	MZ500-34
1N5263A	1N1941
1N5263A	1N1968
1N5263A	1N1995
1N5263A	1N3425
1N5265	1N6019
1N5265A	MZ500-35
1N5266	1N6020
1N5266A	MZ500-36
1N5266A	1N1942
1N5266A	1N1969
1N5266A	1N1996
1N5266A	1N3426
1N5266B	1N670
1N5267	1N6021
1N5267A	MZ500-37
1N5268	CD3174
1N5268	1N6022
1N5268A	MZ500-38
1N5268A	1N1943
1N5268A	1N1970
1N5268A	1N1997
1N5268A	1N3427
1N5270	1N6023
1N5270A	MZ500-39
1N5271	1N6024
1N5271A	MZ500-40
1N5271A	1N1944
1N5271A	1N1971
1N5271A	1N1998
1N5271A	1N3428
1N5271A	1N671
1N5272	1N6025
1N5273	1N6026
1N5273A	1N1945
1N5273A	1N1972
1N5273A	1N1999
1N5273A	1N3429
1N5274	1N6027
1N5276	1N6028
1N5276A	1N1946
1N5276A	1N1973
1N5276A	1N2000
1N5276A	1N3430
1N5276A	1N672
1N5277	1N6029
1N5279	1N6030
1N5279A	1N1947
1N5279A	1N1974
1N5279A	1N2001
1N5279A	1N3431
1N5281	1N6031
1N5281,A,B	UZ3281,A,B
1N5281A	1N3432
1N5283	CL2210
1N5287	CL3310
1N5290	CL4710
1N5293	CL6810
1N5297	CL1020
1N5302	CL1520
1N5306	CL2220
1N5310	CL3320
1N5314	CL4720
1N5335A	525338
1N5341B	1N5118
1N5342A,B	UZ4706,A,B
1N5342B	1N4954
1N5343B	1N4955
1N5344B	1N4956
1N5346B	1N4957
1N5347B	1N4958
1N5348B	1N4959
1N5349B	1N4960
1N5350B	1N4961
1N5352B	1N4962
1N5353B	1N4963
1N5355B	1N4964
1N5357B	1N4965
1N5358B	1N4966
1N5359B	1N4967
1N5361B	1N4968
1N5363B	1N4969
1N5364A	525364
1N5365B	1N4970
1N5365B	1N4971
1N5366B	1N4972
1N5367B	1N4973
1N5368B	1N4974
1N5369B	1N4975
1N5370B	1N4976
1N5371B	1N5122
1N5372B	1N4977
1N5373B	1N4978
1N5374B	1N4979
1N5375B	1N4980
1N5377B	1N4981
1N5378B	1N4982
1N5379B	1N4983
1N5380B	1N4984
1N5381B	1N4985
1N5382B	1N5126
1N5383B	1N4986
1N5384A,B	UZ4116,A,B
1N5384A	1N4987
1N5385B	1N5127
1N5386B	1N4988
1N5387B	1N5128
1N5388A	1N4504
1N5388B	1N4989
1N5391	G2A
1N5391	GP15A
1N5391	GP20A
1N5391	P100A
1N5391	1N2858,A
1N5391	1N4816
1N5391	1N4816GP
1N5391	1N5170
1N5391	1N5171
1N5391	1N5391GP
1N5392	G2B
1N5392	GP15B
1N5392	GP20B
1N5392	MR100
1N5392	P100B
1N5392	SI1
1N5392	1N2859,A
1N5392	1N4817
1N5392	1N4817GP
1N5392	1N5004
1N5392	1N5172
1N5392	1N5392GP
1N5392	10D1
1N5392	20D1
1N5393	G2D
1N5393	GP15D
1N5393	GP20D
1N5393	MR200
1N5393	P100D
1N5393	SI2
1N5393	1N2485
1N5393	1N2860,A
1N5393	1N3082
1N5393	1N3639
1N5393	1N4818
1N5393	1N4818GP
1N5393	1N5005
1N5393	1N5393GP
1N5393	10D2
1N5393	20D2
1N5394	SI3
1N5394	1N5394GP
1N5394	10D3
1N5394	20D3
1N5395	G2G
1N5395	GP15G
1N5395	GP20G
1N5395	MR400
1N5395	P100G
1N5395	SI4
1N5395	1N2486
1N5395	1N2487
1N5395	1N2861,A
1N5395	1N2862,A
1N5395	1N3083
1N5395	1N3640
1N5395	1N4819
1N5395	1N4819GP
1N5395	1N4820
1N5395	1N4820GP
1N5395	1N5006
1N5395	1N5173
1N5395	1N5174
1N5395	1N5395GP
1N5395	10D4
1N5395	20D4
1N5396	SI5
1N5396	1N4821
1N5396	1N4821GP
1N5396	1N5396GP
1N5396	10D5
1N5396	20D5
1N5397	G2J
1N5397	GP15J
1N5397	GP20J
1N5397	MR600
1N5397	P100J
1N5397	SI6
1N5397	1N2488
1N5397	1N2489
1N5397	1N2863,A
1N5397	1N2864,A
1N5397	1N5084
1N5397	1N5641
1N5397	1N4822
1N5397	1N4822GP
1N5397	1N5007
1N5397	1N5175
1N5397	1N5176
1N5397	1N5397GP
1N5397	10D6
1N5397	20D6
1N5398	G2K
1N5398	GP15K
1N5398	GP20K
1N5398	MR800
1N5398	P100K
1N5398	SI7
1N5398	SI8
1N5398	1N3642
1N5398	1N5052
1N5398	1N5053
1N5398	1N5177
1N5398	1N5398GP
1N5398	10D8
1N5398	20D8
1N5399	G2M
1N5399	GP15M
1N5399	GP20M
1N5399	MR1000
1N5399	P100M
1N5399	SI10
1N5399	SI9
1N5399	1N5054
1N5399	1N5178
1N5399	1N5399GP
1N5399	10D10
1N5399	20D10
1N5400	S3A025
1N5401	ER100
1N5401	S3A1
1N5401	40D1
1N5402	HR200
1N5402	S3A2
1N5402	40D2
1N5403	S3A3
1N5404	HR400
1N5404	S3A4
1N5404	40D4
1N5405	S3A5
1N5406	HR600
1N5406	S3A6
1N5406	40D6
1N5407	S3A7
1N5408	S3A8
1N5408	S3A10
1N5409	S3A9
1N5441A	BA138
1N5441A	V7
1N5441A	V907
1N5441A	1N4786
1N5442A	1N4787
1N5443A	BA121
1N5443A	V10
1N5443A	V910
1N5443A	1N4788
1N5444A	V12
1N5444A	V912
1N5445A	1N4789
1N5445A	V915
1N5445A	1N4790
1N5446A	1N4791
1N5446A	1N4800
1N5447A	V20
1N5447A	V920
1N5447A	1N5552
1N5447A	1N5627
1N5448A	1N4792
1N5449A	BA125
1N5449A	V927
1N5449A	1N4793
1N5450A	V933
1N5450A	1N4794
1N5451A	V939
1N5451A	1N4795
1N5452A	1N5628
1N5452A	1N4796
1N5453A	BA150
1N5453A	V956
1N5453A	1N4797
1N5454A	V968
1N5454A	1N4798
1N5454A	1N4813
1N5455A	V982
1N5455A	1N4799
1N5455A	1N4814
1N5456A	V900
1N5456A	1N4815
1N5457A	V82E
1N5457A	1N5946
1N5457A	1N5686
1N5458A	1N5687
1N5461A	1N4091
1N5461A	1N5681
1N5461A	1N5696
1N5462A	1N4802
1N5462A	1N5682
1N5462A	1N5697
1N5463A	1N5683
1N5463A	1N5698
1N5464A	1N5684
1N5464A	1N5699
1N5465A	1N5685
1N5465A	1N5700
1N5467A	V20E
1N5467A	1N5701
1N5468A	1N5702
1N5469A	1N5688
1N5469A	1N5703
1N5470A	1N5689
1N5470A	1N5704
1N5471A	1N5690
1N5471A	1N5705
1N5472A	1N5691
1N5472A	1N5706
1N5473A	V56E
1N5473A	1N5692
1N5473A	1N5707
1N5474A	V68E
1N5474A	1N3947
1N5474A	1N5693
1N5475A	1N5708
1N5475A	1N5694
1N5475A	1N5709
1N5476A	V100E
1N5476A	V900E
1N5476A	1N5695
1N5476A	1N5710
1N5739	PC140
1N5758	MPT20
1N5758	1N5758
1N5758A	1N5758A
1N5759	MPT24
1N5760	MPT28
1N5760	1N5760
1N5760B	1N5751
1N5761	D3202Y
1N5761	G732
1N5761	H732
1N5761	MPT32
1N5761	1N5761
1N5761A	D3202U
1N5761A	1N5761A
1N5762	D30
1N5762	G735
1N5762	H735
1N5762	MPT36
1N5762	1N5762
1N5762A	1N5762A
1N5817	IT8S817
1N5818	IT8S818
1N5819	IT8S819
1N5823	IT8S823
1N5823	VSK520
1N5824	IT8S824
1N5824	VSK530
1N5825	IT8S825
1N5825	VSK540
1N5829	25PQ010
1N5829	25PQ015
1N5829	25PQ020
1N5830	25PQ025
1N5830	25PQ030
1N6267	1N5629
1N6267A	1N5629A
1N6268	1N5630
1N6268A	1N5630A
1N6269	1N5631
1N6269A	1N5631A
1N6270	1N5632
1N6270A	1N5632A
1N6271	1N5633
1N6271A	1N5633A
1N6272	1N5634
1N6272A	1N5634A
1N6273	1N5635
1N6273A	1N5635A
1N6274	1N5636
1N6274A	1N5636A
1N6275	1N5637
1N6275A	1N5637A
1N6276	1N5638
1N6276A	1N5638A
1N6277	1N5639
1N6277A	1N5639A
1N6278	1N5640
1N6278A	1N5640A
1N6279	1N5641
1N6279A	1N5641A
1N6280	1N5642
1N6280A	1N5642A
1N6281	1N5643
1N6281A	1N5643A
1N6282	1N5644
1N6282A	1N5644A
1N6283	1N5555
1N6283	1N5645
1N6283A	MZ5555
1N6283A	1N5556
1N6283A	1N5645A
1N6284	1N5646
1N6284A	1N5646A
1N6287A	MZ5556
1N6289A	MZ5557
1N6289A	1N5557
1N6289A	1N5651A
1N6290	1N5652
1N6290A	1N5652A
1N6291	1N5653
1N6291A	1N5653A
1N6292	1N5654
1N6292A	1N5654A

MOTOR.	Industry Standard No.
1N6293	1N5655
1N6293A	1N5655A
1N6294	1N5656
1N6294A	1N5656A
1N6295	1N5657
1N6295A	1N5657A
1N6296	1N5658
1N6296A	1N5658A
1N6297	1N5659
1N6297A	1N5659A
1N6298	1N5660
1N6298A	1N5660A
1N6299	1N5661
1N6299A	1N5661A
1N6300	1N5662
1N6300A	1N5662A
1N6301	1N5663
1N6301A	1N5663A
1N6302	1N5664
1N6302A	1N5664A
1N6303	N55558
1N6303A	1N5558
1N6303A	1N5665A
1N702A	GLAZ2.6A
1N710A	GLAZ6.8A
1N711A	GLZ7.5A
1N738A	GLZ100A
1N746	DI-746
1N746	MZ92-3.3
1N746	1N5842
1N746A	MLV746A
1N747	MZ92-3.6
1N747	1N5843
1N748	MZ92-3.9
1N748	1N5844
1N749	MZ92-4.3
1N749	1N5845
1N750	MZ92-4.7
1N750	1N5846
1N751	MZ92-5.1
1N751	1N5847
1N752	MZ92-5.6
1N752	1N5848
1N753	MZ92-6.2
1N753	1N5850
1N754	MZ92-6.8
1N754	1N5851
1N755	MZ92-7.5
1N755	1N5852
1N756	MZ92-8.2
1N756	1N5853
1N757	MZ92-9.1
1N757	1N5855
1N758	MZ92-10
1N758	1N5856
1N759	DI-759
1N759	MZ92-12
1N759	1N5858
1N759A	MLV759A
1N763A	GLZ7.0A
1N769A	GLZ24A
1N821	MC6400,MC6401
1N821	1N3500
1N821	1N3553
1N821	N4010
1N821A	N3779
1N821A	1N3780
1N823	MC6402,MC6403
1N823	1N1735
1N823	1N3496
1N823A	1N2765
1N823A	1N3781
1N825	MC6404,MC6405
1N825	PR6105-PR6450
1N825	1N3497
1N825	1N826
1N825A	1N2765A
1N825A	1N3782
1N827	MC6406,MC6407
1N827	PR6105A-PR6450A
1N827	1N3498
1N827	1N828
1N827A	1N3783
1N829	MC6424,MC6425
1N829	1N3499
1N829A	1N3784
1N935	MC6416
1N935A	MC6417
1N936	MC6418
1N936A	MC6419
1N937	MC6420
1N937	MC6428
1N937	PR9110-PR9450
1N937	1N2625
1N937A	MC6421
1N937A	1N2625A
1N937B	MC6422
1N937B	1N2625B
1N938	MC6422
1N938	PR9110A-PR9450A
1N938	1N2626A
1N938A	1N2626A
1N938B	1N2626B
1N939A	MC6423
1N939A	MC6429
1N941	1N3580
1N941	1N1736
1N941A	1N3580A
1N941B	1N3580B
1N942	1N3581
1N942A	1N1736A
1N942A	1N3581A
1N942B	1N3581B
1N943	1N3582
1N943A	1N3582A
1N943B	1N3582B
1N944	1N3583
1N944A	1N3583A
1N944B	1N3583B
1N945	1N3584
1N945A	1N3584A
1N945B	1N3584B
1N957A	DI-957
1N964A	MZ92-13
1N964A	1N5859
1N965A	MZ92-15
1N965A	1N5861
1N966A	MZ92-16
1N966A	1N5862
1N967A	MZ92-18
1N967A	1N5864
1N968A	MZ92-20
1N968A	1N5866
1N969A	MZ92-22
1N969A	1N5867
1N970A	MZ92-24
1N970A	1N5868
1N971A	MZ92-27
1N971A	1N5870
1N972A	MZ92-30
1N972A	1N5872
1N973A	MZ92-33
1N973A	1N5873
1N974A	MZ92-36
1N974A	1N5874
1N975A	MZ92-39
1N975A	1N5875
1N976A	DI-976
1N976A	MZ92-43
1N976A	1N5876
1N977A	MZ92-47
1N977A	1N5877
1N978A	MZ92-51
1N978A	1N5878
1N979A	MZ92-56
1N979A	1N5879
1N980A	MZ92-62
1N980A	1N5881
1N981A	MZ92-68
1N981A	1N5882
1N982A	MZ92-75
1N982A	1N5883
1N983A	MZ92-82
1N983A	1N5884
1N984A	MZ92-91
1N984A	1N5886
1N985A	MZ92-100
1N985A	1N5887
1N986A	MZ92-110
1N986A	1N5888
1N987A	MZ92-120
1N987A	1N5889
1N988A	MZ92-130
1N988A	1N5890
1N989A	MZ92-150
1N989A	1N5892
1N990A	MZ92-160
1N990A	1N5893
1N991A	MZ92-180
1N991A	1N5895
1N992A	MZ92-200
1N992A	1N5897
2N5641	MM1557
2N1132	2N1132
2N1132A	2N1132A
2N1595	MCR914-1
2N1595	MCR914-2
2N1595	S0301M
2N1595	S0501M
2N1595	T1145A0
2N1595	1N1595
2N1595	2N6605
2N1595	2N6606
2N1596	MCR914-3
2N1596	S1001M
2N1596	T1145A1
2N1596	1N1596
2N1596	2N6607
2N1597	MCR914-4
2N1597	S2001M
2N1597	T1145A2
2N1597	1N1597
2N1597	2N6608
2N1598	MCR914-5
2N1598	T1145A3
2N1598	1N1598
2N1599	MCR914-6
2N1599	S4001M
2N1599	T1145A4
2N1599	1N1599
2N1613	1N1613
2N711	1N711
2N711	2N2952
2N1842	C36U
2N1842	2N1842
2N1842A	2N1842A
2N1843	C36P
2N1843	2N1843
2N1843	2N3644
2N1843	2N3654
2N1843A	2N1843A
2N1844	C36A
2N1844	2N1844
2N1844	2N3650
2N1844	2N3655
2N1844A	2N1844A
2N1845	C36Q
2N1845	2N1845
2N1845A	2N1845A
2N1846	C36B
2N1846	2N1846
2N1846	2N2888
2N1846	2N2888
2N1846	2N3656
2N1846A	2N1846A
2N1847	C36I
2N1847	2N1847
2N1847	2N2889
2N1847A	2N1847A
2N1848	C36C
2N1848	2N1848
2N1848	2N3652
2N1848	2N3657
2N1848A	2N1848A
2N1849	C36D
2N1849	2N1849
2N1849	2N3653
2N1849	2N3658
2N1849A	2N1849A
2N1850	C36E
2N1850	2N1850
2N1850A	2N1850A
2N1890	2N1890
2N1893	2N1893
2N1984	1N5985
2N1990	1N5990
2N1991	1N1991
2N2060	2N2060
2N2060	2N2980
2N2060	2N2981
2N2060	2N2982
2N2060	2N3513
2N2060	2N3516
2N2060	2N4878
2N2060	2N4879
2N2060A	2N2060A
2N2060A	2N2060B
2N2060JAN	2N2060JAN
2N2060JTX	2N2060JTX
2N2060TXV	2N2060TXV
2N2102	2N2102
2N2193A	2N2195A
2N2193A	2N2317
2N2193A	2N2389
2N2193A	2N3830
2N2193A	2N3831
2N2218	1N1051
2N2218	1N1074
2N2218	1N1075
2N2218	1N1076
2N2218	1N1077
2N2218	1N1838
2N2218	1N1953
2N2218	1N1983
2N2218	2N2217
2N2218	2N2224
2N2218	2N2236
2N2218	2N2237
2N2218	2N2479
2N2218	2N3332A
2N2218	2N3333A
2N2218	2N3334A
2N2218	2N3335B
2N2218	2N3336A
2N2218	2N3337A
2N2218	2N3338A
2N2218A	2N1987
2N2219	1N3338
2N2219	2N1420
2N2219	2N1566A
2N2219	2N1984
2N2219	2N194B
2N2219	2N2195
2N2219	2N2195A
2N2219	2N2195B
2N2219	2N2395
2N2219	2N2396
2N2219	2N2788
2N2219	2N2863
2N2219	2N2864
2N2219	2N2958
2N2219A	1N1905
2N2219A	2N1944
2N2219A	2N1945
2N2219A	2N1946
2N2219A	2N1986
2N2219A	2N194
2N2219A	2N194A
2N2219A	2N2219A
2N2221	1N1081
2N2221	1N1082
2N2221	1N1103
2N2221	1N1104
2N2221	1N1149
2N2221	1N1150
2N2221	1N1151
2N2221	1N1152
2N2221	1N1153
2N2221	1N1154
2N2221	1N1155
2N2221	1N1156
2N2221	1N1588
2N2221	1N1591
2N2221	1N162
2N2221	1N162A
2N2221	1N163A
2N2221	2N2220
2N2221	2N2221
2N2221	1N3115
2N2221	2N3332
2N2221	2N3333
2N2221	2N3334
2N2221	2N3335
2N2221	2N3336
2N2221	2N3337
2N2221	2N3338
2N2221	2N470
2N2221	2N471
2N2221	2N472
2N2221	2N472A
2N2221	2N473
2N2221	2N474
2N2221	2N475
2N2221	2N475A
2N2221	2N476
2N2221	2N477
2N2221	2N478
2N2221	2N479
2N2221	2N479A
2N2221	2N480
2N2221	2N480A
2N2221	2N715
2N2221	2N716
2N2221	2N717
2N2221	2N731
2N2221	2N842
2N2221	2N844
2N2221	2N988
2N2221	2N989
2N2221A	2N2221A
2N2221A	2N2314
2N2221A	2N2315
2N2221A	2N2677
2N2221A	2N2678
2N2221A	2N2791
2N2221A	2N2909
2N2222	2N1247
2N2222	2N1248
2N2222	2N1249
2N2222	2N1386
2N2222	2N1387
2N2222	2N1388
2N2222	2N1389
2N2222	2N1390
2N2222	2N1592
2N2222	2N1593
2N2222	2N1594
2N2222	2N2222
2N2222	2N2656
2N2222	2N2792
2N2222	2N3116
2N2222	2N843
2N2222	2N845
2N2222	2N2222A
2N2222A	2N2676
2N2222A	2N3704
2N2222A	2N3705
2N2223	2N2223
2N2223A	2N2223A
2N2223A	2N2414
2N2242	2N1005
2N2242	2N1006
2N2270	2N2270
2N2297	2N2297
2N2322	C5U
2N2322	S0301M32
2N2322	2N1875
2N2322	2N1876
2N2322	2N2322
2N2323	C5P
2N2323	S0501M32
2N2323	2N1877
2N2323	2N2323
2N2324	C5A
2N2324	S1001M32
2N2324	2N1878
2N2324	2N2324
2N2324	2N3273
2N2325	C5G
2N2325	2N2325
2N2326	C5B
2N2326	S2001M32
2N2326	2N1879
2N2326	2N1880
2N2326	2N2326
2N2326	2N3274
2N2327	2N2327
2N2328	2N2328
2N2328	2N3275
2N2329	S4001M32
2N2329	2N2329
2N2329	2N3276
2N2368	2N743
2N2369	1N1135
2N2369	2N1135A
2N2369	2N2205
2N2369	2N2369
2N2369	2N3299
2N2369A	2N744
2N2369A	2N1764
2N2369A	2N2369A
2N2369A	2N3227
2N2405	2N2405
2N2453	2N2453
2N2453A	2N2453A
2N2480	2N2480
2N2480A	2N2480A
2N2481	1N1267
2N2481	1N1268
2N2481	1N1269
2N2481	1N1270
2N2481	1N1271
2N2481	1N1272
2N2481	2N2481
2N2484	2N2483
2N2484	2N2484
2N2501	1N1276
2N2501	1N1277
2N2501	1N1278
2N2501	1N1279
2N2501	1N1587
2N2501	1N1590
2N2501	2N2501
2N2501	2N957
2N2540	2N2539
2N2540	2N2540
2N2573	2N2573
2N2574	2N2574
2N2575	2N2575
2N2576	2N2576
2N2577	2N2577
2N2578	2N2578
2N2579	2N2579
2N2604	2N2604
2N2605	2N2605
2N2606	2N2386
2N2608	2N2608
2N2608	2N2844
2N2608	2N3377
2N2608	2N3379
2N2608	2N3578
2N2608	2N5608
2N2608	5U021
2N2609	2N2609
2N2609	2N381
2N2639	2N2639
2N2639	2N2639JAN
2N2639	2N2639JTX
2N2639	2N2639TXV
2N2640	2N2640
2N2641	2N2641
2N2642	2N2642
2N2642	2N2642JAN
2N2642	2N2642JTX
2N2642	2N2642TXV
2N2643	N87300
2N2643	N87301
2N2643	N87302
2N2643	2N2643
2N2644	2N2644
2N2646	MU970
2N2646	2N2646
2N2647	MU971
2N2647	2N2647
2N2647-10	MCR808-10
2N2652	2N2652
2N2652A	2N2652A
2N2720	2N2720
2N2721	2N2721
2N2722	2N2722
2N2723	2N2724
2N2723	2N2725
2N2785	2N2709
2N2800	2N2709
2N2800	2N2800
2N2845	2N1252
2N2845	2N1253
2N2857	A485
2N2857	2N2857
2N2894	2N2894
2N2895	2N2895
2N2895	2N719
2N2895	2N719A
2N2895	2N986
2N2895	2N2896
2N2897	2N2897

MOTOR.	Industry Standard No.
2N2903	2N2903
2N2903	2N3728
2N2903A	2N2903A
2N2903A	2N3729
2N2904	2N1228
2N2904	2N1229
2N2904	2N1230
2N2904	2N1231
2N2904	2N2904
2N2904	2N2927
2N2904A	2N2904A
2N2905	2N1131
2N2905	2N2393
2N2905	2N2394
2N2905	2N2801
2N2905	2N1344
2N2905	2N3671
2N2905A	2N1131A
2N2905A	2N1232
2N2905A	2N1233
2N2905A	2N2905A
2N2905A	2N5189
2N2906	2N1623
2N2906	2N2906
2N2906	2N327
2N2906	2N327A
2N2906	2N327B
2N2906	2N330
2N2906	2N330A
2N2906	2N354
2N2906	2N4384
2N2906	2N4386
2N2906	2N721
2N2906	2N722
2N2906	2N858
2N2906	2N859
2N2906	2N860
2N2906	2N861
2N2906	2N862
2N2906	2N863
2N2906	2N864
2N2906	2N864A
2N2906	2N865
2N2906	2N865A
2N2906	2N866
2N2906	2N867
2N2906	2N923
2N2906	2N924
2N2906	2N925
2N2906	2N926
2N2906	2N927
2N2906	2N928
2N2906	2N978
2N2906A	2N1474
2N2906A	2N1474A
2N2906A	2N1475
2N2906A	2N2906A
2N2907	2N2696
2N2907	2N2837
2N2907	2N2838
2N2907	2N2907
2N2907	2N3136
2N2907	2N3548
2N2907	2N3550
2N2907A	2N1439
2N2907A	2N1440
2N2907A	2N1441
2N2907A	2N1442
2N2907A	2N1443
2N2907A	2N2907A
2N2907A	2N3547
2N2907A	2N3549
2N2907A	2N5763
2N2907A	2N935
2N2907A	2N936
2N2907A	2N937
2N2907A	2N938
2N2907A	2N939
2N2907A	2N940
2N2907A	2N941
2N2907A	2N946
2N2907A	2N947
2N2913	2N2913
2N2913	2N2972
2N2914	2N2914
2N2914	2N2973
2N2915	2N2915
2N2915	2N2915A
2N2915	2N2974
2N2916	2N2916
2N2916	2N2916A
2N2916	2N2975
2N2916	2N4955
2N2917	2N2917
2N2917	2N2976
2N2918	2N2918
2N2918	2N2977
2N2918	2N4956
2N2919	2N2919
2N2919	2N2919A
2N2919	2N2978
2N2919	2N3521
2N2919	2N3522
2N2919	2N3907
2N2919	2N3908
2N2920	2N2920
2N2920	2N2920A
2N2920	2N2936
2N2920	2N2937
2N2920	2N2979
2N2920	2N3680
2N2920	2N4042
2N2920	2N4043
2N2920	2N4044
2N2920	2N4045
2N2920	2N6085
2N2920	2N6086
2N2920	2N6087
2N2920	2N6088
2N2920	2N6089
2N2920	2N6090
2N2920	2N6091
2N2920	2N6092
2N2920	2N6441
2N2920	2N6442
2N2920	2N6443
2N2920	2N6444
2N2920	2N6445
2N2920	2N6446
2N2920	2N6447
2N2920	2N6448
2N2920	2N6521
2N2920	2N6522
2N2944	2N2183
2N2944	2N2944
2N2944	2N3977
2N2945	2N2181
2N2945	2N2182
2N2945	2N2945
2N2945A	2N2944A
2N2945A	2N2945A
2N2946	2N2946
2N2946A	2N2946A
2N2959	2N2959
2N2974	PT2974
2N2975	PT2975
2N2976	NS7303
2N2976	NS7304
2N2976	NS7305
2N2978	PT2978
2N2979	PT2979
2N2979	2N2979A
2N3009	2N3009
2N3011	2N3011
2N3012	2N3012
2N3013	2N3013
2N3013	2N3211
2N3014	2N2710
2N3014	2N3014
2N3014	2N770
2N3014	2N771
2N3014	2N772
2N3014	2N773
2N3014	2N774
2N3014	2N775
2N3014	2N776
2N3014	2N777
2N3014	2N778
2N3019	2N1335
2N3019	2N1336
2N3019	2N1337
2N3019	2N1339
2N3019	2N1340
2N3019	2N1341
2N3019	2N1342
2N3019	2N1508
2N3019	2N1509
2N3019	2N1973
2N3019	2N2192
2N3019	2N2192A
2N3019	2N2192B
2N3019	2N2390
2N3019	2N2438
2N3019	2N2439
2N3019	2N3019
2N3019	2N3107
2N3019	2N3666
2N3019	2N3701
2N3020	2N871
2N3020	2N1206
2N3020	2N1943
2N3020	2N1974
2N3020	2N1975
2N3020	2N1988
2N3020	2N1989
2N3020	2N2086
2N3020	2N2087
2N3020	2N2193
2N3020	2N2193B
2N3020	2N2243
2N3020	2N2243A
2N3020	2N2312
2N3020	2N2316
2N3020	2N2437
2N3020	2N3020
2N3020	2N3108
2N3020	2N3110
2N3020	2N3665
2N3020	2N3678
2N3020	2N497
2N3020	2N698
2N3020	2N870
2N3043	2N3043
2N3043	2N3519
2N3043	2N3520
2N3043	2N5523
2N3043	2N5524
2N3044	2N3044
2N3045	8P2484P
2N3045	2N3045
2N3046	2N3046
2N3047	2N3047
2N3048	2N3048
2N3053	2N3036
2N3053	2N3053
2N3053A	2N3053A
2N3054	2N4910
2N3054	2N4911
2N3055	MJ3055
2N3055	SDP9206
2N3055	SDP9707
2N3055	2N3235
2N3055	2N5054
2N3055	2N5055
2N3055	2N5036
2N3055	2N6037
2N3055	2DC768
2N3055	2SD492
2N3055	2SD878
2N3055A	RC8242
2N3073	2N3073
2N3114	2N2008
2N3114	2N3114
2N3133	2N3133
2N3135	2N3135
2N3137	2N3137
2N3227	2N3227
2N3244	2N3244
2N3245	2N3245
2N3249	2N3249
2N3250	2N1024
2N3250	2N1025
2N3250	2N1026
2N3250	2N1027
2N3250	2N1219
2N3250	2N1223
2N3250	2N2391
2N3250	2N2392
2N3250	2N2424
2N3250	2N2970
2N3250	2N2971
2N3250	2N3250
2N3250	2N3702
2N3250	2N3829
2N3250	2N4034
2N3250	2N2425
2N3250A	2N3250A
2N3251	2N3251
2N3251	2N3703
2N3251	2N4035
2N3251	2N995
2N3251A	2N3251A
2N3252	2N3252
2N3253	2N2868
2N3253	2N3253
2N3287	2N3287
2N3287	2N3288
2N3287	2N3289
2N3287	2N3290
2N3287	2N3295
2N3299	2N3299
2N3300	2N3300
2N3301	2N3301
2N3302	2N3302
2N3303	2N3303
2N3303	2N958
2N3303	2N959
2N3307	2N3307
2N3307	2N3380
2N3308	2N3308
2N3330	U149
2N3330	U168
2N3330	UC805
2N3330	2N2498
2N3330	2N3329
2N3330	2N3332
2N3330	2N3376
2N3330	2N3378
2N3365	U3001
2N3365	U3010
2N3366	U3002
2N3366	U3011
2N3367	U3003
2N3367	U3012
2N3367	2N3068A
2N3375	BLY59
2N3375	2N2876
2N3375	2N3375
2N3393	2N3385
2N3394	2N3382
2N3394	2N3383
2N3394	2N3384
2N3425	2N3425
2N3436	DN3436A
2N3436	FE400
2N3436	FE404
2N3436	U199
2N3436	2N3436
2N3437	DN3437A
2N3437	FE402
2N3437	U198
2N3437	2N3437
2N3437	UC714
2N3438	DN3438A
2N3439	2N3439
2N3440	2N2726
2N3440	2N3440
2N3440	2N4064
2N3441	2N3441
2N3441	2SC489
2N3441	2SD877
2N3442	2N3442
2N3444	2N3444
2N3445	2N3445
2N3446	2N3446
2N3447	2N3447
2N3447	2SC161
2N3447	2SC241
2N3447	2SC244
2N3447	2SC494
2N3447	2SC521
2N3447	2SC646
2N3448	2N5614
2N3448	2N5616
2N3448	2N5618
2N3448	2SC518A
2N3448	2SC520
2N3448	2SC520A
2N3448	2SC521A
2N3448	2SC647
2N3467	2N1238
2N3467	2N1239
2N3467	2N1240
2N3467	2N1241
2N3467	2N3467
2N3467JAN	2N3467JAN
2N3467JTX	2N3467JTX
2N3467JTXV	2N3467JTXV
2N3468	2N3121
2N3468	2N3468
2N3468JAN	2N3468JAN
2N3468JTX	2N3468JTX
2N3468JTXV	2N3468JTXV
2N3485	2N3485
2N3485A	2SC524
2N3485A	2N3485A
2N3486	2N3486
2N3486A	2N3486A
2N3494	2N3494
2N3495	2N1234
2N3495	2N3495
2N3496	2N3496
2N3497	2N2600
2N3497	2N2600A
2N3497	2N3497
2N3498	2N1923
2N3498	2N3498
2N3498	2N498
2N3499	2N3199
2N3499	2N3499
2N3499	2N4960
2N3499	2N4961
2N3499	2N4962
2N3499	2N4963
2N3500	2N1207
2N3500	2N1615
2N3500	2N2472
2N3500	2N2473
2N3500	2N3500
2N3500	2N3923
2N3501	2N3501
2N3502	2N3502
2N3503	2N3503
2N3504	2N3504
2N3505	2N3505
2N3506	2N3506
2N3507	2N3507
2N3508	2N1708
2N3508	2N3519
2N3508	2N3508
2N3509	2N3509
2N3510	2N3510
2N3511	2N3511
2N3546	2N1606
2N3546	2N1607
2N3546	2N1608
2N3546	2N2894A
2N3546	2N3248
2N3546	2N3546
2N3553	BFW47
2N3553	PT8551
2N3553	2N2631
2N3553	2N3118
2N3553	2N3309A
2N3553	2N3553
2N3553	2N5834
2N3553	2N6197
2N3553	2SC597
2N3583	RCA1802
2N3583	SDT4901
2N3583	SDT5952
2N3583	SDT5955
2N3583	2N3583
2N3583	2SC1450
2N3583	2SC483
2N3583	2SC487
2N3583	2SD102
2N3583	2SD150
2N3583	2SD152
2N3583	2SD238
2N3583	2SD92
2N3584	2N3584
2N3584	2SD554
2N3585	SDT4904
2N3585	SDT4905
2N3585	2N3585
2N3585	2SC1031
2N3585	2SC1104
2N3585	2SC1466
2N3585	2SC679
2N3585	2SC825
2N3632	BLY60
2N3632	2N3632
2N3632	2SP733
2N3632	2SC818
2N3632	2SC585
2N3632	2SC635
2N3632	2SC636
2N3634	2N3634
2N3635	2N3635
2N3636	2N3636
2N3637	2N3637
2N3647	2N3647
2N3648	2N3648
2N3668	2N3668
2N3668	2N3669
2N3670	2N3670
2N3677	2N3677
2N3700	2N3700
2N3700	2N760
2N3700	2N760A
2N3712	2N3712
2N3713	MJ480
2N3713	MJ481
2N3713	SDT9307
2N3713	2N3713
2N3713	2N5869
2N3713	2N5873
2N3714	2N3714
2N3714	2N5870
2N3714	2N5874
2N3715	SDT9308
2N3715	2N3715
2N3715	2N5863
2N3715	2N4111
2N3715	2SD163
2N3715JAN	2N3715JAN
2N3715JTX	2N3715JTX
2N3715JTXV	2N3715JTXV
2N3716	SDT9309
2N3716	2N3716
2N3716	2N4113
2N3716	2SD041
2N3716	2SD316
2N3716	2SD369
2N3716JAN	2N3716JAN
2N3716JTX	2N3716JTX
2N3716JTXV	2N3716JTXV
2N3719	SDT3501
2N3719	2N3202
2N3719	2N3719
2N3720	SDT3502
2N3720	2N3203
2N3720	2N3720
2N3724	TI8133
2N3724	TI8134
2N3724	2N3724
2N3725	TI8136
2N3725	2N3725
2N3725	2N5860
2N3725A	2N3725A
2N3726/27	2N3347
2N3726/27	2N3348
2N3726/27	2N3349
2N3726/27	2N3351
2N3726/27	2N3352
2N3727	2N3727
2N3734	2N3724A
2N3734	2N3734
2N3735	2N3735
2N3735JAN	2N3735JAN
2N3735JTX	2N3735JTX
2N3735JTXV	2N3735JTXV
2N3736	2N3736
2N3737	2N2351
2N3737	2N2351A
2N3737	2N3737
2N3738	MJ2251
2N3738	MJ3201
2N3738	2N3738
2N3738	2N4296
2N3738	2N4297
2N3738	2SC1160
2N3738	2SC1161
2N3738	2SC783
2N3738	2SD138
2N3738	2SD156
2N3738	2SD158

MOTOR	Industry Standard No.	MOTOR	Industry Standard No.	MOTOR	Industry Standard No.	MOTOR	Industry Standard No.
2N3738	2SD56	2N3791	2N3791	2N3839	SD1303	2N3993	TP5116
2N3739	MJ3202	2N3791	2N4907	2N3839	2N3839	2N3993	U300
2N3739	MJ400	2N3791	2N4908	2N3839	SD1200	2N3993	U301
2N3739	SE9331	2N3791JAN	2N3791JAN	2N3866	2N3866	2N3993	U304
2N3739	2N3739	2N3791JTX	2N3791JTX	2N3867	SDT3505	2N3993	U305
2N3739	2SC1059	2N3791JTXV	2N3791JTXV	2N3867	SDT3775	2N3993	U306
2N3739	2SC1102	2N3792	2N3792	2N3867	SDT3778	2N3993	UC401
2N3739	2SC1105	2N3792	2N4909	2N3867	2N3867	2N3993	UC450
2N3739	2SC1168	2N3792JAN	2N3792JAN	2N3867JAN	2N3867JAN	2N3993	2N3386
2N3739	2SC1304	2N3792JTX	2N3792JTX	2N3867JTX	2N3867JTX	2N3993	2N3387
2N3739	2SC1391	2N3792JTXV	2N3792JTXV	2N3867JTXV	2N3867JTXV	2N3993	2N3993
2N3739	2SC1456	2N3796	M100	2N3868	SDT3506	2N3993	2N5018
2N3739	2SC2354	2N3796	2N3796	2N3868	SDT3776	2N3993	2N5114
2N3739	2SC515	2N3797	MBM667	2N3868	2N3868	2N3993	2N5115
2N3739	2SC582	2N3797	2N3631	2N3868	2N4314	2N3994	UC451
2N3739	2SC685	2N3797	2N3797	2N3868JAN	2N3868JAN	2N3994	2N3994
2N3739	2SC779	2N3798	2N3798	2N3868JTX	2N3868JTX	2N3994	2N4382
2N3739	2SC782	2N3798	2N3798A	2N3868JTXV	2N3868JTXV	2N3994	2N5019
2N3739	2SC795	2N3799	2N5579	2N3870	MCR2835-3	2N3994	2N5116
2N3739	2SC867	2N3799	2N5580	2N3870	S1035G	2N3997	M101
2N3739	2SD139	2N3799	2N3581	2N3870	2N3870	2N4003	2N6007
2N3739	2SD157	2N3799	2N3582	2N3871	MCR2835-4	2N4012	2N4012
2N3739	2SD159	2N3799	2N3799	2N3871	S2035G	2N4013	2N4013
2N3739	2SD24	2N3799	2N3799A	2N3871	2N3871	2N4014	2N4014
2N3739	2SD324	2N3799	2N4359	2N3872	MCR2835-6	2N4015	2N4015
2N3739	2SD326	2N3806	FT4017	2N3872	S4035G	2N4016	2N4016
2N3739	2SD766	2N3806	FT4018	2N3872	2N3872	2N4026	2N4026
2N3739JAN	2N3739JAN	2N3806	2N4018	2N3873	MCR2835-8	2N4027	2N4027
2N3739JTX	2N3739JTX	2N3806	2N5254	2N3873	S6035G	2N4028	2N4028
2N3739JTXV	2N3739JTXV	2N3807	FT4019	2N3873	2N3873	2N4029	2N4029
2N3740	MJ2253	2N3807	2N3801	2N3896	C135A	2N4030	2N4030
2N3740	2N3740	2N3807	2N4019	2N3896	C38A	2N4031	2N4031
2N3740A	2N3740A	2N3808	FT4021	2N3896	MCR2935-3	2N4032	2N4032
2N3740JAN	2N3740JAN	2N3808	SMT100	2N3896	MCR2935-3	2N4033	2N4033
2N3740JTX	2N3740JTX	2N3808	SMT101	2N3896	S1035H	2N4036	2N4036
2N3740JTXV	2N3740JTXV	2N3808	SMT102	2N3896	2N3896	2N4037	2N4037
2N3741	MJ2254	2N3808	2N5255	2N3897	C135B	2N4040	2N4940
2N3741	2N3741	2N3809	FT4020	2N3897	C38B	2N4072	MM1713
2N3741	2SA616	2N3809	FT4022	2N3897	S38G	2N4072	MM1943
2N3741	2SB502	2N3809	SMT103	2N3897	MCR2935-3	2N4072	MM1945
2N3741	2SB503	2N3809	2N3803	2N3897	MCR2935-4	2N4072	2N4072
2N3741A	2N3741A	2N3810	FT4024	2N3897	S2035H	2N4073	2N4073
2N3741JAN	2N3741JAN	2N3810	SMT104	2N3897	2N3897	2N4073	2N5710
2N3741JTX	2N3741JTX	2N3810	SMT105	2N3898	C135D	2N4091	C413R
2N3741JTXV	2N3741JTXV	2N3810	2N3804	2N3898	C38D	2N4091	CM601
2N3742	2N3742	2N3810	2N4023	2N3898	MCR2935-6	2N4091	CM602
2N3742	2S2279	2N3810	2N4024	2N3898	MCR2935-6	2N4091	CM603
2N3743	2N3743	2N3810	2N4025	2N3898	S4035H	2N4091	CM646
2N3762	2N3762	2N3810A	2N3804A	2N3898	2N3898	2N4091	DN3356A
2N3762JAN	2N3762JAN	2N3810JAN	2N3810JAN	2N3899	C135M	2N4091	DN3356A
2N3762JTX	2N3762JTX	2N3810JTX	2N3810JTX	2N3899	C137M	2N4091	DN3367A
2N3762JTXV	2N3762JTXV	2N3810JTXV	2N3810TXV	2N3899	MCR2935-8	2N4091	FT0655A
2N3763	2N1242	2N3811	FT4023	2N3899	S6035H	2N4091	MFE2098
2N3763	2N1243	2N3811	FT4025	2N3899	2N3899	2N4091	UC250
2N3763	2N3072	2N3811	2N3805	2N3899	2N5204	2N4091	UC751
2N3763	2N3763	2N3811	2N3811	2N3902	DT8401	2N4091	UC752
2N3763JAN	2N3763JAN	2N3811	2N4020	2N3902	DTS402	2N4091	UC753
2N3763JTX	2N3763JTX	2N3811	2N4021	2N3902	IR401	2N4091	UC754
2N3763JTXV	2N3763JTXV	2N3811	2N4022	2N3902	IR402	2N4091	UC755
2N3764	2N3120	2N3811A	2N3805A	2N3902	SDT1056	2N4091	UC756
2N3764	2N3764	2N3811A	2N3806	2N3902	SDT1061	2N4091	2N3640
2N3765	2N3765	2N3811A	2N3807	2N3902	2N3902	2N4091,J	2N4091,J
2N3766	MJ2249	2N3811A	2N3808	2N3903	A5T3903	2N4092	0610
2N3766	MJ3101	2N3811A	2N3809	2N3903	2N3903	2N4092	CM642
2N3766	SDT5901	2N3811A	2N3811A	2N3904	A5T3904	2N4092	CM644
2N3766	SDT5902	2N3811JAN	2N3811JAN	2N3904	EN3904	2N4092	CM645
2N3766	SDT5906	2N3811JTX	2N3811JTX	2N3904	PE9001	2N4092	FT0655B
2N3766	SDT5907	2N3811TXV	2N3811TXV	2N3905	A5T3905	2N4092	TI842
2N3766	2N3766	2N3812	2N3812	2N3905	EN1132	2N4092	U1287
2N3766	2SC490	2N3813	2N3813	2N3905	2N3905	2N4092	U182
2N3766	2SD130	2N3814	2N3814	2N3906	PN3504A	2N4092	U221
2N3766	2SD141	2N3815	2N3815	2N3906	A5T3504	2N4092,J	2N4092,J
2N3766	2SD142	2N3816	2N3816	2N3906	A5T3906	2N4092/3	MFE2097
2N3766	2SD226	2N3816	2N3816A	2N3906	BCX78	2N4093	FT0655C
2N3766	2SD241	2N3817	SP10800	2N3906	BCX79	2N4093	NP510
2N3766	2SD256	2N3817	SP10801	2N3906	EN3502	2N4093	NP511
2N3766	2SD315	2N3817	SP10810	2N3906	EN3504	2N4093	UC201
2N3766	2SD57	2N3817	2N3817	2N3906	EN3906	2N4093	UC251
2N3766	2SD58	2N3817A	2N3817A	2N3906	PN5251A	2N4093	2N5549
2N3766	2SD70	2N3821	D1202	2N3909	U145	2N4093,J	2N4093,J
2N3766	2SD90	2N3821	DNX5	2N3909	U147	2N4103	2N4103
2N3766	2SD91	2N3821	DNX5A	2N3909	U148	2N4117	2N4117
2N3766JAN	2N3766JAN	2N3821	DNX8	2N3909A	2N2499	2N4117A	2N4117A
2N3766JTX	2N3766JTX	2N3821	DNX8A	2N3909A	2N3581	2N4118	2N3454
2N3766JTXV	2N3766JTXV	2N3821	DNX9	2N3924	BFR25	2N4118	2N5457
2N3767	MJ2250	2N3821	DNX9A	2N3924	BFW46	2N4118	2N4118
2N3767	SDT5903	2N3821	U1286	2N3924	2SC1947	2N4118A	2N4118A
2N3767	SDT5908	2N3821	2N3070	2N3926	BLY57	2N4119	2N3453
2N3767	2N3767	2N3821	2N3084	2N3926	2N3926	2N4119	2N3456
2N3767	2SD129	2N3821	2N3085	2N3926	2N4932	2N4119	2N4119
2N3767	2SD143	2N3821	2N3086	2N3926	2N5423	2N4119A	2N4119A
2N3767	2SD144	2N3821	2N3087	2N3926	2SC572	2N4123	A5T4123
2N3767	2SD154	2N3821	2N3088A	2N3926	2SC598	2N4123	2N4123
2N3767	2SD155	2N3821	2N3089A	2N3926	2SC637	2N4124	A5T4124
2N3767	2SD242	2N3821	2N3452	2N3927	BLY58	2N4124	C89011
2N3767	2SD254	2N3821	2N3455	2N3927	2N3927	2N4124	C89017
2N3767	2SD255	2N3821	2N3685	2N3927	2N4933	2N4124	2N4124
2N3767	2SD257	2N3821	2N3686	2N3927	2SC422	2N4124	SE2001
2N3767	2SD28	2N3821	2N3821	2N3927	2SC573	2N4124	SET7095
2N3767	2SD29	2N3821/2	U1284	2N3927	2SC600	2N4124	SET7056
2N3767	2SD291	2N3822	DNX1	2N3927	2SC831	2N4125	A5T4125
2N3767	2SD292	2N3822	DNX1A	2N3927	2SC638	2N4125	BN4125
2N3767JAN	2N3767JAN	2N3822	DNX4	2N3946	2N3946	2N4125	PN3251
2N3767JTX	2N3767JTX	2N3822	DNX4A	2N3946	2N789	2N4125	A5T4126
2N3767JTXV	2N3767JTXV	2N3822	U1281	2N3946	2N790	2N4126	BN4126
2N3771	IR3771	2N3822	U1282	2N3946	2N791	2N4167	C10V
2N3771	MJ3771	2N3822	U1283	2N3946	2N792	2N4167	C11U
2N3772	2N3772	2N3822	UC241	2N3946	2N793	2N4167	C20U
2N3772	IR3772	2N3822	2N3069	2N3947	2N3947	2N4167	MCR2305-1
2N3772	RCS257	2N3822	2N3684	2N3948	2N3948	2N4167	MCR2315-1
2N3772	2N3772	2N3822	2N3822	2N3950	2N3950	2N4167	2N1770
2N3773	IR3773	2N3822	2N5277	2N3959	2N3959	2N4167	2N4167
2N3773	MJ3773	2N3822	2N5543	2N3960	2N3960	2N4168	C10P
2N3773	2N3773	2N3822	2N5544	2N3962	2N3962	2N4168	C11F
2N3773	2SD873	2N3823	KE5103	2N3963	2N3963	2N4168	C15F
2N3789	MJ490	2N3823	KE5104	2N3964	2N3246	2N4168	C20P
2N3789	MJ491	2N3823	KE5105	2N3964	2N3964	2N4168	MCR2305-2
2N3789	2N3171	2N3823	2N3825	2N3965	2N3965	2N4168	MCR2315-2
2N3789	2N3172	2N3824	CM640	2N3966	2N3966	2N4168	2N1600
2N3789	2N3183	2N3824	DNX7	2N3970	NP521	2N4168	2N1771
2N3789	2N3184	2N3824	DNX7A	2N3970	2N3970	2N4168	2N1771A
2N3789	2N3195	2N3824	FT0654A	2N3971	NP523	2N4168	2N4168
2N3789	2N3196	2N3824	FT0654B	2N3971	UC807	2N4169	C10A
2N3789	2N3789	2N3824	2N3824	2N3972	2N3972	2N4169	C11A
2N3789	2N5867	2N3838	FT3838	2N3980	2N3980	2N4169	C15A
2N3789	2N5871	2N3838	2N3838	2N3993	TP5114	2N4169	C20A
2N3789	2SA807	2N3838	2N3838JAN	2N3993	TP5115	2N4169	MCR2305-3
2N3790	2N3173	2N3838	2N3838JTX			2N4169	MCR2315-3
2N3790	2N3185	2N3838	2N3838TXV			2N4169	2N1601
2N3790	2N3197					2N4169	2N1772
2N3790	2N3790					2N4169	2N1772A
2N3790	2N5868					2N4169	2N3269
2N3790	2N5872					2N4169	2N3936
2N3790	2SA808					2N4169	2N4169

MOTOR.	Industry Standard No.	MOTOR.	Industry Standard No.	MOTOR.	Industry Standard No.	MOTOR.	Industry Standard No.
2N4170	C10B	2N422/A	DN5369A	2N4238	2N2041	2N4401	2N6000
2N4170	C10G	2N4220	C6690	2N4238	2N2106	2N4401	2N6002
2N4170	C11B	2N4220	C6691	2N4238	2N2107	2N4401	2N6004
2N4170	C11G	2N4220	PTO654E	2N4238	2N2657	2N4401	2N6006
2N4170	C15B	2N4220	U273,A	2N4238	2N2987	2N4401	2N6010
2N4170	C15G	2N4220	U274,A	2N4238	2N4238	2N4401	2N6012
2N4170	C20B	2N4220	U275	2N4239	2N1506A	2N4402	A5T4402
2N4170	MCR2305-4	2N4220	U275A	2N4239	2N2658	2N4402	A5T2907
2N4170	MCR2315-4	2N4220	UC701	2N4239	2N4239	2N4403	A5T4403
2N4170	2N1602	2N4220	UC703	2N4240	2N4240	2N4403	BC231
2N4170	2N1773	2N4220	2N4220	2N4260	2N4260	2N4403	BC588
2N4170	2N1773A	2N4220A	D1101	2N4261	2N4261	2N4403	BC727
2N4170	2N1774	2N4220A	D1181	2N4338	EN714	2N4403	BC728
2N4170	2N1774A	2N4220A	D1182	2N4338	2N3068	2N4403	BCX75
2N4170	2N3270	2N4220A	D1184	2N4338	2N3071	2N4403	BCX76
2N4170	2N3937	2N4220A	D1185	2N4338	2N3370	2N4403	BN2905
2N4170	2N4170	2N4220A	D1302	2N4338	2N3460	2N4403	PN2904
2N4171	C11C	2N4220A	D1303	2N4338	2N3687	2N4403	PN2904A
2N4171	C11H	2N4220A	DN3070A	2N4338	2N4338	2N4403	PN2906
2N4171	C15C	2N4220A	DN3071A	2N4339	2N3066	2N4403	PN2906A
2N4171	C15H	2N4220A	DN3459A	2N4339	2N3067	2N4403	TI893M
2N4171	C20C	2N4220A	DN3460A	2N4339	2N3088	2N4403	2N6001
2N4171	MCR2305-5	2N4220A	FE304	2N4339	2N3089	2N4403	2N6005
2N4171	MCR2315-5	2N4220A	TN4338	2N4339	2N3369	2N4403	2N6011
2N4171	2N1603	2N4220A	TN4339	2N4339	2N3459	2N4403	2N6013
2N4171	2N1775	2N4220A	TN4340	2N4339	2N4339	2N4403	2N6015
2N4171	2N1775A	2N4220A	TN4341	2N4340	2N3340	2N4404	2N3945
2N4171	2N1776	2N4220A	U1177	2N4341	2N3368	2N4404	2N4404
2N4171	2N1776A	2N4220A	U1181	2N4341	2N3458	2N4404	2N5864
2N4171	2N3271	2N4220A	U1324	2N4342	2N4342	2N4405	2N4405
2N4171	2N3938	2N4220A	U1525	2N4342	2N4343	2N4406	2N4406
2N4171	2N4171	2N4220A	U1714	2N4347	2N4347	2N4406	2N5865
2N4172	C11D	2N4220A	UC110	2N4347	2SC240	2N4407	2N4407
2N4172	C15D	2N4220A	UC120	2N4347	2SC242	2N4409	A5T4409
2N4172	C20D	2N4220A	UC130	2N4347	2SC245	2N4410	A5T4410
2N4172	MCR2305-6	2N4220A	2N3465	2N4347	2SC43	2N4410	ESN870
2N4172	MCR2315-6	2N4220A	2N3466	2N4347	2SC44	2N4410	ESN871
2N4172	2N1604	2N4220A	2N3969	2N4347	2SC492	2N4410	PN1893
2N4172	2N1777	2N4220A	2N3969A	2N4347	2SC493	2N4416	A192
2N4172	2N1777A	2N4220A	2N4220A	2N4347	2SC726	2N4416	BSV22
2N4172	2N2653	2N4220A	2N4867	2N4347	2SC727	2N4416	PTO650
2N4172	2N3272	2N4220A	2N4867A	2N4351	M116	2N4416	ITE4416
2N4172	2N3939	2N4220A	2N4868	2N4351	M117	2N4416	U311
2N4172	2N4172	2N4220A	2N4868A	2N4351	MEM562	2N4416	U312
2N4173	C11E	2N4220A	2N5391	2N4351	MEM562C	2N4416	UC150
2N4173	MCR2305-7	2N4220A	2N5392	2N4351	MEM563C	2N4416	UC150W
2N4173	MCR2315-7	2N4220A*1	PF4339	2N4351	MEM711	2N4416	UC155
2N4173	2N1778	2N4220A*1	PF4340	2N4351	MM2102	2N4416	UC210
2N4173	2N1778A	2N4220A/2A	FE300	2N4351	MPF159	2N4416	UC734
2N4173	2N3940	2N4220A/2A	FE302	2N4351	2N4058	2N4416	2N4416
2N4173	2N4173	2N4220A/2A	U1814	2N4351	2N4059	2N4416	2N4417
2N4174	C11M	2N4221	PTO654C	2N4351	2N4351	2N4416	2N5103
2N4174	C15M	2N4221	PTO654D	2N4351	3N75	2N4416	2N5104
2N4174	MCR2305-8	2N4221	U1321	2N4351	3N175	2N4416	2N5105
2N4174	MCR2315-8	2N4221	UC220	2N4351	3N176	2N4416A	BF980
2N4174	2N2619	2N4221	2N4221	2N4351	3N177	2N4416A	2N4416A
2N4174	2N4174	2N4221	2N4869	2N4352	PI100	2N4427	B1-12
2N4183	MCR1604-1	2N4221A	DN3069A	2N4352	PF704	2N4427	2N4427
2N4183	MCR2604-1	2N4221A	DN3368A	2N4352	HRN836Q	2N4427	2S5421
2N4183	MCR2604L1	2N4221A	LDP603	2N4352	K1504	2N4427	2S6687
2N4183	2N4183	2N4221A	LDP604	2N4352	M114	2N4427	2S0319
2N4184	MCR1604-2	2N4221A	U1180	2N4352	MEM803	2N4427	2S0371
2N4184	MCR2604-2	2N4221A	U1323	2N4352	MEM806	2N4427	2S0821
2N4184	MCR2604L2	2N4221A	UC100	2N4352	MEM806A	2N4428	2N4428
2N4184	2N4184	2N4221A	UC240	2N4352	MEM807	2N4428	2S0651
2N4185	MCR1604-3	2N4221A	UC704	2N4352	MEM807A	2N4440	2N4440
2N4185	MCR2604-3	2N4221A	UC705	2N4352	MEM817	2N4441	MCR3000-2
2N4185	MCR2604L3	2N4221A	2N3968	2N4352	MPF160	2N4441	80508P1
2N4185	2N4185	2N4221A	2N3968A	2N4352	SC1611	2N4441	80508P3
2N4186	MCR1604-4	2N4221A	2N4221A	2N4352	SC1612	2N4441	2N4441
2N4186	MCR2604L4	2N4221A	2N4869A	2N4352	SC1613	2N4442	MCR3000-4
2N4186	2N4186	2N4221A	2N5393	2N4352	SC1614	2N4442	82008P1
2N4187	MCR1604-5	2N4221A	2N5394	2N4352	TIX881	2N4442	82008P3
2N4187	MCR2604-5	2N4221A	2N5395	2N4352	TIX867	2N4442	2N4442
2N4187	MCR2604L5	2N4221A/2A	U1322	2N4352	2N3610	2N4443	MCR3000-6
2N4187	2N4187	2N4222	M4139	2N4352	2N3882	2N4443	84008P1
2N4188	MCR1604-6	2N4222	2N4222	2N4352	2N4267	2N4443	84008P3
2N4188	MCR2604-6	2N4222	D1180	2N4352	2N4352	2N4443	2N4443
2N4188	MCR2604L6	2N4222	D1301	2N4352	2N4353	2N4444	MCR3000-8
2N4188	2N4188	2N4222A	DN3458A	2N4352	3N145	2N4444	86008P
2N4189	MCR1604-7	2N4222A	LDP605	2N4352	3N146	2N4444	86008P3
2N4189	MCR2604-7	2N4222A	2N3967	2N4352	3N149	2N4444	2N4444
2N4189	MCR2604L7	2N4222A	2N3967A	2N4352	3N150	2N4449	2N4449
2N4189	2N4189	2N4222A	2N4222A	2N4352	3N161	2N4453	2N4453
2N4190	MCR1604-8	2N4223	D1183	2N4352	3N163	2N4851	2N4851
2N4190	MCR2604-8	2N4223	2N4223	2N4352	3N164	2N4852	2N4852
2N4190	MCR2604L8	2N4224	D1201	2N4352	3N172	2N4853	2N2160
2N4190	2N4190	2N4224	2N4224	2N4352	3N173	2N4853	2N4853
2N4199	2N4199	2N4231A	MJ4104	2N4352	3N174	2N4854	2N4854
2N4200	2N4200	2N4231A	RC829	2N4360	2N4360	2N4854JAN	2N4854JAN
2N4201	2N4201	2N4231A	RC831	2N4391	CM643	2N4854JTX	2N4854JTX
2N4202	2N4202	2N4231A	SDT9301	2N4391	CM647	2N4854TXV	2N4854TXV
2N4203	2N4203	2N4231A	SDT9304	2N4391	LDP691	2N4855	2N4855
2N4204	2N4204	2N4231A	2N4231	2N4391	U320	2N4856	U222
2N4208	2N4208	2N4231A	2N4231A	2N4391	U321	2N4856	2N4094
2N4208	2N4872	2N4231A	2N6260	2N4391	U322	2N4856,J	2N4856,J
2N4209	2N3304	2N4232A	RC829A	2N4391	2N4391	2N4856A	2N4856A
2N4209	2N4209	2N4232A	RC831A	2N4392	CM650	2N4857	2N4095
2N4209	2N5056	2N4232A	SDT9302	2N4392	LDP692	2N4857,J	2N4857,J
2N4209	2N5057	2N4232A	SDT9305	2N4392	MFE2133	2N4857A	2N4857,J
2N4212	80501M31	2N4232A	2N4232	2N4392	U1715	2N4858,J	2N4858,J
2N4212	2N1869	2N4232A	2N4232A	2N4392	2N4392	2N4858A	2N4858A
2N4212	2N1870	2N4233A	RC829B	2N4393	CM641	2N4859	KE4859
2N4212	2N2009	2N4233A	RC831B	2N4393	LDP693	2N4859,J	2N4859,J
2N4212	2N3555	2N4233A	SDT9303	2N4393	UC200	2N4860	KE4860
2N4212	2N3559	2N4233A	SDT9306	2N4393	2N4393	2N4860,J	2N4860,J
2N4212	2N4212	2N4233A	2N4233	2N4393	2N5396	2N4860A	2N4860A
2N4213	80501M31	2N4233A	2N4233A	2N4398	MJ450	2N4861	KE4861
2N4213	2N1871	2N4233A	2N6261	2N4398	2N4398	2N4861,J	2N4861,J
2N4213	2N2010	2N4234	2N3660	2N4399	2N4399	2N4861A	2N4861A
2N4213	2N3556	2N4234	2N3774	2N4399JAN	2N4399JAN	2N4870	2N4870
2N4213	2N3560	2N4234	2N3778	2N4399JTX	2N4399JTX	2N4871	2N4871
2N4213	2N4213	2N4234	2N3782	2N4399JTXV	2N4399JTXV	2N4877	SDT3421
2N4214	81001M31	2N4234	2N4234	2N4401	A5T2222	2N4877	SDT3422
2N4214	2N1872	2N4235	2N2881	2N4401	BC232	2N4877	SDT3425
2N4214	2N2011	2N4235	2N3661	2N4401	BC387	2N4877	SDT3426
2N4214	2N3557	2N4235	2N3775	2N4401	BC737	2N4877	SDT4451
2N4214	2N3561	2N4235	2N3779	2N4401	BC738	2N4877	SDT4453
2N4214	2N4214	2N4235	2N4235	2N4401	BC873	2N4877	SDT4483
2N4215	2N1873	2N4236	2N3776	2N4401	BCX74	2N4877	SDT5506
2N4215	2N4215	2N4236	2N3780	2N4401	BFR50	2N4877	SDT5507
2N4216	82001M31	2N4236	2N4236	2N4401	BFR51	2N4877	TIP501
2N4216	2N1874	2N4237	2N1479	2N4401	BFR52	2N4877	TIP502
2N4216	2N2012	2N4237	2N1481	2N4401	PN2218	2N4877	2N5334
2N4216	2N3558	2N4237	2N1714	2N4401	PN2218A	2N4890	2N4890
2N4216	2N3562	2N4237	2N1716	2N4401	PN2219	2N4898	MJ3701
2N4216	2N4216	2N4237	2N2038	2N4401	PN2219A	2N4898	MJ3702
2N4217	2N4217	2N4237	2N2040	2N4401	PN2221	2N4898	2N4387
2N4218	2N2013	2N4237	2N4237	2N4401	PN2221A	2N4898	2N4388
2N4218	2N4218	2N4238	2N1480	2N4401	PN3566	2N4898	2N4898
2N4219	84001M31	2N4238	2N1482	2N4401	SB6002	2N4899	MJ3703
2N4219	2N2014	2N4238	2N2033	2N4401	TI890M	2N4899	2N4899
2N4219	2N4219	2N4238	2N2039	2N4401	TI892M	2N4899	2SA613

MOTOR.	Industry Standard No.
2N4900	MJ3704
2N4900	2N4900
2N4900	28A614
2N4901	2N4901
2N4902	2N4902
2N4903	2N4903
2N4904	2N4904
2N4905	2N4905
2N4906	2N4906
2N4912	2N4912
2N4912	28D146
2N4912	28D147
2N4912	28D148
2N4912	28D236
2N4912	28D237
2N4913	2N4913
2N4914	2N4914
2N4915	2N4915
2N4918	2N4918
2N4918	28A496
2N4918	28A779
2N4918	2BA922
2N4918	28B559
2N4918	28B632
2N4919	2N4919
2N4919	28A505
2N4919	28A780
2N4920	2N4920
2N4920	28B526
2N4920	28B527
2N4920	28B528
2N4920	28B548
2N4920	28B549
2N4920	28B631
2N4921	2N4921
2N4921	28C496
2N4921	28D488
2N4922	2N4922
2N4922	28C1517
2N4922	28C2311
2N4922	28D489
2N4923	2N4923
2N4923	28C2024
2N4923	28C495
2N4923	28D556
2N4923	28D557
2N4923	28D558
2N4923	28D490
2N4923	28D600
2N4924	2N4924
2N4925	2N4925
2N4926	2N4926
2N4927	2N4927
2N4928	2N1476
2N4928	2N1477
2N4928	2N4928
2N4929	2N4929
2N4930	2N4930
2N4930JAN	2N4930JAN
2N4930JTX	2N4930JTX
2N4931	2N4931
2N4931JAN	2N4931JAN
2N4931JTX	2N4931JTX
2N4937	2N2802
2N4937	2N2805
2N4937	2N4937
2N4938	2N2803
2N4938	2N2806
2N4939	2N2804
2N4939	2N2807
2N4939	2N4939
2N4941	2N4941
2N4942	2N4942
2N4948	2N4948
2N4949	2N4949
2N4957	NB95312
2N4957	2N4957
2N4957	28A800
2N4959	BPR99
2N4959	MM439
2N4959	2N4958
2N4959	2N4959
2N5016	2N5016
2N5022	2N5022
2N5023	2N5023
2N5031	MM8006
2N5031	NB73412
2N5031	2N5031
2N5031	28C1275
2N5031	28C567
2N5031	28C568
2N5032	MM8007
2N5032	2N3570
2N5032	2N3571
2N5032	2N3572
2N5032	2N5880
2N5032	2N5032
2N5038	MJ5038
2N5038	2N5038
2N5038	2N5386
2N5039	MJ5039
2N5039	SDT12301
2N5039	2N5039
2N5050	SDT5904
2N5050	SDT5905
2N5050	SDT5909
2N5050	SDT5910
2N5050	SDT6901
2N5050	TIP503
2N5050	TIP505
2N5050	2N5050
2N5050	2N6263
2N5050	28C101
2N5050	28C1431
2N5050	28C491
2N5050	28C791
2N5050	28C840
2N5050	28C884
2N5050	28D103
2N5050	28D49
2N5050	28D71
2N5051	SDT5951
2N5051	SDT5954
2N5051	SDT6902
2N5051	TIP504
2N5051	TIP506
2N5051	2N5051
2N5051	2N6264
2N5051	28C840A
2N5051	28D93
2N5052	SDT5953
2N5052	SDT5956
2N5052	SDT6903
2N5052	SDT6904
2N5052	2N5052
2N5052	28C2239
2N5052	28C680
2N5052	28D251
2N5052	28D94
2N5058	2N5058
2N5059	2N5059
2N5060	C103Y
2N5060	T1044
2N5060	T1060
2N5060	2N5060
2N5061	C103YY
2N5061	T1045
2N5061	T1061
2N5061	2N5061
2N5062	C103A
2N5062	T1046
2N5062	T1062
2N5062	2N5062
2N5063	T1063
2N5063	2N5063
2N5064	C103B
2N5064	T1047
2N5064	T1064
2N5064	2N5064
2N5067	2N5067
2N5068	2N5068
2N5069	2N5069
2N5070	2N5070
2N5071	2N5071
2N5086	2N5102
2N5086	A5T2605
2N5086	A5T2609
2N5086	A5T4248
2N5086	A5T4249
2N5086	A5T5086
2N5086	EN3962
2N5086	PN4248
2N5086	PN4249
2N5087	A5T4250
2N5087	A5T5087
2N5087	BC259
2N5087	BC315
2N5087	PN4250
2N5087	PN4250A
2N5089	BC329
2N5090	2N5090
2N5108	2N5108
2N5109	NE74014
2N5109	2N5109
2N5109	28C1253
2N5146	SP5765QD
2N5146	2N5146
2N5160	2N5160
2N5161	2N5161
2N5162	2N5162
2N5164	C33P
2N5164	MCR2818-2
2N5164	MCR2818R2
2N5164	MCR808-2
2N5164	MCR808R2
2N5164	2N5164
2N5165	C33B
2N5165	MCR2818-4
2N5165	MCR2818R4
2N5165	MCR808-4
2N5165	MCR808R4
2N5165	2N5165
2N5166	C33D
2N5166	MCR2818-6
2N5166	MCR2818R6
2N5166	MCR808-6
2N5166	MCR808R6
2N5166	2N5166
2N5167	MCR2818-8
2N5167	MCR2818R8
2N5167	MCR808-8
2N5167	MCR808R8
2N5167	2N5167
2N5168	C37F
2N5168	MCR1305R2
2N5168	MCR1308-2
2N5168	MCR1308R2
2N5168	MCR2918-2
2N5168	MCR2918R2
2N5168	2N5168
2N5169	C37B
2N5169	C37G
2N5169	MCR1305R4
2N5169	MCR1308-4
2N5169	MCR1308R4
2N5169	MCR2918-4
2N5169	MCR2918R4
2N5169	2N5169
2N5170	C37D
2N5170	MCR1305R6
2N5170	MCR1308-6
2N5170	MCR1308-8
2N5170	MCR1308R6
2N5170	MCR2918-6
2N5170	MCR2918R6
2N5170	2N5170
2N5171	C37M
2N5171	MCR1305R8
2N5171	MCR1308R8
2N5171	MCR2918-8
2N5171	MCR2918R8
2N5171	2N5171
2N5179	2N3478
2N5179	2N3600
2N5179	2N5179
2N5180	2N5180
2N5190	MJ33521
2N5190	MJ3482
2N5190	2N5190
2N5190	28D359
2N5190	28D360
2N5190	28D485
2N5191	MJ3483
2N5191	2N5191
2N5191	28D361
2N5191	28D486
2N5192	MJ3484
2N5192	2N5192
2N5192	28D487
2N5193	MJ33571
2N5193	MJ3492
2N5193	2N5193
2N5193	28B525
2N5193	28B529
2N5193	28B572
2N5193	28B575
2N5194	MJ8493
2N5194	2N5194
2N5194	28B524
2N5194	28B573
2N5194	28B576
2N5194	28C2270
2N5195	MJ8494
2N5195	2N5195
2N5195	28B574
2N5195	28B577
2N5208	PN4916
2N5208	2N5208
2N5209	PN4917
2N5210	A5T5209
2N5210	A5T5391
2N5210	A5T5391A
2N5210	A7T5391
2N5210	A7T5391A
2N5210	A8T5391
2N5210	A8T5391A
2N5219	A5T5220
2N5220	A5T5221
2N5223	A5T5223
2N5224	PN5134
2N5225	A5T5225
2N5226	A5T5226
2N5226	EN2894A
2N5227	A5T5227
2N5229	2N5229
2N5230	2N5230
2N5231	2N5231
2N5241	2N5241
2N5245	2N5245
2N5246	2N5246
2N5247	2N5247
2N5265	UC310
2N5265	UC320
2N5265	UC340
2N5265	UC814
2N5265	UC851
2N5265	UC853
2N5265	UC854
2N5265	UC855
2N5265	2M3574
2N5265	2N3277
2N5265	2N5278
2N5265	2N5328
2N5265	2N5573
2N5265	2N5575
2N5265	2N5698
2N5265	2N5020
2N5265	2N5265
2N5266	P1005
2N5266	P1027
2N5266	P1004
2N5267	UC300
2N5267	2N2497
2N5267	2N5267
2N5267/8	2N5267
2N5268	P1028
2N5269	2N5269
2N5270	P1005
2N5270	P1029
2N5271	2N5271
2N5301	2N5301
2N5302	2N3237
2N5302	2N5302
2N5302	28D231
2N5302	28D249
2N5302	28D630
2N5302	28D631
2N5302JAN	2N5302JAN
2N5302JTX	2N5302JTX
2N5302JTXV	2N5302JTXV
2N5303	SDT9701
2N5303	2N5303
2N5303JAN	2N5303JAN
2N5303JTX	2N5303JTX
2N5303JTXV	2N5303JTXV
2N5336	SDT5423
2N5336	SDT5427
2N5336	SDT4452
2N5336	SDT4454
2N5336	SDT5508
2N5336	2N5418
2N5336	2N5419
2N5336	2N5420
2N5336	2N5421
2N5336	28C148
2N5336	2N5150
2N5336	2N5336
2N5337	SDT4455
2N5337	SDT4456
2N5337	SDT5501
2N5337	SDT5502
2N5337	SDT5503
2N5337	SDT5512
2N5337	SDT5513
2N5337	2N4300
2N5337	2N4301
2N5337	2N4305
2N5337	2N4307
2N5337	2N4311
2N5337	2N4877
2N5337	2N5152
2N5337	2N5154
2N5337	2N5335
2N5337	2N5337
2N5337	2N5729
2N5338	SDT5424
2N5338	SDT5428
2N5338	SDT5509
2N5338	2N5338
2N5339	SDT5514
2N5339	2N4309
2N5339	2N5339
2N5345	2N5344
2N5345	2N5345
2N5346	SDTB01
2N5346	SDTB02
2N5346	SDTB05
2N5346	SDTB06
2N5346	SDTB07
2N5346	2N5346
2N5347	SDT12302
2N5347	SDT12303
2N5347	SDT12305
2N5347	SDT12306
2N5347	SDT12307
2N5347	SDT3401
2N5347	SDT3402
2N5347	SDT3403
2N5347	SDT3405
2N5347	SDT3406
2N5347	SDT3407
2N5347	SDT6308
2N5347	SDT6309
2N5347	SDT6310
2N5347	SDT6311
2N5347	SDT6312
2N5347	SDT6313
2N5347	SDT6314
2N5347	SDT6315
2N5347	SDT6316
2N5347	SDT6408
2N5347	SDT6409
2N5347	SDT6410
2N5347	SDT6411
2N5347	SDT6412
2N5347	SDT6413
2N5347	SDT6414
2N5347	SDT6415
2N5347	SDT6416
2N5347	TIP542
2N5347	TIP543
2N5347	2N3996
2N5347	2N3997
2N5347	2N3998
2N5347	2N3999
2N5347	2N4000
2N5347	2N4115
2N5347	2N4116
2N5347	2N4998
2N5347	2N5000
2N5347	2N5002
2N5347	2N5004
2N5347	2N5083
2N5347	2N5084
2N5347	2N5085
2N5347	2N5284
2N5347	2N5285
2N5347	2N5326
2N5347	2N5347
2N5347	2N5477
2N5347	2N5478
2N5347	2N5730
2N5348	SDTB05
2N5348	2N5348
2N5349	SDT3404
2N5349	SDT3408
2N5349	2N5349
2N5349	2N5479
2N5349	2N5480
2N5358	2N5358
2N5359	2N5359
2N5360	2N5360
2N5361	0673
2N5361	0674
2N5361	2N5361
2N5362	2N5362
2N5363	2N5363
2N5364	2N5278
2N5364	2N5364
2N5400	A5T5497
2N5400	A5T5400
2N5400	BC528
2N5401	A5T5401
2N5401	BC530
2N5401	PN4357
2N5401	PN4888
2N5401	PN4889
2N5416	2N5415
2N5416	2N5416
2N5427	SDT5911
2N5427	SDT5912
2N5427	SDT5913
2N5427	SDT7A08
2N5427	SDT7A09
2N5427	SDT5427
2N5428	SDT7A01
2N5428	SDT7A02
2N5428	SDT7A03
2N5428	2N5878
2N5428	2N5202
2N5428	2N5428
2N5428	2N5598
2N5428	2N5600
2N5428	2N5602
2N5428	2N5606
2N5428	2N5610
2N5428	28C1444
2N5428	28D290
2N5428	28D690
2N5429	SDT5914
2N5429	2N5429
2N5430	2N5879
2N5430	2N5430
2N5430	2N5604
2N5430	2N5612
2N5430	2N6500
2N5430	28C1050
2N5430	28C1445
2N5431	28C981
2N5431	2N5431
2N5441	SC2663
2N5441	T6402P
2N5441	T6406B
2N5441	TIC270B
2N5442	2N5441
2N5442	SC266D
2N5442	T6406D
2N5442	TIC270D
2N5442	2N5442
2N5443	SC266M
2N5443	T6402B
2N5443	T6406B
2N5443	T6406M
2N5443	TIC270E
2N5443	2N5443
2N5444	SC265B
2N5444	T6412P
2N5444	TIC270M
2N5444	TIC272B
2N5444	2N5444
2N5445	SC265D
2N5445	TIC272D
2N5445	2N5445
2N5446	SC265M
2N5446	T6412B
2N5446	TIC272E
2N5446	TIC272M
2N5446	2N5446

MOTOR.	Industry Standard No.	MOTOR.	Industry Standard No.	MOTOR.	Industry Standard No.	MOTOR.	Industry Standard No.
2N5457	C94	2N5486	IT109	2N5634	2N3865	2N5681	2N2988
2N5457	C94A	2N5486	ITE4338	2N5634	2N5634	2N5681	2N5681
2N5457	C94B	2N5486	ITE4339	2N5634	2SC1111	2N5682	2N2911
2N5457	C95	2N5486	ITE434	2N5634	2SC1112	2N5682	2N2984
2N5457	C95A	2N5486	ITE4340	2N5634	2SC1115	2N5682	2N2990
2N5457	E101	2N5486	ITE4341	2N5634	2SC1403	2N5682	2N4271
2N5457	E102	2N5486	KE4416	2N5634	2SC1869	2N5682	2N4272
2N5457	E201	2N5486	MPF107	2N5634	2SC2321	2N5682	2N5682
2N5457	E230	2N5486	TI834	2N5634	2SC2337	2N5683	2N5683
2N5457	FE5457	2N5486	TI888	2N5634	2SC2487	2N5683JAN	2N5683JAN
2N5457	ITE4867	2N5486	U1837E	2N5634	2SC2488	2N5683JTX	2N5683JTX
2N5457	KE5684	2N5486	U1994E	2N5634	2SC2489	2N5683JTXV	2N5683JTXV
2N5457	KE5687	2N5486	U315	2N5634	2SC2493	2N5684	2N5684
2N5457	KE4220	2N5486	UC20	2N5634	2SC902	2N5684JAN	2N5684JAN
2N5457	MPF103	2N5486	UC21	2N5634	2SD110	2N5684JTX	2N5684JTX
2N5457	NF5457	2N5486	UC22	2N5634	2SD165	2N5684JTXV	2N5684JTXV
2N5457	NKT80111	2N5486	UC23	2N5634	2SD177	2N5685	2N5575
2N5457	NKT80112	2N5486	UC734E	2N5634	2SD208	2N5685	2N5578
2N5457	NKT80113	2N5486	2N5486	2N5634	2SD218	2N5685	2N5685
2N5457	NPC211N	2N5486	2SK19Y	2N5634	2SD425	2N5685JAN	2N5685JAN
2N5457	NPC212N	2N5486	2SK23	2N5634	2SD432	2N5685JTX	2N5685JTX
2N5457	NPC214N	2N5486	2SK32	2N5635	MM1549	2N5685JTXV	2N5685JTXV
2N5457	NPC215N	2N5539	SDT5504	2N5635	2N4041	2N5686	2N6258
2N5457	WK5457	2N5550	A5T5550	2N5635	2N5675	2N5686	2SD114
2N5457	2N4302	2N5550	BO532	2N5636	MM1550	2N5686JAN	2N5686JAN
2N5457	2SK13	2N5550	BF412	2N5636	2N4040	2N5686JTX	2N5686JTX
2N5457	2SK17GR	2N5550	BB835	2N5636	2N5636	2N5686JTXV	2N5686JTXV
2N5457	2SK170	2N5551	A5T5551	2N5636	2N5774	2N5745	2N5745
2N5457	2SK17R	2N5551	BC533	2N5637	MM1551	2N5745JAN	2N5745JAN
2N5457	3SK20H	2N5551	BF413	2N5637	2N5637	2N5745JTX	2N5745JTX
2N5457	3SK21H	2N5551	PN5965	2N5637	2N6204	2N5745JTXV	2N5745JTXV
2N5458	E202	2N5556	2N5447	2N5638	FE0655A	2N5758	2N5758
2N5458	E231	2N5567	T1020	2N5638	KE3970	2N5758	2SC1618
2N5458	FE3819	2N5567	T4107B	2N5638	KE4091	2N5758	2SC1619
2N5458	FE5458	2N5567	TIC220B	2N5638	NP530	2N5758	2SC1667
2N5458	ITE4868	2N5567	TIC230B	2N5638	TI873	2N5758	2SC664
2N5458	KE3685	2N5567	2N5567	2N5638	U1897E	2N5758	2SC793
2N5458	KE4221	2N5568	T1021	2N5638	2N5949	2N5758	2SC794
2N5458	MPF104	2N5568	T4107D	2N5639	FE0655B	2N5758	2SC962
2N5458	NF4302	2N5568	TIC220D	2N5639	ITE3066	2N5758	2SD116
2N5458	NF4303	2N5568	TIC230D	2N5639	ITE3067	2N5758	2SD119
2N5458	NF4304	2N5568	2N5568	2N5639	ITE3068	2N5758	2SD12
2N5458	NF5458	2N5569	T1022	2N5639	KE3971	2N5758	2SD124
2N5458	NKT80215	2N5569	T4117B	2N5639	KE4092	2N5758	2SD124A
2N5458	NKT80216	2N5569	TIC222B	2N5639	NP52/	2N5758	2SD125
2N5458	NPC213N	2N5569	TIC232B	2N5639	NP531	2N5758	2SD125A
2N5458	NPC216N	2N5569	2N5569	2N5639	TI874	2N5758	2SD131
2N5458	TI858	2N5570	T1023	2N5639	U1898E	2N5758	2SD15
2N5458	WK5458	2N5570	T4117D	2N5639	2N5950	2N5758	2SD16
2N5458	2N4303	2N5570	TIC222D	2N5640	FE0655C	2N5758	2SD180
2N5458	2N4304	2N5570	TIC232D	2N5640	ITE4117	2N5758	2SD188
2N5458	2N5163	2N5570	2N5570	2N5640	ITE4118	2N5758	2SD189
2N5458	2N5458	2N5571	MA05571	2N5640	ITE4119	2N5758	2SD189A
2N5459	C95E	2N5571	T4106B	2N5640	KE3972	2N5758	2SD201
2N5459	C96E	2N5571	TIC240B	2N5640	KE4093	2N5758	2SD247
2N5459	CP2386	2N5571	2N5571	2N5640	KE510	2N5758	2SD26
2N5459	E103	2N5572	MA05572	2N5640	KE511	2N5758	2SD260
2N5459	E203	2N5572	T4106D	2N5640	NP532	2N5758	2SD26A
2N5459	E232	2N5572	TIC240D	2N5640	NP533	2N5758	2SD26B
2N5459	FE5459	2N5572	2N5572	2N5640	TI875	2N5758	2SD335
2N5459	ITE4869	2N5573	MA05573	2N5640	U1899E	2N5758	2SD338
2N5459	KE4222	2N5573	T4116B	2N5640	UC155W	2N5758	2SD339
2N5459	MPF105	2N5573	TIC242B	2N5640	2N5951	2N5758	2SD371
2N5459	NF5459	2N5573	2N5573	2N5640	2N5952	2N5758	2SD379
2N5459	NKT80211	2N5574	MA05574	2N5640	2N5953	2N5758	2SD428
2N5459	NKT80212	2N5574	T4116D	2N5641	A3-28	2N5758	2SD47
2N5459	NKT80213	2N5574	TIC242D	2N5641	B3-28	2N5758	2SD50
2N5459	NKT80214	2N5574	2N5574	2N5641	BLY91A	2N5758	2SD51
2N5459	WK5459	2N5581	2N5581	2N5641	PP9730	2N5758	2SD52
2N5460	MPF151	2N5582	2N2695	2N5641	PP9732	2N5758	2SD597
2N5460	2N5303	2N5582	2N3673	2N5641	SD1220-1	2N5758	2SD68
2N5461	MPF152	2N5582	2N4450	2N5641	SD1242-5	2N5758	2SD73
2N5461	2N5461	2N5582	2N5582	2N5641	2N2347	2N5758	2SD80
2N5462	MPF153	2N5583	MM4500	2N5641	2N3296	2N5758	2SD81
2N5462	2N5462	2N5583	NE71112	2N5641	2N3961	2N5758	2SD82
2N5463	MPF154	2N5583	2N5583	2N5641	2N5641	2N5758	2SD88
2N5463	2N5463	2N5583	2SA711	2N5641	2N5711	2N5759	2N1235
2N5464	MPF155	2N5589	MM1601	2N5641	2N6198	2N5759	2N5759
2N5464	2N5464	2N5589	SD1177	2N5642	MM1558	2N5759	2SC519
2N5465	MPF156	2N5589	SD1256	2N5642	SD1222-5	2N5759	2SC961
2N5465	TI805	2N5589	2N3925	2N5642	SD1244-6	2N5759	2SD202
2N5465	2N5465	2N5589	2N5589	2N5642	2N5642	2N5759	2SD334
2N5471	2N5471	2N5590	MM1602	2N5642	2N5712	2N5759	2SD427
2N5471/3	2N5471/3	2N5590	SD1012	2N5642	2N5713	2N5759	2SD430
2N5471/3	2N3113	2N5590	2N5590	2N5642	2N6199	2N5759	2SD53
2N5472	2N2841	2N5590	2SC1257	2N5643	BLY92A	2N5759	2SD598
2N5472	2N5472	2N5591	MM1603	2N5643	BLY93A	2N5759	2SD67
2N5473	UC330	2N5591	SD1216	2N5643	MM1559	2N5759	2SD673
2N5473	2N2842	2N5591	2N5591	2N5643	SD1224-2	2N5759	2SD674
2N5473	2N5473	2N5629	SDT7734	2N5643	2N5643	2N5760	2N5760
2N5474	U110	2N5629	SDT9702	2N5643	2N5714	2N5760	2N6262
2N5474	2N2606	2N5629	SDT9705	2N5643	2N5994	2N5760	2SC1030
2N5474	U133	2N5629	2N5629	2N5643	2N6200	2N5760	2SC1051
2N5475	2N2607	2N5629	2N6360	2N5643	2SC1689	2N5760	2SC1866
2N5475	2N2843	2N5630	IR6302	2N5644	MM1660	2N5760	2SC2569
2N5475	2N5475	2N5630	MJ6302	2N5644	2N5644	2N5760	2SC519A
2N5476	U112	2N5630	SDT7735	2N5645	C3-12	2N5760	2SC665
2N5476	2N5476	2N5630	SDT9703	2N5645	MM1661	2N5760	2SC897
2N5484	E210	2N5630	SDT9706	2N5645	2N5645	2N5760	2SC898
2N5484	FE0654C	2N5630	2N4348	2N5645	2SC891	2N5760	2SD117
2N5484	FE5484	2N5630	2N5630	2N5646	C12-12	2N5760	2SD118
2N5484	SPB1092	2N5630	2N6302	2N5646	MM1662	2N5760	2SD17
2N5484	U314	2N5631	SDT7736	2N5646	2N5646	2N5760	2SD126
2N5484	U0714A	2N5631	2N5631	2N5646	2SC1081	2N5760	2SD17
2N5484	2N5484	2N5631	2N6259	2N5646	2SC892	2N5760	2SD203
2N5484	2SK17Y	2N5632	SDT9203	2N5646	2SC990	2N5760	2SD26C
2N5484	2SK19BL	2N5632	SDT9208	2N5653	2N5653	2N5760	2SD45
2N5484	2SK19G	2N5632	2N3233	2N5654	2N5654	2N5760	2SD46
2N5485	E211	2N5632	2N3236	2N5655	2N5655	2N5760	2SD60
2N5485	E305	2N5632	2N3864	2N5656	2SD482	2N5760	2SD675
2N5485	FE0654B	2N5632	2N5632	2N5656	2N5656	2N5760	2SD676
2N5485	MPF106	2N5632	2SD111	2N5656	2SD483	2N5760	2SD69
2N5485	NPC108	2N5632	2SD151	2N5657	2N5657	2N5760	2SD728
2N5485	NPC108A	2N5632	2SD164	2N5657	2SD484	2N5760	2SD74
2N5485	SPB1087	2N5632	2SD175	2N5668	KE3823	2N5760	2SD83
2N5485	2N5485	2N5632	2SD176	2N5668	2N5668	2N5777	2N5777
2N5485	2SK35	2N5632	2SD207	2N5669	2N5669	2N5778	2N5778
2N5485	3SK22	2N5632	2SD212	2N5669	TI859	2N5779	2N5779
2N5485	3SK23	2N5632	2SD522	2N5670	2N5670	2N5780	2N5780
2N5485	3SK28	2N5633	SDT9204	2N5670	2N3777	2N5793	2N5793
2N5485	4G2	2N5633	SDT9209	2N5679	2N3781	2N5794	2N5586
2N5485	5G2	2N5633	2N5559	2N5679	2N5679	2N5794	2N5587
2N5486	B114	2N5633	2N5633	2N5680	2SD333	2N5794	2N5794
2N5486	E212	2N5633	2SC2430	2N5680	2N5680	2N5795	2N5795
2N5486	E300	2N5633	2SC2492	2N5681	2N1715	2N5796	2N5117
2N5486	E304	2N5633	2SD769	2N5681	2N1717	2N5796	2N5118
2N5486	FE0654A	2N5633	2SD161	2N5681	2N2108	2N5796	2N5119
2N5486	FE5245	2N5633	2SD213	2N5681	2N2201	2N5796	2N5120
2N5486	FE5246	2N5633	2SD217	2N5681	2N2202	2N5796	2N5122
2N5486	FE5247	2N5633	2SD319	2N5681	2N2204	2N5796	2N5123
2N5486	IT108	2N5633	2SD426			2N5796	2N5124
		2N5633	2SD431			2N5796	2N5125
		2N5633	2SD867				

MOTOR.	Industry Standard No.	MOTOR.	Industry Standard No.	MOTOR.	Industry Standard No.	MOTOR.	Industry Standard No.
2N5796	2N5796	2N5883	2N5883	2N6051JTX	2N6051JTX	2N6081	PT8837
2N5829	2N5829	2N5884	2N5884	2N6051JTXV	2N6051JTXV	2N6081	PT9795A
2N5835	2N5835	2N5885	RC8258	2N6052	IR647	2N6081	2N4127
2N5836	MM8020	2N5885	2N5885	2N6052	PMD11K-100	2N6081	2N5996
2N5836	2N5836	2N5885	2N3559	2N6052	PMD13K-100	2N6081	2N6081
2N5837	MM8021	2N5886	2N5886	2N6052	TIP607	2N6081	2SC1258
2N5837	2N5837	2N5941	MM1632	2N6052	TIP627	2N6082	B25-12
2N5838	MJ3260	2N5942	MM1633	2N6052	TIP647	2N6082	BLX89A
2N5838	SDT1050	2N5943	MM8002	2N6052	2N6052	2N6082	MM1666
2N5838	SDT1055	2N5943	MM8023	2N6052	2SA1045	2N6082	2N4128
2N5838	SDT1060	2N5943	PT3571	2N6052	2SA1046	2N6082	2N6082
2N5838	SVT250-5	2N5943	PT3571A	2N6052	2SB589	2N6082	2SC1297
2N5838	2N5838	2N5943	PT4579	2N6052JAN	2N6052JAN	2N6083	B30-12
2N5838	2SD1195	2N5943	SD1006	2N6052JTX	2N6052JTX	2N6083	MM1667
2N5838	SD1353	2N5943	2N5943	2N6052JTXV	2N6052JTXV	2N6083	PT8864
2N5838	2SD748	2N5943	2SC652	2N6053	MJ4010	2N6083	PT8864A
2N5839	2N5839	2N5943	2SC823	2N6053	TIP625	2N6083	SD1229
2N5840	SDT1051	2N5943	2SC824	2N6053	2N6053	2N6083	2N6083
2N5840	SVT350-5	2N5943	2SC852	2N6053	2SB585	2N6083	2SC1259
2N5840	2N5840	2N5944	BLX67	2N6054	MJ4011	2N6084	B40-12
2N5840	2SD1106	2N5944	C1-12Z	2N6054	TIP626	2N6084	B45-12
2N5840	2SC1152	2N5944	CD5944	2N6054	2N6054	2N6084	MM1668
2N5840	2SC1185	2N5944	PT3537	2N6054	2SB586	2N6084	MM1669
2N5840	2SD1292	2N5944	PT8809	2N6055	MJ4000	2N6084	PT8838
2N5840	2SC792	2N5944	SD1134	2N6055	TIP620	2N6084	2N6084
2N5840	2SC935	2N5944	2N5698	2N6055	2N6055	2N6084	2SC1298
2N5840	2SD198	2N5944	2N5914	2N6055	2N6492	2N6094	MM4020
2N5840	2SD320	2N5944	2N5944	2N6055	2SD502	2N6094	2N6094
2N5840	2SD632	2N5944	2SC1603	2N6055	2SD523	2N6095	MM4021
2N5840	2SD672	2N5944	2SC1808	2N6056	DT81010	2N6095	2N6095
2N5840	2SD800	2N5944	2SC1966	2N6056	IR1010	2N6096	MM4022
2N5841	MM1605	2N5944	3TX620	2N6056	SDM1010	2N6096	2N6096
2N5841	2N5841	2N5944	3TX621	2N6056	TIP621	2N6097	MM4023
2N5841	MM1606	2N5944	3TX820	2N6056	2N4001	2N6097	2N6097
2N5842	MM1607	2N5945	BLX68	2N6056	2N6056	2N6107	RCA1C11
2N5842	2N5842	2N5945	C2-8Z	2N6056	2N6493	2N6107	2N6106
2N5846	MM1608	2N5945	C5-12	2N6056	2N6494	2N6107	2N6107
2N5847	MM1618	2N5945	CD5945	2N6056	2SC1831	2N6107	2SA771
2N5847	PT8851	2N5945	PT8810	2N6056	2SD107	2N6107	2SB707
2N5847	PT8851A	2N5945	SD1135	2N6056	2SD108	2N6107	2SB708
2N5847	SD1069	2N5945	2N5945	2N6056	2SD463	2N6109	2N6108
2N5847	SD1167	2N5945	2SC1967	2N6056	2SD464	2N6109	2N6109
2N5847	2N5689	2N5945	3TX622	2N6056	2SD503	2N6109	2SA770
2N5847	2N5699	2N5945	3TX822	2N6056	2SD524	2N6109	2SB754
2N5847	2N5847	2N5946	BLX53A	2N6056	2SD692	2N6111	2N6110
2N5848	A15-12	2N5946	C10-12A	2N6057	PMD10K-40	2N6111	2N6111
2N5848	PT8852	2N5946	C5-8Z	2N6057	PMD10K-60	2N6114	D5K1
2N5848	PT8852A	2N5946	CD5946	2N6057	2N6057	2N6115	D5K2
2N5848	SD1168	2N5946	PT8811	2N6057	2N6355	2N6116	MPU231
2N5848	2N5690	2N5946	SD1136	2N6057	2N6356	2N6116	2N6116
2N5849	MM1620	2N5946	SD1433	2N6057	2SD504	2N6116	2N6137
2N5849	MM1622	2N5946	2N5915	2N6058	PMD10K-80	2N6117	2N6138
2N5849	MM1646	2N5946	2N5946	2N6058	2N6058	2N6117	MPU232
2N5849	SD1169	2N5947	MM8012	2N6058	2N6357	2N6117	2N6117
2N5849	SD1290	2N5947	PT3570	2N6058	2N6358	2N6118	MPU233
2N5849	2N5691	2N5947	PT4570	2N6058	2SD505	2N6118	2N6118
2N5849	2N5848	2N5947	PT4572A	2N6058JAN	2N6058JAN	2N6121	MJB2482
2N5849	2N5849	2N5947	2N5947	2N6058JTX	2N6058JTX	2N6121	MJB5190
2N5851	MM1510	2N5974	MJ3101	2N6058JTXV	2N6058JTXV	2N6121	NSP5190
2N5852	MM1511	2N5974	2N5974	2N6059	PMD10K-100	2N6121	2N5295
2N5859	2N1409	2N5975	MJ3102	2N6059	PMD12K-100	2N6121	2N5296
2N5859	2N1409A	2N5975	2N5975	2N6059	TIP602	2N6121	2N6121
2N5859	2N1410	2N5975	2SB579	2N6059	2N6059	2N6122	MJB2483
2N5859	2N1410A	2N5976	MJ3103	2N6059	2SC2435	2N6122	MJB5191
2N5859	2N1444	2N5976	2N5976	2N6059	2SC2436	2N6122	NSP5191
2N5859	2N1959	2N5976	2SB580	2N6059	2SD506	2N6122	RCA3054
2N5859	2N2410	2N5976	2SB581	2N6059	2SD628	2N6122	2N5297
2N5859	2N2476	2N5977	MJ3201	2N6059	2SD629	2N6122	2N5298
2N5859	2N2477	2N5977	2N5977	2N6059	2SD803	2N6122	2N6122
2N5859	2N2537	2N5977	2SC932	2N6059JAN	2N6059JAN	2N6123	2SD475
2N5859	2N2538	2N5977	2SD493	2N6059JTX	2N6059JTX	2N6123	MJB5192
2N5859	2N2846	2N5978	MJ3202	2N6059JTXV	2N6059JTXV	2N6123	NSP5192
2N5859	2N2848	2N5978	MJ3203	2N6068	MAC77-1	2N6123	2N5293
2N5859	2N3015	2N5978	2N5978	2N6068	2N6068	2N6123	2N5294
2N5859	2N3512	2N5978	2SD494	2N6068A	2N6068A	2N6123	2N6123
2N5859	2N4046	2N5979	MJ3204	2N6068B	2N6068B	2N6123	2SC1107
2N5859	2N5859	2N5979	2N5979	2N6069	MAC77-2	2N6123	2SC1108
2N5861	2N5861	2N5980	2N5980	2N6069	2N6069	2N6123	2SC1109
2N5862	2N5862	2N5981	2N5981	2N6069A	2N6069A	2N6123	2SC1110
2N5875	MJ2940	2N5982	2N5982	2N6069B	2N6069B	2N6123	2SC789
2N5875	2N5875	2N5983	2N5983	2N6070	MAC77-3	2N6123	2SD476
2N5876	MJ2941	2N5984	2N5984	2N6070	2N6070	2N6123	2SD570
2N5876	2N5876	2N5985	2N5985	2N6070A	T1C206A	2N6124	MJB5193
2N5876	2SA877	2N5986	2N5986	2N6070A	T1C215A	2N6124	NSP5193
2N5877	2SA877	2N5987	2N5987	2N6070A	2N6070A	2N6124	2N6023
2N5877	MJ2840	2N5988	2N5988	2N6070B	2N6070B	2N6124	2SC624
2N5877	2N1487	2N5989	2N5989	2N6071	L2004F71	2N6124	2N6124
2N5877	2N1489	2N5990	2N5990	2N6071	L2004F91	2N6125	MJB5194
2N5877	2N1702	2N5991	2N5991	2N6071	MAC77-4	2N6125	NSP5194
2N5877	2N3232	2N6027	2SD501	2N6071	2N6071	2N6125	2N6025
2N5877	2N5877	2N6027	2N6027	2N6071A	L2004F51	2N6125	2N6026
2N5877	2N6253	2N6028	2N6028	2N6071A	T1C206B	2N6125	2SA490
2N5877	2SD172	2N6029	2N5742	2N6071A	T1C215B	2N6125	2SA670
2N5877	2SD174	2N6029	2N6029	2N6071A	2N6071A	2N6125	2SA671
2N5877	2SD206	2N6030	2N6030	2N6071B	L2004F31	2N6125	2SA755
2N5877	2SD211	2N6031	2N6031	2N6071B	2N6071B	2N6125	2SA768
2N5878	MJ2841	2N6034	2N6034	2N6072	MAC77-5	2N6125	2SB507
2N5878	RCA1B01	2N6035	2N6035	2N6072	2N6072	2N6125	2SB565
2N5878	SDT9202	2N6036	2N6036	2N6072A	2N6072A	2N6126	MJB5195
2N5878	SDT9207	2N6037	2N6037	2N6072B	2N6072B	2N6126	NSP5195
2N5878	2N1488	2N6037	2SD479	2N6073	L4004F79	2N6126	2N6022
2N5878	2N1490	2N6038	2N6038	2N6073	L4004F91	2N6126	2N6126
2N5878	2N5737	2N6038	2SC2324	2N6073	MAC77-6	2N6126	2SA489
2N5878	2N5739	2N6039	2SD480	2N6073	2N6073	2N6126	2SA769
2N5878	2N5878	2N6039	2N6039	2N6073A	L4004F51	2N6126	2SB509
2N5878	2N6254	2N6039	2SD481	2N6073A	T1C206D	2N6126	2SB535
2N5879	2N5879	2N6040	2N6040	2N6073A	T1C215D	2N6126	2SB536
2N5879	2N6246	2N6041	2SB675	2N6073A	2N6073A	2N6126	2SB537
2N5879	2N6469	2N6041	2N6041	2N6073B	L4004F31	2N6126	2SB566
2N5880	RC8618	2N6041	2SB674	2N6073B	2N6073B	2N6126	2SB596
2N5880	2N5880	2N6042	2SB711	2N6074	MAC77-7	2N6129	2SB604
2N5880	2N6247	2N6042	2N6042	2N6074	2N6074	2N6129	2N6129
2N5880	2SA680	2N6042	2SB673	2N6074A	2N6074A	2N6130	RCA1C05
2N5881	MJ2802	2N6042	2SB712	2N6074B	2N6074B	2N6130	2N6130
2N5881	SDT7731	2N6043	D44D3	2N6075	MAC77-8	2N6131	2N6131
2N5881	SDT7732	2N6043	D44D4	2N6075	2N6075	2N6132	2N6132
2N5881	2N5667	2N6043	2N6043	2N6075A	2N6075A	2N6133	RCA1C06
2N5881	2N5881	2N6044	D44D5	2N6075B	2N6075B	2N6133	2N6133
2N5881	2N6470	2N6044	D44D6	2N6077	2N6077	2N6134	2N6134
2N5881	2N6471	2N6044	2N6044	2N6078	2N6078	2N6136	BLX69A
2N5882	RC8617	2N6045	2N6045	2N6080	B2-8Z	2N6136	C25-12
2N5882	SDT7733	2N6045	2SD721	2N6080	B3-12	2N6136	MM1665
2N5882	SDT9704	2N6045	2SD722	2N6080	MM1680	2N6136	MRP602
2N5882	2N3238	2N6045	2SD768	2N6080	PT4537	2N6136	PT8825
2N5882	2N3239	2N6049	2N6049	2N6080	PT8871	2N6136	2N6136
2N5882	2N6240	2N6050	PMD11K-40	2N6080	PT8871A	2N6145	MAC40802
2N5882	2N5882	2N6050	PMD11K-60	2N6080	2N6080	2N6145	2N6145
2N5882	2N5970	2N6050	2N6050	2N6081	B12-12	2N6146	MAC40803
2N5882	2N5971	2N6050	2SB587	2N6081	B15-12	2N6146	2N6146
2N5882	2N6242	2N6051	PMD11K-80	2N6081	B5-8Z	2N6147	MAC40804
2N5882	2SC1777	2N6051	2N6051	2N6081	BLX88A	2N6147	2N6147
2N5882	2SC2432	2N6051	2SB588	2N6081	CD2514	2N6151	MAC10-4
2N5883	2N5741	2N6051JAN	2N6051JAN	2N6081	MM1681		
2N5883	2N5743			2N6081	PT8549		

MOTOR.	Industry Standard No.	MOTOR.	Industry Standard No.	MOTOR.	Industry Standard No.	MOTOR.	Industry Standard No.
2N6151	2N6151	2N6226	TIP544	2N6249	28C1441	2N6306	DTS515
2N6152	MAC10-6	2N6226	2N3174	2N6249	28C1584	2N6306	DTS516
2N6152	2N6152	2N6226	2N3186	2N6249	28C1585	2N6306	DTS517
2N6153	MAC10-8	2N6226	2N5198	2N6249	28C1783	2N6306	KDT515
2N6153	2N6153	2N6226	2N6226	2N6249	28C1785	2N6306	KDT516
2N6154	MAC11-4	2N6226	28A626	2N6249	28C2127	2N6306	KDT517
2N6154	2N6154	2N6226	28A627	2N6249	28C2256	2N6306	RC8579
2N6155	MAC11-6	2N6226	28A657	2N6249	28C2260	2N6306	SDT520
2N6155	2N6155	2N6226	28A658	2N6249	28C2261	2N6306	SDT521
2N6156	MAC11-8	2N6226	28A663	2N6249	28C2262	2N6306	SDT522
2N6156	2N6156	2N6226	28A756	2N6249	28C2428	2N6306	SDT525
2N6157	MAC35-1	2N6226	28A837	2N6249	28C409	2N6306	SDT527
2N6157	MAC35-2	2N6226	28B518	2N6249	28C410	2N6306	SDT530
2N6157	MAC35-3	2N6226	28B531	2N6249	28C940	2N6306	SDT532
2N6157	MAC35-4	2N6226	28B552	2N6249	28D665	2N6306	SDT535
2N6157	MAC37-1	2N6227	TIP545	2N6249	28D753	2N6306	SDT7201
2N6157	MAC37-2	2N6227	2N6227	2N6250	IR515	2N6306	SDT7202
2N6157	MAC37-3	2N6227	28A757	2N6250	IR516	2N6306	SDT7207
2N6157	MAC37-4	2N6227	28B519	2N6250	28C1322	2N6306	SDT7208
2N6157	2N6157	2N6227	28B653	2N6250	28C1586	2N6306	SVT1200-10
2N6158	MAC35-5	2N6227	28B654	2N6250	28C1786	2N6306	SVT1250-10
2N6158	MAC35-6	2N6228	TIP546	2N6250	28D552	2N6306	TIP51
2N6158	MAC37-5	2N6228	2N6228	2N6251	2N6251	2N6306	TIP554
2N6158	MAC37-6	2N6228	28A1063	2N6255	MM1612	2N6306	2N4070
2N6158	2N6158	2N6228	28A656	2N6255	MRF201	2N6306	2N4071
2N6159	MAC35-7	2N6228	28A714	2N6255	SD1174	2N6306	2N5239
2N6159	MAC35-8	2N6228	28A758	2N6255	28C1256	2N6306	2N5804
2N6159	MAC37-7	2N6228	28B506	2N6256	MRF601	2N6306	2N6306
2N6159	MAC37-8	2N6228	28B520	2N6256	2N6256	2N6306	2N6510
2N6159	2N6159	2N6229	28?738	2N6257	MJ6257	2N6306	2N6511
2N6160	MAC36-1	2N6229	2N5740	2N6257	2N6257	2N6306	28C675
2N6160	MAC36-2	2N6229	2N6229	2N6274	KP3946	2N6306	28C676
2N6160	MAC36-4	2N6229	28A980	2N6274	KP3948	2N6306	28C677
2N6160	MAC38-1	2N6229	28B558	2N6274	2N4002	2N6306	28C678
2N6160	MAC38-2	2N6230	28B230	2N6274	2N4003	2N6306	28C759
2N6160	MAC38-3	2N6230	28A1067	2N6274	2N6274	2N6306	28C760
2N6160	MAC38-4	2N6230	28A648	2N6274	2N6278	2N6306	28C886
2N6160	2N5273	2N6230	28A878	2N6275	2N6032	2N6306	28C901
2N6160	2N5806	2N6230	28A981	2N6275	2N6275	2N6306	28C901A
2N6160	2N6160	2N6230	28B530	2N6275	2N6279	2N6306	28D321
2N6161	MAC36-5	2N6230	28B541	2N6275	28D232	2N6306	28D417
2N6161	MAC36-6	2N6230	28B557	2N6276	2N6276	2N6306	28D731
2N6161	MAC38-5	2N6231	2N6231	2N6276	2N6280	2N6306	28D732
2N6161	MAC38-6	2N6231	28A1007	2N6277	2N6033	2N6307	DT8430
2N6161	2N5274	2N6231	28A1064	2N6277	2N6277	2N6307	DTS518
2N6161	2N5807	2N6231	28A1065	2N6277	2N6281	2N6307	FT430
2N6161	2N6161	2N6231	28A1068	2N6282	PMD1600K	2N6307	IR430
2N6162	MAC36-7	2N6231	28A882	2N6282	PMD1601K	2N6307	KDT430
2N6162	MAC36-8	2N6231	28A982	2N6282	PMD16K-40	2N6307	KDT518
2N6162	MAC38-7	2N6231	28B559	2N6282	PMD16K-60	2N6307	MJ3430
2N6162	MAC38-8	2N6231	28B696	2N6282	2N6282	2N6307	SDT430
2N6162	2N5275	2N6233	DT8660	2N6283	PMD1602K	2N6307	SDT536
2N6162	2N5808	2N6233	SDT4902	2N6283	PMD16K-80	2N6307	SDT537
2N6162	2N5809	2N6233	2N5660	2N6283	2N6283	2N6307	SDT540
2N6162	2N6162	2N6233	2N5664	2N6283JAN	2N6283JAN	2N6307	SDT541
2N6163	2N6163	2N6233	2N6233	2N6283JTX	2N6283JTX	2N6307	SDT542
2N6164	2N6164	2N6233	28C1025	2N6283JTXV	2N6283JTXV	2N6307	SDT7204
2N6165	2N6165	2N6233	28C488	2N6284	PMD1603K	2N6307	SDT7209
2N6166	MM1561	2N6233	28D508	2N6284	PMD16K-100	2N6307	SVT300-10
2N6166	SD1019-5	2N6234	82061B	2N6284	2N6284	2N6307	TIP52
2N6166	2N6166	2N6234	SDT4903	2N6284	2N6729	2N6307	TIP555
2N6166	2N6166	2N6234	2N6234	2N6284JAN	2N6284JAN	2N6307	2N6307
2N6167	MCR4018-3	2N6235	DT8663	2N6284JTXV	2N6284JTXV	2N6307	28C758
2N6167	2N6167	2N6235	DT8665	2N6285	PMD1700K	2N6307	28C885
2N6168	MCR4018-4	2N6235	SDT1301	2N6285	PMD17K-40	2N6308	DT8403
2N6168	2N6168	2N6235	SDT1302	2N6285	PMD17K-60	2N6308	DT8409
2N6169	MCR4018-5	2N6235	SDT1303	2N6285	2N6285	2N6308	DT8424
2N6169	MCR4018-6	2N6235	SDT1304	2N6285JTX	2N6284JTX	2N6308	23519
2N6169	2N6169	2N6235	TIP530	2N6286	PMD1702K	2N6308	IR403
2N6170	MCR4018-7	2N6235	2N4298	2N6286	PMD17K-80	2N6308	IR409
2N6170	MCR4018-8	2N6235	2N4299	2N6286	2N6286	2N6308	IR424
2N6170	2N6170	2N6235	2N5651	2N6286JAN	2N6286JAN	2N6308	KDT519
2N6171	MCR4035-3'	2N6235	2N5665	2N6286JTX	2N6286JTX	2N6308	MJ424
2N6171	86420A	2N6235	2N6079	2N6286JTXV	2N6286JTXV	2N6308	SDT424
2N6171	2N6171	2N6235	2N6235	2N6287	PMD1703K	2N6308	SDT545
2N6172	MCR4035-4	2N6235	28C833	2N6287	PMD17K-100	2N6308	SDT546
2N6172	86420D	2N6236	MCR306-1	2N6287	2N6287	2N6308	SDT547
2N6172	2N6172	2N6236	S0505M82	2N6287	2N6693	2N6308	SDT550
2N6173	MCR4035-5	2N6236	S0505M83	2N6287JAN	2N6287JAN	2N6308	SDT551
2N6173	MCR4035-6	2N6236	S2060G	2N6287JTX	2N6287JTX	2N6308	SDT552
2N6173	2N6173	2N6236	S2061Q	2N6287JTXV	2N6287JTXV	2N6308	SDT7205
2N6174	MCR4035-7	2N6236	S2062Q	2N6288	2N6288	2N6308	TIP53
2N6174	MCR4035-8	2N6236	T1039Y	2N6288	2N6289	2N6308	2N6508
2N6174	86420M	2N6236	2N6236	2N6290	RCA1014	2N6508	2N6582
2N6174	2N6174	2N6236	2N6332	2N6290	2N5490	2N6512	RC830
2N6186	MJ6701	2N6237	MCR306-2	2N6290	2N5491	2N6512	RC832
2N6186	SDT3125	2N6237	S0505M82	2N6290	2N5494	2N6512	2N6512
2N6186	SDT3126	2N6237	S0505M83	2N6290	2N5495	2N6513	RC830A
2N6186	2N6186	2N6237	S2060F	2N6290	2N6290	2N6513	RC832A
2N6187	2N4999	2N6237	S2060Y	2N6290	28D844	2N6513	2N6313
2N6187	2N5001	2N6237	S2061F	2N6292	RCA1C10	2N6514	RC830B
2N6187	2N5003	2N6237	S2061Y	2N6292	2N5492	2N6514	RC832B
2N6187	2N5005	2N6237	S2062F	2N6292	2N5493	2N6514	2N6314
2N6187	2N5384	2N6237	S2062Y	2N6292	2N5496	2N6515	2N6315
2N6187	2N5385	2N6237	T1039P	2N6292	2N5497	2N6515	2N6373
2N6187	2N5408	2N6237	2N6237	2N6292	2N6292	2N6515	2N6374
2N6187	2N5410	2N6237	2N6333	2N6292	2N6293	2N6516	2N6372
2N6187	2N6187	2N6238	MCR306-3	2N6294	2N6294	2N6516	2N6495
2N6188	2N6188	2N6238	S1003M82	2N6295	2N6295	2N6517	2N5955
2N6189	2N5286	2N6238	S1003M83	2N6296	2N6296	2N6517	2N5956
2N6189	2N5287	2N6238	S2060A	2N6297	2N6297	2N6517	2N6317
2N6189	2N5409	2N6238	S2061A	2N6298	2N6298	2N6518	28A764
2N6189	2N5411	2N6238	S2062A	2N6299	2N6299	2N6518	2N5954
2N6189	2N6189	2N6238	T1039A	2N6300	2N6300	2N6518	2N6318
2N6190	MJ8101	2N6238	2N6238	2N6300	28C1664	2N6326	28A765
2N6190	SDT3323	2N6238	2N6334	2N6301	2N6301	2N6327	2N6327
2N6190	SDT3327	2N6239	MCR306-4	2N6301	2N6534	2N6327	28C2434
2N6190	2N6190	2N6239	S2003M82	2N6301	28C1894	2N6328	2N6328
2N6191	2N5147	2N6239	S2003M83	2N6301	28C2198	2N6328	28D250
2N6191	2N5149	2N6239	S2060B	2N6301	28D384	2N6328	2N6328
2N6191	2N5151	2N6239	S2062B	2N6301	28D385	2N6329	2N6329
2N6191	2N5153	2N6239	T1039B	2N6303	28D391	2N6330	2N6330
2N6191	2N5404	2N6239	2N6239	2N6303	SDT3503	2N6331	2N6331
2N6191	2N5406	2N6239	2N6335	2N6303	SDT3507	2N6338	MJ7000
2N6191	2N6191	2N6240	MCR306-5	2N6303	SDT3777	2N6338	SDT7603
2N6192	SDT3324	2N6240	MCR306-6	2N6303	2N3204	2N6338	SDT7609
2N6192	SDT3328	2N6240	S2060D	2N6303	2N6303	2N6338	2N5671
2N6192	SDT3504	2N6240	S2061D	2N6304	N86912	2N6338	2N5734
2N6192	2N5405	2N6240	S2062D	2N6304	2N5054	2N6338	2N5929
2N6192	2N6192	2N6240	S4003M82	2N6304	2N6304	2N6338	2N5932
2N6193	SDT3508	2N6240	S4003M83	2N6304	28C1044	2N6338	2N5933
2N6193	2N5407	2N6240	T1039D	2N6304	28C1254	2N6338	2N5936
2N6193	2N6193	2N6240	2N6240	2N6304	28C1260	2N6338	2N6128
2N6211	RC8559	2N6240	2N6337	2N6304	28C988	2N6338	2N6271
2N6211	RC8560	2N6241	S2060M	2N6305	SD5053	2N6338	2N6272
2N6211	TIP507	2N6241	S2061M	2N6305	2N6305	2N6338	2N6273
2N6211	TIP508	2N6241	S2062M	2N6306	DT8310	2N6338	2N6338
2N6211	TIP521	2N6241	2N6241	2N6306	DT8311	2N6338	28D132
2N6211	TIP522	2N6241	RC8564				
2N6211	2N6211	2N6249	SDT7612				
2N6211	28A762	2N6249	TIP538				
2N6212	2N6212	2N6249	2N5264				
2N6213	2N6213	2N6249	2N6249				
		2N6249	28C1436				

MOTOR.	Industry Standard No.
2N6339	SDT7604
2N6339	SDT7610
2N6339	TIP35D
2N6339	TIP515
2N6339	TIP517
2N6339	2N5672
2N6339	2N6339
2N6339	2N6354
2N6339	2N6496
2N6340	TIP35B
2N6340	2N6340
2N6340	2SC1609
2N6340	2SC1818
2N6340	SDT7206
2N6341	SDT7605
2N6341	SDT7611
2N6341	TIP516
2N6341	TIP518
2N6341	2N5931
2N6341	2N5935
2N6341	2N5937
2N6341	2N6341
2N6341	2SC1610
2N6341	2SC1672
2N6341	2SC431
2N6341	2SC432
2N6342	2N6342
2N6342A	SC149B
2N6342A	TIC236B
2N6342A	2N6342A
2N6343	2N6343
2N6343A	SC149C
2N6343A	TIC236C
2N6343A	2N6343A
2N6344	2N6344
2N6344A	SC149M
2N6344A	2N6344A
2N6345	2N6345
2N6345A	2N6345A
2N6346	2N6346
2N6346A	T2850A
2N6346A	T2850B
2N6346A	2N6346A
2N6347	2N6347
2N6347A	T2850D
2N6347A	2N6347A
2N6348	2N6348
2N6348A	2N6348A
2N6349	2N6349
2N6349A	2N6349A
2N6366	MRP415
2N6367	MRP416
2N6367	RF2142
2N6368	MRP417
2N6370	MRP419
2N6370	PT9787
2N6370	PT9787A
2N6370	S10-28
2N6377	2N6377
2N6378	2N6380
2N6378	2N5678
2N6378	2N6378
2N6379	2N6381
2N6379	2N5539
2N6379	2N6379
2N6383	2N6382
2N6383	2N6383
2N6384	TIP600
2N6384	2N6384
2N6385	TIP601
2N6385	2N6385
2N6385	2SD168
2N6385	2SD301
2N6386	D44D1
2N6386	D44D2
2N6386	2N6386
2N6387	D44H2
2N6387	2N6387
2N6388	D44H3
2N6388	RCA1C15
2N6388	2N6388
2N6394	C126P
2N6394	T10126P
2N6394	2N6394
2N6395	C126A
2N6395	T10126A
2N6395	2N6395
2N6396	C126B
2N6396	T10126B
2N6396	2N6396
2N6397	C126D
2N6397	T10126D
2N6397	2N6397
2N6398	C126M
2N6398	T10126M
2N6398	2N6398
2N6399	2N6399
2N6400	C127P
2N6400	2N6400
2N6401	C127A
2N6401	2N6401
2N6402	C127B
2N6402	2N6402
2N6403	C127D
2N6403	2N6403
2N6404	C127M
2N6404	2N6404
2N6405	2N6405
2N6408	2N6408
2N6409	2N6409
2N6410	2N6410
2N6420	MJ3583
2N6420	RCA1B03
2N6420	RC830C
2N6420	2N6420
2N6420	2SA483
2N6420	2SA566
2N6420	2SA652
2N6420	2SA653
2N6420	2SA766
2N6421	MJ3584
2N6421	RC8320C
2N6421	2N6421
2N6422	MJ3585
2N6422	2N6422
2N6423	MJ4240
2N6423	2N6423
2N6424	MJ3738
2N6424	2N6424
2N6425	MJ3739
2N6425	2N6425
2N6430	2N6430
2N6431	2N6431

MOTOR.	Industry Standard No.
2N6432	2N6432
2N6433	2N6433
2N6436	2N6127
2N6436	2N6436
2N6436	2SA1042
2N6436	2SA1044
2N6437	2N6437
2N6438	2N6438
2N6438	2SA1001
2N6438	2SA1002
2N6438	2SA1003
2N6438	2SA1040
2N6438	2SA1041
2N6438	2SA1043
2N6439	C2M50-28R
2N6439	C2M60-28R
2N6439	C50-28
2N6439	J02007A
2N6439	MRP306
2N6439	MRF5178
2N6439	MRP605
2N6439	PH0450D
2N6439	PH0450H
2N6439	PH0550H
2N6439	2N6362
2N6439	2N6439
2N6442	ND5700
2N6442	ND5701
2N6442	ND5702
2N6457	TIP534
2N6486	MJ85983
2N6486	NSP5983
2N6486	TIP73
2N6486	2N6103
2N6486	2N6486
2N6487	MJ85984
2N6487	NSP5984
2N6487	RCA3055
2N6487	TIP73A
2N6487	2N6098
2N6487	2N6099
2N6487	2N6100
2N6487	2N6487
2N6488	MJ85985
2N6488	NSP5985
2N6488	TIP73B
2N6488	2N6101
2N6488	2N6102
2N6488	2N6488
2N6489	MJ85960
2N6489	MJ85980
2N6489	NSP5980
2N6489	TIP74
2N6489	2N6489
2N6490	MJ85981
2N6490	NSP5981
2N6490	TIP74A
2N6490	2N6490
2N6491	MJ85982
2N6491	NSP5982
2N6491	TIP74B
2N6491	2N6491
2N6497	TIP524
2N6497	2N6497
2N6497	2SC2233
2N6497	2SC2516
2N6498	2N6498
2N6499	2SC2536
2N6499	2SD872
2N6501	2N6501
2N6502	2N6502
2N6503	2N6503
2N6504	2N6504
2N6505	2N6505
2N6506	2N6506
2N6507	2N6507
2N6508	2N6508
2N6509	2N6509
2N6521	2N4974
2N6522	2N4975
2N6542	MJ3010
2N6542	MJ3011
2N6542	MJ3012
2N6542	SVT300-5
2N6542	TIP529
2N6542	TIP575B
2N6542	2N5788
2N6542	2N5805
2N6542	2N6542
2N6542	2SC1114
2N6542	2SC1131
2N6543	FM27K380
2N6543	SDT1052
2N6543	SDT1053
2N6543	SDT1054
2N6543	SDT401
2N6543	SDT402
2N6543	SVT400-5
2N6543	SVT450-5
2N6543	TIP575C
2N6543	2N6543
2N6543	2SC1130
2N6543	2SC1156
2N6543	2SC1463
2N6543	2SC2243
2N6543	2SC2247
2N6543	2SC2388
2N6543	2SD171
2N6543	2SD312
2N6543	2SD677
2N6543	2SD749
2N6544	TIP303
2N6544	TIP304
2N6544	TIP535
2N6544	TIP536
2N6544	TIP558
2N6544	TIP559
2N6544	2N5240
2N6544	2N6512
2N6544	2N6514
2N6544	2N6544
2N6544	2N6671
2N6544	BUX82
2N6545	BUX83
2N6545	DTS425
2N6545	IR425
2N6545	MJ425
2N6545	SDT1057
2N6545	SDT1058
2N6545	SDT1059
2N6545	SDT1062
2N6545	SDT1063
2N6545	SDT1064

MOTOR.	Industry Standard No.
2N6545	SDT425
2N6545	SVT350-3
2N6545	SVT400-3
2N6545	SVT450-3
2N6545	TIP305
2N6545	TIP306
2N6545	TIP537
2N6545	TIP54
2N6545	TIP556
2N6545	TIP560
2N6545	TIP561
2N6545	2N5157
2N6545	2N6466
2N6545	2N6467
2N6545	2N6513
2N6545	2N6545
2N6545	2N6583
2N6545	2N6672
2N6545	2N6673
2N6545	2SC2190
2N6545	2SC2244
2N6545	2SC2248
2N6545	2SD265
2N6545	2SD266
2N6545	2SD273
2N6545	2SD274
2N6545	2SD351
2N6545	2SD601
2N6545	2SD802
2N6546	IR518
2N6546	MJ7260
2N6546	RCA8767
2N6546	RCA9113
2N6546	SDT13301
2N6546	TIP531
2N6546	TIP533
2N6546	TIP539
2N6546	TIP562
2N6546	2N5587
2N6546	2N5388
2N6546	2N5389
2N6546	2N6573
2N6546	2N6574
2N6546	2SC1141
2N6546	2SC1434
2N6546	2SC1870
2N6546	2SC2402
2N6546	2SC411
2N6546	2SD262
2N6546JAN	2N6546JAN
2N6546JTX	2N6546JTX
2N6546JTXV	2N6546JTXV
2N6547	BUX80
2N6547	IR519
2N6547	MJ13010
2N6547	MJ7261
2N6547	RCA8767A
2N6547	RCA8767B
2N6547	RCA9113A
2N6547	RCA9113B
2N6547	SDT13302
2N6547	SDT13303
2N6547	TIP532
2N6547	TIP540
2N6547	TIP563
2N6547	2N6547
2N6547	2N6575
2N6547	2SC1140
2N6547	2SC2191
2N6547	2SC2246
2N6547	2SC2429
2N6547	2SD293
2N6547	2SD294
2N6547	2SD310
2N6547	2SD311
2N6547	2SD396
2N6547	2SD429
2N6547JAN	2N6547JAN
2N6547JTX	2N6547JTX
2N6547JTXV	2N6547JTXV
2N6548	NSD152
2N6548	2N6548
2N6548	2SC1226
2N6549	NSD151
2N6549	2N6549
2N6551	RCP701A
2N6551	RCP701B
2N6551	RCP703A
2N6551	RCP703B
2N6551	RCP705
2N6551	RCP705B
2N6551	RCP707
2N6551	RCP707B
2N6551	2N6551
2N6552	NSD104
2N6552	NSD105
2N6552	RCP701C
2N6552	RCP703C
2N6552	2SC1098
2N6553	NSD106
2N6553	RCP135
2N6553	RCP137
2N6553	RCP701D
2N6553	RCP703D
2N6553	2N6553
2N6553	2SC1157
2N6554	RCP700A
2N6554	RCP700B
2N6554	RCP702A
2N6554	RCP702B
2N6554	RCP704
2N6554	RCP704B
2N6554	RCP706
2N6554	RCP706B
2N6554	2N6554
2N6555	NSD204
2N6555	NSD205
2N6555	RCP700C
2N6555	RCP702C
2N6555	2N6555
2N6556	NSD206
2N6556	RCP700D
2N6556	RCP702D
2N6556	2N6556
2N6556	2SA646
2N6556	2SA647
2N6557	NSD130
2N6557	NSD132

MOTOR.	Industry Standard No.
2N6557	RCP111A
2N6557	RCP111B
2N6557	RCP113A
2N6557	RCP113B
2N6557	RCP115B
2N6557	RCP117B
2N6557	RCP135B
2N6557	RCP137B
2N6557	2N6557
2N6557	2SC1519
2N6557	2SC1520
2N6557	2SC1521
2N6558	NSD133
2N6558	NSD134
2N6558	NSD450
2N6558	RCP111C
2N6558	RCP113C
2N6558	RCP131C
2N6558	RCP133C
2N6558	SVT056
2N6558	2N6558
2N6559	NSD135
2N6559	RCP111D
2N6559	RCP113D
2N6559	RCP131D
2N6559	RCP133D
2N6559	2N6177
2N6559	2N6559
2N6656	2N6656
2N6569	MJ2801
2N6569	SDT9201
2N6569	SDT9205
2N6569	SDT9210
2N6569	2N6571
2N6569	2N6569
2N6657	2N6657
2N6576	2N6576
2N6577	2N6577
2N6578	IR642
2N6578	PMD20K-120
2N6578	PMD25K-120
2N6578	TIP622
2N6578	TIP642
2N6578	2N6578
2N6578	2SD1830
2N6578	2SD670
2N6591	NSD123
2N6591	NSD127
2N6591	NSD457
2N6591	RCP115
2N6591	RCP117
2N6591	2N6591
2N6591	2SC1224
2N6592	NSD128
2N6592	RCP131A
2N6592	RCP133A
2N6592	2N6592
2N6593	NSD129
2N6593	NSD458
2N6593	RCP131B
2N6593	RCP133B
2N6593	2N6593
2N6594	MJ2267
2N6594	MJ2901
2N6594	2N6594
2N6603	A500
2N6603	A501
2N6603	AT1845
2N6603	AT1845A
2N6603	AT2645
2N6603	AT2645A
2N6603	BPR49
2N6603	HXTR6104
2N6603	HXTR6105
2N6603	MRP901
2N6603	MRP902
2N6603	NE02107
2N6603	NE02108
2N6603	NE32707
2N6603	NE32708
2N6603	NE41703
2N6603	NE41707
2N6603	NE41708
2N6603	T996
2N6604	A510
2N6604	A511
2N6604	AT1825
2N6604	HXTR2102
2N6604	MRP911
2N6604	MRP912
2N6604	NE02103
2N6609	2N6609
2N6609	2SA971
2N6648	RCA8350
2N6648	TIP8350
2N6648	2N6648
2N6649	RCA8350A
2N6649	TIP605
2N6649	TIP8350A
2N6649	2N6649
2N6650	RCA8350B
2N6650	TIP606
2N6650	TIP8950B
2N6650	2N6650
2N6666	RCA8203
2N6666	2N6666
2N6667	RCA8203A
2N6667	2N6667
2N6668	RCA1016
2N6668	RCA8203B
2N6668	2N6668
2N6681	C30U
2N6681	C31U
2N6681	C32U
2N6681	S0525H
2N6681	2N6681
2N6682	C30F
2N6682	C31F
2N6682	C32F
2N6682	S0525H
2N6682	2N6682
2N6683	C30A
2N6683	C31A
2N6683	S1025H
2N6683	2N6683
2N6684	2N6684
2N6685	C30B
2N6685	C31B
2N6685	C32B
2N6685	S2025H
2N6685	2N6685
2N6686	2N6686
2N6687	C30C

MOTOR.	Industry Standard No.
2N687	C31C
2N687	C32C
2N687	2N687
2N688	C30D
2N688	C31D
2N688	C32D
2N688	84025H
2N688	2N688
2N689	86025H
2N690	86025H
2N690	2N690
2N691	2N691
2N692	2N692
2N697	2N2790
2N697	2N696
2N697	2N697
2N699	2N699
2N703	2N703
2N706	2N2256
2N706	2N706
2N706A	2N1586
2N706A	2N702
2N706A	2N706A
2N706A	2N835
2N706B	2N706B
2N708	2N2242
2N708	2N5210
2N708	2N708
2N718	2N718
2N718A	2N718A
2N720A	2N2509
2N720A	2N720
2N720A	2N720A
2N720A	2N735
2N720A	2N840
2N720A	2N841
2N720A	2N981
2N720A	2N997
2N736	2N736
2N740	2N739
2N740	2N740
2N742	2N1139
2N834	2N1199
2N834	2N1199A
2N834	2N1472
2N834	2N1589
2N834	2N2954
2N834	2N783
2N834	2N848
2N834	2N849
2N834	2N850
2N834	2N851
2N834	2N852
2N834	2N920
2N834	2N921
2N869A	2N1254
2N869A	2N1255
2N869A	2N1256
2N869A	2N1257
2N869A	2N1258
2N869A	2N1259
2N869A	2N1428
2N869A	2N1429
2N869A	2N3209
2N869A	2N3544
2N869A	2N3545
2N869A	2N3576
2N869A	2N726
2N869A	2N869
2N869A	2N869A
2N869A	2N996
2N8979	28D495
2N910	2N910
2N911	2N911
2N914	2N914
2N915	2N915
2N916	2N916
2N918	2N917
2N918	2N918
2N930	2N2510
2N930	2N2529
2N930	2N2530
2N930	2N2531
2N930	2N2532
2N930	2N2533
2N930	2N2534
2N930	2N2610
2N930	2N2693
2N930	2N3706
2N930	2N929
2N930	2N930
2N930A	2N2511
2N930A	2N2586
2N930A	2N3117
2N930A	2N3247
2N956	2N956
2N998	2N998
2N998	2N999
3N124	MM2090
3N124	3N124
3N125	MM2091
3N125	3N125
3N126	MM2092
3N126	SPB1091
3N126	TIX878
3N126	TIX879
3N126	3N126
3N128	MEM554
3N128	MEM554C
3N128	MEM557
3N128	MEM557C
3N128	MEM571C
3N128	MEM660
3N128	3N128
3N128	3N38
3N128	3N39
3N128	3N42
3N128	3N43
3N128	3N152
3N128	3N153
3N128	3N154
3N140	3N140
3N140	3N141
3N140	3N59
3N155	HA2001
3N155	HRN8338D
3N155	HRN8363
3N155	IT1750
3N155	M113
3N155	MEM560
3N155	MEM560C
3N155	MEM804
3N155	MEM814
3N155	3N155
3N155	3N186
3N155A	UC1700
3N155A	3N155A
3N156	MEM511
3N156	MEM511C
3N156	MEM520
3N156	2N4065
3N156	2N4120
3N156	3N156
3N156A	HA2000
3N156A	HA2010
3N157	3N157
3N157	3N184
3N157	3N185
3N157A	K1501
3N157A	K1502
3N157A	3N157A
3N158	M104
3N158	MEM520C
3N158	MEM556
3N158	MEM556C
3N158	3N158
3N158	3N178
3N158	3N179
3N158	3N180
3N158A	3N158A
3N169	3N169
3N170	IT2700
3N170	IT2701
3N170	3N170
3N171	3N171
3N201	MEM564C
3N201	MEM614
3N201	MEM616
3N201	MEM640
3N201	MEM641
3N201	MEM642
3N201	3N187
3N201	3N201
3N201	3N204
3N202	MEM617
3N202	3N202
3N202	3N203
3N202	3N205
3N203	MEM618
3N203	3N206
3N209	3N200
3N209	3N209
3N211	MEM680
3N211	3N211
3N212	MEM681
3N212	3N212
3N213	MEM682
3N213	3N213
4N25	CNY17,18
4N25	CNY21
4N25	FCD825,B
4N25	IL1
4N25	IL5
4N25	OPI2252
4N25	OPI2253
4N25	TLP503
4N25	TLP504
4N25	4N25
4N26	GLI-5
4N26	CQY13
4N26	CQY14
4N26	CQY15
4N26	CQY40,41
4N26	EP2
4N26	H11A10
4N26	H74A1
4N26	OPI2152
4N26	OPI2153
4N26	PC503
4N26	4N26
4N27	FCD,810,A
4N27	FCD836
4N27	IL12
4N27	IL15
4N27	IL16
4N27	IL74
4N27	OPI2151
4N27	SPX26
4N27	SPX28
4N27	TLP501
4N27	4N27
4N28	4N28
4N29	FCD850C,D
4N29	FCD855C,D
4N29	4N29
4N30	4N30
4N31	4N31
4N32	FCD860C,D
4N32	FCD865C,D
4N32	4N32
4N32	4N45
4N32	4N46
4N33	GLI-10
4N33	ILA30
4N33	ILA55
4N33	ILCA2-30
4N33	ILCA2-55
4N33	4N33
4N35	GLI-3
4N35	FCD825C,D
4N35	SPX2
4N35	SPX2E
4N35	SPX35
4N35	SPX36
4N35	SPX37
4N35	SPX4
4N35	SPX5
4N35	SPX6
4N35	4N35
4N36	4N36
4N37	4N37
4N38	GLI-2
4N38	4N38
4N39	4N39
4N40	4N40
5M100ZSB10	MZ5210
5M100ZS5	DBZ3100
5M100ZSB10	UZ222
5M10ZS5	BZX70-C10
5M110ZSB10	MZ222
5M110ZSB10	MZ5222
5M110ZSB10	UZ5222
5M110ZSB20	MZ322
5M110ZSB5	MZ122
5M110ZSB5	UZ122
5M110ZSB5	UZ5122
5M200ZS10	MZ220
5M200ZS10	MZ5220
5M200ZS10	UZ220
5M200ZS10	UZ5220
5M200ZS20	MZ320
5M200ZS5	MZ120
5M200ZS5	UZ120
5M200ZS5	UZ5120
5M200ZSB10	MZ240
5M200ZSB10	MZ5240
5M200ZSB10	UZ240
5M200ZSB10	UZ5240
5M200ZSB20	MZ340
5M200ZSB5	UZ140
5M200ZSB5	UZ5140
5M50ZS10	1N4521
5M6.0ZS5	DBZ3006
5M6.8ZS10	MZ5806
5M6.8ZS10	MZ806
5M6.8ZS10	UZ806
5M6.8ZS20	MZ906
5M6.8ZS5	MZ706
5M6.8ZS5	UZ5706
5M6.8ZS5	UZ706
5M75ZS5	BZX70-C75
5M90ZS10	MZ5890
6	8G307N
10M100ZS10	UZ7210
10M100ZS5	UZ7110
10M100ZZ10	1N2008C
10M100ZZ5	1N2008CA
10M10ZZ10	1N2498C
10M10ZZ5	1N2498CA
10M110ZZ10	1N2009C
10M110ZZ5	1N2009CA
10M11ZZ10	1N2499C
10M11ZZ5	1N2499CA
10M120ZZ10	1N2010C
10M120ZZ5	1N2010CA
10M12ZZ10	1N2500C
10M12ZZ5	1N2500CA
10M130ZZ10	1N2011C
10M130ZZ5	1N2011CA
10M13ZZ10	1N1816C
10M13ZZ5	1N1816CA
10M150ZZ10	1N2012C
10M150ZZ5	1N2012CA
10M15ZZ10	1N1817C
10M15ZZ5	1N1817CA
10M16ZZ10	1N1818C
10M16ZZ5	1N1818CA
10M18ZZ10	1N1819C
10M18ZZ5	1N1819CA
10M20ZZ10	1N1820C
10M20ZZ5	1N1820CA
10M22ZZ10	1N1821C
10M22ZZ5	1N1821CA
10M24ZZ10	1N1822C
10M24ZZ5	1N1822CA
10M27ZZ10	1N1823C
10M27ZZ5	1N1823CA
10M3.3AZ5	10LZ3.3D5
10M30ZZ10	1N1824C
10M30ZZ5	1N1824CA
10M33ZZ10	1N1825C
10M33ZZ5	1N1825CA
10M36ZZ10	1N1826C
10M36ZZ5	1N1826CA
10M39ZZ10	1N1827C
10M39ZZ5	1N1827CA
10M43ZZ10	1N1828C
10M43ZZ5	1N1828CA
10M47ZZ10	1N1829C
10M47ZZ5	1N1829CA
10M51ZZ10	1N1830C
10M51ZZ5	1N1830CA
10M56ZZ10	1N1831C
10M56ZZ5	1N1831CA
10M6.8Z10	UZ7806
10M6.8Z5	UZ7706
10M62ZZ10	1N1832C
10M62ZZ5	1N1832CA
10M68ZZ10	1N1833C
10M68ZZ5	1N1833CA
10M7.5AZ5	10LZ7.5D5
10M7.5Z5	BZY93-C7V5
10M75Z5	BZY93-C75
10M75ZZ10	1N1834C
10M75ZZ5	1N1834CA
10M82ZZ10	1N1835C
10M82ZZ5	1N1835CA
10M91ZZ10	1N1836C
10M91ZZ5	1N1836CA
50M7.5ZS5	BZY91-C7V5
50M75ZS5	BZY91-C75
212B	BCW86

668

Notes

Notes

670

Notes

Notes

672

Notes

_____

_____

_____

_____

_____

_____

_____

_____

_____

_____

_____

_____